The McGraw-Hill Companies

Library of Congress Cataloging in Publication Data

McGraw-Hill concise encyclopedia of science & technology.—6th ed.
v. cm.
Includes bibliographical references and index.
Contents: v. 1. Articles A–L—v. 2. Articles M–Z.
ISBN 978-0-07-161366-8 (set : alk. paper)—
ISBN 978-0-07-161368-2 (v. 1 : alk. paper)—
ISBN 978-0-07-161369-9 (v. 2 : alk. paper)
1. Science—Encyclopedias. 2. Technology—
Encyclopedias. I. McGraw-Hill Publishing Company.
II. Title: Concise encyclopedia of science & technology.
III. Title: McGraw-Hill concise encyclopedia of science and technology.

Q121.M29 2009
503—dc22 2008050987

1 2 3 4 5 6 7 8 9 0 CTP/CTP 0 5 4 3 2 1 0 9

ISBN: P/N 978-0-07-161368-2 of set
P/N 978-0-07-161366-8

MHID: P/N: 0-07-161368-4 of set
0-07-161366-8

This book was printed on acid-free paper.

It was set in Helvetica Black and Souvenir Light by Aptara, New Delhi, India.

China Translation & Printing Service, Ltd., was the printer and binder.

McGRAW-HILL

CONCISE ENCYCLOPEDIA OF SCIENCE & TECHNOLOGY

SIXTH EDITION

New York Chicago San Francisco Lisbon London Madrid Mexico City
Milan New Delhi San Juan Seoul Singapore Sydney Toronto

Contents

Editorial Staff

Mark Licker, Publisher

Jonathan Weil, Senior Staff Editor

David Blumel, Editor

Stefan Malmoli, Editor

Jessa Netting, Editor

Charles Wagner, Manager, Digital Content

Renee Taylor, Editorial Assistant

Editing, Design, and Production Staff

Roger Kasunic, Vice President—Editing, Design, and Production

Frank Kotowski, Jr., Managing Editor

Consulting Editors

Prof. Robert E. Knowlton. *Department of Biological Sciences, George Washington University, Washington, DC.* INVERTEBRATE ZOOLOGY.

Prof. Cynthia K. Larive. *Department of Chemistry, University of California, Riverside.* ANALYTICAL CHEMISTRY.

Prof. Chao-Jun Li. *Canada Research Chair in Green Chemistry, Department of Chemistry, McGill University, Montreal, Quebec, Canada.* ORGANIC CHEMISTRY.

Prof. Donald W. Linzey. *Wytheville Community College, Wytheville, Virginia.* VERTEBRATE ZOOLOGY.

Dr. Philip V. Lopresti. *Retired; formerly, Engineering Research Center, AT&T Bell Laboratories, Princeton, New Jersey.* ELECTRONIC CIRCUITS.

Dr. Dan Luss. *Cullen Professor of Engineering, Department of Chemical Engineering, University of Houston, Texas.* CHEMICAL ENGINEERING.

Prof. Scott M. McLennan. *Department of Geosciences, State University of New York at Stony Brook.* GEOLOGY (PHYSICAL, HISTORICAL, AND SEDIMENTARY).

Prof. Philip L. Marston. *Department of Physics and Astronomy, Washington State University, Pullman.* ACOUSTICS.

Dr. Ramon A. Mata-Toledo. *Professor of Computer Science, James Madison University, Harrisonburg, Virginia.* COMPUTING AND INFORMATION TECHNOLOGY.

Prof. Krzysztof Matyjaszewski. *J. C. Warner Professor of Natural Sciences, Department of Chemistry, Carnegie-Mellon University, Pittsburgh, Pennsylvania.* POLYMER SCIENCE AND ENGINEERING.

Prof. Joel S. Miller. *Department of Chemistry, University of Utah, Salt Lake City.* INORGANIC CHEMISTRY.

Dr. Orlando J. Miller. *Center for Molecular Medicine and Genetics, Wayne State University School of Medicine, Detroit, Michigan.* GENETICS.

Prof. Jay M. Pasachoff. *Director, Hopkins Observatory, Williams College, Williamstown, Massachusetts.* ASTRONOMY.

Prof. David J. Pegg. *Department of Physics, University of Tennessee, Knoxville.* ATOMIC AND MOLECULAR PHYSICS.

Prof. J. Jeffrey Peirce. *Department of Civil and Environmental Engineering, Edmund T. Pratt Jr., School of Engineering, Duke University, Durham, North Carolina.* ENVIRONMENTAL ENGINEERING.

Dr. William C. Peters. *Professor Emeritus, Mining and Geological Engineering, University of Arizona, Tucson.* MINING ENGINEERING.

Prof. Arthur N. Popper. *Department of Biology, University of Maryland, College Park.* NEUROSCIENCE.

Dr. Kenneth P. H. Pritzker. *Pathologist-in-Chief and Director, Head, Connective Tissue Research Group, and Professor, Laboratory Medicine and Pathobiology, University of Toronto, Mount Sinai Hospital, Toronto, Ontario, Canada.* MEDICINE AND PATHOLOGY.

Prof. Justin Revenaugh. *Department of Geology and Geophysics, University of Minnesota, Minneapolis.* GEOPHYSICS.

Dr. Roger M. Rowell. *USDA Forest Service, Forest Products Laboratory, Madison, Wisconsin.* FORESTRY.

Dr. John L. Safko. *Distinguished Professor Emeritus, Physics and Astronomy, Associated Faculty, School of the Environment, University of South Carolina, Columbia.* CLASSICAL MECHANICS.

Dr. Andrew P. Sage. *Founding Dean Emeritus and First American Bank Professor, University Professor, School of Information Technology and Engineering, George Mason University, Fairfax, Virginia.* CONTROL AND INFORMATION SYSTEMS.

Dr. Alfred S. Schlachter. *Advanced Light Source, Lawrence Berkeley National Laboratory, Berkeley, California.* ATOMIC AND MOLECULAR PHYSICS.

Prof. Ivan Schuller. *Department of Physics, University of California-San Diego, La Jolla, California.* CONDENSED-MATTER PHYSICS.

Dr. David Sherman. *Department of Earth Sciences, University of Bristol, United Kingdom.* MINERALOGY.

Dr. Steven A. Slack. *Department of Plant Pathology, Cornell University, Ithaca, New York.* PLANT PATHOLOGY.

Dr. Arthur A. Spector. *Department of Biochemistry, University of Iowa, Iowa City.* BIOCHEMISTRY.

Dr. Bruce A. Stanley. *Director, Scientific Programs, Section of Technology Development and Research Resources H093, The Pennsylvania State University College of Medicine, Hershey.* PHYSIOLOGY.

Prof. Anthony P. Stanton. *Carnegie-Mellon University, Pittsburgh, Pennsylvania.* GRAPHIC ARTS AND PHOTOGRAPHY.

Dr. Trent Stephens. *Professor of Anatomy and Embryology, Idaho State University, Pocatello.* DEVELOPMENTAL BIOLOGY.

Prof. John F. Timoney. *Department of Veterinary Science, University of Kentucky, Lexington.* VETERINARY MEDICINE.

Prof. Antonio J. Torres. *Department of Food Science and Technology, Oregon State University, Corvallis.* FOOD SCIENCE.

Dr. Sally E. Walker. *Associate Professor of Geology and Marine Science, University of Georgia, Athens.* INVERTEBRATE PALEONTOLOGY.

Prof. Pao K. Wang. *Department of Atmospheric and Oceanic Sciences, University of Wisconsin-Madison.* METEOROLOGY AND CLIMATOLOGY.

Prof. Frank M. White. *Professor Emeritus, Department of Mechanical Engineering, University of Rhode Island, Kingston.* FLUID MECHANICS.

Prof. Mary Anne White. *Department of Chemistry, Dalhousie University, Halifax, Nova Scotia, Canada.* MATERIALS SCIENCE AND METALLURGICAL ENGINEERING.

Prof. Thomas A. Wikle. *Department of Geography, and Associate Dean, College of Arts and Sciences, Oklahoma State University, Stillwater.* PHYSICAL GEOGRAPHY.

Preface

For nearly a half century, the *McGraw-Hill Encyclopedia of Science & Technology* has been an indispensable scientific reference work for a broad range of readers, from students to professionals and interested general readers. Found in many thousands of libraries around the world, its 20 volumes authoritatively cover every major field of science. However, the needs of many readers can be served by a concise work offering the same quality of coverage of the full breadth of science and technology. With this in mind, we present the Sixth Edition of the *McGraw-Hill Concise Encyclopedia of Science and Technology*.

In preparing this edition, now in two convenient volumes, the editors condensed the content of the parent work with the objective of retaining the essential information in each article without extensive detail. The original authors, all recognized experts, are identified and their affiliations are included in a complete alphabetical listing of Contributors.

The reader will find 7000 alphabetically arranged entries. Most include cross-references to other articles for background reading or further study. Dual measurement units (U.S. Customary and International System) are used throughout. The outstanding illustrations, one of the hallmarks of the *Concise Encyclopedia*, are now supplemented by 82 full-color plates and maps, illustrating a wide range of scientific and technical areas. The Appendix complements the articles with useful information, such as tables of scientific values, conversion charts, biographies of famous scientists, and more. Extensive bibliographies guide the reader to relevant literature. Finally, the Index provides quick access to specific information the reader needs.

We also invite the reader to go to http://MHEST.com on the Web, where we periodically provide updated information, such as articles, interviews, and multimedia content to enrich the reader's experience.

This concise reference will fill the need for accurate, current scientific and technical information in a convenient format. It can serve as the starting point for research by anyone seriously interested in science, even professionals seeking information outside their own specialty. It should prove to be a much used and much trusted addition to the reader's bookshelf.

MARK LICKER
Publisher

Organization of the Encyclopedia

Alphabetization. The 7000 article titles are sequenced across the two volumes on a word-by-word basis, not letter by letter. Hyphenated words are treated as separate words. In occasional inverted article titles, the comma provides a full stop. The Index is alphabetized on the same principles. Readers can turn directly to the pages for much of their research. An example of sequencing is:

Air	**Air pressure**
Air brake	**Air-traffic control**
Air-cushion vehicle	**Airborne radar**
Air pollution	**Aircraft**
Air pollution, indoor	**Aistopoda**

Cross references. Virtually every article has cross references set in CAPITALS AND SMALL CAPITALS. These references offer the user the option of turning to other articles for related information.

Measurement units. Since some readers prefer the U.S. Customary system while others require the metric or International System of Units (SI), measurements in the Encyclopedia are given in dual units. Insight into the measurement systems is provided by the discussion in the Appendix, which also has handy conversion tables.

Contributors. The authorship of each article is specified at its conclusion. The contributor's full name and affiliation may be found in the "Contributor Affiliations" section at the back of the volume.

Appendix. Every user should explore the variety of succinct information supplied by the Appendix, which includes measurement tables, mathematical notation, fundamental constants, and scientific notation; and a biographical listing of scientists. For users wishing to go beyond the scope of this Encyclopedia, recommended books and journals are listed in the Bibliographies subsection of the Appendix; the titles are grouped by subject area.

Index. The approximately 30,000-entry Index offers the reader the time-saving convenience of being able to quickly locate specific information in the text, rather than approaching the Encyclopedia via article titles only. This elaborate breakdown of the volume's contents assures both the general reader and the professional of efficient use of the *McGraw-Hill Concise Encyclopedia of Science & Technology*.

Table of Contents for Colorplates

The plates are relevant to the subject matter in the respective articles, but may not be specifically mentioned in the text.

Macadamia nut The fruit of a tropical evergreen tree, *Macadamia ternifolia*, native to Queensland and New South Wales and now grown commercially in Australia and Hawaii. The trees bear many small white or pinkish flowers in drooping racemes, each of which may mature from 1 to 20 fruits. These consist of a leathery outer husk (pericarp) which splits along one side at maturity, freeing the very hard-shelled, nearly round seed or nut. Two types of nuts are recognized, the most important commercially having a smooth shell and the other having a rough shell and sometimes referred to another species, *M. integrifolia* (see illustration). Laurence H. MacDaniels

***Macadamia integrifolia.* (*a*) Mature nuts. (*b*) Nuts without husks. (*c*) Nuts in husk showing method of dehiscence. (*R. A. Jaynes, ed., Handbook of North American Nut Trees, Humphrey Press, 1969*)**

Mach number In the flow of a fluid, the ratio of the flow velocity, V, at a given point in the flow to the local speed of sound, a, at that same point. That is, the Mach number, M, is defined as V/a. In a flowfield where the properties vary in time and/or space, the local value of M will also vary in time and/or space. In aeronautics, Mach number is frequently used to denote the ratio of the airspeed of an aircraft to the speed of sound in the freestream far ahead of the aircraft; this is called the freestream Mach number. The Mach number is a convenient index used to define the following flow regimes: (1) subsonic, where M is less than 1 everywhere throughout the flow; (2) supersonic, where M is greater than 1 everywhere throughout the flow; (3) transonic, where the flow is composed of mixed regions of locally subsonic and supersonic flows, all with local Mach numbers near 1, typically between 0.8 and 1.2; and (4) hypersonic, where (by arbitrary definition) M is 5 or greater.

Perhaps the most important physical aspect of Mach number is in the completely different ways that disturbances propagate in subsonic flow compared to that in a supersonic flow. Shock waves are a ubiquitous aspect of supersonic flows. *See* COMPRESSIBLE FLOW; SHOCK WAVE; SONIC BOOM; SUPERSONIC FLIGHT. John D. Anderson, Jr.

Machine A combination of rigid or resistant bodies having definite motions and capable of performing useful work. The term mechanism is closely related but applies only to the physical arrangement that provides for the definite motions of the parts of a machine. For example, a wristwatch is a mechanism, but it does no useful work and thus is not a machine. Machines vary widely in appearance, function, and complexity from the simple hand-operated paper punch to the ocean liner, which is itself composed of many simple and complex machines. *See* MACHINERY; SIMPLE MACHINE. Richard M. Phelan

Machine key Generally, a device used to prevent relative rotation of a shaft and the member to which it is connected, such as the hub of a gear, pulley, or crank. Many types of keys (see illustration) are available, and the choice in any installation depends on such factors as power requirements, tightness of fit, stability of connection, and cost.

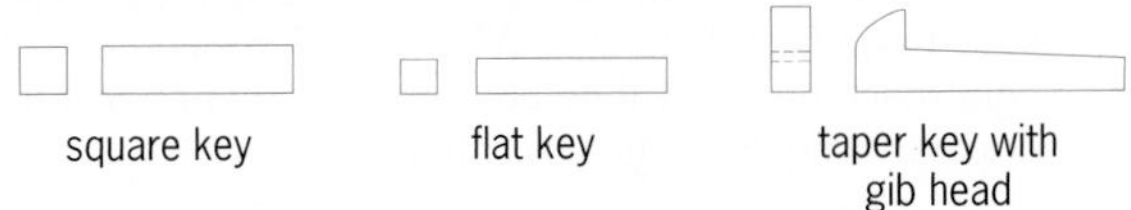

Types of keys. (*After P. H. Black and O. E. Adams, Jr., Machine Design, McGraw-Hill, 3d ed., 1968*)

Square keys are common in general industrial machinery. Flat keys are used where added stability of the connection is desired, as in machine tools. Square or flat keys may be of uniform cross section or they may be tapered. In tapered keys the width is uniform and the height of the key tapers. Tapered keys may have gib heads to facilitate removal. Other types of keys have been developed for special applications. Paul H. Black

Machinery A group of parts arranged to perform a useful function. Normally some of the parts are capable of motion; others are stationary and provide a frame for the moving parts. The terms machine and machinery are so closely related as to be almost synonymous; however, machinery has a plural implication, suggesting more than one machine. Common examples of machinery include automobiles, clothes washers, and airplanes; machinery differs greatly in number of parts and complexity.

Some machinery simply provides a mechanical advantage for human effort. Other machinery performs functions that no human being can do for long-sustained periods. *See* MACHINE; MECHANICAL ENGINEERING; SIMPLE MACHINE. Robert S. Sherwood

Machining An operation that changes the shape, surface finish, or mechanical properties of a material by the application of special tools and equipment. Machining almost always is a process where a cutting tool removes material to effect the desired change in the workpiece. Typically, powered machinery is required to operate the cutting tools. *See* PRODUCTION METHODS.

Although various machining operations may appear to be very different, most are very similar: they make chips. These chips vary in size from the long continuous ribbons produced on a lathe to the microfine sludge produced by lapping or grinding. These chips are formed by shearing away the workpiece material by the action of a cutting tool. Cylindrical holes can be produced in a workpiece by drilling, milling, reaming, turning, and electric discharge machining. Rectangular (or nonround) holes and slots may be produced by broaching, electric discharge machining, milling, grinding, and nibbling. Cylinders may be produced on lathes and grinders. Special geometries, such as threads and gears, are produced with special tooling and equipment utilizing the turning and grinding processes mentioned above. Polishing, lapping, and buffing are variants of grinding where a very small amount of stock is removed from the workpiece to produce a high-quality surface.

In almost every case, machining accuracy, economics, and production rates are controlled by the careful evaluation and selection of tooling and equipment. Speed of cut, depth of cut, cutting-tool material selection, and machine-tool selection have a tremendous impact on machining. In general, the more rigid and vibration-free a machining tool is, the better it will perform. Jigs and fixtures are often used to support the work-piece. Since it relies on the plastic deformation and shearing of the workpiece by the cutting tool, machining generates heat that must be dissipated before it damages the workpiece or tooling. Coolants, which also acts as lubricants, are often used.

To increase the life and speed of cutting tools, they are often coated with a thin layer of extremely hard material such as titanium nitride or zirconium nitride. These materials, which are applied over the cutting edges, provide excellent wear resistance. They are also brittle, so they rely on the toughness of the underlying cutting tool to support them. Coated tools are more expensive than conventional tools, but they can often cut at much higher rates and last significantly longer. When used properly on sufficiently rigid machine tools, they are far more economical than conventional tooling. *See* METAL COATINGS. J. R. Casey Bralla

Mackerel A fish which is a member of the order Perciformes, family Scombridae. There are about 50 carnivorous species found in the middle layer or near the surface of tropical and temperate seas. Mackerel are characterized by a long slender body, pointed head, and large mouth.

Scomber scombrus, the common mackerel, is an important fish commercially. It is a migratory species found on both sides of the North Atlantic. The Pacific mackerel (*Pneumatophorus diego*) is also an important commercial fish but differs from the common mackerel in having a swim bladder. The American Spanish mackerel (*Scomberomorus maculatus*) is a choice food fish. *See* PERCIFORMES. Charles B. Curtin

Macrocyclic compound An organic compound that contains a large ring. In the organic chemistry of alicyclic compounds, a closed chain of 12 carbon (C) atoms is usually regarded as the minimum size for a large ring; crown ethers are similarly defined. Macrocyclic compounds may be a single, continuous thread of atoms, as in cyclododecane [$(CH_2)_{12}$], or they may incorporate more than one strand or other ring systems (subcyclic units) within the macrocycle or macroring. In addition, macrocycles may be composed of aromatic rings that confer considerable rigidity upon the cyclic system. These aromatic rings may be joined together or coupled by spacer units consisting of one or more carbon atoms. *See* AROMATIC HYDROCARBON.

Classes of macrocyclic polyethers. Crown ethers are generally composed of repeating ethylene (CH_2CH_2) units separated by noncarbon atoms such as oxygen (O), nitrogen (N), sulfur (S), phosphorus (P), or silicon (Si). By far, the most common heteroatom present in the macrorings of crowns [X in $(XCH_2CH_2)_n$] is oxygen; but as more intricate structures are prepared, nitrogen, sulfur, phosphorus, silicon, or siloxy residues are becoming much more common.

By adding a third strand to the simple macrocyclic polyethers, three-dimensional compounds based on the crown framework are formed. Typically, two of the oxygen atoms across the ring from each other are replaced by nitrogens, and a third ethyleneoxy chain is attached to them. Known as cryptands, these structures completely encapsulate cations smaller than their internal cavities and strongly bind the most similar in size.

Two crown ether rings may be held together by a crown-ether-like strand to give a bicyclic cryptand. These have sometimes been referred to as ditopic receptors because they possess two distinct binding sites.

Lariat ethers, spherands, calixarenes, cavitands, and carcerands are other types of macrocyclic compounds, all of which are capable of encapsulating "guest" molecules in their interior cavities.

Cyclophane is the name given to macrocyclic compounds that contain organic (usually aromatic) rings as part of a cavity-containing structure. The first such compound was [2.2]-paracyclophane. In it, two benzene rings are joined by ethylene (CH_2CH_2) chains in their para positions. *See* CYCLOPHANE.

Complexation phenomena. It is the ability of these macrocyclic host compounds to complex a variety of guest species that makes these structures interesting. A crown ether can be described as a doughnut with an electron-rich and polar hole and a greasy or lipophilic (hydrophobic) exterior. As a result, these compounds are usually quite soluble in organic solvents but accommodate positively charged species in their holes.

A variety of organic cations have been found to complex with crown ethers and related hosts. It has been suggested that for a host-guest interaction to occur, the host must have convergent binding sites and the guest must have divergent sites. This is illustrated by the interaction between optically active dibinaphtho-22-crown-6 and optically active phenethylammonium chloride. The crown ether oxygen atoms converge to the center of a hole and the ammonium hydrogens diverge from nitrogen. Three complementary O—H—N hydrogen bonds stabilize the complex. In this particular case, different steric interactions between the optically active crown and the enantiomers of the complex permit resolution of the salt.

Other organic cations have also been complexed, either by insertion of the charged function in the crown's polar hole or by less distinct interactions observed in the solid state. *See* COORDINATION CHEMISTRY; COORDINATION COMPLEXES.

Applications. The striking ability of neutral macrocyclic polyethers to complex with alkali and alkaline-earth cations as well as a variety of other species has proved of considerable interest to the chemistry community. Crown ethers may complex the cation associated with an organic salt and cause separation of the ions. In the absence of cations to neutralize them, many anions show considerably enhanced reactivity. *See* ORGANIC REACTION MECHANISM.

One of the important modern developments in synthetic chemistry was the use of the phase-transfer technique. Nucleophiles such as cyanide are often insoluble in media that dissolve organic compounds with which they react. Thus 1-bromooctane may be heated in the presence of sodium cyanide for days with no product formation. When a crown ether is added, two things change. First, solubility is enhanced because the crown wraps about the cation, making it more lipophilic. This, in turn, makes the entire salt more lipophilic. Second, by solvating the cation, the association between cation and anion and the interactions with solvent are weakened, thus activating the anion for reaction. This approach has been used to assist the dissolution of potassium permanganate ($KMnO_4$) in benzene in which solvent permanganate is a powerful oxidizing agent. One striking example of solubilization is the displacement of chloride (Cl^-) by fluoride (F^-) in dimethyl 2-chloroethylene-1,1-dicarboxylate by using the KF complex of dicyclohexano-18-crown-6. In this reaction, a crown provides solubility for an otherwise insoluble or marginally soluble salt. Use of crowns to transfer a salt from the solid phase into an organic phase is often referred to as solid-liquid phase-transfer catalysis. *See* CATALYSIS; PHASE-TRANSFER CATALYSIS.

Since crown ethers and related species complex cations selectively, they can be used as sensors. Crowns have been incorporated into electrodes for this purpose, and crowns having various appended chromophores have been prepared. When a cation is bound within the macroring, a change in electron density is felt in the chromophore. The chromophores are often nitroaromatic residues and therefore highly colored. The color change that accompanies complexation can be easily detected and quantitated. *See* ION-SELECTIVE MEMBRANES AND ELECTRODES. George W. Gokel

Macrodasyida An order of the phylum Gastrotricha. They inhabit marine or brackish waters, seldom fresh waters. Some do not exceed 0.5 mm (0.02 in.) in length, and most are not more than 1–1.5 mm (0.04–0.06 in.); these live in clean to detritus-rich marine sands of littoral or sublittoral areas. All have front and rear groups of adhesive tubes; most also have tubes along their sides. Some have cuticular thickenings or scales or hooks. *Turbanella* and *Tetranchyroderma* are the most common and most abundant of macrodasyids, with numbers of 50–100 per cubic centimeter of sand (800–1600 per cubic inch) not being unusual. *See* GASTROTRICHA. William D. Hummon

Macroevolution Large-scale patterns and processes in the history of life, including the origins of novel organismal designs, evolutionary trends, adaptive radiations, and extinctions. Macroevolutionary research is based on phylogeny, the history of common descent among species. The formation of species and branching of evolutionary lineages mark the interface between macroevolution and microevolution, which addresses the dynamics of genetic variation within populations. Phylogenetic reconstruction, the developmental basis of evolutionary change, and long-term trends in patterns of speciation and extinction among lineages constitute major foci of macroevolutionary studies.

Phylogenetic reconstruction. Phylogenetic relationships are revealed by the sharing of evolutionarily derived characteristics among species, which provides evidence for common ancestry. Shared derived characteristics are termed synapomorphies, and are equated by many systematists with the older concept of homology. Characteristics of different organisms are homologous if they descend, with some modification, from an equivalent characteristic of their most recent common ancestor. Closely related species share more homologous characteristics than do species whose common ancestry is more distant. Species are grouped into clades according to patterns of shared homologies. The clades form a nested hierarchy in which large clades are subdivided into smaller, less inclusive ones, and are depicted by a branching diagram called a cladogram. A phylogenetic tree is a branching diagram, congruent with the cladogram, that represents real lineages of past evolutionary history.

A cladogram or phylogenetic tree is necessary for constructing a taxonomy, but the principles by which higher taxa are recognized remain controversial. The traditional evolutionary taxonomy of G. G. Simpson recognizes higher taxa as units of adaptive evolution called adaptive zones. Species of an adaptive zone share common ancestry, and distinctive morphological or behavioral characteristics associated with use of environmental resources. Higher taxa receive Linnean categorical ranks (genus, family, order, and so forth) reflecting the breadth and distinctness of their adaptive zones. All taxa must have a single evolutionary origin, which means that the taxon must include the most recent common ancestor of all included species. A taxon is monophyletic if it contains all descendants of the group's most recent common ancestor, or paraphyletic if some descendants of the group's most recent common ancestor are excluded because they have evolved a new adaptive zone. For example, evolutionary taxonomy of the anthropoid primates groups the orangutan, gorilla, and chimpanzee in the paraphyletic family Pongidae and the humans in the monophyletic family Hominidae. Although the humans and chimpanzees share more recent common ancestry than either does with the gorilla or orangutan, the chimpanzees are grouped with the latter species at the family level and the humans are placed in a different family because they are considered to have evolved a new adaptive zone. The Hominidae and Pongidae together form a monophyletic group at a higher level (see illustration). *See* SYSTEMATICS.

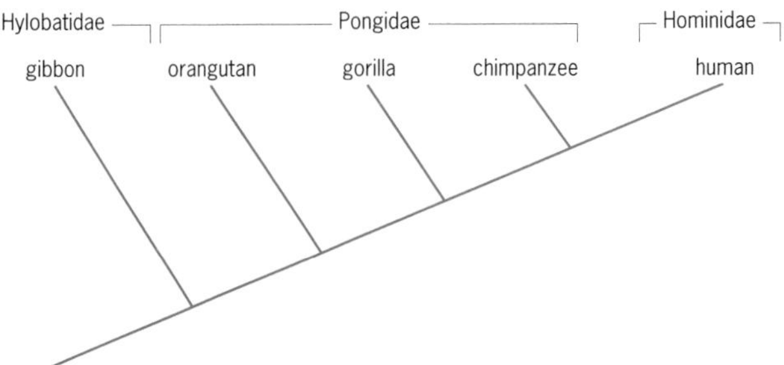

Phylogenetic relationships of anthropoid primates showing traditional family-level taxa. All apes and humans together form a monophyletic group. The family Pongidae is paraphyletic, and therefore considered invalid by cladistic taxonomists. (*After C. P. Hickman, Jr., L. S. Roberts, and A. Larson, Integrated Principles of Zoology, 9th ed., 1993*)

Cladistic taxonomy or phylogenetic systematics accepts only monophyletic taxa because these alone are considered natural units of common descent. Linnean rankings are considered unimportant. Taxa recognized using both the Simpsonian and cladistic taxonomies are standardly used in macroevolutionary analyses of extinction and patterns of diversity through time. The Simpsonian versus cladistic taxonomies often lead to fundamentally different interpretations, however. For example, extinction of a paraphyletic group, such as dinosaurs, would be considered pseudoextinction by cladists because some descendants of the group's most recent common ancestor survive. Birds are living descendants of the most recent common ancestor of all dinosaurs. The dinosaurs as traditionally recognized, therefore, do not form a valid cladistic taxon. *See* AVES; DINOSAURIA; PHYLOGENY.

Developmental processes. Comparative studies of organismal ontogeny are used to find where in development the key features of higher taxa appear and how developmental processes differ between taxa. Evolutionary developmental biologists denote the characteristic body plans of taxa by the term Bauplan. The major characteristics of animal phyla and their developmental and molecular attributes appear to have arisen and stabilized early in the history of life, during the Cambrian Period. Subsequent evolutionary diversification builds upon the Bauplan established early in animal evolution. *See* CAMBRIAN.

Particularly important to the evolutionary diversification of life are historical processes that generate change by altering the timing of organismal development, a phenomenon called heterochrony. Heterochronic changes can produce either paedomorphic or paeramorphic results. Paedomorphosis denotes the retention of preadult characteristics of ancestors in the adult stages of descendants; peramorphosis is the opposite outcome, in which the descendant ontogeny transcends that of the ancestor, adding new features at the final stages. Heterochronic changes can be produced by changing the rates of developmental processes or the times of their onset or termination.

Developmental dissociation occurs when different kinds of heterochronic change alter the development of different parts of the organism independently. Extensive dissociation can fundamentally restructure organismal ontogeny, producing ontogenetic repatterning. However, it is rare that a single heterochronic transformation affects all parts of the organism simultaneously. For most taxa, novel morphologies are produced by a mosaic of different heterochronic processes and by changes in the physical location of developmental events within the organism.

Long-term trends. Traditional Darwinian theory emphasizes natural selection acting on varying organisms within populations as the main causal factor of evolutionary change. Over many generations, the accumulation of favorable variants by natural selection produces new adaptations and new species. Macroevolutionary theory postulates two additional processes analogous to natural selection that act above the species level and on much longer time scales. An evolving lineage ultimately experiences one of two fates, branching speciation or extinction. Lineages that have a high propensity to produce new species and an ability to withstand extinction will dominate evolutionary history.

The higher-level process of differential speciation and extinction caused by the varying characteristics of species or lineages has been called species selection. Because the precise meaning of the term species is controversial, the more neutral terms lineage selection and clade selection are sometimes substituted for species selection. Most species show an evolutionary duration from a few million to approximately 10 million years in the fossil record between geologically instantaneous events of branching speciation. Species selection therefore generally occurs on a time scale of millions of years, rather than the generational time scale of natural selection. Species selection may be the primary factor underlying morphological evolutionary trends at this scale if lineages evolve by punctuated equilibrium, in which most morphological evolutionary change accompanies branching speciation, and species remain morphologically stable between speciational events. *See* SPECIATION.

The fossil record reveals mass extinctions in which enormous numbers of species from many different taxa are lost within a relatively short interval of geological time. Some lineages may be better able to survive mass extinction events than others, and the characteristics that make a lineage prone to survive mass extinction may be very different from those that influence species selection between events of mass extinction. Catastrophic species selection denotes differential survival and extinction of lineages during events of mass extinction as determined by character variation among lineages. Prior to the Cretaceous mass extinction, dinosaur taxa dominated mammalian taxa, whereas mammals survived the mass extinction and then diversified extensively. The characteristics of the ancestral mammals may have permitted them to survive environmental challenges to which dinosaurs were susceptible. *See* EXTINCTION (BIOLOGY); FOSSIL; MAMMALIA; PALEONTOLOGY; PERMIAN.

Because natural selection, species selection, and catastrophic species selection can differ in the biological characteristics they promote, higher-level processes may undo or reverse evolutionary trends arising from lower-level processes. *See* ORGANIC EVOLUTION.

Allan Larson

Macromolecular engineering The process of designing and synthesizing well-defined complex macromolecular architectures. This process allows for the control of molecular parameters such as molecular-weight/molecular-weight distribution, microstructure/structure, topology, and the nature and number of functional groups. In addition, macromolecular engineering is the key to establishing the relationships between the precise molecular architectures and their properties. The understanding of the structure–property interplay is critical for the successful use of these elegantly tailored structures in the design of novel polymeric materials for applications such as tissue engineering, drug delivery, molecular filtration, micro- and optoelectronics, and polymer conductivity. Complex architectures, including star-shaped, branched, grafted, and dendritic-like polymers, have been prepared using living polymerization methods (for which there is no termination step to stop chain growth) such as anionic, cationic, living radical, metal-catalyzed polymerization, or combinations of these methods. *See* POLYMER; POLYMERIZATION.

Nikos Hadjichristidis; Hermis Iatrou; Marinos Pitsikalis

Macroscelidea An order of mammals containing the single family Macroscelididae (elephant shrews). Although formerly thought to be most closely associated with shrews, ungulates, and rabbits and hares, they are now thought to be most closely associated with elephants, aardvarks, golden moles, hyraxes, and sea cows on the basis of DNA investigations. *See* MAMMALIA.

The 15 species of elephant shrews have chubby, soft-furred bodies, long hindlegs, and a long, slender, ratlike tail usually covered with knobbed bristles (see illustration). The fur is brownish or grayish. The eyes are large, the ears are well developed, and the elongated, trunklike snout is flexible and sensitive. The head and body length ranges 95–315 mm and the tail 80–265 mm.

Elephant shrew, *Elephantulus*. (*Photo by Dr. Lloyd Glenn Ingles; © 2001 California Academy of Sciences*)

Elephant shrews are found throughout Africa in a variety of habitats from the deserts of the Sahara to grassy plains, to rocky outcrops and boulder fields, to the well-forested regions of the Congo. They are primarily diurnal, but may be nocturnal during hot weather and on moonlit nights.

Donald W. Linzey

Macrostomorpha More than 300 species of small (mostly 1 mm size) flatworms usually with a ventral mouth situated about one quarter distant from the rostral tip of the body. Macrostomorpha are covered by a single-layered, ciliated epidermis; they move by ciliary gliding and have special duo-gland adhesive papillae, at least in the tail region, which they use for temporary adhesion to and quick release from the substrate (only the marine, pelagic microstomid *Alaurina* does not possess papillae). Many species are found in marine sand and mud, and in various freshwater habitats. These flatworms were formerly members of the Turbellaria-Rhabdocoela, but are now included in the Rhabditophora, which is one of three monophyletic taxa constituting the Platyhelminthes (Acoelomorpha, Catenulida, and Rhabditophora). *See* PLATYHELMINTHES.

Macrostomorpha are subdivided into two taxonomic groups of equal rank: The Haplopharyngida are more than 5 mm (0.2 in.) long. Only two marine species in one family (Haplopharyngidae) are known. The Macrostomida include all other Macrostomorpha. Three families are distinguished: (1) Microstomidae (three genera with 35 species) is known from many marine and freshwater habitats worldwide. (2) Dolichomacrostomidae (35 species in three subfamilies) is known only from marine sediments. (3) Macrostomidae includes most of the 11 genera of widely recognized species. Marine, brackish-water, and freshwater forms are found in all regions of the world. The best-known genus of Macrostomorpha is *Macrostomum*. Reinhard Rieger

Madelung constant A numerical constant α_M in terms of which the electrostatic energy U of a three-dimensional periodic crystal lattice of positive and negative point charges q_+, $-q_-$, N in number, is given by the equation below, where d is

$$U = -\frac{1}{2}\frac{Nq_+q_-}{d}\alpha_M$$

the nearest-neighbor distance between positive and negative charges and N is large. Knowledge of such electrostatic energies as given by the Madelung constant is of importance in the calculation of the cohesive energies of ionic crystals and in many other problems in the physics of solids. See IONIC CRYSTALS.

Madelung constants for some common ionic crystals

Crystal structure	Madelung constant, α_M
Sodium chloride, NaCl	1.7476
Cesium chloride, CsCl	1.7627
Zinc blende, α-ZnS	1.6381
Wurtzite, β-ZnS	1.641
Fluorite, CaF_2	5.0388
Cuprite, CuO_2	4.1155
Rutile, TiO_2	4.816
Anatase, TiO_2	4.800
Corundum, Al_2O_3	25.0312

The Madelung constants for a number of common ionic crystal structures are given in the table. For these cases d is chosen as the nearest-neighbor distance. See CRYSTAL. B. Gale Dick

Magellanic Clouds Two small, irregular galaxies that are close companions of the Milky Way Galaxy. Both are nearby galaxies that are located in the southern sky, not far from the south celestial pole. When viewed without a telescope, they resemble small sections of the Milky Way that might have drifted away from the main arc. The Large Magellanic Cloud (LMC) subtends an angular extent of about 5° in the sky, and the Small Magellanic Cloud (SMC) is about 3° across. Telescopic studies show, however, that each is really much larger than it appears. See MILKY WAY GALAXY.

The Magellanic Clouds are rather small irregular-type galaxies. The Large Magellanic Cloud is at a distance of 160,000 light-years (1.5×10^{18} km or 9×10^{17} mi), and the Small Cloud is about 10% farther away. The explosion of Supernova 1987A in the Large Magellanic Cloud was seen in February 1987. The galaxies are satellites of the Milky Way Galaxy. See SUPERNOVA.

It was shown in 1998, using Hubble Space Telescope measurements, that the oldest globular clusters in the Large Magellanic Cloud have the same ages as the oldest such clusters in the Milky Way Galaxy, indicating that both galaxies must have formed at nearly the same time. The Small Magellanic Cloud may have formed more recently, as no very old clusters have yet been identified among its several hundred star clusters. Both Magellanic Clouds have large numbers of very young stars, most of which are located in stellar associations, each of which contains several hundred recently formed stars in loose aggregates about 200 light-years (2×10^{15} km or 1.2×10^{15} mi) across. See STAR CLUSTERS.

Both objects also are rich in gas, mostly neutral hydrogen gas. In certain areas the gas is heated by nearby bright stars, producing brilliant glowing nebulae. The brightest and biggest is the 30 Doradus nebula, one of the most remarkable objects of the nearby universe, which includes a "nursery" for the formation of supergiant stars. See INTERSTELLAR MATTER; NEBULA; SUPERGIANT STAR. Paul Hodge

Magic numbers The number of neutrons or protons in nuclei which are required to fill major quantum shells. They occur at particle numbers 2, 8, 20, 50, and 82.

In atoms, the electrons that orbit the nucleus fill quantum electron shells at atomic numbers of 2 (helium), 10 (neon), 18 (argon), 36 (krypton), and 54 (xenon). These elements are chemically inert and difficult to ionize because the energies of orbits are grouped in bunches or shells with large gaps between them. In nuclei, an analogous behavior is found; quantum orbits completely filled with neutrons or protons result in extra stability. The neutrons and protons fill their quantum states independently, so that both full neutron and full proton shells can occur as magic nuclei. In a few cases, for example oxygen-16 (${}^{16}_{8}O_8$) and calcium-40 (${}^{40}_{20}Ca_{20}$), doubly magic nuclei have full neutron and proton shells. Between the major shell gaps, smaller subshell gaps cause some extra stabilization and semimagic behavior is found at particle numbers 14, 28, 40, and 64. See ATOMIC STRUCTURE AND SPECTRA; ELECTRON CONFIGURATION.

In very heavy nuclei the Coulomb repulsion between the protons results in a different sequence of states for neutrons and protons and different major shell gaps. For neutrons the magic sequence continues at $N = 126$; the next shell gap is predicted at $N = 184$. For protons the next major shell gap is anticipated at $Z = 114$. The latter shell gaps lie beyond the heaviest nuclei known, but calculations indicate that the extra stability gained by producing nuclei with these particle numbers may result in an island of long-lived superheavy nuclei.

The closing of nuclear quantum shells has many observable consequences. The nuclei are more tightly bound than average, and the extra stability leads to anomalously high abundances of magic nuclei in nature. The full shells require unusually high energies to remove the least bound neutron or proton, and the probability of capturing extra particles is lower than expected. Furthermore, the full shells are spherically symmetric, and the nuclei have very small electric quadrupole moments. Many of these properties were known before the nuclear shell model was developed to account for quantum-level ordering and gaps between major shells. The different shell closures for atomic and nuclear systems reflect the differences between the Coulomb force that binds electrons to nuclei and the strong force that holds the nucleus together. An important component of the strong force in nuclei is the spin-orbit term, which makes the energy of a state strongly dependent on the relative orientation of spin and orbital angular momentum. See ANGULAR MOMENTUM; ELEMENTS, COSMIC ABUNDANCE OF; ISOTOPE; NUCLEAR MOMENTS; NUCLEAR STRUCTURE; STRONG NUCLEAR INTERACTIONS. C. J. Lister

Magma The hot material, partly or wholly liquid, from which igneous rocks form. Besides liquids, solids and gas may be present in magma. Most observed magmas are silicate melts with associated crystals and gas, but some inferred magmas are carbonate, phosphate, oxide, sulfide, and sulfur melts.

Strictly, any natural material which contains a finite proportion of melt (hot liquid) is a magma. However, magmas which contain more than about 60% by volume of solids generally have finite strength and fracture like solids.

Hypothetical, wholly liquid magmas which develop by partial melting of previously solid rock and segregation of the liquid into a volume free of suspended solids and gas are called primary magmas. Hypothetical, wholly liquid magmas which develop by crystallization of a primary magma and isolation of rest liquid free of suspended solids are called parental (or secondary) magmas. Although no unquestioned natural examples of either primary or parental magmas are known, the concepts implied by the definitions are useful in discussing the origins of magmas.

Bodies of flowing lava and natural volcanic glass prove the existence of magmas. Such proven magmas include the silicate magmas corresponding to such rocks as basalt, andesite, dacite, and rhyolite as well as rare carbonate-rich magmas and sulfur melts. Oxide-rich and sulfide-rich magmas are inferred from textural and structural evidence of fluidity as well as mineralogical evidence of high temperature, together with the results of experiments on the equilibrium relations of melts and crystals. *See* IGNEOUS ROCKS; LAVA.

Magma is presumed to underlie regions of active volcanism and to occupy volumes comparable in size and shape to plutons of eroded igneous rocks. However, it is not certain that individual plutons existed wholly as magma at one time. Magma may underlie some regions where no volcanic activity exists, because many plutons appear not to have vented to the surface. *See* PLUTON.

Diverse origins are probable for various magmas. Basaltic magmas because of their high temperatures probably originate within the mantle several tens of kilometers beneath the surface of the Earth. Rhyolitic magmas may originate through crystallization of basaltic magmas or by melting of crustal rock. Intermediate magmas may originate within the mantle or by crystallization of basaltic magmas, by melting of appropriate crustal rock, and also by mixing of magmas or by assimilation of an appropriate rock by an appropriate magma. *See* IGNEOUS ROCKS; VOLCANO.

Alfred T. Anderson

Magnesite A member of the calcite-type carbonates having the formula $MgCO_3$. It forms dolomite [$CaMg(CO_3)_2$] with calcite ($CaCO_3$) in the system $CaCO_3$—$MgCO_3$. Pure magnesite is not common in nature because there exists a complete series of solid solutions between $MgCO_3$ and $FeCO_3$, which is constantly present in magnesite in its natural occurrence. *See* CARBONATE MINERALS; MAGNESIUM.

Magnesite is usually white, but it may be light to dark brown if iron-bearing. The hardness of magnesite is $3^1/_2$ to $4^1/_2$ on the Mohs scale, and the specific gravity is 3.00. *See* HARDNESS SCALES.

Magnesite deposits are of two general types: massive and crystalline. Massive magnesite is an alteration product of serpentine which has been subjected to the action of carbonate waters. Crystalline magnesite is usually found in association with dolomite. It is generally thought to be a secondary replacement of magnesite in preexisting dolomite by magnesium-rich fluids.

Magnesite is an important industrial mineral. Various types of magnesite or magnesia (MgO) are produced by different thermal treatments. The caustic-calcined magnesite or magnesia is used in the chemical industry for the production of magnesium compounds, while dead-burned or sintered magnesite or magnesia is used in refractory materials. Fused magnesia is used as an insulating material in the electrical industry because of its high electrical resistance and high thermal conductivity.

Luke L. Y. Chang

Magnesium A metallic chemical element, Mg, in group 2 of the periodic system. Magnesium is silvery white and extremely light in weight. Because of this lightness combined with alloy strength suitable for many structural uses, magnesium has long been known as industry's lightest structural metal. *See* PERIODIC TABLE.

With a density only two-thirds that of aluminum, magnesium is used in countless applications where weight saving is an important consideration. The metal also has, however, many desirable chemical and metallurgical properties which account for its extensive use in a variety of nonstructural applications.

Table 1. Physical properties of primary magnesium (99.9% pure).

Property	Value
Atomic number	12
Atomic weight	24.305
Atomic volume, cm^3/g-atom	14.0
Crystal structure	Close-packed hexagonal
Electron arrangement in free atoms	(2) (8) 2
Mass numbers of the isotopes	24, 25, 26
Percent relative abundances of ^{24}Mg, ^{25}Mg, ^{26}Mg	77, 11.5, 11.5
Density, g/cm^3 at 20°C	1.738
Specific heat, cal/g/°C at 20°C (1 cal = 4.2 joules)	0.245
Melting point, °C	650
Boiling point, °C	1110 ±10

Table 2. Principal magnesium compounds and uses

Compound	Uses
Magnesium carbonte	Refractories, production of other magnesium compounds, water treatment, fertilizers
Magnesium chloride	Cell feed for production of metallic magnesium, oxychloride cement, refrigerating brines, catalyst in organic chemistry, production of other magnesium compounds, flocculating agent, treatment of foliage to prevent fire and resist fire, magnesium melting and welding fluxes
Magnesium hydroxide	Chemical intermediate, alkali, medicinal
Magnesium oxide	Insulation, refractories, oxychloride and oxysulfate cements, fertilizers, rayon-textile processing, water treatment, papermaking, household cleaners, alkali, pharmaceuticals, rubber filler catalyst
Magnesium sulfate	Leather tanning, paper sizing, oxychloride and oxysulfate cements, rayon delustrant, textile dyeing and printing, medicinal, fertilizer ingredient, livestock-food additive, ceramics, explosives, match manufacture

Magnesium is very abundant in nature, occurring in substantial amounts in many rock-forming minerals such as dolomite, magnesite, olivine, and serpentine. In addition, magnesium is also found in sea water, subterranean brines, and salt beds. It is the third most abundant structural metal in the Earth's crust, exceeded only by aluminum and iron.

Some of the properties of magnesium in metallic form are listed in Table 1. Magnesium is very active chemically. It will actually displace hydrogen from boiling water, and a large number of metals can be prepared by thermal reduction of their salts and oxides with magnesium. The metal will combine with most nonmetals and with practically all acids. Magnesium reacts only slightly or not at all with most alkalies and many organic chemicals, including hydrocarbons, aldehydes, alcohols, phenols, amines, esters, and most oils. As a catalyst, magnesium is useful for promoting organic condensation, reduction, addition, and dehalogenation reactions. It has long been used for the synthesis of complex and special organic compounds by the well-known Grignard reaction. Principal alloying ingredients include aluminum, manganese, zirconium, zinc, rare-earth metals, and thorium. *See* MAGNESIUM ALLOYS

Magnesium compounds are used extensively in industry and agriculture. Table 2 lists the major magnesium compounds and indicates some of their more significant applications.

William H. Gross; Stephen C. Enclesan

Magnesium alloys The most important alloying ingredients used in magnesium alloys are aluminum, zinc, manganese, silicon, zirconium, rare-earth metals, and thorium. The specific gravity of magnesium alloys ranges from 1.74 to 1.83. It has led to a great many structural applications in the aircraft, transportation, materials-handling, and portable-tool and

-equipment industries. Magnesium alloys are commonly used in the form of die castings, which account for 75% of their usage in structural applications. *See* MAGNESIUM. Thomas E. Leontis

Magnet An object or device that produces a magnetic field. Magnets are essential for the generation of electric power and are used in motors, generators, labor-saving electromechanical devices, information storage, recording, and numerous specialized applications, for example, seals of refrigerator doors. The magnetic fields produced by magnets apply a force at a distance on other magnets, charged particles, electric currents, and magnetic materials. *See* GENERATOR; MAGNETIC RECORDING; MOTOR.

Magnets may be classified as either permanent or excited. Permanent magnets are composed of so-called hard magnetic material, which retains an alignment of the magnetization in the presence of ambient fields. Excited magnets use controllable energizing currents to generate magnetic fields in either electromagnets or air-cored magnets. *See* ELECTROMAGNET; FERROMAGNETISM; SUPERCONDUCTIVITY.

The essential characteristic of permanent-magnet materials is an inherent resistance to change in magnetization over a wide range of field strength. Resistance to change in magnetization in this type of material is due to two factors: (1) the material consists of particles smaller than the size of a domain, a circumstance which prevents the gradual change in magnetization which would otherwise take place through the movement of domain wall boundaries; and (2) the particles exhibit a marked magnetocrystalline anisotropy. During manufacture the particles are aligned in a magnetic field before being sintered or bonded in a soft metal or polyester resin. Compounds of neodymium, iron, and boron are used. *See* IRON ALLOYS.

Electromagnets rely on magnetically soft or permeable materials which are well annealed and homogeneous so as to allow easy motion of domain wall boundaries. Ideally the coercive force should be zero, permeability should be high, and the flux density saturation level should be high. Coincidentally the hysteresis energy loss represented by the area of the hysteresis curve is small. This property and high electrical resistance (for the reduction of eddy currents) are required where the magnetic field is to vary rapidly. This is accomplished by laminating the core and using iron alloyed with a few percent silicon that increases the resistivity.

Electromagnets usually have an energizing winding made of copper and a permeable iron core. Applications include relays, motors, generators, magnetic clutches, switches, scanning magnets for electron beams (for example, in television receivers), lifting magnets for handling scrap, and magnetic recording heads. *See* CATHODE-RAY TUBE; CLUTCH; ELECTRIC SWITCH; RELAY.

Special iron-cored electromagnets designed with highly homogeneous fields are used for special analytical applications in, for example, electron or nuclear magnetic resonance, or as bending magnets for particle accelerators. *See* MAGNETIC RESONANCE; PARTICLE ACCELERATOR.

Air-cored electromagnets are usually employed above the saturation flux density of iron (about 2 T); at lower fields, iron-cored magnets require much less power because the excitation currents needed then are required only to generate a small field to magnetize the iron. The air-cored magnets are usually in the form of a solenoid with an axial hole allowing access to the high field in the center. The conductor, usually copper or a copper alloy, must be cooled to dissipate the heat generated by resistive losses. In addition, the conductor and supporting structure must be sufficiently strong to support the forces generated in the magnet. *See* SOLENOID (ELECTRICITY).

In pulsed magnets, higher fields can be generated by limiting the excitation to short pulses (usually furnished by the energy stored in a capacitor bank) and cooling the magnet between pulses. The highest fields are generally achieved in small volumes. A field of 75 T has been generated for 120 microseconds.

Large-volume or high-field magnets are often fabricated with superconducting wire in order to avoid the large resistive power losses of normal conductors. The two commercially available superconducting wire materials are (1) alloys of niobium-titanium, a ductile material which is used for generating fields up to about 9 T; and (2) a brittle alloy of niobium and tin (Nb_3Sn) for fields above 9 T. Practical superconducting wires use complex structures of fine filaments of superconductor that are twisted together and embedded in a copper matrix. The conductors are supported against the electromagnetic forces and cooled by liquid helium at 4.2 K (−452°F). A surrounding thermal insulating enclosure such as a dewar minimizes the heat flow from the surroundings.

Superconducting magnets operating over 20 T have been made with niobium-titanium outer sections and niobium-tin inner sections. Niobium-titanium is used in whole-body nuclear magnetic resonance imaging magnets for medical diagnostics. Other applications of superconducting magnets include their use in nuclear magnetic resonance for chemical analysis, particle accelerators, containment of plasma in fusion reactors, magnetic separation, and magnetic levitation. *See* MAGNETIC LEVITATION; MAGNETIC SEPARATION METHODS; MEDICAL IMAGING; NUCLEAR FUSION; NUCLEAR MAGNETIC RESONANCE (NMR); SUPERCONDUCTING DEVICES.

The highest continuous fields are generated by hybrid magnets. A large-volume (lower-field) superconducting magnet that has no resistive power losses surrounds a water-cooled inner magnet that operates at the highest field. The fields of the two magnets add. Over 35 T has been generated continuously. Simon Foner

Magnet wire Insulated copper or aluminum wire used in the coils of all types of electromagnetic machines and devices. It is single-strand wire insulated with enamel, varnish, cotton, glass, asbestos, or combinations of these. To meet the immense variety of uses and to gain competitive advantage, a great number of kinds of enamel and fiber insulations are widely available. *See* ELECTRICAL INSULATION; MAGNET. Philip L. Alger; C. J. Herman

Magnetic compass A compass depending for its directive force upon the attraction of the Earth's magnetism for a magnet free to turn in any horizontal direction. A compass is an instrument used for determining horizontal direction.

The magnetic compass operates on the principle that like magnetic poles repel each other whereas unlike poles attract each other. The Earth has internal magnetism similar to that which would result from a short, powerful bar magnet at the center of the Earth. The lines of force connecting the two poles are vertical at two points on the surface of the Earth, called magnetic poles, and horizontal along the magnetic equator, a line approximating a great circle nearly midway between the magnetic poles. The horizontal component of the Earth's magnetic field varies from a maximum on or near the magnetic equator to zero at the magnetic poles.

A simple magnetic compass consists of a magnetized needle mounted so as to be free to align itself with the horizontal component of the Earth's field. A magnetic compass is unreliable or inoperative in the vicinities of the magnetic poles. Alton B Moody

Magnetic ferroelectrics Materials that display both magnetic order and spontaneous electric polarization. Research on these materials has enabled considerable advances to be made in understanding the interplay between magnetism and ferroelectricity. The existence of both linear and higher-order coupling terms has been confirmed, and their consequences studied. They have given rise, in particular, to a number of magnetically induced polar anomalies and have even provided an example of a ferromagnet whose magnetic moment per unit volume is totally induced by its coupling via linear terms to a spontaneous electric dipole moment.

Most known ferromagnetic materials are metals or alloys. Ferroelectric materials, on the other hand, are nonmetals by definition. It therefore comes as no surprise to find that there are no known room-temperature ferromagnetic ferroelectrics. In fact, there are no well-characterized materials which are known to be both strongly ferromagnetic and ferroelectric at any temperature.

Somewhat unaccountably, antiferromagnetic ferroelectrics are also comparative rarities in nature. Nevertheless, a few are known, and among them the barium-transition-metal fluorides are virtually unique in providing a complete series of isostructural examples. They have the chemical composition $BaXF_4$ in which X is a divalent ion of one of the 3*d* transition metals, manganese, iron, cobalt, or nickel. These materials are orthorhombic and all spontaneously polar (that is, pyroelectric) at room temperature. For all except the iron and manganese materials, which have a higher electrical conductivity than the others, the polarization has been reversed by the application of an electric field, so that they are correctly classified as ferroelectric. Long-range antiferromagnetic ordering sets in at temperatures somewhat below 100 K (−280°F). Structurally the materials consist of XF_6 octahedra which share corners to form puckered *xy* sheets which are linked in the third dimension *z* by the barium atoms. *See* CRYSTAL.

The importance of these magnetic ferroelectrics is the opportunity they provide to study and to separate the effects of a variety of magnetic and nonmagnetic excitations upon the ferroelectric properties and particularly upon the spontaneous polarization. Measurements are often made via the pyroelectric effect, which is the variation of polarization with temperature. This effect is an extremely sensitive indicator of electronic and ionic charge perturbations in polar materials. Through these perturbations the effects of propagating lattice vibrations (phonons), magnetic excitations (magnons), electronic excitations (excitons), and even subtle structural transitions can all be probed with precision. *See* PYROELECTRICITY.

Of all the X ions present in the series $BaXF_4$, the largest is Mn^{2+}. As the temperature is reduced from room temperature, the fluorine cages contract and eventually the divalent manganous ion becomes too big for its cage, precipitating a complicated structural transition at 250 K (−10°F). One interesting effect of this phase transition is that it produces a lower-temperature phase with a crystal symmetry low enough to support the existence of the linear magnetoelectric effect, a linear coupling between magnetization and polarization. Below the antiferromagnetic transition at 26 K in $BaMnF_4$ this linear coupling produces a canting of the antiferromagnetic sublattices through a very small angle (of order 0.2 degree of arc). The result is a spontaneous, polarization-induced magnetic moment. At low temperatures $BaMnF_4$ is therefore technically a weak ferromagnet, although the resultant magnetic moment is extremely small, and it is more usually referred to as a canted antiferromagnet. This is the only well-categorized example of pyroelectrically driven ferromagnetism. *See* ANTIFERROMAGNETISM; FERROELECTRICS; FERROMAGNETISM; MAGNETISM. Malcolm E. Lines

Magnetic instruments

Magnetic instruments Instruments designed for the measurement of magnetic field strength or magnetic flux density, depending on their principle of operation.

Hall-effect instruments. Often called gaussmeters, these instruments measure magnetic field strength. They have a useful working range from 10 A/m to 2.4 MA/m (0.125 oersted to 30 kilooersteds). When a magnetic field, H_z, is applied in a direction at right angles to the current flowing in a conductor (or semiconductor), a voltage proportional to H_z is produced across the conductor in a direction mutually perpendicular to the current and the applied magnetic field. This phenomenon is called the Hall effect. The output voltage of the Hall probe is proportional to the Hall coefficient, which is a characteristic of the Hall-element material, and is inversely proportional to the thickness of this material. For a sensitive Hall probe, the material is thin with a large Hall coefficient. The semiconducting materials indium arsenide and indium antimonide are particularly suitable. *See* HALL EFFECT.

Fluxgate magnetometer. This instrument is used to measure low magnetic field strengths. It is usually calibrated as a gaussmeter with a useful range of 0.2 millitesla to 0.1 nanotesla (2 gauss to 1 microgauss).

Fluxmeter. This instrument is designed to measure magnetic flux. A fluxmeter is a form of galvanometer in which the torsional control is very small and heavy damping is produced by currents induced in the coil by its motion. This enables a fluxmeter to accurately integrate an emf produced in a search coil when the latter is withdrawn from a magnetic field, almost independently of the time taken for the search coil to be moved. *See* GALVANOMETER; MAGNETIC FLUX.

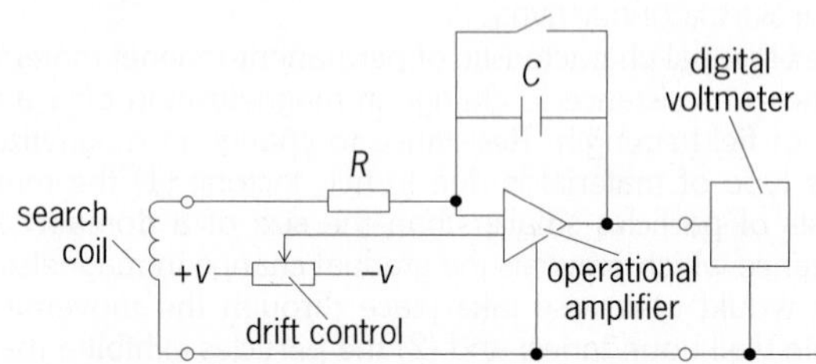

Arrangement of an electronic charge integrator.

Electronic charge intergrators. Often termed an integrator or gaussmeter, an electronic charge integrator, in conjunction with a search coil of known effective area, is used for the measurement of magnetic flux density. Integrators have almost exclusively replaced fluxmeters because of their independence of level and vibration. The instrument (see illustration) consists of a high-open-loop-gain (10^7 or more) operational amplifier with a capacitive feedback and resistive input. *See* OPERATIONAL AMPLIFIER.

Rotating-coil gaussmeter. This instrument measures low magnetic field strengths and flux densities. It comprises a coil mounted on a nonmagnetic shaft remote from a motor mounted at the other end. The motor causes the coil to rotate at a constant speed, and in the presence of a magnetic field or magnetic flux density a voltage is induced in the search coil. The magnitude of the voltage is proportional to the effective area of the search coil and the speed of rotation. *See* MAGNETIC FIELD; MAGNETIC INDUCTION; MAGNETOMETER. A. E. Drake

Magnetic lens

Magnetic lens A magnetic field with axial symmetry capable of converging beams of charged particles of uniform velocity and of forming images of objects placed in the path of such beams. Magnetic lenses are employed as condensers, objectives, and projection lenses in magnetic electron microscopes, as final focusing lenses in the electron guns of cathode-ray tubes, and for the selection of groups of charged particles of specific velocity in velocity spectrographs.

Magnetic lenses may be formed by solenoids or helical coils of wire traversed by electric current, by axially symmetric pole pieces excited by a coil encased in a high-permeability material such as soft iron, or by similar pole pieces excited by permanent magnets. In the last two instances the armatures and pole pieces serve to concentrate the magnetic field in a narrow region about the axis.

Magnetic lenses are always converging lenses. Their action differs from that of electrostatic lenses and glass lenses in that they produce a rotation of the image in addition to the focusing action. For the simple uniform magnetic field within a long solenoid the image rotation is exactly 180°. Thus a uniform magnetic field forms an erect real image of an object on its axis. E. G. Ramberg

Magnetic levitation A method of supporting and transporting objects or vehicles which is based on the physical property that the force between two magnetized bodies is inversely proportional to their distance. By using this magnetic force to counterbalance the gravitational pull, a stable and contactless suspension between a magnet (magnetic body) and a fixed guideway (magnetized body) may be obtained. In magnetic levitation (maglev), also known as magnetic suspension, this basic principle is used to suspend (or levitate) vehicles weighing 40 tons or more by generating a controlled magnetic force. By removing friction, these vehicles can travel at speeds higher than wheeled trains, with considerably improved propulsion efficiency (thrust energy/input energy) and reduced noise. In maglev vehicles, chassis-mounted magnets are either suspended underneath a ferromagnetic guideway (track) or levitated above an aluminum track. *See* MAGNET; MAGNETISM.

In the attraction-type system, a magnet-guideway geometry is used to attract a direct-current electromagnet toward the track. This system, also known as the electromagnetic suspension (EMS) system, is suitable for low- and high-speed passenger-carrying vehicles and a wide range of magnetic bearings. The electromagnetic suspension system is inherently nonlinear and unstable, requiring an active feedback to maintain an upward lift force equal to the weight of the suspended magnet and its payload (vehicle).

In the repulsion-type system, also known as the electrodynamic levitation system (EDS or EDL), a superconducting coil operating in persistent-current mode is moved longitudinally along a conducting surface (an aluminum plate fixed on the ground and acting as the guideway) to induce circulating eddy currents in the aluminum plate. These eddy currents create a magnetic field which, by Lenz's law, oppose the magnetic field generated by the travelling coil. This interaction produces a repulsion force on the moving coil. At lower speeds, this vertical force is not sufficient to lift the coil (and its payload), so supporting auxiliary wheels are needed until the net repulsion force is positive. The speed at which the net upward lift force is positive (critical speed) is dependent on the magnetic field in the airgap and payload, and is typically around 80 km/h (50 mi/h). To produce high flux from the traveling coils, hard superconductors (type II) with relatively high values of the critical field (the magnetic field strength of the coil at 0 K) are used to yield airgap flux densities of over 4 tesla. With this choice, the strong eddy-current induced magnetic field is rejected by the superconducting field, giving a self-stabilizing levitation force at high speeds (though additional control circuitry is required for adequate damping and ride quality). *See* EDDY CURRENT.

Due to their contactless operation, linear motors are used to propel maglev vehicles: linear induction motors for low-speed vehicles and linear synchronous motors for high-speed systems. Operationally they are the unrolled versions of the conventional rotary motors. *See* INDUCTION MOTOR; SYNCHRONOUS MOTOR.

Suspending the rotating part of a machine in a magnetic field may eliminate the contact friction present in conventional mechanical bearings. Magnetic bearings may be based on either attractive or repulsive forces. Although well developed, radial magnetic bearings are relatively expensive and complex, and are used in specialized areas such as vibration dampers for large drive shafts for marine propellers. In contrast, the axial versions of magnetic bearings are in common use in heavy-duty applications, such as large pump shafts and industrial drums. *See* ANTIFRICTION BEARING. P. K. Sinha

Magnetic materials Materials exhibiting ferromagnetism. The magnetic properties of all materials make them respond in some way to a magnetic field, but most materials are diamagnetic or paramagnetic and show almost no response. The materials that are most important to magnetic technology are ferromagnetic and ferrimagnetic materials. Their response to a field H is to create an internal contribution to the magnetic induction B proportional to H, expressed as $B = \mu H$, where μ the permeability, varies with H for ferromagnetic materials. Ferromagnetic materials are the elements iron, cobalt, nickel, and their alloys, some manganese compounds, and some rare earths. Ferrimagnetic materials are spinels of the general composition MFe_2O_4, and garnets, $M_3Fe_5O_{12}$, where M represents a metal. *See* FERRIMAGNETISM; FERROMAGNETISM; MAGNETISM; MAGNETIZATION.

Ferromagnetic materials are characterized by a Curie temperature, above which thermal agitation destroys the magnetic coupling giving rise to the alignment of the elementary magnets (electron spins) of adjacent atoms in a crystal lattice. Below the Curie temperature, ferromagnetism appears spontaneously in small volumes called domains. In the absence of a magnetic field, the domain arrangement minimizes the external energy, and the bulk material appears unmagnetized. *See* CURIE TEMPERATURE.

Magnetic materials are further classified as soft or hard according to the ease of magnetization. Soft materials are used in devices in which change in the magnetization during operation is desirable, sometimes rapidly, as in ac generators and transformers. Hard materials are used to supply a fixed field either to act alone, as in a magnetic separator, or to interact with others, as in loudspeakers and instruments. *See* ELECTRIC ROTATING MACHINERY; ELECTRICAL MEASUREMENTS; GENERATOR; INDUCTOR; LOUDSPEAKER; MAGNETIC SEPARATION METHODS; MICROPHONE; TRANSFORMER. F. E. Luborsky

Magnetic monopoles Magnetically charged particles. Such particles are predicted by various physical theories, but so far all experimental searches have failed to demonstrate their existence.

The fundamental laws governing electricity and magnetism become symmetric if particles exist that carry magnetic charge. Current understanding of electromagnetic physical phenomena is based on the existence of electric monopoles, which are sources or sinks of electric field lines (illus. *a*), and which when set into motion generate magnetic fields. The magnetic field lines produced by such a current have no beginning or end and form closed loops. All magnetic fields occurring in nature can be explained as arising from currents. However, theories of

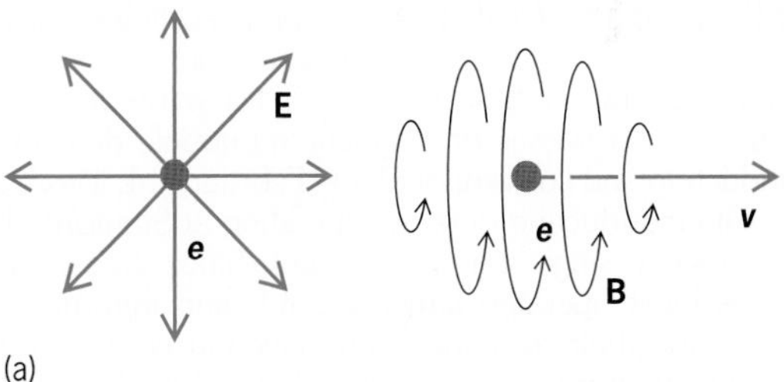

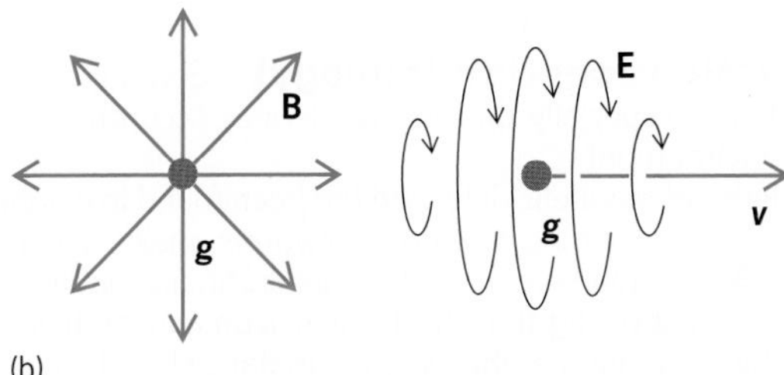

Electric (E) and magnetic (B) field lines generated by monopoles and by their motion with velocity *v*. (*a*) Electric monopole with electric charge *e*. (*b*) Magnetic monopole with magnetic charge *g*.

electromagnetism become symmetric if magnetic charges also exist. These would be sources or sinks of magnetic field and when set into motion would generate electric fields whose lines would be closed without ends (illus. *b*). *See* ELECTRIC FIELD.

In 1931 P. A. M. Dirac found a more fundamental reason for hypothesizing magnetic charges, when he showed that this would explain the observed quantization of electric charge. He showed that all electric and magnetic charges *e* and *g* must obey Eq. (1), where *k* must be an integer and $\hbar$ is Planck's constant

$$eg = k(\tfrac{1}{2}\hbar c) \quad (1)$$

divided by 2π. Equation (1) can be satisfied only if all electric and magnetic charges are integer multiples of an elementary electric charge e_0 and an elementary magnetic charge g_0. Since the size of the elementary electric charge, the charge carried by an electron or proton, is known experimentally, Dirac's equation predicts the size of the elementary magnetic charge to be given by Eq. (2). Since the fine structure constant α is given by Eq. (3), the

$$g_0 = \frac{1}{2}\frac{\hbar c}{e_0} \quad (2)$$

$$\alpha = \frac{e_0^2}{\hbar c} \approx \frac{1}{137} \quad (3)$$

elementary magnetic charge g_0 is about 68.5 times larger than the elementary electric charge e_0. *See* FINE STRUCTURE (SPECTRAL LINES); FUNDAMENTAL CONSTANTS.

In 1983 a successful theoretical unification of the electromagnetic and weak forces culminated in the detection of the W^+, the W^-, and the Z^0 particles predicted by the theory. This success has encouraged the search for a grand unification theory that would include the electroweak force and the nuclear or color force under one consistent description. In 1974 G. 't Hooft and independently A. M. Polyakov showed that magnetically charged particles are necessarily present in all true unification theories (those based on simple or semisimple compact groups). These theories predict the same long-range field and thus the same charge g_0 as the Dirac solution; now, however, the near field is also specified, leading to a calculable mass. The SU(5) model predicts a monopole mass of 10^{16} GeV/c^2, while theories based on supersymmetry or Kaluza-Klein models yield even higher masses up to the Planck mass of 10^{19} GeV/c^2. *See* ELECTROWEAK INTERACTION; FUNDAMENTAL INTERACTIONS; GRAND UNIFICATION THEORIES; QUANTUM GRAVITATION; SUPERGRAVITY; SUPERSYMMETRY.

There are two classes of magnetic monopole detectors, superconducting and conventional. On February 14, 1982, a prototype superconducting detector operating at Stanford University observed a single candidate event. Since then a number of groups have operated larger second- and third-generation detectors, and their combined data have placed a limit on the monopole flux more than 3000 times lower than the value from the data set that included the original event. Thus the possibility that this event was caused by the passage of a magnetic monopole has been largely discounted. Blas Cabrera

Magnetic reception (biology)

Magnetic reception (biology) Sensitivity to magnetic stimuli, especially the very weak ones occurring naturally in the environment.

Evidence of magnetic detection has been found in a variety of invertebrates, including protozoa, flatworms, snails, and insects. In 1968 Martin Lindauer and Herman Martin first published extensive data showing that the Earth's geomagnetic field influences the orientation of the waggle-run dance by which a scout honeybee communicates the distance and direction of a food source to the forager bees. Later, Lindauer and Martin showed that fluctuations of less than 10^{-4} gauss (roughly 1/10,000 of the Earth's field) can influence these bees' behavior. Other investigators found evidence of magnetic detection in other kinds of insects, including termites, beetles, and fruit flies (*Drosophila*).

Most of the evidence for magnetic detection by birds has come from studies of their migratory and homing behavior. Results strongly suggest that birds possess a magnetic compass, that is, they can determine compass bearings from the geomagnetic field. Evidence indicates that birds' sensitivity to magnetic stimuli is roughly similar to the honeybees'. It appears that the tiny fluctuations in the Earth's magnetic field caused by solar flares and other solar disturbances have a detectable effect on birds' navigation. The detection system probably has a narrow range of sensitivity; magnetic fields much stronger or weaker than the Earth's probably cannot be detected. *See* MIGRATORY BEHAVIOR.

Although behavioral effects of magnetic stimuli have been found in many kinds of animals, no one has yet succeeded in conditioning an animal to a magnetic stimulus in the laboratory. There is abundant evidence that the detection process is not quick, usually taking 15 min or more; hence, the flash stimuli presented in most classical conditioning attempts may be undetectable.

The physical mechanism for magnetic detection by living organisms is unknown, though a variety of possibilities have been put forward. William T. Keeton

Magnetic recording

Magnetic recording The technique of storing information as a magnetic pattern on a moving magnetic medium. The medium may be a disk, either flexible (floppy) or rigid, or a tape.

All materials respond in some way to an applied magnetic field, but the term "magnetic material" generally means one that maintains a magnetic polarization in the absence of an applied field. This remanence depends upon the magnetic field history to which the material was exposed. This history can be plotted on a graph of magnetization versus magnetic field, giving rise to hysteresis loops (Fig. 1). If the material is initially completely demagnetized (A in Fig. 1, sometimes referred to as the ac erase state), and the applied field is increased to some intermediate value and then reversed, a minor loop (BCDE) is obtained. At zero magnetic field there is a remanence. If the field is increased to the point that further increases in the field result in no further increases in magnetization, the material is said to be saturated. The remanence, at zero magnetic field, is a function of the

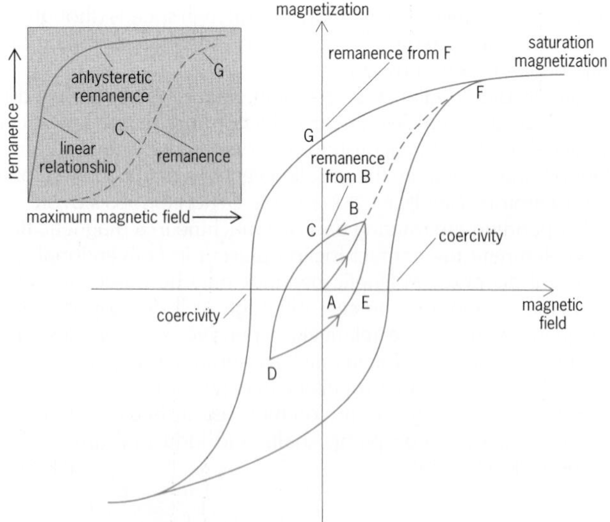

Fig. 1. Typical magnetization curves. At A, material is completely demagnetized. If the magnetic field is increased to B and then reversed, the minor loop BCDE is obtained. The inset shows remanence as a function of maximum field; points C and G correspond to the hysteresis loops shown.

maximum magnetic field to which the material has been exposed. Also, it takes a finite field in the reverse direction to drive the magnetization to zero. This field is called the coercivity of the material, and is an important parameter in magnetic recording. *See* MAGNETIC HYSTERESIS; MAGNETIC MATERIALS; MAGNETIZATION.

The hysteretic behavior of magnetic materials, and in particular, the field dependence of the remanence, is the basis for recording sound. The basic idea is to use the electric current from a microphone to generate a magnetic field (Ampère's law) that magnetizes portions of a magnetic medium in proportion to this current. The resulting magnetic pattern along the medium can then be read back as a voltage induced in a pick-up coil (Faraday's induction law) as the fringing fields from the magnetic medium passby. *See* AMPÈRE'S LAW; FARADAY'S LAW OF INDUCTION; MAGNETIC FIELD.

The writing field associated with a coil can be enhanced by filling the coil with a magnetic material. The reason is that the magnetic flux density generated by an electric current is proportional to the current through a constant of proportionality called the permeability. In free space, this quantity is usually denoted by μ_0. The permeability of a magnetic material, μ, however, is much larger. For example, the nickel-iron alloy permalloy has a relative permeability, μ/μ_0, of the order of 10,000. Therefore, the magnetic flux density inside a permalloy core would be 10,000 times that of an air core.

A high-permeability core also serves to confine the flux density. The field in a very narrow gap in the magnetic material can therefore be relatively large. Thus, recording generally employs an electromagnet with a narrow gap (Fig. 2). The writing and reading element is referred to as the head.

The relationship between the remanence and the field is very nonlinear (Fig. 1), which led to a good deal of distortion in early recorders. However, if an alternating current (ac) is added to the signal current, the resulting remanence, called the anhysteretic remanence, becomes linear at low fields. The amplitude of this bias must be sufficient to produce a field greater than the coercivity, and the bias frequency must be higher than the highest signal frequency. *See* SOUND RECORDING.

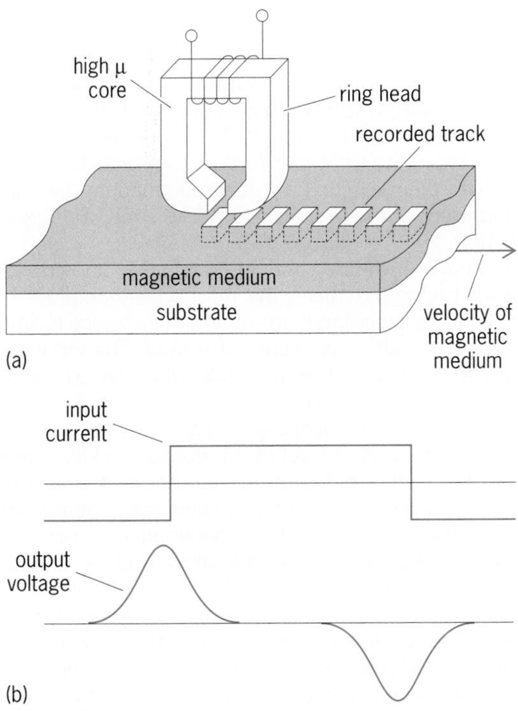

Fig. 2. Writing and reading process. (*a*) Motion of the magnetic medium past the electromagnet in the form of a ring head. (*b*) Variation with time of input current and output voltage.

To extend magnetic recording to video recording with signals as high as 5 MHz requires increasing the speed between the tape and the recording head. Video recording is based on an approach in which a number of heads are mounted on the face of a drum that rotates rapidly in a direction transverse to the direction of the tape motion. *See* TELEVISION.

Magnetic recording was applied to the storage of data in the early 1950s. Data generally means information represented in a digital form, that is, a sequence of 0's and 1's. In the first tape system for data storage, the data were recorded longitudinally along seven tracks. In a disk system, data are stored along concentric tracks. The figure of merit is the areal bit density, which is the product of the linear density along a track and the track density. *See* COMPUTER STORAGE TECHNOLOGY.

Early tapes and disks utilized particulate media in which the magnetic ingredient consisted of microscopic particles of a magnetic oxide. These particles were immersed in a polymeric binder that served to separate the particles from one another and bind them to the substrate. The particles first used were the gamma form of iron oxide. Higher levels of magnetization and coercivity can be obtained in particles made of the ferromagnetic elements, iron and cobalt, and their alloys than is possible with oxide particles. However, metal particles tend to corrode in the atmosphere and to react with binders and so must be passivated at some cost in saturation magnetization. The particles are also difficult to disperse and are much more expensive than particles of iron oxide. Metal particles having coercivities in the range 700–1150 Oe (56–92 kA/m) are used in premium audio tapes, and particles with coercivities of 1350–1550 Oe (107–123 kA/m) are used in 8-mm video tapes.

Metallic films are used on magnetic disks to reduce the thickness of the magnetic medium and retain a large magnetization. Among the magnetic elements (iron, cobalt, and nickel), cobalt has a hexagonal crystalline structure that leads to a large coercivity. Therefore, many metallic media consist of cobalt with additional elements to stabilize the hexagonal phase, and also, for longitudinal recording, to ensure that the hexagonal axis lies in the plane of the film. Robert M. White

Magnetic relaxation The relaxation or approach of a magnetic system to an equilibrium or steady-state condition as the magnetic field is changed. This relaxation is not instantaneous but requires time. The characteristic times involved in magnetic relaxation are known as relaxation times. Relaxation has been studied for nuclear magnetism, electron paramagnetism, and ferromagnetism.

Magnetism is associated with angular momentum called spin, because it usually arises from spin of nuclei or electrons. The spins may interact with applied magnetic fields, the so-called Zeeman energy; with electric fields, usually atomic in origin; and with one another through magnetic dipole or exchange coupling, the so-called spin-spin energy. Relaxation which changes the total energy of these interactions is called spin-lattice relaxation; that which does not is called spin-spin relaxation. (As used here, the term lattice does not refer to an ordered crystal but rather signifies degrees of freedom other than spin orientation, for example, translational motion of molecules in a liquid.) Spin-lattice relaxation is associated with the approach of the spin system to thermal equilibrium with the host material; spin-spin relaxation is associated with an internal equilibrium of the spins among themselves. *See* MAGNETISM; SPIN (QUANTUM MECHANICS). Charles P. Slichter

Magnetic resonance A phenomenon exhibited by the magnetic spin systems of certain atoms whereby the spin systems absorb energy at specific (resonant) frequencies when subjected to alternating magnetic fields. The magnetic fields must alternate in synchronism with natural frequencies of the magnetic system. In most cases the natural frequency is that of precession of the bulk magnetic moment of constituent atoms or nuclei about some magnetic field. Because the natural frequencies are highly specific as to their origin (nuclear magnetism, electron spin magnetism, and so on), the resonant method makes possible the selective study of particular features of interest. For example, it is possible to study weak nuclear magnetism unmasked by the much larger electronic paramagnetism or diamagnetism which usually accompanies it.

Nuclear magnetic resonance (that is, resonance exhibited by nuclei) reveals not only the presence of a nucleus such as hydrogen, which possesses a magnetic moment, but also its interaction with nearby nuclei. It has therefore become a most powerful method of determining molecular structure. The detection of resonance displayed by unpaired electrons, called electron paramagnetic resonance, is also an important application. *See* ELECTRON-NUCLEAR DOUBLE RESONANCE; MAGNETIC RESONANCE; MAGNETISM; NUCLEAR MEGNETIC RESONANCE (NMR).

Charles P. Slichter

Magnetic reversals The Earth's magnetic field has reversed polarity hundreds of times. That is, at different times in Earth's past, a compass would have pointed south instead of north. Recognition that the geomagnetic field has repeatedly reversed polarity played a key role in the revolution that transformed the geological sciences in the 1960s—the acceptance of the theory of plate tectonics. It is generally accepted that the geomagnetic field is generated by motion of electrically conducting molten metal in Earth's outer core. However, the mechanism by which the field decays and reverses polarity remains one of the great unknowns in geophysics. *See* GEODYNAMO; GEOMAGNETISM; PLATE TECTONICS.

The last magnetic field reversal occurred long before humans were aware of the geomagnetic field (780,000 years ago), so it is necessary to study geological records to understand the process by which the field reverses. The ability of rocks to act as fossilized compasses, which record a permanent "memory" of Earth's magnetic field at the time of rock formation, makes them suitable for detailed studies of ancient geomagnetic field behavior. *See* PALEOMAGNETISM; ROCK MAGNETISM. Andrew P. Roberts

Magnetic separation methods All materials possess magnetic properties. Substances that have a greater permeability than air are classified as paramagnetic; those with a lower permeability are called diamagnetic. Paramagnetic materials are attracted to a magnet; diamagnetic substances are repelled. Very strongly paramagnetic materials can be separated from weakly or nonmagnetic materials by the use of low-intensity magnetic separators. Minerals such as hematite, limonite, and garnet are weakly magnetic and can be separated from nonmagnetics by the use of high-intensity separators.

Magnetic separators are widely used to remove tramp iron from ores being crushed, to remove contaminating magnetics from food and industrial products, to recover magnetite and ferrosilicon in the float-sink methods of ore concentration, and to upgrade or concentrate ores. Magnetic separators are extensively used to concentrate ores, particularly iron ores, when one of the principal constituents is magnetic. *See* MECHANICAL SEPARATION TECHNIQUES; ORE DRESSING. Fred D. DeVaney

Magnetic susceptibility The magnetization of a material per unit applied field. It describes the magnetic response of a substance to an applied magnetic field. *See* MAGNETISM; MAGNETIZATION.

All ferromagnetic materials exhibit paramagnetic behavior above their ferromagnetic Curie points. The general behavior of the susceptibility of ferromagnetic materials at temperatures well above the ferromagnetic Curie temperature follows the Curie-Weiss law. The paramagnetic Curie temperature is usually slightly greater than the temperature of transition. *See* CURIE TEMPERATURE; CURIE-WEISS LAW; FERROMAGNETISM.

Most paramagnetic substances at room temperature have a static susceptibility which follows a Langevin-Debye law. Saturation of the paramagnetic susceptibility occurs when a further increase of the applied magnetic field fails to increase the magnetization, because practically all the magnetic dipoles are already oriented parallel to the field. *See* PARAMAGNETISM.

The susceptibility of diamagnetic materials is negative, since a diamagnetic substance is magnetized in a direction opposite to that of the applied magnetic field. The diamagnetic susceptibility is independent of temperature. Diamagnetic susceptibility depends upon the distribution of electronic charge in an atom and upon the energy levels. *See* DIAMAGNETISM.

The susceptibility of antiferromagnetic materials above the Néel point, which marks the transition from antiferromagnetic to paramagnetic behavior, follows a Curie-Weiss law with a negative paramagnetic Curie temperature.

Elihu Abrahams; Frederic Keffer

Magnetic thermometer A thermometer whose operation is based on Curie's law, which states that the magnetic susceptibility of noninteracting (that is, paramagnetic) dipole moments is inversely proportional to absolute temperature. Magnetic thermometers are typically used at temperatures below 1 K (−458°F). The magnetic moments in the thermometric material may be of either electronic or nuclear origin. Generally the magnetic thermometer must be calibrated at one (or more) reference temperatures. *See* ELECTRON; NUCLEAR MOMENTS; PARAMAGNETISM.

At temperatures from a few millikelvins upward, the thermometric material is preferably an electronic paramagnet, typically a nonconducting hydrous rare-earth salt. For higher temperatures, an ion is selected with a large magnetic moment in a crystalline environment with a high density of magnetic ions. In contrast, for low temperature use the magnetic exchange interactions between the magnetic ions should be small, which is accomplished by selecting an ion with a well-localized moment and by maintaining a large separation between the magnetic ions by means of diamagnetic atoms. This is the case in cerium magnesium nitrate (CMN) $[2Ce(NO_3)_3 \cdot 3Mg(NO_3)_3 \cdot 24H_2O]$. Here, the Ce^{3+} ion is responsible for the magnetic moment, which is well localized within the incompletely filled 4*f* shell relatively deep below the outer valence electrons. To reduce the magnetic interactions between the Ce^{3+} ions further, Ce^{3+} may be partly substituted with diamagnetic La^{3+} ions. Lanthanum-diluted CMN has been used for thermometry to below 1 mK. *See* EXCHANGE INTERACTION.

A mutual-inductance bridge, originally known as the Hartshorn bridge, has been the most widely employed measuring circuit for precision thermometry. The bridge is driven by a low-frequency alternating-current source. The inductance at low temperatures consists of two coils, which are as identical as possible. The voltages induced across them by the drive current are compared by means of a high-input-impedance ratio transformer. The output level of this voltage divider is adjusted to equal that of the midpoint between the two coils, using as null indicator a narrow-band preamplifier and a phase-sensitive (lock-in) detector. Thus, without a paramagnetic specimen, the bridge is balanced with the decade divider adjusted at its midpoint, while with the specimen inside one of the coils the change in the divider reading at bridge balance is proportional to the sample magnetization. For high-resolution thermometry it has become standard practice to replace the room-temperature zero detector with a SQUID magnetometer circuit. This also allows the mass of the sample to be reduced from several grams to the 1-mg level. *See* INDUCTANCE MEASUREMENT; SQUID.

Nuclear magnetic moments are smaller by a factor of 10^3 and are used for thermometry only in the ultralow-temperature region. For this the Curie-law behavior is generally sufficient down to the lowest temperatures. The nuclear paramagnetic thermometer loses adequate sensitivity for calibration purposes above 50–100 millikelvins, unless it is operated in a high polarizing field (*H* greater than 0.1 tesla). It can be utilized as a self-calibrating primary thermometer if the spin-lattice relaxation time is measured in parallel with the nuclear Curie susceptibility. Pulsed NMR measurement on the ^{195}Pt isotope in natural platinum metal provides presently the most widely used thermometry at temperatures below 1 mK. In the Curie-susceptibility measuring mode, it has been extended down to 10 μK. *See* LOW-TEMPERATURE THERMOMETRY; MAGNETIC RELAXATION; NUCLEAR MAGNETIC RESONANCE (NMR). Matti Krusius

Magnetic thin films Sheets of magnetic material with thicknesses of a few micrometers or less, used in the electronics industry. Magnetic films can be single-crystal, polycrystalline, amorphous, or multilayered in the arrangement of their atoms. Applications include magnetooptic storage, inductive recording media, magnetoresist sensors, and thin-film heads. *See* COMPUTER STORAGE TECHNOLOGY.

Both ferro- and ferrimagnetic films are used. The ferromagnetic films are usually transition-metal-based alloys. For example, permalloy is a nickel-iron alloy. The ferrimagnetic films, such as garnets or the amorphous films, contain transition metals such as iron or cobalt and rare earths. The ferrimagnetic properties are advantageous in magnetooptic applications where a low overall magnetic moment can be achieved without a significant change in the Curie temperature. *See* FERRIMAGNETISM; FERROMAGNETISM.

The change in electrical properties, such as the electrical resistance, with a magnetic field is used in sensor elements. The most notable of these in semiconductor technology is the magnetoresist head used in disk storage technology. Very large magnetoresist signals (called giant magnetoresistance) are observed in magnetic multilayers and composites containing a magnetic and nonmagnetic material. *See* MAGNETIC MATERIALS; MAGNETISM; MAGNETIZATION. Praveen Chaudhari

Magnetism The branch of science that describes the effects of the interactions between charges due to their motion and spin. These interactions may appear in various forms, including electric currents and permanent magnets. They are described in terms of the magnetic field, although the field hypothesis cannot be tested independently of the electrokinetic effects by which it is defined. The magnetic field complements the concept of the electrostatic field used to describe the potential energy between charges due to their relative positions. Special relativity theory relates the two, showing that magnetism is a relativistic modification of the electrostatic forces. The two together form the electromagnetic interactions which are propagated as electromagnetic waves, including light. They control the structure of materials at distances between the long-range gravitational actions and the short-range "strong" and "weak" forces most evident within the atomic nucleus. *See* ELECTROMAGNETIC RADIATION; RELATIVITY.

The magnetic field can be visualized as a set of lines (Fig. 1) illustrated by iron filings scattered on a suitable surface. The intensity of the field is indicated by the line spacing, and the direction by arrows pointing along the lines. The sign convention is chosen so that the Earth's magnetic field is directed from the north magnetic pole toward the south magnetic pole. The field can be defined and measured in various ways, including the forces on the equivalent magnetic poles, and on currents or moving charges. Bringing a coil of wire into the field, or removing it, induces an electromotive force (emf) which depends on the rate at which the number of field lines, referred to as lines of magnetic flux, linking the coil changes in time. This provides a definition of flux, Φ, in terms of the emf, e, given by Eq. (1) for a coil of N turns wound sufficiently closely to make

$$e = -N\,d\Phi/dt \qquad \text{volts} \tag{1}$$

the number of lines linking each the same. The International System (SI) unit of Φ, the weber (Wb), is defined accordingly as the volt-second. The symbol B is used to denote the flux, or line, density, as in Eq. (2), when the area of the coil is sufficiently

$$B = \Phi/\text{area} \tag{2}$$

small to sample conditions at a point, and the coil is oriented so that the induced emf is a maximum. The SI unit of B, the tesla (T), is the Wb/m^2. The sign of the emf, e, is measured positively in the direction of a right-hand screw pointing in the direction of the flux lines. It is often convenient, particularly when calculating induced emfs, to describe the field in terms of a magnetic vector potential function instead of flux.

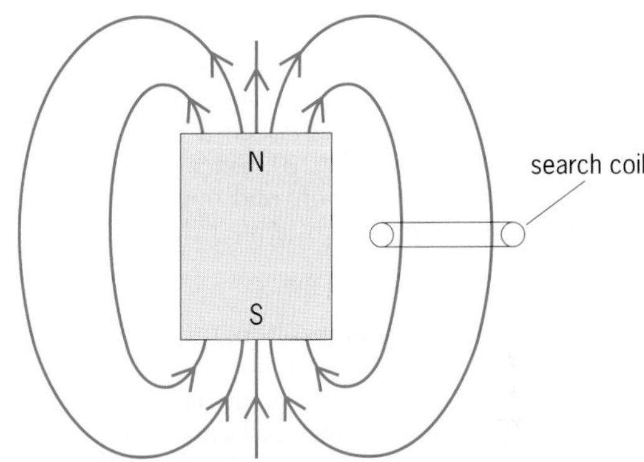

Fig. 1. Magnetic lines of a bar magnet.

Magnetic circuits. The magnetic circuit provides a useful method of analyzing devices with ferromagnetic parts, and introduces various quantities used in magnetism. It describes the use of ferromagnetic materials to control the flux paths in a manner analogous to the role of conductors in carrying currents around electrical circuits. For example, pieces of iron may be used to guide the flux which is produced by a magnet along a path which includes an air gap (Fig. 2), giving an increase in the flux density, B, if the cross-sectional area of the gap is less than that of the magnet. *See* MAGNET; MAGNETIC MATERIALS.

The magnet may be replaced by a coil of N turns carrying a current, i, wound over a piece of iron, or ferromagnetic material, in the form of a ring of uniform cross section. The flux linking each turn of the coil, and each turn of a secondary coil wound separately from the first, is then approximately the same, giving the same induced emf per turn [according to Eq. (1)] when the supply current, i, and hence the flux, Φ, changes in time. The arrangement is typical of many different devices. It provides, for example, an electrical transformer whose input and output voltages are directly proportional to the numbers of turns in the windings. Emf's also appear within the iron, and tend to produce circulating currents and losses. These are commonly reduced by

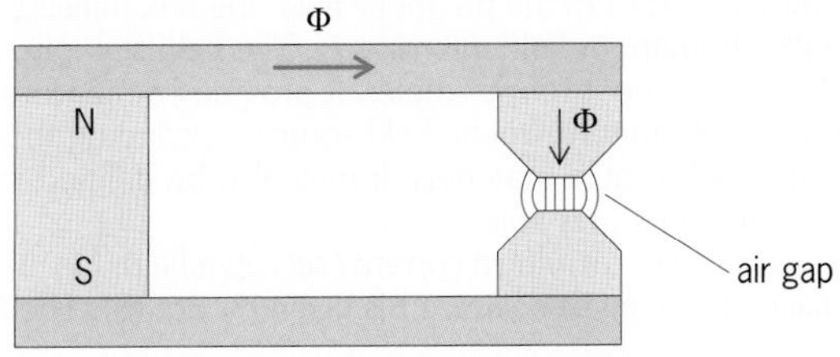

Fig. 2. Magnetic circuit with an air gap.

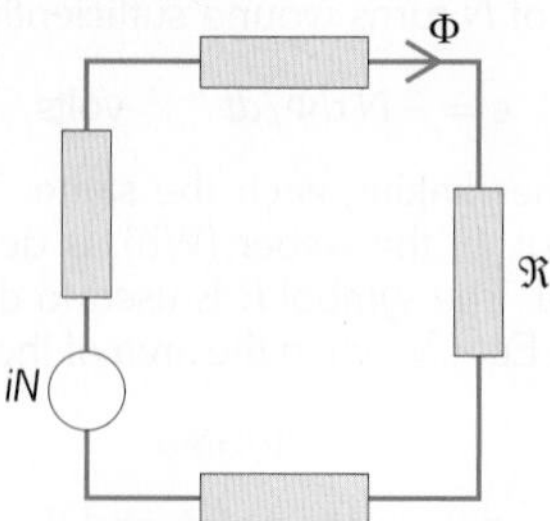

Fig. 3. Circuit analogy. Components of a magnetic circuit carrying a flux Φ analogous to current. The reluctances of the components are analogous to resistance.

dividing the material into thin laminations. *See* EDDY CURRENT; TRANSFORMER.

The amount of flux produced by a given supply current is reduced by the presence of any air gaps which may be introduced to contribute constructional convenience or to allow a part to move. The effects of the gaps, and of different magnetic materials, can be predicted by utilizing the analogy between flux, Φ, and the flow of electric current through a circuit consisting of resistors connected in series (Fig. 3). Since Φ depends on the product, iN, of the winding current and number of turns, as in Eq. (3), the ratio between them, termed the reluctance, $\Re$, is

$$iN = \Phi\Re \tag{3}$$

the analog of electrical resistance. It may be constant or may vary with Φ. The quantity iN is the magnetomotive force (mmf), analogous to voltage or emf in the equivalent electrical circuit. The relationship between the two exchanges the potental and flow quantities, since the magnetic mmf depends on current, i, and the electrical emf on $d\Phi/dt$. Electric and magnetic equivalent circuits are referred to as duals. *See* RELUCTANCE.

Any part of the magnetic circuit of length l, in which the cross section, a, and flux density, B, are uniform has a reluctance given by Eq. (4). This equation parallels Eq. (5) for the resistance, R,

$$\Re = l/(a\mu) \tag{4}$$

$$R = l/(a\sigma) \tag{5}$$

of a conduct of the same dimensions. The permeability, μ, is the magnetic equivalent of the conductivity, σ, of the conducting material. Using a magnet as a flux source (Fig. 2) gives an mmf which varies with the air gap reluctance. In the absence of any magnetizable materials, as in the air gaps, the permeability is given by Eq. (6) in SI units (Wb/A-m). The quantity μ_0 is

$$\mu = \mu_0 = 4\pi \times 10^{-7} \tag{6}$$

sometimes referred to as the permeability of free space. Material properties are described by the relative permeability, μ_r in accordance with Eq. (7). The materials which are important in

$$\mu = \mu_r\mu_0 \tag{7}$$

magnetic circuits are the ferromagnetics and ferrites characterized by large value of μ_r, sometimes in excess of 10,000 at low flux densities.

Magnetic field strength. It is convenient to introduce two different measures of the magnetic field: the flux density, B, and the field strength, or field intensity, H. The field strength, H, can be defined as the mmf per meter. It provides a measure of the currents and other magnetic field sources, excluding those representing polarizable materials. It may also be defined in terms of the force on a unit pole.

A straight wire carrying a current I sets up a field (Fig. 4) whose intensity at a point at distance r is given by Eq. (8). The field of

$$H = \frac{I}{2\pi r} \tag{8}$$

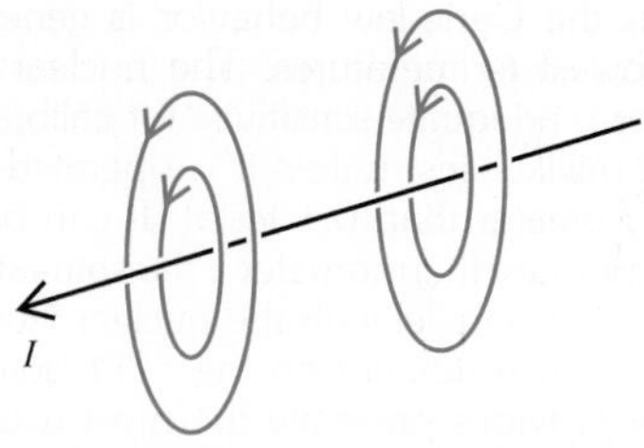

Fig. 4. Magnetic field of a straight wire.

strength, H, like B, is a vector quantity pointing in the direction of rotation of a right-hand screw advancing in the direction of current flow. The intensity of the field is shown by the number of field lines intersecting a unit area. The straight wire provides one example of the circuital law, known as Ampère's law, given by Eq. (9). Here, θ is the angle between H and the element dl any

$$\oint H\cos\theta\, dl = I \tag{9}$$

closed path of summation, or integration, and I is the current which links this path. Choosing a circular path, centered on a straight wire, reduces the integral to $H(2\pi r)$.

A long, straight, uniformly wound coil (Fig. 5), for example, produces a field which is uniform in the interior and zero outside. The interior magnetic field, H, points in the direction parallel to the coil axis. Applying Eq. (14) to the rectangle $pqrs$ of unit length in the axial direction shows that the only contribution is from pq, giving Eq. (10), where n is the number of turns, per unit length,

$$H = In \tag{10}$$

carrying the current, I. The magnetic field strength, H, remains the same, by definition, whether the interior of the coil is empty or is filled with ferromagnetic material of uniform properties. The interior forms part of a magnetic circuit in which In is the mmf per unit length, where mmf is the magnetic analog of electric voltage, or scalar potential, in an electric circuit. The magnetic field strength, H, is the analog of the electric field vector, E, as a measure of potential gradient, pointing down the gradient. The flux density, B, describes the effect of the field, in the sense of the voltage which is induced in a search coil by changes in time [Eq. (1)]. The ratio of H to B is the reluctance of a volume element of unit length and unit cross section in which the field is uniform, so that, from Eq. (4), the two quantities are related by Eq. (11). The permeability, μ, is defined by Eq. (11). The

$$B = \mu H \tag{11}$$

relative permeability, μ_r, of polarizable materials is measured accordingly by subjecting a sample to a uniform field inside a

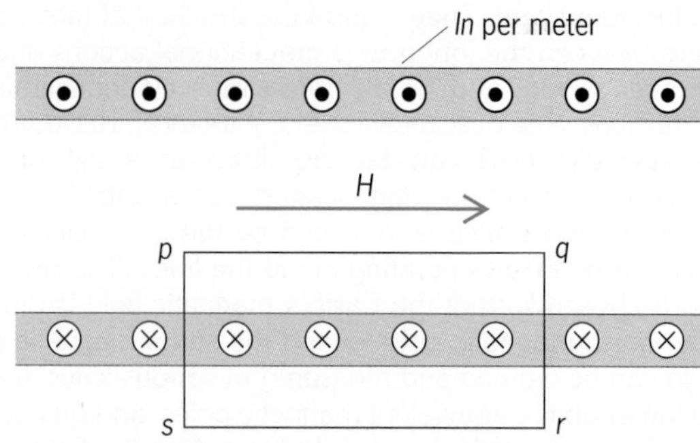

Fig. 5. Cross section of part of a long, straight uniformly wound coil. Black circles indicate current out of the plane of the page, and crosses indicate current into the plane of the page. Rectangle *pqrs* is used to calculate magnetic field strength, *H*, within the coil.

long coil such as that shown in Fig. 5 and using the emf induced in a search coil wound around the specimen to observe the flux in it.

Magnetic flux and flux density. Magnetic flux is defined in terms of the forces exerted by the magnetic field on electric charge. The forces can be described in terms of changes in flux with time [Eq. (1)], caused either by motion relative to the source or by changes in the source current, describing the effect of charge acceleration.

Since the magnetic, or electrokinetic, energy of current flowing in parallel wires depends on their spacing, the wires are subject to forces tending to change the configuration. The force, dF, on an element of wire carrying a current, i, is given by Eq. (12),

$$dF = Bi\,dl \qquad \text{newtons} \tag{12}$$

and this provides a definition of the flux density, B, due to the wires which exert the force. The SI unit of B, called the tesla, or Wb/m^2, is the N/A-m. The flux density, B, equals $\mu_0 H$ in empty space, or in any material which is not magnetizable [Eq. (11)]. An example is the force, F, per meter (length) which is exerted by a long straight wire on another which is parallel to it, at distance r. From Eq. (8), this force is given by Eq. (13), when the wires carry

$$F = \frac{\mu_0 I i}{2\pi r} = 2 \times 10^{-7}\, Ii/r \qquad \text{newtons} \tag{13}$$

currents I and i. The force, F, is accounted for by the electrokinetic interactions between the conduction charges, and describes the relativistic modification of the electric forces between them due to their relative motion.

In general, any charge, q, moving at velocity u is subject to a force given by Eq. (14), where $\mathbf{u} \times \mathbf{B}$ denotes the cross-product

$$\mathbf{f} = q\,\mathbf{u} \times \mathbf{B} \qquad \text{newtons} \tag{14}$$

between vector quantities. That is, the magnitude of $\mathbf{f}$ depends on the sine of the angle θ between the vectors $\mathbf{u}$ and $\mathbf{B}$, of magnitudes u and B, according to Eq. (15). The force on a positive

$$f = quB \sin\theta \tag{15}$$

charge is at right angles to the plane containing $\mathbf{u}$ and $\mathbf{B}$ and points in the direction of a right-hand screw turned from $\mathbf{u}$ to $\mathbf{B}$.

The same force also acts in the axial direction on the conduction electrons in a wire moving in a magnetic field, and this force generates an emf in the wire. The emf in an element of wire of length dl is greatest when the wire is at right angles to the $\mathbf{B}$ vector, and the motion is at right angles to both. The emf is then given by Eq. (16). More generally, u is the component

$$\text{emf} = uB\,dl \tag{16}$$

of velocity normal to $\mathbf{B}$, and the emf depends on the sine of the angle between $\mathbf{dl}$ and the plane containing the velocity and the $\mathbf{B}$ vectors. The sign is given by the right-hand screw rule, as applied to Eq. (15).

Magnetic flux linkage. The magnetic flux linking any closed path is obtained by counting the number of flux lines passing through any surface, s, which is bounded by the path. Stated more formally, the linkage depends on the sum given in Eq. (17), where B_n denotes the component of B in the direction

$$\Phi = \int\!\!\int B_n\,ds \qquad \text{webers} \tag{17}$$

normal to the area element, ds. The rate of change of linkage gives the emf induced in any conducting wire which follows the path [Eq. (1)].

The flux linkage with a coil (Fig. 6) is usually calculated by assuming that each turn of the coil closes on itself, giving a flux pattern which likewise consists of a large number of separate closed loops. Each links some of the turns, so that the two cannot be separated without breaking, or "tearing," either the loop or the turn. The total linkage with the coil is then obtained by adding the contributions from each turn.

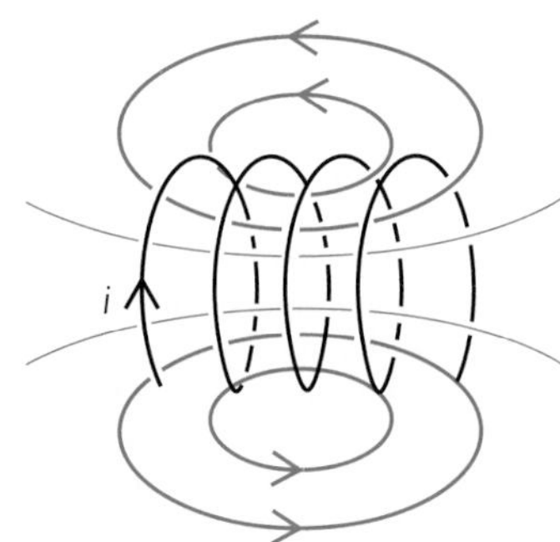

Fig. 6. ***B*** **(magnetic flux) field of a short coil.**

The inductance, L, is a property of a circuit defined by the emf which is induced by changes of current in time, as in Eq. (18).

$$e = -L\,di/dt \tag{18}$$

The SI unit of inductance is the henry (H), or V-s/A. The negative sign shows that e opposes an increase in current (Lenz's law). From Eq. (1) the inductance of a coil of N turns, each linking the same flux, Φ, is given by Eq. (19), so that the henry is also the

$$L = N\Phi/i \qquad \text{henrys} \tag{19}$$

Wb/A. When different turns, or different parts of a circuit, do not link the same flux, the product $N\Phi$ is replaced by the total flux linkage, Φ, with the circuit as a whole.

The mutual inductance, M, between any two coils, or circuit parts, is defined by emf which is induced in one by a change of current in the other. Using 1 and 2 to distinguish between them, the emf induced in coil 1 is given by Eq. (20*a*), where

$$e_1 = -M_{12}\,di_2/dt \tag{20a}$$

$$e_2 = -M_{21}\,di_1/dt \tag{20b}$$

the sign convention is consistent with that used for L, referred to as the self-inductance. Likewise, the emf induced in coil 2 when the roles of the windings are reversed is given by Eq. (20*b*). The interaction satisfies the reciprocity condition of Eq. (21), so that the suffixes may be omitted.

$$M_{21} = M_{12} \tag{21}$$

Magnetostatics. The term "magnetostatics" is usually interpreted as the magnet equivalent of the electrostatic interactions between electric charges. The equivalence is described

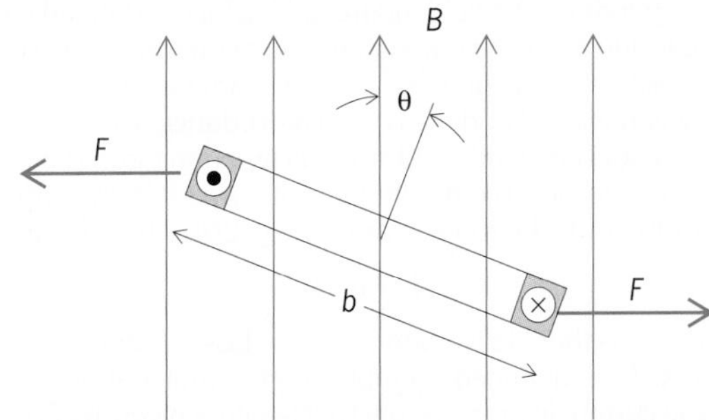

Fig. 7. Cross section of a rectangular current loop placed in magnetic flux density *B*. Equal but opposite forces, *F*, act on opposite sides of the loop carrying current into the plane of the page (indicated by a cross) and out of the plane of the page (indicated by black circle). These forces give rise to a torque on the loop.

most directly in terms of the magnetic pole, since the forces between poles, like those between charges, vary inversely with the square of the separation distance. Although no isolated poles, or monopoles, have yet been observed, the forces which act on both magnets and on coils are consistent with the assumption that the end surfaces are equivalent to magnetic poles.

Magnetic moment. The magnetic moment of a small current loop, or magnet, can be defined in terms of the torque which acts on it when placed in a magnetic flux density, B, which is sufficiently uniform in the region of the loop. For a rectangular loop with dimensions a and b and with N turns, carrying a current, i, equal but opposite forces act on the opposite sides of length a (Fig. 7). The force is $iNBa$ [Eq. (12)], and the torque, given by Eq. (22), depends on the effective distance, $b \sin\theta$, between

$$T = iNBa\,b \sin\theta \quad \text{N-m} \tag{22}$$

the wires. It is proportional to the area ab, and is a maximum when the angle θ between B and the axis of the loop is 90°. A current loop of any other shape can be replaced by a set of smaller rectangles placed edge to edge, and the torques of these added to give the total on the loop. The magnetic moment of any loop of area s is defined as the ratio of the maximum torque to the flux density, so its magnitude is given by Eq. (23). It is a

$$m = iNs \tag{23}$$

vector quantity pointing in the direction of a right-hand screw turned in the direction of current flow. It is expressed in vector cross-product notation by Eq. (24).

$$\mathbf{T} = \mathbf{m} \times \mathbf{B} \tag{24}$$

See TORQUE.

An electron orbiting at frequency f is the equivalent of a current $i = qf$, giving Eq. (25) for the moment, where s is the area

$$m_0 = qfs \tag{25}$$

of the orbit. The permissible values are determined by the quantum energy levels. The electron spin is a quantum state which can likewise be visualized as a small current loop. Atomic nuclei also possess magnetic moments. *See* ELECTRON SPIN; MAGNETON; NUCLEAR MOMENTS.

Magnetic polarization. Materials are described as magnetic when their response to the magnetic field controls the ratio of B to H. The behavior is accounted for by the magnetic moments produced mainly by the electron spins and orbital motions. These respond to the field and contribute to it in a process referred to as magnetic polarization. The effects are greatest in ferromagnetics and in ferrites, in which the action is described as ferrimagnetic. *See* FERRIMAGNETISM; FERRITE; FERROMAGNETISM.

The sources are the equivalent of miniature "Ampèrean currents" whose sum, in any volume element, is equivalent to a loop of current flowing along the surface of the element. The flux density, B, depends on the field intensity, H, which is defined so that its value inside a long ferromagnetic rod of uniform cross section placed inside a long coil (Fig. 5) is the same as in the annular gap between the rod and the coil, in accordance with Eq. (10). If the field is not sufficiently uniform, H can be measured by using a search coil to observe the flux density, $\mu_0 H$, in the gap. The flux density inside the rod is given by Eq. (26), where B_0 denotes

$$B = \mu_r B_0 \tag{26}$$

$\mu_0 H$, and μ_r is the relative permeability [Eqs. (7) and (11)]. The same flux, B, is obtained by replacing the material by a coil in which the current in amperes per unit length is given by Eq. (27).

$$J_s = (B - B_0)/\mu_0 = (1 - 1/\mu_r)B/\mu_0 \tag{27}$$

The magnetic moment, dm, of a volume element of length dz is due to the current flowing over the surface enclosing the area, $dydz$; from Eq. (23), it is given by Eq. (28). The moment per unit volume defines the magnetic polarization, as in Eq. (29).

$$dm = (J_s\,dx)\,dy\,dz \tag{28}$$

$$M = dm/dx\,dy\,dz \tag{29}$$

The polarization, $\mathbf{M}$, is a vector pointing in the direction of $\mathbf{dm}$ with magnitude J_s. The surface current produces an $\mathbf{H}$-like, or $\mathbf{B}/\mu_0$, field which is entirely different from $\mathbf{H}$ in the material. Substituting from Eq. (27) gives Eq. (30). This model of the

$$\mathbf{H} = \frac{\mathbf{B}}{\mu_0} - \mathbf{M} \tag{30}$$

material accounts for the flux field, $\mathbf{B}$, as observed by the voltage induced in a search coil wound around the specimen, and $\mathbf{H}$, becomes an auxiliary quantity representing the sum of the polarization, $\mathbf{M}$, and the magnetizing field, $\mathbf{B}\mu_0$, to which $\mathbf{M}$ responds. The polarization, $\mathbf{M}$, also makes the largest contribution to that field, since the equivalent surface current is in the same direction as the current in the magnetizing coil.

Magnetic hysteresis. The relationship between the flux density, B, and the field intensity, H, in ferromagnetic materials depends on the past history of magnetization. The effect is known as hysteresis. It is demonstrated by subjecting the material to a symmetrical cycle of change during which H is varied continuously between the positive and negative limits $+H_m$ and $-H_m$ (Fig. 8). The path that is traced by repeating the cycle a sufficient number of times is the hysteresis loop. The sequence is counterclockwise, so that B is larger when H is diminishing than when it is increasing, in the region of positive H. The flux density, B_r, which is left when H falls to zero is called the remanence, or retentivity. The magnetically "hard" materials used for permanent magnets are characterised by a high B_r, together with a high value of the field strength, $-H_c$, which is needed to reduce B to zero. The field strength, H_c, is known as the coercive force, or coercivity. Cycling the material over a reduced range in H gives the path in Fig. 8 traced by the broken line, lying inside the larger

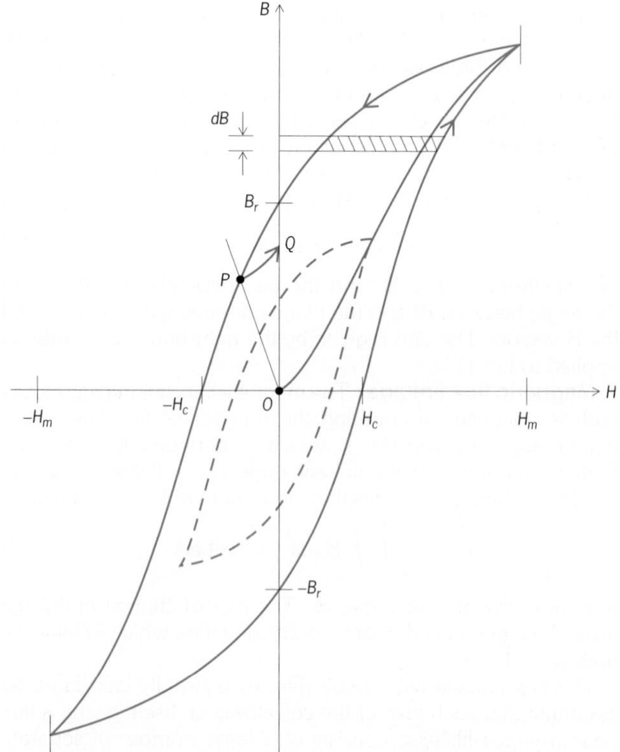

Fig. 8. Hysteresis behavior of a ferromagnetic material.

loop. The locus of the tips of such loops is known as the normal magnetization curve. The initial magnetization curve is the *B-H* relationship which is followed when *H* is progressively increased in one direction after the material has first been demagnetized ($B = H = 0$). C. J. Carpenter

Magnetite A cubic mineral and member of the spinel structure type with composition $[Fe^{3+}]^{IV}[Fe^{2+}Fe^{3+}]^{VI}O_4$. The color is opaque iron-black and streak black, the hardness is 6 (Mohs scale), and the specific gravity is 5.20. The habit is octahedral, but the mineral usually occurs in granular to massive form, sometimes of enormous dimensions. Magnetite is a natural ferrimagnet, but heated above 1072°F (578°C; the Curie temperature) it becomes paramagnetic.

The major magnetic ore of iron, magnetite may be economically important if it occurs in sufficient quantities. The most spectacular ore body occurs at Kiruna in northern Sweden. Other important occurrences are in Norway, Russia, and Canada. *See* IRON; SPINEL. Paul B. Moore

Magnetization The process of becoming magnetized; also the property and in particular the extent of being magnetized. Magnetization has an effect on many of the physical properties of a substance. Among these are electrical resistance, specific heat, and elastic strain. *See* MAGNETOCALORIC EFFECT; MAGNETORESISTANCE; MAGNETOSTRICTION.

The magnetization **M** of a body is caused by circulating electric currents or by elementary atomic magnetic moments, and is defined as the magnetic moment per unit volume of such currents or moments. In the mks (SI) system of units, **M** is measured in webers per square meter. For **M**, 1 weber/m^2 = $10^4/4\pi$ gauss.

The magnetic induction or magnetic flux density **B** is given by the equation below, where **B** and **M** are in webers/m^2, **H**, the

$$\mathbf{B} = \mu_0\mathbf{H} + \mathbf{M} \text{ (mks)}$$

applied magnetic field, is in ampere-turns/m, and μ_0, the permeability of free space, is defined as $4\pi \times 10^{-7}$ henry/m, that is, webers/(ampere-turn)(m). *See* ELECTRICAL UNITS AND STANDARDS.

The topic of magnetization is generally restricted to materials exhibiting spontaneous magnetization, that is, magnetization in the absence of **H**. All such materials will be referred to as ferromagnets, including the special category of ferrimagnets. A ferromagnet is composed of an assemblage of spontaneously magnetized regions called domains. Within each domain, the elementary atomic magnetic moments are essentially aligned, that is, each domain may be envisioned as a small magnet. An unmagnetized ferromagnet is composed of numerous domains, oriented in some fashion.

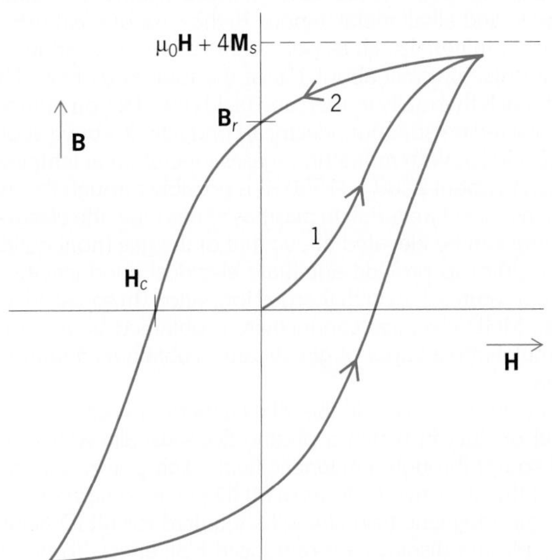

Magnetization or *B-H* curves.

The process of magnetization in an applied field **H** consists of growth of those domains oriented most nearly in the direction of **H** at the expense of others, followed by rotation of the direction of magnetization against anisotropy forces. *See* FERROMAGNETISM.

On removal of the field **H**, some magnetization will remain, called the remanence $\mathbf{M}_r$.

Curves, sometimes called *B-H* curves, are used to describe magnetic materials. They are plotted with **H** as abscissa and with either **M** or **B** as ordinate. In the illustration, $\mathbf{B}_r$ is the remanent induction ($\mathbf{B}_r = \mathbf{M}_r$); $\mathbf{H}_c$ is the coercive force, or reverse field required to bring the induction **B** back to zero; and $\mathbf{M}_s$ is the saturation magnetization, or magnetization when all domains are aligned. The saturation magnetization is equal to the spontaneous magnetization of a single domain, except that it is possible to increase this magnetization slightly by application of an extremely large field. Saturation magnetization is temperature dependent, and disappears completely above the Curie temperature T_c where a ferromagnet changes into a paramagnet. *See* CURIE TEMPERATURE.

The irreversible nature of magnetization is shown most strikingly by the fact that the path of demagnetization does not retrace the path of magnetization—path 2 of the illustration does not retrace path 1. There is a tendency for the magnetization to show hysteresis, that is, to lag behind the applied field, and the loop of the illustration is called a hysteresis loop. Elihu Abrahams; Frederic Keffer; Kenneth V. Manning

Magneto A type of permanent-magnet alternating-current generator frequently used as a source of ignition energy on tractor, marine, industrial, and aviation engines. *See* ALTERNATING-CURRENT GENERATOR.

Modern induction-type magnetos consist of a permanent-magnet rotor and stationary low- and high-tension windings, also called the primary and secondary windings. The energy output of a magneto is obtained as a result of a rapid rate of change of flux through the stationary windings. The primary winding has comparatively few turns and the secondary winding has many thousand turns of fine wire. One end of the secondary winding is connected to an end of the primary winding and grounded to the frame of the magneto. The primary winding is closed on itself through a breaker mechanism actuated by a cam on the magneto shaft. The breaker is mechanically set to interrupt the primary circuit each time the flux through the winding is changing at its greatest rate. The sudden collapse of the primary current induces a very high voltage in the secondary winding. *See* IGNITION SYSTEM; INTERNAL COMBUSTION ENGINE. Robert T. Weil, Jr.

Magnetocaloric effect The reversible change of temperature accompanying the change of magnetization of a ferromagnetic or paramagnetic material. This change in temperature may be of the order of 1°C (2°F), and is not to be confused with the much smaller hysteresis heating effect, which is irreversible. *See* THERMAL HYSTERESIS. Elihu Abrahams; Frederic Keffer

Magnetochemistry The study of magnetic properties of materials by analyzing the interactions between a magnetic field and chemical substances, which include both molecular and extended structures. All substances respond to a magnetic field. Some materials, called permanent magnets, generate magnetic fields.

Magnetism originates in matter from the intrinsic spin of nuclei (protons and neutrons) and electrons, as well as from the orbital motion of electrons. The spins of protons, neutrons, and electrons create magnetic moments. The magnitudes of these moments are related inversely to their masses, so the electronic magnetic moment is about 2000 times larger than that of a proton or neutron. Therefore, macroscopic magnetic behavior arises essentially from electronic properties. The magnetism associated with nuclei is used in nuclear magnetic resonance (NMR) and is very useful for structural information and analysis. *See* ATOMIC STRUCTURE AND SPECTRA; MAGNETIC RESONANCE; NUCLEAR MAGNETIC RESONANCE (NMR); NUCLEAR MOMENTS.

The orbital motion of electrons can also create a magnetic moment if the orbital angular momentum is not quenched. The combination of electron spin and orbital momenta produces a local magnetic moment assigned to each atom in a chemical structure. These local moments can interact with one another and with an external magnetic field, leading to the substance's magnetic behavior and response. If the orbital angular momentum for an atom is quenched, which is often the case, then the local moment comes entirely from the total electron spin. If all electron spins are paired, an atom or ion has no local magnetic moment. Therefore, local magnetic moments primarily arise from unpaired electrons. *See* ELECTRON SPIN; MAGNETISM; MOLECULAR ORBITAL THEORY.

The nature of a material's response to an external magnetic field essentially depends on the number of unpaired electrons. In an external magnetic field, all electrons, as charged particles, will develop trajectories that create an induced magnetic field opposing the applied field, according to Lenz's law. The presence of unpaired electron spins will create local moments that counteract this induced field. If these local moments interact strongly with one another within the material in the absence of the external field, there can be cooperative behavior. *See* LENZ'S LAW.

Diamagnetism is a property of a material which is repelled by an external magnetic field due to the induced motion of electrons. All electron spins are paired giving a net spin of zero. This is the most common behavior for main-group molecules and solids. The susceptibility is independent of temperature and field strength. *See* DIAMAGNETISM.

Paramagnetism is the property of a material which is attracted by an external magnetic field. There are unpaired electrons, which are commonly observed for transition-metal and rare-earth metal complexes. The susceptibility is inversely proportional to temperature but independent of field strength. *See* PARAMAGNETISM.

Ferromagnetism is exhibited by materials that are strongly attracted by an external magnetic field and can produce a permanent magnet. There are unpaired electron spins that are held in parallel alignment between adjacent magnetic atoms. Ferromagnets show paramagnetic behavior above a critical temperature called the Curie temperature. *See* CURIE TEMPERATURE; FERROMAGNETISM.

Antiferromagnetism is exhibited by materials that are weakly attracted by an external magnetic field because there are unpaired electrons that are held in antiparallel alignment between different atoms, and there is no net magnetization. If the external field is oriented parallel to the direction of these local moments (called the easy axis of magnetization), the material would be repelled by the field. Antiferromagnets show paramagnetic behavior above a critical temperature called the Neél temperature. *See* ANTIFERROMAGNETISM.

Ferrimagnetism is a property of materials that are attracted by an external magnetic field as there are unpaired electrons with antiparallel alignment, but there is an overall net magnetization. Ferrimagnets show paramagnetic behavior above a critical temperature. *See* FERRIMAGNETISM.

Superconductors are "perfect" diamagnets with an immeasurably small electrical resistivity. Electrons are coupled into Cooper pairs, and the material completely repels an external magnetic field. Superconductors adopt "normal" behavior above a critical temperature. *See* SUPERCONDUCTIVITY.

Molecular oxygen (O_2) has two unpaired electrons in its ground state and is paramagnetic. Other paramagnetic molecules include NO, NO_2, ClO_2, and ClO_3.

Most organic compounds are diamagnetic, but there are some, called free radicals, with one or two unpaired electrons. Aromaticity assists to stabilize the unpaired electrons. *See* FREE RADICAL.

Metals typically show Pauli paramagnetism. In the presence of an external magnetic field, the energy bands associated with spin-up electrons are separated from the energy bands of the spin-down electrons (Zeeman interaction), resulting in more spin-up than spin-down electrons and an induced, temperature-independent magnetization. This effect is observed for sodium, magnesium, and aluminum (Al becomes superconducting at low temperatures). In poor metals (for example, graphite) and semimetals (such as antimony and bismuth), the induced magnetic field from the trajectories of the conduction electrons in the external magnetic field create a net diamagnetic response called Landau diamagnetism. Some metallic elements, such as iron, cobalt, nickel, gadolinium, and dysprosium, are ferromagnetic, which means they show net magnetization in the absence of a magnetic field. A few elements are antiferromagnetic, such as manganese and chromium. Recent investigations with possible applications to magnetic storage and retrieval involve half-metallic ferromagnets, in which only the spin-up electrons contribute to electronic conduction. Gordon J. Miller

Magnetohydrodynamic power generator A system for the generation of electrical power through the interaction of a flowing, electrically conducting fluid with a magnetic field. As in a conventional electrical generator, the Faraday principle of motional induction is employed, but solid conductors are replaced by an electrically conducting fluid. The interactions between this conducting fluid and the electromagnetic field system through which power is delivered to a circuit are determined by the magnetohydrodynamic (MHD) equations, while the properties of electrically conducting gases or plasmas are established from the appropriate relationships of plasma physics. Major emphasis has been placed on MHD systems utilizing an ionized gas, but an electrically conducting liquid or a two-phase flow can also be employed. *See* ELECTROMAGNETIC INDUCTION; GENERATOR; MAGNETOHYDRODYNAMICS; PLASMA (PHYSICS).

Electrical conductivity in an MHD generator can be achieved in a number of ways. At the heat-source operating temperatures of MHD systems (1300–5000°F or 1000–3000 K), the working fluids usually considered are gases derived from combustion, noble gases, and alkali metal vapors. In the case of combustion gases, a seed material such as potassium carbonate is added in small amounts, typically about 1% of the total mass flow. The seed material is thermally ionized and yields the electron number density required for adequate electrical conductivity above about 4000°F (2500 K). With monatomic gases, operation at temperatures down to about 2200°F (1500 K) is possible through the use of cesium as a seed material. In plasmas of this type, the electron temperature can be elevated above that of the gas (nonequilibrium ionization) to provide adequate electrical conductivity at lower temperatures than with thermal ionization. In so-called liquid metal, MHD electrical conductivity is obtained by injecting a liquid metal into a vapor or gas stream to obtain a continuous liquid phase.

The conversion process in the MHD generator itself occurs in a channel or duct in which a plasma flows usually above the speed of sound through a magnetic field. High power densities are one of the attractive features of MHD power generators.

Under the magnetic field strengths required for MHD generators, the plasma displays a pronounced Hall effect. To permit the basic Faraday motional induction interaction and simultaneously support the resulting Hall potential in the flow direction, a linear channel requires segmented walls comprising alternately electrodes (anode or cathode) and insulators. From an electrical machine viewpoint, both individual cells and the complete generator may be regarded as a gyrator. The optimum loading of the MHD channel is achieved by extracting power from both the Faraday and Hall terminals, and this is most readily accomplished through consolidation of the dc outputs of individual electrode pairs using power electronics. *See* GYRATOR; HALL EFFECT.

For most applications, a superconducting magnet system is needed to provide the 4–6-T field, which is at least twice the value utilized in conventional machines. *See* MAGNET; SUPERCONDUCTING DEVICES.

Improvement of the overall thermal efficiency of central station power plants has been the continuing objective of power engineers. Conventional plants based on steam turbine technology are limited to about 40% efficiency, imposed by a combination of working-fluid properties and limits on the operating temperatures of materials. When combined with a steam turbine system to serve as the high-temperature or topping stage of a binary cycle, an MHD generator has the potential for increasing the overall plant thermal efficiency to around 50%, and values higher than 60% have been predicted for advanced systems. *See* ELECTRIC POWER GENERATION; STEAM TURBINE.

MHD power generation also has important potential environmental advantages. These are of special significance when coal is the primary fuel, for it appears that MHD systems can utilize coal directly without the cost and loss of efficiency resulting from the processing of coal into a clean fuel required by competing systems. William D. Jackson

Magnetohydrodynamics The interaction of electrically conducting fluids with magnetic fields. The fluids can be ionized gases (commonly called plasmas) or liquid metals. Magnetohydrodynamic (MHD) phenomena occur naturally in the Earth's interior, constituting the dynamo that produces the Earth's magnetic field; in the magnetosphere that surrounds the Earth; and in the Sun and throughout astrophysics. In the laboratory, magnetohydrodynamics is important in the magnetic confinement of plasmas in experiments on controlled thermonuclear fusion. Magnetohydrodynamic principles are also used in plasma accelerators for ion thrusters for spacecraft propulsion, for light-ion-beam powered inertial confinement, and for magnetohydrodynamic power generation. *See* COSMIC RAYS; GEOMAGNETISM; ION PROPULSION; MAGNETOHYDRODYNAMIC POWER GENERATOR; MAGNETOSPHERE; NUCLEAR FUSION; PLASMA (PHYSICS); SOLAR WIND; SUN.

The conducting fluid and magnetic field interact through electric currents that flow in the fluid. The currents are induced as the conducting fluid moves across the magnetic field lines. In turn, the currents influence both the magnetic field and the motion of the fluid. Qualitatively, the magnetohydrodynamic interactions tend to link the fluid and the field lines so as to make them move together. *See* ELECTRIC CURRENT; MAGNETIC FIELD.

The generation of the currents and their subsequent effects are governed by the familiar laws of electricity and magnetism. The motion of a conductor across magnetic lines of force causes a voltage drop or electric field at right angles to the direction of the motion and the field lines; the induced voltage drop causes a current to flow as in the armature of a generator.

The currents themselves create magnetic fields which tend to loop around each current element. The currents heat the conductor and also give rise to mechanical ponderomotive forces when flowing across a magnetic field. (These are the forces which cause the armature of an electric motor to turn.) In a fluid, the ponderomotive forces combine with the pressure forces to determine the fluid motion. *See* ELECTRICITY; GENERATOR; MAGNETISM; MOTOR.

Magnetohydrodynamic phenomena involve two well-known branches of physics, electrodynamics and hydrodynamics, with some modifications to account for their interplay. The basic laws of electrodynamics as formulated by J. C. Maxwell apply without any change. However, Ohm's law, which relates the current flow to the induced voltage, has to be modified for a moving conductor. *See* ELECTRODYNAMICS; HYDRODYNAMICS; MAXWELL'S EQUATIONS; OHM'S LAW.

It is useful to consider first the extreme case of a fluid with a very large electrical conductivity. Maxwell's equations predict, according to H. Alfvén, that for a fluid of this kind the lines of the magnetic field move with the material. The picture of moving lines of force is convenient but must be used with care because such a motion is not observable. It may be defined, however, in terms of observable consequences by either of the following statements: (1) a line moving with the fluid, which is initially a line of force, will remain one; or (2) the magnetic flux through a closed loop moving with the fluid remains unchanged.

If the conductivity is low, this is not true and the fluid and the field lines slip across each other. This is similar to a diffusion of two gases across one another and is governed by similar mathematical laws.

As in ordinary hydrodynamics, the dynamics of the fluid obeys theorems expressing the conservation of mass, momentum, and energy. These theorems treat the fluid as a continuum. This is justified if the mean free path of the individual particles is much shorter than the distances that characterize the structure of the flow. Although this assumption does not generally hold for plasmas, one can gain much insight into magnetohydrodynamics from the continuum approximation. The ordinary laws of hydrodynamics can then easily be extended to cover the effect of magnetic and electric fields on the fluid by adding a magnetic force to the momentum-conservation equation and electric heating and work to the energy-conservation equation. Malcolm G. Haines

Magnetometer An instrument that measures the magnitude and/or direction of a magnetic field. A magnetometer can be either a scalar instrument that measures the magnitude of the total field or a vector instrument that measures one or more vector components of the field. Some magnetometers are relative devices that are calibrated with respect to a known field. Others are absolute devices that yield magnetic-field values without calibration. Arrays of magnetometers can be configured as gradiometers that suppress or enhance far-field magnetic sources while enhancing or suppressing near-field sources. Magnetometer technology is available for detecting a range of magnetic fields.

Perhaps the simplest magnetometer is an induction coil that employs the Faraday effect, a voltage induced in a conductor by a time-varying magnetic flux. Induction magnetometers are generally multiturn coils that are either air-wound or wound around a ferromagnetic core to increase flux through the coil. Fixed coils can measure only time-varying fields, but static fields can be measured using moving induction coils that rotate, vibrate, or deform. Induction coils are usually vector sensors that require external calibration. *See* FARADAY EFFECT.

Hall-effect sensors are made from InSb or InAs alloy thin films produced by vacuum deposition or molecular-beam epitaxy. These solid-state vector devices can be fabricated as integrated circuit packages that can measure alternating-current and direct-current magnetic fields. By combining multiple calibrated sensors into a single package, the total magnetic field and magnetic-field gradients can be measured. Hall-effect magnetometers are usually used to measure relatively strong magnetic fields. *See* HALL EFFECT; MAGNETIC FIELD.

The fluxgate (saturable-core) sensor is constructed from a pair of high-magnetic-permeability cores. Wound around each core in opposite directions is a primary field core driven to magnetic saturation by an audio-frequency current. The primary coils are connected in series. A secondary field core wound collectively around both cores is connected to a detector circuit. All fluxgate magnetometers are relative vector instruments that require calibration in a known magnetic field to produce accurate results. Orthogonal sets of fluxgate sensors can be used to measure all three field components and thereby the total field vector.

Two general classes of resonance magnetometers are proton precession magnetometers and optically pumped magnetometers. Both are absolute instruments that measure total field strength without the need for calibration using a known magnetic field.

The cryogenic or SQUID magnetometer uses one or more Josephson junctions as a magnetic-field sensor. A Josephson junction is a zone of weak magnetic coupling (a weak link) between two regions of superconducting material in which current will flow without resistance. A change in the magnetic field applied to the weak link produces a proportional change in magnetic flux within the Josephson junction. The SQUID is the most sensitive magnetometer in general use, capable of measuring flux changes only a small fraction of a flux quantum Φ (2.07 $\times\ 10^{-15}$ weber), but its capabilities are now rivaled by modern spin-polarized alkali vapor magnetometers. *See* JOSEPHSON EFFECT; SUPERCONDUCTIVITY.

Two types of SQUID are in common use. The radio-frequency SQUID employs a single weak link, whereas the direct-current SQUID uses a pair of Josephson junctions. All SQUID magnetometers are relative, vector instruments. The principal advantages of the SQUID magnetometer over proton, optically pumped, and fluxgate magnetometers are sensitivity and frequency response. The principal disadvantage of the SQUID magnetometer is that it must be kept in a superconducting state. *See* SQUID. Michael McWilliams

Magneton A unit of magnetic moment used to describe atomic, molecular, or nuclear magnets. More precisely, one unit, the Bohr magneton, is used at the atomic and molecular levels, and another unit, the nuclear magneton, is used at the nuclear level. Still another unit (which might be called the muon magneton, but is usually not named) is used to describe the magnetic moment of the muon. *See* MAGNETIC MOMENT.

The Bohr magneton μ_B is defined and its value given in Eq. (1), where $-e$ and m are the charge and mass of the electron respectively and $\hbar$ is Planck's constant divided by 2π.

$$\mu_B = \frac{e\hbar}{2m} = (9.274\,009\,15 \pm 0.000\,000\,23) \times 10^{-24}\ \text{joule/tesla} \qquad (1)$$

In Dirac's theory the magnetic moment of the electron is exactly $-\mu_B$, but according to the theory of quantum electrodynamics the electron has a small anomalous magnetic moment. The experimental value of the electron magnetic moment μ_e is given by Eq. (2), in agreement with the prediction of quantum electrodynamics within the errors.

$$\mu_e = -(1.001\,159\,652\,188\,11 \pm 0.000\,000\,000\,000\,74)\mu_B \qquad (2)$$

The unit of magnetic moment to describe the muon is obtained from the Bohr magneton by replacing m in Eq. (1) by the muon mass m_μ. The experimental value of the muon magnetic moment is given in Eq. (3). The deviation of the muon magnetic moment

$$\mu_\mu = (1.001\,165\,920\,67 \pm 0.000\,000\,000\,60)\frac{e\hbar}{2m_\mu} \qquad (3)$$

from its Dirac value can also be accounted for by the theory of quantum electrodynamics. *See* LEPTON.

The nuclear magneton is obtained from the Bohr magneton by replacing m by the proton mass m_p. The value of the nuclear magneton is given in Eq. (4). The nuclear magneton is

$$\mu_N = (5.050\,783\,24 \pm 0.000\,000\,13) \times 10^{-27}\ \text{joule/tesla} \qquad (4)$$

used not only as the unit for the magnetic moment of the proton but also for the neutron and other hadrons and for atomic nuclei. If the proton and neutron were Dirac particles, the proton's magnetic moment would be one nuclear magneton (except for a small correction arising from quantum electrodynamics) and the neutron's magnetic moment would be zero (because the neutron is uncharged). However, the proton and neutron have large anomalous magnetic moments, given in Eqs. (5).

$$\begin{aligned} \mu_p &= (2.792\,847\,356 \pm 0.000\,000\,023)\mu_N \\ \mu_n &= (-1.913\,042\,73 \pm 0.000\,000\,45)\mu_N \end{aligned} \qquad (5)$$

See NEUTRON; NUCLEAR MOMENTS; PROTON.

According to present theory, the proton, neutron, and other hadrons have large anomalous magnetic moments because these particles are not elementary but composite. In the theory of quantum chromodynamics, the principal constituents of a baryon, such as the proton or neutron, are three quarks. *See* BARYON; ELEMENTARY PARTICLE; FUNDAMENTAL CONSTANTS; QUANTUM CHROMODYNAMICS; QUARKS. Don B. Lichtenberg

Magnetooptics That branch of physics which deals with the influence of a magnetic field on optical phenomena. Considering the fact that light is electromagnetic radiation, an interaction between light and a magnetic field would seem quite plausible. It is, however, not the direct interaction of the magnetic field and light that produces the known magnetooptic effects, but the influence of the magnetic field upon matter which is in the process of emitting or absorbing light. *See* COTTON EFFECT; FARADAY EFFECT; MAGNETOOPTIC KERR EFFECT; MAJORANA EFFECT; VOIGT EFFECT; ZEEMAN EFFECT. G. H. Dieke; William W. Watson

Magnetoresistance The change of electrical resistance produced in a current-carrying conductor or semiconductor on application of a magnetic field H. Magnetoresistance is one of the galvanomagnetic effects. It is observed with H both parallel to and transverse to the current flow. The change of resistance usually is proportional to H^2 for small fields, but at high fields it can rise faster than H^2, increase linearly with H, or tend to a constant (that is, saturate), depending on the material. In most nonmagnetic solids the magnetoresistance is positive. *See* GALVANOMAGNETIC EFFECTS.

In semiconductors, the magnetoresistance is unusually large and is highly anisotropic with respect to the angle between the field direction and the current flow in single crystals. When the magnetoresistance is measured as a function of field, it is the basis for the Shubnikov–de Haas effect, much as the field dependence of the magnetization gives rise to the de Haas-van Alphen effect. Measurement of either effect as the field direction changes with respect to the crystal axes serves as a powerful probe of the Fermi surface. Magnetoresistance measurements also yield information about current carrier mobilities. Important to practical applications is the fact that the geometry of a semiconductor sample can generate very large magnetoresistance, as in the Corbino disk. *See* DE HAAS-VAN ALPHEN EFFECT; FERMI SURFACE; SEMICONDUCTOR.

Multilayered structures composed of alternating layers of magnetic and nonmagnetic metals, such as iron/chromium or cobalt/copper, can feature very large, negative values of magnetoresistance. This effect, called giant magnetoresistance, arises from the spin dependence of the electron scattering which causes resistance. When consecutive magnetic layers have their magnetizations antiparallel (antiferromagnetic alignment), the resistance of the structure is larger than when they are parallel (ferromagnetic alignment). Since the magnetic alignment can be changed with an applied magnetic field, the resistance of the structure is sensitive to the field. Giant magnetoresistance can also be observed in a simpler structure known as a spin valve, which consists of a nonmagnetic layer (for example, copper) sandwiched between two ferromagnetic layers (for example, cobalt). The magnetization direction in one of the ferromagnetic layers is fixed by an antiferromagnetic coating on the outside, while the magnetization direction in the other layer, and hence the resistance of the structure, can be changed by an external magnetic field. Films of nonmagnetic metals containing ferromagnetic granules, such as cobalt precipitates in copper, have been found to exhibit giant magnetoresistance as well. *See* ANTIFERROMAGNETISM; FERROMAGNETISM; MAGNETIC FIELD; MAGNETIZATION.

Magnetoresistors, especially those consisting of semiconductors such as indium antimonide or ferromagnets such as permalloy, are important to a variety of devices which detect magnetic fields. These include magnetic recording heads and position and speed sensors. *See* MAGNETIC MATERIALS; MAGNETIC RECORDING.

J. F. Herbst

Magnetosphere A comet-shaped cavity or bubble around the Earth, carved in the solar wind. This cavity is formed because the Earth's magnetic field represents an obstacle to the solar wind, which is a supersonic flow of plasma blowing away from the Sun. As a result, the solar wind flows around the Earth, confining the Earth and its magnetic field into a long cylindrical cavity with a blunt nose. Since the solar wind is a supersonic flow, it also forms a bow shock a few earth radii away from the front of the cavity. The boundary of the cavity is called the magnetopause. The region between the bow shock and the magnetopause is called the magnetosheath. The Earth is located about 10 earth radii from the blunt-nosed front of the magnetopause. The long cylindrical section of the cavity is called the magnetotail, which is on the order of a few thousand earth radii in length, extending approximately radially away from the Sun. *See* SOLAR WIND; SUN.

The magnetosphere has been extensively explored by a number of satellites carrying sophisticated instruments. The satellite observations have indicated that the cavity is not an empty one, but is filled with plasmas of different characteristics. The Earth's dipolar magnetic field is considerably deformed by these plasmas and the electric currents generated by them. *See* VAN ALLEN RADIATION.

All other magnetic planets, such as Mercury, Jupiter, and Saturn, have magnetospheres which are similar in many respects to the magnetosphere of the Earth.

S.-I. Akasofu

Magnetostriction The change of length of a ferromagnetic substance when it is magnetized. More generally, magnetostriction is the phenomenon that the state of strain of a ferromagnetic sample depends on the direction and extent of magnetization. The phenomenon has an important application in devices known as magnetostriction transducers. *See* FERROMAGNETISM.

The magnetostrictive effect is exploited in transducers used for the reception and transmission of high-frequency sound vibrations. Nickel is often used for this application. *See* SONAR; ULTRASONICS.

Elihu Abrahams; Frederic Keffer

Magnetron The oldest of a family of crossed-field microwave electron tubes wherein electrons, generated from a heated cathode, move under the combined force of a radial electric field and an axial magnetic field. By its structure a magnetron causes moving electrons to interact synchronously with traveling-wave components of a microwave standing-wave pattern in such a manner that electron potential energy is converted to microwave energy with high efficiency. Magnetrons have been used since the 1940s as pulsed microwave radiation sources for radar tracking. Because of their compactness and the high efficiency with which they can emit short bursts of megawatt peak output power, they have proved excellent for installation in aircraft as well as in ground radar stations. In continuous operation, a magnetron can produce a kilowatt of microwave power which is appropriate for rapid microwave cooking. *See* ELECTRON TUBE.

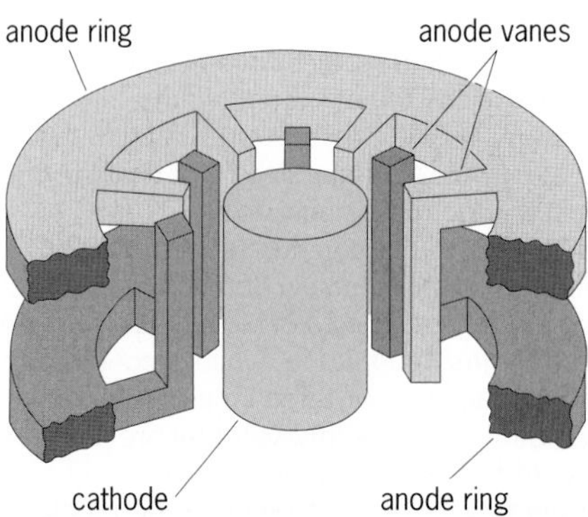

An interdigital-vane anode circuit and cathode, indicating the basic cylindrical geometry of the magnetron. (*After G. D. Sims and I. M. Stephenson, Microwave Tubes and Semiconductor Devices, Blackie and Son, London, 1963*)

The magnetron is a device of essentially cylindrical symmetry (see illustration). On the central axis is a hollow cylindrical cathode. The outer surface of the cathode carries electronemitting materials, primarily barium and strontium oxides in a nickel matrix. Such a matrix is capable of emitting electrons when current flows through the heater inside the cathode cylinder. *See* VACUUM TUBE.

At a radius somewhat larger than the outer radius of the cathode is a concentric cylindrical anode. The anode serves two functions: (1) to collect electrons emitted by the cathode and (2) to store and guide microwave energy. The anode consists of a series of quarter-wavelength cavity resonators symmetrically arranged around the cathode.

A radial dc electric field (perpendicular to the cathode) is applied between cathode and anode. This electric field and the axial magnetic field (parallel and coaxial with the cathode) introduced by pole pieces at either end of the cathode provide the required crossed-field configuration.

R. J. Collier

Magnification A measure of the effectiveness of an optical system in enlarging or reducing an image. For an optical system that forms a real image, such a measure is the lateral magnification m, which is the ratio of the size of the image to the size of the object. If the magnification is greater than unity, it is an enlargement; if less than unity, it is a reduction.

The angular magnification is the ratio of the angles formed by the image and the object at the eye. In telescopes the angular magnification (or, better, the ratio of the tangents of the angles under which the object is seen with and without the lens, respectively) can be taken as a measure of the effectiveness of the instrument.

Magnifying power is the measure of the effectiveness of an optical system used in connection with the eye. The magnifying power of a spectacle lens is the ratio of the tangents of the angles under which the object is seen with and without the lens, respectively. The magnifying power of a magnifier or an ocular is the ratio of the size under which an object would appear when seen through the instrument at a distance of 10 in. or 250 mm (the distance of distinct vision) divided by the object size. *See* LENS (OPTICS); OPTICAL IMAGE.

Max Herzberger

Magnitude (astronomy) The brightness of an astronomical object, expressed on a unique numerical scale. The stellar magnitude scale is logarithmic and is inverted in that fainter objects have numerically larger magnitudes. Although used primarily for stars, the stellar magnitude scale can also be used to express the brightness of the Sun, planets, asteroids, comets, nebulae, galaxies, and even background radiation.

Since the brightness of any object varies with wavelength, many different magnitude scales have been defined corresponding to different spectral regions, bandwidths, and methods of observation. Visual magnitudes, corresponding to the sensitivity of the human eye centered in the yellow part of the spectrum, are usually implied if the type is unspecified.

The star catalog of Hipparchus (about 150 B.C.) is thought to have contained approximately 850 naked-eye stars, classified according to brightness. The 15 or so brightest stars were referred to as stars of the first magnitude, while second-magnitude stars were on the average two or three times fainter, and so on. The scale is logarithmic because intervals that are perceived as equal intervals are, in fact, equal brightness ratios.

Measurements of brightness ratios in the nineteenth century showed that, on average, stars of the sixth magnitude (near the limit of naked-eye vision) were about 100 times fainter than those of the first. On the scale introduced by N. R. Pogson in 1856 and universally adopted, an interval of 5 magnitudes corresponds to a factor of exactly 100, so that each magnitude corresponds to a factor of $\sqrt[5]{100} \approx 2.512 \cdots$. The zero point of the Pogson scale was set so that most stars retained their customary magnitudes.

An attractive feature of the magnitude scale is the ease with which fractional magnitudes can be interpreted. Each change of 1% in the brightness of an object corresponds to a change of 0.01 in the magnitude, and this numerical correspondence holds to good accuracy for changes up to about 30%.

A distinction is made between the apparent magnitude of an object viewed from the Earth and its absolute magnitude, which measures the object's intrinsic luminosity by indicating its apparent magnitude as seen from a standard distance. The absolute magnitude may be defined as the apparent magnitude an object would have if viewed from a distance of 10 parsecs (1 pc = 3.26 light-years = 1.92×10^{13} mi = 3.09×10^{13} km). *See* PARSEC.

Robert F. Wing

Magnolia A genus of trees with large, chiefly white flowers, and simple, entire, usually large alternate leaves. In the winter the twigs may be recognized by their aromatic odor when bruised.

The most important species commercially is *Magnolia acuminata*, commonly called cucumber tree, which grows in the Appalachian and Ozark mountains. The fruit is red when ripe and resembles a small cucumber in shape.

The wood of the magnolia is similar to that of the tulip tree and is rather soft, but it is of such wide natural dimensions that it is valued for furniture, cabinetwork, flooring, and interior finish.

Magnolia species occur naturally in a broad belt in the eastern United States and Central America, with a similar region in eastern Asia and the Himalayas. *See* MAGNOLIALES.

Arthur H. Graves; Kenneth P. Davis

Magnoliales An order of flowering plants consisting of six families, the best known of which are Magnoliaceae (220 species) and Annonaceae (2200 species). The others contain some botanically interesting and peculiar plants (305 species), but none besides Myristicaceae (300 species) are commonly encountered. Previously, many authors have considered these families as among the most primitive of the flowering plants, but in all cases the plants exhibit some highly derived traits. Studies of sequences of deoxyribonucleic acid (DNA) demonstrate that Magnoliales are closely related to Laurales, Piperales, and Winterales, and then more distantly to the monocotyledons.

Some species are of minor economic importance in various parts of the tropics: Annonaceae contain the custard apple, soursop, and sweetsop (*Annona* species), ylang-ylang (an aromatic oil is produced by the flowers of *Canaga odorata*), and *Mkilua fragrans* is the source of a perfume. *Myristica fragrans* (Myristicaeae) is the source of nutmeg and mace, and seeds of several species in this family are used locally to produce consumable oils that are sometimes also used for candle making. Of Magnoliaceae, many species of *Magnolia* (tulip trees) are used as ornamental trees in the temperate and tropical zones, and *Liriodendron* (yellow poplar or green tulip tree) is a commonly planted shade tree throughout the north temperate region. The latter is also a valuable timber-producing genus. *See* LAURALES; PIPERALES; TULIP TREE.

Mark Chase

Magnoliidae A subclass of the class Magnoliopsida (dicotyledons) in the division Magnoliophyta (Angiospermae), the flowering plants. The subclass consists of 8 orders, 39 families, and more than 12,000 species. The Magnoliidae are the most primitive subclass of flowering plants. In general, they have a well-developed perianth, which may or may not be numerous, centripetal stamens, and they are apocarpous. *See* ARISTOLOCHIALES; ILLICIALES; LAURALES; MAGNOLIALES; MAGNOLIOPHYTA; MAGNOLIOPSIDA; NYMPHAEALES; PAPAVERALES; PIPERALES; PLANT KINGDOM; RANUNCULALES.

Arthur Cronquist; T. M. Barkley

Magnoliophyta A division of seed plants consisting of about 250,000 species, which form the bulk and most conspicuous element of the land plants. Often called flowering plants or angiosperms, they have several unique characteristics, the most prominent of which are their reproductive structure, flowers, and covered seeds. The other obvious woody land plants are the gymnosperms, which have cones instead of flowers and have naked seeds. Another trait distinguishing the angiosperms is the presence of double fertilization, which results in the production of stored food (starch or oils) within their seeds. *See* FLOWER.

Angiosperms range from some of the smallest plants known to large forest trees, and they occur in all habitats, including the oceans, where they are only a minor element in most marine ecosystems. Some are capable of growing directly on rock surfaces as well as on the limbs of trees. The angiosperms are usually considered to be the most highly evolved division of the subkingdom Embryobionta. Their highly specialized and relatively efficient conducting tissues, combined with the protection of their ovules in an ovary, give them a competitive advantage over most other groups of land plants in most regions. *See* EMBRYOBIONTA.

The angiosperms may be characterized as vascular plants with roots, stems, and leaves, usually with well-developed vessels in the xylem and with companion cells in the phloem. The central cylinder has leaf gaps or scattered vascular bundles; the ovules are enclosed in an ovary; and the female gametophyte is reduced to a few-nucleate embryo sac without an archegonium. The male gametophyte is reduced to a tiny pollen grain that gives rise to a pollen tube containing a tube nucleus and two sperms; one sperm fuses with the egg in the embryo sac to form a zygote, and the other fuses with two nuclei of the embryo sac to form a triple fusion nucleus that is typically the forerunner of the endosperm of the seed. *See* LEAF; PHLOEM; PLANT REPRODUCTION; POLLEN; ROOT (BOTANY); SEED; STEM; XYLEM.

Among plants with alternation of sporophyte and gametophyte generations, the angiosperms represent the most extreme stage in reduction of the gametophyte, which in effect is reduced to a mere stage in the reproduction of the sporophyte. The pollen grain, with its associated pollen tube, and the embryo sac represent the male and female gametophyte generations; the endosperm is a new structure not referable to either generation; and the remainder of the plant throughout its life cycle is the sporophyte. Many angiosperms can also propagate asexually by means of creeping stems or roots or by other specialized vegetative structures such as bulbils.

It is obvious to biologists that the angiosperms must have evolved from gymnosperms, but beyond this the facts are obscure. They appear in the fossil record early in the Cretaceous Period as obvious angiosperms, without any hint of a connection to any particular group of gymnosperms. Many believe that

(a)

(b)

(c)

(d)

(e)

(f)

(g)

(a) Bat (*Eptesicus fuscus*) (*photo by Lloyd Glenn Ingles; California Academy of Sciences*). (b) North American bison (*Bison bison*) near Soda Butte Creek, Yellowstone National Park (*photo by Jim Peaco; National Park Service*). (c) Caribou (*Rangifer tarandus*) (*photo by Dean Biggins; U.S. Fish and Wildlife Service*). (d) Alpine chamois (*Rupicapra rupicapra*) (*photo by Brent Huffman/Ultimate Ungulate Images*). (e) African bush elephant (*Loxodonta africana*) (*photo by Robert Thomas and Margaret Orr; California Academy of Sciences*). (f) Red panda (*Ailurus fulgens*) (*photo by John White*). (g) Walrus (*Odobenus rosmarus*) (*photo by Donna Dewhurst Collection; U.S. Fish and Wildlife Service*).

among the gymnosperms the seed ferns provide the most likely ancestors. *See* PALEOBOTANY; PINOPHYTA.

The Magnoliophyta consist of two large groups that have not been formally named: the eudicots and the magnoliids. The eudicots are characterized by flowers that are highly organized in terms of the number and orientation of parts, whereas the magnoliids have many parts, without any particular fixed patterns among the parts—except for the monocots, in which the most developed groups, like the eudicots, exhibit developed flowers with highly organized patterns. *See* LILIOPSIDA; MAGNOLIOPSIDA.

Arthur Cronquist; T. M. Barkley; Mark W. Chase

Magnoliopsida One of the two classes of flowering plants which collectively make up the division Magnoliophyta (Angiospermae). The Magnoliopsida, often known as Dicotyledoneae or dicotyledons, embrace 6 subclasses, 64 orders, 318 families, and about 165,000 species.

All of the characters which collectively distinguish the Magnoliopsida from the Liliopsida (monocotyledons) are subject to exception, but in general the Magnoliopsida have two cotyledons and net-veined leaves. The vascular bundles are typically borne in a ring (or cylinder) enclosing a pith. Increase in thickness of stems and roots, after the primary tissues have matured, results from meristematic activity of a cambial layer which passes through the vascular bundles. In about half of the species of the group, the cambium of the stem forms a continuous cylinder which produces a new layer of wood (secondary xylem) and bark (secondary phloem) each growing season for year after year. Such plants become trees or shrubs.

It is widely agreed that the most existing angiosperms belong to the dicotyledons (especially the order Magnoliales) and that most of the characters which distinguish the monocotyledons as a group are derived rather than primitive. *See* ASTERIDAE; CARYOPHYLLIDAE; DILLENIIDAE; HAMAMELIDAE; LILIOPSIDA; MAGNOLIALES; MAGNOLIIDAE; MAGNOLIOPHYTA; PLANT KINGDOM; ROSIDAE.

Arthur Cronquist; T. M. Barkley

Magnon A quantum of a spin wave; an elementary excitation of a magnetic system which is usually long-range-ordered, such as a ferromagnet. *See* ANTIFERROMAGNETISM; FERRIMAGNETISM; FERROMAGNETISM.

In the lowest energy state of a simple ferromagnet, all the magnetic moments of the individual atoms are parallel (say, to the z axis). Each atomic moment derives mainly from the electron spin angular momentum of the atom. In the next-to-lowest energy state (first excited state), the total z component of spin angular momentum, S_z, is reduced by one unit of $\hbar = h/2\pi$, where h is Planck's constant. In the case of a crystalline material, this unit is shared equally by all the spins, each of which lies on a cone (see illustration), precessing at an angular rate ω. These spins form a wave, known as a spin wave, having a repeat distance or wavelength, λ. The wave amplitude (that is, the cone angle) is extremely small, because of the sharing among all the spins whose number N is very large, roughly 10^{23}. Thus, each atom's share of the reduction in S_z, labeled Δ, is only $\hbar/N$, whereas the z component of the atomic spin in the fully aligned state is typically 1–10 times $\hbar$. It follows from simple geometry that the cone half-angle is of order 10^{-11} to 10^{-12} radian. The state with this value of the amplitude is said to be a one-magnon state with wave number $k = 1/\lambda$. If Δ is doubled to $2\hbar/N$, the state is a two-magnon state, and so forth. The integer values of $N\Delta/\hbar$ correspond to the possible changes in S_z being integral multiples of $\hbar$. *See* ELECTRON SPIN; WAVE MOTION.

While the spin waves associated with energy states, that is, stationary states, in crystals vary sinusoidally in space (see illustration), magnons can be associated, instead, with nonstationary states (wave packets) in some situations. Closely analogous to magnons are phonons and photons, quanta of mass-density waves and electromagnetic waves, respectively. *See* ELECTROMAGNETIC WAVE; PHONON; PHOTON; QUANTUM MECHANICS.

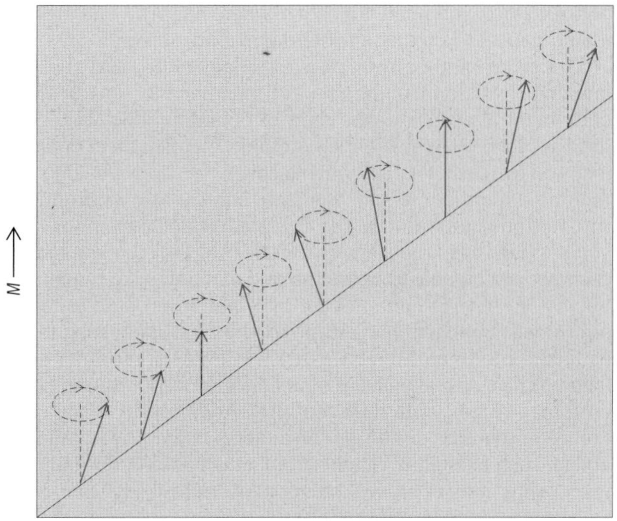

Spin wave in a linear ferromagnetic array of precessing atomic spins of equal magnitude, represented as arrows (vectors) in perspective. The axis of precession is along the vertical direction of the total magnetization, *M*.

Thomas A. Kaplan

Mahogany A hard, red or yellow-brown wood which takes a high polish and is extensively used for furniture and cabinetwork. The West Indies mahogany tree (*Swietenia mahagoni*), a native of tropical regions in North and South America, is a large evergreen tree with smooth pinnate leaves. Together with other species it yields the world's most valuable cabinet wood. In the United States it occurs naturally only in the extreme southern tip of Florida, but it is planted elsewhere in the state as an ornamental and shade tree.

Arthur H. Graves; Kenneth P. Davis

Maillard reaction A nonenzymatic chemical reaction involving condensation of an amino group and a reducing group, resulting in the formation of intermediates which ultimately polymerize to form brown pigments (melanoidins). The reaction was named for the French biochemist Louis-Camille Maillard. It is of extreme importance to food chemistry, especially because of its ramifications in terms of food quality. *See* AMINE; REACTIVE INTERMEDIATES.

There are three major stages of the reaction. The first comprises glycosylamine formation and rearrangement *N*-substituted-1-amino-l-deoxy-2-ketose (Amadori compound). The second phase involves loss of the amine to form carbonyl intermediates, which upon dehydration or fission form highly reactive carbonyl compounds through several pathways. The third phase occurring upon subsequent heating involves the interaction of the carbonyl flavor compounds with other constituents to form brown nitrogen-containing pigments (melanoidins). These are highly desirous compounds in certain foods browned by heating in the presence of oxygen.

The Maillard reaction is considered undesirable in some biological and food systems. The interaction of carbonyl and amine compounds might damage the nutritional quality of proteins by reducing the availability of lysine and other essential amino acids and by forming inhibitory or antinutritional compounds. The reaction is also associated with undesirable flavors and colors in some foods, particularly dehydrated foods. *See* AMINO ACIDS; CARBONYL.

Milton E. Bailey

Maintenance, industrial and production The actions taken to preserve the operation of devices, particularly of electromechanical equipment, to ensure that the devices can perform their intended functions when needed. The field of maintenance science is an interdisciplinary research area that employs techniques from physics, engineering, and decision analysis. Traditionally, the focus of maintenance has been on equipment availability—the ratio of operating time less downtime to total available time. Modern maintenance practices focus on increasing equipment effectiveness, that is, making sure that the equipment is both available and capable of producing superior-quality products. *See* SYSTEMS ENGINEERING.

It has been estimated that up to 50% of all life-cycle equipment costs are attributable to operation and maintenance. Equipment buyers are now requiring better information on time to failure and repair. Suppliers have responded by including such things as failure mode and effects analysis, statistical information on failure times, cost-effective maintenance procedures, and better customer training with their products. Additionally, many companies emphasize design enhancements to improve maintainability, such as built-in diagnostics, greater standardization and modularity, and improved component accessibility. *See* ENGINEERING DESIGN; MACHINE DESIGN.

Maintenance activities can be classified into several broad categories, depending on whether they respond to failures that have occurred or whether they attempt to prevent failures. The simplest and least sophisticated maintenance strategy still used by many companies is reactive maintenance or breakdown maintenance. Equipment is operated until it fails, then repaired or replaced. No effort is expended on activities that monitor the ongoing "health" of the equipment, and maintenance is focused on quick repairs that return the equipment to production as soon as possible. A slightly more sophisticated maintenance strategy is preventive maintenance, also known as calendar-based maintenance. This system involves detailed, planned maintenance activities on a periodic basis, usually monthly, quarterly, semiannually, or annually. As in reactive maintenance, preventive maintenance does not monitor information on equipment status. Rather, it attempts to avoid unplanned failures through planned repairs or replacements. Predictive maintenance is based on an ongoing (continuous or periodic) assessment of the actual operating condition of equipment. The equipment is monitored while in operation, and repair or replacement is scheduled only when measurements indicate that it is required. Predictive maintenance programs seek to control maintenance activities to avoid both unplanned equipment outages and unnecessary maintenance and overhauls. Georgia-Ann Klutke

Malachite A bright-green, basic carbonate of copper [$Cu_2CO_3(OH)_2$]. Malachite is the most stable copper mineral in natural environments in contact with the atmosphere and hydrosphere. It occurs as an ore mineral in oxidized copper sulfide deposits; as a stain on fractures in rock outcrops; as a corrosion product of copper and its alloys (except in industrial-urban environments, where the basic copper sulfate dominates); as suspended particles in streams and in alluvial sediments; and as encrustations on bronze artifacts in seawater and on coccoliths floating in the oceans. It can be distinguished from other green copper minerals by its effervescence in acid. The combination of hardness (3.5–4 on Mohs scale) ideal for carving, color variation in concentric layers, and adamantine-to-silky luster has made malachite a highly prized ornamental stone. Its rare blocky-tabular crystals up to 5 mm (0.2 in.), its pseudomorphs after azurite crystals to 2 cm (0.8 in.), and its more common felty tufts perched on bright blue azurite are eagerly sought by mineral collectors. Malachite is an important copper ore mineral in supergene copper oxide deposits formed by weathering of primary copper sulfide deposits. *See* AZURITE; CARBONATE MINERALS; COPPER. Marco T. Einaudi

Malacostraca The largest and most diversified class of the Crustacea; includes the shrimps, lobsters, crabs, sow bugs, beach hoppers, and their allies. The shell or carapace may be large, small, vestigial, or absent; the tail or abdomen is long or short; the eyes are generally set on movable stalks but may be sessile or even coalesced. Despite this diversity, the unity of the group is demonstrated by the following characteristics which all share. The maximum number of appendages is 19 pairs. The trunk limbs are sharply differentiated into a thoracic series of eight pairs and an abdominal series of six pairs. The female genital duct always opens at the level of the sixth thoracic segment, whereas those of the male open at the level of the eighth. *See* CRUSTACEA.

Malacostraca are divided into three subclasses, the Phyllocarida, Hoplocarida, and Eumalacostraca. Central to the classification of the Eumalacostraca has been the concept of the "caridoid facies" (see illustration). This term refers to a series of

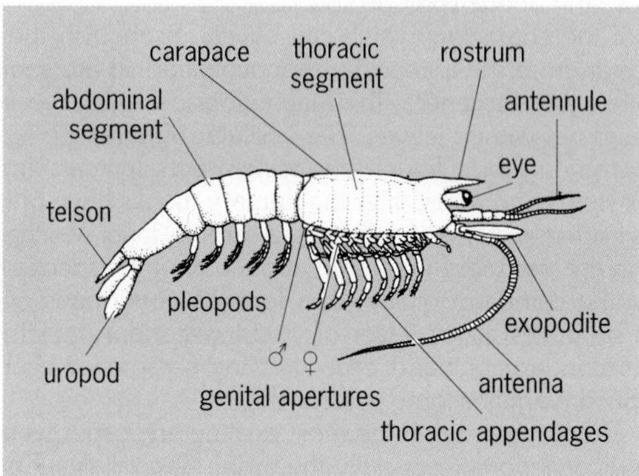

Caridoid facies. (*After E. R. Lankester, ed., A Treatise on Zoology, pt. 7. fasc. 3, A. and C. Black, 1909*)

morphological attributes generally common to the three orders, Syncarida, Peracarida, and Eucarida. *See* EUCARIDA; EUMALACOSTRACA; HOPLOCARIDA; PERACARIDA; PHYLLOCARIDA; SYNCARIDA.

Patsy A. McLaughlin

Malaria A disease caused by members of the protozoan genus *Plasmodium*, a widespread group of sporozoans that parasitize the human liver and red blood cells. Four species can infect humans: *P. vivax*, causing vivax or benign tertian malaria; *P. ovale*, a very similar form found chiefly in central Africa that causes ovale malaria; *P. malariae*, which causes malariae or quartan malaria; and *P. falciparum*, the highly pathogenic causative organism of falciparum or malignant tertian malaria. Malaria is characterized by periodic chills, fever, and sweats, often leading to severe anemia, an enlarged spleen, and other complications that may result in loss of life, especially among infants whose deaths are almost always attributed to falciparum malaria. The infective agents are inoculated into the human bloodstream by the bite of an infected female *Anopheles* mosquito, more than 60 species of which can carry the infection to humans. The disease is found in all tropical and some temperate regions, but it has been eradicated in North America, Europe, and Russia. Despite control efforts, malaria has probably been the greatest single killer disease throughout human history and continues to be a major infectious disease. *See* EPIDEMIC.

The vast reproductive capacity of *Plasmodium* parasites is illustrated by their life cycle, which begins as a series of asexual divisions in human liver and then red blood cells. Transfer of the parasites to the mosquito host depends on the rate of sexual multiplication that begins in the infected human red blood cells and is completed in the mosquito stomach, followed by asexual multiple division of the product of sexual fusion. Clinical malaria usually begins 7–18 days after infection with sporozoites. Red cell infections tend to follow a remarkably synchronous division cycle. The parasite progresses from merozoite to a vegetative phase (trophozoite) to a division stage (schizont), ending with the new generation of merozoites ready to break out in a burst of parasite releases and initiate the chills-fever-sweat phase of the disease. The sequence of chills, fever, and sweats is the result of simultaneous red cell destruction at 48- or 72-h intervals. *See* SPOROZOA.

Chloroquine remains the drug of choice for prevention as well as treatment of vivax, ovale, and malariae malaria. However, most strains of falciparum malaria have become strongly chloroquine-resistant. For prevention of chloroquine-resistant falciparum malaria (and in some areas vivax malaria is now chloroquine-resistant as well) a weekly dose of mefloquine beginning a week before, then during, and for 4 weeks after leaving the endemic area is recommended. Chloroquine-resistant malaria is chiefly treated with the oldest known malaricide, quinine, in the form of quinine sulfate, plus pyrimethamine-sulfadoxine. *See* DRUG RESISTANCE; QUININE.

Failure of earlier efforts to eradicate malaria and the rapid spread of resistant strains of both parasites and their mosquito vectors necessitated renewed interest in prevention of exposure by avoidance of mosquito bites using pyrethrin-treated bednets, coverage of exposed skin during active mosquito periods (usually dawn, dusk, and evening hours), and use of insect-repellent lotions. A balance between epidemiological and immunological approaches to prevention, and the continued development of new drugs for prophylaxis and treatment are recognized as the most effective means to combat one of the most dangerous and widespread threats to humankind from an infectious agent. *See* MEDICAL PARASITOLOGY. Donald Heyneman

Malnutrition Impaired health caused by a dietary deficiency, excess, or imbalance. To support human life, energy (from fat, carbohydrate, and protein), water, and more than 40 different food substances must be obtained from the diet in appropriate amounts. Malnutrition can result from the chronic intake of any of these substances at levels above, as well as below, ranges that are adequate and safe, but commonly the term refers only to deficient intake.

The number of people throughout the world who suffer from nutritional deficiencies as a result of inadequate dietary intake is uncertain, but even the most conservative estimates place that figure at hundreds of millions; many experts consider the actual number to approach 1 billion. Most malnourished people live in developing countries where income, education, and housing are inadequate to buy, transport, store, and prepare food and where nutritional deficiencies are almost always related to poverty. In industrialized countries, chronic conditions of deficient dietary intake occur far less frequently but are reported occasionally among people who are dieting to lose weight, fasting, or on an unusually restrictive ("fad") diet. Pregnant women, infants, and children are most at risk for inadequate dietary intake because their nutritional requirements are relatively high.

Nutritional deficiencies also occur as a result of illness, injury, or alcohol or drug abuse that interferes with appetite; the inability to eat; defective digestion, absorption, or metabolism of food molecules; or disease states that increase nutrient losses. Secondary malnutrition has been observed frequently among medical and surgical patients who are treated in hospitals for prolonged periods of time. Regardless of cause, the effects of malnutrition can range from minor symptoms to severe syndromes of starvation, protein-calorie malnutrition, or single-nutrient deficiencies. *See* METABOLIC DISORDERS.

The chronic intake of energy below the level of expenditure induces rapid losses in body weight and muscle mass accompanied by profound changes in physiology and behavior. Together, these effects cause a starving person to become weak, apathetic, depressed, and unable to work productively and to do whatever is necessary to reverse the malnutrition. The consequences of nutritional deficiencies are seen first in tissues that are growing rapidly. These changes are most evident in the gastrointestinal tract, skin, blood cells, and nervous system as indigestion, malabsorption, skin lesions, anemia, or neurologic and behavior changes. Of special concern is the loss of immune function that accompanies severe malnutrition.

The combined effects of malnutrition and infection in young children are referred to as protein-calorie malnutrition. It classified into two entities, marasmus and kwashiorkor, on the basis of physical appearance and the relative proportions of protein and calories in the diet. Children with the marasmus form appear generally wasted as a result of diets that are chronically deficient in calories as well as protein and other nutrients. Children with kwashiorkor are also very thin but have characteristically bloated bellies due to fluid retention and accumulation of fat in the liver, symptoms attributed to diets relatively deficient in protein. *See* ADIPOSE TISSUE; PROTEIN METABOLISM.

Deficiency conditions due to lack of a single vitamin or mineral occur rarely and usually reflect the lack of the most limiting nutrient in a generally deficient diet. In industrialized countries, single-nutrient deficiencies are most evident in individuals who abuse alcohol or drugs. Classic conditions of deficiency of niacin (pellagra), thiamine (beriberi), vitamin C (scurvy), and vitamin D (rickets) have virtually disappeared as a result of food fortification programs and the development of food distribution systems that provide fresh fruits and vegetables throughout the year. Iron-deficiency anemia also has declined in prevalence, although children in low-income families remain at risk. In developing countries, however, such conditions are still observed among people whose diets depend on one staple food as the major source of calories. A condition of substantial current public health importance is vitamin A deficiency, which is the principal cause of blindness and a major contributor to illness and death among children in developing countries. *See* ANEMIA; VITAMIN. Marion Nestle

Malpighiales One of the largest orders of the rosid eudicotyledons, comprising more than 30 families distributed worldwide. Recent analyses of deoxyribonucleic acid (DNA) sequences, both plastid and nuclear, led to its recognition, even though the group is highly heterogeneous and difficult to characterize. The largest families are Euphorbiaceae (8000 species), Clusiaceae (1400), Malpighiaceae (1100), Flacourtiaceae (900), and Violaceae (850). Most of the order is composed of woody species, many of regional importance as timber and medicines. Several of the smaller families are significant as well, including Salicaceae (used as coppice, and the original source of aspirin) and Rhizophoraceae (the ecologically significant mangroves). *See* MAGNOLIOPHYTA. Mark W. Chase

Malt beverage A fermented beverage produced from grain. Beer is a generic term used to describe alcoholic beverages made from cereal grains, especially barley, in the form of malt. Ale, lager, porter, and stout are different kinds of beer made by recognizably similar processes. The United States is the largest producer of beer in the world.

The manufacture of beer is a complex natural process of three general parts: the preparation of barley by germination, or the malting process; the actual digestion of barley (now malt) starch to produce a solution of sugars (called wort) and the adjustment of flavor with hops, which are the brewhouse processes; and the fermentation of these sugars by yeast to yield alcohol, carbon dioxide gas, and flavor compounds to produce beer. M. J. Lewis

Malvales An order of flowering plants in the core eudicots. The order consists of 10 families and more than 5500 species. The circumscription of the order has been altered greatly in recent years, largely on the basis of deoxyribonucleic acid (DNA) sequence data. Of the five families in the traditional concept

(Bombacaceae, Malvaceae, Sterculiaceae, Tiliaceae, and Elaeocarpaceae), the first four have been combined as Malvaceae, due to lack of monophyly of three of these families in their traditional circumscription; and Elaeocarpaceae are included in Oxalidales. In addition, other families have been transferred to Malvales, including Dipterocarpaceae and Thymelaeaceae. The expanded order is characterized by the presence of mucilage in epidermal cells and cavities; and of palmate leaves, stellate hairs, and numerous stamens with partially fused filaments occur frequently.

Malvaceae are cosmopolitan and include economic crops such as cotton (*Gossypium*), cocoa (*Theobroma*), and durian (*Durio*) and horticultural plants such as hollyhocks (*Althaea*) and rose of Sharon (*Hibiscus*). Dipterocarpaceae are important elements of tropical forests, especially in Southeast Asia, and provide hardwood timbers. Other horticultural genera in the order include *Cistus* and *Helianthemum* (Cistaceae) and *Daphne* (Thymelaeaceae). Several genera of Thymelaeaceae provide fibers used for making paper, and *Bixa* (Bixaceae) is the source of the orange dyestuff anatto. *See* CACAO; COTTON; EUDICOTYLEDONS. Mark W. Chase

Mammalia The class Mammalia has been the dominant group of vertebrates since the extinction of the dinosaurs 65 million years ago. There are over 4200 living species, classified into over 1000 genera, 140 families, and 18 orders. The number of extinct mammals is at least five times that. Most living mammals are terrestrial. However, many groups of mammals moved to the water from land-dwelling ancestors. These included manatees and dugongs (which are distantly related to elephants), otters (which are related to weasels), seals, sea lions, and walruses (which are distantly related to bears), and whales (which are distantly related to even-toed hoofed mammals), as well as numerous extinct groups. Mammals have also taken to the air, with over 920 living species of bats, as well as numerous gliding forms such as the flying squirrels, phalangerid marsupials, and flying lemurs or colugos. Mammals are even more successful at small body sizes, with hundreds of small species of rodents, rabbits, and insectivores.

Mammals are distinguished from all other animals by a number of unique characteristics. These include a body covered with hair or fur (secondarily reduced in some mammals, particularly aquatic forms); mammary glands in the female for nursing the young; a jaw composed of a single bone, the dentary; and three middle ear bones, the incus, malleus, and stapes. All mammals maintain a constant body temperature through metabolic heat. Their four-chambered heart (two ventricles and two atria) keeps the circulation of the lungs separate from that of the rest of the body, resulting in more efficient oxygen transport to the body tissues. They have many other adaptations for their active lifestyle, including specialized teeth (incisors, canines, molars, and premolars) for biting, tearing, and grinding up food for more efficient digestion. These teeth are replaced only once in the lifetime of the animal (rather than continuous replacement, found in other toothed vertebrates). Mammals have a unique set of muscles that allow the jaw to move in many directions for chewing and for stronger bite force. Their secondary palate encloses the internal nasal passage and allows breathing while they have food in the mouth. Ribs (found only in the thoracic region) are firmly attached to the breastbone (sternum), so that expansion of the lung cavity is accomplished by a muscular wall in the abdominal cavity called the diaphragm. *See* CARDIOVASCULAR SYSTEM; DENTITION; EAR (VERTEBRATE); HAIR; LACTATION; MAMMARY GLAND; THERMOREGULATION; TOOTH.

All mammals have large brains relative to their body size. Most mammals have excellent senses, and some have extraordinary senses of sight, smell, and hearing. To accommodate their large brains and more sophisticated development, mammals are born alive (except for the platypus and echidnas, which lay eggs), and may require considerable parental care before they are ready to fend for themselves. Juvenile mammals have separate bony caps (epiphyses) on the long bones, separated from the shaft of the bone by a layer of cartilage. This allows the long bones to grow rapidly while still having a strong, bony articulation at the end. When a mammal reaches maturity, these epiphyses fuse to the shaft, and the mammal stops growing (in contrast to other vertebrates, which grow continuously through their lives). *See* BRAIN; NERVOUS SYSTEM (VERTEBRATE); SKELETAL SYSTEM.

The living mammals are divided into three major groups: the monotremes (platypus and echidnas), which still lay eggs, retain a number of reptilian bones in their skeletons, and have other primitive features of their anatomy and physiology; the marsupials (opossums, kangaroos, koalas, wombats, and their relatives), which give birth to an immature embryo that must crawl into the mother's pouch (marsupium), where it finishes development; and the placentals (the rest of the living mammals), which carry the young through a long gestation until they give birth to relatively well-developed progeny. In addition to these three living groups, there were many other major groups, such as the rodentlike multituberculates, now extinct. The most recent classification of the mammals can be summarized as follows:

Class Mammalia
- Subclass Prototheria (monotremes)
- Subclass Theriiformes
  - Infraclass Holotheria
    - Cohort Marsupialia (marsupials or pouched mammals)
    - Cohort Placentalia (placentals)
    - Magnorder Xenarthra (sloths, anteaters, armadillos)
    - Magnorder Epitheria
      - Grandorder Anagalida (= Glires) (rodents, rabbits, elephant shrews)
      - Grandorder Ferae (carnivores, pangolins, many extinct groups)
      - Grandorder Lipotyphla (hedgehogs, shrews, moles, tenreces, and kin)
      - Grandorder Archonta
        - Order Chiroptera (bats)
        - Order Primates (lemurs, monkeys, apes, humans)
        - Order Scandentia (tree shrews)
      - Grandorder Ungulata (hoofed mammals)
        - Order Tubulidentata (aardvarks)
        - Order Artiodactyla (even-toed hoofed mammals: pigs, hippos, camels, deer, antelopes, cattle, giraffes, pronghorns, and relatives)
        - Order Cete (whales and their extinct land relatives)
        - Order Perissodactyla (odd-toed hoofed mammals: horses, rhinos, tapirs, and extinct relatives)
        - Order Hyracoidea (hyraxes)
        - Order Tethytheria (elephants, manatees, and extinct relatives)

This classification does not list all the extinct groups, which include at least a dozen more ordinal-level taxa. See separate articles on each group. *See* REPRODUCTIVE SYSTEM.

Mammals evolved from the Synapsida, an early branch of the terrestrial amniotes that has been erroneously called the mammallike reptiles. (This name is inappropriate because synapsids were never related to reptiles.) The first undoubted mammals appeared in the Late Triassic (about 210 million years ago), and were tiny insectivorous forms much like living shrews. Through the rest of the age of dinosaurs, a number of different groups evolved over the next 145 million years of the Jurassic and Cretaceous. Most remained tiny, shrewlike animals, hiding from the dinosaurs in the underbrush and coming out mostly at night. The first two-thirds of mammalian history had passed before the dinosaurs became extinct 65 million years ago, and this allowed mammals to emerge from their shadow. Between 65 and

55 million years ago, a rapid adaptive radiation yielded all the living orders of placental mammals and many extinct forms as well. Donald R. Prothero

Mammary gland A unique anatomical structure of mammals that secretes milk for the nourishment of the newborn. The mammary gland contains thousands of milk-producing units called alveoli, each of which consists of a unicellular layer of epithelial cells arranged in a spheroid structure. The alveolar epithelial cells take up a variety of nutrients from the blood that perfuses the outer surface of the alveolar structures. Some of the nutrients are then secreted directly into the alveolar lumen; other nutrients are used to synthesize the unique constituents of milk which are then secreted. Each alveolus is connected to a duct through which milk flows. The ducts from many alveoli are connected via a converging ductal system which opens externally by way of the lactiferous pore.

Surrounding each alveolus and its associated small ducts are smooth muscle cells called myoepithelial cells. These cells contract in response to the posterior pituitary hormone oxytocin; milk is thus forced out of the alveoli, through the ductal system, and out the lactiferous pore for the nourishment of the newborn. The release of oxytocin is a neuroendocrine reflex triggered by the stimulation of sensory receptors by the suckling of the newborn. *See* ENDOCRINE MECHANISMS.

Mammary glands are basically highly modified and specialized sebaceous glands which derive from ectoderm. In the embryo, mammary lines, formed on both sides of the midventral line, mark the location of future mammary glands. Along the mammary lines discrete ectodermal ingrowths, called mammary buds, produce a rudimentary branched system of ducts at birth. In all species (except the monotremes) a nipple or teat develops in concert with the mammary buds. In the most primitive mammal (the duckbill or platypus), which lacks nipples or teats, milk simply oozes out of the two mammary gland areas and is lapped up by the young.

From birth to sexual maturity the mammary gland consists of a nipple and a rudimentary ductal system in both males and females. At the onset of puberty in the female, the enhanced secretion of estrogen causes a further development of the mammary ductal system and an accumulation of lipids in fat cells. After puberty in women, the mammary gland consists of about 85% fat cells and a partially developed ductal system. *See* ESTROGEN.

During pregnancy the mammary gland comes under the influence of estrogen and progesterone which are derived from both the ovary and placenta. These hormones cause a further branching of the ductal system and the development of milk-secreting structures, the alveoli. In humans, approximately 200 alveoli are surrounded by a connective tissue sheath forming a structure called a lobule. About 26 lobules are packaged via another connective tissue sheath into a larger structure called a lobe. Each of 15–20 lobes is exteriorized into the nipple via separate lactiferous pores. *See* PROGESTERONE.

A complement of hormones maximizes the development of the ductal and lobuloalveolar elements in the mammary gland. Optimal ductal growth is attained with estrogen, a glucocorticoid, prolactin, and insulin. Maximal lobuloalveolar growth is obtained with estrogen, progesterone, growth hormone, prolactin, a glucocorticoid, and insulin. During pregnancy estrogen and progesterone stimulate mammary development but inhibit milk production.

During the final third of pregnancy, the alveolar epithelial cells begin secreting a fluid called colostrum. This fluid fills the alveoli and causes a gradual enlargement of the breast or udder. At parturition, the inhibitory influence of estrogen and progesterone is removed, and the gland can secrete milk under the influence of a further complement of hormones including prolactin, a glucocorticoid, insulin, and the thyroid hormones. *See* GLAND; LACTATION; MAMMALIA; MILK; PREGNANCY. James A. Rillema

Mammography The radiological imaging of breast tissue. This procedure is used to identify cancer, preferably when still impalpable. Because of improved resolution in high-contrast film, only minor exposure of skin to x-rays is required, so that national screening programs have been set up in many western industrialized countries.

Mammography depends on the tumor being reflected as a dense focus in contrast to surrounding tissue or less dense glandular and ductal parts of the structure. Discrimination between benign and malignant lesions may sometimes be difficult, but may be aided by increased magnification of the image and by spot compression to confirm otherwise equivocally benign lesions. In addition, ultrasonography may be used to distinguish between cystic (mainly benign) and solid (possibly malignant) masses. Many women with impalpable cancers discovered through mammography have microcalcifications as the diagnostic feature; others are diagnosed by the presence of a distinct mass, architectural asymmetry of the glandular and ductal tissue, or tissue distortion.

Mammography has recognized limitations: the possibility exists that in some postmenopausal women it will fail to reveal cancer that is present. Similarly, in premenopausal women in whom the breast tissue is often dense, mammography can be falsely negative in some women with subsequently proven cancers. However, the ability of mammography to reflect breast changes over time and its value in the surveillance of women with breast cancer treated by conservation techniques (lumpectomy and radiation) has made mammography an invaluable clinical tool. *See* BREAST DISORDERS; CANCER (MEDICINE); MEDICAL ULTRASONIC TOMOGRAPHY; RADIOGRAPHY. Alexander J. Walt

Mandarin A name used to designate a large group of citrus fruits in the species *Citrus reticulata* and some of its hybrids. This group is variable in the character of trees and fruits since the term is used in a general sense to include many different forms, such as tangerines, King oranges, Temple oranges, tangelos (hybrids between grapefruit and tangerine), Satsuma oranges, and Calamondin, presumably a hybrid between a mandarin and a kumquat. *See* KUMQUAT; ORANGE; TANGERINE.

Although tangerines are the most extensively planted of the mandarin group, others, particularly the Temple orange, the Murcott orange, and the tangelos, of which there are several varieties, are important commercial fruits in the United States. *See* FRUIT; FRUIT, TREE; SAPINDALES. Frank E. Gardner

Mandibulata A subphylum of the phylum Arthropoda in the long-established classification scheme which comprises three groups at subphylum level; it is almost certainly not a natural assemblage. Mandibulata, defined as those arthropods possessing mandibles (lateral jaws) and antennae, includes the six classes Insecta, Chilopoda, Diplopoda, Symphyla, Pauropoda, and Crustacea. In fact, neither in their development nor in their functional morphology are the mandibles of crustaceans homologous with the mandibles of insects. The available fossil evidence for primitive crustaceans also supports a different phyletic origin for these appendages. In addition, insects have a single pair of antennae, while crustaceans have two pairs. *See* ARTHROPODA; CHILOPODA; CRUSTACEA; DIPLOPODA; INSECTA; PAUROPODA; SYMPHYLA. W. D. Russell-Hunter

Manganese A metallic element, Mn, atomic number 25, and atomic weight 54.9380 g/mole. Manganese is one of the transition elements of the first long period of the periodic table, falling between chromium and iron. The principal properties of manganese are given in the table. It is the twelfth most abundant element in the Earth's crust (approximately 0.1%) and occurs naturally in several forms, primarily as the silicate ($MnSiO_3$) but also as the carbonate ($MnCO_3$) and a variety of oxides, including pyrolusite (MnO_2) and hausmannite (Mn_3O_4). Weathering of land deposits has led to large amounts of the oxide being washed out to sea, where they have aggregated into the so-called manganese nodules containing 15–30% Mn. Vast deposits, estimated at over 10^{12} metric tons, have been detected on the seabed, and a further 10^7 metric tons is deposited every year. The nodules also contain smaller amounts of the oxides of other metals such as iron (Fe), cobalt (Co), nickel (Ni), and copper (Cu). The economic importance of the nodules as a source of these important metals is enormous. *See* HAUSMANNITE; MANGANESE NODULES; PERIODIC TABLE; PYROLUSITE.

Properties of manganese

Property	Value
Atomic number	25
Atomic weight, g/mole	54.9380
Naturally occurring isotope	^{55}Mn (100%)
Electronic configuration	[Ar]$3d^5 4s^2$
Electronegativity	1.5
Metal radius, picometers	127
Melting point, °C (°F)	1244 ± 3 (2271 ± 5.4)
Boiling point. °C (°F)	1962 (3563)
Density (25 °C or 77°F), g/cm^3 (oz/in.^{3})	7.43 (4.30)
Electrical resistivity, ohm·cm	185×10^{-6}

Manganese is more electropositive than its near neighbors in the periodic table, and consequently more reactive. The bulk metal undergoes only surface oxidation when exposed to atmospheric oxygen, but finely divided metal is pyrophoric.

Manganese is a trace element essential to a variety of living systems, including bacteria, plants, and animals. In contrast to iron (Fe), its neighbor in the periodic table, the exact function of the manganese in many of these systems was determined only recently. The manganese superoxide dismutases have been isolated from bacteria, plants, and animals, and are relatively small enzymes with molecular weights of approximately 20,000. The function of the enzyme is believed to be protection of living tissue from the harmful effects of the superoxide ion (O_2^-), a radical formed from partial reduction of O_2 in the cells of respiring (O_2-utilizing) cells.

The most important biological role yet recognized for manganese is in the enzyme responsible for photosynthetic water oxidation to oxygen in plants and certain photosynthetic bacteria. This reaction represents the source of oxygen gas on the Earth and is therefore responsible for the development of the most common forms of life.

All steels contain some manganese, the major advantage being an increase in hardness, although it also serves as a scavenger of oxygen and sulfur impurities that would induce defects and consequent brittleness in the steel. Manganese even has some use in the electronics industry, where manganese dioxide, either natural or synthetic, is employed to produce manganese compounds possessing high electrical resistivity; among other applications, these are utilized as components in every television set. *See* ELECTROLYSIS; GLASS; METAL; TRANSITION ELEMENTS.

George Christou

Manganese nodules Concentrations of manganese and iron oxides found on the floors of many oceans. The origin of these potato-shaped metal-rich deposits has been elucidated; their complex growth histories are revealed by the textures of nodule interiors shown in the illustration.

Marine manganese nodules from certain regions are significantly enriched in nickel, copper, cobalt, zinc, molybdenum, and other elements so as to make them important reserves for these strategic metals.

Reflected-light photograph of the polished surface of a sectioned manganese nodule showing the complex growth history of the concretionary deposit (diameter 1.6 in. or 4 cm).

Although manganiferous nodules and crusts have been sampled or observed on most sea floors, attention has focused on the nickel-plus-copper-rich nodules (2–3 wt% metals) from the north equatorial Pacific in a belt stretching from southeast Hawaii to Baja California, as well as the high-cobalt nodules from seamounts in the Pacific Ocean. Manganese nodules from the Atlantic Ocean and from higher latitudes in the Pacific Ocean have significantly lower concentrations of the minor strategic metals. However, surveys of the Indian Ocean have revealed metal-enrichment trends comparable to those found in the Pacific Ocean nodules; high Ni + Cu-bearing nodules are found adjacent to the Equator.

Microchemical analyses have revealed that chemical differences exist between the outermost top (exposed to sea water) and bottom (immersed in sediment) layers of manganese nodules. Surfaces buried in underlying sediments are generally higher in Mn, Ni, and Cu contents, compared to the more Fe + Co-rich surfaces exposed to sea water. Episodic rolling-over of a nodule accounts for fluctuating concentrations of Mn, Fe, Ni, Cu, Co, and other metals across sectioned manganese nodules.

Roger G. Burns; Virginia Mee-Burns

Mango A tree (*Mangifera indica*) of the family Anacardiaceae that originated in the Indo-Burma region and is now grown throughout the world. The mango is a medium to large evergreen tree; it produces a dense, round canopy, with leaves which are reddish brown when young and dark green when mature.

In the United States, mangoes are grown only in Florida and on a small scale in Hawaii, usually as backyard trees. Their greatest importance is in India, constituting 75% of the world's area of mango production. The ripe fruit is eaten raw as a dessert or used in the manufacture of juice, jams, jellies, and preserves. Unripe fruit can be made into pickles or chutneys. Mangoes are a good source of vitamins A and C. *See* FRUIT, TREE; SAPINDALES.

Robert M. Warner

Mangrove A taxonomically diverse assemblage of trees and shrubs that form the dominant plant communities in tidal, saline wetlands along sheltered tropical and subtropical coasts. The development and composition of mangrove communities depend largely on temperature, soil type and salinity, duration and frequency of inundation, accretion of silt, tidal and wave energy, and cyclone or flood frequencies. Extensive mangrove communities seem to correlate with areas in which the water temperature of the warmest month exceeds 75°F (24°C), and they are absent from waters that never exceed 75°F (24°C) during the year. Intertidal, sheltered, low-energy, muddy sediments are the most suitable habitats for mangrove communities, and under optimal conditions, forests up to 148 ft (45 m) in height can develop. Where less favorable conditions are found, mangrove communities may reach maturity at heights of only 3 ft (1 m). *See* ECOSYSTEM.

Plants of the mangrove community belong to many different genera and families, many of which are not closely related to one another phylogenetically. However, they do share a variety of morphological, physiological, and reproductive adaptations that enable them to grow in an unstable, harsh, and salty environment. Approximately 80 species of plants belonging to about 30 genera in over 20 families are recognized throughout the world as being indigenous to mangroves. About 60 species occur on the east coasts of Africa and Australasia, whereas about 20 species are found in the Western Hemisphere. At the generic level, *Avicennia* and *Rhizophora* are the dominant plants of mangrove communities throughout the world, with each genus having several closely related species in both hemispheres. At the species level, however, only a few species, such as the portia tree (*Thespesia populnea*), the mangrove fern (*Acrostichum aureum*), and the swamp hibiscus (*Hibiscus tiliaceus*), occur in both hemispheres.

The mangrove community is often strikingly zoned parallel to the shoreline, with a sequence of different species dominating from open water to the landward margins. These zones are the response of individual species to gradients of inundation frequency, waterlogging, nutrient availability, and soil salt concentrations across the intertidal area, rather than a reflection of ecological succession, as earlier studies had suggested. *See* ECOLOGICAL SUCCESSION.

Most plants of the mangrove community are halophytes, well adapted to salt water and fluctuations of tide level. Many species show modified root structures such as stilt or prop roots, which offer support on the semiliquid or shifting sediments, whereas others have erect root structures (pneumatophores) that facilitate oxygen penetration to the roots in a hypoxic environment. Salt glands, which allow excess salt to be extruded through the leaves, occur in several species; others show a range of physiological mechanisms that either exclude salt from the plants or minimize the damage excess salts can cause by separating the salt from the sensitive enzyme systems of the plant. Several species have well-developed vivipary of their seeds, whereby the hypocotyl develops while the fruit is still attached to the tree. The seedlings are generally buoyant, able to float over long distances in the sea and rapidly establish themselves once stranded in a suitable habitat. *See* PLANTS OF SALINE ENVIRONMENTS.

A mangrove may be considered either a sheltered, muddy, intertidal habitat or a forest community. The sediment surface of mangrove communities abounds with species that have marine affinities, including brightly colored fiddler crabs, mound-building mud lobsters, and a variety of mollusks and worms, as well as specialized gobiid fish (mudskippers). The waterways among the mangroves are important feeding and nursery areas for a variety of juvenile finfish as well as crustaceans. Animals with forest affinities that are associated with mangroves include snakes, lizards, deer, tigers, crab-eating monkeys, bats, and many species of birds.

Economically, mangroves are a major source of timber, poles, thatch, and fuel. The bark of some trees is used for tanning materials, whereas other species have food or medicinal value. *See* ECOLOGICAL COMMUNITIES; FOREST MANAGEMENT. Peter Saenger

Manifold (mathematics) A Hausdorff topological space with an n-dimensional atlas of charts for some integer n. The integer n is called the dimension of the manifold. If M is a Hausdorff space, an n-chart on M is a pair (U, φ) with U an open subset of M and φ a homeomorphism of U onto an open subset $\varphi(U)$ of n-dimensional euclidean space $\mathbf{R}^n$. An n-dimensional atlas for M is a system of n-charts (U, φ) such that the union of the sets U is all of M. *See* TOPOLOGY.

The terminology "charts" and "atlases" comes from the geographer's way of viewing the surface of the Earth. The manifold described in this case is the two-dimensional sphere $M = S^2$.

Many mathematicians include an additional restriction in the definition of manifold—sometimes the underlying topological space is connected, sometimes the topology is given by a metric, sometimes both.

With an appropriate additional structure on a manifold, it is possible to speak of differentiability or smoothness of real-valued functions. In this setting, the notion of a smooth mapping between smooth manifolds can be defined, and the differential of a smooth mapping, which is a generalization of the derivative, can be introduced.

Two charts (U_1, φ_1) and (U_2, φ_2) in an atlas for an n-dimensional manifold M are compatible (for defining smoothness) if the mapping $\varphi_2 \circ {\varphi_1}^{-1}$ from the open set $\varphi_1(U_1 \cap U_2)$ in $\mathbf{R}^n$ to the open set $\varphi_2(U_1 \cap U_2)$ is infinitely differentiable and has an infinitely differentiable inverse function. If all the charts in an atlas are compatible with one another, the atlas is said to determine a differentiable structure on M, and M with its atlas is then called a smooth manifold. Two atlases determine the same differentiable structure if all the charts in question are compatible with one another. Smooth manifolds are called also differentiable manifolds, differential manifolds, and C^∞ manifolds.

Alternate definitions of compatibility of charts lead to other classes of manifolds. Thus a manifold with its atlas is said to be piecewise linear (PL) or real analytic if all the mappings $\varphi_2 \circ {\varphi_1}^{-1}$ have the corresponding property.

If the dimension n is even, say $2m$, then the open sets in $\mathbf{R}^n$ can be regarded as open sets in the complex space $\mathbf{C}^m$ The manifold with its atlas is called a complex manifold of dimension m if the mappings $\varphi_2 \circ {\varphi_1}^{-1}$ are analytic functions of several variables. Complex manifolds are used in studying varieties in algebraic geometry. *See* ALGEBRAIC GEOMETRY; COMPLEX NUMBERS; SERIES.

Anthony W. Knapp

Mantophasmatodea An order of Insecta, known as heel-walkers or gladiators, occurring in Africa south of the Equator. Members of the 13 known species are 1–3 cm (0.4–1.2 in.) long, slender, moderately long-legged, and secondarily wingless. The habitus (general appearance) resembles a mixture of grasshopper, stick-insect, and praying mantis (see illustration).

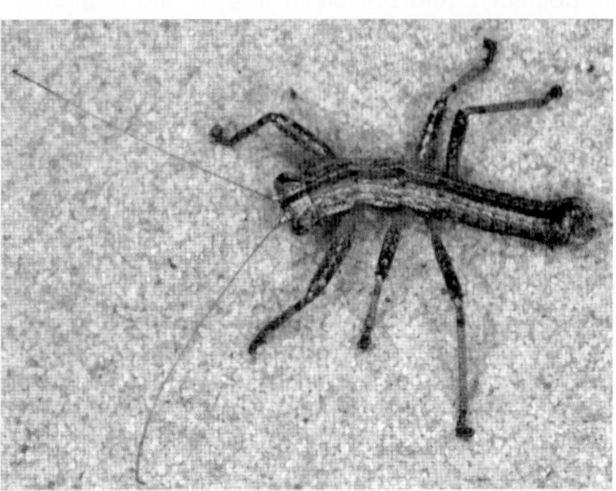

Male of *Karoophasma biedouwensis* from Succulent Karoo biome of western South Africa. (*Photo by M. Picker, University of Cape Town*)

The biting mouthparts are directed downward. The antennae are long and thin; the forelegs have a strong femur and two rows of spines on the tibia. The heel-walkers live singly and prey on insects, which they grasp using their strong, spiny forelegs. The terminal tarsomere (foot segment), with the claws, is usually held elevated (hence the common name).

Four species occur in Namibia (genera *Mantophasma, Sclerophasma, Praedatophasma*, and *Tyrannophasma*), eight in western South Africa (*Austrophasma, Karoophasma, Lobophasma, Hemilobophasma*, and *Namaquaphasma*), and one in Tanzania (*Tanzaniophasma*). Species differ mainly in the structure of the male and female genitalia; *Praedatophasma* and *Tyrannophasma* are conspicuously spinose (hence the common name "gladiator"). Mantophasmatodea inhabit several major biomes of southwestern Africa: the dry Nama Karoo and Succulent Karoo, and the more humid Fynbos. Fossils from the Lower Tertiary were found in Europe (*Raptophasma* and *Adicophasma* in Baltic amber). Klaus-Dieter Klass

Manufactured fiber Any of a number of textile fibers produced from chemical substances of natural origin or synthetic origin; the latter are also known as synthetic fibers. Among the natural sources of manufactured fibers are plant cellulose and protein, rubber, metals, and nonmetallic inorganics. The synthetic fibers are produced from organic intermediates derived from petroleum, coal, and natural gas. *See* NATURAL FIBER; TEXTILE.

With the exceptions of glass and metal fibers, the manufactured fibers are made from very long chainlike molecules called linear polymers. These polymers may be naturally occurring (cellulose from cotton or wood pulp) or may be synthetic (polyester). Irrespective of their chemical nature, fiber-forming polymers must possess the following characteristics: (1) great length—at least 200 monomer units must be joined in a chain; (2) a high degree of intramolecular and intermolecular attraction, whether through primary chemical bonds or other attractive forces; (3) the ability to be oriented along the axis of the fiber; and (4) the ability to form well-ordered crystals or pseudocrystals. All of these parameters are sensitive to the chemical nature of the polymer and the processes of manufacture of the fiber. In turn, they establish the properties of the fiber, such as strength, flexibility, resilience, and abrasion resistance, which contribute to their usefulness in various end uses for apparel, home furnishings, and commercial and industrial applications. *See* POLYMER.

Only a fraction of those substances capable of forming fibers prove to have all of the characteristics necessary for commercial success. The fiber types of major importance in the United States are classified by composition as follows: cellulosic (composed of regenerated cellulose, cellulose diacetate, and cellulose triacetate); synthetic (composed of polyamide, polyester, polyacrylic, polyvinyl, and polyolefin resins); and inorganic (composed of glass and metal).

All of the manufactured fibers are produced according to the same principles: (1) the fiber-forming material must first be made fluid; (2) the fluid is forced under pressure (extruded) through tiny holes into a medium which causes it to solidify; and (3) the solid fibers are further processed to obtain their optimum properties.

Typically one of three procedures is used to produce fibers. In wet spinning, (for example, the production of rayon by the viscose process), the polymer is dissolved in an applicable reagent to form the fluid (dope). The fluid is then pumped through metal plates (spinnerets) containing many small holes into a liquid bath of appropriate composition. A chemical reaction between the spinning dope and the bath causes the fiber to solidify. In dry spinning, the polymer is again dissolved in an appropriate solvent and extruded through a spinneret. However, the liquid bath is replaced by a stream of warm gas (usually air) which evaporates the solvent and allows the polymer to solidify as a filament. Cellulose diacetate and triacetate are produced in this manner. In melt spinning, Nylon, polyester, and the other thermoplastic fibers are produced by melt spinning. No solvents or reagents are required since the polymer can be melted without appreciable decomposition. Thus, the fluid consists of hot molten polymer which, upon extrusion into a stream of cold air, solidifies into a filament. Depending upon the end use, filaments may be produced in various sizes ranging from finer than a human hair to thick bristles for toothbrushes. They may also be produced with different cross-sectional shapes, such as round, lobed, square, or dogbone.

After extrusion, filaments are usually stretched (drawn). Drawing causes an increase in order (crystallinity) by extending the molecules of the fiber so that they pack more closely together, and orients the molecules along the longitudinal axis of the fiber. Higher orientation and increased crystallinity raise the strength of the fiber, decrease its stretch, and improve its elasticity.

Often, the manufactured fibers are textured to improve their comfort properties. Fabrics made from smooth, straight filament yarns are not as comfortable as those made from yarns spun from the shorter natural fibers. Texturing introduces irregularities (crimp) along the length of the filament and leads to bulkier filament yarns which are closer to spun yarns in their performance.

Advances in polymer and fiber technology have led to the development of fibers with exceptionally high temperature resistance and extremely high strength. These properties are desirable in applications such as upholstery and floor coverings in aircraft and other mass-transit vehicles, protective clothing for fire fighters and other emergency personnel, body armor for soldiers and police officers, tire cords, and industrial belting.

Metallic fibers of silver and gold have been used for millennia to decorate fabrics. Today metallic fibers serve useful as well as decorative purposes. These fibers are formed by drawing metal wires through successively finer dies to achieve the desired diameter. Although gold and silver are the easiest to draw, modern methods have allowed the manufacture of steel, tantalum, and zirconium fibers. Because they are electrical conductors, metal fibers have been blended into fabrics to reduce the tendency to develop static electrical charges.

Glass fibers are prepared by the melt spinning of previously formed glass marbles, and the molten filaments are drawn down to very fine dimensions. It is the fineness of the fibers that gives them their flexibility and allows them to be used in textiles. Unfortunately, the fibers are so stiff that when broken they can penetrate human skin. Thus, they are not well suited to use in apparel or upholstery. Glass is widely used in curtains and drapery because of its total resistance to the degrading effects of sunlight, its low cost, and its flame resistance. It provides a nonrotting, nonsettling insulating material for homes and industrial uses. *See* GLASS.

Fiber properties include the physical, mechanical, chemical, biological, and geometrical characteristics of fibers. Some of the more important ones are tensile strength, elongation at break, modulus of elasticity or stiffness, fatigue under repeated stress, resilience or ability to recover from deformation, moisture absorption and wettability, electrostatic properties, friction, color, luster, density, and resistance to light, heat, weathering, abrasion, laundering, mildew, insects, chemicals, and solvents; and finally a number of geometric features, such as diameter, cross-sectional shape, and crimp. Such properties play an important part in determining whether or not the fiber can be made into a fabric that will be wrinkle-resistant, pleasing to the touch, comfortable, easy to clean, durable, and attractive in color, luster, drape, and general appearance. With a knowledge of the physical properties of the available fibers, the textile engineer can choose the best fiber or best blend of several fibers to fit the intended use. The final result, however, is also dependent upon the proper choice and control of additional factors such as the yarn and fabric structure, the weave pattern, and the finishing of the cloth. Ira Block

Manufacturing engineering Engineering activities involved in the creation and operation of the technical and economic processes that convert raw materials, energy, and purchased items into components for sale to other manufacturers or into end products for sale to the public. Defined in this way, manufacturing engineering includes product design and manufacturing system design as well as operation of the factory. More specifically, manufacturing engineering involves the analysis and modification of product designs so as to assure manufacturability; the design, selection, specification, and optimization of the required equipment, tooling, processes, and operations; and the determination of other technical matters required to make a given product according to the desired volume, timetable, cost, quality level, and other specifications. *See* PROCESS ENGINEERING; PRODUCTION.

The formulation of a process plan for a given part has seven aspects: (1) a thorough understanding of processing techniques, their yield and their reliability, precedences, and constraints (both economic as well as technical); (2) the material and tolerances of the part; (3) proper definition of machinability or process data; (4) proper work-holding design of the stock or piece part during the fabrication process, a key consideration in generating piece parts of consistent quality; (5) proper tool selection for the task; (6) the capability of the equipment selected; and (7) personnel skills required and available.

Process planning aids based on computer programs that incorporate a type of spread sheet can be used to reduce significantly the time required to generate individual process plans. For example, systems have been developed that calculate the cycle time for each part as well as the number of tools used per part, the number of unique tools per part set, and the total time for cutting operations per tool type. *See* COMPUTER-AIDED DESIGN AND MANUFACTURING.

In parallel with the definition of process equipment, the manufacturing system designer must determine the most appropriate materials-handling techniques for the transfer of parts from machine to machine of each family of parts. During the manufacture of pieces, the parts are organized by the type of feature desired. The parts are then grouped and manufactured as a family, a method known as group technology. This includes the selection of storage devices appropriate for raw material, work in process, and finished-goods inventory as well as fixtures, gages, and tooling. Materials-handling equipment may be very different for each family, depending on part size and weight, aggregate production volume, part quality considerations during transfer, and ease of loading and unloading candidate machines. Different materials-handling approaches may also be appropriate within individual fabrication systems. *See* MATERIALS-HANDLING EQUIPMENT.

A quality assurance philosophy must be developed that emphasizes process control as the means to assure part conformance rather than emphasizing the detection of part nonconformance as a means of detecting an out-of-control process. The success of any fabrication process is based on rigid work-holding devices that are accurately referenced to the machine, accurate tool sizing, and tool position control. The basic way to determine if these three factors are functioning together acceptably is to measure a feature they produce as they produce it or as soon as possible after that feature is machined. The primary objective of this measurement is to determine that the combination is working within acceptable limits (statistical process control); the fact that the part feature is in conformance to print (conformance to tolerance specified on the print/drawing) is a by-product of a process that is in control. *See* QUALITY CONTROL.

In a modern manufacturing environment an organization's strategies for highly automated systems and the role for workers in these systems are generally based on one of two distinct philosophical approaches. One approach views workers within the plant as the greatest source of error. This approach uses computer-integrated manufacturing technology to reduce the workers' influence on the manufacturing process. The second approach uses computer-integrated manufacturing technology to help the workers make the best product possible. It implies that workers use the technology to control variance, detect and correct error, and adapt to a changing marketplace. *See* COMPUTER-INTEGRATED MANUFACTURING.

The best approach utilizes the attributes of employees in the factory to produce products in response to customer demand. This viewpoint enables the employees to exert some control over the system, rather than simply serving it. The employees can then use the system as a tool to achieve production goals.

As technology and automation have advanced, it has become necessary for manufacturing engineers to gain a much broader perspective. They must be able to function in an integrated activity involving product design, product manufacture, and product use. They also have to consider how the product will be destroyed as well as the efficient recovery of the materials used in its manufacture.

Manufacturing engineers must also be able to use an increasing array of computerized support tools, ranging from process planning and monitoring to total factory simulation—and in some cases, including models of the total enterprise. *See* SIMULATION. J. L. Nevins

Map design The systematic process of arranging and assigning meaning to elements on a map for the purpose of communicating geographic knowledge in a pleasing format. Careful design is crucial to map effectiveness to avoid distorted or inaccurately represented information.

The first design stage involves determining the type of map to be created for the problem at hand. Decisions must be made about the map's spatial format in terms of size and shape, the basic layout, and the data to be represented. In this step, the experience, cultural background, and educational attainment of the intended audience must be considered. The second stage involves the exploration of preliminary ideas through the manipulation of design parameters such as symbols, color, typography, and line weight. In the third step, alternatives are evaluated and may be accepted or rejected. Under some circumstances, prototype maps may be developed for sample readers as a means of evaluating design scenarios. The last step involves the selection of a final design.

Design considerations include the selection of scale (the relationship between the mapping media format and the area being mapped), symbols to represent geographic features, the system of projection (the method used to translate Earth coordinates to flat media), titles, legends, text, borders, and credits. The process of arranging each map element is referred to as map composition. Success in map composition is achieved when design principles are applied to create a pleasing image with a high degree of information content and readability.

Computers now duplicate all capabilities of manual map design. Design iterations can be explored and evaluated faster and with lower cost, compared to manual drafting methods. The digital environment facilitates the independent storage of map elements that can be combined to form composite images. Other computer innovations such as interactive mapping allow a map user to act as map creator in exploring geographic relationships by selecting and tailoring data sets available through software or the World Wide Web. In addition to its benefits to the design process, mapping software has brought challenges to the mapping sciences, including the proliferation of poorly designed maps constructed by persons untrained in cartography. *See* CARTOGRAPHY; COMPUTER GRAPHICS; MAP PROJECTIONS; MAP REPRODUCTION. Thomas A. Wikle

Map projections Systematic methods of transforming the spherical representation of parallels, meridians, and geographic features of the Earth's surface to a nonspherical surface, usually a plane. Map projections have been of concern to cartographers, mathematicians, and geographers for centuries because globes and curved-surface reproductions of the Earth are cumbersome, expensive, and difficult to use for making measurements. Although the term "projection" implies that transformation is accomplished by projecting surface features of a sphere to a flat piece of paper using a light source, most projections are devised mathematically and are drawn with computer assistance. The task can be complex because the sphere and plane are not applicable surfaces. As a result, each of the infinite number of possible projections deforms the geometric relationships among the points on a sphere in some way, with directions, distances, areas, and angular relationships on the Earth never being completely recreated on a flat map.

It is impossible to transfer spherical coordinates to a flat surface without distortion caused by compression, tearing, or shearing of the surface (see illustration). Conceptually, the transformation may be accomplished in two ways: (1) by geometric transfer to some other surface, such as a tangent or intersecting cylinder, cone, or plane, which can then be developed, that is, cut apart and laid out flat; or (2) by direct mathematical transfer to a plane of the directions and distances among points on the sphere. Patterns of deformation can be evaluated by looking at different projection families. Whether a projection is geometrically or mathematically derived, if its pattern of scale variation is like that which results from geometric transfer, it is classed as cylindrical, conic, or in the case of a plane, azimuthal or zenithal. *See* CARTOGRAPHY; TERRESTRIAL COORDINATE SYSTEM.

Cylindrical projections result from symmetrical transfer of the spherical surface to a tangent or intersecting cylinder. True or correct scale can be obtained along the great circle of tangency or the two homothetic small circles of intersection. If the axis of the cylinder is made parallel to the axis of the Earth, the parallels and meridians appear as perpendicular lines. Points on the Earth equally distant from the tangent great circle (Equator) or small circles of intersection (parallels equally spaced on either side of the Equator) have equal scale departure. The pattern of deformation therefore parallel the parallels, as change in scale occurs in a direction perpendicular to the parallels. A cylinder turned 90° with respect to the Earth's axis creates a transverse projection with a pattern of deformation that is symmetric with respect to a great circle through the Poles. Transverse projections based on the Universal Transverse Mercator grid system are commonly used to represent satellite images, topographic maps, and other digital databases requiring high levels of precision. If the turn of the cylinder is less than 90°, an oblique projection results. All cylindrical projections, whether geometrically or mathematically derived, have similar patterns of deformation. *See* GREAT CIRCLE, TERRESTRIAL.

Transfer to a tangent or intersecting cone is the basis of conic projections. For these projections, true scale can be found along one or two small circles in the same hemisphere. Conic projections are usually arranged with the axis of the cone parallel to the Earth's axis. Consequently, meridians appear as radiating straight lines and parallels as concentric angles. Conical patterns of deformation parallel the parallels; that is, scale departure is uniform along any parallel. Several important conical projections are not true conics in that their derivation either is based upon more than one cone (polyconic) or is based upon one cone with a subsequent rearrangement of scale variation. Because conic projections can be designed to have low levels of distortion in the midlatitudes, they are often preferred for representing countries such as the United States.

Azimuthal projections result from the transfer to a tangent or intersecting plane established perpendicular to a right line passing through the center of the Earth. All geometrically developed azimuthal projections are transferred from some point on this line. Points on the Earth equidistant from the point of tangency or the center of the circle of intersection have equal scale

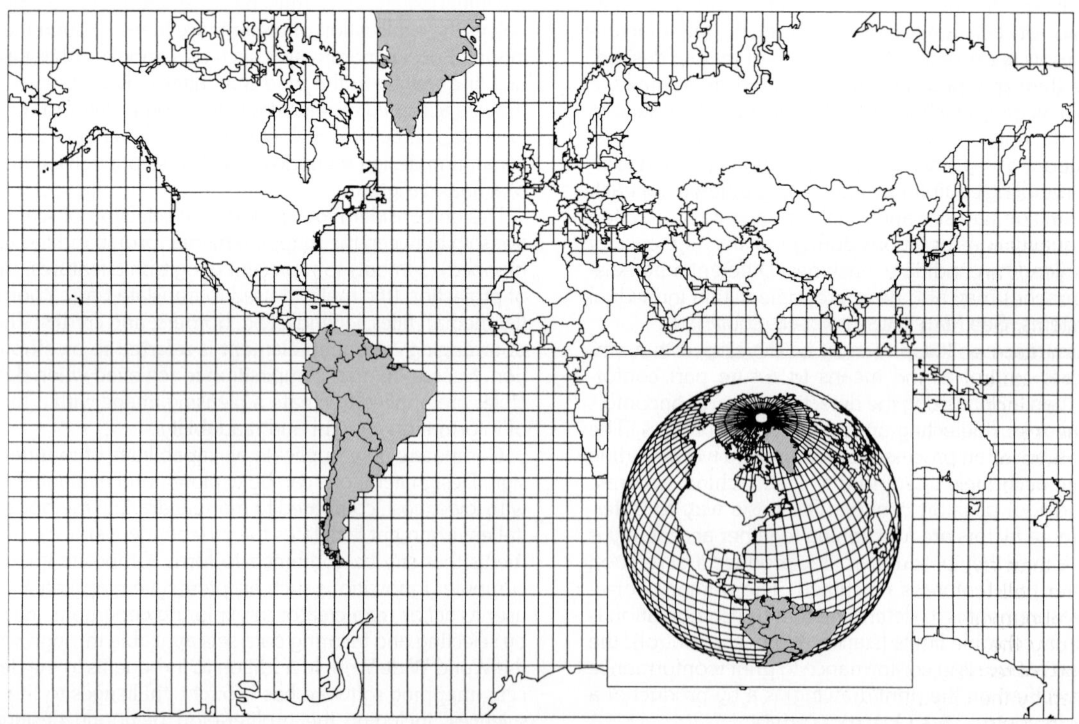

On this Mercator projection, (mathematically derived, cylindrical type), Greenland and South America appear similar in size. The inset map shows that South America is actually about 15 times larger than Greenland.

departure. Hence the pattern of deformation is circular and concentric to the Earth's center. All azimuthal projections, whether geometrically or mathematically derived, have two aspects in common: (1) all great circles that pass through the center of the projection appear as straight lines; and (2) all azimuths from the center are truly displayed. Arthur H. Robinson; Thomas A. Wikle

Maple A genus, *Acer*, of broad-leaved, deciduous trees including about 115 species in North America, Asia, Europe, and North Africa. This genus is characterized by simple, opposite, usually palmately lobed (rarely pinnate) leaves, generally inconspicuous flowers, and a fruit consisting of two long-winged samaras or keys (see illustration).

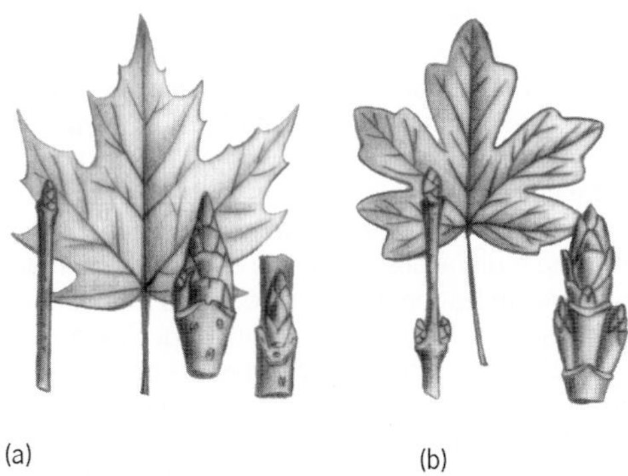

Characteristic maple leaves, twigs, and buds. (*a*) Sugar maple (*Acer saccharum*). (*b*) Hedge maple (*A. campestre*).

The most important commercial species is the sugar or rock maple (*A. saccharum*), called hard maple in the lumber market. This tree grows in the eastern half of the United States and adjacent Canada. It can be recognized by its gray furrowed bark, sharp-pointed scaly winter buds, and symmetrical oval outline of the crown.

Maple is an important source of hardwood lumber. Hard maple is used for flooring, furniture, boxes, crates, woodenware, spools, bobbins, motor vehicle parts, veneer, railroad ties, and pulpwood. It is the source of maple sugar and syrup and is planted as a shade tree. *See* SAPINDALES. Arthur H. Graves; Kenneth P. Davis

Marattiales An order (class Polypodiopsida) of primitive ferns that consists of about seven living genera and 150 species. They exist mostly in tropical regions. These large, coarse, sappy ferns usually have large, highly divided fronds. Stems of living genera (*Angiopteris* and *Marattia*) are short and bulbous, often a foot or more in diameter. Other genera have smaller, mostly horizontal, and somewhat more elongated stems. The internal vascular system (food and water conduction tissue) of the stems and fronds can be quite complex, apparently as a function of size. Marattialeans are included among the primitive ferns primarily because of the way that the sporangia (spore-bearing organs) arise from a mass of cells rather than from a single cell as in more highly evolved ferns. *See* PALEOBOTANY; POLYPODIOPHYTA. Benton M. Stidd

Marble A term applied commercially to any limestone or dolomite taking polish. Marble is extensively used for building and ornamental purposes. *See* DOLOMITE; LIMESTONE.

In petrography the term marble is applied to metamorphic rocks composed of recrystallized calcite or dolomite. Schistosity, often controlled by the original bedding, is usually weak except in impure micaceous or tremolite-bearing types. Calcite (marble) deforms readily by plastic flow even at low temperatures. Therefore, granulation is rare, and instead of schistosity there develops a flow structure characterized by elongation and bending of the grains concomitant with a strong development of twin lamellae. *See* METAMORPHIC ROCKS; MINERALOGY; SCHIST.

Pure marbles attaining 99% calcium carbonate, $CaCO_3$, are often formed by simple recrystallization of sedimentary limestone. Dolomite marbles are usually formed by metasomatism. *See* CALCITE; DOLOMITE; METASOMATISM. T. F. W. Barth

Marchantiales An order of the liverwort subclass Marchantiidae. Characteristic features include the differentiation of upper and lower tissues of the gametophyte, ventral scales, and rhizoids of two kinds. The sporophyte is considerably reduced, and the capsule dehisces irregularly. The archegonia, though dorsal, come to be pendant from an elevated receptacle because of differential growth resulting in decurved margins. The stalks of the receptacles are modified branches often with rhizoids in one, two, or rarely four furrows along their length. The order consists of 12 families grouped in two suborders, the more complex Marchantiineae and the simplified Ricciineae. *See* BRYOPHYTA; MARCHANTIIDAE. Howard Crum

Marchantiidae One of the two subclasses of liverworts (class Hepaticopsida). The gametophytes are ribbonlike or rosette-shaped thalli, usually showing considerable internal tissue differentiation. The rhizoids may be both smooth and internally pegged on the same thalli. Oil bodies, if present, are restricted to scattered cells that lack chloroplasts. The antheridia are usually ovoid, and the archegonia usually consist of six rows of cells. The sporophytes are generally reduced.

The subclass differs from the thallose Metzgeriales of the subclass Jungermanniidae in the internal differentiation of the thallose gametophytes, the smooth and pegged rhizoids, and the variation in the way the capsules dehisce. The subclass is divided into three orders, the Marchantiales, the Monocleales, and the Sphaerocarpales. *See* BRYOPHYTA; JUNGERMANNIIDAE; MARCHANTIALES; METZGERIALES; MONOCLEALES; SPHAEROCARPALES. Howard Crum

Margarine An emulsified fatty food product used as a spread and a baking and cooking fat, consisting of an aqueous phase dispersed in the fat as a continuous phase. Developed originally as a butter substitute, margarine is now considered a food in its own right and is manufactured in forms unknown to butter, such as plastic, soft, or fluid. However, margarine is colored, flavored, fortified with vitamins, and otherwise formulated to have the same or similar taste, appearance, and nutritional value as butter. *See* BUTTER.

Currently, margarine produced in the United States must contain not less than 80% fat. The fats and oils must be edible but may be from any vegetable or animal carcass source, natural or hydrogenated. The required aqueous phase may be water, milk, or solutions of dairy or vegetable protein, and must be pasteurized. Vitamin A must be added to yield a finished margarine with not less than 15,000 international units per pound (0.45 kg). Optional ingredients include salt or potassium chloride for low-sodium diets, nutritive sweeteners, fatty emulsifiers, antioxidants, preservatives, edible colors, flavors, vitamin D, acids, and alkalies. *See* FAT AND OIL (FOOD). Theodore J. Weiss

Marijuana The Spanish name for the dried leaves and flowering tops of the hemp plant, *Cannabis sativa* (Cannabinaceae). The narcotic ingredients allegedly have stimulating effects, and after smoking two or three cigarettes, the smoker often has a feeling of well-being and increased power and ability. After excessive amounts of the drug, illusions are often common, as well as pleasing, fanciful hallucinations. Sometimes the excessive user experiences disorientation and even delirium. *See* HEMP. Perry D. Strausbaugh; Earl L. Core

Marine biological sampling The collection and observation of living organisms in the sea, including the quantitative determination of their abundance in time and space. The biological survey of the ocean depends to a large extent on specially equipped vessels. Sampling in intertidal regions at low tide is one of the few instances where it is possible to observe and collect marine organisms without special apparatus.

A primary aim of marine biology is to discover how ocean phenomena control the distribution of organisms. Sampling is the means by which this aim is accomplished. Traditional techniques use samplers attached to wires lowered over the side of a ship by means of hydraulic winches. These samplers include bottles designed for collecting seawater from particular depths, fine-meshed nets that are towed behind the ship to sieve out plankton and fish, and grabs or dredges that collect animals inhabiting the ocean bottom. These types of gear are relied upon in many circumstances; however, they illustrate some of the problems common to all methods by which the ocean is sampled. First, sampling is never synoptic, which means that it is not possible to sample an area of ocean so that conditions can be considered equivalent at each point. Usually, it is assumed that this is so. Second, there are marine organisms for which there exists no sampling methodology. For example, knowledge of the larger species of squid is confined to the few animals that have been washed ashore, or the rare video recording. A third problem concerns the representativeness of the samples collected. The open ocean has no easily definable boundaries, and organisms are not uniformly distributed. The actual sampling is regularly done out of view of the observer; thus, sampling effectiveness is often difficult to determine. Furthermore, navigational systems are not error-free, so the position of the sample is never precisely known. All developments in methods for sampling the ocean try to resolve one or more of these difficulties by improving synopticity, devising more efficient sampling gear, or devising methods for observation such that more meaningful samples can be obtained.

Direct and remote observation methods provide valuable information on the undersea environment and thus on the representativeness of various sampling techniques. Personnel-operated deep-submergence research vessels (DSRVs) are increasingly being used to observe ocean life at depth and on the bottom, and for determining appropriate sampling schemes. The deep-submergence research vessels are used with cameras and television recording equipment and are also fitted with coring devices, seawater samplers, and sensors of various types. Other cameras are operated unattended at the bottom for months at a time, recording changes occurring there. Scuba diving is playing a larger role, especially in open-ocean areas, and is used to observe marine organisms in their natural habitat as well as to collect the more fragile marine planktonic forms such as foraminifera, radiolaria, and jellyfish. Autonomous underwater vehicles (AUVs) continue to assume greater importance in sampling programs since they can go to greater depths than can divers, and they overcome a limitation in diving in that AUVs can be operated at night and can sample under ice. Optical sensors carried aboard Earth-orbiting satellites can provide images of ocean color over wide areas. Ocean color is related to the turbidity and to the amount of plant material in the seawater. This establishes a means by which sampling programs carried out from ships can be optimized. *See* DIVING; SEAWATER FERTILITY; UNDERWATER PHOTOGRAPHY; UNDERWATER TELEVISION; UNDERWATER VEHICLE.

John Marr

Marine boiler A steam boiler designed to suit the marine environment and generally arranged to supply steam to the main propulsion machinery, ship's service electric generators, feed-pump drivers, and other auxiliary services.

Marine boilers are usually of the two-drum water-tube type with water-cooled furnaces, superheaters, desuperheaters, and heat recovery equipment of the economizer or air-heater type. The majority of ships are fitted with two boilers, although some large passenger ships may have three or more. Some cargo ships are fitted with only one boiler, and in some of these cases a smaller auxiliary boiler may be fitted for emergency or in-port steaming use.

Marine boilers are generally arranged for oil firing. Oil is used extensively because of its simplicity of handling and storing and the fact that it can be stored in spaces that often cannot be used for carrying cargo. *See* BOILER; MARINE MACHINERY; STEAM-GENERATING UNIT.

Robert P. Giblon

Marine conservation The management of marine species and ecosystems to prevent their decline and extinction. As in terrestrial conservation, the goal of marine conservation is to preserve and protect biodiversity and ecosystem function through the preservation of species, populations, and habitats. The importance of conserving marine species and ecosystems is growing as a consequence of human activities. Negative impacts on marine biological systems are caused by such actions as overfishing; overutilization, degradation, and loss of coastal and marine habitats; introduction of nonnative species; and intensification of global climate change, which alters oceanic circulation and disrupts existing trophic relationships. Marine conservation biologists seek to reduce the negative effects of all these actions by conducting directed research and helping to develop management strategies for particular species, communities, habitats, or ecosystems.

A variety of approaches and tools are used in marine conservation. These include population assessment; mitigation, recovery, and restoration efforts; establishment of marine protected areas; and monitoring programs. Many of these approaches overlap with those in terrestrial conservation. However, fundamental differences between terrestrial and marine environments in spatial dimension, habitat type, and organismal life history require that basic conservation techniques be modified for application to the marine environment. *See* BIODIVERSITY; ECOSYSTEM; MARINE ECOLOGY; OCEANOGRAPHY.

Effective management requires knowledge of the size and status of populations. Trends in abundance can be detected through stock assessment methods first developed for marine fisheries and subsequently modified for application to other marine organisms. These methods use estimates of population size, reproduction, survivorship, and immigration to determine whether populations are increasing, decreasing, or stable. Population viability analysis is a specialized statistical assessment in which demographic and environmental information is used to determine the probability that a population will persist in a particular environment for a specified period of time. This method can be used to guide management decisions, and has been used in efforts to manage marine mammals, turtles, seabirds, and other species. *See* ECOLOGICAL COMMUNITIES; ECOLOGICAL METHODS; POPULATION ECOLOGY.

Depleted, threatened, or endangered populations are often subject to mitigation or recovery efforts. The purpose of these efforts is to reduce the immediate threat of extinction or extirpation. This is typically achieved by direct human intervention to increase the size of a population or to prevent further decline in population size. Methods used to achieve recovery for fish and marine invertebrates include reducing fishing quotas, restricting the use of certain types of fishing gear, restricting the seasonal or annual distribution of a fishery, or closing fisheries altogether.

Recovery efforts can be most successful if they are based on multispecies or ecosystem-level management strategies. These strategies take into account positive and negative interactions between species, such as facilitation, competition, and predation. They further take into account interactions between species and their environment. Key to the success of assessment and recovery programs is identification of the appropriate biological unit for conservation (for example, population, subspecies, stock, or evolutionarily significant unit). Maintaining genetic diversity is an important goal of conservation biology, because genetic diversity confers evolutionary potential. Thus, conservation efforts often are aimed at populations that are genetically distinct from other populations of the same species.

Restoration efforts are aimed at returning habitats to an ecologically functional condition, usually consistent with some previous, more pristine condition. *See* ENDANGERED SPECIES; FISHERIES ECOLOGY; MARINE FISHERIES.

Marine protected areas are set aside for the protection or recovery of species, habitats, or ecosystems. They include marine parks, marine reserves, marine sancturaries, harvest refugia, and voluntary or legislated no-take areas. Some marine protected areas allow for consumptive use (such as fishing) or extraction of resources (for example, oil drilling), while others are closed to most human activities.

Monitoring programs are necessary to determine the outcome of specific conservation actions and to guide future conservation decisions. Monitoring programs vary according to the objectives of specific conservation projects but typically include such activities as long-term surveys of population size and status, and the development of mathematical models to help predict specific outcomes. Terrie Klinger

Marine containers Standardized rectangular boxes for the transport of marine cargo. Ocean transportation of cargo containers began in 1956. Containers can be loaded at a remote site, sealed, moved to the pier, and lifted into the ship without the need for intermediate handling. The principal benefits other than reduced pierside handling cost are less cargo damage and reduced pilferage.

Containerization has been the most significant change ever made to cargo transportation by sea. The now ubiquitous container transportation system has made break-bulk shipping almost obsolete on the major trade routes of the world. Even on routes serving minor outports, where cargo volumes have been insufficient to justify the landside infrastructure required for the container system, the phenomenon of containerization is increasingly important.

One aspect of the success of container transportation has centered on the standardization of equipment, which has allowed virtually any container to be carried anywhere in the world—by road, rail, or sea or any combination of these modes—without concern for the ability of the transport facilities to deliver the cargo, safely and untouched, to the ultimate destination, however remote it might be.

The predominant materials of construction are steel and aluminum, although a small number of containers have included reinforced plastic side panels. All are built around a steel frame to provide the requisite strength for lifting and stacking, in vessel holds and on deck, as well as during intermodal portions of the journey.

An increasing number of containers are specially fitted with racks and other devices for dedicated services, of which the transport of motor vehicles and vehicle components is the most common. Although a majority of containers are used for general cargo, an increasing number of specialized boxes carry refrigerated or frozen meats, fish, fruit, and other foods. Refrigerated boxes are also used to carry critical medical supplies such as blood plasma and insulin. These containers are equipped with self-contained refrigeration units powered from the ship's electrical system. Instrumentation systems are used to monitor and control cargo temperature. Tank containers are used to transport a vast array of oils, chemicals, and other liquids. Notable among these liquids are certain potable commodities and cryogenic gases, the latter typically carried in insulated, double-wall tanks. *See* MARINE REFRIGERATION; MERCHANT SHIP. Rod Vulovic

Marine ecology An integrative science that studies the basic structural and functional relationships within and among living populations and their physical-chemical environments in marine ecosystems. Marine ecology draws on all the major fields within the biological sciences as well as oceanography, physics, geology, and chemistry. Emphasis has evolved toward understanding the rates and controls on ecological processes that govern both short- and long-term events, including population growth and survival, primary and secondary productivity, and community dynamics and stability. Marine ecology focuses on specific organisms as well as on particular environments or physical settings. *See* ENVIRONMENT.

Marine environments. Classification of marine environments for ecological purposes is based very generally on two criteria, the dominant community or ecosystem type and the physical-geological setting. Those ecosystems identified by their dominant community type include mangrove forests, coastal salt marshes, submersed seagrasses and seaweeds, and tropical coral reefs. Marine environments identified by their physical-geological setting include estuaries, coastal marine and nearshore zones, and open-ocean-deep-sea regions. *See* DEEP-SEA FAUNA; ECOLOGICAL COMMUNITIES; HYDROTHERMAL VENT; PHYTOPLANKTON; ZOOPLANKTON.

An estuary is a semienclosed area or basin with an open outlet to the sea where fresh water from the land mixes with seawater. The ecological consequences of fresh-water input and mixing create strong gradients in physical-chemical characteristics, biological activity and diversity, and the potential for major adverse impacts associated with human activities. Because of the physical forces of tides, wind, waves, and fresh-water input, estuaries are perhaps the most ecologically complex marine environment. They are also the most productive of all marine ecosystems on an area basis and contain within their physical boundaries many of the principal marine ecosystems defined by community type. *See* ESTUARINE OCEANOGRAPHY; MANGROVE; SALT MARSH.

Coastal and nearshore marine ecosystems are generally considered to be marine environments bounded by the coastal land margin (seashore) and the continental shelf 300–600 ft (100–200 m) below sea level. The continental shelf, which occupies the greater area of the two and varies in width from a few to several hundred kilometers, is strongly influenced by physical oceanographic processes that govern general patterns of circulation and the energy associated with waves and currents. Ecologically, the coastal and nearshore zones grade from shallow water depths, influenced by the adjacent landmass and input from coastal rivers and estuaries, to the continental shelf break, where oceanic processes predominate. Biological productivity and species diversity and abundance tend to decrease in an offshore direction as the food web becomes supported only by planktonic production. Among the unique marine ecosystems associated with coastal and nearshore water bodies are seaweed-dominated communities (for example, kelp "forests"), coral reefs, and upwellings. *See* CONTINENTAL MARGIN; REEF; UPWELLING.

Approximately 70% of the Earth's surface is covered by oceans, and more than 80% of the ocean's surface overlies water depths greater than 600 ft (200 m), making open-ocean–deep-sea environments the largest, yet the least ecologically studied and understood, of all marine environments. The major oceans of the world differ in their extent of landmass influence, circulation patterns, and other physical-chemical properties. Other major water bodies included in open-ocean–deep-sea environments are the areas of the oceans that are referred to as seas. A sea is a water body that is smaller than an ocean and has unique physical oceanographic features defined by basin morphology. Because of their circulation patterns and geomorphology, seas are more strongly influenced by the continental landmass and island chain structures than are oceanic environments.

Within the major oceans, as well as seas, various oceanographic environments can be defined. A simple classification would include water column depths receiving sufficient light to support photosynthesis (photic zone); water depths at which light penetration cannot support photosynthesis and which for all ecological purposes are without light (aphotic zone); and the benthos or bottom-dwelling organisms. Classical oceanography defines four depth zones; epipelagic, 0–450 ft (0–150 m), which is variable; mesopelagic, 450–3000 ft (150–1000 m); bathypelagic, 3000–12,000 ft (1000–4000 m); and abyssopelagic, greater than 12,000 ft (4000 m). These depth strata correspond approximately to the depth of sufficient light penetration to support photosynthesis; the zone in which all light is attenuated; the truly aphotic zone; and the deepest oceanic environments.

Marine ecological processes. Fundamental to marine ecology is the discovery and understanding of the principles that underlie the organization of marine communities and govern their behavior, such as controls on population growth and stability, quantifying interactions among populations that lead to persistent communities, and coupling of communities to form viable ecosystems. The basis of this organization is the flow of energy and cycling of materials, beginning with the capture of radiant solar energy through the processes of photosynthesis and ending with the remineralization of organic matter and nutrients.

Photosynthesis in seawater is carried out by various marine organisms that range in size from the microscopic, single-celled marine algae to multicellular vascular plants. The rate of photosynthesis, and thus the growth and primary production of marine plants, is dependent on a number of factors, the more important of which are availability and uptake of nutrients, temperature, and intensity and quality of light. Of these three, the last probably is the single most important in governing primary production and the distribution and abundance of marine plants. Considering the high attenuation of light in water and the relationships between light intensity and photosynthesis, net autotrophic production is confined to relatively shallow water depths. The major primary producers in marine environments are intertidal salt marshes and mangroves, submersed seagrasses and seaweeds, phytoplankton, benthic and attached microalgae, and—for coral reefs—symbiotic algae (zooxanthellae). On an areal basis, estuaries and nearshore marine ecosystems have the highest annual rates of primary production. From a global perspective, the open oceans are the greatest contributors to total marine primary production because of their overwhelming size.

The two other principal factors that influence photosynthesis and primary production are temperature and nutrient supply. Temperature affects the rate of metabolic reactions, and marine plants show specific optima and tolerance ranges relative to photosynthesis. Nutrients, particularly nitrogen, phosphorus, and silica, are essential for marine plants and influence both the rate of photosynthesis and plant growth. For many phytoplankton-based marine ecosystems, dissolved inorganic nitrogen is considered the principal limiting nutrient for autotrophic production, both in its limiting behavior and in its role in the eutrophication of estuarine and coastal waters. *See* PHOTOSYNTHESIS.

Marine food webs and the processes leading to secondary production of marine populations can be divided into plankton-based and detritus-based food webs. They approximate phytoplankton-based systems and macrophyte-based systems. For planktonic food webs, current evidence suggests that primary production is partitioned among groups of variously sized organisms, with small organisms, such as cyanobacteria, playing an equal if not dominant role at times in aquatic productivity. The smaller autotrophs—both through excretion of dissolved organic compounds to provide a substrate for bacterial growth and by direct grazing by protozoa (microflagellates and ciliates)—create a microbially based food web in aquatic ecosystems, the major portion of autotrophic production and secondary utilization in marine food webs may be controlled, not by the larger organisms typically described as supporting marine food webs, but by microscopic populations.

Macrophyte-based food webs, such as those associated with salt marsh, mangrove, and seagrass ecosystems, are not supported by direct grazing of the dominant vascular plant but by the production of detrital matter through plant mortality. The classic example is the detritus-based food webs of coastal salt marsh ecosystems. These ecosystems, which have very high rates of primary production, enter the marine food web as decomposed and fragmented particulate organics. The particulate organics of vascular plant origin support a diverse microbial community that includes bacteria, flagellates, ciliates, and other protozoa. These organisms in turn support higher-level consumers.

Both pelagic (water column) and benthic food webs in deep ocean environments depend on primary production in the overlying water column. For benthic communities, organic matter must reach the bottom by sinking through a deep water column, a process that further reduces its energy content. Thus, in the open ocean, high rates of secondary production, such as fish yields, are associated with areas in which physical-chemical conditions permit and sustain high rates of primary production over long periods of time, as is found in upwelling regions.

Regardless of specific marine environment, microbial processes provide fundamental links in marine food webs that directly or indirectly govern flows of organic matter and nutrients that in turn control ecosystem productivity and stability. *See* BIOLOGICAL PRODUCTIVITY; ECOLOGY; ECOSYSTEM; SEAWATER FERTILITY.

Richard Wetzel

Marine engine An engine that propels a waterborne vessel. In all except the smallest boats, the engine is but part of an integrated power plant, which includes auxiliary machinery for propulsion engine support, ship services, and cargo, trade, or mission services. Marine engines in common use are diesel engines, steam turbines, and gas turbines. Gasoline engines are widely used in pleasure craft. *See* BOAT PROPULSION; INTERNAL COMBUSTION ENGINE; MARINE MACHINERY.

Diesel engines of all types and power outputs are in use for propulsion of most merchant ships, most service and utility craft, most naval auxiliary vessels, and most smaller surface warships and shorter-range submarines. The diesel engines most commonly used fall into either a low-speed category or the medium- and high-speed category. Low-speed engines are generally intended for the direct drive of propellers without any speed reduction, and therefore are restricted to a range of rotative speeds for which efficient propellers can be designed, generally below 300 revolutions per minute (rpm). The largest engines are rated for power output of over 5000 kW (almost 7500 horsepower) per cylinder at about 100 rpm. Because of their higher rotative speeds, medium- and high-speed engines drive propellers through speed-reduction gears, but they are directly connected for driving generators in diesel-electric installations. Large medium-speed engines are capable of over 1500 kW (2000 hp) per cylinder at about 400 rpm. The upper limit of the medium-speed category, and the start of the high-speed category, is generally placed in the range of 900–1200 rpm. *See* DIESEL ENGINE.

While steam-turbine plants cannot achieve the thermal efficiency of diesel engines, steam turbines of moderately high power levels (above about 7500 kW or 10,000 hp) offer efficient energy conversion from steam, which can in turn be produced by combustion of low-quality fuel oil, coal, or natural gas in boilers, or from a nuclear reactor. For high efficiency, high turbine speeds are required, typically 3000–10,000 rpm, with reduction gearing or electric drive used to achieve low propeller rotative speeds. The combination of turbine and reduction gear or electric drive has usually proven robust and durable, so that most oil-fueled steamships currently in service are held over from an earlier era. Others, more recently built, are capitalizing on the availability, in their trade, of a fuel unsuitable for diesel engines. *See* STEAM TURBINE.

Aircraft-derivative gas turbines have become the dominant type of propulsion engine for medium-sized surface warships, including frigates, destroyers, cruisers, and small aircraft carriers. In all cases the turbines are multishaft, simple-cycle engines, with the power turbine geared to the propeller. In some installations, two to four turbines are the sole means of propulsion; in other cases, one or two turbines provide high-speed propulsion, while diesel engines or smaller gas turbines are used for cruising speeds. Factors favoring the aircraft-derivative gas turbine in this application are low weight, compact dimensions, high power, rapid start and response, standardization of components, and maintenance by replacement. *See* GAS TURBINE.

In the electric drive arrangement, the engine is directly coupled to a generator, and the electricity produced drives an electric motor, which is most often of sufficiently low rotative speed to be directly connected to the propeller shaft. Any number of engine-generator sets may be connected to drive one or more propulsion motors. Electric drive has been used with engines of all types, including low-speed diesels. Advantages of electric drive include flexibility of machinery arrangement, elimination of gear noise, high propeller torque at low speed, and inherent reversing capability. In ships with high electric requirements for cargo, mission, or trade services—for example, passenger ships, tankers with electric-motor-driven cargo pumps, or warships with laser weapons—there is an advantage in integrating propulsion and ship service support through a common electric distribution system. However, electric drive is usually heavier, higher in initial cost, and less efficient than direct or geared drive. Alan L. Rowen

Marine engineering The engineering discipline concerned with the machinery and systems of ships and other marine vehicles and structures. Marine engineers are responsible for the design and selection of equipment and systems, for installation and commissioning, for operation, and for maintenance and repair. They must interface with naval architects, especially during design and construction.

Marine engineers are likely to have to deal with a wide range of systems, including diesel engines, gas turbines, boilers, steam turbines, heat exchangers, and pumps and compressors; electrical machinery; hydraulic machinery; refrigeration machinery; steam, water, fuel oil, lubricating oil, compressed gas, and electrical systems; equipment for automation and control; equipment for fire fighting and other forms of damage control; and systems for cargo handling. Many marine engineers become involved with structural issues, including inspection and surveying, corrosion protection, and repair.

Marine engineers are generally mechanical engineers or systems engineers who have acquired their marine orientation through professional experience, but programs leading to degrees in marine engineering are offered by colleges and universities in many countries. *See* BOAT PROPULSION; MARINE BOILER; MARINE ENGINE; MARINE MACHINERY; MARINE REFRIGERATION; NAVAL ARCHITECTURE; PROPELLER (MARINE CRAFT); SHIP DESIGN; SHIP POWERING, MANEUVERING, AND SEAKEEPING. Alan L. Rowen

Marine fisheries The harvest of animals and plants from the ocean to provide food and recreation for people, food for animals, and a variety of organic materials for industry. Important products in marine capture fisheries include fish; mollusks such as oysters, clams, and squid; and crustaceans such as crabs and shrimps. Marine mammals (whales) and reptiles (turtles) have been important historically in marine landings (the weight of the catch landed at the wharf); and some plants, mostly seaweeds, are harvested in significant amounts. As practiced in recent decades, unregulated or poorly managed fisheries can lead to collapse of major fish stocks. *See* FISHERIES ECOLOGY; FOOD MANUFACTURING. Edward D. Houde

Marine geology The study of the portion of the Earth beneath the oceans. Approximately 70% of the Earth's surface is covered with water. Marine geology involves the study of the sea floor; of the sediments, rocks, and structures beneath the sea floor; and of the processes that are responsible for their formation. The average depth of the ocean is about 3800 m (12,500 ft), and the greatest depths are in excess of 11,000 m (36,000 ft; the Marianas Trench). The study of the sea floor necessitates employing a complex suite of techniques to measure the characteristic properties of the Earth's surface beneath the oceans. Contrary to popular views, only a minority of marine geological investigations involve the direct observation of the sea floor by scuba diving or in submersibles. Instead, most of the ocean floor has been investigated by surface ships using remote-sensing geophysical techniques, and more recently by satellite observations.

The oceanic crust is relatively young, having been formed entirely within the last 200 million years (m.y.), a small fraction of the nearly 5-billion-year history of the Earth. The process of renewing or recycling the oceanic crust is the direct consequence of plate tectonics and sea-floor-spreading processes. It is therefore logical that the geologic history of the sea floor be outlined within the framework of plate tectonic tenets. Where plates move apart, molten lava reaches the surface to fill the voids, creating new oceanic crust. Where the plates come together, oceanic crust is thrust back within the interior of the Earth, creating the deep oceanic trenches. These trenches are located primarily around the rim of the Pacific Ocean. The down-going material can be traced by using the distribution of earthquakes to depths of about 700 km (420 mi). At that level, the character of the subducted lithosphere is lost, and this material is presumably remelted and assimilated with the surrounding upper-mantle material. *See* EARTHQUAKE; GEODYNAMICS; LITHOSPHERE; PLATE TECTONICS.

Mid-oceanic ridges. Most of the ocean floor can be classified into three broad physiographic regions, one grading into the other (see illustration). The approximate centers of the ocean basin are characterized by spectacular, globally encircling mountain ranges, the mid-oceanic ridge (MOR) system, which formed as the direct consequence of the splitting apart of oceanic lithosphere. The detailed morphologic characteristics of these mountain ranges depend somewhat upon the rate of separation of the plates involved. Abyssal hill relief, especially within 500 km (300 mi) of ridge crest, is noticeably rougher on the slow-spreading Mid-Atlantic Ridge than on the fast-spreading East Pacific Rise. The profile of the East Pacific Rise is also broader and shallower than for the Mid-Atlantic Ridge.

The broad cross-sectional shape of this mid-ocean mountain range can be related directly and simply to its age. The depth of the mid-oceanic ridge at any place is a consequence of the steady conduction of heat to the surface and the associated cooling of the oceanic crust and lithosphere. As it cools, contracts, and becomes denser, the oceanic crust plus the oceanic lithosphere sink isostatically (under its own weight) into the more fluid substrate (the asthenosphere). The depth to the top of the oceanic crust is a predictable function of the age of that crust; departures from such depth predictions represent oceanic depth anomalies. These depth anomalies are presumably formed because of processes other than lithospheric cooling, such as intraplate volcanism. The Hawaiian island chain and the Polynesian island

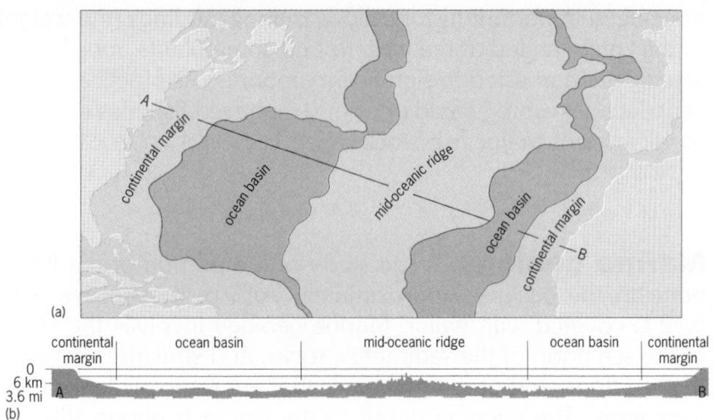

Geology of the North Atlantic Ocean. (*a*) Physiographic divisions of the ocean floor. (*b*) Principal morphologic features along the profile between North America and Africa.

groups are examples of this type of volcanism. *See* ASTHENOSPHERE; MID-OCEANIC RIDGE; OCEANIC ISLANDS; VOLCANOLOGY.

Basins. The deep ocean basins, which lie adjacent to the flanks of the mid-oceanic ridge, represent the older portions of the sea floor that were once the shallower flanks of the ridge (see illustration). The bulk of sediments found on the ocean floor can be broadly classified as terrigenous or biogenic. Terrigenous sediments are derived from drainage of adjacent landmasses and are brought to the sea floor through river systems. This sediment load is sometimes transported across the continental shelves, often utilizing, as pathways, submarine canyons that dissect the shelves, the continental slope, and the continental rise. Biogenic sediments are found in all parts of the ocean, intermixed either with terrigenous sediments or in near "pure form" in those areas inaccessible to terrigenous sedimentation.

Biogenic sediments are composed mostly of the undissolved tests of siliceous and calcareous microorganisms, which settle slowly to the sea floor. This steady so-called pelagic rain typically accumulates at rates of a few centimeters per thousand years. The composition and extent of the input to the biogenic sediment depend upon the composition and abundances of the organisms, which in turn are largely reflective of the water temperature and the available supply of nutrients. The Pacific equatorial zones and certain other regions of deep ocean upwelling are rich in nutrients and correspondingly rich in the microfauna and flora of the surface waters. Such regions are characterized by atypically high pelagic sediment rates. *See* UPWELLING.

Continental margins. The continental margins lie at the transition zone between the continents and the ocean basins and mark a major change from deep to shallow water and from thin oceanic crust to thick continental crust. Rifted margins are found bounding the Atlantic Ocean (see illustration). These margins represent sections of the South American and North American continents that were once contiguous to west Africa and northwest Africa, respectively. These supercontinents were rifted apart 160–200 m.y. ago as the initial stages of sea-floor spreading and the birth of the present Atlantic Ocean sea floor. *See* CONTINENTS, EVOLUTION OF.

Continental margins are proximal to large sources of terrestrial sediments that are the products of continental erosion. The margins are also the regions of very large vertical motions through time. This vertical motion is a consequence of healthy and subsequent cooling of the rifted continental lithosphere and subsidence. During initial rifting of the continents, fault-bound rift basins are formed that serve as sites of deposition for large quantities of sediment. These sedimentary basins constitute significant loads onto the underlying crust, giving rise to an additional component of margin subsidence. The continental margins are of particular importance also because, as sites of thick sediment accumulations (including organic detritus), they hold considerable potential for the eventual formation and concentration of hydrocarbons. As relatively shallow areas, they are also accessible to offshore exploratory drilling and oil and gas production wells. *See* OIL AND GAS, OFFSHORE.

Many sedimentary aprons or submarine fans are found seaward of prominent submarine canyons that incise the continental margins. Studies of these sedimentary deposits have revealed a number of unusual surface features that include a complex system of submarine distributary channels, some with levees. The channel systems control and influence sediment distribution by depositional or erosional interchannel flows. Fans are also effected by major instantaneous sediment inputs caused by large submarine mass slumping and extrachannel turbidity flows. *See* SUBMARINE CANYON.

In contrast to the rifted margins, the continental margins that typically surround the Pacific Ocean represent areas where plates are colliding. As a consequence of these collisions, the oceanic lithosphere is thrust back into the interior of the Earth; the loci of underthrusting are manifest as atypically deep ocean sites known as oceanic trenches. The processes of subducting the oceanic lithosphere give rise to a suite of tectonic and morphologic features characteristically found in association with the oceanic trenches. An upward bulge of the crust is created seaward of the trench that represents the flexing of the rigid oceanic crust as it is bent downward at the trench. The broad zone landward of most trenches is known as the accretionary prism and represents the accumulation of large quantities of sediment that was carried on the oceanic crust to the trench. Because the sediments have relatively little strength, they are not underthrust with the more rigid oceanic crust, but they are scraped off. In effect, they are plastered along the inner wall of the trench system, giving rise to a zone of highly deformed sediments. These sediments derived from the ocean floor are intermixed with sediments transported downslope from the adjacent landmass, thus creating a classic sedimentary melange. *See* CONTINENTAL MARGIN; SEDIMENTOLOGY.

Anomalous features. In addition to the major morphologic and sediment provinces, parts of the sea floor consist of anomalous features that obviously were not formed by fundamental processes of sea-floor spreading, plate collisions, or sedimentation. Examples are long, linear chains of seamounts and islands. Many of these chains are thought to reflect the motion of the oceanic plates over hot spots that are fixed within the mantle. *See* MAGMA; SEAMOUNT AND GUYOT.

The presence of large, anomalously shallow regions known as oceanic plateaus may also represent long periods of anomalous regional magmatic activity that may have occurred either near divergent plate boundaries or within the plate. Alternatively, many oceanic plateaus are thought to be small fragments of continental blocks that have been dispersed through the processes of rifting and spreading, and have subsequently subsided below sea level to become part of the submarine terrain.

Other important features of the ocean floor are the so-called scars represented by fracture zone traces that were formed as part of the mid-oceanic ridge system, where the ridge axis was initially offset. Oceanic crusts on opposite sides of such offsets have different ages and hence they have different crustal depths. A structural-tectonic discontinuity exists across this zone of ridge axis offset known as a transform zone. Although relative plate motion does not occur outside the transform zone, the contrasting properties represented by the crustal age differences create contrasting topographic and subsurface structural discontinuities, which can sometimes be traced for great distances. Fracture zone traces define the paths of relative motion between the two plates involved. Those mapped by conventional methods of marine survey have provided fundamental information that allows rough reconstructions of the relative positions of the

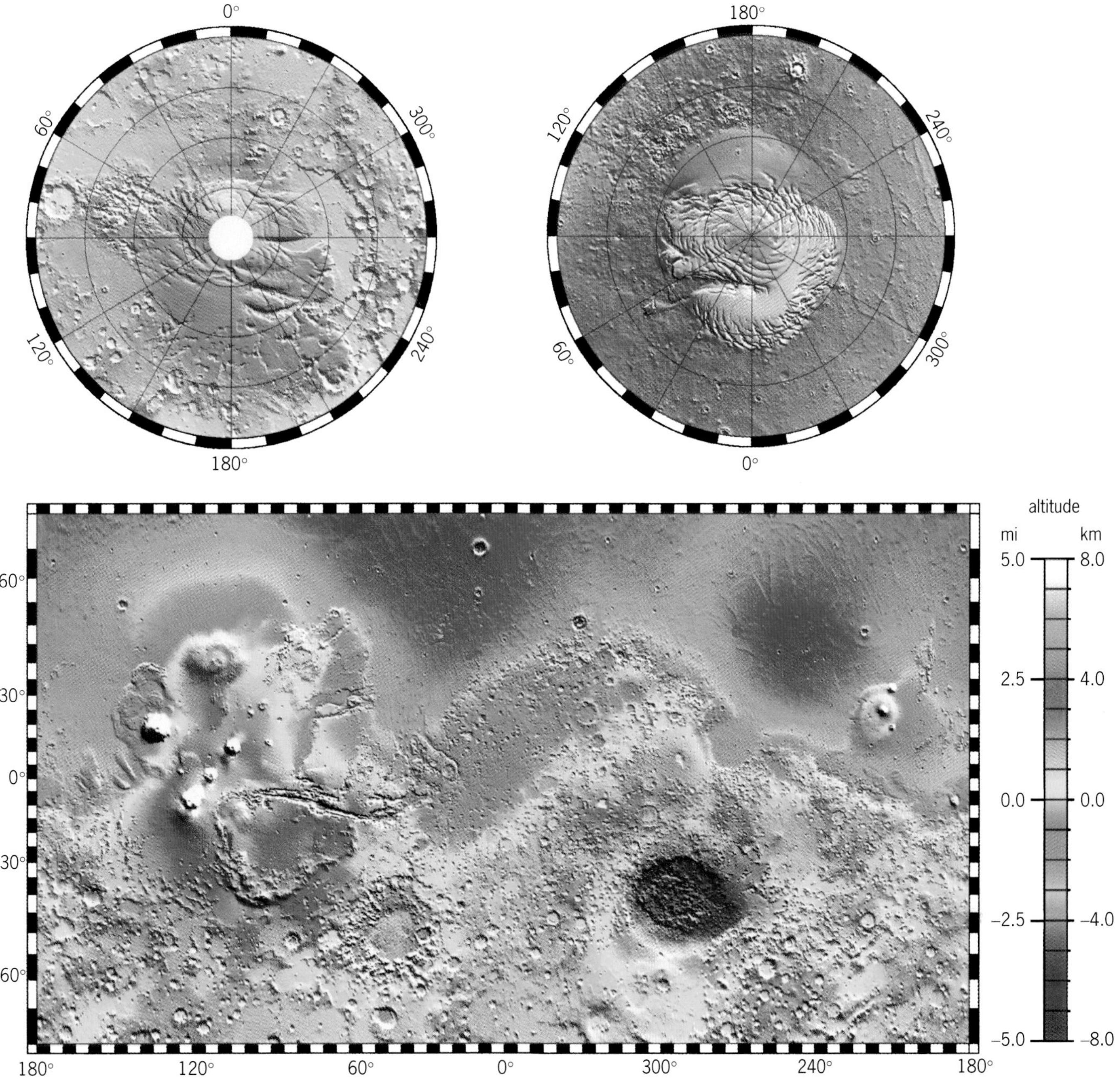

Colorplate 1. (Above) Stereographic projections of Martian topography, which reveal the puzzling dichotomy between the older southern highlands and low-lying northern plains. The large, deep depression centered at 290°W, 40°S is Hellas, the largest known impact basin in the solar system. (*NASA*)

Colorplate 2. (Right) The vast canyon complex called Valles Marineris splits Mars's equatorial region for more than 4000 km (2500 mi).

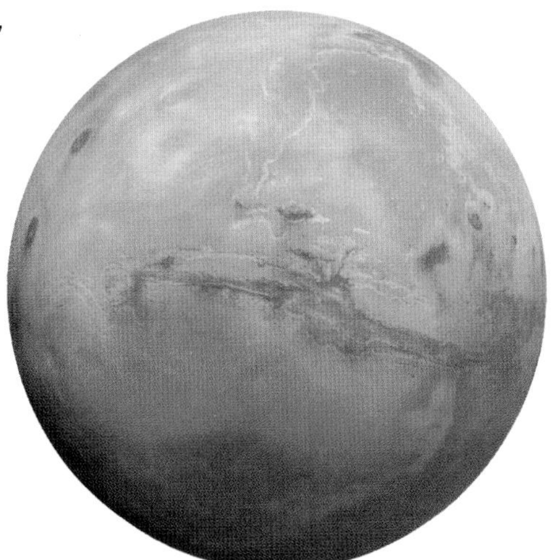

continents and oceans throughout the last 150–200 m.y. The study that deals with the relative motions of the plates is known as plate kinematics. *See* TRANSFORM FAULT.

Marginal seas. The sea-floor features described so far are representative of the main ocean basins and reflect their evolution mostly through processes of plate tectonics. Other, more complicated oceanic regions, typically found in the western Pacific, include a variety of small, marginal seas (back-arc basins) that were formed by the same general processes as the main ocean basins. These regions define a number of small plates whose interaction is also more or less governed by the normal tenets of plate tectonics. One difficulty in studying these small basins is that they are typified by only short-lived phases of evolution. Frequent changes in plate motions interrupt the process, creating tectonic overprints and a new suite of ocean-floor features. Furthermore, conventional methods of analyzing rock magnetism, heat flows, or depths of the sea floor to roughly date the underlying crust do not work well in these small regions. The small dimensions of these seas bring into play relatively large effects of nearby tectonic boundaries and render invalid key assumptions of these analytical techniques. The number of small plates that actually behave as rigid pieces is not well known, but it is probably only 10–20 for the entire world. Dennis E. Hayes

Marine machinery All machinery installed on waterborne craft, including engines, transmissions, shafting, propulsors, generators, motors, pumps, compressors, blowers, eductors, centrifuges, boilers and other heat exchangers, winches, cranes, steering gear, and associated piping, tanks, wiring, and controls, used for propulsion, for ship services, and for cargo, trade, or mission services.

Practically all marine machinery elements have nonmarine counterparts; in some cases, the latter were developed from marine applications, while in other cases specific equipment was "marinized." For marine service, machinery may have to meet higher standards of reliability and greater demands for weight and volume reduction and access for maintenance. Marine machinery must be capable of withstanding the marine environment, which tends toward extreme ambient conditions, high humidity, sea-water corrosion, vibration, sea motions, shock, variable demand, and fluctuating support services. Even higher standards may apply for warship machinery. To improve system reliability, essential equipment may be fitted in duplicate or provided with duplicated or alternative support or control systems, while nonessential equipment may be fitted with bypasses, to permit continued operation of a system following a component failure. Isolation valves or circuit breakers are common, enabling immediate repair.

Machinery on modern ships is highly automated, with propulsion usually directly controlled from the wheelhouse, and auxiliary machinery centrally controlled from an air-conditioned, sound-proofed control room, usually in the engine room. In the typical modern merchant ship (but not in passenger ships), the machinery operates automatically, and the controls are unattended at sea, with engineers called out by alarm in the event of malfunctions.

Propulsion machinery comprises an engine, usually a diesel engine, steam turbine, or gas turbine, with required gearing or other transmission system, and, for steam plants, steam generators. *See* BOAT PROPULSION; MARINE BOILER; MARINE ENGINE; PROPELLER (MARINE CRAFT); SHIP NUCLEAR PROPULSION. Alan L. Rowen

Marine microbiology An independent discipline applying the principles and methods of general microbiology to research in marine biology and biogeochemistry. Marine microbiology focuses primarily on prokaryotic organisms, mainly bacteria. Because of their small size and easy dispersability, bacteria are virtually ubiquitous in the marine environment. Furthermore, natural populations of marine bacteria comprise a large variety of physiological types, can survive long periods of starvation, and are able to start their metabolic activity as soon as a substrate becomes available. As a result, the marine environment, similar to soil, possesses the potential of a large variety of microbial processes that degrade (heterotrophy) but also produce (autotrophy) organic matter. Considering the fact that the marine environment represents about 99% of the biosphere, marine microbial transformations are of tremendous global importance. *See* BIOSPHERE.

Heterotrophic transformations. Quantitatively, the most important role of microorganisms in the marine environment is heterotrophic decomposition and remineralization of organic matter. It is estimated that about 95% of the photosynthetically produced organic matter is recycled in the upper 300–400 m (1000–1300 ft) of water, while the remaining 5%, largely particulate matter, is further decomposed during sedimentation. Only about 1% of the total organic matter produced in surface waters arrives at the deep-sea floor in particulate form. In other words, the major source of energy and carbon for all marine heterotrophic organisms is distributed over the huge volume of pelagic water mass with an average depth of about 3800 m (2.5 mi). In this highly dilute medium, particulate organic matter is partly replenished from dissolved organic carbon by microbial growth, the so-called microbial loop.

Of the large variety of organic material decomposed by marine heterotrophic bacteria, oil and related hydrocarbons are of special interest. Other environmentally detrimental pollutants that are directly dumped or reach the ocean as the ultimate sink by land runoff are microbiologically degraded at varying rates. Techniques of molecular genetics are aimed at encoding genes of desirable enzymes into organisms for use as degraders of particular pollutants.

A specifically marine microbiological phenomenon is bacterial bioluminescence, which may function as a respiratory bypass of the electron transport chain. Free-living luminescent bacteria are distinguished from those that live in symbiotic fashion in light organelles of fishes or invertebrates. *See* BIOLUMINESCENCE.

Photoautotrophs and chemoautotrophs. The type of photosynthesis carried out by purple sulfur bacteria uses hydrogen sulfide (instead of water) as a source of electrons and thus produces sulfur, not oxygen. Photoautotrophic bacteria are therefore limited to environments where light and hydrogen sulfide occur simultaneously, mostly in lagoons and estuaries. In the presence of sufficient amounts of organic substrates, heterotrophic sulfate-reducing bacteria provide the necessary hydrogen sulfide where oxygen is depleted by decomposition processes. Anoxygenic photosynthesis is also carried out by some blue-green algae, which are now classified as cyanobacteria. *See* CYANOBACTERIA; PHOTOSYNTHESIS.

Chemoautotrophic bacteria are able to reduce inorganic carbon to organic carbon (chemosynthesis) by using the chemical energy liberated during the oxidation of inorganic compounds. Their occurrence, therefore, is not light-limited but depends on the availability of oxygen and the suitable inorganic electron source. Their role as producers of organic carbon is insignificant in comparison with that of photosynthetic producers (exempting the processes found at deep-sea hydrothermal vents). The oxidation of ammonia and nitrite to nitrate (nitrification) furnishes the chemically stable and biologically most available form of inorganic nitrogen for photosynthesis. *See* NITROGEN CYCLE.

The generation of methane and acetic acid from hydrogen and carbon dioxide stems from anaerobic bacterial chemosynthesis, and is common in anoxic marine sediments. *See* METHANOGENESIS (BACTERIA).

Marine microbial sulfur cycle. Sulfate is quantitatively the most prominent anion in seawater. Since it can be used by a number of heterotrophic bacteria as an electron acceptor in respiration following the depletion of dissolved oxygen, the resulting sulfate reduction and the further recycling of the reduced sulfur compounds make the marine environment microbiologically distinctly different from fresh water and most soils. The marine anaerobic, heterotrophic sulfate-reducing bacteria are classified in three genera: *Desulfovibrio, Desulfotomaculum*, and *Clostridium*.

The marine aerobic sulfur-oxidizing bacteria fall into two groups: the thiobacilli and the filamentous or unicellular organisms. While the former comprise a wide range from obligately to facultatively chemoautotrophic species (requiring none or some organic compounds), few of the latter have been isolated in pure culture, and chemoautotrophy has been demonstrated in only a few.

Hydrothermal vent bacteria. Two types of hydrothermal vents have been investigated: warm vents (8–25°C or 46–77°F) with flow rates of 1–2 cm (0.4–0.8 in.) per second, and hot vents (260–360°C or 500–600°F) with flow rates of 2 m (6.5 ft) per second. In their immediate vicinity, dense communities of benthic invertebrates are found with a biomass that is orders of magnitude higher than that normally found at these depths and dependent on photosynthetic food sources. This phenomenon has been explained by the bacterial primary production of organic carbon through the chemosynthetic oxidation of reduced inorganic compounds. The chemical energy required for this process is analogous to the light energy used in photosynthesis and is provided by the geothermal reduction of inorganic chemical species. The specific compounds contained in the emitted vent waters and suitable for bacterial chemosynthesis are mainly hydrogen sulfide, hydrogen, methane, and reduced iron and manganese. The extremely thermophilic microorganisms isolated from hydrothermal vents belong, with the exception of the genus *Thermotoga,* to the Archaebacteria. Of eight archaeal genera, growing within a temperature range of about 75–110°C (165–230°F), three are able to grow beyond the boiling point of water, if the necessary pressure is applied to prevent boiling. These organisms are strictly anaerobic. However, unlike mesophilic bacteria, hyperthermophilic marine isolates tolerate oxygen when cooled below their minimum growth temperature. *See* ARCHAEBACTERIA; HYDROTHERMAL VENT. Holger W. Jannasch

Marine mining The process of recovering mineral wealth from sea water and from deposits on and under the sea floor. While mineral resources to the value of trillions of dollars do exist in and under the oceans, their exploitation is not simple. Many environmental problems must be overcome and many technical advances must be made before the majority of these deposits can be mined in competition with existing land resources.

The mineral resources of the marine environment are of three basic types: the dissolved minerals of the ocean waters; the unconsolidated mineral deposits of marine beaches, continental shelf, and deep-sea floor; and the consolidated deposits contained within the bedrock underlying the seas. As with land deposits, the initial stages preceding the production of a marketable commodity include discovery, characterization of the deposit to assess its value and exploitability, and mining, including beneficiation of the material to a salable product. Michael J. Cruickshank

Marine navigation The process of directing a watercraft to a destination in a safe and expeditious manner. From a known present position, a course is determined that avoids dangers, and on this course estimates are made of time schedules. The task is to make periodic adjustments using en route measurements of position or heading and speed.

The method used will depend on the type of vessel and on its role or mission. The devices available range from a simple compass to a host of sophisticated electronic systems. In all cases, the navigator must plan and prepare by setting instruments in order and by checking for predictable current and tidal effects and hazards to navigation en route. This preparation includes having the latest, correct charts and reviewing pertinent sections of sailing directions, tabulations of navigational lights, and tide and current tables.

The methods used to fulfill the requirements of these phases fall under one of the following broad categories of navigation: dead reckoning, piloting, celestial navigation, and electronic navigation. The first three categories have become somewhat standardized; the fourth category has been under constant and innovative development. Electronic navigation is rapidly evolving into integrated systems providing all navigation functions.

Electronic navigation now includes navigation involving any electronic device or instrument. A system that utilizes radio signals to provide lines of position or to fix the position of a craft involves the use of transmitters at accurately defined positions on the surface of the Earth or in satellites. Most common today is the use of satellite navigation using the Global Positioning System (GPS). In the near future new signals will be added to GPS, and the European Union is building a compatible system called Galileo. *See* ELECTRONIC NAVIGATION SYSTEMS.

The Automatic Identification System (AIS) was developed under the guidance of the International Maritime Organization (IMO), primarily to improve maritime safety by assisting the navigation of ships, the protection of the environment, and the operation of vessel traffic services (VTS). AIS serves in a vessel-to-vessel mode for collision avoidance, as a means for coastal states to obtain information about a ship and its cargo, and as a ship-to-shore method for vessel traffic management. The AIS device includes a Global Positioning System (GPS) receiver, a microprocessor, and a very high frequency–frequency-modulation (VHF-FM) transceiver. The microprocessor takes data from the ship's sensors and packages it with the vessel's identification into a digital signal, which is automatically broadcast. When other broadcasts are received, the processor prepares the received data for display.

Vessel traffic services (VTS) have been established in a number of heavily trafficked ports throughout the world in an attempt to reduce the number of collisions and strandings and safeguard the environment. Generally the service provides marine traffic management of an advisory nature, but in an especially hazardous situation it may be necessary for the VTS to exercise emergency control of vessel movements. *See* VESSEL TRAFFIC SERVICE.

Traffic-separation schemes have been established in a number of high-traffic density areas throughout the world, primarily to decrease the risk of collision at sea. A typical scheme consists of the establishment of parallel traffic lanes separated by an intervening buffer zone, analogous to a divided freeway on land.

Today, the navigator must learn to work with complex computer-based systems that integrate several sources of real-time information and compare the results to large quantities of stored hydrography and cartography. The trend is toward more integration with an accompanying assessment of the integrity or quality of separate information sources. In ascending order of sophistication, a brief description of the major shipboard systems follows:

An electronic chart displays on a video screen the same type of hydrographic information that mariners seek in a traditional nautical chart. Electronic charts integrated with a range of information, and with hardware and software that can process a hydrographic database to support decision making, are classified as electronic chart display and information systems (ECDIS). In addition to displaying a real-time picture of the vessel's position in the waterway, an ECDIS manages navigational and piloting information (typically, vessel-route-monitoring, track-keeping, and track-planning information) to support navigational decision making.

Collision avoidance systems (CAS), also known as automatic radar plotting aids (ARPA), of varying degrees of sophistication have been developed to reduce the work load of the navigator and eliminate human error. Typically, such a system consists of a digital computer that receives inputs from the ship's radar, compass, and log, and determines and displays collision threats, and in some installations provides a recommended avoiding action.

Integrated navigation systems integrate the functions of the electronic chart system and collision avoidance systems, and normally some level of automated control of vessel course and speed. The advent of the AIS gives a big technical push towards integration in order to correlate the AIS-derived collision avoidance information with the traditional radar tracks.

There are a variety of intelligent systems deployed aboard automated ship's bridges: piloting expert systems, engineering and vibration expert systems, neural network systems for adaptive and intelligent steering control, and automated intelligent docking systems. *See* EXPERT SYSTEMS; NEURAL NETWORK.

Integrated bridge systems are designed to allow the wheelhouse to function as the operational center for navigational and supervisory tasks aboard the ship. These bridges in many cases become ship's operations centers, incorporating controls and monitors for all essential vessel functions, including navigation, engine control, and communications. Other ship functions such as cargo loading, monitoring, and damage control may be included. *See* NAVIGATION.

Jay Spalding; Alton B Moody; Martha R. Grabowski; Richard Greenspan

Marine refrigeration Marine refrigerating equipment is used for shipboard refrigeration of products as well as for air conditioning the quarters of passengers and crew. Shipboard refrigeration is necessary for the preservation of perishables in transit and foodstuffs to be used by passengers and crew. Marine refrigeration is also used for maintaining certain cargo products in liquid form that would otherwise evaporate when stored at ambient conditions. *See* AIR CONDITIONING; REFRIGERATION.

The common refrigerants that have been in use for years, known as CFCs (chlorofluorocarbons) and HCFCs (hydrochlorofluorocarbons), are still being phased out because they contain chlorine, which depletes Earth's stratospheric ozone layer. Replacements for these refrigerants known as hydrofluorocarbons (HFCs), which are non-ozone-depleting, have been developed. These refrigerants contain fluorine in place of chlorine, and do not destroy the ozone layer. For some systems, the cargo itself is used as the refrigerant, such as when carrying ammonia or petroleum products such as propane or butane.

For refrigerating systems using insulated holds, liquid refrigerant is delivered to cooling coils, where the refrigerant is expanded, and enters the evaporator as a low-pressure liquid. There it evaporates by absorbing heat from the space being cooled. The refrigerant vapor is then drawn from the evaporator by a compressor, which delivers the refrigerant to a condenser as a high-pressure, high-temperature vapor. The condenser, normally a shell-and-tube heat exchanger cooled by seawater flowing through the tubes, removes the heat absorbed in the evaporator and the heat of compression from the refrigerant as it condenses to a high-pressure liquid. The refrigerating cycle is then repeated. For the container system, each container has its own refrigeration system consisting of a compressor, air-cooled condenser, cooling coil, and circulating fan. For liquid cargoes in insulated tanks, the latent heat of vaporization is released as the evaporating liquid changes state to a gas.

John H. Merold

Marine sediments The accumulation of minerals and organic remains on the sea floor. Marine sediments vary widely in composition and physical characteristics as a function of water depth, distance from land, variations in sediment source, and the physical, chemical, and biological characteristics of their environments. The study of marine sediments is an important phase of oceanographic research and, together with the study of sediments and sedimentation processes on land, constitutes the subdivision of geology known as sedimentology. *See* OCEANOGRAPHY.

Traditionally, marine sediments are subdivided on the basis of their depth of deposition into littoral 0–66 ft (0–20 m), neritic 66–660 ft (20–200 m), and bathyal 660–6600 ft (200–2000 m) deposits. This division overemphasizes depth. More meaningful, although less rigorous, is a distinction between sediments mainly composed of materials derived from land, and sediments composed of biological and mineral material originating in the sea. Moreover, there are significant and general differences between deposits formed along the margins of the continents and large islands, which are influenced strongly by the nearness of land and occur mostly in fairly shallow water, and the pelagic sediments of the deep ocean far from land.

Sediments of continental margins. These include the deposits of the coastal zone, the sediments of the continental shelf, conventionally limited by a maximum depth of 330–660 ft (100–200 m), and those of the continental slope. Because of large differences in sedimentation processes, a useful distinction can be made between the coastal deposits on one hand (littoral), and the open shelf and slope sediments on the other (neritic and bathyal). Furthermore, significant differences in sediment characteristics and sedimentation patterns exist between areas receiving substantial detrital material from land, and areas where most of the sediment is organic or chemical in origin.

Coastal sediments include the deposits of deltas, lagoons, and bays, barrier islands and beaches, and the surf zone. The zone of coastal sediments is limited on the seaward side by the depth to which normal wave action can stir and transport sand, which depends on the exposure of the coast to waves and does not usually exceed 66–100 ft (20–30 m); the width of this zone is normally a few miles. The sediments in the coastal zone are usually land-derived. The material supplied by streams is sorted in the surf zone; the sand fraction is transported along the shore in the surf zone, often over long distances, while the silt and clay fractions are carried offshore into deeper water by currents. Consequently, the beaches and barrier islands are constructed by wave action mainly from material from fairly far away, although local erosion may make a contribution, while the lagoons and bays behind them receive their sediment from local rivers.

The types and patterns of distribution of the sediments are controlled by three factors and their interaction: (1) the rate of continental runoff and sediment supply; (2) the intensity and direction of marine transporting agents, such as waves, tidal currents, and wind; and (3) the rate and direction of sea level changes. The balance between these three determines the types of sediment to be found. *See* DELTA; ESTUARINE OCEANOGRAPHY.

On most continental shelves, equilibrium has not yet been fully established and the sediments reflect to a large extent the recent rise of sea level. Only on narrow shelves with active sedimentation are present environmental conditions alone responsible for the sediment distribution. Sediments of the continental shelf and slope belong to one or more of the following types: (1) bio- genic (derived from organisms and consisting mostly of calcareous material); (2) authigenic (precipitated from sea water or formed by chemical replacement of other particles, for example, glauconite, salt, and phosphorite); (3) residual (locally weathered from underlying rocks); (4) relict (remnants of earlier environments of deposition, for example, deposits formed during the transgression leading to the present high sea level stand); (5) detrital (products of weathering and erosion of land, supplied by streams and coastal erosion, such as gravels, sand, silt, and clay).

Much of the fine-grained sediment transported into the sea by rivers is not permanently deposited on the self but kept in suspension by waves. This material is slowly carried across the shelf by currents and by gravity flow down its gentle slope, and is finally deposited either on the continental slope or in the deep sea. If submarine canyons occur in the area, they may intercept these clouds, or suspended material, channel them, and transport them far into the deep ocean as turbidity currents. If the canyons intersect the nearshore zone where sand is transported, they can carry this material also out into deep water over great distances. *See* CONTINENTAL MARGIN; REEF. Tjeerd H. Van Andel

Deep-sea sediments. In general, classifications are difficult to apply because so many deep-sea sediments are widely ranging mixtures of two or more end-member sediment types. However, they can be divided into biogenic and nonbiogenic sediments.

Biogenic sediments, those formed from the skeletal remains of various kinds of marine organisms, may be distinguished according to the composition of the skeletal material, principally either calcium carbonate or opaline silica. The most abundant contributors of calcium carbonate to the deep-sea sediments are the planktonic foraminiferids, coccolithophorids and pteropods. Organisms which extract silica from the sea water and whose hard parts eventually are added to the sediment are radiolaria, diatoms, and to a lesser degree, dilicoflagellates and sponges. The degree to which deep-sea sediments in any area are composed of one or more of these biogenic types depends on the organic productivity of the various organisms in the surface water, the degree to which the skeletal remains are redissolved by sea water while setting to the bottom, and the rate of sedimentation of other types of sediment material. Where sediments are composed largely of a single type of biogenic material, it is often referred to as an ooze, after its consistency in place on the ocean floor.

The nonbiogenic sediment constituents are principally silicate materials and, locally, certain oxides. These may be broadly divided into materials which originate on the continents and are transported to the deep sea (detrital constituents) and those which originate in place in the deep sea, either precipitating from solution (authigenic minerals) or forming from the alteration of volcanic or other materials. The coarser constituents of detrital sediments include quartz, feldspars, amphiboles, and a wide spectrum of other common rock-forming minerals. The finer-grained components also include some quartz and feldspars, but belong principally to a group of sheet-silicate minerals known as the clay minerals, the most common of which are illite, montmorillonite, kaolinite, and chlorite. The distributions of several of these clay minerals have yielded information about their origins on the continents and, in several cases, clues to their modes of transport to the oceans. Pierre E. Biscaye

Maritime meteorology Those aspects of meteorology that occur over, or are influenced by, ocean areas. Maritime meteorology serves the practical needs of surface and air navigation over the oceans. Phenomena such as heavy weather, high seas, tropical storms, fog, ice accretion, sea ice, and icebergs are especially important because they seriously threaten the safety of ships and personnel. The weather and ocean conditions near the air-ocean interface are also influenced by the atmospheric planetary boundary layer, the ocean mixed layer, and ocean fronts and eddies.

To support the analysis and forecasting of many meteorological and oceanographic elements over the globe, observations are needed from a depth of roughly 1 km (0.6 mi) in the ocean to a height of 30 km (18 mi) in the atmosphere. In addition, the observations must be plentiful enough in space and time to keep track of the major features of interest, that is, tropical and extratropical weather systems in the atmosphere and fronts and eddies in the ocean. Over populated land areas, there is a fairly dense meteorological network; however, over oceans and uninhabited lands, meteorological observations are scarce and expensive to make, except over the major sea lanes and air routes. Direct observations in the ocean, especially below the sea surface, are insufficient to make a synoptic analysis of the ocean except in very limited regions. Fortunately, remotely sensed data from meteorological and oceanographic satellites are helping to fill in some of these gaps in data. Satellite data can provide useful information on the type and height of clouds, the temperature and humidity structure in the atmosphere, wind velocity at cloud level and at the sea surface, the ocean surface temperature, the height of the sea, and the location of sea ice. Although satellite-borne sensors cannot penetrate below the sea surface, the height of the sea can be used to infer useful information about the density structure of the ocean interior. *See* REMOTE SENSING.

The motion of the atmosphere and the ocean is governed by the laws of fluid dynamics and thermodynamics. These laws can be expressed in terms of mathematical equations that can be put on a computer in the form of a numerical model and used to help analyze the present state of the fluid system and to forecast its future state. This is the science of numerical prediction, and it plays a very central role in marine meteorology and physical oceanography.

The first step in numerical prediction is known as data assimilation. This is the procedure by which observations are combined with the most recent numerical prediction valid at the time that the observations are taken. This combination produces an analysis of the present state of the atmosphere and ocean that is better than can be obtained from the observations alone. Data assimilation with a numerical model increases the value of a piece of data, because it spreads the influence of the data in space and time in a dynamically consistent way.

The second step is the numerical forecast itself, in which the model is integrated forward in time to predict the state of the atmosphere and ocean at a future time. Models of the global atmosphere and world ocean, as well as regional models with higher spatial resolution covering limited geographical areas, are used for this purpose. In meteorology and oceanography the success of numerical prediction depends on collecting sufficient data to keep track of meteorological and oceanographic features of interest (including those in the earliest stages of development), having access to physically complete and accurate numerical models of the atmosphere and ocean, and having computer systems powerful enough to run the models and make timely forecasts. *See* WEATHER FORECASTING AND PREDICTION. Robert L. Haney

Marjoram The aromatic herb *Majorana hortensis*, a common plant in Mediterranean areas. The spicy camphoraceous odor of marjoram has long been cherished as an addition to a wide variety of foods; in the Middle Ages marjoram was used as an air freshener. Marjoram is in the mint family (Lamiaceae) and is a close relative of European or Greek oregano, with which it is often confused. There is still controversy concerning the proper taxonomic classification of this plant. Some authors place it in the genus *Origanum*, while others continue to separate it into its own genus *Majorana*. *See* OREGANO.

Marjoram is a small perennial (1–2 ft or 30–60 cm tall), and has ovate leaves to 1 in. (2.5 cm) long. The leaves are slightly hairy, as are the erect somewhat woody stems. Majoram flowers are white to very light lavender or pink in color but are very small and usually go unnoticed; the entire flower spike or inflorescence is, however, easily noticeable.

The dried and fresh leaves are used as flavoring for meats (sausage), vegetables, cheeses, poultry stuffing blends, and sauces, especially tomato-based sauces. *See* LAMIALES; SPICE AND FLAVORING. Seth Kirby

Marl A sediment that consists of a mixture of calcium carbonate ($CaCO_3$) and any other constituents in varying proportions. Marls usually are fine-grained, so that most consist of $CaCO_3$ mixed with silt and clay. The dominant carbonate mineral in most marls is calcite ($CaCO_3$), but other carbonate minerals such as aragonite (another form of $CaCO_3$), dolomite [Ca, $Mg(CO_3)_2$], and siderite ($FeCO_3$) may be present. *See* ARAGONITE; CALCITE; CARBONATE MINERALS; DOLOMITE; SIDERITE.

In North America, the name marl may be limited to a lake deposit that is rich in $CaCO_3$, but usually the term is extended to include marine deposits. Deep-sea marls consist of mixtures of clay and the $CaCO_3$ skeletons of microscopic planktonic animals (foraminiferans) and plants (coccoliths and discoasters). Marls deposited in the deeper parts of lakes consist of fine-grained $CaCO_3$, but marls deposited in shallow water may contain $CaCO_3$ in the form of shells of mollusks and fragments of calcareous algae such as *Chara* (stonewort) mixed with fine-grained $CaCO_3$ that is precipitated on the leaves of rooted aquatic vegetation. The $CaCO_3$ from these various sources is then mixed with sand, silt, or clay brought in by streams from the surrounding drainage basin. *See* MARINE SEDIMENTS.

Although the term marl has been applied to sediments with a greatly variable content of $CaCO_3$, strictly speaking the $CaCO_3$ content should range between 30 and 70%. Because of the high $CaCO_3$ content, most marls are light to medium gray, although they can be almost any color. The high $CaCO_3$ content also tends to make dried marl earthy and crumbly.

The indurated rock equivalent of marl is marlstone. Equivalent terms range from calcareous claystone to argillaceous or impure limestone, depending on the amount of $CaCO_3$ that is present. Marlstone is common in marine sequences of all ages, and is particularly common in ancient lake sequences of Tertiary age in the western United States. For example, the famous "oil shale" of the Green River Formation of Wyoming, Colorado, and Utah is not a shale at all but a marlstone in which the dominant carbonate mineral is dolomite. *See* LIMESTONE; OIL SHALE; SEDIMENTARY ROCKS.

Walter E. Dean

Marmot A member of the genus *Marmota* in the squirrel family Sciuridae. Fourteen species occur mainly in mountainous areas in the northern hemisphere. Marmots are large, heavy-bodied rodents attaining weights of 3–7.5 kg (6.6–16.5 lb) and total lengths of 400–820 mm (16–32.8 in.). Fur color varies considerably among the different species. The upper parts may be gray, brown, or yellowish. The underside, although paler, normally differs little in color from the upper parts, although in *M. flaviventris* the belly is yellow. Some species have white markings. *See* RODENTIA; SQUIRREL

Marmots are found in open habitats such as pastures, forest edges, alpine meadows, and steppes. Most of their life is spent in burrows, which they excavate in well-drained soil. Marmots are primarily diurnal and terrestrial, but occasionally climb into shrubs and trees, especially when being pursued. Some are known to be good swimmers. Marmots are herbivorous. Food consists mainly of green vegetation such as grasses and forbs as well as fruits, grains, and legumes.

Most marmots inhabit northern mountainous areas where food is unavailable during the winter, thus making hibernation a necessity. They are true hibernators: they do not store food for winter but subsist entirely on fat reserves stored in their bodies. *See* HIBERNATION AND ESTIVATION.

Donald W. Linzey

Mars The fourth planet outward from the Sun. Mars is the planet most like Earth in terms of geology, climate, and suitability for life. It is also the world that has been most thoroughly studied by interplanetary spacecraft.

Mars has a mean heliocentric distance (semimajor axis) of 1.524 astronomical units, equivalent to 141.6×10^6 mi (227.9×10^6 km). Its orbital eccentricity of 0.093, one of the largest of the major planets, causes Mars's distance to the Sun to vary from 128×10^6 mi (207×10^6 km) at perihelion to 155×10^6 mi (249×10^6 km) at aphelion. Mars revolves around the Sun with a sidereal period of 1.881 years or 686.93 days. The planet comes closest to Earth, reaching opposition, every 780 days on average.

The mean diameter of Mars is 4212 mi (6780 km), or about 53% that of the Earth. The planet has a mass of 7.08×10^{20} tons (6.42×10^{23} kg), about 11% of Earth's. The sidereal period is 24h 37m 22.7s, corresponding to a mean solar day of 24h 39m 35.2s.

Mars appears to the unaided eye as a bright, slightly reddish star. Viewed through a telescope, Mars usually appears as a salmon-hued disk marked by mottled dark regions and white, seasonally variable polar caps.

Interior. Measurements of the planet's gravitational field made by orbiting spacecraft have shown that Mars has a dense core and thus is differentiated into a core, mantle, and crust. The crust, composed of silicate rocks, varies irregularly in thickness from about 22 mi (35 km) near the north pole to 45 mi (70 km) near the south pole. The reason for this gradation is not well understood. The surface exhibits no obvious evidence of terrestrial-style plate tectonism.

At present Mars does not have a global magnetic field of appreciable strength. However, early in its history Mars must have had a more substantial magnetic field, because its near-surface rocks exhibit strong remanent magnetism. *See* ROCK MAGNETISM.

Geology. The Martian surface has been modified extensively by the processes of impact cratering, volcanism, faulting, and fluvial erosion. The terrain in the southern hemisphere is very heavily cratered and thus quite old, having formed some 3.8×10^9 years ago. Much of it stands 1 to 2 mi (1 to 4 km) higher than the planet's mean radius. Although volcanic sources are not common at southern latitudes, surface rocks in the southern hemisphere are predominantly basaltic (silicon-poor) in mineralogy. The northern hemisphere is dominated by vast, lava-covered plains with relatively few impact craters and an average elevation about 3 mi (5 km) below the mean radius. Mineral compositions are typically andesitic (silicon-rich). This hemispheric dichotomy has no well-accepted explanation. The surface also exhibits numerous impact basins.

The Tharsis ridge, or bulge, is an uplifted portion of the surface that stands several kilometers above the mean elevation of the planet. Most of Mars's tectonic features are associated with Tharsis, which affects approximately one-quarter of the entire surface.

The Martian surface exhibits several enormous shield volcanoes. The largest and perhaps the youngest is Olympus Mons, which is nearly 370 mi (600 km) across at its base and stands approximately 16 mi (26 km) above the surrounding terrain. Three other large shield volcanoes—Ascraeus Mons, Pavonis Mons, and Arsia Mons—lie along the nearby Tharsis ridge. In shape and structure, the Martian shield volcanoes bear a strong resemblance to their Hawaiian counterparts. At the summit of each shield is a complex of calderas, collapsed craterlike features that were once vents for lava. *See* VOLCANO.

Perhaps the most spectacular features on the Martian surface are the huge canyons located primarily in the equatorial regions. Valles Marineris, actually a system of canyons, extends for over 3000 mi (5000 km) along the equatorial belt. In places, the canyon complex is as much as 300 mi (500 km) wide and drops to more than 4 mi (6 km) below the surrounding surface. Dwarfing the Grand Canyon, Valles Marineris is comparable in size to the great East African Rift Valley. In general, the walls of the canyon are precipitous and show evidence of slumping and landslide activity.

An astonishing class of features on the Martian surface is a widespread network of channels that bear a very strong

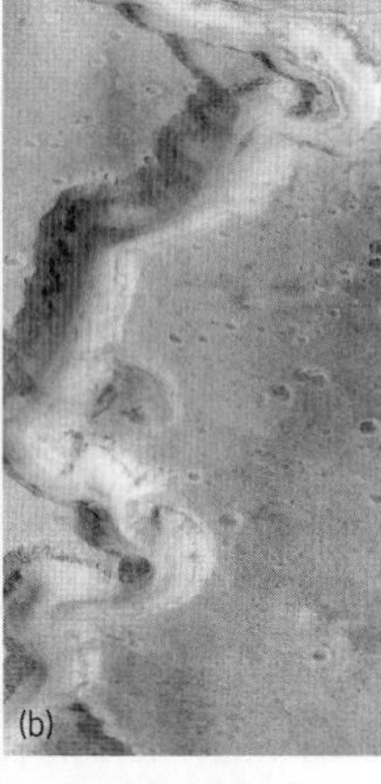

Fig. 1. Nanedi Vallis, a long, water-cut canyon on Mars, which is 1.5 mi (2.5 km) wide. (*a*) Canyon and surrounding terrain. (*b*) Detailed view of area in white rectangle in part a. Rocky outcrops jut from the canyon's walls, and a secondary rivulet is seen in its floor. (*NASA/Malin Space Science Systems*)

resemblance to dry river beds (Fig. 1). Ranging in size from broad, sinuous features nearly 40 mi (60 km) wide to small, narrow networks less than 300 ft (100 m) wide, the channels appear to have been created by water erosion. The largest channels must have been formed by enormous torrents of water, presumably released in some catastrophic manner.

Images from the *Mars Global Surveyor* spacecraft have revealed the existence of more than 120 small gullies cascading down the slopes of hillsides and crater rims that appear to have been carved by flowing water or mud. Because the gullies appear to be extremely fresh, they imply that liquid water in some form has flowed on the Martian surface within the past 10^6 years and perhaps much more recently.

Spacecraft images show that many canyon walls and crater floors exhibit extensive exposures of layered rock, up to 2.5 mi (4 km) thick in some places. Although their stratigraphic age is uncertain, these layers imply that the deposition of sediment was widespread early in Martian history.

Surface exploration. Six spacecraft have landed safely on the Martian surface and returned significant scientific data about its properties. Two American *Viking* landers arrived in 1976. *Mars Pathfinder*, which landed in 1997, carried a small, instrumented rover, *Sojourner*, which was directed from Earth via remote control to several rocks and fine-grained drifts in the lander's immediate vicinity.

The most extensive surface exploration has been conducted by the two Mars Exploration Rovers (MERs), *Spirit* and *Opportunity*, which landed on Mars in early 2004. Each rover was equipped with panoramic and microscopic cameras, three different spectrometers to characterize surface composition, magnets to collect metallic particles, and a grinding tool to expose fresh rock surfaces. Although the MERs had been designed to operate for 3 months, both were still functional in mid-2008, more than $4\frac{1}{2}$ years after their arrival. Both MER landing sites were chosen in the hope of finding evidence for the presence of liquid water at ground level in the geologically recent past. *Spirit* identified an iron oxide called goethite, which forms only in the presence of water. The *Opportunity* landing site exhibited countless millimeter-size concretions, nicknamed blueberries (Fig. 2). These spherules are the source of hematite detected spectroscopically from orbit, which forms only in wet environments. *Opportunity* also found that the outcrops are rich in sulfate minerals, evidence that the rock had been saturated by salt-laden water. The *Phoenix* lander reached Mars on May 25, 2008 to begin a close examination of the surface soil-ice layer and the atmosphere of Mars' northern polar region and to investigate whether the subsurface environment could ever have supported microbial life.

Meteorites from Mars. A small number of meteorites have reached Earth after being violently ejected from Mars during the collision of sizable asteroids or comets. The meteorites are thus samples from Mars itself. The first of these meteorites was recognized in 1982, and 36 distinct falls (some represented by multiple samples) were known by early 2006. They are sometimes collectively termed the SNC meteorites, representing their three subclasses: shergottites, nakhlites, and chassignites. The Martian meteorites contain a distinctive ratio of the isotopes of oxygen found nowhere on Earth, on the Moon, or in other meteorites. Almost all of them crystallized from molten rock 1.3×10^9 years ago, far more recently than the usual 4.6×10^9 years (the age of the solar system) found in all other meteorites. The lone exception to date, designated ALH 84001, crystallized 4.5×10^9 years ago during the planet's infancy. *See* METEORITE.

Atmosphere. The Martian atmosphere is very thin, with surface pressure averaging about 6.5 millibars (650 Pa), or less than 1% of the pressure at Earth's surface. The atmosphere consists principally of carbon dioxide (95.3%) but contains nitrogen, argon, oxygen, and a trace of water vapor, totaling 4.7%. Both carbon dioxide and water form clouds in the Martian atmosphere.

In 2004 a spectrometer aboard *Mars Express*, a spacecraft orbiting the planet, discovered trace amounts of methane in the Martian atmosphere. Methane cannot exist in equilibrium with the other gases, and scientists are unsure of its source.

Localized dust storms appear quite frequently on Mars. Because the Martian atmosphere is extremely thin, wind velocities greater than 90–110 mi/h (40–50 m/s) are needed to set surface dust grains in motion. Some dust storms develop with such intensity that their total extent may be hemipheric or even global.

Water inventory. The seasonal cycle of growth and decay of the bright polar caps has long been taken as evidence of the presence of water on Mars. However, evidence from the *Mariner* spacecraft identified carbon dioxide as the principal constituent of the polar snow, with lesser amounts of water ice also present. The polar caps at the end of winter cover a vast area, but during spring and summer the edges of the caps retreat. The north pole's residual cap consists almost entirely of water ice; although likely to be dominated by water ice, the southern cap is covered with frozen carbon dioxide. If all the water present in the polar caps were distributed uniformly over the surface of Mars, it would produce a layer 70–110 ft (22–33 m) deep.

Fig. 2. Image from the roving spacecraft *Opportunity* showing a rock outcrop at Meridiani Planum strewn with small spherical concretions, nicknamed blueberries, that contain the iron-bearing mineral hematite. This hematite type forms only in the presence of water. The light-colored circle, 2 in. (5 cm) across, was created by the rock abrasion tool on *Opportunity*. (*NASA/USGS*)

Although no spacecraft to date has detected the presence of liquid water directly, there is ample evidence that a great volume of ice lies beneath the surface. A gamma-ray spectrometer on the 2001 *Mars Odyssey* spacecraft has detected areally extensive subsurface layers enriched in hydrogen, whose only plausible source is water ice. A radar instrument on *Mars Express* has detected the likely presence of massive ice reservoirs buried beneath the plains surrounding both polar caps.

Even though at present liquid water cannot exist on the Martian surface, the planet's numerous large flood channels suggest that vast amounts of water flowed on the surface early in Martian history. Circumstantial evidence implies that much of the low-lying northern hemisphere was covered with an ocean roughly 4×10^9 years ago. Although unproven, this hypothesis suggests that 10^7 mi^2 (27×10^6 km^2) of the northern plains were inundated.

However, the surface of Mars exhibits widespread exposures of olivine, an iron-magnesium silicate mineral that readily reacts with water to form other compounds. The frequent occurrence of olivine implies that much of the Martian surface has not been exposed to liquid water for billions of years and argues that Mars has been cold and dry for much of its history.

Possibility of life. The ample evidence that early Mars had an abundance of liquid water, a thicker atmosphere, and a more clement environment has raised speculation that the planet may have once been, or may still be, an abode for primitive life forms. In 1976, the twin *Viking* landers conducted experiments on the Martian surface to test for the existence of life. One of the biological experiments gave a positive result—the release of oxygen from a soil sample when humidified—but there is no consensus as to the source of this reaction.

Twenty years later, a scientific team claimed that it had found evidence of fossilized microbes within a known Martian meteorite (ALH 84001). After years of intensive study, scientists have reached no consensus; the mineral associations within ALH 84001 and other Martian meteorites neither confirm the presence of once-living organisms nor exclude it.

Satellites. Mars has two small satellites, Phobos and Deimos. Neither satellite is massive enough to be gravitationally contracted to a spherical shape. Both satellites are saturated with impact craters. Both satellites have the same low albedo, approximately 0.05, so that surfaces are as dark as the very darkest asteroids. This low reflectivity suggests that the Martian satellites may have originated in the asteroid belt and been captured by Mars long ago. *See* ASTEROID. J. Kelly Beatty

Marsileales A small order of heterosporous, leptosporangiate ferns (division Polypodiophyta) which grow in water or wet places and are rooted to the substrate. The leaves arise on long stalks from the rhizome and typically have floating blades with four leaflets, suggesting a four-leaved clover. The order contains a single family and only 3 genera, about 50 species in all, most of them belonging to the widespread genus *Marsilea* (water clover). *See* POLYPODIOPHYTA. Arthur Cronquist

Marsupialia An order of animals, long considered the only order of the mammalian infraclass Metatheria.

The marsupials are characterized by the presence of a pouch (marsupium) in the female, a skin pocket whose teat-bearing abdominal wall is supported by epipubic bones. The young are born in the embryonic state and crawl unaided to the marsupium, where they attach themselves to the teats and continue their development. In a few species the pouch is vestigial or has disappeared completely.

A wide variety of terrestrial adaptations are found among the 82 genera of living marsupials. Some spectacular examples of evolution convergent upon placental modes of life are exemplified by the Australian marsupial "moles" (*Notoryctes*), "wolves" (*Thylacinus*), and "flying squirrels" (*Petaurus*). For the most part, the adaptive radiation of the marsupials has paralleled that of the placentals, yielding many ecological analogs, such as kangaroos and wallabies, which are the counterparts of the placental deer and antelopes, although these groups differ widely in structure. *See* ANTEATER; EUTHERIA; KANGAROO; KOALA; MAMMALIA; METATHERIA; OPOSSUM. Richard H. Tedford

Marten One of eight groups of carnivorous mammals in the family Mustelidae along with the fisher, weasels, mink, polecats, wolverines, otters, and badgers. The stone marten (*Martes foina*) inhabits coniferous and deciduous woodlands, while the European pine marten (*M. martes*), sable (*M. zibellina*), Japanese marten (*M. melampus*), and American pine marten (*M. americana*) are restricted to northern coniferous forests.

The marten (*Martes*) is a medium-sized, dark-colored, semiarboreal mustelid with a moderately elongated shape, sharply pointed face, rounded ears, and a bushy tail (see illustration). Some members of this genus possess a white throat patch. The throat patch in the American pine marten is reddish-orange, while the yellow-throated marten (*M. flavigula*) and the Nilgiri marten (*M. gwatkinsi*) have yellow throat patches. The fisher (*M. pennanti*) lacks a throat patch. Martens range in head-body length 30–75 cm (12–30 in.) and have a 12–45 cm (5–18 in.) tail. They weigh 0.5–6 kg (1–13 lb). Males are considerably larger than females in all species.

American pine marten, *Martes americana*. (Photo by Gerald and Buff Corsi; © California Academy of Sciences)

Martens are generally solitary. They are among the most agile and graceful members of the weasel family and are able to leap effortlessly from branch to branch. Martens are opportunistic hunters and kill their prey with a bite to the back of the neck. The fur of some martens, especially the sable and fisher, is highly valued. *See* CARNIVORA; FISHER; SABLE. Donald W. Linzey

Maser A device for coherent amplification or generation of electromagnetic waves by use of excitation energy in resonant atomic or molecular systems. "Maser" is an acronym for microwave amplification by stimulated emission of radiation. The device uses an unstable ensemble of atoms or molecules that may be stimulated by an electromagnetic wave to radiate energy at the same frequency and phase as the stimulating wave, thus providing coherent amplification. Amplifiers and oscillators operating on the same principle as the maser exist in many regions of the electromagnetic spectrum. Those operating in the optical region were once called optical masers, but they are now universally called lasers (the "l" stands for "light"). Amplification by maser action is also observed arising naturally from interstellar gases. See COHERENCE; LASER.

Maser amplifiers can have exceptionally low internally generated noise, approaching the limiting effective input power of one-half quantum of energy per unit bandwidth. Their inherently low noise makes maser oscillators that use a narrow atomic or molecular resonance extremely monochromatic, providing a basis for frequency standards. The hydrogen maser, which uses a hyperfine resonance of a gas of hydrogen atoms as the amplification source, is the prime example of this use. Also because of their low noise and consequent high sensitivity, maser amplifiers are particularly useful for reception and detection of very weak signals in radio astronomy, microwave radiometry, and the like. A maser amplifier was used in the experiments that detected the cosmic microwave radiation left over from the big bang that created the universe. See COSMIC BACKGROUND RADIATION; ELECTRICAL NOISE; FREQUENCY MEASUREMENT; RADIO ASTRONOMY; UNCERTAINTY PRINCIPLE.

The quantum theory describes discrete particles such as atoms or molecules as existing in one or more members of a discrete set of energy levels, corresponding to the various possible internal motions of the particle (vibrations, rotations, and so forth). Thermal equilibrium of an ensemble of such particles requires that the number of particles n_1 in a lower energy level 1 be related to the number of particles n_2 in a higher energy level 2 by the Boltzmann distribution, given by the equation below, where E_1

$$\frac{n_1}{n_2} = \exp\frac{(E_2 - E_1)}{kT}$$

and E_2 are the respective energies of the two levels, k is Boltzmann's constant, and T is the absolute (Kelvin) temperature. See BOLTZMANN STATISTICS; QUANTUM MECHANICS.

Particles may be stimulated by an electromagnetic wave to make transitions from a lower energy level to a higher one, thereby absorbing energy from the wave and decreasing its amplitude, or from a higher energy level to a lower one, thereby giving energy to the wave and increasing its amplitude. These two processes are inverses of each other, and their effects on the stimulating wave add together. The upward and downward transition rates are the same, so that, for example, if the number of particles in the upper and lower energy states is the same, the stimulated emission and absorption processes just cancel. For any substance in thermal equilibrium at a positive (ordinary) temperature, the Boltzmann distribution requires that n_1 be greater than n_2 resulting in net absorption of the wave. If n_2 is greater than n_1, however, there are more particles that emit than those that absorb, so that the particles amplify the wave. In such a case, the ensemble of particles is said to have a negative temperature T, to be consistent with the Boltzmann condition. If there are not too many counterbalancing losses from other sources, this condition allows net amplification. This is the basic description of how a maser amplifies an electromagnetic wave. An energy source is required to create the negative temperature distribution of particles needed for a maser. This source is called the pump.

Gas masers. In the first known maser of any kind, the amplifying medium was a beam of ammonia (NH_3) molecules, and the molecular resonance used was the strongest of the rotation-inversion lines, at a frequency near 23.87 GHz (1.26-cm wavelength). Molecules from a pressurized tank of ammonia issued through an array of small orifices to form a molecular beam in a meter-long vacuum chamber. Spatially varying electric fields in the vacuum chamber created by a cylindrical array of electrodes formed a focusing device, which ejected from the beam the molecules in the lower energy level and directed the molecules in the upper energy level into a metal-walled electromagnetic cavity resonator. When the cavity resonator was tuned to the molecular transition frequency, the number of molecules was sufficiently large to produce net amplification and self-sustained oscillation. This type of maser is particularly useful as a frequency or time standard because of the relative sharpness and invariance of the resonance frequencies of molecules in a dilute gas. See CAVITY RESONATOR; MOLECULAR BEAMS.

Solid-state masers. Solid-state masers usually involve the electrons of paramagnetic ions in crystalline media immersed in a magnetic field. At least three energy levels are needed for continuous maser action. The energy levels are determined both by the interaction of the electrons with the internal electric fields of the crystal and by the interaction of the magnetic moments of the electrons with the externally applied magnetic field. The resonant frequencies of these materials can be tuned to a desired condition by changing the strength of the applied magnetic field and the orientation of the crystal in the field. An external oscillator, the pump, excites the transition between levels 1 and 3 [at the frequency $\nu_{31} = (E_3 - E_1)/h$], equalizing their populations. Then, depending on other conditions, the population of the intermediate level 2 may be greater or less than that of levels 1 and 3. If greater, maser amplification can occur at the frequency ν_{21}, or if less, at the frequency ν_{32}. Favorable conditions for this type of maser are obtained only at very low temperature, as in a liquid-helium cryostat. A typical material is synthetic ruby, which contains paramagnetic chromium ions (Cr^{3+}), and has four pertinent energy levels. The important feature of solid-state masers is their sensitivity when used as amplifiers. See PARAMAGNETISM.

Astronomical masers. Powerful, naturally occurring masers have probably existed since the earliest stages of the universe, though that was not realized until a few years after masers were invented and built on Earth. Their existence was first proven by discovery of rather intense 18-cm-wavelength microwave radiation of the free radical hydroxyl (OH) molecule coming from very localized regions of the Milky Way Galaxy.

Masers in astronomical objects differ from those generally used on Earth in that they involve no resonators or slow-wave structures to contain the radiation and so increase its interaction with the amplifying medium. Instead, the electromagnetic waves in astronomical masers simply travel a very long distance through astronomical clouds of gas, far enough to amplify the waves enormously even on a single pass through the cloud. It is believed that usually these clouds are large enough in all directions that a wave passing through them in any direction can be strongly amplified, and hence astronomical maser radiation emerges from them in all directions.

Naturally occurring masers have been important tools for obtaining information about astronomical objects. Since they are very intense localized sources of microwave radiation, their positions around stars or other objects can be determined very accurately with microwave antennas separated by long distances and used as interferometers. This provides information about the location of stars themselves as well as that of the masers often closely surrounding them. The masers' velocity of motion can also be determined by Doppler shifts in their wavelengths. The location and motion of masers surrounding black holes at the centers of galaxies have also provided information on the impressively large mass of these black holes. Astronomical masers often vary in power on time scales of days to years, indicating changing conditions in the regions where they are located. Such masers also give information on likely gas densities, temperature, motions, or other conditions in the rarefied gas of which they are a part. See BLACK HOLE; DOPPLER EFFECT.

Charles H. Townes; James P. Gordon

Masking of sound Interference with the audibility of a sound caused by the presence of another sound. More specifically, the number of decibels (dB) by which the intensity level of a sound (signal) must be raised above its threshold of audibility, to be heard in the presence of a second sound (masker), is called the masking produced by the masker on the signal. The masker and the signal may be identical or may differ in frequency, complexity, or time.

When both the masker and signal are pure tones and the tonal signal and masker have the same frequency, a very low level masker is required to mask the signal, indicating significant masking. As the difference in the frequency between the signal and masker increases, the signal is easier to detect, requiring a higher-level masker to mask the signal. Results from these psychophysical tuning curve measures of masking agree very well with data obtained from single auditory neurons in the auditory periphery, suggesting that tonal masking is mediated by the activity of these peripheral neurons.

The most widely studied complex masking sound is random noise which has energy at all frequencies and is said to be flat if the level for each 1-Hz bandwidth of the noise is the same. When random-flat noise is used to mask a pure tone, only a narrow frequency band (critical band) of the noise centered at the tonal frequency causes masking. If the bandwidth of the noise is narrower than this critical bandwidth, the tone's intensity can be lowered before the tone is masked. If the bandwidth is wider than this critical bandwidth, further widening of the bandwidth causes no changes in the detectability of the signal. The width of the critical band increases proportionally to its center frequency, that is, to the signal frequency. When noise masks speech, either the detectability of speech or speech intelligibility can be measured. The level for speech intelligibility is about 10–14 dB higher than for speech detectability.

Masking can occur when the signal either precedes or follows the masker in time. In backward masking the signal precedes the masker, while in forward masking the signal follows the masker. The physiological basis for this effect, as well as its implications for auditory processing of complex stimuli, is of great interest to the auditory scientist. *See* ACOUSTIC NOISE; HEARING (HUMAN); SOUND. William A. Yost

Masonry Construction of natural building stone or manufactured units such as brick, concrete block, adobe, glass block, or cast stone that is usually bonded with mortar. Masonry can be used structurally or as cladding or paving. It is strong in compression but requires the incorporation of reinforcing steel to resist tensile and flexural stresses. Masonry veneer cladding can be constructed with adhesive or mechanical bond over a variety of structural frame types and backing walls.

Masonry is noncombustible and can be used as both structural and protective elements in fire-resistive construction. It is durable against wear and abrasion, and most types weather well without protective coatings. The mass and density of masonry also provide efficient thermal and acoustical resistance.

Brick, concrete block, and stone are the most widely used masonry materials for both interior and exterior applications in bearing and nonbearing construction. Stone masonry can range from small rubble or units of ashlar (a hewn or squared stone) embedded in mortar, to mechanically anchored thin slabs, to ornately carved decorative elements. Granite, marble, and limestone are the most commonly used commercial building stones. Glass block can be used as security glazing or as elements to produce special daylighting effects. *See* BRICK; CONCRETE; GLASS; GRANITE; MARBLE; LIMESTONE; STONE AND STONE PRODUCTS.

Masonry mortar is made from cement, sand, lime, and water. Masonry grout, a more fluid mixture of similar ingredients, is used to fill hollow cores and cavities and to embed reinforcing steel. Anchors and ties are usually of galvanized or stainless steel. Flashing may be of stainless steel, coated copper, heavy rubber sheet, or rubberized asphalt. *See* GROUT; MORTAR. Christine Beall

Mass The quantitative measure of the inertia of an object, that is, of its resistance to being accelerated. Thus, the more massive an object is, the harder it is to accelerate it (change its state of motion). Isaac Newton described mass as the "quantity of matter" when formulating his second law of motion. He observed that the rate of change of motion is proportional to the impressed force and in the direction of that force. By "motion" he meant the product of the quantity of matter and the velocity. In modern terms Newton's "motion" is known as momentum. The second law is expressed by Eq. (1), where **F** is the applied

$$\frac{d\mathbf{p}}{dt} \propto \mathbf{F} \tag{1}$$

force, t the time, and **p** the momentum. If the mass of the object is constant, then the rate of change of momentum with time is just the product of the mass with the rate of change of the velocity with time, the acceleration **a**. For this situation the second law may be written as Eq. (2).

$$\mathbf{a} \propto \frac{\mathbf{F}}{m} \tag{2}$$

See FORCE; LINEAR MOMENTUM; MOMENTUM.

By suitable choice of units the proportionality becomes an equality as shown in Eq. (3). In particular, the equality holds in

$$\mathbf{a} = \frac{\mathbf{F}}{m} \tag{3}$$

the International System of Units (SI), in which acceleration is measured in meters/second, force is in newtons, and the unit of mass is the kilogram (kg). *See* ACCELERATION; UNITS OF MEASUREMENT.

The primary standard of the unit of mass is the international prototype kilogram kept at Le Bureau International des Poids et Mesures in France. The prototype kilogram is a cylinder of a platinum-iridium alloy, a material chosen for its stability. The standard of mass is kept under conditions specified initially by the first Conférence Général des Poids et Mesures in 1889. Secondary standards are maintained in laboratories around the world and are compared with the mass of the prototype by means of balances whose precision reaches 1 part in 10^8 or better. *See* PHYSICAL MEASUREMENT.

The mass that is described through the second law is known as inertial mass. The inertial mass of any object may be measured by comparing it to the inertial mass of another object. For example, two objects may be made to collide and their accelerations measured. According to Newton's third law of motion, both objects experience forces of equal magnitude. Consequently the ratio of their masses will be related to their accelerations through Eq. (4), where a_1 and a_2 are the magnitudes of accelerations.

$$\frac{m_1}{m_2} = \frac{a_2}{a_1} \tag{4}$$

Thus, if the value of one mass is known, the value of the other mass may be determined by experiment.

The most common way of measuring mass is by weighing, a process that involves the effect of gravity. With a simple two-pan balance, two masses are compared by comparing their weights. The weight of an object on Earth is the gravitational force of attraction exerted on it by the Earth. The response of the balance to the two masses is really a response to their weights. As a result, in commercial and common use, weight is often expressed in units of kilograms instead of the correct units of newtons. *See* GRAVITATION; WEIGHT.

For many years it was thought that mass was a conserved quantity; that is, that it can neither be created nor destroyed. During the eighteenth and nineteenth centuries much of the development of chemistry was based on the conservation of mass. Careful measurements of the mass of constituents of chemical reactions and their products showed that the quantity of matter was unchanged by chemical reactions. However, in the twentieth century, observations and measurements in nuclear and particle physics demonstrated conclusively that mass is not a conserved quantity. For example, radioactive polonium-210 decays into lead-206 by the emission of an energetic alpha particle (a helium nucleus). The combined mass of the lead nucleus and the alpha particle is less than the mass of the original polonium nucleus. The difference in the masses appears as the kinetic energy of the alpha and the lead nucleus. This correspondence between mass and energy was shown by Einstein in Eq. (5).

$$E = mc^2 \tag{5}$$

See CONSERVATION OF MASS.

In the present model for understanding fundamental particles and their interactions, that is, the standard model, the masses of particles arise from the energy of the force fields of their constituents. The masses of the nuclear particles, the protons and neutrons, are thought to arise from the interactions of subnuclear particles called quarks. The mass of a proton, for example, comes at least in part from the energy of the force fields of the quarks. Edwin R. Jones

Mass defect The difference between the mass of an atom and the sum of the masses of its individual components in the free (unbound) state. The mass of an atom is always less than the total mass of its constituent particles; this means, according to Albert Einstein's well-known formula, that an energy of $E = mc^2$ has been released in the process of combination, where m is the difference between the total mass of the constituent particles and the mass of the atom, and c is the velocity of light.

The mass defect, when expressed in energy units, is called the binding energy, a term which is perhaps more commonly used. *See* NUCLEAR BINDING ENERGY. W. W. Watson

Mass–luminosity relation The relation, observed or predicted by theory, between the quantity of matter a star contains (its mass) and the amount of energy generated in its interior (its luminosity). Because of the great sensitivity of the rate of energy production in a stellar interior to the mass of the star, the mass–luminosity relation provides an important test of theories of stellar interiors.

For a family of stars with different masses but with the same mixture of chemical elements uniformly distributed throughout the stellar volumes, there should be a unique mass–luminosity relation. The observed relation is obtained from binary stars for which masses and luminosities can be observationally evaluated. The majority of stars are found along this relation, called the main sequence, and it implies that all stars begin their lives with roughly the same chemical composition. It conforms reasonably well to theory. *See* BINARY STAR; STAR.

There are classes of stars that depart markedly from the mass–luminosity relation for main-sequence stars. Examples are white dwarf stars, which are known to be extremely underluminous for their masses, and are thought to be stars in which all sources of nuclear energy have been exhausted. One of the triumphs of twentieth-century astrophysics was the establishment and exploration of stellar evolution theory, which sees stars evolving off the main sequence as they age, and links an individual star's entire evolutionary roadmap to its initial mass and composition. *See* HERTZSPRUNG-RUSSELL DIAGRAM; WHITE DWARF STAR.

Daniel M. Popper; Gerard van Belle

Mass number The mass number A of an atom is the total number of its nuclear constituents, or nucleons, as the protons and neutrons are collectively called. The mass number is placed before and above the elemental symbol, thus ^{238}U. The mass number gives a useful rough figure for the atomic mass; for example, $^{1}H = 1.00783$ atomic mass units (amu), $^{238}U = 238.051$ amu, and so on. *See* ATOMIC NUMBER. Henry E. Duckworth

Mass spectrometry An analytical technique for identification of chemical structures, determination of mixtures, and quantitative elemental analysis, based on application of the mass spectrometer. Determination of organic and inorganic molecular structure is based on the fragmentation pattern of the ion formed when the molecule is ionized; further, because such patterns are distinctive, reproducible, and additive, mixtures of known compounds may be analyzed quantitatively. Quantitative elemental analysis of organic compounds requires exact mass values from a high-resolution mass spectrometer; trace analysis of inorganic solids requires a measure of ion intensity as well. *See* MASS SPECTROSCOPE.

For analysis of organic compounds the principal methods are electron impact, chemical ionization, field ionization, field desorption, particle bombardment, laser desorption, and electrospray.

In electron impact, when a gaseous sample of a molecular compound is ionized with a beam of energetic (commonly 70-V) electrons, part of the energy is transferred to the ion formed by the collision, as shown in the reaction below.

$$\text{A—B—C} + e \longrightarrow \text{A—B—C}^+ + e + e$$

For most molecules the production of cations is favored over the production of anions by a factor of about 10^4, and the following discussion pertains to cations. The ion corresponding to the simple removal of the electron is commonly called the molecular ion and normally will be the ion of greatest m/e ratio in the spectrum. In the ratio, m is the mass of the ion in atomic mass units (daltons) and e is the charge of the ion measured in terms of the number of electrons removed (or added) during ionization. Occasionally the ion is of vanishing intensity, and sometimes it collides with another molecule to abstract a hydrogen or another group. In these cases an incorrect assignment of the molecular ion may be made unless further tests are applied. Proper identification gives the molecular weight of the sample.

The remaining techniques were devised generally to circumvent the problem of the weak or vanishingly small intensity of a molecular ion.

In chemical ionization the ions to be analyzed are produced by transfer of a heavy particle (H^+, H^-, or heavier) to the sample from ions produced from a reactant gas.

Field ionization and field desorption is used for less volatile material. The sample is ionized when it is in a very high field gradient (several volts per angstrom) near an electrode surface. The molecular potential well is distorted so that an electron tunnels from the molecule to the anode. The ion thus formed is repelled by the anode. Typically, the lifetime of the ion in the mass spectrometer source is much less (10^{-12} to 10^{-9} s) than in electron impact. Because little energy is transferred as internal energy and the ion is removed rapidly, little fragmentation occurs, and the molecular weight is more easily determined.

In electrohydrodynamic ionization, a high electric-field gradient induces ion emission from a droplet of a liquid solution, that is, the sample and a salt dissolved in a solvent of low volatility. An example is the sample plus sodium iodide (NaI) dissolved in glycerol. The spectra include peaks due to cationized molecules of MNa^+, $MNa(C_3H_8O_3)_n{}^+$, and $Na(C_3H_8O_3)_n{}^+$.

A sample heated very rapidly may vaporize before it pyrolyzes. Techniques for heating by raising the temperature of a source probe on which the sample is coated by 200 K (360°F)/s have been developed. Irradiation of an organic sample with laser radiation can move ions of mass up to 1500 daltons into the gas phase for analysis. This technique is the most compatible with analyzers that require particularly low pressures such as ion cyclotron resonance. Time-resolved spectra of surface ejecta are proving to be the most useful kinds of laser desorption spectra available.

A solid sample or a sample in a viscous solvent such as glycerol may have ions sputtered from its surface by bombardment with accelerated electrons, ions, or neutrals. Bombardment by electrons is achieved simply by inserting a probe with the sample directly into the electron beam of an electron impact source (in-beam electron ionization).

In laser desorption, the energy of laser photons may be used to remove an analyte from a surface and ionize it for mass-spectrometric analysis. In virtually all cases involving polar compounds of high molecular weight, the analyte is prepared in or on an organic matrix that is coated on the surface to be irradiated. The matrix material assists in the ionization process; the technique is known as matrix-assisted laser desorption ionization (MALDI). Numerous organic materials have been investigated. Molecular weights in excess of 700,000 have been measured with this technique, and the technique is particularly suited for the molecular-weight determination of large polar biological molecules; for example, enzymes and intact antibodies have been analyzed.

In electrospray ionization, a solution sprayed through a nozzle of very small diameter into a vacuum with an electric field having a gradient of several hundred to a thousand volts per centimeter produces gaseous ions from solutes effectively. Electrospray ionization is the only mass-spectrometric technique that produces a large fraction of multiply charged ions from an organic or biological analyte. Since mass spectrometers measure mass-to-charge ratios of ions, not simply their mass, electrospray ionization has the advantage of permitting ions of very high mass to be analyzed without special mass-analysis instrumentation; for example, an ion of mass 120,000 daltons carrying 60 positive charges appears at mass-to-charge 2000, within the range of many mass analyzers. This technique has been used to measure the masses of ions from molecules of masses up to about 200,000 daltons. Since the distribution of charged species reflects to some extent the degree of protonation in solution, a signal that reflects the folding of a protein, there is evidence that solution conformations of proteins can be studied by electrospray ionization mass spectrometry. Maurice M. Bursey

Measurement using a mass spectrometer takes advantage of the mass dependency in the equations of motion of an ion in an electric or magnetic field. Three common mass analyzers are magnetic-sector, quadrupole, and time-of-flight. An ion traversing the magnetic field of a sector instrument successfully reaches the detector when its mass, m, corresponds to $m/z = r^2B^2/2V$, where z is the charge of the ion, r is the radius of the flight tube, B is the magnetic field strength, and V is the acceleration voltage. A mass spectrum can be recorded by varying B or V. A quadrupole is an array of four parallel rods electronically connected such that a radio frequency (RF) is applied to opposing rods, and the waveforms applied to adjacent rods are out of phase. In addition, a direct-current (DC) component is added to the rods such that one opposing pair has +DC and the other −DC. Ions entering the quadrupole follow complex sinusoidal paths. An ion successfully traversing the rods to the detector is mass-selected by the amplitude of the applied radio frequency and the amount of resolution from adjacent masses by the DC/RF ratio. A spectrum is detected by scanning the RF amplitude and the DC voltage in a fixed ratio. The time-of-flight analyzer is a tube with an acceleration field at one end and a detector at the other. Ions are pulsed and accelerated down the tube, starting a clock. Arrival times at the detector are converted to mass by solving the equation $m/z = 2V(t/L)^2$, where V is the acceleration potential, t is the arrival time, and L is the flight tube length.

The mass analyzers are characterized by their mass resolving power (RP), scan or acquisition rate, and mass range. Sector instruments are usually assembled with an additional energy filter and can be operated up to 100,000 RP. Resolving power is defined as $M/\Delta M$, where M is the mass number of the observed mass and ΔM is the mass difference in two signals with a 10% valley between them. Increased resolution in sectors comes at the sacrifice of sensitivity. To increase resolution, the slits are narrowed on the ion beam. It is typical to lose 90% of the signal going from 1000 to 10,000 RP. Magnetic instruments are available with mass ranges up to 10,000 atomic mass units (amu) for singly charged ions at full acceleration potential. To obtain good quality, data sectors are scanned relatively slowly—a practical rate is 5 s per decade (1000–100 amu). Quadrupoles are usually operated at unit resolution. For example, 100 is resolved from 101, or 1000 from 1001. Quadrupoles can be scanned at 2000 amu/s and are available with mass range up to 4000 amu. Fast scanning and high-gas-pressure tolerance make quadrupoles popular for mass spectrometer–chromatographic hybrid systems. Modern time-of-flight analyzers are capable of 10,000 RP, 50% valley definition of ΔM. Acquisition rates can be 100 spectra per second. The time-of-flight mass range is theoretically unlimited, and singly charged ions over 100,000 amu have been observed. *See* ATOMIC NUMBER; MASS NUMBER. Todd D. Williams

The analysis of solid inorganic samples can be made either by vaporization in a Knudsen cell arrangement at very high temperatures or by volatilization of the sample surface so that particles are atomized and ionized with a high-energy spark (for example, 20,000 eV). The wide range of energies given to the particles requires a double-focusing mass spectrometer for analysis. Detection in such instruments is by photographic plate; exposures for different lengths of time are recorded sequentially, and the darkening of the lines on the plate is related empirically to quantitative composition by calibration charts. The method is useful for trace analysis (parts per billion) with accuracy ranging from 10% at higher concentration levels to 50% at trace levels. Methods for improving accuracy, including interruption and sampling of the ion beam, are under development.

Secondary ion mass spectrometry is most commonly used for surface analysis. A primary beam of ions accelerated through a few kilovolts is focused on a surface; ions are among the products sputtered from this surface, and they may be directly analyzed in a quadrupole filter. Sputtered material may also be analyzed by ionization of the neutrals in an inductively coupled plasma and subsequent mass analysis. This method produces ions with a lower energy spread than spark source mass spectrometry and, since detection then becomes less of a problem, has been supplanting spark source methods. *See* LASER SPECTROSCOPY; SECONDARY ION MASS SPECTROMETRY (SIMS); SPECTROSCOPY. Maurice M. Bursey

Mass spectroscope An instrument used for determining the masses of atoms or molecules found in a sample of gas, liquid, or solid. It is analogous to the optical spectroscope, in which a beam of light containing various colors (white light) is sent through a prism to separate it into the spectrum of colors present. In a mass spectroscope, a beam of ions (electrically charged atoms or molecules) is sent through a combination of electric and magnetic fields so arranged that a mass spectrum is produced (see illustration). If the ions fall on a photographic plate which after development shows the mass spectrum, the instrument is called a mass spectrograph; if the spectrum is allowed to sweep across a slit in front of an electrical detector which records the current, it is called a mass spectrometer.

Mass spectroscopes are used in both pure and applied science. Atomic masses can be measured very precisely. Because of the equivalence of mass and energy, knowledge of nuclear structure and binding energy of nuclei is thus gained. The relative abundances of the isotopes in naturally occurring or artificially produced elements can be determined.

Empirical and theoretical studies have led to an understanding of the relation between molecular structure and the relative abundances of the fragments observed when a complex molecule, such as a heavy organic compound, is ionized. When a high-resolution instrument is employed, the masses of the molecular or fragment ions can be determined so accurately that identification of the ion can frequently be made from the mass alone.

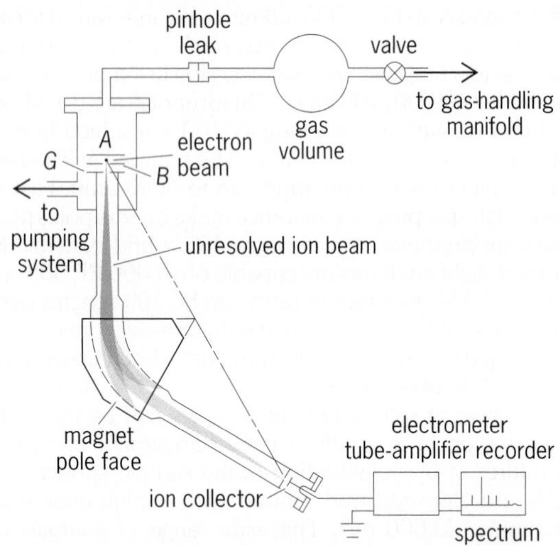

Schematic drawing of mass spectrometer tube. Ion currents are in the range 10^{-10} to 10^{-15} ampere and require special electrometer tube amplifiers for their detection. In actual instruments the radius of curvature of ions in a magnetic field is 4–6 in. (10–15 cm).

Because chemical compounds may have mass spectra as unique as fingerprints, mass spectroscopes are widely used in industries such as oil refineries, where analyses of complex hydrocarbon mixtures are required. *See* BETA PARTICLES. Alfred O. Nier

Mass wasting A generic term for downslope movement of soil and rock, primarily in response to gravitational body forces. Mass wasting is distinct from other erosive processes in which particles or fragments are carried down by the internal energy of wind, running water, or moving ice and snow.

The stability of slope-making materials is lost when their shear strength (or sometimes their tensile strength) is overcome by shear (or tensile) stresses, or when individual particles, fragments, and blocks are induced to topple or tumble. The shear and tensile strength of earth materials depends on their mineralogy and structure. Processes that generally decrease the strength of earth materials include one or more of the following: structural changes, weathering, groundwater, and meteorological changes. Stresses in slopes are increased by steepening, heightening, and external loading due to static and dynamic forces. Processes that increase stresses can be natural or result from human activities. Although other classifications exist, these movements can be conveniently classified according to their velocity into two types: creep and landsliding. *See* SOIL MECHANICS.

Geologically, creep is the imperceptible downslope movement at rates as slow as a fraction of millimeter per year; its cumulative effects are ubiquitously expressed in slopes as the downhill bending of bedded and foliated rock, bent tree trunks, broken retaining walls, and tilted structures. There are two varieties of geologic creep. Seasonal creep is the slow, episodic movement of the uppermost several centimeters of soil, or fractured and weathered rock. It is especially important in regions of permanently frozen ground. Rheologic creep, sometimes called continuous creep, is a time-dependent deformation at relatively constant shear stresses of masses of rock, soil, ice, and snow. This type of creep affects rock slopes down to depths of a few hundred meters, as well as the surficial layer disturbed by seasonal creep. Continuous creep is most conspicuous in weak rocks and in regions where high horizontal stresses (several tens of bars or several megapascals) are known to exist in rock masses at depths of 330 to 660 ft (100 to 200 m).

Landsliding includes all perceptible mass movements. Three types are generally recognized on the basis of the type of movement: falls, slides, and flows. Falls involve free-falling material; in slides the moving mass displaces along one or more narrow shear zones; and in flows the distribution of velocities within the moving mass resembles that of a viscous flow. *See* LANDSLIDE.

Mass wasting is an important consideration in the interaction between humans and the environment. Deforestation accelerates soil creep. Engineering activities such as damming and open-pit mining are known to increase landsliding. On the other hand, enormous natural rock avalanches have buried entire villages and claimed tens of thousands of lives. Alberto S. Nieto

Massif A block of the Earth's crust commonly consisting of crystalline gneisses and schists, the textural appearance of which is generally markedly different from that of the surrounding rocks. Common usage indicates that a massif has limited areal extent and considerable topographic relief. Structurally, a massif may form the core of an anticline or may be a block bounded by faults or even unconformities. In any case, during the final stages of its development a massif acts as a relatively homogeneous tectonic unit which to some extent controls the structures that surround it. Numerous complex internal structures may be present; many of these are not related to its development as a massif but are the mark of previous deformations. Philip H. Osberg

Mastigophora A superclass of the Protozoa also known as the Flagellata. The common morphological flagellate type is spherical to cylindrical on an anteroposterior axis. Despite their common morphological plan, based on the flagellum as a means of locomotion, Flagellata are very diverse in shape, colony formation, internal structure, external shell or test, color, physiology, reproduction, and choice of environment.

The Mastigophora exhibit marked plantlike characteristics, so that texts and references in botany always treat at least some of them. Some workers include all colorless flagellates in the algae; however, flagellates display many distinctly animal features and sometimes are treated as Protozoa. Thus, flagellates are regarded as a link between the plant and animal kingdoms. *See* PHYTAMASTIGOPHOREA; ZOOMASTIGOPHOREA.

Mastigophora possess one common feature, locomotion by means of one or more whiplike protoplasmic extrusions termed flagella. The flagellate body is typically monaxial and elongate to some degree. Cells are practically spherical in *Monas*, but ovoid, cordiform, pyriform, fusiform, acicular, tubular, or flattened cells are more common. Flattened species typically glide along a surface rather than swim. A particular shape is normally maintained by *Ochromonas*, whereas *Mastigamoeba* and some others form pseudopodia, and certain species (*Euglena agilis*) undergo frequent distortions termed euglenoid or metabolic changes of shape.

Structurally the flagellate cell is not simple. The cytoplasm is sometimes quite vacuolated (*Collodictyon or Trepomonas*) and occasionally colored, although color is generally confined to chromatophores when these are present. One or more contractile vacuoles may be located near the flagellar base.

Most flagellates are single-celled but colony formation is frequent. Cells may be naked (*Oicomonas*), enclosed in a thick cellulose cell wall or pellicle (*Euglena spirogyra*), or in a chitinous, calcareous, or silicious test, the lorica or shell (*Trachelomonas, Distephanus speculum*).

Flagellates adjust to wide ranges of pH, osmotic pressure, light, and temperature but optima are found upon investigation. Some flagellate species occur in both fresh and sea water, but cannot be transferred directly from fresh to sea water. Almost any aqueous ecological niche contains Flagellata.

Flagellata are near the base of the food pyramid, probably next to bacteria. Their varied synthetic abilities, fast reproductive rate, and huge numbers in both fresh and salt water compensate for their usual small size. The oceanic blooms which kill fish and other animals are principally flagellates, but green flagellates reaerate polluted water. They often cause tastes and odors in potable water, but they readily attack organic matter in natural water as do bacteria, They include many parasites dangerous to humans and animals and are themselves frequently parasitized. *See* PROTOZOA. James B. Lackey

Mastitis Inflammation of the mammary gland. This condition is most frequently caused by infection of the gland with bacteria that are pathogenic for this organ. It has been described in humans, cows, sheep, goats, pigs, horses, and rabbits. Mastitis causes lactating women to experience pain when nursing the child, it damages mammary tissue, and the formation of scar tissue in the breast may cause disfigurement.

The mammary gland is composed of a teat and a glandular portion. The gland has defensive mechanisms to prevent and overcome infection with bacteria. Nonspecific defense mechanisms include teat duct keratin, lactoferrin, lactoperoxidase, and complement. Specific defense mechanisms are mediated by antibodies and include opsonization, direct lysis of pathogens, and toxin neutralization. Milk contains epithelial cells, macrophages, neutrophils, and lymphocytes. To induce mastitis, a pathogen must first pass through the teat duct to enter the gland, survive the bacteriostatic and bactericidal mechanisms, and multiply. Bacteria possess virulence factors such as capsules and toxins which enable them to withstand these protective mechanisms. *See* LACTATION.

When bacteria multiply within the gland, there is a release of inflammatory mediators and an influx of neutrophils. The severity of mammary infection is classified according to the clinical signs. In humans, infection occurs during lactation, with clinical episodes most frequent during the first 2 months of lactation. In acute puerperal mastitis the tissue becomes hot, swollen, red, and painful, and a fever may be present. In the absence of treatment, this may progress to a pus-forming mastitis, with the development of breast abscesses.

In acute puerperal mastitis of humans, suitable antibiotics are administered by the intravenous or intramuscular route, while in the abscess form surgical drainage is provided in addition to antibacterial therapy. Penicillins, cephalosporins, and erythromycin are administered locally into the infected gland after milking for 1–2 days. Additional antibiotics are given systemically, that is, intravenously or intramuscularly, in severe cases of mastitis, and also to improve bacteriologic cure rates. *See* ANTIBIOTIC. Neil L. Norcross

Matched-field processing A set of related signal processing techniques for remote sensing that involve matching measured signals obtained from an array of sensors to synthesized signals obtained from signal propagation simulations conducted in a model of the actual environment. Here the signals are typically acoustic or electromagnetic waves, but they may be water waves, seismic waves, structural waves, or any other measurable phenomenon that travels in a predictable manner from its source to a distant receiver. Matched-field processing is the extension of elementary sensor-array signal processing techniques for line-of-sight signal propagation in uniform unbounded environments to complicated, but predictable, signal propagation in nonuniform bounded environments. Matched-field processing is superior to these elementary techniques because it can explicitly account for reflection, refraction, and scattering of the signal as it propagates from its source to the sensors. However, this superior performance is possible only when the signal propagation simulations are accurate and the model environment matches the environment in which the signal propagation took place. *See* ACOUSTIC SIGNAL PROCESSING; SIGNAL PROCESSING; SIMULATION; WAVE MOTION.

Remote sensing is common for military, biomedical, surveillance, quality control, and nondestructive evaluation purposes. The primary remote-sensing tasks are typically detection, localization, tracking, and identification of signal sources (or scatterers). These tasks all become more difficult when the environment is complicated and multiple signal-propagation paths are possible between distant signal sources (or scatterers) and the sensors. Such environments occur in applications of radar and sonar where naturally occurring or artificial obstacles and surfaces cause signal echoes and reflections. Matched-field processing is utilized primarily for detection and localization of signal sources, with a greater emphasis on localization. It is the preferred source localization technique for small sources, complicated wave fronts, and environments with known reflection and refraction. *See* RADAR; REMOTE SENSING; SONAR; UNDERWATER SOUND.

The primary limitations of matched-field processing stem from differences between actual signal propagation in the real environment and simulated signal propagation in the model environment. Here, geometrical differences as small as a fraction of a signal wavelength in the placement of a reflecting surface can lead to erroneous source localization, or degradation (or even loss) of the localization peak in the matched-field processing output. Thus, there is a significant burden on the user to know the environment well and to faithfully simulate signal propagation. Yet, even when all the environmental-specification parameters are not known, matched-field processing is still possible by adding parameter searches to the preexisting search space, an extension of matched-field processing called focalization. However, the main computational burden of matched-field processing commonly results from repeated simulation of signal propagation in the model environment as the test source is moved through the search space. Thus, minimizing the number of signal propagation simulations is typically advantageous. David R. Dowling

Material resource planning A formal computerized approach to inventory planning, manufacturing scheduling, supplier scheduling, and overall corporate planning. The material requirements planning (MRP) system provides the user with information about timing (when to order) and quantity (how much to order), generates new orders, and reschedules existing orders as necessary to meet the changing requirements of customers and manufacturing. The system is driven by change and constantly recalculates material requirements based on actual forecast orders. It makes adjustments for possible problems prior to their occurrence, as opposed to traditional control systems which looked at more historical demand and reacted to existing problems. *See* MANUFACTURING ENGINEERING.

The logic of the material requirements planning system is based on the principle of dependent demand, a term describing the direct relationship between demand for one item and demand for a higher-level assembly part or component. For example, the demand for the number of wheel assemblies on a bicycle is directly related to the number of bicycles planned for production; further, the demand for tires is directly dependent on the demand for wheel assemblies. In most manufacturing businesses, the bulk of the raw material and in-process inventories are subject to dependent demand. Dependent demand quantities are calculated, while independent demand items are forecast. Independent demand is unrelated to a higher-level item which the company manufactures or stocks. Generally, independent demand items are carried in finished goods inventory and subject to uncertain end customer demand. Spare parts or replacement requirements for a drill press are an example of an independent demand item.

By use of the computer, material requirements planning is able to manipulate massive amounts of data to keep schedules up to date and priorities in order. The technological advances in computing and processing power, the benefits of on-line capabilities, and reduction in computing cost make computerized manufacturing planning and control systems such as material requirements planning powerful tools in operating modern manufacturing systems productively. *See* INDUSTRIAL ENGINEERING; INVENTORY CONTROL; SYSTEMS ENGINEERING. Larry C. Giunipero

Materials handling The movement, protection, storage, and control of materials and products throughout the process of their manufacture, distribution, consumption, and disposal, as defined by the Material Handling Industry of America. Materials handling is not just about equipment for storage and movement, it also includes controls that direct movement, packaging tools, and techniques that protect goods as they flow from point to point. However, no amount of equipment, automation, or information technology can take the place of a well-designed system that facilitates the steady flow of goods through manufacturing and distribution facilities. *See* INDUSTRIAL FACILITIES.

Materials-handling systems are designed to make sure that materials and goods are delivered to the right place, at the right time, in the right condition, with the correct information (labeling and invoices or other paperwork) in a safe and cost-effective manner as they flow through networks of suppliers and customers. Sometimes materials handling is not considered a value-added activity, but it adds value by putting things where they need to be for use at the right time. Items that are damaged or delivered to the wrong location have less value than those delivered when and where they are needed for further processing or use. This is known as the "time-and-place value" of goods and materials.

Materials handling is typically classified as bulk or discrete handling. Bulk handling deals with the controlled movement and storage of materials, such as coal, grains, liquids, or other types of materials, which flow from point to point and are measured in terms of their volume or mass. Mining operations, chemical processing facilities, and oil refineries are bulk-handling operations. Discrete handling deals with individual items or groups of items packaged together to form what is known as a unit load. Unit loads are stored and moved as if they were a single item. They can be items grouped together in a box, a group of boxes on a pallet, or simply one individual item. Unit-load formation is a fundamental concept of discrete handling. Many criteria influence the decision of how to form unit loads, such as size, weight, available equipment, and how often the goods are to be stored or moved. The design of manufacturing facilities, assembly lines, and warehouses are greatly influenced by the tradeoffs necessary in determining the quantity of a unit load. Examples of discrete handling are most often found in manufacturing, assembly, and warehouses. If you have ever picked up your bags from an airport luggage carousel, you have also experienced discrete-parts materials handling. *See* BULK-HANDLING MACHINES.

In addition to hardware, a materials-handling system has a control system, as well as information-gathering (sensors) and communication components. In a distribution network, manufacturing, warehousing, and transportation execution systems (often collectively referred to as supply-chain execution system components) control the movement (when and where) of goods as they flow through the system. These systems not only track the goods in a facility, many of them also track the equipment resources used to move the goods and the labor used to perform tasks. *See* CONTROL SYSTEMS. Michael Ogle

Materials science and engineering The term "materials science and engineering" emerged in the 1960s to represent how the engineering application of materials was increasingly based on scientific principles, rather than the empiricism. Materials science focuses on understanding how the properties (or characteristics) of materials are based on their atomic- and microscopic-scale structure. The term microstructure refers to this microscopic-scale structure. Materials science grew out of the fields of physical chemistry, polymer chemistry, and condensed-matter physics. *See* PHYSICAL CHEMISTRY; SOLID-STATE PHYSICS.

The engineering of materials involves the processing, testing, and selection of materials for various design applications. Materials engineering grew out of the fields of metallurgy, plastics engineering, and ceramic engineering. *See* METALLURGY.

Most of the elements in the periodic table exhibit atomic bonding that is metallic in nature. The metallic bond involves a mobile "gas" of electrons that produces the electrical conductivity, thermal conductivity, and optical reflectivity characteristic of metals. The crystalline (regular and repeating) arrangement of atoms in metals generally provides attractive mechanical behavior in structural applications. Many alloys are reasonably strong and can be formed into practical shapes. *See* ALLOY; METAL; PERIODIC TABLE.

Polymers are composed of a few elements in the periodic table (usually carbon and hydrogen and possibly a few of the nonmetallic elements such as oxygen, nitrogen, fluorine, and silicon) that exhibit covalent atomic bonding. Polymers are generally good electrical insulators because directional covalent bonding between adjacent atoms usually does not provide "free" electrons for conduction. The use of polymeric insulation on electrical wires is a practical example. Polymers also can be optically transparent (for example, clear plastic wrap for food). The alternative name "plastic" comes from the extensive formability of many polymers during fabrication. *See* DIELECTRIC MATERIALS.

Ceramics can be defined as nonmetallic and inorganic materials. Ceramics are chemical combinations of at least one metallic and at least one nonmetallic element (especially oxygen, with others possibly including carbon, nitrogen, phosphorous, and sulfur). Aluminum oxide (Al_2O_3) is a common example in which the metallic element aluminum is combined with the nonmetallic element oxygen. Although forming amorphous alloys is difficult, forming noncrystalline ceramics, called glasses, is relatively easy, as the crystallization of molten oxides can be a sluggish process. The most common examples are the silicate glasses, based on the mineral silicon dioxide (SiO_2). CaO. *See* CERAMICS; GLASS.

The ionic bond between metallic and nonmetallic atoms is a strong one. The thermal stability associated with the ionic bond makes ceramics resistant to high temperatures. This heat resistance leads to the use of ceramics in refractory applications such as furnaces and boilers.

Metals, polymers, and ceramics, which are fundamentally based on atomic bonding, cover the basic categories of structural materials. Engineers have developed many advanced materials based on the microscopic-scale combinations of them, resulting in a fourth category called composites. A common example is fiberglass, a composite of glass fibers (a few micrometers in diameter) embedded in a polymer matrix. *See* COMPOSITE MATERIAL; METAL MATRIX COMPOSITE; POLYMERIC COMPOSITE.

Materials also can be classified based on electrical conductivity rather than atomic bonding. Electrical classification provides a fifth category of materials—semiconductors. While metals are typically good electrical conductors and polymers and ceramics/glasses are typically electrical insulators, composites tend to have electrical properties that are the average of their individual components. Covalently bonded semiconductors, with intermediate levels of electrical conductivity, have played an increasingly critical role in modern technology. The primary example is silicon, a central component of modern, solid-state electronics. *See* SEMICONDUCTOR.

Nanotechnology involves structures developed on the nanometer (10^{-9} m) scale, rather than the micrometer (10^{-6} m) scale. The resulting nanomaterials can display significantly improved mechanical, electrical, magnetic, thermal, and optical properties, compared to their counterparts on a coarser scale. *See* NANOTECHNOLOGY James F. Shackelford

Maternal behavior The pattern of care given an offspring by its mother. Many species reproduce generation after generation without receiving or providing any parental care. Insects and fish commonly produce vast numbers of offspring that they neither feed nor defend, resulting in the loss of many offspring to predators and to other hazards. Their great numbers, however, ensure that some will survive and reproduce. Also, if the young of a species are self-sufficient at birth or if they mature very rapidly after birth, they can often survive and reproduce with little parental care.

Parental behavior is most highly developed in species that produce only a few offspring at once, mature slowly, and have complex behavior patterns that can be learned only through extended practice. Parental behavior can involve as little as hiding the young and never seeing them again or it can involve years of feeding, defending, and teaching. In most species that care for their young after birth, the female does most or all of the work. The biological explanation for this arises from the fact that a male can father vast numbers of offspring, whereas a female can bear only a few. Each offspring that fails to survive or reproduce can represent a significant proportion of the female's lifetime reproductive output. Thus, the females that pass their genes on to subsequent generations in greatest numbers are not those that bear the most young but those that invest more in caring for the few young they produce. Active care of the young by the male is most common when males of a species limit themselves to one or a few female partners. *See* REPRODUCTIVE BEHAVIOR.

Factors that influence the onset of maternal behavior include physiological and psychological factors related to pregnancy, hormonal influences associated with childbirth, the behavior of the newborn, and cultural factors such as learning and social traditions. Although hormones play necessary roles in recovery from pregnancy and in milk production, there is no evidence that they play a critical role in any other aspect of human maternal behavior. Cultural factors and the behavior of the newborn are the primary cues that prompt human mothers to care for their young. Everett Waters; German Posada

Maternal influence Effects on development that are attributable specifically to the maternal parent or to maternally inherited factors. This influence can take a variety of forms, including maternal-specific inheritance of genetic material (for example, mitochondrial DNA), parent-specific epigenetic modification of genes (genomic imprinting), placement of molecules into the oocyte prior to fertilization, and interactions with the maternal parent during development.

The term "parent-of-origin effects" describes instances where an offspring's phenotype is asymmetrically dependent on the genotypes or phenotypes of its maternal and paternal parents. In most cases where such asymmetry exists, it is the maternal parent that wields the greater influence. This maternal advantage derives in part from the greater size of the oocyte, which means that the mother contributes more cytoplasmic material to the zygote than the father does. This asymmetry is enhanced in species where a part of the offspring's development occurs in physical contact with the mother, as in viviparous animals and many plants. Postnatal care, if present, is often provided preferentially by the mother.

One source of maternal influence is the inheritance of DNA specifically from the mother. Organelles that contain their own DNA, such as mitochondria and chloroplasts, are inherited cytoplasmically. Mitochondria are inherited predominantly, if not exclusively, from the mother. Many of the genes encoded by the mitochondrial DNA are involved in oxidative phosphorylation. A number of maternally transmitted diseases are characterized by oxidative phosphorylation defects; they affect in particular the central nervous system and muscle tissues. *See* DEOXYRIBONUCLEIC ACID (DNA); MITOCHONDRIA..

In some species, asymmetric parental influence derives from differential epigenetic modification of genes inherited by the offspring, or "genomic imprinting." An imprinted gene is one where an allele's pattern of expression differs depending on whether it was inherited from the mother or the father. In the simplest instances, one of the two alleles is active and one is transcriptionally silent. In more complex cases, clusters of imprinted genes are coregulated, and various transcripts are produced from the maternally and paternally derived alleles. The epigenetic modifications involved in imprinting include DNA methylation as well as histone modifications, such as methylation and acetylation.

Imprinted genes have been identified in mammals and angiosperms, where the early part of development occurs in close physical contact with the maternal parent. In these cases, paternally derived genes are selected to place a greater demand for resources on the mother. Maternally derived genes favor more restraint, preserving maternal resources for the mother's other offspring. Many imprinted genes directly influence growth and resource demands during development. At loci where increased expression enhances growth, imprinted genes are paternally expressed and maternally silenced. At growth-suppressing loci, the opposite pattern of expression holds.

A final source of maternal influence results from interactions between the mother and offspring after fertilization. In many plants and in viviparous animals, the early part of development occurs in close physical contact with the mother. In the cases of mammals and angiosperms, specific organs exist (the placenta and endosperm, respectively) that mediate nutrient transfer from mother to offspring. Further maternal influence arises in postnatal care. For instance, mammalian breast feeding involves the transfer of antibodies as well as nutrients to offspring. An offspring's immunity in the earliest stages of life therefore depends on its mother's genotype at various immune system loci, as well as her history of antigen exposure. *See* DEVELOPMENTAL BIOLOGY; DEVELOPMENTAL GENETICS. J. F. Wilkins

Mathematical biology The application of mathematics to biological systems. Mathematical biology spans all levels of biological organization and biological function, from the configuration of biological macromolecules to the entire ecosphere over the course of evolutionary time.

The influence of physics on mathematical biology has been twofold. On the one hand, organisms simply are material systems, and presumably can be analyzed in the same terms as any other material system. Reductionism, the theory that biological processes find their resolution in the particularities of physics, finds its practical embodiment in biophysics. Thus, one of the roots of mathematical biology is what was originally called mathematical biophysics. On the other hand, other early investigations in mathematical biology, such as population dynamics (mathematical ecology), exploited the form of such analyses, such as using differential rate equations, but they expressed their analyses in strictly biological terms. Such approaches were guided by analogy with mathematical physics rather than by reduction to physics and so rest on the form rather than the substance of physics. *See* BIOPHYSICS; PHYSICS.

Both of these approaches are important, especially since organisms possess characteristics that have no obvious counterpart in inorganic systems. As a result, mathematical biology has acquired an independent and unique character. In several important cases, these characteristics have required a reconsideration of physics itself, as in the impact of open systems on classical thermodynamics.

Surrogacy and models. The idea that something can be learned about a system by studying a different system, or surrogate, is central to all science. The relation between a system and its surrogates is embodied in the concept of a model. The basic idea of mathematical biology is that an appropriate formal or mathematical system may similarly be used as a surrogate for a biological system. The use of mathematical models offers possibilities that transcend what can be done on the basis of observation and experiment alone. *See* MODEL THEORY.

For example, morphological differences between related species can be made to disappear by means of relatively simple coordinate transformations of the space in which the forms are embedded. Surrogacy explicitly becomes a matter of intertransformability, or similarity, and what is true for morphology also holds true for other functional relationships that are characteristic of organisms, whether they be chemical, physical, or evolutionary. These assertions of surrogacy and modeling can be restated: closely related implies similar. This is a nontrivial assertion: "closely related" is a metric relation pertaining to genotypes, whereas "similar" is an equivalence relation based on phenotypes. It is the similarity relation between phenotypes that provides the basis for surrogacy. Thus the question immediately arises: given a genotype, how far can it be varied or changed or mutated, and still preserve similarity?

Such questions fall mathematically into the province of stability theory, particularly structural stability. Under very general conditions, there exist many genomes that are unstable (bifurcation points) in the sense that however high a degree of metric approximation is chosen, the associated phenotypes may be dissimilar, that is, not intertransformable. That observation by R. Thom provides the basis for his theory of catastrophes and demonstrates the complexity of the surrogacy relationship. The fundamental importance of such ideas for phenomena of development, for evolution (particularly for macroevolution), and for the extrapolation of data from one species to another, or the relation between health and disease, is evident. *See* CATASTROPHE THEORY; MACROEVOLUTION.

Metaphor. A closely related group of ideas that are characteristic of mathematical biology may be described as metaphoric. One example of a metaphoric approach is the study of brain activities through the application of the properties of neural networks, that is, networks of interconnected boolean (binary-state) switches. Appropriately configured switching networks are known to exhibit behaviors that are analogous to those that characterize the brain, such as learning, memory, and discrimination. That is, networks of neuronlike units can automatically manifest brainlike behaviors and can be regarded as metaphorical brains. Such boolean neural nets also underlie digital computation, a relationship which is explored in the hybrid area of artificial intelligence. The same mathematical formulation of switching networks arises in genetic and developmental phenomena, such as the concept of operon, and in other physiological systems, such as the immune system. *See* ARTIFICIAL INTELLIGENCE; NEURAL NETWORK; OPERON.

Another important example of metaphor in biology is morphogenesis, or pattern generation, through the coupling of chemical reactions with physical diffusion. Chemical reactions tend to make systems heterogeneous, diffusion tends to smooth them out, and combining the two can lead to highly complex behaviors. Since reactions and diffusions typically occur together in biological systems, exploring the general properties of such systems can illuminate pattern generation in general.

Such ideas turn out to be closely related to those of bifurcation and catastrophe and have a profound impact on physics itself, since they are inherently associated with systems that are thermodynamically open and hence completely outside the realm of classical thermodynamics. The behavior of such open systems can be infinitely more complicated than those that are commonly explored in physics. Open systems may possess large numbers of stable and unstable steady states of various types, as well as more complicated oscillatory steady-state behaviors (limit cycles) and still more general behaviors collectively called chaotic. Changes in initial conditions or in environmental circumstances can result in dramatic switching (bifurcations) between these modes of behavior. *See* CHAOS.

Applications. Perhaps the biotechnology that has affected everyone most directly is medicine. Medicine can be regarded as a branch of control theory, geared to the maintenance or restoration of a state of health. It is unique in that the systems needed for control are themselves control systems that are far more intricate and complex than any that can be fabricated. In addition to the light it sheds on the processes needed for control, mathematical biology is indispensable for designing the controls themselves and for assessing their costs, benefits, safety, and efficacy.

In general, the object of any theory of control is to produce an algorithm, or protocol, that will achieve optimal results. Mathematical biology allows one to relate systems of different characters through the exploitation of their mathematical commonalities. Biology has many optimal designs and optimal controls, which are the products of biological evolution through natural selection. The design of optimal therapies in medicine is analogous to the generation of optimal organisms. Thus the mathematical theory appropriate for analyzing one discipline of biology, such as evolution, itself becomes transmuted into a theory of control in an entirely different realm. The same holds true for other biotechnologies, such as the efficient exploitation of biological populations. *See* MATHEMATICAL ECOLOGY. Robert Rosen

Mathematical ecology

The application of mathematical theory and technique to ecology. The earliest studies in ecology were by naturalists interested in organisms and their relationships to the environment. Such investigations continue to this day as a central part of the subject, and have focused attention on understanding the ecological and evolutionary relationships among species. For the most part, such approaches are retrospective, designed to help in understanding how current ecological relationships developed, and to place that development within appropriate evolutionary context. The second major branch is applied ecology, and derives from the need to manage the environment and its resources. Here the necessity for rigorous mathematical treatments is obvious, but the goals are quite different from those in evolutionary ecology. Management and control are the objectives, and the relevant time horizon lies in the future. The focus is no longer simply to derive understanding and explanation; rather, one seeks methods for prediction and algorithms for control.

There has been a dramatic increase in mathematical activity concerning the modeling and control of epidemics, and an increasing recognition of the need to view such problems in their proper ecological context as host-parasite interactions. Researchers are using mathematical models to help to understand the factors underlying disease outbreaks, and to develop methods for control such as vaccination strategies. *See* EPIDEMIOLOGY.

Finally, the need for environmental protection in the face of threats from such competing stresses as toxic substances, acid precipitation, and power generation has led to the development of more sophisticated models that address the responses to stress of community and ecosystem characteristics, for example, succession, productivity, and nutrient cycling. *See* ECOLOGY; ECOLOGY, APPLIED; ENVIRONMENTAL ENGINEERING. Simon A. Levin

Mathematical geography

The branch of geography that examines human and physical activities on the Earth's surface using models and statistical analysis. The primary areas in which mathematical methods are used include the analysis of spatial patterns, the processes that are responsible for creating and modifying these patterns, and the interactions among spatially separated entities.

What sets geographic methods apart from other quantitative disciplines is geography's focus on place and relative location. Latitude and longitude provide an absolute system of recording spatial data, but geographic databases also typically contain large amounts of relative and relational data about places.

Thus, geographers have devoted much effort to accounting for spatial interrelations while maintaining consistency with the assumptions of mathematical models and statistical theory. *See* GEOGRAPHY; MODEL THEORY; STATISTICS.

Spatial pattern methodologies attempt to describe the arrangement of phenomena over space. In most cases these phenomena are either point or area features, though computers now allow for advanced three-dimensional modeling as well. Point and area analyses use randomness (or lack of pattern) as a dividing point between two opposite pattern types—dispersed or clustered.

An important innovation in geographic modeling has been the development of spatial autocorrelation techniques. Unlike conventional statistics, in which many tests assume that observations are independent and unrelated, very little spatial data can truly be considered independent. Soil moisture or acidity in one location, for example, is a function of many factors, including the moisture or acidity of nearby points. Because most physical and human phenomena exhibit some form of spatial interrelationships, several statistical methods, primarily based on the Moran Index, have been developed to measure this spatial autocorrelation. Once identified, the presence and extent of spatial autocorrelation can be built into the specification of geographical models to more accurately reflect the behavior of spatial phenomena. Jonathan C. Comer

Mathematical physics An area of science concerned with the application of mathematical concepts to the physical sciences and the development of mathematical ideas in response to the needs of physics. Historically, the concept of mathematical physics was synonymous with that of theoretical physics. In present-day terminology, however, a distinction is made between the two. Whereas most of theoretical physics uses a large amount of mathematics as a tool and as a language, mathematical physics places greater emphasis on mathematical rigor, and devotes attention to the development of areas of mathematics that are, or show promise to be, useful to physics. The results obtained by pure mathematicians, with no thought to applications, are almost always found to be both useful and effective in formulating physical theories.

Mathematical physics forms the bridge between physics as the description of nature and its structure on the one hand, and mathematics as a construction of pure logical thought on the other. This bridge between the two disciplines benefits and strengthens both fields enormously. *See* MATHEMATICS; PHYSICS; THEORETICAL PHYSICS.

The methods employed in mathematical physics range over most of mathematics, the areas of analysis and algebra being the most commonly used. Partial differential equations and differential geometry, with heavy use of vector and tensor methods, are of particular importance in the formulation of field theories, and functional analysis as well as operator theory in quantum mechanics. Group theory has become an especially valuable tool in the construction of quantum field theories and in elementary-particle physics. There has also been an increase in the use of general geometrical approaches and of topology. For solution methods and the calculation of quantities that are amenable to experimental tests, of particular prominence are Fourier analysis, complex analysis, variational methods, the theory of integral equations, and perturbation theory. *See* ABSTRACT ALGEBRA; COMPLEX NUMBERS; DIFFERENTIAL GEOMETRY; FOURIER SERIES AND TRANSFORMS; GROUP THEORY; INTEGRAL EQUATION; OPERATOR THEORY; TOPOLOGY; VARIATIONAL METHODS (PHYSICS); VECTOR METHODS (PHYSICS). Roger G. Newton

Mathematical software The collection of computer programs that can solve equations or perform mathematical manipulations. The developing of mathematical equations that describe a process is called mathematical modeling. Once these equations are developed, they must be solved, and the solutions to the equations are then analyzed to determine what information they give about the process. Many discoveries have been made by studying how to solve the equations that model a process and by studying the solutions that are obtained.

Before computers, these mathematical equations were usually solved by mathematical manipulation. Frequently, new mathematical techniques had to be discovered in order to solve the equations. In other cases, only the properties of the solutions could be determined. In those cases where solutions could not be obtained, the solutions had to be approximated by using numerical calculations involving only addition, subtraction, multiplication, and division. These methods are called numerical algorithms. These algorithms are often straightforward, but they are usually tedious and require a large number of calculations, usually too many for a human to perform. There are also many cases where there are too many equations to write down. *See* ALGORITHM; NUMERICAL ANALYSIS.

The advent of computers and high-level computer languages has allowed many of the tedious calculations to be performed by a machine. In the cases where there are too many equations, computer programs have been written to manipulate the equations. A numerical algorithm carried out by a computer program can then be applied to these equations to approximate their solutions. Mathematical software is usually divided into two categories: the numerical computation environment and the symbolic computation environment. However, many software packages exist that can perform both numerical and symbolic computation.

Mathematical software that does numerical computations must be accurate, fast, and robust. Accuracy depends on both the algorithm and the machine on which the software is run. Most mathematical software uses the most advanced numerical algorithms. Robustness means that the software checks to make sure that the user is inputting reasonable data, and provides information during the performance of the algorithm on the convergence of the calculated numbers to an answer. Mathematical software packages can approximate solutions to a large range of problems in mathematics, including matrix equations, nonlinear equations, ordinary and partial differential equations, integration, and optimization. Mathematical software libraries contain large collections of subroutines that can solve problems in a wide range of mathematics. These subroutines can easily be incorporated into larger programs.

Early computers were used mainly to perform numerical calculations, while the mathematical symbolic manipulations were still done by humans. Now software is available to perform these mathematical manipulations. Most of the mathematical software packages that perform symbolic manipulations can also perform numerical calculations. Software can be written in the package to perform the numerical calculations, or the calculations can be performed after the symbolic manipulations by putting numbers into the symbolic formulas. Mathematical software that is written to solve a specific problem using a numerical algorithm is usually computationally more efficient than these software environments. However, these software environments can perform almost all the commonly used numerical and symbolic mathematical manipulations. *See* SYMBOLIC COMPUTING.

Parallel computers have more than one processor that can work on the same problem at the same time. Parallel computing allows a large problem to be distributed over the processors. This allows the problem to be solved in a smaller period of time. Many numerical algorithms have been converted to run on parallel computers. *See* COMPUTER PROGRAMMING; CONCURRENT PROCESSING; DIGITAL COMPUTER; DISTRIBUTED SYSTEMS (COMPUTERS); MULTIPROCESSING; SOFTWARE. James Sochacki

Mathematics Mathematics is frequently encountered in association and interaction with astronomy, physics, and other branches of natural science, and it also has deep-rooted affinities to the humanities. It is a realm of knowledge entirely unto itself, and one of considerable scope.

Relation to science. Mathematics is not a branch of natural science itself. It does not deal with phenomena and objects of the external world and their relations to each other but, strictly speaking, only with objects and relations of its own imagery. One can practice meaningful mathematics without being concerned with science at all, and philosophical attempts to reduce all origin of mathematics to utilitarian motives are wholly unconvincing. However, mathematics is the language of science in a deep sense. Mathematics is an indispensable medium by which and within which science expresses, formulates, continues, and communicates itself. And just as the language of true literacy not only specifies and expresses thoughts and processes of thinking but also creates them in turn, so does mathematics not only specify, clarify, and make rigorously workable concepts and laws of science, but also at certain crucial instances becomes an indispensable constituent of their creation and emergence as well.

Creative formulas. A formula is a string of mathematical symbols subject only to certain general rules of composition. To a working mathematician a string of symbols is a formula if it is something worth remembering. Much mathematics is concentrated in and propelled by certain formulas of unusual import.

Foundations—mathematical logic. A prime demand on mathematics is that it be deductively rigorous, and a traditional model for intended rigor is Euclid's presentation of mathematical assertions in theorems. A theorem is a proposition which has been proved, excepting certain first theorems called axioms, which are admitted without proof; and to prove a theorem means to obtain it from other theorems by certain procedures of deduction or inference. It had long been commonplace that each branch of mathematics was based on its own axioms, but during the 19th century, mathematicians arrived at the insight that even the same branch might have alternate axioms. Specifically, there were envisaged alternate versions of two- and three-dimensional geometry, the axiom varied being the axiom on parallels. It was also recognized that a set of axioms becomes mathematically possible if it is logically consistent, that is, if one cannot deduce from the axioms to theorems one of which, as a proposition, is the negation of the other. *See* EUCLIDEAN GEOMETRY.

At the same time certain developments led to the realization that not only the axioms but the rules of inference themselves might be, and even ought to be, subjected to variations. Now if axioms and rules of inferences are both viewed as subject to change, it is customary to speak of a mathematical system or also a formal system, and, of course, an irreducible first requirement is that the system be consistent after the manner just stated. Consistency alone is a somewhat negative property. There is a further property, called completeness, which is more positive, and which, if present, is very welcome. A system is complete if for any proposition which can be formulated it either can be proved that it holds or that its negation holds.

Some of the developments that led to doubt as to whether the traditional rules of inference are inviolate were the following.

1. G. Boole had found in 1854 that the classical Aristotelian connectives "and," "or," "negation of" for propositions follow rules similar to those which the operations addition, multiplication, "the negative of" obey in ordinary algebra (Boole's algebra of propositions); his conclusions took from rules of inference the status of untouchability. *See* BOOLEAN ALGEBRA.

2. G. Cantor, the founder of the theory of sets and operations between them, defined a set (intuitively or naively) as the collection of all objects having a certain property which is verbally expressible. Especially, "the set of all sets" is again a set and it has the peculiarity of being a set which contains itself as one of its elements. But this leads to the following contradictory situation (Russell's paradox): Divide the totality of all possible sets into two categories. A set shall belong to category I if it does not contain itself as an element, and to category II if it does contain itself as an element. Now form the set M whose elements are the sets of category I. It can now be reasoned by deductive steps admissible in Cantor's own theory that the set M cannot belong to either of the two categories, although the original division into categories did assign each set to one of them.

3. In 1904 E. Zermelo formulated the following axiom of choice: Given any family of nonvacuous sets $\{S\}$, no matter how (infinitely) large the family may be, it is possible to choose simultaneously an element $x = x_s$ from each given set S and thus to consider the set M consisting of precisely these elements. By the use of this axiom some striking theorems in classical mathematics could be proved which, without the use of the axiom, seemed to be logically out of reach entirely. Mathematicians began to wonder whether a theorem based on the axiom of choice is indeed valid or, at any rate, whether it has the same level of validity as one without it, and as a consequence, theorems employing the axiom of choice were frequently labeled as such. *See* SET THEORY.

Some of the doubts were resolved eventually; the most striking results are the following ones of K. Gödel. (1) Any consistent mathematical system which is sufficient for classical arithmetic must be incomplete. (2) Any such system remains consistent if one adds to it the axiom of choice, so that working mathematics cannot disprove the axiom of choice. In 1963 P. Cohen showed that the axiom cannot be proved, either. *See* LOGIC.

Constructiveness. Some mathematicians object to mere existence proofs, and they demand that any proof also be constructive. The interpretations of this demand differ widely. Some proofs closely approach what a practical mathematician welcomes; if, for instance, a theorem asserts the existence of a number or a function, then the proof must also embody a procedure for actual computation of the solution, approximately, at least. Other versions are little more than the negative requirement that certain combinations of inference be avoided. There are also views which combine both; the best known among the last is the intuitionist view. It firmly demands a certain kind of constructiveness, which, however, does not necessarily guarantee the calculation by present-day computing machines. However, the actual stricture by which intuitionism became widely known is that proof by contradiction is not admissible. Proof by contradiction is also called proof by double negation, and it is equivalent to the Aristotelian law of the excluded middle. It assumes tentatively that the proposition to be proved is false and from this assumption deduces a contradiction to a previously established theorem.

Space in mathematics. If geometry is the mathematics of space, then, in a superficial sense, all mathematics began with geometry, because apparently it began with measurements of figures: length, area, volume, and size of angles. It did not concern itself with questions of shape but with clarifying and deciding when figures are equal or substantially equal with regard to form. The first true theory of geometry was that of the Greeks, whose primary concern was study of the basic concept of equality of figures—their congruence and similarity—and the Greeks were so determined to dissociate their theory from the preceding phase of merely making measurements that Euclid's extensive work, for instance, avoids to a fault any kind of actual measurements. But, for all its lofty purposes, Greek geometry was too rigid and circumscribed to be able really to cope with the mathematical problem of space. Geometry did not progress further until, with the advent of coordinate systems, introduced by Descartes and his predecessors, a better mathematics of space could be initiated.

If a cartesian coordinate system is etched into two- or three-dimensional euclidean space, then the space becomes a point set, each point being a pair (x^1, x^2) or a triple (x^1, x^2, x^3) of real numbers, and any figure a suitable subset of it. This is a deliberate process of arithmetization of space which unifies space and number at the base. It does not hamper geometry in its task of pursuing problems of shape but instead aids it. In the cartesian plane, two figures are similar if the points of one can be obtained from the points of the other by means of a transformation, Eqs. (1), where Eq. (2) applies and where, for some $\rho < 0$. Eq. (3)

$$\begin{aligned} y^1 &= a^1 + \alpha_1{}^1 x^1 + \alpha_2{}^1 x^2 \\ y^2 &= a^2 + \alpha_1{}^2 x^1 + \alpha_2{}^2 x^2 \end{aligned} \tag{1}$$

$$\alpha_1{}^1 \cdot \alpha_1{}^2 + \alpha_2{}^1 \cdot \alpha_2{}^2 = 0 \tag{2}$$

$$(\alpha_1{}^1)^2 + (\alpha_2{}^1)^2 = (\alpha_1{}^2)^2 + (\alpha_2{}^2)^2 = \rho^2 \tag{3}$$

can be written. The similarity is a congruence if, and only if, $\rho = 1$ (orthogonal transformation). Now this analytic representation of congruence and similarity suggests a geometric examination of the most general linear transformations, Eqs. (1), which are nonsingular, that is, for which the determinant $|\alpha_q{}^p| \neq 0$. They were virtually unknown to the Greeks, although they highlight the axiom of parallels of Euclid's geometry. A one-to-one transformation of the cartesian plane is such a linear transformation if, and only if, it carries a straight line onto a straight line and parallel straight lines into parallel straight lines.

The family of all linear transformations constitutes a transitive group, and the subfamily of orthogonal transformations is already a transitive group. F. Klein made the pronouncement, which is generally accepted, that there arises a geometry on a space if on the space there is given a transitive group of transformations; two figures are considered equal whenever one figure can be carried into the other figure by one of the transformations.

The arithmetization of space led to a purely mathematical creation of n-dimensional space, euclidean and other, for any integer dimension n, by defining its points generally as n-tuples of real numbers $(x^1,\ldots, x^n)$ with suitable definitions for various geometrical relations between such points. The best-known application of this was the four-dimensional space of the theory of relativity, but actually multidimensional geometry had been playing a part in physics before that. If a mechanical system involves M mass points, it was customary in effect to introduce the space of dimension $n = 3M$, whose points are the states of the system, that is, the n-tuples of coordinates $\{x_m{}^1, x_m{}^2, x_m{}^3\}$, $m = 1,\ldots, M$, at any one time point. Also, if there are restraints operative in the system, then the Lagrange-Hamilton theory suitably reduced the dimension of the space by the use of the free parameters of the system instead of the original n coordinates themselves. The use of free parameters spread from mechanical systems to other systems in physics and chemistry, and all so-called equations of state are geared to this. Finally, in quantum theory a state of a system has infinitely many coordinates and the infinitely dimensional space representing it is a Hilbert space. Also, partly under the influence of Hilbert space, mathematicians have become fascinated with infinitely dimensional spaces in general. They are being studied intensively, and large parts of mathematics are being pressed into these new frames of reference.

The arithmetization of space is also reflected in the ever-widening use of graphs and charts. *See* ALGEBRA; CALCULUS OF VECTORS; GEOMETRY; TOPOLOGY. Salomon Bochner

Matrix isolation A technique for providing a means of maintaining molecules in an inert medium at low temperature for spectroscopic study. This method is particularly well suited for preserving reactive species in a solid, inert environment. Absorption (infrared, visible, and ultraviolet), electron-spin resonance, and laser-excitation spectroscopes can be used to examine elusive molecular fragments such as free radicals that may be postulated as important controlling intermediates for chemical transformations used in industrial reactions, high-temperature molecules that are in equilibrium with solids at very high temperatures, weak molecular complexes that may be stable at low temperatures, new reactive molecular species, and molecular ions

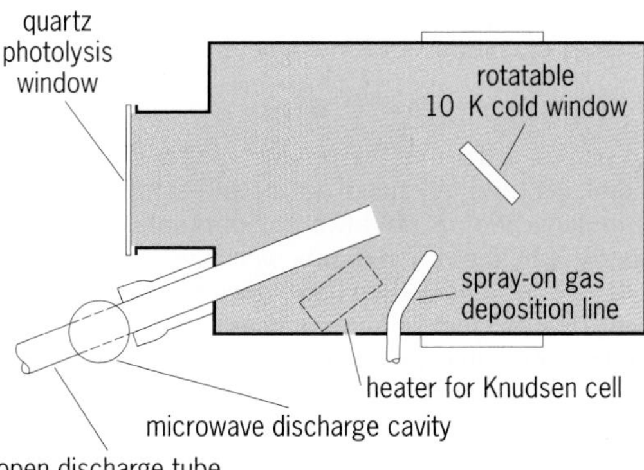

Vacuum-vessel base cross section for matrix photoionization experiments. 10 K = −263°C or −442°F.

that are produced in plasma discharges or by high-energy radiation. The matrix isolation technique enables spectroscopic data to be obtained for reactive molecular fragments, many of which cannot be studied in the gas phase. *See* ELECTRON PARAMAGNETIC RESONANCE (EPR) SPECTROSCOPY; INFRARED SPECTROSCOPY; SPECTROSCOPY.

The experimental apparatus for matrix isolation experiments is designed with the method of generating the molecular transient and performing the spectroscopy in mind. The illustration shows the cross section of a vacuum vessel used for absorption spectroscopic measurements (such as infrared, visible, and ultraviolet). The optical windows must be transparent to the examining radiation. The rotatable cold window is cooled to 4 to 20 K (−452 to −424°F) by closed-cycle refrigeration or liquid helium. The matrix sample is introduced through the spray-on line at rates of 1–5 millimoles per hour; argon is the most widely used matrix gas, although neon, krypton, xenon, and nitrogen are also used. The reactive species can be generated in a number of ways: mercury-arc photolysis of a trapped precursor molecule through the quartz window, evaporation from a Knudsen cell in the heater, chemical reaction of atoms evaporated from the Knudsen cell with molecules deposited through the spray-on line, neodymium-yttrium aluminum garnet (Nd-YAG) laser evaporation of solids, and vacuum-ultraviolet photolysis of molecules deposited from the spray-on line by radiation from discharge-excited atoms flowing through the tube. The cold window is rotated for transmission absorption spectroscopic measurements using ultraviolet-visible or infrared spectrometers. Lester Andrews

Matrix mechanics A formulation of quantum theory in which the operators are represented by time-dependent matrices. *See* MATRIX THEORY.

Matrix mechanics is not useful for obtaining quantitative solutions to actual problems; on the other hand, because it is concisely expressed in a form independent of special coordinate systems, matrix mechanics is useful for proving general theorems. Edward Gerjuoy

Matrix theory The study of matrices and their properties, and of linear transformations on vector spaces, which can be represented by matrices.

A matrix is a rectangular array of numbers, with the numbers that appear in the matrix being called entries. For example, A, given by Eq. (1), is a 2×3 matrix (that is, it has 2 rows and 3

$$A = \begin{bmatrix} 1 & 2 & 3 \\ 4 & 5 & 6 \end{bmatrix} \tag{1}$$

columns). The entry in the ith row and jth column of a matrix is denoted by a_{ij}. For example, in the matrix A, $a_{21} = 4$ and $a_{12} = 2$.

In general, if X is an $m \times n$ matrix and Y is an $n \times q$ matrix, then the product $P = XY$ is an $m \times q$ matrix whose entry in row

i and column j is given by Eq. (2) for $i, \ldots, m$ and $j = 1, \ldots, q$. The number of columns of X must be the same as the number

$$p_{ij} = x_{i1}y_{1j} + x_{i2}y_{2j} + \cdots + x_{in}y_{nj} \tag{2}$$

of rows of Y in order that the product matrix P can be defined. In general, XY and YX need not be the same matrix; that is, matrix multiplication is not always commutative. Multiplication of a matrix X by the $n \times n$ matrix with 1 in row i and column i for $i = 1, \ldots, n$, and 0 elsewhere, does not change the matrix X. Finally, the product of two matrices can have every entry 0 without this being the case for either factor.

Vectors in the plane can be added and multiplied (scaled) by constants. For example, if the vectors u and v and the number c are given by Eq. (3), then $u + v$ is given by Eq. (4) and cu is given by Eq. (5).

$$u = \begin{bmatrix} 1 \\ 2 \end{bmatrix} \quad v = \begin{bmatrix} 3 \\ 4 \end{bmatrix} \quad c = -3 \tag{3}$$

$$u + v = \begin{bmatrix} 1+3 \\ 2+4 \end{bmatrix} = \begin{bmatrix} 4 \\ 6 \end{bmatrix} \tag{4}$$

$$cu = \begin{bmatrix} -3 \cdot 1 \\ -3 \cdot 2 \end{bmatrix} = \begin{bmatrix} -3 \\ -6 \end{bmatrix} \tag{5}$$

There are a number of simple axioms that are satisfied by vector addition and scalar multiplication. A set of vectors V together with a set of numbers F that satisfy these rules is called a vector space over F. The set of numbers F is a field. This means that F has many of the properties associated with rational numbers (fractions), real numbers (decimals), or complex numbers. There are many important mathematical systems that satisfy the axioms for a vector space. Some examples are the two-dimensional space of plane geometry, the three-dimensional space of solid geometry, and polynomials with coefficients from F. *See* FIELD THEORY (MATHEMATICS); LINEAR ALGEBRA.

If V and W are vector spaces over F, then a linear transformation T on V to W is a function that assigns to each vector v in V a unique vector w in W, as in Eq. (6). Moreover, if u and v are

$$Tv = w \tag{6}$$

any vectors in V, and c and d are any scalars in F, then T satisfies Eq. (7).

$$T(cu + dv) = cTu + dTv \tag{7}$$

A basis of the vector space V is an ordered sequence of vectors $v_1, \ldots, v_n$ in V such that any vector v in V can be written in only one way as in Eq. (8) for appropriate scalars $c_1, \ldots, c_n$.

$$v = c_1v_1 + c_2v_2 + \cdots + c_nv_n \tag{8}$$

Once bases $v_1, \ldots, v_n$ and $w_1, \ldots, w_m$ for V and W respectively have been chosen, any linear transformation T on V to W can be completely and uniquely described in terms of the entries in an $m \times n$ matrix A by Eq. (9). In other words, each Tv_j, as a vector

$$Tv_j = a_{1j}w_1 + a_{2j}w_2 + \cdots + a_{mj}w_m \tag{9}$$

$$j = 1, 2, \ldots, n$$

in W, can be expressed uniquely as a sum of scalar multiples of the basis $w_1, \ldots, w_m$. The scalars $a_{1j}, \ldots, a_{mj}$ are the entries in column j of A. This observation is fundamental because it shows that the study of linear transformations is coextensive with the study of matrices.

There are a number of important questions about linear transformations that have been studied extensively. In particular:

1. Given w in W, the problem of determining all vectors v in V (if any) for which Eq. (10) is satisfied.

$$Tv = w \tag{10}$$

2. If $V = W$, the problem of determining all scalars λ and nonzero vectors v for which Eq. (11) is satisfied.

$$Tv = \lambda v \tag{11}$$

The set of all vectors Tv obtained as v varies over V is called the range of T. The range of T is always a vector space. Thus the first problem above is equivalent to determining the range of a linear transformation T. Once bases of V and W have been selected, Eq. (10) is readily seen to be equivalent to solving a system of linear equations of the form of Eqs. (12) for the determination of the numbers $x_1, \ldots, x_n$.

$$\begin{aligned} a_{11}x_1 + \cdots + a_{1n}x_n &= b_1 \\ a_{21}x_1 + \cdots + a_{2n}x_n &= b_2 \\ &\vdots \\ a_{m1}x_1 + \cdots + a_{mn}x_n &= b_m \end{aligned} \tag{12}$$

Finding the numbers λ and vectors v that satisfy Eq. (11) is called the eigenvalue problem; the number λ in (11) is called an eigenvalue of T, and the corresponding vector v in Eq. (11) is called an eigenvector of T. Equation (11) leads to a system similar to Eqs. (12), but $m = n$ and each b_i must be replaced by λx_i for $i = 1, \ldots, n$. Moreover, both the eigenvalue λ and $x_1, \ldots, x_n$ must be determined. Because of the importance of Eqs. (10) and (11) in many areas of applied mathematics, special-purpose computer programs have been developed to deal with them. *See* EIGENFUNCTION; EIGENVALUE (QUANTUM MECHANICS).

Because of their applications in the physical and social sciences, certain special classes of linear transformations and matrices have been studied extensively. If A is an $m \times n$ matrix, then the transpose of A is the $n \times m$ matrix whose rows, in succession, are columns of A written as rows. The transpose of A is denoted by A^T. If A has complex number entries, then A^* is the matrix obtained from A^T by replacing each entry by its complex conjugate. An $n \times n$ matrix A is said to be nonsingular (or invertible) if there is a unique $n \times n$ matrix A^{-1} for which $AA^{-1} = A^{-1}A = I_n$; the matrix I_n is the identity matrix: $a_{ii} = 1$ for $i = 1, \ldots, n$, and $a_{ij} = 0$ if $i \neq j$. If $A^* = A$, then A is hermitian; if $A^* = A^{-1}$, then A is unitary; if $A^*A = AA^*$, then A is normal.

The singular-value decomposition theorem states that any $m \times n$ matrix A can be factored into a product $A = UDV$, in which U is $m \times m$ unitary, V is $n \times n$ unitary, and D is an $m \times n$ matrix whose only (possibly) nonzero entries are $d_{11} \geq \cdots > d_{pp} \geq 0$, where p is the smaller of m and n. The numbers $d_{11}, \ldots, d_{pp}$ are called the singular values of A. The singular-value decomposition theorem is used extensively in solving least-squares problems.

Matrices whose entries are nonnegative real numbers are important in many applications of matrix theory to problems in the social sciences. Matrices with polynomial entries have been studied extensively because of their importance in control theory, systems theory, and other areas of applied mathematics and engineering.

Marvin Marcus

Matter (physics) A term that traditionally refers to the substance of which all bodies consist. Matter in classical mechanics is closely identified with mass. Modern analyses distinguish two types of mass: inertial mass, by which matter retains its state of rest or uniform rectilinear motion in the absence of external forces; and gravitational mass, by which a body exerts forces of attraction on other bodies and by which it reacts to those forces. Expressed in appropriate units, these two properties are numerically equal—a purely experimental fact, unexplained by theory. Albert Einstein made the equality of inertial and gravitational mass a fundamental principle (principle of equivalence), as one of the two postulates of the theory of general relativity. *See* GRAVITATION; INERTIA; MASS; RELATIVITY; WEIGHT.

In quantum mechanics, mass is only one among many properties (quantum numbers) that a particle can have, for example, electric charge, spin, and parity. The nearest quantum-mechanical analogs of traditional matter are fermions, having half-integral values of spin. Forces are mediated by exchange of bosons, particles having integral spins (such as photons). Fermions correspond to classical matter in exhibiting impenetrability (a consequence of the exclusion principle), but the correspondence is only rough. For example, fermions can also be exchanged in interactions (a photon and an electron can exchange an electron), and they also exhibit wavelike behavior. States of classical matter-particles were given by their positions and momenta, but in quantum mechanics it is impossible to assign simultaneous precise positions and momenta to particles. *See* EXCLUSION PRINCIPLE; QUANTUM ELECTRODYNAMICS; QUANTUM MECHANICS; QUANTUM STATISTICS.

Beginning in the 1920s, astronomers accumulated a vast amount of evidence leading to the conclusion that the density of matter in the universe was much larger than that of visible matter. Visible matter is baryonic, consisting almost entirely by mass of protons and neutrons. With the total density of baryonic matter so low, the missing nonbaryonic dark matter would have to be some kind of exotic particle. *See* DARK MATTER; SUPERSYMMETRY; WEAKLY INTERACTING MASSIVE PARTICLE (WIMP).

It was long realized that such exotic matter was still not enough for closure (critical density), however. In 1998, results from studies of distant supernovae indicated strongly that the universe is not merely expanding: the rate of expansion is accelerating. The new view thus suggested was that some kind of dark antigravitational energy was at work, resembleing the cosmological constant that Einstein had proposed as a term in his equations designed to preserve the universe from the possibility that it might expand or contract. *See* ACCELERATING UNIVERSE; COSMOLOGICAL CONSTANT; DARK ENERGY.

In 2003, the *Wilkinson Microwave Anisotropy Probe* (*WMAP*) determined the distribution as follows: 4% of the matter in the universe is baryonic, 23% is dark nonbaryonic matter, and 73% is something else, generally known as dark energy. The *WMAP* results were bolstered by ones from the Sloan Digital Sky Survey (SSDS), also announced in 2003. Several future investigations, expanding and improving on these results, are in preparation. *See* COSMOLOGY; SLOAN DIGITAL SKY SURVEY; UNIVERSE; WILKINSON MICROWAVE ANISOTROPY PROBE.

Dudley Shapere

Matthiessen's rule

An empirical rule which states that the total resistivity of a crystalline metallic specimen is the sum of the resistivity due to thermal agitation of the metal ions of the lattice and the resistivity due to the presence of imperfections in the crystal. This rule is a basis for understanding the resistivity behavior of metals and alloys at low temperatures.

The resistivity of a metal results from the scattering of conduction electrons. Lattice vibrations scatter electrons because the vibrations distort the crystal. Imperfections such as impurity atoms, interstitials, dislocations, and grain boundaries scatter conduction electrons because in their immediate vicinity the electrostatic potential differs from that of the perfect crystal.

Frank J. Blatt

Maxillopoda

A class of Crustacea whose application is gaining moderately widespread use, despite the fact that its validity is not universally accepted by researchers. The constituent subclasses of the Maxillopoda also remain unsettled.

The class Maxillopoda was proposed for those taxa with six thoracic somites (with some exceptions), a well-developed mandibular palp in adults, well-developed maxillules and maxillae adapted for filtering in filter feeders, and the lack of gnathobases on the appendages of the thorax. The Recent taxa included were the Copepoda, Branchiura, Mystacocarida, and provisionally, the Cirripedia. Subsequently the Cirripedia (subdivided into Cirripedia sensu stricto and Ascothoracida) were unequivocally incorporated. The Ostracoda are now included by some but excluded by others. More recent suggestions have included the Tantulocarida, and even the probably noncrustacean Pentastomida, within the Maxillopoda. *See* BRANCHIURA; CIRRIPEDIA; COPEPODA; CRUSTACEA; OSTRACODA; PENTASTOMIDA; TANTULOCARIDA.

Patsy A. McLaughlin

Maxwell's demon

An imaginary being whose action appears to contradict the second law of thermodynamics, which identifies the natural direction of change with the direction of increasing entropy. There has always been a certain degree of discomfort associated with the acceptance of the law, particularly in relation to the time reversibility of physical laws and the role of molecular fluctuations. In 1867, J. C. Maxwell considered, in this connection, the action of "a finite being who knows the paths and velocities of all the molecules by inspection." This being was later referred to as a demon by Lord Kelvin, and the usage has been generally adopted. *See* ENTROPY; THERMODYNAMIC PRINCIPLES; TIME, ARROW OF.

The activity of Maxwell's demon can be modeled by a trapdoor in a partition between two regions full of gas at the same pressure and temperature. The trapdoor needs to be restrained by a light spring to ensure that it is closed unless it is struck by molecules traveling from the left (see illustration). Its hinging is such that molecules traveling from the right cannot open it. The essential point of Maxwell's vision was that molecules striking the trapdoor from the left would be able to penetrate into the right-hand region but those present on the right would not be able to escape back into the left-hand region. Therefore, the initial equilibrium state of the two regions, that of equal pressures, would be slowly replaced by a state in which the two regions acquired different pressures as molecules accumulated in the right-hand region at the expense of the left-hand region. Only a slightly more elaborate mechanical arrangement is needed to change the apparatus to one in which the temperatures of the two regions move apart. In each case, the demonic trapdoor appears to be contriving a change that is contrary to the second law, for an implication of that law is that systems in either mechanical equilibrium (at the same pressure) or thermal equilibrium (at the same temperature) cannot spontaneously diverge from equilibrium.

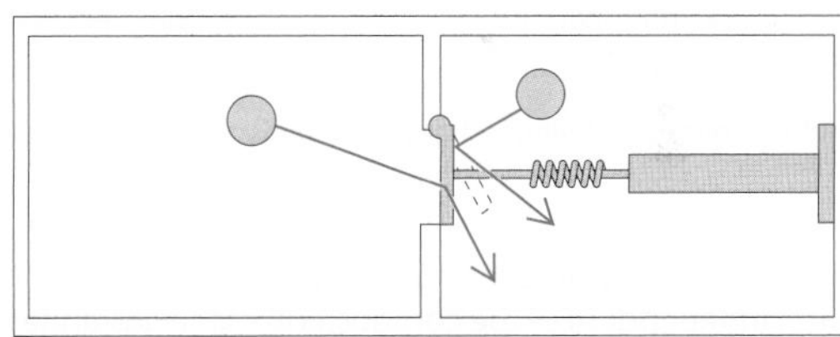

Type of device that emulates mechanically the actions of Maxwell's demon. Molecules traveling to the right can open the trapdoor and enter the right-hand compartment, but those striking it from the right cannot open it, so do not move into the left-hand compartment.

As frequently occurs in science, the resolution of a paradox or the elimination of an apparent conflict with a firmly based law depends on a detailed analysis of the proposed arrangement. Numerous analyses of this kind have shown that the activities of Maxwell's demon do not in fact result in the overthrow of the second law.

P. W. Atkins

Maxwell's equations

Four differential equations proposed by James Clerk Maxwell in 1864 as the basis of the theory of electromagnetic waves. They may be written, in vector notation, as Eqs. (1)–(4), where **D** is the electric displacement, **B** the

$$\nabla \cdot \mathbf{D} = \rho \qquad (1)$$

$$\nabla \cdot \mathbf{B} = 0 \qquad (2)$$

$$\nabla \times \mathbf{E} = -\frac{\delta \mathbf{B}}{\delta t} \qquad (3)$$

$$\nabla \times \mathbf{H} = \mathbf{i} + \frac{\delta \mathbf{D}}{\delta t} \qquad (4)$$

magnetic flux density, **E** the electric field strength or intensity, **H** the magnetic field strength or intensity, ρ the charge density, and **i** the current density.

The first equation states that electric flux lines, if they end at all, will do so on electric charges. The second states that magnetic flux lines never terminate. The third is a form of Faraday's law of induction, which states that the rate of change of the magnetic flux threading a circuit equals the electromotive force or line integral of **E** around the circuit. The fourth integral is based partially on A. M. Ampère's experiments on steady currents which show that the line integral of the magnetic intensity **H** (or $\mathbf{B}/\mu$, where μ is the permeability) around a closed curve equals the current encircled. *See* DISPLACEMENT CURRENT; EQUATION OF CONTINUITY; STOKES' THEOREM. William R. Smythe

McLeod gage

McLeod gage A type of instrument used to measure vacuum by application of the principle of Boyle's law.

A known volume of a gas whose pressure is to be measured is trapped by raising the level of a fluid (mercury or oil) by means of a plunger, by lifting a reservoir, by using pressure, or by tipping the apparatus. As the fluid level is further raised, the gas is compressed into the capillary tube (see illustration). Obeying Boyle's law, the compressed gas now exerts enough pressure to

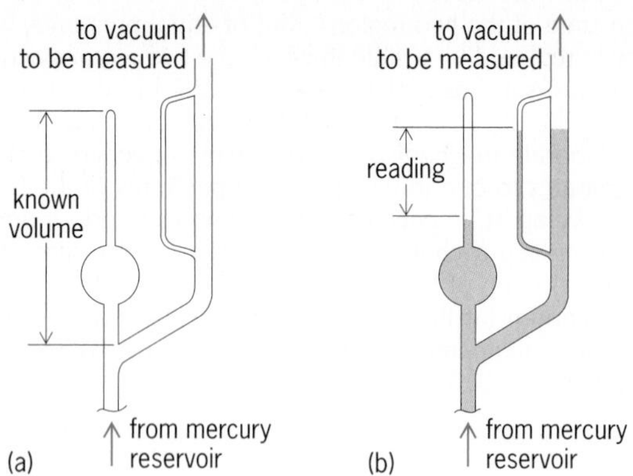

McLeod gage. (*a*) Filling (charging) position. (*b*) Measuring position.

support a column of fluid high enough to read. Readings are somewhat independent of the composition of the gas under pressure. *See* VACUUM MEASUREMENT. Richard Comeau

Mean effective pressure

Mean effective pressure A term commonly used in the evaluation for positive displacement machinery performance which expresses the average net pressure difference in pounds per square inch (psi) on the two sides of the piston in engines, pumps, and compressors. It is also known as mean pressure and is abbreviated as mep or mp.

In an engine (prime mover) it is the average pressure which urges the piston forward on its stroke. In a pump or compressor it is the average pressure which must be overcome, through the driver, to move the piston against the fluid resistance.

The criterion of mep is a vitally convenient device for the evaluation of a reciprocating engine, pump, or compressor design as judged by initial cost, space occupied, and deadweight. *See* COMPRESSOR; DIESEL CYCLE; THERMODYNAMIC CYCLE; VAPOR CYCLE. Theodore Baumeister

Mean free path

Mean free path The average distance traveled between two similar events. The concept of mean free path is met in all fields of science and is classified by the events which take place. The concept is most useful in systems which can be treated statistically, and is most frequently used in the theoretical interpretation of transport phenomena in gases and solids, such as diffusion, viscosity, heat conduction, and electrical conduction. The types of mean free paths which are used most frequently are for elastic collisions of molecules in a gas, of electrons in a crystal, of phonons in a crystal, and of neutrons in a moderator. *See* KINETIC THEORY OF MATTER. W. Dexter Whitehead

Measles

Measles An acute, highly infectious viral disease with cough, fever, and maculopapular rash. It is of worldwide endemicity.

The virus enters the body via the respiratory system, multiplies there, and circulates in the blood. Cough, sneezing, conjunctivitis, photophobia, and fever occur, with Koplik's spots (small red spots containing a bluish-white speck in the center) in the mouth.

A rash appears after 14 days' incubation and persists 5–10 days. Serious complications may occur in 1 out of 15 persons; these are mostly respiratory (bronchitis, pneumonia), but neurological complications are also found. Encephalomyelitis occurs rarely. Permanent disabilities may ensue for a significant number of persons. Measles is one of the leading causes of death among children in the world, particularly in the developing countries.

In unvaccinated populations, immunizing infections occur in early childhood during epidemics which recur after 2–3 years' accumulation of susceptible children. Transmission is by coughing or sneezing. Measles is infectious from the onset of symptoms until a few days after the rash has appeared. Second attacks of measles are very rare. Treatment is symptomatic.

Killed virus vaccine should not be used, as certain vaccinees become sensitized and develop local reactions when revaccinated with live attenuated virus, or develop a severe illness upon contracting natural measles. Live attenuated virus vaccine effectively prevents measles; vaccine-induced antibodies persist for years. *See* BIOLOGICALS. Joseph L. Melnick

Measure

Measure A reference sample used in comparing lengths, areas, volumes, masses, and the like. The measures employed in scientific work are based on the international units of length, mass, and time—the meter, the kilogram, and the second—but decimal multiples and submultiples are commonly employed. Prior to the development of the international metric system, many special-purpose systems of measures had evolved and many still survive, especially in the United Kingdom and the United States. *See* METRIC SYSTEM; PHYSICAL MEASUREMENT; TIME; UNITS OF MEASUREMENT; WEIGHT. Dudley Williams

Measure theory

Measure theory A branch of mathematical analysis connected with the theory of integration. In order to discuss this subject, a formal definition of the term measure must be given.

Let X be an arbitrary set. Let m be a fixed collection of subsets of X satisfying the following conditions:

1. $\phi \in m$. (ϕ is the empty set. The symbol $\in$ indicates that ϕ is an element of m).

2. If $A \in m$, then $A^c \in m$. (A^c is the complement of A. It consists of those elements of X which do not belong to A.)

3. If $A_1, A_2, \ldots \in m$, then (1) is valid. (This set is the intersec-

$$\bigcup_{k=1}^{\infty} A_k \in m \tag{1}$$

tion of $A_1, A_2, \ldots$. It consists of those elements of X which belong to at least one of the sets $A_1, A_2, \ldots$.)

In this situation the collection m is called a σ-algebra. Here the term algebra refers to the various set operations (complementation, union, intersection), and the prefix σ to the fact that countably many such operations can be performed with sets in m and still result in sets in m. For example, the three properties mentioned above imply another property:

4. If $A_1, A_2, \ldots \in m$, then (2) is valid. (This set is the inter-

$$\bigcap_{k=1}^{\infty} A_k \in m \qquad (2)$$

section of $A_1, A_2, \ldots$. It consists of those elements of X which belong to all of the sets $A_1, A_2, \ldots$.)

Now suppose that X is a set and m is a particular σ-algebra of subsets of X. A measure μ is a function which assigns to each set in m a certain nonnegative real number (or $+\infty$) and which satisfies the following conditions:

1. $\mu(\phi) = 0$.
2. If A and $B \in m$ and are disjoint ($A \cap B = \phi$), then Eq. (3) holds.

$$\mu(A \cup B) = \mu(A) + \mu(B) \qquad (3)$$

3. If $A_1, A_2, \ldots \in m$, then Eq. (4) holds.

$$\mu\left(\bigcup_{k=1}^{\infty} A_k\right) \leq \sum_{k=1}^{\infty} \mu(A_k) \qquad (4)$$

Various other properties can be derived from these conditions, such as:

4. If A and $B \in m$ and $A \subset B$ (that is, A is a subset of B), then $[\mu](A) \leq \mu(B)$.
5. If $A_1, A_2, \ldots \in m$ and are mutually disjoint ($A_i \cap A_j = \phi$ if $i \neq j$), then Eq. (5) holds.

$$\mu\left(\bigcup_{k=1}^{\infty} A_k\right) = \sum_{k=1}^{\infty} \mu(A_k) \qquad (5)$$

The last property is of crucial importance and is called the countable additivity property of the measure μ.

Measure theory has a great number of important applications. Undoubtedly, the most important is the application to integration. *See* INTEGRATION. B. Frank Jones, Jr.

Measured daywork A tool used primarily in manufacturing facilities as a control device to measure productive output in relation to labor input within a specific time period. The measurement of the work content is accomplished through the use of time standards which are usually the result of a stopwatch time study, predetermined time standards (methods-time measurement, the work-factor system), or some other form of work-measurement technique designed to measure tasks of labor under normal and average conditions.

A measured daywork plan is similar to incentive pay plans inasmuch as in both plans, time standards are used as a device to measure operator performance and also for various forms of management planning. However, in a measured daywork plan, worker income is based on a fixed hourly rate established by management, and is usually affected only by job classification, shift premiums, and overtime adjustments. Because of the fixed hourly rate in a measured daywork plan, there is little incen- tive for a worker to exceed a normal or standard level of performance or productivity. On the other hand, time standards are more readily acceptable, and become less an item of contention to the employee and bargaining unit (union).

The term daywork as used in industry denotes a fixed hourly rate that is not raised or lowered by varying worker performance levels. The hourly rate for a particular job should be a fair one relative to other jobs in the shop, and should also be comparable to rates of pay for similar jobs in the industrial community.

Once measured labor time standards have been established for shop operations, in addition to evaluating operator performance and identifying labor costs, new-product costs can also be determined prior to release to production, worker-power planning and scheduling can be done, equipment capacity requirements can be identified, and planning and make/buy decisions can be facilitated. *See* PERFORMANCE RATING; PRODUCTIVITY; WAGE INCENTIVES; WORK MEASUREMENT. Dave H. Scott

Mechanical advantage Ratio of the force exerted by a machine (the output) to the force exerted on the machine, usually by an operator (the input). The term is useful in discussing a simple machine, where it becomes a figure of merit. It is not particularly useful, however, when applied to more complicated machines, where other considerations become more important than a simple ratio of forces. *See* EFFICIENCY; SIMPLE MACHINE. Richard M. Phelan

Mechanical alloying A materials-processing method that involves the repeated welding, fracturing, and rewelding of a mixture of powder particles, generally in a high-energy ball mill, to produce a controlled, extremely fine microstructure. The mechanical alloying technique allows alloying of elements that are difficult or impossible to combine by conventional melting methods. In general, the process can be viewed as a means of assembling metal constituents with a controlled microstructure. If two metals will form a solid solution, mechanical alloying can be used to achieve this state without the need for a high-temperature excursion. Conversely, if the two metals are insoluble in the liquid or solid state, an extremely fine dispersion of one of the metals in the other can be accomplished. The process of mechanical alloying was originally developed as a means of overcoming the disadvantages associated with using powder metallurgy to alloy elements that are difficult to combine. *See* CRUSHING AND PULVERIZING; POWDER METALLURGY; SOLID SOLUTION.

Some oxides are insoluble in molten metals. Mechanical alloying provides a means of dispersing these oxides in the metals. Examples are nickel-based superalloys strengthened with dispersed thorium oxide or yttrium oxide (Y_2O_3). These superalloys have excellent strength and corrosion resistance at elevated temperatures, making them attractive candidate materials for use in applications such as jet-engine turbine blades, vanes, and combustors. A number of other potential applications for mechanical alloying material are being explored, including powders for coating applications, alloys of immiscible systems, amorphous alloys, intermetallics, cermets, and organic-ceramic-metallic material systems in general. *See* AMORPHOUS SOLID; CERMET; HIGH-TEMPERATURE MATERIALS; INTERMETALLIC COMPOUNDS; METAL COATINGS.

Liquids or solid immiscible systems are difficult to process by conventional pyrometallurgy; mechanical alloying provides a route to obtain a homogeneous distribution in the solid phase. *See* PYROMETALLURGY. F. H. Froes

Mechanical classification A sorting operation in which mixtures of particles of mixed sizes, and often of different specific gravities, are separated into fractions by the action of a stream of fluid. Water is ordinarily used as the sorting fluid, but other liquids or air or other gases may be used.

The main objective of classification is to separate the particles according to size. This function is identical to that of screening, but classification is applicable to smaller particles, especially those that are undersize. For small particles it is more economical than screening. In classification the oversize and undersize are called sands and slimes, respectively.

Material also may be mechanically classified by specific gravity, a method that separates substances differing in chemical composition. This is called hydraulic separation. Such classification is based on the fact that, in a fluid, particles of the same specific gravity but of different size or shape settle at different constant speeds. Large, heavy, round particles settle faster than small, light, needlelike ones. If the particles also differ in specific gravity, the speed of settling is further affected. This is the basis for the separation of particles by kind rather than by size alone. *See* FLOTATION; MECHANICAL SEPARATION TECHNIQUES; UNIT OPERATIONS. Warren L. McCabe

Mechanical engineering One of several recognized fields of engineering. To grasp the meaning of mechanical engineering, it is desirable to take a close look at what engineering really is. The Engineers' Council for Professional Development has defined engineering as the profession in which a knowledge of the mathematical and physical sciences gained by study, experience, and practice is applied with judgment to develop ways to utilize economically the materials and forces of nature for the progressive well-being of mankind. It is a profession in which study in mathematics and science is blended with experience and judgment for the production of useful things.

Formal training of a mechanical engineer includes mastery of mathematics through the level of differential equations. Training in physical science embraces chemistry, physics, mechanics of materials, fluid mechanics, thermodynamics, statics, and dynamics. *See* ENGINEERING; MACHINERY; TECHNOLOGY.

Robert S. Sherwood

Mechanical impedance For a system executing simple harmonic motion, the mechanical impedance is the ratio of force to particle velocity. If the force is that which drives the system and the velocity is that of the point of application of the force, the ratio is the input or driving-point impedance. If the velocity is that at some other point, the ratio is the transfer impedance corresponding to the two points.

Mechanical impedance is a complex quantity. The real part, the mechanical resistance, is independent of frequency if the dissipative forces are proportional to velocity; the imaginary part, the mechanical reactance, varies with frequency, becoming zero at the resonant and infinite at the antiresonant frequencies of the system. *See* FORCED OSCILLATION; HARMONIC MOTION.

Martin Greenspan

Mechanical separation techniques A group of laboratory and production operations whereby the components of a polyphase mixture are separated by mechanical methods into two or more fractions of different mechanical characteristics. The separated fractions may be homogeneous or heterogeneous, particulate or nonparticulate.

Types of mechanical separator

Materials separated	Separators
Liquid from liquid	Settling tanks, liquid cyclones, centrifugal decanters, coalescers
Gas from liquid	Still tanks, deaerators, foam breakers
Liquid from gas	Settling chambers, cyclones, electrostatic precipitators, impingement separators
Solid from liquid	Filters, centrifugal filters, clarifiers, thickeners, sedimentation centrifuges, liquid cyclones, wet screens, magnetic separators
Liquid from solid	Presses, centrifugal extractors
Solid from gas	Settling chambers, air filters, bag filters, cyclones, impingement separators, electrostatic and high-tension precipitators
Solid from solid	
By size	Screens, air and wet classifiers, centrifugal classifiers
By other characteristics	Air and wet classifiers, centrifugal classifiers, jigs, tables, spiral concentrators, flotation cells, dense-medium separators, magnetic separators, electrostatic separators

The techniques of mechanical separation are based on differences in phase density, in phase fluidity, and in such mechanical properties of particles-as size, shape, and density; and on such particle characteristics as wettability, surface charge, and magnetic susceptibility. Obviously, such techniques are applicable only to the separation of phases in a heterogeneous mixture. They may be applied, however, to all kinds of mixtures containing two or more phases, whether they are liquid-liquid, liquid-gas, liquid-solid, gas-solid, solid-solid, or gas-liquid-solid.

Methods of mechanical separations fall into four general classes: (1) those employing a selective barrier such as a screen or filter cloth; (2) those depending on difference in phase density alone (hydrostatic separators); (3) those depending on fluid and particle mechanics; and (4) those depending on surface or electrical characteristics of particles. A wide variety of separation devices have been devised and are in use. The more important kinds of equipment are listed in the table, grouped according to the phases involved. *See* CENTRIFUGATION; CLARIFICATION; DUST AND MIST COLLECTION; FILTRATION; FLOTATION; MAGNETIC SEPARATION METHODS; MECHANICAL CLASSIFICATION; SCREENING; SEDIMENTATION (INDUSTRY); THICKENING.

Shelby A. Miller

Mechanical vibration The continuing motion, repetitive and often periodic, of a solid or liquid body within certain spatial limits. Vibration occurs frequently in a variety of natural phenomena such as the tidal motion of the oceans, in rotating and stationary machinery, in structures as varied in nature as buildings and ships, in vehicles, and in combinations of these various elements in larger systems. The sources of vibration and the types of vibratory motion and their propagation are subjects that are complicated and depend a great deal on the particular characteristics of the systems being examined. Further, there is strong coupling between the notions of mechanical vibration and the propagation of vibration and acoustic signals through both the ground and the air so as to create possible sources of discomfort, annoyance, and even physical damage to people and structures adjacent to a source of vibration.

Mass-spring-damper system. Although vibrational phenomena are complex, some basic principles can be recognized in a very simple linear model of a mass-spring-damper system (see illustration). Such a system contains a mass M, a spring with spring constant k that serves to restore the mass to a neutral position, and a damping element which opposes the motion of the vibratory response with a force proportional to the velocity of the system, the constant of proportionality being the damping constant c. This damping force is dissipative in nature, and without its presence a response of this mass-spring system would be completely periodic. *See* DAMPING.

Complex systems. The foregoing model of the linear spring-mass-damper system contains within it a number of simplifications that do not reflect conditions of the real world in any obvious way. These simplifications include the periodicity of both the input and, to some extent, the response; the discrete nature of the input, that is, the assumption that it is temporal in nature with no reference to spatial distribution; and the assumption that only a single resonant frequency and a single set of parameters are required to describe the mass, the stiffness, and the damping. The real world is far more complex. Many sources of vibration are not periodic. These include impulsive forces and shock loading, wherein a force is suddenly applied for a very short time to a system; random excitations, wherein the signal fluctuates in time in such a way that its amplitude at any given instant can

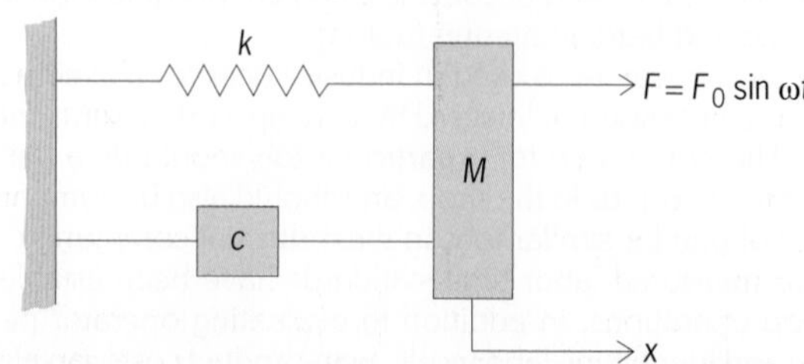

Vibrating linear system (mass-spring-damper) with one degree of freedom.

be expressed only in terms of a probabilistic expectation; and aperiodic motions, wherein the fluctuation in time may be some prescribed nonperiodic function or some other function that is not readily seen to be periodic.

Sources of vibration. There are many sources of mechanical and structural vibration that the engineer must contend with in both the analysis and the design of engineering systems. The most common form of mechanical vibration problem is motion induced by machinery of varying types, often but not always of the rotating variety. Other sources of vibration include: ground-borne propagation due to construction; vibration from heavy vehicles on conventional pavement as well as vibratory signals from the rail systems common in many metropolitan areas; and vibrations induced by natural phenomena, such as earthquakes and wind forces. Wave motion is a source of vibration in mechanical and structural systems associated with offshore structures.

Effect of vibrations. The most serious effect of vibration, especially in the case of machinery, is that sufficiently high alternating stresses can produce fatigue failure in machine and structural parts. Less serious effects include increased wear of parts, general malfunctioning of apparatus, and the propagation of vibration through foundations and buildings to locations where the vibration of its acoustic realization is intolerable either for human comfort or for the successful operation of sensitive measuring equipment. *See* ACOUSTIC NOISE; SOUND; VIBRATION; WEAR.

Clive L. Dym; J. P. Den Hartog

Mechanics

In its original sense, mechanics refers to the study of the behavior of systems under the action of forces. Mechanics is subdivided according to the types of systems and phenomena involved.

An important distinction is based on the size of the system. Those systems that are large enough can be adequately described by the newtonian laws of classical mechanics; in this category, for example, are celestial mechanics and fluid mechanics. On the other hand, the behavior of microscopic systems such as molecules, atoms, and nuclei can be interpreted only by the concepts and mathematical methods of quantum mechanics.

Mechanics may also be classified as nonrelativistic or relativistic mechanics, the latter applying to systems with material velocities comparable to the velocity of light. This distinction pertains to both classical and quantum mechanics.

Finally, statistical mechanics uses the methods of statistics for both classical and quantum systems containing very large numbers of similar subsystems to obtain their large-scale properties. *See* CLASSICAL FIELD THEORY; CLASSICAL MECHANICS; DYNAMICS; FLUID MECHANICS; QUANTUM MECHANICS; STATICS; STATISTICAL MECHANICS.

Bernard Goodman

Mechanism

Classically, a mechanical means for the conversion of motion, the transmission of power, or the control of these. Mechanisms are at the core of the workings of many machines and mechanical devices. In modern usage, mechanisms are not always limited to mechanical means. In addition to mechanical elements, they may include pneumatic, hydraulic, electrical, and electronic elements. In this article, the discussion of mechanism is limited to its classical meaning. *See* MACHINE.

Most mechanisms consist of combinations of a relatively small number of basic components. Of these, the most important are cams, gears, links, belts, chains, and logical mechanical elements. The last include such devices as ratchets, trips, detents, and interlocks. In order to understand how any mechanism works, their degree of freedom, structure, and kinematics must be considered. *See* BELT DRIVE; CAM MECHANISM; CHAIN DRIVE; ESCAPEMENT; GEAR; LINKAGE (MECHANISM); RATCHET.

Degree of freedom is conveniently illustrated for mechanisms with rigid links. The discussion is limited to mechanisms which obey the general degree-of-freedom equation,

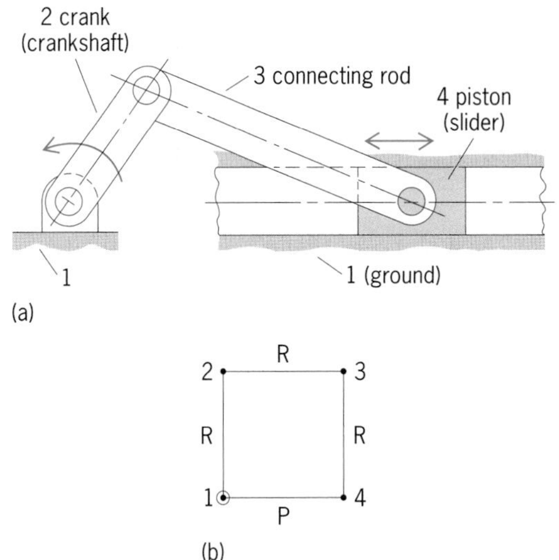

Slider-crank mechanism, (*a*) Mechanism, (*b*) Graph of mechanism. *R* = pin joint; *P* = sliding joint.

$$F = \lambda(l - j - 1) + \sum f_i$$

where F = degree of freedom of mechanism, l = number of links of mechanism, j = number of joints of mechanism, f_i = degree of freedom of relative motion at *i*th joint, σ = summation symbol (summation over all joints), and λ = mobility number (the most common cases are $\lambda = 3$ for plane mechanisms and $\lambda = 6$ for spatial mechanisms). *See* DEGREE OF FREEDOM (MECHANICS).

The kinematic structure of a mechanism refers to the identification of the joint connection between its links. Just as chemical compounds can be represented by an abstract formula and electric circuits by schematic diagrams, the kinematic structure of mechanisms can be usefully represented by abstract diagrams. The structure of mechanisms for which each joint connects two links can be represented by a structural diagram, or graph, in which links are denoted by vertices, joints by edges, and in which the edge connection of vertices corresponds to the joint connection of links; edges are labeled according to joint type, and the fixed link is identified as well. Thus the graph of the slider-crank mechanism of illustration *a* is as shown in illustration *b*. In this figure the circle around vertex 1 signifies that link 1 is fixed.

Kinematics is divided into kinematic analysis (analysis of a mechanism of given dimensions) and synthesis (determination of the proportions of a mechanism for given motion requirements). It includes the investigation of finite as well as infinitesimal displacements, velocities, accelerations and higher accelerations, and curvatures and higher curvatures in plane and three-dimensional motions. *See* KINEMATICS.

The design of mechanisms involves many factors. These include their structure, kinematics, dynamics, stress analysis, materials, lubrication, wear, tolerances, production considerations, control and actuation, vibrations, critical speeds, reliability, costs, and environmental considerations. Modern trends in the design of mechanisms emphasize economical design analysis by means of computer-aided design techniques.

Ferdinand Freudenstein

Mechanoreceptors

Sensory receptors that provide the organism with information about such mechanical changes in the environment as movement, tension, and pressure. In higher animals receptors are actually the only means by which information of the surroundings is gained and by which reactions to environmental changes are started. *See* SENSATION.

Mechanoreceptors are excited by mechanical disturbances of their surroundings through deformation of their structure, through pressure or tension, or through a combination of these. In general, little energy is required for mechanical stimuli to cause a detectable excitation in mechanoreceptors.

From a physical point of view, mechanoreceptors are energy transducers; they convert mechanical into electrical energy, which in turn triggers the nerve impulse. Deformation leads to a sequence of events which may be summarized by the following scheme:

$$\begin{array}{ccccc} \text{Mechanical} & \rightarrow & \text{Generator} & \rightarrow & \text{Nerve impulse} \\ \text{stimulus} & & \text{current} & & \text{(action potential)} \end{array}$$

The generator current is the earliest detectable sign of excitation. The most salient characteristic of the generator current is its graded nature; its amplitude increases continuously, without visible steps, if the stimulus strength is progressively increased. When the generator current reaches a certain critical amplitude, an all-or-nothing potential is discharged in the sense organ which may then propagate as an all-or-nothing nerve impulse along the afferent axon of the receptor. *See* NERVOUS SYSTEM (INVERTEBRATE); NERVOUS SYSTEM (VERTEBRATE). Werner R. Loewenstein

Mecoptera A small order of insects called the scorpion flies. Characteristic of the adult insect is the peculiar prolongation of the head into a beak, which bears chewing mouthparts. They are small to medium in size. The insects either have two pairs of large, net-veined wings of equal size, often with dark areas, or have short and aborted wings. The legs are long and slender. In some species, the male abdomen has a terminal enlargement which is held recurved over the back so that he resembles a scorpion, thus the common name. The Mecoptera are found in moist habitats within densely wooded areas. The adults are omnivorous but feed chiefly on small insects. *See* INSECTA. Bryant E. Rees

Medical bacteriology The study of bacteria that cause human disease. The field encompasses the detection and identification of bacterial pathogens, determination of the sensitivity and mechanisms of resistance of bacteria to antibiotics, the mechanisms of virulence, and some aspects of immunity to infection. *See* VIRULENCE.

The clinical bacteriology laboratory identifies bacterial pathogens present in specimens such as sputum, pus, blood, and spinal fluid, or from swabs of skin, throat, rectal, or urogenital surfaces. Identification involves direct staining and microscopic examination of these materials, and isolation of bacteria present in the material by growth in appropriate media. The laboratory must differentiate bacterial pathogens from harmless bacteria that colonize humans. Species and virulent strains of bacteria can be identified on the basis of growth properties, metabolic and biochemical tests, and reactivity with specific antibodies.

Recent advances in the field of diagnostic bacteriology have involved automation of biochemical testing; the development of rapid antibody-based detection methods; and the application of molecular biology techniques. Once a bacterial pathogen has been identified, a major responsibility of the diagnostic bacteriology laboratory is the determination of the sensitivity of the pathogen to antibiotics. This involves observation of the growth of the bacteria in the presence of various concentrations of antibiotics. The process has been made more efficient by the development of automated instrumentation.

An increasingly serious problem in the therapy of infectious diseases is the emergence of antibiotic-resistant strains of bacteria. An important area of research is the mechanisms of acquisition of antibiotic resistance and the application of this knowledge to the development of more effective antibiotics. *See* ANTIBIOTIC; ANTIGEN-ANTIBODY REACTION; BACTERIAL PHYSIOLOGY AND METABOLISM; BACTERIAL TAXONOMY; IMMUNOCHEMISTRY.

The study of bacterial pathogenesis involves the fields of molecular genetics, biochemistry, cell biology, and immunology. In cases where the disease is not serious and easily treated, research may involve the deliberate infection of human volunteers. Otherwise, various models of human disease must be utilized. These involve experimental infection of animals and the use of tissue cell culture systems. Modern molecular approaches to the study of bacterial pathogenesis frequently involve the specific mutation or elimination of a bacterial gene thought to encode a virulence property, followed by observation of the mutant bacteria in a model system of human disease. In this way, relative contributions of specific bacterial traits to different stages of the disease process can be determined. This knowledge permits the design of effective strategies for intervention that will prevent or cure the disease. *See* BACTERIAL GENETICS.

The presence of specific antibodies is frequently useful in the diagnosis of bacterial diseases in which the pathogen is otherwise difficult to detect. An example is the sexually transmitted disease syphilis; the diagnosis must be confirmed by the demonstration of antibodies specific for *T. pallidum*. *See* ANTIBODY; BIOLOGICALS.

Immunity to some bacteria that survive intracellularly is not mediated by antibodies but by immune effector cells, known as T cells, that activate infected cells to kill the bacteria that they contain. An active area of research is how bacterial components are presented to the immune system in a way that will induce effective cell-mediated immunity. This research may lead to the development of T-cell vaccines effective against intracellular bacterial pathogens. Steve L. Moseley

Medical control systems Physiological and artificial systems that control one or more physiological variables or functions of the human body. Regulation, control processes, and system stability are at the heart of the survival of living organisms, both unicellular and multicellular. In the nineteenth century, C. Bernard concluded that the higher animals, far from being indifferent to their surroundings, must be in close and intimate relation to them. The equilibrium they maintain is the result of compensation established as continually and exactly as if by a very sensitive balance. W. B. Cannon (1929) differentiated the stability properties of biological systems from those of physical systems, and introduced the term homeostasis to describe the steady states in the body that are maintained by complex, coordinated physiological reactions. The condition of homeostasis is achieved either by regulation of supplies (for example, control of blood sugar level) or by regulation of processes (for example, control of body temperature and control of voluntary movements). *See* HOMEOSTASIS.

Medical control systems may be classified into two groups: (1) the physiological control systems in normal or pathological conditions (for example, control of electrolytes, arterial pressure, respiration, body temperature, blood sugar, endocrinal functions, neuromuscular and motor activity, and sensory functions), and (2) the external (artificial) control systems that interface with physiological systems (for example, artificial kidneys or hemodialyzers, blood oxygenators or heart-lung machines used during open-heart surgery, external prosthetics and orthotics, cardiac pacemakers, ventilators, implantable defibrillators, and implantable pumps for drug delivery). For the development and proper functioning of artificial devices, the underlying control mechanisms of the normal and of the disabled physiological systems with which the external devices must interface must be adequately understood. Thus, in its broadest sense, the area of medical control systems encompasses all branches of engineering, mathematical biology, biophysics, physiology, and medicine. *See* BIOMECHANICS; BIOMEDICAL CHEMICAL ENGINEERING; BIOMEDICAL ENGINEERING; CONTROL SYSTEMS; MATHEMATICAL BIOLOGY.

The importance of control systems engineering in medical applications has grown because of the inherent complexity of medical control systems. H. A. Simon's concept of complexity is very appropriate for medical control systems: complex systems are composed of subsystems that in turn have their own subsystems, and so on; and the large number of parts interact in a complicated way so that it is sometimes impossible to infer the properties of the whole from the properties of the parts and their laws of interaction. Indeed, the analytical models developed, using control systems engineering, of the components of a medical system have had limited success in predicting the behavior of the overall system.

Examples of medical control systems include myoelectric prostheses, which are replacement devices for lost limbs; external orthoses, which are used for rehabilitation of patients with acquired disabilities; and implantable devices such as defibrillators and pumps for drug delivery. Numerous other devices, such as cardiac pacemakers, artificial kidneys, heart-lung machines, and artificial ventilators, have been in routine clinical use for many years. Gyan C. Agarwal

Medical imaging A medical specialty that uses x-rays, gamma rays, high-frequency sound waves, and magnetic fields to produce images of organs and other internal structures of the body. In diagnostic radiology the purpose is to detect and diagnose disease, while in interventional radiology, imaging procedures are combined with other techniques to treat certain diseases and abnormalities. *See* RADIOLOGY.

Film x-ray studies, the most common radiologic procedures, are made up of still pictures of the various organs and tissues in the body. In these procedures, x-rays are passed through the body to expose the photographic film that is placed on the opposite side of the body. The changes in film density that result from exposure allow the radiologist to distinguish between normal and abnormal tissue and to diagnose many different disease types. *See* X-RAYS.

Fluoroscopy is a dynamic x-ray imaging technique that produces a moving image over time. It is essential for evaluating organ movement such as the beating of the heart or movement of the diaphragm. The gastrointestinal series and the barium enema are the most common fluoroscopic studies. These procedures begin with the administration of a barium mixture either by ingestion or by an enema that fills the stomach or large intestine. The barium mixture, known as a contrast medium, like dense tissues, blocks the x-ray beam. Fluoroscopy then reveals the location of the barium-coated lining of the stomach and intestine and enables the radiologist to observe as they contract and distend.

Angiography is the radiologic study of blood vessels. Because arteries and veins are not normally visible in conventional x-ray studies, an iodinated compound, which is opaque to the x-ray, must be injected into the bloodstream. An arteriogram is an x-ray study of the arteries; a venogram is an x-ray study of the veins. Arteriography is most often used to show the presence and extent that arteries have become clogged and narrowed by arteriosclerosis, which can lead to strokes and heart attacks.

Computed tomography (CT), also called computed axial tomography (CAT), is a scanning technique that combines computer and x-ray technologies. The computer constructs a two-dimensional anatomic image that represents a cross-sectional slice through the body. Three-dimensional images can be generated by using special computer software. These are especially useful in planning reconstructive orthopedic or plastic surgery. *See* COMPUTERIZED TOMOGRAPHY.

Ultrasound imaging, or sonography, is a diagnostic imaging procedure that uses high frequency sound waves instead of ionizing radiation. During an ultrasound examination, a lightweight transducer is placed on the patient's skin over the part to be imaged. The transducer produces sound waves that penetrate the skin to reach tissues and organs. When the sound waves strike specific tissue surface, echoes are produced. The echoes are detected by the transducer and are then electronically converted into an anatomic image that is displayed on a video screen. The image can also be recorded on film or videotape. Ultrasound imaging is commonly used in obstetrics to monitor the position and development of the fetus and also to detect any fetal abnormalities or problems in the pregnancy. Ultrasound is also used to show problems in other internal structures, including the gallbladder, kidney, and heart. Doppler ultrasound can monitor blood flow through veins and arteries. It is commonly used to study kidney transplants and blood flow to the brain and also to diagnose blocked arteries. *See* MEDICAL ULTRASONIC TOMOGRAPHY.

Magnetic resonance imaging (MRI) is a diagnostic procedure that uses a large, high-strength magnet, radio-frequency signals, and a computer to produce images. The technique of MRI is extremely useful in evaluating diseases of the brain and spine. It is also used to evaluate joints, bone and soft tissue abnormalities, as well as abnormalities of the chest, abdomen, and pelvis. *See* NUCLEAR MAGNETIC RESONANCE (NMR).

Nuclear medicine imaging studies use radioactive compounds called radionuclides of radiopharmaceuticals that emit gamma rays, or some emit beta particles. The chemicals are formulated so that they collect temporarily in the parts of the body to be studied. For most nuclear imaging studies, the radionuclide is injected into the patient and the images are taken with a gamma camera suspended above the patient who lies on a table. The camera detects the gamma rays emitted from the radionuclide in the patient's body and uses this information to produce an image that shows the distribution of the radionuclide within the body. The image is recorded on film and is called a scintigram or scan. Scintigrams of the heart and bone are the two most common nuclear medicine examinations.

The single-photon emission computed tomography (SPECT) examination uses a computer to obtain two-dimensional images that are thin slices of internal organs such as the heart, brain, and liver. The SPECT images can display organs with much greater detail than conventional scintigrams.

Positron emission tomography (PET) is a more refined radiologic technique that is used to study the metabolic activity inside an organ. The technique has been shown to be useful in the study of brain-related disorders, such as epilepsy and Alzheimer's disease, and of the vitality of heart tissue.

Interventional radiology combines imaging procedures with various injection and catheter techniques to treat tumors, blockages, bleeding vessels, and other abnormalities without extensive surgery. Among the more common interventional procedures is angioplasty, which is used to treat blocked or narrowed arteries. *See* RADIOGRAPHY. Michael Lopiano; Raymond Del Fava

Medical information systems Standardized methods of collection, evaluation or verification, storage, and retrieval of data about a patient. The three broad areas of any information system—input, data transformation, and output—suffice to describe an existing system, but they are insufficient to design a new one. To these must be added the action that is expected to take place on the basis of the data output from the system, thereby defining the purpose of the system, and the feedback from such action to the system input, which places the system in a specific medical environment. Crucial to any information system, and especially for a medical system, is the accuracy of the input data, which is of more significance than mere precision. A medical information system is a part of necessary time management that is intended to maximize the amount of data and procedure information in order to make the most accurate clinical decisions. As such, an information management system must be interactive to serve clinical decision purposes.

The medical environment is highly data- and information-intensive. Hospitals have always managed data and information and were pioneers in the use of technologic information systems. With scaled-down technology, medical offices are able to incorporate data and information management in the private-practice environment as well.

Data input is provided by the patient and by other information sources involved. Input information is complex because the patient must be described both cross-sectionally, which is current information, and longitudinally, that is back to childhood and to the parents. Historical information obtainable from the patient is subject to increasing error as the individual goes backward in time. Significant error in the medical information system can arise if imperfect recall is not considered and appropriately discounted. Data from outside sources are subject to the same errors; hence it is valuable to obtain objective data rather than subjective data from either input source.

In a medical information system, input information can be processed through one of two parallel, sometimes mutually exclusive routes—administration and medical service organization. Administration encompasses the admitting clerk, administrative officers, pharmacists, and all those who have no direct patient contact. The service route includes the physician, the nurse, and all others who have direct patient contact. The two routes differ in their method and degree of detail. The passage of such medical information through both routes allows data to reach the output stage. Redundancy, as illustrated by parallel information routes in hospitals, can be a source of economic drain, but a degree of redundancy is necessary for the quality of service required and is a fail-safe feature that is vital to all aspects of quality. Redundancy is particularly important because those in direct contact with the patient are subject to human error or variability.

The output of a medical information system can be viewed as a control system for actions by clinicians. The output, once it has passed through the transformation media, constitutes the patient's medical records and usually includes the history and physical examination data, progress notes, nursing notes, doctor's orders, and summaries and notations of values from physical laboratory data. Items of output information are interrelated to establish a diagnosis of a course of action. Feedback from that stage to the patient, often in the form of interrelations, can be called patient care.

In addition to its primary function of facilitating patient care, a medical information system can also prove useful for investigation and research into disease and therapy, and the administrative portion of the record can aid in planning future improvements in quality of service. *See* DATA COMMUNICATIONS; DATABASE MANAGEMENT SYSTEM; MEDICAL CONTROL SYSTEMS. Cesar A. Caceres

Medical mycology The study of fungi (molds and yeasts) that cause human disease. Fungal infections are classified according to the site of infection on the body or whether an opportunistic setting is necessary to establish disease. Fungal infections that occur in an opportunistic setting have become more common due to conditions that compromise host defenses, especially cell-mediated immunity. Such conditions include acquired immunodeficiency syndrome (AIDS), cancer, and immunosuppressive therapy to prevent transplant rejection or to control inflammatory syndromes. Additionally, opportunistic fungal infections have become more significant as severely debilitated individuals live longer because of advances in modern medicine, and nosocomial (hospital-acquired) fungal infections are an increasing problem. Early diagnosis with treatment of the fungal infection and control of the predisposing cause are essential. *See* OPPORTUNISTIC INFECTIONS.

Antifungal drug therapy is extremely challenging since fungi are eukaryotes, as are their human hosts, leading to problems with toxicity or cross-reactivity with host molecules. Most antifungal drugs target the fungal cell membrane or wall. The "gold standard" for therapy of most severe fungal infections is amphotericin B, which binds to ergosterol, a membrane lipid found in most fungi and some other organisms but not in mammals. Unfortunately, minor cross-reactive binding of amphotericin B to cholesterol in mammalian cell membranes can lead to serious toxicity, especially in the kidney where the drug is concentrated. Recent advances in antifungal therapy include the use of liposomal amphotericin B and newer azoles such as fluconazole and itraconazole, which show reduced toxicity or greater specificity. Conversely, drug resistance in pathogenic fungi is an increasing problem, as it is in bacteria.

Candidiasis is the most common opportunistic fungal infection, and it has also become a major nosocomial infection in hospitalized patients. *Candida albicans* is a dimorphic fungus with a yeast form that is a member of the normal flora of the surface of mucous membranes. In an opportunistic setting, the fungus may proliferate and convert to a hyphal form that invades these tissues, the blood, and other organs. The disease may extend to the blood or other organs from various infected sites in patients who are suffering from a grave underlying disease or who are immunocompromised. Other important opportunistic fungal diseases include aspergillosis, mucormycosis, and cryptococcus.

Healthy persons can acquire disease from certain pathogenic fungi following inhalation of their fungal spores. The so-called deep or systemic mycoses are all caused by different species of soil molds; most infections are unrecognized and produce no or few symptoms. However, in some individuals infection may spread to all parts of the body from the lung, and treatment with amphotericin B or an antifungal azole drug is essential.

Other fungal infections develop when certain species of soil molds are inoculated deep into the subcutaneous tissue, such as by a deep thorn prick or other trauma. A specific type of lesion develops with each fungus as it grows within the tissue. Proper wound hygiene will prevent these infections.

Ringworm, also known as dermatophytosis or tinea, is the most common of all fungal infections. Some species of pathogenic molds can grow in the stratum corneum, the dead outermost layer of the skin. Disease results from host hypersensitivity to the metabolic products of the infecting mold as well as from the actual fungal invasion. Tinea corporis, ringworm of the body, appears as a lesion on smooth skin and has a red, circular margin that contains vesicles. The lesion heals with central clearing as the margin advances. On thick stratum corneum, such as the interdigital spaces of the feet, the red, itching lesions, known as athlete's foot or tinea pedis, become more serious if secondary bacterial infection develops. The ringworm fungi may also invade the hair shaft (tinea capitis) or the nail (onychomycosis). Many pharmaceutical agents are available to treat or arrest such infections, but control of transmission to others is important. *See* FUNGAL INFECTIONS; FUNGI; YEAST. Carlyn Halde; Jon P. Woods

Medical parasitology The study of diseases of humans caused by parasitic agents. It is commonly limited to parasitic worms (helminths) and the protozoa. Current usage places the various nonprotozoan microbes in distinct disciplines, such as virology, rickettsiology, and bacteriology.

Nematodes. The roundworms form an extremely large yet fairly homogeneous assemblage, most of which are free-living (nonparasitic). Some parasitic nematodes, however, may cause disease in humans (zoonosis), and others cause disease limited to human hosts (anthroponosis). Among the latter, several are enormously abundant and widespread. *See* NEMATA (NEMATODA).

The giant roundworm (*Ascaris lumbricoides*) parasitizes the small intestine, probably affecting over a billion people; and the whipworm (*Trichuris trichiura*) infects the human colon, probably affecting a half billion people throughout the tropics. Similarly, the hookworms of humans, *Necator americanus* in the Americas and the tropical regions of Africa and Asia, and *Ancylostoma duodenale* in temperate Asia, the Mediterranean, and Middle East, suck blood from the small intestine and cause major debilitation, especially among the undernourished. The human pinworm (*Enterobius vermicularis*) infects the large intestine of millions of urban dwellers. Most intestinal nematodes, which require a period of egg maturation outside the human host before they are infective, are associated with fecal contamination of soil or food crops and are primarily rural in distribution.

The nonintestinal nematodes are spread by complex life cycles that usually involve bloodsucking insects. One exception is the guinea worm (*Dracunculus medinensis*), a skin-infecting 2–3-ft (0.6–1-m) worm transmitted by aquatic microcrustaceans that are ingested in drinking water that has been contaminated by larvae that escape from the skin sores of infected humans. Such bizarre life cycles are typical of many helminths. Other nematodes of humans include (1) the filarial worms, which are transmitted by mosquitoes and may induce enormously enlarged fibrous masses in legs, arms, or genitalia (elephantiasis), and (2) *Onchocerca volvulus*, which is transmitted by blackflies (genus *Simulium*) and forms microscopic embryos (microfilariae) in the eyes causing high incidence of blindness in Africa and parts of central and northern South America.

A more familiar tissue-infecting nematode of temperate regions is *Trichinella spiralis*, the pork or trichina worm, which is the agent of trichinosis. The tiny spiraled larvae encyst in muscle and can carry the infection to humans and other carnivorous mammals who eat raw or undercooked infected meat.

Trematodes. Parasites of the class Trematoda vary greatly in size, form, location in the human host, and disease produced, but all go through an initial developmental period in specific kinds of fresh-water snails, where they multiply as highly modified larvae of different types. Ultimately, an infective larval stage (cercaria) escapes in large numbers from the snail and continues the life cycle. Each trematode species follows a highly specific pathway from snail to human host, usually by means of another host or transport mechanism. These include the intestinal, liver, blood, and lung flukes. *See* SCHISTOSOMIASIS; TREMATODA.

Cestodes. Tapeworms, the other great assemblage of parasitic flatworms, parasitize most vertebrates, with eight or more species found in humans. Their flat ribbonlike body form consists of a chain of hermaphroditic segments. Like the trematodes, their life cycles are complex, although not dependent on a snail host. The enormous beef tapeworm of humans, *Taenia saginata*, is transmitted by infected beef ("measly beef") from cattle that grazed where human feces containing egg-filled tapeworm segments contaminated the soil. Other tapeworms include the pork, dog, and broad (or fish) tapeworms. *See* CESTODA.

Protozoa. Of the many protozoa that can reside in the human gut, only the invasive strain of *Entamoeba histolytica* causes serious disease. This parasite, ingested in water contaminated with human feces containing viable cysts of *E. histolytica*, can cause the disease amebiasis, which in its most severe form is known as amebic dysentery. Another common waterborne intestinal protozoon is the flagellate *Giardia lamblia*, which causes giardiasis, a mild to occasionally serious or long-lasting diarrhea. *See* GIARDIASIS; PROTOZOA.

Other flagellate parasites infect the human skin, bloodstream, brain, and viscera. The tsetse fly of Africa carries to humans the blood-infecting agents of trypanosomiasis, or African sleeping sickness, *Trypanosoma brucei gambiense* and *T. brucei rhodesiense*. The infection can be fatal if the parasites cross the blood-brain barrier. In Latin America, the flagellate *T. cruzi* is the agent of Chagas' disease, a major cause of debilitation and premature heart disease among those who are poorly housed. The infection is transmitted in the liquid feces of a conenose bug (genus *Triatoma*) and related insects. The infective material is thought to be scratched into the skin or rubbed in the eye, especially by sleeping children. *See* TRYPANOSOMATIDAE.

Another group of parasitic flagellates includes the macrophage-infecting members of the genus *Leishmania*, which are transmitted by blood-sucking midges or sand flies. Cutaneous leishmaniasis is characterized by masses of infected macrophages in the skin, which induce long-lasting dermal lesions of varying form and severity. The broad spectrum of host-parasite interactions is well exemplified by leishmaniases. The various manifestations of the disease are the result of the particular species of agent and vector, the immunological status of the host, the presence or absence of reservoir hosts, and the pattern of exposure.

Two remaining major groups of protozoa are the ciliates and the sporozoans. The former group is largely free-living, with only a single species, *Balantidium coli*, parasitic in humans (and pigs). This large protozoon is found in the large intestine, where it can cause balantidiasis, an ulcerative disease. The sporozoans, on the other hand, are all parasitic and include many parasites of humans. The most important are the agents of malaria. Other disease agents are included in the genera *Isospora, Sarcocystis, Cryptosporidium*, and *Toxoplasma. Pneumocystis*, a major cause of death among persons with acquired immune deficiency syndrome (AIDS), was formerly considered a protozoon of uncertain relationship, but now it is thought to be a member of the Fungi. *See* ACQUIRED IMMUNE DEFICIENCY SYNDROME (AIDS); MALARIA; SPOROZOA.

Toxoplasma gondii, the agent of toxoplasmosis, infects as many as 20% of the world's population. It can penetrate the placenta and infect the fetus if the mother has not been previously infected and has no antibodies. As with most medically important parasites, the great majority of *Toxoplasma* infections remain undetected and nonpathogenic. The parasite primarily affects individuals lacking immune competence—the very young, the very old, and the immunosuppressed. *See* MEDICAL BACTERIOLOGY; MEDICAL MYCOLOGY; PARASITOLOGY; ZOONOSES. Donald Heyneman

Medical ultrasonic tomography A mapping or imaging technique, used to obtain clinically useful information about the structure and functioning of tissues and organs, in which acoustic pulses are emitted from an acoustoelectric transducer, and echoes are received from acoustic impedance discontinuities along the assumed line-of-sight axial propagation path. A number of different modes of operation have emerged, each having areas of usefulness.

The A (amplitude) mode uses acoustic pulse emissions and echo reception along a single line-of-sight axial propagation path, and thus provides a one-dimensional mapping. This mode of operation cannot provide identification of structural features. It is, however, a most accurate method of measuring time delays and, therefore, distances between echo-producing structures or distances of structures from transducers, provided the speed of sound propagation in the medium is known.

The M mode of operation is used to display the movement of time-varying, echo-producing structures by intensity-modulating the trace as it is swept slowly across the oscilloscope screen in a direction at right angles to the fast time-base sweep. This mode of operation is used extensively in diagnosing disorders of the heart. *See* ECHOCARDIOGRAPHY; HEART DISORDERS.

For a two-dimensional picture to be obtained, the line-of-sight propagation path must be scanned and the position and direction of the path monitored and used to form a two-dimensional picture. Typically, the B-mode display is formed by moving the transducer so that the line-of-sight path remains in a single plane. The time-base trace of the cathode-ray oscilloscope screen is moved to correspond, in position and direction, to the ultrasonic line-of-sight propagation path, and echoes are displayed as intensity modulations of the trace.

The C (constant-range) mode provides a two-dimensional image display at constant time delay, and presumed constant distance, from the ultrasonic transducer. The scanning is arranged so that the point at constant depth along the propagation path (beam axis) traverses a plane. *See* MEDICAL IMAGING; ULTRASONICS. Floyd Dunn

Medical waste Any solid waste that is generated in the diagnosis, treatment, or immunization of human beings or animals, in research pertaining thereto, or in the production or testing of biologicals. Since the development of disposable medical products in the early 1960s, the issue of medical waste has confronted hospitals and regulators. Previously, reusable products included items such as linen, syringes, and bandages; they were sterilized or disinfected prior to reuse, and the principal waste product was limited to human pathological tissue.

Most hazardous substances are described by their relevant properties, such as corrosive, poison, or flammable. Medical waste was originally defined in terms of its infectious properties, and thus it was called infectious waste. However, given the difficulty of identifying pathogenic organisms in waste that might cause disease, it has become standard practice to define medical waste by types or categories. While definitions differ somewhat under different regulations, in the United States the Centers for Disease Control and Prevention (CDC) cite four categories of infective wastes that should require special handling and treatment: laboratory cultures and stocks, pathology wastes, blood, and items that possess sharp points such as needles and syringes (sharps). These categories require that the generator of these wastes exercise judgment in identifying the material to be included.

The waste category that has generated a great deal of interest is sharps. Needles and syringes, in particular, pose risks, since the instruments can penetrate into the body, increasing the potential for disease transmission. Improper disposal of these items in the past has been the catalyst for increased regulation and tighter management control.

Treatment of medical waste constitutes a method for rendering it noninfectious prior to disposal in a landfill or other solid-waste site. The treatment technologies currently used for medical waste include incineration, sterilization, chemical disinfection, and microwave, as well as others under development. *See* HAZARDOUS WASTE.

Robert A. Spurgin

Medicine The field of science devoted to healing. Many subdivisions exist and more ramifications appear almost daily. Included in the area of medicine are the clinical specialties of surgery, pediatrics, psychiatry, obstetrics, and others. Internal medicine is the specialization which deals with internal diseases of a nonsurgical nature.

Related to the clinical specialties, particularly in regard to medical education and research, are the basic medical sciences. These include, among others, anatomy, physiology, psychology, pharmacology, biochemistry, and microbiology. Midway between the basic and the clinical sciences lies pathology, the study of the structural and functional alterations caused by diseases or abnormal states.

An important area in all specialties is preventive medicine and public health. This form of medicine supplies a necessary link with the community, state, or large geographic region in matters of prevention, mass treatment, and statistical appraisals of health matters. It is also concerned with socioeconomic factors related to physical and mental well-being. *See* PUBLIC HEALTH.

Socialized medicine is that form which exists under the direct control and financing of the state. The National Health Service of Great Britain is the best-known example, and other systems exist.

Other subdivisions of medicine, with names that are largely self-explanatory, include veterinary, legal, tropical, and military medicine. *See* FORENSIC MEDICINE.

Although medicine is based primarily upon scientific information and method, an important feature is the relationship between the physician and the patient. It is at this point that the necessary scientific background of medicine gives way to the art of healing. *See* SURGERY.

Edward G. Stuart; N. Karle Mottet

Mediterranean Sea The Mediterranean Sea lies between Europe, Asia Minor, and Africa. It is completely landlocked except for the Strait of Gibraltar, the Bosporus, and the Suez Canal. The Mediterranean is conveniently divided into an eastern basin and a western basin, which are joined by the Strait of Sicily and the Strait of Messina.

The total water area of the Mediterranean is 965,900 mi^2 (2,501,000 km^2), and its average depth is 5040 ft (1536 m). The greatest depth in the western basin is 12,200 ft (3719 m), in the Tyrrhenian Sea. The eastern basin is deeper, with a greatest depth of 18,140 ft (5530 m) in the Ionian Sea about 34 mi (55 km) off the Greek mainland. The Atlantic tide disappears in the Strait of Gibraltar. The tides of the Mediterranean are predominantly semidiurnal.

J. Lyman

Meiosis The set of two successive cell divisions that serve to separate homologous chromosome pairs and thus reduce the total number of chromosomes by half. The meiotic process includes two sequential nuclear divisions that must occur prior to the formation of gametes (sperm and eggs). The major purpose of meiosis is the precise reduction in the number of chromosomes by one-half, so a diploid cell can create haploid gametes. Meiosis is therefore a critical component of sexual reproduction. *See* GAMETOGENESIS.

The basic events of meiosis are quite simple. As the cell begins meiosis, each chromosome has already duplicated its DNA and carries two identical copies of the DNA molecule. These are visible as two lateral parts, called sister chromatids, which are connected by a centromere. Homologous pairs of chromosomes are first identified and matched. This process, which occurs only in the first of the two meiotic divisions, is called pairing. The matched pairs are then physically interlocked by recombination, which is also known as exchange or crossing-over. After recombination, the homologous chromosomes separate from each other, and at the first meiotic division are partitioned into different nuclei. The second meiotic division begins with half of the original number of chromosomes. During this second meiotic division, the sister chromatids of each chromosome separate and migrate to different daughter cells. *See* CHROMOSOME.

The patterns by which genes are inherited are determined by the movement of the chromosomes during the two meiotic divisions. It is a fundamental tenet of Mendelian inheritance that each individual carries two copies of each gene, one derived from its father and one from its mother. Moreover, each of that individual's gametes will carry only one copy of that gene, which is chosen at random. The process by which the two copies of a given gene are distributed into separate gametes is referred to as segregation. Thus, if an individual is heterozygous at the *A* gene for two different alleles, *A* and *a*, his or her gametes will be equally likely to carry the *A* allele or the *a* allele, but never both or neither. The fact that homologous chromosomes, and thus homologous genes, segregate to opposite poles at the first meiotic division explains this principle of inheritance. *See* CELL CYCLE.

Meiotic divisions. The two meiotic divisions may be divided into a number of distinct stages. Meiotic prophase refers to the period after the last cycle of DNA replication, during which time homologous chromosomes pair and recombine. The end of prophase is signaled by the breakdown of the nuclear envelope, and the association of the paired chromosomes with the meiotic spindle. The spindle is made up of microtubules that, with associated motor proteins, mediate chromosome movement.

Metaphase I is the period before the first division during which pairs of interlocked homologous chromosomes, called bivalents, line up on the middle of the meiotic spindle. The chromosomes are primarily (but not exclusively) attached to the spindle by their centromeres such that the centromere of one homolog is attached to spindle fibers emanating from one pole, and the

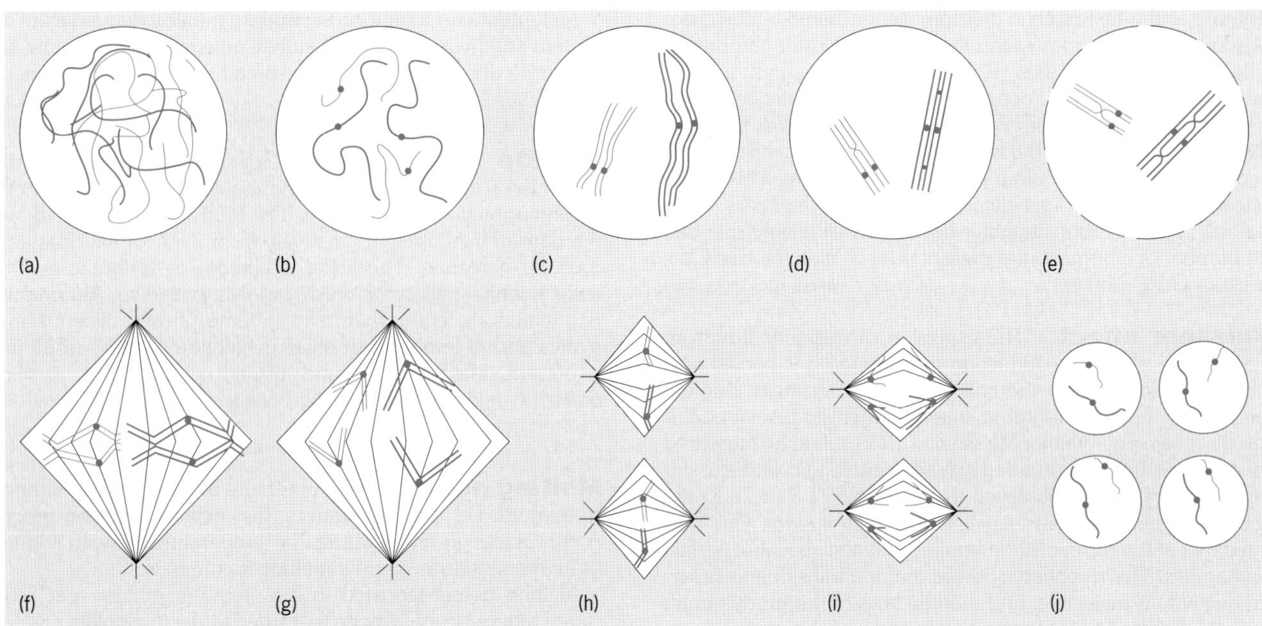

Stages of meiosis. (*a*) Premeiotic interphase. (*b*) Leptotene. (*c*) Zygotene. (*d*) Pachytene. (*e*) Diplotene/diakinesis. (*f*) Metaphase I. (*g*) Anaphase I. (*h*) Metaphase II. (*i*) Anaphase II. (*j*) Telophase II.

centromere of its partner is attached to spindle fibers from the other pole (see illustration). The bivalents are physically held together by structures referred to as chiasmata that are the result of meiotic recombination events. In most meiotic systems, meiosis will not continue until all of the homolog pairs are properly oriented at the middle of the spindle, the metaphase plate.

Anaphase I refers to the point at which homologous chromosome pairs separate and move to opposite poles. This is accomplished by the release of chiasmata. Depending on the organism, there may or may not be a true telophase, or a time in which nuclei reform. In most organisms, the first cell division occurs after the completion of anaphase I.

Following the completion of the first meiotic division, the chromosomes align themselves on a new pair of spindles, with their sister chromatids oriented toward opposite poles. The stage at which each chromosome is so aligned is referred to as metaphase II. In some, but not all, organisms, metaphase II is preceded by a brief prophase II. DNA replication does not occur during prophase II; each chromosome still consists of the two sister chromatids. Nor are there opportunities for pairing or recombination at this stage due to the prior separation of homologs at anaphase I.

The start of anaphase II is signaled by the separation of sister centromeres, and the movement of the two sister chromatids to opposite poles. At telophase II, the sisters have reached opposite poles and the nuclei begin to reform. The second cell division usually occurs at this time. Thus, at the end of the second meiotic division, there will be four daughter cells, each with a single copy of each chromosome.

Details of meiotic prophase. Pairing and recombination occur during the first meiotic prophase. The prophase of the first meiotic division is subdivided into five stages: leptotene, zygotene, pachytene, diplotene, and diakinesis. Homolog recognition, alignment, and synapsis occur during leptotene and zygotene. In the leptotene, initial homolog alignments are made. By zygotene, homologous chromosomes have become associated at various points along their length. These associations facilitate a more intimate pairing that results in the homologous chromosomes lying abreast of a tracklike structure called the synaptonemal complex. The beginning of pachytene is signaled by the completion of a continuous synaptonemal complex running the full length of each bivalent. During diplotene, the attractive forces that mediated homologous pairing disappear, and the homologs begin to repel each other. Luckily, homologs virtually always recombine, and those recombination events can be seen as chiasmata that tether the homologs together. The final stage in meiotic prophase is diakinesis, during which the homologs shorten and condense in preparation for nuclear division.

Recombination. Meiotic recombination involves the physical interchange of DNA molecules between the two homologous chromosomes, thus allowing the creation of new combinations of alleles for genes located on that pair of chromosomes. Mechanistically, recombination involves the precise breakage and rejoining of two nonsister chromatids. The result is the formation of two recombinant chromatids, each of which carries information from both of the original homologs. The number and position of recombination events is very precisely controlled. Exchange occurs only in the gene-rich euchromatin that makes up most of the chromosome arms, never in the heterochromatin that surrounds the centromeres. Moreover, as a result of a process known as interference, the occurrence of one exchange in a given chromosomal region greatly decreases the probability of a second exchange in that region. *See* RECOMBINATION (GENETICS).

Errors of meiosis. The failure of two chromosomes to segregate properly is called nondisjunction. Nondisjunction occurs either because two homologs failed to pair and/or recombine or because of a failure of the cell to properly move the segregating chromosomes on the meiotic spindle. The result of nondisjunction is the production of gametes that are aneuploid, carrying the wrong number of chromosomes. When such a gamete is involved in a fertilization event, the resulting zygote is also aneuploid. Those cases where the embryo carries an extra copy of a given chromosome are said to be trisomic, while those that carry but one copy are said to be monosomic for that chromosome. Most aneuploid zygotes are not viable and result in early spontaneous abortion. There are no viable monosomies for the human autosomes; however, a few types of trisomic zygotes are capable of survival. These are trisomies for the sex chromosomes (XXX, XXY, XYY), trisomy 21 (Down syndrome), trisomy 18, and trisomy 13. *See* CROSSING-OVER (GENETICS).

Meiosis versus mitosis. The fundamental difference between meiosis and mitosis is that sister chromatids do not separate at the first meiotic division; rather, homologous chromosomes separate from each other with their sister chromatids still attached to each other. Recombination is frequent in most meiotic cells; however, it occurs only rarely in mitotic cells, usually as part of DNA repair events. Most critically, DNA synthesis occurs only once within the two meiotic divisions, while there is a complete replication before every mitotic division. This allows mitosis to produce two genetically identical daughter cells, while meiosis produces four daughter cells, each with only one-half the number of chromosomes present prior to meiosis. *See* CELL DIVISION; GENE; MITOSIS. Michelle Y. Walker; R. Scott Hawley

Meissner effect The expulsion of magnetic flux from the interior of a superconducting metal when it is cooled in a magnetic field to below the critical temperature, near absolute zero, at which the transition to superconductivity takes place. It was discovered by Walther Meissner in 1933, when he measured the magnetic field surrounding two adjacent long cylindrical single crystals of tin and observed that at −452.97°F (3.72 K) the Earth's magnetic field was expelled from their interior. This indicated that at the onset of superconductivity they became perfect diamagnets. This discovery showed that the transition to superconductivity is reversible, and that the laws of thermodynamics apply to it. The Meissner effect forms one of the cornerstones in the understanding of superconductivity, and its discovery led F. London and H. London to develop their phenomenological electrodynamics of superconductivity. *See* DIAMAGNETISM; THERMODYNAMIC PRINCIPLES.

The magnetic field is actually not completely expelled, but penetrates a very thin surface layer where currents flow, screening the interior from the magnetic field.

The Meissner effect is subject to limitations. Full diamagnetism is not observed in polycrystalline samples, and the effect is not observed in impure samples or samples with certain geometrics, such as a round flat disk, with the magnetic field parallel to the axis of rotation. *See* SUPERCONDUCTIVITY. Hans W. Meissner

Meitnerium The seventeenth of the synthetic transuranium elements. Element 109 falls in column 9 of the periodic table under the elements cobalt, rhodium, and iridium. It is expected to have chemical properties similar to those of iridium. *See* IRIDIUM; PERIODIC TABLE; TRANSURANIUM ELEMENTS.

Element 109 was discovered in 1982 by a team under P. Armbruster and G. Münzenberg at the Gesellschaft für Schwerionenforschung (GSI) at Darmstadt, Germany. In a sequence of bombardments of bismuth-209 targets with beams of ions of titanium-50, chromium-54, and iron-58, the compound systems $^{259}105$, $^{263}107$, and $^{267}109$ were produced. The decay analysis of the isotopes produced showed in the case of elements 105 and 107 the production of $^{258}105$ and $^{262}107$ by reaction channels in which one neutron is emitted. These isotopes have odd neutron and proton numbers and possess a special stability against spontaneous fission. It was shown that alpha-particle decay dominated the decay chains. Spontaneous fission occurs through a 30% electron capture branch of $^{256}105$ in $^{258}104$. Three decay chains were observed for the three reactions ending by fission of $^{258}104$, and the decay of the first atom of element 109 was observed. *See* ALPHA PARTICLES; DUBNIUM; NUCLEAR FISSION; NUCLEAR REACTION; RADIOACTIVITY; RUTHERFORDIUM.

The single atom of element 109 was produced at a bombarding energy of 299 MeV in the reaction between iron-58 and bismuth-209. A total dose of 7×10^{17} ions was used to bombard thin layers of bismuth during a 250-h irradiation time. Peter Armbruster

Melanterite A mineral having composition $FeSO_4 \cdot 7H_2O$. Melanterite occurs mainly in green, fibrous or concretionary masses, or in short, monoclinic, prismatic crystals. Luster is vitreous, hardness is 2 on Mohs scale, and specific gravity is 1.90.

Melanterite is a common secondary mineral derived from oxidation and hydration of iron sulfide minerals such as pyrite and marcasite. Its occurrence is widespread. It is not an ore mineral. *See* MARCASITE. Edward C. T. Chao

Melilite A complete solid solution series ranging from gehlenite, $Ca_2Al_2SiO_7$, to akermanite, $Ca_2MgSi_2O_7$, often containing appreciable Na and Fe. The Mohs hardness is 5–6, and the density increases progressively from 2.94 for akermanite to 3.05 for gehlenite. The luster is vitreous to resinous, and the color is white, yellow, greenish, reddish, or brown. Akermanite-rich varieties occur in thermally metamorphosed siliceous limestones and dolomites, but more gehlenite-rich ones result if Al is present. Melilites are found instead of plagioclase in silica-deficient, feldspathoid-bearing basalts. *See* SILICATE MINERALS. Lawrence Grossman

Melting point The temperature at which a solid changes to a liquid. For pure substances, the melting or fusion process occurs at a single temperature, the temperature rise with addition of heat being arrested until melting is complete.

Melting points reported in the literature, unless specifically stated otherwise, have been measured under an applied pressure of 1 atm (10^5 pascals), usually 1 atm of air. (The solubility of air in the liquid is a complicating factor in precision measurements.) Upon melting, all substances absorb heat, and most substances expand; consequently an increase in pressure normally raises the melting point. A few substances, of which water is the most notable example, contract upon melting; thus, the application of pressure to ice at 32°F (0°C) causes it to melt. Large changes in pressure are required to produce significant shifts in the melting point.

For solutions of two or more components, the melting process normally occurs over a range of temperatures, and a distinction is made between the melting point, the temperature at which the first trace of liquid appears, and the freezing point, the higher temperature at which the last trace of solid disappears, or equivalently, if one is cooling rather than heating, the temperature at which the first trace of solid appears. *See* PHASE EQUILIBRIUM; SOLUTION; SUBLIMATION; TRIPLE POINT. Robert L. Scott

Membrane distillation A separation method in which a nonwetting, microporous membrane is used with a liquid feed phase on one side of the membrane and a condensing, permeate phase on the other side. Separation by membrane distillation is based on the relative volatility of various components in the feed solution. The driving force for transport is the partial pressure difference across the membrane. Separation occurs when vapor from components of higher volatility passes through the membrane pores by a convective or diffusive mechanism. *See* CONVECTION (HEAT).

Membrane distillation shares some characteristics with another membrane-based separation known as pervaporation, but there also are some vital differences. Both methods involve direct contact of the membrane with a liquid feed and evaporation of the permeating components. However, while membrane distillation uses porous membranes, pervaporation uses nonporous membranes.

Membrane distillation systems can be classified broadly into two categories: direct-contact distillation and gas-gap distillation. These terms refer to the permeate or condensing side of the membrane; in both cases the feed is in direct contact with the membrane. In direct-contact membrane distillation, both sides of the membrane contact a liquid phase; the liquid on the permeate side is used as the condensing medium for the vapors leaving the hot feed solution. In gas-gap membrane distillation, the condensed permeate is not in direct contact with the membrane.

Potential advantages of membrane distillation over traditional evaporation processes include operation at ambient pressures and lower temperatures as well as ease of process scale-up. *See* CHEMICAL SEPARATION TECHNIQUES; MEMBRANE SEPARATIONS.

S. S Kulkarni; Norman N. Li

Membrane mimetic chemistry The study of processes and reactions whose developments have been inspired by the biological membrane. Faithful modeling of the biomembrane is not an objective of membrane mimetic chemistry. Rather, only the essential components of natural systems are recreated from relatively simple, synthesized molecules. (The term membrane mimetic is more restrictive than the term biomimetic. Biomimetic chemistry is directed at the mechanistic elucidation of biochemical reactions and at the development of new compounds modeled on specific biological systems.) *See* CELL MEMBRANES.

Various surfactant aggregate systems have been used in membrane mimetics.

Surfactants (detergents) contain distinct hydrophobic (apolar) and hydrophilic (polar) regions. Depending on the chemical structure of their hydrophilic polar head groups, surfactants can be neutral, positively charged, or negatively charged. *See* DETERGENT.

Aqueous micelles are spherical aggregates, 4–8 nanometers in diameter, formed dynamically from surfactants in water above a characteristic concentration, the critical micelle concentration. *See* MICELLE.

Monomolecular layers are formed by spreading naturally occurring lipids or synthetic surfactants, dissolved in volatile solvents, over water in a trough. The polar head groups of the surfactants are in contact with water, the subphase, while their hydrocarbon tails protrude above it. *See* MONOMOLECULAR FILM.

Other systems used in membrane mimetics are multilayer assemblies (Langmuir-Blodgett films), bilayer lipid membranes, and vesicles prepared by sonication from naturally occurring lipids. *See* SONOCHEMISTRY.

Membrane mimetic chemistry has become a versatile chemical tool. Applications of compartmentalization of reactants in membrane mimetic systems involve altered reaction rates, products, stereochemistries, and isotope distributions. Monolayers and organized multilayers can be employed profitably as molecular electronic devices. Opportunities also exist for using different surfactant aggregates with polymeric membranes for the control and regulation of reverse osmosis and ultrafiltration. *See* SURFACTANT; ULTRAFILTRATION.

Janos H. Fendler

Membrane separations Processes for separating mixtures by using thin barriers (membranes) between two miscible fluids. A suitable driving force across the membrane, for example concentration or pressure differential, leads to the preferential transport of one or more feed components.

Membrane separation processes are classified under different categories depending on the materials to be separated and the driving force applied: (1) In ultrafiltration, liquids and low-molecular-weight dissolved species pass through porous membranes while colloidal particles and macromolecules are rejected. The driving force is a pressure difference. (2) In dialysis, low-molecular-weight solutes and ions pass through while colloidal particles and solutes with molecular weights greater than 1000 are rejected under the conditions of a concentration difference across the membrane. (3) In electrodialysis, ions pass through the membrane in preference to all other species, due to a voltage difference. (4) In reverse osmosis, virtually all dissolved and suspended materials are rejected and the permeate is a liquid, typically water. (5) For gas and liquid separations, unequal rates of transport can be obtained through nonporous membranes by means of a solution and diffusion mechanism. Pervaporation is a special case of this separation where the feed is in the liquid phase while the permeate, typically drawn under subatmospheric conditions, is in the vapor phase. (6) In facilitated transport, separation is achieved by reversible chemical reaction in the membrane. High selectivity and permeation rate may be obtained because of the reaction scheme. Liquid membranes are used for this type of separation. *See* DIALYSIS; ION-SELECTIVE MEMBRANES AND ELECTRODES; OSMOSIS; TRANSPORT PROCESSES; ULTRAFILTRATION.

Norman N. Li; Sudhir S. Kulkarni

Memory The ability to store and access information that has been acquired through experience. Memory is a critical component of practically all aspects of human thinking, including perception, learning, language, and problem solving. *See* PERCEPTION; PROBLEM SOLVING (PSYCHOLOGY).

Stages. The information-processing approach divides memory into three general stages: sensory memory, short-term memory, and long-term memory. Sensory memory refers to the sensations that briefly continue after something has been perceived. Short-term memory includes all of the information that is currently being processed in a person's mind, and is generally thought to have a very limited capacity. Long-term memory is where all the information that may be used at a later time is kept.

A number of interesting facts are known about sensory memory, including the following: (1) sensory memories appear to be associated with mechanisms in the central nervous system rather than at the sensory receptor level, and (2) the amount of attention that a person pays to a stimulus can affect the duration of the sensory memory. Although all of the functions of sensory memory are not understood, one of its most important purposes is to provide people with additional time to determine what should be transferred to the next stage in the memory system, that is, short-term memory.

Information obtained from either sensory memory or long-term memory is processed in short-term memory in order for a person to achieve current goals. In some situations, short-term memory processing simply involves the temporary maintenance of a piece of information, such as remembering a phone number long enough to dial it. Other times, short-term memory can involve elaborate manipulations of information in order to generate new forms. For example, when someone reads $27 + 15$, the person manipulates the symbols in short-term memory in order to come up with the solution. One useful manipulation that can be done in short-term memory is to reorganize items into meaningful chunks. For example, it is a difficult task to keep the letters S K C A U Q K C U D E H T in mind all at once. However, if they are rearranged in short-term memory, in this case reversing them, they can be reduced to a single simple chunk: THE DUCK QUACKS. Short-term memory can accommodate only five to seven chunks at any one time. However, the amount of information contained in each chunk is constrained only by one's practice and ingenuity. In order to increase the amount of information that can be kept in short-term memory at one time, people need to develop specific strategies for organizing that information into meaningful chunks. In addition, many studies have also demonstrated that the transfer of information from short-term to long-term memory is much greater when the information is manipulated rather than simply maintained.

One can keep massive amounts of information in long-term memory. In general, recall from long-term memory simply involves figuring out the heading under which a memory has been filed. Many tricks for effective retrieval of long-term memories involve associating the memory with another more familiar memory that can serve as an identification tag. This trick of using associations to facilitate remembering is called mnemonics. Long-term memory stores related concepts and incidents in close range of one another. This logical association of memories is indicated by subjects' reaction times for identifying various memories. Generally, people are faster at recalling memories if they have recently recalled a related memory. One good way to locate a long-term memory is to remember the general situation under which it was stored. Accordingly, techniques that reinstate the context of a memory tend to facilitate remembering.

Sometimes information may not have been filed in long-term memory in the first place, or if it has, is inaccessible. In these situations, the long-term memory system often fills in the gaps by using various constructive processes. One common component to memory constructions is a person's expectations. Countless studies have also indicated that memories tend to systematically change in the direction of a prior expectation or inference about what is likely to have occurred.

Physiology. A number of physiological mechanisms appear to be involved in the formation of memories, and the mechanisms may differ for short-term and long-term memory. There is both direct and indirect evidence suggesting that short-term memory involves the temporary circulation of electrical impulses around complex loops of interconnected neurons. A number of indirect lines of research indicate that short-term memories are eradicated by any event that either suppresses neural activity (for example, a blow to the head or heavy anesthesia) or causes neurons to fire incoherently (for example, electroconvulsive shock). More direct support for the electric circuit model of short-term memory comes from observing electrical brain activity. By implanting electrodes in the brain of experimental animals, researchers have observed that changes in what an animal is watching are associated with different patterns of circulating electrical activity in the brain. These results suggest that different short-term memories may be represented by different electrical patterns. However, the nature of these patterns is not well understood. *See* ELECTROENCEPHALOGRAPHY.

Long-term memories appear to involve some type of permanent structural or chemical change in the composition of the brain. This conclusion is derived both from general observations of the imperviousness of long-term memories and from physiological studies indicating specific changes in brain composition. Even in acute cases of amnesia where massive deficits in long-term memory are reported, often, with time, all long-term memories return. Similarly, although electroconvulsive therapy is known to eliminate recent short-term memories, it has practically no effect on memories for events occurring more than an hour prior to shocking. Thus the transfer from a fragile short-term memory to a relatively solid long-term memory occurs within an hour. This process is sometimes called consolidation. *See* ELECTROCONVULSIVE THERAPY.

The nature of the "solid" changes associated with long-term memories appears to involve alterations in both the structural (neural connections) and chemical composition of the brain. One study compared the brains of rats that had lived either in enriched environments with lots of toys or in impoverished environments with only an empty cage. The cerebral cortices of the brains of the rats from the enriched environment were thicker, heavier, endowed with more blood vessels, and contained significantly greater amounts of certain brain chemicals (such as the neurotransmitter acetylcholine). Other researchers have observed that brief, high-frequency stimulation of a neuron can produce long-lasting changes in the neuron's communications across synapses.

Researchers believe that different brain structures may be involved in the formation and storage of long-term memories. The hippocampus, thalamus, and amygdala are believed to be critical in the formation of long-term memories. Individuals who have had damage to these structures are able to recall memories prior to the damage, indicating that long-term memory storage is intact; however, they are unable to form new long-term memories, indicating that the long-term memory formation process has been disrupted. It is not known where long-term memories are stored, but they may be localized in the same areas of the brain that participated in the actual learning. *See* BRAIN.

Jonathan Schooler; Elizabeth F. Loftus

Mendelevium A chemical element, Md, atomic number 101, the twelfth member of the actinide series of elements. Mendelevium does not occur in nature; it was discovered and is prepared by artificial nuclear transmutation of a lighter element. Known isotopes of mendelevium have mass numbers from 248 to 258 and half-lives from a few seconds to about 55 days. They are all produced by charged-particle bombardments of more abundant isotopes. The amounts of mendelevium which are produced and used for studies of chemical and nuclear properties are usually less than about a million atoms; this is of the order of a million times less than a weighable amount. Studies of the chemical properties of mendelevium have been limited to a tracer scale. The behavior of mendelevium in ion-exchange chromatography shows that it exists in aqueous solution primarily in the 3 + oxidation state characteristic of the actinide elements. However, it also has a dipositive (2+) and a monopositive (1+) oxidation state. *See* ACTINIDE ELEMENTS; PERIODIC TABLE; TRANSURANIUM ELEMENTS.

Glenn T. Seaborg

Mendelism Fundamental principles governing the transmission of genetic traits, discovered by an Augustinian monk Gregor Mendel in 1856. Mendel performed his first set of hybridization experiments with pea plants. Although the pea plant is normally self-fertilizing, it can be easily crossbred, and grows to maturity in a single season. True breeding strains, each with distinct characteristics, were available from local seed merchants. For his experiments, Mendel chose seven sets of contrasting characters or traits. For stem height, the true breeding strains tall (7 ft or 2.1 m) and dwarf (18 in. or 45 cm) were used. He also selected six other sets of traits, involving the shape and color of seeds, pod shape and color, and the location of flowers on the plant stem.

The most simple crosses performed by Mendel involved only one pair of traits; each such experiment is known as a monohybrid cross. The plants used as parents in these crosses are known as the P_1 (first parental) generation. When tall and dwarf plants were crossed, the resulting offspring (called the F_1 or first filial generation) were all tall. When members of the F_1 generation were self-crossed, 787 of the resulting 1064 F_2 (second filial generation) plants were tall and 277 were dwarf. The tall trait is expressed in both the F_1 and F_2 generations, while the dwarf trait disappears in the F_1 and reappears in the F_2 generation. The trait expressed in the F_1 generation Mendel called the dominant trait, while the recessive trait is unexpressed in the F_1 but reappears in the F_2. In the F_2, about three-fourths of the offspring are tall and one-fourth are dwarf (a 3:1 ratio). Mendel made similar crosses with plants exhibiting each of the other pairs of traits, and in each case all of the F_1 offspring showed only one of the parental traits and, in the F_2, three-fourths of the plants showed the dominant trait and one-fourth exhibited the recessive trait. In subsequent experiments, Mendel found that the F_2 recessive plants bred true, while among the dominant plants one-third bred true and two-thirds behaved like the F_1 plants. *See* DOMINANCE.

Law of segregation. To explain the results of his monohybrid crosses, Mendel derived several postulates. First, he proposed that each of the traits is controlled by a factor (now called a gene). Since the F_1 tall plants produce both tall and dwarf offspring, they must contain a factor for each, and thus he proposed that each plant contains a pair of factors for each trait. Second, the trait which is expressed in the F_1 generation is controlled by a dominant factor, while the unexpressed trait is controlled by a recessive factor. To prevent the number of factors from being doubled in each generation, Mendel postulated that factors must separate or segregate from each other during gamete formation. Therefore, the F_1 plants can produce two types of gametes, one type containing a factor for tall plants, the other a factor for dwarf plants. At fertilization, the random combination of these gametes can explain the types and ratios of offspring in the F_2 generation (see illustration). *See* FERTILIZATION (ANIMAL); GENE.

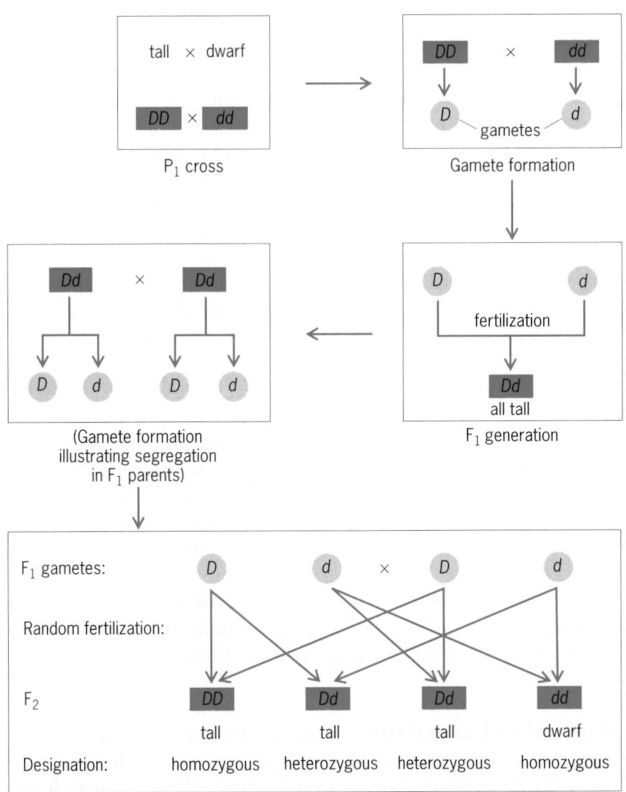

Schematic representation of a monohybrid cross. Pure-bred tall and dwarf strains are crossed, and yield typical 3:1 ratio in the F_2 generation. *D* and *d* represent the tall and dwarf factors (genes), respectively. (*After W. S. Klug and M. R. Cummings, Concepts of Genetics, Charles E. Merrill, 1983*)

Independent assortment. Mendel extended his experiments to examine the inheritance of two characters simultaneously. Such a cross, involving two pairs of contrasting traits, is known as a dihybrid cross. For example, Mendel crossed plants with tall stems and round seeds with plants having dwarf stems and wrinkled seeds. The F_1 offspring were all tall and had round seeds. When the F_1 individuals were self-crossed, four types of offspring were produced in the following proportions: 9/16 were tall, round; 3/16 were tall, wrinkled; 3/16 were dwarf, round; and 1/16 were dwarf, wrinkled. On the basis of similar results in other dihybrid crosses, Mendel proposed that during gamete formation, segregating pairs of factors assort independently of one another. As a result of segregation, each gamete receives one member of every pair of factors [this assumes that the factors (genes) are located on different chromosomes]. As a result of independent assortment, all possible combinations of gametes will be found in equal frequency. In other words, during gamete formation, round and wrinkled factors segregate into gametes independently of whether they also contain tall or dwarf factors. *See* GAMETOGENESIS; MEIOSIS.

It might be useful to consider the dihybrid cross as two simultaneous and independent monohybrid crosses. In this case, the predicted F_2 results are 3/4 tall, 1/4 dwarf, and 3/4 round, 1/4 wrinkled. Since the two sets of traits are inherited independently, the number and frequency of phenotypes can be predicted by combining the two events:

$$3/4\,\text{tall}\begin{cases}3/4\,\text{round} & (3/4)(3/4) = 9/16\,\text{tall, round}\\ 1/4\,\text{wrinkled} & (3/4)(1/4) = 3/16\,\text{tall, wrinkled}\end{cases}$$

$$1/4\,\text{dwarf}\begin{cases}3/4\,\text{round} & (1/4)(3/4) = 3/16\,\text{dwarf, round}\\ 1/4\,\text{wrinkled} & (1/4)(1/4) = 1/16\,\text{dwarf, wrinkled}\end{cases}$$

This 9:3:3:1 ratio is known as a dihybrid ratio and is the result of segregation, independent assortment, and random fertilization. *See* GENETICS.

Michael R. Cummings

Meninges In mammals, the three membranes that cover the brain and spinal cord: the dura mater, the arachnoid membrane, and the pia mater. The outermost, the dura mater, is a tough, fibrous, double-layered structure that is adherent to the skull. The inner layer of the dura mater sends separating sheets between the cerebral hemispheres and between the cerebrum and cerebellum. It also contains large venous sinuses and forms sheaths for nerves leaving the skull. The middle layer, the arachnoid, is a delicate serous layer loosely investing the brain. Below this is the spongy subarachnoid cavity which contains the circulating cerebrospinal fluid. The innermost layer, the pia mater, is a vascular layer which closely follows each convolution of the brain. Together the meninges furnish protection, blood supply, drainage, and cerebrospinal channels for the brain. *See* NERVOUS SYSTEM (VERTEBRATE).

Thomas S. Parsons

Meningitis Inflammation of the meninges. Certain types of meningitis are associated with distinctive abnormalities in the cerebrospinal fluid. With certain types of meningitis, especially bacterial, the causative organism can usually be recovered from the fluid. *See* MENINGES.

Meningeal inflammation in most cases is caused by invasion of the cerebrospinal fluid by an infectious organism. Noninfectious causes also occur. For example, in immune-mediated disorders antigen-antibody reactions can cause meningeal inflammation. Other noninfectious causes of meningitis are the introduction into the cerebrospinal fluid of foreign substances such as alcohol, detergents, chemotherapeutic agents, or contrast agents used in some radiologic imaging procedures. Meningeal inflammation brought about by such foreign irritants is called chemical meningitis. Inflammation also can occur when cholesterol-containing fluid or lipid-laden material leaks into the cerebrospinal fluid from some intracranial tumors.

Bacterial meningitis is among the most feared of human infectious diseases because of its possible seriousness, its rapid progression, its potential for causing severe brain damage, and its frequency of occurrence. Most cases of bacterial meningitis have an acute onset. Common clinical manifestations are fever, headache, vomiting, stiffness of the neck, confusion, seizures, lethargy, and coma. Symptoms of brain dysfunction are caused by transmission of toxic materials from the infected cerebrospinal fluid into brain tissue and the disruption of arterial perfusion and venous drainage from the brain because of blood vessel inflammation. These factors also provoke cerebral swelling, which increases intracranial pressure. Before antibiotics became available, bacterial meningitis was almost invariably fatal. *See* ANTIBIOTIC.

Most types of acute bacterial meningitis are septic-borne in that they originate when bacteria in the bloodstream (bacteremia, septicemia) gain entrance into the cerebrospinal fluid. Meningitis arising by this route is called primary bacterial meningitis. Secondary meningitis is that which develops following direct entry of bacteria into the central nervous system, which can occur at the time of neurosurgery, in association with trauma, or through an abnormal communication between the external environment and the cerebrospinal fluid.

Many viruses can cause meningeal inflammation, a condition referred to as viral aseptic meningitis. The most common viral causes include the enteroviruses, the various herpesviruses, viruses transmitted by arthropods, the human immunodeficiency virus type I (HIV-1), and formerly, the mumps virus. If the virus attacks mainly the brain rather than the spinal cord, the disorder is termed viral encephalitis. *See* ANIMAL VIRUS; ARBOVIRAL ENCEPHALITIDES; ENTEROVIRUS; HERPES.

Fungal, parasitic, and rickettsial meningitis are less common in the United States than are bacterial and viral. These infections are more likely to be subacute or chronic than those caused by bacteria or viruses; in most cases, the meningeal inflammation is associated with brain involvement. An acute form of aseptic meningitis can occur in the spirochetal diseases, syphilis and Lyme disease. *See* LYME DISEASE; MEDICAL MYCOLOGY; MEDICAL PARASITOLOGY; RICKETTSIOSES; SYPHILIS. William E. Bell

Meningococcus A major human pathogen belonging to the bacterial genus *Neisseria*, and the cause of meningococcal meningitis and meningococcemia. The official designation is *N. meningitidis*. The meningococcus is a gram-negative, aerobic, nonmotile diplococcus. It is fastidious in its growth requirements and is very susceptible to adverse physical and chemical conditions.

Humans are the only known natural host of the meningococcus. Transmission occurs by droplets directly from person to person. Fomites and aerosols are probably unimportant in the spread of the organism. The most frequent form of host-parasite relationship is asymptomatic carriage in the nasopharynx.

The most common clinical syndrome caused by the meningococcus is meningitis, which is characterized by fever, headache, nausea, vomiting, and neck stiffness and has a fatality rate of 15% (higher in infants and adults over 60). Disturbance of the state of consciousness quickly occurs, leading to stupor and coma. Many cases also have a typical skin rash consisting of petechiae or purpura. *See* MENINGITIS. Ronald Gold

Menopause The irreversible cessation of regular monthly uterine bleeding in the adult human female, marking the end of her ability to become pregnant. Menopause commonly occurs in the United States between the ages of 47 and 53. It probably occurs because the ovary runs out of eggs and the cyclic rise and fall of brain and ovarian hormones designed to prepare the uterus to receive and nourish pregnancy no longer occur.

Menopause is one event in the climacteric, the period of time during which the reproductive machinery slows down and finally stops. The biochemical hallmark of this period is a reduction in estrogen production by the ovary. Some estrogen continues to be produced by the adrenal gland and the fatty tissues throughout the body, but this amount is very small compared with premenopausal levels. Estrogen has widespread effects on both genital and extragenital systems, and the withdrawal of estrogen accounts for many of the signs and symptoms attributed to menopause, although these are influenced by both hereditary and social factors. Many psychological problems have been attributed to estrogen deprivation, but well-documented proof of those relationships is lacking.

While estrogen can reverse or halt many of the physical changes described, it will not prevent aging or restore reproductive ability. Treatment of menopausal symptoms should be undertaken on an individual basis, with careful discussion of the risks and benefits currently known. *See* ESTROGEN; MENSTRUATION. G. Guzinski

Menstruation Periodic sloughing of the uterine lining in women of reproductive age. Menstrual bleeding indicates the first day of the menstrual cycle, which lasts an average of 27–30 days, although ranges of 21–60 days have been recorded. Menarche, the onset of menstruation, occurs between the ages of 9 and 16. The majority of females begin menstruating at ages 12–14. During the first few years, the duration and intensity of menstrual flow and the total cycle length may be quite variable, but regularity is gradually established. Cessation of menses, or menopause, occurs at an average age of 51, with a range of 42–60 years.

The menstrual cycle consists of cyclic changes in both the ovary and the uterus. These changes are controlled by the interaction of several hormones including follicle-stimulating hormone (FSH) and luteinizing hormone (LH), which are secreted by the anterior pituitary, and the steroid hormones estrogen and progesterone, which are secreted by follicles in the ovary. At the beginning of the cycle, the follicle is stimulated by FSH. In response, it grows and secretes estrogen. The amount of estrogen secretion increases rapidly near the middle of the cycle. Estrogen, in turn, stimulates growth of the uterine lining (mucosa), which becomes thicker and fills with blood vessels. In midcycle, the rapid increase in estrogen causes a massive surge of LH release and a smaller release of FSH from the pituitary. This surge causes ovulation, which is the release of the ovum from the follicle. After ovulation, the follicle undergoes rapid changes and is then called a corpus luteum, which secretes progesterone in response to LH stimulation. Progesterone and estrogen together cause a further thickening of the uterine mucosa, preparing the uterus for pregnancy. If pregnancy does not occur, the corpus luteum degenerates, the uterine mucosa sloughs off, and the cycle begins again.

There is no menstrual bleeding during pregnancy, as the uterine mucosa is needed for the maintenance of pregnancy. This amenorrhea, or lack of normal ovarian function, sometimes continues during nursing. Janice M. Bahr

Mental retardation A developmental disability characterized by significantly subaverage general intellectual functioning, with concurrent deficits in adaptive behavior. The causes are many and include both genetic and environmental factors as well as interactions between the two. In most cases the diagnosis is not formally made until children have entered into school settings. In the preschool years, the diagnosis is more likely to be established by evidence of delayed maturation in the areas of sensory-motor, adaptive, cognitive, social, and verbal behaviors. By definition, evidence of mental retardation must exist prior to adulthood, where vocational limitation may be evident, but the need for supervision or support may persist beyond the usual age of social emancipation.

From the aspect of etiology, mental retardation can be classified by prenatal, perinatal, or postnatal onset. Prenatal causes include genetic disorders, syndromal disorders, and developmental disorders of brain formation. Upward of 700 genetic causes have been suggested as associated with the development of mental retardation. Many environmental influences on the developing fetus, for example, infection, and other unknown errors of development may account for mental retardation.

Perinatal causes include complications at birth, extreme prematurity, infections, and other neonatal disorders. Postnatal causes include trauma, infections, demyelinating and degenerative disorders, consequences of seizure disorders, toxic-metabolic disorders, malnutrition, and environmental deprivation. Often no specific cause can be identified for the mental retardation of a particular individual.

Individuals with mental retardation are typically subclassified in terms of the manifest severity of cognitive disability as reflected by the ratio of mental age to chronological age, or intelligence quotient (IQ). Subaverage intellectual functioning is defined as an IQ score of at least two standard deviations below the mean, or approximately 70 to 75 or below. Mild, moderate, severe, and profound degrees of mental retardation refer to two, three, four, or five standard deviations below the normal IQ for the general population.

Limitations in adaptive behavior must also be demonstrable in order to satisfy diagnostic criteria for mental retardation. This criterion is important because certain artistic or other gifts may not be revealed by formal IQ testing, and different levels of learning difficulty may be accentuated by the demands of specific environments. Outside such environments, an individual may navigate a normal course in life.

A specific genetic or other cause of mental retardation may also predispose to other medical or neurologic conditions. In these circumstances, the comorbid medical conditions may increase the likelihood of emotional or behavioral problems, or contribute to the challenges with which a given child must contend. Thus, the identification of cause can be important in planning for the medical, educational, and treatment needs of a particular individual.

Considerable progress has been made in both prevention and treatment. Diet is a method of treatment following early detection of phenylketonuria; warnings regarding alcohol consumption during pregnancy, lead exposure in infancy, and disease immunization and therapy are measures for prevention of retardation. Advances in prenatal, obstetrical, and neonatal care and genetic counseling have had the effect of reducing the incidence or the severity of various conditions. Energetic training and the application of psychosocial techniques have resulted in improved social performance and adaptive behavior in many persons with mental retardation. Bryan H. King

Menthol A cyclic monoterpene alcohol with melting point 40°C (104°F), specific optical rotation ($[\alpha]_D$) −50°. Menthol is the major component of the essential oil of peppermint. Since menthol has three chiral centers (asymmetric carbons), the structure below is one of eight stereoisomers.

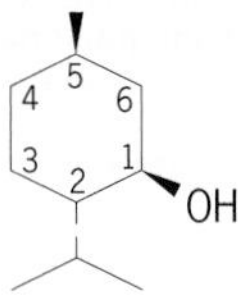

(−)Menthol
1R, 2S, 5R

Menthol has a cooling, slightly anesthetic effect and is used in pharmaceutical preparations as an antitussive and antipruritic agent, and as a decongestant in nasal sprays. Most of the menthol produced is used in dentrifices, mouthwash, skin lotions, and so forth, and as a flavorant in tobacco, chewing gum, and confectionary products. For most of these purposes, the (−)isomer is preferred. *See* TERPENE. James A. Moore

Mercaptan One of a group of organosulfur compounds which are also called thiols or thio alcohols and which have the general structure RSH. Aromatic thiols are called thiophenols, and biochemists often refer to thiols as sulfhydryl compounds. The unpleasant odor of volatile thiols causes them to be classed as stenches, but the odors of many solid thiols are not unpleasant.

Mercaptans (1) form salts with bases, (2) are easily oxidized to disulfides and higher oxidation products such as sulfonic acids, (3) react with chlorine (or bromine) to form sulfenyl chlorides (or bromides), and (4) undergo additions to unsaturated compounds, such as olefins, acetylenes, aldehydes, and ketones. The insoluble mercury salts (mercaptides) are used to isolate and identify mercaptans. *See* ORGANOSULFUR COMPOUND. Norman Kharasch

Merchant ship A power-driven ship employed in commercial transport on the oceans and large inland bodies of water such as the Great Lakes. The relatively small craft used for inland waterway transportation are not commonly referred to as ships.

Commodities transported by water are classified as breakbulk, unitized, or bulk (dry or liquid) cargoes. Generally, water cargo transportation is cheaper per ton-mile than land or air transportation; approximately 90% of the United States overseas trade revenue is waterborne, while the rest is airborne. *See* MARINE CONTAINERS.

A passenger ship, as defined by International Safety of Life at Sea (SOLAS) rules, carries more than 12 passengers on international voyages. Passenger vessels that also transport cargo are called passenger-cargo ships. Amelio M. D'Arcangelo; Robert B. Zubaly

Mercury (element) A chemical element, Hg, atomic number 80 and atomic weight 200.59. Mercury is a silver-white liquid at room temperature (melting point −38.89°C or −37.46°F); it boils at 357.25°C (675.05°F) under atmospheric pressure. It is a noble metal that is soluble only in oxidizing solutions. Solid mercury is as soft as lead. The metal and its compounds are very toxic. With some metals (gold, silver, platinum, uranium, copper, lead, sodium, and potassium, for example) mercury forms solutions called amalgams. *See* AMALGAM; PERIODIC TABLE; TRANSITION ELEMENTS.

In its compounds, mercury is found in the 2+, 1+, and lower oxidation states, for example, $HgCl_2$, Hg_2Cl_2, or $Hg_3(AsF_6)_2$. Often the mercury atoms are doubly covalently bonded, for example, Cl—Hg—Cl or Cl—Hg—Hg—Cl. Some mercury(II) salts, for example, $Hg(NO_3)_2$, or $Hg(ClO_4)_2$, are quite soluble in water and dissociate normally. The aqueous solutions of these salts react as strong acids because of hydrolysis. Other mercury(II) salts, for example, $HgCl_2$ or $Hg(CN)_2$, also dissolve in water, but exist in solution as only slightly dissociated molecules. There are compounds in which mercury atoms are bound directly to carbon or nitrogen atoms, for example, $H_3C—Hg—CH_3$ or $H_3C—CO—NH—Hg—NH—CO—CH_3$. In complex compounds, for example, $K_2(HgI_4)$, mercury often has three or four bonds.

Metallic mercury is used as a liquid contact material for electrical switches, in vacuum technology as the working fluid of diffusion pumps, for the manufacture of mercury-vapor rectifiers, thermometers, barometers, tachometers, and thermostats, and for the manufacture of mercury-vapor lamps. It finds application for the manufacture of silver amalgams for tooth fillings in dentistry. Of importance in electrochemistry are the standard calomel electrode, used as the reference electrode for the measurement of potentials and for potentiometric titrations, and the Weston standard cell.

Mercury is commonly found as the sulfide, HgS, frequently as the red cinnabar and less often as the black metacinnabar. A less common ore is the mercury(I) chloride. Occasionally the mercury ore contains small drops of metallic mercury. *See* CINNABAR.

The surface tension of liquid mercury is 484 dynes/cm, six times greater than that of water in contact with air. Hence, mercury does not wet surfaces with which it is in contact. In dry air metallic mercury is not oxidized. After long standing in moist air, however, the metal becomes coated with a thin layer of oxide. In air-free hydrochloric acid or in dilute sulfuric acid, the metal does not dissolve. Conversely, it is dissolved by oxidizing acids (nitric acid, concentrated sulfuric acid, and aqua regia). Klaus Brodersen

Mercury (planet) The planet closest to the Sun. It is visible to the unaided eye only shortly after sunset or shortly before sunrise, when it is near its greatest angular distance from the Sun (28°). Mercury is the smallest planet since Pluto is no longer classified as a planet. Its diameter is 3031 mi (4878 km), and its mass is 0.055 times the mass of the Earth. Most detailed knowledge of Mercury is derived from data returned by the *Mariner 10* spacecraft, which flew by the planet three times in 1974 and 1975. The coverage and resolution is somewhat comparable to Earth-based telescopic coverage and resolution of the Moon before the advent of space flight. As a consequence, there are still many uncertainties and questions concerning this unusual planet. Mercury represents an end member in solar system origin and evolution because it formed closer to the Sun than any other planet and, therefore, in the hottest part of the solar nebula from which the entire solar system formed. *See* PLUTO; SOLAR SYSTEM.

Mercury has the most eccentric (0.205) and inclined (7°) orbit of any planet in the solar system. Its average distance from the Sun is 3.60×10^7 mi (5.79×10^7 km). The rotation period is 58.646 Earth days, and the orbital period, 87.969. Therefore, Mercury makes exactly three rotations around its axis for every two orbits around the Sun. Thus, a solar day (sunrise to sunrise) lasts two mercurian years (176 Earth days).

Mercury experiences the greatest range in surface temperature (1130°F or 628°C) of any planet or satellite in the solar system because of its proximity to the Sun, its long solar day, and its lack of an insulating atmosphere. Its maximum surface temperature is 800°F (427°C) at perihelion on the equator, hot enough to melt zinc. At night, however, the unshielded surface plunges to below −300°F (−184°C). *See* VENUS.

Mariner 10 discovered an intrinsic dipole magnetic field with a dipole moment equal to about 0.004 that of the Earth. Mercury is the only terrestrial planet besides the Earth with a magnetic field. Although weak compared to that of the Earth, the field has sufficient strength to hold off the solar wind, creating a bow shock and accelerating charged particles from the solar wind. Mercury's dipolar magnetic field is evidence that Mercury currently has an electrically conducting fluid outer core of unknown thickness. *See* GEOMAGNETISM; MAGNETISM; MAGNETOSPHERE.

Mercury's internal structure is unique in the solar system. The planet's mean density is 5.44 g/cm^3 (5.44 times that of water), which is larger than that of any other planet or satellite except Earth (5.52 g/cm^3). Because of Earth's large internal pressures, however, its uncompressed density is only 4.4 g/cm^3 compared to Mercury's uncompressed density of 5.3 g/cm^3. This means that Mercury contains a much larger fraction of iron than any other planet or satellite in the solar system. The iron core must be about 75% of the planet diameter, or 42% of Mercury's volume. It is surrounded by a silicate mantle and crust only about 370 mi (600 km) thick. Earth's core is only 54% of the planet diameter, or just 16% of the total volume. Because Mercury's outer core is at least partly molten at present, a light alloying element in the core must have lowered the melting point and retained a partially molten core over geologic history (4.5×10^9 years); otherwise the core would have solidified long ago. Sulfur is the most reasonable candidate for this alloying element. Mercury probably has between about 0.2 and 7% sulfur in its core. The origin of Mercury and how it acquired such a large percentage of iron is a major unsolved problem. See EARTH

Mercury's atmosphere is very tenuous and is essentially exospheric in that its atoms rarely collide with each other. The atmospheric surface pressure is 10^{12} times less than Earth's. *Mariner 10*'s ultraviolet spectrometer identified hydrogen, helium, oxygen, and argon in the atmosphere, all of which are probably derived largely from the solar wind. Earth-based telescopic observations in 1985 discovered that Mercury is surrounded by a tenuous atmosphere of sodium and potassium that is probably derived from its surface.

The surface of Mercury superficially resembles that of the Moon (see illustration). It is heavily cratered, with large expanses of younger smooth plains (similar to the lunar maria) that fill and surround major impact basins. Unlike the Moon surface, Mercury's heavily cratered terrain is interspersed with large regions of gently rolling intercrater plains, the major terrain type on the planet. Also unlike the Moon surface, a system of thrust faults, unique in the solar system, transects the surface viewed by *Mariner 10*. The largest structure viewed by *Mariner 10* is the 800-mi-diameter (1300-km) Caloris impact basin. Infrared temperature measurements from *Mariner 10* indicate that the surface is a good thermal insulator and, therefore, must be covered with porous soil or rock powder like the lunar regolith. This is expected on a planet whose surface is shattered and stirred by meteorite impacts.

Photomosaic of Mercury as seen by the outgoing *Mariner 10* spacecraft in March 1974. The terminator, the boundary between the lighted and unlighted halves of the planet, runs vertically along the left side of the photograph. The Caloris Basin is on the terminator, slightly above the middle. It is surrounded and filled by younger smooth plains. (*Jet Propulsion Laboratory*)

A new mission to Mercury called *MESSENGER* (MErcury Surface, Space ENvironment, GEochemistry, and Ranging), launched in 2004, is scheduled to orbit the planet in 2011 after two flybys in 2008 and 2009. This mission should answer most of the questions raised by the *Mariner 10* mission and Earth-based observations. The flyby on January 14, 2008 imaged about 21% of Mercury's surface at close range for the first time, and observations were made of Mercury's exosphere, magnetosphere, and surface. *See* PLANET; SPACE PROBE. Robert G. Strom

Mercury-vapor lamp A vapor or gaseous discharge lamp in which the arc discharge takes place in mercury vapor. This lamp is widely used for roadway and all other forms of illumination, and as a source of ultraviolet radiation for industrial applications.

The arc discharge takes place in a transparent tube of fused silica, or quartz, and this quartz tube is usually mounted inside a larger bulb of glass (see illustration). The outer bulb reduces the ultraviolet radiation of the inner arc tube, encloses and protects the mount structure, and can be coated with phosphors that greatly improve the color of the light emitted.

The arc tube has electrode assemblies mechanically sealed into each end, At one end of the arc tube, a smaller starting

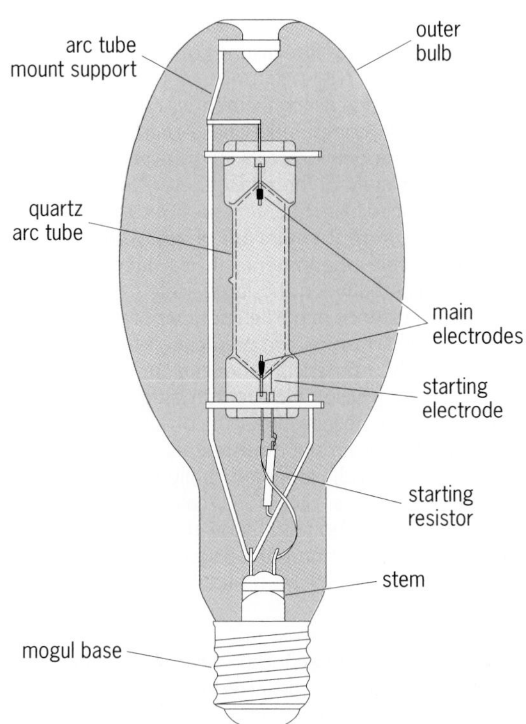

High-pressure mercury-vapor lamp.

electrode is located close to the main electrode. The starting electrode is connected through a high resistance to the opposite electric polarity of the adjacent main electrode. The arc tube itself is filled with argon gas and a small amount of pure mercury before being sealed.

Radiation from the mercury arc is confined to four specific wavelengths in the visible portion of the spectrum and several strong lines in the ultraviolet. The visible radiation from clear mercury lamps provides light with a distinct blue-green appearance and poor color rendition. Red objects, for example, look brown or black, and human skin looks unattractive. For this reason, most mercury lamps have a phosphor color-correcting coating, and clear bulbs are used only where appearance of colors is secondary to some gain in efficiency. *See* ILLUMINATION; LAMP; ULTRAVIOLET LAMP; VAPOR LAMP. T. F. Neubecker

Meridian A line of longitude on Earth or in the sky, perpendicular to the Equator, and extending between the North and South poles.

A meridian of longitude on Earth marks a system of location and timekeeping whose zero was set at Greenwich, England, in the nineteenth century, when it was necessary to regularize time signals in the era of railway transportation. *See* LATITUDE AND LONGITUDE; TIME.

Meridians of celestial longitude are the extensions into space of meridians of longitude on Earth. Each celestial meridian is the half of a great circle that goes between the north and south celestial poles and crosses the celestial equator perpendicularly. The unique line known as the celestial meridian is the great circle going from north pole to south pole through a given observer's zenith, the point directly overhead. It thus marks the north-south line, and celestial objects going through it are said to be at their meridian transits or at their culminations. *See* ASTRONOMICAL COORDINATE SYSTEMS. Jay M. Pasachoff

Meromictic lake A lake whose water is permanently stratified and therefore does not circulate completely throughout the basin at any time during the year. Normally lakes in the Temperate Zone mix completely during the spring and autumn when water temperatures are approximately the same from top to bottom. In meromictic (*mero*, partial; *mixis*, circulation) lakes, there are no periods of overturn or complete mixing because seasonal changes in the thermal gradient either are small or are overridden by the stability of a chemical gradient, or the deeper waters are physically inaccessible to the mixing energy of the wind. Commonly in meromictic lakes, the vertical stratification in density is stabilized by a chemical gradient.

The upper stratum of water in a meromictic lake is mixed by the wind and is called the mixolimnion. The bottom, denser stratum, which does not mix with the water above, is referred to as the monimolimnion. The transition layer between these strata is called the chemocline.

Of the hundreds of thousands of lakes on the Earth, less than 200 have been reported to be meromictic and less than 50 meromictic lakes are reported for North America. Nevertheless, meromictic lakes are widely dispersed across all continents and on islands such as Cuba, the Galápagos and Tasmania. Zige Tangco at 4560 m (15,000 ft) above sea level in the central Tibetan Plateau is the highest known meromictic lake in the world. In North America, meromictic lakes generally are restricted to sheltered basins with proportionally very small surface areas in relation to depth, basins in arid regions, and isolated basins in fiords. Meromixis may occur in lakes with a permanent ice cover. Meromictic lakes frequently contain colored water, which limits penetration of solar radiation. *See* LAKE; LIMNOLOGY.

Gene E. Likens

Merostomata A class of the phylum Arthropoda, subphylum Chelicerata. Merostomes are aquatic chelicerates, characterized by abdominal appendages bearing respiratory organs. Most merostomes are extinct; only the horseshoe crabs, comprising four species and three genera (*Carcinoscorpio* and *Tachypleus* of eastern Asia and *Limulus polyphemus* of eastern North America), survive.

The body of a merostome consists of a prosoma, or head, which lacks antennae, has a pair of compound eyes and a pair of simple median eyes, and bears the chelicerae (pincers) and five pairs of uniramous walking legs with gnathobases for mastication. The opisthosoma, or trunk, consists of 12 or fewer segments which may be freely articulating or partly or entirely fused into a solid shield; the opisthosoma bears the respiratory appendages. The telson (tail) is a solid, usually spikelike, structure. *See* CHELICERATA; LIVING FOSSILS. Niles Eldredge

Mesogastropoda The largest and most diverse order of gastropod mollusks in the subclass Prosobranchia. Mesogastropods are all single-gilled and thus monotocardiac, that is, with only one auricle and asymmetry of other cardiac, renal, and genital structures. Mesogastropods with a wide variety of marine lifestyles are found burrowing in soft muds and living exposed to air near high-tide level. There are also estuarine, freshwater, and terrestrial species in the order. The anatomical asymmetry makes possible a functional separation of genital from renal ducts, which in turn allows the development of internal fertilization, of large eggs, of ovoviviparity, and of true viviparity (all features advantageous for the evolution of nonmarine stocks).

Mesogastropods are structurally diverse, and most classifications divide them into 13 superfamilies, 4 of which number several thousand species and encompass some of the world's most abundant, ubiquitous, and cosmopolitan marine snails. *See* GASTROPODA; PROSOBRANCHIA. W. D. Russell-Hunter

Meso-ionic compound A member of a class of five-membered ring heterocycles (and their benzo derivatives) which possess a sextet of π electrons in association with the atoms composing the ring but which cannot be represented satisfactorily by any one covalent or polar structure.

Two main types, depending formally upon the origin of the electrons in the π system, have been identified; they are exemplified by compounds (**1**) and (**2**). In structure (**1**) the nitrogen and oxygen atoms, 1,3 to each other, are shown as donating two electrons each to the total of eight electrons in the whole π system, whereas in structure (**2**) the two middle nitrogen atoms, 1,2 to each other, are the two-electron donors.

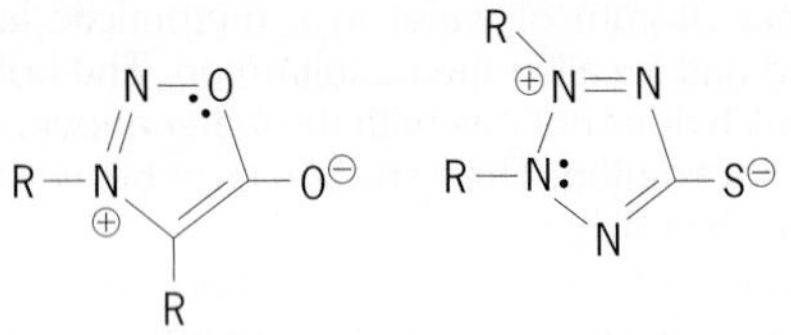

(1) **(2)**

The term "satisfactorily" in the definition refers to the fact that the charge in the ring cannot be associated exclusively with one ring atom. Thus, these compounds are in sharp contrast with other dipolar structures, such as ylides, and such compounds are not considered meso-ionic.

There has been considerable interest in the pharmacological activity of meso-ionic compounds, and derivatives have shown a variety of antibiotic, anthelminthic, antidepressant, and anti- inflammatory properties. *See* HETEROCYCLIC COMPOUNDS.

Jeremiah P. Freeman

Mesometeorology That portion of meteorology comprising the knowledge of intermediate-scale atmospheric phenomena, that is, in the size range of approximately 1–1200 mi (2–2000 km) and with time periods typically, but not always, less than 1 day. Unlike the larger weather systems on synoptic scales (the scales resolved by current weather reporting station networks) which typically produce significant changes over periods of days, most mesoscale phenomena have interdiurnal periods (less than 1 day), and consequently their changes are often more startling. In addition to time and space criteria for defining mesoscale, dynamical considerations can be used.

For observing mesoscale phenomena over midlatitude land masses, the average spacing between atmospheric sounding stations is about 180–360 mi (300–600 km) and soundings are taken twice each day. Consequently, only the largest mesoscale phenomena, with wavelengths greater than about 600 mi (1000 km), are routinely observed (resolved) by this network. Information with higher time-and-space resolution is available from aircraft observations and networks of radar stations, profilers, and satellite imagery. The profiler is a ground-based hybrid observing system of vertically pointing radar and microwave radiometry. The remote-sensing platforms often show that mesoscale weather systems are distinct components of larger synoptic-scale cyclones (low-pressure systems) and anticyclones (high-pressure systems).

Although the satellite imagery and radar data provide extensive areal coverage and clearly reveal the presence of the mesoscale systems, they do not provide measurements of certain atmospheric parameters (such as temperature, moisture, and pressure) in a form in which mesoscale structures and circulations can be readily quantified and understood. Thus, while mesoscale phenomena can be "observed," they cannot be studied and predicted (in the conventional manner) as easily as synoptic-scale systems. *See* METEOROLOGICAL SATELLITES; RADAR METEOROLOGY; WEATHER FORECASTING AND PREDICTION.

Mesoanalysis is the analysis of meteorological data in a manner that reveals the presence and characteristics of mesoscale phenomena. Because sounding stations are so widely spaced, only the largest of mesoscale systems can be resolved by the free-air (sounding) data. On the other hand, the density of surface observing stations is often satisfactory for identifying mesoscale features or circulations. Probably the most common application of mesoanalysis is for forecasting convective (thunderstorm) weather systems. *See* THUNDERSTORM.

Modern weather-forecasting techniques rely heavily upon predictions made from computers. These predictions, commonly called numerical model forecasts, require as input the three-dimensional initial state of the atmosphere. Normally, this initial condition is produced from the previous forecast and the most recent observations from the network of atmospheric soundings. For many mesoscale phenomena, it has been possible to develop relationships between the large-scale (synoptic) environment and the occurrence of particular types of mesoscale events. By using these relationships, the prediction of the synoptic-scale environment by the numerical models is then used to infer the likelihood of specific mesoscale events. Using satellite, radar, and conventional surface observations, the onset of an event is readily detected and appropriate adjustments to local forecasts are implemented. These adjustments usually come in the form of very short-term forecasts, commonly called nowcasts, and typically are valid for only about 3 h. However, depending upon the particular mesoscale phenomena, longer-term (3–12 h) forecasts sometimes are possible. *See* METEOROLOGY; NOWCASTING; STORM DETECTION.

J. Michael Fritsch

Meson Strongly interacting particles are either baryons or mesons. Baryons, such as the proton and neutron, have half-integer spins and are fermions, whereas mesons are bosons with integer spin. The lightest meson is the pion, denoted π, which was originally predicted to exist as the carrier of the powerful force that grips the protons and neutrons within atomic nuclei. *See* BARYON; QUANTUM STATISTICS; SPIN (QUANTUM MECHANICS); STRONG NUCLEAR INTERACTIONS.

Like baryons, mesons are very small (roughly 10^{-15} m in diameter) and consist of even smaller elementary particles: quarks, antiquarks, and gluons. Baryons are fundamental particles of matter, made of quarks; antibaryons are their antimatter analogs and are made of antiquarks. Mesons are neither matter nor antimatter: they are made of quarks and a balancing number of antiquarks (in the simplest case a single quark and a single antiquark). Over 200 examples are known of such *q*–*q*-bar combinations (*q* being the generic symbol for a quark; and *q*-bar, denoted here by $\bar{q}$, for an antiquark). As is the case for baryons, the quarks and antiquarks are bound by the exchange particles of the strong interaction, the gluons. Examples have begun to emerge of mesons that appear to require two quarks and two antiquarks, or even no quarks and antiquarks at all. The former are known as tetraquarks or molecules; and the latter as glueballs, consisting, at least in theory, of just gluons. *See* ANTIMATTER; GLUONS; QUARKS.

Whereas the lightest baryon, the proton, is stable against decay into other particles, all mesons are unstable. Heavy mesons can decay rapidly (in about 10^{-23} s) into lighter mesons so long as the flavor and other properties, such as parity and angular momentum or spin, are preserved. When the lightest meson of a particular flavor is formed, it is nonetheless unstable since its quark and antiquark can mutually annihilate, for example into two photons, an electron and a positron or neutrino, or even gluons. For example, the lightest meson, the pion, occurs either with no electric charge (π^0) or with negative or positive electric charge (π^- or π^+). When the quark and antiquark within the π^0 annihilate, the products are probably two photons; when the π^- or π^+ decay, they leave a neutrino and an electron or positron, or more often a neutrino and a muon (a heavier analog of the electron and positron). The π^0 survives for only 10^{-16} s—small indeed but over a million times longer than the 10^{-23} s of its heavier cousins. The π^- or π^+ survive for 10^{-8} s. These differences in lifetimes are the result of the forces at work. For heavy mesons, the strong force destroys them; for the π^0, it is the electromagnetic force; and for the electrically charged pions, it is the weak force that is responsible. The large differences in lifetimes reflect the differing strengths of the forces. *See* FUNDAMENTAL INTERACTIONS; LEPTON; NEUTRINO.

There are six different kinds, or flavors, of quarks, in three families. The lightest family contains the up (u) and down (d) quarks, and charges of $+^2/_3$ and $-^1/_3$ in units of the electron charge (e), respectively. The corresponding antiquarks, labeled $\bar{u}$ or $\bar{d}$, have the opposite sign of charge to their quark counterparts. Thus, $\bar{u}$ has charge $-^2/_3$ and $\bar{d}$ has $+^1/_3$. *See* FLAVOR.

A $u\bar{d}$ has charge $+1$, and the lightest example is the π^+. The $d\bar{u}$ has charge -1: the π^-. Both the $u\bar{u}$ and $d\bar{d}$ have charge zero. Quantum theory implies that the particular quantum combination $u\bar{u} - d\bar{d}$ corresponds to the π^0; the alternative combination $u\bar{u} + d\bar{d}$ appears in another meson, the electrically neutral η (eta).

Because the difference between the masses of the up and down quarks and the electromagnetic self energy of quarks or antiquarks in a meson are very small compared to meson masses, properties of mesons are to a very good approximation unchanged by replacing an up quark or antiquark with a down quark or antiquark. This equivalence is referred to as isospin symmetry. In this language, the π^+, π^-, or π^0 are referred to collectively as different charge states of the pion. *See* I-SPIN.

The three varieties of pion correspondingly have masses that differ by only some 4%, the charged pions having a mass of 140 MeV/c^2 (where c is the speed of light) and the π^0 mass being 135 MeV/c^2. The η, however, is some four times more massive at 550 MeV/c^2. In part, this is because the η contains (along with $u\bar{u} + d\bar{d}$) a third flavor of quark, the strange quark, accompanied by its antiquark: thus $s\bar{s}$. The strange quark is some 120 MeV/c^2 heavier than an up or down quark, and so an $s\bar{s}$ has added mass relative to pions, which contain only u or d flavors. Indeed, the additional quark flavor leads to three electrically neutral mesons, π^0, η, and η', which testify to the three possibilities, $u\bar{u}$, $d\bar{d}$, and $s\bar{s}$; the η' mass is 960 MeV/c^2.

It is possible for any of the u, d, or s quarks to be attracted to any of the $\bar{u}$, $\bar{d}$, or $\bar{s}$ antiquarks. Mesons that have an s or $\bar{s}$ partnered by a ($\bar{u}$, $\bar{d}$) or (u, d) will have strangeness. Such "strange" mesons are known as kaons and denoted generically by the symbol K. The kaon, containing a light flavor and a strange, has a mass of around 495 MeV/c^2, which is heavier than a pion but lighter than η or η'. *See* STRANGE PARTICLES.

The remaining three flavors of quarks—the charm c, bottom b, and top t quarks—are heavy on the scale of the mass of the pion, K, or η, and η' mesons. These give rise to further families of mesons known as charmonium and bottomonium (a $c\bar{c}$ or $b\bar{b}$ combination) and also mesons that contain unpaired charm or bottom flavors, such as the $D(c\bar{u})$ with mass 1865 MeV/c^2 and $B(b\bar{u})$ at 5279 MeV/c^2. The top quark is very massive, causing it to decay so rapidly that it does not have time to form a meson by binding strongly to other quarks or to an antiquark. Thus toponium is unlikely to occur unless there are other forces at work which are yet unknown. *See* CHARM.

Charmonium mesons with masses greater than 3730 MeV/c^2 decay by the strong interaction into pairs of charmed mesons, $D(c\bar{u})$ and $\bar{D}(\bar{c}u)$. However, there are several examples of charmonium mesons that have masses below this limit and so are prevented by energy conservation from decaying this way. Their dominant decays are by shedding energy in the form of photons or pions, eventually ending up as the etacharm (η_c) or J-psi (J/Ψ), the lightest $c\bar{c}$ mesons (the η_c having $c\bar{c}$ spin totaling 0 and the J/Ψ having spin 1).

The J/Ψ is metastable on the time scale of strong interactions. It decays (after about 10^{-20} s) by its c and $\bar{c}$ mutually annihilating, producing photons or gluons; the gluons do not escape free but rapidly create showers of pions or other particles. *See* J/PSI PARTICLE.

Since the 1980s, a large amount of knowledge about mesons has resulted from experiments where electrons collide with and are annihilated by positrons. If the energy of the collision is the same as that of a vector meson, the vector can be produced directly. This is how the J/Ψ—the vector meson of the charmonium family, $c\bar{c}$, with a mass of 3095 MeV/c^2—was discovered in 1974. The upsilon (Υ), the analogous vector meson of bottomonium, $b\bar{b}$, can also be made this way. *See* UPSILON PARTICLES.

If the total energy of the electron-positron collisions is around 10 GeV, the result is a pair of B mesons, where a b flavor is accompanied by up, down, or strange such that the meson has an overall bottom flavor. The decays of bottom mesons and their antimatter counterparts are expected to give important clues about the origin of the asymmetry between matter and antimatter. *See* ANTIMATTER; ELEMENTARY PARTICLE.

Frank Close

Mesosauria An order of extinct aquatic reptiles, also known as Proganosauria, of latest Carboniferous or earliest Permian time, about 280,000,000 years ago. The best-known genus is *Mesosaurus*. Like many aquatic reptiles, mesosaurs have a very long snout. The teeth are numerous and long, and appear very delicate. They may have served for filter feeding on soft-bodied invertebrates.

Mesosaurs, an early offshoot of the Carboniferous "stem reptiles" or captorhinomorphs, were the first reptiles to invade marine waters, but apparently soon became extinct, leaving no descendants. *See* REPTILIA.

R. L. Carroll

Mesoscopic physics A subdiscipline of condensed-matter physics that focuses on the properties of solids in a size range intermediate between bulk matter and individual atoms or molecules. The size scale of interest is determined by the appearance of novel physical phenomena absent in bulk solids and has no rigid definition; however, the systems studied are normally in the range of 100 nanometers (the size of a typical virus) to 1000 nm (the size of a typical bacterium). Other branches of science, such as chemistry and molecular biology, also deal with objects in this size range, but mesoscopic physics has dealt primarily with artificial structures of metal or semiconducting material which have been fabricated by the techniques employed for producing microelectronic circuits. Thus mesoscopic physics has a close connection to the fields of nanofabrication and nanotechnology. Three categories of new phenomena in such systems are interference effects, quantum size effects, and charging effects. *See* ARTIFICIALLY LAYERED STRUCTURES; NANOSTRUCTURE; QUANTIZED ELECTRONIC STRUCTURE (QUEST); SEMICONDUCTOR HETEROSTRUCTURES.

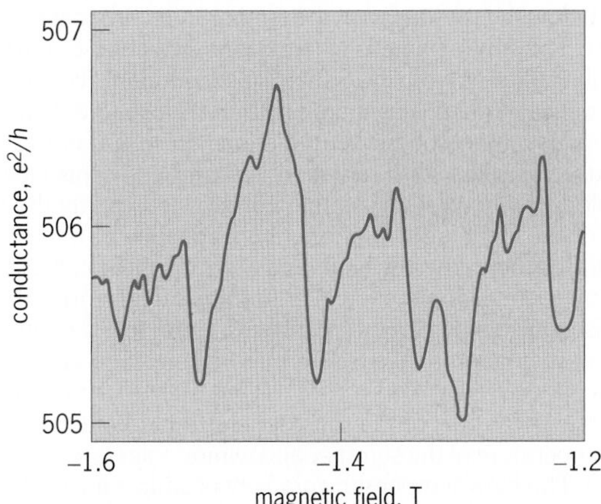

Conductance of 2000-nm gold wire as a function of magnetic field measured at a temperature of 0.04 K. The pattern observed is reproducible over a period of days. (*After R. A. Webb and S. Washbum, Quantum interference fluctuations in disordered materials, Phys. Today, 41(12):46–55, December 1988*)

Interference effects. In the mesoscopic regime, scattering from defects induces interference effects which modulate the flow of electrons. The experimental signature of mesoscopic interference effects is the appearance of reproducible fluctuations in physical quantities. For example, the conductance of a given specimen oscillates in an apparently random manner as a function of experimental parameters (see illustration). However, the same pattern may be retraced if the experimental parameters are cycled back to their original values; in fact, the patterns observed are reproducible over a period of days.

Quantum size effects. Another prediction of quantum mechanics is that electrons confined to a particular region of space may exist only in a certain set of allowed energy levels. The spacing between these levels increases as the confining region becomes smaller. One striking phenomenon which arises from these quantum size effects is the steplike increase of the conductance of electrons flowing through a constriction of several hundred nanometers' width. *See* ENERGY LEVEL (QUANTUM MECHANICS).

Another mesoscopic system that shows quantum size effects consists of isolated islands of electrons that may be formed at the appropriately patterned interface between two different semiconducting materials. The electrons typically are confined to disk-shaped regions termed quantum dots. The confinement of the electrons in these systems changes their interaction with electromagnetic radiation significantly.

Charging effects. Isolated mesoscopic solids such as quantum dots or metallic grains on an insulating substrate also show novel effects associated with the discreteness of the charge on the electron. Devices known as single-electron transistors (SETs) are by far the most sensitive electrometers (instruments for measuring electrical charge) presently known. *See* ELECTROMETER; TRANSISTOR.

A. Douglas Stone

Mesosphere A layer within the Earth's atmosphere that extends from about 50 to 85 km (31 to 53 mi) above the surface. The mesosphere is predominantly characterized by its thermal structure. On average, mesospheric temperature decreases with increasing height.

Temperatures range from as high as 12°C (53°F) at the bottom of the mesosphere to as low as −133°C (−208°F) at its top. The top of the mesosphere, called the mesopause, is the coldest area of the Earth's atmosphere. Temperature increases with increasing altitude above the mesopause in the layer known as the thermosphere, which absorbs the Sun's extreme ultraviolet radiation. In the stratosphere, the atmospheric layer immediately below the mesosphere, the temperature also increases with height. The stratosphere is where ozone, which also absorbs ultraviolet radiation from the Sun, is most abundant. The transition zone between the mesosphere and the stratosphere is called the stratopause. Mesospheric temperatures are comparatively cold because very little solar radiation is absorbed in this layer. Meteorologists who predict weather conditions or study the lowest level of the Earth's atmosphere, the troposphere, often refer to the stratosphere, mesosphere, and thermosphere collectively as the upper atmosphere. However, scientists who study these layers distinguish between them; they also refer to the stratosphere and mesosphere as the middle atmosphere. *See* ATMOSPHERE; METEOROLOGY; STRATOSPHERE; THERMOSPHERE; TROPOSPHERE.

In the lower part of the mesosphere, the difference between the temperature at the summer and winter poles is of order 35°C (63°F). This large temperature gradient produces the north-south or meridional winds that blow from summer to winter. Temperatures in the upper mesosphere are colder in summer and warmer in winter, resulting in return meridional flow from the summer to the winter hemisphere. Although the temperature gradient in the upper part of the mesosphere remains large, additional complications result in wind speeds that are much slower than they are in the lower part of the mesosphere. Winds in the east-west or zonal direction are greatest at mesospheric middle latitudes. Zonal winds blow toward the west in summer and toward the east in winter. Like their meridional counterparts, zonal winds are comparatively strong near the bottom of the mesosphere and comparatively weak near the top. Thus, on average both temperature and wind speed decrease with increasing height in the mesosphere. *See* ATMOSPHERIC GENERAL CIRCULATION.

Meteors which enter the Earth's atmosphere vaporize in the upper mesosphere. These meteors contain significant amounts of metallic atoms and molecules which may ionize. Metallic ions combined with ionized water clusters make up a large part of the D-region ionosphere that is embedded in the upper mesosphere. *See* IONOSPHERE.

The upper mesosphere is also where iridescent blue clouds can be seen with the naked eye and photographed in twilight at high summer latitudes when the Sun lights them up in the otherwise darkening sky. These clouds are called noctilucent clouds (NLC). Noctilucent clouds are believed to be tiny ice crystals that grow on bits of meteoric dust.

Large-scale atmospheric circulation patterns transport tropospheric air containing methane and carbon dioxide from the lower atmosphere into the middle atmosphere. While carbon dioxide warms the lower atmosphere, it cools the middle and upper atmosphere by releasing heat to space. Methane breaks down and contributes to water formation when it reaches the middle atmosphere. If the air is sufficiently cold, the water can freeze and form noctilucent clouds. Temperatures must be below −129°C (−200°F) for noctilucent clouds to form. These conditions are common in the cold summer mesopause region at high latitudes.

Maura Hagan

Mesozoa A division of the animal kingdom sometimes ranked as intermediate between the Protozoa and the Metazoa. These animals are unassignable to any of the better-known phyla, as usually defined. In the absence of proof concerning their relationships, and in view of the disagreement among zoologists relative to their affinities and even with respect to the facts and interpretation of their structure and life cycle, they are treated as a small phylum somewhere between Protozoa and Platyhelminthes. No particular phylogenetic interpretation should be attached to this placement.

The Mesozoa comprise two orders of small, wormlike organisms, the Dicyemida and the Orthonectida. Both are parasitic in marine invertebrates. The body consists of a single layer of ciliated cells enclosing one or more reproductive cells. These body cells are rather constant in number and arrangement for any given species. The internal cells do not correspond to the entoderm of other animals, as they have no digestive function. The life cycles are complex, involving both sexual and asexual generations (metagenesis).

Bayard H. McConnaughey

Mesozoic The middle era of the three major divisions of the Phanerozoic Eon (Paleozoic, Mesozoic, and Cenozoic eras) of geologic time, encompassing an interval from 251 to 65 million years ago (Ma) based on various isotopic-age dates. The Mesozoic Era is known also as the Age of the Dinosaurs and the interval of middle life. The Mesozoic Erathem (the largest recognized time-stratigraphic unit) encompasses all sedimentary rocks, body and trace fossils of organisms preserved, metamorphic rocks, and intrusive and extrusive igneous rocks formed during the Mesozoic Era. *See* GEOCHRONOMETRY.

The Mesozoic Era records dramatic changes in the geologic and biologic history of the Earth. At the beginning of the Mesozoic Era, all the continents were amassed into one large supercontinent, Pangaea. Both the marine and continental biotas were impoverished from the mass extinction that marked the boundary between the Permian and Triassic periods, and the end of the Paleozoic Era. This mass extinction was responsible for the loss of over 90% of the species on Earth. During the Mesozoic Era, many significant events were recorded in the geologic and fossil record of the Earth, including the breakup of Pangaea and the evolution of modern ocean basins by continental drift, the rise of the dinosaurs, the ascension of the angiosperms (flowering plants), the diversification of the insects and crustaceans, and the appearance of the mammals and birds. The end of the Mesozoic Era is marked by a major mass extinction at the Cretaceous-Tertiary boundary that records several meteorite impacts, the extinction of the dinosaurs, the rise to dominance of the mammals, and the beginning of the Cenozoic Era and the life-forms dominant today. *See* CONTINENTAL DRIFT; PLATE TECTONICS.

The Mesozoic Era comprises three periods of geologic time: the Triassic Period (251–200 Ma), the Jurassic Period (200–146 Ma), and the Cretaceous Period (146–65 Ma). These periods are each subdivided into epochs, formal designations of geologic time described as Early, Middle, and Late (except for the Cretaceous, which has no middle epoch designated yet). The packages of rock themselves are subdivided into series designated Lower, Middle, and Upper (except for Cretaceous). Each epoch is subdivided into ages. Likewise, each series is subdivided into stages, which are time-stratigraphic units whose boundaries are based on unconformities, hiatuses, or erosional surfaces, on correlations to a type section (where rocks are first described), or preferably on changes in the biota that depict true measurable time (for example, evolutionary changes). *See* CRETACEONS; JURRASSIC; UNCONFORMITY.

S. T. Hasiotis; R. F. Dubiel

Messier Catalog An early listing of nebulae and star clusters. Although Charles Messier (1730–1817) was primarily a comet hunter, he is remembered for his catalog of objects. As the first serious and systematic comet hunter, Messier learned about the hazards of being misled by the galaxies, nebulae, and star clusters that look like comets. *See* COMET; CRAB NEBULA; GALAXY, EXTERNAL; NEBULA; PLANETARY NEBULA; STAR CLUSTERS.

Messier published three versions of his catalog. Ultimately he listed 103 objects. Today, with the addition of objects that he observed and noted, Messier's catalog lists 110 clusters, nebulae, and galaxies.

As Messier was completing his small catalog, William Herschel was building a far larger one, with thousands of objects seen through his bigger telescopes. That catalog was transformed by John Dreyer into the still more massive, 7000-object *New General Catalogue*, which today is the standard list of distant objects in the sky. *See* ASTRONOMICAL CATALOG.

Although there are other catalogs, Messier's effort remains the most famous. It contains beautiful objects that are spread all over the sky as seen from Messier's home, and consequently can be seen from any location in the Northern Hemisphere and from most in the Southern.

David H. Levy

Metabolic disorders Disorders of metabolism principally involve an imbalance in nucleic acids, proteins, lipids, or carbohydrates. They are usually associated with either a deficiency or excess resulting in an imbalance in a particular metabolic pathway. All metabolic disorders have a genetic background, and some of them are expressed as specific genetic diseases. Other factors affecting metabolism include internal control mechanisms that are superimposed on the genetic background. One of the most important mechanisms is the hormonal control system, which consists of the endocrine, paracrine, and autocrine systems. The second control system that has a significant effect on metabolism is the neural control system. The third control system is the immune control system, which relates to both the endocrine and neural systems. Genetic background, environmental factors, and the three major control mechanisms, in conjunction with age and sex, bring about profound changes in metabolism, which ultimately result in structural and functional alterations. *See* ENDOCRINE MECHANISMS; IMMUNOLOGY; NERVOUS SYSTEM (VERTEBRATE).

Nucleic acids. Abnormalities of nucleic acid metabolism are associated with several diseases, including gout and lupus erythematosus. All genetic disease implies a defect in nucleic acids, and although some genetic diseases are classified as protein abnormalities there is always an inherent defect in the nucleotide. This is either a deficiency, an excess, or a mutation that results in an abnormal protein being formed. Similarly, although lipid storage disease results in an abnormal metabolism of lipids, it is the result of a deficiency of a particular enzyme, which also means a defect in the particular nucleic acid code. In some carbohydrate genetic storage diseases, there is a deficiency of a particular enzyme, which again results from a nucleotide defect. Certain congenital defects that result in malformations of organ systems are the result of either germ cell or somatic cell deficiencies involving differentiation genes composed of nucleic acid.

The genetic disease severe combined immunodeficiency syndrome (SCIDS) results in a deficiency of B lymphocytes, which produce antibody, and T lymphocytes, which are responsible for graft and tumor rejection as a result of their cytotoxic effect. This abnormality is associated with a deficiency of the enzyme adenosine deaminase. *See* ARTHRITIS; CONNECTIVE TISSUE; DISEASE; GOUT; HUMAN GENETICS; NUCLEIC ACID.

Proteins. The diseases associated with protein abnormalities include those associated with increased production of proteins, decreased production of proteins, production of abnormal proteins, and excretion of unusual amounts of amino acids.

Often called macroglobulinemia, hyperproteinemia results in an increase in beta or gamma globulins, but possibly with less total protein. Hyperglobulinemia diseases include multiple myeloma, kala-azar, Hodgkin's disease, lymphogranuloma inguinale, sarcoidosis, cirrhosis, and amyloid disease. They usually involve stem cell lines of the bone marrow macrophages or B cells. *See* CIRRHOSIS; HODGKIN'S DISEASE.

A decrease in the amount of protein (hypoproteinemia) can result from a lack of amino acids for protein synthesis, a metabolic block, or other interference with normal protein synthesis. Increased excretion of protein, particularly in chronic renal disease with a loss of albumin in the urine (albuminuria), is another common cause of hypoproteinemia. Kwashiorkor is the best example of hypoproteinemia resulting from dietary deficiency. In hypogammaglobulinemia and agammaglobulinemia, which may also be classified under hypoproteinemia, the total serum albumin and globulin are not markedly depressed, but the gamma globulins may fall from a normal of 15–20% of total protein to 0.4%.

Many diseases are characterized by abnormal proteins, including multiple myeloma, the hemoglobinopathies, and the various amyloid disturbances. Multiple myeloma, a neoplastic growth of plasma cells, particularly in the bone marrow and lymph nodes, is representative of a disease in which an abnormal protein is produced. Another large group of abnormal proteins includes the hemoglobinopathies. Hemoglobin functions in the transport of oxygen in the blood. Scores of different hemoglobins have been identified, but sickle cell in the United States and thalassemia (Mediterranean anemia) in Europe are two of the most common hemoglobin abnormalities. A third type of abnormal protein is found in amyloid disease, of which there are many variations. *See* AMYLOIDOSIS; HEMOGLOBIN; SICKLE CELL DISEASE.

A fourth group with abnormal protein metabolism is associated with a change in particular amino acids resulting either from an overflow mechanism, where the concentration of amino acids in the serum surpasses the renal threshold of the glomerular membrane, or from defective absorption of amino acids in renal tubules. Tyrosine appears to be one of the most critical amino acids, and its metabolism is related to four key diseases including phenylketonuria, hypothyroidism, albinism, and alkaptonuria. The liver plays a major role in the deamination of amino acids.

Advanced hepatitis and cirrhosis may lead to increased levels of amino acids in the blood and excretion in the urine. Other diseases with amino aciduria that are believed to be the result of defective kidney function include cystinuria (the failure to reabsorb cystine, lysine, arginine, and ornithine), Wilson's disease (a degeneration involving copper metabolism in the liver and brain), Fanconi's syndrome, galactosemia, scurvy, rickets, and lead, cresol, or benzene poisoning. *See* KIDNEY DISORDERS; LIVER DISORDERS; PHENYLKETONURIA; PROTEIN METABOLISM; SICKLE CELL DISEASE.

Lipids. Although lipid stores remain a secondary energy reserve in starvation, the breakdown of lipids associated with diabetes and starvation results in the production of ketone bodies in the urine with general acidosis (ketosis) in the serum. Of concern to many people is an increase in lipids associated with obesity. *See* OBESITY.

Hyperlipemia, an excess of lipid in the blood, is often secondary to uncontrollable diabetes, hypothyroidism, biliary cirrhosis, and lipoid nephrosis. Excess proliferation of fat cells is known as a lipoma, which occasionally becomes malignant, producing a liposarcoma. *See* ARTERIOSCLEROSIS.

In a large group of genetic lipid storage diseases, lipid accumulates because of a disturbance in lipid metabolism that is independent of external stimuli. These genetic diseases are inherent nucleic acid defects that result in abnormal enzymatic proteins, which then result in abnormal lipid metabolism. In these diseases, a large accumulation of lipid appears in many cells, but particularly the reticuloendothelial cells of the lymph nodes, liver, spleen, and bone marrow. Abnormal lipid storage occurs in Niemann-Pick, Gaucher's, and Tay-Sachs diseases.

The heterozygous form of the genetic disease familial hypercholesteremia results in cholesterol levels two to four times normal, and is characterized by arteriosclerotic lesions of the coronary vessels that cause myocardial infarct and death in the 35–50-year age range. In the homozygous condition, with cholesterol levels eight to ten times normal, death occurs usually before the age of 15. The defect resides in the receptor for low-density lipoprotein (LDL). *See* CHOLESTEROL; HEART DISORDERS; LIPID METABOLISM.

Carbohydrates. Abnormal carbohydrate diseases include the genetic diseases that represent a deficiency in nucleotide and eventually protein enzymatic activity. Of the six common carbohydrate storage diseases, two examples are von Gierke's disease, marked by glycogen storage in the heart, and Pompes' disease, in which the carbohydrate is stored in the liver. Variants of carbohydrate disease involve storage of mucopolysaccharides, as in Hurler's disease, and storage of galactose, as in galactosemia.

The most important disease associated with carbohydrate metabolism is diabetes mellitus. *See* CARBOHYDRATE METABOLISM; DIABETES; METABOLISM. Donald W. King

Metabolism All the physical and chemical processes by which living, organized substance is produced and maintained and the transformations by which energy is made available for use by an organism.

In defining metabolism, it is customary to distinguish between energy metabolism and intermediary metabolism, although the two are, in fact, inseparable. Energy metabolism is primarily concerned with overall heat production in an organism, while intermediary metabolism deals with chemical reactions within cells and tissues. In general, the term metabolism is interpreted to mean intermediary metabolism. *See* ENERGY METABOLISM.

Metabolism thus includes all biochemical processes within cells and tissues which are concerned with their building up, breaking down, and functioning. The synthesis and maintenance of tissue structure generally involves the union of smaller into larger molecules. This part of metabolism, the building of tissues, is termed anabolism. The process of breaking down tissue, of splitting larger protoplasmic molecules into smaller ones, is termed catabolism. Growth or weight gain occurs when anabolism exceeds catabolism. On the other hand, weight loss results if catabolism proceeds more rapidly than anabolism, as in periods of starvation, serious injury, or disease. When the two processes are balanced, tissue mass remains the same.

The metabolism of the three major foodstuffs, carbohydrates, fats, and proteins, is intimately interrelated, so any clearcut division of the three is arbitrary and inaccurate. Thus the metabolism of protoplasm is concerned with all three of these foodstuffs. The metabolic pathways of carbohydrates, fats, and proteins cross at many points; thus certain pathways of metabolism are shared in common by fragments of these different classes of foodstuffs.

Some of the metabolic processes of the protoplasm of both plant and animal cells occur along common pathways; carbohydrate metabolism in plants is similar in many details to carbohydrate metabolism in animals. Therefore the study of metabolism in any organism is, in a sense, the study of metabolism in all protoplasm. *See* CARBOHYDRATE METABOLISM; LIPID METABOLISM; PROTEIN METABOLISM. Mary B. McCann

Metadata Data that describe or define all objects that make up a data source. That is, metadata is data about the data.

Widely used within the context of database management systems (DBMS), the term metadata refers to the complete description of all database structures and their constraints (conditions that need to be satisfied by the data stored in the database). This metadata is stored in the data dictionary or catalog. The data dictionary is an integral part of any DBMS environment and can be considered a database in its own right (albeit a system database instead of a user database). For this reason, a DBMS is sometimes referred to as a self-describing collection of integrated data since it holds both the data and the data description. A comprehensive data dictionary contains metadata about all database objects, including connections, users, users' privileges, schemas (external, conceptual, and so on), mappings between the different schemes, and all types of security and integrity constraints. *See* DATABASE MANAGEMENT SYSTEM. Ramon A. Mata-Toledo

Metal An electropositive chemical element. Physically, a metal atom in the ground state contains a partially filled band with an empty state close to an occupied state. Chemically, upon going into solution a metal atom releases an electron to become a positive ion. Consequently in biotic systems metal atoms function prominently in ionic transport and electron exchange. In bulk a metal has a high melting point and a correspondingly high boiling temperature; except for mercury, metals are solid at standard conditions. Direct observation shows a metal to be relatively dense, malleable, ductile, cohesive, highly conductive both electrically and thermally, and lustrous. When crystals of the elements are classified along a scale from plastic to brittle, metals fall toward the plastic end. Furthermore, molten metals mixed with each other over wide ranges of proportions form, upon slowly cooling, homogeneous close-packed crystals. In contrast, a metal mixed with a nonmetal completely combines into a homogeneous crystal only in one or a few discrete stoichiometric proportions.

For detailed discussions of metals, in particular, exceptions to generic behavior, see separate articles on each metal. *See* ELEMENTS; PERIODIC TABLE. Irwin G. Greenfield

Metal, mechanical properties of Commonly measured properties of metals (such as tensile strength, hardness, fracture toughness, creep, and fatigue strength) associated with the way that metals behave when subjected to various states of stress. The properties are discussed independently of theories of elasticity and plasticity, which refer to the distribution of stress and strain throughout a body subjected to external forces.

Stress states. Stress is the internal resistance, per unit area, of a body subjected to external forces. The forces may be distributed over the surface of a body (surface forces) or may be distributed over the volume (body forces); examples of body forces are gravity, magnetic forces, and centrifugal forces. Forces are generally not uniformly distributed over any cross section of the body upon which they act; a complete description of the state of stress at a point requires the magnitudes and directions of the force intensities on each surface of a vanishingly small body surrounding the point. All forces acting on a point may be resolved into components normal and parallel to faces of the body surrounding the point. When force intensity vectors act perpendicular to the surface of the reference body, they are described as normal stresses. When the force intensity vectors are parallel to the surface, they describe a state of shear stress. Normal stresses are positive, when they act to extend a line (tension). Shear stresses always occur in equal pairs of opposite signs.

A complete description of the state of stress requires knowledge of magnitudes and directions of only three normal stresses, known as principal stresses, acting on reference faces at right angles to each other and constituting the bounding faces of a reference parallelepiped. Three such mutually perpendicular planes may always be found in a body acted upon by both normal and shear forces; along these planes there is no shear stress, but on other planes either shear or shear and tensile forces will exist.

The shear stress is a maximum on a plane bisecting the right angle between the principal planes on which act the largest and smallest (algebraic) principal stresses. The largest normal stress in the body is equal to the greatest principal stress. The magnitude and orientation of the maximum shear stress determine the direction and can control the rate of the inelastic shear processes, such as slip or twinning, which occur in metals. Shear stresses also play a role in crack nucleation and propagation, but the magnitude and direction of the maximum normal stress more often control fracture processes in metals capable only of limited plastic deformation.

It is often useful to characterize stress or strain states under boundary conditions of either plane stress (stresses applied only in the plane of a thin sheet) or plane strain (stresses applied to relatively thick bodies under conditions of zero transverse strain). These two extreme conditions illustrate that strains can occur in the absense of stress in that direction, and vice versa.

Tension and torsion. In simple tension, two of the three principal stresses are reduced to zero, so that there is only one principal stress, and the maximum shear stress in numerically half the maximum normal stress. Because of the symmetry in simple tension, every plane at 45° to the tensile axis is subjected to the maximum shear stress. For other kinds of loading, the relationship between the maximum shear stress and the principal stresses are obtained using the same method, with the results depending upon the loading condition.

For example, in simple torsion, the maximum principal stress is inclined 45° to the axis of the bar being twisted. The least principal stress (algebraically) is perpendicular to this, at 45° to the bar axis, but equal to and opposite in sign to the first principal stress—that is, it is compressive. Both of these are in a plane perpendicular to the radial direction, the direction of the intermediate principal stress, which in this case has the magnitude zero. Every free external surface of a body is a principal plane on which the principal stress is zero. In torsion, the maximum shear stress occurs on all planes perpendicular to and parallel with the axis of the twisted bar. But because the principal stresses are equal but of opposite sign, the maximum shear stress is numerically equal to the maximum normal stress, instead of to half of it, as in simple tension. This means that in torsion one may expect more ductility (the capacity to deform before fracture) than in tension. Materials that are brittle (exhibiting little capacity for plastic deformation before rupture) in tension may be ductile in torsion. This is because in tension the critical normal stress for fracture may be reached before the critical shear stress for plastic deformation is reached; in torsion, because the maximum shear stress is equal to the maximum normal stress instead of half of it as in tension, the critical shear stress for plastic deformation is reached before the critical maximum normal stress for fracture.

Tension test. To achieve uniformity of distribution of stress and strain in a tension test requires that the specimen be subjected to no bending moment. This is usually accomplished by providing flexible connections at each end through which the force is applied. The specimen is stretched at a controllable rate, and the force required to deform it is observed with an appropriate dynamometer. The strain is measured by observing the extension between gage marks adequately remote from the ends, or by measuring the diameter and calculating the change in length by using the constancy of volume that characterizes plastic deformation. Diameter measurements are applicable even after necking-down has begun. The elastic properties are seldom determined since these are structure-insensitive.

Yield strength. The elastic limit is rarely determined. Metals are seldom if ever ideally elastic, and the value obtained for the elastic limit depends on the sensitivity of strain measurement. The proportional limit, describing the limit of applicability of Hooke's law of linear dependence of stress on strain, is similarly difficult to determine. Modern practice is to determine the stress required to produce a prescribed inelastic strain, which is called the yield strength.

Tensile strength. Tensile strength, usually called the ultimate tensile strength, is calculated by dividing the maximum load by the original cross-sectional area of the specimen. It is, therefore, not the maximum value of the true tensile stress, which increases continuously to fracture and which is always higher than the nominal tensile stress because the area continuously diminishes. For ductile materials the maximum load, upon which the tensile strength is based, is the load at which necking-down begins. Beyond this point, the true tensile stress continues to increase, but the force on the specimen diminishes. This is because the rate of strain hardening has fallen to a value less than the rate at which the stress is increasing because of the diminution of area.

Yield point. A considerable number of alloys, including those of iron, molybdenum, tungsten, cadmium, zinc, and copper, exhibit a sharp transition between elastic and plastic flow. The stress at which this occurs is known as the upper yield point. A sharp drop in load to the lower yield point accompanies yielding, followed, in ideal circumstances, by a flat region of yield elongation; subsequently, normal strain hardening is observed (see illustration).

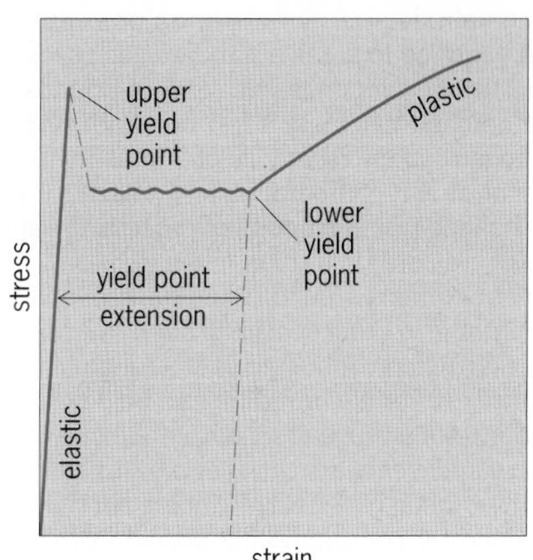

Yield point, mild steel.

Elongation. The tensile test provides a measure of ductility, by which is meant the capacity to deform by extension. The elongation to the point of necking-down is called the uniform strain or elongation because, until that point on the stress-strain curve, the elongation is uniformly distributed along the gage length. The strain to fracture or total elongation includes the extension accompanying local necking. Since the necking extension is a fixed amount, independent of gage length, it is obvious that the total elongation will depend upon the gage length, and will be greater for short gage lengths and less for long gage lengths.

Ductile-to-brittle transition. Many metals and alloys, including iron, zinc, molybdenum, tungsten, chromium, and structural steels, exhibit a transition temperature, below which the metal is brittle and above which it is ductile. The transition temperature very clearly is sensitive to alloy content, but it will vary even for the same material, depending upon such external test conditions as stress state and strain rate, and microstructural variables such as purity and grain size. The ductility transition frequently is accompanied by a change in the mechanism of fracture (as in iron and steels or zinc), but this need not be so.

Notch tensile test. Notch sensitivity in metals cannot be detected by the ordinary tension test on smooth bar specimens. Either a notched sample may be used in a tension test or a notched-bar impact test may be conducted. Notches produce triaxial stresses under the notch root as tensile forces are applied, thereby decreasing the ratio of shear stress to normal stress and increasing the likelihood of fracture. Materials are evaluated by a quantity, notch strength, which is the analog of the ultimate tensile strength in an ordinary tensile test. The notch strength is defined as the maximum load divided by the original cross-sectional area at the notch root.

Compression test. Very brittle metals, or metals used in products which are formed by compressive loading (rolling, forging), often are tested in compression to obtain yield strength or yield point information. Compression test specimens are generally in the form of solid circular cylinders. The ratio of specimen length to diameter is critical in that high ratios increase the likelihood of buckling during a test, thereby invalidating the test results. Proper specimen alignment is important for the same reason. In addition, care must be taken to lubricate specimen ends to avoid spurious effects from friction between the specimen ends and the testing machine. In the case of a metal which fails in compression by a shattering fracture (for example, cast iron), a quantity known as the compressive strength may be reproducibly obtained by dividing the maximum load carried by the specimen by its cross-sectional area. For materials which do not fail in compression by shattering, the compressive strength is arbitrarily defined as the maximum load at or prior to a specified compressive deformation.

Notched-bar impact test. Notched-bar impact tests are conducted to estimate the resistance to fracture of structures which may contain defects. The common procedure is to measure the work required to break a standardized specimen, and to express the results in work units, such as foot-pounds or newton-meters. The notched-bar impact test does not provide design information regarding the resistance of a material to crack propagation. Rather, it is a comparative test, useful for preliminary screening of materials or evaluation of processing variables. The notch behavior indicated in a single test applies only to the specimen size, notch geometry, and test conditions involved and is not generally applicable to other specimen sizes and conditions. The test is most useful when conducted over a range of temperatures so that the ductile-to-brittle transition can be determined.

Notched-bar tests are usually made in either a simple beam (Charpy) or a cantilever beam (Izod) apparatus, in both of which the specimen is broken by a freely swinging pendulum; the work done is obtained by comparing the position of the pendulum before it is released with the position to which it swings after striking and breaking the specimen. In the Izod test, the specimen is held in a vise, with the notch at the level of the top of the vise, and broken as a cantilever beam in bending with the notch on the tension side. In the Charpy test, the specimen is laid loosely on a support in the path of the pendulum and broken as a beam loaded at three points; the tup (striking edge) strikes the middle of the specimen, with the notch opposite the tup, that is, on the tension side. Both tests give substantially the same result with the same specimen unless the material is very ductile, a situation in which there is little interest.

Hardness testing. When the only information that is needed is the comparison of the resistance to deformation of a particular sample or lot with a standard material, indentation hardness tests are used. They are relatively inexpensive and fast. They tell nothing about ductility and little about the relationship between stress and strain, for in making the indent the stress and strain are nonuniformly distributed.

In all hardness tests, a standardized load is applied to a standardized indenter, and the dimensions of the indent are measured. This applies to such methods as scratch hardness testing, in which a loaded diamond is dragged across a surface to produce, by plastic deformation, a furrow whose width is measured, and the scleroscope hardness test, in which an indent is produced by dropping a mass with a spherical tup onto a surface. The dimensions of the indent are proportional to the work done in producing it, and the ratio of the height of rebound to the height from which the tup was dropped serves as an indirect measure of the hardness.

Fatigue. Fatigue is a process involving cumulative damage to a material from repeated stress (or strain) applications (cycles), none of which exceed the ultimate tensile strength. The number of cycles required to produce failure decreases as the stress or strain level per cycle is increased. The fatigue strength or fatigue limit is defined as the stress amplitude which will cause failure in a specified number of cycles. For a few metals, notably steels and titanium alloys, an endurance limit exists, below which it is not possible to produce fatigue failures no matter how often stresses are applied.

Creep and stress rupture. Time-dependent deformation under constant load or stress is measured in a creep test. Creep tests are those in which the deformation is recorded with time, while stress rupture tests involve the measurement of time for fracture to occur. Closely related are stress relaxation tests, in which the decay of load with time is noted for a body under a fixed state of strain. Test durations vary from seconds or minutes to tens of thousand of hours. Appreciable deformation occurs in structural materials only at elevated temperatures, while pure metals may creep at temperatures well below room temperature.

Since creep deformation and rupture time are temperature- and stress-dependent, it is usually necessary to test a material at several stresses and temperatures in order to establish the creep or stress-rupture properties in adequate detail. *See* METAL; METALLURGY.

N. S. Stoloff

Metal-base fuel A fuel containing a metal of high heat of combustion as a principal constituent. High propellent performance in either a rocket or an air-breathing engine is obtained when the heat of combustion of the fuel is high. Chemically, high heats of combustion are attained by the oxidation of the low-atomic-weight metals in the upper left-hand corner of the periodic table. The generally preferred candidates are lithium, beryllium, boron, carbon, magnesium, and aluminum.

The metallized additive can be used in either a liquid or solid propellent. When the pure metal is added to liquid fuels, an emulsifying or gelling agent is employed which maintains the particles in uniform suspension. When used in composite solid rocket propellents, the metal powder is usually mixed with the oxidizer and unpolymerized fuel, and the propellant is then processed in the usual way.

Early compounding of metallized propellents employed the free metal itself. As a result of extensive research in metalloorganic compounds, however, several classes of metallic compounds have been employed. Two major reasons for the use of such compounds are that the solubility of the metal in the fuel can be realized, resulting in a homogeneous propellant, and performance higher than that for the pure metal can be obtained in some cases.

The major classes of metallic compounds of interest as high-performance propellents include the hydrides, amides, and hydrocarbons. Additional classes include mixtures either of two metals or of two chemical groups, such as an amine hydrocarbon.

Most metallic fuels are costly and many, because of particle-size requirements or synthesis in a specific compound, are in limited supply. The combustion gases all produce smoky exhausts which may be objectionable in use. Engine development problems are also increased because of the appearance of smoke and deposits in the engine. David Altman

Metal carbonyl A complex of a transition metal combined with carbon monoxide (CO). In a metal carbonyl, the CO groups form sigma bonds to the metal through lone pairs of electrons on carbon; the metal, in turn, donates electrons to antibonding pi orbitals on CO. In so doing, the metal usually attains the electronic configuration of an inert gas (the 18-electron rule). Thus elemental chromium (Cr^0), with six valence electrons, combines with six molecules of CO, each donating two electrons, to afford chromium hexacarbonyl [$Cr(CO)_6$], in which Cr has 18 valence electrons and is isoelectronic to krypton (Kr). *See* CHEMICAL BONDING; COORDINATION COMPLEXES; VALENCE.

Metal carbonyls usually are electrically neutral and nonpolar, exhibiting many of the physical properties of organic compounds. Typically they are volatile solids or liquids, soluble in common organic solvents but insoluble in water. They are highly toxic.

Anionic and cationic species also are known. Metal carbonyls also may contain one (mononuclear) or more (polynuclear) metal atoms. Polynuclear metal carbonyls may exhibit covalent metal-metal bonds, and in them CO also may function as bridging or capping ligands, forming bonds to two or three metal atoms, respectively. Mononuclear metal carbonyls tend to exhibit geometries expected on the basis of minimization of interligand repulsions (valence shell electron-pair repulsion model); their polynuclear analogs often may be envisioned structurally as arising from the fusion of mononuclear polyhedra. Thus nickel tetracarbonyl [$Ni(CO)_4$], iron pentacarbonyl [$Fe(CO)_5$], and chromium hexacarbonyl [$Cr(CO)_6$] exhibit tetrahedral, trigonal bipyramidal, and octahedral geometries, respectively [structures (**1**)–(**3**)]. The polynuclear manganese carbonyl

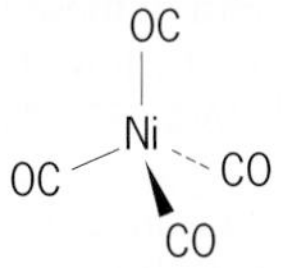

(1) 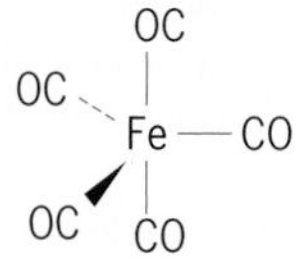 (2) 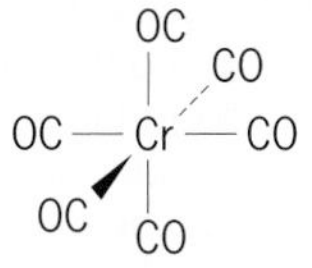 (3)

[$Mn_2(CO)_{10}$] has a structure obtained by joining two octahedra at a vertex [structure (**4**)], and the structure of diiron enneacarbonyl [$Fe_2(CO)_9$] is viewed as two octahedra sharing three bridging carbonyls at a face [structure (**5**)].

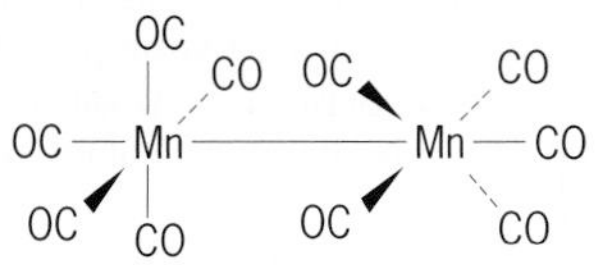

(4) 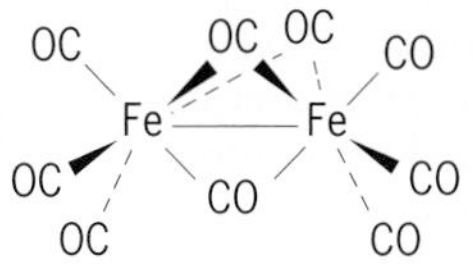 (5)

Metal carbonyls react in solution under relatively mild conditions to form a variety of substitution products, with Lewis bases such as amines, phosphines, sulfides, and nitric oxide and with organic molecules such as alkenes, alkynes, arenes, and other aromatic compounds. Reactions of metal carbonyls can also afford derivatives containing alkyl, acyl, allyl, halide, hydride, carbene, carbyne, and other substituents. They thus are the most important precursors to the synthesis of organotransition metal complexes. Their thermal decomposition can afford pure metals and CO. Gerard R. Dobson

Metal casting A metal-forming process whereby molten metal is poured into a cavity or mold and, when cooled, solidifies and takes on the characteristic shape of the mold. Casting offers several advantages over other methods of metal forming: it is adaptable to intricate shapes, to extremely large pieces, and to mass production; it can provide parts with uniform physical and mechanical properties throughout; and depending on the particular material being cast, the design of the part, and the quantity being produced, it can be more economical.

The two broad categories of metal-casting processes are ingot casting and casting to shape. Ingot castings are produced by pouring molten metal into a permanent or reusable mold. Following solidification, the ingots (bars, slabs, or billets) are processed mechanically into many new shapes. Casting to shape involves pouring molten metal into molds in which the cavity provides the final useful shape, followed by heat treatment and machining or welding, depending upon the specific application.

While design factors are important for producing sound castings with proper dimensions, factors such as the pouring temperature, alloy content, mode of solidification, gas evolution, and segregation of alloying elements control the final structure of the casting and therefore its mechanical and physical properties. Typically, pouring temperatures are selected within 100–300°F (60–170°C) of an alloy's melting point. Exceedingly high pouring temperatures can result in excessive mold metal reactions, producing numerous casting defects.

Almost all metals and alloys used by engineering specialists have at some point been in the molten state and cast. Metallurgists have in general lumped these materials into ferrous and nonferrous categories. Ferrous alloys, cast irons and steels, constitute the largest tonnage of cast metals. Aluminum-, copper-, zinc-, titanium-, cobalt-, and nickel-base alloys are also cast into many forms, but in much smaller quantity than cast iron and steel. Selection of a given material for a certain application will depend upon the physical and chemical properties desired, as well as cost, appearance, and other special requirements. *See* METAL, MECHANICAL PROPERTIES OF; METAL FORMING. Karl Rundman

Metal cluster compound A compound in which two or more metal atoms are bonded to one another. Metal cluster compounds bridge the gap between the solid-state chemistry of the metals—or their lower-valent oxides, chalcogenides, and related salts—and the complexes of the metals in which each metal ion is completely surrounded by and bonded to a set of ligands or ions. The latter group comprises the classical coordination chemistry of metal ions. *See* COORDINATION CHEMISTRY; COORDINATION COMPLEXES; SOLID-STATE CHEMISTRY.

Interest in metal cluster compounds arises from unique features of their chemistry. (1) Cluster compounds provide models for studying fundamental reactions on surfaces. (2) There is a hope that cluster compounds may provide entry to new classes of catalysts that may be tailored to specific syntheses and may thus be more selective than existing processes. (3) The nature of the bonding in cluster compounds is an area wherein experiment and theory are continuously challenging each other. (4) The systematic synthesis of mixed metal clusters may provide for the development of new types of supported catalysts (the discrete clusters are deposited on supports such as alumina, silica, or zeolites). *See* ATOM CLUSTER; NANOSTRUCTURE.

Malcolm H. Chisholm

Metal coatings Thin films of material bonded to metals in order to add specific surface properties, such as corrosion or oxidation resistance, color, attractive appearance, wear resistance, optical properties, electrical resistance, or thermal protection. This article discusses various methods of applying either metallic coatings or nonmetallic coatings, such as vitreous enamel and ceramics, and the conversion of surfaces to suitable reaction-product coatings. For other methods for the protection of metal surfaces *see* CLADDING; ELECTROLESS PLATING; ELECTROPLATING OF METALS; LACQUER; PAINT.

Hot-dipped coatings of low-melting metals provide inexpensive protection to the surfaces of a variety of steel articles. Thoroughly cleaned work is immersed in a molten bath of the coating metal. The coating consists of a thin alloy layer together with relatively pure coating metal that adheres to the work as it is withdrawn from the bath.

Sprayed coating permits the coating of assembled steel structures to obtain corrosion resistance, the building up of worn machine parts for rejuvenation, and the application of highly refractory coatings with melting points in excess of 3000°F (1650°C).

Cementation coatings are surface alloys formed by diffusion of the coating metal into the base metal, producing little dimensional change. Parts are heated in contact with powdered coating material that diffuses into the surface to form an alloy coating, whose thickness depends on the time and the temperature of treatment.

In vapor deposition a thin specular coating is formed on metals, plastics, paper, glass, and even fabrics. Coatings form by condensation of metal vapor originating from molten metal, from high-voltage discharge between electrodes (cathode sputtering), or from chemical means such as hydrogen reduction or thermal decomposition (gas plating) of metal halides.

Immersion coatings are produced either by direct chemical displacement or for thicker coatings by chemical reduction (electroless coating). Metal ions plate out of solution onto the workpiece.

Vitreous enamel coatings are glassy but noncrystalline coatings for attractive durable service in chemical, atmospheric, or moderately high-temperature environments. In wet enameling, a slip is prepared of a water suspension of crushed glass, flux, suspending agent, refractory compound, and coloring agents or opacifiers. The slip is applied by dipping or flow coating; it is then fired at a temperature at which it fuses into a continuous vitreous coating. Dry enameling is used for castings, such as bathtubs. The casting is heated to a high temperature, and then dry enamel powder is sprinkled over the surface, where it fuses. *See* FRIT.

Essentially crystalline, ceramic coatings are used for high-temperature protection above 1100°C (2000°F). The coatings may be formed by spraying refractory materials such as aluminum oxide or zirconium oxide, or by the cementation processes for coatings of intermetallic compounds such as molybdenum disilicide. *See* CERMET.

Surface-conversion coatings provide an insulating barrier of low solubility formed on steel, zinc, aluminum, or magnesium without electric current. The article to be coated is either immersed in or sprayed with an aqueous solution, which converts the surface into a phosphate, an oxide, or a chromate.

Anodic coatings of protective oxide may be formed on aluminum or magnesium by making them the anode in an electrolytic cell. If permanent color is required, the coating is impregnated with a dye before sealing. *See* CORROSION.

William W. Bradley

Powder coating is a process whereby organic polymers such as acrylic, polyester, and epoxies are applied to substrates for protection and beautification. It is essentially an industrial painting process which uses a powdered (25–50-μm particle size) resin rather than the solvent solution. The powders are applied to electrically grounded substrates, usually by means of an electrostatic spray gun. The powder particles are attracted to and adhere to the substrate until it can be transported to an oven, where the powder particles melt, coalesce, flow, and form a smooth coating. Outdoor lawn and patio furniture coated in this process display good weathering and abuse resistance. Powder-coated electrical transformers are insulated electrically and provided with corrosion protection. Powder coatings have also been developed for finishing major appliances and for automotive coatings. *See* SURFACE COATING.

William W. Bradley

Metal forming Manufacturing processes by which parts or components are fabricated from metal stock. In the specific technical sense, metal forming involves changing the shape of a piece of metal. In general terms, however, it may be classified roughly into five categories: mechanical working, such as forging, extrusion, rolling, drawing, and various sheet-forming processes; casting; powder and fiber metal forming; electroforming; and joining processes. *See* DRAWING OF METAL; EXTRUSION; FORGING; METAL CASTING; METAL ROLLING; POWDER METALLURGY; SHEET-METAL FORMING.

Serope Kalpakjian

Metal halide lamp A high-pressure discharge lamp that is enclosed in a quartz envelope containing metal halides, usually iodides, and produces high-efficacy white light. These lamps are widely used for sports stadiums, roadways, commercial interiors, and industrial applications. The singular lamp feature is the compact geometry and high efficacy of nearly white light. *See* LUMINOUS EFFICACY.

The metal halide lamp requires a high voltage in order to start. Once the arc discharge is ignited, the internal vapor pressure begins to increase until it reaches a preset value, usually around 5 atm (500 kilopascals). These lamps are always connected to an auxiliary power supply called the ballast which supplies the proper voltage and current for starting and operating the lamps. The light output gradually increases over approximately 2 min as the various ingredients begin to vaporize and emit light. The light from the arc discharge comes from the metal components of the iodide compounds, which are typically a mixture of sodium, thallium, indium, scandium, dysprosium, and occasionally tin iodide. A combination of these metals produces a pleasing white light of very high efficiency, between 80 and 120 lumens per watt or even higher. *See* ARC DISCHARGE.

Special ingredients can be used inside these lamps to make them suitable for plant growth, photocopying, and scientific usages. Certain types of compact metal halide lamps are ideal for use with movie and slide projection. In these cases, the ingredients chosen (usually indium) are used because they produce a small, intense high-brightness arc. The spectra of these lamps are nearly continuous and approach the spectra found in an incandescent lamp. *See* CINEMATOGRAPHY; INCANDESCENT LAMP; OPTICAL PROJECTION SYSTEMS.

In cases where a more diffuse light source is desired, the outer jacket of the lamp can be coated with a white phosphor in order both to diffuse the light and to change the color of the output and in some cases actually improve the efficiency.

Gilbert H. Reiling

Metal hydrides A compound in which hydrogen is bonded chemically to a metal or metalloid element. The compounds are classified generally as ionic, transition metal, and covalent hydrides. Covalent hydrides are of two subtypes, binary and complex. Certain hydrides have achieved a position of modest industrial importance, but most are of theoretical interest only.

Under extreme conditions such as in electric discharges, many metals form volatile, short-lived transient hydrides of the general

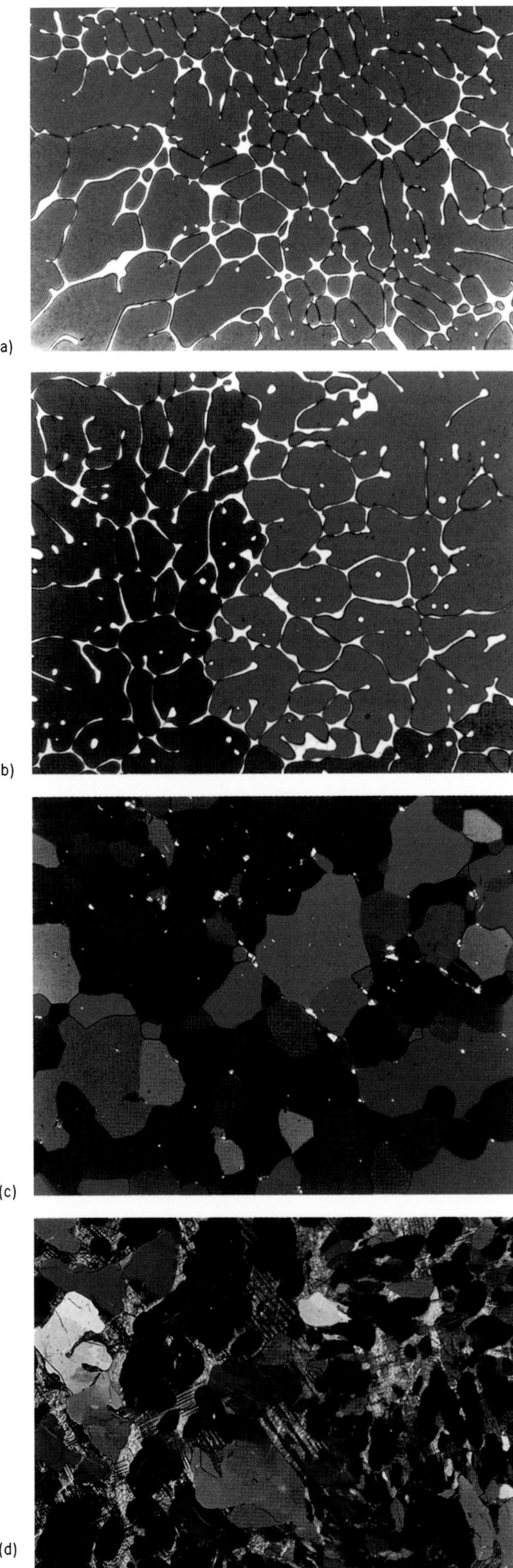

Microstructures of uranium-carbon alloys. (*a*) As-cast $UC_{0.8}$. (*b*) As-cast $UC_{0.8}$ at a different magnification. (*c*) As-cast and heat-treated $UC_{1.5}$, annealed for 60 h at 3128°F (1720°C). (*d*) As-cast $UC_{1.6}$. (*Oak Ridge National Laboratory*)

formula type MH. Although some of these can be prepared experimentally, most are observed only by their spectra. They are important in studying molecular bonding. The action of atomic hydrogen at low temperatures forms surface films of unstable hydrides with many metals. *See* HYDRIDO COMPLEXES; HYDROGEN.

James C. Warf

Metal matrix composite A material in which a continuous metallic phase (the matrix) is combined with another phase (the reinforcement) that constitutes a few percent to around 50% of the material's total volume. In the strictest sense, metal matrix composite materials are not produced by conventional alloying. This feature differentiates most metal matrix composites from many other multiphase metallic materials, such as pearlitic steels or hypereutectic aluminum-silicon alloys. *See* ALLOY.

The particular benefits exhibited by metal matrix composites, such as lower density, increased specific strength and stiffness, increased high-temperature performance limits, and improved wear-abrasion resistance, are dependent on the properties of the matrix alloy and of the reinforcing phase. The selection of the matrix is empirically based, using readily available alloys; and the major consideration is the nature of the reinforcing phase.

A large variety of metal matrix composite materials exist. The reinforcing phase can be fibrous, platelike, or equiaxed (having equal dimensions in all directions); and its size can also vary widely, from about 0.1 to more than 100 micrometers. Matrices based on most engineering metals have been explored, including aluminum, magnesium, zinc, copper, titanium, nickel, cobalt, iron, and various aluminides. This wide variety of systems has led to an equally wide spectrum of properties for these materials and of processing methods used for their fabrication. Reinforcements used in metal matrix composites fall in five categories: continuous fibers, short fibers, whiskers, equiaxed particles, and interconnected networks.

Composite properties depend first and foremost on the nature of the composite; however, certain detailed microstructural features of the composite can exert a significant influence on its behavior. Physical properties of the metal, which can be significantly altered by addition of a reinforcement, are chiefly dependent on the reinforcement distribution. A good example is aluminum-silicon carbide composites, for which the presence of the ceramic increases substantially the elastic modulus of the metal without greatly affecting its density. However, the level of improvement depends on the shape and alignment of the silicon carbide. Also, it depends on the processing of the reinforcement: for the same reinforcement shape (continuous fibers), microcrystalline polycarbosilane-derived silicon carbide fibers yield much lower improvements than do crystalline β-silicon carbide fibers. Other properties, such as the strength of metal matrix composites, depend in a much more complex manner on composite microstructure. The strength of a fiber-reinforced composite, for example, is determined by fracture processes, themselves governed by a combination of microstructural phenomena and features. These include plastic deformation of the matrix, the presence of brittle phases in the matrix, the strength of the interface, the distribution of flaws in the reinforcement, and the distribution of the reinforcement within the composite. Consequently, predicting the strength of the composite from that of its constituent phases is generally difficult. *See* BRITTLENESS; PLASTIC DEFORMATION OF METAL.

The combined attributes of metal matrix composites, together with the costs of fabrication, vary widely with the nature of the material, the processing methods, and the quality of the product. In engineering, the type of composite used and its application vary significantly, as do the attributes that drive the choice of metal matrix composites in design. For example, high specific modulus, low cost, and high weldability of extruded aluminum oxide particle-reinforced aluminum are the properties desirable for bicycle frames. High wear resistance, low weight, low cost, improved high-temperature properties, and the possibility for incorporation in a larger part of unreinforced aluminum are the considerations for design of diesel engine pistons. *See* COMPOSITE MATERIAL; HIGH-TEMPERATURE MATERIALS.

Mel M. Schwartz

Metal rolling Reducing or changing the cross-sectional area of a workpiece by the compressive forces exerted by rotating rolls. The original material fed into the rolls is usually an ingot from a foundry. The largest product in hot rolling is called a bloom; by successive hot- and then cold-rolling operations the bloom is reduced to a billet, slab, plate, sheet, strip, and foil, in decreasing order of thickness and size. The initial breakdown of the ingot by rolling changes the coarse-grained, brittle, and porous structure into a wrought structure with greater ductility and finer grain size.

A schematic presentation of the rolling process, in which the thickness of the metal is reduced as it passes through the rolls, is shown in illustration *a*. The speed at which the metal moves during rolling changes, as shown in illustration *b*, to keep the volume rate of flow constant throughout the roll gap. Hence, as the thickness decreases, the velocity increases; however, the surface speed of a point on the roll is constant, and there is therefore relative sliding between the roll and the strip. The normal pressure distribution on the roll and hence on the strip is of the form shown in illustration *c*. Because of its particular shape this pressure distribution is known as the friction hill.

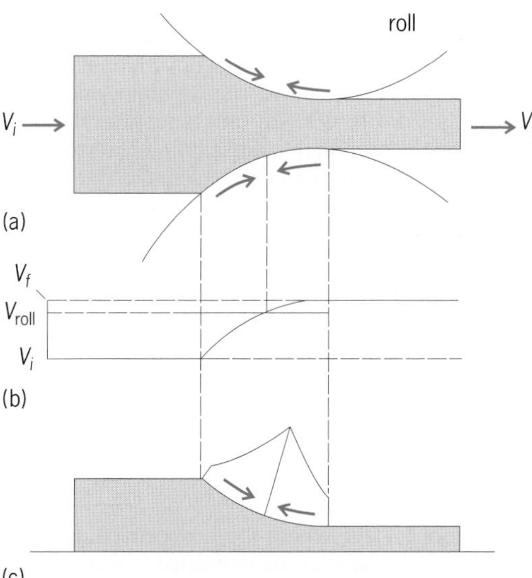

The rolling process, (*a*) Direction of friction forces in the roll gap. (*b*) Velocity distribution, (*c*) Normal pressure acting on the strip in the roll gap. V_i = initial velocity, V_f = final velocity, V_{roll} = velocity during rolling operation.

A great variety of roll arrangements and equipment are used in rolling. The proper reduction per pass in rolling depends on the type of material and other factors; for soft, nonferrous metals, reductions are usually high, while for high-strength alloys they are small. Requirements for roll materials are mainly strength and resistance to wear. Common roll materials are cast iron, cast steel, and forged steel.

Serope Kalpakjian

Metallic glasses Metals and metallic alloys having an amorphous structure on the atomic scale. Although the word "glass" is commonly used to refer to the familiar transparent oxide glasses (such as silicate glasses used to make windowpanes), in a more general sense a glass is any solid obtained from a liquid that does not crystallize upon cooling. As a result,

Mechanical properties of some amorphous alloys, together with properties of common crystalline engineering alloys

	Density, g/cm^3 (lb/in.^{3})	Young's modulus, GPa (10^6 psi)	Poisson's ratio	Yield strength, MPa (10^3 psi)	Tensile strength, MPa (10^3 psi)	Percent elongation	Fracture toughness, MPa · m$^{1/2}$ (ksi · in.$^{1/2}$)
			Amorphous alloys				
$Zr_{41}Ti_{14}Cu_{12.5}Ni_{10}Be_{22.5}$	6.12 (0.221)	101 (14.6)	0.34	1900 (275)	1900 (275)	2*	25 (23)
$Mg_{65}Cu_{25}Tb_{10}$	3.98 (0.144)	48 (7.0)	0.31	700 (100)	700 (100)	1.5*	2 (1.8)
$Pt_{57.5}Cu_{14.7}Ni_{5.3}P_{22.5}$	15.0 (0.541)	95 (14)	0.42	1400 (200)	1470 (213)	1.5*	90 (83)
			Crystalline alloys				
Steel (4340, quenched and tempered)	7.85 (0.283)	207 (30)	0.30	1620 (235)	1760 (255)	12	50 (46)
Titanium (Ti-6Al-4V, solutionized and aged)	4.43 (0.160)	114 (16.5)	0.34	1100 (160)	1170 (170)	10	55 (50)
Aluminum (7075-T6)	2.80 (0.101)	71 (10)	0.33	505 (73)	572 (83)	11	24 (22)

*Elastic deformation; under tensile loading, most metallic glasses fail without measurable plastic deformation.

a glass retains an atomic-scale structure in which the atoms are more or less randomly arranged, similar to that of the liquid state. Because most metals crystallize quickly, special care must be taken with processing and alloy design to produce a metallic glass. Metallic glasses produced directly by quenching the liquid are quite similar in terms of structure and properties to amorphous alloys (of the same chemical composition) produced by other means, such as physical vapor deposition, irradiation, mechanical alloying, or solid-state reaction. *See* AMORPHOUS SOLID; CRYSTAL; GLASS; GLASS TRANSITION.

Metallic glasses are also notable for their lack of structure over length scales longer than the atomic. Conventional crystalline alloys have microstructural features such as grains and precipitates with typical sizes of 0.1–100 μm; these features make fabrication of very small components difficult and lead to anisotropic mechanical properties that can be problematic. Because metallic glasses have no structural features on these length scales, they should be well suited for highly miniaturized devices, such as for micro-electro-mechanical systems (MEMS). The lack of microstructure also means that metallic glasses have low ability to suppress vibrations (damping capacity). *See* MICRO-ELECTRO-MECHANICAL SYSTEMS (MEMS).

At low stresses and at temperatures well below the glass transition, metallic glasses display isotropic linear elastic behavior in which the strain ε, Eq. (1), is directly proportional to the stress

$$\varepsilon = \frac{\sigma}{E} \tag{1}$$

σ through Hooke's law, where the proportionality constant E (Young's modulus) describes the resistance to elastic (reversible) deformation, or stiffness. The disordered atomic structure of a metallic glass enables slight rearrangements of the atomic positions in response to an applied load, which would not be possible if the material were crystalline. This has the effect of reducing the stiffness slightly, so that E for metallic glasses is typically 20–30% lower than that for a crystalline alloy of the same composition. Distortion of the atomic bonds requires energy, and the elastic strain energy (per unit volume of material), given by Eq. (2),

$$U_{\text{el}} = \frac{1}{2}\sigma\varepsilon = \frac{\sigma^2}{2E} \tag{2}$$

is called the resilience of the material. The high yield stress of metallic glasses, together with low modulus, means that metallic glasses can sustain large elastic strains of 2–3% (compared with 0.1–0.2% for a typical crystalline metal) and have exceptionally large values of resilience (see table). *See* HOOKE'S LAW; STRESS AND STRAIN; YOUNG'S MODULUS.

At elevated temperatures, approaching the glass transition temperature, the strength of a metallic glass drops and the deformation becomes homogeneous instead of being concentrated into shear bands. If heated above the glass transition, a metallic alloy becomes a fluid, the viscosity of which can be controlled over a wide range of temperature. Because the stresses required to cause deformation are low, either of these states can be useful for processing metallic glasses into complex shapes. *See* METAL; MECHANICAL PROPERTIES OF.

Like crystalline metals, metallic glasses are electrically and thermally conductive, due to the presence of conduction electrons. However, the disordered atomic structure and the high alloy content mean that metallic glasses are not especially good conductors, with an electrical resistivity on the order of $10^{-6}\,\Omega \cdot$m, as compared to about $10^{-8}\,\Omega \cdot$ m for copper at room temperature. However, the resistivity of metallic glasses does not vary strongly with temperature (except near absolute zero, where some amorphous alloys become superconducting). The presence of conduction electrons also allows metallic glasses to scatter and absorb incident light, giving metallic glasses the shiny luster typical of metals. Thus, unlike common oxide glasses, metallic glasses are not transparent to visible light. *See* ELECTRICAL CONDUCTIVITY OF METALS.

Because the early metallic glasses could only be produced in limited shapes (such as thin ribbons and foils), applications were mostly limited to those making use of the advantageous magnetic properties of iron-based ferromagnetic alloys. For instance, these alloys can be used to make high-efficiency cores for electrical transformers, and the large magnetostriction is useful for making antishoplifting tags. The more recent development of bulk glass-forming alloys has opened up a much wider range of applications that make use of the unique mechanical properties of metallic glasses. In particular, the ability to soften the glass by heating it above the glass transition temperature makes it possible to form complex shapes and castings with thin sections. This, together with the high strength and stiffness (compared to polymers), makes metallic glasses attractive for applications such as cases for cellular telephones and other electronic devices. High strength is related to high hardness. This, together with the ability to form fine features, makes metallic glasses well suited for precision knife-edges, such as scalpels. Also, the high resilience and low damping capacity mean that metallic glasses make good springs. One example is the use of zirconium-based metallic glasses as golf club heads. Todd C. Hufnagel

Metallocenes Bis-cyclopentadienyl derivatives of transition metals whose bonding involves overlap of *ns*, $(n-1)d$, and *np* orbitals of the metal with molecular orbitals of appropriate symmetry of each cyclopentadienyl ring. The resulting complexes often possess two parallel rings (sandwich structure), but in some cases, for example those involving the titanium subgroup of metals, the rings are canted (see illustration). Metals in

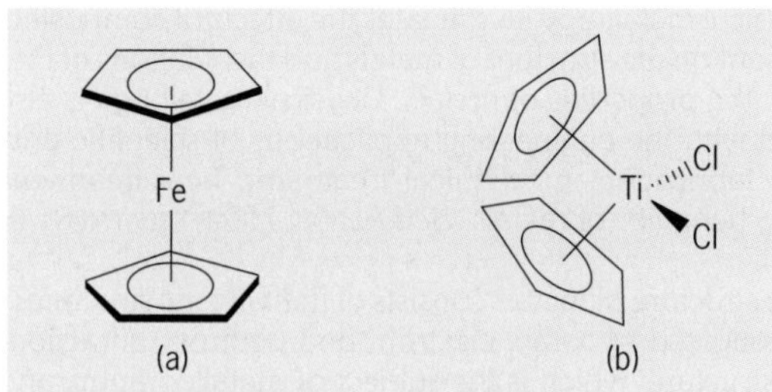

Metallocene structures. (*a*) Staggered sandwich structure of ferrocene. (*b*) Canted cyclopentadienyl ring structure of titanocene dichloride. The distribution of the ligands about the Ti atom is tetrahedral.

the periodic table commonly known to form metallocene complexes are titanium, zirconium, hafnium, vanadium, chromium, molybdenum, tungsten, manganese, iron, ruthenium, osmium, cobalt, rhodium, and nickel. *See* COORDINATION COMPLEXES.

The reactions of metallocenes can be divided into two classes: the first is typified by the iron triad, and comprises essentially the reactions of aromatic molecules; the second consists of the reactions of the other metallocenes where the 18-electron rare-gas configuration is not found. Reactions in these latter systems often lead to a product where the 18-electron rule is obeyed.

Ferrocene is a very electron-rich system, and undergoes electrophilic substitution with great rapidity. For example, acylation proceeds about 10^6 times faster than that of benzene under similar conditions. Ferrocene also undergoes several other typical aromatic substitution reactions besides acylation, including sulfonation, dimethylaminomethylation (Mannich reaction), metalation, and the like. Bis substitution tends to factor a product where each ring is monosubstituted, although several cases are known where two substituents are introduced into one ring. Ferrocene is oxidized and deactivated under conditions for nitration and halogenation.

Uses of metallocenes include reaction of chromocene with alumina to make a polymerization catalyst for ethylene. Ferrocene and some alkyl-substituted ferrocenes have been used as moderators in high-temperature combustions such as occur in solid rocket fuels. A cyclopentadienyl complex, $CH_3C_5H_4Mn(CO)_3$, briefly replaced tetraethyllead as an octane booster and anti-knock agent in liquid fuels. *See* ORGANOMETALLIC COMPOUND.

Donald W. Slocum

Metallochaperones A family of proteins that shuttle metal ions to specific sites within a cell. The target sites for metal delivery include a number of metalloenzymes, or proteins that bind metal ions, such as copper, zinc, or iron, and use these ions as cofactors to carry out essential biochemical reactions. Metallochaperones escort the ion to a specific intracellular location and facilitate incorporation of the metal into designated metalloenzymes. *See* BIOINORGANIC CHEMISTRY; CELL (BIOLOGY).

The bulk of current knowledge on metallochaperones is restricted to copper, although it is reasonable to assume that a distinct class of proteins is responsible for the incorporation of other metal ion cofactors into metalloenzymes. Among the metallochaperones that have been studied in detail are a family of three copper chaperones. These molecules operate in eukaryotic (nucleated) cells to direct copper to distinct intracellular locations: the mitochondria, the secretory pathway, and the cytosol. The first copper chaperone identified, COX17, is a small protein that specifically directs copper to the mitochondria. The copper delivered by COX17 is inserted into the metalloenzyme cytochrome oxidase, needed for respiration. A second copper chaperone identified was ATX1, which carries copper to the secretory pathway, a cellular compartment that functions to shuttle proteins toward the cell surface. The metal delivered by ATX1 is incorporated into copper enzymes destined for the cell surface or the extracellular milieu. The most recently identified copper chaperone is CCS, which specifically delivers copper to a single metalloenzyme, superoxide dismutase. This copper-requiring enzyme is located in the soluble cytosolic compartment of the cell and acts to detoxify harmful reactive oxygen species. *See* MITOCHONDRIA.

Intracellular copper is normally present at exquisitely low levels, and activation of copper enzymes is wholly dependent upon copper chaperones. Copper not only is an essential nutrient but also is quite toxic to living cells, and elaborate detoxification mechanisms prevent the free metal ion from accumulating to any substantial degree. The copper-requiring metalloenzymes cannot compete for these vanishingly low levels of available metal, explaining the requirement for the copper metallochaperones.

Valeria Cizewski Culotta

Metallography The study of the structure of metals and alloys by various methods, especially light and electron microscopy. Light microscopy of metals is conducted with reflected light on surfaces suitably prepared to reveal structural features. The method is often called optical microscopy or light optical microscopy. A resolution of about 200 nanometers and a linear magnification of at most 2000× can be obtained. Electron microscopy is generally carried out by the scanning electron microscope (SEM) on specimen surfaces or by the transmission electron microscope (TEM) on electron-transparent thin foils prepared from bulk materials. Magnifications can range from 10× to greater than 1,000,000×, sufficient to resolve individual atoms or planes of atoms.

Metallography serves both research and industrial practice. Light microscopy has long been a standard method for observing the morphology of phases resulting from industrial processes that involve phase transformations, such as solidification and heat treatment, and plastic deformation and annealing. Microscopy, both light and electron, is also indispensable for the analysis of the causes of service failures of components and products.

In light microscopy, microstructural features observed in photomicrographs include the size and shape of the grains (crystals) in single-phase materials (see illustration), the structure of alloys containing more than one phase such as steel, the effects of deformation, microcracking, and the effects of heat treatment. Other structural features investigated by light microscopy include the morphology and size of precipitates, compositional inhomogeneities (microsegregation), microporosity, corrosion, thickness and structure of surface coatings, and microstructure and defects in welds.

Photomicrographs of typical microstructures of annealed brass (70% Cu–30% Zn). ***(Courtesy of W. R. Johnson)***

The electron microscope offers improved depth of field and higher resolution than the light microscope, as well as the possibility of in-place spectroscopy techniques. The scanning electron microscope images the surface of a material, while the transmission electron microscope reveals internal microstructure. Images produced by the scanning electron microscope are generally easier to interpret; in addition, the instrument operates at lower voltages, offers lower magnification, and requires less specimen preparation than is necessary for the transmission electron microscope. Consequently it is important to view a specimen with light microscopy and often with the scanning electron microscope before embarking on transmission electron microscopy.

However there are some disadvantages. Electron microscope specimens are viewed under vacuum, the instruments cost significantly more than light microscopes, electron beam damage is always a danger, and representative sampling becomes more difficult as the magnification increases. *See* ELECTRON MICROSCOPE.

The ionizing nature of electron irradiation means that x-ray spectrometry and electron spectrometry, both powerful tools in their own right, can be performed in both scanning electron microscopy and transmission electron microscopy. The various signals detected spectroscopically can also be used to form images of the specimen, which reveal elemental distribution among other information. In particular, the characteristic x-ray signal can be detected and processed to map the elemental distribution quantitatively on a micrometer scale in the scanning electron microscope and a nanometer scale in the transmission electron microscope. Electron spectroscopic signals permit not only elemental images to be formed but also images that reveal local changes in bonding, dielectric constant, thickness, band gap, and valence state. *See* ELECTRON SPECTROSCOPY; METALLURGY.

David A. Thomas; David B. Williams

Metalloid An element which exhibits the external characteristics of a metal but behaves chemically both as a metal and as a nonmetal. Arsenic and antimony, for example, are hard crystalline solids that are definitely metallic in appearance. They may, however, undergo reactions that are characteristic of both metals and nonmetals. However, only when this dualistic chemical behavior is very marked and the external appearance metallic is the element commonly called a metalloid. *See* METAL; NONMETAL.

Francis J. Johnston

Metallurgy The technology and science of metallic materials. Metallurgy as a branch of engineering is concerned with the production of metals and alloys, their adaptation to use, and their performance in service. As a science, metallurgy is concerned with the chemical reactions involved in the processes by which metals are produced and the chemical, physical, and mechanical behavior of metallic materials.

The field of metallurgy may be divided into process metallurgy (production metallurgy, extractive metallurgy) and physical metallurgy. In this system metal processing is considered to be a part of process metallurgy and the mechanical behavior of metals a part of physical metallurgy.

Process metallurgy, the science and technology used in the production of metals, employs some of the same unit operations and unit processes as chemical engineering. These operations and processes are carried out with ores, concentrates, scrap metals, fuels, fluxes, slags, solvents, and electrolytes. Different metals require different combinations of operations and processes, but typically the production of a metal involves two major steps. The first is the production of an impure metal from ore minerals, commonly oxides or sulfides, and the second is the refining of the reduced impure metal, for example, by selective oxidation of impurities or by electrolysis. *See* ELECTROMETALLURGY; HYDROMETALLURGY; IRON METALLURGY; ORE DRESSING; PYROMETALLURGY; STEEL MANUFACTURE.

Physical metallurgy investigates the effects of composition and treatment on the structure of metals and the relations of the structure to the properties of metals. Physical metallurgy is also concerned with the engineering applications of scientific principles to the fabrication, mechanical treatment, heat treatment, and service behavior of metals. *See* ALLOY; HEAT TREATMENT (METALLURGY).

The structure of metals consists of their crystal structure, which is investigated by x-ray, electron, and neutron diffraction, their microstructure, which is the subject of metallography, and their macrostructure. Crystal imperfections, which provide mechanisms for processes occurring in solid metals, are investigated by x-ray diffraction and metallographic methods, especially electron microscopy. The microstructure is determined by the constituent phases and the geometrical arrangement of the microcrystals (grains) formed by those phases. Macrostructure is important in industrial metals. It involves chemical and physical inhomogeneities on a scale larger than microscopic. Examples are flow lines in steel forgings and blowholes in castings. *See* METALLOGRAPHY; X-RAY DIFFRACTION.

Phase transformations occurring in the solid state underlie many heat-treatment operations. The thermodynamics and kinetics of these transformations are a major concern of physical metallurgy. Physical metallurgy also investigates changes in the structure and properties resulting from mechanical working of metals.

For more information on metallurgy and some associated techniques see articles on individual metals and their metallurgy. *See* ELECTROPLATING OF METALS; METAL COATINGS; METAL FORMING.

Michael B. Bever

Metameres The successive subdivisions along the length of the body axis in bilaterally symmetrical animals; also called somites or segments. Common examples are the muscles and spinal nerves in the human body and in the body and tail of many mammals, snakes and lizards, salamanders, and fishes. It also occurs in other chordates, and in arthropods and annelid worms. It never involves reproductive organs, and thus differs from strobilization in tapeworms and certain jellyfish. This serial repetition of parts (metamerism or segmentation) arises either from a bilateral series of coelomic pouches which form the segmental muscles, kidneys, and body cavities of lower forms, or from mesoblastic somites which form the skeletal and muscular segments of vertebrates. Repetitive features of the nervous system are acquired secondarily through the influence of mesodermal metameres upon adjoining ectodermal tissues. *See* ANIMAL SYMMETRY; COELOM; MUSCULAR SYSTEM; NEURULATION.

Howard L. Hamilton

Metamict state The state of a special class of amorphous materials that were initially crystalline. W. C. Broegger first used the term *metamikte* in 1893 to describe minerals that were optically isotropic with a "glasslike" fracture but still retained well-formed crystal faces. In 1914 A. Hamburg correctly attributed the transition from the periodic, crystalline state to the aperiodic, metamict state as induced by alpha-decay damage. In minerals, this damage is the result of the decay of naturally occurring radionuclides and their daughter products in the uranium and thorium (^{238}U, ^{235}U, and ^{232}Th) decay series. A wide variety of complex oxides, silicates, and phosphates are reported as occurring in the metamict state. All of these structures can accommodate uranium and thorium. *See* ALPHA PARTICLES; AMORPHOUS SOLID; RADIOACTIVITY.

The presence of uranium and thorium distinguishes metamict minerals from other naturally occurring amorphous materials that have not experienced this radiation-induced transformation. Lanthanide elements are also common (in some cases over 50 wt %) and water of hydration may be high (up to 70 mol %).

The radiation damage caused by the alpha-decay event is the result of two separate but simultaneous processes: (1) An alpha particle with an energy of approximately 4.5 MeV and a range of 10,000 nanometers dissipates most of its energy by ionization; however, at low velocities near the end of its track, it displaces several hundred atoms, creating Frenkel defect pairs. (2) The alpha-recoil atom with an energy of approximately 0.09 MeV and a range of 10 to 20 nm produces several thousand atomic displacements, creating tracks of disordered material. These two damaged areas are separated by thousands of unit cell distances and have different effects on the crystalline structure. Local point defects cause an increase in the distortion; therefore, there is an increase in the strain in the structure. Alpha-recoil tracks create regions of aperiodic material that at high enough alpha-decay doses (usually 10^{24} to 10^{25} alpha-decay events/m^3) overlap and finally lead to the metamict state. The former causes broadening of x-ray diffraction maxima and an increase in unit cell volume (and a decrease in the density); the latter causes a decrease in diffraction peak intensities. The radiation-induced transition from the crystalline to the metamict state occurs over a narrow range of alpha-decay dose (10^{24}–10^{25} alpha decays/m^3), which corresponds to 0.1 to 1.0 displacements per atom (dpa). *See* CRYSTAL DEFECTS; RADIATION DAMAGE TO MATERIALS.

A renewed interest in the metamict state has been stimulated by concern for the long-term stability of crystalline materials (nuclear waste forms) that will serve as hosts for actinides (for example, plutonium, americium, curium, and neptunium). Various crystalline materials (phases) may appear in a single waste form; each phase may or may not suffer radiation damage. For some nuclear waste-form phases, the radiation-induced transformation to the metamict state has been stimulated by doping phases with highly radioactive plutonium-238 or curium-244.

Rodney C. Ewing

Metamorphic rocks Preexisting rock masses in which new minerals, textures, or structures are formed at higher temperatures and greater pressures than those normally present at the Earth's surface. *See* IGNEOUS ROCKS; SEDIMENTARY ROCKS.

Two groups of metamorphic rocks may be distinguished; cataclastic rocks, formed by the operation of purely mechanical forces; and recrystallized rocks, or the metamorphic rocks properly so called, formed under the influence of metamorphic pressures and temperatures. Cataclastic rocks are mechanically sheared and crushed. They represent products of dynamometamorphism, or kinetic metamorphism. Chemical and mineralogical changes generally are negligible. The rocks are characterized by their minute mineral grain size. Each mineral grain is broken up along the edges and is surrounded by a corona of debris or strewn fragments. *See* METAMORPHISM.

Metamorphic rocks, properly so called, are recrystallized rocks. The laws of recrystallization are not the same as those of simple crystallization from a liquid, because the crystals can develop freely in a liquid, but during recrystallization the new crystals are encumbered in their growth by the old minerals. Consequently, the structures which develop in metamorphic rocks are distinctive and of great importance, because in many ways they reflect the physiochemical environment of recrystallization and thereby the genesis and history of the metamorphic rock.

The metamorphic minerals may be arranged in an idioblastic series (crystalloblastic series) in their order of decreasing force of crystallization as follows: (1) sphene, rutile, garnet, tourmaline, staurolite, kyanite; (2) epidote, zoisite; (3) pyroxene, hornblende; (4) ferrogmagnesite, dolomite, albite; (5) muscovite, biotite, chlorite; (6) calcite; (7) quartz, plagioclase; and (8) orthoclase, microcline. Crystals of any of the listed minerals tend to assume idioblastic outlines at surfaces of contact with simultaneously developed crystals of all minerals of lower position in the series.

Igneous magma at high temperature may penetrate into sedimentary rocks, it may reach the surface, or it may solidify in the form of intrusive bodies (plutons). Heat from such bodies spreads into the surrounding sediments, and because the mineral assemblages of the sediments are adjusted to low temperatures, the heating-up will result in a mineralogical and textural reconstruction known as contact metamorphism. *See* CONTACT AUREOLE; PLUTON.

The effects produced do not depend only upon the size of the intrusive. Other factors are amount of cover and the closure of the system, composition and texture of the country rock, and abundance of gaseous and hydrothermal magmatic emanations. The heat conductivity of rocks is so low that gases and vaporous emanations become chiefly responsible for transportation and transfer of heat into the country rock.

Crystalline schists, gneisses, and migmatites are typical products of regional metamorphism and mountain building. If sediments accumulate in a slowly subsiding geosynclinal basin, they are subject to down-warping and deep burial, and thus to gradually increasing temperature and pressure. They become sheared and deformed, and a general recrystallization results. However, subsidence into deeper parts of the crust is not the only reason for increasing temperature. It is not known what happens at the deeper levels of a live geosyncline, but obviously heat from the interior of the Earth is introduced regionally and locally, partly associated with magmas, partly in the form of "emanations" following certain main avenues, determined by a variety of factors. From this milieu rose the lofty mountain ranges of the world, with their altered beds of thick sediments intercalated with tuffs, lava, and intrusives, all thrown into enormous series of folds and elevated to thousands of feet. Thus were born the crystalline schists with their variants of gneisses and migmatites. *See* OROGENY.

Well-defined series of mineral facies have been singled out. Sedimentary rocks of the lowest metamorphic grade have recrystallized to give rocks of the zeolite facies. At slightly higher temperatures the greenschist facies develops—chlorite, albite, and epidote being characteristic minerals. A higher degree of metamorphism produces the epidote-amphibolite facies, and a still higher degree the true amphibolite facies in which hornblende and plagioclase mainly take the place of chlorite and epidote. Representative of the highest regional metamorphic grade is the granulite facies, in which most of the stable minerals are water-free, for example, pyroxenes and garnets. Any sedimentary unit will recrystallize according to the rules of the several mineral facies, the complete sequence of events being a progressive change of the sediment by deformation, recrystallization, and alteration in the successive stages: greenschist facies→epidote-amphibolite facies→amphibolite facies→granulite facies. *See* FACIES (GEOLOGY); GRANULITE.

The normal continental crust is entirely made up of metamorphic rocks; where thermal, mechanical, and geochemical equilibrium prevails, there are only metamorphic rocks. Border cases of this normal situation occur in the depths where ultrametamorphism brings about differential melting and local formation of magmas. When equilibrium is restored, these magmas congeal and recrystallize to (metamorphic) rocks. At the surface, weathering processes oxidize and disintegrate the rocks superficially and produce sediments as transient products. Thus the cycle is closed; petrology is without a break. All rocks that are found in the continental crust were once metamorphites. *See* AMPHIBOLITE; GREISEN; MARBLE; MIGMATITE; PHYLLITE; QUARTZITE; SCAPOLITE; SERPENTINITE; SOAPSTONE.

T. F. W. Barth; Robert C. Newton

Metamorphism The alterations and transformations in preexisting rock masses effected by temperature and pressure, but excluding changes produced by weathering and sedimentation. The changes may include the production of new minerals, structures, or textures, or all three. They give a distinctive new character to the rock as a whole, but they do not involve the loss of individuality of a rock mass, such as changes brought about by fusion. Quantitatively, the metamorphic rocks, including gneisses and migmatites, are the most important group of rocks in the crust of the continents. *See* GNEISS; METAMORPHIC ROCKS; MIGMATITE.

Different kinds of metamorphism may be defined according to genetic criteria, such as the geologic processes that were assumed to have caused the metamorphism, or the physical and chemical conditions that appear to have been predominant in determining the course of metamorphism. Using these criteria three general kinds of metamorphism are noted below.

1. Dislocation, mechanical, or dynamic metamorphism is the result of pressure (or stress) along dislocations in the Earth's crust. The deformed rocks commonly show marked zones of extremely fine-grained rocks, such as mylonites, whose structures are determined by crushing and movement of the grains without important recrystallization of old, or growth of new, minerals. This type of metamorphism is local and restricted in occurrence.

2. Contact or thermal metamorphism occurs in response to increased temperature induced by adjacent intrusions of magma. Chemical reconstitution of the rocks is due to magmatic exhalation; other conditions, such as confining pressure, exert subordinate influence.

3. Regional metamorphism, the most widespread type, is brought about by an increase in both temperature and pressure in orogenic regions, which are vast segments of the crust represented by the folded mountain ranges. Heat and pressure are mainly consequences of downwarping and deep burial. Pressure is also generated by shearing stresses accompanying the orogenic movements. *See* OROGENY. T. F. W. Barth; R. C. Newton

A fluid phase plays an important role during metamorphism as an agent of heat and mass transfer. The presence of a static film of fluid around mineral grains greatly facilitates chemical reactions because the fluid film speeds the movement of matter from reactant to product minerals. Flowing fluid carries substantial quantities of materials in solution that may be precipitated far from their source. Heated rocks are cooled more rapidly, and cool rocks are more quickly heated, by flowing fluids than would otherwise be possible by heat conduction. Fluids of metamorphic rocks consist primarily of water (H_2O), variable amounts of carbon dioxide (CO_2) and methane (CH_4), and minor quantities of hydrogen sulfide (H_2S), carbon monoxide (CO), hydrogen (H_2), and sulfur dioxide (SO_2). Douglas Rumble III

Metamorphosis A pronounced change in both the internal and external morphology of an animal that takes place in a short amount of time, triggered by some combination of external and internal cues. The extent of morphological change varies considerably among species. Even when morphological changes are relatively slight, metamorphosis typically brings about a pronounced shift in habitat and lifestyle. The precise morphological, physiological, and biochemical changes that constitute metamorphosis; the neural, hormonal, and genetic mechanisms through which those changes are controlled; and the ecological consequences of those changes and when they take place continue to be studied in a wide variety of animals. The hormonal and genetic control of metamorphosis has been best examined in a few species of insect, amphibian, and fish (such as flounder), but other aspects of metamorphosis have been investigated for other insect, amphibian, and fish species as well as for marine invertebrates and, indeed, representatives of essentially every animal phylum.

Amphibians exhibit extensive tissue remodeling during metamorphosis, including resorption of the tail musculature and skeletal system; major reconstruction of the digestive tract; degeneration of larval skin and pronounced alteration in skin chemical composition; growth of the hind and fore limbs; degeneration of the gills and associated support structures; shifts in mode of nitrogen excretion, from ammonia to urea; alteration in visual system biochemistry; replacement of larval hemoglobin with adult hemoglobin; and differential growth of the cerebellum. *See* AMPHIBIA.

Metamorphosis among insects is associated primarily with wing development. Bristletails and other species that do not develop wings and are not descended from winged ancestors exhibit no pronounced metamorphosis. Metamorphosis is most dramatic among holometabolous species, which pass through a distinctive and largely inactive pupal stage; in such species, all of the transformations separating the larval morphology and physiology from that of the adult take place in the pupa. Wings, compound eyes, external reproductive parts, and thoracic walking legs develop from discrete infolded pockets of tissue (imaginal discs) that form during larval development. *See* INSECTA.

The most dramatic metamorphic changes in fish are seen among flounder and other flatfish: in such species, during metamorphosis a symmetrical fish larva becomes an asymmetrical adult, with both eyes displaced to the dorsal surface. The transformation of leptocephalus larvae into juvenile eels is also dramatic; such transformation includes a shift in the position of the urinary and digestive tracts from posterior to anterior. *See* EEL; PLEURONECTIFORMES.

The control of metamorphosis among crabs, barnacles, gastropods, bivalves, bryozoans, echinoderms, sea squirts, and other marine invertebrates is poorly understood, partly due to the very small size of the larvae—they rarely exceed 1 mm in length, and most are less than 0.5 mm. The larvae of some marine invertebrate species are triggered to metamorphose by specific substances associated with adults of the same species, or with the algae or animals on which they prey. *See* ANNELIDA; BIVALVIA; CRAB; DECAPODA (CRUSTACEA); ECHINODERMATA; GASTROPODA; MOLLUSCA.

Among insects, the timing of metamorphosis is influenced by environmental factors such as temperature, humidity, photoperiod, pheromone production by neighboring individuals, and the nutritional quality of the diet. In a number of species, larvae can undergo developmental arrest (a diapause) in response to unfavorable environmental conditions, so that metamorphosis can be delayed for many months or even years. The hormonal basis for such effects has been at least partly worked out for a number of insect species.

Among marine invertebrates and in at least some fish species, there is also considerable flexibility in the timing of metamorphosis. At some point in the development of marine invertebrates and apparently also in the development of some coral reef fishes, individual larvae become "competent" to metamorphose. It is not yet clear what makes larvae competent; the development of external receptor cells, or the completion of specific neural pathways, or the activation of hormonal systems or their receptors are likely possibilities. *See* ENDOCRINE MECHANISMS; ENDOCRINE SYSTEM (INVERTEBRATE); INVERTEBRATE EMBRYOLOGY. Jan A. Pechenik

Metasomatism The process by which the bulk chemical composition of a rock is changed from some previous state by the introduction of components from an external source. In contrast with metamorphism, where rocks are converted to a new set of minerals with little or no change in bulk composition, metasomatism involves the import and export of chemical components through the agency of a chemically active fluid.

Clastic rocks and mafic-to-felsic igneous rocks react in similar ways with metasomatic fluids by exchange of alkalies, alkaline earths, and hydrogen. Hydrolytic alteration dominates lower-temperature metasomatic processes. Hydrolysis involves hydrogen-ion metasomatism—exchange of hydrogen ion for potassium, sodium, and calcium.

Metasomatism of aluminum-poor rocks carbonate and ultramafic rocks generally involves addition of silica, metals, and alumina.

Metasomatism is best developed in environments characterized by extreme physical and chemical gradients and high fluid flux. At the centimeter scale, chemical contrasts along shale-limestone contacts lead to diffusive exchange of components on heating during regional or contact metamorphism. At the kilometer scale in mid-ocean rifts, island arcs, and continental-margin plutonic arcs, metasomatism results from emplacement of magma at depths of a few kilometers and infiltration of hot, saline, aqueous fluids through fractured rocks. At global scales, metasomatism accompanies mass fluxing between the crust and the mantle, such as on emplacement of mantle plumes into the lower crust or subduction of oceanic crust into the mantle. *See* ASTHENOSPHERE; EARTH CRUST; MAGMA; METAMORPHISM.

Marco T. Einaudi

Metastable state In quantum mechanics, a state that is not truly stationary but is almost stationary.

In practice, especially in atomic and nuclear physics applications, the designation metastable state usually is reserved for states whose lifetimes are unusually long. For example, the excited states of atoms usually decay with the emission of a single photon, in a time of the order of 10^{-8} s. However, the necessity for angular momentum and parity conservation forces the second excited state ($2S_{1/2}$) of atomic hydrogen to decay by simultaneous emission of two photons; consequently, the lifetime is increased to an estimated value of 0.15 s. Thus, the $2S_{1/2}$ state of atomic hydrogen is usually termed metastable, but most other hydrogenic states are not. Similarly, emission of a gamma-ray photon by an excited nucleus usually occurs in 10^{-13} s or less; however, the lifetime of one excited state of the ^{113}In nucleus, the state that customarily is termed metastable, is about 100 min. Since radiative transition probabilities for emission of photons generally decrease rapidly with decreasing frequency, a low-lying atomic or nuclear excited state may have a lifetime longer than most excited states of atoms and nuclei and yet not be metastable in the practical sense just described, because photon emission from the state may not be hindered by any general requirement or selection rule, such as is invoked for the $2S_{1/2}$ state of hydrogen. *See* EXCITED STATE; NUCLEAR ISOMERISM; RADIOACTIVITY.

Edward Gerjuoy

Metatheria An infraclass of therian mammals including a single order, the Marsupialia. The Metatheria are distinguished from the Eutheria (the placental mammals) by numerous characters. The braincase is small, the angular process of the mandible is inflected, and a pair of marsupial bones articulates with the pelvis. Almost all living marsupials have a pouch on the belly of the female in which the young are carried after birth. The early marsupials were unable to compete with the more progressive later placental forms and died out except in South America and Australia, where they were isolated by water barriers. *See* EUTHERIA; MAMMALIA; THERIA.

D. Dwight Davis; Frederick S. Szalay

Metazoa The monophyletic group of eukaryotic organisms comprising all multicellular animals, but not those in which their cells form colonies, such as choanoflagellates. In colonies, all the cells need to feed because they are not in contact with each other, such that nutrients cannot be transported between them. In multicellular organisms, cells are engaged in a more specialized division of labor because nutrients can be transported between them. Therefore metazoans have specific characters related to cell recognition, adhesion, communication, and skeletal elements responsible for maintenance of the shape of the whole body and its organs. The need for cell-to-cell communication via neurotransmitters evolved into one of the most intriguing systems in all organisms, the nervous system. Other specific characteristics of metazoans are connected to sexual reproduction, such as the origin and structure of haploid gametes, fertilization, and development from the zygote to the adult organism. For most recent authors, Metazoa is a synonym of Animalia.

Metazoans are divided into about 35 major lineages that represent unique body plans. Each one of these lineages (named phyla; phylum in singular) can be defined by certain unique characteristics not found in any other phyla. The simplest metazoans are the sponges (phylum Porifera) without a well-defined axis of symmetry, few cell types, and without true tissues. With a few exceptions, the remaining metazoans have well-defined axes of symmetry. *See* PORIFERA.

Complex metazoans have tissues and organs, including muscle cells that allowed the animals to become independent from the substrate. This motility probably triggered an arms race of predators and prey escape mechanisms. The same race is probably responsible for the evolution of increased body complexity, size, and development of jaws and skeletons. *See* ANIMAL KINGDOM; SKELETAL SYSTEM.

Metazoans evolved during the Precambrian (Proterozoic) in the primitive oceans. Most preserved Precambrian animals had simple body plans, probably similar to those of extant sponges and cnidarians. It is believed that most of the extant animal body plans also originated later during the Precambrian times.

Gonzalo Giribet

Meteor The luminous streak lasting seconds or fractions of a second and seen at night when a solid, natural body plunges into the Earth's (or another planet's) atmosphere. The entering object is called a meteoroid and, if any of it survives atmospheric passage, the remainder is called a meteorite. Cosmic dust particles (with masses of micrograms) entering the atmosphere and leaving very brief, faint trails are called micrometeors, with the surviving pieces known as micrometeorites. If the apparent brightness of a meteor exceeds that of the planet Venus as seen from Earth, it is called a fireball, and when a bright meteor is seen to explode, it is called a bolide. *See* METEORITE; MICROMETEORITE.

Under normal, clear atmospheric conditions and dark skies (no moonlight or artificial lights), an observer will see an average of five meteors per hour. The spatial distribution of meteoroid orbits relative to the Sun and the circumstances of their intersections with the moving Earth are responsible for pronounced variations in meteor rates.

The average meteor seen by the unaided eye starts with a meteoroid velocity of 18 mi/s (30 km/s) and leaves a luminous trail from 67 to 50 mi (110 to 80 km) high. The meteor trails are rapidly expanding columns of atoms, ions, and electrons dislodged from the meteoroid by collisions with air molecules, and can be excited to temperatures of several thousand degrees Celsius. For a time after trail formation, the free electrons are dense enough to reflect radio waves in the very high frequency range, and therefore can be used to transmit radio messages. *See* RADIO-WAVE PROPAGATION.

Under the right circumstances, particularly with high-power ultrahigh-frequency (UHF) radars, the ionization right around and moving with the meteoroid itself is seen. This is known as the head echo, and a determination of its velocity is the most accurate way to determine radar meteor speeds. *See* RADAR; RADAR ASTRONOMY.

The Earth moves around the Sun with an average speed of 18 mi/s (30 km/s). According to the laws of celestial mechanics, if a meteoroid comes from beyond the solar system, its velocity at the Earth's distance from the Sun must be greater than 26 mi/s (42 km/s). If such a meteoroid hits the Earth head-on, indications of preatmospheric speeds in excess of 45 mi/s (72 km/s) would be observed. The fact that the vast majority of observed meteoroids have orbits with Earth-approaching velocities of less than 45 mi/s indicates that most of these are comet and asteroid fragments, and are therefore long-term members of the solar system. However, in the 1980s and 1990s, a combination of spacecraft and high-power radar observations indicated that hypervelocity micrometeoids do indeed exist with seeming interstellar dust connections. *See* INTERSTELLAR MATTER.

A combination of the meteoroid's and Earth's velocities of travel around the Sun make the meteor itself seem to originate from a specific direction in the sky called the radiant. If there are numerous meteoroids in nearly the same orbit (sometimes incorrectly called meteor streams), the Earth sweeps them up at specific times of the year and a so-called meteor shower is observed. Meteor showers are named after the constellation or single star in the sky from which they appear to radiate. While shower meteoroids are really moving nearly parallel through space and result in nearly parallel meteor trails, the effects of perspective make the meteors appear to diverge from the radiant. Meteors that cannot be shown to be associated with a known shower are termed sporadic meteors.

A number of meteor showers have been observed to be in orbits that are similar to those traveled by known comets. Thus an association between shower meteors and comets has gradually become a firmly entrenched concept. There are numerous theoretical scenarios where vaporization of the more volatile cometary ices ejects small solid particles from the surface of the nucleus. A fair proportion of these fragments, particularly the smaller dust-sized ones, escape and take up their own orbits as meteoroids. Cometary nuclei have been known to split into two or more pieces and, when this occurs, it is likely that particles larger than dust size are released as well.

The strategy of photographic or electronic measurements is to place at least two cameras 10–52 mi (15–85 km) apart over a known baseline, but arranged to examine the same volume of space at a height of about 56 mi (90 km). Each camera has a rotating shutter so that the meteor trail consists of a line of bright dashes. Meteor imaging is one of the most difficult areas of astronomical detection, even with ultrafast cameras. Meteor spectroscopy is even more difficult since the light is spread out over areas hundreds of times larger than the meteor trail itself. *See* ASTRONOMICAL IMAGING; ASTRONOMICAL PHOTOGRAPHY; ASTRONOMICAL SPECTROSCOPY.

Radio and radar observations depend on the fact that the initial ion-electron densities in a meteor trail are considerably higher than the average for the ionosphere at an altitude of 56 mi (90 km). For a very high frequency (VHF) or somewhat lower-frequency radar system, the maximum reflected signal occurs when the meteor trail is at right angles to the outgoing wave, with head echoes rarely seen. At ultrahigh frequencies (UHF), radar reflections from the head-echo predominate. From these, high-accuracy radial velocities are determined directly, using the Doppler effect. *See* DOPPLER EFFECT; RADIO ASTRONOMY.

The parent comet of the Leonid stream, 1866 I, made another of its periodic (33-year intervals) approaches to the Sun in 1999. With the appearance of the comet, it was expected that the strong meteor storm that happened last in 1966 would again make a brief but spectacular appearance. However, perturbations by the outer planets (particularly by Neptune) once again played a significant role in this shower's behavior. The perturbations moved a number of thin meteoroid streams produced by the comet many orbit periods in the past into intersection range of the Earth. This produced a unique succession of strong peaks covering a span of seven years (1996–2003). The scientific yield of this extended display was much more than anyone had hoped. While meteor trails had been previously recorded from the space shuttle and other spacecraft, in 1997 the first above-atmosphere, far-ultraviolet spectrum of a bright meteor was recorded during the Leonid shower. Spectra obtained from the ground also yielded new information. The large number of fireballs in the Leonid streams enabled many details of the ablation processes at higher than average meteoroid incoming velocities to be recorded with high-speed cameras. David D. Meisel

Meteorite A naturally occurring solid object from interplanetary space that survives impact on a planetary surface. While in space, the object is called a meteoroid, and a meteor if it produces light or other visual effects as it passes through a planetary atmosphere. Explosive surface impacts by large meteorites are believed to have created the plethora of craters on the solid planets and moons of the solar system. *See* METEOR; MICROMETEORITE.

A meteorite seen to strike a surface is known as a fall, whereas a meteorite discovered by chance is known as a find. In both cases, meteorites are named after their geographic places of recovery.

The major classification of meteorites is into chondrites and nonchondrites (or nonchondritic meteorites), and major classes of the nonchondrites are achondrites, stony-irons, and irons, in recognition of their compositions that are dominated by silicate minerals and iron-nickel alloys either alone or as admixtures. Within each of the major categories, detailed classifications are based on distinctive mineralogical and chemical compositions and physical structures.

Meteorites represent the most ancient rocks known. Their ages, as determined by radiometric dating, extend to more than 4.5×10^9 years, which is thought to be near the time of solar system formation. As samples of primordial material, stony meteorites known as chondrites are studied for clues about how the solar system formed. In contrast, achondrites, stony-irons, and irons are samples of melt products formed during processing of solid material in planetary or preplanetary bodies. *See* DATING METHODS; SOLAR SYSTEM.

Asteroids are believed to be the sources of most meteorites. In 1982, however, it was conclusively demonstrated that a small achondrite found in Antarctica in 1981 was from the Moon. Even more exciting is the prospect that 34 (as of early 2006) closely related achondrites (24 shergottites, seven nakhlites, two chassignites, and one orthopyroxenite), from various recovery locations around the world, are from Mars; one of them contains trapped gases that are nearly identical to those measured for the Martian atmosphere by the Viking lander in 1976. *See* ASTEROID; MARS; MOON. James L. Gooding

Chondrites. Chondrites are the most abundant sort of meteorite, over 15,000 being known. They constitute about 82% of all meteorites. Chondrites include three main categories: ordinary, carbonaceous (C), and enstatite (E). Chondrites are so-named because nearly all contain small (generally 0.5–2 mm) beadlike chondrules. Only CI chondrites are essentially chondrule-free; their matrix texture and chemical composition closely relate them to the other chondrule-containing carbonaceous chondrites.

Ordinary chondrites constitute 94% of all classified chondrites. Their main constituent minerals are olivine [$(Mg,Fe)_2SiO_4$], low-calcium pyroxene [$(Mg,Fe)SiO_3$], plagioclase [$(Na,Ca)Al(Si,Al)Si_2O_3$], iron-nickel (Fe-Ni) metal, and troilite (FeS). They may contain silicate glass. *See* FELDSPAR; OLIVINE; PYROXENE; PYRRHOTITE.

Chondrules and chondrite matrixes consist of the same minerals: the difference is textural. Chondrule minerals crystallized within molten droplets, and show a variety of shapes consistent with a molten origin involving very rapid (minute-scale) heating and cooling. Matrix minerals are small and granular.

Carbonaceous chondrites constitute less than 5% of chondrite falls and are the least evolved meteorites. The same major chemical elements compose all chondrites, but carbonaceous chondrites can contain substantial carbon, hydrogen, and nitrogen, present at trace levels in ordinary chondrites. In addition to chondrules, they contain many other types of inclusions. Most important, the constituent minerals of chondrules and inclusions in carbonaceous chondrites differ in composition and kind from those composing the surrounding matrix.

Enstatite chondrites make up not more than 2% of all chondrites and consist of 60–80% enstatite (Fs0, that is, $MgSiO_3$) with 10–30% metal and 5–15% troilite. Enstatite chondrites range from those with many distinct chondrules to those having nearly unrecognizable chondrules. *See* ENSTATITE.

Chondrites contain components from two genetic environments: nebular (those formed from dispersed materials in space) and planetary (those formed within parent bodies). Carbonaceous chondrites contain the most obvious nebular components. Refractory mineral inclusions in them formed by direct condensation from a hot gas cloud surrounding the primitive Sun. During late-formation stages, as temperatures in the nebular gas fell, lower-temperature minerals formed and accreted as matrix, into asteroidal bodies together with refractory minerals. Most (about 90%) were not subsequently metamorphosed at high temperatures in their parent bodies, which would have erased their primitive characteristics. No consensus exists about the nebular condensation and accretion histories of ordinary and enstatite chondrite parent materials into asteroid-sized bodies but accretion (planetary) processes are broadly understood.

Michael E. Lipschutz

Chondrules are the most abundant particles in chondrites, with diameters up to about 1 mm (0.04 in.). They are distinguished from other particles in the chondrite matrix by textural and compositional criteria, being iron-magnesium silicate spherules with igneous textures.

Chondrules formed either by the melting of rock, dustballs, or free dust grains, or by the condensation of gas, and possibly by all of these mechanisms. It seems that chondrules must form in less than 1% of the nebula in order to achieve the nonnebular gas composition. This local rather than nebula-wide formation is consistent with the relatively rapid cooling rates of chondrules.

Formation of a solar system involves numerous sources of energy that could have been responsible for chondrule melting. Chondrules might have been formed by heating when shock waves propagated through the solar disk encountered concentrations of chondrule precursors. This process is the mechanism whose temperature distributions best explain the relative proportions of different kinds of chondrules, but how these shocks were generated needs to be established. Roger H. Hewins

Achondrites. These are stony meteorites that have few, if any, chondrules and differ chemically from chondrites. They constitute about 8% of all meteorite falls and 1% of all finds. Although achondrites can be divided into several distinct groups based on chemical and isotopic composition, they are generally believed, based on aspects of their textures and composition, to have formed as the result of igneous processes on asteroidal or planetary bodies. Much of the interest in these meteorites derives from the fact that they provide clues into the nature of igneous processes and planetary differentiation early in the history of the solar system on planetary bodies outside the Earth–Moon system and on bodies presumed to be much smaller than the Earth and Moon.

The eucrites, howardites, and diogenites—often collectively referred to as the basaltic achondrites—are the most abundant achondritic meteorites. They appear to be samples of a series of related igneous rocks and of regolith breccias composed of fragments of these igneous rocks. They define a coherent group in terms of their oxygen isotope compositions, suggesting they are closely related. With ages near 4.5 billion years, they are products of igneous activity from the earliest history of planetary bodies in the solar system.

It is clear that igneous processes as they are known from study of terrestrial and lunar rocks were active on small bodies very soon after their formation. The heat source for such igneous activity is still under investigation, but it could be the decay of the aluminium isotope ^{26}Al or perhaps heating by electric currents induced in small planets by the passage of an intense solar wind associated with a very active early Sun (T-Tauri phase). The reflectance spectrum of the surface of the asteroid 4 Vesta closely resembles those of eucritic meteorites, and it has been suggested that this could be the source of the basaltic achondrites, although there are dynamical difficulties with such a source. *See* BASALT; IGNEOUS ROCKS; SOLAR WIND.

Shergottites, nakhlites, and chassignites (often referred to as the SNC group) are rare meteorites. Their young crystallization ages ($\sim 1.3 \times 10^9$ years), plus the similarity of the bulk compositions of the shergottites to that of the Martian soil as determined by the Viking landers, first led to the suggestion that these meteorites could be derived from Mars. It is difficult to conceive of a heat source for endogenous igneous activity (these meteorites have no features resembling known impact melts) on an asteroidal parent body at about 1.3×10^9 years ago; and given the limited choice of available larger planets, Mars seemed the most likely choice. The similarity of relative noble gas and nitrogen abundances and isotopic ratios in the Martian atmosphere and shock-produced glass in one shergottite provide strong support for this hypothesis. The very low paleomagnetic intensities of the shergottites are also consistent with a Martian origin. It is still a subject of controversy whether fragments of sufficient size to explain measured cosmic-ray exposure ages could be ejected more or less intact from Mars by impact and subsequently delivered to Earth. It is generally accepted that the question of whether or not these meteorites are from Mars will be resolved only after a sample-return mission to Mars. *See* PALEOMAGNE-TISM.

Edward Stolper

Iron and stony-iron meteorites. Iron and stony-iron meteorites are made largely or partly of metallic iron-nickel and come from deep inside asteroids that were melted around 4560 million years ago. In most of these bodies, the metal formed dense cores that were surrounded by silicate mantles. Iron and stony-iron meteorites are much stronger than other meteorites, so they can survive longer in space. At least 12 of the 20 craters on Earth under 1.2 km (0.75 mi) in diameter were made by iron or stony-iron meteoroids.

Iron meteorites are pieces of once molten metallic cores and pools in asteroids that were fragmented by impacts after they cooled slowly. About 700 different iron meteorites have been identified; 40 were seen to fall, and the rest fell during the last million years. The smallest iron meteorites, which weigh only 5–30 g (0.18–1.1 oz), were found in Antarctica and are aerodynamically shaped like certain tektites. The largest single iron meteorite weighs about 60 metric tons (66 tons) and still lies in Namibia. *See* TEKTITE.

Nine much larger iron masses hit the Earth during the last million years, forming numerous craters 50 to 1200 m (150 to 3900 ft) in diameter. However, the surviving fragments from these meteoroids weigh less than a ton. The largest crater, which is in Arizona, was formed about 50,000 years ago by the impact of a meteoroid measuring about 50 m (150 ft) across. The impact released energy equivalent to about 20–60 megatons of TNT.

The chemical and mineralogical evidence shows that iron meteorites formed from molten pools of metal that solidified and then cooled over many millions of years. This evidence is consistent with an origin for iron meteorites in the cores of asteroids that melted and differentiated. When an asteroid is partly melted, iron-nickel and iron sulfide, being denser than the associated silicates, will begin to sink to the center. With sufficient heating, a core of molten sulfur-rich metal will form. Since most iron meteorites contain no silicates and most achondrites have only trivial amounts of metal, it is likely that metallic cores are the source of many iron meteorites.

There are two major types of stony-iron meteorites: pallasites and mesosiderites. They are among the rarest groups of meteorites, together constituting 1% of meteorite falls. About 50 specimens of each type have been found. Edward R. D. Scott

Isotopic anomalies in meteorites. Primitive meteorites show a large variety of isotopic anomalies, that is, deviations from the average solar system composition (called "normal" composition) that cannot be explained by chemical fractionation and radioactive decay taking place today. They are rather of presolar origin or the result of irradiation from an early active Sun or in interplanetary space. These anomalies provide information about the nucleosynthetic sources of the material that formed the solar system. *See* ISOTOPE.

It is well established that carbon and all the heavier elements are produced in stars by nuclear processes (stellar nucleosynthesis). These elements are ejected into the interstellar medium either from explosions of massive stars (supernovae) or as stellar winds and planetary nebulae from low-to-intermediate mass stars at the end of their evolution. Stars of different masses and ages produce elements with very different isotopic ratios. The solar system was formed from material originating from many stars. However, until the early 1970s it was believed that all this material was thoroughly homogenized into a mix of uniform isotopic composition in a hot solar nebula. Isotopic measurements of samples from the Earth, the Moon, and various meteorites seemed to confirm this belief. Exceptions were well-understood excesses of certain isotopes that were the daughters of long-lived (they still exist today) radioactive isotopes or isotopic fractionation effects as the result of physical and chemical processes. *See* NUCLEOSYN- THESIS.

This belief in isotopic homogeneity was shattered by the discovery of isotopic anomalies in several elements in 1973; and today, thanks to increases in the precision of isotopic analysis and the capability to measure small samples, a plethora of isotopic anomalies are known in meteorites. The largest anomalies are found in small samples, where the effects are not diluted by isotopically normal material.

Isotopic anomalies in meteorites and interplanetary dust particles (IDPs) can be divided into four classes.

1. *Mass-dependent fractionation caused by physicochemical processes (diffusion, evaporation, condensation, and chemical reactions)*. Certain physical and chemical processes can also lead to non-mass-dependent fractionation that might mimic effects of nuclear origin.

2. *Effects from the decay of radioisotopes*. In addition to effects from the decay of long-lived isotopes, meteorites show also the effects of short-lived, now extinct isotopes.

3. *Nuclear effects reflecting nucleosynthetic processes in stellar sources*. These effects are found in samples that formed in the solar system but inherited some nucleosynthetic components, as well as in presolar grains, bona fide stardust, that formed in stellar atmospheres and thus represent the isotopic compositions of their parent stars.

4. *Effects due to irradiation of material by an early active Sun or irradiation of meteorites by galactic and solar cosmic rays*. These effects provide information on the exposure history of meteorites, both on their parent bodies and in interplanetary space during their travel between their asteroidal sources and the Earth. Ernst Zinner

Meteorite impact. The process of impact cratering was of fundamental importance for the accumulation of planets in the early solar system, the formation of planetary landscapes, and the Archean geology of the Earth. In addition, meteorite impacts are implicated in the Moon's origin and the extinction of the dinosaurs.

The precise outcome of a planetary collision depends on the size of the meteorite and conditions on the target planet. Small meteorites striking planets such as the Earth or Venus dissipate most of their energy in the atmosphere and do not strike the surface at high speed. In general, if the mass of the meteorite is small compared to the mass of atmospheric gases displaced during its entry, it will not create an impact crater. On airless bodies such as the Moon, there seems to be no lower limit on impact crater size: Craters as small as a few micrometers in diameter have been discovered on the lunar rocks. On Earth, the atmosphere prevents stony meteorites or comets from making craters smaller than a few kilometers in diameter, and even iron meteorites cannot make high-speed impact craters smaller than a few hundred meters in diameter.

When a large meteorite does penetrate a planet's atmosphere, it initiates a series of swift but orderly processes that eventually create a characteristic landform, an impact crater. Three principal stages are recognized in this process.

1. The meteorite first plunges into the surface rocks at high speed, compressing the underlying rocks and converting its initial kinetic energy into both heat and kinetic energy of the surface rocks. The high pressures produce a series of characteristic mineralogical changes in the surrounding rocks that often permit verification of the impact origin of a suspected crater. The duration of this compression stage is short, however, lasting only as long as it takes the meteorite to travel a distance equal to its own diameter.

2. Subsequently, the pent-up pressures in the compressed rocks create an explosion, blasting aside the surrounding rocks as a strong shock wave radiates away from the impact site. The nearly hemispherical shock wave from the impact expands and weakens as time passes, leaving behind outward-moving rock debris that eventually excavate the crater. The maximum depth of excavation is only about 10% of the crater's diameter. The immediate result of crater excavation is called the transient crater. This is a relatively deep, steep-walled crater that begins to collapse as soon as it forms.

3. Small transient craters are quickly filled by a lens of broken rock that forms from debris that slides down from the rim and pools at the bottom of the crater. Such bowl-shaped craters floored by broken rock are called simple craters.

In larger craters the floor rises as the rim sinks, producing craters with central mounds that are thinly veneered with broken and melted rock. The rims of such craters are scalloped and terraced with great blocks of slumped rock. Still larger craters exhibit circular mountainous rings instead of central peaks.

The very largest impact structures, particularly on the Moon, are surrounded by inward-facing, roughly circular (but often incomplete) mountain rings that probably formed well outside the crater cavity by a process of inward flow and slumping in the fluid asthenosphere beneath the crater. They are termed multi-ring basins. These enormous structures dominate the Moon's surface and form the principal stratigraphic markers on that body. H. J. Melosh

Meteorological instrumentation Devices that measure or estimate properties of the Earth's atmosphere. Meteorological instruments take many forms, from simple mercury thermometers and barometers to complex observing systems that remotely sense winds, thermodynamic properties, and chemical constituents over large volumes of the atmosphere.

Weather station measurements provide a description of conditions near the ground. In addition to the average regional conditions, these measurements also provide local information on mesoscale phenomena such as cold fronts, sea breezes, and disturbed conditions resulting from nearby thunderstorms. Traditional thermodynamic instruments are mechanical or heat-conductive devices relying on the expansion and contraction of metallic and nonmetallic liquids or solid materials as a function of temperature, pressure, and humidity. Among these are the mercury, alcohol, and bimetallic thermometers for measurement of temperature, mercury and metallic bellow (aneroid) barometers for measurement of pressure, human hair hygrometers, and wet/dry-bulb thermometers (called psychrometers) for measurement of relative humidity. Mercury barometers are simply weighing devices that balance the mass of the atmospheric column against the mass of a mercury column. On average, a column of atmosphere weighs the same as 76 cm (29.92 in.) of mercury. Psychrometers measure humidity by means of the wet-bulb depression technique. A moist thermometer is cooled by evaporation when relative humidity is less than 100%. The temperature difference between wet and dry thermometers is referred to as the wet-bulb depression, a well-known function of relative humidity at standard airflow speeds. A related method of humidity measurement is the chilled mirror technique (dewpointer). A polished surface is cooled to the temperature of water vapor saturation, at which point the cooled surface becomes fogged. Dewpoint saturation uniquely defines humidity at a known temperature and pressure. *See* BAROMETER; DEW POINT; HYGROMETER; PSYCHROMETER; TEMPERATURE MEASUREMENT; THERMOMETER.

Precipitation measurement devices may be described as precision buckets, which measure the depth or weight of that which falls into them. These gages work best for rainfall, but they are also used in an electrically heated mode for weighing snow. Rulers are routinely used for measurement of snow depth. Time-resolved measurements of rainfall are traditionally made by counting quantum amounts (0.01 in. or 0.25 mm) of rain with a small, mechanically controlled tipping bucket located beneath a large collecting orifice. Modern rain measuring is sometimes performed along short paths via drop-induced scintillations of infrared radiation, which is emitted by a laser. When the raindrop size distribution is needed, optical-shadowing spectrometers are employed, as are momentum-measuring impact distrometers, devices that measure the number density versus the size distribution of raindrops or other hydrometeors. *See* SNOW GAGE.

Wind measurements are performed by anemometers, some of which use wind-driven spinning cups for wind speed determination. Vanes are used in conjunction with cups for indication of wind direction. Alternatively, three-axis propeller anemometers may be employed to provide orthogonal components of the three-dimensional wind vector. Many hybrids of these basic approaches continue to be successfully employed. Fast-response sonic anemometers employ ultrasound transmission, where the apparent propagation speed of sound is measured. The difference between this measured speed and the actual speed for a fluid at rest is the wind speed. Such measurements are made on a time scale of 0.01 s and are used to determine the fluxes of momentum, water vapor, sensible heat, and other scalars in the planetary boundary layer. *See* ANEMOMETER; WIND MEASUREMENT.

Balloon-borne vertical profiles or soundings of temperature, humidity, and winds are central to computerized (numerical) weather prediction. Such observations are made simultaneously or synoptically worldwide on a daily basis. The temperature and humidity sensors are lightweight expendable versions of traditional surface station instruments. Balloon drift during ascent provides the wind measurement. The preferred method of tracking these rawinsondes is to use global navigation aid systems such as Omega, Loran-C, and the Global Positioning System. Parachute-borne dropsondes are often released from aircraft in data-sparse regions. *See* LORAN; SATELLITE NAVIGATION SYSTEMS.

Remote sensing, principally via electromagnetic radiation, is a mainstay of modern meteorology. Such devices typically operate in the optical, infrared, millimeter-wave, microwave, and high-frequency radio regions of the electromagnetic spectrum. Passive radiometers typically operate at infrared and microwave frequencies; they are used for estimates of temperature, water vapor, cloud heights, cloud liquid water mass, and trace-gas concentrations. These observations are made from the ground, aircraft, and satellites, usually measuring naturally emitted radiation. Radarlike, active remote-sensing devices are among the most powerful tools available to meteorology. Collectively, these instruments are capable of measuring kinematic, microphysical, chemical, and thermodynamic properties of the troposphere at high spatial and temporal resolution. Active meteorological remote sensors are principally deployed on land, ships, and aircraft platforms, as well as aboard satellites. Unlike passive instruments, active remote sensors can precisely resolve the distance at which a measurement is located.

At optical frequencies, lidars measure conditions in relatively clear air. Capabilities include determining the properties of tenuous clouds; determining concentrations of aerosol, ozone, and water vapor; and measuring winds through the Doppler frequency-shift effect. Millimeter-wave radars are used to probe opaque, nonprecipitating clouds. Polarimetric and Doppler techniques reveal hydrometeor type, water mass, and air motions. *See* LIDAR.

The best-known meteorological remote sensor is the microwave weather radar. In addition to measuring rainfall and tracking movement of storms, powerful and sensitive meteorological radars can measure detailed flow fields in and around storms by using hydrometeors, insects, and blobs of water vapor as reflective targets. These radars can also distinguish between rain, hail, and snow. When Doppler measurements are combined with the atmospheric equations of motion, thermodynamic perturbation fields, such as buoyancy, are revealed inside violent convective storms. At ultrahigh and very high radio frequencies, radars known as wind profilers measure the mean wind as a function of height in the clear and cloudy air. Superior to infrequent weather balloons, radio wind profiling methods permit continuous measurement of winds with regularity and high accuracy. When radio wind profilers are colocated with acoustic transponders, the speed of sound is easily measured through radar tracking of the acoustic wave. This permits the computation of atmospheric density and temperature profiles, on which the speed of sound is strongly dependent. *See* DOPPLER RADAR; METEOROLOGY; RADAR METEOROLOGY; REMOTE SENSING. Richard E. Carbone

Meteorological optics The study of optical phenomena occurring in the atmosphere. Many light effects can be seen by looking skyward, and all of them, resulting from the interaction of light with the atmosphere, lie in the province of atmospheric optics or meteorological optics. The subject also includes the effect of light waves too long or too short to be detected by the human eye—light-type radiation in the infrared or ultraviolet regions of the spectrum. Light interacts with the different components of the atmosphere by a variety of physical processes, the most important being scattering, reflection, refraction, diffraction, absorption, and emission. *See* ABSORPTION OF ELECTROMAGNETIC RADIATION; ATMOSPHERE; OPTICS; REFLECTION OF ELECTROMAGNETIC RADIATION; SCATTERING OF ELECTROMAGNETIC RADIATION. Robert Greenler

Meteorological radar A remote-sensing device that transmits and receives microwave radiation for the purpose of detecting and measuring weather phenomena. Radar is an acronym for radio detection and ranging. Today, many types of sophisticated radars are used in meteorology, ranging from Doppler radars, which are used to determine air motions (for example, to detect tornadoes), to multiparameter radars, which

Parabolic dish antenna of the CSU-CHILL 11-cm multiparameter Doppler radar operating at Colorado State University. The antenna is housed within a large, inflatable radome. (*P. Kennedy, Colorado State University*)

provide information on the phase (ice or liquid), shape, and size of hydrometeors. Airborne Doppler radars play a vital role in meteorological research. Radars are also used to detect hail, estimate rainfall rates, probe the clear-air atmosphere to monitor wind patterns, and study the electrification processes in thunderstorms that generate lightning discharges.

Commonly used, pulsed Doppler radar operates in the microwave region, with standard wavelengths of 10, 5, and 3 cm, referred to as S-, C-, and X-band radars, respectively. The electromagnetic radiation is focused into a narrow beam by illuminating a parabolic dish reflector with microwave energy provided by the radar transmitter. S-band radars require the use of large antennas (see illustration) to generate a narrow beam of microwave energy; transmit high power (peak power of 1 megawatt); and suffer relatively little attenuation as the radar beam passes through regions of heavy rain and hail. X-band radars use much smaller antennas to achieve similar narrow beams, and are highly portable. However, X-band radars suffer from attenuation when used to probe precipitation, which significantly limits their range. Attenuation results when the radar energy is either absorbed by the raindrops or reemitted from the raindrops in directions other than toward the radar. *See* ANTENNA (ELECTROMAGNETISM); DOPPLER RADAR; MICROWAVE.

A pulsed Doppler radar typically emits 1000 electromagnetic pulses per second. These individual pulses are typically 1 microsecond (10^{-6} s) in duration. The Doppler radar provides information on the target's velocity, either toward or away from the radar when viewed along the radar beam. The Doppler shift, which is measured as a small difference between the frequency of the transmitted pulse and the frequency of the energy backscattered to the radar, provides a measure of the scatterer's radial motion. Scatterers in the case of meteorological radar include raindrops, ice particles (snowflakes), hailstones, and even insects, providing clear air returns. A Doppler radar also detects the amplitude of the backscattered signal, which can be used as a measure of storm intensity and as a means of estimating rainfall rates. *See* PRECIPITATION (METEOROLOGY); SCATTERING OF ELECTROMAGNETIC RADIATION; STORM; STORM DETECTION.

Dual-wavelength radar transmits electromagnetic energy at two wavelengths, and it also receives energy at both wavelengths. Typically, S- and X-band wavelengths are used. Dual-wavelength techniques were originally proposed to detect large hail. At S-band, hail is usually a Rayleigh target, whereas at X-band, hail is considered a Mie scatterer. Since radar energy is scattered in various directions by a Mie target, the power returned at the X-band wavelength is reduced relative to that at S-band. The presence of large hail is interpreted on the basis of the ratio of backscattered power at X-band to that at S-band. Dual-wavelength techniques are also used to estimate rainfall rates by comparing the backscattered power at a nonattenuating wavelength (S-band) to that at attenuating wavelengths (X-band). *See* HAIL.

Dual-polarization radar is able to transmit and receive both horizontally and vertically polarized radiation (the polarization of a radar beam is defined by the orientation of the electric field vector that comprises an electromagnetic wave). These radars are now used in meteorological research and have largely superseded dual-wavelength radars. A suite of multiparameter variables is being used to infer information on particle phase (ice or water), size, orientation, and shape.

Radar operating at a wavelength of 2 cm is in low Earth orbit (350 km or 210 m above the surface) and is used for mapping tropical precipitation. Understanding the amount and distribution of tropical rainfall is crucial for better understanding the Earth's climate. This space-borne radar and associated satellite was jointly developed by the United States (NASA) and Japan (National Space Development Agency); it is known as the *Tropical Rainfall Measuring Mission* (*TRMM*) satellite. Space-borne radar presents many challenging problems, including cost, size constraints, reliability issues, and temporal sampling. It is obviously impossible to continuously sample every precipitating cloud in the tropics from radar orbiting the Earth. But the *TRMM* satellite will help scientists develop a statistical distribution of rain rates within a certain area, and calculate the probability of a specific rain rate occurring. Based on this information, it will be possible to generate monthly mean rain amounts within areas of 10^5 km^2. Such information will be vital for the verification of climate models. *See* SATELLITE METEOROLOGY.

Radars used to probe the clear air, or regions devoid of clouds, are known as profilers. A profiler is essentially a Doppler radar that operates at much longer wavelengths compared to weather radar. Wavelengths of 6 m, 70 cm, and 33 cm are commonly used. In the case of a profiler, the reflected power is not only from hydrometeors but also from gradients in the index of refraction of air, which are caused by turbulent motions in the atmosphere. These turbulent motions in turn cause small fluctuations in air temperature and moisture content, which also change the index of refraction. A profiler can determine the airflow in the cloud-free atmosphere, roughly up to 10 km above the Earth's surface. Optical radars, called lidars, use lasers as the radiation source. At these short wavelengths (0.1–10 μm), the laser beam is scattered by small aerosol particles and air molecules, allowing air motions to be determined, especially in thin, high tropospheric clouds and in the Earth's boundary layer (approximately the lowest 1 km or 0.6 mi of the Earth's atmosphere). *See* AEROSOL; HYDROMETEOROLOGY; LASER; LIDAR.

In the late 1990s, the weather radars used by the National Weather Service to provide warnings of impending severe weather were updated from antiquated WSR-57 and WSR-74 noncoherent radars to NEXRADs (Next Generation Weather Radars). NEXRAD (WSR-88D) radars are state-of-the-art Doppler radars operating at a wavelength of 10 cm. Using NEXRAD's Doppler capability, weather forecasters are able to warn the public sooner of approaching tornadoes and other severe weather. In severe storms, a mesocyclone first develops within the storm. The mesocyclone may be 10 km (6 mi) or more wide, and represents a deep rotating column of air within the storm. Severe and long-lasting tornadoes are often associated with mesocyclones. The mesocyclone is readily detected by a Doppler radar such as NEXRAD. The entire continental United States is covered by the NEXRAD network, consisting of more than 100 radars. NEXRADs along the Gulf Coast, in Florida, and along the eastern seaboard provide warning information on land-falling hurricanes. About 60 of the nation's busiest airports are also equipped with Doppler radars. These radars (operating at a wavelength of 5 cm, known as Terminal Doppler Weather Radars) provide weather-related warnings to air-traffic controllers and pilots. One particularly dangerous weather condition is wind shear, which often occurs as a microburst or intense downdraft. Microbursts can severely affect the flight of landing and departing aircraft, and have been identified as a factor in many aircraft accidents. *See* RADAR; RADAR METEOROLOGY; TORNADO; WEATHER FORECASTING AND PREDICTION.

Steven A. Rutledge

Meteorological rocket A small rocket system used for extending observations of the atmosphere above feasible limits for balloon-borne and telemetering instruments. Synoptic exploration of the middle-atmospheric circulation (20–95 km or 12–60 mi altitude) through use of these systems (also known as rocketsondes) matured in the 1960s into a highly productive source of information on atmospheric structure and dynamics. Many thousands of small meteorological rockets have been launched in a coordinated investigation of the wind field and the temperature and ozone structures in the middle atmosphere region at 25–55 km (16–34 mi) altitude. These data produced dramatic changes in the scientific view of this region of the atmosphere, with a resulting alteration of the structural concepts into an atmospheric model that is primarily characterized by intense dynamics.

The development of the small meteorological rocket began in 1959 with a rocketsonde system known as ARCAS. The maximum altitude reached by this system was about 60 km (37 mi). A less wind-sensitive rocketsonde system, the Loki-Datasonde (PWN-8B), replaced the ARCAS system during the early 1970s. It was soon replaced with the Super Loki-Datasonde (PWN-11D). The PWN-11D rocketsonde motor burns for 2 s before separation from its inert dart and payload, which are thereby propelled to about 80 km (50 mi) altitude, where the payload is ejected. The payload consists of a small bead thermistor temperature sensor attached to a radio transmitter that sends the temperature data to a ground receiver, and a Starute parachute. The meteorological measurements are made during payload descent. At launch, the Super Loki-Datasonde has an overall weight of 31 kg (68 lb), and its length is approximately 4 m (13 ft).

Synoptic-scale circulation systems in the upper atmosphere are demonstrated by rocketsonde data to be very obviously keyed to the geographic and orographic structures of the Earth's surface. In winter, oceanic regions characteristically have poleward extensions of ridges of high pressure, and continental regions have shifty troughs of low pressure extending equatorward over them. This intimate relationship between the surface and 50 km (31 mi) is most likely the direct result of turbulent energy transport in the vertical direction, and a total understanding of the entire atmospheric system cannot be realized until these factors are incorporated. See ATMOSPHERIC GENERAL CIRCULATION; STRATOSPHERE; UPPER-ATMOSPHERE DYNAMICS.

Willis L. Webb; Francis J. Schmidlin

Meteorological satellites Satellites dedicated to the observation of meteorological phenomena and atmospheric or surface properties used for weather forecasting. Operational meteorological satellites provide routine observations of weather conditions as well as an ever expanding range of environmental properties, such as aerosol, dust and ash clouds from volcanic eruptions, ozone, and land vegetation cover. For this reason, they are known in the United States as operational environmental satellites. See METEOROLOGY; SATELLITE (SPACECRAFT); SATELLITE METEOROLOGY.

Optical imaging sensors. The first recognized application of orbital observation was the visual exploitation of cloud images associated with weather systems. Recent instruments, such as the Advanced Very High Resolution Radiometer (AVHRR) on *NOAA* satellites, use a variety of quantitative applications, such as remote sensing of sea surface temperature, monitoring changes in land vegetation, and discriminating between different kinds of clouds. There is a pervasive trend to increase the number of spectral bands in imaging sensors, from 5 channels in the current AVHRR to 36 channels in the experimental Moderate-resolution Imaging Spectroradiometer (MODIS) developed by NASA. These channels sample the full spectrum of backscattered solar radiation in the visible, near-infrared, and longwave infrared, and a good part of the emitted terrestrial radiation spectrum (thermal infrared). This multiplicity of spectral bands allows the detection of a wide variety of features, from aerosols and smoke in the atmosphere to chlorophyll in the ocean. See CLOUD; REMOTE SENSING; TERRESTRIAL RADIATION; WEATHER.

Except for observing polar regions, or providing meteorological support to operations in remote locations worldwide, the ideal platforms for cloud imaging are those in geosynchronous equatorial orbit, also known as geostationary orbit, at the precise altitude (35,900 km) where the orbital period matches the period of rotation of the Earth, so that the satellite appears to hover over a fixed location at the Equator. The international system of four to six geostationary meteorological satellites provides uninterrupted visibility of the global tropics and midlatitudes (up to 60° north and south at the satellite longitude) with the ability to monitor fast-developing weather systems that often are the most dangerous. The sharpness of cloud images (1-km picture elements in the visible), as well as the ability to scan the same scene repeatedly at time intervals as short as 5 minutes, allow for tracking the apparent motion of clouds, deducing wind velocity, and instantaneously assessing the strength of developing storms, a valuable capability in warm climate regions. See EARTH ROTATION AND ORBITAL MOTION; TROPICAL METEOROLOGY.

Imaging microwave radiometers. Also interesting is the detection of diverse atmospheric properties and surface features using multifrequency microwave radiometers with small antenna beams. Water molecule absorption of microwave radiation emitted by the ocean provides an accurate estimation of total precipitable water in the atmospheric column. Microwave radiation emitted by the relatively homogeneous moist atmosphere below is scattered in a recognizable way by waterdrops and ice particles in rain clouds, thus providing an indirect means to estimate precipitation rates. Microwave radiation contrast discriminates ice floes from open ocean water, and wet from dry soil. Microwave radiometry enables diagnostics of sea state and wind strength over the surface of the ocean, or the sea surface temperature. The principal design constraint of imaging microwave radiometers is the diffraction limit of the sensor—large apertures are desirable, but bulky antennas are a problem because mechanical scanning is needed to preserve radiometric accuracy. In order to achieve reasonably small footprints, microwave sensors are currently deployed in low Earth orbit. See MICROWAVE; PRECIPITATION (METEOROLOGY); RADIOMETRY.

Sounding sensors. The retrieval of temperature profile and water vapor information from spectral data is a difficult and not a fully determined mathematical problem. The solutions are highly sensitive to spectral resolution and small errors in radiometric measurements. The latest Atmospheric Infra-Red Sounder (AIRS) instrument developed by NASA is expected to yield temperature profiles as accurate as balloon measurements, 1°C within each successive 1-km-thick layer of the lower atmosphere. See HYDROMETEOROLOGY; INFRARED RADIATION.

Atmospheric sounders operate in the thermal infrared, using the absorption bands of carbon dioxide molecules (3.7–4.9 μm and 13–15 μm), and in the microwave spectrum, using the 54-GHz absorption band of oxygen. Emitted radiation is much weaker and atmospheric sounders correspondingly less sensitive in the microwave region. However, nonprecipitating clouds are largely transparent to such relatively long wavelengths, thus allowing all-weather albeit less accurate observations.

Measurements of temperature and moisture are used mainly to update numerical weather prediction computations that forecast the circulation of the global atmosphere several days in advance. For this quantitative application, a delay of a few hours is immaterial but homogeneous global coverage is essential. Thus, atmospheric sounders are principally deployed on Sun-synchronous polar orbits. The parameters of these circular low Earth orbits are selected from a discrete set of altitudes (800–1000 km) and inclinations (retrograde quasi-polar) that allow the orbital plane to drift by about 1° of longitude per day and match the change in Sun-Earth direction. Thus, a Sun-synchronous satellite crosses the Equator at (nearly) the same local time on every successive orbit.

Active sensors. Orbital systems are now powerful enough to probe the atmospheric medium or the surface with beams of electromagnetic radiation generated in space. The first operational sensor of this kind was a coarse radar or scatterometer that measured microwave radiation backscattered by the ocean surface. Backscatter is sensitive to surface roughness and thus provides a measurement of vector wind speed over the ocean (as well as a coarse all-weather mapping of sea ice).

Various radar altimeters have been used to map the changing topography of the ocean surface (principally to reconstruct the oceanic circulation from measured altitude gradients).

Higher-frequency experimental radar and lidar systems are being tested to profile the distribution and optical properties of aerosol and cloud ice particles and waterdrops. *See* LIDAR; METEOROLOGICAL RADAR; RADAR METEOROLOGY. Pierre Morel

Meteorology A discipline involving the study of the atmosphere and its phenomena. Meteorology and climatology are rooted in different parent disciplines, the former in physics and the latter in physical geography. They have, in effect, become interwoven to form a single discipline known as the atmospheric sciences, which is devoted to the understanding and prediction of the evolution of planetary atmospheres and the broad range of phenomena that occur within them. The atmospheric sciences comprise a number of interrelated subdisciplines. *See* CLIMATOLOGY.

Atmospheric dynamics (or dynamic meteorology) is concerned with the analysis and interpretation of the three-dimensional, time-varying, macroscale motion field. It is a branch of fluid dynamics, specialized to deal with atmospheric motion systems on scales ranging from the dimensions of clouds up to the scale of the planet itself. The activity within dynamic meteorology that is focused on the description and interpretation of large-scale (greater than 1000 km or 600 mi) tropospheric motion systems such as extratropical cyclones has traditionally been referred to as synoptic meteorology, and that devoted to mesoscale (10–1000 km or 6–600 mi) weather systems such as severe thunderstorm complexes is referred to as mesometeorology. Both synoptic meteorology and mesometeorology are concerned with phenomena of interest in weather forecasting, the former on the day-to-day time scale and the latter on the time scale of minutes to hours. *See* DYNAMIC METEOROLOGY; MESOMETEOROLOGY.

The complementary field of atmospheric physics (or physical meteorology) is concerned with a wide range of processes that are capable of altering the physical properties and the chemical composition of air parcels as they move through the atmosphere. It may be viewed as a branch of physics or chemistry, specializing in processes that are of particular importance within planetary atmospheres. Overlapping subfields within atmospheric physics include cloud physics, which is concerned with the origins, morphology, growth, electrification, and the optical and chemical properties of the droplets within clouds; radiative transfer, which is concerned with the absorption, emission, and scattering of solar and terrestrial radiation by aerosols and radiatively active trace gases within planetary atmospheres; atmospheric chemistry, which deals with a wide range of gas-phase and heterogeneous (that is, involving aerosols or cloud droplets) chemical and photochemical reactions on space scales ranging from individual smokestacks to the global ozone layer; and boundary-layer meteorology or micrometeorology, which is concerned with the vertical transfer of water vapor and other trace constituents, as well as heat and momentum across the interface between the atmosphere and the underlying surfaces and their redistribution within the lowest kilometer of the atmosphere by motions on scales too small to resolve explicitly in global models. Aeronomy is concerned with physical processes in the upper atmosphere (above the 50-km or 30-mi level). *See* AERONOMY; ATMOSPHERIC CHEMISTRY; ATMOSPHERIC ELECTRICITY; ATMOSPHERIC GENERAL CIRCULATION; ATMOSPHERIC WAVES, UPPER SYNOPTIC; CLOUD PHYSICS; METEOROLOGICAL OPTICS; MICROMETEOROLOGY; RADIATIVE TRANSFER; TERRESTRIAL RADIATION.

Although atmospheric dynamics and atmospheric physics in some circumstances can be successfully pursued as separate disciplines, important problems such as the development of numerical weather prediction models and the understanding of the global climate system require a synthesis. Physical processes such as radiative transfer and the condensation of water vapor onto cloud droplets are ultimately responsible for the temperature gradients that drive atmospheric motions, and the motion field, in turn, determines the evolving, three-dimensional setting in which the physical processes take place.

The atmospheric sciences cannot be completely isolated from related disciplines. On time scales longer than a month, the evolution of the state of the atmosphere is influenced by dynamic and thermodynamic interactions with the other elements of the climate system, that is, the oceans, the cryosphere, and the terrestrial biosphere. A notable example is the El Niño-Southern Oscillation phenomenon in the equatorial Pacific Ocean, in which changes in the distribution of surface winds force anomalous ocean currents; the currents can alter the distribution of sea-surface temperature, which in turn can alter the distribution of tropical rainfall, thereby inducing further changes in the surface wind field. On a time scale of decades or longer, the cycling of chemical species such as carbon, nitrogen, and sulfur between these same global reservoirs also influences the evolution of the climate system. Human activities represent an increasingly significant atmospheric source of some of the radiatively active trace gases that play a role in regulating the temperature of the Earth. *See* BIOSPHERE; MARITIME METEOROLOGY; TROPICAL METEOROLOGY.

Throughout the atmospheric sciences, prediction is a unifying theme that sets the direction for research and technological development. Prediction on the time scale of minutes to hours is concerned with severe weather events such as tornadoes, hail, and flash floods, which are manifestations of intense mesoscale weather systems, and with urban air-pollution episodes; day-to-day prediction is usually concerned with the more ordinary weather events and changes that attend the passage of synoptic-scale weather systems such as extratropical cyclones; and seasonal prediction is concerned with regional climate anomalies such as drought or recurrent and persistent cold air outbreaks. Prediction on still longer time scales involves issues such as the impact of human activity on the temperature of the Earth, regional climate, the ozone layer, and the chemical makeup of precipitation. *See* CLIMATE MODELING; DROUGHT; HAIL; TORNADO.

The evolution of the atmospheric sciences from a largely descriptive field to a mature, quantitative physical science discipline is apparent in the development of vastly improved predictive capabilities based upon the numerical integration of specialized versions of the Navier-Stokes equations, which include sophisticated parametrizations of physical processes such as radiative transfer, latent heat release, and microscale motions. The so-called numerical weather prediction models have largely replaced the subjective and statistical prediction methods that were widely used as a basis for day-to-day weather forecasting. The state-of-the-art numerical models exhibit significant skill for forecast intervals as long as about a week. *See* NAVIER-STOKES EQUATION.

A distinction is often made between weather prediction, which is largely restricted to the consideration of dynamic and physical processes internal to the atmosphere, and climate prediction, in which interactions between the atmosphere and other elements of the climate system are taken into account. The importance and complexity of these interactions tend to increase with the time scale of the phenomena of interest in the forecast. Weather prediction involves shorter time frames (days to weeks), in which the information contained in the initial conditions is the dominant factor in determining the evolution of the state of the atmosphere; and climate prediction involves longer time frames (seasons and longer), for boundary forcing is the dominant factor in determining the state of the atmosphere.

Atmospheric prediction has benefited greatly from major advances in remote sensing. Geostationary and polar orbiting satellites provide continuous surveillance of the global distribution of cloudiness, as viewed with both visible and infrared imagery. These images are used in positioning of features such as cyclones and fronts on synoptic charts. Cloud motion vectors derived from consecutive images provide estimates of winds in regions that have no other data. Passive infrared and microwave sensors aboard satellites also provide information on the distribution of sea-surface temperature, sea state, land-surface vegetation, snow and ice cover, as well as vertical profiles of temperature and moisture in cloud-free regions. Improved ground-based radar imagery and vertical profiling devices provide detailed coverage of convective cells and other significant mesoscale features over land areas. Increasingly sophisticated data assimilation schemes are being developed to incorporate this variety of information into numerical weather prediction models on an operational basis. *See* ATMOSPHERE; CLIMATIC PREDICTION; CYCLONE; FRONT; RADAR METEOROLOGY; SATELLITE METEOROLOGY; WEATHER FORECASTING AND PREDICTION. John M. Wallace

Methane The simplest compound of carbon and hydrogen, CH_4. At room temperature, methane is a gas less dense than air. The gas liquefies at −164°C (−263°F) and solidifies at −183°C (−297°F). It is not very soluble in water. Methane is combustible, and mixtures of about 5–15% in air are explosive. Complete combustion of methane produces carbon dioxide and water. Methane is not toxic when inhaled, but it can produce suffocation by reducing the concentration of oxygen. *See* ALKANE.

Methane is widely distributed in nature. It is the principal component of natural gas. The combustible gas found in coal mines is chiefly methane. Marsh gas, which is produced under water by anaerobic bacterial decomposition of plant and animal matter, is also methane. Some plants emit methane, and many animals emit methane as a by-product of food digestion. Methane is one of the most powerful greenhouse gases in the atmosphere, being more than 60 times as effective as carbon dioxide. *See* COALBED METHANE; GREENHOUSE EFFECT; NATURAL GAS.

In the chemical industry, methane is a raw material for the manufacture of methanol (CH_3OH), formaldehyde (CH_2O), nitromethane (CH_3NO_2), chloroform (CH_3Cl), carbon tetrachloride (CCl_4), and some halogenated hydrocarbons.

Bassam Z. Shakhashiri; Rodney Schreiner

Methanogenesis (bacteria) The microbial formation of methane, which is confined to anaerobic habitats where occurs the production of hydrogen, carbon dioxide, formic acid, methanol, methylamines, or acetate—the major substrates used by methanogenic microbes (methanogens). In fresh-water or marine sediments, in the intestinal tracts of animals, or in habitats engineered by humans such as sewage sludge or biomass digesters, these substrates are the products of anaerobic bacterial metabolism. Methanogens are terminal organisms in the anaerobic microbial food chain—the final product, methane, being poorly soluble, anaerobically inert, and not in equilibrium with the reaction which produces it.

Two highly specialized digestive organs, the rumen and the cecum, have been evolved by herbivores to delay the passage of cellulose fibers so that microbial fermentation may be complete. In these organs, large quantities of methane are produced from hydrogen and carbon dioxide or formic acid by methanogens. From the rumen, an average cow may belch 26 gallons (100 liters) of methane per day.

Methanogens are the only living organisms that produce methane as a way of life. The biochemistry of their metabolism is unique and definitively delineates the group. Two reductive biochemical strategies are employed: an eight-electron reduction of carbon dioxide to methane or a two-electron reduction of a methyl group to methane. All methogens form methane by reducing a methyl group. The major energy-yielding reactions used by methanogens utilize substrates such as hydrogen, formic acid, methanol, acetic acid, and methylamine. Dimethyl sulfide, carbon monoxide, and alcohols such as ethanol and propanol are substrates that are used less frequently. *See* ARCHAEA; BACTERIAL PHYSIOLOGY AND METABOLISM; METHANE. Ralph S. Wolfe

Methanol The first member of the homologous series of aliphatic alcohols, with the formula CH_3OH. It is produced commercially from a mixture of carbon monoxide (CO) and hydrogen (H_2). Methanol is a highly flammable liquid, boiling point 64.7°C (149°F), and is miscible with water and most organic liquids. It is a highly poisonous substance; sublethal amounts can cause permanent blindness. *See* ALCOHOL.

Methanol is one of the major industrial organic chemicals. Its major derivatives are methyl tertiary butyl ether (MTBE), formaldehyde, and acetic acid. Other derivatives and uses include chloromethanes, methyl methacrylate, methylamines, dimethyl terephthalate, solvents (such as glycol methyl ethers), antifreeze, and fuels. James A. Moore

Methods engineering A technique used by industrial engineers to improve productivity and quality and to reduce costs in both direct and indirect operations of manufacturing and service organizations. Methods engineering is applicable in any enterprise requiring human effort. It can be defined as the systematic procedure for subjecting all direct and indirect operations to close scrutiny in order to introduce improvements that will make work easier to perform while maintaining or improving quality, and will allow work to be done more smoothly, in less time, with less energy, effort, and fatigue, and with less investment per unit. The ultimate objective of methods engineering is increasing profits, but it is also important in improving worker health and safety.

The terms methods engineering, operation analysis, and work design are frequently used synonymously. Methods engineering entails analysis work at two different times during the history of a product. Initially, the methods engineer is responsible for designing and developing the various work centers where the product will be produced. Next, he or she continually restudies the work centers to find a better way to produce the product and/or improve its quality. More recently, this second analysis has been called reengineering or lean manufacturing. *See* LEAN MANUFACTURING; OPERATIONS RESEARCH; PRODUCTIVITY.

Methods engineers use a systematic procedure. To develop a work center, to produce a product, or to provide a service. This procedure is outlined below. Note that Steps 6 and 7 are not strictly part of a methods study but are necessary in a fully functioning work center.

1. *Select the project.* Typically, projects selected represent either new or existing products that have a high cost of manufacture and a low profit. Also, products that have quality problems and difficulty in meeting competition are logical methods engineering projects.

2. *Get and present the data.* Assemble all the important facts relating to the product or service. These include drawings and specifications, quantity requirements, delivery requirements, and projections about the anticipated life of the product or service. Once all the important information has been acquired, record it in an orderly form for study and analysis. The development of process charts at this point is very helpful.

3. *Analyze the data.* Utilize the primary approaches to operations analysis to decide which alternative will produce the best product or service. These primary approaches include purpose of operation, design of part, tolerances and specifications, materials, manufacturing process, setup and tools, working conditions, materials handling, plant layout, and principles of motion economy.

4. *Develop the ideal method.* Select the best procedure for each operation, inspection, and transportation by considering the various constraints associated with each alternative, including productivity, ergonomic, and health and safety implications.

5. *Present and install the method.* Explain the proposed method in detail to those responsible for its operation and maintenance. Consider all details of the work center to assure that the proposed method will provide the results anticipated.

6. *Develop a job analysis.* Make a job analysis of the installed method to assure that the operator or operators are adequately selected, trained, and rewarded.

7. *Establish time standards.* Establish a fair and equitable standard for the installed method.

8. *Follow up the method.* At regular intervals, audit the installed method to determine if the anticipated productivity and quality are being realized, if costs were correctly projected, and if further improvements can be made. Andris Freivalds

Metric system A system of measurement units used in scientific work and for everyday applications by most countries, with the notable exception of the United States. A prime advantage of the metric system is its international acceptance as a standard of measurement, providing a common measurement language for most of the world's population.

Table 1. SI base units

Quantity	Unit	Symbol
Length	meter	m
Mass	kilogram	kg
Time	second	s
Electric current	ampere	A
Thermodynamic temperature	kelvin	K
Amount of substance	mole	mol
Luminous intensity	candela	cd

The meter-kilogram-second (MKS) and centimeter-gram-second (cgs) systems were major variations of the metric system. Both are now superseded by the SI-metric system adopted in 1960 by the 11th General Conference on Weights and Measures (CGPM). The new SI-metric system, given the official abbreviation SI for Système International d'Unités, is also called the modernized metric system.

Qualities of the metric system include being decimal and coherent. The decimal nature makes it easy to manipulate these units with the decimal number system used throughout the world. Coherency is a quality of the derivation of SI units from the base units of the system. There is a one-to-one relationship between the equations of physics and the SI units for the quantities expressed in an equation. Metric units and symbols can be manipulated algebraically in the same manner as the physical quantities they represent. There is no need for either multipliers or conversion factors, as is often the case with nonmetric systems of units.

Table 2. SI prefixes

Prefix	Symbol	Multiplication factor	
yotta	Y		10^{24}
zetta	Z		10^{21}
exa	E		10^{18}
peta	P		10^{15}
tera	T		10^{12}
giga	G	1,000,000,000	10^{9}
mega	M	1,000,000	10^{6}
kilo	k	1000	10^{3}
hecto	h	100	10^{2}
deka	da	10	10^{1}
deci	d	0.1	10^{-1}
centi	c	0.01	10^{-2}
milli	m	0.001	10^{-3}
micro	μ	0.000,001	10^{-6}
nano	n	0.000,000,001	10^{-9}
pico	p		10^{-12}
femto	f		10^{-15}
atto	a		10^{-18}
zepto	z		10^{-21}
yocto	y		10^{-24}

The SI-metric system has seven base units from which all other units are derived (Table 1). These seven units are independent quantities, six of which are now defined in terms of reproducible experimental procedures. Only the kilogram is still defined using an artifact, a platinum-iridium cylinder kept under glass at the International Bureau of Weights and Measures (BIPM) in France, copies of which are available to other countries.

Twenty SI prefixes, from plus to minus the 24th power of 10, cover a range of measurement from the extremely large to the extremely small (Table 2). The prefixes are spaced in increments of 10 to the power 3, except for additional powers of ±1 and ±2. Prefixes for other powers are no longer to be used. The use of the proper SI prefix with a quantity allows values to stay between 1 and 1000.

Derived units in the metric system are formed by combining base and other derived units in a manner similar to the algebraic relationships that link the corresponding quantities. Likewise,

Table 3. SI derived units with special names and symbols

Quantity	Name	Symbol	Expression in terms of other SI units
Angle, plane	radian*	rad	m/m = 1
Angle, solid	steradian*	sr	$m^2/m^2 = 1$
Celsius temperature	degree Celsius	°C	K
Electric capacitance	farad	F	C/V
Electric charge, quantity of electricity	coulomb	C	A·s
Electric conductance	siemens	S	A/V
Electric inductance	henry	H	Wb/A
Electric potential difference, electromotive force	volt	V	W/A
Electric resistance	ohm	Ω	V/A
Energy, work, quantity of heat	joule	J	N·m
Force	newton	N	$kg{\cdot}m/s^2$
Frequency (of a periodic phenomenon)	hertz	Hz	1/s
Illuminance	lux	lx	lm/m^2
Luminous flux	lumen	lm	cd·sr
Magnetic flux	weber	Wb	V·s
Magnetic flux density	tesla	T	Wb/m^2
Power, radiant flux	watt	W	J/s
Pressure, stress	pascal	Pa	N/m^2
Activity (referred to a radionuclide)	becquerel	Bq	1/s
Absorbed dose, specific energy imparted, kerma	gray	Gy	J/kg
Dose equivalent, ambient dose equivalent, directional dose equivalent, personal dose equivalent, organ dose equivalent	sievert	Sv	J/kg
Catalytic activity	katal	kat	mol/s

*The radian and steradian, previously classified as supplementary units, are dimensionless derived units that may be used or omitted in expressing the values of physical quantities.

Table 4. Units in use with SI

Quantity	Unit	Symbol	Value in SI units
Time	minute	min	1 min = 60 s
	hour	h	1 h = 60 min = 3600 s
	day	d	1 d = 24 h = 86,400 s
	week, month, etc.		
Plane angle	degree*	°	$1^\circ = (\pi/180)$ rad
	minute*	′	$1' = (1/60)^\circ = \pi/10{,}800$ rad
	second*	″	$1'' = (1/60)' = (\pi/648{,}000)$ rad
	revolution, turn	r	$1\ \mathrm{r} = 2\pi$ rad
Area	hectare	ha	$1\ \mathrm{ha} = 1\ \mathrm{hm}^2 = 10^4\ \mathrm{m}^2$
Volume	liter†	L, l	$1\ \mathrm{L} = 1\ \mathrm{dm}^3 = 10^{-3}\ \mathrm{m}^3$
Mass	metric ton, tonne	t	$1\ \mathrm{t} = 10^3$ kg

*Decimal degrees should be used for division of degrees, except for fields such as astronomy and cartography.
†The symbol L is preferred for use in the United States.
Metric ton or tonne is restricted to commercial usage.

the symbols for those derived units are obtained by use of the same relationships. Examples of the many derived units without special names are speed, acceleration, and density. There are, however, 22 SI derived units with special names (Table 3).

Some other units are in use with SI. These include units that are already used worldwide for time measurement, and for plane angles in degrees, minutes, and seconds (Table 4). In addition, there are specially named units for area, volume, and mass that arose from a combination of need and common usage. *See* PHYSICAL MEASUREMENT. Donald W. Hillger

Metzgeriales An order of liverworts in the subclass Jungermanniidae. Twelve families make up the Metzgeriales, which are also known as the Anacrogynae because of archegonia produced behind the growing apex. The gametophyte plant body is flat, elongated, and usually thallose, with no tissue differentiation or surface pores; less commonly there is a stem with two rows of leaves. Capsules dehisce by valves. *See* BRYOPHYTA; JUNGERMANNIIDAE. Howard Crum

Mica Any one of a group of hydrous aluminum silicate minerals with platy morphology and perfect basal (micaceous) cleavage.

The most common micas are muscovite $[KAl_2(AlSi_3O_{10})(OH)_2]$, paragonite $[NaAl_2(AlSi_3O_{10})(OH)_2]$, phlogopite $[K(Mg,Fe)_3(AlSi_3O_{10})(OH)_2]$, biotite $[K(Fe,Mg)_3(AlSi_3O_{10})(OH)_2]$, and lepidolite $[K(Li,Al)_{2.5\text{-}3.0}(Al_{1.0\text{-}0.5}Si_{3.0\text{-}3.5}O_{10})(OH)_2]$. Calcium (Ca), barium (Ba), rubidium (Rb), and cesium (Cs) can substitute for sodium (Na) and potassium (K); manganese (Mn), chromium (Cr), and titanium (Ti) for magnesium (Mg), iron (Fe), and lithium (Li); and fluorine (F) for hydroxyl (OH). The three major species, muscovite, biotite, and phlogopite, are widely distributed rock-forming minerals, occurring as essential constituents in a variety of igneous, metamorphic, and sedimentary rocks and in many mineral deposits.

Mica is commonly found as small flakes or lamellar plates without a crystal outline. Muscovite and biotite sometimes occur in thick books, tabular prisms with a hexagonal outline that can be up to several feet across. The prominent basal cleavage is a consequence of the layered crystal structure. Thin cleavage sheets of micas, particularly muscovite and phlogopite, are flexible, elastic, tough, and translucent to transparent (isinglass). They have low electrical and thermal conductivity and high dielectric strength.

Micas have Mohs hardnesses of 2–3 and specific gravities of 2.8–3.2. Upon heating in a closed tube, they evolve water. They have a vitreous-to-pearly luster. Muscovite is colorless to pale shades of brown, green, or gray. Paragonite is colorless to pale yellow. Phlogopite is pale yellow to brown. Biotite is dark green, brown, or black. Lepidolite is most often pale lilac, but it can also be colorless, pale yellow, or pale gray. *See* BIOTITE; HARDNESS SCALES.

Commercial mica is of two main types: sheet, and scrap or flake. Sheet muscovite, mostly from pegmatites, is used as a dielectric in capacitors and vacuum tubes in electronic equipment. Lower-quality muscovite is used as an insulator in home electrical products such as hot plates, toasters, and irons. Scrap and flake mica is ground for use in coatings on roofing materials and waterproof fabrics, and in paint, wallpaper, joint cement, plastics, cosmetics, well drilling products, and a variety of agricultural products. *See* CAPACITOR; ELECTRIC INSULATOR; SILICATE MINERALS. Lawrence Grossman; Steven Simon

Mice and rats The names associated with a great number of species of mammals in a number of different families of the order Rodentia. The rodents constitute about 42% of the known mammalian species of the world. Most of the smaller rodents are called rats or mice. There is no fundamental taxonomic difference between mice and rats; they differ only in size. (Smaller species are generally called mice, larger ones rats.) Both mice and rats are often in the same family, and there are mice and rats in a number of different rodent families. *See* MAMMALIA; RODENTIA.

Some of the rats are of great economic significance to humans, particularly the Norway rat, *Rattus norvegicus*, and the black or roof rat, *R. rattus*. In addition to harboring many diseases transmissible to humans, such as bubonic plague, endemic typhus, rat-bite fever, and infectious jaundice, infested rats can transmit trichinosis to swine. *See* PLAGUE.

Some species of mice and rats are used for research in biology and medicine. In addition to their use in studying the mechanisms of genetics, they are important in the study of carcinogenesis, effects of drugs, and virology. They are also important experimental animals in studying cell physiology, such as for cell and tissue culture research, and in animal behavior.

There is much current discussion and disagreement concerning the higher classification of rodents. The Rodentia are divided into five suborders, with mice and rats in four of the five (Sciuromorpha, Castorimorpha, Myomorpha, Hystricomorpha).

The Sciuromorpha are squirrel-like rodents, including the dormice (Family Gliridae). They are one of the oldest living families of rodents, appearing in Eocene deposits, and they are widespread in the Old World, their 28 species occurring in much of Eurasia and Africa below the Sahara. Dormice are squirrel-like, but unlike most squirrels are generally nocturnal. They are generally smaller than squirrels, but larger than mice. The fat or edible dormouse of Europe, *Glis glis*, is about 140–240 mm (5–9 in.) in total length and weighs about 70–180 g (2.5–6.3 oz).

The suborder Castorimorpha comprises the single family Heteromyidae (kangaroo rats and mice and pocket mice), with 6 genera and 60 species. The 19 species of kangaroo rats (*Dipodomys*) occur in desert regions of North America. They have elongate hindlegs and hop rather than run, thus resembling a kangaroo. These animals do not usually drink water, but obtain fluid from food, an adaptation for water balance in desert environments.

The suborder Myomorpha includes the families Dipodidae, Platacanthomyidae, Spalacidae, Callomyscidae, Nesomyidae, Cricetidae, and Muridae. The Dipodidae includes three subfamilies: the jumping mice, birchmice, and jerboas. The major families of mice and rats are Muridae with 730 species and Cricetidae with 130 genera and 681 species. The Cricetidae or "New World rats and mice" include the hamsters, voles, lemmings, bog lemmings, and muskrats. The members of the Muridae are the Old World rats and mice. Except for some introduced species, all occur in the Old World. Most are in the subfamily Murinae, which includes *Mus* and *Rattus*, which have been introduced into much of the rest of the world. The common house mouse is one of the oldest known species of domestic rodent pests. They begin to breed at 3 months of age, have a gestation of about 3 weeks, and have 4–6 litters of four to eight young per year. Adults have a pointed snout, compact body about 3 in. (8 cm) long, and an equally long tail. Of the 44 species of *Mus* known, only one species occurs in the United States, and it has become wild in some parts of the country. The familiar white laboratory mouse is an albino form of the house mouse and is used extensively in laboratory research. The genus *Rattus* contains about 137 species with many varieties and races throughout the world.

The Hystricomorpha are referred to as the South American rodents, although some of the species occur in Africa.

John O. Whitaker, Jr.

Micelle A colloidal aggregate of a unique number (50 to 100) of amphipathic molecules, which occurs at a well-defined concentration called the critical micelle concentration. In polar media such as water, the hydrophobic part of the amphiphiles forming the micelle tends to locate away from the polar phase while the polar parts of the molecule (head groups) tend to locate at the polar micelle solvent interface. A micelle may take several forms, depending on the conditions and composition of the system, such as distorted spheres, disks, or rods (see illustration). Micelles are formed in nonpolar media such as benzene, where the amphiphiles cluster around small water droplets in the system, forming an assembly known as a reversed micelle.

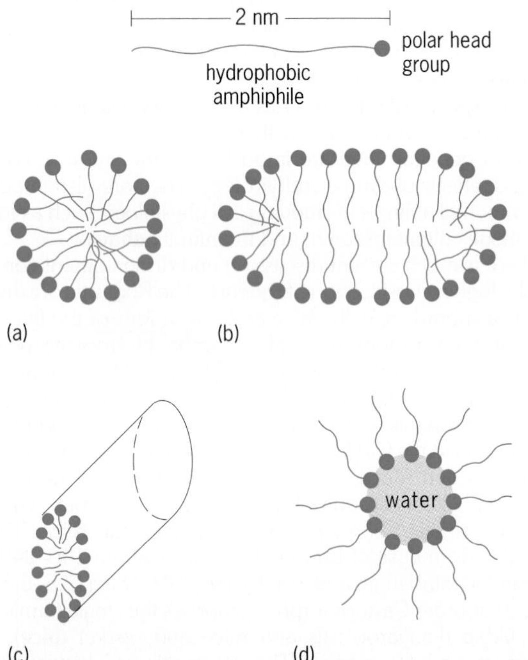

Form of an amphiphile and several forms of micelle: *(a)* spherical, *(b)* disk, *(c)* rod, and *(d)* reversed.

Micellar systems have the unique property of being able to solubilize both hydrophobic and hydrophilic compounds. They are used extensively in industry for detergency and as solubilizing agents. *See* DETERGENT; SOAP.

J. K. Thomas

Microbial ecology The study of interrelationships between microorganisms and their living and nonliving environments. Microbial populations are able to tolerate and to grow under varying environmental conditions, including habitats with extreme environmental conditions such as hot springs and salt lakes. Understanding the environmental factors controlling microbial growth and survival offers insight into the distribution of microorganisms in nature, and many studies in microbial ecology are concerned with examining the adaptive features that permit particular microbial species to function in particular habitats.

Within habitats some microorganisms are autochthonous (indigenous), filling the functional niches of the ecosystem, and others are allochthonous (foreign), surviving in the habitat for a period of time but not filling the ecological niches. Because of their diversity and wide distribution, microorganisms are extremely important in ecological processes. The dynamic interactions between microbial populations and their surroundings and the metabolic activities of microorganisms are essential for supporting productivity and maintaining environmental quality of ecosystems. Microorganisms are crucial for the environmental degradation of liquid and solid wastes and various pollutants and for maintaining the ecological balance of ecosystems—essential for preventing environmental problems such as acid mine drainage and eutrophication. *See* ECOSYSTEM; EUTROPHICATION.

The various interactions among microbial populations and between microbes, plants, and animals provide stability within the biological community of a given habitat and ensure conservation of the available resources and ecological balance. Interactions between microbial populations can have positive or negative effects, either enhancing the ability of populations to survive or limiting population densities. Sometimes they result in the elimination of a population from a habitat. *See* RHIZOSPHERE.

The transfer of carbon and energy stored in organic compounds between the organisms in the community forms an integrated feeding structure called a food web. Microbial decomposition of dead plants and animals and partially digested organic matter in the decay portion of a food web is largely responsible for the conversion of organic matter to carbon dioxide. *See* BIOMASS; FOOD WEB.

Only a few bacterial species are capable of biological nitrogen fixation. In terrestrial habitats, the microbial fixation of atmospheric nitrogen is carried out by free-living bacteria, such as *Azotobacter*, and by bacteria living in symbiotic association with plants, such as *Rhizobium* or *Bradyrhizobium* living in mutualistic association within nodules on the roots of leguminous plants. In aquatic habitats, cyanobacteria, such as *Anabaena* and *Nostoc*, fix atmospheric nitrogen. The incorporation of the bacterial genes controlling nitrogen fixation into agricultural crops through genetic engineering may help improve yields. Microorganisms also carry out other processes essential for the biogeochemical cycling of nitrogen. *See* BIOGEOCHEMISTRY; NITROGEN CYCLE; NITROGEN FIXATION.

The biodegradation (microbial decomposition) of waste is a practical application of microbial metabolism for solving ecological problems. Solid wastes are decomposed by microorganisms in landfills and by composting. Liquid waste (sewage) treatment uses microbes to degrade organic matter, thereby reducing the biochemical oxygen demand (BOD). *See* ESCHERICHIA; SEWAGE TREATMENT; WATER PURIFICATION.

Ronald M. Atlas

Microbiology The multidisciplinary science of microorganisms. The prefix micro generally refers to an object sufficiently small that a microscope is required for visualization. In the seventeenth century, Anton van Leeuwenhoek first documented observations of bacteria by using finely ground lenses. Bacteriology, as a precursor science to microbiology, was based on Louis Pasteur's pioneering studies in the nineteenth century, when it was demonstrated that microbes as minute simple living organisms were an integral part of the biosphere involved in fermentation and disease. Microbiology matured into a scientific discipline when students of Pasteur, Robert Koch, and others sustained microbes on various organic substrates and determined that microbes caused chemical changes in the basal nutrients to derive energy for growth. Modern microbiology continued to evolve from bacteriology by encompassing the identification, classification, and study of the structure and function of a wide range of microorganisms including protozoa, algae, fungi, viruses, rickettsia, and parasites as well as bacteria. The comprehensive range of organisms is reflected in the major subdivisions of microbiology, which include medical, industrial, agricultural, food, and dairy. *See* ALGAE; BACTERIOLOGY; BIOTECHNOLOGY; FUNGI; IMMUNOLOGY; INDUSTRIAL MICROBIOLOGY; MEDICAL BACTERIOLOGY; MEDICAL MYCOLOGY; MEDICAL PARASITOLOGY; MICROSCOPE; PROTOZOA; RICKETTSIOSES; VIRUS. Edward W. Voss, Jr.

Microbiota (human) Microbial flora harbored by normal, healthy persons. In a healthy human, internal tissues (such as brain, blood, cerebrospinal fluid, and muscle) are normally free of microorganisms. Conversely, surface tissues (such as skin and mucous membranes) are constantly in contact with environmental microorganisms and are readily colonized by certain microbial species. The mixture of microorganisms regularly found at any anatomical site is referred to as the normal microbiota, the indigenous microbial population, the microflora, or the normal flora. Because bacteria make up most of the normal microbiota, they are emphasized over the fungi (mainly yeasts) and protozoa. *See* BACTERIA; FUNGI; MICROBIAL ECOLOGY; PROTOZOA; YEAST.

The adult human is covered with approximately 2 m^2 (21.5 ft^2) of skin. It has been estimated that this surface area supports about 10^{12} bacteria. Commensal microorganisms living on the skin can be either resident (normal) or transient microbiota. Resident organisms normally grow on or in the skin. Their presence becomes fixed in well-defined distribution patterns. Those microorganisms that are temporarily present are transients. Transients usually do not become firmly entrenched and are unable to multiply.

The large intestine, or colon, has the largest microbial community in the body. The colon can be viewed as a large fermentation vessel, and the microbiota consist primarily of anaerobic, gram-negative, nonsporing bacteria and gram-positive, spore-forming or nonsporing rods. Not only are the vast majority of microorganisms anaerobic, but many different species are present in large numbers. Besides the many bacteria in the colon, the yeast *Candidia albicans* and certain protozoa may occur as harmless commensals. John P. Harley

Microcline Triclinic potassium feldspar, $KAlSi_3O_8$, that usually contains a few percent sodium feldspar (Ab = $NaAlSi_3O_8$) in solid solution. Its hardness is 6; specific gravity, 2.56; mean refractive index, 1.52; color, white (green varieties are called amazon stone or amazonite). Microcline is found in some relatively high-grade regional metamorphic rocks, but is much more common in pegmatities, granites, and related plutonic igneous rocks. In the last, it often occurs as a microcline perthite, containing exsolved low albite intergrowths. *See* FELDSPAR; PERTHITE. Paul H. Ribbe

Microcomputer A digital computer whose central processing unit consists of a microprocessor, a single semiconductor integrated circuit chip. Once less powerful than larger computers, microcomputers are now as powerful as the minicomputers and superminicomputers of just several years ago. This is due in part to the growing processing power of each successive generation of microprocessor, plus the addition of mainframe computer features to the chip, such as floating-point mathematics, computation hardware, memory management, and multiprocessing support. *See* INTEGRATED CIRCUITS; MICROPROCESSOR; MULTIPROCESSING.

Microcomputers are the driving technology behind the growth of personal computers and workstations. The capabilities of today's microprocessors in combination with reduced power consumption have created a new category of microcomputers: hand-held devices. Some of these devices are actually general-purpose microcomputers: They have a liquid-crystal-display (LCD) screen and use an operating system that runs several general-purpose applications. Many others serve a fixed purpose, such as telephones that provide a display for receiving text-based pager messages and automobile navigation systems that use satellite-positioning signals to plot the vehicle's position. *See* LIQUID CRYSTALS; MOBILE RADIO; RADIO PAGING SYSTEMS; SATELLITE NAVIGATION SYSTEMS.

The microprocessor acts as the microcomputer's central processing unit (CPU), performing all the operations necessary to execute a program (see illustration).

A memory subsystem uses semiconductor random-access memory (RAM) for the temporary storage of data or programs. The memory subsystem may also have a small secondary memory cache that improves the system's performance by storing frequently used data objects or sections of program code in special high-speed RAM.

The graphics subsystem consists of hardware that displays information on a color monitor or LCD screen: a graphics memory buffer stores the images shown on the screen, digital-to-analog convertors (DACs) generate the signals to create an image on an analog monitor, and possibly special hardware accelerates the drawing of two- or three-dimensional graphics. (Since LCD screens are digital devices, the graphics subsystem sends data to the screen directly rather than through the DACs.) *See* DIGITAL-TO-ANALOG CONVERTER.

The storage subsystem uses an internal hard drive or removable media for the persistent storage of data.

The communications subsystem consists of a high-speed modem or the electronics necessary to connect the computer to a network.

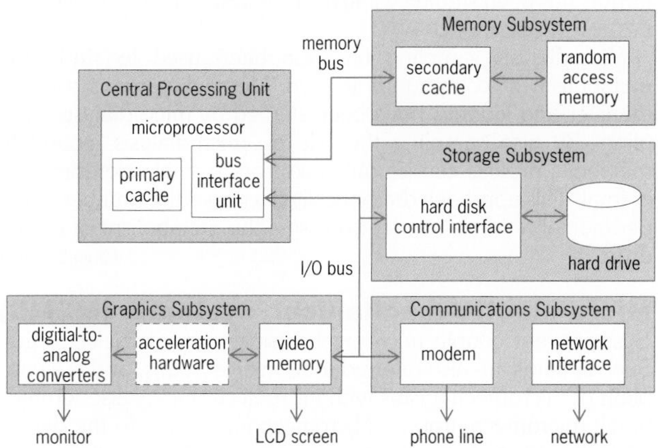

Elements of a microcomputer. The various subsystems are controlled by the central processing unit. Some designs combine the memory bus and bus input/output into a single system bus. The graphics subsystem may contain optional graphics acceleration hardware.

Microcomputer software is the logic that makes microcomputers useful. Software consists of programs, which are sets of instructions that direct the microcomputer through a sequence of tasks. A startup program in the microcomputer's ROM initializes all of the devices, loads the operating system software, and starts it. All microcomputers use an operating system that provides basic services such as input, simple file operations, and the starting or termination of programs. While the operating system used to be one of the major distinctions between personal computers and workstations, today's personal computer operating systems also offer advanced services such as multitasking, networking, and virtual memory. All microcomputers exploit the use of bit-mapped graphics displays to support windowing operating systems. *See* OPERATING SYSTEM; SOFTWARE. Tom Thompson

Microdialysis sampling An approach for sampling the extracellular space of essentially any tissue or fluid compartment in the body. Continuous sampling can be performed for long periods with minimal perturbation to the experimental animal. Microdialysis provides a route for sampling the extracellular fluid without removing fluid, and administering compounds without adding fluid. The resulting sample is clean and amenable to direct analysis.

Microdialysis sampling is performed by implanting a short length of hollow-fiber dialysis membrane at the site of interest. The fiber is slowly perfused with a sampling solution (the perfusate) having an ionic composition and pH that closely matches the extracellular fluid of the tissue being sampled. Low-molecular-weight compounds in the extracellular fluid diffuse into the fiber and are swept to a collection vial for subsequent analysis. The system is analogous to an artificial blood vessel that can deliver compounds and remove the resulting metabolites. *See* DIALYSIS; MEMBRANE SEPARATIONS.

Microdialysis is a diffusion-controlled process. The perfusion rate through the probe is generally in the range of 0.5 to 5.0 ml/min. At this flow rate, there is no net flow of liquid across the dialysis membrane. The driving force for mass transport is the concentration gradient between the extracellular fluid and the fluid in the probe. *See* DIFFUSION; TRANSPORT PROCESSES.

The greatest use of microdialysis sampling has been in the neurosciences. Microdialysis probes can be implanted in specific brain regions of conscious animals in order to correlate neurochemical activity with behavior. Most studies have focused on determining dopamine or the other monoamine neurotransmitters. *See* NEUROBIOLOGY.

Microdialysis probes have been implanted in the skin of experimental animals and humans to determine the transdermal delivery of drugs from ointments. Delivery of anticancer drugs to tumors has been studied using microdialysis. *See* DRUG DELIVERY SYSTEMS; PHARMACOLOGY.

Microdialysis sampling has also been used to study the metabolism of compounds in vivo. Metabolic organs such as the liver and kidneys have been studied by microdialysis sampling. By also sampling the bile by microdialysis, complete metabolic profiles can be obtained from a single experimental animal. This approach dramatically decreases the number of experimental animals needed to assess the metabolism of a new drug. Craig E. Lunte

Micro-electro-mechanical systems (MEMS) Systems that couple micromechanisms with microelectronics. Such systems are also referred to as microsystems, and the coupling of micromechanisms with microelectronics is also termedbreak micromechatronics. Micromechanics refers to the design and fabrication of micromechanisms that predominantly involve mechanical components with submillimeter dimensions and corresponding tolerances of the order of 1 micrometer or less. The types of systems encompassed by MEMS represent the need for transducers that act between signal and information processing functions, on the one hand, and the mechanical world, on the other. This coupling of a number of engineering areas leads to a highly interdisciplinary field that is commensurately impacting nearly all branches of science and technology in fields such as biology and medicine, telecommunications, automotive engineering, and defense. Ultimately, realization of a "smart" MEMS may be desired for certain applications whereby information processing tasks are integrated with transduction tasks, yielding a device that can autonomously sense and accordingly react to the environment. *See* TRANSDUCER.

Motivating factors behind MEMS include greater independence from packaging shape constraints due to decreased device size. In addition, the advantages of repeatable manufacturing processes as well as economic advantages can follow from batch fabrication schemes such as those used in integrated circuit processing, which has formed the basis for MEMS fabrication. Many technical and manufacturing trade-offs, however, come into play in deciding whether an integrated approach is beneficial. In some cases, the device design with the greatest utility is based on a hybrid approach, where mechanical processing and electronic processing are separated until a final packaging step. Two broad categories of devices follow from the transduction need addressed by MEMS: the input transducer or microsensor, and the output transducer or microactuator. *See* INTEGRATED CIRCUITS.

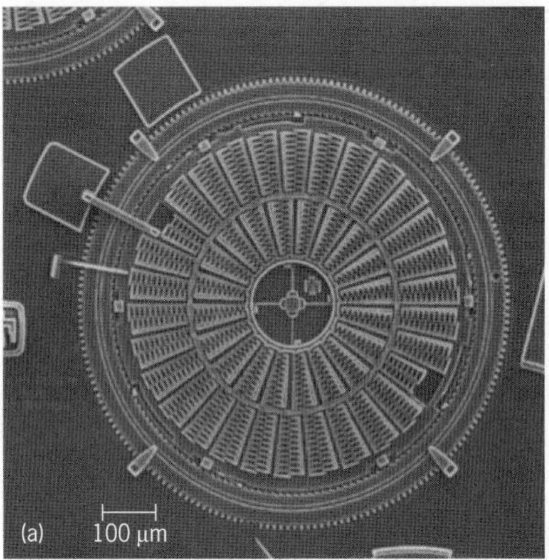

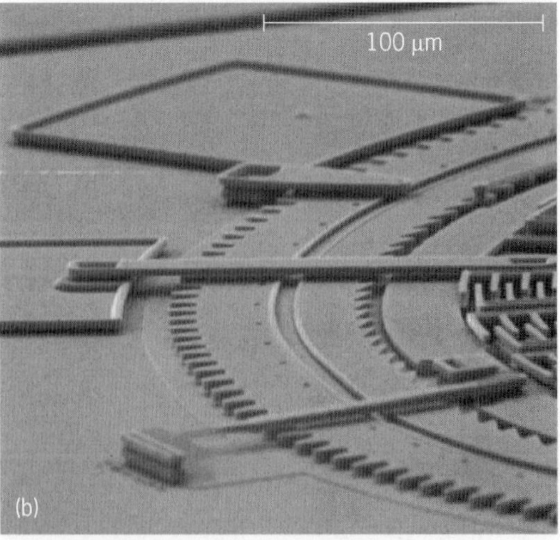

Fig. 1. Torsional ratcheting actuator fabricated by surface micromachining. (*a*) Overview. (*b*) Close-up. (*J. Jakubczala, Sandia National Laboratories*)

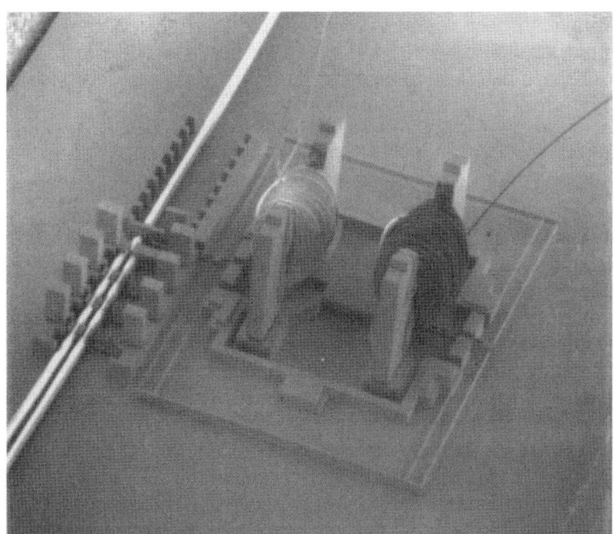

Fig. 2. Magnetic 1 × 2 optical fiber switch fabricated by deep-x-ray lithography. Total device size is approximately 4 mm × 4 mm. (*Henry Guckel, University of Wisconsin*)

Microfabrication technology. The development of process tools and materials for MEMS is the pivotal enabler for integration success. A material is chosen and developed for its mechanical attributes and patterned with a process amenable to co-electronic fabrication. Two basic approaches to patterning a material are used. Subtractive techniques pattern via removal of unwanted material, while additive techniques make use of temporary complementary molds within which the resulting structure conforms. Both approaches use a mask to transfer a pattern to the desired material. For batch processes, this step typically occurs via photolithography and may itself entail several steps. The basic process is to apply a photoresist, a light-sensitive material, and use a photomask to selectively expose the photoresist in the desired pattern. A solvent chemically develops the photoresist-patterned image, which then may be used as a mask for further processing.

Subtractive processing is accomplished via chemical etching. Wet etching occurs in the liquid phase, and dry etching or gas-phase etching may occur in a vapor phase or plasma. *See* PLASMA (PHYSICS).

A primary microfabrication technology that has been used for most commercial devices is bulk micromachining, which is the process of removing, or etching, substrate material. The important aspect of precision bulk micromachining is etch directionality. The two limiting cases are isotropic, or directionally insensitive, and anisotropic, or directionally dependent, up to the point of being unidirectional.

An alternative processing approach to bulk microfabrication was driven by the desire to reduce the fraction of the substrate area that had to be devoted to the mechanical components, thereby allowing a larger number of device dies per wafer. The approach, termed surface micromachining (SMM), realizes mechanical structures by depositing and patterning mechanical material layers in conjunction with sacrificial spacer material layers.

Applications. A highly successful device that is fabricated with both bulk and surface micromachining is the integrated pressure transducer. The process sequence uses surface micromachining techniques to form a polysilicon-plate-covered cavity. Application areas include air pressure sensing in automobile engines, environmental monitoring, and blood pressure sensing. Similar processing has resulted in the integration of surface-micromachined polysilicon inertial reference proof masses with microelectronic processing, yielding single-chip force-feedback accelerometers. *See* ACCELEROMETER.

The use of surface micromachining technology to implement microactuators has resulted in steerable micromirror arrays with as many as 1024 × 768 pixels on a chip. These arrays have revolutionized digital display technology. Further electrostatic microactuator designs are possible and may be extremely intricate, such as a torsional ratcheting actuator fabricated with five polysilicon levels (Fig. 1). These types of devices are suited for a variety of micropositioning applications. Processing based on deep-x-ray lithography has been used to produce precision magnetic microactuators. One such microactuator directly switches a single-mode optical fiber in a 1 × 2 switch configuration (Fig. 2).

Todd R. Christenson

Microfluidics Technology that involves manipulating fluids in structures in which at least one linear dimension is less than a millimeter. Currently, the smallest dimension of common microfluidic elements is 10–500 micrometers. The most mature and commercially successful microfluidic systems are the print heads found in common inkjet printers. More recent efforts have been aimed at developing microfluidic systems for performing chemical processes. The principal target applications of these chemical systems are as portable analyzers of chemicals and biomolecules [proteins and deoxyribonucleic acid (DNA)] for biomedical applications and for defense against chemical and biological weapons. Some chemical syntheses have also been performed in microfluidic devices.

Flow characteristics. Experimental and theoretical attention has been paid to the possibility that fluid behavior in microchannels deviates from that found in macroscopic flows. For flows of simple liquids, such as water and organic solvents, the same general equations (Navier-Stokes equations) that govern macroscopic flows, with the no-slip condition at solid boundaries (meaning that the fluid is stationary relative to the solid at the boundary) are valid on scales above tens of nano meters. *See* BOUNDARY-LAYER FLOW; FLUID-FLOW PRINCIPLES; NAVIER-STOKES EQUATION.

While the basic governing equations are the same, there are several general features that distinguish flows in microstructures from flows in common macroscopic systems (such as water faucets and coffee cups). As the dimension of a flow decreases, the importance of forces that act on the volume of the fluid, such as inertia and gravity, diminishes relative to that of forces that act at the surfaces, such as viscous friction and surface tension.

Challenges. Microfluidic technology is still in its infancy. Optimal designs for pumps, valves, injectors, and so forth must still be invented and characterized, and ideal materials and methods of microfabrication must be developed. Future applications of microfluidic devices must also be conceived of and explored. For example, microfluidic systems could act as implants to deliver drugs and monitor physiological parameters. *See* FLUID MECHANICS.

Abraham D. Stroock

Microlithography The formation of small three-dimensional relief images on the surface of a substrate for subsequent transfer of this pattern into the substrate itself, as used in such applications as semiconductor fabrication. The fabrication of an integrated circuit (IC) requires a variety of physical and chemical processes performed on a semiconductor (for example, silicon) substrate. In general, the various processes used to make an IC fall into three categories: film deposition, patterning, and semiconductor doping. Films of both conductors (such as polysilicon, aluminum, and more recently copper) and insulators (various forms of silicon dioxide, silicon nitride, and others) are used to connect and isolate transistors and their components. Selective doping of various regions of silicon allows the conductivity of the silicon to be changed with the application of voltage. By creating structures of these various components, millions of transistors can be built and wired together to form the complex circuitry of a modern microelectronic device. Fundamental to

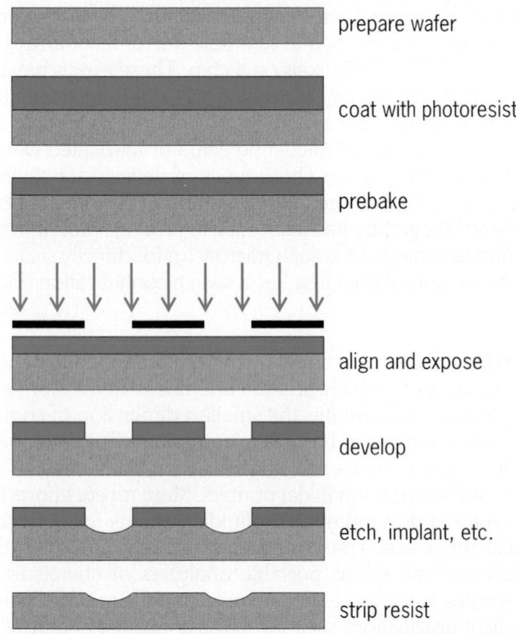

Example of a typical sequence of lithographic processing steps (with no postexposure bake in this case), illustrated for a positive resist.

all of these processes is lithography. *See* INTEGRATED CIRCUITS; SEMICONDUCTOR.

The word lithography, from Greek, literally means writing on stones. In the case of semiconductor lithography, our stones are silicon wafers and we are writing our patterns in a light-sensitive polymer called a photoresist. To build the complex structures that make up a transistor and the many wires that connect the millions of transistors of a circuit, lithography and etch pattern transfer steps are repeated at least 10 times, but more typically are done 20–30 times in order to make one circuit. Each pattern being printed on the wafer is aligned to the previously formed patterns, and slowly the conductors, insulators, and selectively doped regions are built up to form the final device. *See* POLYMER; TRANSISTOR.

Optical microlithography is a photographic process by which a photoresist is exposed and developed to form three-dimensional images on the substrate. The general sequence of processing steps for a typical optical lithography process is as follows: substrate preparation, photoresist spin coat, prebake, exposure, postexposure bake, development, and postbake. The resist is removed in the final operation of the lithographic process, after the pattern has been transferred into the underlying layer. This sequence is shown in the illustration, and is generally performed on several tools linked together into a contiguous unit called a lithographic cluster. Chris A. Mack

Micromanipulation The technique or practice involving manipulation of objects too small to be easily seen with the unaided eye. When a microscope is used to allow the operator to visually guide the microtools used in the manipulation, the technique is called micrurgy. When an object that is not extremely small needs to be positioned with extreme precision, as when aiming a laser to make a surgical lesion, the technique is called micropositioning.

To manipulate microscopic objects, it is necessary to use a tool called a micromanipulator, which allows relatively coarse hand movements to execute proportionately slower and smaller movements of a probe or microtool. Micromanipulators are used to probe and test integrated circuit chips, to align fiber-optic communication cables, as well as to accomplish techniques in biological research.

The simplest and most common micromanipulators consist of three orthogonal mechanical slides whose position is controlled by a threaded screw or a rack and pinion. Such devices are relatively inexpensive and adequate for magnifications up to 150 power. However, the imperfections of simple machines are revealed by magnifications greater than 150 power.

There has been increasing use of electrically powered micromanipulators. The three types are direct-current-motorized, stepper-motorized, and piezoelectric. Direct-current-motor-driven micromanipulators may be operated by remote control with a joystick or single-step push button, which allows for precise control of pulse length. It is easy to adjust the ratio of movement reduction electrically. At highest magnification, however, the slight vibration of the motors begins to interfere with visualization, making direct-current-motor-driven micromanipulators best suited to medium magnifications. Micromanipulators with stepper-motor drives are the only micromanipulators capable of fast, direct movements to a specific set of coordinates, and can be directed by a computer over a preplanned route. Specialized piezoelectric micromanipulators, called cell penetrators, provide movements over a very short range of 1–10 μm. They move very abruptly but are excellent for puncturing small and hard-to-penetrate cell membranes. Charles W. Scouten

Micrometeorite A submillimeter extraterrestrial particle that has survived entry into the atmosphere without melting. Meteoroids are natural interplanetary objects that orbit the Sun, and they range in size from small dust grains to objects that are miles in diameter. Particles below 0.04 in. (1 mm) in diameter are considered micrometeoroids, and the micrometeoroids that enter the atmosphere without melting are called micrometeorites. Micrometeorites survive entry without severe heating because they are small and they totally decelerate from cosmic velocity at high altitudes near 55 mi (90 km). Most of the mass of extraterrestrial matter that annually collides with the Earth is in the micrometeoroid size range, a total of about 10^4 tons (10^7 kg), but only a small fraction survives as micrometeorites. Usually only the particles smaller than 0.1 mm survive as true unmelted micrometeorites, although the survival of an individual micrometeorite depends on entry velocity, angle of entry, melting point, and density as well as size. *See* METEORITE.

Micrometeorites are of particular interest because they are samples of comets and asteroids, small primitive bodies that have survived without major change since the earliest history of the solar system. Some of these particles are generated by collisions in the asteroid belt, while others are released from comets when these bodies approach the Sun and ice volatilization releases dust grains and propels them into space. Once released from a parent comet or asteroid, particles survive only for a few thousand to a hundred thousand years, depending on size, before they are either destroyed or collide with a planet. Particles are destroyed either when they collide with other particles or when they spiral into the Sun because of the Poynting-Robertson drag, an effect of sunlight that causes the orbits of small particles to decay. During exposure in space, the small particles accumulate large amounts of helium implanted by the solar wind, and they also are riddled with radiation damage tracks produced by solar cosmic rays, high-energy particles accelerated from solar flares. *See* ASTEROID; COMET; COSMIC RAYS; SOLAR WIND; SUN.

The collection and laboratory analysis of micrometeorites provide an important source of information on the nature of materials in comets and asteroids. Most micrometeorites are collected in the stratosphere with aircraft such as the U2, which is capable of flying at an altitude of 12 mi (20 km) where terrestrial particles as large as 10 μm are rare. Micrometeorites are collected from the stratosphere by direct impact onto sticky plates that are extended from aircraft wings into the ambient airstream. After a cumulative exposure of many hours, the plates are returned to a clean room where the microscopic particles are picked off with needles and placed onto mounts where they can be studied by electron microscopes, mass spectrometers, and other instruments. The collection of micrometeorites in the stratosphere is usually limited to the size range from 2 to 100 μm in diameter. Most particles larger than this limit melt to form cosmic spherules during atmospheric entry and are not true micrometeorites. *See* COSMIC SPHERULES. Donald E. Brownlee

Micrometeorology The study of small-scale meteorological processes associated with the interaction of the atmosphere and the Earth's surface. The lower boundary condition for the atmosphere and the upper boundary condition for the underlying soil or water are determined by interactions occurring in the lowest atmospheric layers. Momentum, heat, water vapor, various gases, and particulate matter are transported vertically by turbulence in the atmospheric boundary layer and thus establish the environment of plants and animals at the surface. These exchanges are important in supplying energy and water vapor to the atmosphere, which ultimately determine large-scale weather and climate patterns. Micrometeorology also includes the study of how air pollutants are diffused and transported within the boundary layer and the deposition of pollutants at the surface.

In many situations, atmospheric motions having time scales between 15 min and 1 h are quite weak. This represents a spectral gap that provides justification for distinguishing micrometeorology from other areas of meteorology. Micrometeorology studies phenomena with time scales shorter than the spectral gap (time scales less than 15 min to 1 h and horizontal length scales less than 2–10 km or 1–6 mi). Some phenomena studied by micrometeorology are dust devils, mirages, dew and frost formation, evaporation, and cloud streets. *See* AIR POLLUTION; ATMOSPHERE; MESOMETEOROLOGY.

Much of the early understanding of micrometeorology was obtained by studying conditions in large, flat, uniform areas that are relatively simple situations. Micrometeorologists have turned their attention to more complex situations that represent conditions over more of the Earth's surface. The micrometeorology of complex terrain, that is, hills and mountains, is important for air pollution in many towns and cities and for visibility in national parks and for locating wind generators. Another interest is the study of micrometeorology in areas of widely varied surface conditions. For instance, several different crops, dry unirrigated lands, lakes, and rivers may be located near one another. In these cases it is important to understand how the micrometeorology associated with each of these surfaces interacts to produce the overall heat and moisture fluxes of the region so that these areas can be correctly included in weather and climate forecast computer programs. *See* CLIMATOLOGY; MOUNTAIN METEOROLOGY; WEATHER FORECASTING AND PREDICTION.

Microscale meteorological features are too small to be observed by the standard national and international weather observing network. Generally, micrometeorological phenomena must be studied during specific experiments by using specially designed instruments. Instruments used to study turbulent fluxes must be able to respond to very rapid fluctuations. Special cup anemometers are made from very light materials, and high-quality bearings are used to minimize drag. Other anemometers use the speed of sound waves or measure the temperature of heated wires to measure wind. Tiny thermometers are used, so that time constants are short. Instruments are usually placed on towers or in aircraft, or are suspended in packages from tethered balloons. Instruments have been developed that can measure turbulence remotely. Wind speed and boundary-layer convection can be measured with Doppler radar, lidar devices using lasers, and sodar (sound detection and ranging) using sound waves. *See* ATMOSPHERIC ACOUSTICS; LIDAR; METEOROLOGICAL INSTRUMENTATION; METEOROLOGICAL RADAR; METEOROLOGY. Steven A. Stage

Micrometer A precision instrument used to measure small distances and angles. A common use is on a machinist's caliper, as in the illustration. *See* CALIPER.

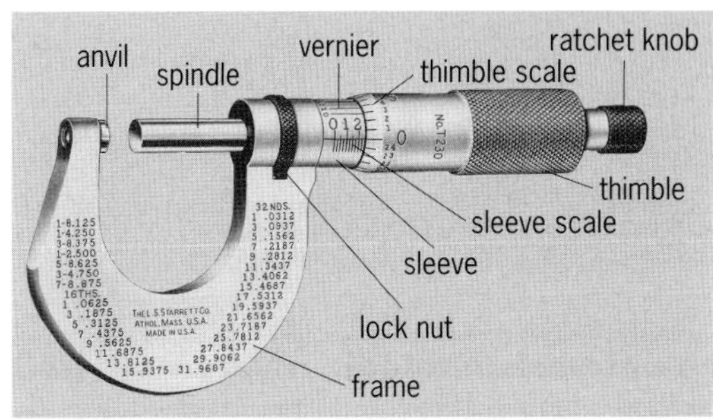

Machinist's outside caliper with micrometer reading 0.250 in. (*L. S. Sfarrett Co.*)

The spindle of the caliper is an accurately machined screw, which is rotated by the thimble or the ratchet knob until the object to be measured is in contact with both spindle and anvil. The ratchet slips after correct pressure is applied, ensuring consistent, accurate gaging. The number 1 on the sleeve represents 0.1 in.; the smallest divisions are 0.025 in. A vernier scale allows accurate reading to 0.0001 in. *See* VERNIER. Frank H. Rockett

Micro-opto-electro-mechanical systems (MOEMS) A class of microsystems that combine the functions of optical, mechanical, and electronic components in a single, very small package or assembly. MOEMS devices can vary in size from several micrometers to several millimeters. MOEMS may be thought of as an extension of micro-electro-mechanical systems (MEMS) technology by the provision of some optical functionality. This optical functionality may be in the form of moving optical surfaces such as mirrors or gratings, the integration of guided-wave optics into the device, or the incorporation of optical emitters or detectors into the system. The term may be confused with micro-opto-mechanical systems (MOMS), which more properly refers to microsystems that do not include electronic functions at the microsystem location. MOEMS is a rapidly growing area of research and commercial development with great potential to impact daily life. The basic concept is the miniaturization of combined optical, mechanical, and electronic functions into an integrated assembly, or monolithically integrated substrate, through the use of micromachining processes derived from those used by the microelectronics industry. These processes, utilizing microlithography and various etch (subtractive) or deposition (additive) steps on a planar substrate, enable the production of extremely precise shapes, structures, and patterns in various materials. *See* INTEGRATED CIRCUITS; INTEGRATED OPTICS; MICRO-ELECTRO-MECHANICAL SYSTEMS (MEMS); MICRO-OPTO-MECHANICAL SYSTEMS (MOMS).

The microsystems realized by these techniques can have many unique capabilities. The miniaturization that is realized is useful in itself, allowing the systems to be utilized as sensors or actuators in environments that were not previously accessible, including inside living organisms, in hand-held instruments, or in small spacecraft. The miniaturization also allows for high-speed operation of the system, as the operating speed of mechanical systems is related to their inertial and frictional properties as well as the actuating forces. Optomechanical systems have been historically constrained in this area because of the mass required for stable optical elements and the extremely precise alignment requirements of most opto-mechanical systems, which limits the forces that can be tolerated for rapid motion. In the more integrated forms of MOEMS, the systems are prealigned by the precise fabrication processes, eliminating one of the more expensive aspects of assembling conventional optical systems. The miniaturization along with the scalability of microfabrication processes allows the development of massively parallel opto-mechanical systems, with millions of moving parts, that would not be possible in conventional technologies. MOEMS can incorporate detection and drive electronics in close proximity to provide improvements in signal-to-noise ratio for sensors and simplified interfaces for actuated systems. Ultimately, these electronics may be monolithically integrated in some technologies. Because of the production volumes achievable with micromachining techniques, MOEMS are potentially much less expensive than their conventional counterparts. Mial E. Warren

Micro-opto-mechanical systems (MOMS)

Miniaturized optome-chanical devices or assemblies that are typically formed using micromachining techniques that borrow heavily from the microelectronics industry. The term may be used to distinguish devices and microsystems that combine optical and mechanical functions without the use of internal electronic devices or signals. Systems that use electronic devices as part of the microsystem may be referred to as MOEMS (micro-opto-electro-mechanical systems). In some cases, these terms may be used synonymously. A related area is MEMS (micro-electro-mechanical systems), in which electronic and mechanical functions are combined in a miniature device or system, but not necessarily implementing optical functions. The progress of MOMS technology has been greatly enabled by the simultaneous development of microelectronics and optical fiber-based telecommunications technology. *See* INTEGRATED CIRCUITS; MICRO-ELECTRO-MECHANICAL SYSTEMS (MEMS); MICRO-OPTO-ELECTRO-MECHANICAL SYSTEMS (MOEMS); OPTICAL COMMUNICATIONS.

Advantages. Although similar in concept to MOEMS technologies, MOMS has unique advantages for some applications. The use of only optical energy and signals gives MOMS an inherent immunity to electromagnetic interference (EMI) that is important for applications in electrically noisy or high-voltage environments. The absence of semiconductor electronic devices greatly increases the high-temperature tolerance of the system. MOMS devices can be designed to work immersed in liquids, which is of great importance for chemical sensing and biomedical applications. The fact that the power and signal sources can be remotely provided via an optical fiber, allowing the sensor to be passive, is of great utility and reduces the impact of a MOMS sensor on its local environment. MOMS can be used safely in flammable and explosive environments, making them uniquely valuable in the petrochemical industry.

Applications. Some examples of MOMS technology include optical pressure transducers—microphones or hydrophones that have a thin mechanical membrane that is one surface in a Fabry-Perot interferometer formed by the reflection from the membrane surface and the reflection from the end of the fiber. (A similar arrangement for sensing vibration is shown in the illustration.) Other versions have a planar optical waveguide on the surface of a sensitive membrane that is one arm of a two-beam Mach-Zehnder interferometer. Another example is an accelerometer in which a small mass is suspended from flexure attachments to the substrate. Optical fibers are positioned with a small gap in which the moving mass can interrupt the transfer of light from one fiber to another to modulate the light intensity transmitted through the fibers. One of the most well developed MOMS applications is optical sensing of the position of small cantilevers used in scanning tip microscopy processes such as atomic force microscopy. *See* ACCELEROMETER; MICROPHONE; PRESSURE TRANSDUCER; SCANNING TUNNELING MICROSCOPE. Mial E. Warren

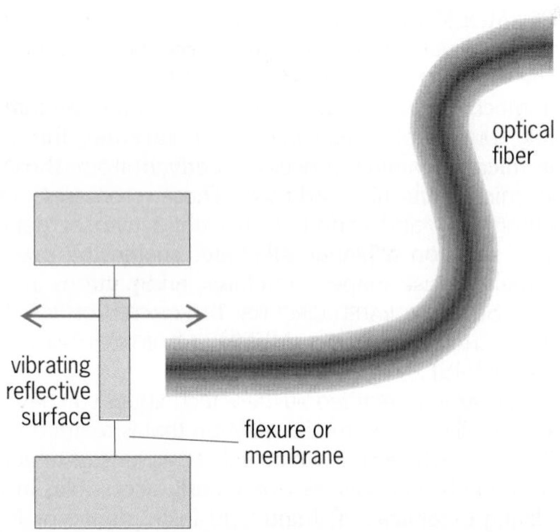

A simplified MOMS sensor using a reflecting surface on a flexible mount or membrane and the end face of an optical fiber to form an optical interferometer that can sense vibration. The vibration of the flexible membrane allows the reflecting surface to move, changing the resonance wavelength of the interferometer, modulating the intensity of the light that is reflected back into the interferometer. At the other end of the interferometer is a light source and a detector.

Micropaleontology

A branch of paleontology dealing with the fossilized microscopic organic remains (microfossils) of the geologic past, their structure, biology, phylogenetic relations, and distribution in space and time. The study of these microfossils has become an independent scientific field largely because:

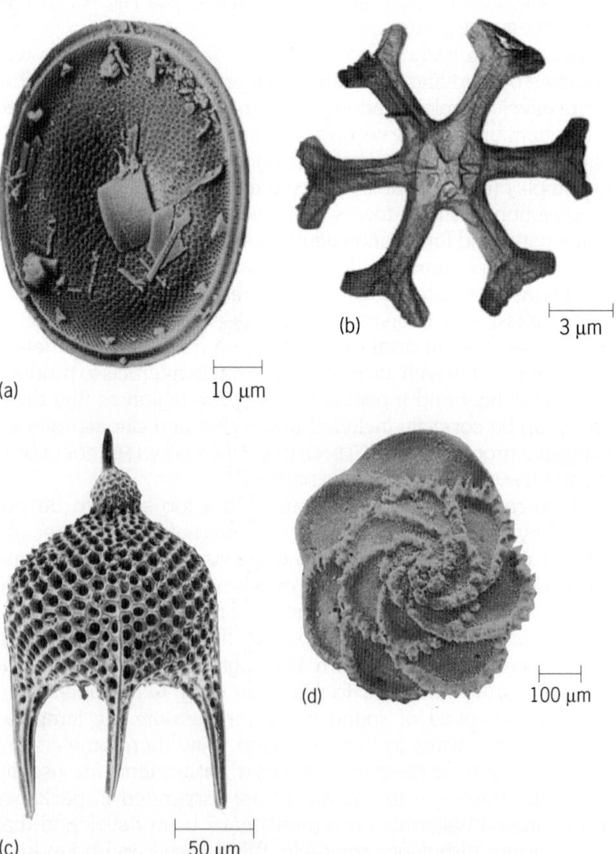

Representative microfossils. (*a*) Diatom; (*b*) asterolith, a calcareous nannoplankton; (*c*) radiolarian; and (*d*) planktonic foraminiferan.

(1) The size of these fossils requires special methods for collection and examination. (2) Their abundance in geologic formations makes it possible to analyze their spatial distribution and the rates of morphological changes during the course of evolution by means of statistical methods which can be used only under exceptional circumstances in the study of larger fossils. (3) Microfossils have become indispensable tools in certain branches of applied geology, especially in the exploration for oil-bearing strata, because countless numbers of these minute fossils may be obtained from small pieces of subsurface rock recovered from drill holes. (4) The diversity of microfossils, their wide spatial distribution in varied environments, and their distinctive steps in evolution and the ease of studying them have contributed to make micropaleontology one of the most actively studied branches of the earth sciences.

The material subjected to micropaleontological studies forms a spectrum from primitive plants to advanced vertebrates (see illustration). The only prerequisite for organisms to become the subject of micropaleontological studies is their possession of resistant skeletal components ensuring their preservation in sedimentary strata as fossilized remains even after biological, chemical, or mechanical processes have destroyed the organisms' soft parts.

Most major groups of organisms incorporate, besides organic compounds, hard resistant materials that serve for structural support or protection. The more common substances found among the microfossils are calcium carbonate, silicon dioxide (or silica), calcium phosphate in the form of the mineral apatite (typical of bones and teeth), sporonine (principal constituent of pollen and spore walls), and various complex organic compounds.

Tsunemasa Saito

Microphone An electroacoustic device containing a transducer which is actuated by sound waves and delivers electric signals proportional to the sound pressure. Microphones are usually classified with respect to the transducer principle used. Their directional characteristics are also of interest, that is, the voltage output as a function of the direction of incidence for constant sound pressure. *See* DIRECTIVITY; SOUND; SOUND PRESSURE; TRANSDUCER.

In addition to directional characteristics, some other important characteristics of microphones include open-circuit sensitivity, equivalent noise level, dynamic range, and vibration sensitivity.

Open-circuit sensitivity is defined as the ratio of open-circuit output voltage and sound pressure. The pressure sensitivity refers to the actual pressure acting upon the diaphragm of the microphone, while the free-field sensitivity refers to the pressure that existed in the sound field before insertion of the microphone. Pressure sensitivity and free-field sensitivity are equal at low frequencies. Sensitivities are measured in volts/pascal (V/Pa).

Equivalent noise level is equal to the level of a sound pressure which generates an output voltage of the microphone corresponding to its inherent A-weighted noise voltage. It is measured in dB(A).

Dynamic range is defined as the range of sound pressure levels in decibels (dB) extending from the equivalent noise level to the level where the nonlinear distortion reaches 3%.

Vibration sensitivity is defined as the ratio of the output voltage of the microphone as a result of acceleration of its case to the magnitude of the acceleration. Vibration sensitivities are measured in volts/g, where g is the acceleration of the Earth's gravity, or in volts/(m/s^2).

Electrostatic (condenser) microphones. These consist of a fixed electrode (the backplate), a movable electrode (the diaphragm), and an air gap between the electrodes. To decrease the acoustic stiffness of the airgap, which is generally about 20 to 30 micrometers (0.8 to 1.2 mils) thick, the backplate is often perforated with holes connecting the air gap to a larger air cavity. The diaphragm is a thin [typically 4 to 6 μm thick (0.16 to 0.24 mil)] foil under mechanical tension. *See* CAPACITANCE; CAPACITOR; ELECTRICAL IMPEDANCE.

Condenser microphones are renowned for their excellent acoustic qualities such as flat frequency response, high sensitivity, large dynamic range, and small vibration sensitivity. Also important is their suitability for miniaturization, with the smallest units having dimensions of only about 0.12 × 0.12 × 0.08 in. (3 × 3 × 2 mm). They can be designed as precision instruments and as such are widely used in measurement and in high-fidelity sound production. *See* HEARING AID; MAGNETIC RECORDING; SOUND RECORDING; TELEPHONE.

Piezoelectric microphones. These consist of a material having piezoelectric properties. A deformation of the material leads to the generation of a voltage which corresponds to the deformation. Piezoelectric materials can be crystals, polycrystalline ceramics, or semicrystalline polymers. The best-known piezoelectric crystals are quartz and ammonium dihydrogen phosphate (ADP). Representative of polycrystalline ceramics are lead zirconate titanate (PZT) and barium titanate, which are initially electrostrictive; they have to be poled, that is, exposed to a high electric field at elevated temperatures, to become piezoelectric. An example of a semicrystalline polymer is poly(vinylidenefluoride) [PVDF]. It is also made piezoelectric by poling. *See* ELECTRET; ELECTROSTRICTION; PIEZOELECTRICITY.

Well-designed piezoelectric microphones have acceptable quality. A drawback is the relatively high vibration sensitivity. They are still in occasional use in telephones in some countries and are also employed in the near-ultrasonic range at frequencies up to about 100 kHz.

Dynamic microphones. These consist of a conductor located in the gap of a permanent magnet. Motion of the conductor produces a voltage proportional to its velocity. In the moving-coil microphone the coil, often referred to as voice coil, is connected to a diaphragm actuated by the sound waves. Motion of the coil induces a voltage proportional to its velocity. To obtain a frequency-independent sensitivity, the coil must respond to the sound pressure with frequency-independent velocity. This is accomplished by resistance-controlling the system: the acoustical resistance is made larger in magnitude than the acoustical reactance due to the mass of the diaphragm and coil and due to the compliance of the suspension. A silk cloth or a piece of felt placed behind the voice coil is used for this purpose. In modern moving-coil microphones, the diaphragm is made of a plastic film. The impedance of the voice coil is typically 200 to 1000 ohms. *See* ACOUSTIC IMPEDANCE.

Dynamic microphones are relatively complicated systems. If well designed, they are of good quality. Drawbacks are the difficulties encountered in miniaturization and the relatively high vibration sensitivity. Moving-coil microphones are still widely used in high-fidelity, radio, television, and concert applications. In many other areas they have been replaced by electret-based condenser microphones.

Magnetic microphones. These consist of a diaphragm connected to an armature which, when vibrating, varies the reluctance in a magnetic field. The variation in reluctance leads to a variation in the magnetic flux through a surrounding coil and therefore to an induced voltage. This voltage is proportional to the velocity of the armature. To obtain a frequency-independent sensitivity, the velocity of the armature in response to the sound pressure must be independent of frequency. As in dynamic microphones, this is accomplished by resistance-controlling the system, for example, by placing an acoustic resistance behind the diaphragm.

Magnetic microphones are relatively complicated and have poor frequency response and high vibration sensitivity. While never extensively used, they have now disappeared completely. However, the magnetic principle is still used in telephone receivers and in earphones employed in hearing aids. *See* EARPHONES.

Silicon (MEMS) microphones. Microelectronic processing methods allow fabrication of batch-processed, high-performance sensors. Among these methods are doping, deposition, oxidation, lithography, and, as key technologies, various etching processes. The last are usually subdivided into dry and wet methods. The dry processes, such as plasma, sputter, ion, and laser etching, allow one to fabricate holes with walls perpendicular to the surface of the substrate. The wet methods, subdivided into isotropic and anisotropic processes, usually result in curved walls or walls extending under a certain angle to the surface. An important variant of the etching processes is sacrificial-layer etching where a buried layer of material is etched out through holes in the covering layer. The etching technologies enable fabrication of membranes, holes, pits, and recesses as required in acoustic sensors. The methods in their entirety, if used to make three-dimensional mechanical structures, are referred to as micromachining, and the resulting devices are called microelectromechanical systems, or MEMS. *See* MICRO-ELECTRO-MECHANICAL SYSTEMS (MEMS).

Acoustic MEMS sensors or silicon microphones utilizing the condenser and the piezoelectric principles have been built since the mid-1980s. In addition, new concepts of transducer design, such as the modulation of the drain current of a field-effect transistor or the modulation of light propagation in an optical waveguide by the sound waves, have been realized in silicon. Progress has been such that silicon microphones based on the condenser principle are now commercially available. *See* FIBER-OPTIC SENSOR; TRANSISTOR.

Silicon microphones have several advantages as compared to conventional microphones. They can be made considerably smaller with membrane areas of only about 1 mm^2, as opposed to about 5 mm^2 for the smallest conventional transducers. They also have very low vibration sensitivity due to the use of thin diaphragms. They are thus not susceptible to pickup originating from vibrations due to walking or other motions or caused by vibration sources such as motors in cassette recorders or camcorders. Furthermore, they can be produced together with proper signal-processing electronics on the same chip with the same semiconductor methods. Finally, they can be made inexpensively through batch-processing techniques.

Gerhard M. Sessler

Microprocessor A device that integrates the functions of the central processing unit (CPU) of a computer onto one semiconductor chip or integrated circuit (IC). In essence, the microprocessor contains the core elements of a computer system, its computation and control engine. Only a power supply, memory, peripheral interface ICs, and peripherals (typically input/output and storage devices) need be added to build a complete computer system. *See* COMPUTER PERIPHERAL DEVICES.

A microprocessor consists of multiple internal function units. A basic design has an arithmetic logic unit (ALU), a control unit, a memory interface, an interrupt or exception controller, and an internal cache. More sophisticated microprocessors might also contain extra units that assist in floating-point match calculations, program branching, or vector processing (see illustration).

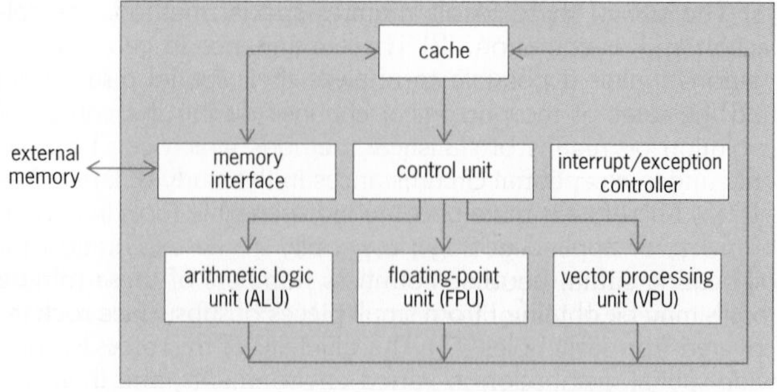

A microprocessor consists of multiple independent function units. The memory interface fetches instructions from, and writes data to, external memory. The control unit issues one or more instructions to other function units. These units process the instructions in parallel to boost performance.

The ALU performs all basic computational operations: arithmetic, logical, and comparisons.

The control unit orchestrates the operation of the other units. It fetches instructions from the on-chip cache, decodes them, and then executes them. Each instruction has the control unit direct the other function units through a sequence of steps that carry out the instruction's intent. The execution path taken by the control unit can depend upon status bits produced by the arithmetic logic unit or the floating-point unit (FPU) after the instruction sequence completes. This capability implements conditional execution control flow, which is a critical element for general-purpose computation. *See* BIT.

The memory interface enables the microprocessor to maintain two-way communication with off-chip semiconductor memory, which stores programs and data. This interface typically supports memory reads and writes in blocks of words (the number of bits that the processor operates on at one time). The block size facilitates burst data transfers to and from the chip's internal cache. *See* SEMICONDUCTOR MEMORIES.

The interrupt or exception controller enables the microprocessor to respond to requests from the external environment or to error conditions by allowing interruptions of the ongoing operation. An interrupt might be an external peripheral requesting service, while an exception typically consists of a floating-point math error or an unrecognized instruction. The interrupt controller can prioritize and selectively handle these interrupts.

The internal cache is an on-chip memory storage area that holds recently used data values or instruction sequences that are likely to be used again in the near future. Since this information is already on-chip, it can be accessed rapidly, thereby accelerating the computation rate. Items not in the cache can take several or more extra operations to access, which significantly degrades the computation rate. Software writers often organize a program's code and data structures so that the most frequently used elements often occupy the cache, thus maintaining a high level of computational throughput. *See* COMPUTER STORAGE TECHNOLOGY; COMPUTER SYSTEMS ARCHITECTURE.

The design of instruction sets (the commands that produce basic work when executed by the microprocessor) often influences the design of the microprocessor itself. Instruction sets—and as a consequence, the microprocessor architecture—are of two types: reduced instruction set computers (RISC) and complex instruction set computers (CISC). Because of the limits of early computer technology, most computers were by necessity RISC machines. Since most of the software was written in assembly language (that is, a programming language that represented the program's intent in actual machine instructions), there was a drive to build instruction sets of greater sophistication and complexity. These new CISC instruction sets made assembly language programming easier, but they also made it difficult to build high-speed computer hardware. First, CISC instructions were harder to decode. In addition, since CISC instructions involved long and complex operation sequences, they incurred a major cost by requiring more complicated logic to implement. Second, such instructions were also difficult to interrupt or abort if an exception occurred. Finally, such instructions usually carried many data dependencies that made it more difficult to support advanced architectural techniques. By returning to a RISC design, much faster computers can be built. In fact, an enhancement in performance by a factor of 2 to 3 has been attributed to this simple organizational change. To achieve these efficiencies, most of the RISC microprocessor's function units must be kept as busy as possible. This requires optimizing compilers that can translate a program's high-level source code and then reorder the resulting low-level instructions in such a way as to ensure the high throughput. *See* COMPUTER PROGRAMMING; PROGRAMMING LANGUAGES.

Microprocessors are found in virtually every consumer product that requires electric power, such as microwave ovens, automobiles, video recorders, cellular telephones, digital cameras, and hand-held computers. High-performance microprocessors implement the servers that store and distribute Web content, such as streaming audio and video, desktop computers, and the high-speed network switches that constitute the Web's infrastructure. More modest-powered microprocessors are at the heart of notebook computers and electronic games. Low-power microprocessors provide the control and flow logic of hand-held devices, digital cameras, cellular and cordless phones, pagers, and the diagnostic and pollution control of automobile engines. *See* INTERNET; VIDEO GAMES; WIDE-AREA NETWORKS; WORLD WIDE WEB. Tom Thompson

Micropygoida An order of regular echinoids belonging to the Diadematacea, established for the two recognized species of *Micropyga*. They have an aulodont lantern with grooved teeth. Test plating is imbricate, and ambulacra are composed of trigeminate compound plates in which upper and lower elements are reduced to demiplates. Pore pairs are biserially arranged in ambulacral columns, and there are unique umbrellalike aboral tube feet. *Micropyga* is a deep-water echinoid, found between 480 and 4290 ft (150 and 1340 m) depth in the Indo-West Pacific. There are no fossils that can be placed in this taxon with certainty, but it is possible that the Upper Jurassic *Pedinothuria* may belong here. *See* ECHINODERMATA. Andrew B. Smith

Microradiography The process of producing enlarged images of the interior of thin, usually small specimens by penetration of low-energy (0.1–10 keV) x-rays. The magnification can be obtained geometrically during the exposure by subsequent enlargement of the initial image by optical or electronic means, or by a combination of both processes. As with radiography on other size scales, microradiography shows the spatial distribution of mass and elemental composition of the sample. If a pulsed (flash) source or time-gated detector is employed, radiographs also provide stop-motion details of fast-changing objects. Microradiography has numerous applications in biology, material science, the characterization of fabricated microstructures, and assessment of plasma-driven compression of thermonuclear fuel.

Microradiography is largely synonymous with x-ray microscopy, both techniques being concerned with producing enlarged images of opaque objects by using x-rays. However, x-ray microscopy of defects in crystals can be performed by using diffraction, rather than simple absorption, in a technique termed x-ray topography. Microradiography by x-ray absorption is complementary, or related to a variety of other techniques for characterization of microstructures. *See* X-RAY DIFFRACTION; X-RAY MICROSCOPE.

There are three classes of sources employed for microradiography: electron-impact devices, storage ring sources of synchrotron radiation, and multimillion-degree plasmas. The first microradiography measurements, and most studies since then, have been done with electron-impact sources. These contain an electron-emitting cathode and an anode in a high vacuum, between which a high voltage is applied. *See* X-RAY TUBE.

High-energy (of the order of gigaelectronvolts) electrons orbiting within an evacuated toroid, called a storage ring, produce intense continuum radiation which is tightly collimated and polarized. Such synchrotron radiation is emitted in nanosecond pulses at MHz rates, and can be 1000 or more times as intense as the x-rays from electron-impact sources. The use of synchrotron radiation for microradiography began in 1977 and has rapidly expanded. *See* SYNCHROTRON RADIATION.

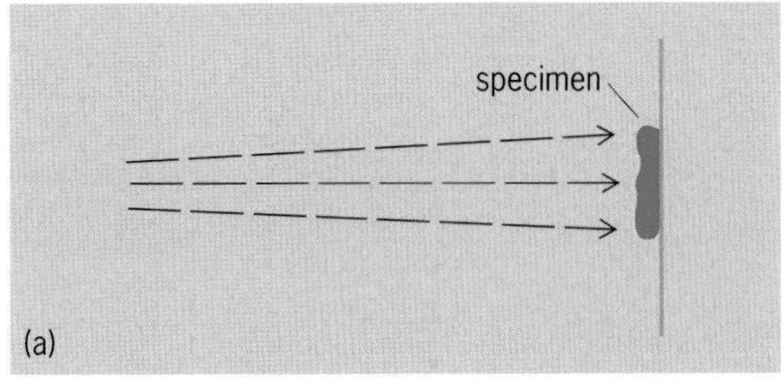

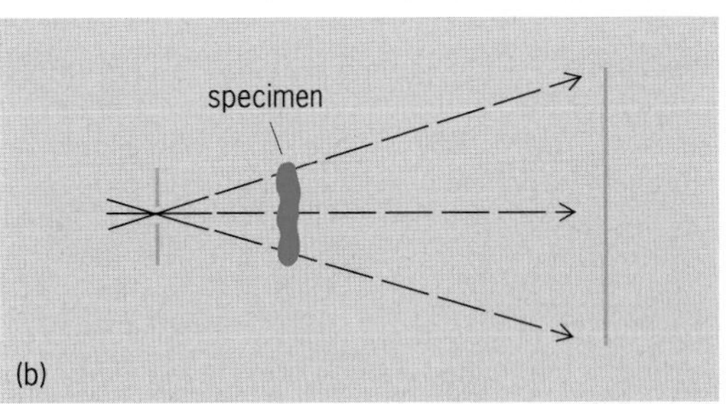

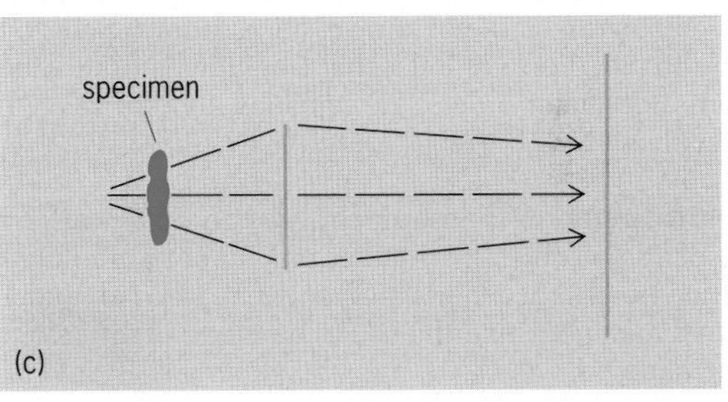

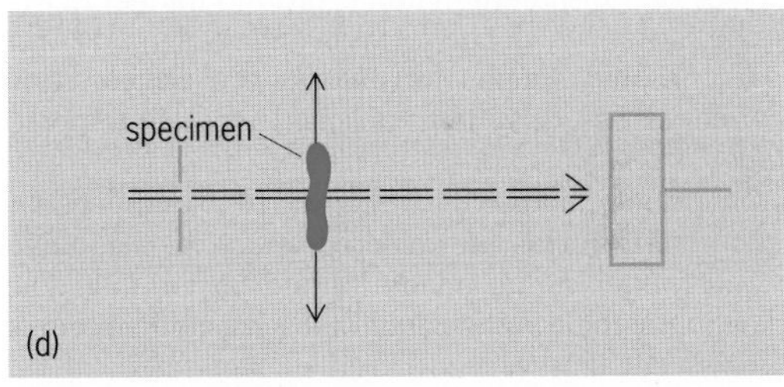

Schematics of techniques for x-ray microradiography. X-ray paths are indicated by broken lines. (*a*) Contact. (*b*) Projection. (*c*) Imaging. (*d*) Scanning.

Very high temperature plasmas can be heated by high-power lasers or electrical discharges. They are typically 10 μm to 1 mm in size. Multimillion-degree plasmas emit uncollimated line and continuum spectra predominantly in the soft x-ray region, below a few kiloelectronvolts, with pulse lengths of 0.1–100 ns. Stop-motion microradiographs were first made with plasma x-radiation in 1980. *See* PLASMA (PHYSICS).

X-ray micrographs may be obtained by four geometrically different techniques (see illustration). Two of them, the contact and the projection methods, have been in use for decades. However, they have undergone significant development because of the availability of bright synchrotron radiation and plasma radiation sources. The other two techniques, the true imaging and the scanning methods, require the bright sources.

Microradiography was developed in order to observe fine details in the interior of opaque natural (for example, biological) and manufactured (for example, metallurgical) samples. Optical microscopy has long been available to observe the surfaces of opaque specimens, or the interior of clear samples, with spatial resolutions approaching 200 nm. Early interest in microradiography waned for two reasons: weak sources required long exposure times, and electron beam techniques which provide submicrometer details were developed. The availability of bright soft x-ray sources, especially synchrotron radiation, has led to a resurgence of interest in microradiography, especially for cellular samples. Sub-100-nm resolution of live specimens is needed to understand biological structures with sizes between those observable with light microscopes (down to about 200 nm) and the molecular level (about 1 nm) probed by diffraction from ordered arrays of molecules. Microradiography of inorganic natural, manufactured, and dynamic structures remains of interest. *See* ELECTRON MICROSCOPE; OPTICAL MICROSCOPE. David J. Nagel

Microsauria A diverse order of small, extinct amphibians, known only from the Pennsylvanian and lower Permian of North America and Europe. They range from obligatorily aquatic, perennibranchiate genera with lateral-line canal grooves, to fully terrestrial lizardlike forms. Several families include long-bodied, possibly burrowing, species. Limbs are always retained, and the tail is never specialized as a swimming organ. Microsaurs are recognized by the possession of a broad, strap-shaped occipital condyle, and no more than a single bone in the temporal series. The trunk vertebrae are spool-shaped.

The specific origin of the group remains unknown. Although intermediate forms are not known, microsaurs appear to be the most probable group of Paleozoic amphibians from which two modern amphibian orders, the apodans and the salamanders, have evolved. *See* AMPHIBIA; LEPOSPONDYLI. Robert L. Carroll

Microscope An instrument used to obtain an enlarged image of a small object. The image may be seen, photographed, or sensed by photocells or other receivers, depending upon the nature of the image and the use to be made of the information of the image.

A simple microscope, hand lens, or magnifier usually is a round piece of transparent material, ground thinner at the edge than at the center, which can form an enlarged image of a small object. Commonly, simple microscopes are double convex or planoconvex lenses, or systems of lenses acting together to form the image.

The compound microscope utilizes two lenses or lens systems. One lens system forms an enlarged image of the object and the second magnifies the image formed by the first. The total magnification is then the product of the magnifications of both lens systems (see illustration).

The typical compound microscope consists of a stand, a stage to hold the specimen, a movable body-tube containing the two lens systems, and mechanical controls for easy movement of the body and the specimen. The lens system nearest the specimen is called the objective; the one nearest the eye is called the eyepiece or ocular. A mirror is placed under the stage to reflect light into the instrument when the illumination is not built into the stand. For objectives of higher numerical aperture than 0.4, a condenser is provided under the stage to increase the illumination of the specimen. Various optical and mechanical attachments may be added to facilitate the analysis of the information in the doubly enlarged image. *See* ELECTRON MICROSCOPE; FLUORESCENCE MICROSCOPE; INTERFERENCE MICROSCOPE; LENS (OPTICS); OPTICAL MICROSCOPE; PHASE-CONTRAST MICROSCOPE; REFLECTING MICROSCOPE; X-RAY MICROSCOPE. Oscar W. Richards

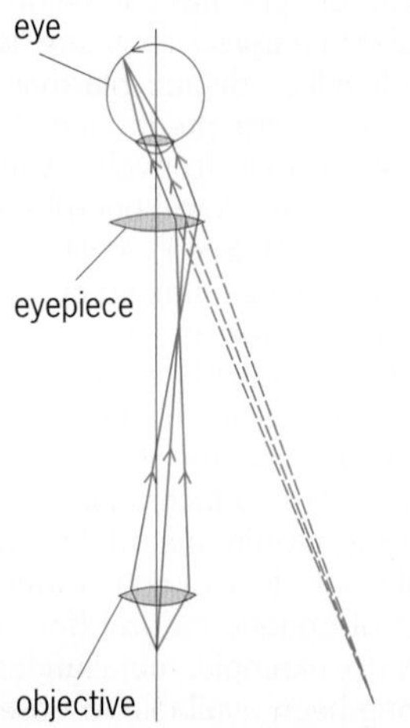

Compound microscope diagram. (*After F. A. Jenkins and H. E. White, Fundamentals of Optics, 4th ed., McGraw-Hill, 1976*)

Microsensor A very small sensor with physical dimensions in the submicrometer to millimeter range. A sensor is a device that converts a nonelectrical physical or chemical quantity, such as pressure, acceleration, temperature, or gas concentration, into an electrical signal. Sensors are an essential element in many measurement, process, and control systems, with countless applications in the automotive, aerospace, biomedical, telecommunications, environmental, agricultural, and other industries. The stimulus to miniaturize sensors lies in the enormous cost benefits that are gained by using semiconductor processing technology, and in the fact that microsensors are generally able to offer a better sensitivity, accuracy, dynamic range, and reliability, as well as lower power consumption, than their larger counterparts.

Mechanical microsensors form perhaps the largest family of microsensors because of their widespread availability. Microsensors have been produced to measure a wide range of mechanical properties, including force, pressure, displacement, acceleration, rotation, and mass flow. Force sensors generally use a sensing element that converts the applied force into the deformation of the elastic element.

Applications for chemical and biochemical microsensors are environmental monitoring and medicine. Applications in the medical industry may involve monitoring blood, urine, and breath, which contain a wealth of information about the patient's state of health. Only a few such devices now exist. Examples include a glucose biochemical microsensor and ion-selective field-effect devices used to measure blood pH. The use of microsensors to gather medical diagnostic information is an attractive proposition, and eventually there may even be implanted microsensors to diagnose health problems, using smell-sensitive array devices. *See* BIOELECTRONICS.

Andrew C. Pike; Chris J. Welham; Julian W. Gardner

Microsporidea A class of Cnidospora characterized by the production of minute spores with a single intrasporal or one or two intracapsular filaments and a single sporoplasm. The spore membrane of these protozoans is usually a single piece. Microsporidians are mainly intracellular parasites of arthropods and fishes. Microsporida is the only order of this class. *See* CNIDOSPORA. Ross F. Nigrelli

Microtechnique The art of preparing objects for examination under the microscope and of preserving objects so prepared. Few objects yield useful information if examined without such preparation, which may involve, in addition to preliminary preservation, hardening, rendering transparent, selective coloration of parts, and cutting into thin slices.

The four types of microscope slide commonly made are wholemounts, smears, squashes, and sections. The last three methods are merely devices to make thinner, or smaller, objects unsuitable for the first method. In all four methods, the objects are permanently preserved in a mounting medium between a glass slide, about 0.04 in. (1 mm) thick, and a glass cover slip 0.04 to 0.008 in. (1 to 0.2 mm) thick. Preliminary fixation and preservation, staining, and the final mounting media are common to all four types. Peter Gray

Microwave Electromagnetic energy with wavelengths in free space ranging roughly from 0.3 to 30 cm. Corresponding frequencies range from 1 to 100 GHz. Frequency and wavelength are related by $f\lambda = c$, where f is the frequency, λ is the free-space wavelength, and c is the velocity of light in vacuum, approximately 3×10^8 m/s. *See* ELECTROMAGNETIC RADIATION; FREQUENCY (WAVE MOTION); WAVE MOTION; WAVELENGTH.

Characteristic transmission media for microwaves are hollow-pipe waveguides, where the cross-sectional dimensions are of the order of the wavelength and thus are of convenient size. Coaxial transmission lines are also used, however, especially in the lower-frequency bands, and various stripline techniques are used on microwave integrated circuits. Resonant cavities are commonly used as circuit elements, and radiation or reception of the energy is typically by horns, parabolic reflectors, or arrays. *See* ANTENNA (ELECTROMAGNETISM); CAVITY RESONATOR; COAXIAL CABLE; ELECTROMAGNETIC WAVE TRANSMISSION; TRANSMISSION LINES; WAVEGUIDE.

Generation. For most applications, microwaves are generated in electronic devices that produce oscillations at microwave frequencies. The devices may be single-frequency or tunable, and continuous-wave (cw) or pulsed. Vacuum-tube generators include klystrons, magnetrons, and backward-wave oscillators; solid-state generators include tunnel diodes, Gunn diodes, IMPATT diodes, transistor oscillators, masers, and harmonic generators using varactor diodes. The vacuum-tube generators are used to produce higher powers, which can be as much as thousands of kilowatts. Solid-state generators were formerly limited in power to a few watts, but their power capabilities are continually increasing and now may reach hundreds of watts. *See* GYROTRON; KLYSTRON; MAGNETRON; MICROWAVE SOLID-STATE DEVICES; MICROWAVE TUBE; TRAVELING-WAVE TUBE.

Circuit elements. Physical elements which produce specific effects on microwaves are called circuit elements.

The most common method of microwave transmission within a system is through hollow circular or rectangular metal tubes of uniform cross section called waveguides. The microwave energy is confined within these tubes and guided along them.

Various forms of stripline are used in interconnecting components on a dielectric or semiconductor substrate when microwave devices are integrated. One important example is the microstrip, in which a metallic strip or ribbon is placed on a thin dielectric, which is in turn backed by a conducting ground plane. Striplines have more losses than hollow-pipe waveguides but are generally used over very short distances.

Filters are needed in communication or information-processing systems for blocking of high frequencies (low-pass), blocking of low frequencies (high-pass), elimination of undesired bands (band elimination), or passing of desired bands while attenuating others (band pass). All of these may be made for microwaves by adding periodic perturbations, such as posts, irises, diaphragms, or dimensional variations, to the wave-guide or other transmission system. *See* ELECTRIC FILTER; MICROWAVE FILTER.

A thin sheet of plastic can be used to alter the amplitude or phase of microwaves. If the sheet is coated with powdered carbon with appropriate electrical conductivity and placed in a waveguide with the lossy material parallel to the lines of electrical intensity, it will absorb microwave power. Variable attenuation can be achieved by mechanically inserting more or less of the lossy strip into the path of the wave. *See* ATTENUATION (ELECTRICITY).

A phase shifter changes the phase of a microwave without changing its amplitude. It can be constructed in the same manner as an attenuator without the lossy material.

An attenuator which has a very large loss and is closed at one end is called a termination; it absorbs all the power transmitted into it, reflecting none.

The most common microwave detector is a silicon diode designed for high frequencies and mounted in a waveguide or a stripline. The diode rectifies the microwave signal, producing an average current which can be indicated by a direct-current meter connected between the diode terminals. If the microwave signal is modulated in amplitude, the modulation will appear in the output current. *See* AMPLITUDE MODULATION; SEMICONDUCTOR DIODE.

The bolometer is a detector which absorbs microwave power, causing a temperature increase and a corresponding change in resistance. The bolometer does not respond fast enough to detect high-frequency modulation. It is often used as one arm of a resistance bridge circuit in microwave power meters. *See* BOLOMETER; MICROWAVE POWER MEASUREMENT.

A transmitting antenna takes microwave power from a waveguide and converts it into a plane wave that propagates through space to a distant receiving antenna. Two important characteristics of antennas are efficiency and directivity, efficiency being the ratio of the power delivered into space to the power available in the waveguide. High directivity is accomplished by large antennas which focus the microwave energy in the same way a searchlight focuses a beam of light.

The gyrator is a lossless, nonreciprocal, two-port circuit which has 180° more phase shift in one direction than in the other. This principle is used in the broadband microwave circulator. The three-port circulator has the property that all the power into port 1 exits at port 2, all the power into port 2 exits at port 3, and all the power into port 3 exits at port 1. The nonreciprocal phase shift is achieved in a magnetic ferrite placed in the waveguide junction and magnetized with an external permanent magnet. An isolator is a circulator with one port terminated, resulting in a circuit which transmits power in one direction and not the other. The input circuit is thus isolated from the output circuit. When the terminated port is internal, the isolator appears to be a two-port element. *See* GYRATOR.

A varactor is a solid-state diode whose capacitance changes with applied voltage. Varactors are used as harmonic generators to obtain microwave power efficiently from lower-frequency sources such as quartz crystal-controlled oscillators in the 10–100-MHz range. Varactors are also used in up-converters and down-converters performing the same functions as similar circuits using resistive diodes, but more efficiently and at the expense of narrower bandwidth. *See* OSCILLATOR; VARACTOR.

A microwave amplifier converts a low-power input signal to a higher-power output signal while preserving one or more characteristics. A linear amplifier preserves the amplitude, frequency, and phase of the input signal. When a linear amplifier is overloaded, it becomes saturated: the output amplitude tends to remain constant and the envelope of the input signal becomes distorted.

High power output is achieved with klystron and traveling-wave tube (TWT) amplifiers, both of which can be operated in the linear or saturated modes. Moderate power can be achieved with transistor amplifiers, and these are continually being extended in their frequency range of usefulness. Very low noise levels are achieved in the maser and the parametric amplifier, both of which require power at a single frequency to pump the active element. *See* AMPLIFIER; MASER; PARAMETRIC AMPLIFIER.

Microwave integrated circuits. Microwave integrated circuits (MICs) are of two types, hybrid MICs and monolithic microwave integrated circuits (MMICs). In the hybrid circuits, some or all of the active and passive devices are added to a dielectric or semiconducting substrate and interconnected by striplines as discussed above. For MMICs, all of the active and passive devices of the functional unit are formed on the substrate, and interconnected by striplines through a variety of microfabrication techniques, including photolithography, epitaxy, ion implantation, etching, diffusion, sputtering, and evaporation. *See* INTEGRATED CIRCUITS.

Receiver. The first active element in nearly all microwave receivers is a silicon diode operated as a down-converter. In this type of receiver, a strong, continuous-wave local oscillator signal is used to pump the diode over its nonlinear resistance range. In this manner, the local oscillator and the input signal are mixed, shifting the input signal down to an intermediate frequency, which is the difference between the frequencies of the local oscillator signal and the received signal. Intermediate frequencies of a few tens of megahertz are common. Frequency, phase, or amplitude modulation on the received signal appears in the detector output at the intermediate frequency. A bandpass intermediate-frequency amplifier, providing most of the gain of the receiver, follows the detector, after which a demodulator converts the modulation on the intermediate-frequency signal to usable form, for example, an audio or a television signal. *See* RADIO RECEIVER.

Transmitter. The main components of a microwave transmitter are a microwave power source, a modulator, and, if necessary, a power amplifier. The modulation can be done directly at microwave frequencies or it can be performed at intermediate frequency and shifted to the microwave frequency in an up-converter, which is very much like a down-converter.

Propagation. In free space, microwaves travel in straight lines as do optical waves. Near the Earth, however, the atmosphere has an index of refraction which normally decreases with distance above the Earth and causes the wave to travel in a circular path which bends slightly toward the Earth. Microwaves are reflected and refracted by objects just as are optical waves. *See* MICROWAVE OPTICS.

Occasionally during the summer, atmospheric conditions cause microwaves transmitted from an antenna to travel to a receiver via two or more paths. These waves interfere at the receiver and may cause large decreases in the received signal amplitude. This phenomenon, called multipath fading, is a serious problem in microwave transmission parallel to the surface of the Earth. Other atmospheric conditions can result in what is known as earth-bulge fading. When microwaves are directed well above the horizon, neither of these two problems occurs. Thus satellite microwave systems do not suffer from either multipath or earth-bulge fading.

At frequencies above about 10 GHz, rain absorbs microwave energy, resulting in large signal losses. Both satellite and point-to-point microwave systems are seriously affected by rain attenuation. For most frequencies the attenuation of microwaves by the Earth's atmosphere is very small. *See* RADIO-WAVE PROPAGATION.

Applications. Areas in which microwave radiation is applied include radar, communications, radiometry, medicine, physics, chemistry, and cooking food.

Radar is used in military applications, commercial aviation, remote sensing of the atmosphere, and astronomy. The high antenna directivity and the excellent propagation characteristics of microwaves in the atmosphere make this the preferred band for radar applications. Microwaves are also used in electronic countermeasures to radar. *See* ELECTRONIC WARFARE; RADAR.

There is at least 100 times as much frequency space available for communications in the microwave band as in the entire spectrum below microwaves. In addition, the high directivity obtainable at microwave frequencies allows reuse of these frequencies many times in the same area, a practice not possible at lower frequencies. The high directivity also makes possible communication to satellites and deep-space probes. *See* COMMUNICATION SATELLITE; RADIO SPECTRUM ALLOCATIONS; SPACE COMMUNICATIONS.

All objects, including liquids and gases, emit electromagnetic radiation in the form of noise, the amount of the noise being proportional to the absolute temperature of the object. A noise temperature can be assigned to the object corresponding to the amount of noise radiating from it. A microwave radiometer is a sensitive receiver which measures the noise power received by an antenna; from this measurement, the noise temperature of the source object can be determined. Radiometers are used extensively for remote sensing. Microwave radiometers are used to study astronomical sources of noise and to observe planets from deep space probes. *See* PASSIVE RADAR; RADIOMETRY; REMOTE SENSING.

Applications of microwaves in medicine include (1) thermography, the measurement of tissue temperature; (2) hyperthermia, microwave heating used in the treatment of cancer and in the treatment of hypothermic subjects; and (3) biomedical imaging, the use of microwaves to study the structure of tissue beneath the skin. *See* RADIOLOGY.

Physics and chemistry. Microwave energy is used in large particle accelerators to accelerate charged particles such as electrons and protons to very high energies and cause them to collide. Knowledge of the structure of matter is also obtained from microwave spectroscopy. *See* MICROWAVE SPECTROSCOPY; PARTICLE ACCELERATOR.

Microwave energy is absorbed in most foods and has been found to be a source of quick, uniform heating or cooking. Microwave ovens based upon this principle are now widely used. Microwaves are also used for the industrial heating of foodstuffs and other materials. Clyde L. Ruthroff; J. R. Whinnery

Microwave filter A two-port component used to provide frequency selectivity in satellite and mobile communications, radar, electronic warfare, metrology, and remote-sensing systems operating at microwave frequencies (1 GHz and above). Microwave filters perform the same function as electric filters at lower frequencies, but differ in their implementation because circuit dimensions are on the order of the electrical wavelength at microwave frequencies. Thus, in the microwave regime,

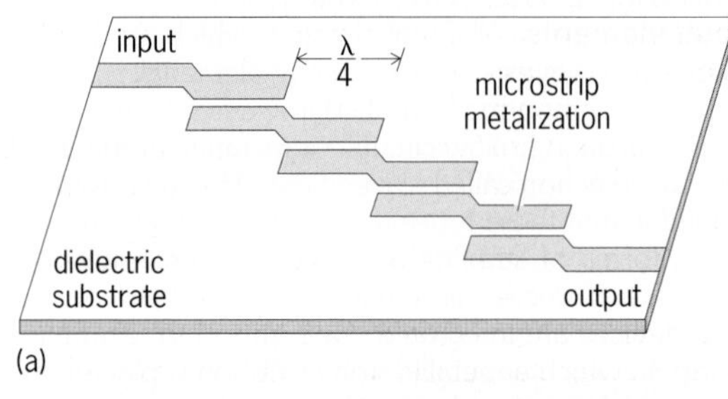

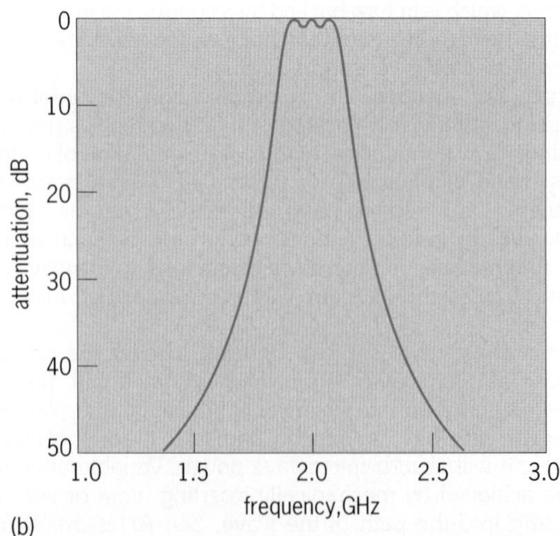

Third-order (four-section) microwave band-pass filter. (*a*) Layout of the parallel coupled stripline filter. (*b*) Calculated frequency response of the filter, with center frequency of 2 GHz and 0.5-dB equal-ripple passband.

distributed circuit elements such as transmission lines must be used in place of the lumped-element inductors and capacitors used at lower frequencies. This can make microwave filter design more difficult, but it also introduces a variety of useful coupling and transmission effects that are not possible at lower frequencies.

The majority of modern microwave filters are designed by using the insertion-loss method, whereby the amplitude response of the filter is approximated by using network synthesis techniques that have been extended to accommodate microwave distributed circuit elements. A general four-step procedure is followed: determination of filter specifications, design of a low-pass prototype filter, scaling and transforming the filter, and implementation (conversion of lumped elements to distributed elements).

Microwave filters are implemented in many ways. Waveguide cavity band-pass filters have very low insertion loss, making them preferred for frequency multiplexing in satellite communication systems. Coaxial low-pass filters, made with sections of coaxial line with varying diameters, are compact and inexpensive. Planar filters in microstrip or stripline form (see illustration) are important for integration with hybrid or monolithic microwave integrated circuits. While planar filters are usually more cost effective than waveguide versions, their insertion loss is usually greater. Computer-aided design procedures are used in the synthesis of more sophisticated amplitude and phase responses, and active microwave devices (field-effect transistors) are used to provide filters with gain or tunable response characteristics. *See* COAXIAL CABLE; COMPUTER-AIDED DESIGN AND MANUFACTURING; ELECTRIC FILTER; MICROWAVE SOLID-STATE DEVICES; TRANSMISSION LINES; WAVEGUIDE. David M. Pozar

Microwave free-field standards The means for setting up electromagnetic fields of precisely determined intensity at microwave frequencies in unbounded regions of space. Such standards are used to evaluate field probes and antennas for measuring field strength and power density. The standardization of these devices is necessary before they are used for determining the performance of radar and communications systems or for assessing such systems for health and safety risks or electromagnetic compatibility. *See* ELECTROMAGNETIC RADIATION; MICROWAVE.

Antennas used in free-field standards are usually either half-wave dipoles or wave-guide horns. The half-wave dipole is a collinear device with a length of approximately one-half of the free-space wavelength of the radiated wave. The gain and pattern characteristics of a pyramidal horn can be calculated to accuracies of about ± 2 dB; however, the reflections at the throat and aperture discontinuities significantly influence the gain-frequency characteristic and calibration is usually necessary. *See* WAVEGUIDE.

The ideal environment for making measurements on antennas is a large unobstructed volume which is free of reflecting objects and electromagnetically interfering signals—that is, a free-space condition. A practical solution is to use an anechoic chamber to set up simulated free-space conditions in a bounded environment. Low reflection of electromagnetic signals from the walls of such a chamber is achieved by the use of an electromagnetic wave absorbent layer covering all of the reflecting surfaces within the room or chamber, the outer shell of which is a metallic structure to give shielding against interfering signals encroaching on the test region.

Clear open sites can be used at low frequencies, or for high-gain antennas where far-field measurements require such large distances that enclosed or anechoic environments are impractical. However, since electromagnetic waves are strongly reflected by a ground plane, this effect must be accommodated, and one of two courses can be adopted. One is to do measurements as far above the ground as possible. The alternative is to make use of the ground plane by working close to it—if necessary, enhancing its reflection with a metal ground plane or grid—and by making allowance for the reflection in the analysis of performance. *See* ANECHOIC CHAMBER.

Probably the single most important parameter of a standard antenna, when considered for metrological applications, is the boresight or maximum gain, and a number of methods of determining this parameter have been devised. The three-antenna method involves measurement of the transmission between two polarization-matched antennas. Only the product of the antenna gains can be obtained from this measurement; however, with three antennas the measured combinations will yield the gain of each antenna uniquely. The extrapolation method involves determining the transmission characteristic between two antennas as the transmission path is increased through about 4 to 10 Rayleigh distances. With good metrology it is possible to get a sufficiently accurate characterization to allow extrapolation to the true far-field range. Antenna metrology has increasingly concentrated on near-field scanning techniques with the objective of improving antenna characterization. In this method a probe antenna is used to sample the magnitude and phase, for orthogonal polarizations, of the radiated fields over a well-defined surface, which can be a plane, a cylinder, or a sphere, a few wavelengths from the antenna under test.

The essential requirement for the calibration of devices for measuring power flux density or field strength is the creation of a substantially plane wave of known power density which encompasses the effective aperture of the device to be tested. This is effected by launching a known power through an antenna or transverse-electromagnetic cell of known characteristics and calculating the field strength or power density from the appropriate equation. Ralph W. Yell

Microwave landing system (MLS) An all-weather aircraft landing-guidance system that operates at microwave frequencies and provides deviations from the landing runway centerline using a time-referenced scanning beam (TRSB) technique. The MLS was standardized in 1988 and approved for use in international civil aviation until at least the year 2020. MLS is used to support low-visibility instrument precision approach and landing operations in North America and Europe. In addition to the fixed-base MLS equipment design, a compact mobile microwave landing system (MMLS) equipment design exists. The instrument landing system (ILS) is also standardized internationally and approved for use indefinitely as countries implement their transition to new technologies. Standards for a third landing system, the Global Navigation Satellite System (GNSS), based primarily on Global Positioning System (GPS) technology, exist. Multimode receivers enable an aircraft to conduct an instrument approach using ILS, MLS, or GNSS. *See* INSTRUMENT LANDING SYSTEM (ILS); SATELLITE NAVIGATION SYSTEMS.

The operating frequencies for MLS lie in a portion of the C band (5030–5091 MHz) designated for use in aeronautical telecommunications. This frequency choice allows a 12-ft (3.6-m) antenna to generate the 1° beamwidth pattern needed to exclude most reflections.

As with ILS, the MLS equipment is sited near the primary runway, with the azimuth transmitter (lateral guidance) and distance-measuring equipment (DME) transponder located near the runway stop end, and the elevation transmitter (vertical guidance) located alongside the runway near landing threshold. This type of siting is referred to as a split-site configuration. With this geometry, the approach course and glide path, generated by the ground equipment, are monitored at the landing runway. Also, the aircraft lateral and vertical displacements due to guidance errors become vanishingly small as the runway is approached and the angular guidance converges to its origin. For special-purpose applications, such as providing guidance to a heliport-only facility, all the MLS equipment may be sited at one common location, known as a collocated site configuration. See DISTANCE-MEASURING EQUIPMENT.

Unlike ILS, the 50 times higher frequency of the MLS allows generation of narrow beams by relatively small equipment. Because of this 50:1 scale factor, a 1° beamwidth antenna for MLS requires a 12-ft (3.6-m) antenna, while for ILS a 600-ft (180-m) antenna would be required. See ANTENNA (ELECTROMAGNETISM).

The large coverage volume of MLS is provided by scanning the narrow beams clockwise then counterclockwise for azimuth functions and up then down for elevation functions. This scanning is electronically controlled at a precise rate of 20,000°/s and fills a lateral sector of 60° (maximum) on each side of the runway centerline and a vertical sector of 30° (maximum). The angular position of the aircraft is decoded by the airborne receiver, which measures the time elapsed between successive passages of azimuth or elevation beams.

The antennas typically used are phased arrays where beam scanning is accomplished by a stored set of commands which, at the appropriate time in the transmission sequence, are directed to variable signal delay devices (phase shifters) associated with each radiating element of the array. See AIR NAVIGATION.; ELECTRONIC NAVIGATION SYSTEMS. Douglas B. Vickers; Michael F. DiBenedetto

Microwave measurements A collection of techniques particularly suited for development of devices and monitoring of systems where physical size of components varies from a significant fraction of an electromagnetic wavelength to many wavelengths. See MICROWAVE.

Virtually all microwave devices are coupled together with a transmission line having a uniform cross section. The concept of traveling electromagnetic waves on that transmission line is fundamental to the understanding of microwave measurements. See MICROWAVE TRANSMISSION LINES.

At any reference plane in a transmission line there are considered to exist two independent traveling electromagnetic waves moving in opposite directions. One is called the forward or incident wave, and the other the reverse or reflected wave. The electromagnetic wave is guided by the transmission line and is composed of electric and magnetic fields with associated electric currents and voltages. Any one of these parameters can be used in considering the traveling waves, but the measurements in the early development of microwave technology made principally on the voltage waves led to the custom of referring only to voltage. One parameter in very common use is the voltage reflection coefficient Γ, which is related to the incident, V_i, and reflected, V_r voltage waves by Eq. (1).

$$\Gamma = \frac{V_r}{V_i} \qquad (1)$$

Impedance. The voltage reflection coefficient Γ is related to the impedance terminating the transmission line and to the impedance of the line itself. If a wave is launched to travel in only one direction on a uniform reflectionless transmission line of infinite length, there will be no reflected wave. The input impedance of this infinitely long transmission line is defined as its characteristic impedance Z_0. An arbitrary length of transmission line terminated in an impedance Z_0 will also have an input impedance Z_0. See ELECTRICAL IMPEDANCE.

If the transmission line is terminated in the arbitrary complex impedance load Z_L, the complex voltage reflection coefficient Γ_L at the termination is given by Eq. (2).

$$\Gamma = \frac{Z_L - Z_0}{Z_L + Z_0} \qquad (2)$$

Even when there is no unique expression for Z_L and Z_0 such as in the case of hollow uniconductor waveguides, the voltage reflection coefficient Γ has a value because it is simply a voltage ratio. In general, the measurement of microwave impedance is the measurement of Γ. Both amplitude and phase of Γ can be

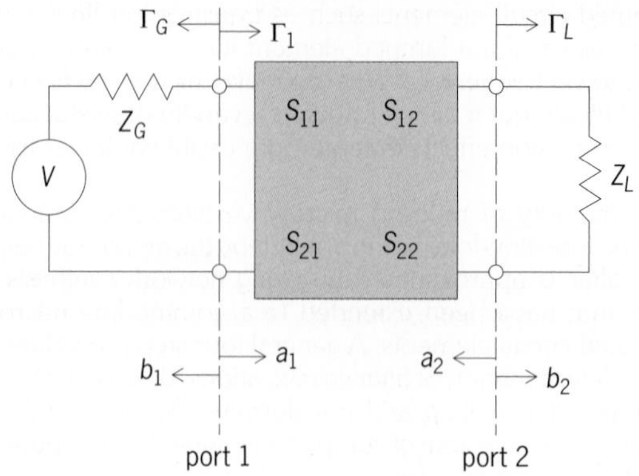

A two-port inserted between a load and a generator. S_{nm} are the scattering coefficients of the two-port.

measured by direct probing of the voltage standing wave set up along a transmission line by the two opposed traveling waves, but this is a slow technique. Directional couplers have been used for many years to perform much faster swept frequency measurement of the magnitude of Γ, and more recently the use of automatic network analyzers under computer control has made possible rapid, accurate measurements of amplitude and phase of Γ over very broad frequency ranges. See DIRECTIONAL COUPLER.

Power. A required increase in microwave power is expensive whether it be the output from a laboratory signal generator, the power output from a power amplifier on a satellite, or the cooking energy from a microwave oven. To minimize this expense, absolute power must be measured. Most techniques involve conversion of the microwave energy to heat energy which, in turn, causes a temperature rise in a physical body. This temperature rise is measured and is approximately proportional to the power dissipated. The whole device can be calibrated by reference to low-frequency electrical standards and application of appropriate corrections. See RADIOMETRY.

The power sensors are simple and can be made to have a very broad frequency response. A power meter can be connected directly to the output of a generator to measure available power P_A, or a directional coupler may be used to permit measurement of a small fraction of the power actually delivered to the load.

Scattering coefficients. While the measurement of absolute power is important, there are many more occasions which require the measurement of relative power which is equivalent to the magnitude of voltage ratio and is related to attenuation. Also there arises frequently the need to measure the relative phase of two voltages. Measurement systems having this capability are referred to as vector network analyzers, and they are used to measure scattering coefficients of multi-port devices. The concept of scattering coefficients is an extension of the voltage reflection coefficient applied to devices having more than one port. The most simple is a two-port. Its characteristics can be specified completely in terms of a 2 × 2 scattering matrix, the coefficients of which are indicated in the illustration. The incident voltage at the reference plane of each port is defined as a, and the reflected voltage is b. Voltages a and b are related by matrix equation (3), where (S_{nm}) is the scattering matrix of the junction. Writing Eq. (3) out for a two-port device gives Eqs. (4) and (5).

$$(b_n) = (S_{nm})(a_m) \qquad (3)$$

$$b_1 = S_{11}a_1 + S_{12}a_2 \qquad (4)$$

$$b_2 = S_{21}a_1 + S_{22}a_2 \qquad (5)$$

Examination of Eq. (4) shows, for example, that S_{11} is the voltage reflection coefficient looking into port 1 if port 2 is terminated with a Z_0 load ($a_2 = 0$). See MATRIX THEORY.

Heterodyne. The heterodyne principle is used for scalar attenuation measurements because of its large dynamic range and for vector network analysis because of its phase coherence. The microwave signal at frequency f_s is mixed with a microwave local oscillator at frequency f_{LO}, in a nonlinear mixer. The mixer output signal at frequency $f_s - f_{LO}$ is a faithful amplitude and phase reproduction of the original microwave signal but is at a low, fixed frequency so that it can be measured simply with low-frequency techniques. One disadvantage of the heterodyne technique at the highest microwave frequencies is its cost. Consequently, significant effort has been expended in development of multiport network analyzers which use several simple power detectors and a computer analysis approach which allows measurement of both relative voltage amplitude and phase with reduced hardware cost. *See* HETERODYNE PRINCIPLE.

Noise. Microwave noise measurement is important for the communications field and radio astronomy. The measurement of thermal noise at microwave frequencies is essentially the same as low-frequency noise measurement, except that there will be impedance mismatch factors which must be carefully evaluated. The availability of broadband semiconductor noise sources having a stable, high, noise power output has greatly reduced the problems of source impedance mismatch because an impedance-matching attenuator can be inserted between the noise source and the amplifier under test. *See* ELECTRICAL NOISE; ELECTRICAL NOISE GENERATOR; MICROWAVE NOISE STANDARDS.

Use of computers. The need to apply calculated corrections to obtain the best accuracy in microwave measurement has stimulated the adoption of computers and computer-controlled instruments. An additional benefit of this development is that measurement techniques that are superior in accuracy but too tedious to perform manually can now be considered. For a discussion of attenuation in microwave circuits *see* ATTENUATION (ELECTRICITY). Richard F. Clark

Microwave noise standards Electrical noise generators which produce calculable noise intensities at microwave frequencies, and which are used to calibrate other noise sources by using comparison methods. Noise standards are based upon the blackbody or thermal radiator and generate noise power according to Planck's radiation law. The practical realization of a blackbody in the microwave region consists of a microwave absorber with unity absorptivity. This can be achieved by using a transmission line terminated in its characteristic impedance, or in microwave terminology a matched termination. *See* HEAT RADIATION; MICROWAVE TRANSMISSION LINES; TRANSMISSION LINES.

The range of sources which require calibration and the desire to obtain low uncertainties dictate that microwave thermal noise standards are required with temperatures both above and below the ambient temperature. Sources have been developed with temperatures in the range from 4 to 1300 K (−452 to 1900°F). The low temperatures are normally achieved by immersion of the matched termination in a cryogenic liquid of which liquid nitrogen (77 K or −321°F) is the most common. Standards for measurement of high-temperature sources have the termination in a heated oven. A transition section supports the temperature gradient from the thermal termination to the ambient temperature output which connects to the measurement system. *See* MICROWAVE; MICROWAVE MEASUREMENTS; RADIOMETRY. Malcolm W. Sinclair

Microwave optics The study of those properties of microwaves which are analogous to the properties of light waves in optics. The fact that microwaves and light waves are both electromagnetic waves, the major difference being that of frequency, already suggests that their properties should be alike in many respects. But the reason microwaves behave more like light waves than, for instance, very low-frequency waves for electrical power (50 or 60 Hz) is primarily that the microwave wavelengths are usually comparable to or smaller than the ordinary physical dimensions of objects interacting with the waves.

As is the case with light, a beam of microwaves propagates along a straight line in a perfectly homogeneous infinite medium. This phenomenon follows directly from a general solution of the wave equation in which the direction of a wave normal does not change in a homogeneous medium. *See* WAVE EQUATION.

With some modification the laws of reflection and refraction can be applied to the propagation of microwaves inside a dielectric-filled metallic waveguide. Another interesting application is associated with the microwave analog of total internal reflection in optics. A properly designed dielectric rod (without metal walls) can serve as a waveguide by totally reflecting the elementary plane waves. Still another case of interest is that of a microwave lens. *See* ANTENNA (ELECTROMAGNETISM); REFLECTION OF ELECTROMAGNETIC RADIATION; REFRACTION OF WAVES; WAVEGUIDE.

In an analogous manner to light, a microwave undergoes diffraction when it encounters an obstacle or an opening which is comparable to or somewhat smaller than its wavelength. *See* MICROWAVE. C. K. Jen

Microwave solid-state devices Semiconductor devices used for the detection, generation, amplification, and control of electromagnetic radiation with wavelengths from 30 cm to 1 mm (frequencies from 1 to 300 GHz). The number and variety of microwave semiconductor devices, used for wireless and satellite communication and optoelectronics, have increased as new techniques, materials, and concepts have been developed and applied. Passive microwave devices, such as *pn* and PIN junctions, Schottky barrier diodes, and varactors, are primarily used for detecting, mixing, modulating, or controlling microwave signals. Step-recovery diodes, transistors, tunnel diodes, and transferred electron devices (TEDs) are active microwave devices that generate power or amplify microwave signals. *See* MICROWAVE; SEMICONDUCTOR DIODE.

Typical high-frequency semiconductor materials include silicon (Si), germanium (Ge), and compound semiconductors, such as gallium arsenide (GaAs), indium phosphide (InP), silicon germanium (SiGe), silicon carbide (SiC), and gallium nitride (GaN). In general, the compound semiconductors work best for high-frequency applications due to their higher electron mobilities. *See* GALLIUM; GERMANIUM; SEMICONDUCTOR; SILICON.

Passive devices. A PIN (*p*-type/intrinsic/*n*-type) diode is a *pn* diode that has an undoped (intrinsic) region between the *p*- and *n*-type regions. The use of an intrinsic region in PIN diodes allows for high-power operation and offers an impedance at microwave frequencies that is controllable by a lower frequency or a direct-current (DC) bias. The PIN diode is one of the most common passive diodes used at microwave frequencies. PIN diodes are used to switch lengths of transmission line, providing digital increments of phase in individual transmission paths, each capable of carrying kilowatts of peak power. PIN diodes come in a variety of packages for microstrip and stripline packages, and are used as microwave switches, modulators, attenuators, limiters, phase shifters, protectors, and other signal control circuit elements. *See* JUNCTION DIODE.

A Schottky barrier diode (SBD) consists of a rectifying metal-semiconductor barrier typically formed by deposition of a metal layer on a semiconductor. The SBD functions in a similar manner to the antiquated point contact diode and the slower-response *pn*-junction diode, and is used for signal mixing and detection. The point contact diode consists of a metal whisker in contact with a semiconductor, forming a rectifying junction. The SBD is more rugged and reliable than the point contact diode. The SBD's main advantage over *pn* diodes is the absence of minority carriers, which limit the response speed in switching applications and the high-frequency performance in mixing and detection applications. SBDs are zero-bias detectors. Frequencies to 40 GHz

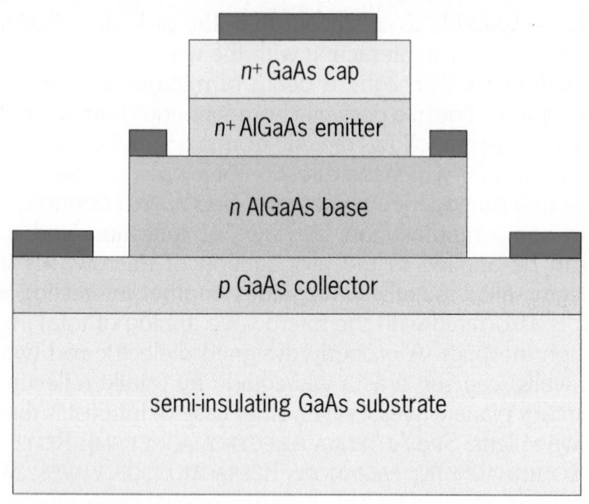

Fig. 1. Materials composition for a heterojunction bipolar transistor (HBT).

are available with silicon SBDs, and GaAs SBDs are used for higher-frequency applications. *See* SCHOTTKY EFFECT.

The variable-reactance (varactor) diode makes use of the change in capacitance of a *pn* junction or Schottky barrier diode, and is designed to be highly dependent on the applied reverse bias. The capacitance change results from a widening of the depletion layer as the reverse-bias voltage is increased. As variable capacitors, varactor diodes are used in tuned circuits and in voltage-controlled oscillators. For higher-frequency microwave applications, silicon varactors have been replaced with GaAs. Typical applications of varactor diodes are harmonic generation, frequency multiplication, parametric amplification, and electronic tuning. Multipliers are used as local oscillators, low-power transmitters, or transmitter drivers in radar, telemetry, telecommunication, and instrumentation. *See* VARACTOR.

Active devices. Transistors are the most widely used active microwave solid-state devices. At very high microwave frequencies, high-frequency effects limit the usefulness of transistors, and two-terminal negative resistance devices, such as transferred-electron devices, avalanche diodes, and tunnel diodes, are sometimes used. Two main categories of transistors are used for microwave applications: bipolar junction transistors (BJTs) and field-effect transistors (FETs). In order to get useful output power at high frequencies, transistors are designed to have a higher periphery-to-area ratio using a simple stripe geometry. The area must be reduced without reducing the periphery, as large area means large interelectrode capacitance. For high-frequency applications the goal is to scale down the size of the device. Narrower widths of the elements within the transistor are the key to superior high-frequency performance. *See* TRANSISTOR.

A BJT consists of three doped regions forming two *pn* junctions. These regions are the emitter, base, and collector in either an *npn* or *pnp* arrangement. Silicon *npn* BJTs have an upper cutoff frequency of about 25 GHz (varies with manufacturing improvements). The cutoff frequency is defined as the frequency at which the current amplification drops to unity as the frequency is raised. The primary limitations to higher frequency are base and emitter resistance, capacitance, and transit time. To operate at microwave frequencies, individual transistor dimensions must be reduced to micrometer or submicrometer size. To maintain current and power capability, various forms of internal paralleling on the chip are used. Three of these geometries are interdigitated fingers that form the emitter and base, the overlaying of emitter and base stripes, and the matrix approach. Silicon BJTs are mainly used in the lower microwave ranges. Their power capability is quite good, but in terms of noise they are inferior to GaAs metal semiconductor field-effect transistors (MESFETs) at frequencies above 1 GHz and are mainly used in power amplifiers and oscillators. They may also be used in small-signal microwave amplifiers when noise performance is not critical. *See* ELECTRICAL NOISE.

Heterojunction bipolar transistors (HBTs) have been designed with much higher maximum frequencies than silicon BJTs. HBTs are essentially BJTs that have two or more materials making up the emitter, base, and collector regions (Fig. 1). In HBTs, the major goal is to limit the injection of holes into the emitter by using an emitter material with a larger bandgap than the base. The difference in bandgaps manifests itself as a discontinuity in the conduction band or the valence band, or both. For *npn* HBTs, a discontinuity in the valence band is required. In general, to make high-quality heterojunctions, the two materials should have matching lattice constants. For very thin layers, lattice matching is not absolutely necessary as the thin layer can be strained to accommodate the crystal lattice of the other material. Fortunately, the base of a bipolar transistor is designed to be very thin and thus can be made of a strained layer material. Combinations such as AlGaAs/InGaAs and Si/SiGe are possible. *See* BAND THEORY OF SOLIDS; ELECTRICAL CONDUCTIVITY OF METALS; HOLE STATES IN SOLIDS; SEMICONDUCTOR HETEROSTRUCTURES.

Field-effect transistors (FETs) operate by varying the conductivity of a semiconductor channel through changes in the electric field across the channel. The three basic forms of FETs are the junction FET (JFET), the metal semiconductor FET (MESFET), and the metal oxide semiconductor FET (MOSFET). All FETs have a channel with a source and drain region at each end and a gate located along the channel, which modulates the channel conduction (Fig. 2). Microwave JFETs and MESFETs work by channel depletion. The channel is *n*-type and the gate is *p*-type for JFETs and metal for MESFETs. FET structures are well suited for microwave applications because all contacts are on the surface to keep parasitic capacitances small. The cutoff frequency is mainly determined by the transit time of the electrons under the gate; thus short gate lengths (less than 1 μm) are used.

Power devices consist of a number of MESFETs in parallel with air bridges connecting the sources. GaAs MESFET devices are used in low-noise amplifiers (LNAs), Class C amplifiers, oscillators, and monolithic microwave integrated circuits. The performance of a GaAs FET is determined primarily by the gate width and length. The planar structure of a MESFET makes it straightforward to add a second gate which can be used to control the amplification of the transistor. Dual-gate MESFETs can be used as mixers (with conversion gain) and for control purposes. Applications include heterodyne mixers and amplitude modulation of oscillators. *See* AMPLIFIER; HETERODYNE PRINCIPLE; MIXER; OSCILLATOR.

The MOSFET has a highly insulating silicon dioxide (SiO_2) layer between the semiconductor and the gate; however, silicon MOSFETs are not really considered microwave transistors. Compared with the GaAs MESFET, MOSFETs have lower electron mobility, larger parasitic resistances, and higher noise levels. Also, since the silicon substrate cannot be made semi-insulating, larger parasitic capacitances result. MOSFETs therefore do not perform very well above 1 GHz. Below this frequency, MOSFETs find application mainly as radio-frequency (RF) power amplifiers.

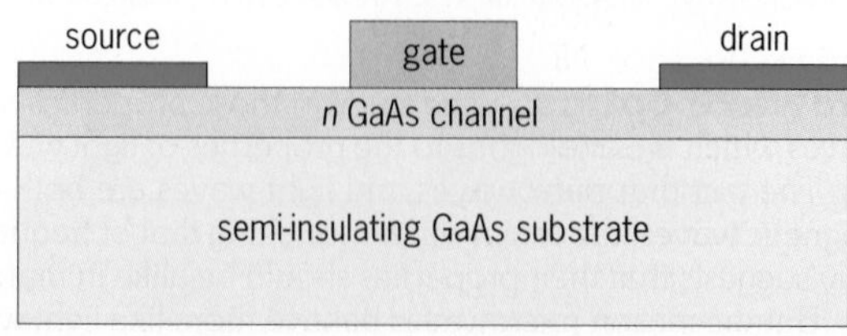

Fig. 2. Gallium arsenide metal semiconductor field-effect transistor (MESFET).

A disadvantage of the MESFET is that the electron mobility is degraded since electrons are scattered by the ionized impurities in the channel. By using a heterojunction consisting of *n*-type AlGaAs with undoped GaAs, electrons move from the AlGaAs to the GaAs and form a conducting channel at the interface. The electrons are separated from the donors and have the mobility associated with undoped material. A heterojunction transistor made in this fashion has many different names: high electron mobility transistor (HEMT), two-dimensional electron gas FET (TEGFET), modulation-doped FET (MODFET), selectively doped heterojunction transistor (SDHT), and heterojunction FET (HFET). The HEMT has high power gain at frequencies of 100 GHz or higher with low noise levels.

A monolithic microwave integrated circuit (MMIC) can be made using silicon or GaAs technology with either BJTs or FETs. For high-frequency applications, GaAs FETs are the best choice. A MMIC has both the active and passive devices fabricated directly on the substrate. MMICs are typically used as low-noise amplifiers, as mixers, as modulators, in frequency conversion, in phase detection, and as gain block amplifiers. Silicon MMIC devices operate in the 100-MHz to 3-GHz frequency range. GaAs FET MMICs are typically used in applications above 1 GHz.

Active microwave diodes. Active microwave diodes differ from passive diodes in that they are used as signal sources to generate or amplify microwave frequencies. These include step-recovery, tunnel, Gunn, avalanche, and transit time diodes, such as impact avalanche and transit-time (IMPATT), trapped plasma avalanche triggered transit-time (TRAPATT), barrier injection transit-time (BARITT), and quantum well injection transit time (QWITT) diodes.

A step recovery diode is a special PIN type in which charge storage is used to produce oscillations. When a diode is switched from forward to reverse bias, it remains conducting until the stored charge has been removed by recombination or by the electric field. A step recovery diode is designed to sweep out the carriers by an electric field before any appreciable recombination has taken place. Thus, the transition from the conducting to the nonconducting state is very fast, on the order of picoseconds. Because of the abrupt step, this current is rich in harmonics, so these diodes can be used in frequency multipliers. *See* FREQUENCY MULTIPLIER.

For microwave power generation or amplification, a negative differential resistance (NDR) characteristic at microwave frequencies is necessary. NDR is a phenomenon that occurs when the voltage (*V*) and current (*I*) are 180° out of phase. NDR is a dynamic property occurring only under actual circuit conditions; it is not static and cannot be measured with an ohmmeter. Transferred electron devices (TEDs), such as Gunn diodes, and avalanche transit-time devices use NDR for microwave oscillation and amplification. TEDs and avalanche transit-time devices today are among the most important classes of microwave solid-state devices. *See* NEGATIVE-RESISTANCE CIRCUITS.

The tunnel diode uses a heavily doped abrupt *pn* junction resulting in an extremely narrow junction that allows electrons to tunnel through the potential barrier at near-zero applied voltage. This results in a dip in the current-voltage (*I-V*) characteristic, which produces NDR. Because this is a majority-carrier effect, the tunnel diode is very fast, permitting response in the millimeter-wave region. Tunnel diodes produce relatively low power. The tunnel diode was the first semiconductor device type found to have NDR. *See* TUNNEL DIODE; TUNNELING IN SOLIDS.

Avalanche diodes are junction devices that produce a negative resistance by appropriately combining impact avalanche breakdown and charge-carrier transit time effects. Avalanche breakdown in semiconductors occurs if the electric field is high enough for the charge carriers to acquire sufficient energy from the field to create electron-hole pairs by impact ionization. The avalanche diode is a *pn*-junction diode reverse-biased into the avalanche region. By setting the DC bias near the avalanche threshold, and superimposing on this an alternating voltage, the diode will swing into avalanche conditions during alternate half-cycles. The hole-electron pairs generated as a result of avalanche action make up the current, with the holes moving into the *p* region, and the electrons into the *n* region. The carriers have a relatively large distance to travel through the depletion region. At high frequencies, where the total time lag for the current is comparable with the period of the voltage, the current pulse will lag the voltage. By making the drift time of the electrons in the depletion region equal to one-half the period of the voltage, the current will be 180° out of phase. This shift in phase of the current with respect to the voltage produces NDR, so that the diode will undergo oscillations when placed in a resonant circuit.

A Gunn diode is typically an *n*-type compound semiconductor, such as GaAs or InP, which has a conduction band structure that supports negative differential mobility. Although this device is referred to as a Gunn diode, after its inventor, the device does not contain a *pn* junction and can be viewed as a resistor below the threshold electric field (E_{thres}). For applied voltages that produce electric fields below E_{thres}, the electron velocity increases as the electric field increases according to Ohm's law. For applied voltages that produce electric fields above E_{thres}, conduction band electrons transfer from a region of high mobility to low mobility, hence the general name "transferred electron device." Beyond E_{thres}, the velocity suddenly slows down due to the significant electron transfer to a lower mobility band producing NDR. For GaAs, E_{thres} is about 3 kV/cm. The Gunn effect can be used up to about 80 GHz for GaAs and 160 GHz for InP. Two modes of operation are common: nonresonant bulk (transit-time) and resonant limited space-charge accumulation (LSA). *See* ELECTRIC FIELD.

Impact avalanche and transit-time diodes (IMPATTs) are NDR devices that operate by a combination of carrier injection and transit time effects. There are several versions of IMPATT diodes, including simple reverse-biased *pn* diodes, complicated reverse-biased multidoped *pn* layered diodes, and reverse-biased PIN diodes. The IMPATT must be connected to a resonant circuit. At bias turn-on, noise excites the tuned circuit into a natural oscillation frequency. This voltage adds algebraically across the diode's reverse-bias voltage. Near the peak positive half-cycle, the diode experiences impact avalanche breakdown. When the voltage falls below this peak value, avalanche breakdown ceases. A 90° shift occurs between the current pulse and the applied voltage in the avalanche process. A further 90° shift occurs during the transit time, for a total 180° shift which produces NDR. An IMPATT oscillator has higher output power than a Gunn equivalent. However, the Gunn oscillator is relatively noise-free, while the IMPATT is noisy due to avalanche breakdown.

A trapped plasma avalanche triggered transit-time (TRAPATT) diode is basically a modified IMPATT diode in which the holes and electrons created by impact avalanche ionization multiplication do not completely exit from the transit domain of the diode during the negative half-cycle of the microwave signal. These holes and electrons form a plasma which is trapped in the diode and participates in producing a large microwave current during the positive half-cycle.

A barrier injection transit-time diode (BARRITT) is basically an IMPATT structure that employs a Schottky barrier formed by a metal semiconductor contact instead of a *pn* junction to create similar avalanche electron injection.

A variety of approaches have been investigated to find alternative methods for injecting carriers into the drift region without relying on the avalanche mechanism, which is inherently noisy. Quantum well injection transit-time diodes (QWITT) employ resonant tunneling through a quantum well to inject electrons into the drift region. The device structure consists of a single GaAs quantum well located between two AlGaAs barriers in series with a drift region of made of undoped GaAs. This structure is then placed between two n^+-GaAs regions to form contacts.

Laurence P. Sadwick

Microwave spectroscopy The study of the interaction of matter and electromagnetic radiation in the microwave region of the spectrum. See SPECTROSCOPY.

The interaction of microwaves with matter can be detected by observing the attenuation or phase shift of a microwave field as it passes through matter. These are determined by the imaginary or real parts of the microwave susceptibility (the index of refraction). The absorption of microwaves may also trigger a much more easily observed event like the emission of an optical photon in an optical double-resonance experiment or the deflection of a radioactive atom in an atomic beam. See MOLECULAR BEAMS.

At room temperature, the relative population difference between the states involved in a microwave transition is a few percent or less. The population difference can be close to 100% at liquid helium temperatures, and microwave spectroscopic experiments are often performed at low temperatures to enhance population differences and to eliminate certain line-broadening mechanisms. The population differences between the states involved in a microwave transition can also be enhanced by artificial means. When the molecules or atoms with inverted populations are placed in an appropriate microwave cavity, microwave oscillations will build up spontaneously in the cavity through maser (microwave amplification by stimulated emission of radiation) action. See MASER.

The magnetic dipole and electric quadrupole interactions between the nuclei and electrons in atoms and molecules can lead to energy splittings in the microwave region of the spectrum. Thus, microwave spectroscopy has been used extensively for precision determinations of spins and moments of nuclei. See HYPERFINE STRUCTURE; MOLECULAR BEAMS; NUCLEAR MOMENTS.

The rotational frequencies of molecules often fall within the microwave range, and microwave spectroscopy has contributed a great deal of information about the moments of inertia, the spin coupling mechanisms, and other physical properties of molecules. See MOLECULAR STRUCTURE AND SPECTRA.

The magnetic resonance frequencies of electrons in fields of a few thousand gauss (a few tenths of a tesla) lie in the microwave region. Thus, microwave spectroscopy is used in the study of electron-spin resonance or paramagnetic resonance. See ELECTRON PARAMAGNETIC RESONANCE (EPR) SPECTROSCOPY; MAGNETIC RESONANCE.

The cyclotron resonance frequencies of electrons in solids at magnetic fields of a few thousand gauss (a few tenths of a tesla) lie within the microwave region of the spectrum. Microwave spectroscopy has been used to map out the dependence of the effective mass on the electron momentum.

For other application See ATOMIC CLOCK; COSMIC BACKGROUND RADIATION; RADIO ASTRONOMY.

William Happer

Microwave tube A high-vacuum tube designed for operation in the frequency region from approximately 3000 to 300,000 MHz. Two considerations distinguish a microwave tube from vacuum tubes used at lower frequencies: the dimensions of the tube structure in relation to the wavelength of the signal that it generates or amplifies, and the time during which the electrons interact with the microwave field. See VACUUM TUBE.

In the microwave region wavelengths are in the order of centimeters; resonant circuits are in the forms of transmission lines that extend a quarter of a wavelength from the active region of the microwave tube. With such short circuit dimensions the internal tube structure constitutes an appreciable portion of the circuit. For these reasons a microwave tube is made to form part of the resonant circuit. Leads from electrodes to external connections are short, and electrodes are parts of surfaces extending through the envelope directly to the external circuit that is often a coaxial transmission line or cavity. See CAVITY RESONATOR; TRANSMISSION LINES.

At microwaves the period of signal is in the range of 0.001-1 nanosecond. Only if transit time is less than a quarter of the signal period do significant numbers of electrons exchange appreciable energy with the signal field. Transit time is reduced in several ways. Electrodes are closely spaced and made planar in configuration, and high interelectrode voltages are used.

Tubes designed by the foregoing principles are effective for wavelengths from a few meters to a few centimeters. At shorter wavelengths different principles are necessary. To obtain greater exchange of energy between the electron beam and the electromagnetic field several alternative designs have proved practical.

Instead of collecting the electron beam at a plate formed by the opposite side of the resonant circuit, the beam is allowed to pass into a field-free region before reacting further with an external circuit. The electron cloud can be deflected by a strong static magnetic field so as to revolve and thereby react several times with the signal field before reaching the plate. See KLYSTRON; MAGNETRON.

Instead of producing the field in one or several resonant circuits, the field can be supported by a distributed structure along which it moves at a velocity comparable to the velocity of electrons in the beam. The electron beam is then directed close to this structure so that beam and field interact over an extended interval of time. See MICROWAVE; TRAVELING-WAVE TUBE.

Frank H. Rockett

Mictacea A proposed order of the Peracarida established for two small crustacean species, *Hirsutia bathyalis* and *Mictocaris halope*. The two species share many features common to other peracaridans but differ sufficiently to justify their assignment to a distinct order with two monotypic families. Common peracaridan features include a brood pouch formed by basal lamellae of the perepods (oöstegites) in the female; a small movable process (lacinia mobilis) on the mandible; free thoracic somites not fused to a carapace shield; a single maxilliped of typical peracarid form; and partially immobile pereopodal basal segments.

Mictocaris is a cave-dwelling species, whereas *Hirsutia* has been found only in soft muddy sediment of the deep sea. Nothing is known about the feeding habits of these crustaceans, but with its spined first pereopod and fossorial second, *Hirsutia* is thought to be carnivorous. In contrast, the feeding appendages in *Mictocaris* resemble those of thermosbaenaceans, which scrape food particles from the substrate.

Mictacea appear to be most closely related to the Thermosbaenacea, Spelaeogriphacea, and Mysidacea. See CRUSTACEA; MYSIDA; PERACARIDA; SPELAEOGRIPHACEA; THERMOSBAENACEA.

Patsy A. McLaughlin

Mid-Oceanic Ridge A largely interconnected system of broad submarine rises totaling at least 60,000–80,000 km (37,000–50,000 mi) long, the precise length depending on what is included and how it is measured. Thus the Mid-Oceanic Ridge is the longest mountain range system on the planet. The origin of the Mid-Oceanic Ridge is intimately connected with plate tectonics. Wherever plates move apart sufficiently far and fast for oceanic crust to form in the void between them, a branch of the Mid-Oceanic Ridge will be created. The plate boundary of the Mid-Oceanic Ridge comprises an alternation of spreading centers (or axes or accreting plate boundaries) interrupted or offset by a range of different discontinuities, the most prominent of which are transform faults. As the plates move apart, new oceanic crust is formed along the spreading axes, and the ideal transform fault zones are lines along which plates slip past each other and where oceanic crust is neither created nor destroyed. See PLATE TECTONICS; TRANSFORM FAULT.

Separation of plates causes the hot upper mantle to rise along the spreading axes of the Mid-Oceanic Ridge; partial melting of this rising mantle generates magmas of basaltic composition that segregate from the mantle and rise in a narrow zone at the axis of the Mid-Oceanic Ridge to form the oceanic crust. The partially molten mantle "freezes" to the sides and bottoms of the diverging plates to form the mantle lithosphere that, together with the overlying "rind" of oceanic crust, comprises the lithospheric plate. At the axis of the Mid-Oceanic Ridge the underlying column of crust and mantle is hot and thermally expanded; this thermal expansion explains why the Mid-Oceanic Ridge is a ridge. With time, a column of crust plus mantle lithosphere cools and shrinks as it moves away from the ridge axis as part of the plate. The gentle regional slopes of the Mid-Oceanic Ridge (typically from 3 to 50 parts per thousand near the axis, and decreasing smoothly toward the flanks) therefore represent the combined effects of sea-floor spreading (divergent plate motion) and thermal contraction. *See* EARTH CRUST; LITHOSPHERE.

The height and thermal contraction rate of the ridge crest are relatively independent of the rate of sea-floor spreading; thus, the width and regional slopes of the Mid-Oceanic Ridge depend primarily on the rate of plate separation (spreading rate). Where the plates are separating at 2 cm (0.8 in.) per year, the Mid-Oceanic Ridge has five times the regional slope but only one-fifth the width of a part of the ridge forming where the plates are separating at 10 cm (4 in.) per year (for example, the northern East Pacific Rise). One consequence of the relation between the width and plate separation rate of the Mid-Oceanic Ridge is that more ocean water is displaced, thereby raising sea level, during times of globally faster plate motion.

The modern Mid-Oceanic Ridge is typically 1000–4000 km (600–2500 mi) wide, depending on the rate of plate separation and other factors. Actually, the ridge as a feature of thermal expansion has no sharp outer edge; the plate continues to cool and contract gradually and at ever-decreasing rates. However, the outer edge may be defined functionally as that line or zone beyond which the sea floor, deepening from the axis of the Mid-Oceanic Ridge, ceases to deepen further. Several processes can affect the location of the ridge's outer edge. Where postulated plumes of hot mantle material rise under the plates away from the axis of the Mid-Oceanic Ridge, the crust and mantle lithosphere become reelevated by as much as 1000–2000 m (3300–6600 ft). Whether formed by such a plume or not, an example of such a midplate swell is the Bermuda Rise. *See* HOT SPOTS (GEOLOGY).

The axis of the Mid-Oceanic Ridge—that is, the active plate boundary between two separating plates—is a narrow zone only a few kilometers wide, characterized by frequent earthquakes, intermittent volcanism, and scattered clusters of hydrothermal vents where seawater, percolating downward and heated by proximity to hot rock, is expelled back into the ocean at temperatures as high as 350°C (660°F). Surrounding such vents are deposits of hydrothermal minerals rich in metals, as well as exotic animal communities including, in some vent fields, giant tubeworms and clams. Sulfate-reducing bacteria, rather than photosynthesizing plants, are the base of the food chain at the vents. *See* HYDROTHERMAL VENT; MARINE GEOLOGY; VOLCANO.

Peter R. Vogt

Middle-atmosphere dynamics The motion of that portion of the atmosphere that extends in altitude roughly from 10 to 100 km (6 to 60 mi). The Earth's climate is determined by a balance between incoming solar and outgoing Earth thermal radiative energy, both of which must necessarily pass through the middle atmosphere. The lower portion, the stratosphere, contains many greenhouse gases (ozone, water vapor, carbon dioxide, methane, nitrous oxide, chlorofluorocarbons, and others); and it is predicted to cool at the same time as the lower atmosphere is warmed by the greenhouse effect. The middle atmosphere is also a focus for effects of emissions from proposed commercial fleets of stratospheric aircraft. In addition to the chemistry involved, dynamical transport modeling and measurements are needed to predict the widespread transport of these important trace gases and emissions over the globe. *See* ATMOSPHERE; STRATOSPHERE; TERRESTRIAL RADIATION.

The stratosphere constitutes the lower part of the middle atmosphere, from about 10 to 50 km (6 to 30 mi) altitude; from about 50 to 80 km (30 to 48 mi) or so lies the mesosphere. The location of the base of the stratosphere (called the tropopause) depends on meteorological conditions, varying on average from about 10 km (6 mi) in altitude at the poles to about 16 km (10 mi) at the Equator.

Atmospheric gravity waves result from combined gravitational and pressure gradient forces. Typical characteristics are transverse polarization, vertical wavelengths of 0.1–0 km (0.06–6 mi), horizontal wavelengths of 1–100 km (0.6–60 mi) or more, and periods in the range of 5 min to several hours. These waves may be excited by airflow over orography (mountains) as standing lee waves, by growing clouds, and by large-scale storm complexes in the lower atmosphere; and then they propagate up into the middle atmosphere.

Planetary-scale Rossby waves are large and slowly moving waves affected by the Coriolis effect due to the Earth's rotation. Rossby waves are common at middle latitudes in winter, where they can propagate up into the middle atmosphere from excitation regions below. Near the Equator, hybrid Rossby-gravity waves and also Kelvin waves (a special class of eastward-propagating internal gravity waves having no north-south velocity component) have been observed in the middle atmosphere. *See* CORIOLIS ACCELERATION.

A variety of global-scale normal-mode oscillations are also found in the middle atmosphere, prominent examples being wave 1, westward-moving waves with periods of about 5 and 16 days, and wave 3, a westward-moving feature with a period of about 2 days. Another observed oscillation in the middle atmosphere has been found with periods in the range 1–2 months (propagating up from the lower atmosphere).

Waves resulting from fluid dynamical instabilities are also observed: medium-scale (waves 4–7) eastward-moving waves, which are actually the tops of tropospheric storm systems, can dominate the circulation of the summer Southern Hemisphere lower stratosphere. The medium-scale waves have periods of 10–20 days. *See* DYNAMIC INSTABILITY; DYNAMIC METEOROLOGY.

J. L. Stanford

Midnight sun The phenomenon occurring when the Sun does not set, but only approaches the horizon at midnight. The effect occurs near the time of the summer solstice, on June 21, for latitudes north of the Arctic Circle. The same effect occurs near the time of the winter solstice, on December 21, for latitudes south of the Antarctic Circle. (Here "summer" and "winter" refer to the Northern Hemisphere; seasons are reversed in the Southern Hemisphere.)

The Earth orbits the Sun on a plane called the ecliptic. The Earth's Equator is inclined with the ecliptic by 23°26′. As a result, the North and South poles are in turn inclined toward the Sun for 6 months. Close to the summer solstice, on June 21, the Northern Hemisphere reaches its maximum inclination toward the Sun and the Sun illuminates all the polar area down to latitude +66°34′. As seen from the polar area, the Sun does not set, but only reaches its lowest altitude in the north at midnight. Latitude +66°34′ defines the Arctic Circle, which is the southernmost latitude in the Northern Hemisphere where the midnight sun can be observed near the summer solstice. However, atmospheric refraction raises objects on the horizon by about 34′, and the midnight sun can therefore be seen for a few days from locations 80 km (50 mi) south of the Arctic Circle. Observers at latitudes above the Arctic Circle see the midnight sun higher above the northern horizon, or correspondingly, see the midnight sun before the summer solstice. See EARTH ROTATION AND ORBITAL MOTION; ECLIPTIC; METEOROLOGICAL OPTICS; REFRACTION OF WAVES; SEASONS. Pekka Parviainen

Migmatite Rocks originally defined as of hybrid character due to intimate mixing of older rocks (schist and gneiss) with granitic magma. Now most plutonic rocks of mixed appearance, regardless of how the granitic phase formed, are called migmatites. Commonly they appear as veined gneisses.

Several modes of origin have been proposed. (1) Granitic magma may be intercalated between thin layers of schist (lit-par-lit injection) to form a banded rock called injection gneiss. (2) The granitic magma may form in place by selective melting of the rock components. (3) The granitic layers may develop by metamorphic differentiation (redistribution of minerals in solid rock by recrystallization). (4) The granitic layers may represent selectively replaced or metasomalized portions of the rock. See METAMORPHISM; METASOMATISM. Carleton A. Chapman

Migratory behavior Regularly occurring, oriented seasonal movements of individuals of many animal species. The term migration is used to refer to a diversity of animal movements, ranging from short-distance dispersal and one-way migration to round-trip migrations occurring on time scales from hours (the vertical movements of aquatic plankton) to years (the return of salmon to their natal streams following several years and thousands of kilometers of travel in the open sea).

Many temperate zone species, including many migrants, are known to respond physiologically to changes in the day length with season (photoperiodism). For example, many north temperate organisms are triggered to come into breeding condition by the interaction between the lengthening days in spring and their biological clocks (circadian rhythms). Similar processes, acting through the endocrine system, bring animals into migratory condition.

To perform regular oriented migrations, animals need some mechanism for determining and maintaining compass bearings. Animals use many environmental cues as sources of directional information. Work with birds has shown that species use several compasses.

Many species of vertebrates and invertebrates possess a time-compensated Sun compass. With such a system, the animal can determine absolute compass directions at any time of day; that is, its internal biological clock automatically compensates for the changing position of the Sun as the Earth rotates during the day. Many arthropods, fish, salamanders, and pigeons can perceive the plane of polarization of sunlight, and may use that information to help localize the Sun even on partly cloudy days.

Only birds that migrate at night have been shown to have a star compass. Unlike the Sun compass, it appears not to be linked to the internal clock. Rather, directions are determined by reference to star patterns which seem to be learned early in life.

Evidence indicates that several insects, fish, a salamander, certain bacteria, and birds may derive directional information from the weak magnetic field of the Earth. See MAGNETIC RECEPTION (BIOLOGY).

Many kinds of animals show the ability to return to specific sites following a displacement. The phenomenon can usually be explained by familiarity with landmarks near "home" or sensory contact with the goal. For example, salmon are well known for their ability to return to their natal streams after spending several years at sea. Little is known about their orientation at sea, but they recognize the home stream by chemical (olfactory) cues in the water. The young salmon apparently imprint on the odor of the stream in which they were hatched. Current evidence indicates that birds imprint on or learn some feature of their birthplace, a prerequisite for them to be able to return to that area following migration. On its first migration, a young bird appears to fly in a given direction for a programmed distance. Upon settling in a wintering area, it will also imprint on that locale and will thereafter show a strong tendency to return to specific sites at both ends of the migratory route.

Only in birds can an unequivocal case be made for the existence of true navigation, that is, the ability to return to a goal from an unfamiliar locality in the absence of direct sensory contact with the goal. This process requires both a compass and the analog of a map. Present evidence suggests that the map is not based on information from the Sun, stars, landmarks, or magnetic field. Other possibilities such as olfactory, acoustic, or gravitational cues are being investigated, but the nature of the navigational component of bird homing remains the most intriguing mystery in this field. Kenneth P. Able

Military aircraft Aircraft that are designed for highly specialized military applications. Fixed-wing aircraft, rotary-wing aircraft, free-flight balloons, and blimps have all been used in both crewed and crewless flight modes for military purposes. See DRONE.

Bombers are usually characterized by relatively long range, low maneuverability, and large weapon-carrying capability. The use of aerial refueling by tanker aircraft gives most bombers a global range. Bombers may be equipped to deliver conventional or nuclear weapons in day, night, or adverse weather.

Aircraft that provide airborne early warning and control include a variant of the Russian Ilyushin IL-76 and the United States E-3A. Increasing numbers of the airborne early warning and control variant of the Russian Ilyushin IL-76 aircraft (NATO codename Mainstay) have been produced for early warning against low-altitude penetration and for air battle management. The U.S. Air Force Airborne Warning and Control System (AWACS), designated E-3A, is a versatile surveillance, command, and control center, designated to provide battle management in the conduct of air warfare.

Unlike bombers and AWACS aircraft, fighters are relatively short-range, highly maneuverable, fast aircraft, designed to destroy enemy aircraft and to attack ground targets. They can carry machine guns, cannons, rockets, guided missiles, and bombs, depending upon the mission. They may be interceptor fighters, designed to shoot down enemy airplanes or missiles during day, night, or adverse weather conditions. Other fighters may be designated for close-in attack of mobile enemy ground forces to provide close support for friendly ground troops. Some fighters, called fighter-bombers, can carry conventional or nuclear weapons several hundreds of kilometers behind enemy lines to strike priority ground targets.

The United States reconnaissance program provides capabilities to meet many peacetime and wartime information collection requirements. Reconnaissance resources include strategic, tactical standoff, and penetration aircraft systems that are flexible and responsive. Reconnaissance aircraft carry photographic, infrared, radar, and television sensors. These aircraft may be specially designed or may be modified from a basic fighter or bomber type. Some are equipped with special electronic gear for such purposes as submarine detection; others serve as picket planes for early warning of an enemy approach.

Transport aircraft provide dedicated logistic support to all types of military operations. Transport aircraft carry troops and war supplies. Many are adaptations of airplanes used by commercial airlines. The aerial tanker is a special-purpose transport aircraft. Fighters, bombers, and helicopters can refuel from tankers while in flight by means of special probe and drogue fittings. Another special-purpose transport is a gunship. This aircraft is equipped with rapid-fire weapons for saturation attack on ground targets.

Helicopters deserve special mention as military aircraft. They are unexcelled for rescue work and for delivery of people and material to otherwise inaccessible areas. Some helicopters are armed and serve as attack aircraft, providing gun and rocket fire against ground targets. Other helicopters deliver assault troops to advanced combat areas and supply them with ammunition and other needs. *See* HELICOPTER.

Special-purpose research aircraft are occasionally designed, assembled, and tested in order to experiment with advanced aerodynamic, structural, avionic, or propulsion concepts that must be validated before they can be applied to other aircraft designs. Research aircraft are usually well instrumented, with performance data telemetered on radio-frequency data links to ground stations located at the test ranges where they are flown. *See* AIRCRAFT TESTING. Robert J. Strohl

Military satellites Artificial satellites used for a variety of military purposes. Of approximately 4000 satellites successfully launched between 1957 and 1999, about 50% have been either specifically military or usable for military purposes. Major functions of military satellites include communications, positioning and navigation, meteorology, reconnaissance and surveillance, early warning, remote sensing, geodesy, and research.

Although only certain satellites are used continuously for military purposes, all communication satellites, including commercial, may find use during conflict. All contain the necessary equipment to transmit a signal over great distances to assist in the command, control, administration, and logistic support of military forces. Military communication satellites differ from commercial satellites only in that they contain specialized components, certain capabilities, and multiple redundant systems designed to make them less vulnerable and more effective in a hostile environment. *See* COMMUNICATIONS SATELLITE.

Military forces must be able to quickly and precisely determine their position on the ground, in the air, or at sea. The Navstar Global Positioning System (GPS) is the most accurate and reliable satellite navigation system available. It includes 25 spacecraft in semisynchronous (12-h) orbits inclined at 55° to the Equator at 11,600 mi (18,700 km) altitude. Of these, 21 are operational and 4 are spares. The inclined orbits provide worldwide coverage, including the Poles. Russia also maintains global navigation satellite systems. Its Tsikada/Nadezhda low-Earth-orbit system functions similar to the United States' decommissioned Transit system. In addition, Russia operates the GLONASS navigation system. Similar to GPS, the system is less complex, but its satellites have proven less reliable than the United States' version. Receivers are available that will accept navigational data from either GPS or GLONASS. *See* SATELLITE NAVIGATION SYSTEMS.

From orbit, it is possible to obtain a wide-field-of-view image of the Earth, its cloud formations, and their movements. This meteorological information is valuable for military planning and operations.

The *Television Infrared Observation Satellites* (*TIROS*) have, for decades, traveled in Sun-synchronous, low Earth orbits providing images of cloud cover, snow, ice, and the sea surface. The Defense Meteorological Satellite Program (DMSP) consists of several satellites in low Earth, Sun-synchronous, polar orbits at an altitude of 517 mi (833 km), spaced to provide complete coverage of the Earth at various times of the day and night.

Russia operates the Meteor system of satellites. As many as five satellites are in low Earth orbit similar to *TIROS* with an orbital inclination of 81.2°. They image in the infrared spectrum. Several other countries maintain a meteorological satellite capability that provides useful information for military operations, although not specifically designed for military use. *See* METEOROLOGICAL SATELLITES.

Military reconnaissance and surveillance satellites offer near-real-time unrestricted access over almost any area on Earth. Operating in many parts of the electromagnetic spectrum, they can be used to observe weapons development and deployment of forces, and to provide warning of attack by ground forces as well as targeting intelligence, technical intelligence on enemy capabilities, electronic intelligence, and bomb damage assessment.

Early warning satellites provide information on missile launch and nuclear detonation that give governments time to make strategic military decisions.

Remote-sensing satellites afford a unique view of Earth, providing vital information to military forces. The images produced by these satellites are used to conduct routine reconnaissance, analyze waterways, assist in exercise and strike planning, and provide up-to-date maps for forces deploying to unfamiliar areas. Important information is gathered not only in the visible spectrum but in other bands of the spectrum. *See* REMOTE SENSING.

Geodesy. Geodesy is the study of the Earth's size and shape. Geodetic data are important to the military in that the data affect position determination, navigation, map making, and a variety of other missions. Almost all satellites can be used for geodesy, provided their position in space can be accurately determined by optical or electronic means from the Earth.

In addition to the missions discussed above, there have been hundreds of military research and technology spacecraft, as well as thousands of experimental investigations for military purposes on space vehicles launched by all space-faring nations.

Daniel F. Moorer, Jr.

Milk The lacteal secretion, practically free from colostrum, obtained by the complete milking of one or more healthy cows and containing not less than 8.25% milk solids (not fat) and not less than 3.25% milk fat. Among mammals, humans utilize milk as a source of food. The dairy cow supplies the vast majority of milk for human consumption, particularly in the United States; however, milk from goats, water buffalo, and reindeer is also consumed in other countries. Without qualification, the general term milk refers to cow's milk.

Average composition of milk is 87.2% water, 3.7% fat, 3.5% protein, 4.9% lactose, and 0.7% ash. Whole milk and skim milk are classified as excellent sources of calcium, phosphorus, and riboflavin because 10% of the daily nutritional requirement is supplied by not over 100 kcal (420 kilojoules). These two beverages are also classified as good sources of protein and thiamine; and whole milk is a good source of vitamin A. To be classified as good, the source must contribute 10% of a nutrient in not over 200 kcal (840 kilojoules). Milk is a good source of protein rich in all the essential amino acids.

Processing. Most raw milk collected at farms is pumped from calibrated and refrigerated stainless steel tanks into tank trucks for delivery to processing plants. The actual processing of raw milk begins with either separation or clarification. These machines are essentially similar except that in the clarifier the cream and skim milk fractions are not separated. Many processers have units called standardizer-clarifiers which separate only a small fraction of the fat from the raw whole milk; the amount of fat removed can be regulated. This facilitates the production of milk of standard fat content even though that in the raw product may vary.

Milk is rendered free of pathogenic bacteria by pasteurization. This is accomplished in a manner so that every particle of milk is heated to a specified temperature and held at that temperature for a specified time. *See* PASTEURIZATION.

Fat globules in fluid milk products are broken by homogenization into sizes that are 2 micrometers or less and thus are relatively unaffected by gravitational forces. The U.S. Public Health Service specifies that the fat content of the upper 6 in. (100 m]) of a quart of homogenized milk that has been undisturbed for 48 h cannot differ by more than 10% from that of the remainder.

Most milk is fortified with 400 international units (IU) of vitamin D per quart, and some skim milk is fortified also with 2000 IU of vitamin A per quart. These vitamins as concentrates are added either by automatic dispensing into a continuous flow of milk prior to pasteurization or as a single quantity in a batch operation.

Products. Many fermented or cultured products are produced from milk. These fermentations require the use of bacteria that ferment lactose or milk sugar.

Cultured buttermilk consists of skim milk or low-fat milk which is pasteurized at 180°F (82°C) for 30 min, cooled to 72°F (22°C), and inoculated with an active starter culture containing *Streptococcus lactis* and *Leuconostoc citrovorum*. The mixture is incubated at 70°F (21°C) and cooled when acidity is developed to approximately 0.8%. This viscous product is then agitated, packaged, and cooled. The desired flavor is created by volatile acids and diacetyl; the latter is produced by *L. citrovorum*.

One of the oldest fermented milks known is yogurt. Yogurt is prepared using whole or low-fat milk with added nonfat milk solids. The milk is heated to approximately 180°F (82°C) for 30 min, homogenized, cooled to 115°F (46°C), inoculated with an active culture, and packaged. Yogurt cultures are mixtures of *Streptococcus thermophilus* and *Lactobacillus bulgaricus* in a 1:1 ratio. Balance of these organisms in the culture is important for production of a quality product.

Concentrated and dried milk products. To reduce costs of transportation and handling, either part or all of the water is removed from milk. Moreover, the partly dehydrated milk can either be sterilized or dried to permit unrefrigerated storage for prolonged periods. Many different milk products (such as dry whole milk, evaporated milk, and condensed milk) are produced for these specific reasons. The composition of some of these is controlled by standards of identity and some by request of the commercial buyer. Robert L. Bradley, Jr.

Milky Way Galaxy The large disk-shaped aggregation of stars, gas, and dust in which the solar system is located. The term "Milky Way" is used to refer to the diffuse band of light visible in the night sky emanating from the Milky Way Galaxy. Although the two terms are frequently used interchangeably, Milky Way Galaxy, or simply the Galaxy, refers to the physical object rather than its appearance in the night sky, while Milky Way is used to refer to either.

Structure and contents. The Milky Way Galaxy contains about 2×10^{11} solar masses of visible matter. Roughly 96% is in the form of stars, and about 4% is in the form of interstellar gas. The gas both inside the stars and in the interstellar medium is primarily hydrogen and helium with a small admixture of all of the heavier atoms. The mass of dust is about 1% of the interstellar gas mass and is an insignificant fraction of the total mass of the Galaxy. Its presence, however, limits the view from the Earth in the plane of the Galaxy to a small fraction of the Galaxy's diameter in most directions. *See* INTERSTELLAR MATTER.

The Milky Way Galaxy contains four major structural subdivisions: the nucleus, the bulge, the disk, and the halo. The Sun is located in the disk about half way between the center and the indistinct outer edge of the disk of stars. The currently accepted value of the distance of the Sun from the galactic center is 8.5 kiloparsecs, although some measurements suggest that the distance may be as small as 7 kpc. *See* PARSEC.

The nucleus of the Milky Way is a region within a few tens of parsecs of the geometric center and is totally obscured at visible wavelengths. The nucleus is the source of very energetic activity detected by means of radio waves and infrared radiation.

At the galactic center, there is a very dense cluster of hot stars observed by means of its infrared radiation. In 1997, astronomers confirmed the existence of a black hole with a mass of about 2.5 million times the mass of the Sun at the position of an unresolved source of radio emission known as Sgr A* in the middle of the central star cluster. The black hole appears to be the dynamical center of the Milky Way. *See* BLACK HOLE; INFRARED ASTRONOMY; RADIO ASTRONOMY.

The bulge is a spheroidal distribution of stars centered on the nucleus which extends to a distance of about 3 kpc from the center. It contains a relatively old population of stars, very nearly as old as the Milky Way itself. Direct imaging with infrared satellites has demonstrated that the bulge is actually an elongated barlike structure with a length about two to three times its width. The Milky Way is thus classified as a barred spiral galaxy, a classification that includes about half of all disk-shaped galaxies. *See* STELLAR POPULATION.

The disk is a thin distribution of stars and gas orbiting the nucleus of the Galaxy. The disk of stars begins near the end of the bar and can be identified to about 16 kpc from the center of the Galaxy; the disk of gas can be identified to about twice this distance, about 35 kpc from the center. The faint, low-mass stars make up most of the mass of the disk. There is also a thick disk of stars and gas. The thin disk of stars contains most of the mass and has a thickness relative to its diameter similar to that of a commercial compact disk. The disk is the location of the spiral arms that are characteristic of most disk-shaped galaxies, as well as most of the present-day star formation.

The halo is a rarefied spheroidal distribution of stars nearly devoid of the interstellar gas and dust that surrounds the disk. The stars found in the halo are the oldest stars in the Galaxy. The stars are found individually as "field" stars as well as in globular clusters: spherical clusters of up to about a million stars with very low abundances of elements heavier than helium. The extent of the halo is not well determined, but globular clusters with distances of about 40 kpc from the center have been identified. Dynamical evidence suggests that the halo contains nonluminous matter in some unknown form, commonly referred to as dark matter. The dark matter contains most of the mass of the Galaxy, dominating even that in the form of stars. *See* DARK MATTER; STAR CLUSTERS.

Current evidence suggests that the ages of the oldest stars in the Milky Way are within about 10% of the age of the universe as a whole; thus parts of the Milky Way must have formed early in the history of the universe, about 12 billion years ago. There is increasing evidence that the Milky Way formed as a result of the coalescence of small galaxies and protogalaxies, objects with the masses of small dwarf galaxies that are thought to have been among the first objects to form in the universe. The coalescence would have proceeded rapidly at first and then more irregularly as relatively large pieces merged to form the Milky Way. According to this picture, the formation of the Milky Way is not yet complete, and both small galaxies (such as the Magellanic clouds and the Sagittarius Dwarf) and starless clouds of gas are continuing to rain in on the Milky Way. *See* GALAXY FORMATION AND EVOLUTION. Leo Blitz

Milleporina An order of the class Hydrozoa of the phylum Cnidaria. These are the "stinging corals" of shallow tropical seas. Their structure is similar to that of hydroids except for the

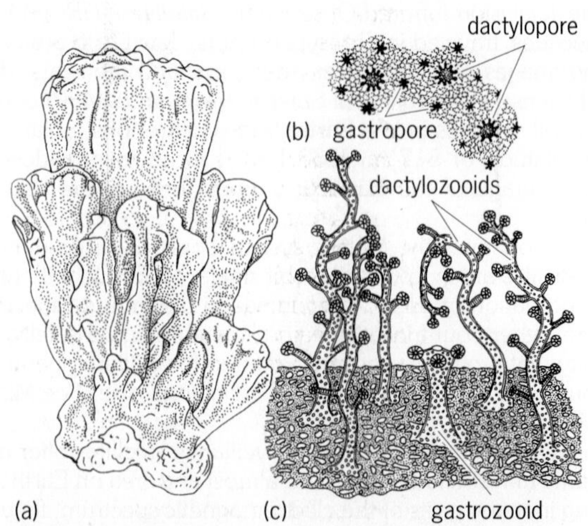

Milleporina. (*a*) Piece of dry *Millepora*, showing typical flabellate shape. (*b*) Same, magnified, showing pores. (*c*) Polyps of *Millepora*. (*After L. H. Hyman, The Invertebrates, vol. 1, McGraw-Hill, 1940*)

addition of a calcareous exoskeleton (illus. *a*). Because of this skeleton, they resemble true corals (Anthozoa). The skeleton is covered by a thin layer of tissue, is penetrated by interconnecting tubes, and is perforated by tiny holes through which the bodies of the "coral animals," or polyps, are extended (illus. *b*).

The polyps are of two types (illus. *c*): nutritive gastrozooids with tentacles and mouth; and protective polyps, which are long and armed with stinging cells but have no mouth. Millepores produce medusae, or jellyfish, in which sex cells develop. *See* HYDROZOA.

Sears Crowell

Millerite A mineral having composition NiS and crystallizing in the hexagonal system. Millerite usually occurs in hair-like tufts and radiating groups of slender to capillary crystals. The hardness is 3–3.5 (Mohs scale) and the specific gravity is 5.5. The luster is metallic and the color pale brass yellow. Millerite is found in many localities in Europe, notably in Germany and Czechoslovakia. In the United States it is found with pyrrhotite at the Gap Mine, Lancaster County, Pennsylvania; with hematite at Antwerp, New York; and in geodes in limestone at Keokuk, Iowa. In Canada large cleavable masses are mined as a nickel ore in Lamotte Township, Quebec.

Cornelius S. Hurlbut, Jr.

Millet A common name applied to at least five related members of the grass family grown for their edible seeds: foxtail millet (*Setaria italica*), proso millet (*Panicum miliaceum*), pearl or cat-tail millet (*Pennisetum typhoideum*), Japanese barnyard millet (*Echinochloa frumentacea*), raggee or finger millet (*Eleusine coracana*), and koda millet (*Paspalum scrobiculatum*).

As a crop for human food, pearl millet is grown widely in the tropics and subtropics in regions of limited rainfall where there is a growing season of 90 to 120 days. The naked seeds are yellowish to whitish in color and about the size of wheat grain. The dried grain is usually pulverized to make a meal or flour and then cooked in soups, in porridge, or as cakes.

Howard B. Sprague

Mineral A naturally occurring solid phase with a restricted chemical composition and a definite atomic arrangement. This is really an idealized definition insofar as some minerals (especially those that form at low temperature) have poorly defined atomic structures. A few solid geological materials (such as coal and obsidian) are not minerals because they have no ordered crystal structure and a completely variable composition. Rocks are usually made up of one or more minerals. In a thermodynamic sense, a mineral is a separate phase that is made up of one or more components. For example, the mineral olivine [$(Fe,Mg)_2SiO_4$] can be viewed as being made up of fayalite (Fe_2SiO_4) and forsterite (Mg_2SiO_4) in a solid solution.

Classification and systematics. At present, the International Mineralogical Association recognizes over 4000 known minerals. Most of these are very rare, and only a few minerals are common enough to be geologically significant. As developed from the scheme of J. D. Dana (1848), minerals are viewed as nominally ionic compounds (salts) and are classified by the nature of the anion. The most abundant elements in the Earth's crust are oxygen (O), silicon (Si), and aluminum (Al). These elements combine to form a variety of anionic units based on SiO_4 and AlO_4 tetrahedra. Such aluminosilicate minerals are, by far, the most volumetrically important in the Earth's crust. The silicate minerals are classified by the degree of polymerization of the $(Si,Al)O_4$ tetrahedra. The SiO_4^{4-} tetrahedra can polymerize by sharing oxygen atoms to form [$Si_2O_7^{6-}$] dimers (sorosilicates), $[(Si_2O_6)^{4-}]_n$ single chains, $[(Si_8O_{22}(OH)_2^{14-}]_n$ double chains, $[(Al,Si)_4O_{10}(OH)_2^{6-}]_n$ sheets, and three-dimensional [$Si_{2-n}Al_nO_4^{n-}$] frameworks. *See* SILICATE MINERALS.

There are many nonsilicate minerals based on anions such as sulfide, oxide, hydroxide, sulfate, and carbonate, but by volume they are not as significant as silicates in the Earth's crust. Nearly all important ore minerals are nonsilicates. Moreover, the second most abundant mineral in the Earth's lower mantle is the oxide ferropericlase [$(Mg,Fe)O$].

Occurrence and formation. Igneous rocks are those that form from a molten state. Partial melting of upper mantle rocks (peridotite) will yield a melt with a basaltic composition. As this melt cools, minerals such as olivine [$(Mg,Fe)_2SiO_4$] and plagioclase feldspar [$(Ca,Na)(AlSi)AlSi_2O_8$] form. If the crystals that form can be separated from the melt, the melt composition will evolve to become more silica rich, and from it minerals such as amphibole, mica, and potassium feldspar may form. *See* AMPHIBOLE; BASALT; FELDSPAR; IGNEOUS ROCKS; MAGMA; MICA; PERIDOTITE; PETROLOGY.

When rocks are exposed to the Earth's surface, minerals are broken down by reactions with water and carbon dioxide. These weathering reactions give rise to clay minerals such as montmorillonite and kaolinite. These minerals always occur as colloids and cannot form large crystals. *See* CLAY MINERALS; COLLOID; KAOLINITE; MONTMORILLONITE; SEDIMENTARY ROCKS; WEATHERING PROCESSES.

Some minerals, especially sulfides, are deposited by hydrothermal solutions. These are usually H_2S-NaCl-rich fluids that are able to dissolve metals such as lead (Pb), zinc (Zn), and copper (Cu) by the formation of metal-chloride complexes. Hydrothermal fluids may have originated by exsolution from magma, or they may result from seawater and ground water that has circulated in a geothermal regime. As these fluids react with minerals such as feldspar ($KAlSi_3O_8$) or calcite ($CaCO_3$) the pH rises and metal sulfides will precipitate from solution either in vein deposits or as disseminated sulfide crystals in the host rock. Subsequent alteration of primary sulfide minerals by oxidized acidic ground water will yield a complex variety of sulfate and oxide minerals. *See* HYDROTHERMAL ORE DEPOSITS; ORE AND MINERAL DEPOSITS; SULFIDE AND ARSENIDE MINERALS.

Relatively soluble minerals such as halite (NaCl) and gypsum ($CaSO_4 \cdot 2H_2O$) can form in evaporite deposits. Marine evaporates result when seawater is trapped in isolated basins. Smaller evaporite deposits can form on continents when drainage basins feed into extremely arid environments. Evaporation of such waters yields unusual minerals such as nitrates and borates. *See* BORATE MINERALS; GYPSUM; HALITE; NITRATE MINERALS; SALINE EVAPORITES.

Some minerals are formed by biological processes. The most important is aragonite ($CaCO_3$), which is formed by mollusks and planktonic organisms such as foraminfera and coccolithophorids. Aragonite dissolves and reprecipitates as the more stable polymorph calcite ($CaCO_3$) which makes up the vast deposits of limestone, a major rock in the Earth's crust. Iron(III) and manganese(IV) oxides are often formed by Iron- and manganese-oxidizing bacteria. Other bacteria use these oxides, along with sulfate minerals, as electron acceptors during respiration, leading to the formation of various iron(II) oxide and sulfide minerals such as pyrite (FeS_2). The mineral apatite [$Ca_5(PO_4)_3(F,Cl,OH)$] is the inorganic phase in bone and teeth. *See* APATITE; ARAGONITE; CALCITE; LIMESTONE.

Economic aspects. A few minerals are used as gemstones. Most economically important metals (such as copper, lead, and zinc) occur as sulfide minerals. Iron ore minerals are oxides such as hematite (Fe_2O_3) and magnetite (Fe_3O_4). The primary ores of aluminum are oxides and oxide hydroxide minerals [for example, diaspore (AlOOH)]. *See* MAGNETITE; OXIDE AND HYDROXIDE MINERALS.

Industrial minerals are those that are economically useful in their natural state. Clay minerals such as kaolinite [$Al_2Si_2O_5(OH)_4$] are used to make china, pottery, bricks, and tile. Gypsum [$CaSO_4 \cdot 2H_2O$] is the main component of plaster and plasterboard. Corundum (Al_2O_3) is a common abrasive. Because of its fibrous habit, chrysotile [$Mg_3Si_2O_5(OH)_4$] is the main mineral of asbestos. The mineral talc [$Mg_3Si_4O_{10}(OH)_2$] is a lubricant owing to its layer structure held together by weak van der Waals bonds. *See* ASBESTOS; MINERALOGY; TALC.

David M. Sherman

Bibliography. H. Blatt et al., *Petrology: Igneous, Sedimentary, and Metamorphic*, W. H. Freeman, 3d ed., 2005; C. Klein, *Manual of Mineral Science*, 22d ed., Wiley, 2001; H. Paquet et al. (eds.), *Soils and Sediments: Mineralogy and Geochemistry*, Springer, 1998.

Mineralogy The study of the crystalline phases (chemical compounds or pure elements) that make up the Earth and other rocky bodies in the solar system—that is, the terrestrial planets (Mercury, Venus, and Mars), meteorites, asteroids, and planetary satellites. In universities, nearly all mineralogists are found in earth science or geology departments.

Before the twentieth century, the science of mineralogy was concerned with identifying new minerals and determining their chemical compositions and physical properties. Investigations of the chemical compositions of minerals played a central role in the early development of chemistry and atomic theory. One of the most striking aspects of minerals is that they often occur as well-formed crystals. This motivated the hypothesis that the external symmetry reflected an internal symmetrical arrangement of atoms. The development of x-ray diffraction in 1915 enabled mineralogists to determine the internal crystal structures of minerals (that is, how the atoms are arranged). Once the chemical compositions and crystal structures of minerals were known, it became possible to systematically determine the thermodynamic properties of minerals and understand their stability as a function of pressure, temperature, and composition. This enabled geologists to understand the stabilities of mineral assemblages and the origins of the different kinds of igneous and metamorphic rocks. *See* CRYSTAL STRUCTURE; GEOLOGY; MINERAL; PETROLOGY; X-RAY DIFFRACTION.

By the 1960s, the crystal structures of most of the rock-forming minerals had been determined by x-ray diffraction. New techniques, however, were used in pursuit of a deeper understanding of the structural chemistry of minerals. Transmission electron microscopy, developed in 1961, has revealed the nanoscale (10–1000 Å) structures of minerals that result from twinning, dislocations, and phase transformations. Starting in the 1960s, a variety of spectroscopic methods (such as Mössbauer, nuclear magnetic resonance, infrared, and Raman) were used to understand how atoms are ordered among the crystallographic sites in minerals and to identify the nature of defects and impurities. *See* ELECTRON MICROSCOPE; SPECTROSCOPY.

With the advent of increasingly powerful computers, mineralogists began investigating the structures and properties of minerals using methods in computational chemistry. In the 1980s and 1990s, these calculations were largely based on classical models for interatomic interactions. In recent years, calculations of mineral stabilities and properties have been done using first-principles calculations based on either the Hartree-Fock approximation or density functional theory. *See* COMPUTATIONAL CHEMISTRY; QUANTUM CHEMISTRY.

Mineralogy is concerned with understanding the physics and chemistry of minerals insofar as they control processes in geochemistry and geophysics. The boundaries between mineralogy, petrology, and geochemistry have largely disappeared. *See* GEOCHEMISTRY; GEOPHYSICS; MINERAL. David M. Sherman

Minimal principles In the treatment of physical phenomena, it can sometimes be shown that, of all the processes or conditions that might occur, the ones actually occurring are those for which some characteristic physical quantity assumes a minimum value. These processes or conditions are known as minimal principles. The application of minimal principles provides a powerful method of attacking certain problems that would otherwise prove formidable if approached directly from first principles.

One simple minimal principle asserts that the state of stable equilibrium of any mechanical system is the state for which the potential energy is a minimum. Other general theorems of classical dynamics that are related to minimal principles are Hamilton's principle and the principle of least action. *See* HAMILTON'S PRINCIPLE; LEAST-ACTION PRINCIPLE. Dudley Williams

Minimal surfaces A branch of mathematics belonging to the calculus of variations, differential geometry, and geometric measure theory. A surface, interface, or membrane is called minimal when it has assumed a geometric configuration of least area among those configurations into which it can readily deform. Soap films spanning wire frames or compound soap bubbles enclosing volumes of trapped air are common examples. *See* DIFFERENTIAL GEOMETRY; MEASURE THEORY.

Geometrically, the mean curvature of a surface S at a point is the difference between the maximum upward curvature there and the maximum downward curvature; in particular, a surface of zero mean curvature has such principal curvatures equal and opposite and hence typically appears "saddle-shaped." It turns out that S is a minimal surface; that is, it cannot be perturbed to less area leaving its boundary fixed, provided the mean curvature is zero at each of its points; such a surface could occur, for example, as a soap film spanning a wire frame. The corresponding minimal surface equation is partial differential equation. If, alternatively, S were part of a soap bubble enclosing trapped air, the mean curvature of S would be proportional to the difference in air pressure between the two sides of S. In the calculus of variations, typically area-minimizing properties of minimal surfaces are emphasized. In differential geometry, minimal surfaces are defined as surfaces of zero mean curvature; surfaces of constant mean curvature are also extensively studied.

The two-dimensional surface is dominant in determining shape whenever the energy of a system is changed significantly by a displacement or a change in area of the surface. Such surfaces include the interfaces between crystals in a typical rock or metal, the film of soapy water between the air cells in a soap froth, the membrane separating the cells in living tissue, and the cracks separating basalt columns. Minimization of surface area plays a role in determining the shape of many living organisms. *See* FOAM; GRAIN BOUNDARIES. Frederick J. Almgren, Jr.

Mining The taking of minerals from the earth, including production from surface waters and from wells. Usually the oil and gas industries are regarded as separate from the mining industry. The term mining industry commonly includes such functions as exploration, mineral separation, hydrometallurgy, electrolytic reduction, and smelting and refining, even though these are not actually mining operations. *See* HYDROMETALLURGY; METALLURGY; ORE DRESSING.

Mining is broadly divided into three basic methods: opencast, underground, and fluid mining. Opencast mining is done either from pits or gouged-out slopes or by surface mining, which involves extraction from a series of successive parallel trenches. Dredging is a type of surface mining, with digging done from barges. Hydraulic mining uses jets of water to excavate material.

Underground mining involves extraction from beneath the surface, from depths as great as 10,000 ft (3 km), by any of several methods.

Fluid mining is extraction from natural brines, lakes, oceans, or underground waters; from solutions made by dissolving underground materials and pumping to the surface; from underground oil or gas pools; by melting underground material with hot water and pumping to the surface; or by driving material from well to well by gas drive, water drive, or combustion. Most fluid mining is done by wells. In one experimental type of well mining, insoluble material is washed loose by underground jets and the slurry is pumped to the surface. *See* COAL MINING; OPEN-PIT MINING; PETROLEUM ENGINEERING; PLACER MINING; SOLUTION MINING; SURFACE MINING; UNDERGROUND MINING.

The activities of the mining industry begin with exploration, which, since accidental discoveries or surficially exposed deposits are no longer sufficient, has become a complicated, expensive, and highly technical task. After suitable deposits have been found and their worth proved, development, or preparation for mining, is necessary. For opencast mining, this involves stripping off overburden; and for underground mining, the sinking of shafts, driving of adits and various other underground openings, and providing for drainage and ventilation. For mining by wells, drilling must be done. For all these cases, equipment must be provided for such purposes as blasthole drilling, blasting, loading, transporting, hoisting, power transmission, pumping, ventilation, storage, or casing and connecting wells. Mines may ship their crude products directly to reduction plants, refiners, or consumers, but commonly, concentrating mills are provided to separate useful from useless (gangue) minerals. *See* PROSPECTING.

A unique feature of mining is the circumstance that mineral deposits undergoing extraction are "wasting assets," meaning that they are not renewable as are other natural resources. This depletability of mineral deposits requires that mining companies must periodically find new deposits and constantly improve their technology to stay in business. Depletion means that the supplies of any particular mineral, except those from oceanic brine, must be drawn from ever-lower-grade sources. Evan Just

Mink A semiaquatic, carnivorous mammal in the family Mustelidae. The American mink (*Mustela vison*) is found in Alaska and Canada and throughout most of the United States. The range extends southward in the western United States to northern New Mexico, northern Nevada, and central California. This species was deliberately introduced into the former Soviet Union, and escaped animals have established populations throughout Europe.

American mink (*Mustela vison*).

The mink has a long, slender, weasel-like body and a tail that is bushy at the tip. It is short-legged and almost uniformly dark brown (see illustration). Long, glistening guard hairs partially conceal the soft, luxurious underfur. Five toes are present on each foot, and the hindfeet may be slightly webbed. Adult mink are 530–660 mm (20–26 in.) in total length, including a 175–225-mm (7–9-in.) tail. They normally weigh 0.5–1.4 kg (1–3 lb).

American mink are found mainly along the banks of streams, rivers, ponds, and lakes and in swamps and marshes. Forested, log-strewn, or brushy areas are preferred. Mink are mainly nocturnal. They are excellent swimmers and divers. Mink feed primarily on mice, muskrats, rabbits, birds, frogs, fish, and crayfish. *See* CARNIVORA; MARTEN; WEASEL. Donald W. Linzey

Miocene The second subdivision of the Tertiary Period (Eocene, Miocene, and Pliocene) by Charles Lyell in 1833; the fourth in a more modern sevenfold subdivision (epochs) of the Cenozoic Era; and the first epoch of the Neogene Period (which includes in successive order the Miocene, Pliocene, Pleistocene, and Holocene). The Miocene represents the interval of time from the end of the Oligocene to the beginning of the Pliocene and the rocks (series) formed during this epoch. *See* CENOZOIC; HOLOCENE; OLIGOCENE; PLEISTOCENE; PLIOCENE; TERTIARY.

The Miocene spans the time interval between 23.8 and 5.32 million years ago (Ma) based on integrated astronomical and radioisotopic dating. The Miocene/Pliocene boundary is located in Sicily, just above a major unconformity separating the youngest late Miocene (Messinian) deposits (of the Great Terminal Miocene Salinity Crisis) and the overlying white chalks of the Zanclean. *See* UNCONFORMITY.

Major orogenic and volcanic events characterize the Miocene. Plate-tectonic motions, originating in the Mesozoic, resulted in the gradual dismemberment of the Tethyan Ocean and the upthrusting of the Alpine-Himalayan orogenic belt in three major phases: the late Eocene (about 40 Ma) and the early (21–17 Ma) and mid-late Miocene (10–7 Ma). Along the eastern margins of the Pacific Ocean, the ocean crust was subducted under the North and South American continents, giving rise to major orogenic movements stretching from the Aleutians to Tierra del Fuego. The Andes range was thrust up during the later part of the Miocene. The Pacific Coast developed as a result of westward drift of North America over, and partial consumption by, the Farallon plate and collision with the Farallon Ridge. Only two relatively minor plates remain as remnants of the Farallon plate: the Juan de Fuca and Cocos plates between Mexico and Alaska. The plate margin was bounded by transform faults rather than a subduction zone, and northwestern propagation of a major transform fault issuing from the Cocos plate formed the Gulf of California in the late Miocene, and its continued extension northward is familiar to residents of the west coast as the San Andreas Fault System. The latter was responsible for the formation of many of the off- and onshore basins of southern California, some of which contain prolific petroleum resources. Subduction of the Pacific plate at the Middle America Trench during the late Paleogene and Neogene resulted in arc magmatism and eventual uplift of the Central American Isthmus into a series of archipelagos in the late Miocene (about 7 Ma) and eventual fusion into a continuous land bridge in the early Pliocene (about 3 Ma) that resulted in the separation of the Atlantic and Pacific oceans and concomitant disruption in marine faunal communities as well as transcontinental migration of vertebrate animals in the Great American Faunal Interchange. *See* OROGENY; PLATE TECTONICS; SUBDUCTION ZONES; TRANSFORM FAULT.

Ocean circulation essentially assumed its modern form during the Miocene as enhanced refrigeration in the form of growth of the Antarctic Ice Sheet plunged the Earth inexorably deeper into an icehouse state, although there were some details that were completed during the succeeding Pliocene and Pleistocene epochs. An ice cap has been present on Antarctica, at least intermittently, since at least the early Oligocene (about 34 Ma). The opening of the Drake Passage between South America and Antarctica took place during the latest Oligocene–early Miocene (about 25–23 Ma), allowing the unhindered circulation of ocean currents around the Antarctic continent. The development of the Circum-Antarctic Current thermally isolated high southern latitude waters and the continent of Antarctica from the warmer, low-latitude waters and resulted in the replacement of calcareous oozes (comprising planktonic foraminifera and calcareous nannoplankton) by biosiliceous oozes (diatoms and radiolarians). *See* GLACIAL EPOCH.

During the Miocene, life assumed much of its modern aspect. The spread of grasses and weeds throughout this epoch, but particularly in the late Miocene, and concomitant reduction in and thinning of forests reflected the global Neogene cooling as the Earth entered deeper into an ice house state. In this environment, snakes, frogs, and murids (rats, mice) expanded in diversity and habitat; songbirds reflect the expansion of seed-bearing herbs and, like frogs, the concomitant diversification of insects, many of which are found entombed in middle Miocene amber from the Dominican Republic. Grazing animals (elephants, rodents, horses, camelids, and rhinos, for example) developed high crowned teeth to resist significant wear caused by silicon fragments in the developing grasses. Some animals assumed gigantic proportions such as *Baluchi- therium*, a Eurasian rhino that stood 16 ft (5 m) at the shoulders, and the tallest camel known, a giraffelike form that was over 12 ft (3.5 m) tall.

The relatively free interchange between Eurasia and Africa between 18 and 12 Ma appears to have come to an end in the late Miocene, after which (about 8 Ma) the hominoids of Eurasia and Africa appear to have followed separate and independent lines of evolution: the pongids in Asia, and the panids and hominoids leading eventually to the true hominids in (predominantly East) Africa. This scenario has been linked, in turn, with the development of the East African Rift (and its northward extension into the Red Sea and Gulf of Suez), which would have served as a geographic barrier allowing independent evolution toward forest (panid) and savannah (hominid) adapted forms. With the late Miocene change in climate (7–5 Ma) to cooler, drier conditions and the spread of open savannah and grasslands, monkeys came to dominate the African forest at the expense of dryomorphs. There is a gap in the terrestrial fossil record during this interval of time, and it is only in the early Pliocene (about 4 Ma) that the story of human evolution resumes with the discovery of the earliest true hominids (australopithecines) in East Africa, about 1 million years older than the australopithecine footprints of Laeotili and the skeletons of Lucy and other australopithecines at Hadar in Ethiopia at about 3 million years. *See* AUSTRALOPITHECINE; FOSSIL HUMANS.

In the marine realm, major radiation of mammals including walruses, seals, sea lions, and whales occurred during the early Miocene. On the sea floor, large bivalve mollusks of the scallop family thrived in the early Miocene, and a distinct horizon of large pectenids occurs in lower Miocene rocks of Europe and North America and in corresponding levels in the deposits of the Paratethyan Sea in east-central Europe and at least as far to the east as Iran, attesting to an interval of global climatic amelioration. Among the protozoa, planktonic foraminifera experienced a major radiation in the early and middle Miocene following the drastic reduction in diversity during the middle and late Eocene, some 15–20 million years earlier. Mangroves and coral reefs flourished in a circumequatorial belt spanning the Indo-Pacific and Caribbean regions, but the latter were eliminated from the Mediterranean during the terminal Miocene Salinity Crisis, never to return with early Pliocene flushing from the Atlantic. *See* FORAMINIFERIDA; MANGROVE; MOLLUSCA; REEF. W. A. Berggren

Mira The first star recognized to have a periodic brightness variation. Mira (officially designated Omicron Ceti, in the constellation Cetus, the Whale) was discovered in 1596 by David Fabricius (or Fabricus). Mira is the prototype of an entire class of Mira-type pulsating long-period variables. Although it once resembled the Sun, Mira has evolved into a cool red giant star that is at the end of its life. *See* GIANT STAR; STELLAR EVOLUTION.

Contracting and expanding every 332 days, Mira typically varies in visual brightness from about magnitude 3.4 (brightest) to about 9.3 (faintest). However, an individual maximum can sometimes be as bright as second magnitude, while at other times the maximum may reach barely fifth magnitude. Mira is the brightest long-period variable because it is intrinsically bright and only 420 light-years from the Earth. Angular diameter measurements such as those from the Very Large Telescope (VLT) Interferometer show that the diameter of Mira varies from about 330 to 400 times the diameter of the Sun during its pulsational cycle, and has an average size larger than the orbit of Mars. *See* MAGNITUDE (ASTRONOMY).

Mira is also known as a symbiotic variable: a long-period variable star with a close hot compact companion star. The companion, VZ Ceti (or Mira B), forms a close visual double with Mira (or Mira A; the pair is often known as Mira AB). VZ Ceti was also discovered in 1959 by G. van Biesbroeck to be a variable star in its own right. It is a hotter, bluer, burned-out star, called a white dwarf, surrounded by material captured from Mira's wind. *See* BINARY STAR; SYMBIOTIC STAR; WHITE DWARF STAR.

The *Chandra X-ray Observatory* has observed Mira and its companion. The *Chandra* image shows. Mira A losing matter from its upper atmosphere via a stellar wind. The image shows a faint bridge of this material between the two stars. *See* STAR; VARIABLE STAR. Arne Henden; Janet Akyüz Mattei

Mirage A name for a variety of unusual images of distant objects seen as a result of the bending of light rays in the atmosphere during abnormal vertical distribution of air density. If the air closer to the ground is much warmer than the air above, the rays are bent in such a way that they enter the observer's eyes along a line lower than the direct line of sight. The object is then seen below the horizon, the inferior mirage. If the air closer to the ground is much colder than the air above, the rays are bent in the opposite direction, arriving at the observer's eyes above the line of sight; the object then seems to be elevated or floating in the air, the superior mirage. Mirages can be seen most frequently along an overheated highway surface; the inferior mirage of the sky gives the impression of water reflection over a wet pavement, which disappears upon a closer viewing. Zdenek Sekeva

Mirror optics The use of plane or curved reflecting surfaces for the purpose of reverting, directing, or forming images. An optical surface which specularly reflects the largest fraction of the incident light is called a reflecting surface. Such surfaces are commonly fabricated by polishing of glass, metal, or plastic substrates, and then coating the surface of the substrate with a thin layer of metal, which may be covered in addition by a single or multiple layers of thin dielectric films. The law of reflection states that the incident and reflected rays will lie in the plane containing the local normal to the reflecting surface and that the angle of the reflected ray from the normal will be equal to the angle of the incident ray from the normal. *See* GEOMETRICAL OPTICS.

The formation of images in the plane mirrors is easily understood by applying the law of reflection. The illustration shows the formation of the image of a point formed by a plane mirror. Each of the reflected rays appears to come from a point image located a distance behind the mirror equal to the distance of the object point in front of the mirror. The face of the observer can be considered as a set of points, each of which is imaged by the plane mirror. Since the observer is viewing the facial image from

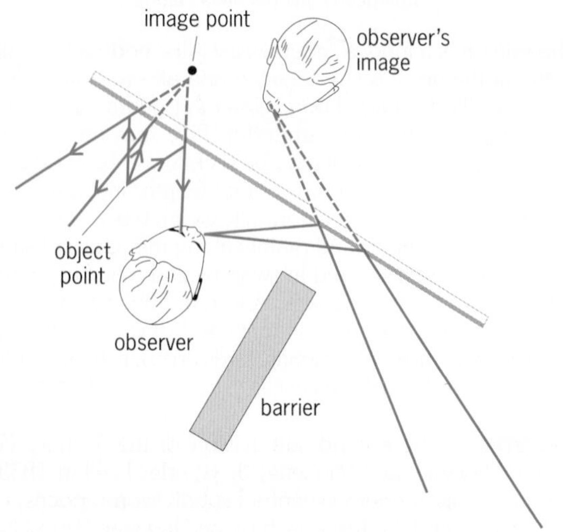

Formation of images by a plane mirror.

the object side of the mirror, the face will appear to be reversed left for right in the virtual image formed by the mirror. The illustration also indicates the redirection of light by a plane mirror, in that a viewer who cannot observe the object point directly can observe the virtual image of the point formed by the mirror. A simple optical device which is based on this principle is the simple mirror periscope, which uses two mirrors to permit viewing of scenes around an obstacle. *See* PERISCOPE.

A curved mirror, either spherical or conic in form, will produce a real or virtual image in much the same manner as a lens, but generally with reduced aberrations. There will be no chromatic aberrations since the law of reflection is independent of the color or wavelength of the incident light. *See* ABERRATION (OPTICS); OPTICAL IMAGE.

Both concave and convex spherical mirrors are commonly encountered. Convex mirrors are commonly used as wide-angle rearview mirrors in automobiles or on trucks. A common application of concave mirrors is the magnifying shaving mirror frequently found in bathrooms.

A spherical mirror will form an image which is not perfect, except for particular conjugate distances. The use of a mirror which has the shape of a rotated conic section, such as a parabola, ellipsoid, or hyperboloid, will form a perfect image for a particular set of object-image conjugate distances and will have reduced aberrations for some range of conjugate relations. The most familiar applications for conic mirrors are in reflecting tele- scopes. *See* OPTICAL PRISM; OPTICAL SURFACES; OPTICAL TELESCOPE; REFLECTION OF ELECTROMAGNETIC RADIATION; TELESCOPE. R. R. Shannon

Misophrioida A copepod order of free-living crustaceans made up of 34 species in 16 genera and 3 families. *Misophria pallida* was the first species described, in 1865, from a sample taken close to the sea floor in shallow water. A second genus and species, *Benthomisophria palliata*, was established in 1909 for a bathypelagic species from the North Atlantic Ocean. The remaining species have been described since 1964. Many of the recently described species have been recovered during the exploration of anchialine caves, which are coastal marine water bodies with subsurface hydrologic connections to the sea. Anchialine caves are one category of crevicular habitat in coastal marine geological formations. *See* COPEPODA; CRUSTACEA.

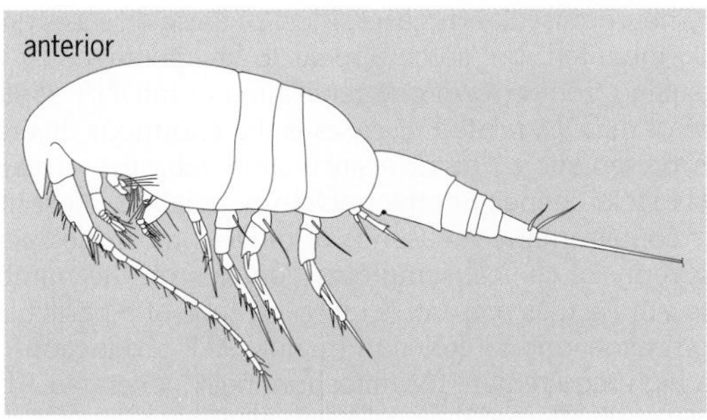

Side view of typical misophrioidan.

Adult females are 0.6–5.7 mm (0.02–0.23 in.) long and males 0.6–3.7 mm (0.02–0.15 in.); the length range includes small species living close to the sea floor and the larger, bathypelagic species. The cuticle of the adult body is distinctly ornamented. The second thoracic somite, bearing the first swimming leg, articulates with the first somite. However, a posterior extension of the cuticle of the first thoracic somite completely covers the second thoracic somite so that these two somites appear unarticulated (see illustration). The last two thoracic somites are associated with the posterior part of the body (the podoplean architecture). The internal anatomy of *B. palliata* includes a heart, but the naupliar (larval) eye, a normal feature of adult copepods, is absent.

Frank D. Ferrari

Mississippian The fifth period of the Paleozoic Era. The Mississippian System (referring to rocks) or Period (referring to time during which these rocks were deposited) is employed in North America as the lower (or older) subdivision of the Carboniferous, as used on other continents. The name Mississippian is derived from rock exposures on the banks of the Mississippi River between Illinois and Missouri.

The limits of the Mississippian Period are radiometrically dated. Its start (following the Devonian) is dated as 345–360 million years before the present (Ma). Its end (at the start of the next younger North American period, the Pennsylvanian) is dated as 320–325 Ma. The duration of the Mississippian is generally accepted as 40 million years (m.y.). Biochronologic dating within the Mississippian, based on a combination of conodont (phosphatic teeth of eel- or hagfish-like primitive fish), calcareous foraminiferan, and coral zones permits a relative time resolution to within about 1 m.y. *See* CARBONIFEROUS; PENNSYLVANIAN.

The Mississippian is divided, in ascending order, into the Lower Mississippian, comprising the Kinderhookian and Osagean, and the Upper Mississippian, comprising the Meramecian and Chesterian. Kinderhookian, Osagean, Meramecian, and Chesterian are used in North America as series (for rocks) and as stages (for time). In Illinois, Valmeyeran is commonly used for the Osagean and Meramecian combined.

During much of Mississippian time, the central North American craton (stable part of the continent) was the site of an extensive marine carbonate platform on which mainly limestones and some dolostones and evaporites were deposited. This platform extended either from the present Appalachian Mountains or Mississippi Valley to the present Great Basin. The craton was covered by shallow, warm, tropical epicontinental seas that had maximum depths of only about 60 m (200 ft) at the shelf edge.

Mississippian North America was subjected to a number of sea-level rises, associated with transgressions, and sea-level falls, associated with regressions. The two major rises, which were probably eustatic (referring to worldwide change of sea level), took place during the late Kinderhookian and at the start of the middle Osagean zone. These sea-level rises caused progradation or seaward migration of carbonate platforms as organisms produced buildups that maintained their niches relative to sea level. The rises also caused stratification of the water column in deeper basins, so that bottom conditions became deficient to lacking in oxygen. *See* ANOXIC ZONES; BASIN.

Crinoids were probably the most abundant biota in Mississippian seas, but are only uncommonly used for correlation because most specific identifications require study of calyxes, which generally disarticulated after death. Corals are probably the most widely preserved Mississippian megafossil group, and they are one of the most useful biochronologic tools for constructing biostratigraphic zones for carbonate-platform rocks. The earliest Mississippian is characterized mainly by solitary corals and tubular colonial corals known as *Syringopora*, which had survived the late Frasnian mass extinction. Other colonial corals began a gradual return late in the Kinderhookian. The Late Mississippian was a heyday for reef-building colonial corals and large solitary corals. *See* CORALLINALES; CRINOIDEA; REEF.

Forests flourished during the Mississippian, and tree trunks, plant stems, roots, and spores occur commonly in terrestrial and peritidal rocks, particularly in coal beds. *Lepidodendron* trunks and *Stigmaria* roots are among the best-known plant remains. *See* GEOLOGIC TIME SCALE; PALEONTOLOGY; PALEOZOIC.

Charles A. Sandberg

Mistletoe The name given to several species of the mistletoe family (Loranthaceae). The true mistletoe of Europe is *Viscum album*, and among the early nations this was an important ceremonial plant, which probably accounts for the origin of the custom of kissing under the mistletoe. In the United States, the common representative of the group is *Phoradendron* *flavescens.* All of the mistletoes are green hemiparasites; that is, they obtain water and minerals from the host plant but manufacture their own food. *See* SANTALALES.

Perry D. Strausbaugh; Earl L. Core

Mitochondria Specialized organelles of all eukaryotic cells that use oxygen (see illustration). Often called the powerhouses of the cell, mitochondria are responsible for energy generation by the process of oxidative phosphorylation. In this process, electrons produced during the oxidation of simple organic compounds are passed along a chain of four membrane-bound enzymes (the electron transport or respiratory chain), finally reacting with and reducing molecular oxygen to water. The movement of the electrons releases energy that is used to build a gradient of protons across the membrane in which the electron transport chain is situated. Like a stream of water that drives the turbines in a hydroelectric plant, these protons flow back through adenosine triphosphate (ATP) synthase, a membrane-bound enzyme that acts as a molecular turbine. Rotation of part of ATP synthase results in storage of energy in the form of ATP, the universal energy currency of the cell.

Besides their role in energy generation, mitochondria house numerous enzymes that carry out steps essential to metabolism. Defects in mitochondrial assembly or function generally have serious consequences for survival of the cell. In humans, mitochondrial dysfunction is the underlying cause of a wide range of degenerative diseases, with energy-demanding cells such as those of the central nervous and endocrine systems, heart, muscle, and kidney being most severely affected.

Mitochondria are bounded by two concentric membranes referred to as the outer and the inner. This creates two distinct compartments, the matrix and the intermembrane space. The outer membrane consists of a bilayer containing about 80% lipid. It is freely permeable to molecules smaller than about 5000 daltons. The inner membrane is also a lipid bilayer. It is extremely rich in protein (about 75%) and is impermeable to even the smallest of ions. The inner membrane contains the enzymes of the electron transport chain and the ATP synthase, together with a set of transporter proteins that regulate the movement of metabolites in and out of the matrix space. Mitochondria of cells that depend on a high level of ATP production are usually extensively folded to produce structures called cristae (see illustration). These greatly increase the surface area of the inner membrane, allowing many more copies of the enzymes of oxidative phosphorylation.

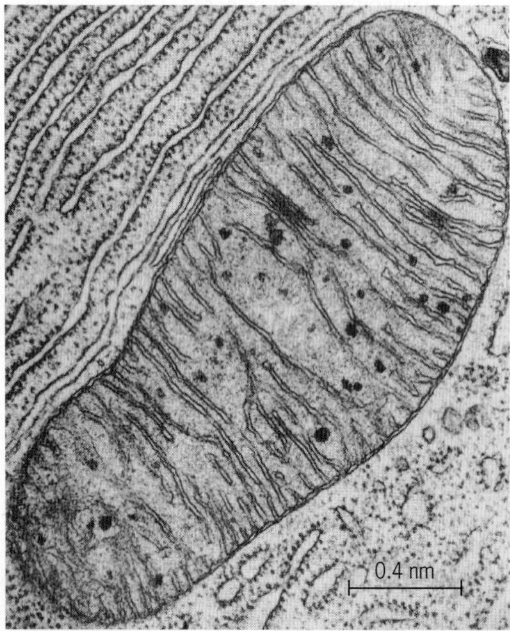

Electron micrograph of a thin section through the pancreas of a bat, showing a typical mitochondrion in profile. Note how the cristae are formed by extensive folding of the inner membrane. (*Courtesy of K. R. Porter*)

The intermembrane space contains enzymes capable of using some of the ATP that is transported out of the matrix to phosphorylate other nucleotides. The matrix space is packed with a hundred or so water-soluble proteins that form a sort of semisolid gel. They include enzymes of the tricarboxylic acid (Krebs) cycle and enzymes required for the oxidation of pyruvate and fatty acids, comprising steps in the biosynthesis or degradation of amino acids, nucleotides, and steroids.

Both mitochondria and chloroplasts contain DNA and the machinery necessary to express the information stored there. In both cases, the DNAs are relatively small and simple compared with DNA in the nucleus. While chloroplast DNA tends to be very similar in size in all organisms examined, mtDNA varies widely in complexity, from 16–18 kilobases in metazoa to upward of 2000 kilobases in some higher plants. In the case of higher-plant mtDNAs, some extra sequences appear to have been picked up from chloroplast and nuclear DNAs. MtDNA is generally circular, although some linear exceptions are known among the yeasts, algae, and protozoa.

The information content of most mtDNA is limited. This means that most of the several hundred proteins found in these organelles are encoded by genes located in nuclear DNA. These proteins are synthesized in the cytosol and subsequently transported specifically to the respective organelles. The contributions of the two genetic systems are usually closely coordinated, so that cells synthesize organelles of more or less constant composition.

Many organisms, including humans, show uniparental inheritance of mitochondrial genes because one parent contributes more cytoplasm to the zygote than the other. In humans, it is the egg cell provided by the mother that contributes the cytoplasm. Human mitochondrial genes are thus inherited maternally.

Recent years have seen a growing interest in human diseases that result from mitochondrial dysfunction. A number of these result from mutations in mtDNA. Others are linked to nuclear genes, whose mutation disturbs oxidative phosphorylation or impairs mitochondrial assembly. Mutations in mtDNA are remarkably frequent and lead to a wide range of degenerative, mainly neuromuscular diseases. Most of these diseases are maternally inherited, but some appear to be spontaneous, possibly resulting from error-prone replication of mtDNA. A striking feature of mtDNA-related diseases is the enormous diversity in clinical presentation. This diversity is attributable to two main factors: (1) Heterogeneity in the mtDNA population. Most human tissues contain many thousands of mtDNA molecules per cell. The severity of clinical symptoms depends on the number of mutated molecules present. (2) Dependence of a particular cell type on mitochondrial function (mainly ATP production). Cells with a high requirement for mitochondrially generated ATP are more severely affected than cells with alternative sources of ATP.

Besides their role in metabolism and energy-linked processes, mitochondria have recently been identified as important players in the initiation of apoptosis (programmed cell death). On one hand, the mitochondrial outer membrane houses a number of members of the Bcl-2 family of apoptosis regulatory proteins. On the other hand, release of certain mitochondrial proteins from the intermembrane space is instrumental in activating specialized proteases called caspases. These catalyze a degradative cascade in the cytoplasm that eventually ends in cell death.

Mammalian mtDNAs accumulate mutations at high rates and evolve correspondingly fast (up to 12–15 times faster than single-copy genes in nuclear DNA and up to 100 times faster for rRNA and tRNA genes). This behavior reflects both a high incidence of mutations and a high probability of their fixation. The first is probably related to oxidative damage to mtDNA by oxygen free radicals produced as by-products of electron transfer through the respiratory chain. The second has been attributed to the lack of efficient DNA repair (mitochondria lack nucleotide-excision repair) and to a relatively high tolerance of many mitochondrial gene products to mutational change.

The rapid rate of sequence evolution of mammalian mtDNAs makes these genomes highly sensitive indicators of recent evolutionary relationships. Unlike their nuclear counterparts, mtDNAs do not undergo recombination during sexual transmission and are strictly maternally inherited. Sequence changes in mtDNA therefore provide a clear record of the history of the female lineages through which this DNA has been transmitted.

Les A. Grivell

Mitosis The series of visible changes that occur in the nucleus and chromosomes of non-gamete-producing plant and animal cells as they divide. During mitosis, the replicated genes, packaged within the nucleus as chromosomes, are precisely distributed into two genetically identical daughter nuclei (see illustration). The series of events that prepares the cell for mitosis is known as the cell cycle. When viewed in the context of the cell cycle, the definition of mitosis is often expanded to include cytokinesis, the process by which the cell cytoplasm is partitioned during cell division. *See* CELL CYCLE.

Chromosome segregation is mediated in all nonbacterial cells (that is, eukaryotes) by the transient formation of a complex structure known as the mitotic spindle. During mitosis in most higher plants and animals, the nuclear membrane surrounding the replicated chromosomes breaks down, and the spindle is formed in the region previously occupied by the nucleus (open mitosis). In lower organisms, including some protozoa and fungi, the spindle is formed and functions entirely within the nucleus which remains intact throughout the process (closed mitosis). *See* CHROMOSOME.

All spindles are bipolar structures, having two ends or poles. In animal cells, each spindle pole contains an organelle, the centrosome, onto which the spindle focuses. The polar regions of plant spindles lack centrosomes and, as a result, are much broader. In animals the bipolar nature of the spindle is established by the separation of the centrosomes, which is critical for a successful mitosis; the presence of only one pole produces a monopolar spindle in which chromosome segregation is inhibited. The presence of more than two poles produces multipolar spindles which distribute the chromosomes unequally among three or more nuclei. Centrosomes are duplicated during interphase near the time that the DNA is replicated, but then act as a single functional unit until the onset of mitosis. In plants, and during meiosis in some animals, the two spindle poles are organized by the chromosomes and by molecular motors that order randomly nucleated microtubules into parallel bundles. *See* CENTROSOME; DEOXYRIBONUCLEIC ACID (DNA); PLANT CELL.

Microtubules are the primary structural components of the mitotic spindle and are required for chromosome motion. These 25-nanometer-diameter, hollow, tubelike structures are formed from the polymerization of protein subunit dimers composed of alpha and beta tubulin. During interphase, microtubules are distributed throughout the cytoplasm, where they serve to maintain cell shape and also function as polarized roadways for transporting

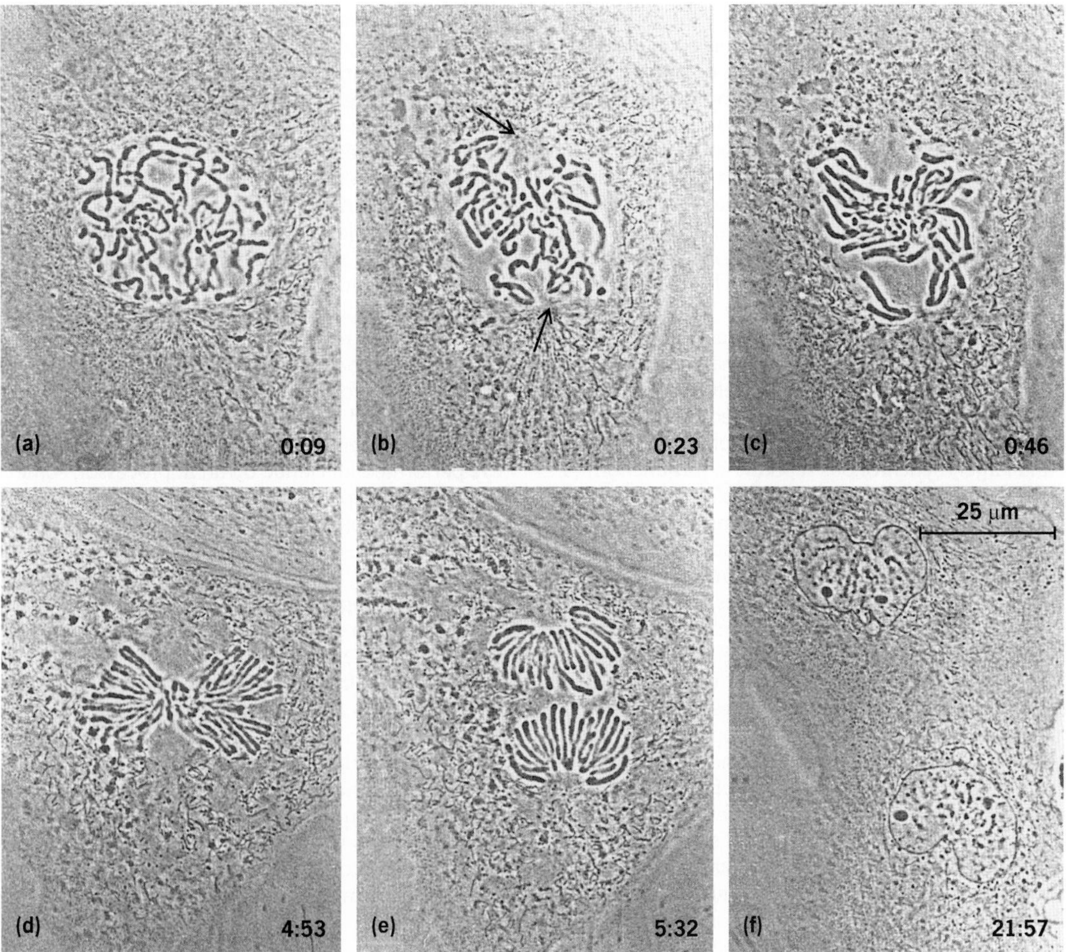

Selected phase-contrast light micrographs showing changes in chromosome position during mitosis in a living newt lung epithelial cell (elapsed time in hours and minutes). (*a*) Late prophase. (*b*) Prometaphase. (*c*) Mid-prometaphase. (*d*) Metaphase. (*e*) Anaphase. (*f*) Telophase.

organelles and cell products. As the cell enters mitosis, the cytoplasmic microtubule network is disassembled and replaced by the mitotic spindle. The microtubules in animal cells originate from the centrosome that, like the chromosomes, was inherited during the previous mitosis where it functioned as a spindle pole. The motion associated with microtubules is mediated by several families of molecular motors which bind to and move along the wall of the microtubule. *See* CYTOSKELETON.

As mitosis begins, each replicated chromosome consists of two identical sister chromatids that are joined along their length. In most cells, chromosomes possess a unique region of highly condensed chromatin (DNA plus protein), known as the centromere, which forms an obvious constriction on the chromosome, referred to as the primary constriction. Spindle microtubules attach to a small specialized structure on the surface of the centromere known as the kinetochore. Fragments of chromosomes lacking a kinetochore do not move poleward; it is always the kinetochore that leads in the poleward motion of the chromosome. Kinetochores in most plant cells are shaped like a ball, whereas those in animal cells resemble a plate. The centromere region of each replicated chromosome contains two sister kinetochores, one attached to each chromatid, that lie on opposite sides of the primary constriction.

Once initiated, mitosis is a continuous process that, depending on the temperature and organism, requires several minutes to many hours to complete. Traditionally it has been subdivided into five consecutive stages that are distinguished primarily by chromosome structure, position, and behavior. These stages are prophase, prometaphase, metaphase, anaphase, and telophase. In prophase, the first visible sign of an impending mitosis occurs within the nucleus as the DNA-containing chromatin begins to condense into chromosomes. By late prophase the individual chromosomes are well defined and the nucleoli begin to dissipate. Near this time the cytoplasmic network of microtubules breaks down, and is replaced in animal cells by two radial astral arrays of microtubules growing from the two spindle poles (centrosomes). The breakdown of the nuclear envelope initiates the prometaphase stage of mitosis. During this stage the chromosomes attach to the separating centrosomes as their kinetochores capture dynamically unstable astral microtubules that are randomly probing the cytoplasm. Once both sister kinetochores are attached to the spindle, so that each kinetochore is attached to different and opposing poles, the chromosome is considered to be bioriented. Then, it undergoes a complex series of motions, termed congression, that ultimately position its centromere on the spindle equator midway between the poles. When the last chromosome becomes positioned on the spindle equator, the cell is considered to be in metaphase of mitosis. The sudden and largely synchronous separation (disjunction) of sister chromatids initiates the anaphase stage of mitosis. During this time the once highly organized spindle begins to disassemble and the poles themselves begin to move farther apart. During telophase, the groups of separated sister chromosomes, which are now positioned near their respective poles, begin to swell and stick to one another. As this occurs, a nuclear envelope is deposited on the surface of the decondensing chromatin, and a Golgi apparatus reforms in association with each of the two centrosomes. New microtubule-based structures, known in animals as stem bodies and in plants as phragmoplasts, also form between the now-separated daughter nuclei. The phragmoplasts are responsible for directing the construction of a new cell wall that ultimately partitions the dividing plant cell into two separate (but connected) entities. In animals the stem bodies act as a template and catalyst for forming and stabilizing the cytokinetic furrow, which constricts the cell into two independent daughters. At the end of telophase the centrosome inherited by each daughter cell nucleates another interphase complex of cytoplasmic microtubules. *See* CELL (BIOLOGY); CELL DIVISION; CELL NUCLEUS.

Conly L. Rieder

Mitteniales An order of true mosses (subclass Bryidae) found in Australia and New Zealand. The order consists of a single species. *Mittenia plumula*, which is adapted for growth in caves or cavelike places. Branches of the persistent protonema consist of spherical cells which reflect light from a backing of chloroplasts, thus providing a glow. The stems are simple and erect, and the leaves are 2–4-ranked and oblong-lingulate and blunt or apiculate. The midrib ends above the midleaf, and the cells are short and smooth.

The order is remarkably similar to the Schistostegales of North Temperate distribution in protonema and habitat, but the other features of gametophyte and sporophyte are quite different. A distinctive feature is the double peristome with 32 segments derived in part from the outermost cell walls of the endothecium (that is, the inner portion of the embryonic capsule). *See* BRYIDAE; BRYOPHYTA; BRYOPSIDA; SCHISTOSTEGALES.

Howard Crum

Mixer A device with two or more signal inputs and one common output. The two primary classes are linear (additive) and nonlinear (multiplicative) mixers. Linear mixers are used to add or blend together two or more signals, nonlinear mixers mainly to shift the spectrum (center frequency) of one signal by the frequency of a second signal.

Linear mixing is the process of combining signals additively, such as the summing of audio signals in a recording studio. This operation can be accomplished passively by simply using a resistive summing network. Although this approach appears very economical, there is a loss in signal strength and an interaction of the signal amplitudes as the gains are adjusted.

Inexpensive integrated circuits have improved this application dramatically. Operational amplifiers of reasonably high quality that will eliminate the adjustment interactions and also provide gain are readily available. The input signals are summed into the virtual ground summing node at the input of the operational amplifier. There is a sign change in the output, but that is a small drawback compared to the advantage of having the virtual ground provided by the operational amplifier. *See* AMPLIFIER; INTEGRATED CIRCUITS; LINEARITY; OPERATIONAL AMPLIFIER.

Perhaps the most familiar application of nonlinear mixers is in radio and television receivers. They are widely used in such applications as amplitude modulation (AM) and demodulation, frequency demodulation, phase detection, frequency multiplication, and single-sideband (SSB) generation. The incoming information to a receiver has been transmitted and received at a frequency far too high to permit efficient amplification and processing. Therefore the signal is translated or frequency-shifted or heterodyned by a mixer to a lower frequency, known as the intermediate frequency (IF), where amplification and processing are performed efficiently by an IF processor, sometimes referred to as the IF strip. *See* AMPLITUDE-MODULATION DETECTOR; AMPLITUDE MODULATOR; FREQUENCY-MODULATION DETECTOR; FREQUENCY MODULATOR; FREQUENCY MULTIPLIER; PHASE-ANGLE MEASUREMENT; RADIO RECEIVER; SINGLE SIDEBAND; TELEVISION RECEIVER.

A second application of a nonlinear mixer is frequency synthesis, where a stable but not easily changed signal at a high frequency is made tunable by mixing it with an easily tunable signal at a low frequency, which, perhaps, can be varied in precise increments of any size. The utility of the method is limited by the ability to filter or separate one frequency term from another, thereby determining the minimum practical value of low frequency for the application.

A mixer is an integral part of an AM-radio integrated circuit which contains virtually all AM-radio functions except filters. A particular type of mixer, the quadrature detector, is included in the frequency-modulation (FM)-radio integrated circuit.

Stanley A. White

Mixing A common operation to effect distribution, intermingling, and homogeneity of matter. Actually the operation is

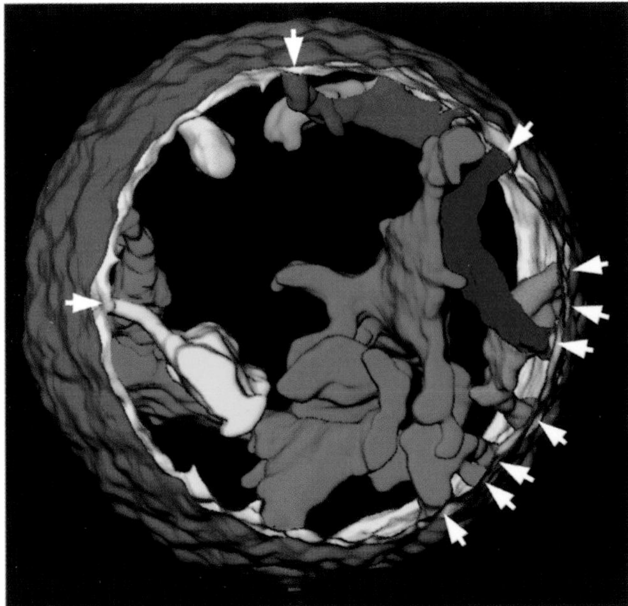

Colorplate 1. Three-dimensional reconstruction of a rat liver mitochondrion produced by electron microscopic tomography. In contrast to standard transmission electron microscopy in which samples are cut as thin as possible (typically 50–80 nm), this technique depends on registration of multiple images from a specimen that is thick enough to contain a representative portion of the object of interest. During imaging, the sample is tilted around one or more axes to generate different projections that are later digitally combined to produce a full three-dimensional view. This reconstruction shows the outer membrane (red), peripheral inner membrane (yellow), and a few selected cristae (various colors). Arrows point to tubular regions of the inner membrane that connect intracristal compartments to the periphery and to each other. The mitochondrion is 1.5 μm in diameter and 0.5 μm thick. (*Courtesy of Dr. Carmen Mannella, Resource for the Visualization of Biological Complexity, Wadsworth Center, Albany, NY*)

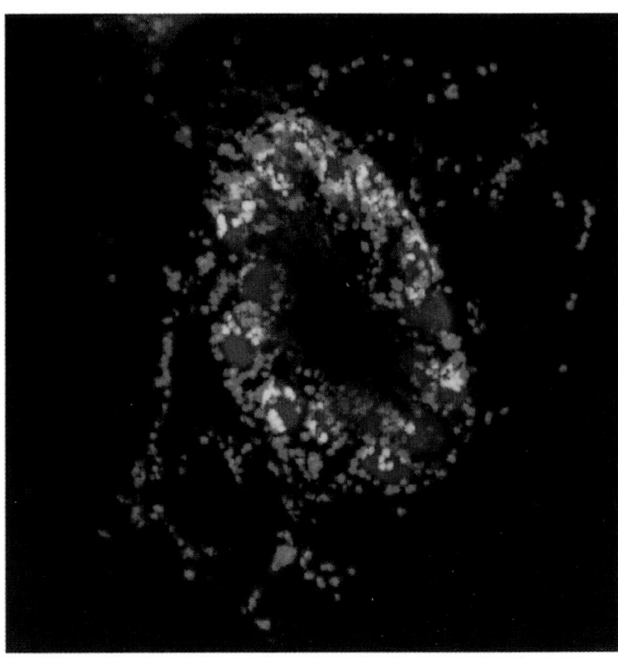

Colorplate 2. Guard cells of a transgenic tobacco plant expressing green fluorescent protein (GFP) targeted to mitochondria. Chloroplasts show red chlorophyll autofluorescence and mitochondria, green protein fluorescence. (*Image provided by Rainer Köhler and Maureen Hanson, Cornell University*)

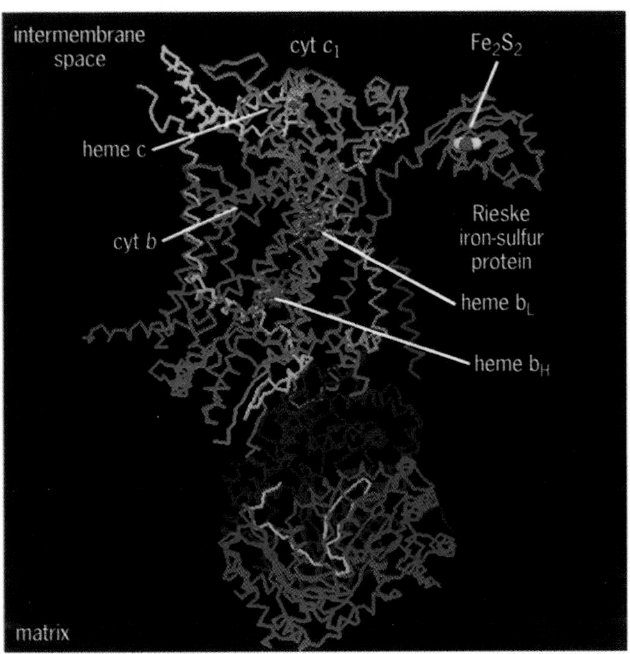

Colorplate 3. Three-dimensional structure of bovine cytochrome bc_1 complex (ubiquinol-cytochrome *c* oxidoreductase) deduced from x-ray crystallography. The complex consists of two monomer units that combine to form a stable dimer. The molecule shown is one of the monomers. It contains 11 subunits with a total molecular mass of about 240 kilodaltons. Cytochrome *b*, the Rieske iron-sulfur protein, and cytochrome c_1 carry the redox centers necessary for electron transport. The complex catalyzes a two-cycle oxidation of ubiquinol produced by the action of NADH dehydrogenase (complex I). This reaction generates a semiquinone as intermediate and leads to the translocation across the membrane of two protons for every electron transferred to cytochrome *c*.

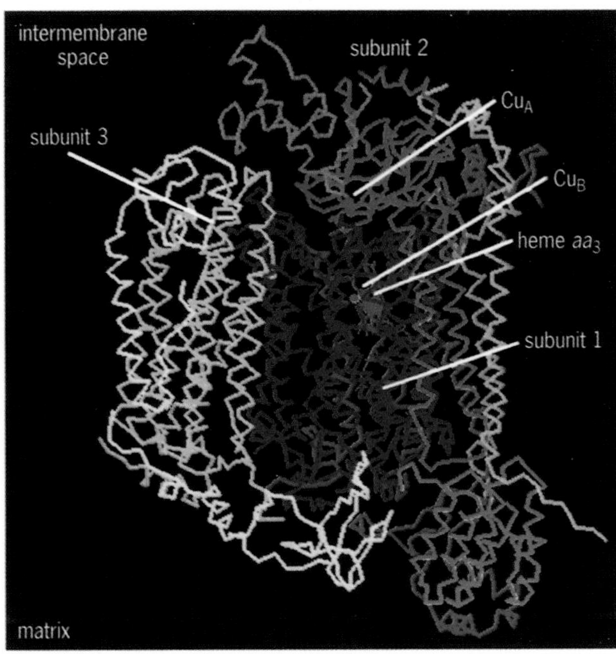

Colorplate 4. Three-dimensional structure of bovine cytochrome *c* oxidase deduced from x-ray crystallography. Like cytochrome bc_1, cytochrome *c* oxidase forms a dimer, of which the monomer unit is shown. This consists of 13 subunits with a total molecular weight of approximately 200 kilodaltons. Cytochrome c reduced by the action of cytochrome bc_1 donates two electrons to the complex whose functional core is formed by the three mitochondrially encoded subunits Cox1, Cox2, and Cox3. Cox2 contains a mixed valence copper center (Cu_A) and is the first component to receive the electrons. Cox1 contains two heme groups (heme aa_3) and a bimetallic copper group (Cu_B) which together form the active site of the enzyme.

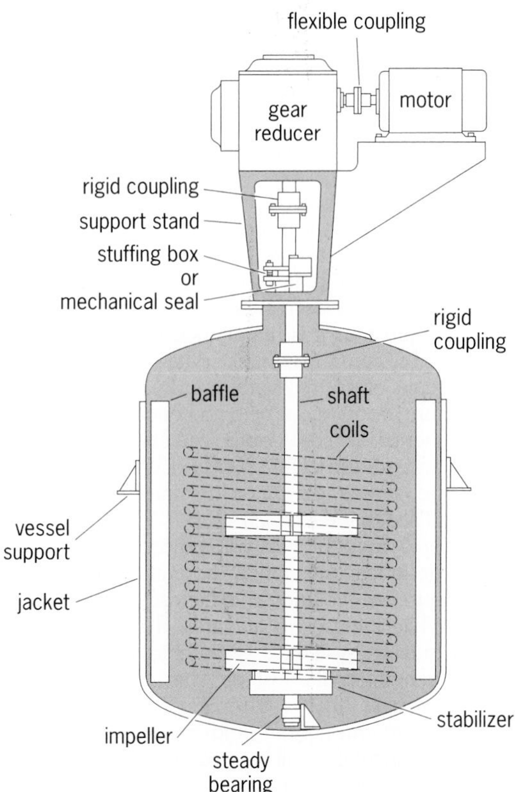

Typical impeller-type liquid mixer. (*After V. W. Uhl and J. B. Gray, Mixing: Theory and Practice, vol. 2, Academic Press, 1967*)

called agitation, with the term mixing being applicable when the goal is blending, that is, homogeneity. Other processes, such as reaction, mass transfer (includes solubility and crystallization), heat transfer, and dispersion, are also promoted by agitation. The type, extent, and intensity of agitation determine both the rates and adequacy of a particular process result. The agitation is accomplished by a variety of equipment.

Most liquid mixing is done by rotating impellers in vertical cylindrical vessels. A typical impeller-type liquid mixer with a variety of features is shown in the illustration. The internal features, including the vessel itself, are considered as a whole, that is, as the agitated system. The forces applied by the impeller develop overall circulation or bulk flow. Superimposed on this flow pattern, there is molecular diffusion, and if turbulence is present, also turbulent eddies. These provide micromixing. Solids, granular to powder, are mixed in a variety of contrivances.

Solids of different density and size are mixed in tumblers (a double cone turning end on end) or with agitators (a helical ribbon rotating in a horizontal trough). The duration of mixing is an important additional variable because classification and separation often occur after attainment of the desired distribution if the operation is carried on too long. Vincent W. Uhl

Mizar The second-magnitude star that marks the bend in the handle of the Big Dipper, ζ Ursae Majoris. Mizar was the first telescopic visual binary to be discovered (1650), the first ever to be photographed (1857), and the first star discovered to be a spectroscopic binary (1889). Its distance to the Sun is 24 parsecs (7.5×10^{14} km or 4.7×10^{14} mi). *See* BINARY STAR; CONSTELLATION; URSA MAJOR.

Mizar (the Horse) forms a wide visual binary with the fourth-magnitude star Alcor (the Rider), at a separation of 12′. The two stars have nearly the same proper motion, and the pair is visible to the naked eye. Mizar itself is a close visual pair, with bluish-white components of apparent magnitude 2.27 and 3.95, approximately 14″ apart. The spectrum of the primary star, Mizar A, shows periodic doubling of the absorption lines due to the Doppler effect, caused by the motion of two stars around their common center of mass. The orbital period of this spectroscopic binary is 20.5 days, and the stars have been resolved with an interferometer, the mean angular separation being only 0.012″. Mizar B is also a spectroscopic binary with a period of 176 days. *See* DOPPLER EFFECT; SPECTRAL TYPE. David W. Latham

Mobile communications Radio communication in which one or both ends of the communication path are movable. The term "mobile" refers to movement of the radio rather than association with a vehicle (for example, hand-held portable radios are included in the definition). The movement need not occur during the communications. The Federal Communications Commission (FCC) licenses and regulates non-federal-government radio activity in the United States, while the National Telecommunications and Information Administration (NTIA) oversees federal-government uses. Other countries have similar agencies. International coordination is afforded through the International Telecommunication Union (ITU) and international treaty.

The category of private land mobile services encompasses users who lease or purchase radio equipment for personal use. Examples are public safety, special emergency, industrial, land transportation, and location radio services. Spectrum over a wide range of radio frequency bands is allocated; for example, low band (30–50 MHz), high band (150–174 MHz), and 900-MHz band (896–901 MHz, paired with 935–940 MHz). Dispatch is the normal mode of operation; that is, all members of the group hear all communications. To accomplish this, high-power, high-site, base station repeaters are generally used so that a single site covers the entire area of interest. Coverage radius varies with frequency band, local terrain, and permissible power levels, but values on the order of 20 mi (32 km) are commonplace. Where areas to be covered are even larger (for example, statewide police systems) or where coverage reliability must be greater than that possible from a single site (for example, for ambulance communications), multiple sites can simulcast the communications. Current technology allows for data exchanges, vehicle location, and secure digitized voice. *See* RADIO SPECTRUM ALLOCATIONS.

Specialized mobile radio (SMR) is a type of mobile radio service in which individual users with business interests are licensed to operate their mobile, portables, and control stations on channel pairs repeated by specialized mobile radio base stations. Full interconnection to the public switched telephone network (PSTN) is possible. To boost the spectrum efficiency of specialized mobile radios relative to shared repeaters already in use, the FCC required that channels be trunked. Trunking in the context of radio systems means not only sharing equipment but sharing frequencies as well. Trunking channels means that when a user wishes to place a call it can be served by any one of the channel pairs that is available.

Although paging is primarily a one-way radio system, two-way operation with such functions as page acknowledgment and short message reply is available. Some types of paging receivers display digits and letters (alphanumeric displays) that allow the calling party's number and a brief message to be displayed, and a message operator becomes unnecessary. Since display of paging messages involves little information, thousands of users can share a paging channel, thus making the system extremely spectrum efficient. Other types of paging receivers provide for brief voice messages following the alert (tone and voice). *See* RADIO PAGING SYSTEMS.

Cellular technology allows hundreds of thousands of users to be handled in a single metropolitan area. Rather than link into the telephone system from a single high-power, high site that covers the entire metropolitan area, users are linked via many low-power, low sites. A single low site, of course, can cover only a limited area, termed a cell, but many low sites taken together can cover the entire metropolitan area. Spectrum efficiency stems from reusing the same frequency at all sites that are sufficiently separated. To further limit interference caused by such frequency reuse, each cell may be divided into sectors, and directive antenna patterns may be used.

An attractive feature of cellular radio is the ability to vary the size of cells in accordance with user density; hence, cell size can increase away from city centers. To sustain the reuse pattern with mixed cell sizes, power levels are tailored to produce comparable signal levels at all cell boundaries. Also, as more customers are added, radio channels can be created to serve them by constructing new base stations (hence, new cells) in geographical locations between existing cells. This concept is called cell splitting. Geographical coverage of the system can be expanded as well by constructing new base stations on the periphery of the existing system and assigning frequencies consistent with the original reuse pattern.

Automatic, continuous coverage as users move across cell boundaries is provided by the cell hand off feature of cellular (also termed handover and automatic link transfer). Calls in need of handoff are recognized by monitoring call quality and comparing it to some required threshold. Handoff control procedures for first-generation analog frequency-modulation systems are also in operation.

Despite the improvements in analog cellular operation, schemes that digitally encode speech and suitably modulate the carrier have been adopted in second-generation (2G) cellular systems. Such schemes handle data traffic as well as voice traffic. *See* MODULATION.

European digital technology (GSM, or Global System for Mobile Communications) uses 124 carriers of 200 kHz each over 25 MHz of spectrum allocated for each direction of communication (890–915 MHz and 935–960 MHz). Each 200-kHz carrier uses time-division multiple access (TDMA) with eight user time slots per carrier; hence, eight simultaneous conversations can be supported. Several standards for digital cellular also exist in the United States and Canada. *See* MULTIPLEXING AND MULTIPLE ACCESS.

The great demand for cellular phones and related wireless services has been addressed to some extent by the addition of new spectrum, by the introduction of narrow-band and digital cellular systems, and by cell splitting where practical. Acknowledging that these techniques for increasing capacity would quickly be exhausted, most countries have allocated additional spectrum for mobile and portable communications. Since these frequency bands have much more spectrum that those previously allocated to cellular service, a greater variety of services are possible.

A spectrum of 120 MHz in the 1850–1910 and 1930–1990 MHz bands was allocated for licensed personal communication system (PCS) operation in the United States, and 20 MHz of spectrum in the 1920–1930 MHz band for unlicensed operation, split evenly between voice (isochronous) and data (asynchronous) applications. The spectrum allocation for licensed PCS operation is divided into six frequency blocks, three of which contain 30 MHz of spectrum, and the other three 10 MHz. It is thus possible in a given region to have as many as six competing service providers, in addition to the two 900-MHz cellular service providers.

In response to the ever-increasing demand for higher capacity and new services, the ITU defined the standards for third-generation systems, also known as International Mobile Telecommunications 2000 (IMT-2000). The ITU specified minimum bit rates of 144 kilobits per second (kbps) in mobile (outdoor) environments and 2 megabits per second (Mbps) in fixed (indoor) environments.

The IMT-2000 initiative gave birth to the Third-Generation Partnership Project (3GPP) for the evolution of GSM networks and the Third-Generation Partnership Project-2 (3GPP2) for the evolution of code-division multiple access (CDMA) networks. The bands 1885–2025 MHz and 2110–2200 MHz are intended for use, on a worldwide basis, by administrations wishing to implement IMT-2000.

With the increased interest in Internet usage in particular and data applications in general, extensions of the 3G standards have been developed to increase performance by taking advantage of the distinct nature of data as compared to voice traffic. For instance, introducing delay can be very annoying in conversational speech, and thus it is maintained under strict control in voice-optimized systems. In contrast, data packets can be subject to variable delays of up to several hundred milliseconds to exploit better channel conditions with no noticeable degradation in the user experience. The data-optimized evolutions of the third-generation standards are CDMA2000 1xEV-DO (Evolution Data Optimized) and High Speed Down Link Packet Access (HSDPA) for W-CDMA (Wideband CDMA).

With the deployment of third-generation systems well underway, system designers and standard organizations have turned their attention to the evolution of systems "beyond 3G" and to 4G proper. While there is no universal definition of what will constitute a true 4G system, it is possible to extrapolate from the previous generational transitions that a key element will be a significant increase (that is, a tripling or at a least doubling) in system capacity, throughput, and average bit rates. It is expected that such increases will be made possible in large part by advanced antenna technologies.

A well-known multiple-antenna technique is antenna diversity, where multiple antennas are used to maximize the chances that the radio channel condition to at least one of the antennas will be satisfactory. In the technique of selection diversity, the signal from the best antenna is selected. The optimum approach, known as maximal ratio combining, is to perform a sum of all received signals, weighting each in proportion to its signal-to-noise ratio, thus emphasizing the better-quality signals.

Wireless systems can benefit greatly from using antennas that focus radio energy in the geographical direction of interest. Modern electronics make it possible to coordinate the signals from a set of antenna elements to generate, and even steer, a beam in the desired direction, by properly phasing the signals associated with each antenna element. The beam can be configured in reception as well and in transmission.

With sufficient antenna elements, it is possible to form more than a single beam. In this case, multiple messages intended for users in different directions can be associated with correspondingly many beams, generated by the appropriate beamforming weights for multiple beams in each of the desired directions.

Link capacity can be greatly increased with a promising technique using multiple antennas at the transmitter and the receiver, known as multiple-input, multiple-output (MIMO). In MIMO systems, the transmitter sends multiple simultaneous messages in the same channel bandwidth without increasing the transmitted power. In a scattering environment, the receiver can process the signals at multiple receive antennas and separate the multiple transmitted messages, thereby creating multiple parallel virtual channels.

There are many scenarios where mobile communications services are needed, yet no network infrastructure, such as base station transceivers, is available. In this case, satellite-based mobile services are extremely valuable. The use of geostationary Earth orbit (GEO) satellites, where the orbit is chosen such that the satellite appears stationary with respect to Earth, simplifies the Earth station system design. However, geostationary orbits are at such a large distance from Earth that large transmission delays are introduced. More recently, lower orbits have been used to reduce delay in medium Earth orbit (MEO) and low Earth orbit (LEO) systems. These systems must provide for user terminal handoffs, as the serving satellite disappears below the horizon and another satellite appears above the horizon, and many satellites are needed to ensure coverage at all times. *See* COMMUNICATIONS SATELLITE. Reinaldo A. Valenzuela

Mode of vibration A characteristic manner in which vibration occurs. In a freely vibrating system, oscillation is restricted to certain characteristic frequencies; these motions are called normal modes of vibration.

An ideal string, for example, can vibrate as a whole with a characteristic frequency $f = (1/2L)\sqrt{T/m}$, where L is the length of string between rigid supports, T the tension, and m the mass per unit length of the string. The displacements of different parts of the string are governed by a characteristic shape function. The frequency of the second mode of vibration is twice that of the first mode. Similarly, modes of higher order have frequencies that are integral multiples of the fundamental frequency.

Because the frequencies are in the ratios 1:2:3. . . , the modes of vibration of an ideal string are properly called harmonics. Not all vibrating bodies have harmonic modes of vibration, however. *See* HARMONIC (PERIODIC PHENOMENA); VIBRATION. Robert W. Young

Model theory The body of knowledge that concerns the fundamental nature, function, development, and use of formal models in science and technology. In its most general sense, a model is a proxy. A model is one entity used to represent some other entity for some well-defined purpose. Examples of models include: (1) An idea (mental model), such as the internalized model of a person's relationships with the environment, used to guide behavior. (2) A picture or drawing (iconic model), such as a map used to record geological data, or a solids model used to design a machine component. (3) A verbal or written description (linguistic model), such as the protocol for a biological experiment or the transcript of a medical operation, used to guide and improve procedures. (4) A physical object (scale model, analog model, or prototype), such as a model airfoil used in the wind-tunnel testing of a new aircraft design. (5) A system of equations and logical expressions (mathematical model or computer simulation), such as the mass- and energy-balance equations that predict the end products of a chemical reaction, or a computer program that simulates the flight of a space vehicle. Models are developed and used to help hypothesize, define, explore, understand, simulate, predict, design, or communicate some aspect of the original entity for which the model is a substitute.

Formal models are a mainstay of every scientific and technological discipline. Social and management scientists also make extensive use of models. Indeed, the theory of models and modeling cannot be divorced from broader philosophical issues that concern the origins, nature, methods, and limits of human knowledge (epistemology) and the means of rational inquiry (logic and the scientific method). *See* LOGIC; SCIENTIFIC METHODS.

Models are usually more accessible to study than the system modeled. Changes in the structure of a model are easier to implement, and changes in the behavior of a model are easier to isolate, understand, and communicate to others. A model can be used to achieve insight when direct experimentation with the actual system is too dangerous, disruptive, or demanding. A model can be used to answer questions about a system that has not yet been observed or built, or even one that cannot be observed or built with present technologies.

Specific models developed in different disciplines may differ in subject, form, and intended use. However, basic concepts such as model description, validation, simplification, and simulation are not unique to any particular discipline. Model theory seeks a formal logical and axiomatic understanding of the underlying concepts that are common to all modeling endeavors.

General and mathematical systems theory have stimulated many of the important developments in model theory. Mathematical models are particularly useful, because of the large body of mathematical theory and technique that exists for the study of logical expressions and the solution of equations. The power and accessibility of digital computers have increased the use and importance of mathematical models and computer simulation in all branches of modern science and technology. A great variety of programming languages and applications software are now available for modeling, computational analysis, and system simulation. *See* DIGITAL COMPUTER; SIMULATION; SYSTEMS ANALYSIS; SYSTEMS ENGINEERING. K. Preston White, Jr.

Modeling languages Graphical or textual languages that are used to build models by describing the qualities with which the models are concerned. A model represents something and includes some but not all qualities of what it represents. When models are used in systems and software engineering, their important qualities are structure, behavior, communication, and requirements.

Need for models. Systems and software engineering are both characterized by a combination of large information content with a need for different people (system stakeholders) to understand at least some system aspects. During the development of systems, developers can see only a small part of the complete system at a time. They will then need models to assist them in understanding the context of their own work and the requirements on it. Models are useful prior to system development to study requirements and design alternatives, and to ensure that all stakeholders have a common understanding of why a system should be built or modified. They are also useful during and after development of a system to assist developers and users to navigate the system's structure and to understand the system's behavior. *See* MODEL THEORY.

Obviously, software is a subset of systems engineering. However, since building the right software requires a correct understanding of the software's environment, it is advisable to include this environment in software engineering models. *See* SOFTWARE ENGINEERING; SYSTEMS ENGINEERING.

Unified Modeling Language diagrams. The Unified Modeling Language (UML™) has emerged as a standard for software engineering. This standard includes a set of diagrams:

Use case diagram. This diagram shows the interaction between a user and a software system.

Class diagram. This diagram is basically an enhanced entity-relationship diagram where the entities are classes with names, attributes, and actions.

Sequence diagram. Message sequence charts have a long tradition for visualizing the interaction between concurrent processes communicating by way of messages.

Collaboration diagram. This diagram basically contains the same information as the sequence diagram.

Package diagram. This diagram shows dependencies between major system components.

State diagram. This diagram visualizes behavior.

Activity diagram. This diagram is a development from flow charts.

Deployment diagram. This diagram shows the distribution of software objects on hardware nodes.

Component diagram. This diagram shows dependencies between components and can also be drawn to show aggregation (components included in other components).

Other modeling languages. While the UML is the emerging standard for modeling languages, there are several alternatives, including mathematical formal languages, Petri nets, and formalized natural language.

Mathematical formal languages can be used to create formal specifications, which can then be checked automatically for compliance with a completed software system.

Petri nets give a graphical model of how concurrent processes communicate and influence each other through the transmission of "tokens."

A formalized natural language is obtained by simplifying a programming language for modeling use while keeping the "reserved words" and including variables of defined types. *See* NATURAL LANGUAGE PROCESSING; PROGRAMMING LANGUAGES.

Ingmar Ogren

Modem A device that converts the digital signals produced by terminals and computers into the analog signals that telephone circuits are designed to carry. Despite the availability of several all-digital transmission networks, the analog telephone network remains the most readily available facility for voice and data transmission. Since terminals and computers transmit data using digital signaling, whereas telephone circuits are designed to transmit analog signals used to convey human speech, a device is required to convert from one to the other in order to transmit data over telephone circuits. The term modem is a contraction of the two main functions of such a unit, modulation and demodulation. The device is also called a data set. *See* INTEGRATED SERVICES DIGITAL NETWORK (ISDN): MODULATION.

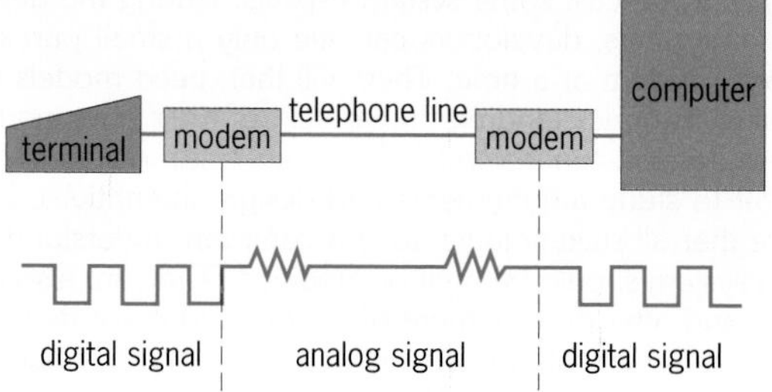

Signal conversion performed by modems. A modem converts a digital signal to an analog tone (modulation) and reconverts the analog tone into its original digital signal (demodulation).

In its most basic form a modem consists of a power supply, transmitter, and receiver. The power supply provides the voltage necessary to operate the modem's circuitry. The transmitter section contains a modulator as well as filtering, wave-shaping, and signal control circuitry that converts digital pulses (often input as a direct-current signal with one level representing a digital one and another level a digital zero) into analog, wave-shaped signals that can be transmitted over a telephone circuit. The receiver section contains a demodulator and associated circuitry that is used to reverse the modulation process by converting the received analog signals back into a series of digital pulses (see illustration). *See* DATA COMMUNICATIONS; DEMODULATOR; ELECTRIC FILTER; ELECTRICAL COMMUNICATIONS; ELECTRONIC POWER SUPPLY; MODULATOR; WAVE-SHAPING CIRCUITS.

Gilbert Held

Modulation A technique employed in telecommunications transmission systems whereby an electromagnetic signal (the modulating signal) is encoded into one or more of the characteristics of another signal (the carrier signal) to produce a third signal (the modulated signal), whose properties are matched to the characteristics of the medium over which it is to be transmitted. The encoding preserves the original modulating signal in that it can be recovered from the modulated signal at the receiver by the process of demodulation. The main purpose of modulation is to overcome any inherent incompatibilities between the electromagnetic properties of the modulating signal and those of the transmission medium. Of primary importance in this respect is the spectral distribution of power in the modulating signal relative to the passband of the medium. Modulation provides the means for shifting the power of the modulating signal to a part of the frequency spectrum where the medium's transmission characteristics, such as its attenuation, interference, and noise level, are favorable. *See* ELECTROMAGNETIC WAVE TRANSMISSION; RADIO-WAVE PROPAGATION.

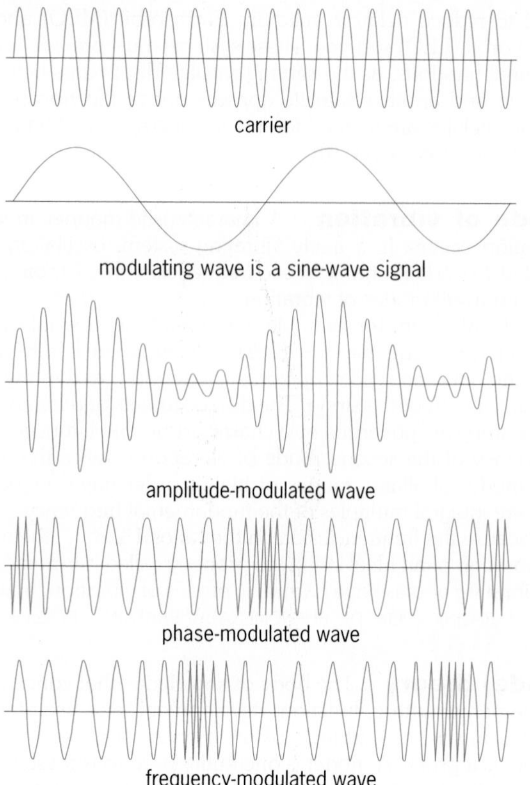

Amplitude, phase, and frequency modulation of a sine-wave carrier by a sine-wave signal. (*After H. S. Black, Modulation Theory, Van Nostrand, 1953*)

Two forms of modulation are generally distinguished, although they have many properties in common: If the modulating signal's amplitude varies continuously with time, it is said to be an analog signal and the modulation is referred to as analog. In the case where the modulating signal may vary its amplitude only between a finite number of values and the change may occur only at discrete moments in time, the modulating signal is said to be a digital signal and the modulation is referred to as digital.

In most applications of modulation the carrier signal is a sine wave, which is completely characterized by its amplitude, its frequency, and its phase relative to some point in time. Modulating the carrier then amounts to varying one or more of these parameters in direct proportion to the amplitude of the modulating signal. In analog modulation systems, varying the amplitude, frequency, or phase of the carrier signal results in amplitude modulation (AM), frequency modulation (FM), or phase modulation (PM), respectively. Since the frequency of a sine wave expressed in radians per second equals the derivative of its phase, frequency modulation and phase modulation are sometimes subsumed under the general term "angle modulation" or "exponential modulation." The illustration shows an example of an unmodulated sine-wave carrier signal and the signal resulting from modulating its amplitude, phase, or frequency with the amplitude of an analog modulating signal, which is also taken to be a sine wave.

If the modulating signal is digital, the modulation is termed amplitude-shift keying (ASK), frequency-shift keying (FSK), or phase-shift keying (PSK), since in this case the discrete amplitudes of the digital signal can be said to shift the parameter of the carrier signal between a finite number of values. For a

Moiré pattern representing the ionic bond. (G. Oster, *Mount Sinai School of Medicine*)

modulating signal with only two amplitudes, "binary" is sometimes added before these terms.

Digital modulating signals with more than two amplitudes are sometimes encoded into both the amplitude and phase of the carrier signal. For example, if the amplitude of the modulating signal can vary between four different values, each such value can be encoded as a combination of one of two amplitudes and one of two phases of the carrier signal. Quadrature amplitude modulation (QAM) is an example of such a technique.

In certain applications of modulation the carrier signal, rather than being a sine wave, consists of a sequence of electromagnetic pulses of constant amplitude and time duration, which occur at regular points in time. Changing one or the other of these parameters gives rise to three modulation schemes known as pulse-position modulation (PPM), pulse-duration modulation (PDM), and pulse-amplitude modulation (PAM), in which the time of occurrence of a pulse relative to its nominal occurrence, the time duration of a pulse, or its amplitude are determined by the amplitude of the modulating signal. *See* PULSE MODULATION.

Hermann J. Helgert

Modulator A device that combines an electrical information signal with a periodic electrical carrier signal for efficient transmission. The modification of a usually sinusoidal electrical carrier signal (or simply carrier) for the purpose of transmitting information which is carried on a second electrical modulating signal or information signal is called modulation. The device or circuit that performs the modulation is called a modulator. The modulator varies the carrier in amplitude, frequency, phase, or some combination in order to imbed the information for efficient transmission. During the process of modulation, information is shifted in frequency from baseband to the carrier frequency for efficient transmission. *See* CARRIER (COMMUNICATIONS).

The information signals usually are of analog origin but may be converted to digital format for more exact processing. Transmission is always analog or continuous. Robust and versatile digital signal modulation of any type may be implemented directly with microprocessors or digital components (such as multipliers and adders) by mechanizing the modulation algorithms directly. The precisely controlled modulated digital signal is transformed into an analog signal by a digital-to-analog converter (DAC). Finally, by means of a linear amplifier, its power is increased to a level required for successful radio-frequency (RF) transmission. Because linear RF amplifiers are notoriously inefficient, an alternative lower-cost approach to AM and quadrature amplitude modulation (QAM) is used to first generate the baseband analog signals and then impress them on the RF carriers. High-quality data resolution is sacrificed in exchange for economy. *See* DIGITAL-TO-ANALOG CONVERTER.

Amplitude modulation is a multiplication process. The multiplication function called for in simple and inexpensive analog amplitude modulation is achieved economically by varying, via the modulating signal, the applied power-supply voltage to the output RF amplifier that follows the carrier-frequency oscillator. This can be done by simply transformer-coupling the modulating signal voltage into the amplifier's power-supply output line. *See* AMPLITUDE MODULATOR.

A simple quadrature amplitude modulator can be implemented by inserting a second signal path from the reference carrier-frequency oscillator output consisting of a 90° phase shifter followed by a second linear amplifier whose power-supply output line is varied by a second transformer-coupled modulating signal voltage. A QAM signal is formed by passively summing the outputs of the two amplifiers delivering amplitude-modulated sine and cosine (quadrature) carriers.

By using analog or digital signals to amplitude-modulate each quadrature carrier, a QAM transmitter can generate an information-bearing carrier varying in both amplitude and phase. For digital signals, a so-called *M*-ary (that is, each data symbol can be represented by one of *M* possible states in the complex plane) digital QAM signal occupies the same bandwidth as a PSK signal (discussed below), but the power efficiency is greater.

In a frequency modulator, the operating frequency of the carrier-frequency oscillator can be determined by the reactance of a tuned *LC* (inductor-capacitor) circuit. The oscillator frequency is determined by $1/\sqrt{LC}$, so varying either element in the oscillator's tuning circuit varies the frequency. A voltage-controlled capacitor called a varactor can be used in the circuit. The modulating voltage is applied directly to the varactor in order to produce an FM signal. *See* VARACTOR.

Transmission of digital data by frequency modulation is called frequency-shift keying (FSK). The frequency of the transmitted signal is determined by the digitized information signal. *M*-ary FSK employs *M* distinct frequencies. *See* FREQUENCY MODULATOR.

In phase modulation, the signal to be transmitted is varied in phase according to the information signal. Transmission of digital data by phase-angle modulation is called phase-shift keying (PSK). PSK employs a single frequency. The phase of the transmitted signal is determined by the digitized information signal. Each data symbol in an *M*-ary PSK system may be represented by a point on a unit circle because the amplitude is held constant. The points are separated in angle (phase) by $360°/M$. *See* MODULATION; PHASE MODULATOR.

Stanley A. White

Mohair The long, lustrous hair of the Angora goat, which originated in the area around Ankara (Angora), Turkey. Mohair is a smooth, strong, durable, and resilient fiber. It enhances softness and luster in fabrics. Mohair absorbs dye evenly and brilliantly, retains color well, and permits unusual decorative effects. It is mainly used as an apparel fiber but may be used in upholstery, draperies, wigs, hairpieces, and rugs. Leather produced from the skin is useful for gloves, purses, and novelties. *See* WOOL.

Clair E. Terrill

Moho (Mohorovicić discontinuity) The level in the Earth where the velocity of sonic waves first increases rapidly or discontinuously to a value between 7.6 and 8.6 km/s (4.7 and 5.3 mi/s). A. Mohorovicić discovered this boundary while investigating seismograms of the Zagreb (now capital of Croatia) earthquake of October 8, 1909. He recognized that low-velocity waves traveling directly from the earthquake source were overtaken at large distances by refracted waves traveling through the deeper, high-velocity layer. Modern determinations of the depth and nature of the Moho are commonly made in seismic refraction studies that use artificial seismic sources, such as explosions, rather than earthquakes. This method allows identification of the wave traveling in the high-velocity medium (P_n) and a wide-angle reflection (P_mP) from the boundary. The Moho is generally assumed to mark the boundary between the crust and mantle, although this need not always be the case. Despite drilling attempts in the early 1960s, the Moho has not been directly sampled. Knowledge of the oceanic and continental Moho is based on interpretation of geophysical data, as well as geological interpretation of ophiolites (believed to be sections of oceanic crust and mantle uplifted onto land). *See* EARTH INTERIOR; OPHIOLITE.

David M. Fountain; Maya Tolstoy

While the definition of Moho in the continents is based solely on seismic refraction observations, the term has been utilized to describe a range of observations pertaining to the oceanic environment. In the strictest sense, Moho refers to the transition boundary where material velocities, as determined from seismic refraction methods, exceed 5 mi/s (8 km/s). From geological studies of ocean crust and ophiolites (ancient oceanic sections subsequently emplaced on continents), oceanic crust is understood to originate from partial melting of mantle that upwells and decompresses in response to sea-floor spreading. The eruption and intrusion of this melt, and the formation of Moho, occurs in a very narrow zone at the axis of sea-floor spreading. Different mantle conditions can lead to different volumes of melting, accounting for most of the observed variations in crustal thickness (depth to the Moho). Very slow spreading centers, while rare, are expected to form somewhat thinner crust, largely reflecting the magma-starved nature of these areas. The Moho is a proxy for the transition from crustal materials (for example, basalts, diabase dikes, or gabbros) formed by mafic melts extracted from the mantle to the ultramafic residual mantle (for example, peridotite) that has remained at depth. *See* BASALT; DOLERITE; GABBRO; PETROLOGY; PLATE TECTONICS. Carolyn Z. Mutter; Maya Tolstoy

Moiré pattern When one family of curves is superposed on another family of curves, a new family called the moirè pattern appears.

To produce moirè patterns, the lines of the overlapping figures must cross at an angle of less than about 45°. The moirè lines are then the locus of points of intersection. The illustration shows the

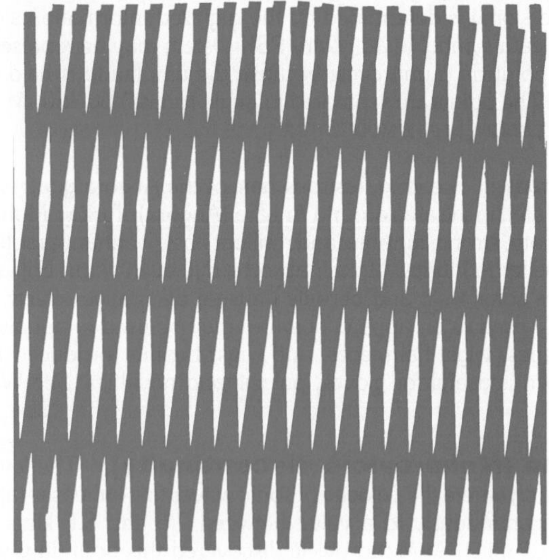

Two simple gratings crossed at a small angle.

case of two identical figures of simple gratings of alternate black and white bars of equal spacing. When the figures are crossed at 90°, a checkerboard pattern with no moirè effect is seen. At crossing angles of less than 45°, however, one sees a moirè pattern of equispaced lines, the moirè fringes. The spacing of the fringes increases with decreasing crossing angle. This provides one with a simple method for measuring extremely small angles (down to 1 second of arc). As the angle of crossing approaches zero, the moirè fringes approach 90° with respect to the original figures.

Even when the spacings of the original figures are far below the resolution of the eye, the moirè fringes will still be readily seen. This phenomenon provides a means of checking the fidelity of a replica of a diffraction grating. *See* DIFFRACTION GRATING.

Moiré techniques are widely used in the stress analysis of metals, in the examination of large optical surfaces, in investigating aberrations of lenses, and in determining a refractive index gradient (for example, that of sugar molecules diffusing into water). Gerald Oster

Moisture-content measurement Measurement of the ratio or percentage of water present in a gas, a liquid, or a solid (granular or powdered) material. Nearly all materials contain free water, the relative amount being dependent upon the physical and chemical properties of the material. The primary purpose of determining and maintaining moisture contents within specified limits can usually be traced to economic factors, trade practices, or legal requirements.

Moisture content has a number of synonymous terms, many of which are specific to certain industries, types of product, or material. The water content in solid, granular, or liquid materials is usually referred to as moisture content on either the wet or dry basis; the wet basis is common to most industries. Specifically, moisture content on the wet basis refers to the quantity of water per unit weight or volume of the wet material. A weight basis is preferred. The textile industry uses the dry basis for moisture content of textile fibers. Often referred to as regain moisture content, the dry basis or regain refers to the quantity of water in a material expressed as a percentage of the weight of the bone-dry (thoroughly dried) material.

The moisture content in air is referred to as humidity, either absolute or relative. Absolute humidity is the number of pounds of water vapor associated with 1 lb (0.5 kg) of dry air, also called just humidity, Relative humidity is the ratio, usually expressed as a percentage, of the partial pressure of water vapor in the actual atmosphere to the vapor pressure of water at the prevailing temperature. Relative humidity is customarily reported by the U.S. Weather Bureau because it essentially describes the degree of saturation of the air. However, air which is saturated (100% RH) at 50°F (10°C) is quite dry (19% RH) when heated to 100°F (38°C). A changing basis of this type is not convenient for many purposes such as computations used in air conditioning, combustion, or chemical processing; therefore absolute units, such as dew point or grains of water per pound of dry air, are more acceptable. Dew point is the temperature at which a given mixture of air and water vapor is saturated with water vapor. *See* DEW POINT; HUMIDITY.

Gases. The measurement of water content in gases and mixtures of air and gases is important in industry. A number of commercially manufactured instruments are available for these measurements; their principles of operation include condensation, used in dew- or fog-point indicators; dimensional change, used by hygrometers; thermodynamic equilibrium, used by wet-bulb psychrometers; and absorption methods, which serve as the basic principle for gravimetric and electric conductivity or dielectric types. *See* HYGROMETER; PSYCHROMETER; PSYCHROMETRICS.

The importance of humidity in relation to personal comfort is well known. The air conditioning industry produces equipment to maintain comfortable conditions of temperature and humidity. Considerable industrial air conditioning is also done for process reasons. Control of humidity is also important in the preservation of materials, especially those which are hygroscopic, and in the storage of food products.

Liquids and solids. The development of instrumentation for the measurement and control of moisture in liquids and solids has been due in a large part to the great need that exists in many processes where the control of a precise moisture is critical. The desirability of a specific moisture content in a product during its preliminary manufacturing process is often required. In general, however, the rigid control of moisture content occurs most frequently in the final product to assure its quality and the fulfillment of legal or trade practices for the individual product.

Instruments suitable for the measurement of moisture content may be classified as periodic and continuous. In general, only those instruments offering continuous measurement are practical for the automatic control of moisture content in a product. The periodic instrument types are generally automated versions of conventional laboratory moisture-analysis procedures.

Moisture measuring instruments may also be classified by operating principle. Those instruments employing electrical conductivity (either dc or ac), absorption of electromagnetic energy (radio-frequency regions), electrical capacitance (dielectric constant change), and infrared energy radiations are more readily adapted to continuous measurements inasmuch as the response of these instruments to moisture changes is very fast. Those instruments employing automatic oven drying, chemical titrations, equilibrium hygrometric methods, distillation methods, and so forth are usually of the intermittent, or periodic, type. Lee E. Cuckler

Mole (chemistry) A unit (symbolized mol) used to measure the amount of material in a chemical sample. The mole is defined by international agreement as the amount of substance (chemical amount) of a chemical system that contains as many molecules or entities as there are atoms in 12 g of carbon-12 (^{12}C). When the mole is used, the elementary entities need not be molecules, but they must always be specified. They may be atoms, molecules, ions, electrons, or specified groups of such particles.

Three obvious ways of measuring the amount of material in a given sample are to measure the mass of the sample, to measure the volume, or to count the number of molecules in the sample. Although it is more difficult to devise an experiment to count molecules, this third way of measuring amount is of special interest to chemists because molecules react in simple rational proportions (for example, one molecule of A may react with one, or two, or three molecules of B, and so forth). However, to count molecules is inconvenient in practice because the numbers are so large. For any chemical, a mass of 1 kilogram of the sample contains a large number of molecules, of the order 10^{23}–10^{24}. The mole is defined so that 1 mole of any substance always contains the same number of molecules. This number approximately 6.02×10^{23}, and is known as the Avogadro number. The mole is a more convenient unit in which to measure the amount of a chemical than counting the number of molecules, and it has the same advantages. *See* AVOGADRO NUMBER.

The amount of substance (chemical amount) of a sample, n, may be determined in practice by one of three methods.

The value of n (the amount of substance) may be determined from the mass m by dividing by the molar mass M of the sample, as in Eq. (1). If m is expressed in g and M in g/mol, then the value of n will be obtained in mol.

$$n = \frac{m}{M} \qquad (1)$$

For a gas, the value of n may be determined from the volume V, pressure p, and absolute temperature T by using the ideal gas equation (2), where R is the gas constant ($R = 8.3145$

$$n = \frac{pV}{RT} \qquad (2)$$

J K^{-1} mol^{-1}). If pV is expressed in (N m^{-2}) $\times$ (m^3) = J, and RT in J mol^{-1}, then pV/RT gives the value of n in mol.

For a solution, the amount of solute (or the amount concentration of solution) is frequently determined by titration: if ν_A molecules of A react with ν_B molecules of B in the titration, then at the end point the amount of A used (n_A) is related to the amount of B (n_B) by Eq. (3), so that if one is known the other may be determined.

$$n_A = \frac{\nu_A}{\nu_B} n_B \qquad (3)$$

See TITRATION.

The concentration of a solution may be recorded as (mass of solute)/(volume of solution), in units gram/liter; or as (chemical amount of solute)/(volume of solution) in units mol/liter. Because of the proportionality of chemical amount to number of molecules, the latter is the more useful measure of concentration and is generally used in chemistry and biochemistry. *See* CONCENTRATION SCALES. Ian. M. Mills

Mole (zoology) A mammal belonging to the order Soricomorpha (previously Insectivora), family Talpidae (the true moles). There are 39 species in 17 genera of talpids distributed on all continents except Australia. *See* INSECTIVORA; MAMMALIA.

Moles are small, subterranean mammals that are highly specialized for burrowing. The eyes and ears are much reduced, and they have a long pointed snout and short soft fur that can lie either way. The powerful forelimbs are adapted for digging, having very powerful muscles and very large strong claws. Moles have sharp pointed teeth. The mole's body is stout and cylindrical with a short neck. Moles are solitary animals and rarely come above-ground. Earthworms are a staple food of many moles, and they are often stored for later use. There is usually one litter per year, which is consistent with their underground mode of existence; that is, they have few predators. The young approach the size of the adults and begin to fend for themselves when about 4 weeks old.

European common mole (*Talpa europaea*).

The main North American moles—*Condylura* (1 species), *Scalopus* (1), *Parascalops* (1), and *Scapanus* (3)—fall into a related group, the subfamily Scalopinae, separate from the Old World moles, but this group also includes one species from China, *Scapanulus oweni*. The rest of the moles occur mostly in the Old World and are in the subfamily Talpinae, although one genus and species from this group, *Neurotrichus gibbsii*, occurs on the west coast of North America. The genus *Talpa* of Eurasia contains nine species. The European mole (*T. europaea*; see illustration), found in northern and central Asia as well as Europe, is a rather typical representative of the group and spends most of its life underground. John O. Whitaker, Jr.

Molecular adhesion The tendency of dissimilar solids or liquids to cling together as a result of the interatomic forces which they exert upon each other across their common interface. Some factors affecting molecular adhesion are the physical state of the materials, the composition and the topology of the surfaces in contact, the temperature, and the presence of foreign materials such as adsorbed gases on one or both surfaces.

Mutual forces of attraction, electrical in origin, exist between virtually all atoms and molecules. It is these forces which, under proper conditions of temperature and pressure, compel isolated atoms or molecules to condense into solids and liquids. On average, these forces are approximately equally strong in all directions and result in an atom or molecule achieving its lowest possible energy, that which is thermodynamically favored, when it is surrounded on all sides by other atoms. The presence of a real surface breaks the three-dimensional continuity of the substance, leaving atoms at the surface with an unfulfilled capacity for bonding. It is this capacity for bonding that is responsible for the phenomena of surface tension, adsorption of gases, and adhesion. *See* ADSORPTION; COHESION (PHYSICS); INTERMOLECULAR FORCES; SURFACE TENSION.

Molecular adhesion plays an especially significant role in the deposition of thin films on substrates of dissimilar composition. A particularly important example is the use of such films to form conducting paths—microscopic wires only a few tens of nanometers thick—in integrated circuits. These films must both adhere well to the underlying substrate and produce contacts with appropriate electrical properties. At present the understanding of adhesion is not sufficiently detailed to be able to predict such properties in advance of experiments. *See* INTEGRATED CIRCUITS; SPUTTERING. Robert A. Weller

Molecular anthropology The study of primate phylogeny and human evolution through the genetic information in the deoxyribonucleic acid (DNA) of genomes and in the proteins that genes encode. The first studies in molecular anthropology used immunological and biochemical methods to obtain information from proteins on the degrees of genetic similarity of humans and other primates. These results not only placed chimpanzees and gorillas closest to humans rather than to orangutans but also indicated that the very close kinship between chimpanzees and gorillas was not any closer than the relation of each to humans. Subsequent studies that extracted genetic information directly from DNA extended this original finding. Indeed, the accumulating comparative DNA sequence data provide mounting evidence that the closest genetic kinship is between chimpanzees and humans rather than chimpanzees and gorillas.

The results gathered in DNA studies of primate phylogeny challenge the traditional anthropological view that humans are very different from all other animals. DNA results show that, genetically, humans are only slightly remodeled apes. Humans share with their closest relatives, the chimpanzees and bonobos (pygmy chimpanzees), more than 98.3% identity in typical noncoding DNA and probably about 99.5% identity in the active coding sequences of functional nuclear genes. Humans share about 98.0% identity in nuclear genomic DNA with gorillas, 96.5% identity with orangutans, and 95.0% identity with the most distant ape relatives, the gibbons and siamangs. Apes and humans share with the other branch of catarrhines, the Old World monkeys, about 92% identity in nuclear genomic DNA, and with the platyrrhines, the New World monkeys, about 87% identity. Even with nonanthropoid primates, the tarsiers and strepsirhines (lemurs and loriforms, such as bushbabies), the anthropoids (platyrrhines and catarrhines) share a DNA identity in the range of 76–71%. *See* DEOXYRIBONUCLEIC ACID (DNA).

Traditional primate classifications use the vague concept of grades of evolutionary advancement to place the smaller-brained primates in the suborder Prosimii (the primitive grade) and the larger-brained primates in the suborder Anthropoidea (the advanced grade). Moreover, on viewing humans as the most advanced primates, traditional primate classifications have humans as the sole living members of family Hominidae, while the great apes of Africa (chimpanzees, bonobos, gorillas) and Asia (orangutans) are members of subfamily Ponginae of family Pongidae. In contrast, a strictly objective view based on molecular evidence, but also congruent morphological evidence from both living and fossil primates, not only places apes with humans within the family Hominidae but also within this family places chimpanzees and bonobos with humans in the genus *Homo*. *See* APES; FOSSIL APES; FOSSIL HUMANS; FOSSIL PRIMATES; MAMMALIA.

After the divergence of *Homo* (*Homo*) from *Homo* (*Pan*) *paniscus* (bonobos, or pygmy chimpanzees), humankind's emergence was marked by mutations (such as DNA sequence changes) that spread to fixation in the ancestors of all modern humans. These mutations are the human-specific factors which distinguish the human species genetically from all other species. Ongoing evolution involves mutations that have not spread to fixation, either because they occurred too recently or because natural selection has maintained a polymorphic state. These mutations occur at frequencies that occasionally differ from one human group to another. They account for the genetic diversity found in the human species. Extensive comparative data now exist on the genetic diversity due to mitochondrial genetic variants (which arise from mutations in the DNA carried by mitochondria). The species-wide distribution of mitochondrial haplotypes in present-day human populations compared to the distribution in chimpanzees reveals that humans show less DNA diversity than chimpanzees. Human populations in different ethnic groups (European Norwegians and South African hottentots) and even within the same group share 85–90% of the total species diversity. The human genetic diversity due to mutations in nuclear DNA (the DNA carried by the chromosomes of cells) also shows this same pattern in which most of the variation within the human species as a whole is contained within single populations. These findings argue that all living humans are biologically not only members of the same species but also the same subspecies and even the same race, the human race. A further inference from the data on mitochondrial genetic diversity is that the birth of the human race took place in Africa about 200,000 years ago. *See* MITOCHONDRIA. Morris Goodman

Molecular beams Well-directed streams of atoms or molecules in vacuum. Utilization of molecular beams is a cornerstone technique in the investigation of molecular structure and interactions. Molecular beams are usually formed at sufficiently low particle density for the interaction of one beam molecule with another to be negligible. This ensemble of truly isolated molecules is available for the spectroscopic study of molecular energy levels using photon probes from the radio-frequency to optical portions of the electromagnetic spectrum. Some of the best-determined fundamental knowledge of physics comes from spectroscopic molecular-beam experiments. Beyond this, beams can be applied as probes of the multifaceted nature of gases, plasmas, surfaces, and even the structure of solids. An application intermediate in complexity is the study of molecular interactions by means of two colliding beams, where one might be a beam of charged particles such as ions or electrons. *See* SCATTERING EXPERIMENTS (ATOMS AND MOLECULES).

One simple means of forming a beam is to permit gas from an enclosed chamber to escape through a small orifice into a second chamber maintained at high vacuum by means of large pumps (illus. *a*). A useful number of molecules passes forward along the horizontal axis of the apparatus. A well-collimated beam is then formed by requiring that those molecules entering the test chamber where an experiment is to be performed pass not only through the orifice but also through a second small hole separating the collimating and test chambers.

If higher velocities are desired, a charge-exchange beam system can be used. In this scheme (illus. *b*), ions are produced by

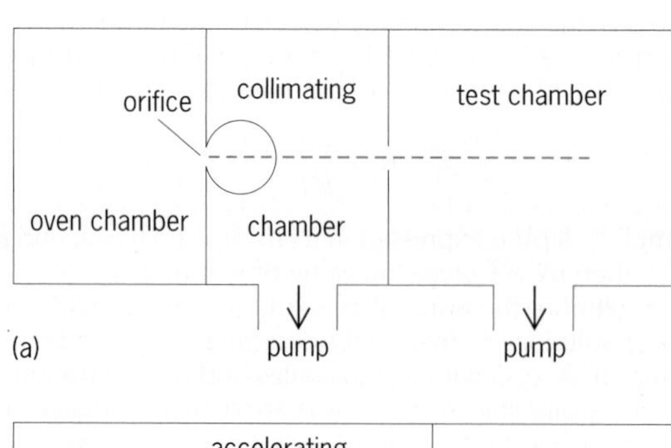

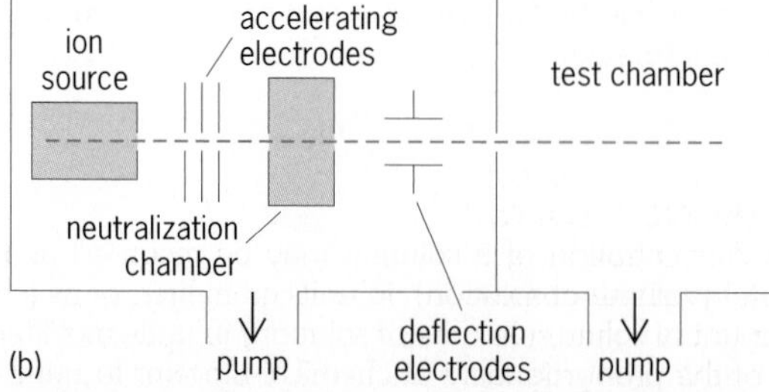

Schematic diagrams of systems for producing molecular beams. (*a*) Conventional oven-beam system. (*b*) Charge-exchange beam system.

some ionizing process such as electron impact on atoms within a gas discharge. Since the ions are electrically charged, they can be accelerated to the desired velocity and focused into a beam using electric or magnetic fields. The last step in neutrally charged beam formation is to pass the ions through a neutralizing gas where electrons from the gas molecules are transferred to the beam ions in charge-exchange molecular collisions. *See* ION SOURCES.

Much of molecular spectroscopy involves the absorption or emission of light by molecules in a gas sample. The frequency of the light photon is proportional to the separation of molecular energy levels involved in the spectroscopic transition. However, the molecule density in typical gas samples is so high that the energy levels are slightly altered by collisions between molecules, with the transition frequency no longer characteristic of the free molecule. The use of low-density molecular beams with their sensitive detection techniques can reduce this collision alteration problem, with the result that atomic properties can be measured to accuracies of parts per million or even better. If the very simplest atoms or molecules are employed, the basic electromagnetic interactions holding the component electrons and nuclei together can be precisely studied. This is of great importance to fundamental physics, since theoretical understanding of electromagnetic interactions through quantum electrodynamics represents the most successful application of quantum field theory to elementary particle physics problems. *See* QUANTUM ELECTRODYNAMICS; QUANTUM FIELD THEORY.

The development of tunable, strong laser sources of single-frequency light beams has added another dimension to molecular-beam experiments. With laser radiation resonantly tuned to excite a molecule from its normal ground state to one of its infinite number of vibrationally, rotationally, and electronically excited states, the number of possible studies and applications of excited molecular beams becomes enormous. *See* LASER SPECTROSCOPY; MOLECULAR STRUCTURE AND SPECTRA; NUCLEAR STRUCTURE.

James E. Bayfield

Molecular biology The study of structural and functional properties of biological systems, pursued within the context of understanding the roles of the various molecules in living cells and the relationship between them. Molecular biology has its roots in biophysics, genetics, and biochemistry. A prime focus of the field has been the molecular basis of genetics, and with the demonstration in the mid-1940s that deoxyribonucleic acid (DNA) is the genetic material, emphasis has been on structure, organization, and regulation of genes. Initially, molecular biologists restricted their studies to bacterial and viral systems, largely because of their genetic and biochemical simplicity. *Escherichia coli* has been extensively examined because of its limited number of cellular functions and the corresponding restricted amount of genetic information encoded in the bacterial chromosome. Simple eukaryotic cells, such as protozoa and yeast, offer similar advantages and also have been studied. For these same reasons, bacteriophage and animal viruses have provided molecular biologists with the ability to study the structural and functional properties of molecules in intact cells. However, a series of conceptual and technological developments occurred rapidly during the late 1970s that permitted molecular biologists to approach a broad spectrum of plant and animal cells with experimental techniques. One of the major factors has been the development and applications of genetic engineering. Recombinant DNA technology allowed the isolation and selective modification of specific genes, thereby reducing both their structural and functional complexity and facilitating the study of gene expression in higher cells. The concepts and techniques used by molecular biologists have been rapidly and effectively employed to resolve numerous cellular, biological, and biochemical problems—becoming routine at both the basic and applied levels.

The recognition of DNA as the genetic material coupled with the discovery that genes reside in chromosomes resulted in an intensive effort to map genes to specific chromosomes. Initially genes were assigned to chromosomes on the basis of correlations between modifications in cellular function, particularly biochemical defects, and the addition, loss, or modification of specific chromosomes. *See* CHROMOSOME ABERRATION; MUTATION.

A major breakthrough was the development of somatic cell genetics. This is an approach in which, for example, human and hamster cells are fused, resulting in a hybrid cell initially containing the complement of human and hamster chromosomes. As the cells grow and divide in culture, the hamster chromosomes are retained while there is a progressive loss of human chromosomes. By correlating the loss of human biological or biochemical traits with the loss of specific human chromosomes, a number of human genes have been successfully mapped. *See* SOMATIC CELL GENETICS.

The development of methods for isolating genes and for determining the genetic sequences of the DNA in which the genes are encoded, led to rapid advances in gene mapping at several levels of resolution. Localization of specific genes to chromosomes is routinely carried out with cloned genes as probes. Further information about the segment of a chromosome in which a specific gene resides can be obtained by directly determining the DNA sequences of both the gene itself and the surrounding region.

Chromosome localization of specific genes has numerous applications at both the basic and clinical levels. At the basic level, knowledge of the positions of various genes provides insight into potentially functional relationships. At the clinical level, chromosome aberrations are now routinely used in prenatal diagnosis of an extensive series of human genetic disorders, and several chromosomal modifications have been linked to specific types of cancer. Knowledge of genetic defects at the molecular level has permitted the development of diagnostic procedures that in some instances, such as sickle cell anemia, are based on a single nucleotide change in the DNA.

Recombinant DNA. Recombinant DNA technology has provided molecular biology with an extremely powerful tool. In broad terms, applications of recombinant DNA technology can be divided into four areas—biomedical, basic biological, agricultural, and industrial. Biomedical applications include the elucidation of the cellular and molecular bases of a broad spectrum of diseases, as well as both diagnostic and therapeutic applications in clinical medicine.

In a strictly formal sense, the term recombinant DNA designates the joining or recombination of DNA segments. However, in practice, recombinant DNA has been applied to a series of molecular manipulations whereby segments of DNA are rearranged, added, deleted, or introduced into the genomes of other cells.

The ability to manipulate or "engineer" genetic sequences is based on several developments.

1. *Methods for breaking and rejoining DNA*. The precise breaking and rejoining of DNA has been made possible by the discovery of restriction endonucleases, enzymes that have the ability to recognize specific DNA sequences and to cleave the double helix precisely at these sites. Also important are the ability to join fragments of DNA together with the enzyme DNA ligase, and the techniques to determine the nucleotide sequence of genes and thereby confirm the identity and location of structural and regulatory sequences.

2. *Carriers for genetic sequences*. Bacterial plasmids, that is, circular double-stranded DNA molecules that replicate extrachromosomally, have been modified so that they can serve as efficient carriers for segments of DNA, complete genes, regions of genes, or sequences contained within several different genes. Bacteriophage and animal viruses, retroviruses, and bovine papilloma virus have also been successfully utilized as DNA

carriers. These carriers are referred to as cloning vectors. Host cells in which vectors containing cloned genes can replicate range from bacteria to numerous other cells, including normal, transformed, and malignant human cells.

3. *Introduction of recombinant DNA molecules.* Genetic sequences in the form of isolated DNA fragments, or chromosomes, or of DNA molecules cloned in plasmid vectors can be introduced into host cells by a procedure referred to as transfection or DNA-mediated gene transfer—a technique that renders the cell membrane permeable by a brief treatment with calcium phosphate, thereby facilitating DNA uptake. Genes cloned in viruses can also be introduced by infection of host cells.

4. *Selection of cells containing cloned sequences.* Bacterial cells containing plasmids with cloned genes can be detected by selective resistance or sensitivity to antibiotics. In addition, the presence of introduced genes in bacterial, plant, or animal cells can be assayed by a procedure known as nucleic acid hybridization.

5. *Amplification.* Amplification of genetic sequences cloned in bacterial plasmids is efficiently achieved by treatment of host cells with antibiotics which suppress replication of the bacterial chromosome, yet do not interfere with replication of the plasmid with its cloned gene. Sequences cloned in bacterial or animal viruses are often amplified by virtue of the ability of the virus to replicate preferentially. *See* GENE AMPLIFICATION.

6. *Expression.* Expression of cloned human genes can be mediated by regulatory sequences derived from the natural gene, from exogenous genes, or by host cell sequences.

Two clinically important genes, human insulin and human growth hormone, have been cloned and introduced into bacteria under conditions where biologically active hormones can be produced.

Progress has been made in applications of recombinant DNA technology to the resolution of agricultural problems, especially for the improvement of both crops and livestock. *See* ADENOHYPOPHYSIS HORMONE; BREEDING (ANIMAL); BREEDING (PLANT); GENETIC ENGINEERING; INSULIN.

Biophysical analysis. Understanding of the structural properties of molecules and the interaction between molecules that constitute biologically important complexes has been facilitated by biophysical analysis. For example, developments in the resolution offered by techniques such as electron microscopy, x-ray diffraction, and neutron scattering have provided valuable insight into the structure of chromatin, the protein-DNA complex which constitutes the genome of eukaryotic cells. These techniques have also provided clues about modifications in chromatin structure that accompany functional changes. One possible application of biophysical analysis is the diagnosis of human disorders by adaptation of nuclear magnetic resonance for tissue and whole body evaluation of soft tissue tumors, blood flow, and cardiac function. *See* ELECTRON MICROSCOPE; NUCLEAR MAGNETIC RESONANCE (NMR); X-RAY DIFFRACTION.

Flow of molecular information. Information for all cellular activities is encoded in DNA; selective elaboration of this information is prerequisite to meeting both structural and biochemical requirements of the cell. In this regard, there are three major areas of investigations by molecular biologists: (1) the composition, structure, and organization of chromatin, the protein-DNA molecular complex in which genetic information is encoded and packaged; (2) the molecular events associated with the expression of genetically encoded information so that specific cellular biochemical requirements can be met; and (3) the molecular signals that trigger the expression of specific genes and the types of communication and feedback operative to monitor and mediate gene control. *See* CHROMOSOME; DEOXYRIBONUCLEIC ACID (DNA); GENE; GENETIC CODE; NUCLEIC ACID.

Gary S. Stein; Janet L. Stein

Molecular chaperone A specialized cellular protein that binds nonnative forms of other proteins and assists them to reach a functional conformation, in most cases through the expenditure of adenosine triphosphate (ATP). Originally identified by their increased abundance after heat shock, chaperone proteins in general bind to exposed hydrophobic surfaces of nonnative proteins, preventing them from forming intermolecular interactions that lead to irreversible multimolecular aggregation. Thus, the role of chaperone proteins under conditions of stress, such as heat shock, is to protect proteins by binding to incipiently misfolded conformations, preventing aggregation; then, following return of normal conditions, they allow refolding to occur, associated with protein release. *See* ADENOSINE TRIPHOSPHATE (ATP); PROTEIN.

Chaperones also play essential roles in folding under normal conditions. Proteins in general contain in their primary amino acid sequences all the information necessary for proper folding. Under cellular conditions, however, "off-pathway" misfolding becomes a major competing reaction, particularly for multidomain proteins, potentially lodging them in kinetic traps (local energetic minima that are separated by energetic barriers from the global energetic minimum typical of the native state). Chaperones provide kinetic assistance to the folding process, binding such trapped, misfolded states through exposed hydrophobic surfaces and expending energy of ATP to place the protein back on a productive folding path. Thus, chaperones are not true folding catalysts; that is, they do not accelerate on-pathway folding. Rather, as kinetic assistants, they serve to smooth out the energy landscape, thus improving the overall rate and extent of productive folding in vivo. One of the most extensively investigated roles of chaperones is to assist in the proper folding of newly translated proteins in the cytosol. *See* CYTOPLASM.

In addition to roles in protection against stress and in de novo cytosolic folding, molecular chaperones participate in the following: 1. Maintaining precursor forms of endoplasmic reticulum and mitochondrial proteins in unfolded, extended states in the eukaryotic cytosol, enabling the precursors to be recognized and translocated into these organelles. 2. Enabling translocation of proteins into the endoplasmic reticulum and mitochondria, by action at the trans (inside) aspect of the organellar membrane. 3. Assisting folding of proteins imported into mitochondria. 4. Assisting folding of some proteins imported into the endoplasmic reticulum. 5. Binding near-native forms of signal transduction kinases and nuclear receptors in the cytosol. 6. Functioning as ATP-driven enzymes to recognize appropriately tagged (for example, ubiquitinated) proteins, unfold them, and then bind and translocate them for degradation into coaxially associated proteolytic cylinders such as the proteasome. *See* CELL (BIOLOGY); ENDOPLASMIC RETICULUM; MITOCHONDRIA; PROTEIN DEGRADATION.

Arthur Horwich; Wayne Fenton

Molecular cloud A large and relatively dense cloud of cold gas and dust in interstellar space from which new stars are born. Molecular clouds consist primarily of molecular hydrogen (H_2) gas, with temperatures in the range 10–100 K. Molecular hydrogen is not directly observable under most conditions in molecular clouds. Therefore, almost all current knowledge about the properties of molecular clouds has been deduced from observations of trace constituents, mostly simple molecules such as carbon monoxide (CO), which have strong emission lines in the centimeter-, millimeter-, and submillimeter-wavelength portions of the electromagnetic spectrum. The majority of clouds lie in a broad Molecular Ring encircling the galactic center with an inner radius of about 3 kiloparsecs and an ill-defined outer radius extending to beyond 20 kpc. *See* RADIO ASTRONOMY.

Molecular clouds are the principal sites of ongoing star formation. Therefore, they tend to be associated with young stars and star-forming regions. The nearest star-forming clouds are found in the constellations Ophiuchus, Taurus, and Perseus, at distances of 125, 140, and 300 parsecs (1 parsec $= 3 \times 10^{13}$ km or 2×10^{13} mi), where the nearest regions of active low- and intermediate-mass star formation are found. These cloud complexes have masses ranging from several thousand to perhaps over 10,000 times the mass of the Sun. However, most of the molecular gas in the Milky Way Galaxy is concentrated into giant clouds with masses more than 100,000 times the mass of the Sun. The nearest giant molecular clouds are located at a distance of 460 parsecs toward the constellation Orion where, over the last 10^7 years, they gave birth to tens of thousands of stars, including several dozen relatively rare high-mass stars. *See* ORION NEBULA; PROTOSTAR.

Most molecular clouds have temperatures of only 10 K. Molecular clouds are orders of magnitude more dense than the general interstellar medium, with gas densities ranging from about 10 molecules per cubic centimeter on large scales to over 10^6 molecules per cubic centimeter in cloud cores. The sizes of individual clouds range from less than 0.1 parsec for small clouds and dense cores to over 100 parsecs for giant molecular clouds. In addition to star-forming molecular clouds concentrated toward the plane of the Milky Way, there are many smaller and lower-density molecular clouds visible most clearly far away from the galactic plane, with the closest ones only about 50 parsecs from the Sun.

Molecular clouds have a very complex internal structure consisting of clumps and filaments of dense gas surrounded by interclump gas of much lower density. Individual clumps usually have supersonic internal motions with a velocity of several kilometers per second. The powerful outflows produced by young stars during the first 100,000 years of their existence may be a major source of these chaotic motions. Magnetic fields which thread molecular clouds may play a role in the longevity of turbulent motions and may support clouds against gravitational collapse.

About 100 different chemical species have been so far identified within molecular clouds, indicating that there is a rich chemistry taking place. *See* INTERSTELLAR EXTINCTION. John Bally

Molecular electronics The use of molecules as active electronic devices. As contemporary integrated-circuit technology approaches its limits, molecular electronic devices are being pursued not just as a potential successor to electronic systems but also as a method for introducing new functionality or interfaces.

The advantages of molecular device integration into complex systems are an active area of research, as contemporary integrated circuits struggle with power dissipation, materials properties, and interconnect problems. Molecular components offer the possibility of self-assembly processing and different functionality (as well as alternatives to charge-based computation), and represent the ultimate endpoint of any electronic system. Another application issue is precision, mainly of the interface to the active molecules. At the current stage of structural uncertainty, one sees fluctuations in expected device behavior due to this uncertainty. Advances in the field, as well as most other electronics approaches on this length scale, await the development of a truly planar (or 3D) atomic-scale technology with a precision better than 0.1 nm as well as structural characterization tools at this size scale. Mark A. Reed

Molecular isomerism The property of compounds (isomers) which have the same molecular formula but different physical and chemical properties. The difference in properties is caused by a difference in molecular structure (that is, molecular architecture). A typical example is dimethyl ether, CH_3OCH_3, a chemically quite inert gas which condenses at −24°C, and ethyl alcohol, CH_3CH_2OH, a liquid of substantial chemical reactivity which boils at 78°C; both compounds have the molecular formula C_2H_6O.

Isomers may be classified as constitutional isomers or stereoisomers. Constitutional isomers differ in constitution or connectedness, relating to the question as to which atoms are linked to which others and how. Dimethyl ether and ethanol (Fig. 1) are constitutional isomers. In dimethyl ether each carbon is connected to three hydrogen atoms and the one oxygen atom; the two carbon atoms are thus equivalent. In ethyl alcohol (ethanol) one carbon is linked to three hydrogen atoms and the other carbon; the second carbon is linked to the first carbon, two hydrogens, and the oxygen atom which, in turn, is linked to the sixth hydrogen atom; the two carbon atoms are not equivalent. Stereoisomers, in contrast, have the same constitution but differ in the three-dimensional array of the atoms in space, called configuration.

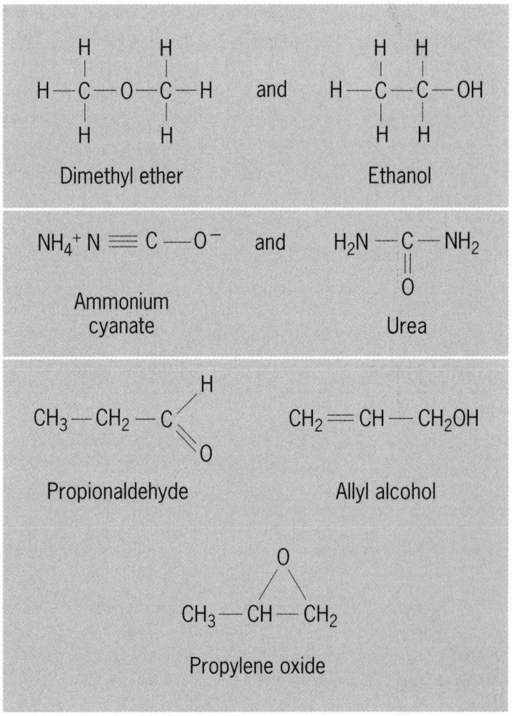

Fig. 1. Functional isomers.

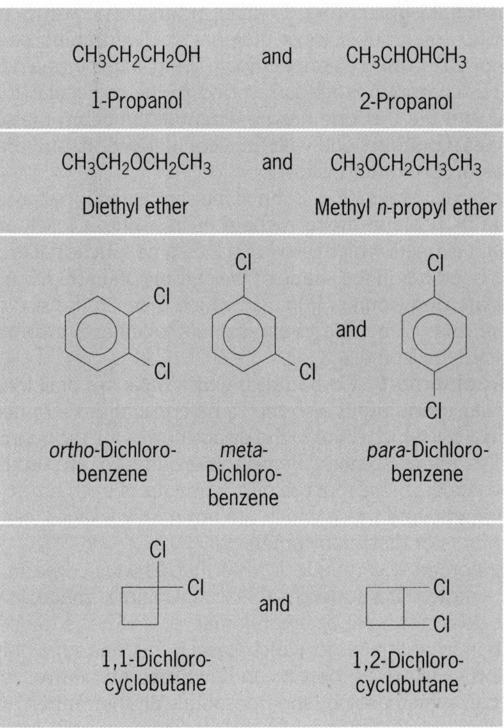

Fig. 2. Positional isomers.

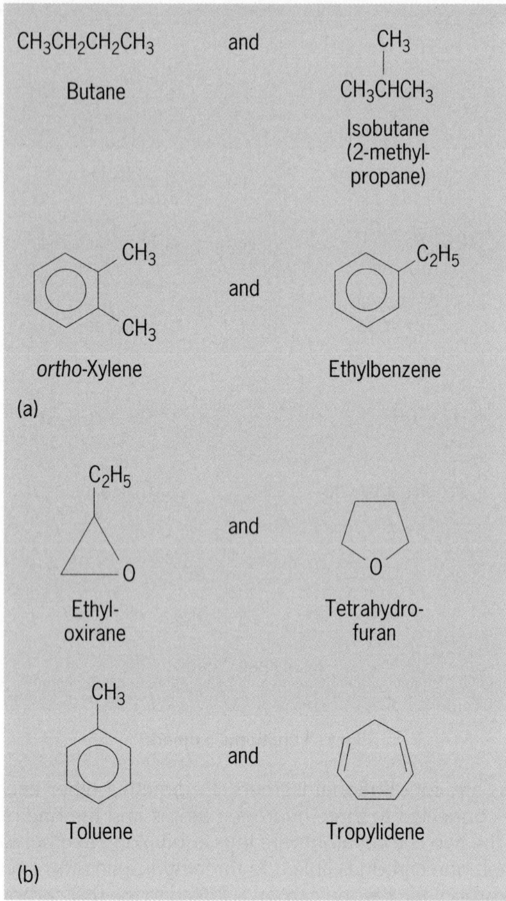

Fig. 3. Skeletal isomers. (*a*) Chain. (*b*) Ring.

Constitutional isomers. Constitutional isomers have been subdivided into functional isomers, positional isomers, and chain isomers.

Functional isomers (Fig. 1) differ in functional group, that is, the group (or groups) most material in determining chemical behavior. In the third example shown in Fig. 1 (propionaldehyde) the three compounds all correspond to the molecular formula C_3H_6O, but the first one has an aldehyde function, the second combines a double bond with an alcohol function, and the third one has an epoxide function.

Positional isomers (Fig. 2) have the same functional group but differ in its position along a chain or in a ring. Closely related are chain isomers which also have the same functional group or groups but differ in the shape of the carbon chain (Fig. 3*a*); quite similar are ring isomers (Fig. 3*b*) which differ in the size of one or more rings. Ring and chain isomers together are sometimes called skeletal isomers.

Stereoisomers. Compounds which have not only the same molecular formula but also the same constitution (connectivity of atoms) but which differ in the disposition of the atoms in space are called Stereoisomers. Stereoisomers, in turn, are subdivided into two types: those that are mirror images of each other, called enantiomers, and those which are not mirror images, called diastereomers or diastereoisomers.

Enantiomers are unique in that they always come in pairs. Either a molecule is superposable with its mirror image, in which case it does not have an enantiomer, or it is not superposable with its mirror image, in which case it has one and only one enantiomer (since an object can have only one mirror image). Molecules which are not superposable with their mirror images are called chiral; those which are so superposable are called achiral. Enantiomers are much more alike than are other sets of isomers (constitutional isomers or diastereomers); thus they have the same melting point, boiling point, free energy, spectral properties, x-ray diffraction pattern, and so on.

Diastereomers have the same constitution but different spatial arrangement and are not mirror images. They resemble constitutional isomers in that there may be more than two isomers in a set and that their physical, energetic, and spectral properties are generally quite distinct. *See* CONFORMATIONS; ANALYSIS; STEREOCHEMISTRY; TAUTOMERISM.

Ernest L. Eliel

Molecular machine A molecular device is an assemblage of a discrete number of molecular components (that is, a supramolecular structure) designed to achieve a specific function. Each molecular component performs a single act, while the entire supramolecular structure performs a more complex function, which results from the cooperation of the various molecular components. Molecular devices operate via electronic or nuclear rearrangements. Like any device, they need energy to operate and signals to communicate with the operator. The extension of the concept of a device, so common on a macromolecular level, to the molecular level is of interest not only for basic research but also for the growth of nanoscience and nanotechnology. *See* NANOTECHNOLOGY; SUPRAMOLECULAR CHEMISTRY.

A molecular machine is a particular type of molecular device in which the component parts can display changes in their relative positions as a result of some external stimulus. Such molecular motions usually result in changes of some chemical or physical property of the supramolecular system, resulting in a “readout” signal that can be used to monitor the operation of the machine. The reversibility of the movement, that is, the possibility to restore the initial situation by means of an opposite stimulus, is an essential feature of a molecular machine. Although there are a number of chemical compounds whose structure or shape can be modified by an external stimulus (for example, photoinduced cis-trans isomerization processes), the term “molecular machines” is used only for systems showing large-amplitude movements of molecular components.

The human body can be viewed as a very complex ensemble of molecular-level machines that power motions, repair damage, and orchestrate an inner world of sense, emotion, and thought. Among the most studied natural molecular machines are those based on proteins such as myosin and kynesin, whose motions are driven by adenosine triphosphate (ATP) hydrolysis. One of the most interesting molecular machines of the human body is ATP synthase, a molecular-level rotatory motor. In this machine, a proton flow through a membrane spins a wheellike molecular structure and the attached rodlike species. This changes the structure of catalytic sites, allowing uptake of adenosine diphosphate (ADP) and inorganic phosphate, their reaction to give ATP, and then the release of the synthesized ATP. *See* ADENOSINE TRIPHOSPHATE (ATP).

An artificial molecular machine performs mechanical movements analogous to those observed in artificial macroscopic machines (for example, tweezers, piston/cylinder, and rotating rings). Analogously to what happens for macroscopic machines, the energy to make molecular machines work (that is, the stimulus causing the motion of the molecular components of the supramolecular structure) can be supplied as light, electrical energy, or chemical energy. In most cases, the machinelike movement involves two different, well-defined and stable states, and is accompanied by on/off switching of some chemical or physical signal [absorption and emission spectra, nuclear magnetic resonance (NMR), redox potential, or hydronium ion (H_3O^+) concentration]. For this reason, molecular machines can also be regarded as bistable devices for information processing (Fig. 1). *See* ACID AND BASE.

The interest in molecular machines arises not only from their mechanical movements but also from their switching aspects. Computers are based on sets of components constructed by the

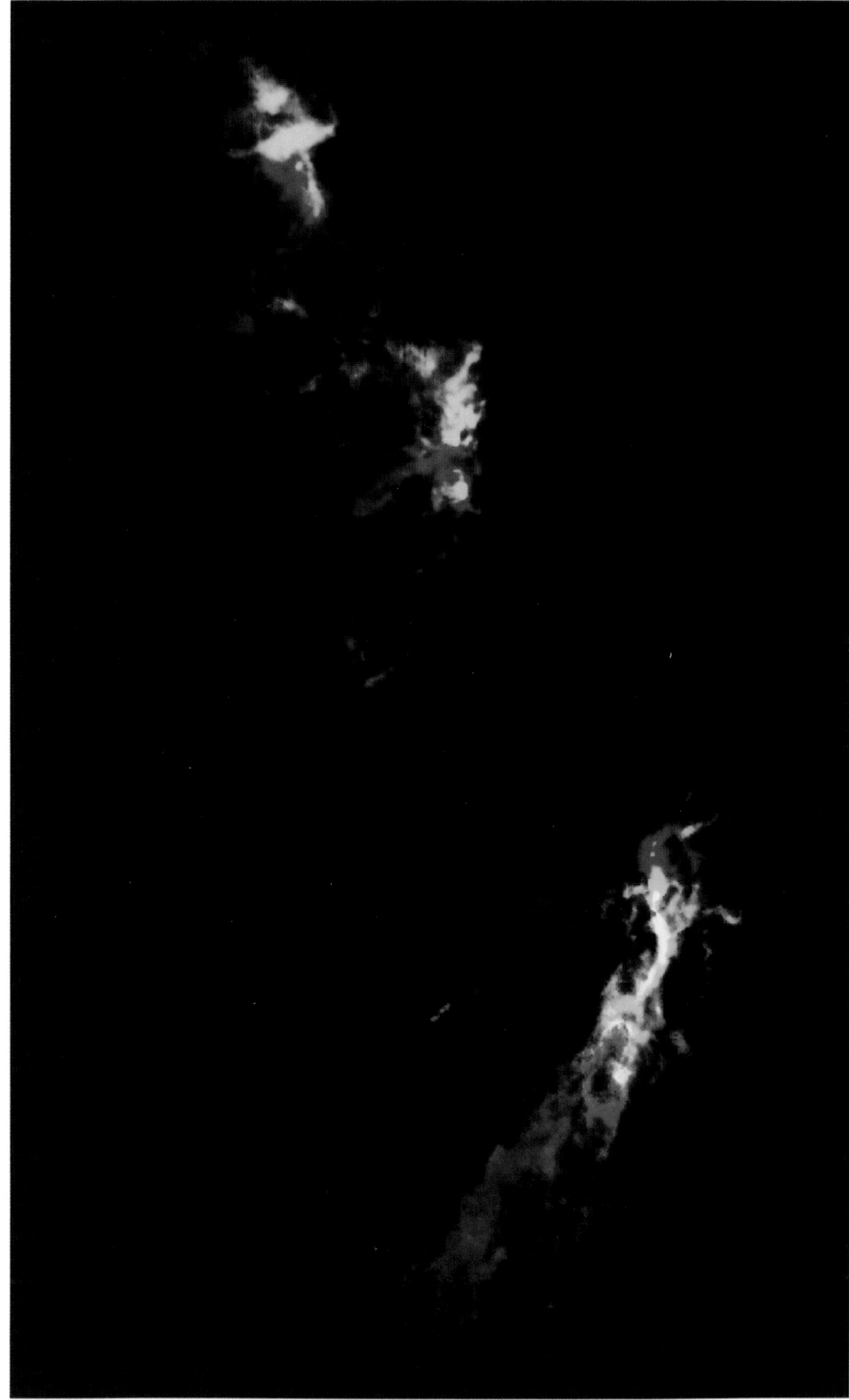

Emission of 2.6-mm-wavelength (110-GHz) radiation of ^{13}CO from two giant molecular clouds in the constellation Orion. The vertical extent of this image is about 10°, which corresponds to about 80 parsecs at the 460-parsec distance to these clouds. Colors show the motion of the emitting gas along the line of sight to the Earth as measured by the Doppler effect. Red indicates the presence of gas moving away from the Earth at about 3 km/s (2 mi/s) with respect to the mean velocity of the cloud, shown in green. Blue indicates the presence of gas approaching the Earth at 3 km/s (2 mi/s) with respect to mean cloud velocity. These motions are superimposed on the overall recession of the Orion molecular cloud complex from the stars near the Sun at about 9 km/s (6 mi/s). (*John Bally*)

top-down approach. This approach, however, is now close to its intrinsic limitations. A necessary condition for further miniaturization to increase the power of information processing and computation is the bottom-up construction of molecular-level components capable of performing the functions needed (chemical computer). The molecular machines described above operate according to a binary logic and therefore can be used for switching processes at the molecular level. It has already been shown that suitable designed machinelike systems can be employed to perform complex functions such as multipole switching, plug/socket connection of molecular wires, and XOR logic operation. *See* LOGIC. V. Balzani

Molecular mechanics An empirical computational method that provides structural, energetic, and property information about molecules. One simple way to consider molecules is as a collection of balls (atoms) held together by springs (bonds)—the basis of many molecular model kits. Molecular mechanics offers a way to model the behavior of matter mathematically in this manner. Mechanical spring-based theory begins with a fundamental assumption that matter consists of atoms and that the potential energy of a collection of atoms can be defined for every set of positions. The collection of atoms is treated as a mechanical system moving within this potential energy, just as a clockwork's motions are determined by its spring potentials. Molecular mechanics methods are a natural outgrowth of concepts of bonding between atoms in molecules and van der Waals forces between nonbonded atoms. In 1930, T. L. Hill proposed the first such potential energy function made up of a simple representation of this type, including van der Waals interactions together with stretching and bending deformation functions. The idea was to minimize this energy function, leading to information about the structure and steric energy in congested molecules. Subsequent studies established more elaborate formulations of the potential-energy functional form, and were used to understand increasingly more detail about molecular systems, such as the equilibrium structure, energetics, thermodynamic properties, and vibrational spectra. Further work in the 1940s verified the theory for more complex reactions. *See* CHEMICAL BONDING; ENERGY; VAN DER WAALS EQUATION.

Unlike quantum-mechanical approaches, electrons are not explicitly included in molecular mechanics calculations. This is possible due to the Born-Oppenheimer approximation, which states that the electronic and nuclear motions can be uncoupled from one another and considered separately. Molecular mechanics assumes that the electrons in a molecule find their optimum distribution, and approaches chemical problems from the standpoint of the nuclear structure. A molecule from this perspective is considered to be a collection of masses that are interacting with each other via nearly ideal harmonic forces—hence, the analogy to the ball-and-spring model. Potential energy functions are used to describe the interactions between nuclei. With judicious parameterization, the electronic system is implicitly taken into account. The resulting energy function of structure, Energy $= f$ (nuclear positions), is presumed to have a minimum corresponding to the most stable equilibrium geometry. Any deviation of the model from the ideal molecular geometry will correspond to an increase in energy. *See* NUCLEAR STRUCTURE; QUANTUM MECHANICS.

Molecular mechanics methods treat molecules as collections of particles held together by simple forces. The various types of forces are described in terms of simple individual potential functions, each summed over all respective atoms involved in that force. The total of all such contributions then constitutes the overall potential energy, or steric energy, of the molecular system. The resulting potential energy landscape of a molecule is also referred to as an empirical force field, since the derivative of the potential energy determines the forces acting on the atoms when in motion. *See* MOLECULAR STRUCTURE AND SPECTRA.

Knowledge of the structure and properties of a series of molecules or understanding of very large macromolecular structures, such as enzymes and proteins, are problems particularly well suited for study using molecular mechanics. In fact, the mechanical molecular model formulation was developed out of a need to describe molecular structure and properties at a time when relatively little computer power was available, requiring the resulting method to be as practical as possible. The application to molecular systems with thousands of atoms, even with the advent of computational power, still mandates a more practical method, such as offered by molecular mechanics force-field methods. A variety of mathematical and computational approaches have been used over the years. This continues to be an area of active research. *See* COMPUTATIONAL CHEMISTRY.

Because molecular mechanics–based methods tend to be computationally very efficient, application is practical for conformational analysis studies to predict the most likely conformations for highly flexible molecules. In structural biology, this is quite important in the protein-folding problem, the goal of which is to predict the three-dimensional structure of a protein from the linear sequence of amino acids. Another wide use of molecular mechanics is in studies involving a small molecule binding into an active site, called docking studies. Estimating correct three-dimensional atomic structures of complexes between proteins and ligands is an important component of the drug-design process in the pharmaceutical industry. Basic aspects of ligand–protein interactions, categorized under the general term molecular recognition, are concerned with the specificity as well as the stability of the ligand to bind into a protein. *See* MOLECULAR RECOGNITION. Kim K. Baldridge; Celine Amoreira

Molecular orbital theory A quantum-mechanical model concerned with the description of the discrete energy levels associated with electrons in molecules. One useful way to generate such levels is to assume that the molecular orbital wave function (ψ_j) may be written as a simple weighted sum of the constituent atomic orbitals (χ_i) [Eq. (1)]; this is called

$$\psi_j = \Sigma c_{ij} \chi_i \quad (1)$$

the linear combination of atomic orbitals approximation. The c_{ij} coefficients may be determined numerically by substitution of Eq. (1) into the Schrödinger equation and application of the variational theorem. The theorem states that an approximate wave function will always be an upper bound to the true energy; thus minimization of the energy of the system given by the wave function of Eq. (1) will provide the best values of c_{ij}. Once the wave function is known, its associated energy may be calculated. The energies of the occupied orbitals in molecules may be probed by using photoelectron spectroscopy, which gives a good check on the accuracy of the theory. There are some simple concepts that contribute to a qualitative understanding of these molecular orbital energy levels and hence an insight into chemical bonding in molecules. They may be illustrated with reference to the hydrogen molecule. *See* CHEMICAL BONDING; ELECTRON SPECTROSCOPY; QUANTUM MECHANICS; SCHRÖDINGER'S WAVE EQUATION.

First, the basis orbitals (χ_i) used in the expansion of Eq. (1) can usefully be restricted to include the valence orbitals only. For molecular hydrogen (H_2) the 1*s* orbitals on the two hydrogen atoms are then the only two orbitals to be included. Second, since hydrogen atoms are chemically identical, any observable characteristic whose value might be computed with Eq. (1) must be the same for both atoms. This leads to the requirement that $c^2_{1j} = c^2_{2j}$, where the labels 1,2 refer to hydrogen atoms 1 and 2. As a consequence $c_{1j} = \pm c_{2j}$.

When the signs of the two coefficients are the same, the two hydrogen orbitals are mixed in phase; when they are different, the two hydrogen orbitals are mixed out of phase. When the atomic orbitals are mixed in phase, electron density is built up

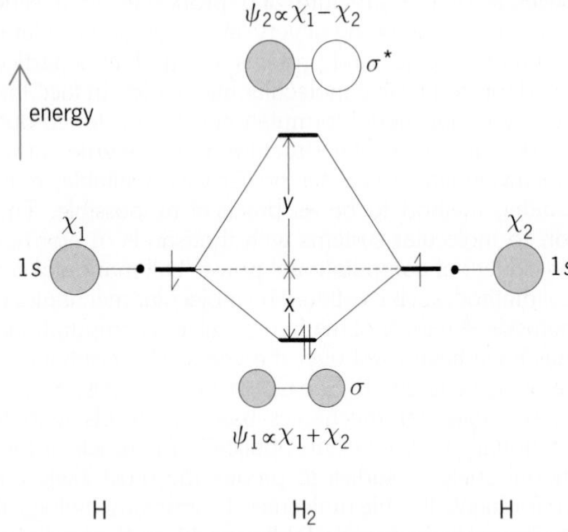

Fig. 1. Molecular orbital diagram of H_2.

between the two hydrogen nuclei and the potential energy of the nuclei and electrons is lowered. In fact, a reduction of kinetic energy also occurs. An electron lying in the molecular orbital is then of lower energy than an electron associated with an isolated hydrogen 1*s* orbital. It is called a bonding orbital. The increase in electron density between the two nuclei is the electronic "glue" holding the nuclei together. When the atomic orbitals are mixed out of phase, the opposite behavior occurs. Electron density is removed from the region between the two nuclei, resulting in an increase of both potential and kinetic energy of the electrons. An electron lying in such a molecular orbital would experience an energetic destabilization relative to an electron associated with an isolated hydrogen 1*s* orbital.

Such a molecular orbital is called an antibonding orbital. Figure 1 shows this information as a molecular orbital diagram. The shading convention of the orbitals has been adopted to indicate the in-phase and out-of-phase mixing of the basis orbitals. Just as the energy levels of atoms are filled in an Aufbau process, so the orbitals of the molecule may be analogously filled up with electrons, each level accommodating two electrons of opposite spin. In H_2 there are two electrons to be accounted for. They lie in the bonding orbital, and the stabilization energy relative to two isolated hydrogen atoms (the bond energy) is $2x$. Antibonding orbitals are invariably destabilized more than their bonding counterparts are stabilized. This is shown in Fig. 1 by making $y > x$. With four electrons to be accommodated in this collection of orbitals (this would correspond to the hypothetical case of the He_2 molecule), one electron pair resides in the bonding orbital and one pair in the antibonding orbital. Since $y > x$, this molecule is less stable relative to two isolated helium atoms, and as a result the molecule does not exist as a stable entity. He_2^+, however, with only three electrons is known.

The size of the interaction energy associated with two atomic orbitals (x in Fig. 1) is controlled by the extent of their spatial overlap. This overlap integral clearly depends upon the internuclear separation. The equilibrium bond length in the hydrogen molecule (and indeed in all molecules) is then a balance between the attractive forces associated with bonding orbital formation and the electrostatic repulsion between the nuclei. Such a molecular parameter is amenable to numerical calculation.

The description of the bonding in the H_2 molecule using this model is one where two electrons occupy a bonding orbital and give rise to a simple two-center–two-electron bond traditionally written as H—H. Since the electron density associated with the bonding orbital is cylindrically and symmetrically located about the H—H axis, this bond is called a σ bond. In Fig. 1 the bonding orbital is labeled with a σ and the corresponding antibonding orbital with a σ^*.

Ideas similar to those above are readily extended to diatomic molecules from the first row of the periodic table, such as N_2 and O_2, where the valence orbitals to be considered are the one 2*s* and the three 2*p* orbitals of the atoms. The 2*s* orbitals lie deeper in energy than the triply degenerate 2*p* orbitals. The atomic 2*s* orbitals form bonding and antibonding orbitals ($s\sigma$ and $s\sigma^*$) just as in the case of elemental hydrogen described above, but the behavior of the 2*p* orbitals is a little different. Here there are three possible types of interaction between the *p* orbitals on one center and those on the other. The end-on overlap of two *p* orbitals gives rise to a σ interaction (Fig. 2*a*), and the sideways overlap of two *p* orbitals gives rise to a π interaction (Fig. 2*b*). The interaction in Fig. 2*c* can be ignored since the overlap between the two orbitals in this orientation can be seen to be identically zero. The result is a σ-bonding orbital and a σ-antibonding orbital ($p\sigma$ and $p\sigma^*$), and a pair of π-bonding and a pair of π-antibonding orbitals (π and π^*). A larger interaction energy is associated with $p\sigma$ compared to $p\pi$, due to the larger σ overlap compared to π overlap in Fig. 2.

Filling these orbitals with electrons allows comment on the stability of the resulting diatomics. The molecule Li_2 $(s\sigma)^2$ is known and, like H_2, may be written as Li—Li to emphasize the single, two-center, two-electron bond between the nuclei. The molecule Be_2 which would have the configuration $(s\sigma)^2(s\sigma^*)^2$ is unknown since, just as in He_2, $s\sigma^*$ is destabilized more than $s\sigma$ is stabilized relative to an atomic 2*s* level. If the molecular orbital bond order is written as expression (2), then the bond order in Li_2 is one but

$$\text{Molecular bond order} = \left(\begin{matrix}\text{number of bonding}\\ \text{electron pairs}\end{matrix}\right) - \left(\begin{matrix}\text{number of antibonding}\\ \text{electron pairs}\end{matrix}\right) \quad (2)$$

the bond order in Be_2 is zero.

By filling up the molecular orbital levels derived from the 2*p* orbitals, the bond order associated with the other diatomics may be generated: $B_2(1)$, $C_2(2)$, $N_2(3)$, $O_2(2)$, $F_2(1)$, and $Ne_2(0)$. All of these species are known except Ne_2, which is predicted, like He_2 and Be_2, to have a zero bond order and therefore not to exist as a stable molecule. The molecular orbital bond orders for the three best-known diatomics are consistent with their traditional formulation as N≡N, O=O, and F—F. N_2, for example, would be described as having one σ and two π bonds. With the configuration $(s\sigma)^2(s\sigma^*)^2(p\sigma)^2-(\pi)^4(\pi^*)^2$, there are four bonding pairs of electrons and two antibonding pairs giving rise to a net bond order of two. The pair of π^* orbitals is only doubly occupied, whereas there is space for four electrons. Hund's rules (which for the electronic ground state maximize the number of

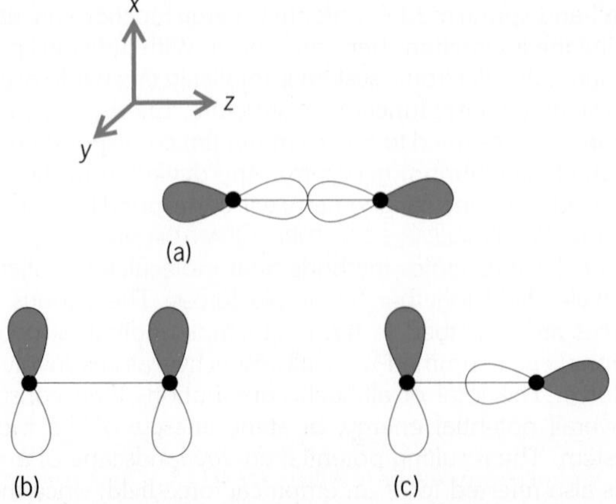

Fig. 2. Possible orientations of the *p* orbitals on adjacent atomic centers. (*a*) End-on overlap. (*b*) Sideways overlap. (*c*) Zero overlap.

electrons with parallel spins) identify the lowest-energy arrangement as the one where each of the degenerate π^* components is singly occupied, the spins of the two electrons being parallel. Unpaired electrons give rise to paramagnetic behavior, and gaseous oxygen is indeed paramagnetic. Jeremy K. Burdett

Molecular pathology A discipline that deals with the origins and mechanisms of diseases at their most fundamental level, that of macromolecules such as deoxyribonucleic acid (DNA) and protein, in order to provide precise diagnoses and discover possible avenues for treatment. It is interdisciplinary, including infectious disease, oncology, inherited genetic disease, and legal issues such as parentage determination or forensic identity testing. While a variety of biophysical and biochemical techniques can be applied to study the molecular basis of disease, antibodies and nucleic acid probes are two of the principal approaches. *See* ANTIBODY; NUCLEIC ACID; ONCOLOGY.

When monoclonal antibodies are either tagged to permit their detection or immobilized on a chromatographic column to purify their specific target molecule, they serve as powerful tools for analyzing pathologic processes. A monoclonal antibody conjugated to an enzyme generating a colored reaction product is the basis of the enzyme-linked immunosorbent assay (ELISA), which is widely applied in many diagnostic tests. Autoantibodies from the sera of individuals with autoimmune diseases are often used as highly specific reagents for understanding the nature of these diseases and the role that the affected molecules or organelles normally perform in the cell. These autoantibodies can be detected in tissues by using fluorescently tagged antibodies directed against human immunoglobulins. *See* AUTOIMMUNITY; MONOCLONAL ANTIBODIES.

In an analogous manner to immunohistochemistry, traditional histopathology can be enhanced by using in situ hybridization. With this technique, an infectious agent such as a virus or a specific messenger ribonucleic acid (mRNA) can be localized within a specific cell or tissue.

Diseases often result from germline or somatic mutations in the individual's DNA, such as are seen in sickle cell disease or cancer, respectively. These abnormalities can be detected by using two basic techniques of molecular genetics: the Southern blot and the polymerase chain reaction (PCR). For the Southern blot, high-molecular-weight DNA isolated from a specimen (most commonly peripheral-blood white cells) is digested by using an appropriate restriction endonuclease. The resulting fragments are then separated by gel electrophoresis and transferred to a nylon membrane, which is incubated with a solution containing a specific, labeled probe, also in single-stranded form. Probe-target hybrids formed by annealing of their complementary sequences can be detected by autoradiography or a colorimetric reaction. The Southern blot technique can detect DNA polymorphisms, mutations, or the presence of viral, bacterial, or specific sex chromosomes. In the polymerase chain reaction, short oligonucleotide primers flank the specific gene region or RNA sequence to be amplified and are combined with the target specimen and free nucleotides, which are synthesized into new DNA. An automated thermal cycler repeatedly alters the temperature to denature the target DNA, to allow the primers to reanneal to the target, and then to synthesize the product. This amplified polymerase chain reaction product is typically detectable as a band in a gel.

Understanding diseases at the genetic level has several advantages. Even when the gene's protein product is not expressed, definitive diagnoses can be made by using DNA-based techniques. Diseases that are similar clinically (phenotypically) can, in fact, be due to different mutations (genotypes) within a single gene or due to mutations in different genes, often related in an enzyme complex or as a portion of a group of structural proteins. Another advantage of molecular diagnosis is the ability to detect phenotypically normal carriers of genetic diseases in order to provide information for appropriate genetic counseling and prenatal diagnosis. *See* HUMAN GENETICS; MUTATION; PATHOLOGY. Mark A. Lovell

Molecular physics The study of the physical properties of molecules. Molecules possess a far richer variety of physical and chemical properties than do isolated atoms. This is attributable primarily to the greater complexity of molecular structure, as compared to that of the constituent atoms. Molecules also possess additional energy modes because they can vibrate; that is, the constituent nuclei oscillate about their equilibrium positions and rotate when unhindered. These modes give rise to additional spectroscopic properties, as compared to those of an atom; molecular spectroscopy in the optical, infrared, and microwave regions is one of the physical chemist's most powerful means of identifying and understanding molecular structure. Molecular spectroscopy has also given rise to the rapidly growing field of molecular astronomy.

Molecular physics is primarily concerned with the study of properties of isolated molecules, as contrasted to the more general study of molecular reactions, which is the domain of physical chemistry. Such properties, in addition to the broad field of spectroscopy, include electron affinities (for the formation of molecular negative ions); polarizabilities (the "distortability" of the molecule along its various symmetry axes by external electric fields); magnetic and electric multipole moments, attributable to the distributions of electric charge; currents and spins of the molecule; and the (nonreactive) interactions of molecules with other molecules, atoms, and ions. *See* INFRARED SPECTROSCOPY; INTERMOLECULAR FORCES; MICROWAVE SPECTROSCOPY; MOLECULAR BEAMS; MOLECULAR STRUCTURE AND SPECTRA; SPECTROSCOPY. Benjamin Bederson

Molecular recognition The ability of biological and chemical systems to distinguish between molecules and regulate behavior accordingly. How molecules fit together is fundamental in disciplines such as biochemistry, medicinal chemistry, materials science, and separation science. A good deal of effort has been expended in trying to evaluate the underlying intermolecular forces. The weak forces that act over short distances (hydrogen bonds, van der Waals interactions, and aryl stacking) provide most of the selectivity observed in biological chemistry and permit molecular recognition. The recognition event initiates behavior such as replication in nucleic acids, immune response in antibodies, signal transduction in receptors, and regulation in enzymes. Most studies of recognition in organic chemistry have been inspired by these biological phenomena. It has been the task of bioorganic chemistry to develop systems capable of such complex behavior with molecules that are comprehensible and manageable in size, that is, with model systems. *See* ANTIBODY; CHEMORECEPTION; ENZYME; HYDROGEN BOND; INTERMOLECULAR FORCES; NUCLEIC ACID; SYNAPTIC TRANSMISSION.

The advantage of cyclic structures lies in their ability to restrict conformation or flexibility. A rigid matrix of binding sites, that is, preorganized sites, is usually associated with high selectivity in binding. A flexible matrix tends to accept several binding partners. Although sacrificing selectivity, this has the advantage of transmitting conformational information and is relevant to biological signaling events.

Macrocyclic (crown) ethers can bind and transport ions and imitate biological processes involving macrolides. Large ring structures that are lined with oxygen present an inner surface which is complementary to the spherical outer surface of positively charged ions.

Cyclophane-type structures offer considerable rigidity because of the aromatic nuclei. Binding forces between host and guest are largely hydrophobic. A typical system is a cyclophane-naphthalene complex [structure (**1**)], in which a naphthalene

(1)

guest is bound by a water-soluble cyclophane derivative. Other macrocyclic structures include the cyclodextrins and hybrid structures assembled from macrocyclic subunits. *See* AROMATIC HYDROCARBON; COORDINATION COMPLEXES; NAPHTHALENE.

Because the encircling of larger, more complex molecules with macrocycles poses structural problems, other molecular shapes have been explored. Cleft molecules offer advantages in this regard. The principle underlying these systems involves the shape of the small organic target molecules: convex in surface and bearing functional groups that diverge from their centers. Accordingly, designing a trap for such targets requires molecules of a concave surface in which functional groups converge. This complementarity is also a feature of the immune system: the "hot spots" of an antigen tend to be convex, whereas the binding sites of the antibody are concave.

Systems featuring a cleft have been developed to bind adenine derivatives and other heterocyclic systems through chelation, as shown in (**2**).

R = ribose or deoxyribose

(2)

See CHELATION.

Apart from the abstract questions concerning articulation of molecules, some practical applications in the pharmaceutical industry may be envisioned. Many of the target structures are biologically active, and the use of synthetic sequestering agents for metabolic substrates can represent a novel approach to biochemical methods and drug delivery. Julius Rebek, Jr.

Molecular sieve Any one of the crystalline metal aluminosilicates belonging to a class of minerals known as zeolites. An important characteristic of the zeolites is their ability to undergo dehydration with little or no change in crystal structure. The dehydrated crystals are honeycombed with regularly spaced cavities interlaced by channels of molecular dimensions which offer a very high surface area for the absorption of foreign molecules.

The basic formula for all crystalline zeolites can be represented as

$$M_{2/n}O:Al_2O_3:xSiO_2:yH_2O$$

Molecular sieve type-A crystal model. Dark spheres represent the included cations, and light spheres the SiO_4 or AlO_4 tetrahedrons.

where M represents a metal ion and n its valence. The crystal structure consists basically of a three-dimensional framework of SiO_4 and AlO_4 tetrahedrons (see illustration). The tetrahedrons are cross-linked by the sharing of oxygen atoms, so that the ratio of oxygen atoms to the total of silicon and aluminum atoms is equal to 2. The electrovalence of the tetrahedrons containing aluminum is balanced by the inclusion of cations in the crystal. One cation may be exchanged for another by the usual ion-exchange techniques. The size of the cation and its position in the lattice determine the effective diameter of the pore in a given crystal species.

The properties of molecular sieves as adsorbents which distinguish them from nonzeolitic adsorbents are (1) the relatively strong coulomb fields generated by the adsorption surface and (2) the uniform pore size; the pore size is controlled, in a given crystal species, by the associated cation.

The basic characteristics of molecular sieves are utilized commercially in several production and research applications. Their absorption properties make them useful for drying, purification, and separations of gases and liquids. Conversely, molecular sieves can be preloaded with chemical agents, which are thereby isolated from the reactive system in which they are dispersed until released from the adsorbent either thermally or by displacement by a more strongly adsorbed compound. They are also used as cation exchange media and as novel catalysts and catalyst supports. *See* ADSORPTION; GAS CHROMATOGRAPHY; ION EXCHANGE; ZEOLITE. R. L. Mays

Molecular simulation A tool for predicting entirely computationally many useful functional properties of systems of interest in the chemical, pharmaceutical, materials, and related industries. Included are thermodynamic, thermochemical, spectroscopic, mechanical, and transport properties, and morphological information (such as location and shape of binding sites on a biomolecule and crystal structure).

The two main molecular simulation techniques are molecular dynamics and Monte Carlo simulation, both of which are rooted in classical statistical mechanics. Given mathematical models for the internal structure of each molecule (the intramolecular potential which describes the energy of each conformation of the molecule) and the interaction between molecules (the intermolecular potential which describes the energy associated with molecules being in a particular conformation relative to each other), classical statistical mechanics provides a formalism for predicting properties of a macroscopic collection of such molecules based on statistically averaging over the possible microscopic states of the system as it evolves under the rules of classical mechanics. Thus, the building blocks are molecules, the dynamics are described by classical mechanics, and the key concept is statistical averaging. In molecular dynamics, the microscopic states of the system are generated by solving the classical equations of motion as a function of time (typically over a period limited to tens of nanoseconds). Thus, one can observe the relaxation of a system to equilibrium (provided the time for the relaxation falls within the time accessible to molecular dynamics simulation), and so molecular dynamics permits the calculation of transport properties which at the macroscopic scale describe the relaxation of a system in response to inhomogeneities. In Monte Carlo simulation, equilibrium configurations of systems are generated stochastically according to the probabilities rigorously known from classical statistical mechanics. Thus, Monte Carlo simulation generates equilibrium states directly (which has many advantages, including bypassing configurations which are not characteristic of equilibrium but which may be difficult to escape dynamically) and so can be used to study equilibrium configurations of systems which may be expensive or impossible to access via molecular dynamics. The drawback of Monte Carlo simulation is that it cannot yield the kind of dynamical response information that leads directly to transport properties. *See* CHEMICAL DYNAMICS; COMPUTATIONAL CHEMISTRY; MONTE CARLO METHOD; SIMULATION; STOCHASTIC PROCESS.

Computational quantum chemistry and molecular simulation methods can be used to predict properties that once were only accessible experimentally, resulting in several significant applications in basic and industrial research. These applications include providing estimates of properties for systems for which little or no experimental data are available, which is especially useful in the early stages of chemical process design; yielding insight into the molecular basis for the behavior of particular systems, which is very useful in developing engineering correlations, design rules, or quantitative structure-property relations; and providing guidance for experimental studies by identifying the interesting systems or properties to be measured. Peter J. Cummings

Molecular structure and spectra Until the advent of quantum theory, ideas about the structure of molecules evolved gradually from analysis and interpretation of the facts of chemistry. Chemists developed the concept of molecules as built from atoms in definite proportions, and identified and constructed (synthesized) a great variety of molecules. Later, when the structure of atoms as built from nuclei and electrons began to be understood with the help of quantum theory, a beginning was made in explaining why atoms can combine in definite ways to form molecules; also, infrared spectra began to be used to obtain information about the dimensions and the nuclear motions (vibrations) in molecules. However, a fundamental understanding of chemical binding and molecular structure became possible only by application of the present form of quantum theory, called quantum mechanics. This theory makes it possible to obtain from the spectra of molecules a great deal of information about the nature of molecules in their normal as well as excited states, and about dissociation energies and other characteristics of molecules.

Molecular sizes. The size of a molecule varies approximately in proportion to the numbers and sizes of the atoms in the molecule. Simplest are diatomic molecules. These may be thought of as built of two spherical atoms of radii r and r', flattened where they are joined. The equilibrium value R_e of the distance R between their nuclei is then smaller than the sum of the atomic radii. However, the nuclei of atoms in two different molecules cannot normally approach more closely than a distance $r + r'$; r and r' are called the van der Waals radii of the atoms.

To describe a polyatomic molecule, one must specify not merely its size but also its shape or configuration. For example, carbon dioxide (CO_2) is a linear symmetrical molecule, the O—C—O angle being 180°. The H—O—H angle in the nonlinear water (H_2O) molecule is 105°. Many molecules which are essential for life contain thousands or even millions of atoms. Proteins are often coiled or twisted and cross-linked in curious ways which are important for their biological functioning.

Dipole moments. Most molecules have an electric dipole moment. In atoms, the electron cloud surrounds the nucleus so symmetrically that its electrical center coincides with the nucleus, giving zero dipole moment; in a molecule, however, these coincidences are disturbed, and a dipole moment usually results.

Thus, when the atoms of HCl come together, there is some shifting of the H-atom electron toward the Cl. A complete shift would give H^+Cl^-, which would constitute an electric dipole of magnitude eR_e, where e is the electronic charge. But in fact the dipole moment is only $0.17eR_e$. This is because the actual electronic shift is only fractional. *See* ELECTRONEGATIVITY.

Molecular polarizability. In the preceding consideration of dipole moments, the discussion has been in terms of atoms and molecules free from external forces. An electric field pulls the electrons of an atom or molecule in one direction and pushes the nuclei in the opposite direction. This action creates a small induced dipole moment, whose magnitude per unit strength of the field is called the polarizability.

Molecular energy levels. The stationary states of motion of nuclei and electrons in a molecule, or of electrons in an atom, are restricted by quantum mechanics to special forms with definite energies. (Nonstationary states, which vary in the course of time, are constructed by mixing stationary states of different energies.) The state of lowest energy is called the ground state; all others are excited states. In analogy to water levels in a river, the energies of the stationary states are called energy levels. Excited states exist only momentarily, following an electrical or other stimulus. *See* ENERGY LEVEL (QUANTUM MECHANICS); QUANTUM CHEMISTRY; QUANTUM MECHANICS.

Excitation of an atom consists of a change in the state of motion of its electrons. Electronic excitation of molecules can also occur, but alternatively or additionally, molecules can be excited to discrete states of vibration and rotation.

The total internal energy of a molecule is a sum of several terms: the largest is the electronic energy, E_{el}, followed by the vibrational energy, E_v. The total energy can be written as Eq. (1).

$$E = E_{el} + E_v + (E_r + E_{fs} + E_{hfs} + E_{ext}) \quad (1)$$

Both the electronic energy E_{el} and vibration energy E_v can be discrete or continuous. The quantities E_r, E_{fs}, and E_{hfs} denote rotational, fine-structure, and hyperfine-structure energies, respectively. The last two appear as small or minute splittings of the rotation levels. Fine-structure splittings are caused by any unpaired electrons, and hyperfine splittings result from the presence of nuclear spins. The spacings ΔE of adjacent discrete levels of each type are usually in the order given in Eq. (2).

$$\Delta E_{el} \gg \Delta E_v \gg \Delta E_r \gg \Delta E_{fs} \gg \Delta E_{hfs} \quad (2)$$

The E_{ext} term in Eq. (1) refers to additional fine structure which appears on subjecting molecules to external magnetic fields (Zeeman effect) or electric fields (Stark effect). *See* FINE STRUCTURE (SPECTRAL LINES); HYPERFINE STRUCTURE; STARK EFFECT; ZEEMAN EFFECT.

Polyatomic molecules have much more complicated patterns of vibrational and (usually) of rotational energy levels than diatomic molecules.

Molecular spectra. Radiation emitted by a molecule initially in an upper state level E' as it makes a transition to a lower state level E'' obeys the Einstein-Bohr equation (3). Molecular emis-

$$h\nu = E' - E'' \quad (3)$$

sion spectra accompany jumps in energy from higher to lower

levels; absorption spectra accompany transitions from lower to higher levels.

Molecular spectra can be classified as fine-structure or low-frequency spectra, rotation spectra, vibration-rotation spectra, and electronic spectra. *See* ELECTRON PARAMAGNETIC RESONANCE (EPR) SPECTROSCOPY; MAGNETIC RESONANCE; MICROWAVE SPECTROSCOPY; MOLECULAR BEAMS; SPECTROSCOPY.

Transitions between energy levels differing only in rotational state give rise to pure rotation spectra. These typically consist of a sequence of lines spaced almost equidistantly and lying in the far infrared or the microwave region.

Spectra involving only vibrational and rotational state changes lie mainly in the infrared. Each band consists of two sets of rotational lines, one on each side of a central frequency, which is called the band origin. Vibration absorption bands of liquids and solutions are widely used in chemical analysis. *See* INFRARED SPECTROSCOPY.

Electronic band spectra are the most general type of molecular spectra. For any one electronic transition, the spectrum consists typically of many vibrational bands. *See* ATOMIC STRUCTURE AND SPECTRA; INTERMOLECULAR FORCES; MOLECULAR WEIGHT; RAMAN EFFECT; RESONANCE (MOLECULAR STRUCTURE); SCATTERING EXPERIMENTS (ATOMS AND MOLECULES); VALENCE.

Robert S. Mulliken; U. Fano; Peter Bernath

Lasers and synchrotron sources. Modern sources of radiation have changed the field of molecular spectroscopy enormously. Ultrafast pulsed lasers can offer time resolutions of a few femtoseconds (10^{-15} s) and very high peak powers of the order of terawatts (10^{12} W). Laser linewidths of a few hundred hertz for continuous-wave semiconductor lasers operating at 445 terahertz (674 nm) are readily achieved so that the spectral resolution is limited by Doppler or pressure broadening of the molecular lines of the gas sample. Special laser techniques such as saturation spectroscopy are required to take advantage of these exceedingly narrow laser linewidths. *See* LASER; LASER SPECTROSCOPY.

Synchrotron radiation also has many advantages for molecular spectroscopy. It has long been used at short wavelengths, for example, in the photoionization of molecules and surfaces and in x-ray crystallography. More recently synchrotron radiation has been used also for infrared and far-infrared spectroscopy. The far-infrared radiation emitted from the edge of a bending magnet that turns the electron beam is many orders of magnitude brighter than that from a traditional thermal or arc source. In this context, brightness means power (watts) emitted per unit frequency interval, per unit emitting area, and per unit solid angle. The high brightness of the radiation can be used to improve the signal-to-noise ratio or the sensitivity of a typical infrared absorption measurement. Alternatively the high brightness can be used for very fast time resolution or for imaging.

Synchrotron radiation (like a laser) can be focused to a very small diffraction-limited spot (similar in size to the wavelength of the radiation) without excessive loss of radiation. This small spot can be used for hyperspectral imaging of biological or heterogeneous industrial samples, meaning that each pixel of a two-dimensional slice of a sample has an associated spectrum. *See* SYNCHROTRON RADIATION.

Peter Berrath

Molecular weight The sum of the atomic weights of all atoms making up a molecule. Actually, what is meant by molecular weight is molecular mass. The use of this expression is historical, however, and will be maintained. The atomic weight is the mass, in atomic mass units, of an atom. It is approximately equal to the total number of nucleons, protons and neutrons composing the nucleus. Since 1961 the official definition of the atomic mass unit (amu) has been that it is 1/12 the mass of the carbon-12 isotope, which is assigned the value 12.000 exactly. *See* ATOMIC MASS; ATOMIC MASS UNIT; RELATIVE ATOMIC MASS; RELATIVE MOLECULAR MASS.

A mole is an amount of substance containing the Avogadro number, N_A, approximately 6.022×10^{23}, of molecules or atoms. Molecule, in this definition, is understood to be the smallest unit making up the characteristic compound. Originally, the mole was interpreted as that number of particles whose total mass in grams was numerically equivalent to the atomic or molecular weight in atomic mass units, referred to as gram-atomic or gram-molecular weight. This is how the above value for N_A was calculated. As the ability to make measurements of the absolute masses of single atoms and molecules has improved, however, modern metrology is tending to alter its approach and define the Avogadro number as an exact quantity, thereby changing slightly the definition of the atomic mass unit and removing the need to define atomic weight with respect to a particular isotopic species. The latest and most accurate value for the Avogadro number is $6.02214179(30) \times 10^{23}$ mol^{-1}. *See* AVOGADRO NUMBER; MOLE (CHEMISTRY).

As the masses of all the atomic species are now well known, masses of molecules can be determined once the composition of the molecule has been ascertained. Alternatively, if the molecular weight of the molecule is known and enough additional information about composition is available, such as the basic atomic constituents, it is possible to begin to assemble structural information about the molecule. Thus, the determination of the molecular weight is one of the first steps in the analysis of an unknown species. Given the increasing emphasis on the study of biologically important molecules, particular attention has been focused on the determination of molecular weights of larger and larger units. There are a number of methods available, and the one chosen will depend on the size and physical state of the molecule. All processes are physical macroscopic measurements and determine the molecular weight directly. Connection to the absolute mass scale is straightforward by using the Avogadro number, although, for extremely large molecules, this connection is often unnecessary or impossible, as the accuracy of the measurements is not that good. The main function of molecular weight determination of large molecules is elucidation of structure.

Molecular weight determination of materials which are solid or liquid at room temperature is best achieved by taking advantage of one of the colligative properties of solutions, boiling-point elevation, freezing-point lowering, or osmotic pressure, which depend on the number of particles in solution, not on the nature of the particle. The choice of which to use will depend on a number of properties of the substance, the most important of which will be the size. All require that the molecule be small enough to dissolve in the solution but large enough not to participate in the phase change or pass through a semipermeable membrane. Freezing-point lowering is an excellent method for determining molecular weights of smaller organic molecules, and osmometry, as the osmotic pressure determination is called, for determining molecular weights of larger organic molecules, particularly polymeric species. Boiling-point elevation is used less frequently. *See* POLYMER.

The basis of all the methods involving colligative properties of solutions is that the chemical potentials of all phases must be the same. (Chemical potential is the partial change in energy of a system as matter is transferred into or out of it. For two systems in contact at equilibrium, the chemical potentials for each must be equal.) *See* CHEMICAL EQUILIBRIUM; CHEMICAL THERMODYNAMICS.

Another measurement from which molecular weights can be obtained is based on the scattering of light from the molecule. A beam of light falling on a molecule will induce in the molecule a dipole moment which in its turn will radiate. The interference between the radiated beam and the incoming beam produces an angular dependence of the scattered radiation which depends on the molecular weight of the molecule. This occurs whether the molecule is free or in solution. While the theory for this effect is complicated and varies according to the size of the molecule, the general result for molecules whose size is considerably less than that of the wavelength λ of the radiation (less than $\lambda/50$) is given by the equation below; $I(\theta)$ is the intensity of radiation at

$$\frac{I(\theta)}{I_0} = \text{constant } (1 + \cos^2\theta)\ Mc$$

angle θ, I_0 the intensity of the incoming beam, M the molecular weight, and c the concentration in grams per cubic centimeter of the molecule. If the molecules are much larger than $\lambda/50$ (about 9 nanometers for visible light), this relationship in this simple form is no longer valid, but the method is still viable with appropriate adjustments to the theory. In fact, it can be used in its extended version even for large aggregates. *See* SCATTERING OF ELECTROMAGNETIC RADIATION. C. Denise Caldwell

Molecule The smallest unit of a compound, which consists of atoms bonded together in a unique arrangement. A diatomic molecule consists of two atoms linked together by chemical bonds. Examples include the oxygen and nitrogen of the air (O_2 and N_2, respectively) and the neurotransmitter nitric oxide (NO). A polyatomic molecule consists of more than two linked atoms. Examples include water (H_2O), carbon dioxide (CO_2), ethanol (C_2H_5OH), and the vast molecule of deoxyribonucleic acid (DNA) with thousands of atoms.

The bonds that hold the atoms together are covalent. That is, they consist of pairs of electrons that are shared by the neighboring atoms. This type of bonding is in contrast to ionic bonding, in which there is complete transfer of one or more electrons from one atom to another and large numbers of the resulting ions (charged atoms) clump together, as in sodium chloride (NaCl). Some chemists refer to the smallest unit of an ionic compound (in the case of sodium chloride, a sodium ion, Na^+, and a chloride ion, Cl^-) as constituting a "molecule" of the compound, but this usage is uncommon and the term "formula unit" is preferred. *See* CHEMICAL BONDING; IONIC CRYSTALS.

Many molecules are organic. That is, they contain at least one atom and commonly many atoms of carbon. Molecules that do not contain carbon, with a few exceptions, are inorganic. The principal exception to this general classification is carbon dioxide, which is treated as an honorary inorganic compound. The term "organic" arose when it was thought erroneously that such compounds could be produced only by living organisms.

The bonds between atoms may be single, double, or triple (in rare cases, quadruple), in which neighboring atoms share one, two, or three electron pairs. Broadly speaking, the greater the multiplicity of bonds, the more tightly bonded the atoms.

Molecules have not only a characteristic atomic composition but also, because the atoms are linked together in a characteristic array, a characteristic shape. Thus, H_2O is an angular molecule in which the two OH bonds make an angle of about 104° to each other, and CO_2 is linear (all three atoms in a straight line). The four bonds that carbon commonly forms (as in methane, CH_4) are typically arranged in a tetrahedral array with bond angle close to 109°.

The tetrahedral geometry of the bonds to carbon applies only when all the bonds are single. Moreover, neighboring groups of atoms are free to rotate around a single bond, which acts as a miniature axle. As a result, a polyatomic molecule consisting of many singly bonded atoms should not be thought of as a rigid framework, but as ceaselessly writhing and twisting into different shapes or "conformations." In contrast, the presence of a multiple bond confers structural rigidity because the neighboring groups of atoms joined by a double bond are not free to rotate relative to one another.

Double bonds between carbon atoms play important roles in physiology and industry. The primary act of vision, for instance, can be traced to the properties of double bonds. The industrial importance of double bonds between carbon atoms lies in the reactivity they confer on organic molecules (in contrast to the inertness conferred on nitrogen). One aspect of this reactivity is the ability to combine small double-bonded organic molecules into long chains. That is, certain molecules act as monomers that may be polymerized to form the plastics characteristic of the modern world. *See* PHOTORECEPTION; POLYMER. Peter Atkins

Mollusca A major phylum of the animal kingdom comprising an extreme diversity of external body forms (oysters, clams, chitons, snails, slugs, squid, and octopuses among others), all based on a remarkably uniform basic plan of structure and function. The phylum name is derived from *mollis*, meaning soft, referring to the soft body within a hard calcareous shell, which is usually diagnostic. Soft-bodied mollusks make extensive use of ciliary and mucous mechanisms in feeding, locomotion, and reproduction. Most molluscan species are readily recognizable as such.

The Mollusca constitute a successful phylum; there are probably over 110,000 living species of mollusks, a number second only to that of the phylum Arthropoda, and more than double the number of vertebrate species. More than 99% of living molluscan species belong to two classes: Gastropoda (snails) and Bivalvia. Ecologically, these two classes can make up a dominant fraction of the animal biomass in many natural communities, both marine and fresh-water.

Classification. The phylum Mollusca is divided into seven distinct extant classes, three of which (Gastropoda, Bivalvia, and Cephalopoda) are of major significance in terms both of species numbers and of ecological bioenergetics, and one extinct class. An outline of their classification follows.

Class Monoplacophora (mainly fossil; but
  one living genus *Neopilina*)
Class Aplacophora
  Subclass Neomeniomorpha
  Subclass Chaetodermomorpha
Class Polyplacophora
Class Scaphopoda
Class Rostroconchia (fossil only)
Class Gastropoda
  Subclass Prosobranchia
    Order: Archaeogastropoda
      Mesogastropoda
      Neogastropoda
  Subclass Opisthobranchia
    Order: Bullomorpha (or Cephalaspidea)
      Aplysiomorpha (or Anaspidea)
      Thecosomata
      Gymnosomata
      Pleurobranchomorpha
        (or Notaspidea)
      Acochlidiacea
      Sacoglossa
      Nudibranchia (or Acoela)
  Subclass Pulmonata
    Order: Systellommatophora
      Basommatophora
      Stylommatophora
Class Bivalvia (or Pelecypoda)
  Subclass Protobranchia
  Subclass Lamellibranchia
    Order: Taxodonta
      Anisomyaria
      Heterodonta
      Schizodonta
      Adapedonta
      Anomalodesmata
  Subclass Septibranchia

Class Cephalopoda
Subclass Nautiloidea
Subclass Ammonoidea (fossil only)
Subclass Coleoidea
Order: Belemnoidea
Sepioidea
Teuthoidea
Vampyromorpha
Octopoda

Functional morphology. The unique basic plan of the Mollusca involves the different modes of growth and of functioning of the three distinct regions of the molluscan body (see illustration). These are the head-foot with some nerve concentrations, most of the sense organs, and all the locomotory organs; the visceral mass (or hump) containing organs of digestion, reproduction, and excretion; and the mantle (or pallium) hanging from the visceral mass and enfolding it and secreting the shell. In its development and growth, the head-foot shows a bilateral symmetry with an anterioposterior axis of growth. Over and around the visceral mass, however, the mantle-shell shows a biradial symmetry, and always grows by marginal increment around a dorsoventral axis. It is of considerable functional importance that a space is left between the mantle-shell and the visceral mass forming a semi-internal cavity; this is the mantle cavity or pallial chamber within which the typical gills of the mollusk, the ctenidia, develop. This mantle cavity is almost diagnostic of the phylum; it is primarily a respiratory chamber housing the ctenidia, but with alimentary, excretory, and genital systems all discharging into it.

In looking at any mollusk, it is important to realize that whatever the shape of the shell, it is always underlain by the mantle, a fleshy fold of tissues which has secreted it. The detailed structure of the shell and of the mantle edge (with three functionally distinct lobes) is also consistent throughout the Mollusca. The shell is made up of calcium carbonate crystals enclosed in a meshwork of tanned proteins. It is always in three layers: the outer periostracum, the prismatic layer and the innermost or nacreous layer.

Each of the eight classes of the Mollusca has a characteristic body form and shell shape. Two classes are enormous (Gastropoda and Bivalvia), one of moderate extent (Cephalopoda), the others being minor by comparison. The Gastropoda constitute a diverse group with the shell usually in one piece. This shell may be coiled as in typical snails—that is, helicoid or turbinate—or it may form a flattened spiral, or a short cone as in the limpets, or it may be secondarily absent as in the slugs. Most gastropods are marine, but many are found in fresh waters and on land; in fact, they are the only successful nonmarine mollusks.

The Bivalvia are a more uniform group, with the shell in the form of two calcareous valves united by an elastic hinge ligament. Mussels, clams, and oysters are familiar bivalves. The group is mainly marine with a few genera in estuaries and in fresh waters. There can be no land bivalves since their basic functional organization is as filter feeders. The third major group, the Cephalopoda, includes the most active and most specialized mollusks. There is a chambered, coiled shell in *Nautilus* and in many fossil forms; this becomes an internal structure in cuttlefish and squids, and is usually entirely absent in octopods.

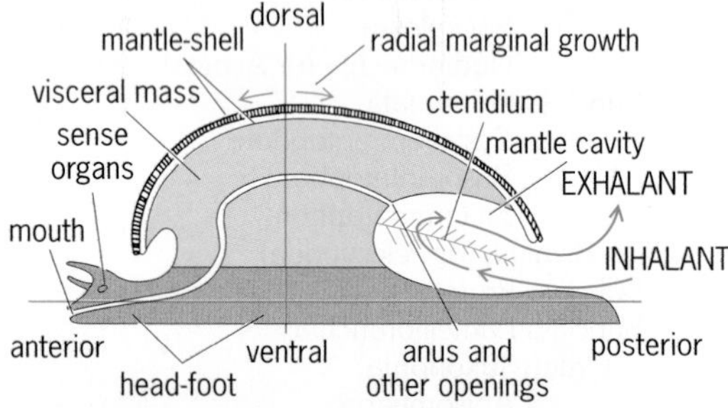

Generalized model of a stem mollusk (or archetype) in side view. There are three distinct regions in the molluscan body: head-foot, visceral mass, and mantle-shell. Water circulation through the mantle cavity, gills (ctenidia), and pallial complex is from ventral inhalant to dorsal exhalant. (*After W. D. Russell-Hunter, A Life of Invertebrates, Macmillan, 1979*)

A diversity of gill patterns have evolved in the major molluscan groups, paralleling the evolution of the mantle-shell patterns. The more advanced gastropods show reduction from a pair of aspidobranch ctenidia to a single one, and from that to a one-sided pectinibranch ctenidium (or comb gill), and subsequently to no gill at all in the pulmonate snails. The bivalves show enlargement of gill leaflets to longer filaments and their subsequent folding into the true lamellibranch condition, used in filter feeding. The gills in the cephalopods, while still structurally homologous, are modified with new skeletal elements to resist the stresses of water pumping by muscles.

Besides the gills, the other organs of the mantle cavity (termed collectively the pallial complex) again show morphological and functional consistency throughout the main groups of the Mollusca. The ctenidia form a curtain functionally dividing the mantle cavity into an inhalant part (usually ventral) containing the osphradia (pallial sense organs which sample the incoming water), and an exhalant part (usually dorsal) containing hypobranchial glands and both the anus and the openings of the kidney and genital ducts.

The cardiac structures of mollusks are also closely linked to the pallial complex. If there is a symmetrical pair of ctenidia, there will be a symmetrical pair of auricles on either side of the muscular ventricle of the heart; if one ctenidium, one auricle; if four ctenidia, four auricles. Note that body fluids in mollusks are almost all blood, just as body cavities are almost all hemocoel.

The respiratory pigment is usually hemocyanin in solution, so that neither circulatory efficiency nor blood oxygen-carrying capacity is high. However, mollusks are mostly sluggish animals with low metabolic (and hence respiratory) rates.

Uniquely molluscan is the use of cilia in "sorting surfaces," which can segregate particles into different size categories and send them to be disposed of in different ways in several parts of the organism. In a simpler type of sorting surface, the epithelium is thrown into a series of ridges and grooves, the cilia in the grooves beating along them and the cilia on the crests of the ridges beating across them. Thus, fine particles impinging on the surface can be carried in the direction of the grooves, while larger particles are carried at right angles. Such sorting surfaces occur both externally on the feeding organs and internally in the gut of many mollusks. For example, on the labial palps of bivalves, they are used to separate the larger sand grains (which are rejected) from the smaller microorganisms which then pass to the mouth.

The range in levels of complexity of molluscan nervous systems is comparable to that found in the phylum Chordata. The four-strand nervous system with one pair of tiny ganglia found in chitons is not dissimilar to the neural plan in turbellarian flatworms. In contrast, the nervous system and sense organs of a cephalopod like an octopus are equaled and exceeded only by those of some birds and mammals. In the majority of mollusks the nervous system is in an intermediate condition. In mollusks other than cephalopods, the main effectors controlled by the nervous system are cilia and mucous glands. In fact, apart from the muscles which withdraw it into its shell, the typical mollusk is a slow-working animal with little fast nervous control or quick reflexes. In the brain of modern cephalopods, paired ganglia have been fused into a massive structure, with over 300 million neurons and extensive "association" centers providing considerable mnemic and learning capacities. *See* OCTOPUS.

In all primitive mollusks, the sexes are separate, and external fertilization follows the spawning of eggs and sperm into the sea.

In more advanced mollusks, eggs are larger (and fewer), fertilization may become internal (with complex courtship and copulatory procedures), and larval stages may be sequentially suppressed. A remarkably large number of mollusks (including many higher snails) are hermaphroditic. Although some are truly simultaneous hermaphrodites, many more show various kinds of consecutive sexuality. Most often the male phase occurs first, and these species are said to show protandric hermaphroditism.

Distributional ecology. Mollusks are largely marine. The extensive use of ciliary and mucous mechanisms in feeding, locomotion, reproduction, and other functions demands a marine environment for the majority of molluscan stocks. Apart from a small number of bivalve genera living in brackish and freshwaters, all nonmarine mollusks are gastropods.

Despite the soft, hydraulically moved bodies and relatively permeable skins typical of all mollusks, some snails are relatively successful as land animals, although they are largely limited to more humid habitats. The primary physiological requirements for life on land concern water control, conversion to air breathing, and temperature regulation.

In the sea, all classes of mollusks are found, and all habitats have mollusks. Protobranchiate bivalves are found at depths of over 30,000 ft (9000 m). Although ecologically cephalopod mollusks are limited to the sea, there are sound reasons for claiming modern cephalopods as the most highly organized invertebrate animals. The functional efficiencies of jet propulsion and of massive brains in squid, cuttlefish, and octopuses have not been paralleled in their other physiological systems.

In addition to the extreme diversity of external body form exhibited by different mollusks, they show a remarkable diversity in their ecological distribution and life styles. However, the basic molluscan plan of structure and function always remains recognizable. *See* APLACOPHORA; BIVALVIA; CEPHALOPODA; GASTROPODA; LAMELLIBRANCHIA; MONOPLACOPHORA; POLYPLACOPHORA; SCAPHOPODA.

W. D. Russell-Hunter

Molybdenum

Molybdenum A chemical element, Mo, atomic number 42, and atomic weight 95.96, in the periodic table in the triad of transition elements that includes chromium (atomic number 24) and tungsten (atomic number 74). Research has revealed it to be one of the most versatile chemical elements, finding applications not only in metallurgy but also in paints, pigments, and dyes; ceramics; electroplating; industrial catalysts; industrial lubricants; and organometallic chemistry. Molybdenum is an essential trace element in soils and in agricultural fertilizers. Molybdenum atoms have been found to perform key functions in enzymes (oxidases and reductases), with particular interest being directed toward its role in nitrogenase, which is employed by bacteria in legumes to convert inert nitrogen (N_2) of the air into biologically useful ammonia (NH_3). *See* NITROGEN FIXATION; PERIODIC TABLE.

Molybdenum is widely distributed in the Earth's crust at a concentration of 1.5 parts per million by weight in the lithosphere and about 10 parts per billion in the sea. It is found in at least 13 minerals, mainly as a sulfide [molybdenite (MoS_2)] or in the form of molybdates [for example, wulfenite ($PbMoO_4$) and magnesium molybdate ($MgMoO_4$)].

Although molybdenum is closer to chromium in atomic weight and atomic number, its chemical behavior is usually very similar to that of tungsten, which has nearly the same atomic radius. (This is due to the so-called lanthanide contraction in which atomic radii decrease for elements 57 to 71 found in the period between molybdenum and tungsten.) *See* CHROMIUM; LANTHANIDE CONTRACTION; TUNGSTEN.

Molybdenum atoms contain six valence electrons ($4d^55s^1$), which are employed with great versatility in forming compounds and complexes in which electronic configurations vary from d^0 (no *d* electrons in oxidation state + 6) to d^8 (8 *d* electrons in oxidation state -2). The $+6$ state is preferred, but all states from -2 to $+6$ are known. States usually exhibit a variety of coordination numbers (4 to 9), and include polynuclear complexes and metal-metal bonds in metallic clusters with two to six metal atoms in their metallic cores. Molybdenum forms a very large number of compounds with oxygen. Low-valent molybdenum [for example, $Mo(CO)_6$ and Mo_2, Mo_3, and Mo_6 clusters] has a very rich organometallic chemistry, including clusters that are being studied as models for molybdenum metal surfaces that catalyze organic reactions employed in industrial syntheses and oil refining. The ability of molybdenum atoms to vary oxidation state, coordination number, and coordination geometry and to form metal-metal bonds in clusters accounts in part for the large number of industrial catalysts and biological enzymes in which Mo atoms are found at the active site for catalysis. *See* CHEMICAL BONDING; COORDINATION CHEMISTRY; ELECTRON CONFIGURATION.

Physical properties of molybdenum metal

Property	Value
Density	10.22 g/cm^3 (5.911 oz/in.^{3})
Heat of vaporization	491 kJ/mol
Heat of fusion	28 kJ/mol
Specific heat	0.267 J/g°C
Thermal conductivity	1.246 J/s/cm^2/cm°C (200°C) 0.923 J/s/cm^2/cm°C (2200°C)
Electrical conductivity	34% International Copper Standard
Electrical resistivity	5.2 microhm-cm, 20°C 78.2 microhm-cm, 2525°C
Magnetic susceptibility	0.93×10^{-6} emu, 25°C 1.11×10^{-6} emu, 1825°C
Mean linear expansion coefficient	6.65×10^{-6}/°C, 20–1600°C
Modulus of elasticity	0.324 N/m^2
Lattice parameter	0.314767 nm (body-centered cube)

Molybdenum is a high-melting silver-gray metal, strong even at high temperatures, hard, and resistant to corrosion (see table). It also exhibits high conductivity, a high modulus of elasticity, high thermal conductivity, and a low coefficient of expansion. Its major use is in alloy steels, for example, as tool steels ($\leq$10% molybdenum), stainless steel, and armor plate. Up to 3% molybdenum is added to cast iron to increase strength. Up to 30% molybdenum may be added to iron-, cobalt-, and nickel-based alloys designed for severe heat- and corrosion-resistant applications. It may be used in filaments for light bulbs, and it has many applications in electronic circuitry. *See* ALLOY; IRON ALLOYS; STAINLESS STEEL.

Molybdenum trioxide, molybdates, sulfo-molybdates, and metallic molybdenum are found in thousands of industrial catalysts used in oil refining, ammonia synthesis, and industrial syntheses of organic chemicals. Monomeric molybdenum(IV) in aqueous solution is a powerful catalyst for the reduction of inert oxo-anions such as perchlorate (ClO_4^-) or nitrate (NO_3^-) as well as other oxidized nonmetals such as azide ion (N_3^-) and dinitrogen (N_2). The trinuclear cation $MO_3O_4^{4+}$ is inert, unreactive, and noncatalytic. *See* CATALYSIS; HOMOGENEOUS CATALYSIS.

The molybdenum enzymes comprise two major categories. The first category contains the single, highly important enzyme nitrogenase, which is responsible for biological nitrogen fixation. The second category contains all other known molybdenum enzymes, which are crucial for the metabolism of bacteria, plants, and animals, including humans.

Edward I. Stiefel

Molybdenum alloys

Molybdenum alloys Solid solutions of molybdenum and other metals. Molybdenum is classified as a refractory metal by virtue of its high melting point (2623°C or 4750°F), and many of its applications result from its strength at high temperatures. A number of other physical and mechanical properties make it attractive for use in a wide variety of applications. Molybdenum is used extensively as electrodes in electric-boost furnaces because it erodes very slowly and does not contaminate the glass bath. Its high-temperature strength allows it to support significant structural loads imposed during operation of the furnaces. *See* MOLYBDENUM; REFRACTORY.

Four main classes of commercial molybdenum-base alloys exist. The most common of the carbide-strengthened alloys is known as TZM, containing about 0.5% titanium, 0.08% zirconium, and 0.03% carbon. Other alloys in this class include TZC (1.2% titanium, 0.3% zirconium, 0.1% carbon), MHC (1.2% hafnium, 0.05% carbon), and ZHM (1.2% hafnium, 0.4% zirconium, 0.12% carbon). The high-temperature strength imparted by these alloys is their main reason for existence. Both TZM and MHC have found application as metalworking tool materials. Their high-temperature strength and high thermal conductivity make them quite resistant to the collapse and thermal cracking that are common failure mechanisms for tooling materials.

Tungsten and rhenium are the two primary solid-solution alloys. The most common compositions are 30% tungsten (Mo-30W), 5% rhenium (Mo-5Re), 41% rhenium (Mo-41Re), and 47.5% rhenium (Mo-50Re). With the exception of the Mo-30W alloy which is available as a vacuum-arc-cast product, these alloys are normally produced by powder metallurgy. The tungsten-containing alloys find application as components in systems handling molten zinc, because of their resistance to this medium. They were developed as a lower-cost, lighter-weight alternative to pure tungsten and have served these applications well over the years. The 5% rhenium alloy is used primarily as thermocouple wire, while the 41% and 47.5% alloys are used in structural aerospace applications. *See* RHENIUM; TUNGSTEN.

The beneficial effects of solid-solution hardening and dispersion hardening found in the carbide-strengthened alloys have been combined in the HWM-25 alloy (25% tungsten, 1% hafnium, 0.07% carbon). This alloy offers high-temperature strength greater than that of carbide-strengthened molybdenum, but it has not found wide commercial application because of the added cost of tungsten and the expense of processing the material.

Dispersion-strengthened alloys rely exclusively on powder metallurgy manufacturing techniques. This allows the production of fine stable dispersions of second phases that stabilize the wrought structure against recrystallization, resulting in a material having improved high-temperature creep strength as compared to pure molybdenum. Once recrystallization occurs, the dispersoids also stabilize the interlocked recrystallized grain structure. This latter effect produces significant improvements in the ductility of the recrystallized material.

The potassium- and silicon-doped alloys such as MH (150 ppm potassium, 300 ppm silicon) and KW (200 ppm potassium, 300 ppm silicon, 100 ppm aluminum) are the oldest of this category; they are analogs to the doped tungsten alloys in common use for tungsten lamp filament. John A. Shields

Moment of inertia A measure of the "opposition to rotation" around a fixed axis that a rigid body presents to a torque applied about that axis. The concept is used in analyzing the dynamics of systems in which rotation occurs, for example, in attitude control of spacecraft. It extends naturally into quantum mechanics and the notion of quantization of rotational energy.

The simplest system to illustrate moment of inertia is the rigid rotor lying in a horizontal plane. This is a rigid massless rod of length r, pivoted at the origin of a coordinate system, at the end of which is a point mass m. The state of the system is specified by one quantity: the angle coordinate ϕ. The angular velocity ω is given by Eq. (1). It follows from Newton's first law of motion that

$$\omega \equiv \frac{d\phi}{dt} \tag{1}$$

if no external forces are applied, ω remains constant. However, if a torque of magnitude L is applied to the system, centered on the origin, then from Newton's second law of motion it can be shown that L is related to the angular acceleration by Eq. (2). For

$$L = mr^2 \frac{d\omega}{dt} \tag{2}$$

any given system, the factor preceding the derivative is defined as the moment of inertia, usually symbolized by the letter I. Thus, the angular acceleration is given by Eq. (3). Evidently, for the rigid rotor, the moment of inertia is given by Eq. (4).

$$\frac{d\omega}{dt} = \frac{L}{I} \tag{3}$$

$$I_{\text{rotor}} = mr^2 \tag{4}$$

See ACCELERATION; VELOCITY; TORQUE.

For a rigid system of N mass particles lying in the plane, the total moment of inertia is simply the sum of the individual moments of inertia, that is, the sum given by Eq. (5).

$$I_{\text{total}} = \sum_{i=1}^{n} m_i r_i^2 = \sum_{i=1}^{n} I_t \tag{5}$$

The radius of gyration k of a body is defined as the radius at which all its mass could be concentrated to yield the same moment of inertia; that is, k is given by Eq. (6).

$$k \equiv \sqrt{\frac{I}{m}} \tag{6}$$

The angular momentum of a rotating system is given by Eq. (7), and its kinetic energy of rotation is given by Eq. (8).

$$J = I\omega \tag{7}$$

$$T_{\text{rotor}} = \tfrac{1}{2} I \omega^2 \tag{8}$$

See ANGULAR MOMENTUM; ENERGY.

These principles apply equally well to rigid bodies, or to systems of rigidly connected masses in three dimensions. The situation is more complex because not only can one choose any line in space as an axis of rotation, but the body itself may have any orientation with respect to that axis. However, analysis of rotation dynamics is greatly simplified for such systems in two ways. *See* RIGID-BODY DYNAMICS.

The first simplification is by the parallel-axis theorem. Assume that a rigid body of mass M is spun about an axis that passes through its center of mass and for which the moment of inertia is I_{COM}. Now consider a new axis, parallel to the first but shifted to a perpendicular distance R away. The parallel-axis theorem states that the new moment of inertia is given by Eq. (9).

$$I' = I_{\text{COM}} + MR^2 \tag{9}$$

See CENTER OF MASS.

The second simplification arises from the fact that no matter what shape the body may have, only six quantities are needed to allow full analysis. These are the components of the matrix called the moment of inertia tensor $\mathbf{I}$, given by Eq. (10), where the terms

$$\mathbf{I} = \begin{pmatrix} I_{xx} & I_{xy} & I_{xz} \\ I_{xy} & I_{yy} & I_{yz} \\ I_{xz} & I_{yz} & I_{zz} \end{pmatrix} \tag{10}$$

on the diagonals are the moments of inertia for rotation about the x, y, and z axes, respectively, and the off-diagonal quantities are the products of inertia.

It can be shown that the matrix can always be transformed, through rotation of the assumed coordinate axes, such that the products of inertia are reduced tozero. The resulting axes are known as the principal axes, corresponding to which are the principal moments of inertia, I_a, I_b, and I_c. Andrej Tenne-Sens

Momentum A foundational quantity in physics that is always conserved in isolated systems, defined in classical mechanics as the product of mass and velocity. An equivalent definition that is better suited to modern physics is that momentum is the rate of change of energy with respect to velocity: this allows extending the conservation laws to include entities of zero mass. *See* CLASSICAL MECHANICS.

The word momentum by itself usually means linear momentum, the vector quantity obtained by multiplying (or scaling) the particle's velocity vector by its mass. The linear momentum of a system is the sum of the linear momenta of the particles composing it. By Newton's second law of motion, a particle's linear momentum increases with time in direct proportion to the applied force. *See* DYNAMICS; FORCE; LINEAR MOMENTUM; MOTION; NEWTON'S LAWS OF MOTION.

A particle's angular momentum is a vector quantity defined as its moment of (linear) momentum about a specified axis. The angular momentum of a system is the sum of the angular momenta (about the same axis) of the particles composing it. A system's angular momentum increases with time in direct proportion to the total applied moment of force, or torque, about the axis. *See* ANGULAR MOMENTUM; ROTATIONAL MOTION; TORQUE.

The concept of momentum is of fundamental importance because in isolated physical systems (that is, ones not acted on by some outside influence) both linear and angular momentum do not change with time: they are said to be conserved. (This statement also applies to total system energy.) The law of conservation of linear momentum comes from the fact that the laws of physics do not depend on where the system is located (the universe is homogeneous). The law of conservation of angular momentum comes from the fact that the laws of physics do not depend on the orientation of the system (the universe is isotropic). The two conservation laws of momentum along with the law of conservation of energy are the three foundational axioms of classical physics. *See* CONSERVATION LAWS (PHYSICS); CONSERVATION OF ENERGY; CONSERVATION OF MOMENTUM; KINEMATICS; MECHANICS; SYMMETRY LAWS (PHYSICS).

When velocities approach the speed of light, observed mass increases and Newton's laws of motion must give way to Einstein's more accurate laws of relativity. Nevertheless, the same form of law of conservation of linear momentum remains valid if the relativistic increase of mass is taken into account. Relativity views space and time united into a single entity, called spacetime, and correspondingly unites momentum and energy into momentum-energy, or momenergy. *See* RELATIVITY. Andrej Tenne-Sens

Monazite A rare mineral that incorporates the light rare-earth elements (lanthanum, cerium, praseodymium, neodymium, promethium, samarium, europium, gadolinium) and also yttrium. Monazite has a general formula of (La,Ce, Nd)PO_4, but Pr, Sm, Eu, Gd, and Y substitute for La, Ce, and Nd in solid solution in minor amounts. The dominant rare-earth element in a particular monazite is denoted by the atomic suffix, such as monazite-(Ce) in which cerium exists in amounts greater than other rare-earth atoms. Monazite-(Ce), monazite-(La), and monazite-(Nd) are officially recognized by the International Mineralogical Association. *See* CERIUM; MINERAL; RARE-EARTH ELEMENTS; YTTRIUM.

The atomic arrangement of monazite is formed of a packing arrangement of (PO_4) tetrahedra and distorted (REO_9) polyhedra, where RE = the rare-earth elements in the particular monazite mineral. The arrangement is formed of chains of alternating phosphate tetrahedra and RE polyhedra, parallel to the *c* axis. Monazite is similar in structure and chemistry to the tetragonal mineral xenotime, Y(PO_4), that selectively incorporates the heavy rare-earth elements. *See* PHOSPHATE MINERALS.

Monazite is variably green, yellow, brown, or red-brown, and rarely occurs in crystals large enough to discern with the unaided eye. Mohs hardness is 5–5.5, and the specific gravity is 4.6–5.5, varying with substitution of different elements.

Monazite is one of the main ore minerals for the rare-earth elements that are used in the manufacture of television and computer screens, fluorescent light bulbs, and highly efficient batteries, among other industrial applications.

John M. Hughes; John Rakovan

Mongoose Any of a group of 38 species of carnivorous mammals in the family Herpestidae. Mongooses inhabit a broad band along both sides of the Equator from western Africa to Southeast Asia. Although native to southern Asia, the East Indies, Africa, and Madagascar, one genus (*Herpestes*) was introduced into Spain and Portugal as well as to other parts of Europe and to many islands around the world including the West Indies, Fiji, and the Hawaiian Islands.

Banded mongoose, *Mungos mungo*. (*Photo by Gerald and Buff Corsi; © California Academy of Sciences*)

Mongooses are long slender carnivores with long faces, small rounded ears, short legs, and long tapering bushy tails. Most have long coarse, usually grizzled fur and lack body markings, although a few have stripes (see illustration). Color ranges from dark gray through brown to yellowish or reddish. Most species have a large anal sac containing at least two glandular openings. Adult mongooses have a head–body length of 24–60 cm (9.5–23 in.), a tail length of 19–44 cm (7.5–17 in.), and a weight ranging from 320 g to 5 kg (11 oz to 11 lb).

Mongooses, particularly the genus *Herpestes*, have been introduced into many regions of the world. Originally introduced to help control populations of rodents and venomous snakes in sugarcane plantations, they quickly multiplied and spread to become pests. They have destroyed not only the rats and snakes but also harmless reptiles, birds, and mammals. They have contributed to the extinction or endangerment of many desirable species of wildlife, as well as preying upon domestic poultry and consuming eggs. Now the importation or possession of mongooses is forbidden by law in some countries, including the United States and Australia. *See* CARNIVORA; SCENT GLAND. Donald W. Linzey

Monhysterida An order of nematodes in which, generally, the stoma is funnel shaped and lightly cuticularized; however, in some families the stoma is spacious and heavily cuticularized, and is armed with protrusible teeth. The amphids vary from simple spirals to circular forms. Usually the second and third circlets of cephalic sensilla are combined, but in some taxa the third circlet of four distinct setae is separate. The normal pattern of distribution is often disrupted by numerous cervical setae. The near-cyclindrical esophagus is sometimes swollen posteriorly. The cuticle may be smooth or may have annuli or ornamentation. When the annuli are distinct, the somatic setae may be long and in four to eight longitudinal rows. The female gonads are outstretched and either single or paired.

There are three monhysterid superfamilies: The Linhomoeoidea and Siphonolaimoidea are primarily marine forms; feeding habits among the groups are unknown. The Monhysteroidea are free-living nematodes found in all environments, from marine waters to fresh waters and soil; feeding habits are unknown. *See* NEMATA (NEMATODA). Armand R. Maggenti

Monkey An adaptive or evolutionary grade among the primates, represented by members of two of the three modern anthropoid superfamilies. The New World, platyrrhine monkeys (Ateloidea) and Old World, catarrhine forms (Cercopithecoidea) probably reached a monkey level of adaptation independently some time after their separation from a common ancestor, perhaps 45 million years ago. The term monkey is not indicative of taxonomic or phylogenetic relationship: the closest relatives of the cercopithecoids are not the ateloid monkeys but the Old World apes and humans. *See* APES; FOSSIL APES; FOSSIL PRIMATES; PRIMATES.

The Ateloidea comprise two extant families, while the living Cercopithecoidea are considered today to comprise only one family, with two subfamilies. A modern classification of the Anthropoidea follows:

Hyporder Anthropoidea
  Infraorder Platyrrhini
    Superfamily Ateloidea (New World or platyrrhine monkeys)
      Family Atelidae
        Subfamily Atelinae (howler and spider monkeys)
        Subfamily Pitheciinae (saki, owl, and titi monkeys)
      Family Cebidae
        Subfamily Cebinae (capuchin and squirrel monkeys)
        Subfamily Callitrichinae (marmosets and tamarins)
      Family Branisellidae (extinct early ateloids)
  Infraorder Catarrhini (Old World anthropoids)
    Parvorder Eucatarrhini (modern catarrhines)
      Superfamily Hominoidea (gibbons, great apes, and humans)
      Superfamily Cercopithecoidea
        Family Cercopithecidae (Old World or catarrhine monkeys)
          Subfamily Cercopithecinae (cheek-pouched monkeys: macaques, baboons, geladas, mangabeys, and guenons)
          Subfamily Colobinae (leaf eaters: langurs and colobus)
          Subfamily Victoriapithecinae (extinct early cercopithecids)
    Parvorder Eocatarrhini (archaic catarrhines)
        Family Propliopithecidae (early archaic catarrhines)
        Family Pliopithecidae (later archaic catarrhines)
  Infraorder Paracatarrhini (extinct early anthropoids)
        Family Oligopithecidae (extinct archaic anthropoids)
        Family Parapithecidae (extinct Egyptian monkeys)

Monkeys are hard to characterize as a group because of their great diversity, and because much of the discussion reflects a comparison with the apes. Both monkeys and apes contrast with the prosimian grade in that they are typically large, diurnal animals that live in social groups. Monkeys differ from apes in their possession of a tail, a smaller brain, quadrupedal pronograde (with long axis of body horizontal) posture, and a usually longer face. They are generally smaller than apes, but large monkeys outweigh gibbons. Like almost all primates, monkeys are pentadactyl, with nails rather than claws on the digits in most cases. They have pectoral mammary glands and well-developed vision. Monkeys are primarily vegetarian and inhabit forested tropical or subtropical regions of Africa, Asia, and South America. The differences between the New and Old World monkeys are summarized in the table.

Old World species are found throughout all the warmer regions of the Eastern Hemisphere except Australia and Madagascar. Many of the familiar monkeys are included in this family, such as the rhesus macaque, Barbary "ape," proboscis monkey, baboon, and mandrill.

The New World monkeys, or ateloids, occupy forested areas from southern Mexico to Argentina. They are divided into two main groups, or families (their major characteristics are given in the table). All are arboreal, including a few with prehensile tails; there is no living form, or any evidence of a fossil form, that has come to the ground habitually. Familiar monkeys include

Major contrasts between New and Old World monkeys and special features of each

New World species (Ateloidea)	Old World species (Cercopithecoidea)
Nose platyrrhine (nasal septum wide, nostrils opening to sides)	Nose catarrhine (septum narrow, nostrils opening downward)
Tail long, prehensile only in atelines and *Cebus*	Tail short to long, nonprehensile
3 premolar teeth in each quadrant	2 premolars in each quadrant
24 deciduous, 36 permanent teeth (4 fewer in Callitrichinae), I 2/2 C 1/1 P 3/3 M 3/3/(M 2/2 in Callitrichinae)	20 deciduous, 32 permanent teeth, I 2/2 C 1/1 P 2/2 M 3/3
No ischial callosities	Ischial callosities present
Jaws and teeth lightly built in Cebidae; more robust in Atelidae, with deep lower jaw	Cheek pouches in Cercopithecinae; sacculated stomach in Colobinae
Opening on skull to internal ear circular but flat ("ringlike")	Opening extended into tube
Fingers and toes with curved nails (clawlike in Callitrichinae)	All nails tending to be flattened
Big toe opposable, thumb not fully so and sometimes reduced in Cebidae	Thumb and big toe opposable, thumb reduced in Colobinae

marmosets, capuchins, titis, sakis and uakaris, howler monkeys, and spider and wooly monkeys. Eric Delson

Monocleales An order of liverworts of the subclass Marchantiidae, consisting of a single genus (*Monoclea*). The gametophyte is among the largest of all liverworts; its thallus consists of homogeneous cells except for scattered oil cells. The rhizoids are smooth and both thick-walled and thin-walled. The sex organs are grouped but not elevated: the antheridia are sunken in cavities and grouped into receptacles, while the archegonia are enclosed in groups by involucres. The very long, massive seta considerably elevates the capsules which dehisce spoonlike by one slit. The lobing of spore mother cells is unique in the subclass Marchantiidae. *See* BRYOPHYTA; MARCHANTIIDAE. Howard Crum

Monoclonal antibodies Antibody proteins that bind to a specific target molecule (antigen) at one specific site (antigenic site). In response to either infection of immunization with a foreign agent, the immune system generates many different antibodies that bind to the foreign molecules. Individual antibodies within this polyclonal antibody pool bind to specific sites on a target molecule known as epitopes. Isolation of an individual antibody within the polyclonal antibody pool would allow biochemical and biological characterization of a highly specific molecular entity targeting only a single epitope. Realization of the therapeutic potential of such specificity launched research into the development of methods to isolate and continuously generate a supply of a single lineage of antibody, a monoclonal antibody (mAb).

In 1974, W. Köhler and C. Milstein developed a process for the generation of monoclonal antibodies. In their process, fusion of an individual B cell (or B lymphocyte), which produces an antibody with a single specificity but has a finite life span, with a myeloma (B cell tumor) cell, which can be grown indefinitely in culture, results in a hybridoma cell. This hybridoma retains desirable characteristics of both parental cells, producing an antibody of a single specificity that can grow in culture indefinitely.

Generation of monoclonal antibodies through the hybridoma process worked well with B cells from rodents but not with B cells from humans. Consequently, the majority of the first monoclonal antibodies were from mice. When administered into humans as therapeutic agents in experimental tests, the human immune system recognized the mouse monoclonal antibodies as foreign agents, causing an immune response, which was sometimes severe. Although encouraging improvements in disease were sometimes seen, this response made murine (mouse) antibodies unacceptable for use in humans with a functional immune system.

Fueled by advances in molecular biology and genetic engineering in the late 1980s, efforts to engineer new generations of monoclonal antibodies with reduced human immunogenicity have come to fruition. Today there are a number of clonal antibodies approved for human therapeutic use in the United States.

Characterization of the structure of antibodies and their genes laid the foundation for antibody engineering. In most mammals, each antibody is composed of two different polypeptides, the immunoglobulin heavy chain (IgH) and the immunoglobulin light chain (IgL). Comparison of the protein sequences of either heavy of light antibody chain reveals a portion that typically varies from one antibody to the next, the variable region, and a portion that is conserved, the constant region. A heavy and a light chain are folded together in an antibody to align their respective variable and constant regions. The unique shape of the cofolded heavy- and light-chain variable domains creates the variable domain of the antibody, which fits around the shape of the target epitope and confers the binding specificity of the antibody.

Mice genetically engineered to produce fully human antibodies allow the use of established hybridoma technology to generate fully human antibodies directly, without the need for additional engineering. These transgenic mice contain a large portion of human DNA encoding the antibody heavy and light chains. Inactivation of the mouse's own heavy- and light-chain genes forces the mouse to use the human genes to make antibodies. Current versions of these mice generate a diverse polyclonal antibody response, thereby enabling the generation and recovery of optimal monoclonal antibodies using hybridoma technology.

Disease areas that currently are especially amenable to antibody-based treatments include cancer, immune dysregulation, and infection. Depending upon the disease and the biology of the target, therapeutic monoclonal antibodies can have different mechanisms of action. A therapeutic monoclonal antibody may bind and neutralize the normal function of a target. For example, a monoclonal antibody that blocks the activity of the of protein needed for the survival of a cancer cell causes the cell's death. Another therapeutic monoclonal antibody may bind and activate the normal function of a target. For example, a monoclonal antibody can bind to a protein on a cell and trigger an apoptosis signal. Finally, if a monoclonal antibody binds to a target expressed only on diseased tissue, conjugation of a toxic payload (effective agent), such as a chemotherapeutic or radioactive agent, to the monoclonal antibody can create a guided missile for specific delivery of the toxic payload to the diseased tissue, reducing harm to healthy tissue. *See* ANTIBODY; ANTIGEN; GENETIC ENGINEERING; IMMUNOLOGY. Frances M. Brodsky

Monocotyledons This group of flowering plants (angiosperms), with one seed leaf, was previously thought to be one of the two major categories of flowering plants (the other group is dicotyledons). However, deoxyribonucleic acid (DNA) studies have revealed that, although they do constitute a group of closely related families, they are closely related to the magnoliids, with which they share a pollen type with a single aperture. The eudicots are much more distantly related. In general, monocots can also be recognized by their parallel-veined leaves and three-part flowers. Their roots have disorganized vascular bundles, and if they are treelike (yuccas, aloes, dracaenas) their wood is unusually structured. Among the important monocots are grasses (including corn, rice, and wheat), lilies, orchids, palms, and sedges. *See* DICOTYLEDONS; EUDICOTYLEDONS; FLOWER; GRASS CROPS; LILIALES; MAGNOLIOPHYTA. Mark W. Chase; Michael F. Fay

Monogenea A class of the Platyhelminthes which are ectoparasites of the gills, skin, and orifices of fishes and, less frequently, of the esophageal tracts and bladders of amphibians and turtles. They have conspicuous anterior and posterior holdfasts, the latter usually armed. The terminal genitalia are frequently sclerotized. The group is characterized by sexual reproduction, direct development, and a single host in the life cycle.

The most widely used classification employs two subclasses, the Monophisthocotylea, in which the posthaptor is without discrete multiple suckers or clamps, and the Polyopisthocotylea, with suckers or clamps on the posthaptor.

Body shapes of the various genera are distinctive, sometimes bizarre, as in *Vallisia*, which is sickle-shaped. Paired external suckers or buccal cavity suckers and adhesive glands occur anteriorly. The posterior holdfast is either solid and armed with central anchors and marginal hooks (Gyrodactyloidea), sucker-shaped with anchors and hooks (Capsaloidea; see illustration), or solid and bearing suckers or clamps (Polyopisthocotylea).

Monogenea usually have direct development involving simple metamorphosis from the ciliated larval stage to the nonciliated juvenile. Juvenile anchors and hooks may be retained or replaced by adult suckers or clamps. Cross-fertilization or, perhaps less frequently, self-fertilization of hermaphroditic individuals

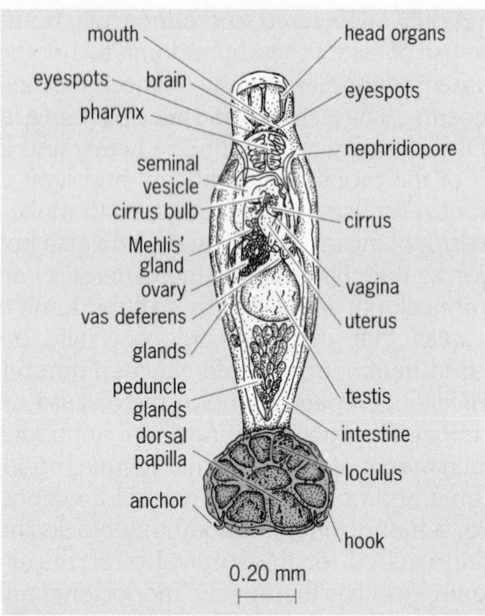

A monogeneid of the superfamily Capsaloidea, *Heterocotyle aetobatis* from the spotted eagle ray, ventral view.

resulting in egg capsules which hatch on the host or in its environment is most common. *See* PLATYHELMINTHES. William J. Hargis

Monogononta A class of the phylum Rotifera which contains the majority of species in this invertebrate class. The organisms of this order are characterized by the presence of a single gonad in both males and females. There is a striking degree of sexual dimorphism, with the males being small and degenerate. The order is made up of three suborders: Ploima, Flosculariacea, and Collothecacea.

In suborder Ploima there is an exceptional diversity of form, varying from soft-bodied wormlike rotifers to species with variously ornamented, loricate shells (see illustration). Most of the free-swimming benthonic and pelagic rotifers belong to this suborder. Locomotion is by the ciliated corona.

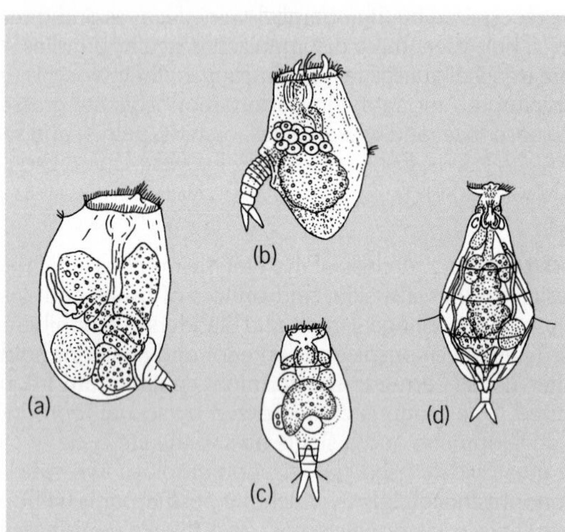

Ploimates. (*a*) *Asplanchnopus* sp. (*b*) *Ploesoma* sp. (*c*) *Euchlanis dilata.* (*d*) *Notomata copeus.*

The suborder Flosculariacea contains the spectacular sessile rotifers formerly known as melicertaceans of the family Flosculariidae, as well as a number of equally notable free-swimming forms included in the family Testudinellidae.

The suborder Collothecacea contains but a single family, the Collothecidae, made up of five genera. Most species of Collothecidae are sessile, and many are encased in gelatinous tubes. *See* ROTIFERA. Elbert H. Ahlstrom

Monomolecular film A film one molecule thick. It is often referred to as a monolayer. Films that form at surfaces or interfaces are of special importance. Such films may reduce friction, wear, or corrosion, or may stabilize emulsions, foams, and solid dispersions. Thin films on water surfaces can reduce evaporation losses, though the spreading of thin films on such surfaces can represent a serious environmental problem. In some instances, the film spreads by itself and is essentially insoluble in the substrate (an insoluble monolayer). In other cases, the film molecules have a low substrate solubility and concentrate at the surface (a Gibbs monolayer).

In many cases, the film molecules are amphipathic, having a dual nature. On water surfaces, polar groups can anchor the film molecules, preventing evaporation, while the nonpolar sections prevent them from going into solution. On solid surfaces, polar interactions, or even chemisorption, may resist film removal and provide an essentially new surface. Based on the design of the film molecules, the newly created surface can be tailored to a number of different tasks. There is a wide range of applications of such films, but two areas stand out. On solid surfaces, the formation of multilayers from monolayers can lead to the creation of new materials, often having properties significantly different from that of typical three-dimensional (3D) solids. On water surfaces, the greatest number of studies relate to biological systems, with an emphasis on biological membranes. Many of the earlier studies were on long-chain fatty acids and related substances at the air–water interface. Using typical membrane lipids, such as long-chain phospholipids, cholesterol, and membrane proteins, the monolayer constitutes an approximation to one-half of a biological membrane, and has provided considerable information on membrane behavior at the molecular level. *See* CELL MEMBRANES; MEMBRANE MIMETIC CHEMISTRY; POLAR MOLECULE. D. Allan Cadenhead; George L. Gaines, Jr.

Mononchida An order of nematodes having a full complement of cephalic sensilla on the lips in two circlets of 6 and 10. The amphids are small and cuplike, and are located just posterior to the lateral lips; the amphidial aperture is either slit-like or ellipsoidal. The stoma is globular and heavily cuticularized, and is derived primarily from the cheilostome. The stoma bears one or more massive teeth that may be opposed by denticles in either transverse or longitudinal rows. The esophagus is cylindrical conoid, with a heavily cuticularized luminal lining. The excretory system is atrophied. Males have ventromedial supplements and paired spicules. The gubernaculum may possess lateral accessory pieces. Females have one or two ovaries. Caudal glands and a spinneret are common; however, they may be degenerate or absent.

There are three mononchid superfamilies: The Mononchoidea contain some of the most common and easily recognized free-living nonparasitic nematodes that occur in soils and fresh waters throughout the world. The closely related Bathyodontoidea are inhabitants of soil or fresh water and prey on small microorganisms. The nonparasitic Monochuloidea comprise both soil and fresh-water species, all of which are predators of microfauna. *See* NEMATA (NEMATODA). Armand R. Maggenti

Monoplacophora A class of the phylum Mollusca. Although fossil monoplacophorans had been known since the end of the nineteenth century, it was the discovery of a living species in deep water off Costa Rica in the 1950s that led to universal acceptance of the class. Monoplacophorans are bilaterally symmetrical, univalved mollusks that vanish from the fossil record at the end of the Paleozoic, about 240 million years ago. The living species, such as *Neopilina galatheae*, are rare and inhabit deep water, which may explain their absence from Mesozoic and Cenozoic rocks. Living monoplacophorans have a limpet-shaped shell, a circular foot attached by pairs of retractor muscles, and several gills on each side of the body. *See* GASTROPODA; MOLLUSCA; POLYPLACOPHORA. Bruce Runnegar

Monopulse radar Radar capable of estimating target position based on the return from a single pulse. In many radars, precise angular position is estimated by conically scanning a single beam around the initial coarse angle estimate; the orderly amplitude variation of echoes during such scanning provides the refinement. Such measurement is limited, however, by pulse-to-pulse fluctuations in echo strength, a property quite common in radar targets.

Monopulse radars use antennas that provide a local cluster of simultaneous beams (instead of scanning just one beam) to make the same precise angle estimate with each pulse transmitted. Since angle information is contained in each return, fluctuations in echo strength do not significantly degrade the measurement. Monopulse radars with mechanically positioned antennas address only a single target, and average the measurements over many pulses for improved accuracy. Radars using electronic beam steering in stationary phased-array antennas may make such a measurement in a single-pulse "dwell," doing so on dozens of targets, returning to each several times a second if necessary. Robert T. Hill

Monorail A distinctive type of materials-handling machine that provides an overhead, normally horizontal, fixed path of travel in the form of a trackage system and individually propelled hand or powered trolleys which carry their loads suspended freely with an intermittent motion. Because monorails operate over fixed paths rather than over limited areas, they differ from overhead-traveling cranes, and they should not be confused with such overhead conveyors as cableways. *See* BULK-HANDLING MACHINES; MATERIALS-HANDLING EQUIPMENT. Arthur M. Perrin

Monosaccharide A class of simple sugars containing a chain of 3–10 carbon atoms in the molecule, known as polyhydroxy aldehydes (aldoses) or ketones (ketoses). They are very soluble in water, sparingly soluble in ethanol, and insoluble in ether. The number of monosaccharides known is approximately 70, of which about 20 occur in nature. The remainder are synthetic. The existence of such a large number of compounds is due to the presence of asymmetric carbon atoms in the molecules. Aldohexoses, for example, which include the important sugar glucose, contain no less than four asymmetric atoms, each of which may be present in either D or L configuration. The number of stereoisomers rapidly increases with each additional asymmetric carbon atom.

A list of the best-known monosaccharides is given below:

Trioses: $CH_2OH \cdot CHOH \cdot CHO$, glycerose (glyceric aldehyde)
$CH_2OH \cdot CO \cdot CH_2OH$, dihydroxy acetone

Tetroses: $CH_2OH \cdot (CHOH)_2 \cdot CHO$, erythrose
$CH_2OH \cdot CHOH \cdot CO \cdot CHO$, erythrulose

Pentoses: $CH_2OH \cdot (CHOH)_3 \cdot CHO$, xylose, arabinose, ribose
$CH_2OH \cdot (CHOH)_2 \cdot CO \cdot CH_2OH$, xylulose, ribulose

Methyl pentoses (6-deoxyhexoses):
$CH_4(CHOH)_4 \cdot CHO$, rhamnose, fucose

Hexoses: $CH_2OH \cdot (CHOH)_4 \cdot CHO$, glucose, mannose, galactose
$CH_2OH \cdot (CHOH)_3 \cdot CO \cdot CHOH$, fructose, sorbose

Heptoses: $CH_2OH \cdot (CHOH)_5 \cdot CHO$, glucoheptose, galamannoheptose
$CH_2OH \cdot (CHOH)_4 \cdot CO \cdot CH_2OH$, sedoheptulose, mannoheptulose

Aldose monosaccharides having 8, 9, and 10 carbon atoms in their chains have been synthesized. *See* CARBOHYDRATE; KETONE; OPTICAL ACTIVITY; STEREOCHEMISTRY. William Z. Hassid

Monosodium glutamate The single sodium salt of glutamic acid used in foods to accentuate flavors. It is also known as MSG. Molecular structure is represented below.

```
          H   H   H
          |   |   |
HO — C — C — C — C — C — ONa
     ||   |   |   |   ||
     O    H   H   N   O
                 / \
                H   H
```

The crystal form available in commerce is the monohydrate, with structure as represented plus one molecule of water of hydration.

Originally produced from seaweed in the Orient, it is now made principally from cereal glutens, such as those of wheat, corn, and soybeans, from solutions evolved in the manufacture of beet sugar, and by microbiological fermentation of carbohydrates. The two raw materials used for the greater proportion of commercial production are wheat gluten and desugared beet-sugar molasses.

Monosodium glutamate is recognized as a standard of identity ingredient in several commercial food preparations. Its principal use is in the preparation of canned and dried soups, but it also enters into the production of some meat, vegetable, fowl, and fish products. It is the so-called secret ingredient used by many of the famous restaurant and hotel chefs. Paul V. D. Manning

Monotremata The single order of the mammalian subclass Prototheria. Two living families, the Tachyglossidae and the Ornithorhynchidae, make up this unusual order of quasimammals, or mammallike reptiles.

The Tachyglossidae comprise the echidnas (spiny anteaters), which have relatively large brains with convoluted cerebral hemispheres. The known genera, *Tachyglossus* and *Zaglossus*, are terrestrial, feeding on termites, ants, and other insects. They are capable diggers, both to obtain food and to escape enemies. Like hedgehogs, they can erect their spines and withdraw their limbs when predators threaten. Commonly one egg, but occasionally two or even three, is laid directly into the marsupium (pouch) of the mother where it is incubated for up to 10 days. Species of *Tachyglossus* live in rocky areas, semideserts, open forests, and scrublands. They are found in Australia, Tasmania, New Guinea, and Salawati Island. Species of *Zaglossus* are found in mountainous, forested areas.

The duck-billed platypus, constituting the Ornithorhynchidae, has a relatively small brain with smooth cerebral hemispheres. The young have calcified teeth, but in the adult these are replaced by horny plates which form around the teeth in the gums. The snout is duck-billed. The semiaquatic platypus is a capable swimmer, diver, and digger. Two eggs are usually laid by the female into a nest of damp vegetation. After incubating the eggs for about 10 days the female leaves, returning only when the eggs are hatched. The platypus is found in Australia and Tasmania in almost all aquatic habitats. *See* MAMMALIA; PROTOTHERIA. Frederick S. Szalay

Monsoon meteorology The study of the structure and behavior of the atmosphere in those areas of the world that have monsoon climates. In lay terminology, monsoon connotes the rains of the wet summer season that follows the dry winter. However, for mariners, the term monsoon has come to mean the seasonal wind reversals.

In true monsoon climates, both the wet summer season that follows the dry winter and the seasonal wind reversals should occur. Winds from cooler oceans blow toward heated continents in summer, bringing warm, unsettled, moisture-laden air and the season of rains, the summer monsoon. In winter, winds from the cold heartlands of the continents blow toward the oceans, bringing dry, cool, and sunny weather, the winter monsoon.

Based on these criteria, monsoon climates of the world include almost all of the Eastern Hemisphere tropics and subtropics, which is about 25% of the surface area of the Earth. The areas of maximum seasonal precipitation straddle or are adjacent to the Equator. Two of the world's areas of maximum precipitation (heavy rainfall) are within the domain of the monsoons: the central and south African region, and the larger south Asia-Australia region. The monsoon surface winds emanate from the cold continents of the winter hemisphere, cross the Equator, and flow toward and over the hot summer-hemisphere landmasses.

India presents the classic example of a monsoon climate region, with an annual cycle that brings southwesterly winds and heavy rains in summer (the Indian southwest monsoon) and northeasterly winds and dry weather in winter (the northeast winter monsoon).

Like all weather systems on Earth, monsoons derive their primary source of energy from the Sun. About 30% of the Sun's energy that enters the top of the atmosphere is transmitted back to space by cloud and surface reflections. Little of the remainder is absorbed directly by the clear atmosphere; it is absorbed at the Earth's surface according to a seasonal cycle. The opposition of seasons in the Northern and Southern hemispheres leads to a slow movement of surface air across the Equator from winter hemisphere to summer hemisphere, forced by horizontal pressure gradients and vertical buoyancy forces resulting from differential seasonal heating. Such a seasonally reversing rhythm is most pronounced in the monsoon regions. *See* ALBEDO; ATMOSPHERE; HEAT BALANCE; INSOLATION; METEOROLOGY; TROPICAL METEOROLOGY. Jay S. Fein

Monstrilloida A copepod order of crustaceans containing 122 species in five genera, *Cymbasoma*, *Guanabaraenia*, *Monstrilla*, *Monstrillopsis*, and *Thaumatohessia*; Monstrillidae is the only family. *See* COPEPODA; CRUSTACEA.

Adult females (1.5–2.5 mm or 0.07–0.11 in.) and males (1.0–1.5 mm or 0.04–0.07 in.) of the Monstrilloida are free-swimming and are often collected in plankton samples taken near the surface of shallow marine waters. The body has a podoplean architecture (the last two thoracic somites with the posterior part of the body), although the broader anterior part merges indistinctly with the narrower posterior part. Although antenna 1, the four swimming legs, and legs 5 and 6 are present, the feeding limbs, which include antenna 2, mandible, maxilla 1, maxilla 2, and the maxilliped, fail to form at this stage, and adults are assumed not to feed. An embryo hatches as a feeble-swimming, simple nauplius (a crustacean larval stage) with well-developed antenna 1, antenna 2, and mandible. This nauplius penetrates the body of a polychaete or a mollusk; a hooklike structure on the mandibular endopod facilitates initial attachment. The next phase of the life cycle is passed within the blood system of the host as a sac-like form with several processes; this endoparasitic phase cannot be identified as a copepod. The monstrilloid leaves its host as an immature copepodid, apparently the fifth copepodid, and molts once to the adult. Monstrilloids appear to be found throughout the world's oceans. Frank D. Ferrari

Monte Carlo method A technique for estimating the solution, *x*, of a numerical mathematical problem by means of an artificial sampling experiment. The estimate is usually given as the average value, in a sample, of some statistic whose mathematical expectation is equal to *x*. In many of the useful applications, the mathematical problem itself arises in a problem of probability in physics or other sciences, operational research, image analysis, general statistics, mathematical economics, or econometrics. The importance of the method arises primarily from the need to solve problems for which other methods are more expensive or impracticable, and from the increased importance of all numerical methods because of the development of the electronic digital computer.

The main advantage of Monte Carlo is that other methods can be more costly or impracticable. A familiar example is the estimation of the probability of winning a game of pure chance: Sometimes the only reasonably simple method of estimation is to play the game several times. There are also numerical problems that can be solved by deterministic methods but can be more simply solved approximately by the Monte Carlo method. Sometimes poor approximations are satisfactory because the aim is merely to determine the strategic variables of a problem. This is likely to be a fruitful technique in mathematical economics.

Another situation where a poor approximation is satisfactory occurs when there is available an iterative method of calculation, that is, a method of successive approximation, which converges closely to the right answer in a reasonable time provided that the first trial solution is not too far from the truth. The Monte Carlo method may then perhaps be used for obtaining a first trial solution. Modern Monte Carlo techniques are themselves usually iterative.

Sometimes the expense of a Monte Carlo method does not increase as fast as that of other methods when the dimensionality of a problem is increased. This seems to be true for multiple integration when it cannot be done analytically, and for the solution of Schrödinger's equation for several particles. *See* SCHRÖDINGER'S WAVE EQUATION.

The main disadvantage of some Monte Carlo methods is that for each extra decimal place required, it is necessary to multiply the sample size by 100. Thus, to calculate π to five decimal places by throwing a needle would require about 10^{10} throws, or 1 throw per second for about 300 years. Irving J. Good

Month Any of several units of time based on the revolution of the Moon around Earth.

The calendar month is one of the 12 arbitrary periods into which the calendar year is divided. *See* CALENDAR.

The synodic month, the period of the lunar phases, is the average period of revolution of the Moon with respect to the Sun, the same as the average interval between successive full moons. Its duration is 29.531 days. *See* PHASE (ASTRONOMY).

The tropical month is the period required for the mean longitude of the Moon to increase 360°, or 27.322 days.

The sidereal month, 7 s longer than the tropical month, is the average period of revolution of the Moon with respect to a fixed direction in space.

The anomalistic month, 27.555 days in duration, is the average interval between closest approaches of the Moon to Earth. The variation in the Moon's distance from the Earth causes a variation in the apparent size of the Moon and thus in the duration of solar eclipses.

The nodical month, 27.212 days in duration, is the average interval between successive northward passages of the Moon across the ecliptic, points known as nodes. Since eclipses can occur only when the Sun and Moon are near such nodes, this period is also known as a draconic month, after the Chinese mythical dragon that supposedly ate the Sun to cause a solar eclipse. *See* ECLIPSE; MOON; TIME. Gerald M. Clemence; Jay M. Pasachoff

Montmorillonite A group name for all clay minerals with an expanding structure, except vermiculite, and also a specific mineral name for the high alumina end member of the group. *See* CLAY MINERALS; VERMICULITE.

Montmorillonite clays have wide commercial use. The high colloidal, plastic, and binding properties make them especially in demand for bonding molding sands and for oil-well drilling muds. They are also widely used to decolorize oils and as a source of petroleum cracking catalysts. *See* CLAY.

Members of the montmorillonite group of clay minerals vary greatly in their modes of formation. Alkaline conditions and the presence of magnesium particularly favor the formation of these minerals. Several important modes of occurrence are in soils, in bentonites, in mineral veins, in marine shales, and as alteration products of other minerals. Recent sediments have a fairly high montmorillonite content. *See* BENTONITE; MARINE SEDIMENTS.

Floyd M. Wahl; Ralph E. Grim

Moon The Earth's natural satellite. United States and Soviet spacecraft have obtained lunar data and samples, and American astronauts have orbited, landed upon, and roved upon the Moon.

The Earth and Moon now make one revolution about their barycenter, or common center of mass (a point about 4670 km from the Earth's center), in $27^d\ 7^h\ 43^m\ 11.6^s$. This sidereal period is slowly lengthening, and the distance (now about 60.27 earth radii) between centers of mass is increasing, because of tidal friction in the oceans of the Earth.

The Moon's present orbit is inclined about 5° to the plane of the ecliptic. As a result of differential attraction by the Sun on the Earth-Moon system, the Moon's orbital plane rotates slowly relative to the ecliptic (the line of nodes regresses in an average period of 18.60 years) and the Moon's apogee and perigee rotate slowly in the plane of the orbit (the line of apsides advances in a period of 8.850 years). Looking down on the system from the north, the Moon moves counterclockwise. It travels along its orbit at an average speed of nearly 0.6 mi/s (1 km/s) or about 1 lunar diameter per hour.

As a result of the Earth's annual motion around the Sun, the direction of solar illumination changes about 1° per day, so that lunar phases do not repeat in the sidereal period given above but in the synodic period, which averages $29^d\ 12^h\ 44^m$.

When the lunar line of nodes coincides with the direction to the Sun and the Moon happens to be near a node, eclipses can occur. *See* ECLIPSE.

The relation between the Moon's shape and its mass distribution is very important to theories of lunar origin and the history of the Earth-Moon system. By radio altimetry, Apollo confirmed that the Moon's surface on the far side is higher on the average than the near side; that is, the center of mass is offset from the center of figure. The offset is about 2 km (1.2 mi) toward the Earth. These observations suggest that the Moon's crust is thicker on the far side than on the near side. The *Clementine* mission in 1994 extended measurements to nearly the whole Moon and revealed the depth of a huge basin on the southern far side.

Aristarchus-Harbinger region of the Moon, photographed from the *Apollo 15* spacecraft in lunar orbit, with the craters Aristarchus and Herodotus and Schroeter's Valley, the largest sinuous rille on the Moon. The impact crater Aristarchus, about 25 mi (40 km) in diameter and more than 2.5 mi (4 km) deep, lies at the edge of a mountainous region that shows evidence of volcanic activity. (*NASA*)

The Moon's small size and low mean density result in surface gravity too low to hold a permanent atmosphere, and therefore it was to be expected that lunar surface characteristics would be very different from those of Earth. However, the bulk properties of the Moon are also quite different—the density alone is evidence of that. The Moon is too small to have compressed its silicates into a metallic phase by gravity; therefore, if it has a dense core at all, the core should be of nickel-iron. Available data suggest that the Moon's iron core may have a diameter of at most a few hundred kilometers.

As can be seen from the Earth with the unaided eye, the Moon has two major types of surface: the dark, smooth maria and the lighter, rougher highlands. Photography by spacecraft shows that, for some unknown reason, the Moon's far side consists mainly of highlands. Both maria and highlands are covered with craters of all sizes. Numerous different types of craters can be recognized. Most prominent at full moon are the bright ray craters whose grayish ejecta appear to have traveled for hundreds of miles across the lunar surface. Observers have long recognized that some erosive process has been and may still be active on the Moon. Bombardment of the airless Moon by meteoritic matter and solar particles, and extreme temperature cycling, are now considered the most likely erosive agents, but local internal activity is also a possibility.

The lunar mountains, though very high (26,000 ft or 8000 m), are not extremely steep, and lunar explorers see rolling rather than jagged scenery. Though widespread networks of cracks are visible, there is no evidence on the Moon of the great mountain-building processes seen on the Earth.

Basins on the Moon's near side, namely, Imbrium, Serenitatis, and Crisium, appear fully flooded. These were maria created by giant impacts, followed by subsidence of the ejecta and (probably much later) upwelling of lava from inside the Moon. Examination of small variations in Lunar Orbiter motions has revealed that each of the great circular maria is the site of a positive gravity anomaly (excess mass). The old argument about impact versus vulcanism as the primary agent in forming the lunar relief appears to be entering a new, more complicated phase with the confirmation of extensive flooding of impact craters by lava on the Moon's near side, while on the far side, where the crust is thicker, the great basins remain mostly empty.

In some of the Moon's mountainous regions bordering on the maria are found sinuous rilles (see illustration). These winding valleys were shown in Lunar Orbiter pictures to have an exquisite fineness of detail. No explanation for them yet offered has proved entirely convincing.

The Moon seems to be totally covered, to a depth of at least tens of meters, by a layer of rubble and soil with very peculiar optical and thermal properties. This layer is called the regolith. The observed optical and radio properties all point to a highly porous or underdense structure for at least the top few millimeters of the lunar surface material. A dark-gray, fine soil appears to mantle the entire Moon, softening most surface contours and covering everything except occasional fields of rocks. This soil, with a slightly cohesive character like that of damp sand and a chemical composition similar to that of some basic silicates on the Earth, is a product of the radiation, meteoroid, and thermal environment at the lunar surface.

James D. Burke

Moose An even-toed ungulate (Artiodactyla) which is a member of the deer family, Cervidae. *Alces alces* is the largest member of the family and ranges in the boreal forested areas throughout North America and in northern Eurasia. The moose is known as the elk in Europe and is believed by some authorities to be a race of the American moose (*A. americana*). The legs are long, making the animal well-adapted for its feeding habits of wading for aquatic plants and browsing on trees and bushes. During the rutting season in the early fall, the male gathers a number of cows together, and mating takes place. After a gestation period of about 37 weeks, one or two calves are born. *See* ARTIODACTYLA. Charles B. Curtin

Moraxella A genus of bacteria that are parasites of mucous membranes. Subgenus *Moraxella* is characterized by gram-negative rods that are often very short and plump, frequently resembling a coccus, and usually occurring in pairs. Subgenus *Branhamella* has gram-negative cocci occurring as single cells or in pairs with the adjacent sides flattened. They are usually harmless parasites of humans and other warm-blooded animals and are generally considered not to be highly pathogenic. Most species may be opportunistic pathogens in predisposed or debilitated hosts.

There are presently six species in the subgenus *Moraxella*: *M.* (*M.*) *lacunata* (also known as *Diplobacillus moraxaxenfeld* and *liquefaciens*), *M.* (*M.*) *bovis*, *M.* (*M.*) *nonliquefaciens* (also known as *Bacillus duplex nonliquefaciens*), *M.* (*M.*) *atlantae*, *M.* (*M.*) *phenylpyruvica*, and *M.* (*M.*) *osloensis*. The different species are recognized on the basis of phenotypic properties, including liquefaction of coagulated serum, hemolysis of human blood in blood agar media, nitrate reduction, phenylalanine deaminase activity, urease activity, and growth on mineral salts medium with ammonium ion and acetate as the sole carbon source.

The subgenus *Branhamella* presently contains four species: *M.* (*B.*) *catarrhalis*, *M.* (*B.*) *caviae*, *M.* (*B.*) *ovis*, and *M.* (*B.*) *cuniculi*. The different species are recognized on the basis of hemolysis of human blood in blood agar media, and nitrate and nitrite reduction, among other properties.

Moraxella (*M.*) *lacunata*, the type species of the subgenus *Moraxella*, was a significant causative agent of human conjunctivitis and keratitis in the past but is only rarely isolated at present. Infectious keratoconjunctivitis in cattle, called pinkeye, is caused by *M.* (*M.*) *bovis*. *Moraxella* (*M.*) *nonliquefaciens* is considered to be a well-established parasite of humans and rarely causes disease, but it has been associated with endophthalmitis and pneumonitis with pulmonary abscess. *Moraxella* (*M.*) *osloensis*, usually a harmless parasite, has been frequently associated with such human infections as osteomyelitis, endocarditis, septicemia, meningitis, stomatitis, and septic arthritis.

Moraxella (*B.*) *catarrhalis*, the type species of *Branhamella*, is the only species of this subgenus recovered from humans. The organism is considered to be a well-adapted parasite but has been judged the etiologic agent of middle-ear infection, maxillary sinus infection, bronchitis, tracheitis, conjunctivitis, pneumonia, otitis media of infants, respiratory disease in the compromised host, septicemia, meningitis, and endocarditis. The remaining *Branhamella* species (*caviae*, *ovis*, and *cuniculi*) are parasites of guinea pigs, sheep, cattle, and rabbits.

Moraxella species are susceptible to most antimicrobial agents with the exception of the lincomycins. Their usually high susceptibility to the penicillins is a feature that separates them from most other gram-negative rods. *See* CLINICAL MICROBIOLOGY. Gerald Gilardi

Mordant A substance or combination of substances that facilitates the fixing of a dye to a fiber. A mordant enables the production of a more permanent and often deeper color. Metallic salts or hydroxides are most frequently used as mordants.

Certain mordants act directly on the fiber, making it more susceptible to the dye. Fabrics are then pretreated with the mordant before exposure to the dye. Other mordants function through the formation of a complex with the dye. The complex acts as the dyeing agent. Mordant and dye in this case are exposed simultaneously to the fabric. *See* DYE; DYEING. Francis Johnston

Mormonilloida A copepod order of crustaceans containing four species. *Mormonilla minor* and *M. phasma* were described in the late 1800s; *M. polaris* and *M. atlantica* were described in the early 1900s. The status of a fifth species in a second genus, *Corynuropis tenuicaudatus*, remains uncertain. *See* COPEPODA; CRUSTACEA.

Lengths range from 0.2–1.7 mm (0.01–0.07 in.) for adult females to 0.9–1.4 mm (0.04–0.06 in.) for adult males. The second thoracic somite, bearing the first swimming leg, articulates with the first, and the last two thoracic somites are associated with the posterior part of the body (podoplean architecture). The only internal organ system that has been studied is the musculature; it is simpler than the musculature of copepods from other orders. Species of Mormonilloida are deep-water, free-swimming copepods reported from the Arctic, eastern Indian, and eastern North Atlantic oceans, and the Mediterranean, Red, and Arabian seas. Phylogenetic relationships of Mormonilloida to the other copepod orders remain uncertain. Frank D. Ferrari

Morphogenesis The development of form and pattern in animals. Animals have complex shapes and structural patterns that are faithfully reproduced during the embryonic development of each generation. Morphogenesis is a higher outcome of the process of differentiation, defined as the progressive structural and functional diversification and specialization of cells, and recognizable by specific molecular markers and morphological and histological organization. Morphogenesis is accomplished via a complex series of individual and group cell movements called morphogenetic movements. The cell differentiations that lead to these movements are preterminal differentiations that result in coordinated changes in individual cell shape, adhesion, and motility, as well as production of extracellular matrix. The starting point for these events is a single cell, the fertilized egg, or zygote, that divides repeatedly to form a multicellular embryo capable of carrying out the series of patterned differentiations leading to the morphogenetic movements that shape the embryo. *See* ANIMAL GROWTH; DEVELOPMENTAL BIOLOGY; EMBRYONIC DIFFERENTIATION; MOLECULAR BIOLOGY.

Cell differentiation involves the differential expression of genes in the nuclear DNA that encode proteins specifying the structure and function of each cell type. During cleavage, nuclei divide equivalently so that all cells of the embryo receive the total complement of genes contained in the DNA of the zygote nucleus. However, cells in different regions of the embryos of many animals contain cytoplasm which differs in composition. The cytoplasmic composition of a cell controls which genes will be expressed in each region of the embryo, resulting in a patterned differentiation. Initial differences in cells arise because of the positions they occupy: cells on the outside of the early mammalian embryo become trophoblast cells, whereas cells on the inside become the inner cell mass from which the embryo develops. In all developing multicellular organisms, interactions between these early cell types result in new patterns of gene activity that further increase the diversity of cell types. By means of cascades of such interactions, all of the different cell types of an animal body gradually emerge in the proper spatial patterns. *See* CELL DIFFERENTIATION; CLEAVAGE (DEVELOPMENTAL BIOLOGY); DEOXYRIBONUCLEIC ACID (DNA).

The structural patterns of animals and their parts exhibit polarity; that is, they display structural differences along anteroposterior (AP), dorsoventral (DV), and right-left axes. In many animals, the axial polarities of the embryo are established during oogenesis and/or the period between fertilization and the first cleavage, by the localization of maternal determinants into specific parts of the egg. For example, in the nematode worm *Caenorhabditis elegans*, the sperm nucleus moves from its locus of entry to the nearest end of the oblong egg. This end of the egg becomes the posterior pole of the AP axis, and the opposite end becomes the anterior pole. *See* FERTILIZATION (ANIMAL); OOGENESIS; OVUM.

Once axial polarities are established, the pattern of structural differentiation along those axes is elaborated. In *C. elegans*, patterning relies heavily on further segregation of maternal determinants as cleavage proceeds. In vertebrates, patterning relies mainly on cell interactions. The first step in patterning is to induce the formation of the mesoderm. At the mid-blastula stage, when the embryo consists of several thousand cells, an inductive interaction takes place between the endoderm cells and the cells derived from the gray crescent region of the egg, causing the latter to become mesoderm cells. Genes are then activated that confer the ability to organize the AP and DV pattern of the mesoderm and initiate the morphogenetic movements of gastrulation. Cells form the circular blastopore of the gastrula, through which the prospective endoderm, mesoderm, and endoderm migrate to form three concentric layers, with ectoderm on the outside, endoderm on the inside, and mesoderm in between. The ectoderm develops into the epidermis and nervous system, the endoderm into the organs of the alimentary tract, and the mesoderm into muscles, skeleton, heart, kidneys, and connective tissue. *See* BLASTULATION; GASTRULATION; GERM LAYERS.

Prior to gastrulation, the prospective organ regions of the ectoderm and endoderm are not yet determined, as shown by the fact that they are unable to differentiate in isolation, and that they can develop in ways different from their normal fate when transplanted elsewhere in the embryo. Mesodermal organ regions, however, are highly self-organizing under these conditions. Ectodermal and endodermal organ regions become determined during and after gastrulation by the inductive action of the mesoderm. For example, dorsal mesoderm normally invaginates and stretches out along the dorsal midline where it differentiates as notochord and trunk muscles. The ectoderm overlying the dorsal mesoderm differentiates as the central nervous system. *See* FATE MAPS (EMBRYOLOGY).

Once induced, organ regions of vertebrate embryos can themselves induce additional organs from undetermined tissue. For example, the retina and iris of the eye develop from a vesicle growing out of the forebrain. This vesicle induces a lens from the overlying head ectoderm, and the lens then induces the cornea from head ectoderm to complete the eye. By means of such cascades of inductive interactions, all the organs of the body are blocked out. *See* EMBRYONIC INDUCTION.

Once determined, an organ region constitutes a developmental system, called a morphogenetic field, which specifies the detailed pattern of cell differentiation within the organ. Cells differentiate in patterns dictated by their relative positions within the field. It is generally accepted that graded molecular signals, or cues, are the basis of this positional information. It is proposed that the source of the signals is a set of boundary cells which define the limits of the field. All the cells of a field thus derive their positional information from a common set of boundary cells.

Most fields become inactive after the pattern they specify begins to differentiate; the ability to form normal organs after removal or interchange of cells is then lost. However, in some animals, such as salamanders, the fields of certain organs can be reactivated by loss of a part, even in the adult. The missing part is then redeveloped in a process called regeneration. *See* REGENERATIVE BIOLOGY.

Each cell type of an animal is characterized by a specific set of gene-encoded proteins which determine the structure, shape, mobility, and function of the cell. In addition, different kinds of cells secrete molecules of protein and protein complexed with carbohydrate which constitute equally specific types and patterns of extracellular matrix. The matrix stabilizes tissue and organ structure, guides migrating cells to their proper locations, and is a medium through which cell interactions take place. Cell interactions take place at cell surfaces, the molecular composition of which is distinct from one cell type to another, allowing them to recognize one another. These differences are reflected in varying degrees of adhesivity between different kinds of cells, and between cells and different kinds of extracellular matrix. Differential adhesivity is the property upon which cell migration, clustering, and rearrangement is based, and is thus important in creating the conditions for cell–cell and cell–matrix interactions. *See* CELL ADHESION.

Another important mechanism for the development of complex patterns and shapes is the differential growth of organs and their parts. Differential growth is evident as soon as embryonic organ regions begin to develop. During the early part of their development, the growth rates of organs are controlled by intrinsic factors. At later stages of development, including postnatal life, organ growth is largely under the control of hormones secreted by cells of the endocrine glands. When growth ceases at maturity, the relative sizes of the body parts constitute distinctive features of individual and species-specific form.

Apoptosis, or cell death by suicide, is an important feature in the refinement of final cell number and shape of developing organs and appendages. Striking examples of the role of apoptosis in developing vertebrates are the elimination of excess neurons in the spinal cord, and foot development in ducks and chickens. Ducks have webbed toes while chickens do not. This difference results because as the chicken leg bud grows and forms the toes, the cells between the developing digits undergo apoptosis. Various grafting and culturing experiments have indicated that it is a cell's relative position in the limb bud which establishes its fate to die at some later time in development. Apoptosis is either triggered by a signal from other cells or is cell-autonomous, but is preventable if a rescue signal is available. Thus within each cell, there are genes that protect against apoptosis and genes that carry out the death sentence. David L. Stocum

Mortar A binding agent used in construction of clay brick, concrete masonry, and natural stone masonry walls and, to much less extent, landscape pavements. Modern mortars are improved versions of the lime and sand mixtures historically used in building masonry walls. *See* BRICK; MASONRY.

Masonry mortar is composed of one or more cementitious materials, such as masonry cement or portland cement and lime, clean sand, and sufficient water to produce a plastic, workable mixture.

Mortars are closely related to concrete but, like grout, generally do not contain coarse aggregate. Mortars function with the same calcium silicate-based chemistry as concrete and grouts, bonding with masonry units into a contiguous, weatherproof surface in the process. Masonry cement or portland cement-lime mortars can be formulated to address job-specific requirements including setting time, rate of hardening, water retentivity, and extended workability. *See* CEMENT; CONCRETE; GROUT; LIME (INDUSTRY). John Melander

Mosaicism The condition in which more than one genetically distinct population of cells coexists within one individual. The term mosaic is used for an individual composed of two or more cell lines of different genetic or chromosomal constitutions, where the cell lines originate from one zygote. Mosaicism is a common phenomenon in both the plant and animal kingdoms; it may originate at any state in the course of ontogeny, and in any tissue in which cells proliferate. Cells with a particular genetic constitution form a clone, and this may appear as a mosaic spot embedded in a background containing cells of different genotype. The several types of mosaicism include chromosomal mosaicism, mosaicism due to mitotic recombination or gene mutation, and functional mosaicism.

In chromosomal mosaicism, abnormal segregation of chromosomes in the course of mitosis results in progeny cells with irregular numbers of chromosomes. Chromosome nondisjunction brings about aneuploidy, a condition in which the cell bears

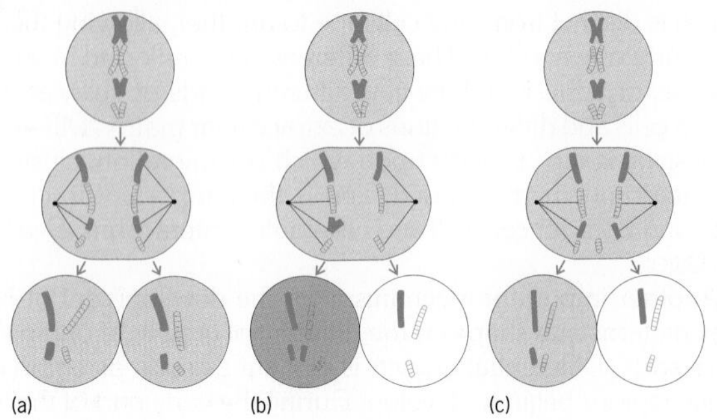

Chromosomal mosaicism. (*a*) Normal mitosis with two pairs of homologous chromosomes. (*b*) Somatic nondisjunction resulting in aneuploid daughter cells: one bears an additional chromosome (trisomy) and the other lacks one chromosome (monosomy). (*c*) Chromosome elimination leading to monosomy.

more or fewer chromosomes than normal (see illustration). Such a condition is often lethal to the cell due to genetic imbalance, especially when more than one large autosome is involved. In humans, trisomic mosaicism for several autosomes has been reported, and trisomy 21, that is, mosaicism for Down syndrome, is the most common. *See* CHROMOSOME ABERRATION; DOWN SYNDROME.

The elimination of complete chromosomes or chromosome fragments is a frequent source of mosaicism. Loss of chromosomes or chromosome fragments can occur spontaneously (with a low frequency), but it can also be induced by several chemicals, and by x-rays and gamma rays.

Elimination of sex chromosomes is often the source of sex mosaicism. In humans most of the XX/XY mosaics are hermaphrodites (persons with both ovarian and testicular tissues). *See* HUMAN GENETICS.

In plants, elimination of chromosomes or chromosome segments that code for the synthesis of chlorophyll or other colored substances results in a colorless cell, and the clonal descendants of this cell form a lightly colored or noncolored patch on the leaves, flowers, or fruits of the plant. This type of pattern is called variegation.

In mitotic recombination, the exchange of parts of homologous chromosomes often results in cells with a different genetic constitution from that of the mother and the surrounding cells. Progeny of the daughter cells, when properly marked genetically, can be recognized after cell proliferation in both somatic and germ cells. Mitotic recombination (also called somatic crossing-over) is a relatively rare event, but has been known to be the source of spontaneous mosaicism in *Drosophila* and some other insects, in the nematode *Caenorhabditis elegans*, in mice, and in a number of plant species (soybean, tobacco). The frequency of somatic crossing-over can be greatly increased by ultraviolet, x, and gamma rays and also by chemical mutagens and carcinogens. *See* CROSSING-OVER (GENETICS); MUTAGENS AND CARCINOGENS; RECOMBINATION (GENETICS).

In some cells that are heterozygous for mutations, the wild-type alleles may lose their activities as a consequence of a new mutation and thus mosaicism can develop. Mutagens and carcinogens are known to induce such mutations and, indeed, these substances can be screened on the basis of this behavior.

Although the genetic constitution is the same (XX) in each of the cells of female eutherian mammals, they show mosaicism for the majority of X-linked genes. This functional mosaicism is due to X-chromosome inactivation. Maternally and paternally derived X chromosomes are turned off randomly during embryogenesis, and consequently recessive X-linked genes can be expressed despite their heterozygous condition. For example, the calico cat is heterozygous for black and orange alleles of an X-linked coat color gene. Due to X-chromosome inactivation, calico cats exhibit an irregular patchwork of black and orange sectors, with each sector representing a clone of cells derived from one cell. *See* GENETICS; SEX-LINKED INHERITANCE. Janos Szabad

Mosquito Any member of the family Culicidae in the insect order Diptera. Mosquitoes are holometabolous insects and all larval stages are aquatic. Adults are recognized by their long proboscis for piercing and sucking, and characteristic scaled wing venation. This is a relatively large group of well-known flies with nearly 3000 species in 34 genera reported in the world. There are 13 genera and 167 recognized species of mosquitoes in North America north of Mexico. Almost 75% of these species belong to three genera: *Aedes* (78 species), *Culex* (29 species), and *Anopheles* (16 species).

Adult females lay their eggs on or near water. Most larvae, or wrigglers, feed on algae and organic debris that they filter from the water with their oral brushes, although certain genera may be predaceous and feed on other mosquito larvae. Larvae go through three molts and four instars before pupation. Pupae, or tumblers, are active but nonfeeding stages in which metamorphosis to the adult stage occurs. Both larvae and pupae usually breathe through air tubes at the surface of the water.

Adult male mosquitoes are relatively short-lived, and do not suck blood, but feed primarily on nectar and other plant juices. Females also feed on nectar as their primary energy source, but they require a blood meal for egg production in most species. Some mosquito species are very host-specific, blood-feeding only on humans, birds, mammals, or even reptiles and amphibians, although many species will feed on any available host.

Mosquitoes are of major importance in both human and veterinary medicine. They can cause severe annoyance and blood loss when they occur in dense populations, and they act as vectors of three important groups of disease-causing organisms: *Plasmodium*, the protozoan parasite that produces malaria; filarial worms, parasitic nematodes causing elephantiasis in humans and heartworm disease in canines; and arboviruses, which are the causative agents of yellow fever, dengue fever, LaCrosse encephalitis, St. Louis encephalitis, western equine encephalomyelitis, eastern and Venezuelan equine encephalitis, and several other viral diseases. Human malaria is transmitted exclusively by *Anopheles*, filariasis by *Culex*, *Anopheles*, and *Aedes*, and arboviruses primarily by *Culex* and *Aedes* species. *See* ARBOVIRAL ENCEPHALITIDES; HEARTWORMS; INSECTA; MALARIA; MEDICAL PARASITOLOGY; YELLOW FEVER. Bruce M. Christensen

Mössbauer effect Recoil-free gamma-ray resonance absorption. The Mössbauer effect, also called nuclear gamma resonance fluorescence, has become the basis for a type of spectroscopy which has found wide application in nuclear physics, structural and inorganic chemistry, biological sciences, the study of the solid state, and many related areas of science.

The fundamental physics of this effect involves the transition (decay) of a nucleus from an excited state of energy E_e to a ground state of energy E_g with the emission of a gamma ray of energy E_γ. If the emitting nucleus is free to recoil, so as to conserve momentum, the emitted gamma ray energy is $E_\gamma = (E_e - E_g) - E_r$, where E_r is the recoil energy of the nucleus. Tlie magnitude of E_r is given classically by the relationship $E_r = E_\gamma{}^2/2mc^2$, where m is the mass of the recoiling atom and c is the speed of light. Since E_r is a positive number, the E_γ will always be less than the difference $E_e - E_g$, and if the gamma ray is now absorbed by another nucleus, its energy is insufficient to promote the transition from E_g to E_e.

In 1957 R. L. Mössbauer discovered tnat if the emitting nucleus is held by strong bonding forces in the lattice of a solid, the whole lattice takes up the recoil energy, and the mass in the recoil energy equation given above becomes the mass of the whole lattice. Since this mass typically corresponds to that of 10^{10} to 10^{20} atoms, the recoil energy is reduced by a factor

Three mosaics from *Clementine* spacecraft of the same scene over the crater Tycho, 53 mi (85 km) in diameter. (*a*) Simple ratio image. (*b*) False, multicolor version, which allows different rock units to be distinguished. (*c*) "Stretched" color mosaic, exaggerating the blue reflectance of the central peak. (*U.S. Geological Survey*)

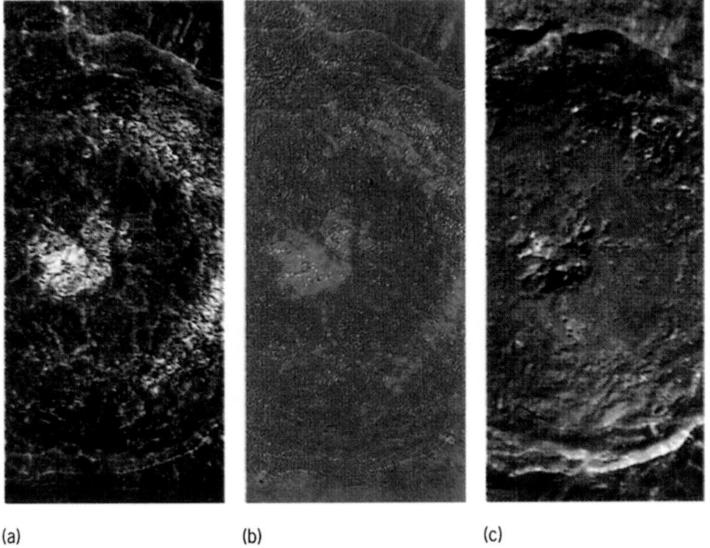

(a) (b) (c)

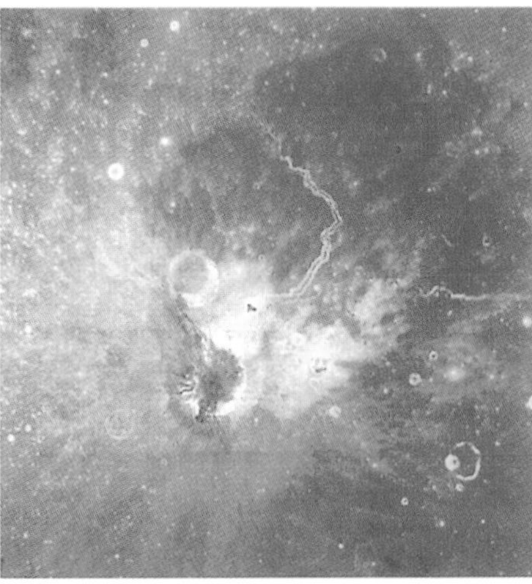

False-color mosaic from *Clementine* spacecraft of the Aristarchus Plateau. Blues are fresh highland materials, deep red is dark pyroclastic (ash) deposits, yellow is outcrop of fresh basalt (in crater and rille walls), and reddish purples are more lava flows. Sinuous rille is Schröter's Valley, a large lava channel on the plateau. (*U.S. Geological Survey*)

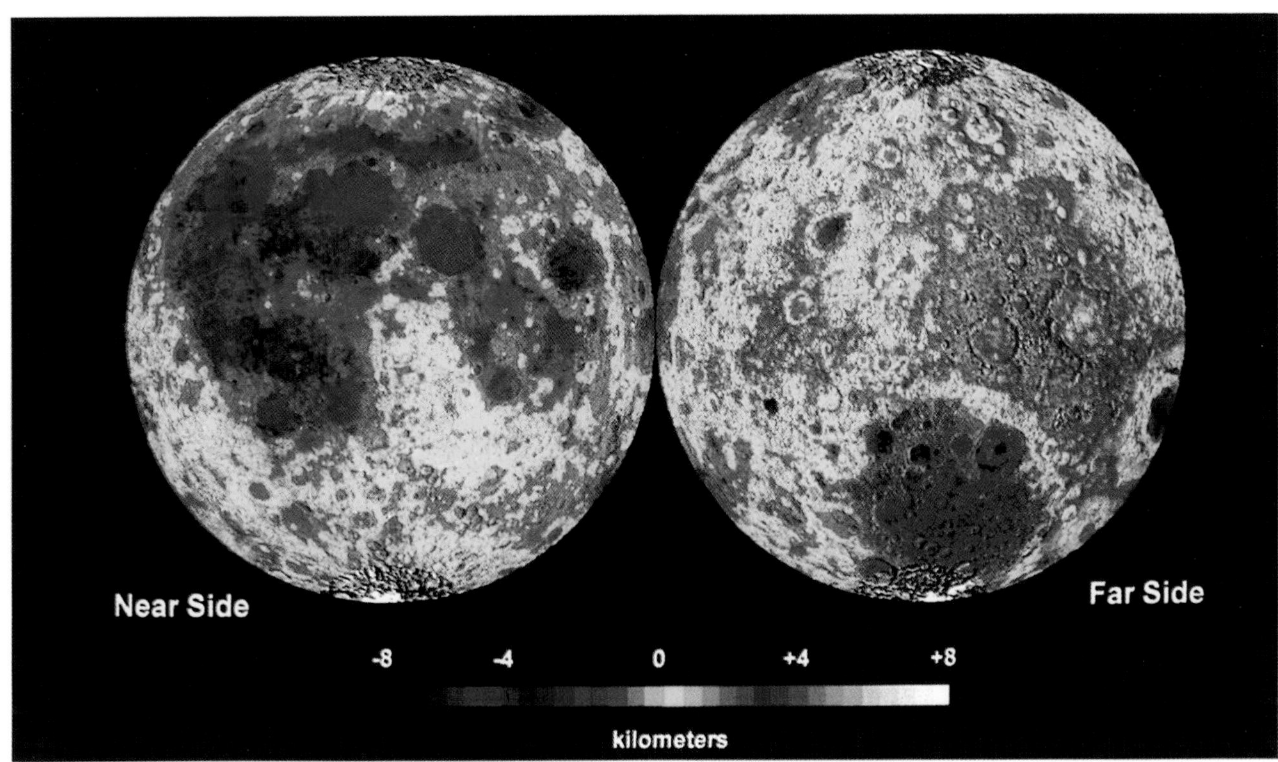

Topographic map of the near and far sides of the Moon, according to *Clementine* laser altimetry. The color bar indicates depths (purple) and heights (red). The far side shows the South Pole–Aitken Basin, 1500 mi (2500 km) in diameter. (*Lunar and Planetary Institute*)

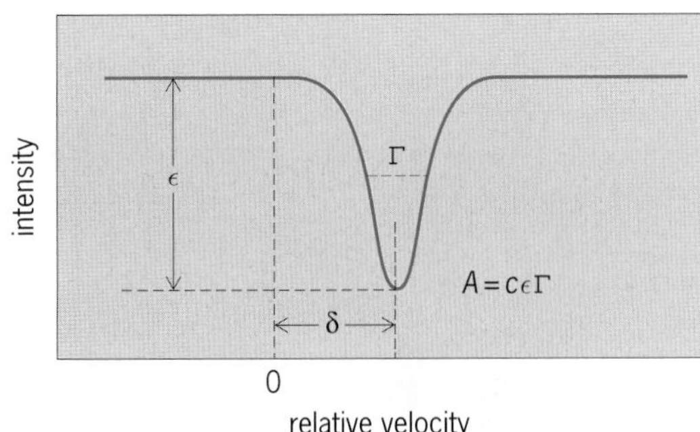

Mössbauer spectrum of an absorber which gives an unsplit resonance line. The spectrum Is characterized by a posi- tion δ, a line width Γ, and an area *A* related to the effect magnitude ε.

of 10^{-10} to 10^{-20}, with the important result that $E_r \approx 0$ so that $E_\gamma = E_e - E_g$; that is, the emitted gamma-ray energy is exactly equal to the difference between the nuclear ground-state energy and the excited-state energy. Consequently, absorption of this gamma ray by a nucleus which is also firmly bound to a solid lattice can result in the "pumping" of the absorber nucleus from the ground state to the excited state. *See* ENERGY LEVEL (QUANTUM MECHANICS); EXCITED STATE; GAMMA RAYS; GROUND STATE.

In a typical Mössbauer experiment the radioactive source is mounted on a velocity transducer which imparts a smoothly varying motion (relative to the absorber, which is held stationary), up to a maximum of several centimeters per second, to the source of the gamma rays. These gamma rays are incident on the material to be examined (the absorber). Some of the gamma rays are absorbed and reemitted in all directions, while the remainder of the gamma rays traverse the absorber and are registered in an appropriate detector.

A typical display of a Mössbauer spectrum, which is the result of many repetitive scans through the velocity range of the transducer, is shown in the illustration. In certain nuclides the Mössbauer resonance line displays splitting that arises from the coupling of the nuclear electric quadrupole moment with the electric field gradient or of the nuclear magnetic dipole moment with the magnetic field at the nucleus, providing information on the magnitude of these interactions.

Mössbauer effect experiments have been used to elucidate problems in a very wide range of scientific disciplines. Applications include the measurement of nuclear magnetic and quadrupole moments and of excited-state lifetimes involved in the nuclear decay process; study of the chemical consequences of nuclear decay; study of the nature of magnetic interactions in iron-containing alloys and of the dependence of the magnetic field in these alloys on various parameters; study of the effects of high pressure on chemical properties of materials; investigation of the relationship between chemical composition and structure on the one hand and the superconductive transition on the other; investigation of the structure of compounds; and study of the structure and bonding properties of metal atoms in complex biological molecules. Rolfe H. Herber

Motion If the position of a material system as measured by a particular observer changes with respect to time, that system is said to be in motion with respect to the observer. Absolute motion, then, has no significance, and only relative motion may be defined; what one observer measures to be at rest, another observer in a different frame of reference may regard as being in motion. *See* FRAME OF REFERENCE; RELATIVE MOTION.

The time derivatives of the various coordinates used to specify the system may be used to prescribe the motion at any instant of time. How the motion develops in subsequent instants is then determined by the laws of motion. In classical dynamics it is supposed that in principle the motion and configuration of the system may be specified to an arbitrary precision, although in quantum mechanics it is recognized that the measurement of the one disturbs the other.

The most general theory of motion that has yet been developed is quantum field theory, which combines both quantum mechanics and relativity theory, as well as the experimentally observed fact that elementary particles can be created and annihilated. *See* DEGREE OF FREEDOM (MECHANICS); DYNAMICS; EULER'S EQUATION OF MOTION; HAMILTON'S EQUATIONS OF MOTION; HARMONIC MOTION; KINEMATICS; LAGRANGE'S EQUATIONS; NEWTON'S LAWS OF MOTION; OSCILLATION; PERIODIC MOTION; QUANTUM FIELD THEORY; QUANTUM MECHANICS; RECTILINEAR MOTION; RELATIVITY; ROTATIONAL MOTION. Herbert C. Corben; Bernard Goodman

Motivation The intentions, desires, goals, and needs that determine human and animal behavior. An inquiry is made into a person's motives in order to explain that person's actions.

Different roles have been assigned to motivational factors in the causation of behavior. Some have defined motivation as a nonspecific energizing of all behavior. Others define it as recruiting and directing behavior, selecting which of many possible actions the organism will perform. The likely answer is that both aspects exist. More specific determinants of action may be superimposed on a dimension of activation or arousal that affects a variety of actions nonselectively. The situation determines what the animal does; arousal level affects the vigor, promptness, or persistence with which the animal does it.

Early drive theorists saw motivated behavior as adjunct to physiological mechanisms of homeostasis, that is, the mechanisms by which the body regulates internal variables such as temperature, blood sugar level, and the volume and concentration of body fluids. Thus, motivated behavior forms part of a negative-feedback loop, an arrangement characteristic of regulatory systems.

However, the homeostatic model faces difficulties. First, not all "basic biological drives" work this way. Second, motivated behavior can be influenced by external as well as internal factors. Since these external influences are not coupled with the animal's internal state, they can lead to behavior that does not promote homeostasis and may even threaten it. To add to the complexity, internal and external factors are not independent and additive; rather they interact with each other. In such cases, internal influences affect behavior by setting the animal's responsiveness to certain external signals. The interaction occurs in the opposite direction as well: external signals can affect internal state. Third, especially in humans, vigorous and persistent goal-directed behavior can occur in the absence of any physiological need. *See* HOMEOSTASIS.

Even relatively simple motives can be influenced by much more than the existing internal and external situation. They respond to potential or expected factors, as registered by cognitive apparatus. Even relatively simple motives such as hunger and thirst are responsive to cognitive factors. *See* THIRST.

To a hungry rat, food becomes a goal. The rat will make various responses, including arbitrary learned ones or operants, that lead to contact with food. A rat can be trained to do whatever else is necessary (within its capabilities) to attain its goal. It is this flexibility of goal-directed behavior that justifies the concept of motivation. If an animal will do whatever is necessary to obtain food, it must want food. Internal factors then may act by setting the goal status of environmental commodities: the effect of hunger is to make food a goal.

There is a question as to how behavior can be guided by a state or event (goal attainment) that does not yet exist. Modern approaches to this question lean heavily on cognitive concepts. Mammals, birds, and even some insects can represent to themselves a nonexistent state of affairs. They can represent what a goal object is (search images): a chimpanzee may show behavioral signs of surprise if a different food is substituted for the usual one. They can represent where it is (cognitive maps): a digger wasp remembers the location of its nest relative to arbitrary landmarks, and will fly to the wrong place if the landmarks are moved.

If this idea is generalized, motivated behavior can be thought of as guided by a feedback control system with a set point. A set point establishes a goal state which the control system seeks to bring about. Behavior is controlled, not by present external or internal stimuli alone, but by a comparison between the existing state of affairs and a desired state of affairs, that is, the set point or goal, registered or specified within the brain. The animal then acts to reduce the difference between the existing and the desired state of affairs.

This way of looking at motivation helps bridge the gap between simple motives in animals and complex ones in humans. If to be motivated is to do whatever is necessary to bring about an imagined state of affairs, then human motives can literally be as complex, and be projected as far into the future, as human imaginations permit. *See* COGNITION.

Another approach to motivation comes from ethology, which has formed links with cognitive psychology. The broken-wing display of the piping plover provides an example. If a predator approaches a nest with eggs, the parent bird may behave as if injured (hence easy prey) and thus lead the intruder away from the nest. This action pattern is characteristic of the species and unlearned in its gross topography; yet the bird monitors the intruder's behavior and modulates the display accordingly. It may approach more closely and intensify the display if the intruder is not at first diverted from its path. Thus a species-typical action pattern can be used in ways suggestive of purpose and goal direction: the bird modifies it as necessary to promote the goal of diverting the intruder. *See* ETHOLOGY.

Motivation and emotion are closely related. Indeed, it has been argued that emotions are the true motivators and that other factors internal, situational, and cognitive take hold of behavior by way of the emotions they evoke. In the simplest case, pleasure and displeasure have been recognized for centuries as having motivational force. In more complex cases, the role of cognitive operations, such as how an individual feels about an event, as well as what is done about it, can depend heavily on how an individual thinks about it.

The culture in which an individual is raised has a powerful effect on how the individual behaves. It has been argued that culture teaches its members what to believe are the consequences of a specific action (cognitive), and how the individuals should feel about those consequences or about the actions themselves (emotional/motivational). Douglas G. Mook

Motor A machine that converts electrical into mechanical energy. Motors that develop rotational mechanical motion are most common, but linear motors are also used. A rotary motor delivers mechanical power by means of a rotating shaft extending from one or both ends of its enclosure (see illustration). The shaft is attached internally to the rotor. Shaft bearings permit the rotor to turn freely. The rotor is mounted coaxially with the stationary part, or stator, of the motor. The small space between the rotor and stator is called the air gap, even though fluids other than air may fill this gap in certain applications.

In a motor, practically all of the electromechanical energy conversion takes place in the air gap. Commercial motors employ magnetic fields as the energy link between the electrical input and the mechanical output. The air-gap magnetic field is set up by current-carrying windings located in the rotor or the stator, or by a combination of windings and permanent magnets. The magnetic field exerts forces between the rotor and stator to produce the mechanical shaft torque; at the same time, in accord with Faraday's law, the magnetic field induces voltages in the windings. The voltage induced in the winding connected to the

Cutaway view of a single-phase induction motor. (*Emerson Motor Division*)

electrical energy source is often called a countervoltage because it is in opposition to the source voltage. By its magnitude and, in the case of alternating-current (ac) motors, its phase angle, the countervoltage controls the flow of current into the motor's electrical terminals and hence the electrical power input. The physical phenomena underlying motor operation are such that the power input is adjusted automatically to meet the requirements of the mechanical load on the shaft. *See* ELECTROMAGNETIC INDUCTION; MAGNET; WINDINGS IN ELECTRIC MACHINERY.

Both the rotor and stator have a cylindrical core of ferromagnetic material, usually steel. The parts of the core that are subjected to alternating magnetic flux are built up of thin sheet steel laminations that are electrically insulated from each other to impede the flow of eddy currents, which would otherwise greatly reduce motor efficiency. The windings consist of coils of insulated copper or aluminum wire or, in some cases, heavy, rigid insulated conductors. The coils may be placed around pole pieces, called salient poles, projecting into the air gap from one of the cores, or they may be embedded in radial slots cut into the core surface facing the air gap. In a slotted core, the core material remaining between the slots is in the form of teeth, which should not be confused with magnetic poles. *See* EDDY CURRENT.

Direct-current (dc) motors usually have salient poles on the stator and slotted rotors. Polyphase ac synchronous motors may have salient poles on the rotor and slotted stators. Rotors and stators are both slotted in induction motors. Permanent magnets may be inserted into salient pole pieces, or they may be cemented to the core surface to form the salient poles.

The windings and permanent magnets produce magnetic poles on the rotor and stator surfaces facing each other across the air gap. If a motor is to develop torque, the number of rotor poles must equal the number of stator poles, and this number must be even because the poles on either member must alternate inpolarity (north, south, north, south) circularly around the air gap. George McPherson, Jr.

Motor-generator set A motor and one or more generators, with their shafts mechanically coupled, used to convert the voltage or frequency of an available power source to another desired frequency or voltage. The motor of the set is selected so that it operates from the available power supply; the generators are designed to provide the desired output voltage or frequency.

The principal advantage of a motor-generator set over other conversion systems is the flexibility offered by the use of separate machines for each function. Since energy conversion is employed twice, electrical to mechanical and back to electrical, the efficiency of this system is lower than that in most other conversion methods. *See* GENERATOR; MOTOR. Arthur R. Eckels

Motor systems Those portions of nervous systems that regulate and control the contractile activity of muscle and the secretory activity of glands. Muscles and glands are the two types of organ by which an organism reacts to its environment; together they constitute the machinery of behavior. Cardiac muscle and some smooth muscle and glandular structures can function independently of the nervous system but in a poorly coordinated fashion. Skeletal muscle activity, however, is entirely dependent on neural control. Destruction of the nerves supplying skeletal muscles results in paralysis, an inability to move. The somatic motor system includes those regions of the central nervous system involved in controlling the contraction of skeletal muscles in a manner appropriate to environmental conditions and internal states. *See* GLAND; MUSCLE.

Skeletal muscle. The nerve supply to skeletal muscles of the limbs and trunk is derived from large nerve cells called motoneurons, whose cell bodies are located in the ventral horn of the spinal cord. Muscles of the face and head are innervated by motoneurons in the brainstem. The axons of the motoneurons traverse the ventral spinal roots (or the appropriate cranial nerve roots) and reach the muscles via peripheral nerve trunks. In the muscle, the axon of every motoneuron divides repeatedly into many terminal branches, each of which innervates a single muscle fiber. The region of innervation, called the neuromuscular junction or motor end plate, is a secure synaptic contact between the motoneuron terminal and the muscle fiber membrane. *See* SYNAPTIC TRANSMISSION.

Since synaptic transmission at the neuromuscular junction is very secure, an action potential in the motoneuron will produce contraction of every muscle fiber that it contacts. For this reason, the motoneuron and all the fibers it innervates form a functional unit called the motor unit. The number of muscle fibers in a single motor unit may be as small as six (for intrinsic eye muscles) or over 700 (for motor units of large limb muscles). In general, muscles involved in delicate rapid movements have fewer muscle fibers per motor unit than large muscles concerned with gross movements.

Components of skeletal motor system. Motoneurons are activated by nerve impulses arriving through many different neural pathways. Some of their neural input originates in peripheral receptor organs located in the muscles themselves, or in receptors in skin or joints. Many muscle receptors discharge in proportion to muscle length or tension; such receptors have relatively potent connections to motoneurons, either direct monosynaptic connections or relays via one or more interneurons. Similarly,

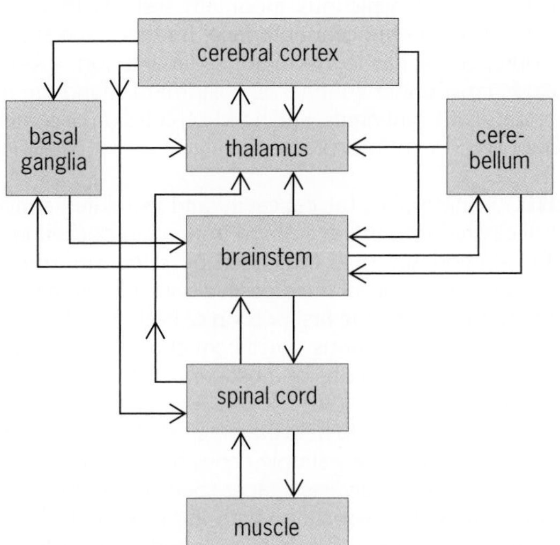

Schematic diagram of the major components of the vertebrate motor systems. Arrows indicate the main neural connections between regions.

stimulation of skin and joints, particularly painful stimulation, can strongly affect motoneurons. Such simple segmental pathways constitute the basis for spinal reflexes. The other major source of input to motoneurons arises from supraspinal centers. The illustration shows the main nervous system centers involved in controlling the input to motoneurons.

Segmental circuits. At the spinal level, the motoneurons and muscles have a close reciprocal connection. Afferent connections from receptors in the muscles return sensory feedback to the same motoneurons which contract the muscle. Connections to motoneurons of synergist and antagonist muscles are sufficiently potent and appropriately arranged to subserve a variety of reflexes. In animals with all higher centers removed, these segmental circuits may function by themselves to produce simple reflex responses. Under normal conditions, however, the activity of segmental circuits is largely controlled by supraspinal centers. Descending tracts arise from two major supraspinal centers: the cerebral cortex and the brainstem.

Brainstem. The brainstem, which includes the medulla and pons, is a major and complex integrating center which combines signals descending from other higher centers, as well as afferent input arising from peripheral receptors. The descending output from brainstem neurons affects motor and sensory cells in the spinal cord. Brainstem centers considerably extend the motor capacity of an animal beyond the stereotyped reflex reactions mediated by the spinal cord. In contrast to segmental reflexes, these motor responses involve coordination of muscles over the whole body. Another major motor function of the brainstem is postural control, exerted via the vestibular nuclei of the ear.

Besides neurons controlling limb muscles, the brainstem also contains a number of important neural centers involved in regulating eye movements. These include motor neurons of the eye muscles and various types of interneurons that mediate the effects of vestibular and visual input on eye movement.

Cerebellum. Another important coordinating center in the motor system is the cerebellum, an intricately organized network of cells closely interconnected with the brainstem. The cerebellum receives a massive inflow of sensory signals from peripheral receptors in muscles, tendons, joints, and skin, as well as from visual, auditory, and vestibular receptors. Higher centers, particularly the cerebral cortex, also provide extensive input to the cerebellum via pontine brainstem relays. The integration of this massive amount of neural input in the cerebellum somehow serves to smooth out the intended movements and coordinate the activity of muscles. Without the cerebellum, voluntary movements become erratic, and the animal has difficulty accurately terminating and initiating responses. The output of the cerebellum affects primarily brainstem nuclei, but it also provides important signals to the cerebral cortex.

Basal ganglia. At another level of motor system are the basal ganglia. These massive subcortical nuclei receive descending input connections from all parts of the cerebral cortex. Their output projections send recurrent information to the cerebral cortex via the thalamus, and their other major output is to brainstem cells.

Cerebral cortex. At the highest level of the nervous system is the cerebral cortex, which exerts control over the entire motor system. The cerebral cortex performs two kinds of motor function: certain motor areas exert relatively direct control over segmental motoneurons, via a direct corticospinal pathway, the pyramidal tract, and also through extrapyramidal connections via supraspinal motor centers. The second function, performed in various cortical association areas, involves the programming of movements appropriate in the context of sensory information, and the initiation of voluntary movements on the basis of central states. Cortical language areas, for example, contain the circuitry essential to generate the intricate motor patterns of speech. Limb movements to targets in extrapersonal space appear to be programmed in parietal association cortex. Such cortical areas involved in motor programming exert their effects via corticocortical connections to the motor cortex, and by descending connections to subcortical centers, principally basal ganglia and brainstem.

As indicated in the illustration, the motor centers are all heavily interconnected, so none really functions in isolation. In fact, some of these connections are so massive that they may form functional loops, acting as subsystems within the motor system. For example, most regions of the cerebral cortex have close reciprocal interconnections with underlying thalamic nuclei, and the corticothalamic system may be considered to form a functional unit. Another example is extensive connection from cerebral cortex to pontine regions of the brainstem, controlling cells that project to the cerebellum, which in turn projects back via the thalamus to the cerebral cortex. Such functional loops are at least as important in understanding motor coordination as the individual centers themselves. *See* BRAIN; NERVOUS SYSTEM (VERTEBRATE). Eberhard E. Fetz

Mountain A feature of the Earth's surface that rises high above its base and has generally steep slopes and a relatively small summit area. Commonly the features designated as mountains have local heights measurable in thousands of feet, lesser features of the same type being called hills, but there are many exceptions. *See* HILL AND MOUNTAIN TERRAIN.

Mountains rarely occur as isolated individuals. Instead they are usually found in roughly circular groups or massifs, such as the Olympic Mountains of northwestern Washington, or in elongated ranges, like the Sierra Nevada of California. An array of linked ranges and groups, such as the Rocky Mountains, the Alps, or the Himalayas, is a mountain system. North America, South America, and Eurasia possess extensive cordilleran belts, within which the bulk of their higher mountains occur. *See* CORDILLERAN BELT; MASSIF; MOUNTAIN SYSTEMS.

As a rule, mountains represent portions of the Earth's crust that have been raised above their surroundings by upwarping, folding, or buckling, and have been deeply carved by streams or glaciers into their present surface form. Some individual peaks and massifs have been constructed upon the surface by outpourings of lava or eruptions of volcanic ash. *See* OROGENY. Edwin H. Hammond

Mountain meteorology The effects of mountains on the atmosphere, ranging over all scales of motion, including very small (such as turbulence), local (for instance, cloud formations over individual peaks or ridges), and global (such as the monsoons of Asia and North America).

The most readily perceived effects of a mountain, or even of a hill, are related to the blocking of air flow. When there is sufficient wind, the air either goes around the obstacle or over it, causing waves in the flow similar to those in a river washing over a boulder. Since ascending air cools by adiabatic expansion, the saturation point of water vapor may be reached in such waves as they form over an obstacle, and a cloud then forms in the ascending branch of the wave motion. Such a cloud dissipates in the descending branch where adiabatic warming takes place. The shapes and amplitudes of these lee waves (they form over and to the lee of mountains) depend not only on the thermal stability and on the vertical wind shear in the overlying atmosphere but also on the shape of the underlying terrain. *See* CLOUD; CLOUD PHYSICS; WAVE (PHYSICS).

On a grander scale, mountain ranges, such as the Sierras of North and South America, place an obstacle in the path of the westerly winds (that is, winds from the west), which generally prevail in middle latitudes. Such a blockage tends to generate a high-pressure region upwind from the mountains (this may be viewed as air piling up as it prepares to jump the hurdle), and a low-pressure area downwind. Thus, there is a stronger push against the mountains on the high-pressure western side than on the low-pressure eastern side. The net effect is the slowing down of the atmospheric flow (mountain torque). *See* TORQUE.

Less subtle than mountain torque effects are the large-scale meanders that develop in the global flow patterns once they have been perturbed, mainly by the North and South American Andes and by the Plateau of Tibet and its Himalayan mountain ranges. These meanders in the large-scale flow are known as planetary waves. They appear prominently in the pressure patterns of hemispheric or global weather maps. *See* WEATHER MAP.

The major monsoon circulations interact with the global circulation, shaped in part by sea-surface temperature anomalies in the equatorial Pacific. The various aspects of mountain meteorology, therefore, have to be viewed within the larger picture. There is a continuous interaction between the weather effects on all space and time scales generated by the mountains and the weather patterns that prevail elsewhere on the Earth. *See* METEOROLOGY. Elmar R. Reiter

Mountain systems Long, broad, linear to arcuate belts in the Earth's crust where extreme mechanical deformation and thermal activity have been (or are being) concentrated.

Mountain systems in the general sense occur both on continents and in ocean basins, but the geological properties of the systems in continental as opposed to oceanic settings are distinctly different. The mechanical strain in classical, continental mountain systems is expressed in the presence of major folds, faults, and intensive fracturing and cleavage. Thermal effects are in the form of vast volcanic outpourings, intruded bodies of igneous magma, and metamorphism. Uplift and deformation in young mountain systems are conspicuously displayed in the physiographic forms of topographic relief. Where mountain building is presently taking place, the dynamics are partly expressed in warping of the land surface and significant shallow or deep earthquake activity. Locations of ancient mountain systems in continental regions now beveled flat by erosion are clearly disclosed by the presence of highly deformed, intruded, and metamorphosed rocks.

Two basic classes of oceanic mountain systems exist. A world-encircling oceanic rift mountain system has been built along the extensional tectonic boundary between plates diverging at rates of 0.8–2.4 in. or 2–6 cm per year from the mid-oceanic ridges. This rift mountain system is exposed to partial view in Iceland. The second type, island arc mountain systems, occur in oceanic basins where the crust dives downward at trench sites, thus underthrusting adjacent oceanic crust. *See* MARINE GEOLOGY.

The classical, conspicuous mountain systems of the Earth occur at the continent/ocean interface, for this is the site where plate convergence has led to major sedimentation, subduction of oceanic crust under continents, collision of island arc mountain systems with continents, and head-on collision of continents. *See* OROGENY; PLATE TECTONICS. George H. Davis

Mouth The oral or buccal cavity and its related structures. The oral cavity forms in the embryo from an in-pocketing of the skin, the stomodeum; it is thus lined by ectoderm and is not, properly speaking, part of the digestive tract. Functionally, however, the mouth forms the first portion of both the digestive and respiratory systems. Various special structures are found in, or associated with, the mouths of most vertebrates. *See* DIGESTIVE SYSTEM; RESPIRATORY SYSTEM.

Teeth may be present to help grasp or grind food. In most vertebrates they are relatively simple cones but in some, especially mammals, they are of diverse shapes. *See* DENTITION; TOOTH.

Various glands are associated with the mouth. These are of infrequent occurrence in fish but are found in most tetrapods. Humans have three pairs of salivary glands: the parotid, submaxillary, and sublingual. In forms such as some snakes salivary glands may produce a poison used to subdue prey.

Other structures also vary greatly. Most tetrapods have a mobile tongue attached to the floor of the mouth, but few fish do. The structure of the roof of the mouth, or palate, is quite different in different groups. *See* PALATE; TONGUE.

In mammals, including humans, the margins of the lips mark the junction between the outer skin and the inner mucous lining of the oral cavity. The mucosa of the mouth forms the lining and the gums surrounding the teeth and covers the surface of the tongue. The roof of the mammalian mouth consists of the hard palate and, behind this, the soft palate which merges into the oropharynx. The lateral walls consist of the distensible cheeks, and the floor is formed principally by the tongue and the soft tissues that lie between the two sides of the lower jaw, or mandible.

The posterior limit of the oral cavity of mammals is marked by the fauces, an aperture which leads to the pharynx. On either side of the fauces are two muscular arches covered by mucosa, the glossopalatine and pharyngopalatine arches; between them lie masses of lymphoid tissue, the tonsils. Suspended from the posterior portion of the soft palate is the soft retractable uvula. *See* MOUTH DISORDERS; TONSIL. Thomas S. Parsons

Mouth disorders The mouth, or oral cavity, which comprises the lips, tongue, teeth, gums, and related structures, is subject to a large number of disease processes. Periodontitis, an inflammatory disease of the tissues supporting and surrounding the teeth, and dental decay are the most common diseases; together they account for almost all tooth loss. Other diseases of the mouth can be classified as cysts; diseases of the salivary glands; keratotic, inflammatory, ulcerative, and proliferative lesions; oral infections; and an unusual form of rapidly destructive periodontitis observed in some patients with acquired immune deficiency syndrome (AIDS).

The oral cavity is made up of many varieties of tissue, each of which can give rise to neoplasms. Thus, several types of malignant and benign neoplasms can be found in and around the mouth. The benign tumors most often seen are fibromas, papillomas, and hemangiomas. Oral squamous-cell carcinoma (epidermoid carcinoma) accounts for over 90% of all malignancies of the mouth. Its most common site is the lower lip, and it is more common in males than in females. *See* ONCOLOGY.

Gingivitis and periodontitis, as well as dental caries, are a result of oral bacterial infections. Mycotic infection by *Candida albicans* is recognized clinically by the presence of red, inflamed mucosa bearing colonies of white fungi with the appearance of milk curds. The disease is called candidiasis or thrush, and it occurs most commonly in children, in adults after extensive antibiotic therapy, and in some patients with acquired immune deficiency syndrome (AIDS). *See* ACQUIRED IMMUNE DEFICIENCY SYNDROME (AIDS); PERIODONTAL DISEASE.

Infection of the mouth and lips by the virus herpes simplex type I is possibly the most common human viral infection, next to the common cold. Initially, the lesions appear as small blebs or blisters which rupture to form open ulcers. The ulcers are extremely painful and can involve the entire mouth. *See* HERPES.

Roy C. Page

Moving-target indication A method of presenting pulse-radar echoes in a manner that discriminates in favor of moving targets and suppresses stationary objects. Moving-target indication (MTI) is almost a necessity when moving targets are being sought over a region from which the ground clutter echoes are very strong. The most common presentation of the output of a radar with MTI is a plan-position indicator (PPI) display. The moving targets appear as bright echoes, while ground clutter is suppressed. *See* RADAR. John M. Carroll

Mucilage A naturally occurring, high-molecular-weight (200,000 and up), organic plant product of unknown detailed structure. The term is loosely used, often interchangeably with the term gum. Chemically, mucilage is closely allied to gums and pectins but differs in certain physical properties. Although gums swell in water to form sticky, colloidal dispersions and pectins gelatinize in water, mucilages form slippery, aqueous colloidal dispersions. Mucilages are formed in normal plant growth within the plant by mucilage-secreting hairs, sacs, and canals, but they are not found on the surface as exudates as a result of bacterial or fungal action after mechanical injury, as are gums. Mucilages occur in nearly all classes of plants in various parts of the plant, usually in relatively small percentages, and are not infrequently associated with other substances, such as tannins. The chief industrial sources of mucilages are Icelandic and Irish moss, linseed, locust bean, slippery elm bark, and quince seed. *See* ADHESIVE; GUM; PECTIN. Elbert H. Hadley

Muffler A device used to attenuate sound while also allowing fluid (usually gas) to flow through it; also known as silencer in British usage. Mufflers are extensively used to reduce the intake and exhaust noise from pumps, fans, compressors, and internal combustion engines. Although active noise control techniques are emerging, most mufflers continue to use passive silencing methods. Passive mufflers are categorized as reactive or dissipative based on their primary method of attenuation. Reactive mufflers reflect sound back toward the noise source, and dissipative mufflers use porous materials to absorb the sound.

Reactive mufflers reflect acoustic waves at locations where a duct expands, contracts, or branches. Often a combination of reactive elements such as expansion chambers, resonators, and flow reversals is used. Reactive mufflers can be designed to provide better low-frequency attenuation than a dissipative muffler of similar size. Also, reactive mufflers can be used in harsh environments that dissipative or active mufflers might not withstand. In most cases, reactive mufflers are best suited for low-to-moderate frequencies, where acoustic wavelengths are larger than any cross dimension of the muffler. At these frequencies, mufflers can exhibit resonance or broadband attenuation behavior. *See* RESONANCE (ACOUSTICS AND MECHANICS).

Dissipative mufflers use absorptive materials that dissipate the acoustic energy into heat. A variety of porous media can be used for absorption, with fibrous materials such as fiberglass being common. The linings and baffles can be flat, contoured, constructed from layers of different materials, or mixed and matched for a particular application. Absorptive materials may face challenges due to harsh conditions such as high temperatures and potential clogging from particulate-laden flows. Dissipative mufflers are best suited for moderate-to-high frequencies, since absorption is less effective at low frequencies. At frequencies where the absorptive materials are effective, the attenuation is broadband, and the passbands exhibited by reactive mufflers are reduced or eliminated. Compared to reactive mufflers of similar size, dissipative mufflers can have higher attenuation (except at resonances for the reactive muffler) and lower pressure drop. At higher frequencies, where the acoustic wavelength is smaller than the duct width, the attenuation of a dissipative muffler may decrease considerably. *See* SOUND ABSORPTION.

Active mufflers attenuate unwanted noise by adding sound to counteract it. The disturbances add algebraically, resulting in a cancellation of the unwanted sound. An active muffler consists of sensors (such as microphones), a controller, and actuators (such as loudspeakers). The controller unit processes the signals from the sensor, and computes an appropriate signal for the actuator. Numerous control systems and strategies exist, and are under continuous development. Active mufflers are best suited for low frequencies where the sound field is relatively simple. The effectiveness of active mufflers has been demonstrated for a number of situations, but several challenges are the topic of ongoing research. There is a need for rugged sensors and actuators that can withstand high temperatures and harsh environments. Also, high-intensity disturbances at low frequencies require large-displacement, high-power actuators. *See* ADAPTIVE SOUND CONTROL.

Ahmet Selamet

Mugiliformes An order in the class Actinopterygii, subdivision Euteleostei, superorder Acanthopterygii. The Mugiliformes comprise 17 genera and about 72 species, all in the single family Mugilidae, the mullets (also known as grey mullets). They are presently deemed subperciforms and are the only order in Mugilomorpha, one of three series making up the Acanthopterygii. *See* ACTINOPTERYGII; PERCIFORMES.

The following characters distinguish the family: The head and body sectors are more or less round in cross section and the tail sector is moderately compressed; the dorsal fins are well separated, the first has four weak spines, and the second 8 to 10 soft rays; the pectoral fins are high on the body; the pelvic fins are subabdominal, each with one spine and five rays; the anal fin has two or three spines and 7 to 11 soft rays; an adipose eyelid is usually present; the mouth is small and terminal; teeth are small or absent; gill rakers are long; the stomach is usually muscular (gizzardlike), and the intestine is long and coiled; scales of adults are usually ctenoid (or cycloid in juveniles); and the lateral line is absent or very faint. Adult mullet species range from 15 cm (6 in.) to 1.2 m (4 ft) in maximum length, but most species are about 30 to 70 cm (11.8 to 27.5 in.) long [see illustration].

Largescale mullet. (*Photo courtesy of John E. Randall*)

Mullets inhabit coastal marine and brackish-water environments in all tropical and temperate seas, as well as freshwater. Some marine species enter freshwater only temporarily and may ascend great distances up rivers, such as the striped mullet (*Mugil cephalus*), a common and economically important species that occurs practically worldwide in warm waters. Most species form schools, some quite large.

Herbert Boschung

Mulberry A genus (*Morus*) of trees characterized by milky sap and simple, often lobed, alternate leaves. White mulberry (*M. alba*) was introduced into the United States from China during the 19th century as a source of food for silkworms. The silkworm project was unsuccessful, but the trees remained and are common in cities and on the borders of forests. Red mulberry (*M. rubra*) grows in the eastern half of the United States and in southern Ontario. The wood is used for fence posts, furniture, interior finish, agricultural implements, and barrels.

Arthur H. Graves; Kenneth P. Davis

Mule A hybrid sired by a male ass (*Equus asinus*) out of a female horse (*E. caballus*). The opposite cross, very seldom made, produces the hinny (a hybrid between a stallion and a female ass). The mule and the hinny are usually sterile, but two authenticated cases of mules producing living progeny are known. Male mules, often called horse mules, are almost always castrated to make them more tractable as work animals. Mules are noted for their endurance, surefootedness, and ability to stand hard work in hot weather. They can safely be self-fed in lots or corrals, whereas horses cannot. Usually steady and free from nervous excitability, mules can be handled by inexperienced or careless farm labor.

John M. Kays

Multiaccess computer A computer system in which computational and data resources are made available simultaneously to a number of users. Users access the system through terminal devices, normally on an interactive or conversational basis. A multiaccess computer system may consist of only a single central processor connected directly to a number of terminals (that is, a star configuration), or it may consist of a number of processing systems which are distributed and interconnected with each other as well as with the user terminals.

The primary purpose of multiaccess computer systems is to share resources. The resources being shared may be simply the data-processing capabilities of the central processor, or they may be the programs and the data bases they utilize. The earliest examples of the first mode of sharing are the general-purpose, time-sharing, computational services. Examples of the latter mode are airlines reservation systems in which it is essential that all ticket agents have immediate access to current information.

System components. The major hardware components of a multiaccess computer system are terminals or data entry/display devices, communication lines to interconnect the terminals to the central processors, a central processor, and on-line mass storage. Terminals may be quite simple, providing only the capabilities for entering or displaying data, or they may have an appreciable amount of "local intelligence" to support simple operations like editing of the displayed text without requiring the involvement of the central processor. The interconnecting communication lines can be provided by utilizing the common-user telephone system or by obtaining leased, private lines from the telephone company or a specialized carrier.

System operating requirements. A multiaccess system must include the following functional capabilities: (1) multiline communications capabilities that will support simultaneous conversations with a reasonably large number of remote terminals; (2) concurrent execution of a number of programs with the ability to quickly switch from executing the program of one user to executing that of another; (3) ability to quickly locate and make available data stored on the mass storage devices while at the same time protecting such data from unauthorized access.

The ability of a system to support a number of simultaneous sessions with remote users is an extension of the capability commonly known as multiprogramming. In order to provide such service, certain hardware and software features should be available in the central processor. Primary among these is the ability to quickly switch from executing one program to another while protecting all programs from interference with one another.

Memory sharing is essential to the efficient operation of a multiaccess system. A popular memory management technique is the utilization of paging. The program is broken into a number of fixed-size increments called pages. Similarly, central memory is divided into segments of the same size called page frames. (Typical sizes for pages and page frames are 512 to 4096 bytes.) Under the concept known as demand paging, only those pages that are currently required by the program are loaded into central memory.

Software capabilities. The control software component of most interest to an interactive user is the command interpreter. This routine interacts directly with users, accepting requests for service and translating them into the internal form required by the remainder of the operating system, as well as controlling all interaction with the system.

The capability to page the memory as outlined above can be utilized to provide users with the impression that each has available a memory space much larger than is actually assigned. Such a system is said to provide a virtual memory environment. Similarly, the ability of the operating system to quickly change context from one executing program to another will result in users' receiving the impression that each has an individual processor. *See* DIGITAL COMPUTER. Philip H. Enslow, Jr.

Multidimensional scaling Any of a number of procedures for the fitting of a selected geometric model to a body of multivariate data. In the original form advanced by Warren S. Torgerson, multidimensional scaling dealt with judgments about stimuli, with its major aim being the derivation of psychological dimensions presumed to underlie the judgments, and the determination of scale values for stimuli along the derived dimensions. Classical multidimensional scaling was a significant methodological advance in the characterization of complex stimulus domains.

With the advent of multidimensional scaling and related techniques, psychological inquiry benefited in two important ways. First, the data collection procedures associated with the new techniques did not require prior specification of constituent dimensions before scaling could take place. As a consequence, it was necessary to instruct the subjects in the specific dimensions to be considered in making judgments. Since most multidimensional scaling procedures made use of judgments on the metadimensions of similarity of preference rather than judgments on specific psychological dimensions (such as asking whether something is good or bad), the elicited responses were considered to be less artificially constrained by the investigator, and hence more likely to reflect the true nature of the subject's psychological experience. Second, the multidimensional scaling procedures allowed all relevant stimulus dimensions to be extracted in the same operation and to exhibit their relative strengths and interrelationships within the confines of a single model. This feature thus freed researchers of the requirement of specifying dimensions in advance. These two features, rightly or wrongly, served to promote the idea that multidimensional scaling would allow "recovery" of the cognitive structure underlying the judgments.

Fields other than psychology also found applications for multidimensional scaling. Through the use of metadimensions such as proximity or dominance, analysis could be extended to well-defined surrogates, such as indices of overlap or confusion, measures of net migration, rankings of dominance, or coefficients of interaction. The procedure then allowed psychological scaling methodology to be applied in nonpsychological sciences and permitted results to be interpreted, not in terms of psychological processes, but in terms of processes appropriate for the particular field of application. Richard Degerman

Multilevel control theory An approach to the control of large-scale systems based on (1) decomposition of the complex overall control problem into simpler and more easily managed subproblems and (2) coordination of the subproblems so that overall system objectives and constraints are satisfied.

The controllers are organized in a multilevel hierarchical structure according to three basic criteria: functional, plant, and temporal decomposition. Functional and temporal decompositions are often classified as multilayer or vertical structures; plant decomposition is often classified as a multilevel or horizontal structure.

In the functional decomposition approach, the overall control problem is partitioned into a nested set of generic control functions such as the regulatory (or direct control) function, optimizing control function, the adaptive control function, and the self-organizing function.

In the plant decomposition multilevel approach, the controlled system (plant) is partitioned into subsystems along lines of weak interaction. In a two-level control hierarchy, each subsystem has its own (first-level) controller which acts to satisfy local objectives and constraints. A second-level controller (coordinator) influences the actions of the local controllers to compensate for subsystem interactions so that overall objectives and constraints are satisfied.

In the temporal decomposition approach, the control or decision-making problem is partitioned into subproblems based on the different time scales relevant to the associated action functions. These time scales reflect such factors as the response time of the plant, the bandwidth characteristics of the disturbance inputs, and trade-off considerations relating the benefit of control action to its cost. At each layer of the hierarchy, decisions and control actions determined at higher layers (corresponding to lower-frequency events) are treated as constants; disturbances and actions associated with lower layers are treated as noise to be represented by their respective mean values. The temporal hierarchy may embrace a broad spectrum of control and decision-making activities, including process control, production control, scheduling, and planning functions. Time scales may range from seconds to years. *See* ADAPTIVE CONTROL; OPTIMAL CONTROL THEORY; PROCESS CONTROL; PRODUCTION ENGINEERING.

Hierarchical control has become accepted technology in many industries. It is often implicitly embedded in the design of the system, providing the conceptual framework for integrated (or plant-wide) systems control. Motivating considerations are improved productivity, operating efficiency, product quality, or other economic-based objectives. Hierarchical control has been applied to various nonindustrial systems, such as robotics and air traffic control systems. *See* AIR-TRAFFIC CONTROL; CONTROL SYSTEMS; DISTRIBUTED SYSTEMS (CONTROL SYSTEMS). Irving Lefkowitz

Multimedia technology Computer-based, interactive applications having multiple media elements, including text, graphics, animations, video, and sound. Multimedia technology refers to both the hardware and software used to create and run such systems.

The mode of delivery for each application depends on the amount of information that must be stored, the privacy desired, and the potential expertise of the users. Applications that require large amounts of data are usually distributed on CD-ROMs, while personal presentations might be made directly from a computer using an attached projector. Advertising and some training materials are often placed on the WWW for easy public access. Museums make use of multimedia kiosks with touch screens and earphones. *See* COMPACT DISK; INTERNET; VIDEO GAMES; WORLD WIDE WEB.

Multimedia products may be created and run on the commonly used computer environments. Multimedia system users may employ a variety of input devices in addition to the keyboard and mouse, such as joysticks and trackballs. Touch screens provide both input and display capabilities and are often the choice when potentially large numbers of novices may use the system. Other display devices include high-resolution monitors and computer projectors. Generally the abundance of graphics and video in multimedia applications requires the highest resolution and deepest color capacity possible in display devices.

Input devices for the creation of multimedia applications include graphics tablets, which are pressure-sensitive surfaces for drawing with special pens; digital cameras, which take pictures electronically; and scanners, which convert existing pictures and graphics into digital form. Other hardware devices, such as a video card and video digitizing board, are required both to create and to play digital video elements.

The hardware for incorporating sound elements into multimedia systems includes microphones, voice-recognition systems, sound chips within the computer, and speakers, which come in a wide variety of forms with varying capabilities and quality. *See* SPEECH RECOGNITION.

The future of multimedia technology is dependent upon the evolution of the hardware. As storage devices get faster and larger, multimedia systems will be able to expand, and increased use of DVD should result in improved quality. Rising network speeds will increase the possibility of delivering multimedia applications over the WWW. Currently, Virtual Reality Modeling Language (VRML) is used for some WWW applications and may drastically expand the multimedia experience. Virtual reality is becoming more realistic and will stretch the multimedia experience to envelop the user. The one certainty in multimedia technology is that it will continue to change, to be faster, better, and more realistic. *See* VIRTUAL REALITY.

Pauline K. Cushman; Robert A. Kolvoord

Multimeter An instrument designed to measure electrical quantities. A typical multimeter can measure alternating- and direct-current potential differences (voltages), current, and resistance, with several full-scale ranges provided for each quantity. Sometimes referred to as a volt-ohm meter (VOM), it is a logical development of the electrical meter, providing a general-purpose instrument. Many kinds of special-purpose multimeters are manufactured to meet the needs of such specialists as telephone engineers and automobile mechanics testing ignition circuits. *See* AMMETER; CURRENT MEASUREMENT; OHMMETER; RESISTANCE MEASUREMENT; VOLTAGE MEASUREMENT; VOLTMETER.

Multimeters originated when all electrical measuring instruments used analog techniques. They were generally based on a moving-coil indicator, in which a pointer moves across a graduated scale. Accuracy was typically limited to about 2%, although models achieving 0.1% were available. Analog multimeters are still preferred for some applications. For most purposes, digital instruments are now used. In these, the measured value is presented as a row of numbers in a window. Inexpensive hand-held models perform at least as well as a good analog design. High-resolution multimeters have short-term errors as low as 0.1 part per million (ppm) and drift less than 5 ppm in one year. Many digital multimeters can be commanded by, and send their indications to, computers or control equipment.

R. B. D. Knight

Multiple cropping Planting two or more species in the same field in the same year. Preserved through history to maintain biological, economic, and nutritional diversity, multiple-species systems still are used by the majority of the world's farmers, especially in developing countries. Where farm size is small and the lack of capital has made it difficult to mechanize and expand, farm families that need a low-risk source of food and income often use multiple cropping. These systems maintain a green and growing crop canopy over the soil through much of the year, the total season depending on rainfall and temperature. Systems with more than one crop frequently make better use of total sunlight, water, and available nutrients than is possible with a single crop. The family has a more diverse supply of food and more than one source of income, with both spread over much of the year.

Multiple-cropping patterns are described by the number of crops per year and the intensity of crop overlap. Double cropping or triple cropping signifies systems with two or three crops planted sequentially with no overlap in growth cycle. Intercropping indicates that two or more crops are planted at the same time, or at least planted so that significant parts of their growth cycles overlap. Relay cropping describes the planting of a second crop after the first crop has flowered; in this system there still may be some competition for water or nutrients. When a crop is harvested and allowed to regrow from the crowns or root systems, the term ratoon cropping is used. Sugarcane, alfalfa, and sudangrass are commonly produced in this way, while the potential exists for such tropical cereals as sorghum and rice. Mixed cropping, strip cropping, associated cropping, and alternative cropping represent variations of these systems. *See* AGRICULTURAL SCIENCE (PLANT); AGRICULTURAL SOIL AND CROP PRACTICES; AGRICULTURE; AGRONOMY.

Charles A. Francis

Multiple proportions, law of This law states that, when two elements combine together to form more than one compound, the weights of one element that unite with a given weight of the other are in the ratio of small whole numbers. The law can be illustrated by the composition of the five oxides of nitrogen. One gram of nitrogen is combined with 2.85 g of oxygen in nitrogen pentoxide, N_2O_5; with 2.28 g in nitrogen dioxide, NO_2; with 1.71 g in nitrogen trioxide, N_2O_3; with 1.14 g in nitric oxide, NO; and with 0.57 g in nitrous oxide, N_2O. These numbers are in the simple ratio of 5:4:3:2:1. *See* DEFINITE COMPOSITION, LAW OF.

Thomas C. Waddington

Multiple sclerosis A neuromuscular disorder that characteristically involves the destruction of myelin, the insulating material around nerve fibers. The onset of the disease is unusual in persons under 15 or over 60 years of age, and peak incidence is found in people in their 20s and 30s. Multiple sclerosis affects females more frequently than males by approximately 2:1. Distribution is worldwide, but there is an unusual relationship to latitude, with a much higher incidence at northern latitudes than near the Equator.

In multiple sclerosis, only the central nervous system is affected, but both incoming and outgoing processes may be disrupted. Common initial symptoms reflect this underlying disease mechanism. They include blindness in one eye due to disruption of the conduction of the nerve impulse through the optic nerve; weakness of one side of the body due to impairment of the downstream signals from the motor areas of the cerebral cortex through the spinal cord; difficulties with coordination related to problems with cerebellar function; and disturbances in sensation, such as tingling and numbness in an arm or a leg that is related to dysfunction of incoming sensory signals. Multiple sclerosis is a progressive disease, so that over time there is often an accumulation of new symptoms and problems.

In its classic form, the disease spreads both temporally and anatomically. Temporally, there may be a series of acute attacks, but between attacks a person may recover fully and remain well for some time. Anatomically, areas of disruption of myelin (demyelination) are scattered throughout the nervous system and spinal cord. Thus, symptoms depend upon what part of the nervous system is affected at any given time. The course of the disease is unpredictable.

The basic cause of the disease is not known. There clearly are genetic factors in that the incidence of the disease is 20 times higher in first-degree relatives than in the general population. However, other factors must be involved, including infection (presumably viral) or possibly an immunological mechanism. The prevailing hypothesis combines these two possible etiological mechanisms to suggest that some type of viral infection occurs early in life to alter the patient's immune system. Thus, the activity and progression of the disease are related to altered immune functions within the central nervous system. In keeping with this hypothesis, therapy is aimed at altering the immune status of the affected individual. *See* HUMAN GENETICS; IMMUNOGENETICS; NERVOUS SYSTEM DISORDERS.

Guy M. McKhann

Multiplexing and multiple access In telecommunications, multiplexing refers to a set of techniques that enable the sharing of the usable electromagnetic spectrum of a telecommunications channel (the channel passband) among multiple users for the transfer of individual information streams. It is assumed that the user information streams join at a common access point to the channel. The term "multiple access" is usually applied to multiplexing schemes by which multiple users who are geographically dispersed gain access to the shared telecommunications facility or channel. Various methods of multiplexing and multiple access are in common use.

In frequency-division multiplexing (FDM) and frequency-division multiple access (FDMA), the passband of a channel is shared among multiple users by assigning distinct and nonoverlapping sections of the electromagnetic spectrum within the passband to individual users. The information stream from a particular user is encoded into a signal whose energy is confined to the part of the passband assigned to that user. *See* RADIO SPECTRUM ALLOCATION.

Time-division multiplexing (TDM) and time-division multiple access (TDMA) permit a user access to the full passband of the channel, but only for a limited time, after which the access right is assigned to another user. Normally the access rights are assigned in a cyclical order to the competing users. However, statistical time-division multiplexing assigns time on the channel on a demand basis, which typically increases the number of users who may be accommodated on the same channel, but may result in delays in accessing the channel during periods when the demand exceeds the supply.

In code-division multiple access (CDMA), all users are assigned the entire passband of the channel and are permitted to transmit their information streams simultaneously. To maintain the ability to recover the individual signals at the receiver, at the transmitter each signal has impressed on it a characteristic signature.

Space-division multiple access (SDMA) refers to the use of the same portion of the electromagnetic spectrum over two or more spatially distinct transmission paths. In most applications of space-division multiple access, the paths are formed by multibeam antennas, in which each beam is directed toward a different geographic area. *See* ANTENNA (ELECTROMAGNETISM).

In wavelength-division multiplexing (WDM) schemes, transmission systems that employ the optical portion of the electromagnetic spectrum, such as those using fiber-optic cables as the transmission medium, share the total available passband of the medium by assigning individual information streams to signals of different wavelengths or "colors." *See* OPTICAL COMMUNICATIONS; OPTICAL FIBERS.

Polarization refers to the direction or geometric orientation of the electric field vector of an electromagnetic field. Polarization-division multiplexing and polarization-division multiple access assign electric fields of different polarization to individual channels or users. *See* ELECTRICAL COMMUNICATIONS; POLARIZATION OF WAVES; POLARIZED LIGHT. Hermann J. Helgert

Multiplication One of the fundamental operations of arithmetic and algebra. The symbol × is commonly employed in arithmetic to denote multiplication. Because of its resemblance to the letter *x*, it is rarely used in algebra, where multiplication is frequently denoted by a dot (as in $a \cdot b$) or, most often, merely by juxtaposition of letters (for example, *ab*), Multiplication of numbers (real or complex) is associative, $a(bc) = (ab)c$; commutative, $ab = ba$; and distributive with respect to addition, $a(b + c) = ab + ac$; but the term has been extended to denote binary operations on many other kinds of objects, and these operations need not possess all the properties of ordinary multiplication listed above (for example, multiplication of matrices is not commutative). *See* ADDITION; ALGEBRA; DIVISION; SUBTRACTION. Leonard M. Blumenthal

Multipole radiation Standard patterns of radiation distribution about their source. The term radiation applies primarily to the transport of energy by acoustic, elastic, electromagnetic, or gravitational waves, and extends to the transport of atomic or subatomic particles (as represented by quantum-mechanical wave functions). *See* ELECTROMAGNETIC RADIATION; GRAVITATIONAL RADIATION; QUANTUM MECHANICS; SOUND; WAVE MOTION.

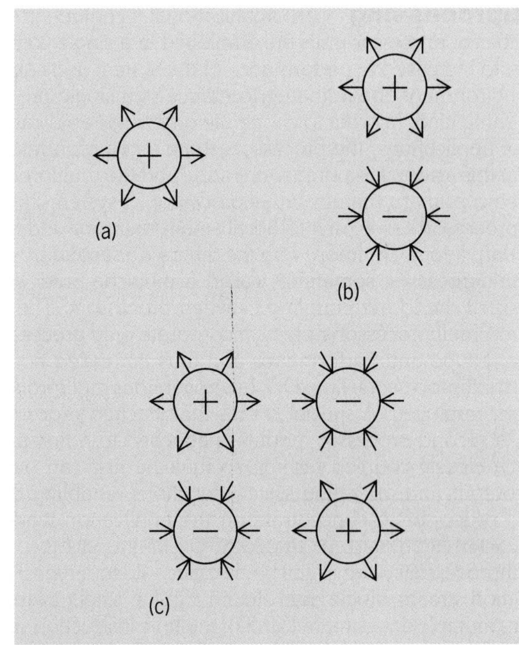

Static electric potentials generated by fixed multipoles. (*a*) Monopole ($l = 0$). (*b*) Dipole ($l = 1$). (*c*) Quadrupole ($l = 2$).

Each multipole pattern reflects the source's geometrical shape (or the shape of a source component). These geometrical features stand out clearly for the static electric potentials generated by fixed charges as shown by the small set of monopole, dipole, and quadrupole charges (see illustration), elements of all multipoles being named (in terms of powers of 2) 2^l-poles, with l equal to any nonnegative integer. A monopole ($l = 0$) acoustic wave radiates from a perfectly spherical bubble with oscillating radius; higher multipoles would arise from bubble distortions. So-called transverse waves, elastic or electromagnetic (including light), have only $l \geq 1$ components, gravitational waves only $l \geq 2$. The angular distributions, in azimuth (φ) and colatitude (θ), of 2^l-pole waves have amplitudes distributed in directions (θ, φ) in proportion to the spherical harmonic functions $Y_l^m(\theta, \varphi)$. The index *m* is a positive or negative integer whose absolute value is equal to less than *l*. *See* COORDINATE SYSTEMS; DIPOLE; SPHERICAL HARMONICS.

The multipolarity index *l* also represents the number of angular momentum quanta $\hbar$ (Planck's constant divided by 2π) radiated together with each energy quantum $h\nu$ (phonon, photon, graviton, and so forth). Detection and measurement of received energy quanta, together with measurement of their detection rate and mapping of their directional distribution, generally serve to diagnose the mechanics of the radiation source. Energy and momentum conservation underlie this analysis; so does the conservation of angular momentum which states that the initial angular momentum of the source equals the vector sum of the final angular momentum of the source and the angular momentum of the radiation. The quantitative implications of this vector relation are studied by the branch of quantum theory called angular momentum algebra. The balancing of parity, that is, of each variable's sign reversal (or persistence) under reflection through the source's center, also contributes to the analysis of experimental data. Further, more complex angular-momentum considerations play a role in the analysis of the behavior of spin-carrying particles. *See* ANGULAR MOMENTUM; CONSERVATION LAWS (PHYSICS); GRAVITON; PHONON; SELECTION RULES (PHYSICS); SPIN (QUANTUM MECHANICS). Ugo Fano

Multiprocessing An organizational technique in which a number of processor units are employed in a single computer system to increase the performance of the system in its application environment above the performance of a single processor of the same kind. In order to cooperate on a single application or class of applications, the processors share a common resource. Usually this resource is primary memory, and the multiprocessor is called a primary memory multiprocessor. A system in which each processor has a private (local) main memory and shares secondary (global) memory with the others is a secondary memory multiprocessor, sometimes called a multicomputer system because of the looser coupling between processors. The more common multiprocessor systems incorporate only processors of the same type and performance and thus are called homogeneous multiprocessors; however, heterogeneous multiprocessors are also employed. A special case is the attached processor, in which a second processor module is attached to a first processor in a closely coupled fashion so that the first can perform input/output and operating system functions, enabling the attached processor to concentrate on the application workload. *See* COMPUTER STORAGE TECHNOLOGY; OPERATING SYSTEM.

Multiprocessor systems may be classified into four types: single instruction stream, single data stream (SISD); single instruction stream, multiple data stream (SIMD); multiple instruction stream, single data stream (MISD): and multiple instruction stream, multiple data stream (MIMD). Systems in the MISD category are rarely built. The other three architectures may be distinguished simply by the differences in their respective instruction cycles:

In an SISD architecture there is a single instruction cycle; operands are fetched in serial fashwion into a single processing unit before execution. Sequential processors fall into this category.

An SIMD architecture also has a single instruction cycle, but multiple sets of operands may be fetched to multiple processing units and may be operated upon simultaneously within a single instruction cycle. Multiple-functional-unit, array, vector, and pipeline processors are in this category. *See* SUPERCOMPUTER.

In an MIMD architecture, several instruction cycles may be active at any given time, each independently fetching instructions and operands into multiple processing units and operating on them in a concurrent fashion. This category includes multiple processor systems in which each processor has its own program control, rather than sharing a single control unit.

MIMD systems can be further classified into throughput-oriented systems, high-availability systems, and response-oriented systems. The goal of throughput-oriented multiprocessing is to obtain high throughput at minimal computing cost in a general-purpose computing environment by maximizing the number of independent computing jobs done in parallel. High-availability multiprocessing systems are generally interactive, often with never-fail real-time online performance requirements.

The goal of response-oriented multiprocessing (or parallel processing) is to minimize system response time for computational demands. *See* COMPUTER SYSTEMS ARCHITECTURE; CONCURRENT PROCESSING; FAULT-TOLERANT SYSTEMS; REAL-TIME SYSTEMS.

Peter C. Patton

Multituberculata An extinct order in the class Mammalia, subclass Allotheria, comprising a major group of early mammals, ranging from Late Jurassic to Late Eocene (about 155 to 35 million years ago). The group is best known from North America, Mongolia, and Europe. Most multituberculates were mouselike in size, although the North American Paleocene *Taeniolabis* was larger, probably closer to the woodchuck in its proportions. Multituberculates appear primarily to have been terrestrial creatures, but some were probably arboreal (living in trees) and others fossorial (adapted for digging). The Eocene decline and extinction of multituberculates may reflect competition from small placental mammals, especially primates and rodents. *See* ARCHAIC UNGULATE; RODENTIA.

The most primitive multituberculates are traditionally classified in the suborder Plagiaulacoidea (Late Jurassic to Early Cretaceous). Advanced multituberculates are traditionally classified in the suborders Ptilodontoidea and Taeniolabidoidea (both Late Cretaceous to Eocene).

Although clearly mammalian and often characterized as rodentlike, multituberculates are unlike rodents or other living mammals in several important features. The dentition is distinctive and, with associated jaw fragments, it is all that is known for most species. Functional studies using living mammals as analogues suggest that multituberculate incisors were used in grasping, the bladelike premolars in cutting, and the molars in grinding food. Microscopic wear patterns on multituberculate tooth enamel, however, show that jaw motion differed radically from that in other mammals: during chewing, the mandible moved only in the vertical plane, not transversely, and was strongly retracted during the power stroke. These studies also imply that most multituberculates were probably omnivores, although species having gnawing incisors were likely specialized herbivores feeding on tough plant tissues. Multituberculate relationships to other mammals are unclear. *See* ALLOTHERIA; DENTITION; MAMMALIA; THERIA.

Richard C. Fox; Thomas Martin

Multivariable control The control of systems characterized by multiple inputs, which are usually referred to as the controls; or by multiple outputs, which are often the measured variables and the variables to be controlled (see illustration); or by both multiple inputs and multiple outputs (MIMO). Automobiles, chemical processing and manufacturing plants, aircraft and aerospace vehicles, biological systems, and the national economy are examples of multivariable systems which require and receive some form of regulation or control, be it mathematically contrived or not. Control systems are based on mathematical models of the behavior to be controlled, and the methodologies rely on sound mathematical theories.

Most of the mathematical control theory which has been developed for multivariable systems has assumed, as a starting point, some form of known linear, time-invariant, continuous, and causal dynamical mathematical model for the system to be controlled. In that model the manipulated inputs and the outputs to be controlled have been identified. Such a mathematical description of the dynamic behavior of variables of interest may be derived directly from first physical principles and then simplified, for example via linearization of a nonlinear description around an operating point.

The control objectives to be achieved in the multivariable (MIMO) case include those control objectives which are generally sought in the scalar, that is, single input/single output (SISO) case, and also depend on the specific application being considered. In particular, stability is usually a primary concern in the design of any control system, along with various measures of robustness to uncertainties in the plant model.

While controller failure in the scalar case can lead to catastrophic failure of the overall system, such need not be true in

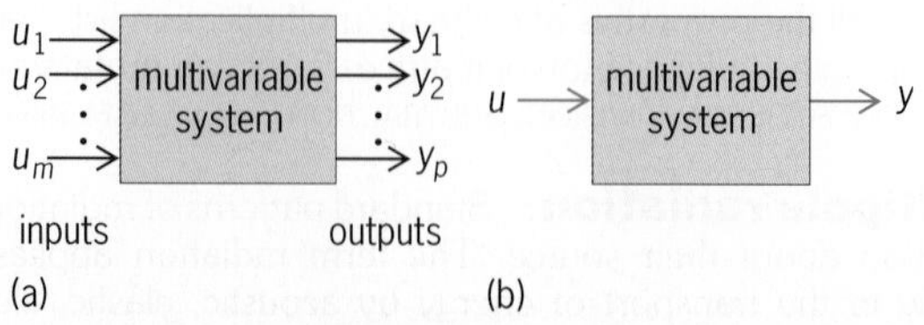

Two multivariable system representations. (*a*) Input and output variables shown separately. (*b*) Input and output variables shown in vector form.

the multivariable case due to the interactions between various input/output pairs. More specifically, in certain applications it may be possible to design a multivariable controller "driven" by all of the outputs which performs satisfactorily even when feedback information from one or more outputs is lost; or the same control objectives may be attained even when some of the control inputs are not available. The integrity of such a multivariable controller would be much better than that of its scalar counterpart. *See* CONTROL SYSTEM STABILITY.

In many control applications, a primary objective is that of tracking, or ensuring that the system output or outputs track a desired input or inputs with little or no steady-state errors. In addition, simultaneous regulation of external disturbances is also often desired; that is, ensuring that any external disturbances affect the plant outputs as little as possible. These performance objectives are to be attained with adequate robustness to uncertainty in the plant and the environment. It is also obvious that simplicity of controller design is a highly desirable design objective in both the scalar and multivariable cases. One final design objective that is unique to the multivariable case is that of minimizing or eliminating the interaction between loops, which is often referred to as decoupling; that is, perfect decoupling would imply a controlled system where each input affects one and only one output or, otherwise stated, a system whose compensated transfer matrix is diagonal and nonsingular. *See* CONTROL SYSTEMS; PROCESS CONTROL. Panos J. Antsaklis; William A. Wolovich

Multivibrator A form of electronic circuit that employs positive feedback to cross-couple two devices so that two distinct states are possible, for example, one device ON and the other device OFF, and in which the states of the two devices can be interchanged either by use of external pulses or by internal capacitance coupling. When the circuit is switched between states, transition times are normally very short compared to the ON and OFF periods. Hence, the output waveforms are essentially rectangular in form.

Multivibrators may be classified as bistable, monostable, or astable. A bistable multivibrator, often referred to as a flip-flop, has two possible stable states, each with one device ON and the other OFF, and the states of the two devices can be interchanged only by the application of external pulses. A monostable multivibrator, sometimes referred to as a one-shot, also has two possible states, only one of which is stable. If it is forced to the opposite state by an externally applied trigger, it will recover to the stable state in a period of time usually controlled by a resistance-capacitance (*RC*) coupling circuit. An astable multivibrator has two possible states, neither of which is stable, and switches between the two states, usually controlled by two *RC* coupling time constants. The astable circuit is one form of relaxation oscillator, which generates recurrent waveforms at a controllable rate.

Symmetrical bistable multivibrator. In bistable multivibrators, either of the two devices in a completely symmetrical circuit may remain conducting, with the other nonconducting, until the application of an external pulse. Such a multivibrator is said to have two stable states.

The original form of bistable multivibrator made use of vacuum tubes and was known as the Eccles-Jordan circuit, after its inventors. It was also called a flip-flop or binary circuit because of the two alternating output voltage levels. The junction field-effect transistor (JFET) circuit (Fig. 1) is a solid-state version of the Eccles-Jordan circuit. Its resistance networks between positive and negative supply voltages are such that, with no current flowing to the drain of the first JFET, the voltage at the gate of the second is slightly negative, zero, or limited to, at most, a slightly positive value. The resultant current in the drain circuit of the second JFET causes a voltage drop across the drain load resistor; this drop in turn lowers the voltage at the gate of the first JFET to a sufficiently negative value to continue to reduce the drain current to zero. This condition of the first device OFF and the second ON will be maintained as long as the circuit remains undisturbed. *See* TRANSISTOR.

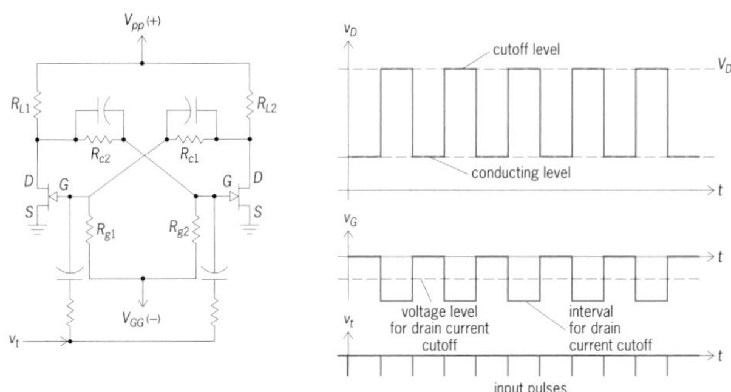

Fig. 1. Bistable multivibrator with triggering, gate, and drain waveforms shown for one transistor.

If a sharp negative pulse is applied to the gate of the ON transistor, its drain current decreases and its drain voltage rises. A fraction of this rise is applied to the gate of the OFF transistor, causing some drain current to flow. The resultant drop in drain voltage, transferred to the gate of the ON transistor, causes a further rise at its drain. The action is thus one of positive feedback, with nearly instantaneous transfer of conduction from one device to the other. There is one such reversal each time a pulse is applied to the gate of the ON transistor. Normally pulses are applied to both transistors simultaneously so that whichever device is ON will be turned off by the action. The capacitances between the gate of one transistor and the drain of the other play no role other than to improve the high-frequency response of the voltage divider network by compensating for the input capacitances of the transistors and thereby improving the speed of transition.

A bipolar transistor counterpart of the JFET bistable multivibrator uses *npn* bipolar transistors. The base of the transistor corresponds to the gate, the emitter to the source, and the collector to the drain. Although waveforms are of the same polarity and the action is roughly similar to that of the JFET circuit, there are important differences. The effective resistance of the base-emitter circuit, when it is forward-biased and being used to control collector current, is much lower than the input gate resistance of the JFET when the latter resistance is used to control drain current (a few thousand ohms compared to a few megohms). This fact must be taken into account when the divider networks are designed. If *pnp* transistors are used, all voltage polarities and current directions are reversed.

Unsymmetrical bistable circuits. Bistable action can be obtained in the emitter- or source-coupled circuit with one of the set of cross-coupling elements removed (Fig. 2). In this case, regenerative feedback necessary for bistable action is obtained by the one remaining common coupling element, leaving one emitter or gate free for triggering action. Biases can be adjusted such that device 1 is ON, forcing device 2 to be OFF. In this case, a pulse can be applied to the free input in such a direction as to reverse the states. Alternatively, device 1 may initially be OFF with device 2 ON. Then an opposite polarity pulse is required to reverse states. Such an unsymmetrical bistable circuit, historically referred to as the Schmitt trigger circuit, finds widespread use in many applications.

Monostable multivibrator. A monostable or one-shot multivibrator has only one stable state. If one of the normally active devices is in the conducting state, it remains so until an external pulse is applied to make it nonconducting. The second device is thus made conducting and remains so for a duration dependent upon *RC* time constants within the circuit itself. Monostable multivibrators are available commercially in integrated chip form. *See* INTEGRATED CIRCUITS.

Astable multivibrator. The astable multivibrator has capacitance coupling between both of the active devices and therefore

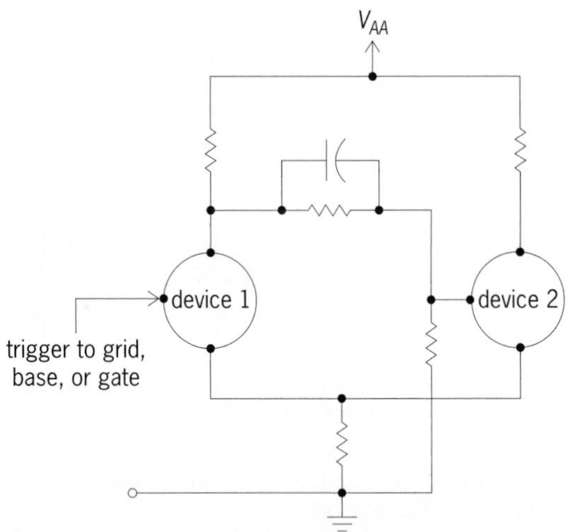

Fig. 2. Unsymmetrical bistable multivibrator.

has no permanently stable state. Each of the two devices functions in a manner similar to that of the capacitance-coupled half of the monostable multivibrator. It will therefore generate a periodic rectangular waveform at the output with a period equal to the sum of the OFF periods of the two devices.

Astable multivibrators, although normally free-running, can be synchronized with input pulses recurrent at a rate slightly faster than the natural recurrence rate of the device itself. If the synchronizing pulses are of sufficient amplitude, they will bring the internal waveform to the conduction level at an earlier than normal time and will thereby determine the recurrence rate.

Logic gate multivibrators. Multivibrators may be formed by using two cross-coupled logic gates, with the unused input terminals used for triggering purposes. The bistable forms of such circuits are usually referred to as flip-flops. *See* LOGIC CIRCUITS.

Glenn M. Glasford

Mumps An acute contagious viral disease, characterized chiefly by enlargement of the parotid glands (parotitis).

Besides fever, the chief signs and symptoms are the direct mechanical effect of swelling on glands or organs where the virus localizes. One or both parotids may swell rapidly, producing severe pain when the mouth is opened. In orchitis, the testicle is inflamed but is enclosed by an inelastic membrane and cannot swell; pressure necrosis produces atrophy, and if both testicles are affected, sterility may result. The ovary may enlarge, without sequelae.

An attenuated live virus vaccine can induce immunity without parotitis. It is recommended particularly for adults exposed to infected children, for students in boarding schools and colleges, and for military troops. Joseph L. Melnick

Muonium An exotic atom, Mu or (μ^+e^-), formed when a positively charged muon (μ^+) and an electron are bound by their mutual electrical attraction. It is a light, unstable isotope of hydrogen, with a muon replacing the proton. Muonium has a mass 0.11 times that of a hydrogen atom due to the lighter mass of the muon, and a mean lifetime of 2.2 microseconds, determined by the spontaneous decay of the muon ($\mu^+ \rightarrow e^+\nu_e \overline{\nu}_\mu$). Muonium is formed when beams of μ^+ produced in particle accelerators are stopped in certain nonmetallic targets.

Since muonium is a system consisting only of leptons, it serves as a testing ground for the theory of quantum electrodynamics (QED), which describes the electromagnetic interaction between particles. Muonium chemistry and muonium spin rotation (MSR) are two developing subfields which seek to understand the chemical and physical behavior of a light hydrogen isotope in matter and to probe the structure of materials. *See* POSITRONIUM; QUANTUM ELECTRODYNAMICS. Patrick O. Egan

Muscle The tissue in the body in which cellular contractility has become most apparent. Almost all forms of protoplasm exhibit some degree of contractility, but in muscle fibers specialization has led to the preeminence of this property. In vertebrates three major types of muscle are recognized: smooth, cardiac, and skeletal.

Smooth muscle. Smooth muscle, also designated visceral and sometimes involuntary, is the simplest type. These muscles consist of elongated fusiform cells which contain a central oval nucleus. The size of such fibers varies greatly, from a few micrometers up to 0.02 in. (0.5 mm) in length. These fibers contract relatively slowly and have the ability to maintain contraction for a long time. Smooth muscle forms the major contractile elements of the viscera, especially those of the respiratory and digestive tracts, and the blood vessels. Smooth muscle fibers in the skin regulate heat loss from the body. Those in the walls of various ducts and tubes in the body act to move the contents to their destinations, as in the biliary system, ureters, and reproductive tubes.

Smooth muscle is usually arranged in sheets or layers, commonly oriented in different directions. The major physiological properties of these muscles are their intrinsic ability to contract spontaneously and their dual regulation by the autonomic nerves of the sympathetic and parasympathetic systems. *See* AUTONOMIC NERVOUS SYSTEM.

Cardiac muscle. Cardiac muscle has many properties in common with smooth muscle; for example, it is innervated by the autonomic system and retains the ability to contract spontaneously. Presumably, cardiac muscle evolved as a specialized type from the general smooth muscle of the circulatory vessels. Its rhythmic contraction begins early in embryonic development and continues until death. Variations in the rate of contraction are induced by autonomic regulation and by many other local and systemic factors.

The cardiac fiber, like smooth muscle, has a central nucleus, but the cell is elongated and not symmetrical. It is a syncytium, a multinuclear cell or a multicellular structure without cell walls. Histologically, cardiac muscle has cross-striations very similar to those of skeletal muscle, and dense transverse bands, the intercalated disks, which occur at short intervals. *See* HEART (VERTEBRATE).

Skeletal muscle. Skeletal muscle is also called striated, somatic, and voluntary muscle, depending on whether the description is based on the appearance, the location, or the innervation. The individual cells or fibers are distinct from one another and vary greatly in size from over 6 in. (15 cm) in length to less than 0.04 in. (1 mm). These fibers do not ordinarily branch, and they are surrounded by a complex membrane, the sarcolemma. Within each fiber are many nuclei; thus it is actually a syncytium formed by the fusion of many precursor cells.

The transverse striations of skeletal muscle form a characteristic pattern of light and dark bands within which are narrower bands. These bands are dependent upon the arrangement of the two sets of sliding filaments and the connections between them. *See* MUSCLE PROTEINS; MUSCULAR SYSTEM. Walter Bock

Muscle proteins Specialized proteins in muscle cells that are the building blocks of the structures constituting the moving and regulatory machinery of muscle. The moving machinery comprises myofilaments that are discernible by electron microscopy. These myofilaments are of two kinds, myosin and actin, and their regular arrangement within the cell gives the striated pattern to skeletal muscle fibers (see illustration). It is recognized that the sliding of the two sets of filaments relative to each other is the molecular basis of muscle contraction. To understand the ultimate mechanism that causes the movement

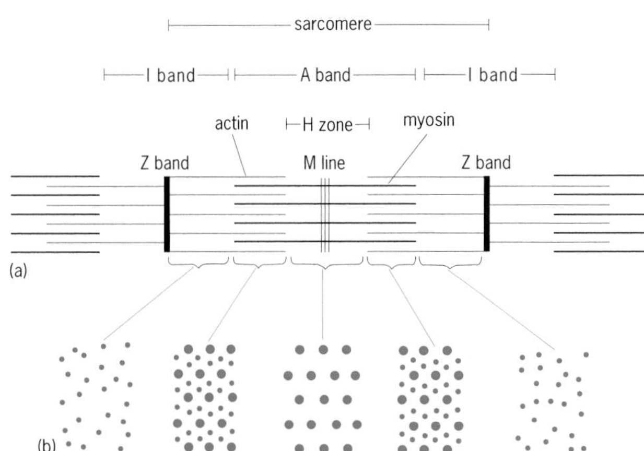

Origin of striations in muscle. (*a*) Banding pattern of a sarcomere, showing the arrangement of myosin and actin molecules. (*b*) Cross-sectional appearance of sarcomere at different points along its length. (*After R. Craig, The structure of the contractile filaments, in A. G. Engle and B. Q. Banker, eds., Myology, McGraw-Hill, 1986*)

of these filaments relative to each other, it is necessary to consider the features of the individual molecules that make up these filaments. Practically all nonmuscle cells, although lacking the filaments of muscle, contain proteins similar to those found in muscle; these proteins are likely to be involved in cell motility and in determining properties of cell membranes. *See* CELL MEMBRANES.

The banding pattern of muscle fibers produced by the regular arrangement of myosin and actin is further enhanced by the Z membranes, or disks, which appear as dark bands when the myofibril is viewed longitudinally and which separate adjacent sarcomeres. There is now considerable evidence for elastic connections occurring between the Z disks and interacting with the myosin filaments. The main constituent of these filaments is titin, a protein with a mass of about 3×10^6 daltons; it is also known as connectin.

Molecules of myosin, amounting to about 60% of the total muscle protein, are arranged in filaments occupying the central zone of each segment (sarcomere) of the fibril, the A band. Myosin is an elongated molecule whose overall length is about 150 nm. It is made up of two intertwined heavy peptide chains (molecular weight of about 2×10^5) whose ends form two separate globular structures. The intertwined portion has a high α-helical content and forms a rigid rod. Myosin functions as an ATP-cleaving enzyme, adenosine triphosphatase (ATPase). Each myosin head has an ATP binding site as well as an actin-binding site. The ATPase activity of myosin itself is low in media that contain magnesium ion, but if actin is added to myosin at low ionic strength in the presence of magnesium ion, considerable activation of ATPase takes place. This actin-activated ATPase reaction, which furnishes energy of contraction and the interaction of myosin with actin, is an essential element of the contraction mechanism. *See* ADENOSINE TRIPHOSPHATE (ATP).

In the monomeric form, actin is a bilobular-shaped, nearly globular unit with a diameter of about 6 nm, a height of about 4 nm, and a molecular mass of 4.2×10^4 daltons. Actin, as isolated in the monomeric form, contains a tightly bound atom of magnesium and one bound ATP molecule in a cleft between the two lobes.

The interaction of actin and myosin in striated muscle is regulated by the cellular calcium level through a protein complex attached at regular intervals to the thin filament. Two protein complexes associated with the actin filaments are tropomyosin and troponin. The mechanism by which the calcium ion–induced changes in the thin filament lead to activation of muscle contraction is still not fully understood. Earlier views regarded the attachment of myosin heads to actin as the regulated step. In contrast, later biochemical studies of myosin ATPase activated by regulated actin in the test tube suggest that activation involves a shift among states of myosin attached to actin, from one of weak binding to one of strong binding. Time-resolved x-ray diffraction on live muscle shows that the first structural changes in the thin filaments may involve a movement of tropomyosin, but the link between the movement and the changes in actin–myosin interaction require further clarification. *See* CALCIUM METABOLISM; MUSCLE. John Gergely

Muscovite A mineral of the mica group with an ideal composition of $KAl_2(AlSi_3)O_{10}(OH)_2$. Sometimes it is referred to as a white mica or potash mica.

Physical properties include specific gravity 2.76–2.88, hardness on the Mohs scale 2–2.5, and luster vitreous to pearly. Thin sheets are flexible and may be colorless, with books (thick crystals) translucent, yellow, brown, reddish, or green. Muscovite occurs commonly in all the major rock types, in igneous rocks (granites, pegmatites, and hydrothermal alteration products), in metamorphic rocks (slates, phyllites, schists and gneisses), and in sedimentary rocks (sandstones and other clastic rocks). As larger flakes, muscovite is used as an electrical insulator, both for its dielectric properties and for its resistance to heat. Ground muscovite is used for fireproofing, as an additive to paint to provide a sheen and for durability, as a filler, and for many other applications. *See* MICA; SILICATE MINERALS. Stephen Guggenheim

Muscular dystrophy A group of muscle diseases that are hereditary and characterized by progressive muscle weakness and wasting.

The muscular dystrophies are primary diseases of the muscle cells characterized by progressive degeneration and replacement by fibrous tissue, resulting in progressive muscle weakness. In some types of muscular dystrophies, the disease appears to be restricted to the skeletal muscles alone (facioscapulohumeral muscular dystrophy, limb-girdle muscular dystrophy), and in others skeletal-muscle involvement is a part of a more generalized process, with abnormalities in other organ systems as well (Duchenne's muscular dystrophy, myotonic dystrophy). These features as well as the differing patterns of inheritance indicate that the various muscular dystrophies are different diseases with different genetic and biochemical abnormalities underlying them.

The gene for the Duchenne and the Becker muscular dystrophy has been identified. This gene produces the muscle protein dystrophin, which is absent in Duchenne's dystrophy and qualitatively altered in Becker's dystrophy. Although the gene for myotonic dystrophy has not been identified, it has been found to be closely linked to genetic markers on chromosome 19. *See* HUMAN GENETICS; MUSCLE PROTEINS.

Duchenne's muscular dystrophy is the most rapidly progressive form of muscular dystrophy. It affects boys before the age of 4, and is characterized initially by progressive weakness of the hip muscles with difficulty in rising from the floor or chair and in climbing stairs. This is accompanied by enlargement of the calf muscles, which are infiltrated by fat and fibrous tissue (pseudohypertrophy). Weakness of the muscles of the upper arms and shoulder muscles follows.

Becker's muscular dystrophy is also characterized by calf pseudohypertrophy but is much more slowly progressive.

Myotonic dystrophy is very slowly progressive and affects the muscles of the face, neck, and hands. It usually begins in early adulthood. In addition to progressive weakness and wasting of the affected muscles, these individuals also exhibit myotonia, that is, a delayed relaxation of a muscle after forceful contraction. Mental retardation, frontal balding, cataracts, and gonadal degeneration are common.

Treatment remains largely symptomatic. *See* MUSCULAR SYSTEM DISORDERS.

S. M. Sumi; Thomas D. Bird

Muscular system The muscular system consists of muscular cells, the contractile elements with the specialized property of exerting tension during contraction, and associated connective tissues. The three morphologic types of muscles are voluntary muscle, involuntary muscle, and cardiac muscle. The voluntary, striated, or skeletal muscles are involved with general posture and movements of the head, body, and limbs. The involuntary, nonstriated, or smooth muscles are the muscles of the walls of hollow organs of the digestive, circulatory, respiratory, and reproductive systems, and other visceral structures. Cardiac muscle is the intrinsic muscle tissue of the heart. *See* MUSCLE.

Anatomy. Muscle groups are particularly distinct in elasmobranchs and other primitive fishes, and they are generally defined on the basis of their embryonic origin in these animals. Two major groups of skeletal muscles are recognized, somatic (parietal) muscles, which develop from the myotomes, and branchiomeric muscles, which develop in the pharyngeal wall from lateral plate mesoderm. The somatic musculature is subdivided into axial muscles, which develop directly from the myotomes and lie along the longitudinal axis of the body, and appendicular muscles, which develop within the limb bud from mesoderm derived phylogenetically as buds from the myotomes.

The vertebrate muscular system is the largest of the organ systems, making up 35–40% of the body weight in humans. The movement of vertebrates is accomplished exclusively by muscular action, and muscles play the major role in transporting materials within the body. Muscles also help to tie the bones of the skeleton together and supplement the skeleton in supporting the body against gravity. *See* SKELETAL SYSTEM.

Most of the axial musculature is located along the back and flanks of the body, and this part is referred to as trunk musculature. But anteriorly the axial musculature is modified and assigned to other subgroups. Certain of the occipital and neck myotomes form the hypobranchial muscles, and the most anterior myotomes form the extrinsic ocular muscles.

The hypaxial musculature of tetrapods can be subdivided into three groups: (1) a subvertebral (hyposkeletal) group located ventral to the transverse processes and lateral to the centra of the vertebrae, (2) the flank muscles forming the lateral part of the body wall, and (3) the ventral abdominal muscles located on each side of the midventral line. The subvertebral musculature assists the epaxial muscles in the support and movement of the vertebral column. Most of the flank musculature takes the form of broad, thin sheets of muscle that form much of the body wall and support the viscera. The midventral hypaxial musculature in all tetrapods consists of the rectus abdominis, a longitudinal muscle on each side of the midline that extends from the pelvic region to the anterior part of the trunk.

The hypobranchial musculature extends from the pectoral girdle forward along the ventral surface of the neck and pharynx to the hyoid arch, chin, and into the tongue. It is regarded as a continuation of part of the hypaxial trunk musculature.

Limb muscles are often classified as intrinsic if they lie entirely within the confines of the appendage and girdle, and extrinsic if they extend from the girdle or appendage to other parts of the body. In fishes, movements of the paired fins are not complex or powerful and the appendicular muscles in the strictest sense are morphologically simple. In terrestrial vertebrates, the limbs become the main organs for support and locomotion, and the appendicular muscles become correspondingly powerful and complex. The muscles are too numerous to describe individually, but they can be sorted into dorsal and ventral groups, because tetrapod muscles originate embryonically in piscine fashion from a dorsal and a ventral premuscular mass within the limb bud. In general, the ventral muscles, which also spread onto the anterior surface of the girdle and appendage, act to protract and adduct the limb and to flex its distal segments; the dorsal muscles, which also extend onto the posterior surface of the girdle and appendage, have the opposite effects (retraction, abduction, and extension). The limb muscles also serve as flexible ties or braces that can fix the bones at a joint and support the body.

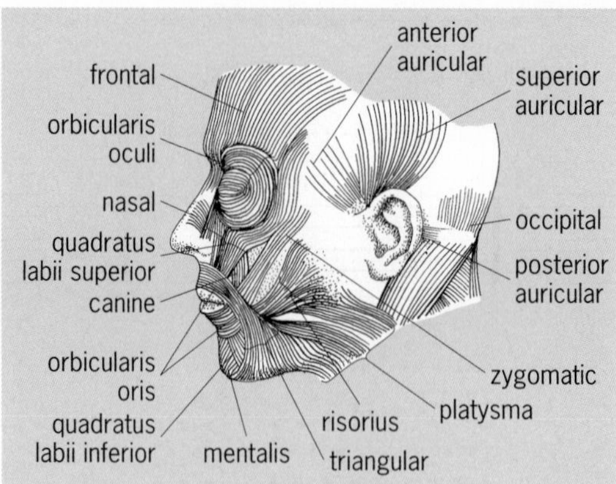

Fig. 1. Human facial muscles. (*After H. W. Rand, The Chordates, Blakiston, 1950*)

Flight in birds has entailed a considerable modification of the musculature of the pectoral region. As one example, the ventral adductor muscles are exceedingly large and powerful, and the area from which they arise is increased by the enlargement of the sternum and the evolution of a large sternal keel. Not only does a ventral muscle, the pectoralis, play a major role in the downstroke of the humerus, but a ventral muscle, the supracoracoideus, is active in the upstroke as well.

In a number of terrestrial vertebrates, particularly amniotes, certain of the more superficial skeletal muscles of the body have spread out beneath the skin and inserted into it. These may be described as integumentary muscles. Integumentary muscles are particularly well developed in mammals and include the facial muscles (Fig. 1) and platysma, derived from the hyoid musculature, and often a large cutaneous trunci. The last is derived from the pectoralis and latissimus dorsi and fans out beneath the skin of the trunk. The twitching of the skin of an ungulate is caused by this muscle.

Muscle mechanics. Many of the bones serve as lever arms, and the contractions of muscles are forces acting on these arms (Fig. 2). The joint, of course, is the fulcrum and it is at one end of the lever. The length of the force arm is the perpendicular distance from the fulcrum to the line of action of the muscle; the length of the work arm is the perpendicular distance from the fulcrum to the point of application of the power generated in the lever. Compactness of the body and physiological properties of the muscle necessitates that a muscle attach close to the fulcrum; therefore, the force arm is considerably shorter than the work arm. Most muscles are at a mechanical disadvantage, for they must generate forces greater than the work to be done, but an advantage of this is that a small muscular excursion can induce a much greater movement at the end of the lever. *See* SKELETAL SYSTEM.

Slight shifts in the attachments of a muscle that bring it toward or away from the fulcrum, and changes in the length of the work arm, can alter the relationship between force and amount or speed of movement.

In general, the force of a muscle is inversely related to the amount and speed of movement that it can cause. Certain patterns of the skeleton and muscles are adapted for extensive, fast movement at the expense of force, whereas others are adapted for force at the expense of speed. In the limb of a horse, which is adapted for long strides and speed, the muscles that move

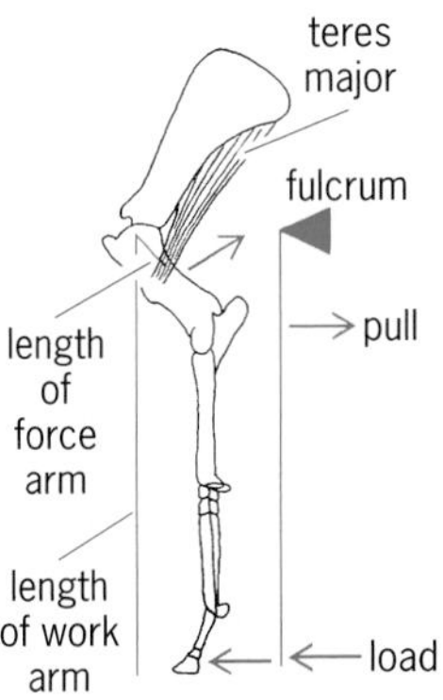

Fig. 2. A typical vertebrate lever system.

the limb insert close to the fulcrum and the appendage is long. This provides a short force arm but a very long work arm to the lever system (Fig. 2). In the front leg of a mole, which is adapted for powerful digging, the distance from the fulcrum to the insertion of the muscles is relatively greater and the length of the appendage is less, with the result that the length of the force arm is increased relative to the length of the work arm. Warren F. Walker

Muscular system disorders Disorders affecting skeletal (voluntary) muscle. The normal functioning of the skeletal muscle is dependent not only on the integrity of the muscle fibers themselves, but also on that of the motor cortex, the pyramidal tract, and the extrapyramidal system (including the cerebellum). It also depends on innervation by the motoneurons of the brainstem and the spinal cord. In addition, the proper functioning of the other organ systems, such as the endocrine system, and variations in the concentration of various electrolytes may also affect muscle function.

Damage to the motor cortex or the pyramidal tract produces the type of weakness seen in humans after a stroke or spinal cord injury. Although the paralyzed limb may initially be flaccid (hypotonic), spasticity (hypertonia) eventually develops. Despite the weakness, muscle atrophy is usually not striking. When the extrapyramidal system or the cerebellum is the site of damage, instead of weakness there are uncontrolled movements, difficulty with coordination, or both. In either of these situations there is no characteristic change in the muscle, either grossly or microscopically. At most, atrophy of type 2 fibers is seen.

Motoneuron damage. With damage to the spinal motoneuron or its axon, there is flaccid weakness of the muscle with proportionate wasting. Direct involvement of the spinal motoneuron was typically seen in poliomyelitis, but is now seen more commonly in the progressive spinomuscular atrophies of infancy and childhood, and in amyotrophic lateral sclerosis (ALS) in adults. Spontaneous twitching of groups of muscle fibers (fasciculation) innervated by the same motoneuron (motor unit) is frequently seen in these disorders. *See* POLIOMYELITIS.

Progressive diseases of unknown cause may occur in infancy (Werdnig-Hoffmann disease) or later in childhood (Kugelberg-Wielander disease). Both diseases result from degeneration of the motoneurons and appear to be inherited in an autosomal recessive pattern. In the infantile form, the baby is often floppy from birth with generalized weakness, a poor cry, and difficulty in sucking and breathing. Many children succumb in early childhood. In the later childhood forms, the rate of progression is slower and the outlook better. In these children weakness is more marked in the proximal muscles of the limbs.

Amyotrophic lateral sclerosis is a progressive disease of unknown cause, in which both the brainstem and spinal motoneurons, as well as the corticospinal tracts, undergo degeneration. This is a relatively rapid progressive disease, with death usually occurring within 3 years of diagnosis due to swallowing difficulty and respiratory failure. Although some cases of ALS appear to be inherited, most cases occur sporadically.

Neuromuscular junction. The most common example of disease at the neuromuscular junction is myasthenia gravis. Other, less common diseases are the myasthenic Eaton-Lambert syndrome and botulism. *See* MYASTHENIA GRAVIS.

Myopathy. Abnormalities of the muscle itself (myopathy) obviously result in muscle weakness, and muscle diseases fall into two large groups: those with a genetic basis and those which are nongenetic.

Congenital myopathies and the muscular dystrophies constitute the genetic muscle diseases. Congenital myopathies are characterized by a generalized weakness which is present at birth. The weakness is usually not progressive and often improves with time. Muscle fiber necrosis, a characteristic of the muscular dystrophies, is not seen in the congenital myopathies. *See* MUSCULAR DYSTROPHY.

Nongenetic or acquired diseases are all characterized by rapidly progressive weakness of the proximal muscles of the limbs, and so resemble the limb-girdle form of muscular dystrophy in the distribution of weakness. They are often included as inflammatory myopathies. Individuals with these disorders have difficulty arising from a recumbent or sitting position, in climbing stairs, and in lifting heavy objects onto a shelf. They also often have tender and painful muscles, may be febrile, and may have other manifestations of a systemic illness.

Metabolic diseases. A number of metabolic diseases have been associated with muscle symptoms. These include thyroid diseases and certain endocrine diseases, particularly Cushing's syndrome, which are either spontaneous or secondary to therapeutically administered adrenocorticosteroid hormones. No specific histologic changes have been described in the muscle in these disorders. Muscular symptoms are also seen with some of the glycogen storage diseases. Muscle weakness and paralysis may also be associated with alterations in the level of serum potassium.

Myotonia. A delayed relaxation of the muscle after forceful contraction (myotonic) is another symptom of muscle disease. This phenomenon is a characteristic feature of myotonic dystrophy and of congenital myotonia. Myotonia is usually present from early life and is often associated with muscle hypertrophy, but muscle weakness is not a feature of this disease. The muscle shows no definite histologic changes. *See* MUSCULAR SYSTEM. S. Mark Sumi

Mushroom A macroscopic fungus with a fruiting body (also known as a sporocarp). Approximately 14% (10,000) described species of fungi are considered mushrooms. Mushrooms grow aboveground or underground. They have a fleshy or nonfleshy texture. Many are edible, and only a small percentage are poisonous.

Mushrooms reproduce via microscopic spheres (spores) that are roughly comparable to the seeds of higher plants. Spores are produced in large numbers on specialized structures in or on the fruiting body. Spores that land on a suitable medium absorb moisture, germinate, and produce hyphae that grow and absorb nutrients from the substratum. If suitable mating types are present and the mycelium (the threadlike filaments or hyphae that become interwoven) develops sufficiently to allow fruiting, the life cycle will continue. In nature, completion of the life cycle is dependent on many factors, including temperature, moisture and nutritional status of the substratum, and gas exchange capacity of the medium.

Fewer than 20 species of edible mushrooms are cultivated commercially. The most common cultivated mushroom is *Agaricus bisporus*, followed by the oyster mushroom (*Pleurotus* spp.). China is the leading mushroom-producing country; Japan leads the world in number of edible species cultivated commercially.

Mushrooms may be cultivated on a wide variety of substrates. They are grown from mycelium propagated on a base of steam-sterilized cereal grain. This grain and mycelium mixture is called spawn, which is used to seed mushroom substrata.

Mushrooms contain digestible crude protein, all essential amino acids, vitamins (especially provitamin D-2), and minerals; they are high in potassium and low in sodium, saturated fats, and calories. Although they cannot totally replace meat and other high-protein food in the diet, they can be considered an important dietary supplement and a health food.

Fungi have been used for their medicinal properties for over 2000 years. Although there remains an element of folklore in the use of mushrooms in health and medicine, several important drugs have been isolated from mushroom fruiting bodies and mycelium. The best-known drugs obtained are lentinan from *L. edodes*, grifolin from *Grifola frondosa*, and krestin from *Coriolus versicolor*. These compounds are protein-bound polysaccharides or long chains of glucose, found in the cell walls, and function as antitumor immunomodulatory drugs. *See* FUNGI; MEDICAL MYCOLOGY. D. Martínez-Carrera

Musical acoustics The branch of acoustics that deals with the generation of sound by musical instruments, the transmission of sound to the listener, and the perception of musical sound. A main research activity in musical acoustics is the study of the way in which musical instruments vibrate and produce sound. The most common way of classifying musical instruments is according to the nature of the primary vibrator, into string instruments, wind instruments, and percussion instruments. The vibrations of a plucked string, a struck membrane, or a blown pipe can be described in terms of normal modes of vibration. Determining the normal modes of a complex vibrator is often termed modal analysis. Much of the progress in understanding how musical instruments generate sound is due to new methods of modal analysis, such as holographic interferometry and experimental modal testing. *See* CAVITY RESONATOR; INTERFEROMETRY; MODE OF VIBRATION; VIBRATION.

In the case of most percussion, plucked string, and struck string instruments, the player delivers energy to the primary vibrator (string, membrane, bar, or plate) and thereafter has little control over the way it vibrates. In the case of wind and bowed string instruments, however, the continuing flow of energy is controlled by feedback from the vibrating system. In brass and reed woodwinds, pressure feedback opens or closes the input valve. In flutes or flue organ pipes, however, the input valve is flow-controlled. In bowed string instruments, pulses on the string control the stick-slip action of the bow on the string.

Four attributes are frequently used to describe musical sound: loudness, pitch, timbre, and duration. Each of the subjective qualities depends on one or more physical parameters that can be measured. Loudness, for example, depends mainly on sound pressure but also on the spectrum of the partials and the physical duration. Pitch depends mainly on frequency, but also shows lesser dependence on sound pressure and envelope. Timbre includes all the attributes by which sounds with the same pitch and loudness are distinguished. Relating the subjective qualities of sound to the physical parameters is a central problem in psychoacoustics, and musical acousticians are concerned with this same problem as it applies to musical sound. *See* PSYCHOACOUSTICS.

Sound pressure level is measured with a sound level meter and is generally expressed on a logarithmic scale of decibels (dB) using an appropriate reference level and weighting network. From measurements of the sound pressure level at different frequencies, it is possible to calculate a subjective loudness, expressed in sones, which describes the sensation of loudness heard by an average listener. Musicians prefer to use dynamic markings ranging from *ppp* (very soft) to *fff* (very loud). *See* DECIBEL; LOUDNESS; SOUND; SOUND PRESSURE.

Pitch is defined as that attribute of auditory sensation in terms of which sounds may be ordered on a scale extending from low to high. Pitch is generally related to a musical scale where the octave, rather than the critical bandwidth, is the "natural" pitch interval. *See* PITCH; SCALE (MUSIC).

Timbre is defined as that attribute of auditory sensation in terms of which a listener can judge two sounds similarly presented and having the same loudness and pitch as dissimilar. Timbre depends primarily on the spectrum of the sound, but it also depends upon the waveform, the sound pressure, the frequency location of the spectrum, and the temporal characteristics of the sound. It has been found impossible to construct a single subjective scale of timbre (such as the sone scale of loudness); multidimensional scales have been constructed. The term "tone color" is often used to refer to that part of timbre that is attributable to the steady-state part of the tone, but the time envelope (and especially the attack) has been found to be very important in determining timbre as well.

Another subject relating to the perception of music is combination tones. When two tones that are close together in frequency are sounded at the same time, beats generally are heard, at a rate that is equal to their frequency difference. When the frequency difference Δf exceeds 15 Hz or so, the beat sensation disappears, and a roughness appears. As Δf increases still further, a point is reached at which the "fused" tone at the average frequency gives way to two tones, still with roughness. The respective resonance regions on the basilar membrane are now separated sufficiently to give two distinct pitches, but the excitations overlap to give a sense of roughness. When the separation Δf exceeds the width of the critical band, the roughness disappears, and the two tones begin to blend. *See* EAR (VERTEBRATE).

Pythagoras of ancient Greece is considered to have discovered that the tones produced by a string vibrating in two parts with simple ratios such as 2:1, 3:2, or 4:3 sound harmonious. These ratios define the so-called perfect intervals of music, which are considered to have the greatest consonance. Other consonant intervals in music are the major sixth ($f_2/f_1 = 5/3$), the major third ($f_2/f_1 = 5/4$), the major sixth ($f_2/f_1 = 8/5$), and the minor third ($f_2/f_1 = 6/5$). Why are some intervals more consonant than others? H. Helmholtz concluded that dissonance (the opposite of consonance) is greatest when partials of the two tones produce 30 to 40 beats per second (which are not heard as beats but produce roughness). The more the partials of one tone coincide in frequency with the partials of the other, the less chance of roughness. This explains why simple frequency ratios define the most consonant intervals. More recent research has concluded that consonance is related to the critical band. If the frequency difference between two pure tones is greater than a critical band, they sound consonant; if it is less than a critical band, they sound dissonant. The maximum dissonance occurs when Δf is approximately 1/4 of a critical band, which agrees reasonably well with Helmholtz's criterion for tones around 500 Hz. Thomas D. Rossing

Musical instruments Instruments for producing musical sounds have long been classified as woodwinds, brass, percussion, or strings; to these must be added electrical and electronic instruments. In a sense, all these instruments implement and extend the capability of the original musical instrument, the singing voice. The classes mentioned are useful in grouping instruments in a general way for the kinds of sounds they produce, even though woodwind instruments are not necessarily made of wood, nor are brass instruments always made of metal.

Woodwind instruments. Woodwind instruments are distinguished primarily by the fact that the effective length of the vibrating air column is shortened by opening lateral side holes in succession. Two distinctly different means of generating the sound are employed. For the flute, and its half-size version the piccolo, the player blows across the embouchure hole near one end in such away as to cause periodic puffs of air to enter the tube;

after a turbulent turning these puffs excite the air column longitudinally. This method of excitation leaves the tube acoustically open in the sense that the contained air vibrates much as it does in a simple tube with both ends open to the atmosphere.

For the double-reed oboe or bassoon, the player holds between the lips a pair of thin reeds (pieces of cane appropriately thinned, shaped, and bound together) that beat against each other to change the player's breath to puffs of air. For clarinets and saxophones, a single reed attached to a mouthpiece by a ligature functions in a similar way. The portion of the mouthpiece (the lay) against which the reed beats must be appropriately curved; the character of the sound is modified somewhat by the volume of the mouthpiece as well as by the shape and material of the reed.

For both the single- and double-reed instruments, the reeds vibrate under the influence of sound waves reflected back from the distant end of the air column and allow the puffs of air to enter when the sound pressure within the instrument is large. Thus (in contrast to the flute) the air vibration at the reed end is that associated with an acoustically closed end.

Brass instruments. The typical brass instrument consists of a cup-shaped mouthpiece, a slightly tapered mouthpipe, cylindrical tubing including valves, and a roughly hyperbolic bell. Puffs of air are introduced by the player via vibrating lips stretched over the mouthpiece. The action is comparable to that of the clarinet, in that the mouthpiece end is nearly closed acoustically. The length of the air column is increased by tubing switched in by use of valves, either piston or rotary: a common arrangement is such that the first valve lowers the intonation by two semitones, the second by one semitone, and the third by three semitones. For a given length of tubing, different tones are produced by tensioning the lips to excite different modes of vibration whose frequencies are approximately in the ratios 2:3:4:5:6:8.

Percussion instruments. Instruments such as the timpani (kettledrums) and xylophone are called percussion instruments because the sound is initiated by a blow. Two kinds of sound producers are involved: a membrane under tension, associated with a cavity that can influence the frequency of vibration, as in the case of the timpani; and a rigid bar or plate vibrating transversely, whose frequency is little affected by any resonator that may be attached. Some percussion instruments give a well-defined sound that excites a sensation of definite pitch, such as does a church bell; others, such as drums, cymbals, and triangles, are useful primarily for rhythm effects.

Stringed instruments. For the guitar and harp, strings are set into vibration by plucking; for the other stringed instruments the vibration is usually initiated and maintained by bowing. The frequency of vibration is primarily established by the length, tension, and mass per unit length of the string. A string vibrates not only at the lowest (fundamental) frequency, but also at the same time at higher frequencies which tend toward integer multiples of the fundamental frequency. The sound radiated from the instrument is thus complex.

The radiation of sound from a stringed instrument is enhanced by a resonator consisting of an almost closed air cavity. Some of the energy of the vibrating string is transmitted via the bridge to the walls of the cavity. In a carefully constructed violin the resonances of the air cavity and its vibrating walls are distributed in frequency in order to afford a relatively uniform response throughout the playing range of the instrument.

Keyboard instruments. Instruments such as the celesta, pipe organ, accordion, and piano are usually put in a group called keyboard instruments, because the respective vibrating bars, pipes, reeds, and strings in these instruments are selected by use of keys in a keyboard. The celesta and piano could also be described as percussion instruments, because hammers strike the bars and strings; the pipe organ and the accordion, with its wind-driven free reeds, are wind instruments. By its multiple keyboards (and pedal board) the pipe organ puts under the control of a single player thousands of sources whose distinctive sounds can be reproduced on command. Robert W. Young

Electrical, electronic, and software instruments. Electrical musical instruments produce electrical tone signals for amplification (and loudspeaker listening) without using tuned mechanical vibrators or air columns. Early electrical instruments predated electronics, generating their tone signals by electromagnetic or electrostatic rotating machinery. Later, most of the instruments became electronic, producing their tone waves from vacuum-tube circuits, then transistor circuits, and now integrated electronic circuits. In the 1980s, electronic instruments transitioned from analog to digital, offering much more precise tuning (and alternative tunings) and repeatability of configuration. *See* INTEGRATED CIRCUITS.

Many different means of sound generation have been used during the evolution of electronic musical instruments. In concept at least, the most simple electronic method would be to substitute directly for each conventional mechanical (string, reed) or acoustical (resonant air column) sound source an electronic circuit source or software algorithm that generates the same complex tone wave. However, in different parts of the musical scale the waveform typically differs for the same instrument. Moreover, keyboard instruments require that a number of tones be produced at the same time. Consequently, the substitution method is less feasible economically than other methods described below. *See* WAVEFORM.

The additive synthesis method synthesizes each desired complex tone by adding together an appropriate number of simple (sine) waves at harmonically related frequencies, with the relative amplitudes of the harmonic waves adjustable to produce the desired resultant waveform. Additive synthesis has been used extensively in the field of computer music since at least the 1960s, and it has been the basis of a number of limited-production hardware devices. In current music synthesizers, however, there are typically more efficient ways than additive synthesis to achieve desired sounds, and mass-produced electronic musical instruments are generally not based on additive synthesis. However, the concept of additive synthesis remains important, and there are closely related descendents, such as sinusoidal modeling and group additive synthesis.

Subtractive synthesis is similar in principle to that of speech (and singing) sound formation within the human voice system. At each desired musical frequency an electronic circuit produces a standard basic tone waveform (such as a sawtooth wave), which already contains a very large number of harmonics at known relative amplitudes. Then a variety of electric or electronic filters is provided in the instrument wherein switch selection, individually or collectively, converts the basic tone signals into the desired musical tone waveforms. Subtractive synthesis is somewhat the opposite of additive synthesis, since it practically subtracts undesired harmonics instead of adding desired harmonics. Subtractive synthesis has been the basis for most voice synthesizers for over half a century. *See* ELECTRIC FILTER; FUNCTION GENERATOR; WAVE-SHAPING CIRCUITS.

The first all-digital synthesizer was based on frequency-modulation (FM) synthesis. Simple FM is carried out using two digital oscillators, with the output of one adding to the frequency (or phase) control of the other. In additive synthesis, two oscillators can provide only two harmonics. In FM synthesis, two oscillators can provide any number of harmonics. It is even possible for the output of a digital oscillator to modulate its own phase, giving so-called feedback FM. A major advantage of FM is the synthesis of a wide variety of complex sounds from an extremely simple digital algorithm. Disadvantages of FM are its extreme sensitivity to its control parameters, and the fact that many parameter settings sound artificial. FM synthesis remains a valuable complexity-reduction technique in the context of additive synthesis. *See* FREQUENCY MODULATION.

Sampling synthesis can refer to any synthesis method based on playing back digitally recorded and manipulated sounds. Today, ROM-based wavetable synthesizers dominate the field, from single-chip devices used in notebook computers to the most expensive synthesizers. With enough digital recordings of an instrument, it is possible to achieve any sound with sampling synthesis. However, approaching full expressive fidelity for most instruments, particularly bowed strings and solo woodwinds, requires prohibitive amounts of memory, system development, and control complexity.

A more recent approach to the synthesis of acoustic musical instruments, also made possible by digital technology, is known as physical modeling synthesis. In this approach, the algorithms are based directly on the mathematical physics of the instrument. One variant of physical modeling synthesis, which is especially effective for synthesizing string and wind instruments, is waveguide synthesis. Waveguide synthesis explicitly simulates traveling waves on a string or inside a bore or horn using digital delay lines. The theory is analogous to that of sampled (discrete-time) electric transmission lines. Filters are used in conjuction with digital waveguides to simulate losses due to bridge motion, horn radiation, air absorption, and the like.

Daniel W. Martin; Julius O. Smith III

Musk-ox An even-toed ungulate, *Ovibos moschatus*, which is a member of the family Bovidae in the mammalian order Artiodactyla. This single species is the northernmost representative of the family, ranging through the tundra areas and snowfields of Canada and Alaska, as well as Greenland.

The musk-ox derives its name from the musky odor it emits. It is a stoutly built animal, stands about 4 ft (1.2 m) high at the shoulder, and may weigh up to 700 lb (320 kg; see illustration).

Musk-ox (*Ovibos moschatus*). (*Brent Huffman/Ultimate Ungulate Images*)

It has a coat of long dense hair that is resistant to the extreme cold of the windswept treeless tundra. These ruminants are herbivores. They do not hibernate and are usually found in herds of 20–100 animals huddled together for warmth. As protection against its natural enemy, the wolf, the musk-ox will form a circle with the young inside. Since a cow produces a single calf every 2 years, the numbers have been so reduced that they are now protected by the Canadian government. *See* ARTIODACTYLA.

Charles B. Curtin

Muskeg A term derived from Chippewan Indian for "grassy bog." In North America ecologists apply diverse usage, but most include peat bogs or tussock meadows, with variable woody vegetation such as spruce or tamarack. Plant remains accumulate when trapped in the water or in media such as sphagnum moss which inhibit decay. The peat might accrue indefinitely or might approach a steady state of raised bogs, blanket bogs, or forest if input became balanced by erosion or by loss as methane and CO_2.

Organic terrain occurs on every continent. The largest expanses of it are in Russia (especially western Siberia) and Canada. Typical bog or spruce-larch muskeg develops in cool temperature and subarctic to arctic lands. In the subtropics and the tropics there is considerable organic terrain, as in Paraguay, Uruguay, and Guyana in South America. Thus, despite differences in climate or floristics, the phenomenon of peat formation persists if local aeration or biochemical conditions hinder decay. Despite climatic and biotic differences, gross peat structure categories seem comparable the world over. *See* BOG; PEAT.

Norman W. Radforth

Muskmelon The edible fruit of *Cucumis melo*, belonging to the gourd family, Cucurbitaceae, as do other vine crops such as cucumber, watermelons, pumpkin, and squash. The muskmelon appears to be indigenous to Africa. *See* CANTALOUPE; HONEY DEW MELON; PERSIAN MELON; VIOLALES.

The plants are annual, trailing vines, with three to five runners. The runners produce short fruiting branches, which bear the perfect flowers and later the fruits. Muskmelons maturing on the vine without becoming overripe are superior in quality to those harvested immature. The sugar content, flavor, and texture of the fresh flesh improves very rapidly as the fruit approaches maturity. When mature, the melon is sweet, averages 6 to 8% sugar, and has a slight to distinctly musky odor and flavor, depending upon cultivar and environment. The flesh is rich in potassium, in vitamin C and, when deep orange, also in vitamin A.

Frank W. Zink

Muskrat A large aquatic rodent, *Ondatra zibethicus*, in the family Cricetidae and in the vole subfamily (Arvicolinae). It is known as the common muskrat (see illustration). It is dark reddish-brown, with a long, essentially naked, scaly, laterally compressed tail. The pelage is dense and shiny, and consists of a thick coat of underfur over which lies a covering of long, glossy guard hairs. The ears and eyes are relatively small for an animal of this size. The hindfeet bear webbed toes and are much larger than the front feet. The muskrat received its name because of its inguinal glands, which produce a musky odor. *See* MAMMALIA; RODENTIA; SCENT GLAND.

The muskrat is found only in North America, but is common over much of the United States and Canada where suitable habitat is present. Marshes and other wet areas with an abundance of emergent vegetation, especially cattails, are the preferred habitat of muskrats, but the species also occurs along streams and ditches and about lakes and ponds. The muskrat is primarily nocturnal but is often active by day, especially in spring and fall. It spends most of the daytime in burrows or houses that it constructs in or near water. The houses are the most conspicuous

Common muskrat, *Ondatra zibethicus*. (*Photo by Dr. Lloyd Glenn Ingles; © California Academy of Sciences*)

indicators of the muskrat's presence on a marsh. Most muskrat houses are in water 1–2 ft (0.3–0.6 m) deep and are constructed of emergent and submerged vegetation.

Muskrats are primarily vegetarian but will eat animal matter, even carrion, under certain conditions. Probably, most emergent plants are eaten. Animal foods include winter-killed fishes, frogs, crayfishes, mussels, and other dead animals. John O. Whitaker

Mustard Any one of a number of annual crucifer species of Asiatic origin belonging to the plant order Capparales. Mustards eaten as greens are *Brassica juncea*, *B. juncea* var. *crispifolia*, and *B. hirta*. Table mustard and oils are obtained from *B. nigra*. Important production centers for mustard greens are in the South, where the crop is popular. Montana and the West Coast states are important sources of mustard seed. *See* CAPPARALES.

H. John Carew

Mutagens and carcinogens A mutagen is a substance or agent that induces heritable change in cells or organisms. A carcinogen is a substance that induces unregulated growth processes in cells or tissues of multicellular animals, leading to cancer. Although mutagen and carcinogen are not synonymous terms, the ability of a substance to induce mutations and its ability to induce cancer are strongly correlated. Mutagenesis refers to processes that result in genetic change, and carcinogenesis (the processes of tumor development) may result from mutagenic events.

A mutation is any change in a cell or in an organism that is transmitted to subsequent generations. Mutations can occur spontaneously or be induced by chemical or physical agents. The cause of mutations is usually some form of damage to DNA or chromosomes that results in some change that can be seen or measured. However, damage can occur in a segment of DNA that is a noncoding region and thus will not result in a mutation. Mutations may or may not be harmful, depending upon which function is affected. They may occur in either somatic or germ cells. Mutations that occur in germ cells may be transmitted to subsequent generations, whereas mutations in somatic cells are generally of consequence only to the affected individual.

Not all heritable changes result from damage to DNA. For example, in growth and differentiation of normal cells, major changes in gene expression occur and are transmitted to progeny cells through changes in the signals that control genes that are transcribed into ribonucleic acid (RNA). It is possible that chemicals and radiation alter these processes as well. When such an effect is seen in newborns, it is called teratogenic and results in birth defects that are not transmitted to the next generation. However, if the change is transmissible to progeny, it is a mutation, even though it might have arisen from an effect on the way in which the gene is expressed. Thus, chemicals can have somatic effects involving genes regulating cell growth that could lead to the development of cancer, without damaging DNA.

Cancer arises because of the loss of growth control by derangement of regulatory signals. Included in the phenotypic consequences of mutations are alterations in gene regulation brought about by changes either in the regulatory region or in proteins involved with coordinated cellular functions. Altered proteins may exhibit novel interactions with target substrates and thereby lose the ability to provide a regulatory function for the cell or impose altered functions on associated molecules. Through such a complex series of molecular interactions, changes occur in the growth properties of normal cells leading to cancer cells that are not responsive to normal regulatory controls and can eventually give rise to a visible neoplasm or tumor. While mutagens can give rise to neoplasms by a process similar to that described above, not all mutagens induce cancer and not all mutational events result in tumors.

The identification of certain specific types of genes, termed oncogenes, that appear to be causally involved in the neoplastic process has helped to focus mechanistic studies on carcinogenesis. Oncogenes can be classified into a few functionally different groups, and specific mutations in some of the genes have been identified and are believed to be critical in tumorigenesis. Tumor suppressor genes or antioncogenes provide a normal regulatory function; by mutation or other events, the loss of the function of these genes may release cells from normal growth-control processes, allowing them to begin the neoplastic process. *See* ONCOGENES; ONCOLOGY.

There are a number of methods and systems for identifying chemical mutagens. Mutations can be detected at a variety of genetic loci in very diverse organisms, including bacteria, insects, cultured mammalian cells, rodents, and humans. Spontaneous and induced mutations occur very infrequently, the estimated rate being less than 1 in 10,000 per gene per cell generation. This low mutation rate is probably the result of a combination of factors that include the relative inaccessibility of DNA to damaging agents and the ability of cellular processes to repair damage to DNA.

Factors that contribute to the difficulty in recognizing substances that may be carcinogenic to humans include the prevalence of cancer, the diversity of types of cancer, the generally late-life onset of most cancers, and the multifactorial nature of the disease process. Approximately 50 substances have been identified as causes of cancer in humans, but they probably account for only a small portion of the disease incidence. *See* CANCER (MEDICINE); HUMAN GENETICS; MUTATION; RADIATION BIOLOGY.

Raymond W. Tennant

Mutation An abrupt, heritable change in genes or chromosomes manifested by changes in the phenotype (the appearance) of an organism. It is theoretically preferable to define mutations as changes in deoxyribonucleic acid (DNA) sequences, but the classical definition remains the operational definition in most circumstances. *See* DEOXYRIBONUCLEIC ACID (DNA).

The word mutation has two common meanings, one being the process and the other the product (the altered gene or chromosome carries a mutation). The process is also called mutagenesis. An organism bearing a mutation is called a mutant. An agent that induces a mutation is called a mutagen.

The study of mutation has long occupied a central position in genetics. Mutations are the ultimate sources of variability upon which evolution acts, despite being random changes that are far more likely to harm than to improve a complicated and highly evolved organism. Mutation has consistently been the most telling probe into the nature of the gene, and understanding of most aspects of biology has benefited from studies of the properties of mutant organisms. Mutation is also an important component of disease, either causing it directly (for example, through birth defects) or predisposing humans to a vast array of disorders that together constitute a substantial fraction of illnesses. Finally, deliberate selection of mutant plants and animals for economic or esthetic purposes has long been practiced, and has grown into an important aspect of genetics. With advances in molecular genetics, it is now possible to construct specific mutations at will, rather than merely selecting among an array of random mutations for the infrequent useful ones. *See* BREEDING (ANIMAL); BREEDING (PLANT).

Chromosome mutations, which alter sections composed of many DNA base pairs, consist of partial losses (deletions or deficiencies), rearrangements, and additions. Most deletions that remove many genes are highly deleterious. Rearrangements may be less deleterious if they shuffle genes about but do not interrupt them or relocate them to sites where they cannot function well. They involve either inversions (simple reversals of an internal segment of a chromosome) or translocations (transfer of a segment of a chromosome to a new location). Even when not directly deleterious, rearrangements lead to anomalies of genetic recombination; a common secondary consequence in humans is sterility. Addition mutations are of two types, duplications and insertions. Duplications usually consist of tandem repeats of a segment of a chromosome, and may range from innocuous to lethal depending upon their location and extent. Insertions occur through the movement of special DNA sequences (transposons) that range from hundreds to thousands of DNA base pairs in length. *See* CHROMOSOME; CHROMOSOME ABERRATION; TRANSPOSONS.

Gene mutations affect only a single gene and consist of intragenic chromosomal mutations, additions or deletions of one or a few base pairs, base-pair substitutions (point mutations), and complex mutations comprising simultaneously arising clusters of any of the above. The severity of a gene mutation depends on its individual nature and on the importance of the affected gene, and can range from innocuous to lethal.

Because the genetic code employs consecutive sets of three DNA base pairs to specify consecutive amino acids in proteins, the addition or deletion of multiples of three base pairs leads to the addition or deletion of one or more amino acids. However, the addition or deletion of one or two base pairs (or any nonmultiple of three) shifts the reading frame, so that everything from that point onward is read out of its normal frame, with drastic consequences for that particular gene. These occurrences are called frameshift mutations. *See* GENETIC CODE.

Although important mechanisms of mutagenesis undoubtedly remain to be discovered, many, and perhaps most, of the predominant mechanisms are now known, at least in outline. Genome mutations that alter the number of chromosomes in a cell usually result from the faulty distribution of chromosomes during mitotic or meiotic cell divisions. The fault probably often lies in the systems of spindle fibers that segregate daughter chromosomes into daughter cells; chemicals (such as colchicine) that interfere with such fibers induce aneuploidy at high frequencies. Chromosome mutations can be formed by chromosome breakage followed by incorrect patterns of rejoining. Many agents, including ionizing radiations and numerous chemicals, can induce chromosome breaks. In addition to events triggered by breaks, however, deletions and duplications are triggered by anomalies of genetic recombination between similar but nonhomologous DNA sequences. *See* MEIOSIS; MITOSIS; RADIATION BIOLOGY; RECOMBINATION (GENETICS).

Cells direct a number of repair operations against anomalies of DNA structure that would otherwise result in death or mutation. While some of these repair mechanisms circumvent death, others actually cause mutations. Most DNA damage is detected by excision repair systems before it can interfere with replication. These systems enzymatically remove the damaged bases and their associated sugar-phosphate backbones; the resulting gap is then filled in by DNA repair synthesis, using the complementary strand as a template.

With the growing recognition that many chemicals are mutagenic, and that many humans suffer from mutations of ancient or recent origin, much attention has been directed to discovering the mutagenic components of the environment. Many artificial chemicals are mutagenic, although these are still only a small percent of the total; and as a part of their ongoing war against predators, most plants elaborate chemicals that are mutagenic for at least some organisms. *See* MUTAGENS AND CARCINOGENS.

While the search for ways to prevent unwanted mutation continues vigorously, other searches seek methods to introduce specific, predetermined, desirable mutations into organisms of economic importance. By using the methods of molecular genetics, it is now usually possible to introduce specific mutations into specific genes in order, for instance, to increase the production of some marketable product. A more distant possibility is intervention in the human germ line in order, for instance, to cure specific genetic diseases. *See* GENETIC ENGINEERING; GENETICS; HUMAN GENETICS.

John W. Drake

Mutualism An interaction between two species that benefits both. Individuals that interact with mutualists experience higher sucess than those that do not. Hence, behaving mutualistically is advantageous to the individual, and it does not require any concern for the well-being of the partner. At one time, mutualisms were thought to be rare curiosities primarily of interest to natural historians. However, it is now believed that every species is involved in one or more mutualisms. Mutualisms are thought to lie at the root of phenomena as diverse as the origin of the eukaryotic cell, the diversification of flowering plants, and the pattern of elevated species diversity in tropical forests.

Mutualisms generally involve an exchange of substances or services that organisms would find difficult or impossible to obtain for themselves. For instance, *Rhizobium* bacteria found in nodules on the roots of many legume (bean) species fix atmospheric nitrogen into a form (NH_3) that can be taken up by plants. The plant provides the bacteria with carbon in the form of dicarboxylic acids. The carbon is utilized by the bacteria as energy for nitrogen fixation. Consequently, leguminous plants often thrive in nitrogen-poor environments where other plants cannot persist. Another well-known example is lichens, in which fungi take up carbon fixed during photosynthesis of their algae associates. *See* NITROGEN FIXATION.

A second benefit offered within some mutualisms is transportation. Prominent among these mutualisms is biotic pollination, in which certain animals visit flowers to obtain resources and return a benefit by transporting pollen between the flowers they visit. A final benefit is protection from one's enemies. For example, ants attack the predators and parasites of certain aphids in exchange for access to the aphids' carbohydrate-rich excretions (honeydew).

Another consideration about mutualisms is whether they are symbiotic. Two species found in intimate physical association for most or all of their lifetimes are considered to be in symbiosis. Not all symbioses are mutualistic; symbioses may benefit both, one, or neither of the partners.

Mutualisms can also be characterized as obligate or facultative (depending on whether or not the partners can survive without each other), and as specialized or generalized (depending on how many species can confer the benefit in question).

Two features are common to most mutualisms. First, mutualisms are highly variable in time and space. Second, mutualisms are susceptible to cheating. Cheaters can be individuals of the mutualist species that profit from their partners' actions without offering anything in return, or else other species that invade the mutualism for their own gain.

Mutualism has considerable practical significance. Certain mutualisms play central roles in humans' ability to feed the growing population. It has been estimated that half the food consumed is the product of biotic pollination. *See* ECOLOGY; PLANT PATHOLOGY.

Judith L. Bronstein

Myasthenia gravis A disease resulting from an abnormality in neuromuscular transmission, characterized by a fluctuating degree of muscle weakness. The weakness is usually aggravated by activity, and there is partial or complete restoration of strength after a period of rest or the administration of anticholinesterase medications.

It has been shown that the basic defect in myasthenia gravis is a reduction in the number of acetylcholine receptor sites in the postsynaptic membrane of the neuromuscular junction. It has also been shown that many myasthenic patients have immunoglobulins in the serum that partially block acetylcholine receptors. *See* AUTOIMMUNITY; IMMUNOGLOBULIN.

Abnormalities have been demonstrated in the thymus gland and skeletal muscle in myasthenia gravis. There is an increased incidence of thymoma in myasthenia gravis, and in those without a thymoma, hyperplasia of the germinal centers is a common finding in the thymus gland.

Although the disease affects young women more commonly, usually in the third decade, it can occur in either sex at any age. In the majority of persons, weakness affects muscles of head, neck, and limbs (generalized myasthenia), but in some the weakness is restricted to the muscles of the eyes (ocular myasthenia), in which case the disease is usually benign.

The standard treatment for myasthenia gravis has been the use of longer-acting anticholinesterase agents; thymectomy and immunosuppressive drugs are reserved for those patients with generalized myasthenia that does not respond sufficiently to these agents.

S. Mark Sumi

Mycobacterial diseases Diseases caused by mycobacteria, a diffuse group of acid-fast, rod-shaped bacteria in the genus *Mycobacterium*. Some mycobacteria are saprophytes, while others can cause disease in humans. The two most important species are *M. tuberculosis* (the cause of tuberculosis) and *M. leprae* (the cause of leprosy); other species have been called by several names, particularly the atypical mycobacteria or the nontuberculous mycobacteria. *See* LEPROSY; TUBERCULOSIS.

These bacteria are classified according to their pigment formation, rate of growth, and colony morphology. The most commonly involved disease site is the lungs. Nontuberculous mycobacteria are transmitted from natural sources in the environment, rather than from person to person, and thus are not a public health hazard. Nontuberculous mycobacteria have been cultured from various environmental sources.

The diagnosis of disease caused by nontuberculous mycobacteria can be difficult, since colonization or contamination of specimens may be present rather than true infection.

Pulmonary disease resembling tuberculosis is a most important manifestation of disease caused by nontuberculous mycobacteria. The symptoms and chest x-ray findings are similar to those seen in tuberculosis. *Mycobacterium kansasii* and *M. avium intracellulare* are the most common pathogens. The disease usually occurs in middle-aged men and women with some type of chronic coexisting lung disease. The pathogenic mechanisms are obscure. Pulmonary infections due to *M. kansasii* can be treated successfully with chemotherapy. The treatment of pulmonary infections due to *M. avium intracellulare* complex is difficult.

Chronic infection involving joints and bones, bursae, synovia, and tendon sheaths can be caused by various species.

Localized abscesses due to *M. fortuitum* or *M. chelonei* can occur after trauma, after surgical incision, or at injection sites. The usual treatment is surgical incision. The most common soft tissue infection is caused by *M. marinum*, which may be introduced, following an abrasion or trauma, from handling fish or fish tanks, or around a swimming pool. Treatment is surgical. *Mycobacterium ulcerans* causes a destructive skin infection in tropical areas of the world. It is treated by wide excision and skin grafting.

Disseminated *M. avium intracellulare* is one of the opportunistic infections seen in the acquired immune deficiency syndrome (AIDS). In individuals with AIDS, the organism has been cultured from lung, brain, cerebrospinal fluid, liver, spleen, intestinal mucosa, and bone marrow. No treatment has yet been effective in this setting. *See* ACQUIRED IMMUNE DEFICIENCY SYNDROME (AIDS).

George M. Lordi; Lee B. Reichman

Mycology The study of organisms of the fungal lineage, including mushrooms, boletes, bracket or shelf fungi, powdery mildew, bread molds, yeasts, puffballs, morels, stinkhorns, truffles, smuts, and rusts. Fungi, in the traditional sense, are an ecological grouping of organisms found in every ecological niche. Mycologists estimate that there are 1.5 million species of fungi, with only 70,000 species now described. Fungi typically have a filamentous-branched somatic structure surrounded by thick cell walls; these structures are known as hyphae. *See* FUNGI; MUSHROOM.

The names adopted for major taxa (groups) of fungi and fungi-like organisms are as follows:

Eumycota (true fungi)
  Chytridiomycota (Chytridiomycetes)
  Zygomycota (Zygomycetes, Trichomycetes)
  Ascomycota (Archiascomycetes, Hemiascomycetes, Euascomycetes)
  Basidiomycota (Urediniomycetes, Ustilaginomycetes, Hymenomycetes)
  Deuteromycota (Agonomycetes, Balstomycetes, Coelomycetes, Hyphomycetes)
Pseudomycota (fungi-like organisms)
  Oomycota (Oomycetes)
  Hyphochytriomycota (Hyphochytriomycetes)
  Plasmodiophoromycota (Plasmodiophoromycetes)
Myxomycota (slime molds)
  Myxomycota (Mycomycetes)

Fungi are heterotrophic, meaning they cannot make their own food as plants do. They are osmotrophic or nutrition-absorptive, obtaining their food by releasing enzymes into their environment to break down complex organic compounds. Fungi play a significant role in nature as decomposers of wood and wood products, and in forest ecosystems fungi release nutrients back to soil. They reproduce through spores produced sexually and asexually. *See* FOREST ECOSYSTEM; FUNGAL ECOLOGY; WOOD DEGRADATION.

Shung-Chang Jong

Mycoplasmas The smallest prokaryotic microorganisms that are able to grow on cell-free artificial media. Their genome size is also among the smallest recorded in prokaryotes, about 5×10^8 to 10^9 daltons. The mycoplasmas differ from almost all other prokaryotes in lacking a rigid cell wall and in their incapability to synthesize peptidoglycan, an essential component of the bacterial cell wall.

Taxonomically, the mycoplasmas are assigned to a distinct class, the Mollicutes, containing two orders, Mycoplasmatales and Acholeplasmatales. The distinction between the orders is based primarily on differences in nutritional criteria: members of the Mycoplasmatales require cholesterol or other sterols for growth whereas those of the second order do not. The main criteria used for the subdivision of the orders into families and genera are shown:

Class: Mollicutes
  Order I: Mycoplasmatales (sterol required for growth; $NADH_2$ oxidase localized in cytoplasm)
    Family I: Mycoplasmataceae (genome size approximately 5×10^8 daltons)
      Genus I: *Mycoplasma* (70 species; do not hydrolyze urea)
      Genus II: *Ureaplasma* (2 species; hydrolyze urea)
    Family II: Spiroplasmataceae (helical organisms; genome size approximately 10^9 daltons)
      Genus I: *Spiroplasma* (4 species)
  Order II: Acholeplasmatales (sterol not required for growth; $NADH_2$ oxidase localized in membrane; genome size approximately 10^9 daltons)
    Family I: Acholeplasmataceae
      Genus I: *Acholeplasma* (9 species)
  Genus of uncertain taxonomic position
    Anaeroplasma (2 species)

The term mycoplasmas is generally used as the vernacular or trivial name for all members of the class Mollicutes, irrespective of the classification in aparticular genus. *See* PROKARYOTAE.

The mycoplasmas are almost ubiquitous in nature. Several species areimportant pathogens of humans, animals, and plants, while others constitute part of the normal microbial flora of, for example, the upper respiratory and lowerurogenital tracts of humans. *Mycoplasma pneumoniae* was found to be the cause of cold agglutinin-associated primary atypical pneumonia. This disease is particularly frequent in the 5–15-year age group; it is probably endemical most all over the world and often reaches epidemic proportions at intervals of 4 to 5 years.

Mycoplasmas are generally highly resistant to benzyl penicillin and other antibiotics which act by interfering with the biosynthesis of peptidoglycan. They are usually susceptible to antibiotics that specifically inhibit protein synthesis in prokaryotes, such as tetracyclines and chloramphenicol. Susceptibility to other antibiotics, such as erythromycin and other macrolides, is variable. *See* ANTIBIOTIC; BACTERIAL PHYSIOLOGY AND METABOLISM; PLANT PATHOLOGY; PNEUMONIA. E. A. Freundt

Mycorrhizae Dual organs of absorption that are formed when symbiotic fungi inhabit healthy absorbing organs (roots, rhizomes, or thalli) of most terrestrial plants and many aquatics and epiphytes.

Mycorrhizae appear in the earliest fossil record of terrestrial plant roots. Roughly 80% of the nearly 10,000 plant species that have been examined are mycorrhizal. Present-day plants that normally lack mycorrhizae are generally evolutionarily advanced. It has been inferred that primitive plants evolved with a symbiosis between fungi and rhizoids or roots as a means to extract nutrients and water from soil. The degree of dependence varies between species or groups of plants. In absolute dependence, characteristic of perennial, terrestrial plants, the host requires mycorrhizae to survive. Some plants are facultative; they may form mycorrhizae but do not always require them. This group includes many of the world's more troublesome weeds. A minority of plant species characteristically lack mycorrhizae, so far as is known, including many aquatics, epiphytes, and annual weeds.

The three major types of mycorrhizae differ in structural details but have many functions in common. The fungus colonizes the cortex of the host root and grows its filaments (hyphae) into surrounding soil from a few centimeters to a meter or more. The hyphae absorb nutrients and water and transport them to host roots. The fungi thus tap far greater volumes of soil at a relatively lower energy cost than the roots could on their own. Moreover, many, if not all, mycorrhizal fungi produce extracellular enzymes and organic acids that release immobile elements such as phosphorus and zinc from clay particles, or phosphorus and nitrogen bound in organic matter. The fungi are far more physiologically capable in extracting or recycling nutrients in this way than the rootlets themselves.

Mycorrhizal fungi are relatively poorly competent in extracting carbon from organic matter. They derive energy from host-photosynthesized carbohydrates. Hosts also provide vitamins and other growth regulators that the fungi need.

The major types are ectomycorrhizae, vesicular-arbuscular mycorrhizae, and ericoid mycorrhizae. Ectomycorrhizae are the most readily observed type. Ectomycorrhizal hosts strongly depend on mycorrhizae to survive. Relatively few in number of species, they nonetheless dominate most forests outside the tropics. Vesicular-arbuscular mycorrhizae (sometimes simply termed arbuscular mycorrhizae) form with the great majority of terrestrial herbaceous plant species plus nearly all woody perennials that are not ectomycorrhizal. Vesicular-arbuscular mycorrhizal hosts range from strongly mycorrhiza-dependent, especially the woody perennials, to faculative, as are many grasses.

Ericoid mycorrhizae are restricted to the Ericales, the heath order. The hosts are strongly mycorrhiza-dependent. Though relatively few in number, heath species dominate large areas around the world and are common understory plants in many forests. Other mycorrhiza types include those special for the Orchidaceae (orchids) and Gentianaceae (gentians). *See* ASCOMYCOTA; ERICALES; ZYGOMYCOTA.

The succession of plants from pioneering through seral to climax communities is governed by availability of mycorrhizal propagules. When catastrophic fire, erosion, or clearcutting reduce the availability of mycorrhizal fungi in the soil, plants dependent on those fungi will have difficulty becoming established. Each mycorrhizal fungus has its own array of physiological characteristics. Some are especially proficient at releasing nutrients bound in organic matter, some produce more effective antibiotics or growth regulators than others, and some are more active in cool, hot, wet, or dry times of year than others. Healthy plant communities or crops typically harbor diverse populations of mycorrhizal fungal species. This diversity, evolved over a great expanse of time, is a hallmark of thriving ecosystems. Factors that reduce this diversity also reduce the resilience of ecosystems.

Mycorrhizal inoculation of plants in nurseries, orchards, and fields has succeeded in many circumstances, resulting in improved survival and productivity of the inoculated plants. Inoculation with selected fungi is especially important for restoring degraded sites or introducing exotics. Because ectomycorrhizal fungi include many premier edibles such as truffles, seedlings can also be inoculated to establish orchards for production of edible fungi. *See* FOREST SOIL; FUNGI. James M. Trappe

Mycotoxin Any of the mold-produced substances that may be injurious to vertebrates upon ingestion, inhalation, or skin contact. The diseases they cause, known as mycotoxicoses, need not involve the toxin-producing fungus. Diagnostic features characterizing mycotoxicoses are the following: the disease is not transmissible; drug and antibiotic treatments have little or no effect; in field outbreaks the disease is often seasonal; the outbreak is usually associated with a specific foodstuff; and examination of the suspected food or foodstuff reveals signs of fungal activity.

The earliest recognized mycotoxicoses were human diseases. Ergotism, or St. Anthony's fire, results from eating rye infected with *Claviceps purpurea*. Yellow rice disease, a complex of human toxicoses, is caused by several *Penicillium islandicum* mycotoxins. World attention was directed toward the mycotoxin problem with the discovery of the aflatoxins in England in 1961. The aflatoxins, a family of mycotoxins produced by *Aspergillus flavus* and *A. parasiticus*, can induce both acute and chronic toxicological effects in vertebrates. Aflatoxin B_1, the most potent of the group, is toxic, carcinogenic, mutagenic, and teratogenic. Major agricultural commodities that are often contaminated by aflatoxins include corn, peanuts, rice, cottonseed, and various tree nuts. *See* AFLATOXIN; ERGOT AND ERGOTISM. Alex Ciegler; Maren Klich

Myiasis The infestation of vertebrates by the larvae, or maggots, of numerous species of flies. These larvae may invade different parts of the bodies of these animals or may appear externally. The Diptera of medical and veterinary importance are largely confined to the families Oestridae, Calliphoridae, and Sarcophagidae.

In cutaneous myiasis, the larvae are found in or under the skin. There may be a migration of some species of these larvae through host tissues, resulting in a swelling with intense itching. Such a condition is known as larva migrans, or creeping eruption, and may require surgical treatment.

Intestinal myiasis in humans is usually the result of accidentally swallowing the eggs or larvae of these flies. It occurs commonly in many herbivores who ingest the eggs when feeding on contaminated herbage. The larvae settle in the stomach or intestinal tract of the animal host.

Cavity, or wound, myiasis occurs when the larvae invade natural orifices, such as the nasopharynx, vulva, and sinuses, or artificial openings such as wounds. External myiasis includes infestation by those maggots which are blood feeders. *See* DIPTERA; MEDICAL PARASITOLOGY. Charles B. Curtin

Myliobatiformes An order of batoid fishes (subclass Elasmobranchii), the typical members of which are described by the following characteristics: The disc is strongly depressed and varies from oval longitudinally to much broader than long; the tail is well marked off from the body sector, very short to long and whiplike, and equipped with a poisonous spine in some species; the pectoral rays are either continuous along the side of the head or separate from the head and modified to form rostral lobes or finlike rostral appendages (cephalic fins); the dorsal fin, if present, is near the base of the tail; and development is ovoviviparous. Most species are tropical or subtropical, although some species occur in warm temperate and cool temperate zones. The usual habitat is shallow shore waters and upper continental and insular slopes. *See* BATOIDEA; ELASMOBRANCHII.

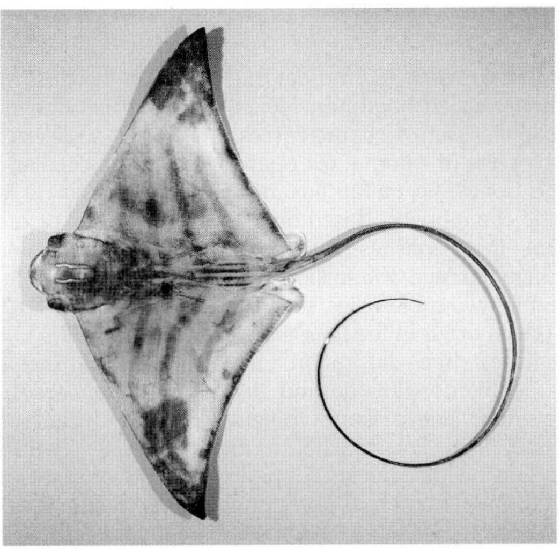

***Aetomylaeus nichofii*, a banded eagle ray. (*Photo courtesy of J. E. Randall*)**

Recent studies on the phylogeny of batoids have led to some major changes in their classification, one of which results in the inclusion of the platyrhinids and *Zanobatus* in Myliobatiformes. The most recently proposed classification includes three suborders: Platyrhinoidei, Zanobatoidei, and Myliobatoidei. The suborder Myliobatoidei is the most important with a number of notable families, including the Hexatrygonidae (sixgill stingrays), Plesiobatidae (deepwater stingrays), Urolophidae (round stingrays), Urotrygonidae (American round stingrays), Dasyatidae (whiptail stingrays), Gymnuridae (butterfly rays), and Myliobatidae (eagle rays [see illustration], including the subfamily Mobulinae [devil rays]). Urolophids, urotrygonids, and dasyatids, collectively known as stingarees, are potentially dangerous to humans. The piercing tail spine delivers a poison that causes immediate and excruciating pain, followed by swelling and most often a serious infection. Devil rays are famous for their large sizes and spectacular leaps out of the water. These rays comprise two genera: *Manta*, with one species, *Manta birostris*, commonly known as the giant manta or giant devil ray; and *Mobula*, with nine species. Herbert Boschung

Mylonite A rock that has undergone significant modification of original textures by predominantly plastic flow due to dynamic recrystallization. Mylonites form at depth beneath brittle faults in continental and oceanic crust, in rocks from quartzofeldspathic to olivine-pyroxenite composition. Mylonites were once confused with cataclasites, which form by brittle fracturing, crushing, and comminution. Microstructures that develop during mylonitization vary according to original mineralogy and modal compositions, temperature, confining pressure, strain, strain rate, applied stresses, and presence or absence of fluids.

At low to moderate metamorphic grades, mylonitization reduces the grain size of the protolith and commonly produces a very fine-grained, well-foliated rock with a pronounced linear fabric defined by elongate minerals. Lineations may be weak or absent, however, in high-strain zones that lack a significant rotational component. At high metamorphic grades, grain growth during mylonitization can produce a net increase in grain size, and the term mylonitic gneiss is used where there is a preserved or inferred undeformed protolith. *See* METAMORPHIC ROCKS. Carol Simpson

Myodocopida An order of marine organisms that forms an important part of the class Ostracoda (subphylum Crustacea) and comprises three suborders. It has a long geological history that extends back at least to the Early Silurian. However, the lightly calcified carapaces of the species in this order are a factor in their having a sparse fossil record. Indeed, of the sixteen families of myodocopids, three are known only from modern marine environments. Species of those families that have a fossil record are quite rare and discontinuous in their stratigraphical distribution. Myodocopids are only abundant as fossils from organically rich shale that was deposited in deep, anoxic environments, especially those of the Devonian seas. Such environments were devoid of benthic (bottom-dwelling) life, but the valves and carapaces of the nektonic (free-swimming) myodocopids, both immature forms (termed instars) and adults, sank into such environments and were preserved.

The myodocopids are typically much larger than other ostracodes and may be more than 1 cm long, although adults of most species are unlikely to be longer than 3 mm. As is characteristic of ostracodes, the carapace of the myodocopids comprises two valves that are joined along the animal's dorsum by a hinge and ligament. Among the myodocopids, the anterior portions of both valves are marked by a characteristic notch through which the setaceous antennae or antennules protrude. The appendages of all ostracodes are homologous with those of other crustaceans and, as is characteristic of arthropods in general, have been highly modified during their evolution. The appendages of the myodocopids are especially well adapted for swimming. Sexual dimorphism is not as pronounced among the myodocopids as it is among most smaller, more heavily calcified ostracodes.

Myodocopids are not exclusively planktonic as is often assumed. They are typically strong swimmers, but many are nektobenthic, swimming in proximity to the substrate. Others, however, live near the surface of the open ocean and are not associated with the environments of deposition that lie far beneath them on the ocean floor, where their valves and carapaces will ultimately be deposited. A fascinating characteristic of the myodocopids is the ability of many species to secrete bioluminescent material into the water. This bioluminescence functions as highly elaborate courtship displays that are species-specific and facilitate recognition of potential mates. *See* BIOLUMINESCENCE; CRUSTACEA; OSTRACODA. Roger A. Kaesler

Myricales An order of flowering plants, division Magnoliophyta (Angiospermae), in the subclass Hamamelidae of the class Magnoliopsida (dicotyledons). The order consists of the single family Myricaceae, with about 50 species. Within its subclass the order is marked by its simple, resinous-dotted, aromatic leaves and unilocular ovary with two styles and a single ovule. The plants are trees or shrubs, and the flowers are much reduced and borne in catkins. The fruit is a small, waxy-coated drupe or nut. Several species of *Myrica* are occasionally cultivated as ornamentals. *See* FLOWER; HAMAMELIDAE; MAGNOLIOPSIDA. Arthur Cronquist

Myrtales An order of flowering plants in the core eudicots. The order consists of 10 families and approximately 9300 species. The two largest families are Melastomataceae (approximately 4500 species) and Myrtaceae (approximately 3000 species). Thymelaeaceae are excluded in recent concepts of the order, being related instead to families of Malvales.

Myrtales are chiefly tropical, but Onagraceae and Penaeaceae are predominantly temperate. Myrtales usually have opposite, simple, entire leaves and perigynous to epigynous flowers with a compound pistil and most commonly axile placentation. The seeds have little or no endosperm. The stamens are normally numerous, and many species have tetramerous flowers. Vascular bundles characteristically have internal phloem, which is otherwise rare in the rosid dicots.

Economic crops in Myrtaceae include spice trees such as allspice (*Pimenta*) and cloves (*Syzygium*), and the timber trees *Eucalyptus*. Other important economic crops in the order include evening primrose (*Oenothera*, Onagraceae) and pomegranate (*Punica*, Lythraceae). *See* EUDICOTYLEDONS. Michael F. Fay

Mysida An order of free-swimming, shrimplike crustaceans, commonly known as opossum shrimps, belonging to the subclass Eumalacostraca. They occur in vast numbers in coastal and open oceanic regions of the world. The mysidan adult body length generally averages about 0.6 in. (15 mm), and most species are distributed in shallow coastal and shelf waters of the oceans. A few live in the surface layers of the ocean bottoms. In addition, some species have invaded freshwaters, including specialized forms strictly confined to caves. *See* CRUSTACEA; EUCARIDA; EUMALACOSTRACA.

Mysida are among the most primitive of the peracarid eumalacostracans. The carapace generally covers the entire thorax, and it is not fused to the underlying segments. The thoracic limbs are biramous (having two branches), possessing both a well-developed medial endopod and a lateral natatory (adapted to swimming) exopod; the thoracopods generally bear well-developed gills on the base of the limbs. Generally the pleopods are well developed. Mysidans feed on algae, detritus, and zooplankton. They pass through up to 12 molts before achieving sexual maturity. *See* PERACARIDA.

Most species of mysids form aggregations. These are of different types for different purposes. The functions of these aggregations, except for breeding, are not always clear, although protection of the population from predators is important. Aggregations occur in many coastal mysids, especially in estuarine or sandy beach habitats. Swarms of mysids in coastal waters are exploited commercially in tropical and subtropical regions of the world, and are used to make shrimp paste and sauces.

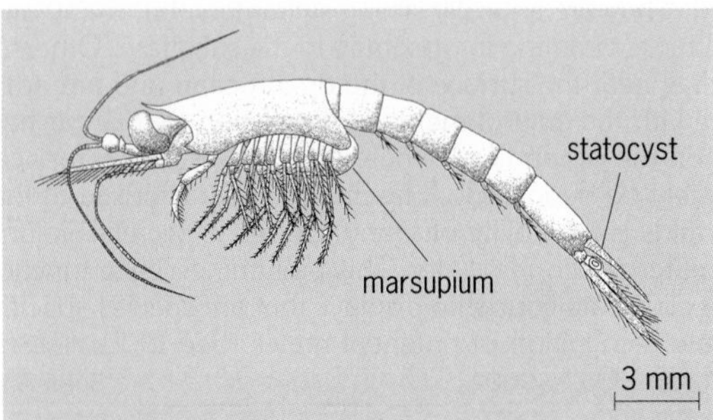

***Boreomysis arctica*, a member of the Mysida showing the marsupium and the paired uropods with a statocyst in the inner branch or endopod. (*From W. M. Tattersall and O. S. Tattersall, 1951*)**

The order Mysida currently contains almost 1100 species in some 168 genera ascribed to four families: Petalophthalmidae, Mysidae (by far the largest family), Lepidomysidae, and Stygiomysidae. Except for the deep-sea petalophthalmids (six species) and the cave-dwelling lepidomysids (nine species), all mysidans have statocysts in the uropods of their tail fan (see illustration), a feature peculiar to mysidans. John Mauchline; Frederick R. Schram

Mystacocarida A subclass of primitive Crustacea. The three species are in the genus *Derocheilocaris*. The body is wormlike and about 0.2 in. (5 mm) long. The cephalothorax bears first antennae, second antennae, mandibles, first maxillae, and second maxillae. Maxillipeds are on a separate segment, and four additional free thoracic segments bear platelike appendages. The six abdominal segments are without appendages, but large caudal rami are present, The labrum is enormous, mouthparts and nervous system are primitive, and the genital pore is on the thorax. *See* CRUSTACEA. Robert W. Pennak

Myxiniformes An order of Craniata in the class Myxini; commonly called hagfishes. Myxiniforms can be identified by the following set of characteristics: an eel-like scaleless body; 1 to 16 pairs of external gill openings; one semicircular canal; and no bones, no paired fins, no neuromasts, and no lens or extrinsic eye muscles. Based on a very scanty fossil record, it is probable that hagfishes have remained little changed since the Pennsylvanian age 300 million years ago. The order comprises one family, two subfamilies, seven genera, and about 70 species, all of which occur in the temperate seas of the world.

Hagfishes (see illustration) have a low fin continuous from the middorsum, around the tail, and forward on the midventrum, but it is not differentiated as dorsal, caudal, and anal fins. The maximum total length is 18–116 cm, depending on the species. Copious amounts of slime are excreted through numerous pores on the ventrolateral sides of the body. The slime can aid in escaping the grip of a predator as well as clogging its tail and causing suffocation. The eyes are degenerate and not visible externally. Hagfishes are benthic, burrowing in mud that overlays substrates ranging in particle sizes from silt to rocks. They feed on soft-bodied invertebrates, as well as the remains of crabs, shrimps, and fishes. Tough items such as the carcass of a large vertebrate require a technique called knotting. The hagfish grasps the flesh with its teeth and then throws its body into a simple overhand knot. Waves of muscular contractions move the knot toward the head, providing a mechanical means of tearing away the flesh. The same technique is used to

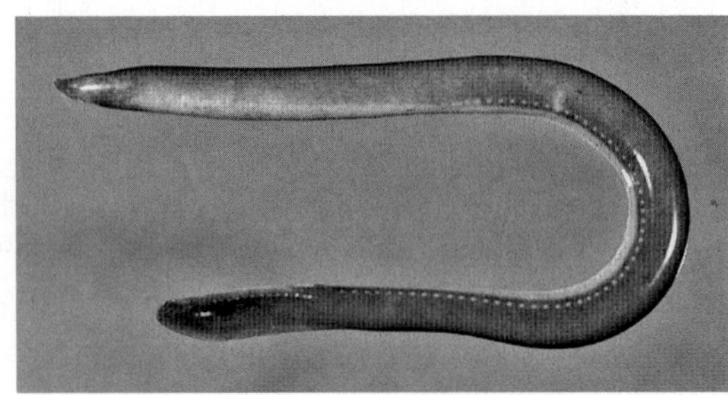

***Myxine glutinosa*. (*Courtesy of Donald Flescher*)**

clear the body of entrapment in its own slime, to avoid capture, and to escape from traps. Herbert Boschung

Myxomycota Organisms that are classified in the kingdom Fungi and given the class name Myxomycetes, following the rules of botanical nomenclature; or classified in the kingdom Protista at various taxonomic ranks, as class Mycetozoa, following the rules of zoological nomenclature. Evolutionary origins are controversial, but many now believe, based on DNA sequencing techniques, that the Myxomycetes diverge early on the tree of life in the region where other protists are found.

The class consists of 3 subclasses, 6 orders, approximately 57 genera, and 600 species. Subclasses Ceratiomyxomycetidae, Myxogastromycetidae, and Stemonitomycetidae are distinguished by the type of sporophore development, type of plasmodium, and method of bearing spores. The various orders, families, genera, and species are distinguished by characteristics of the fruiting bodies such as spore color, peridium, capillitium, calcium carbonate, or columella.

Myxomycetes begin to appear in May and fruit throughout the summer until October in the north temperate regions. Many species are universally distributed and live in moist and dark places on decaying organic matter. Some species are restricted to more specialized habitats.

Spores are released from the fruiting bodies when disturbed and fall onto the substratum where, when water is present, they germinate and release protoplasts. The protoplasts may develop into either a myxamoeba or a flagellated swarm cell, both of which are haploid and behave like gametes (sex cells). The haploid (monoploid) gametes fuse in pairs forming diploid zygotes, which then divide mitotically without subsequent cell division, resulting in the formation of a multinucleated, free-living mass of unwalled protoplasm called the plasmodium. The diploid plasmodium is representative of the slime stage, and hence the common names sometimes used for this group of organisms include plasmodial, acellular, or true slime molds. The plasmodia ingest food as particulate matter (usually bacteria) by engulfment and are capable of growing to over 70 cm in diameter.

The separate stages in their life cycle make myxomycetes ideal organisms to study basic biological problems, ranging through protoplasmic streaming, the mitotic cycle, morphogenesis, aging, and cell division in cancerous cells. *See* EUMYCOTA. Harold W. Keller

Myxosporida An order of the protozoan class Myxosporidea (subphylum Cnidospora). It is characterized by the production of spores with one or more valves and polar capsules, and by possession of a single sporoplasm with or without an iodinophilous vacuole. Myxosporidians are mainly parasites of fishes. They infect all parts of the body, including the heart and brain, and often induce considerable pathological changes in the host tissue.

Infection begins with the ingestion of the spore by a host fish. The digestive fluids cause the polar filaments to be extruded, and at the same time the sporoplasm is released from the spore (see illustration). The sporoplasm, or amebula, reaches the specific site of infection directly through the gut wall or by way of the bloodstream. The amebula becomes a trophozoite when it starts feeding on the host tissues. The trophozoite then goes through a series of nuclear divisions and, by a process of budding, gives rise to a number of cells, each of which eventually develops into a sporont. A sporont is a monosporoblast if one spore is produced and a pansporoblast if two or more spores are formed. The sporont undergoes a series of nuclear divisions, in which the number of nuclei produced will determine the number of spores and polar capsules to be formed. That is, in every spore one nucleus is involved in the formation of each valve and each polar capsule. Two nuclei become the gametic nuclei, which then fuse

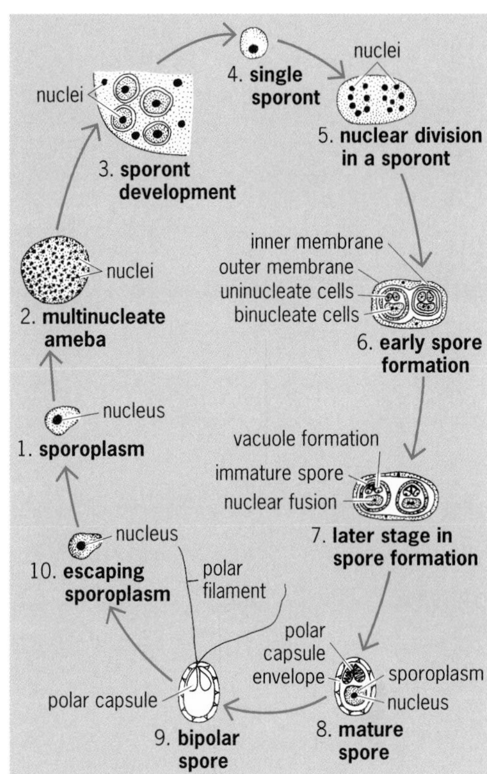

Life cycle of *Myxobolus*, a fish parasite.

to form the zygotic nucleus of the sporoplasm. *See* CNIDOSPORA; MYXOSPORIDEA. Ross F. Nigrelli

Myxosporidea A class of the protozoan subphylum Cnidospora. Members of this class, which includes the orders Myxosporida, Actinomyxida, and Helicosporida, are parasites in fish, a few amphibians, and invertebrates. The Myxosporida are divided into two suborders, the Unipolarina and Bipolarina.

Unipolarina is characterized by spores with one to six (never five) polar capsules located at the anterior end, except in some genera in which the capsules are widely separated or located in the central part of the spore but in which the polar filament is attached near the anterior end. The Unipolarina contains nine or more families.

Bipolarina, containing a single family with three genera, is characterized by the presence of one capsule at or near each end of a fusiform or ellipsoid spore. *See* ACTINOMYXIDA; CNIDOSPORA; HELICOSPORIDA; MYXOSPORIDA. Ross F. Nigrelli

Myzostomida Small, soft-bodied marine worms associated with echinoderms, mainly crinoids. They are found in all oceans from subtidal to over 3000 m (9840 ft) depth. These worms are often considered to be an order of Polychaeta. The Myzostomida include about 180 species within eight families. Most of them are ectocommensals of crinoids (that is, they live on their outer body surface without affecting them), but some are parasites of crinoids, asteroids, or ophiuroids that infest the gonads, coelom, integument, or digestive system. The body plan of most myzostomids is singular and differs from the regular body plan of polychaetes as they are incompletely segmented, parenchymous, acoelomate organisms with chaetae. *See* POLYCHAETA.

For most myzostomids, the body consists of an anterior cylindrical introvert (also called proboscis) and a flat, oval or disk-like trunk. The introvert is extended when the individual feeds, but it is retracted into an anteroventral pouch of the trunk most of the time. The trunk ranges from a few millimeters to 3 cm (1.2 in.) long. Igor Eeckhaut

N

Nailing The driving of nails in a manner that will position and hold two or more members, usually of wood, in a desired relationship to each other. The contact pressures between the surfaces of the nails and the surrounding wood fibers hold the nails in position. Some types of nails are shown in the illustration.

Factors that determine the strength and efficiency of a nailed joint are (1) the type of wood, (2) the nail used, (3) the conditions under which the nailed joint is used, and (4) the number of nails. In general, hard, dense woods hold nails better than soft woods. The better the resistance of a nail to direct withdrawal from a piece of wood, the tighter the joint will remain. To increase resistance to withdrawal or loosening, nails may be coated, etched, spirally grooved, annularly grooved, or barbed, as illustrated.

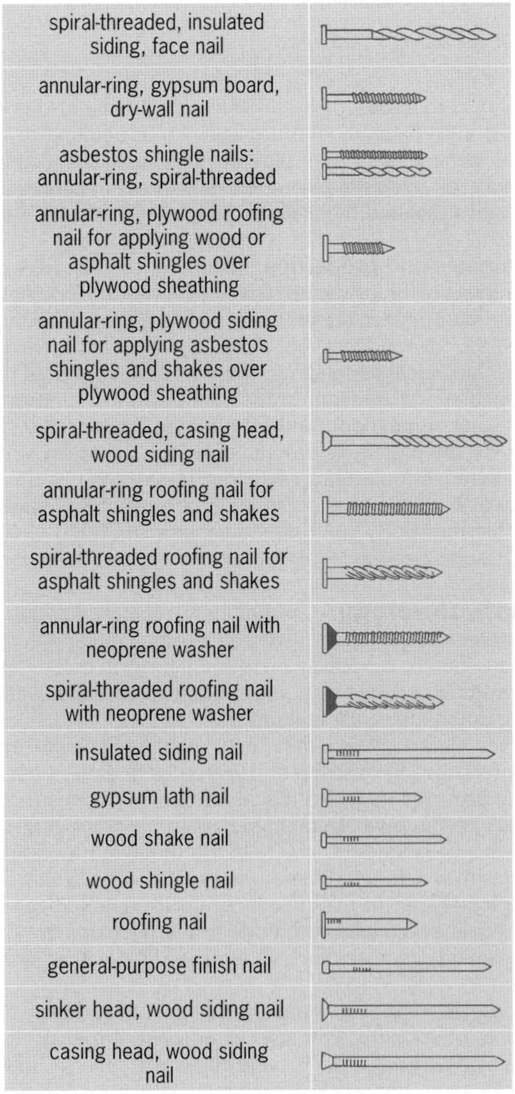

spiral-threaded, insulated siding, face nail
annular-ring, gypsum board, dry-wall nail
asbestos shingle nails: annular-ring, spiral-threaded
annular-ring, plywood roofing nail for applying wood or asphalt shingles over plywood sheathing
annular-ring, plywood siding nail for applying asbestos shingles and shakes over plywood sheathing
spiral-threaded, casing head, wood siding nail
annular-ring roofing nail for asphalt shingles and shakes
spiral-threaded roofing nail for asphalt shingles and shakes
annular-ring roofing nail with neoprene washer
spiral-threaded roofing nail with neoprene washer
insulated siding nail
gypsum lath nail
wood shake nail
wood shingle nail
roofing nail
general-purpose finish nail
sinker head, wood siding nail
casing head, wood siding nail

Special- and general-purpose nails.

Blunt-pointed nails are often used to prevent the wood from splitting. Using nails of a smaller diameter also tends to prevent splitting but requires a greater number of nails per joint. Beeswax is sometimes applied to nail points to make them drive more easily, but it also reduces the holding power of the nail. Alan H. Tuttle

Najadales An order of aquatic and semiaquatic flowering plants, division Magnoliophyta (Angiospermae), in the subclass Alismatidae of the class Liliopsida (monocotyledons). The order consists of 10 families and a little more than 200 species. The Potamogetonaceae, with about 100 species, are the largest family of the order, and the name Potamogetonales is sometimes used instead of Najadales for the group. The Najadales are Alismatidae in which the perianth, when present, is not differentiated into evident sepals and petals. Usually the flowers are not individually subtended by bracts. The Zosteraceae of this order are unique among flowering plants in that they grow submersed in the ocean, albeit in shallow water near the shore. *Zostera marina*, or eelgrass, is a common member of the family. *See* ALISMATIDAE; FLOWER; LILIOPSIDA; MAGNOLIOPHYTA; PLANT KINGDOM.

Arthur Cronquist; T. M. Barkley

Nanochemistry The study of the synthesis and characterization of materials in the nanoscale size range (1 to 10 nanometers). These materials include large organic molecules, inorganic cluster compounds, and metallic or semiconductor particles. The synthesis of nanoscale inorganic materials is important because the small size endows these particles with unusual structural and optical properties that may find application in catalysis and electrooptical devices. Approaches to the synthesis of these materials have focused on constraining the reaction environment through the use of surface-bound organic additives, porous glasses, zeolites, clays, or polymers. The use of synthetic approaches that are inspired by the biological processes result in the deposition of inorganic materials such as bones, shells, and teeth (biomineralization). This biomimetic approach involves the use of assemblies of biological molecules that provide nanoscale reaction environments in which inorganic materials can be prepared in an organized and controlled manner. Examples of biological assemblies include phospholipid vesicles and the polypeptide micelle of the iron storage protein, ferritin. *See* MICELLE.

Vesicles are bounded by an organic membrane that provides a spatial limit on the size of the reaction volume. If a chemical reaction is undertaken in this confined space that leads to the formation of an inorganic material, the size of the product will also be constrained to the dimensions of the organic host structure. Provided that the chemical and physical conditions are not too severe to disrupt the organic membrane, these supramolecular assemblies may have advantages over inorganic hosts such as clays and zeolites because the chemical nature of the organic surface can be systematically modified so that controlled reactions can be accomplished. *See* SUPRAMOLECULAR CHEMISTRY.

One problem encountered with the use of phospholipid vesicles is their sensitivity to changes in temperature and ionic strength. Procedures have been developed in which the biomolecular cage of the iron storage protein, ferritin, has been used as a nanoscale reaction environment for the synthesis of inorganic materials. In the simplest approach the native iron oxide core is transformed into another material by chemical reaction within the protein shell. Stephen Mann

Nanoparticles Synthetic particles that range from 1 to 100 nanometers in diameter. Semiconductor nanoparticles around 1–20 nm in diameter are often called quantum dots, nanocrystals, or Q-particles. These particles possess short-range structures that are essentially the same as the bulk semiconductors, yet have optical or electronic properties that are dramatically different from the bulk properties. The confinement of electrons within a semiconductor nanocrystal results in a shift of the band gap to higher energy with smaller crystalline size. This effect is known as the quantum size effect. In the strong confinement regime, the actual size of the semiconductor particle determines the allowed energy levels and thus the optical and electronic properties of the material. *See* MESOSCOPIC PHYSICS.

Due to their finite, small size and the high surface-to-volume ratio, nanoparticles often exhibit novel properties. These properties can ultimately lead to new applications, ranging from catalysis, ceramics, microelectronics, sensors, pigments, and magnetic storage, to drug delivery and biomedical applications. Research in this area is motivated by the possibility of designing nanostructured materials that possess novel electronic, optical, magnetic, mechanical, photochemical, and catalytic properties. Such materials are essential for technological advances in photonics, quantum electronics, nonlinear optics, and information storage and processing. *See* NANOTECHNOLOGY. M. Samy El-Shall

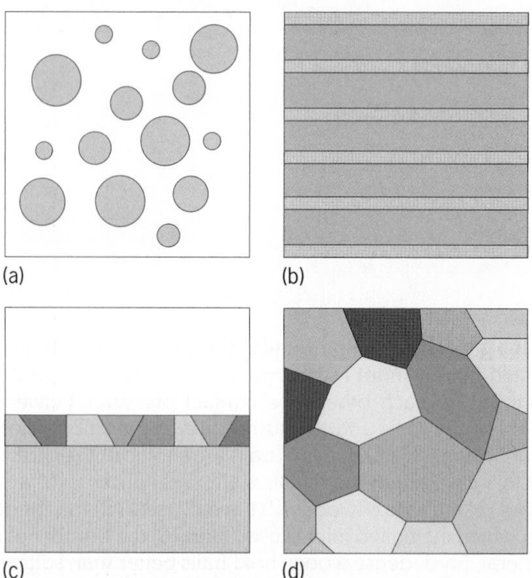

Schematic of four basic types of nanostructured materials, classified according to integral modulation dimensionality. (*a*) Dimensionality 0: clusters of any aspect ratio from 1 to infinity. (*b*) Dimensionality 1: multilayers. (*c*) Dimensionality 2: ultrafine-grained overlayers (coatings) or buried layers. (*d*) Dimensionality 3: nanophase materials. (*After R. W. Siegel, Nanostructured materials: Mind over matter, Nanostruct. Mater., 3:1–18, 1993*)

Nanostructure A material structure assembled from a layer or cluster of atoms with size of the order of nanometers. Interest in the physics of condensed matter at size scales larger than that of atoms and smaller than that of bulk solids (mesoscopic physics) has grown rapidly since the 1970s, owing to the increasing realization that the properties of these mesoscopic atomic ensembles are different from those of conventional solids. As a consequence, interest in artificially assembling materials from nanometer-sized building blocks arose from discoveries that by controlling the sizes in the range of 1–100 nm and the assembly of such constituents it was possible to begin to alter and prescribe the properties of the assembled nanostructures. *See* MESOSCOPIC PHYSICS.

Nanostructured materials are modulated over nanometer length scales in zero to three dimensions. They can be assembled with modulation dimensionalities of zero (atom clusters or filaments), one (multilayers), two (ultrafine-grained overlayers or coatings or buried layers), and three (nanophase materials), or with intermediate dimensionalities (see illustration).

Multilayers and clusters. Multilayered materials have had the longest history among the various artificially synthesized nanostructures, with applications to semiconductor devices, strained-layer superlattices, and magnetic multilayers. Recognizing the technological potential of multilayered quantum heterostructure semiconductor devices helped to drive the rapid advances in the electronics and computer industries. A variety of electronic and photonic devices could be engineered by utilizing the low-dimensional quantum states in these multilayers for applications in high-speed field-effect transistors and high-efficiency lasers, for example. Subsequently, a variety of nonlinear optoelectronic devices, such as lasers and light-emitting diodes, have been created by nanostructuring multilayers. *See* ARTIFICIALLY LAYERED STRUCTURES; LIGHT-EMITTING DIODE; SEMICONDUCTOR HETEROSTRUCTURES; TRANSISTOR.

The advent of beams of atom clusters with selected sizes allowed the physics and chemistry of these confined ensembles to be critically explored, leading to increased understanding of their potential, particularly as the constituents of new materials, including metals, ceramics, and composites of these materials. A variety of carbon-based clusters (fullerenes) have also been assembled into materials of much interest. In addition to effects of confinement, interfaces play an important and sometimes dominant role in cluster-assembled nanophase materials, as well as in nanostructured multilayers. *See* ATOM CLUSTER; CERAMICS; FULLERENE.

Synthesis and properties. A number of methods exist for the synthesis of nanostructured materials. They include synthesis from atomic or molecular precursors (chemical or physical vapor deposition, gas condensation, chemical precipitation, aerosol reactions, biological templating), from processing of bulk precursors (mechanical attrition, crystallization from the amorphous state, phase separation), and from nature (biological systems). Generally, it is preferable to synthesize nanostructured materials from atomic or molecular precursors, in order to gain the most control over a variety of microscopic aspects of the condensed ensemble; however, other methodologies can often yield very useful results. *See* VAPOR DEPOSITION. Richard W. Siegel

Nanotechnology Techniques and products involving nanometer-scale structures, with dimensions ranging from 1 to 100 nanometers, especially those that transform matter, energy, and information using nanometer-scale components with precisely defined molecular features.

In the late 1980s, the term nanotechnology entered widespread use to describe anticipated technologies based on the use of molecule-based machine systems designed to build complex products with atomic precision. Since the mid-1990s, usage has broadened to embrace instruments, processes, and products in which key dimensions are in the 1–100-nm range. Technologies that fit this definition are extremely diverse, but many could potentially contribute to the development of new products and processes such as advanced molecular manufacturing.

Nanoscale technologies are extremely diverse, rapidly changing, and often only tenuously connected. Products include nanoscale particles, fibers, and films of diverse materials and structures; nanoscale lithographic structures for electronics (many integrated circuits now qualify); structures formed by spontaneous molecular aggregation (self-assembly); and solids containing nanoscale grains or pores. The means and materials used to produce nanoscale and nanotextured structures often have little in common, and their applications range from stain-resistant clothing to state-of-the-art electronics. Many nanotechnologies are a continuation of preexisting fields under a new label. What they share (particularly toward the lower end of the 1–100-nm range) is the emergence of novel properties, relative to the corresponding bulk materials, associated with surface and quantum effects, together with a distinctive set of instruments and computational modeling techniques. Grouping these diverse nanotechnologies together has fostered a vibrant cross-fertilization of disciplines. *See* MOLECULAR SIMULATION; MONOMOLECULAR FILM; NANOPARTICLES.

Looking forward, the metrics of complexity and scale define the chief frontiers. In small structures, precision has already reached the atomic limit. Examples include quantum dots, engineered biomolecular objects, self-assembled molecular structures, and sections of carbon nanotubes. For systems built with atomic precision, scale limits complexity. Great complexity is possible, even in nanosystems of microscopic scale. For example, a cubic micrometer of a typical material contains roughly 10^{11} atoms; with generalized atomic control on that scale, a cubic micrometer could contain roughly 10^9 distinct functional components. *See* CARBON NANOTUBES; NANOSTRUCTURE; SUPRAMOLECULAR CHEMISTRY.

Although complex systems with precise molecular features cannot be made with existing techniques, certain nanosystems can be designed and analyzed. Systems based on mechanical (rather than electronic) degrees of freedom are particularly tractable. These are of special interest, because programmable nanoscale mechanical systems could be used to produce atomically precise structures of arbitrary complexity. The development of productive nanosystems is a key strategic objective.

K. Eric Drexler

Naphtha Any one of a wide variety of volatile hydrocarbon mixtures. They are sometimes obtained from coal tar but are more often derived from petroleum. Physical properties vary widely. The initial boiling point may be as low as 27°C (80°F), and end points may reach 260°C (500°F). Boiling ranges are sometimes as narrow as 11°C (20°F) or as wide as 110°C (200°F). Products sold as naphthas find their greatest use as solvents, thinners, or carriers.

There is a fairly sharp differentiation between aliphatic and aromatic naphthas. Aliphatic naphthas are relatively low in odor and toxicity and tend, also, to be low in solvent power. The aro- matic naphthas are highly solvent. Their main components are toluene and xylenes; benzene is less desirable because of the extreme toxicity of its vapors. *See* PETROLEUM PRODUCTS.

J. K. Roberts

Narcotic A drug which diminishes the awareness of sensory impulses, especially pain, by the brain. This action makes narcotics useful therapeutically as analgesics. While they are the most powerful pain-relieving agents available, their use is complicated by a number of undesirable side actions. *See* ANALGESIC.

All of the generally used narcotics are in some way related to opium, and the term opiate is sometimes used interchangeably with the term narcotic. Opium is a gummy exudate obtained from the unripe seed capsules of the opium poppy. Crude opium contains over a dozen alkaloids, all of which have been isolated and identified as to their structural chemistry. From this knowledge chemists have developed a number of synthetic chemical compounds, some of which have important advantages over the naturally occurring alkaloids. Therapeutically important natural alkaloids are morphine, codeine, and papaverine. Among the important synthetic narcotics are meperidine (Demerol), dihydromorphine (Dilaudid), oxymorphone (Numorphan), alphaprodine (Nisentil), anileridine (Leritine), piminodine (Alvodine), levorphanol (Levo-Dromoran), methadone (Dolophine), and phenazocine (Prinadol). *See* ALKALOID; OPIATES; POPPY.

Nalorphine (Nalline) is a narcotic antagonist and is used in the treatment of acute overdosage from narcotics; it is dangerous to drug addicts. Heroin is a highly addicting narcotic, and is so dangerous in this regard that the drug has been completely banned by both federal and state laws under all circumstances.

Pharmacology. The pharmacology of narcotics is generally similar to that of morphine, the principal narcotic used for its analgesic effects. Differences among them lie in the potency of their action and in the degree and variety of the side actions which they produce. Effects are those of analgesia, accompanied by a state of euphoria characterized by drowsiness and a change of mood from anxiety and tension to calmness and equanimity. It should be remembered that whatever narcotic is used, the effects are dose-related, and in higher doses all narcotics produce deep sleep and eventually general depression of all brain functions. Death from overdosage is due to depression of the respiratory centers with resultant failure of respiration.

The predominant pharmacological effect of morphine (and the other narcotics) is on the central nervous system. From the standpoint of its medicinal use, its most important action is relief of pain. Along with its valuable medicinal use morphine produces a great many undesirable side actions; the most frequent are depressed respiratory activity, the production of nausea and vomiting, and the inhibition of defecation and urination.

Drug dependence. All narcotics have the potential for producing dependence and addiction when used repeatedly over a period of time. Drug dependence results from compulsive, continued use of the drug, and is characterized by one or more of the following conditions: habituation, tolerance, or addiction.

Like any other habit pattern, habitual use of a drug can develop. Common examples are the use of nicotine in the form of cigarettes, or caffeine in the form of coffee or tea. Such habituation is generally regarded as innocuous.

Repeated ingestion of a drug in which the effect produced by the original dose no longer occurs results in tolerance. To produce the original effect, it is necessary to increase the dose.

When the body develops a dependence for the drug, addiction occurs. If the drug is suddenly stopped after a period of frequent use, a withdrawal syndrome develops, which is characterized by physical pain and widespread body reactions. The addict comes to dread the development of such painful and distressing reactions, and is trapped into continuing the drug.

All narcotics can produce habituation, tolerance, and addiction to a greater or less degree. Addiction to codeine is relatively rare but possible. Addiction to heroin develops rapidly, and this narcotic is therefore exceedingly dangerous.

James M. Dille

Native elements Those elements which occur in nature uncombined with other elements. Aside from the free gases of the atmosphere there are about 20 elements that are found as minerals in the native state. These are divided into metals, semimetals, and nonmetals. Gold, silver, copper, and platinum are the most important metals and each of these has been found abundantly enough at certain localities to be mined as an ore. Rarer native metals are others of the platinum group, lead, mercury, tantalum, tin, and zinc. Native iron is found sparingly both as terrestrial iron and meteoric iron.

The native semimetals can be divided into (1) the arsenic group, including arsenic, antimony, and bismuth; and (2) the tellurium group, including tellurium and selenium.

The native nonmetals are sulfur, and carbon in the forms of graphite and diamond. Native sulfur is the chief industrial source of that element.

Cornelius S. Hurlbut, Jr.

Natrolite A fibrous or needlelike mineral belonging to the zeolite family of silicates. Most commonly it is found in radiating fibrous aggregates. The hardness is 5–5$^1/_2$ on Mohs scale, and the specific gravity is 2.25. The mineral is white or colorless with a vitreous luster that inclines to pearly in fibrous varieties. The chemical composition is $Na_2(Al_2Si_3O_{10}) \cdot 2H_2O$, but some potassium is usually present substituting for sodium.

Natrolite is a secondary mineral found lining cavities in basaltic rocks. Its outstanding locality in the United States is at Bergen Hill, New Jersey. *See* ZEOLITE. Clifford Frondel; Cornelius S. Hurlbut, Jr.

Natural fiber A fiber obtained from a plant, animal, or mineral. The commercially important natural fibers are those cellulosic fibers obtained from the seed hairs, stems, and leaves of plants; protein fibers obtained from the hair, fur, or cocoons of animals; and the crystalline mineral asbestos. Until the advent of the manufactured fibers near the beginning of the twentieth century, the chief fibers for apparel and home furnishings were linen and wool in the temperate climates and cotton in the tropical climates. However, with the invention of the cotton gin in 1798, cheap cotton products began to replace the more expensive linen and wool until by 1950 cotton accounted for about 70% of the world's fiber production. Despite the development of new fibers based on fossil fuels, cotton has managed to maintain its position as the fiber with the largest production volume, although its use has fallen. *See* COTTON; MANUFACTURED FIBER; WOOL.

The natural fibers may be classified by their origin as cellulosic (from plants), protein (from animals), and mineral. The plant fibers may be further ordered as seed hairs, such as cotton; bast (stem) fibers, such as linen from the flax plant; hard (leaf) fibers, such as sisal; and husk fibers, such as coconut. The animal fibers are grouped under the categories of hair, such as wool; fur, such as angora; or secretions, such as silk. The only important mineral fiber is asbestos, which because of its carcinogenic nature has been banned from consumer textiles. *See* TEXTILE. Ira Block

Natural gas A combustible gas that occurs beneath the surface of the earth, often found in conjunction with petroleum deposits. Its main use is for fuel, but it is also used to make carbon black, certain chemicals, and liquefied petroleum gas. *See* LIQUEFIED PETROLEUM GAS (LPG).

Natural gas consists predominantly of methane, but also contains a mixture of hydrocarbons such as ethane, propane, and pentanes. Carbon dioxide, nitrogen, helium, and hydrogen sulfide may also be present. The types of natural gas vary according to composition and can be dry or lean gas (mostly methane), wet gas (considerable amounts of other hydrocarbons), sour gas (much hydrogen sulfide), sweet gas (little hydrogen sulfide), residue gas (higher hydrocarbons having been extracted), and casinghead gas (derived from an oil well by extraction at the surface). *See* METHANE.

Natural gas occurs on every continent. Wherever oil has been found, a certain amount of natural gas is also present. In addition, natural gas is associated with coal. Natural-gas hydrates are solids composed of water molecules forming a rigid lattice of cages (clathrates), with most of the cages containing a molecule of natural gas. Methane hydrate is the dominant natural-gas hydrate. Methane hydrates are found in polar and other continental shelves, and in deep-ocean outer continental margins in all oceans, including polar oceans. *See* CLATHRATE COMPOUNDS; COALBED METHANE; HYDRATE; PETROLEUM GEOLOGY.

Natural gas comes from reservoirs, usually in sedimentary rock lying below the earth surface at depths varying from a few hundred feet to several miles. Commonly, it forms a gas cap, or mass of gas, entrapped between liquid petroleum and an impervious capping rock layer in a petroleum reservoir. Under conditions of greater pressure, it is mixed with or dissolved in crude oil. Successful exploitation involves drilling, producing, gathering, processing, transporting, and metering the use of the gas. Before it is commercially distributed, natural gas usually is processed to remove propane, butane, and nonhydrocarbon gases such as hydrogen sulfide. Processed natural gas has no distinct odor, so an odorant is added since an undetected leak could result in an explosion or asphyxiation. Transmission pipelines move the gas from processing centers to the market, where distribution pipelines carry it to consumers. Natural gas is also distributed worldwide in liquid form (liquefied natural gas, LNG), which is produced by chilling the gas to below its boiling point. *See* LIQUEFIED NATURAL GAS (LNG); OIL AND GAS FIELD EXPLOITATION; OIL AND GAS WELL DRILLING; PETROLEUM RESERVOIR ENGINEERING.

Michael A. Adewumi; Michel T. Halbouty

Natural language processing Computer analysis and generation of natural language text. The goal is to enable natural languages, such as English, French, or Japanese, to serve either as the medium through which users interact with computer systems such as database management systems and expert systems (natural language interaction), or as the object that a system processes into some more useful form such as in automatic text translation or text summarization (natural language text processing).

In the computer analysis of natural language, the initial task is to translate from a natural language utterance, usually in context, into a formal specification that the system can process further. Further processing depends on the particular application. In natural language interaction, it may involve reasoning, factual data retrieval, and generation of an appropriate tabular, graphic, or natural language response. In text processing, analysis may be followed by generation of an appropriate translation or a summary of the original text, or the formal specification may be stored as the basis for more accurate document retrieval later. Given its wide scope, natural language processing requires techniques for dealing with many aspects of language, in particular, syntax, semantics, discourse context, and pragmatics.

The first aspect of natural language processing, and the one that has perhaps received the most attention, is syntactic processing, or parsing. Syntactic processing is important because certain aspects of meaning can be determined only from the underlying structure and not simply from the linear string of words. A second phase of natural language processing, semantic analysis, involves extracting context-independent aspects of a sentence's meaning. Given that most natural languages allow people to take advantage of discourse context, their mutual beliefs about the world, and their shared spatio-temporal context to leave things unsaid or say them with minimal effort, the purpose of a third phase of natural language processing, contextual analysis, is to elaborate the semantic representation of what has been made explicit in the utterance with what is implicit from context. A fourth phase of natural language processing, pragmatics, takes into account the speaker's goal in uttering a particular thought in a particular way—what the utterance is being used to do.

Bonnie Webber

Nautiloidea A group of externally shelled cephalopods, represented by the two living genera, *Nautilus* and the more recently defined *Allonautilus*. The formal designation of this group as a subclass is now generally used only for those externally shelled cephalopods that resemble *Nautilus* in having completely coiled shells (thus the subclass includes *Nautilus* and *Allonautilus*). In living forms, the basic structural plan includes a shell consisting of a septate phragmocone, a living chamber, and a siphuncle. In fossil nautiloids, this simple pattern is modified in great variety with respect to shell form and size, structure and size of the siphuncle, and the large number of devices to counteract the buoyancy of the phragmocone. The shape of fossil nautiloids may deviate in many ways from the simple *Nautilus* model.

Fossil nautiloids are found on all continents, including Antarctica. Especially noteworthy are the rich Ordovician and Silurian faunas of North America, northern Europe, and China. Coiled nautiloids almost certainly moved around by jet propulsion like *Nautilus* and lived close to the sea floor, at moderate to intermediate depths in many different environments. Other types may have included agile swimmers as well as slow-moving benthic adaptations. *See* CEPHALOPODA. Curt Teichert

Naval architecture An engineering discipline concerned with the design of ships, boats, drill rigs, submarines, and other floating or submerged craft. The naval architect creates the initial overall concept for a new ship, integrates the work of other specialists as the ship design is developed, and is specifically responsible for the hull and superstructure shape, general arrangements, structural design, weights and centers calculations, stability analysis, hydrodynamic performance assessments, propeller and rudder design, and the arrangement and outfit of all living and working spaces, other than machinery spaces. The naval architect's ally, the marine engineer, is responsible for the design of the propulsion plant, the electric plant, and other ship machinery and mechanical systems, including the so-called distributive systems: electric cabling, piping, and ventilation system ducting. The marine engineer also is responsible for ship control systems, including propulsion and electric plant controls and the steering system. *See* FERRY; HYDROFOIL CRAFT; ICEBREAKER; MERCHANT SHIP; NAVAL SURFACE SHIP; OIL AND GAS, OFFSHORE; SUBMARINE.

In the past, naval architecture was as much an art as a science, but research, coupled with advances in computer-aided design, has greatly enhanced the scientific basis of the profession. Naval architecture is a specialized form of mechanical engineering, as is marine engineering. Thus the education of naval architects is very similar in content to that of mechanical engineers, and the same types of degree programs are offered. Some colleges and universities combine naval architecture and marine engineering education and offer a combined degree. *See* MARINE ENGINEERING; MECHANICAL ENGINEERING. Peter A. Gale

Naval armament A general term that covers the ordnance and control systems used by naval ships and aircraft. It includes a wide spectrum of weapons designed for use against targets in the air, on land or sea, or under the ocean surface. The spectrum of weaponry used by naval forces runs from small arms to nuclear warheads and includes weapons that are intended for use against a particular type of target as well as general-purpose weapons.

Naval armament may be air-, surface-, or submarine-launched. It can be categorized as tactical or strategic, or by its intended primary target: surface attack, air defense, or antisubmarine. Many weapons can be used against different types of targets. Naval weaponry includes guns, guided missiles, rockets, bombs, depth charges, torpedoes, and mines.

Guided missiles. In the years since World War II, guided missiles have taken first place among families of naval weapons. Naval missiles may be adaptable to multiple launch modes: from ship, submarine, and aircraft. Modern missiles are more compact, saving critical space and weight, and their guidance systems have steadily become more sophisticated. Shipboard launchers can handle two or three different weapons, eliminating the need for separate launchers.

An *Ohio*-class missile submarine carries 24 Trident fleet ballistic missiles (Fig. 1), developed to replace the earlier Polaris and Poseidon. *See* SUBMARINE.

Standard, the Navy's principal air defense missile, replaced the first-generation Tartar, Terrier, and Talos. A supersonic solid-fuel

Fig. 1. Trident missile being fired from a submerged submarine. (*U.S. Navy*)

weapon, it is produced in medium-range (MR) and extended-range (ER) versions.

Sea Sparrow is an antiaircraft adaptation of the airborne Sparrow III missile, developed as a relatively uncomplicated basic point-defense missile system (BPDMS) to protect ships without Standard missiles.

Tomahawk, a long-ranged land attack cruise missile, was used in the Gulf War and in Kosovo. Capable of attacking targets at a range up to 1000 mi (1600 km), Tomahawk has greatly increased the striking power of the surface warship, which at one time was thought to have been relegated to a subsidiary role by the aircraft carrier. It is also used by aircraft; submarines can carry them in torpedo tubes, and some submarines have been armed with vertical tube launchers.

Harpoon is a long-range antiship missile, originally designed as an air-to-surface weapon but now used in surface ships and submarines as well.

Antisubmarine weapons. ASROC (antisubmarine rocket), launched by surface warships, was originally designed to carry either a nuclear depth charge or a homing torpedo. All nuclear ASROC warheads were taken out of service by 1989. ASROC is an unguided rocket carrying a Mark 46 homing torpedo. Aimed by shipboard computers using target information obtained by sonar, the rocket is fired from a launcher and follows a ballistic trajectory to the target's predicted position. Torpedo and rocket then separate; the torpedo, slowed by a drag parachute, lands in the water and seeks the target. *See* ANTISUBMARINE WARFARE.

Rockets. Naval rockets, as distinguished from guided missiles, are unguided weapons carrying explosive warheads. Their light weight, in proportion to explosive payload, and lack of recoil allow them be used by attack planes and helicopters. *See* ROCKET PROPULSION.

Torpedoes. Torpedoes travel underwater on their own power to attack the vulnerable hulls of surface ships and submarines. Modern naval torpedoes are fast, far-ranging, and armed with a powerful explosive warhead. Torpedoes may be homing (guiding themselves acoustically to the target); nonhoming (following a preset course); or wire-guided (controlled by

Fig. 2. Mark 45 lightweight 5-in. (127-mm) gun mount. (*U.S. Navy*)

signals from the firing ship, transmitted through a trailing wire). They can be launched from surface ships, submarines, or aircraft. Homing torpedoes are used as payload by the ASROC system. Methods for countering the homing torpedo, like the weapons themselves, have been worked on since World War II. It remains a highly effective weapon, and will probably continue in service for a long time. *See* ACOUSTIC TORPEDO.

Guns. Though missiles are widely used by ships and aircraft, guns remain significant naval weapons. Missiles are superior for most long-range attack missions and for defense against supersonic planes and missiles at high altitudes and long ranges; the opposite, however, is often true for such missions as shore bombardment, fire support of land forces, and defense against small attack craft. Renewed attention has been given, both in the United States and in other countries, to lighter guns, with high rates of fire, and to quick-reaction control systems for close-in defense against aircraft and missiles in combination with short-range antiaircraft missiles (Fig. 2).

Bombs. These are free-falling weapons, unlike missiles, which are self-propelled. Bombs take many shapes and sizes, from small antitank and antipersonnel bomblets dispensed from a larger shell or bomb, to heavy weapons designed for blast effect. Most planes and helicopters carry arms externally to accommodate weapon-mix versatility and to keep aircraft size and weight down. High aircraft speeds led to development of streamlined, low-drag bombs. Bombs can be "dumb," that is, uncontrolled, or "smart." Smart bombs have guidance systems and movable control surfaces, and their trajectory can be adjusted to steer them toward a target.

Mines. A mine is a thin-cased, non-self-propelled weapon filled with high explosive and placed underwater, where it is designed to explode when struck, or closely approached, by a ship. Mines can be contact type (fired by actually striking the hull of a passing ship) or influence type (detonated by the close approach of a ship). An influence mine may be magnetic (actuated by a ship's magnetic field), acoustic (actuated by the underwater sound that a ship generates), or pressure (actuated by the change in water pressure caused by a ship's passage). It may also be fired by a combination of these influences. Influence mines are thus much harder to sweep than contact mines. Mines are planted by submarines or aircraft; some navies also use surface minelayers.

John C. Reilly

Naval surface ship

A surface ship designed primarily for use in warfare, either to operate in direct combat or to provide support to other ships engaged therein. Naval ships can therefore be categorized as either combatants or noncombatants (auxiliaries), both with unique design characteristics.

Fig. 1. Aegis guided-missile cruiser, *Ticonderoga* CG 47 class. USS *Cape St. George* is shown. (*U.S. Navy photo*)

Combatant ships. This category includes battle ships, cruisers, destroyers, frigates, aircraft carriers, amphibious warfare ships, mine warfare ships, and patrol ships.

After World War II, all the surviving battleships of the U.S. Navy were disposed of except for the four most modern ships, which were retained in a decommissioned (standby or "mothball") status. In 1995 all four battleships were stricken from the naval register, but Congress later directed that two be retained as mobilization assets.

The primary purpose of modern cruisers is to provide antiair and antisubmarine protection to aircraft carriers and other friendly forces. These highly capable multimission ships can also operate independently. Equipped with antiair, antisubmarine, antiship, and land attack missiles, they are self-contained offensive units in their own right. The dominant cruisers in the world are the U.S. CG 47 *Ticonderoga* class of Aegis guided-missile cruisers (Fig. 1) designed in the 1970s.

If battleships and cruisers are the heavyweights and middleweights, then destroyers are the lightweights with a "go-anywhere, do-anything" outlook. They have tremendous firepower for their size and rely on high speed, dash, and maneuverability. Current destroyers tend either to be multipurpose or to emphasize specific warfare areas. An example of a multipurpose destroyer is the U.S. DDG 51 *Arleigh Burke* class of Aegis guided-missile destroyers. An example of a specialized destroyer is the U.S. DD 963 *Spruance* class designed primarily to detect and destroy submarines.

The smallest and most numerous surface combatants in the U.S. Navy are the *Oliver Hazard Perry* (FFG 7) class of multi-purpose frigates. Designed in the 1970s, this 50-ship class was expected to protect merchant and military convoys. The speed of FFG-7 is 27–28 knots (13–14 m/s) since carrier operations were not a primary mission. By means of sonars, an embarked ASW helicopter, and shipboard torpedo tubes, it can detect and destroy submarines.

The largest warships in the world are aircraft carriers (Fig. 2), which through the aircraft they support can project power at great distances. An aircraft carrier is a floating mobile air station. Its flight deck provides the runways, its island is the control tower, the large hangar below the flight deck is the garage that contains maintenance and repair shops, and deep in the hull are storage tanks for aviation fuel and magazines for aviation ordnance. These ships, which can operate and maintain up to 80 aircraft, require a length of almost 1100 ft (335 m), a displacement of nearly 100,000 tons (101,600 metric tons), and a crew of 6000.

The ships used to transport, land, support, and control assault troops collectively constitute the amphibious warfare force. Despite the success of Allied amphibious operations in World War II,

Colorplate 1. Trifid Nebula (M20), an H II region in Sagittarius, and a reflection nebula appearing adjacent to it. The Trifid is the red region divided by absorbing dust lanes. The reflection nebula is the blue object at the top and is many light-years distant from the Trifid. (*Todd Boronson/ NOAO/AURA/NSF*)

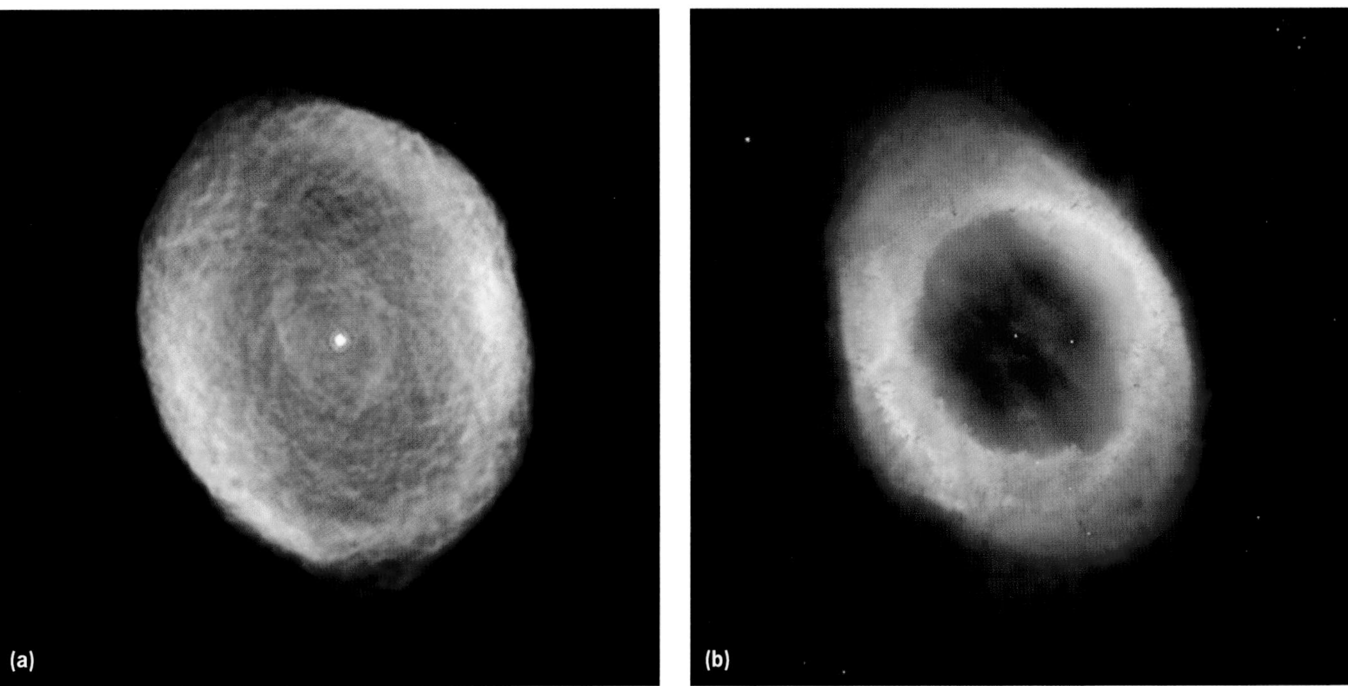

Colorplate 2. *Hubble Space Telescope* images of the planetary nebulae (*a*) IC 418 and (*b*) NGC 6720. The color of the nebulae is the result of emissions from different atomic lines. (*Hubble Heritage Team, AURA/STScI/NASA*)

Fig. 2. Nuclear-powered aircraft carrier, *Nimitz* CVN 68 class. USS *Theodore Roosevelt* (CVN 71) is shown. (*U.S. Navy photo*)

the need for improvement was recognized. The first major breakthrough came with the advent of the helicopter and its successful adaptation for landing assault troops (called vertical assault). The second was the development of the high-speed (40 knots; 19 m/s) landing craft, air-cushioned (LCAC). The employment of these two new vehicle types permits landing ships to remain offshore, over the horizon, during amphibious assaults. All current United States amphibious warfare ships have both flight decks for operating helicopters and well-decks for operating LCACs and other craft.

Landing craft are used to ferry tanks, vehicles, equipment, ammunition, general cargo, and personnel directly onto the beach from the landing ships offshore.

Amphibious vehicles are capable of being launched directly into the ocean from landing ships, proceeding to the beach, and then moving inland.

Minelayer ships are built in varying sizes, but can be characterized by a mine stowage system, and rails for moving the mines, and dropping them off the stern or side of the ship. Minesweepers can be broadly categorized as having either a hunter role (locate and mark mines) or a hunter/killer role (locate and destroy). Mines can be located by variable-depth sonar and later destroyed by divers or by means of a mine neutralization vehicle (MNV). The MNV is a crewless minisubmarine that lays an explosive charge near the mine, backing off before detonation. Mines can also be located by the minesweeper towing a mechanical sweep that cuts the cable between the bottom anchor and the mine. After the mine surfaces, it is destroyed by divers or gunfire. Mines can also be detonated by towing magnetic and acoustic cables that trigger them. Minesweepers are often built of wood or composite materials to reduce their magnetic signature.

Patrol ships are small ships that augment conventional surface forces in coastal areas and restricted seas. Their primary mission is coastal patrol and interdiction surveillance—an important aspect of littoral operations. These ships also provide full mission support for Navy SEALs and other special operations forces.

Noncombatant ships. Naval auxiliary ships provide services and support naval operations. Floating configurations that provide services but are not ships are called service craft. Service craft include floating dry-docks, harbor tugs, berthing barges, diving support boats, and fuel barges.

Included in the auxiliary category are the tenders for submarines, and the ships that replenish the fleet with supplies of oil, stores, ammunition, and combat support items. Also included are oceangoing salvage rescue ships, acoustic research ships, oceanographic research ships, surveying ships, hospital ships, ocean surveillance ships, cable repair ships, oceangoing tugs, marine prepositioning ships, and even experimental submarines.

Barry Tibbitts

Navier-Stokes equation

Navier-Stokes equation A partial differential equation which describes the conservation of linear momentum for a linearly viscous (newtonian), incompressible fluid flow. In vector form, this relation is written as Eq. (1), where ρ is fluid density,

$$\rho\left[\frac{\partial \mathbf{V}}{\partial t} + (\mathbf{V}\cdot\nabla)\mathbf{V}\right] = -\nabla p + \rho\mathbf{g} + \mu\nabla^2\mathbf{V} \qquad (1)$$

$\mathbf{V}$ is fluid velocity, p is fluid pressure, $\mathbf{g}$ is the gravitational acceleration, μ is fluid viscosity, ∇ is the del or grad operator, and ∇^2 is the laplacian operator. The equation is named after its two principal developers, French engineer C. L. M. H. Navier (1823) and Irish scientist George G. Stokes (1845). When coupled with the conservation of mass relation, $\nabla \cdot \mathbf{V} = 0$, Eq. (1) can be solved for the space-time distribution of $\mathbf{V}$ and p in a given region of viscous fluid flow. Typical boundary conditions are (1) the knowledge of the velocity and pressure in the far field, and (2) the no-slip condition at solid surfaces (fluid velocity equals solid velocity). *See* CALCULUS OF VECTORS; GRADIENT OF A SCALAR; LAPLACIAN; NEWTONIAN FLUID; VISCOSITY.

Equation (1) has been successfully applied to the prediction of both laminar flow and disorderly turbulent fluid flows. In the latter case, the evaluation of the equation's adequacy has been limited to a comparison between numerical approximations to solutions of Eq. (1) and measured time-average values and higher statistics. A significant limitation of Navier-Stokes theory is the lack of any proof regarding uniqueness or existence of solutions of Eq. (1) for given boundary and initial conditions.

The primary dimensionless parameter which governs Eq. (1) is the Reynolds number, given by Eq. (2), where L is a characteristic

$$\mathrm{Re} = \frac{\rho V L}{\mu} \qquad (2)$$

body dimension. For small $\mathrm{Re} \ll 1$, Eq. (1) can be simplified by neglecting the left-hand side, resulting in a linear approximation called Stokes flow, or creeping flow, for which many solutions are known. *See* CREEPING FLOW; LUBRICATION; REYNOLDS NUMBER.

For large $\mathrm{Re} \gg 1$, viscous effects are often confined to a thin boundary layer near solid surfaces, with the remaining flow being nearly inviscid. *See* BOUNDARY-LAYER FLOW.

Frank M. White; Arthur E. Bryson, Jr.; A. Gordon L. Holloway

Navigation

Navigation The process of directing the movement of a craft from one place to another. Navigation involves position, direction, distance, time, and speed.

The process of keeping track of a craft's location by measuring and applying progress from a previous position is called dead reckoning. The location of a craft relative to external reference points such as landmarks or aids to navigation is called piloting. Radio navigation involves determining distances or directions to radio transmitters. Celestial navigation involves the use of celestial bodies. *See* CELESTIAL NAVIGATION; DEAD RECKONING; PILOTING.

The craft to be navigated may be a ship, small marine craft, land vehicle, aircraft, missile, spacecraft, or any moving object requiring direction or capable of being directed, even an animal or bird. The characteristics of the craft have a significant influence upon the type of navigation and the equipment used. Size, mission, weight and space limitations, and economic factors are important considerations.

Anything used in navigation, whether aboard the craft or external to it, is properly termed a navigational aid. Thus, in addition to onboard navigational equipment, the term includes such external aids as natural landmarks, prominent buildings, or other structures. Although sometimes used synonymously with "navigational aid," the expression "aid to navigation" is generally restricted to an object or device, external to the craft, established expressly to assist navigation. In this restricted sense, aids to navigation for mariners consist of buoys, beacons, lighthouses, lightships, and navigation sound and electronic transmitters including LORAN and navigation satellites. Aids to navigation for aviators consist primarily of radio ranges and beacons and radio position-fixing transmitters including navigation satellites. *See* BUOY; ELECTRONIC NAVIGATION SYSTEMS; HYPERBOLIC NAVIGATION SYSTEM; LIGHTHOUSE; MARINE NAVIGATION. Alton B Moody

Neandertals A group of late archaic humans from Europe, the Near East, and central Asia that immediately preceded the first modern humans in those regions. The Neandertals (also spelled Neanderthals) are included by some within the species *Homo sapiens*, recognizing their close affinities to modern humans; others place them in their own species, *Homo neanderthalensis*, emphasizing the differences between them and modern humans.

The first recognized Neandertal remains were found in the Neander Valley near Düsseldorf, Germany, in 1856. Since then the remains of several hundred Neandertals have been discovered. Since the Neandertals were the first humans to bury their dead, a number of largely complete skeletons are preserved, providing detailed knowledge of their biology. *See* EARLY MODERN HUMANS.

In the early twentieth century, when Neandertals were the only archaic humans known, they were reconstructed as semihuman, dull-witted, and brutish. Hence their popular image was that of the archetypical cavemen. They are now recognized as relatively recent members of the human lineage; they lived between about 125,000 and 36,000 years ago (and as late as 30,000 years ago in certain isolated regions), as compared with earlier members of the genus *Homo* who extend back more than 2 million years. The Neandertals share many features with modern humans both anatomically and behaviorally. Yet, a number of important contrasts between them and more recent humans are recognized.

Physically, the Neandertals were about the same height as most modern humans, on the average 5 ft 5 in. (166 cm), but they were much more heavily built. They had heavy necks, broad and muscular shoulders, and extremely muscular arms, hands, and legs. Estimates of their strength show them to have been about as strong as very athletic modern humans. Their leg bones show a marked thickening of their shafts, indicative of both marked strength and endurance—a necessary part of their survival.

The Neandertals are known for their long, low braincases and their projecting faces with large brows and prominent noses. Their brains were larger than those of modern humans. The large brain size was due in part, as with early modern humans, to their large body masses. The length and lowness of their braincases was due to relatively slow brain growth during infancy. There is no evidence that they were less intelligent than modern humans, only that their behavioral system was less elaborate.

The position of the Neandertals in modern human ancestry remains controversial. Whatever the extent to which Neandertals can be claimed to be ancestors of modern humans, they represent the most recent phase of premodern humans, one in which people were less efficient than modern humans at hunting and gathering, and compensated for their cultural limitations with biological attributes such as tremendous strength, large front teeth, and thermal adaptations. Yet they exhibited the beginnings of many of the attributes of modern humans. They were very successful for about 100,000 years, but they were eventually replaced by humans who were better able to exploit their environments. *See* FOSSIL HUMANS. Erik Trinkaus; Steven Churchill

Nearshore processes Processes that shape the shore features of coastlines and begin the mixing, sorting, and transportation of sediments and runoff from land. In particular, the processes include those interactions among waves, winds, tides, currents, and land that relate to the waters, sediments, and organisms of the nearshore portions of the continental shelf. The nearshore extends from the landward limit of storm-wave influence, seaward to depths where wave shoaling begins. *See* COASTAL LANDFORMS.

The energy for nearshore processes comes from the sea and is produced by the force of winds blowing over the ocean by the gravitational attraction of Moon and Sun acting on the mass of the ocean, and by various impulsive disturbances at the atmospheric and terrestrial boundaries of the ocean. These forces produce waves and currents that transport energy toward the coast. The configuration of the landmass and adjacent shelves modifies and focuses the flow of energy and determines the intensity of wave and current action in coastal waters. Rivers and winds transport erosion products from the land to the coast, where they are sorted and dispersed by waves and currents.

In temperate latitudes, the dispersive mechanisms operative in the nearshore waters of oceans, bays, and lakes are all quite similar, differing only in intensity and scale, and are determined primarily by the nature of the wave action and the dimensions of the surf zone. The most important mechanisms are the orbital motion of the waves, the basic mechanism by which wave energy is expended on the shallow sea bottom, and the currents of the nearshore circulation system that produce a continuous interchange of water between the surf zone and offshore areas. The dispersion of water and sediments near the coast and the formation and erosion of sandy beaches are some of the common manifestations of nearshore processes.

Erosional and depositional nearshore processes play an important role in determining the configuration of coastlines. Whether deposition or erosion will be predominant in any particular place depends upon a number of interrelated factors: the amount of available beach sand and the location of its source; the configuration of the coastline and of the adjoining ocean floor; and the effects of wave, current, wind, and tidal action. The establishment and persistence of natural sand beaches are often the result of a delicate balance among a number of these factors, and any changes, natural or anthropogenic, tend to upset this equilibrium. *See* DEPOSITIONAL SYSTEMS AND ENVIRONMENTS; EROSION. Douglas L. Inman

Nebula The term "nebula" was originally used to refer to any fixed, extended, and usually fuzzy luminous celestial object, including stellar systems, such as galaxies and star clusters. According to the modern definition, nebulae are gaseous objects usually located within the Milky Way Galaxy, although with increasingly powerful telescopes gaseous nebulae can now be observed in external galaxies. *See* GALAXY, EXTERNAL; MILKY WAY GALAXY; STAR CLUSTERS.

Because of space observations, the study of gaseous nebulae has undergone a renaissance. The extension of observations into the x-ray, ultraviolet, infrared, millimeter, and submillimeter wavelengths has revealed a much richer makeup of nebulae, containing almost all states of matter. Although the term "gaseous nebulae" is used, in fact they consist of ions, atoms, molecules, and solid particles.

Types. Gaseous nebulae can be divided into three main types. Those that radiate brightly in the visible are called emission nebulae. Those that are detected through their effects on the obscuration of background stars are called dark nebulae or absorption nebulae. Nebulae that do not self-radiate in visible light but reflect light from nearby stars are called reflection nebulae. Examples of emission nebulae include the Great Orion Nebula, which is a region of active star formation, and planetary nebulae and supernova remnants, which are objects associated with dying stars. *See* ORION NEBULA.

New stars are formed out of concentrations of dust and gas in the interstellar medium. New stars often form in groups. For clusters of newborn stars containing hot stars, the ultraviolet light from the stars is able to ionize the surrounding gas, making them luminous. In astronomical nomenclature, they are called H II regions, refering to the fact that the hydrogen atoms in the nebulae are in ionized form, with the electron detached from the proton.

Planetary nebulae are ejected by very old red giants. Since they are continuously expanding, they have short lifetimes lasting only tens of thousands of years. After the ejection, the remnant of the original red giant becomes a hot, blue star that eventually evolves to become a white dwarf. The ultraviolet light from the star can ionize the ejecta, leading to the emission of a variety of atomic lines in different colors. Planetary nebulae have well-defined morphologies and have sizes of fractions of a light-year. *See* GIANT STAR; PLANETARY NEBULA; WHITE DWARF STAR.

The detonation of a star in a supernova event causes the ejection of the outer layers into the surrounding interstellar medium. In early stages, as in the Crab Nebula, the radiating material consists of ejecta from the star. In the later stages, this rapidly moving material is slowed down as it mixes with the surrounding dust and gas of the interstellar medium. Heating by shock waves causes the material to radiate optically. Supernova remnants are also characterized by their x-ray emission arising from the rarefied gas behind the shock front, and by synchrotron radiation at radio frequencies. *See* CRAB NEBULA; SUPERNOVA.

The optical spectrum of a reflection nebula is made up of the continuous spectrum of the illuminating stars. However, the color of a reflection nebula is often bluer than the stellar spectrum because the dust particles scatter more efficiently in the blue. The best-known example is the Pleiades reflection nebula, which is illuminated by the brightest stars in the Pleiades cluster. *See* PLEIADES.

Dark clouds represent a concentration of solid dust particles that obscure starlight. By preventing light from stars located behind the cloud from reaching us, they appear as dark patches in the sky. Many dark clouds have no strong heating sources in the form of central stars, but are heated primarily by external nearby stars or cosmic rays. For this reason, the gas temperature inside dark clouds can be very low (about 10 K or −442°F). On a smaller scale, there are compact, dense clouds known as Bok globules. They are often spherical in shape and have densities higher than dark clouds. *See* GLOBULE.

Many dark clouds also contain molecules that can be detected through their rotational transitions in the millimeter wavelengths. Interstellar clouds that are heavy in molecular content are called molecular clouds. Molecular clouds that are bright in the infrared and molecular line emissions are called giant molecular clouds. Their dust and gas temperatures, inferred from infrared continuum and molecular-line emissions respectively, are in the range of 50 to 100 K (−370 to −280°F). Such relatively high temperatures suggest that there must be heating sources inside the cloud. These internal sources can be identified by near-infrared imaging and are believed to be newly formed massive (OB) stars. *See* MOLECULAR CLOUD.

Composition. Gaseous nebulae are made up of free-flying atoms in the gaseous state. These atoms can be in the neutral (uncharged) state, or in an ionized state. Molecular gas is also present in nebulae. Although most molecules consist of two, three, or four atoms, very complex molecules with over a dozen atoms have also been seen.

Also present in nebulae are solid-state particles, which are generally referred to as "dust" in astronomical literature. The most common inorganic dust consists of silicates, which are made up of atoms of oxygen, silicon, magnesium, and iron. Organic carbonaceous dust, consisting of primarily carbon and hydrogen, is also widely seen. *See* INTERSTELLAR MATTER.

Spectral characteristics. The different states of matter in a nebula can be traced by their different radiation mechanisms. Atoms and ions can be identified by their atomic electronic transitions, molecules by their vibrational or rotational transitions, and solid particles by their lattice vibrational modes. Since different atoms radiate in different specific frequencies, they can be uniquely identified by spectroscopic observations.

The densities in gaseous nebulae are lower than the best vacuum that can be achieved in the laboratory. Consequently, the excitation of atoms and molecules and their subsequent radiation are very different from usual mechanisms observed in the terrestrial environment. Sun Kwok

Nectarine A smooth-skinned, fuzzless form of peach, *Prunus persica*. The nectarine's lack of pubescence is a simple recessive genetic characteristic. Classically, the fruits were thought of as being somewhat smaller, softer, and richer in flavor than those of the peach. More recently developed cultivars, however, approximate fresh-market peaches in size and firmness but are not usually superior in flavor.

California is practically the sole commercial producer of nectarines. There is a considerable number of plantings in irrigated areas in south-central Washington. *See* FRUIT; FRUIT, TREE; PEACH; ROSALES. L. F. Hough; Catherine H. Bailey

Nectridea An order of mostly aquatic lepospondyl amphibians known from Carboniferous and Permian rocks of North America, Europe, and North Africa. They were small, usually less than 20 in. (50 cm) in length, and outwardly newtlike with short trunks and long tails. Limbs were small but well developed. Carpal and tarsal bones were rarely ossified. Their vertebrae exhibit the one-piece centrum characteristic of lepospondyls, but are distinct in bearing spatulate neural spines (see illustration) with crenulated dorsal edges. Three nectridean families are recognized: Urocordylidae, Keraterpetontidae, and Scincosauridae. *See* AMPHIBIA; LEPOSPONDYLI. C. F. Wellstead

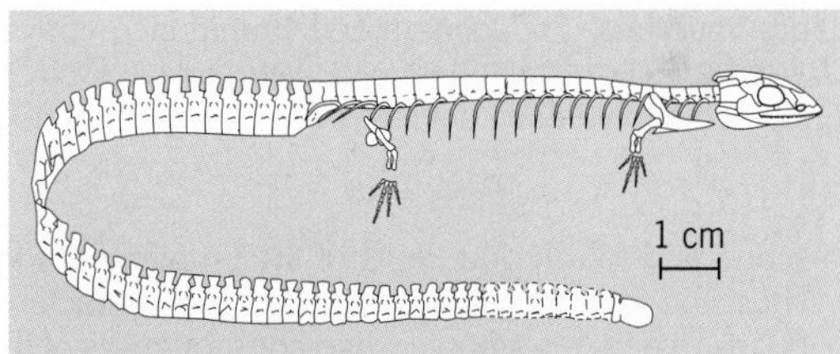

Urocordylid nectridean *Ptyonius marshii*. (*After R. L. Carroll, Vertebrate Paleontology and Evolution, W. H. Freeman, 1988*)

Negative ion An atomic or molecular system having an excess of negative charge. Negative ions, also called anions, are formed in attachment processes in which an additional electron is captured by an atom, molecule, or cluster. They can also be formed when a molecule or cluster dissociates. Doubly charged negative ions, also called dianions, have also been observed in the case of molecules and clusters. Here, two additional electrons have become attached to the neutral systems. Negative ions are destroyed in a controlled manner in detachment processes and, in the case of molecular ions or clusters, dissociation processes, when the ion interacts with photons, electrons, heavy particles, or external fields. Experimental studies of negative ions involve measurements of cross sections for detachment and dissociation. *See* SCATTERING EXPERIMENTS (ATOMS AND MOLECULES).

Negative ions were first reported in the early days of mass spectrometry. It was soon learned that even a small concentration of such weakly bound, negatively charged systems had an appreciable effect on the electrical conductivity of gaseous discharges. Negative ions now play a major role in a number of areas of physics and chemistry involving weakly ionized gases and plasmas. Applications include accelerator technology, mass spectrometry, injection heating of thermonuclear plasmas, material processing, and the development of tailor-made gaseous dielectrics. In nature, negative ions are known to be present in tenuous plasmas such as those found in astrophysical and aeronomical environments. The absorption of radiation by negative hydrogen ions in the solar photosphere, for example, determines the Sun's spectral distribution. *See* ION; ION SOURCES; PLASMA (PHYSICS). David J. Pegg

Negative-resistance circuits Electronic circuits or devices that, over some range of voltage v and current i, satisfy Eq. (1) for equivalent resistance R_{eq} (where the voltage and current polarities are defined in Fig. 1*a*). They are used as building blocks in designing circuits for a wide range of applications, including amplifiers, oscillators, and memory elements. *See* AMPLIFIER; ELECTRICAL RESISTANCE; OSCILLATOR.

$$R_{eq} = \frac{dv}{di} < 0 \qquad (1)$$

An ideal negative resistor would have the voltage-current relationship (transfer characteristic) shown in Fig. 1*b*, and thus satisfy Ohm's law with a negative value for the resistance. However, the same effect can generally be obtained with any circuit (or physical device) whose voltage-current curve contains a region of negative slope. Figure 1*c*, for example, shows transfer characteristics typical of a tunnel diode and a neon bulb, which can be operated in the negative-resistance regions indicated. *See* OHM'S LAW; RESISTOR.

Common generalizations of the negative-resistance idea include negative capacitors, negative inductors, and frequency-dependent negative resistors. Some of the circuits used to implement them are negative impedance converters, negative impedance inverters, and generalized immittance converters. *See* CAPACITOR; ELECTRICAL IMPEDANCE; IMMITTANCE; INDUCTOR.

The power dissipated in a device, given by Eq. (2), is negative

$$P_{DISS} = vi \qquad (2)$$

in the second and fourth quadrants of the v-i plane of Figs. 1*b* and *c*. Thus, the ideal negative resistor whose characteristic is shown in Fig. 1*b* generates power. Two consequences of this are that an active circuit (a circuit containing a power supply) is required to implement the ideal characteristic of Fig. 1*b* but is not necessary for the small-signal negative resistances of Fig. 1*c*; and that for any practical circuit, the characteristic curve must eventually fold over into the power-dissipating quadrants, as shown in Fig. 2*a* or *b*. If the curve did not fold but just continued forever, it would be possible to extract an infinite amount of power from the device.

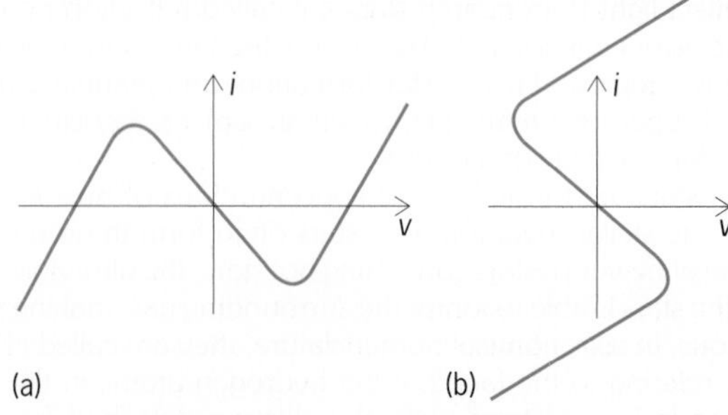

Fig. 2. Large-signal behavior of a negative resistance having a finite internal power supply. (*a*) Voltage-controlled resistance. (*b*) Current-controlled resistance.

The two types of curve of Fig. 2 correspond to an important dichotomy in types of negative resistance. The N-shaped curve of Fig. 2*a* allows current to be a single-valued function of voltage (but not vice versa), and circuits with this behavior are therefore called voltage-controlled negative resistors. Dually, the S-shaped curve of Fig. 2*b*, for which Eq. (3) is appropriate, describes a current-controlled negative resistor. The tunnel-diode characteristic of Fig. 1*c* can be seen to be voltage-controlled, while the neon tube is current-controlled.

$$v = f(i) \qquad (3)$$

If the terminals of a current-controlled negative resistor are open-circuited, then $i = 0$ and there is a unique solution $v = f(0)$. The voltage-controlled circuit, however, can have any of three voltages in this situation (the three intersections of the N with the horizontal axis). Dually, the S-curve gives a device with multiple equilibrium states when short-circuited. When the dynamic behavior of these circuits is accounted for, it is found that some of these equilibria are stable and some are unstable. These stability considerations are essential to designing a negative-resistance circuit for a particular application.

Negative resistors can be implemented by using amplifiers in positive-feedback configurations. Figure 3*a* shows how a voltage amplifier with a gain of 2 can be used to simulate a grounded negative resistor, and Fig. 3*b* shows an operational-amplifier implementation of the same idea. *See* OPERATIONAL AMPLIFIER.

In the practical case of a clipping amplifier, which has the input-output characteristics shown in Fig. 3*c*, the resulting large-signal voltage-current behavior of the simulated resistor is as shown in the figure. This is a voltage-controlled resistor.

The best-known negative-resistance device is the tunnel diode. It is very useful because the phenomenon that it exploits is a quantum-mechanical effect that happens much more rapidly than most others in electronics.

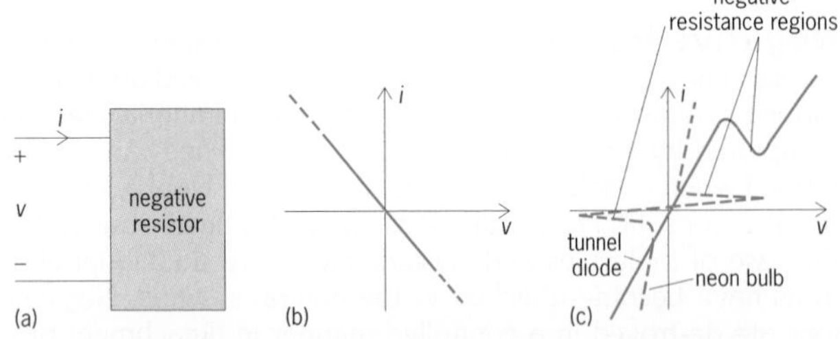

Fig. 1. Characteristics of negative resistors. (*a*) Definition of voltage (*v*) and current (*i*) polarities. (*b*) Voltage-current transfer characteristic of an ideal negative resistor. (*c*) Transfer characteristic of practical physical devices with negative-resistance regions: a tunnel diode and a neon bulb (not to the same scale).

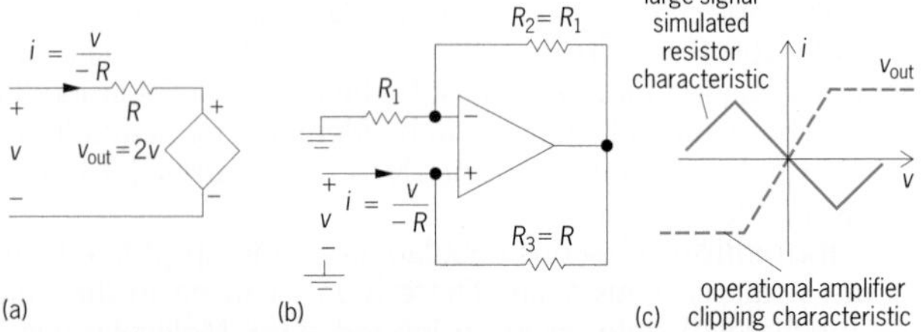

Fig. 3. Active circuits that simulate a negative resistance. (*a*) Circuit that uses an ideal voltage amplifier with a gain of 2. (*b*) Circuit that uses an operational amplifier. (*c*) The operational-amplifier clipping characteristic and resulting large-signal voltage-current characteristic of the simulated negative resistor.

A tunnel diode consists of two very heavily doped regions of a semiconducting material with a very abrupt junction between them. These regions, like any crystalline material, can contain electrons only with energies in certain bands. One side of the junction is doped to have a generous supply of electrons in a certain band of energies, while the other side has a great many vacancies (holes) for electrons in another band. As the applied voltage increases, the bands of electrons and holes on the two sides of the junction start to slide past one another, and eventually their region of overlap starts to decrease. Since quantum tunneling can occur only from an electron in the "supply" to a vacancy at the same energy, this reduction in overlap reduces the amount of charge flowing. Thus, an increasing voltage produces decreasing current, for a negative differential resistance like that shown in Fig. 1*c*. *See* BAND THEORY OF SOLIDS; SEMICONDUCTOR RECTIFIER.

A number of other quantum electronic devices have been developed that also have negative-resistance characteristics. In particular, devices have been constructed that have two barriers (instead of the single barrier created by the tunnel-diode junction) and make use of resonant tunneling, where the spacing between the barriers creates a resonance for electrons at certain frequencies. This resonance, in turn, enhances the rate of tunneling. These devices are claimed to be useful at terahertz (10^{12} Hz) frequencies. *See* SEMICONDUCTOR HETEROSTRUCTURES; TUNNEL DIODE; TUNNELING IN SOLIDS. Martin Snelgrove

Negative temperature The property of a thermodynamical system which satisfies certain conditions and whose thermodynamically defined absolute temperature is negative. The essential requirements for a thermodynamical system to be capable of negative temperature are: (1) the elements of the thermodynamical system must be in thermodynamical equilibrium among themselves in order for the system to be described by a temperature at all; (2) there must be an upper limit to the possible energy of the allowed states of the system; and (3) the system must be thermally isolated from all systems which do not satisfy both requirements (1) and (2); that is, the internal thermal equilibrium time among the elements of the system must be short compared to the time during which appreciable energy is lost to or gained from other systems.

The second condition must be satisfied if negative temperatures are to be achieved with a finite energy. Most systems do not satisfy this condition; for example, there is no upper limit to the possible kinetic energy of a gas molecule. Systems of interacting nuclear spins, however, have the characteristic that under suitable circumstances they can satisfy all three of the conditions, in which case the nuclear spin system can be at negative absolute temperature. *See* KINETIC THEORY OF MATTER; STATISTICAL MECHANICS.

The transition between positive and negative temperatures is through infinite temperature, not absolute zero; negative absolute temperatures should therefore not be thought of as colder than absolute zero, but as hotter than infinite temperature. *See* ABSOLUTE ZERO; TEMPERATURE. Norman F. Ramsey

Nemata (Nematoda) A phylum of unsegmented worms. A classification of nematodes follows:

Phylum Nemata
- Class: Adenophorea
  - Subclass: Enoplia
    - Order: Enoplida
      - Oncholaimida
      - Tripylida
      - Isolaimida
      - Mononchida
      - Dorylaimida
      - Stichosomida
  - Subclass: Chromadoria
    - Order: Araeolaimida
      - Chromadorida
      - Desmoscolecida
      - Desmodorida
      - Monhysterida
- Class: Secernentea
  - Subclass: Rhabditia
    - Order: Rhabditida
      - Strongylida
  - Subclass: Spiruria
    - Order: Spirurida
      - Ascaridida
  - Subclass: Diplogasteria
    - Order: Diplogasterida
      - Tylenchida

Description. The Nemata are unsegmented or pseudosegmented (any superficial annulation limited to the cuticle) bilaterally symmetrical worms with a basically circular cross section. The body is covered by a noncellular cuticle. The cylindrical body is usually bluntly rounded anteriorly and tapering posteriorly. The body cannot be easily divided into head, neck, and trunk or tail, although a region posterior to the anus is generally referred to as the tail. The oral opening is terminal (rarely subterminal) and followed by the stoma, esophagus, intestine, and rectum which opens through a subterminal anus. Females have separate genital and digestive tract openings. In males the tubular reproductive system joins posteriorly with the digestive tract to form a cloaca. The sexes are separate and the gonads may be paired or unpaired. Females may be oviparous or ovoviviparous.

Adult nematodes are extremely variable in size, ranging from less than 0.012 in. (0.3 mm) to over 26 ft (8 m). Nematodes are generally colorless except for food in the intestinal tract or for those few species which have eyespots.

Life cycle. Reproduction among nematodes is either amphimictic or parthenogenetic (rarely hermaphroditic). After the completion of oogenesis the chitinous egg shell is formed and a waxy vitelline membrane forms within the egg shell; in some nematodes the uterine cells deposit an additional outermost albuminoid coating. Upon deposition or within the female body, the egg proceeds through embryonation to the eellike first- or second-stage larva, but following eclosion the larva proceeds through four molts to adulthood. This represents a direct life cycle, but among parasites more diversity occurs.

Distribution. Nemata comprise the third largest phylum of invertebrates, being exceeded only by Mollusca and Arthropoda. In sheer numbers of individuals they exceed all other metazoa. As parasites of animals they exceed all other helminths combined. Nematodes have been recovered from the deepest ocean floors to the highest mountains, from the Arctic to the Antarctic, and in soils as deep as roots can penetrate. Armand R. Maggeati

Nematicide A type of chemical used to kill plant-parasitic nematodes. Nematicides may be classed as soil fumigants or soil amendments, space fumigants, surface sprays, or dips. Soil treatments are commonly used because most plant-pathogenic species spend part or all of their life cycle in the soil, in or about the roots of plants. Nematicides may be liquids, gases, or solids, but on a field scale, liquids are most practical. *See* NEMATA; PESTICIDE. Dewey J. Raski

Nematomorpha A phylum of worms that was formerly considered to be a class of the phylum Aschelminthes; commonly called the hairworms, and closely allied to the nematodes. The adults are free-living in aquatic habitats, while the juveniles are parasitic in arthropods. The nematomorphs are found all over the world. They are divided into two classes, the Nectonematoidea and Gordioidea, with a total of 225 species. *See* NEMATA (NEMATODA).

The body is long and slender with a maximum length of 5 ft (1.5 m) and a diameter of 0.02–0.12 in. (0.5–3 mm). The females are longer than the males. The posterior end may be rounded with a terminal cloaca, or it may form two or three lobes in a forklike structure. The body color is yellowish, brown, or almost black. The body wall consists of three layers: an outer, rather thick fibrous cuticle; an epidermis consisting of a single layer of cells; and innermost, a muscle layer with longitudinal fibers only.

The body cavity extends the length of the body. It may be filled with tissue so that only minor spaces are left around the digestive system and the gonads.

The sexes are always separate, and the gonads are paired and stringlike extending the length of the body. During copulation the male coils itself around the female and places a drop of sperm near the cloacal opening of the female. The sperm cells actively enter the seminal receptacle. The eggs are laid in water in strings, and the adults die after egg laying. When hatched, the larvae swim to an aquatic arthropod. They penetrate the body wall of the host by means of their characteristic proboscis, which is armed with hooks and three long stylets. The gradual development in the host lasts some months without any metamorphosis. When they are mature, the worms leave the host. Bent J. Muus

Nematophytales An enigmatic group of fossil plants, in mid-Silurian to lower Upper Devonian rocks, composed of intertwined, branching tubes of two sizes: 10–50 micrometers in diameter, and 1–10 micrometers. Although they are referred to the algae by some authors, the occurrence of this group in inland swamps, coastal plain deposits, and marine deposits close to shore indicates that they were terrestrial organisms, unrelated to any known groups, perhaps at an intermediate level between algae and bryophytes. *See* PALEOBOTANY. Harlan P. Banks

Nemertea A phylum of bilaterally symmetrical, nonsegmented, ribbonlike worms. They have an eversible proboscis and a complete digestive tract with an anus. There is no coelom or body cavity, and the mesenchyme or parenchyma and the muscle fibers fill the area between the ciliated epidermis and the cellular lining of the digestive tract. Many species are brightly colored, sometimes having stripes or transverse bars.

The nemerteans are the simplest animals with a circulatory system. There are two lateral blood vessels and in some a third, unpaired dorsal vessel. The blood consists of a colorless fluid which may contain blood cells of several types. In species in which the blood is colored, the pigment is present in the cells. There is no heart, but the walls of the principal vessels may be contractile.

The nervous system has a pair of cerebral ganglia forming the brain as well as two longitudinal nerve cords and many smaller nerves. The ganglia and lateral cords may contain unusually large neurochord cells. In the epidermis there are scattered sensory nerve cells, probably tactile. A few to many simple eyes, or ocelli, may be present in front of the cerebral ganglia. There are no special respiratory organs; respiration occurs through the body surface. Nemerteans are usually either male or female, but a few individuals have both sex organs. Fertilization occurs outside the body in many species but may be internal in certain forms.

The nemerteans are mostly marine, bottom-dwelling worms, found in greatest numbers along the coasts of northern temperate regions. They live under stones, among the tangled masses of plants, in sand, mud, or gravel, and sometimes form mucus-lined tubes. A few are pelagic, freshwater, or terrestrial. Certain species are commensal with other animals.

The phylum Nemertea, containing about 1250 known species, is divided into two classes, Anopla and Enopla. *See* ANOPLA; ENOPLA. Arthur G. Humes; J. B. Jennings

Neodymium A metallic chemical element, Nd, atomic number 60, atomic weight 144.24. Neodymium belongs to the rare-earth group of elements. The naturally occurring element includes the six isotopes. The oxide, Nd_2O_3, is a light-blue powder. It dissolves in mineral acids to give reddish-violet solutions. *See* PERIODIC TABLE; RARE-EARTH ELEMENTS.

The salts have found application in the ceramic industry for coloring glass and for glazes. The glass is particularly useful in goggles used by glass blowers, since it absorbs the intense yellow D line of sodium present in the flame. The element has found commercial application in the manufacture of lasers. Frank H. Spedding

Neogastropoda The most highly specialized order in the subclass Prosobranchia (phylum Mollusca, class Gastropoda). Neogastropods have simplified pallial and cardiac structures involving complete separation of genital from renal organs, and a "half-gill" (that is, a one-sided comb-shaped or pectinibranch ctenidium) with its axis and major blood vessels fused to the mantle wall. The order comprises mainly marine carnivores and carrion feeders, all with a long extensible proboscis bearing a flesh-tearing radula. More efficient hydrodynamically with their simplified mantle cavity and fused ctenidial axis, neogastropods are not limited to clean waters over hard substrata (as are the archaeogastropods) but have successfully invaded all areas of the seashore and sea bottom, whether covered with sand, silt, or mud.

Neogastropods occur in all depths of the world's oceans from the tropics to polar waters, and there are at least 6000 species, mostly in four important superfamilies. The larger whelks of the superfamily Buccinacea are found from the shallow sublittoral and continental shelves down to depths of 9800 ft (3000 m). The flesh of many whelk species provides human food, and almost all species have been used in commercial longline fisheries as resilient and attractive bait. The smaller tingles, dog whelks, and oyster drills of the superfamily Muricacea are the abundant neogastropod predators in inshore and intertidal waters. A third important superfamily, Volutacea, encompasses more beautiful, much collected shells. The most specialized neogastropods are the tropical toxoglossans (superfamily Conacea) belonging to the families Conidae, or cone shells, and Terebridae, or auger shells. Both groups have been prized by shell collectors for centuries.

Despite their large number of species and diversity of habitats, the neogastropods show more anatomical uniformity (efficient mantle cavity, inhalant siphon with chemoreceptive osphradium, extensible proboscis, stenoglossan radula, and simple carnivore gut) than is found in any of the other major orders of gastropods. *See* GASTROPODA; MOLLUSCA; PROSOBRANCHIA. W. D. Russell-Hunter

Neognathae One of the two recognized superorders making up the subclass Neornithes of the class Aves. They are characterized as flying birds with fully developed wings and sternum with a keel, caudal vertebrae fused into a pygostyle, and absence of teeth in both jaws, or modifications of these conditions in secondary flightless birds.

This superorder includes all living birds and all known fossil birds since the Late Cretaceous; only the ancestral Jurassic *Archaeopteryx* and the specialized Cretaceous *Hesperonis* and its allies do not belong to the Neognathae. *See* ARCHAEORNITHES; AVES; ODONTOGNATHAE; RATITES. Walter Bock

Neognathostomata A superorder of Echinoidea, subclass Euechinoidea. These invertebrates are characterized by having a rigid, exocyclic test and a lantern or jaw apparatus developed sometime during the life history and usually persisting into the adult stage. The included orders are the Clypeasteroida, Cassiduloida, Neolampadoida, and Oligopygoida. *See* ECHINODERMATA; ECHINOIDEA; EUECHINOIDEA. Howard B. Fell

Neogregarinida An order of the protozoan subclass Gregarinia, class Telosporea, subphylum Sporozoa. All gregarines are parasites of the digestive tract and body cavity of invertebrates or lower chordates; their large, mature trophozoites (vegetative stages) live outside the host'scells. The Neogregarinida are thought to be relatively advanced gregarines which live in insects. There are only about 29 species of about 12 genera, and 4 families. *See* GREGARINIA. Norman D. Levine

Neolampadoida A group of small, deep-water cassiduloid echinoids with neotenous characteristics, treated as an order by some workers; possibly polyphyletic. The presence of bourrelets and phyllodes, the elongate first ambulacral plates, the undifferentiated tuberculation, and undifferentiated posterior interambulacral plating all indicate their relationships lie with cassiduloids. The only character shared by members of this group is the lack of petals (they have simple ambulacral pores only). Other characteristics, such as apical disc plating, are varied, indicating at least two independent origins from shallow-water cassiduloids.

There are seven genera, each monospecific. Five are living today and are usually found at depths of 430–1280 ft (135–400 m). A Miocene species and an Upper Eocene species are also known. *See* ECHINODERMATA. Andrew B. Smith

Neolithic The period of prehistoric culture whose basic defining attributes are the emergence of agriculture, animal domestication, and sedentary farmsteads or villages. This definition has evolved over the last century from the original characterization of this period based on the appearance of polished stone axes. By 1865, when John Lubbock published *Prehistoric Times*, two types of Stone Age had been recognized in Europe: *période de la pierre taillée* (period of chipped stone implements) and *période de la pierre polie* (period of polished stone implements). Lubbock termed the former Palaeolithic and the latter Neolithic. Subsequently, it was realized that the definition of this period based on a single artifact type was spurious, since Neolithic peoples also continued to make chipped stone tools. A more comprehensive view developed that saw the Neolithic as characterized by pottery manufacture, agriculture, livestock, and settled villages, but without the use of metals. Thus the Neolithic formed the final Stone Age precursor to the Bronze Age and the Iron Age in the classic northern European prehistoric sequence, which was soon extended throughout most of Eurasia. *See* PALEOLITHIC.

As the term Neolithic is presently used, it refers specifically to prehistoric societies in Europe, Asia, and northern Africa that derived the majority of their diet from agriculture and livestock and that lived in sedentary communities, either dispersed farmsteads or villages, but that did not yet know the use of alloyed metals. Outside of this area, the term Neolithic is rarely used, although clearly in most other parts of the world there was also a transition from hunting and gathering to agriculture at some time in prehistory. For the purposes of this article, "Neolithic" will be expanded to include a global consideration of the origins and dispersal of agriculture and sedentary life, what might be called a neolithic cultural pattern.

With the exception of the dog, which was domesticated by late Pleistocene hunter-gatherers in many parts of the world, the domestication of plants appears to have preceded that of animals. Although many parts of the world can claim to have been the location of the domestication of some food plants, the regions where the crop species which underlie the major agricultural systems of the world were domesticated are fairly well delimited. The three foci of domestication include the "Fertile Crescent" area of the Near East (wheat and barley), the Huanghe and Yangzi valleys of China (millet and rice), and Mesoamerica (maize and beans), while the broader areas of where important domestication also occurred include northern and Andean South America (potatoes and other root crops), northern sub-Saharan Africa (sorghum), and southeast Asia and many Pacific islands (various tree and root crops). Other research has added another area to this list: the indigenous domestication in eastern North America of a complex of weedy plants that include chenopod, sunflower, and marsh elder.

Just as wheat and barley were the founder crops of Near Eastern agriculture, sheep and goat can be considered the founder animals of ungulate domestication around 9000 years ago. Cattle and pigs appear to have been domesticated around 8000 years ago, with evidence pointing toward Anatolia as the likely location. The pig and the chicken were the most important domesticated animals in the Chinese Neolithic, supplemented soon after with the water buffalo.

Not only are the domestication of plants and animals and the establishment of farming settlements the hallmarks of the Neolithic, but also these provided the platform for subsequent cultural developments. The Neolithic in the Old World and its New World parallels were periods of dramatic change in human society, laying the foundation for the socially stratified societies that followed. Peter Bogucki

Neomeniomorpha A subclass of creeping, vermiform mollusks in the class Aplacophora. They are covered by a spicular integument and recognized by the presence of a ventral groove within which lies a narrow foot and by the absence of on oral shield. Neomenioids range in size from less than 0.08 to 12 in. (2 mm to 300 mm) and are found from subtidal areas to the abyss, at depths over 16,000 ft (5000 m). There are 23 families with 70 genera and 193 species worldwide.

Neomenioids creep by means of their ciliated foot along a track of sticky mucus produced from a ciliated, reversible pedal pit at the anterior end of the pedal groove. Anterior to the pedal pit, the head end is held above the substratum and freely moved.

All neomenioids are hermaphroditic. A barrel-shaped, nonfeeding larva called a pericalymma either is brooded or swims by means of a ciliated cellular test within which the animal develops; metamorphosis through loss or resorption of the test occurs within 10 days. *See* APLACOPHORA; MOLLUSCA. Amelie H. Scheltema

Neon A gaseous chemical element, Ne, with atomic number 10 and atomic weight 20.180. Neon is a member of the family of noble gases. The only commercial source of neon is the Earth's atmosphere, although traces of neon are found in natural gas, minerals, and meteorites. *See* INERT GASES; PERIODIC TABLE.

Considerable quantities of neon are used in high-energy physics research. Neon fills spark chambers used to detect the passage of nuclear particles. Liquid neon can be utilized as a refrigerant in the temperature range about 25 to 40 K (−416 to −387°F). Neon is also used in some kinds of electron tubes, in

Physical properties of neon

Property	Value
Atomic number	10
Atomic weight (atmospheric neon only)	20.180
Melting point, °C	−248.6
Boiling point at 1 atm pressure, °C	−246.1
Gas density at 0°C and 1 atm pressure, g/liter	0.8999
Liquid density at its boiling point, g/ml	1.207
Solubility in water at 20°C, ml neon (STP)/1000 g water at 1 atm partial pressure neon	10.5

Geiger-Müller counters, in spark-plug test lamps, and in warning indicators on high-voltage electric lines. A very small wattage produces visible light in neon-filled glow lamps; such lamps are used as economical night and safety lights. *See* NEON GLOW LAMP.

Neon is colorless, odorless, and tasteless; it is a gas under ordinary conditions. Some of the other properties of neon are given in the table. Neon does not form any chemical compounds in the ordinary sense of the word; there is only one atom in each molecule of gaseous neon. Arthur W. Francis

Neon glow lamp A low-wattage lamp often used as an indicator light or as an electronic circuit component. The neon lamp usually consists of a pair of electrodes sealed within a bulb containing neon gas at a low pressure. Some of the smaller bulbs are equipped with wire leads that are connected directly into the electrical supply circuit; others are equipped with conventional bases that vary with the size of the lamps (see illustration).

Some examples of glow lamps.

Electrodes sealed in a neon atmosphere will emit electrons if a sufficient voltage difference is impressed across them. In glow lamps the electrodes are usually treated to emit electrons freely. With a sufficiently high voltage between electrodes, the velocity of electron flow is high enough to ionize the neon nearest the negative electrode (cathode). The neon then emits a reddish-orange glow similar to the color of neon sign tubing. With direct current the glow is restricted to the immediate vicinity of the negative electrode. With alternating current, both electrodes act alternately as cathodes, and the glow appears alternately at both surfaces. At usual frequencies, the alternations occur so rapidly that both electrodes appear to glow constantly. In dc circuits, the voltage across the electrodes may be reduced significantly, once the lamp has started, without causing the lamp to go out. Alfred Makulec

Neornithes The subclass of Aves that contains all of the known birds other than those placed in the Archaeornithes. Comprising more than 30 orders, both fossil and living, its members are characterized by a bony, keeled sternum with fully developed powers of flapping flight (secondarily lost in a number of groups); a short tail with the caudal vertebrae fused into a single platelike pygostyle to which all tail feathers attach; a large fused pelvic girdle with a reversed pubis which is fused to a large synsacrum; and a large brain and eyes contained within a fused braincase. The jaws are specialized into a beak covered with a horny rhamphotheca; the upper jaw is kinetic, being either prokinetic or rhynchokinetic. Prokinesis refers to a bending zone at the base of the upper jaw, and rhynchokinesis to one within the upper jaw. A few fossil groups still possess teeth, but most fossil and all Recent birds have lost teeth. *See* ARCHAEORNITHES.

The Neornithes contains two superorders, the Odontognathae and the Neognathae. The Odontognathae, alternately known as the Odontornithes, may be an artificial group. Its members, which include the Cretaceous fossil orders Hesperornithiformes and Ichthyornithiformes, are united only by the presence of teeth in all species. The Neognathae contains the remaining modern birds, which have lost the teeth, and includes 26 orders. *See* AVES; NEOGNATHAE; ODONTOGNATHAE. Walter J. Bock

Neoteny A phenomenon among some salamanders, in which larvae of large size, while still retaining the gills and other larval features, become sexually mature, mate, and produce fertile eggs. In certain lakes of Mexico, only the neotenous larvae are present and are called axolotls. Neoteny occurs in certain species of the family Ambystomidae. Tracy I. Storer

Nepenthales An order of flowering plants, division Magnoliophyta (Angiospermae), in the subclass Dilleniidae of the class Magnoliopsida (dicotyledons). The order (also known as Sarraceniales) consists of 3 well-marked small families: the Droseraceae, with about 90 species; the Nepenthaceae, with about 80 species; and the Sarraceniaceae, with only about 16. The plants characteristically grow in waterlogged soils which are deficient in available nitrogen. They are herbs or shrubs with alternate, simple leaves that are modified for catching insects, from which they absorb nitrogenous nutrients. The pitcher plants (*Sarracenia*), sundew (*Drosera*), and Venus' flytrap (*Dionaea muscipula*) are well-known members of the order. *See* MAGNOLIOPSIDA. Arthur Cronquist

Neper A unit of attenuation used in transmission-line theory. On a uniform transmission line having waves traveling in only one direction, the magnitudes of voltage E and of current I decrease with distance x traveled, as given by Eq. (1), where

$$\frac{E}{E_0} = \frac{I}{I_0} = \epsilon^{-\alpha x} \qquad (1)$$

E_0, I_0, and α are constants. The attenuation in nepers between the points where E_0, and I_0 are measured and where E and I are measured is given by Eq. (2), in which ln denotes the natural (or

$$\alpha x = \ln \frac{E_0}{E} = \ln \frac{I_0}{I} \qquad (2)$$

napierian) logarithm. One neper equals 8.686 dB, the decibel being the practical unit of attenuation. *See* DECIBEL; TRANSMISSION LINES. Edward W. Kimbark

Nepheline A mineral of variable composition: in its purest state, $NaAlSiO_4$; often nearly $Na_3K(AlSiO_4)_4$; but generally (Na, K, □, Ca, Mg, Fe^{2+}, Mn,Ti)$_8$(Al, Si, Fe^{3+})$_{16}$O$_{32}$, where □ represents vacant crystallographic sites, and Ca, Mg, Fe^{2+}, Mn, Ti, and Fe^{3+} are usually present in only minor or trace amounts. The most important variations in nepheline composition are due to crystalline solution of $KAlSiO_4$ (the mineral kalsilite), and substitution of □ for K. *See* SILICATE MINERALS.

The salient physical properties of nepheline are: a Mohs scale hardness of 5.5–6.0; a specific gravity between 2.56 and 2.67; a typically dark gray, light gray, or white color, but it can also be colorless (nepheline is colorless in petrographic thin section); and a vitreous or greasy luster. Nepheline occurs as simple hexagonal prisms or, more commonly, as isolated shapeless grains or irregular polycrystalline masses.

Nepheline is the most abundant feldspathoid mineral; it occurs in a wide variety of SiO_2-deficient (quartz-free) and alkali-rich volcanic, plutonic, and metamorphic rocks. In volcanic rocks, nepheline occurs chiefly as a primary mineral in phonolites, kenytes, and melilite basalts, and it is the characteristic mineral of nephelinites.

Both "pure" (processed) nepheline and nepheline syenite are used as raw materials for the manufacture of glass, various ceramic materials, alumina, pottery, and tile. *See* FELDSPATHOID; IGNEOUS ROCKS; NEPHELINITE. James G. Blencoe

Images of the Neptune system from *Voyager 2. (Right)* The planet, with cloud features including a large dark oval near the western limb (the left edge) and a second dark spot with a bright core near the terminator (the lower right edge). (*Below*) Composite picture of Triton showing the illuminated southern hemisphere, with patches of dark, windblown material directed away from the subsolar point. The "cantaloupe" terrain northward of the equator (which runs left to right in this picture) stretches away into the darkness. It is crossed by intersecting ridges and valleys. The absence of impact craters shows that this surface is relatively young. (*NASA*)

Nephelinite A dark-colored, aphanitic (very finely crystalline) rock of volcanic origin, composed essentially of nepheline (a feldspathoid) and pyroxene. *See* KALSILITE.

The texture is usually porphyritic with large crystals (phenocrysts) of augite and nepheline in a very fine-grained matrix. Augite phenocrysts may be diopsidic or titanium-rich and may be rimmed with soda-rich pyroxene (aegirine-augite). Microscopically the matrix is seen to be composed of tiny crystals or grains of nepheline, augite, aegirite, and sodalite with occasional soda-rich amphibole, biotite, and brown glass.

Nephelinite and related rocks are very rare. They occur as lava flows and small, shallow intrusives. A great variety of these feldspathoidal rocks is displayed in Kenya. *See* FELDSPATHOID; IGNEOUS ROCKS; LEUCITE. Carleton A. Chapman

Neptune The outermost of the four giant planets. Neptune is a near twin of Uranus in size, mass, and composition. Its discovery in 1846 within a degree from the theoretically predicted position was one of the great achievements of celestial mechanics. *See* CELESTIAL MECHANICS.

Through a small telescope, Neptune appears as a tiny greenish disk. Its linear equatorial diameter of 30,775 mi (49,528 km) is very similar to that of Uranus. The mass of Neptune is 17.15 times the mass of Earth, corresponding to a mean density of 1.64, somewhat above that of its sister planet. This suggests that the proportion of heavy elements is somewhat greater in Neptune than in Uranus. *See* URANUS.

Most of what is known about Neptune is the result of the flyby of the planet by the *Voyager 2* spacecraft in August 1989. The cloud features included a large dark oval (about the size of Earth), reminiscent of Jupiter's Great Red Spot, as well as the white clouds of condensed methane whose brilliant contrast with the blue-green atmosphere made them visible from Earth. Unlike the Great Red Spot, Neptune's dark oval proved to have a short lifetime. By following the clouds over several weeks, scientists were able to deduce the presence of currents at different latitudes, with the high-latitude winds faster than those near the equator. Storm systems on Neptune can cross latitude lines, moving toward the equator as the dark oval did before disappearing. On Jupiter and Saturn, such latitudinal motion does not occur.

This circulation pattern resembles that of Uranus, despite the different inclinations of the rotational axes of the two planets (that of Neptune is 29.6°, while that of Uranus is 97.9°, and that of Earth is 23.5°), and the fact that Neptune has an internal energy source that releases some 2.7 times the amount of heat absorbed from the Sun, while Uranus has no excess internal heat. The explanation may lie in similar atmospheric opacities. However, the difference in internal heat may be responsible for the greater cloud activity on Neptune than on Uranus.

The atmosphere of Neptune, like those of the other giant planets, is composed predominantly of hydrogen and helium. The relative abundance of methane is enhanced slightly more than on Uranus, between 25 and 40 times the value corresponding to solar abundances of the elements.

The orientation of Neptune's magnetic field is surprisingly similar to that of Uranus. It can be represented by a bar magnet inclined at an angle of 46.8° with respect to the axis of rotation and offset by 0.55 planetary radius. This field has trapped a plasma of ionized and neutral gases in the planet's magnetosphere.

Before the *Voyager* encounter, only two satellites of Neptune were known, both in highly irregular orbits. Triton was discovered visually by W. Lassell in 1846. It is moving in a retrograde direction around Neptune with a period of 5.9 days in a nearly circular orbit. Nereid was found in 1949 as a result of a photographic search by G. P. Kuiper. It has the most eccentric orbit of any known satellite. In sharp contrast to these two bodies, the six satellites discovered by the *Voyager* cameras all have very regular orbits: in the plane of the planet's equator and nearly circular. They are all close to the planet. In 2002 and 2003, five irregular satellites were discovered from Earth at large distances from the planet. The search for additional satellites continues. *See* ELLIPSE; ORBITAL MOTION; SATELLITE (ASTRONOMY).

Triton has a tenuous atmosphere containing nitrogen, methane, and carbon monoxide. These three gases plus carbon dioxide and water are also present as frozen ices on the satellite's surface. *Voyager* revealed that this remarkable object has a diameter of only 1681 mi (2705 km), making it considerably smaller than the Earth's Moon (2086 mi or 3476 km). The surface temperature of Triton is -391 ± 7°F (38 ± 4 K). The size of this satellite, as well as the temperature and composition of its surface, makes Triton very similar to Pluto. *See* PLUTO.

The *Voyager* cameras showed that there are three well-defined, complete rings around Neptune, accompanied by a sheet of material that itself constitutes a broad ring. The outermost of these, the Adams ring, contains three concentrated clumps of material known as arcs. A fourth, much smaller arc is barely visible in the best *Voyager* image.

The confinement of these narrow rings is commonly assumed to require the presence of small shepherding satellites. Galatea and Despina orbit, respectively, just inside the outer two narrow rings, but the corresponding outer shepherds have not been found. Similarly, the persistence of the three arcs within the outer ring remains an enigma. Tobias C. Owen

Neptunium A chemical element, symbol Np, atomic number 93. Neptunium is a member of the actinide or 5*f* series of elements. It was synthesized as the first transuranium element in 1940 by bombardment of uranium with neutrons to produce neptunium-239. The lighter isotope ^{237}Np, a long-lived alpha emitter with half-life 2.14×10^6 years, is particularly important chemically. *See* PERIODIC TABLE.

Neptunium metal is ductile, low-melting (637°C or 1179°F), and in its alpha form is of high density, 20.45 g/cm^3 (11.82 oz/in.^{3}). The chemistry of neptunium may be said to be intermediate between that of uranium and plutonium. Neptunium metal is reactive and forms many binary compounds, for example, with hydrogen, carbon, nitrogen, phosphorus, oxygen, sulfur, and the halogens. *See* ACTINIDE ELEMENTS; NUCLEAR CHEMISTRY; TRANSURANIUM ELEMENTS. Robert A. Penneman

Nerve A group of nerve fibers coursing together as a bundle in the peripheral nervous system. The individual fibers are covered by Schwann cells, many of which contain large amounts of myelin, which makes the nerve appear shiny white. The nerve fibers with their Schwann cell sheaths are held together by connective tissue. In most nerves, some of the fibers are sensory (carrying information to the central nervous system) and some are motor (carrying information from the central nervous system to peripheral glands and muscles). When both sensory and motor fibers are in a nerve, it is called a mixed nerve.

In the central nervous system (brain and spinal cord) a group of nerve fibers running together is called a tract. Glial cells, not Schwann cells, form the sheaths of tract fibers, and there is no connective tissue holding the bundle together. Whereas most nerves are mixed, there is functional segregation in the central nervous system so that most tracts have only one functional type of fiber. *See* MOTOR SYSTEMS; NERVOUS SYSTEM (VERTEBRATE).

Douglas B. Webster

Nervous system (invertebrate) All multicellular organisms have a nervous system, which may be defined as assemblages of cells specialized by their shape and function to act as the major coordinating organ of the body. Nervous tissue underlies the ability to sense the environment, to move and react to stimuli, and to generate and control all behavior of the organism. Compared to vertebrate nervous systems, invertebrate systems are somewhat simpler and can be more easily analyzed. Invertebrate nerve cells tend to be much larger and fewer in number than those of vertebrates. They are also easily accessible and less complexly organized; and they are hardy and amenable to revealing experimental manipulations. However, the rules governing the structure, chemistry, organization, and function of nervous tissue have been strongly conserved phylogenetically. Therefore, although humans and the higher vertebrates have unique behavioral and intellectual capabilities, the underlying physical-chemical principles of nerve cell activity and the strategies for organizing higher nervous systems are already present in the lower forms. Thus neuroscientists have taken advantage of the simpler nervous systems of invertebrates to acquire further understanding of those processes by which all brains function. *See* NERVOUS SYSTEM (VERTEBRATE).

Invertebrate and vertebrate nerve cells differ more in quantity, or degree, than in qualitative features. Aside from differences in size and numbers, the most striking difference is that invertebrate neurons have a unipolar shape, whereas most vertebrate neurons are multipolar. An additional general contrast between invertebrate and vertebrate nervous systems is that invertebrates tend to have more neurons displaced to the periphery (outside the central nervous system) and to perform more integrative and processing functions in the periphery. Vertebrates perform almost all their integration within the central nervous system, using interneurons. Invertebrate nervous systems also seem to have a greater potential for regrowth, regeneration, or repair after damage than do vertebrate nerve cells. Many invertebrates continue to add new nerve cells to their ganglia with age; vertebrates, in general, do not. Only vertebrate neurons have myelin sheaths, a specialized wrapping of glial membrane around axons, increasing their conduction speed. Invertebrates tend to enhance conduction velocity by using giant axons, particularly for certain escape responses. James E. Blankenship; Becky Houck

Nervous system (vertebrate) A coordinating and integrating system which functions in the adaptation of an organism to its environment. An environmental stimulus causes a response in an organism when specialized structures, receptors, are excited. Excitations are conducted by nerves to effectors which act to adapt the organism to the changed conditions of the environment.

Comparative morphology. The brain of all vertebrates, including humans, consists of three basic divisions: prosencephalon, mesencephalon, and rhombencephalon (Fig. 1). The individual divisions or patterns of the brain do not function separately to bring about a final response; rather, each pattern acts on a common set of connections in the spinal cord.

Spinal patterns are the final common patterns used by all higher brain pathways to influence all organs of the body. These reflexes are divided into two basic patterns: the monosynaptic arc and the multisynaptic arc. The monosynaptic arc, or myotatic reflex, maintains tonus and posture in vertebrates and consists of two neurons, a sensory and a motor neuron.

The multisynaptic arc, or flexor reflex, is the pattern by which an animal withdraws a part of its body from a noxious stimulus. Both sensory neurons and internuncial neurons send information to brain centers. Coordinated limb movement is based on a connective pattern of neurons at the spinal level.

The structure of the spinal cord and its connections are basically similar among all vertebrates. The major evolutionary changes in the spinal cord have been the increased segregation of cells and fibers of a common function from cells and fibers of other functions and the increase in the length of fibers which connect brain centers with spinal centers. *See* POSTURAL EQUILIBRIUM.

The rhombencephalon of the brain is subdivided into a roof, or cerebellum, and a floor, or medulla oblongata. The medulla is similar to the spinal cord and is divided into a dorsal sensory

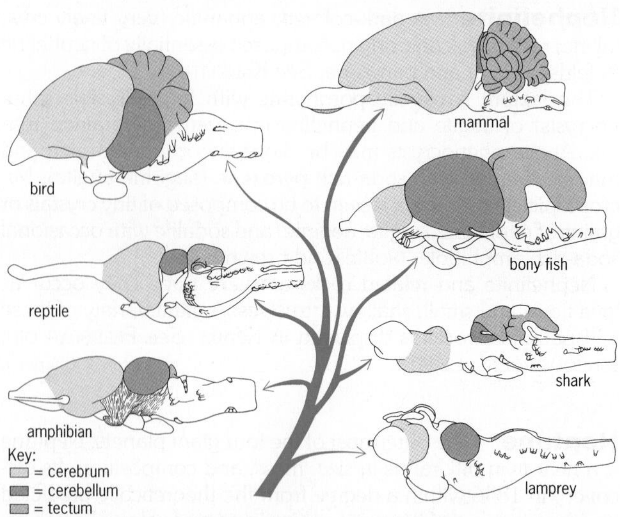

Fig. 1. Lateral views of several vertebrate brains showing evolutionary relationships.

region and a ventral motor region. It is an integrating and relay area between higher brain centers and the spinal cord. In addition to these nuclei and their connections, the medulla consists of both ascending and descending pathways to and from higher brain centers. The same basic connections occur throughout vertebrates.

In mammals, the cerebellum does not initiate movement; it only times the length of muscle contractions and orders the sequence in which muscles should contract to bring about a movement. The command to initiate a movement is received from the cerebral cortex (Fig. 2). Similarly, the cerebral cortex receives information regarding limb position and state of muscular contraction to ensure that its commands can be carried out by the cerebellum.

The mesencephalon is divided into a roof or optic tectum and a floor or tegmentum. The tegmentum contains the nuclei of the oculomotor and trochlear cranial nerves and a rostral continuation of the sensory nucleus of the trigeminal cranial nerve.

In the evolution of vertebrates, the prosencephalon develops as two major divisions, the diencephalon and the telencephalon. The diencephalon retains the tubular form and serves as a relay and integrating center for information passing to and from the telencephalon and lower centers. The telencephalon is divided into a pair of cerebral hemispheres and an unpaired telencephalon medium.

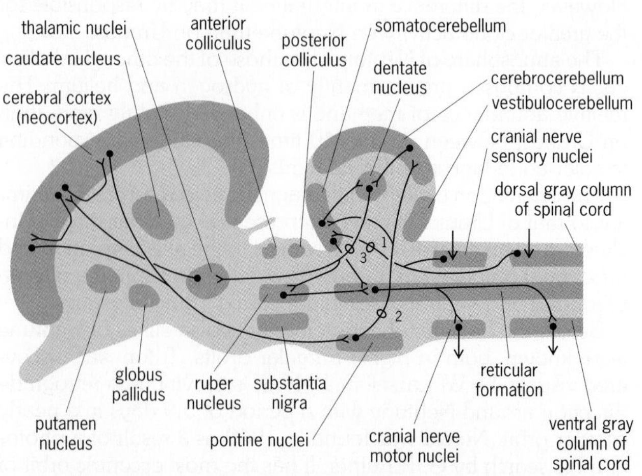

Fig. 2. Mammalian brain in sagittal section. Cerebellar patterns: tract 1, posterior cerebellar peduncle; 2, middle cerebellar peduncle; 3, anterior cerebellar peduncle.

There are three divisions of the diencephalon in all vertebrates: an epithalamus which forms the roof of the neural tube, a thalamus which forms the walls of the neural tube, and a hypothalamus which forms the floor of the neural tube. The epithalamus and hypothalamus are primarily concerned with autonomic functions such as homeostasis. The thalamus is subdivided into dorsal and ventral regions. The dorsal region relays and integrates sensory information, and the ventral thalamus relays and integrates motor information. *See* HOMEOSTASIS; INSTINCTIVE BEHAVIOR.

The telencephalon is the most complex brain division in vertebrates. It is divided into a roof, or pallium, and a floor, or basal region. The pallium is divided into three primary divisions: a medial PI or hippocampal division, a dorsal PII or general pallial division, and a lateral PII division, often called the pyriform pallium.

The most striking change in the telencephalon of land vertebrates involves the PIIIa component. In mammals, it has proliferated with the PIIb component of the dorsal pallium to produce the mammalian neocortex. In all land vertebrates except amphibians, the PIIb and the PIIIa components, along with the corpus striatum (BI and BII), are the highest centers for the analysis of sensory information and motor coordination. The PI, PIIa, PIIIb, BIII, and posterior parts of BI and BII form part of the limbic system which is concerned with behavioral regulation.

R. Glenn Northcutt

Comparative histology. The nervous system is composed of several basic cell types, including nerve cells called neurons, interstitial cells called neurolemma (cells of Schwann), satellite cells, oligodendroglia, and astroglia; and several connective-tissue cell types, including fibroblasts and microglia, blood vessels, and extracellular fluids.

Each neuron possesses three fundamental properties, involving specialized capacity to react to stimuli, to transmit the resulting excitation rapidly to other portions of the cell, and to influence other neurons, muscle, or glandular cells. Each neuron consists of a cell body (soma), one to several cytoplasmic processes called dendrites, and one process called an axon. Cell bodies vary from about 7 to more than 70 micrometers in diameter; each contains a nucleus and several cytoplasmic structures, including Nissl (chromophil) granules, mitochondria, and neurofibrils. The cell body is continuously synthesizing new cytoplasm, especially protein, which flows down the cell processes. The dendrites range from a fraction of a millimeter to a few millimeters in length. An axon may range from about a millimeter up to many feet in length. The site where two neurons come into contact with each other and where influences of one neuron are transmitted to the other neuron is called a synapse. Neurotransmitters are secreted across the presynaptic membrane into the synaptic cleft where they may excite (excitatory synapse) or inhibit (inhibitory synapse) the postsynaptic membrane. *See* BIOPOTENTIALS AND IONIC CURRENTS; SENSATION; SYNAPTIC TRANSMISSION.

There are three layers of connective tissue membranes, the meninges, covering the brain and spinal cord: the inner, pia mater; the middle layer, the arachnoid; and the outermost, the dura mater. Between the pia mater and the arachnoid is the subarachnoid space; this space and the ventricular cavities within the brain are filled with an extracellular fluid, the cerebrospinal fluid. *See* MENINGES. Charles Noback

Comparative embryology. The anlage of the nervous system is formed in the outer germ layer, the ectoderm, although some later contributions are also obtained from the middle germ layer, the mesoderm. In most vertebrates a neural plate is formed, which later folds into a neural groove, then closes to form a neural tube. The formation of neural tissue within the ectoderm is due to inductive influences from underlying chordomesodermal structures. *See* DEVELOPMENTAL BIOLOGY; EMBRYONIC INDUCTION; NEURAL CREST.

When the neural tube is developing, a segmentation of the central nervous system occurs by the formation of transverse bulges, neuromeres. At the time of neuromeric segmentation, the brain is subdivided into the so-called brain vesicles by local widenings of its lumen. In the rostral end more or less well-developed hemispheres are formed; in the middle of the brain anlage the mesencephalic bulge develops; and behind the latter the walls of the tube thicken into cerebellar folds. In this way the brain anlage is divided into five sections: the telencephalon, diencephalon, mesencephalon, metencephalon, and myelencephalon, and its cavity is divided into the rudiments of the adult ventricles.

In spite of the extraordinary variation in adult morphology of the vertebrate brain in different species, the early phases of development are essentially similar. The spinal cord remains as a comparatively slightly differentiated tube.

The cranial or cerebral nerves are the peripheral nerves of the head that are related to the brain. Twelve pairs of cranial nerves have been distinguished in human anatomy and these nerves have been numbered rostrally to caudally as follows:

I. Olfactory nerve, fila olfactoria
II. Optic nerve, fasciculus opticus
III. Oculomotor nerve
IV. Trochlear nerve
V. Trigeminal nerve, in most vertebrates divided into three branches: ophthalmic, maxillary, and mandibular
VI. Abducens nerve
VII. Facial nerve
VIII. Statoacoustic nerve
IX. Glossopharyngeal nerve
X. Vagus nerve
XI. Accessory nerve
XII. Hypoglossal nerve

The spinal ganglia are formed from the neural crest which grows out like a continuous sheet from the dorsal margin of the neural tube and is secondarily split up into cell groups, the ganglia, by a segmentating influence from the somites. Fibers grow out from the ganglionic cells and form the sensory fibers of the spinal nerves. Motor nerve fibers emerge from cells situated in the ventral horns of the spinal cord. The ventral motor fibers and the dorsal sensory fibers fuse to form a common stem, which is again laterally divided into branches, innervating the corresponding segment of the body.

The ganglia of the sympathetic nervous system develop ventrolateral to the spinal cord as neural crest derivatives. At first a continual column of sympathetic nerve cells is formed; it later subdivides into segmental ganglia.

The parasympathetic system is made up of preganglionic fibers emanating as general visceromotor fibers from the brain and from the sacral cord segments. Cells migrate to form the peripheral ganglia along them. *See* AUTONOMIC NERVOUS SYSTEM.

Bengt Kallen

Nervous system disorders A satisfactory classification of diseases of the nervous system should include not only the type of reaction (congenital malformation, infection, trauma, neoplasm, vascular diseases, and degenerative, metabolic, toxic, or deficiency states) but also the site of involvement (meninges, peripheral nerves or gray or white matter of the spinal cord, brainstem, cerebellum, and cerebrum). To these may be added various other correlates, such as age and sex. The nerve cell may be damaged primarily, as in certain infections, but much more commonly the nerve cell is damaged secondarily as the result of metabolic or vascular diseases affecting other important organs, such as the heart, lungs, liver, and kidneys.

Malformation. The central nervous system develops as a hollow neural tube by the fusion of the crests of the neural groove, beginning in the cervical area and progressing rostrally and caudally, the last points to close being termed the anterior and posterior neuropores. If the anterior neuropore fails to close (about 24 days of fetal age), anencephaly develops. The poorly organized brain is exposed to amniotic fluid and becomes necrotic and hemorrhagic, with death usually within hours after birth. *See* CONGENITAL ANOMALIES.

If the posterior neuropore fails to close (about 26 days of fetal age), the lumbosacral neural groove is exposed to amniotic fluid. The nervous tissue becomes partially necrotic and incorporated in a scar. Such a meningomyelocele is readily infected unless buried surgically within a few hours after birth. In addition, in about 95% of such infants hydrocephalus occurs, which usually can be adequately treated by shunting the ventricular fluid into the venous system or peritoneal cavity.

Other developmental disorders of the nervous system may appear as hypoplasia or hyperplasia (decrease or increase in growth of cells, respectively) or as a destruction of otherwise normally developing tissues. Rapidly growing tissues such as the embryonic nervous system are generally rather easily damaged by many toxic agents. The time of onset and the extent of repair rather than the nature of the agent determine the resulting pattern of abnormal development. *See* BEHAVIORAL TOXICOLOGY.

Infection. Infections of the nervous system may occur through a defect in the normal protective coverings caused by certain congenital malformations, as mentioned above, but also through other defects as the result of trauma, especially penetrating wounds or fractures opening into the paranasal sinuses or mastoid air cells. Subsequent infection of the nervous system may be the major complication of such "open head" injuries.

Infections may also spread directly from adjacent structures, as from mastoiditis, sinusitis, osteomyelitis, or subcutaneous abscesses. Such infections usually spread along venous channels producing epidural abscess, subdural empyema, leptomeningitis, and brain abscess. All of these infections are characteristically caused by pyogenic (pus-forming) bacteria. Other pyogenic bacteria may metastasize by way of the bloodstream from more distant infections, such as bacterial endocarditis, pneumonia, and enteritis.

Infections of the nervous system must be treated promptly as medical emergencies. The diagnosis is easily established by spinal puncture; the microorganisms can be visualized with special stains.

Many other microorganisms can infect the nervous system: *Mycobacterium tuberculosis* (the organism causing tuberculosis), *Treponema pallidum* (the organism causing syphilis), several fungi and rickettsiae, and many viruses.

Viral infections vary widely geographically, generally related to the necessity for intermediate hosts and vectors (animal reservoirs) by which the virus is spread. Poliomyelitis, now largely prevented by effective vaccination of most children, is primarily an intestinal infection which occasionally spreads to the nervous system, infecting and destroying motor nerve cells, thereby producing weakness of certain muscles. Herpes zoster has a similar preference for infecting sensory nerve cells and producing an acute skin eruption in the distribution of the affected sensory cells. Herpes simplex is closely related to herpes zoster, resides in the trigeminal or sacral sensory nerve cells, and intermittently produces eruptions in the distribution of these cells: "fever blisters" in and around the mouth in type I herpes, or similar blisters in the genital area in type II herpes. The latter is increasingly being recognized as a venereal disease. Rabies virus also affects certain nerve cells in the temporal lobe of the brain, as well as in the cerebellum, and is transmitted through the saliva of animals that bite other animals or humans; rabies is the single exception to the rule that immunization must precede infection to be effective, and the immunization must begin promptly after the bite. *See* ANIMAL VIRUS; HERPES; POLIOMYELITIS.

Inflammation. Certain viruses frequently produce a meningitis in humans from whose cerebrospinal fluid the virus is relatively easily grown. Other viruses, such as measles and varicella, occasionally produce meningitis or encephalomyelitis, but the cerebrospinal fluid does not contain the virus. *See* MENINGITIS.

Allergy to one's own tissue elements is an interesting possibility that has evoked many experimental approaches. Two human diseases, multiple sclerosis, a demyelinating disease affecting the central nervous system, and the Landry-Guillain-Barré syndrome, a demyelinating disease affecting the peripheral nervous system, are considered likely candidates to be related to experimental allergic encephalomyelitis and experimental allergic neuritis, respectively. *See* AUTOIMMUNITY; MULTIPLE SCLEROSIS.

Vascular disease. Vascular diseases of the nervous system are commonly called strokes, a term which emphasizes the suddenness of onset of neurological disability. Such a cataclysmic onset is characteristic of vascular diseases, since the nerve cell can function without nutrients for only a matter of seconds and will die if not renourished within several minutes. *See* VASCULAR DISORDERS.

Two main types of hemorrhage occur: hemorrhage into the subarachnoid space from rupture of an aneurysm (a focal weakening and dilatation) of a large artery; and hemorrhage into the brain from rupture of an aneurysm of a small artery or arteriole. Both types of hemorrhage occur more commonly in hypertensive adults. *See* ARTERIOSCLEROSIS; HEMORRHAGE; HYPERTENSION.

Nerve cells require oxygen and glucose for functional activity, and can withstand only brief periods of hypoxia or hypoglycemia. Even a few seconds of hypoxia can block the nerve cell's function, and more than 10 min is almost certainly fatal to most nerve cells. Transient ischemic attacks may result, with temporary impairment of blood flow to a part of the brain and consequent focal neurological dysfunction. These attacks may also be successfully treated with drugs or surgery and the disastrous major stroke prevented. Myocardial infarction, postural hypotension, and stenosis or narrowing of the carotid or vertebral arteries greater than 60% are common causes of cerebral ischemia. If the ischemia is not rapidly reversed, the neurons undergo selective necrosis; if the ischemia is more severe or prolonged, the glia and blood vessels in the gray matter also undergo necrosis; and if the ischemia is still more severe or prolonged, all the gray and white matter in the ischemic zone becomes necrotic, a condition known as cerebral infarction or encephalomalacia. One of the common ways the brain reacts to small or large hemorrhages or ischemic episodes is by swelling. Such swelling itself may be fatal within a few days to a week or so by a process known as transtentorial herniation, compressing the brainstem, where there are important neural circuits for vital functions, such as breathing and maintenance of blood pressure.

Degenerative and other diseases. Degenerative, metabolic, toxic, and deficiency states include the largest numbers of both common and rare diseases of the nervous system. Since neurons in the brain may be destroyed after birth and cannot be replaced, mental deterioration, deafness and blindness, incoordination and adventitious movements, and other neurologic signs that are so typical of these disorders are generally not reversible even if the basic metabolic defect can be corrected. Advances have been made in the early diagnosis and treatment of several diseases usually manifest in infancy with mental retardation. Three examples are phenylpyruvic oligophrenia (phenylketonuria or PKU), which is treatable with a phenylalanine-deficient diet; galactosemia, requiring a galactose-free diet also as early as possible to avoid cataracts and mental retardation; and cretinism, which requires treatment with thyroid. *See* HUNTINGTON'S DISEASE; METABOLIC DISORDERS; PARKINSON'S DISEASE; PHENYLKETONURIA.

Neoplasm. Neoplasms of the nervous system can be divided into primary and metastatic, the primary into gliomas and others, and the metastatic into bronchogenic and others. These four groups each account for about 25% of all intracranial neoplasms. *See* TUMOR. Ellsworth C. Alvord, Jr.; Cheng-mei Shaw

Network theory

The systematizing and generalizing of the relations between the variables flowing in and across elements within an electrical network. To be precise, certain terms are introduced. *See* ALTERNATING-CURRENT CIRCUIT THEORY.

Elements. The elements of a network model are resistance, inductance, and capacitance (the passive elements) and sources of energy (the active elements), which may be either independent sources or controlled, that is, dependent, sources. An independent-voltages source produces a voltage across its terminals that is not dependent on any current or voltage, although it may be a function of time, as in the case of an alternating source; an independent-current source carries current that is independent of all voltages or currents but may be a function of time. *See* CAPACITANCE; ELECTRICAL RESISTANCE; ELECTROMAGNETIC INDUCTION; ELECTROMOTIVE FORCE (EMF); INDUCTANCE.

A number of definitions, together with theorems that relate them, are taken from the mathematical subject of topology. Two or more elements are joined at a node (*see* illustration). If three or more elements are connected together at a node, that node is called a junction. (The term major node may be used instead of junction; topological terminology varies among authors.) An element extends from one node to another. A branch of a network extends from one junction to another and may consist of one element or several elements connected in series. A loop, or circuit, is a single closed path for current. A mesh, or window, is a loop with no interior branch. *See* TOPOLOGY.

The illustration shows a network with 13 elements, of which 12 are passive and 1 is active. Nodes are indicated by dots; of the 9 nodes in the figure, 6 are junctions. There are 10 branches of which 5 come together at a single node or junction at the bottom of the figure.

Equation (1) is an extension of Ohm's law, where I is the

$$V = IZ \tag{1}$$

current through an element, Z is the impedance of that element, and V is the voltage or potential difference between the nodes that terminate that element; this applies to every passive element of an electrical network. The number of elements (active or passive) in a network may be designated as E. *See* ELECTRICAL IMPEDANCE; OHM'S LAW.

At every node of an electrical network the sum of currents entering that node is zero. Equation (2) expresses Kirchhoff's

$$\Sigma I = 0 \tag{2}$$

current law. An equation of this form can be written for each node, but in a fully connected network one of these equations

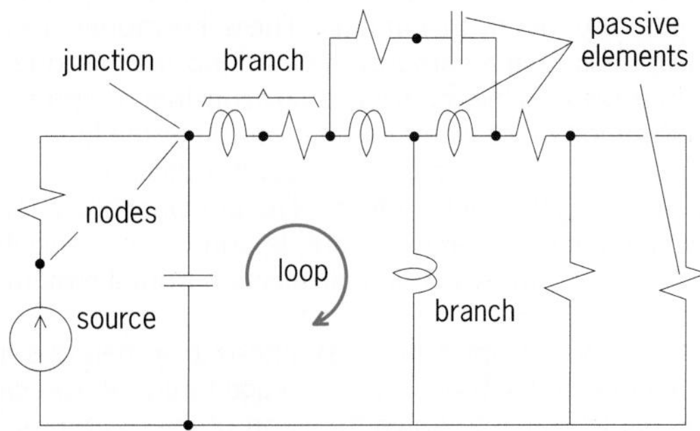

Parts of a network. (*After H. H. Skilling, Electrical Engineering Circuits, 2d ed., Wiley, 1965*)

can be derived from the others; hence the number of independent node equations is one less than the number of nodes. The number of independent node equations, called N, equals in a fully connected network the number of nodes minus one. *See* KIRCHHOFF'S LAWS OF ELECTRIC CIRCUITS.

Around every loop of an electrical network the sum of the voltages across the elements is zero. Equation (3) expresses Kirch-

$$\Sigma V = 0 \tag{3}$$

hoff's voltage law. A network such as that shown in the illustration can have many possible loops and hence many equations of this form, but only a limited number are independent. If L is the number of independent loops, topology gives Eq. (4), from which L can be computed, E and N having been

$$E = N + L \tag{4}$$

counted. Thus in the illustration there are 13 elements and 9 nodes, hence 8 independent nodes, so that there are 13 − 8 or 5 independent loops; that is, $L = 5$.

If a network is planar (flat) and fully connected, the number of independent loops is equal to the number of meshes, or windows. The illustration shows such a network, and the number of meshes is obviously 5, so again $L = 5$.

Branch equations. There are E elements in a network. Suppose that all impedances are known. There is a voltage across each element, and there is a current through each element; hence there are $2E$ voltages and currents to be known. One equation is provided by each element; either the element is a source for which voltage or current is given, or it is an impedance for which there is a relation given by Ohm's law in the form of Eq. (1). Hence there are E equations from the elements. From Eqs. (2) and (3) the nodes and loops provide $N + L = E$. Thus there are $2E$ equations relating voltages and currents.

In an actual solution for current and voltage in a network, it is probably desirable to reduce the number of equations by combining elements that are in series. This can reduce the number of elements to the number of branches and the number of nodes to the number of junctions; the network is then described by $2B$ branch equations.

These branch equations are easy to write but tedious to solve, unless a computer is used. Two modifications have been devised, however, that eliminate a great deal of the labor and reduce the number of equations from $2B$ to either L, the number of independent loops, or N, the number of independent nodes, as will be described in the following paragraphs.

Although branch equations can be written for networks containing either linear or nonlinear elements, the solution is more difficult for nonlinear networks. A linear network is one that gives rise to linear systems of equations, which are subject to special methods of solution, and for which the principle of superposition applies, allowing the use of loop or node equations. In a linear system the values of resistance, inductance, and capacitance are constant with respect to voltage and current, and a controlled source produces a voltage or a current that is proportional to another voltage or current. *See* SUPERPOSITION THEOREM (ELECTRIC NETWORKS).

Fortunately, many electrical networks are linear or are nearly enough linear to be so considered, at least in the useful range of operation or in a piecewise linear fashion.

Loop equations. The ingenuity of the loop method lies in the selection of currents to be determined. It is necessary to find only as many independent currents as there are independent loops, instead of finding as many different currents as there are branches.

Nodal equations. Loop equations are written based on the concept of loop currents. This makes it unnecessary to give any attention to Kirchhoff's current law, for loop currents necessarily add to zero at every node, and Kirchhoff's current law is automatically satisfied. The loop-current concept therefore reduces the number of equations that must be solved simultaneously from the $2B$ equations of the branch method to L, the number of independent loops, which is usually about one-fourth as many.

In the node-equation method, the simplifying concept is the idea of measuring voltage from all the nodes of the network to one particular node that is called the reference node, or the datum node. This makes it unnecessary to give any attention to Kirchhoff's voltage law. It is only necessary to satisfy Kirchhoff's current law at each node, for the voltage law is automatically satisfied. Thus the number of simultaneous equations is reduced to the number of independent nodes N, a number much smaller than $2B$ and comparable with L.

Whether the node method or the loop method is the more convenient depends on the network. Some networks have fewer loops than nodes, and some have fewer nodes than loops. Other factors also affect the relative convenience. Hugh Hildreth Skilling

Neural crest A strip of ectodermal material in the early vertebrate embryo inserted between the prospective neural plate and epidermis. After closure of the neural tube the crest cells migrate into the body and give rise to parts of the neural system: the main part of the visceral cranium, the mesenchyme, the chromaffin cells, and pigment cells. The true nature of the neural crest eluded recognition for many years because this primary organ has a temporary existence; its cells and derivatives are difficult to analyze when dispersed throughout the body. The fact that mesenchyme arises from this ectodermal organ was directly contrary to the doctrine of the specificity of the germ layers.

Neural crest no doubt exists, with similar qualities, in all vertebrate groups, including the cyclostomes. It has been most thoroughly studied in amphibians and the chick. *See* GERM LAYERS.

Sven Hörstadius

Neural network An information-processing device that consists of a large number of simple nonlinear processing modules, connected by elements that have information storage and programming functions. The field of neural networks is an emerging technology in the area of machine information processing and decision making. The main thrusts are toward highly innovative machine and algorithmic architectures, radically different from those that have been employed in conventional digital computers. The information-processing elements and components of neural networks, inspired by neuroscientific studies of the structure and function of the human brain, are conceptually simple. Three broad categories of neural-network architectures have been formulated which exhibit highly complex information-processing capabilities. Several generic models have been advanced which offer distinct advantages over traditional digital-computer implementation. Neural networks have created an unusual amount of interest in the engineering and industrial communities by opening up new research directions and commercial and military applications. *See* NEUROBIOLOGY.

Automated information processing is achieved by means of modules that in general involve four functions: input/output (getting in and out of the machine), processing (executing prescribed specific information-handling tasks), memory (storing information), and connections between different modules providing for information flow and control. Neural networks contain a very large number of simple processing modules. This contrasts with traditional digital computers, which contain a small number of complex processing modules that are rather sophisticated in the sense that they are capable of executing very large sets of prescribed arithmetic and logical tasks (instructions). In conventional digital computers, the four functions listed above are carried out by separate dedicated machine units. In neural networks information storage is achieved by components which at the same time effect connections between distinct machine units. These key distinctions between the neural-network and the digital computer architectures are of a fundamental nature and have major implications in machine design and in machine utilization.

The information-processing properties of neural networks depend mainly on two factors: the network topology (the scheme used to connect elements or nodes together), and the algorithm (the rules) employed to specify the values of the weights connecting the nodes. While the ultimate configuration and parameter values are problem-specific, it is possible to classify neural networks, on the basis of how information is stored or retrieved, in four broad categories: neural networks behaving as learning machines with a teacher; neural networks behaving as learning machines without a teacher; neural networks behaving as associative memories; and neural networks that contain analog as well as digital devices and result in hybrid-machine implementations that integrate complex continuous dynamic processing and logical functions. Within these four categories, several generic models have found important applications, and still others are under intensive investigation.

Neural-network research is developing a new conceptual framework for representing and utilizing information, which will result in a significant advance in information epistemology. Communication technology is based on the notions of coding and channel capacity (bits per second), which provide the conceptual framework for information representation appropriate to machine-based communication. Neural-network systems (biological or artificial) do not store information or process it in the way that conventional digital computers do. Specifically, the basic unit of neural-network operation is not based on the notion of the instruction but on the connection. The performance of a neural network depends directly on the number of connections per second that it effects, and thus its performance is better understood in terms of its connections-per-second (CPS) capability. *See* INFORMATION THEORY. Nicholas DeClaris

Neurobiology Study of the development and function of the nervous system, with emphasis on how nerve cells generate and control behavior. The major goal of neurobiology is to explain at the molecular level how nerve cells differentiate and develop their specific connections and how nerve networks store and recall information. Ancillary studies on disease processes and drug effects in the nervous system also provide useful approaches for understanding the normal state by comparison with perturbed or abnormal systems. The functions of the nervous system may be studied at several levels: molecular, subcellular (organelle), cellular, simple multicellular interacting systems, complex systems, and higher functions (whole animal behavior). *See* BIOPOTENTIALS AND IONIC CURRENTS; MEMORY; MOTOR SYSTEMS; NERVOUS SYSTEM (INVERTEBRATE); NERVOUS SYSTEM (VERTEBRATE); NERVOUS SYSTEM DISORDERS; NEURON; SENSE ORGAN; SYNAPTIC TRANSMISSION. James R. Bamburg; Michael D. Brown

Neurohypophysis hormone Either of two peptide hormones secreted by the neurohypophysis, or posterior lobe of the pituitary gland, in humans. These hormones, oxytocin and vasopressin, each comprise nine amino acid residues. Vasopressin is responsible for arterial vasoconstriction (pressor action) and inhibition of water excretion through the kidneys (antidiuretic action), and has a weak effect on contraction of smooth muscle including that of the uterus. The principal action of oxytocin is stimulation of smooth muscle contraction, specifically that of the uterine muscle, and milk ejection from the mammary gland.

Oxytocin and vasopressin are synthesized in neurons in the hypothalamus and subsequently packaged into neurosecretory granules, which migrate down the axon of the neuron and are stored in the posterior lobe of the pituitary gland, from where they are secreted into the systemic circulation. These hormones are also secreted directly from the hypothalamus into the third ventricle and into the hypothalamo-hypophysial portal circulation of the anterior pituitary gland. *See* NEUROSECRETION.

Major stimuli controlling the release of vasopressin include changes in osmolality of the blood, alterations in blood volume, and psychogenic stimuli such as pain, fear, and apprehension. Stimuli evoking release of oxytocin include nipple stimulation or suckling, and stretching of the cervix and vagina (Ferguson reflex).

Oxytocin probably plays an important role in the onset of labor and delivery (parturition) in primates. During lactation, significant amounts of oxytocin are released by the mother during suckling. When there is total destruction of the pituitary or the neurohypophysis, diabetes insipidus may occur. *See* DIABETES; HORMONE; PITUITARY GLAND. M. Yusoff Dawood

Neuroimmunology The study of basic interactions among the nervous, endocrine, and immune systems during development, homeostasis, and host defense responses to injury. In its clinical aspects, neuroimmunology focuses on diseases of the nervous system, such as myasthenia gravis and multiple sclerosis, which are caused by pathogenic autoimmune processes, and on nervous system manifestations of immunological diseases, such as primary and acquired immunodeficiencies. *See* AUTOIMMUNITY; IMMUNOLOGICAL DEFICIENCY.

Neuroimmune interactions are dependent on the expression of at least two structural components: immunocytes must display receptors for nervous system-derived mediators, and the mediators must be able to reach immune cells in concentrations sufficient to alter migration, proliferation, phenotype, or secretory or effector functions. More than 20 neuropeptide receptors have been identified on immunocompetent cells.

It has been found that stimuli derived from the nervous system could affect the course of human disease. The onset or progression of tumor growth, infections, or chronic inflammatory diseases, for example, could be associated with traumatic life events or other psychosocial variables such as personality types and coping mechanisms. More direct indications of the influence of psychosocial factors on immune function have been provided by findings that cellular immunity can be impaired in individuals who are exposed to unusually stressful situations, such as the loss of a close relative. *See* CELLULAR IMMUNOLOGY.

During responses to infection, trauma, or malignancies, cells of the immune system produce some cytokines in sufficiently high quantities to reach organs that are distant from the site of production. These cytokines are known to act on the nervous system. Fever is the classic example of changes in nervous system function induced by products of the immune system; interleukin 1, which is produced by monocytes after stimulation by certain bacterial products, binds to receptors in the hypothalamus and evokes changes via the induction of prostaglandins. Interleukin 1 also induces slow-wave sleep. Both fever and sleep may be regarded as protective behavioral changes. *See* ENDOCRINE SYSTEM (VERTEBRATE); IMMUNOLOGY; NERVOUS SYSTEM (VERTEBRATE); NEUROSECRETION. Martin Lotz; Wietse Kuis; Peter Villiger

Neuron A nerve cell: the functional unit of the nervous system. Structurally, the neuron is made up of a cell body or soma and one or more long processes: a single axon and dendrites. The cell body contains the nucleus and usual cytoplasmic organelles with an exceptionally large amount of rough endoplasmic reticulum, called Nissl substance in the neuron. The longest cell process is the axon, which is capable of transmitting propagated nerve impulses. There may be none, one, or many dendrites composing part of a neuron. If there is no dendrite, it is a unipolar neuron; with one dendrite, it is a bipolar neuron; if there is more than one dendrite, it is a multipolar neuron. In most neurons only the axon propagates nerve impulses; the dendrites and somas are also irritable but do not propagate nerve impulses. *See* NERVOUS SYSTEM (VERTEBRATE). Douglas B. Webster

Neuroptera An order of delicate insects having endopterygote development, chewing mouthparts, and soft bodies. Included are the insects commonly termed lacewings, ant lions, dobsonflies, and snake flies. The order consists of about 25 families and is widely distributed.

The adults have long, slender antennae and usually four similar wings, although the front pair is generally slightly larger than the hind pair. The adults of most species are strongly attracted to lights. The larvae are aggressive predators. The larvae of lacewings are especially destructive to aphids, scale insects, and mites. *See* ENDOPTERYGOTA; INSECTA. Frank M. Carpenter

Neurosecretion The synthesis and release of hormones by neurons. Such neurons are called neurosecretory cells, and their products are often called neurohormones. Like conventional (that is, nonglandular or ordinary) neurons, neurosecretory cells are able to receive signals from other neurons. But unlike ordinary neurons that have cell-to-cell communication over short distances at synapses, neurosecretory cells release their product into an extracellular space that may be at some distance from the target cells. In an organism with a circulatory system, the neurohormones are typically sent by the vascular route to their target, whereas in lower invertebrates that lack an organized circulatory system the neurohormones apparently simply diffuse from the release site to the target. It is now clear that the nervous and endocrine systems interact in many ways, as in the suckling reflex of mammals (where the hormone oxytocin, a neurohormone, elicits milk ejection and is reflexly released in response to nerve impulses generated by stimulation of the nipples), and neurosecretory cells form a major link between them. *See* ENDOCRINE MECHANISMS; ENDOCRINE SYSTEM (INVERTEBRATE); ENDOCRINE SYSTEM (VERTEBRATE).

It has been shown that peptides or low-molecular-weight proteins as well as amines, such as octopamine and dopamine, are released from neurosecretory cells into the circulatory systems of various animals, where they function as neurohormones. In classical neurosecretory cells, the secreted material is synthesized in the cell body by the rough-surfaced endoplasmic reticulum and subsequently packaged in the form of membrane-bounded granules by the Golgi apparatus, and is then typically transported along the axon to the axonal terminals, where it is stored until released. The release of neurohormones from axonal terminals into an extracellular space is triggered when the electrical activity (action potential) that is propagated by the axon enters the neurosecretory terminals. Calcium ions are essential for neurohormone release. *See* BIOPOTENTIALS AND IONIC CURRENTS; ENDOPLASMIC RETICULUM; GOLGI APPARATUS.

Neurohormones have a wide variety of functions. The role of the vertebrate hypothalamo-neurohypophysial system has been especially well elucidated. The pars nervosa is the site of release of vasopressin (also called the antidiuretic hormone) and oxytocin, and the median eminence is the release site for several hypothalamic neurohormones that regulate the adenohypophysis, the nonneural portion of the pituitary gland. *See* ADENOHYPOPHYSIS HORMONE; NERVOUS SYSTEM (VERTEBRATE); NEUROHYPOPHYSIS HORMONE; PITUITARY GLAND. Milton Fingerman

Neurulation The process by which the vertebrate neural tube is formed. The primordium of the central nervous system is the neural plate, which arises at the close of gastrulation by inductive action of the chorda-mesoderm on the overlying ectoderm. The axial mesodermal substratum causes the neural ectoderm to thicken into a distinct plate across the dorsal midline and influences both its size and shape. Its shieldlike appearance, broader anteriorly and narrower posteriorly, presages the areas of brain and spinal cord, respectively. The lateral edges of the neural plate then rise as neural folds which meet first at the level of the future midbrain, above the dorsal midline, then fuse anteriorly and posteriorly to form the neural tube. The body ectoderm becomes confluent above the closing neural tube and separates from it. Upon closure, the cells (known as neural crest cells) which occupied the crest of the neural folds leave the roof of the tube and migrate through the mesenchyme to all parts of the embryo, forming diverse structures. The neural tube thus formed gives rise to the brain and about half of the spinal cord. The remainder of the neural tube is added by the tail bud, which proliferates a solid nerve cord that secondarily hollows into a tube. *See* NERVOUS SYSTEM (VERTEBRATE); NEURAL CREST. Howard L. Hamilton

Neutral currents Exchange currents which carry no electric charge and mediate certain types of electroweak interactions. The discovery of the neutral-current weak interactions and the agreement of their experimentally measured properties with the theoretical predictions were of great significance in establishing the validity of the Weinberg-Salam model of the electroweak forces.

The electroweak forces come in three subclasses: the electromagnetic interactions, the charged-current weak interactions, and the neutral-current weak interactions. The electromagnetic interaction is mediated by an exchanged photon γ. Since the photon carries no electric charge, there is no change in charge between the incoming and the outgoing particles. The charged-current weak interaction is mediated by the exchange of a charged intermediate boson, the W^+, and thus, for example, an incoming neutral lepton such as the ν_μ is changed into a charged lepton, the μ^-. In the neutral-current weak interactions, the exchanged intermediate boson, the Z^0, carries no electric charge (hence the name neutral-current interaction), and thus for example, an incident neutral lepton, such as the ν_μ, remains an outgoing neutral ν_μ. *See* ELECTRON; INTERMEDIATE VECTOR BOSON; LEPTON; NEUTRINO; PHOTON.

The neutral-current interactions were experimentally discovered in 1973, and have since been extensively studied, in neutrino scattering processes. Very important information about the properties of the neutral currents have been obtained by studying the interference effects between the electromagnetic and the neutral-current weak interactions in the scattering of polarized electrons on deuterium. Parity violating effects in atomic physics processes due to the neutral weak currents have been observed, and predicted parity-violating nuclear effects have been searched for. *See* ELEMENTARY PARTICLE; FUNDAMENTAL INTERACTIONS; PARITY (QUANTUM MECHANICS); SYMMETRY LAWS (PHYSICS); WEAK NUCLEAR INTERACTIONS. Charles Baltay

Neutralization reaction (immunology) A procedure in which the chemical or biological activity of a reagent or a living organism is inhibited, usually by a specific neutralizing antibody. As an example, the lethal or the dermonecrotic actions of diphtheria toxin on animals may be completely neutralized by an equivalent amount of diphtheria antitoxin.

Antibodies to bacterial, snake-venom, and other enzyme preparations regularly precipitate them from solution so that the supernates are devoid of enzyme activity; however, the neutralization of activity in the precipitate may range from complete to negligible. *See* IMMUNOLOGY; NEUTRALIZING ANTIBODY; SEROLOGY.

Henry P. Treffers

Neutralizing antibody An antibody that reduces or abolishes some biological activity of a soluble antigen or of a living microorganism. Thus, diphtheria antitoxin is a neutralizing antibody that, in adequate amounts, abolishes the pathological effects of diphtheria toxin in animals. This is only one characteristic; the other general properties of the antibody are those of the immunoglobulin family (IgG, IgA, or IgM) to which it belongs. *See* ANTIBODY; IMMUNOGLOBULIN. Henry P. Treffers

Neutrino An elusive elementary particle that interacts with matter principally through the weak nuclear force. Neutrinos are electrically neutral spin-$^1/_2$ fermions with left-handed helicity. Many weak interaction processes (interactions that involve the weak force), such as radioactive nuclear beta decay and thermonuclear fusion, involve neutrinos. Present experimental knowledge is consistent with neutrinos being point particles that have no internal constituents. Neutrinos are classified as neutral leptons, where leptons are defined as elementary particles that interact with the electroweak (electromagnetic and weak nuclear) and gravitational forces but not with the strong nuclear force. *See* ELEMENTARY PARTICLE; FUNDAMENTAL INTERACTIONS; HELICITY (QUANTUM MECHANICS); LEPTON; SPIN (QUANTUM MECHANICS); WEAK NUCLEAR INTERACTIONS.

Because the role of gravitational forces is negligible in nuclear and particle interactions and because neutrinos have zero electric charge, neutrinos have the unique property that they interact almost completely via the weak nuclear force. Consequently, neutrinos can be used as sensitive probes of the weak force. As such, neutrino beams at particle accelerators have been employed to study charge-changing (charged current) and charge-preserving (neutral current) weak interactions. However, the extreme weakness (compared to the electromagnetic and strong forces) and short range (of the order of 10^{-18} m) of the weak interaction have made determination of many neutrino properties extremely difficult.

Currently, three distinct flavors (or types) of neutrinos are known to exist: the electron neutrino (ν_e), the muon neutrino (ν_μ), and the tau neutrino (ν_τ). Each neutrino flavor is associated with a corresponding charged lepton, the electron (e), muon (μ), and tau (τ) particle. The electron, muon, and tau neutrinos (or their antiparticles) have been observed in experiments. Based on the observation of interactions involving neutrinos, the lepton flavor families, which comprise the charged and neutral leptons and their antiparticles (e^-, ν_e, e^+, $\bar{\nu}_e$; μ^-, ν_μ, μ^+, $\bar{\nu}_\mu$; τ^-, ν_τ, τ^+, $\bar{\nu}_\tau$), obey laws of conservation of lepton number. These empirical laws state that the number of leptons minus antileptons does not change, both within a flavor family and overall. *See* ELECTRON; SYMMETRY LAWS (PHYSICS).

The existence of neutrino oscillations (a phenomenon whereby neutrinos change their flavors during the flight from a neutrino source to a detector), seen clearly in observations of atmospheric and solar neutrinos, shows that neutrinos have tiny finite masses which are many orders of magnitude smaller than the masses of their charged lepton counterparts, and also shows that the physical neutrinos do not have pure mass states (quantum-mechanical states) but contain mixtures of two or more neutrino mass states. This mixing indicates that the empirical laws of lepton number conservation are not exact and that they are violated in some physical processes. It is not known whether neutrinos have magnetic or electric dipole moments. Yoichiro Suzuki

Neutrino astronomy The detection and study of neutrinos to learn about astronomical objects and the universe. These neutral, weakly interacting particles come almost without any disruption straight from their sources, traveling at very close to the speed of light. A low-energy neutrino in flight would not notice a barrier of lead 50 light-years thick. Neutrino light would provide a wondrous new view of the universe.

Neutrinos in the universe. Neutrinos were made in huge numbers at the time of the big bang. Like the cosmic background radiation, they now possess little kinetic energy as a result of the expansion of the universe. The problem with observing these relic neutrinos is that probability of a neutrino interacting within a detector decreases with the square of the neutrino's energy, for low energies. Nobody has been able to detect these lowest-energy neutrinos, and prospects are not good for doing so. *See* BIG BANG THEORY; COSMIC BACKGROUND RADIATION.

Stellar neutrinos. Neutrinos also originate in the nuclear fusion in stars. The Sun close by produces a huge flux of neutrinos, which have been detected in five experiments. However, observations of stellar neutrinos are limited to the Sun. Just as the sky is dark at night despite all the stars, the Sun far outshines all the rest of the cosmos in numbers of detectable neutrinos. *See* SOLAR NEUTRINOS.

Supernovae. On February 23, 1987, two detectors in deep mines in the United States (the IMB experiment) and Japan (the Kamiokande experiment) recorded a total of 19 neutrino interactions over a span of 13 seconds. Two and a half hours later, astronomers in the Southern Hemisphere saw the first supernova to be visible with the unaided eye since 1604. Many deductions followed about the nature of neutrinos, such as limits on mass, charge, gravitational attraction, and magnetic moment.

Supernovae of the gravitational-collapse type occur when elderly massive stars run out of nuclear fusion energy and can no longer resist the force of gravity. The neutrinos carry off most of the in-fall energy. Much can be learned from the final stages of stellar evolution, not only about the process of stellar collapse to a neutron star or black hole (the latter if the progenitor is very massive) but also about properties of neutrinos. Four underground detectors have significant capability for supernova detection from the Milky Way Galaxy. From historical records and from observations of distant spiral galaxies, the rate of supernovae in the Milky Way Galaxy is expected to be between one and five per century. Thus experimentalists may have to wait a long time before the next observation, and there is no way of predicting when it will occur. *See* SUPERNOVA.

High-energy cosmic neutrinos. Higher-energy neutrinos must be made in many of the most luminous and energetic objects in the universe, such as active galactic nuclei and gamma-ray bursters. Two things make prospects brighter in the near future for higher-energy neutrino astronomy than for lower energies: (1) the interaction probability for neutrinos goes up with energy, and (2) the consequences of neutrino interaction with a target (Earth or detector) become more detectable as the energy release is greater. The favored method is to detect muons produced by neutrinos. These charged particles produce Cerenkov radiation, a short flash of light detectable at tens of meters distance by photomultipliers in clear water or ice. *See* CERENKOV RADIATION; PHOTOMULTIPLIER.

View of the ANTARES Project from the ocean bottom. The optical detectors consist of modules of a cluster of photomultipliers and electronics, spaced along vertical buoy strings. Spherical floats at the top of each detector string keep the string close to vertical; anchors and releases are at the bottom. Fiber-optic cables go from each string to a junction box, which is serviced by the submarine at the far side of the array. A cable descends the slope from shore in the background. Scales are exaggerated. (*ANTARES Collaboration*)

High-energy neutrino telescopes. Neutrino detectors must be placed deep underground or underwater to escape the backgrounds caused by the rain of cosmic rays upon the atmosphere. The lead project, DUMAND (Deep Underwater Muon and Neutrino Detector), was canceled in 1995, but made great headway in pioneering techniques, studying backgrounds, exploring detector designs, and stimulating interest in astrophysical neutrinos.

Two projects similar to DUMAND are under way in the Mediterranean, the more developed NESTOR (Neutrino Experimental Submarine Telescope with Oceanographic Research) Project, and the ANTARES (Astronomy with a Neutrino Telescope and Abyss Environmental Research) Project (see illustration). These projects employ basically the same method of bottom-anchored cables, with photomultipliers protected in spherical glass pressure housings, as developed for DUMAND. A different type of neutrino telescope, the AMANDA (Antarctic Muon and Neutrino Detector Array) Project, is under construction in ice at the South Pole. John G. Learned

Neutron An elementary particle having approximately the same mass as the proton, but lacking a net electric charge. It is indispensable in the structure of the elements, and in the free state it is an important reactant in nuclear research and the propagating agent of fission chain reactions. Neutrons, in the form of highly condensed matter, constitute the substance of neutron stars. *See* NEUTRON STAR.

Neutrons and protons are the constituents of atomic nuclei. The number of protons in the nucleus determines the chemical nature of an atom, but without neutrons it would be impossible for two or more protons to exist stably together within nuclear dimensions, which are of the order of 10^{-13} cm. The protons, being positively charged, repel one another by virtue of their electrostatic interactions. The presence of neutrons weakens the electrostatic repulsion, without weakening the nuclear forces of cohesion. In light nuclei the resulting balanced, stable configurations contain protons and neutrons in almost equal numbers, but in heavier elements the neutrons outnumber the protons; in ^{238}U, for example, 146 neutrons are joined with 92 protons. Only one nucleus, ^{1}H, contains no neutrons. For a given number of protons, neutrons in several different numbers within a restricted range often yield nuclear stability—and hence the isotopes of an element. *See* ISOTOPE; NUCLEAR STRUCTURE; PROTON.

Free neutrons have to be generated from nuclei, and since they are bound therein by cohesive forces, an amount of energy equal to the binding energy must be expended to get them out. Nuclear machines, such as cyclotrons and electrostatic generators, induce many nuclear reactions when their ion beams strike target material. Some of these reactions release neutrons, and these machines are sources of high neutron flux. Neutrons are released in the act of fission, and nuclear reactors are unexcelled as intense neutron sources. *See* NUCLEAR BINDING ENERGY; NUCLEAR FISSION.

Neutrons occur in cosmic rays, being liberated from atomic nuclei in the atmosphere by collisions of the high-energy primary or secondary charged particles. They do not themselves come from outer space. *See* COSMIC RAYS; DELAYED NEUTRON.

Having no electric charge, neutrons interact so slightly with atomic electrons in matter that energy loss by ionization and atomic excitation is essentially absent. Consequently they are vastly more penetrating than charged particles of the same energy. The main energy-loss mechanism occurs when they strike nuclei. The most efficient slowing-down occurs when the bodies that are struck in an elastic collision have the same mass as the moving bodies; hence the most efficient neutron moderator is hydrogen, followed by other light elements: deuterium, beryllium, and carbon. The great penetrating power of neutrons imposes severe shielding problems for reactors and other nuclear machines, and it is necessary to provide walls, usually of concrete, several feet in thickness to protect personnel. The currently accepted health tolerance levels for an 8-h day correspond for fast neutrons to a flux of 20 neutrons/(cm^2)(s) or 130 neutrons/(in.^{2})(s); for slow neutrons, 700/(cm^2)(s) or 4500/(in.^{2})(s). On the other hand, fast neutrons are useful in some kinds of cancer therapy. *See* RADIATION DAMAGE TO MATERIALS; RADIATION INJURY (BIOLOGY); RADIATION SHIELDING; RADIOLOGY.

Free neutrons are radioactive, each transforming spontaneously into a proton, an electron (β^- particle), and an antineutrino. This instability is a reflection of the fact that neutrons are slightly heavier than hydrogen atoms. The neutron's rest mass is 1.0086649 atomic mass units on the unified mass scale (1.67493 $\times$ 10^{-24} g), as compared with 1.0078250 atomic mass units for the hydrogen atom.

Neutrons are, individually, small magnets. This property permits the production of beams of polarized neutrons, that is, beams of neutrons whose magnetic dipoles are aligned predominantly parallel to one direction in space. The magnetic moment is -1.913042 nuclear magnetons. *See* MAGNETON; NUCLEAR MOMENTS; NUCLEAR ORIENTATION; SPIN (QUANTUM MECHANICS).

Despite its overall neutrality, the neutron does have an internal distribution of electric charge, as has been revealed by scattering experiments. On a still finer scale, the neutron can also be presumed to have a quark structure in analogy of that of the proton. *See* QUANTUM CHROMODYNAMICS; QUARKS.

When neutrons are completely slowed down in matter, they have a maxwellian distribution in energy that corresponds to the temperature of the moderator with which they are in equilibrium. The de Broglie wavelength of these ultracold neutrons is greater than 50 nm, which is so much larger than interatomic distances in solids that they interact with regions of a surface rather than with individual atoms, and as a result they are reflected from polished surfaces at all angles of incidence. Ultracold neutrons are important in basic physics and have applications in studies of surfaces and of the structure of inhomogeneities and magnetic domains in solids. *See* ELEMENTARY PARTICLE; NEUTRON DIFFRACTION; THERMAL NEUTRONS.

Arthur H. Snell

Neutron diffraction

Neutron diffraction The phenomenon associated with the interference processes which occur when neutrons are scattered by the atoms within solids, liquids, and gases. The use of neutron diffraction as an experimental technique is relatively new compared to electron and x-ray diffraction, since successful application requires high thermal-neutron fluxes, which can be obtained only from nuclear reactors. These diffraction investigations are possible because thermal neutrons have energies with equivalent wavelengths near 0.1 nanometer and are therefore ideally suited for interatomic interference studies.

In the scattering of neutrons by atoms, there are two important interactions. One is the short-range, nuclear interaction of the neutron with the atomic nucleus. This interaction produces isotropic scattering because the atomic nucleus is essentially a point scatterer relative to the wavelengths of thermal neutrons. Strong resonances associated with the scattering process prevent any regular variation of the nuclear scattering amplitudes with atomic number. The other important process for the scattering of neutrons by atoms is the interaction of the magnetic moment of the neutron with the spin and orbital magnetic moments of the atom. *See* SCATTERING EXPERIMENTS (ATOMS AND MOLECULES); SCATTERING EXPERIMENTS (NUCLEI).

Since the nuclear scattering amplitudes for neutrons do not vary uniformly with atomic number, there are certain types of chemical structures which can be investigated more readily by neutron diffraction than by x-ray diffraction. Moreover, since neutron scattering is a nuclear process, when the scattering amplitude of an element is not favorable for a particular investigation, it is frequently possible to substitute an enriched isotope which has scattering characteristics that are markedly different. The most significant application of neutron diffraction in chemical crystallography is the structure determination of composite crystals which contain both heavy and light atoms, and the most important compounds in this general classification are the hydrogen-containing substances.

The interaction of the magnetic moment of the neutron with the orbital and spin moments in magnetic atoms makes neutron scattering a unique tool for the study of a wide variety of magnetic phenomena, because information is obtained on the magnetic properties of the individual atoms in a material. This interaction depends on the size of the atomic magnetic moment and also on the relative orientation of the neutron spin and of the atomic magnetic moment with respect to the scattering vector and with respect to each other. Consequently, detailed information can be obtained on both the magnitude and orientation of magnetic moments in any substance which displays magnetic properties.

The investigation of antiferromagnetic and ferrimagnetic substances is one of the most important applications of the neutron diffraction technique, because detailed information on the magnetic configuration in these systems cannot be obtained by other methods.

One of the most important uses of inelastic neutron scattering is the study of thermal vibrations of atoms about their equilibrium positions, because lattice vibration quanta, or phonons, can be excited or annihilated in their interactions with low-energy neutrons. The measurements provide a direct determination of the dispersion relations for the normal vibrational modes of the crystal and do not require the large corrections necessary in similar x-ray investigations. These measured dispersion relations furnish the best experimental information available on interatomic forces that exist in crystals. *See* ELECTRON DIFFRACTION; MAGNON; NEUTRON SPECTROMETRY.

Michael K. Wilkinson

Neutron optics

Neutron optics The general class of experiments designed to emphasize the wavelike character of neutrons. Like all elementary particles, neutrons can be made to display wavelike, as well as particlelike, behavior. They can be reflected and refracted, and they can scatter, diffract, and interfere, like light or any other type of wave. Many classical optical effects, such as Fresnel diffraction, have been performed with neutrons, including even those involving the construction of Fresnel zone plates. *See* DIFFRACTION; INTERFERENCE OF WAVES; REFLECTION OF ELECTROMAGNETIC RADIATION; REFRACTION OF WAVES; SCATTERING OF ELECTROMAGNETIC RADIATION; WAVE (PHYSICS).

The typical energy of a neutron produced by a moderated nuclear reactor is about 0.02 eV, which is approximately equal to the kinetic energy of a particle at about room temperature (80°F or 300 K), and which corresponds to a wavelength of about 10^{-10} m. This is also the typical spacing of atoms in a crystal, so that solids form natural diffraction gratings for the scattering of neutrons, and much information about crystal structure can be obtained in this way. However, the wavelike properties of neutrons have been confirmed over a vast energy range from 10^{-7} eV to over 100 MeV. *See* NEUTRON DIFFRACTION.

Neutrons, being uncharged, can be made to interfere over large spatial distances, since they are relatively unaffected by the stray fields in the laboratory that deflect charged particles. This property has been exploited by using the neutron interferometer. This device is made possible by the ability to grow essentially perfect crystals of up to 4 in. (10 cm). The typical interferometer is made from a single perfect crystal cut so that three parallel "ears" are presented to the neutron beam. This allows the incident beam to be split and subsequently recombined coherently. *See* COHERENCE; INTERFEROMETRY; SINGLE CRYSTAL.

One of the most significant experiments performed with the interferometer involved rotating the interferometer about the incident beam so that one neutron path was higher than the other, creating a minute gravitational potential difference (of 10^{-9} eV) between the paths. This was sufficient to cause a path difference of 20 or so wavelengths between the beams. This remains the only type of experiment that has ever seen a quantum-mechanical interference effect due to gravity. It also verifies the extension of the equivalence principle to quantum theory (although in a form more subtle than its classical counterpart). *See* GRAVITATION; RELATIVITY.

Many noninterferometer experiments have also been done with neutrons. In one experiment, resonances were produced in transmitting ultracold neutrons (energy about 10^{-7} eV) through several sheets of material. This is theoretically similar to seeing the few lowest states in a square-well potential in the Schrödinger equation. *See* NEUTRON; QUANTUM MECHANICS.

Daniel M. Greenberger

Neutron spectrometry A generic term applied to experiments in which neutrons are used as the probe for measuring excited states of nuclides and for determining the properties of these states. The term neutron spectroscopy is also used. The strength of the interaction between a neutron and a target nuclide can vary rapidly as a function of the energy of the incident neutron, and it is different for every nuclide. At particular neutron energies the interaction strength for a specific nuclide can be very strong; these narrow energy regions of strong interactions are called resonances (see illustration). The strength of the interaction, expressing the probability that an interaction of a given kind will take place, can be considered as the effective cross-sectional area σ presented by a nucleus to an incident neutron.

Neutron spectroscopy can be carried out by two different techniques (or a combination): (1) by the use of a time-pulsed neutron source which emits neutrons of many energies simultaneously, combined with the time-of-flight technique to measure the velocities of the neutrons; this time-of-flight technique can be used for neutron measurements from 10^{-3} eV to about 200 MeV; (2) by the use of a beam of nearly monoenergetic neutrons whose energy can be varied in small steps approximately equal to the energy spread of the neutron beam; however, useful "monoenergetic" neutron sources are not available from about 10 eV to about 10 keV.

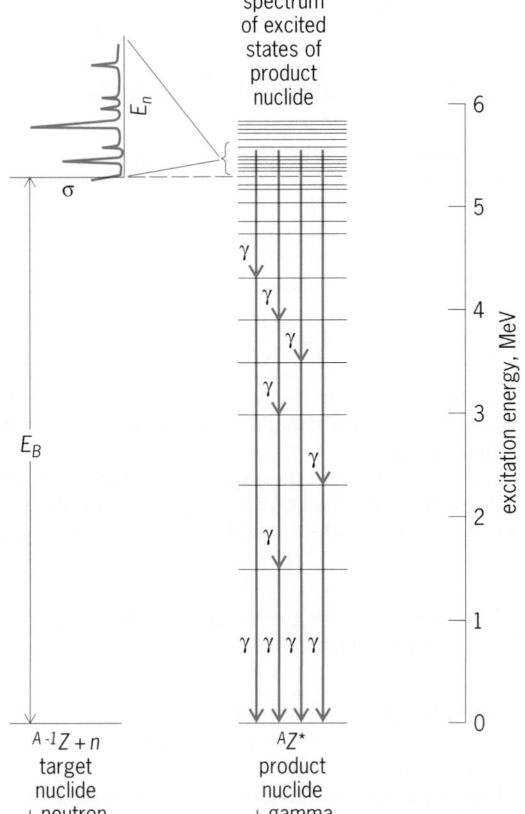

Energy-level diagram for the product nucleus $^AZ^*$ with mass number *A* and charge number *Z*. The asterisk emphasizes that the product nucleus is in an excited state, from which it returns to ground state by emitting gamma (γ) rays. Excitation energy is the sum of the energy of the neutron E_n, and the binding energy E_B of the neutron which has been added to the target nuclide.

Neutron spectroscopy has yielded a mass of valuable information on nuclear systematics for almost all nuclides. The distribution of the spacings between nuclear levels and the average of these spacings have provided valuable tests for various nuclear theories. The properties of these levels, that is, the probabilities that they decay by neutron or gamma-ray emission, or by fission, and the averages and distribution of these probabilities have stimulated much theoretical effort.

In addition, knowledge of neutron cross sections is fundamental for the optimum design of thermal fission power reactors and fast neutron breeder reactors, as well as fusion power reactors now in the conceptual stage. Cross sections are needed for nuclear fuel materials such as ^{235}U or ^{239}Pu, for fertile materials such as ^{238}U, for structural materials such as iron and chromium, for coolants such as sodium, for moderators such as beryllium, and for shielding materials such as concrete. *See* NUCLEAR STRUCTURE; REACTOR PHYSICS.

John A. Harvey

Neutron star A star containing about $1\frac{1}{2}$ solar masses of material compressed into a volume approximately 6 mi (10 km) in radius. (1 solar mass equals 4.4×10^{33} lbm or 2.0×10^{33} kg.) Neutron stars are one of the end points of stellar evolution and are the final states of stars that begin their lives with considerably more mass than the Sun. The density of neutron star material is 10^{14} to 10^{15} times the density of water and exceeds the density of matter in the nuclei of atoms. Neutron stars are pulsars (pulsating radio sources) if they rotate sufficiently rapidly and have strong enough magnetic fields. *See* PULSAR; STELLAR EVOLUTION.

Neutron stars play a role in astrophysics which extends beyond their status as strange, unusual types of stellar bodies. The interior of a neutron star is a cosmic laboratory in which matter is compressed to densities which are found nowhere else in the universe. Precise measurements of the rotation of neutron stars can probe the behavior of matter at such densities. Neutron stars in double-star systems can emit x-rays when matter flows toward the neutron star, swirls around it, and heats up. Neutron stars are almost certainly formed in supernova explosions. A few pulsars are found in double-star systems, and careful timing of the pulses they emit can test Einstein's general theory of relativity. *See* BINARY STAR; GRAVITATION; RELATIVITY; SUPERNOVA.

Measured values of masses of neutron stars in double star systems range from 1.4 to 1.8 solar masses. The highest mass that a neutron star can have is about 2 solar masses. More massive neutron stars might possibly exist if current ideas about the behavior of neutron star matter turn out to be wrong. Objects that end their lives with masses higher than this neutron star mass limit will become black holes. *See* BLACK HOLE.

Most of the interior of a neutron star consists of matter which is almost entirely composed of neutrons. In the bulk of the star, this matter is in a superfluid state, where circulation currents can flow without resistance. This material is under pressure, since it must be able to support the tremendous weight of the overlying layers at each point in the neutron star. This pressure, called degeneracy pressure, is caused by the close packing of the neutrons rather than by the motion of the particles. As a result, neutron stars can be stable no matter what the internal temperature is, because the pressure that supports the star is independent of temperature. *See* SUPERFLUIDITY.

Harry L. Shipman

New Zealand A landmass in the Southern Hemisphere, bounded by the South Pacific Ocean to the north, east, and south and the Tasman Sea to the west, with a total land area of 103,883 mi^2 (269,057 km^2). The exposed landmass represents about one-quarter of a subcontinent, with three-quarters submerged. This long, narrow, mountainous country, oriented northeast to southwest, consists of two main islands, North Island and South Island, surrounded by a much greater area of crust submerged to depths reaching 1.2 mi (2 km).

South Island lowlands are either alluvial plains as in Otago, Southland, and Nelson, or glacial outwash fans as in Westland and Canterbury. North Island lowlands such as Hawke's Bay, Wairarapa, and Manawatu are alluvial; the Waikato, Hauraki, and Bay of Plenty lowlands occupy structural basins that contain large volumes of reworked volcanic debris from the central volcanic region. The alluvial lowlands of both main islands form the most agriculturally productive areas of the country. *See* PLAINS.

The climate of New Zealand is influenced by three main factors: a location in latitudes where the prevailing airflow is westerly; an oceanic environment; and the mountain chains, which modify the weather systems as they pass eastward, causing high rainfalls on windward slopes and sheltering effects to leeward.

Weather is determined mostly by series of anticyclones and troughs of low pressure that produce alternating periods of settled and variable conditions. Westerly air masses are occasionally replaced by southerly airstreams, which bring cold conditions with snow in winter and spring to areas south of 39°S, and northerly tropical maritime air, which brings warm humid weather to the north and east coasts. *See* METEOROLOGY.

Rainfall on land is 16–470 in. (400–12,000 mm) per year, with the highest rainfall being on the western windward slopes of the mountains, and the lowest on the eastern basins in the lee of the Southern Alps in Central Otago and south Canterbury. Annual rain days are at least 130 for most of North Island, but on South Island the totals are far more variable, with over 200 occurring in Fiordland, 180 on the west coast, and fewer than 80 in Central Otago. Summer droughts are relatively common in Northland, and in eastern regions of both islands. *See* DROUGHT; PRECIPITATION (METEOROLOGY).

Droughts, springtime air frosts, and hailstorms are the major common climatic hazards for the farming industry, but floods associated with prolonged intense rainstorms are the major general hazard.

The economy is heavily dependent on the natural resources soil, water, and plants. New Zealand has few exploitable minerals, but possesses a climate generally favorable for agriculture, pastoral farming, renewable forestry, and tourism. With a small population (3.4 million), much of its manufacturing is concerned with processing produce from the land and surrounding seas, and supplying the needs of those industries.

Because of its high relief and its location on an active crustal plate boundary in the zone of convergence between Antarctic air masses and tropical air masses, New Zealand is prone to high-intensity and high-frequency natural hazards—earthquakes, volcanic eruptions, large and small landslides, and floods. M. J. Selby

Newcastle disease A viral infection that affects the digestive, intestinal, and respiratory tracts and the neurological system of birds. The causative agent is an enveloped ribonucleic acid (RNA) virus that is classified as a paramyxovirus. *See* PARAMYXOVIRUS.

Newcastle disease occurs in five forms based on a virulence in chickens ranging from inapparent infection to severe disease and death. Viscerotropic-velogenic Newcastle disease causes a very severe infection, producing hemorrhagic lesions in the intestinal tract and high mortality. The neurotropic-velogenic type is also highly lethal and produces neurologic and respiratory signs in infected birds. The mesogenic form causes an acute respiratory or neurologic infection that may be lethal only in young birds. The lentogenic type is a mild or inapparent respiratory infection of chickens. The last group includes the viruses causing inapparent or asymptomatic infections of the digestive tract.

The wide susceptibility of avian species to infection with Newcastle disease has complicated control. Newcastle disease is spread worldwide by the international transportation of live birds disseminating the virus. Control of and protection from Newcastle disease can be achieved by the correct use of vaccines. Lentogenic and some mesogenic strains are used to produce vaccines that can be administered by aerosol, intranasal drops, or intramuscular injection, or as an additive to the drinking water. *See* ANIMAL VIRUS. Mary Lynne Vickers

Newtonian fluid A fluid whose stress at each point is linearly proportional to its strain rate at that point. The concept was first deduced by Isaac Newton and is directly analogous to Hooke's law for a solid. All gases are newtonian, as are most common liquids such as water, hydrocarbons, and oils. *See* HOOKE'S LAW; STRESS AND STRAIN.

A simple example, often used for measuring fluid deformation properties, is the steady one-dimensional flow $u(y)$ between a fixed and a moving wall (see illustration). The no-slip condition

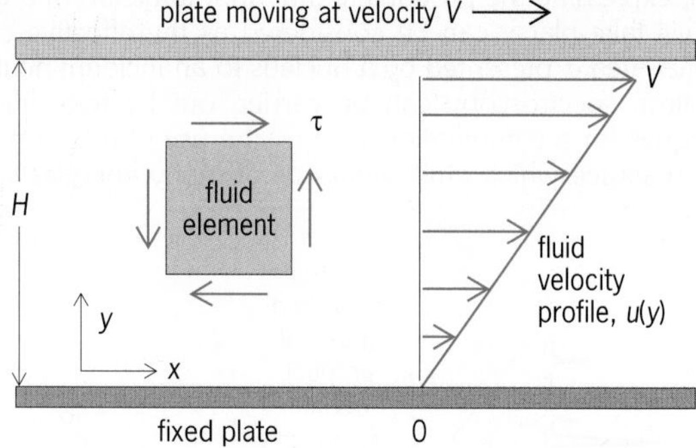

A fluid sheared between two plates. The resulting strain rate equals *V*/*H*.

at each wall forces the fluid into a uniform shear strain rate ϵ, given by Eq. (1), which is induced by a uniform shear stress τ.

$$\varepsilon = \frac{\partial u}{\partial y} = \frac{V}{H} \quad (1)$$

Here V is the speed of the moving wall, H is the perpendicular distance between the walls, and u is the fluid velocity at distance y from the fixed wall.

If the fluid is newtonian, the experimental plot of τ versus ε will be a straight line. The constant of proportionality is called the viscosity μ of the fluid, as stated in Eq. (2). The viscosity co-

$$\tau = \mu\varepsilon \quad (2)$$

efficients of common fluids vary by several orders of magnitude. *See* FLUID MECHANICS; FLUIDS; VISCOSITY. Frank M. White

Newton's laws of motion Three fundamental principles which form the basis of classical, or newtonian, mechanics. They are stated as follows:

First law: A particle not subjected to external forces remains at rest or moves with constant speed in a straight line.

Second law: The acceleration of a particle is directly proportional to the resultant external force acting on the particle and is inversely proportional to the mass of the particle.

Third law: If two particles interact, the force exerted by the first particle on the second particle (called the action force) is equal in magnitude and opposite in direction to the force exerted by the second particle on the first particle (called the reaction force).

The newtonian laws have proved valid for all mechanical problems not involving speeds comparable with the speed of light and not involving atomic or subatomic particles. *See* DYNAMICS; FORCE.

Dudley Williams

Niacin A vitamin also known as nicotinic acid, a component of the vitamin B complex. It is a white water-soluble powder stable to heat, acid, and alkali, with the structure shown below.

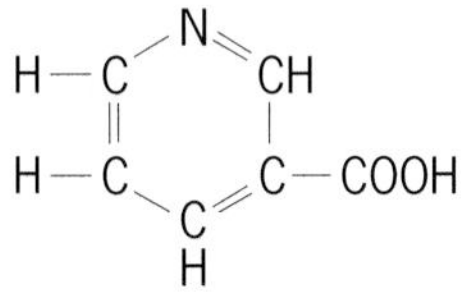

```
          N
       ⁄     ⸌
H—C           CH
   ‖           |
H—C           C—COOH
     ⸌      ⁄
        C
        |
        H
```

Many animals, including humans, are capable of synthesizing niacin in varying degrees from the amino acid tryptophan. Niacin is widely distributed in foods. Yeasts, wheat germ, and meats, particularly organ meats, are rich sources of the vitamin. Some foods such as milk are relatively poor sources of niacin but contain generous quantities of tryptophan.

Niacin-deficiency disease is known as pellagra and is particularly prevalent among the poor people whose diet is largely corn. The recommended dietary allowance for niacin is 6.6 niacin equivalents per 1000 kcal. *See* VITAMIN.

Stanley N. Gershoff

Nickel A chemical element, Ni, atomic number 28, a silver-white, ductile, malleable, tough metal. The atomic mass of naturally occurring nickel is 58.693. *See* PERIODIC TABLE.

Nickel consists of five natural isotopes having atomic masses of 58, 60, 61, 62, 64. Seven radioactive isotopes have also been identified, having mass numbers of 56, 57, 59, 63, 65, 66, and 67.

Most commercial nickel goes into stainless steel and other corrosion-resistant alloys. Nickel is also important in coins as a replacement for silver. Finely divided nickel is used as a hydrogenation catalyst. *See* NICKEL ALLOYS.

Nickel is a fairly plentiful element, making up about 0.008% of the Earth's crust and 0.01% of the igneous rocks. Appreciable quantities of nickel are present in some kinds of meteorite, and large quantities are thought to exist in the Earth's core. Two important ores are the iron-nickel sulfides, pentlandite and pyrrhotite $(Ni,Fe)_xS_y$; the ore garnierite, $(Ni,Mg)SiO_3 \cdot nH_2O$, is also commercially important. Nickel occurs in small quantities in plants and animals. It is present in trace amounts in sea water, petroleum, and most coal.

Nickel metal is of moderate strength and hardness (3.8 on Mohs scale). When viewed as very small particles, nickel appears black. The density of nickel is 8.90 times that of water at 20°C (68°F). Nickel melts at 1455°C (2651°F) and boils at 2840°C (5144°F). Nickel is only moderately reactive. It resists alkaline corrosion and does not burn in the massive state, although fine nickel wires can be ignited. Nickel is above hydrogen in the electrochemical series, and it dissolves slowly in dilute acids, releasing hydrogen. In metallic form nickel is a moderately strong reducing agent.

Nickel is usually dipositive in its compounds, but it can also exist in the oxidation states 0, 1+, 3+, and 4+. Besides the simple nickel compounds, or salts, nickel forms a variety of coordination compounds or complexes. Most compounds of nickel are green or blue because of hydration or other ligand bonding to the metal. The nickel ion present in water solutions of simple nickel compounds is itself a complex, $[Ni(H_2O)_6]^{2+}$.

William E. Cooley

Nickel alloys Combinations of nickel with other metals. Nickel-base alloys may be melted in open-hearth, electric-arc, or induction furnaces in air, under inert gas, or in vacuum. Casting may also be done under these same ambient conditions.

Nickel 211 and Duranickel alloy 301 are essentially binary alloys with 4.75% manganese and 4.5% aluminum, respectively. A characteristic use of nickel 211 is as wire for sparkplug electrodes. Duranickel alloy 301 is well suited to the manufacture of springs and diaphragms.

Monel alloy 400 contains about two-thirds nickel and one-third copper and is the oldest of the commercial nickel-base alloys, dating from about 1905. Nickel-chromium binary alloys are used primarily in specialty high-temperature service. Other nickel alloys contain combinations of other metals in varying quantities. *See* NICKEL.

E. N. Skinner; Gaylord Smith

Nickel metallurgy The extraction and refining of nickel from its ores. Nickel's properties of strength, toughness, and resistance to corrosion have been used to advantage in alloys since ancient times. Although nickel ranks twenty-fourth in order of abundance of the elements, there are relatively few nickel deposits of commercial importance. Nickel ores are of two generic types, sulfides and laterites. Explorations of the ocean bottoms have revealed vast deposits of manganese oxide nodules which contain significant values of nickel, copper, and cobalt.

Selection of processes for nickel extraction is largely determined by the type of ore to be treated. Sulfide ores are amenable to concentration by such methods as flotation or magnetic separation. The state of combination of nickel in the lateritic ores usually precludes such enrichment, thus requiring treatment of the total ore.

Sulfide ores are first crushed and ground to liberate the mineral values and then subjected to froth flotation or magnetic separation to concentrate the valuable constituents and reject the gangue or rock fraction. The nickel concentrate is treated by pyrometallurgical processes. The major portion undergoes partial roasting in multihearth or fluidized-bed furnaces to eliminate about half of the sulfur and to oxidize the associated iron. The hot calcine, plus flux, is smelted in natural gas-coal-fired reverberatory furnaces operating at about 1200°C to produce a furnace matte, enriched in nickel, and a slag for discard. The furnace matte is transferred to converters and blown with air in the presence of more flux to oxidize the remaining iron and associated sulfur, yielding Bessemer matte containing nickel, copper, cobalt, small amounts of precious metals, and about 22% sulfur. The molten Bessemer matte is cast into 25-ton (22.5-metric ton) molds in which it undergoes controlled slow cooling. After crushing and grinding, the metallics are removed magnetically and treated in a refining complex for recovery of metal values.

The bulk of the nickel originating from lateritic ores is marketed as ferronickel. The process employed is basically simple and involves drying and preheating the ore usually under reducing conditions. The hot charge is then further reduced and melted in an electric-arc furnace, and the crude metal is refined and cast into ferronickel pigs. A substantial amount of nickel is produced from lateritic ores by the nickel sulfide matte technique. In this process the ore is mixed with gypsum or other sulfur-containing material such as high-sulfur fuel oil, followed by a reduction and smelting operation to form matte. The molten furnace matte is upgraded in either conventional or top-blown rotary converters to a high-grade matte, which can be further refined by roasting and reduction to a metallized product. *See* NICKEL; PYROMETALLURGY.

Alexander Illis

Nicotinamide adenine dinucleotide (NAD) An organic coenzyme and one of the most important components of the enzymatic systems concerned with biological oxidation-reduction reactions. It is also known as NAD, diphosphopyridine nucleotide (DPN), coenzyme I, and codehydrogenase I. NAD is found in the tissues of all living organisms. *See* COENZYME.

The nicotinamide, or pyridine, portion of NAD can be reduced chemically or enzymatically with the formation of reduced or hydrogenated NAD (NADH). NAD functions as the immediate oxidizing agent for the oxidation, or dehydrogenation, of various organic compounds in the presence of appropriate dehydrogenases, which are specific apoenzymes, or protein portions of the enzyme. In the dehydrogenase reactions one hydrogen atom is transferred from the substrate to NAD, while another is liberated as hydrogen ion.

NAD and its reduced form, NADH, serve to couple oxidative and reductive processes and are constantly regenerated during metabolism. Hence, they serve as catalysts and NAD is referred to as a coenzyme. In some enzymatic reactions a different coenzyme, triphosphopyridine nucleotide, is required. Dehydrogenases are generally quite specific with respect to the coenzyme which they can utilize. *See* ENZYME; NICOTINAMIDE ADENINE DINUCLEOTIDE PHOSPHATE (NADP). Michael Doudoroff

Nicotinamide adenine dinucleotide phosphate (NADP) A coenzyme and an important component of the enzymatic systems concerned with biological oxidation-reduction systems. It is also known as NADP, triphosphopyridine nucleotide (TPN), coenzyme II, and codehydrogenase II. The compound is similar in structure and function to nicotinamide adenine dinucleotide (NAD). It differs structurally from NAD in having an additional phosphoric acid group esterified at the 2′ position of the ribose moiety of the adenylic acid portion. In biological oxidation-reduction reactions the NADP molecule becomes alternately reduced to its hydrogenated form (NADPH) and reoxidized to its initial state. *See* CARBOHYDRATE METABOLISM; COENZYME; ENZYME. Michael Doudoroff

Nicotine alkaloids Alkaloids found in various species of the genus *Nicotiana*. The species most often used for the production of tobacco because of its high level of nicotine is

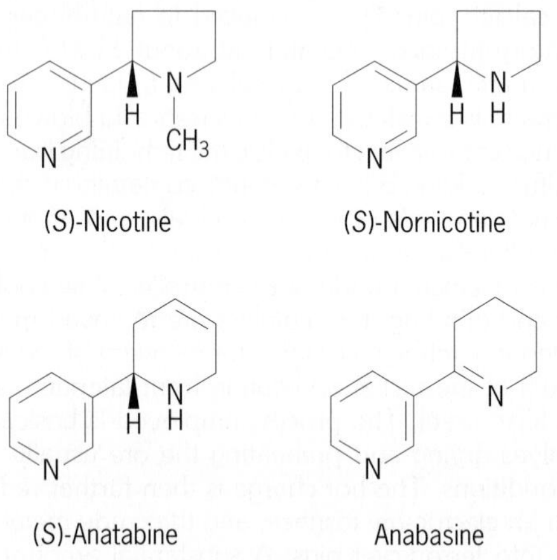

(*S*)-Nicotine

(*S*)-Nornicotine

(*S*)-Anatabine

Anabasine

Structures of nicotine and some other alkaloids found in tobacco.

N. tabacum, which is cultivated in many parts of the world for the preparation of cigarettes, cigars, and pipe tobacco. Nicotine is the most abundant alkaloid in *N. tabacum*, occurring to the extent of 2–8% based on the dry weight of the cured leaf. Other alkaloids that are found in this species are nornicotine, anabasine, and anatabine. The chemical structures of these alkaloids are shown in the illustration. *See* ALKALOID; TOBACCO. Edward Leete

Niobium A chemical element, Nb, atomic number 41 and atomic weight 92.906. In the United States this element was originally called columbium. The metallurgists and metals industry still use this older name. *See* PERIODIC TABLE.

Most niobium is used in special stainless steels, high-temperature alloys, and superconducting alloys such as Nb_3Sn. Niobium is also used in nuclear piles.

Niobium metal has a density of 8.6 g/cm^3 (5.0 oz/in.^{3}) at 20°C (68°F), a melting point of 2468°C (4474°F), and a boiling point of 4927°C (8900°F). Metallic niobium is quite inert to all acids except hydrofluoric, presumably owing to an oxide film on the surface. Niobium metal is slowly oxidized in alkaline solution. It reacts with oxygen and the halogens upon heating to form the oxidation state V oxide and halides, with nitrogen to form NbN, and with carbon to form NbC, as well as other elements such as arsenic, antimony, tellurium, and selenium.

The oxide Nb_2O_5, melting point 1520°C (2768°F), dissolves in fused alkali to yield a soluble complex niobate, $Nb_6O_{19}{}^{8-}$. Normal niobates such as $NbO_4{}^{3-}$ are insoluble. The oxide dissolves in hydrofluoric acid to give ionic species such as $NbOF_5{}^{2-}$ and $NbOF_6{}^{3-}$, depending on the fluoride and hydrogen-ion concentration. The highest fluoro complex which can exist in solution is $NbF_6{}^{-}$. Edwin M. Larsen

Nippotaeniidea An order of tapeworms of the subclass Cestoda. The few known species are intestinal parasites of Eurasian fresh-water fishes. The head bears a single terminal sucker. The segmental anatomy shows relationships to the Pseudophyllidea and Cyclophyllidea. The life history is unknown. It is probable that this order is related to the proteocephalids. *See* CESTODA; CYCLOPHYLLIDEA; PSEUDOPHYLLIDEA. Clark P. Read

Niter A potassium nitrate mineral with chemical composition KNO_3. Niter crystallizes in the orthorhombic system, generally in thin crusts and delicate acicular crystals; it occurs in massive, granular, or earthy forms. It is brittle; hardness is 2 on Mohs scale; specific gravity is 2.109. The luster is vitreous, and the color and streak are colorless to white. *See* NITRATE MINERALS.

Niter is commonly found, usually in small amounts, as a surface efflorescence in arid regions and in caves and other sheltered places. Niter occurs associated with soda niter in the desert regions of northern Chile, and in similar occurrences in Italy, Egypt, Russia, the western United States, and elsewhere. George Switzer

Nitrate minerals These minerals are few in number and with the exception of soda niter are of rare occurrence. Normal anhydrous and hydrated nitrates occurring as minerals are soda niter, $NaNO_3$; niter, KNO_3: ammonia niter, NH_4NO_3; nitrobarite, $Ba(NO_3)_2$; nitrocalcite, $Ca(NO_3)_2 \cdot 4H_2O$; and nitromagnesite, $Mg(NO_3)_2 \cdot 6H_2O$. In addition there are three known naturally occurring nitrates containing hydroxyl or halogen, or compound nitrates. They are gerhardtite, $Cu_2(NO_3)(OH)_3$; buttgenbachite, $Cu_{19}(NO_3)_2Cl_4(OH)_{32}\dot{m}\,3H_2O$; and darapskite, $Na_3(NO_3)(SO_4) \cdot H_2O$. *See* NITER; SODA NITER.

The natural nitrates are for the most part readily soluble in water. For this reason they occur most abundantly in arid regions, particularly in South America along the Chilean coast. *See* FERTILIZER; NITROGEN. George Switzer

Nitration A process in which a nitro group ($—NO_2$) becomes chemically attached to a carbon, oxygen, or nitrogen atom in an organic compound. A hydrogen or halogen atom is often replaced by the nitro group. Three general reactions summarize nitration chemistry:

1. C nitration, in which the nitro group attaches itself to a carbon atom [reaction (1)].

$$\rangle C—H + HNO_3 \longrightarrow \rangle C—NO_2 + H_2O \qquad (1)$$

2. O nitration (an esterification reaction), in which an O-N bond is formed to produce a nitrate [reaction (2)].

$$\searrow\!\!-C-OH + HNO_3 \longrightarrow \;\searrow\!\!-C-O + NO_2 + H_2O \qquad (2)$$

3. N nitration, in which a N-N bond is formed [reaction (3)].

$$>NH + HNO_3 \longrightarrow\; >N-NO_2 + H_2O \qquad (3)$$

Aromatics, alcohols, glycols, and amines are generally nitrated with mixed acids via an ionic reactions. The mixture includes nitric acid, a strong acid such as sulfuric acid which acts as a catalyst, and a small amount of water. With sulfuric acid, nitric acid is ionized to the nitronium ion, NO_2^+, which is the nitrating agent.

Propane is commercially nitrated in relatively large amounts using nitric acid in gas-phase free-radical reactions at temperatures of about 380–420°C. Nitric acid decomposes at these temperatures to produce nitrogen dioxide radicals (actually a mixture of $\cdot NO_2$ and $\cdot ONO$) and a hydroxy radical ($\cdot OH$). In the free-radical reaction, about 35–40% of the nitric acid reacts to form four C_1-C_3 nitroparaffins; C-C bonds are broken during the nitration. The remaining nitric acid acts mainly as an oxidizing agent to form aldehydes, alcohols, carbon monoxide, carbon dioxide, water, and small amounts of other oxidized materials. Commercially, an adiabatic reactor is used, and the heat of reaction is employed to preheat and vaporize the nitric acid feed (containing water).

The product stream from free-radical nitrations is a condensation of mixed nitroparaffins. This liquid mixture is washed to remove the aldehydes, and then is distilled to recover each of the four nitroparafins—nitromethane, nitroethane, 1-nitropropane, and 2-nitropropane. The unreacted propane is recovered, combined with the feed propane, and returned to the reactor. The oxides of nitrogen are converted back to nitric acid; carbon monoxide, carbon dioxide, and water are discarded.

In the classical Victor Meyer process, an organic halide (often a bromide) is reacted with silver nitrite to produce a nitrohydrocarbon and silver halide. In a modified process, sodium nitrite, dissolved in a suitable solvent, is substituted for the more expensive silver nitrite. The desired nitroalkanes are produced in high yields by these processes, whereas they are produced in rather low yields in free-radical nitrations.

Nitrations can also often be performed by addition reactions using unsaturated hydrocarbons with nitric acid or nitrogen dioxide. Lyle F. Albright

Nitric acid A strong mineral acid having the formula HNO_3. Pure nitric acid is a colorless liquid with a specific gravity of 1.52 at 25°C (77°F); it freezes at −47°C (−53°F). Nitric acid is used in the manufacture of ammonium nitrate and phosphate fertilizers, nitro explosives, plastics, dyes, and lacquers. The principal commercial process for the manufacture of nitric acid is the Ostwald process, in which ammonia, NH_3, is catalytically oxidized with air to form nitrogen dioxide, NO_2. When the dioxide is dissolved in water, 60% nitric acid is formed. Production of 90–100% nitric acid is based on processes such as the reaction of sulfuric acid with sodium nitrate (an older method of nitric acid manufacture), dehydration of 60% acid, and oxidation of nitrogen dioxide in a solution of dilute nitric acid. *See* AMMONIA; NITROGEN. Francis J. Johnston

Nitric oxide An important messenger molecule in mammals and other animals. It can be toxic or beneficial, depending upon the amount and where in the body it is released. Initial research into the chemistry of nitric oxide (NO) was motivated by its production in automobile emissions and other combustion processes, which results in photochemical smog and acid rain. In the late 1980s, researchers in immunology, cardiovascular pharmacology, neurobiology, and toxicology discovered that nitric oxide is a crucial physiological messenger molecule. Nitric oxide is now thought to play a role in blood pressure regulation, control of blood clotting, immune defense, digestion, neuronal signaling, the senses of sight and smell, and possibly learning and memory. Underproduction or unregulated overproduction of nitric oxide may also contribute to disease processes such as diabetes, artherosclerosis, stroke, hypertension, carcinogenesis, multiple sclerosis, transplant rejection, damage associated with reperfusion in ischemic (oxygen-deprived) tissues, impotence, septic shock, and long-term depression. *See* IMMUNOLOGY; NEUROBIOLOGY.

Most cellular messengers are large, unreactive biomolecules that make specific contacts with their targets. In contrast, nitric oxide is a small molecule that can diffuse freely throughout biological tissues, reacting with target molecules at remote sites that are as far away as ~300 micrometers from its site of generation, a distance corresponding to 10–20 cell lengths. Nitric oxide is a free radical—that is, it contains an unpaired electron. However, unlike most free radicals, it is remarkably unreactive toward other compounds, and its chemistry is limited primarily to bond-forming combination with other radicals and coordination to metal centers. The chemical inertness of NO extends to the biological components of cells, where its direct reactions are limited primarily to coordination of transition metals such as iron centers in heme (iron porphyrin–containing) proteins and reaction with the diradical O_2, O_2-derived free radicals, and organic free radicals that are formed upon oxidation of various biomolecules. *See* FREE RADICAL; NITRO AND NITROSO COMPOUNDS; NITROGEN; NITROGEN OXIDES; PORPHYRIN.

Nitric oxide is produced in the body by enzymes called nitric oxide synthases, which convert the amino acid L-arginine to nitric oxide and L-citrulline. There are three recognized types of nitric oxide synthase, often designated nNOS, eNOS, and iNOS for the neuronal, endothelial, and inducible forms, respectively. Both nNOS and eNOS are constitutive—that is, they are always present in cells—while iNOS is normally not present but is synthesized in response to biochemical signals generated within the cellular environment. Nitric oxide synthases are structurally organized into two domains called the reductase and oxygenase domains. The reductase domain contains two redox-active flavin cofactors (FAD and FMN) that deliver electrons one at a time from the physiological electron donor (NADPH) to the oxygenase domain that contains the active site for arginine oxidation. The oxygenase domain contains a biologically unique combination of heme and tetrahydrobiopterin (H_4B) cofactors. Oxidation of arginine to citrulline plus NO requires two cycles of oxidation with O_2. Recent structural and kinetic evidence strongly suggests that H_4B plays essential roles in both cycles to deliver electrons to reactive oxo-heme intermediates, thereby directing substrate oxidation and minimizing formation of unwanted reactive oxygen species, such as superoxide ion (O_2^-) and nitroxyl anion (NO^-), as side products. A calcium-calmodulin binding site is located between the two domains. The constitutive enzymes, nNOS and eNOS, are activated by Ca^{2+}-calmodulin binding at this site, which in turn is regulated by the amount of Ca^{2+} in the medium. In contrast, Ca^{2+}-calmodulin binds so tightly to iNOS that this enzyme is effectively "turned on" under all physiological conditions. The practical consequence of this difference in binding affinities is that the constitutive enzymes generate only low levels of NO, whereas, once its formation has been induced, iNOS generates relatively massive amounts of NO. These differences in enzymatic activity are consistent with the presumed functions of the enzymes, which are secondary messenger generation and cytotoxin production, respectively. *See* AMINO ACIDS; ENZYME. James K. Hurst; Judith N. Burstyn; Mark F. Reynolds

Nitrile One of a group of organic chemical compounds of general formula RC≡N. A nitrile is named from the acid to which it can be hydrolyzed by adding the suffix -onitrile to the acid stem, for example, acetonitrile from acetic acid. An alternative system names the group attached to CN, thus CH_3CN is also named methyl cyanide. In more complex structures the CN group is named as a substituent, cyano.

Industrially, nitriles are formed by heating carboxylic acids with ammonia and a dehydration catalyst under pressure. For the preparation of acrylonitrile, which is used on a large scale in the plastics industry, a vapor-phase catalytic ammoxidation of propylene has been developed. *See* ACRYLONITRILE; AMINE. Paul E. Fanta

Nitro and nitroso compounds Nitro compounds are derivatives of organic hydrocarbons having one or more —NO_2 groups with nitrogen-to-carbon bonding. They differ from the oxygen-linked nitrites, which are esters. The group lacks enough electrons to form double bonds with both oxygens. However, both oxygens react alike; hence the bond is regarded as a resonance hybrid of single and double bonds.

Aromatic nitro compounds have been used chiefly as dye intermediates, explosives, and pharmaceuticals. They are formed readily by the reaction of aromatic compounds with nitric acid; H is replaced by the —NO_2 group, for example,

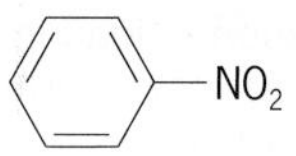

NO_2

Aliphatic nitro compounds are prepared with difficulty and have grown in importance only since the development of vapor-phase nitration of hydrocarbons with nitric acid vapors at 420°C (788°F).

Nitroso compounds contain the —NO group attached to carbon or nitrogen. Many are unstable intermediates, for example, nitrosobenzene formed during the reduction of nitrobenzene. *See* NITRATION. Allen L. Hanson

Nitroaromatic compound A member of the class of organic compounds in which the nitro group (—NO_2) is attached directly to the cyclic, aromatic nucleus. The prototypal compound is nitrobenzene. It is prepared by the reaction of benzene with nitric acid in the presence of sulfuric acid, as shown in the reactions below. The most significant use of nitrobenzene

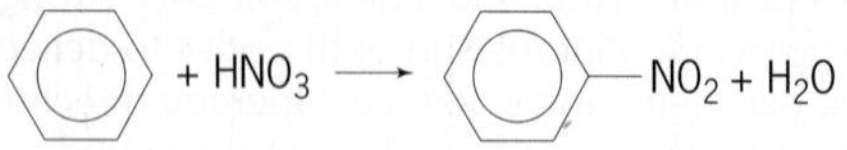

$$\text{Benzene} + HNO_3 \longrightarrow \text{Nitrobenzene}(-NO_2) + H_2O$$

is in the manufacture of aniline [structure (**1**)]. About 97% of

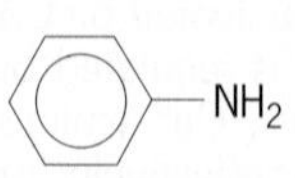

NH_2

(**1**)

the nitrobenzene produced in the United States is converted to aniline, which is used in the manufacture of plastics, rubber additives, dyes, drugs, and other products. *See* BENZENE.

Second to nitrobenzene in commercial importance are the mononitrotoluenes, particularly the ortho and para isomers, (**2**) and (**3**), respectively. Reaction of toluene with a mixture of

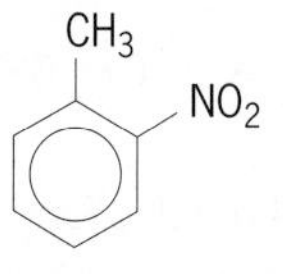

CH_3 NO_2 (**2**) CH_3 NO_2 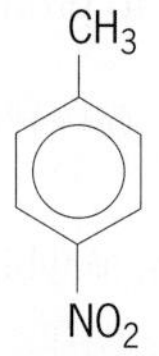 (**3**)

nitric and sulfuric acid at about 40°C (104°F) gives a high yield of a mixture of the isomers, which are separated by a combination of fractional distillation and crystallization. The nitrotoluenes are important intermediates in the preparation of dyes, rubber chemicals, and agricultural chemicals.

2,4,6-Trinitrotoluene (TNT; **4**) is a military explosive that is sta-

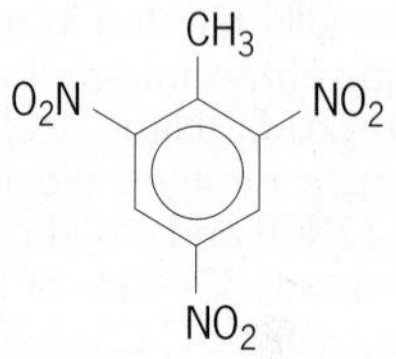

CH_3 O_2N NO_2 NO_2

(**4**)

ble, nonhygroscopic, and relatively insensitive to impact, friction, shock, and electric spark. It is produced by nitration of toluene in successive stages at progressively higher temperatures and concentrations of acid.

Although literally thousands of other aromatic ring compounds, including the heterocyclics, have been converted to their nitro derivatives, few such compounds have achieved any significant industrial importance. *See* AROMATIC HYDROCARBON; NITRATION. Paul E. Fanta

Nitrogen A chemical element, N, atomic number 7, atomic weight 14.0067. Nitrogen, a gas under normal conditions, is the lightest element of periodic group 5 (nitrogen family). *See* PERIODIC TABLE.

At standard temperature and pressure, elemental nitrogen exists as a gas with a density of 1.25046 g/liter. This value indicates that the molecular formula is N_2. Some physical properties of elemental nitrogen are listed in Table 1.

Elemental nitrogen has a low reactivity toward most common substances at ordinary temperatures. At high temperatures, molecular nitrogen, N_2, reacts with chromium, silicon, titanium, aluminum, boron, beryllium, magnesium, barium, strontium, calcium, and lithium (but not the other alkali metals) to form nitrides; with O_2 to form NO; and at moderately high temperatures and pressures in the presence of a catalyst, with hydrogen to form ammonia. Above 1800°C (3300°F), nitrogen, carbon, and hydrogen combine to form hydrogen cyanide.

Table 2 lists the principal classes of inorganic nitrogen compounds. Thus, in addition to the typical oxidation states of the family (−3, +3, and +5), nitrogen forms compounds with a variety of additional oxidation states. *See* AMINE;

Table 1. Properties of nitrogen

Property	Value
Heat of transformation (α–β)	54.71 cal/mole
Heat of fusion	172.3 cal/mole
Heat of vaporization	1332.9 cal/mole
Critical temperature	126.26 ± 0.04 K
Critical pressure	33.54 ± 0.02 atm
Density: α form	1.0265 g/ml at −252.6°C
β form	0.8792 g/ml at −210.0°C
Liquid	1.1607–0.0045T(T = abs temp)

Consulting Editors

Dr. Milton B. Adesnik. *Department of Cell Biology, New York University School of Medicine, New York.* CELL BIOLOGY.

Prof. Eugene A. Avallone. *Consulting Engineer; Professor Emeritus of Mechanical Engineering, City College of the City University of New York.* MECHANICAL AND POWER ENGINEERING.

Prof. William P. Banks. *Department of Psychology, Pomona College, Claremont, California.* GENERAL AND EXPERIMENTAL PSYCHOLOGY.

Dr. Phillip Barak. *Department of Soil Science, University of Wisconsin-Madison.* CROP SCIENCE.

Prof. Ted Barnes. *Physics Division, Oak Ridge National Laboratory, Oak Ridge, Tennessee.* ELEMENTARY PARTICLE PHYSICS.

Dr. Paul Barrett. *Department of Palaeontology, The Natural History Museum, London, United Kingdom.* VERTEBRATE PALEONTOLOGY.

Prof. S. H. Benabdallah. *Department of Mechanical Engineering, Royal Military College of Canada, Kingston, Ontario.* MECHANICAL ENGINEERING.

Prof. Ray Benekohal. *Department of Civil and Environmental Engineering, University of Illinois at Urbana-Champaign.* TRANSPORTATION ENGINEERING.

Prof. Carrol Bingham. *Department of Physics, University of Tennessee, Knoxville.* NUCLEAR AND ELEMENTARY PARTICLE PHYSICS.

Michael Bosworth. *Vienna, Virginia.* NAVAL ARCHITECTURE AND MARINE ENGINEERING.

Robert D. Briskman. *Technical Executive, Sirius Satellite Radio, New York.* TELECOMMUNICATIONS.

Prof. Richard O. Buckius. *Department of Mechanical & Industrial Engineering, University of Illinois at Urbana-Champaign.* MECHANICAL ENGINEERING.

Dr. Robyn J. Burnham. *Department of Geological Sciences, University of Michigan, Ann Arbor.* PALEOBOTANY.

Dr. Mark W. Chase. *Molecular Systematics Laboratory, Royal Botanic Gardens, Kew, Richmond, Surrey, United Kingdom.* PLANT TAXONOMY.

Prof. Wai-Fah Chen. *Dean, College of Engineering, University of Hawaii at Manoa.* CIVIL ENGINEERING.

Dr. John F. Clark. *Director, Graduate Studies, and Professor, Space Systems, Spaceport Graduate Center, Florida Institute of Technology, Satellite Beach.* SPACE TECHNOLOGY.

Prof. J. John Cohen. *Department of Immunology, University of Colorado School of Medicine, Denver.* IMMUNOLOGY.

Prof. Mark Davies. *Department of Mechanical & Aeronautical Engineering, University of Limerick, Ireland.* AERONAUTICAL ENGINEERING AND PROPULSION.

Prof. Peter J. Davies. *Department of Plant Biology, Cornell University, Ithaca, New York.* PLANT PHYSIOLOGY.

Dr. M. E. El-Hawary. *Associate Dean of Engineering, Dalhousie University, Halifax, Nova Scotia, Canada.* ELECTRICAL POWER ENGINEERING.

Barry A. J. Fisher. *Director, Scientific Services Bureau, Los Angeles County Sheriff's Department, California.* FORENSIC SCIENCE.

Dr. Peter L. Forey. *Department of Palaeontology, The Natural History Museum, London.* ANIMAL SYSTEMATICS AND EVOLUTION.

Dr. Richard L. Greenspan. *The Charles Stark Draper Laboratory, Cambridge, Massachusetts.* NAVIGATION.

Dr. Timothy A. Haley. *Food Science Department, Purdue University, West Lafayette, Indiana.* FOOD SCIENCE.

Dr. Lisa Hammersley. *Department of Geology, California State University, Sacramento.* PETROLOGY.

Prof. John P. Harley. *Department of Biological Sciences, Eastern Kentucky University, Richmond.* MICROBIOLOGY.

Prof. Terry Harrison. *Department of Anthropology, Paleoanthropology Laboratory, New York University, New York.* ANTHROPOLOGY AND ARCHEOLOGY.

Dr. Ralph E. Hoffman. *Yale Psychiatric Institute, Yale University School of Medicine, New Haven, Connecticut.* PSYCHIATRY.

Dr. Gary Hogg. *Professor of Industrial Engineering, Arizona State University, Tempe.* INDUSTRIAL AND PRODUCTION ENGINEERING.

Prof. A. Gordon L. Holloway. *Chair, Department of Mechanical Engineering, University of New Brunswick, Fredericton, Canada.* FLUID MECHANICS.

Dr. Hong Hua. *Director of 3D Visualization and Imaging System Lab, College of Optical Sciences, The University of Arizona, Tucson.* ELECTROMAGNETIC RADIATION AND OPTICS.

Dr. George Hudler. *Department of Plant Pathology, Cornell University, Ithaca, New York.* PLANT PATHOLOGY.

Dr. S. C. Jong. *Senior Staff Scientist and Program Director, Mycology and Protistology Program, American Type Culture Collection, Manassas, Virginia.* MYCOLOGY.

Dr. Edwin Kan. *Associate Professor, School of Electrical and Computer Engineering, Cornell University, Ithaca, New York.* PHYSICAL ELECTRONICS.

Dr. Peter M. Kareiva. *Lead Scientist, Pacific Western Conservation Region, The Nature Conservancy, Seattle, Washington.* ECOLOGY AND CONSERVATION.

Dr. Bryan P. Kibble. *National Physical Laboratory, Teddington, Middlesex, England.* ELECTRICITY AND ELECTROMAGNETISM.

Prof. Robert E. Knowlton. *Department of Biological Sciences, George Washington University, Washington, DC.* INVERTEBRATE ZOOLOGY.

Prof. Cynthia K. Larive. *Department of Chemistry, University of California, Riverside.* ANALYTICAL CHEMISTRY.

Prof. Chao-Jun Li. *Canada Research Chair in Green Chemistry, Department of Chemistry, McGill University, Montreal, Quebec, Canada.* ORGANIC CHEMISTRY.

Prof. Donald W. Linzey. *Wytheville Community College, Wytheville, Virginia.* VERTEBRATE ZOOLOGY.

Dr. Philip V. Lopresti. *Retired; formerly, Engineering Research Center, AT&T Bell Laboratories, Princeton, New Jersey.* ELECTRONIC CIRCUITS.

Dr. Dan Luss. *Cullen Professor of Engineering, Department of Chemical Engineering, University of Houston, Texas.* CHEMICAL ENGINEERING.

Prof. Scott M. McLennan. *Department of Geosciences, State University of New York at Stony Brook.* GEOLOGY (PHYSICAL, HISTORICAL, AND SEDIMENTARY).

Prof. Philip L. Marston. *Department of Physics and Astronomy, Washington State University, Pullman.* ACOUSTICS.

Dr. Ramon A. Mata-Toledo. *Professor of Computer Science, James Madison University, Harrisonburg, Virginia.* COMPUTING AND INFORMATION TECHNOLOGY.

Prof. Krzysztof Matyjaszewski. *J. C. Warner Professor of Natural Sciences, Department of Chemistry, Carnegie-Mellon University, Pittsburgh, Pennsylvania.* POLYMER SCIENCE AND ENGINEERING.

Prof. Joel S. Miller. *Department of Chemistry, University of Utah, Salt Lake City.* INORGANIC CHEMISTRY.

Dr. Orlando J. Miller. *Center for Molecular Medicine and Genetics, Wayne State University School of Medicine, Detroit, Michigan.* GENETICS.

Prof. Jay M. Pasachoff. *Director, Hopkins Observatory, Williams College, Williamstown, Massachusetts.* ASTRONOMY.

Prof. David J. Pegg. *Department of Physics, University of Tennessee, Knoxville.* ATOMIC AND MOLECULAR PHYSICS.

Prof. J. Jeffrey Peirce. *Department of Civil and Environmental Engineering, Edmund T. Pratt Jr., School of Engineering, Duke University, Durham, North Carolina.* ENVIRONMENTAL ENGINEERING.

Dr. William C. Peters. *Professor Emeritus, Mining and Geological Engineering, University of Arizona, Tucson.* MINING ENGINEERING.

Prof. Arthur N. Popper. *Department of Biology, University of Maryland, College Park.* NEUROSCIENCE.

Dr. Kenneth P. H. Pritzker. *Pathologist-in-Chief and Director, Head, Connective Tissue Research Group, and Professor, Laboratory Medicine and Pathobiology, University of Toronto, Mount Sinai Hospital, Toronto, Ontario, Canada.* MEDICINE AND PATHOLOGY.

Prof. Justin Revenaugh. *Department of Geology and Geophysics, University of Minnesota, Minneapolis.* GEOPHYSICS.

Dr. Roger M. Rowell. *USDA Forest Service, Forest Products Laboratory, Madison, Wisconsin.* FORESTRY.

Dr. John L. Safko. *Distinguished Professor Emeritus, Physics and Astronomy, Associated Faculty, School of the Environment, University of South Carolina, Columbia.* CLASSICAL MECHANICS.

Dr. Andrew P. Sage. *Founding Dean Emeritus and First American Bank Professor, University Professor, School of Information Technology and Engineering, George Mason University, Fairfax, Virginia.* CONTROL AND INFORMATION SYSTEMS.

Dr. Alfred S. Schlachter. *Advanced Light Source, Lawrence Berkeley National Laboratory, Berkeley, California.* ATOMIC AND MOLECULAR PHYSICS.

Prof. Ivan Schuller. *Department of Physics, University of California-San Diego, La Jolla, California.* CONDENSED-MATTER PHYSICS.

Dr. David Sherman. *Department of Earth Sciences, University of Bristol, United Kingdom.* MINERALOGY.

Dr. Steven A. Slack. *Department of Plant Pathology, Cornell University, Ithaca, New York.* PLANT PATHOLOGY.

Dr. Arthur A. Spector. *Department of Biochemistry, University of Iowa, Iowa City.* BIOCHEMISTRY.

Dr. Bruce A. Stanley. *Director, Scientific Programs, Section of Technology Development and Research Resources H093, The Pennsylvania State University College of Medicine, Hershey.* PHYSIOLOGY.

Prof. Anthony P. Stanton. *Carnegie-Mellon University, Pittsburgh, Pennsylvania.* GRAPHIC ARTS AND PHOTOGRAPHY.

Dr. Trent Stephens. *Professor of Anatomy and Embryology, Idaho State University, Pocatello.* DEVELOPMENTAL BIOLOGY.

Prof. John F. Timoney. *Department of Veterinary Science, University of Kentucky, Lexington.* VETERINARY MEDICINE.

Prof. Antonio J. Torres. *Department of Food Science and Technology, Oregon State University, Corvallis.* FOOD SCIENCE.

Dr. Sally E. Walker. *Associate Professor of Geology and Marine Science, University of Georgia, Athens.* INVERTEBRATE PALEONTOLOGY.

Prof. Pao K. Wang. *Department of Atmospheric and Oceanic Sciences, University of Wisconsin-Madison.* METEOROLOGY AND CLIMATOLOGY.

Prof. Frank M. White. *Professor Emeritus, Department of Mechanical Engineering, University of Rhode Island, Kingston.* FLUID MECHANICS.

Prof. Mary Anne White. *Department of Chemistry, Dalhousie University, Halifax, Nova Scotia, Canada.* MATERIALS SCIENCE AND METALLURGICAL ENGINEERING.

Prof. Thomas A. Wikle. *Department of Geography, and Associate Dean, College of Arts and Sciences, Oklahoma State University, Stillwater.* PHYSICAL GEOGRAPHY.

A crystal of barium sodium niobate shown generating the second harmonic of an infrared laser beam. The infrared beam is made visible by using a card treated with an infrared-sensitive phosphor. Barium sodium niobate is one of a number of new nonlinear crystals which have been developed for nonlinear optical applications. (*Bell Telephone Laboratories*)

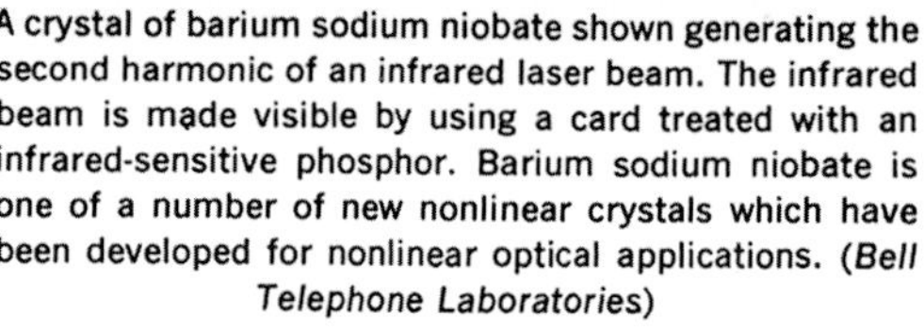

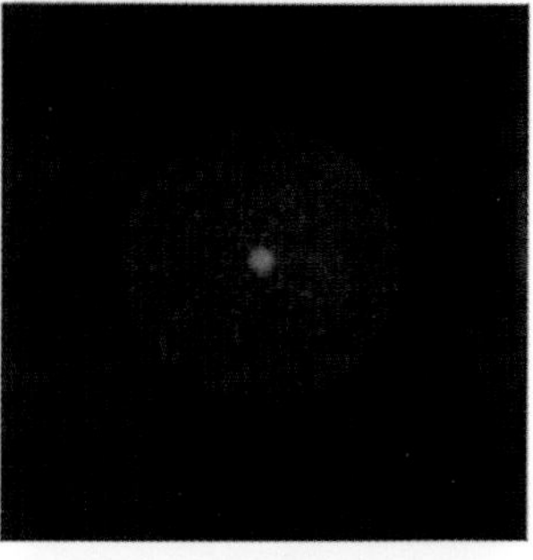

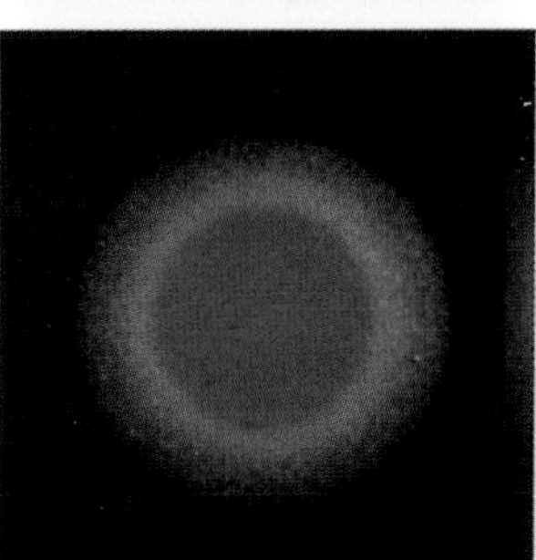

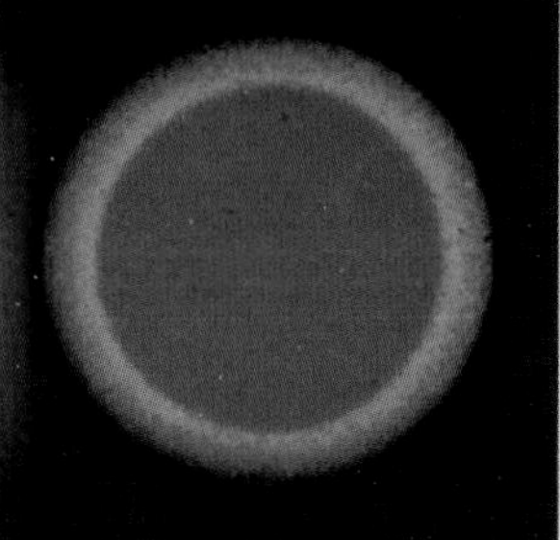

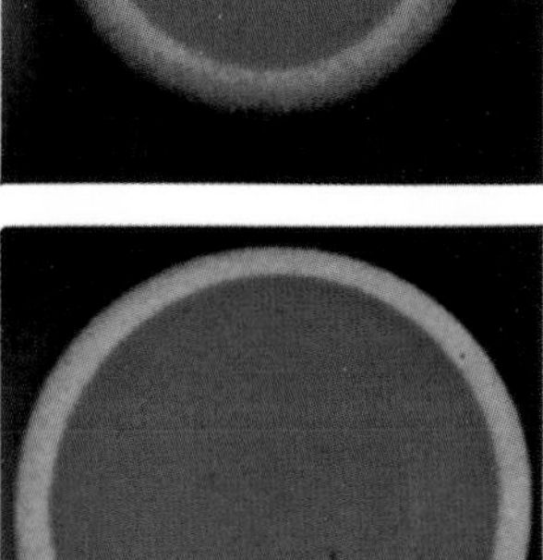

Parametric fluorescence from a crystal of $LiNbO_3$ when irradiated by the 4880-A line of an argon-ion laser for temperatures of 350, 300, 250, 200, and 150°C (top to bottom) of the $LiNbO_3$ crystal. Radiation in the infrared is also emitted but is not visible in the photographs. (*Courtesy of S. E. Harris, M. K. Oshman, and R. L. Byer*)

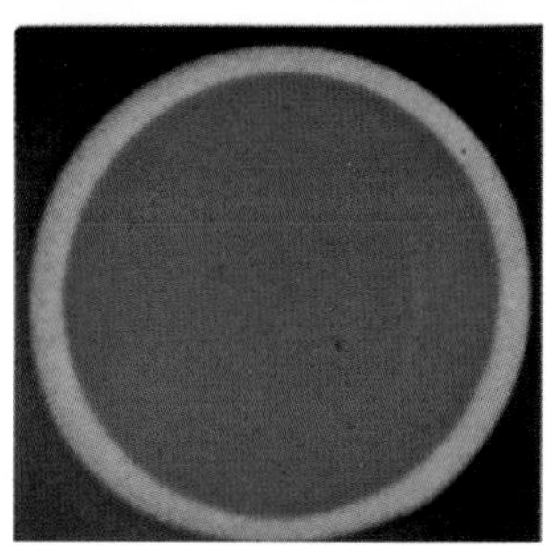

Coherent anti-Stokes Raman scattering from a cell of benzene. Red and yellow laser beams enter cell from left. Beam spots on screen at right correspond to incident lasers (center red, yellow), first-order anti-Stokes Raman (green), and second-order Stokes Raman (red spot at left). (*Courtesy of A. Compaan, S. C. Chandra, and E. Wiener-Avnear, Kansas State University*)

Table 2. Compounds of nitrogen

Oxidation state	Examples
+5	N_2O_5, HNO_3, nitrates, NO_2X
+4	$N_2O_4 \rightleftharpoons 2NO_2$
+3	N_2O_3, HNO_2, nitrites, NOX, NX_3
+2	NO, Na_2NO_2, nitrohydroxylamates
+1	N_2O, $H_2N_2O_2$, hyponitrites
0	N_2
−1/3	HN_3, acids
−1	NH_2OH, hydroxylammonium salts
−2	NH_2NH_2, hydrazinium salts, hydrazides
−3	NH_3, ammonium salts, amides, imides, nitrides

AMMONIA; HYDRAZINE; NITRIC ACID; NITROGEN COMPLEXES; NITROGEN OXIDES.

Molecular nitrogen is the principal constituent of the atmosphere (78% by volume of dry air), in which its concentration is a result of the balance between the fixation of atmospheric nitrogen by bacterial, electrical (lightning), and chemical (industrial) action, and its liberation through the decomposition of organic materials by bacteria or combustion. In the combined state, nitrogen occurs in a variety of forms. It is a constituent of all proteins (both plant and animal) as well as of many other organic materials. Its chief mineral source is sodium nitrate.

The methods for the preparation of elementary nitrogen may be grouped into two classes, separation from the atmosphere and decomposition of nitrogen compounds. The industrial method for the production of nitrogen is the fractional distillation of liquid air. Nitrogen containing about 1% argon and traces of other inert gases may be obtained by the chemical removal of oxygen, carbon dioxide, and water vapor from the atmosphere by appropriate chemical reagents.

Because the importance of nitrogen compounds in agriculture and chemical industry, much of the industrial interest in elementary nitrogen has been in processes for converting elemental nitrogen into nitrogen compounds. The principal methods for doing this are the Haber process for the direct synthesis of ammonia from nitrogen and hydrogen, the electric are process, which involves the direct combination of N_2 and O_2 to nitric oxide, and the cyanamide process. Nitrogen is also used for filling bulbs of incandescent lamps and, in general, wherever a relatively insert atmosphere is required. Harry H. Sisler

Nitrogen complexes Compounds containing the dinitrogen molecule, N_2, bound to a metal (also called dinitrogen complexes). Outstanding in their ability to form coordination compounds with nitrogen are a number of metals which belong to the group 7 transition metal family. For each metal of this group, several nitrogen complexes have been identified. Nitrogen complexes of these metals occur in low oxidation states, such as Co(I) or Ni(O). The other ligands present in these complexes besides N_2 are usually of a type known to stabilize low oxidation states; phosphines appear to be particularly prominent bonding partners in this respect. The illustration shows the structure of a typical N_2 complex, elucidated by a crystal structure determination. The N-N bond axis in this complex is aimed, within the limits of experimental error, directly toward the position of the metal atom. The Co-N_2 bond length, 0.18 nanometer, is within the normal range of comparable metal-ligand bonds. *See* COORDINATION CHEMISTRY.

Even in most favorable cases the binding of the dinitrogen molecule to the metal is fairly labile; all the compounds lose their nitrogen on mild heating. Some of the nitrogen complexes are only metastable to loss of dinitrogen even at room temperature; accordingly, they cannot be obtained by direct uptake of gaseous nitrogen. In the synthesis of these metastable complexes, hydrazine or azide compounds serve as a source of nitrogen

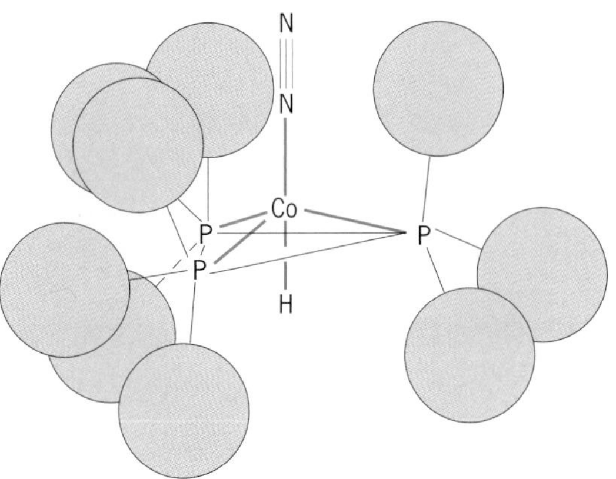

Structure of a coordination compound with N_2 (circles represent phenyl groups).

molecules within the coordination sphere of the metal. Addition of other coordinating agents to the nitrogen complexes usually results in a displacement of N_2 from the metal. The cobalt compound in the illustration exchanges its N_2 ligand quite reversibly for other ligand molecules, such as NH_3 and $H_2C{=}CH_2$. Whereas these ligands are easily displaced again by an excess of N_2, an irreversible exchange occurs with carbon monoxide. The bulky organic groups on the phosphine ligands are likely to interfere with the approach to the metal of all but the slimmest ligands and thereby help the "thin" dinitrogen molecule to maintain or regain its position on the metal in competition with most other ligands. Hans Brintzinger

Nitrogen cycle The collective term given to the natural biological and chemical processes through which inorganic and organic nitrogen are interconverted. It includes the process of ammonification, ammonia assimilation, nitrification, nitrate assimilation, nitrogen fixation, and denitrification.

Nitrogen exists in nature in several inorganic compounds, namely N_2, N_2O, NH_3, NO_2^-, and NO_3^-, and in several organic compounds such as amino acids, nucleotides, amino sugars, and vitamins. In the biosphere, biological and chemical reactions continually occur in which these nitrogenous compounds are converted from one form to another. These interconversions are of great importance in maintaining soil fertility and in preventing pollution of soil and water.

An outline showing the general interconversions of nitrogenous compounds in the soil-water pool is presented in the illustration. There are three primary reasons why organisms metabolize nitrogen compounds: (1) to use them as a nitrogen source, which means first converting them to NH_3, (2) to use certain nitrogen compounds as an energy source such as in the oxidation of NH_3 to NO_2^- and of NO_2^- to NO_3^-, and (3) to use certain nitrogen compounds (NO_3^-) as terminal electron acceptors under conditions where oxygen is either absent or in limited supply. The reactions and products involved in these three metabolically different pathways collectively make up the nitrogen cycle.

There are two ways in which organisms obtain ammonia. One is to use nitrogen already in a form easily metabolized to ammonia. Thus, nonviable plant, animal, and microbial residues in soil are enzymatically decomposed by a series of hydrolytic and other reactions to yield biosynthetic monomers such as amino acids and other small-molecular-weight nitrogenous compounds. These amino acids, purines, and pyrimidines are decomposed further to produce NH_3 which is then used by plants and bacteria for biosynthesis, or these biosynthetic monomers can be used directly by some microorganisms. The decomposition process is called ammonification.

The second way in which inorganic nitrogen is made available to biological agents is by nitrogen fixation (this term is

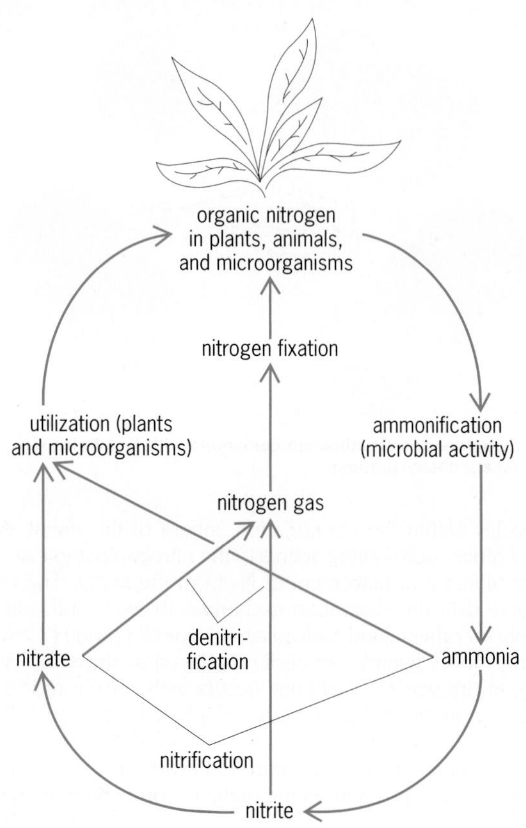

Diagram of the nitrogen cycle.

maintained even though N_2 is now called dinitrogen), a process in which N_2 is reduced to NH_3. Since the vast majority of nitrogen is in the form of N_2, nitrogen fixation obviously is essential to life. The N_2-fixing process is confined to prokaryotes (certain photosynthetic and nonphotosynthetic bacteria). The major nitrogen fixers (called diazotrophs) are members of the genus *Rhizobium*, bacteria that are found in root nodules of leguminous plants, and of the cyanobacteria (originally called blue-green algae). *See* NITROGEN FIXATION. Leonard E. Mortenson

Nitrogen excretion In the quest for sufficient food energy to meet caloric requirements, animals ingest more nitrogen, largely as amino acids, than they require. Accordingly, the excess nitrogen ingested must be excreted in some form. Through the action of a series of related enzymes called transaminases, virtually all metabolic nitrogen can be transferred to α-ketoglutaric acid to form glutamic acid. Under the influence of glutamic dehydrogenase, glutamic acid may be oxidized by the coenzyme diphosphopyridine nucleotide (DPN) with the reformation of α-ketoglutaric acid plus ammonia.

Aquatic animals in general simply excrete the ammonia as such, typically by diffusion across the gills or body surface. Other animals, notably amphibians and mammals, first convert ammonia into urea. To accomplish the transformation, advantage is taken of the enzymically catalyzed metabolic sequence by which the amino acid arginine is synthesized from ornithine, a sequence common to almost all living forms. It is adapted for urea synthesis by mammals, reptiles, and other forms by the additional presence in liver of the enzyme arginase, which catalyzes the hydrolysis of arginine to urea plus ornithine, which is then available for recycling.

Terrestrial animals, notably insects, arachnids, reptiles, and birds, convert the ammonia into the purine xanthine, which is then oxidized to form uric acid, the form in which the ammonia is ultimately eliminated. This mechanism is an adaptation to embryonic development in an egg with closed shell, in which the accumulation of ammonia would be toxic and the formation of urea would cause osmic difficulties. *See* EXCRETION; PROTEIN METABOLISM. Philip Handler; Bradley T. Scheer

Nitrogen fixation The chemical or biological conversion of atmospheric nitrogen (N_2) into compounds which can be used by plants, and thus become available to animals and humans. In the 1990s, chemical and biological processes together contributed about 260 million tons (230 million metric tons) of fixed nitrogen per year globally. Industrial production of nitrogen fertilizer accounted for about 85 million tons (80 million metric tons) of nitrogen per year, while spontaneous chemical processes, such as lightning, ultraviolet irradiation, and combustion, leading to the synthesis of nitrogen oxides from O_2 and N_2, may have accounted for 44 million tons (40 million metric tons) per year. The remainder, roughly half of the global input of newly fixed nitrogen, arose from biological processes. World agriculture, which is very dependent on nitrogen fixation, is increasingly reliant on chemical nitrogen sources. *See* NITROGEN.

Chemical fixation. Three chemical processes for fixing atmospheric nitrogen have been developed. All require considerable thermal or electrical energy and yield different products. In arc processes, which are now rarely used, air is passed through an electric arc and about 1% nitric oxide is formed, which can be chemically converted to nitrates. In the cyanamide process, which is now obsolete, heating calcium carbide in nitrogen generates calcium cyanamide, which when moistened hydrolyzes to urea and ammonia. In the widely used Haber process, hydrogen (generated by heating natural gas) is mixed with nitrogen (from air), and burned to yield a nitrogen-hydrogen mixture. The nitrogen-hydrogen mixture is compressed (10–80 megapascals) and heated (200–700°C or 390–1300°F) in the presence of a metal oxide catalyst to give ammonia. The Haber process is the major source of ammonia used for fertilizer. *See* AMMONIA; CYANAMIDE; ELECTROCHEMICAL PROCESS; FERTILIZER; HIGH-PRESSURE PROCESSES.

Biological fixation. Only prokaryotes—bacteria, archaea, and cyanobacteria (earlier called blue-green algae)—fix nitrogen. Nitrogen-fixing microbes, called diazotrophs, fall into two main groups, free-living and symbiotic. *See* ARCHAEBACTERIA; BACTERIA; CYANOBACTERIA; PROKARYOTAE.

The free-living diazotrophs are subclassified. Aerobic diazotrophs, of which there are over 50 genera, including *Azotobacter*, methane-oxidizing bacteria, and cyanobacteria, require oxygen for growth and fix nitrogen when oxygen is present. *Azotobacter*, some related bacteria, and some cyanobacteria fix nitrogen in ordinary air, but most members of this group fix nitrogen only when the oxygen concentration is low. Free-living diazotrophs, which fix nitrogen only when oxygen is absent or vanishingly low, are widespread. The genera *Bacillus* and *Klebsiella* include many strains of this type, and representatives of symbiotic diazotrophs behave in this way as well. *See* ALGAE; BACTERIAL PHYSIOLOGY AND METABOLISM.

The best-known symbiotic bacteria belong to the genus *Rhizobium*. Species of *Rhizobium*, or related genera, such as *Bradyrhizobium* and *Sinorhizobium*, colonize the roots of leguminous plants and stimulate the formation of nodules within which they fix nitrogen microaerobically. Both plants and bacteria show specificity; for example, certain types of plants require special strains of rhizobia. Some types of rhizobium, such as *Bradyrhizobium*, can fix nitrogen in the absence of plant tissue, but require low oxygen, though most rhizobia fix nitrogen only within the nodules. *See* SOIL MICROBIOLOGY.

The enzymes responsible for nitrogen fixation are called nitrogenases. The most common nitrogenase consists of two proteins, one large containing molybdenum, iron, and inorganic sulfur (the MoFe-protein or dinitrogenase), the other smaller containing iron and inorganic sulfur (the Fe-protein or dinitrogenase reductase). Nitrogenase reduces one molecule of N_2 to two of ammonia (NH_3), a reaction which is accompanied by the conversion of 16 molecules of adenosine triphosphate (ATP) to adenosine diphosphate (ADP) and the release of one molecule of H_2 as a by-product. Nitrogenase is irreversibly destroyed by air, so all aerobic diazotrophs have developed means of restricting access of oxygen to the active enzyme. John Postgate

Nitrogen-fixing trees Trees that utilize nitrogen fixation. Nitrogen-fixing tree species occur in all three subfamilies of the Leguminosae (Fabaceae). Most of the mimosoid group (such as *Acacia, Albizia, Leucaena*) and virtually all of the papilionoid group (including *Dalbergia, Gliricidia, Robinia*) fix nitrogen, but nodulation and nitrogen fixation in the caesalpinioid group (such as *Chamaecrista*) is very restricted. However, in some genera, for example the acacias, not all the species form nitrogen-fixing symbioses. Most legume trees are found in the tropics and subtropics, but some grow and are of importance in temperate climates, for example the horticultural laburnums or black locust (*Robinia pseudoacacia)*, which is used extensively in rehabilitation of derelict land. *See* LEGUME; NITROGEN FIXATION.

The ability to fix atmospheric nitrogen into a form that can be used for plant growth is confined to bacteria and cyanobacteria. Plants fix nitrogen only by virtue of associations with these simple organisms. The best-known associations are the symbioses of *Rhizobium* bacteria with agricultural legumes, such as clovers, peas, and beans. Rhizobia stimulate the formation of root nodules, which provide a specialized environment within which high rates of nitrogen fixation can occur. Nitrogen fixation not only supports plant growth independent of mineral nitrogen in the soil but also can improve soil nitrogen status as plant residues, notably leaves and fine roots, decay and are mineralized. *See* BACTERIA; CYANOBACTERIA.

Many species of trees and shrubs other than legumes form nitrogen-fixing root nodules not with rhizobia but with a filamentous bacterium, *Frankia*, or with nitrogen-fixing cyanobacteria, such as *Nostoc*. Symbiosis with cyanobacteria produces a large, coralloid mass of nodules on surface roots of some cycads. They are restricted to Central and South America, South Africa, and Australasia, where they are a source of fixed nitrogen for natural ecosystems. *Frankia* forms nitrogen-fixing nodules on the roots of trees and shrubs in eight families other than legumes, most being native to temperate regions. Species nodulated by *Frankia* are known as actinorhizal plants. They are very important ecologically for their ability to colonize and to improve the nitrogen status of denuded and degraded soils. In temperate regions, the most widely utilized actinorhizal tree species belong to the genus *Alnus* (the alders) of the Betulaceae. Their main use now is in land reclamation.

Most species of nitrogen-fixing trees belong to the Leguminosae and have diverse uses. Timber and pulpwood species include *Dalbergia* (rosewood) and *Albizia* (ipil ipil), which is an important component of sustainable forestry in the tropics. Many species are used for fuel wood, and trees able to grow in deforested nitrogen-deficient soils of arid regions are very important in this respect socially, for example some *Acacia* and *Prosopis* (mesquite) species. The multifaceted properties of nitrogen-fixing trees ensure them a favored role in agroforestry systems, particularly on impoverished lands, where their minimal requirement for fertilizer application and the high nitrogen content of residues are of great value for intercropping practices. C. T. Wheeler

Nitrogen oxides Chemical compounds of nitrogen and oxygen. Nitrogen and oxygen do not combine when mixed directly (as in air), but they do combine during chemical reactions of compounds containing them. A number of nitrogen oxides can be isolated which differ from one another in the numbers of nitrogen and oxygen atoms present in each molecule.

Oxides of nitrogen and their properties

Name	Stoichiometric formula	Melting point, °C (°F)	Boiling point, °C (°F)
Nitrous oxide (dinitrogen monoxide)	N_2O	−90.8 (−131)	−88.5 (−127.3)
Nitric oxide (nitrogen monoxide)	NO	−163.6 (−262.5)	−151.7 (241.0)
Dinitrogen trioxide	N_2O_3	−103 (−155)	3.5 (38.3)
Dinitrogen tetroxide (⇌ nitrogen dioxide)	N_2O_4 (⇌ NO_2)	−11.2 (11.8)	21.2 (70.2)
Dinitrogen pentoxide	N_2O_5	41 (106)	

The table gives data for the five nitrogen oxides which are well established.

Nitrous oxide and nitric oxide. When inhaled, nitrous oxide has anesthetic effects; in small amounts it produces mild hysteria and hence is sometimes called laughing gas. It is colorless, is the least reactive of the oxides, and dissolves in water without chemical reaction. Some nitric oxide is formed in an electric arc, as in the technical production of nitric acid.

With oxygen or air, nitric oxide is rapidly converted to nitrogen dioxide. Nitric oxide is colorless and is soluble in water without reaction. It is an important messenger molecule in animals. It is one of the few "odd" molecules which contain an odd number of electrons. As an odd molecule, it has the ability to lose or gain one electron, thus giving the electrically charged ions NO^+ and NO^-. The important nitrosyl compounds contain these ions.

Trioxide. Dinitrogen trioxide exists pure only in the solid state. It is the anhydride of nitrous acid; when the oxide is dissolved in an alkaline solution, nitrite ion is produced.

Dioxide and tetroxide. The position of the equilibrium between nitrogen dioxide and dinitrogen tetroxide depends upon temperature and physical state. Dinitrogen tetroxide reacts readily with water to give an equimolecular mixture of nitrous and nitric acids. As temperature is raised, the nitrous acid decomposes to nitric acid and nitric oxide. These reactions are important in the technical production of nitric acid by catalytic oxidation of ammonia. Dinitrogen tetroxide is an oxidizing agent comparable in strength to bromine, and is employed as such in the lead-chamber process for sulfuric acid. In organic chemistry the tetroxide finds use as a special oxidizing agent (for example, in the production of sulfoxides and phosphine oxides) and as a nitrating agent.

Pentoxide. Solid dinitrogen pentoxide readily volatilizes, and the molecular type of structure found in the gaseous state is observed also in solutions of the oxide in low dielectric solvents such as carbon tetrachloride and chloroform. Sodium metal reacts with the liquid oxide, liberating nitrogen dioxide and forming sodium nitrate. Gaseous dinitrogen pentoxide decomposes readily, and is a strong oxidizing agent. With water it is converted to nitric acid. *See* NITRIC OXIDE; NITROGEN; OXYGEN. Cyril C. Addison

Nitroparaffin Any derivative of an aliphatic hydrocarbon that contains one or more —NO_2 groups bonded via nitrogen to the carbon framework. Nitroparaffins are also known as nitroalkanes.

Low-molecular-weight nitroparaffins are prepared via the vapor-phase nitration of alkanes at >400°C (750°F). However, the process is not generally satisfactory for higher-molecular-weight nitroparaffins because of polynitration and chain cleavage. The direct nitration of propane is used commercially to prepare nitromethane (boiling point 101°C or 214°F), nitroethane (bp 114°C or 237°F), 1-nitropropane (bp 131°C or 268°F), and 2-nitropropane (bp 120°C or 248°F). Nitroparaffins are prepared in the laboratory by the reaction of nitrite salts with alkyl bromides or iodides, from the oxidation of amines or oximes by using peroxycarboxylic acids, and by the chain homologation of simple nitroparaffins. *See* NITRATION.

Nitromethane, nitroethane, and the nitropropanes are useful solvents with high dielectric constants that readily dissolve many polymers. In addition, these simple nitroparaffins are versatile intermediates for the synthesis of specialty chemicals.

A. G. M. Barrett

Nobelium A chemical element, No, atomic number 102. Nobelium is a synthetic element produced in the laboratory. It decays by emitting an alpha particle, that is, a doubly charged helium ion. Only atomic quantities of the element have been produced to date. Nobelium is the tenth element heavier than uranium to be produced synthetically. It is the thirteenth member of the actinide series, a rare-earth-like series of elements. *See* ACTINIDE ELEMENTS; PERIODIC TABLE; RADIOACTIVITY; RARE-EARTH ELEMENTS; TRANSURANIUM ELEMENTS.

Paul R. Fields

Noeggerathiales An incompletely known and poorly defined group of vascular plants whose geologic range extends from Upper Carboniferous to Triassic. Their taxonomic status and position in the plant kingdom are uncertain since morphological evidence (because of the paucity of the fossil record) does not make it possible to place the group confidently in any recognized major subdivision of the vascular plants. The Noeggerathiales have been proposed in the evolutionary scheme for vascular plants. *See* PALEOBOTANY; PLANT KINGDOM.

Elso S. Barghoorn

Noise measurement The process of quantitatively determining one or more properties of acoustic noise. In noise assessment and control studies, knowledge of the physical properties of the undesirable sound is the initial step toward understanding the situation and what should be done to reduce or eliminate a noise problem.

The most common measures of noise are of the magnitude and frequency content of the noise sound pressure, time-averaged or as a function of time. Of increasing interest are metrics of sound quality (that may include both physical and psychoacoustic factors), such as loudness, pitch strength, and fluctuation strength. To characterize the noise output of a source, sound power level may be determined. To locate a source or to quantify propagation paths, sound intensity level may be measured.

Essentially all noise measurements are performed using electronic equipment. An electroaoustic transducer (a microphone in air and other gases; a hydrophone in water and other liquids) transforms sound (usually the sound pressure) at the point of observation into a corresponding electrical signal. This electrical signal is then operated on by analog or digital means with devices such as signal conditioners, filters, and detectors to determine the value (or values) of interest. This value is then given on an indicating meter or digital display. *See* ELECTRIC FILTER; HYDROPHONE; MICROPHONE; TRANSDUCER.

For noise measurements in air, a sound-level meter is the most commonly used instrument. The simplest sound-level meter comprises a microphone, a frequency-weighting filter, a root-mean-square (rms) detector, and logarithmic readout of the sound pressure level in decibels relative to 20 micropascals. Standard frequency weightings, designated A and C, were originally developed to approximate human response to noise at low and high levels, respectively, but now are used as specified in standards and legislation without regard to their origin. More sophisticated sound-level meters incorporate standardized octave-band or fractional-octave-band filters, and provide additional metrics and analysis capabilities. *See* DECIBEL; LOUDNESS.

Selection of a noise metric and measurement instrument generally depends upon the application. Four classes of application are: (1) discrete source emission, (2) hearing conservation, (3) outdoor environmental noise, and (4) indoor room noise. Within these classes there are found additional subclasses, measurements for enforcement of legal limits often being done differently than those for comfort, aircraft sonic boom being measured differently than power plant noise, and so on.

Joseph Pope

Nomograph A graphical relationship between a set of variables that are related by a mathematical equation or law. The fundamental principle involved in the construction of a nomographic or alignment chart consists of representing an equation containing three variables, $f(u, v, w) = 0$, by means of three scales in such a manner that a straight line cuts the three scales in values of u, v, and w, satisfying the equation. The cutting line is called the isopleth or index line. Numbers may be quickly and easily read from the scales of such a chart even by one unfamiliar with the construction of the chart and the equation involved. The illustration shows such an example. Assume that it is desired to find the value of E when $D = 2$ and $Q = 50$. Lay a straightedge through 50 on the Q scale and through 2 on the D scale and read 11.8 at its intersection with the E scale. As another example, it might be desired to know what value or values of D should be used if E and Q are required to be 10 and 60, respectively. A straightedge through $E = 10$ and $Q = 60$ cuts the D scale in two points, $D = 2.8$ and $D = 9.4$. This is equivalent to finding two positive roots of the cubic equation $D^3 - 10D^2 + 56.25 = 0$. It is assumed that $g = 32$ ft/s^2 in this equation.

Raymond D. Douglass

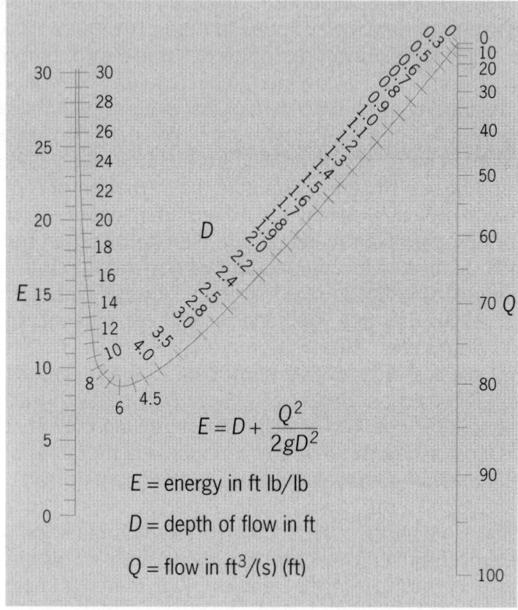

Nomograph for energy content of a rectangular channel with uniform flow.

Nondestructive evaluation Nondestructive evaluation (NDE) is a technique used to probe and sense material structure and properties without causing damage. It has become an extremely diverse and multidisciplinary technology, drawing on the fields of applied physics, artificial intelligence, biomedical engineering, computer science, electrical engineering, electronics, materials science and engineering, mechanical engineering, and structural engineering. Historically, NDE techniques have been used almost exclusively for detection of macroscopic defects (mostly cracks) in structures which have been manufactured or placed in service. Using NDE for this purpose is usually referred to as nondestructive testing (NDT).

A developing use of NDE methods is the nondestructive characterization (NDC) of materials properties (as opposed to revealing flaws and defects). Characterization typically sets out to establish absolute or relative values of material properties such as mechanical strength (elastic moduli), thermal conductivity or diffusivity, optical properties, magnetic parameters, residual strains, electrical resistivity, alloy composition, the state of cure in polymers, crystallographic orientation, and the degree of crystalline perfection. Nondestructive characterization can also be used for a variety of other specialized properties that are relevant to some aspect of materials processing in production, including determining how properties vary with the direction within the material, a property called anisotropy.

Much effort has been directed to developing techniques that are capable of monitoring and controlling (1) the materials production process; (2) materials stability during fabrication, transport, and storage; and (3) the amount and rate of degradation during the postfabrication in-service life for both components and structures. Real-time process monitoring for more efficient real-time process control, improved product quality, and increased reliability has become a practical reality. *See* MATERIALS SCIENCE AND ENGINEERING.

Visual inspection is the oldest and most versatile NDE tool. In visual inspection, a worker examines a material using only eyesight. The liquid (or dye) penetrant visual method uses brightly colored liquid dye to penetrate and remain in very fine surface cracks after the surface is cleaned of residual dye. The magnetic particle visual method requires that a magnetic field be generated inside a ferromagnetic test object. Flux leakage occurs where there are discontinuities on the surface. Magnetic particles (dry powder or a liquid suspension) are captured at the leakage location and can be readily seen with proper illumination.

The eddy current method uses a probe held close to the surface of a conducting test object. X-rays provide a varied and powerful insight into material, but they are somewhat limited for use in the field. The acoustic emission technique typically uses a broadband piezoelectric transducer to listen for acoustic noise. The thermography technique uses a real-time "infrared camera," much like a home camcorder, except that it forms images using infrared photons instead of visible ones. Contact ultrasonics technique is the workhorse of traditional and mature NDE technology. It uses a transducer held in contact with a test object to launch ultrasonic pulses and receive echoes. *See* EDDY CURRENT; X-RAY DIFFRACTION.

Many noncontact measurements have been developed that enhance the mature technologies. These include noncontact ultrasonic transducers that involve laser ultrasonics, electromagnetic acoustic transducers, and air- or gas-coupled transducers. Thermal wave imaging uses a main laser beam to scan the surface of the object to be examined. Electronic speckle pattern interferometry is a noncontact, full-field optical technique for high-sensitivity measurement of extremely small displacements in an object's surface. "Speckle" refers to the grainy appearance of an optically rough surface illuminated by a laser. Development of microwave techniques are under way in such diverse applications as ground-penetrating radar for land-mine detection, locating delaminations in highway bridge decks, and monitoring the curing process in polymers. *See* INTERFEROMETRY; LASER; TRANSDUCER; ULTRASONICS. John M. Winter, Jr.

Noneuclidean geometry A system of geometry based upon a set of axioms different from those of the euclidean geometry based on the three-dimensional space of experience. Noneuclidean geometries, especially riemannian geometry, are useful in mathematical physics. *See* DIFFERENTIAL GEOMETRY; EUCLIDEAN GEOMETRY; PROJECTIVE GEOMETRY; RIEMANNIAN GEOMETRY. John De Cicco

K. F. Gauss is credited with discovering an "elliptic" noneuclidean geometry in which there are no parallels, but the other euclidean axioms are satisfied. One way to model it is to call each diameter of a sphere a "point," and call each plane through the center (intersecting the sphere in a great circle) a "line." Then each two points determine a unique line and each two lines determine one point.

Another type of noneuclidean geometry, called hyperbolic geometry, was discovered about 1830 by J. Bolyai and N. I. Lobachevski. Through a given point P not on a line l, there are many lines that do not meet l. This geometry may be modeled by defining as "lines" those circular arcs that meet a large circle C at right angles and restricting the term "points" to those within the absolute circle C. Then two points determine a unique line, but there are many lines through a point P not on a line l that do not intersect l. Since consistent geometries exist in which the parallel postulate does not hold, this postulate is independent of the others. J. Sutherland Frame

According to whether the geometry is elliptic, euclidean, or hyperbolic, the sum of the angles of a geodesic triangle is greater than π, equal to π, or less than π.

Elliptic geometry of two dimensions may be represented upon a sphere in euclidean space of three dimensions. On the other hand, hyperbolic geometry of two dimensions can be depicted on a pseudosphere in euclidean space of three dimensions. The pseudosphere is obtained by revolving the tractrix about its asymptote. John De Cicco

Nonlinear acoustics The study of amplitude-dependent acoustical phenomena. The amplitude dependence is due to the nonlinear response of the medium in which the sound propagates, and not to the nonlinear behavior of the sound source. According to the linear theory of acoustics, increasing the level of a source by 10 dB results in precisely the same sound field as before, just 10 dB more intense. Linear theory also predicts that only frequency components radiated directly by the source can be present in the sound field. These principles do not hold in nonlinear acoustics. *See* AMPLITUDE (WAVE MOTION); LINEARITY; NONLINEAR PHYSICS.

The extent to which nonlinear acoustical effects are strong or even significant depends on the competing influences of energy loss, frequency dispersion, geometric spreading, and diffraction. When conditions are such that nonlinear effects are strong, acoustic signals may experience substantial waveform distortion and changes in frequency content as they propagate, and shock waves may be present. Nonlinear acoustical effects occur in gases, liquids, and solids, and they are observed over a broad range of frequencies. Shock waves present in sonic booms and thunder claps are in the audio frequency range. Principles of nonlinear acoustics form the basis for procedures at megahertz frequencies used in medical ultrasound and nondestructive evaluation of materials. Nonlinearity can also induce changes in nonfluctuating properties of the medium. These include acoustic streaming, which is the steady fluid flow produced by the absorption of sound, and radiation pressure, which results in a steady force exerted by sound on its surroundings. *See* ACOUSTIC RADIATION PRESSURE; BIOMEDICAL ULTRASONICS; NONDESTRUCTIVE EVALUATION; SHOCK WAVE; SONIC BOOM; THUNDER.

The principal feature that distinguishes nonlinear acoustics from nonlinear optics is that most acoustical media exhibit only weak dispersion, whereas media in which nonlinear optical effects arise exhibit strong dispersion. Dispersion is the dependence of propagation speed on frequency. In optical media, strong nonlinear wave interactions require that phase-matching conditions be satisfied, which can be accomplished only for several frequency components at one time. In contrast, all frequency components in a sound wave propagate at the same speed and are automatically phase-matched, which permits strong nonlinear interactions to occur among all components in the frequency spectrum. *See* NONLINEAR OPTICS.

Acoustic streaming is a nonlinear effect because the velocity of the flow depends quadratically on the amplitude of the sound, and the flow is not predicted by linear theory. Absorption due to viscosity and heat conduction results in a transfer of momentum from the sound field to the fluid. This momentum transfer manifests itself as steady fluid flow.

Acoustic streaming produced in sound beams is enhanced considerably when shocks develop. Shock formation generates a frequency spectrum rich in higher harmonics. Because thermoviscous absorption increases quadratically with frequency, attenuation of the wave, and therefore the streaming velocity, increases markedly following shock formation. Streaming is also generated in acoustic boundary layers formed by standing waves in contact with surfaces. Measurements of acoustic streaming have been used to determine the bulk viscosity coefficients of fluids. Thermoacoustic engines and refrigerators are adversely affected by heat transport associated with streaming. *See* THERMOACOUSTICS.

Phase conjugation refers to wavefront reversal, also called time reversal, at a single frequency. The latter terminologies more clearly describe this procedure. A waveform is captured by a phase conjugation device and reversed in such a way that it propagates back toward the source in the same way that it propagated toward the conjugator. Sound that is radiated from a point source and propagates through an inhomogeneous medium that introduces phase distortion in the wave field is thus retransmitted by the conjugator in such a way as to compensate for the phase distortion and to focus the wave back on the point source.

Phase conjugation is used to compensate for phase distortion in applications involving imaging and retargeting of waves on sources. The most successful techniques for acoustical phase conjugation are based on modulation of acoustical properties of a material that captures the incident sound wave. The modulation is twice the frequency of the incident sound wave, and it is induced by an electric field applied to piezoelectric material, or a magnetic field applied to magnetostrictive material. Often the modulated property of interest is the sound speed in the material. When the incident wave at frequency f propagates through a medium in which the sound speed fluctuates at frequency $2f$, parametric interaction generates a wave at the difference frequency f that propagates backward as though reversed in time. *See* MAGNETOSTRICTION; OPTICAL PHASE CONJUGATION; PIEZOELECTRICITY.

Phenomena associated with nonlinear acoustics have proved useful in both diagnostic and therapeutic applications of biomedical ultrasound. A very significant breakthrough in diagnostic imaging, especially for echocardiography and abdominal ultrasound imaging, is based on second-harmonic generation. Medical ultrasound imaging is performed at frequencies of several megahertz. Images constructed from the backscattered second-harmonic component have substantially reduced clutter and haze associated with the propagation of ultrasound through the outer layers of skin, which is the primary cause of phase aberrations. In another technique, microbubbles are injected into the bloodstream to enhance echoes backscattered from blood flow. The microbubbles are fabricated to make them resonant at diagnostic imaging frequencies, and they become strongly nonlinear oscillators when excited by ultrasound. Imaging is based on echoes at harmonics of the transmitted signal. Frequencies backscattered from the microbubbles differ from those in echoes coming from the surrounding tissue, which highlights the locations of the microbubbles and therefore of the blood flow itself.

A notable therapeutic application is lithotripsy, which refers to the noninvasive disintegration of kidney stones and gallstones with focused shock waves. Nonlinear acoustical effects in lithotripsy are associated not only with propagation of the shock wave but also with the generation of cavitation activity near the stones. Radiation of shock waves due to the collapse of cavitation bubbles is believed to be the dominant cause of stone breakup. An emerging therapeutic application, high-intensity focused ultrasound (HIFU), utilizes the heat dissipated by shock waves that develop in beams of focused ultrasound. The heating is so intense and localized that the potential exists for noninvasive cauterization of internal wounds and removal of tumors and scar tissue. *See* CAVITATION; ULTRASONICS. Mark F. Hamilton

Nonlinear control theory

A control system involves a plant and a controller. Plants are objects as diverse as a satellite, a distillation column, a robot arm, and a colony of bacteria. After measuring actual outputs of the plant, the controller computes signals that are applied at the inputs to the plant to achieve desired outputs. The design of controllers must be based upon mathematical models of plants, which are in most realistic situations composed of nonlinear differential and difference equations. A standard approach is to linearize the equations and use the powerful methods available for the design of linear control systems. *See* DIFFERENTIAL EQUATION; LINEAR SYSTEM ANALYSIS; OPTIMAL CONTROL (LINEAR SYSTEMS).

When the controlled outputs are allowed to have large deviations from the desired steady-state values, a linearized model will cease to describe the plant accurately, thereby causing erroneous results in the design. Linearized design models also fail in those important situations where nonlinearities are introduced into a controller to achieve a desired performance, generally at reduced cost. Typical examples of nonlinear controllers are on-off relays for temperature regulation in heating and cooling systems, switching elements in robot manipulators, and jet thrusters for the attitude control of space vehicles.

Unlike linear systems, there is no general theory for nonlinear systems. Nonlinear control theory is fragmented into areas centered around those classes of systems that are most prominent in applications.

A simple nonlinear control system is the inverted pendulum (illus. *a*), where it is required to keep the pendulum in an upright position by applying the torque $u(t)$ at its base. The pendulum can be considered as a generic model for a rocket booster balanced on top of a gimbaled thruster engine (illus. *b*), and a controlled robot arm (illus. *c*). Newton's second law of motion for the pendulum is Eq. (1), where m is the mass of the bob,

$$m\dot{\theta}(t) = \frac{mg}{\ell}\sin\theta(t) + u(t) \qquad (1)$$

ℓ is the length of the pendulum, and g is the acceleration due to gravity. By choosing the angular position $\theta = x_1$ and angular velocity $\dot{\theta} = x_2$ of the bob as the state of the system, Eq. (1) can

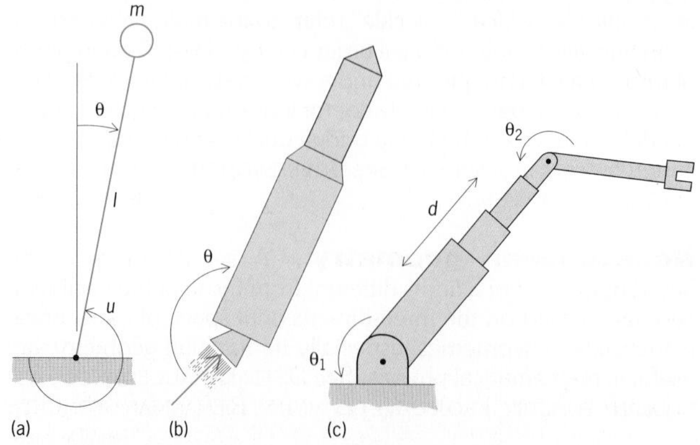

Nonlinear systems. (*a*) Inverted pendulum, which serves as a model for (*b*) a rocket booster and (*c*) a robot arm.

be rewritten as two first-order state equations (2).

$$\begin{aligned} \dot{x}_1 &= x_2 \\ \dot{x}_2 &= \frac{g}{\ell}\sin x_1 + u \end{aligned} \qquad (2)$$

See PENDULUM.

If the torque is absent ($u = 0$), the upright position of a motionless pendulum ($\dot{x}_1 = 0$, $\dot{x}_2 = 0$) is an equilibrium state: $x_1 = 0$, $x_2 = 0$. If the pendulum is slightly perturbed at this state, it will fall down and keep oscillating around the other equilibrium point at $x_1 = \pi$ and $x_2 = 0$, where again, $\dot{x}_1 = 0$ and $\dot{x}_2 = 0$.

To keep the pendulum in the upright position in physically realistic situations involving perturbations, the torque u is chosen to be a suitable function ϕ of the state x_1, x_2, as in Eq. (3). That

$$u = \phi(x_1, x_2) \qquad (3)$$

u should be a feedback control law in terms of the state is one of the most important results of control theory. Dragoslav D. Šiljak

Nonlinear optical devices Devices that use the fact that the polarization in any real medium is a nonlinear function of the optical field strength to implement various useful functions. The nonlinearities themselves can be grouped roughly into second-order and third-order. Materials that possess inversion symmetry typically exhibit only third-order nonlinearities, whereas materials without inversion symmetry can exhibit both second- and third-order nonlinearities. *See* CRYSTALLOGRAPHY; ELECTRIC SUSCEPTIBILITY; ELECTROMAGNETIC RADIATION; POLARIZATION OF DIELECTRICS.

Second-order devices. Devices based on the second-order nonlinearity involve three-photon (or three-wave) mixing. In this process, two photons are mixed together to create a third photon, subject to energy- and momentum-conservation constraints. Different names are ascribed to this mixing process, depending upon the relative magnitudes of the energies of the three photons. *See* CONSERVATION OF ENERGY; CONSERVATION OF MOMENTUM.

When the two beginning photons are of equal energy or frequency, the mixing process gives a single photon with twice the energy or frequency of the original ones. This mixing process is called second-harmonic generation. Second-harmonic generation is used often in devices where photons of visible frequency are desired but the available underlying laser system is capable of producing only infrared photons. For example, the neodymium-doped yttrium-aluminum-garnet (Nd:YAG) laser produces photons in the infrared with a wavelength of 1.06 micrometers. These photons are mixed in a crystal with a large second-order nonlinearity and proper momentum-conservation characteristics to yield green second-harmonic photons of 0.532-μm wavelength. Under different momentum-conservation constraints, a similar interaction can take place between two photon fields of different frequency, resulting in photons whose energy or frequency is the sum of those of the original photons. This process is called sum-frequency mixing. *See* LASER.

Optical parametric oscillation/amplification occurs when one of the two initial photons has the largest energy and frequency of the three. A high-energy photon and a low-energy photon mix to give a third photon with an energy equal to the difference between the two initial photons. If initially the third field amplitude is zero, it is possible to generate a third field whose frequency is not previously present; in this case the process is called optical parametric oscillation. If the third field exists but at a low level, it can be amplified through the optical parametric amplification process. *See* PARAMETRIC AMPLIFIER.

Third-order devices. Devices based on the third-order nonlinearity involve a process called four-photon (or four-wave) mixing. In this process, three photons are mixed together to create a fourth photon, subject to energy- and momentum-conservation constraints. The four-photon mixing nonlinearity is responsible for the existence of so-called self-action effects where the refractive index and absorption coefficient of a light field are modified by the light field's own presence, for third-harmonic generation and related processes, and for phase-conjugation processes.

In a medium with a third-order nonlinearity, the refractive index and absorption coefficient of a light field present in the medium are modified by the strength of the light intensity. Because the field effectively acts on itself, this interaction is termed a self-action effect. The momentum-conservation constraints are automatically satisfied because of the degenerate frequencies involved in the interaction. Such an interaction manifests itself by changing the total absorption experienced by the light field as well as by changing the velocity of propagation of the light field. *See* ABSORPTION OF ELECTROMAGNETIC RADIATION; REFRACTION OF WAVES.

There are many devices based on the self-action effects. A reverse saturable absorber becomes more opaque because of the nonlinear absorption (also called two-photon adsorption) that it manifests. Refractive-index changes can be used to change the transmission characteristics of resonant cavities and other structures by modifying the effective optical path length (the product of actual structure length times the effective refractive index for the structure) and shifting the cavity resonances to other frequencies. Several nonlinear optical switches have been proposed based upon this resonance-shifting phenomenon. *See* CAVITY RESONATOR; OPTICAL BISTABILITY.

In a third-harmonic generation process, three photons of like energy and frequency are mixed to yield a single photon with three times the energy and frequency of the initial photons. Applications of third-harmonic generation are typically in the areas of frequency upconversion.

Phase-conjugation devices make use of a property that third-order media possess whereby energy- and frequency-degenerate photons from two counterpropagating fields are mixed with an incoming photon to yield a photon with exactly the opposite propagation direction and conjugate phase. This phase-conjugate field will pass out of the nonlinear optical device in exactly the direction opposite to the incoming field. Such devices are used in phase-conjugate mirrors, mirrors which have the ability to cancel phase variation in a beam due to, for example, atmospheric turbulence. *See* ADAPTIVE OPTICS; OPTICAL PHASE CONJUGATION.

The suitability of available nonlinear optical materials is a critical factor in the development of nonlinear optical devices. For certain applications, silica glass fibers may be used. Because of the long propagation distances involved in intercontinental transmission systems, the small size of the optical nonlinearity in silica is not a drawback. Other key materials are semiconductors [such as gallium arsenide (GaAs), zinc selenide (ZnSe), and indium gallium arsenide phosphide (InGaAsP)], certain organic polymeric films, hybrid materials such as semiconductor-doped glasses, and liquid crystals. *See* NONLINEAR OPTICS; OPTICAL MATERIALS. David R. Andersen

Nonlinear optics A field of study concerned with the interaction of electromagnetic radiation and matter in which the matter responds in a nonlinear manner to the incident radiation fields. The nonlinear response can result in intensity-dependent variation of the propagation characteristics of the radiation fields or in the creation of radiation fields that propagate at new frequencies or in new directions. Nonlinear effects can take place in solids, liquids, gases, and plasmas, and may involve one or more electromagnetic fields as well as internal excitations of the medium. Most of the work done in the field has made use of the high powers available from lasers. The wavelength range of interest generally extends from the far-infrared to the vacuum ultraviolet, but some nonlinear interactions have been observed at wavelengths extending from the microwave to the x-ray ranges. *See* LASER.

Nonlinear materials. Nonlinear effects of various types are observed at sufficiently high light intensities in all materials. It is convenient to characterize the response of the medium mathematically by expanding it in a power series in the electric and magnetic fields of the incident optical waves. The linear terms in such an expansion give rise to the linear index of refraction, linear absorption, and the magnetic permeability of the medium, while the higher-order terms give rise to nonlinear effects. *See* ABSORPTION OF ELECTROMAGNETIC RADIATION; REFRACTION OF WAVES.

In general, nonlinear effects associated with the electric field of the incident radiation dominate over magnetic interactions. The even-order dipole susceptibilities are zero except in media which lack a center of symmetry, such as certain classes of crystals, certain symmetric media to which external forces have been applied, or at boundaries between certain dissimilar materials. Odd-order terms can be nonzero in all materials regardless of symmetry. Generally the magnitudes of the nonlinear susceptibilities decrease rapidly as the order of the interaction increases. Second- and third-order effects have been the most extensively studied of the nonlinear interactions, although effects up to order 30 have been observed in a single process. In some situations, multiple low-order interactions occur, resulting in a very high effective order for the overall nonlinear process. For example, ionization through absorption of effectively 100 photons has been observed. In other situations, such as dielectric breakdown or saturation of absorption, effects of different order cannot be separated, and all orders must be included in the response. *See* ELECTRIC SUSCEPTIBILITY; POLARIZATION OF DIELECTRICS.

Stimulated scattering. Light can scatter inelastically from fundamental excitations in the medium, resulting in the production of radiation at a frequency that is shifted from that of the incident light by the frequency of the excitation involved. The difference in photon energy between the incident and scattered light is accounted for by excitation or deexcitation of the medium. Some examples are Brillouin scattering from acoustic vibrations; various forms of Raman scattering involving molecular rotations or vibrations, electronic states in atoms or molecules, lattice vibrations or spin waves in solids, spin flips in semiconductors, and electron plasma waves in plasmas; Rayleigh scattering involving density or entropy fluctuations; and scattering from concentration fluctuations in gases. *See* SCATTERING OF ELECTROMAGNETIC RADIATION.

At the power levels available from pulsed lasers, the scattered light experiences exponential gain, and the process is then termed stimulated, in analogy to the process of stimulated emission in lasers. In stimulated scattering, the incident light can be almost completely converted to the scattered radiation. Stimulated scattering has been observed for all of the internal excitations listed above. The most widely used of these processes are stimulated Raman scattering and stimulated Brillouin scattering.

Self-action and related effects. Nonlinear polarization components at the same frequencies as those in the incident waves can result in effects that change the index of refraction or the absorption coefficient, quantities that are constants in linear optical theory. For example, propagation through optical fibers can involve several nonlinear optical interactions. Self-phase modulation resulting from the nonlinear index can be used to spread the spectrum, and subsequent compression with diffraction gratings and prisms can be used to reduce the pulse duration. The shortest optical pulses, with durations of the order of 6 femtoseconds, have been produced in this manner. Linear dispersion in fibers causes pulses to spread in duration and is one of the major limitations on data transmission through fibers. Dispersive pulse spreading can be minimized with solitons, which are specially shaped pulses that propagate long distances without spreading. They are formed by a combined interaction of spectral broadening due to the nonlinear refractive index and anomalous dispersion found in certain parts of the spectrum. *See* SOLITON.

Coherent effects. Another class of effects involves a coherent interaction between the optical field and an atom in which the phase of the atomic wave functions is preserved during the interaction. These interactions involve the transfer of a significant fraction of the atomic population to an excited state. As a result, they cannot be described with the simple perturbation expansion used for the other nonlinear optical effects. Rather they require that the response be described by using all powers of the incident fields. These effects are generally observed only for short light pulses, of the order of several nanoseconds or less. In one interaction, termed self-induced transparency, a pulse of light of the proper shape, magnitude, and duration can propagate unattenuated in a medium which is otherwise absorbing.

Other coherent effects involve changes of the propagation speed of a light pulse or production of a coherent pulse of light, termed a photon echo, at a characteristic time after two pulses of light spaced apart by a time interval have entered the medium. Still other coherent interactions involve oscillations of the atomic polarization, giving rise to effects known as optical nutation and free induction decay. Two-photon coherent effects are also possible.

Nonlinear spectroscopy. The variation of the nonlinear susceptibility near the resonances that correspond to sum- and difference-frequency combinations of the input frequencies forms the basis for various types of nonlinear spectroscopy which allow study of energy levels that are not normally accessible with linear optical spectroscopy.

Nonlinear spectroscopy can be performed with many of the interactions discussed earlier. Multiphoton absorption spectroscopy can be performed by using two strong laser beams, or a strong laser beam and a weak broadband light source. If two counterpropagating laser beams are used, spectroscopic studies can be made of energy levels in gases with spectral resolutions much smaller than the Doppler limit. Nonlinear optical spectroscopy has been used to identify many new energy levels with principal quantum numbers as high as 150 in several elements. *See* RESONANCE IONIZATION SPECTROSCOPY; RYDBERG ATOM.

Many types of four-wave mixing interactions can also be used in nonlinear spectroscopy. The most widespread of these processes, termed coherent anti-Stokes Raman spectroscopy (CARS), offers the advantage of greatly increased signal levels over linear Raman spectroscopy for the study of certain classes of materials.

Phase conjugation. Optical phase conjugation is an interaction that generates a wave that propagates in the direction opposite to a reference, or signal, wave, and has the same spatial variations in intensity and phase as the original signal wave, but with the sense of the phase variations reversed. Several nonlinear interactions are used to produce phase conjugation.

Optical phase conjugation allows correction of optical distortions that occur because of propagation through a distorting medium. This process can be used for improvement of laser-beam quality, optical beam combining, correction of distortion because of mode dispersion in fibers, and stabilized aiming. It can also be used for neural networks that exhibit learning properties. *See* NEURAL NETWORK; OPTICAL PHASE CONJUGATION.

John F. Reintjes

Photorefractive effect. The photorefractive effect occurs in many electrooptic materials. A change in the index of refraction in a photorefractive medium arises from the redistribution of charge that is induced by the presence of light. Charge carriers that are trapped in impurity sites in a photorefractive medium are excited into the material's conduction band when exposed to light. The charges migrate in the conduction band until they become retrapped at other sites. The charge redistribution produces an electric field that in turn produces a spatially varying index change through the electrooptic effect in the material. Unlike most other nonlinear effects, the index change of the photorefractive effect is retained for a time in the absence of the light and thus may be used as an optical storage mechanism. Storage times range from milliseconds to months or years, depending upon the material and the methods employed. *See* TRAPS IN SOLIDS.

Photorefractive materials are often used for holographic storage. In this case, the index change mimics the intensity interference pattern of two beams of light. Over 500 holograms have been stored in the volume of a single crystal of iron-doped lithium niobate. *See* HOLOGRAPHY.

Photorefractive materials are typically sensitive to very low light levels. The photorefractive effect is, however, extremely slow by the standards of optical nonlinearity. Because of their sensitivity, photorefractive materials are increasingly used for image and optical-signal processing applications. *See* IMAGE PROCESSING; NONLINEAR OPTICAL DEVICES. Dana Z. Anderson

Nonlinear physics The study of situations where, in a general sense, cause and effect are not proportional to each other; in other words, if the measure of what is considered to be the cause is doubled, the measure of its effect is not simply twice as large. Many examples have been known in physics for a long time, and they seemed well understood. Over the last few decades, however, physicists have noticed that this lack of proportionality in some of the basic laws of physics often leads to unexpected complications, if not to outright contradictions. Thus, the term nonlinear physics refers more narrowly to these developments in the understanding of physical reality.

Linearity in nonlinear systems. When a large number of particles starts out in a condition of stable equilibrium, the result of small external forces is well-coordinated vibrations of the whole collection, for example, the vibrations of a violin string, or of the electric current in an antenna. Each collective motion acts like an independent oscillator, each with its own frequency. In more complicated systems, many vibrational modes can be active simultaneously without mutual interference. A large dynamical system is, therefore, described in terms of its significant degrees of freedom, thought to be only loosely coupled. The motion of any part of the whole becomes multiperiodic; for example, a water molecule has bending and stretching vibrations with different frequencies, and both are coupled with the rotational motion of the whole molecule. *See* ANTENNA (ELECTROMAGNETISM); DEGREE OF FREEDOM (MECHANICS); MOLECULAR STRUCTURE AND SPECTRA; VIBRATION.

Failure of perturbation theory. H. Poincaré discovered at the end of the nineteenth century that for many problems this perturbation theory is not entirely satisfactory. He showed, in the case of the Moon's motion around the Earth, that the disturbance by the Sun is strong enough that this standard mathematical procedure fails. The main culprits are resonances, which occur when the frequencies of different degrees of freedom are combined through their nonlinear coupling. A key nonperturbative phenomenon is known to engineers as phase lock: When different frequencies arise in simple multiples of one another, the whole dynamical system falls into a dynamical trap; and for a continuous range of initial conditions, the interaction changes the frequencies of the individual degrees of freedom sufficiently to "lock" the motion into the resonance. *See* ASTEROID; PHASE-LOCKED LOOPS; RESONANCE (ACOUSTICS AND MECHANICS).

KAM theorem. In the 1950s, A. N. Kolmogoroff provided a first account of how the addition of a weak coupling generates chaotic regions and islands in phase space. This problem was later worked out in detail by V. Arnold and J. Moser to yield the KAM theorem. This theorem gives detailed information about the loss of the regular structure as the strength of the coupling increases. It does not say anything, however, about the trajectories in the newly created areas of chaotic behavior. These further investigations are the main goal of such fields as chaos or complexity. The impact of Poincaré's general arguments and the KAM theorem reaches into every area of nonlinear physics. The oldest among the areas is hydrodynamics, where the phenomenon of turbulent flow has so far resisted any effective control. This is what makes weather prediction so difficult. Signal propagation along the nerves and transmission of pulses through synaptic connections are other well-known nonlinear processes. *See* CHAOS; NEUROBIOLOGY; NONLINEAR ACOUSTICS; NONLINEAR OPTICS. Martin C. Gutzwiller

Nonlinear programming The area of applied mathematics and operations research concerned with finding the largest or smallest value of a function subject to constraints or restrictions on the variables of the function. Nonlinear programming is sometimes referred to as nonlinear optimization.

A useful example concerns a power plant that uses the water from a reservoir to cool the plant. The heated water is then piped into a lake. For efficiency, the plant should be run at the highest possible temperature consistent with safety considerations, but there are also limits on the amount of water that can be pumped through the plant, and there are ecological constraints on how much the lake temperature can be raised. The optimization problem is to maximize the temperature of the plant subject to the safety constraints, the limit on the rate at which water can be pumped into the plant, and the bound on the increase in lake temperature.

The nonlinear programming problem refers specifically to the situation in which the function to be minimized or maximized, called the objective function, and the functions that describe the constraints are nonlinear functions. Typically, the variables are continuous; this article is restricted to this case.

Researchers in nonlinear programming consider both the theoretical and practical aspects of these problems. Theoretical issues include the study of algebraic and geometric conditions that characterize a solution, as well as general notions of convexity that determine the existence and uniqueness of solutions. Among the practical questions that are addressed are the mathematical formulation of a specific problem and the development and analysis of algorithms for finding the solution of such problems.

The general nonlinear programming problem can be stated as that of minimizing a scalar-valued objective function $f(x)$ over all vectors x satisfying a set of constraints. The constraints are in the form of general nonlinear equations and inequalities. Mathematically, the nonlinear programming problem may be expressed as below, where $\mathbf{x} = (x_1, x_2, \ldots, x_n)$ are the variables of the problem,

$$\begin{aligned} &\text{minimize } f(\mathbf{x}) \text{ with respect to } \mathbf{x} \\ &\text{subject to: } g_i(\mathbf{x}) \leq 0, \quad i = 1, 2, \ldots, m \\ &\qquad\qquad h_j(\mathbf{x}) = 0, \quad j = 1, 2, \ldots, p \end{aligned}$$

f is the objective function, $g_i(\mathbf{x})$ are the inequality constraints, and $h_j(\mathbf{x})$ are the equality constraints. This formulation is general in that the problem of maximizing $f(\mathbf{x})$ is equivalent to minimizing $-f(\mathbf{x})$ and a constraint $g_i(\mathbf{x}) \geq 0$ is equivalent to the constraint $-g_i(\mathbf{x}) \leq 0$.

Since general nonlinear equations cannot be solved in closed form, iterative methods must be used. Such methods generate a sequence of approximations, or iterates, that will converge to a solution under specified conditions. Newton's method is one of the best-known methods and is the basis for many of the fastest methods for solving the nonlinear programming problem. Paul T. Boggs

Nonmetal The elements are conveniently, but arbitrarily, divided into metals and nonmetals. The nonmetals do not conduct electricity readily, are not ductile, do not have a complex refractive index, and in general have high ionization potentials.

If the periodic table is divided diagonally from upper left to lower right, all the nonmetals are on the right-hand side of the diagonal. Examples of elements which do not fit neatly into this useful but arbitrary classification are tin, which exists in two allotropic modifications, one definitely metallic and the other with many properties of a nonmetal, and tellurium and antimony. Such elements are called metalloids. *See* METAL; METALLOID; PERIODIC TABLE.

Thomas C. Waddington

Non-newtonian fluid A fluid that departs from the classic linear newtonian relation between stress and shear rate. In a strict sense, a fluid is any state of matter that is not a solid, and a solid is a state of matter that has a unique stress-free state. A conceptually simpler definition is that a fluid is capable of attaining the shape of its container and retaining that shape for all time in the absence of external forces. Therefore, fluids encompass a wide variety of states of matter including gases and liquids as well as many more esoteric states (for example, plasmas, liquid crystals, and foams). *See* FLUIDS; FOAM; GAS; LIQUID; LIQUID CRYSTALS; PLASMA (PHYSICS).

A newtonian fluid is one whose mechanical behavior is characterized by a single function of temperature, the viscosity, a measure of the "slipperiness" of the fluid. For the example of Fig. 1, where a fluid is sheared between a fixed plate and a moving plate, the viscosity is given by Eq. (1). Thus, as the vis-

$$\text{Viscosity} = \frac{\text{force/area}}{\text{velocity/height}} \tag{1}$$

cosity of a fluid increases, it requires a larger force to move the top plate at a given velocity. For simple, newtonian fluids, the viscosity is a constant dependent on only temperature; but for non-newtonian fluids, the viscosity can change by many orders of magnitude as the shear rate (velocity/height in Fig. 1) changes. Typically, the viscosity (η) of these fluids is given as a function of the shear rate ($\dot{\gamma}$). A common dependence for this function is given in Fig. 2. For other non-newtonian fluids, the viscosity might increase as the shear rate increases (shear-thickening fluids). *See* NEWTONIAN FLUID; VISCOSITY.

Many of the fluids encountered in everyday life (such as water, air, gasoline, and honey) are adequately described as being newtonian, but there are even more that are not. Common examples include mayonnaise, peanut butter, toothpaste, egg whites, liquid soaps, and multigrade engine oils. Other examples such as molten polymers and slurries are of considerable technological importance. A distinguishing feature of many non-newtonian fluids is that they have microscopic or molecular-level structures that can be rearranged substantially in flow. *See* PARTICLE FLOW; POLYMER.

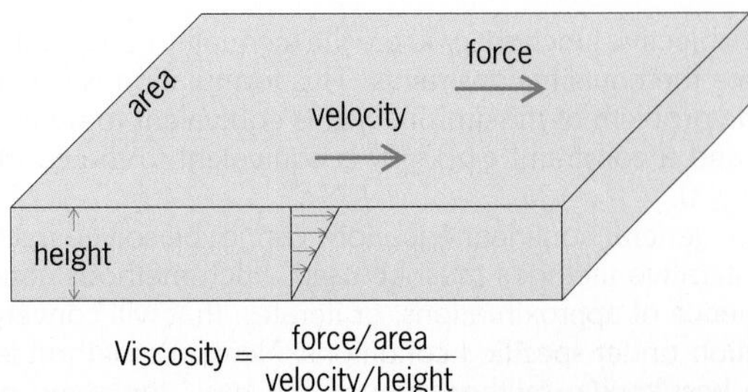

Fig. 1. Steady shear flow of a fluid between a fixed plate and a parallel plate, illustrating the concept of viscosity.

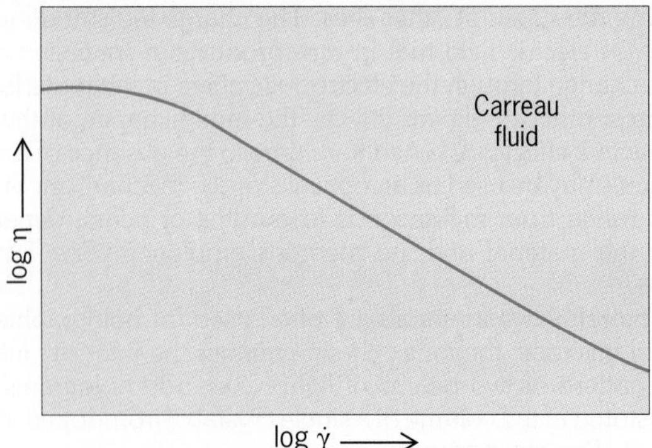

Fig. 2. Typical dependence of the viscosity (η) on shear rate ($\dot{\gamma}$) for a non-newtonian fluid (Carreau model).

Our intuitive understanding of how fluids behave and flow is built primarily from observations and experiences with newtonian fluids. However, non-newtonian fluids display a rich variety of behavior that is often in dramatic contrast to these expectations. For example, an intuitive feel for the slipperiness of fluids can be gained from rubbing them between the fingers. Furthermore, the slipperiness of water, experienced in this way, is expected to be the same as the slipperiness of automobile tires on a wet road. However, the slipperiness (viscosity) of many non-newtonian fluids changes a great deal depending on how fast they move or the forces applied to them.

Intuitive expectations for how the surface of a fluid will deform when the fluid is stirred (with the fluid bunching up at the wall of the container) are also in marked contrast to the behavior of non-newtonian fluids. When a cylindrical rod is rotated inside a container of a newtonian fluid, centrifugal forces cause the fluid to be higher at the wall. However, for non-newtonian fluids, the normal stress differences cause the fluid to climb the rod; this is called the Weissenberg effect. Intuitive understanding about the motion of material when the flow of a fluid is suddenly stopped, for example, by turning off a water tap, is also notably at odds with the behavior of non-newtonian fluids. *See* CENTRIFUGAL FORCE.

A non-newtonian fluid also displays counterintuitive behavior when it is extruded from an opening. A newtonian fluid tapers to a smaller cross section as it leaves the opening, but the cross section for a non-newtonian fluid first increases before it eventually tapers. This phenomenon is called die swell. *See* NOZZLE.

When a newtonian fluid is siphoned and the fluid level goes below the entrance to the siphon tube, the siphoning action stops. For many non-newtonian fluids, however, the siphoning action continues as the fluid climbs from the surface and continues to enter the tube. This phenomenon is called the tubeless siphon.

Perhaps the most striking behavior of non-newtonian fluids is a consequence of their viscoelasticity. Solids can be thought of as having perfect memory. If they are deformed through the action of a force, they return to their original shape when the force is removed. This happens when a rubber ball bounces; the ball is deformed as it hits a surface, but the rubber remembers its undeformed spherical shape. Recovery of the shape causes the ball to bounce back. In contrast, newtonian fluids have no memory; when a force is removed, they retain their condition at the time the force is removed (or continue moving as the result of inertia). When a newtonian fluid is dropped onto a surface, it does not bounce. Non-newtonian fluids are viscoelastic in the sense that they have fading memory. If a force is removed shortly after it is applied, the fluid will remember its undeformed shape and return toward it. However, if the force is applied on the fluid for a long time, the fluid will eventually forget its undeformed shape. If a sample of a non-newtonian fluid is dropped onto a surface, it will bounce like a ball. However, if the fluid is simply placed on the surface, it will flow smoothly. Viscoelasticity is frequently the cause of many of the secondary flows that are observed for non-newtonian fluids. These are fluid motions that are small for newtonian fluids (for example, swirling motions) but can become dominant for non-newtonian fluids. *See* ELASTICITY.

Analysis of fluid flow operations is typically performed by examining local conservation relations—conservation of mass, momentum (Newton's second law), and energy. This analysis requires material-specific information (for example, the relation between density, pressure, and temperature) that is collectively known as constitutive relations. The science devoted to obtaining suitable constitutive equations for description of the behavior of non-newtonian fluids is called rheology. The most important constitutive equation for fluid mechanics is that relating the stress in the fluid to the kinematics of the motion (that is, the velocity, the derivatives of the velocity with respect to position, and the time history of the velocity).

Although the non-newtonian behavior of many fluids has been recognized for a long time, the science of rheology is, in many respects, still in its infancy, and new phenomena are constantly being discovered and new theories proposed. Advancements in computational techniques are making possible much more detailed analyses of complex flows and more sophisticated simulations of the structural and molecular behavior that gives rise to non-newtonian behavior. Engineers, chemists, physicists, and mathematicians are actively pursuing research in rheology, particularly as more technologically important materials are found to display non-newtonian behavior. *See* FLUID MECHANICS; RHEOLOGY.

John M. Wiest

Nonsinusoidal waveform The representation of a wave that does not vary in a sinusoidal manner. Electric circuits containing nonlinear elements, such as iron-core magnetic devices, electrifying devices, and transistors, commonly produce nonsinu-soidal currents and voltages. When these are repetitive functions of time, they are called nonsinusoidal electric waves. Oscillograms, tabulated data, and sometimes mathematical functions for segments of such waves are often used to describe the variation through one cycle. The term cycle corresponds to 2π electrical radians and covers the period, which is the time interval T in seconds in which the wave repeats itself.

These electric waves can be represented by a constant term, the average or dc component, plus a series of harmonic terms in which the frequencies of the harmonics are integral multiples of the fundamental frequency. The fundamental frequency f_1, if it does exist, has the time span $T = 1/f_1$ seconds for its cycle. The second-harmonic frequency f_2 then will have two of its cycles within T seconds, and so on.

The series of terms stated above is known as a Fourier series and can be expressed in the form of the equation below, where $y(t)$, plotted over acycle of the fundamental, gives the shape of

$$y(t) = B_0 + C_1 \sin(\omega t + \phi_1) + \cdots + C_n \sin(n\omega t + \phi_n) + \cdots$$
$$= \sum_{n=0}^{\infty} C_n \sin(n\omega t + \phi_n)$$

the nonsinusoidal wave. The radian frequency of the fundamental is $\omega = 2\pi f_1$, and n is an integer. C_1 is the amplitude of the fundamental ($n = 1$), and succeeding C_n's are the amplitudes of the respective harmonics having frequencies corresponding to $n = 2, 3, 4$, and so on, with respect to the fundamental. The phase angle of the fundamental with respect to a chosen time reference axis is ϕ_1, and the succeeding ϕ_n's are the phase angles of the respective harmonics. *See* FOURIER SERIES.

The equation for $y(t)$, in general, includes an infinite number of terms. In practice, the first several terms usually yield an approximate result sufficiently accurate for portrayal of the actual wave. The degree of accuracy desired in representing faithfully the actual wave determines the number of terms that must be used in any computation.

Burtis L. Robertson; W. S. Pritchett

Nonstoichiometric compounds Chemical compounds in which the relative number of atoms is not expressible as the ratio of small whole numbers, hence compounds for which the subscripts in the chemical formula are not rational (for example, $Cu_{1.987}S$). Sometimes they are called berthollide compounds to distinguish them from daltonides, in which the ratio of atoms is generally simple. Nonstoichiometry is a property of the solid state and arises because a fraction of the atoms of a given kind may be (1) missing from the regular structure (for example, $Fe_{1-\delta}O$), (2) present in excess over the requirements of the structure (for example, $Zn_{1+\delta}O$), or (3) substituted by atoms of another kind (for example, $Bi_2Te_{3\pm\delta}$). The resulting materials are generally of variable composition, intensely colored, metallic or semiconducting, and different in chemical reactivity from the parent stoichiometric compounds from which they are derived.

Nonstoichiometry is best known in the binary compounds of the transition elements, particularly the hydrides, oxides, chalcogenides, pnictides, carbides, and borides. It is also well represented in the so-called insertion or intercalation compounds, in which a metallic element or neutral molecule has been inserted in a stoichiometric host. Nonstoichiometric compounds are important in some solid-state devices (such as rectifiers, thermoelectric generators, and photodetectors) and are probably formed as chemical intermediates in many reactions involving solids (for example, heterogeneous catalysis and metal corrosion).

The simplest way to classify nonstoichiometric compounds is to consider which element is in excess and how this excess is brought about. A classification scheme largely based on this distinction but which also includes some examples of ternary systems is as follows.

Binary compounds:

- I. Metal nonmetal ratio greater than stoichiometric
  - (*a*) Metal in excess, for example, $Zn_{1+\delta}O$
  - (*b*) Missing nonmetal, for example $UH_{3-\delta}$, $WO_{3-\delta}$
- II. Metal: nonmetal ratio less than stoichiometric
  - (*a*) Metal-deficient, for example, $Co_{1-\delta}O$
  - (*b*) Nonmetal in excess, for example, $UO_{2+\delta}$
- III. Deviations on both sides of stoichiometry, for example, $TiO_{1\pm\delta}$

Ternary compounds (insertion compounds):

- IV. Oxide "bronzes," for example, $M_\delta WO_3$, $M_\delta V_2O_5$
- V. Intercalation compounds, for example, $K_{1.5+\delta}MoO_3$, $LL_\delta TiS_2$

Excluded from consideration are the recognized impurity materials, such as $Na_{1-2x}Ca_xCl$, which are best considered as conventional solid solutions wherein ions of one kind and perhaps vacancies have replaced an equivalent number of ions of another kind.

Michell J. Sienko

Noradrenergic system A neuronal system that is responsible for the synthesis, storage, and release of the neurotransmitter norepinephrine. Norepinephrine, also known as noradrenalin, consists of a single amine group and a catechol nucleus (a benzene ring with two hydroxyl groups) and is therefore referred to as a monoamine or catecholamine. It exists in both the central and peripheral nervous systems. Norepinephrine is the primary neurotransmitter released by the sympathetic nervous system, which mediates the "fight or flight" reaction, preparing the body for action by affecting cardiovascular function, gastrointestinal motility and secretion, bronchiole dilation, glucose metabolism, and so on. Within the central nervous system, norepinephrine has been associated with several brain functions, including sleep, memory, learning, and emotions.

After synthesis, the majority of norepinephrine is transported into synaptic vesicles in the nerve terminals, where it remains until needed. When the nerve terminal is activated by depolarization, calcium flows into it, leading to the release of norepinephrine into the synaptic cleft. Once released into the synaptic cleft, norepinephrine is free to bind to specific receptors located on the presynaptic or postsynaptic terminal, which initiates a chain of events (the effector system) in the target cell that can be mediated by a number of different second messenger systems. The exact effect is determined by the identity of the receptor activated. *See* EPINEPHRINE; SYMPATHETIC NERVOUS SYSTEM; SYNAPTIC TRANSMISSION.

Termination of norepinephrine occurs by a reuptake mechanism in the presynaptic membrane. Once transported back into the presynaptic terminal, norepinephrine can be stored in vesicles for future use or enzymatically degraded by monoamine oxidase.

Certain medications achieve their effect by altering various stages of synthesis, storage, release, and inactivation of norepinephrine. The behavioral manifestations of these alterations have led to a better understanding of norepinephrine's role in various psychiatric disorders. *See* AFFECTIVE DISORDERS; MONOAMINE OXIDASE; PSYCHOPHARMACOLOGY; SCHIZOPHRENIA; STRESS.

Michelle Mynlieff; Dennis S. Charney; Alan Breier; Steven Southwick

Normal (mathematics) A term generically synonymous with perpendicular, which often refers specifically to a line that goes through a point *P* of a curve *C* and is perpendicular to the tangent to *C* at *P*. If curve *C* is not a plane curve, all normal lines of *C* at point *P* on *C* lie in a plane, the normal plane of *C* at *P*. *See* ANALYTIC GEOMETRY.

Leonard M. Blumenthal

North America The third largest continent, extending from the narrow isthmus of Central America to the Arctic Archipelago. The physical environments of North America, like the rest of the world, are a reflection of specific combinations of the natural factors such as climate, vegetation, soils, and landforms. *See* CONTINENT.

Location. North America covers 9,400,000 mi^2 (24,440,000 km^2) and extends north to south for 5000 mi (8000 km) from Central America to the Arctic. It is bounded by the Pacific Ocean on the west and the Atlantic Ocean on the east. The Gulf of Mexico is a source of moist tropical air, and the frozen Arctic Ocean is a source of polar air. With the major mountain ranges stretching north-south, North America is the only continent providing for direct contact of these polar and tropical air masses, leading to frequent climatically induced natural hazards such as violent spring tornadoes, extreme droughts, subcontinental floods, and winter blizzards, which are seldom found on other continents. *See* AIR MASS; ARCTIC OCEAN; ATLANTIC OCEAN; GULF OF MEXICO; PACIFIC OCEAN.

Geologic structure. The North American continent includes (1) a continuous, broad, north-south-trending western cordilleran belt stretching along the entire Pacific coast; (2) a northeast-southwest-trending belt of low Appalachian Mountains paralleling the Atlantic coast; (3) an extensive rolling region of old eroded crystalline rocks in the north-central and northeastern part of the continent called the Canadian Shield; (4) a large, level interior lowland covered by thick sedimentary rocks and extending from the Arctic Ocean to the Gulf of Mexico; and (5) a narrow coastal plain along the Atlantic Ocean and the Gulf of Mexico. These broad structural geologic regions provide the framework for the natural regions of this continent and affect the location and nature of landform, climatic, vegetation, and soil regions.

Canadian Shield. Properly referred to as the geological core of the continent, the exposed Canadian Shield extends about 2500 mi (4000 km) from north to south and almost as much from east to west. The rest of it dips under sedimentary rocks that overlap it on the south and west. The Canadian Shield consists of ancient Precambrian rocks, over 500 million years old, predominantly granite and gneiss, with very complex structures indicating several mountain-building episodes. It has been eroded into a rolling surface of low to moderate relief with elevations generally below 2000 ft (600 m). Its surface has been warped into low domes and basins, such as the Hudson Basin, in which lower Paleozoic rocks, including Ordovician limestones, have been preserved. Since the end of the Paleozoic Era, the Shield has been dominated by erosion. Parts of the higher surface remain at about 1500–2000 ft (450–600 m) above sea level, particularly in the Labrador area. The Shield remained as land throughout the Mesozoic Era, but its western margins were covered by a Cretaceous sea and by Tertiary terrestrial sediments derived from the Western Cordillera. *See* CRETACEOUS; ORDOVICIAN; MESOZOIC; PALEOZOIC; PRECAMBRIAN; TERTIARY.

The entire exposed Shield was glaciated during the Pleistocene Epoch, and its surface was intensely eroded by ice and its meltwaters, erasing major surface irregularities and eastward-trending rivers that were there before. The surface is now covered by glacial till, outwash, moraines, eskers, and lake sediments, as well as drumlins formed by advancing ice. A deranged drainage pattern is evolving on this surface with thousands of lakes of various sizes. *See* DRUMLIN; ESKER; GLACIAL EPOCH; GLACIATED TERRAIN; MORAINE; PLEISTOCENE; TILL.

The Canadian Shield extends into the United States as Adirondack Mountains in New York State, and Superior Upland west of Lake Superior.

Southeastern Coastal Plain. The Southeastern Coastal Plain is geologically the youngest part of the continent, and it is covered by the youngest marine sedimentary rocks. This flat plain, which parallels the Atlantic and Gulf coastline, extends for over 3000 mi (4800 km) from Cape Cod, Massachusetts, to the Yucatán Peninsula in Mexico. It is very narrow in the north but increases in width southward along the Atlantic coast and includes the entire peninsula of Florida. As it continues westward along the Gulf, it widens significantly and includes the lower Mississippi River valley. It is very wide in Texas, narrows again southward in coastal Mexico, and then widens in the Yucatán Peninsula and continues as a wide submerged plain, or a continental shelf, into the sea. *See* COASTAL PLAIN.

Extending from Cape Cod, Massachusetts, to Mexico and Central America, the Coastal Plain is affected by a variety of climates and associated vegetation. While a humid, cool climate with four seasons affects its northernmost part, subtropical air masses affect the southeastern part, including Florida, and hot and arid climate dominates Texas and northern Mexico; Central America has hot, tropical climates.

Varied soils characterize the Coastal Plain, including the fertile alluvial soils of the Mississippi Valley. Broadleaf forests are present in the northeast, citrus fruits grow in Florida, grasslands dominate the dry southwest, and tropical vegetation is present on Central American coastal plains.

Eastern Seaboard Highlands. Between the Southeastern Coastal Plain and the extensive interior provinces lies a belt of mountains that, by their height and pattern, create a significant barrier between the eastern seaboard and the interior of North America. These mountains consist of the Adirondack Mountains and the New England Highlands.

The Adirondack Mountains are a domal extension of the Canadian Shield, about 100 mi (160 km) in diameter, composed of complex Precambrian rocks. The New England Highlands consist of a north-south belt of mountains east of the

Hudson Valley, including the Taconic mountains in the south and the Green mountains in the north, and continuing as the Notre Dame Mountains along the St. Lawrence Valley and the Chic-Choc Mountains of the Gaspé Peninsula. The large area of New England east of these mountains is an eroded surface of old crystalline rocks culminating in the center as the White Mountains, with their highest peak of the Presidential Range, Mount Washington, reaching over 6200 ft (1880 m). This area has been intensely glaciated, and it meets the sea in a rugged shoreline. Nova Scotia and Newfoundland have a similar terrain.

New England is a hilly to mountainous region carved out of ancient rocks, eroded by glaciers, and covered by glacial moraines, eskers, kames, erratics, and drumlins, with hundreds of lakes scattered everywhere. It has a cool and moist climate with four seasons, thin and acid soils, and mixed coniferous and broadleaf forests.

Appalachian Highlands. The Appalachian Highlands are traditionally considered to consist of four parts: the Piedmont, the Blue Ridge Mountains, the Ridge and Valley Section, and the Appalachian Plateau. These subregions are all characterized by different geologic structures and rock types, as well as different geomorphologies.

The northern boundary of the entire Appalachian System is an escarpment of Paleozoic rocks trending eastward along Lake Erie, Lake Ontario, and the Mohawk Valley. The boundary then swings south along Hudson River Valley and continues southwestward along the Fall Line to Montgomery, Alabama. The western boundary trends northeastward through Cumberland Plateau in Tennessee, and up to Cleveland, Ohio, where it joins the northern boundary. Together with New England, this region forms the largest mountainous province in eastern United States.

Interior Domes and Basins Province. The southwestern part of the Appalachian Plateau, overlain mainly by the Mississippian and Pennsylvanian sedimentary rocks, has been warped into two low structural domes called the Blue Grass and Nashville Basins, and a structural basin, drained by the Green River; its southern fringe is called the Pennyroyal Region. The Interior Dome and Basin Province is contained roughly between the Tennessee River in the south and west and the Ohio River in the north.

There is no boundary on the east, because the domes are part of the same surface as the Appalachian Plateau. However, erosional escarpments, forming a belt of hills called knobs, clearly mark the topographic domes and basins. The northern dome, called the Blue Grass Basin or Lexington Plain, has been eroded to form a basin surrounded by a series of inward-facing cuesta escarpments. The westernmost cuesta reaches about 600 ft (180 m) elevation while the central part of the basin lies about 1000 ft (300 m) above sea level, which is higher than the surrounding hills. This gently rolling surface with deep and fertile soils exhibits some solutional karst topography. *See* FLUVIAL EROSION LANDFORMS.

Ozark and Ouachita Highlands. The Paleozoic rocks of the Pennyroyal Region continue Westward across southern Illinois to form another dome of predominantly Ordovician rocks, called the Ozark Plateau. This dome, located mainly in Missouri and Arkansas, has an abrupt east side, and a gently sloping west side, called the Springfield Plateau. Its surface is stream eroded into hilly and often rugged topography that is developed mainly on limestones, although shales, sandstone, and chert are present. Much residual chert, eroded out of limestone, is present on the surface. There are some karst features, such as caverns and springs. In the northeast, Precambrian igneous rocks protrude to form the St. Francois Mountains, which reach an elevation of 1700 ft (515 m).

Central Lowlands. One of the largest subdivisions of North America is the Central Lowlands province which is located between the Appalachian Plateau on the east, the Interior Domes and Basins Province and the Ozark Plateau on the south, and the Great Plains on the west. It includes the Great Lakes section and the Manitoba Lowland in Canada. This huge lowland in the heart of the continent (whose elevations vary from about 900 ft or 270 m above sea level in the east and nearly 2000 ft or 600 m in the west) is underlain by Paleozoic rocks that continue from the Appalachian Plateau and dip south under the recent coastal plain sediments; meet the Cretaceous rocks on the west; and overlap the crystalline rocks of the Canadian Shield on the northeast.

The present surface of nearly the entire Central Lowlands, roughly north of the Ohio River and east of the Missouri River, is the creation of the Pleistocene ice sheets. When the ice formed and spread over Canada, and southward to the Ohio and Missouri rivers, it eroded much of the preexisting surface. During deglaciation, it left its deposits over the Canadian Shield and the Central Lowlands.

The Central Lowlands are drained by the third longest river system in the world, the Missouri-Mississippi, which is 3740 mi (6000 km) long. This mighty river system, together with the Ohio and the Tennessee, drains not only the Central Lowlands but also parts of the Appalachian Plateau and the Great Plains, before it crosses the Coastal Plain and ends in the huge delta of the Mississippi. The river carries an enormous amount of water and alluvium and continues to extend its delta into the Gulf. In 1993 it reached a catastrophic level of a hundred-year flood, claimed an enormous extent of land and many lives, and created an unprecedented destruction of property. This flood again alerted the population to the extreme risk of occupying a river floodplain. *See* FLOODPLAIN; RIVER.

Great Plains. The Great Plains, which lie west of the Central Lowlands, extend from the Rio Grande and the Balcones Escarpment in Texas to central Alberta in Canada. On the east, they are bounded by a series of escarpments, such as the Côteau du Missouri in the Dakotas. The dry climate with less than 20 in. (50 cm) of precipitation, and steppe grass vegetation growing on calcareous soils, help to determine the eastern boundary of the Great Plains. On the west, the Great Plains meet the abrupt front of the Rocky Mountains, except where the Colorado Piedmont and the lower Pecos River Valley separate them from the mountains.

The Great Plains region shows distinct differences between its subsections from south to north. The southernmost part, called the High Plains or Llano Estacado, and Edwards Plateaus are the flattest. While Edwards Plateau, underlain by limestones of the Cretaceous age, reveals solutional karst features, the High Plains have the typical Tertiary bare cap rock surface, devoid of relief and streams.

The central part of the Great Plains has a recent depositional surface of loess and sand. The Sand Hills of Nebraska form the most extensive sand dunes area in North America, covering about 24,000 mi^2 (62,400 km^2). They are overgrown by grass and have numerous small lakes. The loess region to the south provides spectacular small canyon topography. *See* DUNE.

The northern Great Plains, stretching north of Pine Ridge and called the Missouri Plateau, have been intensly eroded by the western tributaries of the Missouri River into river breaks and interfluves. In extreme cases, badlands were formed, such as those of the White River and the Little Missouri.

The terrain of the Canadian Great Plains consists of three surfaces rising from east to west: the Manitoba, Saskatchewan, and Alberta Prairies developed on level Creteceous and Tertiary rocks. Climatic differences between the arid and warm southern part and the cold and moist northern part have resulted in regional differences. The eastern boundary of the Saskatchewan Plain is the segmented Manitoba Escarpment, which extends for 500 mi (800 km) northwestward, and in places rises 1500 ft

(455 m) above the Manitoba Lowland. Côteau du Missouri marks the eastern edge of the higher Alberta Plain.

Western Cordillera. The mighty and rugged Western Cordilleras stretch along the Pacific coast from Alaska to Mexico. There are three north-south-trending belts: (1) Brooks Range, Mackenzie Mountains, and the Rocky Mountains to the north and Sierra Madre Oriental in Mexico; (2) Interior Plateaus, including the Yukon Plains, Canadian Central Plateaus and Ranges, Columbia Plateau, Colorado Plateau, and Basin and Range Province stretching into central Mexico; and (3) Coastal Mountains from Alaska Range to California, Baja California, and Sierra Madre Occidental in Mexico.

This subcontinental-size mountain belt has the highest mountains, greatest relief, roughest terrain, and most beautiful scenery of the entire continent. It has been formed by earth movements resulting from the westward shift of the North American lithospheric plate. The present movements, and the resulting devastating earthquakes along the San Andreas fault system paralleling the Pacific Ocean, are part of this process. *See* CORDILLERAN BELT; PLATE TECTONICS.

This very high, deeply eroded and rugged Rocky Mountains region comprises several distinct parts: Southern, Middle, and Northern Rockies, plus the Wyoming Basin in the United States, and the Canadian Rockies. The Southern Rockies, extending from Wyoming to New Mexico, include the Laramie Range, the Front Range, and Spanish Peaks with radiating dikes on the east; Medicine Bow, Park, and Sangre de Cristo ranges in the center; and complex granite Sawatch Mountains and volcanic San Juan Mountains of Tertiary age on the west. Most of the ranges are elongated anticlines with exposed Precambrian granite core, and overlapping Paleozoic and younger sedimentary rocks which form spectacular hogbacks along the eastern front. There are about 50 peaks over 14,000 ft (4200 m) high, while the Front Range alone has about 300 peaks over 13,000 ft (3940 m) high. The southern Rocky Mountains, heavily glaciated into a beautiful and rugged scenery with permanent snow and small glaciers, form a major part of the Continental Divide.

The interior Plateaus and Ranges Province of the Western Cordillera lies between the Rocky Mountains and the Coastal Mountains. It is an extensive and complex region. It begins in the north with the wide Yukon Plains and Uplands; narrows into the Canadian Central Plateaus and Ranges; widens again into the Columbia Plateau, Basin and Range Province, and Colorado Plateau; and finally narrows into the Mexican Plateau and the Central American isthmus.

The coastal Lowlands and Ranges extend along the entire length of North America and include Alaskan Coast Ranges, Aleutian Islands, Alaska Range, Canadian Coast Ranges, and a double chain of the Cascade Mountains and Sierra Nevada on the east, and Coast Ranges on the west, separated by Puget Sound, Willamette Valley, and Great Valley of California. These ranges continue southward as Lower California Peninsula, Baja California, and Sierra Madre Occidental in Mexico.

The basin-and-range type of terrain of the southwest United States continues into northern Mexico and forms its largest physiographic region, the Mexican Plateau. This huge tilted block stands more than a mile above sea level—from about 4000 ft (1200 m) in the north, it rises to about 8000 ft (2400 m) in the south. The Mexican Plateau is separated from the Southern Mexican Highlands (Sierra Madre del Sur) by a low, hot and dry Balsas Lowland drained by the Balsas River. To the east of the Southern Highlands lies a lowland, the Isthmus of Tehuantepec, which is considered the divide between North and Central America. Here the Pacific and Gulf coasts are only 125 mi (200 km) apart. The lowlands of Mexico are the coastal plains. The Gulf Coastal Plain trends southward for 850 mi from the Rio Grande to the Yucatan Peninsula. It is about 100 mi (160 km) wide in the north, just a few miles wide in the center, and very wide in the Yucatan Peninsula. Barrier beaches, lagoons, and swamps occur along this coast. The Pacific Coastal Plains are much narrower and more hilly. North-south-trending ridges of granite characterize the northern part, and islands are present offshore. Toward the south, sandbars, lagoons, and deltaic deposits are common.

East of the Isthmus of Tehuantepec begins Central America with its complex physiographic and tectonic regions. This narrow, mountainous isthmus is geologically connected with the large, mountainous islands of the Greater Antilles in the Caribbean. They are all characterized by east-west-trending rugged mountain ranges, with deep depressions between them. One such mountain system begins in Mexico and continues in southern Cuba, Puerto Rico, and the Virgin Islands. North of this system, called the Old Antillia, lies the Antillian Foreland, consisting of the Yucatán Peninsula and the Bahama Islands. Central American mountains are bordered on both sides by active volcanic belts. Along the Pacific, a belt of young volcanoes extends for 800 mi (1280 km) from Mexico to Costa Rica. Costa Rica and Panama are mainly a volcanic chain of mountains extending to South America. Nicaragua is dominated by a major crustal fracture trending northwest-southeast. Barbara Zakrzewska Borowiecka

North Pole

North Pole That end of the Earth's axis which points toward the North Star, Polaris (Alpha Ursae Minoris). It is the geographical pole where all meridians converge, and should not be confused with the north magnetic pole, which is in the Canadian Archipelago. The North Pole's location falls near the center of the Arctic Sea.

The North Pole has phenomena unlike any other place except the South Pole. For 6 months the Sun does not appear above the horizon, and for 6 months it does not go below the horizon. As there is a long period (about 7 weeks) of continuous twilight before March 21 and after September 23, the period of light is considerably longer than the period of darkness. Van H. English

North Sea

North Sea A flooded portion of the northwest continental margin of Europe occupying an area of over 200,000 mi^2 (500,000 km^2). The North Sea has extensive marine fisheries and important offshore oil and gas reserves. In the south, its depth is less than 150 ft (50 m), but north of 58° it deepens gradually to 600 ft (200 m) at the top of the continental slope. A band of deep water down to 1200 ft (400 m) extends around the south and west coast of Norway and is known as the Norwegian Trench.

The nontidal residual current circulation of the southern North Sea is mainly determined by wind velocity, but in the north, well-defined non-wind-driven currents have been identified, especially in the summer. Two of these currents bring in water from outside the North Sea; one flows through the channel between Orkney and Shetland (the Fair Isle current), and the other follows the continental slope north of Shetland and merges with the Fair Isle current southwest of Norway before entering the Skagerrak. The north-flowing Norwegian coastal current provides the exit route for North Sea waters, and is formed from the waters of these two major inflows and from other much smaller inputs such as river runoff, the English Channel, and the Baltic Sea.

There is a rich diversity of zooplankton within the North Sea. Copepods are of particular importance in the food web. There are a wide range of fish stocks in the North Sea and adjacent waters and, in terms of species exploited by commercial fisheries, they constitute the richest area in the northeast Atlantic. The commercially important stocks exploited for human consumption include cod, haddock, whiting, pollock, plaice, sole, herring, mackerel, lobster, prawn, and brown shrimp (*Crangon crangon*). A number of stocks are used for fishmeal and oil; these stocks include sand eel, Norway pout, blue whiting, and sprat. *See* COPEPODA; MARINE ECOLOGY; ZOOPLANKTON. H. D. Dooley

Nose The nasal cavities and the structures surrounding and associated with them. The nose functions primarily as the organ of smell and in most tetrapods also assumes a respiratory function, forming the anterior end of the air passage through which air is drawn in and in which it is warmed and moistened.

In humans the nasal cavities are triangular openings that pass from the external nares back to the dorsal part of the pharynx (see illustration). The lateral walls are composed principally of portions of the ethmoid and sphenoid bones and projections of three turbinate bones, or conchae, on each side. The floor of the nose is formed by the palate, which is also the roof of the mouth. The nasal cavities are lined with respiratory epithelium, which also lines the paranasal sinuses. The latter are cavities in the frontal, ethmoid, sphenoid, and maxillary bones which communicate with the nasal passages. The external nose consists of the two nasal bones that form the bony bridge and two pairs of lower nasal cartilages. These together with the tightly adherent skin determine the individual shape and size of the human nose. *See* OLFACTION.

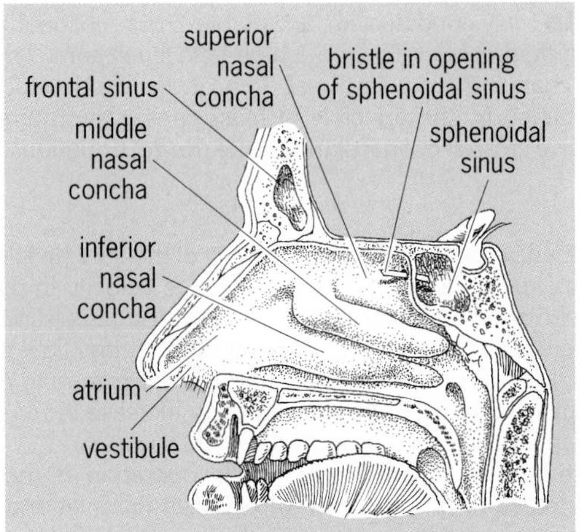

Human nose. (*After W. J. Hamilton et al., Textbook of Human Anatomy, Macmillan, 1956*)

Thomas S. Parsons

Nose cone The forward portion of a spacecraft that is designed for atmospheric entry. Nose cones are utilized for intercontinental ballistic missiles and spacecraft such as Apollo and space shuttles. The nose cone is required to withstand heating encountered during atmospheric entry, maintain the structural integrity of the spacecraft, prevent overheating of the payload, and usually maintain the aerodynamic characteristics of the spacecraft.

Even for a properly designed shape, it is inevitable that some fraction of the spacecraft's initial kinetic energy will finally reach the nose cone in the form of heat. The design of the heat shield for the nose cone is a complex procedure, which is highly dependent on the heating level. There are a variety of surface-protection or cooling systems which have been used. Generally these systems consist of heat sinks of various types: the absorption of heat by virtue of a material's sensible heat capacity, latent heat capacity, or chemical heat capacity. That is, heat absorption is accomplished by a temperature rise, a phase change, or a chemical reaction. Aerodynamic lift is employed by such vehicles as reusable space shuttles to lower heating rates so that the nose cone material can radiate away much of the incident heating.

Ablation is used to provide surface protection. The designer can divert heat from the spacecraft by allowing the nose cone's outer layer of material to melt, vaporize, or sublime. While large ablation rates provide excellent thermal protection, the resulting change in profile due to surface recession can adversely change the aerodynamic characteristics of the spacecraft. The designer must account for this change. *See* SPACE SHUTTLE; SPACECRAFT STRUCTURE.

Philip R. Nachtsheim

Notacanthoidei A suborder of Albuliformes consisting of two families, Halosuridae (halosaurs) and Notacanthidae (spiny eels). They are also known as Lyopomi and Heteromi, respectively. The body of these fishes is elongate and tapers posteriorly. The caudal fin is absent or essentially absent; pectoral fins are high on the body; pelvic fins are abdominal; and the anal fin is long (see illustration). Some species have photophores (light emitting organs), a characteristic of many deep-sea fishes. The halosaurs differ from spiny eels in lacking spines in the dorsal fin and having gill membranes that are completely separate rather than joined or partly joined.

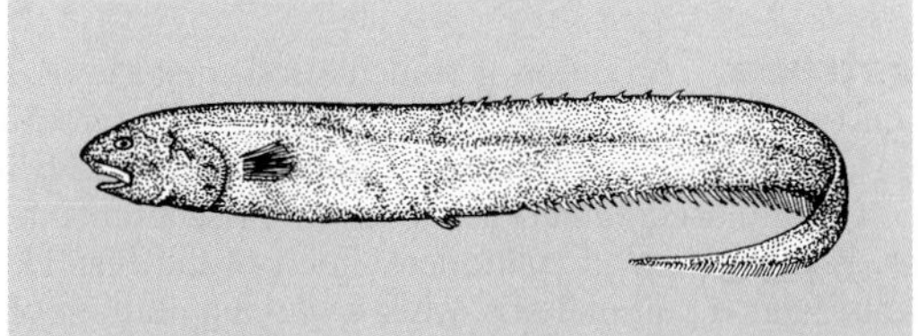

Spiny eel (*Notacanthus nasus*). (*After D. S. Jordan and B. W. Evermann, The Fishes of North and Middle America, U.S. Nat. Mus. Bull. 47, 1900*)

The Notacanthoidei is a small group with a history extending back to the Upper Cretaceous; it includes 2 families, 6 Recent genera, and about 25 species. The species inhabit deep seas worldwide, in waters between 125 and 4900 m (413 and 16,170 ft). They are like true eels (Anguilliformes) in lacking a firm suspension of the pectoral girdle from the skull, but some have fin spines similar to those of perciform fishes. *See* ALBULIFORMES; ANGUILLIFORMES; PHOTOPHORE GLAND.

Reeve M. Bailey; Herbert Boschung

Nothosauria An extinct clade (evolutionary lineage) of eosauropterygian reptiles (that is, a suborder of Eosauropterygia) that also includes pachypleurosaurs (suborder Pachypleurosauria) and pistosaurs (suborder Pistosauroidea). Representatives of these groups are known from uppermost Lower Triassic to Upper Triassic marine deposits in Europe, the Near and Middle East, Northern Africa, China, and North America. Triassic nothosaurs and their relatives are typically restricted to near-shore habitats or shallow epicontinental seas. Some derived members of the pistosaurs constitute the sister group (closest relatives) to plesiosaurs, a clade that greatly diversified during the Jurassic and Cretaceous. *See* PLACODONTIA; PLESIOSAURIA; SAUROPTERYGIA.

Nothosaurs range approximately 3–13 ft (1–4 m) in length. The neck is long, with at least 15 cervical vertebrae. Like aquatic lizards and crocodiles, early nothosaurs probably swam primarily by lateral undulation of the trunk and tail, with the limbs held to the side to reduce drag. Advanced nothosaurs show modifications of the front limbs to act as paddles; the rear limbs are somewhat reduced.

Nothosaurs (and placodonts) are distinguished from other aquatic reptiles by the closure of the openings in the palate that occur in more primitive reptiles, and by the covering of the base of the braincase. As in plesiosaurs (and placodonts), part of the scapula is located on the outside of the clavicle, a reversal of the relationship of these bones in other reptiles. More advanced members of the nothosaurs evolved a narrow and strongly depressed skull with a much-elongated rostrum (elongated snout) that was furnished with strongly procumbent, fanglike teeth in its anterior part. These pincer-jaws allowed the capture of fish and cephalopod prey with rapid lateral strikes of the head.

Robert L. Carroll; Olivier C. Rieppel

Notomyotida A small and distinctive order of valvatacean Asteroidea with long, straight-sided arms and relatively small disks. The skeleton is differentiated into marginals, actinals, and abactinals, and there is a clear calycinal ring. Both supra- and inframarginals are present and are offset or alternate rather than vertically aligned as they are in other asteroids. Digestive and reproductive organs are confined to the disk and most proximal parts of the arms and do not extend well into the arms as in other valvataceans.

Only a single family, Benthopectinidae, is included. Extant benthopectinids are specialist deep-water suspension feeders that live at depths usually between 1000 and 10,000 ft (300 and 3000 m). There are eight genera and 80 species, distributed in all the world's major oceans. *See* ASTEROIDEA; ECHINODERMATA.

Andrew B. Smith

Notostraca An order of branchiopod crustaceans, sometimes called tadpole shrimps. Generally they range from about 20 mm (0.4 in.) to (exceptionally) about 90 mm (3.5 in.) in length. The multisegmented trunk, up to 44 segments in some species, is elongate and cylindrical. Each of the first 11 trunk segments bears a pair of limbs, while a varying number of more posterior segments bear up to six pairs of smaller limbs per segment—a very unusual situation. The trunk terminates in a telson that bears a pair of slender, segmented, caudal filaments. Notostracans feed on a variety of small organisms that are seized by the trunk limbs and passed forward to the mouthparts, but they also collect and eat detritus with the same limbs.

Some species are bisexual, but in some parts of their range some are self-fertilizing hermaphrodites. The highly resistant eggs, which can withstand desiccation if necessary, hatch as nauplii, but in some cases this stage is transient, molting almost at once to a more advanced stage. The two extant genera, *Triops* and *Lepidurus*, are essentially worldwide in distribution and occur mostly in temporary waters. *See* BRANCHIOPODA.

Geoffrey Fryer

Notoungulata An order of dominant, hoofed herbivores of the Cenozoic of South America that are abundantly represented in Paleocene through Pleistocene nonmarine sedimentary rocks of that continent. Diverging from a primitive condylarth ancestry at an early date, they radiated into a wide diversity of forms, some of which were convergent with Northern Hemisphere ungulates.

Notoungulates were characterized by a skull with an expanded temporal region due to the presence of a large sinus in the squamosal and no postorbital bar. The feet were primitive, with five toes (three or two in some advanced forms), and the weight was borne mainly by the third digit. *See* EUTHERIA.

Richard H. Tedford

Nova The sudden brightening of a previously inconspicuous star. The name, short for nova stella (new star), formerly included objects now classified as supernovae and as other kinds of cataclysmic variables. Classical novae now include only those events where the energy source is hydrogen fusion (burning) on the surface of a white dwarf in a close binary system and the white dwarf is not destroyed in the process.

A handful of novae are discovered each year in the Milky Way Galaxy, and the total rate is probably 20–50 per year. A comparable number are found in other, nearby galaxies. The system consists of a normal, hydrogen-burning star in a close orbit (periods of a few days or less) around a white dwarf or degenerate star. A stream of gas flows from the normal star into a disk around the white dwarf and then accretes onto its surface. Hydrogen gradually builds up there until it is hot and dense enough for nuclear burning, normally with carbon, oxygen, neon, or magnesium from the white dwarf itself acting as a catalyst. Any nuclear fuel ignited under degenerate conditions explodes, because energy released does not cause the gas to expand, so temperature rises rapidly. *See* BINARY STAR; WHITE DWARF STAR.

Novae brighten in a few days and fade in months to years. The peak brightness is more than 100 times the solar luminosity, and the total energy release more than 10^{45} ergs (10^{38} joules). Novae recur every 10^4–10^5 years. *See* CATACLYSMIC VARIABLE; LIGHT CURVES; VARIABLE STAR.

Virginia Trimble

Nozzle A conduit with a variable cross-sectional area in which a fluid accelerates into a high-velocity stream. The effect of the changing cross-sectional area on the fluid velocity can be explained by the principle of mass conservation applied to successive cross-sectional planes of the nozzle. Equation (1) must

$$\dot{m} = \rho V A \tag{1}$$

be satisfied, where ρ is the mass density of the fluid [in kilograms per cubic meter (kg/m^3)], V is the average velocity in the cross section [in meters per second (m/s)], A is the cross-sectional area [in square meters (m^2)], and $\dot{m}$ is the rate of mass flow through the nozzle [in kilograms per second (kg/s)]. Decreasing A along the length of the nozzle must result in an increase in ρV since $\dot{m}$ is the same at every cross section. *See* FLUID FLOW.

An important parameter of nozzle operation is the difference in pressure [in pascals (Pa)] between the inlet and outlet. Higher pressure differences push the fluid to higher velocities and achieve higher mass flow rates for a given nozzle size.

For liquid nozzles it can be assumed that ρ is constant and therefore V increases as A decreases and vice versa. Liquid nozzles, such as those on fire hoses, are called converging because the area decreases along the length of the nozzle to increase the speed. Typical liquid nozzles have a simple conical shape and are designed to a specific ratio of inlet to outlet areas.

In the case of gas nozzles the gas density can change dramatically as a result of the pressure reduction between the inlet and outlet of the nozzle. At very high gas speeds this effect is so significant that the basic shape of the nozzle must change to a converging-diverging form (*see* illustration). The diverging portion is necessary to accommodate the expansion of the gas as it accelerates to lower pressure.

An understanding of the operation of the converging-diverging nozzle requires knowledge of the Mach number, which is the ratio of the gas speed to the speed of sound in the gas c; $Ma = V/c$. A subsonic flow has $Ma < 1$, a sonic flow has $Ma = 1$, and a supersonic flow has $Ma > 1$. The pressure, temperature, and speed of air flowing in a nozzle are related to Mach number and profile of nozzle cross-sectional area. If the flow in the converging portion of the nozzle is subsonic, the flow at the throat can at most be sonic, and supersonic flow can occur only in the diverging portion. To achieve supersonic flow in the diverging section, the reservoir pressure must be sufficient to achieve sonic flow at the throat. The fall of air temperature with increasing Ma is a direct result of the gas expansion. *See* MACH NUMBER; SOUND.

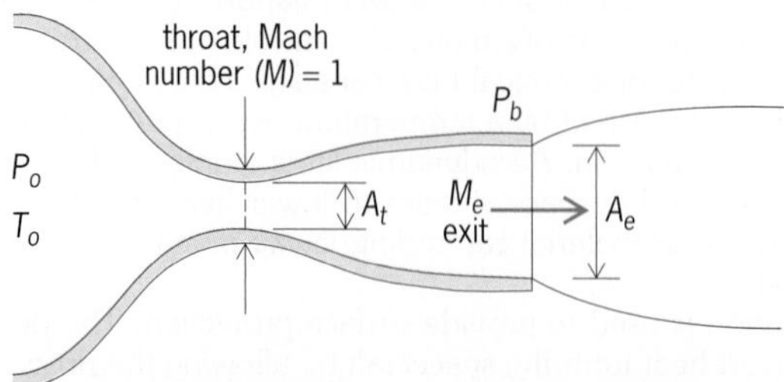

Typical convergent-divergent nozzle with a jet plume. P_0, T_0 = pressure and temperature upstream of the nozzle; A_t = area at the throat; P_b = back pressure; M_e, A_e = Mach number and area at exit.

A nozzle can be used for a variety of purposes. It is an indispensable piece of equipment in many devices employing fluid as a working medium. The reaction force that results from the fluid acceleration may be employed to propel a jet aircraft or a rocket. In fact, most military jet aircraft employ the simple convergent conical nozzle, with adjustable conical angle, as their propulsive device. If the high-velocity fluid stream is directed to turn a turbine blade, it may drive an electric generator or an automotive vehicle. High-velocity streams are produced inside a wind tunnel so that the conditions of flight of a missile or an aircraft may be simulated for research purposes. The nozzle must be carefully designed in this case to provide uniformly flowing fluid with the desired velocity, pressure, and temperature at the test section of the wind tunnel. Nozzles are used to disperse fuel into an atomized mist, such as that in diesel engines, for combustion purposes. Nozzles may also be used as metering devices for gas or liquid. *See* ATOMIZATION; FLOW MEASUREMENT; IMPULSE TURBINE; INTERNAL COMBUSTION ENGINE; ROCKET PROPULSION; TURBINE PROPULSION; WIND TUNNEL. Wen L. Chow; A. Gordon L. Holloway

Nuclear battery A battery that converts the energy of particles emitted from atomic nuclei into electric energy. Two basic types have been developed: (1) A high-voltage type, in which a beta-emitting isotope is separated from a collecting electrode by a vacuum or a solid dielectric, provides thousands of volts but the current is measured in picoamperes (pA); (2) a low-voltage type gives about 1 volt with current in microamperes (μA).

In the high-voltage type, a radioactive source is attached to one electrode, emitting charged particles. The source might be strontium-90, krypton-85, or hydrogen-3 (tritium), all of which are pure beta emitters. An adjacent electrode collects the emitted particles. A vacuum or solid dielectric separates the source and the collector electrodes. The principal use of the high-voltage battery is to maintain the voltage of a charged capacitor. The current output of the radioactive source is sufficient for this purpose.

Three different concepts have been employed in the low-voltage type of nuclear batteries: (1) a thermopile, (2) the use of an ionized gas between two dissimilar metals, and (3) the two-step conversion of beta energy into light by a phosphor and the conversion of light into electric energy by a photocell. *See* BATTERY. Jack Davis; L. Rozeanu; Kenneth Franzese

Nuclear binding energy The amount by which the mass of an atom is less than the sum of the masses of its constituent protons, neutrons, and electrons expressed in units of energy. This energy difference accounts for the stability of the atom. In principle, the binding energy is the amount of energy which was released when the several atomic constituents came together to form the atom. Most of the binding energy is associated with the nuclear constituents (protons and neutrons), or nucleons, and it is customary to regard this quantity as a measure of the stability of the nucleus alone. *See* NUCLEAR STRUCTURE.

A widely used term, the binding energy (BE) per nucleon, is defined by the equation below, where ${}_ZM^A$ represents the mass of

$$\text{BE/nucleon} = \frac{[ZH + (A - Z)n - {}_ZM^A]c^2}{A}$$

an atom of mass number A and atomic number Z, H and n are the masses of the hydrogen atom and neutron, respectively, and c is the velocity of light. The binding energies of the orbital electrons, here practically neglected, are not only small, but increase with Z in a gradual manner; thus the BE/nucleon gives an accurate picture of the variations and tends in nuclear stability.

The binding energy, when expressed in mass units, is known as the mass defect, a term sometimes incorrectly applied to quantity $M - A$, where M is the mass of the atom. *See* MASS DEFECT. Henry E. Duckworth

The term binding energy is sometimes also used to describe the energy which must be supplied to a nucleus in order to remove a specified particle to infinity, for example, a neutron, proton, or alpha particle. A more appropriate term for this energy is the separation energy. This quantity varies greatly from nucleus to nucleus and from particle to particle. For example, the binding energies for a neutron, a proton, and a deuteron in ^{16}O are 15.67, 12.13, and 20.74 MeV, respectively, while the corresponding energies in ^{17}O are 4.14, 13.78, and 14.04 MeV, respectively. The usual order of neutron or proton separation energy is 7–9 MeV for most of the periodic table. D. H. Wilkinson

Nuclear chemical engineering The branch of chemical engineering that deals with the production and use of radioisotopes, nuclear power generation, and the nuclear fuel cycle. A nuclear chemical engineer requires training in both nuclear and chemical engineering. As a nuclear engineer, he or she should be familiar with the nuclear reactions that take place in nuclear fission reactors and radioisotope production, with the properties of nuclear species important in nuclear fuels, with the properties of neutrons, gamma rays, and beta rays produced in nuclear reactors, and with the reaction, absorption, and attenuation of these radiations in the materials of reactors. *See* BETA PARTICLES; GAMMA RAYS; NEUTRON; NUCLEAR FUELS.

As a chemical engineer, he or she should know the properties of materials important in nuclear reactors and the processes used to extract and purify these materials and convert them into the chemical compounds and physical forms used in nuclear systems. *See* CHEMICAL ENGINEERING; NUCLEAR REACTOR.

Aspects of nuclear reactors of concern to nuclear chemical engineers include production and purification of the uranium dioxide fuel, production of the hafnium-free zirconium tubing used for fuel cladding, and control of corrosion and radioactive corrosion products by chemical treatment of coolant. A chemical engineering aspect of heavy-water reactor operation is control of the radioactive tritium produced by neutron activation of deuterium. Aspects of liquid-metal fast-breeder reactors of concern to nuclear chemical engineers include fabrication of the mixed uranium dioxide-plutonium dioxide fuel, purity control of sodium coolant to prevent fouling and corrosion, and reprocessing of irradiated fuel to recover plutonium and uranium for recycle. *See* NUCLEAR FUEL CYCLE; PLUTONIUM; URANIUM. Manson Benedict

Nuclear chemistry An interdisciplinary field that, in general, encompasses the application of chemical techniques to the solution of problems in nuclear physics. The discovery of the naturally occurring radioactive elements and of nuclear fission are classical examples of the work of nuclear chemists.

Although chemical techniques that are employed in nuclear chemistry are essentially the same as those in radiochemistry, these fields may be distinguished on the basis of the aims of the investigation. Thus, a nuclear chemist utilizes chemical techniques as a tool for the study of nuclear reactions and properties, whereas a radiochemist utilizes the radioactive properties of certain substances as a tool for the study of chemical reactions and properties. For the application of radioactive tracers to chemical problems *see* RADIOCHEMISTRY.

For the chemical effects of radiation on various systems *see* RADIATION CHEMISTRY.

The chemical identification of radioactive nuclides and the determination of their nuclear properties has been one of the major activities of nuclear chemists. Such studies have produced an extensive array of radioisotopes, and present studies are concerned mainly with the more difficult identification of nuclides of very short half-life. Nuclear chemical investigations led to the discovery of the synthetic radioactive elements which do not have any stable isotopes and are not formed in the natural radioactive series (technetium, promethium, astatine, and the transuranium elements). Other major areas of nuclear chemistry include studies of nuclear structure and spectroscopy and of the probability and mechanisms of various nuclear reactions. *See* NUCLEAR FISSION; NUCLEAR REACTION; NUCLEAR STRUCTURE. Ellis P. Steinberg

Nuclear engineering The branch of engineering that deals with the production and use of nuclear energy and nuclear radiation. The multidisciplinary field of nuclear engineering is studied in many universities. In some it is offered in a special nuclear engineering department; in others it is offered in other departments, such as mechanical or chemical engineering. Primarily, nuclear engineering involves the conception, development, design, construction, operation, and decommissioning of facilities in which nuclear energy or nuclear radiation is generated or used.

Examples of facilities include nuclear power plants; nuclear propulsion reactors used for the propulsion of ships and submarines; space nuclear reactors, used to power satellites, probes, and vehicles; nuclear production reactors, which produce fissile or fusile materials used in nuclear weapons; nuclear research reactors, which generate neutrons and gamma rays for scientific research and medical and industrial applications; gamma cells, which are used for sterilizing medical equipment and food and for manufacturing polymers; particle accelerators, which produce nuclear radiation for use in medical and industrial applications; and nuclear waste repositories. *See* NUCLEAR MEDICINE; NUCLEAR POWER; NUCLEAR REACTOR; PARTICLE ACCELERATOR; RADIOACTIVE WASTE MANAGEMENT; SHIP NUCLEAR PROPULSION; SPACE POWER SYSTEMS; SUBMARINE.

Many nuclear engineers are also involved in the research and development of future fusion power plants—plants that will be based on the fusion reaction for generating nuclear energy. Many challenging engineering problems are involved, including the development of technologies for heating the fusion fuel to hundreds of millions of degrees; confining this ultrahot fuel; and compressing fusion fuel to many thousand times their natural solid density. *See* NUCLEAR FUSION. Ehud Greenspan

Nuclear explosion An explosion whose energy is produced by a nuclear transformation, either fission or fusion. *See* NUCLEAR FISSION; NUCLEAR FUSION.

The energy of a nuclear explosion is usually stated in terms of the mass of trinitrotoluene (TNT) which would provide the same energy. The complete fissioning of 1 kg of uranium or plutonium would be equivalent to 17,000 metric tons of TNT (17 kilotons); 2 lb would be equivalent to 17,000 short tons. The indicated yield-to-mass ratio of 1.7×10^7 cannot be realized, largely because of the ancillary equipment necessary to assemble the nuclear components into an explosive configuration in the very short time required.

Though the size of a typical nuclear explosion is appalling, the most significant feature of a nuclear device derives from its yield-to-weight ratio. The first nuclear weapons (1945) weighed about 5 tons, but with yields of 15 to 20 kT their yield-to-weight ratio was, nevertheless, close to 4000 times larger than that of previous weapons. By 1960 the United States had developed a weapon with a yield of about 1 megaton (MT) in a weight of about 1 ton for use in an intercontinental missile. With this, the yield-to-weight ratio was raised to 10^6.

Although weapons with yields of up to 15 and 60 MT were fired by the United States and Soviet Union respectively, about 1960 the main interest of the major nuclear powers focused on adapting nuclear devices for delivery by missile carriers, and this called for smaller weapons with so-called moderate yields of tens or hundreds of kilotons, up to a few megatons. *See* MISSILE.

The damage mechanism of a conventional explosion is blast—the pressure, or shock, wave transmitted in the surrounding medium. The blast wave from a nuclear explosion is similar, except for the great difference in scale. A nuclear explosion also produces several kinds of effects not experienced with ordinary explosives.

About one-third of the energy of the explosion is distributed as thermal radiation on line-of-sight trajectories. Exposure to 5–10 cal/cm^2 (2–4×10^5 J/m^2) of thermal radiation energy in the short time (a second, or so) during which the thermal pulse is delivered will ignite many combustible materials (fabrics, paper, dry leaves, and so forth). It will cause serious flash burns on exposed skin. Such energy levels will be delivered in clear air to 0.3–0.4 mi (0.5–0.6 km) by an explosion of 1 kT. At Hiroshima, burn injuries alone would have been fatal to almost all persons in the open without protection out to a little over 1 mi (1.6 km), and burns serious enough to require treatment were experienced at distances greater than 2 mi (3.2 km).

During the few tens of seconds before the fireball rises away from the point at which the explosion occurred, it provides an intense source of gamma rays. The radiation emitted during this interval is referred to as prompt radiation. The remaining radioactivity, which is swept upward from the scene of the explosion to the altitude at which the fireball stops rising, and some of which ultimately returns to the surface, constitutes the residual radiation. *See* ALPHA PARTICLES; NEUTRON; RADIOACTIVITY.

At Hiroshima a dose equal to or greater than 450 rads (4.5 grays) extended to almost 1 mi (1.6 km). About half of the persons exposed to this dose in a short time will die within a few weeks, so that persons in this area who were not protected by heavy building walls experienced severe hazard from radiation. This is about the same distance for severe hazards from blast and thermal effects. The prompt radiation exposure falls off more rapidly with distance than the blast effect which, in turn, falls more rapidly than the intensity of thermal radiation. For an explosion much larger than 15 or 20 kT, the hazard range from thermal radiation or blast will be larger than that from prompt radiation, and prompt radiation will be a relatively unimportant effect. For much smaller yields, this order of importance will be reversed. J. Carson Mark

Nuclear fission An extremely complex nuclear reaction representing a cataclysmic division of an atomic nucleus into two nuclei of comparable mass. This rearrangement or division of a heavy nucleus may take place naturally (spontaneous fission) or under bombardment with neutrons, charged particles, gamma rays, or other carriers of energy (induced fission). Although nuclei with mass number A of approximately 100 or greater are energetically unstable against division into two lighter nuclei, the fission process has a small probability of occurring, except with the very heavy elements. Even for these elements, in which the energy release is of the order of 200 megaelectronvolts, the lifetimes against spontaneous fission are reasonably long. *See* NUCLEAR REACTION.

Liquid-drop model. The stability of a nucleus against fission is most readily interpreted when the nucleus is viewed as being analogous to an incompressible and charged liquid drop with a surface tension. Long-range Coulomb forces between protons act to disrupt the nucleus, whereas short-range nuclear forces, idealized as a surface tension, act to stabilize it. The degree of stability is then the result of a delicate balance between the relatively weak electromagnetic forces and the strong nuclear forces. Although each of these forces results in potentials of several hundred megaelectronvolts, the height of a typical barrier against fission for a heavy nucleus, because they are of opposite sign but do not quite cancel, is only 5 or 6 MeV. Investigators have used this charged liquid-drop model with great success in describing the general features of nuclear fission and also in reproducing the total nuclear binding energies. *See* NUCLEAR BINDING ENERGY; NUCLEAR STRUCTURE; SURFACE TENSION.

Shell corrections. The general dependence of the potential energy on the fission coordinate representing nuclear elongation

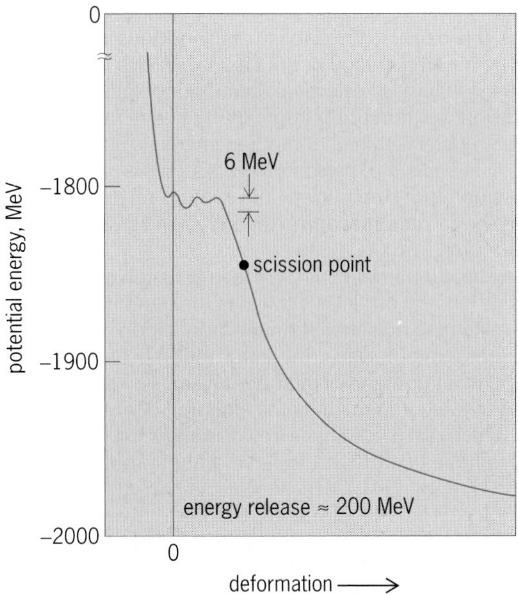

Fig. 1. Plot of the potential energy in MeV as a function of deformation for the nucleus ^{240}Pu. (*After M. Bolsteli et al., New calculations of fission barriers for heavy and superheavy nuclei, Phys. Rev., 5C:1050–1077, 1972*)

or deformation for a heavy nucleus such as ^{240}Pu is shown in Fig. 1. The expanded scale used in this figure shows the large decrease in energy of about 200 MeV as the fragments separate to infinity. It is known that ^{240}Pu is deformed in its ground state, which is represented by the lowest minimum of −1813 MeV near zero deformation. This energy represents the total nuclear binding energy when zero of potential energy is the energy of the individual nucleons at a separation of infinity. The second minimum to the right of zero deformation illustrates structure introduced in the fission barrier by shell corrections, that is, corrections dependent upon microscopic behavior of the individual nucleons, to the liquid-drop mass. Although shell corrections introduce small wiggles in the potential-energy surface as a function of deformation, the gross features of the surface are reproduced by the liquid-drop model. Since the typical fission barrier is only a few megaelectronvolts, the magnitude of the shell correction need only be small for irregularities to be introduced into the barrier. This structure is schematically illustrated for a heavy nucleus by the double-humped fission barrier in Fig. 2, which represents the region to the right of zero deformation in Fig. 1 on an expanded scale. The fission barrier has two maxima and a rather deep minimum in between. For comparison, the single-humped liquid-drop barrier is also schematically illustrated. The transition in the shape of the nucleus as a function of deformation is schematically represented in the upper part of the figure.

Experimental consequences. The observable consequences of the double-humped barrier have been reported in numerous experimental studies. In the actinide region more than 30 spontaneously fissionable isomers have been discovered between uranium and berkelium, with half-lives ranging from 10^{-11} to 10^{-2} s. These decay rates are faster by 20 to 30 orders of magnitude than the fission half-lives of the ground states, because of the increased barrier tunneling probability (Fig. 2). Several cases in which excited states in the second minimum decay by fission are also known. Normally these states decay within the well by gamma decay; however, if there is a hindrance in gamma decay due to spin, the state (known as a spin isomer) may undergo fission instead.

Fission probability. The cross section for particle-induced fission $\sigma(y, f)$ represents the cross section for a projectile y to react with a nucleus and produce fission, as shown by the equation below. The quantities $\sigma_R(y)$, Γ_f, and Γ_t are the total reaction

$$\sigma(y, f) = \sigma_R(y)(\Gamma_f / \Gamma_t)$$

across sections for the incident particle y, the fission width, and the total level width, respectively, where $\Gamma_t = \Gamma_f + \Gamma_n + \Gamma_y + \cdots$ is the sum of all partial-level widths. All the quantities in the above equation are energy-dependent.

When the incoming neutron has low energy, the likelihood of reaction is substantial only when the energy of the neutron is such as to form a compound nucleus in one or another of its resonance levels. The requisite sharpness of the "tuning" of the energy is specified by the total level width Γ. The nuclei ^{233}U, ^{235}U, and ^{239}Pu have a very large cross section to take up a slow neutron and undergo fission because both their absorption cross section and their probability for decay by fission are large. The probability for fission decay is high because the binding energy of the incident neutron is sufficient to raise the energy of the compound nucleus above the fission barrier. The very large, slow neutron fission cross sections of these isotopes make them important fissile materials in a chain reactor. *See* CHAIN REACTION (PHYSICS); REACTOR PHYSICS.

Postscission phenomena. After the nuclear fragments are separated, they are further accelerated as the result of the large Coulomb repulsion. The initially deformed fragments collapse to their equilibrium shapes, and the excited primary fragments lose energy by evaporating neutrons. After neutron emission, the fragments lose the remainder of their energy by gamma radiation, with a lifetime of about 10^{-11} s. The variation of neutron yield with fragment mass is directly related to the fragment excitation energy. Minimum neutron yields are observed for nuclei near closed shells because of the resistance to deformation of nuclei with closed shells. Maximum neutron yields occur for fragments that are "soft" toward nuclear deformation.

After the emission of the prompt neutrons and gamma rays, the resulting fission products are unstable against β-decay For example, in the case of thermal neutron fission of ^{235}U, each fragment undergoes on the average about three β-decays before it settles down to a stable nucleus. For selected fission products (for example, ^{87}Br and ^{137}I) β-decay leaves the daughter nucleus with excitation energy exceeding its neutron binding energy. The resulting delayed neutrons amount, for thermal neutron fission

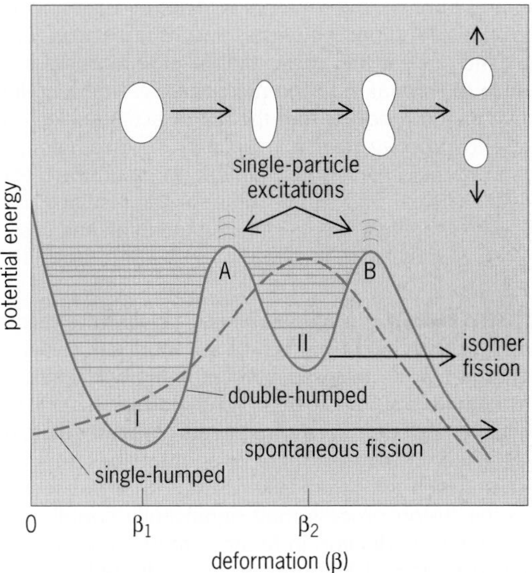

Fig. 2. Schematic plots of single-humped fission barrier of liquid-drop model and double-humped barrier Introduced by shell corrections. Humps at *A* and *B* result in minima in potential energy at deformation of β_1 and β_2. States in these wells are designated class I and class II, respectively. (*After J. R. Huizenga, Nuclear fission revisited, Science, 168:1405–1413, 1979*)

of ^{235}U, to about 0.7% of all the neutrons given off in fission. Though small in number, they are quite important in stabilizing nuclear chain reactions against sudden minor fluctuations in reactivity. *See* DELAYED NEUTRON; NEUTRON; THERMAL NEUTRONS.

John R. Huizenga

Nuclear fuel cycle The nuclear fuel cycle typically involves the following steps: (1) finding and mining the uranium ore; (2) refining the uranium from other elements; (3) enriching the uranium-235 content to 3–5%; (4) fabricating fuel elements; (5) interim storage and cooling of spent fuel; (6) reprocessing of spent fuel to recover uranium and plutonium (optional); (7) fabricating recycle fuel for added energy production (optional); (8) cooling of spent fuel or reprocessing waste, and its eventual transport to a repository for disposal in secure long-term storage. *See* NUCLEAR FUELS; URANIUM.

Steps 6 and 7 are used in Britain, France, India, Japan, and Russia. They are no longer used in the United States, which by federal policy has been restricted to a "once through" fuel cycle, meaning without recycle. Belgium, China, France, Germany, Japan, and Russia, with large and growing nuclear power capacities, use recycled plutonium. Disposal of highly enriched uranium from nuclear weapons is beginning to be undertaken by blending with natural or depleted uranium to make the 3–5% low-enrichment fuel. Similarly, MOX (mixed oxides) fuel capability can be used to dispose of plutonium stockpiled for nuclear weapons. This option is being planned in Europe and Russia, and is beginning to be considered in the United States. *See* PLUTONIUM.

Nuclear reactors produce energy using fuel made of uranium slightly enriched in the isotope ^{235}U. The basic raw material is natural uranium that contains 0.71% ^{235}U (the only naturally occurring isotope that can sustain a chain reaction). The other isotopes of natural uranium consist of ^{238}U, part of which converts to plutonium-239, during reactor operation. The isotope ^{239}Pu also sustains fission, typically contributing about one-third of the energy produced per fuel cycle. *See* NUCLEAR FISSION; NUCLEAR REACTOR.

Various issues revolve around the type of nuclear fuel cycle chosen. For instance, the question is still being argued whether "burning" weapons materials in recycle reactors is more or less subject to diversion (that is, falling into unauthorized hands) than storing and burying these materials. Another issue involves the composition of radioactive wastes and its impact on repository design. The nuclear fuel cycles that include reprocessing make it possible to separate out the most troublesome long-lived radioactive fission products and the minor actinide elements that continue to produce heat for centuries. The remaining waste decays to radiation levels comparable to natural ore bodies in about 1000 years. The shorter time for the resulting wastes to decay away simplifies the design, management, and costs of the repository. *See* ACTINIDE ELEMENTS.

Edwin L. Zebroski

Nuclear fuels Materials whose ability to release energy derives from specific properties of the atom's nucleus. In general, energy can be released by combining two light nuclei to form a heavier one, a process called nuclear fusion; by splitting a heavy nucleus into two fragments of intermediate mass, a process called nuclear fission; or by spontaneous nuclear decay processes, which are generically referred to as radioactivity. Although the fusion process may significantly contribute to the world's energy production in future centuries and although the production of limited amounts of energy by radioactive decay is a well-established technology for specific applications, the only significant industrial use of nuclear fuel so far utilizes fission. Therefore, the term nuclear fuels generally designates nuclear fission fuels only. *See* NUCLEAR BATTERY; NUCLEAR FISSION; NUCLEAR FUSION; NUCLEAR POWER; RADIOACTIVITY AND RADIATION APPLICATIONS.

Large releases of energy through a fission or a fusion reaction are possible because the stability of the nucleus is a function of its size. The binding energy per nucleon provides a measure of the nucleus stability. By selectively combining light nuclei together by a fusion reaction or by fragmenting heavy nuclei by a fission reaction, nuclei with higher binding energies per nucleon can be formed. The result of these two processes is a release of energy. The fissioning of one nucleus of uranium releases as much energy as the oxidation of approximately 5×10^7 atoms of carbon. *See* NUCLEAR BINDING ENERGY.

Many heavy elements can be made to fission by bombardment with high-energy particles. However, only neutrons can provide a self-sustaining nuclear fission reaction. Upon capture of a neutron by a heavy nucleus, the latter may become unstable and split into two fragments of intermediate mass. This fragmentation is generally accompanied by the emission of one or several neutrons, which can then induce new fissions. Only a few long-lived nuclides have been found to have a high probability of fission: ^{233}U, ^{235}U, and ^{239}Pu. Of these nuclides, only ^{235}U occurs in nature as 1 part in 140 of natural uranium, the remainder being mostly ^{238}U. The other nuclides must be produced artificially: ^{233}U from ^{232}Th, and ^{239}Pu from ^{238}U. The nuclides ^{233}U, ^{235}U, and ^{239}Pu are called fissile materials since they undergo fission with either slow or fast neutrons, while ^{232}Th and ^{238}U are called fertile materials. The latter, however, can also undergo the fission process at low yields with energetic neutrons; therefore, they are also referred to as being fissionable.

The term nuclear fuel applies not only to the fissile materials, but often to the mixtures of fissile and fertile materials as well. Using a mixture of fissile and fertile materials in a reactor allows capture of excess neutrons by the fertile nuclides to form fissile nuclides. Depending on the efficiency of production of fissile elements, the process is called conversion or breeding. Breeding is an extreme case of conversion corresponding to a production of fissile material at least equal to its consumption. *See* NUCLEAR FUEL CYCLE; NUCLEAR REACTOR.

David Franklin; Albert Machiels

Nuclear fuels reprocessing Nuclear fuels are reprocessed for military or civilian purposes. In military applications, reprocessing is applied to extract fissile plutonium from fuels that are designed and operated to optimize production of this element. In civilian applications, reprocessing is used to recover valuable uranium and transuranic elements that remain in fuels discharged from electricity-generating nuclear power plants, for subsequent recycle in freshly constituted nuclear fuel. This military-civilian duality has made the development and application of reprocessing technology a sensitive issue worldwide and necessitates stringent international controls on reprocessing operations. It has also stimulated development of alternative processes to produce less plutonium and more uranium (or transuranic elements), so that the proliferation of nuclear weapons is held in check. *See* NUCLEAR POWER; PLUTONIUM; TRANSURANIUM ELEMENTS; URANIUM.

Nuclear fuel is removed from civilian power reactors due to chemical, physical, and nuclear changes that make it increasingly less efficient for heat generation as its cumulative residence time in the reactor core increases. The fissionable material in the fuel is not depleted; however, the buildup of fission product isotopes (with strong neutron-absorbing properties) tends to decrease the nuclear reactivity of the fuel. *See* NUCLEAR FISSION; NUCLEAR FUELS; NUCLEAR REACTOR.

A typical composition of civilian reactor spent fuel at discharge is 96% uranium, 3% fission products, and 1% transuranic elements (generally as oxides, because most commercial nuclear fuel is in the form of uranium oxide). The annual spent fuel output from a 1.2-gigawatt electric power station totals approximately 33 tons (30 metric tons) of heavy-metal content. This spent fuel can be discarded as waste or reprocessed to recover the uranium and plutonium that it contains (for recycle in fresh fuel elements). The governments of France, the United Kingdom, Russia, and China actively support reprocessing as a means for the management of highly radioactive spent fuel and as a source of fissile material for future nuclear fuel supply. The United States forbids the reprocessing of civilian reactor fuel for plutonium recovery and is the only one of the five declared nuclear weapons states with complete fuel recycling capabilities that actively opposes commercial fuel reprocessing.

Decisions to reprocess are not made on economic grounds only, making it difficult to evaluate the economic viability of reprocessing in various scenarios. In the ideal case, a number of factors must be considered, including: (1) cost of uranium/U_3O_8; (2) cost of enrichment; (3) cost of fuel fabrication; (4) cost of reprocessing; (5) waste disposal cost; and (6) fissile content of spent fuel.

The once-through fuel cycle (that is, direct disposal/no reprocessing) is favored when fuel costs and waste disposal costs are low and reprocessing costs are high. However, technological advancements and escalating waste disposal costs can swing the balance in favor of reprocessing. *See* ACTINIDE ELEMENTS; NUCLEAR FUEL CYCLE; RADIOACTIVE WASTE MANAGEMENT.

The technology of reprocessing nuclear fuel was created as a result of the Manhattan Project during World War II, with the purpose of plutonium production. Early reprocessing methods were refined over the years, leading to a solvent extraction process known as PUREX (plutonium uranium extraction). The PUREX process is an aqueous method that has been implemented by several countries and remains in operation on a commercial basis. A nonaqueous reprocessing method known as pyroprocessing was developed in the 1990s as an alternative to PUREX. It has not been deployed commercially, but promises greatly decreased costs and reduced waste volumes, with practically no secondary wastes or low-level wastes being generated. It also has the important attribute of an inability to separate pure plutonium from irradiated nuclear fuel. *See* SOLVENT EXTRACTION.

Both the PUREX process and the pyroprocess can be used in a waste management role in support of a once-through nuclear fuel cycle if the economics of this application are favorable. The PUREX process can be operated with a low decontamination factor for plutonium. The pyroprocess can place the transuranic elements in the salt waste stream that leads to a glass-ceramic waste form. Both systems are effective in placing the fission products and actinide elements present in spent nuclear fuel into more durable waste forms that can be safely disposed in a high-level waste repository. James J. Laidler

Nuclear fusion

One of the primary nuclear reactions, the name usually designating an energy-releasing rearrangement collision which can occur between various isotopes of low atomic number. *See* NUCLEAR REACTION.

Interest in the nuclear fusion reaction arises from the expectation that it may someday be used to produce useful power, from its role in energy generation in stars, and from its use in the fusion bomb. Since a primary fusion fuel, deuterium, occurs naturally and is therefore obtainable in virtually inexhaustible supply, solution of the fusion power problem would permanently solve the problem of the present rapid depletion of chemically valuable fossil fuels. The lack of radioactive waste products from the fusion reaction is another argument in its favor as opposed to the fission of uranium. *See* HYDROGEN BOMB; NUCLEAR FISSION.

In a nuclear fusion reaction the close collision of two energy-rich nuclei results in a mutual rearrangement of their nucleons (protons and neutrons) to produce two or more reaction products, together with a release of energy. The energy usually appears in the form of kinetic energy of the reaction products, although when energetically allowed, part may be taken up as energy of an excited state of a product nucleus. In contrast to neutron-produced nuclear reactions, colliding nuclei, because they are positively charged, require a substantial initial relative kinetic energy to overcome their mutual electrostatic repulsion so that reaction can occur. This required relative energy increases with the nuclear charge Z, so that reactions between low-Z nuclei are the easiest to produce. The best known of these are the reactions between the heavy isotopes of hydrogen, deuterium (D), and tritium (T). *See* DEUTERIUM; TRITIUM.

Nuclear fusion reactions can be self-sustaining if they are carried out at a very high temperature. That is to say, if the fusion fuel exists in the form of a very hot ionized gas of stripped nuclei and free electrons termed a plasma, the agitation energy of the nuclei can overcome their mutual repulsion, causing reactions to occur. This is the mechanism of energy generation in the stars and in the fusion bomb. It is also the method envisaged for the controlled generation of fusion energy. *See* PLASMA (PHYSICS).

The cross sections (effective collisional areas) for many of the simple nuclear fusion reactions have been measured with high precision. It is found that the cross sections generally show broad maxima as a function of energy and have peak values in the general range of 0.01 barn (1 barn $= 10^{-28}$ m^2) to a maximum value of 5 barns, for the deuterium-tritium reaction. The energy releases of these reactions can be readily calculated from the mass difference between the initial and final nuclei or determined by direct measurement.

Some of the important simple fusion reactions, their reaction products, and their energy releases are:

$$\begin{aligned} D + D &\rightarrow {}^3He + n + 3.25\ \text{MeV} \\ D + D &\rightarrow T + p + 4.0\ \text{MeV} \\ T + D &\rightarrow {}^4He + n + 17.6\ \text{MeV} \\ {}^3He + D &\rightarrow {}^4He + p + 18.3\ \text{MeV} \\ {}^6Li + D &\rightarrow 2{}^4He + 22.4\ \text{MeV} \\ {}^7Li + p &\rightarrow 2{}^4He + 17.3\ \text{MeV} \end{aligned}$$

If it is remembered that the energy release in the chemical reaction in which hydrogen and oxygen combine to produce a water molecule is about 1 eV per reaction, it will be seen that, gram for gram, fusion fuel releases more than a million times as much energy as typical chemical fuels. Richard F. Post

Nuclear hormone receptors

Specialized proteins in the nucleus that are involved in the regulation of gene activity in response to hormones and other signals. The nuclear hormone receptors represent one of the largest family of proteins that directly regulate transcription in response to hormones and other ligands (chemical entities that bind to and activate the receptor).

Nuclear hormone receptors have been identified in many species ranging from *Caenorhabditis elegans* (worms) to humans. Nearly 50 distinct nuclear receptor genes have been identified in the human genome. These include receptors for the steroid hormones (glucocorticoids, mineralocorticoids, estrogens, progestins, and androgens), thyroid hormones, vitamin D, and retinoic acid. An even larger number of nuclear receptor proteins have been found for which no known hormone or ligand has yet been identified. These proteins have been termed orphan receptors. Orphan receptors hold considerable promise, as they provide the first clues toward the identification of novel regulatory molecules and new drug therapies. *See* GENE; HORMONE.

Although each of the nuclear receptors mediates distinct biological effects, their activities at the molecular level are remarkably similar. Nuclear receptors contain two functional modules that characterize this family of proteins: the deoxyribonucleic acid (DNA)–binding domain and the hormone-binding or ligand-binding domain (see illustration). In order to regulate gene transcription, nuclear receptors must first bind specific DNA sequences adjacent to their target genes. These sequences are referred to as hormone response elements (HREs).

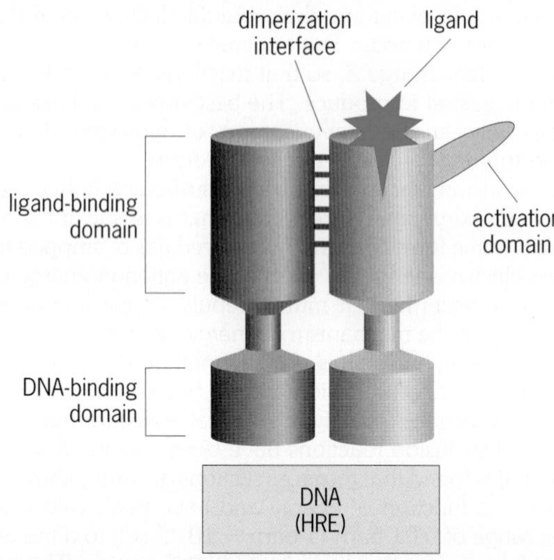

Schematic illustration of a nuclear receptor dimer bound to DNA. The DNA-binding domains are shown contacting the hormone response element (HRE). The receptor at the right is shown bound to its ligand. In addition to binding ligand, the ligand-binding domain contains subdomains for transcriptional activation and dimerization.

The DNA-binding domain is the portion of the receptor that binds to these sequences.

While the DNA-binding domain selects the target genes, it is the hormone- or ligand-binding domain that determines which hormones activate these genes. The ligand-binding domain contains a cavity or pocket that specifically recognizes the cognate ligand. Thus, the ligand-binding domain senses the presence of specific ligands in the body, and in the presence of that ligand it allows the receptor to activate gene transcription. *See* DEOXYRIBONUCLEIC ACID (DNA); NUCLEIC ACID.

In the nucleus, DNA is wrapped around histone proteins in a structure known as chromatin. DNA that is in a more compact chromatin structure is less transcriptionally active. Indeed, one way in which nuclear receptors activate transcription is to modulate chromatin structure. An important advance in understanding nuclear receptors has been the realization that these receptors interact with a variety of coactivators and corepressor proteins. Barry Marc Forman

Nuclear isomerism The existence of excited states of atomic nuclei with unusually long lifetimes. If the lifetime of a specific excited state is unusually long, compared with the lifetimes of other excited states in the same nucleus, the state is said to be isomeric. The definition of the boundary between isomeric and normal decays is arbitrary, and the term is therefore used loosely. *See* EXCITED STATE; PARITY (QUANTUM MECHANICS); SPIN (QUANTUM MECHANICS).

The predominant decay mode of excited nuclear states is by γ-ray emission. The rate at which this process occurs is determined largely by the spins, parities, and excitation energies of the decaying state and of those to which it is decaying. In particular, the rate is extremely sensitive to the difference in the spins of initial and final states and to the difference in excitation energies. Both extremely large spin differences and extremely small energy differences can result in a slowing of the γ-ray emission by many orders of magnitude, resulting in some excited states having unusually long lifetimes and therefore being termed isomeric.

In addition to spin isomers, two other types of isomers have been identified. The first of these arises from the fact that some excited nuclear states represent a drastic change in shape of the nucleus from the shape of the ground state. In many cases this extremely deformed shape displays unusual stability, and states with this shape are therefore isomeric. A particularly important class of these shape isomers is observed in the decay of heavy nuclei by fission, and the study of such fission isomers has been the subject of intensive effort. *See* NUCLEAR FISSION.

A more esoteric form of isomer has also been observed, the so-called pairing isomer which results from differences in the microscopic motions of the constituent nucleons in the nucleus. A state of this type has a quite different character from the ground state of the nucleus, and is therefore also termed isomeric. *See* NUCLEAR STRUCTURE. Russell Betts

Nuclear magnetic resonance (NMR) A phenomenon exhibited when atomic nuclei in a static magnetic field absorb energy from a radio-frequency field of certain characteristic frequencies. Nuclear magnetic resonance is a powerful analytical tool for the characterization of molecular structure, quantitative analysis, and the examination of dynamic processes. It is based on quantized spectral transitions between nuclear Zeeman levels of stable isotopes, and is unrelated to radioactivity. *See* ZEEMAN EFFECT.

The format of nuclear magnetic resonance data is a spectrum that contains peaks referred to as resonances. The resonance of an isotope is distinguished by the transition frequency of the nucleus. The intensity of the resonance is directly proportional to the number of nuclei that produce the signal. Although the majority of nuclear magnetic resonance spectra are measured for samples in solution or as neat liquids, it is possible to measure nuclear magnetic resonance spectra of solid samples. Nuclear magnetic resonance is a nondestructive technique that can be used to measure spectra of cells and living organisms.

Nuclear magnetic properties. The nuclei of most atoms possess an intrinsic nuclear angular momentum. The classical picture of nuclear angular momentum is a spherical nucleus rotating about an axis in a manner analogous to the rotation of the Earth. Nuclear angular momentum, like most other atomic quantities, can be expressed as a series of quantized levels.

The transitions that give rise to the nuclear magnetic resonance spectrum are produced when the nuclei are placed in a static magnetic field. The external or applied magnetic field defines a geometric axis, denoted as the z axis. The external magnetic field orients the z components of nuclear angular momentum and the magnetic moment with respect to the z axis. The nuclear magnetic resonance spectrum is produced by spectral transitions between different spin states and is therefore dependent on the value of the nuclear spin. Nuclei for which the nuclear spin value is 0 have a magnetic spin-state value of 0. Therefore, these nuclei do not give rise to a nuclear magnetic resonance spectrum under any circumstances. Many important elements in chemistry have zero values of nuclear spin and are inherently nuclear magnetic resonance inactive, including the most abundant isotopes of carbon-12, oxygen-16, and sulfur-32. Nuclear magnetic resonance spectra are measured most often for nuclei that have nuclear spin values of $^1/_2$. Chemically important spin-$^1/_2$ nuclei include the isotopes hydrogen-1, phosphorus-31, flourine-19, carbon-13, nitrogen-15, silicon-29, iron-57, selenium-77, cadmium-113, silver-107, platinum-195, and mercury-199. Nuclear magnetic resonance spectra of nuclei that have nuclear spin values of greater than $^1/_2$, known as quadrupolar nuclei, can be measured; but the measurements are complicated by faster rates of nuclear relaxation. Nuclei that fall into this class include the isotopes hydrogen-2, lithium-6, nitrogen-14, oxygen-17, boron-11, chlorine-35, sodium-23, aluminum-27, and sulfur-33.

The transition frequency in nuclear magnetic resonance depends on the energy difference of the spin states. The intensity of the resonance is dependent on the population difference of the two spin states, which in turn depends directly on the magnitude of the energy difference. For spin-$^1/_2$ nuclei, ΔE is the difference in energy between the + and − values of the magnetic spin state. In contrast to other spectroscopic methods, the

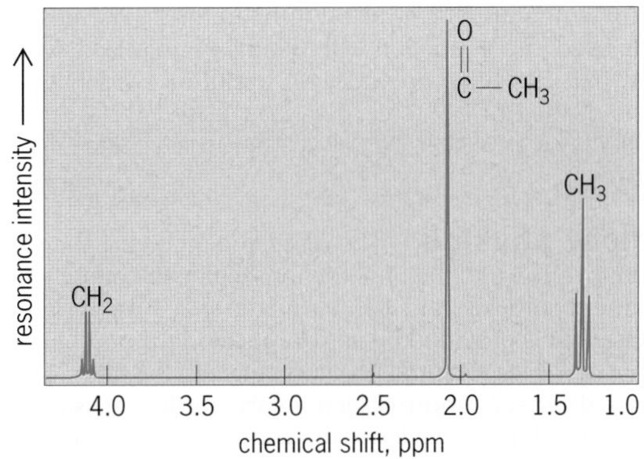

Proton spectrum of ethyl acetate ($CH_3COOH_2CH_3$). Chemical shift is relative to the protons of tetramethylsilane. The integrated intensity of the resonances corresponds to the relative number of protons which give rise to each resonance.

nuclear magnetic resonance frequency is variable, depending on the strength of the magnet used to measure the spectrum. Larger magnetic fields produce a greater difference in energy between the spin states. This translates into a larger difference in the population of the spin states and therefore a more intense resonance in the resulting nuclear magnetic resonance spectrum.

Individual nuclei of the same isotope in a molecule have transition frequencies that differ depending on their chemical environment. This phenomenon, called chemical shift, occurs because the effective magnetic field at a particular nucleus in a molecule is less than the applied magnetic field due to shielding by electrons.

An example of resonance chemical shift is observed in the ^1H nuclear magnetic resonance spectrum of ethyl acetate ($CH_3COOCH_2CH_3$; see illustration). Resonances at several frequencies are observed in this spectrum. The methylene (CH_2) protons are affected by the electron-withdrawing oxygen atoms of the neighboring ester group, and as a result the chemical shift of the methylene proton resonance is significantly different from the chemical shift of the resonances of the protons of the methyl (CH_3) groups of ethyl acetate. The two methyl groups are in different chemical environments and therefore give rise to resonances that have different chemical shifts. Because of the dependence of the transition frequency of a nucleus on its chemical environment, chemical shift is diagnostic of the functional group containing the nucleus of interest. Nuclear magnetic resonance spectroscopy is a frequently employed tool in chemical synthesis studies, because the nuclear magnetic resonance spectrum can confirm the chemical structure of a synthetic product.

In most nuclear magnetic resonance spectra, rather than frequency units, the spectra are plotted in units of chemical shift expressed as part per million (ppm). The ppm scale calculates the ratio of the resonance frequency in hertz (Hz) to the Larmor frequency of the nucleus at the magnetic field strength of the measurement. The ppm scale allows direct comparison of nuclear magnetic resonance spectra acquired by using magnets of differing field strength. This unit of nuclear magnetic resonance chemical shift should not be confused with the concentration units of ppm (mg/kg), often referred to by analytical chemists in the field of trace analysis.

The chemical shift (either in hertz or ppm) of a resonance is assigned relative to the chemical shift of a standard reference material. The nuclear magnetic resonance community has agreed to arbitrarily set the chemical shift of certain standard compounds to 0 ppm. For ^1H and ^{13}C nuclear magnetic resonance, the accepted standard is tetramethylsilane, which is defined to have a chemical shift of 0 ppm. However, any molecule with a resonance frequency in the appropriate chemical shift region of the spectrum that does not overlap with the resonances of the sample can be employed as a chemical shift reference. The use of a chemical shift reference compound other than the accepted standard is particularly common for nuclei that have a very large chemical shift range such as fluorine-19 or selenium-77, since it may be impossible to measure the spectrum of the accepted chemical shift reference compound and the analyte of interest simultaneously because of their very different frequencies.

In addition to the differences in chemical shift, resonances in a given spectrum, for example for ethyl acetate (see illustration), also differ in the number of signals composing the resonance detected for each group of protons. Instead of one resonance resulting from a single transition between two spin states, the methylene group (CH_2) resonance is actually a quartet composed of four lines. Similarly, although the acetate CH_3 resonance is a single resonance, the ethylene group (CH_3) resonance consists of three lines. This splitting of some resonances, called spin-spin or scalar coupling, arises from interactions between nuclei through their bonding electrons rather than through space. Scalar coupling is a short-range interaction that usually occurs for nuclei separated by one to three bonds.

The effects of spin-spin coupling can be removed from the spectrum by application of a low-strength magnetic field to one of the coupled spins. The effect of this secondary field is to equalize the population of the coupled transitions and to remove the effects of spin-spin coupling. Decoupling can be homonuclear or heteronuclear. Homonuclear decoupling is useful for assigning the resonances of coupled spin systems in the spectrum of a compound or complex mixture. For example, in ethyl acetate, decoupling at the frequency of the methylene group (CH_2) protons would remove the effects of spin-spin coupling, collapsing the ethylene group (CH_3) resonance into a singlet and confirming the resonance assignments. Cynthia K. Larive

Nuclear medicine A subspecialty of medicine based on the use of radioactive substances in medical diagnosis, treatment, and research. Nuclear medicine makes it possible to examine regional physiology and biochemistry in ways that at times surpass the perception of surgeons during an operation or pathologists during an autopsy. Imaging methods make it possible to measure regional as well as overall organ function, and to portray the results in the form of functional or biochemical pictures of the body in health and in disease. Such pictures enhance the information about structure that is obtained by other imaging methods, such as computerized tomography (CT) or magnetic resonance imaging (MRI), often providing unique, objective evidence of disease long before structural changes are seen. *See* COMPUTERIZED TOMOGRAPHY; NUCLEAR MAGNETIC RESONANCE (NMR); RADIOACTIVE TRACER.

In a typical examination, a radioactive molecule is injected into an arm vein, and its distribution at specific time periods afterward is imaged in certain organs of the body or in the entire body. The images are created by measuring the gamma-ray photons emitted from the organs or regions of interest within the body. Nuclear medicine imaging procedures differ from ordinary x-rays in that the gamma rays are emitted from the body rather than transmitted across the body, as in the case of x-rays. As in most modern imaging, the principle of tomography is used, that is, the person is viewed by radiation detectors surrounding the body, or by rotation of a gamma camera around the body. Such procedures include single-photon emission computed tomography (SPECT), based on the use of iodine-123 or technetium-99m, and positron emission tomography (PET), based on the use of carbon-11 and fluorine-18. *See* RADIOISOTOPE (BIOLOGY).

The nature of the injected material, called a radiopharmaceutical, determines the information that will be obtained. In most cases, either blood flow or biochemical processes within an organ or part of an organ are examined. The essence of a nuclear medicine examination is measurement of the regional chemistry of a living human body.

In some diseases, radiation can be used to produce a biological effect. An example is the use of radioactive iodine to treat hyperthyroidism or cancer of the thyroid. The effects of treatment can be assessed with nuclear medicine techniques as well. *See* MEDICAL IMAGING; RADIATION THERAPY; RADIOLOGY.

Henry N. Wagner, Jr.

Nuclear molecule A quasistable entity of nuclear dimensions formed in nuclear collisions and comprising two or more discrete nuclei that retain their identities and are bound together by strong nuclear forces. Whereas the stable molecules of chemistry and biology consist of atoms bound through various electronic mechanisms, nuclear molecules do not form in nature except possibly in the hearts of giant stars; this simply reflects the fact that all nuclei carry positive electrical charges, and that under all natural conditions the long-range electrostatic repulsion prevents nuclear components from coming within the grasp of the short-range attractive nuclear force which could provide molecular binding. But in energetic collisions this electrostatic repulsion can be overcome. *See* NUCLEAR STRUCTURE.

D. Allan Bromley

Nuclear moments Intrinsic properties of atomic nuclei: electric moments result from deviations of the nuclear charge distribution from spherical symmetry; magnetic moments are a consequence of the intrinsic spin and the rotational motion of nucleons within the nucleus. The classical definitions of the magnetic and electric multipole moments are written in general in terms of multipole expansions. *See* NUCLEAR STRUCTURE; SPIN (QUANTUM MECHANICS).

In special cases nuclear moments can be measured by direct methods involving the interaction of the nucleus with an external magnetic field or with an electric field gradient produced by the scattering of high-energy charged particles. In general, however, nuclear moments manifest themselves through the hyperfine interaction between the nuclear moments and the fields or field gradients produced by either the atomic electrons' currents and spins, or the molecular or crystalline electronic and lattice structures. *See* HYPERFINE STRUCTURE.

Noémie Koller

Nuclear orientation The directional ordering of an assembly of nuclear spins I with respect to some axis in space. Under normal conditions nuclei are not oriented; that is, all directions in space are equally probable. For a system of nuclear spins with rotational symmetry about an axis, the degree of orientation is completely characterized by the relative populations a_m of the $2I + 1$ magnetic sublevels m $(= I, I - 1, \ldots, -I)$.

Nuclear orientation can be achieved in various ways. The most obvious way is to modify the energies of the $2I + 1$ magnetic sublevels so as to remove their degeneracy and thereby change the populations of these sublevels. The spin degeneracy can be removed by a magnetic field interacting with the nuclear magnetic dipole moment, or by an inhomogeneous electric field interacting with the nuclear electric quadrupole moment. Significant differences in the populations of the sublevels can be established by cooling the nuclear sample to low temperatures. This means of producing nuclear orientation is called the static method. In contrast, there is the dynamic method, which is related to optical pumping in gases. There are other ways to produce oriented nuclei; for example, in a nuclear reaction such as the capture of polarized neutrons (produced by magnetic scattering) by unoriented nuclei. *See* DYNAMIC NUCLEAR POLARIZATION; OPTICAL PUMPING.

Oriented nuclei have been used to measure nuclear properties, for example, magnetic dipole and electric quadrupole moments, spins, parities, and mixing ratios of nuclear states. Oriented nuclei have been used to examine some of the fundamental properties of nuclear forces, for example, nonconservation of parity in the weak interaction. Measurement of hyperfine fields, electric-field gradients, and other properties relating to the environment of the nucleus have been made by using oriented nuclei. Nuclear orientation thermometry is one of the few sources of a primary temperature scale at low temperatures. Oriented nuclear targets used in conjunction with beams of polarized and unpolarized particles have proved very useful in examining certain aspects of the nuclear force. *See* LOW-TEMPERATURE THERMOMETRY; NUCLEAR MOMENTS; NUCLEAR STRUCTURE; PARITY (QUANTUM MECHANICS).

Harvey Marshak

Nuclear physics The discipline involving the structure of atomic nuclei and their interactions with each other, with their constituent particles, and with the whole spectrum of elementary particles that is provided by very large accelerators. The nuclear domain occupies a central position between the atomic range of forces and sizes and those of elementary-particle physics, characteristically within the nucleons themselves. As the only system in which all the known natural forces can be studied simultaneously, it provides a natural laboratory for the testing and extending of many fundamental symmetries and laws of nature. Containing a reasonably large, yet manageable number of strongly interacting components, the nucleus also occupies a central position in the universal many-body problem of physics. *See* ATOMIC NUCLEUS; ATOMIC STRUCTURE AND SPECTRA; ELEMENTARY PARTICLE; SYMMETRY LAWS (PHYSICS).

Nuclear physics is unique in the extent to which it merges the most fundamental and the most applied topics. Its instrumentation has found broad applicability throughout science, technology, and medicine; nuclear engineering and nuclear medicine are two very important areas of applied specialization. *See* NUCLEAR ENGINEERING; NUCLEAR RADIATION (BIOLOGY); RADIOLOGY.

Nuclear chemistry, certain aspects of condensed matter and materials science, and nuclear physics together constitute the broad field of nuclear science; outside the United States and Canada elementary particle physics is frequently included in this more general classification. *See* ANALOG STATES; COSMIC RAYS; FUNDAMENTAL INTERACTIONS; ISOTOPE; NUCLEAR CHEMISTRY; NUCLEAR FISSION; NUCLEAR FUSION; NUCLEAR ISOMERISM; NUCLEAR MOMENTS; NUCLEAR REACTION; NUCLEAR REACTOR; NUCLEAR SPECTRA; NUCLEAR STRUCTURE; PARTICLE ACCELERATOR; PARTICLE DETECTOR; RADIOACTIVITY; SCATTERING EXPERIMENTS (NUCLEI); WEAK NUCLEAR INTERACTIONS.

D. Allan Bromley

Nuclear power Power derived from fission or fusion nuclear reactions. More conventionally, nuclear power is interpreted as the utilization of the fission reactions in a nuclear power reactor to produce steam for electric power production, for ship propulsion, or for process heat. Fission reactions involve the breakup of the nucleus of high-mass atoms and yield an energy release which is more than a millionfold greater than that obtained from chemical reactions involving the burning of a fuel. Successful control of the nuclear fission reactions utilizes this intensive source of energy. *See* NUCLEAR FISSION.

Fission reactions provide intensive sources of energy. For example, the fissioning of an atom of uranium yields about 200 MeV, whereas the oxidation of an atom of carbon releases only 4 eV. On a weight basis, this 50×10^6 energy ratio becomes about 2.5×10^6. Uranium consists of several isotopes, only 0.7% of which is uranium-235, the fissile fuel currently used in reactors. Even with these considerations, including the need to enrich the fuel to several percent uranium-235, the fission reactions are attractive energy sources when coupled with abundant and relatively cheap uranium ore.

Although the main process of nuclear power is the release of energy in the fission process which occurs in the reactor, there are a number of other important processes, such as mining and waste disposal, which both precede and follow fission. Together they constitute the nuclear fuel cycle. *See* NUCLEAR FUEL CYCLE.

Power reactors include light-water-moderated and -cooled reactors (LWRs), including the pressurized-water reactor (PWR) and the boiling-water reactor (BWR). The high-temperature gas-cooled reactor (HTGR), and the liquid-metal-cooled fast breeder reactor (LMFBR) have reached a high level of development but are not used for commercial purposes. *See* NUCLEAR REACTOR.

Critics of nuclear power consider the radioactive wastes generated by the nuclear industry to be too great a burden for society to bear. They argue that since the high-level wastes will contain highly toxic materials with long half-lives, such as a few tenths of one percent of plutonium that was in the irradiated fuel, the safekeeping of these materials must be assured for time periods longer than social orders have existed in the past. Nuclear proponents answer that the time required for isolation is much shorter, since only 500 to 1000 years is needed before the hazard posed by nuclear waste falls below that posed by common natural ore deposits in the environment. *See* RADIOACTIVE WASTE MANAGEMENT.

Nuclear power facilities present a potential hazard rarely encounted with other facilities; that is, radiation. A major health hazard would result if, for instance, a significant fraction of the core inventory of a power reactor were released to the atmosphere. Such a release of radioactivity is clearly unacceptable, and steps are taken to assure it could never happen. These include use of engineered safety systems, various construction and design codes, regulations on reactor operation, and periodic maintenance and inspection. Frank J. Rahn

Nuclear quadrupole resonance A selective absorption phenomenon observable in a wide variety of polycrystalline compounds containing nonspherical atomic nuclei when placed in a magnetic radio-frequency field. Nuclear quadrupole resonance (NQR) is very similar to nuclear magnetic resonance (NMR), and was originated as an inexpensive (no stable homogeneous large magnetic field is required) alternative way to study nuclear moments. It later gained a modest popularity. *See* MAGNETIC RESONANCE; NUCLEAR MAGNETIC RESONANCE (NMR).

In the simplest case, for example, ^{35}Cl in solid Cl_2, NQR is associated with the precession of the angular momentum I (and the nuclear magnetic dipole moment μ) of the nucleus, depicted in the illustration as a flat ellipsoid of rotation, around the

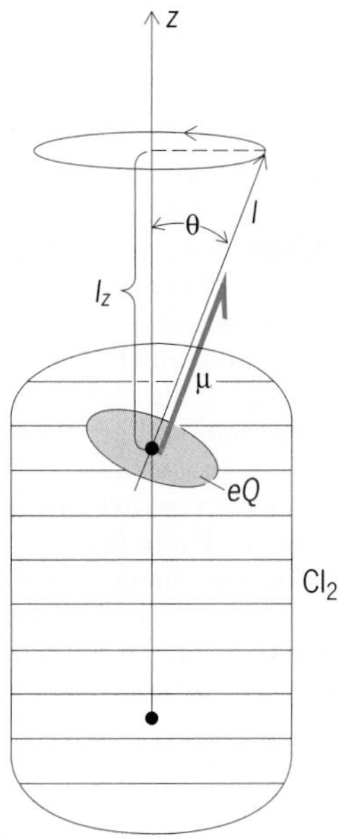

Interaction of ^{35}Cl nucleus with the electric field of a Cl_2 molecule.

symmetry axis (taken as the z axis) of the Cl_2 molecule fixed in the crystalline solid. The precession, with constant angle θ between the nuclear axis and symmetry axis of the molecule, is due to the torque which the inhomogeneous molecular electric field exerts on the nucleus of electric quadrupole moment eQ. The absorption occurs classically when the frequency of the rf field and that of the precessing motion of the angular momentum coincide.

NQR spectra have been observed in the approximate range 1–1000 MHz. Most of the NQR work has been on molecular crystals. For such crystals the coupling constants found do not differ very much from those measured for the isolated molecules in microwave spectroscopy. The most precise nuclear information which may be extracted from NQR data are quadrupole moment ratios of isotopes of the same element. If values for the axial gradient of the molecular electric field can be estimated from atomic fine structure data, then fair values of the quadrupole moment may be obtained. However, it has also proved very productive to use the quadrupole nucleus as a probe of bond character and orientation and crystalline electric fields and lattice sites, and extensive data have been accumulated in this area. *See* MICROWAVE SPECTROSCOPY. Hans Dehmelt

Nuclear radiation All particles and radiations emanating from an atomic nucleus due to radioactive decay and nuclear reactions. Thus the criterion for nuclear radiations is that a nuclear process is involved in their production. The term was originally used to denote the ionizing radiations observed from naturally occurring radioactive materials. These radiations were alpha rays (energetic helium nuclei), beta rays (negative electrons), and gamma rays (electromagnetic radiation with wavelength much shorter than visible light). *See* ALPHA PARTICLES; BETA PARTICLES; GAMMA RAYS.

Nuclear radiations have traditionally been considered to be of three types based on the manner in which they interact with matter as they pass through it. These are the charged heavy particles with masses comparable to that of the nuclear mass (for example, protons, alpha particles, and heavier nuclei), electrons (both negatively and positively charged), and electromagnetic radiation. For all of these, the interactions with matter are considered to be primarily electromagnetic. The behavior of mesons and other particles is intermediate between that of the electron and heavy charged particles.

A striking difference in the absorption of the three types of radiations is that only heavy charged particles have a range. That is, a monoenergetic beam of heavy charged particles, in passing through a certain amount of matter, will lose energy without changing the number of particles in the beam. Ultimately, they will be stopped after crossing practically the same thickness of absorber. For electromagnetic radiation (gamma rays) and neutrons, on the other hand, the absorption is exponential. The difference in behavior reflects the fact that charged particles are not removed from the beam by individual interactions, whereas gamma radiation photons (and neutrons) are removed. Electrons exhibit a more complex behavior. *See* ELECTRON; NUCLEAR REACTION. Dennis G. Kovar

Nuclear radiation (biology) Nuclear radiations are used in biology because of their common property of ionizing matter. This makes their detection relatively simple, or makes possible the production of biological effects in any living cell.

Ionizing radiation is any electromagnetic or particulate radiation capable of producing ions, directly or indirectly, in its passage through matter.

All ionizing radiations produce biological changes, directly by ionization or excitation of the atoms in the molecules of biological entities, such as in chromosomes, or indirectly by the formation of active radicals or deleterious agents, through ionization and excitation, in the medium surrounding the biological entities. Ionizing radiation, having high penetrating power, can reach the most vulnerable part of a cell, an organ, or a whole organism, and is thus very effective. In terms of the energy absorbed per unit mass of a biological entity in which an effect is produced, some ionizing radiations are more effective than others. The relative biological effectiveness (RBE) depends in fact on the density of ionization (also termed the specific ionization or linear energy transfer, LET) along the path of the ionizing particle rather than on the nature of the particle itself. Relative biological effectiveness depends also on many other factors. *See* LINEAR ENERGY TRANSFER (BIOLOGY); RADIATION BIOLOGY.

The medical uses of nuclear radiations may be divided into three distinct classes:

1. The radiations, which are principally x-rays, are used to study the anatomical configuration of body organs, usually for the purpose of detecting abnormalities as an aid in diagnosis.
2. The radiations are used for therapeutic purposes to produce biological changes in such tissues as tumors.
3. The radiations are used as a simple means of tracing a suitable radioactive substance through different steps in its course through the body, in the study of some particular physiological process. *See* RADIOLOGY. Gioacchino Failla; Edith H. Quimby

The radiations emitted by radioactive isotopes of the various elements are used in biological research. The most useful ones in biological research are the isotopes of the elements which are important in metabolism and in the structural materials of cells. These include carbon, hydrogen, sulfur, and phosphorus. In addition, the radioactive metals like cobalt-60, radium, and others can be used to produce radiations for external application to cells and tissues. Most of the isotopes mentioned emit beta particles when they decay, and a few emit gamma rays. Therefore, they can be easily detected by various means. These procedures are sometimes used in the study of the movement of elements or their compounds in plants and animals. They are frequently used for tracing the sequence of reactions in metabolism. *See* BETA PARTICLES; GAMMA RAYS; RADIOISOTOPE (BIOLOGY).

In addition to the radiations emitted, some of the elements change to a different element when they decay. Phosphorus-32 changes to sulfur, sulfur-35 changes to chlorine, and tritium (hydrogen-3) changes to helium when they decay by the emission of an electron, a beta particle. Therefore, in addition to the radiation produced, the transmutation of the element affects the molecule and the cell of which it is a part. In other experiments, the decay of phosphorus-32 has been used to give information on the nature and importance of the phosphorus-containing molecules to the survival or reproduction of a cell or virus particle. Nuclear radiations have also proved useful in many studies of the nature and the mechanisms of the effects of radiations on cells and cell constituents. J. Herbert Taylor

Nuclear reaction A process that occurs as a result of interactions between atomic nuclei when the interacting particles approach each other to within distances of the order of nuclear dimensions ($\simeq 10^{-12}$ cm). While nuclear reactions occur in nature, understanding of them and use of them as tools have taken place primarily in the controlled laboratory environment. In the usual experimental situation, nuclear reactions are initiated by bombarding one of the interacting particles, the stationary target nucleus, with nuclear projectiles of some type, and the reaction products and their behaviors are studied.

Types of nuclear interaction. As a generalized nuclear process, consider a collision in which an incident particle strikes a previously stationary particle, to produce an unspecified number of final products. If the final products are the same as the two initial particles, the process is called scattering. The scattering is said to be elastic or inelastic, depending on whether some of the kinetic energy of the incident particle is used to raise either of the particles to an excited state. If the product particles are different from the initial pair, the process is referred to as a reaction.

The most common type of nuclear reaction, and the one which has been most extensively studied, involves the production of two final products. Such reactions can be observed, for example, when deuterons with a kinetic energy of a few megaelectronvolts are allowed to strike a carbon nucleus of mass 12. Protons, neutrons, deuterons, and alpha particles are observed to be emitted, and reactions (1)–(4) are responsible. In these

$$ {}^{2}_{1}\mathrm{H} + {}^{12}_{6}\mathrm{C} \rightarrow {}^{2}_{1}\mathrm{H} + {}^{12}_{6}\mathrm{C} \qquad (1) $$

$$ {}^{2}_{1}\mathrm{H} + {}^{12}_{6}\mathrm{C} \rightarrow {}^{1}_{1}\mathrm{H} + {}^{13}_{6}\mathrm{C} \qquad (2) $$

$$ {}^{2}_{1}\mathrm{H} + {}^{12}_{6}\mathrm{C} \rightarrow {}^{0}_{1}n + {}^{13}_{7}\mathrm{N} \qquad (3) $$

$$ {}^{2}_{1}\mathrm{H} + {}^{12}_{6}\mathrm{C} \rightarrow {}^{4}_{2}\mathrm{He} + {}^{10}_{5}\mathrm{B} \qquad (4) $$

equations the nuclei are indicated by the usual chemical symbols; the subscripts indicate the atomic number (nuclear charge) of the nucleus, and the superscripts the mass number of the particular isotope. These reactions are conventionally written in the compact notation ${}^{12}\mathrm{C}(d,d){}^{12}\mathrm{C}$, ${}^{12}\mathrm{C}(d,p){}^{13}\mathrm{C}$, ${}^{12}\mathrm{C}(d,n){}^{13}\mathrm{N}$, and ${}^{12}\mathrm{C}(d,\alpha){}^{10}\mathrm{B}$, where d represents deuteron, p proton, n neutron, and α alpha particle. In each of these cases the reaction results in the production of an emitted light particle and a heavy residual nucleus. If the residual nucleus is formed in an excited state, it will subsequently emit this excitation energy in the form of gamma rays or, in special cases, electrons. The residual nucleus may also be a radioactive species, in which case it will undergo further transformation in accordance with its characteristic radioactive decay scheme. *See* RADIOACTIVITY.

Nuclear cross section. In general one is interested in the probability of occurrence of the various reactions as a function of the bombarding energy of the incident particle. The measure of probability for a nuclear reaction is its cross section. Consider a reaction initiated by a beam of particles incident on a region which contains N atoms per unit area (uniformly distributed), and where I particles per second striking the area result in R reactions of a particular type per second. The fraction of the area bombarded which is effective in producing the reaction products is R/I. If this is divided by the number of nuclei per unit area, the effective area or cross section $\sigma = R/IN$. This is referred to as the total cross section for the specific reaction, since it involves all the occurrences of the reaction. The dimensions are those of an area, and total cross sections are expressed in either square centimeters or barns (1 barn $= 10^{-24}$ cm^2). The differential cross section refers to the probability that a particular reaction product will be observed at a given angle with respect to the beam direction. Its dimensions are those of an area per unit solid angle (for example, barns per steradian).

Reaction mechanism. Various reaction models have been extremely successful in describing certain classes or types of nuclear reaction processes. In general, all reactions can be classified according to the time scale on which they occur, and the degree to which the kinetic energy of the incident particle is converted into internal excitation of the final products. A large fraction of the reactions observed has properties consistent with those predicted by two reaction mechanisms which represent the extremes in this general classification. These are the mechanisms of compound nucleus formation and direct interaction.

Compound nucleus formation is envisioned to take place in two distinct steps. In the first step the incident particle is captured by (or fuses with) the target nucleus, forming an intermediate or compound nucleus which lives a long time ($\simeq 10^{-16}$ s) compared to the approximately 10^{-22} s it takes the incident particle to travel past the target. During this time the kinetic energy of the incident particle is shared among all the nucleons, and all memory of the incident particle and target is lost. The compound nucleus is always formed in a highly excited unstable state, is assumed to approach themodynamic equilibrium involving all or most of the available degrees of freedom, and will decay, as the second step, into different reaction products, or through so-called exit channels. The essential feature of the compound nucleus formation or fusion reaction is that the probability for a specific reaction depends on two independent probabilities: the probability for forming the compound nucleus, and the probability for decaying into that specific exit channel.

Some reactions have properties which are in striking conflict with the predictions of the compound nucleus hypothesis. Many of these are consistent with the picture of a mechanism where no long-lived intermediate system is formed, but rather a fast mechanism where the incident particle, or some portion of it, interacts with the surface, or some nucleons on the surface, of the target nucleus. These direct reactions are assumed to involve only a very small number of the available degrees of freedom. Most direct reactions are of the transfer type, where one or more nucleons are transferred to or from the incident particle as it passes the target, leaving the two final partners either in their ground states or in one of their many excited states. Such transfer reactions are generally referred to as stripping or pickup reactions, depending on whether the incident particle has lost or acquired nucleons in the reaction.

Inelastic scattering is also a direct reaction. Whereas the states preferentially populated in transfer reactions are those of specific single-particle or shell-model structure, the states preferentially excited in inelastic scattering are collective in nature. *See* NUCLEAR STRUCTURE; SCATTERING EXPERIMENTS (NUCLEI).

Dennis G. Kovar

Nuclear reactor A system utilizing nuclear fission in a controlled and self-sustaining manner. Neutrons are used to fission the nuclear fuel, and the fission reaction produces not only energy and radiation but also additional neutrons. Thus a neutron chain reaction ensues. A nuclear reactor provides the assembly of materials to sustain and control the neutron chain reaction, to appropriately transport the heat produced from the fission reactions, and to provide the necessary safety features to cope with the radiation and radioactive materials produced by its operation. *See* CHAIN REACTION (PHYSICS); NUCLEAR FISSION.

Nuclear reactors are used in a variety of ways as sources for energy, for nuclear irradiations, and to produce special materials by transmutation reactions. The generation of electrical energy by a nuclear power plant makes use of heat to produce steam or to heat gases to drive turbogenerators. Direct conversion of the fission energy into useful work is possible, but an efficient process has not yet been realized to accomplish this. Thus, in its operation the nuclear power plant is similar to the conventional coal-fired plant, except that the nuclear reactor is substituted for the conventional boiler as the source of heat.

The rating of a reactor is usually given in kilowatts (kW) or megawatts-thermal [MW(th)], representing the heat generation rate. The net output of electricity of a nuclear plant is about one-third of the thermal output. Significant economic gains have been achieved by building improved nuclear reactors with outputs of about 3300 MW(th) and about 1000 MW-electrical [MW(e)]. *See* ELECTRIC POWER GENERATION; NUCLEAR POWER.

Fuel and moderator. The fission neutrons are released at high energies and are called fast neutrons. The average kinetic energy is 2 MeV, with a corresponding neutron speed of 1/15 the speed of light. Neutrons slow down through collisions with nuclei of the surrounding material. This slowing-down process is made more effective by the introduction of materials of low atomic weight, called moderators, such as heavy water (deuterium oxide), ordinary (light) water, graphite, beryllium, beryllium oxide, hydrides, and organic materials (hydrocarbons). Neutrons that have slowed down to an energy state in equilibrium with the surrounding materials are called thermal neutrons, moving at 0.0006% of the speed of light. The probability that a neutron will cause the fuel material to fission is greatly enhanced at thermal energies, and thus most reactors utilize a moderator for the conversion of fast neutrons to thermal neutrons. *See* NEUTRON; THERMAL NEUTRONS.

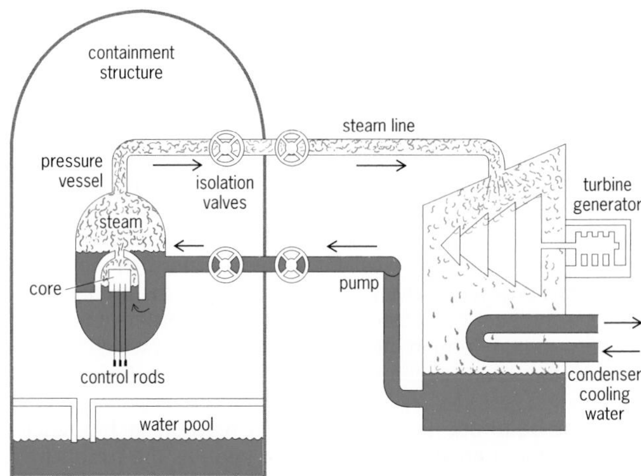

Boiling-water reactor. (*Atomic Industrial Forum, Inc.*)

With suitable concentrations of the fuel material, neutron chain reactions also can be sustained at higher neutron energy levels. The energy range between fast and thermal is designated as intermediate. Fast reactors do not have moderators and are relatively small.

Only three isotopes—uranium-235, uranium-233, and plutonium-239—are feasible as fission fuels, but a wide selection of materials incorporating these isotopes is available.

Heat removal. The major portion of the energy released by the fissioning of the fuel is in the form of kinetic energy of the fission fragments, which in turn is converted into heat through the slowing down and stopping of the fragments. For the heterogeneous reactors this heating occurs within the fuel elements. Heating also arises through the release and absorption of the radiation from the fission process and from the radioactive materials formed. The heat generated in a reactor is removed by a primary coolant flowing through it.

Reactor coolants. Coolants are selected for specific applications on the basis of their heat-transfer capability, physical properties, and nuclear properties.

Water has many desirable characteristics. It was employed as the coolant in many of the first production reactors, and most power reactors still utilize water as the coolant. In a boiling-water reactor (BWR; see illustration), the water boils directly in the reactor core to make steam that is piped to the turbine. In a pressurized-water reactor (PWR), the coolant water is kept under increased pressure to prevent boiling. It transfers heat to a separate stream of feed water in a steam generator, changing that water to steam.

For both boiling-water and pressurized-water reactors, the water serves as the moderator as well as the coolant. Both light water and heavy water are excellent neutron moderators, although heavy water (deuterium oxide) has a neutron-absorption cross section approximately 1/500 that for light water that makes it possible to operate reactors using heavy water with natural uranium fuel. The high pressure necessary for water-cooled power reactors determines much of the plant design. *See* NUCLEAR REACTION.

Gases are inherently poor heat-transfer fluids as compared with liquids because of their low density. This situation can be improved by increasing the gas pressure; however, this introduces other problems and costs. Helium is the most attractive gas (it is chemically inert and has good thermodynamic and nuclear properties) and has been selected as the coolant for the development of high-temperature gas-cooled reactor (HTGR) systems, in which the gas transfers heat from the reactor core to a steam generator. The British advanced gas reactor (AGR), however, uses carbon dioxide (CO_2). Gases are capable of operation at extremely high temperature, and they are being considered for special process applications and direct-cycle gas-turbine applications.

The alkali metals, in particular, have excellent heat-transfer properties and extremely low vapor pressures at temperatures of interest for power generation. Sodium is attractive because of its relatively low melting point (208°F or 98°C) and high heat-transfer coefficient. It is also abundant, commercially available in acceptable purity, and relatively inexpensive. It is not particularly corrosive, provided low oxygen concentration is maintained. Its nuclear properties are excellent for fast reactors. In the liquid-metal fast breeder reactor (LMFBR), sodium in the primary loop collects the heat generated in the core and transfers it to a secondary sodium loop in the heat exchanger, from which it is carried to the steam generator in which water is boiled to make steam.

Plant balance. The nuclear chain reaction in the reactor core produces energy in the form of heat, as the fission fragments slow down and dissipate their kinetic energy in the fuel. This heat must be removed efficiently and at the same rate it is being generated in order to prevent overheating of the core and to transport the energy outside the core, where it can be converted to a convenient form for further utilization. The energy transferred to the coolant, as it flows past the fuel element, is stored in it in the form of sensible heat and pressure and is called the enthalpy of the fluid. In an electric power plant, the energy stored in the fuel is further converted to kinetic energy through a device called a prime mover which, in the case of nuclear reactors, is predominantly a steam turbine. Another conversion takes place in the electric generator, where kinetic energy is converted into electric power as the final energy form to be distributed to the consumers through the power grid and distribution system. *See* ENTHALPY; GENERATOR; PRIME MOVER; STEAM TURBINE.

Fluid flow and hydrodynamics. Because heat removal must be accomplished as efficiently as possible, considerable attention must be given to fluid-flow and hydrodynamic characteristics of the system. *See* FLUID FLOW; HYDRODYNAMICS.

The heat capacity and thermal conductivity of the fluid at the temperature of operation have a fundamental effect upon the design of the reactor system. The heat capacity determines the mass flow of the coolant required. The fluid properties (thermal conductivity, viscosity, density, and specific heat) are important in determining the surface area required for the fuel—in particular, the number and arrangement of the fuel elements. These factors combine to establish the pumping characteristics of the system because the pressure drop and coolant temperature rise in the core are directly related. *See* CONDUCTION (HEAT); HEAT CAPACITY; VISCOSITY.

Thermal stress. The temperature of the reactor coolant increases as it circulates through the reactor core. Fluctuations in power level or in coolant flow rate result in variations in the temperature rise. A reactor is capable of very rapid changes in power level, particularly reduction in power level, which is a safety feature of the plant. Reactors are equipped with mechanisms (reactor scram systems) to ensure rapid shutdown of the system in the event of leaks, failure of power conversion systems, or other operational abnormalities. Therefore, reactor coolant systems must be designed to accommodate the temperature transients that may occur because of rapid power changes. In addition, they must be designed to accommodate temperature transients that might occur as a result of a coolant system malfunction, such as pump stoppage.

Coolant system components. The development of reactor systems has led to the development of special components for reactor component systems. Because of the hazard of radioactivity, leak-tight systems and components are a prerequisite to safe, reliable operation, and maintenance. Special problems are introduced by many of the fluids employed as reactor coolants.

More extensive component developments have been required for sodium, which is chemically active and is an extremely poor lubricant. Centrifugal pumps employing unique bearings and seals have been specially designed. Sodium is an excellent electrical conductor and, in some special cases, electromagnetic-type pumps have been used. These pumps are completely sealed, contain no moving parts, and derive their pumping action from electromagnetic forces imposed directly on the fluid. *See* CENTRIFUGAL PUMP; ELECTROMAGNETIC PUMP.

Core design. A typical reactor core for a power reactor consists of the fuel element rods supported by a grid-type structure inside a vessel.

Structural materials employed in reactor systems must possess suitable nuclear and physical properties and must be compatible with the reactor coolant under the conditions of operation. The most common structural materials employed in reactor systems are stainless steel and zirconium alloys. Zirconium alloys have favorable nuclear and physical properties, whereas stainless steel has favorable physical properties. Aluminum is widely used in low-temperature test and research reactors; zirconium and stainless steel are used in high-temperature power reactors. Zirconium is relatively expensive, and its use is therefore confined to applications in the reactor core where neutron absorption is important. *See* ALUMINUM; STAINLESS STEEL; ZIRCONIUM.

Reactors maintain a separation of fuel and coolant by cladding the fuel. The cladding is designed to prevent the release of radioactivity from the fuel. The cladding material must be compatible with both the fuel and the coolant.

The cladding materials must also have favorable nuclear properties. The neutron-capture cross section is most significant because the unwanted absorption of neutrons by these materials reduces the efficiency of the nuclear fission process. Aluminum is a very desirable material in this respect; however, its physical strength and corrosion resistance in water decrease very rapidly above about 300°F (149°C).

Zirconium has favorable neutron properties, and in addition is corrosion-resistant in high-temperature water. It has found extensive use in water-cooled power reactors. Stainless steel is used for the fuel cladding in fast reactors, in some light-water reactors for which neutron captures are less important.

Control. A reactor is critical when the rate of production of neutrons equals the rate of absorption in the system. The control of reactors requires the continuing measurement and adjustment of the critical condition. The neutrons are produced by the fission process and are consumed in a variety of ways, including absorption to cause fission, nonfission capture in fissionable materials, capture in fertile materials, capture in structure or coolant, and leakage from the reactor to the shielding. A reactor is subcritical (power level decreasing) if the number of neutrons produced is less than the number consumed. The reactor is supercritical (power level increasing) if the number of neutrons produced exceeds the number consumed. *See* REACTOR PHYSICS.

Reactors are controlled by adjusting the balance between neutron production and neutron consumption. Normally, neutron consumption is controlled by varying the absorption or leakage of neutrons; however, the neutron generation rate also can be controlled by varying the amount of fissionable material in the system.

The reactor control system requires the movement of neutron-absorbing rods (control rods) in the reactor under carefully controlled conditions. They must be arranged to increase reactivity (increase neutron population) slowly and under good control. They must be capable of reducing reactivity, both rapidly and slowly.

The control drives can be operated by the reactor operator or by automatic control systems. Reactor scram (rapid reactor shutdown) can be initiated automatically by a wide variety of system scram-safety signals, or it can be started by the operator depressing a scram button in the control room.

Control drives are electromechanical or hydraulic devices that impart in-and-out motion to the control rods. They are usually equipped with a relatively slow-speed reversible drive system for normal operational control. Scram is usually effected by a high-speed overriding drive accompanied by disconnecting the main drive system.

Applications. Reactor applications include mobile, stationary, and packaged power plants; production of fissionable fuels (plutonium and uranium-233) for military and commercial applications; research, testing, teaching-demonstration, and experimental facilities; space and process heat; dual-purpose design; and special applications. The potential use of reactor radiation or radioisotopes produced for sterilization of food and other products, steam for chemical processes, and gas for high-temperature applications has been recognized. *See* NUCLEAR FUEL CYCLE; NUCLEAR FUELS REPROCESSING; RADIOACTIVITY AND RADIATION APPLICATIONS; SHIP NUCLEAR PROPULSION. Frank J. Rahn

Nuclear spectra The distribution of the intensity of particles (or radiation) emitted in a nuclear process as a function of energy. The nuclear spectrum is a unique signature of the process.

For example, when very slow neutrons (with speeds less than 0.5% of the speed of light) hit nitrogen nuclei, there is a high probability that they will be captured and that the nuclear system which is formed will emit a set of gamma rays (electromagnetic radiation) of very precise energies. The 24 gamma rays have energies ranging from 1.68 to 10.83 MeV, and their relative intensities are well known. A spectrum of these gamma rays, that is, the number of gamma rays having a particular energy, versus that energy can provide a unique signature of the presence of nitrogen. An application is the passing of a beam of slow neutrons through luggage at an airport: the presence of unusual amounts of nitrogen indicates that a plastic explosive may be present. This testing is nondestructive: relatively few neutrons are needed to produce the characteristic spectrum, and the luggage and its contents are not harmed. *See* GAMMA RAYS; NONDESTRUCTIVE EVALUATION.

Measurements. The methods used to measure nuclear spectra depend on the nature of the particles (radiation) involved. The most accurate energy measurements are those of gamma rays. Gamma-ray spectra can be measured by determining the energy deposited by the gamma rays in a crystal, often made of sodium iodide, containing thallium impurities [NaI(Tl)], or of germanium, containing lithium impurities [Ge(Li)]. In a NaI(Tl) detector, the gamma-ray energy is transferred to electrons within the crystal, and these charged particles in turn produce electromagnetic radiation with frequencies in the visible range. The crystal is surrounded by detectors (photomultipliers) that are sensitive to the visible light. The intensity of the signal in the photomultipliers is proportional to the energy of the gamma rays that entered the NaI(Tl) crystal. The signal pulse is amplified electronically, and the pulse heights (pulse sizes) are displayed in a pulse-height multichannel analyzer in a histogram. Usually the number of pulses having a certain height (strength) is plotted versus the height. What results is a plot showing the number of gamma rays having a certain energy versus the energy of the gamma rays, a spectrum. *See* GAMMA-RAY DETECTORS; PHOTOMULTIPLIER.

Neutron spectra are often determined by measuring their velocities. This is done by a time-of-flight technique in which an electronic timer measures the time interval between the emission of the neutron from a nucleus and its arrival at a detector a known distance away. This measurement uniquely determines the velocity, and thus the kinetic energy, of the neutrons. *See* NEUTRON SPECTROMETRY; TIME-OF-FLIGHT SPECTROMETERS.

Measurements of nuclear spectra involving charged particles, such as pions, protons and alpha particles, are often made by determining their momenta (mass × velocity) and then calculating the corresponding kinetic energy. Momentum measurements are made by passing the beam of charged particles through a region in which a magnetic field exists. A magnetic field that is constant in time will not cause a change in a charged particle's speed, but it will cause a charged particle to deviate in its path. *See* PARTICLE ACCELERATOR.

Modern magnetic spectrometers use sophisticated counter telescopes and multiwire proportional counters, which permit not only the registering of the particles characterized by a certain value of the radius of curvature (and therefore of momentum) but enable the particular particle (proton, alpha particle, or whatever) that caused the signal to be identified. Contemporary magnetic spectrometer systems not only utilize complex arrangements of magnetic fields, detectors, and electronics but also generally require powerful computers to monitor and analyze the results. *See* PARTICLE DETECTOR. Fay Ajzenberg-Selove

Nuclear structure At the center of every atom lies a small, dense nucleus, which carries more than 99.97% of the atomic mass in less than 10^{-12} of its volume. The nucleus is a tightly bound system of protons and neutrons which is held together by strong forces that are not normally perceptible in nature because of their extremely short range. The small size, strong forces, and many particles in the nucleus result in a highly complex and unique quantal system that at present defies exact analysis. The study of the nucleus and the forces that hold it together constitute the field of nuclear structure physics. *See* ATOMIC STRUCTURE AND SPECTRA; NEUTRON; PROTON; QUANTUM MECHANICS; STRONG NUCLEAR INTERACTIONS.

The protons of the nucleus, being positively charged, generate a spherically symmetric electric field in which the atomic electrons orbit. The cloud of negatively charged atomic electrons normally balances the positive nuclear charge, making the atom electrically neutral. The atomic number of protons is usually denoted by Z and the number of neutrons, which are electrically neutral, by N. The total number of protons and neutrons (or nucleons) is the mass number $A = Z + N$. Isotopes have the same atomic number, Z, and hence are forms of the same chemical element, having the same chemical properties, but they differ in neutron number; isotones have a common number of neutrons, N, and isobars have the same mass number, A. *See* ISOBAR (NUCLEAR PHYSICS); ISOTONE; ISOTOPE.

Nuclei have masses less than the sum of the constituents, the missing mass ΔM being accounted for by the binding energy ΔMc^2 (where c is the speed of light), which holds the nuclear system together. The characteristic energy scale is in megaelectronvolts (1 MeV = 1.6×10^{-13} joule). The internuclear forces generate an attractive potential field which holds the nucleus together and in which the nucleons orbit in highly correlated patterns. The volume of nuclei increases approximately linearly with mass number A, and the radius is roughly $R = 1.2 \times 10^{-15} \cdot A^{1/3}$ m. *See* NUCLEAR BINDING ENERGY.

Size, shape, and density distributions. A variety of sophisticated techniques have been developed for precise estimates of the nuclear charge distribution, including electron scattering, the study of muonic atoms, and the laser spectroscopy of hyperfine atomic structure. An overall picture of the nuclear charge distributions emerges. The nuclear charge density saturates in the interior and has a roughly constant value in all but the lightest nuclei. The nucleus has a diffuse skin which is of nearly constant thickness.

Many nuclei are found to have nonspherical shapes. Unlike the atom, which has a spherically symmetric Coulomb field generated by the nucleus, the nuclear field is composed of a complicated superposition of short-range interactions between nucleons, and the most stable nuclear shape is the one that minimizes the energy of the system. In general, it is not spherical, and the nuclear shape is most simply described by a multipole power series, the most important term of which is the nuclear quadrupole moment. A positive quadrupole moment reflects the elongation of nuclei into a prolate or football-like shape, while a negative value reflects an oblate shape like that of Earth. *See* NUCLEAR MOMENTS.

An accurate determination of nuclear matter distributions, that is, the distribution of both protons and neutrons in nuclei, is harder to precisely ascertain.

Nuclear masses and binding energies. The variation of average binding energy with mass number is approximated by the Bethe-Weizsacker mass formula, which is noteworthy for its simplicity in reproducing the overall binding energy systematics. The formula is developed by modeling the nucleus on a liquid drop. By analogy with a drop of liquid, there is an attractive volume term, which depends on the number of particles; a repulsive surface-tension term; and a term due to the mutual Coulomb repulsion of protons, which is responsible for the decrease in binding energy for heavy nuclei. The model is spectacularly successful in reproducing the overall trends in nuclear binding energies, masses, and the energetics of nuclear fission, and in predicting the limits of stability where neutrons and protons become unbound. As in the case of predicting a mean nuclear shape, a comparison of the prediction of the Bethe-Weizsacker mass formula to measured masses shows periodic fluctuations with both N and Z, which are due to the quantum shell effects. *See* NUCLEAR FISSION.

Nuclear excited states. The small nuclear size and tightly bound nature impose very restrictive constraints on the orbits that protons and neutrons can undergo inside the system. Thus, each nucleus has a series of quantum states that particles can occupy. The Pauli principle requires that each particle have a unique set of quantum labels. Each nuclear state can then be filled with four particles: protons with internal angular momentum "up" and "down," and likewise two neutrons. *See* ANGULAR MOMENTUM; ENERGY LEVEL (QUANTUM MECHANICS); EXCLUSION PRINCIPLE; PARITY (QUANTUM MECHANICS); QUANTUM NUMBERS; QUARKS; SPIN (QUANTUM MECHANICS).

A nucleus is most stable when all of its nucleons occupy the lowest possible states without violating this occupancy rule. This is called the nuclear ground state. During nuclear collisions the protons and neutrons can be excited from their most bound states and promoted to higher-lying unoccupied states. The process is usually very short-lived and the particles deexcite to their most stable configuration on a time scale of the order of 10^{-12} s. The energy is usually released in the form of gamma rays of well-defined energy corresponding to the difference in energy of the initial and final nuclear states. Occasionally, gamma decay is not favored because of angular momentum selection rules, and long-lived nuclear isomers result. *See* GAMMA RAYS; NUCLEAR ISOMERISM; NUCLEAR SPECTRA; SELECTION RULES (PHYSICS).

Nuclear models. The detailed categorization of the excitation of protons and neutrons allows a mapping of the excited states of each nucleus and determination of its quantum numbers. These data are the essential information required for development of detailed models that can describe the motion of nucleons inside nuclei. Unlike atomic molecules, where rotational, vibrational, and single-particle degrees of freedom involve different time scales and energies, the nucleus is highly complex, with rotation, vibration, and single-particle degrees of freedom being excited at similar energies and often strongly mixed. *See* MOLECULAR STRUCTURE AND SPECTRA.

The measurement of static electric and magnetic moments of nuclear states and of dynamic transition moments has provided a great deal of information. Electric moments have revealed a variety of enhanced collective modes of excitation, including elongated, flattened, and pear-shaped nuclear shapes. Magnetic moments have provided detailed information on the differences between excitations involving neutrons (negative moments) and protons (positive moments).

For atoms, the solutions of the Schrödinger equation with a Coulomb potential lead to a reasonable prediction of the energies of quantized atomic states, as well as their spins, parities, and moments. Attempts to make the same progress for nuclei, using a variety of spherically symmetric geometric potentials of nuclear dimensions, failed to reproduce known properties until it was realized by M. G. Meyer and independently by J. H. Jensen in 1949 that an additional spin-orbit potential was required to reproduce the known sequence of nuclear states. A potential of this form binds states having the internal spin of the nucleons parallel to its orbital angular momentum more tightly than when they are antiparallel. The ensuing sequence of quantum shell gaps and subgaps was then correctly reproduced. The shell model has evolved rapidly, and its domain of applicability has widened from the limited regions of sphericity near doubly magic nuclei to encompass most light nuclei with $A < 60$ as well as enlarged regions around shell closures.

As valence particles are added to a closed core of nucleons, the mutual residual interactions can act coherently and polarize the nuclear system away from sphericity. The polarization effects are strongest when both valence protons and neutrons are involved. The deformed nuclear potential can then undergo collective rotation, which generally involves less energy than other degrees of freedom and thus dominates the spectrum of strongly deformed nuclei.

Nuclei undergo collective vibrations about both spherical and deformed shapes. The degree of softness of these vibrations is characterized by the excitation energy required to populate states. The distinguishing feature of vibrational excited states is that they are grouped in nearly degenerate angular momentum multiplets, each group being separated by a characteristic phonon energy.

It has been a goal of nuclear structure studies to develop models that incorporate all of the features described above in order to produce a unified nuclear picture. The development of generalized nuclear models has relevance to other fields of physics. There are many isotopes that will never be accessible in the laboratory but may exist in stars or may have existed earlier in cosmological time. The evolution of generalized models greatly increases the power to predict nuclear behavior and provides information that is required for cosmological calculations. *See* COSMOLOGY; NUCLEOSYNTHESIS.

Nuclei at high excitation energies. As nuclei are excited to ever higher excitation energies, it is anticipated that shell effects will be slowly replaced by statistical, or chaotic, behavior. The number of states per megaelectronvolt of excitation energy with each spin and parity rise exponentially with increasing excitation energy until the levels become sufficiently close that they overlap and mix strongly and so become a continuum of states.

Toward the top of this energy regime, new modes of nuclear collectivity become accessible. Giant resonance states can be excited that involve compression and oscillation of the nuclear medium with vibrations of the protons and neutrons in phase (isoscalar) or beating against each other (isovector). The excitation and decay of these giant resonances can provide information about shapes of nuclei at high excitation and about the compressibility of nuclear matter. Results from giant resonance studies indicate that the shell effect persists high into the region previously thought to be statistical. *See* GIANT NUCLEAR RESONANCES.

The semiclassical statistical and hydrodynamic behavior of hot nuclear matter and its experimental, theoretical, and astrophysical aspects are of great interest at the highest nuclear energies. The influence of compression and heat on the density of nuclear matter is being investigated in order to measure a nuclear equation of state in analogy with the properties of a classical fluid. It has been suggested that the nuclear matter may undergo phase changes under compression, with high-density condensates possibly providing a new metastable state. At the highest densities and temperatures, the nucleons themselves are forced to overlap and merge, leading to a plasma of quarks and gluons that are the nucleonic constituents. *See* QUARK-GLUON PLASMA; RELATIVISTIC HEAVY-ION COLLISIONS. C. J. Lister

Nucleation The formation within an unstable, supersaturated solution of the first particles of precipitate capable of spontaneous growth into large crystals of a more stable solid phase. These first viable particles, called nuclei, may either be formed from solid particles already present in the system (heterogeneous nucleation) or be generated spontaneously by the supersaturated solution itself (homogeneous nucleation). *See* SUPERSATURATION.

Nucleation is significant in analytical chemistry because of its influence on the physical characteristics of precipitates. Processes occurring during the nucleation period establish the rate of precipitation, and the number and size of the final crystalline particles. *See* COLLOID; PRECIPITATION (CHEMISTRY). David H. Klein; Louis Gordon

Nucleic acid An acidic, chainlike biological macromolecule consisting of multiply repeated units of phosphoric acid, sugar, and purine and pyrimidine bases. Nucleic acids as a class are involved in the preservation, replication, and expression of hereditary information in every living cell. There are two types of nucleic acid: deoxyribonucleic acid (DNA) and ribonucleic acid (RNA). Edward Johnson

Nucleon The collective name for a proton or a neutron. These subatomic particles are the principal constituents of atomic nuclei and therefore of most matter in the universe. The proton and neutron share many characteristics. They have the same intrinsic spin, nearly the same mass, and similar interactions with other subatomic particles, and they can transform into one another by means of the weak interactions. Hence it is often useful to view them as two different states or configurations of the same particle, the nucleon. Nucleons are small compared to atomic dimensions and relatively heavy. Their characteristic size is of order 1/10,000 the size of a typical atom, and their mass is of order 2000 times the mass of the electron.

The proton and neutron differ chiefly in their electromagnetic properties. The proton has electric charge $+1$, the opposite of the electron, while the neutron is electrically neutral. They have significantly different intrinsic magnetic moments. Because the neutron is slightly heavier than the proton, roughly 1 part in 1000, the neutron is unstable, decaying into a proton, an electron, and an antineutrino with a characteristic lifetime of approximately 900 s. Although some unified field theories predict that the proton is unstable, no experiment has detected proton decay.

The complex forces between nucleons and the discovery during the 1950s of many similar subatomic particles led physicists to suggest that nucleons might not be fundamental particles. During the late 1960s and 1970s, inelastic electron and neutrino scattering experiments indicated that nucleons are composed of pointlike particles with spin $^1/_2$ and electric charges that are fractions of the charge on the electron. Particles with similar properties, named quarks, had been hypothesized in the early 1960s to explain other regularities among the properties of hadrons. In the early 1970s, it became clear that nucleons and other hadrons are indeed bound states of quarks. *See* HADRON; NUCLEAR STRUCTURE.

Quarks are believed to be fundamental particles without internal structure. The proton consists of two up-type quarks and one down-type quark (*uud*), while the neutron consists of *ddu*. Quarks are bound into nucleons by strong forces carried by gluons. The nucleon contains ambient gluon fields in somewhat the same way that the atom contains ambient electromagnetic fields. Because quarks and gluons are much less massive than the nucleon itself, their motion inside the nucleon is relativistic, making quark-antiquark pair creation a significant factor. Thus the nucleon contains fluctuating quark-antiquark pairs in addition to quarks and gluons. The theory of quark-gluon interactions is known as quantum chromodynamics (QCD), in analogy to the quantum theory of electrodynamics (QED). *See* ELEMENTARY PARTICLE; GLUONS; NEUTRON; PROTON; QUANTUM CHROMODYNAMICS; QUANTUM ELECTRODYNAMICS; QUARKS. Robert L. Jaffe

Nucleoprotein A generic term for any member of a large class of proteins associated with nucleic acid molecules. Nucleoprotein complexes occur in all living cells and viruses, where they play vital roles in deoxyribonucleic acid (DNA) replication, transcription, ribonucleic acid (RNA) processing, and protein synthesis.

Classification of nucleoproteins depends primarily upon the type of nucleic acid involved—DNA or RNA—and on the biological function of the associated proteins. Deoxyribonucleoproteins (complexes of DNA and proteins) constitute the genetic material of all organisms and many viruses. They function as the chemical basis of heredity and are the primary means of its expression and control. Most of the mass of chromosomes is made up of DNA and proteins whose structural and enzymatic activities are required for the proper assembly and expression of the genetic information encoded in the molecular structure of the nucleic acid. *See* DEOXYRIBONUCLEIC ACID (DNA).

Ribonucleoproteins (complexes of RNA and proteins) occur in all cells as part of the machinery for protein synthesis. This complex operation requires the participation of messenger RNAs (mRNAs), amino acyl transfer RNAs (tRNAs), and ribosomal RNAs (rRNAs), each of which interacts with specific proteins to form functional complexes called polysomes, on which the synthesis of new proteins occurs. *See* RIBONUCLEIC ACID (RNA).

In simpler life forms, such as viruses, most of the mass of the viral particle is due to its nucleoprotein content. The material responsible for the hereditary continuity of the virus may be DNA or RNA, depending on the type of virus, and it is usually enveloped by one or more proteins which protect the nucleic acid and facilitate infections. *See* BACTERIOPHAGE; VIRUS.

The genetic complexity of higher organisms reflects a corresponding complexity in their DNA sequences. A typical human diploid nucleus, for example, contains enough DNA that, if fully extended, would be over 1 m (3.3 ft) long. However, the ability to compact this amount of DNA into a nucleus 0.01 mm in diameter is accomplished routinely by cells. The reduction in size is largely due to interactions between the DNA and sets of small basic proteins called histones. All somatic cells of higher organisms contain five major histone classes, all of which are characterized by a high content of the basic amino acids arginine, lysine, and histidine (which are positively charged to interact with negatively charged DNA). The characterization of histones was originally based on their chromatographic or solubility properties, but emphasis is now placed on details of their structure and their capacity to organize DNA. The major classes are called H1, H2A, H2B, H3, and H4. Timothy J. Moss; Vincent G. Allfrey; Lori L. Wallrath

Nucleosome The fundamental histone-containing structural subunit of eukaryotic chromosomes. In most eukaryotic organisms, nuclear deoxyribonucleic acid (DNA) is complexed with an approximately equal mass of histone protein. The nucleosome is organized so that the DNA is exterior and the histones interior. The DNA makes two turns around a core of eight histone molecules, forming a squat cylinder 11 nanometers in diameter and 5.5 nm in height. A short length of linker or spacer DNA connects one nucleosome to the next, forming a nucleosomal chain that has been likened to a beaded string. This basic structure is found in all forms of chromatin (the DNA-protein complex that forms eukaryotic chromosomes), from the dispersed type in nondividing nuclei to the compact chromosomes visible in mitotic and meiotic nuclei. Nucleosomes have been found in all eukaryotic organisms examined, the only exceptions being some sperm nuclei where protamines replace histones during spermatogenesis, and the dinoflagellate algae, which lack histones as major nuclear components. *See* Chromosome; Deoxyribonucleic acid (DNA).

A chain of adjacent nucleosomes is approximately sixfold shorter than the DNA it contains. Moreover, chains of nucleosomes can self-assemble into thicker fibers in which the DNA packing ratio approaches 35:1. These observations, and the lack of any obvious catalytic activity, led to the assumption that the primary function of the nucleosome consists of organizing and packing DNA. More recently, however, nucleosomes have also been shown to have important gene regulatory functions.

C. L. F. Woodcock

Nucleosynthesis Theories of the origin of the elements involve synthesis with charged and neutral elementary particles (neutrons, protons, neutrinos, photons) and other nuclear building blocks of matter, such as alpha particles. The theory of nucleosynthesis comprises a dozen distinct processes, including big bang nucleosynthesis, cosmic-ray spallation in the interstellar medium, and static or explosive burning in various stellar environments (hydrogen-, helium-, carbon-, oxygen-, and silicon-burning, and the, *s*-, *r*-, *p*-, γ-, and ν-processes). Acceptable theories must lead to an understanding of the cosmic abundances observed in the solar system, stars, and the interstellar medium. Hydrogen and helium constitute about 98% of the total element content by mass and more than 99.8% by number of atoms. There is a rapid decrease with increasing nuclear mass number A, although the abundances of iron-group elements like iron and nickel is remarkably large. *See* Elements, cosmic abundance of.

Observations of the expanding universe and of the 3-K background radiation indicate that the universe originated in a primordial event known as the big bang about 15×10^9 years ago. Absence of stable mass 5 and mass 8 nuclei preclude the possibility of synthesizing the major portion of the nuclei of masses greater than 4 in those first few minutes of the universe, when the density and temperature were sufficiently high to support the necessary nuclear reactions to synthesize elements. *See* Big bang theory; Cosmic background radiation; Cosmology.

The principal source of energy in stars is certainly nuclear reactions which release energy by fusion of lighter nuclei to form more massive ones. There is considerable evidence that nucleosynthesis has been going on in stars for billions of years. Observations show that the abundance ratio of iron and heavier elements to hydrogen decreases with increasing stellar age. The oldest known stars in the disk of the Milky Way Galaxy exhibit a ratio 10,000 times smaller than in the Sun. This low ratio is understood on the basis of element synthesis occurring in previous-generation stars that evolved to the point of exploding as supernovae, thus enriching the interstellar medium with the nuclei that were synthesized during their lifetimes and by the explosive synthesis that accompanies the ejection of the stellar envelope. Later-generation stars were then formed from enriched gas and dust. *See* Milky Way Galaxy; Stellar evolution; Stellar population; Supernova.

Hydrogen burning, the first process of nucleosynthesis, converts hydrogen to helium. In stars of 1.2 or less solar masses, this process occurs via the proton-proton chain. A similar reaction chain was also responsible for much of the element synthesis during the big bang, producing the bulk of deuterium and helium, and some of the ^{7}Li, observed today. In more massive stars where the central temperatures exceed 2×10^7 K, hydrogen burning is accomplished through proton captures by carbon, nitrogen, and oxygen nuclei, in the carbon-nitrogen-oxygen (CNO) cycles, to form ^{4}He. The product (ash) of hydrogen burning is helium, but much of the helium produced is consumed in later stages of stellar evolution or is locked up forever in stars of lower mass that never reach the temperatures required to ignite helium burning. The observed abundances of some carbon, nitrogen, oxygen, and fluorine nuclei are attributed to hydrogen burning in the CNO cycles. *See* Carbon-nitrogen-oxygen cycles; Nuclear fusion; Proton-proton chain.

When the hydrogen fuel is exhausted in the central region of the star, the core contracts and its temperature and density increase. Helium, the ash of hydrogen burning, cannot be burned immediately due to the larger nuclear charge of helium ($Z = 2$) producing a much higher Coulomb barrier against fusion. When the temperature eventually exceeds about 10^8 K, helium becomes the fuel for further energy generation and nucleosynthesis. The basic reaction in this thermonuclear phase is the triple-alpha process in which three ^{4}He nuclei (three alpha particles) fuse to form ^{12}C, a carbon nucleus of mass 12 (atomic mass units). Capture of an alpha particle by ^{12}C then forms ^{16}O, symbolically written as $^{12}\text{C} + {}^4\text{He} \rightarrow {}^{16}\text{O} + \gamma$, or simply $^{12}\text{C}(\alpha,\gamma)^{16}\text{O}$, where γ represents energy released in the form of electromagnetic radiation. Other reactions that are included in helium burning are $^{16}\text{O}(\alpha,\gamma)^{20}\text{Ne}$, $^{20}\text{Ne}(\alpha,\gamma)^{24}\text{Mg}$, $^{14}\text{N}(\alpha,\gamma)^{18}\text{F}$, and $^{18}\text{O}(\alpha,\gamma)^{22}\text{Ne}$. Fluorine-18, produced when ^{14}N captures an alpha particle, is unstable and decays by emitting a positron (e^+) and a neutrino (ν) to form ^{18}O [in short, $^{14}\text{N}(\alpha,\gamma)^{18}\text{F}(e^+,\nu)^{18}\text{O}$]. Because there is likely to be ^{13}C in the stellar core if hydrogen burning proceeded by the carbon-nitrogen-oxygen cycles, the neutron-producing reaction $^{13}\text{C}(\alpha,n)^{16}\text{O}$ should also be included with the helium-burning reactions. The neutrons produced by this reaction are probably responsible for the bulk of *s*-process nucleosynthesis (discussed below). Helium burning is probably responsible for much of the ^{12}C observed in the cosmic abundances, although in more massive stars the later burning stages will consume the ^{12}C produced earlier by helium burning. *See* Nuclear reaction.

Upon exhaustion of the helium supply, if the star has an initial mass of at least 8 solar masses, gravitational contraction of the stellar core can lead to a temperature exceeding 5×10^8 K, where it becomes possible for two ^{12}C nuclei to overcome their high mutual Coulomb-repulsion barrier and fuse to form ^{20}Ne, ^{23}Na, and ^{24}Mg through reactions such as $^{12}\text{C}(^{12}\text{C},\alpha)^{20}\text{Ne}$, $^{12}\text{C}(^{12}\text{C},p)^{23}\text{Na}$, and $^{12}\text{C}(^{12}\text{C},\gamma)^{24}\text{Mg}$. Carbon burning can produce a number of nuclei with masses less than or equal to 28 through further proton and alpha-particle captures.

Carbon burning is followed by a short-duration stage, sometimes referred to as neon burning, in which ^{20}Ne disintegrates by the reaction $^{20}\text{Ne}(\gamma,\alpha)^{16}\text{O}$. The alpha particle released is then captured by a remaining ^{20}Ne nucleus to produce ^{24}Mg; thus, the effective neon burning reaction is $^{20}\text{Ne} + {}^{20}\text{Ne} \rightarrow {}^{16}\text{O} + {}^{24}\text{Mg}$. The eventual result is that most of the carbon from helium burning becomes oxygen, which supplements the original oxygen formed in helium burning. This stage is followed by the fusion of oxygen nuclei at much higher temperatures. (Temperatures greater than 10^9 K are required for ^{16}O nuclei to overcome their mutual Coulomb barrier.) Some relevant reactions for oxygen burning are $^{16}\text{O}(^{16}\text{O},\alpha)^{28}\text{Si}$, $^{16}\text{O}(^{16}\text{O},p)^{31}\text{P}$, and $^{16}\text{O}(^{16}\text{O},\gamma)^{32}\text{S}$. Nuclei of masses up to $A = 40$ may be produced in this phase through proton, neutron, and alpha-particle captures.

Silicon burning commences when the temperature exceeds about 3×10^9 K. In this phase, photodisintegration of ^{28}Si and other intermediate-mass nuclei around $A = 28$ produces copious supplies of protons, neutrons, and alpha particles. These particles capture on the seed nuclei left from previous burning stages and thus produce new isotopes up to mass 60, resulting in the buildup of the abundance peak near $A = 56$. Because the binding energy per nucleon is a peak for the iron-group (A near 60) nuclei, further fusion reactions no longer release energy. Silicon burning is therefore the end stage of stellar burning. Production of nuclei with mass higher than A near 60 therefore occurs by special processes that are ancillary to the mainline stellar burning stages.

Because neutrons are neutral particles, their capture is not affected by the Coulomb barrier that inhibits charged-particle reactions. If the number of neutrons per seed nucleus is small, so that time intervals between neutron captures are long compared to the beta-decay lifetimes of unstable nuclei that are formed, the *s*-process (slow process) takes place. The seed nuclei are predominantly in the iron peak, but the abundances of low-mass nuclei are also affected by neutron-capture processing. In the *s*-process, if neutron capture produces a nucleus that is unstable (due to an excess number of neutrons), the newly formed nucleus undergoes beta decay to a stable isobar by emitting an electron and an antineutrino. The resulting nucleus eventually captures another neutron, and the capture-decay step repeats. The presence of free neutrons leads to a chain of capture-decay events that drives the abundances along a unique *s*-process path that zigzags along a single line in the nuclear *Z-N* diagram near the low of beta-stable nuclei (valley of beta stability). Given enough neutrons, the *s*-process synthesizes nuclei of masses up to 209, when alpha decay becomes a deterrent to further buildup by neutron capture. *See* RADIOACTIVITY.

The *r*-process occurs when a large neutron flux allows rapid neutron capture, so that seed nuclei capture many neutrons before undergoing beta decay. The rapid neutron capture takes the nuclei far away from the valley of beta stability, into the regime of extremely neutron-rich nuclei. The time scale for the *r*-process is very short, 1–100 s, and the abundance of free neutrons is very large. These conditions are found deep inside the interiors of exploding massive stars, supernovae. The *r*-process can synthesize nuclei all the way into the transuranic elements.

The *p*-process produces heavier elements on the proton-rich side of the beta valley. The *p*-nuclei are blocked from formation by stable nuclei produced by either the *r*- or *s*-process. The major task for *p*-process theory is thus to find ways, other than beta decay, to process the more abundant *r*- and *s*-nuclei into the less abundant *p*-nuclei. Two possible mechanisms are: radiative proton capture, and gamma-induced neutron, proton, or alpha-particle removal reactions. In both cases the temperature should be in excess of $2\text{–}3 \times 10^9$ K.

The neutrino (ν) flux emitted from a cooling proto neutron star alters the yields of explosive nucleosynthesis from type II supernovae. Inelastic scattering of neutrinos (all flavors) off abundant nuclei excites states that can decay via single or multiple nucleon emission. The ν-process is probably responsible for significant contributions to the synthesis in nature of about a dozen isotopes. While the neutrino interaction cross section with matter is extremely small (about 10^{-44} cm^2), the high neutrino energies and the large number flux close to the collapsing iron core of a massive star lead to significant synthesis of nuclei, either directly or indirectly. *See* NEUTRINO.

The bulk of the light elements lithium, beryllium, and boron found in the cosmic abundance curve cannot have survived processing in stellar interiors because they are readily destroyed by proton capture. Although some ^{7}Li originated in the big bang, primordial nucleosynthesis cannot be responsible for the bulk of the ^{7}Li in existence. Spallation of more abundant nuclei such as carbon, nitrogen, and oxygen by high-energy protons and alpha particles can account for the low-abundance nuclides ^{6}Li, ^{9}Be, ^{10}B, ^{11}B, and for some ^{7}Li. The canonical process is spallation of carbon, nitrogen, and oxygen nuclei in the interstellar medium by fast light particles, such as alpha particles and protons, which are abundant in the gas between the stars. These high-energy particles are referred to as cosmic rays, leading to the term "cosmic-ray spallation (CRS) process." *See* COSMIC RAYS.

Dieter H. Hartmann; Bradley S. Meyer

Nucleotide A cellular constituent that is one of the building blocks of ribonucleic acids (RNA) and deoxyribonucleic acid (DNA). In biological systems, nucleotides are linked by enzymes in order to make long, chainlike polynucleotides of defined sequence. The order or sequence of the nucleotide units along a polynucleotide chain plays an important role in the storage and transfer of genetic information. Many nucleotides also perform other important functions in biological systems. Some, such as adenosine triphosphate (ATP), serve as energy sources that are used to fuel important biological reactions. Others, such as nicotinamide adenine dinucleotide (NAD) and coenzyme A (CoA), are important cofactors that are needed to complete a variety of enzymatic reactions. Cyclic nucleotides such as cyclic adenosine monophosphate (cAMP) are often used to regulate complex metabolic systems. Chemically modified nucleotides such as fluoro-deoxyridine monophosphate (Fl-dUMP) contain special chemical groups that are useful for inactivating the normal function of important enzymes. These and other such compounds are widely used as drugs and therapeutic agents to treat cancer and a variety of other serious illnesses. *See* COENZYME; CYCLIC NUCLEOTIDES; NICOTINAMIDE ADENINE DINUCLEOTIDE (NAD).

Nucleotides are generally classified as either ribonucleotides or deoxyribonucleotides. Both classes consist of a phosphorylated pentose sugar that is linked via an *N*-glycosidic bond to a purine or pyrimidine base. The combination of the pentose sugar and the purine or pyrimidine base without the phosphate moiety is called a nucleoside. *See* PURINE; PYRIMIDINE.

Ribonucleosides contain the sugar D-ribose, whereas deoxyribonucleosides contain the sugar 2-deoxyribose. The four most common ribonucleosides are adenosine, guanosine, cytidine, and uridine. The purine ribonucleosides, adenosine and guanosine, contain the nitrogenous bases adenine and guanine, respectively. The pyrimidine ribonucleosides, cytidine and uridine, contain the bases cytosine and uracil, respectively. Similarly, the most common deoxyribonucleosides include deoxyadenosine, deoxyguanosine, deoxycytidine, and thymidine, which contains the pyrimidine base thymine. Phosphorylation of the ribonucleosides or deoxyribonucleosides yields the corresponding ribonucleotide or deoxyribonucleotide. *See* DEOXYRIBONUCLEIC ACID (DNA); ENZYME; NUCLEIC ACID; RIBONUCLEIC ACID (RNA).

E. Patrick Groody

Nuclide A species of atom that is characterized by the constitution of its nucleus, in particular by its atomic number *Z* and its neutron number $A - Z$, where *A* is the mass number. The total number of stable nuclides is approximately 275. About a dozen radioactive nuclides are found in nature, and hundreds of others have been created artificially.

Henry E. Duckworth

Nudibranchia An order of the gastropod subclass Opisthobranchia containing about 2500 living species of carnivorous sea slugs. They occur in all the oceans, at all depths, but reach their greatest size and diversity in warm shallow seas.

In this, the largest order of the Opisthobranchia, there is much evidence indicating polyphyletic descent from a number of long-extinct opisthobranch stocks. In all those lines which persist to the present day, the shell and operculum have been discarded in the adult form. In many of them the body has quite independently become dorsally papillate. In at least two suborders these dorsal papillae have acquired the power to nurture nematocysts derived from their coelenterate prey so as to use them for the nudibranch's own defense. In other cases such papillae (usually called cerata) have independently become penetrated by lobules of the adult digestive gland, or they may contain virulent defensive glands or prickly bundles of dagger-like calcareous spicules. *See* OPISTHOBRANCHIA.

T. E. Thompson

Number theory The study of the properties and relationships between integers and other special types of numbers. There are many sets of positive integers of particular interest, such as the primes and the perfect numbers. Number theory, of ancient and continuing interest for its intrinsic beauty, also plays a crucial role in computer science, particularly in the area of cryptography.

Elementary number theory. This part of number theory does not rely on advanced mathematics, such as complex analysis and ring theory. The basic notion of elementary number theory is divisibility. An integer d is a divisor of n, written $d \mid n$, if there is an integer t such that $n = dt$. A prime number is a positive integer that has exactly two positive divisors, 1 and itself. The ten smallest primes are 2, 3, 5, 7, 11, 13, 17, 19, 23, and 29. Euclid (around 300 BC) proved that there are infinitely many primes by showing that if the only primes were 2, 3, 5, . . . p, a prime not in this list could be found by taking a prime factor of the number shown in Eq. (1). Primes are the building blocks

$$N = (2 \cdot 3 \cdot 5 \cdots p) + 1 \qquad (1)$$

of the positive integers. The fundamental theorem of arithmetic, established by K. F. Gauss in 1801, states that every positive integer can be written as the product of prime factors in exactly one way when the order of the primes is disregarded.

A perfect number is a positive integer equal to the sum of its positive divisors other than itself. L. Euler showed that $2^{n-1}(2^n - 1)$ is perfect if and only if $2^n - 1$ is prime. If $2^n - 1$ is prime, then n itself must be prime. Primes of the form $2^p - 1$ are known as Mersenne primes after M. Mersenne, who studied them in the seventeenth century. As of October 2008, 46 Mersenne primes were known, the largest being $2^{43,112,609} - 1$, a number with 12,978,189 decimal digits.

If $a - b$ is divisible by m, then a is called congruent to b modulo m, and this relation is written $a \equiv b \pmod{m}$. This relation between integers is an equivalence relation and defines equivalence classes of numbers congruent to each other, called residue classes. Congruences to the same modulus can be added, subtracted, and multiplied in the same manner as equations. However, when both sides of a congruence are divided by the same integer d, the modulus m must be divided by gcd(d, m). [The greatest common divisor of the positive integers a and b, written gcd(a, b), is the largest integer that divides both a and b. Two integers with greatest common divisor equal to 1 are called coprime.] There are m residue classes modulo m. The number of classes containing only numbers coprime to m is denoted by $\phi(m)$, where $\phi(m)$ is called the Euler phi function.

In the third century B.C., Eratosthenes showed how all primes up to an integer n can be found when only the primes p up to $\sqrt{n}$ are known. It is sufficient to delete from the list of integers, starting with 2, the multiples of all primes up to $\sqrt{n}$. The remaining integers are all the primes not exceeding n.

Single equations or systems of equations in more unknowns than equations, with restrictions on solutions such as that they must all be integral, are called diophantine equations, after Diophantus who studied such equations in ancient times. A wide range of diophantine equations have been studied. For example, the diophantine equation has (2) infinitely many solutions

$$x^2 + y^2 = z^2 \qquad (2)$$

in integers. These solutions are known as pythagorean triples, since they correspond to the lengths of the sides of right triangles where these sides have integral lengths. All solutions of this equation are given by $x = t(u^2 - v^2)$, $y = 2tuv$, and $z(u^2 + v^2)$, where t, u, and v are positive integers.

Perhaps the most notorious diophantine equation is Eq (3).

$$x^n + y^n = z^n \qquad (3)$$

Fermat's last theorem states that this equation has no solutions in integers when n is an integer greater than 2 where $xyz \neq 0$. Establishing Fermat's last theorem was the quest of many mathematicians over 200 years. In the 1980s, connections were made between the solutions of this equations and points on certain elliptic curves. Using the theory of elliptic curves, A. Wiles completed a proof of Fermat's last theorem based on these connections in 1995.

Algebraic number theory. Attempts to prove Fermat's last theorem led to the development of algebraic number theory, a part of number theory based on techniques from such areas as group theory, ring theory, and field theory. Gauss extended the concepts of number theory to the ring $R[i]$ of complex numbers of the form $a + bi$, where a and b are integers. Ordinary primes $p \equiv 3 \pmod{4}$ are also prime in $R[i]$, but $2 = -i(1 + i)^2$ is not prime, nor are primes $p \equiv 1 \pmod{4}$ since such primes split as $p = (a + bi)(a - bi)$. More generally, an algebraic number field $R(\theta)$ of degree n is generated by the root θ of a polynomial equation $f(x) = 0$ of degree n with rational coefficients. A number α in this field is called an algebraic integer if it satisfies an algebraic equation with integer coefficients with initial coefficient 1. The algebraic integers in an algebraic number field form an integral domain. But, prime factorization may not be unique; for example, in $R[\sqrt{-5}]$, $21 = 3 \cdot 7 = (1 + 2\sqrt{-5}) \cdot (1 - 2\sqrt{-5})$ where each of the four factors in the two products is prime. To restore unique factorization, the concept of ideals is needed, as shown by E. E. Kummer and J. W. R. Dedekind. *See* RING THEORY.

Analytic number theory. There are many important results in number theory that can be established by using methods from analysis. For example, analytic methods developed by G. F. B. Riemann in 1859 were used by J. Hadamard and C. J. de la Vallée Poussin in 1896 to prove the famous prime number theorem. This theorem, first conjectured by Gauss about 1793, states that $\pi(x)$, the number of primes not exceeding x, behaves as shown in Eq. (4). These methods of Riemann are based on

$$\lim_{x \to \infty} \frac{\pi(x)}{(x/\log x)} = 1 \qquad (4)$$

$\zeta(s)$, the function defined by Eq. (5), where $s = \sigma + it$ is a com-

$$\zeta(s) = \sum_{n=1}^{\infty} \frac{1}{n^s} = \prod_p \frac{1}{1 - p^{-s}} \qquad (5)$$

plex variable; the series in this equation is convergent for $\sigma > 1$. Via an analytic continuation, this function can be defined in the whole complex plane. It is a meromorphic function with only a simple pole of residue 1 at $s = 1$. However, the fundamental theorem of arithmetic can be used to show that this series equals the product over all primes p shown in Eq. (5). It can be shown that $\zeta(s)$ has no zeros for $\sigma = 1$; this result and the existence of a pole at $s = 1$ suffice to prove the prime number theorem. Many additional statements about $\pi(x)$ have been proved. Riemann's work contains the still unproved so-called Riemann hypothesis: all zeros of $\zeta(s)$ have a real part not exceeding $^1/_2$. *See* COMPLEX NUMBERS AND COMPLEX VARIABLES.

Diophantine approximation. A real number x is called rational if there are integers p and q such that $x = p/q$; otherwise x is called irrational. The number $b^{1/m}$ is irrational if b is an integer which is not the mth power of an integer (for example, $\sqrt{2}$ is irrational). A real number x is called algebraic if it is the root of a monic polynomial with integer coefficients; otherwise x is called transcendental. The numbers e and π are transcendental. That π is transcendental implies that it is impossible to square the circle. *See* CIRCLE; E (MATHEMATICS).

The part of number theory called diophantine approximation is devoted to approximating numbers of a particular kind by numbers from a particular set, such as approximating irrational numbers by rational numbers with small denominators. A basic result is that, given an irrational number x, there exist infinitely many fractions h/k that satisfy the inequality (6), where c is any

$$\left| x - \frac{h}{k} \right| < \frac{1}{ck^2} \qquad (6)$$

positive number not exceeding $\sqrt{5}$. However, when c is greater than $\sqrt{5}$, there are irrational numbers x for which there are only finitely many such h/k.

In 1851, J. Liouville showed that transcendental numbers exist; he did so by demonstrating that the number x given by Eq. (7) has the property that, given any positive real number m,

$$x = \sum_{j=1}^{\infty} 10^{-j!} \qquad (7)$$

there is a rational number h/k that satisfies Eq. (8).

$$\left| x - \frac{h}{k} \right| < \frac{1}{k^m} \qquad (8)$$

See ALGEBRA; NUMBERING SYSTEMS; ZERO. Kenneth H.Rosen

Numbering systems A numbering system is a systematic method for representing numbers using a particular set of symbols. The most commonly used numbering system is the decimal system, based on the number 10, which is called the basis or radix of the system. The basis tells how many different individual symbols there are in the system to represent numbers. In the decimal system these symbols are the digits 0, 1, 2, 3, 4, 5, 6, 7, 8, 9. The range of these numbers varies from 0 to (10 − 1). This is a particular case of a more general rule: Given any positive basis or radix N, there are N different individual symbols that can be used to write numbers in that system. The range of these numbers varies from 0 to $N - 1$.

In the computer and telecommunication fields, three of the most frequently used numbering systems are the binary (base 2), the octal (base 8), and the hexadecimal (base 16). The binary system has only two symbols: 0 and 1. Either of these symbols can be called a binary digit or a bit. The octal system has eight symbols: 0, 1, 2, 3, 4, 5, 6, 7. The hexadecimal system has 16 symbols: 0, 1, 2, 3, 4, 5, 6, 7, 8, 9, A, B, C, D, E, F. A stands for 10, B for 11, C for 12, D for 13, E for 14, and F for 15. The reason for choosing single letters to represent numbers higher than 9 is to keep all individual symbols single characters. *See* BIT.

All the numbering systems mentioned so far are positional systems. That is, the value of any symbol depends on its position in the number. For example, the value of 2 in the decimal number 132 is that of two units, whereas its value in decimal 245 is that of two hundreds. In the decimal system, the rightmost position of a number is called the ones (10^0) place, the next position from the right is called the tens (10^1) place, the next position the hundreds (10^2) place, and so on. Observe that the powers increase from the right. The power of the rightmost digit is zero, the power of the next digit is one, the power of the next digit is two, and so on. These powers are sometimes called the weight of the digit.

Conversion to decimal numbers. In any positional system, the decimal equivalent of a digit in the representation of the number is the digit's own value, in decimal, multiplied by a power of the basis in which the number is represented. The sum of all these powers is the decimal equivalent of the number. The corresponding powers of each of the digits can be better visualized by writing superscripts beginning with 0 at the rightmost digit, and increasing the powers by 1 in moving toward the left digits of the number.

Decimal numbers to other bases. The conversion of a given decimal number to another basis r ($r > 0$) is carried out by initially dividing the given decimal number by r, and then successively dividing the resulting quotients by r until a zero quotient is obtained. The decimal equivalent is obtained by writing the remainders of the successive divisions in the opposite order in which they were obtained.

Binary to hexadecimal or octal and vice versa. The table shows the decimal numbers 1 through 15 written in binary, octal, and hexadecimal. Since each four-bit binary number corresponds to one and only one hexadecimal digit and vice versa, the hexadecimal system can be viewed as a shorthand notation of the binary system. Similar reasoning can be applied to the octal system. This one-to-one correspondence between the symbols of the binary system and the symbols of the octal and hexadecimal system provides a method for converting numbers between these bases.

To convert binary numbers to hexadecimal, the following procedure may be used: (1) Form four-bit groups beginning from the rightmost bit of the number. If the last group (at the leftmost position) has fewer than four bits, add extra zeros to the left of the bits in this group to make it a four-bit group. (2) Replace each four-bit group by its hexadecimal equivalent. A process that is almost the reverse of the previous procedure can be used to convert from hexadecimal to binary. However, there is no need to add extra zeros to any group since each hexadecimal number will always convert to a group with four binary bits.

A similar process can be followed to convert a binary number to octal, except that in this case three-bit groups must be formed. Individual octal numbers will always convert to groups with three binary bits. Ramon A. Mata-Toledo

The first 15 integers in binary, octal, hexadecimal, and decimal notation

Binary	Octal	Hexadecimal	Decimal
0001	1	1	1
0010	2	2	2
0011	3	3	3
0100	4	4	4
0101	5	5	5
0110	6	6	6
0111	7	7	7
1000	10	8	8
1001	11	9	9
1010	13	A	10
1011	14	B	11
1100	14	C	12
1101	15	D	13
1110	16	E	14
1111	17	F	15

Numerical analysis The development and analysis of computational methods (and ultimately of program packages) for the minimization and the approximation of functions, and for the approximate solution of equations, such as linear or nonlinear (systems of) equations and differential or integral equations. Originally part of every mathematician's work, the subject is now often taught in computer science departments because of the tremendous impact which computers have had on its development. Research focuses mainly on the numerical solution of (nonlinear) partial differential equations and the minimization of functions.

Numerical analysis is needed because answers provided by mathematical analysis are usually symbolic and not numeric; they are often given implicitly only, as the solution of some equation, or they are given by some limit process. A further complication is provided by the rounding error which usually contaminates every step in a calculation (because of the fixed finite number of digits carried).

Even in the absence of rounding error, few numerical answers can be obtained exactly. Among these are (1) the value of a piece-wise rational function at a point and (2) the solution of a (solvable) linear system of equations, both of which can be produced in a finite number of arithmetic steps. Approximate answers to all other problems are obtained by solving the first few in a sequence of such finitely solvable problems. A typical example is provided by Newton's method: A solution c to a nonlinear equation $f(c) = 0$ is found as the limit

$$c = \lim_{n\to\infty} x_n,$$

with x_{n+1} being a solution to the linear equation

$$f(x_n) + f'(x_n)(x_{n+1} - x_n) = 0,$$

that is, $x_{n+1} = x_n - f(x_n)/f'(x_n)$, $n = 0, 1, 2, \ldots$ Of course, only the first few terms in this sequence $x_0, x_1\ x_2, \ldots$ can ever be calculated, and thus one must consider when to break off such a solution process and how to gauge the accuracy of the current approximation.

An otherwise satisfactory computational process may become useless, because of the amplification of rounding errors. A computational process is called stable to the extent that its results are not spoiled by rounding errors. The extended calculations involving millions of arithmetic steps now possible on computers have made the stability of a computational process a prime consideration.

Interpolation and approximation. Polynomial interpolation provides a polynomial p of degree n or less which uniquely matches given function values $f(x_0), \ldots, f(x_n)$ at corresponding distinct points $x_0, \ldots, x_n$. The interpolating polynomial p is used in place of f, for example in evaluation, integration, differentiation, and zero finding. Accuracy of the interpolating polynomial depends strongly on the placement of the interpolation points, and usually degrades drastically as one moves away from the interval containing these points (that is, in case of extrapolation). *See* EXTRAPOLATION.

When many interpolation points (more than 5 or 10) are to be used, it is often much more efficient to use instead a piecewise polynomial interpolant or spline. Suppose the interpolation points above are ordered, $x_0 < x_1 < \cdots x_n$. Then the cubic spline interpolant to the above data, for example, consists of cubic polynomial pieces, with the ith piece defining the interpolant on the interval $[x_{i-1}, x_i]$ and so matched with its neighboring piece or pieces that the resulting function not only matches the given function values (hence is continuous) but also has a continuous first and second derivative.

Interpolation is but one way to determine an approximant. In full generality, approximation involves several choices: (1) a set P of possible approximants, (2) a criterion for selecting from P a particular approximant, and (3) a way to measure the approximation error, that is, the difference between the function f to be approximated and the approximant p, in order to judge the quality of approximation.

Solution of linear systems. Solving a linear system of equations is probably the most frequently confronted computational task. It is handled either by a direct method, that is, a method which obtains the exact answer in a finite number of steps, or by an iterative method, or by a judicious combination of both. Analysis of the effectiveness of possible methods has led to a workable basis for selecting the one which best fits a particular situation.

Direct methods require a number of operations which increases with the cube of the number of unknowns. Some types of problems arise wherein the matrix of coefficients is sparse, but the unknowns may number several thousand; for these, direct methods are prohibitive in computer time required. One frequent source of such problems is the finite difference treatment of partial differential equations. A significant literature of iterative methods exploiting the special properties of such equations is available. For certain restricted classes of difference equations, the error in an initial iterate can be guaranteed to be reduced by a fixed factor, using a number of computations that is proportional to $n \log n$, where n is the number of unknowns. Since direct methods require work proportional to n^3, it is not surprising that as n becomes large, iterative methods are studied rather closely as practical alternatives.

Differential equations. Classical methods yield practical results only for a moderately restricted class of ordinary differential equations, a somewhat more restricted class of systems of ordinary differential equations, and a very small number of partial differential equations. The power of numerical methods is enormous here, for in quite broad classes of practical problems relatively straightforward procedures are guaranteed to yield numerical results, whose quality is predictable. *See* DIFFERENTIAL EQUATION.

Carl de Boor

Numerical representation (computers)

Numerical data in a computer are written in basic units of storage made up of a fixed number of consecutive bits. The most commonly used units in the computer and communication industries are the byte (8 consecutive bits), the word (16 consecutive bits), and the double word (32 consecutive bits). A number is represented in each of these units by setting the bits according to the binary representation of the number. By convention the bits in a byte are numbered, from right to left, beginning with zero. Thus, the rightmost bit is bit number 0 and the leftmost bit is number 7. The rightmost bit is called the least significant bit, and the leftmost bit is called the most significant bit. Higher units are numbered also from right to left. In general, the rightmost bit is labeled 0 and the leftmost bit is labeled $(n - 1)$, where n is the number of bits available. *See* BIT.

Since each bit may have one of two values, 0 or 1, n bits can represent 2^n different unsigned numbers. The range of these nonnegative integers varies from 0 to $2^n - 1$. To represent positive or negative numbers, one of the bits is chosen as the sign bit. By convention, the leftmost bit (or most significant bit) is considered the sign bit. A value of 0 in the sign bit indicates a positive number, whereas a value of 1 indicates a negative one. A similar convention is followed for higher storage units, including words and double words. Various conventions exist for representing integers and real numbers.

Ramon A. Mata-Toledo

Numerical taxonomy

The grouping by numerical methods of taxonomic units based on their character states. The application of numerical methods to taxonomy, dating back to the rise of biometrics in the late nineteenth century, has received a great deal of attention with the development of the computer and computer technology. Numerical taxonomy provides methods that are objective, explicit, and repeatable, and is based on the ideas first put forward by M. Adanson in 1963. These ideas, or principles, are that the ideal taxonomy is composed of information-rich taxa based on as many features as possible, that a priori every character is of equal weight, that overall similarity between any two entities is a function of the similarity of the many characters on which the comparison is based, and that taxa are constructed on the basis of diverse character correlations in the groups studied. *See* TAXONOMY.

In the early stages of development of numerical taxonomy, phylogenetic relationships were not considered. However, numerical methods have made possible exact measurement of evolutionary rates and phylogenetic analysis. Furthermore, rapid developments in the techniques of direct measurement of the homologies of deoxyribonucleic acid (DNA), and ribonucleic acid (RNA), between different organisms now provide an estimation of "hybridization" between the DNAs of different taxa and, therefore, possible evolutionary relationships. Thus, research in numerical taxonomy often includes analyses of the chemical and physical properties of the nucleic acids of the organisms, the data from which are correlated with phenetic groupings established by numerical techniques. *See* PHYLOGENY; TAXONOMIC CATEGORIES.

Rita R. Colwell

Nummulites

A genus of unicellular shelled protoctist (protozoa) of the class Foraminifera (order Rotaliida, family

***Nummulites* in a borehole core through the early Eocene El Garia Formation of the Hasdrubal Gas Field, offshore Tunisia. The large individual developed from gametes; the smaller specimens were likely produced asexually. (*Photo courtesy of Simon Beavington-Penney*)**

Nummulitidae). The shells of *Nummulites* and other large, internally complex foraminifers commonly occurred in rock-forming abundances on continental shelves and around oceanic islands throughout subtropical and tropical regions during the Cenozoic. Nummulitic limestones have been used as building materials around the Mediterranean; the pyramids of Egypt were constructed of blocks of Eocene age. *Nummulites* shells are useful stratigraphic and paleoenvironmental indicators; the Paleogene was formerly known as the Nummulitic Period. *See* LIMESTONE.

The shells, also known as tests, are composed of finely perforate calcite; may be discoidal, lenticular, or globular in shape; and can be as much as 6 in. (15 cm) in diameter, though sizes of 1–20 mm are more common. Shells are planispirally enrolled whorls of tiny undivided chambers (see illustration). Modern representatives of the Nummulitidae are benthic (bottom-dwelling) foraminifers that harbor diatom (single-celled alga) symbionts. The symbionts live within the host cytoplasm and assist the host in metabolism and calcification by providing photosynthetically produced carbon. Because the diatoms require sunlight to photosynthesize, the foraminifers must live at depths to which some sunlight can penetrate. Reproduction is through alternation of sexual and asexual generations. *See* FORAMINIFERIDA.

Alfred R. Loelich, Jr.; Pamela Hallock

Nursing The application of principles from the basic sciences, social sciences, and humanities to assist healthy and sick individuals and their families or other caring persons in performing those activities that contribute to the individuals' physical and mental well-being and that they would perform unaided if able to do so. Nursing includes providing physical and emotional care, promoting comfort, serving as patient advocates, assisting in rehabilitative efforts, teaching self-care and health promotion activities, and administering treatments prescribed by a licensed physician or dentist. Patient-care activities are conceived and coordinated so as to help individuals gain independence as rapidly as possible or maintain an optimal level of function. When multiple health-care providers are involved, nursing coordinates patient-care efforts to improve the quality of care.

Nursing practice is conducted in a variety of settings, including hospitals, community facilities, private homes, nursing homes, schools, industry, physician's offices, the military, and civil service arenas. Standards for nursing practice and licensure are governed by state nurse practice acts and are directed by professional nursing organizations.

Two types of nurses are legally recognized in the United States: the registered professional nurse and the licensed practical nurse. Licensed practical or vocational nurses (LPNs or LVNs) are trained to perform uncomplicated patient-care tasks in hospitals or other health-care facilities under the aegis of registered nurses or physicians. *See* MEDICINE.

Judith A. Vessey

Nut (engineering) In mechanical structures, an internally threaded fastener. Plain square and hexagon nuts for bolts and screws are available in three degrees of finish: unfinished, semi-finished, and finished. There are two standard weights: regular and heavy. For specific applications, there are other standard forms such as jam nut, castellated nut, slotted nut, cap nut, wing nut, and knurled nut.

Wing and knurled nuts are designed for applications where a nut is to be tightened or loosened by using finger pressure only. *See* SCREW FASTENER.

Warren J. Luzadder

Nut crop culture The cultivation of plants, primarily trees, that produce nuts. The term nut is used loosely. Generally, a nut is defined as any edible fruit or seed enclosed in a hard shell. Botanically, a nut is a hard, indehiscent, one-seeded (nut), pericarp (shell) generally resulting from a compound ovary of a flower. Indehiscent means the shell does not split open spontaneously when ripe. Examples are chestnuts, filberts, and acorns. Technically, a nut is a dry edible fruit consisting of a kernel or seed enclosed in a woody shell. Only a fraction of the 80 fruits and seeds designated as nuts fit this description. The peanut is notable because the plant is a herbaceous annual legume in which the nuts are analogous to peapods that mature underground. *See* FLOWER.

Although most nuts are spherical, they can vary considerably in size, shape, shell, and hull. Popular edible nuts range in size from the small sunflower seed to the large coconut. Nuts can be round (macadamias), elongated (pistachios and almonds), kidney shaped (cashews), or triangular (brazil nuts), or have the distinctive double-lobed kernel of the walnut and pecan.

Plants producing nut crops are very diverse in their botanical classification, type, and climatic and cultural requirements. Cultural methods differ according to the nature of the plant, its soil requirements, the climate, available labor, and other factors. As with most orchard crops, commercially cultivated nut trees are now grafted onto rootstocks that allow better adaptation to the soil or climate, and disease or pest resistance. Relative to other tree fruits nut crop culture offers some distinct advantages in the ability to use machinery rather than hand labor for harvesting. *See* AGRICULTURAL MACHINERY.

The edible part of most nuts, the kernel, is high in fat (30–70%) and low in carbohydrates (generally under 25%). Although nuts contain high-quality protein, they are considered primarily a source of fat and are rarely a dietary staple in modern cultures. The most common nuts, such as walnuts, almonds, pecans, and cashews, are consumed primarily as snack food or as a confectionery additive, less frequently in savory foods. Other nuts, such as the betel nuts, are masticatories, that is, nuts that are chewed but not consumed. Not all nuts are edible or chewable. With increasing world population and diversity in diets, commercially produced nut crops have become economically important. *See* ALMOND; BRAZIL NUT; CASHEW; CHESTNUT; COCONUT; COLA; FILBERT; MACADAMIA NUT; PEANUT; PECAN; PINE NUT; PISTACHIO; WALNUT.

Louise Ferguson

Nutation (astronomy and mechanics) In mechanics, a bobbing motion that accompanies the precession of a spinning rigid body, such as a top. In simple precession, the axis of a top with a fixed point of contact sweeps out a cone, whose axis is the vertical direction. In the general motion, the angle between the axis of the top and the vertical varies with time. This motion of the top's axis, bobbing up and down as it precesses, is known as nutation. *See* RIGID-BODY DYNAMICS.

Astronomical nutation refers to irregularities in the precessional motion of rotating bodies. A well-studied example is the irregularities in the precessional rotation of the equinoxes caused by the varying torque applied to the Earth by the Sun and Moon. Astronomical nutation should not be confused with nutation as defined in mechanics; the latter is present even if the source of the torques is unvarying. *See* CELESTIAL MECHANICS; PRECESSION OF EQUINOX. Vernon D. Barger; Ray E. Bolz; John L. Safko

Nutmeg A delicately flavored spice obtained from the nutmeg tree (*Myristica fragrans*), a native of the Moluccas, or Spice Islands. The tree is a dark-leafed evergreen, and is a member of the nutmeg family (Myristicaceae). The golden-yellow, mature fruits resemble apricots (see illustration). They gradually lose moisture and when completely ripe, the husk (pericarp) splits open, exposing the shiny brown seed which is the nutmeg of commerce. Nutmeg oil is used in medicine, perfumery, and dentifrices, and in the tobacco industry.

Mature nutmeg (*Myristica fragrans*) fruits. (*USDA*)

Perry D. Strausbaugh; Earl L. Core

Nutrition The science of nourishment, including the study of the nutrients that each organism must obtain from its environment to maintain life and health and to reproduce. Although each kind of organism has its distinctive needs, a far-reaching biochemical unity in nature has been discovered which gives vastly more coherence to the whole subject. Many nutrients, such as amino acids, minerals, and vitamins, needed by higher organisms may also be needed by the simplest forms of life—single-celled bacteria and protozoa. The recognition of this fact has made possible highly important developments in biochemistry.

Mammals need for their nutrition (aside from water and oxygen) a highly complex mixture of more than 40 chemical substances, including amino acids; carbohydrates; certain lipids; fibers; a great variety of minerals, including several that are required only in minute amounts, commonly referred to as trace minerals; and vitamins. *See* AMINO ACIDS; CARBOHYDRATE METABOLISM; LIPID METABOLISM; PROTEIN METABOLISM; VITAMIN.

Early workers in human nutrition focused on the minimum amounts needed to prevent or cure acute deficiency diseases, such as scurvy and beriberi. Since that time, the Recommended Dietary Allowances and Adequate Intakes (RDAs and AIs, collectively called Dietary Reference Intakes) in the United States and similar recommendations in other countries include consideration of biochemical criteria of adequacy. They also include approximate adjustments for age, sex, and pregnancy and lactation, along with rough estimates for some other sources of individual variation. However, statistical data needed to adequately assess individual variations are not yet available for any nutrient.

Interests have shifted toward what may be more nearly optimal nutritional intakes based on the amounts needed to promote health (not merely to avoid disease or biochemical deficiency), longevity, and resistance to chronic disorders, including cardiovascular disease, cancer, hypertension, and diabetes. Increasing evidence indicates that nutrients also protect against environmental pollutants and some human birth defects that formerly were not believed to be nutrition-related.

For modern humans, the problems of suboptimal nutrition have increased with the advent and extensive consumption of technologically derived, refined foods. In advanced western nations, more than half of the dry weight and energy content of food supplies derives from purified sugars, separated fats, alcohol, and milled grains. These nonwhole "foods" have lost most or all of the nutrients present in the whole foods from which they derive. Their excessive use facilitates various kinds of malnutrition and overconsumption that do not occur readily with whole foods. Modern dietary guidelines and nutrition education focus substantially on partially replacing nonwhole foods with whole grains, legumes, low-fat meats and dairy products, fish, vegetables, fruits, and nuts that retain their natural biochemical unity. It is clear that improper nutrition may produce or contribute to almost every type of illness. Nutritional and medical research are yielding important advances in using improved nutrition to prevent, cure, and ameliorate disease and illness. *See* DISEASE; MALNUTRITION; METABOLIC DISORDERS. Roger J. Williams; Donald R. Davis

Nymphaeales An order of flowering plants that have previously been included in the same larger grouping as Magnoliales, the Magnoliidae. Deoxyribonucleic acid (DNA) sequence studies have demonstrated that Nymphaeales as previously defined contain two families, Ceratophyllaceae (water hornwort) and Nelumbonaceae (lotus), that are not closely related to the others. Remaining in the order is the family Nymphaeaceae, the waterlilies, from which a small group of tropical plants, Cabombaceae, are split by some scientists. Nymphaeaceae contain nearly 100 species of fresh-water aquatics that are typically found in river and lake systems throughout the world. The ovaries of these plants are filled with mucilage, which mediates pollen tube growth from the stigmas to the ovules, and they have either inappertuate or monosulcate pollen. A spectacular plant is the Amazonian water lily (*Victoria*), which has leaves up to 15 ft (5 m) in diameter. The water lily family has been shown by DNA analyses to be one of the oldest lineages of flowering plants and distantly related to all others as well. The family members are relics of the early diversification of the flowering plants. *See* EUMAGNOLIIDS; PLANT KINGDOM; POLLEN. Mark Chase

Oak A genus (*Quercus*) of trees, some of which are shrubby, with about 200 species, mainly in the Northern Hemisphere. About 50 species are native in the United States. All oaks have scaly winter buds, usually clustered at the ends of the twigs, and single at the nodes. The fruit is a nut (acorn). The leaves are simple and usually lobed.

Oaks furnish the most important hardwood lumber in the United States. Principal uses are for charcoal, barrels, building construction, flooring, railroad ties, mine timbers, boxes, crates, vehicle parts, ships, agricultural implements, caskets, woodenware, fence posts, piling, and veneer. Oak is also used for pulp and paper products. *See* FAGALES. Arthur H. Graves; Kenneth P. Davis

Oasis An isolated fertile area, usually limited in extent and surrounded by desert. The term was initially applied to small areas in Africa and Asia typically supporting trees and cultivated crops with a water supply from springs and from seepage of water originating at some distance. However, the term has been expanded to include areas receiving moisture from intermittent streams or artificial irrigation systems. Thus the floodplains of the Nile and Colorado rivers can be considered vast oases, as can arid areas irrigated by humans. *See* DESERT.

Oases are restricted to climatic regions where precipitation is insufficient to support crop production. Such regions may be classified as extremely arid (annual rainfall less than 2 in. or 50 mm), arid (annual rainfall less than 10 in. or 250 mm), and semiarid (rainfall less than 20 in. or 500 mm). Many African and Asian oases are in extremely arid areas. Most oases are found in warm climates. Oasis soils are weakly developed, high in organic matter but often saline, and have been strongly affected by human occupation. William G. McGinnies

Oats An agricultural crop grown for its grain and straw in most countries of the temperate zones of the world. In the major oat-growing states of the midwestern United States (Iowa, North Dakota, South Dakota, Minnesota, and Wisconsin) the crop is raised for grain, whereas in the Southern states (Texas, Oklahoma, and Georgia) it is used for pasture or a combination of pasture and grain. About 90% of the annual oat grain production is used for animal feeds, and about 10% is processed into food for humans, for example, oatmeal and other cereal products. In general, oats are a cool-season crop which requires a moist climate. They grow well on both light and heavy soils if sufficient moisture and fertility nutrients are available. *See* ANIMAL FEEDS; CEREAL.

The fifteen species of oats in the genus *Avena* are divided into three groups on the basis of chromosome number: 14, 28, or 42. Within the 42-chromosome cultivated species there is wide variation among varieties for all plant traits. Oats belong to the Graminae (grass) family; thus the oat plant forms a crown at the soil surface from which a fibrous root system penetrates the soil. Culms usually grow 2–5 ft (0.6–1.5 m) tall, and they are terminated with inflorescences called panicles. Each panicle usually bears 10–75 spikelets on its numerous branches. A spikelet is enclosed by two papery glumes and bears two or three florets, each with an ovary, two stigmas, and three anthers enclosed in a lemma and palea (see illustration). In most varieties the lemma and palea adhere to the oat seed after threshing. A trait used to determine market grade of oats is the color of the lemma, which may be white, yellow, gray, brown, red, or black. The major trait that distinguishes wild from cultivated oats is seed shattering. In cultivated species the seed attachment is persistent, and it can be separated from the panicle only by threshing. *See* CYPERALES; FLOWER; GRASS CROPS; INFLORESCENCE.

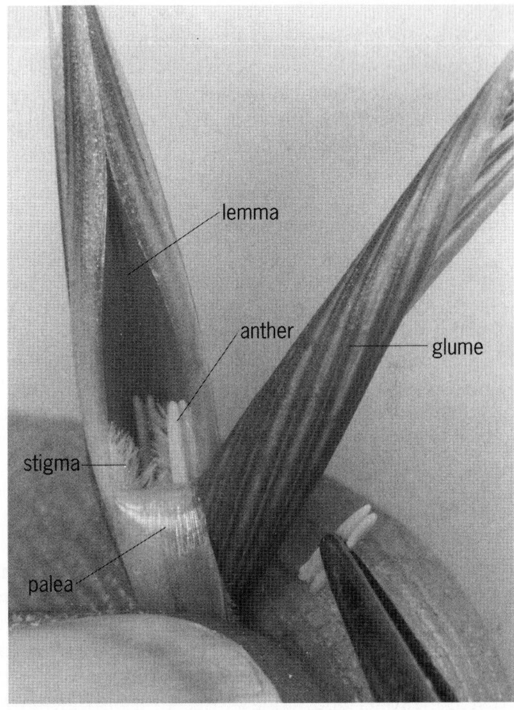

An oat flower; two oat grains are shown at the lower right.

The world collection of oats, maintained by the U.S. Department of Agriculture, contains more than 14,000 lines of 42-chromosome types. These represent lines from wild species and from varieties produced at breeding stations. The collection represents a vast range of genetic types that can be used for varietal improvement. Kenneth J. Frey

The milling of oats is less complex than wheat milling and has many similarities to rice or barley milling operations because there is limited fractionation of the kernel. The oat grain is covered with a coarse, adhering hull which must be removed prior to production of ingredients or consumer foods. Oats as received at the mill house are termed green oats and must be cleaned to remove foreign seeds and trash. Clean, sound oats are heated slowly prior to hull removal. The green oats have active lipolytic enzymes (lipases) which will catalyze hydrolysis of triglycerides and yield free fatty acids. The heating, drying, or roasting procedures inactivate lipases, facilitate hull removal, and impart a distinctive roasted flavor to the oat product. Roasted oats are air-cooled and size-graded prior to dehulling. The products of the dehuller are primarily the whole kernel or groat and the fiber hull which are readily separated by air aspiration. The low-density oat hulls possess particularly high levels of fiber and pentosans which are suitable feedstock for industrial production of furfural (an important chemical used in nylon manufacturing) through high-temperature acid hydrolysis and dehydration. Whole, cleaned oat groats are not frequently available in the commercial food market. Selected large-sized groats are excellent for puffing into ready-to-eat cereals.

Oat cutting and flaking procedures provide more extensive utilization of groats in the form of rolled oats. Oat flour is obtained from further reduction and sieving or hammer milling of the whole groat or flaked product. This high-protein flour is frequently used in the formulations of ready-to-eat cereals and many prepared baby foods. Composite flours blended from oat flour and other cereals providing high protein content and extended shelf life have been proposed as suitable for world feeding programs. *See* FOOD MANUFACTURING. Mark A. Uebersax

Obesity The presence of excess body fat. The great prevalence of this condition, its severe consequences for physical and mental health, and the difficulty of treating it make the prevention of obesity a major public health priority.

Obesity is most often defined in terms of body weight relative to height, since both height and weight are easily measured. Obesity is considered to begin at a weight-for-height that is 20–30% above desirable weight, with this desirable weight taken as the midpoint of ranges of weight associated with the greatest longevity in studies of life-insured individuals. In population surveys, obesity is defined as a body weight that meets or exceeds the 85th percentile of the Body Mass Index (BMI), an index of weight-for-height that correlates well with body fat content. *See* ADIPOSE TISSUE.

The prevalence of obesity increases with age, is higher in women than men, and is highest among the poor and minority groups. Obesity increases the likelihood of high blood cholesterol, high blood pressure, and diabetes, and therefore of the diseases for which such conditions are risk factors—coronary heart disease, stroke, and kidney disease. It also increases the likelihood of gallbladder disease and cancers of the breast and uterus. Thus, obesity increases overall mortality rates, and it does so in proportion to the degree and duration of overweight. Individuals who become obese at the earliest ages are at highest risk of premature mortality. Distribution of excess fat to the upper body rather than the lower body may also increase risk.

The causes of most cases of obesity are poorly understood. At the simplest level, obesity results from an excess of energy (caloric) intake over expenditure, but this statement does not explain why some individuals can eat as much as they like without gaining weight while others remain overweight despite constant dieting. Studies of genetically obese animals and those with damage to the part of the brain called the hypothalamus suggest that individuals may balance body weight around a "setpoint" that is maintained—without conscious control—by variations in metabolic rate in response to caloric intake. Variations in the prevalence of obesity among population groups suggest a genetic basis for the condition. The complexity of body-weight regulatory mechanisms suggests that obesity is not due to a single cause but, like other chronic diseases, is multifactorial in origin. Specific inherited differences that might influence setpoints include differences in nearly every anatomic, neurologic, and biochemical factor known to affect food intake and utilization, energy metabolism, and energy expenditure. *See* ENDOCRINE MECHANISMS; ENERGY METABOLISM; METABOLIC DISORDERS.

Because the causes of obesity are incompletely understood, it is difficult to formulate effective treatment strategies. Studies suggest that programs combining diet and exercise help obese individuals lose more weight and maintain losses longer than either program does separately. *See* FOOD; NUTRITION. Marion Nestle

Object-oriented programming A computer-programming methodology that focuses on data items rather than processes. Traditional software development models assume a top-down approach. A functional description of a system is produced and then refined until a running implementation is achieved. Data structures (and file structures) are proposed and evaluated based on how well they support the functional models.

The object-oriented approach focuses first on the data items (entities, objects) that are being manipulated. The emphasis is on characterizing the data items as active entities which can perform operations on and for themselves. It then describes how system behavior is implemented through the interaction of the data items.

The essence of the object-oriented approach is the use of abstract data types, polymorphism, and reuse through inheritance.

Abstract data types define the active data items described above. A traditional data type in a programming language describes only the structure of a data item. An abstract data type also describes operations that may be requested of the data item. It is the ability to associate operations with data items that makes them active. The abstract data type makes operations available without revealing the details of how the operations are implemented, preventing programmers from becoming dependent on implementation details. The definition of an operation is considered a contract between the implementor of the abstract data type and the user of the abstract data type. The implementor is free to perform the operation in any appropriate manner as long as the operation fulfills its contract. Object-oriented programming languages give abstract data types the name class.

Polymorphism in the object-oriented approach refers to the ability of a programmer to treat many different types of objects in a uniform manner by invoking the same operation on each object. Because the objects are instances of abstract data types, they may implement the operation differently as long as they fulfill the agreement in their common contract.

A new abstract data type (class) can be created in object-oriented programming simply by stating how the new type differs from some existing type. A feature that is not described as different will be shared by the two types, constituting reuse through inheritance. Inheritance is useful because it replaces the practice of copying an entire abstract data type in order to change a single feature.

In the object-oriented approach, a class is used to define an abstract data type, and the operations of the type are referred to as methods. An instance of a class is termed an object instance or simply an object. To invoke an operation on an object instance, the programmer sends a message to the object. John J. Shilling

Obolellida A small extinct order of articulated brachiopods that ranges in age from Early to Middle Cambrian and includes the earliest known calcitic brachiopods.

Phylum Brachiopoda
  Subphylum Rhynchonelliformea
    Class Obolellata
      Order Obolellida
        Superfamily Obolelloidea
          Family Obolellidae
            5 genera (Lower Cambrian)
          Family Trematobolidae
            3 genera (Early-Middle Cambrian)

Obolellids have a biconvex, calcite, impunctate (lacking holes) shell with an elongated oval shape and a laminar secondary layer. The order includes forms that have primitive articulation of the ventral and dorsal valves—consisting of paired ventral denticles (hinge teeth), where the pedicle (fleshy stalk) attaches to the shell, and dorsal sockets along the internal posterior margin (hinge line)—and forms that lack denticles, such as the genus *Obolella*. The ventral valve (formerly named pedicle valve) has a well-defined, low and relatively short flat shelf (interarea) at the posterior. The pedicle opening (delthyrium) is located between the valves, and is either uncovered, as in *Obolella*, or closed off by convex plates (pseudodeltidium). Members of this group were presumably epifaunal and sessile. *See* BRACHIOPODA; RHYNCHONELLIFORMEA. Christian Emig

Observatory, astronomical A telescope or telescopes, their protective enclosures (if any), support and headquarters buildings, and the staff of astronomers, engineers, technicians, and other support personnel. The telescopes can be optical or infrared (reflecting or refracting) inside a corotating dome, or radio dishes without enclosures.

The on-site support building contains the control room with the computers and control electronics to operate and point the telescope, as well as the data acquisition computers and electronics for detector instruments. It houses laboratories for testing and calibrations, a machine shop for emergency repair and fabrication of parts, and storage for inactive instruments or telescope optics. These areas are often incorporated into a single building along with the telescope and dome. Since telescopes are generally located at remote sites to maximize their usefulness, the administration and support offices are often located in a separate headquarters building miles away in a town or university.

Optical and infrared observatories are the most common types of observatory. They are often located in remote mountains to minimize the effects of contaminating artificial lights and atmospheric blurring. Some countries place their largest telescopes at the best sites in the world such as the 13,800-ft-high (4200-m) volcanic mountain Mauna Kea in Hawaii (see illustration) and La Palma in the Canary Islands.

Optical observatories study planets, stars, nebulae, and galaxies. A large aperture is needed to collect the faint light of these sources. Sizes typically range from 0.5 to 10 m (20–400 in.). The smaller telescopes can use glass lenses, the largest refractor being the University of Chicago 1.0-m (40-in.) Yerkes. For the largest telescopes, engineering and manufacturing constraints dictate the reflecting design. The 9.8-m (386-in.) Keck Telescope is the world's largest reflector. The Very Large Telescope Project seeks to combine the light from four 8.2-m (323-in.) telescopes to effectively have the light-gathering capability of a 16-m (630-in.) telescope.

Most optical observatories can also observe in the near-infrared region. Infrared observatories have been optimized to work farther into the longer (thermal) infrared. Since glass absorbs infrared light, all infrared telescopes are of the reflecting type. Special design techniques must be used, as the telescope itself glows in the infrared.

Radio observatories have become an essential complement to optical observatories. With their longer wavelengths, the radio telescopes can see through cosmic clouds and measure the properties (temperature, pressure, chemical composition, and velocities) of gases which pervade the universe. Since the wavelengths are so long, the telescopes must be large to achieve a high resolution (the ability to discriminate between two nearby sources), and radio dishes are typically tens to hundreds of meters wide.

The longer radio wavelengths allow the use of interferometers, where the signal from several small radio telescopes can be mathematically combined to yield results as if they had been collected from a very large dish. The Very Large Array radio observatory in Socorro, New Mexico, has twenty-seven 25-m (82-ft) antennas arranged in a Y pattern, which can yield the resolution of a single telescope 36 km (22 mi) wide. The Very Long Baseline Array uses ten 25-m (82-ft) antennas spread from Hawaii to the Virgin Islands, allowing the resolution of a telescope dish that distance across. *See* RADIO TELESCOPE.

Gemini North Observatory in Mauna Kea. (*a*) Telescope and interior of dome. (*b*) Exterior of dome. Vent gates surrounding the dome allow air to flow throughout the dome, thereby minimizing distortion of images due to swirling. (*Neelon Crawford, Polar Fine Arts and Gemini Observatory*)

Crewless balloons have been used to hoist telescopes above the absorbing atmosphere for gamma-ray, x-ray, and ultraviolet observations. Their lower cost (compared to space missions) is offset by the short duration of the observations. The National Aeronautics and Space Administration (NASA) successfully operated the Kuiper Airborne Observatory for 20 years. Space observatories have amply proven their worth despite their initial high cost and limited mission lifetimes. *See* SATELLITE (ASTRONOMY).

An astronomical instrument is the working heart of an observatory. The telescope exists solely to collect and funnel light into the instrument. Astronomical instruments range from a simple eyepiece for direct viewing (with different magnifications), to a camera for imaging on film or electronic detectors, to a spectrograph which records the wavelength distribution of light energy (analogous to a rainbow). The charge-coupled device (CCD) has now virtually supplanted the photographic plate. For radio telescopes the instrument is an amplifier that isolates and boosts the weak celestial electric signal. *See* ASTRONOMICAL SPECTROSCOPY; CAMERA; CHARGE-COUPLED DEVICES; SPECTROGRAPH.

Besides observing light (electromagnetic radiation), astronomers are mapping the universe via neutrinos, particles that can transverse matter with a minimum of interaction. Gravitational wave observatories attempt to detect ripples (gravitational waves or gravitons) in the fabric of space-time itself. *See* GRAVITATIONAL WAVES; NEUTRINO ASTRONOMY; TELESCOPE. John Hamilton

Obsessive-compulsive disorder A type of anxiety disorder (commonly referred to as OCD) characterized by recurrent, persistent, unwanted, and unpleasant thoughts (obsessions) or repetitive, purposeful ritualistic behaviors that the person feels driven to perform (compulsions). A cardinal feature of this disorder is an awareness of the irrationality or excess of the obsessions and compulsions accompanied by an inability to control them.

Typical compulsions include an irresistible urge to wash (particularly the hands) or clean, to check doors to confirm that they are locked, to return repeatedly to appliances to make sure they are turned off, to touch, to repeat, to count, to arrange, or to save. Typical obsessions include overconcern about dirt and contamination, fear of acting on violent or aggressive impulses, feeling overly responsible for the safety of others, abhorrent religious (blasphemous) and sexual intrusions, and inordinate concern with arrangement or symmetry. Obsessions may accompany compulsions, or compulsions may occur alone.

The difference between obsessive-compulsive disorder and milder forms of obsession or compulsion seen in otherwise healthy people is that for the sufferer the obsessions or compulsions cause marked distress, are time-consuming, and significantly interfere with the person's normal routine, occupational functioning, usual social activities, and relationships with others.

Onset in adolescence occurs in about a third of cases. In another third symptoms appear in early adulthood, and in the last third they start later in life. If not treated appropriately, the disorder is often chronic, with waxing and waning of symptoms.

Obsessive-compulsive disorder is generally resistant to traditional psychotherapy, which has tried to trace the condition to conflicts of early childhood. An effective mode of psychotherapy is behavioral therapy, in which the patients are gradually exposed to their feared or triggering situation but are prevented from performing accompanying compulsions. This approach, which focuses on treating the symptoms rather than trying to understand their origin, seems to be more effective in treating the ritualistic behavior (compulsions) than the pervasive thoughts (obsessions). Obsessive-compulsive disorder is also refractory to most drugs used to treat anxiety, depression, and psychosis. However, it often eases with medications that affect the brain's serotonergic system, such as clorimipramine, fluvoxamine, and fluoxetine.

The specific response of patients with obsessive-compulsive disorder to serotonergic drugs, their hypersensitivity to activation of the serotonergic system, and the distinct functional anatomy differences found in those patients suggest a biological cause for this disorder. In this regard, obsessive-compulsive disorder represents a shift from a psychological to a neurobiological approach in the study of anxiety disorders. *See* ANXIETY DISORDERS; NEUROTIC DISORDERS; SEROTONIN. Joseph Zohar

Obsidian A volcanic glass, usually of rhyolitic composition, formed by rapid cooling of viscous lava. The color is jet-black because of abundant microscopic, embryonic crystal growths (crystallites) which make the glass opaque except on thin edges. Iron oxide dust may produce red or brown obsidian.

Obsidian usually forms the upper parts of lava flows. Well-known occurrences are Obsidian Cliffs in Yellowstone Park, Wyoming; Mount Hekla, Iceland; and the Lipari Islands off the coast of Italy. *See* IGNEOUS ROCKS; VOLCANIC GLASS.

Carleton A. Chapman

Occultation The temporary blocking from view of one celestial body by another. The occulting body is the one closer to the observer, and can be a planet, moon, ring system, or other body, usually in the solar system. The occulted body is smaller in apparent, projected size and is usually a distant star, although it can also be a spacecraft radio signal, as in the case of the *Voyager* spacecraft radio occultations at the outer planets, or another solar system body. Examples of occultations are that of a star by a planet, a lunar occultation of a star, and an occultation by Pluto of its satellite Charon. Although a solar eclipse is not usually thought of in these terms, this event is actually an occultation of the Sun by the Moon. *See* ECLIPSE; PLANET; SATELLITE (ASTRONOMY); SOLAR SYSTEM.

Observations of occultations can reveal information about the physical size of the blocking body, the structure of its atmosphere (bound, in the case of planetary atmospheres, or unbound in the case of comets), or the structure of its rings. Occultations of stars by the Moon are used primarily to study the structure of the occulted stars, quantities such as their projected size, and limb darkening, though the figure of the lunar limb used to make predictions of Baily's beads at eclipses is based on lunar occultations.

Although occultations provide a wealth of information about solar system objects, the observation of an occultation is primarily a monitoring of stellar brightness over time. In fact, the best data are obtained when the occulting solar system body is as close to invisible as is possible. Observations of stellar occultations typically require fast sampling rates, monitoring the stellar brightness several times each second. Such observations are possible with photoelectric photometers and charge-coupled devices (CCDs) designed for such rapid readout. The quality of the resulting light curve depends on the stellar brightness and the brightness of the background. The resulting light curve, or plot of stellar brightness against time, shows the star at full level, then a drop in brightness when star is occulted, and then a rise of the signal back to the full level. *See* ASTRONOMICAL IMAGING; CHARGE-COUPLED DEVICES.

Analysis of stellar occultation data involves converting the time series light curve into a spatial scan, using knowledge of the geometry of the occulting body and the observer. These calculations rely heavily on solar system ephemerides; for occultations by small bodies in the outer solar system, such as Pluto, Triton, comets, and Kuiper Belt objects, the accuracy of event predictions are constrained by the accuracy of their ephemerides. To compensate for these uncertainties in the data analysis, typically two or more light curves from different observing sites are required, since the circular silhouette has different-length chords at different parts of the surface. These are then combined and the ephemeris uncertainties are removed. *See* EPHEMERIS.

Amanda S. Bosh

Ocean One of the major subdivisions of the interconnected body of salt water that occupies almost three-quarters of the Earth's surface. Earth is the only planet in the solar system whose surface is covered with significant quantities of water. Of the nearly 1.4 billion cubic kilometers of water found either on the surface or in relatively accessible underground supplies, more than 97% is in the oceans. *See* OCEANOGRAPHY.

Oceans and the seas that connect them cover some 73% of the surface of the Earth, with a mean depth of 3729 m (12,234 ft) (table). More than 70% of the oceans have a depth between 3000 and 6000 m (10,000 and 20,000 ft). Less than 0.2% of the oceans have depths as great as 7000 m (23,000 ft).

The oceans are cold and salty. Some 50% have a temperature between 0 and 2°C (32 and 36°F) and a salinity between 34.0 and 35.0. To a high degree of approximation, a salinity of 34 is the equivalent of 34 grams of salt in a kilogram of seawater. Water with a temperature above a few degrees Celsius is confined to a relatively thin surface layer of the ocean. *See* SEAWATER.

Ocean basin characteristics

	Area, km^2	Volume, km^3	Mean depth, m
Pacific	181,344,000	714,410,000	3940
Atlantic	94,314,000	337,210,000	3575
Indian	74,118,000	284,608,000	3840
Arctic	12,257,000	13,702,000	1117
Total	362,033,000	1,349,929,000	3729

Ocean salinity is primarily controlled by the balance of precipitation, river runoff, and evaporation of water at the sea surface. The highest salinities are found in major evaporation basins with little rainfall or river runoff, such as the Red Sea. The lowest salinities are found near the mouths of major rivers such as the Amazon. *See* RED SEA.

Nearly all elements known to humankind have been found dissolved in seawater, and those that have not are assumed to be present. However, all but a few are found in very small amounts. Sodium chloride accounts for some 85% of the dissolved salts, and an additional four ions (sulfate, magnesium, calcium, and potassium) bring the total to more than 99.3%. The ratio of ions is remarkably constant from one ocean to another and from top to bottom of each.

The oceans are continually transporting excess heat (warm water) from the tropics toward the Poles and returning colder water toward the tropics. This process of moving excess heat from lower (south of 40°) to higher (north of 40°) latitudes is shared approximately equally by the oceans and the atmosphere. A significant part of the ocean heat exchange process is carried out by the major ocean currents, the "named" currents such as the Gulf Stream, Brazil Current, California Current, and Kuroshio. These currents are primarily driven by the winds, and there is considerable similarity in their pattern from one ocean basin to another. *See* GULF STREAM; KUROSHIO.

The average winds over the North and South Atlantic as well as the North and South Pacific oceans come out of the west (westerlies) at the middle latitudes and from the east at the lower latitudes (trade winds). The frictional drag of these winds on the surface of the water imparts a spin or torque to the surface of the ocean, clockwise in the Northern Hemisphere and counterclockwise in the Southern Hemisphere. The major exception is the Indian Ocean north of the Equator, where the circulation is strongly influenced by the winds of the seasonal monsoon. *See* ATLANTIC OCEAN; CORIOLIS ACCELERATION; EQUATORIAL CURRENTS; INDIAN OCEAN; OCEAN CIRCULATION; PACIFIC OCEAN. John A. Knauss

Ocean circulation The general circulation of the ocean. The term is usually understood to include large-scale, nearly steady features, such as the Gulf Stream, as well as current systems that change seasonally but are persistent from one year to the next, such as the Davidson Current, off the northwestern United States coast and the equatorial currents in the Indian Ocean. A great number of energetic motions have periods of a month or two and horizontal scales of a few hundred kilometers—a very low-frequency turbulence, collectively called eddies. Energetic motions are also concentrated near the local inertial period (24 h, at 30° latitude) and at the periods associated with tides (primarily diurnal and semidiurnal). *See* TIDE.

The greatest single driving force for currents, as for waves, is the wind. Furthermore, the ocean absorbs heat at low latitudes and loses it at high latitudes. The resultant effect on the density distribution is coupled into the large-scale wind-driven circulation. Some subsurface flows are caused by the sinking of surface waters made dense by cooling or high evaporation. *See* OCEAN WAVES.

Except in western boundary currents, and in the Antarctic Circumpolar Current, the system of strong surface currents is restricted mainly to the upper 330–660 ft (100– 200 m) of the sea. The mid-latitude anticyclonic gyres, however, are coherent in the mean well below 3300 ft (1000 m). The average speeds of the open-ocean surface currents remain mostly below 0.4 knot (20 cm/s). Exceptions to this are found in the western boundary currents, such as the Gulf Stream, and in the Equatorial Currents of the three oceans, all of which have velocities of 2–4 knots (1–2 m/s).

The deep circulation results in part from the wind stress and in part from the internal pressure forces which are maintained by the budgets of heat, salt, and water. Both groups of forces are dependent upon atmospheric influences. Apart from Coriolis and frictional forces, the topography of the sea bottom exercises a decisive influence on the course of deep circulation.

The deep circulation in marginal seas depends largely on the climate of the region, whether arid or humid. Under the influence of an arid climate, evaporation is greater than precipitation. The marginal sea is therefore filled with relatively salty water of a high density. Its surface lies at a lower level than that of the neighboring ocean. Examples of this type are the Mediterranean Sea, Red Sea, and Persian Gulf. The deep circulation of marginal seas in humid climates shows a different pattern. The level of the sea is higher than in the neighboring ocean. Therefore, the surface water with its lower density and accordingly its lower salinity flows outward, and the relatively salty ocean water of higher density flows over the sill into the marginal sea. Examples of this circulation are the Baltic Sea with the shallow Darsser and Drogden rises, the Norwegian and Greenland fiords, and the Black Sea with its entrance through the Bosporus. *See* BLACK SEA; FIORD; MEDITERRANEAN SEA.

The deep circulation in the oceans is more difficult to perceive than the circulation in the marginal seas. In addition to the internal pressure forces, determined by the distribution of density and the piling up of water by the wind, there are also the influences of Coriolis forces and large-scale turbulence. There are areas in tropical latitudes in which the surface water, as a result of strong evaporation, has a relatively high density. In thermohaline convection, the water sinks while flowing horizontally until it reaches a density corresponding to its own, and then spreads out horizontally. In this way the colder and deeper levels of the oceans take on a layered structure consisting of the so-called bottom water, deep water, and intermediate water. *See* ATLANTIC OCEAN; PACIFIC OCEAN. Wilton Sturges

Wherever oceanographers have made long-term current and temperature measurements, they have found energetic fluctuations with periods of several weeks to several months. These low-frequency fluctuations (compared to tides) are caused by oceanic mesoscale eddies which are in many respects analogous to the atmospheric mesoscale pressure systems that form weather. Like the weather, mesoscale eddies often dominate the instantaneous current, and are thought to be an integral part of the ocean's general circulation.

Eddies occur in virtually all oceans and seas, but their amplitude varies greatly from place to place. The largest amplitudes are found on the western sides of the oceans in conjunction with the strongest ocean currents (the Gulf Stream in the North Atlantic, the Kuroshio in the North Pacific) and near the Equator. Much weaker eddies are found in the ocean interior, distant from major currents. This consistent pattern of eddy amplitude suggests that instabilities of western boundary currents are an important source of eddy energy. Atmospheric forcing by variable winds can also generate eddies, and is probably most important at low latitudes where the horizontal scales of the oceanic eddies best match the scales of the atmospheric forcing. James F. Price

Ocean waves The irregular moving bumps and hollows on the ocean surface. Winds blowing over the ocean, in addition to producing currents, create surface water undulations called waves or a sea. The characteristics of these waves (or the state of the sea) depend on the speed of the wind, the length of time that it has blown, the distance over which it has blown, and the depth of the water. If the wind dies down, the waves that remain are called a dead sea.

Surface waves. Ocean surface waves are propagating disturbances at the atmosphere-ocean interface. They are the most familiar ocean waves. Surface waves are also seen on other bodies of water, including lakes and rivers. *See* WAVE MOTION IN LIQUIDS.

A simple sinusoidal wave train is characterized by three attributes: wave height (H), the vertical distance from trough to crest; wavelength (L), the horizontal crest-to-crest distance; and wave period (T), the time between passage of successive crests past a fixed point. The phase velocity ($C = L/T$) is the speed of propagation of a crest. For a given ocean depth (h), wavelength increases with increasing period. The restoring force for these surface waves is predominantly gravitational. Therefore, they are known as surface gravity waves, unless their wavelength is shorter than 1.8 cm (0.7 in.), in which case surface tension provides the dominant restoring force.

Surface gravity waves may be classified according to the nature of the forces producing them. Tides are ocean waves induced by the varying gravitational influence of the Moon and Sun. They have long periods, usually 12.42 h for the strongest constituent. Storm surges are individual waves produced by the wind and dropping barometric pressure associated with storms; they characteristically last several hours. Earthquakes or other large, sudden movements of the Earth's crust can cause waves, called tsunamis, which typically have periods of less than an hour. Wakes are waves resulting from relative motion of the water and a solid body, such as the motion of a ship through the sea or the rapid flow of water around a rock. Wind-generated waves, having periods from a fraction of a second to tens of seconds, are called wind waves. Like tides, they are ubiquitous in the ocean, and continue to travel well beyond their area of generation. The ocean is never completely calm. *See* STORM SURGE; TIDE; TSUNAMI.

The growth of wind waves by the transfer of energy from the wind is not fully understood. At wind speeds less than 1.1 m/s (2.5 mi/h), a flat water surface remains unruffled by waves. Once generated, waves gain energy from the wind by wave-coupling of pressure fluctuations in the air just above the waves. For waves traveling slower than the wind, secondary, wave-induced airflows shift the wave-induced pressure disturbance downwind so the lowest pressure is ahead of the crests. This results in energy transfer from the wind to the wave, and hence growth of the wave.

If a constant wind blows over a sufficient length of ocean, called the fetch, for a sufficient length of time, a wave field develops whose statistical characteristics depend only on wind velocity.

Because of viscosity, surface waves lose energy as they propagate, short-period waves being dampened more rapidly than long-period waves. Waves with long periods (typically 10 s or more) can travel thousands of kilometers with little energy loss. Such waves, generated by distant storms, are called swell.

When waves propagate into an opposing current, they grow in height. For example, when swell from a Weddell Sea storm propagates northeastward into the southwestward-flowing Agulhas Current off South Africa, high steep waves are formed. Many large ships in this region have been severely damaged by such waves.

Because actual ocean waves consist of many components with different periods, heights, and directions, occasionally a large number of these components can, by chance, come in phase with one another, creating a freak wave with a height several times the significant wave height of the surrounding sea. According to linear theory, waves with different periods propagate with different speeds in deep water, and hence the wave components remain in phase only briefly. But nonlinear effects are bound to be significant in a large wave. In such a wave, the effects of nonlinearity can compensate for those of dispersion, allowing a solitary wave to propagate almost unchanged. Consequently, a freak wave can have a lifetime of a minute or two. *See* SOLITON.

Mark Wimbush

Internal waves. Internal waves are wave motions of stably stratified fluids in which the maximum vertical motion takes place below the surface of the fluid. The restoring force is mainly due to gravity; when light fluid from upper layers is depressed into the heavy lower layers, buoyancy forces tend to return the layers to their equilibrium positions. In the oceans, internal oscillations have been observed wherever suitable measurements have been made. The observed oscillations can be analyzed into a spectrum with periods ranging from a few minutes to days. At a number of locations in the oceans, internal tides, or internal waves having the same periodicity as oceanic tides, are prominent.

Internal waves are important to the economy of the sea because they provide one of the few processes that can redistribute kinetic energy from near the surface to abyssal depths. When they break, they can cause turbulent mixing despite the normally stable density gradient in the ocean. Internal waves are known to cause time-varying refraction of acoustic waves because the sound velocity profile in the ocean is distorted by the vertical motions of internal waves. Internal waves have been found by recording fluctuating currents in middepths by moored current meters, by acoustic backscatter Doppler methods, and by studies of the fluctuations of the depths of isotherms as recorded by instruments repeatedly lowered from shipboard or by autonomous instruments floating deep in the water.

Internal waves are thought to be generated in the sea by variations of the wind pressure and stress at the sea surface, by the interaction of surface waves with each other, and by the interaction of tidal motions with the rough sea floor. Charles S. Cox

Oceanic islands Islands rising from the deep sea floor. Oceanic islands range in size from mere specks of rock or sand above the reach of tides to large masses such as Iceland (39,800 mi^2 or 103,000 km^2). Excluded are islands that have continental crust, such as the Seychelles, Norfolk, or Sardinia, even though surrounded by ocean; all oceanic islands surmount volcanic foundations. A few of these have active volcanoes, such as on Hawaii, the Galápagos islands, Iceland, and the Azores, but most islands are on extinct volcanoes. On some islands, the volcanic foundations have subsided beneath sea level, while coral reefs growing very close to sea level have kept pace with the subsidence, accumulating thicknesses of as much as 5000 ft (1500 m) of limestone deposits between the underlying volcanic rocks and the present-day coral islands. *See* REEF; VOLCANO.

Oceanic islands owe their existence to volcanism that began on the deep sea floor and built the volcanic edifices, flow on flow, up to sea level and above. The highest of the oceanic islands is Hawaii, where the peak of Mauna Kea volcano reaches 14,000 ft (4200 m). Most volcanic islands are probably built from scratch in less than 10^6 years, but minor recurrent volcanism may continue for millions of years after the main construction stage. *See* VOLCANOLOGY.

Islands in regions of high oceanic fertility are commonly host to colonies of sea birds, and the deposits of guano have been an important source of phosphate for fertilizer. On some islands, for example, Nauru in the western equatorial Pacific, the original guano has been dissolved and phosphate minerals reprecipitated in porous host limestone rocks. The principal crop on most tropical oceanic islands is coconuts, exploited for their oil content, but some larger volcanic islands, with rich soils and abundant water supplies, are sites of plantations of sugarcane and pineapple. Atoll and barrier-reef islands have very limited water supplies, depending on small lenses of ground water, augmented by collection of rainwater. *See* ATOLL; ISLAND BIOGEOGRAPHY; REEF.

Edward L. Winterer

Oceanographic vessels Research vessels designed to collect quantitative data from the sea's surface, its depths, the sea floor, and the overlying atmosphere. Their primary purpose is to carry scientists and increasingly sophisticated equipment to and from study sites on the ocean's surface, and in some cases below the surface. The ships must have the ability to lower and retrieve instruments by using winches and wires. The ship's equipment and instrumentation must determine precisely the location on the sea surface, and provide suitable communication, data gathering, archiving, and computational facilities for the scientific party.

The requirements list includes seakeeping (sea-kindliness, a measure of a ship's response to severe seas; and station-keeping, the ability of a ship to maintain its fixed location on the sea surface); work environment; endurance (range, days at sea); scientific complement (number of researchers accommodated); operating economy and scientific effectiveness; subdued acoustical characteristics; payload (scientific storage, weight handling); speed; and ship control. These requirements often conflict, necessitating compromise.

Ships typically can be considered in three major groups based on use: general purpose (classical biological, physical, chemical, geological, and ocean engineering research, or a combination); dedicated special purpose (hydrographic survey, mapping, geophysical, or fisheries); or unique (deep-sea drilling, crewed spar buoy, or support of submersible operations). They can be used simply as delivery and support systems for exploratory devices, such as floats and bottom landers, as well as crewless remote operating vehicles (ROVs—tethered, powered, surface-controlled robots), or autonomous underwater vehicles (AUVs—freely operating robots, using computer programmed guidances). *See* MARINE ENGINEERING; NAVAL ARCHITECTURE; OCEANOGRAPHY. J. J. Griffin; J. F. Bash

Oceanography The science of the sea; including physical oceanography, marine chemistry, marine geology, and marine biology. The need to know more about the impact of marine pollution and possible effects of the exploitation of marine resources, together with the role of the ocean in possible global warming and climate change, means that oceanography is an important scientific discipline. Improved understanding of the sea has been essential in such diverse fields as fisheries conservation, the exploitation of underwater oil and gas reserves, and coastal protection policy, as well as in national defense strategies. The scientific benefits include not only improved understanding of the oceans and their inhabitants, but important information about the evolution of the Earth and its tectonic processes, and about the global environment and climate, past and present, as well as possible future changes. *See* CLIMATE HISTORY; COASTAL ENGINEERING; MARINE MINING; MARINE SEDIMENTS; MARITIME METEOROLOGY; OIL AND GAS, OFFSHORE.

The traditional basis of modern oceanography is the hydrographic station. Hydrographic studies are still carried out at regular intervals, with the research vessel in a specific position. Seawater temperature, depth, and salinity can be measured continuously by a probe towed behind the ship. The revolution in electronics has provided not only a new generation of instruments for studying the sea but also new ways of collecting and analyzing the data they produce. Computers are employed in gathering and processing data in all fields, and are also used in the creation of mathematical models to aid in understanding. Much information can also be gained by remote sensing using satellites, which are also a valuable navigational aid. These provide data on sea surface temperature and currents, and on marine productivity. Satellite altimetry gives information on wave height and winds and even bottom topography (because this affects sea level). Deep-sea cameras and submersibles now permit visual evidence of creatures in remote depths. *See* HYDROGRAPHY; OCEANOGRAPHIC VESSELS; MARINE ECOLOGY; REMOTE SENSING; SATELLITE NAVIGATION SYSTEMS; SEAWATER.

Since the early 1900s, all recorded ocean depths have been incorporated in the General Bathymetric Chart of the Ocean. The amount of data available increased greatly with the introduction of continuous echo sounders; subsequently, side-scan sonar permitted very detailed topographical surveys to be made of the ocean floor. The features thus revealed, in particular the midocean ridges (spreading centers) and deep trenches (subduction zones), are integral to the theory of plate tectonics. An important discovery made toward the end of the twentieth century was the existence of hydrothermal vents, where hot mineral-rich water gushes from the Earth's interior. The deposition of minerals at these sites and the discovery of associated ecosystems make them of potential economic as well as great scientific interest. *See* ECHO SOUNDER; HYDROTHERMAL VENT; MARINE GEOLOGY; MID-OCEANIC RIDGE; PLATE TECTONICS; SONAR; SUBDUCTION ZONES. Margaret Deacon

Octane number A standard laboratory measure of a fuel's ability to resist knock during combustion in a spark-ignition engine, A single-cylinder four-stroke engine of standardized design is used to determine the knock resistance of a given fuel by comparing it with that of primary reference fuels composed of varying proportions of two pure hydrocarbons, one very high in knock resistance and the other very low. A highly knock-resistant isooctane (2,2,4-trimethylpentane, C_8H_{18}) is assigned a rating of 100 on the octane scale, and normal heptane (C_7H_{16}), with very poor knock resistance, represents zero on the scale. Octane number is defined as the percentage of isooctane required in a blend with normal heptane to match the knocking behavior of the gasoline being tested. *See* SPARK KNOCK; OCTANE.

For fuels with a rating higher than 100 octane, the rating is usually obtained by determining the amount of tetraethyllead compound that needs to be added to pure isooctane to match the knock resistance of the test fuel. John C. Lane

Octocorallia (Alcyonaria) A subclass of Anthozoa. These cnidarians are benthic as adults, living attached to firm substrata or burrowed into soft sediments, from the intertidal zone to great depths. As in all anthozoans, the adult form is a polyp—a cylindrical organism that has, at its free end, the single body opening, the mouth, which is surrounded by eight tentacles. They are colonial, with a few possible exceptions.

Each of the eight tentacles has many small or elongate side branches (pinnules) oriented perpendicular to the axis of the hollow, flexible tentacle. The opposite end of the polyp is attached to the substratum or, more commonly in alcyonarians, emerges from the tissue that unites the members of the colony. Sexual reproduction typically results in a planula larva, which eventually metamorphoses into a polyp. Asexual (vegetative) reproduction typically involves budding a new polyp from an existing one or from the tissue connecting members of the colony (the coenenchyme); in a colony, the resulting polyps remain attached to one another.

All alcyonarians have a type of polyp known as the autozooid. The internal body space (the gastrovascular cavity) is divided by eight longitudinally oriented sheets of tissue known as mesenteries. The best-developed pair of mesenteries (the asulcal pair) bears large, heavily ciliated filaments. The other six mesenteries bear filaments having many gland cells. Gametes form and mature in the mesenteries, which rupture to release them when they are ripe; some kinds of alcyonarians are hermaphroditic and some are gonochoric (with separate sexes). Typically gametes are spawned freely into the sea, where egg and sperm meet and the planula develops; in some alcyonarians, however, embryos are brooded internally or on the surface of the colony, so a free-swimming stage of the life cycle is lacking.

Three orders are recognized. The two that have long been recognized are Helioporacea, the blue corals, and Pennatulacea, the sea pens and sea pansies. All other alcyonarians belong to order Alcyonacea. *See* ANTHOZOA; CNIDARIA. Daphne G. Fautin

Octopoda An order of the class Cephalopoda (subclass Coleoidea), characterized by eight appendages that encircle the mouth, a saclike body, and an internal shell that is much modified or reduced from that of its ancestors. One or two rows of suckers without chitinous rings occur along the eight highly flexible contractile arms. The approximately 200 species of octopods include shallow-water forms like the common octopus, *Octopus vulgaris*; open-ocean species like the paper argonaut, *Argonauta argo*; and deep-sea forms with fins like the flapjack devilfish, *Opisthoteuthis californica*. Two suborders divide the Octopoda into those with paddle-shaped fins on the body and tendrillike cirri on the arms (Cirrata) and those without fins or cirri (Incirrata). *See* CEPHALOPODA; COLEOIDEA. Clyde F. E. Roper

Octopus Any species in the subclass Coleoidea, order Polypoidea, of the cephalopod mollusks or, more particularly, any of the approximately 120 species of the genus *Octopus*. Widely distributed in coastal areas of all oceans, such octopods have a saclike body without any skeletal support and eight arms of equal length which bear suckers. They are predacious carnivores, crawling on the sea bottom and lurking in submarine caves and crevices. They can swim by jet propulsion but rarely do so, except as an escape mechanism. The body ranges in size from 1.2 to 14 in. (3 to 35 cm) across. Prey, such as living crabs, are captured by the arms with their suckers and pressed in toward the mouth, which is armed with both a chitinous parrotlike beak and a powerful radular apparatus. The saliva of *Octopus* contains a neurotoxin.

Rapid and patterned color changes are possible in octopuses, cuttlefish, and their allies because their chromatophores (unlike those of other invertebrates) are miniature bags of pigment expanded by the pull of muscles under nervous control. As well as being used in concealment and for terrorizing displays, rapid color changes along with tactile caresses form part of the courtship behavior of such cephalopods. The sense organs of octopuses are complex and efficient in discrimination; they include eyes, olfactory pits, chemotactile and mechanotactile sensillae on the arms, and statocysts capable of detecting direction and angle of acceleration as well as static posture. *See* CEPHALOPODA; MOLLUSCA; OCTOPODA. W. D. Russell-Hunter

Odonata An order of the class Insecta, commonly known as dragonflies and damselflies (see illustration). Odonates generally are conspicuous insects with strong flying ability and a complex range of behaviors. They occur on all continents except Antarctica, although more species occur near the Equator than at higher latitudes. The larvae are aquatic, inhabiting all types of freshwater, including streams, rivers, ponds, and marshes. The adults fly over or near these localities, though some species may disperse and are occasionally found far from water. *See* INSECTA.

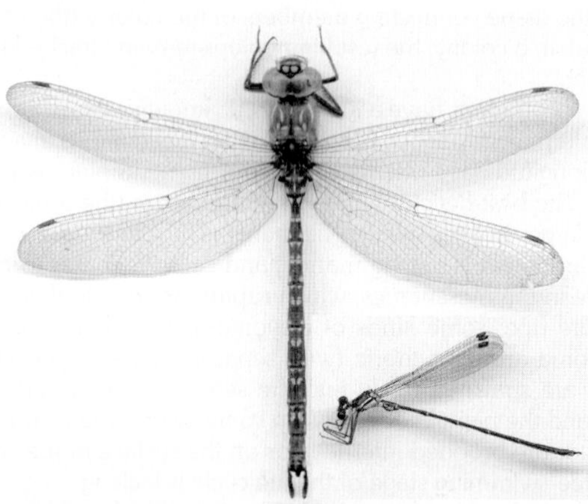

An aeshnid dragonfly (note the distinctive wing venation) and a lestid damselfly.

Adult Odonata range in size from small (with a wingspan of 15 mm, or 0.6 in.) to very large (with a wingspan greater than 150 mm, or 6 in.). They have large compound eyes on either a globular or transversely elongate head. The antennae are short and bristlelike. There are two pairs of elongate, membranous wings, which often are clear or transparent; however, in some species, the wings have patches or bands of color. The wings have a distinctive venation, with five main longitudinal veins that originate from the base and are connected by a network of many crossveins.

Odonata contains more than 6000 species worldwide, and is divided into three suborders: the Anisoptera, or dragonflies; the Zygoptera, or damselflies; and an intermediate suborder called the Anisozygoptera. The term dragonfly is also used for the order as a whole. All the commonly encountered species belong to either Anisoptera or Zygoptera. There are three general stages in the life history: the egg, the larva (sometimes incorrectly called a nymph), and the adult. Eggs are laid into emergent or submerged vegetation or in or on waterside vegetation, or they are dropped on to the water surface, where they sink to the bottom.

Common enemies of the larvae include other predacious aquatic invertebrates, wading birds, and fish. Small larvae also are at risk of attack from larger larvae of their own or other species. Birds prey on the adults and some are taken by other odonates. Birds, reptiles, frogs, and fish prey heavily on newly emerged and ovipositing adults. Odonata are wholly beneficial insects, which (in large numbers) can ameliorate outbreaks of aquatic pest species. *See* ENTOMOLOGY, ECONOMIC. John Trueman

Oedogoniales An order of filamentous fresh-water green algae (Chlorophyceae) with unique morphological features including (1) an elaborate method of cell division that results in the accumulation of apical caps, (2) zoospores and antherozoids with a subapical crown of flagella, and (3) a highly specialized type of oogamy. There is a single family, Oedogoniaceae, comprising three genera. *See* CHLOROPHYCEAE.

Oedogonium has the largest number of species (several hundred) and is the most common of the three genera. Its unbranched filaments are initially attached by a holdfast cell to submerged vegetation, stones, or wood, usually in permanent ponds or pools, but at maturity they may form free-floating masses. The cells are cylindrical, each containing a reticulate chloroplast with numerous pyrenoids.

Vegetative multiplication by fragmentation is common. In asexual reproduction, zoospores with a subapical crown of up to 120 flagella are formed singly within a cell. In sexual reproduction, there is a highly specialized interplay between female and male elements. The egg is a metamorphosed protoplast of an enlarged spherical cell (oogonium). Antherozoids, with a subapical crown of about 30 flagella, are produced in pairs or tetrads in very small discoid antheridia. The two types of sex organs may occur on the same filament (homothallic species), or on different filaments (heterothallic species), but in either case certain species have an indirect development of antherozoids. These species, termed nannandrous in distinction to those with direct development (macrandrous), form short cylindrical cells which initially appear like antheridia, but in which the protoplast metamorphoses into a single swarmer bearing a subapical crown of flagella. Paul C. Silva; Richard L. Moe

Ohmmeter A portable instrument for measuring relatively low values of electrical resistance. The range of resistance measured is typically from 0.1 microhm to 1999 ohms (Ω).

The ohmmeter solves quickly and easily a variety of measurement problems, including measuring the resistance of cladding and tracks on printed circuit boards, electrical connectors, and switch and relay contacts, as well as determining the quality of ground-conductor continuity and bonding, cables, bus-bar joints, and welded connector tags. *See* RESISTANCE MEASUREMENT.

A. Douglas Skinner

Ohm's law The statement that the current I flowing in an electrical circuit is often, to a very good approximation, proportional to the voltage V of its source; that is, $V = IZ$. If the current is steady direct current (DC), Z equals the total resistance of the circuit, R.

Ohm's law, although closely obeyed in most metallic and some other conductors, is an approximation, and does not have the same physical status as, say, Maxwell's equations. *See* MAXWELL'S EQUATIONS.

If Z or R is independent of I to a sufficient approximation, the circuit and the elements in it are said to be linear and network theory can be applied. If V and I are time-dependent, alternating-current circuit theory applies to their frequency components. The current is derived from the voltage by considering not only the resistances but also the vector impedances composed of self- and mutual inductances and capacitances together with their losses. These latter components combine into the total impedance Z, and are often also linear to a sufficient approximation. *See* ALTERNATING CURRENT; CAPACITANCE; ELECTRICAL IMPEDANCE; INDUCTANCE; NETWORK THEORY.

This 'law' was first described by Georg Simon Ohm in 1827 as a result of his experiments with metallic conductors. *See* ELECTRICAL RESISTANCE; ELECTRICAL RESISTIVITY; RESISTANCE MEASUREMENT; SEMICONDUCTOR; SKIN EFFECT (ELECTRICITY); THERMAL CONDUCTION IN SOLIDS.

Bryan P. Kibble

Oichnus A morphologically distinct type of trace fossil left by boring organisms in hard biogenic substrates. *Oichnus* Bromley 1981 is a formal taxonomic name with the rank of ichnogenus (a trace fossil genus). *Oichnus* is a Latin-derived term that literally means a trace (ichnus) shaped like the letter "O"; it denotes the small circular to subcircular holes (borings) found primarily in fossilized skeletal remains of invertebrate animals, such as mollusk shells, ostracode valves, or echinoid tests.

Oichnus includes several ichnospecies (trace fossil species). Seven have been named formally so far, including the type ichnospecies *O. simplex* Bromley 1981, as well as six other forms: *O. paraboloides* Bromley 1981, *O. ovalis* Bromley 1991, *O. asperus* Nielsen and Nielsen 1991, *O. coronatus* Nielsen and Nielsen 1991, *O. gradates* Nielsen and Nielsen 1991, and *O. excavatus* Donovan and Jagt 2002. *See* TRACE FOSSILS.

Most widespread and well-studied fossil types of *Oichnus* borings, such as *O. simplex* and *O. paraboloides*, are attributed typically (based on extrapolation from modern ecosystems) to predatory and/or ectoparasitic organisms that drilled holes in external skeletons of live victims. Because of its high informative value and frequent occurrence in the fossil record, *Oichnus* provides arguably some of the best direct paleontological evidence for ecological interactions between ancient marine predators (or parasites) and their prey (or hosts). *Oichnus* borings occur continuously through the entire fossil record of marine macroorganisms, since the dawn of metazoan animals in the Late Proterozoic Era (approximately 600 million years ago) until present day. *See* FOSSIL; PALEOECOLOGY; PALEONTOLOGY; PREDATOR-PREY INTERACTIONS.

Michal Kowalewski

Oil analysis Analysis of petroleum, or crude oil, to determine its value in modern refinery operations. In addition, procedures have been developed for analysis of lubricating oil.

Typical products derived from crude oil

Product	Carbon atom range	Boiling point range, °F (°C)
Gas	C_1 and C_2	
Liquefied petroleum gas	C_3 and C_4	
Gasolines	C_4 to C_{10}	59–370 (15–190)
Kerosines	C_9 to C_{15}	300–540 (150–280)
Middle distillates	C_{12} to C_{20}	390–640 (200–340)
Gas oils	C_{20} to C_{45}	640–1040 (340–560)
Bottoms	Unvaporized residua	

For refinery operations, oil analysis, or assay, must provide the refinery planner with the data needed to predict yields, qualities, and operating costs for a wide variety of refinery operating conditions and product demands. In a refinery, crude oil is distilled and separated into products according to the boiling points of the crude oil components (see table). *See* PETROLEUM; PETROLEUM PRODUCTS.

A crude assay follows much the same procedure. The oil is distilled and separated into up to 40 narrow-boiling-range cuts. Each cut is then subjected to a variety of tests sufficient to characterize it for the products the cut could be included in. The refinery planners may then calculate yield and qualities of any product by blending the yields and qualities of the cuts that are included in the product.

Petroleum consists primarily of compounds of carbon and hydrogen containing from 1 to about 60 carbon atoms. Carbon atoms in natural petroleum occur in straight and branched chains (paraffins), in single or multiple saturated rings (cycloparaffins or naphthenes), and in cyclic structures of the aromatic type such as benzene, naphthalene, and phenanthrene. Cyclic structures may have attached to them side chains of paraffinic carbons. In lubricating oil it is usual to have naphthene rings built onto the aromatic rings and side chains attached. In products produced by cracking in the refinery, olefins or compounds with carbon-carbon double bonds not in aromatic rings are also found. The high-boiling fractions of petroleum contain increasing amounts of oxygen, nitrogen, and sulfur compounds, as well as traces of organic compounds of metals such as vanadium, nickel, and iron.

Three types of tests are used to characterize a crude oil and its narrow-boiling-range cuts: (1) tests for physical properties such as specific gravity, refractive index, freeze temperature, vapor pressure, octane number, and viscosity; (2) tests for specific chemical species such as sulfur, nitrogen, metals, and total paraffins, naphthenes, and aromatics; (3) tests for determining actual chemical composition.

Jack Rees

Analysis of used lubricating oils is part of the maintenance program for many types of engines and industrial equipment. To analyze a lubricating oil, a representative sample of the lubricating oil flowing in the system at normal operating temperature is collected in a clear container and labeled with pertinent data. The sample is then delivered to the testing laboratory as soon as possible. Tests applied to used engine oils include viscosity, fuel dilution, water, insolubles, and spectrochemical or spectrographic analysis. Wear-metal concentrations are determined in parts per million (ppm); these metals may include silver, aluminum, chromium, copper, iron, nickel, lead, and tin, in addition to silicon. An increase in any of these concentrations may indicate abnormal engine wear. *See* ENGINE; INTERNAL COMBUSTION ENGINE; SPECTROSCOPY.

Donald L. Anglin

Oil and gas, offshore For many years, petroleum companies stopped at the water's edge or sought and developed oil and gas accumulations only in inland waters or shallow seas bordering onshore producing areas. Exploration deeper under the sea, and production from the continental shelves beyond territorial limits, did not begin in earnest until the world's increasing demand for petroleum energy sources, coupled with a lessening return from land drilling, provided the incentives for the huge investments needed for drilling in the open sea. *See* CONTINENTAL MARGIN; OIL AND GAS WELL DRILLING; PETROLEUM RESERVES.

Today offshore oil exploration and production is a worldwide industry. By the early 1990s, offshore sources accounted for 30% of worldwide crude oil production and 14% of worldwide natural gas. Until now, most offshore production came from reservoirs located under the continental shelf, in water depths up to 600 ft (180 m). Spurred by technology and a need to find additional secure sources of energy, exploration and production are now moving even farther from shore and into the deeper waters of the continental slopes. Exploration wells have been drilled in water deeper than 9000 ft (2750 m), and hydrocarbons are being produced from offshore fields in waters deeper than 5000 ft (1500 m).

There is a sound geologic basis for the petroleum industry turning to the continental shelves and slopes. Favorable sediments and structures exist beneath the present seas of the world in geologic settings that have proven highly productive onshore. In fact, the subsea geologic similarity, or in some cases superiority, to geologic conditions on land has been a vital factor in the expansion of the world's investment in offshore exploration and production. *See* MARINE GEOLOGY.

Significant exploration is taking place in water depths up to 10,000 ft (3050 m) in the Gulf of Mexico, offshore Brazil, offshore West Africa, and in the northern Atlantic off the coasts of Norway, Ireland, and the United Kingdom. Oil and natural gas fields in waters as deep as 5500 ft (1675 m) are already on-stream in the Gulf of Mexico and offshore Brazil. In the North Atlantic west of the Shetland Islands, where weather conditions are particularly severe, production has started from reservoirs in water depths up to 1500 ft (450 m). *See* PETROLEUM GEOLOGY.

The underwater search has been made possible only by vast improvements in offshore technology. Drillers first took to sea with land rigs mounted on barges towed to location and anchored, or with fixed platforms accompanied by a tender ship. As the search for oil and natural gas advanced worldwide and farther away from shore, types of exploration rigs evolved which could move easily between locations and operate in a wide range of water depths.

The move into the open and often hostile sea has required not only the development of drilling vessels but also a host of auxiliary equipment and techniques. An entire industrial complex has developed to serve the offshore industry, including construction of fixed platform structures from which the majority of the world's offshore oil and gas production is presently drilled and produced.

The United States Gulf coast, where a large percentage of the world's offshore drilling has taken place, is regularly hit by hurricanes that damage structures in their path, and a number of platforms were lost early on due to high wind and waves. However, improved understanding of the environment, as well as more advanced materials and structural analysis techniques, has significantly reduced failures due to environmental forces. The major causes of accident are now due to human error, process hazards, and transportation to and from offshore facilities by helicopter. *See* HURRICANE.

Improved reservoir management and further discoveries have increased production from onshore fields, where costs are less than for offshore. However, the world's increasing demand for petroleum energy continues to force the search for new reserves into even deeper waters and more remote corners of the world. G. R. Schoonmaker; J. A. Turley; Cyril Arney

Oil and gas field exploitation In the petroleum industry, a field is an area underlain without substantial interruption by one or more reservoirs of commercially valuable oil or gas, or both. A single reservoir (or group of reservoirs which cannot be separately produced) is a pool. Several pools separated from one another by barren, impermeable rock may be superimposed one above another within the same field. Pools have variable areal extent. Any sufficiently deep well located within the field should produce from one or more pools. However, each well cannot produce from every pool, because different pools have different areal limits.

Development of a field includes the location, drilling, completion, and equipment of wells necessary to produce the commercially recoverable oil and gas in the field.

General considerations. Oil and gas production necessarily are intimately related, since approximately one-third of the gross gas production in the United States is produced from wells that are classified as oil wells. However, the naturally occurring hydrocarbons of petroleum are not only liquid and gaseous but may even be found in a solid state, such as asphaltite and some asphalts.

Where gas is produced without oil, the production problems are simplified because the product flows naturally throughout the life of the well and does not have to be lifted to the surface. However, there are sometimes problems of water accumulations in gas wells, and it is necessary to pump the water from the wells to maintain maximum, or economical, gas production. The line of demarcation between oil wells and gas wells is not definitely established. Most gas wells produce quantities of condensable vapors, such as propane and butane, that may be liquefied and marketed for fuel, and the more stable liquids produced with gas can be utilized as natural gasoline.

Production methods in producing wells. The common methods of producing oil wells are (1) natural flow; (2) pumping with sucker rods; (3) gas lift; (4) hydraulic subsurface pumps; (5) electrically driven centrifugal well pumps; and (6) swabbing.

However, most wells are not self-flowing and various lifting methods must be employed. Approximately 90% of the wells made to produce by some artificial lift method in the United States are equipped with sucker-rod–type pumps. In these the pump is installed at the lower end of the tubing string and is actuated by a string of sucker rods extending from the surface to the subsurface pump. The two common variations are mechanical and hydraulic long-stroke pumping. Other lifting mechanisms are the gas lift, hydraulic subsurface pumps, swabs, bailers, jet pumps, and sonic pumps. *See* OIL AND GAS WELL COMPLETION.

Production instruments. The commoner and more important instruments required in petroleum production operations are the following:

1. Gas meters, which are generally of the orifice type, are designed to record the differential pressure across the orifice, and the static pressure.

2. Recording subsurface pressure gages small enough to run down 2-in. (3-cm) ID (inside diameter) tubing are used extensively for measuring pressure gradients down the tubing of flowing wells, recording pressure buildup when the well is closed in, and measuring equilibrium bottom-hole pressures.

3. Subsurface samplers designed to sample well fluids at various levels in the tubing are used to determine physical properties.

4. Oil meters of various types are utilized to meter crude oil flowing to or from storage.

5. Dynamometers are used to measure polished-rod loads.

6. Liquid-level gages and controllers are used. They are similar to those used in other industries, but with special designs for closed lease tanks. Roy L. Chenault

Oil and gas storage Storage, usually in great quantities, of crude oil and natural gas after production from natural reservoirs. Large amounts of refined products are stored as well. Storage is necessary to meet seasonal and other fluctuations in demand; for efficient operation of producing equipment, pipelines, tankers, and refineries; and for emergency use.

Crude oil and refined products. Oil from producing wells is first collected in welded-steel, bolted-steel, or wooden tanks of 100 bbl (16 m^3) or greater capacity. These tanks, upright cylinders with low-pitched conical roofs, provide temporary storage while the oil is awaiting shipment. Several tanks grouped together are a tank battery. Assemblages of large steel tanks, known as tank farms, are used for more permanent storage at pipeline pump stations, points where tankers load and unload, and refineries.

For offshore producing fields a number of unique storage systems have been designed. In several instances old tankers have been adapted for storage, and barges have been constructed especially for offshore storage use.

To minimize vaporization losses, lease tanks are sometimes equipped to hold several ounces pressure. At large-capacity storage sites, special tanks are generally used. Tanks with lifter or floating roofs are used to store crude oil, motor gasoline, and less volatile natural gasoline. Motor and natural gasolines are also stored in spheroid containers. Spherical containers are used for more volatile liquids, such as butane. Horizontal cylindrical containers are used for propane and butane storage. Refrigerated insulated tank systems enabling propane to be stored at a lower pressure are also in use.

Large quantities of volatile liquid-petroleum products, including propane and butane, are stored in underground caverns dissolved in salt formations and in mined caverns, gas reservoirs, and water sands. Refrigerated propane is also being stored in excavations in frozen earth and in underground concrete tanks.

Natural gas. Natural gas is stored in low-pressure surface holders, buried high-pressure pipe batteries and bottles, depleted or partially depleted oil and gas reservoirs, water sands, and several types of containers at extremely low temperature (−258°F or −161°C) after liquefaction. Low-pressure holders, which store relatively small volumes of gas, basically use either a water or a dry seal, and variations of each type exist.

In the United States gas pipeline and utility companies store large quantities of natural gas in underground reservoirs. In most cases these reservoirs are located near market areas. Underground storage permits greatly increased pipeline utilization, resulting in lower transportation costs and reduced gas cost to the consumer. Underground storage is the only economical method of storing large enough quantities of gas to meet the seasonal fluctuations in pipeline loads.

In operating storage reservoirs only a portion of the stored gas, called working gas, is normally withdrawn. The remaining gas, called cushion gas, stays in the reservoir to provide the necessary pressure to produce the storage wells at desired rates. In aquifer storages some water returns to help maintain the reservoir pressure. In aquifer storages the original hydrostatic pressure must be exceeded in order to push the water back.

Storage of liquefied natural gas throughout the world is in connection with shipment of liquefied natural gas by tanker, and is located at the loading and unloading ends of the tanker runs as well as at peak sharing facilities operated by gas pipeline and local utility companies. Storage is in insulated metal tanks, buried concrete tanks, or frozen earth excavations. In two projects using frozen earth excavations, excessive boil-off of the liquefied gas led to replacement with insulated metal tanks. *See* LIQUEFIED NATURAL GAS (LNG); OIL AND GAS FIELD EXPLOITATION; PETROLEUM; PIPELINE.

Peter G. Burnett

Oil and gas well completion

The operations that prepare a well bore for producing oil or gas from the reservoir. The goal of these operations is to optimize the flow of the reservoir fluids into the well bore, up through the producing string, and into the surface collection system. *See* OIL AND GAS FIELD EXPLOITATION; OIL AND GAS WELL DRILLING.

The well bore is lined (cased) with steel pipe, and the annulus between well bore and casing is filled with cement. Properly designed and cemented casing prevents collapse of the well bore and protects fresh-water aquifers above the oil and gas reservoirs from becoming contaminated with oil and gas and the oil reservoir brine. Similarly, the oil and gas reservoir is prevented from becoming invaded by extraneous water from aquifers that were penetrated above or below the productive reservoir. *See* AQUIFER.

The nature of the reservoir, evaluated from a core analysis, cuttings, or logs, or from experience with like productive formations, determines the type of completion to be used. In a barefoot completion, the casing is set just above the producing formation, and the latter is drilled out and produced with no pipe set across it. Such a completion can be used for hard rock formations which are not friable and will not slough, and when there are no opportunities for producing from another, lower reservoir. Set-through and perforated completions are also employed for relatively well-consolidated formations from which the potential for sand production is small. However, the perforated completion is used when a long producing interval must be prevented from collapse, when multiple intervals are to be completed in the one borehole, or when intervening water sands within the oil-producing interval are to be shut off and the oil-saturated intervals selectively perforated. *See* WELL LOGGING.

A string of steel tubing is lowered into the casing string and serves as the conduit for the produced fluids. The tubing may be hung from the well-head or supported by a packer set above the producing zone. The packer is used when it is desirable to isolate the casing string from the produced fluids because of the latter's pressure, temperature, or corrosivity, or when such isolation may improve production characteristics.

The tops of wells from which fluids flow as a result of the indigenous reservoir energy are equipped with a manifold known as the Christmas tree. However, only some reservoirs have sufficient pressure and sufficient gas in solution (which is released at the lower pressure existing in the well bore and therefore lowers the effective density of the fluid in the tubing) to permit natural flow to the surface. The reservoir fluids from other reservoirs and, after pressure depletion, even from those which initially flowed must be brought to the surface by one of several methods of artificial lift.

Excessive water production increases the cost of oil production since energy must be expended in lifting the water to the surface. Water production may also jeopardize the production of oil and gas by saturating the oil-productive interval with water. Such damage is more likely to occur in low-pressure formations or formations which contain water-sensitive clays that swell in an excess of water. Water exclusion may be effected by the application of cements of various types. If it is determined that water is entering from the lower portion of a producing sand in a relatively shallow, low-pressure well, a cement plug may be placed in the bottom of the hole so that it will cover the oil-water interface of the reservoir. This technique is called laying in a plug and may be accomplished by placing the cement with a dump-bottom bailer on a wire line or by pumping cement down the drill pipe or tubing. For deeper, higher-pressure, or more troublesome wells, a squeeze method is used. Squeeze cementing is the process of applying hydraulic pressure to force a cement into an exposed formation or through openings in the casing or liner. It is also used for repairing casing leaks.

Production may be impaired from a well bore as a result of drilling-mud invasion or of accumulation of clays and fine silts carried by the producing fluids to the borehole, or the lithology of the formation itself may have a naturally low permeability to reservoir fluids. Since the permeability to fluids of the formation within the first few feet of the well bore has an exponential effect on limiting the influx of fluid, the productivity of a well can frequently be increased manyfold by increasing the permeability of this element of the reservoir or removing the skin just at the face of the producing interval. This is accomplished by acidization and fracturing, and in some instances by the use of surfactants, solvents, and explosives. Todd M. Doscher; R. E. Wyman

Oil and gas well drilling The drilling of holes for exploration and extraction of crude oil and natural gas. Deep holes and high pressures are characteristics of petroleum drilling not commonly associated with other types of drilling. In general, it becomes more difficult to control the direction of the drilled hole as the depth increases, and additionally, the cost per foot of hole drilled increases rapidly with the depth of the hole. Drilling-fluid pressure must be sufficiently high to prevent blowouts but not high enough to cause fracturing of the borehole. Formation-fluid pressures are commonly controlled by the use of a high-density clay-water slurry, called drilling mud. The chemicals used in drilling mud can be expensive, but the primary disadvantage in the use of drilling muds is the relatively low drilling rate which normally accompanies high bottom-hole pressure. Drilling rates can often be increased by using water to circulate the cuttings from the hole; when feasible, the use of gas as a drilling fluid can lead to drilling rates as much as 10 times those attained with mud. Drilling research has the objectives of improving the utilization of current drilling technology and the development of improved drilling techniques and tools.

The hole direction must be controlled within permissible limits in order to reach a desired target at depths as great as 25,000 ft (7600 m). Inclined layers of rocks with different hardnesses tend to cause the direction of drilling to deviate; consequently, deep holes are rarely truly straight and vertical. The drilling rate generally increases as additional drill-collar weight is applied to the bit by adjusting the pipe tension at the surface. However, crooked-hole tendency also increases with higher weight-on-bit. A so-called packed-hole technique has been used to reduce the tendency to hole deviation. One version of this technique makes use of square drill collars that nearly fill the hole on the diagonals but permit fluid and cuttings to circulate around the sides. This procedure reduces the rate at which the hole direction can change.

In mountainous terrain, it is difficult to construct well locations over each subsurface drilling target, and from offshore drilling platforms it is necessary to drill many wells from a single surface location. For these situations, technology has been developed that permits wells to be drilled directionally from the single surface location to the desired subsurface point.

Advances in technology and the need to accomplish special objectives have led to drilling horizontal wells in deep oil or gas reservoirs. The angle of the well is successively built up in order to reach the ultimate horizontal course of the well. The program for drilling the well is developed before hand, based on the technology for actually increasing the angle and then continuing the drilling at the desired angle.

Novel drilling methods include studies of rock failure by mechanical, thermal, hydraulic, fusion and vaporization, and chemical means. Jet piercing is widely used for drilling very hard, spallable rocks, such as taconite. Other methods include the use of electric arc, laser, plasma, spark, and ultrasonic drills. *See* OIL AND GAS WELL COMPLETION; PETROLEUM GEOLOGY; ROCK MECHANICS; TURBODRILL; WELL LOGGING. Todd M. Doscher

Oil burner A device for converting fuel oil from a liquid state into a combustible mixture. A number of different types of oil burners are in use for domestic heating. These include sleeve burners, natural-draft pot burners, forced-draft pot burners, rotary wall flame burners, and air-atomizing and pressure-atomizing gun burners. The most common and modern type that handles 80% of the burners used to heat United States homes is the pressure-atomizing-gun-type burner shown in the illustration.

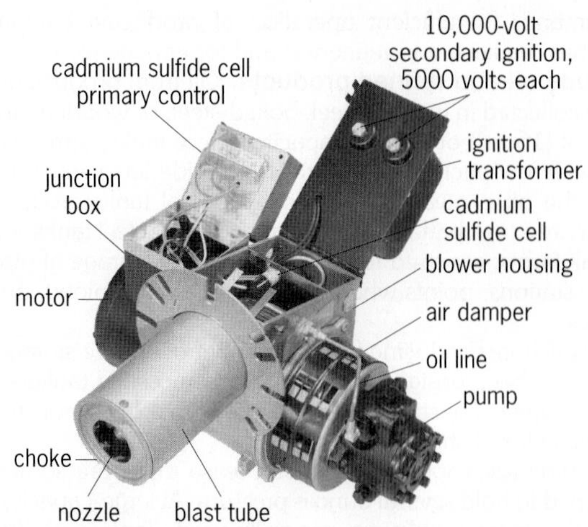

An oil burner of the pressure-atomizing type. (*Automatic Burner Corp.*)

The sleeve burner, commonly known as a range burner because of its use in kitchen ranges, is the simplest form of vaporizing burner. The natural-draft pot burner relies on the draft developed by the chimney to support combustion. A modification of this burner is the forced-draft pot burner which supplies its own air for combustion and does not rely totally on the chimney. The rotary wall flame burners have mechanically assisted vaporization. The gun-type burner uses a nozzle to atomize the fuel so that it becomes a vapor, and burns easily when mixed with air.

The oil burner is used for a wide assortment of heating, air conditioning, and processing applications. Oil burners heat commercial buildings such as hospitals, schools, and factories. Air conditioners using the absorption refrigeration system have been developed and fired with oil burners. Oil burners are used to produce CO_2 in greenhouses to accelerate plant growth. They also produce hot water for many commercial and industrial applications. *See* AIR COOLING; COMFORT HEATING; HOT-WATER HEATING SYSTEM. Robert A. Kaplan

Oil furnace A combustion chamber in which oil is the heat-producing fuel. Fuel oils, having from 18,000 to 20,000 Btu/lb (42–47 megajoules/kg), which is equivalent to 140,000 to 155,000 Btu/gal (39–43 megajoules/liter), are supplied commercially. The lower flash-point grades are used primarily in domestic and other furnaces without preheating. Grades having higher flash points are fired in burners equipped with preheaters. *See* FUEL OIL.

Domestic oil furnaces with automatic thermostat control usually operate intermittently, being either off or operating at maximum capacity. *See* OIL BURNER. Frank H. Rockett

Oil sand A loose to consolidated sandstone or a porous carbonate rock, impregnated with a heavy asphaltic crude oil, too viscous to be produced by conventional methods; also known as tar sand or bituminous sand.

Oil sands are distributed throughout the world but the largest proven accumulation occurs in Alberta, Canada (the Athabasca deposit). A large accumulation appears to be present in the Orinoco Basin in Venezuela, and far smaller deposits occur in Russia, the United States, Madagascar, Albania, Trinidad, and Romania. G. Ronald Gray

Oil shale A sedimentary rock containing solid, combustible organic matter in a mineral matrix. The organic matter, often called kerogen, is largely insoluble in petroleum solvents, but decomposes to yield oil when heated. Although "oil shale" is used as a lithologic term, it is actually an economic term referring to the rock's ability to yield oil; oil shale appears to be the cheapest source after natural petroleum for large amounts of liquid fuels. No real minimum oil yield or content of organic matter can be established to distinguish oil shale from sedimentary rocks. Additional names given to oil shales include black shale, bituminous shale, carbonaceous shale, coaly shale, cannel shale, cannel coal, lignitic shale, torbanite, tasmanite, gas shale, organic shale, kerosine shale, coorongite, maharahu, kukersite, kerogen shale, algal shale, and "the rock that burns." *See* KEROGEN.

The world's oil shale deposits represent a tremendous store of fossil energy. It has been estimated that the organic matter in sedimentary rocks contains 1.2×10^{16} tons (1.1×10^{16} metric tons) of organic carbon, nearly 1000 times that found in coals. Although part of that organic carbon has matured to produce oil and gas, most of it is still oil shale. Unfortunately, most of this tremendous resource is not well known. Oil shales occur on every continent in sediments ranging in age from Cambrian to Tertiary. Estimates for the total oil resource in shales of all grades reached 1.75×10^{15} barrels. Just 1% of that total shale oil represents more oil than the world is expected to produce as natural petroleum (2×10^{12} bbl). Oil shale represents a tremendous supply of liquid fuels.

Although the oil potential of the world's oil shales is great, commercial production of this oil has been considered uneconomic. Oil shales are lean ores, producing only limited amounts of oil which historically has been low in price. Mining and heating 1 ton of relatively rich oil shale yielding 25 gal/ton produces only 0.6 bbl of oil.

Shale oil is produced from the organic matter in oil shale when the rock is heated in the absence of oxygen (destructive distillation). This heating process is called retorting, and the equipment that is used to do the heating is known as a retort. The rate at which the oil is produced depends upon the temperature at which the shale is retorted. Most references report retorting temperatures as being about 500°C (930°F). John Ward Smith; Howard B. Jensen

Oilfield waters Waters of varying mineral content which are found associated with petroleum and natural gas or have been encountered in the search for oil and gas. They are also called oil field brines, or brines. They include a variety of underground waters, usually deeply buried, and have a relatively high content of dissolved mineral matter. These waters may be (1) present in the pore space of the reservoir rock with the oil or gas, (2) separated by gravity from the oil or gas and thus lying below it, (3) at the edge of the oil or gas accumulation, or (4) in rock formations which are barren of oil and gas. Brines are commonly defined as water containing high concentrations of dissolved salts. Potable or fresh waters usually are not considered oil field waters but may be encountered, generally at shallow depths, in areas where oil and gas are produced.

Probably the most important geological use of oil field water analyses is their application to the quantitative interpretation of electrical and neutron well logs, particularly micrologs. *See* PETROLEUM GEOLOGY; WELL LOGGING. Prestar McGain

Okra A warm-season annual, *Hibiscus esculentus*, of Ethiopian origin. Okra, also called gumbo, is grown for its immature pods (see illustration), which are generally used for preparing soups but are also eaten as a freshly cooked vegetable. It is a member of the order Malvales and is related to cotton. Georgia, Florida, and Louisiana are important producing states. *See* MALVALES. H. John Carew

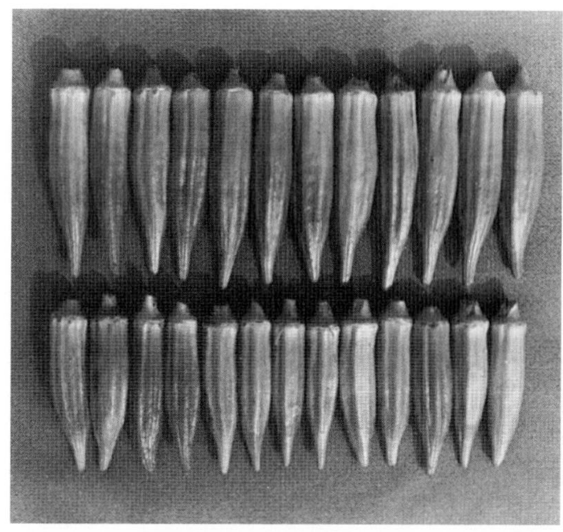

Okra pods. (*Asgrow Seed Co., subsidiary of The Upjohn Co.*)

Olbers' paradox The riddle of why is the sky is dark at night. This celebrated riddle originated in the sixteenth century. In 1823, Wilhelm Olbers presented it in the simplest terms: In an infinite universe, populated everywhere with stars, a line of sight in any direction when extended out into space must ultimately intercept the surface of a star. Hence stars should cover the entire sky. And if all stars are sunlike, the sky at every point should blaze as brightly as the disk of the Sun. Olbers has been credited incorrectly with the discovery of the riddle; he did, however, express it in this lucid form, and showed that the riddle still holds even when stars are irregularly distributed in clusters.

Modern calculations confirm that most lines of sight do not intercept stars but extend back to the beginning of the universe. Light travels at approximately 300,000 km/s (186,000 mi/s), and a static universe $10–20 \times 10^9$ years old is not old enough for starlight to reach the Earth from regions sufficiently distant for visible stars to cover the sky. If the sky at night is dark in a static universe, then obviously in an expanding universe of the same age the night sky is even darker. In the big bang universe, $10–20 \times 10^9$ years old, we cannot see sufficient stars to cover the sky. Instead, on looking out in space, we look far back in time to the beginning of the universe and see in all directions the big bang covering the sky. The expansion of the universe has reduced the incandescent glare of the big bang and redshifted its radiation into the invisible infrared. *See* BIG BANG THEORY; COSMIC BACKGROUND RADIATION; UNIVERSE. Edward Harrison

Olfaction One of the chemical senses, specifically the sense of smell. Olfaction registers chemical information in organisms ranging from insects to humans, including marine organisms. For terrestrial animals, its stimuli comprise airborne molecules. The typical stimulus is an organic chemical with molecular weight below 300 daltons. A few inorganic chemicals can also stimulate olfaction, notably hydrogen sulfide, ozone, ammonia, and the halogens.

The anatomy of olfactory structures and the neurophysiology of olfaction differ significantly among different animal groups. For examples, insect olfactory receptors exist within sensory hairs on the antennae. The olfactory organ of fishes resides typically in tubular chambers on either side of the mouth. In terrestrial vertebrates, the olfactory receptors reside within a sac or cavity more or less similar to the human nasal cavity. The olfactory mucosa patch in the cavity characteristically contains millions of receptor cells, though in some olfactory-dominated mammals, such as the dog and rabbit, it contains tens of millions. The location of

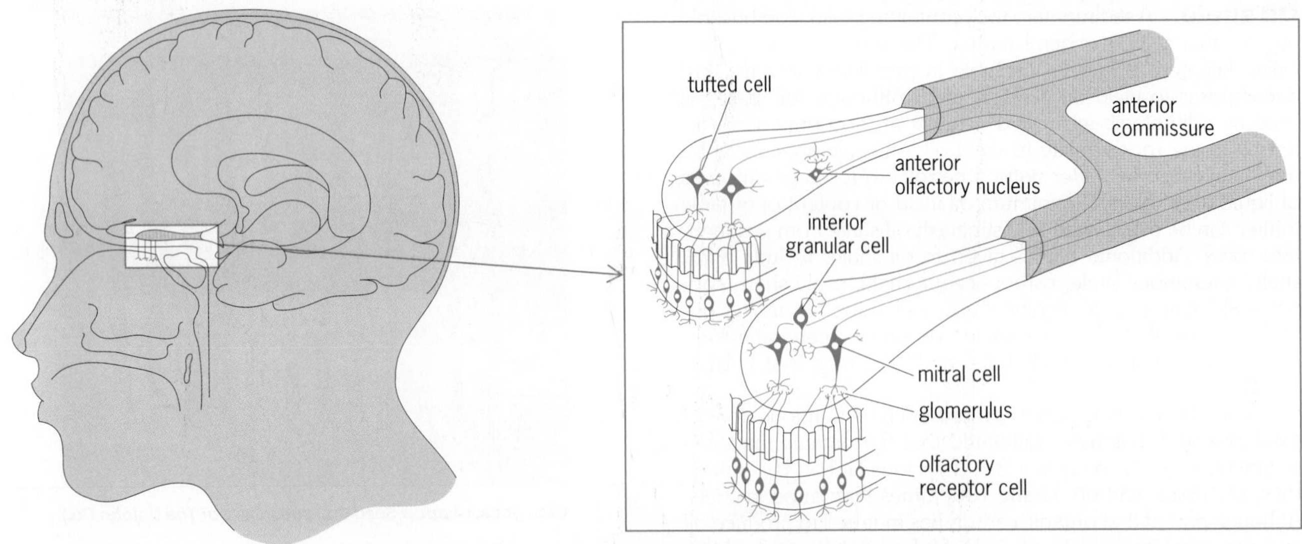

Location of the olfactory bulbs at the interior surface of the human brain and their connections via the anterior commissure. (*After D. Ottoson, Physiology of the Nervous System, Oxford University Press, 1983*)

the olfactory mucosa relative to air currents in the cavity plays some role in the ongoing olfactory vigilance of the organism. In the human the mucosa sits out of the main airstream. During quiet breathing eddy currents may carry just enough stimulus to evoke a sensation, whereupon sniffing will occur. Sniffing amplifies the amount of stimulus reaching the receptors by as much as tenfold.

Reception of the chemical stimulus and transduction into a neural signal apparently occur on the olfactory receptor cilia. The ciliary membrane contains receptor protein molecules that interact with stimulating molecules through reversible binding. Vertebrate receptor cells show broad tuning, that is, they respond to many odorants.

Adjacent points in the mucosa generally project to adjacent points in the olfactory bulb of the brain (see illustration). The synapses between the incoming olfactory nerve fibers and the second-order cells, mitral cells, occur in basketlike structures called glomeruli. On average, a glomerulus receives about 1000 receptor cell fibers for each mitral cell. The location of cells within the bulb seems to play a role in encoding odor quality: each odorant stimulates a more or less unique spatial array.

The central neural pathways of the olfactory system have a complexity unmatched among the sensory systems. One pathway carries information to the pyriform cortex (paleocortex of the temporal lobe), to a sensory relay in the thalamus (dorsomedial nucleus), and to the frontal cortex (orbitofrontal region). This pathway seems rather strictly sensory. Another pathway carries information to the pyriform cortex, the hypothalamus, and other structures of the limbic system. The latter have much to do with the control of emotions, feeding, and sex. The strong affective and motivational consequences of olfactory stimulation seem compatible with projections to the limbic system and with the role of olfaction in certain types of physiological regulation. In many vertebrate species, reception of pheromones occurs via an important accessory olfactory organ, known as the vomeronasal organ, which characteristically resides in the hard palate of the mouth or floor of the nasal cavity. *See* PHEROMONE.

Human olfactory sensitivity varies from odorant to odorant over several orders of magnitude. A common range of thresholds for materials used in fragrances and flavors is 1 to 100 parts per 10^9 parts of air. Thresholds gathered from various groups of human subjects permit certain generalities about how the state of the organism affects olfaction. For instance, persons aged 70 and above are about tenfold less sensitive than young adults. Males and females have about equal sensitivity, except perhaps in old age, where females are more sensitive. Persons with certain medical disorders, such as multiple sclerosis, Parkinson's disease, paranasal sinus disease, Kallmann's syndrome, and olfactory tumors, exhibit decreased sensitivity (hyposmia) or complete absence of sensitivity (anosmia).

Above its threshold, the perceived magnitude of an odor changes by relatively small amounts as concentration increases. A tenfold increment in concentration will cause, on average, about a twofold change in perceived magnitude. The perceived magnitude of an odor is often greatly influenced by olfactory adaptation, a process whereby during continuous short-term exposure to a stimulus its perceived magnitude falls to about one-third of its initial value.

The stimuli for olfaction are commonly complex, that is, they are mixtures. Such products as coffee, wine, cigarettes, and perfumes contain at least hundreds of odor-relevant constituents. Only rarely does the distinctive quality of a natural product, such as a vegetable, arise from only a single constituent. A chemical analysis of most products will not usually allow a simple prediction of odor intensity or quality. One general rule, however, is that the perceived intensity of the mixture falls well below the sum of the intensities of the unmixed components.

General notions about the properties that endow a molecule with its quality have spawned more than two dozen theories of olfaction, including various chemical and vibrational theories. Most modern theories hold that the key to quality lies in the size and shape of molecules, with some influence of chemical functionality. For molecules below about 100 daltons, functional group has obvious importance: for example, thiols smell skunky, esters fruity, amines fishy-uriny, and carboxylic acids rancid. For larger molecules, the size and shape of the molecule seem more important. Shape detection is subtle enough to enable easy discrimination of some optical isomers. Progressive changes in molecular architecture along one or another dimension often lead to large changes in odor quality. No current theory makes testable predictions about such changes. *See* CHEMICAL SENSES; CHEMORECEPTION.

William Cain

Oligocene The third oldest of the seven geological epochs of the Cenozoic Era. It corresponds to an interval of geological time (and rocks deposited during that time) from the close of the Eocene Epoch to the beginning of the Miocene Epoch. The most recent geological time scales assign an age of 34 to 24 million years before present (m.y. B.P.) to the Oligocene Epoch. *See* CENOZOIC; EOCENE; MIOCENE.

An important event that characterizes the Oligocene Epoch was the development of extensive glaciation on the continent of Antarctica. Prior to that time, the world was largely ice-free through much of the Mesozoic and early Tertiary. A significant amount of ice is now known to have existed on the Antarctic continent since at least the beginning of the Oligocene, when the Earth was ushered into its most recent phase of ice-house conditions. This in turn created revolutions in the global climatic and hydrographic systems, with important repercussions for the marine and terrestrial biota. The changes include steepened latitudinal and vertical thermal gradients affecting major fluctuations in global climates, and the shift in the route of global dispersal of marine biota from an ancestral equatorial Tethys seaway, which had become severely restricted by Oligocene time, to the newly initiated circum-Antarctic circulation. *See* GLACIAL EPOCH.

Worldwide, the epoch represents an overall regressive sequence when there was a drawdown of global sea level, with relatively deeper, marine facies in the Early Oligocene and shallower-water to nonmarine facies in the late Oligocene. *See* FACIES (GEOLOGY).

Due to accentuated thermal gradients and seasonality in the Oligocene, marine biotic provinces became more fragmented. Extreme climates, with greater diurnal and seasonal temperature contrasts, are held responsible for reduced diversities in marine plankton. The Oligocene was characterized by transitional faunal features between the Paleogene and the Neogene.

Rhinocerids, tapirs, and wild boarlike hog species with strong incisors and large canines appeared. *Hyaenodon* was an Oligocene carnivore with strong canines and sharp molars, much like those of the modern cat species. The horses that had first appeared in the Eocene continued to increase in size and became three-toed, typified by *Mesohippus*. Elephants made their first appearance near the Eocene-Oligocene boundary and developed a short trunk and two pairs of tusks. An early simian, *Propliopithecus*, made its first appearance in Oligocene, and is considered ancestral to the modern family of gibbons. The general uniformity of mammalian fauna in the Oligocene suggests that the widespread regressions of the sea most likely resulted in land bridges that reconnected some of the Northern Hemispheric landmasses, which may have led to transmigrations of some families of mammals between North America, Asia, and Africa. Birds had achieved some of their modern characteristics, and at least 10 modern genera had already made their appearance by the close of Oligocene time. *See* AVES; GEOLOGIC TIME SCALE; MAMMALIA; PALEOECOLOGY; PERISSODACTYLA. Bilal U. Haq

Oligochaeta A class or subclass (depending on the classification system used) of worms, including the earthworms, within the phylum Annelida. These animals exhibit both external and internal segmentation—that is, furrows in the body wall and transverse partitions (septa) dividing the coelom into chambers. They usually possess chaetae (setae), or bristles, made of chitin, which are not borne on parapodia and are few in number (compared to the Polychaeta). Oligochaetes are hermaphroditic. The gonads are few in number and situated in the anterior part of the body, the male gonads being anterior to the female gonads. The gametes are discharged through the oviducts and sperm ducts. At maturity, a portion of the body wall is swollen into a secretory area called the clitellum. There is no larval stage during the development of oligochaetes. *See* POLYCHAETA.

The oligochaetes are primarily freshwater and burrowing terrestrial animals. About 200 species are marine, mostly occurring in the intertidal zone. They range in size from aquatic species that are 0.02 in. (0.5 mm) long to the giant Australian earthworm (*Megascolides*), which is up to 3 m (10 ft) long. About 25 families and over 3000 species are presently recognized.

Oligochaetes are cylindrical, elongated animals with the anterior mouth usually overhung by a fleshy lobe, the prostomium, and the anus located terminally. The body plan is that of a tube within a tube. The chaetae, usually a pair of ventrolateral bundles, are borne on most segments. Other external features are the pores of the reproductive systems opening on certain segments, the openings of the excretory organs (metanephridia), and in many earthworms dorsal pores which open externally from the coelom. Movement in terrestrial forms is accomplished by alternate contractions of the muscle layers, which produce a peristaltic motion in which the chaetae aid forward progression by gripping the substrate. Other, smaller muscles protract and retract the chaetae and are associated with the intestine and reproductive organs.

The oligochaetes have been used in studies of anatomy, physiology, regeneration, and metabolic gradients. Some aquatic forms are important in studies of stream pollution as indicators of organic contamination. Earthworms are important in turning over the soil and reducing vegetable material into humus. It is likely that fertile soil furnishes a suitable habitat for earthworms, rather than being a result of their activity. *See* ANNELIDA.

Perry C. Holt; Robert Knowlton

Oligoclase A plagioclase feldspar with composition in the range $Ab_{90}An_{10}$ to $Ab_{70}An_{30}$, where Ab represents the composition of albite, $NaAlSi_3O_8$, and An represents the composition of anorthite, $CaAl_2Si_2O_8$. The diagnostic properties are hardness on Mohs scale, 6–6.5; density, 2.65 g/cm^3; and color usually white or colorless, transparent to translucent. The presence of minute, mutually parallel inclusions of hematite (Fe_2O_3) causes a golden play of color in the variety of oligoclase called aventurine or sunstone. Oligoclase is triclinic. The mineral is common in igneous rocks and metamorphic rocks. *See* ALBITE; ANORTHITE; FELDSPAR; HARDNESS SCALE; IGNEOUS ROCKS; METAMORPHIC ROCKS.

Dana T. Griffen

Oligonucleotide A single-stranded short polymer of deoxyribonucleic acid (DNA) or ribonucleic acid (RNA). Although the number of nucleic acid monomers (known as nucleotides) in the polymer varies widely, practical uses for oligonucleotides ranging in size from 4 to over 100 monomers have been reported.

Each nucleotide in the nucleic acid polymer is composed of three parts: a five-carbon sugar, a nitrogenous base, and a phosphate group. The five-carbon sugar (ribose or 2′-deoxyribose) is covalently attached to the nitrogenous base through a glycosidic linkage to create a nucleoside. If the sugar is ribose, the resulting oligonucleotide is known as an RNA oligonucleotide, or oligoribonucleotide; if the sugar is 2′-deoxyribose, the resulting polymer is called either a DNA oligonucleotide or an oligodeoxyribonucleotide. There are two classes of bases: purines (adenine and guanine) and pyrimidines (thymine, cytosine, and uracil). Guanine, cytosine, adenine, and uracil are the four bases found in RNA oligonucleotides; DNA oligonucleotides contain guanine, cytosine, adenine, and thymine. The phosphate molecule covalently links the 5′ hydroxyl group of one nucleoside to the 3′ hydroxyl group of the next nucleoside. This creates a phosphodiester bond between successive sugar molecules, forming the backbone of the nucleic acid strand. *See* DEOXYRIBONUCLEIC ACID (DNA); NUCLEIC ACID; RIBONUCLEIC ACID (RNA).

Oligonucleotides are synthesized from high-molecular-weight DNA or RNA in a manner that defines the order, or sequence, of purines and pyrimidines. This synthesis is most often carried out chemically using automated instruments, but may also be carried out enzymatically using RNA or DNA polymerase. The basic chemical structure of the oligonucleotide can be modified in a number of ways to enhance desired properties. For instance, purines or pyrimidines can be modified by alkylation (addition or substitution of an alkyl group) or deamination (removal of an amino group) to alter their interaction with other nucleic acids.

Oligonucleotides are commonly used to bind to other nucleic acids or proteins through base-specific hydrogen bonds. Generally, they are used to form antiparallel duplexes, in which the sugar-phosphate backbones run in opposite directions (one strand runs in the 5′- to 3′-direction and the other runs 3′ to 5′). Oligonucleotides may also bind, in a sequence-specific manner, to a nucleic acid duplex to form three-stranded structures known as a triplexes. Oligonucleotides used to form triplexes with genomic DNA have been used experimentally to inhibit gene expression and to create site-directed mutations. Perhaps the most common current use of chemically synthesized oligonucleotides is as part of a nucleic acid amplification technique called the polymerase chain reaction (PCR). *See* GENE AMPLIFICATION.

The therapeutic use of DNA oligonucleotides to treat diseases in humans has taken two main forms. The first is used in clinical trials as antiviral or anticancer therapeutic agents. Oligonucleotides act by serving as a specific inhibitor of messenger mRNA translation inside cells. The effect of the oligonucleotide is to reduce the level of proteins (encoded by mRNAs) that allow a virus to propagate or a cancer cell to divide in an unregulated way. A second therapeutic use for DNA oligonucleotides is to activate the immune system. The sequence cytosine-guanine (CpG) is usually methylated in vertebrate, but not bacterial, DNA. DNA oligonucleotides with unfamiliar cytosine methylation patterns provide a signal in humans to initiate cellular processes that activate both innate and acquired immunological responses. CpG immune modulators are currently being studied as agents to treat a number of human diseases. *See* GENE; GENETIC CODE; GENETIC ENGINEERING; IMMUNITY. Daniel Weeks; John Dagle

Oligopygoida An order of irregular echinoids in the superorder Neognathostomata resembling clypeasteroids but lacking the accessory ambulacral pores characteristic of that group. Oligopygoids have well-developed petals, and there are characteristic small demiplates present below the petals. The apical disk is monobasal and the mouth oval and usually deeply sunken. Oligopygoids have a lantern, which closely resembles that of clypeasteroids, and their lantern muscle-attachment structures are a mixture of ambulacral and interambulacral processes.

There are two genera, *Oligopygus* and *Haimea*, containing about 25 species, all from the middle and upper Eocene of the Caribbean and Gulf of Mexico regions. They were probably infaunal deposit feeders like present-day laganiids. *See* ECHINODERMATA; NEOGNATHOSTOMATA. Andrew B. Smith

Oligosaccharide A carbohydrate molecule composed of 3–20 monosaccharides (simple sugars). Generally, free oligosaccharides do not constitute a significant proportion of naturally occurring carbohydrates. Most carbohydrates that occur in nature are in the form of monosaccharides (such as blood sugar, or glucose), disaccharides (such as table sugar, or sucrose, and milk sugar, or lactose), and polysaccharides (such as starch and glycogen, polyglucose molecules, or chitin). *See* GLUCOSE; LACTOSE; MONOSACCHARIDE; POLYSACCHARIDE.

The monosaccharides of multiple sugar units such as oligosaccharides are connected with each other through bonds called glycosidic linkages. They are linked primarily to other sugars and to other molecules through aldehyde or ketone reducing groups.

Most naturally occurring oligosaccharides are linked either to proteins (glycoproteins) or to lipids (glycolipids). Glycoconjugates are present in essentially all life forms and particularly in cell membranes and cell secretions. Many hormones are glycoproteins, and an increasing number of enzymes have been shown to have sugars attached. Antigenic properties of the human red blood cell ABO blood group system are determined by glycolipid oligosaccharides. In fact, all the major protein components of blood serum, with the exception of serum albumin, are glycoproteins. *See* BLOOD GROUPS; CELL MEMBRANES; GLYCOLIPID; GLYCOPROTEIN.

Many changes in the structures of oligosaccharides of glycoconjugates have been detected in cancer cells. Changes or differences in oligosaccharide structures are generally the result of differences in biosynthetic pathways or of degradative pathways. An understanding of glycoconjugates in normal biological systems and in certain disease states is currently of great importance. Don M. Carlson

Olive Olive fruits, which are produced by a small to medium-sized evergreen tree (*Olea europaea*), can be eaten, after processing, as table olives, or can be extracted for oil that is used on salads, for cooking, for body lotions, or for medicinal purposes. The olive tree is a historically ancient cultivated plant, having been domesticated by early civilizations in the eastern Mediterranean regions. Olive culture later spread to all the Mediterranean countries and subsequently to South America, California, South Africa, and Australia.

Olive fruits are harvested for table olives in the autumn, when the fruits change from green to straw or to a slightly red color. Raw fruits contain a bitter glucoside which makes them inedible, but treatment with an alkali such as sodium hydroxide (lye) neutralizes the bitterness. The lye must subsequently be leached out of the fruits with water. In another method most of the bitterness can be removed by leaching with salt water. A lactic acid fermentation process is widely used in the Mediterranean countries to preserve the olives.

For oil production, the fruits are harvested in midwinter when they have become black and have reached their maximum oil content—15 to 25% of the fresh weight, depending on the variety. In producing olive oil, the freshly harvested fruits including pits are ground, after which this material is placed in burlap or cloth bags and placed under high pressure. Oil and water are extracted and transferred to tanks, where the oil rises to the top and is removed. *See* FAT AND OIL (FOOD). Hudson T. Hartmann

Olivine It is generally accepted that the Earth's upper mantle consists mainly of olivine, an orthorhombic silicate with the composition $(Mg_{1.8},Fe_{0.2})SiO_4$, together with some pyroxene and garnet. The natural occurrence of two high-pressure forms (polymorphs) of olivine—orthorhombic wadsleyite, and cubic ringwoodite (with a spinel structure)—was predicted from high-pressure experiments and was later confirmed by meteorite investigations. The names olivine, wadsleyite, and ringwoodite refer only to naturally occurring compositions $[(Mg,Fe)_2SiO_4]$.

Because of their abundance in the Earth's mantle, knowledge of physical and chemical properties of olivine, wadsleyite, and ringwoodite is of great geophysical importance. Until recently, many of these properties had to be inferred from theoretical considerations and from experiments on chemical analogs, which transform at lower pressures. With the development of new experimental apparatus capable of generating very high pressures and temperatures (multianvil press and diamond anvil cell), a growing number of experimental studies are being performed on phases of natural composition.

Experimentally determined thermodynamic phase equilibria data indicate that in the Earth's mantle olivine transforms to wadsleyite, then to ringwoodite, and finally to compositions of magnesiowüstite plus perovskite. Estimated transformation pressures correspond closely to the discontinuities of seismic velocities at, respectively, 246, 322, and 417 mi (410, 520, and 670 km) depth in the mantle. Ljuba Kerschofer

Omphacite The pale to bright green monoclinic pyroxene found in eclogites and related rocks. Omphacites are essentially members of the solid solution series between jadeite $(NaAlSi_2O_6)$

and diopside ($CaMgSi_2O_6$). The density ranges from 3.16 to 3.43 g/cm^3; hardness is 5–6.

Omphacite is stable only at the relatively high pressures of the blueschist and eclogite facies of metamorphism, where it is associated with minerals such as glaucophane and lawsonite or pyropic garnet, respectively. In such environments it also occurs in veins, either on its own or with quartz. *See* GLAUCOPHANE; PYROXENE.

Timothy J. B. Holland

Oncholaimida An order of nematodes comprising the single superfamily Oncholaimoidea. These nematodes are principally marine and brackish-water forms with alleged predaceous and carnivorous feeding habits. The external amphidial aperture is an oval or a widened ellipse. Generally, the stoma is armed with one dorsal tooth and two subventral teeth, and its wall may be further fortified with transverse rows of small denticles. The stoma is divisible into the cheilostome (secondary blastocoel invagination) and the esophastome (primary blastocoel invagination). In some species the stoma of the adult male is collapsed or indistinct. The cephalic sensilla are in two whorls: one is circumoral and composed of six papilliform sensilla; the second combines the ancestral two whorls of six and four into a single whorl of ten setiform sensory organs. In some forms these sensilla are papilliform. The cylindrical-to-conoid esophagus may exhibit a series of muscular bulbs posteriorly. The cuticle is generally smooth, and over the length of the body there are scattered sensory setae or papillae. *See* NEMATA (NEMATODA).

Armand R. Maggenti

Oncofetal antigens Small proteins or carbohydrates (capable of stimulating an immune response) that are present in normal fetal tissues during early development but also are abnormally expressed by adult tumors. Oncofetal antigens are different from tumor-associated antigens, which are found only in tumors and have no role in development. These antigens, primarily glycoprotein in nature and produced by cancerous cells, are the products of one or more genes that normally are expressed only during fetal development and then are repressed in adult life. Production of these proteins in association with cancer as a result of activation of control genes, by unknown mechanisms, has been termed retrogenetic expression. Advances in tumor immunology have enabled the use of oncofetal antigens in both the diagnosis and treatment of cancer. In particular, fetal antigens in adult body fluids can serve as tumor markers for the detection of early oncogenic processes and can aid in monitoring the efficiency of cancer treatment as well as in the development of new treatment modalities.

The two best-known oncofetal antigens are α-fetoprotein and carcinoembryonic antigen (CEA); however, several lesser-known oncofetal antigens, such as pancreatic oncofetal antigen, have been described.

Alpha fetoprotein is a substance produced by the liver of a healthy fetus. The exact function of this protein is unknown. However, the physiochemical properties of α-fetoprotein are similar to those of albumin, the major normal serum component in the adult. Although α-fetoprotein in human blood gradually disappears after birth, it never disappears entirely. It may reappear in liver disease or in tumors of the liver, ovaries, or testicles. The α-fetoprotein test is used to screen people at high risk for these conditions. After a cancerous tumor is removed, an α-fetoprotein test can monitor the progress of treatment. Continued high α-fetoprotein levels suggest the cancer is growing. *See* ALPHA FETOPROTEIN; CANCER (MEDICINE).

The carcinoembryonic antigen of human digestive system cancer has been the most studied oncofetal antigen. Originally thought to be a specific antigen of the fetal digestive tract as well as cancer of the colon, carcinoembryonic antigen is now known to occur normally in feces and secretions from the pancreas and bile ducts. It also appears in the plasma due to cigarette smoking and a diverse group of neoplastic and non-neoplastic conditions, including cancers of the colon, pancreas, stomach, lung, and breast; alcoholic cirrhosis; pancreatitis; inflammatory bowel disease; and rectal polyps. The carcinoembryonic antigen levels in the blood are one of the factors that doctors consider when determining the prognosis, or most likely outcome, of cancer. The carcinoembryonic antigen test is ordered for patients with known cancers, commonly cancer of the gastrointestinal system. In general, a higher carcinoembryonic antigen level predicts a more severe disease, one that is less likely to be curable. *See* ANTIGEN; ONCOLOGY.

Zoë Cohen; T. Ming Chu

Oncogenes Genes that contribute to the conversion of a normal cell into a cancerous cell. Oncogenes can derive from cellular genes that undergo mutations that alter their expression or activity, or from viruses that carry oncogenes within their genome and are transferred into the cell by infection. *See* ANIMAL VIRUS; CANCER (MEDICINE); GENE; ONCOLOGY.

Most, if not all, neoplasms (cancers) arise when individual cells within the body suffer irreversible genetic damage that leads to unrestrained cell growth, a block in the normal process of differentiation, or programmed cell death (apoptosis). Since genes consist of segments of deoxyribonucleic acid (DNA) that are linked in a specific order along each chromosome, any agent or process that breaks the DNA or alters the individual chemical subunits of the DNA can cause genetic damage.

Genetic lesions in neoplastic cells can affect two classes of genes. The first genes identified were called oncogenes, and the mutations that alter these genes occur on only one of the two similar chromosomes in a cell. The unaffected chromosome retains the normal cellular gene, whereas the mutant form of the gene on the affected chromosome overrides the normal gene function and promotes neoplastic growth. In contrast to oncogenes, mutations in cancer cells can also inactivate or even delete genes whose function is required for normal cell growth. This second class of genes, called tumor suppressors, appears to restrain cell growth or prevent the accumulation of mutations, and a cell remains normal as long as at least one copy of the tumor suppressor gene is functional.

Oncogenic mutations in cellular genes are never inherited from parent to child, presumably because the mutations are so detrimental to normal development that an embryo with the mutation would not survive. Instead, these mutations arise during the lifetime of an individual through spontaneous mutation or exposure to environmental carcinogens, both of which damage DNA. *See* DEOXYRIBONUCLEIC ACID (DNA); MUTAGENS AND CARCINOGENS; MUTATION; TUMOR.

Cellular genes have highly specific functions and patterns of expression in normal cells, and the mutations that convert a normal gene into an oncogene within a tumor cell can alter either of these properties. Four major types of oncogene-forming lesions are commonly found in tumor cells: (1) Point mutations can change individual bases within the segment of the gene that encodes a protein, and the resulting mutant protein has enhanced or altered activity that contributes to cancer cell growth. (2) Chromosomes can break near specific genes and rejoin in such a way that an oncogene is created through an abnormally high level of gene expression or by the loss of the gene's normal mode of regulation. These lesions are referred to as chromosomal translocations. (3) Oncogenes can be created by the localized amplification of small chromosomal domains, which leads to extra copies of the gene within the amplified segment and elevated levels of oncogene-encoded ribonucleic acid (RNA) and protein. (4) Viruses or mobile genetic elements can be inserted into the chromosome near a specific gene and alter the expression of the gene by imposing new regulatory signals, a process called insertional mutagenesis. *See* CHROMOSOME ABERRATION.

Some viruses contain oncogenes that are part of the viral genetic material and not derived from the cell. Although most human cancer is not associated with infectious agents such as viruses, some cancers have been shown to be caused at least in part by viruses. The most common is cervical carcinoma, where the causative agent is human papilloma virus (HPV). *See* TUMOR VIRUSES.

Normal cells divide or remain quiescent as the result of diverse extracellular signals (such as growth factors and cell-cell interactions). These signals are interpreted by receptors and cytoplasmic factors, and eventually the signals lead to reprogramming of gene expression in the nucleus. Thus, it is logical that the majority of oncogenes encode proteins that are components of this signal transduction pathway and they inappropriately activate the normal growth signals. Many oncogene-encoded proteins localize to the plasma membrane and are enzymes that phosphorylate other proteins (protein kinases), altering the activity of these substrate proteins. Other oncogenic proteins function as regulatory subunits of enzymes, and the mutations signal the enzymes to be active continuously rather than in a regulated manner. The oncogene-encoded proteins that localize to the nucleus are largely DNA-binding proteins that regulate specific sets of cellular genes involved in growth control. Finally, oncogenes can function not by promoting cell growth but by blocking apoptosis. Since many cells within the body (especially blood cells) are programmed to die after a limited lifespan, overexpression of a protein that blocks this programmed cell death can also contribute to cancerous growth. *See* CELL SENESCENCE; GENETIC ENGINEERING. Michael Cole

Oncology The study of cancer. There are five major areas of oncology: etiology, prevention, biology, diagnosis, and treatment. As a clinical discipline, it draws upon a wide variety of medical specialties; as a research discipline, oncology also involves specialists in many areas of biology and in a variety of other scientific areas. Oncology has led to major progress in the understanding not only of cancer but also of normal biology.

Cancer defies simple definition. It is a disease that develops when the orderly relationship of cell division and cell differentiation becomes disordered. In cancer, dividing cells seem to lose the capacity to differentiate, and they acquire the ability to invade through basement membranes and spread (metastasize) to many areas of the body through the bloodstream or lymphatics. Cancer is usually clonal, that is, it develops initially in a single cell. That abnormal cell then produces progeny that may behave rather heterogeneously. Some progeny continue to divide, some develop the capacity to metastasize, and some develop resistance to therapeutic agents. This single cell and its progeny, if unchecked, typically lead to the death of the host. *See* CANCER (MEDICINE).

Causes of cancer. Cancer is generally thought to result from one or more permanent genetic changes in a cell. In some cells a single mutational event can lead to neoplastic transformation, but for most tumors it appears that carcinogenesis is a multistep process. Although some rare congenital conditions lead to cancer in infancy, the vast majority of human cancers arise as a result of the complex interplay between genetic and environmental factors. Without question, there are forms of cancer clearly related to particular environmental exposures; it is equally clear, however, that these factors act on a genetic substrate that may be either susceptible or resistant to the development of cancer.

The emergence of cancer appears to involve the accumulation of genetic damage in a target tissue. Such complex, sequential genetic changes specific to tissues appear to underlie the progression to cancer. Multistep progression is quite complicated to study in experimental systems. Much work has focused on the identification, isolation, and characterization of oncogenes, which have the ability to transform non-neoplastic cells into cancer cells. More than 50 bona fide or putative oncogenes have been characterized and mapped throughout the human genome. *See* ONCOGENES; TUMOR.

Environmental factors involved in the development of cancers can be chemical, physical, or biological carcinogenic agents. At least three stages occur in the natural history of cancer development from environmental factors. The first stage is initiation, which is a specific alteration in the deoxyribonucleic acid (DNA) of a target cell; environmental agents may act by inducing expression of oncogenes. The second phase, promotion, involves the reversible stimulation of expansion of the initiated cell or the reversible alteration of gene expression in that cell or its progeny. Because promotion is thought to be reversible, it is a target for prevention. The final phase of carcinogenesis is progression. It is characterized by the development of aneuploidy and clonal variation in the tumor; these in turn result in invasiveness, metastasis, and growth advantage. *See* MUTAGENS AND CARCINOGENS.

Cancer prevention. An obvious starting point for cancer prevention is avoidance of environmental agents that contribute to carcinogenesis. Eliminating use of tobacco and alcohol would reduce cancer mortality by more than 30%. Skin cancer prevention is a simple matter of using the available tools (clothing and sunscreens) to protect the skin from Sun damage. The role of diet in cancer prevention is controversial. Epidemiologic evidence suggests a particularly strong link between a high-fat, high-calorie, low-fiber diet and an increased risk of colon cancer. Calorie restriction and exercise may reduce the carcinogenicity of carcinogenic agents, but there is no evidence that any dietary supplement will prevent cancer. *See* NUTRITION.

Cancer biology. The study of cancer biology picks up where cancer etiology leaves off, namely, at the point where the tumor has developed into a clonal cluster of autonomously proliferating cells. The pathological correlate of this stage of tumor development is carcinoma in situ; a condition in which no tissue destruction is evident, but atypical-appearing cancer cells are present at their site of origin. The transition from carcinoma in situ to locally invasive cancer is accompanied by dissolution of the basement membrane, penetration of tumor cells through the membrane and into the supportive tissues, and disruption of the supportive tissues. Expansion of the primary tumor in locally invasive cancer is always accompanied by the development of blood vessels, which are often defective and easily invaded by individual and clumped tumor cells. The tumor cells can also invade regional blood vessels and lymphatics and circulate throughout the body, attaching to endothelium in a distant organ site, inducing retraction of the endothelium, and becoming attached to the endothelial basement membrane. Once attached to the basement membrane, the tumor cells are covered over by the endothelial cells and effectively separated from the flow of blood. Local dissolution of the basement membrane then occurs, allowing the tumor to completely spread into the tissue and reestablish a blood flow in the breached vessel. As it grows, more blood vessel development nourishes the enlarging tumor.

During metastasis, tumor cells must overcome host defenses. They have various mechanisms to do so. For example, they produce new cell surface receptors to facilitate basement membrane and matrix binding; make new enzymes such as collagenases, serine proteases, metalloproteinases, cysteine proteinases, and endoglycosidases to facilitate their invasiveness; and secrete motility factors to enable them to move through the holes and pathways created by their enzymes. They avoid detection by the immune system through a variety of techniques. Unlike animal tumors, most human tumors are poorly immunogenic. Tumor cells often produce factors that are immunosuppressive. An unexplained feature of metastasis is the propensity of certain tumor types to spread to specific organs. *See* CELLULAR IMMUNOLOGY; HISTOCOMPATIBILITY.

Tumor detection. There are two major strategies to detect tumors at the earliest possible stage in their history: responding to the seven warning signals of cancer and screening populations at high risk. The seven danger signals of cancer are (1) unusual bleeding or discharge, (2) a lump or thickening in the breast or elsewhere, (3) a sore that does not heal, (4) change in bowel or bladder habits, (5) persistent hoarseness or cough, (6) persistent indigestion or difficulty in swallowing, and (7) change in a wart or mole.

Diagnosis. The diagnosis of cancer depends on the careful examination of biopsy material. Cancers arising in tissues having ectodermal or endodermal origins are generally called carcinomas; those derived from glands are called adenocarcinomas. Cancers arising in tissues derived from mesoderm are called sarcomas; those of lymphohematopoietic origin are lymphomas and leukemias. The cardinal microscopic features of malignancy are anaplasia, invasion, and metastasis.

Once a diagnosis of cancer is made, it is critical to determine the extent to which the disease has spread. This is called staging. It is distinct from grading, which is an assessment of histologic atypia performed with a microscope. Staging entails performing a careful physical examination, various radiographic studies, and perhaps surgical procedures (biopsies, endoscopies) to examine those sites to which a particular tumor type is most likely to spread. Often the results of such staging tests determine the nature and extent of therapy.

Treatment. There are four major approaches to cancer treatment: surgery, radiation therapy, chemotherapy, and biological therapy. These modalities are often used together with additive or synergistic effects. Surgery and radiation therapy are most effective in curing localized tumors and together result in the cure of about 50% of all newly diagnosed cases. Once the cancer has spread to regional nodes or distant sites, it is generally incurable with the use of local therapies alone. Systemic administration of a combination of chemotherapeutic agents may cure another 15–18% of all patients. *See* CHEMOTHERAPY AND OTHER ANTINEOPLASTIC DRUGS. Dan L. Longo

Onion A cool-season biennial, *Allium cepa*, of Asiatic origin and belonging to the plant order Liliales. The onion is grown for its edible bulbs.

Related species are leek (*A. porrum*), garlic (*A. sativum*), Welsh onion (*A. fistulosum*), shallot (*A. ascalonicum*), and chive (*A. schoenoprasum*).

Onion varieties (cultivars) are classified mainly according to pungency (mild or pungent) and use (dry bulbs or green bunching). Bulbs may be white, red, or yellow. Varieties differ markedly in their keeping quality and in their response to length of day. Hybrid varieties, with increased disease resistance, longer storage life, and improved quality, are rapidly displacing older varieties. Texas, New York, and California are important producing states. *See* LILIALES. H. John Carew

Ontogeny The developmental history of an organism from its origin to maturity. It starts with fertilization and ends with the attainment of an adult state, usually expressed in terms of both maximal body size and sexual maturity. Fertilization is the joining of haploid gametes (a spermatozoon and an ovum, each bearing half the number of chromosomes typical for the species) to form a diploid zygote (with a full chromosome number), a new unicellular living being. The gametes are the link between one generation and the next: the fusion of male and female gametes is the onset of a new ontogenetic cycle. Many organisms die shortly after sexual reproduction, whereas others live longer and generations are overlapped. Species are usually perceived as consisting mostly of adults, but in most cases the majority of their representation in the environment is as intermediate ontogenetic stages. *See* ANIMAL REPRODUCTION; FERTILIZATION (ANIMAL).

Zygotes undergo a series of asexual reproductions (via mitoses). In unicellular organisms, mitoses usually lead to the formation of new independent cells (or sometimes to polynucleated cells, or to colonies of identical cells), deriving from a first sexually derived individual, so forming a clone of genetically identical individuals. In multicellular organisms, the products of the asexual reproductions starting with the first division of the zygote are initially clones of one another but remain connected, and eventually differentiate to form an individual. Clonation of individuals occurs even in humans, when the first results of asexual reproduction of the zygote separate from each other, leading to twin formation. *See* MITOSIS.

The ontogeny of a multicellular organism involves segmentation (or cleavage): the zygote divides into two, four, etc., cells which continue to divide. These cells are initially similar to the zygote, although smaller in size. They soon start to differentiate from their ancestors, acquiring special features, and forming specific tissue layers and, eventually, organs. These processes lead to the formation and growth of an embryo. Embryos can develop either freely, or within egg shells, or within the body of one parent; they can grow directly into juveniles (as in humans) or into larvae (with an indirect development, as in insects). *See* ANIMAL GROWTH; EMBRYOLOGY.

Juveniles are similar to adults but are smaller and not sexually mature. Their ontogeny continues until they reach a maximal size and reproductive ability. Usually ontogeny is interrupted at adulthood, but some organisms can grow throughout their life, so that ontogeny ends with death.

Ferdinando V. Boero; Jean Bouillon; Stefano Piraino

Onychophora The only living animal phylum with true lobopods (annulate, saclike legs with internal musculature). There are about 70 known living species in two families, Peripatopsidae and Peripatidae. These terrestrial animals are frequently referred to as Peripatus. Onychophora comprise a single class or order of the same name. They were once considered a missing link between annelid worms and arthropods, but are best considered to be aligned with the arthropods.

They have a cylindrical body, 0.5–6 in. (1.4–15 cm) long, with one antennal pair, an anterior ventral mouth, and 14–43 pairs of stubby, unsegmented legs ending in walking pads and paired claws. Mandibles are present as modified tips of the first appendage pair. The body surface has a flexible chitinous cuticle. The body wall has three layers of smooth muscle, as in annelids, but the coelom is reduced to gonadal and nephridial cavities; the body cavity has an arthropodlike partitioned hemocoel; the heart is tubular with metameric ostia; and the nephridia are segmental. Gas exchange takes place by means of tracheae; spiracles are minute and numerous, located between skin folds. Slow locomotion is effected by legs and body contractions; the animals can squeeze into very tight spaces. The eyes, located at the antennal base, are the direct type with a chitinous lens and retinal layer. The sexes are separate; the testes and ovaries are paired; and the genital tracts open though the posterior ventral pore. Onychophora are oviparous, ovoviviparous, or viviparous.

The Onychophora are predatory, feeding on small invertebrates. They are largely nocturnal, occurring in humid habitats in forests. Stewart B. Peck

Onychopoda A specialized order of brachiopod crustaceans formerly included in the order Cladocera. The body is up to about 12 mm (0.5 in.) in length, but much of the length of the longest species is made up by a caudal process.

The head and thorax are short, as is the abdomen in some species, but in others it is drawn out into a long caudal process. A carapace is present but is reduced to a dorsal brood pouch, leaving the body naked. A large median compound eye occupies much of the head.

Onychopods swim actively by means of their antennae—the antennules are small and sensory—and seize their food with their four pairs of grasping trunk limbs. Most are predators, but detritus is also eaten by some species. The mandibles are stoutly denticulate.

Reproduction is mostly by parthenogenesis (males are unknown in some species from the Caspian Sea); eggs and young are carried in the brood pouch. Sexual reproduction gives rise to freely shed resistant eggs that overwinter in temperate zone species. Onychopods occur in the sea and fresh water and are worldwide in distribution, but fresh-water species occur only in the Holarctic temperate zone. There is a remarkable group of endemic species in the Ponto-Caspian region. *See* BRACHIOPODA.

Geoffrey Fryer

Onyx Banded chalcedonic quartz, in which the bands are straight and parallel, rather than curved, as in agate. Unfortunately, in the colored-stone trade, gray chalcedony dyed in various solid colors such as black, blue, and green is called onyx, with the color used as a prefix. Because the color is permanent, the fact that it is the result of dyeing is seldom mentioned.

The natural colors of true onyx are usually red or brown with white, although black is occasionally encountered as one of the colors. When the colors are red-brown with white or black, the material is known as sardonyx; this is the only kind commonly used as a gemstone. Its most familiar gem use is in cameos and intaglios. *See* CAMEO; CHALCEDONY. Richard T. Liddicoat, Jr.

Oogenesis The generation of ova or eggs, the female gametes. Primordial germ cells, once they have populated the gonads, proliferate and differentiate into sperm (in the testis) or ova (in the ovary). The decision to produce either spermatocytes or oocytes is based primarily on the genotype of the embryo. In rare cases, this decision can be reversed by the hormonal environment of the embryo, so that the sexual phenotype may differ from the genotype. Formation of the ovum most often involves substantial increases in cell volume as well as the acquisition of organellar structures that adapt the egg for reception of the sperm nucleus, and support of the early embryo. In histological sections, the structure of the oocyte often appears random but as the understanding of its chemical and structural organization increases, an order begins to emerge. *See* OVUM; SPERMATOGENESIS.

Among lower vertebrates and invertebrates, mitotic divisions of the precursor cells, the oogonia, continue throughout the reproductive life of the adult; thus extremely large numbers of ova are produced. In the fetal ovary of mammals, the oogonia undergo mitotic divisions until the birth of the fetus, but a process involving the destruction of the majority of the developing ova by the seventh month of gestation reduces the number of oocytes from millions to a few hundred. Around the time of birth, the mitotic divisions cease altogether, and the infant female ovary contains its full complement of potential ova. At puberty, the pituitary hormones, follicle stimulating hormone (FSH), and luteinizing hormone (LH) stimulate the growth and differentiation of the ova and surrounding cells (see illustration). *See* MITOSIS.

One important feature of oocyte differentiation is the reduction of the chromosome complement from the diploid state of the somatic cells to the haploid state of gametes. Fusion with the haploid genome of the sperm will restore the normal diploid number of chromosomes to the zygote. The meiotic divisions which reduce the chromosome content of the oocyte occur after the structural differentiation of the oocyte is complete, often only after fertilization. Unlike the formation of sperm, in which the two divisions of meiosis produce four equivalent daughter cells, the cytoplasm of the oocyte is divided unequally, so that three polar bodies with reduced cytoplasm and one oocyte are the final products. Generally, each fertilized oocyte produces a single embryo, but there are exceptions. Identical twins, for example, arise from the same fertilized egg. *See* FERTILIZATION; MEIOSIS.

The provision of nutrients for the embryo is a major function of the egg, and this is accomplished by the storage of yolk in the cytoplasm. Yolk consists of complex mixtures of proteins (vitellins), lipids, and carbohydrates in platelets, which are

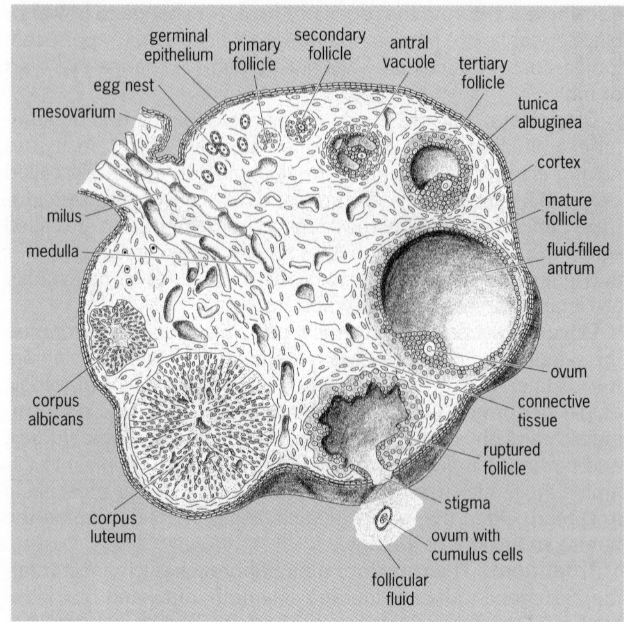

Three-dimensional view of the cyclic changes in the mammalian ovary.

membrane-surrounded packets dispersed throughout the egg cytoplasm (ooplasm). The amount of yolk in an egg correlates with the nutritional needs of the embryo. Although the eggs of mammals are extremely small as compared to the fetus, the bulk of the nutrition is supplied by the placenta; yolk is required only until implantation in the uterine wall.

Egg cytoplasm also contains large stores of ribonucleic acid (RNA) in the form of ribosomal, messenger, and transfer RNA. These RNAs direct the synthesis of proteins in the early embryo, and may have a decisive influence on the course of development. The mechanism by which the RNA is supplied to the egg is the basis for a major classification of ovary types. Panoistic ovaries, in which the egg nucleus is responsible for the production of all the stored RNA in the ooplasm, are typical of vertebrates, primitive insects, and a number of invertebrates. The amounts of RNA produced during the meiotic prophase in such ovaries are much larger than those produced by a somatic cell, and thus special mechanisms seem to be involved in the synthetic process. *See* DEOXYRIBONUCLEIC ACID (DNA); RIBONUCLEIC ACID (RNA). Spencer J. Berry

Oolite A deposit containing spheroidal grains with a mineral cortex, most commonly calcite or aragonite, accreted around a nucleus formed primarily of shell fragments or quartz grains. The term ooid is applied to grains less than 0.08 in. (2 mm) in diameter, and the term pisoid to those greater than 0.08 in. (2 mm). Accretionary layering (growth banding) is usually developed clearly. A flattened or elongate shape may occur if the nucleus shows that form. Ooids formed on nuclei of shells and shell fragments and composed of fine, radial calcite are cemented by coarse, clear calcite. Growth banding is visible in most ooids. The pisoids are composed of many thin layers of very small, tangential (lighter layers) and radial (darker layers) aragonite crystals. These pisoids are cemented with fibrous aragonite.

Ooids are primarily marine, forming in agitated shallow, warm waters. Under those conditions, the ooids are kept intermittently moving, so accretion occurs on all sides. Some ooids and most pisoids form in nonmarine environments, such as hypersaline and fresh-water lakes, hot springs, caves, caliche soils, and some rivers. *See* ARAGONITE; CALCITE. Philip A. Sandberg

Oomycota A class of fungi in the subdivision Mastigomycotina. They comprise a group of heterotropic, funguslike organisms that are classified with the zoosporic fungi (Mastigomycotina) but in reality are related to the heterokont algae. They are distinguished from other zoosporic fungi by the presence of biflagellate zoospores. Some taxa are nonzoosporic. Asexual reproduction involves the release of zoospores from sporangia; in some taxa the sporangium germinates with outgrowth of a germ tube. Sexual reproduction occurs when an oogonial cell is fertilized by contact with an antheridium, resulting in one or more oospores.

Oomycota are cosmopolitan, occurring in fresh and salt water, in soil, and as terrestrial parasites of plants. Many species can be grown in pure culture on defined media. There are five orders: The Saprolegniales and Leptomitales are popularly known as water molds. Some species are destructive fish parasites. Species of Rhipidiales grow in stagnant or polluted waters. Many Lagenidiales are parasites of invertebrates and algae. The Peronosporales are primarily plant parasites attacking the root, stem, or leaf, and include some of the more destructive plant pathogens. *See* EUMYCOTA; FUNGI. Donald J. S. Barr

Opal A natural hydrated form of silica. Opal is a relatively common mineral in its nongem form, which is known as common opal and lacks the play of color for which gem, or precious, opal is known. All opal is of relatively simple chemical composition, $SiO_2 \cdot nH_2O$. The hardness of opal on the Mohs hardness scale ranges from 5 to 6, the specific gravity from 2.25 to 1.99, and the refractive index from 1.455 to 1.435. *See* HARDNESS SCALES; SILICATE MINERALS.

The color of common opal ranges from transparent, glassy, and colorless to white and bluish white. Common pigmenting agents, such as iron, produce yellow, brown, red, and green colors, and frequently several colors in a single specimen. Precious opal has a play of color that is the result of white light being diffracted by the relatively regular internal array of silica spheres. Because opal is a hydrous mineral, certain opals from specific geologic occurrences may crack because of water loss. Therefore, considerable care is required in the polishing and handling of opal.

Several trade terms are used to describe the appearance of precious opal based on transparency, body color, and the type of play of color. Some of these terms are black opal, which is translucent to almost opaque, with dark gray to black body color, with play of color; fire opal, which is transparent to semitransparent, with yellow, orange, red, or brown body color and with or without play of color; harlequin or mosaic opal, in which the play of color occurs in distinct, broad, angular patches; and matrix opal, which consists of thin seams of high-quality gem opal in a matrix. *See* GEM. Peter J. Darragh; Cornelis Klein

Opalescence The milky iridescent appearance of a dense transparent medium when the system (or medium) is illuminated by polychromatic radiation in the visible range, such as sunlight. Slight changes in the rainbowlike color of the system can occur, depending on the scattering angle, that is, the angle between the directions of incident radiation and of observation.

Opalescence is a general term which applies to the optical phenomenon of intense scattering in the visible range of the electromagnetic radiation by a system with strong local optical inhomogeneities. The iridescence, or rainbowlike display of interference of colors, arises because the intensity of scattered light is approximately proportional to the reciprocal fourth power of the wavelength of incident light (Rayleigh's law). *See* SCATTERING OF ELECTROMAGNETIC RADIATION. Benjamin Chu

Open channel A natural or artificial conveyance through which liquid (typically water) having a free surface moves. The free surface is the interface with a gas (usually the atmosphere), along which the pressure is constant. The liquid is accelerated or decelerated in the flow direction due to an imbalance between the driving gravity force and the viscous boundary resistance force (friction). Such flows occur naturally in rivers, streams, and estuaries as a part of the hydrologic process of surface runoff and artificially in free-surface conduits for the transport of water for irrigation, water supply, drainage, flood control, and other useful purposes. In contrast to full pipe flow, the free surface introduces an additional freedom into the description of open-channel flow, which is the position of the free surface itself as it adjusts to the imposed flow conditions. *See* CANAL; PIPE FLOW; RIVER; VISCOSITY.

The effect of gravity as the driving force in open-channel flows is manifested first as a body force (weight) with a component in the direction of motion, parallel to the sloping channel bottom, and second as a gradient (spatial variation) in the pressure with flow depth. Open-channel flows exhibiting changes in depth or velocity at a fixed point over time are classified as unsteady. Flows exhibiting spatial changes in depth or velocity in the flow direction, or the lack thereof, result in a classification as nonuniform or uniform flow, respectively. As a special case, steady uniform flow does not change with time and has the property of constant depth and velocity in the flow direction so that the streamwise gravity force is exactly balanced by the boundary resistance force exerted on the flowing fluid. Artificial channels are often designed for this condition. Terry W. Sturm

Open circuit A condition in an electric circuit in which there is no path for current between two points; examples are a broken wire and a switch in the open, or off, position. *See* CIRCUIT (ELECTRICITY).

Open-circuit voltage is the potential difference between two points in a circuit when a branch (current path) between the points is open-circuited. Open-circuit voltage is measured by a voltmeter which has a very high resistance (theoretically infinite). Clarence F. Goodheart

Open-pit mining The process of extracting beneficial minerals by surface excavations. Open pit mining is a type of surface excavation which often takes the shape of an inverted cone; the shape of the mine opening varies with the shape of the mineral deposit. Other types of surface mining are specific to the type and shape of the mineral deposit. *See* COAL MINING; PLACER MINING; SURFACE MINING.

The open pit mine, like any other mining operation, must extract the product minerals at a positive economic benefit. All costs of producing the product, including excavation, beneficiation, processing, reclamation, environmental, and social costs, must be paid for by the sales of the mineral product. A mineral that is in sufficient concentration to meet or exceed these economic constraints is called ore. The terms ore body and ore deposit are used to refer to the natural occurrence of an economic mineral deposit. *See* ORE AND MINERAL DEPOSITS.

Ore bodies occur as the result of natural geologic occurrences. The geologic events that lead to the concentration of a mineral into an ore deposit are generally complex and rare. If those events placed the deposit sufficiently near the surface, open pit mining may be viable.

Material encountered during the mining process that has little or no economic value is called waste or overburden. One important economic criterion for open pit mining is the amount of overlying waste which must be removed to extract the ore. The ratio of the amount of waste to the amount of ore is referred to as the strip ratio. In general, the lower the strip ratio, the more likely an ore body is to be mined by open pit methods.

Modern open pit mining utilizes large mechanical equipment to remove the ore and waste from the open pit excavation. The amount of equipment and its type and size depend on the characteristics of the ore and waste and the required production capacity. In general, there are four basic unit operations common to most open pit mining operations. These are drilling, blasting, loading, and hauling.

Waste material that is generated during the course of mining at most mines must be discarded as economically as possible without jeopardizing future mining activities but while respecting environmental regulations. Two types of waste material are generated at most mining operations: waste rock and overburden from the mine, and tailings—the waste material from the processing plant after treatment of the ore. *See* LAND RECLAMATION.

Computer software is available to assist the mining engineer in ore reserve estimation with the application of geostatistics, mine planning and design, and production and maintenance monitoring and reporting. With the help of high-speed computers the engineering and production staff can evaluate aspects of the mining activities, which allows a more efficient and economical extraction of the mineral commodity. *See* MINING; OPERATIONS RESEARCH; OPTIMIZATION. Herb Welhener; John M. Marek

Operating system The software component of a computer system that is responsible for managing and coordinating activities and sharing the resources of the computer. The operating system (OS) acts as a host for application programs that are run on the machine. (In computer jargon, application programs are said to run on top of the operating system.) As a host, one of the functions of an operating system is to handle the details of the hardware operation. This relieves application programs from having to manage these details and makes it easier to write applications. Almost all computers use an operating system, such as hand-held computers (including personal data assistants or PDAs), laptop and desktop computers, supercomputers, and even modern video game consoles. *See* COMPUTER ARCHITECTURE.

Users may interact with the operating system by typing commands or using a graphical user interface (GUI, commonly pronounced "gooey"). The purpose of a GUI is to make it easier to interact with objects in the operating system, such as files, folders, and application programs. The GUI represents these objects as icons and provides easy methods for moving and copying objects, launching and switching between applications, scrolling through windows, and so on. The GUI is generally thought of as part of the operating system; however, most operating systems implement it as a specialized application program or service that runs on top of the operating system. The Microsoft Windows® operating system and Apple's MacOS® each have their own distinctive GUIs that run for the most part outside the operating system. Other operating systems, such as Linux, use GUIs based on the X Window System, which similarly runs on top of the OS. *See* COMPUTER PROGRAMMING; HUMAN-COMPUTER INTERACTION.

Operating systems offer a number of services to application programs and users. Applications access these services through application programming interfaces (APIs) or system calls. By invoking these interfaces (which is usually done by making a special subroutine call in the program), the application can request a service from the operating system, pass parameters, and receive the results of the operation.

Modern operating systems provide the capability of running multiple application programs simultaneously, which is referred to as multiprogramming or multitasking. Each program running is represented by a process in the operating system. The operating system provides an execution environment for each process by sharing the hardware resources so that each application does not need to be aware of the execution of other processes.

The main memory of a computer (known as random access memory, or RAM) is a finite resource. All running applications require the use of memory to store the program's instructions and data while the application is executing. The operating system is responsible for sharing the memory among the currently running processes. When a user initiates an application, the operating system decides where to place it in memory and may allocate additional memory to the application if it requests it. Most operating systems use capabilities in the hardware to prevent one application from overwriting the memory of another. This provides security and prevents applications from interfering with one another. The operating system can also allow portions of memory to be shared among processes. This provides a way for processes on the computer to communicate efficiently. Shared memory is an example of an interprocess communication (IPC) facility. *See* COMPUTER STORAGE TECHNOLOGY.

The details of device management are left to the operating system. The operating system provides a set of APIs to the applications for accessing I/O devices in a consistent and relatively simple manner regardless of the specifics of the underlying hardware. The operating system will generally use a software component called a device driver to control an I/O device. A device driver is typically provided by the manufacturer when the I/O device is purchased and is configured into the operating system when the device is first installed. This allows the operating system to be upgraded to support new devices as they become available. *See* COMPUTER PERIPHERAL DEVICES.

Operating systems provide security by preventing unauthorized access to the computer's resources. Many operating systems also prevent computer users from accidentally or intentionally interfering with each other. The security policies that an operating system enforces range from none in the case of a video game console, to simple password protection for hand-held and desktop computers, to very elaborate schemes for use in high-security environments. The security policies of an operating system can control who can log into the machine, which users can read or modify which files, which users can install software or reconfigure the operating system, which applications can use the network, which users can terminate applications, and so forth. *See* COMPUTER SECURITY. Curt Schimmel

Operational amplifier A voltage amplifier that amplifies the differential voltage between a pair of input nodes. For an ideal operational amplifier (also called an op amp), the amplification or gain is infinite.

Most existing operational amplifiers are produced on a single semiconductor substrate as an integrated circuit. These integrated circuits are used as building blocks in a wide variety of applications. *See* INTEGRATED CIRCUITS.

Although an operational amplifier is actually a differential-input voltage amplifier with a very high gain, it is almost never used directly as an open-loop voltage amplifier in linear applications for several reasons. First, the gain variation from one operational amplifier to another is quite high and may vary by ± 50% or more from the value specified by the manufacturer. Second, other nonidealities such as the offset voltage make it impractical to stabilize the dc operating point. Finally, performance characteristics such as linearity and bandwidth of the open-loop operational amplifier are poor. In linear applications, the operational amplifier is almost always used in a feedback mode.

A block diagram of a classical feedback circuit is shown in illus. *a*. The transfer characteristic, often termed the feedback gain A_f of this circuit,is given by Eq. (1). In the limiting case,

$$\frac{X_o}{X_i} = A_f = \frac{A}{1 + A\beta} \tag{1}$$

as A becomes very large, the feedback gain is approximated by Eq. (2).

$$A_f \simeq \frac{1}{\beta} \tag{2}$$

See FEEDBACK CIRCUIT.

An operational amplifier is often used for the amplifier designated A in this block diagram. Since A_f in the limiting case is

Basic circuits. (*a*) Classical feedback circuit. (*b*) Operational amplifier symbol typically used in circuit diagrams.

independent of A, the exact gain characteristics of the operational amplifier become unimportant provided the gain is large. Although linear applications of the operational amplifier extend well beyond the simple feedback block diagram of illus. *a*, the applications invariably involve circuit structures with feedback that make the characteristics of the circuit nearly independent of the exact characteristics of the operational amplifier. Such circuits are often termed active circuits.

The commonly used operational amplifier symbol is shown in illus. *b*. In this circuit, the output voltage is related to the gain A of the operational amplifier by Eq. (3), where A is very large

$$V_0 = A(V^+ - V^-) \qquad (3)$$

and the input currents I^+ and I^- are nearly zero. *See* AMPLIFIER; CIRCUIT (ELECTRONICS).

Peter M. VanPeteghem

Operations research

The application of scientific methods and techniques to decision-making problems. A decision-making problem occurs where there are two or more alternative courses of action, each of which leads to a different and sometimes unknown end result. Operations research is also used to maximize the utility of limited resources. The objective is to select the best alternative, that is, the one leading to the best result.

To put these definitions into perspective, the following analogy might be used. In mathematics, when solving a set of simultaneous linear equations, one states that if there are seven unknowns, there must be seven equations. If they are independent and consistent and if it exists, a unique solution to the problem is found. In operations research there are figuratively "seven unknowns and four equations." There may exist a solution space with many feasible solutions which satisfy the equations. Operations research is concerned with establishing the best solution. To do so, some measure of merit, some objective function, must be prescribed.

In the current lexicon there are several terms associated with the subject matter of this program: operations research, management science, systems analysis, operations analysis, and so forth. While there are subtle differences and distinctions, the terms can be considered nearly synonymous. *See* SYSTEMS ENGINEERING.

Methodology. The success of operations research, where there has been success, has been the result of the following six simply stated rules: (1) formulate the problem; (2) construct a model of the system; (3) select a solution technique; (4) obtain a solution to the problem; (5) establish controls over the system; and (6) implement the solution.

The first statement of the problem is usually vague and inaccurate. It may be a cataloging of observable effects. It is necessary to identify the decision maker, the alternatives, goals, and constraints, and the parameters of the system. A statement of the problem properly contains four basic elements that, if correctly identified and articulated, greatly eases the model formulation. These elements can be combined in the following general form: "Given (the system description), the problem is to optimize (the objective function), by choice of the (decision variable), subject to a set of (constraints and restrictions)."

In modeling the system, one usually relies on mathematics, although graphical and analog models are also useful. It is important, however, that the model suggest the solution technique, and not the other way around.

With the first solution obtained, it is often evident that the model and the problem statement must be modified, and the sequence of problem-model-technique-solution-problem may have to be repeated several times. The controls are established by performing sensitivity analysis on the parameters. This also indicates the areas in which the data-collecting effort should be made.

Implementation is perhaps of least interest to the theorists, but in reality it is the most important step. If direct action is not taken to implement the solution, the whole effort may end as a dust-collecting report on a shelf.

Mathematical programming. Probably the one technique most associated with operations research is linear programming. The basic problem that can be modeled by linear programming is the use of limited resources to meet demands for the output of these resources. This type of problem is found mainly in production systems, but is not limited to this area. *See* LINEAR PROGRAMMING.

Stochastic processes. A large class of operations research methods and applications deals with stochastic processes. These can be defined as processes in which one or more of the variables take on values according to some, perhaps unknown, probability distribution. These are referred to as random variables, and it takes only one to make the process stochastic.

In contrast to the mathematical programming methods and applications, there are not many optimization techniques. The techniques used tend to be more diagnostic than prognostic; that is, they can be used to describe the "health"of a system, but not necessarily how to "cure" it. *See* PROBABILITY; QUEUEING THEORY; STOCHASTIC PROCESS.

Scope of application. There are numerous areas where operations research has been applied. The following list is not intended to be all-inclusive, but is mainly to illustrate the scope of applications: optimal depreciation strategies; communication network design; computer network design; simulation of computer time-sharing systems; water resource project selection; demand forecasting; bidding models for offshore oil leases; production planning; classroom size mix to meet student demand; optimizing waste treatment plants; risk analysis in capital budgeting; electric utility fuel management; optimal staffing of medical facilities; feedlot optimization; minimizing waste in the steel industry; optimal design of natural-gas pipelines; economic inventory levels; optimal marketing-price strategies; project management with CPM/PERT/GERT; air-traffic-control simulations; optimal strategies in sports; optimal testing plans for reliability; optimal space trajectories. *See* DECISION THEORY; GERT; INVENTORY CONTROL; PERT.

William G. Lesso

Operator theory

At one level of abstraction an operator is simply a function whose arguments and values are real- (or complex-) valued functions of one or more real variables; in more naive terms an operator is a rule for converting such real- (or complex-) valued functions into others. The following are simple examples: (i) the operator which takes each differentiable real-valued function of one variable into its derivative; (ii) the operator which takes each twice-differentiable function f of one variable into expression (1); (iii) the operator which takes each twice-differentiable function f of three variables into expression (2); and (iv) the operator which takes the continuous function f of one real variable into the function g where relation (3) holds.

$$\left(\frac{df}{dx}\right)^2 + x^2\frac{d^2f}{dx^2} \qquad (1)$$

$$\frac{\partial^2 f}{\partial x^2} + \frac{\partial^2 f}{\partial y^2} + \frac{\partial^2 f}{\partial z^2} \qquad (2)$$

$$g(x) \equiv \int_0^1 \sqrt{x+y}\, f(y)dy \qquad (3)$$

Since an operator is a function, the usual functional notation is applicable. $L(f)$ may be used to denote the result of operating on f with the operator L. The set of all functions f for which $L(f)$ is defined is called the domain of L, and the set of all functions g such that $L(f) = g$ for some f in the domain of L is called the range of L. It is obvious that solving a differential or integral equation is equivalent (in many ways) to solving an operator equation $L(f) = g$, where g and L are given and it is required to find f. Moreover, the operator concept can be very useful both in theory and practice, producing a great variety of illuminating insights. *See* DIFFERENTIAL EQUATION; INTEGRAL EQUATION.

In large part the fruitfulness of the operator concept can be traced to two sources. One of these is the possibility of adding and multiplying operators in such a way that many, though not all, of the laws of ordinary algebra hold. The other is the fact that the ranges and domains of operators behave in many respects like ordinary space and, indeed, may be regarded as contained in infinite dimensional generalizations of the familiar three-dimensional space of solid geometry. This makes it possible to think of an operator as a geometrical transformation and to exploit one's spatial intuition. George W. Mackey

Operon A group of distinct genes that are expressed and regulated as a unit. Each operon is a deoxyribonucleic acid (DNA) sequence that contains at least two regulatory sites, the promoter and the operator, and the structural genes that code for specific proteins (see illustration). The promoter (*p*) site is the location at which ribonucleic acid (RNA) polymerase binds to the operon. RNA polymerase moves down the operon catalyzing the synthesis of a messenger RNA (mRNA) molecule with a sequence that is complementary to DNA. This process is called transcription. The mRNA is used as a template by ribosomes to synthesize the proteins coded for by the structural genes (in the original DNA) in a process called translation. This mRNA is referred to as polycistronic because its sequence directs the synthesis of more than one protein. The operator (*o*) site is located between the *p* site and the beginning of the coding region for the first structural gene. It is at this site that molecules called repressors can bind to the DNA and block RNA polymerase from transcribing the DNA, thus shutting off the operon. Some systems can be derepressed by the addition of small molecules called effectors, which bind to the repressor protein and cause a conformational (shape) change that makes it no longer able to bind to the DNA at the operator site. *See* DEOXYRIBONUCLEIC ACID (DNA); RIBONUCLEIC ACID (RNA).

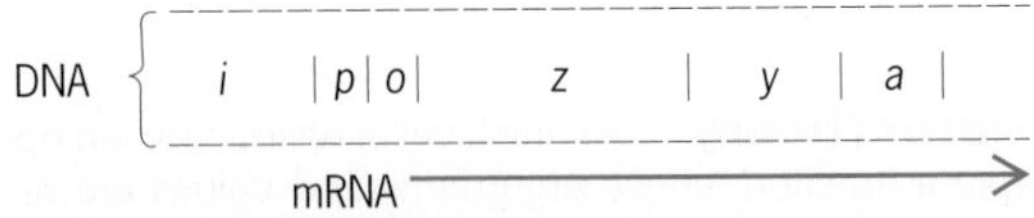

The lactose (*lac*) operon from *Escherichia coli: z, y,* and *a* are structural genes; *i* is the *lac* repressor gene; *p* is the promoter site; and *o* is the operator site. The arrow indicates length and direction of mRNA synthesis.

Activation is believed to arise from the binding of a protein immediately adjacent to the promoter. The protein provides additional locations with which RNA polymerase can interact; the extra interactions result in an increased amount of polymerase binding to the promoter. Activators are more frequently involved in the regulation of genes in eukaryotes than in prokaryotes.

Once RNA polymerase begins transcribing a gene, it continues making RNA until a termination site is reached. Antiterminators are proteins that prevent termination at certain sites. In the presence of these antiterminators, RNA polymerase continues along the genome and transcribes the genes following the termination site until a different class of termination site is encountered.

Attenuation is the premature termination of the mRNA translation. Although the exact mechanism of attenuation has not been determined, it is thought that attenuation is due to the formation of a translation termination site in mRNA. *See* GENE. Douglas H. Ohlendorf

Ophioglossales An order of the class Polypodiopsida known as the adder's-tongue ferns. It is a small group with only 3 genera and about 80 species. Two genera, *Ophioglossum* and *Botrychium*, are widely distributed in tropical and temperate regions and have about the same number of species; the third genus, *Helminthostachys*, is represented by a single species confined to southeastern Asia and Polynesia. These are considered the most primitive of the present-day ferns. No fossils have been reported for this group.

The chromosome number is high in the species of *Ophioglossum*. *Ophioglossum petiolatum*, a tropical species, has a chromosome count of over 1000, the largest number observed in a naturally occurring species of vascular plants. This group is distinguished from other ferns by the arrangement of the sporogenous tissue in the characteristic fertile spike of the sporophyte. The leaves are erect or merely bent over in bud and not circinate. The gametophyte is a small, nongreen, fleshy, subterranean saprophyte, associated with an endophytic fungus. The group appears to be an evolutionary dead end. *See* FUNGI; LEAF; POLYPODIALES; POLYPODIOPSIDA; PTEROPSIDA. Paul A. Vestal

Ophiolite A distinctive assemblage of mafic plus ultramafic rocks generally considered to be fragments of the oceanic lithosphere that have been tectonically emplaced onto continental margins and island arcs. An ophiolite is a formation made up of an association of typical rocks in a clearly defined sequence. A complete idealized ophiolite sequence from bottom to top includes (1) an ultramafic tectonite complex composed mostly of multilayered, deformed harzburgite, dunite, and minor chromitite; (2) a plutonic complex of layered mafic-ultramafic cumulates at the base, grading upward to massive gabbro, diorite, and possibly plagiogranite; (3) a mafic sheeted-dike complex; (4) an extrusive section of massive and pillow lavas, pillow breccias, and intercalated pelagic sediments; and (5) a top layer of abyssal or bathyal sediments, which may include ribbon chert, red pelagic limestone, metalliferous sediments, volcanic breccias, or pyroclastic deposits. Most ophiolites lack complete sections, and are dismembered and fragmented. Their estimated original thickness is variable, ranging from about 2 km (1.2 mi) to more than 8 km (5 mi). *See* EARTH CRUST; LITHOSPHERE.

Ophiolites typically occur in collisional mountain belts or island arcs and define a suture zone marking the boundary where two plates have welded together. The ophiolite complex is interpreted as evidence for a closed marginal ocean or back-arc basin. Throughout the world, ophiolites occur as long narrow belts, up to 10 km (6 mi) wide, that can extend more than 1000 km (600 mi) in length, in two distinct geographic settings. (1) Those in the Alpine-Mediterranean region, Tethyan ophiolite, were formed in small ocean basins that were surrounded by older, attenuated continental crust. (2) Those in western North America and the Circum-Pacific (Cordilleran) region seem to have formed in inter-arc basins. The Cordilleran ophiolites, such as the Trinity ophiolite and the Coast Range ophiolite of California, are generally incomplete, metamorphosed, or dismembered, but they commonly form the basement rocks for many North American continental margin terranes. *See* BASIN; CONTINENTAL MARGIN; STRUCTURAL GEOLOGY; GEODYNAMICS.

Ophiolites represent new oceanic crust formed in a variety of spreading environments, including oceanic ridge, back-arc basin, and island arcs above a subduction zone, and subsequently emplaced onto the continents. Their occurrence along plate sutures marks the sites of ancient tectonic interaction between oceanic and continental crust. Ophiolites provide the best opportunity for geologists to study the ocean floor on land; they also offer vertical sections in addition to horizontal distributions. Moreover, ophiolite formations record the ages of oceanic fragments that escaped, disappearing into subduction zones. Juhn Liou; Shige Maruyama; Yoshi Ogasawara

Ophiuroidea A class of the Asterozoa, known as the brittle stars, in which the arms are usually clearly demarcated from a central disk and perform whiplike locomotor movements, and the tube feet are nonsuctorial sensory tentacles. In all existing ophiuroids the ambulacral plates fuse together in pairs to form articulating joints termed vertebrae, and the ambulacral groove is converted into an internal epineural canal.

There are about 1900 extant species referred to 230 genera, arranged to form 3 orders: Oegophiurida, Phrynophiurida, and Ophiurida. There is also one Paleozoic order, the Stenurida.

Ophiuroids are usually five-armed, with a few species being regularly six- or seven-armed. In some Euryalae the arms may branch repeatedly; these are the so-called basket stars. Tropical species are often patterned in contrasting colors, but most ophiuroids tend to match their environment. Some species are luminescent, although not constantly so. Andrew B. Smith

Ophiuroids occur in all the oceans from low-tide level downward, often in dense populations which number millions to the hectare. Six families range below a depth of 2 mi (3.2 km); the genera *Ophiura*, *Amphiophiura*, and *Ophiacantha* range below 4 mi (6.4 km). The shallow-water forms hide among algae, under stones, or within sponges or bury the disk in sand or mud, leaving only the arms protruding. Deep-water forms lie in or on the bottom material or adhere to corals or cidarids. Howard B. Fell

Opiates Drugs derived from opium, the dried juice of the oriental poppy seed. The pharmacologically active substances, which constitute approximately 25% of the extract, are the alkaloids morphine, codeine, and papaverine. The newer synthetic compounds which resemble morphine in their action are called opioids.

The principal effect of opium and opioids is to relieve pain. Even today morphine remains the best analgesic. It also assuages anxiety and causes slight drowsiness, relaxation, and a euphoric state of mind. These psychic effects are so agreeable that many troubled individuals seek solace by ingesting, smoking, or injecting opiates. *See* MORPHINE.

Codeine has an action similar to morphine, but its analgesic effects are less. Papaverine has almost no analgesic action, and is used as an antispasmodic to relieve vascular spasm and undesirable contraction of smooth muscle. *See* ANALGESIC; NARCOTIC. Raymond D. Adams

Opiliones The harvestmen or daddy longlegs, an order of the class Arachnida; sometimes known as Phalangida. About 4500 species are known. They are common in temperate and tropical climates. Most are red, brown, or black and 5–20 mm (0.2–0.8 in.) long, although a few are only 1–2 mm (0.04–0.08 in.). Some large tropical species have bright iridescent colors and elaborate spines.

The cephalothorax (prosoma) is broadly attached to the segmented abdomen (opisthosoma). The center of the cephalothoracic shield (carapace) bears a tubercle with a simple *eye* on each side. Many have scent glands opening on the sides which produce repellent fluids with strong odors, possibly phenols or quinones. The six pairs of appendages include relatively small chelate chelicerae (jaws), leglike (usually) pedipalps, and four pairs of legs. The legs may be very long and slender, the distal ends being able to wrap around plant stalks.

Harvestmen are predators of small invertebrates or may scavenge or eat decaying vegetation. Respiration is by means of tracheae (thin tubes), which open through spiracles on the abdomen; there may be additional spiracles on the legs.

The group is divided into three suborders: the mitelike Cyphophthalmi; the tropical Laniatores with strong pedipalps, and some species adapted for cave life; and the long-legged Palpatores, the most common opilionids in temperate areas. *See* ARACHNIDA. H. W. Levi

Opisthobranchia A subclass in the class Gastropoda containing about 4000 living species, arranged in nine orders, including the herbivorous Aplysiomorpha (sea hares) and Sacoglossa and the carnivorous Thecosomata (sea butterflies) and Nudibranchia (sea slugs). Primitive members of many of the orders show adaptations for burrowing beneath sand or mud; more advanced members are always active surface-living or pelagic forms. This trend is accompanied by a decrease in the importance of the shell and operculum for passive defense. These are replaced by more dynamic chemical (some species secrete decinormal sulfuric acid through the skin if annoyed), physical (daggerlike calcareous epidermal spicules), or biological (redirected nematocysts derived from coelenterate prey) defensive mechanisms.

The adult shells of the primitive opisthobranchs living today are often strongly developed and sometimes colorful; a more typical opisthobranch shell is fragile, inflated, and egg-shaped. In rather more advanced forms, the external shell has a very wide gape, and in animals like *Berthella* the widely gaping shell is wholly internal, covered by the mantle. The most varied shells are found in the Sacoglossa. In the highest sacoglossans, some bullomorphs and aplysiomorphs, and in all the nudibranchs, the true shell is completely lost after larval metamorphosis. *See* GASTROPODA; MOLLUSCA; NUDIBRANCHIA; SACOGLOSSA. T. E. Thompson

Opisthocomiformes A small order of birds that contains a single family, the Opisthocomidae. Its one species, the hoatzin, is restricted to South America. In the absence of definite evidence of its relationship to other avian orders, the Opisthocomiformes is best considered as a distinct order that evolved in South America since its isolation from the rest of Gondwanaland sometime in the late Mesozoic.

The hoatzin is a unique bird, often considered rather reptilian because of its nearly unfeathered young, which leaves the nest soon after birth to crawl around trees by using its clawed wings as well as its feet. The hoatzin has an anisodactyl perching foot with three anterior toes and a well-developed hallux (first toe). Hoatzins are adept swimmers, capable of surviving a fall into water. They are medium-sized birds of dark brown plumage streaked with white; a distinct crest of reddish-brown feathers tops the head, which is small and has a short stout bill. The wings are long and rounded, the legs are short with strong toes, and the long hallux is on the same level as the anterior toes. Hoatzins are weak fliers, but they are almost strictly arboreal, clambering about in trees to feed on leaves and fruit. Hoatzins are found in forests of northern South America, especially along rivers and streams.

The hoatzins are known in the fossil record only by the Miocene *Hoazinoides* from Colombia, a typical form which reveals nothing about the evolutionary history of the order. *See* AVES. Walter J. Bock

Opossum A New World marsupial belonging to the family Didelphidae. Didelphidae consists of one North American species, the Virginia opossum (*Didelphis virginiana*), and 65 Central and South American species. Opossums are primarily nocturnal and may be either arboreal or terrestrial. *See* MARSUPIALIA.

The Virginia opossum *D. virginiana* ranges from southern Canada to Costa Rica. It is a medium-sized mammal with long, rather coarse, grayish-white fur. Individuals have large, thin, leathery, naked ears and a long, scaly, sparsely haired tail. The nose is pink, the eyes are black, and the ears are bluish-black. Adult Virginia opossums are approximately 600–875 mm in length, including a 275–350 mm tail. Adults usually weigh between 1.3 and 4.5 kg.

Opossums are omnivorous, eating a wide variety of foods but preferring animal matter during all seasons. They are solitary mammals and inhabit a variety of habitats from grasslands to forests. Opossums readily climb, using their prehensile tail extensively for balance and support. These mammals are probably best known for "playing possum," a temporary state of catatonia. This phenomenon, which is a passive defensive tactic that apparently has survival value, seems to be a nonvoluntary (reflex) action and probably is not deliberate or willful. Donald W. Linzey

Opportunistic infections Infections that cause a disease only when the host's immune system is impaired. The classic opportunistic infection never leads to disease in the normal host. The protozoon *Pneumocystis carinii* infects nearly everyone at some point in life but never causes disease unless the immune system is severely depressed. The most common immunologic defect associated with pneumocystosis is acquired immune deficiency syndrome (AIDS). *See* ACQUIRED IMMUNE DEFICIENCY SYNDROME (AIDS).

A compromised host is an individual with an abnormality or defect in any of the host defense mechanisms that predisposes that person to an infection. The altered defense mechanisms or immunity can be congenital, that is, occurring at birth and genetically determined, or acquired. Congenital immune deficiencies are relatively rare. Acquired immunodeficiencies are associated with a wide variety of conditions such as (1) the concomitant presence of certain underlying diseases such as cancer, diabetes, cystic fibrosis, sickle cell anemia, chronic obstructive lung disease, severe burns, and cirrhosis of the liver; (2) side effects of certain medical therapies and drugs such as corticosteroids, prolonged antibiotic usage, anticancer agents, alcohol, and nonprescribed recreational drugs; (3) infection with immunity-destroying microorganisms such as the human immunodeficiency virus that leads to AIDS; (4) age, both old and young; and (5) foreign-body exposure, such as occurs in individuals with prosthetic heart valves, intravenous catheters, and other indwelling prosthetic devices.

Virtually any microorganism can become an opportunist. The typical ones fall into a number of categories and may be more likely to be associated with a specific immunologic defect. Examples include (1) gram-positive bacteria: both *Staphylococcus aureus* and the coagulase-negative *S. epidermidis* have a propensity for invading the skin and as well as catheters and other foreign implanted devices; (2) gram-negative bacteria: the most common is *Escherichia coli* and the most lethal is *Pseudomonas aeruginosa*; these pathogens are more likely to occur in cases of granulocytopenia (granulocyte deficiency, as occurs in leukemia or chemotherapy); (3) acid-fast bacteria: *Mycotuberculum tuberculosis* is more likely to reactivate in the elderly and in those individuals with underlying malignancies and AIDS; (4) protozoa: defects in cell-mediated immunity, such as AIDS, are associated with reactivated infection with *Toxoplasma gondii* and *Cryptosporidium*; (5) fungi: *Cryptococcus neoformans* is a fungus that causes meningitis in individuals with impaired cell-mediated immunity such as AIDS, cancer, and diabetes; *Candida albicans* typically causes blood and organ infection in individuals with granulocytopenia. *See* CELLULAR IMMUNOLOGY; ESCHERICHIA; MEDICAL MYCOLOGY; STAPHYLOCOCCUS; TUBERCULOSIS.

The first step in treatment of opportunistic infections involves making the correct diagnosis, which is often difficult as many of the pathogens can mistakenly be thought of as benign. The second step involves administration of appropriate antimicrobial agents. As a third step, if possible, the underlying immune defect needs to be corrected. *See* IMMUNOLOGICAL DEFICIENCY; INFECTION; MEDICAL BACTERIOLOGY. Robert Murphy

Opsonin A term used in serology and immunology to refer to a substance that enhances the phagocytosis of bacteria by leukocytes. Opsonin is generally synonymous with the bacteriotropin of F. Neufeld and coworkers (1904–1905), a relatively thermostable antibody, increased in amount during specific immunization, that renders the corresponding bacterium more susceptible to phagocytosis. There is evidence that this action can be promoted to some extent by antibody alone, but that it is substantially increased by the further addition of the thermolabile complement system. *See* AGGLUTINATION REACTION; ANTIBODY; LYTIC REACTION; NEUTRALIZATION REACTION (IMMUNOLOGY); PHAGOCYTOSIS; PRECIPITIN; SERUM. Henry P. Treffers

Optical activity The effect of asymmetric compounds on polarized light. To exhibit this effect, a molecule must be nonsuperimposable on its mirror image, that is, must be related to its mirror image as the right hand is to the left hand. An optically active compound and its mirror image are called enantiomers or optical isomers (see illustration). Enantiomers differ only in their geometric arrangements; they have identical chemical and physical properties. The right-handed and left-handed forms of a molecule can be distinguished only by their optical activity or by their interactions with other asymmetric molecules. Optical activity can be used to probe other aspects of molecular geometry, as well as to identify which enantiomer is present and its purity.

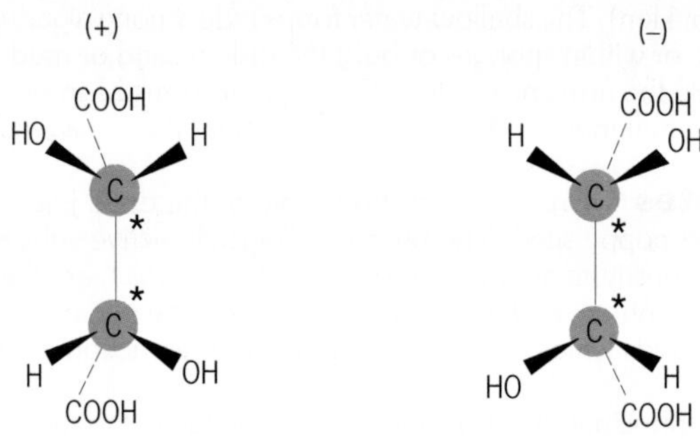

Enantiomers of tartaric acid.

The physical basis of optical activity is the differential interaction of asymmetric substances with left versus right circularly polarized light. If solids and substances in strong magnetic fields are excluded, optical activity is an intrinsic property of the molecular structure and is one of the best methods of obtaining structural information from a sample in which the molecules are randomly oriented. The relationship between optical activity and molecular structure results from the interaction of polarized light with electrons in the molecule. Thus the molecular groups that contribute most directly to optical activity are those that have mobile electrons which can interact with light. Such groups are called chromophores, since their absorption of light is responsible for the color of objects. For example, the chlorophyll chromophore makes plants green. *See* FARADAY EFFECT; POLARIZED LIGHT; STEREOCHEMISTRY.

Optical activity is measured by two methods, optical rotation and circular dichroism. The optical rotation method depends on the different velocities of left and right circularly polarized light beams in the sample. The velocities are not measured directly, but both beams are passed through the sample simultaneously. This is equivalent to using plane-polarized light. The differing velocities of the left and right circularly polarized components yield a rotation of the plane of polarization. Circular dichroism is the difference in absorption of left and right circularly polarized light. Since this difference is about a millionth of the absorption of either polarization, special techniques are needed to determine it accurately. Circular dichroism is reported as a difference in absorption, or as an ellipticity (a measure of the elliptical polarization of the emergent beam). Vincent Madison

Optical bistability A phenomenon exhibited by certain resonant optical structures whereby it is possible to have two stable steady transmission states for the device, depending upon the history of the input. Such a bistable device may be useful for optical computing elements because of its memory characteristics. The bistability can result from the intrinsic properties of the optical device or from some external feedback such as an electrical voltage supplied by another device. This second type, extrinsic or hybrid optical bistability, is not true optical bistability.

Optical bistability is an inherently steady-state phenomenon, and typically any cycling of the device through its hysteresis cycle must be done adiabatically; that is, changes in the propagating light amplitude, envelope phase, and profile must occur sufficiently slowly that their impact on the evolution of the system may be neglected. This requirement imposes some rather severe frequency-response limitations on the use of intrinsically bistable devices in optical circuits. The two primary types of intrinsic optical bistability, each arising from a distinct physical mechanism, are absorptive bistability and refractive bistability. *See* ADIABATIC PROCESS; HYSTERESIS.

Absorptive optical bistability is based upon coupling the feedback mechanism inherent in an optical cavity with an absorbing nonlinear optical medium in which the absorption coefficient decreases with increasing light intensity (a saturable absorber). The basic theory of operation is: the saturable absorber is placed in the cavity, and the cavity is resonantly pumped. For low light intensities, the transmission coefficient for the cavity is small because of the presence of the highly absorbing medium inside the cavity. As the pump intensity is increased, the absorption of the nonlinear medium decreases. Finally, for some threshold pump intensity, the cavity switches into a high transmission state, because the absorption coefficient is reduced sufficiently that the intrinsic cavity feedback mechanism dominates. The threshold is very sharp because, when the cavity is in a highly transmittive state, the built-up intensity inside the cavity becomes very large compared to the pump intensity (due to the feedback) and effectively bleaches virtually all of the absorption in the nonlinear medium. The intense pump is then largely transmitted, although some energy is stored in the cavity to bleach the absorber. *See* ABSORPTION OF ELECTROMAGNETIC RADIATION; LASER; OPTICAL PUMPING.

This device exhibits two characteristics that constrain its usefulness in particular applications. (1) The device is based on an absorption mechanism, so the energy absorbed from the pump light must be dissipated in the bistable element or heat-sinked elsewhere. (2) It is highly frequency-sensitive because its operation is based on the switching characteristics of a resonant cavity.

Refractive optical bistability is based on coupling the feedback mechanism inherent in an optical cavity with a nonlinear optical medium that exhibits a change in the refractive index as a function of light intensity. The nonlinear refractive medium is placed inside the optical cavity, and the cavity is pumped slightly off-resonance so that the transmission coefficient is small compared to unity. However, a small amount of light intensity does exist inside the cavity, and changes the effective optical path length inside the cavity by inducing change in the refractive index of the nonlinear medium. As the pump intensity is increased, this change in the effective path length becomes larger, until at some point the cavity switches into, and possibly past, resonance. The transmission coefficient switches abruptly to a value close to unity, and the built-up intensity inside the cavity increases abruptly. If the pump intensity is increased further, it is possible to switch the cavity through a second resonance, with an additional threshold in the transmission coefficient. *See* REFRACTION OF WAVES.

The most common implementation scheme for a bistable optical device is the nonlinear Fabry-Perot etalon. The device is typically fabricated from a semiconductor, and consists of a slab of material of approximately 1 micrometer thickness. On each surface of the semiconductor, a highly reflective coating may be deposited to increase the bandwidth of the Fabry-Perot cavity. The choice of a proper nonlinear material is based upon the operating wavelength and the temporal response time desired, and possibly other considerations. Typically for applications in the far-infrared, near-infrared, and visible wavelengths, the proper materials are indium antimonide (InSb), gallium arsenide (GaAs), and zinc selenide (ZnSe), respectively. *See* INTERFEROMETRY.

David R. Andersen

Optical coherence tomography Optical coherence tomography (OCT) is a recently developed, noninvasive technique for imaging subsurface tissue structure with micrometer-scale resolution. The principles of time gating, optical sectioning, and optical heterodyning are combined to allow cross-sectional imaging. Depths of 1–2 mm (0.04–0.08 in.) can be imaged in turbid tissues such as skin or arteries; greater depths are possible in transparent tissues such as the eye. Optical coherence tomography complements other imaging modalities commonly used to image subsurface tissue structure, including ultrasound and confocal microscopy.

Principles of operation. In a typical optical coherence tomography system (see illustration), light from a broadband, near-infrared source and a visible aiming beam is combined and coupled into one branch of a fiber-optic Michelson interferometer. Broadband sources include superluminescent diodes, fiber amplifiers, and femtosecond pulse lasers in the wavelength range of 800–1550 nanometers. The light is split into two fibers using a 2 × 2 coupler, one leading to a reference mirror and the second focused into the tissue. Light reflects off the reference mirror and is recoupled into the fiber leading to the mirror. Concurrently, light is reflected from index-of-refraction mismatches in the tissue and recoupled into the fiber leading to the tissue. Reflections result from changes in the index of refraction within the structure of the tissue, for instance between intercellular fluid and collagen fibers. Light that has been back-reflected from the tissue and light from the reference arm recombine within the 2 × 2 coupler.

Because the broadband source has a short coherence length, only light which has traveled very close to the same time (or optical path length) in the reference and tissue arms will interfere constructively and destructively. By changing the length of the reference arm, reflection sites at various depths in the tissue can be sampled. The depth resolution of the optical coherence tomography system is determined by the effectiveness of this time gating and hence is inversely proportional to the bandwidth of the source. An optical detector in the final arm of the Michelson interferometer detects the interference between the reference and tissue signals. During optical coherence tomography imaging, the reference-arm mirror is scanned at a constant velocity, allowing depth scans (analogous to ultrasound A-scans) to be made. Either the tissue or the interferometer optics is mounted

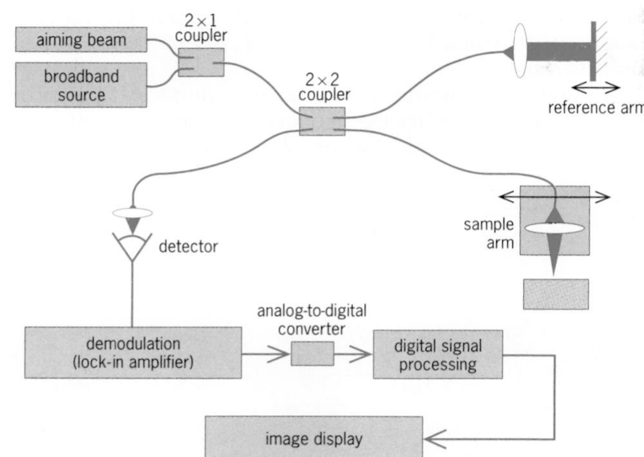

Typical optical coherence tomography system, based on a broadband source and fiber-optic Michelson interferometer.

on a stage so that the beam can be scanned laterally across the tissue to build up two- and three-dimensional images, pixel by pixel.

Variations. Several instruments have been built based on variations of the basic optical coherence tomography system. For instance, polarization-sensitive optical coherence tomography uses polarization-altering optics in the arms of the interferometer to determine the sample birefringence from the magnitude of the back-reflected light. Optical coherence microscopy uses a system of high numerical aperture to achieve resolutions comparable to confocal microscopy but with increased depth of penetration. Color Doppler optical coherence tomography (CDOCT) is an augmentation capable of simultaneous blood flow mapping and spatially resolved imaging.

Applications. Optical coherence tomography can be used to probe the structure of any accessible tissue. Noninvasive studies of the eye and skin are being performed, and a commercial device has been developed for retinal imaging. Using optical coherence tomography, parameters such as eye length can be accurately measured, and the cross-section images of the retina give a clear and quantifiable assessment of retinal separation and macular degeneration, among other pathologies. In skin, the morphology of normal skin layers and components, and disorders such as psoriasis, can be imaged. *See* COHERENCE; COMPUTERIZED TOMOGRAPHY; CONFOCAL MICROSCOPY; DOPPLER EFFECT; FIBER-OPTICS IMAGING; HETERODYNE PRINCIPLE; INTERFEROMETRY; MEDICAL ULTRASONIC TOMOGRAPHY; ULTRASONICS. Jennifer Kehlet Barton

Optical communications

Optical communications The transmission of speech, data, video, and other information using light in the frequency range below the infrared limit of the visible spectrum. An optical communications link consists of a data or information source, a modulated laser transmitter, an optically transparent transmission medium, possibly an amplifier, and a photodetector receiver that recovers the transmitted data (see illustration). The predominant medium is optically transparent fiber; in this form, the technology is called fiber-optic communications.

Light by nature is an electromagnetic wave and it follows the laws of waves, such as refraction, diffraction, interference, and polarization. It is also a particle and exerts pressure; this is particularly noticeable when a satellite enters the Sun's light and starts drifting due to pressure exerted on it. The smallest quantity of light is known as a photon, which, according to atomic physics, is generated when an atom from an excited state makes a transition to either a less excited state or to its nonexcited (ground) state. The frequencies used in optical communications are much higher than those used for radio, television, and microwaves, and lower than those of visible light. Thus, light used in optical communications is at frequencies below the infrared limit of visible light. *See* DIFFRACTION; ELECTROMAGANETIC RADITION; INFRARED RADIATION; INTERFERENCE OF WAVES; LIGHT; PHOTON; POLARIZATION OF WAVES; RADIATION PRESSURE; REFRACTION OF WAVES; WAVE MOTION.

The amount of information that a wave can carry depends on its frequency; the higher the frequency, the more the information. Thus, frequencies in the optical spectrum can carry orders-of-magnitude more information than radio and microwave frequencies. For example, the frequency of light is of the order of 3×10^8 MHz as compared with radio and television frequencies that are in the kilohertz-to-megahertz range. Therefore, optical communication is the preferred communications technology for very high capacity transmission, since it is capable of delivering an unprecedented amount of information per second compared with other communications technologies. *See* BANDWIDTH REQUIREMENTS (COMMUNICATIONS); INFORMATION THEORY.

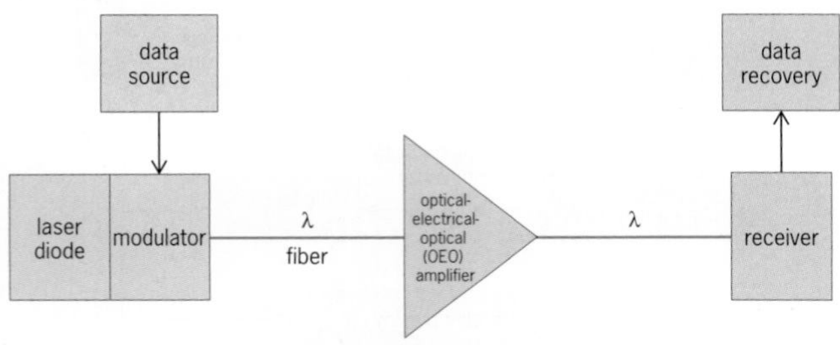

Block diagram of a simplified optical communications link.

Basic components. For optical communications to be effective, certain basic components are required to construct the communications link. These, in addition to couplers, are a laser transmitter, a modulator, the optical fiber, and a photodetector receiver (see illustration).

The laser (light amplification by stimulated emission of radiation) used in optical communications is a semiconductor device. When a voltage is applied to it, a strong field is generated within it that excites specific atoms. When excited atoms are stimulated, they emit light at a specific frequency (or wavelength), which is determined by the material, its geometry, and the applied field. Typical lasers that are used in optical communications generate a continuous optical (yet invisible) beam, and are made with compound materials that consist of the elements indium, gallium, arsenic, and phosphorus. *See* COHERENCE; LASER.

To carry information, the continuous optical beam that is generated by the laser must be modulated. That is, the characteristics of the laser light must be altered by impressing on it a stream of digital data or a string of ones and zeros. An optical device that modulates the continuous laser beam is known as a modulator. In optical communications, a modulator may be thought of a fast shutter that acts on the continuous beam. When the shutter opens it allows light to pass through, and when it closes it blocks light; that is, the optical shutter affects the optical power of the beam. This modulation method is known as on-off keying. *See* MODULATION; MODULATOR.

The fiber used in optical communications consists of two silica layers, the core and the cladding. What distinguishes the two is the refractive index. The core includes elements, known as dopants, such that its refractive index is higher than the refractive index of the cladding. This refractive index variation helps to keep light within the core in which it is transmitted. The cladding is surrounded by nonsilica layers that are added for strength and for fiber-type identification. *See* OPTICAL FIBERS.

Photodetectors are diode semiconductor devices that generate one or more electron-hole pairs for each photon that enters the biased *p-n* junction. When modulated light impinges on the photodetector device, the generated electron-hole pairs constitute an electric current known as photocurrent. The generated photocurrent converts the optical digital signal to an electrical digital signal, which is passed through filters and a resistive load to convert it to voltage pulses. The two photodetector types used predominantly in optical communications are, the *p*-intrinsic-*n* (PIN) diode and the avalanche photodiode (APD). *See* OPTICAL DETECTORS; PHOTOVOLTAIC CELL; SEMICONDUCTOR DIODE.

Wavelength-division multiplexing. Initial optical communications networks used a single wavelength at 1310 nm as the carrier frequency, and at modulation rates 2.5 Gbps or less. Soon thereafter, a second wavelength at 1550 nm was added that was suitable for longer fiber lengths. The protocol used was called synchronous optical network (SONET) in the United States, and synchronous digital hierarchy (SDH) in Europe. SONET and SDH, although very similar, differ enough, as a result of specific local needs for services and demographics, to create two different specification standards. As bandwidth requirements increased, another modulation rate was added at 10 Gbps, and more recently another at 40 Gbps.

Despite this incremental and dramatic increase in data rate, bandwidth demand exceeded bandwidth deliverability by a single fiber; this situation is known as capacity exhaust. To meet the ever-increasing bandwidth demand, a need arose for additional fiber, which is very costly. However, at about the same time (in the early 1990s) optical and photonic developments came to fruition that could be applied cost-effectively to optical communications. These developments allowed more than one optical wavelength to be multiplexed in the same fiber, giving birth to a technology called wavelength-division multiplexing (WDM). Providing a multiplicity of wavelengths, or optical channels, in a single fiber, solved the capacity exhaust problem and alleviated the need for additional fibers and networks. *See* MULTIPLEXING AND MULTIPLE ACCESS.

The optical technology that was required in basic optical communications (laser, fiber, and photodiode) was now enhanced with more components to make WDM possible. Key WDM components are optical multiplexers, optical demultiplexers, and optical amplifiers. Other components include couplers, splitters, filters, equalizers, compensators, and optical add-drop multiplexers.

Optical switching and routing. Optical networks consist of a mesh of nodes that are interconnected with optical fiber. Nodes receive traffic from one fiber and, based on the predetermined route, they pass it onto another fiber that connects with the next node in the mesh, and so on. This is known as a mesh topology, which is used in the optical networks that connect cities across the United States, also known as the backbone.

To be able to pass traffic from one fiber to another, a node must support an optical switching function. One such technology uses lithium niobate ($LiNbO_3$) solid-state devices that, by the application of a voltage, couple an optical signal from one waveguide to another, similar to a train switching tracks. This technology is fast but attenuates the signal and is not suitable for very large switches. Another technology uses micromirrors. This technology is also known as micro-electro-mechanical systems (MEMS). Tilting the mirrors by applying an electrostatic field will redirect the optical beam in the desired direction, and thus a switching function for many optical channels is accomplished. *See* ELECTROOPTICS; MICRO-ELECTRO-MECHANICAL SYSTEMS (MEMS); MICRO-OPTO-ELECTRO-MECHANICAL SYSTEMS (MOEMS); OPTICAL MODULATORS.

Protection. As in previous communication networks, the network protection strategy in optical networks is very important. The mesh topology has excellent protection because, if one fiber is cut (fiber cuts happen frequently when digging roads and fields through which fiber is laid), then the switching nodes detect the fiber failure and, by using sophisticated algorithms and control messages, find another route on to which to move the affected traffic. Another topology of interest is the ring, which is used in many metropolitan and large campus applications, and is thus named the Metro ring. The Metro ring consists of a single fiber, a dual fiber, or a quad fiber. Optical add-drop nodes are placed on the ring to remove and add traffic to be sent to its destination. Clearly, a single ring does not exhibit good protection against fiber cuts unless it is a bidirectional ring; that is, traffic flows in both directions of the ring.

Free-space optical (FSO) communications. Installing fiber in a busy city is a rather lengthy process because over 1 or 2 km distance many rights of way must be obtained. However, shooting a laser beam from the top of one building to the top of another building that is much easier. *See* COMMUNICATIONS SATELLITE; MICROWAVE; MOBILE RADIO; RADIO-WAVE PROPAGATION.

FSO technology may also include multiple wavelengths, thus utilizing WDM technology.

Laser beams may be used as intersatellite links. Satellites form a network of their own, communicating with each other via the laser beams. S. V. Kartalopoulos

Optical detectors Devices that respond to incident ultraviolet, visible, or infrared electromagnetic radiation by giving rise to an output signal, usually electrical. Based upon the manner of their interaction with radiation, they fall into three categories. Photon detectors are those in which incident photons change the number of free carriers (electrons or holes) in a semiconductor (internal photoeffect) or cause the emission of free electrons from the surface of a metal or semiconductor (external photoeffect, photoemission). Thermal detectors respond to the temperature rise of the detecting material due to the absorption of radiation, by changing some property of the material such as its electrical resistance. Detectors based upon wave-interaction effects exploit the wavelike nature of electromagnetic radiation, for example by mixing the electric-field vectors of two coherent sources of radiation to generate sum and difference optical frequencies. *See* NONLINEAR OPTICAL DEVICES.

The most widely used photon effects are photoconductivity, the photovoltaic effect, and the photoemissive effect. Photoconductivity, an internal photon effect, is the decrease in electrical resistance of a semiconductor caused by the increased numbers of free carriers produced by the absorbed radiation. *See* PHOTOCONDUCTIVE CELL.

The photovoltaic effect, also an internal photoeffect, occurs at a *pn* junction in a semiconductor or at a metal-semiconductor interface (Schottky barrier). Absorbed radiation produces free hole-electron pairs which are separated by the potential barrier at the *pn* junction or Schottky barrier, thereby giving rise to a photovoltage. This is the principle employed in a solar cell. *See* PHOTODIODE; PHOTOVOLTAIC CELL; PHOTOVOLTAIC EFFECT; SEMICONDUCTOR DIODE; SOLAR CELL.

The photoemissive effect, also known as the external photoeffect, is the emission of an electron from the surface of a metal or semiconductor (cathode) into a vacuum or gas due to the absorption of a photon by the cathode. The photocurrent is collected by a positively biased anode. Internal amplification of the photoexcited electron current can be achieved by means of secondary electron emission at internal structures (dynodes). Such a vacuum tube is known as a photomultiplier. Internal amplification by means of an avalanche effect in a gas is employed in a Geiger tube. *See* GEIGER-MÜLLER COUNTER; LIGHT AMPLIFIER; PHOTOELECTRIC DEVICES; PHOTOEMISSION; PHOTOMULTIPLIER; PHOTOTUBE.

Semiconductors are key to the development of most photon detectors. These materials are characterized by a forbidden energy gap which determines the minimum energy that a photon must have to produce a free hole-electron pair in an intrinsic photoeffect. Since the energy of a photon is inversely proportional to its wavelength, the minimum energy requirement establishes a long-wavelength limit of an intrinsic photoeffect. It is also possible to produce free electrons or free holes by photoexcitation at donor or acceptor sites in the semiconductor; this is known as an extrinsic photoeffect. Here the long-wavelength limit of the photoeffect is determined by the minimum energy (ionization energy) required to photoexcite a free electron from a donor site or a free hole from an acceptor site. *See* SEMICONDUCTOR.

The choice of materials also plays a role in thermal detectors. The most widely used thermal detector is a bolometer, that is, a temperature-sensitive resistor in the form of a thin metallic or semiconductor film (although superconducting films are also used). Incident electromagnetic radiation absorbed by the film causes its temperature to rise, thereby changing its electrical resistance. The change in resistance is measured by passing a current through the film and measuring the change in voltage. Materials with a high temperature coefficient of resistance are desired for bolometers, a criterion which usually favors semiconductors over metals. *See* BOLOMETER. Paul W. Kruse

Optical fibers Transparent strands of glass, plastic, or other flexible material used for light delivery. The light delivery can be for the purpose of signal (data) transmission, light amplification, image transmission, or energy transmission. The light can be confined by total internal reflection, by mirrorlike reflection from an internal coated surface, or by a suitably designed set of interior air holes formed parallel to the axis of the fiber. Optical fibers are, in their most common form, a type of dielectric guide for electromagnetic waves (waveguide). In certain forms, fibers are also termed lightguides. *See* Reflection of electromagnetic radiation.

In the most common form, the optical fiber consists of a core of material with a refractive index higher than the surrounding cladding. If a light source such as a laser is directed into the fiber, the light travels (propagates) along the fiber and may be detected at the other end. Information may be encoded on the light by on-off keying (a binary system in which on = 1, off = 0), frequency modulation, or phase modulation. If the wavelength is chosen correctly, fibers may transmit light over very long distances of 100 km (60 mi) or more. For even longer distances, specialty fibers are fabricated with rare-earth ions, which enable optical amplification. Fiber amplifiers enable both undersea and terrestrial communication over great distances. A fiber amplifier which is equipped with suitable end mirrors may also function as a laser source. *See* Fiber-optic circuit; Frequency modulation; Modulation; Optical communications; Phase modulation; Refraction of waves.

Fused silica (SiO_2) is the most common fiber material. Plastic fibers are used for low-cost, short-distance optical links. Fluoride-based glasses are required for transmission in the mid-infrared region of the optical spectrum.

There are five basic types of optical fibers (see illustration). Propagation in these lightguides is most easily understood by ray optics, although the wave or modal description must be used for exactness. In a multimode, stepped-refractive-index-profile fiber (illus. *a*), the number of rays or modes of light which are guided, and thus the amount of light power coupled into the lightguide, is determined by the core size and the core-cladding refractive index difference. Such fibers are limited to short distances for information transmission due to pulse broadening.

A graded-index multimode fiber (illus. *b*), where the core refractive index varies across the core diameter, is used to minimize pulse broadening due to intermodal dispersion. Since light travels more slowly in the high-index region of the fiber relative to the low-index region, significant equalization of the transit time for the various modes can be achieved to reduce pulse broadening. This type of fiber is suitable for intermediate-distance, intermediate-bit-rate transmission systems. For both fiber types, light from a laser or light-emitting diode can be effectively coupled into the fiber. *See* Laser; Light-emitting diode.

A single-mode fiber (illus. *c*) is designed with a core diameter and refractive index distribution such that only one fundamental mode is guided, thus eliminating intermodal pulse-broadening effects. Material and waveguide dispersion effects cause some pulse broadening, which increases with the spectral width of the light source. These fibers are best suited for use with a laser source in order to efficiently couple light into the small core of the waveguide and to enable information transmission over long distances at very high bit rates. *See* Waveguide.

There are two fiber structures that confine light without total internal reflection. Microstructured optical fibers (illus. *d*), a category which includes photonic bandgap fibers, confine light in a small core region surrounded by micrometer-size air holes. Such fibers provide useful and interesting applications in nonlinear optics, optical switching, and frequency conversion. The confinement of an ultrashort laser pulse within such a waveguide can result in the generation of a supercontinuum, or a very large range of optical frequencies. A hollow-core optical fiber (illus. *e*) confines light through the use of an interior coating (metal or multilayer dielectric) of high reflectance. Such fibers are used in cases of very high laser power or for wavelengths at which transparent fiber materials are not available.

The attenuation or loss of light intensity is an important property of the lightguide since it limits the achievable transmission distance, and is caused by light absorption and scattering. Optical fibers based on silica glass have an intrinsic transmission window at near-infrared wavelengths with extremely low losses. Glass fibers, intrinsically brittle, are coated with a protective plastic to preserve their strength. *See* Optical materials.

Suzanne R. Nagel; Thomas G. Brown

cross section | index profile | light path | input pulse | output pulse

(a)

(b)

(c)

(d)

(e)

Types of optical fiber designs. (*a*) Multimode, stepped-refractive-index-profile. (*b*) Multimode, graded-index-profile. (*c*) Single-mode, stepped-index. Graded-index is possible. (*d*) Microstructured optical fiber. (*e*) Hollow-core fiber for high-power laser delivery.

Optical flat A disk of high-grade quartz glass approximately $^3/_4$ in. (2 cm) thick, having at least one side ground and polished with a deviation in flatness usually not exceeding 50 nanometers all over, and a surface quality of 5 microfinish or less. When two surfaces of this quality are placed lightly together

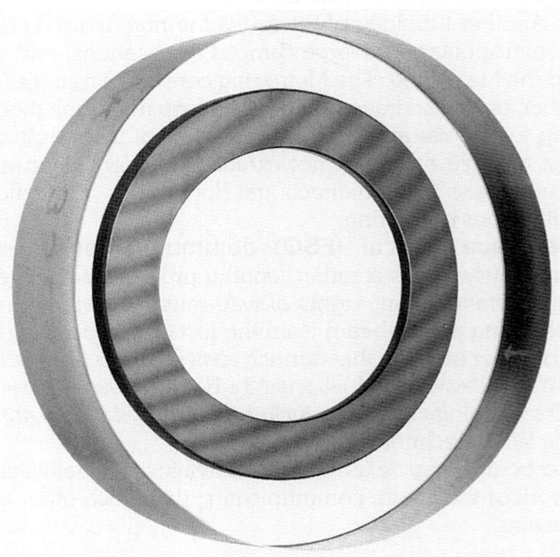

Optical flat being used to determine flatness of seal ring. Interference bands on seal ring face show lines of constant depth. (*Van Keuren Co.*)

so that the air is not wrung out from between them, they are separated by a film of air and actually touch at only one point. This point is the vertex of a wedge of air separating the two pieces.

If parallel beams of light pass through the flat, part will be reflected against the surface being inspected, while part will be reflected directly back through the flat. Because the distance between the surfaces is constantly increasing along the angle, the beams reflected from the flat and the beams reflected from the workpiece will alternately reinforce and interfere with each other, producing a pattern of alternate light and dark bands (see illustration). Each succeeding full band from a point of contact means the distance between surfaces is one wavelength thicker. If the light is relatively monochromatic, the wavelength is known. Red with a wavelength of 295 nm is commonly used. Thus a definite relationship is established between lineal measurement and light waves. Optical flats are used for two general purposes, determination of surface contour and comparison of lineal measurement. Rush A. Bowman

Optical guided waves Optical-frequency electromagnetic waves confined within an optical waveguide, a structure designed to carry such waves from one place to another somewhat as a pipe carries water. Optical waveguides confine light by the method of total internal reflection. (The terms optical and light are used here in the broadest sense to include visible and near-infrared electromagnetic radiation.) The demonstrations of the first semiconductor laser and the first low-loss glass optical fiber initiated a technological revolution. *See* REFLECTION OF ELECTROMAGNETIC RADIATION.

Because of the high data rates that can be achieved, the transmission of information in the form of optical guided waves confined within an optical-fiber waveguide has become the preferred method for the telecommunications industry. The optical fiber consists of two concentric glass cylinders. The inner core region is made of a glass that has a slightly higher index of refraction than the outer cladding region, as required for total internal reflection to occur. Core diameters in the range of 4–50 micrometers are used routinely to carry near-infrared optical radiation. Pulses of light with wavelengths between 0.8 and 1.55 μm are injected into the fiber on the near end. The presence or absence of a pulse is interpreted as a one or a zero by a receiver on the far end, typically many miles (kilometers) away. *See* OPTICAL COMMUNICATIONS; OPTICAL FIBERS.

Devices such as semiconductor lasers make use of optical guided waves over distances much shorter than those spanned by optical fibers. In a semiconductor laser, a planar waveguide also serves as an electrically driven amplifying medium. Gallium arsenide (GaAs), aluminum gallium arsenide (AlGaAs), or indium gallium arsenide phosphide (InGaAsP) are common examples of semiconductor laser materials. When mirrors are formed on the ends of the semiconductor waveguide, typically a few hundred micrometers apart, light travels back and forth between those mirrors just as with any laser, but in the form of optical guided waves. *See* LASER.

Optical waveguides are the basic constituents of an emerging technology known variously as integrated optics, integrated optoelectronics, or photonic integrated circuits. This technology integrates optical and electronic components into or on an optical waveguide. The aim is to process and manipulate light while it is trapped as optical guided waves within the confines of the optical waveguide. *See* INTEGRATED OPTICS; WAVEGUIDE. Dennis G. Hall

Optical image The image formed by the light rays from a self-luminous or an illuminated object that traverse an optical system. The image is said to be real if the light rays converge to a focus on the image side and virtual if the rays seem to come from a point within the instrument (see illustration).

The optical image of an object is given by the light distribution coming from each point of the object at the image plane of an optical system. The ideal image of a point according to geometrical optics is obtained when all rays from an object point unite in a single image point. However, diffraction theory teaches that even in this case the image is not a point but a minute disk. *See* DIFFRACTION.

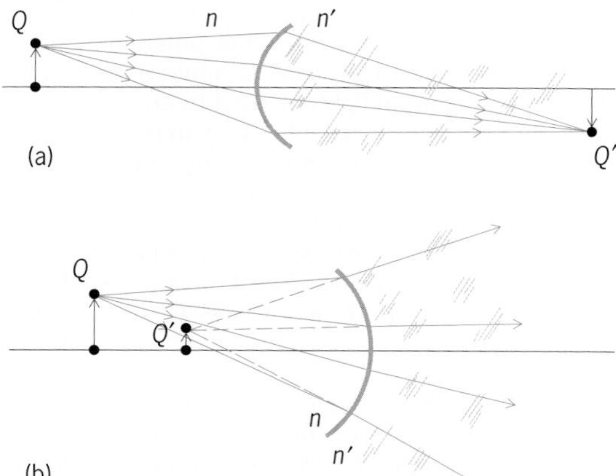

Optical images. (*a*) Real image. Rays leaving object point *Q* and passing through the refracting surface separating media *n* and *n*′ are brought to a focus at the image point *Q*′. (*b*) Virtual image. Rays leaving *A* and refracted by the concave surface separating *n* and *n*′ appear to be coming from the virtual image point *Q*′. As the rays are diverging, they cannot be focused at any point. (*Modified from F. A. Jenkins and H. E. White, Fundamentals of Optics, 4th ed., McGraw-Hill, 1976*)

From the standpoint of geometrical optics, if this most desirable type of image formation cannot be achieved, the next best objective is to have the image free from all but aperture errors (spherical aberration). In this case the light distribution in the image plane is still circular, resembling the point image; there is a true coordination of object point and image, although the image may be slightly unsharp. If the aperture errors are small, or if the image is viewed from a distance, such an image formation may be very satisfactory. *See* ABERRATION (OPTICS).

Asymmetry and deformation errors may be very disturbing if not held in check, because the light distribution of the image of a point in this case has a decidedly undesirable shape. Max Herzberger

Optical information systems Systems that use light to process information. Optical information systems or processors consist of one or several light sources; one- or two-dimensional planes of data such as film transparencies, various lenses, and other optical components; and detectors. These elements can be arranged in various configurations to achieve different data-processing functions. As light passes through various data planes, the light distribution is spatially modulated proportional to the information present in each plane. This modulation occurs in parallel in one or two dimensions, and the processing is performed at the speed of light. Optical processors offer various advantages compared to other technologies: data travels at the speed of light; all data in one-dimensional and two-dimensional arrays are operated on in parallel; multiple planes of data can be processed in parallel by various multiplexing schemes; it is possible to have large numbers of interconnections with no interaction (which is not possible with electrical connections); and power dissipation is less and size and weight can be less for optical processors than for their electronic counterparts. *See* CONCURRENT PROCESSING; MULTIPLEXING AND MULTIPLE ACCESS.

In practice, the processing speed is limited by the rate at which data can be introduced into the system and the rate at which processed data (produced on output detector arrays) can be analyzed. The reusable real-time spatial light modulators used to produce new input data, filters, interconnections, and so forth, are the major components required for these optical information-processing systems to realize their full potential. Spatial light modulators convert electrical input data into a form suitable for spatially modulating input light, or react to an optical input and generate a different optical output. The manipulation of the light passing through the system is controlled by spatial light modulators, lenses, holographic optical elements, computer-generated holograms, or fiber optics. Four major application areas are image processing, signal processing, computing and interconnections, and neural networks. *See* HOLOGRAPHY; IMAGE PROCESSING; NEURAL NETWORK; OPTICAL FIBERS; OPTICAL MODULATORS.

David Casasent

Optical isolator A device that is interposed between two systems to prevent one of them from having undesired effects on the other, while transmitting desired signals between the systems by optical means. Optical isolators are used for both electrical systems and optical systems such as lasers.

An optical isolator for electrical systems is a very small four-terminal electronic circuit element that includes in an integral package a light emitter, a light detector, and, in some devices, solid-state electronic circuits. The emitting and detecting devices are so positioned that the majority of the emission from the emitter is optically coupled to the light-sensitive area of the detector. The device is also known as an optoisolator, optical-coupled isolator, and optocoupler. The device is housed in an integral opaque package so that the only optical emission impinging on the detector is that produced by the emitter. This configuration of components can perform as a solid-state electronic transformer or relay, since an electronic input signal causes an electronic output signal without any electrical connection between the input and the output terminals.

Optical isolators are used in electrical systems to protect humans or machines when high-voltage or high-power equipment is being controlled. In addition, optical isolators are used in electronic circuit design in situations where two circuits have large voltage differences between them and yet it is necessary to transfer small electrical signals between them without changing the basic voltage level of either. Robert D. Compton

The need for optical isolation has broadened considerably since the advent of lasers. It is often necessary to prevent light from reentering the laser, irrespective of any electrical consideration. One example is a small laser followed by high-power laser amplifiers. If the powerful amplified light reenters the small (master oscillator) laser, it can destroy it. Another example is a frequency-stabilized laser, whose oscillation frequency is perturbed by reentering (injected signal) light.

A polarizer-plus-quarterwave-plate isolator prevents laser light from reentering the laser when the light is scattered back by specular reflectors. This device cannot ensure isolation if there is diffuse reflection or if polarization-altering (birefringent) optics are encountered. Another limitation of this isolator is that the transmitted light is circularly polarized. *See* BIREFRINGENCE; POLARIZED LIGHT.

In contrast to the quarter-wave polarizer isolator, the Faraday isolator can provide truly one-way transmission irrespective of polarization changes from the exit side if an exit polarizer (which passes light that has undergone the Faraday rotation after passing though the entrance polarizer) is used in addition to the entrance polarizer. For example, it isolates against diffuse reflections and any light source on the exit side. *See* FARADAY EFFECT.

The isolation properties of an acoustooptic deflector are based on the fact that light deflected by it is shifted in frequency by an amount equal to the acoustic frequency. The reflected beam, passing through the deflector a second time, is again shifted in frequency by the same amount and in the same sense if the deflector is operated in the Bragg mode. Hence, the reflected light that is returned to the laser is shifted in frequency by an amount $2f$, where f is the frequency of the acoustic wave. Provided the frequency of the light returned to the laser is not close to any resonant frequency of the laser cavity, it will not perturb the laser and will simply be reflected from the output mirror. *See* ACOUSTOOPTICS.

Stephen F. Jacobs

Optical materials All substances used in the construction of devices or instruments whose function is to alter or control electromagnetic radiation in the ultraviolet, visible, or infrared spectral regions. Optical materials are fabricated into optical elements such as lenses, mirrors, windows, prisms, polarizers, detectors, and modulators. These materials serve to refract, reflect, transmit, disperse, polarize, detect, and transform light. The term "light" refers here not only to visible light but also to radiation in the adjoining ultraviolet and infrared spectral regions. At the microscopic level, atoms and their electronic configurations in the material interact with the electromagnetic radiation (photons) to determine the material's macroscopic optical properties such as transmission and refraction. These optical properties are functions of the wavelength of the incident light, the temperature of the material, the applied pressure on the material, and in certain instances the external electric and magnetic fields applied to the material. *See* ATOMIC STRUCTURE AND SPECTRA; DISPERSION (RADIATION); ELECTROMAGNETIC RADIATION; ELECTROOPTICS; INFRARED RADIATION; LENS (OPTICS); LIGHT; MAGNETOOPTICS; MIRROR OPTICS; OPTICAL DETECTORS; OPTICAL MODULATORS; OPTICAL PRISM; POLARIZED LIGHT; REFLECTION OF ELECTROMAGNETIC RADIATION; REFRACTION OF WAVES; ULTRAVIOLET RADIATION.

There is a wide range of substances that are useful as optical materials. Most optical elements are fabricated from glass, crystalline materials, polymers, or plastic materials. In the choice of a material, the most important properties are often the degree of transparency and the refractive index, along with each property's spectral dependency. The uniformity of the material, the strength and hardness, temperature limits, hygroscopicity, chemical resistivity, and availability of suitable coatings may also need to be considered. *See* HARDNESS SCALES; STRENGTH OF MATERIALS.

Glass technology provided the foundation for classical optical elements, such as lenses, prisms, and filters. Glasses developed for use in the visible region have internal transmittances of over 99% throughout the wavelength range of 380–780 nanometers. However, the silicate structure in glasses limits their transmission to about 2.5 micrometers in the infrared. Chalcogenide glasses, heavy-metal fluoride glasses, and heavy-metal oxide glasses extend this transmission to 8–12 μm. *See* COLOR FILTER.

Advances in the process for manufacturing optical fibers led to the present fiber-optic communication systems that operate in the near-infrared region with windows at wavelengths of 850, 1310, 1550, and 1625 nm. An advanced fiber-optic system, LEAF (Large Effective Area Fiber), was designed to minimize nonlinearities by spreading the optical power over large areas. *See* OPTICAL COMMUNICATIONS; OPTICAL FIBERS; VAPOR DEPOSITION.

The use of photolithography for printing integrated circuits has necessitated the improvement in the transmission of glasses for the ultraviolet region. Fused silica, which transmits to about 180 nm, is well suited for the lithography in the ultraviolet region. However, the crystalline material calcium fluoride, which transmits into the ultraviolet region to about 140 nm, outperforms any glass in printing microchips using fluorine excimer lasers. Deep-ultraviolet applications of fused-silica glasses include high-energy lasers, spacecraft windows, blanks for large astronomical mirrors, optical imaging, and cancer detection using ultraviolet-laser-induced autofluorescence. *See* FLUORESCENCE; INTEGRATED CIRCUITS; TELESCOPE.

The need for an inexpensive, unbreakable lens that could be easily mass-produced precipitated the introduction of plastic optics in the mid-1930s. Although the variety of plastics suitable for precision optics is limited compared to glass or crystalline materials, plastics are often preferred when difficult or unusual shapes, lightweight elements, or economical mass-production techniques are required.

The softness, inhomogeneity, and susceptibility to abrasion intrinsic to plastics often restrict their application. Haze (which is the light scattering due to microscopic defects) and birefringence (resulting from stresses) are inherent to plastics. Plastics also exhibit large variations in the refractive index with changes in temperature. Shrinkage resulting during the processing must be considered. *See* BIREFRINGENCE; PHOTOELASTICITY.

Organic synthetic polymers are emerging as key materials for information technologies. Polymers often have an advantage over inorganic materials because they can be designed and synthesized into compositions and architectures not possible with crystals, glasses, or plastics. They are manufactured to be durable, optically efficient, reliable, and inexpensive. Many uses of polymers in photonic and optoelectronic devices have emerged, including light-emitting diodes, liquid-crystal–polymer photodetectors, polymer-dispersed liquid-crystal devices (for projection television), optical-fiber amplifiers doped with organic dyes (rhodamine), organic thin-film optics, and electrooptic modulators. *See* LIGHT-EMITTING DIODE; LIQUID CRYSTALS.

Although most of the early improvements in optical devices were due to advancements in the production of glasses, the crystalline state has taken on increasing importance. Historically, the naturally occurring crystals such as rock salt, quartz, and fluorite plus suitable detectors permitted the first extension of visible optical techniques to harness the invisible ultraviolet and infrared rays. Synthetic crystal-growing techniques have made available single crystals such as lithium fluoride (of special value in the ultraviolet region, since it transmits at wavelengths down to about 120 nm), calcium fluoride, and potassium bromide (useful as a prism at wavelengths up to about 25 μm in the infrared). Many alkali-halide crystals are important because they transmit into the far-infrared. *See* CRYSTAL GROWTH; CRYSTAL STRUCTURE; SINGLE CRYSTAL.

Following the invention of the transistor, germanium and silicon ushered in the use of semiconductors as infrared optical elements or detectors. Polycrystalline forms of these semiconductors could be fabricated into windows, prisms, lenses, and domes by casting, grinding, and polishing. Compound semiconductors such as gallium arsenide (GaAs), ternary compounds such as gallium aluminum arsenide ($Ga_{1-x}Al_xAs$), and quaternary compounds such as indium gallium arsenide phosphide (InGaAsP) now serve as lasers, light-emitting diodes, and photodetectors. *See* SEMICONDUCTOR.

Single crystals are indispensable for transforming, amplifying, and modulating light. Birefringent crystals serve as retarders, or wave plates, which are used to convert the polarization state of the light. In many cases, it is desirable that the crystals not only be birefringent, but also behave nonlinearly when exposed to very large fields such as those generated by intense laser beams. A few examples of such nonlinear crystals are ammonium dihydrogen phosphate (ADP), potassium dihydrogen phosphate (KDP), beta barium borate (BBO), lithium borate (LBO), and potassium titanyl phosphate (KTP). *See* CRYSTAL OPTICS; NONLINEAR OPTICS.

Other optical materials are the liquid crystals used in displays as light valves, materials used in erasable optical disks for computers and in liquid cells (Kerr cells), laser dyes, dielectric multilayer films, filter materials, and the many metals (aluminum, gold, beryllium, and so forth) and alloys that are important as coating materials. *See* COMPUTER STORAGE TECHNOLOGY; KERR EFFECT; OPTICAL RECORDING. James Steve Browder

Optical microscope An instrument used to obtain an enlarged image of a small object. In general, a compound microscope consists of a light source, a condenser, an objective, and an ocular or eyepiece, which can be replaced by a recording device such as a photoelectric tube or a photographic plate. The optical microscope is limited by the wavelengths of the light used and by the materials available for manufacturing the lenses.

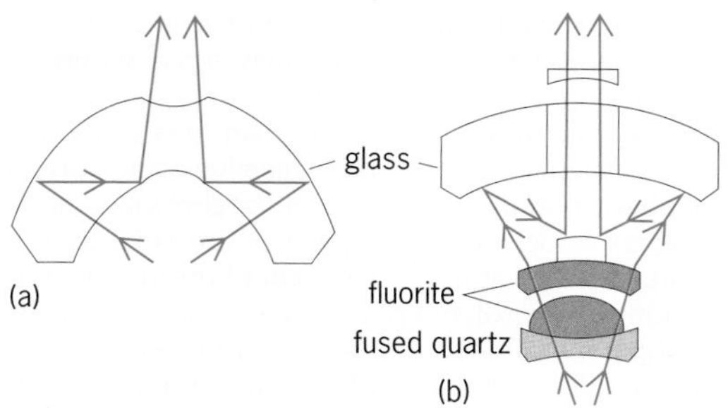

Fig. 1. Two types of catadioptric objective. (*a*) Maksutov type. (*b*) 53X, NA 0.72, ultraviolet objective, designed by Gray. Glass elements in the latter serve purely as reflectors. (*Photographic Service Department, Kodak Research Laboratory*)

Magnifying power. The quality and design of the lens system determines the magnifying power, details of image formation, and color correcting capabilities of a light microscope. The magnifying power of a compound microscope is the product of the magnification of the objective and the magnifying power of the eyepiece. The latter is computed like that of any magnifier. The magnification of the objective is equal to the distance from the second focal point to the image formed by the objective, divided by the focal length. An objective of 18-mm (0.7-in.) focal length thus has a power of 10×. It is customary to specify objectives in terms of magnifying power instead of focal length. The distance mentioned is called the optical tube length (generally 180 mm or 7 in.), and is to be distinguished from the mechanical tube length, which is the length of the mechanical tube itself. *See* MAGNIFICATION.

Catadioptric systems. Catadioptric systems have been developed for microscopes. Their great advantage is their comparatively small chromatic aberration. Pure mirror systems have no color aberrations. In catadioptric systems, therefore, it is customary to assign all the power to the mirror or mirrors, keeping the refracting system nearly afocal (Fig. 1*a*). The chromatic errors of the entire system remain small, and the refracting part can be used to correct the remaining monochromatic errors. All microscopic work in the ultraviolet region is done with catadioptric systems (Fig. 1*b*).

Condensers. An external auxiliary lens is used to condense the light from a light source so that the object is brightly and uniformly illuminated. The usual purpose of a condenser system is to make sure that as much light as possible coming from the object goes through an optical system. Condensers are used in macroscopic projection, in which an illuminated film or slide is imaged with the help of a projection objective or magnifier. In microscope systems, they are used to direct the light from a light source so that the rays from any object point fill most of the entrance pupil. A condenser system is usually arranged to image the light source onto the entrance pupil of the optical system (Köhler illumination). The condenser is generally corrected for spherical aberration, color, and sine condition, although the requirements are slightly different than in an image-forming system. Max Herzberger;

Light microscope. The mirror, condenser, oculars, and body tube of the light microscope are frequently known as the optical train. The stand, stage, and adjustments comprise the mechanical part of the microscope.

A mirror is usually attached to the substage of the microscope to reflect light along the optic axis of the microscope. When no condenser is used, the concave mirror is used because it concentrates more light on the specimen; a plane mirror is used with a condenser.

Objectives vary from a simple doublet lens to complex corrected lens systems. Achromatic objectives are corrected for spherical aberration in one color and for chromatic aberrations in two colors. Apochromatic objectives are corrected to focus three colors together and the spherical aberration is minimized for two colors. The resolving power of an objective, the least distance at which two objects can be seen to be separate, is equal to the wavelength of light λ divided by the sum of the numerical apertures of the condenser and objective used. The larger the numerical aperture, the greater is the resolving power. Objectives are described also by the equivalent focal length. Objectives of shorter focal length have less depth of field, less working distance, and greater magnification.

Photomicrographic objectives are designed to produce a flat image with little distortion. For convenience, two to five objectives can be mounted on a revolving nosepiece to be parfocal and parcentric, so that the specimen remains almost in focus at the center of the field as the objectives are changed.

The commonly used Huygenian ocular has a fairly flat field with marked pincushion distortion. Compensating oculars complete the color correction for apochromatic objectives and have less distortion, but they do have curvature of field. *See* EYEPIECE.

The monocular body tube may be of adjustable length. American microscopes are designed for a mechanical tube length of 160 mm (6.3 in.) and a cover-glass thickness of 0.18 mm (0.007 in.). The draw tube is lengthened for thinner and shortened for thicker cover glasses to correct for the spherical aberrations from cover glasses of incorrect thickness.

Binocular bodies are designed for the use of both eyes. Most binocular bodies use prisms to reflect one-half of the light to each eye. Because each eye sees the same field, these binocular bodies do not give stereoscopic vision. The binocular is often longer than the monocular body and the proper tube length is maintained with a compensating lens.

Inverted microscope. The inverted microscope has the body of the microscope, including the objective and the ocular, below the stage and the illumination above the stage for transmitted light. The inverted microscope is especially useful for the examination of surfaces. Large and awkward specimens can be moved over the stage more readily than with the usual microscope. The inverted microscope is also useful for microdissection and the observation of hanging-drop preparations and is convenient for observing chemical reactions, melting-point determinations, and photomicrography.

Comparison microscope. The comparison microscope is an arrangement of two microscopes connected by a special viewing ocular so that the field of one microscope is seen at one side of a vertical dividing line and the field of the other microscope on the opposite side of the dividing line; or it may be a projection type of microscope in which the image is compared with a template or known pattern.

Dissecting microscope. Dissecting microscopes are of two types. The simplest is a magnifying glass mounted on a support above a glass plate, used for the dissection of materials.

The more usual dissecting microscope, often called a Greenough microscope, is a stereoscopic microscope composed of two separate microscopes fastened together and used as a single unit on one stand (Fig. 2). This is a truly stereoscopic instrument because the right eye sees the specimen from the right side and the left eye from the left side. Prisms are usually included in the body tube to erect the image; thus movements of the specimen are direct and are not reversed as with the monobjective microscope.

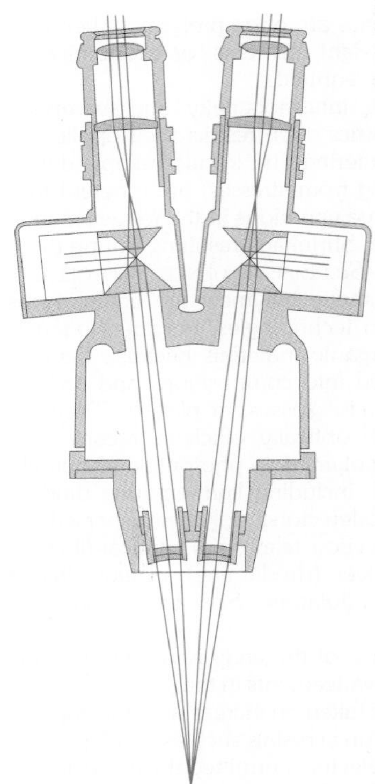

Fig. 2. Diagram of light rays as they pass through a binocular biobjective microscope.

Metallurgical microscope. The metallurgical microscope is a laboratory microscope with a focusing stage and a vertical illuminator, used primarily for the examination of metal surfaces.

Near-infrared microscopy. Near-infrared microscopy is an optical method that can be used for studying a variety of materials that are opaque in transmitted visible light (400–700 nm) yet translucent in the near infrared (700–1200 nm). The method utilizes the near-infrared optical microscope, a device for the conversion of the near-infrared image to a visible image.

Lawrence A. Harris

Near-field optical microscopy. A fundamental law of optics, the so-called diffraction limit, states that two objects can be imaged as separate entities only if their distance is larger by about one-half the wavelength of visible light, which ranges from 400 to 700 nm. As a consequence, conventional optical microscopy is restricted to a resolution of about 200 nm, and this is not enough for many important observations. The scanning near-field optical microscope (NSOM or SNOM) circumvents the diffraction limit. In contrast to nonoptical methods of surpassing this limit, such as electron microscopy, it can provide information on such qualities as color, luster, transmissivity, and birefringence, which are sensitive indicators of material composition and status. Furthermore, it operates at ambient conditions, a prerequisite for observations of living organisms. *See* MICROSCOPE. D. W. Pohl

Optical modulators

Optical modulators Devices that serve to vary some property of a light beam. The direction of the beam may be scanned as in an optical deflector, or the phase or frequency of an optical wave may be modulated. Most often, however, the intensity of the light is modulated.

Rotating or oscillating mirrors and mechanical shutters can be used at relatively low frequencies (less than 10^5 Hz). However, these devices have too much inertia to operate at much higher frequencies. At higher frequencies it is necessary to take advantage of the motions of the low-mass electrons and atoms in liquids or solids. These motions are controlled by modulating the applied electric fields, magnetic fields, or acoustic waves in phenomena known as the electrooptic, magnetooptic, or acoustooptic effect, respectively. *See* ACOUSTOOPTICS; ELECTROOPTICS; KERR EFFECT; MAGNETOOPTICS. Ivan P. Kaminow

Optical phase conjugation A process that involves the use of nonlinear optical effects to precisely reverse the direction of propagation of each plane wave in an arbitrary beam of light, thereby causing the return beam to exactly retrace the path of the incident beam. The process is also known as wavefront reversal or time-reversal reflection. The unique features of this phenomenon suggest widespread application to the problems of optical beam transport through distorting or inhomogeneous media. Although closely related, the field of adaptive optics will not be discussed here. *See* ADAPTIVE OPTICS.

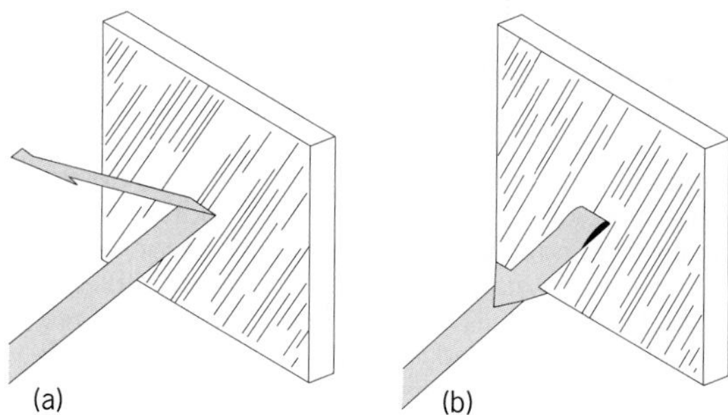

Comparison of reflections (*a*) from a conventional mirror and (*b*) from an optical phase conjugator. (*After V. J. Corcoran, ed., Proceedings for the International Conference on Laser '78 for Optical and Quantum Electronics, STS Press, McLean, Virginia, 1979*)

Optical phase conjugation is a process by which a light beam interacting in a nonlinear material is reflected in such a manner as to retrace its optical path. As the illustration shows, the image-transformation properties of this reflection are radically different from those of a conventional mirror. The incoming rays and those reflected by a conventional mirror (illus. *a*) are related by reversal of the component of the wave vector $\vec{k}$ which is normal to the mirror surface. Thus a light beam can be arbitrarily redirected by adjusting the orientation of a conventional mirror. In contrast, a phase-conjugate reflector (illus. *b*) inverts the vector quantity $\vec{k}$ so that, regardless of the orientation of the device, the reflected conjugate light beam exactly retraces the path of the incident beam. This retracing occurs even though an aberrator (such as a piece of broken glass) may be in the path of the incident beam. Looking into a conventional mirror, one would see one's own face, whereas looking into a phase-conjugate mirror, one would see only the pupil of the eye.

These new and remarkable image-transformation properties (even in the presence of a distorting optical element) open the door to many potential applications in areas such as laser fusion, atmospheric propagation, fiber-optic propagation, image restoration, real-time holography, optical data processing, nonlinear microscopy, laser resonator design, and high-resolution nonlinear spectroscopy. *See* HOLOGRAPHY; LASER; MIRROR OPTICS; NONLINEAR OPTICS; OPTICAL COMMUNICATIONS; OPTICAL FIBERS. Robert A. Fisher; Barry J. Feldman

Optical prism A simple component, made of a light-refracting and transparent material such as glass and bounded by two or more plane surfaces at an angle, that is used in optical devices, especially to change the direction of light travel, to accomplish image rotation or inversion, and to disperse light into its constituent colors. Once light enters a prism, it can be reflected one or more times before it exits the prism.

A variety of prisms can be classified according to their function. Some prisms, such as the dove prism, can be used to rotate an image and to change its parity. Image inversion by prisms in traditional binoculars is a typical application. Some prisms take advantage of the phenomenon of total internal reflection to deviate light, such as the right-angle prism and the pentaprism used in single lens reflex cameras. A thin prism is known as an optical wedge; it can be used to change slightly the direction of light travel, and therefore it can be used in pairs as an alignment device. Optical wedges are also use in stereoscopic instruments to allow the viewer to observe the three-dimensional effect without forcing the eyes to point in different directions. A variable wedge can be integrated into a commercial pair of binoculars to stabilize the line of sight in the presence of the user's slight hand movements. Other prisms such as corner-cubes can be used to reflect light backward, and are fabricated in arrays for car and bicycle retroreflectors. *See* BINOCULARS; MIRROR OPTICS; PERISCOPE; RANGEFINDER (OPTICS); REFLECTION OF ELECTROMAGNETIC RADIATION.

An important application of a prism is to disperse light. When light enters at an angle to the face of a prism, it is refracted. Since the index of refraction depends on the wavelength, the light is refracted at different angles and therefore it is dispersed into a spectrum of colors. The blue color is refracted more than the red. When light reaches the second face of the prism, it is refracted again and the initial dispersion can be added to or canceled, depending on the prism angle. A combination of prisms in tandem can increase the amount of light dispersion. Dispersing prisms have been used in monochromators and spectroscopic instruments. With two prisms of different materials, it is possible to obtain light deviation without dispersion (an achromatic prism) or dispersion without deviation. *See* DISPERSION (RADIATION); REFRACTION OF WAVES; SPECTROSCOPY. Jose M. Sasian

Optical projection systems Optical projection is the process whereby a real image of a suitably illuminated object is formed by an optical system in such a manner that it can be viewed, photographed, or otherwise observed, Essential equipment in an optical projection system consists of a light source, a condenser, an object holder, a projection lens, and (usually) a screen on which the image is formed (Fig. 1). For some important applications of optical projection *see* CINEMATOGRAPHY.

The luminance of the image in the direction of observation will depend upon (1) the average luminance of the image of the light source as seen through the projection lens from the image point under consideration, (2) the solid angle subtended by the exit pupil of the projection lens at this image point, and (3) the reflective or transmissive characteristics of the screen. Usually it is desirable to have this luminance as high as possible. Therefore, with a given screen, tens, and projection distance, the best arrangement is to have the light source imaged in the projection lens, with its image filling the exit pupil as completely and as uniformly as possible.

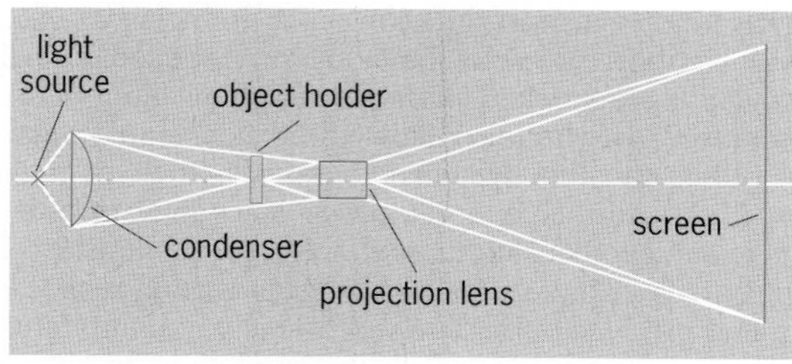

Fig. 1. A simple optical projection system.

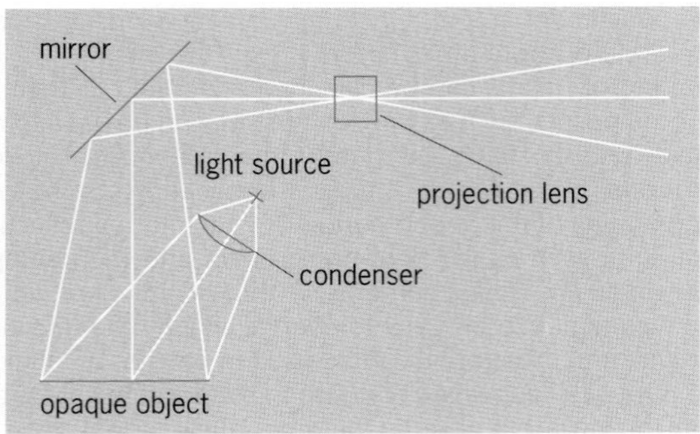

Fig. 2. An epidiascope, or system for projecting an image of an opaque object.

The object is placed between the condenser and the projection lens. If transparent, it can be inserted directly in the light beam; however, it should be positioned, and the optical system should be so designed that it does not vignette (cut off) any of the image of the light source in the projection lens. If the object is opaque, an arrangement known as an epidiascope (Fig. 2) is used. Armin J. Hill

Optical pulses Bursts of electromagnetic radiation of finite duration. Optical pulses are used to transmit information or to record the chronology of physical events. The simplest example is the photographic flash. This was probably first developed by early photographers who used flash powder that, when ignited, produced a short burst of intense light. This was followed by the flash lamp, in which a tube filled with an inert gas such as xenon is excited by a brief electrical pulse. A great advance in the creation of short optical pulses came with the invention of the laser. Lasers are now the most common and effective way of generating a variety of short optical pulses, of different durations, energies, and wavelengths. *See* LASER; STROBOSCOPIC PHOTOGRAPHY.

Pulses of millisecond (10^{-3} s) duration are very simply generated by mechanically modulating a constant light source such as a lamp or a continuous-wave laser. This can be done, for example, by placing a rotating disk with holes in it in front of the light source. Shorter laser pulses, of microsecond (10^{-6} s) or nanosecond (10^{-9} s) duration, are generated by using a technique known as Q-switching. A modulating device is incorporated inside the laser cavity that allows the buildup of the laser radiation inside the cavity and then switches it out in an instant. The modulating device is usually controlled by external electrical pulses. Semiconductor diode lasers, which are used to transmit information (voice or data) over a fiber-optic cable, are pumped by electricity and can be directly pulsed by applying to them a pulsed electrical signal. *See* OPTICAL COMMUNICATIONS; OPTICAL FIBERS.

Ultrashort laser pulses, with durations of the order of picoseconds (1 ps = 10^{-12} s) or femtoseconds (1 fs = 10^{-15} s), are generated by using a general principle known as mode locking, whereby several frequency modes of the laser structure are made to resonate simultaneously and with a well-orchestrated relationship so as to form a short-duration pulse at the laser output.

Pulses as short as 11 fs have been produced directly by a passively mode-locked titanium:sapphire laser. The titanium:sapphire laser has also allowed the extension of ultrashort optical pulses to other wavelength ranges, such as the near-infrared (2–10 μm). Dye lasers, based on organic dyes in solution, have achieved durations as short as 27 fs. Ultrashort diode laser pulses have been obtained by active and passive mode locking and produce pulses as short as a few hundred femtoseconds. They are more commonly operated so as to give rise to pulses in the picosecond range, appropriate for optical communication systems.

The generation of ultrashort laser pulses has been motivated by the quest for ever better resolution in the study of the temporal evolution and dynamics of physical systems, events, and processes. Such laser pulses are capable of creating snapshots in time of many events that occur on the atomic or molecular scale, a technique known as time-resolved spectroscopy. This stroboscopic aspect of ultrashort laser pulses is their most important scientific application and is used in physics, engineering, chemistry, and biology. For example, ultrashort pulses can excite and take snapshots of molecular vibrations and deformations. They can track the passage of charge carriers through a microscopic semiconductor device. This ability to understand the dynamics of the more elemental building blocks of nature can in turn make it possible to build ever faster devices for use in information processing and information transmission, in addition to providing a better understanding of the physical world. *See* LASER PHOTOCHEMISTRY; LASER SPECTROSCOPY; OPTICAL INFORMATION SYSTEMS; ULTRAFAST MOLECULAR PROCESSES. Philippe C. Becker

Optical pumping The process of causing strong deviations from thermal equilibrium populations of selected quantized states of different energy in atomic or molecular systems by the use of optical radiation (that is, light of wavelengths in or near the visible spectrum), called the pumping radiation.

Optical pumping is vital for light amplification by stimulated emission in an important class of lasers. For example, the action of the ruby laser involves the fluorescent emission of red light by a transition from an excited level E_2 to the ground level E_1. In this case E_2 is relatively high above E_1 and the equilibrium population of E_2 is practically zero. Amplification of the red light by laser action requires that number of atoms N_2 exceed N_1 (population inversion). The inversion is accomplished by intense green and violet light from an external source which excites the chromium ion in the ruby to a band of levels, E_3 above E_2. From E_3 the ion rapidly drops without radiation to E_2, in which its lifetime is relatively long for an excited state. Sufficiently intense pumping forces more luminescent ions into E_2 by way of the E_3 levels than remain in the ground state E_1, and amplification of the red emission of the ruby by stimulated emission can then occur. *See* LASER. William West

Optical recording The process of recording signals on a medium through the use of light, so that the signals may be reproduced at a subsequent time. Photographic film has been widely used as the medium, but in the late 1970s development of another medium, the so-called optical disk, was undertaken. The introduction of the laser as a light source greatly improves the quality of reproduced signals. The pulse-code modulation (PCM) techniques make it possible to obtain extremely high-fidelity reproduction of sound signals in optical disk recording systems.

Optical film recording. Optical film recording is also termed motion picture recording or photographic recording. A sound motion picture recording system consists basically of a modulator for producing a modulated light beam and a mechanism for moving a light-sensitive photographic film relative to the light beam and thereby recording signals on the film corresponding to the electrical signals. A sound motion picture reproducing system is basically a combination of a light source, an optical system, a photoelectric cell, and a mechanism for moving a film carrying an optical record by means of which the recorded photographic variations are converted into electrical signals of approximately similar form.

In laser-beam film recording, an optical film system utilizes a laser as a light source, a combination of an acoustooptical modulator (AOM) and an acoustooptical deflector (AOD) instead of a galvanometer. A 100-kHz pulse-width modulation (PWM) circuit converts the audio input signal into a PWM signal. The laser beam is made to continuously scan the sound track area

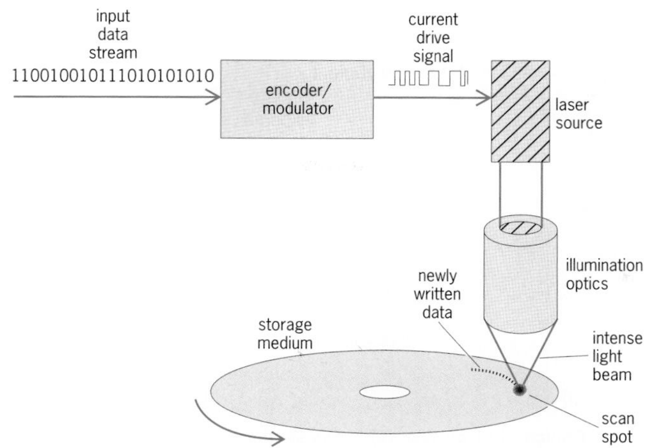

Fig. 1. Recording process for a simple optical medium. Writing data into the recording layers involves modulating an intense laser beam as the layers move under the scan spot.

at right angles to the direction of the film transport. This is done by means of the acoustooptical deflector, which in turn is driven by a 100-kHz sawtooth signal. Simultaneously, the laser beam is pulse-width-modulated by means of the acoustooptical modulator, which is driven by a 100-kHz PWM signal. The scanning signal and the pulse-width-modulated signal combine and generate the variable-area sound track exposure on the film. The traces of successive scans are fused into a pattern of variable-area recording. H. Date

Optical data storage. Optical data storage involves placing information in a medium so that, when a light beam scans the medium, the reflected light can be used to recover the information. There are many forms of storage media, and many types of systems are used to scan data.

In the recording process (Fig. 1), an input stream of digital information is converted with an encoder and modulator into a drive signal for a laser source. The laser source emits an intense light beam that is directed and focused into the storage medium with illumination optics. As the medium moves under the scanning spot, energy from the intense scan spot is absorbed, and a small localized region heats up. The storage medium, under the influence of the heat, changes its reflective properties. Since the light beam is modulated in correspondence to the input data stream, a circular track of data marks is formed as the medium rotates. After every revolution, the path of the scan spot is changed slightly in radius to allow another track to be written.

In readout of the medium (Fig. 2), the laser is used at a constant output power level that will not heat the medium beyond its thermal writing threshold. The laser beam is directed through a beam splitter into the illumination optics, where the beam is focused into the medium. As the data to be read pass under the scan spot, the reflected light is modulated. The modulated light is collected by the illumination optics and directed by the beam splitter to the servo and data optics, which converge the light onto detectors. The detectors change the light modulation into current modulation that is amplified and decoded to produce the output data stream.

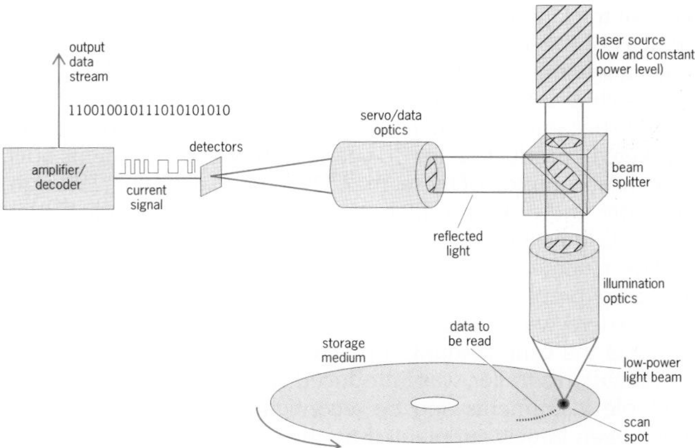

Fig. 2. Readout of an optical medium. Low-power laser beam illuminates the recording layers, and modulation of the reflected light is observed with the detectors. Beam splitter serves to direct a portion of the reflected light to detectors.

Optical media can be produced in several different configurations. The most common configuration is the single-layer disk, such as the compact disk (CD), where data are recorded in a single storage layer. A substrate provides mechanical support for the storage layer. The substrate also provides a measure of contamination protection, because light is focused through the substrate and into the recording layer. Dust particles on the surface of the substrate only partially obscure the focused beam, so enough light can penetrate for adequate signal recovery. *See* COMPACT DISK.

In order to increase data capacity of the disk, several layers can be used. Each layer is partially transmitting, which allows a portion of the light to penetrate throughout the thickness of the layers. The scan spot is adjusted by refocusing the illumination optics so that only one layer is read out at a time.

Data can also be recorded in volumetric configurations. As with the multiple-layer disk, the scan spot can be refocused throughout the volume of material to access information. Volumetric configurations offer the highest efficiency for data capacity, but they are not easily paired with simple illumination optics.

The final configuration is to place the information on a flexible surface, such as ribbon or tape. As with magnetic tape, the ribbon is pulled under the scan spot and data are recorded or retrieved. Flexible media have about the same capacity efficiency as volumetric storage. The advantage of a flexible medium over a volumetric medium is that no refocusing is necessary. The disadvantage is that a moderately complicated mechanical system must be used to move the ribbon.

There are several types of optical storage media. The most popular media are based on pit-type, magnetooptic, phase-change, and dye-polymer technologies. CD and digital versatile disc (DVD) products use pit-type technology. Erasable disks using magnetooptic (MO) technology are popular for workstation environments. Compact-disk-rewritable (CD-RW) products [also known as compact-disk-erasable (CD-E)] use phase-change technology, and compact-disk-recordable (CD-R) products use dye-polymer technology. CD and DVD products are read-only memories (ROMs); that is, they are used for software distribution and cannot be used for recording information. CD-R products can be used for recording information, but once the information is recorded, they cannot be erased and reused. Both CD-RW and MO products can be erased and reused. Tom D. Milster; Glenn T. Sincerbox

Optical rotatory dispersion The change in rotation as a function of wavelength experienced by linearly polarized light as it passes through an optically active substance. *See* OPTICAL ACTIVITY.

In all materials the rotation varies with wavelength. The variation is caused by two quite different phenomena. The first accounts in most cases for the majority of the variation in rotation and should not strictly be termed rotatory dispersion. It depends on the fact that optical activity is actually circular birefringence. In other words, a substance which is optically active transmits right circularly polarized light with a different velocity from left circularly polarized light.

In addition to this pseudodispersion which depends on the material thickness, there is a true rotatory dispersion which depends on the variation with wavelength of the indices of refraction for right and left circularly polarized light. *See* POLARIZED LIGHT. Bruce H. Billings

For wavelengths that are absorbed by the optically active sample, the two circularly polarized components will be absorbed to differing extents. This unequal absorption is known as circular dichroism. Circular dichroism causes incident linearly polarized light to become elliptically polarized. *See* ABSORPTION.

Optical rotatory dispersion and circular dichroism are closely related, just as are ordinary absorption and dispersion. If the entire optical rotatory dispersion spectrum is known, the circular dichroism spectrum can be calculated, and vice versa.

In order for a molecule (or crystal) to exhibit circular birefringence and circular dichroism, it must be distinguishable from its mirror image. An object that cannot be superimposed on its mirror image is said to be chiral, and optical rotatory dispersion and circular dichroism are known as chiroptical properties.

Most biological molecules have one or more chiral centers and undergo enzyme-catalyzed transformations that either maintain or reverse the chirality at one or more of these centers. Still other enzymes produce new chiral centers, always with a high specificity. These properties account for the fact that optical rotatory dispersion and circular dichroism are widely used in organic and inorganic chemistry and in biochemistry. *See* ENZYME; STEREOCHEMISTRY.

In the absence of magnetic fields, only chiral substances exhibit optical rotatory dispersion and circular dichroism. In a magnetic field, even substances that lack chirality rotate the plane of polarized light, as shown by M. Faraday. Magnetic optical rotation is known as the Faraday effect, and its wavelength dependence is known as magnetic optical rotatory dispersion. In regions of absorption, magnetic circular dichroism is observable. *See* FARADAY EFFECT. R. W. Woody

Optical surfaces Interfaces between different optical media at which light is refracted or reflected. From a physical point of view, the basic elements of an optical system are such things as lenses and mirrors. However, from a conceptual point of view, the basic elements of an optical system are the refracting or reflecting surfaces of such components. Surfaces are the basic elements of an optical system because they are the elements that affect the light passing through the system. Every wavefront has its curvature changed on passing through each surface so that the final set of wavefronts in the image space may converge on the appropriate image points. Also, the aberrations of the system depend on each surface, the total aberrations of the system being the sum of the aberrations generated at the individual surfaces. *See* ABERRATION (OPTICS); REFLECTION OF ELECTROMAGNETIC RADIATION; REFRACTION OF WAVES.

Optical systems are designed by ray tracing, and refraction at an optical surface separating two media of different refractive index is the fundamental operation in the process. The transfer between two surfaces is along a straight line if, as is usually the case, the optical media are homogeneous. The refraction of the ray at a surface results in a change in the direction of the ray. This change is governed by Snell's law.

The vast majority of optical surfaces are spherical in form. This is so primarily because spherical surfaces are much easier to generate than nonspherical, or aspheric, surfaces. Moreover, lens systems seldom need aspherics because the aberrations can be controlled by changing the shape of the component lenses without changing their function in the system, apart from modifying the aberrations. Also, many lens components can be included in a lens system in order to control the aberrations. *See* LENS (OPTICS).

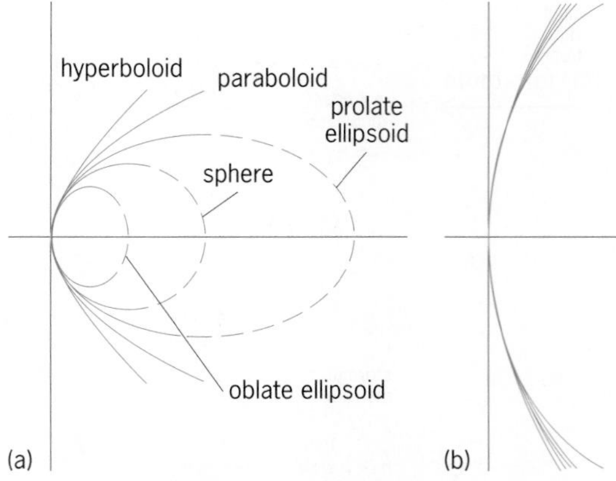

Conics of revolution. (*a*) Cross sections of entire surfaces. (*b*) Cross sections of portions near the optical axis.

On the other hand, mirror systems usually require aspheric surfaces. Unlike lenses, where the shape can be changed to modify the aberrations, mirrors cannot be changed except by introducing aspheric surfaces. Mirror systems are further constrained by the fact that only a few mirrors, usually two, are used in a system because each successive mirror occludes part of the beam going to the mirror preceding it. *See* MIRROR OPTICS.

The most common form of rotationally symmetric surface is the conic of revolution. The departure of conic surfaces from spherical form is shown in the illustration. The classical virtue of the conics of revolution for mirrors is the fact that light from a point located at one focus of the conic is perfectly imaged at the other focus. If these conic foci are located on the axis of revolution, the mirror is free of spherical aberration for such conjugate points. *See* CONIC SECTION. Roland V. Shack

Optical tracking systems Multipurpose instruments used to make measurements on a remote object, often an airborne vehicle. These systems are used to provide two basic types of data: accurate measurement of the position, velocity, and other motion parameters of the target; and information about the target, such as images or optical spectra.

Many optical tracking systems are used for weapon system qualification. A system tracking a towed aerial target can accurately determine the miss distance of an interceptor missile. One tracking a military aircraft can measure its rate of climb or the release characteristics of its ordnance.

Another common use of optical tracking systems is for range safety at missile and rocket launch sites. A remotely operated optical tracking system can provide real-time position data and images of the launch vehicle, assuring that all technical parameters are within specification without risk to personnel.

The cinetheodolite was developed in 1952. These instruments quickly became standard equipment at test ranges. Various models were in production for 30 years, with over 400 systems built. The most common of these, the EOTS-F, had a 7.5-in. (190-mm) aperture and a 35-mm high-speed film camera. Nearly 100 of these systems are still in use.

The cinetheodolites were extremely accurate, but required specially prepared sites and had a single optical configuration. There was a need for a more versatile instrument, and in 1963 the Cine Sextant, a precision two-axis tracking system mounted on a trailer, was introduced. These systems could carry multiple instruments and be reconfigured for each test. They have been largely supplanted by the Compact Tracking Mount (CTM) and the KINETO tracking mount (see illustration). These somewhat smaller mounts have all-electric drives instead of the hydraulic/electric combination on the Cine Sextant and as a result are considerably less expensive. Over 150 KINETOs have

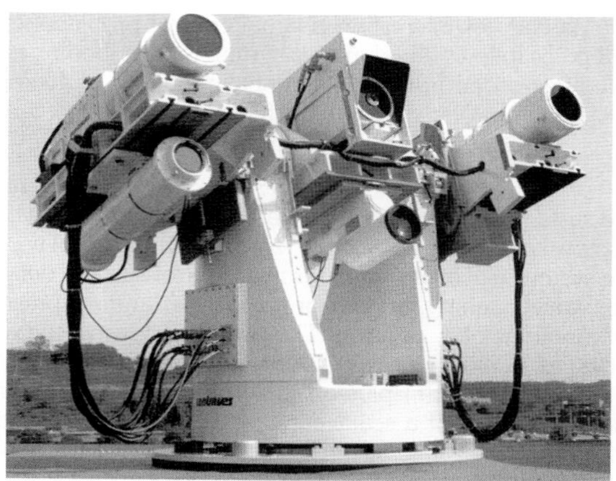

KINETO tracking mount. (*L-3 Communications Corp.*)

been produced, making it the most common system in use at test ranges.

Optical tracking systems use long-focal-length, telephoto-type lenses. The system resolution is the angular extent of the smallest discernible features. The angular velocity is a measure of how fast the system rotates. A jet flying at the speed of sound 1 mi (1.6 km) from the tracking system requires a line-of-sight rate of 12°/second. A KINETO tracking mount can move at eight times that rate.

Optical tracking systems have two basic modes of operation, feedback tracking and ephemeris tracking. With feedback tracking, the position of the target in the field of view of the tracking sensor provides information that corrects the trajectory of the system. Feedback tracking can be as simple as an operator moving a joystick based on what he or she sees on a screen, or it can employ an automatic video tracker (AVT), a specialized image-processing computer that measures the target position in the sensor field of view. The most common type of AVT is called a centroid tracker because it determines the outline of the target and then calculates its center.

An ephemeris is a mathematical prediction of the path of a celestial body, the usual means of tracking a satellite. Most satellites are visible only when the Sun is shining on them and the angle between the Sun and the observer is favorable. By accurately calibrating the pointing of the tracking system and commanding it from the same equations that determine the motion of the satellite, the line of sight can be directed correctly without feedback. *See* EPHEMERIS.

Historically, the instrument of choice for optical tracking systems has been the high-speed film camera. Digital cameras have largely replaced film cameras on the ranges, partly due to the desire for real-time data and partly due to the expense and hazard of film processing. An increasing number of optical tracking systems employ infrared cameras. These sensors do not have the same resolution as visible-band cameras, but have the advantage of being able to track objects that are not sunlit or that have poor contrast. It is often necessary to measure the range to a target. This is accomplished by adding a radar ranging system to the optical tracking system or by incorporating a laser rangefinder. Most laser rangefinders now meet standards for eye safety and can easily be added to a KINETO or other configurable tracking system. *See* CAMERA; INFRARED IMAGING DEVICES; LASER; RADAR.

James E. Kimbrell

Optics Narrowly, the science of light and vision; broadly, the study of the phenomena associated with the generation, transmission, and detection of electromagnetic radiation in the spectral range extending from the long-wave edge of the x-ray region to the short-wave edge of the radio region. This range, often called the optical region or the optical spectrum, extends in wavelength from about 1 nanometer to about 1 millimeter. *See* GEOMETRICAL OPTICS; METEOROLOGICAL OPTICS; PHYSICAL OPTICS; VISION.

The discoveries of the experimentalists of the early seventeenth century formed the basis of the science of optics. The statement of the law of refraction, the development of the astronomical telescope, observations of diffraction, and the principles of the propagation of light all came in this relatively short period. The publication of Isaac Newton's *Opticks* in 1704, with its comprehensive and original studies of refraction, dispersion, interference, diffraction, and polarization, established the science.

In the early nineteenth century many productive investigators established the transverse-wave nature of light. The relationship between optical and magnetic phenomena led to the crowning achievement of classical optics—the electromagnetic theory of J. C. Maxwell. Maxwell's theory, which holds that light consists of electric and magnetic fields propagated together through space as transverse waves, provided a general basis for the treatment of optical phenomena. In particular, it served as the basis for understanding the interaction of light with matter and, hence, as the basis for treatment of the phenomena of physical optics. *See* ELECTROMAGNETIC RADIATION; LIGHT; MAXWELL'S EQUATIONS.

In the twentieth century optics has been in the forefront of the revolution in physical thinking caused by the theory of relativity and especially by the quantum theory.

The science of optics finds itself in a position that is satisfactory for practical purposes but less so from a theoretical standpoint. The theory of Maxwell is sufficiently valid for treating the interaction of high-intensity radiation with systems considerably larger than those of atomic dimensions. The modern quantum theory is adequate for an understanding of the spectra of atoms and molecules and for the interpretation of phenomena involving low-intensity radiation, provided one does not insist on a very detailed description of the process of emission or absorption of radiation. However, a general theory of relativistic quantum electrodynamics valid for all conditions and systems has not been worked out.

The development of the laser has been an outstanding event in the history of optics. The theory of electromagnetic radiation from its beginnings was able to comprehend and treat the properties of coherent radiation, but the controlled generation of coherent monochromatic radiation of high power was not achieved in the optical region until the work of C. H. Townes and A. L. Schawlow in 1958 pointed the way. Many achievements in optics, such as holography and interferometry over long paths, have resulted from the laser. *See* HOLOGRAPHY; INTERFEROMETRY; LASER.

Richard C. Lord

Optimal control (linear systems) A branch of modern control theory that deals with designing controls for linear systems by minimizing a performance index that depends on the system variables. Under some mild assumptions, making the performance index small also guarantees that the system variables will be small, thus ensuring closed-loop stability. *See* CONTROL SYSTEM STABILITY.

Classical control theory applies directly in the design of controls for systems that have one input and one output. Complex modern systems, however, have multiple inputs and outputs. Examples include aircraft, satellites, and robot manipulators, which, though nonlinear, can be linearized about a desired operating point. Modern control theory was developed, beginning about 1960, for these multivariable systems. It is characterized by a state-space description of systems, which involves the use of matrices and linear algebra, and by the use of optimization techniques to determine the control policies. These techniques facilitate the use of modern digital computers, which lead to an interactive approach to design of controls for complex systems. *See* DIGITAL COMPUTER; LINEAR ALGEBRA; MATRIX THEORY; MULTIVARIABLE CONTROL; OPTIMIZATION.

The multivariable state-variable description is of the form of Eqs. (1), where $u(t)$ is an m-dimensional control input vector and

$$x = Ax + Bu \qquad z = Hx \tag{1}$$

$z(t)$ is a performance output vector for which there are specified performance requirements. The state $x(t)$ is an internal variable of dimension n that describes the energy storage properties of the system. Matrices A and B describe the system dynamics and are determined by a physical analysis using, for example, Newton's laws of motion. They may in general be time-varying. Matrix H is chosen to select the variables of importance as the performance outputs. In the case of nonlinear systems, the description of Eq. (1) results when the system is linearized about a desired operating point. *See* LINEAR SYSTEM ANALYSIS; NONLINEAR CONTROL THEORY.

Traditionally, modern control-system design assumes that all the states $x(t)$ are available for feedback so that the control input is of the form of Eq. (2), where $K(t)$ is an $m \times n$ feedback gain

$$u = -Kx \tag{2}$$

matrix, generally time-varying. Substituting the control of Eq. (2) into the system of Eq. (1) yields the closed-loop system given in Eqs. (3). The control design problem is to choose the mn entries

$$\dot{x} = (A - BK)x \qquad z = Hx \tag{3}$$

of the feedback matrix K to yield a desired closed-loop behavior of the performance output $z(t)$.

To obtain satisfactory performance of the system of Eq. (1), a quadratic performance index of the form of Eq. (4) may be

$$J = {}^1/_2 x^T(T)S(T)x(T) + {}^1/_2 \int_0^T (x^T Qx + u^T Ru)\,dt \tag{4}$$

chosen, where $[0, T]$ is the time interval of interest and the symmetric weighting matrices $S(T)$, Q, and R are design parameters that are selected to obtain the required performance. They must be chosen so that $x^T(T)S(T)x(T) \geq 0$ and $x^T Qx \geq 0$ for all $x(t)$, and $u^T Ru > 0$ for all $u(t)$. That is, Q and $S(T)$ should be positive semidefinite while R should be positive definite. Frank L. Lewis

Optimal control theory An extension of the calculus of variations for dynamic systems with one independent variable, usually time, in which control (input) variables are determined to maximize (or minimize) some measure of the performance (output) of a system while satisfying specified constraints. Theory is conveniently divided into two parts: optimal programming, where the control variables are determined as functions of time for a specified initial state of the system, and optimal feedback control, where the control variables are determined as functions of the current state of the system.

Examples of optimal control problems are: (1) determining paths of vehicles between two points to minimize fuel or time, and (2) determining feedback control logic for vehicles or industrial processes to keep them near a desired operating point in the presence of disturbances with acceptable control magnitudes.

Dynamic systems are conveniently divided into two categories: continuous dynamic systems, where the control and state variables are functions of a continuous independent variable, such as time or distance, and discrete dynamic systems, where the independent variable changes in discrete increments. Many discrete systems are discretized versions of continuous systems; the discretization is often made so that (1) the system can be analyzed or controlled by digital computers (or both), or (2) measurements of continuous outputs are made at discrete intervals of time (sampled-data systems) in order to share data transmission channels.

A large class of interesting optimal control problems can be described as follows: the dynamic system is described by a set of coupled first-order ordinary differential equations of the form of Eq. (1), where x (the state vector) represents the state variables

$$\dot{x} = f(x, u, t) \tag{1}$$

of the system (such as position, velocity, temperature, voltage, and so on); u (the control vector) represents the control variables of the system (such as motor torque, control-surface deflection angle, valve opening, and so on); t is time; and f represents a set of functions of x, u, and t. The performance index J, which one desires to minimize, is a scalar function of the final time t_f and the final state $x(t_f)$ plus an integral from the initial time t_0 to the final time t_f of a scalar function of the state and control vectors and time, as given in Eq. (2). Possible constraints in-

$$J = \varphi[x(t_f), t_f] = \int_{t_0}^{t_f} L[x(t), u(t), t]\,dt \tag{2}$$

clude: (a) specified vector functions constraining the initial time t_0 and the initial state $x(t_0)$, as in Eq. (3); (b) specified vector

$$\alpha[x(t_0), t_0] = 0 \tag{3}$$

functions constraining the final time and final state, as in Eq. (4);

$$\psi[x(t_f), t_f] = 0 \tag{4}$$

(c) specified vector functions constraining the control variables [inequality (5)]; (d) specified vector functions constraining both

$$C[u(t), t] \leq 0 \tag{5}$$

control and state variables [inequality (6)]; and (e) specified vec-

$$CS[x(t), u(t), t] \leq 0 \tag{6}$$

tor functions constraining only the state variables [inequality (7)].

$$S[x(t), t] \leq 0 \tag{7}$$

Constraints a and b are equality constraints, whereas constraints c, d, and e are inequality constraints. *See* LINEAR SYSTEM ANALYSIS.

Classes of problems. There are several important classes of optimal control problems.

Bang-bang control is a type of control in which the control variables are either at their maximum or at their minimum values, never in between. This type of control is optimal, for example, in certain minimum-fuel problems where the control variables enter linearly into system equations (1), the performance index in Eq. (2), and the constraints in Eq. (5). *See* NONLINEAR CONTROL THEORY.

Bang-zero-bang control is a type of control in which the control values are at their maximum values, zero, or their minimum values. This type of control is optimal, for example, in certain minimum-fuel problems where the control variables enter linearly into system equations (1) and the constraints in Eq. (5), and the performance index depends on the magnitude (but not the sign) of the control variables.

Linear-quadratic problems are problems in which system equations (1) are linear and the performance index in Eq. (2) is quadratic in the states x and the control u.

Neighboring optimum feedback control (NOFC) about a nominal optimal path is a viable alternative to a "complete" optimal feedback control solution. NOFC constitutes an approximation to the optimal feedback control solution in a region neighboring the nominal optimal path.

Stochastic optimal control theory. This takes into consideration, in an ensemble average sense, random disturbances on the dynamic system, random initial conditions, and random errors in measurements. A control design is sought that maximizes (or minimizes) the performance criterion on the average and satisfies the constraints within certain tolerances, also on the average. *See* ESTIMATION THEORY; STOCHASTIC CONTROL THEORY; STOCHASTIC PROCESS.

Differential games. These are two-sided optimal control problems. Typically, control system A tries to minimize the performance criterion, while control system B tries to maximize it. This may lead to a minimax feedback strategy in which both systems do their best, taking into account the other controller's intelligence in seeking the opposite goal. *See* CONTROL SYSTEMS; DECISION THEORY; GAME THEORY. Arthur E. Bryson, Jr.

Optimization

Optimization The design and operation of systems or processes to make them as good as possible in some defined sense. The approaches to optimizing systems are varied and depend on the type of system involved, but the goal of all optimization procedures is to obtain the best results possible (again, in some defined sense) subject to restrictions or constraints that are imposed. While a system may be optimized by treating the system itself, by adjusting various parameters of the process in an effort to obtain better results, it generally is more economical to develop a model of the process and to analyze performance changes that result from adjustments in the model. In many applications, the process to be optimized can be formulated as a mathematical model; with the advent of high-speed computers, very large and complex systems can be modeled, and optimization can yield substantially improved benefits.

Optimization is applied in virtually all areas of human endeavor, including engineering system design, optical system design, economics, power systems, water and land use, transportation systems, scheduling systems, resource allocation, personnel planning, portfolio selection, mining operations, blending of raw materials, structural design, and control systems. Optimizers or decision makers use optimization in the design of systems and processes, in the production of products, and in the operation of systems.

The first step in modern optimization is to obtain a mathematical description of the process or the system to be optimized. A mathematical model of the process or system is then formed on the basis of this description. Depending on the application, the model complexity can range from very simple to extremely complex. An example of a simple model is one that depends on only a single nonlinear algebraic function of one variable to be selected by the optimizer (the decision maker). Complex models may contain thousands of linear and nonlinear functions of many variables. As part of the procedure, the optimizer may select specific values for some of the variables, assign variables that are functions of time or other independent variables, satisfy constraints that are imposed on the variables, satisfy certain goals, and account for uncertainties or random aspects of the system.

System models used in optimization are classified in various ways, such as linear versus nonlinear, static versus dynamic, deterministic versus stochastic, or time-invariant versus time-varying. In forming a model for use with optimization, all of the important aspects of the problem should be included, so that they will be taken into account in the solution. The model can improve visualization of many interconnected aspects of the problem that cannot be grasped on the basis of the individual parts alone. A given system can have many different models that differ in detail and complexity. Certain models (for example, linear programming models) lend themselves to rapid and well-developed solution algorithms, whereas other models may not. When choosing between equally valid models, therefore, those that are cast in standard optimization forms are to be preferred. *See* MODEL THEORY.

The model of a system must account for constraints that are imposed on the system. Constraints restrict the values that can be assumed by variables of a system. Constraints often are classified as being either equality or inequality constraints. The types of constraints involved in any given problem are determined by the physical nature of the problem and by the level of complexity used in forming the mathematical model.

Constraints that must be satisfied are called rigid constraints. Physical variables often are restricted to be nonnegative; for example, the amount of a given material used in a system is required to be greater than or equal to zero. Rigid constraints also may be imposed by government regulations or by customer-mandated requirements. Such constraints may be viewed as absolute goals.

In contrast to rigid constraints, soft constraints are those constraints that are negotiable to some degree. These constraints can be viewed as goals that are associated with target values. The amount that the goal deviates from its target value could be considered in evaluating trade-offs between alternative solutions to the given problem.

When constraints have been established, it is important to determine if there are any solutions to the problem that simultaneously satisfy all of the constraints. Any such solution is called a feasible solution, or a feasible point in the case of algebraic problems. The set of all feasible points constitutes the feasible region.

If no feasible solution exists for a given optimization problem, the decision maker may relax some of the soft constraints in an attempt to create one or more feasible solutions; a class of approaches to optimization under the general heading of goal programming may be employed to relax soft constraints in a systematic way to minimize some measure of maximum deviations from goals.

A key step in the formulation of any optimization problem is the assignment of performance measures (also called performance indices, cost functions, return functions, criterion functions, and performance objectives) that are to be optimized. The success of any optimization result is critically dependent on the selection of meaningful performance measures. In many cases, the actual computational solution approach is secondary. Ways in which multiple performance measures can be incorporated in the optimization process are varied. Donald A. Pierre

Oral glands

Oral glands Glands located in the mouth that secrete fluids to moisten and lubricate the mouth and food and may initiate digestive activity, and that may perform other specialized functions. Fishes and aquatic amphibians have only solitary mucus-secreting cells in the epithelium of the mouth cavity. Multicellular glands first appeared in land animals to keep the mouth moist and make food easier to swallow. Some glands of terrestrial amphibians have a lubricative secretion; others serve to make the tongue sticky for use in catching insects. Some frogs secrete a serous fluid that contains ptyalin, a digestive enzyme. The oral glands of reptiles are much the same, but are more distinctly grouped. In poisonous snakes and the single poisonous lizard, the Gila monster, certain oral glands of the serous type are modified to form venom. Also many of the lizards have glands that are mixed in character, containing both mucous and serous cells. Oral glands are poorly developed in crocodilians and sea turtles. Birds bolt their food, yet grain-eaters have numerous glands, some of which secrete ptyalin.

All mammals except aquatic forms are well supplied with oral glands. There are numerous small glands, such as the labial glands of the lips, buccal glands of the cheeks, lingual glands of the tongue, and palatine glands of the palate. Besides these, there are larger paired sets in mammals that are quite constant from species to species and are commonly designated as salivary glands (see illustration). The parotid gland, near each ear, discharges into the vestibule. The submaxillary or submandibular gland lies along the posterior part of the lower jaw; its duct opens well forward under the tongue. The sublingual gland lies

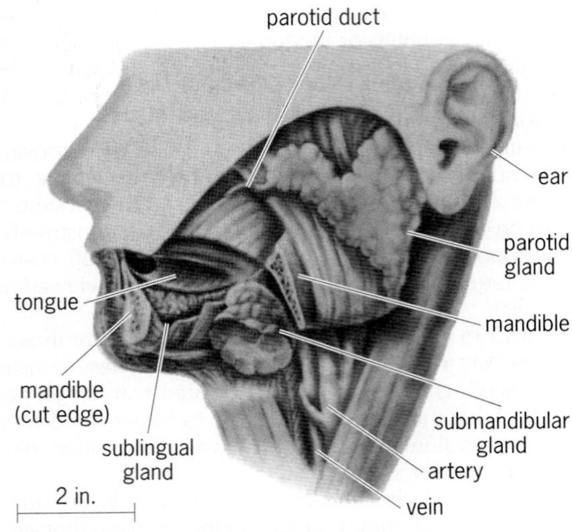

Salivary glands, shown by a partial dissection of the head. (*After J. C. Brash, ed., Cunningham's Textbook of Anatomy, 9th ed., Oxford, 1951*)

in the floor of the mouth. It is really a group of glands, each with its duct. *See* GLAND. Leslie B. Arey

Orange The sweet orange (*Citrus sinensis*) is the most widely used species of citrus fruit and commercially is the most important. The sour or bitter oranges, of lesser importance, are distinct from sweet oranges and are classified as a separate species, *C. aurantium*. The United States is the largest producer of oranges, followed by Spain, Italy, and Brazil. The orange is also a major crop in several other countries.

Sweet orange fruit is consumed fresh or as frozen or canned juice. A large portion of the crop, particularly in the United States, is used as frozen concentrate. After the juice is extracted, the peel and pulp are used for cattle feed. Peel oil is used in perfumes and flavoring, and citrus molasses is used as a livestock feed.

The sweet orange tree is a moderately vigorous evergreen with a rounded, densely foliated top. The fruits are round or somewhat elongate and usually orange-colored when ripe. They can be placed in four groups: the common oranges, acid-less oranges, pigmented oranges, and navel oranges. They may also be distinguished on the basis of early midseason, and late maturity. *See* FRUIT; FRUIT, TREE. R. K. Soost

Orbital motion In astronomy the motion of a material body through space under the influence of its own inertia, a central force, and other forces. Johann Kepler found empirically that the orbital motions of the planets about the Sun are ellipses. Sir Isaac Newton, starting from his laws of motion, proved that an inverse-square gravitational field of force requires a body to move in an orbit that is a circle, ellipse, parabola, or hyperbola.

Two bodies revolving under their mutual gravitational attraction, but otherwise undisturbed, describe orbits of the same shape about a common center of mass. The less massive body has the larger orbit. In the solar system, the Sun and Jupiter have a center of mass just outside the visible disk of the Sun. For each of the other planets, the center of mass of Sun and planet lies within the Sun.

For this reason, it is convenient to consider only the relative motion of a planet of mass m about the Sun of mass M as though the planet had no mass and moved about a center of mass $M + m$. The orbit so determined is exactly the same shape as the true orbits of planet and Sun about their common center of mass, but it is enlarged in the ratio $(M + m)/M$. *See* CENTER OF MASS; PLANET.

Orbital velocity v of a planet moving in a relative orbit about the Sun may be expressed by Eq. (1) where a is the semimajor

$$v^2 = G(M + m)\left(\frac{2}{r} - \frac{1}{a}\right) \tag{1}$$

axis, and r is the distance from the planet to the Sun. In the special case of a circular orbit, $r = a$, and the expression becomes Eq. (2). When the eccentricity of an orbit is exactly unity, the length

$$v^2 = \frac{G(M + m)}{a} \tag{2}$$

of the major axis becomes infinite and the ellipse degenerates into a parabola. The expression for the velocity then becomes Eq. (3). This parabolic velocity is referred to as the velocity of

$$v^2 = G(M + m)\left(\frac{2}{r}\right) \tag{3}$$

escape, since it is the minimum velocity required for a particle to escape from the gravitational attraction of its parent body. *See* ESCAPE VELOCITY.

Eccentricities greater than unity occur with hyperbolic orbits. Because in a hyperbola the semimajor axis a is negative, hyperbolic velocities are greater than the escape velocity.

Parabolic and hyperbolic velocities seem to be observed in the motions of some comets and meteors. Aside from the periodic ones, most comets appear to be visitors from cosmic distances, as do about two-thirds of the fainter meteors. It is possible that many of the "parabolic" comets are actually moving in elliptical orbits of extremely long period. The close approach of one of these visitors to a massive planet, such as Jupiter, could change the velocity from parabolic to elliptical if retarded, or from parabolic to hyperbolic if accelerated. It is possible that many of the periodic comets, especially those with periods under 9 years, have been captured in this way. *See* COMET; GRAVITATION; PERTURBATION (ASTRONOMY). Raynor L. Duncombe

Orchid Any member of the orchid family (Orchidaceae), among the largest families of plants, estimated to contain up to 35,000 species. Orchids are monocots; their flowers have inferior ovaries, three sepals, and three petals. They are distinguished by the differentiation of one petal into a labellum, and the fusion of pistil and stamens into the column. Pollen is usually contained in pollinia, that is, bundles that are removed intact by pollinators, usually insects or sometimes birds. Self-pollination and asexual reproduction without fertilization also occur. The combination of lip and column structure, flower color, fragrance, and other factors may limit the range of pollinators. Differing pollination mechanisms often provide barriers to cross-pollination between related species. Each flower can produce large quantities of seeds, with numbers in the millions in some tropical species. Seeds are minute, with undifferentiated embryo and no endosperm. Germination and establishment depends on symbiotic mycorrhizae that provide nutrients and water.

Orchids occur on all continents except Antarctica; they range from arctic tundra and temperate forest and grassland to tropical rainforest, where as epiphytes they reach their greatest abundance and diversity. Vanilla is obtained from seed pods of some species of *Vanilla*, and the beauty and mystique of many orchids make them important horticultural subjects. *See* MYCORRHIZAE; ORCHIDALES; VANILLA. Charles J. Sheviak

Orchidales An order of flowering plants, division Magnoliophyta (Angiospermae), in the subclass Liliidae of the class Liliopsida (monocotyledons). The order consists of four families: the Orchidaceae (see illustration), with perhaps 15,000–20,000 species; the Burmanniaceae, with about 130 species;

Eastern American species of moccasin flower (*Cypripedium acaule*). (*U.S. Forest Service photograph by R. Dale Sanders*)

the Cordiaceae, with 9 species; and the Geosiridaceae, with 1 species. The Orchidales are mycotrophic, sometimes nongreen Liliidae with very numerous tiny seeds that have an undifferentiated embryo and little or no endosperm. The ovary is always inferior and apparently lacks the septal nectaries found in many Liliopsida although other kinds of nectaries are present. *See* FLOWER; LILIIDAE; LILIOPSIDA; MAGNOLIOPHYTA; ORCHID.

Arthur Cronquist; T. M. Barkley

Ordovician The second-oldest period in the Paleozoic Era. The Ordovician is remarkable because not only did one of the most significant Phanerozoic radiations of marine life take place (early Middle Ordovician), but also one of the two or three most severe extinctions of marine life occurred (Late Ordovician). The early Middle Ordovician radiation of life included the initial colonization of land. These first terrestrial organisms were nonvascular plants. Vascular plants appeared in terrestrial settings shortly afterward. *See* GEOLOGIC TIME SCALE.

The rocks deposited during this time interval (these are termed the Ordovician System) overlie those of the Cambrian and underlie those of the Silurian. The Ordovician Period was about 7×10^7 years in duration, and it lasted from about 5.05×10^8 to about 4.35×10^8 years ago.

The Ordovician System is recognized in nearly all parts of the world, including the peak of Mount Everest, because the groups of fossils used to characterize the system are so broadly delineated. Biogeographic provinces limited the distribution of organisms in the past to patterns similar to those of modern biogeographic provinces. Three broadly defined areas of latitude—the tropics, the midlatitudes (approximately 30–60°S), and the Southern Hemisphere high latitudes—constitute the biogeographic regions. Provinces may be distinguished within these three regions based upon organismal associations unique to each province.

Early Ordovician environmental conditions in most areas were similar to those of the Late Cambrian. Accordingly, Early Ordovician life was similar to that of the latter part of the Cambrian. Trilobites were the prominent animal in most shelf sea environments. Long straight-shelled nautiloids, certain snails, a few orthoid brachiopods, sponges, small echinoderms, algae, and bacteria flourished in tropical marine environments. Linguloid brachiopods and certain bivalved mollusks inhabited cool-water, nearshore environments.

Middle Ordovician plate motions were acompanied by significant changes in life. On land, nonvascular, mosslike plants appeared in wetland habitats. Vascular plants appeared slightly later in riverine habitats. The first nonvascular plants occurred in the Middle East on Gondwanan shores. The Middle Ordovician radiation of marine invertebrates is one of the most extensive in the record of Phanerozoic marine life. Corals, bryozoans, several types of brachiopods, a number of crinozoan echniderms, conodonts, bivalved mollusks, new kinds of ostracodes, new types of trilobites, and new kinds of nautiloids suddenly developed in tropical marine environments. As upwelling conditions formed along the plate margins, oxygen minimum zones—habitats preferred by many graptolites—expanded at numerous new sites. Organic walled microfossils (chitinozoans and acritarchs) radiated in mid- to high-latitude environments. Ostracoderms (jawless, armored fish) radiated in tropical marine shallow-shelf environments. These fish were probably bottom detritus feeders. *See* PALEOECOLOGY.

The latest Ordovician stratigraphic record suggests that glacial ice melted relatively quickly, accompanied by a relatively rapid sea-level rise in many areas. Some organisms—certain conodonts, for example—did not endure significant extinctions until sea levels began to rise and shelf sea environments began to expand. *See* PALEOCEANOGRAPHY; STRATIGRAPHY. W. B. N. Berry

Ore and mineral deposits Ore deposits are naturally occurring geologic bodies that may be worked for one or more metals. The metals may be present as native elements, or, more commonly, as oxides, sulfides, sulfates, silicates, or other compounds. The term ore is often used loosely to include such nonmetallic minerals as fluorite and gypsum. The broader term, mineral deposits, includes, in addition to metalliferous minerals, any other useful minerals or rocks. Minerals of little or no value which occur with ore minerals are called gangue. Some gangue minerals may not be worthless in that they are used as by-products; for instance, limestone for fertilizer or flux, pyrite for making sulfuric acid, and rock for road material.

Mineral deposits that are essentially as originally formed are called primary or hypogene. The term hypogene also indicates formation by upward movement of material. Deposits that have been altered by weathering or other superficial processes are secondary or supergene deposits. Mineral deposits that formed at the same time as the enclosing rock are called syngenetic, and those that were introduced into preexisting rocks are called epigenetic.

The distinction between metallic and nonmetallic deposits is at times an arbitrary one since some substances classified as nonmetals, such as lepidolite, spodumene, beryl, and rhodochrosite, are the source of metals. The principal reasons for distinguishing nonmetallic from metallic deposits are practical ones, and include such economic factors as recovery methods and uses.

Most mineral deposits are natural enrichments and concentrations of original material produced by different geologic processes. Economic considerations, such as the amount and concentration of metal, the cost of mining and refining, and the market value of the metal, determine whether the ore is of commercial grade. To be of commercial grade, for example, the following metals must be concentrated in the amounts indicated: aluminum, about 30%; copper, 0.7–10%; lead, 2–4%; zinc, 3–8%; and gold, silver, and uranium, only a small fraction of a percent of metal. *See* GEOCHEMICAL PROSPECTING; MINERAL

A. F. Hagner

Ore dressing Treatment of ores to concentrate their valuable constituents (minerals) into products (concentrate) of smaller bulk, and simultaneously to collect the worthless material (gangue) into discardable waste (tailing). The fundamental operations of ore-dressing processes are the breaking apart of the associated constituents of the ore by mechanical means (severance) and the separation of the severed components (beneficiation) into concentrate and tailing, using mechanical or physical methods which do not effect substantial chemical changes.

Comminution is a single- or multistage process whereby ore is reduced from run-of-mine size to that size needed by the beneficiation process. The process is intended to produce individual particles which are either wholly mineral or wholly gangue, that is, to produce liberation. Since the mechanical forces producing fracture are not susceptible to detailed control, a class of particles containing both mineral and gangue (middling particles) are also produced. Comminution is divided into crushing (down to 6- to 14-mesh) and grinding (down to micrometer sizes).

Screening is a method of sizing whereby graded products are produced, the individual particles in each grade being of nearly the same size. In beneficiation, screening is practiced for two reasons: as an integral part of the separation process, for example, in jigging; and to produce a feed of such size and size range as is compatible with the applicability of the separation process. *See* SCREENING.

Beneficiation consists of two fundamental operations: the determination that an individual particle is either a mineral or a gangue particle (selection); and the movement of selected particles via different paths (separation) into the concentrate and tailing products. When middling particles occur, they will either be selected according to their mineral content and then caused to report as concentrate or tailing, or be separated as a third product (middling), which is reground to achieve further liberation. *See* FLOTATION; LEACHING; MECHANICAL SEPARATION TECHNIQUES.

Separation is achieved by subjecting each particle of the mixture to a set of forces which is usually the same irrespective of the nature of the particles excepting for the force based upon the discriminating property. This force may be present for both mineral and gangue particles but differing in magnitude, or it may be present for one type of particle and absent for the other. As a result of this difference, separation is possible, and the particles are collected in the form of concentrate or tailing.

Magnetic separation utilizes the force exerted by a magnetic field upon magnetic materials to counteract partially or wholly the effect of gravity. Thus under the action of these two forces, different paths are produced for the magnetic and nonmagnetic particles. *See* MAGNETIC SEPARATION METHODS. Menelaos D. Hassialis

Orectolobiformes

Orectolobiformes An order of sharks known as the carpet sharks. Members share the following characteristics: two dorsal fins without spines; a short snout that leaves the mouth terminal or almost terminal; and each nostril connected to the mouth by a deep groove, the anterior margin having a well-developed barbel. The classification hierarchy is:

Class Chondrichthyes
  Subclass Elasmobranchii
    Superorder Euselachii
      Order Orectolobiformes

The order consists of about 31 species in 14 genera and 7 families and is represented by at least one family in the seas of the world. Among them is the nurse shark (Family Gingymostomatidae). This order also includes the whale shark (Family Rhincodontidae). The whale shark reaches a length of at least 12 m (40 ft), making it the world's largest cold-blooded animal. *See* CHONDRICHTHYES; ELASMOBRANCHII. Herbert Boschung

Oregano

Oregano A herb, also known as wild marjoram. The dried leaves of several species of aromatic plants are known as oregano; thus oregano is a common name for a general flavor and aroma rather than the name of a specific plant.

European (*Origanum vulgare*) and Greek (*O. hervacleoticum*) oregano are both in the mint family (Lamiaceae). Mexican oregano is obtained primarily from plants of *Lippia graveolens*. These small aromatic shrubs in the verbena family grow wild in Mexico. Origanum oil used in perfumery is steam-distilled primarily from Spanish oregano, *Thymus capitatus*. *See* LAMIALES.

European oregano can be distinguished by its strong piquant character and tall growth with dark, broad leaves; it is a perennial erect herb 2–3 ft tall (0.6–1 m) with pubescent stems, ovate dark green leaves, and white or purple flowers. Native to southern Europe, southwest Asia, and the Mediterranean countries, European oregano is usually found growing in the dry, rocky, calcareous soils of the mountain regions. Greece, Italy, Spain, Turkey, and the United States are the primary sources of European oregano.

Dried oregano leaves are used as a culinary herb in meat and sausage products, salads, soups, Mexican foods, and barbeque sauces. The essential oil of oregano is used in food products, cosmetics, and liqueurs. *See* MARJORAM; SPICE AND FLAVORING. Seth Kirby

Organic chemistry

Organic chemistry The study of the structure, preparation, properties, and reactions of carbon compounds. The term organic was early applied to compounds derived from plant and animal sources. These substances from living systems were usually distillable liquids or low-melting solids and were flammable, in contrast to metals, salts, and oxides from mineral sources. Until about 1830 it was held by some that organic compounds contained some special quality, or vital force. This notion was dispelled, but the term organic remained and became broadened to include carbon compounds in general. *See* CARBON.

Structure. The structures of organic compounds are described by a molecular framework of carbon atoms on which substituents may be located at various points.

Structures can be represented in several ways, as illustrated for the three-carbon alcohol 2-propanol [structure (**1**)] and the cyclic ketone 2-methyl-3-cyclohexenone (**2**). The expanded structure

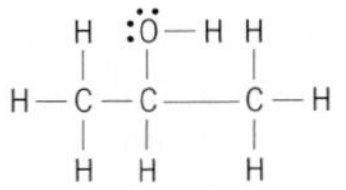

(**1a**) (**1b**) CH_3CHCH_3 with OH (**1c**) (**2**)

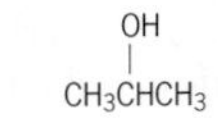

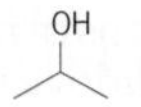

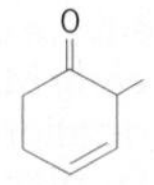

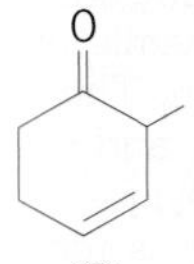

(**2**)

of 2-propanol (**1a**) shows all bonds and electron pairs, including unshared electrons on oxygen. More compact and convenient is the condensed structure (**1b**) in which the C—C and C—H bonds are implied. In the bond-line convention (**1c**), all C—C bonds are indicated by a line, as shown for 2-propanol. Carbon atoms are not shown explicitly, but rather are implied at the ends of each line segment, together with enough hydrogen atoms to complete the tetravalency at each carbon. The bond-line convention is particularly convenient for cyclic structures such as 2-methyl-3-cyclohexenone; each vertex and the end of each line segment represents a carbon and appropriate number of hydrogens. *See* STRUCTURAL CHEMISTRY; VALENCE.

A functional group is an atom other than carbon or a multiple bond, such as the hydroxyl group (OH) of 2-propanol, the

Principal organic functional groups

Composed class	Group	Structure
Alkene	Double bond	$>C{=}C<$
Alkyne	Triple bond	$-C{\equiv}C-$
Alcohol	Hydroxyl	$-OH$
Amine	Amino	$-NH_2(-NR_2)$*
Aldehyde	Carbonyl	$-\overset{O}{\overset{\parallel}{C}}H$
Ketone	Carbonyl	$-\overset{O}{\overset{\parallel}{C}}R$
Acid	Carboxyl	$-\overset{O}{\overset{\parallel}{C}}OH$
Ester	Alkoxycarbonyl	$-\overset{O}{\overset{\parallel}{C}}OR$
Amide	Carbamoyl	$\overset{O}{\overset{\parallel}{C}}N<$
Nitrile	Cyano	$-C{\equiv}N$
Azide	Azido	$-N{=}N{=}N$
Nitro		$-NO_2$
Sulfide		$-S-$
Sulfoxide		$-\overset{O}{\overset{\parallel}{S}}-$
Sulfonic acid		$-SO_3H$

*R = any carbon group, for example, CH_3.

double bond (C=C), or the carbonyl group (C=O) of 2-methyl-3-cyclohexenone. The group defines a class of compounds and is the point at which characteristic reactions occur, for example, oxidation, reduction, or addition of an electrophilic or nucleophilic reagent. Some of the principal functional groups are shown in the table. *See* ELECTROPHILIC AND NUCLEOPHILIC REAGENTS.

The fact that there can be two or more compounds, known as isomers, with the same molecular composition was one of the key points in development of a structural theory. One type of isomerism, structural or constitutional, is illustrated by the two isomers that have the formula C_4H_{10}, butane (**3*a***) and isobutane (2-methylpropane; **3*b***). The number of possible structural isomers becomes enormous in larger molecules.

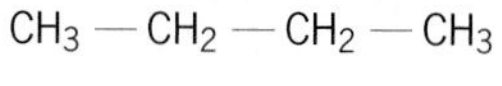

$CH_3—CH_2—CH_2—CH_3$ (3*a*)

$CH_3—CH(CH_3)—CH_3$ (3*b*)

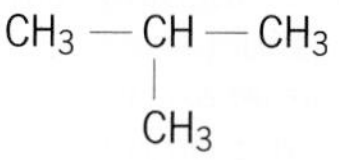

See MOLECULAR ISOMERISM.

Several three-dimensional representations of butane, showing the tetrahedral geometry of the carbon atoms, are given in structures (**4**). As indicated in these structures, butane can exist in

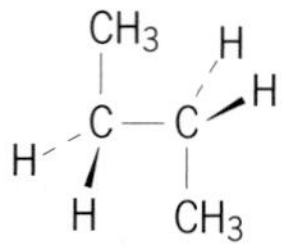

(4*a*) 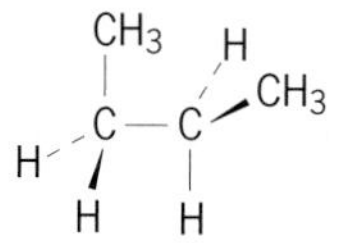 (4*b*) 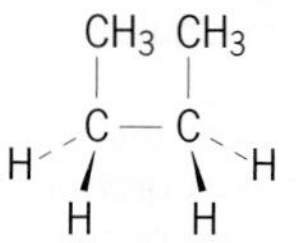 (4*c*)

several forms, called conformations, which differ in the relative positions of the carbon atoms, and thus the overall shape of the molecule. However, the barrier to rotation around the central C—C bond is so low that these individual conformational isomers are not separable, and butane is thus a single compound. *See* CONFORMATIONAL ANALYSIS.

In an alkene, rotation around the C=C bond does not occur, and 2-butene, for example, exists as two isomeric compounds, cis (Z) and trans (E) [**5*a*** and **5*b***, respectively].

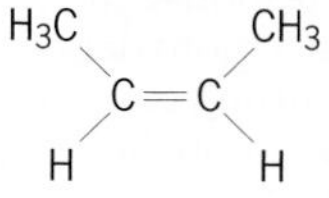

(5*a*) 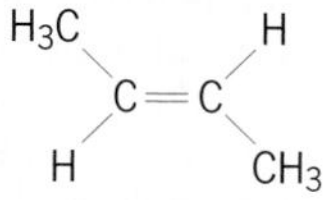 (5*b*)

Stereoisomers are compounds that have the same bond sequence but differ in the spatial array of the bonds. When a carbon atom is bonded to four unlike atoms or groups, the tetrahedral geometry of carbon causes the atom to be dissymmetric or chiral. A compound with a chiral atom can exist in two isomeric forms, known as enantiomers. The relative positions of all atoms is identical in the two enantiomers, but they differ in handedness, a characteristic of an asymmetric object and its nonsuperposable mirror image, as in structures (**6**) of 1-chlorobutane.

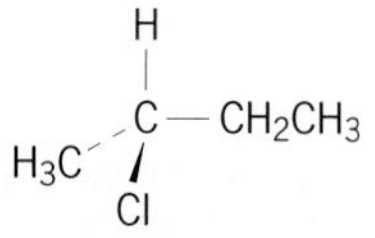

(6*a*) 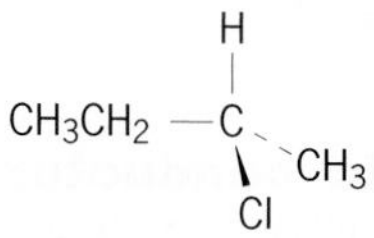 (6*b*)

See STEREOCHEMISTRY.

When two chiral centers are present, two stereoisomers can arise from each enantiomer. Thus enantiomer (**7*a***) of chlorobutane can lead to isomeric structures (**7*b***) and (**7*c***), in which the

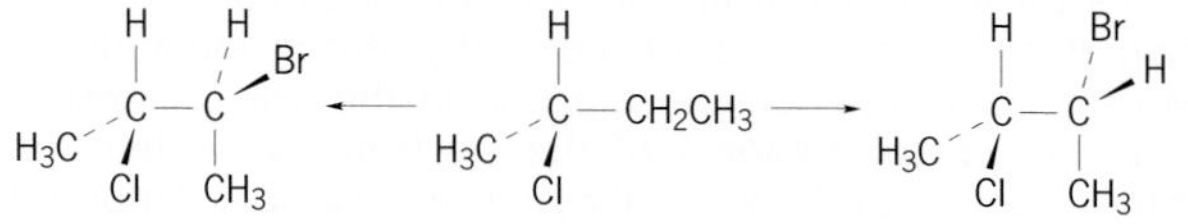

(7*b*) (7*a*) (7*c*)

relative positions of the atoms is not identical. In this case, the isomers are known as diastereoisomers. With *n* chiral centers, there can be 2^n stereoisomers.

Acyclic compounds. The simplest organic compound is methane (CH_4). It is the first member of the homologous series of alkanes, in which successive compounds differ by an additional $—CH_2—$ group (CH_3CH_3, $CH_3CH_2CH_3$, and so forth). *See* ALKANE; METHANE.

Higher alkanes, $CH_3(CH_2)_nCH_3$ ($n = 3–20$), and also branched isomers and cyclic hydrocarbons are the principal components of petroleum. These compounds have no reactive functional groups.

Both acyclic and cyclic carbon frameworks can contain multiple bonds; oxygen, nitrogen, and sulfur atoms; and other functional groups listed in the table.

Carbocyclic compounds. The two large groups of compounds with rings containing only carbon are alicyclic and aromatic. The parent hydrocarbons in the former series are cycloalkanes and in the latter, benzene. The structure of benzene is a planar six-numbered ring with six electrons in a delocalized array. *See* AROMATIC HYDROCARBON; BENZENE.

Heterocyclic compounds. A nitrogen, oxygen, or sulfur atom can take the place of carbon in either alicyclic or aromatic rings. The most numerous and important hetrocyclic compounds are those with nitrogen in a five- or six-membered aromatic system.

Synthesis reactions. The preparation of compounds occupies much of the effort of organic chemistry, and is the principal business of the chemical industry. The manufacture of drugs, pigments, and polymers entails the preparation of organic compounds on a scale of thousands to billions of kilograms per year, and there is constant research to develop new products and processes. Synthesis of new substances is carried out for many purposes beyond the goal of a commercial product. A compound of a specified structure may be needed to test a mechanistic proposal or to evaluate a biochemical response such as inhibition of an enzyme. Synthesis may provide a more dependable and less expensive source of a naturally occurring compound; moreover, a synthetic approach permits variations in the structure that may lead to enhanced biological activity.

The term synthesis usually implies a planned sequence of steps leading from simple starting compounds to a desired end product. Each of these steps involves a reaction that may lead to formation of a C—C bond or to the introduction, alteration, or removal of a functional group. Progress in synthesis depends on the availability of a wide range of reactions that bring about these changes in good yield, with a minimum of interfering by-products. An integral part of synthesis is the development of new methods and reagents that are selective for a desired transformation, and, very importantly, proceed with control of the stereochemistry. *See* ASYMMETRIC SYNTHESIS; ORGANIC SYNTHESIS.

James A. Moore

Organic conductor An organic substance with low electrical resistance. Two major classes of organic conductors are charge-transfer compounds and conducting polymers.

Charge-transfer compounds. The search for organic conductors in the early 1970s led to the observation of metallic-like electrical conduction in well-ordered molecular crystals and the discovery of many new phenomena such as the stabilization of charge-density waves and spin-density waves, new mechanisms for electronic transport, organic superconductivity, and new states of matter produced under strong magnetic fields. Most of these phenomena are due to the low dimensionality (one or two dimensions) of the electron gas in the charge-transfer compounds where they were observed. Among these properties, superconductivity has created much interest since the zero-resistance state is now observed at temperatures as high as 10 K (–442°F) in certain organic superconductors. *See* SUPERCONDUCTIVITY.

Charge-transfer compounds are two-component materials containing anionic and cationic species originating by charge transfer between donor and acceptor entities; these may be two organic molecules or an organic molecule with an inorganic ion. Tetrathiafulvalene-tetracyanoquinodimethane (TTF-TCNQ) is the prototype of charge-transfer organic crystals. Its crystal structure exhibits piled-up segregated columns of donor TTF and acceptor TCNQ molecules (illus. *a*). In the solid state, the amount of charge transferred from donor to acceptor is determined by the overall crystal stability.

Another class of organic conductors is exemplified by radical cation salts such as $(TMTSF)_2X$ where the organic molecule is tetramethyltetraselenafulvalene and X is an inorganic anion. In this class of materials (Bechgaard salts) the molecules display a zigzag packing along the stacking axis (illus. *b*), where one positive charge (hole) is shared between two organic molecules.

The strong overlap between electron clouds of neighboring molecules along the stacks spreads the partially filled molecular electronic states into an energy band 0.5–1 eV wide. This bandwidth is large enough to allow electron delocalization among all molecules on a given stack and to promote electrical conduction similar to that in metal crystals. *See* BAND THEORY OF SOLIDS; CONDUCTION (ELECTRICITY); DELOCALIZATION; MOLECULAR ORBITAL THEORY.

D. Jérome

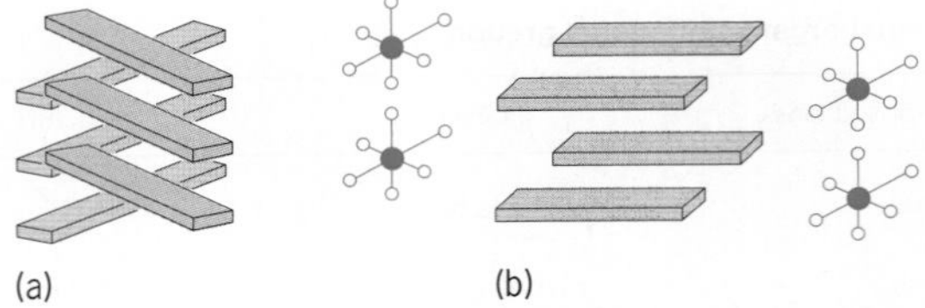

Stacking of charge-transfer compounds. (*a*) Segregated stacking of TTF-TCNQ-like materials. (*b*) Typical zigzag stacking of the $(TMTSF)_2X$ series.

Conducting polymers. Polymeric materials are typically considered as insulators. However, research since the late 1970s has led to the discovery of polymeric materials with extremely high conductivity, approaching that of copper. The prospect of materials combining the properties of plastics and metals or semiconductors has led to a search for applications, made attractive because improved polymers no longer suffer from such drawbacks as low stability, processing difficulties, and brittleness. Most conducting polymers can be switched reversibly between conductive and nonconductive states, with the result that their conductivities can span an enormous range. This switching is accomplished through oxidation-reduction (redox) chemistry, the conductivity being sensitive to the degree of oxidation of the polymer backbone. This property distinguishes conducting polymers from metals and semiconductors and is the basis of many existing and potential applications. In addition, certain polymers become conducting upon oxidation or reduction and thus can exhibit *p*- or η-type conduction. *See* OXIDATION-REDUCTION; POLYMER.

In a conducting polymer an oxidant removes electrons from the π-electron system of the polymer, creating radical cations that, at high concentrations, dimerize to form cation pairs known as bipolarons. Charge-balancing counterions are concomitantly incorporated between polymer chains. The overall process is referred to as doping, and the counteranion (or countercation in the case of reduction) is the dopant.

Considerable progress has been made on the theory of important parameters such as oxidation potentials, band gaps, and band widths, often with good agreement with experiment. Such work is important for the design of new conductive polymers with specific properties. *See* SOLID-STATE CHEMISTRY.

D. Jérome; Gary E. Wnek

Organic evolution The modification of living organisms during their descent, generation by generation, from common ancestors. Organic, or biological, evolution is to be distinguished from other phenomena to which the term evolution is often applied, such as chemical evolution, cultural evolution, or the origin of life from nonliving matter. Organic evolution includes two major processes: anagenesis, the alteration of the genetic properties of a single lineage over time; and cladogenesis, or branching, whereby a single lineage splits into two or more distinct lineages that continue to change anagenetically.

Anagenesis consists of change in the genetic basis of the features of the organisms that constitute a single species. Populations in different geographic localities are considered members of the same species if they can exchange members at some rate and hence interbreed with each other, but unless the level of interchange (gene flow) is very high, some degree of genetic difference among different populations is likely to develop. The changes that transpire in a single population may be spread to other populations of the species by gene flow. *See* SPECIES CONCEPT.

Almost every population harbors several different alleles at each of a great many of the gene loci; hence many characteristics of a species are genetically variable. All genetic variations ultimately arise by mutation of the genetic material. Broadly defined, mutations include changes in the number or structure of the chromosomes and changes in individual genes, including substitutions of individual nucleotide pairs, insertion and deletion of nucleotides, and duplication of genes. Many such mutations alter the properties of the gene products (ribonucleic acid and proteins) or the timing or tissue localization of gene action, and consequently affect various aspects of the phenotype (that is, the morphological and physiological characteristics of an organism). Whether and how a mutation is phenotypically expressed often depends on developmental (epigenetic) events. *See* GENE; GENETIC CODE; MUTATION; RIBONUCLEIC ACID (RNA).

Natural selection is a consistent difference in the average rate at which genetically different entities leave descendants to subsequent generations; such a difference arises from differences in fitness (that is, in the rate of survival, reproduction, or both). In fact, a good approximate measure of the strength of natural selection is the difference between two such entities in their rate of increase. The entities referred to are usually different alleles at a locus, or phenotypically different classes of individuals in the population that differ in genotype. Thus selection may occur at the level of the gene, as in the phenomenon of meiotic drive, whereby one allele predominates among the gametes produced by a heterozygote. Selection at the level of the individual organism, the more usual case, entails a difference in the survival and reproductive success of phenotypes that may differ at one locus or at more than one locus. As a consequence of the difference in fitness, the proportion of one or the other allele increases in subsequent generations. The relative fitness of different genotypes usually depends on environmental conditions.

Different alleles of a gene that provides an important function do not necessarily differ in their effect on survival and reproduction; such alleles are said to be neutral. The proportion of two neutral alleles in a population fluctuates randomly from generation to generation by chance, because not all individuals in the population have the same number of surviving offspring. Random fluctuations of this kind are termed random genetic drift. If different alleles do indeed differ in their effects on fitness, both genetic drift and natural selection operate simultaneously. The deterministic force of natural selection drives allele frequencies toward an equilibrium, while the stochastic (random) force of genetic drift brings them away from that equilibrium. The outcome for any given population depends on the relative strength of natural selection (the magnitude of differences in fitness) and of genetic drift (which depends on population size).

The great diversity of organisms has come about because individual lineages (species) branch into separate species, which continue to diverge. This splitting process, speciation, occurs when genetic differences develop between two populations that prevent them from interbreeding and forming a common gene pool. The genetically based characteristics that cause such reproductive isolation are usually termed isolating mechanisms. Reproductive isolation seems to develop usually as a fortuitous by-product of genetic divergence that occurs for other reasons (either by natural selection or by genetic drift). *See* SPECIATION.

A frequent consequence of natural selection is that a species comes to be dominated by individuals whose features equip them better for the environment or way of life of the species. Such features are termed adaptations. Although many features of organisms are adaptive, not all are, and it is a serious error to suppose that species are capable of attaining ideal states of adaptation. Some characteristics are likely to have developed by genetic drift rather than natural selection, and so are not adaptations; others are side effects of adaptive features, which exist because of pleiotropy or developmental correlations.

Higher taxa are those above the species level, such as genera and families. A taxon such as a genus is typically a group of species, derived from a common ancestor, that share one or more features so distinctive that they merit recognition as a separate taxon. The degree of difference necessary for such recognition, however, is entirely arbitrary: there are often no sharp limits between related genera, families, or other higher taxa, and very often the diagnostic character exists in graded steps among a group of species that may be arbitrarily divided into different higher taxa. Moreover, a character that in some groups is used to distinguish higher taxa sometimes varies among closely related species or even within species. In addition, the fossil record of many groups shows that a trait that takes on very different forms in two living taxa has developed by intermediate steps along divergent lines from their common ancestor; thus the inner ear bones of mammals may be traced to jaw elements in reptiles that in turn are homologous to gill arch elements in Paleozoic fishes.

The characteristics of a species evolve individually or in concert with certain other traits that are developmentally or functionally correlated. Because of this mosaic pattern of evolution, it is meaningful to speak of the rate of evolution of characters, but not of species or lineages as total entities. Thus in some lineages, such as the so-called living fossils, many aspects of morphology have evolved slowly since the groups first came into existence, but evolution of their deoxyribonucleic acid and amino acid sequences has proceeded at much the same rate as in other lineages. Every species, including the living fossils, is a mixture of traits that have changed little since the species' remote ancestors, and traits that have undergone some evolutionary change in the recent past. The history of life is not one of progress in any one direction, but of adaptive radiation on a grand scale: the descendants of any one lineage diverge as they adapt to different resources, habitats, or ways of life, acquiring their own specialized features as they do so. There is no evidence that evolution has any goal, nor does the mechanistic theory of evolutionary processes admit of any way in which genetic change can have a goal or be directed toward the future. However, for life taken as a whole, the only clearly discernible trend is toward ever-increasing diversity.

Douglas J. Futuyma

Organic geochemistry The study of the abundance and composition of naturally occurring organic substances, their origins and fate, and the processes that affect their distributions on Earth and in extraterrestrial materials. These activities share the common need for identification, measurement, and assessment of organic matter in its myriad forms.

Organic geochemistry was born from a curiosity about the organic pigments extractable from petroleum and black shales. It developed with extensive investigations of the chemical characteristics of petroleum and petroleum source rocks as clues to their occurrence and formation, and now encompasses a broad scope of activities within interdisciplinary areas of earth and environmental science. This range of studies recognizes the potential of geological records of organic matter to help characterize sedimentary depositional environments and to provide evidence of ancient life and indications of evolutionary developments through the Earth's history. Organic geochemistry includes determinations of anthropogenic contaminants amid the natural background of organic molecules and the assessment of their environmental impact and fate. Marine organic geochemistry addresses and interprets aquatic processes involving carbon species. It involves investigations of the chemical character of particulate and dissolved organic matter, evaluation of oceanic primary production including the factors (light, temperature, nutrient availability) that influence the uptake of carbon dioxide (CO_2), the composition of marine organisms, and the subsequent processing of organic constituents through the food web. Organic geochemistry extends to broader biogeochemical issues, such as the carbon cycle, and the effects of changing carbon dioxide levels, especially efforts to use geochemical data and proxies to help constrain global climate models. Examination of the organic chemistry of meteorites and lunar materials also falls within its compass, and as a critical part of the quest for remnants of life on Mars, such extraterrestrial studies are now regaining the prominence they held in the 1970s during lunar exploration. *See* COSMOCHEMISTRY; GEOCHEMISTRY.

Global inventories of carbon. Carbon naturally exists as oxidized and reduced forms in carbonate carbon and organic matter. The major reservoir of both forms of carbon on Earth is the geosphere. It contains carbonate minerals deposited as sediments and organic matter accumulated from the remains of dead organisms. Estimates of the size of the geological reservoir of carbon vary within the range of 5 to 7 × 10^{22} g, of which 75% is carbonate carbon and 25% is organic carbon. The amounts of carbon contained in living biota (5 × 10^{17} g), dissolved in the ocean (4 × 10^{19} g), and present in atmospheric gases (7 × 10^{17} g) are miniscule compared to the quantity of organic carbon buried in the rock record. The importance of buried organic matter extends beyond its sheer magnitude; it includes the fossil fuels—coal, natural gas, and petroleum—that supply 85% of the world's energy. *See* BIOGEOCHEMISTRY; CARBON; CARBON DIOXIDE; CARBONATE MINERALS; COAL; FOSSIL FUEL; NATURAL GAS; PETROLEUM; SEDIMENTARY ROCKS.

Sedimentary organic matter. The vast amounts of organic matter contained in geological materials represent the accumulated vestiges of organisms amassed over the expanse of geological time. Yet, survival of organic cellular constituents of biota into the rock record is the exception rather than the norm. Only a small portion of the carbon fixed by organisms during primary production, especially by photosynthesis, escapes degradation as it settles through the water column and eludes microbial alteration during subsequent incorporation and assimilation into sedimentary detritus. *See* BIODEGRADATION.

Sedimentary organic matter can be divided operationally into solvent-extractable bitumen and insoluble kerogen. Bitumens contain a myriad of structurally distinct molecules, especially hydrocarbons, which can be individually identified (such as by gas chromatography-mass spectrometry) although they may be present in only minute quantities (nanograms or picograms). The range of components includes many biomarkers that retain structural remnants inherited from their source organisms, which attest to their biological origins and subsequent geological fate. *See* BITUMEN; KEROGEN.

Biomarkers are individual compounds whose chemical structures carry evidence of their origins and history. Recognition of the specificity of biomarker structures initially helped confirm that petroleum was derived from organic matter produced by biological processes. Of the thousands of individual petroleum components, hundreds reflect precise biological sources of organic matter, which distinguish and differentiate their disparate origins. The diagnostic suites of components may derive from individual families of organisms, but contributions at a species level can occasionally be recognized. Biomarker abundances and distributions help to elucidate sedimentary environments, providing evidence of depositional settings and conditions. They also reflect sediment maturity, attesting to the progress of the successive, sequential transformations that convert biological precursors into geologically occurring products. Thus, specific biomarker characteristics permit assessment of the thermal history of individual rocks or entire sedimentary basins. *See* BASIN.

Carbon isotopes. Carbon naturally occurs as three isotopes: carbon-12 (^{12}C), carbon-13 (^{13}C), and radiocarbon (^{14}C). Temporal excursions in the ^{13}C values of sediment sequences can reflect perturbations of the global carbon cycle. Radiocarbon is widely employed to date archeological artifacts, but the sensitivity of its measurement also permits its use in exploration of the rates of biogeochemical cycling in the oceans. This approach permits assessment of the ages of components in sediments, demonstrating that bacterial organic matter is of greater antiquity than components derived from phytoplankton sources. *See* ISOTOPE; MARINE SEDIMENTS; PALEOCEANOGRAPHY; RADIOCARBON DATING.

Simon C. Brassell

Organic nomenclature A system by which a unique and unambiguous name or other designation is assigned to a given organic molecular structure. A set of rules adopted by the International Union of Pure and Applied Chemistry (IUPAC) is the basis for a standardized name for any organic compound. Common or nonsystematic names are used for many compounds that have been known for a long time. The latter names have the advantage of being short and easily recognized, as in the examples below. However, in contrast to systematic names, common or trivial names do not convey information from which the structure can be written by reference to prescribed rules.

In the IUPAC system, a name is formed by combination of a parent alkyl chain or ring with prefixes and suffixes to denote substituents. For aliphatic compounds, the parent name is a stem that denotes the longest straight chain in the structure with the ending-ane. The first four members of the alkane series are methane, ethane, propane, and butane. From five carbons on, the names follow the Greek numerical roots, for example, pentane and hexane. Changing the ending-ane to -yl gives the name of the corresponding radical or group; for example, CH_3CH_2— is the ethyl radical or ethyl group. Branches on an alkyl chain are indicated by the name of a radical or group as a prefix. The location of a branch or other substituent is indicated by number; the chain is numbered from whichever end results in the lowest numbering. *See* ALKANE.

Double or triple bonds are indicated by the endings-ene or -yne, respectively. The configuration of the chain at a double bond is denoted by *E*- when two similar groups are on opposite sides of the plane bisecting the bond, and *Z*- when they are on the same side. *See* ALKENE; ALKYNE.

A compound containing a functional group is named by adding to the parent name a suffix characteristic of the group. If there are two or more groups, a principal group is designated by suffix and the other(s) by prefix.

The same general principles apply to cyclic compounds. For alicyclic rings, the prefix cyclo- is followed by a stem indicating the number of carbon atoms in the ring, as is illustrated in the structure below. In bicyclic compounds, the total number of carbons in the ring system is prefixed by bicyclo- and numbers in brackets which indicate the number of atoms in each connecting chain.

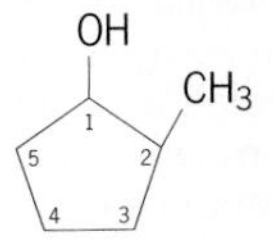

OH
CH_3
1 2 3 4 5

2-Methylcyclopentanol

The names of aromatic hydrocarbons have the ending -ene, which denotes a ring system with the maximum number of noncumulative double bonds. Each of the simpler polycyclic hydrocarbons has a different parent name. To number the positions in a polycyclic aromatic ring system, the structure must first be oriented with the maximum number of rings arranged horizontally and to the right. The system is then numbered in clockwise sequence starting with the atom in the most counterclockwise position of the upper right-hand ring, omitting atoms that are part of a ring fusion. *See* AROMATIC HYDROCARBON.

Basis of nomenclature of heterocyclic compounds

Heteroatom	Prefix	Ring size	Suffix
O	ox(a)-	4	-ete
S	thi(a)-	5	-ole
N	az(a)-	6	-ine
		7	-epine

Systematic names for rings containing a heteroatom (O, S, N) are based on a combination of a prefix denoting the heteroatom(s) and a suffix denoting the ring size, as indicated in the table. *See* CHEMICAL SYMBOLS AND FORMULAS; HETEROCYCLIC COMPOUNDS; ORGANIC CHEMISTRY.

James A. Moore

Organic photochemistry A branch of chemistry that deals with light-induced changes of organic material. Because it studies the interaction of electromagnetic radiation and matter, photochemistry is concerned with both chemistry and physics. In considering photochemical processes, therefore, it is also necessary to consider physical phenomena that do not involve strictly chemical changes, for example, absorption and emission of light, and electronic energy transfer.

Because several natural photochemical processes were known to play important roles (namely, photosynthesis in plants, process of vision, and phototropism), the study of organic photochemistry started very early in the twentieth century. However, the breakthrough occurred only after 1950 with the availability of commercial ultraviolet radiation sources and modern analytical instruments for nuclear magnetic resonance (NMR) spectroscopy and (gas) chromatography.

In general, most organic compounds consist of rapidly interconvertible, nonseparable conformers, because of the free rotation about single bonds. Each conformer has a certain energy associated with it, and its own electronic absorption spectrum. This is one cause of the broad absorption bands produced by organic compounds in solution. The equilibrium of the conformers may be influenced by the solvent and the temperature. According to the Franck-Condon principle, which states that promotion of an electron by the absorption of a photon is much faster than a single vibration, each conformer will have its own excited-state configuration. *See* CONFORMATIONAL ANALYSIS; MOLECULAR STRUCTURE AND SPECTRA.

Because of the change in the pi-bond order upon excitation of the substrate and the short lifetime of the first excited state, the excited conformers are not in equilibrium and each yields its own specific photoproduct. Though different conformers may lead to the same photoproduct and one excited conformer may lead to several photoproducts, a change in solvent, temperature, or wavelength of excitation influences the photoproduct composition. This is especially true with small molecules; larger molecules with aromatic groups are less sensitive for small wavelength differences.

The influence of wavelength is also important when the primary photoproduct also absorbs light and then gives rise to another photoreaction. Excitation with selected wavelengths or monochromatic light by use of light filters or a monochromator, respectively, may then be profitable for the selective production of the primary product. Similarly, irradiation at a low temperature is helpful in detecting a primary photoproduct that is unstable when heated (thermolabile). *See* COLOR FILTER.

Wim. H. Laarhoven

Organic reaction mechanism A complete, step-by-step account of how a reaction of organic compounds takes place. A fully detailed mechanism would correlate the original structure of the reactants with the final structure of the products and account for changes in structure and energy throughout the progress of the reaction. It would also account for the formation of any intermediates and the rates of interconversions of all of the various species. Because it is not possible to detect directly all of these details, evidence for a reaction mechanism is always indirect. Experiments are designed to produce results that provide logical evidence for (but can never unequivocally prove) a mechanism. For most organic reactions, there are mechanisms that are considered to be well established based on bodies of experimental evidence. Nevertheless, new data often become

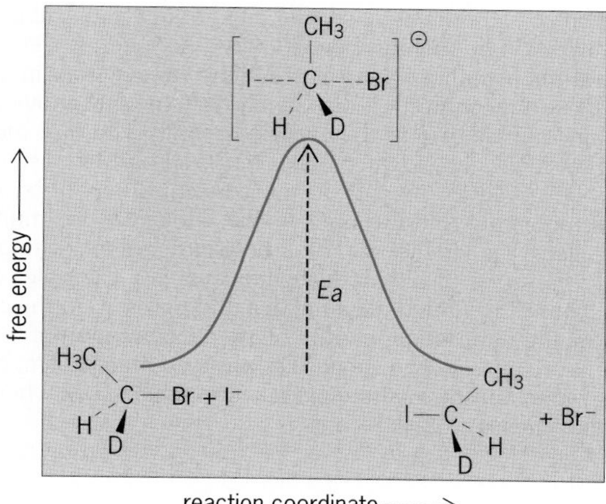

Potential energy–reaction coordinate diagram for a typical nucleophilic substitution reaction that proceeds by the S_N2 mechanism. E_a = activation energy.

available that provide further insight into new details of a mechanism or that occasionally require a complete revision of an accepted mechanism.

Classification of organic reactions. The description of an organic reaction mechanism typically includes designation of the overall reaction (for example, substitution, addition, elimination, oxidation, reduction, or rearrangement), the presence of any reactive intermediates (that is, carbocations, carbanions, free radicals, radical ions, carbenes, or excited states), the nature of the reagent that initiates the reaction (such as electrophilic or nucleophilic), the presence of any catalysis (such as acid or base), and any specific stereochemistry. For example, reaction (1) would be

$$(H_3C)(H)(D)C{-}Br + I^- \longrightarrow I{-}C(CH_3)(D)(H) + Br^- \qquad (1)$$

described as a concerted nucleophilic substitution of an alkyl halide that proceeds with inversion of stereochemistry. A reaction that proceeds in a single step, without intermediates, is described as concerted or synchronous. Reaction (1) is an example of the S_N2 mechanism (substitution, nucleophilic, bimolecular).

Potential energy diagrams. A common method for illustrating the progress of a reaction is the potential energy diagram, in which the free energy of the system is plotted as a function of the completion of the reaction (see illustration).

The reaction coordinate is intended to represent the progress of the reaction, and it may or may not correlate with an easily observed or measurable feature. In reaction (1), the reaction coordinate could be considered to be the increasing bond length of the carbon-bromine (C-Br) bond as it is broken, or the decreasing separation of C and iodine (I) as they come together to form a bond. In fact, a complete potential energy diagram should illustrate the variation in energy as a function of both of these (and perhaps several other relevant structural features), but this would require a three- dimensional (or higher) plot.

Besides identifying the energy levels of the original reactants and the final products, the potential energy diagram indicates the energy level of the highest point along the reaction pathway, called the transition state. Because the transition state represents the highest energy that the molecules must attain as they proceed along the reaction pathway, the energy level of the transition state is a key indication of how easily the reaction can occur. Features that tend to make the transition state more stable (lower in energy) make the reaction more favorable. Such stabilizing features could be intramolecular, such as electron donation or withdrawal by substituents, or intermolecular, such as stabilization by solvent. *See* CHEMICAL BONDING; ENERGY.

Kinetics. Another way to illustrate the various steps involved in a reaction mechanism is as a kinetic scheme that shows all of the individual steps and their rate constants. The S_N2 mechanism is a single step, so the kinetics must represent that step; the rate is observed to depend on the concentrations of both the organic substrate and the nucleophile. However, for multistep mechanisms the kinetics can be a powerful tool for distinguishing the presence of alternative pathways. For example, when more highly substituted alkyl halides undergo nucleophilic substitution, the rate is independent of the concentration of the nucleophile. This evidence suggests a two-step mechanism, called the S_N1 mechanism, as shown in reaction scheme (2), where the k terms represent rate constants.

$$(CH_3)_3C\text{—Br} \underset{k_{-1}}{\overset{k_1}{\rightleftharpoons}} (CH_3)_3C^{\oplus} + Br^{\ominus} \qquad (2a)$$

$$(CH_3)_3C^{\oplus} + I^{\ominus} \xrightarrow{k_2} (CH_3)_3C\text{—I} \qquad (2b)$$

The S_N1 mechanism accomplishes the same overall nucleophilic substitution of an alkyl halide, but does so by initial dissociation of the leaving group (Br^-) to form a carbocation, step (2*a*). The nucleophile then attaches to the carbocation to form the final product, step (2*b*). Alkyl halides that have bulky groups around the carbon to be substituted are less likely to be substituted by the direct S_N2 mechanism, because the nucleophile encounters difficulty in making the bond to the inaccessible site (called steric hindrance). If those alkyl groups have substituents that can support a carbocation structure, generally by electron donation, then the S_N1 mechanism becomes preferable.

A crucial feature of a multistep reaction mechanism is the identification of the rate-determining step. The overall rate of reaction can be no faster than its slowest step. In the S_N1 mechanism, the bond-breaking reaction (2*a*) is typically much slower than the bond-forming reaction (2*b*). Hence, the observed rate is the rate of the first step only. Thus, kinetics can distinguish the S_N1 and S_N2 mechanisms, as shown in Eqs. (3) and (4),

$$\text{Rate} = k\,[\text{RX}]\,[\text{Nu}] \quad \text{for an } S_N2 \text{ mechanism} \qquad (3)$$

$$\text{Rate} = k\,[\text{RX}] \quad \text{for an } S_N1 \text{ mechanism} \qquad (4)$$

where R is an alkyl group, X is a halogen or other leaving group, Nu is a nucleophile, and the terms in the brackets represent concentrations.

A more complete description of the S_N1 mechanism was recognized when it was observed that the presence of excess leaving group [for example, Br^- in reaction (2)] can affect the rate (called the common ion rate depression). This indicated that the mechanism should include a reverse step [k_{-1} in reaction step (2*a*)] in which the leaving group returns to the cation, regenerating starting material. In this case, the rate depends in a complex manner on the competition of nucleophile and leaving group for reaction with the carbocation. *See* CHEMICAL DYNAMICS; REACTIVE INTERMEDIATES; STERIC EFFECT (CHEMISTRY).

Activation parameters. The temperature dependence of the rate constant provides significant information about the transition state of the rate-determining step. The Arrhenius equation (5) expresses that dependence in terms of an exponential function of temperature and an activation energy, E_a; A is called the Arrhenius or preexponential factor, and R is the gas constant. *See* GAS; LOGARITHM.

$$k = Ae^{-E_a/RT} \qquad (5)$$

The activation energy represents the energy difference between the reactants and the transition state, that is, the amount of energy that must be provided in order to proceed along the reaction pathway successfully from reactant to product.

Stereochemistry. Careful attention to the stereochemistry of a reaction often provides crucial insight into the specific orientation of the molecules as they proceed through the reaction mechanism. The complete inversion of stereochemistry observed in the S_N2 mechanism provides evidence for the backside attack of the nucleophile. Alkyl halides that undergo substitution by the S_N1 mechanism do not show specific stereochemistry, since the loss of the leaving group is completely uncorrelated with the bonding of the necleophile.

In addition reactions, possible stereochemical outcomes are addition of the new bonds to the same or opposite sides of the original pi bond, called syn and anti addition, respectively. The anti addition of bromine to double bonds provides evidence

$$Br_2 + \text{cyclohexene} \longrightarrow \text{bromonium ion } (Br^{\oplus}) + Br^{\ominus} \longrightarrow \text{trans-1,2-dibromocyclohexane} + \text{trans-1,2-dibromocyclohexane} \qquad (6)$$

for the intermediacy of a bridged bromonium ion, as shown in reaction (6).

Carl C. Wamser

Organic synthesis The making of an organic compound fro simpler starting materials. Organic synthesis plays an important role by allowing for the creation of specific molecules for scientific and technological investigations.

The heart of organic synthesis is designing synthetic routes to a molecule. The simplest synthesis of a molecule is one in which the target molecule can be obtained by submitting a readily available starting material to a single reaction that converts it to the desired target molecule. However, in most cases the synthesis is not that straightforward; in order to convert a chosen starting material to the target molecule, numerous steps that add, change, or remove functional groups, and steps that build up the carbon atom framework of the target molecule may need to be done.

A systematic approach for designing a synthetic route to a molecule is to subject the target molecule to an intellectual exercise called a retrosynthetic analysis. This involves an assessment of each functional group in the target molecule and the overall carbon atom framework in it; a determination of what known reactions form each of those functional groups or that build up the necessary carbon framework as a product; and a determination of what starting materials for each such reaction are required. The resulting starting materials are then subjected to the same retrosynthetic analysis, thus working backward from the target molecule until starting materials are derived.

The retrosynthetic analysis of a target molecule usually results in more than one possible synthetic route. It is therefore necessary to critically assess each derived route in order to chose the single route that is most feasible and most economical. The safety of each possible synthetic route (the toxicity and reactivity hazards associated with the reactions involved) is also considered when assessing alternative synthetic routes to a molecule.

Selectivity is an important consideration in the determination of a synthetic route to a target molecule. Stereoselectivity refers to the selectivity of a reaction for forming one stereoisomer of a product in preference to another. Stereoselectivity cannot be achieved for all organic reactions; the nature of the mechanism of some reactions may not allow for the formation of one particular configuration of a chiral (stereogenic) carbon center or

Colorplate 1. Closeup view of the Trapezium stars obtained with the Planetary Camera on the Hubble Space Telescope. The field-of-view shown is 35 arcseconds of each side. Most faint stars have tails pointing away from the brightest star. These low-mass stars are surrounded by circumstellar disks (proplyds) which are being evaporated by the intense radiation field of the hot star. (*John Bally, Dave Devine, and Ralph Sutherland*)

Colorplate 2. Hubble Space Telescope image showing the inner portion of the Orion Nebula in the light of doubly ionized oxygen (blue), atomic hydrogen (green), and singly ionized nitrogen (red). (*C. R. O'Dell and S. K. Wong, Rice University; NASA*)

TABLE 1. Examples of some functional-group interconversions

General equation for the reaction*	Net transformation (name)
$R_1R_2R_3C{-}X + Nu^- \longrightarrow Nu{-}CR_1R_2R_3 + X^-$ (X = Cl, Br, I, or OSO_2R; Nu = OH, OR, CN, NR_2, others)	Alkyl halide to various functional groups (alcohols, ethers, nitriles, amines, others)
$R_1R_2CH{-}CXR_3R_4$ + base (such as CH_3O^-) $\longrightarrow R_1R_2C{=}CR_3R_4$	Alkyl halide to alkene (elimination)
$ROH + HX \longrightarrow RX$ (X = Cl, Br, I)	Alcohol to alkyl halide
$R_1R_2CH{-}C(OH)R_3R_4 + H_2SO_4 \longrightarrow R_1R_2C{=}CR_3R_4$	Alcohol to alkene (dehydration)
$R_1CH(OH)R_2 \xrightarrow{CrO_3\text{-pyridine}} R_1C({=}O)R_2$	Oxidation of alcohol to ketone or aldehyde
$R_1YH + R_2C({=}O)X \longrightarrow R_2C({=}O)OR_1$ (Y = O or N; X = OH, Cl, others)	Alcohol and carboxylic acid derivative to ester (esterification); amine and carboxylic acid derivative to amide
$R_1R_2C{=}CR_3R_4 + RCO_3H \longrightarrow$ epoxide of $R_1R_2C{-}CR_3R_4$ (O bridging)	Alkene to epoxide (epoxidation)
$R_1R_2C{=}CR_3R_4 + H_2 \xrightarrow[\text{(or other catalyst)}]{Pd} H{-}CR_1R_2{-}CR_3R_4{-}H$	Alkene to alkane (hydrogenation)
$R_1C({=}O)R_2 + NaBH_4 \longrightarrow R_1CH(OH)R_2$	Reduction of ketone or aldehyde to alcohol
$R_1COOR_2 \xrightarrow[\text{2. } H_2O \text{ workup}]{\text{1. LiAlH4}} R_1CH_2OH + R_2OH$	Reduction of ester to two alcohols
$R_1COOR_2 \xrightarrow{H_2O\text{, acid or base}} R_1COOH + R_2OH$	Ester to carboxylic acid and alcohol (ester hydrolysis)
$R{-}CN \xrightarrow[\text{2. } H_2O \text{ workup}]{\text{1. LiAlH4}} RCH_2NH_2$	Reduction of nitrile to amine
Benzene $+ E^+ \longrightarrow$ C_6H_5E (E = Br, NO_2, R, RCO, others)	Benzene to substituted benzene (electrophilic aromatic substitution)

*R = any organic group (alkyl, aryl, alkenyl) or a hydrogen atom. Nu = nucleophile.

one particular geometry (cis versus trans) for a double bond or ring. When stereoselectivity can be achieved, it requires that the reaction proceed via a geometrically defined transition state and that one or both of the reactants possess a particular geometrical shape during the reaction. For example, if one or both of the reactants is chiral, the absolute configuration of the newly formed stereogenic carbon center can be selected for in many reactions. *See* ASYMMETRIC SYNTHESIS; ORGANIC REACTION MECHANISM; STEREOCHEMISTRY.

Chemoselectivity is the ability of a reagent to react selectively with one functional group in the presence of another similar functional group. An example of a chemoselective reagent is a reducing agent that can reduce an aldehyde and not a ketone. In cases where chemoselectivity cannot be achieved, the functional

TABLE 2. Examples of some carbon-carbon bond-forming reactions

General equation for the reaction	Name of reaction
$R_1X + Mg \longrightarrow R_1MgX \xrightarrow[2.\ H^+]{1.\ R_2COR_3} R_1-C(OH)(R_2)-R_3$ (X = Cl, Br, I)	Grignard
$R_1X \xrightarrow[2.\ CuI]{1.\ Li} (R_1)_2CuLi \xrightarrow{R_2X} R_1-R_2$ (X = Cl, Br, I)	Gilman
$XC(=O)CH_2R_1 \xrightarrow[2.\ R_2CHO;\ 3.\ H^+]{1.\ (C_3H_7)_2NLi} XC(=O)CH(R_1)CH(OH)R_2$ (X = R, RO, NR_2, others)	Aldol addition
$XCH_2CO-Y + RCHO \xrightarrow[2.\ H_2O\ workup]{1.\ Zn} RCH(OH)-CH_2CO-Y$ (X = Cl, Br, or I; Y = R, RO, NR_2, others)	Reformatsky
$XC(=O)C(R_1)=C(R_3)R_2 + H_2C(COOR)_2 \xrightarrow{NaOR} XC(=O)CH(R_1)C(R_2)(R_3)CH(COOR)_2$ (X = R, RO, NR_2, others)	Michael addition
$XC(=O)CH_2R_1 + R_2CHO \xrightarrow[2.\ H_3O^+,\ heat]{1.\ NaOCH_3} XC(=O)C(R_1)=CHR_2$ (X = R, RO, NR_2, others)	Aldol condensation
$R_1COOR_2 + R_3CH_2COOR_2 \xrightarrow{NaOR_2} R_1C(=O)CH(R_1)C(=O)OR_2$	Claisen condensation
$R_1CH_2Br \xrightarrow[2.\ BuLi]{1.\ PPh_3} R_1-CH=PPh_3 \xrightarrow{3.\ R_2CHO} R_1CH=CHR_2$	Wittig
$2RCHO \xrightarrow[2.\ H_2O\ workup]{1.\ Ti(0)} R-CH(OH)-CH(OH)-R$	Pinacol coupling
$BrCH_2CH_2CH_2CH_2CH=CH_2 \xrightarrow[Bu_3SnH]{free\text{-}radical\ initiator}$ cyclopentane$-CH_3$	Free-radical cyclization
R-substituted diene + $CH_2=CHY \xrightarrow{heat}$ cyclohexene bearing Y and R (Y = COOR, COR, CN, others)	Diels-Alder
$H_2C=C(R_2)CH_2CH_2CH=CHR_1 \xrightarrow{heat}$ product with R_2, CH_2, R_1	Cope rearrangement

group that should be prevented from participating in the reaction can be protected by converting it to a derivative that is unreactive to the reagent involved. The usual strategy employed to allow for such selective differentiation of the same or similar groups is to convert each group to a masked (protected) form which is not reactive but which can be unmasked (deprotected) to yield the group when necessary.

A large variety of organic reactions that can be used in syntheses are known. They can be categorized according to whether they feature a functional group interconversion or a carbon-carbon bond formation.

Functional group interconversions (Table 1) are reactions that change one functional group into another functional group. A functional group is a nonhydrogen, non-all-singly-bonded carbon atom or group of atoms. Included in functional group interconversions are nucleophilic substitution reactions, electrophilic additions, oxidations, and reductions. *See* COMPUTATIONAL CHEMISTRY; ELECTROPHILIC AND NUCLEOPHILIC REAGENTS; OXIDATION-REDUCTION; OXIDIZING AGENT; SUBSTITUTION REACTION.

Carbon-carbon bond-forming reactions (Table 2) feature the formation of a single bond or double bond between two carbon atoms. This is a particularly important class of reactions, as the basic strategy of synthesis—to assemble the target molecule from simpler, hence usually smaller, starting materials—implies that most complex molecules must be synthesized by a process that builds up the carbon skeleton of the target by using one or more carbon-carbon bond-forming reactions. Robert D. Walkup

Organoactinides Organometallic compounds of the actinides—elements 90 and beyond in the periodic table. Both the large sizes of actinide ions and the presence of 5*f* valence orbitals are unique features which differ distinctly from most, if not all, other metal ions.

Organometallic compounds have been prepared for all actinides through curium (element 96), although most investigations have been conducted with readily available and more easily handled natural isotopes of thorium (Th) and uranium (U). Organic groups (ligands) which bind to actinide ions include both π- and σ-bonding functionalities. The importance of this type of compound reflects the ubiquitous character of metal-carbon two-electron sigma bonds in both synthesis and catalysis. *See* CATALYSIS.

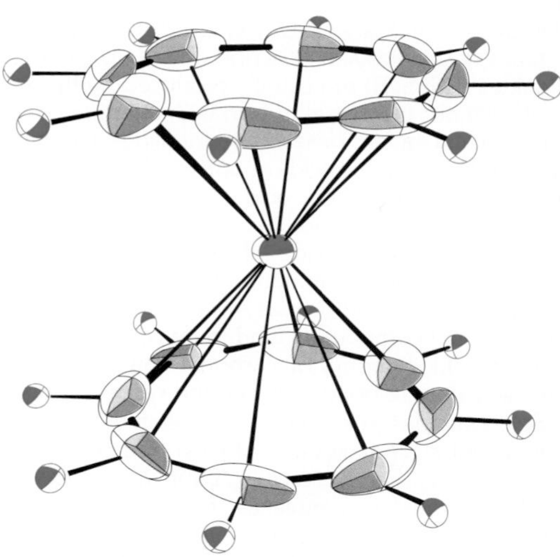

Molecular structure of $U(C_8H_8)_2$ determined by single-crystal x-ray diffraction. (*After K. O. Hodgson and K. N. Raymond, Inorg. Chem., 12:458, 1973*)

The moleculaor structures of a number of organoactinides have been determined by single-crystal x-ray and neutron-diffraction techniques. In almost all cases the large size of the metal ion gives rise to unusually high (as compared to a transition-metal compound) coordination numbers. That is, a greater number of ligands or ligands with greater spatial requirements can be accommodated within the actinide coordination sphere. The sandwich complex bis(cyclooctatetraenyl)-uranium (uranocene), an example of this latter type, is shown in the illustration. *See* ACTINIDE ELEMENTS; COORDINATION CHEMISTRY; METALLOCENES; NEUTRON DIFFRACTION; ORGANOMETALLIC COMPOUND; X-RAY DIFFRACTION. Tobin J. Marks

Organometallic compound A member of a broad class of compounds whose structures contain both carbon (C) and a metal (M). Although not a required characteristic of organometallic compounds, the nature of the formal carbon-metal bond can be of the covalent, ionic, or π-bound type.

The term organometallic chemistry is essentially synonymous with organotransition-metal chemistry; it is associated with a specific portion of the periodic table ranging from groups 3 through 11, and also includes the lanthanides. *See* CHEMICAL BONDING; LIGAND; PERIODIC TABLE; TRANSITION ELEMENTS.

From the perspective of inorganic chemistry, organometallics afford seemingly endless opportunities for structural variations due to changes in the metal coordination number, alterations in ligand-metal attachments, mixed-metal cluster formation, and so forth. From the viewpoint of organic chemistry, organometallics allow for manipulations in the functional groups that in unique ways often result in rapid and efficient elaborations of carbon frameworks for which no comparable direct pathway using non-transition organometallic compounds exists.

In moving across the periodic table, the early transition metals have seen relatively limited use in synthesis, with two exceptions: titanium (Ti) and zirconium (Zr).

Titanium has an important role in a reaction known as the Sharpless asymmetric epoxidation, where an allylic alcohol is converted into a chiral, nonracemic epoxy alcohol with excellent and predictable control of the stereochemistry [reaction (1)]. The

R¹, R², R³, OH — Allylic alcohol → R¹, R², O, R³, OH — Optically active epoxy alcohol (1)

significance of the nonracemic product is that the reaction yields a single enantiomer of high purity.

There are many applications utilizing the Sharpless asymmetric synthesis. Examples of synthetic targets that have relied on this chemistry include riboflavin (vitamin B_2) and a potent inhibitor of cellular signal transduction known as FK-506. *See* ASYMMETRIC SYNTHESIS; TITANIUM.

Below titanium in group 4 in the periodic table lies zirconium. Most modern organozirconium chemistry concerns zirconium's ready formation of the carbenelike complex [Cp_2Zr:], which because of its mode of preparation is more accurately thought of as a π complex (2). Also important are reactions of the zirconium chloride hydride, $Cp_2Zr(H)Cl$, commonly referred to as Schwartz's reagent, with alkenes and alkynes, and the subsequent chemistry of the intermediate zirconocenes. *See* METALLOCENES.

When the complex Cp_2ZrCl_2 is exposed to two equivalents of ethyl magnesium bromide EtMgBr, the initially formed Cp_2ZrEt_2 loses a molecule of ethane (C_2H_6) to produce the complexed zirconocene [Cp_2Zr: (**1**)], as shown in reaction (2). Upon intro-

$$Cp_2ZrCl_2 \xrightarrow[\text{tetrahydrofuran}]{\text{2EtMgBr}} [Cp_2Zr:] \equiv Cp_2Zr \text{—}\!\!\begin{matrix} CH_2 \\ \| \\ CH_2 \end{matrix} \quad (2)$$

(**1**)

duction of another alkene or alkyne, a zirconacyclopentane or zirconacyclopentene is formed, respectively [reaction (3)], where

$$\underset{(\mathbf{1})}{Cp_2Zr \text{—}\!\!\begin{matrix} CH_2 \\ \| \\ CH_2 \end{matrix}} \xrightarrow[\text{tetrahydrofuran}]{R\text{—}\equiv\text{—}R} Cp_2Zr\text{(cyclopentene, R, R)} \quad (3)$$

the structure above the arrow indicates that the chemistry applies to either alkenes (no third bond) or alkynes (with third bond)]. These are reactive species that can be converted to many useful derivatives resulting from reactions such as insertions, halogenations, and transmetalation/quenching. When the preformed complex (**1**) is treated with a substrate containing both an alkene and alkyne, a bicyclic zirconacene results that can ultimately yield polycyclic products (for example, pentalenic acid, a likely intermediate in the biosynthesis of the antibiotic pentalenolactone). *See* REACTIVE INTERMEDIATES.

Among the group 6–8 metals, chromium (Cr), molybdenum (Mo), and tungsten (W) have been extensively utilized in the synthesis of complex organic molecules in the form of their electrophilic Fischer carbene complexes, which are species having, formally, a double bond between carbon and a metal. They are normally generated as heteroatom-stabilized species bearing a "wall" of carbon monoxide ligands (**2**).

CO, CO, OMe, OC, Cr, OC, R, CO

(2)

Most of the synthetic chemistry has been performed with chromium derivatives, which are highly electrophilic at the carbene center because of the strongly electron-withdrawing carbonyl (CO) ligands on the metal. Many different types of reactions are characteristic of these complexes, such as α alkylation, Diels-Alder cycloadditions of α,β-unsaturated systems, cyclopropanation with electron-deficient olefins, and photochemical extrusions/cycloadditions. The most heavily studied and applied in synthesis, however, is the Dötz reaction, which has been applied in the production of antitumor antibiotics in the anthracycline and aureolic acid families. *See* DIELS-ALDER REACTION; METAL CARBONYL.

Groups 9–11 contain transition metals that have been the most widely used not only in terms of their abilities to effect C—C bond formations but also organometallic catalysts for some of the most important industrial processes. These include cobalt (Co), rhodium (Rh), palladium (Pd), and copper (Cu). Bruce H. Lipshutz

Organophosphorus compound One of a series of derivatives of phosphorus that have at least one organic (alkyl or aryl) group attached to the phosphorus atom linked either directly to a carbon atom or indirectly by means of another element (for example, oxygen). The mono-, di-, and trialkylphosphines (and their aryl counterparts) can be regarded formally as the parent compounds of all organophosphorus compounds.

Considering the large number of organic groups that may be joined to phosphorus as well as the incorporation of other elements in these materials, the number of combinations is practically unlimited. A vast family in itself is composed of the heterocyclic phosphorus molecules, in which phosphorus is one of a group of atoms in a ring system.

Some organophosphorus compounds have been used as polymerization catalysts, lubricant additives, flameproofing agents, plant growth regulators, and insecticides. Organophosphorus compounds were made during World War II for use as chemical warfare agents in the form of nerve gases (Sarin, Trilon 46, Soman, and Tabun). *See* PHOSPHORUS. Sheldon E. Cremer

Organoselenium compound One of a group of compounds that contain both selenium (Se) and carbon (C) and frequently other elements as well, for example, halogen, oxygen (O), sulfur (S), or nitrogen (N). Organoselenium compounds have become common in organic chemistry laboratories, where they have numerous applications, particularly in the area of organic synthesis. Organoselenium compounds formally resemble their sulfur analogs and may be classified similarly. For instance, selenols, selenides, and selenoxides are clearly related to thiols, sulfides, and sulfoxides. Despite structural similarities, however, sulfur and selenium compounds are often strikingly different with respect to their stability, properties, and ease of formation. *See* ORGANOSULFUR COMPOUND; SELENIUM.

Selenols have the general formula RSeH, where R represents either an aryl or alkyl group. They are prepared by the alkylation of hydrogen selenide (H_2Se) or selenide salts, as well as by the reaction of Grignard reagents or organolithium compounds with selenium, as in reactions (1).

$$\underset{\text{Selenol}}{\text{RSeH}} \xleftarrow[2.\ H_3O^+]{1.\ Se} \underset{\substack{\text{Grignard}\\\text{reagent}}}{\text{RMgX}} \text{ or } \underset{\substack{\text{Organolithium}\\\text{compound}}}{\text{RLi}} \qquad H_2Se\ (\text{or } HSe^-) \xrightarrow{RX} \quad (1)$$

Selenols are stronger acids than thiols. They and their conjugate bases are powerful nucleophiles that react with alkyl halides (R′X; R′ = alkyl group) or similar electrophiles to produce selenides (RSeR′); further alkylation yields selenonium salts, as shown in reactions (2), where R″ represents an alkyl group that

$$\underset{\text{Selenol}}{\text{RSeH}} \xrightarrow{R'X} \underset{\text{Selenide}}{\text{RSeR}'} \xrightarrow{R''X} \underset{\text{Selenonium salt}}{\text{R—}\overset{R''}{\overset{|}{\text{Se}^+}}\text{—R}' + X^-} \quad (2)$$

may be different from the alkyl group R′. Both acyclic and cyclic selenides are known. *See* ALKYLATION; ELECTROPHILIC AND NUCLEOPHILIC REAGENTS; GRIGNARD REACTION; REACTIVE INTERMEDIATES.

Diselenides (RSeSeR) are usually produced by the aerial oxidation of selenols; they react in turn with chlorine or bromine, yielding selenenyl halides or selenium trihalides. Diselenides are easily oxidized to seleninic acids or anhydrides by reagents such as hydrogen peroxide or nitric acid.

Selenoxides are readily obtained from the oxidation of selenides with hydrogen peroxide (H_2O_2) or similar oxidants. Selenoxides undergo facile elimination to produce olefins and selenenic acids, a process known as syn-elimination.

Selenocarbonyl compounds tend to be considerably less stable than their thiocarbonyl or carbonyl counterparts because of the weaker double bond between the carbon and selenium atoms. Selenoamides, selenoesters and related compounds can be isolated, but selenoketones (selones) and selenoaldehydes are highly unstable. *See* ALDEHYDE; AMIDE; ESTER; KETONE.

The charge-transfer complexes formed between certain Organoselenium donor molecules and appropriate acceptors such as tetracyanoquinodimethane are capable of conducting electric current. Selenium-containing polymers are also of interest as organic conductors. *See* COORDINATION COMPLEXES; ORGANIC CONDUCTOR.

Selenium is an essential trace element, and its complete absence in the diet is severely detrimental to human and animal health. The element is incorporated into selenoproteins such as glutathione peroxidase, which acts as a natural antioxidant. *See* BIOINORGANIC CHEMISTRY; COORDINATION CHEMISTRY; PROTEIN. Thomas G. Back

Organosilicon compound One of a group of compounds in which silicon (Si) is bonded to an organic functional group (R) either directly or indirectly via another atom. Formally, all organosilanes can be viewed as derivatives of silane (SiH_4) by the substitution of hydrogen (H) atoms. The most common substituents are methyl (CH_3; Me) and phenyl (C_6H_5; Ph) groups. However, tremendous diversity results with cyclic structures and the introduction of heteroatoms. *See* SILICON.

Organosilicon compounds are not found in nature and must be prepared in the laboratory. The ultimate starting material is sand (silicon dioxide, SiO_2) or other inorganic silicates, which make up over 75% of the Earth's crust. The useful properties of silicone polymers were identified in the 1940s; widespread interest in Organosilicon chemistry followed.

The chemistry of organosilanes can be explained in terms of the fundamental electronic structure of silicon and the polar nature of its bonds to other elements. Silicon appears in the third row of the periodic table, immediately below carbon (C) in group IV, and has many similarities with carbon. However, it is the fundamental differences between carbon and silicon that make silicon so useful in organic synthesis and of such great theoretical interest. *See* CARBON.

In the majority of organosilicon compounds, silicon follows the octet rule and is 4-coordinate. This trend can be explained by the electronic configuration of atomic silicon ($3s^2 3p^2 3d^0$) and the formation of sp^3-hybrid orbitals for bonding. Unlike carbon ($2s^2 2p^2$), however, silicon is capable of expanding its octet and can form 5- and 6-coordinate species with electronegative substituents, such as the 6-coordinate octahedral dianion, $[Me_2SiF_4]^{2-}$. *See* COORDINATION CHEMISTRY; ELECTRON CONFIGURATION; VALENCE.

Silicon is a relatively electropositive element that forms polar covalent bonds ($Si^{\delta+}$—$X^{\delta-}$) with carbon and other elements, including the halogens, nitrogen, and oxygen. The strength and reactivity of silicon bonds depend on the relative electronegativities of the two elements. For example, the strongly electronegative elements fluorine (F) and oxygen (O) form bonds that have tremendous thermodynamic stabilities. *See* CHEMICAL THERMODYNAMICS; ELECTRONEGATIVITY.

Before 1981, multiply bonded silicon compounds were identified only as transient species both in solution and in the gas phase. However, when tetramesityldisilene ($Ar_2Si{=}SiAr_2$, where Ar = 2,4,6-trimethylphenyl) was isolated, it was found to be a crystalline, high-melting-point solid having excellent stability in the absence of oxygen and moisture.

Numerous reactive species have been generated and characterized in organosilicon chemistry. Silylenes are divalent silicon species (SiR_2, where R = alkyl, aryl, or hydrogen), analogous to carbenes in carbon chemistry. The dimerization of two silylene units gives a disilene. Silicon-based anions, cations, and radicals are important intermediates in the reactions of silicon. *See* FREE RADICAL.

The direct process for the large-scale preparation of organosilanes has provided a convenient source of raw materials for the development of the silicone industry. The process produces a mixture of chloromethylsilanes from elemental silicon and methyl chloride in the presence of a copper (Cu) catalyst, as in the reaction below.

$$\text{Me—Cl} + \text{Si} \xrightarrow{\text{Cu catalyst}} Cl_2SiMe_2 + \text{other } Cl_nSiMe_{4-n}$$

In spite of concentrated research efforts in this area, the synthetic methods available for the controlled formation of silicon-carbon bonds remain limited to only a few general reaction types. These include the reaction of organometallic reagents with silanes, catalytic hydrosilylation of multiple bonds, and reductive silylation.

The role of silicon in organic synthesis is quite extensive, and chemists exploit the unique reactivity of organosilanes to accomplish a wide variety of transformations. Silicon is usually introduced into a molecule to perform a specific function and is then removed under controlled conditions.

Polysilanes are organosilicon compounds that contain either cyclic arrays or linear chains of silicon atoms. The isolation and characterization of both cyclosilanes (with up to 35 silane units) and high-molecular-weight linear polysilanes (with up to 3000 silane units) have demonstrated that silicon is capable of extended chain formation (catenation). Polysilanes contain only silicon-silicon bonds in their backbone, which differentiates them from polysiloxanes (silicones), which contain alternating silicon and oxygen repeat units. *See* SILICONE RESINS.

Polysilanes have found applications as photoresists in microlithography, charge carriers in electrophotography, and photoinitiators for vinyl polymerization. Polysilanes also function as preceramic polymers for the manufacture of silicon carbide fibers.

Howard Yokelson

Organosulfur compound A member of a class of organic compounds with any of several dozen functional groups containing sulfur (S).

Sulfur is an element of the third row of the periodic table; it is larger and less electronegative than oxygen, which lies above it in the second row. Compounds with an expanded valence shell, that is, compounds bonding to as many as six ligands around sulfur, are therefore possible, and a broad range of compounds can be formed. Moreover, sulfur has a much greater tendency than oxygen to undergo catenation to give chains with several atoms linked together through S—S bonds. *See* CHEMICAL BONDING; PERIODIC TABLE; STRUCTURAL CHEMISTRY; VALENCE.

The structures and names of representative types of organosulfur compounds are shown in the table. Some compounds and groups are named by using the prefix thio to denote replacement of oxygen by sulfur. The prefix thia can be used to indicate that one or more —CH_2— groups have been replaced by sulfur, as in 2,7-dithianonane [$CH_3S(CH_2)_4SCH_2CH_3$].

Thiols and sulfides are sulfur counterparts of alcohols and ethers, respectively, and can be prepared by substitution reactions analogous to those used for the oxygen compounds. Sulfonium salts are obtained by further alkylation of sulfides.

Although thiols and alcohols are structurally analogous, there are significant differences in the properties of these two groups. Hydrogen bonding of the type —S–H—S— is very weak compared to —O—H–O—, and thiols are thus more volatile and have lower boiling points than the corresponding alcohols; for

Some types of organosulfur compounds and groups

Structure	Name
RSH	Thiol (mercaptan)
RSR	Sulfide (thioether)
RSSR	Disulfide
RSSSR	Trisulfide (trisulfane)
$R\overset{+}{S}R\ X^-$ with R on S	Sulfonium salt
$R_2C{=}S$	Thioketone
$RN{=}C{=}S$	Isothiocyanate
RC(=O)SR	Thiolate ester (thioic acid S-ester)
RC(=S)OR	Thionoate ester
RCS_2R	Dithioate ester
RSOH	Sulfenic acid
RSCl	Sulfenyl chloride
RSOR	Sulfoxide
RS(=NR)R	Sulfimide
RSO_2H	Sulfinic acid
RSO_2R	Sulfinate ester
$R_2S{=}R_2$	Sulfonium ylide (sulfurane)
RSO_2R	Sulfone
RSO_3H	Sulfonic acid
RSO_2NH_2	Sulfonamide
RSO_2Cl	Sulfonyl chloride
$ROSO_3R$	Sulfate ester

example, methanethiol (CH_3SH) has a boiling point of 5.8°C (42.4°F) compared to 65.7°C (150.3°F) for methanol (CH_3OH).

Thiols form insoluble precipitates with heavy-metal ions such as lead or mercury. Both thiols and sulfides are extremely malodorous compounds, recalling the stench of rotten eggs (hydrogen sulfide). However, traces of these sulfur compounds are an essential component of the distinctive flavors and aromas of many vegetables, coffee, and roast meat. *See* MAILLARD REACTION; MERCAPTAN; SPICE AND FLAVORING.

Thiocarbonyl compounds contain a carbon-sulfur double bond (C=S). Thiocarbonyl compounds (thiones) are much less common than carbonyl compounds (C=O bond). Simple thioaldehydes or thioketones have a strong tendency to form cyclic trimers, polymers, or other products.

Sulfides can be oxidized sequentially to sulfoxides and sulfones, containing the sulfinyl (—SO—) and sulfonyl (—SO_2—) groups, respectively, as in the reaction below.

$$\mathrm{RSR'} \longrightarrow \underset{\text{Sulfoxide}}{\mathrm{RS(=O)R'}} \longrightarrow \underset{\text{Sulfone}}{\mathrm{RS(=O)_2R'}}$$

Dimethyl sulfoxide (DMSO) is available in large quantities as a by-product of the Kraft sulfite paper process. It is useful as a polar solvent with a high boiling point and as a selective oxidant and reagent in organic synthesis. *See* DIMETHYL SULFOXIDE.

Compounds containing the sulfonyl group include sulfones, sulfonyl chlorides, sulfonic acids, and sulfonamides. The sulfonyl group resembles a carbonyl in the acidifying effect on an α-hydrogen. The diaryl sulfone unit is the central feature of polysulfone resins, used in some high-performance plastics. Sulfonic acids are obtained by oxidation of thiols or by sulfonation. Sulfonamides, prepared from the chlorides, were the mainstay therapeutic agents in infections until the advent of antibiotics; they are still used for some conditions. *See* POLYSULFONE RESINS; SULFONAMIDE; SULFONIC ACID.

A number of proteins and metabolic pathways in systems of living organisms depend on the amino acid cysteine and other sulfur compounds. In many proteins, for example, in the enzyme insulin, disulfide bonds formed from the —SH groups of cysteine units are an essential part of the structure. The —SH groups of cysteine also play a role in the metal-sulfur proteins that mediate electron-transport reactions in respiration and photosynthesis. *See* INSULIN; PHOTOSYNTHESIS.

The coenzyme lipoic acid is a cyclic disulfide that functions together with the coenzyme thiamine diphosphate to accept electrons and undergo reduction of the —S—S— bond in the oxidative decarboxylation of pyruvic acid. Two other major pathways in metabolism, the transfer of acetyl groups and of methyl groups, are mediated by organosulfur compounds. Acetyl transfer, a key step in lipid and carbohydrate metabolism, occurs by way of thioesters. *See* CARBOHYDRATE METABOLISM; COENZYME; LIPID METABOLISM; THIAMINE.

Sulfur is present in numerous other compounds found in natural sources. Petroleum contains variable amounts of sulfur, both as simple thiols and sulfides, and also heterocyclic compounds such as benzothiophene. Removal of these is an important step in petroleum refining. *See* PETROLEUM; PETROLEUM PROCESSING AND REFINING.

Several sulfur-containing compounds from natural sources have important pharmacological properties. Examples are the β-lactam antibiotics penicillin, cephalosporin, and thienamycin, and the platelet anticoagulating factor ajoene from garlic, produced by a series of complex enzymatic reactions from allicin. *See* ANTIBIOTIC; HETEROCYCLIC COMPOUNDS; ORGANIC CHEMISTRY; SULFUR.

James A. Moore

Oriental vegetables

Oriental vegetables are very important in Asian countries, but are considered as minor crops in the United States and Europe. However, in recent years there has been an increased interest in these crops because of their unusual flavors and textures and in some cases their high nutritional values. Some of the more common ones are described below.

Chinese cabbage, celery cabbage, napa, or pe-tsai (*Brassica campestris*, pekinensis group; *B. rapa*, pekinensis group; *B. pekinensis*) belongs to the mustard (Cruciferae) family, and is a biennial leafy plant but is grown as an annual. The harvested part is a head which is composed of broad crinkled leaves with a very wide, indistinct, white midrib. The outer leaves are pale green, and the inner leaves of the head are blanched. It is a good salad vegetable because of the mild flavor and crisp texture.

There are many varieties of pak choy, bok choi, Chinese mustard, or celery mustard (*Brassica campestris*, chinensis group; *B. rapa*, chinensis group; *B. chinensis*). The crop is grown as an annual. Pak choy does not form a head like most varieties of Chinese cabbage, but forms a celerylike stalk of tall, dark green leaves, with prominent white veins and long white petioles (see illustration). The base of the petiole may be expanded and spoon-shaped; the blade of the leaf is smooth, and not crinkled like that of Chinese cabbage.

The large, long, white radish (*Raphanus sativus*, longipinnatus group) is often called oriental winter radish, daikon, Chinese winter radish, lobok, and lob paak. Radish is a dicotyledonous herbaceous plant grown for its long, enlarged roots.

Edible podded peas, China peas, sugar peas, or snow peas (*Pisum sativum*, macrocarpon group) belong to the Leguminosae family. The primitive forms of pea are slightly bitter and have a tough seed coat, which allows for long dormant periods. In the garden pea the pod swells first as the seeds enlarge, but in the edible podded pea the immature seeds bulge the pod, which demarks the developing seeds. Unlike the regular garden peas which have tough fibery seed pods, the edible podded peas were selected for the tender pods and not for the seeds.

Yard-long bean or asparagus bean (*Vigna sinensis*, sesquipedelis group) belongs to the Leguminosae family. It is an annual climbing plant. It is a relative of the cow pea (*V. unguiculata*).

Pak choy or Chinese mustard (*Brassica campestris*). (*University of California Agricultural Experiment Station*)

Mung bean or green gram (*Phaseolus aureus*) sprouts have been used by the Chinese for their remarkable healing qualities for thousands of years. Only recently has the Western world recognized the value of sprouted leguminous seeds (mung, soy, and alfalfa sprouts). The unsprouted mung bean seeds contain negligible amounts of vitamin C (ascorbic acid), whereas the sprouted seeds contain 20 mg per 100 g of sprouts, which is as high as tomato juice.

Jicama (*Pachyrrizus erosus*), also called the yam bean, is indigenous to Mexico and Central America. It belongs to the pea (Leguminosae) family. The crop is grown for its enlarged turnip-shaped root, which is eaten raw or cooked, has a crisp texture, and is sweetish in taste.

Chinese winter melon, wax gourd, winter gourd, white gourd, ash gourd, Chinese preserving melon, ash pumpkin, or tung kwa (*Benincasa hispida; B. cerifera*) is a viny annual cucurbit. The flesh of mature fruits is used in making Chinese soups. It can be eaten raw or made into sweet preserves similar to citron or watermelon rind preserves.

Balsam pear, alligator pear, bitter gourd, bitter melon, bitter cucumber, or fu kwa (*Momordica charantia*) is an annual herbaceous vine. The heart-shaped to cylindrical fruits are extremely bitter; some of the bitterness is removed before cooking by peeling and steeping in salt water. Immature fruits are less bitter than mature ones.

Chinese okra or angled loofa (*Luffa acutangula*) is called zit kwa by the Chinese; it is a close relative of *L. cylindrica*, known as loofa, sponge gourd, or dishcloth gourd. Chinese okra is an annual climbing vine grown for the immature fruits. The thoroughly mature fruits can be made into vegetable sponge.

Water spinach, water convolvulus, swamp cabbage, kang kong, or shui ung tsoi (*Ipomoea aquatica*) is a perennial semi-aquatic plant grown for its long, tender shoots. It is a relative of the sweet potato (*I. batatas*), but does not produce an enlarged root.

M. Yamaguchi

Orion The Hunter, a prominent constellation in the evening winter sky (see illustration). Orion is perhaps the easiest constellation to identify in the sky, because of the three bright stars in a line that form the belt of the mythical figure. The cool, red supergiant star Betelgeuse glows brightly as Orion's shoulder, above which is an upraised club, and the hot, blue star Rigel marks Orion's heel. From the leftmost star of the belt dangles several faint stars and the Horsehead Nebula. The faint stars and the Orion Nebula, M42 (the 42nd object in Messier's eighteenth-century catalogue), one of the most prominent and beautiful emission nebulae and a nursery for star formation, make up Orion's sword. Four bright stars close together (within about 1 light-year) make the Trapezium, which provides the energy to make the Orion Nebula glow. *See* BETELGEUSE; CONSTELLATION; ORION NEBULA; RIGEL.

Jay M.Pasachoff

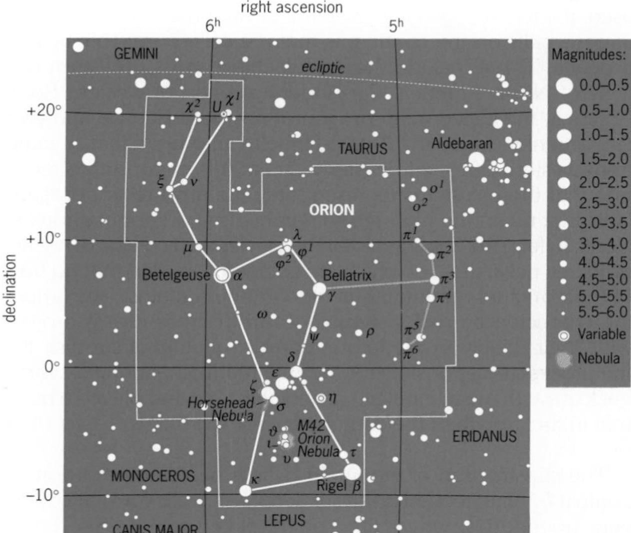

Modern boundaries of the constellation Orion, the Hunter. The celestial equator is 0° of declination, which corresponds to celestial latitude. Right ascension corresponds to celestial longitude, with each hour of right ascension representing 15° of arc. Apparent brightness of stars is shown with dot sizes to illustrate the magnitude scale, where the brightest stars in the sky are 0th magnitude or brighter and the faintest stars that can be seen with the unaided eye at a dark site are 6th magnitude. (*Wil Tirion*)

Orion Nebula The brightest emission nebula in the sky, designated M42 in Messier's catalog. The Great Nebula in Orion consists of ionized hydrogen and other trace elements (see illustration). The nebula belongs to a category of objects known as H II regions (the Roman numeral II indicates that hydrogen is in the ionized state), which mark sites of recent massive star formation. Located in Orion's Sword at a distance of 460 parsecs or 1500 light-years (8.8×10^{15} mi or 1.4×10^{16} km), the Orion Nebula consists of dense plasma, ionized by the ultraviolet radiation of a group of hot stars less than 100,000 years old known as the Trapezium cluster. The nebula covers an area slightly smaller than the full moon and is visible with the aid of binoculars or a small telescope. *See* ORION.

In addition to the Trapezium cluster of high-mass stars, the Orion Nebula contains about 700 low-mass stars packed into an unusually small volume of space. As the Orion Nebula evolves,

Visual-wavelength view of the Orion Nebula obtained by taking a time exposure on a photographic plate with the 4-m-diameter (158-in.) telescope on Kitt Peak, Arizona (*National Optical Astronomy Observatories*).

the cluster is expected to expand and possibly dissolve into the smooth background of stars in the Milky Way.

Observations with the Hubble Space Telescope have shown that many low-mass stars in the Orion Nebula are surrounded by disks of dense gas and dust. These so-called proplyds (derived from the term "proto planetary disks") appear to be rapidly losing their mass as they are irradiated by intense ultraviolet radiation produced by high-mass stars in the nebula. However, the proplyds may eventually evolve into planetary systems if most of the solids in these disks have already coagulated into centimeter-sized objects which are resistant to ablation by the radiation field.

A star-forming molecular cloud core (known as OMC1 for Orion Molecular Cloud 1) is hidden behind the Orion Nebula by a shroud of dust. This cloud core is only a small part of a giant molecular cloud (the Orion A cloud), 100,000 times more massive than the Sun. *See* INTERSTELLAR MATTER; MOLECULAR CLOUD; NEBULA. John Bally

Ornamental plants Plants used for ornamental purposes. The use of ornamental plants developed early in history for it is known that primitive peoples planted them near their dwellings. Gardening in various forms has been a part of the cultures of all known great civilizations. Although increased urbanization has often reduced the areas which can be devoted to ornamentals, nevertheless there is a growing interest in them and the sums spent for plants and horticultural supplies are quite large. New plants from various parts of the world are constantly being introduced by individuals and by botanic gardens and arboreta.

Ornamentals have been selected from virtually the whole kingdom of plants. Ground covers include turfgrasses, herbaceous annuals, biennials or perennials, and some scandent shrubs or vines. The woody structural landscape plants include vines (deciduous and evergreen), shrubs (deciduous and evergreen), deciduous trees, broad-leaved evergreen trees, gymnosperms and cyads, bamboos, and palms.

Plants for special habitats or uses include lower plants such as mosses, selaginellas or horsetails, ferns and tree ferns, sword-leaved monocots, bromeliads and epiphytic plants, aquatic plants, cacti and succulents, and bulbous and cormous plants. Interior decorative plants in a large measure are shade plants from the tropics.

The best choice of an ornamental plant for a given situation depends on the design capabilities of the plant, the ecological requirements, and various horticultural factors. The plant must fulfill the purposes of the designer regarding form, mass, scale, texture, and color. It must also have the ability to adjust well to the ecology of the site in such factors as temperature range, humidity, light or shade, water, wind, and various soil factors, including drainage, aeration, fertility, texture, and salinity or alkalinity. Some important horticultural factors are growth rate, ultimate size and form, predictability of performance, and susceptibility to insects, diseases, deer, gophers, snails, and other pests. The hardiness of the plant is often important, and pruning and litter problems may influence choice. *See* ARBORETUM; BOTANICAL GARDENS; FLORICULTURE. Vernon T. Stoutemyer

Ornithischia One of two constituent clades of Dinosauria (the other being Saurischia). Ornithischian dinosaurs are characterized by the possession of numerous anatomical features, notably the configuration of the pelvis. In ornithischian hips, the pubis bone (which points downward and forward in the majority of other reptile groups) has rotated posteriorly to lie parallel to the ischium. Other features that characterize ornithischians include the presence of ossified tendons that form a complex lattice arrangement supporting the vertebral column, a single midline bone called the predentary that lies at the front end of and connects the two lower jaws, and a palpebral bone that traverses the orbit.

During the Late Triassic–Early Jurassic interval, ornithischians were relatively rare components of dinosaur faunas; however, the group began to radiate spectacularly in terms of species richness, abundance, and morphological diversity during the Middle Jurassic. From the Middle Jurassic onward, ornithischians attained a global distribution and became the dominant terrestrial vertebrates of the Cretaceous. Almost all ornithischians were herbivorous, though a few primitive forms may have been omnivores.

Phylogenetic analyses of ornithischian interrelationships indicate that the clade can be subdivided into three major lineages (Thyreophora, Marginocephalia, and Ornithopoda), along with a small number of primitive forms that do not fall within any of these groupings. Ornithischian evolution was dominated by several major themes: experimentation with stance (quadrupedality vs. bipedality), the appearance of ever more complex feeding mechanisms (such as alterations to teeth and skull morphology), and the development of sociality (gregarious living, parental care, and intra- and interspecific communication).

Thyreophora (the "armored" dinosaurs) consists of two clades: Stegosauria and Ankylosauria. Both of these groups were quadrupedal and characterized by the possession of numerous osteoderms (bony plates) that were embedded within the skin.

Marginocephalia are united by the possession of a distinct shelf of bone at the rear of the skull and consist of two groups: Pachycephalosauria ("bone-headed" dinosaurs) and Ceratopsia ("horned" dinosaurs, including *Triceratops*).

Ornithopoda is a conservative but very successful group that retained the bipedal pose of primitive ornithischians. All ornithopods possessed sophisticated grinding jaw mechanisms, allowing them to process vegetation as efficiently as living grazers and browsers. One subgroup of ornithopods, the hadrosaurs ("duck-billed" dinosaurs), had teeth arrayed in dense "batteries," consisting of multiple columns and rows of functional teeth that provided a large continuous surface area for chewing. Hadrosaurs were also among the most social of dinosaurs. *See* ARCHOSAURIA; CRETACEOUS; DINOSAURIA; JURASSIC; SAURISCHIA. Paul M. Barrett

Orogeny The process of mountain building. As traditionally used, the term orogeny refers to the development of long, mountainous belts on the continents that are called orogenic belts or orogens. These include the Appalachian and Cordilleran orogens of North America, the Andean orogen of western South America, the Caledonian orogen of northern Europe and eastern Greenland, and the Alpine-Himalayan orogen that stretches from western Europe to eastern China. It is important to recognize that these systems represent only the most recent orogenic belts that retain the high-relief characteristic of mountainous regions. In fact, the continents can be viewed as a collage of ancient orogenic belts, most of which are so deeply eroded that no trace of their original mountainous topography remains. By comparing characteristic rock assemblages from more recent orogens with their deeply eroded counterparts, geologists surmise that the processes responsible for mountain building today extended back through most (if not all) of geologic time and played a major role in the growth of the continents. *See* CONTINENTS, EVOLUTION OF.

The construction of mountain belts is best understood in the context of plate tectonics theory. Orogenic belts form at convergent boundaries, where lithosphere plates collide. *See* LITHOSPHERE; PLATE TECTONICS.

There are two basic kinds of convergent plate boundaries, leading to the development of two end-member classes of orogenic belts. Oceanic subduction boundaries are those at which oceanic lithosphere is thrust (subducted) beneath either continental or oceanic lithosphere. The process of subduction leads to partial melting near the plate boundary at depth, which is manifested by volcanic and intrusive igneous activity in the overriding plate. Where the overriding plate consists of oceanic lithosphere, the result is an intraoceanic island arc, such as the Japanese islands. Where the overriding plate is continental, a continental arc is formed. The Andes of western South America is an example. *See* MARINE GEOLOGY; OCEANIC ISLANDS; SUBDUCTION ZONES.

The second kind of convergent plate boundary forms when an ocean basin between two continental masses has been completely consumed at an oceanic subduction boundary and the continents collide. Continent collisional orogeny has resulted in some of the most dramatic mountain ranges on Earth; a good example is the Himalayan orogen, which began forming roughly 50 million years ago when India collided with the Asian continent. Because the destruction of oceanic lithosphere at subduction boundaries is a prerequisite for continental collision, continent collisional orogens contain deformational features and rock associations developed during arc formation as well as those produced by continental collision. Kip Hodges

Orpiment A mineral having composition As_2S_3. Crystals are small, tabular, and rarely distinct; the mineral occurs more commonly in foliated or columnar masses. The hardness is 1.5–2 (Mohs scale) and the specific gravity is 3.49. The luster is resinous and pearly on the cleavage surface; the color is lemon yellow. Orpiment is found in Romania, Peru, Japan, and Russia. In the United States it occurs at Mercer, Utah; Manhattan, Nevada; and in deposits from geyser waters in Yellowstone National Park. *See* ARSENIC. Cornelius S. Hurlbut, Jr.

Orthida An extinct order of brachiopods that lived in shallow-shelf seas during the Paleozoic. It contains the oldest members of the Rhynchonelliformea and is the stem group for all other brachiopods in this subphylum. Orthids are characterized by unequally biconvex valves containing distinct ribs and a strophic (straight) hinge line with associated well-developed interareas. Openings in the interareas provided space for a fleshy protrusion (pedicle) that was used for attachment to the substrate.

Orthids were sessile, attached, epifaunal suspension feeders. They arose in the Early Cambrian but remained low in diversity. Beginning in the Early Ordovician, orthid brachiopods began to increase dramatically in diversity, taking part in the great Ordovician radiation of marine animal life. During this taxonomic diversification, orthids also became the dominant member of marine epifaunal communities and helped to establish rhynchonelliform brachiopods as one of the main components of the Paleozoic evolutionary fauna. Orthids suffered high extinction in the Late Ordovician mass extinction and never returned to their earlier Ordovician diversity levels. Orthids were finally extinguished in the late Permian mass extinction. *See* ARTICULATA (ECHINODERMATA); BRACHIOPODA; RHYNCHONELLIFORMEA. Mark E. Patzkowsky

Orthoclase Potassium feldspar (Or = $KAlSi_3O_8$) that usually contains up to 30 mole % albite (Ab = $NaAlSi_3O_8$) in solid solution. Its hardness is 6; specific gravity, 2.57–2.5, depending on Ab content; mean refractive index, 1.52; color, white to dull pink or orange-brown. Some orthoclases may be intergrown with relatively pure albite which exsolved during cooling from a high temperature in pegmatites, granites, or granodiorites. This usually is ordered low albite, but in rare cases it may show some degree of Al,Si disorder, requiring it to be classified as analbite or high albite. If exsolution is detectable by eye, the Or-Ab composite mineral is called perthite; if microscopic examination is required to distinguish the phases, it is called microperthite; and if exsolution is detectable only by x-ray diffraction or electron optical methods, it is called cryptoperthite. Orthoclase is optically monoclinic. Its structure averaged over hundreds of nanometers may be monoclinic, but its true symmetry is triclinic. *See* ALBITE; CRYSTAL STRUCTURE; PERTHITE. Paul H. Ribbe

Orthogonal polynomials A special case of orthogonal functions that arise in many physical problems (often as the solutions of differential equations), in the study of distribution functions, and in certain other situations where one approximates fairly general functions by polynomials. *See* PROBABILITY.

Each set of orthogonal polynomials is defined with respect to a particular averaging procedure. The average value of a suitable function f is denoted by $E(f)$. An example is shown in Eq. (1). In general an averaging procedure has the form shown in Eq. (2),

$$E\{f\} = \frac{1}{2}\int_{-1}^{1} f(x)\,dx \tag{1}$$

$$E\{f\} = \int_{-\infty}^{\infty} f(x)\,d\sigma(x) \tag{2}$$

a Stieltjes integral, where σ is a distribution function, that is, an increasing function with $\sigma(-\infty) = 0$ and $\sigma(+\infty) = 1$.

Two functions f and g are said to be orthogonal with respect to a given averaging procedure if $E\{f\bar{g}\} = 0$ where the bar denotes complex conjugation. By the system of orthogonal polynomials associated with the averaging procedure is meant a sequence $P_0, P_1, P_2 \ldots$ of polynomials P_n having exact degree n, which are mutually orthogonal, that is, $E\{P_m\bar{P}_n\} = 0$ for $m \neq n$. This last condition is equivalent to the statement that each P_n is orthogonal to all polynomials of degree less than n. Thus P_n has the form $P_n(x) = a_0 + a_1x + a_2x^2 + \cdots + a_nx^n$ where $a_n \neq 0$ and is subject to the n conditions $E\{x^k P_n\} = 0$ for $k = 0, 1, \ldots, n - 1$. This gives n linear equations in the $n + 1$ coefficients of P_n, leaving one more condition, called a normalization, to be imposed. The method of normalization differs in different references. *See* POLYNOMIAL SYSTEMS OF EQUATIONS. Carl S. Herz

Orthonectida An order of Mesozoa. The orthonectids parasitize various marine invertebrates as multinucleate plasmodia. The plasmodia multiply by fragmentation. Eventually they give rise asexually, by polyembryony, to sexual males and females. Commonly only one sex arises from a given plasmodium.

These sexually mature forms escape as minute ciliated organisms. Structurally they are composed of a single layer of ciliated epithelial cells surrounding an inner mass of sex cells. The ciliated cells are disposed in rings around the body.

After insemination the eggs develop in the female and form ciliated larvaea. When liberated, these larvae invade new individuals of their host and then disaggregate, liberating germinal cells which give rise to new plasmodia. *See* MESOZOA. Bayard H. McConnaughey

Orthoptera An order that includes over 20,000 species of terrestrial insects, including most of the "singing" insects, some of the world's largest insects, and some well-known pests. Most species of Orthoptera (from "orthos," meaning straight, and "pteron," meaning wing) have enlarged hind legs adapted for jumping. These include grasshoppers and locusts (in the suborder Caelifera, a mainly diurnal group); and the crickets, katydids (bush-crickets), New Zealand weta, and allied families (suborder Ensifera, which comprises mainly nocturnal species). Orthoptera share with other orthopteroid insects, such as mantids and stick insects (now in separate orders Mantodea and Phasmatodea), gradual metamorphosis, chewing mouthparts, and two pairs of wings, the anterior pair of which is usually thickened and leathery and covers the fanwise folded second pair. Wings are reduced or absent in many species. Characters that define the Orthoptera as a natural group (the inclusive set of all species stemming from a common ancestor) are the jumping hind legs, small and well-separated hind coxae (basal leg segments), a pronotum with large lateral lobes, and important molecular (genetic) characters.

Food habits range from omnivorous to strictly carnivorous or herbivorous. Habitats are nearly all terrestrial, including arctic-alpine tundra and tropical areas with aquatic floating plants. Female orthopterans usually lay their eggs into soil or plant material. There are no parasitic species, but a few crickets live as cleptoparasitic "guests" in ant nests. *See* INSECTA.

Males of many species are outstandingly noisy or musical. Their songs typically function in obtaining mates. In some species females move to the singing male, and in others a female answering song is involved in pair formation. Song frequencies can range from the audible to the very high ultrasonic. Grasshoppers (Acridoidea) stridulate by rubbing the hind femora against the outer wings, whereas crickets (Gryllidae), katydids (Tettigoniidae), and humped-winged crickets (Haglidae) rapidly rub the forewings together. Some wingless species stridulate with the overlapping edges of abdominal segments or the mandibles. *See* ANIMAL COMMUNICATION; PHONORECEPTION.

Darryl T. Gwynne; Robert B. Willey

Orthorhombic pyroxene A group of minerals having the general chemical formula $XYSi_2O_6$, in which the Y site contains Fe or Mg and the X site contains Fe, Mg, Mn, or a small amount of Ca (up to about 3%). The end members of this solid solution series are enstatite ($Mg_2Si_2O_6$) and ferrosilite ($Fe_2Si_2O_6$). Names used for intermediate members of the series are enstatite, bronzite, hypersthene, and orthoferrosilite.

Many of the physical and optical properties of orthopyroxene are strongly dependent upon composition, and especially upon the Fe-Mg ratio. In hand specimens, orthopyroxene can be distinguished from amphibole by its characteristic 88° cleavage angles, and from augite by color—augite is typically green to black, while orthopyroxene is more commonly brown, especially on slightly weathered surfaces.

Orthopyroxene is a widespread mineral in metamorphic rocks. It is characteristic of granulite facies metamorphism, in both mafic and feldspathic gneisses. Orthopyroxene occurs in many basalts and gabbros, particularly those of tholeiitic composition, and many meteorites, but is notably absent from most alkaline igneous rocks. The greatest abundance of orthopyroxene is in ultramafic rocks, especially those in large layered intrusions. *See* PYROXENE.

Robert J. Tracy

Orthotrichales An order of the true mosses (subclass Bryidae) consisting of five families and 23 genera. The plants grow in mats or tufts in relatively exposed places, on trunks of trees and on rock. They are rather freely branched and prostrate, or may be sparsely forked and erect-ascending; the habit is more or less pleurocarpous, but sporophytes may be produced at the ends of leading shoots or branches. The leaves are generally oblong and broadly pointed, with a strong midrib. The capsules are often ribbed, and the peristome, if present, has a poor development of endostome. The calyptrae are often hairy. *See* BRYIDAE; BRYOPHYTA; BRYOPSIDA.

Howard Crum

Osage orange The genus *Maclura* of the mulberry family, with one species, *M. pomifera*. This tree may attain a height of 60 ft (18 m) and has yellowish bark, milky sap, simple entire leaves, strong axillary thorns, and aggregate green fruit about the size and shape of an orange. It is planted for hedges and as an ornament, especially in the eastern United States. The wood is used for fence posts and fuel and as a source of a yellow dye. It has also been used for archery bows, hence one of its common names, bowwood. *See* FOREST AND FORESTRY; TREE.

Arthur H. Graves; Kenneth P. Davis

Oscillation Any effect that varies in a back-and-forth or reciprocating manner. Examples of oscillation include the variations of pressure in a sound wave and the fluctuations in a mathematical function whose value repeatedly alternates above and below some mean value.

The term oscillation is for most purposes synonymous with vibration, although the latter sometimes implies primarily a mechanical motion. The alternating current and the associated electric and magnetic fields are referred to as electric (or electromagnetic) oscillations.

If a system is set into oscillation by some initial disturbance and then left alone, the effect is called a free oscillation. A forced oscillation is one in which the oscillation is in response to a steadily applied periodic disturbance.

Any oscillation that continually decreases in amplitude, usually because the oscillating system is sending out energy, is spoken of as a damped oscillation. An oscillation that maintains a steady amplitude, usually because of an outside source of energy, is undamped. *See* ANHARMONIC OSCILLATOR; DAMPING; FORCED OSCILLATION; HARMONIC OSCILLATOR; MECHANICAL VIBRATION; OSCILLATOR; VIBRATION.

Joseph M. Keller

Oscillator An electronic circuit that generates a periodic output, often a sinusoid or a square wave. Oscillators have a wide range of applications in electronic circuits: they are used, for example, to produce the so-called clock signals that synchronize the internal operations of all computers; they produce and decode radio signals; they produce the scanning signals for television tubes; they keep time in electronic wristwatches; and they can be used to convert signals from transducers into a readily transmitted form.

Oscillators may be constructed in many ways, but they always contain certain types of elements. They need a power supply, a frequency-determining element or circuit, a positive-feedback circuit or device (to prevent a zero output), and a nonlinearity (to define the output-signal amplitude). Different choices for these elements give different oscillator circuits with different properties and applications.

Oscillators are broadly divided into relaxation and quasilinear classes. Relaxation oscillators use strong nonlinearities, such as switching elements, and their internal signals tend to have sharp edges and sudden changes in slope; often these signals are square waves, trapezoids, or triangle waves. The quasilinear oscillators, on the other hand, tend to contain smooth sinusoidal signals because they regulate amplitude with weak nonlinearities. The type of signal appearing internally does not always determine the application, since it is possible to convert between sine and square waves. Relaxation oscillators are often simpler to design and more flexible, while the nearly linear types dominate when precise control of frequency is important.

Relaxation oscillators. Illustration *a* shows a simple operational-amplifier based relaxation oscillator. This circuit can be understood in a number of ways (for example, as a negative-resistance circuit), but its operation can be followed by studying the signals at its nodes (illus. *b*). The two resistors, labeled *r*, provide a positive-feedback path that forces the amplifier output to saturate at the largest possible (either positive or negative) output voltage. If v_+, for example, is initially slightly greater than v_-, then the amplifier action increases v_o, which in turn further increases v_+ through the two resistors labelled *r*. This loop continues to operate, increasing v_o until the operational amplifier saturates at some value V_{max}. [An operational amplifier ideally follows Eq. (1), where A_v is very large, but is restricted to output

$$v_o = A_v(v_+ - v_-) \quad (1)$$

levels $|v_o| \leq V_{max}$.] For the purposes of analyzing the circuit, the waveforms in the illustration have been drawn with the assumption that this mechanism has already operated at time 0 and that the initial charge on the capacitor is zero. *See* AMPLIFIER; OPERATIONAL AMPLIFIER.

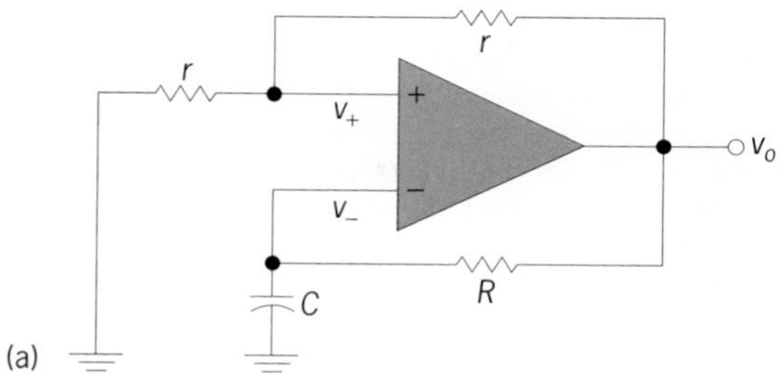

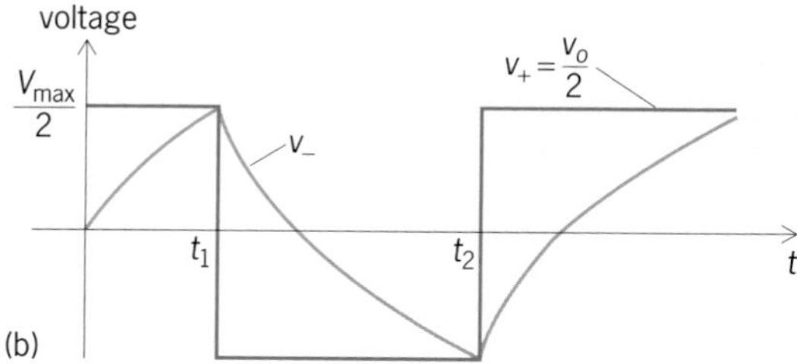

Simple operational-amplifier relaxation oscillator. (*a*) Circuit diagram. (*b*) Waveforms.

Capacitor C will now slowly change from v_o through resistor R, toward V_{max}, according to Eq. (2). Up until time t_1, this process

$$v_- = V_{max}(1 - e^{-t/RC}) \tag{2}$$

continues without any change in the amplifier's output because $v_+ > v_-$, and so $v_o = V_{max}$. At t_1, however, $v_+ = v_-$ and v_o will start to decrease. This causes v_+ to drop, and the positive-feedback action now drives the amplifier output negative until $v_o = -V_{max}$. Capacitor C now discharges exponentially toward the new output voltage until once again, at time t_2, $v_+ = v_-$, and the process starts again. The period of oscillation for this circuit is $2RC \ln 3$.

The basic elements of an oscillator that were mentioned above are all clearly visible in this circuit. Two direct-current power supplies are implicit in the diagram (the operational amplifier will not work without them), the RC circuit sets frequency, there is a resistive positive-feedback path that makes the mathematical possibility $v_o(t) = 0$ unstable, and the saturation behavior of the amplifier sets the amplitude of oscillation at the output to $\pm V_{max}$.

Relaxation oscillators that have a low duty cycle—that is, produce output pulses whose durations are a small fraction of the overall period—are sometimes called blocking oscillators because their operation is characterized by an "on" transient that "blocks" itself, followed by a recovery period.

Inverters (digital circuits that invert a logic signal, so that a 0 at the input produces a 1 at the output, and vice versa) are essentially voltage amplifiers and can be used to make relaxation oscillators in a number of ways. A circuit related to that of the illustration uses a loop of two inverters and a capacitor C to provide positive feedback, with a resistor R in parallel with one of the inverters to provide an RC charging time to set frequency. This circuit is commonly given as a simple example, but there are a number of problems with using it, such as that the input voltage to the first gate sometimes exceeds the specified limits for practical gates. A more practical digital relaxation oscillator, called a ring oscillator, consists simply of a ring containing an odd number N (greater than 1) of inverters. *See* LOGIC CIRCUITS.

Sine-wave oscillators. Oscillators in the second major class have their oscillation frequency set by a linear circuit, and their amplitudes set by a weak nonlinearity.

A simple example of a suitable linear circuit is a two-component loop consisting of an ideal inductor [whose voltage is given by Eq. (3), where i is its current] and a capacitor [whose

$$v = L\frac{di}{dt} \tag{3}$$

current is given by Eq. (4)], connected in parallel. These are said

$$i = C\frac{dv}{dt} \tag{4}$$

to be linear elements because, in a sense, output is directly proportional to input, for example, doubling the voltage v across a capacitor also doubles dv/dt and therefore doubles i. The overall differential equation for a capacitor-inductor loop can be written as Eq. (5). Mathematically this has solutions of the form of

$$i + LC\frac{d^2i}{dt^2} = 0 \tag{5}$$

Eq. (6), where $\omega = 1/LC$ [which means that the circuit oscillates

$$i = A \sin(\omega t + \phi) \tag{6}$$

at a frequency $1/(2\pi LC)$] and A and ϕ are undefined. They are undefined precisely because the elements in the circuit are linear and do not vary with time: any solution (possible behavior) to the equation can be scaled arbitrarily or time-shifted arbitrarily to give another. Practically, A and ϕ are determined by weak nonlinearities in a circuit. *See* DIFFERENTIAL EQUATION; LINEARITY.

Equation (5) is a good first approximation to the equation describing a pendulum, and so has a long history as an accurate timekeeper. Its value as an oscillator comes from Galileo's original observation that the frequency of oscillation ($\omega/2\pi$) is independent of the amplitude A. This contrasts sharply with the case of the relaxation oscillator, where any drift in the amplitude (resulting from a threshold shift in a comparator, for instance) can translate directly into a change of frequency. Equation (5) also fundamentally describes the operation of the quartz crystal that has replaced the pendulum as a timekeeper; the physical resonance of the crystal occurs at a time constant defined by its spring constant and its mass. *See* HARMONIC MOTION; HARMONIC OSCILLATOR; PENDULUM.

Frequency locking. If an external signal is injected into an oscillator, the natural frequency of oscillation may be affected. If the external signal is periodic, oscillation may lock to the external frequency, a multiple of it, or a submultiple of it, or exhibit an irregular behavior known as chaos. *See* CHAOS.

This locking behavior occurs in all oscillators, sometimes corrupting intended behavior (as when an oscillator locks unintentionally to a harmonic of the power-line frequency) and sometimes by design. An important example of an oscillator that exploits this locking principle is the human heart. Small portions of heart muscle act as relaxation oscillators. They contract, incidentally producing an output voltage that is coupled to their neighbors. For a short time the muscle then recovers from the contraction. As it recovers, it begins to become sensitive to externally applied voltages that can trigger it to contract again (although it will eventually contract anyway). Each small section of heart muscle is thus an independent oscillator, electrically coupled to its neighbors, but the whole heart is synchronized by the frequency-locking mechanism. *See* CARDIAC ELECTROPHYSIOLOGY.

Martin Snelgrove

Oscillatory reaction A chemical reaction in which some composition variable of a chemical system exhibits regular periodic variations in time or space. It is a basic tenet of chemistry that a closed system moves inexorably toward an unchanging state called chemical equilibrium. That motion can be described by the monotonic increase of entropy if the system is isolated, and by the monotonic decrease of Gibbs free energy if the system is constrained to constant temperature and pressure. *See* CHEMICAL EQUILIBRIUM; ENTROPY.

The species taking part in a chemical reaction can be classified as reactants, products, or intermediates. The concentrations of reactants decrease. Intermediates are formed by some steps and destroyed by others. If there is only one intermediate, and if its concentration is always much less than the initial concentrations of reactants, this intermediate attains a stable steady state in which the rates of formation and destruction are virtually equal. Some oscillations require at least two intermediates which interact in such a way that the steady state of the total system is unstable to the minor fluctuations present in any collection of molecules. The concentrations of the intermediates may then oscillate regularly, although the oscillations must disappear before the inevitable monotonic approach to equilibrium.

The systems whose chemistries are best understood all involve an element that can exist in several different oxidation states. An example is the so-called Belousov-Zhabotinsky reaction. A strong oxidizing agent (bromate) attacks an organic substrate (such as malonic acid), and the reaction is catalyzed by a metal ion (such as cerium) that can exist in two different oxidation states.

As long as bromide ion (Br) is present, it is oxidized by bromate (BrO_3^-), as in reaction (1).

$$BrO_3^- + 2Br^- + 3H^+ \rightarrow 3HOBr \qquad (1)$$

When bromide ion is almost entirely consumed, the cerous ion (Ce^{3+}) is oxidized, as in reaction (2). Reaction (2) is inhibited by

$$BrO_3^- + 4Ce^{3+} + 5H^+ \rightarrow HOBr + 4Ce^{4+} + 2H_2O \qquad (2)$$

Br^-, but when the concentration of bromide has been reduced to a critical level, reaction (2) accelerates autocatalytically until bromate is being reduced by Ce^{3+} many times as rapidly as it is by Br^- when reaction (3) is dominant.

The hypobromous acid (HOBr) brominates the organic substrate to form bromomalonic acid (BrMA), as in reaction (3).

$$2Ce^{4+} + BrMA \rightarrow 2Ce^{3+} + Br^- + \text{oxidized organic matter} \qquad (3)$$

Reaction (3) creates the bromide ion necessary to shut off fast reaction (2) and throw the system back to dominance by slow reaction (1).

As other redox oscillators become understood, they fit the same pattern of a slow reaction destroying a species that inhibits a fast reaction that can be switched on autocatalytically; the fast reaction then generates conditions to produce the inhibitor again. *See* OXIDATION-REDUCTION. Richard M. Noyes

Oscilloscope An electronic measuring instrument which produces a display showing the relationship of two or more variables. In most cases it is an orthogonal (x,y) plot with the horizontal axis being a linear function of time. The vertical axis is normally a linear function of voltage at the signal input terminal of the instrument. Because transducers of many types are available to convert almost any physical phenomenon into a corresponding voltage, the oscilloscope is a very versatile tool that is useful for many forms of physical investigation. *See* TRANSDUCER.

The oscillograph is an instrument that performs a similar function but provides a permanent record. The light-beam oscillograph used a beam of light reflected from a mirror galvanometer which was focused onto a moving light-sensitive paper. These instruments are obsolete. The mechanical version, in which the galvanometer drives a pen which writes on a moving paper chart, is still in use, particularly for process control. *See* GALVANOMETER; GRAPHIC RECORDING INSTRUMENTS.

Oscilloscopes are one of the most widely used electronic instruments because they provide easily understood displays of electrical waveforms and are capable of making measurements over an extremely wide range of voltage and time. Although a very large number of analog oscilloscopes are in use, digitizing oscilloscopes (also known as digital oscilloscopes or digital storage oscilloscopes) are preferred, and analog instruments are likely to be superseded. *See* ELECTRONIC DISPLAY.

An analog oscilloscope, in its simplest form, uses a linear vertical amplifier and a time base to display a replica of the input signal waveform on the screen of a cathode-ray tube (CRT). The screen is typically divided into 8 vertical divisions and 10 horizontal divisions. Analog oscilloscopes may be classified into nonstorage oscilloscopes, storage oscilloscopes, and sampling oscilloscopes.

Analog nonstorage oscilloscopes are the oldest and most widely used type. Except for the cathode-ray tube, the circuit descriptions also apply to analog storage oscilloscopes. A typical oscilloscope might have a bandwidth of 150 MHz, two main vertical channels plus two auxiliary channels, two time bases (one usable for delay), and a cathode-ray-tube display area; and it might include on-screen readout of some control settings and measurement results. A typical oscilloscope is composed of five basic elements: (1) the cathode-ray tube and associated controls; (2) the vertical or signal amplifier system with input terminal and controls; (3) the time base, which includes sweep generator, triggering circuit, horizontal or x-amplifier, and unblanking circuit; (4) auxiliary facilities such as a calibrator and on-screen readout; and (5) power supplies.

Digital techniques are applied to both timing and voltage measurement in digitizing oscilloscopes. A digital clock determines sampling instants at which analog-to-digital converters obtain digital values for the input signals. The resulting data can be stored indefinitely or transferred to other equipment for analysis or plotting. *See* VOLTAGE MEASUREMENT; WAVEFORM DETERMINATION.

In its simplest form a digitizing oscilloscope comprises six basic elements: (1) analog vertical input amplifier; (2) high-speed analog-to-digital converter and digital waveform memory; (3) time base, including triggering and clock drive for the analog-to-digital converter and waveform memory; (4) waveform reconstruction and display circuits; (5) display, generally, but not restricted to, a cathode-ray tube; (6) power supplies and ancillary functions. In addition, most digitizing oscilloscopes provide facilities for further manipulation of waveforms prior to display, for direct measurements of waveform parameters, and for connection to external devices such as computers and hard-copy units.

Higher measurement accuracy is available from digitizing oscilloscopes. The first decision to be made in choosing an oscilloscope is whether this or any of the other properties exclusive to the digitizing type are essential. If not, the option of an analog design remains. The selected instrument must be appropriate for the signal under examination. It must have enough sensitivity to give an adequate deflection from the applied signal, sufficient bandwidth, adequately short rise time, and time-base facilities capable of providing a steady display of the waveform. An analog oscilloscope needs to be able to produce a visible trace at the sweep speed and repetition rate likely. A digitizing oscilloscope must have an adequate maximum digitizing rate and a sufficiently long waveform memory. R. B. D. Knight

Osmium A chemical element, Os, atomic number 76, atomic weight 190.23. The element is a hard white metal of rare natural occurrence, usually found in nature alloyed with other platinum metals. *See* METAL; PERIODIC TABLE; PLATINUM.

Physical properties of the element, which is found as seven naturally occurring isotopes, are given in the table. The metal is exceeded in density only by iridium. Osmium is a very hard metal and unworkable, and so it must be used in cast form or fabricated by powder metallurgy. Osmium is a third-row transition element and has the electronic configuration $[Xe](4f)^{14}(5d)^6(6s)^2$; in the periodic table it lies below iron (Fe) and ruthenium (Ru). In

Principal properties of osmium

Property	Value
Density, g/cm^3	22.6
Naturally occurring isotopes (% abundance)	184 (0.018)
	186 (1.59)
	187 (1.64)
	188 (13.3)
	189 (16.1)
	190 (26.4)
	192 (41.0)
Ionization enthalpy, kJ/mol: 1st 2d	840
	1640
Oxidation states	−1 to VIII
Most common	IV, VI, VIII
Ionic radius, Os^{4+}, nm	0.078
Melting point, °C (°F)	3050 (5522)
Boiling point, °C (°F)	5500 (9932)
Specific heat, cal/g·°C	0.032
Crystal structure	Hexagonal close-packed
Lattice constant *a* at 25°C, nm *c*/*a* at 25°C	0.27341
	0.15799
Thermal neutron capture cross section, barns	15.3
Thermal conductivity, 0–100°C, (cal · cm)/(cm^2 · s · °C)	0.21
Linear coefficient of thermal expansion at 20–100°C, (μin./in./°C)	6.1
Electrical resistivity at 0°C, $\mu\Omega$-cm	8.12
Temperature coefficient of electrical resistance, 0–100°C/°C	0.0042
Young's modulus at 20°C, lb/in.^{2}, static	81×10^6

powder form the metal may be attacked by the oxygen in air at room temperature, and finely divided osmium has a faint odor of the tetraoxide. In bulk form it does not oxidize in air below 500°C (750°F), but at higher temperatures it yields OsO_4. It is attacked by fluorine or chlorine at 100°C (212°F). It dissolves in alkaline oxidizing fluxes to give osmates ($OsO_4{}^{2-}$). *See* ELECTRON CONFIGURATION; IRIDIUM; IRON; RUTHENIUM.

The chemistry of osmium more closely resembles that of ruthenium than that of iron. The high oxidation states VI and VIII ($OsO_4{}^{2-}$ and OsO_4) are much more accessible than for iron. *See* OXIDATION-REDUCTION.

Osmium forms many complexes. In water, osmium complexes with oxidation states ranging from II to VIII may be obtained. Oxo compounds, which contain Os═O, are very common and occur for oxidation states IV to VIII. Although OsO_4 is tetrahedral in the gas phase and in noncomplexing solvents such as dichloromethane (CH_2Cl_2), it tends to be six-coordinate when appropriate ligands are available; thus, in sodium hydroxide (NaOH) solution, dark purple $OsO_4(OH)_2{}^{2-}$ is formed from OsO_4. Similarly, $OsO_2(OH)_4{}^{2-}$ is formed by addition of hydroxide to osmate anion. Analogous reactions with ligands such as halides, cyanide, and amines give osmyl derivatives like the cyanide-deduct $OsO_2(CN)_4{}^{2-}$, in which the trans dioxo group (O═Os═O) is retained.

Osmium tetraoxide, a commercially available yellow solid (melting point 40°C or 104°F), is used commercially in the important *cis*-hydroxylation of alkenes and as a stain for tissue in microscopy. It is poisonous and attacks the eyes. Osmium metal is catalytically active, but it is not commonly used for this purpose because of its high price. Osmium and its alloys are hard and resistant to corrosion and wear (particularly to rubbing wear). Alloyed with other platinum metals, osmium has been used in needles for record players, fountain-pen tips, and mechanical parts. *See* ALLOY; STAIN (MICROBIOLOGY); TRANSITION ELEMENTS.

Carol Crevtz

Osmoregulatory mechanisms Physiological mechanisms for the maintenance of an optimal and constant level of osmotic activity of the fluid within and around the cells, considered to be most favorable for the initiation and maintenance of vital reactions in the cell and for maximal survival and efficient functioning of the entire organism.

The actions of osmoregulatory mechanisms are, first, to impose constraints upon the passage of water and solute between the organism and its surroundings and, second, to accelerate passage of water and solute between organism and surroundings. The first effect requires a change of architecture of membranes in that they become selectively permeable and achieve their purpose without expenditure of energy. The accelerating effect, apart from requiring a change of architecture of cell membranes, requires expenditure of energy and performance of useful osmotic work. Thus substances may be moved from a region of low to a region of higher chemical activity. Such movement can occur in opposition to the forces of diffusion of an electric field and of a pressure gradient, all of which may act across the cell membrane. It follows that there must be an energy source which is derived from the chemical reactions of cellular metabolism and that part of the free energy so generated must be stored in molecules which are driven across the membrane barrier. Active transport is the modern term for such processes. *See* CELL MEMBRANES; OSMOSIS.

William A. Brodsky; T. P. Schilb

Osmosis The transport of solvent through a semipermeable membrane separating two solutions of different solute concentration. The solvent diffuses from the solution that is dilute in solute to the solution that is concentrated.

The flow of liquid through such a barrier may be stopped by applying pressure to the liquid on the side of higher solute concentration. The applied pressure required to prevent the flow of solvent across a perfectly semipermeable membrane is called the osmotic pressure and is a characteristic of the solution. The walls of cells in living organisms permit the passage of water and certain solutes, while preventing the passage of other solutes, usually of relatively high molecular weight. These walls act as selectively permeable membranes, and allow osmosis to occur between the interior of the cell and the surrounding media. *See* EDEMA; OSMOREGULATORY MECHANISMS; SOLUTION.

Francis J. Johnston

Ostariophysi A very large group of fishes containing 27% of all known fishes of the world and accounting for 64% of the freshwater species. Although the perciforms dominate modern seas, the ostariophysans command preeminent faunal importance in freshwaters. *See* PERCIFORMES.

The classification hierarchy is as follows:

Class Actinopterygii
- Subclass Neopterygii
  - Division Teleostei
    - Subdivision Euteleostei
      - Superorder Ostariophysi
        - Series Anotophysi
          - Order Gonorhynchiformes
        - Series Otophysi
          - Order Cypriniformes
          - Order Characiformes
          - Order Siluriformes
          - Order Gymnotiformes

Several characteristics alone unite the Ostariophysi: modification of the anterior three to five vertebrae as a protective encasement for a series of bony ossicles that form a chain connecting the swim bladder with the inner ear, the so-called Weberian apparatus; multicellular horn tubercles, in the form of breeding and nuptial tubercles or pearl organs; minute unicellular horny projections called unculi; and an alarm substance (pheromone) that can be released by injured tissue to elicit a fright reaction. Other identifying characteristics include abdominal pelvic fins, if present, usually with many rays; a pelvic girdle which does not have contact with the cleithra; and a more or less horizontal pectoral fin placed low on the side. Scales are the cycloid type, which may be modified into bony plates, and branchiostegal rays are variable in number and arrangement. *See* EAR (VERTEBRATE); SWIM BLADDER. Herbert Boschung

Osteichthyes A group consisting of the classes Actinopterygii (rayfin fishes) and Sarcopterygii (lungfishes and lobefins, or coelacanths, but excluding tetrapods). The Sarcopterygii and Actinopterygii are well-defined phyletic lines which had evolved by the Middle Devonian, and many members of the two taxa persist today.

The osteichthyans (bony fishes) include most of the Recent fishes. They differ from the living Agnatha in having jaws, paired nostrils, true teeth, paired pelvic and pectoral fins supported by girdles (unless lost secondarily), three semicircular canals, and bony scales (unless lost or modified). Many fossil agnathans possess bony scales but differ from the higher fishes in the abovementioned features. Separation of the osteichthyans from the Paleozoic Placodermi and Acanthodii is more difficult because these three groups agree in most basic vertebrate features. The osteichthyans contrast with the Chondrichthyes in having a bony skeleton (although some Recent bony fishes possess a largely cartilaginous skeleton), a swim bladder (at least primitively), a true gill cover, and mesodermal scales (sharks possess dermal denticles, or placoid scales) which are sometimes modified or lost. Fertilization is usually external, but if it is internal, the intromittent organ is not derived from pelvic-fin claspers. Most often a modified anal fin or a fleshy tube or sheath functions in sperm transfer. *See* COPULATORY ORGAN; SCALE (ZOOLOGY); SWIM BLADDER.

The Recent osteichthyan fauna includes 2 classes, 45 orders, 435 families, 4076 genera, and 24,497 species. A classification of extant and selected fossil osteichthyans follows. Equivalent names are given in parentheses.

Osteichthyes
  Class Actinopterygii
    Subclass Cladistia
      Order Polypteriformes (Brachiopterygii)
    Subclass Chondrostei
      Order Acipenseriformes
    Subclass Neopterygii
      Order Lepisosteiformes (in part, Semionotiformes)
        Amiiformes
      Division Teleostei
        Subdivision Osteoglossomorpha
          Order Hiodontiformes
            Osteoglossiformes
        Subdivision Elopomorpha
          Order Elopiformes
            Albuliformes
            Anguilliformes (Apodes)
            Saccopharyngiformes
        Subdivision Clupeomorpha
          Order Clupeiformes
        Subdivision Euteleostei
          Superorder Ostariophysi
            Series Anotophysi
              Order Gonorhynchiformes
            Series Otophysi
              Order Cypriniformes
                Characiformes
                Siluriformes (Nematognathi)
                Gymnotiformes
          Superorder Protacanthopterygii
            Order Argentiniformes,
              Osmeriformes
              Salmoniformes
              Esociformes (Haplomi, Esocae)
          Superorder Stenopterygii
            Order Stomiiformes (Stomiatiformes)
              Ateleopodiformes
          Superorder Cyclosquamata
            Order Aulopiformes
          Superorder Scopelomorpha
            Order Myctophiformes
          Superorder Lampridiformes (Lampriformes)
            Order Lampriformes
          Superorder Polymixiomorpha
            Order Polymixiiformes
          Superorder Paracanthopterygii
            Order Percopsiformes
              Gadiformes
              Ophidiiformes
              Batrachoidiformes (Haplodoci)
              Lophiiformes
          Superorder Acanthopterygii
            Series Mugilomorpha
              Order Mugiliformes
            Series Atherinomorpha
              Order Atheriniformes
                Beloniformes
                Cyprinodontiformes (Microcyprini)
            Series Percomorpha
              Order Stephanoberyciformes (Xenoberyces, in part)
                Beryciformes
                Zeiformes
              Order Gasterosteiformes
                Suborder Gasterosteoidei (Thoracostei)
                Suborder Syngathoidei (Solenichthys)
              Order Synbranchiformes (Symbranchii)
                Scorpaeniformes
                Perciformes
                  Suborder Percoidei
                    Elassomatoidei
                    Labroidei
                    Zoarcoidei
                    Notothenioidei
                    Trachinoidei
                    Pholidichthyoidei
                    Blennioidei
                    Icosteoidei
                    Gobiesocoidei (Xenopterygii)
                    Callionymoidei
                    Gobioidei
                    Kurtoidei
                    Acanthuroidei
                    Scombrolabracoidei
                    Scombroidei
                    Stromateoidei
                  Suborder Anabantoidei (Labyrinthici, in part)
                    Channoidei (Ophiocephaliformes)
                    Caproidei

Order Pleuronectiformes
(Heterostomata)
Tetraodontiformes
(Plectognathi)
Class Sarcopterygii
Subclass Coelacanthimorpha
Order Coelacanthiformes
Subclass Dipnotetrapodomorpha
Infraclass Dipnoi (Dipneusti)
Order Ceratodontiformes

Reeve M. Bailey; Herbert Boschung

The Osteichthyes evolved from some group among the primitive, bony gnathostome fishes. The oldest osteichthyans do not appear until the Lower Devonian, when lobefins and lungfishes enter the paleontological record. The osteichthyans became well represented in the Middle Devonian, at which time the rayfin fishes, lobefin fishes, and lungfishes were well differentiated. The ancestral stock of the osteichthyans may be sought among the other major groups of gnathostomes, the Placodermi and Acanthodii, but the still-fragmentary character of the early fossil record prevents clear elucidation of the origin and initial history of the group. Although the beginnings of the osteichthyans are shrouded in uncertainty, advancements in paleontology have laid to rest the belief that they evolved from that other great class of modern fishes, the Chondrichthyes. *See* ACANTHODII; CHONDRICHTHYES; PLACODERMI.

John G. Maisey

Osteoglossiformes An order of teleost fishes consisting of two monophyletic clades, suborders Osteoglossoidei and Notopteroidei. The Osteoglossoidei consists of one family, the Osteoglossidae (bonytongues and butterfly fishes), which occur in freshwaters of tropical South America, Africa, and Southeast Asia to northern Australia. The Notopteroidei consist of three extant families, Notopteridae (Old World knifefishes and featherfin knifefishes, or featherbacks), Mormyridae (elephantfishes; see illustration), and Gymnarchidae (gymnarchid eel). The featherbacks are native to central tropical Africa, India, and the Malay Archipelago. The elephantfish family, having a long proboscislike snout in many species, is the most species-rich osteoglossiform taxon but is limited to tropical Africa and the Nile River. The family Gymnarchidae is represented by a single species, *Gymnarchus niloticus*, of tropical Africa including the upper Nile. *See* TELEOSTEI.

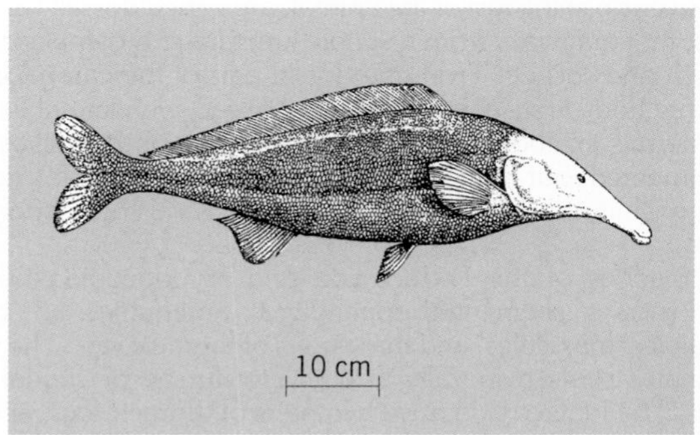

African elephantnose (*Mormyrus proboscirostris*). (*After G. A. Boulenger, Catalogue of Fresh Water Fishes of Africa in the British Museum, Natural History, vol. 1, 1909*)

Osteoglossiformes have the primary bite between the well-toothed tongue and the strongly toothed parasphenoid and certain pterygoid bones in the roof of the mouth. The mouth is bordered by a small premaxilla, which is fixed to the skull, and the maxilla. The anterior part of the intestine passes posteriorly to the left of the esophagus and stomach, whereas in all but a very few fishes the intestine pass to the right. A unique feature is the presence of paired bony rods at the base of the second gill arch.

Osteoglossiforms represent one of the basal stocks among the teleosts. Recent members may be classified in 2 suborders, 6 families, 29 genera, and about 217 species. All inhabit freshwater, and all are tropical. Numerically Osteoglossiformes is dominated by the Mormyridae. Of the roughly 198 species, many are valued as food. Mormyrids and gymnarchids are of especial interest in that they are electrogenic. Modified muscles in the caudal peduncle generate and emit a continuous electric pulse. The discharge frequency varies, being low at rest and high when the fish is under stress. The mechanism operates like a radar device, since the fish is alerted whenever an electrical conductor enters the electromagnetic field surrounding it. The receptor mechanism is not fully understood, but the brain of mormyrids (especially the cerebellum) is the largest of all fishes. *See* ACTINOPTERYGII; ELECTRIC ORGAN (BIOLOGY).

Osteoglossiforms vary greatly in size, shape, and biology. Their trophic biology is greatly diverse, the various species being either piscivores, insectivores, filter feeders, or omnivores. The well-vascularized swim bladder can act as a lung, which is especially helpful in oxygen-deficient habitats.

Reeve M. Bailey; Herbert Boschung

Osteolepiformes Fossil lobe-finned (sarcopterygian) fishes of Devonian to Permian age (395–270 million years ago), the ancestral group for the land vertebrates, or "tetrapods." *See* SARCOPTERYGII; TETRAPODA.

The term "osteolepiforms" is applied to a range of fossil fishes, found worldwide in Devonian to Permian strata, that appear to be closely related to tetrapods. After decades of debate about their precise position in the vertebrate family tree, a consensus has emerged that they occupy the bottom part of the tetrapod stem lineage, that is, the evolutionary line that leads toward land vertebrates from the last common ancestor we share with our nearest living fish relatives, the lungfishes. Immediately above the osteolepiforms in the stem lineage, just below the most primitive tetrapods, we find the transitional fossil Panderichthys. *See* ANIMAL EVOLUTION; PANDERICHTHYS.

General characteristics of the osteolepiforms include an intracranial joint running vertically through the skull behind the eyes, which allowed the front part of the head to move slightly up and down relative to the back part; a pair of internal nostrils, or choanae, connecting the nose to the palate; and paired fin skeletons in which one can recognize equivalents of the tetrapod limb bones humerus, radius, ulna (forelimb/pectoral fin) and femur, tibia, fibula (hindlimb/pelvic fin). The similarities between osteolepiform fin skeletons and tetrapod limb skeletons are quite specific, extending to the detailed pattern of muscle attachments on the bones. However, despite their tetrapod-like anatomy, osteolepiforms retained the overall body form of "normal" open-water fishes and do not seem to have been adapted for an amphibious life. Osteolepiforms are known from both marine and nonmarine environments.

The tristichopterids (or eusthenopterids) are the most advanced osteolepiforms. They had narrow, sharply pointed heads

The tristichopterid *Eusthenopteron foordi*, approximately 50–70 cm in length. (*Reprinted from E. Jarvik, Basic Structure and Evolution of Vertebrates, vol. 1, pp. xx, © 1980, with permission from Elsevier*)

and, in particular, long snouts. These features are shared with *Panderichthys* and tetrapods, and may reflect a change from suction feeding to a "snapping" mode of prey capture. The best-known tristichopterid, and arguably the most thoroughly studied of all fossil fishes, is *Eusthenopteron foordi* from the early Late Devonian of Quebec, Canada (see illustration). Per E. Ahlberg

Osteoporosis A metabolic bone disease in which the amount of bone tissue is reduced sufficiently to increase the likelihood of fracture. Fractures of the vertebrae, femur (hip), and wrist are the most common osteoporotic fractures, but other bones such as the ribs, upper arm, and pelvis may also fracture.

Although low bone mass is the major factor in osteoporotic fractures, there may also be qualitative and architectural changes in bone with aging that lead to increased fragility. Osteoporosis can be primary or secondary. Primary osteoporosis occurs independently of other causes. The secondary osteoporoses result from identifiable causes, such as exogenous cortisone administration, Cushing's disease, hyperparathyroidism, hyperthyroidism, hypogonadism, multiple myeloma, prolonged immobilization, alcoholism, anorexia nervosa, and various gastrointestinal disorders. Primary osteoporosis occurring in children is called juvenile osteoporosis; that occurring in premenopausal women and middle-aged or young men is known as idiopathic osteoporosis. Osteoporosis, which is found in older persons, can be classified as postmenopausal (type I) or involutional (type II) osteoporosis. *See* ALCOHOLISM; ANOREXIA NERVOSA; GASTROINTESTINAL TRACT DISORDERS; METABOLIC DISORDERS.

Goals should include prevention of both the underlying disorder (the disease) and its effects (osteoporotic fractures). Secondary osteoporoses are managed by eliminating the underlying disorder. To prevent primary osteoporosis, good health-related behavior during childhood and young adulthood has been suggested as the most important factor. Such behavior includes avoiding cigarette smoking and excess alcohol intake, maintaining a normal body weight, and maintaining optimal dietary intake of calcium. It has been suggested that the recommended daily allowance (RDA) for calcium in the postmenopausal period and the recommended intake of vitamin D in the aged should be increased. At menopause, the possibility of estrogen replacement therapy should be considered for women who are at high risk for osteoporosis. Fracture prevention should be a lifelong effort. During childhood and the premenopausal years, a maximal peak bone mass should be developed through weight-bearing exercise. Exercise to improve coordination and flexibility and to maintain good posture is also useful. *See* ESTROGEN; NUTRITION; VITAMIN D.

The diagnosis of primary osteoporosis is made in the presence of either low bone mass or a characteristic fracture that cannot be attributed to some other cause.

The goals of treatment are rehabilitation and minimization of the risk of future fractures. Goals for rehabilitation include restoring a positive outlook on life, treating depression if it exists, increase of physical activity, restoring independence, relieving pain, restoring muscle mass, and improving posture. Various medications, including estrogen and calcitonin, can maintain bone mass. *See* AGING; BONE; BONE DISORDERS. John F. Aloia

Osteostraci An order of extinct jawless vertebrate fishes, also called Cephalaspida, known from the Middle Silurian to Upper Devonian of Europe, Asia, and North America. They were mostly small, about 2 in. to 2 ft (5 to 60 cm) in length. The head and part of the body were encased in a solid armor of bone, and the posterior part of the body and the tail were covered with thick scales. Some early forms lacked paired fins, though most possessed flaplike pectoral fins. One or two dorsal fins were present. Their depressed shape and the position of the eyes on the top of the head suggest that Osteostraci were bottom dwellers (see illustration). The underside of the throat region was covered by small plates and there was a small mouth in front. Robert H. Denison

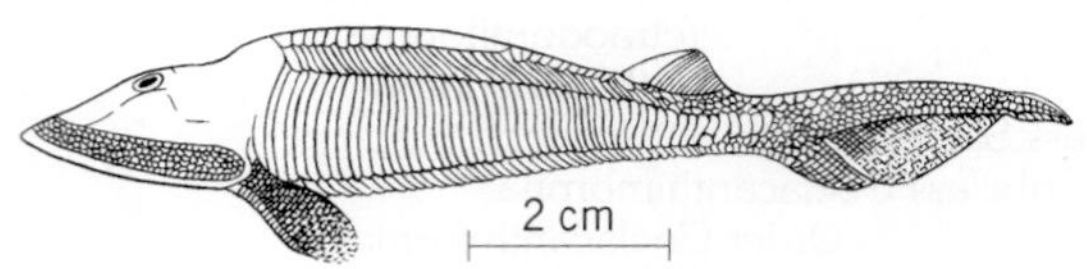

***Hemicyclaspis*, a cephalaspid, a Lower Devonian jawless vertebrate. (*After E. H. Colbert, Evolution of the Vertebrates, Wiley, 1955*)**

Ostracoda A major taxon of the Crustacea containing small bivalved animals 0.004–1.4 in. (0.1–33 mm) long, with most between 0.04 and 0.08 in. (1 and 2 mm). They inhabit aquatic environments in nearly all parts of the world. Semiterrestrial species have been described from moss and leaf-litter habitats in Africa, Madagascar, Australia, and New Zealand, and from vegetable debris of marine origin in the Kuril Archipelago. Of the more than 2000 species extant, none is truly parasitic and most are free-living. However, a few fresh-water and marine forms live commensally on other animals. Most ostracodes are scavengers, some are herbivorous, and a few are predacious carnivores. Exceptional biological features are known; there are myodocopine ostracodes that produce bioluminescence, and some species of podocopines form a secretion from spinning glands to enable them to climb polished surfaces.

Researchers disagree on the hierarchical classification of the Ostracoda. It is recognized as a distinct class by some and a subclass within the Maxillopoda by others. Consequently, the subdivisions within the Ostracoda may be ranked as subclasses, superorders, or orders. Six major subdivisions are currently recognized: Bradoriida, Phosphatocopida, Leperditicopida, Paleocopa, Myodocopa, and Podocopa. The first four taxa are extinct. The Myodocopa are further subdivided into the Myodocopida and Halocyprida, and the Podocopa into Platycopida and Podocopida. All fresh-water Ostracoda belong to the Podocopida. *See* MAXILLOPODA; PODOCOPA.

Knowledge of the morphology of the Ostracoda is based primarily on extensive study of many species. The two valves, sufficient to enclose the rest of the animal, are joined dorsally along a hinge, that may vary from a simple juncture to a complex series of teeth and sockets. From the dorsal part of the carapace, the elongate body is suspended as a pliable sac, with lateral flaps of hypodermis extending between the lamellae of the valves. It is unsegmented, but reinforced by chitinous processes for rigidity in the vicinity of the appendages. There is no true abdominal region.

Ostracodes of the Podocopida and Myodocopida possess seven pairs of segmented appendages: antennules, antennae, mandibles, maxillules, and three pairs of thoracic legs. The body terminates posteroventrally in a pair of furcae or caudal processes. The Platycopida have two pairs of thoracic legs, and the Halocyprida frequently only one pair. Appendages have muscles within them and are connected by musculature to the valves and to a centrally located chitinous structure, the endoskeleton. They are usually specialized for particular functions, such as swimming, walking, food gathering, mastication, and cleaning the interior of the carapace.

Ostracodes have separate sexes, although many species are parthenogenetic and lack males. Most ostracodes lay their eggs on the substrate or on vegetation, but some transfer them to the posterior space within the carapace, where they hatch and the young brood is retained for a time. Patsy A. McLaughlin

Ostracoderm A popular name applied to several groups of extinct jawless vertebrates (fishes). Most of them were covered with an external skeleton or armor of bone, from which is derived their name, meaning "shell-skinned" . They are known from the Ordovician, Silurian, and Devonian periods, and thus include the earliest known vertebrates. *See* JAWLESS VERTEBRATES.

Robert H. Denison

Otter A member of the family Mustelidae, which also includes weasels, mink, ferrets, martens, sables, fishers, badgers, and wolverines. Otters are found worldwide except in Australia, Madagascar, and on the islands of the South Pacific. Most inhabit freshwater lakes and rivers. The sea otter (*Enhydra lutris*), however, inhabits the waters and shores of the northern Pacific Ocean from southern California to the Kurile Islands.

North American river otter (*Lontra canadensis*). (*Alan and Sandy Carey/Getty Images*)

All otters are much alike no matter where they live. The North American otter (*Lontra canadensis*) has a lithe, muscular body; a broad, flat head; small ears; and a long, powerful, tapering tail which serves as a rudder (see illustration). The ears and nostrils are capable of being closed when the otter is underwater. The limbs are short, but the strong hind feet are large, and the five toes are broadly webbed. The short, dense fur is dark brown and is impervious to water. Perianal scent glands are present in all species except the sea otter.

Otters are skillful predators. They usually prey on sluggish varieties of fish, but may also feed on trout, frogs, crayfish, crabs, ducks, muskrats, and young beavers. Otters catch fish with their forepaws, then rip them apart with their teeth. An otter's den is usually a hole in the bank of a stream or lake and has an underwater entrance. A second entrance for ventilation is hidden in bushes on the bank. Sea otters rarely come to land, preferring to live in great beds of floating kelp, a type of brown seaweed. *See* CARNIVORA; MAMMALIA.

Donald W. Linzey

Otto cycle The basic thermodynamic cycle for the prevalent automotive type of internal combustion engine. The engine uses a volatile liquid fuel (gasoline) or a gaseous fuel to carry out the theoretic cycle shown in the illustration. The cycle consists of two isentropic (reversible adiabatic) phases interspersed between two constant-volume phases. The theoretic cycle should not be confused with the actual engine built for such service as automobiles, motor boats, aircraft, lawn mowers, and other small self-contained power plants.

The thermodynamic working fluid in the cycle is subjected to isentropic compression, phase 1–2; constant-volume heat addition, phase 2–3; isentropic expansion, phase 3–4; and constant-volume heat rejection (cooling), phase 4–1.

The Otto cycle is represented in many millions of engines utilizing either the four-stroke principle or the two-stroke principle. Evidence indicates that actual Otto engines offer peak

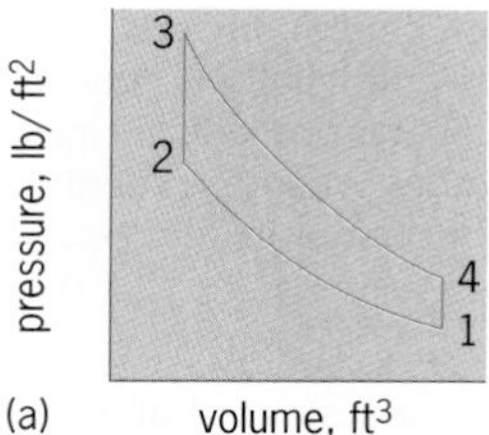

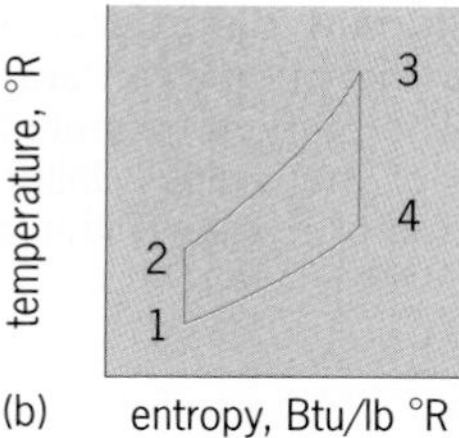

Diagrams of (*a*) pressure-volume and (*b*) temperature-entropy for Otto cycle.

efficiencies (25±%) at compression ratios of 15±. Above this ratio, efficiency falls. The most probable explanation is that the extreme pressures associated with high compression cause increasing amounts of dissociation of the combustion products. This dissociation, near the beginning of the expansion stroke, exerts a more deleterious effect on efficiency than the corresponding gain from increasing compression ratio. *See* BRAYTON CYCLE; CARNOT CYCLE; DIESEL CYCLE; INTERNAL COMBUSTION ENGINE; THERMODYNAMIC CYCLE.

Theodore Baumeister

Ovarian disorders A variety of neoplastic and nonneoplastic disorders that occur in the ovary. Ovarian neoplasms are of greater diversity in histologic appearance and biologic behavior than for any other organ. The nonneoplastic disorders include physiologic cysts, pregnancy luteomas, and polycystic ovarian disease (Stein-Leventhal syndrome). The ovary can also be a site of metastasis from malignant tumors originating in the genital tract, breast, and gastrointestinal tract. *See* CANCER (MEDICINE).

As ovarian enlargement due to tumors occurs, compression of pelvic and abdominal structures produces vague symptoms such as constipation, pelvic discomfort, a feeling of heaviness, and frequent urination. Pain can be an initial symptom of both benign and malignant ovarian disorders. The symptoms of malignant ovarian disorders include abdominal pain and swelling, bloating, heartburn, nausea, and anorexia.

An examination of the pelvis is the most crucial component in the diagnosis and evaluation of ovarian disorders. This often allows the physician to differentiate between disorders of the ovary and other pelvic structures such as the uterus, rectum, and bladder.

Benign disorders. Physiologic cysts occur in response to cyclic stimulation of the ovary by hormones. Simple cysts in women of reproductive age which are greater than about 1.2 in. (3 cm) can often be treated with oral contraceptives. Persistence of the cyst for 2 months warrants surgical exploration with removal of the cyst, leaving the ovary intact.

Polycystic ovarian syndrome is caused by abnormal regulation of the hypothalamic-pituitary-ovarian axis. In this condition, the ovaries are bilaterally enlarged with multiple follicular cysts. Therapy consists of administering oral contraceptives, progesterone preparations, or clomiphene.

Endometriosis is characterized by ectopic endometrial glandular and stromal tissue that often involves the ovary as well as other pelvic structures. Ovarian endometriosis often produces a cystic mass filled with old blood termed a chocolate cyst. The two most common symptoms of endometriosis are pelvic pain and infertility. This can be treated medically with agents that suppress ovulation, or surgically by removing the ovary.

Neoplasms of the ovary can originate from epithelial, stromal, or germ cells, and produce a variety of symptoms. In a reproductive-age woman who has not completed childbearing or desires conservative therapy, removal of the cyst or one ovary is adequate treatment. In the individual who has completed childbearing or is peri- or postmenopausal, removal of both ovaries is acceptable.

Malignant disorders. As with benign tumors, malignant ovarian neoplasms can arise from epithelial, stromal, or germ cells. The primary treatment is surgery; procedures include removal of the ovaries, fallopian tubes, uterus, omentum, and other tumor masses within the abdominal cavity. *See* ONCOLOGY; OVARY; TUMOR. David L. Tait

Ovary

Ovary A part of the reproductive system of all female vertebrates. Although not vital to individual survival, the ovary is vital to perpetuation of the species. The function of the ovary is to produce the female germ cells or ova, and in some species to elaborate hormones that assist in regulating the reproductive cycle.

The ovaries develop as bilateral structures in all vertebrates, but adult asymmetry is found in certain species of all vertebrates from the elasmobranchs to the mammals. The ovary of all vertebrates functions in essentially the same manner. However, ovarian histology of the various groups differs considerably. Even such a fundamental element as the ovum exhibits differences in various groups. *See* OVUM.

The mammalian ovary is attached to the dorsal body wall. The free surface of the ovary is covered by a modified peritoneum called the germinal epithelium. Just beneath the germinal epithelium is a layer of fibrous connective tissue. Most of the rest of the ovary is made up of a more cellular and more loosely arranged connective tissue (stroma) in which are embedded the germinal, endocrine, vascular, and nervous elements.

The most obvious ovarian structures are the follicles and the corpora lutea. The smallest, or primary, follicle consists of an oocyte surrounded by a layer of follicle (nurse) cells. Follicular growth results from an increase in oocyte size, multiplication of the follicle cells, and differentiation of the perifollicular stroma to form a fibrocellular envelope called the theca interna. Finally, a fluid-filled antrum develops in the granulosa layer, resulting in a vesicular follicle.

The cells of the theca interna hypertrophy during follicular growth and many capillaries invade the layer, thus forming the endocrine element that is thought to secrete estrogen. The other known endocrine structure is the corpus luteum, which is primarily the product of hypertrophy of the granulosa cells remaining after the follicular wall ruptures to release the ovum. Ingrowths of connective tissue from the theca interna deliver capillaries to vascularize the hypertrophied follicle cells of this new corpus luteum; progesterone is secreted here. *See* ESTROGEN; ESTRUS; MENSTRUATION; PROGESTERONE. Kenneth L. Duke

Overvoltage

Overvoltage The difference between the electrical potential of an electrode or cell under the passage of current and the thermodynamic value of the electrode or cell potential under identical experimental conditions in the absence of electrolysis; it is also known as overpotential. Overvoltage is expressed in volts, often in absolute value; it is a measure of the rates of the different processes associated with an electrode reaction.

An understanding of the factors that contribute to the overvoltage is important in the operation of practical electrochemical systems. In batteries, the overvoltage plays a significant role in the available voltage and power. In large-scale industrial electrolysis, overvoltage is a major factor in determining the energy efficiency of a process, and hence, the cost of electricity. *See* DECOMPOSITION POTENTIAL; ELECTROCHEMICAL PROCESS; ELECTRODE POTENTIAL; ELECTROLYSIS.

Since the overvoltage is governed by kinetic considerations, all of the experimental conditions that can affect the rate of an electrolytic reaction are of importance. These include concentration of electrolyzed substance, temperature, composition of solvent and electrolyte, nature of the electrode surface, mode of mass transfer, and the current density (current per unit area of electrode). A rapid reaction occurs with a small overvoltage (a few millivolts). A slow reaction requires a large overvoltage (a few volts).

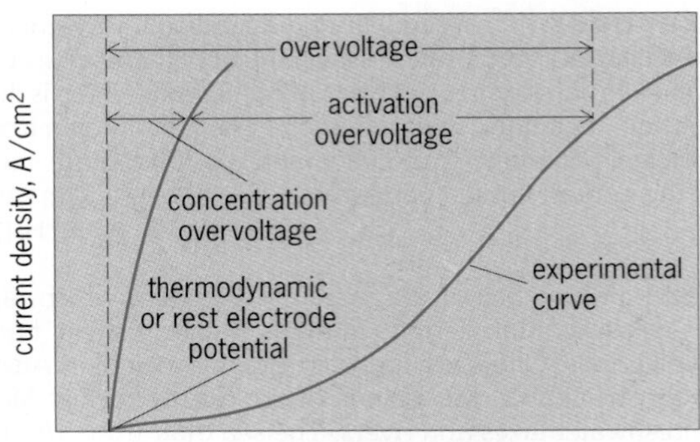

Current density-potential curve.

The rates of electrode reactions are frequently determined from current density-potential curves (see illustration). Since the departure of the electrode or cell potential from the thermodynamic value upon passage of current is sometimes termed polarization, such curves are also known as polarization curves. These curves are obtained by measurements with three-electrode electrolytic cells. The current density that flows through the electrode of interest (the working electrode) is adjusted with an external direct-current power supply, and the potential of this electrode is measured with respect to a reference electrode whose potential is known and fixed. The measured potential contains a contribution, known as the ohmic drop, that results from the flow of current through the solution resistance between the working electrode and the reference electrode. This drop is minimized by placing these electrodes close together and using various experimental approaches. The thermodynamic potential is obtained from available data for the electrode reaction of interest, corrected for the concentration of the reaction species in the solution under the experimental conditions. It is also sometimes given by the working electrode potential in the electrolysis cell when no current flows (the rest potential). The overvoltage can be read directly from the current-potential curve, as shown in the illustration. *See* ELECTRODE; REFERENCE ELECTRODE.

The total overvoltage can be decomposed into different components, which are assigned to different sources of rate limitations, for example, concentration overvoltage, activation overvoltage, reaction overvoltage, and crystallization overvoltage.

Concentration overvoltage occurs when the concentration of the reactants or products at the electrode surface are different from those in the bulk solution. These differences arise because the electroactive reactant is consumed, and products are produced by the passage of current.

Activation overvoltage arises from slowness in the rate of the electron transfer reaction at the electrode surface. To drive an electrode reaction at a given rate (current density), it is necessary to overcome an energy barrier, the energy of activation for the reaction. The additional energy (that is, beyond the thermodynamic requirements) needed to overcome this barrier is provided by the electrical energy supplied to the cell in the form of an increase in the applied potential. The magnitude of this activation overvoltage often depends upon the nature of the electrode material.

Reaction overvoltage arises when a chemical reaction is associated with the overall electrode reaction. For example, if the electroactive substance is generated from the major reactant by a chemical reaction that precedes the electron transfer, its concentration at the electrode surface will be governed by the rate of this reaction. This preceding reaction will thus affect the potential at which the electrode reaction occurs. Slow steps in the formation of nuclei and the crystal lattice, for example, in the electroplating of a metal, can lead to nucleation and crystallization overvoltages, respectively. Allen J. Bard

Ovum The egg or female sex cell. Strictly speaking, the term refers to this cell when it is ready for fertilization, but it is often applied to earlier or later stages. Confusion is avoided by using qualifying adjectives such as immature, ripe, mature, fertilized, or developing ova. The mature ova are generally spheroidal and large. The number of ova produced at one time varies in different animals, from millions in many marine animals that spawn into the surrounding sea water to about a dozen or less in mammals in which adaptations for internal nourishment of the developing embryo and care of the young are highly developed.

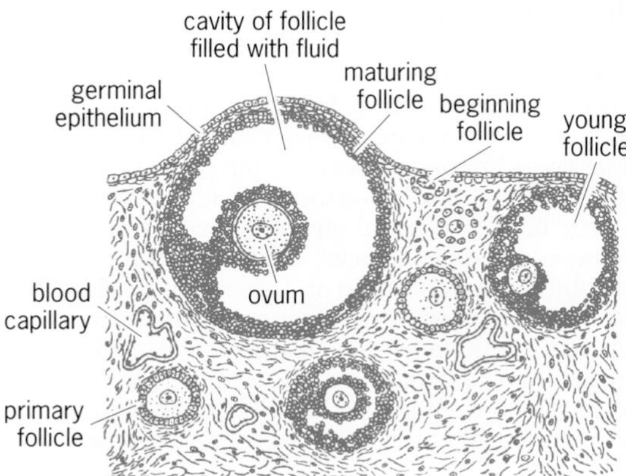

Section of a mammalian ovary.

In the ovary the immature ovum is associated with follicle cells through which it receives material for growth. In mammals, as the egg matures, these cells arrange themselves into a structure known as the Graafian, or vesicular, follicle, consisting of a large fluid-filled cavity into which the ovum, surrounded by several layers of cells, projects from the layer of follicle cells that constitutes the inner wall (see illustration). The fluid contains estrogenic female sex hormone secreted by cells in an intermediate layer of the follicular wall.

Yolk, or deutoplasm, is essentially a food reserve in the form of small spherules, present to a greater or lesser extent in all eggs. It accounts largely for the differences in size of eggs. Eggs are classified according to the distribution of yolk. In the isolecithal type there is a nearly uniform distribution through the cytoplasm, as in most small eggs. The yolk in telolecithal eggs is increasingly concentrated toward one pole, as in the large eggs of fish, amphibians, reptiles, and birds. Centrolecithal, or centrally located, yolk occurs in eggs of insects and cephalopod mollusks. *See* GAMETOGENESIS; OOGENESIS. Albert Tyler; Howard L. Hamilton

Oxalidales An order of flowering plants (angiosperms) in the eurosid I group of the rosid dicots. The order is previously unrecognized in classifications of the angiosperms but is indicated by numerous studies of DNA sequences. Oxalidales consist of five small families: Cephalotaceae (one species), Connaraceae (300 species of tropical trees and vines), Cunoniaceae (250 species of trees and shrubs mostly from the Southern Hemisphere), Elaeocarpaceae (350 species of trees and shrubs from the Southern Hemisphere and Asian tropics), and Oxalidaceae (350 species, mostly in *Oxalis*, mostly herbs that are found throughout the world). Oxalidales are heterogeneous in their morphological traits. Many species of the order are locally economically important, producing timbers and fruits, including zebrawood (*Connarus*, Connaraceae), star fruit (*Averrhoa*, Oxalidaceae), and lightwood (*Ceratopetalum* and *Eucryphia*, Cunoniaceae). *Oxalis* (Oxalidaceae) has some species that are grown as ornamentals and several that are noxious introduced weeds. *See* MAGNOLIOPHYTA; MAGNOLIOPSIDA; ORNAMENTAL PLANTS; PITCHER PLANT; ROSIDAE; WEEDS. Mark W. Chase

Oxidation process A process in which oxygen is caused to combine with other molecules. The oxygen may be used as elemental oxygen, as in air, or in the form of an oxygen-containing molecule which is capable of giving up all or part of its oxygen. Oxidation in its broadest sense, that is, an increase in positive valence or removal of electrons, is not considered here if oxygen itself is not involved. *See* OXIDATION-REDUCTION.

Most oxidations occur with the liberation of large amounts of energy in the form of either heat, light, or electricity. The stable ultimate products of oxidation are oxides of the elements involved. These oxidations occur in nature as corrosion, decay, and respiration and in the deliberate burning of matter such as wood, petroleum, sulfur, or phosphorus to oxides of the constituent elements.

The principal variables to be considered and controlled in any partial oxidation are temperature, pressure, reaction time (or contact time), nature of catalyst, if any, mole ratio of oxidizing agent, and whether the substance to be oxidized is to be kept in the liquid or vapor phase. Only a narrow range of conditions unique to each substance being oxidized and each product desired will give satisfactory yields. It is also essential to maintain conditions outside the range of spontaneous ignition, to avoid explosive mixtures or the accidental accumulation of unstable peroxides, and to choose materials which not only can resist the environmental conditions but also which do not have adverse catalytic effects or otherwise interfere with the desired reaction. *See* COMBUSTION. I. E. Levine

Oxidation-reduction An important concept of chemical reactions which is useful in systematizing the chemistry of many substances. Oxidation can be represented as involving a loss of electrons by one molecule and reduction as involving an absorption of electrons by another. Both oxidation and reduction occur simultaneously and in equivalent amounts during any reaction involving either process.

Oxidation number. The oxidation state is a concept which describes some important aspects of the state of combination of the elements. An element in a given substance is characterized by a number, the oxidation number, which specifies whether the element in question is combined with elements which are more electropositive or more electronegative than it is. It further specifies the combining capacity which the element exhibits in a particular combination. A scale of oxidation numbers is defined by assigning to an oxygen atom in an ion such as SO_4^{2-} the value of 2−. That for sulfur as 6+ then follows from the requirement that the sum of the oxidation numbers of all the atoms add up to the net charge on the species. The value of 2− for oxygen is not chosen arbitrarily. It recognizes that oxygen is more electronegative than sulfur, and that when it reacts with other elements it seeks to acquire two more electrons, by sharing or outright transfer from the electropositive partner, so as to complete a stable valence shell of eight electrons.

Although oxidation number is in some respects similar to valence, the two concepts have distinct meanings. In the substance H_2, the valence of hydrogen is 1 because each H makes a single bond to another H, but the oxidation number is 0, because the hydrogen is not combined with a different element. *See* VALENCE.

When the oxidation number of an atom in a species is increased, the process is described as oxidation, no matter what reagent produces it; when a decrease in oxidation number takes place, the process is described as reduction, again without regard to the identity of the reducing agent. The term oxidation has been generalized to imply combination of an element with an element more electronegative than itself.

Reactions. In an oxidation-reduction reaction, some element decreases in oxidation state and some element increases in oxidation state. The substances containing these elements are defined as the oxidizing agents and reducing agents, and they are said to be reduced and oxidized, respectively. The processes in question can always be represented formally as involving electron absorption by the oxidizing agent and electron donation by the reducing agent. For example, reaction (1) can be regarded as the sum of the two partial processes, or half-reactions, (2) and (3). Similarly, reaction (4) consists of the two half-reactions (5) and (6), with half-reaction (5) being taken five times to balance the electron flow from reducing agent to oxidizing agent.

$$2Fe^{3+} + 2I^- \rightarrow 2Fe^{2+} + I_2 \quad (1)$$

$$2I^- \rightarrow I_2 + 2e^- \quad (2)$$

$$2Fe^{3+} + 2e^- \rightarrow 2Fe^{2+} \quad (3)$$

$$16H^+ + 2MnO_4{}^- + 10I^- \rightarrow 8H_2O + 2Mn^{2+} + 5I_2 \quad (4)$$

$$2I^- \rightarrow I_2 + 2e^- \quad (5)$$

$$16H^+ + 2MnO_4{}^- + 10e^- \rightarrow 2Mn^{2+} + 8H_2O \quad (6)$$

Each half-reaction consists of an oxidation-reduction couple; thus, in half-reaction (6) the reducing agent and oxidizing agent making up the couple are manganous ion, Mn^{2+}, and permanganate ion, $MnO_4{}^-$, respectively; in half-reaction (5) the reducing agent is I^- and the oxidizing agent is I_2. The fact that $MnO_4{}^-$ reacts with I^- to produce I_2 means that $MnO_4{}^-$ in acid solution is a stronger oxidizing agent than is I_2. Because of the reciprocal relation between the oxidizing agent and reducing agent comprising a couple, this statement is equivalent to saying that I^- is a stronger reducing agent than Mn^{2+} in acid solution. Reducing agents may be ranked in order of tendency to react, and this ranking immediately implies an opposite order of tendency to react for the oxidizing agents which complete the couples. In the list below some common oxidation-reduction couples are ranked in this fashion:

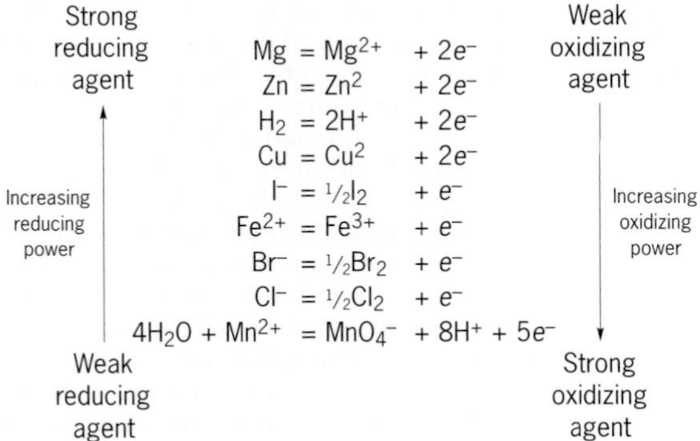

See ELECTROCHEMICAL SERIES; ELECTRONEGATIVITY; OXIDIZING AGENT.

Henry Taube

Oxide A binary compound of oxygen with another element. Oxides have been prepared for essentially all the elements except the noble gases. Often, several different oxides of a given element can be prepared; a number exist naturally in the Earth'scrust and atmosphere: silicon dioxide (SiO_2) in quartz; aluminum oxide ($A1_2O_3$) in corundum; iron oxide (Fe_2O_3) in hematite; carbon dioxide (CO_2) gas; and water (H_2O).

Most elements will react with oxygen at appropriate temperature and oxygen pressure conditions, and many oxides may thus be directly prepared. Most metals in massive form react with oxygen only slowly at room temperatures because the first thin oxide coat formed protects the metal. The oxides of the alkali and alkaline-earth metals, except for beryllium and magnesium, are porous when formed on the metal surface, and they provide only limited protection to the continuation of oxidation, even at room temperatures. Gold is exceptional in its resistance to oxygen, and its oxide (Au_2O_3) must be prepared by indirect means. The other noble metals, although ordinarily resistant to oxygen, will react at high temperatures to form gaseous oxides.

Oxides may be classified as acidic or basic according to the character of the solution resulting from their reactions with water. The nonmetal oxides generally form acid solutions and the metal oxides generally form alkaline solutions. *See* ACID AND BASE; EQUIVALENT WEIGHT; OXYGEN.

Russell K. Edwards

Oxide and hydroxide minerals Mineral phases containing only oxide or hydroxide anions in their structures. By volume, oxide and hydroxide minerals comprise only a small fraction of the Earth's crust. However, their geochemical and petrologic importance cannot be overstated. Oxide and hydroxide minerals are important ores of metals such as iron, aluminum, titanium, uranium, and manganese. Oxide and hydroxide minerals occur in all geological environments. Some form as primary minerals in igneous rocks, while others form as secondary phases during the weathering and alteration of silicate and sulfide minerals. Some oxide and hydroxide minerals are biogenic; for example, iron(III) and manganese(IV) hydroxides and oxides often result from bacterial oxidation of dissolved Fe^{2+} and Mn^{2+} in low-temperature aqueous solutions. *See* HYDROXIDE; MINERAL; ORE AND MINERAL DEPOSITS; OXIDE; SILICATE MINERALS; WEATHERING PROCESSES.

Iron and manganese hydroxide minerals often occur as nanocrystalline or colloidal phases with high, reactive surface areas. Adsorption of dissolved aqueous ions onto colloidal iron and manganese oxides plays a major role in the fate of micronutrients and heavy metals in soil and ground water and the trace-element chemistry of the oceans. Much current research is focused on measuring the thermodynamics and kinetics of

Table 1. Summary of important simple oxide minerals; classified by structure

Mineral	Formula
X_2O Oxides	
Cuprite	Cu_2O
Argentite	Ag_2O
NaCl Structure	
Periclase	MgO
Manganosite	MnO
Wustite	FeO
Other MO Oxides	
Tenorite	CuO
Zincite	ZnO
Corundum Structure	
Corundum	Al_2O_3
Eskaolite	Cr_2O_3
Hematite	Fe_2O_3
Ilmenite	$Fe^{2+}Ti^{4+}O_3$
Rutile Structure	
Rutile	TiO_2
Cassiterite	SnO_2
Pyrolusite	MnO_2
Spinel Structure	
Spinel	$MgAl_2O_4$
Magnetite	Fe_3O_4
Chromite	$FeCr_2O_4$
Franklinite	$ZnFe_2O_4$
Ulvospinel	$Fe^{2+}(Fe^{2+}Ti^{4+})O_4$
Hausmannite	Mn_3O_4 (distorted)
Fluorite Structure	
Uraninite	UO_2
Thorianite	ThO_2

metal adsorption by colloidal Fe-Mn hydroxides and oxides in the laboratory. In anoxic sedimentary environments, bacteria may use iron(III) and manganese(IV) hydroxide minerals as electron acceptors. Consequently, these minerals may facilitate the biodegradation of organic pollutants in soil and ground water. *See* ADSORPTION; COLLOID.

To a first approximation, the bonding in oxide minerals can be viewed in the ionic (electrostatic) model. According to Pauling's rules, the coordination number of a metal cation (such as Mg^{2+}, Al^{3+}, and Ti^{4+}) is determined by the radius of the cation relative to that of the oxide anion (O^{2-}). This allows one to predict the structures of some of the simpler oxide minerals (see table). Cations with similar ionic radii (such as Mg^{2+} and Fe^{2+}) are able to substitute for each other and form a solid solution. Sulfide minerals (which are more covalent) show little solid solution. The simple ionic radii arguments will fail when electronic configurations preclude the spherical symmetry of the cations. The ions Cu^{2+} and Mn^{3+}, with nine and four d electrons, tend to adopt distorted coordination environments because of the Jahn-Teller effect. Also, d-electron configurations can give rise to large octahedral site-preference energies that cause small cations, such as Mn^{4+}, to always adopt octahedral coordination. The magnetic and semiconducting properties of transition-metal oxide minerals, such as magnetite (Fe_3O_4), give useful geophysical signatures for subsurface exploration. *See* IONIC CRYSTALS; JAHN-TELLER EFFECT; PROSPECTING; SOLID-STATE CHEMISTRY; STRUCTURAL CHEMISTRY; VALENCE. David M. Sherman

Oxidizing agent A participant in a chemical reaction that absorbs electrons from another reactant. In the process a component atom of this substance undergoes a decrease in oxidation number. In this action as an oxidizing agent, the substance undergoes reduction.

A measure of the effectiveness of a reagent as an oxidizing agent is its reduction potential. This is, in electrochemical terms, the equivalent of the free-energy change for the reduction process. The element with the highest reduction potential (and, therefore, the strongest oxidizing agent) is fluorine, F_2. The practical effectiveness of a given oxidizing (or reducing) agent will depend upon both the thermodynamics and the available kinetic pathway for the reaction process. *See* CHEMICAL THERMODYNAMICS.

Substances that are widely used as oxidizing agents in chemistry include ozone (O_3), permanganate ion (MnO_4^-), nitric acid (HNO_3), as well as oxygen itself. Organic chemists have empirically developed combinations of reagents to carry out specific oxidation steps in synthetic processes. The action of molecular oxygen as an oxidizing agent may be made more specific by photochemical excitation to an excited singlet electronic state. F. J. Johnson

Oxime One of a group of chemical substances with the general formula RR′C=N—OH, where R and R′ represent any carbon group or hydrogen. Oximes are derived from aldehydes (aldoximes, RHC=NOH) and ketones (ketoximes, RR′C=NOH, where R and R′ are not hydrogen). Oximes and oxime ethers (RR′C=N—OR″) have important pharmaceutical and synthetic applications. The oxime (and oxime ether) functional group is incorporated into many organic medicinal agents, including some antibiotics, for example, gemifloxacin mesylate; and pralidoxime chloride and obidoxime chloride are used in the treatment of poisoning by organophosphate insecticides malathion and diazinon. *See* ALDEHYDE; KETONE.

Hydroxylamine (H_2NOH) reacts readily with aldehydes or ketones to give oximes. The rate of the reaction of hydroxylamine with acetone is greatest at pH 4.5. Oximes are formed by nucleophilic attack of hydroxylamine at the carbonyl carbon (C=O) of an aldehyde or ketone to give an unstable carbinolamine intermediate, as in the reaction below. Since the breakdown of the carbinolamine intermediate to an oxime is acid-catalyzed, the rate of this step is enhanced at low pH. If the pH is too low, however, most of the hydroxylamine will be in the nonnucleophilic protonated form (NH_3OH^+), and the rate of the first step will decrease. Thus, in oxime formation the pH has to be such that there is sufficient free hydroxylamine for the first step and enough acid so that dehydration of the carbinolamine is facile. *See* REACTIVE INTERMEDIATES.

One of the best-known reactions of oximes is their rearrangement to amides. This reaction, the Beckmann rearrangement, can be done with a variety of reagents [such as phosphorus pentachloride (PCl_5), concentrated sulfuric acid (H_2SO_4), and perchloric acid ($HClO_4$)] that induce the rearrangement by converting the oxime hydroxyl group into a group of atoms that easily departs in a displacement reaction (a good leaving group) by either protonation or formation of a derivative. The industrial synthesis of ϵ-caprolactam is done by a Beckmann rearrangement on cyclohexanone oxime. ϵ-Caprolactam is polymerized to the polyamide known as nylon 6, which is used in tire cords. *See* ORGANIC SYNTHESIS; POLYAMIDE RESINS; TIRE. James E. Johnson

Oximetry Any of various methods for determining the oxygen saturation of blood. Oximeters are instruments that measure the saturation of blood in arteries by transilluminating intact tissue, thus eliminating the need to puncture vessels. Aside from the major advantage of being noninvasive and therefore free of risk, oximeters permit long-term monitoring in a variety of clinical settings, giving a continuous record of oxygen saturation.

When dark venous blood passes through the pulmonary circulation or is exposed to air, it becomes bright red. In severe lung disease, the hemoglobin in the blood fails to become adequately oxygenated in its passage through the lungs and the mucous membranes lose their characteristic pink color. The mucous membranes then appear dusky or blue, a physical sign known as cyanosis. Oximeters determine oxygen saturation by making use of this color change; they compare photoelectrically the differences in light absorption between oxygenated and reduced hemoglobin. *See* HEMOGLOBIN.

The original ear oximeters used a single wavelength of visible red light to monitor oxyhemoglobin levels. Later models used visible red and green light, allowing a distinction to be made between the effects of changes in hemoglobin concentration and saturation. Modern two-wavelength oximeters are known as pulse oximeters. Incident light is emitted by two alternating light-emitting diodes—one in the visible red region (660 nm) and one in the infrared region (940 nm)—to determine the ratio of oxyhemoglobin to deoxyhemoglobin. Small light-emitting diodes are readily incorporated into a variety of probes; sites for transillumination include the earlobe, finger, toe, bridge of the nose and, in neonates, palms of the hand or foot. Oximeters are used widely in modern hospital settings, where their ability to reflect oxygen saturation changes within 1–2 s can alert caregivers to the presence of life-threatening episodes of hypoxemia (inadequate oxygenation). In addition to monitoring the critically ill, oximeters are used routinely in the operating room to safeguard individuals having routine anesthetics. Kenneth R. Chapman; A. S. Rebuck

Oxygen A gaseous chemical element, O, atomic number 8, and atomic weight 15.9994. Oxygen is of great interest because it is the essential element both in the respiration process in most living cells and in combustion processes. It is the most abundant element in the Earth's crust. About one-fifth (by volume) of the air is oxygen. *See* PERIODIC TABLE.

Oxygen is separated from air by liquefaction and fractional distillation. The chief uses of oxygen in order of their importance are (1) smelting, refining, and fabrication of steel and other metals; (2) manufacture of chemical products by controlled oxidation; (3) rocket propulsion; (4) biological life support and medicine;

Properties of oxygen

Property	Value
Atomic number	8
Atomic weight	15.9994
Triple point (solid, liquid, and gas in equilibrium)	−218.80°C (−139.33°F)
Boiling point at 1 atm pressure	−182.97°C (−119.4°F)
Gas density at °C and 10^5 Pa pressure, g/liter	1.4290
Liquid density at the boiling point, g/ml	1.142
Solubility in water at 20°C, oxygen (STP) per 1000 g water at 10^5 Pa partial pressure of oxygen	30

and (5) mining, production, and fabrication of stone and glass products.

Uncombined gaseous oxygen usually exists in the form of diatomic molecules, O_2, but oxygen also exists in a unique triatomic form, O_3, called ozone. *See* OZONE.

Under ordinary conditions oxygen is a colorless, odorless, and tasteless gas. It condenses to a pale blue liquid, in contrast to nitrogen, which is colorless in the liquid state. Oxygen is one of a small group of slightly paramagnetic gases, and it is the most paramagnetic of the group. Liquid oxygen is also slightly paramagnetic. Some data on oxygen and some properties of its ordinary form, O_2, are listed in the table. *See* PARAMAGNETISM.

Practically all chemical elements except the inert gases form compounds with oxygen. Most elements form oxides when heated in an atmosphere containing oxygen gas. Many elements form more than one oxide; for example, sulfur forms sulfur dioxide (SO_2) and sulfur trioxide (SO_3). Among the most abundant binary oxygen compounds are water, H_2O, and silica, SiO_2, the latter being the chief ingredient of sand. Among compounds containing more than two elements, the most abundant are the silicates, which constitute most of the rocks and soil. Other widely occurring compounds are calcium carbonate (limestone and marble), calcium sulfate (gypsum), aluminum oxide (bauxite), and the various oxides of iron which are mined as a source of iron. Several other metals are also mined in the form of their oxides. Hydrogen peroxide, H_2O_2, is an interesting compound used extensively for bleaching. *See* HYDROGEN PEROXIDE; OXIDATION-REDUCTION; OXIDE; PEROXIDE; WATER.

Arthur W. Francis, Sr.; Lawrence M. Sayre

Oxygen toxicity A toxic effect in a living organism caused by a species of oxygen. Oxygen has two aspects, one benign and the other malignant. Those organisms that avail themselves of the enormous metabolic advantages provided by dioxygen (O_2) must defend themselves against its toxicity. The complete reduction of one molecule of O_2 to two of water (H_2O) requires four electrons; therefore, intermediates must be encountered during the reduction of O_2 by the univalent pathway. The intermediates of O_2 reduction, in the order of their production, are the superoxide radical ($O_2{}^-$), hydrogen peroxide (H_2O_2), and the hydroxyl radical (HO·). *See* OXYGEN; SUPEROXIDE CHEMISTRY.

The intermediates of oxygen reduction, rather than O_2, itself, are probably the primary cause of oxygen toxicity. It follows that defensive measures must deal with these intermediates. The superoxide radical is eliminated by enzymes that catalyze the following reaction.

$$O_2^- + O_2^- + 2H^+ \leftrightharpoons H_2O_2 + O_2$$

These enzymes, known as superoxide dismutases, have been isolated from a wide variety of living things.

Hydrogen peroxide (H_2O_2) must also be eliminated, and this is achieved by two enzymatic mechanisms. The first of these is the dismutation of H_2O_2 into water and oxygen, a process catalyzed by catalases. The second is the reduction of H_2O_2 into two molecules of water at the expense of a variety of reductants, a process catalyzed by peroxidases. *See* ENZYME.

The multiplicity of superoxide dismutases, catalases, and peroxidases, and the great catalytic efficiency of these enzymes, provides a formidable defense against $O_2{}^-$ and H_2O_2. If these first two intermediates of O_2 reduction are eliminated, the third (HO·) will not be produced. No defense is perfect, however, and some HO· is produced; therefore its deleterious effects must be minimized. This is achieved to a large extent by antioxidants, which prevent free-radical chain reactions from propagating. *See* ANTIOXIDANT; CHAIN REACTION (CHEMISTRY); FREE RADICAL; PEPTIDE.

The apparent comfort in which aerobic organisms live in the presence of an atmosphere that is 20% O_2 is due to a complex and effective system of defenses against this peculiar gas. Indeed, these defenses are easily overwhelmed, and overt symptoms of oxygen toxicity become apparent when organisms are exposed to 100% O_2. For example, a rat maintained in 100% O_2 will die in 2 to 3 days.

Irwin Fridovich

Oxymonadida An order of class Zoomastigophorea in the phylum Protozoa. These are colorless flagellate symbionts in the digestive tract of the roach *Cryptocercus* and of certain termites. They are xylophagous; that is, they ingest wood particles taken in by the host. Seven or more genera of medium or large size have been identified, the organisms varying from pyriform to ovoid in shape. At the anterior end a pliable neck-like rostrum attaches the organism to the host intestinal wall, but they are sometimes free. They can be either uni- or multinucleate. *See* ZOOMASTIGOPHOREA.

James B. Lackey

Ozone A powerfully oxidizing allotropic form of the element oxygen. The ozone molecule contains three atoms (O_3). Ozone gas is decidedly blue, and both liquid and solid ozone are an opaque blue-black color, similar to that of ink. *See* OXYGEN.

Some properties of ozone are given in the table. Ozone has a characteristic, pungent odor familiar to most persons because ozone is formed when electrical apparatus produces sparks in air. Ozone is irritating to mucous membranes and toxic to human beings and lower animals.

Some properties of ozone

Property	Value
Density of the gas at 0°C, 1 atm pressure	2.154 g/liter
Density of the liquid	
−111.9 °C	1.354 g/ml
−183°C	1.573 g/ml
Boiling point at 1 atm pressure	−111.9°C
Melting point of the solid	−192.5°C

Ozone is a more powerful oxidizing agent than oxygen, and oxidation with ozone takes place with evolution of more heat and usually starts at a lower temperature than when oxygen is used. In the presence of water, ozone is a powerful bleaching agent, acting more rapidly than hydrogen peroxide, chlorine, or sulfur dioxide. *See* OXIDIZING AGENT.

Ozone is utilized in the treatment of drinking-water supplies. Odor- and taste-producing hydrocarbons are effectively eliminated by ozone oxidation. Iron and manganese compounds which discolor water are diminished by ozone treatment. Compared to chlorine, bacterial and viral disinfection with ozone is up to 5000 times more rapid.

Ozone occurs to a variable extent in the Earth's atmosphere. Near the Earth's surface the concentration is usually 0.02–0.03 ppm in country air, and less in cities except when there is smog. At vertical elevations above 13 mi (20 km), ozone is formed by photochemical action on atmospheric oxygen. Maximum concentration of 5×10^{12} molecules/cm^3 (more than

1000 times the normal peak concentration at Earth's surface) occurs at an elevation of 19 mi (30 km). Arthur W. Francis

Ozonolysis A process which uses ozone to cleave unsaturated organic bonds. Generally, ozonolysis is conducted by bubbling ozone-rich oxygen or air into a solution of the reactant. The reaction is fast at moderate temperatures. Intermediates are usually not isolated but are subjected to further oxidizing conditions to produce acids or to reducing conditions to form alcohols or aldehydes. An unsymmetrical olefin is capable of yielding two different products whose structures are related to the groups substituted on the olefin and the position of the double bond.

Before World War I, ozonolysis was applied commercially to the preparation of vanillin from isoeugenol. The only modern application of the technique in the United States is in the manufacture of azelaic and pelargonic acids from oleic acid. *See* ALKENE; OZONE. Robert K. Barnes

P

Pacific islands A geographic designation that includes thousands of mainly small coral and volcanic islands scattered across the Pacific Ocean from Palau in the west to Easter Island in the east. Island archipelagos off the coast of the Asian mainland, such as Japan, Philippines, and Indonesia, are not included even though they are located within the Pacific Basin. The large island constituting the mainland of Papua New Guinea and Irian Jaya is also excluded, along with the continent of Australia and the islands that make up Aotearoa or New Zealand. The latter, together with the Asian Pacific archipelagos, contain much larger landmasses, with a greater diversity of resources and ecosystems, than the oceanic islands, commonly labelled Melanesia, Micronesia, and Polynesia. *See* AUSTRALIA; NEW ZEALAND; OCEANIC ISLANDS.

The great majority of these islands are between 4 and 4000 mi^2 (10 and 10,000 km^2) in land surface area. The three largest islands include the main island of New Caledonia (6220 mi^2 or 16,100 km^2), Viti Levu (4053 mi^2 or 10,497 km^2) in Fiji, and Hawaii (4031 mi^2 or 10,440 km^2) the big island in the Hawaiian chain. When the 80-mi (200-km) Exclusive Economic Zones are included in the calculation of surface area, some Pacific island states have very large territories. These land and sea domains, far more than the small, fragmented land areas per se, capture the essence of the island world that has meaning for Pacific peoples. *See* EAST INDIES.

Oceanic islands are often classified on the basis of the nature of their surface lithologies. A distinction is commonly made between the larger continental islands of the western Pacific, the volcanic basalt island chains and clusters of the eastern Pacific, and the scattered coral limestone atolls and reef islands of the central and northern Pacific.

It has been suggested that a more useful distinction can be drawn between plate boundary islands and intraplate islands. The former are associated with movements along the boundaries of the great tectonic plates that make up the Earth's surface. Islands of the plate boundary type form along the convergent, divergent, or tranverse plate boundaries, and they characterize most of the larger island groups in the western Pacific. These islands are often volcanically and tectonically active and form part of the Pacific so-called Ring of Fire, which extends from Antarctica in a sweeping arc through New Zealand, Vanuatu, Bougainville, and the Philippines to Japan.

The intraplate islands comprise the linear groups and clusters of islands that are thought to be associated with volcanism, either at a fixed point or along a linear fissure. Volcanic island chains such as the Hawaii, Marquesas, and Tuamotu groups are classic examples. Others, which have their volcanic origins covered by great thickness of coral, include the atoll territories of Kiribati, Tuvalu, and the Marshall Islands. Another type of intraplate island is isolated Easter Island, possibly a de- tached piece of a mid-ocean ridge. The various types of small islands in the Pacific are all linked geologically to much larger structures that lie below the surface of the sea. These structures contain the answers to some puzzles about island origins and locations, especially when considered in terms of the plate tectonic theory of crustal evolution. *See* MARINE GEOLOGY; MID-OCEANIC RIDGE; PLATE TECTONICS; SEAMOUNT AND GUYOT; VOLCANO.

The climate of most islands in the Pacific is dominated by two main forces: ocean circulation and atmospheric circulation. Oceanic island climates are fundamentally distinct from those of continents and islands close to continents, because of the small size of the island relative to the vastness of the ocean surrounding it. Because of oceanic influences, the climates of most small, tropical Pacific islands are characterized by little variation through the year compared with climates in continental areas.

The major natural hazards in the Pacific are associated either with seasonal climatic variability (especially cyclones and droughts) or with volcanic and tectonic activity. *See* CLIMATE HISTORY.

Richard D. Bedford

Pacific Ocean The Pacific Ocean has an area of 6.37×10^7 mi^2 (1.65×10^8 km^2) and a mean depth of 14,000 ft (4280 m). It covers 32% of the Earth's surface and 46% of the surface of all oceans and seas, and its area is greater than that of all land areas combined. Its mean depth is the greatest of the three oceans and its volume is 53% of the total of all oceans. Its greatest depths in the Marianas and Japan trenches are the world's deepest, more than 6 mi (10 km).

The two major wind systems driving the waters of the ocean are the westerlies which lie about 40–50° lat in both hemispheres (the "roaring forties") and the trade winds from the east which dominate in the region between 20°N and 20°S. These give momentum directly to the west wind drift (flow to the east) in high latitudes and to the equatorial currents which flow to the west. At the continents there is flow of water from one system to the other and huge circulatory systems result. *See* OCEAN CIRCULATION; SOUTHEAST ASIAN WATERS.

The swiftest flow (greater than 2 knots) is found in the Kuroshio Current near Japan. It forms the northwestern part of a huge clockwise gyre whose north edge lies in the west wind drift centered at about 40°N, whose eastern part is the south-flowing California Current, and whose southern part is the North Equatorial Current.

Equatorward of 30° lat heat received from the Sun exceeds that lost by reflection and back radiation, and surface waters flowing into these latitudes from higher latitudes (California and Peru currents) increase in temperature as they flow equatorward and turn west with the Equatorial Current System. They carry heat poleward and transfer part of it to the high-latitude cyclones along the west wind drift. The temperature of the equatorward currents along the eastern boundaries of the subtropical anticyclones is thus much lower than that of the currents of their western boundaries at the same latitudes. The highest temperatures (more than 82°F or 28°C) are found at the western end of the equatorial region. Along the Equator itself somewhat lower temperatures are found. The cold Peru Current contributes to its eastern end, and there is apparent upwelling of deeper, colder water at the Equator.

Upwelling also occurs at the edge of the eastern boundary currents of the subtropical anticyclones. When the winds blow strongly equatorward (in summer) the surface waters are driven offshore, and the deeper colder waters rise to the surface and further reduce the low temperatures of these equatorward-flowing currents. *See* UPWELLING.

The limiting temperature in high latitudes is that of freezing. Ice is formed at the surface at temperatures slightly less than 30°F (−1°C) depending upon the salinity; further loss of heat is retarded by its insulating effect. The ice field covers the northern and eastern parts of the Bering Sea in winter, and most of the Sea of Okhotsk, including that part adjacent to Hokkaido (the north island of Japan). Summer temperatures, however, reach as high as 43°F (6°C) in the northern Bering Sea and as high as 50°F (10°C) in the northern part of the Sea of Okhotsk. *See* BERING SEA.

Pack ice reaches to about 62°S from Antarctica in October and to about 70°S in March, with icebergs reaching as far as 50°S. *See* ICEBERG; SEA ICE.

Surface waters in high latitudes are colder and heavier than those in low latitudes. As a result, some of the high-latitude waters sink below the surface and spread equatorward, mixing mostly with water of their own density as they move, and eventually become the dominant water type in terms of salinity and temperature of that density over vast regions.

The most conspicuous water masses formed in the Pacific are the Intermediate Waters of the North and of the South Pacific, which on the vertical sections include the two huge tongues of low salinity extending equatorward beneath the surface from about 55°S and from about 45°N. The southern tongue is higher in salinity and density and lies at a greater depth. Joseph L. Reid

Packet switching A software-controlled means of directing digitally encoded information in a communication network from a source to a destination, in which information messages may be divided into smaller entities called packets. Switching and transmission are the two basic functions that effect communication on demand from one point to another in a communication network, an interconnection of nodes by transmission facilities. Each node functions as a switch in addition to having potentially other nodal functions such as storage or processing.

Switched (or demand) communication can be classified under two main categories: circuit-switched communication and store-and-forward communication. Store-and-forward communication, in turn, has two principal categories: message-switched communication (message switching) and packet-switched communication (packet switching).

In circuit switching, an end-to-end path of a fixed bandwidth (or speed) is set up for the entire duration of a communication or call. The bandwidth in circuit switching may remain unused if no information is being transmitted during a call. In store-and-forward switching, the message, either as a whole or in parts, transits through the nodes of the network one node at a time. The entire message, or a part of it, is stored at each node and then forwarded to the next.

In message switching, the switched message retains its integrity as a whole message at each node during its passage through the network. For very long messages, this requires large buffers (or storage capacity) at each node. Also, the constraint of receiving the very last bit of the entire message before forwarding its first bit to the next node may result in unacceptable delays. Packet switching breaks a large message into fixed-size, small packets and then switches these packets through the network as if they were individual messages. This approach reduces the need for large nodal buffers and "pipelines" the resources of the network so that a number of nodes can be active at the same time in switching a long message, reducing significantly the transit delay. One important characteristic of packet switching is that network resources are consumed only when data are actually sent.

All public packet networks require that terminals and computers connecting to the network use a standard access protocol. Interconnection of one public packet network to others is carried out by using another standardized protocol.

Packet-switched networks using satellite or terrestrial radio as the transmission medium are known as packet satellite or packet radio networks, respectively. Such networks are especially suited for covering large areas for mobile stations, or for applications that benefit from the availability of information at several locations simultaneously.

Asynchronous transfer mode (ATM) is a type of packet switching that uses short, fixed-size packets (called cells) to transfer information. The ATM cell is 53 bytes long, containing a 5-byte header for the address of the destination, followed by a fixed 48-byte information field. The rather short packet size of ATM, compared to conventional packet switching, represents a compromise between the needs of data communication and those of voice and video communication, where small delays and low jitter are critical for most applications.

Data communication (or computer communication) has been the primary application for packet networks. Computer communication traffic characteristics are fundamentally different from those of voice traffic. Data traffic is usually bursty, lasting from several milliseconds to several minutes or hours. The holding time for data traffic is also widely different from one application to another. These characteristics of data communication make packet switching an ideal choice for most applications. The principal motivation for ATM is to devise a unified transport mechanism for voice, still image, video, and data communication. *See* DATA COMMUNICATIONS.

The Internet is a global network, the largest packet-switched network in the world. The specific packet-switching protocol used by the Internet is TCP/IP (Transmission Control Protocol/Internet Protocol). *See* INTERNET. Pramode K. Verma

Packing A seal usually used for high pressure as in steam and hydraulic applications. The motion between parts may be infrequent as in valve stems, or continual as in pump or engine piston rods. There is no sharp dividing line between seals and packing; both are dynamic pressure resistors under motion. Diverse materials are used for packing such as impregnated fiber, rubber, cork, or asbestos compounds. In packings, it is necessary that the surface finish of the contacting metal part be smooth for long life of the material. *See* PRESSURE SEAL. Paul H. Black

Pain Pain, especially in its acute form, is usually a reflection of a tissue-damaging or potentially tissue-damaging stimulus. There is a transmission system that conveys this information to the central nervous system. This phenomenon is called nociception. Pain is more complex than other sensory systems such as vision or hearing because it not only involves the transfer of sensory information to the nervous system, but produces suffering which then leads to aversive corrective behavior. In certain disease states, defects in the transmission system can of themselves generate false information to the nervous system, as though tissue damage were occurring in the periphery. An example of this is phantom limb pain, in which the individual often has a crushing type of pain in a foot that has been amputated.

Acute pain such as occurs with broken bones and other significant injuries is almost inevitably accounted for by the phenomenon of nociception and is probably a purely neurophysio-logical event. However, the more pain becomes a chronic phenomenon, the more such influences as psychological factors and behavior become part of the expression of pain.

Acute pain is a useful warning system. There are specific nerve paths for conducting this sensation (see illustration). Pain receptors in the skin and other tissues are nerve terminals which lack any special characteristics, and they are probably triggered by a chemical stimulus when potential tissue damage occurs. There appear to be two types of terminals: one responds to many types

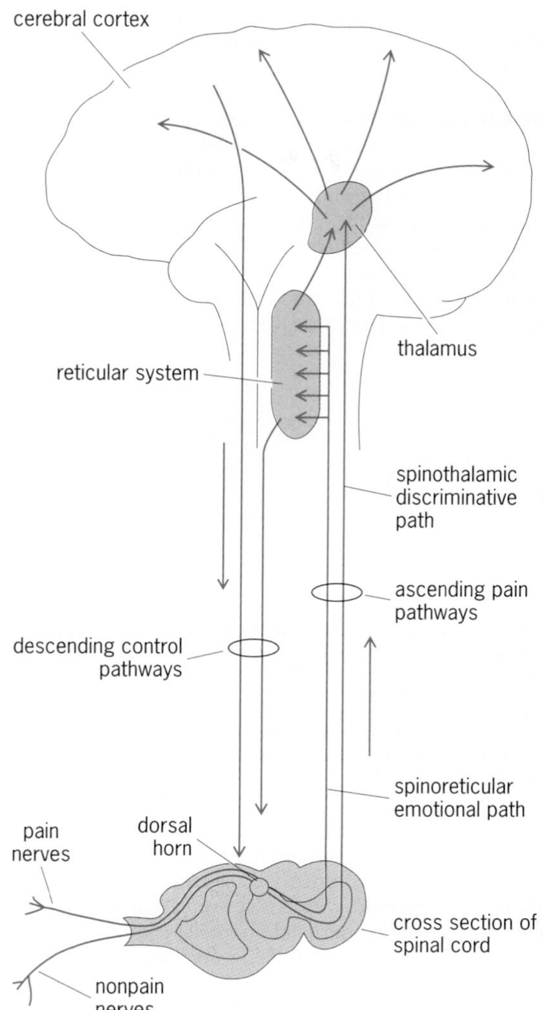

Neurophysiology of incoming pain. Sensation from peripheral receptors travels along specific pain nerves, and is modulated throughout the spinal cord and brain.

of painful stimuli, whereas the other specifically responds to either mechanical or thermal energy. When the terminals are stimulated, the pain (that is, nociception message) is carried along specific small sensory fibers called A-delta and C fibers. The A-delta fibers are larger and transmit the "first pain" or "fast pain" The smaller C fibers transmit a secondary dull continuous pain. These nerve fibers were traditionally believed to enter the spinal cord through the dorsal root, but it now seems that many also enter through the ventral root into the spinal cord.

Having entered the spinal cord, these fibers relay in the dorsal horn of the spinal gray matter, an area of considerable regulation and modulation of the incoming pain stimulus which is influenced by other incoming sensory stimuli; that is, touch or pressure sensations can suppress the transmission of signals in the small pain fibers. This helps to explain why when a person is hurting, the pain can be reduced by rubbing the affected part, and this phenomenon forms the basis of some of the treatment strategies of stimulation-produced analgesia. In addition, the incoming pain signal in the spinal cord is also modulated by descending signals from the brain. At times of anxiety, these pain signals may be augmented. From these relay stations in the dorsal horn, the pain signal is carried by two nerve paths up to the brain. The classical pathway is the spinothalamic tract, on the side of the spinal cord opposite to the incoming stimulus, and this leads to the posterior part of the thalamus in the brainstem, and from there nerve paths radiate the pain sensation to many parts of the cerebral cortex, where the pain is appreciated. In addition to this direct path, there is also a diffuse ascending path known as the spinoreticular tract which relays to many of the basal ganglia in the brain, and from there to areas of the brain connected with motivational and affective behavior such as the hippocampus and the cingulate gyrus. It is possible that narcotic analgesics exert some of their action on this ascending spinoreticular tract because these drugs tend to reduce the suffering aspects of pain, but still preserve many of the discriminative qualities so that individuals can still feel the pain, but it does not bother them so much. *See* ANALGESIC; NARCOTIC.

Certain parts of the brainstem around the central canal appear to exert a strong inhibitory effect on incoming pain signals. Stimulation of these areas probably releases endorphins, which are morphinelike substances produced by the body and liberated at various sites on the incoming pain path to suppress these signals. *See* ENDORPHINS. Terence M. Murphy

Paint and coatings Substances applied to other materials to change the surface properties, such as color, wear, and chemical or scratch resistance, without changing the bulk properties. The terms paint and coatings often are used interchangeably. However, it has become a common practice to use coatings as the broader term and to restrict paints to the familiar architectural and household coatings and sometimes to maintenance coatings for doors and windows, bridges, and tanks. Another common term for paint and coatings is finish.

Coatings are used for (1) decoration, (2) protection, or (3) some functional purpose. The low-gloss paint used on the ceiling of a room fulfills not only a decorative need but also functions to reflect and diffuse light for even illumination. The exterior coating on an automobile adds beauty, while protecting it from rusting. The public commonly thinks of house paint when talking about coatings; however, all kinds of coatings are important, as they make essential contributions to most high-tech fields.

Coatings may be described by their appearance (for example, clear, pigmented, metallic, or glossy) and their function (such as decorative, corrosion protection, abrasion protection, skid resistance). Coatings may be distinguished as organic or inorganic, although there is overlap. For example, many coatings consist of inorganic pigments dispersed in an organic matrix (binder).

The binder in an inorganic coating is a metal salt. Inorganic coatings also may contain organic additives such as lubricants. Electroplated copper, nickel, and zinc coatings are example.

Organic coatings are complex mixtures of chemical substances that can be grouped into four broad categories: (1) binders, (2) volatile components, (3) pigments, and (4) additives.

Binders form the continuous film that adheres to the substrate, binds together the other substances in the coating, and provides an adequately hard outer surface. The binder governs, largely, the properties of the dried or cured film.

Volatile components, such as solvents, water, or both, are included in a majority of coatings. They play a major role in the process of applying coatings; that is, they make the coating fluid enough for application and they evaporate during and after application.

Pigments are finely divided insoluble solids that are dispersed in the vehicle and remain suspended in the binder after film formation. Generally, the purpose of pigments is to provide color and opacity. However, they also have substantial effects on application characteristics and on film properties. *See* PIGMENT (MATERIAL).

Additives are all the materials that are included in small quantities to modify some property of the coating. Almost all coating formulations contain some type of additive. Some additives help to stabilize the coating before use, while others broaden its application properties or improve its durability. Jamil Baghdachi

Palaeacanthocephala An order of Acanthocephala, the adults of which are parasitic worms found in fishes, aquatic birds, and mammals. They have the following characteristics.

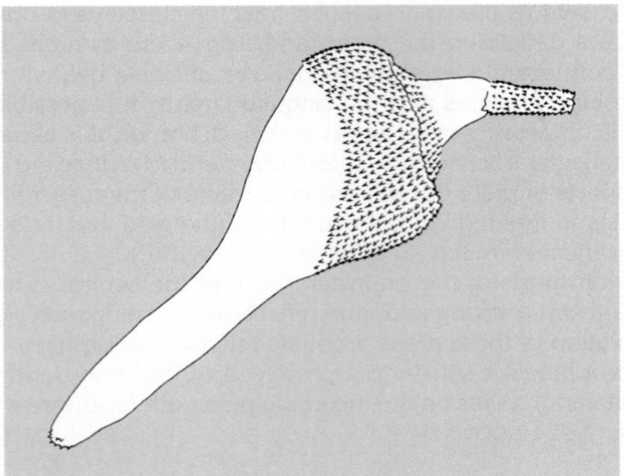

Corynosoma reductum. (*After H. J. Van Cleave, Acanthocephala of North American Mammals, University of Illinois Press, 1953*)

The nuclei of the hypodermis are fragmented and the chief lacunar vessels are lateral. The males have usually 2–7 cement glands. The ligament sac in the female breaks down so that the eggs develop in the body cavity. Proboscis hooks occur in long rows and spines are present on the body of some species. Species which commonly occur in vertebrates are *Leptorhynchoides thecatus* and *Corynosoma* (see illustration). *See* ACANTHOCEPHALA.

Donald V. Moore

Palaeoisopus A peculiar arthropod, evidently related to the Pycnogonida and represented by a number of well-preserved fossils from the Devonian Hunsruck shales. It was formerly considered by a number of paleontologists to have an anterior jointed proboscis and a bulbous terminal abdomen, but studies of material under ultraviolet light have demonstrated that the bulbous abdomen is in fact a pair of robust, well-developed chelae (see illustration). Furthermore,

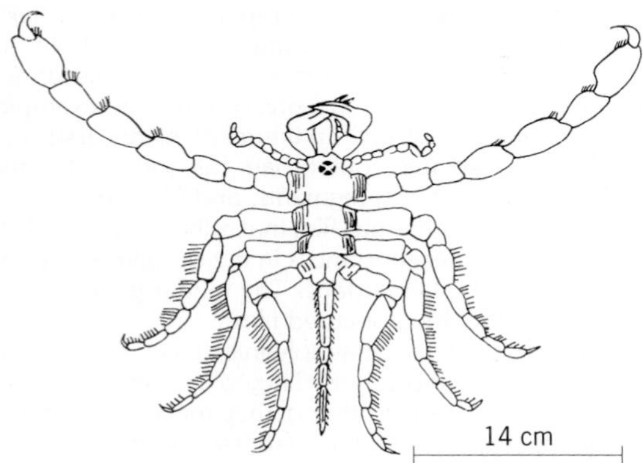

Palaeoisopus problematicus.

additional appendages (palps and ovigers) not apparent in previously studied material have been discerned, bringing its complement of anterior appendages into agreement with that of the Pycnogonida. *See* PYCNOGONIDA.

Joel W. Hedgpeth

Palaeonemertini A rarer order of the class Anopla in the phylum Rhynchocoela, characterized by an unarmed proboscis, a thin gelatinous dermis, and either a two-layered or three-layered body musculature. Many members show primitive features, such as a peripherally located nervous system and the absence of ocelli, ciliated grooves, and intestinal diverticula. Cerebral organs, if present, are generally simple. *See* ANOPLA; NEMERTEA.

J. B. Jennings

Palaeonisciformes A large, extinct assemblage of primitive ray-finned (actinopterygian) fishes known from the Silurian to Cretaceous, which were most numerous in Carboniferous and Permian times. They are known as fossils from all continents except Antarctica. Approximately 40 families are recognized but collectively they are not a natural group; some are genealogically more closely related to derived ray-finned fishes (holosteans and teleosts) than to other palaeonisciforms. *See* ACTINOPTERYGII; HOLOSTEI; TELEOSTEI.

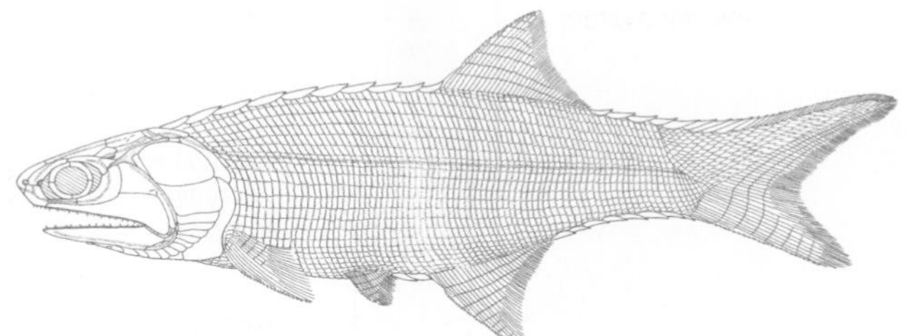

Mimia toombsi (Gardiner & Bartram, 1977), Late Devonian, Gogo, Australia. One of the earliest palaeonisciforms, reaching 15 cm (6 in.) long, and known by complete fossils. (*Reprinted with permission from B. G. Gardiner, The relationships of the palaeoniscid fishes, a review based on new specimens of Mimia and Moythomasia from the Upper Devonian of Western Australia, Bull. Brit. Mus. (Nat. His.), Geology, 37(4):173–428, 1984*)

Primitive palaeonisciforms are mostly small [5–30 cm (2–12 in.) long], with a streamlined body covered with thick, shiny scales and head bones (see illustration). The scales are rhomboid and covered with a substance called ganoine. The ganoine consists of a superficial enamellike layer overlying dentine. In turn, the ganoine overlies a thin bone layer—the latter is all that remains in the scales of the vast majority of modern bony fishes. Over most of the body, each scale abuts with its neighbor above and below by means of a peg-and-socket joint. This type of scale, found as isolated remains, is the first evidence of ray-finned fishes in late Silurian rocks, and it gives its name to the group (palaeonisciform literally means "ancient scale").

The body tapers to a gently upturned, long, asymmetrical tail, in which all the fin rays are inserted along the ventral edge; there is a single dorsal fin placed midway along the back. The head tapers to a blunt snout, and the large eyes (placed well forward) suggest that palaeonisciforms relied on sight for capturing prey, such as small crustaceans and other fishes, as well as to avoid predators. The upper jaw bone (maxilla) is rigidly attached to the other cheek bones so that it is incapable of independent movement. Therefore, the only way to open the mouth is to raise the head at the same time as dropping the lower jaw.

Classification of palaeonisciforms is complicated because most are known only from flattened and fragmentary fossils. Although many separate groups, often called families, are recognized, the relationships of these families to each other and to other fishes are not well understood. Three modern fish groups must have originated from palaeonisciform fishes: the Polypterifomes, or bichirs; the Acipenseriformes, or sturgeons and paddlefishes; as well as fishes classified as holosteans and teleosteans. *See* ACIPENSERIFORMES; OSTEICHTHYES; POLYPTERIFORMES.

Peter L. Forey

Palaeospondylus A tiny fossil fish found only in rocks of Middle Devonian age within a small geographic region of Caithness, Scotland, and principally at one site, Achanaras Quarry, where it is surprisingly abundant. One of the most enigmatic of all fossil fishes, *Palaeospondylus gunni* was discovered in 1890. Its original environment was a large, deep, freshwater lake system with a broad diversity of fishes. Unusually for a tiny fish, *Palaeospondylus* is typically found in sediments associated with the deepest part of the lake.

Specimens of *Palaeospondylus* never exceed 6 cm (2.36 in.) in length. Usually, the head, vertebral column, and a dorsal tail fin are preserved, along with occasional traces of limb girdles. (*Palaeospondylus* means "ancient vertebrae.") The front of the head includes a strange structure identified as the nasal apparatus or jaws. A characteristic feature of the back of the head is a pair of elongated, backwardly projecting dorsal ribs. Past attempts to identify *Palaeospondylus* with any known fishes, living or fossil, have always failed. Keith Thomson

Palate The roof of the mouth in those vertebrates whose mouth cavity and nasal passages are wholly or partially separate.

The palate of mammals consists of two portions. The hard palate, more anterior in position, underlies the nasal cavity, whereas the soft palate hangs like a curtain between the mouth and nasal pharynx. The hard palate has an intermediate layer of bone. The oral surface of the hard palate is a mucous membrane, covered with a stratified squamous epithelium.

The soft palate is a backward continuation from the hard palate. Its free margin connects on each side with two folds of mucous membrane, the palatine arches, enclosing a palatine tonsil. In the midline the margin extends into a fingerlike projection named the uvula. The oral side of the soft palate continues as the covering of the hard palate, and the submucosa contains pure mucous glands.

Besides separating the nasal passages from the mouth, the hard palate is a firm plate, against which the tongue crushes and manipulates food. The soft palate, at rest, is pendant. In sucking, swallowing, or vomiting it is raised to separate the oral from the nasal portion of the pharynx. The closing action also occurs in speech, except for certain consonants requiring nasal resonance. *See* SPEECH. Leslie B. Arey

Paleobiochemistry The study of chemical processes used by organisms that lived in the geological past. Most information on the nature of life in the geological past comes from the study of fossils; a record of biochemical processes that occurred can be found in the organic molecules of sedimentary rocks and fossils. The organic matter in fossil fuel deposits and finely dispersed in shales and limestones represents the debris of cells which have been chemically altered to a more stable form. A comparison of the molecular structure of these preserved organic compounds with that of components of living cells enables the researcher to identify similarities and dissimilarities between past and present biochemistry.

Paleobiochemical studies have shown that a number of the common chemical processes used by living organisms today have been in use for a very great length of time. Paleobiochemical techniques are also used on materials returned from extraterrestrial sources to determine whether life exists outside of Earth. *See* PALEOECOLOGY; PALEONTOLOGY. Thomas C. Hoering

Paleobotany The study of fossil plants of the geologic past. A paleobotanist is a plant historian who carefully pieces together the geologic history of the plant kingdom. Other organisms, including fungi and various types of microscopic plankton, are also studied by paleobotanists. Paleobotany is a branch of paleontology that requires a knowledge of both plant biology (botany) and the geological sciences. *See* BOTANY; FOSSIL; PALEONTOLOGY; PLANT KINGDOM.

Materials used by paleobotanists to reconstruct plant life through geologic time include fossilized remains preserved in the rock layers of the Earth. Materials such as fossil leaves, seeds, fragments of wood, fruits, and flowers are used to interpret the biology and evolution of ancient plants. In addition, the products and distribution of plants are recorded in the rock record in the form of coal, resins, various chemicals, and other substances produced by plants. Some plant fossils, such as pollen grains and spores, are studied by palynologists; palynology has been especially important in mineral and petroleum exploration and in correlating rock layers that are widely separated geographically. From all of these materials, paleobotanists attempt to reconstruct the habit, structure, and biology of plants that grew on the Earth millions of years ago. *See* PALYNOLOGY.

Paleobotany not only involves the collection, description, reconstruction, and naming of fossil plants but is also concerned with the evolution of major groups, relationships that exist between fossil and living forms, how ancient plants functioned and reproduced, what type of environment they lived in, how they were fossilized, and many other biological and geological topics.

Plants are preserved in a variety of ways, and various combinations of physical and chemical processes are involved at the time of preservation. For a plant part to become fossilized, several unique circumstances must occur. First, the item must be close to a site where sediments are accumulating; second, it must be rapidly buried. Rapid burial is necessary to ensure that the biological activities of various microbes do not destroy the tissues of the plant. *See* BIODEGRADATION.

Plant parts are best preserved in very fine grained silts and shales, which are the lithified muds of ancient deposits. Such sediments generally yield excellent fossils because the small grain size preserves minute details of the leaf; coarser-grained sediments such as sands generally do not reproduce delicate features. Plant parts composed of thick-walled cells or those having fibers within their tissues have a better chance of being preserved (fossilization potential) than more delicate tissues composed of thin-walled cells. The aboveground parts of most plants are covered by a layer of cutin or wax that is generally resistant to decay and aids in the preservation of some tissues. In addition, the site of burial is critical as to whether fossil remains will ever be discovered.

Fossil plants have provided an enormous body of knowledge about the evolution of plants from the earliest Precambrian unicellular forms to the complex multicellular flowering plants used for food and shelter today. Tracing plants through geologic time has provided the basis for examining how and when the first plants became adapted to a terrestrial environment when the angiosperms first appeared, why seeds evolved, and why some plants acquired the ability to produce wood. Once plants are identified, the composition of the community and ecosystem can be examined, including the types of interactions that existed with other organisms. Another set of questions that can be answered by the paleobotanist relates to biostratigraphy. For example, plant fossils have been used to correlate rock layers that are widely separated geographically. Thus, where the geologic time range of a plant or assemblage of plants has been determined in rocks whose relative age is known, the presence of the same plants in other rocks will indicate an equivalent age. *See* STRATIGRAPHY. Thomas N. Taylor

Paleoceanography The study of the history of the ocean with regard to circulation, chemistry, biology, and patterns of sedimentation. The source of information is largely the biogenous deep-ocean sediments, so the field may be considered a branch of sedimentology or paleontology. However, there are also strong links to geophysics, marine geochemistry, and mathematical modeling. Geophysical sciences are called upon for reconstruction of geography (position of continents, horizontal motion of the ocean floor), topography (changing depth patterns in the ocean, general subsidence of any given piece of sea floor), and dating (radioisotopes, magnetic patterns on the sea floor and in sediments). Geochemical analyses deliver information on sediment composition (stable and unstable isotopes; major and minor components such as carbonate, opal, and trace elements). Such information is useful in correlating sedimentary sequences and, when combined with geochemical arguments, yields insights about the dynamics of carbon and nutrient cycles. Mathematical modeling introduces a strong quantitative element. It draws upon the knowledge reservoir of modern oceanography and climatology. *See* CLIMATOLOGY; GEOCHEMISTRY; GEOPHYSICS; PALEONTOLOGY; SEDIMENTOLOGY.

The study of marine sediments on land is as old as geology itself. Modern paleoceanography is set apart by the study of sediments recovered from the ocean, especially the deep ocean, and by the use of concepts developed by oceanographers (controls on ocean currents and upwelling, vertical stratification, heat budget, nutrient and carbon cycles, pelagic biogeography, and water masses). Important early studies were on cores raised by the United States cable ship *Lord Kelvin* (1936) in the North Atlantic. The glacial debris zones noted in the *Kelvin* cores are now commonly referred to as Heinrich layers; they are witness to sporadic input of iceberg armadas during the last glacial epoch. *See* GLACIAL EPOCH.

Comparisons of climate-related changes between major ocean basins became possible through the systematic recovery of long cores by the circumglobal Swedish Deep Sea Expedition (1947–1949) on the research vessel *Albatross*. The expedition retrieved cores up to 15 m (45 ft) long, with records reaching back 500,000–1,000,000 years. Many fundamental paleoceanographic concepts were established by the geologists who analyzed these cores, including the role of trade winds in promoting glacial-age upwelling, the changing supply of North Atlantic Deep Water, the cyclicity of climatic change in the late Quaternary, and large-scale shifts in biogeographic boundaries. *See* CLIMATE MODELING; PALEOCLIMATOLOGY; QUATERNARY.

A quantum jump in paleoceanographic research resulted from the initiation of deep-sea drilling using the research vessel *Glomar Challenger* (1968). Enormous blank regions on the world's map suddenly became accessible for detailed exploration far back into geologic time, that is, into the Early Cretaceous. Highlights of the first decade of drilling results include the documentation of cooling steps as the planet moved into the present ice age; the reconstruction of long-term fluctuations in the carbonate compensation depth; the documentation of large-scale salt deposition in an isolated Mediterranean basin; and the discovery of temporary anoxic conditions in the Cretaceous deep sea. *See* CRETACEOUS. Wolfgang H. Berger; Gerold Wefer

Paleocene The oldest of the seven geological epochs of the Cenozoic Era, and the oldest of the five epochs that make up the Tertiary Period. The Paleocene Epoch represents an interval of geological time (and rocks deposited during that time) from the end of the Cretaceous Period to the beginning of the Eocene Epoch. Recent revisions of the geological time scales place the Paleocene Epoch between 65 to 55 million years before present. *See* CENOZOIC; EOCENE; GEOLOGIC TIME SCALE; TERTIARY.

The close of the Cretaceous Period was characterized by the disappearance of many terrestrial and marine animals and plants. The dawn of the Cenozoic in the Paleocene Epoch saw the establishment of new fauna and flora that have evolved into modern biota.

Modern schemes of the Paleocene subdivide it into Lower and Upper series, and their formal equivalents, the Danian and Selandian stages. Some authors prefer to use a threefold subdivision of the Paleocene, adding the Thanetian at the top. The older, Danian lithofacies generally tend to be calcium carbonate-rich (pure chalk in the Danian type area), whereas the younger, Selandian and Thanetian facies have greater land-derived components and are more siliciclastic (sand, sandstone, marl). *See* CHALK; FACIES (GEOLOGY); MARL; SAND; SANDSTONE.

Several major tectonic events that began in the Mesozoic continued into the Paleocene. For example, the Laramide Orogeny that influenced deformation and uplift in the North American Rocky Mountains in the Mesozoic continued into the Paleocene. *See* OROGENY.

The establishment of deeper connections between the North and South Atlantic in the Paleocene facilitated enhanced deep-water flow from the northern to the southern basin. In the south, the Drake Passage between South America and Antarctica was still closed, although Australia had already separated from Antarctica by Paleocene time. The lack of circum-Antarctic flow precluded the geographic isolation of Antarctica and the development of cold deep water from a southern source. *See* PALEOCEANOGRAPHY; PALEOGEOGRAPHY.

Terrestrial floras and faunas corroborate the peak warming in the latest Paleocene and early Eocene and suggest that the warm tropical-temperate belt may have been twice its modern latitudinal extent. The temperate floral and faunal elements extended to 60°N, which has been used as an argument to invoke a very low angle of inclination of the Earth's rotational axis in the Paleocene-Eocene. Alternatively, the mild, equable polar climates and well-adapted physiological responses of plants and animals of those times to local conditions may be enough to explain the presence of a rich vertebrate fauna on Ellesmere Island in arctic Canada. *See* CLIMATE HISTORY; PALEOBOTANY; PALEOCLIMATOLOGY.

The Paleocene Epoch began after a meteorite struck the Earth, causing massive extinctions at the end of Cretaceous and decimating a large percentage of the terrestrial and marine biota. In the oceans, all ammonites, genuine belemnites, rudistids, most species of planktonic foraminifera and nannoplankton, and marine reptiles disappeared at the close of the Cretaceous Period. Even though some groups, such as squids, octopus, nautilus, and a few species of marine plankton, survived, the genetic pool was relatively small at the dawn of the Tertiary Period. The recovery of the marine biota was, however, fairly rapid after the mid-Paleocene due to overall transgressing seas and ameliorating climates. By the late Paleocene, the biota was well on its way to the high diversification of the Eocene. The end of the Paleocene Epoch saw marked changes in deep-water circulation of the world ocean that resulted in a massive extinction of the benthic marine species. *See* EXTINCTION (BIOLOGY).

On land the large dinosaurs, which had been on the decline for over 20 million years, died out at the close of the Cretaceous Period. However, smaller reptiles, including alligators and crocodiles, and some of the land flora escaped extinction and continued into the Paleocene. The Paleocene saw the first true radiation of mammals. The mammals of this epoch were characteristically primitive and small in size (50 cm or 20 in. or less). As the continent of Australia became more isolated geographically, its mammalian fauna, such as the marsupials, became sequestered and more specialized. *See* DINOSAUR; MAMMALIA; PALEONTOLOGY. Bilal U. Haq

Paleoclimatology The study of ancient climates. Climate is the long-term expression of weather; in the modern world, climate is most noticeably expressed in vegetation and soil types and characteristics of the land surface. To study ancient climates, paleoclimatologists must be familiar with various disciplines of geology, such as sedimentology and paleontology, and with climate dynamics, which includes aspects of geography and atmospheric and oceanic physics. Understanding the history of the Earth's climate system greatly enhances the ability to predict how it might behave in the future. *See* CLIMATOLOGY.

Information about ancient climates comes principally from three sources: sedimentary deposits, including ancient soils; the past distribution of plants and animals; and the chemical composition of certain marine fossils. These are all known as proxy indicators of climate (as opposed to direct indicators, such as temperature, which cannot be measured in the past). In addition, paleoclimatologists use computer models of climate that have been modified for application to ancient conditions. *See* GEOLOGY; PALEONTOLOGY.

Like modern climatologists, paleoclimatologists are concerned with boundary conditions, forcing, and response. Boundary conditions are the limits within which the climate system operates. The boundary conditions considered by paleoclimatologists depend on the part of Earth history that is being studied. For the recent past, that is, the last few million years, boundary conditions that can change on short time scales are considered, for example, atmospheric chemistry. For the more distant past, paleoclimatologists must also consider boundary conditions that change on long time scales. Geographic features—that is, the positions of the continents, the location and orientation of major mountain ranges, the positions of shorelines, and the presence or absence of epicontinental seaways—are important for understanding paleoclimatic patterns. Forcing is a change in boundary conditions, such as continental drift, and response is how forcing changes the climate system. Forcing and response are cause and effect in paleoclimatic change. *See* CONTINENTAL DRIFT; CONTINENTS, EVOLUTION OF; PALEOGEOGRAPHY; PLATE TECTONICS.

Proxy indicators of paleoclimate are abundant in the geologic record. Important sedimentary indicators forming on land are coal, eolian sandstone (ancient sand dunes), evaporites (salt), tillites (ancient glacial deposits), and various types of paleosols (ancient soils), such as bauxite (aluminum ore) and laterite (some iron ores). Coals may form where conditions are favorable for growth of plants and accumulation and preservation of peat, conditions that are partly controlled by climate, especially seasonality of rainfall. *See* BAUXITE; COAL; PALEOSOL.

Fossil indicators provide information about climate mostly by their distribution (paleobiogeography), although a few specific types of fossils may be indicative of certain climatic conditions. The latter are usually fossils from the younger part of the geologic record and are closely related to modern species that have narrow environmental tolerances. Another type of information available for documenting paleoclimatic patterns and change is stable isotope geochemistry of fossils and certain types of sedimentary rock. Many elements that are used by organisms to make shells, teeth, and stems occur naturally in several different forms, known as isotopes. The most climatically useful isotopes are those of oxygen (O). Although the effects of temperature change and ice volume change can be difficult to distinguish, the analysis of oxygen isotopes has provided a powerful quantitative tool for the study of both long-term temperature change and the history of the polar ice caps. *See* FOSSIL; ISOTOPE.

A great deal of research in paleoclimatology has been devoted to understanding the causes of climatic change, and the overriding conclusion is that any given shift in the paleoclimatic history of the Earth was brought about by multiple factors operating in concert. The most important forcing factors for paleoclimatic variation are changes in paleogeography and atmospheric chemistry and variations in the Earth's orbital parameters. *See* ATMOSPHERIC CHEMISTRY; BIOGEOCHEMISTRY; EARTH ROTATION AND ORBITAL MOTION; PRECESSION OF EQUINOXES. Judith Totman Parrish; Eric J. Barron

Paleocopa An extinct order of the crustacean class Ostracoda; also called Palaeocopida. The order is divided into nine superfamilies, one of which, Barychilinacea, is only tentatively assigned to the order. A principal feature of the species in the order is their long, straight hinge that extends along the dorsal margin of the carapace and joins the two valves together. As is true of most benthic ostracodes, the palaeocopes lack a frontal opening through which to extend their walking legs. Unlike modern ostracodes, however, they have no calcified inner lamella, and their muscle-scar patterns, which are quite useful in the taxonomy of many other groups of ostracodes, are very poorly known. *See* OSTRACODA.

The carapaces of many palaeocopes are marked by ornamentation in the form of lobes and sulci. Appendages of the palaeocopes are largely unknown except for a few rather poorly preserved specimens.

One of the most remarkable morphological features of many species of palaeocope ostracodes is that their sexual dimorphism is quite pronounced and is carried far beyond the rather simple sexual dimorphism that characterizes Ostracoda. In typical instances, the males resemble the instars (immature forms), and the females have developed strongly modified morphology that is associated with reproduction—especially the development of pouches for carrying eggs and brooding the young.

The study of the ontogeny of the palaeocopes presents paleontologists with a unique opportunity to learn about the pathways and mechanisms of evolution. The palaeocopes lend themselves to this sort of study because they molt, have a predetermined number of growth stages, and have pronounced sexual dimorphism that allows one to determine the sex of individuals and the sex ratios of populations. Thus, research on the palaeocopes can reveal the pathways their evolution has followed in a way that is possible for few other kinds of fossil organisms.

As is true of other benthic ostracodes, the paleocopes do not have a planktonic larval stage. As a result, they are quite limited in their means of dispersal, and few species are biogeographically widespread. Roger L. Kaesler

Paleoecology Ecology of prehistoric times, extending from about 10,000 to about 3.5×10^9 years ago. Although the principles of paleoecology are the same as those underlying modern ecology, the two fields actually differ greatly. Paleoecology is a historical science that must rely on empirical data from fossils and their enclosing sedimentary rocks to make inferences about past conditions. Experimental approaches and direct measurement of environmental parameters, which are critical components of modern ecology, are generally impossible in paleoecology. Furthermore, distortion and loss of information during fossilization means that fossil assemblages and distributions are rarely congruent with living communities. Hence, the resolution of ancient ecosystems must remain relatively imprecise. The lack of precision is compensated for by the fact that paleoecology deals with processes occurring over vast spans of time that are unavailable to modern ecology. Long-term changes in communities (replacement) may be discerned and related to patterns of environmental change. More significantly, overall patterns of ecological change in the global biosphere may be documented; evolutionary paleoecology focuses on recognition and interpretation of long-term ecological trends that have been critical in shaping evolution. *See* ECOLOGY; PALEOCLIMATOLOGY; PALEOGEOGRAPHY; PALEONTOLOGY.

Among the goals of paleoecology are the reconstruction of ancient environments (primarily depositional environments), the inference of modes of life for ancient organisms from fossils, the recognition of recurring groupings of ancient organisms that define relicts of communities (paleocommunities), the reconstruction of the interactions of organisms with their environments and with each other, and the documentation of large-scale and long-term patterns of stasis or change in ecosystems. *See* ECOSYSTEM.

To reconstruct ancient marine environments, many different parameters must be inferred, such as temperature, water salinity, oxygen levels, nutrient concentrations, and water movements and depth. In this regard, paleoecology interfaces directly with the fields of sedimentology and stratigraphy, including study of modern depositional environments. *See* DEPOSITIONAL SYSTEMS AND ENVIRONMENTS; STRATIGRAPHY.

One of the most useful, but also potentially misused, aspects of paleoecological application is known as taxonomic uniformitarianism. This concept relies on studies of modern organisms to determine limiting environmental factors, such as salinity tolerance, temperature preference, or depth ranges. Fossils of the same or closely related species are then inferred to have had similar environmental preferences, and their occurrence is judged to indicate that particular strata were deposited under a comparable range of environmental conditions. Such an approach is valid for very closely related organisms in relatively recent geologic time.

Paleoautecology, the interpretation of modes of life (broadly, niches) of ancient organisms, involves a multidisciplinary approach. Although ancient modes of life cannot be determined completely, paleoecologists can often assign fossils to generalized guilds in terms of types of feeding, substrate preference, and degree of activity. A thorough understanding of the biology of closest modern analogs is particularly important in any attempt to reconstruct paleoautecology. *See* LIVING FOSSILS.

The fossil record contains highly biased remnants of past communities or paleocommunities. Paleocommunities are generally recognized as recurring associations of fossil species. Multivariate statistical techniques such as cluster analysis and ordination analysis are commonly employed to aid in discerning the recurrent groupings of fossil species, or persistent gradients of species composition. Such analyses are based upon field studies in which data on the presence, absence, or relative abundance of fossil taxa have been recorded in a large number of samples, typically from many stratigraphic levels.

Communities and paleocommunities are not static entities in time, but undergo important structural changes on at least three different time scales: succession, replacement, and evolution. Because it operates on a very short time scale, from decades to centuries, ecological succession can be resolved only in a few fossil samples. Longer-term changes in community composition, encompassing thousands of years, are not truly succession, but instead record allogenic effects such as sea level or climate variations. These changes are properly termed community replacement, and involve wholesale migration or restructuring of communities at particular locations due to changing environments. On a scale of millions of years, communities show evolutionary changes because their component species have evolved. Organisms evolve within the context of other organisms, not in a vacuum. There is substantial fossil evidence to indicate increasing complexity of organism interactions through time. This escalation in the intensity of predatory interactions, for example, may have important implications for evolutionary change. *See* ECOLOGICAL COMMUNITIES; ECOLOGICAL SUCCESSION. Carlton E. Brett

Paleogeography The geography of the ancient past. Paleogeographers study the changing positions of the continents and the ancient extent of land, mountains, and shallow-sea and deep-ocean basins. The Earth's geography changes because its surface is in constant motion due to plate tectonics. The continents move at rates of 2–10 cm/yr (0.75–4 in./yr). Though this may seem slow, over millions of years continents can travel across the globe. As the continents move, new ocean basins form, mountains rise and erode, and sea level rises and falls. Paleogeographic maps are necessary in order to understand global climatic change, migration routes, oceanic circulation, mountain building, and the formation of many of the Earth's natural resources, including oil and gas. *See* BASIN; CONTINENTS, EVOLUTION OF; GEOGRAPHY; MID-OCEANIC RIDGE; PALEOCLIMATOLOGY; PLATE TECTONICS; SUBDUCTION ZONES.

In the late Precambrian the continents were colliding to form supercontinents, and the Earth was locked in a major ice age. About 1100 million years ago (Ma), the supercontinent of Rodinia was assembled. Rodinia split into halves approximately 750 Ma, opening the Panthalassic Ocean. By the end of the Precambrian three continents came together to form the supercontinent of Gondwana(land). This major continent-continent collision is known as the Pan-African orogeny. *See* OROGENY; PRECAMBRIAN; PROTEROZOIC; SUPERCONTINENT.

The supercontinent that formed at the end of the Precambrian Era, approximately 600 Ma, had already begun to break apart by the beginning of the Paleozoic Era. Gondwana, which was considerably larger than any of the other continents, stretched from the Equator to the south. *See* ORDOVICIAN; PALEOZOIC.

By the end of the Paleozoic Era, the continents had collided to form the supercontinent of Pangea. Centered on the Equator, Pangea stretched from the South Pole to the North Pole. Though the supercontinent that formed at the end of the Paleozoic Era is called Pangea (literally, "all land,"), this supercontinent probably did not include all the landmasses that existed at that time.

The supercontinent of Pangea did not rift apart all at once, but in three main episodes. The first episode of rifting began in the Middle Jurassic, about 180 Ma when North America rifted away from northwest Africa, opening the Central Atlantic. *See* JURASSIC.

The second phase in the breakup of Pangea began in the Early Cretaceous, about 140 Ma. Gondwana continued to fragment as South America separated from Africa, opening the South Atlantic, and India together with Madagascar rifted away from Antarctica and the western margin of Australia, opening the Eastern Indian Ocean. *See* CRETACEOUS.

The third and final phase in the breakup of Pangea took place during the early Cenozoic. North America and Greenland split away from Europe, and Antarctica released Australia. Australia, like India some 50 million years earlier, moved rapidly northward on a collision course with Southeast Asia. *See* CENOZOIC.

About 18,000 years ago, all of Antarctica and much of North America, northern Europe, and the mountainous regions of the world were covered by glaciers and great sheets of ice. These ice sheets melted approximately 10,000 years ago, giving rise to familiar geographic features such as Hudson's Bay, the Great Lakes, the English Channel, and the fiords of Norway.

Christopher R. Scotese

Paleoindian The oldest archeological cultures of the New World, the ancestors of modern Native Americans, are termed Paleoindian. These colonizing populations of the Americas were *Homo sapiens sapiens* who arrived during the late Pleistocene (Ice Age) from Asia, though precisely when, and whether in a single or multiple pulses of migration, are not yet known.

The presumed entryway was the Bering Straits, which emerged as dry land during glacial periods. The Bering Land Bridge (or Beringia) existed most recently between 25,000 and 11,000 years before present (B.P.), during the last major episode of the Pleistocene. The conditions that linked Siberia to Alaska may have simultaneously hindered migration south from Alaska. Groups headed in that direction had two potential routes (broadly defined): down the Pacific coast, or via the continental interior along the eastern flank of the Rocky Mountains. There is no evidence yet as to which routes might have been taken.

So far, the earliest archeologically confirmed dates put human groups in the Lena Basin and Lake Baikal region of northeast Asia at about 39,000 years B.P., in subarctic Siberia by 25,000 years B.P., but not in western Beringia (such as Kamchatka) until 14,000 years B.P. Humans were in eastern Beringia (Alaska) soon after 12,000 years B.P., and present south of the ice sheets in North America by at least 11,500 years B.P.—the latter represented by the Clovis culture. Yet, the earliest accepted archeological evidence puts human groups in South America earlier still, by at least 12,500 years B.P. at the site of Monte Verde, Chile.

There are no obvious historical or technological affinities between Clovis and the Monte Verde materials, suggesting that the two may represent populations with distinct archeological traditions and separate migratory pulses: a later one (Clovis) that came south through the ice-free corridor soon after it became viable for travel, and an earlier population that perhaps moved along the Pacific coast and reached South America without, so far at least, any traces being found in North America.

Clovis is a widespread entity that first appears on the western Plains and southwest at 11,500 years B.P. and in eastern North

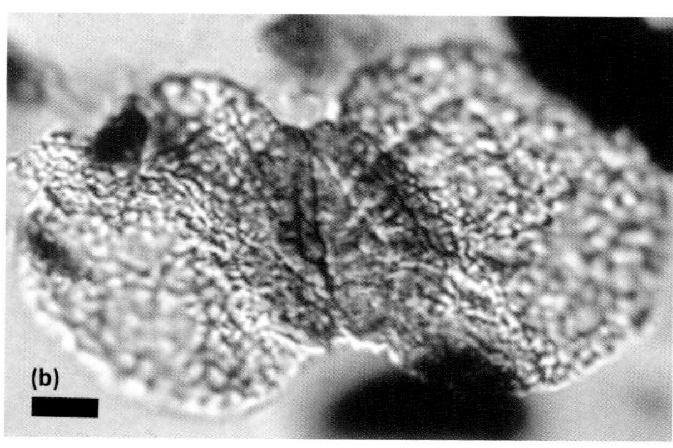

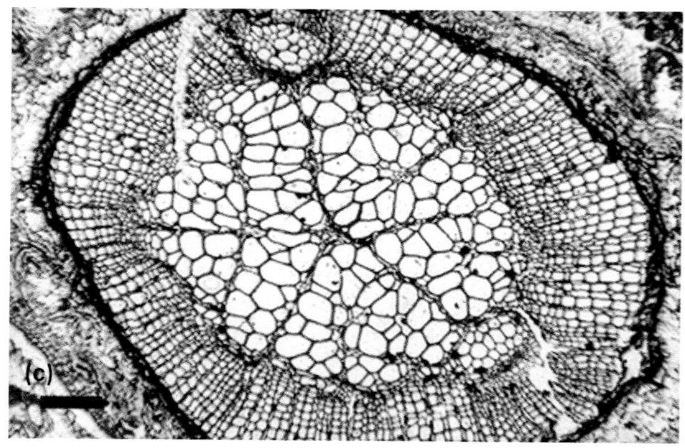

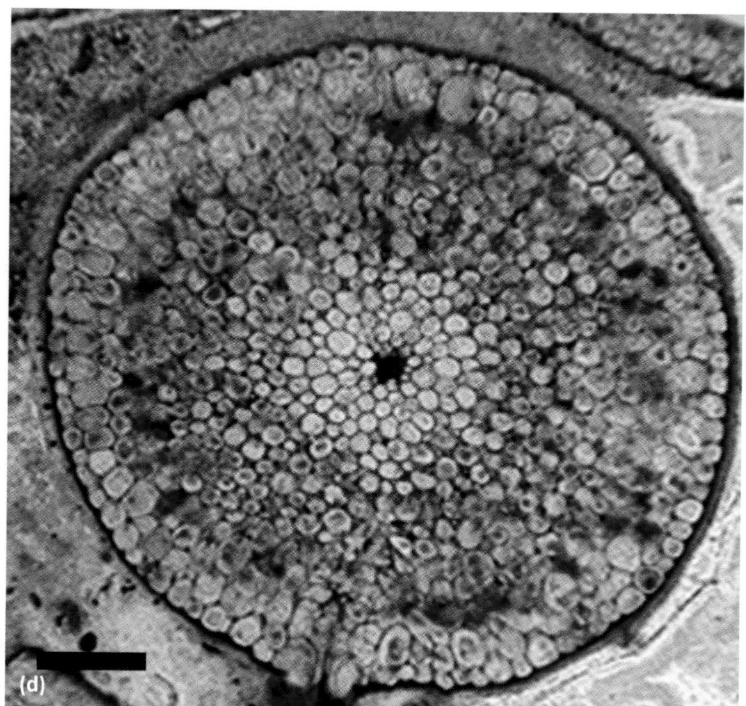

Paleobotany

(*a*) Compression specimen of the Triassic leaf type *Dicroidium*. Bar scale: 10 cm.

(*b*) Permian pollen grain macerated from Antarctic coal. Bar scale: 10 μm.

(*c*) Cross section of a permineralized stem from a Carboniferous coal ball. Bar scale: 0.2 mm.

(*d*) Thin section of a petrified Rhynie chert stem. Bar scale: 1.0 mm.

(*e*) Several *Glossopteris* leaves from the Permian of Antarctica. Leaves of this type were initially some of the first evidence that at one time all the southern continents were physically linked into a major continental plate called Gondwana. Bar scale: 1.5 cm.

America at 10,600 years B.P. That Clovis and related groups apparently expanded across the continent in what may have been less than 1000 years is all the more remarkable given that they spread at a time of geologically rapid environmental and climatic change. Yet Clovis groups seemingly coped with such adaptive challenges with ease: their stylistically distinctive projectile points and tool kits—often including bifacial knives, a variety of unifacial scrapers, occasional blades and flake tools, and (more rarely) bone and ivory implements—are surprisingly similar across the continent. These were highly mobile groups who relied on high-quality stone often obtained from geological sources hundreds of kilometers from the sites where the stone was used and discarded. Their rapid radiation, broadly similar tool kits, and long-distance movement bespeak a cultural "founders effect," suggesting their access to large areas of North America was largely unrestricted. Clovis subsistence, it appears, more often involved less risky and smaller prey—and presumably plants, though remains of such are rarely preserved in the archeological record of this period.

Although the timing varies by area, by 10,500 years B.P. the Clovis tradition was replaced by regional Paleoindian variants, which generally have reduced settlement mobility (relative to Clovis), and include new technologies, prey-specific strategies for hunting and processing, increasing use of local resources, and distinctive stylistic elements and functional artifact forms.

The South American Paleoindian record, by contrast, does not evince any artifact forms that dominate the archeological landscape as Clovis does. Instead, this period is marked by more diverse unifacial and bifacial stone tool technologies, often made of stone acquired locally (and not necessarily of superior quality), and includes forms such as bolas—modified spherical stones used in slings or as hand missiles. Projectile points tend to be less common in assemblages here than in North America, and show considerable stylistic variety. South American Paleoindians utilized a wide range of animals, and early on even made occasionally heavy use of plants.

Once the founding population dispersed across South America (over an unknown length of time), subgroups became geographically isolated relatively quickly. From the earliest known site at 12,500 years B.P. (Monte Verde) until the end of the Pleistocene (10,000 years B.P.), there is a continuing diversification in tool forms and technology, evidently reflecting less mobility, increasing heterogeneity and regional mosaics in culture and adaptations, and less expansive social networks and territories. All told, it is a different trajectory from the one that unfolded in North America—testimony that the earliest colonization of the two continents, though ultimately derived from the same northeast Asian source, may have taken place at different times under very different circumstances. *See* ARCHEOLOGY; PLEISTOCENE; PREHISTORIC TECHNOLOGY. David J. Meltzer

Paleolithic The prehistoric period when people made stone tools exclusively by chipping or flaking. John Lubbock proposed and defined the term Paleolithic, or Old Stone Age, in 1865, and also defined a subsequent stage, Neolithic or New Stone Age, during which some stone tools were formed by polishing or grinding. Later archeologists altered these definitions; to many today the Paleolithic is the period during which human beings lived entirely by hunting and gathering, while the Neolithic is the following interval during which plant and animal domestication was introduced. To other archeologists, the Paleolithic is simply a time interval, roughly equivalent to the Pleistocene Epoch, while the Neolithic comprises the early part of the succeeding Holocene (or Recent) Epoch. *See* GEOLOGIC TIME SCALE; NEOLITHIC.

It is impossible, however, to devise a rigorous, global definition of the Paleolithic or any other cultural stage, because artifact technology and economic practices have changed independently at different times in different parts of the world. Hence, Paleolithic will be used here informally to refer to the time interval between the earliest appearance of stone tools, more than 2.5 million years before present (m.y. B.P.) and the end of the last glacial period, 12,000–10,000 years B.P.

The oldest artifacts found so far come from sites in Ethiopia, Kenya, and Tanzania where they are dated to between 2.6 and 1.6 m.y. B.P. They comprise crude flakes and the modified pebbles and stone chunks from which the flakes were struck. Approximately 1.6–1.5 m.y. B.P., at least some people in East Africa, most likely a *Homo erectus* type, began to manufacture the bifacially flaked tools known to archeologists as hand axes. The flaking of the first hand axes was crude. As time passed, however, their flaking tended to become more refined, and the thinness and bilateral symmetry of some later hand axes may reflect esthetic as well as functional considerations. By 1 m.y. B.P. hand-ax makers had spread through most of Africa and the Near East, and by 900,000–600,000 years B.P. they had reached Europe.

The Early Stone Age/Lower Paleolithic apparently persisted until sometime between 200,000 and 130,000 years B.P., the exact time perhaps depending on the region. In sub-Saharan Africa the Early Stone Age was succeeded by the Middle Stone Age, while in North Africa and Eurasia the Lower Paleolithic was followed by the Middle Paleolithic, also commonly known as the Mousterian, after the cave of Le Moustier, southwestern France, where rich Middle Paleolithic levels were first excavated in the 1860s. Most Middle Stone Age/Middle Paleolithic assemblages lack hand axes. The principal Middle Stone Age/Middle Paleolithic tools are well-made stone flakes, often modified by edge flaking ("retouch") into types called sidescrapers, knives, denticulates (serrate-edged pieces), and so forth. In Europe, Middle Paleolithic people were the Neandertals, *H. (sapiens) neanderthalensis*. Neandertals also lived in the Near East during the early Middle Paleolithic. Neandertal occupation of the region apparently overlapped with that of some of the first anatomically modern people (*H. sapiens sapiens*) who disappear from the archaeological record for some time, but then return and supplant the Neandertals after ?70,000–60,000 years B.P. Outside of Europe and the Near East, bones representing Middle Paleolithic/Middle Stone Age people are rare, very fragmentary, or both. *See* NEANDERTALS.

In Europe, the Near East, and North Africa, the Middle Paleolithic was followed by the Upper Paleolithic, perhaps 50,000 years B.P. in the Near East, adjacent North Africa, and eastern Europe, and beginning around 40,000 years B.P. in western Europe. The time difference within Europe may reflect the movement of technology, people, or both from east to west. The Upper Paleolithic and Later Stone Age are difficult to characterize artifactually, because wherever they are well known, they exhibit great variability in time and space. The amount of artifactual change through time and space during the Upper Paleolithic/Later Stone Age far exceeds that in earlier periods and suggests an ability to innovate that earlier people did not exhibit. Conventionally, the Upper Paleolithic is said to end with the end of the Last Ice Age, 12,000–10,000 years B.P., although Upper Paleolithic/Later Stone Age artifact types and economic practices continued for many millennia throughout much of Eurasia and sub-Saharan Africa. *See* PREHISTORIC TECHNOLOGY.

Throughout the long Paleolithic time span, all human beings lived by hunting and gathering wild resources. Only from the very end of the Paleolithic, about 12,000–10,000 years B.P., is there evidence that some people domesticated animals, plants, or both. *See* DOMESTICATION (ANTHROPOLOGY).

The oldest reasonably secure evidence for human use of fire comes from the famous Peking man (*H. erectus*) site of Zhoukoudian in north China, tentatively dated to 500,000–240,000 years ago. More equivocal evidence for older or equally old controlled use of fire has been found in Kenya, South Africa, and Europe. Unequivocal fireplaces are found in many sites occupied by European Neandertals and their near-modern African contemporaries after 127,000 years B.P.

Concentrations of rocks or other debris that may mark the positions of ancient structures such as wind breaks and flimsy tent-like shelters have been uncovered at several Lower Paleolithic sites in Africa and across Eurasia, but the oldest widely accepted "ruins" date from the Middle Paleolithic. However, it is not until 50,000–40,000 years ago that evidence for housing becomes compelling, with such features as fireplaces and substantial wall supports being commonplace.

The oldest undeniable evidence for burial of the dead comes from the Middle Paleolithic of Europe and the Near East. The deceased were Neandertals, and their skeletons sometimes exhibit pathologies or deformities that must have incapacitated the owners long before death. Perhaps these people could only have survived with care and aid from other members of their group. Upper Paleolithic people also buried their dead, and the pathological or senile state of some skeletons again suggests a characteristically human care for sick or aged comrades. In addition, unlike Middle Paleolithic graves, Upper Paleolithic ones often contain special objects, such as hundreds of ivory beads, carved pendants, or other body ornaments, perhaps the deceased's personal belongings or items to facilitate the transition to an afterlife. Upper Paleolithic graves, together with Upper Paleolithic art, provide the oldest available evidence for the intangible part of culture called ideology or religion. Richard G. Klein; Anne Pike-Tay

Paleomagnetism The study of the direction and intensity of the Earth's magnetic field through geologic time. Paleomagnetism is an important tool in measuring the past movements of the Earth's tectonic plates. By studying ancient magnetic field directions recorded in rocks, scientists learn how the plates moved relative to the Earth's spin axis and relative to one another. The calibrated history of geomagnetic polarity reversals provides a basis for the temporal correlation of rocks on a local to global geographic scale, called magnetostratigraphy. *See* PLATE TECTONICS; ROCK MAGNETISM; STRATIGRAPHY.

At all points on the Earth's surface, the geomagnetic field is represented by a vector of specific length and direction. Two features of the geomagnetic field are particularly useful. First, when averaged over 10^4 to 10^5 years, the field can be sufficiently represented by a dipole magnet located at the Earth's center with its north-south axis aligned with the rotation axis. In this model, the time-averaged magnetization vector from a rock sequence magnetized on the Equator would be inclined at 0° with respect to local horizontal and would point toward the north geographic pole. If the same rock sequence were magnetized at the north pole, the time-averaged magnetization vector would be inclined vertically downward. For all points in between, the relationship between latitude and inclination is $\tan I = 2 \tan \lambda$, where I is inclination and λ is latitude. Thus, if the magnetization vectors of 10^8-year-old rocks now at 30°N have a mean inclination of 0°, these rocks (and probably the tectonic plate upon which they lie) have moved 30° northward since the magnetization was acquired. Longitudinal motion is undetectable. For example, rocks formed anywhere on the Equator will acquire a magnetization with 0° inclination, and subsequent motion of the rocks along the Equator results in no latitudinal change. On time scales shorter than about 10^5 years, the Earth's field at a given point exhibits considerable change in both direction and intensity. These variations, termed secular variations, can be useful for detailed stratigraphic correlation over a limited geographical range and time interval.

A second useful feature of the geomagnetic field is its period polarity reversals. The time-averaged axis of the central dipole remains parallel to the rotational axis before and after the reversal, but the north-seeking pole becomes the south-seeking pole and vice versa. Polarity changes from normal (N) to reverse (R) occur over 3000–5000 years and are effectively an instantaneous geological markers for stratigraphic correlation. Complete N-R-N or R-N-R sequences occur on time scales on the order of 10^5 to 10^7 years, making the geomagnetic polarity time scale a very powerful chronological tool.

Many rocks acquire remanent magnetizations when they form. These magnetizations are usually parallel to the direction of the ambient magnetic field at the time. Magnetic minerals in igneous rocks, such as basalt or gabbro, acquire a thermoremanent magnetization (TRM) as they cool below the Curie point T_C of the mineral [T_C for magnetite, Fe_3O_4, is 1076°F (580°C) and for hematite, Fe_2O_3, 1238°F (670°C)]. This equilibrium TRM is later "frozen" in the rock at the blocking temperature T_B, whereupon the magnetization vector is no longer in equilibrium with the Earth's magnetic field and can theoretically remain stable for periods longer than 10^{11} years. Blocking temperatures for many magnetite-bearing rocks range from 930 to 1080°F (500 to 580°C). If a preexisting rock is later heated to temperatures near or exceeding T_B, it may acquire a new magnetization upon cooling. *See* CURIE TEMPERATURE; IGNEOUS ROCKS.

When sediment accumulates in a basin, magnetic grains sink through the water column and accumulate at the sediment-water interface together with other sedimentary grains. There, they may become aligned with the ambient magnetic field, and a detrital remanent magnetization (DRM) is the result. As sediment is transformed into sedimentary rock by diagenesis and lithification, a previously acquired DRM can be modified by compaction and dewatering, and an additional chemical remanent magnetization (CRM) may be acquired by the rock. CRM is acquired by the formation of new magnetic minerals as they grow through a critical grain size (about 0.2 micrometer for magnetite, Fe_2O_3). Chemical remanent magnetization may also be acquired at higher temperatures during alteration of plutonic, volcanic, and metamorphic rocks. *See* SEDIMENTARY ROCKS.

An original TRM, DRM, or CRM acquired by a rock when it formed may be partially or completely overprinted by a younger TRM or CRM in a subsequent geologic event such as metamorphism or hydrothermal alteration. Rocks may also be overprinted by viscous remanent magnetization (VRM), the statistical thermal realignment of the magnetization of smaller grains parallel to a younger field direction. At the surface, lightning strikes create large electric currents that can impart a spurious isothermal remanent magnetization (IRM). Various demagnetization techniques are employed to selectively remove these undesired magnetizations, isolating the older magnetizations residing in the rock. *See* GEOMAGNETISM. Michael McWilliams

Paleontology The study of animal history as recorded by fossil remains. The fossil record includes a very diverse class of objects ranging from molds of microscopic bacteria in rocks more than 3×10^9 years old to unaltered bones of fossil humans in ice-age gravel beds formed only a few thousand years ago. Quality of preservation ranges from the occasional occurrence of soft parts (skin and feathers, for example) to barely decipherable impressions made by shells in soft mud that later hardened to rock. *See* FOSSIL; MICROPALEONTOLOGY.

The most common fossils are hard parts of various animal groups. Thus the fossil record is not an accurate account of the complete spectrum of ancient life but is biased in overrepresenting those forms with shells or skeletons. Fossilized worms are extremely rare, but it is not valid to make the supposition that worms were any less common in the geologic past than they are now. *See* EDIACARAN BIOTA.

The data of paleontology consist not only of the parts of organisms but also of records of their activities: tracks, trails, and burrows. Even chemical compounds formed only by organisms can, if extracted from ancient rocks, be considered as part of the fossil record. Artifacts made by people, however, are not termed fossils, for these constitute the data of the related science of archeology, the study of human civilizations. *See* ARCHEOLOGY; PALEOBIOCHEMISTRY.

Paleontology lies on the boundary between two disciplines, biology and geology. *See* BIOLOGY; GEOLOGY.

Geological aspects. A major task of any historical science, such as geology, is to arrange events in a time sequence and to describe them as fully as possible.

Fossils only tell that a rock is older or younger than another; they do not give absolute age. The decay of radioactive minerals may provide an age in years, but this method is expensive and time-consuming, and cannot always be applied since most rocks lack suitable radioactive minerals. Correlation by fossils remains the standard method for comparing ages of events in different areas. *See* INDEX FOSSIL; STRATIGRAPHY.

The physical appearance and climate of the Earth during a given period of the geologic past can be described from compilation and analysis of the data which is obtained through studies of the habitats of extant fauna, the geographic distribution of fossils, and the climatic preferences of ancient forms of life. *See* PALEOCLIMATOLOGY; PALEOECOLOGY; PALEOGEOGRAPHY.

Biological aspects. The most fundamental fact of paleontology is that organisms have changed throughout earth history and that each geological period has had its characteristic forms of life. An evolutionist has two major interests: first, to know how the process of evolution works; this is accomplished by studying the genetics and population structure of modern organisms; second, to reconstruct the events produced by this process, that is, to trace the history of life. Any modern animal group is merely a stage, frozen at one moment in time, of a dynamic, evolving lineage. Fossils give the only direct evidence of previous stages in these lineages. Horses and rhinoceroses, for example, are very different animals today, but the fossil history of both groups is traced to a single ancestral species that lived early in the Cenozoic Era. From such evidence, a tree of life can be constructed whereby the relationships among organisms can be understood. *See* ANIMAL EVOLUTION.

Stephen J. Gould

Paleopathology

Paleopathology The study of ancient diseases and their origins. Paleopathology is especially important in the understanding of the origins, prevalence, and spread of infectious diseases, including how humans have contributed to the spread of disease and how they can overcome it. *See* EPIDEMIOLOGY; INFECTIOUS DISEASE; PATHOLOGY.

Hypothesis testing of populations has contributed to the field of paleopathology, as has application of macroscopic (visual) examination, routine x-ray, computerized tomography (CAT) scans, magnetic resonance imaging (MRI), electron microscopy, and immunologic, chemical, and mass spectrophotometry techniques to skeletons, soft tissue, and even scat (animal droppings). *See* COMPUTERIZED TOMOGRAPHY; MAGNETIC RESONANCE; X-RAY DIFFRACTION.

The scientific method in paleopathology is based upon comparison of archeologic or paleontologic findings with individuals documented to have the disease. To this end the following basic tenets are observed: (1) Tissue must be adequately preserved to allow recognition of disease and distinguish possible pseudopathology or postdeath artifact. (2) The manifestations of a disease must be sufficiently stable across generations to allow comparison of ancient with modern disease. (3) Analysis of entire skeletons is more accurate than analysis of isolated bones. (4) Analysis of afflicted populations (paleoepidemiology) is more accurate than analysis of isolated skeletons.

The range of diagnostic methods used in paleopathology is extensive. Skeletal remains are visually examined to identify occurrence and nature of alterations, mapping their skeletal distribution. Internal structure can then be assessed, preferably by a nondestructive technique. Even fossils are not simply casts of external surfaces, but have a visualizable internal structure.

Mummies provide an additional source of information. Rehydration of mummy tissue allows standard soft tissue histology, providing information often transcending that available through study of bones. Anthropologic study of artifacts such as daggers that sometimes accompany mummies and skeletons has also contributed to the understanding of ancient lifestyles and the diseases which impacted them. *See* HISTOLOGY.

Some of the diseases recognized in living individuals for which evidence has also been found in ancient life forms include tuberculosis, leprosy, arthritis, cancer, and various parasitic diseases. *See* ARTHRITIS; CANCER (MEDICINE); LEPROSY; PARASITOLOGY; TUBERCULOSIS.

Bruce M. Rothschild

Paleoseismology

Paleoseismology The study of geological evidence for past earthquakes. This scientific discipline has contributed greatly to modern understanding of the nature of earthquakes. The patterns of earthquakes, in both space and time, evolve over centuries and millennia and cannot be discovered by modern instruments. Knowledge of these patterns is important for understanding the physics of earthquakes and for forecasting future destructive earthquakes.

In certain natural environments, the features related to ancient earthquakes are preserved in the landforms and superficial layers of the Earth's surface. Geologists use this paleoseismological evidence to extend the short historical and instrumental record of earthquakes into ancient centuries and millennia. Such paleoseismological studies have clarified the earthquake record of many parts of the world, including the midcontinent and east coast of the United States, northern Africa, southern Europe, China, Japan, Indonesia, and New Zealand.

The geological preservation of ancient earthquakes has enabled scientists to compare modern earthquakes with those of the past. In 1983, for example, a sparsely populated region of Idaho was struck by a magnitude-7.3 earthquake. Subsequent investigations revealed a fresh, 30-km-long (18-mi) fault scarp running along the western base of the lofty Lost River Range. Inspection of the fresh escarpment revealed that it is surmounted by a more subdued, vegetated escarpment of nearly identical length and height. Excavations across this ancient fault scarp showed that it had formed about 5000 years earlier during an event very similar to the 1983 earthquake. This is one of several examples of what paleoseismologists call a characteristic earthquake: apparently, some earthquakes are nearly identical repetitions of their predecessors. *See* EARTHQUAKE; SEISMOLOGY.

Kerry Sieh

Paleosol

Paleosol A soil of the past, that is, a fossil soil. Paleosols are most easily recognized when they are buried by sediments. They also include surface profiles that are thought to have formed under very different conditions from those now prevailing, such as the deeply weathered tropical soils of Tertiary geological age that are widely exposed in desert regions of Africa and Australia. Such profiles are generally known as relict paleosols. Those that can be shown to have been buried and then uncovered by erosion are known as exhumed paleosols. The main problem in defining the term paleosol comes from defining what is meant by soil, a term that has very different meanings for agronomists, engineers, geologists, and soil scientists. Soil can be considered distinct from sediment in that it forms in place, but soil need not necessarily include traces of life. At its most general level, soil is material forming the surface of a planet or similar body and altered in place from its parent material by physical, chemical, or biological processes. *See* SOIL.

Paleosols are especially abundant in volcanic, alluvial, and eolian sedimentary sequences. Along with the fossils, sedimentary structures, and volcanic rocks found in such deposits, paleosols provide an additional line of evidence for ancient environments during times between eruptions and depositional events. *See* PALEOCLIMATOLOGY; SEDIMENTOLOGY.

Gregory J. Retallack

Paleozoic A major division of time in geologic history, extending from about 540 to 250 million years ago (Ma). It is the earliest era in which significant numbers of shelly fossils are found, and Paleozoic strata were among the first to be studied in detail for their biostratigraphic significance.

The Paleozoic Era is divided into six systems; from oldest to youngest they are Cambrian, Ordovician, Silurian, Devonian, Carboniferous, and Permian. The Carboniferous is subdivided into two subsystems, the Mississippian and the Pennsylvanian which, in North America, are considered systems by many geologists. The Silurian and Devonian systems are closer to international standardization than others; all the series and stage names and lower boundaries have been agreed upon, and most have been accepted. *See* CAMBRIAN; CARBONIFEROUS; DEVONIAN; ORDOVICIAN; PERMIAN; SILURIAN.

Because Alpine and Appalachian mountain chains were among the first studied in detail, orogenies were first named there. In eastern North America, mountain-building effects during the early Paleozoic were ascribed to the Taconic orogeny (Middle and Late Ordovician); middle Paleozoic events were assigned to the Acadian orogeny (Middle and Late Devonian); and late Paleozoic movements were called Appalachian (more accurately Alleghenian) for Permian and, perhaps, Triassic events. *See* DATING METHODS; ISOTOPE; OROGENY; PLATE TECTONICS; UNCONFORMITY.

The major changes in lithofacies during the Paleozoic were also effected by biotic evolution through the era. Limestone facies became more abundant and more diversified in the shallow warm seas as calcium-fixing organisms became more diverse and more widespread. Sediment input from the land was modified as plants moved from the seas to the low coastal plains and, eventually, to the higher ground during the Devonian. Primitive vertebrates evolved during the Cambro-Ordovician, but true fishes and sharks did not flourish until the Devonian. Amphibians invaded the land during the Late Devonian and early Carboniferous at about the same time that major forests began to populate the terrestrial realm. These changes produced an entirely new suite of nonmarine facies related to coal formation, and the Carboniferous was a time of formation of major coal basins on all continental plates.

Major cycles of cold and warm climates were overlaid on depositional and evolutionary patterns, producing periods of continental glaciation when large amounts of the Earth's water were tied up in ice during the Late Ordovician, the Late Devonian, and the late Permian. During the earliest and latest of these periods, icesheets were concentrated in the Southern Hemisphere on a single large Paleozoic continental mass—Gondwana. *See* DEPOSITIONAL SYSTEMS AND ENVIRONMENTS; FACIES (GEOLOGY); PALEOCLIMATOLOGY.

The Paleozoic featured a single southern landmass (Gondwana) for most of the era. This megaplate moved relatively sedately northward during this entire time interval (540–250 Ma) and always contained the magnetic and geographic south poles. Consequently, many of the facies and biologic provinces in the Gondwanan region were influenced by the cooler marine realms and continental and mountain glaciers in nearly every Paleozoic period. Most of the tectonic action that produced major periods of collision, mountain building, carbonate platform building, back-arc fringing troughs with their distinctive faunas and lithofaces, and formation of coal basins and evaporites took place in the Northern Hemisphere. These pulsations produced combinations of Laurentian (North American), Euro-Baltic, Uralian, Siberian, and Chinese plates a various times during the Paleozoic; and these combined units, in turn, moved slowly across the latitudes, producing climatic change; lithofacies changed in response to both the climate and the plate tectonics. *See* PALEOGEOGRAPHY.

There were fewer and simpler life forms in the Cambrian—often termed the Age of Trilobites. All groups of invertebrates and plants became more numerous through geologic time. For example, 7 major invertebrate animal groups at the beginning of the Cambrian doubled to 14 by the end of the period, 20 by the end of the Ordovician, 23 at the end of the Devonian, and 25 at the end of the Paleozoic. The pattern for plant diversification, although starting later, is similar. Three simple plant groups became 5 by the end of the Silurian, 7 at the end of the Devonian, and 13 at the end of the Paleozoic. The vertebrates also diversified very slowly. From one or two groups in the Cambro-Ordovician (conodonts are now considered primitive vertebrates), the number of major kinds rose to 6 at the end of the Devonian and 8 at the end of the Paleozoic. *See* BIOGEOGRAPHY; GEOLOGIC TIME SCALE; INDEX FOSSIL; PALEOECOLOGY; STRATIGRAPHY; TRILOBITA.

J. Thomas Dutro, Jr.

Palladium A chemical element, Pd, atomic number 46. A transition metal, palladium occurs in combination with platinum (Pt) and is the second most abundant platinum-group metal, accounting for 38% of the reserves of these metals. *See* PERIODIC TABLE; PLATINUM.

Palladium is soft and ductile and can be fabricated into wire and sheet. The metal forms ductile alloys with a broad range of elements. Palladium is not tarnished by dry or moist air at ordinary temperatures. At temperatures from 350 to 790°C (660 to 1450°F) a thin protective oxide forms in air, but at temperatures from 790°C (1450°F) this film decomposes by oxygen loss, leaving the bright metal. In the presence of industrial sulfur-containing gases a slight brownish tarnish develops; however, alloying palladium with small amounts of iridium or rhodium prevents this action. Important physical properties of palladium are given in the table. *See* ALLOY; METAL.

At room temperature, palladium is resistant to nonoxidizing acids such as sulfuric acid, hydrochloric acid, hydrofluoric acid, and acetic acid. The metal is attacked by nitric acid, and a mixture of nitric acid and hydrochloric acid is a solvent for the metal. Palladium is also attacked by moist chlorine (Cl) and bromine (Br). *See* ELECTROPLATING OF METALS; NONSTOICHIOMETRIC COMPOUNDS.

The major applications of palladium are in the electronics industry, where it is used as an alloy with silver for electrical contacts or in pastes in miniature solid-state devices and in integrated circuits. Palladium is widely used in dentistry as a substitute for gold. Other consumer applications are in automobile exhaust catalysts and jewelry. *See* INTEGRATED CIRCUITS.

Physical properties of palladium

Property	Value
Atomic weight	106.4
Naturally occurring isotopes (percent abundance)	102 (0.96) 104 (10.97) 105 (22.23) 106 (27.33) 108 (26.71) 110 (11.81)
Crystal structure	Face-centered cubic
Thermal neutron capture cross section, barns	8.0
Density at 25°C (77°F), g/cm^3	12.01
Melting point, °C (°F)	1554 (2829)
Boiling point, °C (°F)	2900 (5300)
Specific heat at 0°C (32°F), cal/g	0.0584
Thermal conductivity,(cal·cm)(cm^2·s·°C)	0.18
Linear coefficient of thermal expansion, (μin./in./)/°C	11.6
Electrical resistivity at 0°C (32°F), $\mu\Omega$-cm	9.93
Young's modulus, lb/in.^{2}, static, at 20°C (68°F)	16.7×10^6
Atomic radius in metal, nm	0.1375
Ionization potential, eV	8.33
Binding energy, eV	3.91
Pauling electronegativity	2.2
Oxidation potential, V	−0.92

Palladium supported on carbon or alumina is used as a catalyst for hydrogenation and dehydrogenation in both liquid- and gas-phase reactions. Palladium finds widespread use in catalysis because it is frequently very active under ambient conditions, and it can yield very high selectivities. Palladium catalyzes the reaction of hydrogen with oxygen to give water. Palladium also catalyzes isomerization and fragmentation reactions. *See* CATALYSIS.

Halides of divalent palladium can be used as homogeneous catalysts for the oxidation of olefins (Wacker process). This requires water for the oxygen transfer step, and a copper salt to reoxidize the palladium back to its divalent state to complete the catalytic cycle. *See* HOMOGENEOUS CATALYSIS; TRANSITION ELEMENTS. D. Max Roundhill

Palpigradi An order of rare arachnids comprising 21 known species from tropical and warm temperate regions. American species occur in Texas and California. All are minute, whitish, eyeless animals, varying from 0.027 to 0.112 in. (0.68 to 2.8 mm) in length, that live under stones, in caves, and in other moist, dark places. The elongate body terminates in a slender, multisegmented flagellum set with setae. In a curious reversal of function, the pedipalps, the second pair of head appendages, serve as walking legs. The first pair of true legs, longer than the others and set with sensory setae, has been converted to tactile appendages which are vibrated constantly to test the substratum. *See* ARACHNIDA. Willis J. Gertsch

Palynology The study of pollen grains and spores, both extant and extinct, as well as other organic microfossils. Although the origin of the discipline dates back to the seventeenth century, when modern pollen was first examined microscopically, the term palynology was not coined until 1944. The term palynology is used by both geologists and biologists. Palynologists use a range of sophisticated methodologies and instruments in studying both paleopalynological and neopalynological problems, but the utilization of modern microscopy is fundamental in both subdisciplines.

Palynologists study microscopic bodies generally known as palynomorphs. These include an array of organic structures, each consisting of a highly resistant wall component. Examples include acritarchs and chitinozoans (microfossils with unknown affinities), foraminiferans (protists), scolecodonts (tooth and mouth parts of marine annelid worms), fungal spores, dinoflagellates, algal spores, and spores and pollen grains of land plants. Spores and pollen grains are reproductive structures and play a paramount role in the life history of land plants. *See* MICROPALEONTOLOGY; PLANT REPRODUCTION; POLLEN.

Neopalynology focuses on several subdisciplines, including taxonomy, genetics, and evolution; development, functional morphology, and pollination; aeropalynology; and melissopalynology. Aeropalynology is the study of pollen grains and spores that are dispersed into the atmosphere. Melissopalynologists analyze bee pollen loads and the pollen component within honeys. *See* POLLINATION.

The main fields of study within paleopalynology are paleobotany; past vegetation and climate reconstruction; geochronology and biostratigraphy; and petroleum and natural gas exploration. *See* FOSSIL; INDEX FOSSIL; PALEOBOTANY; POSTGLACIAL VEGETATION AND CLIMATE. Jeffrey M. Osborn

Pancreas A composite gland in most vertebrates, containing both exocrine cells—which produce and secrete enzymes involved in digestion—and endocrine cells, arranged in separate islets which elaborate at least two distinct hormones, insulin and glucagon, both of which play a role in the regulation of metabolism, and particularly of carbohydrate metabolism. *See* CARBOHYDRATE METABOLISM; PANCREAS DISORDERS.

The pancreas is a more or less developed gland connected with the duodenum. It can be considered as an organ which is characteristic of vertebrates. The pancreas of mammals shows large variations. The extremes are the unique, massive pancreas of humans, and the richly branched organ of the rabbit. Usually, the main duct, the duct of Wirsung, opens into the duodenum very close to the hepatic duct. In humans, the pancreas weighs about 70 g (2.5 oz). It can be divided into head, body, and tail. Accessory pancreases are frequently found anywhere along the small intestine, in the wall of the stomach, and in Meckel's diverticulum. *See* DIGESTIVE SYSTEM.

The pancreatic parenchyma is formed by two elements; one is the exocrine tissue of which the secretion empties into the pancreatic ducts and ultimately into the duodenum; the latter is the endocrine islands whose secretions enter the blood vessels. The acini are the part of the exocrine pancreas that produces the enzymes of the pancreatic juice. The digestive enzymes of the pancreatic juice are peptidases, lipases, esterases, amylase, and nucleases. The endocrine portion of the pancreas consists of cellular masses called the islets (or islands) of Langerhans scattered throughout the exocrine portion. In the adult human their number is estimated to range from 200,000 to 2,300,000 and their diameter varies from 30 to 300 micrometers. The islets are demarcated from the surrounding tissue by an irregular, thin layer of connective tissue. The two main cell types are the alpha or A cells and the beta or B cells, which exist usually in the proportion of 1:4. Another cell type, less abundant (about 5%), is the delta or D cell. S. Haumont

The alpha and beta cells in the islets of Langerhans are the sources of two hormones, insulin from the beta, and glucagon, also known as the hyperglycemic factor, from the alpha cells. The former is a hormone which influences carbohydrate metabolism, enabling the organism to utilize sugar. The latter accelerates the conversion of liver glycogen, the form in which carbohydrate is stored in liver and muscles until needed by the body, into glucose, the principal sugar used by the body to meet its energy requirements. Thus, glucagon elevates the blood sugar level, and its effects are the opposite of those of insulin, so that the two hormones together maintain the sugar metabolism of the body in balance. When the level of sugar in the blood becomes too low, the secretion of glucagon is stimulated. *See* DIABETES; GLUCAGON; INSULIN. Choh Hao Li

Pancreas disorders The pancreas is affected by a variety of congenital and acquired diseases. Because of the dual functional role, the diseases of the exocrine portion of the pancreas will be separated from the endocrine lesions in this discussion.

The most frequent congenital lesion of the pancreas is more appropriately designated as a developmental abnormality—ectopic or aberrant pancreas. Ectopic pancreas can be found anywhere within the gastrointestinal tract, but is more frequent in the stomach and duodenum.

Cystic fibrosis (mucoviscidosis) is a systemic disease in which mucus secretion is altered so that a viscid mucus is produced. The disease is inherited as a mendelian recessive. Cystic fibrosis affects all exocrine glands, including the acinar portion of the pancreas. Production of altered mucus leads to dilation of the exocrine ducts (cystic), destruction of acinar tissue, and replacement of the destroyed tissue by fibrous connective tissue (fibrosis). The islets are not affected by this disease. Elevation in secretion of sodium and chloride in sweat is also common.

Acute hemorrhagic pancreatitis is a serious disease of unknown etiology which causes sudden liberation of activated pancreatic enzymes that digest the pancreatic parenchyma. The digestive process leads to dissolution of fat and production of calcium soaps. In addition, rupture of pancreatic vessels occurs with resultant hemorrhage and shock. This disease is associated with biliary tract disease, especially gallstones (cholelithiasis), alcoholism, hyperlipidemia, and hypercalcemia. *See* ALCOHOLISM; GALLBLADDER DISORDERS.

Chronic pancreatitis, perhaps better designated chronic relapsing pancreatitis, is a condition in which recurrent episodes of pancreatitis occur without the production of symptoms or with the production of mild symptoms. Destruction of the pancreatic tissue, with repair by fibrosis, calcification, and cyst formation, is frequent.

Diabetes mellitus is the principal disease associated with the endocrine portion of the pancreas. Two clinical forms of the disease are recognized—insulin-dependent diabetes mellitus and non-insulin-dependent diabetes mellitus. While many factors are involved in the causation of this disease, basically the disease is a result of the failure of the beta cells of the pancreas to produce appropriate kinds and amounts of insulin to meet metabolic needs. *See* DIABETES.

Tumors of the pancreas can be either benign or malignant. They affect both the endocrine and exocrine portions of the pancreas. Benign tumors of the exocrine pancreas are extremely rare. Malignant tumors of the exocrine pancreas arise most frequently from the pancreatic ducts. Acinar carcinomas also exist but are very rare. Exocrine pancreatic carcinomas are very malignant tumors. Islet cell lesions are quite rare but may be associated with increased hormone production. The tumors can be single or multiple, benign or malignant, and they can form anywhere in the pancreas. Hyperfunction of the islets of Langerhans can result in three distinct clinical syndromes: hyperinsulinism and hypoglycemia, the Zollinger-Ellison syndrome (gastrinoma), and multiple endocrine neoplasia. *See* ONCOLOGY; PANCREAS.

H. Thomas Norris

Panda The family Ailuropodidae contains two species of pandas—the giant panda (*Ailuropoda melanoleuca*) and the lesser or red panda (*Ailurus fulgens*). Until relatively recently, the giant panda had been classified in the family Ursidae with the bears, and the red panda had been included in the family Procyonidae along with raccoons, ringtails, and coatis.

Giant pandas are an endangered species, surviving only in small isolated populations in China. They inhabit a narrow zone of bamboo forest from 1200 to 3300 m (3800 to 10,600 ft) in elevation. Giant pandas have massive heads and bodies with an unmistakable black and white fur pattern. The body is whitish except for the ears, eye spots, nose, limbs, and shoulders which are black. The limbs are relatively short, and each forepaw possesses a "pseudothumb," or sixth toe, which is an adaptation for stripping bamboo leaves from the stalk. Adults are 150–180 cm (5-6 ft) long with a 10–15 cm (4–6 in.) tail. They weigh 75–110 kg (165–242 lb). Their primary food consists of the sprouts, stems, and leaves of bamboo, although they occasionally consume bulbs, grasses, insects, and rodents. Reproduction in captivity is poor, although artificial insemination and hand raising of young are increasingly successful.

Red pandas have long, soft, thick rusty to chestnut-brown fur on their dorsal surface. Small dark-colored patches are present beneath each eye, and the muzzle, lips, cheeks, and edges of the ears are white. The backs of the ears, the limbs, and the underparts are dark reddish-brown to black. The claws are sharp and semiretractile. The bushy nonprehensile tail has 12 inconspicuous dark brown rings on a reddish background. The head is rounded with a short snout, and the ears are large and pointed. Adults are 79–110 cm (31–44 in.) in total length, including a 28–48 cm (11–19 in.) tail. They weigh 3–6 kg (7–13 lb). This species inhabits the mountain forests of China and the southeastern mountainsides of the Himalayas ranging in elevation from 2200 to 4800 m (7000 to 15,000 ft). Preferred habitat consists of giant rhododendron, oak, and bamboo forests. The principal food is bamboo, but the diet is supplemented by grasses, acorns, fruits, roots, lichens, insects, eggs, young birds, and small rodents. A "pseudothumb" is present to facilitate the handling of bamboo leaves and stems. *See* CARNIVORA; MAMMALIA. Donald W. Linzey

Pandanales An order of monocotyledons, the composition of which only recently has been revealed by deoxyribonucleic acid (DNA) sequence studies of four genes. Included are four families. Pandanaceae (800 species; the screw pine family), Cyclanthaceae (230 species; the Panama hat family), Stemonaceae (35 species), and Velloziaceae (200 species). Pandanaceae are often lianas or large herbs from the Old World; Cyclanthaceae are herbs or lianas from the New World tropics; Stemonaceae are herbs or lianas of the Old World (but with one species in the southeastern United States); and Velloziaceae are herbs or small shrubs of Africa and particularly South America (with one genus in southwest China). Cyclanthaceae, Pandanaceae, and Stemonaceae have flower parts in twos or fours, which is unusual among monocotyledons, in which threes are most common.

Several species in Pandanales are economically important. The leaves of Pandanaceae are fibrous and used for making rope and roofing, and the fruits are eaten in many areas. Cyclanthaceae leaves are fibrous and have similar uses, including the manufacture of Panama hats. *See* ARECIDAE; LILIOPSIDA; MAGNOLIOPHYTA; PLANT KINGDOM. Mark W. Chase

Panderichthys A fossil lobe-finned (sarcopterygian) fish from rock strata of the Baltic region dating to the Middle–Late Devonian Period (about 372–368 million years ago), which provides key evidence about the evolutionary transition from fish to land vertebrates (tetrapods). As the most tetrapod-like fish known, it reveals the earliest steps of the transformation that adapted the fish body plan to terrestrial life. *See* SARCOPTERYGII; TETRAPODA.

Panderichthys fits into the tetrapod phylogeny (family tree) above the osteolepiform fishes and immediately below the first tetrapods such as *Ichthyostega* and *Acanthostega*. It shares certain tetrapod-like characteristics with the osteolepiforms, such as a pair of internal nostrils (choanae) on the palate, and paired fin skeletons that contain recognizable equivalents of the major tetrapod limb bones, such as the humerus and femur. However, whereas osteolepiforms still had the body form and fins of "normal" open-water fishes, *Panderichthys* looked quite different (see illustration): it had a crocodile-like head with a long snout and eyes raised on top under distinct bony "eyebrows"; a body that is slightly flattened from top to bottom; and a tail carrying only a simple fin fringe, rather than the separate dorsal, caudal, and anal fins of osteolepiforms. All these features are shared with early tetrapods. The internal anatomy reveals additional tetrapod features that are not present in osteolepiforms. However, unlike tetrapods, *Panderichthys* retained well-developed gills, and its paired appendages were still fins with fin rays, not limbs with toes. *See* ICHTHYOSTEGA; OSTEOLEPIFORMES.

Panderichthys can be envisioned as a specialist shallow-water predator of deltaic environments, which used its fins to crawl through water too shallow to support its body, and perhaps also

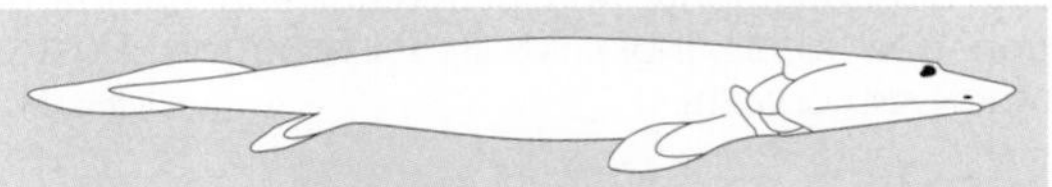

***Panderichthys rhombolepis*, whole-body reconstruction based on specimens from Lode, Latvia. Total length approximately 1.3 m (4.3 ft). (*Reprinted by permission from Nature, P. E. Ahlberg and A. R. Milner, The origin and early diversification of tetrapods, 368:507–514, © 1994, Macmillan Publishers Ltd.*)**

for short excursions over land from one channel to the next. It is possible that it exploited the intertidal zone, emerging to feed on fishes stranded on mudflats and in tidal pools. *See* ANIMAL EVOLUTION; DEVONIAN; FOSSIL. Per E. Ahlberg

Panel heating and cooling A system in which the heat-emitting and heat-absorbing means is the surface of the ceiling, floor, or wall panels of the space which is to be environmentally conditioned. The heating or cooling medium may be air, water, or other fluid circulated in air spaces, conduits, or pipes within or attached to the panel structure. For heating only, electric current may flow through resistors in or on the panels. *See* ELECTRIC HEATING.

Heat energy is transmitted from a warmer to a cooler mass by conduction, convection, and radiation. The output from heating surfaces comprises both radiation and convection components in varying proportions. In panel heating systems, especially the ceiling type, the radiation component predominates. *See* RADIANT HEATING.

When a panel system is used for cooling, the dew-point temperature of the ambient air must remain below the surface temperature of the heat-absorbing panels to avoid condensation of moisture on the panels. Panel cooling effectively prevents the disagreeable feeling of cold air blown against the body and minimizes the occurrence of summer colds. *See* COMFORT HEATING; HOT-WATER HEATING SYSTEM. Erwin L. Weber; Richard Koral

Pantodonta An extinct order of relatively large placental mammals represented by Paleocene-Eocene fossils from western Europe, North America, and eastern Asia. Pantodonts were an early evolutionary experiment in large-bodied herbivory by primitive placental mammals. They first appeared in Asia during the early Paleocene and disappeared during the middle Eocene, leaving no descendants. With the possible exception of the most primitive pantodonts, all were herbivores, and pantodonts were either the largest or among the largest mammals of their time. The adaptive radiation of pantodonts was diverse and encompassed mammals as different as small [1 kg (2 lb) or less in body mass], arboreal herbivores, and large [650 kg (1430 lb)], ground-sloth–like, terrestrial herbivores. *See* EOCENE; PALEOCENE.

Pantodonts are unique among placental mammals in having upper third and fourth premolar tooth crowns in which there are V-shaped crests. All pantodonts were obligate quadrupeds, with four nearly equal-sized limbs. Most pantodonts had large tusks and long, heavy tails. When compared to living mammals, many pantodonts would have looked somewhat like bears, pigs, or small hippos. *See* DENTITION; MAMMALIA.

The most primitive pantodonts are assigned to the family Bemalambdidae. All other pantodonts belong to the suborder Eupantodonta, which diverged early into the superfamilies Pantolambdodontoidea and Pantolambdoidea. The pantolambdodontoids include the Asian family Pantolambdodontidae, the North American family Titanoideidae, and the South American genus *Alcidedorbignyia*. The pantolambdoids include the North American families Pantolambdidae and Barylambdidae and the more cosmopolitan (known from North America and Eurasia) family Coryphodontidae. Spencer G. Lucas

Pantograph A four-bar parallel linkage, with no links fixed, used as a copying device for generating geometrically similar figures, larger or smaller in size, within the limits of the mechanism. In the illustration the curve traced by point T will be similar to that generated by point S. This similarity results because points T and S will always lie on the straight line $\overline{OTS}$; triangles $\overline{OBS}$ and $\overline{TCS}$ are always similar because lengths $\overline{OB}$, $\overline{BS}$, $\overline{CT}$, and $\overline{CS}$ are constant and $\overline{OB}$ is always parallel to $\overline{CT}$. Distance $\overline{OT}$ always maintains a constant proportion to distance $\overline{OS}$ because of the similarity of the above triangles. Numerous

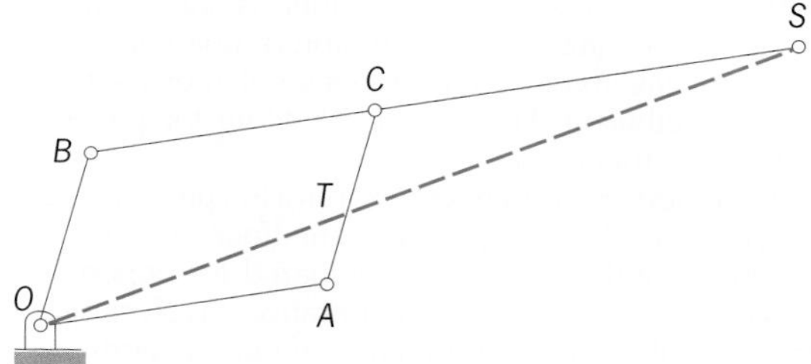

Similar triangles of a pantograph.

modifications of the pantograph as a copying device have been made. *See* FOUR-BAR LINKAGE.

A second use of the pantograph geometry is seen in the collapsible parallel linkage used on electric locomotives and rail cars to keep a current-collector bar or wheel in contact with an overhead wire. Two such congruent linkages in planes parallel to the train's motion are affixed securely on the top of the locomotive with joining horizontal members perpendicular to each other. The uppermost member collects the current, and powerful springs thrust the configuration upward with sufficient pressure normally to make low-resistance contact from wire to collector. Douglas P. Adams

Papaverales An order of flowering plants, division Magnoliophyta (Angiospermae), subclass Magnoliidae of the class Magnoliposida (dicotyledons). The order consists of only two families: Papaveraceae with some 200 species, and Fumariaceae with about 400 species. Within its subclass, the order is marked by its syncarpous gynoecium, parietal placentation, and only two (seldom three) sepals. Most of the species are herbaceous, and many of them contain isoquinoline alkaloids similar to those in the order Ranunculales. The Papaveraceae, with regular flowers, numerous stamens, and a well-developed latex system, include the poppies (*Papaver* and related genera; see illustration), bloodroot (*Sanguinaria*), and celandine (*Chelidonium*). *Papaver somniferum* is the source of opium. The Fumariaceae, with four or six

Oriental poppy (*Papaver orientale*) of the family Papaveraceae and the order Papaverales. (*John H. Gerard, National Audubon Society*)

stamens, irregular flowers that usually have some of the petals spurred or saccate, and no latex system, include the bleeding heart (*Dicentra spectabilis*) and some other common ornamentals. *See* MAGNOLIIDAE; MAGNOLIOPHYTA; MAGNOLIOPSIDA; POPPY; RANUNCULALES. Arthur Cronquist; T. M. Barkley

Paper A flexible web or mat of fibers isolated from wood or other plants materials by the operation of pulping. Nonwovens are webs or mats made from synthetic polymers, such as high-strength polyethylene fibers, that substitute for paper in large envelopes and tote bags.

Paper is made with additives to control the process and modify the properties of the final product. The fibers may be whitened by bleaching, and the fibers are prepared for papermaking by the process of refining. Stock preparation involves removal of dirt from the fiber slurry and mixing of various additives to the pulp prior to papermaking. Papermaking is accomplished by applying a dilute slurry of fibers in water to a continuous wire or screen; the rest of the machine removes water from the fiber mat. The steps can be demonstrated by laboratory handsheet making, which is used for process control.

Although paper has numerous specialized uses in products as diverse as cigarettes, capacitors, and counter tops (resin-impregnated laminates), it is principally used in packaging (~50%), printing (~40%), and sanitary (~7%) applications.

Material of basis weight greater than 200 g/m^2 is classified as paperboard, while lighter material is called paper. Production by weight is about equal for these two classes. Paperboard is used in corrugated boxes; corrugated material consists of top and bottom layers of paperboard called linerboard, separated by fluted corrugating paper. Paperboard also includes chipboard (a solid material used in many cold-cereal boxes, shoe boxes, and the backs of paper tablets) and food containers.

Mechanical pulp is used in newsprint, catalog, and other short-lived papers; they are only moderately white, and yellow quickly with age because the lignin is not removed. A mild bleaching treatment (called brightening) with hydrogen peroxide or sodium dithionite (or both) masks some of the color of the lignin without lignin removal. Paper made with mechanical pulp and coated with clay to improve brightness and gloss is used in 70% of magazines and catalogs, and in some enamel grades. Bleached chemical pulps are used in higher grades of printing papers used for xerography, typing paper, tablets, and envelopes; these papers are termed uncoated wood-free (meaning free of mechanical pulp). Coated wood-free papers are of high to very high grade and are used in applications such as high-quality magazines and annual reports; they are coated with calcium carbonate, clay, or titanium dioxide.

Like wood, paper is a hygroscopic material; that is, it absorbs water from, and also releases water into, the air. It has an equilibrium moisture content of about 7–9% at room temperature and 50% relative humidity. In low humidities, paper is brittle; in high humidities, it has poor strength properties.

The heaviest grades of papers, such as chipboard, are made on multiformer (cylinder) machines that form three to eight layers of fiber mats. These fiber mats are combined prior to pressing and drying. The lightest grades of paper, tissues, cannot withstand numerous felt transfers and are dried on very large Yankee dryers.

Paper may be smoothed against a series of rolls made from metal or rubbery material to impart smoothness or gloss. Paper may also be coated with a paintlike material to give it high brightness and gloss. In addition, numerous other converting operations may be performed on paper. Christopher J. Biermann

Paprika A type of pepper, *Capsicum annuum* (order Polemoniales), with nonpungent flesh, grown for its long red fruit. It is of American origin, but is most popular in Hungary and adjacent countries. Seeds are removed from the mature fruit, and the flesh is dried and ground to prepare the dry condiment commonly referred to as paprika. California is the only important producing state in the United States. *See* PEPPER; SOLANALES.

H. John Carew

Parablastoidea A small extinct class of relatively advanced blastozoan echinoderms containing five genera (three named), ranging from the middle Early Ordovician to the early Late Ordovician in eastern Canada; northeastern, eastern, south-central, and western United States; south and north Wales; and near St. Petersburg, western Russia. Parablastoids have a bud-shaped theca or body with well-developed pentameral symmetry. A stem with one-piece columnals attached the theca to the sea floor, suggesting that parablastoids were attached, medium- to high-level, suspension feeders. Some parablastoids are found in bank or reef deposits, suggesting they may have been adapted for rough-water conditions. Although they converged on blastoids in thecal design and way of life, parablastoids had differences in their plating, ambulacra, and respiratory structures that indicate a separate origin and evolutionary history. This justifies assigning parablastoids and blastoids to different blastozoan classes. *See* BLASTOIDEA; ECHINODERMATA; ORDOVICIAN. James Sprinkle

Parabola A member of the class of curves that are intersections of a plane with a cone of revolution. It is obtained (see illustration) when the cutting plane is parallel to an element of the cone. *See* CONIC SECTION.

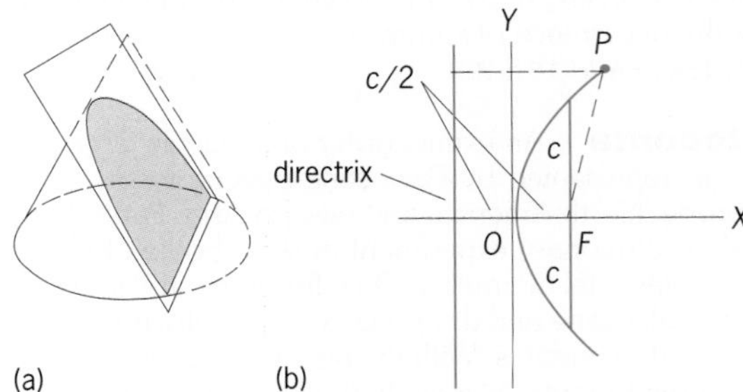

Parabola as (*a*) conic section and (*b*) locus of points.

In analytic geometry the parabola is defined at the locus of points (in a plane) equally distant from the fixed point F (focus) and a fixed line (directrix) not through the point. It is symmetric about the line through F perpendicular to the directrix.

The curve has numerous other properties of interest in both pure and applied mathematics. For example, the trajectory of an artillery shell, assumed to be acted upon only by the force of gravity, is a parabola.

Archimedes found the area bounded by an arc of a parabola and its chord; for example, the area bounded by the parabola $y^2 = 2cx$ and its latus rectum (the chord through F perpendicular to the axis) is $^2/_3c^2$. Leonard M. Blumenthal

Parachute A flexible, lightweight structure, generally intended to retard the passage of an object through the atmosphere by materially increasing the resistive surface. Parachutes have continued to be the simplest and cheapest devices for the deceleration of payloads, people, and vehicles since their first recorded use in 1797. Comprising cloth and suspension lines, their construction is far simpler than that of aircraft. As a result, parachute construction and design is a fairly mature art. However, their very "softness" makes their aerodynamics much more complicated.

Types. Parachutes come in two basic shapes: a near-hemispherical cup or a wing. Most are made of nylon fabric. When fully inflated, some hemispherical parachutes have a slight conical shape, while in others slots, vents, or concentric gaps are cut to enhance stability. Winglike parachutes, also commonly known as parafoils or ram-air parachutes, are made of bags or cells sewn together to form a wing. They maintain this shape via the wind entering inlets cut on the leading edge of the parachute, thereby pressurizing each cell.

Deployment and inflation stages. All parachutes are packed into a small container prior to actual use, and therefore require hardware that will extract the parachute out of its container and deploy it into the wind. The initial deployment stages of hemispherical parachutes and parafoils are similar. Both begin with the extraction of the bag containing the folded parachute from the harness-container assembly (or vehicle). The extraction is normally carried out with either a lanyard attached to the aircraft (that is, the so-called static line), or with a "pilot" or "extraction" chute, usually a small hemispherical parachute that is previously deployed into the airstream by a latch and ejection spring mechanism. Inflation is typically characterized by several stages whose duration depends on the specific design of the parachute (or parafoil) and on whether the parachute is reefed. Reefing devices are designed to limit the rate of canopy expansion during the early phase of inflation, allowing enough time for the parachute-payload system to reduce its speed with the help of the small but sufficient drag provided by the partially opened parachute.

Technology development. New applications for parachutes constantly require advances in design and packaging. Many current systems have been in use (with minor updates) since the 1950s. Refinements or outright new designs provide more stability while providing the most drag, doing so in harsher environments and at less cost, weight, and pack volume.

The availability of supercomputers has led to the development of computer models that track the interactions of the airflow with the parachute cloth and lines. These models are still approximate in the way that they reproduce fabric crumpling and unsteady turbulent flows, but they are good enough to yield inflation sequences that are very close to reality. Jean Potvin

Paracrinoidea

A medium-sized class of extinct blastozoan echinoderms containing about 17–19 genera, ranging from the late Early Ordovician to the Early Silurian throughout North America and in northern Europe. Paracrinoids have a lens-shaped to globular body or theca, a short often curved stem, and 2–5 ambulacra or erect arms bearing smaller uniserial appendages with a single row of ossicles. Three features distinguish paracrinoids from most other blastozoan echinoderms: (1) they have asymmetrical, uniserial, recumbent ambulacra or erect arms typically with smaller, erect, uniserial appendages branching off the right side looking adorally (toward the mouth) and a covered food groove extending up the left side to the mouth; (2) in many paracrinoids, the summit has shifted so that the anal opening (and not the mouth) lies on the thecal axis opposite the stem facet; and (3) three basal plates, two somewhat larger and one smaller (usually on the left anterior side), form the stem facet at the thecal base.

Paracrinoids were low- to medium-level suspension feeders either held some distance up in the water column by the curved stem or sitting recumbent on the sea floor tethered in place with a short buried stem. A few paracrinoid genera show external growth lines on the larger thecal plates, indicating they were blastozoan and not crinozoan echinoderms. *See* BLASTOIDEA; ECHINODERMATA; EOCRINOIDEA; ORDOVICIAN. James Sprinkle

Paraffin

A term used variously to describe either a waxlike substance or a group of compounds. The former use pertains to the high-boiling residue obtained from certain petroleum crudes. It is recovered by freezing out on a cold drum and is purified by crystallization from methyl ethyl ketone. Paraffin wax is a mixture of 26- to 30-carbon alkane hydrocarbons; it melts at 52–57°C (126–135°F). Microcrystalline wax contains compounds of higher molecular weight and has a melting point as high as 90°C (190°F). The name paraffin was formerly used to designate a group of hydrocarbons—now known as alkanes. *See* ALKANE. Allen L. Hanson

Parainfluenza virus

A member of the genus *Paramyxovirus* of the family Paramyxoviridae which is associated with a variety of respiratory illnesses. The virus particles range in size from 90 to 200 nanometers, agglutinate red blood cells, and (like the influenza viruses) contain a receptor-destroying enzyme. They differ from the influenza viruses in their large size, their possession of the larger ribonucleoprotein helix characteristic of the paramyxoviruses, their tendency to lyse as well as agglutinate erythrocytes, and their generally poor growth in eggs. *See* COMPLEMENT-FIXATION TEST; EMBRYONATED EGG CULTURE; PARAMYXOVIRUS; TISSUE CULTURE.

Four subgroups are known, designated parainfluenza 1, 2, 3, and 4. Types 1, 2, and 3 are distributed throughout the world, but thus far type 4 has been found only in the United States. Parainfluenza 1 and 3 are ubiquitous endemic agents producing infections all through the year. Types 2 and 4 occur more sporadically. With all of the parainfluenza viruses, most primary infections take place early in life. About half of the first infections with parainfluenza 1, about two-thirds of those with parainfluenza 2, and three-fourths of those with parainfluenza 3 produce febrile illnesses. The target organ of type 3 is the lower respiratory tract, with first infections frequently resulting in bronchial pneumonia, bronchiolitis, or bronchitis. Type 1 is the chief cause of croup, but the other types have also been incriminated, to the extent that one-half of all cases of croup can be shown to be caused by parainfluenza viruses. *See* ANIMAL VIRUS; INFLUENZA; VIRUS CLASSIFICATION. Joseph L. Melnick; M. E. Reichmann

Parainsecta

A class of hexapod (six-legged) arthropods consisting of the orders Collembola and Protura. Class Parainsecta (also known as Ellipura) is one of several classifications of those hexapods, which, while similar to insects, are sufficiently different that most researchers believe they deserve separate class status.

Collembola and Protura have frequently been lumped with the Diplura (Entotrophi) into one class (Entognatha) on the basis of their mouthparts, which are inside the head capsule rather than exposed as in the Insecta. However, some recent morphological and paleontological studies support a closer link between insects and Diplura. In addition, there are a number of features found in Protura and Collembola that are not found in Diplura. These include the developmental process: Both Collembola and Protura have epimorphic development (that is, development of the young is completed in the egg). Two unique features of the Parainsecta are the presence of a ventral midline or groove and coxal vesicles (fluid-filled sacs, generally associated with leg bases) that are limited to the first abdominal segment. In the Collembola these vesicles are fused into a single ventral tube. The digestive tract lacks anterior or posterior enlargements, which are often present in Diplura and primitive insects. Other distinguishing features of the Parainsecta are the absence of styli and unpaired pretarsal claws.

Both Collembola and Protura contain very small animals, generally less than 4 mm long. Protura are exclusively, and Collembola primarily, soil and litter inhabitants. Genetic evidence concerning relationships is at present confusing. In addition, differences of interpretation concerning the phylogenetic importance of various morphological features leave the status of the Parainsecta ambiguous. *See* ARTHROPODA; DIPLURA; INSECT PHYSIOLOGY; INSECTA; PROTURA. Kenneth Christiansen

Parallax (astronomy)

The apparent angular displacement of a celestial object due to a change in the position of the observer. With a baseline of known length between two observations, the distance to the object can be determined directly.

The rotation of the Earth or the linear separation of two points on its surface can be used to establish distances within the solar system. The parallax determined is scaled to the equatorial radius of the Earth, which is equal to 6378.1366 km (3963.1903 mi). At the mean distance of the Moon, this baseline subtends an angle of 57′03″, and at the mean distance to the Sun it amounts to 8.794143″. This latter distance is defined as the astronomical unit and serves as a measure of distances within the solar system. One astronomical unit is 149,597,870.691 ± 0.006 km (92,955,807.267 ± 0.004 mi) in length, and its high precision results from tracking interplanetary space probes. *See* ASTRONOMICAL UNIT; EARTH ROTATION AND ORBITAL MOTION.

The astronomical unit is the baseline for the measure of stellar parallaxes or distances and ultimately every other distance in the universe outside the solar system. Observations made from the Earth in its orbit on opposite sides of the Sun are scaled to the astronomical unit. The stellar parallax is given in units of arc-seconds and is by definition the reciprocal of the distance in parsecs. One parsec is the distance at which one astronomical unit subtends an angle of one second of arc, and equals 206,264.8 astronomical units or 3.2616 light-years. *See* LIGHT-YEAR; PARSEC.

In 1989 the European Space Agency launched the first space satellite entirely devoted to astrometry, named *HIPPARCOS* (from *High Precision Parallax Collecting Satellite*, and also honoring the ancient astronomer Hipparchus). In 1993 the ESA completed its mission by obtaining positions, parallaxes, and proper motions for 118,322 stars listed in its final catalog. The accuracy averages about 0.0015″ for stars brighter than the completeness limit of about the 8th magnitude, and rises to perhaps twice that amount at the ultimate limiting magnitude of 12. This precision is about equal to that of the best of current ground-based work for the brighter stars and less for faint stars. The advantage of *HIPPARCOS* lies in the much greater numbers of stars, including all the bright ones, with quality parallaxes. The contributions to stellar knowledge from this satellite are enormous. While the *Hipparcos Catalogue* is the preferred source for most stars brighter than 9th magnitude, the fourth edition of the *Yale Parallax Catalogue*, published in 1995, is the only source for parallaxes of stars fainter than that limit, as well as for stars requiring long time duration such as for the determination of the masses of the stars. *See* BINARY STAR; HERTZSPRUNG-RUSSELL DIAGRAM.

In 1990 the National Aeronautics and Space Administration (NASA) launched the *Hubble Space Telescope* (*HST*) with astrometric capabilities in its imaging cameras and fine guidance sensors, the later with accuracy reaching 200 microarcseconds. A relatively small number of high-accuracy parallaxes have been produced for stars of special calibration importance, such as Cepheids, planetary nebulae, and novae. *See* HUBBLE SPACE TELESCOPE.

Distance estimates are possible based on the assumption that the apparent mean space motion of a large number of stars with similar characteristics is a reflection of the peculiar motion of the Sun, derived from previous observations. Two related methods apply; the first equates the mean drift of the stars to the Sun's motion vector. The second requires measurements of radial velocities as well as proper motions, and assumes that the radial velocity distribution at the solar apex and antapex (the directions toward and away from which the Sun is moving, respectively) matches that in the transverse velocities along the direction normal to that of the assumed solar motion. These methods are referred to as secular and statistical parallax, respectively, although the terms are sometimes confused.

Several methods for the establishment of distance are frequently and incorrectly labeled parallax. These methods, such as spectroscopic or photometric parallax, involve a comparison of an object to another of known distance and like luminosity, and scaling the distance as necessary. *See* ASTROMETRY.

William F. van Altena

Parallel circuit An electric circuit in which the elements, branches (elements in series), or components are connected between two points with one of the two ends of each component connected to each point. The illustration shows a simple parallel circuit. In more complicated electric networks one or more branches of the network may be made up of various combinations of series or series-parallel elements. *See* CIRCUIT (ELECTRICITY).

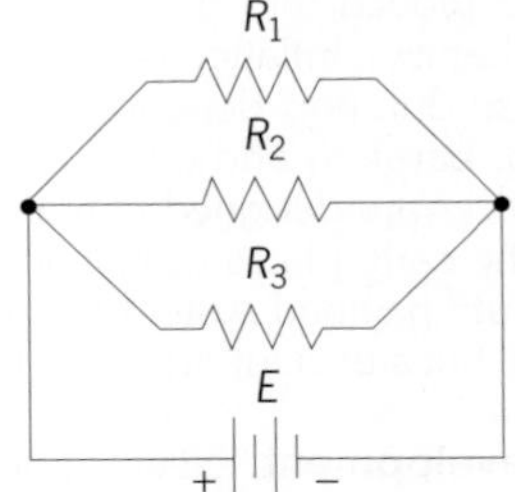

Schematic of a parallel circuit. *E* is a battery; R_1, R_2, and R_3 are resistors.

In a parallel circuit the potential difference (voltage) across each component is the same. However, the current through each branch of the parallel circuit may be different. For example, the lights and outlets in a house are connected in parallel so that each load will have the same voltage (120 volts) but each load may draw a different current (0.5 ampere in a 60-watt lamp and 10 amperes in a toaster).

Clarence F. Goodheart

Paramagnetism A property exhibited by substances which, when placed in a magnetic field, are magnetized parallel to the field to an extent proportional to the field (except at very low temperatures or in extremely large magnetic fields). Paramagnetic materials always have permeabilities greater than 1, but the values are in general not nearly so great as those of ferromagnetic materials. Paramagnetism is of two types, electronic and nuclear.

The following types of substances are paramagnetic:

1. All atoms and molecules which have an odd number of electrons. According to quantum mechanics, such a system cannot have a total spin equal to zero; therefore, each atom or molecule has a net magnetic moment which arises from the electron spin angular momentum. Examples are organic free radicals and gaseous nitric oxide.

2. All free atoms and ions with unfilled inner electron shells and many of these ions when in solids or in solution. Examples are transition, rare-earth, and actinide elements and many of their salts. This includes ferromagnetic and antiferromagnetic materials above their transition temperatures. For a discussion of these materials *See* ANTIFERROMAGNETISM; FERRIMAGNETISM; FERROMAGNETISM.

3. Several miscellaneous compounds including molecular oxygen and organic biradicals.

4. Metals. In this case, the paramagnetism arises from the magnetic moments associated with the spins of the conduction electrons and is called Pauli paramagnetism.

Relatively few substances are paramagnetic. Aside from the Pauli paramagnetism found in metals, the most important paramagnetic effects are found in the compounds of the transition and rare-earth elements which have partially filled 3*d* and 4*f* electron shells respectively.

Electronic paramagnetism arises in a substance if its atoms or molecules possess a net electronic magnetic moment. The magnetization arises because of the tendency of a magnetic field to orient the electronic magnetic moments parallel to itself.

Nuclear paramagnetism arises when there is a net magnetic moment due to the magnetic moments of the nuclei in a substance. Nuclear magnetic moments are about 10^3 times smaller than electron magnetic moments. As a result, nuclear paramagnetism produces effects 10^6 times smaller than electron paramagnetic or diamagnetic effects. *See* DIAMAGNETISM; MAGNETIC RESONANCE; NUCLEAR MOMENTS. Elihu Abrahams; Frederic Keffer

Parameter An auxiliary variable, functions of which give the coordinates of a curve or surface. The coordinates of a curve are functions of one parameter. A curve in 3-space has parametric equations (1).

$$x = f(t) \quad y = g(t) \quad z = h(t) \tag{1}$$

The coordinates of a surface are functions of two parameters, shown in Eqs. (2).

$$x = f(u, v) \quad y = g(u, v) \quad z = h(u, v) \tag{2}$$

An arbitrary constant in an equation is also called a parameter. Variations in the values of the parameter generate a system of equations which may represent a family of curves or surfaces. Such families are called one-parameter, two-parameter, and so on, according to the number of independent parameters. *See* PARAMETRIC EQUATION. Louis Brand

Parametric amplifier A highly sensitive low-noise amplifier for ultrahigh-frequency and microwave radio signals, utilizing as the active element an inductor or capacitor whose reactance is varied periodically at another microwave or ultrahigh frequency. A varactor diode is most commonly used as the variable reactor. Amplification of weak signal waves occurs through a nonlinear modulation or signal-mixing process which produces additional signal waves at other frequencies. This process may provide negative-resistance amplification for the applied signal wave and increased power in one or more of the new frequencies which are generated. *See* VARACTOR.

There are several possible circuit arrangements for obtaining useful parametric amplification. The two most common are the up-converter and the negative-resistance amplifier. In both types, the pump frequency is normally much higher than the input-signal frequency. In the up-converter, a new signal wave is generated at a higher power than the input wave. In the negative-resistance device, negative resistance is obtained for the input-signal frequency, causing an enhancement of signal power at the same frequency. *See* NEGATIVE-RESISTANCE CIRCUITS.

The most important advantage of the parametric amplifier is its low level of noise generation. The parametric amplifier finds its greatest use as the first stage at the input of microwave receivers where the utmost sensitivity is required. Its noise performance has been exceeded only by the maser. Maser amplifiers are normally operated under extreme refrigeration using liquid helium at about 4 K above absolute zero (−452°F). The parametric amplifier does not require such refrigeration but in some cases cooling to very low temperatures has been used to give improved noise performance that is only slightly poorer than the maser. *See* AMPLIFIER; MASER. M. E. Hines

Parametric array A virtual array of sources of sound in a fluid (more fully termed a parametric acoustic array) that is formed by two collinear acoustic waves propagating in the same direction, which interact because of the nonlinearity (parameter) of the fluid medium to generate new acoustic fields. The array is virtual because it is formed in the fluid medium away from the devices responsible for the primary field components. If the primary interacting field is composed of two essentially monochromatic waves, with frequencies ν_1 and ν_2, then the new fields will form at the sum and difference frequencies, $\nu_1 \pm \nu_2$. If ν_1 and ν_2 are slightly different, then the difference frequency, $\nu_1 - \nu_2$, will be quite small. Since absorption of sound in fluids generally increases rapidly with frequency, this difference-frequency wave will outlast the primary- and sum-frequency waves. In addition to suffering less absorption and propagating further, the difference-frequency wave possesses two remarkable properties: most of its energy is concentrated in a narrow anglar sector in the direction of propagation of the primary waves, and it has no sidelobes. Parametric arrays can be used for both directional transmission and directional reception of sound. *See* DIRECTIVITY; SOUND; SOUND ABSORPTION.

Many applications have been pursued underwater because of the special properties of the parametric array. The narrowness of the parametric beam is exploited in several applications requiring high angular resolution, for example, measurement of small patches of the sea surface and sea floor, and volumetric inhomogeneities. Objects lying on or buried in the sea bed are detected by means of parametric arrays. These include, for example, hazardous waste, including munitions dumped in the Baltic Sea and mines in many shallow coastal areas. Techniques of imaging and mapping with parametric sonar are used in these applications, as well as in marine archeology, resulting in the discovery of sunken vessels. The acoustic transmission loss has been measured to characterize sediments. Parametric sonar is used for subbottom profiling and seismic analysis, as in oil and gas exploration. *See* GEOPHYSICAL EXPLORATION; SEA-FLOOR IMAGING; SEISMOLOGY; SONAR.

Beamwidth and bandwidth properties of parametric arrays as transmitters are exploited in underwater communications and in fish-swimbladder-resonance absorption spectroscopy. In the second application, effects of excited, breathing-mode resonances of fish swimbladders on transmitted signals are measured, with the aim of determining the numerical density of such fish. Kenneth G. Foote

Parametric equation A type of mathematical equation used, typically, to represent curves in a plane or in space of three dimensions, In principle, however, there is no limitation to any particular number of dimensions. A parameter is actually an independent variable. In elementary analytic geometry a curve in the xy plane is often studied, in the first instance, as the locus of an equation $y = F(x)$ or $G(x,y) = 0$. The form $y = F(x)$ is not adequate for the complete representation of certain curves, whereas the form $G(x,y) = 0$ may be adequate. The circle $x^2 + y^2 - 16 = 0$ affords an example. But the form $G(x,y) = 0$ is not always convenient. The parametric form $x = f(t)$, $y = g(t)$ is often the most convenient; moreover, it is often the naturally occurring form of representation of the curve. For the circle $x^2 + y^2 - 16 = 0$, one possible parametric representation is $x = 4 \cos t$ and $y = 4 \sin t$.

A pair of equations $x = f(t)$, $y = g(t)$, where f and g are continuous functions defined for some interval of values of t, for example, $a \leq t \leq b$, is said to define a parametric curve. If one thinks of t as time, the equations define the motion of the point (x, y) as t increases from a to b. Clearly the path can cross itself, double back on itself, or the point may even remain motionless.

A parametric surface in space of three dimensions is defined by $x = f(u,v)$, $y = g(u,v)$, $z = h(u,v)$, where f, g, h are continuous functions of the two parameters, u,v. *See* ANALYTIC GEOMETRY; CALCULUS; PARTIAL DIFFERENTIATION. Angus E. Taylor

Paramo A biological community, essentially a grassland, covering extensive high areas in equatorial mountains of the Western Hemisphere. Geographically, paramos are limited to the Northern Andes and adjacent mountains. Paramos occur in alpine regions above timberline and are controlled by a complex of climatic and soil factors peculiar to mountains near the Equator. The richly diverse flora and the fauna of the paramos are adapted to severely cold, mostly wet conditions. Humans have found some paramos suitable for living and use. Harriet G. Barclay

Paramyxovirus A group of viruses that belong to the genus *Paramyxovirus* of the family Paramyxoviridae. The family includes two other genera, *Morbillivirus* and *Pneumovirus*. The genus *Parainfluenza* includes viruses such as mumps and Newcastle disease (in fowl and humans). Measles virus is a member of the genus *Morbillivirus*, and respiratory syncytial virus belongs to the genus *Pneumovirus*. Related members exist in nonhuman species; human measles virus is related to canine distemper and bovine rinderpest virus. Simian and bovine parainfluenza viruses also are known. Like influenza viruses, the paramyxoviruses are ribonucleic acid (RNA)–containing viruses and possess an ether-sensitive lipoprotein envelope. *See* ANIMAL VIRUS; MEASLES; MUMPS; NEWCASTLE DISEASE; PARAINFLUENZA VIRUS; RESPIRATORY SYNCYTIAL VIRUS; VIRUS CLASSIFICATION. Joseph L. Melnick; M. E. Reichmann

Paranoia A mode of thought, feeling, and behavior characterized centrally by false persecutory beliefs, more specifically referred to as paranoidness. Commonly associated with these core persecutory beliefs are properties of suspiciousness, fearfulness, hostility, hypersensitivity, rigidity of conviction, and an exaggerated sense of self-reference. These properties are evident with varying degrees of intensity and duration.

The paranoid mode can be triggered at either biological or psychological levels. Common precipitating biological causes are brain trauma or tumor, thyroid disorder, cerebral arteriosclerosis, and intoxication with certain drugs, including alcohol, amphetamines, cocaine, other psychostimulants, and hallucinogens such as mescaline or lysergic acid diethylamide (LSD). They can produce disordered activity of central dopaminergic and noradrenergic pathways. At the psychological level, triggering causes include false arrest, birth of a deformed child, social isolation, deafness, and intensely humiliating experiences. *See* NORADRENERGIC SYSTEM.

The paranoid mode is resistant to modification by psychotherapeutic or pharmacological methods. Acute psychotic states of paranoidness accompanied by high levels of anxiety are usually responsive to neuroleptic medication. *See* PSYCHOPHARMACOLOGY. Kenneth Mark Colby

Parasexual cycle A series of events, discovered in filamentous fungi, which lead to genetic recombination outside the standard sexual cycle. Processes occurring in certain filamentous fungi were termed the parasexual cycle by Guido Pontecorvo. There are three essential steps in the cycle: heterokaryosis; fusion of unlike haploid nuclei in the heterokaryon to yield heterozygous diploid nuclei; and recombination and segregation at mitosis by two independent processes, mitotic crossing-over and haploidization. The features and operation of the cycle were first demonstrated in the homothallic ascomycete *Aspergillus nidulans*.

Some or all of the features of the parasexual cycle have been demonstrated in many species of fungi. Diploid strains find use in formal genetic analysis and in biochemical research. Mitotic segregation is used for genetic analysis both in species which have a sexual cycle and in imperfect species which have no known sexual cycle. In the former, mitotic analysis is a valuable additional analytical tool; in the latter, it provides the only approach to formal genetic analysis and planned breeding. *See* BREEDING (PLANT); FUNGI; GENETICS; RECOMBINATION (GENETICS); SOMATIC CELL GENETICS. J. Alan Roper

Parasitology The scientific study of parasites and of parasitism. Parasitism is a type of symbiosis and is defined as an intimate association between an organism (parasite) and another, larger species of organism (host) upon which the parasite is metabolically dependent. Implicit in this definition is the concept that the host is harmed, while the parasite benefits from the association. Although technically parasites, pathogenic bacteria and viruses, and fungal and insect parasites of plants are traditionally outside the field of parasitology.

Parasites often cause important diseases of humans and animals. For this reason, parasitology is an active field of study; advances in biotechnology have raised expectations for the development of new drugs, vaccines, and other control measures. However, these expectations are dampened by the inherent complexity of parasites and host-parasite relationships, the entrenchment of parasites and vectors in their environments, and the vast socioeconomic problems in the geographical areas where parasites are most prevalent.

The ecological and physiological relationships between parasites and their hosts constitute some of the most impressive examples of biological adaptation known. Much of classical parasitology has been devoted to the elucidation of one of the most important aspects of host-parasite ecological relationships: the dispersion and transmission of parasites to new hosts.

Parasite life cycles range from simple to highly complex. Simple life cycles (transmission from animal to animal) are direct and horizontal with adaptations that include high reproduction rates, and the production of relatively inactive stages (cysts or eggs) that are resistant to environmental factors such as desiccation, ultraviolet radiation, and extreme temperatures. The infective stages are passively consumed when food or water is contaminated with feces that contain cysts. The cysts are then activated in the gut by cues such as acidity to continue their development. Other direct-transmission parasites, such as hookworms, actively invade new hosts by penetrating the skin. Physiologically more complicated are those life cycles that are direct and vertical, with transmission being from mother to offspring. The main adaptation of the parasite for this type of life cycle is the ability to gain access to the fetus or young animal through the ovaries, placenta, or mammary glands of the mother.

Many parasites have taken advantage of the food chain of free-living animals for transmission to new hosts. During their life cycle, these parasites have intermediate hosts that are the normal prey of their final hosts. Parasites may ascend the food chain by utilizing a succession of progressively larger hosts, a process called paratenesis. *See* FOOD WEB.

Vectors are intermediate hosts that are not eaten by the final host, but rather serve as factories for the production of more parasites and may even carry them to new hosts or to new environments frequented by potential hosts. Blood-sucking arthropods such as mosquitoes and tsetse flies are well-known examples. After acquiring the parasite from an infected host, they move to another host, which they bite and infect. Snails are important vectors for two-host trematodes (flukes), which increase their numbers greatly in the snail by asexual reproduction. The stages that leave the snail may infect second intermediate hosts that are eaten by carnivorous final hosts, may encyst on vegetation that is eaten by herbivorous hosts, or in the case of the blood flukes (schistosomes) may swim to and directly penetrate the final host.

Metabolic dependency is the key to parasitism, and parasites employ many ways to feed off their hosts. The simplest is exhibited by the common intestinal roundworm, *Ascaris*, which consumes the host's intestinal contents. Parasites require from their hosts not only energy-yielding molecules but also basic monomers for macromolecular synthesis and essential cofactors for these synthetic processes. Many examples of the specific absence of key parts of energy-yielding or biosynthetic pathways in parasites are known, and these missing enzymes, cofactors, or intermediates are supplied by the host. Tapeworms are more complex than *Ascaris* in nutritional requirements from the host. They lack a gut, but their surface actively takes up, by facilitated diffusion or active transport, small molecules such as amino acids and simple sugars.

Parasites, by coevolving with their hosts, have the ability to evade the immune response. The best-known evasive tactic is antigenic variation, as found in African trypanosomes, which have a complicated genetic mechanism for producing alternative forms of a glycoprotein that virtually cover the entire parasite. By going through a genetically programmed sequence of variant surface glycoproteins, the trypanosome population in a host stays one step ahead of immunity and is not eliminated. Other possible immune escape mechanisms in parasites have been discovered and probably cooperate to prolong parasite survival.

Parasites are not altogether exempt from the effects of immunity. Rather than completely eliminating parasites, the immune system more often functions to control their populations in the host. Thus a balance is achieved between hosts and parasites that have lived in long evolutionary association, with both surviving through compromise. Enhancing these particular antiparasite mechanisms and neutralizing the parasite's evasion mechanisms would tip the balance in favor of the host. *See* MEDICAL PARASITOLOGY; POPULATION ECOLOGY. Raymond T. Damian

Parasympathetic nervous system A portion of the autonomic system. It consists of two neuron chains, but differs from the sympathetic nervous system in that the first neuron has a long axon and synapses with the second neuron near or in the organ innervated. In general, its action is in opposition to that of the sympathetic nervous system, which is the other part of the autonomic system. It cannot be said that one system, the sympathetic, always has a excitatory role and the other, the parasympathetic, an inhibitory role; the situation depends on the organ in question. However, it may be said that the sympathetic system, by altering the level at which various organs function, enables the body to rise to emergency demands encountered in flight, combat, pursuit, and pain. The parasympathetic system appears to be in control during such pleasant periods as digestion and rest. The alkaloid pilocarpine excites parasympathetic activity while atropine inhibits it. *See* AUTONOMIC NERVOUS SYSTEM; SYMPATHETIC NERVOUS SYSTEM. Douglas B. Webster

Parathyroid gland An endocrine organ usually associated with the thyroid gland and possessed by all vertebrates except the fishes. In response to lowered serum calcium concentration, a hormone is produced which promotes bone destruction and inhibits the phosphorus-conserving activity of the kidneys. *See* THYROID GLAND.

In humans, there are typically four glands situated as shown in the illustration; however, the number varies between three and six, with four appearing about 80% of the time. Variations in the positioning of the glands along the craniocaudal axis occur but, excepting parathyroid III which may occasionally be found upon the anterior surface of the trachea, the relation to the posterior surface of the thyroid is rarely lost. William E. Dossel

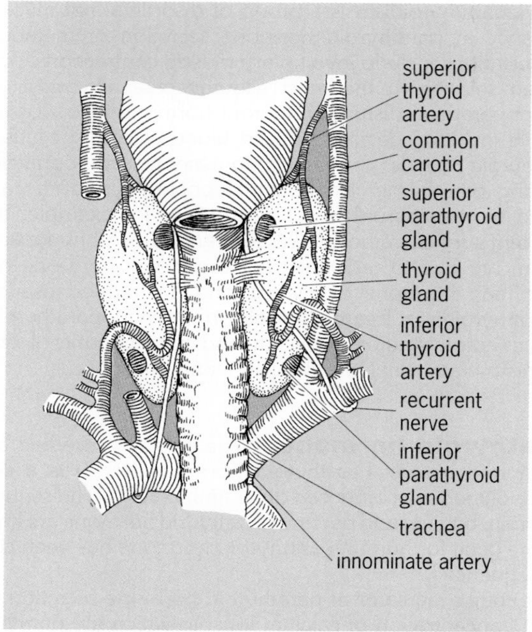

Common positions of human parathyroid glands on the posterior aspect of the thyroid. (***After W. H. Hollinshead, Anatomy of the endocrine glands, Surg. Clin. N. Amer., 21(4):1115–1140, 1952***)

The parathyroid glands are essential for the regulation of calcium and phosphate concentrations in the extracellular fluids of amphibians and higher vertebrates. Parathyroid hormone has two major target organs, bone and kidney. It acts on bone in several ways. Short-term changes include a rapid uptake of bone fluid calcium into osteoblast cells, which in turn pump the calcium into the extracellular fluids. Long-term effects include increased activity and number of osteoclasts, bone cells which act to break down bone matrix and release calcium from bone. All of these effects result in increased blood calcium values. *See* BONE; CALCIUM METABOLISM.

Parathyroid hormone inhibits the renal reabsorption of phosphate, thus increasing the urinary output of phosphate. Phosphate reabsorption across the renal tubule is dependent upon sodium transport, and parathyroid hormone interferes with this sodium-dependent phosphate transport in the proximal tubule. Another important effect of parathyroid hormone on the kidney is to increase the renal reabsorption of calcium, thus reducing the loss of calcium in the urine and conserving calcium in the body. *See* ENDOCRINE SYSTEM (VERTEBRATE); KIDNEY; PARATHYROID HORMONE; VITAMIN D. Nancy B. Clark

Parathyroid gland disorders Disorders involving excessive or deficient blood levels of parathyroid hormone caused by abnormal functioning of the parathyroid gland. Parathyroid hormone is responsible for keeping the concentration of calcium in blood within a narrow normal range. If the blood calcium concentration falls, the parathyroid glands respond by secreting hormone which tends to increase the concentration of calcium. Parathyroid hormone acts directly on bone and kidney, and indirectly on the intestine, to increase the concentration of calcium in blood. It also acts on the kidney to increase excretion of phosphate in the urine, causing a lowering of the concentration of phosphorus in blood. *See* PARATHYROID GLAND.

Hyperparathyroidism is a group of disorders characterized by excessive parathyroid hormone secretion. Primary hyperparathyroidism is defined as increased parathyroid hormone secretion despite elevated blood calcium. Symptoms and signs range from overt loss of calcium from bone and formation of calcium stones in the kidney to nonspecific manifestations such as weakness and fatigue. The disorder occurs most frequently in women after menopause. There are also inherited forms of the disease. Treatment of primary hyperparathyroidism involves surgical removal of the abnormal gland(s). Secondary hyperparathyroidism is defined as excessive secretion of parathyroid hormone in response to reduction in blood calcium caused by kidney or intestinal disorders or vitamin D deficiency. Treatment is directed at the underlying disorder (for example, kidney transplantation for kidney failure). *See* CALCIUM METABOLISM; VITAMIN D.

Hypoparathyroidism is a group of disorders that involves a deficiency in parathyroid hormone secretion or metabolism. Manifestations include involuntary muscle contractions, or generalized seizures in the most extreme cases. In most forms of hypoparathyroidism, parathyroid hormone is low or undetectable in blood despite reduced blood calcium because the parathyroid glands are unable to respond appropriately by secreting parathyroid hormone. Hypoparathyroidism can be caused by parathyroid gland destruction, for example, by inadvertent surgical removal. Certain rare forms of hypoparathyroidism are due to resistance to parathyroid hormone action rather than hormone deficiency, and are termed pseudohypoparathyroidism. Treatment of all forms of hypoparathyroidism involves administration of calcium and large amounts of vitamin D to restore a normal blood calcium and phosphorus.

Allen M. Speigel

Parathyroid hormone The secretory product of the parathyroid glands. Parathyroid hormone (PTH) is a single-chain polypeptide composed of 84 amino acids. The sequences of human, bovine, and porcine parathyroid hormone are known, and the gene for human parathyroid hormone has been cloned and sequenced.

The major regulator of parathyroid hormone secretion is the serum concentration of calcium ions, to which the parathyroid cells are exquisitely sensitive. Only a limited amount of parathyroid hormone is stored in secretory granules, so that a hypocalcemic stimulus must ultimately influence biosynthesis as well as secretion of the hormone. Parathyroid secretory protein is a large, acidic glycoprotein which is stored and cosecreted with parathyroid hormone in roughly equimolar amounts; the biological function of parathyroid secretory protein is unknown.

Parathyroid hormone is responsible for the fine regulation of serum calcium concentration on a minute-to-minute basis. This is achieved by the acute effects of the hormone on calcium resorption in bone and calcium reabsorption in the kidney. The phosphate mobilized from bone is excreted into the urine by means of the hormone's influence on renal phosphate handling. Parathyroid hormone also stimulates calcium absorption in the intestine, this being mediated indirectly by 1,25-dihydroxyvitamin D. Thus, a hypocalcemic stimulus of parathyroid hormone secretion results in an increased influx of calcium from three sources (bone, kidney, and intestine), resulting in a normalization of the serum calcium concentration without change in the serum phosphate concentration. *See* CALCIUM METABOLISM; PARATHYROID GLAND; PARATHYROID GLAND DISORDERS; THYROCALCITONIN; VITAMIN D.

Arthur E. Broadus

Parazoa A name proposed for a subkingdom of animals which includes the sponges. Erection of a separate subkingdom for the sponges implies that they originated from protozoan ancestors independently of all other Metazoa. This theory is supported by the uniqueness of the sponge body plan and by peculiarities of fertilization and development. Much importance is given to the fact that during the development of sponges with parenchymella larvae, the flagellated external cells of the larva take up an internal position as choanocytes after metamorphosis, whereas the epidermal and mesenchymal cells arise from what was an internal mass of cells in the larva. These facts suggest that either the germ layers of sponges are reversed in comparison with those of other Metazoa or the choanocytes cannot be homologized with the endoderm of other animals. Either interpretation supports the wide separation of sponges from all other Metazoa to form the subkingdom Parazoa or Enantiozoa. On the other hand, there are cogent arguments in favor of the basic similarity of the development of sponges and other Metazoa. *See* CALCAREA; DEMOSPONGIAE; METAZOA; PORIFERA.

Willard D. Hartman

Parenchyma A ground tissue chiefly concerned with the manufacture and storage of food. The primary functions of plants, such as photosynthesis, assimilation, respiration, storage, secretion, and excretion—those associated with living protoplasm—proceed mainly in parenchymal cells. Parenchyma is frequently found as a homogeneous tissue in stems, roots, leaves, and flower parts. Other tissues, such as sclerenchyma, xylem, and phloem, seem to be embedded in a matrix of parenchyma; hence the use of the term ground tissue with regard to parenchyma is derived. The parenchymal cell is one of the most frequently occurring cell types in the plant kingdom. *See* PLANT ANATOMY; PLANT PHYSIOLOGY.

Typical parenchyma occurs in pith and cortex of roots and stems as a relatively undifferentiated tissue composed of polyhedral cells that may be more or less compactly arranged and show little variation in size or shape. The mesophyll, that is, the tissue located between the upper and lower epidermis of leaves, is a specially differentiated parenchyma called chlorenchyma because its cells contain chlorophyll in distinct chloroplastids.

This chlorenchymatous tissue is the major locus of photosynthetic activity and consequently is one of the more important variants of parenchyma. Specialized secretory parenchymal cells are found lining resin ducts and other secretory structures. *See* PHOTOSYNTHESIS; SECRETORY STRUCTURES (PLANT).

Robert L. Hulbary

Pareto's law A law (sometimes called the 20–80 rule) describing the frequency distribution of an empirical relationship fitting the skewed concentration of the variate-values pattern. The phenomenon wherein a small percentage of a population accounts for a large percentage of a particular characteristic of that population is an example of Pareto's law. When the data are plotted graphically, the result is called a maldistribution curve. To take a specific case, an analysis of a manufacturer's inventory might reveal that less than 15% of the component part items account for over 90% of the total annual usage value.

The mathematics required to calculate and graph the curve of Pareto's law is simple arithmetic. It should be noted, however, that the calculations need not be done in all cases. It may suffice to merely make a rough approximation of a situation in order to determine whether or not Pareto's law is present and whether benefits may subsequently accrue.

Vincent M. Altamuro

Parity (quantum mechanics) A physical property of a wave function which specifies its behavior under simultaneous reflection of all spatial coordinates through the origin, that is, when x is replaced by $-x$, y by $-y$, and z by $-z$. If the single-particle wave function ψ satisfies Eq. (1), it is said to have even parity. If, on the other hand, Eq. (2) holds, the wave function is said to have odd parity. These two expressions can be combined in Eq. (3), where $P = \pm 1$ is a quantum number, parity, having

$$\psi(x,y,z) = \psi(-x,-y,-z) \tag{1}$$

$$\psi(x,y,z) = -\psi(-x,-y,-z) \tag{2}$$

$$\psi(x,y,z) = P\psi(-x,-y,-z) \tag{3}$$

only the two values $+1$ (designated as even parity) and -1 (odd parity). More precisely, parity is defined as the eigenvalue of the operation of space inversion. Parity is a concept that has meaning only for fields or waves and therefore has meaning only in classical field theory or in quantum mechanics. *See* QUANTUM MECHANICS.

The conservation of parity follows from the inversion symmetry of space, that is, the invariance of the Schrödinger equation $H\psi = E\psi$ (the wave equation satisfied by the wave function ψ) to the inversion of space coordinates, $\mathbf{r} \rightarrow -\mathbf{r}$. The parity (or inversion) operator, which changes $\mathbf{r}$ to $-\mathbf{r}$, has the alternative interpretation that the coordinate values remain unchanged but the coordinate axes are inverted; that is, the positive x axis of the new frame points along the old negative x axis, and similarly for y and z. If the original frame was right-handed, then the new frame is left-handed. [A cartesian coordinate system (frame, for short) is called right-handed if it is possible to place the right hand at the origin and point the thumb and first and second fingers along the positive x, y, and z axes, respectively.] Thus, parity would be conserved if the statement of physical laws were independent of the handedness of the coordinate system that was being used. Of course, the fact that most people are right-handed is not a physical law but an accident of evolution; there is nothing in the relevant laws of physics which favors a right-handed over a left-handed human. The same holds for optically active organic compounds, such as the amino acids. However, the statement that the neutrino is left-handed is a physical law. *See* NEUTRINO.

All the strong interactions between hadrons (for example, nuclear forces) and the electromagnetic interactions are symmetrical to inversion, so that parity is conserved by these interactions. As far as is known, only the weak interactions fail to conserve parity. Thus parity is not conserved in the weak decays of elementary particles (including beta decay of nuclei); in all other processes the weak interactions play a small role, and parity is very nearly conserved. Likewise, in energy eigenstates, weak interactions can be neglected to a very good approximation, and parity is very nearly a good quantum number, so that each atomic, nuclear, or hadronic state is characterized by a definite value of parity, and its conservation in reactions is an important principle. *See* FUNDAMENTAL INTERACTIONS; WEAK NUCLEAR INTERACTIONS.

One of the selection rules which follows from parity conservation is the following: A spin zero boson cannot decay sometimes into two π mesons and sometimes into three π mesons, because these final states have different parities, even and odd respectively. But the positive K meson is observed to have both these decay modes, originally called the θ and the τ mesons, respectively, but later shown by the identity of masses and lifetimes to be decay modes of the same particle. This τ-θ puzzle was the first observation of parity nonconservation. In 1956, T. D. Lee and C. N. Yang made the bold hypothesis that parity also is not conserved in beta decay. They reasoned that the magnitude of the beta-decay coupling is about the same as the coupling which leads to decay of the K meson, and so these decay processes may be manifestations of a single kind of coupling. Also, there is a very natural way to introduce parity nonconservation in beta decay, namely, by assuming a restriction on the possible states of the neutrino (two-component theory). They pointed out that no beta-decay experiment had ever looked for the spin-momentum correlations that would indicate parity nonconservation; they urged that these correlations be sought.

In the first experiment to show parity nonconservation in beta decay, the spins of the beta-active nuclei cobalt-60 were polarized with a magnetic field at low temperature; the decay electrons were observed to be emitted preferentially in directions opposite to the direction of the ^{60}Co spin. The magnitude of this correlation shows that the parity-nonconserving and parity-conserving parts of the beta interaction are of equal size, substantiating the two-component neutrino theory.

It was at first somewhat disconcerting to find parity not conserved, for that seemed to imply a handedness of space. But this is not really the situation; the saving thing is that anti-^{60}Co decays in the opposite direction. Thus, after all, there is nothing intrinsically left-handed about the world, just as there is nothing intrinsically positively charged about nuclei. What really exists here is a correlation between handedness and sign of charge.

Charles J. Goebel

Parkinson's disease A progressive degenerative disorder of the nervous system belonging to a group of conditions called motor system disorders. Parkinson's disease typically begins in middle age with an average age of onset of 59 years, but it can also begin in the 30s or 40s as well as in the 70s or later. Males are affected slightly more commonly than females. Parkinson's disease results in the loss of brain cells that produce dopamine. The cause is unknown but, increasingly, specific environmental and genetic factors are being evaluated for their role in the disease. Typically Parkinson's disease is sporadic, but it can occur in familial clusters. *See* NERVOUS SYSTEM (VERTEBRATE).

The cardinal features of Parkinson's disease are tremor, rigidity, akinesia (absence of movement)–bradykinesia (slowness of movement), and altered posture (these features compose the acronym TRAP). This tetrad of features is not present in every patient. The tremor of Parkinson's disease primarily occurs at rest when the limb is in repose (and is therefore called a rest tremor), and lessens when an action is initiated. Classically, the tremor begins on one side of the body, and later affects the other side as well. Rigidity or stiffness can be manifested as reduced arm swing on one side of the body, resulting in some patients mistakenly suspecting that they had a stroke. Typically, most disabling is the akinesia-bradykinesia. As a result, there is slowed gait with reduced stride length as well as difficulty getting out of low chairs and turning in bed. The individual's posture becomes stooped with flexion at the hips, elbows, and wrists. Patients develop balance problems over time, which may result in falls.

Other features appear that impact quality of life, including fluctuations in mobility (called motor fluctuations) that occur throughout the day and difficulty initiating walking called freezing of gait. With the development of motor fluctuations, patients lose the ability to function independently due to slowness from the unpredictable response to medication therapy as well as progressive balance difficulties. Cognitive changes, which may include a form of dementia that is distinct from Alzheimer's disease, develop as the disease advances. Typically, individuals with Parkinson's disease do not die of the disease per se, but succumb to the complications related to infections or other conditions that come with their progressive immobility. *See* ALZHEIMER'S DISEASE; MOTOR SYSTEMS.

The underlying pathology in Parkinson's disease is the progressive loss of dopamine cells that reside in the brainstem, deep within the base of the brain. Within the brainstem, the dopamine-producing nerve cells are clustered in an area called the substantia nigra because of the black pigmentation within the cells. These cells die off prematurely due to the presence of genetic and/or environmental factors. Once the loss of the dopamine nerve cell pool reaches 50–80%, patients begin to experience symptoms. *See* BRAIN; DOPAMINE.

To date there is no treatment that reverses or slows the course of the disease, and treatment is primarily directed at alleviating symptoms by restoring dopamine signaling in the brain. One common strategy is to give the dopamine precursor levodopa (L-dopa). It is combined with another medication called carbidopa so that less of the L-dopa is metabolized by the liver and more L-dopa is available in the brain for dopamine production. Other medications which simulate the action of dopamine, and are therefore called dopamine agonists, are often used separately or with L-dopa to help manage symptoms. With treatment, symptoms are often reasonably well managed for the first 5–10 years. After this period of stable response to treatment, patients experience an increasing number of hours of impaired movement as the disease progresses.

Paul Tuite

Parsec A unit of measure of astronomical distances. One parsec is equivalent to 3.084×10^{13} kilometers, or 1.916×10^{13} miles. There are 3.26 light-years in 1 parsec. The parsec is defined as the distance at which the semimajor axis of Earth's orbit around the Sun (1 astronomical unit) subtends 1 second of arc. Thus, because the angle is small, the equation below holds.

$$\frac{1 \text{ astronomical unit}}{1 \text{ parsec}} = 1 \text{ second} = \frac{1}{206{,}265}$$

A parsec is then 206,265 astronomical units. At a distance of 1 parsec, the parallax is 1 second of arc. The nearest star is about 1.3 parsecs distant; the farthest known galaxy is several billion parsecs. See PARALLAX (ASTRONOMY). Jesse L. Greenstein

Parsley A biennial, *Petroselinum crispum*, of European origin belonging to the plant order Umbellales. Parsley is grown for its foliage and is used to garnish and flavor foods. It contains large quantities of vitamins A and C. Two types, plain-leafed and curled, are grown for their foliage; Hamburg parsley (*P. crispum* var. *tuberosum*), also called turnip-rooted parsley, is grown for its edible parsniplike root. See APIALES. H. John Carew

Parsnip A hardy biennial, *Pastinaca sativa*, of Mediterranean origin belonging to the plant order Umbellales. The parsnip is grown for its thickened taproot and is used primarily as a cooked vegetable. Exposure of mature roots to low temperatures, not necessarily freezing, improves the quality of the root by favoring the conversion of starch to sugar. See APIALES. H. John Carew

Partial differentiation A mathematical operation performed on functions of more than one variable. In this article only two or three variables are considered; however, the principles apply to functions of n variables, for any positive integer $n > 1$. If $z = f(x,y)$, the partial derivative $\partial z/\partial x$ is defined as the derivative of $f(x,y)$ with respect to x, y being regarded as fixed; that is,

$$\frac{\partial z}{\partial x} = \lim_{h\to 0} \frac{f(x+h,\, y) - f(x,\, y)}{h}$$

Another notation for $\partial z/\partial x$ is $f_1(x,y)$. The other first partial derivative is $\partial z/\partial y$, also written $f_2(x,y)$. For values at particular points the notation is

$$\left(\frac{\partial z}{\partial x}\right)_{(a,b)} = f_1(a,\, b)$$

In the case of a function of three variables, $f(x,y,z)$, the expression is

$$\frac{\partial f}{\partial z} = f_3(x,\, y,\, z)$$

The second derivatives of $f(x,y)$ are given by

$$f_{11}(x,\, y) = \frac{\partial}{\partial x}\left(\frac{\partial f}{\partial x}\right) \qquad f_{12}(x,\, y) = \frac{\partial}{\partial y}\left(\frac{\partial f}{\partial x}\right)$$

$$f_{21}(x,\, y) = \frac{\partial}{\partial y}\left(\frac{\partial f}{\partial y}\right) \qquad f_{22}(x,\, y) = \frac{\partial}{\partial y}\left(\frac{\partial f}{\partial y}\right)$$

It can happen that $f_{12}(x,\, y) \neq f_{21}(x,y)$, but this will not happen in common practice, especially with elementary functions. If f_1, f_2, f_{12}, f_{21} are defined in neighborhood of (a,b), and if f_{12}, f_{21} are continuous at $(a,\, b)$, then $f_{12}(a,b) = f_{21}(a,b)$. In addition, there are more delicate theorems relating to this matter.

The notion of the differentiability of a function is fundamental in the theory of partial differentiation. The requirement that $f(x,y)$ be differentiable is not the same as the requirement that $f_1,(x,y)$ and $f_2(x,\, y)$ both exist; it is a more inclusive requirement. The geometric meaning of f being differentiable at (a,b) is that the surface defined by $z = f(x,y)$ has a tangent plane not parallel to the z axis when $x = a$, $y = b$. In analytic terms the condition is that if

$$\epsilon = f(a+h,\, b+k) - f(a,b) - f_1(a,b)h - f_2(a,b)k$$

then

$$\lim_{(h,k)\to(0,0)} \frac{\epsilon}{|h| + |k|} = 0$$

A sufficient condition that f be differentiable at (a,b) is that the partial derivatives f_1, f_2 be defined at all points near (a,b), and continuous at (a,b).

The prime importance of the differentiability concept is that the differentiability property is needed in proving the chain rule for functions of several variables. This rule asserts that a differentiable function of a differentiable function is differentiable, and the rule tells how to compute partial derivatives of the composite function. For example, if $x = f(s,t)$, $y = g(s,t)$, where f and g are differentiable, and if $z = F(x,y)$, where F is differentiable, then the composite function is $G(s,t) = F[f(s,t),g(s,t)]$. Then $z = G(s,t)$ is differentiable as a function of s and t, and

$$\frac{\partial G}{\partial s} = \frac{\partial F}{\partial x}\frac{\partial f}{\partial s} + \frac{\partial F}{\partial y}\frac{\partial g}{\partial s}$$

$$\frac{\partial G}{\partial t} = \frac{\partial F}{\partial x}\frac{\partial f}{\partial t} + \frac{\partial F}{\partial y}\frac{\partial g}{\partial t}$$

These equations, expressing the formal part of the chain rule, are often written in the form

$$\frac{\partial z}{\partial s} = \frac{\partial z}{\partial x}\frac{\partial x}{\partial s} + \frac{\partial z}{\partial y}\frac{\partial y}{\partial s}$$

$$\frac{\partial z}{\partial t} = \frac{\partial z}{\partial x}\frac{\partial x}{\partial t} + \frac{\partial z}{\partial y}\frac{\partial y}{\partial t}$$

See CALCULUS; DIFFERENTIATION; PARAMETRIC EQUATION. Angus E. Taylor

Particle accelerator An electrical device that accelerates charged atomic or subatomic particles to high energies. The particles may be charged either positively or negatively. If subatomic, the particles are usually electrons or protons and, if atomic, they are charged ions of various elements and their isotopes throughout the entire periodic table of the elements.

Accelerators that produce various subatomic particles at high intensity have many practical applications in industry and medicine as well as in basic research. Electrostatic generators, pulse transformer sets, cyclotrons, and electron linear accelerators are used to produce high levels of various kinds of radiation that in turn can be used to polymerize plastics, provide bacterial sterilization without heating, and manufacture radioisotopes which are utilized in industry and medicine for direct treatment of some illnesses as well as research. They can also be used to provide high-intensity beams of protons, neutrons, heavy ions, pi mesons, or x-rays that are used for cancer therapy and research. The x-rays used in industry are usually produced by arranging for accelerated electrons to strike a solid target. However, with the advent of electron synchrotron storage rings that produce x-rays in the form of synchrotron radiation, many new industrial applications of these x-rays have been realized, especially in the field of solid-state microchip fabrication and medical diagnostics. See ISOTOPIC IRRADIATION; RADIATION BIOLOGY; RADIATION CHEMISTRY; RADIOACTIVITY AND RADIATION APPLICATIONS; RADIOGRAPHY; RADIOISOTOPE; RADIOLOGY; SYNCHROTRON RADIATION.

Particle accelerators fall into two general classes—electrostatic accelerators that provide a steady dc potential, and varieties of accelerators that employ various combinations of time-varying electric and magnetic fields.

Electrostatic accelerators. Electrostatic accelerators in the simplest form accelerate the charged particle either from the source of high voltage to ground potential or from ground potential to the source of high voltage. All particle accelerations are carried out inside an evacuated tube so that the accelerated particles do not collide with air molecules or atoms and may follow trajectories characterized specifically by the electric fields utilized for the acceleration. The maximum energy available from this kind of accelerator is limited by the ability of the evacuated tube to withstand some maximum high voltage.

Time-varying field accelerators. In contrast to the high-voltage-type accelerator which accelerates particles in a continuous stream through a continuously maintained increasing

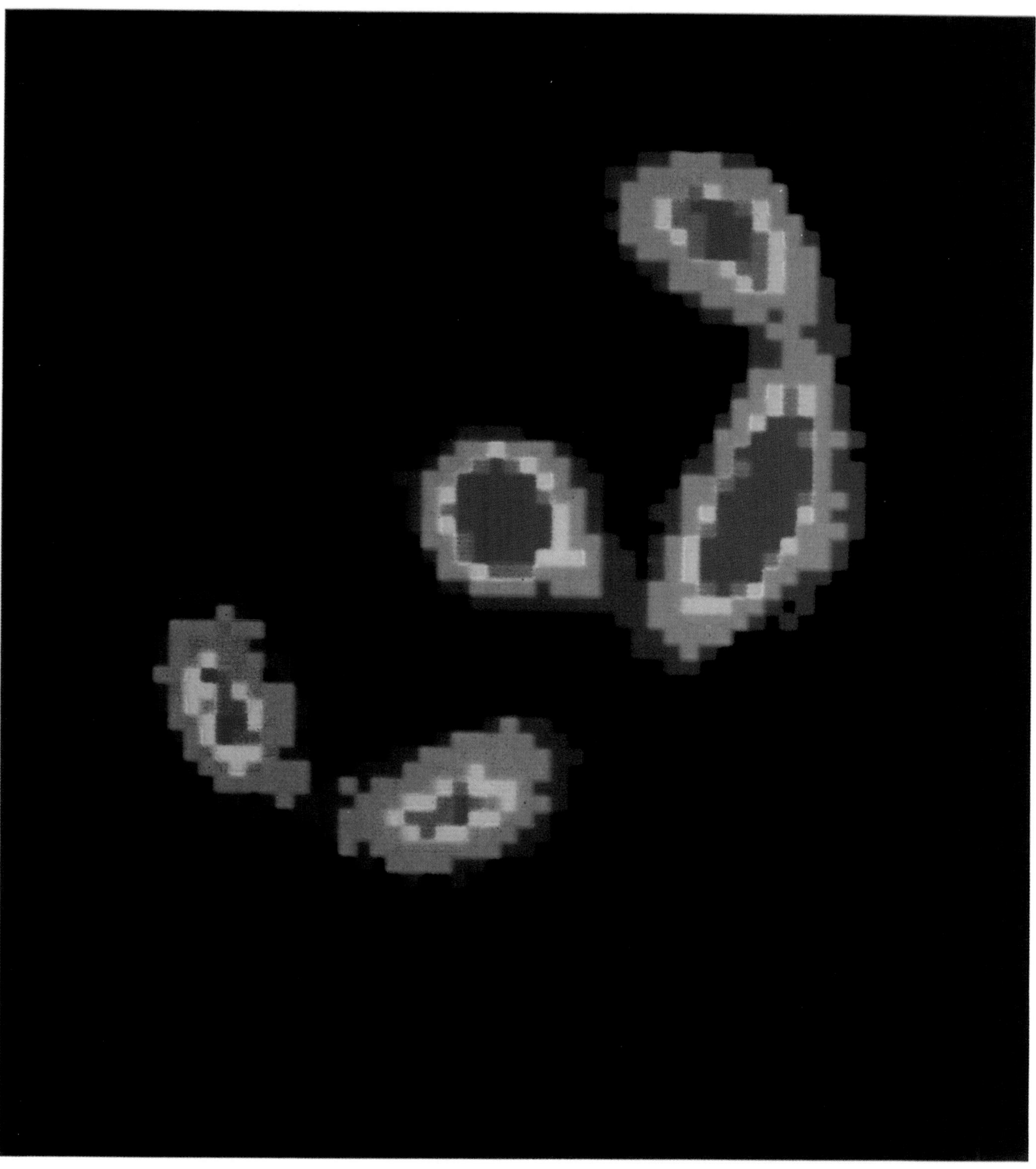

False-color image of individual ions in a particle trap. The ions have been cooled to the point where they form an ordered array called an ion crystal. This array has six ions, one of which is not visible in the image. Five ions lie in a plane at the vertices of a regular pentagon; a sixth ion lies at the center of the pentagon. The five visible ions are strongly fluorescing mercury ions. The nonfluorescing ion at one corner of the pentagon may be a heavier isotope of mercury or a molecular ion such as $HgOH^+$.

potential, the time-varying accelerators must necessarily accelerate particles in small discrete groups or bunches.

An accelerator that varies only in electric field and does not use any magnetic guide or turning field is customarily referred to as a linear accelerator or linac. In the simplest version of this kind of accelerator, the electrodes that are used to attract and accelerate the particles are connected to a radio-frequency (rf) power supply or oscillator so that alternate electrodes are of opposite polarity. In this way, each successive gap between adjacent electrodes is alternately accelerating and decelerating. If these acceleration gaps are appropriately spaced to accommodate the increasing velocity of the accelerated particle, the frequency can be adjusted so that the particle bunches are always experiencing an accelerating electric field as they cross each successive gap. In this way, modest voltages can be used to accelerate bunches of particles indefinitely, limited only by the physical length of the accelerator construction.

All conventional (but not superconducting) research linacs usually are operated in a pulsed mode because of the extremely high rf power necessary for their operation. The pulsed operation can then be adjusted so that the duty cycle or amount of time actually on at full power averages to a value that is reasonable in cost and practical for cooling. This necessarily limited duty cycle in turn limits the kinds of research that are possible with linacs; however, they are extremely useful (and universally used) as pulsed high-current injectors for all electron and proton synchrotron ring accelerators. Superconducting linear accelerators have been constructed that are used to accelerate electrons and also to boost the energy of heavy ions injected from electrostatic machines. These linacs can easily operate in the continuous-wave (cw) rather than pulsed mode, because the rf power losses are only a few watts.

The Continuous Electron Beam Accelerator Facility (CEBAF) uses two 400-MeV superconducting linacs to repeatedly accelerate electrons around a racetracklike arrangement where the two linacs are on the opposite straight sides of the racetrack and the circular ends are a series of recirculation bending magnets, a different set for each of five passes through the two linacs in succession. The continuous electron beam then receives a 400-MeV acceleration on each straight side or 0.8 GeV per turn, and is accelerated to a final energy of 4 GeV in five turns and extracted for use in experiments. The superconducting linacs allow for continuous acceleration and hence a continuous beam rather than a pulsed beam. This makes possible many fundamental nuclear and quark structure measurements that are impossible with the pulsed electron beams from conventional electron linacs. *See* SUPERCONDUCTING DEVICES.

As accelerators are carried to higher energy, a linac eventually reaches some practical construction limit because of length. This problem of extreme length can be circumvented conveniently by accelerating the particles in a circular path maintained by either static or time-varying magnetic fields. Accelerators utilizing steady magnetic fields as guide paths are usually referred to as cyclotrons or synchrocyclotrons, and are arranged to provide a steady magnetic field over relatively large areas that allow the particles to travel in an increasing spiral orbit of gradually increasing size as they increase in energy. *See* MAGNETISM.

Practical limitations of magnet construction and cost have kept the size of circular proton accelerators with static magnetic fields to the vicinity of 100 to 1000 MeV. For even higher energies, up to 400 GeV per nucleon in the largest conventional (not superconducting) proton synchrotron in operation, it is necessary to vary the magnetic field as well as the electric field in time. In this way the magnetic field can be of a minimal practical size, which is still quite extensive for a 980-GeV accelerator (6500 ft or 2000 m in diameter). This circular magnetic containment region, or "racetrack," is injected with relatively low-energy particles that can coast around the magnetic ring when it is at minimum field strength. The magnetic field is then gradually increased to stay in step with the higher magnetic rigidity of the particles as they are gradually accelerated with a time-varying electric field.

Superconducting magnets. The study of the fundamental structure of nature and all associated basic research require an ever increasing energy in order to allow finer and finer measurements on the basic structure of matter. Since the voltage-varying and magnetic-field-varying accelerators also have limits to their maximum size in terms of cost and practical construction problems, the only way to increase particle energies even further is to provide higher-varying magnetic fields through superconducting magnet technology, which can extend electromagnetic capability by a factor of 4 to 5. Large superconducting cyclotrons and superconducting synchrotrons are in operation. *See* MAGNET.

Storage rings. Beyond the limit just described, the only other possibility is to accelerate particles in opposite directions and arrange for them to collide at certain selected intersection regions around the accelerator. The main technical problem is to provide adequate numbers of particles in the two colliding beams so that the probability of a collision is moderately high. Such storage ring facilities are in operation for both electrons and protons. Besides storing the particles in circular orbits, the rings can operate initially as synchrotrons and accelerate lower-energy injected particles to much higher energies and then store them for interaction studies at the beam interaction points.

Large proton synchrotrons have been used as storage-ring colliders by accelerating and storing protons in one direction around the ring while accelerating and storing antiprotons (negative charge) in the opposite direction. The proton and antiproton beams are carefully programmed to be in different orbits as they circulate in opposite directions and to collide only when their orbits cross at selected points around the ring where experiments are located. The antiprotons are produced by high-energy proton collisions with a target, collected, stored, cooled, and eventually injected back into the synchrotron as an antiproton beam.

Electron-positron synchrotron accelerator storage rings have been in operation for many years in the basic study of particle physics, with energies ranging from 2 GeV + 2 GeV to 104 GeV + 104 GeV. The by-product synchrotron radiation from many of these machines is used in numerous applications. However, the synchrotron radiation loss forces the machine design to larger and larger diameters, characterized by the Large Electron Positron Storage Ring (LEP) at CERN, near Geneva, Switzerland (closed down in 2000), which was 17 mi (27 km) in circumference. Harvey E. Wegner

In proton-proton colliders, the two counterrotating beams are of the same charge, but moving in opposite directions. Each beam must be provided with a separate aperture with opposite magnetic field direction in order to follow the same curved trajectory. This can be achieved with two rings of magnets (above or below one another, or side by side), or the magnets can be built to contain the two apertures of opposite magnetic field direction.

The LEP tunnel at CERN is being used for installation of the Large Hadron Collider (LHC), a proton-proton collider which will have an energy of 7 TeV per beam, or 70 times of energy of the electron ring, and a luminosity of $10^{34}/\mathrm{cm}^{-2}\mathrm{s}^{-1}$. Thus, when it comes into operation, it will be the foremost collider for high-energy physics research. There will be over 1232 dipole magnets, each 49 ft (15 m) long and operating at a nominal magnetic field of 8.33 T, and 392 main quadrupole magnets, each 11 ft (3.25 m) long and operating at a nominal gradient of 223 T/m (resulting in a peak field at the coils of 6.85 T). The magnets are of a unique two-in-one design, with both beam apertures in one cold iron mass and cryostat. The magnets will operate at 1.9 K (−456°F). The cooling of the magnets will be done with superfluid helium. Helen T. Edwards; Oliver Brüning

Linear electron-positron colliders. While the LHC is pushing the energy frontier for proton-proton collisions, electron-positron colliders with beam energies of 0.25–2.5 TeV are considered useful to complement the experimental program of the

LHC. However, because of the power loss due to synchrotron radiation, a storage ring with energy significantly higher than the 104 GeV achieved with LEP is not considered reasonable.

This limitation can be overcome with a linear collider, in which an electron linear accelerator and a positron linear accelerator collide their beams heads on. The absence of magnetic bending of the beam orbits eliminates the synchrotron radiation losses and the achievable energy is given by the overall length times the effective accelerating field. Since the accelerated electrons and positrons pass the interaction point only once and are dumped downstream of the interaction point, beam spot size at the interaction point has to be kept very small to reach the required luminosities.

The Stanford Linear Collider (SLC), operated from 1989 to 1998 at Stanford University, Palo Aro, California, with beam energy of 50 GeV, demonstrated the feasibility of the linear collider scheme. SLC, however, was a kind of hybrid between linear and circular collider. Both beams were accelerated simultaneously in a single linear accelerator, separated with a magnet after acceleration, and collided after appropriate bending in opposite directions by magnetic arcs.

Several approaches to linear colliders are being studied. They are based on superconducting as well as normal-conducting rf technology for the accelerating structures. Superconducting rf technology makes possible high efficiency in transforming electrical power into beam power, but imposes a fundamental limit on the accelerating field at about 50 MV/m. Hans Braun; Daniel Schulte

Particle detector A device used to detect and measure radiation characteristically emitted in atomic, molecular, or nuclear processes, including photons (light, including visible light, gamma rays, and x-rays), lightweight charged particles (electrons or positrons), nuclear constituents (neutrons, protons, and heavier ions), energetic neutral particles (fast-moving atoms, for example), and subnuclear constituents such as mesons. The device can also be known as a radiation detector. Since human senses generally do not respond to these types of radiation (with the obvious exception of visible light), detectors are essential tools for the discovery of radioactive minerals, for all studies of the structure of matter at the atomic, nuclear, and subnuclear levels, and for protection from the effects of radiation. They have also become important practical tools in the analysis of materials using the techniques of neutron activation and x-ray fluorescence analysis. *See* ACTIVATION ANALYSIS; ELEMENTARY PARTICLE; NUCLEAR REACTION; NUCLEAR SPECTRA; PARTICLE ACCELERATOR; PROSPECTING; RADIOACTIVITY; X-RAY FLUORESCENCE ANALYSIS.

A convenient way to classify radiation detectors is according to their mode of use: (1) For detailed observation of individual photons or particles, a pulse detector is used to convert each such event (that is, photon or particle) into an electrical signal. (2) To measure the average rate of events, a mean-current detector, such as an ion chamber, is often used. Radiation monitoring and neutron flux measurements in reactors generally fall in this category. Sometimes, when the total number of events in a known time is to be determined, an integrating version of this detector is used. (3) Position-sensitive detectors are used to provide information on the location of particles or photons in the plane of the detector. (4) Track-imaging detectors image the whole three-dimensional structure of a particle's track. The output may be recorded by immediate electrical readout or by photographing tracks as in the bubble chamber. (5) The time when a particle passes through a detector or a photon interacts with it is measured by a timing detector. Such information is used to determine the velocity of particles and when observing the time relationship between events in more than one detector. *See* TIME-OF-FLIGHT SPECTROMETERS.

The ionization produced by a charged particle is the effect commonly employed in a particle detector. In the basic type of gas ionization detector, an electric field applied between two electrodes separates and collects the electrons and positive ions produced in the gas by the radiation to be measured. Multiwire proportional chambers and spark chambers are position-sensitive adaptations of gas detectors. The signal division or time delay that occurs between the ends of an electrode made of resistive material is sometimes used to provide position sensitivity in gas and semiconductor detectors. Track-imaging detectors rely on a secondary effect of the ionization along a particle's track to reveal its structure. *See* IONIZATION CHAMBER.

Another type of position-sensitive detector uses a microchannel plate or an array of microchannel plates at an amplifier, with an imaging anode. This technology is used in many practical applications, such as in image intensifiers and night-vision devices. *See* CHANNEL ELECTRON MULTIPLIER; IMAGE TUBE (ASTRONOMY).

In a semiconductor detector, a solid replaces the gas. The "insulating" region (depletion layer) of a reverse-biased *pn* junction in a semiconductor is employed. Since solids are approximately 1000 times denser than gases, absorption of radiation can be accomplished in relatively small volumes. A less obvious but fundamental advantage of semiconductor detectors is the fact that much less energy is required ($\sim$3 eV) to produce a hole-electron pair than that required ($\sim$30 eV) to produce an ion electron pair in gases. *See* CRYSTAL COUNTER; JUNCTION DETECTOR.

In addition to producing free electrons and ions, the passage of a charged particle through matter temporarily raises electrons in the material into excited states. When these electrons fall back into their normal state, light may be emitted and detected as in the scintillation detector. *See* SCINTILLATION COUNTER.

Neutral particles, such as neutrons, cannot be detected directly by ionization. Consequently, they must be converted into charged particles by a suitable process and then observed by detecting the ionization caused by these particles.

Although ionization detectors dominate the field, a number of detector types based on other radiation-induced effects are used. Notable examples are (1) transition radiation detectors, which depend on the x-rays and light emitted when a particle passes through the interface between two media of different refractive indices; (2) track detectors, in which the damage caused by charged particles in plastic films and in minerals is revealed by etching procedures; (3) thermo- and radiophotoluminescent detectors, which rely on the latent effects of radiation in creating traps in a material or in creating trapped charge; and (4) Cerenkov detectors, which depend on measurement of the light produced by passage of a particle whose velocity is greater than the velocity of light in the detector medium. *See* CERENKOV RADIATION; PARTICLE TRACK ETCHING; TRANSITION RADIATION DETECTORS.

Another type of detector widely used in atomic and molecular physics is a Channeltron or microchannel-plate detector. High gain (amplification) is achieved by multiplication in secondary electron emission as a single incident particle initiates an electron cascade process.

The very large detector systems used in relativistic heavy-ion experiments and in the detection of the products of collisions of charged particles at very high energies, typically at the intersection region of storage rings, deserve special consideration. These detectors are frequently composites of several of the basic types of detectors discussed above and are designed to provide a detailed picture of the multiple products of collisions at high energies. The complete detector system may occupy a space tens of feet in extent and involve tens or hundreds of thousands of individual signal processing channels, together with large computer recording and analysis facilities. Fred S. Goulding; Alfred S. Schlacter

Particle flow Particle flow is important to many industrial processes, including the pneumatic conveying of solids, transport of solids in liquids (slurries), removal of particulates from gas streams for pollution control, combustion of pulverized coal, and drying of particulates in the food and pharmaceutical industries. The vast majority of flow problems in industrial design involve the flow of gas or liquids with suspended solids. *See* DRYING; FLUIDIZED-BED COMBUSTION; PIPELINE.

A key parameter in fluid-particle flows is the Stokes number, which is the ratio of the response time of a particle to a time characteristic of a flow system. Particle response time is the time that a particle takes to respond to a change in carrier flow velocity. If the Stokes number is small (say, less than 0.1), the particles have sufficient time to respond to the change in fluid velocity, so the particle velocity approaches the fluid velocity. However, if the Stokes number is large (say, greater than 10), the particles have little time to respond to the varying fluid velocity and the particle velocity shows little change.

The relative concentration of the particles in the fluid is referred to as loading. The loading may be defined in several ways, such as the ratio of particle mass flow to fluid mass flow. Many industrial applications involve highly loaded particle flows.

If the particle loading is small, the fluid will affect the particle properties (velocity, temperature, and so forth), but the particles will not influence the fluid properties. This is referred to as one-way coupling. If the conditions are such that there is a mutual interaction between the particles and fluid, the flow is two-way-coupled. Two-way coupling effects are reduced with increasing Stokes number because the particles undergo less acceleration.

If the particle motion is controlled by the action of the fluid on the particle, the flow is termed dilute. But if the particle concentration is sufficiently high, the particles will collide with each other and their motion will be dependent on particle-particle collisions; the flow is then regarded as dense. *See* FLUID FLOW; PARTICULATES.

Clayton T. Crowe

Particle track etching A technique of selective chemical etching to reveal tracks of heavy nuclear particles in a wide variety of solid substances. Developed in order to see fossil particle tracks in extraterrestrial materials, the technique finds application in many fields of science and technology.

An etchable track is produced if the charged particle has a sufficiently high radiation-damage rate and if the damaged region in the solid is permanently localized. Thus only highly ionizing particles are detectable; only nonconductors record tracks; and radiation-sensitive plastics can detect lighter particles than can radiation-insensitive minerals and glasses. The conical shape of the etched track depends on the ratio of the rate of etching along the track to the bulk etching rate of the solid.

The lunar surface, meteorites, and other objects exposed in space have been irradiated by charged particles from a variety of sources in the Sun and the Galaxy. Comparison of fossil particle tracks in lunar rocks and meteorites with spacecraft measurements of present-day radiations has established that solar flares and galactic cosmic rays have not changed over the last 2×10^7 years—the typical time a lunar rock exists before being shattered by impacting interplanetary debris.

Studies of tracks in a piece of glass from the *Surveyor 3* spacecraft after a 2.6-year exposure on the lunar surface, and of tracks in plastic detectors exposed briefly above the Earth's atmosphere in rockets, have led to the surprising discovery that the Sun preferentially ejects heavy elements in its flares rather than an unbiased sample of its atmosphere. The existence of galactic cosmic rays with atomic number greater than 30 was discovered in 1966 when fossil particle tracks were first studied in meteorites. Several particles heavier than uranium have been detected, indicating that cosmic rays originate in sources where synthesis has proceeded explosively beyond uranium. *See* COSMIC RAYS.

Unique advantages of etched-track detectors in nuclear and elementary particle physics are their ability to distinguish heavy-particle events in a large background of lightly ionizing radiation and their ability to detect individual rare events by a specialized technique such as electric-spark scanning or ammonia penetration through etched holes. These advantages have permitted such advances as the measurement of very long fission half-lives and the discovery of ternary fission. *See* NUCLEAR FISSION; TRANSURANIUM ELEMENTS.

The spontaneous fission of ^{238}U, present as a trace-element purity, gives tracks that can be used to date terrestrial samples ranging from rocks to human artifacts. Because fission tracks are erased in a particular mineral at a well-defined temperature, one can use the apparent fission-track ages as a function of distance from the heat source to measure the thermal (tectonic) history of regions. *See* FISSION TRACK DATING.

Filters are produced by irradiating thin plastic sheets with fission fragments and then etching holes to the desired size. Uses include biological research, wine filtration, and virus sizing. A uranium exploration method relies on a survey of radon emanation, as measured by alpha-particle tracks in plastic detectors, to locate promising locations in which to drill. Plastic detectors are also used in conjunction with a beam of high-energy heavy ions to take radiographs of cancer patients that reveal details not detectable in x-rays.

P. Buford Price

Particle trap A device used to confine charged or neutral particles where their interaction with the wall of a container must be avoided. Electrons or protons accelerated to energies as high as 1 teraelectronvolt (10^{12} electronvolts) are trapped in magnetic storage rings in high-energy collision studies. Other forms of magnetic bottles are designed to hold dense hot plasmas of hydrogen isotopes for nuclear fusion. At the other end of the energy spectrum, ion and atom traps can store isolated atomic systems at temperatures below 1 millikelvin. Other applications of particle traps include the storage of antimatter such as antiprotons and positrons (antielectrons) for high-energy collision studies or low-energy experiments. *See* ANTIMATTER; NUCLEAR FUSION; PARTICLE ACCELERATOR; PLASMA (PHYSICS); POSITRON.

Charged-particle traps. Charged particles can be trapped in a variety of ways. An electrostatic (Kingdon) trap is formed from a thin charged wire. The ion is attracted to the wire, but its angular momentum causes it to spiral around the wire in a path with a low probability of hitting the wire.

A magnetostatic trap (magnetic bottle) is based on the fact that a charged particle with velocity perpendicular to the magnetic field lines travels in a circle, whereas a particle moving parallel to the field is unaffected by it. In general, the particle has velocity components both parallel and perpendicular to the field lines and moves in a helical spiral. In high-energy physics, accelerators and storage rings also use magnetic forces to guide and confine charged particles. A tokamak has magnetic field lines configured in the shape of a torus, confining particles in spiral orbits. This type of bottle is used to contain hot plasmas in nuclear fusion studies. Another type of bottle uses a magnetic mirror.

The radio-frequency Paul trap uses inhomogeneous radio-frequency electric fields to confine particles, forcing them to oscillate rapidly in the alternating field (see illustration). If the amplitude of oscillation (micromotion) is small compared to the trap dimensions, the trap may be thought of as increasing the (kinetic) energy of the particle in a manner that is a function of the particle position. The particle moves to the position of minimal energy and is therefore attracted to the center of the trap where the oscillating electric fields are weakest. At the center of the trap, the fields are exactly zero, and a single, cold ion or electron trapped there is essentially at rest with almost no micromotion.

The Penning trap, with the same electrode configuration as the Paul trap, uses a combination of static electric and magnetic fields instead of oscillating electric fields.

Neutral-particle traps. Uncharged particles such as neutrons or atoms are manipulated by higher-order moments of the charge distribution such as the magnetic or electric dipole moments.

Magnetic traps of neutral particles use the fact that atoms usually have a magnetic dipole moment on which the gradient of a magnetic field exerts a force. The atom can be in a state whose

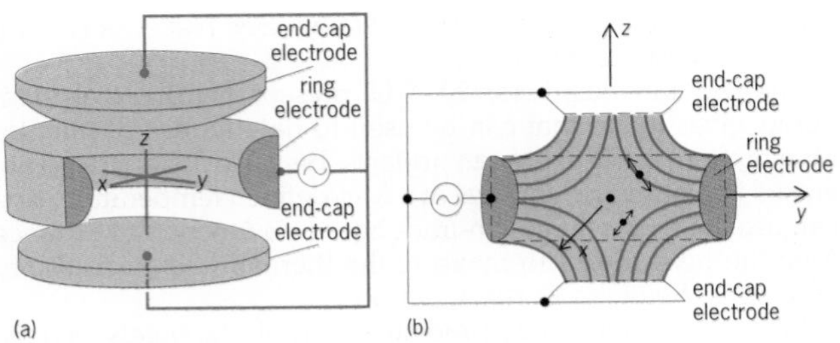

Radio-frequency Paul trap consisting of two end caps and a ring electrode. (*a*) Cutaway view (*after G. Kamas, ed., Time and Frequency Users's Manual, National Bureau of Standards Technical Note 695, 1977*). (*b*) Cross section, showing the amplitude of the instantaneous oscillations for several locations in the trap.

magnetic energy increases or decreases with the field strength, depending on whether the moment is antiparallel or parallel to the field. A magnetic field cannot be constructed with a local maximum in a current-free region, but a local minimum is possible, allowing particles seeking a weak field to be trapped.

Laser traps use the strong electric fields of the laser beam to induce an electric dipole moment on the atom. A laser field tuned below the atomic resonance polarizes the atom in phase with the driving field; the instantaneous dipole moment points in the same direction as the field. Thus the energy of the atom is lowered if it is in a region of high laser intensity. The high-intensity trapping region is formed simply by focusing the beam of a laser. *See* LASER.

Magnetooptic hybrid traps use, instead of the dipole forces induced by the laser field, the scattering force that arises when an atom absorbs photons. An inhomogeneous magnetic field separates the magnetic substates of an atom in a position-dependent manner. These states interact differently with circularly polarized light. It is possible to arrange a combination of laser beams with proper polarizations to create net scattering forces that drive the atom into the region of zero magnetic field. Such a trap requires much lower laser intensities and weaker magnetic fields. *See* LASER COOLING.

Steven Chu

Particulates Solids or liquids in a subdivided state. Because of this subdivision, particulates exhibit special characteristics which are negligible in the bulk material. Normally, particulates will exist only in the presence of another continuous phase, which may influence the properties of the particulates. A particulate may comprise several phases. The table categorizes particulate systems and relates them to commonly recognized designations. *See* ALLOY; EMULSION; FOAM; GEL.

Fine-particle technology deals with particulate systems in which the particulate phase is subject to change or motion, and is concerned with those particles which are tangible to human senses, yet small compared to the human environment—particles that are larger than molecules but smaller than gravel. Fine particles are in abundance in nature (as in rain, soil, sand, minerals, dust, pollen, bacteria, and viruses) and in industry (as in paint pigments, insecticides, powdered milk, soap, powder, cosmetics, and inks). Particulates are involved in such undesirable forms as fumes, fly ash, dust, and smog and in military strategy in the form of signal flares, biological and chemical warfare, explosives, and rocket fuels.

Many of the characteristics of particulates are influenced to a major extent by the particle size. For this reason, particle size has been accepted as a primary basis for characterizing particulates. However, with anything but homogeneous spherical particles, the measured "particle size" is not necessarily a unique property of the particulate but may be influenced by the technique used. Consequently, it is important that the techniques used for size analysis be closely allied to the utilization phenomenon for which the analysis is desired.

Types of particulate systems

System			Hydrosol	Aerosol	Powder
Continuous phase		Solid	Liquid	Gas	None (or gas)
Dispersed or particulate phase	Gas	Sponge	Foam	—	—
	Liquid	Gel	Emulsion	Mist Spray Fog Rain	—
	Solid	Alloy	Slurry Suspension	Fume Dust Snow Hail	Single phase Multi-phase (ores, flour)

Size is generally expressed in terms of some representative, average, or effective dimension of the particle. The most widely used unit of particle size is the micrometer (μm). Another common method is to designate the screen mesh that has an aperture corresponding to the particle size. The screen mesh normally refers to the number of screen openings per unit length or area; several screen standards are in general use.

Particulate systems are often complex. Primary particulates may exist as loosely adhering (as by van der Waals forces) particles called floes or as strongly adhering (as by chemical bonds) particulates called agglomerates. Primary particles are those whose size can only be reduced by the forceful shearing of crystalline or molecular bonds. *See* CHEMICAL BONDING; INTERMOLECULAR FORCES.

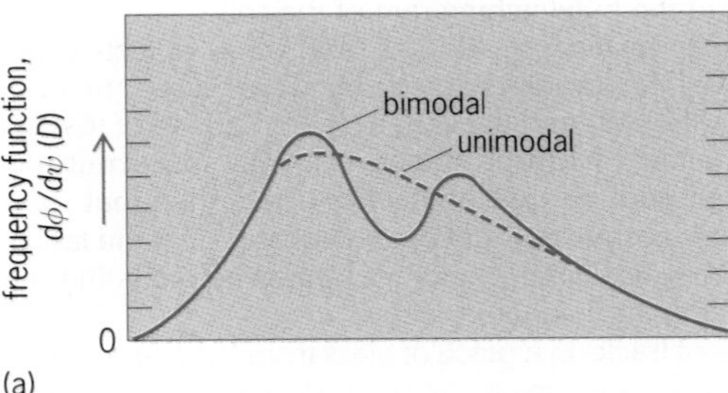

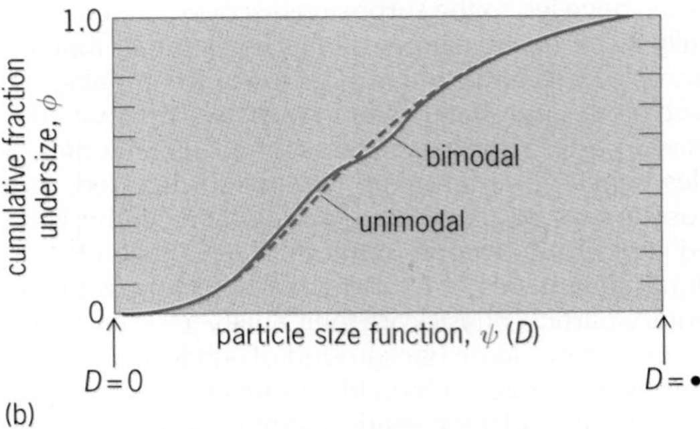

Methods for representing size distribution. (*a*) Frequency distribution. (*b*) Cumulative distribution.

Mechanical dispersoids are formed by comminution, decrepitation, or disintegration of larger masses of material, as by grinding of solids or spraying of liquids, and usually involve a wide distribution of particle sizes. Condensed dispersoids are formed by condensation of the vapor phase (or crystallization of a solution) or as the product of a liquid- or vapor-phase reaction; these are usually very fine and often relatively uniform in size. Condensed dispersoids and very fine mechanical dispersoids generally tend to flocculate or agglomerate to form loose clusters of larger particle size.

Most real systems are composed of a range of particle sizes. The two common general methods for representing size distribution graphically are shown in the illustration. The frequency distribution (illus. *a*) gives the fraction of particles $d\phi$ (on whatever basis desired) that lie in a given narrow size range dD as a function of the average size of the range (or of some function of the average size). A cumulative distribution (illus. *b*) is the integral of the frequency curve. It gives the fraction ϕ of the particles that are smaller or larger than a given size D. *See* INTEGRATION; STATISTICS.

If a particle suspended in a fluid is acted upon by a force, it will accelerate to a terminal velocity at which the resisting force due to fluid friction just balances the applied force. If a particle falls under the action of gravity, this velocity is known as the terminal gravitational settling velocity.

Particles suspended in a fluid partake of the molecular motion of the suspending fluid and hence acquire diffusional characteristics analogous to those of the fluid molecules. This random zigzag motion of the particles, commonly known as brownian motion, is obvious under the microscope for particles smaller than 1 μm. *See* BROWNIAN MOVEMENT. Charles E. Lapple

Paschen-Back effect An effect on spectral lines obtained when the light source is placed in a very strong magnetic field, first explained by F. Paschen and E. Back in 1921. In such a field the anomalous Zeeman effect, which is obtained with weaker fields, changes over to what is, in a first approximation, the normal Zeeman effect. The term "very strong field" is a relative one, since the field strength required depends on the particular lines being investigated. It must be strong enough to produce a magnetic splitting that is large compared to the separation of the components of the spin-orbit multiplet. *See* ATOMIC STRUCTURE AND SPECTRA; ZEEMAN EFFECT. F. A. Jenkins; W. W. Watson

Passeriformes The largest and most diverse order of birds, which is found worldwide, including the northernmost tip of Antarctica, most oceanic islands, and all terrestrial habitats. Over 5500 species of passerine birds (of which about 4400 species are Oscines, the true songbirds) are known, which is just over half of the known species of Recent birds. The most closely related orders may be other land birds, such as the Coraciiformes and the Piciformes. *See* CORACIIFORMES; PICIFORMES.

The Passeriformes are divided into four suborders. The composition of most families within the suborders and their relationships to one another are still much disputed. The Eurylaimi are Old World tropical and subtropical birds and may be an artificial group. The Furnarii and the Tyranni appear to be South American in origin and are still Neotropical in distribution (Central and South America), except for about 30 species of tyrant flycatchers and one species of cotingas which occur in North America. The area of origin of the Oscines is not known, but it may be Old World Gondwanaland after the separation of South America.

The perching birds are small to medium-sized, ravens (*Corvus corax*) being the largest. The wings are short to medium in length and vary from rounded to pointed. A few species, including lyrebirds, scrubbirds, and New Zealand wrens, are almost flightless. The tail varies from nearly absent to long. The bill is widely variable in shape. Passeriforms have legs of short-to-medium length that are usually strong. The four toes show the usual avian anisodactyl arrangement, three in front and a well-developed hallux behind. Most forms can walk or climb well, or both. Plumage varies widely, from all black to mostly white and from bright colors and bold patterns to cryptic coloration. Feeding habits and food choices show wide variation. Most species eat insects or small animals; however, some groups have become specialized on a diet of nectar, seeds, fruits, small vertebrates, or even leaves and the waxy covering of berries for part of the year.

Song is important to most perching birds for species recognition and courtship. The song in young Oscines is learned from adults. The pair bond is usually strong, with both sexes of most species incubating and caring for the young, which remain in the nest until they are able to fly. Many arctic and cold-temperature passerine species migrate to warmer areas for the cold months. Some of these migratory flights measure several thousand miles. *See* AVES; MIGRATORY BEHAVIOR. Walter J. Bock

Pasteurella A genus of gram-negative, nonmotile, nonsporulating, facultatively anaerobic coccobacillary to rod-shaped bacteria which are parasitic and often pathogens in many species of mammals, birds, and reptiles. It was named to honor Louis Pasteur in 1887. Genetic studies have shown that *Pasteurella*, together with *Haemophilus* and *Actinobacillus*, constitute a family, Pasteurellaceae.

The genus contains at least 10 species. *Pasteurella multocida* causes hemorrhagic septicemia in various mammals and fowl cholera, and is occasionally transmitted to humans, mainly in rural areas. Human pasteurellosis may include inflammation in bite and scratch lesions, infections of the lower respiratory tract and of the small intestine, and generalized infections with septicemia and meningitis. *Pasteurella canis* and *P. stomatis* may cause similar, though generally less severe, infections in humans after contact with domestic or wild animals. Although drug-resistant *Pasteurella* strains have been encountered, human *Pasteurella* infections are as a rule readily sensitive to the penicillins and a variety of other chemotherapeutic agents. *See* ANTIBIOTIC; DRUG RESISTANCE. Walter Mannheim

Pasteurellosis A variety of infectious diseases caused by the coccobacilli *Pasteurella multocida* and *P. haemolytica*; the term also applies to diseases caused by any *Pasteurella* species. All *Pasteurella* species occur as commensals in the upper respiratory and alimentary tracts of their various hosts. Although varieties of some species cause primary disease, many of the infections are secondary to other infections or result from various environmental stresses. *Pasteurella* species are generally extracellular parasites that elicit mainly a humoral immune response. Several virulence factors have been identified. *See* VIRULENCE.

Pasteurella multocida is the most prevalent species of the genus causing a wide variety of infections in many domestic and wild animals, and humans. It is a primary or, more frequently, a secondary pathogen of cattle, swine, sheep, goats, and other animals. As a secondary invader, it is often involved in pneumonic pasteurellosis of cattle (shipping fever) and in enzootic or mycoplasmal pneumonia of swine. It is responsible for a variety of sporadic infections in many animals, including abortion, encephalitis, and meningitis. It produces severe mastitis in cattle and sheep, and toxin-producing strains are involved in atrophic rhinitis, an economically important disease of swine. Hemorrhagic septicemia, caused by capsular type B strains, has been reported in elk and deer in the United States.

All strains of *P. haemolytica* produce a soluble cytotoxin (leukotoxin) that kills various leukocytes of ruminants, thus lowering the primary pulmonary defense. It is the principal cause of the widespread pneumonic pasteurellosis of cattle. Other important diseases caused by certain serotypes of *P. haemolytica* are mastitis of ewes and septicemia of lambs.

All of the *Pasteurella* species can be isolated by culturing appropriate clinical specimens on blood agar. Multiple drug resistance is frequently encountered. Treatment is effective if initiated early. Among the drugs used are penicillin and streptomycin, tetracyclines, chloramphenicol, sulphonamides, and some cephalosporins. Sound sanitary practices and segregation of affected animals may help limit the spread of the major pasteurelloses. Live vaccines and bacterins (killed bacteria) are used for the prevention of some. *See* PASTEURELLA. G. R. Carter

Pasteurization The treatment of foods or beverages with mild heat, irradiation, or chemical agents to improve keeping quality or to inactivate disease-causing microorganisms. Originally, Louis Pasteur observed that spoilage of wine and beer could be prevented by heating them a few minutes at 122–140°F (50–60°C). Today pasteurization as a thermal treatment is applied to many foods. In foods consumed directly, destruction of pathogens to protect consumer health is paramount, while in products without public health hazards, control of spoilage microorganisms is primary. In fermentation processes, the raw material may be pasteurized to eliminate microorganisms that produce abnormal end products, or the final product may be heated to stop the fermentation at the desired level.

Milk and dairy products probably represent the most widespread use of pasteurization. Several time-temperature combinations have been approved as equivalent: 145°F (63°C) for 30 min; 161°F (72°C) for 15 s; 191°F (89°C) for 1 s; 194°F (90°C) for 0.5 s; 201°F (94°C) for 0.1 s; 204°F (96°C) for 0.05 s; or 212°F (100°C) for 0.01 s. These precise heat treatments are based on the destruction of the rickettsia *Coxiella burnetii*, which is considered the most heat-resistant nonsporeforming pathogen found in milk. Absolute control of the thermal treatment is essential for safety. Pasteurization of milk has successfully eliminated the spread of diseases such as diphtheria, tuberculosis, and brucellosis through contaminated milk. *See* DAIRY MACHINERY; FOOD MANUFACTURING; MALT BEVERAGE; MILK; WINE. Francis F. Busta

Patent Common designation for letters patent, which is a certificate of grant by a government of an exclusive right with respect to an invention for a limited period of time. A United States patent confers the right to exclude others from making, using, or selling the patented subject matter in the United States and its territories. Portions of those rights deriving naturally from it may be licensed separately, as the rights to use, to make, to have made, and to lease. Any violation of this right is an infringement.

An essential substantive condition which must be satisfied before a patent will be granted is the presence of patentable invention or discovery. To be patentable, an invention or discovery must relate to a prescribed category of contribution, such as process, machine, manufacture, composition of matter, plant, or design. In the United States there are different classes of patents for different members of these categories. Donald W. Banner

Paterinida A small extinct order of inarticulate brachiopods that ranges in age from Early Cambrian to Middle Ordovician. The shell is chitinophosphatic in composition, and its outline is circular or elliptical. The ventral (pedicle) valve is more convex than the dorsal (brachial) valve. The pedicle was absent or emerged between the valves. The muscle scars are unusual compared to other inarticulates, and form narrow triangular tracks radiating from the posterior extremity of each valve. Except for their shell composition and lack of articulation, the Paterinida resemble articulate brachiopods. Members were presumably epifaunal and sessile. *See* BRACHIOPODA; INARTICULATA. Merrill W. Foster

Pathogen Any agent capable of causing disease. The term pathogen is usually restricted to living agents, which include viruses, rickettsia, bacteria, fungi, yeasts, protozoa, helminths, and certain insect larval stages. *See* DISEASE.

Pathogenicity is the ability of an organism to enter a host and cause disease. The degree of pathogenicity, that is, the comparative ability to cause disease, is known as virulence. The terms pathogenic and nonpathogenic refer to the relative virulence of the organism or its ability to cause disease under certain conditions. This ability depends not only upon the properties of the organism but also upon the ability of the host to defend itself (its immunity) and prevent injury. The concept of pathogenicity and virulence has no meaning without reference to a specific host. For example, gonococcus is capable of causing gonorrhea in humans but not in lower animals. *See* MEDICAL MYCOLOGY; MEDICAL PARASITOLOGY; PLANT PATHOLOGY; PLANT VIRUSES AND VIROIDS; VIRULENCE. Daniel N. Lapedes

Pathology The study of the etiologies, mechanisms, and manifestations of disease. Techniques and knowledge gained from other disciplines, including anatomy, physiology, microbiology, biochemistry, and histology, are utilized. The information obtained from the study of pathology is necessary prior to developing methods with which to control and prevent disease.

With the light microscope it became possible to correlate the observed signs and symptoms in an individual with cellular changes. In its early stages pathology was very descriptive. Diseases were understood and categorized in part, by how gross and microscopic anatomy was altered. In the last half of the nineteenth century, by using this approach to pathology, coupled with microbiological techniques, it was learned that the major causes of human death were biotic agents: protozoans, bacteria, viruses, and fungi. Infectious diseases took a heavy toll in human lives. Better sanitation and public health measures were instrumental in controlling these diseases, and the production of antibiotics and immunization procedures further reduced their importance. It is now apparent that all diseases reflect changes at the molecular level. Scientists are beginning to understand what these biochemical alterations are in some diseases.

There are many branches of pathology. Divisions are made depending upon focus of interest. Clinical pathology is concerned with diagnosis of disease. As medicine has expanded, subspecialties such as surgical pathology and neuropathology have developed. Experimental pathology attempts to study disease mechanisms under controlled conditions. General pathology covers all areas, but in less detail, and serves in medical education.

A relatively new area of pathology is environmental pathology, which deals with disease processes resulting from physical and chemical agents. At present, the leading causes of death have environmental agents as the known or suspected major etiologic factors; these diseases include heart disease, atherosclerosis, and cancer. It is believed that with understanding, many such diseases, like those produced in response to biotic agents, can be brought under control. *See* DISEASE. N. Karle Mottet; Carol Quaife

Pathotoxin A chemical of biological origin, other than an enzyme, that plays an important causal role in a plant disease. Most pathotoxins are produced by plant pathogenic fungi or bacteria, but some are produced by higher plants, and one has been reported to be the product of an interaction between a plant and a bacterial pathogen. Some pathogen-produced pathotoxins are highly selective in that they cause severe damage and typical disease symptoms only on plants susceptible to the pathogens that produce them. Others are nonselective and are equally toxic to plants susceptible or resistant to the pathogen involved. A few pathotoxins are species-selective, and are damaging to many but not all plant species. In these instances, some plants resistant to the pathogen are sensitive to its toxic product. *See* PLANT PATHOLOGY. Harry Wheeler

Pattern formation (biology) The mechanisms that ensure that particular cell types differentiate in the correct location within the embryo and that the layers of cells bend and grow in the correct relative positions. Pattern formation is one of four processes that underlie development, the others being growth, cell diversification, and morphogenesis. *See* ANIMAL GROWTH; CELL DIFFERENTIATION; MORPHOGENESIS; PLANT GROWTH.

Pattern formation is the creation of a predictable arrangement of cell types in space during embryonic development. The types of patterns of cell types found in animals and plants can be conveniently described as simple or complex. Simple patterns involve the spatial arrangement of identical or equivalent structures such as bristles on the leg of a fly, hairs on a person's head, or leaves on a plant. Such equivalent patterns are thought to be produced by mechanisms that are the same or very similar in the fly and the plant. Complex patterns are those that are made up of parts that are not equivalent to one another. In the vertebrate limb, for example, the structure of the arm is different at each level, with one bone (humerus) in the upper arm, two bones (radius and ulna) in the lower arm, and a complex set of bones making up the wrist and the hand. How are such nonequivalent parts patterned during development? The theoretical framework that allows a basis for understanding how such patterns arise is called positional information. Two stages exist in the positional information framework. First, a cell must become aware of its position within a developing group, or field, of cells. This specification of cellular position requires a mechanism by which each cell within a field can obtain a unique value or address. The second component is the interpretation of the positional address by a cell to manifest a particular cell type by the expression of a particular set of genes. *See* DEVELOPMENTAL BIOLOGY; EMBRYONIC DIFFERENTIATION.

Nigel Holder

Pauropoda A class, and perhaps the most obscure group, of the Myriapoda. They are pale creatures, no more than 0.04–0.08 in. (1–2 mm) in length, inhabiting damp situations in leaf litter, under bark, stones, and debris, and in humus and similar detritus. Apparently very widely distributed as a class, they have been undiscovered only in deserts and in the arctic and antarctic regions.

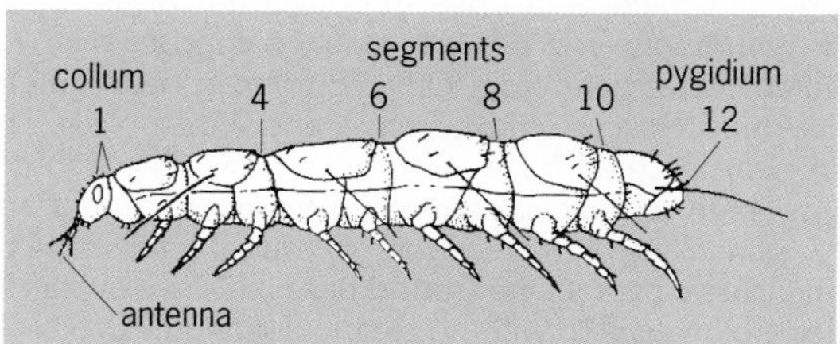

Pauropus silvaticus. **(*After R. E. Snodgrass, A Textbook of Arthropod Anatomy, Cornell University Press, 1952*)**

Like millipedes, they are progoneate and have one pair of maxillae, and their trunk segments display a certain degree of amalgamation. Their peculiar bifurcate antennae and adult complement of 12 trunk segments with 9 pairs of functional legs are distinctive within the myriapod complex (see illustration). All pauropods lack eyes, spiracles, tracheae, and a circulatory system. The class currently consists of 4 families; there are over 500 species known.

Ralph E. Crabill

Pavement An artificial surface laid over the ground to facilitate travel. A pavement's ability to support loads depends primarily upon the magnitude of the load, how often it is applied, the supporting power of the soil underneath, and the type and thickness of the pavement structure. Before the necessary thickness of a pavement can be calculated, the volume, type, and weight of the traffic (the traffic load) and the physical characteristics of the underlying soil must be determined.

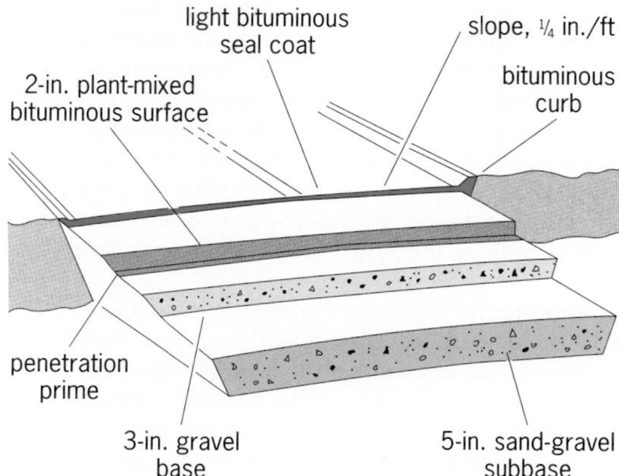

Flexible pavement design for a city collector street with maximum traffic load of 5 tons (4.5 metric tons) per axle. Right-of-way is 60 ft (18 m) wide and the pavement width is 38 ft (11.6 m). Berms or boulevards at the sides are sloped in order to drain toward the street. 1 in. = 2.5 cm.

Once the grading operation has been completed and the subgrade compacted, construction of the pavement can begin. Pavements are either flexible or rigid. Flexible pavements, which are composed of aggregate (sand, gravel, or crushed stone) and bituminous material (see illustration), have less resistance to bending than do rigid pavements, which are made of concrete. Both types can be designed to withstand heavy traffic. Selection of the type of pavement depends, among other things, upon (1) estimated construction costs; (2) experience of the highway agency doing the work with each of the two types; (3) availability of contractors experienced in building each type; (4) anticipated yearly maintenance costs; and (5) experience of the owner in maintenance of each type. *See* CONCRETE; HIGHWAY ENGINEERING.

Archie N. Carter

Pawl The driving link or holding link of a ratchet mechanism, also called a click or detent. In the illustration the driving pawl at *A*, forced upward by lever *B*, engages the teeth of the ratchet wheel and rotates it counterclockwise. Holding pawl *C* prevents clockwise rotation of the wheel when the pawl at *A* is making its return stroke, Pawl and ratchet are an open, upper pair.

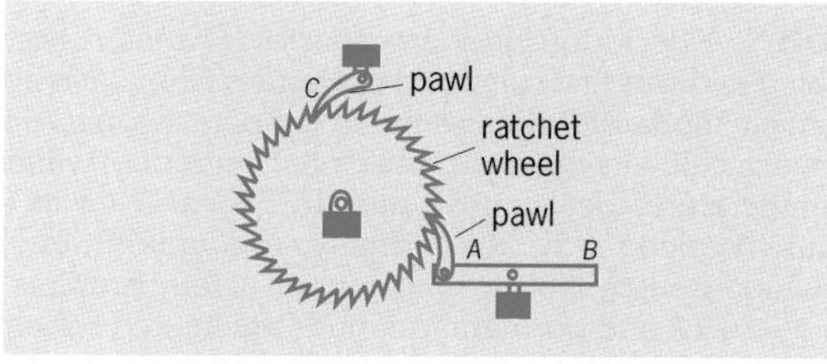

Holding and driving pawls with a ratchet wheel.

Driving and holding pawls likewise engage rack teeth on the plunger of a ratchet lifting jack, such as those supplied with automobiles. A ratchet wheel with a holding pawl only, acting as a safety brake, is fastened to the drum of a capstan, winch, or other powered hoisting device.

A double pawl can drive in either direction or be easily reversed in holding. A cam pawl prevents the wheel from turning clockwise by a wedging action while permitting free counterclockwise rotation. This technique is used in the automobile hill-holder to prevent the vehicle from rolling backward. *See* ESCAPEMENT.

Douglas P. Adams

Paxillosida An order of sea stars and members of the class Asteroidea. The name is derived from the club-shaped plates, or paxillae, that form the sea star's upper skeletal surface and that have tiny spinelets or granules covering their tips. Paxillosida encompasses six families, the largest being the Astropectinidae, Luidiidae, and Porcellanasteridae. The Ctenodiscidae, Goniopectinidae, and Radiasteridae are represented by comparatively few members. Astropectinids and luidiids are primarily predators of mollusks and other echinoderms. The former are found over a wide range of depths, whereas the latter live in relatively shallow water. Porcellanasterids are deep-water asteroids that swallow sediment in bulk as they bury themselves to a level just below the surface.

In addition to the presence of paxillae, paxillosidans are characterized by a number of unusual features that have been thought to indicate a primitive phylogenetic position among living asteroids. Although the tube feet of most asteroids have suckered disks, those of most paxillosidans are pointed. The digestive system of most asteroids is relatively complex and complete, terminating in an anus, whereas that of the paxillosidans is simple, saclike, and lacking an anus in some members. Most asteroids can extrude their stomach during feeding, but that ability is limited in paxillosidans. A brachiolarian larval stage has been recognized in the development of most asteroids, yet that stage is thought to be absent from paxillosidans. *See* ASTEROIDEA; ECHINODERMATA. Daniel B. Blake

Pea The pea is one of the oldest cultivated crops. It is a native to western Asia from the Mediterranean Sea to the Himalaya Mountains. It appears to have been carried to Europe as early as the time of the lake dwellers of prehistoric times. Peas were introduced into China from Persia about A.D. 400; they were introduced into the United States in very early Colonial days.

Garden peas (*Pisum sativum*) have wrinkled seed coats at maturity when dry; field peas (*P. arvense*) have a smooth seed coat. Both types are annual leafy plants. Each leaf bears three pairs of leaflets and ends in a slender tendril. Five to nine round seeds are enclosed in a pod about 3 in. (7.5 cm) long. Seed color varies from white to cream, green, yellow, or brown. Smooth-seeded varieties may be harvested fresh for freezing or canning, or harvested dry as edible peas. Dry peas may be split or ground and prepared in various ways, such as for split-pea soup.

Wisconsin, Washington, Minnesota, Oregon, Illinois, New York, Pennsylvania, Utah, and Idaho lead in the production of peas harvested green. Kenneth J. Morrison

Peach A deciduous fruit tree species (*Prunus persica*) that originated and was first cultivated in western China. It is adapted to relatively moderate climates in the temperate zone. Although most peach cultivars require a substantial amount of winter chilling (temperatures between 32 and 45°F, or 0–7°C) to ensure adequate breaking of winter dormancy and uniform budbreak, peach wood is susceptible to winter injury at temperatures below −15°F (−25°C) and dormant fruit buds are injured by temperatures below 0°F (−18°C). Consequently, commercial cultivation is limited to lower latitudes in the temperate zone or to higher latitudes where large bodies of water have a moderating influence on climate. The principal peach-growing regions in North America, ranked in order of commercial production, are central California, Georgia and the Carolinas, the mid-Atlantic region, the Great Lakes region, and the Pacific northwestern region. Other important peach-growing regions in the world include Italy, southern France, Spain, Japan, China, Argentina, southern Brazil, Chile, South Africa, and southeastern Australia. *See* FRUIT; FRUIT, TREE; ROSALES.

Peach cultivars can vary greatly and are usually distinguished by their fruit types. Peach fruits are covered with short epidermal trichomes called fuzz (smooth-skinned peaches are called nectarines) and at maturity are usually yellow or white with a red blush. The internal flesh is also yellow or white. Clingstone cultivars have a relatively firm flesh that adheres to the pit at maturity, and are primarily used for canning. Freestones usually have a softer flesh that separates from the pit at fruit maturity, and are primarily used for the fresh market, freezing, and drying. Theodore M. De Jong

Peanut A self-pollinated, one- to six-seeded legume which is cultivated throughout the tropical and temperate climates of the world. The oil, expressed from the seed, is of high quality, and a large percentage of the annual world production is used for this purpose. In the United States some 65% goes into the cleaned and shelled trade, the end products of which are roasted or salted peanuts, peanut butter, and confections. *See* ROSALES.

Botanically, peanuts may be divided into three main types, Virginia, Spanish, and Valencia, based on branching order and pattern and the number of seeds per pod. The USDA Marketing Standards includes an additional type, Runner, which refers to the small-seeded Virginia type produced in Georgia and Alabama. *See* SEED.

The peanut's most distinguishing characteristic is the yellow flower, which resembles a butterfly (papilionaceous) and is borne above ground. The pod, a one-loculed legume, splits under pressure along a longitudinal ventral suture. Pod size varies, and seed weight varies from 0.08 to 0.2 oz (0.2 to 5 g). The number of seeds per pod usually is two in the Virginia type, two or three in the Spanish, and three to six in the Valencia. *See* LEGUME. Astor Perry

Pear Any of approximately 20 species of deciduous tree fruits in the genus *Pyrus*. About half of the species are native to Europe, North Africa, and the Middle East around the Mediterrean Sea; the others are native to Asia. Pear culture is documented to have started as early as 1100 B.C. Pears are best adapted to temperate climates with warm, dry summers and cold winters. They require winter cold to break the dormant period but are injured by temperatures below −10 to −15°F (−23 to −26°C). Commercial pear production in the United States is concentrated in the interior valleys of California, Oregon, and Washington.

Nearly all United States pear production is of the European pear, *P. communis*. The Bartlett variety comprises over 75% of the United States pear crop. Other European pear varieties include d'Anjou, Bosc, Comice, Seckel, and Winter Nelis. The European pear is noted for its soft, juicy flesh. The skin color is medium green to yellow, depending on fruit maturity and the variety. Skin texture can be smooth or rough. Fruit shape ranges from the classic pear shape (round base with narrow neck) to a rounded oblong shape with no clearly defined neck area.

The crisp-fleshed Asian pear, *P. pyrifolia*, is the second most popular type of pear grown worldwide. The Asian pear is characterized by a crisp, juicy flesh that has a gritty texture, and has been referred to as the sand pear. The fruit shape is round, the skin color is generally yellow to amber at maturity, and the skin texture is smooth or corky. *Pyrus communis* and *P. pyrifolia* hybrids, such as Kieffer and Leconte, have been developed for limited commercial use in the southeastern United States. *Pyrus ussuriensis* has also been selected for Asian pear varieties. The snow pear, *P. nivalis*, is produced in Europe for cider and perry (a fermented liquor). *See* FRUIT; FRUIT, TREE. Kathleen M. Williams

Pearl Any mollusk-formed calcareous concretion that displays an orient and is lustrous. There are two major groups of bivalved mollusks in which gem pearls may form: the salt-water pearl oyster (*Pinctada*), and a number of genera of freshwater clams. Usually, jewelers refer to salt-water pearls as Oriental pearls, regardless of their place of discovery, and to those from freshwater bivalves as freshwater pearls.

Between the body mass and the valves of the mollusk extends a curtainlike tissue called the mantle. In order for a pearl to form, a tiny object such as a parasite or a grain of sand must work through the mantle. When this happens, secretion of nacre around the invading object builds a pearl within the body of the mollusk. Whole pearls form within the body mass of the mollusk, in contrast to blister pearls, which form as protrusions on the inner surface of the shell. Edible oysters produce lusterless concretions, but never pearls.

The substitute for natural pearls, to which the name cultured pearl has been given, is usually made by inserting a large bead into a mollusk to be coated with nacre. Richard T. Liddicoat, Jr.

Peat A dark-brown or black residuum produced by the partial decomposition and disintegration of mosses, sedges, trees, and other plants that grow in marshes and other wet places. Forest-type peat, when buried and subjected to geological influences of pressure and heat, is the natural forerunner of most coal. Moor peat is formed in relatively elevated, poorly drained moss-covered areas, as in parts of Northern Europe. *See* COAL; HUMUS. Gilbert H. Cady

Pebble mill A tumbling mill that grinds or pulverizes materials without contaminating them with iron. Because the pebbles have lower specific gravity than steel balls, the capacity of a given size shell with pebbles is considerably lower than with steel balls. The lower capacity results in lower power consumption. The shell has a nonmetallic lining to further prevent iron contamination, as in pulverizing ceramics or pigments (see illustration). Selected hard pieces of the material being ground can be used as pebbles to further prevent contamination. *See* TUMBLING MILL. Ralph M. Hardgrove

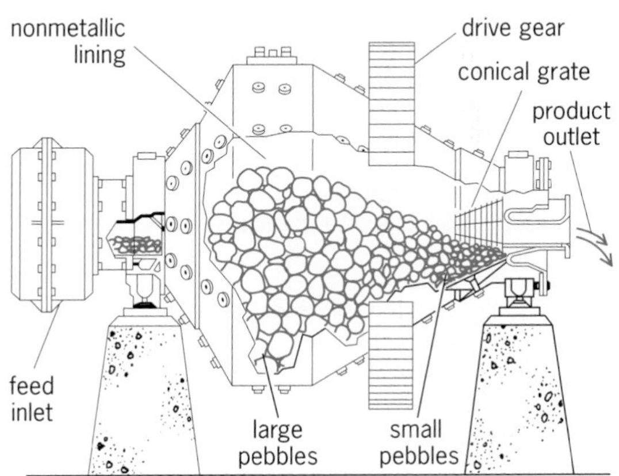

Diagrammatic sketch of a conical pebble mill.

Pecan A large tree (*Carya illinoensis*) of the family Juglandaceae, and the nut from this tree. Native to valleys of the Mississippi River and tributaries as far north as Iowa, to other streams of Texas, Oklahoma, and northern and central Mexico, this nut tree has become commercially important throughout the southern and southwestern United States and northern Mexico. Ralph H. Sharpe

Pectin A group of polysaccharides occurring in the cell walls and intercellular layers of all land plants. They are extractable with hot water, dilute acid, or ammonium oxalate solutions. Pectins are precipitated from aqueous solution by alcohol and are commercially used for their excellent gel-forming ability.

Commercially, the primary source of pectin is the peel of citrus fruits such as lemon and lime, although orange and grapefruit may be used. A secondary source is apple pomace and sunflower heads.

Pectin is widely used in the food industry, principally in the preparation of gels. It is used as a base for jelly and as a stabilizer in some dairy products and frozen desserts, such as sherbet, and also as edible protective coatings for sausages, almonds, candied dried fruit, and soft dates. *See* GEL; POLYSACCHARIDE. Roy L. Whistler; James R. Daniel

Pectolite A mineral inosilicate with composition $Ca_2NaSi_3O_8(OH)$. The hardness is 5 on Mohs scale, and the specific gravity is 2.75. The mineral is colorless, white, or gray with a vitreous to silky luster. Pectolite is found in the United States at Paterson, Bergen Hill, and Great Notch, New Jersey. *See* SILICATE MINERALS. Cornelius S. Hurlbut, Jr.

Pediculosis Human infestation with lice. There are two biological varieties of the human louse, *Pediculus humanus*, var. *capitis* and var. *corporis*, each showing a strong preference for a specific location on the human body. *Pediculus humanus capitis* colonizes the head and *P. h. corporis* lives in the body-trunk region. These lice are wingless insects which are ectoparasites. Their mouthparts are modified for piercing skin and sucking blood. The terminal segments of their legs are modified into clawlike structures which are utilized to grasp hairs and clothing fibers.

Lice are important vectors of human diseases. Their habit of sucking blood and their ability to crawl rapidly from one human to another transmit such diseases as typhus (rickettsial) and relapsing fever (spirochetal). The body fluids and feces of infected lice transmit these diseases. Richard Sudds

Pedinoida An order of Diadematacea, making up those genera which possess solid spines and a rigid test. The ambulacra show typical diadematoid structure, and the tubercles are noncrenulate. The single known family, Pedinidae, includes 15 genera, ranging from the Late Triassic onward, though only one genus, *Caenopedina*, survives today. *See* DIADEMATACEA. Howard B. Fell

Pedology Defined narrowly, a science that is concerned with the nature and arrangement of horizons in soil profiles; the physical constitution and chemical composition of soils; the occurrence of soils in relation to one another and to other elements of the environment such as climate, natural vegetation, topography, and rocks; and the modes of origin of soils. Pedology so defined does not include soil technology, which is concerned with uses of soils.

Broadly, pedology is the science of the nature, properties, formation, distribution, and function of soils, and of their response to use, management, and manipulation. The first definition is widely used in the United States and less so in other countries. The second definition is worldwide. *See* SOIL; SOIL MECHANICS. Roy W. Simonson

Pegasidae The sea moths or sea dragons, a small family of peculiar actinopterygian fishes also known as the Hypostomides. The body is encased in a broad, bony framework anteriorly and has bony rings posteriorly, simulating the seahorses and pipefishes (see illustration). Unlike those fishes, which have the mouth at the tip of the produced snout, sea moths have enlarged nasal bones which form a rostrum that projects well forward of the small, toothless mouth. The greatly expanded horizontal pectoral fin belies its appearance and does not function in aerial gliding. There is no swim bladder. *See* GASTEROSTEIFORMES.

There are two genera, *Pegasus* and *Eurypegasus*, and five species. They live amidst vegetation on Indo-Pacific shores from East Africa to Japan, Australia, and Hawaii. They rarely exceed

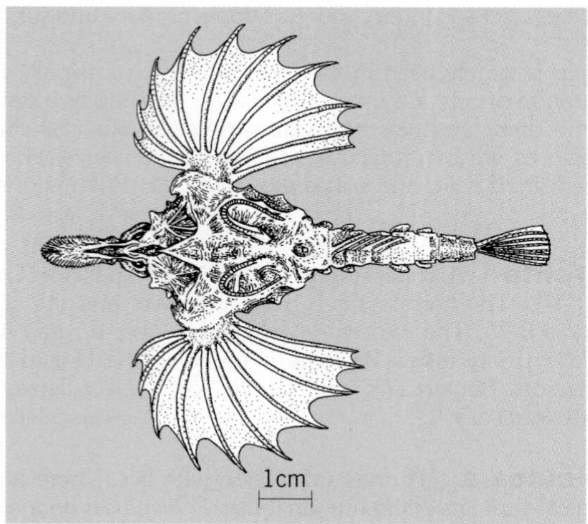

Sea moth (*Pegasus draconis*). (*After D. S. Jordan and J. O. Snyder, vol. 24, Leland Stanford University Contributions to Biology, 1901*)

4 in. (10 cm) in length. There is no fossil record. *See* ACTINOPTERYGII.

Reeve M. Bailey

Pegasus The Flying Horse, a giant northern constellation (see illustration). It is noticeable for a huge square of stars (of which one is now considered to be in neighboring Andromeda), the Great Square of Pegasus. Each side is half again

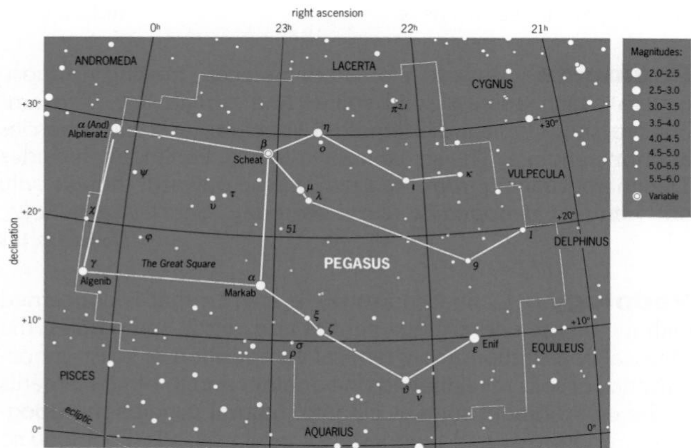

Modern boundaries of the constellation Pegasus, the Flying Horse. The celestial equator is 0° of declination, which corresponds to celestial latitude. Right ascension corresponds to celestial longitude, with each hour of right ascension representing 15° of arc. Apparent brightness of stars is shown with dot sizes to illustrate the magnitude scale, where the brightest stars in the sky are 0th magnitude or brighter and the faintest stars that can be seen with the unaided eye at a dark site are 6th magnitude. (*Wil Tirion*)

larger than a fist held at the end of the observer's outstretched arm. The Great Square is marked by Alpheratz (now alpha Andromedae), Scheat (beta Pegasi), Markab (alpha Pegasi), and Algenib (gamma Pegasi). Pegasus is prominent in the autumn evening sky. *See* ANDROMEDA; CONSTELLATION.

Jay M. Pasachoff

Pegmatite Exceptionally coarse-grained and relatively light-colored crystalline rock composed chiefly of minerals found in ordinary igneous rocks. Extreme variations in grain size also are characteristic, and close associations with dominantly fine-grained aplites are common. Pegmatites are widespread and very abundant where they occur, especially in host rocks of Precambrian age, but their aggregate volume in the Earth's crust is small. Many pegmatites have been economically valuable as sources of clays, feldspars, gem materials, industrial crystals, micas, silica, and special fluxes, as well as beryllium, bismuth, lithium, molybdenum, rare-earth, tantalumniobium, thorium, tin, tungsten, and uranium minerals. *See* APLITE; IGNEOUS ROCKS.

Essential minerals (1) in granitic pegmatities are quartz, potash feldspar, and sodic plagioclase; (2) in syenitic pegmatites, alkali feldspars with or without feldspathoids; and (3) in diorite and gabbro pegmatities, soda-lime or lime-soda plagioclase. Varietal minerals such as micas, amphiboles, pyroxenes, black tourmaline, fluorite, and calcite further characterize the pegmatites of specific districts. Accessory minerals include allanite, apatite, beryl, garnet, magnetite, monazite, tantalite-columbite, lithium tourmaline, zircon, and a host of rarer species.

Richard H. Jahns

Pelecaniformes A small order of diverse aquatic, mainly marine, fish-eating birds that includes the pelicans, boobies, and cormorants. The members of the order, found worldwide, are very different, and some researchers believe that this "order" is an artificial group; however, all members are characterized by several unique features (for example, a bare throat pouch). The five suborders and nine families of the order are listed below.

Order Pelecaniformes
- Suborder Phaethontes
  - Family: Prophaethontidae (fossil; Eocene of England)
  - Phaethontidae (tropic birds; 3 species; oceanic pantropical)
- Suborder Odontopterygia
  - Family Pelagornithidae (fossil; lower Eocene to Miocene; worldwide marine)
- Suborder Pelecani
  - Family Pelecanidae (pelicans; 8 species; worldwide)
- Suborder Sulae
  - Family: Sulidae (boobies, gannets; 9 species; oceanic worldwide)
  - Plotopteridae (fossil; Oligocene to Miocene; North Pacific)
  - Phalacrocoracidae (cormorants; 33 species; worldwide)
  - Anhingidae (anhingas; 4 species; freshwater pantropical)
- Suborder Fregatae
  - Family Fregatidae (frigate birds; 5 species; oceanic worldwide)

The pelecaniforms are medium-sized to large marine birds that are characterized by a foot with four toes united in a common web (totipalmate). The legs are short and stout or weak, and they are used for swimming and perching. The birds are poor walkers, but all living forms—with the exception of the flightless Galápagos cormorant—are excellent fliers, particularly the frigate birds. The bill varies widely in size and shape, from the short stout bill of tropical birds to the long hooked bill of frigate birds and the long, flat, flexible bill of pelicans. All species have a bare throat pouch; the largest such pouch is found in the pelicans, which use it as a fishing net. Pelecaniform plumage is black, gray, or white. All pelecaniforms feed on fish, crustaceans, and squid that are caught by diving from the air or the water surface, are taken from the water surface, or are stolen from other birds.

Walter J. Bock

Pelmatozoa A division of the Echinodermata made up of those forms which are anchored to the substrate during at least a part of the life history. Formerly treated as a formal unit of classification with the rank of subphylum, pelmatozoans are now realized to be a heterogeneous assemblage of forms with similar habits but dissimilar ancestry, their common features having arisen by convergent evolution. Most pelmatozoan echinoderms are members of the subphylum Crinozoa, but some echinozoans also exhibit a sedentary, anchored life, with modifications for such existence. *See* CRINOZOA; ECHINODERMATA; ECHINOZOA; ELEUTHEROZOA. Howard B. Fell

Peltier effect A phenomenon discovered in 1834 by J. C. A. Peltier, who found that at the junction of two dissimilar metals carrying a small current the temperature rises or falls, depending upon the direction of the current. In view of experiments, which establish that the rate of intake or output of heat is proportional to the magnitude of the current, it can be shown that an electromotive force resides at a junction. Electromotive forces of this type are called Peltier emf's. *See* SEEBECK EFFECT; THERMOELECTRICITY; THOMSON EFFECT. John W. Stewart

Pelycosauria A paraphyletic grade of fossil (extinct) synapsids that is ancestral to later forms, such as the therapsids and mammals. The earliest pelycosaur-grade synapsids are known from fragmentary material recovered from the Middle Pennsylvanian of North America, approximately 310 million years in age (Ma). By Late Pennsylvanian and Early Permian times (305–270 Ma), however, synapsids were the dominant terrestrial vertebrates and had diversified into herbivorous, insectivorous, and carnivorous groups.

Paleontologists currently recognize six family-level pelycosaur clades (Caseidae, Edaphosauridae, Eothyrididae, Ophiacodontidae, Sphenacodontidae, and Varanopidae) that encompass approximately 30 genera. Ecologically, large-bodied sphenacodontids, such as the sailback *Dimetrodon*, were the earliest land-living vertebrates adapted to prey upon similarly sized animals. Caseids and edaphosaurids, by contrast, were among the first terrestrial herbivores. Pelycosaur-grade synapsids are not known after the Middle Permian (260 Ma). *See* MAMMALIA; SYNAPSIDA; THERAPSIDA. Christian A. Sidor

Pendulum A rigid body mounted on a fixed horizontal axis, about which it is free to rotate under the influence of gravity. The period of the motion of a pendulum is virtually independent of its amplitude and depends primarily on the geometry of the pendulum and on the local value of g, the acceleration of gravity. Pendulums have therefore been used as the control elements in clocks, or inversely as instruments to measure g.

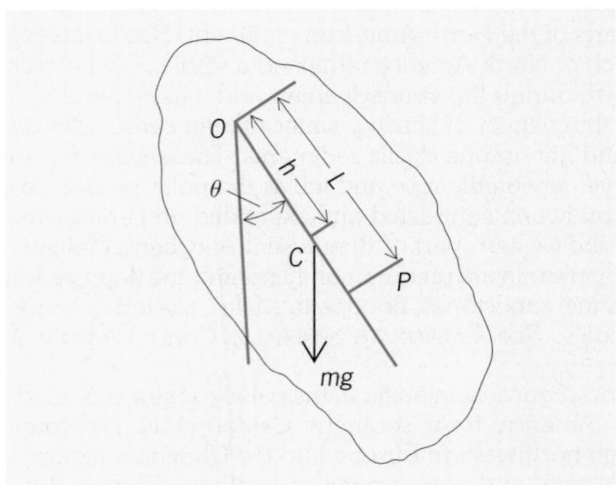

Schematic diagram of a pendulum. *O* represents the axis, *C* is the center of mass, and *P* the center of oscillation.

Motion. In the schematic representation of a pendulum shown in the illustration, O represents the axis and C the center of mass. The line OC makes an instantaneous angle θ with the vertical. In rotary motion of any rigid body about a fixed axis, the angular acceleration is equal to the torque about the axis divided by the moment of inertia I about the axis. If m represents the mass of the pendulum, the force of gravity can be considered as the weight mg acting at the center of mass C.

If the amplitude of motion is small, the motion is simple harmonic. The period T, time for a complete vibration (for example, from the extreme displacement right to the next extreme displacement right), is given by Eq. (1).

$$T = 2\pi\sqrt{I/mgh} \tag{1}$$

See HARMONIC MOTION.

The actual form of a pendulum often consists of a long, light bar or a cord that serves as a support for a small, massive bob. The idealization of this form into a point mass on the end of a weightless rod of length L is known as a simple pendulum. An actual pendulum is sometimes called a physical or compound pendulum. In a simple pendulum the lengths h and L become identical, and the moment of inertia I equals mL^2. Equation (1) for the period becomes Eq. (2).

$$T = 2\pi\sqrt{L/g} \tag{2}$$

Center of oscillation. Equation (2) can be used to define the equivalent length of a physcal pendulum. Comparison with Eq. (1) shows that Eq. (3) holds. The point P on line OC of the

$$L = I/mh \tag{3}$$

illustration, whose distance from the axis O equals L, is called the center of oscillation. Points O and P are reciprocally related to each other in the sense that if the pendulum were suspended at P, O would be the center of oscillation.

Types. Kater's reversible pendulum is designed to measure g, the acceleration of gravity. It consists of a body with two knife-edge supports on opposite sides of the center of mass as at O and P (and with at least one adjustable knife-edge). If the pendulum has the same period when suspended from either knife-edge, then each is located at the center of oscillation of the other, and the distance between them must be L, the length of the equivalent simple pendulum. The value for g follows from Eq. (2).

The ballistic pendulum is a device to measure the momentum of a bullet. The pendulum bob is a block of wood into which the bullet is fired. The bullet is stopped within the block and its momentum transferred to the pendulum. This momentum is determined from the amplitude of the pendulum swing. *See* BALLISTICS.

The spherical pendulum is a simple pendulum mounted on a pivot so that its motion is not confined to a plane. The bob then moves over a spherical surface. A Foucault pendulum is a spherical pendulum suspended so that its plane of oscillation is free to rotate. Its purpose is to demonstrate the rotation of the Earth. *See* FOUCAULT PENDULUM; SCHULER PENDULUM. Joseph M. Keller

Penicillin One of the beta-lactam antibiotics, all of which possess a four-ring beta-lactam structure fused with a five-membered thiazolidine ring. These antibiotics are nontoxic and kill sensitive bacteria during their growth stage by the inhibition of biosynthesis of their cell wall mucopeptide. *See* PLANT CELL.

The antibiotic properties of penicillin were first recognized by A. Fleming in 1928 from the serendipitous observation of a mold, *Penicillium notatum*, growing on a petri dish agar plate of a staphylococcal culture. The mold produced a diffuse zone which lysed the bacterial cells. Commercial production of penicillin came from the pioneer work of E. Chain and

H. W. Florey in 1938. Penicillin (as penicillin G) was made available to the allied troops in Europe in the latter part of World War II.

Penicillin is produced from the fungal culture *P. chrysogenum* that was isolated from a moldy cantaloupe. The biosynthesis of penicillin is known in detail, and all the enzymes involved in the formation of this secondary metabolite have been isolated and purified.

The fermented penicillin G and penicillin V are susceptible to destruction by an enzyme (beta-lactamase) produced by certain bacteria which makes them resistant. The penicillins methicillin, oxacillin, nafcillin, cloxacillin and dicloxacillin are resistant to hydrolysis by beta-lactamases and are used to treat staphylococcal infections. Cloxacillin and dicloxacillin are used orally. Ampicillin and amoxicillin are penicillins with extended spectra, and they are effective against many gram-negative bacteria. They are used mainly orally against streptococci and other respiratory-tract pathogens, including *Haemophilus influenzae*, in the treatment of sinusitis, bronchitis, and pneumonia. They are used extensively in pediatrics and against *Listeria monocytogenes* and *Salmonella* spp. *See* ANTIBIOTIC; DRUG RESISTANCE; SALMONELLOSES; STREPTOCOCCUS.

Penicillin G, the most commonly fermented penicillin, is produced by the addition of a precursor, phenylacetic acid, to the growing culture. Use of phenoxyacetic acid as a precursor produces penicillin V. Both penicillins are recovered by extraction into organic solvents at acid pH, and precipitation as their potassium or sodium salt. Penicillin G is generally given by injection against penicillin-sensitive streptococci such as pneumococci (meningitis), and in treatment of endocarditis and gonorrhea. Penicillin V is acid stable and is usually given orally. It is effective in the treatment of upper respiratory infections and periodontal work. *See* GONORRHEA; MENINGITIS.

D. A. Lowe

Penis The male organ of copulation, or phallus. In mammals the penis consists basically of three elongated masses of erectile tissue. The central corpus spongiosum (corpus urethrae) lies ventral to the paired corpora cavernosa. The urethra runs along the underside of the spongiosum and then normally rises to open at the expanded, cone-shaped tip, the glans penis, which fits like a cap over the end of the penis. Loose skin encloses the penis and also forms the retractable foreskin, or prepuce.

Erection of the penis is caused by nervous stimulation resulting in engorgement of the spiral helicine arteries and the plentiful venous sinuses of the organ. In most mammals other than Primates the penis is retracted into a sheath when not in use.

In submammalian forms the penis is not as well developed. Crocodilians, turtles, and some birds have a penis basically like that of mammals, lying in the floor of the cloaca. When erected, it protrudes from the cloaca and functions in copulation. Other vertebrates lack a penis, although various functionally comparable organs may be developed such as the claspers on the pelvic fins of sharks and the gonopodia on the anal fins of certain teleost fishes. *See* COPULATORY ORGAN.

Thomas S. Parsons

Pennatulacea An order of the cnidarian subclass Alcyonaria (Octocorallia), commonly called the sea pens. These animals lack stolons and live with their bases embedded in the soft substratum of the sea. The colony consists of a distal rachis bearing many polyps and a polypless proximal peduncle, whose terminal end sometimes expands to form a bladder. The colony of *Pennatula* looks like a feather being formed of numerous secondary polyps which arise from leaf-shaped lateral expansions of the very elongated primary axial or terminal polyp. In the other form of colony the polyps arise directly from the primary one, as in *Veretillum* (see illustration), *Renilla*, and *Cavernularia*. The colony has a horny unbranched axial skeleton composed of pennatulin and some calcium carbonate and phosphate, but *Cavernularia*, including the luminous species, has a rudimentary one and *Renilla* lacks a skeleton. *See* OCTOCORALLIA (ALCYONARIA).

Kenji Atoda

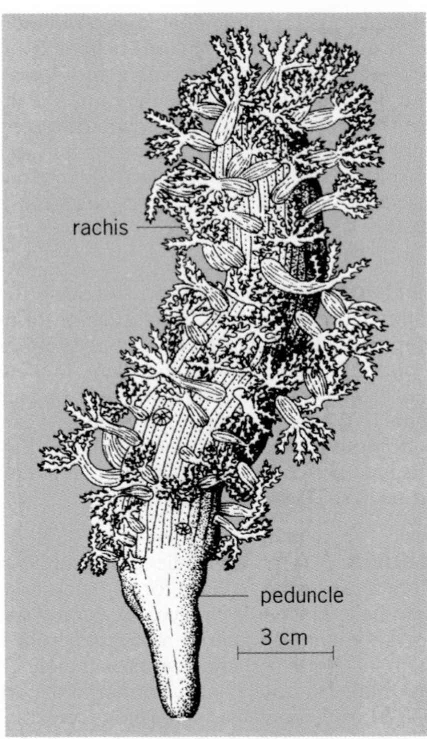

Veretillum cynomorium.

Pennsylvanian A major division of late Paleozoic time, considered either as an independent period or as the younger subperiod of the Carboniferous. In North America, the Pennsylvanian has been widely recognized as a geologic period and derives its name from a thick succession of mostly nonmarine, coal-bearing strata in Pennsylvania. Radiometric ages place the beginning of the period at approximately 320 million years ago and its end at about 290 million years ago. In northwestern Europe, strata of nearly equivalent age are commonly designated as Upper Carboniferous and in eastern Europe as Middle and Upper Carboniferous. *See* CARBONIFEROUS.

In North America, the Pennsylvanian Period was characterized by the progressive growth and enlargement of the Alleghenian-Ouachita-Marathon orogenic belt, which formed as the northwestern parts of the large continent Gondwana (mainly northwestern Africa, the area that is now Florida, and northern South America) collided against and deformed the eastern and southern parts of the North American continent. *See* OROGENY.

Much of North America remained a stable, low-lying cratonic platform during the Pennsylvanian and was covered by a relatively thin veneer of shallow-water marine carbonates and marine and nonmarine clastic sediments. These were deposited as sea level repeatedly rose and fell as the polar glaciers of southern Gondwana contracted and expanded. In Pennsylvania and along the western part of the present Allegheny Plateau, Pennsylvanian strata are predominantly nonmarine deposits made up of channel sandstones, floodplain shales, siltstones, sandstones, and coals. *See* CARBONATE MINERALS; COAL; CRATON; MARINE SEDIMENTS.

During Pennsylvanian time, the paleoequator extended across North America from southern California to Newfoundland, through northwestern Europe into the Ukrainian region of eastern Europe, and across parts of northern China. This was a time of extensive coal deposition in a tropical belt that appears to have included areas from 15 to 20° north and south of the paleoequator. Coal of this age is abundant and relatively widespread and has great economic importance.

Petroleum is commonly trapped in nearshore marine deposits of Pennsylvanian age, particularly in carbonate banks near the edge of shelves, in longshore bars and beaches, in reefs and mounds, and at unconformities associated with transgressive-regressive shore lines. Many of these traps contribute significantly to petroleum production. *See* PETROLEUM.

Pennsylvanian paleogeography changed significantly during the period as the supercontinent Pangaea gradually was formed by the joining together of Gondwana and Laurasia. North America and northern Europe, which had been combined into the continent Laurasia since the late Silurian, and South America and northwestern Africa, which formed the northern part of the continent of Gondwana, came together along the Ouachita–Southern Appalachian–Hercynian geosyncline. The result was an extensive orogeny, or mountain-building episode, which supplied the vast amounts of sediments that make up most of the Pennsylvanian strata in the eastern and midwestern parts of the United States. *See* PALEOGEOGRAPHY.

Evidence in the form of well-developed tree rings, less diverse fossil floras and faunas, and glacial deposits indicates that temperate and glacial conditions were common in nonequatorial climatic belts during Pennsylvanian time. Climatic fluctuations during the period caused significant increases and decreases in the amount of water that was temporarily stored in the glaciers in Gondwana and contributed to eustatic changes of sea level. *See* PALEOCLIMATOLOGY. Charles A. Ross; June R. P. Ross

Pentamerida An extinct order of Brachiopoda that lived from the Lower Cambrian to Upper Devonian (540 to 375 million years ago), with a peak during the Silurian Period. Pentamerides are the largest brachiopods in the Lower Paleozoic, with adults ranging from 1 to 18 cm (0.4 to 7 in.) in length, and the order forms part of the subphylum Rhynchonelliformea (previously termed the Articulata). *See* BRACHIOPODA; RHYNCHONELLIFORMEA.

The two valves, always composed of calcite ($CaCO_3$), are convex, with a larger ventral valve and a smaller dorsal valve, articulated at the posterior and which opened anteriorly during feeding. Externally, most pentamerides are smooth, apart from concentric growth line increments, but some have radial ribs (costae), which may be smooth or sinuous in cross section. Internally, the space between the two pentameride valves is divided by variably developed walls (septa), with at least one septum in the ventral valve and two in the dorsal valve, thus dividing the internal space into several divisions. In some genera, the septa unite anteriorly; and in the posterior part of the shell, the ventral valve septum divides to form a platform termed the spondylium, which was used as an attachment area for the muscles which opened and closed the valves. The form and disposition of the soft parts are not well known.

Within the order Pentamerida, there are currently 214 genera, which have been described from Paleozoic rocks of all continents. The order is subdivided into the suborder Syntrophiidina, with 85 genera ranging from the Early Cambrian to the Early Devonian, and the suborder Pentameridina, with 129 genera ranging from the Late Ordovician to the Late Devonian. *See* ORDOVICIAN; PALEOGEOGRAPHY. L. Robin M. Cocks

Pentastomida A subclass of bloodsucking arthropods, parasitic in the respiratory organs of vertebrates, that frequently are referred to as the Linguatulida or tongue worms. The adults are vermiform, with a short cephalothorax and an elongate, annulate abdomen that may be cylindrical or flattened.

The subclass is divided into two orders: the Cephalobaenida, a more primitive group, and the Porocephalida, a more specialized one. The first has six-legged larvae and the other four-legged larvae. The mitelike form of the larvae, with short stumpy legs, demonstrates relationship to the arthropods. Characteristic arthropod features include the presence of (1) jointed appendages in the larvae; (2) stigmata or breathing pores in the body wall; (3) specialized reproductive organs, especially those of the male; and (4) ecdysis or molting of larvae and nymphs. More than 50 species have been described.

Human infection occurs frequently in Africa and Asia, where humans are accidental intermediate hosts of the nymphal form. The liver is a common site of infection, and large numbers of larvae may produce serious and even fatal effects. *See* ARTHROPODA; CEPHALOBAENIDA; POROCEPHALIDA. Horace W. Stunkard

Pepper The garden pepper, *Capsicum annuum* (family Solanaceae), is a warm-season crop originally domesticated in Mexico. It is usually grown as an annual, although in warm climates it may be perennial. This species includes all peppers grown in the United States except for the "Tabasco" pepper (*C. frutescens*), grown in Louisiana. Other cultivated species, *C. chinense*, *C. baccatum*, and *C. pubescens*, are grown primarily in South America. Some 10–12 strictly wild species also occur in South America. Peppers are grown worldwide, especially in the more tropical areas, where the pepper is an important condiment.

Sweet (nonpungent) peppers, harvested fully developed but still green, are widely used in salads or cooked with other foods. Perfection pimento, harvested red ripe, is used for canning. Paprika is made from ripe red pods of several distinct varieties; the pods are dried and ground. *See* PAPRIKA; PIMENTO.

The ripe color of most varieties is red, a few varieties are orange-yellow, and in Latin America brown-fruited varieties are common. Nutritionally, the mature pepper fruit has three to four times the vitamin C content of an orange, and is an excellent source of vitamin A. *See* ASCORBIC ACID; VITAMIN A. Paul G. Smith

Peppermint The mint species *Mentha piperita* (family Lamiaceae), a sterile interspecific hybrid believed to have occurred in nature from the hybridization of fertile *M. spicata.* Peppermint oil is obtained by steam distillation from the partially dried hay. The main uses of peppermint oil are to flavor chewing gum, confectionary products, toothpaste, mouthwashes, medicines, and as a carminative in certain medical preparations for the alleviation of digestive disturbances. Merritt J. Murray

Pepsin A proteolytic enzyme found in the gastric juice of mammals, birds, reptiles, and fish. It is formed from a precursor, pepsinogen, which is found in the stomach mucosa. Pepsinogen is converted to pepsin either by hydrochloric acid, naturally present in the stomach, or by pepsin itself. *See* ENZYME.

Pepsin is prepared commercially from the glandular layer of fresh hog stomachs. It is a part of the crude preparation known as rennet, which is used to curdle milk in preparation for cheese manufacture. Pepsin is also used for a variety of other applications in food manufacturing; to modify soy protein and gelatin, thereby providing whipping qualities; to modify vegetable proteins for use in nondairy snack items; to make precooked cereals into instant hot cereals; and to prepare animal and vegetable protein hydrolysates for use in flavoring foods and beverages. Myron Solberg

Peptide A compound that is made up of two or more amino acids joined by covalent bonds which are formed by the elimination of a molecule of H_2O from the amino group of one amino acid and the carboxyl group of the next amino acid. Peptides larger than about 50 amino acid residues are usually classified as proteins. Glutathione is the most abundant peptide in mammalian tissue. Hormones such as oxytocin (8), vasopressin (8), glucagon (29), and adrenocorticotropic hormone (39) are peptides whose structures have been deduced; in parentheses are the numbers of amino acid residues for each peptide.

For each step in the biological synthesis of a peptide or protein there is a specific enzyme or enzyme complex that catalyzes each reaction in an ordered fashion along the biosynthetic route. However, it is noteworthy that, although the biological synthesis of proteins is directed by messenger RNA on cellular structures called ribosomes, the biological synthesis of peptides does not require either messenger RNA or ribosomes. *See* AMINO ACIDS; PROTEIN; RIBONUCLEIC ACID (RNA); RIBOSOMES. James M. Manning

Peracarida A large crustacean superorder generally classified within the phylum Arthropoda, subphylum Crustacea, class Malacostraca, subclass Eumalacostraca. The Peracarida are defined as having the following characteristics: a telson (posteriormost body segment) without appendages; one pair (rarely two or three pairs) of maxillipeds (thoracic appendages modified as mouthparts); mandibles (primary jaws) with articulated accessory processes in adults, between the molar and incisor teeth, called the lacinia mobilis; a carapace, when present, that is not fused with posterior thoracic segments and that is usually reduced in size; basalmost leg segments of the thorax with unique, thin, flattened ventral plates (oöstegites) that enclose a ventral brood pouch (marsupium) in all orders except the Thermosbaenacea (which use the carapace to brood embryos); and having the young hatch as a prejuvenile stage called a manca, lacking the last pair of legs (that is, no free-living larvae occur in Peracarida). The roughly 21,500 species of peracarids are divided among nine orders: Mysida, Lophogastrida, Cumacea, Tanaidacea, Mictacea, Spelaeogriphacea, Thermosbaenacea, Isopoda, and Amphipoda. *See* CRUSTACEA; MALACOSTRACA.

The peracarids are an extremely successful group of crustaceans and are known from many habitats. Although most are marine, many also occur on land and in freshwater, and several species live in hot springs at temperatures of 30–50°C (86–122°F). Aquatic forms include planktonic as well as bottom-dwelling species at all depths. The group includes the most successful terrestrial crustaceans—the pillbugs and sowbugs of the order Isopoda (suborder Oniscidea)—and a few amphipods that have invaded land and live in damp leaf litter and gardens. Peracarids range in size from tiny interstitial (living among sand grains) forms only a few millimeters long to planktonic amphipods over 12 cm (4.7 in.) long and deep-water bottom-dwelling isopods that grow to 50 cm (20 in.) in length. Peracarids exhibit all sorts of feeding strategies, including carnivory, herbivory, and detritivory, and a number of them, especially isopods and amphipods, are symbionts (parasites or commensals). *See* DEEP-SEA FAUNA.

The largest order of Peracarida is the Isopoda, characterized by a complete lack of a carapace, a heart located primarily in the abdomen, and biphasic molting (posterior region molts before anterior region). The isopods comprise about 10,000 described marine, freshwater, and terrestrial species. The suborder Oniscidea includes about 5000 species that have invaded land (pillbugs and sowbugs). *See* ISOPODA.

Like the isopods, the order Amphipoda is characterized by the lack of a carapace, but the heart is thoracic and molting is not biphasic. The roughly 8000 species of amphipods have invaded most marine and many freshwater habitats and constitute a large portion of the biomass in many marine regions. The principal amphipod suborder is Gammaridea. Richard C. Brusca

Percent A ratio comparison of two quantities expressed by using 100 equal parts, or hundredths; symbolized %. There are three major uses of percent: part of a whole, rate, and comparison of any two quantities.

Part of a whole. The basic idea of percent is as a ratio that shows a part of a whole. The technical name for the whole is base.

If 89 out of 100 problems are correct, the part-whole comparison shows 89/100 or 89% correct. If the whole is not already divided into equal parts, an equivalent ratio to 100 is found.

Certain ratios are easy to express as hundredths. When 3 baskets are made in basketball out of 10 attempts, an equivalent ratio using 100 is found from which the percent is obvious: $3/10 = 30/100 = 30\%$.

For 35 hits out of 126 times at bat, the ratio 35/126 is not easily expressed as a ratio using 100. Hundredths will be more obvious by dividing 35 by 126, and then reading the number of hundredths to find the percent: $35 \div 126 \approx 0.278$ or 27.8%. The percent is obtained by moving the decimal point two places to the right.

Rate. While percent always means a comparison to 100, percent can show a rate of so many per 100, not so many out of 100. A sales tax of 6% means a rate of 6 cents for each dollar, and this 6 cents is in addition to the dollar. The tax amount can often be calculated mentally by multiplying the 6% rate by the number of dollars.

Interest paid or interest received is done by using rates. Simple interest at a rate of 6% means $6 per $100 for a full year. If interest is calculated monthly on the unpaid balance and the yearly rate is 18%, the monthly rate is approximately $18\% \div 12$ or about 1.5%.

With compound interest, the amount of interest is added each compounding period, and this total amount is subject to compounding for the next period. For example, $1.00 invested at 8% will be worth 108% of $1.00 at the end of one year, or $1.08. The worth at the end of the second year is 108% of $1.08, or $(1.08)^2$. At this same rate, after 10 years the compounded value is $(1.08)^{10} = 2.1589247$, or $2.15, using 1.08 as a factor for 10 times.

Comparing any two quantities. Percent is used to compare any two quantities, but special care must be given to the base for the comparison. Comparison of city A with a population of 42,000 people and city B with 67,000 people will depend on the base. A compared to B is 42,000/67,000, 62.7%; A is 62.7% of B. B compared to A is 67,000/42,000, 1.595, or 159.5%; B is 159.5% of A. *See* ARITHMETIC. Joseph N. Payne

Perception Those subjective experiences of objects or events that ordinarily result from stimulation of the receptor organs of the body. This stimulation is transformed or encoded into neural activity (by specialized receptor mechanisms) and is relayed to more central regions of the nervous system where further neural processing occurs. Most likely, it is the final neural processing in the brain that underlies or causes perceptual experience, and so perceptionlike experiences can sometimes occur without external stimulation of the receptor organs, as in dreams.

In contemporary psychology, interest generally focuses on perception or the apprehension of objects or events, rather than simply on sensation or sensory process. While no sharp line of demarcation between these topics exists, it is fair to say that sensory qualities are generally explicable on the basis of mechanisms within the receptor organ, whereas object and event perception entails higher-level activity of the brain. *See* HEARING (HUMAN); SENSATION; VISION.

Since objects or events are not experienced only through vision, the term perception obviously applies to other sense modalities as well. Certainly things and their movement may be experienced through the sense of touch. Such experiences derive from receptors in the skin (tactile perception), but more importantly, from the positioning of the fingers with respect to one another when an object is grasped, the latter information arising from receptors in the muscles and joints (haptic or tactual perception). The position of the parts of the body are also perceived with respect to one another whether they are stationary (proprioception) or in motion (kinesthesis), and the position of the body is experienced with respect to the environment through receptors sensitive to gravity such as those in the vestibular apparatus in the inner ear. Auditory perception yields recognition of the location of sound sources and of structures such as melodies and speech. Other sense modalities such as taste (gustation), smell (olfaction), pain, and temperature provide sensory qualities but not perceptual structures as do vision, audition, and touch, and thus are usually dealt with as sensory processes. *See* OLFACTION; PAIN; PROPRIOCEPTION.

Constancy. By and large, these perceptual properties of objects remain remarkably constant despite variations in distance, slant, and retinal locus caused by movements of the observer. This fact, referred to as perceptual constancy, is perhaps the hallmark of perception and more than any other, serves to characterize the field of perception.

Examples of perceptual constancy are: size (except at very great distances, an object appears the same size whether seen nearby or far away, although the size of its image on the retina can be very different); shape (a circle seen from the side is perceived as a circle, although it appears as an ellipse on the retina); orientation (objects appear to keep the same orientation in space, independently of the orientation of the observer's head); and position (a fixed object remains perceived as stationary even when its image on the retina moves because of eye or head movements).

Motion perception. Perceived movement cannot simply be explained by the motion of an object's retinal image since image motion caused by observer or eye movement does not lead to perceived object movement. Moreover, an object tracked by smooth-pursuit eye movements will appear to move, although in that case there is essentially no motion of the object's image over the retina. Similarly, an afterimage will appear to move during eye movement even in a completely darkened room. Where ordinarily the movement of the retinal image caused by the moving eye is computed to signify "no object motion," thus yielding position constancy (since the image motion and eye motion are equal in magnitude), the same computational rule must signify "object motion" in the case of the afterimage.

Form perception. Form perception means the experience of a shaped region in the field. Recognition means the experience that the shape is familiar. Identification means that the function or meaning or category of the shape is known. For those who have never seen the shape before, it will be perceived but not recognized or identified. For those who have, it will be perceived as a certain familiar shape and also identified. Recognition and identification obviously must be based on past experience, which means that through certain unknown processes, memory contributes to the immediate experience that one has, giving the qualities of familiarity and meaning.

The figure of a 4 in Fig. 1*a* is seen as one unit, separate from other units in the field, even if these units overlap. This means that the parts of the figure are grouped together by the perceptual system into a whole, and these parts are not grouped with the parts of other objects. This effect is called perceptional organization. There are other problems about form perception that remain to be unraveled. For example, the size of a figure can vary, as can its locus on the retina or even its color or type of contour, without affecting its perceived shape (Fig. 2).

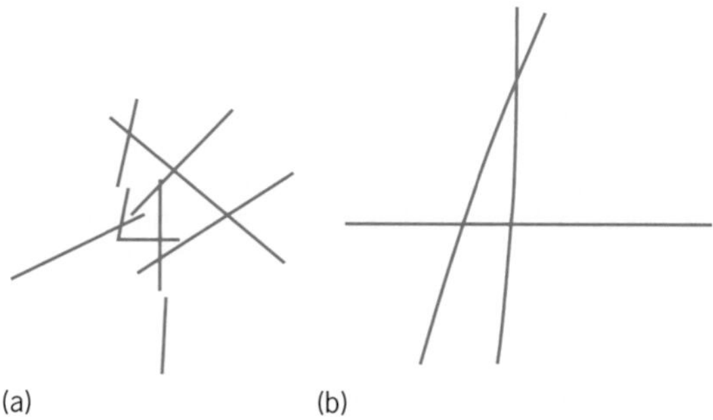

Fig. 1. Perceptual organization. (*a*) The figure of a four is immediately and spontaneously perceived despite the presence of other overlapping and adjacent lines. (*b*) The four, although physically present, is not spontaneously perceived and is even difficult to see when one knows it is there.

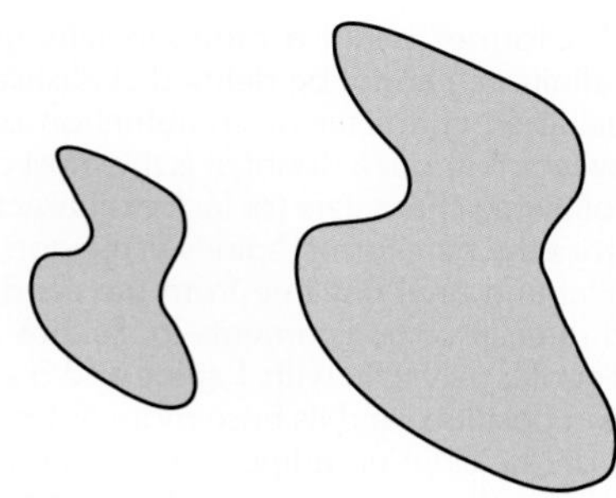

Fig. 2. Transposition of form; the two shapes clearly look the same despite the difference in size.

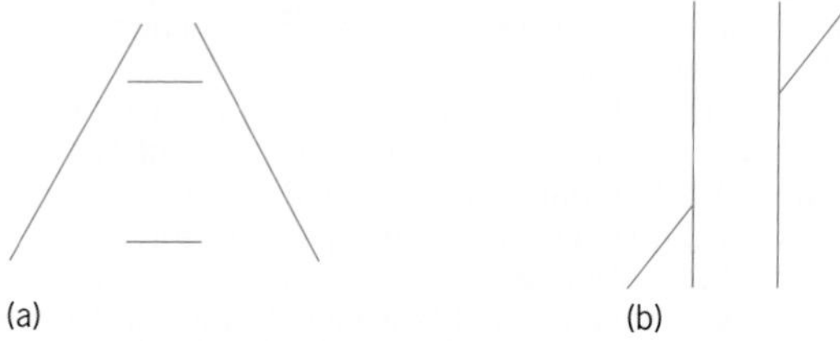

Fig. 3. Geometrical illusions. (*a*) The Ponzo illusion in which the two horizontal lines of equal length appear unequal. (*b*) The Poggendorff illusion in which the two oblique line segments are aligned with one another (that is, are collinear) but appear to be misaligned.

A further fact about form perception is that it is dependent upon orientation. It is a commonplace observation that printed or written words are difficult to read when inverted, and faces look very odd or become unrecognizable when upside down. Simple figures also look different when their orientation is changed: a square looks like a diamond when tilted by 45°.

Geometrical illusions. Related to the topic of form perception is the misperception of the size or direction of parts of figures that constitutes many of the geometric illusions. In an illusion figure, one particular part is perceived to be either longer or shorter than another part, although they are objectively equal (Fig. 3*a*); or the direction of a contour is perceived to be different from that of another contour although they are the same (Fig. 3*b*). For reasons still not understood, the background or context of the rest of the figure affects these parts.

Innate or learned? A central problem is whether the perception of properties such as form and depth or the achievement of veridical perception as in the constancies is innately determined or is based on past experience. By "innate" it is meant that the perception is the result of evolutionary adaptation and thus is present at birth or when the necessary neural maturation has occurred. By "past experience" it is meant that the perception in question is the end result of prior exposure to certain relevant patterns or conditions, a kind of learning process. Despite centuries of discussion of this problem, and considerable experimental work, there is still no final answer to the question. It now seems clear that certain kinds of perception are innate, but equally clear that past experience also is a determining factor. *See* INTELLIGENCE.

Irvin Rock

Perciformes The largest order of fishes, comprising 20 suborders, 160 families, about 1540 genera, and over 10,000 species, which is about 36% of all fish species. Eighty percent of the perciforms are marine species; the perciforms are dominant in the oceans. *See* OSTEICHTHYES.

The order Perciformes is not a monophyletic group, and it evades clear definition, cannot be defined cladistically, and has no unique specialized character or combination of specialized characters; however, without a doubt, it is the most diverse order of fishes. The following characters (or lack of characters) can suffice to characterize the perciforms: spines in the dorsal, anal, and pelvic fins; dorsal and anal fins free from the caudal fin; pelvic girdle attached directly or by ligaments to cleithra or coracoids of the pectoral girdle; pelvic fin with 1 spine and 5 soft rays, thoracic or jugular in position, and its base more or less vertical and placed well up on the side; no adipose fin; upper jaw bordered largely or entirely by premaxillae; no orbitosphenoid, mesocoracoid, epipleural, and epicentral bones; physoclistic (closed) swim bladder; double nostrils on each side of the snout; ctenoid scales; and caudal fin with 17 principal rays (15 branched). *See* SCALE (ZOOLOGY); SWIM BLADDER.

The Perciformes first appeared in the Upper Cretaceous, experienced tremendous diversification, and throughout the Cenozoic became the dominant fish group in the marine environment, as well as having several large groups in freshwater environs. Many of the basic structural types, as well as most major perciform derivatives, such as the Pleuronectiformes and Tetraodontiformes, were present in the Eocene. It is in the shore areas, the offshore banks, the coral reefs, the coastal beaches and lagoons, and the intertidal zone that the perciforms have attained their greatest diversity. *See* PISCES (ZOOLOGY).

Ichthyologists are rapidly extending knowledge of phyletic relationships in this great order of fishes, so it is inevitable that classification is changing and controversial. Some 52 families of perciforms have a single genus, 23 have a single species (that is, are monotypic), and 21 have 100 or more species. Three suborders—Percoidei, Labroidei, and Gobioidei—account for over three-quarters of the species. The eight largest families—Gobiidae, Cichlidae, Serranidae, Labridae, Blenniidae, Pomacentridae, Apogonidae, and Sciaenidae—together account for 5479 species, or about 55% of the perciform fishes.

Herbert Boschung

Percopsiformes An order of the Paracanthopterygii, a fish superorder of questionable monophyly. Percopsiforms are identifiable by the following combination of characters: nonprotractile premaxilla; tooth-bearing ectopterygoid and palatine bones; absence of a suborbital shelf, orbitosphenoid, and basisphenoid bones; weak spines usually present in the dorsal fin; and pelvic fins, if present, behind the pectoral fins. There are three families, seven genera, and nine species, all confined to freshwaters of North America. *See* ACTINOPTERYGII; OSTEICHTHYES; TELEOSTEI.

The family Percopsidae was well represented in Eocene times and presently represented by one genus and two species: *Percopsis omiscomaycus* (trout-perch; see illustration) and *P. transmontana* (sand roller). The trout-perch occurs in lakes, deep flowing pools of creeks, and rivers from Alaska eastward to Quebec and southward to the Great Lakes and Mississippi River basins to West Virginia, Kentucky, Illinois, Missouri, North Dakota, and Montana. It attains a maximum length of 20 cm (8 in.). Sand rollers are usually found near vegetation over sand in quiet backwaters and pool margins of small to large streams of the Columbia River drainage in Idaho, Washington, and Oregon. They attain a maximum length of 9.6 cm (3.8 in.).

Aphredoderidae (pirate perch) is a monotypic family represented by *Aphredoderus sayanus*, which ranges in lowlands of Atlantic Slope streams from Long Island to Florida, the Gulf Slope west to the Brazos River drainage in Texas, the Mississippi Basin to Michigan, and disjunctly to Lakes Erie and Ontario, avoiding Appalachia. It feeds aggressively at night, primarily on isopods and amphipods, and attains a maximum total length of about 13 cm (5.2 in.).

Amblyopsidae (cavefishes) is a family of five genera and six species known from unglaciated southern and eastern regions of the United States. Some species are totally blind, usually without pigment, appearing white or pinkish, and are restricted to limestone caves.

Trout-perch (*Percopsis omiscomaycus*). (*Photo by Konrad P. Schmidt*)

Herbert Boschung

Performance rating A procedure for determining the value for a factor which will adjust the measured time for an observed task performance to a task time that one would expect of a trained operator performing the task, utilizing the approved method and performing at normal pace under specified workplace conditions. Normal time (ultimately subjectively based) is the time that a trained worker requires to perform the specified task under defined workplace conditions, employing the assumed philosophy of "a fair day's work for a fair day's pay."

The performance rating process is concerned with determining normal pace during the work portion of an average day and must, therefore, consider the fatigue recovery aspects of allowance (nonwork) times occurring during the day. The following two equations relate factors in determining how much time a worker will be allowed per unit of output:

$$\text{Standard time} = \text{normal time} \times \text{allowances}$$
$$\text{Normal time} = \text{observed time} \times \text{rating factor}$$

If the observed time for a task is adjusted by the performance rating factor to determine normal time, and allowance time is added for nonwork time, the standard time will represent the allowed time per unit of production.

The most commonly employed rating technique throughout the history of stopwatch time study, including the present, is referred to as pace rating. A properly trained employee of average skill is time-studied while performing the approved task method under specified work conditions. Rating consists only of determining the relative pace (speed) of the operator in relation to the observer's concept of what normal pace should be for the observed task, including consideration of expected allowances to be applied to the standard. *See* HUMAN-FACTORS ENGINEERING; METHODS ENGINEERING; WORK MEASUREMENT.

Philip E. Hicks

Performing arts medicine A subspecialty in medicine that deals with problems specific to the activities of dancers, musicians, and vocalists. It is an outgrowth of the fields of sports medicine and occupational medicine. Performing artists have medical problems similar to those of many athletes in that their activities involve certain parts of the body repetitively performing a specific motion. Athletes and performance artists often practice at least 3–5 h each day on their own and then may have team practice or rehearsal and finally actual competition or a performance. The vast majority of injuries in performers can be attributed to overuse. Poor technique often plays an important role in overuse injuries: poor body positioning and inefficient use of abdominal musculature adversely affect dancers, singers, and musicians. Inappropriate choice of dance style, musical instrument, or vocal repertoire also may lead to overuse injuries.

Each field of the performing arts tends to have its own characteristic injuries, which afflict the parts of the musculoskeletal system subjected to the greatest stress. Tendinitis in dancers occurs most commonly in the rectus femoris tendon in the thigh

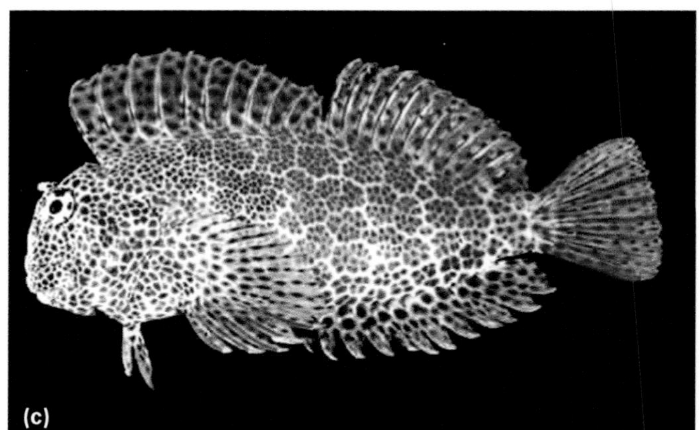

Perciformes

(*a*) Yellowspotted grouper (*Epinephelus timorensis*).

(*b*) Bluespotted wrasse (*Anampses caeruleopunctatus*).

(*c*) Leopard (or shortbodied) blenny (*Exallias brevis*).

(*d*) Halfspotted goby (*Asterropteryx semipunctatus*).

(*e*) Whitespotted surgeonfish (*Acanthurus guttatus*).

(*All photos by Richard C. Wass; Checklist of the Fishes from the National Park of American Samoa/NPS*)

and the Achilles tendon in the leg. Many dancers also suffer from pain related to mechanical stress on the kneecap, a consequence of jumping and deeply bending the knees. Tendinitis in musicians most commonly occurs in the extensor tendons of the wrist and fingers, the flexor tendons of the fingers, and the rotator cuff muscles in the shoulder, producing pain with use of the upper extremities. Neck pain is also common. Singers are prone to overuse injuries of the vocal cords. These injuries take the form of swelling (laryngitis), nodules, and even hemorrhage. *See* SPORTS MEDICINE. Carol C. Teitz

Pericycle As commonly defined, the outer boundary of the stele of plants. Originally it was interpreted as a band of cells between the phloem and the innermost layer (endodermis) of the cortex. Such pericycle is commonly found in roots and, in lower vascular plants, also in stems. In higher vascular plants, however, a distinct layer of cells may not be present between the phloem and the cortex. The pericycle, if present, may be composed of parenchyma or sclerenchyma cells with relatively thin or heavily thickened walls. It may be one to several layers in radial dimensions.

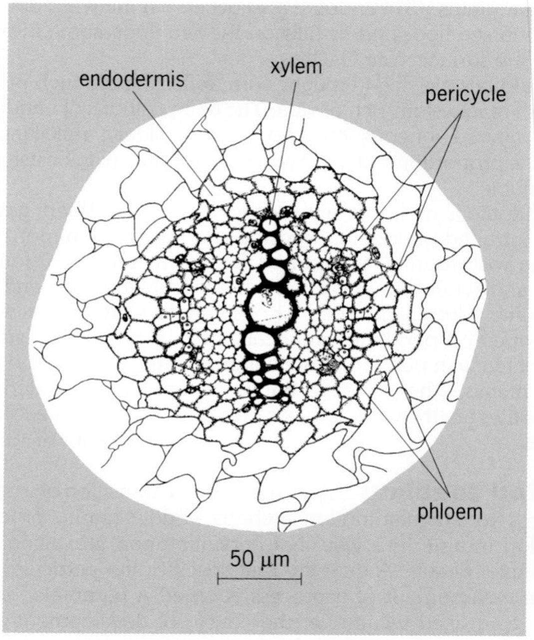

Transection of central part of sugarbeet. (*From K. Esau, Hilgardia, 9(8), 1935*)

Primordia of branch roots commonly arise in the pericycle in seed plants, most frequently outside the xylem ridges (see illustration). The first cork cambium may also arise in the pericycle of those roots that have secondary vascular tissues. In roots, a part of the vascular cambium itself (that outside the primary xylem ridges) originates from pericycle cells. *See* CORTEX (PLANT); ENDODERMIS; LATERAL MERISTEM; PARENCHYMA; PHLOEM; ROOT (BOTANY); SCLERENCHYMA; STEM; XYLEM. Vernon I. Cheadle

Pericyclic reaction Concerted (single-step) processes in which bond making and bond breaking occur simultaneously (but not necessarily synchronously) via a cyclic (closed-curve) transition state. Although a given reaction may appear formally to be pericyclic, it cannot be assumed to be a concerted process. In each case, the detailed mechanism of the reaction must be established experimentally. Pericyclic reactions can be promoted either by heat or by light; the stereochemistry of the reaction is determined by the mode of activation employed and the number of electrons that are delocalized in the transition state. *See* CHEMICAL BONDING; PHYSICAL ORGANIC CHEMISTRY; STEREOCHEMISTRY.

Four types of pericyclic reactions that are frequently encountered in organic chemistry are electrocyclic processes, cycloadditions, sigmatropic shifts, and cheletropic reactions.

Electrocyclic processes are reactions that involve their cyclization across the termini of a conjugated π-system with concomitant formation of a new σ-bond or the microscopic reverse. The sequence of steps involved in the forward reaction must be the same, in the reverse order, as that in the reverse direction when the forward and reverse reactions are carried out under identical conditions. This statement is known as the principle of microscopic reversibility.

The effect of the mode of activation upon the stereochemistry of an electrocyclic process is shown in the reaction (1), where

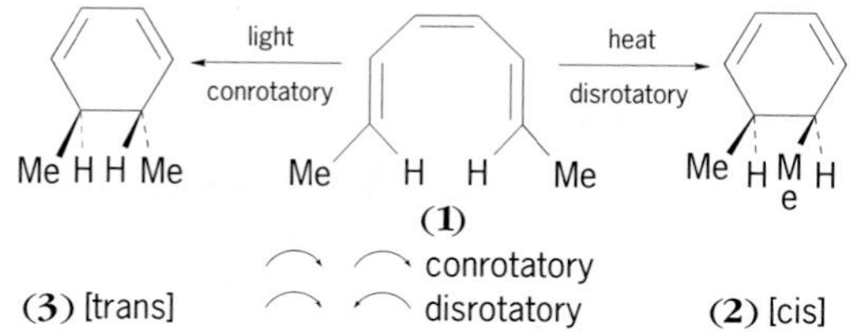

(1)

Me = methyl, for the hexatrienecyclohexadiene interconversion (a six-electron electrocyclic process). Thus, when *trans,cis,trans*-2,4,6-octatriene [structure (**1**)] is heated, disrotatory motion of the two terminal 2*p* orbitals occurs; that is, they rotate in opposite directions thereby resulting in exclusive formation of *cis*-5,6-dimethylcyclohexa-l,3-diene (**2**). The corresponding photochemical process results in conrotatory motion of the termini in structure (**1**); that is, the two terminal 2*p* orbitals rotate in the same direction thereby yielding *trans*-5,6-dimethylcyclohexa-l,3-diene (**3**) exclusively.

Cycloadditions occur when two (or more) π-electron systems react under the influence of heat or light to form a cyclic compound with concomitant formation of two new σ-bonds that join the termini of the original π-systems. The stereochemistry of this reaction is classified with respect to the two molecular planes of the reactants. Thus, if σ-bond formation occurs from the same face of the molecular plane across the termini of one of the component π-systems, the reaction is said to be suprafacial on that component. If instead σ-bond formation occurs from opposite faces of the molecular plane, the reaction is said to be antarafacial on that component. This distinction is illustrated in reaction (2) for two thermal processes, where the symbol $\neq$ indicates the structure of the transition state. Reaction (2*a*) shows the Diels-Alder [4 + 2] cycloaddition of butadiene (**4**) to ethylene (**5**), a six-electron pericyclic reaction in which additions across the termini of the diene (four-electron component) and dienophile (two-electron component) both occur suprafacially. Reaction (2*b*) shows a [14 + 2] cycloaddition in which σ-bond formation occurs suprafacially on the two-electron component [tetracyanoethylene (**6**)] and antarafacially on the fourteen-electron component [heptafulvalene (**7**)].

Sigmatropic shifts involve migration of a σ-bond that is flanked at either (or both) ends by conjugated π-systems. Either one or both ends of the σ-bond may migrate to a new location within the one or more flanking π-systems.

Cheletropic reactions involve extrusion of a fragment via concerted cleavage of two σ-bonds that terminate at a single atom or the reverse process. Cheletropic fragmentations may be either linear or nonlinear [reaction 3].

R. B. Woodward and R. Hoffmann introduced an application of molecular orbital theory that permits prediction of rates and products of pericyclic reactions. They utilized symmetry properties of molecular orbitals to estimate relative energies of diastereoisomeric transition states for structurally similar pericyclic reactions.

In an alternative theoretical approach to understanding pericyclic reactions, the transition state is examined directly, and

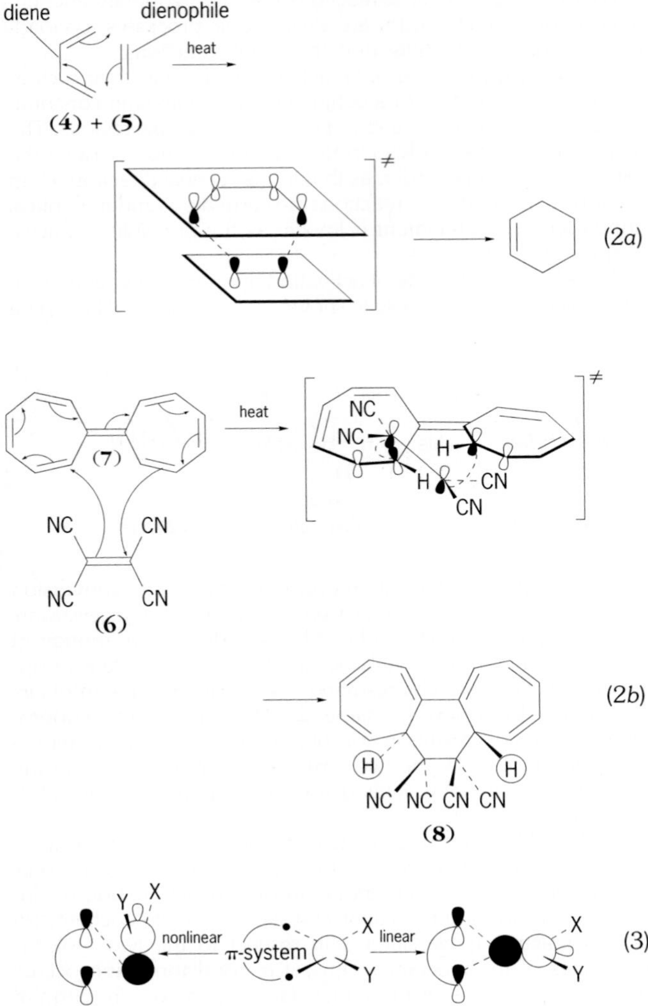

attempts to estimate the degree of electronic stabilization (allowedness) or destabilization (forbiddenness) inherent in that transition state are made. One such approach emphasizes the importance of frontier orbitals (highest-occupied-lowest- unoccupied molecular orbitals) in determining the course of a pericyclic reaction. *See* DELOCALIZATION; DIELS-ALDER REACTION; ELECTRON CONFIGURATION; MOLECULAR ORBITAL THEORY; ORGANIC REACTION MECHANISM; WOODWARD-HOFFMANN RULE. Alan P. Marchand

Periderm A group of tissues which replaces the epidermis in the plant body. Its main function is to protect the underlying tissues from desiccation, freezing, heat injury, mechanical destruction, and disease. Although periderm may develop in leaves and fruits, its main function is to protect stems and roots. The fundamental tissues which compose the periderm are the phellogen, phelloderm, and phellem.

The phellogen is the meristematic portion of the periderm and consists of one layer of initials. These exhibit little variation in form, appearing rectangular and somewhat flat in cross and radial sections, and polygonal in tangential sections.

The phelloderm cells are phellogen derivatives formed inward. The number of phelloderm layers varies with species, season, and age of the periderm. In some species, the periderm lacks the phelloderm altogether. The phelloderm consists of living cells with photosynthesizing chloroplasts and cellulosic walls.

The phellem, or cork, cells are phellogen derivatives formed outward. These cells are arranged in tiers with almost no intercellular spaces except in the lenticel regions. After completion of their differentiation, the phellem cells die and their protoplasts disintegrate. The cell lumens remain empty, excluding a few species in which various crystals can be found. The remarkable impermeability of the suberized cell walls is largely due to their impregnation with waxes, tannins, cerin, friedelin, and phellonic and phellogenic acids.

Lenticels are loose-structured openings that develop usually beneath the stomata and that facilitate gas transport through the otherwise impermeable layers of phellem. *See* BARK; SCLERENCHYMA. Y. Waisel; H. Wilcox

Peridotite A rock consisting of more than 90% of millimeter-to-centimeter-sized crystals of olivine, pyroxene, and hornblende, with more than 40% olivine. Other minerals are mainly plagioclase, chromite, and garnet. Much of the volume of the Earth's mantle probably is peridotite.

Peridotites have three principal modes of occurrence corresponding approximately to their textures: (1) Peridotites with well-formed olivine crystals occur mainly as layers in gabbroic complexes. (2) Peridotite nodules in alkaline basalts and diamond pipes generally have equigranular textures, but some have irregular grains. (3) Peridotite also occurs on the walls of rifts in the deep sea floor and as hills on the sea floor, some of which reach the surface. *See* GABBRO.

Peridotites are rich in magnesium, reflecting the high proportions of magnesium-rich olivine. The compositions of peridotites from layered igneous complexes vary widely, reflecting the relative proportions of pyroxenes, chromite, plagioclase, and amphibole.

Peridotite is an important rock economically. Where granites have intruded peridotite, asbestos and talc are common. Pure olivine rock (dunite) is quarried for use as refractory foundry sand and refractory bricks used in steelmaking. Serpentinized peridotite is locally quarried for ornamental stone. Tropical soils developed on peridotite are locally ores of nickel. The sulfides associated with peridotites are common ores of nickel and platinoid metals. The chromite bands commonly associated with peridotites are the world's major ores of chromium. *See* IGNEOUS ROCKS. Alfred T. Anderson, Jr.

Period doubling A scenario for the transition of a natural process from regular motion to chaos. Various natural processes develop in time in a way that depends upon prevailing environmental details. A quantity that specifies the particular state of the environment of a process is called a parameter, and is taken as a fixed value over the course of development of the process.

It is a frequent natural occurrence for a process to have a regular and easily describable motion for some range of parameters, but to have complex, irregular, and difficult-to-describe motions for other ranges of parameters. In the context of fluid flow, the latter circumstance is termed turbulence. In a more general context it is called chaos (which includes fluid turbulence but presages an underlying generality). *See* FLUID FLOW; TURBULENT FLOW.

Sometimes, as the environmental parameters are varied, a process may systematically exhibit more irregular motions, turning over into chaotic motion beyond some parameter value. In analogy to the phenomenology of phase transitions, this circumstance is termed a transition to chaos. There are a variety of qualitatively different transitions to chaos, each termed a scenario. Period doubling is one frequently encountered scenario leading to chaos for which a full theoretical account exists. Since it occurs in a wide variety of processes of significantly divergent physical characters (for example, fluid-flow, chemical reactions, and electronic devices), it is sensible to consider it as a phenomenon in its own right. *See* PHASE TRANSITIONS.

In order to observe this scenario, it is sufficient that all but one parameter is held fixed. Over some range of this varied parameter (it shall be defined to increase over the range of investigation) the motion is observed to be periodic. Above a certain value of the parameter the motion grows more complicated (a bifurcation has occurred): after the amount of time T for which the motion exactly repeated itself just prior to the bifurcation, the motion now slightly fails to do so, exactly repeating, however, after another T seconds. That is, the period has doubled from T to $2T$. As the parameter is further increased, the error to repeat after the first half of the new period systematically increases. A still further increase of parameter produces another bifurcation resulting in a new doubling of the period: the motion slightly fails to repeat after two roughly periodic cycles, exactly doing so after four. As the parameter is further increased, there are successive period-doubling bifurcations, more and more closely spaced in parameter value until at a critical value the doubling has occurred an infinite number of times, so that the motion is now no longer periodic and hence of a more complex character than had yet been encountered. Unpredictably complex motions occur for values of the parameter above its critical value, although ranges of parameter still exist for which the system exhibits new periodic motions. Indeed any period-doubling system exhibits the same sequence of truly chaotic motion and interspersed periodicities as its parameter increases. Thus there is a strong degree of qualitatively universal behavior for all systems experiencing this scenario.

However, there is also a precise quantitative universality. That is, without knowing the system (or its equations) essentially all measurable quantities can be predicted: By looking at the data alone, it would not be possible to guess the physical system responsible for that data. Thus, reminiscent of thermodynamics, questions can be posed and answered in a general manner that bypasses the specific mechanisms governing any particular system.

Mitchell J. Feigenbaum

Periodic motion Any motion that repeats itself identically at regular intervals. If $x(t)$ represents the displacement of any coordinate of the system at time t, a periodic motion has the property defined by the following equation for every value of the variable time t. The fixed time interval T between repeti-

$$x(t + T) = x(t)$$

tions, or the duration of a cycle, is known as the period of the motion.

The motion of the escapement mechanism of a watch, the motion of the Earth about the Sun, and the more complicated motion of the crankshaft, piston rods, and pistons in an engine running at uniform speed are all examples of periodic motion.

The vibration of a piano string after it is struck is a damped periodic motion, not strictly periodic according to the definition. Although the motion very nearly repeats itself, and with a fixed repetition time, each successive cycle has a slightly smaller amplitude. *See* DAMPING; HARMONIC MOTION; VIBRATION; WAVE MOTION.

Joseph M. Keller

Periodic table A list of chemical elements arranged along horizontal rows in increasing atomic number. It is organized such that the vertical columns consist of elements with remarkably similar properties. The first column, known as the alkali metals (albeit with hydrogen, a nonmetal on top), contains elements with just one outer (valence) electron. The last column has completely filled valence orbitals leading to chemically inert elements called the noble gases. The position of elements in the periodic table provides a powerful method of classifying not only the physical properties of elements but also their expected properties in molecules and solids. *See* ALKALI METALS; ATOMIC NUMBER; ELECTRON CONFIGURATION; INERT GASES; VALENCE.

KEY: Atomic number → 79; Symbol → Au; Atomic weight → 196.97

1	2	3	4	5	6	7	8	9	10	11	12	13	14	15	16	17	18
1 H 1.0079																	2 He 4.0026
3 Li 6.941	4 Be 9.012											5 B 10.811	6 C 12.011	7 N 14.007	8 O 15.999	9 F 18.998	10 Ne 20.180
11 Na 22.990	12 Mg 24.305											13 Al 26.982	14 Si 28.086	15 P 30.974	16 S 32.065	17 Cl 35.453	18 Ar 39.948
19 K 39.098	20 Ca 40.078	21 Sc 44.956	22 Ti 47.867	23 V 50.942	24 Cr 51.996	25 Mn 54.938	26 Fe 55.845	27 Co 58.933	28 Ni 58.693	29 Cu 63.546	30 Zn 65.39	31 Ga 69.723	32 Ge 72.64	33 As 74.922	34 Se 78.96	35 Br 79.904	36 Kr 83.80
37 Rb 85.468	38 Sr 87.62	39 Y 88.906	40 Zr 91.224	41 Nb 92.906	42 Mo 95.94	43 Tc (98)	44 Ru 101.07	45 Rh 102.91	46 Pd 106.42	47 Ag 107.87	48 Cd 112.41	49 In 114.82	50 Sn 118.71	51 Sb 121.76	52 Te 127.60	53 I 126.90	54 Xe 131.29
55 Cs 132.91	56 Ba 137.33	57–71	72 Hf 178.49	73 Ta 180.95	74 W 183.84	75 Re 186.21	76 Os 190.23	77 Ir 192.22	78 Pt 195.08	79 Au 196.97	80 Hg 200.59	81 Tl 204.38	82 Pb 207.2	83 Bi 208.98	84 Po (209)	85 At (210)	86 Rn (222)
87 Fr (223)	88 Ra (226)	89–103	104 Rf (261)	105 Db (262)	106 Sg (266)	107 Bh (264)	108 Hs (277)	109 Mt (268)	110 Ds (271)	111 Rg (272)	112 Uub (285)	113	114	115	116	117	118

Lanthanides	57 La 138.91	58 Ce 140.12	59 Pr 140.91	60 Nd 144.24	61 Pm (145)	62 Sm 150.36	63 Eu 151.96	64 Gd 157.25	65 Tb 158.92	66 Dy 162.50	67 Ho 164.93	68 Er 167.26	69 Tm 168.93	70 Yb 173.04	71 Lu 174.97
Actinides	89 Ac (227)	90 Th 232.04	91 Pa 231.04	92 U 238.03	93 Np (237)	94 Pu (244)	95 Am (243)	96 Cm (247)	97 Bk (247)	98 Cf (251)	99 Es (252)	100 Fm (257)	101 Md (258)	102 No (259)	103 Lr (262)

Periodic table of the elements. Atomic weights are those of the most commonly available long-lived isotopes on the 1999 IUPAC Atomic Weights of the Elements. A value in parentheses denotes the mass number of the longest-lived isotope.

The periodic table dates back to around 1870 when the Russian chemist D. Mendeleev used the similarities in chemical reactivity attributed to different elements to group them according to increasing atomic mass.

The modern periodic table is divided into 18 columns called groups or families (see illustration). Elements in each family tend to have similar properties. In column 1, each alkali metal is soft, relatively low-melting, and highly reactive toward air and water. Column 2 contains the alkaline earth metals, which have higher melting points and are less reactive. Columns 3–12 are filled by the transition metals, which are shiny and good conductors of both heat and electricity. Columns 13–18 are often discussed along with columns 1 and 2, and collectively they are known as the main group or representative elements. Column 15, headed by nitrogen, is known as the pnicogens; column 16, beginning with oxygen, as the chalcogens; column 17, starting with fluorine, as the halogens; and column 18, starting with helium, as the noble gases. *See* ALKALINE-EARTH METALS; TRANSITION ELEMENTS.

The horizontal rows of the periodic table are called periods. Atomic mass generally increases from left to right across a period, while atomic size generally decreases. The decrease in size is due to incomplete screening of the positive nuclear charge by the valence electrons, which causes the outer electron shells to contract. Other properties follow periodic trends, including the ionization potential (the energy needed to remove an electron), electron affinity (the energy released on accepting an electron), and electronegativity (the ability of an atom in a compound to attract electron density). *See* ATOMIC MASS; ATOMIC STRUCTURE AND SPECTRA; ELECTRON AFFINITY; ELECTRONEGATIVITY; IONIZATION POTENTIAL.

After element 57 (lanthanum) comes a series of 14 metallic elements numbered 58–71 with very closely related properties. Below the lanthanides are 14 more metallic elements (90–103) called the actinides. *See* ACTINIDE ELEMENTS; LANTHANIDE CONTRACTION; RARE-EARTH ELEMENTS.

Each box in the periodic table contains a one- or two-letter symbol representing a different element such as C for carbon (6) or Sg for Seaborgium (106) [see illustration]. The number in the upper left corner is the atomic number indicating how many protons are in the atom's nucleus. The atomic mass generally appears below the symbol indicating the average mass observed for that element.

The elements beyond 92 (uranium) do not occur naturally and are produced using nuclear reactions. Elements beyond 100 are not particularly useful, since they generally undergo rapid nuclear decay by emitting radiation. *See* TRANSURANIUM ELEMENTS.

Many periodic tables include a stair-step line separating metals from the metalloids and nonmetals. Nonmetals are found to the right of the metals. Metalloids straddle the stair-step line and often have properties in between metals and nonmetals. With intermediate conductivities, elements such as silicon form important semiconductors used in computer chips and solar cells. *See* METAL; METALLOID; NONMETAL; SEMICONDUCTOR. Richard B. Kaner

Periodontal disease An inflammatory lesion caused by bacteria affecting the tissues housing the roots of the teeth. The disease, sometimes called pyorrhea, increases in prevalence and severity with increasing age, and it is the principal cause of tooth loss in adult humans throughout the world. When only the gum tissue or gingiva is affected, the disease is called gingivitis, but when the process extends into the deeper structures it is known as periodontitis. The diseased tissues appear abnormally red and slightly swollen, and they tend to bleed, sometimes profusely, when the teeth are brushed. In some cases the gums may become thickened and scarred, and they may recede, exposing the root surface. As the disease advances, the attachment of the gum to the tooth is lost, creating a periodontal pocket, a large portion of the gum is destroyed, and the bone surrounding the roots is resorbed. The teeth become loose, abscesses form, and extraction is required.

Both gingivitis and periodontitis are caused by bacteria that form plaques on the surfaces of the teeth at the gingival sulcus or pocket. These plaques may contain 250 or more separate microbial species. Plaques of any microbial composition can cause gingivitis, but specific bacteria appear to be necessary for induction of periodontitis. Among the bacteria involved in periodontitis are various species of *Porphyromonas*, *Bacteroides*, *Actinobacillus*, *Eikenella*, *Fusobacterium*, *Wolinella*, and other less well-characterized species. Spirochetes are present in active lesions, but their role remains unclear. The bacteria extend apically along the interface between the tooth root and the gingival tissue and causes periodontal pockets to form.

The principal features of the pathogenesis of periodontitis have been described. The lesions begin as an acute inflammatory response followed by a dense accumulation of lymphoid cells. There is a net loss of collagen in the area nearest the junctional epithelium and periodontal pocket, with scarring and fibrosis of the connective tissues at more distant sites. The junctional epithelium is converted into an ulcerated pocket epithelium, the alveolar bone housing the tooth roots is resorbed, and the periodontal ligament is destroyed. Products released by infiltrating leukocytes, including prostaglandins, interleukins and collagenase, and other hydrolytic enzymes, are involved in tissue destruction.

Bacterial colonization and extension activate several host defense mechanisms. The most effective of these is the accumulation of functional neutrophilic granulocytes between the surface of the plaque and the gingival tissue. These cells tend to counter and limit microbial extension. The bacteria appear to invade the periodontal connective tissues, where they induce immunopathologic and other destructive inflammatory reactions in the host, and these lead, in major part, to the observed tissue destruction. Periodontal destruction is episodic, with periods of exacerbation characterized by highly acute inflammation, followed by periods of quiescence.

Although bacteria are essential for induction of the disease, predisposing factors are also important, though their elucidation is not complete. Individuals who manifest functionally abnormal neutrophilic granulocytes or monocytes are unusually susceptible to the severe early onset forms of periodontitis. The leukocyte abnormality appears to be genetically transmitted. The early-onset forms have been designated as prepubertal, juvenile, and rapidly progressive periodontitis; adult periodontitis has a later onset and does not seem to be related to leukocyte abnormalities. Some persons with acquired immune deficiency syndrome (AIDS) manifest a highly destructive, unique form of periodontitis. Other predisposing conditions include unusually stressful situations and periods of hormone imbalance occurring at puberty, during pregnancy, and in some women taking birth control drugs.

Good daily oral hygiene practices, including vigorous brushing of the exposed surfaces of the teeth and use of dental floss, interproximal brushes, and other devices to clean between the teeth, constitute the most effective measures to prevent periodontal disease. Basic ingredients of treatment of existing disease include bringing the infection under control and establishing conditions which preclude reinfection. All of the microbial deposits must be removed from the crown and root surfaces. In individuals with severe forms of periodontitis, these procedures may be supplemented by use of antibiotics either systematically or directly into the pocket. These procedures usually lead to reduction of the inflammation and to some shrinkage in the gums, but the periodontal pockets remain. Based on the traditional view that treated pockets may become reinfected and the disease may continue to spread, surgical treatment may be performed with the aim of reducing pocket depth and restoring normal tissue contours. Alternatively, regenerative procedures use various grafting materials, including freeze-dried decalcified bone or bone substitutes and guided tissue regeneration. To perform guided tissue regeneration, flaps are opened in the gingival tissue and the root surfaces are thoroughly cleaned; a porous membrane is placed around the tooth, covering the bone defect with or without placing grafts, and flaps covering the membranes are sutured into place. The membrane permits the wound site to become populated with cells having the capacity to generate new bone, cementum, and periodontal attachment. *See* TOOTH DISORDERS.

Roy C. Page

Perischoechinoidea A subclass of Echinoidea lacking stability in the number of columns of plates that make up the ambulacra and interambulacra. The ambulacral columns vary from 2 to 20, the interambulacral from 1 to 14. There are two orders, Bothriocidaroida and Echinocystitoida. *See* ECHINODERMATA; ECHINOIDEA.

Howard B. Fell

Periscope An optical instrument that permits viewing along a displaced or deflected axis, providing an observer with the view from a position which may be inaccessible or dangerous. Periscopes range in complexity from the simple unit-power tank periscope to the complex multielement submarine periscope.

The tank periscope, intended to protect the user from bullets, employs a pair of plane, parallel, reflecting surfaces (either mirrors or prisms), so arranged in a mount that the path of light through the instalment forms a crude letter *Z*. If powers greater than unity are desired or if the periscope is to be used for sighting, a terrestrial telescope can be added to the periscope. *See* TELESCOPE.

In the submarine periscope, it is necessary to employ a telescope system having a wide field of view and uniform illumination across a field which can be fitted into a long, narrow tube whose length-to-diameter ratio may be 50 or greater. This is achieved by utilizing a plurality of lenses so spaced along the length of the tube as to cause the incoming principal rays from the edge of the field to be deviated from side to side within the tube (see illustration). In general, the greater the number of lenses, the wider the field of view.

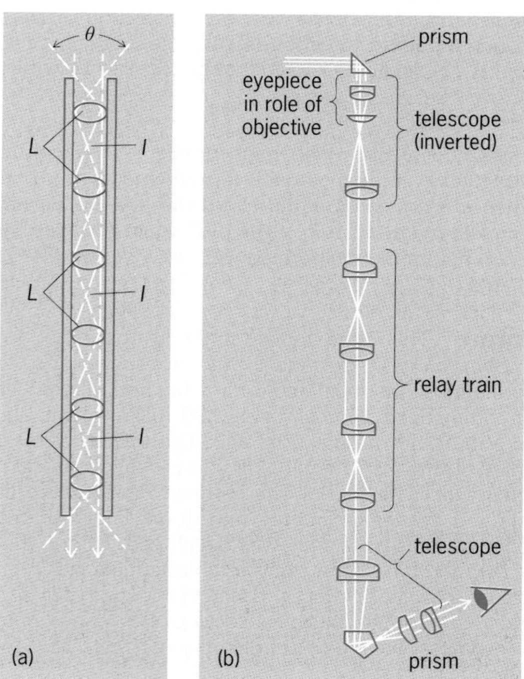

Periscopic relay train. (*a*) Showing lenses *L*, inversions *i*, and angle of view θ. (*b*) Between a pair of facing telescopes in a submarine periscope.

Various modifications of the basic optical systems described here are employed as viewing periscopes in military aircraft and as viewing devices in particle accelerators and nuclear reactors. The cystoscope and endoscope are slender, sometimes mechanically flexible periscopes used for visual examination and photography of body cavities inaccessible to direct observation; an entirely different basis for the design of such instruments is in the use of bundles of optical fibers. *See* OPTICAL FIBERS.

Edward K. Kaprelian

Perissodactyla An order of herbivorous, odd-toed, hoofed mammals, including the living horses, zebras, asses, tapirs, rhinoceroses, and their extinct relatives. They are defined by a number of unique specializations, but the most diagnostic feature is their feet. Most perissodactyls have either one or three toes on each foot, and the axis of symmetry of the foot runs through the middle digit.

The perissodactyls are divided into three groups: the Hippomorpha (horses and their extinct relatives); the Titanotheriomorpha (the extinct brontotheres); and the Moropomorpha (tapirs, rhinoceroses, and their extinct relatives). *See* RHINOCEROS; TAPIR.

Perissodactyls originated in Asia some time before 57 million years ago (Ma). By 55 Ma, the major groups of perissodactyls had differentiated, and migrated to Europe and North America. Before 34 Ma, the brontotheres and the archaic tapirs were the largest and most abundant hoofed mammals in Eurasia and North America. After these groups became extinct, horses and rhinoceroses were the most common perissodactyls, with a great diversity of species and body forms. Both groups were decimated during another mass extinction about 5 Ma, and today only five species of rhinoceros, four species of tapir, and a few species of horses, zebras, and asses cling to survival in the wild. The niches of large hoofed herbivores have been taken over by the ruminant artiodactyls, such as cattle, antelopes, deer, and their relatives.

Most extinct horses were browsers and ate soft, leafy vegetation, but all living horses are grazers, using their sharp incisors and mobile lips to crop low-growing grasses. The only common wild horse, the plains zebra, lives in large herds (up to 100 individuals) and migrates over large areas of grasslands in search of food. However, desert-dwelling asses and Grevy's zebra live in small herds, with a stallion guarding a small harem of mares. Most species of wild horses, including the Grevy's and mountain zebras, all species of onagers and asses, and Przewalski's horse (an ancestor of domesticated horses), are nearly extinct in the wild.

The earliest moropomorphs, such as *Homogalax*, from strata about 55 million years old, are virtually indistiguishable from the earliest horses. From this unspecialized ancestry, a variety of archaic tapirlike animals diverged. Most retained the simple leaf-cutting teeth characteristic of tapirs and, like brontotheres, died out about Ma when their forest habitats shrank. Only the modern tapirs, with their distinctive long proboscis, survive in the jungles of Central and South America (three species), and southeast Asia (one species). All are stocky, piglike beasts with short stout legs, oval hooves, and a short tail. They have no natural defenses against large predators (such as jaguars or tigers), so they are expert at fleeing through dense brush and swimming to make their escape.

Rhinoceroses have been highly diverse and successful throughout the past 50 million years. They have occupied nearly every niche available to a large herbivore, from dog-sized running animals, to several hippolike forms, to the largest land mammal that ever lived—the 18-ft-tall (6-m), 44,000-lb (20,000-kg) *Paraceratherium*. Between 20 and 5 Ma, rhinos diversified into several browsing (leaf-eating) lineages, and hippolike grazing lineages, and browser-grazer pairs of rhinos were found all over the grasslands of Eurasia, Africa, and North America. The mass extinction event that occurred about 5 Ma wiped out North American rhinos and decimated most of the archaic rhino lineages in the Old World. During the ice ages, woolly rhinos and their relatives were common all over Eurasia. Their only surviving descendant is the endangered Sumatran rhinoceros. Only a few hundred individuals still live in the mountainous jungles of Sumatra. Four other species of rhino survive in Asia and Africa, but all are on the brink of extinction because of heavy poaching for their horns. *See* MAMMALIA. Donald R. Prothero

Peritoneum The membranous lining of the coelomic, especially the abdominal, cavity, which surrounds most of the organs. It is composed mainly of flattened epithelial cells that produce a small amount of watery, or serous, fluid. In the embryo the coelomic wall is lined by this membrane, which continues over the developing viscera so that they are suspended and supported by the reflected peritoneum, principally from the dorsal body wall, but also from the ventral body wall in the region of the liver. *See* EPITHELIUM; FETAL MEMBRANE. Thomas S. Parsons

Peritonitis Inflammation of the peritoneum. The condition may be caused by infectious organisms or foreign substances introduced into the abdominal cavity. The small amount of serous fluid normally present as a lubricant acts as an excellent culture medium for bacterial growth and also as a means of spreading invading materials. The source of such substances or organisms is commonly a gastrointestinal inflammation, especially if perforation has occurred. Appendicitis, peptic ulcer, cancer of the bowel, gallbladder disease, and dysentery are common sources of infection that may produce peritonitis, as well as blood-borne forms of tuberculosis and pneumonia. *See* PERITONEUM. Edward G. Stuart; N. Karle Mottet

Peritrichia A specialized subclass of the class Ciliatea composed of a large group of unusual-looking ciliate protozoans. Many are sessile and stalked, while some form colonies which may reach a large size. A number are attached as ectocommensals to a variety of animals and plants. A free-swimming stage in the life cycle, indispensable for distribution, is known as the telotroch. It is a small, mouthless form equipped with a single girdle of posteriorly located locomotor cilia. This is quite unlike the morphology of the mature, sedentary form, which is an inverted bell form atop a long stalk.

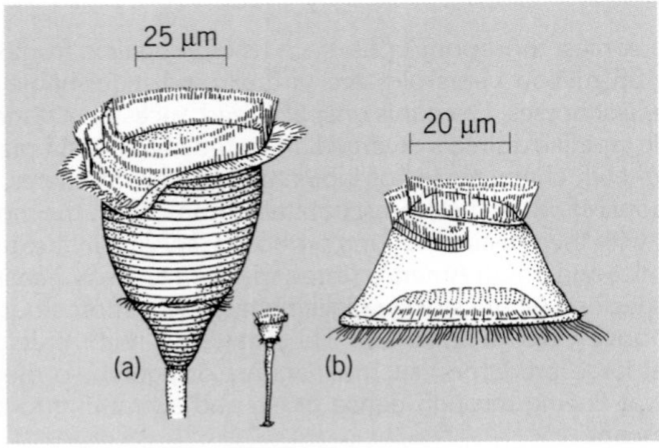

(*a*) *Vorticella*, a stalked peritrich. (*b*) *Trichodina*, a mobile peritrich.

Vorticella (illustration *a*) and *Epistylis* are probably the best-known stalked forms. The former is a solitary ciliate, the latter a colony builder. *Trichodina* (illustration *b*) belongs to the group of mobile peritrichs. *See* CILIATEA. John O. Corliss

Permafrost Perennially frozen ground, occurring wherever the temperature remains below 32°F (0°C) for several years, whether the ground is actually consolidated by ice or not and regardless of the nature of the rock and soil particles of which the earth is composed. Perhaps 25% of the total land area of the Earth contains permafrost; it is continuous in the polar regions and becomes discontinuous and sporadic toward the Equator. During glacial times permafrost extended hundreds of miles south of its present limits in the Northern Hemisphere.

Temperature of permafrost at the depth of no annual change, about 30–100 ft (10–30 m), crudely approximates mean annual air temperature. It is below 23°F (−5°C) in the continuous zone, between 23–30°F (−5 and −1°C) in the discontinuous zone, and above 30°F (−1°C) in the sporadic zone. Temperature gradients vary horizontally and vertically from place to place and from time to time.

Ice is one of the most important components of permafrost, being especially important where it exceeds pore space. Physical properties of permafrost vary widely from those of ice to those of normal rock types and soil. The cold reserve, that is, the number of calories required to bring the material to the melting point and melt the contained ice, is determined largely by moisture content.

Permafrost develops today where the net heat balance of the surface of the Earth is negative for several years. Much permafrost was formed thousands of years ago but remains in equilibrium with present climates. Permafrost eliminates most groundwater movement, preserves organic remains, restricts or inhibits plant growth, and aids frost action. It is one of the primary factors in engineering and transportation in the polar regions. Frederick E. Nelson

Permeance The reciprocal of reluctance in a magnetic circuit. It is the analog of conductance (the reciprocal of resistance) in an electric circuit, and is given by Eq. (1), where **B** is

$$P_m = \frac{\text{magnetic flux}}{\text{magnetomotive force}} = \frac{\int\int \mathbf{B} \cdot d\mathbf{S}}{\oint \mathbf{H} \cdot d\mathbf{l}} \qquad (1)$$

the magnetic flux density, **H** is the magnetic field strength, and the integrals are respectively over a cross section of the circuit and around a path within it. *See* CONDUCTANCE.

From Eq. (1), it can be shown that Eq. (2) is valid, where *A* is

$$P_m = \mu A/l \qquad (2)$$

the cross-sectional area of the magnetic circuit, *l* its length, and μ the permeability. If the material is ferromagnetic, as is often the case, then μ is not constant but varies with the flux density and the complete magnetization curve of *B* against *H* may have to be used to determine the permeance. *See* MAGNETIC MATERIALS; RELUCTANCE. A. Earle Bailey

Permian The name applied to the last period of geologic time in the Paleozoic Era and to the corresponding system of rock formations that originated during that period. The Permian Period commenced approximately 290 million years ago and ceased about 250 million years ago. The system of rocks that originated during this interval of time is widely distributed on all the continents of the world. The Permian Period was a time of variable and changing climates, and during much of this time latitudinal climatic belts were well developed. During the latter half of Permian time, many long-established lineages of marine invertebrates became extinct and were not immediately replaced by new fossil-forming lineages. Rocks of Permian age contain many resources, including petroleum, coal, salts, and metallic ores. *See* LIVING FOSSILS.

During the Permian Period, several important changes took place in the paleogeography of the world. The joining of Gondwana to western Laurasia, which had started during the

Carboniferous, was completed during Wolfcampian time (earliest Permian). The addition of eastern Laurasia (Angara) to the eastern edge of western Laurasia finished during Artinskian time (middle to latest early Permian) and completed the assembly of the supercontinent Pangaea. The climatic effects of these changes were dramatic. Instead of having a circumequatorial tropical ocean, such as during the middle Paleozoic, a large landmass with several high chains of mountains extended from the South Pole across the southern temperate, the tropical, and into the north temperate climatic belts. One very large world ocean, Panthalassa and its western tropical branch, the Tethys, occupied the remaining 75% of the Earth's surface, with a few much smaller cratonic blocks, island arcs, and atolls. *See* CONTINENTAL DRIFT; CONTINENTS, EVOLUTION OF; PALEOGEOGRAPHY.

Most marine invertebrates of the Early Permian were continuations of well-established phylogenetic lines of middle and late Carboniferous ancestry. During early Permian time, these faunas were dominated by brachiopods, bryozoans, conodonts, corals, fusulinaceans, and ammonoids. The Siberian traps, an extensive outflow of very late Permian basalts and other basic igneous rocks (dated at about 250 million years ago), are considered by many geologists as contributing to climatic stress that resulted in major extinctions of many animal groups, particularly the shallow-water marine invertebrates. The end of the Permian is also associated with unusually sharp excursions in values of the carbon-12 isotope (^{12}C) in organic material trapped in marine sediments, suggesting major disruption of the ocean chemistry system.

Terrestrial faunas included insects which showed great advances over those of the Carboniferous Coal Measures. Several modern orders emerged, among them the Mecoptera, Odonata, Hemiptera, Trichoptera, Hymenoptera, and Coleoptera. *See* INSECTA.

Of the vertebrates, labyrinthodont amphibians were common and varied; however, reptiles showed the greatest evolutionary radiation and the most significant advances. Reptiles are found in abundance in the lower half of the system in Texas and throughout most of the upper part of the system in Russia and also are common in Gondwana sediments. Of the several Permian reptilian orders, the most significant was the Theriodonta. These reptiles carried their bodies off the ground and walked or ran like mammals. Unlike most reptiles, their teeth were varied—incisors, canines, and jaw teeth as in the mammals—and all the elements of the lower jaw except the mandibles showed progressive reduction. Most of the known theriodonts are from South Africa and Russia. *See* PALEOZOIC; REPTILIA; THERAPSIDA.

Charles A. Ross; June R. P. Ross

Permineralization A fossilization, or taphonomic, process characterized by the deposition of a mineral or mineraloid (an amorphous "gel mineral") in the pore spaces of an organism's body structures. Pores in bone, shell, or plant material may be impregnated by minerals or mineraloids precipitated from ground water, lake or stream water, or ocean water. Minerals commonly involved in permineralization are quartz, calcite, pyrite, siderite (iron carbonate), and apatite (calcium phosphate). Opal, a mineraloid, is also a common agent of permineralization. Normally, the structural parts of organisms undergo little, if any, shape change as the pores are filled, leading in many examples to exquisite preservation of the fine structural details of bodily remains. *See* FOSSIL; TAPHONOMY.

Permineralization can occur in two ways. One way is by the simple infilling of pore spaces. This often occurs in the permineralization of bone (notably dinosaur bone) by silica, or the formation of a coal ball (a subspherical mass containing mineral matter embedded with plant material) by carbonate mineralization of peat. The fossil retains its original shape because its pore spaces were filled before sediment compaction occurred. The second type of permineralization, called petrifaction, involves the filling of pore spaces together with replacement of organic matter or skeletal parts by mineral or mineraloid matter. Large pores and smaller spaces within and between cells of the tissue are filled first. Simultaneously or shortly afterward, the structural framework of the organism's remains, usually organic tissues, decays or dissolves. Finally, to ensure preservation of delicate structures in physically undistorted condition, replacement of the lost material must occur before sediment is compacted. A common form of petrifaction is that of fossil wood: as infilling of pore spaces occurs, cellulose of the woody material decays or dissolves and is quickly replaced by silica or another substance. *See* COAL BALLS; PALEOBOTANY; PETRIFACTION.

Loren E. Babcock

Permittivity A property of a dielectric medium that determines the forces that electric charges placed in the medium exert on each other. If two charges of q_1 and q_2 coulombs in free space are separated by a distance r meters, the electrostatic force F newtons acting upon each of them is proportional to the product of the charges and inversely proportional to the square of the distance between them. Thus, F is given by Eq. (1), where

$$F = \frac{q_1 q_2}{4\pi\epsilon_0 r^2} \qquad \text{newtons} \qquad (1)$$

$1/(4\pi\epsilon_0)$ is the constant of proportionality, having the magnitude and dimensions necessary to satisfy Eq. (1). This condition leads to a value for ϵ_0, termed the permittivity of free space, given by Eq. (2), where c is the velocity of light in vacuum.

$$\epsilon_0 = \frac{1}{4\pi 10^{-7} c^2} \simeq 8.8542 \times 10^{-12} \qquad \text{farads/meter} \qquad (2)$$

If now the charges are placed in a dielectric medium that is homogeneous and isotropic, the force on each of them is reduced by a factor ϵ_r, where ϵ_r is greater than 1. This dimensionless scalar quantity is termed the relative permittivity of the medium, and the product $\epsilon_0\epsilon_r$ is termed the absolute permittivity ϵ of the medium.

A consequence is that if two equal charges of opposite sign are placed on two separate conductors, then the potential difference between the conductors will be reduced by a factor ϵ_r when the conductors are immersed in a dielectric medium compared to the potential difference when they are in vacuum. Hence a capacitor filled with a dielectric material has a capacitance ϵ_r times greater than a capacitor with the same electrodes in vacuum would have. Except for exceedingly high applied fields, unlikely normally to be reached, ϵ_r is independent of the magnitude of the applied electric field for all dielectric materials used in practice, excluding ferroelectrics. *See* CAPACITANCE; CAPACITOR; FERROELECTRICS.

James H. Calderwood

Perovskite A minor accessory mineral, formula $CaTiO_3$, occurring in basic rocks. Perovskite has given its name to a large family of materials, synthetic and natural, crystallizing in similar structures. The crystal structure is ideally cubic, with a framework of corner-sharing octahedra, containing titanium (Ti) or other relatively small cations surrounded by six oxygen (O) or fluorine (F) anions. Within this framework are placed calcium (Ca) or other large cations, surrounded by twelve anions. Tilting of the octahedra and other distortions often lower the symmetry from cubic, giving the materials important ferroelectric properties and decreasing the coordination of the central cation. This flexibility gives the structure the ability to incorporate ions of different sizes and charges. Substitution of niobium (Nb), cerium (Ce), and other rare-earth elements in natural calcium titanate ($CaTiO_3$) is common and can make perovskite an ore for these elements. *See* COORDINATION CHEMISTRY; CRYSTAL STRUCTURE; RARE-EARTH ELEMENTS.

A number of synthetic perovskites are of major technological importance. Barium titanate ($BaTiO_3$) and lead zirconate-titanate (PZT) ceramics form the basis of a sizable industry in ferroelectric and piezoelectric materials crucial to transducers, capacitors, and electronics. Lanthanum chromate ($LaCrO_3$) and related materials find applications in fuel cells and high-temperature electric heaters. *See* CERAMICS; FUEL CELL; SOLID-STATE CHEMISTRY. Alexandra Navrotsky

Peroxide A chemical compound which contains the peroxy (—O—O—) group, which may be considered to be a derivative of hydrogen peroxide (HOOH). An organic (or inorganic) peroxide is one in which some organic (or inorganic) substituent has replaced one or both hydrogens. Peroxides are used in such diverse reactions as oxidation, synthesis, polymerization, and oxygen generation. Inorganic peroxides include persulfates, hydrogen peroxide (H_2O_2), sodium peroxide, bivalent metal peroxides, and H_2O_2 addition compounds. Organic peroxides include peroxyacetic acid, dibenzoyl peroxide, and cumene peroxide. *See* HYDROGEN PEROXIDE; OXIDIZING AGENT; OXYGEN. W. A. Pryor

Peroxisome An intracellular organelle found in all eukaryotes except the archezoa (original lifeforms). In electron micrographs, peroxisomes appear round with a diameter of 0.1–1.0 micrometer, although there is evidence that in some mammalian tissues peroxisomes form an extensive reticulum (network). They contain more than 50 characterized enzymes and perform many biochemical functions, including detoxification. *See* CELL ORGANIZATION; ENZYME.

Peroxisomes are important for lipid metabolism. In humans, the β-oxidation of fatty acids greater than 18 carbons in length occurs in peroxisomes. In yeast, all fatty acid β-oxidation occurs in peroxisomes. Peroxisomes contain the first two enzymes required for the synthesis of plasmalogens. Peroxisomes also play important roles in cholesterol and bile acid synthesis, purine and polyamine catabolism, and prostaglandin metabolism. In plants, peroxisomes are required for photorespiration. *See* LIPID METABOLISM; PHOTORESPIRATION.

A number of recessively inherited peroxisomal disorders have been described and grouped into three categories. Group I is the most severe and is characterized by a general loss of peroxisomal function. Many of the enzymes normally localized to the peroxisome are instead found in the cytosol. Among the diseases found in group I are Zellweger syndrome, neonatal adrenoleukodystrophy, and infantile Refsum disease. Patients with these disorders usually die within the first years after birth and exhibit neurological and hepatic (liver) dysfunction, along with craniofacial dysmorphism (malformation of the cranium and the face). Groups II and III peroxisomal disorders are characterized by a loss of peroxisomal function less severe than in group I. Richard A. Rachubinski

Peroxisome proliferator-activated receptor A member of the nuclear hormone receptor superfamily of transcription factors, similar in both structure and function to the steroid, thyroid, and retinoid receptors. Peroxisome proliferator-activated receptors (PPARs) respond to specific factors by altering gene expression in a cell-, developmental-, sex-specific manner. PPARs were initially characterized as being activated by peroxisome proliferators (PPs), a large group of hypolipidemic agents and pollutants. Subsequently, it was shown that PPARs respond to endogenous fatty acids and control a variety of target genes involved in lipid homeostasis. Further, PPARs were shown to play a key role in the response to antidiabetic drugs. The multifaceted responses of PPARs are actually carried out by three subtypes of PPARs expressed in different tissues. Currently, the subfamily has been defined as PPARα, PPARβ (also called PPARδ and NUC1), and PPARγ, each with a possibility of different ligands, target genes, and biological roles. *See* GENE; LIPID; LIPID METABOLISM.

Many peroxisome proliferators and fatty acids are PPAR activators. Similar to other steroid hormone receptors, PPARs are ligand-activated transcription factors that control gene expression by interacting with specific DNA response elements (PPREs) located upstream of responsive genes. The ultimate response of a cell to PPAR activators is the sum total of the genes being regulated in that cell. Both PPARα and PPARγ play key roles in regulating fatty acid metabolism. The result of PPARα activation in hepatocytes and certain other tissues is an increase in the peroxisomal and mitochondrial oxidation of fatty acids. The array of genes regulated by PPARγ in adipocytes is indicative of fatty acid accumulation.

The potency of various chemicals to activate PPARs is subtype-specific, and the expression of PPARα, β, and γ varies widely from tissue to tissue. In numerous cell types from ectodermal, mesodermal, or endodermal origin, PPARs are coexpressed, although their concentration relative to each other varies widely. PPARα is highly expressed in hepatocytes (liver cells), cardiomyocytes (heart cells), enterocytes (intestinal cells), and the proximal tubule cells of kidney. PPARβ is expressed ubiquitously and often at higher levels than PPARα and γ. PPARγ is expressed predominantly in adipose tissue, colon epithelium, and macrophages. The distinct tissue distribution suggests that the PPAR subtypes play different biological roles. In fact, of the different subtypes, PPARα predominates in hepatic lipid metabolism and PPARγ plays a pivotal role in adipogenesis and immune responses. There is also evidence to support the role of PPARs in cell growth and differentiation. The biological role of PPARβ has not been clearly established. *See* CELL DIFFERENTIATION. John P. Vanden Heuvel

Peroxynitrite A nitrogen oxyanion containing an O—O peroxo bond that is a structural isomer of the nitrate ion. These species are generally distinguished as $ONOO^-$ and NO_3^-, respectively. Other names for peroxynitrite include pernitrite and peroxonitrite; the systematic name recommended by the International Union of Pure and Applied Chemistry (IUPAC) is oxoperoxonitrate (1−). Energy calculations indicate that there are two stable conformations of $ONOO^-$, for which all of the atoms lie in a plane with the peroxo O—O and N═O bonds forming dihedral angles of approximately 0° (cis isomer) or 180° (trans isomer) [notation (1)].

$$O{=}N{-}O{-}O^- \ (\text{cis}) \rightleftharpoons O{=}N{-}O{-}O^- \ (\text{trans}) \quad (1)$$

See CHEMICAL BONDING.

Peroxynitrite is formed in nitrate salts or nitrate-containing solutions when exposed to ionizing radiation or ultraviolet light. Solutions can also be prepared by a variety of chemical reactions, including the reaction of hydrogen peroxide with nitrous acid (2); reaction of the hydroperoxide anion with organic and inorganic nitrosating agents (3); reaction of ozone with the azide ion (4); or, apparently, reaction of O_2 with compounds capable of generating the nitroxyl anion (NO^-) [5]. These preparations

$$HOOH + HNO_2 \rightarrow ONOOH + H_2O \quad (2)$$

$$HOO^- + RONO \rightarrow ROH + ONOO^- \quad (3)$$

$$2O_3 + N_3^- \rightarrow ONOO^- + N_2O + O_2 \quad (4)$$

$$2O_2 + NH_2OH + 2OH^- \rightarrow ONOO^- + HO_2^- + 2H_2O \quad (5)$$

invariably contain unreacted materials or decomposition products, particularly nitrite ion, which can significantly modulate the peroxynitrite chemical reactivity. Peroxynitrite is also formed in radical-radical coupling reactions, notably superoxide ($\cdot O_2^-$)

with nitric oxide ($\cdot$NO) [reaction (6)], and hydroxyl radical with nitrogen dioxide [reaction (7)].

$$\cdot NO + \cdot O_2^- \rightarrow ONOO^- \quad (6)$$

$$\cdot OH + \cdot NO_2 \rightarrow ONOOH \quad (7)$$

See SUPEROXIDE CHEMISTRY.

Peroxynitrite has been isolated as the tetramethylammonium salt by carrying out reaction (6) in liquid ammonia. Formation of peroxynitrite in both solids and solutions is indicated by the appearance of yellow coloration, which is due to tailing of intense near-ultraviolet absorption bands into the visible region.

Peroxynitrite is a powerful oxidant that has been shown to react with a wide variety of inorganic and organic reductants. Interest in these reactions has been greatly stimulated by recognition that $\cdot$NO and $\cdot O_2^-$ radicals are generated in the bloodstream, neuronal tissues, and phagocytic cells of animals in sufficient quantities to form peroxynitrite [reaction (6)]. Correspondingly, major roles for this powerful oxidant have been proposed both in diseases and tissue damage associated with oxidative stress, and in natural cellular defense mechanisms against microbial infection. *See* BIOINORGANIC CHEMISTRY. James K. Hurst

Perpetual motion The expression perpetual motion, or perpetuum mobile, arose historically in connection with the quest for a mechanism which, once set in motion, would continue to do useful work without an external source of energy or which would produce more energy than it absorbed in a cycle of operation. This type of motion, now called perpetual motion of the first kind, involves only one of the three distinct concepts presently associated with the idea of perpetual motion.

Perpetual motion of the first kind refers to a mechanism whose efficiency exceeds 100%. Clearly such a mechanism violates the now firmly established principle of conservation of energy, in particular that statement of the principle of conservation of energy embodied in the first law of thermodynamics. (Indeed, the first law of thermodynamics is sometimes stated as "A perpetuum mobile of the first kind cannot exist."). *See* CONSERVATION OF ENERGY.

Perpetual motion of the second kind refers to a device that extracts heat from a source and then converts this heat completely into other forms of energy, a process which satisfies the principle of conservation of energy. A dramatic scheme of this type would be an ocean liner, which extracts heat from the nearly limitless oceanic source and then uses this heat for propulsion. This type of perpetual motion is, however, precluded by the second law of thermodynamics which is sometimes stated as "A perpetuum mobile of the second kind cannot exist."

The third type of perpetual motion is, in contrast to the two types described above wherein useful output was the goal, merely a device which can continue moving forever. It could result in actual systems if all mechanisms by which energy is dissipated could be eliminated. Since experience indicates that dissipative effects in mechanical systems can be reduced, by lubrication in the case of friction, for example, but not eliminated, mechanical perpetual motion of the third kind can be approximated but never achieved. An example of a genuine case of this kind occurs in a superconductor. If a direct current is caused to flow in a superconducting ring, this current will continue to flow undiminished in time without application of any external force. *See* SUPERCONDUCTIVITY; THERMODYNAMIC PRINCIPLES. K. L. Kliewer

Perseus A northern constellation named for a Greek hero (see illustration). Perseus and adjacent Cassiopeia are prominently located in the northern Milky Way. *See* CASSIOPEIA.

The double cluster, a pair of star clusters known as h and χ Persei, are fine objects for small telescopes. Algol, an eclipsing binary star known as the Demon Star, shows a cycle of regular dimming by a factor of 6 with a period of 2 days 21 hours on

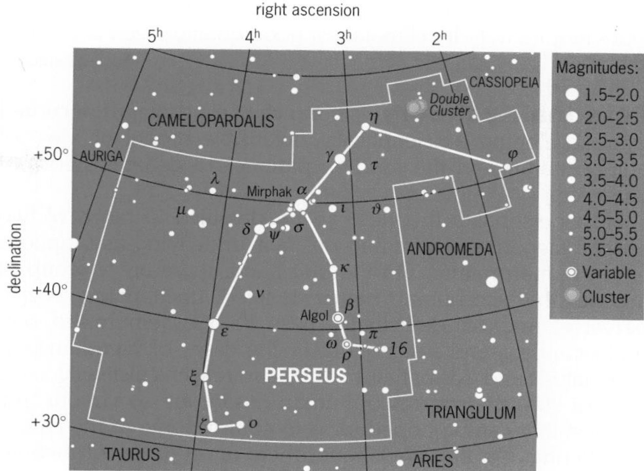

Modern boundaries of the constellation Perseus, a Greek hero. The celestial equator is 0° of declination, which corresponds to celestial latitude. Right ascension corresponds to celestial longitude, with each hour of right ascension representing 15° of arc. Apparent brightness of stars is shown with dot sizes to illustrate the magnitude scale, where the brightest stars in the sky are 0th magnitude or brighter and the faintest stars that can be seen with the unaided eye at a dark site are 6th magnitude. (*Wil Tirion*)

a predictable schedule. *See* BINARY STAR; ECLIPSING BINARY STAR; STAR CLUSTERS.

Each August 12, the Perseid meteor shower appears to come from this part of the sky. *See* CONSTELLATION; METEOR. Jay M. Pasachoff

Persian melon A long-season cultivar of muskmelon, *Cucumis melo*, of the gourd family, Cucurbitaceae. The fruit is round and without sutures; it has dark-green skin and thin, abundant netting. The flesh is deep orange, very thick and firm, and distinctly sweet in flavor. The flesh is very rich in potassium, vitamin A, and vitamin C. *See* MUSKMELON; VIOLALES. Oscar A. Lorenz

Persimmon A deciduous fruit tree species, *Diospyros kaki*. Persimmons originated in the subtropical regions of China but were cultivated more extensively in Japan, the current leader in world production. Persimmons have been introduced to a number of temperate zone countries; however, they have not attained substantial popularity, and only small commercial plantings exist in the United States (primarily in California), Italy, Brazil, and Israel. Persimmon cultivars adapt to a wide climatic range. Although they have a low chilling requirement, they can tolerate temperatures as low as 5°F (−15°C) if they are dormant. However, areas with late spring and early fall temperatures below 27°F (−3°C) should be avoided, as young growth and maturing fruit will be damaged. Mean annual temperatures averaging 57–59°F (14–15°C) are required for good growth and quality. Multiple, selective harvests are done in late fall to ensure that the fruits achieve the desirable deep orange-red color.

Persimmons are clonally propagated by budding onto seedling rootstocks. Seed extracted from mature fruits of *D. lotus*, *D. kaki*, and *D. virgiana* are germinated in the greenhouse in the fall. Seedlings are transplanted in outdoor nursery rows when temperatures are over 55°F (13°C). For horticultural purposes, persimmons are classified as astringent or nonastringent. Astringent persimmons have water-soluble tannins in the flesh that decrease as the fruit softens to ripeness. They are conical in shape. Nonastringent persimmons are firm when ripe, as their soluble tannins decrease with pollination. They have an oblate shape. Astringent cultivars produce best in cool climates, whereas nonastringent types require hot, dry climates for good quality. *See* FRUIT, TREE.

Louise Ferguson

Personality theory A branch of psychology concerned with developing a scientifically defensible model or view of human nature—in the modern parlance, a general theory of behavior.

Most personality theories can be classified in terms of two broad categories, depending on their underlying assumptions about human nature. On the one hand, there are a group of theories that see human nature as fixed, unchanging, deeply perverse, and self-defeating. These theories emphasize self-understanding and resignation; in the cases of Freudian psychoanalysis and existentialism, they also reflect a distinctly tragic view of life—the sources of human misery are so various that the best that can be hoped for is to control some of the causes of suffering. On the other hand, there are a group of theories that see human nature as plastic, flexible, and always capable of growth, change, and development. Human nature is basically benevolent; therefore bad societies are the source of personal misery. Social reform will produce human happiness if not actual perfection. These theories emphasize self-expression and self-actualization—in the cases of Carl Rogers and Abraham Maslow, they reflect a distinctly optimistic and romantic view of life.

Robert Hogan

PERT An acronym for program evaluation and review technique; a planning, scheduling, and control procedure based upon the use of time-oriented networks which reflect the interrelationships and dependencies among the project tasks (activities). The major objectives of PERT are to give management improved ability to develop a project plan and to properly allocate resources within overall program time and cost limitations, to control the time and cost performance of the project, and to replan when significant departures from budget occur.

The basic requirements of PERT, in its time or schedule form of application, are:

1. All individual tasks required to complete a given program must be visualized in a clear enough manner to be put down in a network composed of events and activities. An event denotes a specified program accomplishment at a particular instant in time. An activity represents the time and resources that are necessary to progress from one event to the next.

2. Events and activities must be sequenced on the network under a logical set of ground rules.

3. Time estimates can be made for each activity of the network on a three-way basis. Optimistic (minimum), most likely (modal), and pessimistic (maximum) performance time figures are estimated by the person or persons most familiar with the activity involved. The three-time estimates are used as a measure of uncertainty of the eventual activity duration.

4. Finally, critical path and slack times are computed. The critical path is that sequence of activities and events on the network that will require the greatest expected time to accomplish. Slack time is the difference between the earliest time that an activity may start (or finish) and its latest allowable start (or finish) time, as required to complete the project on schedule.

5. The difference between the pessimistic (b) and optimistic (a) activity performance times is used to compute the standard deviation ($\hat{\sigma}$) of the hypothetical distribution of activity performance times [$\hat{\sigma} = (b - a)/6$]. The PERT procedure employs these expected times and standard deviations (σ^2 is called variance) to compute the probability that an event will be on schedule, that is, will occur on or before its scheduled occurrence time.

In the actual utilization of PERT, review and action by responsible managers is required, generally on a biweekly basis, concentrating on important critical path activities. A major advantage of PERT is the kind of planning required to create an initial network. Network development and critical path analysis reveal interdependencies and problem areas before the program begins that are often not obvious or well defined by conventional planning methods.

Joseph J. Moder

Perthite Any of the oriented intergrowths of potassium- and sodium-rich feldspars, $(K,Na)AlSi_3O_8$, whose proportions are determined in part by the initial composition of the alkali feldspar from which they exsolved and whose physical properties are thus somewhat variable. The early stages of perthite formation from homogeneous, usually monoclinic $(K,Na)AlSi_3O_8$ may be observed experimentally by high-magnification electron microscopy.

If the final K- and Na-rich lamellae or particles are submicroscopic, the composite feldspar is called cryptoperthite. If the particles are small enough and the feldspar relatively clear, Rayleigh-type scattering of light may occur, giving rise to the beautiful blue-to-whitish luster of the semiprecious gem called moonstone. If coarsening has progressed to the micrometer scale and can be seen on a polarizing microscope, the composite is called microperthite; and if the two feldspars are visible to the eye in hand specimen, it is called perthite or macroperthite. Often the albite phase will appear as white veins or blotchy patches against a colored K-rich phase, which may be green to blue microcline (amazonite) or dull pink to orange-brown orthoclase. *See* ALBITE; ANORTHITE; ANORTHOCLASE; FELDSPAR; MICROCLINE; ORTHOCLASE.

Paul H. Ribbe

Perturbation (astronomy) Departure of a celestial body from the trajectory it would follow if moving only under the action of a single central force. Perturbations may be caused by either gravitational or nongravitational forces.

Planetary orbits are subject to two classes of disturbances: secular, or long-term, perturbations; and periodic, or relatively short-term, perturbations. Secular perturbations, so called because they are either progressive or have excessively long periods, arise because of the relative orientation of the orbits in space. They cause slow oscillatory changes of eccentricities and inclinations about their mean values with accompanying changes in the motions of the nodes and perihelia. Periodic perturbations arise from the relative positions of the planets in their orbits. When the disturbed and disturbing planets are aligned on the same side of the Sun, the perturbation reaches a maximum, and reduces to minimum when alignment is reached on opposite sides of the Sun.

The motions of planetary satellites, natural and artificial, reflect both gravitational and nongravitational perturbations. The centrifugal force arising from the rotation of a planet causes a deformation or oblateness of figure. In such a case the central mass does not attract as if it were concentrated at its center. For a close satellite the principal perturbation arises from the attraction of this equatorial bulge.

Raynor L. Duncombe

Perturbation (mathematics) A modification in the mathematical structure of a problem changing the problem from one that can be solved exactly, the unperturbed problem, to one, the perturbed problem, for which it is usually possible to obtain only an approximate solution. The methods employed for this purpose form perturbation theory. These methods attempt to express the solution of the perturbed problem in terms of the properties of the solutions of the unperturbed problem.

Examples of perturbation problems can be found in nearly every branch of mathematics and physics, and in astronomy. The simplest case occurs in ordinary algebra. Suppose that the roots of the equation $f(x) = 0$ are known (the unperturbed problem), and that the roots of the equation $f(x) + \epsilon g(x) = 0$ are to be found (the perturbed problem). The parameter ϵ measures the size of the perturbation. Another set of examples occurs in linear differential equations and in particle dynamics. Possible perturbations include changes in the forces considered to be acting on the particle as well as changes in initial conditions. See PERTURBATION (ASTRONOMY).

Several examples occur in partial differential equations. One physical realization occurs in the theory of wave propagation where the perturbations can be changes in the index of refraction, changes in initial conditions, or changes in the nature or shape of the surfaces encountered by the waves. All of these changes can occur separately or concurrently. The first of these changes is called a volume perturbation, the second a perturbation of initial conditions, and the third a perturbation of boundary conditions. Similar examples can be taken from quantum mechanics, where the volume perturbation corresponds to a change in the hamiltonian, and perturbation of initial conditions to quantum mechanical time-dependent perturbation theory. Other partial differential equations of physics, such as the Laplace equation, the diffusion equation, and the equations of hydrodynamics, furnish further examples. See PERTURBATION (QUANTUM MECHANICS).

All of these problems are linear and can therefore be cast into an equation of the form $A\psi = \lambda\psi$, where ψ is the unknown quantity, λ is a constant, and A is an operator involving among other possibilities differentiation and integration. The quantity σ may be a scalar, a vector, or more generally a matrix quantity. When solutions can be obtained for only special values of λ, the eigenvalues, the equation is called the eigenvalue equation, and the associated problem is called the eigenvalue problem. The operator A contains the perturbation: that is, A equals $A_0 + \epsilon A_1$, where A_0 is the unperturbed operator and ϵA_1, the perturbing term.

Herman Feshbach

Perturbation (quantum mechanics) An expansion technique useful for solving complicated quantum-mechanical problems in terms of solutions for simple problems. Perturbation theory in quantum mechanics provides an approximation scheme whereby the physical properties of a system, modeled mathematically by a quantum-mechanical description, can be estimated to a required degree of accuracy. Such a scheme is useful because very few problems occurring in quantum mechanics can be solved analytically. Consequently an approximation technique must be employed in order to give an approximate analytic solution or to provide suitable algorithms for a numerical solution. Even for problems which admit an exact analytic solution, the exact solution may be of such mathematical complexity that its physical interpretation is not apparent. For these situations, perturbation techniques are also desirable.

Here the discussion of the application of perturbation techniques to quantum mechanics is limited to the domain of nonrelativistic quantum theory. Applications of a similar but mathematically more intricate nature have also been made in quantum electrodynamics and quantum field theory. See QUANTUM ELECTRODYNAMICS; QUANTUM FIELD THEORY; QUANTUM MECHANICS.

Perturbation theory is applied to the Schrödinger equation, $H\Psi = (H_0 + \lambda V)\Psi = i\hbar(\partial/\partial t)\Psi$ [where $\hbar$ is Planck's constant h divided by 2π, and $(\partial/\partial t)$ represents partial differentiation with respect to the time variable t], for which the exact hamiltonian H is split into two parts: the approximate (unperturbed) time-independent hamiltonian H_0 whose solutions of the corresponding Schrödinger equation are known analytically, and the perturbing potential λV. The basic idea is to expand the exact solution Ψ in terms of the solution set of the unperturbed hamiltonian H_0 by means of a power series in the coupling constant λ. Such a procedure is expected to be successful if the system characterized by the unperturbed hamiltonian closely resembles that characterized by the exact hamiltonian. Supposedly the differences are not singular in character, but change as a continuous function of the parameter λ.

Perturbation theory is used in two contexts to provide information about the state of the system, which in quantum mechanics is determined by the wave function Ψ. If λV is time-independent, an objective may be to find the stationary states of the system Ψ_n whose time dependence is given by $\exp(-iE_n t/\hbar)$, where $i = \sqrt{-1}$ and E_n represents the energy of the stationary state labeled by n. If λV is either time-independent or time-dependent, an objective may be to find the time evolution of a state which at some specified time was a stationary state of the unperturbed hamiltonian. The perturbing potential is then considered as causing transitions from the original state to other states of the unperturbed hamiltonian, and application of time-dependent perturbation theory provides the probability of such transitions. See PERTURBATION (MATHEMATICS).

David M. Fradkin

Pesticide A material useful for the mitigation, control, or elimination of plants or animals detrimental to human health or economy. Algicides, defoliants, desiccants, herbicides, plant growth regulators, and fungicides are used to regulate populations of undesirable plants which compete with or parasitize crop or ornamental plants. Attractants, insecticides, miticides, acaricides, molluscicides, nematocides, repellants, and rodenticides are used principally to reduce parasitism and disease transmission in domestic animals, the loss of crop plants, the destruction of processed food, textile, and wood products, and parasitism and disease transmission in humans.

Some pesticides are obtained from plants and minerals. Examples include the insecticides cryolite, a mineral, and nicotine, rotenone, and the pyrethrins which are extracted from plants. A few pesticides are obtained by the mass culture of microorganisms. Two examples are the toxin produced by *Bacillus thuringiensis*, which is active against moth and butterfly larvae, and the so-called milky disease of the Japanese beetle produced by the spores of *B. popilliae*. Most pesticides, however, are products which are chemically manufactured. Two outstanding examples are the insecticide DDT and the herbicide 2,4-D.

Concern over the undesirable effects of pesticides on nonpest organisms culminated in laws to prevent exposure of either humans or the environment to unreasonable hazard from pesticides through rigorous registration procedures. The purpose of regulations are to classify pesticides for general or restricted use as a function of acute toxicity, to certify the qualifications of users of restricted pesticides, to identify accurately and label pesticide products, and to ensure proper and safe use of pesticides. Recommendations as to the product and method of choice for control of any pest problem—weed, insect, or varmint—are best obtained from county or state agricultural extension specialists.

George F. Ludvik

Sophisticated methods of pest control are continually being developed. Highly specific synthetic insect hormones are being developed. In an increasing number of pest situations, a natural predator of an insect has been introduced, or conditions are maintained that favor the propagation of the predator. The numbers of the potential pest species are thereby maintained below a critical threshold. An insect control program in which use of insecticides is only one aspect of a strategy based on ecologically sound measures is known as integrated pest management. See AGRICULTURAL CHEMISTRY; CHEMICAL ECOLOGY; FUNGISTAT AND FUNGICIDE; HERBICIDE; INSECT CONTROL, BIOLOGICAL; INSECTICIDE.

Robert W. Risebrough

Petalite A rare pegmatitic mineral with composition $LiAlSi_4O_{10}$. Its economic significance is markedly disproportionate to the number of its occurrences. It is the only basic raw material suitable for production of a group of materials known as crystallized glass ceramics (melt-formed ceramics). These extremely fine-grained substances are based on a keatite-type structure (stuffed silica derivative). Among the desirable properties of such submicroscopic aggregates are their exceedingly low thermal expansion and high strength, making them suitable for use in cooking utensils and telescopic mirror blanks.

The color is white, pink, pale green, or gray to black. Hardness is $6^{1}/_{2}$ on Mohs scale. Petalite is mined from a large lithium-rich pegmatite at Bikita, Zimbabwe, the only major world source.

E. William Heinrich

Petrifaction A mechanism by which the remains of extinct organisms are preserved in the fossil record. In petrifactions (though chiefly in plants rather than animals) the original shape and topography of the tissues, and occasionally even minute cytological details, are retained relatively undeformed.

The term petrifaction was adopted as a scientific term before knowledge existed of the geochemical mechanism or processes involved. It was formerly widely believed that in the formation of a petrifaction the organic matter of the organism or tissue was replaced molecule by molecule with mineral material entering in solution in percolating groundwater. It is now evident that what actually happens is that the mineral fills cell lumena and the intermicellar interstices of cell walls with insoluble salts depositing from solution. Petrifaction is hence a form of mineral emplacement or embedding, by which the organic residues are filled with solid substance which infiltrates in solution. The most common substances involved in petrifactions are silica, SiO_2, and calcium carbonate, $CaCO_3$ (calcite). Occasionally phosphate minerals, pyrite, hematite, and other less common minerals make up all or part of the petrifaction matrix. *See* FOSSIL; PALEOBOTANY; PETRIFIED FORESTS.

Elso S. Barghoorn

Petrified forests Concentrations of appreciable numbers of petrified tree trunks exposed at the earth surface, standing upright in the position of growth, lying prostrate, or both. They typically consist of the mineralized remains of woody plants and sometimes are called fossil forests. *See* FOSSIL; PETRIFACTION.

Although petrified wood is relatively common, petrified forests are quite rare because special circumstances, such as rapid burial of living or dead trees, are necessary for them to form. Recent studies indicate that only about two dozen are presently known to science. The known forests are scattered from the Antarctic to the Arctic but are most common in the midlatitudes of the Northern Hemisphere. They range in age from the Devonian (about 390 million years old) to the Neogene (about 3 million years old), but they do not become common until the Mesozoic about 248 million years ago. *See* PALEOBOTANY.

Petrified forests are significant because they contain the remains of past life—particularly trees—usually preserved in or close to the terrestrial setting that they once inhabited and sometimes even in the position of growth. Also they often are associated with the remains of coexisting invertebrates and vertebrates, which permits a more complete reconstruction of ancient environments. Furthermore, they help establish the paleoclimates of the time and demonstrate the profound climatic changes that have taken place in the past. *See* CLIMATE MODIFICATION; PALEOCLIMATOLOGY.

Sidney R. Ash

Petrochemical Any of the chemicals derived from petroleum or natural gas. The definition of petrochemicals has been broadened to include the whole range of aliphatic, aromatic, and naphthenic organic chemicals, as well as carbon black and such inorganic materials as sulfur and ammonia.

Petrochemicals are made or recovered from the entire range of petroleum fractions, but the bulk of petrochemical products are formed from the lighter (C_1–C_4) hydrocarbon gases as raw materials. These materials generally occur in natural gas, but they are also recovered from the gas streams produced during refinery operations, especially cracking. Refinery gases are particularly valuable because they contain substantial amounts of olefins that, because of their double bonds, are much more reactive then the saturated (paraffin) hydrocarbons. Also important as raw materials are the aromatic hydrocarbons (benzene, toluene, and xylene) that are obtained from various refinery product streams. For example, catalytic reforming processes convert nonaromatic hydrocarbons to aromatic hydrocarbons by dehydrogenation and cyclization. *See* PETROLEUM; PETROLEUM PRODUCTS.

Thermal cracking processes (such as coking) are focused primarily on increasing the quantity and quality of gasoline and other liquid fuels, but also produce gases, including lower-molecular-weight olefins such as ethylene ($CH_2{=}CH_2$), propylene ($CH_3CH{=}CH_2$), and butylenes (butenes, $CH_3CH{=}CHCH_3$ and $CH_3CH_2CH{=}CH_2$). Catalytic cracking is a valuable source of propylene and butylene, but it is not a major source of ethylene, the most important of the petrochemical building blocks. *See* CRACKING; ETHYLENE.

The starting materials for the petrochemical industry are obtained from crude petroleum in one of two ways. They may be present in the raw crude oil and are isolated by physical methods, such as distillation or solvent extraction; or they are synthesized during the refining operations. Unsaturated (olefin) hydrocarbons, which are not usually present in natural petroleum, are nearly always manufactured as intermediates during the refining sequences. *See* DISTILLATION; PETROLEUM PROCESSING AND REFINING; SOLVENT EXTRACTION.

The main objective in producing chemicals from petroleum is the formation of a variety of well-defined chemical compounds, including (1) chemicals from aliphatic compounds; (2) chemicals from olefins; (3) chemicals from aromatic compounds; (4) chemicals from natural gas; (5) chemicals from synthesis gas (carbon monoxide and hydrogen); and (6) inorganic petrochemicals.

A significant proportion of the basic petrochemicals are converted into plastics, synthetic rubbers, and synthetic fibers. These materials, known as polymers, are high-molecular-weight compounds made up of repeated structural units. The major polymer products are polyethylene, polyvinyl chloride, and polystyrene, all derived from ethylene, and polypropylene, derived from propylene. Major raw material sources for synthetic rubbers include butadiene, ethylene, benzene, and propylene. Among synthetic fibers the polyesters, which are a combination of ethylene glycol and terephthalic acid (made from xylene), are the most widely used. They account for about one-half of all synthetic fibers. The second major synthetic fiber is nylon, its most important raw material being benzene. Acrylic fibers, in which the major raw material is the propylene derivative acrylonitrile, make up most of the remainder of the synthetic fibers. *See* MANUFACTURED FIBER; POLYACRYLATE RESIN; POLYAMIDE RESINS; POLYESTER RESINS; POLYMER; POLYMERIZATION; POLYOLEFIN RESINS; POLYURETHANE RESINS; POLYVINYL RESINS; RUBBER.

An inorganic petrochemical is one that does not contain carbon atoms; typical examples are sulfur (S), ammonium sulfate [$(NH_4)_2SO_4$], ammonium nitrate (NH_4NO_3), and nitric acid (HNO_3). Of the inorganic petrochemicals, ammonia is by far the most common. Ammonia is produced by the direct reaction of hydrogen with nitrogen, with air being the source of nitrogen. Refinery gases, steam reforming of natural gas (methane) and naphtha streams, and partial oxidation of hydrocarbons or higher-molecular-weight refinery residual materials (residua, asphalt) are the sources of hydrogen. The ammonia is used predominantly for the production of ammonium nitrate (NH_4NO_3) as well as other ammonium salts and urea (H_2HCONH_2) that are major constituents of fertilizers. *See* AMMONIA; AMMONIUM SALT; FERTILIZER; UREA.

James G. Speight

Petrofabric analysis The systematic study of the fabrics of rocks, generally involving statistical study of the orientations and distribution of large numbers of fabric elements. The term fabric denotes collectively all the structural or spatial characteristics of a rock mass. The fabric elements are classified into two groups: (1) megascopic features, including bedding, schistosity, foliation, cleavage, faults, joints, folds, and mineral lineations; and (2) microscopic features, including the shapes, orientations, and mutual arrangement of the constituent mineral crystals (texture) and of internal structures (twin lamellae, deformation bands, and so on) inside the crystals.

The aim of fabric analysis is to obtain as complete and accurate a description as possible of the structural makeup of the rock mass with a view to elucidating its kinematic history. The fabric of a sedimentary rock, for example, may retain evidence of the mode of transport, deposition, and compaction of the sediment in the size, shape, and disposition of the particles; similarly, that of an igneous rock may reflect the nature of the flow or of gravitational segregation of crystals and melt during crystallization. The fabrics of deformed metamorphic rocks (tectonites) have been most extensively studied by petrofabric techniques with the objective of determining the details of the history of deformation and recrystallization. *See* STRUCTURAL GEOLOGY; STRUCTURAL PETROLOGY.

John M. Christie

Petrography The description of rocks with goals of classification and interpretation of origin. Most schemes for the classification of rocks are based on the size of grains and the proportions of various minerals. Interpretations of origin rely on field relations, structure, texture, and chemical composition as well as sizes and proportions of different kinds of grains. The names of rocks are based on the sizes and relative proportions of different minerals; boundaries between the names are arbitrary. The conditions of formation of a rock can be estimated from the types and textures of its constituent minerals.

The description of rocks begins in the field with observation of the shape and structure of bodies of rock at the scale of centimeters to kilometers. The geometrical relations between and structures within mappable rock units are generally the domain of field geology, but are simply rock descriptions at a reduced scale.

A petrographer can correctly name most rocks in which most crystals are larger than about 0.04 in. (1 mm) simply by examining the rock with a 10-power magnifying lens. Rocks with smaller grains require either microscopical examination or chemical analysis for proper classification.

Sizes, shapes, and orientations of grains and voids are the most important features of a rock relevant to its origin. The same features also affect density, porosity, permeability, strength, and magnetic behavior. It is also essential to know the identity, abundance, and compositions of minerals constituting the grains in order to name a rock and infer its conditions of formation.

Petrographers study organic as well as inorganic objects, and petrographic analyses are useful to both paleontologists and petroleum geologists. The quality of coal is revealed with polarizing and reflecting light microscopes. Inclusions of petroleum and brine in crystals of silicates and salt in rocks help scientists infer how petroleum formation is connected with cementation and other modifications of buried sediments. *See* PALEONTOLOGY; PETROLEUM GEOLOGY.

Petrographers also study synthetic objects. The textures of metals and alloys are scrutinized by petrographers in order to understand what makes these materials strong and resistant to corrosion. Flaws in glasses and ceramics are revealed by microscopical and polarizing techniques. Fragments of minerals and rocks in some pottery can help point to its source and help trace prehistoric routes of trade. The industrial, agricultural, and natural sources of particles in the air and water may be established from petrographic study. *See* MINERALOGY; PETROLOGY.

Alfred T. Anderson, Jr

Petrolatum A smooth, semisolid blend of mineral oil with waxes crystallized from the residual type of petroleum lubricating oil. The wax molecules are microneedles and hold a large amount of oil in a gel. Petrolatums are useful because they cling, lubricate, and resist both moisture and oxidation. They serve as lubricants in baking and candymaking; as carriers in polishes, cosmetics, and ointments; as rust preventives; as waterproofing agents for paper; and in other uses calling for an inert greaselike material.

J. K. Roberts

Petroleum Unrefined, or crude, oil is found underground and under the sea floor, in the interstices between grains of sandstone and limestone or dolomite (not in caves). Petroleum is a mixture of liquids varying in color from nearly colorless to jet black, in viscosity from thinner than water to thicker than molasses, and in density from light gases to asphalts heavier than water. It can be separated by distillation into fractions that range from light color, low density, and low viscosity to the opposite extreme. In places where it has oozed from the ground, its volatile fractions have vaporized, leaving the dense, black parts of the oil as a pool of tar or asphalt (such as the Brea Tar Pits in California). Much of the world's crude oil is today produced from drilled wells. *See* PETROLEUM ENGINEERING.

Petroleum consists mostly of hydrocarbon molecules. The four main classes of hydrocarbons are paraffins (also called alkanes), olefins (alkenes), cycloparaffins (cycloalkanes), and aromatics. Olefins are absent in crude oil but can be formed in certain refining processes. The simplest hydrocarbon is one carbon atom bonded to four hydrogen atoms (chemical formula CH_4), and is called methane. *See* ALKANE; PARAFFIN.

Petroleum usually contains all of the possible hydrocarbon structures except alkenes, with the number of carbon atoms per molecule going up to a hundred or more. These fractions include compounds that contain sulfur, nitrogen, oxygen, and metal atoms. The proportion of compounds containing these atoms increases with increasing size of the molecule.

Asphaltic molecules contain many cyclic compounds in which the rings contain sulfur, nitrogen, or oxygen atoms; these are called heterocyclic compounds. An example is pyridine. *See* ASPHALT AND ASPHALTITE.

It is generally agreed that petroleum formed by processes similar to those which yielded coal, but was derived from small animals rather than from plants. Dead organisms have been buried in mud over millions of years. Further layers deposited over these mud layers have in some cases reached a thickness of thousands of feet, and compacted the layers beneath them, until the mud has become shale rock. The mud layers were heated and compressed by the layers above. The bodies of the organisms in the mud were decomposed and converted into fatty liquids and solids. Heating these fatty materials over a very long time caused their molecules to break into smaller fragments and combine into larger ones, so the original range of molecular size was spread greatly into the range found in crude oil. Bacteria were usually present, and helped remove oxygen from the molecules and turned them into hydrocarbon compounds. The great pressure of the overlying rock layers helped to force the oil out of the compacted mud (shale) layers into less compacted limestone, dolomite, or sandstone layers next to the shale layers. *See* DOLOMITE; LIMESTONE; ORGANIC GEOCHEMISTRY; PETROLEUM GEOLOGY; SANDSTONE; SEDIMENTOLOGY; SHALE.

At depths greater than about 25,000 ft (7620 m), the temperature is so high that the oil conversion processes go all the way to natural gas and soot. Natural gas formed by the conversion processes is now also found over a variety of depths which do not indicate the depth and temperature of their origin. *See* NATURAL GAS.

The oil formed by the natural thermal and bacterial processes was squeezed out of the compacting mud layers into sandstone or limestone layers and migrated upward in tilted layers. Tectonic processes caused such uptilting and bulging of layers to form ridges and domes. When the ridges and domes were covered by shale already formed, the pores of the shale were too tiny to let the oil through, so the shale acted as a sealing cap. When the oil could not rise farther, it was trapped. Porous rock in such a structure that contains oil or gas is called an oil or gas reservoir.

The recovery from typical reservoirs is not as high as might be thought. Multiple-layer reservoirs will typically contain oil-bearing layers with a wide range of permeability. When recovery from the highest-permeability layers is as complete as it can be, the low-permeability layers will usually have been only slightly depleted, despite all efforts to improve the recovery. Despite recovery efforts, half or more of the oil originally present in oil reservoirs is still in them. *See* PETROLEUM ENHANCED RECOVERY; PETROLEUM RESERVES; PETROLEUM RESERVOIR ENGINEERING. Elmond L. Claridge

Heavy oil and tar sand oil (bitumen) are petroleum hydrocarbons found in sedimentary rocks. They are formed by the oxidation and biodegradation of crude oil, and occur in the liquid or semiliquid state in limestones, sandstones, or sands. *See* BITUMEN.

These oils are characterized by their viscosity; however, density (or API gravity) is also used when viscosity measurements are not available. Heavy oils contain 3 wt % or more sulfur and as much as 200 ppm vanadium. Titanium, zinc, zirconium, magnesium, manganese, copper, iron, and aluminum are other trace elements that can be found in these deposits. Their high naphthenic acid content makes refinery processing equipment vulnerable to corrosion. *See* OIL AND GAS FIELD EXPLOITATION. Ender Okandan

Petroleum engineering The technologies used for the exploitation of crude oil and natural gas reservoirs. It is usually subdivided into the branches of petrophysical, geological, reservoir drilling, production, and construction engineering. After an oil or gas accumulation is discovered, technical supervision of the reservoir is transferred to the petroleum engineering group, although in the exploration phase the drilling and petrophysical engineers have played a role in the completion and evaluation of the discovery.

By the use of down-hole logging tools and of laboratory analysis of cores made during the drilling operation, the petrophysical engineer estimates the porosity, permeability, and oil content of the reservoir rock that has been sampled at the drill site. *See* WELL LOGGING.

The geological engineer, using the petrophysical data, the seismic surveys conducted during the exploration operations, and an analysis of the regional and environmental geology, develops inferences concerning the lateral continuity and extent of the reservoir. *See* PETROLEUM GEOLOGY.

The reservoir engineer, using the initial studies of the petrophysicist and geological engineers together with the early performance of the wells drilled into the reservoir, attempts to assess the producing rates (barrels of oil or millions of cubic feet of gas per day) that individual wells and the entire reservoir are capable of sustaining. One of the major assignments of the reservoir engineer is to estimate the ultimate production that can be anticipated from both primary and enhanced recovery from the reservoir. *See* PETROLEUM ENHANCED RECOVERY; PETROLEUM RESERVES; PETROLEUM RESERVOIR ENGINEERING.

The drilling engineer has the responsibility for the efficient penetration of the earth by a well bore, and for cementing of the steel casing from the surface to a depth usually just above the target reservoir. The drilling engineer or another specialist, the mud engineer, is in charge of the fluid that is continuously circulated through the drill pipe and back up to surface in the annulus between the drill pipe and the bore hole.

The production engineer, upon consultation with the petrophysical and reservoir engineers, plans the completion procedure for the well. This involves a choice of setting a liner across the formation or perforating a casing that has been extended and cemented across the reservoir, selecting appropriate pumping techniques, and choosing the surface collection, dehydration, and storage facilities. *See* OIL AND GAS WELL COMPLETION.

Major construction projects, such as the design and erection of offshore platforms, require the addition of civil engineers to the staff of petroleum engineering departments, and the design and implementation of natural gasoline and gas processing plants require the addition of chemical engineers. *See* OIL AND GAS, OFFSHORE; PETROLEUM. Todd M. Doscher; R. E. Wyman

Relational databases and advanced computer graphics are used in petroleum exploration. There is a heavy emphasis on facile gathering of data and extraction of selected items to provide effective displays and interpretations. In general, petroleum computing can be viewed on three levels: geological computing, geophysical computing, and engineering applications. Geological computing trends have focused on database and spatial system configurations, with specialty applications such as cross-section balancing or geochemical modeling. Geophysical computing tends to be computer-intensive; interpretive installations are, like all interactive workstation environments, driven by graphics. Engineering applications are also computer-intensive; they are generally classified as either simulation or process types. Brian Robert Shaw

Petroleum enhanced recovery Technology to increase oil recovery from a porous formation beyond that obtained by conventional means. Conventional oil recovery technologies produce an average of about one-third of the original oil in place in a formation. Conventional technologies are primary or secondary. Primary technologies rely on native energy, in the form of fluid and rock compressibility and natural aquifers, to produce oil from the formation to wells. Secondary technologies supplement the native energy to drive oil to producing wells by injecting water or low-pressure gas at injection wells. The target of enhanced recovery technologies is that large portion of oil that is not recovered by primary and secondary means. *See* PETROLEUM ENGINEERING.

Many of the challenges encountered by secondary technologies are identical to those encountered by enhanced recovery technologies. Those challenges include reducing residual oil saturation, improving sweep efficiency, fitting the technology to the reservoir heterogeneities, and minimizing up-front and operating costs.

Residual oil remains trapped in a porous rock after the rock has been swept with water, gas, or any other recovery fluid. The residual oil saturation is the percentage of the pore space occupied by the residual oil. The residual oil saturation depends on the pore size distribution and connectivity, the interfacial tension between a recovery agent and the oil, the relative wettability of the rock surfaces with respect to the recovery agent and the oil, the viscosity of the fluids, and the rate at which the fluids are moving through the rock.

The sweep efficiency specifies that portion of a reservoir that is contacted by a recovery fluid. Sweep efficiency increases with volume of injected fluid. It also depends on the pattern of injection and production wells in a formation, on the mobility of the oil and the recovery fluid, and on heterogeneities in the formation.

A wide variety of processes have been considered for enhancing oil recovery: thermal processes, high-pressure gas processes, and chemical processes. Specifically, low residual oil saturation can be obtained by selecting a recovery fluid that provides a very low interfacial tension between the oil and the fluid. With very low interfacial tension, the capillary number is large. And high sweep efficiency can be obtained by selecting a recovery agent with low mobility or by increasing the mobility of the oil. Richard L. Christiansen

Petroleum geology The practice of utilizing geological principles and applying geological concepts to the discovery and recovery of petroleum. Related fields in petroleum discovery include geochemistry and geophysics. The related areas in petroleum recovery are petroleum and chemical engineering. *See* CHEMICAL ENGINEERING; GEOCHEMISTRY; GEOPHYSICS.

Petroleum occurs in a liquid phase as crude oil and condensate, and in a gaseous phase as natural gas. The phase is dependent on the kind of source rock from which the petroleum was formed and the physical and thermal environment in which it exists. Most petroleum occurs at varying depths below the ground surface, but generally petroleum existing as a liquid (crude oil) is found at depths of less than 20,000 ft (6100 m) while natural gas is found both at shallow depths and at depths exceeding 30,000 ft (9200 m). In some cases, oil may seep to the surface, forming massive deposits of oil or tar sands. Natural gas also seeps to the surface but escapes into the atmosphere, leaving little or no surface trace. *See* NATURAL GAS; OIL SAND; PETROLEUM.

Most petroleum is found in sedimentary basins in sedimentary rocks, although many of the 700 or so sedimentary basins of the world contain no known significant accumulations. Several conditions must exist for the accumulation of petroleum: (1) There must be a source rock, usually high in organic matter, from which petroleum can be generated. (2) There must be a mechanism for the petroleum to move, or migrate. (3) A reservoir rock with voids to hold petroleum fluids must exist. (4) The reservoir must be in a configuration to constitute a trap and be covered by a seal—any kind of low-permeability or dense rock formation that prevents further migration. If any of these conditions do not exist, petroleum either will not form or will not accumulate in commercially extractable form. *See* BASIN; SEDIMENTARY ROCKS.

The aim of petroleum geologists is to find traps or accumulations of petroleum. The trap not only must be defined but must exist where other conditions such as source and reservoir rocks occur.

To locate these traps, the geologist must rely on subsurface information and data gathered by drilling exploratory wells and data obtained by geophysical surveying. These data, once interpreted, are used to construct maps, cross sections, and models that are used to infer or to actually depict subsurface configurations that might contain petroleum. Such depictions are prospects for drilling. *See* GEOPHYSICAL EXPLORATION; OIL AND GAS WELL DRILLING.

Oil and gas must be trapped in an individual reservoir in sufficient quantities to be commercially producible. Worldwide, 25% of all oil discovered so far is contained in only ten fields, seven of which are in the Middle East. Fifty percent of all oil discovered to date is found in only 50 fields.

Most of the large and fairly obvious fields in the United States have been discovered, except those possibly existing in frontier or lightly explored areas such as Alaska and the deep waters offshore. Few areas of the world remain entirely untested, but many areas outside the United States are only partly explored, and advanced techniques have yet to be deployed in the recovery of oil and gas found so far. *See* PETROLEUM RESERVES.

Greater efforts in petroleum geology along with petroleum engineering are being made to increase recovery from existing fields. Of all oil discovered so far, it is estimated that there will be recovery of only 35% on the average. Recovering some part of this huge oil resource will require geological reconstruction of reservoirs, a kind of very detailed and small-scale exploration. These reconstructions and models have allowed additional recovery of oil that is naturally movable in the reservoir. If the remaining oil is immobile because it is too viscous or because it is locked in very small pores or is held by capillary forces, techniques must be used by the petroleum geologist and the petroleum engineer to render the oil movable. William L. Fisher

Petroleum microbiology Those aspects of microbiology in which crude oil, refined petroleum products, or pure hydrocarbons serve as nutrients for the growth of microorganisms or are altered as a result of their activities. Applications of petroleum microbiology include oil pollution control, enhanced oil recovery, microbial contamination of petroleum fuels and oil emulsions, and conversion of petroleum hydrocarbons into microbial products.

Many species of bacteria, fungi, and algae have the enzymatic capability to use petroleum hydrocarbons as food. Biodegradation of petroleum requires an appropriate mixture of microorganisms, contact with oxygen gas, and large quantities of utilizable nitrogen and phosphorus compounds and smaller amounts of other elements essential for the growth of all microorganisms. Part of the hydrocarbons are converted into carbon dioxide and water and part into cellular materials, such as proteins and nucleic acids. The requirement for a mixture of different microorganisms arises from the fact that petroleum is composed of a wide variety of different groups of hydrocarbons, whereas any specific microorganism is highly specialized with regard to the type of hydrocarbon it can digest. The bacterial genera that contain the most frequently isolated hydrocarbon degraders are *Pseudomonas*, *Acinetobacter*, *Flavobacterium*, *Brevibacterium*, *Corynebacterium*, *Arthrobacter*, *Mycobacterium*, and *Nocardia*. The fungal genera that contain oil utilizers include *Candida*, *Cladosporium*, *Rhodotorula*, *Torulopsis*, and *Trichosporium*. *See* BIODEGRADATION.

Oil pollution results from natural hydrocarbon seeps, accidental spills, and intentional discharge of oily materials into the environment. Once the oil is released and comes into contact with water, air, and the necessary salts, microorganisms present in the environment begin the natural process of petroleum biodegradation. If this process did not occur, the world's oceans would soon become completely covered with a layer of oil. The reason that oil spills become a pollution problem is that the natural microbial systems for degrading the oil become temporarily overwhelmed. *See* WATER POLLUTION.

The largest potential application of petroleum microbiology is in the field of enhanced oil recovery. Microbial products, as well as viable microorganisms, have been used as stimulation agents to enhance oil recovery from petroleum reservoirs. Xanthan, a polysaccharide produced by *Xanthomonas campestris*, is used as a waterflood thickening agent in oil recovery. Emulsan, a lipopolysaccharide produced by a strain of *Acinetobacter calcoaceticus*, stabilizes oil-in-water emulsions. A number of other microbial products are being tested for potential application in enhanced oil recovery processes. Field tests have indicated that injection of viable microorganisms with their nutrients into petroleum reservoirs can lead to enhanced oil recovery, presumably due to production of carbon dioxide gas, acids, and surfactants. *See* PETROLEUM ENHANCED RECOVERY.

A variety of valuable materials, such as amino acids, carbohydrates, nucleotides, vitamins, enzymes, antibiotics, citric acid, long-chain dicarboxylic acids, and biomass, can be produced by microbial processes using petroleum hydrocarbons as substrates. The main advantage of using hydrocarbons as substrates is their lower cost. Also, certain products, such as tetradecane-1,14-dicarboxylic acid, a raw material for preparing perfumes, are synthesized in higher yields on hydrocarbon than on carbohydrate substrates.

The most active area of research and development in petroleum microbiology since the mid-1960s has been in the large-scale production and concentration of microorganisms for animal feed and human food. Dried microbial cells are collectively referred to as single-cell protein. In spite of its advantages, single-cell protein has not yet played a significant role in providing protein for animal feed or human consumption. However, many scientists are optimistic about its potential. The ability of microorganisms to utilize petroleum also has its detrimental aspects, particularly with respect to the deterioration of petroleum fuels, asphalt coatings, and oil emulsions used with cutting machinery. All hydrocarbons become contaminated if they come into contact with water during storage.

See Bacterial physiology and metabolism; Corrosion; Industrial microbiology. Eugene Rosenberg

Petroleum processing and refining

Petroleum processing and refining The separation of petroleum into fractions and the treating of these fractions to yield marketable products. Petroleum is a mixture of gaseous, liquid, and solid hydrocarbon compounds that occurs in sedimentary rock deposits throughout the world. In the crude state, petroleum has little value but, when refined, it provides liquid fuels (gasoline, diesel fuel, aviation fuel), solvents, heating oil, lubricants, and the distillation residuum asphalt, which is used for highway surfaces and roofing materials. *See* Petroleum; Petroleum products.

Crude petroleum (oil) is a mixture of compounds with different boiling temperatures that can be separated into a variety of fractions (see table). Since there is a wide variation in the composition of crude petroleum, the proportions in which the different fractions occur vary with origin. Some crude oils have higher proportions of lower-boiling components, while others have higher proportions of residuum (asphaltic components).

Petroleum fractions and their uses*

Fraction	Boiling range °C	Boiling range °F	Uses
Fuel gas	−160 to −40	−260 to −40	Refinery fuel
Propane	−40	−40	Liquefied petroleum gas (LPG)
Butane(s)	−12 to −1	11–30	Increases volatility of gasoline, advantageous in cold climates
Light naphtha	−1 to 150	30–300	Gasoline components, may be (with heavy naphtha) reformer feedstock
Heavy naphtha	150–205	300–400	Reformer feedstock, with light gas oil, jet fuels
Gasoline	−1 to 180	30–355	Motor fuel
Kerosine	205–260	400–500	Fuel oil
Stove oil	205–290	400–550	Fuel oil
Light gas oil	260–315	500–600	Furnace and diesel fuel components
Heavy gas oil	315–425	600–800	Feedstock for catalytic cracker
Lubricating oil	>400	>750	Lubrication
Vacuum gas oil	425–600	800–1100	Feedstock for catalytic cracker
Residuum	>600	>1100	Heavy fuel oil, asphalts

*From J. G. Speight (ed.), *The Chemistry and Technology of Petroleum*, 3d ed., Marcel Dekker, New York, 1999.

Petroleum processing and refining involves a series of steps by which the original crude oil is converted into products with desired qualities in the amounts dictated by the market. In fact, a refinery is essentially a group of manufacturing plants that vary in number with the variety of products in the mix. Refinery processes must be selected and products manufactured to give a balanced operation; that is, crude oil must be converted into products according to the demand for each. For example, the manufacture of products from the lower-boiling portion of petroleum automatically produces a certain amount of higher-boiling components. If the latter cannot be sold as, say, heavy fuel oil, these products will accumulate until refinery storage facilities are full. To prevent such a situation, the refinery must be flexible and able to change operations as needed. This usually means more processes, such as thermal processes to change excess heavy fuel oil into gasoline with coke as the residual product, or vacuum distillation processes to separate heavy oil into lubricating oil stocks and asphalt.

Distillation. In a petroleum distillation unit, a tower is used for fractionation. The feedstock of crude oil flows through one or more pipes arranged within a large furnace where it is heated to a temperature at which a predetermined portion of the feed changes into vapor. The heated feed is introduced into a fractional distillation tower where the nonvolatiles or liquid portions pass downward to the bottom of the tower and are pumped away, while the vapors pass upward through the tower and are fractionated into gas oils, kerosine, and naphthas.

Vacuum distillation is used in petroleum refining to separate the less volatile products, such as lubricating oils, from petroleum without subjecting the high-boiling products to cracking conditions. Operating pressure for vacuum distillation is usually 50– 100 mm of mercury (6.7–13.3 kilopascals) [atmospheric pressure = 760 mm of mercury]. By this means, a heavy gas oil that has a boiling range in excess of 315°C (600°F) at atmospheric pressure may be obtained at temperatures of around 150°C (300°F); and lubricating oil, having a boiling range in excess of 370°C (700°F) at atmospheric pressure may be obtained at temperatures of 250–350°C (480–660°F). Atmospheric and vacuum distillation are major parts of refinery operations, and no doubt will continue to be used as the primary refining operation.

Thermal processes. One of the earliest conversion processes used in the petroleum industry was the thermal decomposition of higher-boiling materials into lower-boiling products. This process is known as thermal cracking. The majority of the thermal cracking processes use temperatures of 455–540°C (850–1005°F) and pressures of 100–1000 psi (690–6895 kPa). For example, the feedstock (reduced crude) is preheated by direct exchange with the cracking products in the fractionating columns. Cracked gasoline and heating oil are removed from the upper section of the column. Light and heavy distillate fractions are removed from the lower section and are pumped to separate heaters. Higher temperatures are used to crack the more stable light distillate fraction. The streams from the heaters are combined and sent to a soaking chamber where additional time is provided to complete the cracking reactions. The cracked products are then separated in a low-pressure flash chamber where a heavy fuel oil is removed as bottoms. The remaining cracked products are sent to fractionating columns. The thermal cracking of higher-boiling petroleum fractions to produce gasoline is now virtually obsolete. The antiknock requirements of modern automobile engines together with the different nature of crude oils (compared to those of 50 years ago) has reduced the ability of the thermal cracking process to produce gasoline on an economic basis. *See* Distillation column.

Visbreaking (viscosity breaking) is a mild thermal cracking operation that can be used to reduce the viscosity of residua to allow the products to meet fuel oil specifications. Alternatively, the visbroken residua can be blended with lighter product oils to produce fuel oils of acceptable viscosity. By reducing the viscosity of the residuum, visbreaking reduces the amount of light heating oil that is required for blending to meet fuel oil specifications.

Delayed coking is a thermal process for converting residua into lower-boiling products, such as gases, naphtha, fuel oil, gas oil, and coke. It is a semicontinuous process in which the heated charge is transferred to large soaking (or coking) drums, which provide the long residence time needed to allow the cracking reactions to proceed to completion. The feedstock is introduced into a product fractionator where it is heated and the lighter fractions are removed as a side streams. Gas oil, often the major product of a coking operation, serves primarily as a feedstock for catalytic cracking units. The coke obtained is typically used as fuel; but specialty uses, such as electrode manufacture, and production of chemicals and metallurgical coke are also possible, increasing the value of the coke. For these uses, the coke may require treatment to remove sulfur and metal impurities. *See* Coke; Cracking; Fuel oil; Naphtha.

Catalytic cracking is basically the same as thermal cracking, but differs by the use of a catalyst, which directs the course of the cracking reactions to produce more of the desired higher-octane hydrocarbon products. Catalytic cracking is regarded as the modern method for converting high-boiling petroleum fractions, such as gas oil, into gasoline and other low-boiling fractions. The usual commercial process involves contacting a gas oil faction with an active catalyst at a suitable temperature, pressure, and residence time so that a substantial part (>50%) of the gas oil is converted into gasoline and lower-boiling products, usually in a single-pass operation. *See* GASOLINE; OCTANE NUMBER.

Hydroprocesses. The use of hydrogen in thermal processes was perhaps the single most significant advance in refining technology during the twentieth century. The process uses the principle that the presence of hydrogen during a thermal reaction of a petroleum feedstock will terminate many of the coke-forming reactions and enhance the yields of the lower-boiling components, such as gasoline, kerosine, and jet fuel. *See* HYDROGENATION.

Destructive hydrogenation (hydrogenolysis or hydrocracking) is characterized by the conversion of the higher-molecular-weight constituents in a feedstock to lower-boiling products. Such treatment requires severe processing conditions and the use of high hydrogen pressures to minimize the polymerization and condensation reactions that lead to coke formation. *See* HYDROCRACKING.

Nondestructive hydrogenation is used for improving product quality without appreciable alteration of the boiling range. Nitrogen, sulfur, and oxygen compounds undergo reaction with the hydrogen, forming ammonia, hydrogen sulfide, and water, respectively. Unstable compounds that might lead to the formation of gums or insoluble materials are converted to more stable compounds.

James G. Speight

Petroleum products Petroleum products are those fractions derived from petroleum that have commercial value as a bulk product. Petrochemicals, in contrast, are individual chemicals, derived from bulk fractions, that are used as the basic building blocks of the chemical industry. Gases and liquid fuels are currently the main products of the petroleum industry (see table). However, other products, such as lubricating oils, waxes, and asphalt, have also added to the value of petroleum resources. *See* PETROCHEMICAL; PETROLEUM.

Petroleum products are hydrocarbon compounds, containing combinations of hydrogen and carbon with various molecular forms. Many compounds occur naturally. Other compounds are created by commercial processes for altering one combination to form another. Each combination has its unique set of chemical and physical properties. Specifications for petroleum products are based on properties such as density and boiling range to assure that a petroleum product can perform its intended task. *See* PETROLEUM PROCESSING AND REFINING.

Commercial names and uses for major petroleum products

Crude oil cuts	Refinery blends	Consumer products
Gases	Still gases	Fuel gas
	Propane/butane	Liquefied petroleum gas (LPG)
Light/heavy naphtha	Motor fuel	Gasoline
	Aviation turbine, Jet-B	Jet fuel (naphtha type)
Kerosine	Aviation turbine, Jet-A	Jet fuel (kerosine type)
	No. 1 fuel oil	Kerosine (range oil)
Light gas oil	Diesel	Auto and tractor diesel
	No. 2 fuel oil	Home heating oil
Heavy gas oil	No. 4 fuel oil	Commercial heating oil
	No. 5 fuel oil	Industrial heating oil
	Bright stock	Lubricants
Residuals	No. 6 fuel oil	Bunker C oil
	Heavy residual	Asphalt
	Coke	Coke

Natural gas is predominantly methane (CH_4), which has the lowest boiling point and least complex structure of all hydrocarbons. Natural gas from an underground reservoir, when brought to the surface, may contain other, higher-boiling-point hydrocarbons, and is often referred to as wet gas. Wet gas is processed to remove the entrained hydrocarbons that are higher-boiling than methane. The high-boiling hydrocarbons that are isolated and liquefied are called natural gas condensates. *See* METHANE; NATURAL GAS.

Gasoline (motor fuel) is a complex mixture of hydrocarbons that boils below 200°C (390°F) and is intended for most spark-ignition engines (such as those used in passenger cars, light-duty trucks, motorcycles, and motorboats). The properties of gasoline are intended to satisfy the requirements of smooth and clean burning, easy ignition in cold weather, minimal evaporation in hot weather, and stability during long storage periods. *See* GASOLINE; INTERNAL COMBUSTION ENGINE.

Petroleum naphtha is a generic term applied to refined, partly refined, or unrefined petroleum products. Naphthas are prepared by several methods, including (1) fractionation of distillates or crude petroleum, (2) solvent extraction, (3) hydrogenation of distillates, (4) polymerization of unsaturated (olefinic) compounds, and (5) alkylation processes. The naphtha may also be a combination of product streams from more than one process. The main uses of petroleum naphthas fall into the general areas of (1) solvents (diluents) for paints, (2) dry-cleaning solvents, (3) solvents for cutback asphalts, (4) solvents in rubber industry, and (5) solvents for industrial extraction processes. Turpentine, the traditional solvent for paints, has been almost completely replaced by the cheaper and more abundant petroleum naphthas. *See* NAPHTHA; SOLVENT.

Kerosine is essentially a distillation fraction of petroleum. The quantity and quality of the kerosine vary with the type of crude oil; some crude oils yield excellent kerosine, while others produce kerosine that requires substantial refining. Kerosine is a very stable product, and additives are not required to improve the quality. Apart from the removal of excessive quantities of aromatics, kerosine fractions may need only a lye (alkali) wash if hydrogen sulfide is present. Kerosine is used as a fuel for heating and cooking, jet engines, and lamps, for weed burning, and as a base for insecticides. *See* KEROSINE.

Diesel fuel is a distillate product that has a higher boiling point than gasoline (or naphtha) but that also must self-ignite easily. This is determined through the cetane rating, derived from the reference fuel *n*-cetane. Cetane number is a measure of the tendency of a diesel fuel to knock in a diesel engine. The scale is based upon the ignition characteristics of two hydrocarbons, *n*-hexadecane (cetane) and 2,3,4,5,6,7,8-heptamethylnonane. Diesel fuel oil is essentially the same as furnace fuel oil, but the proportion of cracked gas oil is usually less since the high aromatic content of the cracked gas oil reduces the cetane value of the diesel fuel. *See* CETANE NUMBER; DIESEL ENGINE; DIESEL FUEL.

Domestic fuel oil is used primarily in the home, and includes kerosine, stove oil, and furnace fuel oil. Stove oil is a straight-run (distilled) fraction from crude oil, whereas other fuel oils are usually blends of two or more fractions. The straight-run fractions available for blending into fuel oils are heavy naphtha, light and heavy gas oil, and residua. Cracked fractions such as light and heavy gas oil from catalytic cracking, cracking coal tar, and fractionator bottoms from catalytic cracking may also be used as blends to meet the specifications of the different fuel oils.

Heavy fuel oil includes a variety of oils, ranging from distillates to residual oils that must be heated to 260°C (500°F) or higher before they can be used. In general, heavy fuel oil consists of residual oil blended to suit specific needs and to meet designed specifications.

Heavy fuel oil usually contains residuum that is mixed (cut back) to a specified viscosity with gas oils and fractionator bottoms. For some industrial purposes where flames or flue gases contact the product (ceramics, glass, heat treating, open hearth furnaces), the fuel oil must be blended to have a minimum specified sulfur content. *See* FUEL OIL.

Asphalt is a residuum that cannot be distilled even under the highest vacuum since the temperatures required to volatilize the residuum promote the formation of coke. Asphalts have complex chemical and physical compositions that usually vary with the source of the crude oil. *See* ASPHALT AND ASPHALTITE.

Petroleum coke is the residue left by the noncatalytic destructive distillation (thermal decomposition with simultaneous removal of distillate) of petroleum residua. The coke formed in catalytic cracking operations is usually nonrecoverable because it adheres to the catalyst employed as fuel for the process. The composition of the coke varies with the source of the crude oil, but in general, large amounts of high-molecular-weight complex hydrocarbons (rich in carbon but correspondingly poor in hydrogen) make up a high proportion. Petroleum coke is employed for a number of purposes, but the major use is in the manufacture of carbon electrodes for aluminum refining, which requires a high-purity carbon (that is, low in ash and sulfur-free). In addition, petroleum coke is employed in the manufacture of carbon brushes, silicon carbide abrasives, and structural carbon (such as pipes and Rashig rings), as well as in the manufacture of calcium carbide (CaC_2) from which acetylene is produced. *See* COKE.

James G. Speight

Petroleum reserves Proved reserves are the estimated quantities of crude oil liquids which with reasonable certainty can be recovered in future years from delineated reservoirs under existing economic and operating conditions. Thus, estimates of crude oil reserves do not include synthetic liquids which at some time in the future may be produced by converting coal or oil shale, nor do reserves include fluids which may be recovered following the future implementation of a supplementary or enhanced recovery scheme.

Indicated reserves are those quantities of petroleum which are believed to be recoverable by already implemented but unproved enhanced oil recovery processes or by the application of enhanced recovery processes to reservoirs similar to those in which such recovery processes have been proved to increase recovery.

Thus, crude oil reserves can be called upon in the future with a high degree of certainty, subject of course to the limitations placed on production rate by fluid flow within the reservoir and the capacity of the individual producing wells and surface facilities to handle the produced fluids. It is important to bear in the mind the distinction between resources and reserves. The former term refers to the total amount of oil that has been discovered in the subsurface, whereas the latter refers to the amount of oil that can be economically recovered in the future. The ratio of the ultimate recovery (the sum of currently proved reserves and past production) to the resource or original oil in place is the anticipated recovery efficiency. *See* NATURAL GAS; PETROLEUM; PETROLEUM ENHANCED RECOVERY.

Todd M. Doscher

Petroleum reservoir engineering The technology concerned with the prediction of the optimum economic recovery of oil or gas from hydrocarbon-bearing reservoirs. It is an eclectic technology requiring coordinated application of many disciplines: physics, chemistry, mathematics, geology, and chemical engineering. Originally, the role of reservoir engineering was exclusively that of counting oil and natural gas reserves. The reserves—the amount of oil or gas that can be economically recovered from the reservoir—are a measure of the wealth available to the owner and operator. It is also necessary to know the reserves in order to make proper decisions concerning the viability of downstream pipeline, refining, and marketing facilities that will rely on the production as feedstocks.

The scope of reservoir engineering has broadened to include the analysis of optimum ways for recovering oil and natural gas, and the study and implementation of enhanced recovery techniques for increasing the recovery above that which can be expected from the use of conventional technology.

The amount of oil in a reservoir can be estimated volumetrically or by material balance techniques. A reservoir is sampled only at the points at which wells penetrate it. By using logging techniques and core analysis, the porosity and net feet of pay (oil-saturated interval) and the average oil saturation for the interval can be estimated in the immediate vicinity of the well. The oil-saturated interval observed at one location is not identical to that at another because of the inherent heterogeneity of a sedimentary layer. It is therefore necessary to use statistical averaging techniques in order to define the average oil content of the reservoir (usually expressed in barrels per net acre-foot) and the average net pay. The areal extent of the reservoir is inferred from the extrapolation of geology and fluid content as well as the drilling of dry holes beyond the productive limits of the reservoir. The definition of reservoir boundaries can be heightened by study of seismic surveys, particularly 3D surveys, and analysis of pressure buildups in wells after they have been brought on production. *See* PETROLEUM GEOLOGY; WELL LOGGING.

The overall recovery of crude oil from a reservoir is a function of the production mechanism, the reservoir and fluid parameters, and the implementation of supplementary recovery techniques. In general, recovery efficiency is not dependent upon the rate of production except for those reservoirs where gravity segregation is sufficient to permit segregation of the gas, oil, and water. Where gravity drainage is the producing mechanism, which occurs when the oil column in the reservoir is quite thick and the vertical permeability is high and a gas cap is initially present or is developed on producing, the reservoir will also show a significant effect of rate on the production efficiency. Reservoir engineering expertise, together with geological and petrophysical engineering expertise, is being used to make very detailed studies of the production performance of crude oil reservoirs in an effort to delineate the distribution of residual oil and gas in the reservoir, and to develop the necessary technology to enhance the recovery. *See* PETROLEUM ENHANCED RECOVERY. Todd M. Doscher

Well testing broadly refers to the diagnostic tests run on wells in petroleum reservoirs to determine well and reservoir properties. The most important well tests are called pressure transient tests and are conducted by changing the rate of a well in a prescribed way and recording the resulting change in pressure with time.

The information obtained from pressure transient tests includes estimates of (1) unaltered formation permeability to the fluid(s) produced in the well; (2) altered (usually reduced) permeability near the well caused by drilling and completion practices; (3) altered (increased) permeability near the well created by deliberately stimulating the well by injecting either an acid that dissolves some of the formation or a high-pressure fluid that creates fractures in the formation; (4) distances to flow barriers located in the area drained by the well; and (5) average pressure in the area drained by the well. In addition, some testing programs may confirm hypothesized models of the reservoir, including important variations of formation properties with distance or location of gas/oil, oil/water, or other fluid/fluid contacts.

Pressure transient tests are usually interpreted by comparing the observed pressure-time response to the predicted response by a mathematical model of the well/reservoir system. Graphical techniques are used to calculate permeability. More sophisticated graphical techniques involve matching changes in pressure to preplotted analytical solutions (type-curve matching). Regression analysis is used to match observed pressure-time data to mathematical models. Although analytical solutions are being found for more and more complex reservoir models each year, many reservoirs are still so complex that their behavior cannot be described accurately by analytical solutions. In such cases, finite-difference approximations to the governing flow equations can be used in commercial reservoir simulators, the reservoir properties treated as unknowns, and properties found that fit the observed data well. W. John Lee

Reservoir behavior can be simulated using models that have been constructed to have properties similar either to an ideal geometric shape of constant properties or to the shape and varying properties of a real (nonideal) oil or gas reservoir. *See* MODEL THEORY; SIMULATION.

For application to petroleum reservoirs, it is necessary to predict the simultaneous flow behavior of more than one fluid phase having different properties (water, gas, and crude oil). The permeability, the relative permeability, and the density and viscosity of each phase constitute its transport properties for calculating its flow. The relative permeability is a factor for each phase (oil, water, gas) which, when multiplied by the permeability for a single phase such as water, will give the permeability for the given phase. It varies with the volume fraction of the pore space occupied by the phase, called the saturation of the given phase. Generally, the relative permeability of the water phase depends only on its own saturation, and likewise for the gas phase. The relative permeability of the oil phase is a function of the saturations of both gas and water phases. *See* FLUID FLOW; FLUID-FLOW PRINCIPLES; FLUID MECHANICS. Elmond L. Claridge

Petrology The study of rocks, their occurrence, composition, and origin. Petrography is concerned primarily with the detailed description and classification of rocks, whereas petrology deals primarily with rock formation, or petrogenesis. A petrological description includes definition of the unit in which the rock occurs, its attitude and structure, its mineralogy and chemical composition, and conclusions regarding its origin. *See* MINERALOGY; PETROGRAPHY; ROCK. William Ingersoll Rose, Jr.

One aim of mineralogy and petrology is to decipher the history of igneous and metamorphic rocks. Detailed study of the field geology, the structures, the petrography, the mineralogy, and the geochemistry of the rocks is used as a basis for hypotheses of origin. The conditions at depth within the Earth's crust and mantle, the processes occurring at depth, and the whole history of rocks once deeply buried are deduced from the study of rocks now exposed at the Earth's surface. One approach used to test hypotheses so developed is experimental petrology; the term experimental minerals refers to similar studies involving minerals rather than rocks (mineral aggregates). *See* IGNEOUS ROCKS; METAMORPHIC ROCKS.

The experimental petrologist reproduces in the laboratory the conditions of high pressure and high temperature encountered at various depths within the Earth's crust and mantle where the minerals and rocks were formed. By suitable selection of materials the petrologist studies the chemical reactions that actually occur under these conditions and attempts to relate these to the processes involved in petrogenesis. Peter J. Wyllie

Petromyzontiformes An order of jawless fishes also known as the Hyperoartii or commonly as the lampreys. It is the only order of the class Petromyzontida. Characteristics include the following: eellike body; no bones; no paired fins; one or two dorsal fins; diphycercal (isocercal) tail (that is, with symmetrical upper and lower lobes) in adults, hypocercal (that is, with a larger lower lobe) in ammocoetes (larval lampreys); no barbels; teeth on oral disc and tongue; two semicircular canals; seven pairs of external lateral gill openings; well-developed eyes in adults, and rudimentary eyes in ammocoetes, usually laterally placed; single median nostril; spiral valve and cilia in intestinal tract; small cerebellum; separate sexes; and oviparous, with small and numerous eggs. The ammocoetes undergo radical metamorphosis, and all species spawn in freshwater. *See* JAWLESS VERTEBRATES.

Lampreys are either parasitic or nonparasitic, and it is believed that nonparasitic species are derived from a parasitic species. The nonparasitic lampreys are usually called brook lampreys. Brook lampreys are confined to freshwater and spend most of their lifetime (usually about 3 to 5 years) as ammocoetes. The ammocoetes lack teeth, oral disc, and functional eyes. As filter feeders, they passively extract microorganisms from the surroundings, and they use little energy in the process. Also, by burrowing into the substrate, they are protected from predators.

Parasitic lampreys, by contrast, may be freshwater or anadromous (ascending into freshwater streams from the sea to spawn), but most species are limited to freshwater; however, no parasitic freshwater lampreys are known from the Southern Hemisphere. Parasitic lampreys have a prolonged adult life, and gonadal development is slower. Rasplike teeth rake the flesh of a host fish, allowing the lamprey to feed on its blood and tissue fluids, which is kept liquid by enzymes (lamphedrins) secreted by the lamprey's buccal glands. Some species are serious predators on commercially important fishes, with the notorious sea lamprey (*Petromyzon marinus*) being a good example. It is an anadromous species that occurs in Atlantic drainages of the United States, Canada, Iceland, and Europe, including the Mediterranean Sea, and is landlocked in the Great Lakes and other large lakes of that region. It is by far the largest lamprey, attaining a total length of 120 cm (4 ft), whereas all other lamprey species vary from about 12 to 76 cm (4.8 to 30 in.). *See* PARASITOLOGY; PREDATOR-PREY INTERACTIONS.

Lampreys are nearly antitropical (found in both hemispheres but not in equatorial regions) and are absent entirely from the broad belt around the world between the Tropic of Cancer and the Tropic of Capricorn. Except for the fossil family Mayomyzontidae, a single family, Petromyzontidae, comprising eight genera and 34 species, occurs in the Northern Hemisphere, essentially north of 30°N. Only two genera and four species in two families, Geotriidae and Mordaciidae, occur in the Southern Hemisphere. Herbert Boschung

Pewter A tarnish-resistant alloy of lead and tin always containing appreciably more than 63% tin. Other metals are sometimes used with or in place of the lead; among them are copper, antimony, and zinc. Pewter is commonly worked by spinning and it polishes to a characteristic luster. Because pewter work hardens only slightly, pewter products can be finished without intermediate annealing. Early pewter, with high lead content, darkened with age. With less than 35% lead, pewter was used for decanters, mugs, tankards, bowls, dishes, candlesticks, and canisters. The lead remained in solid solution with the tin so that the alloy was resistant to the weak acids in foods. Frank H. Rockett

pH An expression for the effective concentration of hydrogen ions in solution. The activity of hydrogen ions or, more correctly, hydronium ions, which are hydrated hydrogen ions $H(H_2O)_n^+$, affects the equilibria and kinetics of a wide variety of chemical and biochemical reactions. Because these effects are activity-dependent, it is extremely important to distinguish between the hydrogen-ion concentration and activity. The concentration, or total acidity, is obtained by titration and corresponds to the total concentration of hydrogen ions available in a solution, that is, free, unbound hydrogen ions as well as hydrogen ions associated with weak acids. The hydrogen-ion activity refers to the effective concentration of unassociated hydrogen ions, the form that directly affects physicochemical reaction rates and equilibria. This activity is therefore of fundamental importance in many areas of science and technology. The relationship between hydrogen-ion activity (a_{H^+}) and concentration (C) is given by Eq. (1), where

$$a_{H^+} = \gamma C \qquad (1)$$

the activity coefficient γ is a function of the total ionic strength (concentration) of the solution and approaches unity as the ionic strength approaches zero; that is, the difference between the activity and the concentration of hydrogen ion diminishes as the solution becomes more dilute. *See* ACTIVITY (THERMODYNAMICS); CHEMICAL EQUILIBRIUM; HYDROGEN ION.

The effective concentration of hydrogen ions in solution is expressed in terms of pH, which is the negative logarithm of the hydrogen-ion activity [Eq. (2)]. Because of the negative log-

$$pH = -\log_{10} a_{H^+} \quad (2)$$

arithmic (exponential) relationship, the more acidic a solution, the smaller the pH value. The pH of a solution may have little relationship to the titratable acidity of a solution that contains weak acids or buffering substances; the pH of a solution indicates only the free hydrogen-ion activity. If total acid concentration is to be determined, an acid-base titration must be performed. *See* ACID AND BASE; BUFFERS (CHEMISTRY); TITRATION.

Two methods, electrometric and chemical indicator (optical), are used for measuring pH. The more commonly used electrometric method is based on measurement of the difference between the pH of a test solution and that of a standard solution. The pH scale is defined by a series of reference buffer solutions that are used to calibrate the pH measurement system. The instrument measures the potential difference developed between the pH electrode and a reference electrode of constant potential. The difference in potential obtained when the electrode pair is removed from the standard solution and placed in the test solution is converted to the pH value. In the indicator method, the pH value is obtained by simple visual comparison of the color of pH-sensitive dyes to standards (for example, color charts) or by use of calibrated optical readout devices (photometers), often in combination with fiber-optic sensors. *See* ELECTRODE; REFERENCE ELECTRODE.

Richard A. Durst

pH regulation (biology) The processes operating in living organisms to preserve a viable acid-base state. In higher animals, much of the body substance (60–70%) consists of complex solutions of inorganic and organic solutes. For convenience, these body fluids can be subdivided into the cellular fluid (some two-thirds of the total) and the extracellular fluid. The latter includes blood plasma and interstitial fluid, the film of fluid that bathes all the cells of the body. For normal function, the distinctive compositions of these various fluids are maintained within narrow limits by a process called homeostasis. A crucial characteristic of these solutions is pH, an expression representing the concentration (or preferably the activity) of hydrogen ions, $[H^+]$, in solution. The pH is defined as $-\log [H^+]$, so that in the usual physiological pH range of 7 to 8, $[H^+]$ is exceedingly low, between 10^{-7} and 10^{-8} *M*. Organisms use a variety of means to keep pH under careful control, because even small deviations from normal pH can disrupt living processes. *See* HOMEOSTASIS; PH.

The most accessible and commonly studied body fluid is blood, and it, therefore, provides the most information on pH regulation. Blood pH in humans, and in mammals generally, is about 7.4. This value indicates that blood is slightly alkaline, because neutrality, the condition in which the concentration of hydrogen ions $[H^+]$ equals the concentration of hydroxyl ions $[OH^-]$, is pH 6.8 at mammalian body temperature of 98°F (37°C). The pH within cells, including the red blood cells, is typically lower by 0.2–0.6 unit, and is thus close to neutrality. In most animals other than warm-blooded mammals, blood pH deviates from the familiar value of 7.4. The major reason is that body temperature has an important influence on pH regulation. Consequently, animals that experience significant changes in body temperature have no single normal pH at which they regulate, but rather a series of values depending on body temperature.

Blood pH regulation is necessary because metabolic and ingestive processes add acidic or basic substances to the body and can displace pH from its proper value. For true regulation, active physiological mechanisms are required that can alter the acid-base composition of the blood in a controlled fashion. It is conventional to identify these control mechanisms with their effects on the principal buffer of the extracellular fluid, carbon dioxide (CO_2). Carbon dioxide is produced by cellular metabolism and distributes readily throughout the body because of its high solubility and rapid diffusion. In solution, CO_2 is hydrated to carbonic acid (H_2CO_3) which dissociates almost completely to H^+ and bicarbonate ions $[HCO_3^-]$. Dissolved CO_2 can be identified with a particular partial pressure of CO_2 (PCO_2). To regulate blood pH, organisms have mechanisms to independently control PCO_2 and $[HCO_3^-]$.

The cells, while benefitting from the stability afforded by whole body mechanisms (respiratory and ionic), also have local means for their own pH regulations. An acute acid load on a cell, whether from its intrinsic metabolism or from an external source, is dealt with first by the cell's chemical buffering capacity, a capacity that exceeds that of the blood by severalfold. Other cellular mechanisms include the conversion of organic acids to neutral compounds through metabolic transformations, and the transfer of acid equivalents from the primary cell fluid, the cytoplasm, into cellular organelles. *See* BIOPOTENTIALS AND IONIC CURRENTS; CHEMIOSMOSIS; ENZYME.

Donald C. Jackson

Phaeophyceae A class of plants, commonly called brown algae, in the chlorophyll *a-c* phyletic line (Chromophycota). Brown algae occur almost exclusively in marine or brackish water, where they are attached to rocks, wood, sea grasses, or other algae. Approximately 265 genera and 1500 species are recognized, arranged in about 15 orders. *See* ALGAE; CHROMOPHYCOTA.

Phaeophyceae are characterized primarily by biochemical and ultrastructural features. The cells are typically uninucleate and contain one or more chloroplasts with or without pyrenoids. Photosynthetic pigments include chlorophyll *a* and *c*, β-carotene, and several xanthophylls, principally fucoxanthin. The simplest thallus exhibited by Phaeophyceae is generally considered to be an erect, unbranched or branched, uniseriate filament arising from a prostrate filamentous base. Many Phaeophyceae are crustose or bladelike. Complex thalli differentiated into macroscopic organs are produced by kelps and rockweeds.

The geographic distribution of Phaeophyceae is bimodal. Kelps are most abundant and diverse on surf-swept rocky shores of the North Pacific, but they form an ecologically important vegetation belt in the lower intertidal and upper subtidal zones on all cold-water shores except Antarctica. Rockweeds are similarly abundant on cold-water shores, forming conspicuous belts in the upper intertidal zone. They also form extensive stands in salt marshes in the Northern Hemisphere. Tropical waters support a diverse array of Dictyotales and members of the fucalean family Sargassaceae.

Paul C. Silva; Richard L. Moe

Phagocytosis A mechanism by which single cells of the animal kingdom, such as smaller protozoa, engulf and carry particles into the cytoplasm. It differs from endocytosis primarily in the size of the particle rather than in the mechanism; as particles approach the dimensions and solubility of macromolecules, cells take them up by the process of endocytosis.

Cells such as the free-living amebas or the wandering cells of the metazoa often can "sense" the direction of a potential food source and move toward it (chemotaxis). If, when the cell contacts the particle, the particle has the appropriate chemical composition, or surface charge, it adheres to the cell. The cell responds by forming a hollow, conelike cytoplasmic process around the particle, eventually surrounding it completely. Although the particle is internalized by this sequence of events, it is still enclosed in a portion of the cell's surface membrane and thus isolated from the cell's cytoplasm. The combined particle and membrane package is referred to as a food or phagocytic vacuole. *See* VACUOLE.

Ameboid cells of the metazoa also selectively remove foreign particles, bacteria, and other pathogens by phagocytosis. After the foreign particle or microorganism is trapped in a vacuole inside the macrophage, it is usually digested. To accomplish this, small packets (lysosomes) of lytic proenzymes are introduced into the phagocytic vacuole, where the enzymes are then dissolved and activated. *See* LYSOSOME. Philip W. Brandt

Phanerozoic The uppermost eonothem (a time-stratigraphic unit composed of rocks formed during an eon of geologic time) of the Standard Global Chronostratigraphic Scale. The Phanerozoic Eonothem is the physical stratigraphic representation of the Phanerozoic Eon, a geologic time unit, embracing the Paleozoic, Mesozoic, and Cenozoic eras. Phanerozoic time, which extends from 542 million years ago to the present, is estimated to be slightly less than 12% of the Earth's history. *See* GEOLOGIC TIME SCALE; STRATIGRAPHIC NOMENCLATURE; STRATIGRAPHY.

The Phanerozoic essentially represents the most recent cycle of supercontinent breakup, assembly, and subsequent breakup. Following the dispersal of Pannotia, beginning in the latest Proterozoic, rifted crustal pieces (Gondwana, Laurentia, Baltica, Siberia, Avalonia, and others) reassembled as the supercontinent Pangea in the late Paleozoic. Pangea split into the modern continents beginning in the mid-Mesozoic, and has continued to undergo rifting and dispersal through the Cenozoic. *See* PALEOGEOGRAPHY; SUPERCONTINENT.

The Phanerozoic Eonothem is characterized by a good visible fossil record in sedimentary strata. Through much of the Phanerozoic, global sea level was at positions higher than its present position, and continental shelves were flooded with widespread shallow seas that have left a rich marine fossil record. Sea level was near or below its present position in parts of the late Paleozoic, Mesozoic, and late Cenozoic, and these are intervals for which a good terrestrial fossil record is also available. *See* CENOZOIC; FOSSIL; MESOZOIC; PALEOZOIC. Loren E. Babcock

Pharetronidia A subclass of sponges of the class Calcarea in which the skeleton is formed of quadriradite spicules cemented together into a compact network, or includes an aspiculous massive basal skeleton formed of irregular calcitic spherulites. Spicules shaped like tuning forks are characteristically present, and free triradiates, quadriradiates, and diactinal spicules may occur. *See* CALCAREA. Willard D. Hartman

Pharmaceutical chemistry The chemistry of drugs and of medicinal and pharmaceutical products. The important aspects of pharmaceutical chemistry are as follows:

1. Isolation, purification, and characterization of medicinally active agents and materials from natural sources used in treatment of disease and in compounding prescriptions.

2. Synthesis of medicinal agents not known from natural sources, or the synthetic duplication, for reasons of economy, purity, or adequate supply, of substances first known from natural sources.

3. Semisynthesis of drugs, whereby natural substances are transformed by means of comparatively simple steps into products which possess more favorable therapeutic or pharmaceutical properties.

4. Determination of the derivative or form of a medicinal agent which exhibits optimum medicinal activity and at the same time lends itself to stable formulation and elegant dispensing.

5. Determination of incompatibilities, chemical and biological, between the various ingredients of a prescription.

6. Establishment of safe and practical standards, with respect to both dosage and quality, to assure uniform and therapeutically reliable forms for all medication.

7. Improvement and promotion of the use of chemical agents for prevention of illness, alleviation of pain, cure of disease, and search for new therapeutic agents, particularly where no satisfactory remedy now exists. Walter H. Hartung

Pharmaceuticals testing Techniques used to determine that pharmaceuticals conform to specified standards of identity, strength, quality, and purity. Conformance to these standards assures pharmaceuticals which are safe and efficacious, of uniform potency and purity, and of acceptable color, flavor, and physical appearance. Pharmaceuticals are medicinal products which are prescribed by medical doctors and dispensed through pharmacies and hospitals. They are usually taken orally or by parenteral injection.

Standards or specifications and their attendant procedures are designed to provide desired characteristics and acceptable tolerances for all raw materials, intermediates, and finished products. These standards thus provide an objective determination of whether pharmaceuticals are properly constituted. Two components are vital to the makeup of such standards: appropriate analytical procedures to permit a comprehensive examination, and a list of specifications to define the acceptable limits for each property tested. Standards are established by the pharmaceutical manufacturer, official compendiums (*U.S. Pharmacopeia* and the *National Formulary*), and regulations promulgated by the U.S. Food and Drug Administration (FDA).

The steps in the production cycle for pharmaceuticals must be rigidly and uniformly controlled so that each phase is completely accurate. The four control phases are generally designated as raw materials, manufacturing procedures, finished product testing, and control of identity.

Raw materials are usually referred to as components and are purchased on specifications. Physical specifications include such characteristics as bulk density, mesh size, color, odor, extraneous contamination such as fibers, and homogeneity. Chemical specifications usually include such characteristics as chemical or physiological potency, melting point, boiling range, optical rotation, moisture, heavy-metals content, chemical identity, solubility, and presence of chemical contaminants. Samples are taken upon receipt of a specific batch of raw material. The samples are tested to ensure conformance to each specification. Only after the raw material has been checked against each of the specifications can it be approved for use in pharmaceuticals.

Pharmaceutical manufacturers must make products in conformance with the Good Manufacturing Practices as prescribed by the FDA. These regulations provide criteria on the following: buildings, equipment, personnel, components, master-formula and batch-production records, production and control procedures, product containers, packaging and labeling, laboratory controls, distribution records, stability, and complaint files. The quality-control system must provide regular and continuous use of all reasonable procedures, methods, and operations that are necessary to ensure uniform safety and effectiveness of the pharmaceutical.

Finished pharmaceuticals must conform to appropriate standards of identity, strength, quality, and purity. Accordingly each batch of a pharmaceutical must satisfy five requirements: conformance with (1) the label claim for potency, (2) homogeneity standards, (3) standards of pharmaceutical elegance (the physical appearance of the dosage units), (4) identity specifications, and (5) regulatory standards if they are applicable to the specific pharmaceutical.

Identity is the final requirement in pharmaceutical testing. Identity techniques guarantee proper labeling, that is, that the right product is in the right bottle with the right label. To maintain the identity of the product, extensive checks are made throughout the manufacturing operation, including the use of duplicate label tags on all bulk goods, and very rigid controls are applied to printing, storage, and application of labels on finished pharmaceuticals. *See* BIOASSAY; QUALITY CONTROL. W. Brooks Fortune

Pharmacognosy The general biology, biochemistry, and economics of nonfood natural products of value in medicine, pharmacy, and other health professions. The products studied are of biologic origin, either plant or animal. Pharmacognosy literally means knowledge of drugs, as do pharmacology and pharmacy. The center of interest in pharmacology, however, is on the mode of action of drugs. In pharmacy major attention is directed toward provision of suitable dosage forms, their production and distribution. Pharmacognosy is restricted to natural products with attention centered on sources of drugs, plant and animal, and on the biosynthesis and identity of their pharmacodynamic constituents.

Organs, or occasionally entire plants or animals, are dried or frozen for preservation and are termed crude drugs. They may be used medicinally in essentially this form or as sources of mixtures or of chemicals obtained by processes of extraction. Mixtures obtained by exudation from living plants include such drugs as opium, turpentine, and acacia. Processes of extraction are required to obtain such mixtures as peppermint oil (steam distillation), podophyllum resin (percolation), and parathyroid extract (solution).

Pure chemicals may be extracted from a crude drug (for example, the glycoside digitoxin from digitalis or the hormone insulin from pancreas), from a mixture obtained by exudation (for example, the alkaloid morphine from opium), or from an extracted mixture (for example, the terpene menthol from peppermint oil).

Vitamins as a class of natural products are within the scope of pharmacognosy, although many are obtained commercially by laboratory synthesis. Included also are antibiotics and biolog-icals (serums, vaccines, and diagnostic biological products).

Development of synthetic drugs related chemically to the active constituent of a natural product has frequently followed investigation of native use of the natural product as drug or poison. The objective of such development is usually to produce a drug having fewer undesirable side effects while retaining the useful therapeutic action. *See* PHARMACEUTICAL CHEMISTRY; PHARMACOLOGY; PHARMACY. Richard A. Deno

Pharmacology The science of detection and measurement of the effects of drugs or other chemicals on biological systems. The effect of chemicals may be beneficial (therapeutic) or harmful (toxic). The pure chemicals or mixtures may be of natural origin (plant, animal, or mineral) or may be synthetic compounds.

The broad area covered may be conveniently divided into a number of categories: chemotherapy, the use of chemicals to destroy invading organisms such as bacteria and molds in or on the host; pharmacotherapy, the use of drugs to restore or replace normal function in various tissue cells, organs, or integrated units; pharmacodynamics, studies on the mechanism of action of drugs which may utilize physiological, biochemical, or electrical techniques; toxicology, the study of the poisonous effects of chemicals; psychopharmacology, the study of the effects of chemicals on the behavior of humans or animals; biochemical pharmacology, the effects of chemicals on biochemical reactions in living systems, and the effects of these systems on the chemicals, that is, their metabolism; structure-activity relationship, relationship of biological activity to chemical structure and molecular properties; and clinical pharmacology, the study and evaluation of the effects of drugs in humans. *See* CHEMOTHERAPY AND OTHER NEOPLASTIC DRUGS; PATHOLOGY; TOXICOLOGY. Charles J. Kensler

Pharmacy The health profession concerned with the discovery, development, production, and distribution of drugs. Drugs are substances (other than devices) used to diagnose, prevent, cure, or relieve the symptoms of disease. For relations to closely allied fields. *See* MEDICINE; PHARMACEUTICAL CHEMISTRY; PHARMACOGNOSY; PHARMACOLOGY.

General practice is carried on in exclusive prescription pharmacies, semiprofessional pharmacies, and drug stores. It consists of compounding and dispensing drugs on order of the physician, dentist, or veterinarian; serving as consultant on drugs to the health professions and to the public; and selling other health supplies such as antiseptics, bandages, and home remedies.

A hospital pharmacy includes special administrative features, provision of drugs for nursing stations, manufacturing of pharmaceutical preparations, teaching of nurses and medical and pharmacy interns, service to the hospital committee on pharmacy and therapeutics, preparation and revision of a hospital formulary, and monitoring the drug regimen of the individual patient (clinical pharmacy). The pharmacist may have charge of investigational drugs, radioactive pharmaceuticals, medical and surgical sterile supplies, and gaseous drugs for inhalation therapy. Richard A. Deno

Pharynx A chamber at the oral end of the vertebrate alimentary canal, leading to the esophagus. In adult humans it is divided anteriorly by the soft palate into a nasopharynx and an oropharynx, lying behind the tongue but anterior to the epiglottis; there is also a retropharyngeal compartment, posterior to both epiglottis and soft palate. The nasopharynx receives the nasal passages and communicates with the two middle ears through auditory tubes. The retropharynx leads to the esophagus and to the larynx, and the paths of breathing and swallowing cross within it. *See* ESOPHAGUS; LARYNX; PALATE. William W. Ballard

Phase (astronomy) The changing fraction of the disk of an astronomical object that is illuminated, as seen from some particular location. The monthly phases of the Moon are a familiar example (see illustration). When the Sun is approximately on the far side of the Moon as seen from Earth (conjunction), the dark side of the Moon faces the Earth and there is a new moon. The phase waxes, beginning with crescent phases, as an increasing fraction of the illuminated face of the Moon is seen. At quadrature, when half the visible face of the Moon is illuminated, the phase is called the first-quarter moon, since the Moon is now one-quarter of the way through its cycle of phases. The waxing moon continues through its gibbous phases until it is in opposition; the entire visible face of the Moon is illuminated, the full moon. During the full moon, the Moon and the Sun are on opposite sides of the Earth, a configuration known as a syzygy. Then the Moon wanes, going through waning gibbous,

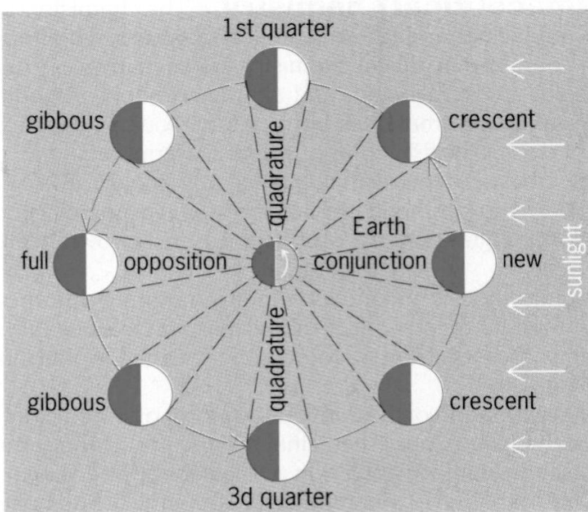

From Earth, different fractions of the illuminated half of the Moon are seen at different times as the Moon goes through a 29.53-day cycle of phases.

third-quarter, and waning crescent phases until it is new again. The cycle of moon phases takes approximately 29.53 days and explains the origin of the word month. *See* MOON.

Galileo discovered the phases of the planet Venus when he observed the sky with his telescope in 1610. Giovanni Zupus discovered the phases of the planet Mercury in 1639. Because of the angle at which the outer planets are seen from Earth, and because of their great distance, they do not appear to go through phases as seen from Earth. Jay M. Pasachoff

Phase (periodic phenomena) The fractional part (portion) of a period through which the time variable of a periodic quantity (alternating electric current, vibration) has moved, as measured at any time instant from an arbitrary time origin. In the case of a sinusoidally varying quantity, the time origin is usually assumed to be the last point at which the value of the quantity was zero, passing from a negative to a positive value.

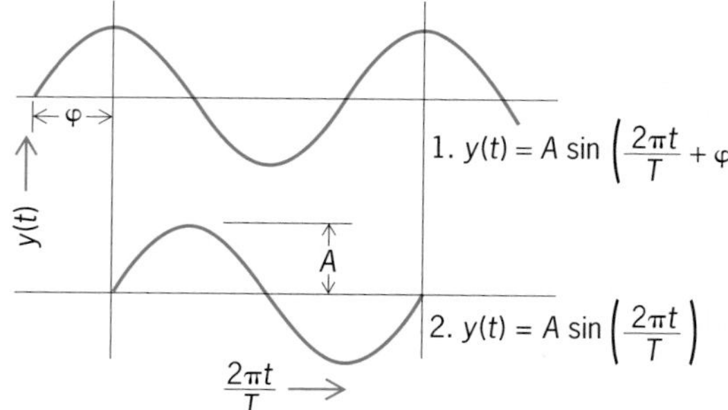

An illustration of the meaning of phase for a sinusoidal wave. The difference in phase between waves 1 and 2 is φ and is called the phase angle. For each wave, *A* is the amplitude and *T* is the period.

In comparing the phase relationships at a given instant between two time-varying quantities, the phase of one is usually assumed to be zero, and the phase of the other is described, with respect to the first, as the fractional part of a period through which the second quantity must vary to achieve a zero of its own (see illustration). In this case, the fractional part of the period is usually expressed in terms of angular measure, with one period being equal to 360° or 2π radians. *See* PHASE-ANGLE MEASUREMENT; SINE WAVE. William J. Galloway

Phase-angle measurement Measurement of the time delay between two periodic signals. The phase difference between two sinusoidal waveforms that have the same frequency and are free of a dc component can be conveniently described as shown in the illustration. It can be seen that the phase angle can be considered as a measure of the time delay between two periodic signals expressed as a fraction of the wave period. This fraction is normally expressed in units of angle, with a full cycle corresponding to 360°. For example, in the illustration, where the voltage v_1 passes through zero $\frac{1}{8}$ cycle before a second voltage v_2, it leads by 360°/8 or 45°. Phase angle is usually defined from the fundamental component of each waveform; therefore distortion of either or both signals can give rise to errors, the extent of which depends on the nature of the distortion and the method of measurement. *See* DISTORTION (ELECTRONIC CIRCUITS).

Many phase-measuring devices are based on the use of zero-crossing detectors. The time at which each signal crosses the zero-voltage axis is determined, usually by means of a squaring-up circuit (for example, an overdriven amplifier) followed by a high-speed comparator. This produces, in each channel, a trigger pulse that is used to drive a bistable flip-flop. The output

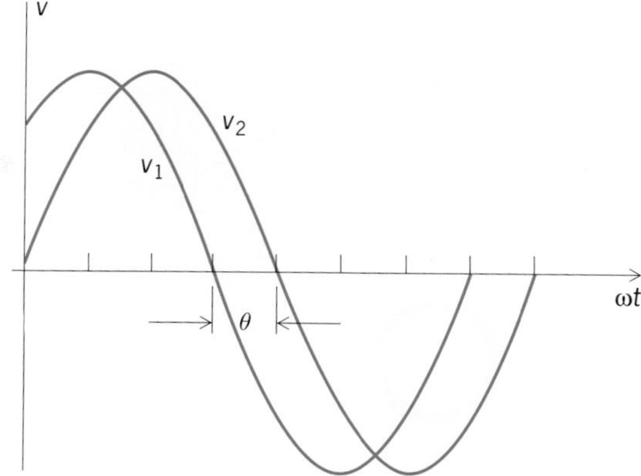

Phase angle θ between voltages v_1 and v_2.

of the flip flop is a rectangular wave, the duty cycle of which is proportional to the phase difference between the input signals. If this signal is integrated by means of a suitable filter, a dc voltage is produced that is an analog representation of the phase angle. This voltage is then displayed on a panel meter (analog or digital) suitably scaled in degrees or radians. Instrumentation using this principle is capable of measuring phase differences to approximately ±0.05° over a wide range of amplitudes and frequencies. *See* AMPLIFIER; COMPARATOR; ELECTRIC FILTER; MULTIVIBRATOR; SWITCHING CIRCUIT; WAVE-SHAPING CIRCUITS.

Conventional phase meters have an upper frequency limit of a few hundred kilohertz. This limit is imposed mainly by the ability of the arrangement consisting of a comparator and a flip-flop to maintain a clean and precise rectangular waveform under conditions of high-speed operation. In order to measure phase angle at frequencies between about 100 kHz and several gigahertz, it is necessary to down-convert the radio-frequency signals to a frequency that can be handled correctly by the phase meter. At microwave frequencies, instruments such as slotted lines, air lines, and vector network analyzers are also used for phase-angle measurements. *See* MICROWAVE MEASUREMENTS. J. Hurll

Phase-contrast microscope A microscope used for making visible differences in phase or optical path in transparent or reflecting specimens. It is one of the most important instruments available for studying living cells and is widely used in biological and medical research.

The essential features of a phase-contrast microscope are shown in the illustration. The practical problem is to find some way of separating the incident or direct light from that diffracted by the object. This is done by placing a diaphragm D of easily recognizable shape, such as an annulus, at the front focal plane of the substage condenser C. Light from each point of the focal plane passes as a parallel pencil of rays through the specimen S and is brought to a focus at the rear focal plane P of the objective O. Thus, on removing the eyepiece, an image of the annulus will be seen at the back of the objective lens. This image corresponds to the incident light. In addition, when a specimen is present, some light is diffracted by it and spreads out to fill the whole of the back lens of the objective. Thus, apart from the small area of overlap over the image of the annulus, the direct and diffracted waves are essentially separated at the plane P. A phase plate is now inserted at this level. This can be a transparent disk with an annular groove of such dimensions that it coincides exactly with the image of the diaphragm D. All the direct light now passes through the groove in the phase plate, whereas the diffracted light passes mainly outside the groove. Since the diffracted light has to pass through a greater thickness of transparent material than the direct light, a phase difference, depending on the refractive index of the phase-plate material

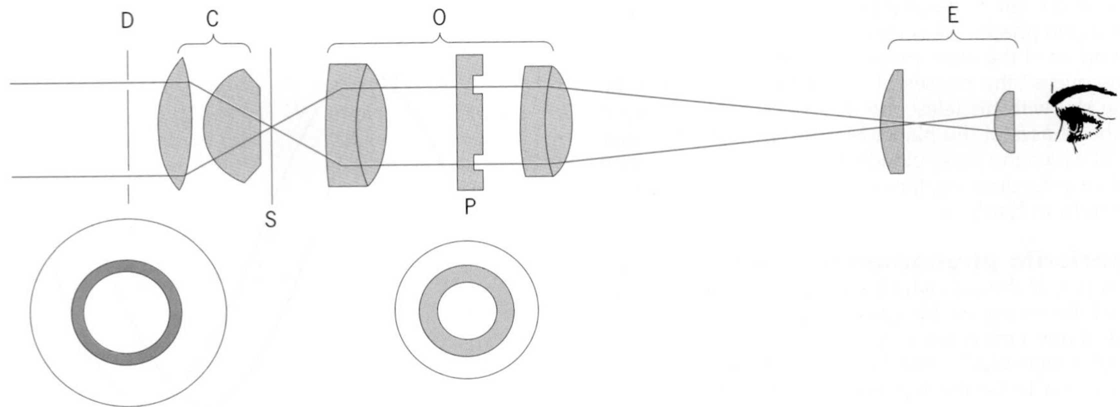

Diagram of a phase-contrast microscope.

and on the thickness of the groove, is introduced between them. If this phase difference is about one-quarter of a wavelength, the basic conditions for phase contrast will have been achieved. If the phase plate is made to retard the incident wave by a quarter of a wavelength, the crests and troughs of the two waves will coincide, giving a resultant of greater amplitude. Refractile details will appear bright (negative contrast) instead of dark (positive contrast).

The phase-contrast microscope is the routine instrument for the examination of living cells because it is possible to study the cell structure under excellent optical conditions and with no loss in resolving power. The method is also useful for the study of unstained tissue sections and has found considerable use for the comparison of material in the electron and optical microscopes. *See* ELECTRON MICROSCOPE; MICROSCOPE; OPTICAL MICROSCOPE.

Robert Barer

Phase equilibrium A general field of physical chemistry dealing with the various situations in which two or more phases (or states of aggregation) can coexist in thermodynamic equilibrium with each other, with the nature of the transitions between phases, and with the effects of temperature and pressure upon these equilibria. Many superficial aspects of the subject are largely qualitative, for example, the empirical classification of types of phase diagrams; but the basic problems always are susceptible to quantitative thermodynamic treatment, and in many cases, statistical thermodynamic methods can be applied to simple molecular models.

Thermodynamics requires that when two phases, α and β, are free to exchange heat, mechanical work, and matter (chemical species), the temperature T, the pressure P, and the chemical potential (partial molar free energy) μ_i(of each particular component i must be equal in both phases at equilibrium. Algebraically, equilibrium exists when $T_\alpha = T_\beta$, $P_\alpha = P_\beta$, $\mu_{i,\alpha} = \mu_{i,\beta}$, and $\mu_{j,\alpha} = \mu_{j,\beta}$.

These conditions of thermal, mechanical, and material equilibrium need not all be present if the equilibrium between phases is subject to inhibiting restrictions. Thus, for a solution of a nonvolatile solute in equilibrium with the solvent vapor, the condition of equality of solute chemical potentials $\mu_{2,\alpha} = \mu_{2,\beta}$ need not apply, since there can be no solute molecules in the vapor phase. Similarly, in osmotic equilibria, in which solvent molecules can pass through a semipermeable membrane, whereas solute molecules cannot, $\mu_{1,\alpha} = \mu_{1,\beta}$ and $T_{1,\alpha} = T_{2,\beta}$, but the solute chemical potentials μ_2 are unequal, as are the pressures on opposite sides of the membrane. *See* OSMOSIS; SOLUTION.

If a system consists of P phases and C distinguishable components, there are $C + 2$ thermodynamic variables (C chemical potentials μ_i, plus the temperature and pressure) which are interrelated by an equation for each phase. Since there are P independent equations relating the $C + 2$ variables, only $F = C + 2 - P$ variables need be fixed to define completely the state of the system at equilibrium; the other variables are then beyond control. This relation for the number of degrees of freedom F, or variance, is called the phase rule. It has proved to be a powerful tool in interpreting and classifying types of phase equilibria.

When chemical changes may occur in the system, the number of components C is the number of independent components whose amounts can be varied by the experimenter; this is equal to the total number of chemical species present less the number of independent chemical equilibria between them.

An invariant system has no degrees of freedom ($F = 0$), for which the number of phases $P = C + 2$. For a one-component system, such an invariant point is a triple point at which three phases coexist at a single temperature and pressure only; for a two-component system, a quadruple point (four phases) would be invariant. *See* TRIPLE POINT.

Robert L. Scott

Phase inverter A circuit having the primary function of changing the phase of a signal by 180°. The phase inverter is most commonly employed as the input stage for a push-pull amplifier. Therefore, the phase inverter must supply two voltages of equal magnitude and 180° phase difference. A variety of circuits are available for the phase inversion. *See* PUSH-PULL AMPLIFIER.

Overall fidelity of a phase inverter and push-pull amplifier can be adversely affected by improper design of the phase inverter. The principal design requirement is that frequency response of one input channel to the push-pull amplifier be identical to the frequency response of the other channel.

The simplest form of phase-inverter circuit is a transformer with a center-tapped secondary. Careful design of the transformer assures that the secondary voltages are equal. The transformer forms a good inverter when the inverter must supply power to the input of the push-pull amplifier. The transformer inverter has several disadvantages. It usually costs more, occupies more space, and weighs more than a transistor circuit. Furthermore, some means must be found to compensate for the frequency response of the transformer, which may not be as uniform as that which can be obtained from solid-state circuits. *See* TRANSFORMER.

An amplifier that provides two equal output signals 180° out of phase is called a paraphase amplifier. If coupling capacitors can be omitted, the simplest paraphase amplifier is shown in the illustration. Approximately the same current flows through R_L and R_E, and therefore if R_L and R_E are equal, the ac output voltages from the collector and from the emitter are equal in

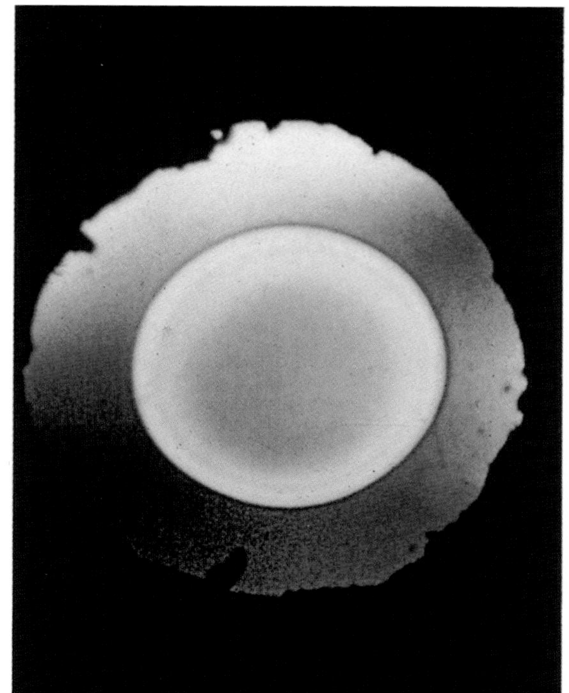

(a)

Phase equilibrium is the coexistence of two or more phases in thermodynamic equilibrium with each other; such equilibria are affected by temperature and pressure. Photomicrographs magnified approximately 500X, using polarized light (crossed Nicols), of (*a*) single crystal and (*b*) 11 crystals of benzene I coexisting with liquid benzene, at about 680 atmospheres and at room temperature; (*c*) potassium nitrate under pressure, where the central portion is potassium nitrate IV (~3000 atmospheres) and surrounding it is potassium nitrate II (~2500 atmospheres). Benzene I and potassium nitrate II and IV are dense forms of the solids favored at these high pressures. Photomicrographs are of specimens contained in a diamond-anvil high-pressure cell. (*National Institute of Standards and Technology*)

(b)

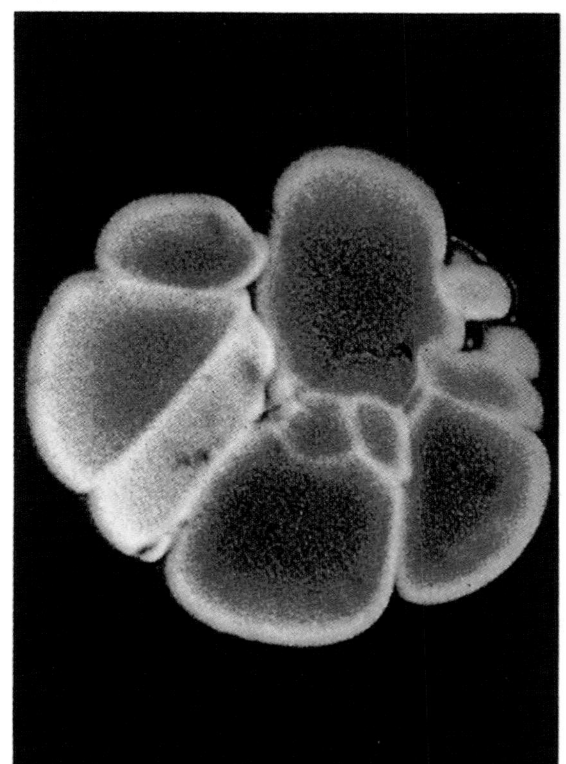

(c)

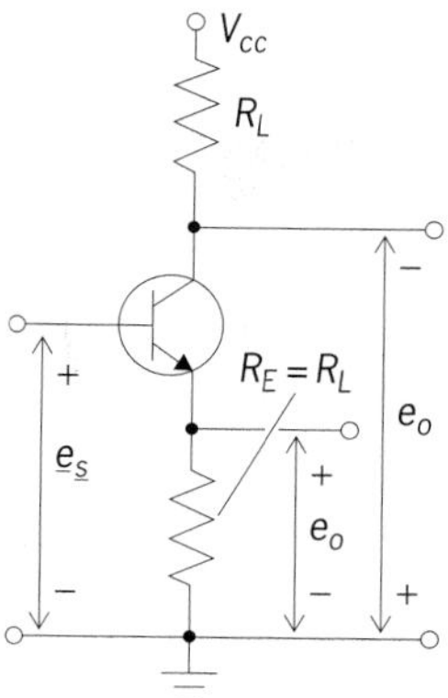

Single-transistor inverter. e_s **= signal voltage;** e_o **= output voltage;** V_{cc} **= collector supply voltage;** R_L **= load resistance;** R_E **= emitter resistance.**

magnitude and 180° out of phase. *See* PHASE (PERIODIC PHENOMENA); PHASE-ANGLE MEASUREMENT. Harold F. Klock

Phase-locked loops Electronic circuits for locking an oscillator in phase with an arbitrary input signal. A phase-locked loop (PLL) is used in two fundamentally different ways: (1) as a demodulator, where it is employed to follow (and demodulate) frequency or phase modulation, and (2) to track a carrier or synchronizing signal which may vary in frequency with time. When operating as a demodulator, the PLL may be thought of as a matched filter operating as a coherent detector. When used to track a carrier, it may be thought of as a narrowband filter for removing noise from the signal and regenerating a clean replica of the signal. *See* DEMODULATOR; ELECTRIC FILTER.

The basic components of a phase-locked loop are shown in the illustration. The input signal is a sine or square wave of arbitrary frequency. The voltage-controlled oscillator (VCO) output signal is a sine or square wave of the same frequency as the input, but the phase angle between the two is arbitrary. The output of the phase detector consists of a direct-current (dc) term, and components of the input frequency and its harmonics. The low-pass filter removes all alternating-current (ac) components, leaving the dc component, the magnitude of which is a function of the phase angle between the VCO signal and the input signal. If the frequency of the input signal changes, a change in phase angle between these signals will produce a change in the dc control voltage in such a manner as to vary the frequency of the VCO to track the frequency of the input signal.

The most widespread use of phase-locked loops is undoubtedly in television receivers. Synchronization of the horizontal oscillator to the transmitted sync pulses is universally accomplished with a PLL. The color reference oscillator is often synchronized with a phase-locked loop. Phase-locked loops are also used as frequency demodulators. They have been applied to stereo decoders made on silicon monolithic integrated circuits.

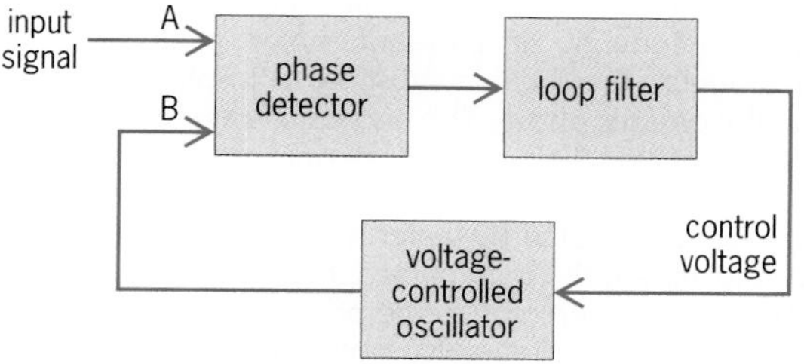

Phase-locked loop. R_1 **and** R_2 **are resistors;** C_1 **and** C_2 **are capacitors.**

High-performance amplitude demodulators may be built using phase-lock techniques. *See* AMPLITUDE-MODULATION DETECTOR.

Thomas B. Mills

Phase modulation A technique used in telecommunications transmission systems whereby the phase of a periodic carrier signal is changed in accordance with the characteristics of an information signal, called the modulating signal. Phase modulation (PM) is a form of angle modulation. For systems in which the modulating signal is digital, the term "phase-shift keying" (PSK) is usually employed. *See* ANGLE MODULATION.

In typical applications of phase modulation or phase-shift keying, the carrier signal is a pure sine wave of constant amplitude, represented mathematically as Eq. (1), where the constant A is

$$c(t) = A \sin \theta(t) \tag{1}$$

its amplitude, $\theta(t) = \omega t$ is its phase, which increases linearly with time, and $\omega = 2\pi f$ and f are constants that represent the carrier signal's radian and linear frequency, respectively.

Phase modulation varies the phase of the carrier signal in direct relation to the modulating signal $m(t)$, resulting in Eq. (2),

$$\theta(t) = \omega t + km(t) \tag{2}$$

where k is a constant of proportionality. The resulting transmitted signal $s(t)$ is therefore given by Eq. (3).

$$s(t) = A \sin [\omega t + km(t)] \tag{3}$$

At the receiver, $m(t)$ is reconstructed by measuring the variations in the phase of the received modulated carrier.

Phase modulation is intimately related to frequency modulation (FM) in that changing the phase of $c(t)$ in accordance with $m(t)$ is equivalent to changing the instantaneous frequency of $c(t)$ in accordance with the time derivative of $m(t)$. *See* FREQUENCY MODULATION.

Among the advantages of phase modulation are superior noise and interference rejection, enhanced immunity to signal fading, and reduced susceptibility to nonlinearities in the transmission and receiving systems. *See* DISTORTION (ELECTRONIC CIRCUITS); ELECTRICAL INTERFERENCE; ELECTRICAL NOISE.

When the modulating signal $m(t)$ is digital, so that its amplitude assumes a discrete set of values, the phase of the carrier signal is "shifted" by $m(t)$ at the points in time where $m(t)$ changes its amplitude. The amount of the shift in phase is usually determined by the number of different possible amplitudes of $m(t)$. In binary phase-shift keying (BPSK), where $m(t)$ assumes only two amplitudes, the phase of the carrier differs by 180°. An example of a higher-order system is quadrature phase-shift keying (QPSK), in which four amplitudes of $m(t)$ are represented by four different phases of the carrier signal, usually at 90° intervals. *See* MODULATION. Hermann J. Helgert

Phase-modulation detector A device that recovers or detects a signal from a phase-modulated carrier. A phase-modulating (PM) signal operating on a sinusoidal carrier can be recovered in a variety of ways by phase-modulation detection (or demodulation). *See* CARRIER; DEMODULATOR; MODULATION; PHASE MODULATION.

A phase-locked loop (PLL) is a closed-loop servo that comprises a phase detector for comparing the waveforms of the input sinusoidal signal with the output of a frequency-controlled oscillator whose frequency is responsive to an input-control voltage. The simple PLL is a very good phase-modulation detector of analog signals. *See* OSCILLATOR; PHASE-LOCKED LOOPS; SERVOMECHANISM.

Phase-shift demodulation recovery always requires a reference waveform for measuring phase shift. A superior alternative approach to obtaining the reference waveform is the two-phase PLL, called the Costas loop, which takes advantage of the fact that the data bits are ± 1, so the squared data values are always one. The oscillator outputs are both sine and cosine at the input carrier frequency with a slight phase error, θ. The input signal is demodulated with respect to both sine and cosine outputs of the oscillator, and low-pass-filtered, and then the two resulting data-modulated signals are multiplied together to obtain a data-independent oscillator frequency-control input signal that is proportional to the sine of double the phase difference between the input signal and the oscillator output.

Recovery of differential binary phase-shift keyed (DBPSK) signals is accomplished by delaying the signal by one symbol time, then correlating the input and output of the delay element. A correlator simply multiplies two signals together and integrates over their product waveform; just multiply and average.

Recovery of multiple phase-shift keyed (MPSK) signals requires coherent detection, that is, correlation with all the M possible symbols (M phase-shifted sinusoids), and selecting the largest value during each symbol time.

Recovery of quadrature amplitude-modulation (QAM) signals is a different situation because we are attempting to decode simultaneously both amplitude and phase modulation. The first step is to correlate the input waveform with the sine and cosine of the carrier over one symbol time. We are not looking for the largest output from the two channels; rather, the outputs from these two channels will be applied to digital-to-analog converters (DACs) to recover amplitude information. The data sets are interleaved to produce the desired digital output. *See* AMPLITUDE MODULATOR.

Stanley A. White

Phase modulator A device that shifts the phase of a sinusoidal carrier signal, $c(t)$, in proportion to the instantaneous amplitude of the information signal, $s(t)$. Since the definition of frequency is the rate of change of phase, phase modulator and phase modulation are very closely related to frequency modulator and frequency modulation.

Either analog or digital information may be modulated onto the sinusoidal carrier, but the methods may differ. Sets of bits that comprise a digital information unit, called a symbol, can have only a finite number of well-defined possible values (usually an integer power of 2), so the digital phase-modulating signal is constrained to a specific set of values. This modulating process is known as phase-shift keying, or PSK. *See* ANGLE MODULATION; FREQUENCY MODULATION; MODULATION; PHASE MODULATION.

An analog phase-modulating signal can take on an infinite number of values, and is an excellent choice for transmitting any continuous signal, such as audio, by continuous phase modulation (CPM). The phase-modulated carrier is of the form $A \cos [2\pi f_c t + k_p s(t)]$, where k_p is the phase-modulation index in radians, and $s(t)$ is the amplitude-normalized information signal, where $|s(t)| \leq 1$.

Consider the most general CPM case. Suppose that we wish to shift that unmodulated sinusoidal carrier through a dynamically changing phase angle $\theta(t)$, where $\theta(t) = k_p s(t)$. Then $A \cos [2\pi f_c t + \theta(t)] = A \cos (2\pi f_c t) \cos [\theta(t)] - A \sin (2\pi f_c t) \sin [\theta(t)]$. In the continuous case, if $\theta(t)$ is constrained to be small, $\cos[\theta(t)] \approx 1$ and $\sin [\theta(t)] \approx \theta(t)$, and therefore the phase-shifted sinusoid is very nearly equal to $A \cos (2\pi f_c t) - \theta(t)\, A \sin (2\pi f_c t)$. This narrow-band phase modulation can therefore be achieved with a simple 90° phase shift of the cosine to generate the sine, multiplication of that sine by $\theta(t)$, and subtraction of the product from the initial cosine.

Wideband phase modulation of a continuous signal offers no such shortcut. We have to form both $\sin (\theta)$ and $\cos (\theta)$ as well as the two products, or rely on frequency multiplication. In frequency multiplication, the signals are produced at a frequency $1/N$th of the target frequencies, and then scaled up in frequency by a factor of N by using a nonlinear device, called a frequency multiplier, to generate a rich set of harmonics, followed by a narrow bandpass filter to preserve only the desired frequency components. *See* FREQUENCY MULTIPLIER.

For transmitting digital data, the simplest form of PSK is binary, or BPSK. We can generalize to M-bit PSK, or MPSK. The two-state BPSK data are transmitted by a plain carrier signal for a logical "1" and inverting of its phase for a logical zero. The inverse convention may also be used, but both carry a hidden cost at the receiver or detector because in order to demodulate phase information one must have a reliable reference phase available from a phase-"unshifted" reference signal. Differential binary phase-shift keying (DBPSK) at the transmitter avoids the data recovery problem at the receiver by encoding not the phase of the absolute data, but the change in phase from one clock cycle to the next. *See* PHASE-MODULATION DETECTOR.

We can double the data rate over BPSK by transmitting bit pairs, and we can combine both amplitude coding and phase coding to improve bandwidth efficiency and again double the data rate. Quadrature PSK (QPSK) encodes data in pairs of bit pairs to signal a phase shift of 0, $\pm\pi/2$, or π and amplitudes of ± 1 and ± 3. Mechanization of a phase shift of π can be accomplished either by a phase-shift network of by modulating the signal onto a quadrature carrier, that is, a second carrier that is orthogonal to (or in quadrature with or 90° out of phase with) the first carrier. Other MPSK schemes are implemented by quadrature amplitude modulation (QAM).

Stanley A. White

Phase rule A relationship used to determine the number of state variables F, usually chosen from among temperature, pressure, and species compositions in each phase, which must be specified to fix the thermodynamic state of a system in equilibrium. It was derived by J. Willard Gibbs. The phase rule (in the absence of electric, magnetic, and gravitational phenomena) is given by the equation below, where C is the number

$$F = C - P - M + 2$$

of chemical species present at equilibrium, P is the number of phases, and M is the number of independent chemical reactions. Here phase is used to indicate a homogeneous, mechanically separable portion of the system, and the term independent reactions refers to the smallest number of chemical reactions which, upon forming various linear combinations, includes all reactions which occur among the species present. The number of independent state variables F is referred to as the degrees of freedom or variance of the system. *See* CHEMICAL EQUILIBRIUM; CHEMICAL THERMODYNAMICS; PHASE EQUILIBRIUM; THERMODYNAMIC PROCESSES.

Stanley I. Sandler

Phase-transfer catalysis A process in which the rate of a reaction occurring in a two-phase organic-water system is enhanced by addition of a compound that helps transfer the water soluble reactant across the interface to the organic phase.

An important factor, which contributes to the slowness of many organic reactions, is the lack of homogeneity of the reaction mixture. This is particularly the case with nucleophilic substitution reactions such as the reaction below, where RX is an organic

$$RX + Nu^- \rightarrow RNu + X^-$$

reagent and Nu^+ is the nucleophilic reagent, The nucleophilic reagent is frequently an inorganic anion, which is soluble in water in which the organic substrate is insoluble, but is insoluble in the organic phase. The encounter rate between Nu^- and RX is consequently low, as they can only meet at the interface of the heterogeneous system. The water-soluble anion is also frequently highly solvated by water molecules, which stabilize the anion and thus reduce its nucleophilic reactivity. These problems have been overcome in the past by the use of polar aprotic solvents, which will dissolve both the organic and inorganic reagents, or by the use of homogeneous mixed-solvent systems, such as water:ethanol or water:dioxan. *See* ELECTROPHILIC AND NUCLEOPHILIC REAGENTS.

Phase-transfer catalysis involves the transportation of the inorganic anion, Nu^- , from the aqueous phase into the organic phase by the formation of a nonsolvated ion-pair with a cationic phase-transfer catalyst, Q^+. With highly lipophilic catalysts, the reactive ion pair $[Q^+Nu^-]$ is formed at the interface between the aqueous and organic phases, followed by rapid transportation into the bulk of the organic phase. The rate of the reaction is enhanced, as the encounter rate of the nucle-ophile, Nu^-, with the organic reagent, RX, in the single phase will be significantly higher than at the interface, Moreover, as the anion is transferred without water of solvation, its nucleophilic reactivity can be considerably higher in the organic phase than in the aqueous phase. Rate enhancements of greater than 10^7 have thus been observed. *See* CATALYSIS; HETEROGENEOUS CATALYSIS; HOMOGENEOUS CATALYSIS; QUATERNARY AMMONIUM SALTS; STEREOCHEMISTRY. R. Alan Jones

Phase transitions Changes of state brought about by a change in an intensive variable (for example, temperature or pressure) of a system. Some familiar examples of phase transitions are the gas-liquid transition (condensation), the liquid-solid transition (freezing), the normal-to-superconducting transition in electrical conductors, the paramagnet-to-ferromagnet transition in magnetic materials, and the superfluid transition in liquid helium. Further examples include transitions involving amorphous or glassy structures, spin glasses, charge-density waves, and spin-density waves. *See* AMORPHOUS SOLID; CHARGE-DENSITY WAVE; METALLIC GLASSES; SPIN-DENSITY WAVE; SPIN GLASS; SUPERCONDUCTIVITY; SUPERFLUIDITY.

Typically the phase transition is brought about by a change in the temperature of the system. The temperature at which the change of state occurs is called transition temperature (usually denoted by T_c). For example, the liquid-solid transition occurs at the freezing point.

The two phases above and below the phase transition can be distinguished from each other in terms of some ordering that takes place in the phase below the transition temperature. For example, in the liquid-solid transition, the molecules of the liquid get "ordered" in space when they form the solid phase. In a paramagnet, the magnetic moments on the individual atoms can point in any direction (in the absence of an internal magnetic field), but in the ferromagnetic phase the moments are lined up along a particular direction, which is then the direction of ordering. Thus in the phase above the transition, the degree of ordering is smaller than in the phase below the transition. One measure of the amount of disorder in a system is its entropy, which is the negative of the first derivative of the thermodynamic free energy with respect to temperature. When a system possesses more order, the entropy is lower. Thus at the transition temperature the entropy of the system changes from a higher value above the transition to some lower value below the transition. *See* ENTROPY; FERROMAGNETISM; PARAMAGNETISM.

This change in entropy can be continuous or discontinuous at the transition temperature. In other words, the development of order in the system at the transition temperature can be gradual or abrupt. This leads to a convenient classification of phase transitions into two types, namely, discontinuous and continuous.

Discontinuous transitions involve a discontinuous change in the entropy at the transition temperature. A familiar example of this type of transition is the freezing of water into ice. As water reaches the freezing point, order develops without any change in temperature. Thus there is a discontinuous decrease in the entropy at the freezing point. This is characterized by the amount of latent heat that must be extracted from the water for it to be "ordered" into the solid phase (ice). Discontinuous transitions are also called first-order transitions.

In a continuous transition, entropy changes continuously, and hence the growth of order below T_c is also continuous. There is no latent heat involved in a continuous transition. Continuous transitions are also called second-order transitions. The paramagnet-to-ferromagnet transition in magnetic materials is an example of such a transition.

The degree of ordering in a system undergoing a phase transition can be made quantitative in terms of an order parameter. At temperatures above the transition temperature the order parameter has a value zero, and below the transition it acquires some nonzero value. For example, in a ferromagnet the order parameter is the magnetic moment per unit volume (in the absence of an externally applied magnetic field). It is zero in the paramagnetic state since the individual magnetic moments in the solid may point in any random direction. Below the transition temperature, however, there exists a preferred direction of ordering, and as the temperature is decreased below T_c, more and more individual magnetic moments start to align along the preferred direction of ordering, leading to a continuous growth of the magnetization or the macroscopic magnetic moment per unit volume in the ferromagnetic state. Thus the order parameter changes continuously from zero above to some nonzero value below the transition temperature. In a first-order transition, the order parameter would change discontinuously at the transition temperature. D. J. Sellmyer; S. S. Jaswal

Phase velocity The velocity of propagation of a pure sine wave of infinite extent. In one dimension, for example, the form of the disturbance for such a wave is given by $y(x, t) = A \sin [2\pi(x/\lambda - t/T)]$. Here x is the position at which the disturbance $y(x, t)$ exists at time t, λ is the wavelength, T is the period which is related to the wave frequency by $T = 1/f$, and A is the disturbance amplitude. The argument of the sine function is called the phase. The phase velocity is the speed with which a point of constant phase can be said to move. Thus $x/\lambda - ft =$ constant, so the phase velocity v_p is given by $dx/dt = v_p = \lambda f$. This is the basic relationship connecting phase velocity, wavelength, and frequency. *See* PHASE (PERIODIC PHENOMENA); SINE WAVE; WAVE MOTION.

The phase velocity for waves in a medium is determined in part by intrinsic properties of the medium. For all mechanical waves in elastic media, the square of the phase velocity is proportional to the ratio of the appropriate elastic property of the medium to the appropriate inertia property. The phase velocity of electromagnetic waves depends upon the medium as well. In vacuum, the phase velocity c is given by $c^2 = 1/\epsilon_0\mu_0 \approx 9 \times 10^{16}\ m^2/s^2$, where ϵ_0 and μ_0 are respectively the permittivity and permeability of the vacuum. Phase velocity may also depend upon the mode of wave propagation—in general, upon the frequency of the wave. Waves of different frequencies will travel at different speeds, resulting in a phenomenon called dispersion. *See* ELECTROMAGNETIC RADIATION; LIGHT; WAVE EQUATION; YOUNG'S MODULUS. S. A. Williams

Phasmatodea The order of insects comprising the leaf insects and the stick insects, also known as walking sticks. There are about 2800 species of phasmids (Phasmatodea is alternatively known as Phasmida) cataloged into six families, the majority of which live in the tropics. All are herbivores, and most have a strong resemblance to twigs and sticks with long slender bodies and legs. Phasmatodea species have cryptic coloration: either brown like a stick or green like a leaf. Many species have knobs, spines, and lobes to camouflage themselves, thus avoiding predation from visual predators such as birds and spiders. Members of the family Timemidae (=Phyllidae) are leaf mimics; their bodies and legs are flattened and marked to resemble leaves. *See* INSECTA; PROTECTIVE COLORATION.

Phasmids range in size from tiny *Timema* males from North America that are only 1.2 cm (0.5 in.) long to giants of Malaysia in the genus *Phobeticus* that can be 0.5 m (1.65 ft) long. Some species are winged and some wingless. In the majority of species, females drop eggs from where they are feeding high among the plant leaves. The eggs with hard cases fall to the ground like the seeds they resemble, and remain hidden among the leaf litter. Phasmids hatch out as perfect miniatures of their parents.

Mary Morgan-Richards

Phenocryst A relatively large crystal embedded in a finer-grained or glassy igneous rock. The presence of phe-nocrysts gives the rock a porphyritic texture. Phenocrysts are represented most commonly by feldspar, quartz, biotite, hornblende, pyroxene, and olivine. Strictly speaking, phenocrysts crystallize from molten rock material (lava or magma). They commonly represent an earlier and slower stage of crystallization than does the matrix in which they are embedded. *See* IGNEOUS ROCKS; PORPHYROBLAST.

Carleton A. Chapman

Phenol The simplest member of a class of organic compounds possessing a hydroxyl group attached to a benzene ring or to a more complex aromatic ring system. Phenol itself, C_6H_5OH, may also be called hydroxybenzene or carbolic acid. Pure phenol is a colorless solid melting at 42°C (108°F), moderately soluble in water, and weakly acidic (p*K* 9.9).

Phenol has broad biocidal properties, and dilute aqueous solutions have long been used as an antiseptic. At higher concentrations phenol causes severe skin burns; it is a violent systemic poison. *See* ANTISEPTIC.

Phenol has the structure shown. Simple substituted phenols, such as the three isomeric chlorophenols, are named as indicated, using the ortho (*o*), meta (*m*), and para (*p*) prefixes. In more highly substituted phenols the positions of substitution are indicated by numbers (as in 2,4-dichlorophenol). Compounds with more than one hydroxyl group per aromatic ring are known as polyhydric phenols, and include catechol, resorcinol, hydroquinone, phloroglucinol, and pyrogallol. Until World War I phenol was essentially a natural coal tar product. However, synthetic methods have replaced extraction from natural sources. There are many possible syntheses.

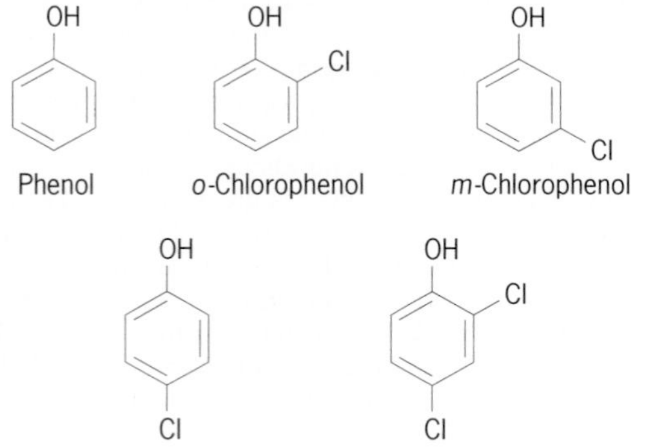

Phenol   *o*-Chlorophenol   *m*-Chlorophenol

p-Chlorophenol   2,4-Dichlorophenol

Phenol is one of the most versatile and important industrial organic chemicals. It is the starting point for many diverse products used in the home and industry. A partial list includes: nylon, epoxy resins, surface active agents, synthetic detergents, plasticizers, antioxidants, lube oil additives, phenolic resins (with formaldehyde, furfural, and so on), polyurethanes, aspirin, dyes, wood preservatives, herbicides, drugs, fungicides, gasoline additives, inhibitors, explosives, and pesticides. *See* PHENOLIC RESIN.

Robert I. Stirton; Martin Stiles

Phenolic resin One of the condensation products of phenols or phenolic derivatives with aldehydes such as formaldehyde and furfural. The phenol-formaldehyde resins, developed commercially between 1905 and 1910, were the first truly synthetic polymers and have found wide usage. They are characterized by low cost, dimensional stability, high strength, and resistance to aging.

Phenolic resins can be cast from syrupy intermediates or molded from B-stage solid resins. Laminated products can be produced by impregnating fiber, cloth, wood, and other materials with the resin. An important type of phenolic resin product is rigid foam. Cured phenolic plastics are rigid, hard, and resistant to chemicals (except strong alkali) and to heat.

Some of the uses for phenolic resins are for making precisely molded articles, such as telephone parts, for manufacturing strong and durable laminated boards, or for impregnating fabrics, wood, or paper. Phenolic resins are also widely used as adhesives, as the binder for grinding wheels, as thermal insulation panels, as ion-exchange resins, and in paints and varnishes. *See* ADHESIVE; ION EXCHANGE; PHENOL; PLASTICS PROCESSING; POLYMERIZATION.

John A. Manson

Phenylketonuria An inborn error of metabolism in which affected individuals lack the liver enzyme needed to metabolize phenylalanine, an amino acid essential for normal growth and development. If untreated, affected individuals may become severely mentally retarded, become microcephalic, have behavioral problems, develop epilepsy, or show other signs of neurological impairment. Phenylketonuria (PKU) is inherited as an autosomal recessive trait and is found in all ethnic groups but most frequently in individuals of northern European descent. Its incidence is about 1 per 14,000 births in the United States.

Newborn screening programs for phenylketonuria have been successful in identifying most cases within a few weeks of birth. A low phenylalanine diet with restriction of proteins, if initiated early in infancy, can prevent the development of severe intellectual and neurological handicaps that would otherwise occur in virtually all untreated cases. *See* PROTEIN METABOLISM.

Felix de la Cruz

Pheromone A substance that acts as a molecular messenger, transmitting information from one member of a species to another member of the same species. A distinction is made between releaser pheromones, which elicit a rapid, behavioral response, and primer pheromones, which elicit a slower, developmental response and may pave the way for a future behavior.

Communication via pheromones is common throughout nature, including some eukaryotic microorganisms such as fungi that exchange vital chemical signals. The cellular slime molds form large aggregations of amebas which unite to form a sorocarp made up of a long, slender stalk that supports a spore-containing fruiting body. A pheromone is responsible for the aggregation. In several species of algae, relatively simple hydrocarbons act as sperm attractants.

By far the largest number of characterized pheromones come from insect species. In social insects, such as termites and ants, there may be as many as a dozen different types of messages that are used to coordinate the complex activities which must be carried out to maintain a healthy colony. These activities might require specialized pheromones such as trail pheromones (to lead to a food source), alarm pheromones (recruiting soldiers to the site of an enemy attack), or pheromones connected with reproductive behavior. Much less is known about mammalian pheromones because mammalian behavior is more difficult to study. There are, however, a small number of well-characterized mammalian pheromones from pigs, dogs, hamsters, mice, and marmosets.

There is great potential for controlling the behavior of a given species by manipulating its natural chemical signals. For example, pheromones have been used to disrupt the reproduction of certain insect pests. This approach can lead to reduced use of pesticides as well as advances in the control of both agricultural pests and disease vectors. *See* CHEMICAL ECOLOGY; CHEMORECEPTION; INSECT CONTROL, BIOLOGICAL; SOCIAL INSECTS. J. Meinwald

Philosophy of science The subfield of philosophy that treats fundamental questions pertaining to science. What is science, as contrasted with technological innovation? What are its objectives and methods? Are there aims or goals that are proprietary to science. Does science make genuine progress, and if so how? What kind of a world does science as we know it testify to? These questions are as urgent today as they were in the days of Aristotle, who originated the first theory of science, as well as engaged in some of the very earliest scientific inquiries and defined the very first disciplinary boundaries. *See* PHYSICAL SCIENCE; SCIENCE; SCIENTIFIC METHODS; TECHNOLOGY.

Science is a relative newcomer on the evolutionary scene, while technological innovation is considerably older. What is perhaps indisputable is that science grew up alongside philosophy. It responded favorably to the ministrations of those cultures that promoted and protected arenas (however exclusive of women, foreigners, and the underclasses) for disputation of ideas. The concept of an open society plays a large role in the philosophies of the most important twentieth-century figures in the philosophy of science: Karl Popper and Thomas Kuhn. Science requires an integrated and highly interactive community. No individual thinker can attain the vision for scientific innovation unless he or she will submit to standing upon the shoulders of those giants who came before—and this is increasingly true as history marches forward. And so we are left with a tension between innovation and deference: a culture of disputation is the most fertile ground for science; but no long-term scientific enterprise can flourish without accumulation of insight. In the face of these facts, the questions of how exactly science makes progress—still a central open question in the philosophy of science—becomes more urgent than ever.

The traditional cumulative conception of scientific knowledge (that scientific knowledge grows simply through accumulation or accretion), accepted without debate up until the nineteenth century, was effectively challenged in the twentieth century by Karl Popper and Thomas Kuhn, in quite similar ways. They argued convincingly that the foundations of scientific thinking (sometimes discussed as the logic of science, or the rationality of science) is a logic of disputation—as Popper put it, a logic of conjectures and refutations, a logic of challenge and reply. No scientific hypothesis is immune to challenge. And so, far from being a predominantly antiquarian exercise in maintaining old knowledge and increasing it when possible, the primary scientific exercise is better characterized as disputation of old orthodoxy and proposals for its replacement with a challenger, when the challenger can stand up to further challenges. *See* HYPOTHESIS.

The cumulative view of scientific progress, still quite common today, was an important ingredient in the optimism of the thinkers—G. W. F. Hegel and Jean Jacques Rousseau. But if (as philosophers currently believe) science does not make progress in this straightforward way, is there any reason to think that human society itself can be progressive? In what sense can science—as distinct from mere technological innovation—be part of the solution to the problems faced by human societies today? What can science bring to the negotiation table when the high stakes associated with public policies crafted in the twenty-first century and beyond are involved? Can science help us understand ourselves and our impact upon our fragile planet? And perhaps most importantly, can science help us craft policies and polities that will effectively safeguard the future of the Earth and the diversity of its life. These are now, as they have always been, open questions for philosophers of science. Mariam Thalos

Phlebitis An inflammation of a vein. Individuals with phlebitis typically experience tenderness, redness, and hardness along the course of the vein. The cause of the inflammation may be related to injury of the vein or infection. The presence of varicose veins and the long-term use of indwelling intravenous catheters or irritating intravenous solutions place individuals at risk of developing phlebitis. In addition, those with certain diseases, including systemic lupus erythematosus, vasculitis, or malignancy, are at increased risk. Two varieties of phlebitis are recognized: phlebothrombosis and thrombophlebitis.

Phlebothrombosis is a condition in which a blood clot develops within an inflamed vein. As the clot enlarges, it may detach and travel to the lung, becoming a pulmonary embolism. Thrombophlebitis begins with an inflammatory reaction in the vein wall. When the lining of the vein is damaged, three reactions influence the development of thrombosis. Initially, damage to the lining results in adherence of white blood cells, coagulation, and a loss of the lining's nonthrombogenic characteristics. Subsequently, the deep lining of the vein is exposed, bringing it into contact with blood and allowing platelets to adhere and aggregate. Finally, the exposed lining and activated platelets result in changes in coagulation, causing more platelets to interact with deep-lining structures. These factors are influenced by the velocity of blood flow in the affected area. *See* EMBOLISM; THROMBOSIS.

Symptomatic thrombophlebitis usually results in a clot which is firmly adherent to the vein wall with a decreased risk of embolizing. Some individuals may develop symptoms suggestive of deep venous thrombosis such as pain and swelling, and should undergo noninvasive ultrasound examination of the deep veins of the leg. In the absence of deep venous thrombosis, the goal of treatment of superficial phlebitis is symptomatic relief. Analgesics, warm compresses and elevation of the affected limb may be beneficial. Late effects of phlebitis include damage to the vein wall and destruction of the venous valves or obliteration of the vein. When the deep veins of the lower extremity are involved, many individuals develop chronic venous insufficiency and its associated morbidity. *See* CIRCULATION; INFLAMMATION. Lazar J. Greenfield; Mary C. Proctor

Phloem The principal food-conducting tissue in vascular plants. Its conducting cells are known as sieve elements, but phloem may also include companion cells, parenchyma cells, fibers, sclereids, rays, and certain other cells. As a vascular tissue, phloem is spatially associated with xylem, and the two together form the vascular system. *See* XYLEM.

Sieve elements differ from phloem parenchyma cells in the structure of their walls and to some extent in the character of their protoplasts. Sieve areas, distinctive structures in sieve element walls, are specialized primary pit fields in which there may be numerous modified plasmodesmata. Plasmodesmata are strands of cytoplasm connecting the protoplasts of two contiguous cells. These strands are often surrounded by callose, a carbohydrate material, that appears to form rapidly in plants when they are placed under stress.

Typical sieve cells are long elements in which all the sieve areas are of equal specialization, though sieve areas may be more numerous in some walls than in others. In contrast, a sieve-tube member has some sieve areas more specialized than others; that is, the pores, or modified plasmodesmata, are larger in some sieve areas. Parts of the walls containing such sieve areas are called sieve plates.

Companion cells are specialized parenchyma cells that occur in close ontogenetic and physiologic association with sieve tube members. Some sieve-tube members lack companion cells. The precise functional relationship between these two kinds of cells is unknown.

Parenchyma cells in the phloem occur singly or in strands of two or more cells. They store starch, frequently contain tannins or crystals, commonly enlarge as the sieve elements become obliterated, or may be transformed into sclereids or cork cambium cells.

Phloem fibers vary greatly in length (from less than 0.04 in. or 1 mm in some plants to 20 in. or 50 cm in the ramie plant). The secondary walls are commonly thick and typically have simple pits, but may or may not be lignified. Michael A. Walsh

Phlogopite A mineral with an ideal composition of $KMg_3(AlSi_3)O_{10}(OH)_2$. Phlogopite belongs to the mica mineral group. It has been occasionally called bronze mica. Phlogopite is a trioctahedral mica, where all three possible octahedral cation sites are occupied by magnesium (Mg). The magnesium octahedra, $Mg(O,OH)_6$, form a sheet by sharing edges. As in all micas, tetrahedra are located on either side of the octahedral sheet, which may be occupied by aluminum (Al) or silicon (Si). Adjacent tetrahedra share corners to form a two-dimensional network of sixfold rings, thus producing a tetrahedral sheet. Two opposing tetrahedral sheets and the included octahedral sheet form a 2:1 layer. Potassium (K) ions are located between adjacent tetrahedral sheets in the interlayer region.

Specific gravity is 2.86, hardness on the Mohs scale is 2.5–3.0, and luster is vitreous to pearly. Thin sheets are flexible. Color is yellow brown, reddish brown, or green, and thin sheets are transparent. Thermal stability varies greatly with composition, with iron or fluorine substitutions reducing or increasing stability, respectively. Weathering of phlogopite may produce vermiculite. *See* HARDNESS SCALES; VERMICULITE; WEATHERING PROCESSES.

Phlogopite occurs in marbles produced by the metamorphism of siliceous magnesium-rich limestones or dolomites and in ultrabasic rocks, such as peridotites and kimberlites. *See* DOLOMITE; LIMESTONE; PERIDOTITE.

Phlogopite is used chiefly as an insulating material and for fireproofing. It has high dielectric properties and high thermal stability. *See* MICA; SILICATE MINERALS. Stephen Guggenheim

Phobia An intense irrational fear that often leads to avoidance of an object or situation. Phobias (or phobic disorders) are common (for example, fear of spiders, or arachnophobia; fear of heights, or acrophobia) and usually begin in childhood or adolescence. Psychiatric nomenclature refers to phobias of specific places, objects, or situations as specific phobias. Fear of public speaking, in very severe cases, is considered a form of social phobia. Social phobias also include other kinds of performance fears (such as playing a musical instrument in front of others; signing a check while observed) and social interactional fears (for example, talking to people in authority; asking someone out for a date; returning items to a store). Individuals who suffer from social phobia often fear a number of social situations. Although loosely regarded as a fear of open spaces, agoraphobia is actually a phobia that results when people experience panic attacks (unexpected, paroxysmal episodes of anxiety and accompanying physical sensations such as racing heart, shortness of breath).

The origin of phobias is varied and incompletely understood. Most individuals with specific phobias have never had anything bad happen to them in the past in relation to the phobia. In a minority of cases, however, some traumatic event occurred that likely led to the phobia. It is probable that some common phobias, such as a fear of snakes or a fear of heights, may actually be instinctual, or inborn. Both social phobia and agoraphobia run in families, suggesting that heredity plays a role. However, it is also possible that some phobias are passed on through learning and modeling.

Phobias occur in over 10% of the general population. Social phobia may be the most common kind, affecting approximately 7% of individuals. When persons encounter the phobic situation or phobic object, they typically experience a phobic reaction consisting of extreme fearfulness, physical symptoms (such as racing heart, shaking, hot or cold flashes, or nausea), and cognitive symptoms (particularly thoughts such as "I'm going to die" or "I'm going to make a fool of myself"). These usually subside quickly when the individual is removed from the situation. The tremendous relief that escape from the phobic situation provides is believed to reinforce the phobia and to fortify the individual's tendency to avoid the situation in the future.

Many phobias can be treated by exposure therapy: the individual is gradually encouraged to approach the feared object and to successively spend longer periods of time in proximity to it. Cognitive therapy is also used (often in conjunction with exposure therapy) to treat phobias. It involves helping individuals to recognize that their beliefs and thoughts can have a profound effect on their anxiety, that the outcome they fear will not necessarily occur, and that they have more control over the situation than they realize.

Medications are sometimes used to augment cognitive and exposure therapies. For example, beta-adrenergic blocking agents, such as propranolol, lower heart rate and reduce tremulousness, and lead to reduced anxiety. Certain kinds of antidepressants and anxiolytic medications are often helpful. It is not entirely clear how these medications exert their antiphobic effects, although it is believed that they affect levels of neurotransmitters in regions of the brain that are thought to be important in mediating emotions such as fear. *See* NEUROTIC DISORDERS. Murray B. Stein

Phoenicopteriformes The flamingos, a small monotypic order of wading birds that includes the family Phoenicopteridae, which has five species in one genus *Phoenicopterus* found worldwide in tropical to temperate marine and fresh waters; four species are found in South America, of which three live

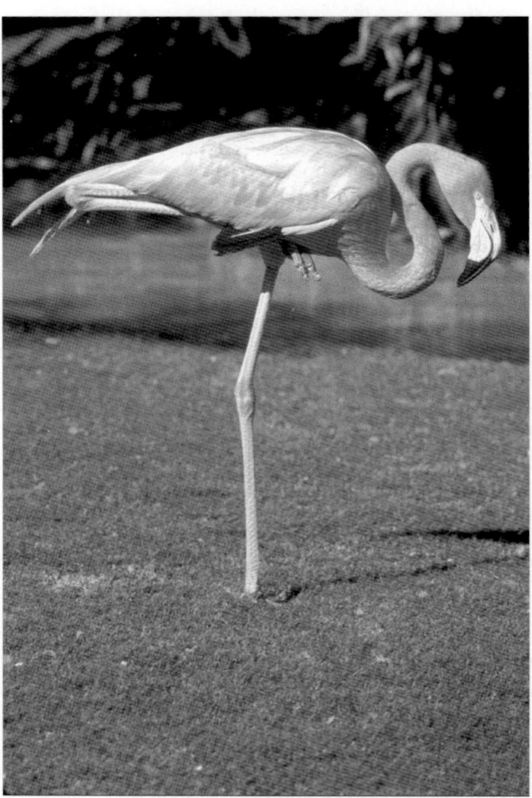

Greater flamingo (*Phoenicopterus ruber*). (*Photo by Dr. Lloyd Glenn Ingles; California Academy of Sciences*)

in the Andes. The flamingos were formally included in the Ciconiiformes and are still placed there by many researchers; others have advocated a close relationship with either the Anseriformes or the Charadriiformes. None of these classifications, however, are strongly supported by available evidence, and so it is best to place these bizarre birds in a unique order. *See* ANSERIFORMES; CHARADRIIFORMES; CICONIIFORMES.

The earliest definite flamingo fossil is *Juncitarsus*, from the middle Eocene of Wyoming; it possesses several primitive traits for the family. Beginning in the late Oligocene, modern flamingos are found in the fossil records from all areas of the world.

Flamingos (see illustration) are long-legged, long-necked wading birds with a thick bill which is sharply bent downward at the midpoint. The three anterior toes are webbed, possibly for walking on soft substrates. The long and broad wings enable the bird to fly well. The adult plumage is pink to light red, with black flight feathers. Flamingos are gregarious, often congregating in flocks in excess of 1 million in the rift valley lakes of Africa. They breed monogamously in large colonies, and both parents incubate the one or two eggs in a nest formed of a stout pillar of mud 1 ft (0.3 m) high on a mud flat. *See* AVES. Walter J. Bock

Pholidota An order of mammals comprising the living pangolins, or scaly anteaters, and their poorly known fossil predecessors. All living pangolins are assigned to the genus *Manis*. They are found in Africa south of the Sahara and in southeastern Asia, including certain islands of the East Indies.

Pangolins feed principally on termites and ants. The elongate tubular skull without teeth, long protrusive tongue, small eyes with heavy eyelids, thick skin, strong legs, five-toed feet with large claws, and large tail enable these unique animals to rip open ant nests and termite dens and devour the animals therein. The greatest peculiarity of animals in the genus *Manis* is a covering of all but the undersides of the body by an armor of large imbricating dermal horny scales. Living pangolins are frequently characterized as being animated pine cones. The position and number of hairs in relation to the scales are peculiar to each modern species. *See* MAMMALIA. Donald E. Savage

Phonetics The science that deals with the production, transmission, and perception of spoken language. At each level, phonetics overlaps with some other sciences, such as anatomy, physiology, acoustics, psychology, and linguistics. In each case, phonetics focuses on phenomena relevant to the study of spoken language.

Speech is normally produced by exhaling air from the lungs through the vocal tract. The vocal tract extends from the larynx through the pharynx and the oral cavity to the lips. If the velum (soft palate) is not raised, the air also passes through the nasal cavities. The shape and size of the oral cavity can be varied by the movement of active articulators: tongue, lips, and velum. *See* PALATE.

Phoneticians usually describe speech sounds with reference to their point (or place) of articulation and their manner of articulation. The point of articulation of a sound is the place of maximum constriction within the vocal tract. The great majority of sounds are produced by moving some part of the tongue toward some region on the roof of the mouth. Exceptions are articulations involving lips and those sounds in which the vocal folds serve as articulators.

At most of these points of articulation, sounds can be produced with several manners of articulation. One way to classify manners of articulation refers to the degree of stricture employed in producing the sound. Sounds produced with complete constriction of the vocal tract are stops, or plosives. If the closure is incomplete, but the articulators are brought close enough so that the air passing between them is set into turbulent motion, the resultant sounds are fricatives or spirants. If the articulators are approximated, but the constriction remains large enough so that air can pass through without friction, the sounds are called approximants—vowellike sounds functioning as consonants. Most of these consonant sounds can be voiced or voiceless; vowels are normally voiced. The terms "voiced" and "voiceless" refer to the presence and the absence of vocal fold vibration.

Acoustic phonetics deals with the manner in which the spoken message is encoded in the sound waves. According to the generally accepted source-filter theory of speech acoustics, sound is generated at a source (which for phonated speech is constituted by the vibrating vocal folds) and passed through the vocal tract. The opening and closing of the vocal folds create a succession of condensations and rarefactions of air molecules—variations in air pressure—and transform kinetic energy into acoustic energy. The sound wave generated at the glottis can be considered, for practical purposes, a complex periodic wave, and as such it contains energy at frequencies that are multiples of the fundamental frequency (harmonics).

The vocal tract acts as a filter, transmitting more energy at those frequencies that correspond to the resonances of the vocal tract than at other frequencies. Energy concentrations at the resonance frequencies of the vocal tract are referred to as formants.

In principle, the source and filter are independent of each other; consider the fact that the same vowel can be sung at different fundamental frequencies (pitches), and different vowels can be produced at the same pitch. The sound wave can be described by specifying its fundamental frequency, amplitude, and spectrum.

The subject matter of phonetics is not limited to the production and perception of vowels and consonants; of equal importance are such prosodic and suprasegmental aspects of spoken language as duration, fundamental frequency, and intensity, as they determine such linguistically relevant phenomena as tone and intonation, stress and emphasis, and the signaling of various boundaries—boundaries of morphemes and words, phrases, clauses, and sentences. *See* SPEECH. Ilse Lehiste

Phonolite A light-colored, aphanitic (not visibly crystalline) rock of volcanic origin, composed largely of alkali feldspar, feldspathoids (nepheline, leucite, sodalite), and smaller amounts of dark-colored (mafic) minerals (biotite, soda amphibole, and soda pyroxene). Phonolite is chemically the effusive equivalent of nepheline syenite and similar rocks. Rocks in which plagioclase (oligoclase or andesine) exceeds alkali feldspar are rare and may be called feldspathoidal latite. *See* FELDSPATHOID; MAGMA.

Phonolites are rare and highly variable rocks. They occur as volcanic flows and tuffs and as small intrusive bodies (dikes and sills). They are associated with trachytes and a wide variety of feldspathoidal rocks. *See* IGNEOUS ROCKS; TRACHYTE. Carleton A. Chapman

Phonon A quantum of vibrational energy in a solid or other elastic medium. This vibrational energy can be transported by elastic waves. The energy content of each wave is quantized. For a wave of frequency f, the energy is $(N + \frac{1}{2})hf$, where N is an integer and h is Planck's constant. Apart from the zero-point energy, $\frac{1}{2}hf$, there are N quanta of energy hf. In elastic or lattice waves, these quanta are called phonons. Quantization of energy is not related to the discreteness of the lattice, and also applies to waves in a continuum. *See* QUANTUM MECHANICS; WAVE MOTION.

The concept of phonons closely parallels that of photons, quanta of electromagnetic wave energy. The indirect consequences of quantization were established for phonons just as for photons in the early days of quantum mechanics—for example, the decrease of the specific heat of solids at low temperatures. Direct evidence that the energy of vibrational modes is changed one phonon at a time came much later than that for photons—for example, the photoelectric effect—because phonons exist only within a solid, are subject to strong attenuation and scattering, and have much lower quantum energy than optical or x-ray photons. *See* PHOTOEMISSION; SPECIFIC HEAT OF SOLIDS.

Like photons, phonons can be regarded as particles, each of energy *hf* and momentum proportional to the wave vector of the elastic or lattice wave. Such a particle can be said to transport energy, thus moving with a velocity equal to the group velocity of the underlying wave. *See* LATTICE VIBRATIONS; PHOTON.

Paul G. Klemens

Phonoreception The perception of sound by animals through specialized sense organs. A sense of hearing is possessed by animals belonging to two divisions of the animal kingdom: the vertebrates and the insects. The sense is mediated by the ear, a specialized organ for the reception of vibratory stimuli. Such an organ is found in all except the most primitive vertebrates, but only in some of the many species of insects. The vertebrate and insect types of ear differ in evolutionary origin and in their modes of operation, but both have attained high levels of performance in the reception and discrimination of sounds. *See* SOUND.

The vertebrate ear is a part of the labyrinth, located deep in the bone or cartilage of the head, one ear on either side of the brain. A complex assembly of tubes and chambers contains a membranous structure which bears within it a number of sensory endings of different kinds. Beginning with the amphibians, which are the earliest vertebrates to spend a considerable portion of their lives on land, there appears a special mechanism, the middle ear, whose function is the transmission of aerial vibrations to the sensory endings of the inner ear. All the vertebrates above the fishes, and certain of the fishes as well, have some type of sound-facilitative mechanism. *See* EAR (VERTEBRATE); HEARING (VERTEBRATE).

The group of invertebrates which has received the most attention has been the insects. Other arthropods, such as certain crustaceans and spiders, have also been found to be sensitive to sound waves. The insect ear consists of a superficial membrane of thin chitin with an associated group of sensilla called scolophores. These ears are found in most species of katydids, crickets, grasshoppers, cicadas, waterboatmen, mosquitoes, and nocturnal and spinner moths. The occur in different places in the body: on the antennae of mosquitoes, on the forelegs of katydids and crickets, on the metathorax of cicadas and waterboatmen, and on the abdomen of grasshoppers. Probably these differently situated organs represent separate evolutionary developments, through the association of a thinned-out region of the body wall with sensilla that are found extensively in the bodies of insects and that by themselves seem to serve for movement perception.

The insects mentioned above are noted for their production of stridulatory sounds made by rubbing the edges of the wings together, or a leg against a wing, or by other means. These sounds are produced by the males and serve for enticing the females in mating. A striking adaptation is that shown by mosquitoes: The ear of the male mosquito is sensitive only to a narrow range of frequencies around 380 Hz, and this frequency is the one which is produced by the wings of the female in flight. If the ear of the male mosquito is made nonfunctional, the mosquito fails to find a mate.

Ernest G. Wever

Phoresy A relationship between two different species of organisms in which the larger, or host, organism transports a smaller organism, the guest. It is regarded as a type of commensalism in which the relationship is limited to transportation of the guest.

Charles B. Curtin

Phoronida A small phylum of marine animals; in the recent past, they have been grouped with Brachiopoda and Bryozoa into the Lophophorata. The Phoronida phylum or class has no intermediate hierarchical level until the generic level. Two genera, *Phoronis* and *Phoronopsis*, are recognized with respectively seven and three well-defined species.

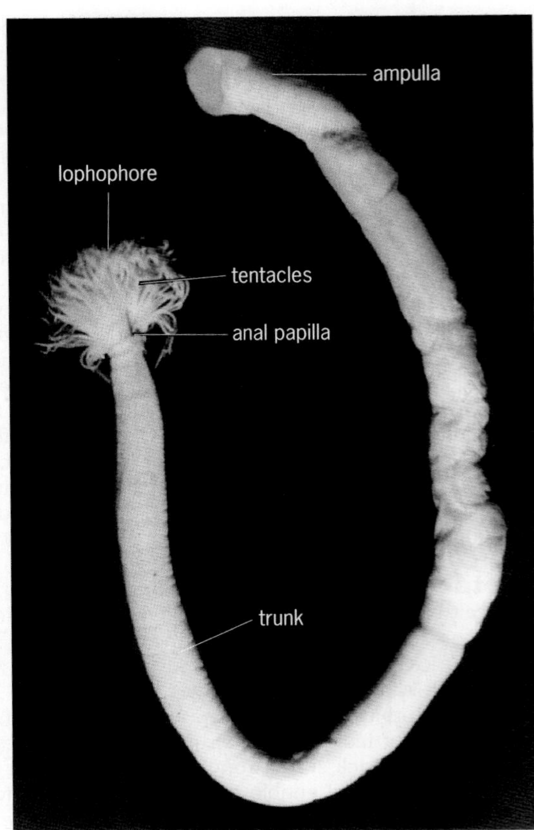

Phoronopsis harmeri **removed from its tube. Length is about 8 in. (20 cm).**

Phoronids may occur in vertical tubes embedded in soft sediments (sand, mud, or fine gravel) or form tangled masses of many individuals, buried in or encrusting limestone rocks and shells of dead mollusks. Phoronids secrete characteristic rigid tubes consisting of layers of chitin to which particles and debris adhere. Phoronids are found in all oceans (except the polar seas) at depths ranging from the intertidal zone to about 400 m (1300 ft).

The body is elongate, ranging in length from about 0.1 to 18 in. (2.5 mm to more than 45 cm), and bears a terminal, bilaterally symmetrical crown of tentacles (the lophophore) that surrounds the mouth (see illustration). The anus occurs at the level of the mouth and is borne on a papilla immediately outside the lophophore. The digestive tract is therefore U-shaped, with the mouth and anus opening close together at one end. The tentacles vary in number from 11 to over 1500 and have complex arrays of cilia that create a current which carries food particles to the mouth and absorbs oxygen for respiration. Associated with the mouth is a ciliated flap of tissue known as the epistome that occurs along the inner row of tentacles. *See* LOPHOPHORE.

Phoronids include both dioecious animals and hermaphrodites. Phoronids reproduce asexually by transverse fission or sexually by internal fertilization. Three types of developmental patterns occur: (1) brooding in the parental tube, (2) brooding on nidamental glands in the lophophore until the larval stage is reached, (3) shedding fertilized embryos into the sea water that grow into a characteristic ciliated, free-swimming pelagic larva, the *Actinotrocha* or actinotroch.

Christian C. Emig; Joan R. Marsden

Phosphatase One of a large group of enzymes, having various chemical structures, that catalyze the hydrolysis of phosphoric acid esters or anhydrides. The phosphatases are a complex group, some of them having highly specific functions. In general, phosphatases promote the degradation of organic phosphorus compounds into inorganic phosphate and an organic moiety. In addition, phosphatases readily bind phosphates and transport them to cells. *See* ENZYME; PHOSPHORUS.

Phosphatases may be intracellular or extracellular. Intracellular phosphatases are found in blood plasma, cartilage, intestinal mucosa, spleen, liver, kidney, and plant tissues; they are found in the cytoplasm or periplasm, or attached to some cells. Intracellular phosphatase production of inorganic phosphates from organic phosphates is crucial to cellular metabolism and the growth and development of organisms. Soil and water contain a large amount of extracellular phosphatases, in either free or immobilized forms, which are important in the biogeochemical cycling of phosphorus. *See* BIOGEOCHEMISTRY; CELL (BIOLOGY).

Phosphatases are adaptive enzymes, so their production is stimulated by a deficiency in inorganic phosphorus. In contrast, the presence of inorganic phosphates inhibits both phosphatase activity and synthesis in various tissues. In soils and aquatic systems, production of extracellular phosphatases is stimulated by the presence of organic matter containing phosphorus. The activity of extracellular phosphatases is maximum in plant root zones. Bacteria, fungi, yeast, soil invertebrates, mycorrhizae, and all plants produce extracellular phosphatases. *See* BACTERIA; FUNGI; MYCORRHIZAE; RHIZOSPHERE; ROOT (BOTANY); SOIL. H. K. Pant

Phosphate metabolism Organic phosphate compounds are present in the structural units of every animal cell, and inorganic phosphate is associated with calcium in bone and teeth. The total phosphorus in the adult human body is about 12 g/kg, with only 1.4 g/kg present in the soft tissues and the remainder in mineralized tissue in the form of apatite crystals. Blood phosphate plays an important role in regulating neutrality, and it is in equilibrium with both bone and cellular organic phosphates. The blood level is held relatively constant by regulating phosphate excretion by the kidney. This control is primarily mediated by action of parathyroid hormone. Vitamin D enhances the entry of phosphate into bone. Phosphate plays an important role in absorption of sugars from the intestine and reabsorption of glucose from the kidney. *See* PARATHYROID HORMONE; VITAMIN D.

The central role of phosphates in life processes is indicated by their occurrence in ribonucleic acid (RNA) and deoxyribonucleic acid (DNA). Through the formation of lecithins, phosphates are involved in fat metabolism. Phosphates play a major role in the conservation and transfer of energy, particularly of the energy produced in the tricarboxylic acid cycle (Krebs cycle), in glycolysis, and in the pentose shunt. They do so by participating in many phosphorylation and transphosphorylation reactions involving sugars and other organic compounds. *See* CARBOHYDRATE METABOLISM; CHROMOSOME; CITRIC ACID CYCLE; LIPID METABOLISM; NUCLEIC ACID.

In phosphorylation reactions, compounds such as phosphocreatine (PC) and adenosine triphosphate (ATP) are formed which are capable of yielding relatively large amounts of free energy (5000–11,000 cal or 21–46 kilojoules per mole) when the phosphate bonds are broken by hydrolysis. ATP and PC have a central role in energy storage and transfer in all tissues.

Phosphorus-containing coenzyme systems include the pyridine (nicotinamide) and the riboflavin nucleotide systems concerned with oxidation-reduction reactions; coenzyme A, the functional form of pantothenic acid, concerned with transacetylation, acylation, and condensation reactions; the diphosphothiamine system concerned with decarboxylation; and pyridoxal phosphate concerned with transamination. *See* BIOCHEMISTRY; BIOLOGICAL OXIDATION; COENZYME; ENERGY METABOLISM. Morton K. Schwartz

Phosphate minerals Any naturally occurring inorganic salts of phosphoric acid, $H_3[PO_4]$. All known phosphate minerals are orthophosphates. There are over 150 species of phosphate minerals, and their crystal chemistry is often very complicated. Phosphate mineral paragenesis can be divided into three categories: primary phosphates (crystallized directly from a melt or fluid), secondary phosphates (derived from the primary phosphates by hydrothermal activity), and rock phosphates (derived from the action of water upon buried bone material, skeletons of small organisms, and so forth). *See* MINERAL; PHOSPHATE. Paul B. Moore

Phospholipid A lipid that contains one or more phosphate groups. Like fatty acids, phospholipids are amphipathic in nature; that is, each molecule consists of a hydrophilic (having strong affinity for water) portion and a hydrophobic (lacking affinity for water) portion. Due to the amphipathic nature and insolubility in water, phospholipids are ideal compounds for forming biological membranes. Medically, phospholipids are used to form therapeutic liposomes—microscopic spheres enclosed by phospholipid membranes and containing specific drugs for delivery into the body. The phospholipid sphere fuses with the plasma membrane of the target cell, and the drug inside the liposome is emptied into the cell. Some liposomes are preferentially recognized by certain organs in the body, thus facilitating the targeted delivery of drugs. *See* LIPID; LIPOSOMES.

There are two classes of phospholipids: those that have a glycerol backbone and those that contain sphingosine (synthesized from serine and palmitic acid). Both classes are present in the biological membrane. Phospholipids that contain a glycerol backbone are called phosphoglycerides (or glycerophospholipids), the most abundant class of phospholipids found in nature. Phosphatidylcholine (lecithin) is the most abundant type of phosphoglyceride in eukaryotes, whereas phosphatidylethanolamine is the major phosphoglyceride found in bacteria. Other naturally occuring phosphoglycerides include phosphatidylserine, phosphatidylinositol, phosphatidylglycerol, and cardiolipin. The structural diversity within each type of phosphoglyceride is due to variability in chain length and degree of saturation of the fatty acid ester groups.

Sphingomyelin is the major sphingosine-containing phospholipid in eukaryotic cells. Its general structure consists of a fatty acid attached to sphingosine by an amide linkage, producing a ceramide.

The biological membrane is a barrier that separates the inside of a cell from its outside environment. It also serves as a barrier to separate cellular contents into different compartments. A bilayer membrane is formed spontaneously when phospholipids are dispersed in an aqueous solution. In this bilayer structure, phospholipids are arranged in two leaflets with the hydrophobic tails facing each other, and the hydrophilic ends exposed to the aqueous medium. The shape of the bilayer is determined by the asymmetrical distribution of phospholipid types, the chain length of each phospholipid, as well as the degree of saturation of the acyl groups in the phospholipid. Individual phospholipid molecules are able to move freely in the lateral plane of the bilayer but not in the transverse plane (flip-flop). Small uncharged molecules are able to diffuse through the bilayer structure, but the permeability of larger or charged molecules is restricted. The arrangement of phospholipid molecules into a bilayer in an aqueous medium follows the laws of thermodynamics and represents the structural basis for the formation of all biological membranes. Selected protein molecules, which are responsible for other membrane functions such as providing channels for large or polar molecules to enter or exit the cell, are embedded in the phospholipid bilayer. *See* CELL MEMBRANES.

For a long time, phospholipids were regarded as merely building blocks for the biological membrane. It was discovered in the mid-1970s, however, that phospholipids participate extensively in the transduction of biological signals across the membrane. For example, a special form of phosphoglyceride, 1-alkyl-2-acetyl-glycero-3-phosphocholine, acts as a very powerful signaling molecule. It causes the aggregation and degranulation of blood platelets, and is known as platelet-activating factor (PAF).

The production of signal molecules is largely through the action of phospholipases. Phospholipases are enzymes responsible for the degradation (via hydrolysis) of phosphoglycerides. The most common type is phospholipase A_2 (PLA_2), which is found in all tissues, in pancreatic juice, as well as in snake and bee venoms. Patrick Choy

Phosphorescence A delayed luminescence, that is, a luminescence that persists after removal of the exciting source. It is sometimes called afterglow.

This original definition is rather imprecise, because the properties of the detector used will determine whether or not there is an observable persistence. There is no generally accepted rigorous definition or uniform usage of the term phosphorescence. In the literature of inorganic luminescent systems, some authors define phosphorescence as delayed luminescence whose persistence time decreases with increasing temperature. According to this usage, luminescence whose persistence time is independent of temperature is called fluorescence regardless of the length of the afterglow; a temperature-independent afterglow of long duration is called simply a slow fluorescence, which implies that the atomic or molecular transition involved is forbidden to a greater or lesser degree by the spectroscopic selection rules. The most common mechanism of phosphorescence in photoconductive inorganic systems, however, occurs when electrons or holes, set free by the excitation process and trapped at lattice defects, are expelled from their traps by the thermal energy in the system and recombine with oppositely charged carriers with the emission of light. *See* HOLE STATES IN SOLIDS; SELECTION RULES (PHYSICS).

In the organic literature the term phosphorescence is reserved for the forbidden luminescent transition from a metastable energy state M to the ground state G, while the afterglow corresponding to the M→ E→ G process (where E is a higher energy state) is called delayed fluorescence. *See* FLUORESCENCE; LIGHT; LUMINESCENCE. Clifford G. Klick; James H. Schulman

Phosphorus A chemical element, P, atomic number 15, atomic weight 30.9738. Phosphorus forms the basis of a very large number of compounds, the most important class of which are the phosphates. For every form of life, phosphates play an essential role in all energy-transfer processes such as metabolism, photosynthesis, nerve function, and muscle action. The nucleic acids which among other things make up the hereditary material (the chromosomes) are phosphates, as are a number of coenzymes. Animal skeletons consist of a calcium phosphate. *See* PERIODIC TABLE; PHOSPHATE.

About three-quarters of the total phosphorus (in all of its chemical forms) used in the United States goes into fertilizers. Other important uses are as builders for detergents, nutrient supplements for animal feeds, water softeners, additives for foods and Pharmaceuticals, coating agents for metal-surface treatment, additives in metallurgy, plasticizers, insecticides, and additives for petroleum products. *See* DETERGENT; WATER SOFTENING.

Of the nearly 200 different phosphate minerals, only one, fluorapatite, $Ca_5F(PO_4)_3$, is mined chiefly from large secondary deposits originating from the bones of dead creatures deposited on the bottom of prehistoric seas and from bird droppings on ancient rookeries. *See* PHOSPHATE MINERALS.

Research in phosphorus chemistry indicates that there may be as many compounds based on phosphorus as on carbon. In organic chemistry it has been customary to group the various chemical compounds based on carbon into families which are called homologous series. This can also be done in the chemistry of phosphorus compounds, even though many phosphorus-based families are incomplete. The best known of the families of compounds based on phosphorus is the group of chain phosphates. Phosphate salts consist of cations, such as sodium, along with chain anions, such as $(P_nO_{3n+1})^{(n+2)-}$, which may have 1–1,000,000 phosphorus atoms per anion.

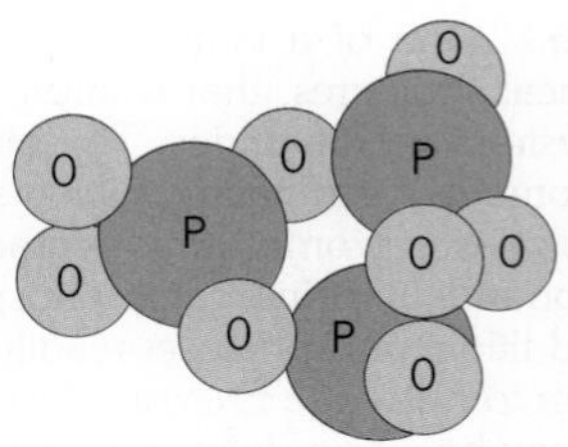

Fig. 1. Ring phosphate anion, $(P_3O_9)^{3-}$.

The phosphates are based on phosphorus atoms tetrahedrally surrounded by oxygen atoms, with the lowest member of the series being the simple PO_4^{3-} anion (the orthophosphate ion). The family of chain phosphates is based on a row of alternating phosphorus and oxygen atoms in which each phosphorus atom remains in the center of a tetrahedron of four oxygen atoms. There is also a closely related family of ring phosphates, a member of which, the trimetaphosphate, is shown in Fig. 1.

An interesting structural characteristic of many known phosphorus compounds is the formation of cagelike structures. Such cagelike molecules are exemplified by white phosphorus, P_4, and one of the phosphorus pentoxides, P_4O_{10} (Fig. 2). Network structures are also common; for example, black phosphorus crystals in which the atoms are bonded together in the form of vast, corrugated planes (Fig. 3).

In the majority of its compounds, phosphorus is chemically bonded to four neighboring atoms. There is a large number of compounds in which one of the four neighboring atoms is absent, and in which its place is taken by an unshared pair of electrons. There are also a few compounds in which there are five or six neighboring atoms bonded to the phosphorus. These compounds are very reactive and tend to be unstable.

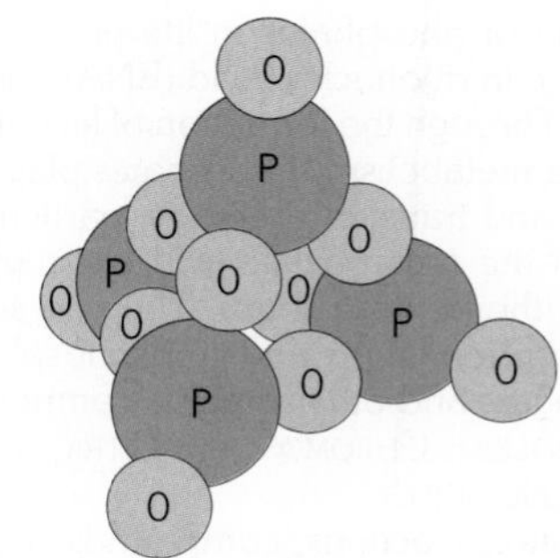

Fig. 2. Phosphorus pentoxide, P_4O_{10}, in vapor state.

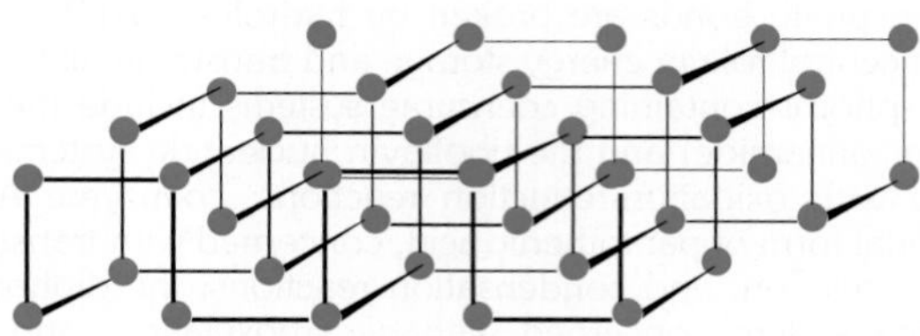

Fig. 3. Black phosphorus, P_n.

During the 1960s and 1970s a large number of organic-phosphorus compounds were prepared. Most of these chemical structures involve three or four neighboring atoms bonded to the phosphorus, but stable structures having two, five, or six neighboring atoms per phosphorus are also known. *See* ORGANOPHOSPHORUS COMPOUND; PHOSPHATE.

Essentially all of the phosphorus used in commerce is in the form of phosphates. The majority of phosphatic fertilizers consist of highly impure monocalcium or dicalcium orthophosphate, $Ca(H_2PO_4)_2$ and $CaHPO_4$. These phosphates are salts of orthophosphoric acid. *See* FERTILIZER.

The phosphorus compound of major biological importance is adenosine triphosphate (ATP), which is an ester of sodium tripolyphosphate, widely employed in detergents and water-softening compounds. Practically every reaction in metabolism and photosynthesis involves the hydrolysis of this tripolyphosphate to its pyrophosphate derivative, called adenosine diphosphate (ADP). John R. Van Wazer

Photoacoustic spectroscopy A technique for measuring small absorption coefficients in gaseous and condensed media, involving the sensing of optical absorption by detection of sound. It is frequently called optoacoustic spectroscopy.

During the transmission of optical radiation through a sample (gas, liquid, or solid), the absorption of radiation by the sample can be measured by several techniques. The straightforward detection technique requires a measurement of the optical radiation level with and without the sample in the optical path. The transmitted power P_{out} and the incident power P_{in} are related through the equation below, where α is the absorption

$$P_{out} = P_{in}e^{-\alpha l}$$

coefficient and l is the length of the absorber. With this technique, the minimum measurable value of αl is of the order of 10^{-4} unless special precautions have been taken to stabilize the source of radiation.

Optoacoustic detection is a calorimetric method where no direct detection of optical radiation is carried out but, instead, a measurement is made of the power absorbed by the medium from the incident radiation. The optoacoustic signal is proportional to the incident power and the absorption-length product αl. Thus, for given sources of noise from the detection transducers, the signal-to-noise ratio improves as the incident energy is increased.

If optical radiation is amplitude-modulated at an audio frequency, the absorption of such radiation by a gaseous medium that has been confined in a cell with appropriate optical windows for the entrance and exit of the radiation, and nonradiative relaxation of the medium, will cause a periodic variation in the temperature of the column of the irradiated gas. Such a periodic rise and fall in temperature gives rise to a corresponding periodic variation in the gas pressure at the audio frequency. The audio-frequency pressure fluctuations (that is, sound) are efficiently detected using a sensitive gas-phase microphone.

The capability of measuring extremely small absorption coefficients and correspondingly small concentrations of the absorption gases has many applications, including high-resolution spectroscopy of isotopically substituted gases, excited states of molecules and forbidden transitions, and pollution detection. The pollution measurements have demonstrated that the optoacoustic spectroscopy technique in conjunction with tunable lasers can be routinely used for on-line real-time in-place detection of undesirable gaseous constituents at sub-parts-per-billion levels. *See* LASER; LASER SPECTROSCOPY.

A very sensitive calorimetric spectroscopic technique has been developed for the study of weak absorption in liquids and solids. This technique uses a pulsed tunable laser for excitation and a submerged piezoelectric transducer, in the case of a liquid, or a contacted piezoelectric transducer, in the case of a solid, for the detection of the ultrasonic signal generated due to the absorption of the radiation and its subsequent conversion into a transient ultrasonic signal (see illustration). Because of the capability of measuring very small fractional absorptions, the technique is clearly applicable to the area of monitoring water pollution, impurity detection in thin semiconductor wafers, transmission studies of ultrapure glasses (used in optical fibers for optical communications), and so forth. *See* ABSORPTION. C. K. N. Patel

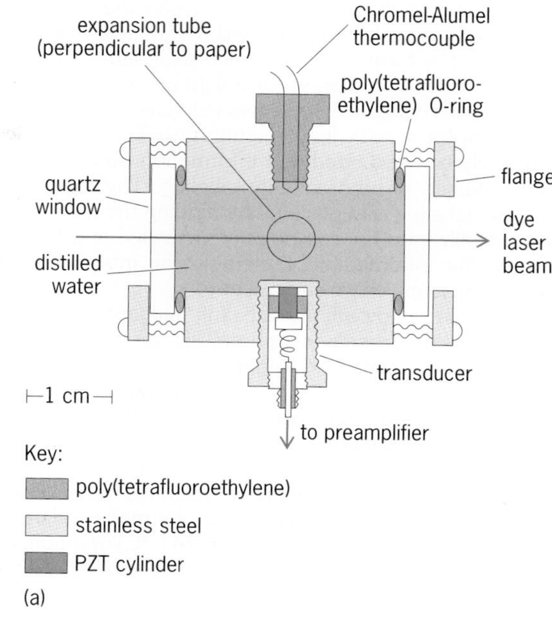

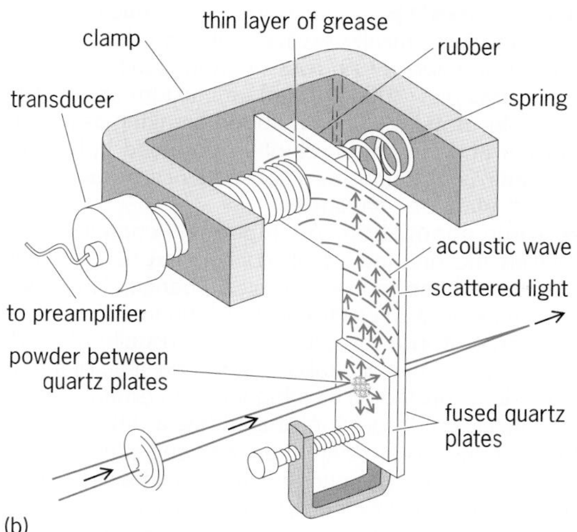

Arrangement for pulsed-laser (*a*) immersed and (*b*) contacted piezoelectric transducer optoacoustic spectroscopy.

Photoaffinity labeling A process by which a macromolecule can be labeled at or near its binding or active site. The method can be applied to proteins, nucleic acids, and lipids by chemists who need to identify the parts of these molecules that are significant for particular biological functions. X-ray crystallography can frequently determine the complete structure of proteins and sometimes of nucleic acids, although the method does not necessarily highlight the reactive groups in a biomolecule. Nuclear magnetic resonance spectroscopy is increasingly useful in illuminating such structures. *See* NUCLEAR QUADRUPOLE RESONANCE; X-RAY CRYSTALLOGRAPHY.

Many important biological processes involve the formation of a complex between a biopolymer and a chemical reagent; for example, enzymes form complexes with their substrates on the way to catalysis, and antibodies form tight complexes with their antigens. Many biochemical receptors (such as hormone receptors) are complexed with lipid membranes. These complexes cannot usually be crystallized, but information about them can often be obtained by photoaffinity labeling. The idea behind photoaffinity labeling is to place in the binding site a compound that is essentially inert but can be photoactivated at will to yield a highly reactive intermediate. Preferably, this intermediate will react with almost any chemical structure.

The successful photoaffinity labeling reagents include diazirines that yield carbenes on photolysis, arylazides that are photochemically decomposed to nitrenes (highly reactive compounds that contain univalent nitrogen), and ketones that yield free radicals on irradiation. Special structural features (such as trifluromethyl substituents) can enhance the activity of the reagents.

In order to identify the products of reactions, many of which occur only in low yield, the reagent must be made radioactive. This is usually accomplished with carbon-14 (^{14}C) or tritium (^{3}H), radioactive isotopes of relatively low energy that nevertheless can easily be traced.

F. H. Westheimer

Photochemistry The study of chemical reactions of molecules in electronically excited states produced by the absorption of infrared (700–1000 nanometers), visible (400–700 nm), ultraviolet (200–400 nm), or vacuum ultraviolet (100–200 nm) light. Bond making and bond breaking as well as electron transfer and ionization are often observed in both organic and inorganic compounds as a consequence of such excitation. *See* ELECTROMAGNETIC RADIATION.

Electronic absorption. An important generalization sometimes called the first law of photochemistry is that only light that is absorbed can induce chemical change. The absorption of a photon induces an electronic transition in which an electron originally present in a molecular orbital, usually a bonding or nonbonding molecular orbital of the ground state of the absorbing molecule, is promoted to a higher-lying orbital. The excited state produced by absorption of light has a different electronic structure than its ground-state precursor and can reasonably be regarded as an isomeric species with distinct and characteristic chemical and physical properties. *See* PHOTON.

Most organic molecules exist as ground-state singlets in which all electrons are paired. Because photoexcitation causes the promotion of only a single electron, two singly occupied orbitals are produced upon excitation. If the electronic transition takes place without a spin inversion, these two electrons have opposite spins, and a singlet excited state is produced. The number of unpaired electrons in a molecule determines its multiplicity: a molecule with no unpaired spins is a singlet; one with one unpaired spin is a doublet; one with two unpaired spins is a triplet; one with three unpaired spins is a quartet, and so forth. If an electronic transition were to take place with a spin inversion, the two singly occupied orbitals would be populated by electrons with parallel spins, producing a triplet excited state. Spin restrictions forbid spin inversion during excitation, and only singlet-singlet electronic transitions are easily observed spectroscopically. After excitation, however, a change in state multiplicity can take place by a process called intersystem crossing. The facility of intersystem crossing is influenced by the magnitude of spin-orbital coupling, which can be enhanced by the presence of a heavy atom (an atom in the third row or below of the periodic table), either bound to the absorbing molecule or present externally as solvent. *See* PERIODIC TABLE; TRIPLET STATE.

Transitions. A chromophore is that part of the molecule that accounts for its absorption of light and its photochemical activity. The absorption corresponding to a particular chromophore depends on the type of transition involved in that particular excitation. The promotion of an electron from a π-bonding molecular orbital to a π-antibonding orbital is referred to as a π,π^* (read pi to pi star) transition. Such transitions are frequently encountered in alkenes, alkynes, aromatic molecules, and other unsaturated compounds. Because the spatial overlap of π and π^* orbitals is substantial, such a transition typically has high oscillator strength and a large extinction coefficient (absorptivity). Promotion of an electron from a nonbonding molecular orbital to a π-antibonding orbital, referred to as an n,π^* transition, involves orbitals that are nearly orthogonal; and it takes place only inefficiently; that is, it has a low oscillator strength and a small extinction coefficient. Such transitions are often encountered in compounds containing carbon-heteroatom or heteroatom-heteroatom double bonds. Because nonbonding molecular orbitals lie at higher energy than bonding ones, n,π^* transitions are of lower energy than the corresponding π,π^* transitions. Both n,π^* and π,π^* transitions are usually found in the ultraviolet region of the electromagnetic spectrum. Transitions involving sigma (σ) bonds (for example $n\sigma^*$ transitions in amines, alcohols, ethers, and alkyl halides and σ,σ^* transitions in alkanes) are usually encountered at the high-energy end of the ultraviolet spectrum or in the vacuum ultraviolet region. *See* CHEMICAL BONDING; ULTRAVIOLET RADIATION.

Each allowed transition of a compound registers as a band in the absorption spectrum, with the intensity of the transition (measured by its extinction coefficient) being governed by the operative selection rules. The transition intensity of a given absorption is measured by integrating over the whole absorption band. The resulting integrated absorption coefficient is directly proportional to the oscillator strength of the transition. The oscillator strength, a measure of the allowedness of an electric dipole transition compared to that of a free electron oscillating in the three dimensions, is directly related to an experimentally measured value, the extinction coefficient (ε). Beer's law is given by Eq. (1), where A is the observed absorbance, ε is the extinction

$$A = \varepsilon bc \tag{1}$$

coefficient, b is the path length (in centimeters) of the cell used for the measurement, and c is the molar concentration of the absorbing species. This law is used to correlate the observed absorbance with the extinction coefficient and concentration of the absorbing species. *See* ABSORPTION OF ELECTROMAGNETIC RADIATION.

Photophysics. The excited state produced by absorption of a photon is not generally a stable species. After a characteristic lifetime that can vary from femtoseconds (10^{-15} s) to hours, the excited molecule will either relax to its ground-state precursor or undergo a chemical transformation. The term photophysics is used to describe nonreactive relaxation processes, which include radiative (taking place with the emission of light) and nonradiative (taking place without the emission of light) pathways.

The energies of the lowest singlet and triplet excited states (relative to the ground state) can be obtained from the longest wavelength band of the fluorescence and phosphorescence spectra, respectively. This band is called a 0,0 band to indicate a transition between the lowest vibrational levels of the lowest-lying states. Singlet and triplet energies can also be determined indirectly by measuring quenching efficiencies. The shift between the 0,0 bands for absorption and emission in a single molecule is called its Stokes shift. A small Stokes shift is usually observed when the excited state has a geometry similar to the ground state. A Jablonski diagram (see illustration) is often used to graphically depict the relationship between competing photophysical processes.

Quantum yield, or quantum efficiency, is defined as the number of molecules participating in a given photophysical process or reaction divided by the number of photons absorbed. The

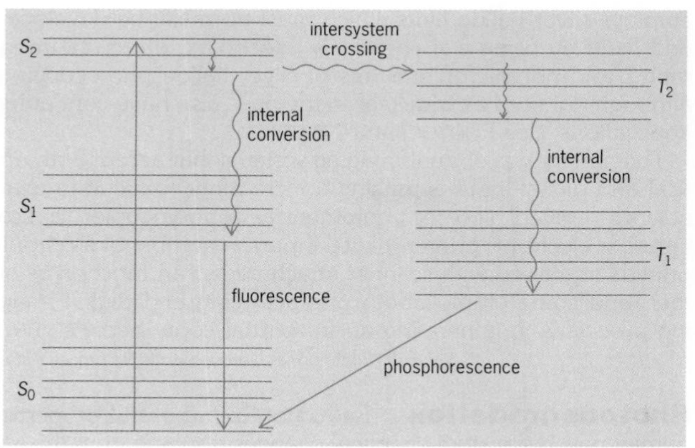

Jablonski diagram. Solid arrows represent radiative processes; and wavy arrows nonradiative processes. S terms = singlet states; T terms = triplet states.

quantum yield ranges between zero and one for photoreactions induced by a single photon; values larger than one are indicative of a chain process in which product is formed in a repeating, dark cycle initiated by the photoexcitation. For a photochemical reaction, the number of molecules participating in the reaction is determined spectroscopically or chromatographically as a chemical yield per volume unit per time. The number of photons absorbed is obtained by measuring with a radiometer the light flux per volume unit per time or by employing a chemical actinometer, a known chemical reaction for which the quantum yield is known and accepted as a standard. *See* QUANTUM CHEMISTRY.

Energy transfer. The process by which an excited state molecule, M*, in an excited singlet or triplet state transfers all or part of its excitation energy to a reaction partner or quencher, Q, is called energy transfer or quenching when the molecule of interest is M [reaction (2)]. This same process is called sensitization when the molecule of interest is Q. In the latter case, M is called the sensitizer. Energy transfer permits an exception to the first law of photochemistry in that Q* is produced without having absorbed the incident light.

$$M^* + Q \rightarrow M + Q^* \quad (2)$$

For energy transfer to take place, an incident wavelength must be chosen so that M is primarily excited, producing an excited state M* whose energy lies above that of Q*. Symmetry selection rules require that all energy transfer events preserve spin multiplicity. Thus, if M* is an excited singlet and Q is a ground-state singlet, M will be produced as a ground-state singlet and Q* as an excited singlet. If M* is an excited triplet and Q is a ground-state singlet, M will be produced as a ground-state singlet and Q* as an excited triplet.

Photochemical mechanisms. As in all studies of mechanisms of chemical reactions, determining the structure of all products is the first step in the specification of a photochemical reaction. Spectroscopic (nuclear magnetic resonance spec- troscopy, electron spin resonance spectroscopy, infrared spectroscopy, mass spectroscopy, x-ray analysis, absorption spectroscopy) and chromatographic (gas, liquid, or thin-layer chromatography) techniques are used to establish product structure and to determine product yields. Monitoring the effect of solvent polarity on reaction rate, the retention or loss of optical activity during the reaction, the positions of isotopic labels, and the success of intermediate trapping experiments can distinguish step-wise chemical reactions (those that proceed through one or more intermediates) from concerted reactions (those that proceed without intermediates). In addition to these mechanistic approaches, the identity and lifetime of the reactive excited state (singlet, triplet, and so forth) and the quantum yields for both product formation and for other competing photophysical processes are required for a full photochemical mechanistic characterization. Time-resolved flash photolysis and pulse radiolysis measurements can, in addition, be used sometimes for direct spectroscopic detection of absorptive or emissive intermediates encountered in a photochemical mechanism, as well as for their kinetic characterization. The addition of specific reactive quenchers or traps, conducting a photoreaction in a low-temperature matrix in which diffusion processes are stopped, and sensitization experiments are effective means for assigning the observed transient absorptions or emissions. The energetics of a well-defined photochemical reaction can be obtained by photoacoustic calorimetry measurements. *See* CHROMATOGRAPHY; MATRIX ISOLATION; PHOTOLYSIS; SPECTROSCOPY. Marye Anne Fox

Photoclinometer

A term applied to directional surveying instruments which record photographically the direction and magnitude of well deviations from the vertical. Two instruments of this type are in wide use, the Schlumberger photoclinometer and the Surwell clinograph. Both instruments record a series of deviation measurements on one trip into and out of the well. From this series of data it is possible to plot quite accurately the course of the well.

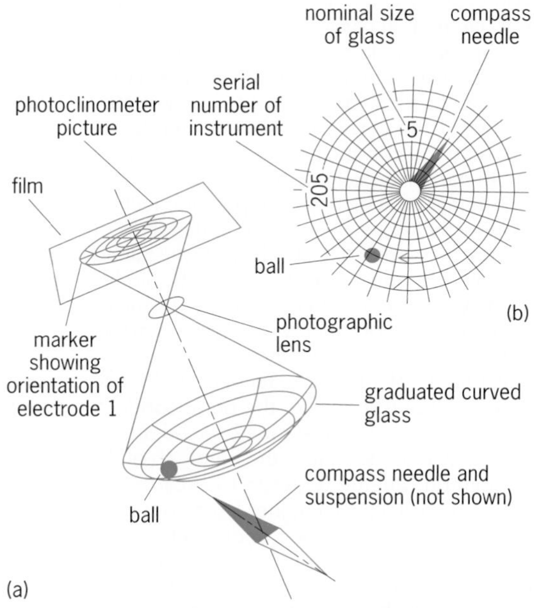

Diagram of a Schlumberger photoclinometer. (*a*) Principal features. (*b*) Type of record obtained. (*Schlumberger Well Surveying Corp.*)

In the Schlumberger photoclinometer (see illustration) the deviation from the vertical is indicated by a small metal ball which rolls in a transparent glass bowl graduated in circular degrees. The direction of the deviation in azimuth is indicated by a magnetic compass. With the instrument suspended by an electrical cable, the positions of the compass and steel ball are photographed on a 35-mm film by operation of electrical controls at the surface. Correlation of the pictures with the depths at which they are taken yields a measure of the magnitude and direction of deviation of the hole as a function of depth.

The Surwell clinograph also operates electrically but is powered by batteries contained in the instrument. The deviation from the vertical is indicated by a box level gage and the direction in azimuth by a gyroscopic compass, permitting its use inside steel pipe. *See* SURVEYING. Holbrook G. Botset

Photoconductive cell

A device for detecting electromagnetic radiation (photons) by variation of the electrical conductivity of a substance (a photoconductor) upon absorption of the radiation by this substance. During operation the cell is connected in series with an electrical source and current-sensitive meter, or in series with an electrical source and resistor. Current in the cell, as indicated by the meter, is a measure of the photon intensity, as is the voltage drop across the series resistor. Photoconductive cells are made from a variety of semiconducting materials in the single-crystal or polycrystalline form. *See* PHOTOCONDUCTIVITY; PHOTOELECTRIC DEVICES. Sebastian R. Borrello

Photoconductivity The increase in electrical conductivity caused by the excitation of additional free charge carriers by light of sufficiently high energy in semiconductors and insulators. Effectively a radiation-controlled electrical resistance, a photoconductor can be used for a variety of light- and particle-detection applications, as well as a light-controlled switch. Other major applications in which photoconductivity plays a central role are television cameras (vidicons), normal silver halide emulsion photography, and the very large field of electrophotographic reproduction. *See* OPTICAL DETECTORS; OPTICAL MODULATORS; PARTICLE DETECTOR; PHOTOGRAPHY; TELEVISION CAMERA TUBE.

Although all insulators and semiconductors may be said to be photoconductive, that is, they show some increase in electrical conductivity when illuminated by light of sufficiently high energy to create free carriers, only a few materials show a large enough change, that is, show a large enough photosensitivity, to be practically useful in applications of photoconductors.

Since the electrical conductivity σ of a material is given by the product of the carrier density, its charge, and its mobility, an increase in the conductivity can be formally due to either an increase in carrier density or an increase in mobility. Although cases are found in which both types of effects are observable, photoconductivity in single-crystal materials is due primarily to an increase in earner density. In polycrystalline materials, on the other hand, where transport may be limited by potential barriers between the crystalline grains, an increase in mobility due to photoexcitation effects on these intergrain barriers may dominate the photoconductivity.

The variation of photoconductivity with photon energy is called the spectral response of the photoconductor. Spectral response curves typically show a fairly well-defined maximum at a photon energy close to that of the bandgap of the material, that is, the minimum energy required to excite an electron from a bond in the material into a higher-lying conduction band where it is free to contribute to the conductivity. This energy ranges from 3.7 eV, in the ultraviolet, for zinc sulfide (ZnS) to 0.2 eV, in the infrared, for cooled lead selenide (PbSe).

Another major characteristic of a photoconductor of practical concern is the rate at which the conductivity changes with changes in photoexcitation intensity. If a steady photoexcitation is turned off at some time, for example, the length of time required for the current to decrease to $1/e$ of its initial value is called the decay time of photoconductivity, t_d. The magnitude of the decay time is determined by the lifetime π and by the density of carriers trapped in imperfections as a result of the previous photoexcitation, which must now also be released in order to return to the thermal equilibrium situation. *See* PHOTOCONDUCTIVE CELL. Richard H. Bube

Photocopying processes Processes that use light to generate copies directly from original paper documents. Light is employed to examine an original document and to detect the presence or absence of an image. In most cases, light reflected from the original subject directly exposes the medium used to produce an image on a copy. However, in an increasing number of cases, the light is converted into an electrical signal, which is later converted back to light to expose the copying medium.

Xerography, the most popular of the photocopying processes, relies on photoconductors and toner powders to create copies. The toner is fused to copy paper by heat or pressure. Microfilming employs silver halide films which must normally be developed and fixed by using wet chemicals. Electrofax, diazo, thermography, and nonmicrofilm forms of silver halide photocopying, although not used as widely as in the past, also have continuing applications. *See* PHOTOGRAPHIC MATERIALS.

The emergence of dual-imaging systems that accept both optical and digital input is making it increasingly difficult to characterize a given piece of equipment as a photocopier (optical input) or electronic printer (digital input). A number of electronic printers equipped with scanner attachments can function as either, and some employ both photocopying and digital imaging processes in generating an individual copy. *See* PRINTING.
Robert I. Edelman; J. Gordon Jarvis; Thomas Destree

Photodegradation Reduction in the useful properties of materials because of chemical changes resulting from the absorption of light. The chemical changes can include bond scission (especially of the molecular backbone), color formation, cross-linking, and chemical rearrangements. All organic materials can photodegrade, but the process has greatest practical relevance for polymers where scission of the polymer backbone is particularly important. Photodegradations of polymers in the absence of oxygen (photolysis) or using wavelengths shorter (more energetic) than those at the Earth's surface (<280 nanometers) have been studied extensively, but only the more practical situation of polymers exposed to terrestrial sunlight (or its equivalent) in air is discussed in this article.

Although all organic polymers can be degraded by light, the rate of degradation varies enormously from polymer to polymer, and is also dependent on the incident wavelengths. Light containing ultraviolet (UV; shorter-wavelength) components is much more destructive than visible light, so that polymers exposed indoors, behind window glass (transmitting >330 nm), will degrade much more slowly than samples exposed outdoors.

For many aromatic polymers, such as polyester and the aramids, in which the polymer itself is the chromophore (light-absorbing group), backbone scission results predominantly from this direct absorption of light energy. For many other polymers, including polyolefins, and polyvinyl chloride where only impurities absorb energy from sunlight, scission of a chemical bond by light to give free radicals is followed by reaction of these highly reactive free radicals with atmospheric oxygen. *See* FREE RADICAL.

Although numerous organic materials will undergo photodegradation, hydrocarbon polymers are particularly vulnerable because their useful properties depend entirely on their high molecular weights, in the tens or hundreds of thousands. Anything that reduces the molecular weight of polymeric systems will alter the characteristics of these systems and limit their service life. In fact, the scission of as few as one carbon-carbon bond in a thousand in a polymer molecule can completely destroy its useful physical properties. This sensitivity is not observed in lower-molecular-weight substances such as liquid hydrocarbons.

A general approach to reducing the rates of photodegradation for all types of polymers is the use of low levels of additives. These additives, known as photostabilizers or UV stabilizers, are effective at fractions of a weight percent. *See* PHOTOCHEMISTRY; POLYMER; STABILIZER (CHEMISTRY). D. M. Wiles; D. J. Carlsson

Photodetachment microscopy Detachment is the removal of an electron from a negative ion, which produces a neutral atom or molecule and a free electron. Photodetachment is detachment in which the necessary energy is provided by light, more precisely by a photon. Photodetachment microscopy consists in observing accurately the motion of the free electron produced when this process takes place in the presence of an external electric field. In such a field, the electron undergoes constant acceleration, as in free fall, and its possible trajectories are parabolas. With the magnitude of the initial electron velocity set by the photon energy, every point within reach on the detection plane can be reached by two, and only two, such parabolas. The ability of a single electron to follow several trajectories simultaneously is an essential feature of quantum mechanics, which is revealed visually in this case by the observation of an electron interference pattern. The sensitivity of the pattern to variations of the electron's ejection energy makes it possible to measure that energy with an accuracy 1000 times better than by classical techniques. Photodetachment microscopy is used to measure the electron affinities of atoms and molecules with unprecedented accuracy. *See* ELECTRON; ELECTRON AFFINITY; INTERFERENCE OF WAVES; NEGATIVE ION; PHOTON; QUANTUM MECHANICS; SUPERPOSITION PRINCIPLE.

Photodetachment. Lasers have become a popular tool for the excitation of atoms, molecules, and their ions. Photons emitted by a laser are essentially monoenergetic, since the laser wavelength can be set very precisely. This makes it possible to selectively excite the electrons bound in atoms and molecules to higher energy levels. If the photon energy is sufficiently high, it can eject an electron from an atom or molecule to form a positive ion. This process is called photoionization. Negative ions are systems that are weakly bound due to the fact that the outermost electron moves in the short-range field of a neutral atom or molecule. The binding energies of these electrons in negative ions are significantly smaller than those in atoms or molecules. Photodetachment is a process analogous to photoionization that ejects an electron from a negative ion. *See* LASER; PHOTOIONIZATION.

Photodetachment spectrometry. The most important parameter that comes out of the fitting of the experimental data with a theoretical propagator formula is the exact energy ε at which the electron is ejected. The energy of the detached electron, ε, can be determined with an accuracy to less than 1 μeV. This makes the photodetachment microscope the most accurate electron spectrometer ever built, as far as the absolute uncertainty is concerned. This technique has been applied to measurements of the electron affinities of fluorine, oxygen, silicon, and sulfur, with an accuracy of a few microelectronvolts. Christophe Blondel

Photodiode A semiconductor two-terminal component with electrical characteristics that are light-sensitive. All semiconductor diodes are light-sensitive to some degree, unless enclosed in opaque packages, but only those designed specifically to enhance the light sensitivity are called photodiodes.

Most photodiodes consist of semiconductor *pn* junctions housed in a container designed to collect and focus the ambient light close to the junction. They are normally biased in the reverse, or blocking, direction; the current therefore is quite small in the dark. When they are illuminated, the current is proportional to the amount of light falling on the photodiode. *See* JUNCTION DIODE.

Photodiodes are used both to detect the presence of light and to measure light intensity. *See* PHOTOELECTRIC DEVICES.

W. R. Sittner

Photoelasticity An experimental technique for the measurement of stresses and strains in material objects by means of the phenomenon of mechanical birefringence. Photoelasticity is especially useful for the study of objects with irregular boundaries and stress concentrations, such as pieces of machinery with notches or curves, structural components with slits or holes, and materials with cracks. The method provides a visual means of observing overall stress characteristics of an object by means of light patterns projected on a screen or photographic film. Photoelasticity is generally used to study objects stressed in two planar directions (biaxial), but with refinements it can be used for objects stressed in three spatial directions (triaxial). *See* BIREFRINGENCE.

Isochromatic fringe pattern for plate with hole. (*From M. M. Frocht, Photoelasticity, vol. 2, copyright © 1948 by John Wiley and Sons, Inc.; used with permission*)

When a stressed model is subjected to monochromatic polarized light in a polariscope the birefringence of the model causes the light to emerge refracted into two orthogonal planes. Because the velocities of light propagation are different in each direction, there occurs a phase shifting of the light waves.

When the waves are recombined with the polariscope, regions of stress where the wave phases cancel appear black, and regions of stress where the wave phases combine appear light. Therefore, in models of complex stress distribution, light and dark fringe patterns (isochromatic fringes) are projected from the model (see illustration). These fringes are related to the stresses. *See* POLARIZED LIGHT.

When white light is used in place of monochromatic light, the relative retardation of the model causes the fringes to appear in colors of the spectrum. White light is often used for demonstration, and monochromatic light is used for precise measurements. *See* STRESS AND STRAIN. William Zuk

Photoelectric devices Devices which give an electrical signal in response to visible, infrared, or ultraviolet radiation. They are often used in systems which sense objects or encoded data by a change in transmitted or reflected light. Photoelectric devices which generate a voltage can be used as solar cells to produce useful electric power. The operation of photoelectric devices is based on any of the several photoelectric effects in which the absorption of light quanta liberates electrons in or from the absorbing material. *See* PHOTOVOLTAIC EFFECT; SOLAR CELL.

Photoconductive devices are photoelectric devices which utilize the photo-induced change in electrical conductivity to provide an electrical signal. Photoemissive systems have also been used in photoelectric applications. These vacuum-tube devices utilize the photoemission of electrons from a photocathode and collection at an anode. *See* PHOTOCONDUCTIVE CELL; PHOTOEMISSION.

Many photoelectric systems now utilize silicon photodiodes or phototransistors. These devices utilize the photovoltaic effect, which generates a voltage due to the photoab-sorption of light quanta near a *pn* junction. Modern solid-state integrated-circuit fabrication techniques can be used to create arrays of photodiodes which can be used to read printed information. *See* PHOTODIODE ; PHOTOTRANSISTOR. Richard A. Chapman

Photoemission The ejection of electrons from a solid (or less commonly, a liquid) by incident electromagnetic radiation. Photoemission is also referred to as the external photoelectric effect. The visible and ultraviolet regions of the electromagnetic spectrum are most often involved, although the infrared and x-ray regions are also of interest. Photoemission has several distinguishing experimental features: (1) the process of photon

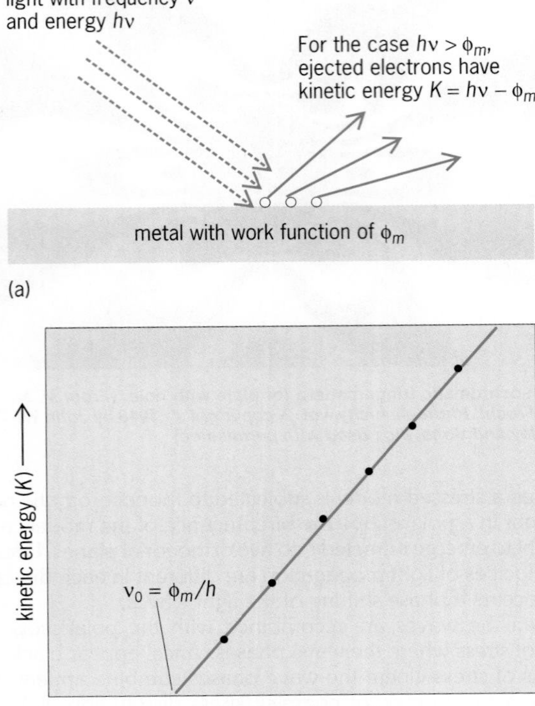

Photoelectric experiment, (*a*) Photoelectric effect with incident light of energy $h\nu$ and work function of ϕ_m. For the case of $h\nu > \phi_m$, electrons are ejected with kinetic energy of $K = h\nu - \phi_m$. (*b*) Typical relationship of kinetic energy *K* of electrons with the frequency ν of the incident light. For the case of $\nu < \nu_0$, no electrons are ejected from the surface of the metal. For the case of $\nu > \nu_0$, electrons are ejected with a certain kinetic energy.

absorption and photoelectron generation is instantaneous, that is, there is no detectable time lag; (2) at a given frequency the number of photoelectrons ejected per second is proportional to the intensity of the incident radiation; and (3) the kinetic energies of the photoelectrons depend on the incident photon frequency and the work function of the surface, but are independent of the incident intensity.

It is important to distinguish between the external and internal photoelectric effects. In the case of the external photoelectric effect, the photoelectron is ejected from the surface to the vacuum energy level. It is the fundamental process in photomultiplier tubes and vacuum photodiodes, which are referred to as photoemissive devices. In the case of the internal photoelectric effect, the absorbed photon is converted into photogenerated electron-hole pairs in semiconductors. It is the fundamental process enabling semiconductor photodetectors and solar cells. *See* OPTICAL DETECTORS; PHOTOCONDUCTIVE CELL; PHOTOCONDUCTIVITY; PHOTODIODE; PHOTOMULTIPLIER; PHOTOVOLTAIC CELL; PHOTOVOLTAIC EFFECT; SOLAR CELL.

The fundamental physical process of photoemission stems from the duality of light as both waves and particles. One of the most important concepts of the quantization of light into particles called photons emerged from the measurements of blackbody thermal radiation, for which Max Planck successfully provided the correct explanation by using discrete light energy in his analysis in 1900. By extending Planck's idea of quantization of electromagnetic radiation, Albert Einstein in 1905 proposed a theory to successfully explain the photoelectric effect. *See* DUALITY (PHYSICS); PHOTON; QUANTUM MECHANICS.

A schematic of the photoelectric experiment is shown in the illustration. It is based on shining incident light onto the surface of a metal and measuring the kinetic energy *K* of the electrons ejected from the surface of the metal, as shown in illus. *b*.

The experiment reveals several important discoveries about the photoelectric effect:

1. If the frequency of incident light is smaller than ν_0, no electron can be ejected from the surface of the metal (regardless of how high the incident light intensity is). Note that $\nu_0 = \phi_m/h$, where ϕ_m corresponds to the work function of the metal and *h* is Planck's constant.
2. If the frequency of incident light is higher than ν_0, electrons will be ejected instanteously (regardless of how low the incident light intensity is).
3. For $\nu > \nu_0$, the number of ejected electrons is proportional to the light intensity.
4. The kinetic energy *K* of the ejected electrons depends on the frequency/energy of the incident light, as expressed by $K = h\nu - \phi_m$.

In 1905, Einstein provided the explanation for the photoelectric effect by treating light as a particle with quantized energy of $h\nu$ (in agreement with Planck's work). This quantized light particle is called a quantum or photon. *K* is the maximum kinetic energy of the ejected electrons; it is also referred to as the Einstein maximum energy. *See* HEAT RADIATION; PLANCK'S CONSTANT.

L. Apker; Nelson Tansu

Photoferroelectric imaging The process of storing an image in a ferroelectric material by utilizing either the intrinsic or extrinsic photosensitivity in conjunction with the ferroelectric properties of the material. Specifically, photoferroelectric (PFE) imaging is a process of storing photographic images or other optical information in transparent lead lanthanum zirconate titanate (PLZT) ceramics. The photoferroelectric imaging device consists simply of a thin flat plate (about 0.01 in. or 0.2–0.3 mm thick) of optically polished PLZT ceramic with transparent conductive indium-tin oxide (ITO) electrodes sputter-deposited on the two major surfaces. The image to be stored is exposed onto one of the ITO electroded surfaces by using near-ultraviolet illumination in the intrinsic photosensitivity region (corresponding to a bandgap energy of approximately 3.35 eV) of the PLZT. Simultaneously, a voltage pulse is applied across the electrodes to switch the ferroelectric polarization from one stable remanent state to another. Images are stored both as spatial distributions of light-scattering centers in the bulk of the PLZT and as surface deformation strains which form a relief pattern of the image on the exposed surface. Both the light scattering and surface strains are related to spatial distributions of ferroelectic domain orientations introduced during the image-storage process. These spatial distributions correspond to brightness variations in the image to which the PLZT is exposed. The stored image may be viewed directly or it may be projected onto a screen by using either transmitted or reflected light.

Important potential applications of photoferroelectric imaging are temporary image storage and display. Various types of image processing, including image contrast enhancement, are also offered by the capability of switching from a positive to a negative stored image in discrete steps. *See* ELECTRONIC DISPLAY; FERROELECTRICS; PHOTOCONDUCTIVITY.

Cecil E. Land

Photogrammetry The practice of obtaining surveys by means of photography. The camera commonly is airborne with its axis vertical, but oblique and horizontal (ground-based) photographs also are applicable. Data reduction is accomplished by stereoscopic line-of-sight geometry with use of both analytical and analog methods. *See* AERIAL PHOTOGRAPH; REMOTE SENSING; SURVEYING.

Typical automated stereoplotting system. (*Lockwood, Kessler, and Bartlett, Inc.*)

In vertical aerial surveys adjacent photos are overlapped. The two images of the same terrain are then superimposed for three-dimensional viewing by human operators or automated sensors.

In a widely used analog procedure the two photos are placed in the projectors of a stereoplotting instrument. With the aid of visible ground-control points the photos are oriented to the relative positions they had at the instants of exposure. In a typical automated stereoplotting system (see illustration) scanning devices substitute for human eyes to sense model-surface slope and thus to control servomechanisms that raise and lower the plotting table and translate it along a succession of closely spaced parallel horizontal dimensions, or ground profiles. Robert H. Dodds

Photographic materials The light-sensitive recording materials of photography, that is, photographic films, plates, and papers. They consist primarily of a support of plastic sheeting, glass, or paper, respectively, and a thin, light-sensitive layer, commonly called the emulsion, in which the image will be formed and stored. The material will usually embody additional layers to enhance its photographic or physical properties.

Film support, for many years made mostly of flammable cellulose nitrate, is now exclusively made of slow-burning "safety" materials, usually cellulose triacetate or polyester terephthalate, which are manufactured to provide thin, flexible, transparent, colorless, optically uniform, tear-resistant sheeting. Film supports usually range in thickness from 0.0025 to 0.009 in. (0.06 to 0.23 mm) and are made in rolls up to 60 in. (1.5 m) wide and 6000 ft (1800 m) long. *See* ESTER; POLYESTER RESINS.

Glass is the predominant substrate for photographic plates, though methacrylate sheet, fused quartz, and other rigid materials are sometimes used. Plate supports are selected for optical clarity and flatness. Thickness, ranging usually from 0.04 to 0.25 in. (1 to 6 mm), is increased with plate size as needed to resist breakage and retain flatness. *See* GLASS.

Photographic paper is made from bleached wood pulp of high α-cellulose content, free from ground wood and chemical impurities. It is often coated with a suspension of baryta (barium sulfate) in gelatin for improved reflectance and may be calendered for high smoothness. Fluorescent brighteners may be added to increase the appearance of whiteness. *See* PAPER.

Most emulsions are basically a suspension of silver halide crystals in gelatin. The crystals, ranging in size from 2.0 to less than 0.05 micrometers, are formed by precipitation by mixing a solution of silver nitrate with a solution containing one or more soluble halides in the presence of a protective colloid. During manufacture, the emulsion is ripened to control crystal size and structure. Chemicals are added in small but significant amounts to control speed, image tone, contrast, spectral sensitivity, keeping qualities, fog, and hardness; to facilitate uniform coating; and, in the case of color films and papers, to participate in the eventual formation of dye instead of metallic silver images upon development. The gelatin, sometimes modified by the addition of synthetic polymers, is more than a simple vehicle for the silver halide crystals. It interacts with the silver halide crystals during manufacture, exposure, and processing and contributes to the stability of the latent image. *See* EMULSION; GELATIN; SILVER.

The silver halides (and silver behenates) are normally sensitive only to x-radiation and to ultraviolet, violet, and blue wavelengths, but they can be made sensitive to longer wavelengths by adding special dyes, predominantly polymethines, to the emulsion. The process is known as spectral sensitizing to distinguish it from the chemical sensitizing used to raise the overall or inherent sensitivity of the grains. *See* PHOTOGRAPHY. Robert D. Anwyl

Photography The process of forming stable or permanent visible images directly or indirectly by the action of light or other forms of radiation on sensitive surfaces. Traditional photography uses the action of light to cause changes in a film of silver halide crystals in which development converts exposed silver halide to (nonsensitive) metallic silver. Following exposure in a camera or other device, the film or plate is developed, fixed in a solution that dissolves the undeveloped silver halide, washed to remove the soluble salts, and dried. Printing from the original, if required, is done by contact or optical projection onto a second emulsion-coated material, and a similar sequence of processing steps is followed. Digital photography captures images directly with an electronic photosensor. *See* PHOTOGRAPHIC MATERIALS.

Photography is practiced on a professional level for portraiture and for various commercial and industrial applications, including the preparation of photographs for advertising, illustration, display, and record-keeping. Press photography is for newspaper and magazine illustrations of topical events and objects. Photography is used at several levels in the graphic arts to convert original photographs or other illustrations into printing plates for high-quality reproduction in quantity. Industrial photography includes the generation and reproduction of engineering drawings, high-speed photography, schlieren photography, metallography, and many other forms of technical photography which can aid in the development, design, and manufacture of various products. Aerial photography is used for military reconnaissance and mapping, civilian mapping, urban and highway planning, and surveys of material resources. Biomedical photography is used to reveal or record biological structures, often of significance in medical research, diagnosis, or treatment. Photography is widely applied to preparing projection slides and other displays for teaching through visual education. *See* PRINTING; SCHLIEREN PHOTOGRAPHY.

Photography is one of the most important tools in scientific and technical fields. It extends the range of vision, allowing records to be made of things or events which are difficult or impossible to see because they are too faint, too brief, too small, or too distant, or associated with radiation to which the eye is insensitive. Technical photographs can be studied at leisure, measured, and stored for reference or security. The acquisition and interpretation of images in scientific and technical photography usually requires direct participation by the scientist or skilled technicians. Robert D. Anwyl

Digital still photography has experienced exponential growth. This is primarily due to a series of technical improvements and price reductions that have made the technology more appealing to the mass market and professionals alike. Parallel growth has been taking place in digital video, and the most recent cameras can be used to capture high-resolution still images or to make lower-resolution video recordings. In addition, the use of digital photography has been increasing in a host of other imaging applications such as surveillance, astronomy, and medical imaging.

Digital photography has advantages over film photography in productivity, quality control, and profitability. Instead of shooting, processing, printing, and scanning, digital photography allows the capture of images directly as digital files, thus saving time and materials.

Digital photography is based on the photoelectric effect of silicon. When photons strike silicon, they excite electrons from the valence band to the conduction band. The number of electrons freed is proportional to the size of the light-sensing element, the amount of light exposure, and the wavelength of the light. Silicon light-sensing elements are arranged in a grid format with each photosite comprising one picture element (pixel). The electrons resulting from light exposure are collected and measured from the pixel array. The charge voltages are converted from analog to digital (A/D) format and amplified to create digital picture files. The cameras have random-access-memory (RAM) buffers to hold images while they are being downloaded or recorded on disk.

Two types of sensors—charge-coupled device (CCD) and complementary metal-oxide semiconductor (CMOS)—are used for the image-sensing elements in digital cameras. CCDs produce higher-quality images and are used in most digital cameras. However, CMOS sensors have inherent size and cost advantages over CCDs and are making inroads on the low end of the image sensor market. *See* CHARGE-COUPLED DEVICES; SILICON.

CCD photosites are light-receptive over most of the pixel surface, with photodiodes collecting electrons based on the intensity of exposure. CCD sensors have wide dynamic ranges and high light sensitivity, and can operate at 30 frames/second for video applications. The CCD architecture makes it impractical to integrate other electronics onto the same chip with the light sensor. CCD cameras require supplemental integrated circuits to support the imaging function. *See* INTEGRATED CIRCUITS.

CMOS sensors have the advantage that many imaging functions are manufactured on the same chip with the light-sensing elements. This leads to smaller, less expensive product designs.

The arrangements of CCDs in digital cameras are divided between linear sensors that scan the image plane and area arrays that capture the entire image with a single exposure. Linear arrays are capable of capturing higher-resolution images, but they are restricted to still-life photography and are not compatible with strobe lighting, which is commonly used in photo studios. Area arrays, by contrast, produce lower-resolution images, but they can be used for live action photography and are strobe-light-compatible.

Capturing color images requires that three separate signals (blue, green, and red) be generated from a scene. This is generally accomplished by placing separation filters in the light path during exposure. Three approaches are used: filter wheels, dyed photo sites, and silicon filtering. Anthony Stanton

Like conventional cameras, digital cameras come in compact, single-lens reflex, and large-format varieties. Low-resolution compacts are useful for producing classified advertisements and tend to have relatively simple optics, image-sensing electronics, and controlling software. Digital cameras are often based on existing single-lens reflex camera designs with the addition of CCD backs and storage subsystems. *See* CAMERA. Thomas Destree

Photoionization The ejection of one or more electrons from an atom, molecule, or positive ion following the absorption of one or more photons. The process of electron ejection from matter following the absorption of electromagnetic radiation has been under investigation for over a century. The earliest measurements involved the irradiation of metal surfaces by ultraviolet radiation. The theoretical interpretation of this phenomenon, known as the photoelectric effect, played an important role in establishing quantum mechanics. It was shown that, contrary to classical ideas, energy exchanges between radiation and matter are mediated by integral numbers of photons. The ejection of electrons from a metal requires that the photon energy exceed the binding energy of electrons in the metal (that the wavelength of light be sufficiently small), whatever the photon flux. *See* PHOTOEMISSION.

Photons (light) can interact with atoms or molecules which are free, that is, not bound to a solid or liquid. In the gas phase the photoeffect is called either photoionization (atoms, molecules, and their positive ions) or photodetachment (atomic and molecular negative ions). *See* PHOTODETACHMENT SPECTROSCOPY.

Photoionization involves a radiative bound-free transition from an initial state consisting of n photons and an atom, molecule, or ion in a bound state to a final continuum state consisting of a residual ion (or an atom in the case of photodetachment) and m free electrons, that is, the reaction below. This

$$nh\nu + X \rightarrow X^{m+} + me^-$$

means that one or more photons (n) interact with a target atom or ion, resulting in the emission of one or more electrons (m).

In the simplest atomic photoionization process a single electron is ejected from an atom following the absorption of a single photon. Each mode of photoionization defines a final-state channel that is characterized by the energy and angular momentum of the outgoing electron as well as the excitation state of the residual ion. Since the photoionization process is endoergic, each channel has a well-defined threshold energy below which the channel is energetically closed. The threshold photon energy for a particular channel is equal to the binding energy of the electron that is to be ejected plus the excitation energy, if any, of the residual ion.

Above threshold, the energy carried off by the outgoing electron represents the balance between the energy supplied by the photon and the binding energy of the electron plus the excitation energy of the residual ion (neglecting the small recoil of the heavy ion). A photoelectron energy spectrum is characterized by a discrete set of peaks at different photon energies, each peak being associated with a particular state of the residual ion. Information on the excitation state of the ion following photoionization can also be obtained by monitoring the fluorescence emitted in the subsequent radiative decay of the state. One of the earliest applications of photoionization measurements was the investigation of the structure of atoms by determining the binding energies of both outer- and inner-shell electrons by means of photoelectron spectroscopy. *See* ATOMIC STRUCTURE AND SPECTRA; ELECTRON SPECTROSCOPY. David J. Pegg; Alfred S. Schlachter

Photoluminescence A luminescence excited in a body by some form of electromagnetic radiation incident on the body. The term photoluminescence is generally limited to cases in which the incident radiation is in the ultraviolet, visible, or infrared regions of the electromagnetic spectrum.

Photoluminescence may be either a fluorescence or a phosphorescence, or both. Energy can be stored in certain luminescent materials by subjecting them to light or some other exciting agent, and can be released by subsequent illumination of the material with light of certain wavelengths. This type of photoluminescence is called stimulated photoluminescence. *See* FLUORESCENCE; LUMINESCENCE; PHOSPHORESCENCE.

Clifford C. Klick; James H. Schulman

Photolysis The chemical decomposition of matter due to absorption of incident light. For example, illumination of microcrystals of silver bromide embedded in gelatin results in formation of metallic silver, and is the basis of the photographic process. *See* PHOTOGRAPHIC MATERIALS.

Numerous metal complexes, azides, nitrides, and sulfides, and most organometallic compounds undergo decomposition upon illumination, often with concomitant evolution of a gaseous product. In the presence of water and oxygen, illumination of many semiconductors results in their corrosion; for example, an aqueous suspension of cadmium sulfide (CdS) undergoes rapid decomposition upon irradiation with sunlight. The outcome of these photochemical processes can often be controlled by addition of adventitious materials. For example, illumination of cadmium sulfide in aqueous solution containing hydrogen sulfide (H_2S) and colloidal platinum results in evolution of hydrogen gas. Here, the semiconductor photosensitizes (or photocatalyzes) decomposition of hydrogen sulfide into its elements, and the reaction can be used to remove sulfides from industrial waste. *See* COORDINATION COMPLEXES; SEMICONDUCTOR.

Many ketones, for example, acetone, abstract hydrogen atoms from adjacent organic matter under illumination. According to the circumstances, the resultant free radicals may be used to initiate polymerization of a monomer or cause decomposition of a plastic film. Both reactions have important commercial applications, and photoinitiators are commonly used for emulsion paints, inks, polymers, explosives, fillings for teeth and for development of photodegradable plastics. Photolysis of carbonyl compounds, released into the atmosphere by combustion of fossil fuels, is responsible for the onset of photochemical smog. Many other types of photochemical transformation of organic molecules are known, including isomerization of unsaturated bonds, cleavage of carbon-halogen bonds, olefin addition reactions, halogenation of aromatic species, hydroxylation, and oxygenation processes. Indeed, photochemistry is often used to produce novel pharmaceutical products that are difficult to synthesize by conventional methods. *See* FREE RADICAL; PHOTODEGRADATION; SMOG.

The most important photochemical reaction is green plant photosynthesis. Here, chlorophyll that is present in the leaves absorbs incident sunlight and catalyzes reduction of carbon dioxide to carbohydrate. *See* PHOTOSYNTHESIS.

Excitation of an organic molecule results in spontaneous generation of the singlet excited state. In most molecules, this highly unstable excited singlet state may undergo an intersystem-crossing process that results in population of the corresponding (less energetic) excited triplet state, in competition to fluorescence. The excited triplet state, because of spin restriction rules, retains a significantly longer lifetime than is found for the corresponding excited singlet state, and may be formed in high yield.

Almost without exception, these triplet states react quantitatively with molecular oxygen (O_2) present in the system via a triplet energy-transfer process. The resulting product is singlet molecular oxygen. This species is a potent and promiscuous reactant, and it is responsible for widespread damage to both synthetic and natural environments. Indeed, plants and photosynthetic bacteria contain carotenoids to protect the organism against attack by singlet oxygen. The same species is known to be responsible, at least in part, for photodegradation of paint, plastic, fabric, colored paper, and dyed wool. Secondary reactions follow from attack on a substrate by singlet oxygen, resulting in initiation of chain reactions involving free radicals. However, modern technological processes have evolved in which singlet molecular oxygen is used to destroy unwanted organic matter, such as tumors, viruses, and bacteria, in a controlled and specific manner. In photodynamic therapy a dye is injected into a tumor and selectively illuminated with laser light. The resultant singlet oxygen destroys the tumor. Similar methodology can be used to produce photoactive soap powders, bleaches, bactericides, and pest-control reagents. *See* CHAIN REACTION (CHEMISTRY); FLUORESCENCE; TRIPLET STATE. A. Harriman

Photometer An instrument used for making measurements of light, or electromagnetic radiation, in the visible range. In general, photometers may be divided into two classifications: laboratory photometers, which are usually fixed in position and yield results of high accuracy; and portable photometers, which are used in the field or outside the laboratory and yield results of lower accuracy. Each class may be subdivided into visual (subjective) photometers and photoelectric (objective or physical) photometers. These in turn may be grouped according to function, such as photometers to measure luminous intensity (candelas or candlepower), luminous flux, illumination (illuminance), luminance (photometric brightness), light distribution, light reflectance and transmittance, color, spectral distribution, and visibility. Visual photometric methods have largely been supplanted commercially by physical methods, but because of their simplicity, visual methods are still used in educational laboratories to demonstrate photometric principles. *See* ILLUMINANCE; LUMINANCE; LUMINOUS FLUX; LUMINOUS INTENSITY. G. A. Horton

Photometry That branch of science which deals with measurements of light (visible electromagnetic radiation) according to its ability to produce visual sensation. Specifically, photometry deals with the attribute of light that is perceived as intensity, while the related attribute of light that is perceived as color is treated in colorimetry. *See* COLOR; COLORIMETRY.

The purely physical attributes of light such as energy content and spectral distribution are treated in radiometry. Sometimes the word photometry is used to denote measurements that have nothing to do with human vision, but this is a mistake according to modern usage. Such measurements are properly referred to as radiometry, even if they are performed in the visible spectral region. *See* RADIOMETRY.

The relative visibility of a fixed power level of monochromatic electromagnetic radiation varies with wavelength over the visible spectral region (400–700 nanometers). The relative visibility of radiation also depends upon the illumination level that is being observed. The cone cells in the retina determine the visual response at high levels of illumination, while the rod cells dominate in the dark-adapted eye at very low levels (such as starlight). Cone-controlled vision is called photopic, and rod-controlled vision is called scotopic, while the intermediate region where both rods and cones play a role is called mesopic. *See* VISION.

Originally, photometry was carried out by using the human visual sense as the detector of light. As a result, photometric measurements were subjective. In order to put photometric measurements on an objective basis, and to allow convenient electronic detectors to replace the eye in photometric measurements, the Commission Internationale de l'Eclairage (CIE; International Commission on Illumination) has adopted two relative visibility functions as standards. These internationally accepted functions are called the spectral luminous efficiency functions for photopic and scotopic vision, and are denoted by $V(\lambda)$ and $V'(\lambda)$, respectively. *See* LUMINOUS EFFICIENCY.

Thus photopic and scotopic (but not mesopic) photometric quantities have objective definitions, just as do the purely physical quantities. However, there is a difference. The purely physical quantities are defined in terms of physical laws, whereas the photometric quantities are defined by convention. In recognition of this difference the photometric quantities are called psychophysical quantities.

According to the International System of Units, SI, the photometric units are related to the purely physical units through a defined constant called the maximum spectral luminous efficacy. This quantity, which is denoted by K_m, is the number of lumens per watt at the maximum of the $V(\lambda)$ function. K_m is defined in SI to be 683 lm/W for monochromatic radiation whose wavelength is 555 nanometers, and this defines the photometric units with which the photometric quantities are to be measured.

At various times, the photometric units have been defined in terms of the light from different standard sources, such as candles made according to specified procedures, and blackbodies at the freezing point of platinum. According to these definitions, K_m was a derived, rather than defined, quantity. *See* ILLUMINATION; LIGHT; PHYSICAL MEASUREMENT; UNITS OF MEASUREMENT. Jon Geist

Photomorphogenesis The regulatory effect of light on plant form involving growth, development, and differentiation of cells, tissues, and organs. Morphogenic influences of light on plant form are quite different from light effects that nourish the plant through photosynthesis, since the former usually occur at much lower energy levels than are necessary for photosynthesis. While photosynthetic processes are energy-transducing, converting light energy into chemical energy for running the machinery of the plant, light serves only as a trigger in photomorphogenesis, frequently resulting in energy expenditure orders of magnitude larger than the amount required to induce a given response. Photomorphogenic processes determine the nature and direction of a plant's growth and thus play a key role in its ecological adaptations to various environmental changes. *See* PHOTOSYNTHESIS.

Morphogenically active radiation is known to control seed and spore germination, growth and development of stems and leaves, lateral root initiation, opening of the hypocotyl or epicotyl hook in seedlings, differentiation of tracheary elements in the stem, and form changes in the gametophytic phase of ferns, to mention but a few of such known phenomena. Many nonmorphogenic processes in plants are also basically controlled by light independent of photosynthesis. Among these are chloroplast movement, biochemical reactions involved in the synthesis of flavonoids, anthocyanins, chlorophyll, and carotenoids, and leaf movements in certain legumes. A large number of these processes are affected only by wavelengths of light at the red end of the spectrum. The similarities in the action spectra for such different systems led to the concept that these divergent sets of developmental and physiological processes might be controlled by a single pigment. This has proved to be the case, the pigment being phytochrome. *See* CHLOROPHYLL; LEAF; PHYTOCHROME; ROOT (BOTANY); SEED; STEM.

Others of these processes, such as rapid stem and internode growth inhibition, directional organ growth (phototropism), and stomatal aperture regulation, are controlled primarily by wavelengths in the near-ultraviolet (UV-A) and blue regions of the spectrum. Although many of their action spectra are similar, suggesting a single pigment as with responses controlled by phytochrome, at least two distinct classes of nonphytochrome pigments, the cryptochromes and the phototropins, are now known to control these latter responses. *See* PLANT GROWTH; PLANT PIGMENT.

Emmanuel Linscum; Winslow R. Briggs

Photomultiplier A very sensitive vacuum-tube detector of light or radiant flux containing a photocathode which converts the light to photoelectrons; one or more secondary-electron-emitting electrodes or dynodes which amplify the number of photoelectrons; and an output electrode or anode which collects the secondary electrons and provides the electrical output signal. It is also known as a multiplier phototube. Because of the very large amplification provided by the secondary-emission mechanism, and the very short time variation associated with the passage of the electrons within the device, the photomultiplier is applied to the detection and measurement of very low light levels, especially if very high speed of response is required.

The illustration is a schematic of a typical photomultiplier and shows its operation. Light incident on a semitransparent photocathode located inside an evacuated envelope causes photoelectron emission from the opposite side of the photocathode. The efficiency of the photoemission process is called the quantum efficiency, and is the ratio of emitted photoelectrons to incident photons (light particles). Photoelectrons are directed by an accelerating electric field to the first dynode, where from 3 to 30 secondary electrons are emitted for each incedent electron, depending upon the dynode material and the applied voltage. These secondaries are directed to the second dynode, where the process is repeated and so on until the multiplied electrons from the last dynode are collected by the anode.

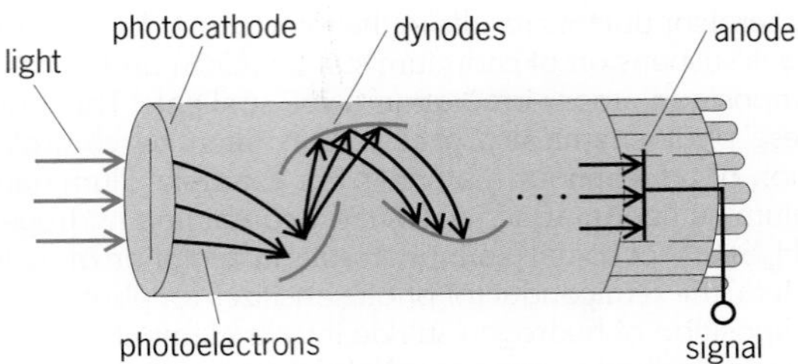

Schematic of a photomultiplier. (*After P. W. Engstrom, Photomultipliers—then and now, RCA Eng., 24(1):18–26, June–July 1978*)

A typical photomultiplier may have 10 stages of secondary emission and may be operated with an overall applied voltage of 2000 V. In most photomultipliers the focusing of the electron streams is done by electrostatic fields shaped by the design of the electrodes. Some special photomultipliers designed for very high speed utilize crossed electrostatic and magnetic fields which direct the electrons in approximate cycloidal paths between electrodes.

Ever since their invention, photomultipliers have been found useful in low-level photometry and spectrometry. The most important applications of photomultipliers are related to scintillation counting, and these form the basis of a whole science of tracer chemistry that has been applied to agriculture, medicine, and industrial problems. In medicine, photomultiplier-scintillator combinations are used in the gamma-ray camera, computerized tomography, and the positron scanner. *See* COMPUTERIZED TOMOGRAPHY; GAMMA-RAY DETECTORS; NUCLEAR MEDICINE; RADIOLOGY; SCINTILLATION COUNTER.

Ralph W. Engstrom

Photon An entity that can be loosely described as a quantum of energy of electromagnetic radiation. According to classical electromagnetic theory, an electromagnetic wave can transfer arbitrarily small amounts of energy to matter. According to the quantum theory of radiation, however, the energy is transferred in discrete amounts. The energy of a photon is the product of Planck's constant and the frequency of the electromagnetic field. In addition to energy, the photon possesses momentum and also possesses angular momentum corresponding to a spin of unity. The interaction of radiation with matter involves the absorption, scattering, and emission of photons. Consequently, the energy interchange is inherently quantized. *See* ANGULAR MOMENTUM; ENERGY; MOMENTUM; SPIN (QUANTUM MECHANICS).

For many purposes, the photon behaves like a particle of zero rest mass moving at the speed of light. The particlelike nature of the photon is vividly exhibited by the photoelectric effect, predicted by A. Einstein, in which light is absorbed in a metal, causing electrons to be ejected. An electron absorbs a photon, gaining its energy. In leaving the metal, it loses energy because of interactions with the surface; the energy loss equals the product of the so-called work function of the surface and the charge of the electron. The final kinetic energy of the electron therefore equals the energy of the incident photon minus this energy loss. *See* PHOTOEMISSION.

A second demonstration of the particlelike behavior of photons is provided by the scattering of an x-ray photon from an electron bound in an atom. The electron recoils because of the momentum of the photon, thereby gaining energy. As a result, the frequency, and hence the wavelength of the scattered x-ray, is altered. If the x-ray is scattered through a certain angle, the wavelength is shifted by an amount determined by this scattering angle and the mass of an electron, according to the laws of conservation of energy and momentum. *See* COMPTON EFFECT.

From a more fundamental view, the photon is the quantum of excitation of a single mode of a radiation field. The dynamical equations for the electric and magnetic energy in such a field are identical to those of a harmonic oscillator. According to quantum theory, the allowed energies of a harmonic oscillator are given by $E = (j + 1/2)hf$, where h is Planck's constant, f is the frequency of the oscillator, and the quantum number $j = 0, 1, 2, \ldots$, describes the state of excitation of the oscillator. This quantum relation was first postulated by M. Planck for the material oscillators in the walls of a thermal enclosure in order to obtain the correct form for the density of radiation in a thermal field, but it was quickly applied by Einstein to describe the state of the radiation field itself. In this picture, j describes the number of photons in the field. *See* HARMONIC OSCILLATOR; QUANTUM ELECTRODYNAMICS; QUANTUM MECHANICS. Daniel Kleppner

Photoperiodism The growth, development, or other responses of organisms to the length of night or day or both. Photoperiodism has been observed in plants and animals, but not in bacteria (prokaryotic organisms), other single-celled organisms, or fungi.

A true photoperiodism response is a response to the changing duration of day or night. Responses to the quantity of light energy in photoperiodism are secondary complications. Some species respond to increasing day lengths and decreasing night lengths (for example, by forming flowers or developing larger gonads); this is called a long-day response. Other species may exhibit the same response, or the same species may respond in some different way, to decreasing days and increasing nights; this is a short-day response. Sometimes a response is independent or nearly independent of day length and is said to be day-neutral.

Plant responses. There are many plant responses to photoperiod. These include development of reproductive structures in lower plants (mosses) and in flowering plants; rate of flower and fruit development; stem elongation in many herbaceous species as well as coniferous and deciduous trees (usually a long-day response and possibly the most widespread photoperiodism response in higher plants); autumn leaf drop and formation of winter dormant buds (short days); development of frost hardiness (short days); formation of roots on cuttings; formation of many underground storage organs such as bulbs (onion, long days), tubers (potato, short days), and storage roots (radish, short days); runner development (strawberry, long days); flower formation (strawberry, short days); balance of male to female flowers or flower parts (especially in cucumbers); aging of leaves and other plant parts; and even such obscure responses as the formation of foliar plantlets (such as the minute plants formed on edges of *Bryophyllum* leaves), and the quality and quantity of essential oils (such as those produced by jasmine plants). Note that a single plant, for example, the strawberry, might be a short-day plant for one response and a long-day plant for another response.

Animal responses. There are also many responses to photoperiod in animals, including control of several stages in the life cycle of insects (for example, diapause) and the long-day promotion in birds of molting, development of gonads, deposition of body fat, and migratory behavior. Even feather color may be influenced by photoperiod (as in the ptarmigan). In several mammals the induction of estrus and spermatogenic activity is controlled by photoperiod (sheep, goat, snowshoe hare), as is fur color in certain species (snowshoe hare). Growth of antlers in American elk and deer can be controlled by controlling day length. Increasing day length causes antlers to grow, whereas decreasing day length causes them to fall off. *See* MIGRATORY BEHAVIOR.

Seasonal responses. Response to photoperiod means that a given manifestation will occur at some specific time during the year. Response to long days (shortening nights) normally occurs during the spring, and response to short days (lengthening nights) usually occurs in late summer or autumn. Since day length is accurately determined by the Earth's rotation on its tilted axis as it revolves in its orbit around the Sun, detection of day length and its rate of change provides an extremely accurate means of determining the season at a given latitude. Such other environmental factors as temperature and light levels also vary with the seasons but are clearly much less dependable from year to year.

Mechanisms. It has long been the goal of researchers on photoperiodism to understand the plant or animal mechanisms that account for the responses. Light must be detected, the duration of light or darkness must be measured, and this time measurement must be metabolically translated into the observed response: flowering, stem elongation, gonad development, fur color, and so forth. Many of the basic mechanisms appear to differ not only between plants and animals, but among different species as well, although the roles (synchronization, anticipation, and so on) are similar in all organisms that exhibit photoperiodism.

In 1938 Karl Hamner and James Bonner investigated whether day or night was more important in photoperiodism. Among other things, they interrupted the day with a short period of darkness and the night with a brief interval of light. Darkness during the day had essentially no effect on short-day cocklebur plants, but the plants completely failed to flower when the long dark period was interrupted by light. Other workers found that long-day plants were promoted in their flowering by a night interruption. Again, the opposite nature of short-day and long-day responses is demonstrated. It was concluded from these experiments that day length was of secondary importance to night length, but the important role of photoperiod as well as darkness is now recognized.

Orange-red wavelengths (660 nm) used as a night interruption are by far the most effective part of the spectrum in inhibition of short-day responses, and the effects of orange-red light can be reversed by subsequent exposure of plants to light of somewhat longer wavelengths (730 nm), called far-red light. With some long-day plants, irradiation with far-red light and blue light are more effective in promotion of flowering than irradiation with orange-red light. These observations led to discovery in the 1950s of the phytochrome pigment system, which plays an extremely important role in plant responses to light, including in photoperiodism. *See* PHYTOCHROME.

In photoperiodism of short-day plants, an optimum response is usually obtained when the phytochromes are in the far-red-receptive form during the day and the red-receptive form during the night. Although normal daylight contains a balance of red and far-red wavelengths, the red-receptive forms are most sensitive, and so the phytochromes under normal daylight conditions are driven mostly to the far-red receptive form. At dusk this form is changed metabolically, and the red-receptive form builds up. It is apparently this shift in the forms of phytochromes that initiates measurement of the dark period. This is how a plant "sees." When the far-red-sensitive form of the pigment is abundant, the plant "knows" it is in the light; the red-sensitive form (or lack of far-red form) provides a biochemical indication that it is in the dark.

The measurement of time—durations of the day or night—is the very essence of photoperiodism. The discovery of a biological clock in living organisms was strongly confirmed in the late 1920s. It was shown that the movement of leaves on a bean plant (from horizontal at noon to vertical at midnight) continued uninterruptedly for several days, even when plants were placed in total darkness and at a constant temperature, and that the time between given points in the cycle (such as the most vertical leaf position) was almost but not exactly 24 h. In the case of bean leaves, it was about 25.4 h. Many other cycles have now been found with similar characteristics in virtually all groups of plants and animals. This is strong evidence that the clocks are internal and not driven by some subtle daily change in the environment. Such rhythms are called circadian (from Latin *circa* = approximately, and *dies* = day).

Circadian rhythms usually have period lengths that are remarkably temperature-insensitive, which is also true of time measurement in photoperiodism. Furthermore, the rhythms are normally highly sensitive to light, which may shift the cycle to an extent that depends upon when the light is given. Thus, daily rhythms in nature are normally synchronized with the daily cycle as the Sun rises and sets each day. Their circadian nature appears only when they are allowed to manifest themselves under constant conditions of light (or darkness) and temperature, so that their free-running periods can appear. Many lines of evidence show that the photoperiodism clock is dependent on the circadian clock. *See* BIOLOGICAL CLOCKS. Frank B. Salisbury

Photophore gland A highly modified integumentary gland which arises from an epithelial invagination into the dermis. It becomes cut off from its site of origin and develops into a luminous organ composed of a lens and a light-emitting gland, at the back of which is a pigmented reflector of probably dermal-cell origin. These luminous bodies occur in deep-sea teleosts and elasmobranchs which live in areas of total darkness. *See* EPITHELIUM; GLAND. Olin E. Nelsen

Photoreception The process of absorption of light energy by plants and animals and its utilization for biologically important purposes. In plants photoreception plays an essential role in photosynthesis and an important role in orientation. Photoreception in animals is the initial process in vision. *See* PHOTOSYNTHESIS; TAXIS; VISION.

The photoreceptors of animals are highly specialized cells or cell groups which are light-sensitive because they contain pigments which are unstable in the presence of light of appropriate wavelengths. These light-sensitive receptor pigments absorb radiant energy and then undergo physicochemical changes, which lead to the initiation of nerve impulses that are conducted to the central nervous system. *See* EYE (INVERTEBRATE); EYE (VERTEBRATE). Verner J. Wulff

Photorespiration Light-dependent carbon dioxide release and oxygen uptake in photosynthetic organisms caused by the fixation of oxygen instead of carbon dioxide during photosynthesis. When the reactions of photosynthesis use oxygen instead of carbon dioxide, phosphoglycolate is formed and there is a loss of carbon from the photosynthetic pathway. Phosphoglycolate inhibits photosynthesis if it is allowed to accumulate in the plant. The reactions of photorespiration break down phosphoglycolate and recover 75% of the carbon for the photosynthetic reaction sequence. The remaining 25% of the carbon is released as carbon dioxide. Photorespiration reduces the rate of photosynthesis in plants in three ways: carbon dioxide is released; energy is diverted from photosynthetic reactions to photorespiratory reactions; and competition between oxygen and carbon dioxide reduces the efficiency of the important photosynthetic enzyme ribulose-bisphosphate (RuBP) carboxylase. There is no known function of the oxygenation reaction; most scientists believe it is an unavoidable side reaction of photosynthesis, but it is possible that photorespiration improves nitrate reduction. *See* CARBON DIOXIDE; OXYGEN; PHOTOSYNTHESIS.

Photorespiration reduces the potential photosynthetic rate by as much as 50%. Since photosynthesis provides the material necessary for plant growth, photorespiration inhibits plant growth by reducing the net rate of carbon dioxide assimilation (photosynthesis). Plants grow faster and larger when photorespiration is reduced by either raising the carbon dioxide concentration or lowering the oxygen concentration in the air surrounding a plant. Most of the beneficial effects on plant growth achieved by increasing CO_2 may result from the reduced rate of photorespiration. *See* PLANT GROWTH.

Photorespiration is a process that is wasteful to the plant and may not be essential for plant survival. It is, therefore, surprising that a mutation eliminating photorespiration has not arisen. Many scientists believe that this indicates that oxygenation and the subsequent photorespiration are unavoidable side reactions of photosynthesis.

There are, however, some plants that avoid photorespiration by actively accumulating carbon dioxide inside the cells that have RuBP carboxylase/oxygenase. Many cacti do this by taking up carbon dioxide at night and then releasing it during the day to allow normal photosynthesis. These plants are said to have crassulacean acid metabolism (CAM). Another group of plants, including corn (*Zea mays*), take up carbon dioxide and attach it to a molecule in one part of the leaf, and then transport it to another part of the leaf for release and fixation by normal photosynthesis. The compound used to transport the carbon dioxide has four carbon atoms: thus these plants are called C_4 plants. Plants that have no mechanism for accumulating carbon dioxide produce the three-carbon compound phosphoglycerate first instead of the C_4 molecules that are seen first in C_4 photosynthesis. These plants are therefore called C_3 plants. Most species of plants are C_3. *See* PHYSIOLOGICAL ECOLOGY (PLANT); PLANT RESPIRATION. Thomas D. Sharkey

Photosphere The apparent, visible surface of the Sun. The photosphere is a gaseous atmospheric layer a few hundred miles deep with a diameter of 864,000 mi (1,391,000 km; usually considered the diameter of the Sun) and an average temperature of approximately 5800 K (10,500°F). Radiation emitted from the photosphere accounts for most of the solar energy flux at the Earth.

Convective cells give the photosphere a granular appearance with bright cells (hot rising gas) surrounded by dark inter- granular lanes (cool descending gas). A typical granule is approximately 600 mi (1000 km) in diameter. Measurements of horizontal velocity reveal a larger convective pattern, the supergranulation; the horizontal motion of individual granules reveals intermediate-scale convective flows. *See* SUN. Stephen L. Keil

Photosynthesis Literally, the synthesis of chemical compounds using light. The term photosynthesis, however, is used almost exclusively to designate one particularly important natural process: the use of light in the manufacture of organic compounds (primarily certain carbohydrates) from inorganic materials by chlorophyll- or bacteriochlorophyll-containing cells. This process requires a supply of energy in the form of light, since its products contain much more chemical energy than its raw materials. This is clearly shown by the liberation of energy in the reverse process, namely the combustion of organic material with oxygen, which takes place during respiration. *See* PLANT RESPIRATION; RESPIRATION.

Among chlorophyll-containing plant and algal cells, and in cyanobacteria (formerly known as blue-green algae), photosynthesis involves the oxidation of water (H_2O) to produce oxygen molecules, which are then released into the environment. This is called oxygenic photosynthesis. In contrast, bacterial photosynthesis does not involve O_2 evolution (production). In this case, other electron donors, such as H_2S, are used instead of H_2O. This process is called anoxygenic photosynthesis.

The light energy absorbed by the pigments of photosynthesizing cells, especially by chlorophyll or bacteriochlorophyll pigments, is efficiently converted into stored chemical energy. Together, the two aspects of photosynthesis—the conversion of inorganic matter into organic matter, and the conversion of light energy into chemical energy—make it the fundamental process of life on Earth: it is the ultimate source of all living matter and of almost all of the energy of life.

The net overall chemical reaction of oxygenic photosynthesis (by plants, algae, and cyanobacteria) is shown in the equation below, where $\{CH_2O\}$ stands for a carbohydrate (sugar). The

$$H_2O + CO_2 + \text{light energy} \xrightarrow[\text{enzymes}]{\text{chlorophyll}} \{CH_2O\} + O_2$$

photochemical reaction in photosynthesis belongs to the type known as oxidation-reduction, with CO_2 acting as the oxidant (electron acceptor) and water as the reductant (electron donor). The unique characteristic of this particular oxidation-reduction is that it is energetically unfavorable; that is, it converts chemically stable materials into chemically unstable products. Light energy is used to make this "uphill" reaction possible. A considerable part of the light energy used for this process is stored as chemical energy.

Photosynthesis is a complex multistage process that consists of nearly a hundred physical processes and chemical reactions. To make this complex process more understandable, it is useful to divide it into four temporal stages. Each phase is based roughly on the time scale in which it occurs. These phases are (1) photon absorption and energy transfer processes in antennas (or antenna chlorophylls, molecules which collect light quanta); (2) primary electron transfer in photochemical reaction centers; (3) electron transport and adenosine triphosphate (ATP) formation; and (4) carbon fixation and export of stable products.

The photosynthetic process takes place within certain structural elements found in plant cells. All algae, as well as all higher plants, contain pigment-bearing subcellular organelles called chloroplasts. Under the electron microscope, all chloroplasts show a layered structure with alternating lighter and darker layers of roughly 0.01 μm in thickness. These layers are membranes that contain proteins. These membranes are called thylakoid membranes (thylakoid stands for a membrane sac). The proteins bind all of the chlorophyll. The membranes are the sites of the first three phases of photosynthesis. *See* CELL PLASTIDS.

The photochemical apparatus is less complex in cyanobacteria. These cells are prokaryotes and therefore lack a nucleus and organelles such as chloroplasts and mitochondria. The early phases of photosynthesis take place on thylakoid membranes, which extend throughout the interior of the cell.

Experiments suggest that plants contained two pigment systems. One (called photosystem I, or PSI; sensitizing reaction I) is primarily composed of chlorophyll *a*; the other (called photosystem II, or PSII; sensitizing reaction II) is also composed of chlorophyll *a*, but includes most of chlorophyll *b* or other auxiliary pigments. These other auxiliary pigments include red and blue pigments, called phycobilins, in red algae and cyanobacteria, for example, and the brown pigment fucoxanthol in brown algae and diatoms. Efficient photosynthesis requires the absorption of an equal number of quanta in PSI and in PSII. This ensures that within both systems excitation energy is absorbed by the antenna system and partitioned to each photosystem, where the energy drives the chemical reactions. *See* CHLOROPHYLL; PLANT PIGMENT. Robert E. Blankenship; Govindjee

The light-dependent conversion of radiant energy into chemical energy as adenosine triphosphate (ATP) and reduced nicotinamide adenine dinucleotide phosphate (NADPH) serves as a prelude to the utilization of these compounds for the reductive fixation of CO_2 into organic molecules. Such molecules, broadly designated as photosynthates, are usually but not invariably in the form of carbohydrates such as glucose polymers or sucrose, and form the base for the nutrition of all living things, as well as serving as the starting material for fuel, fiber, animal feed, oil, and other compounds used by people. Collectively, the biochemical processes by which CO_2 is assimilated into organic molecules are known as the photosynthetic dark reactions, not because they must occur in darkness but because, in contrast to the photosynthetic light reactions, light is not required.

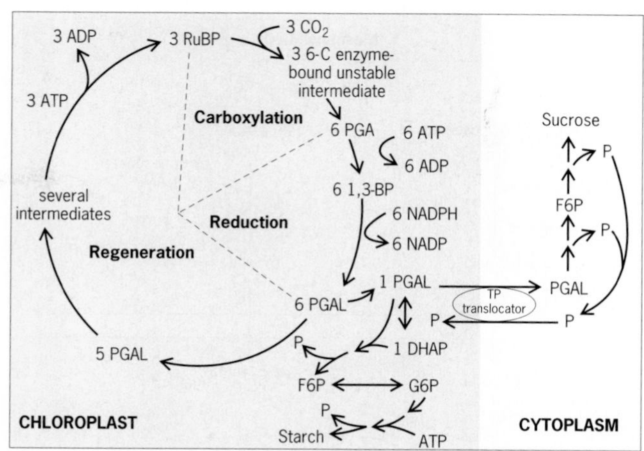

Fig. 1. Schematic outline of the C_3 (Calvin-Benson-Bassham) CO_2 assimilation cycle showing the partitioning of assimilated carbon into starch within the chloroplast [via the phosphorylated 6-C intermediates, fructose-6-phosphate (F6P) and glucose-6-phosphate (G6P)], and the assimilate efflux across the chloroplast inner envelope to the cytoplasm leading to sucrose synthesis.

C_3 photosynthesis. The essential details of C_3 photosynthesis can be seen in Fig. 1. The entire cycle can be separated into three phases—carboxylation, reduction, and regeneration. Since the smallest intermediate in the cycle consists of three carbons, we will start with three molecules of CO_2. During the initial carboxylation phase, these three molecules of CO_2 are combined with three molecules of the five-carbon compound ribulose 1,5-bisphosphate (RuBP) in a reaction catalyzed by the enzyme RuBP carboxylase/oxygenase (rubisco) to form three molecules of an intermediate, unstable enzyme-bound six-carbon compound. These unstable molecules are hydrolyzed by the enzyme into six molecules of the three-carbon compound phosphoglyceric acid (PGA). These products of the carboxylation phase, the six (three-carbon) PGA molecules, are phosphorylated by six molecules of ATP (releasing ADP to be used for photophosphorylation via the light reactions) to form six 1,3-bisphosphoglycerate (1,3-BP) molecules. The resulting compounds are reduced (that is, in the reduction phase of the C_3 cycle) by the NADPH formed in photosynthetic light reactions to form six molecules of the three-carbon compound phosphoglyceraldehyde (PGAL). PGAL is isomerized to form another three-carbon compound, dihydroxyacetone phosphate (DHAP). The rest of the C_3 photosynthetic cycle (the regeneration phase) involves enzymatic steps that allow regeneration of RuBP, the initial carboxylation substrate. One molecule of PGAL is made available for combination with DHAP isomerized from a second PGAL (requiring a second "turn" of the Calvin-Benson-Bassham cycle wheel) to form a six-carbon sugar. The other five PGAL molecules, through a complex series of enzymatic reactions, are rearranged into three molecules of RuBP, which can again be carboxylated with CO_2 to continue the cycle. The net product of two "turns" of the cycle, a six-carbon sugar (G6P or F6P), is formed either within the chloroplast in a pathway leading to starch (a polymer of many glucose molecules) or externally in the cytoplasm in a pathway leading to sucrose (condensed from two six-carbon sugars, glucose and fructose). *See* PLANT METABOLISM.

C_4 photosynthesis. Initially, the C_3 cycle was thought to be the only route for CO_2 assimilation, although it was recognized by plant anatomists that some rapidly growing plants (such as maize, sugarcane, and sorghum) possessed an unusual organization of the photosynthetic tissues in their leaves (Kranz morphology). It was then demonstrated that plants having the Kranz anatomy utilized an additional CO_2 assimilation route now known as the C_4-dicarboxylic acid pathway (Fig. 2).

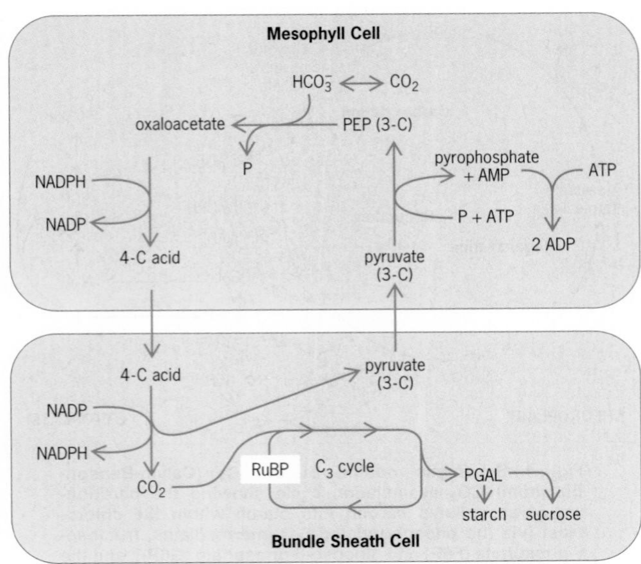

Fig. 2. Schematic outline of the C_4 carbon dioxide assimilation process in the two cell types of a NADP-ME-type plant.

Carbon dioxide enters a mesophyll cell where it is combined (in the form of bicarbonate) with the three-carbon compound phosphoenolpyruvate (PEP) via the enzyme PEP carboxylase to form a four-carbon acid, oxaloacetate, which is reduced to malic acid or transaminated to aspartic acid. The four-carbon acid moves into bundle sheath cells where the acid is decarboxylated and the CO_2 reassimilated via the C_3 cycle. The resulting three-carbon compound, pyruvic acid, moves back into the mesophyll cell and is transformed into PEP (at the cost of 2 ATP molecules) via the enzyme pyruvate phosphate dikinase located in the mesophyll chloroplasts to complete the cycle.

C_4 metabolism is classified into three types, depending on the primary decarboxylation reaction used with the four-carbon acid in the bundle sheath cells. The majority of C_4 species (exemplified by sugarcane, maize, crabgrass, and sorghum) are of type 1 (see below), and employ NADP-malic enzyme (NADP-ME) for decarboxylation. NAD-malic enzyme (NAD-ME) C_4 plants (type 2) include amaranthus, atriplex, millet, pigweed, and purslane. Type 3 C_4 types use phosphoenolpyruvate carboxykinase (PCK) for decarboxylation and include *Panicum* grasses. The decarboxylases are also located in different intracellular compartments as indicated:

1. NADP-ME type,

$$\text{NADP}^+ + \text{malic acid} \xrightarrow{\text{NADP-malic enzyme (chloroplasts)}} \text{pyruvic acid} + CO_2 + \text{NADPH}$$

2. NAD-ME type,

$$\text{NAD}^+ + \text{malic acid} \xrightarrow{\text{NAD-malic enzyme (mitochondria)}} \text{pyruvic acid} + CO_2 + \text{NADH}$$

3. PCK type,

$$\text{Oxaloacetic acid} + \text{ATP} \xrightarrow{\text{phosphoenolpyruvate carboxykinase (cytosol)}} \text{PEP} + CO_2 + \text{ADP}$$

CAM photosynthesis. Under arid and desert conditions, where soil water is in short supply, transpiration during the day when temperatures are high and humidity is low may rapidly deplete the plant of water, leading to desiccation and death. By keeping stomata closed during the day, water can be conserved, but the uptake of CO_2, which occurs entirely through the stomata, is prevented. Desert plants in the Crassulaceae, Cactaceae, Euphorbiaceae, and 15 other families have evolved, apparently independently of C_4 plants, a similar strategy of concentrating and assimilating CO_2 by which the CO_2 is taken in at night when the stomata open; water loss is low because of the reduced temperatures and correspondingly higher humidities. First studied in plants of the Crassulaceae, the process has been called crassulacean acid metabolism (CAM).

In contrast to C_4, where two cell types usually cooperate, the entire CAM process occurs within an individual cell; the separation of C_4 and C_3 is thus temporal rather than spatial. At night, CO_2 combines with PEP through the action of PEP carboxylase, resulting in the formation of oxaloacetic acid and its conversion into malic acid. The PEP is formed from starch or sugar via the glycolytic route of respiration. Thus, there is a daily reciprocal relationship between starch (a storage product of C_3 photosynthesis) and the accumulation of malic acid (the terminal product of nighttime CO_2 assimilation.

Gerald A. Berkowitz; Archie R. Portis, Jr.; Govindjee

Phototransistor A semiconductor device with electrical characteristics that are light-sensitive. Phototransistors differ from photodiodes in that the primary photoelectric current is multiplied internally in the device, thus increasing the sensitivity to light. *See* PHOTODIODE; TRANSISTOR.

Some types of phototransistors are supplied with a third, or base, lead. This lead enables the phototransistor to be used as a switching, or bistable, device. The application of a small amount of light causes the device to switch from a low current to a high current condition. *See* PHOTOELECTRIC DEVICES. W. R. Sittner

Phototube An electron tube comprising a photocathode and an anode mounted within an evacuated glass envelope through which radiant energy is transmitted to the photocathode. A gas phototube contains, in addition, argon or other inert gas which provides amplification of the photoelectric current by partial ionization of the gas. The photocathode emits electrons when it is exposed to ultraviolet, visible, or near-infrared radiation. The anode is operated at a positive potential with respect to the photocathode. *See* ELECTRICAL CONDUCTION IN GASES; ELECTRON TUBE.

A phototube responds to radiation over a limited range of the spectrum that is determined by the photocathode material. Radiant sensitivity is the photoelectric current emitted per unit of incident monochromatic radiant power. *See* PHOTOEMISSION.

Quantum efficiency, or photoelectron yield, is the number of electrons emitted per incident photon. For photometric applications a useful parameter is luminous sensitivity: the photoelectric current per lumen incident from a specified source of light. A source commonly used is a tungsten-filament lamp operated at a color temperature of 4700°F (2870 K). *See* INCANDESCENCE; LUMINOUS FLUX; PHOTON.

Photocathodes are semiconductors which contain one or more of the alkali metals sodium, potassium, rubidium, or cesium chemically combined with bismuth, antimony, or silver oxide. The cathode surface contains a critical excess of the alkali metal which enhances photoelectric emission by decreasing the affinity of the surface for electrons. Negative affinity for electrons is achieved with the gallium arsenide:cesium (GaAs:Cs) and indium gallium arsenide:cesium (InGaAs:Cs) photocathodes used in photomultipliers. Phototubes also emit electrons thermionically at ambient temperatures. This "dark current," observed in the absence of all irradiance, increases almost exponentially with temperature. Thermionic emission from the cesium antimonide (CsSb) photocathode is about 10^{-15} A/cm^2 at 68°F (20°C). *See* PHOTOMULTIPLIER; SEMICONDUCTOR.

Vacuum phototubes are used as detectors of radiant energy in the spectral range from 200 to 1100 nanometers. Since the photoelectric current is directly proportional to the intensity of the radiation, these tubes are used in radiometers, photometers, and colorimeters. By virtue of their narrow pulse response, vacuum phototubes are also used to measure the intensity of very short pulses of light generated by lasers and visible nuclear radiation. Gas phototubes can be used in light-operated relays and for the reproduction of sound from motion picture film, although their response to intensity-modulated light is limited to frequencies below 15 kHz. Vacuum as well as phototubes have been replaced in many applications by semiconductor photodiodes and photovoltaic cells. *See* COLORIMETRY; LASER; PHOTODIODE; PHOTOMETER; PHOTOVOLTAIC CELL; RADIOMETRY. James L. Weaver

Photovoltaic cell A device that detects or measures electromagnetic radiation by generating a current or a voltage, or both, upon absorption of radiant energy. Specially designed photovoltaic cells are used for power generation, as in solar batteries or solar cells, and for sensitive detection of electromagnetic radiation in radiometry, optical communications, spectroscopy, and other applications. An important advantage of the photovoltaic cell in these particular applications is that no separate bias supply is needed—the device generates a signal (voltage or current) simply by the absorption of radiation.

Most photovoltaic cells consist of a semiconductor *pn* junction or Schottky barrier in which electron-hole pairs produced by absorbed radiation are separated by the internal electric field in the junction to generate a current, a voltage, or both, at the device terminals. Under open-circuit conditions (current $I = 0$) the terminal voltage increases with increasing light intensity, and under short-circuit conditions (voltage $V = 0$) the magnitude of the current increases with increasing light intensity. When the current is negative and the voltage is positive, the photovoltaic cell delivers power to the external circuit. In this case, if the source of radiation is the Sun, the photovoltaic cell is referred to as a solar battery or solar cell. When a photovoltaic cell is used as a photographic exposure meter, it produces a current proportional to the light intensity, which is indicated by a low-impedance galvanometer or microammeter. For use as sensitive detectors of infrared radiation, specially designed photovoltaic cells can be operated with either low-impedance (current) or high-impedance (voltage) amplifiers, although the lowest noise and highest sensitivity are achieved in the current or short-circuit mode. Another mode of operation of a *pn* junction diode as a photodetector involves the application of a reverse bias voltage to the diode. In this case, the photogenerated current is directly proportional to the incident power, and the diode is said to be operated in the photodiode mode rather than the photovoltaic mode. *See* EXPOSURE METER; JUNCTION DIODE; OPTICAL DETECTORS; PHOTODIODE; PHOTOELECTRIC DEVICES; PHOTOVOLTAIC EFFECT; RADIOMETRY; SEMICONDUCTOR; SEMICONDUCTOR DIODE; SOLAR CELL. Gregory E. Stillman

Photovoltaic effect The conversion of electromagnetic radiation into electric power through absorption by a semiconducting material. Devices based on this effect serve as power sources in remote terrestrial locations and for satellites and other space applications. Photovoltaic powered calculators and other consumer electronic products are widely available, and solar photovoltaic automobiles and aircraft have been demonstrated.

The basic requirements for the photovoltaic effect are (1) the absorption of photons through the creation of electron-hole pairs in a semiconductor; (2) the separation of the electron and hole so that their recombination is inhibited and the electric field within the semiconductor is altered; and (3) the collection of the electrons and holes, separately, by each of two current-collecting electrodes so that current can be induced to flow in a circuit external to the semiconductor itself.

There are many approaches to achieving these three requirements simultaneously. A very common approach for separating the electrons from the holes is to use a single-crystal semiconductor, for example, silicon, into which a *pn* junction has been diffused. Silicon is often chosen because its optical band gap permits the absorption of a substantial portion of solar photons via the generation of electron-hole pairs. The fabrication of such a device structure causes a local transfer of negative charges from the *n* layer into the *p* layer, bending the conduction and valence bands in the vicinity of the *p-n* boundary, and thereby creating a rectifying junction. Electrons generated in the *p* region can lower their energy by migrating into the *n* region, which they will do by a random walk process in the electric-field-free region far from the junction, or by drift induced by the electric field in the junction region. Holes created in the *n* region, conversely, lose energy by migrating into the *p* region. Thus the presence of such a junction leads to the spontaneous spatial separation of the photogenerated carriers, thereby inducing a voltage difference between current-carrying electrodes connected to the *p* and *n* regions. This process will continue until the difference in potential between the two electrodes is large enough to flatten the bands in the vicinity of the junction, canceling out the internal electric field existing there and so eliminating the source of carrier separation. The resulting voltage is termed the open-circuit voltage, and approximates the built-in voltage associated with the *pn* junction in the dark, a value which cannot exceed the band gap of the semiconductor. *See* HOLE STATES IN SOLIDS; SEMICONDUCTOR; SEMICONDUCTOR DIODE.

In the limit when the device is short-circuited by the external circuit, no such buildup of potential can occur. In this case, one electron flows in the external circuit for each electron or hole which crosses the junction, that is, for each optically generated electron-hole pair which is successfully separated by the junction. The resulting current is termed the short-circuit current and, in most practical photovoltaic devices, approaches numerically the rate at which photons are being absorbed within the device. Losses can arise from the recombination of minority carriers (for example, electrons in the *p*-type region, holes in the *n*-type region) with majority carriers. *See* ELECTRON-HOLE RECOMBINATION.

For a photovoltaic device to generate power, it is necessary to provide a load in the external circuit which is sufficiently resistive to avoid short-circuiting the device. In this case, the voltage will be reduced compared to the open-circuit voltage because a continuing requirement exists for carrier separation at the junction; thus some band bending and its associated internal field must be retained.

Various multiple-layered device configurations based on doped and undoped alloys of amorphous silicon have been developed for photovoltaic devices used in applications ranging from solar watches and calculators to remote power generators. The photovoltaic effect in these devices is particularly intriguing since it is possible to build up so-called tandem devices by stacking one device electrically and optically in series above another. In addition to the increased voltage and concomitant reduction in the required current-carrying capability of electrode grid structures, such devices permit, in principle, an increased efficiency of solar photovoltaic energy conversion. *See* SOLAR CELL. John P. de Neufville

Phthiraptera An order of small wingless insects, permanently parasitic on mammals and birds. The Phthiraptera are one of approximately 30 major groups of insects and can be divided into four suborders, three of which (the Amblycera, Ischnocera, and Rhynchophthirina) are known as chewing or biting lice, and the fourth (the Anoplura) as sucking lice. Amblycera and Ischnocera are found on both mammals and birds, whereas species of Rhynchophthirina and Anoplura are confined to mammals. In total there are approximately 5000 known species, of which just under 70% are recorded from a single host species. *See* ANOPLURA; INSECTA; PARASITOLOGY.

Lice spend their entire life cycle in the microhabitat provided by the hosts' skin, fur, or feathers, and are highly adapted to this environment. The louse body is dorsoventrally flattened, and their legs are modified for clinging to feathers or fur. Their adult size ranges from 3 to 11 mm (0.1 to 0.4 in.). Coloration varies from pale white through shades of yellow and brown to black. Patterns are evident on some species, and cryptic coloration is often employed to match the coloration of the host. Chewing lice have mandibles that have been variously modified in each of the suborders. The mandibles are involved in cutting up the hair or feather material before it is digested, and play a vital secondary role in anchoring the louse to the host. In most species of sucking lice, the mandibles have been completely lost. Instead, sharp stylets protrude from the head and are used to pierce small vessels located close to the skin's surface. Anoplura draw up blood using these stylets—hence their name sucking lice.

The present distribution of lice essentially mirrors that of their hosts, although true geographic distributions of lice are known to occur within the range of certain host species. Consequently lice are found worldwide on every continent and in virtually every habitat occupied by mammals and birds. With the exception of the human body louse, all species complete their entire life cycle from egg to adult on the body of their host.

Lice are important pests of domesticated mammals and birds. The human body louse (*Pediculus humanus humanus*) is the vector of *Rickettsia prowazekii*, which causes louse-borne typhus. The human head louse (*Pediculus humanus capitis*) is commonplace in the developed world, and infests millions of schoolchildren every year. The only other louse known to infect humans is the pubic louse (*Pthirus pubis*), which is normally transmitted through sexual contact. *See* PEDICULOSIS; RICKETTSIOSES; SEXUALLY TRANSMITTED DISEASES.

Vincent S. Smith

Phycobilin Any member of a class of intensely colored pigments found in some algae that absorb light for photosynthesis. Phycobilins are structurally related to mammalian bile pigments, and they are unique among photosynthetic pigments in being covalently bound to proteins (phycobiliproteins). In at least two groups of algae, phycobiliproteins are aggregated in a highly ordered protein complex called a phycobilisome.

Phycobilins occur only in three groups of algae: cyanobacteria (blue-green algae), Rhodophyta (red algae), and Cryptophyceae (cryptophytes), and are largely responsible for their distinctive colors, including blue-green, yellow, and red. Five different phycobilins have been identified to date, but the two most common are phycocyanobilin, a blue pigment, and phycoerythrobilin, a red pigment. In the cell, these pigments absorb light maximally in the orange (620-nanometers) and green (550-nm) portion of the visible light spectrum, respectively. A blue-green light (495-nm) absorbing pigment, phycourobilin, is found in some cyanobacteria and red algae. A yellow light (575-nm) absorbing pigment, phycobiliviolin (also called cryptoviolin), is apparently found in all cryptophytes but in only a few cyanobacteria. A fifth phycobilin, which absorbs deep-red light (697 nm), has been identified spectrally in some cryptophytes, but its chemical properties are unknown. *See* CRYPTOPHYCEAE; CYANOPHYCEAE; RHODOPHYCEAE.

Phycobilins are associated with the photosynthetic light-harvesting system in chloroplasts of red algae and cryptophytes and with the photosynthetic membranes of cyanobacteria, which lack chloroplasts. Phycobilins are covalently bound to a water-soluble protein that aggregates on the surface of the photosynthetic membrane. All other photosynthetic pigments (for example, chlorophylls and carotenoids) are bound to photosynthetic membrane proteins by hydrophobic attraction. Phycobiliprotein can constitute a major fraction of an alga. In some cyanobacteria, phycobiliproteins can account for more than 50% of the soluble protein and one-quarter of the dry weight of the cell. *See* CELL PLASTIDS.

Phycobilins are photosynthetic accessory pigments that absorb light efficiently in the yellow, green, orange, or red portion of the light spectrum, where chlorophyll *a* only weakly absorbs. Light energy absorbed by phycobilins is transferred with greater than 90% efficiency to chlorophyll *a*, where it is used for photosynthesis. *See* CHLOROPHYLL; PHOTOSYNTHESIS.

Todd M. Kana

Phylactolaemata A class of the phylum Bryozoa. Phylactolaemates have lophophores (crowns of tentacles surrounding the mouth) which are markedly U-shaped (or rarely nearly circular but still kidney-shaped) in basal outline, and relatively short wide zooecia (exoskeletal living chambers); these animals dwell only in freshwater. *See* BRYOZOA; LOPHOPHORE.

Phylactolaemate colonies are either encrusting threadlike networks of relatively isolated zooecia with solid chitinous uncalcified walls, or small to large masses of gelatinous material in which the individual zooids are embedded side by side without definite separating zooecial walls. Stolons (elongated projections of the body wall) are not present.

Only about 50–100 phylactolaemate species exist, all classified in a single order, the Plumatellida. Exclusively freshwater, phylactolaemate colonies occur on shallowly submerged tree branches, roots, floating leaves, pond weeds, stones, and debris (like automobile tires), in streams, ponds, and lakes, both natural and artificial. The phylactolaemates may have evolved relatively recently from ctenostomes, although some workers have suggested that phylactolaemates might be very primitive ectoprocts surviving as evolutionary relics. *See* CTENOSTOMATA; FRESHWATER ECOSYSTEM.

Roger J. Cuffey

Phyllite A type of metamorphic rock formed during low-grade metamorphism of clay-rich sediments called pelites. Phyllites are very fine grained rocks with a grain size barely visible in a hand specimen. They have a well-developed planar element called cleavage defined by alignment of mica grains and interlayering of quartz-rich and mica-rich domains. Typically, mica grains show the greater alignment, although other mineral components (quartz, carbonate, and feldspars) may show a preferred shape orientation. Where all minerals of a particular type show the same degree of alignment and the fabric is well developed throughout the rock, the fabric is termed a penetrative fabric. Cleavage surfaces in phyllites have a glittery, lustrous sheen due to light reflecting off grains of chlorite and muscovite. The mineralogy of phyllites is dependent on chemical composition; typical minerals in phyllites are chlorite, muscovite, and quartz. Other minerals that may be present in phyllites formed during low-grade metamorphism include chlorotoid, garnet (rarely), sodium-mica, and sulfide minerals. *See* CHLORITE; MUSCOVITE; QUARTZ.

Phyllite is found in most regionally metamorphosed terranes in the world, including the Appalachians of eastern North America, the Scottish Highlands, and the Alps. *See* METAMORPHIC ROCKS.

Matthew W. Nyman

Phyllocarida A subclass of the crustacean class Malacostraca containing the extant order Leptostraca and the fossil order Archiostraca. The Phyllocarida has a long fossil record, and many early fossil taxa were referred to this subclass. However, studies of presumed phyllocarids from the Burgess Shale have shown that only the archiostracans agree with the definition of the Phyllocarida.

Phyllocarids are distinct from other malacostracan crustaceans because of two other characteristics considered to reflect the primitive condition, which strengthen the hypothesis of early separation from the main evolutionary line. The first is the presence of a bivalve carapace. The second is an abdomen consisting of seven fully formed somites and terminating in a telson that bears caudal rami. *See* CRUSTACEA; LEPTOSTRACA; MALACOSTRACA.

Patsy A. McLaughlin

Phylogeny The genealogical history of organisms, both living and extinct. Phylogeny represents the historical pattern of relationships among organisms which has resulted from the actions of many different evolutionary processes. Phylogenetic relationships are depicted by branching diagrams called

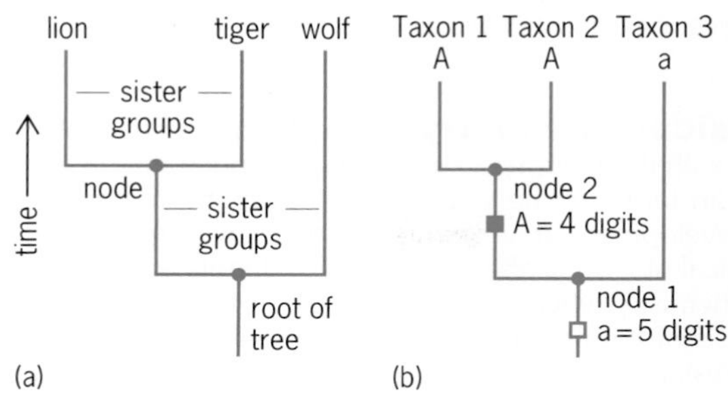

Phylogenetic trees. (*a*) A tree representing a hierarchical pattern of sister-group relationships (those taxa descended from a common ancestor). (*b*) Relationships are determined by identifying derived characters; in this case the condition of five digits is primitive, whereas the loss of a digit is derived and unites taxa 1 and 2.

cladograms, or phylogenetic trees. Cladograms show relative affinities of groups of organisms called taxa. Such groups of organisms have some genealogical unity, and are given a taxonomic rank such as species, genera, families, or orders. For example, two species of cats—say, the lion (*Panthera leo*) and the tiger (*Panthera tigris*)—are more closely related to each other than either is to the gray wolf (*Canis latrans*). The family including all cats, Felidae, is more closely related to the family including all dogs, Canidae, than either is to the family that includes giraffes, Giraffidae. The lion and tiger, and the Felidae and Canidae, are called sister taxa because of their close relationship relative to the gray wolf, or to the Giraffidae, respectively.

Cladograms thus depict a hierarchy of relationships among a group of taxa (illus. *a*). Branch points, or nodes, of a cladogram represent hypothetical common ancestors (not specific real ancestors), and the branches connect descendant sister taxa. If the taxa being considered are species, nodes are taken to signify speciation events. The goal of the science of cladistics, or phylogenetic analysis, is to discover these sister-group (cladistic) relationships and to identify what are termed monophyletic groups—two or more taxa postulated to have a single, common origin.

The acceptance of a cladogram depends on the empirical evidence that supports it relative to alternative hypotheses of relationship for those same taxa. Evidence for or against alternative phylogenetic hypotheses comes from the comparative study of the characteristics of those taxa. Similarities and differences are determined by comparison of the anatomical, behavioral, physiological, or molecular [such as deoxyribonucleic acid (DNA) sequences] attributes among the taxa. A statement that two features in two or more taxa are similar and thus constitute a shared character is, in essence, a preliminary hypothesis that they are homologous; that is, the taxa inherited the specific form of the feature from their common ancestor. However, not all similarities are homologs; some are developed independently through convergent or parallel evolution, and although they may be similar in appearance, they had different histories and thus are not really the same feature. In cladistic theory, shared homologous similarities are either primitive (plesiomorphic condition) or derived (apomorphic condition), whereas nonhomologous similarities are termed homoplasies (or sometimes, parallelisms or convergences). This distinction over concepts and terminology is important because only derived characters constitute evidence that groups are actually related.

As evolutionary lineages diversify, some characters will become modified. Examples include the enlargement of forelimbs or the loss of digits on the hand. Thus, during evolution the foot of a mammal might transform from a primitive condition of having five digits to a derived form with only four digits (illus. *b*). Following branching at node 1, the foot in one lineage undergoes an evolutionary modification involving the loss of a digit (expressed as character state A). A subsequent branching event then produced taxa 1 and 2, which inherited that derived character. The lineage leading to taxon 3, however, retained the primitive condition of five digits (character state a). The presence of the shared derived character, A, is called a synapomorphy, and identifies taxa 1 and 2 as being more closely related to each other than either is to taxon 3. Distinguishing between the primitive and derived conditions of a character within a group of taxa (the ingroup) is usually accomplished by comparisons to groups postulated to have more distant relationships (outgroups). Character states that are present in ingroups but not outgroups are postulated to be derived. Systematists have developed computer programs that attempt to identify shared derived characters (synapomorphies) and, at the same time, use them to construct the best phylogenetic trees for the available data.

Knowledge of phylogenetic relationships provides the basis for classifying organisms. A major task of the science of systematics is to search for monophyletic groups. Some groups, such as birds and mammals, are monophyletic; that is, phylogenetic analysis suggests they are all more closely related to each other than to other vertebrates. However, other traditional groups, such as reptiles, have been demonstrated to be nonmonophyletic (some so-called reptiles, such as dinosaurs and their relatives, are more closely related to birds than they are to other reptiles such as snakes). Classifications based on monophyletic groups are termed natural classifications. Phylogenies are also essential for understanding the distributional history, or biogeography, of organisms. Knowing how organisms are related to one another helps the biogeographer to decipher relationships among areas and to reconstruct the spatial histories of groups and their biotas. *See* ANIMAL EVOLUTION; BIOGEOGRAPHY; SYSTEMATICS; TAXONOMIC CATEGORIES.

Joel Cracraft

Phymosomatoida An order of regular sea urchins, class Echinoidea, characterized by imperforate tubercles, complex ambulacral compounding in which one or more elements are occluded from the perradial suture, and a stirodont lantern with unfused epiphyses. They comprise two families. They first appeared in the Lower Jurassic and are probably paraphyletic, since they include the ancestors of camarodonts. The two extant genera are each known from a single species: *Glyptocidaris*, from a depth of 30–490 ft (10–150 m) around northen Japan, and *Stomechinus*, a common inhabitant of rocky shores around the Indo-West Pacific. They are epifaunal grazers. *See* ECHINODERMATA; ECHINOIDEA.

Andrew Smith

Physical anthropology The subfield of anthropology that deals with human and nonhuman primate evolution, the biological bases of human behavior, and human biological variability and its significance. Some refer to the field as biological anthropology to signal the close links with other biological sciences. The term physical anthropology is largely an American and British invention; in most European and many other countries physical anthropologists are the only anthropologists, while persons who study behavioral aspects of the human condition are known as archeologists, ethnologists, linguists, or prehistorians. *See* ANTHROPOLOGY; ARCHEOLOGY; FORENSIC ANTHROPOLOGY.

Paleoanthropology. Paleoanthropology is the multidisciplinary study of human evolution as evidenced by fossils, artifacts, and their geological and burial site contexts. Physical anthropologists organize expeditions and direct excavations that lead to discoveries of fossil Hominidae, and then engage in the painstaking repair and reconstruction of specimens, their anatomical description, comparison with other specimens, and placement in hominoid phylogeny.

Morphological paleoanthropologists must have detailed knowledge of human and other primate anatomy and the principles of taxonomy in order to restore and interpret their discoveries. In addition to traditional anatomical descriptions and measurements, it is requisite that the variations in samples be presented and, when possible, be tested for statistical significance. Because of the fragmentary nature of many fossils, one of the most difficult problems is to decide whether the new discoveries belong with previously described species or represent new undescribed ones. *See* DATING METHODS; FOSSIL; FOSSIL HUMANS; PALEOECOLOGY.

Paleoprimatology. Physical anthropologists also look to nonhuman primates for clues to human physical history and status as mammals, and for analogies to the behavior and cognitive abilities of human ancestors. Like paleoanthropologists, paleoprimatologists employ methods of other paleobiologists to collect, describe, and interpret fossil specimens phylogenetically and functionally. Contextual information, including the manner in which specimens were deposited and their possible alteration over time, is vital. *See* FOSSIL PRIMATES.

Recent primates, including humans, are the current end products of evolution. Anatomical studies of modern primates are essential to model the morphologies of fossil forms and to reveal their singularity. Carefully controlled comparative studies on extant primates hold the greatest promise for modeling the functional morphology, physiology, and habitat preferences of extinct forms. Advanced techniques and concepts from rehabilitation medicine, kinesiology, orthopedics, orthodontics, radiology, and neurology have been used to establish a truly functional morphology of the primates. *See* PRIMATES.

Molecular anthropology. The explosion of molecular biology that followed the cracking of the genetic code attracted physical anthropologists who wished to test hypotheses about the propinquity of humans with the apes, and the relationships of other primates to one another and to other creatures. Some have endeavored to find "molecular clocks" that could tell when species diverged from one another. Assuming that genetic changes occur at fairly steady rates and given a few well-dated fossils, the time when the living species may have branched from one another can be estimated. *See* MOLECULAR ANTHROPOLOGY.

Primate behavior and ecology. Although firmly rooted in comparative psychology, behavioral primatology has also become a major section of physical anthropology; anthropologists have contributed mainly through field studies. An original goal of primate field studies by anthropologists was to find predictable relationships between specific habitats and the patterns of sociality of the primates that inhabit them. Because of their close genetic relationships to humans, chimpanzees have emerged as the most popular model for early hominid sociality. But theorists emphasizing different aspects of their behavior arrive at markedly different models for early hominid behavior. *See* BEHAVIORAL ECOLOGY; SOCIOBIOLOGY.

Human variation. The term human variation is rapidly replacing its historical predecessor "race" in anthropology because the latter carries so much negative connotation. Many scientists believe that the concept of race should be abandoned. Instead, researchers should simply record the gene frequencies and biological traits of human populations that are otherwise identified only by their geographic localities. This genotypic and phenotypic information would be interpreted in terms of historical and proximate selective forces in each environment.

According to the modern perspective, people should no longer be labeled Caucasoid, Mongoloid, Negroid, Australian, Nordic, Black, Brown, White, and so forth. While this may cause problems in communication as anthropologists attempt to apply their knowledge practically to forensic and medical problems, in the long run humanity will benefit from the exclusion of racial typologies. *See* HUMAN BIOLOGICAL VARIATION; HUMAN GENETICS.

Russell H. Tuttle

Physical chemistry The branch of chemistry that deals with the interpretation of chemical phenomena and properties in terms of the underlying physical processes, and with the development of techniques for their investigation. The term chemical physics is often employed to denote a branch of physical chemistry where the emphasis is on the interpretation and analysis of the physical properties of individual molecules and bulk systems, instead of their reactions. Theoretical chemistry is another major branch, where the emphasis is on the calculation of the properties of molecules and systems, and which used the techniques of quantum mechanics and statistical thermodynamics. It is convenient to regard physical chemistry as dealing with three aspects of matter: its equilibrium properties, structure, and ability to change.

Equilibrium properties. The study of matter in a state of equilibrium constitutes the field of chemical thermodynamics. In particular, chemical thermodynamics provides a technique for discussing the response of a system to a change in the external conditions (such as the shift in the boiling and freezing point of either a pure substance or a mixture when the applied pressure is changed, or when the composition of the mixture is modified), and for rationalizing the energy changes that occur in the course of a chemical reaction. The branch of thermodynamics dealing with the latter is called thermochemistry. Chemical thermodynamics also provides a framework for the determination of the maximum amount of work that may be generated by a system undergoing a specified change, and it therefore provides a way of establishig bounds for the efficiencies of a variety of devices, including engines, refrigerators, and electrochemical cells. Thermodynamics is used in chemistry to assess the position of equilibrium of a chemical reaction (that is, how far it will proceed), and to determine what conditions are necessary in order to optimize the yield of a particular product. The branch of chemical thermodynamics dealing with ionic reactions occurring in the presence of electrodes constitutes the field of equilibrium electrochemistry. *See* CHEMICAL EQUILIBRIUM; CHEMICAL THERMODYNAMICS; ELECTROCHEMISTRY; ENTHALPY; ENTROPY; FREE ENERGY; THERMOCHEMISTRY.

Structure. The principal role of quantum mechanics in chemistry is in the discussion of atomic and molecular structure, and in the interpretation of spectroscopic data. In the branch of physical chemistry known as computational quantum chemistry, interest centers on the numerical solution of the Schrödinger equation in order to obtain wave functions and geometries of molecules. Computational quantum chemistry is so developed that it is capable of being used to map the changes in the structures of molecules while they are in the course of reaction, when atoms and groups of atoms are being transferred from one molecule to another. *See* QUANTUM CHEMISTRY; SCHRÖDINGER'S WAVE EQUATION.

Spectroscopic techniques are used not only to identify molecules present in a sample, but also to determine their shape, size, and electron distribution. The techniques fall into four categories: absorption spectroscopy, emission spectroscopy, Raman spectroscopy, and resonance techniques. *See* ELECTRON PARAMAGNETIC RESONANCE (EPR) SPECTROSCOPY; ELECTRON SPECTROSCOPY; MOLECULAR STRUCTURE AND SPECTRA; MÖSSBAUER EFFECT; NUCLEAR MAGNETIC RESONANCE (NMR); PHOTOCHEMISTRY; RAMAN EFFECT; SPECTROSCOPY.

Techniques for the investigation of molecular structure based on diffraction depend on the observation of the direction through which radiation and particles are scattered when they impinge on a sample. Other techniques for investigating structure include the electric and magnetic properties of molecules, in particular, the determination of electric polarizabilities and dipole moments, magnetic properties, and the properties based on optical birefringenece, such a optical activity and the Faraday effect. *See* X-RAY DIFFRACTION.

Structural properties and thermodynamic properties are brought together by statistical thermodynamics. This major theoretical procedure gives a way of predicting the thermodynamic properties of assemblies of molecules in terms of their individual energy levels.

Physical and chemical change. The third major branch of physical chemistry is concerned with change: physical change and chemical change. In particular, it is concerned with the rate of change. Physical change includes the diffusion of one substance into another, or the migration of ions in an electrode solution. The application of thermodynamics to change in general constitutes the field of nonequilibrium thermodynamics. *See* GAS; TRANSPORT PROCESSES.

Chemical change may be studied at a variety of levels. Empirical chemical kinetics is the study of reactions in order to determine how their rates depend on the concentrations of the participants in the reaction and on the conditions, mainly the temperature. Investigation of the time dependence of reactions yields a detailed picture of the sequence of molecular transformations involved in a complex chemical reaction. *See* CHEMICAL DYNAMICS; SHOCK TUBE; ULTRAFAST MOLECULAR PROCESSES.

An important extension of chemical kinetics is to the reactions that occur on surfaces; these are the processes involved in heterogeneous catalysis. A special application of surface chemistry is to the stability of colloidal suspensions of species in fluids, and another is to the processes that occur at the interface between an electrode and the solution in which it is immersed. *See* ADSORPTION; COLLOID; HETEROGENEOUS CATALYSIS; SURFACE PHYSICS.

P. W. Atkins

Physical geography The study of the Earth's surface features and associated processes. Physical geography aims to explain the geographic patterns of climate, vegetation, soils, hydrology, and landforms, and the physical environments that result from their interactions. Physical geography merges with human geography to provide a synthesis of the complex interactions between nature and society.

The basic content of physical geography comprises a number of areas of specialization. Climatology, the scientific study of climates, concerns the total complex of weather conditions at a given location over an extended time period; it deals not only with average conditions but with extremes and variations. Geomorphology is the interpretive description and explanation of landforms and the fluvial, glacial, coastal, and eolian process that operate on them. The forms, processes, and patterns within the biosphere, including vegetation and animal distributions, are studied as biogeography. With strong ties to fluvial geomorphology, geographic hydrology concerns the scientific study of water from the aspects of distribution, movement, and utilization. Soil geography, with emphasis on the origin, characteristics, classification, and utilization potential of soils, provides an area of specialization with links to land use. Ultimately, the physical geography of a region is understood through an integration of the multiple aspects.

John E. Oliver

Physical law A term that designates four different concepts: (1) objective pattern (or natural regularity), (2) formula purporting to represent an objective pattern, (3) law-based rule (or uniform procedure), and (4) principle concerning any of the preceding.

For example, Newton's second law of motion, $ma = F$, is a law of type 2. It represents, to a good approximation, the actual behavior (law of type 1) of medium-size particles moving slowly relative to the speed of light. Alternative laws of motion, such as the relativistic and quantum-mechanical ones, are different laws of type 2 representing the same objective pattern or law of type 1 to even better approximations. One of the rules (laws of type 3) associated with Newton's second law of motion is: In order to set in motion a stationary particle, exert a force on it. Another is: In order to stop a moving particle, exert on it a force in the opposite direction. An example of a law of type 4 is: Newton's laws of motion are invariant under a Galileo transformation. *See* NEWTON'S LAWS OF MOTION.

A physical law of type 1, or objective pattern, is a constant relation among two or more properties of a physical entity. In principle, any such pattern can be conceptualized in different ways, that is, as alternative laws of type 2. The history of theoretical physics is to a large extent a sequence of laws of type 2. Every one of these is hoped to constitute a more accurate representation of the corresponding objective pattern or law of type 1, which is assumed to be constant and, in particular, untouched by human efforts to grasp it. Likewise, the history of engineering is to some extent a sequence of laws of type 3, or law-based rules of action, of which there are least two for every law of type 2. As for the laws of type 4, or laws of laws, they are of two kinds: scientific and philosophical. The general covariance principle is of the first kind, whereas the hypothesis that all events are lawful is a philosophical thesis. Unlike the former, whose truth can be checked, the principle of lawfulness is irrefutable. *See* ENGINEERING; THEORETICAL PHYSICS.

Not all formulas are called physical laws. For example, the regularities found by curve fitting are called empirical formulas. In physics a formula is called a law if and only if it meets the following conditions: it is part of a theory, and it has been satisfactorily confirmed by measurement or experiment at least within a certain domain (for example, for small mass densities or high field intensities). Thus, the basic assumptions of all the standard physical theories are laws, and so are their logical consequences. In particular, the usual variational principles, such as Hamilton's, are basic laws. However, the equations of motion and field equations entailed by such principles are derived laws (theorems); so are the conservation laws entailed by the equations of motion and field equations. However, the distinction between basic and derived laws is contextual: what is a principle in one theory may be a theorem in another. For example, Newton's second law of motion is a theorem in analytical dynamics, and the first principle of thermodynamics is a theorem of statistical mechanics. *See* CONSERVATION LAWS (PHYSICS); CURVE FITTING; HAMILTON'S PRINCIPLE; PHYSICAL THEORY; SCIENTIFIC METHODS; STATISTICAL MECHANICS; THERMODYNAMIC PRINCIPLES; VARIATIONAL METHODS (PHYSICS).

Mario Bunge

Physical measurement The determination of the size of a physical quantity by comparison with a standard. All physical measurement is ultimately based on measurement unit standards defined by the SI-metric system. Metric units are the foundation from which all other measurement units are derived, even the conventional nonmetric units used by the United States. *See* METRIC SYSTEM; UNITS OF MEASUREMENT.

The SI-metric system consists of seven base units representing length, mass, time, electric current, thermodynamic temperature, amount of substance, and luminous intensity (Table 1).

Length (meter). One of the most basic measurements is that of length. The first attempt at a scientific and logical set of

Table 1. SI-metric base units

Quantity	Unit	Symbol
Length	meter	m
Mass	kilogram	kg
Time	second	s
Electric current	ampere	A
Thermodynamic temperature	kelvin	K
Amount of substance	mole	mol
Luminous intensity	candela	cd

Table 2. Definitions of the SI-metric base units

Unit	Year	Definition*
Meter	1793	1/10,000,000 of the distance from the pole to the equator.
	1795	Provisional meter bar constructed in brass.
	1799	Definitive prototype meter bars constructed in platinum.
	1889	International prototype meter bar in platinum-iridium, cross-section X.
	1906	1,000,000/0.643,846,96 wavelengths in air of the red line of the cadmium spectrum.
	1960	1,650,763.73 wavelengths in vacuum of the radiation corresponding to the transition between levels $2p_{10}$ and $5d_5$ of the krypton-86 atom.
	1983	**The meter is the length traveled by light in vacuum during 1/299,792,458 of a second.**
Kilogram	1901	**The kilogram is the unit of mass; it is equal to the mass of the international prototype kilogram.**
Second	1956	The fraction 1/31,556,925.9746 of the tropical year for 1900 January 0 at 12 hours ephemeris time.
	1967	**The second is the duration of 9,192,631,770 periods of the radiation corresponding to the transition between two hyperfine levels of the ground state of the cesium-133 atom.**
Ampere	1948	**The ampere is that constant current which, if maintained in two straight parallel conductors of infinite length, of negligible circular cross section, and placed 1 meter apart in vacuum, would produce between these conductors a force equal to 2 × 10^{-7} newton per meter of length.**
Kelvin	1967	**The kelvin, unit of thermodynamic temperature, is the fraction 1/273.16 of the thermodynamic temperature of the triple point of water.**
Mole	1971	**The mole is the amount of substance of a system which contains as many elementary entities as there are atoms in 0.012 kilogram of carbon-12.**
Candela	1979	**The candela is the luminous intensity, in a given direction, of a source that emits monochromatic radiation of frequency 540 × 10^{12} hertz and that has a radiant intensity in that direction of 1/683 watt per steradian.**

*Present definition is in boldface.

measures was made in 1793, when the meter (metre) was defined as one ten-millionth of the distance from the pole to the equator of the Earth. The meter so defined was first standardized as the distance between two lines etched on a platinum-iridium bar. Copies of this prototype meter were distributed as standards of length to various countries signing on to the Treaty of the Metre in 1875. As technology improved, it was possible to devise more precise definitions of the meter, and in particular, a definition was sought that was reproducible by physical experiment, and not defined solely by an artifact. When the SI-metric system was formally established in 1960, the definition of the meter utilized a transition between two quantum levels of the krypton-86 atom (Table 2). That was followed in 1983 by a definition of the meter that fixed the speed of light as a constant and defined the meter as the length traveled in a very precise interval of time. This was possible because the second, as the base unit of time (discussed below), is defined even more precisely than the meter. *See* LIGHT.

Although the definition of the meter has changed several times over the years, the length of the meter has not changed, but the precision by which it is measured has improved. *See* LENGTH.

Mass (kilogram). Another very basic measure is that of mass, commonly called weight. (The distinction between mass and weight is an important one. In the metric system, mass and weight are distinct quantities and are assigned different unit names.) The kilogram is the only base unit of the SI-metric system whose name contains a prefix. That is because the gram originated as the unit of mass in the centimeter-gram-second (cgs) metric system, being defined as the mass of 1 cubic centimeter of water at standard temperature and pressure conditions. The kilogram remains defined by a platinum-iridium cylinder kept in a vault at the BIPM along with the original prototype meter, and is the only unit still defined by an artifact (Table 2). Competing studies are underway to create a definition of the kilogram, like that of the meter, which is reproducible in laboratories. *See* MASS; WEIGHT.

Time (second). The second originated with the Babylonian system of measuring time that is still used today, a system of 24-hour days divided into 60 minutes of 60 seconds each. However, the day cannot be used to precisely define the second, as the length of the day is not constant, nor is the length of the year. An early attempt to make the second more precise resulted in the second being defined in terms of a specific year, the tropical year 1900 (Table 2).

As with other SI base units, a more precise definition of the second was sought, one based on a physically reproducible experiment. The result is that the second is currently defined in terms of a transition between hyperfine levels of the ground state of the cesium-133 atom (Table 2).

Electric current (ampere). The "international" electrical units, the ampere and ohm, were standardized around the turn of the twentieth century. However, "absolute" definitions of the electrical units were desired. Of the various electrical units, the ampere was chosen as an SI-metric base unit, since it can be defined in terms of mechanical units using basic principles of physics (Table 2). This ties together mechanical and electrical units into a unified system. The realization of the ampere, however, is difficult and time-consuming, and this limits the relative uncertainty in the value of the ampere. *See* CURRENT MEASUREMENT; ELECTRICAL UNITS AND STANDARDS.

Thermodynamic temperature (kelvin). The SI-metric unit of thermodynamic temperature is the kelvin, the basis for an absolute temperature scale. The definition of the kelvin depends on the triple point of water, a fundamental fixed point in the temperature scale to which is assigned the temperature of 273.16 K (0.01°C). The kelvin unit is then defined as a fraction of the difference between the temperature of the triple point of water and absolute zero (Table 2). The reason for this definition was to make the size of the unit kelvin equal to the degree Celsius, which is defined as 1/100 of the difference between the melting temperature of ice (0°C) and the boiling temperature of water (100°C). *See* ABSOLUTE ZERO; TRIPLE POINT.

The kelvin scale is related to the degree Celsius scale through the relationship degrees Celsius = kelvin − 273.15 K, the constant being the melting temperature of ice (Table 3). Temperatures on this scale are called kelvins, not degrees kelvin. Also, kelvin is not capitalized, and the symbol (capital K) stands alone with no degree symbol. *See* TEMPERATURE; TEMPERATURE MEASUREMENT.

Amount of substance (mole). There is a need to define an amount of substance, as in chemical reactions. A standard amount of chemical element or compound was at one time called atomic weight or molecular weight, or more correctly atomic or molecular mass. These amounts were in fact masses relative

Table 3. Some baseline temperatures in two temperature scales

	kelvins, K	degrees Celsius, °C
Boiling point of water	373.15	100.
Melting point of ice	273.15	0.
Absolute zero	0.	−273.15

to the atomic weight (or mass) of oxygen, which was agreed to be 16, associated with the number of protons and neutrons in the oxygen atom. Oxygen, however, has a slightly variable mixture of isotopes ranging from 16 to 18, to which chemists associated the number 16. But physicists associated 16 with the isotope of oxygen of mass 16 only. This duality of definition, a source of confusion, ended when physicists and chemists agreed to assign a value of 12, exactly, to the relative atomic mass of the isotope of carbon with mass number 12, carbon-12. In SI-metric the amount of substance, now called a mole, is defined as a fixed mass of the carbon-12 isotope (Table 2). *See* ATOMIC MASS; ATOMIC MASS UNIT; MOLE (CHEMISTRY); MOLECULAR WEIGHT.

Luminous intensity (candela). The final SI-metric base unit is used in the area of photometry. Originally a "new candle" was defined as the luminance of a Planck (blackbody) radiator at the temperature of freezing platinum. This unit was given a new international name, the candela, in 1948. However, difficulties in realizing this unit at the high temperatures specified in the definition led to a new definition of the candela in 1979 (Table 2). *See* CANDLEPOWER; LUMINOUS INTENSITY; PHOTOMETRY.

Other units and measurements. All other measurement units can be traced back to SI-metric base units or combinations of SI-metric base and derived units, through the logical principles of the metric system for creating derived units. Even nonmetric units that are still in common use are defined in terms of SI-metric units.

In particular, the units of the metric system became legal for use in the United States in 1866. In 1893 the Mendenhall Order redefined conventional units (such as the yard and the pound) in terms of the metric system, rather than maintaining measurement standards for those units.

The United States is the only remaining major holdout of nonmetric usage. However, the process of conversion was made official in 1975 with the passage of the Metric Conversion Act. That Act was strengthened in 1988 to designate the metric system as the preferred system of units for the U.S. government. As a result, the metric transition, although voluntary, continues as more products and services adopt metric units in order to come in line with the dominant use of metric in the rest of the world.

Donald W. Hillger

Physical optics

The study of the interaction of electromagnetic waves in the optical range with material systems. The optical range of wavelengths may be taken as the range from about 1 nanometer to about 1 millimeter.

The explanation of the absorption, reflection, scattering, polarization, and dispersion of light by a material medium in terms of the properties of the atoms and molecules making up the medium is the objective of physical optics. In the course of seeking this objective, physicists have found that optical investigations are powerful methods of determining the structures of atoms and molecules and of large systems composed thereof. *See* ABSORPTION; ATOMIC STRUCTURE AND SPECTRA; CRYSTAL OPTICS; DIFFRACTION; DISPERSION (RADIATION); ELECTROMAGNETIC RADIATION; ELECTROOPTICS; FARADAY EFFECT; FLUORESCENCE; INTERFERENCE OF WAVES; LASER; LIGHT; MAGNETOOPTICS; MOLECULAR STRUCTURE AND SPECTRA; POLARIZED LIGHT; REFLECTION OF ELECTROMAGNETIC RADIATION; REFRACTION OF WAVES; SCATTERING OF ELECTROMAGNETIC RADIATION; SPECTROSCOPY.

Richard C. Lord

Physical organic chemistry

A branch of science concerned with the scope and limitations of the various rules, effects, and generalizations in use in organic chemistry by means of physical and mathematical methods. It includes, but is not limited to, the dynamics and energetics of organic chemical transformations, transient intermediates in these reactions, rate comparisons between families of reactions, dynamic stereochemistry, conservation of orbital symmetry, the least-motion principle, the isomer number for a given elemental composition, conformational analysis, nonexistent compounds, aromaticity, tautomerism, strain and steric hindrance, and the double-bond rule. Spectroscopy is the main tool employed, with nuclear magnetic resonance being the most widely used spectroscopic technique. With the advent of modern fast computers, computational chemistry has also become an important tool. *See* NUCLEAR MAGNETIC RESONANCE (NMR); SPECTROSCOPY.

Physical organic chemistry is traditionally distinguished from, yet totally intertwined with, synthetic organic chemistry, which deals with the question of how to obtain desired products from available compounds. This distinction can be illustrated with a diagram (see illustration) showing how the energy might vary during a chemical reaction in which the reactant R yields a product P ($R \rightarrow P$). Whereas the synthetic organic chemist will be interested primarily in the practical problem of how to convert R into P, the physical organic chemist studies the curve or curves connecting R and P as well as the structure and physical properties at all extrema, including R and P. However,the demarcation between synthetic organic chemistry and physical organic chemistry is not sharp. Physical organic chemists have contributed greatly to the understanding of the chemistry of hydrocarbons and their derivatives and have enhanced the repertoire of the synthetic organic chemists. In turn, synthetic organic chemists have made possible the construction of the custom-made, often intricate molecules that physical organic chemists use for their studies. The efforts of both groups, moreover, have made possible the birth of such new fields as molecular biochemistry and computational chemistry. *See* COMPUTATIONAL CHEMISTRY; MOLECULAR BIOLOGY.

Chemical reaction mechanisms. The diagram shown in the illustration is also useful in discussions of the dynamics of chemical reactions. It is an attempt to portray how the atoms in the reactant molecule R may move in space to their final positions in the product molecule P, and how the potential energy of the system would vary as a function of these positions. A complete correlation would be multidimensional; what is normally shown is a cross section in which the maximum potential energy is in fact a minimum (saddle point). While an essentially infinite number of pathways between R and P can be imagined and followed, the vast majority of the molecules will in practice use the one that makes the least demand on energy to reach the next maximum; this pathway is known as the reaction mechanism. The maxima (T terms in the illustration) are known as transition states, and the minima (I terms in the illustration) as intermediates.

If a reaction has a single transition state (and, hence, no intermediate), it is known as concerted; alternatively, it is step-wise. A stepwise reaction is simply a succession of concerted steps in which the intermediates are not isolated. *See* FREE RADICAL; ORGANIC REACTION MECHANISM; REACTIVE INTERMEDIATES; PHOTOCHEMISTRY.

Chemical kinetics. The most important route to quantitative information about a reaction is the study of its kinetics. This must begin with an experimental determination of the rate law: the expression that shows how the rate of formation of product $d[P]/dt$ (or loss of reactant: $-d[R]/dt$) depends on the concentration of all species involved in the reaction other than the solvent. The differential might be found to equal $k[R_1]$, $k[R_1]\,[R_2]$, $k[R_1]^2$, and so on; k is known as the rate constant, and the reaction is described as first-order, second-order, and so forth, depending

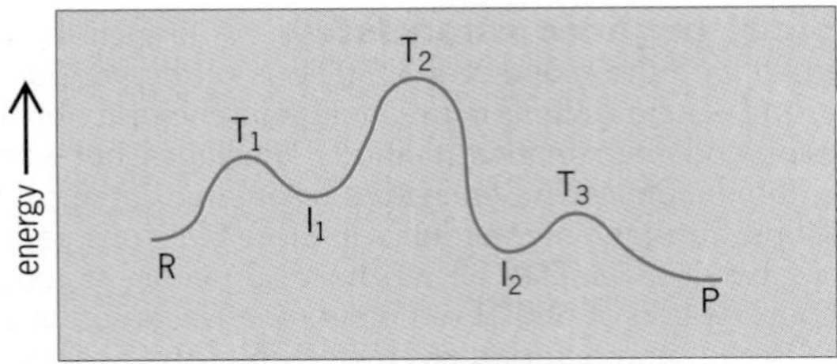

Energy profile of an organic reaction R → P. T terms indicate transition states and I terms indicate intermediates.

on the total number of concentration terms. One important feature is that the reaction order equals the sum of all molecules that have participated in the formation of the transition state (T_2 in the illustration). Thus, for a reaction to be concerted, it is necessary that the order equal the sum of all reactant molecules involved in the stoichiometry. *See* CATALYSIS; CHEMICAL DYNAMICS; PERICYCLIC REACTION.

Stereochemistry. This is also a powerful tool in physical organic chemistry. Experimental work has demonstrated that when a chiral compound such as $(-)HCR_1R_2X$ is converted into optically active HCR_1R_2Y by direct displacement with Y, the product obtained has an optical rotation that is opposite to that exhibited by the same material if it is produced in two steps, via initial displacement by A to give intermediate HCR_1R_2A as shown in the following reaction scheme. This result demonstrates

$$Y + HCR_1R_2{-}X \xrightarrow{\text{direct displacement}} Y{-}CR_2R_1H$$

$$HCR_1R_2{-}X \xrightarrow[-X]{A} A{-}CR_2R_1H \xrightarrow[-A]{Y} HCR_1R_2{-}Y \quad \text{(two steps)}$$

that displacement reactions occur with inversion; the reagent approaches in front, and the leaving group departs in the back. *See* OPTICAL ACTIVITY; STEREOCHEMISTRY.

Isomers. The isomer number has perhaps been the most important organizing principle in organic chemistry since its inception. Simply put, it means that for every elemental composition and molecular mass the isomer number can be predicted by writing all possible sequences of the atoms present, obeying the valence numbers of the atoms: four for carbon, three for nitrogen, two for oxygen, one for hydrogen and the halogens, and so forth. Once reliable atomic weights became available so that elemental compositions could be reported with confidence, this simple rule proved remarkably successful. *See* CHEMICAL BONDING; ORGANIC CHEMISTRY; VALENCE.

The need to specify molecular mass has proved more troublesome: it requires a definition of the concept of a molecule. Such definitions usually refer to covalent bonds as the entities that hold the atoms together, to rule out ionic species such as sodium chloride as candidates.

Among the extra compounds, none have affected organic chemistry more drastically than the stereoisomers. It turns out that for all but the simplest compounds a given sequence of the atoms may represent two, more than two, or even many more isomers. *See* CONFORMATIONAL ANALYSIS; MOLECULAR ISOMERISM.

Many compounds that are considered to be nonexistent, even though they are allowed by the simple rules of isomer numbers, in fact are transient intermediates in various reactions. Sometimes they can be detected spectroscopically, but cannot be isolated. There are instances in which neither of two isomers can be isolated, but mixtures of the two can. In other words, the barrier between the two is low, and the equilibrium constant is close to unity. An example is acetoacetic ester, which normally contains about 15% of the enol isomer. Such isomers are known as tautomers. *See* TAUTOMERISM.

William J. LeNoble

Physical science

Physical science The fields of inquiry to which the general designation science may be appropriately applied are broadly divided into social science and natural science. The latter is further subdivided into biology and physical science. Physical science is generally considered to include astronomy, chemistry, geology, mineralogy, meteorology, and physics. These overlap more or less, as illustrated by astrophysics, chemical physics, physical chemistry, and geophysics. There is overlap, likewise, between the physical and biological sciences, as seen in biochemistry, biophysics, virology, and the close relation between geology and paleontology. The boundaries implied in all such classifications are artificial and consist of regions where one field shades into another. *See* ASTRONOMY; BIOLOGY; CHEMISTRY; GEOLOGY; METEOROLOGY; MINERALOGY; PHYSICS; SCIENCE.

Joel H. Hildebrand

Physical theory

Physical theory A physical theory usually involves the attempt to explain a certain class of physical phenomena by deducing them as necessary consequences of other phenomena regarded as more primitive and less in need of explanation. The value of a theory depends on both the success with which it coordinates a wide range of presently known facts and its fertility in suggesting places to look for presently unknown phenomena.

Percy W. Bridgman; Gerald Holton

Physical vapor deposition

Physical vapor deposition Production of a film of material often on a heated surface and in a vacuum. Physical vapor deposition technology is used in a variety of applications. Coatings are produced from a wide range of materials, including metals, alloys, compound, cermets, and composites. The two basic processes for physical vapor deposition are evaporation deposition and sputter deposition. In evaporation, thermal energy converts a solid or liquid target material to the vapor phase. In sputtering, the target is biased to a negative potential and bombarded by positive ions of the working gas from the plasma, which knock out the target atoms and convert them to vapor by momentum transfer. *See* ALLOY; CERMET; COMPOSITE MATERIAL; METAL.

Physical vapor deposition processes consist of three major steps: generation of the depositing species, transport of the species from source to substrate, and film growth on the substrate. The processes are versatile because the steps occur sequentially and can be controlled independently.

The first step involves generating the depositing species by evaporation using resistance, induction, electron-beam, or laser-beam heating, or by sputtering using direct-current or radio-frequency plasma generation. In the second step involves transport from source to substrate. The third step is film growth on the substrate.

Alloy deposition by evaporation can be accomplished in three ways: coevaporation, evaporation from a single source, and flash evaporation. Coevaporation involves the use of a number of sources corresponding to the number of elements in the alloy. Its advantage is the ability to control the evaporation rate from each source. Its disadvantage is that the alloy composition varies along the plane of the substrate. Evaporation from a single source operates under steady-state conditions where the composition and volume of the liquid pool on top of a solid rod are kept constant. In the flash evaporation process, pellets of an alloy are dropped onto a very hot strip and are vaporized completely, thus maintaining the composition of the alloy in the deposit. This process works very well for elements with high vapor pressures.

Alloy deposition by sputtering is analogous to the evaporation method. In cosputtering, several sputter cathodes or sputter guns are used to generate the various vapor species. The same composition variation in the plane of the substrate occurs as in evaporation. An alloy can be sputtered from an alloy target if the modified surface composition is inversely proportional to the sputtering yield of the various elements. This is accomplished by a conditioning period of the target to allow progress from the alloy composition to the modified surface composition. *See* SPUTTERING.

Coatings provided by vapor deposition have many applications. For optical functions they are used as reflective or transmitting coatings for lenses, mirrors, and headlamps; for energy transmission and control they are used in glass, optical, or selective solar absorbers and in heat blankets; for electrical and magnetic functions they are used for resistors, conductors, capacitors, active solid-state devices, photovoltaic solar cells, and magnetic recording devices; for mechanical functions they are used for solid-state lubricants, tribological coatings for wear and erosion resistance in cutting and forming tools and other engineering surfaces; and for chemical functions they are used to provide resistance to chemical and galvanic corrosion in high-temperature oxidation, corrosion, and catalytic applications as well as to afford protection in the marine environment. *See* VACUUM METALLURGY. Rointan F. Bunshah

Physics Formerly called natural philosophy, physics is concerned with those aspects of nature which can be understood in a fundamental way in terms of elementary principles and laws. In the course of time, various specialized sciences broke away from physics to form autonomous fields of investigation. In this process physics retained its original aim of understanding the structure of the natural world and explaining natural phenomena.

The most basic parts of physics are mechanics and field theory. Mechanics is concerned with the motion of particles or bodies under the action of given forces. The physics of fields is concerned with the origin, nature, and properties of gravitational, electromagnetic, nuclear, and other force fields. Taken together, mechanics and field theory constitute the most fundamental approach to an understanding of natural phenomena which science offers. The ultimate aim is to understand all natural phenomena in these terms. *See* CLASSICAL FIELD THEORY; MECHANICS; QUANTUM FIELD THEORY.

The older, or classical, divisions of physics were based on certain general classes of natural phenomena to which the methods of physics had been found particularly applicable. The divisions are all still current, but many of them tend more and more to designate branches of applied physics or technology, and less and less inherent divisions in physics itself. The divisions or branches, of modern physics are made in accordance with particular types of structures in nature with which each branch is concerned.

In every area physics is characterized not so much by its subject-matter content as by the precision and depth of understanding which it seeks. The aim of physics is the construction of a unified theoretical scheme in mathematical terms whose structure and behavior duplicates that of the whole natural world in the most comprehensive manner possible. Where other sciences are content to describe and relate phenomena in terms of restricted concepts peculiar to their own disciplines, physics always seeks to understand the same phenomena as a special manifestation of the underlying uniform structure of nature as a whole. In line with this objective, physics is characterized by accurate instrumentation, precision of measurement, and the expression of its results in mathematical terms.

For the major areas of physics and for additional listings of articles in physics *see* ACOUSTICS; ASTROPHYSICS; ATOMIC PHYSICS; BIOPHYSICS; CLASSICAL MECHANICS; ELECTRICITY; ELECTROMAGNETISM; ELEMENTARY PARTICLE; FLUID MECHANICS; HEAT; LOW-TEMPERATURE PHYSICS; MOLECULAR PHYSICS; NUCLEAR PHYSICS; OPTICS; SOLID-STATE PHYSICS; STATISTICAL MECHANICS; THEORETICAL PHYSICS; THERMODYNAMIC PRINCIPLES. William G. Pollard

Physiological acoustics The study of specific responses that may occur in the ear or elsewhere along the central auditory pathways, following presentation of an appropriate stimulus at any level of the auditory system. Such responses may be recorded with the aid of various techniques which may be mechanical, electrical, optical, and so forth. The specific stimulus for the ear is acoustic energy. Experimentally, signals with well-defined parameters are used. The approach employed by physiological acoustics thus is purely analytical. This is in contrast to the holistic approach employed by psychoacoustics, which lends itself well to experiments on human subjects. Systematic physiological experiments can be performed only in animals, but differences between humans and other mammals are mainly in degree, not in principle. *See* EAR; HEARING (HUMAN); PSYCHOACOUSTICS. Juergen Tonndorf

Physiological ecology (animal) A discipline that combines the study of physiological processes, the functions of living organisms and their parts, with ecological processes that connect the individual organism with population dynamics and community structure. *See* POPULATION ECOLOGY.

Physiological ecologists focus on whole-animal function and adjustments to ever-changing environments, in both laboratory and field. Short-term behavioral adjustments and longer-term physiological adjustments tend to maximize the fitness of animals, that is, their capacity to survive and reproduce successfully. Among the processes that physiological ecologists study are temperature regulation, energy metabolism and energetics, nutrition, respiratory gas exchange, water and osmotic balance, and responses to environmental stresses. These environmental stresses may include climate variation, nutrition, disease, and toxic exposure. For instance, climate affects animal heat and mass balances, and such changes affect body temperature regulation. Behavioral temperature regulation (typically, avoidance of temperature extremes) modifies mass and energy intake and expenditure, and the difference between intake and expenditure provides the discretionary mass for growth and reproduction. Mortality risk (survivorship) also depends on temperature-dependent behavior, which determines daily activity. Activity time constrains the time for foraging and habitat selection, which in turn influence not only mortality risk but also community composition. Animals are similarly constrained in their discretionary mass and energy by reduction in nutrition, which decreases absorbed food, and by disease and toxins, which may elevate the costs to maintain a higher body temperature (fever). *See* BEHAVIORAL ECOLOGY; HOMEOSTASIS. Warren Porter

Physiological ecology (plant) The branch of plant science that seeks physiological (mechanistic) explanations for ecological observations. Emphasis is placed on understanding how plants cope with environmental variation at the physiological level, and on the influence of resource limitations on growth, metabolism, and reproduction of individuals within and among plant populations, along environmental gradients, and across different communities and ecosystems. The responses of plants to natural, controlled, or manipulated conditions above and below ground provide a basis for understanding how the features of plants enable their survival, persistence, and spread. Information gathered is often used to identify the physiological and morphological features of a plant that permit adaptation to different sets of environmental conditions.

The environments that plants occupy are often subject to variation or change. The ecophysiological characteristics of these plants must be able to accommodate this or the plants face extinction. Given the right conditions, ample time, and genetic variation among a group of interbreeding individuals, plant populations and species can evolve to accommodate marked ecological change or habitat heterogeneity. If evolutionary changes in physiology or morphology occur on a local or regional scale, populations within a single species may diverge in their characteristics. Separate ecological races (ecotypes) arise in response to an identifiable, set of environmental conditions. Ecotypes are genetically distinct and are particularly well suited to the local or regional environment they occupy. Such ecotypes can often increase the geographical range and amplitude of environmental conditions that the species occupies or tolerates. Ecotypes may also occur as a series of populations arrayed over a well-defined environmental gradient called an ecocline. In contrast, if ecotypes are not present, some plant species may still be able to accommodate a wide range of growth conditions through morphological and physiological adjustments, by acclimation to a single factor (such as light) or acclimatization to a complex suite of factors which define the entire habitat. Acclimatization can occur when individuals from several different regions or populations are grown in a common location and adjust, physiologically or morphologically, to this location. Acclimation and acclimatization can therefore be defined as the ability of a single genotype (individual) to express multiple phenotypes (outward appearances) in response to variable growing conditions. Neither requires underlying genetic changes, though some genetic change might occur which could mean that the response seen may itself evolve. Acclimation and acclimatization may also be called phenotypic plasticity. *See* PLANT EVOLUTION.

Studies of metabolic rates in relation to environmental conditions within populations, ecotypes, or species provide a way to measure the tolerance limits expressed at different scales. These data in turn help identify the scales at which different adaptations are expressed, and enhance an understanding of the evolution of physiological processes. Combining observations and measurements from the field with those obtained in laboratory and controlled environment experiments can help identify which conditions may be most influential on plant processes and therefore what may have shaped the physiological responses seen. Laboratory and controlled environment (common garden) experiments also assist in helping identify how much of the variation expressed in a particular metabolic process can be assigned to a particular environmental factor and how much to the plants themselves and the genetic and developmental plasticity they possess. *See* ECOLOGY; ECOSYSTEM; PLANT PHYSIOLOGY.

T. E. Dawson

Phytamastigophorea A class of the subphylum Sarcomastigophora, also known as the Phytomastigina. These are the plant flagellates which contain chlorophyll and other pigments, but colorless forms are also included. Encystment is frequent among phytoflagellates, cyst composition being one method of determining relationships for some colorless species.

The Phytamastigophorea include 10 orders: Chrysomonadida, Silicoflagellida, Coccolithophora, Heterochlorida, Cryptomonadida, Dinoflagellida, Ebriida, Euglenida, Chloromonadida, and Volvocida. See articles on these groups. *See* SARCOMASTIGOPHORA.

James B. Lackey

Phytoalexins Antibiotics produced by plants in response to microorganisms. Plants use physical and chemical barriers as a first line of defense. When these barriers are breached, however, the plant must actively protect itself by employing a variety of strategies. Plant cell walls are strengthened, and special cell layers are produced to block further penetration of the pathogen. These defenses can permanently stop a pathogen when fully implemented, but the pathogen must be slowed to gain time.

The rapid defenses available to plants include phytoalexin accumulation, which takes a few hours, and the hypersensitive reaction, which can occur in minutes. The hypersensitive reaction is the rapid death of plant cells in the immediate vicinity of the pathogen. Death of these cells is thought to create a toxic environment of released plant components that may in themselves interfere with pathogen growth, but more importantly damaged cells probably release signals to surrounding cells and trigger a more comprehensive defense effort. Thus, phytoalexin accumulation is just one part of an integrated series of plant responses leading from early detection to eventual neutralization of a potentially lethal invading microorganism.

The tremendous capacity of plants to produce complex chemical compounds is reflected in the structural diversity of phytoalexins. Each plant species produces one or several phytoalexins, and the types of phytoalexins produced are similar in related species. The diversity, complexity, and toxicity of phytoalexins may provide clues about their function. The diversity of phytoalexins may reflect a plant survival strategy. That is, if a plant produces different phytoalexins from its neighbors, it is less likely to be successfully attacked by pathogens adapted to its neighbor's phytoalexins. Diversity and complexity, therefore, may reflect the benefits of using different deterrents from those found in other plants. *See* PLANT PATHOLOGY.

Arthur R. Ayers

Phytochrome A pigment that controls most photomorphogenic responses in higher plants. Mechanisms have evolved in plants that allow them to adapt their growth and development to more efficiently seek and capture light and to tailor their life cycle to the climatic seasons. These mechanisms enable the plant to sense not only the presence of light but also its intensity, direction, duration, and spectral quality. Plants thus regulate important developmental processes such as seed germination, growth direction, growth rate, chloroplast development, pigmentation, flowering, and senescence, collectively termed photomorphogenesis.

To perceive light signals, plants use several receptor systems that convert light absorbed by specific pigments into chemical or electrical signals to which the plants respond. This signal conversion is called photosensory transduction. Pigments used include cryptochrome, a blue light-absorbing pigment; an ultraviolet light-absorbing pigment; and phytochrome, a red/far-red light-absorbing pigment.

Phytochrome consists of a compound that absorbs visible light (chromophore) bound to a protein. The chromophore is an open-chain tetrapyrrole closely related to the photosynthetic pigments found in the cyanobacteria and similar in structure to the circular tetrapyrroles of chlorophyll and hemoglobin. Phytochrome is one of the most intensely colored pigments found in nature, enabling phytochrome in seeds to sense even the dim light present well beneath the surface of the soil and allowing leaves to perceive moonlight. *See* CHLOROPHYLL; HEMOGLOBIN.

Phytochrome can exist in two stable photointerconvertible forms, P_r or P_{fr}, with only P_{fr} being biologically active. Absorption of red light (near 666 nanometers) by inactive P_r converts it to active P_{fr}, while absorption of far-red light (near 730 nm) by active P_{fr} converts phytochrome back to inactive P_r. Plants frequently respond quantitatively to light by detecting the amount of P_{fr} produced. As a result, the amount of P_{fr} must be strictly regulated nonphotochemically by precisely controlling both the synthesis and degradation of the pigment. *See* ABSORPTION OF ELECTROMAGNETIC RADIATION.

Phytochrome has a variety of functions in plants. Initially, production of P_{fr} is required for many seeds to begin germination. This requirement prevents germination of seeds that are buried too deep in the soil to successfully reach the surface. In etiolated (dark-grown) seedlings, phytochrome can measure an increase in light intensity and duration through the increased formation of P_{fr}. Light direction also can be deduced from the asymmetry of P_{fr} levels from one side of the plant to the other. Different phytochrome responses vary in their sensitivity to P_{fr}; some require very low levels of P_{fr} (less than 1% of total phytochrome) to elicit a maximal response, while others require almost all of the pigment to be converted to P_{fr}. Thus, as the seedling grows toward the soil surface, a cascade of photomorphogenic responses are induced, with the more sensitive responses occurring first. This chain of events produces a plant that is mature and photosynthetically competent by the time it finally reaches the surface. Production of P_{fr} also makes the plant aware of gravity, inducing shoots to grow up and roots to grow down into the soil. *See* PLANT MOVEMENTS; SEED.

In light-grown plants, phytochrome allows for the perception of daylight intensity, day length, and spectral quality. Intensity is detected through a measurement of phytochrome shuttling between P_r and P_{fr}; the more intense the light, the more interconversion. This signal initiates changes in chloroplast morphology to allow shaded leaves to capture light more efficiently. If the light is too intense, phytochrome will also elicit the production of pigments to protect plants from photodamage.

Temperate plants use day length to tailor their development, a process called photoperiodism. How the plant measures day length is unknown, but it involves phytochrome and actually measures the length of night. *See* PHOTOPERIODISM.

Finally, phytochrome allows plants to detect the spectral quality of light, a form of color vision, by measuring the ratio of P_r to P_{fr}. When a plant is grown under direct sun, the amounts of red and far-red light are approximately equal, and the ratio of P_r to P_{fr} in the plant is about 1:1. Should the plant become shaded by another plant, the P_r/P_{fr} ratio changes dramatically to 5:1 or greater. This is because the shading plant's chlorophyll absorbs much of the red light needed to produce P_{fr} and absorbs almost none of the far-red light used to produce P_r. For a shade-intolerant plant, this change in P_r/P_{fr} ratio induces the plant to grow taller, allowing it to grow above the canopy.

It is not known how phytochrome elicits the diverse array of photomorphogenic responses, but the regulatory action must result from discrete changes in the molecule following photoconversion of P_r to P_{fr}. These changes must then start a chain of events in the photosensory transduction chain leading to the photomorphogenic response. Many photosensory transduction chains probably begin by responding to P_{fr} or the P_r/P_{fr} ratio and branch off toward discrete end points. *See* PHOTOMORPHOGENESIS.

Richard D. Vierstra

Phytoplankton Mostly autotrophic microscopic algae which inhabit the illuminated surface waters of the sea, estuaries, lakes, and ponds. Many are motile. Some perform diel (diurnal) vertical migrations, others do not. Some nonmotile forms regulate their buoyancy. However, their locomotor abilities are limited, and they are largely transported by horizontal and vertical water motions.

A great variety of algae make up the phytoplankton. Diatoms (class Bacillariophyceae) are often conspicuous members of marine, estuarine, and freshwater plankton. Dinoflagellates (class Dinophyceae) occur in both marine and freshwater environments and are important primary producers in marine and estuarine environments. Coccolithophorids (class Haptophyceae) are also marine primary producers of some importance. They do not occur in freshwater.

Even though marine and freshwater phytoplankton communities contain a number of algal classes in common, phytoplankton samples from these two environments will appear quite different. These habitats support different genera and species and groups of higher rank in these classes. Furthermore, freshwater plankton contains algae belonging to additional algal classes either absent or rarely common in open ocean environments. These include the green algae (class Chlorophyceae), the euglenoid flagellates (class Euglenophyceae), and members of the Prasinophyceae.

The phytoplankton in aquatic environments which have not been too drastically affected by human activity exhibit rather regular and predictable seasonal cycles. Coastal upwelling and divergences, zones where deeper water rises to the surface, are examples of naturally occurring phenomena which enrich the mixed layer with needed nutrients and greatly increase phytoplankton production. In the ocean these are the sites of the world's most productive fisheries. *See* EUTROPHICATION.

Robert W. Holmes

Phytotronics Research using whole plants and conducted under controlled environmental conditions to determine responses to a single or known combination of environmental elements. Originally, the term phytotronics was used to identify research conducted specifically in phytotrons where controlled plant growth units are available for simultaneous use. The name phytotronics is also often applied to any research conducted with whole plants in a controlled environment plant growth chamber or room.

Henry Hellmers

Piciformes A large order of land birds, second in number of species only to the Passeriformes, that is found throughout the world except for the Australian region and is concentrated in tropical areas. The Piciformes is divided into the following three suborders and eight families; the family Picidae is the largest, with 204 species of woodpeckers.

Order Piciformes
- Suborder Galbulae
  - Family: Primobucconidae (fossil; Eocene of North America)
  - Galbulidae (jacamars; 17 species; Neotropics)
  - Bucconidae (puffbirds; 3 species; Neotropics)
- Suborder Zygodactyli
  - Family Zygodactylidae (fossil; Miocene of Germany)
- Suborder Pici
  - Family: Capitonidae (barbets; 81 species; pantropical)
  - Ramphastidae (toucans; 33 species; Neotropics)
  - Indicatoridae (honeyguides; 16 species; Africa and southern Asia)
  - Picidae (woodpeckers; 204 species; worldwide except the Australasian region)

The piciforms are small to medium-sized, hole-nesting land birds. The bill is short to medium long, straight, and strong, and the wings are of medium length and rounded. The legs are short and strong, with the strong toes arranged in a zygodactylous (yoke) pattern, with two toes forward and two toes back. The tail may have stiffened feathers serving as a prop when climbing. The plumage, which varies greatly in hue, is frequently brightly colored and boldly patterned. Piciforms are good fliers and can easily perch and climb, but they walk poorly. Most species feed on insects; toucans, on the other hand, are primarily fruit eaters. All breed in cavities dug in earthen banks or hollowed out in trees. The eggs are incubated by both sexes, and both parents care for the unfeathered young, which remain in the nest. Except for a few species of woodpeckers, the piciforms are nonmigratory. *See* AVES.

Walter J. Bock

Picornaviridae A viral family made up of the small (18–30 nanometer) ether-sensitive viruses that lack an envelope and have a ribonucleic acid (RNA) genome. The name is derived from "pico" meaning very small, and RNA for the nucleic acid type. Picornaviruses of human origin include the following subgroups: enteroviruses (polioviruses, coxsackieviruses, and echoviruses) and rhinoviruses. There are also picornaviruses of lower animals (for example, bovine foot-and-mouth disease, a rhinovirus). *See* ANIMAL VIRUS; COXSACKIEVIRUS; ECHOVIRUS; ENTEROVIRUS; FOOT-AND-MOUTH DISEASE; POLIOMYELITIS; RHINOVIRUS.

Joseph L. Melnick

Picrite The term picrite has been used with several different meanings. It is generally considered to include certain medium- to fine-grained ultramafic igneous rocks composed chiefly of olivine with smaller amounts of pyroxene, hornblende, and plagioclase feldspar (labradorite).

Its feldspar content is slightly higher than that of peridotite and lower than that of gabbro. Certain analcite-bearing types, associated with teschenite, have also been included under the term picrite. A characteristic feature is poikilitic texture in which large pyroxene or hornblende crystals enclose numerous small grains of olivine. *See* GABBRO; PERIDOTITE.

Picrite is rare and is found in small sills and dikes at or very near the surface. It may also occur in the lower portions of basaltic lava flows where olivine and pyroxene crystals have accumulated under the influence of gravity. *See* IGNEOUS ROCKS.

Carleton A. Chapman

Pictorial drawing A view of an object (actual or imagined) as it would be seen by an observer who looks at the object either in a chosen direction or from a selected point of view. Pictorial sketches often are more readily made and more clearly understood than are front, top, and side views of an object. Pictorial drawings, either sketched freehand or made with drawing instruments, are frequently used by engineers and architects to convey ideas to their assistants and clients. *See* DESCRIPTIVE; GEOMETRY; ENGINEERING DRAWING.

In making a pictorial drawing, the viewing direction that shows the object and its details to the best advantage is chosen. The resultant drawing is orthographic if the viewing rays are considered as parallel, or perspective if the rays are considered as meeting at the eye of the observer. Perspective drawings provide the most realistic, and usually the most pleasing, likeness when compared with other types of pictorial views.

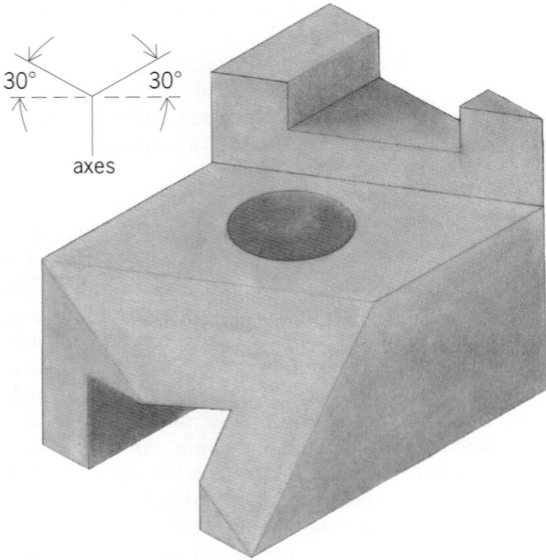

Fig. 1. Isometric drawing; measurements along each axis are made with the same scale.

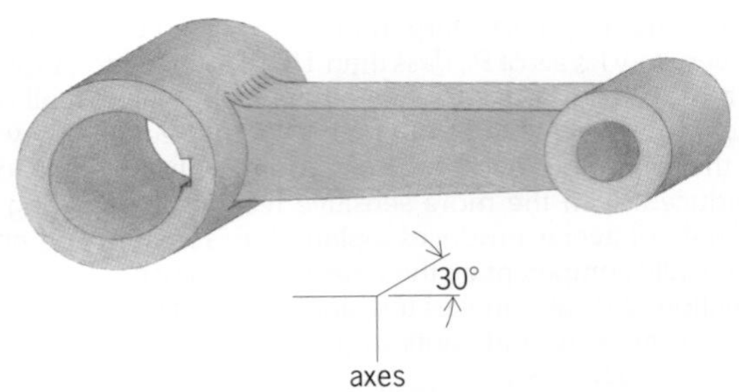

Fig. 2. Oblique pictorial drawing.

Several types of nonperspective pictorial views can be sketched, or drawn with instruments. In the isometric pictorial, the direction of its axes and all measurements along these axes are made with one scale (Fig. 1). Oblique pictorial drawings, while not true orthographic views, offer a convenient method for drawing circles and other curves in their true shape (Fig. 2). In order to reduce the distortion in an oblique drawing, measurements along the receding axis may be foreshortened. When they are halved, the method is called cabinet drawing. Charles J. Baer

Picture tube A cathode-ray tube, or CRT, used as a television display. A cathode-ray tube has an electron gun, electromagnetic deflection yoke, shadow mask or grill assembly, and a phosphor screen that is the light emitter (see illustration). All of this is contained in a glass envelope. Television picture tubes are largely being replaced by flat-panel displays, such as liquid-crystal displays.

An amplified waveform from the input video signal is sent to the cathode in the gun. There are three cathodes used in a color CRT, one for each phosphor color. As the gun is heated, it emits electrons at a current density in relation to the input signal. Grids in the gun shape and accelerate the beam toward the

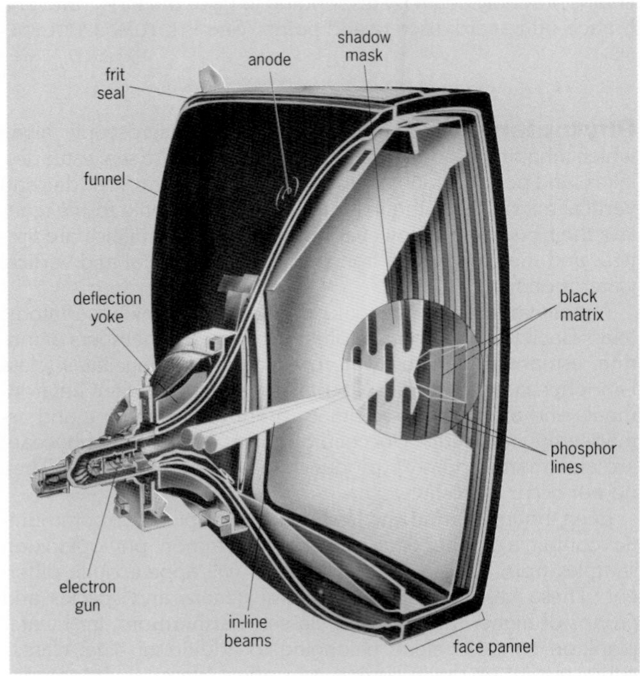

Components of a color picture tube.

deflection yoke area, while magnetic fields generated by the yoke move the beam over the CRT face as a raster scan. The beam passes through the shadow mask and lands on the appropriate phosphor, causing it to emit the desired color.

The CRT is made of several parts, all of which are mounted inside a glass envelope, sometimes also called the bulb. Normally this is made up of a funnel piece and face panel. The separate face panel, made of optical glass, allows for the coating of the phosphors and the mounting of the shadow mask. The entire assembly is evacuated of atmospheric gases, and small amount of material, called a getter, is used to absorb the residual gas molecules inside the envelope.

The conversion of the electronic signals to light is an electrochemical process, whereby a phosphor material lights when an electron beam energizes it. The phosphors are applied to the face panel as a screen of lines or dots, on top of which a film (mirror) of aluminum is vacuum-evaporated. The reflection of light by the aluminum mirror increases the picture brightness and improves picture contrast by preventing stray light from illuminating the back of the phosphor screen. For color CRTs, three different phosphors are used to generate red, green, and blue light. When these three primary colors are viewed, they are perceived as many colors depending on the balance between the primary colors. *See* CATHODE-RAY TUBE; TELEVISION; PHOSPHORESCENCE.

Gary Mandle

Piezoelectricity Electricity, or electric polarity, resulting from the application of mechanical pressure on a dielectric crystal. The application of a mechanical stress produces in certain dielectric (electrically nonconducting) crystals an electric polarization (electric dipole moment per cubic meter) which is proportional to this stress. If the crystal is isolated, this polarization manifests itself as a voltage across the crystal, and if the crystal is short-circuited, a flow of charge can be observed during loading. Conversely, application of a voltage between certain faces of the crystal produces a mechanical distortion of the material. This reciprocal relationship is referred to as the piezoelectric effect. The phenomenon of generation of a voltage under mechanical stress is referred to as the direct piezoelectric effect, and the mechanical strain produced in the crystal under electric stress is called the converse piezoelectric effect. *See* POLARIZATION OF DIELECTRICS.

The necessary condition for the piezoelectric effect is the absence of a center of symmetry in the crystal structure. Of the 32 crystal classes, 21 lack a center of symmetry, and with the exception of one class, all of these are piezoelectric. Hydrostatic pressure produces a piezoelectric polarization in the crystals of those 10 classes that show pyroelectricity in addition to piezoelectricity. *See* CRYSTALLOGRAPHY; PYROELECTRICITY.

Molecular theory. Quantitative theories based on the detailed crystal structure are very involved. Qualitatively, however, the piezoelectric effect is readily understood for simple crystal structures. The illustration shows this for a particular cubic crystal, zincblende (ZnS). Every Zn ion is positively charged and is located in the center of a regular tetrahedron *ABCD*, the corners of which are the centers of sulfur ions, which are negatively charged. When this system is subjected to a shear stress in the *xy* plane, the edge *AB*, for example, is elongated, and the edge *CD* of the tetrahedron becomes shorter. Consequently, these edges are no longer equivalent, and the Zn ion will be displaced along the *z* axis, thus giving rise to an electric dipole moment. The dipole moments arising from different octahedrons sum up because they all have the same orientation with respect to the axes *x*, *y*, and *z*.

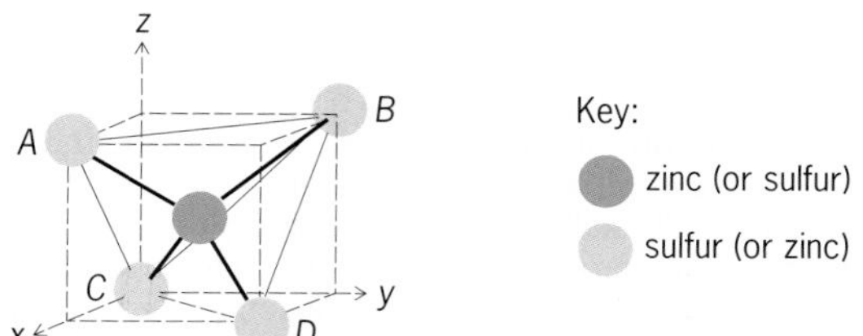

Tetrahedral structure of zincblende, ZnS. Only part of unit cell is shown. Size of circles has no relation to size of ions.

Applications. The sharp resonance curve of a piezoelectric resonator makes it useful in the stabilization of the frequency of radio oscillators. Quartz crystals are used almost exclusively in this application. In vacuum-tube oscillators, the crystal generally is part of the feedback circuit. Selective band-pass filters with low losses can be built by using piezoelectric resonators as circuit elements. A synthetic piezoelectric crystal which is often substituted for quartz in this application is ethylenediamine tartrate (EDT). *See* QUARTZ CLOCK.

Piezoelectric materials are used extensively in transducers for converting a mechanical strain into an electrical signal. Such devices include microphones, phonograph pickups, vibration-sensing elements, and the like. The converse effect, in which a mechanical output is derived from an electrical signal input, is also widely used in such devices as sonic and ultrasonic transducers, headphones, loudspeakers, and cutting heads for disk recording. *See* MICROPHONE; ULTRASONICS. H. Granicher

Piezoelectric materials. The principal piezoelectric materials used commercially are crystalline quartz and rochelle salt, although the latter is being superseded by other materials, such as barium titanate. Quartz has the important qualities of being a completely oxidized compound (silicon dioxide), and is almost insoluble in water. Therefore, it is chemically stable against changes occurring with time. It also has low internal losses when used as a vibrator. Rochelle salt has a large piezoelectric effect, and is thus useful in acoustical and vibrational devices where sensitivity is necessary, but it decomposes at high temperatures (131°F or 55°C) and requires protection against moisture. Barium titanate provides lower sensitivity, but greater immunity to temperature and humidity effects. Other crystals that have been used for piezoelectric devices include tourmaline, ammonium dihydrogen phosphate (ADP), and EDT. *See* QUARTZ.

Frank D. Lewis

Pigeonite Monoclinic pyroxenes of the general formula $(Mg,Fe)SiO_3$ having some augite in solid solution. Pigeonite bears the same relation to the orthorhombic pyroxenes as augite does to the diopside-hedenbergite series. Pigeonite is the orthorhombic pyroxene equivalent in the volcanic rocks. *See* AUGITE; DIOPSIDE; ORTHORHOMBIC PYROXENE; PYROXENE.

George W. De Vore

Pigment A finely divided material which contributes to optical and other properties of paint, finishes, and coatings. Pigments are insoluble in the coating material, whereas dyes dissolve in and color the coating. Pigments are mechanically mixed with the coating and are deposited when the coating dries. Their physical properties generally are not changed by incorporation in and deposition from the vehicle. Pigments may be classified according to composition (inorganic or organic) or by source (natural or synthetic). However, the most useful classification is by color (white, transparent, or colored) and by function. Special pigments include anticorrosive, metallic, and luminous pigments. *See* DYE; PAINT AND COATINGS. C. R. Martinson; C. W. Sisler

Pigmentation A property of biological materials that imparts coloration. Hence, pigmentation determines the quantity and quality of reflected visible light. The characteristics of light returning from living matter are a function of its chemical and physical properties and, therefore, are not only due to pigments proper but can be of structural origin (for example, due to reflection, scattering, or interference) as well.

Pigments are essential constituents of the living world. Their contribution to the evolution and maintenance of life, and its manifold expressions, is most evident in the role of chlorophylls and the associated carotenoids of certain bacteria and most plants. These pigments harvest solar light energy for utilization in the photosynthesis of organic material from inorganic precursors. *See* CAROTENOID; CHLOROPHYLL; PHOTOSYNTHESIS.

The outermost structures on the animal skin are pigmented for many reasons, for example, to reduce the animal's visibility against a colored background or to provide optical signals to the other sex or to other species. Conspicuously pigmented flowers attract pollinators, and colored fruits are easily found by animals, which eat them and then disperse the undigested seeds.

The role of pigments in communication depends on the ability of organisms to discriminate between different regions of the solar spectrum. In animals with eyes, this is accomplished by differently colored visual pigments contained in specialized receptor cells. Microorganisms, fungi, and plants also have special pigment systems that permit these organisms to move or grow toward, or away from, light (positive and negative phototaxis and phototropism, respectively). *See* PLANT MOVEMENTS.

Since most organisms are totally dependent on light—at least indirectly—elaborate pigment systems have evolved which tune metabolic and activity patterns to the daily pattern of light and dark, and to the changes in the relative lengths of day and night in the course of a year. The phytochrome of plants and the pigments of the eye or of extraretinal photoreceptor organs of many vertebrates and invertebrates are typical representatives of pigments that correlate biological activity with light-dark cycles (photoperiodism). *See* COLOR VISION; PHOTOPERIODISM; PHOTORECEPTION.

In the examples listed above, pigments mediate, in various ways, the beneficial actions of light. Absorbed solar light energy may, however, also have detrimental effects by causing undesirable or even destructive reactions. Pigmentations can provide a light-absorbing shield that protects the tissue below from such potentially damaging radiation of the Sun. *See* INTEGUMENT; SKIN.

Peter H. Homann

Pike Any of five species of fishes which compose the family Esocidae in the order Esociformes, known by a variety of names such as pickerel and muskellunge. These fishes are voracious predators with an elongated beaklike snout and sharp teeth. The head is partly scaled and the body is covered with cycloid scales that have deeply scalloped edges. The body is cylindrical and compressed; thus, these fishes are well adapted for rapid movements as they dart after prey. They prey upon each other, as well as other fishes, amphibians, small aquatic birds and mammals, and rats. All species are edible but are considered second-rate game fishes. *See* ESOCIFORMES; OSTEICHTHYES; PREDATOR-PREY INTERACTIONS; SCALE (ZOOLOGY).

Charles B. Curtin

Pile foundation A special type of foundation that enables a structure to be supported by a layer of soil found at any depth below the ground surface. A pile foundation comprises two basic structural elements, the pile and the pile cap. A pile cap is a structural base, similar to a spread footing, that supports a structural column, wall, or slab, except that it bears on a single pile or group of piles. A pile can be described as a structural stilt hammered into the ground. Each pile carries a portion of the pile cap load and transfers it to the soil in the vicinity of the pile tip, located at the bottom of the pile (see illustration).

The pile and pile cap configuration has provided the basic design solution to the difficult problem of obtaining deep foundation support below areas where poor soil conditions prevail. Poor soil conditions may be difficult to excavate through, and are incapable of supporting structural loads. They are typically characterized by the presence of a soft, compressible layer of clay, high ground-water levels, loosely filled soils, uncontrolled landfills, boulders, abandoned underground structures, and natural bodies of water. By supporting a structure on piles in lieu of spread footings, any adverse soil condition may be virtually bypassed, and adequate foundation support can be obtained at any depth, without the need to perform deep excavation, dewater, and install temporary sheeting and bracing.

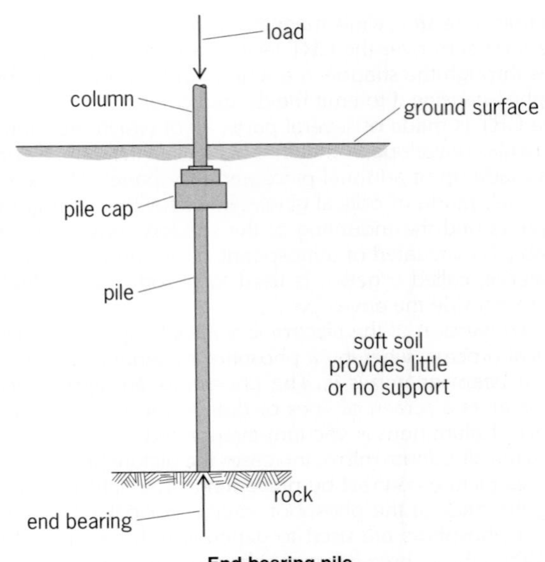

End-bearing pile.

Piles are available in a variety of sizes, shapes, and materials that enable a particular type of pile foundation to be viable both economically and structurally. Principal materials are timber, concrete, and steel.

Pile foundations are used to support marine structures and offshore platforms, since they are located over bodies of water. On land, pile foundations are used primarily in locations where poor soil conditions exist.

Anthony J. Mazzo

Pilot production The production of a product, process, or piece of equipment on a simulated factory basis. In mass-production industries where complicated products, processes, or equipment are being developed, a pilot plan often leads to the presentation of a better product to the customer, lower development and manufacturing costs, more efficient factory operations, and earlier introduction of the product. Following the engineering development of a product, process, or complicated piece of equipment and its one-of-a-kind fabrication in the model shop, it becomes desirable and necessary to "prove out" the development on a simulated factory basis. *See* PRODUCT DESIGN; PRODUCTION ENGINEERING; QUALITY CONTROL.

James E. Woodall

Piloting The form of navigation in which position is determined relative to external reference points, usually fixed points on the Earth. It is the oldest form of navigation. The following remarks are formulated in the context of marine navigation. However, the use of landmarks to navigate terrestrial vehicles and aircraft is well established. Some authors refer to this means of navigating aircraft as pilotage. With the development of electronic aids, piloting techniques were extended far from shore. However, the term "piloting" is generally associated with nearness of land, where tidal and other currents may be strong, shoals and other underwater obstructions may be in near proximity, and maneuvering room is limited when other vessels are encountered. Thus it is not unusual for ships to employ the services of a local expert, called a pilot, to assist in the navigation of the vessel while it enters or leaves port.

A conspicuous object, structure, or light that serves as an indicator for establishing the position of a craft or otherwise assisting in its safe navigation is called a mark. To be useful, not only must a mark be identified, but its position must be known accurately. Artificial marks designed and erected specifically to assist the navigator are called aids to navigation and include beacons, both lighted and unlighted, lighthouses, buoys, both lighted and unlighted, and lightships. Unlighted aids are called daymarks. *See* BUOY; LIGHTHOUSE.

In addition to visible aids to navigation, bottom topography can be of assistance in locating the position of a vessel. Sound signals transmitted through water or air may be used for navigation. Electronic beacons and positioning systems have been established at a number of places to assist in navigation. *See* ELECTRONIC NAVIGATION SYSTEMS.

The measurements made for piloting purposes are of direction, distance, differential distance between two points, and distance to bottom.

Bearings are usually measured (1) by noting when two objects are in range (directly in line); (2) by means of a suitable attachment to a compass or compass repeater; (3) by pelorus, a compasslike instrument without directive properties; or (4) electronically, by radio direction finder, by radar, or by the indication of the receiver-indicator of an electronic system of navigation. Distance is generally measured by radar. The difference in distance from the ship to two points is usually measured electronically by means of the receiver-indicator of a hyperbolic navigation system. Depth measurement is usually made by an echo sounder. *See* DIRECTION-FINDING EQUIPMENT; ECHO SOUNDER; HYPERBOLIC NAVIGATION SYSTEM; RADAR.

Traditionally, position by piloting has been determined by means of lines of positions, each indicating a series of possible positions of the craft at the time of measurement. A measured bearing provides a straight line of position (actually part of a great circle) passing through the object sighted. A measured distance provides a circular line of position with the object as the center and the distance as the radius. A measured differential distance provides a hyperbolic line of position. A position, called a fix, is usually determined by crossing two or more lines of position taken simultaneously or nearly so.

Digitization of a nautical chart provides the information needed for display of an electronic facsimile of a published chart. Selective determination of features to be shown makes possible the tailoring of the facsimile to individual requirements. If a radar image is added to the display, the position of the vessel relative to its surroundings is immediately apparent. A suitable electronic positioning system such as the Global Positioning System (GPS) or differential GPS (DGPS) can be used to display a symbol indicating the position of the vessel, thus providing a check on accuracy of the radar data. Other information can be added as desired. The combined display is called an electronic chart display information system (ECDIS). *See* CELESTIAL NAVIGATION; DEAD RECKONING; MARINE NAVIGATION; NAVIGATION; POLAR NAVIGATION; SATELLITE NAVIGATION SYSTEMS. Alton B Moody

Piltdown man The scientifically most successful fraud in the history of anthropology. Between 1910 and 1915, Charles Dawson, a lawyer and amateur prehistorian, claimed to have recovered craniodental remains of a fossil human ancestor, along with primitive stone tools and mammalian fossils, from a gravel pit near Piltdown in Sussex, southern England. The skull fragments included a partial braincase with relatively thick bone, but otherwise the skull was like modern humans in having a relatively large braincase, steep forehead, and poorly developed bony ridge above the eye socket. The lower canine was apelike in being quite large, but it had a humanlike wear pattern. The fragmentary lower jaw was missing many of the critical parts useful for distinguishing humans from apes, such as the articular condyle and the chin region, but it was generally similar to that of great apes. The molar teeth had flat wear resembling humans. Overall, the skull provided an excellent intermediate stage or "missing link" between apes and humans, suggesting that brain expansion had preceded facial specialization during the course of human evolution. *See* FOSSIL HUMANS.

The Piltdown remains were published in 1912 and were named *Eoanthropus dawsoni* (that is, Dawson's dawn man, in honor of its discoverer). This sensational discovery caused a great deal of excitement among the scientific community and the public at large. During the 1920s and 1930s, *Eoanthropus* was considered pivotal to understanding human evolution. However, with a growing number of fossil finds from Europe and Asia, and especially from Pliocene sites in Africa, the Piltdown finds became increasingly difficult to reconcile with the rest of the fossil record. Eventually, logic led to the hypothesis of a deliberate hoax. Chemical tests showed that none of the human bones could be more than a few centuries old (this was later confirmed in 1959 with the advent of carbon-14 dating). Also, on reexamination, signs of forging were evident. The cheek teeth had been filed unnaturally flat, not worn by chewing, and the canine had also been reduced in size by filing. The thickened skull bones were from a modern human with Paget's disease (which is characterized by an increase in bone turnover). The mandible could be identified as that of an orangutan. The inevitable conclusion was that all of the evidence was fabricated and that it had been planted at Piltdown as part of a carefully orchestrated and deliberate hoax. The identity of the perpetrator of the hoax has been the subject of much speculation. Few people doubt that Dawson was involved in the deception, but most believe that someone with greater expertise probably assisted him. *See* ARCHEOLOGICAL CHRONOLOGY. T. Harrison; W. W. Howells

Pimento A type of pepper, *Capsicum annuum*, grown for its thick, sweet-fleshed red fruit. A member of the plant order Polemoniales, pimento is of American origin, and gets its name from the Spanish word designating all sweet peppers. In the United States, however, the term pimento generally refers to the heart-shaped varieties (cultivars) grown for canning and used for stuffing olives and flavoring foods. Georgia is the only important pimento-producing state. *See* PEPPER; SOLANALES. H. John Carew

Piña A fiber, also known as pineapple fiber, obtained from the large leaves of the pineapple plant grown in tropical countries. This natural fiber is white and especially soft and lustrous. In the Philippine Islands, it is woven into piña cloth, which is soft, durable, and resistant to moisture. Piña is also used in making coarse grass cloth and for mats, bags, and clothing. *See* NATURAL FIBER; PINEAPPLE. M. David Potter

Pinales An order of the class Pinopsida (=Coniferopsida), of the division Pinophyta (Gymnospermae), with about 50 genera and 600 species still living. All are woody plants, as shrubs or trees, and are often the principal trees of the forests worldwide. Pine, spruce, fir, hemlock, cedar, larch, juniper, cypress, yew, redwood, big tree, kauri, podocarpus, araucaria, and others are all part of this order. The big tree (*Sequoia gigantea*) of California is the largest plant, reaching a height of 330 ft (100 m) and a diameter of 33 ft (10 m), and being over 3500 years old. Leaves are usually needlelike or scalelike, with a few exceptions such as *Agathis* (kauri), *Podocarpus*, *Phyllocladus*, and *Araucaria*. Most species are evergreen, bearing their leaves year-round. Modern conifers form some of the most extensive forests in recent times, mainly occurring in the temperate regions or mountainous regions of the subtropics. The families of extant conifers are Araucariaceae, Pinaceae, Taxodiaceae, Cupressaceae, Podocarpaceae, Taxaceae, and Cephalotaxaceae. The phylogenetic relations among these families are practically unknown. *See* PINOPHYTA; PINOPSIDA.

The cones are unisexual. In some species, both male and female cones are borne on the same plant (monoecious); in others, on separate plants (dioecious). Usually, a year intervenes between pollination and fertilization, and another year between fertilization and embryo formation. *See* REPRODUCTION (PLANT).

The conifers are a principal source of lumber and pulp for paper and wood products. Turpentine, tar, resin, and essential oils are some by-products. The group yields little food for humans; some seeds are edible (pine, pinyon, or piñon nuts). *See* PINE NUT; PINE TERPENE.

Thomas A. Zanoni

Pinch effect A name given to manifestations of the magnetic self-attraction of parallel electric currents having the same direction. The effect at modest current levels of a few amperes can usually be neglected, but when current levels approach a million amperes such as occur in electrochemistry, the effect can be damaging and must be taken into account by electrical engineers. The pinch effect in a gas discharge has been the subject of intensive study, since it presents a possible way of achieving the magnetic confinement of a hot plasma (a highly ionized gas) necessary for the successful operation of a thermonuclear or fusion reactor.

The law of attraction which describes the interaction between parallel electric currents was discovered by A. M. Ampère in 1820. For a cylindrical wire of radius r meters carrying a total surface current of I amperes, it manifests itself as an inward pressure on the surface (Fig. 1) given by $I^2/2 \times 10^7 \pi r^2$ pascals. For the electric currents of normal experience, this force is small and passes unnoticed, but it is significant that the pressure increases with the square of the current, I^2. For example, at 25,000 amperes the pressure amounts to about 1 atm (100 kilopascals) for a wire of 1-cm radius, but at 10^6 amperes the pressure is about 1600 atm or about 12 tons in.$^{-2}$ (160 megapascals).

There are a number of ways in which the magnetic field of a fusion reactor can be arranged around the plasma to hold it together, and one of these methods is the pinch effect. A fusion reactor using this type of confinement would ideally be a toroidal tube in which the confined plasma would carry a large electric current induced in it by magnetic induction from a transformer core passing through the major axis of the torus. The current would have the double function of ohmically heating the plasma and compressing the plasma toward the center of the tube.

Characteristically, as can be shown by high-speed photography, the pinch forms at the inner surface of a discharge tube wall and contracts radially inward, forming an intense line, the pinch, on the axis; the pinch rebounds slightly; the contracted discharge rapidly develops necks and kinks; and in a few microseconds all structure is lost in an apparently turbulent glowing gas which fills the tube. Thus, the pinch turns out to be unstable, and plasma confinement is soon lost by contact with the wall. The cause of the instability is easily seen qualitatively: The pinch confinement can be described as being caused by the magnetic field lines encircling the pinch which are stretched longitudinally but which are in compression transversely (Fig. 2). For a uniform cylindrical pinch, the magnetic pinch pressure is everywhere equal to the outward plasma pressure, but at a neck or on the inward side of a kink, the magnetic field lines crowd together, creating a higher magnetic pressure than the outward gas pressure. Consequently, the neck contracts still further, the kink cuts in on the concave side and bulges out on the convex side, and both perturbations grow. The instability has a disastrous effect on the confinement time.

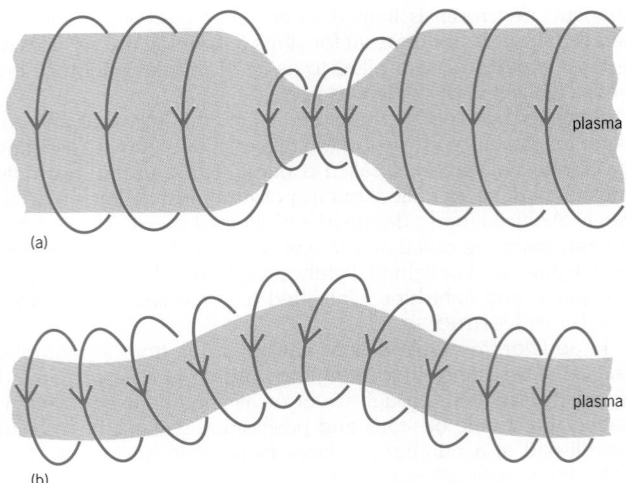

Fig. 2. Instability. (*a*) Sausage type. (*b*) Kink type.

The term theta pinch has come into wide usage to denote an important plasma confinement system which relies on the repulsion of oppositely directed currents and which is thus not in accord with the original definition of the pinch effect (self-attraction of currents in the same direction). Plasma confinement systems based on the original pinch effect are known as Z pinches.

Tokamak is essentially a low-density, slow Z pinch in a torus with a very strong longitudinal field. The helical magnetic field lines, resultant from the externally applied field and that of the pinch, do not close, that is, do not complete one revolution of the minor axis in going around the major axis of the torus once. This is known theoretically to prevent the growth of certain helical distortions of the plasma. The performance of tokamak experiments has raised the possibility of achieving a net power balance. *See* NUCLEAR FUSION.

James A. Phillips

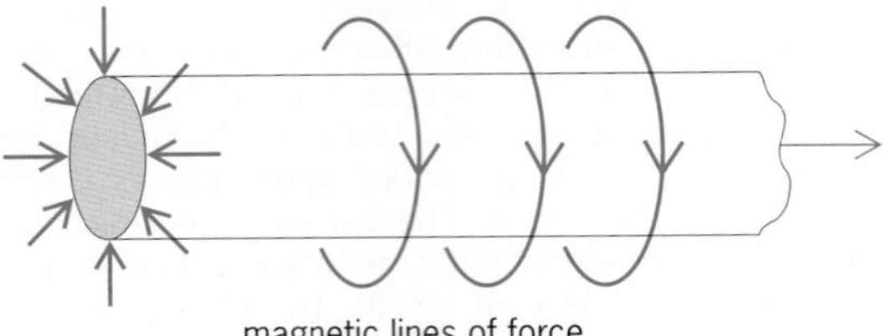

Fig. 1. Pinch pressure on a current-carrying conductor. Arrows at left show direction of pinch pressure.

Pine The genus *Pinus*, of the pine family, characterized by evergreen leaves, usually in tight clusters (fascicles) of two to five, rarely single. There are about 80 known species distributed throughout the Northern Hemisphere. Botanically the leaves are of two kinds: (1) a scalelike form, the primary leaf, which subtends a much shortened and eventually deciduous shoot bearing (2) the secondary leaves or needles. The wood of pines is easily recognized by the numerous resin ducts and by the characteristic resinous odor. *See* PINALES; PINE NUT.

Arthur H. Graves; Kenneth P. Davis

Pine nut The edible seed of more than a dozen species of evergreen cone-bearing trees in the genus *Pinus*, native to the temperate zone of the Northern Hemisphere. The important nut-producing species are the stone pine (*P. pinea*) of southern Europe; the Swiss stone pine (*P. cembra*), native to the Swiss Alps and eastward through Siberia to Mongolia; and the pinon pine (*P. cembroides* var. *edulis*) of the arid regions of the southwestern United States. The seeds or nuts, variable in size according to species, are borne in cones which take 3–4 years to develop. *See* PINE.

Laurence H. MacDaniels

Pine terpene A major component of the essential oils obtained from various *Pinus* species. The principal terpenes of the oil of southern pines [longleaf pine (*P. palustris*) and slash pine (*P. caribaea*)] are α- and β-pinene, whose structures are shown below.

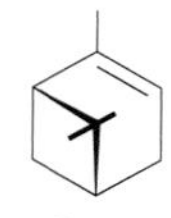

α-Pinene β-Pinene

See ESSENTIAL OIL; PINE.

Gum turpentine (gum spirits) is the volatile fraction of the oleoresin that exudes from cuts made in the trunks of live trees. The resin is collected and distilled by a process that yields about 20% turpentine, mainly α- and β-pinene, and 70% rosin; it was the basis of the original naval stores industry.

Wood turpentine is obtained by steam distillation from stumps and other logging residues. The volatile material in this case consists of about 50% turpentine and 30–40% of higher-boiling-point alcohols; the latter fraction is known as pine oil. The bulk of the wood turpentine and pine oil produced by modern industrial processes is a by-product of the sulfate wood-pulping process (sulfate turpentine).

Important uses of turpentine or the purified pinenes derived from turpentine are in terpene resins, as a thinner in paints and varnishes, and as a starting material in the synthesis of other commercially valuable terpenes.

Pine oil is a mixture of monoterpene alcohols, mainly α-terpineol, obtained in large amounts mixed with wood turpentine or sulfate turpentine. The term pine oil is also used to designate the essential oil of various species of pine.

Much of the pine oil of commerce is prepared synthetically by acid-catalyzed hydration of α-pinene. This process involves a complex series of reactions that occur via cationic intermediates.

The composition of industrial-grade pine oil is approximately 65% α-terpineol, 20–25% of other monoterpene alcohals, and 10–15% hydrocarbons. Pine oil has surfactant and emulsifying properties and is also a disinfectant. Most of the pine oil manufactured is used in the manufacture of cleansers and textile penetrants. *See* SURFACTANT; TERPENE; WOOD CHEMICALS.

James A. Moore

Pineal gland An endocrine gland located in the brain which secretes melatonin, is strongly regulated by light stimuli, and is an important component of the circadian timing system. The pineal gland is virtually ubiquitous throughout the vertebrate animal kingdom. In nonmammalian vertebrates, it functions as a photoreceptive third eye and an endocrine organ. In mammals, it serves as an endocrine organ that is regulated by light entering the body via the eyes. Despite extensive species variation in anatomy and physiology, the pineal gland generally serves as an essential component of the circadian system which allows animals to internally measure time and coordinate physiological time-keeping with the external environment. *See* BIOLOGICAL CLOCKS; BRAIN.

The pineal gland is an unpaired organ attached by a stalk to the roof of the diencephalon. In frogs and lizards, one component of the pineal complex (the frontal organ or parietal eye) projects upward through the skull to lie under the skin; in all other vertebrates the pineal is located beneath the roof of the skull. Across evolution, cells within the pineal gland have progressed from classic photoreceptor cells in the earliest vertebrates, to rudimentary photoreceptors in birds, to classic endocrine cells in mammals. *See* PHOTORECEPTION; SENSE ORGAN.

In mammals, nerve fibers extend from a variety of sources in the brain to the pineal gland. The best studied of these neural inputs is through the retinohypothalamic tract, which extends from the eyes to the pineal gland in mammals. Originating in the retina, the majority of the retinohypothalamic fibers project to or around the bilateral suprachiasmatic nuclei in the hypothalamus. These nuclei serve as endogenous oscillators with period lengths close to 24 h. Thus, the suprachiasmatic nuclei function as pacemakers for the circadian system, which regulates daily physiological and behavioral rhythms. From the suprachiasmatic nuclei there are short projections to the paired paraventricular hypothalamic nuclei, and then long descending axons project from these nuclei to synapse on preganglionic sympathetic neurons in the upper thoracic spinal cord. These sympathetic neurons then extend out of the central nervous system to the superior cervical ganglia in the neck region. From there, postganglionic sympathetic axons reenter the cranium and ultimately innervate the pineal gland.

In mammals, information about environmental light and darkness is relayed from the eye to entrain circadian neural activity of the suprachiasmatic nuclei. In turn, the suprachiasmatic nuclei synchronize circadian rhythms in the pineal gland through its sympathetic innervation. One of the best-studied rhythms in the pineal gland is the biosynthesis of the hormone melatonin. Pinealocytes also have the necessary enzymes for converting tryptophan into a larger family of indole compounds, and numerous polypeptides have been localized in the pineal gland. The biological functions of these other pineal indole and peptide constituents are currently unknown.

In all vertebrate species studied, high levels of melatonin are produced and secreted during the night, while low levels are released during the day. The melatonin circadian rhythm is produced by the endogenous pacemaking activity of the suprachiasmatic nuclei, while the entrainment of this rhythm is coordinated by signals of light and darkness relayed from the eyes. Day length or photoperiod can influence the duration that melatonin production is elevated during the night. This represents a seasonal effect of light on the pineal gland. Specifically, in the summer when days are longer and nights are shorter, the duration of increased nocturnal melatonin secretion is shorter than during the winter when nights are longer. This effect of photoperiod length influencing the duration of nighttime melatonin rise has been documented in many species, including humans.

There is extensive species diversity in the capacity of melatonin to regulate physiology. Numerous species, ranging from insects to mammals, have yearly cycles of activity, morphology, reproduction, or development which are responsive to seasonal changes in day length (photoperiodism). Among many species that breed seasonally, melatonin has been shown to be a potent regulator of the reproductive axis in both males and females. The effects of melatonin on the regulation of circadian physiology have been elucidated in many vertebrate species, including humans. In addition, melatonin has been studied in different species for its influence on retinal physiology, sleep, body temperature regulation, immune function, and cardiovascular regulation.

George C. Brainard

Pineapple A low-growing perennial plant, indigenous to the Americas. The cultivated varieties (cultivars) belong to the species *Ananas sativus* of the plant order Bromeliales.

The edible portion of the pineapple develops from a mass of ovaries on a fleshy flower stock having persistent bracts (see illustration). On the cultivated types, the flowers are usually abortive. The leaves are long and swordlike and usually rough-edged. Commercial plantings bear fruit at the age of 12–20 months, and may continue to be productive for as much as 8–10 years. *See* BROMELIALES.

Pineapple (*Ananas sativus*), fruit and leaves. (*USDA*)

The major producing area is Hawaii, where special methods of culture and harvesting have been developed. Pineapples are also grown in the West Indies and other tropical areas, and to a limited extent in southern Florida.

Pineapples are consumed fresh in considerable quantity, but because of distance from markets and the problems of transporting fresh fruit, most of the crop is canned as sliced pineapple or as juice. *See* FRUIT. J. Harold Clarke

Pinnipeds Carnivorous mammals of the suborder Pinnipedia, which includes 32 species of seals, sea lions, and walrus in three families. All species of the order are found along coastal areas from the Antarctic to the Arctic regions. Although many species have restricted distribution, the group as a whole has worldwide distribution.

Pinnipeds are less modified both anatomically and behaviorally for marine life than are other marine mammals. Each year they return to land to breed, and they have retained their hindlimbs. They are primarily carnivorous mammals, with fish supplying the basic diet. They also eat crustaceans, mollusks, and in some instances sea birds. Their body is covered with a heavy coat of fur, the limbs are modified as flippers, the eyes are large, the external ear is small or lacking, and the tail is absent or very short.

Odobenidae. The single species of this family, the walrus (*Odobenus rosmarus*) [see illustration], is large, growing to 10 ft (3 m) and weighing 3000 lb (1350 kg). It has no external ears but does have a distinct neck region. The upper canines of both sexes are prolonged as tusks which can be used defensively. Walrus populations are about equally distributed in shallow waters of the Atlantic and Pacific oceans in the Arctic polar ice regions. The walrus feeds on marine invertebrates and fish.

Walrus (*Odobenus rosmarus*). (*Photo from Donna Dewhurst Collection/U.S. Fish and Wildlife Service*)

Otariidae. This family, the eared seals, includes the sea lions and fur seals, which are characterized by an external ear. The neck is longer and more clearly defined than that of true seals (Phocidae) and the digits lack nails. The California sea lion (*Zalophus californianus*) occurs along the Pacific coast, and is commonly seen in zoos and performing in circuses. The southern sea lion (*Otaria byronia*) is found around the Galápagos Islands and along the South American coast. Steller's sea lion (*Eumetopias jubatus*) is a large species, the male sometimes reaching a length of 13 ft (4 m) and a weight of about 1300 lb (600 kg), and is found along the Pacific coast. Minor species are the Australian sea lion (*Neophoca cinerea*) and the New Zealand species (*Phocarctos hookeri*).

Phocidae. This is the largest family of pinnipeds, the true seals. It includes the monk seals, elephant seals, common seals, and other less well-known forms. The family is unique in that the digits have nails, the soles and palms are covered with hair, and the necks are very short. Most species live in marine habitats; however, the Caspian seal (*Pusa caspica*) lives in brackish water. The only freshwater seal is the Baikal seal (*P. sibirica*). It is estimated that there are between 40,000 and 100,000 Baikal seals in Lake Baikal, Russia. These animals are hunted for their valuable fur. One of the best-known species is the Atlantic gray seal (*Halichoreus grypus*), a species found in the North Atlantic along the coasts of Europe, Iceland, and Greenland. *See* CARNIVORA; MAMMALIA. Charles B. Curtin

Pinophyta One of the two divisions of the seed plants, comprising about 600 to 700 species extant on all continents except Antarctica. The most familiar and common representatives are the evergreen, cone-bearing trees of the Pinales. Because the ovules (young seeds) are exposed directly to the air at the time of pollination, the Pinophyta are commonly known as the gymnosperms, in contrast to the other division of flowering plants, the angiosperms (division Magnoliophyta), which have the ovules enclosed in an ovary. The division Pinophyta consists of three classes: Ginkgoopsida, Cycadopsida, and Pinopsida. *See* CYCADOPSIDA; GINKGOOPSIDA; MAGNOLIOPHYTA; PINOPSIDA; PLANT KINGDOM. Thomas A. Zanoni

Pinopsida The largest and most important class of the division Pinophyta (Gymnospermae), the other classes being Ginkgoopsida and Cycadopsida. There are two orders: the Cordaitanthales, with three extinct families, and the Pinales (Coniferales), with six extinct families and seven families with some extant genera.

The living Pinopsida are woody plants; most are trees with a central axis and excurrent branches. Leaves are simple, alternate, or opposite or in whorls, scalelike or needlelike or rarely planar. The wood lacks vessels and usually has resin canals. Male reproductive structures are aggregated on microsporophylls directly attached to the cone axis. The ovules are borne in compound cones or singly or paired at the end of a stalk (Taxaceae). The main seed plane is tangential to the cone axis (if the seed scale is regarded as a modified dwarf shoot). The embryo has two or more cotyledons. *See* CYCADOPSIDA; GINKGOOPSIDA; PINALES.

Thomas A. Zanoni

Pionium The electromagnetically bound or Coulomb-bound pion-pion system, the $\pi^+\pi^-$ atom, also called $A_{2\pi}$. Nowadays, this doubly exotic system is called pionium. In some older literature the term pionium was used for atoms consisting of only one pion and a muon or an electron. *See* ELEMENTARY PARTICLE; HADRONIC ATOM; LEPTON; MESON.

The first $A_{2\pi}$ atoms were observed in 1993. In the DIRAC (DI-meson Relativistic Atom Complex) experiment at CERN (Conseil Européen pour la Recherche Nucléaire) in Geneva, Switzerland, a 24-GeV proton beam from the Proton Synchrotron is directed onto a thin nickel target to produce pionium in a proton-nuclear reaction. In 2001, the DIRAC collaboration was able to collect and analyze more than 6000 $A_{2\pi}$ breakups. This sample allowed the $A_{2\pi}$ lifetime to be determined to be 2.9 fs (1 femtosecond $= 10^{-15}$ s) within a statistical error of 15%. *See* PARTICLE ACCELERATOR.

The strong pion-pion interaction in $A_{2\pi}$ is described by low-energy quantum chromodynamics (QCD). Theory predicts that pionium decays, mainly into two neutral pions, and has a lifetime of (2.9 ± 0.1) fs. A lifetime measurement such as that of the DIRAC experiment will check sensitively the understanding of chiral symmetry breaking in QCD. *See* QUANTUM CHROMODYNAMICS.

Juerg Schacher

Pipe flow Conveyance of fluids in closed circular ducts. Flow in closed conduits is probably the most common way of transporting fluids. Crude oil and its components are moved through pipes in a refinery. Water in the home is transported through tubing. Heated and conditioned air is distributed to all parts of a dwelling in circular or rectangular ducts. *See* PIPELINE.

Flow in a closed conduit (circular or otherwise) can be either laminar or turbulent. In laminar flow, the fluid particles move smoothly through the duct in layers called laminae. A fluid particle in one layer stays in that layer. In turbulent flow, flowing fluid particles move tortuously about the cross section, resulting in an effective mixing action. Eddies and vortices are responsible for the mixing, which does not occur in laminar flow. Turbulent flow exists at much higher flow rates than laminar flow.

The criterion for distinguishing between laminar and turbulent flow is this observed mixing action. When injected into a laminar flow in a duct, a dye moves downstream in a threadlike line. When injected into a turbulent flow, a dye disperses quickly. Experiments have shown that laminar flow exists when the dimensionless Reynolds number, $\mathrm{Re} = VD/\nu$, is less than 2100; here V is the average velocity, D is the inside diameter of the pipe or tube, and ν is the kinematic viscosity (a property of the fluid). *See* FLUID FLOW; LAMINAR FLOW; REYNOLDS NUMBER; TURBULENT FLOW.

The energy loss experienced by the fluid is manifested as a pressure drop Δp, which is found in terms of a dimensionless friction factor f as shown in Eq. (1). Here, the density ρ is a

$$\Delta p = \frac{fL}{D}\frac{\rho V^2}{2} \qquad (1)$$

property of the fluid, L is the length over which the pressure drop occurs, D is the inside diameter of the pipe or tube, V is the average velocity, and the quantity $\rho V^2/2$ is the kinetic energy of the flow per unit volume. For laminar flow through a circular duct, the friction factor is found in terms of the Reynolds number, as indicated in Eq. (2). For turbulent flow, the friction factor is

$$f = \frac{64}{\mathrm{Re}} \qquad \left(\begin{array}{c}\text{laminar flow,}\\ \text{circular duct}\end{array}\right) \qquad (2)$$

dependent upon the wall roughness, the fluid properties, the average velocity, and the pipe diameter; that is, $f = f(V, D, \rho, \mu, \varepsilon)$, in which ε is a measure of the absolute roughness of the conduit wall, having the dimension of length. Values of the roughness ε have been measured for many commercial pipe materials. The friction factor for turbulent flow may be obtained from a Moody diagram in which the friction factor f is graphed as a function of the Reynolds number Re, with the relative roughness ε/D as an independent parameter. The Moody diagram is a result of many flow rate and pressure drop measurements made on commercial pipe and tube materials.

William S. Janna

Pipeline A line of piping and the associated pumps, valves, and equipment necessary for the transportation of a fluid. Major uses of pipelines are for the transportation of petroleum, water (including sewage), chemicals, foodstuffs, pulverized coal, and gases such as natural gas, steam, and compressed air. Pipelines must be leakproof and must permit the application of whatever pressure is required to force conveyed substances through the lines. Pipe is made of a variety of materials and in diameters from a fraction of an inch up to 30 ft (9 m). Principal materials are steel, wrought and cast iron, concrete, clay products, aluminum, copper, brass, cement and asbestos (called cement-asbestos), plastics, and wood.

Pipe is described as pressure and nonpressure pipe. In many pressure lines, such as long oil and gas lines, pumps force substances through the pipelines at required velocities. Pressure may be developed also by gravity head, as for example in city water mains fed from elevated tanks or reservoirs.

Nonpressure pipe is used for gravity flow where the gradient is nominal and without major irregularities, as in sewer lines, culverts, and certain types of irrigation distribution systems.

Design of pipelines considers such factors as required capacity, internal and external pressures, water- or airtightness, expansion characteristics of the pipe material, chemical activity of the liquid or gas being conveyed, and corrosion.

Leslie M. McClellan

Piperales A small order of flowering plants (3600 species) in the eumagnoliid group, which is composed of three anomalously woody vines (shrubs) or herbaceous families—the pipeworts (Aristolochiaceae), the black pepper family (Piperaceae), and the lizard's tail family (Saururaceae). The last two families have reduced flowers in dense spikelike flower stems, and the first has medium-sized to enormous flowers that often trap insects for a period before releasing them, covered with pollen.

Black pepper comes from *Piper nigrum* and betel nuts from *P. betle*. Several species of *Aristolochia* have medicinal properties, and some genera in each of these families are commonly grown ornamentals in the temperate zones or house plants, such as *Asarum* (wild ginger), *Peperomia* (pepper elders), and *Houttuynia*. *See* EUMAGNOLIIDS; LAURALES; MAGNOLIALES; MONOCOTYLEDONS.

Mark Chase

Pisces (constellation) The Fishes, a large zodiacal constellation in the northern sky. None of the stars in the constellation is very bright, but many are joined in the shape of

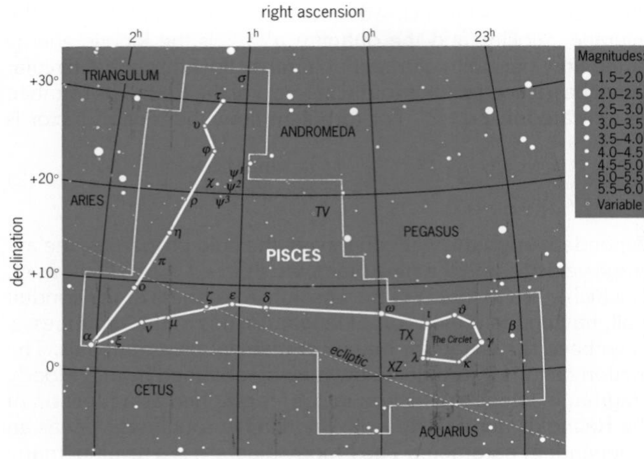

Modern boundaries of the constellation Pisces, the Fishes. The celestial equator is 0° of declination, which corresponds to celestial latitude. Right ascension corresponds to celestial longitude, with each hour of right ascension representing 15° of arc. Apparent brightness of stars is shown with dot sizes to illustrate the magnitude scale, where the brightest stars in the sky are 0th magnitude or brighter and the faintest stars that can be seen with the unaided eye at a dark site are 6th magnitude. (*Wil Tirion*)

a large V, making them easy to notice. The constellation also contains the Circlet of stars. *See* CONSTELLATION; ZODIAC.

Jay M. Pasachoff

Pisces (zoology) A term that embraces all fishes and fishlike vertebrates. The Pisces include four well-defined groups that merit recognition as classes: the Agnatha or jawless fishes, the most primitive; the Placodermi or armored fishes, known only as Paleozoic fossils; the Chondrichthyes or cartilaginous fishes; and the Osteichthyes or bony fishes. *See* CHONDRICHTHYES; JAWLESS VERTEBRATES; OSTEICHTHYES; PLACODERMI.

Reeve M. Bailey

Pistachio A tree, *Pistacia vera*, of the Anacardiaceae family. It is native to central Asia and has been grown for its edible nuts throughout recorded history in various countries of the Mediterranean region. Extensive areas in California were planted with pistachios in the 1970s, and the first commercial nut crop was harvested in 1977. *See* SAPINDALES.

The pistachio tree, relatively slow-growing, reaches a height and spread of 20–25 ft (6–8 m). It thrives under long, hot summers with low humidity, but needs moderately cold winters to satisfy its chilling requirement. Pistachio is deciduous and has imparipinnate leaves, most often 2-paired. It is dioecious, and both staminate and pistillate inflorescences are panicles that may have 150 or more individual flowers. They lack petals and nectaries and, consequently, are wind-pollinated. The fruit, a semidry drupe, is borne on 1-year-old wood in clusters similar to grapes and matures in September. The hull (exocarp and mesocarp) at that time slips easily from the shell (endocarp) which has already dehisced, exposing the kernel. Pistachio kernels contain only 5–10% sugars, but their protein and oil content of about 20 and 40%, respectively, make them high in food value. *See* NUT CROP CULTURE.

Julian C. Crane

Pitch The psychological property of sound characterized by highness or lowness. Pitch is one of the two major auditory attributes of simple sounds, the other being loudness.

A simple sound source, such as a tuning fork, produces an acoustic wave that approximates a perfect sinusoid, and the pitch of a sinusoid wave is almost completely determined by its frequency. Many sounds, however, are complex and contain a number of sinusoidal components. Complex sounds often appear to have a strong pitch, which is the frequency of a sinusoid that appears to match the complex sound. Hence, a tuning fork that vibrates at about 440 Hz will have a pitch very nearly equal to the note A above middle C on the piano. Loudness is determined by the amplitude of the sound vibrations. *See* LOUDNESS; TUNING FORK; WAVE (PHYSICS).

A sequence of different sounds having definite pitches produces a musical tune, making pitch extremely important in music. Practically all musical conventions recognize that doubling the frequency of vibration produces a particular pitch interval, known as an octave. *See* MUSICAL ACOUSTICS; SCALE (MUSIC).

The human auditory system can hear frequencies in the range of 20–20,000 Hz. For frequencies between 100 and 4000 Hz, sinusoidal sounds have a clear pitch. Beyond these limits, the pitch of sound is not distinct. Sounds below 100 Hz may be described as rumbles, while those above 4000 Hz may be described as shrill and squeaky. The ability to detect changes in pitch is remarkably acute. The just-detectable change in frequency is about 0.3% for the midfrequency range. Frequency changes are best detected when the sound is loud. Weaker sounds require greater changes in frequency to be detectable. *See* AUDIOMETRY; HEARING (HUMAN).

David M. Green

Pitcher plant Any member of the families Sarraceniaceae and Nepenthaceae. In these insectivorous plants the leaves form deep cups or pitchers in which water collects. Visiting insects, falling into this water, are drowned and digested by the action of enzymes secreted by cells located in the walls of the pitcherlike structures of these plants. Often these plants climb by tendrils. The end of a tendril may develop into a pitcher, which captures and digests insects. *See* INSECTIVOROUS PLANTS; NEPENTHALES.

Perry D. Strausbaugh; Earl L. Core

Pitchstone A natural glass with dull or pitchy luster and generally brown, green or gray color. It is extremely rich in microscopic, embryonic crystal growths (crystallites) which may cause its dull appearance. The water content of pitchstone is high and generally ranges from 4 to 10% by weight. Pitchstone is formed by rapid cooling of molten rock material (lava or magma) and occurs most commonly as small dikes or as marginal portions of larger dikes. *See* IGNEOUS ROCKS; VOLCANIC GLASS.

Carleton A. Chapman

Pith The central zone of tissue of an axis in which the vascular tissue is arranged as a hollow cylinder. Pith is present in most stems and in some roots. Stems without pith rarely occur in angiosperms but are characteristic of psilopsids, lycopsids, *Sphenophyllum*, and some ferns. Roots of some ferns, many monocotyledons, and some dicotyledons include a pith, although most roots have xylem tissue in the center.

Pith is composed usually of parenchyma cells often arranged in longitudinal files. This arrangement results from predominantly transverse division of pith mother cells near the apical meristem. *See* PARENCHYMA; ROOT (BOTANY); STEM.

H. Weston Blaser

Pitot tube A device to measure the stagnation pressure due to isentropic deceleration of a flowing fluid. In its original form it was a glass tube bent at 90° and inserted in a stream flow, with its opening pointed upstream. Water rises in the tube a distance, h, above the surface, and if friction losses are negligible, the velocity of the stream, V, is approximately 2gh, where g is the acceleration of gravity. However, there is a significant measurement error if the probe is misaligned at an angle α with respect to the stream. For an open tube, the error is about 5% at $\alpha \approx 10°$.

The misalignment error of a pitot tube is greatly reduced if the probe is shielded, as in the Kiel-type probe. The Kiel probe is accurate up to $\alpha \approx 45°$.

The modern application is a pitot-static probe, which measures both the stagnation pressure, with a hole in the front, and the static pressure in the moving stream, with holes on the sides. A pressure transducer or manometer records the difference between these two pressures. Pitot-static tubes are generally unshielded and must be carefully aligned with the flow to carry out accurate measurements. *See* BERNOULLI'S THEOREM.

When used with gases, estimate of the stream velocity is only valid for a low-speed or nearly incompressible flow, where the stream velocity is less than about 30% of the speed of sound of the fluid. At higher velocities, estimate of the stream velocity must be replaced with a Bernoulli-type theory, which accounts for gas density and temperature changes. If the gas stream flow is supersonic, or the stream velocity is greater than the speed of sound of the gas, a shock wave forms in front of the probe and the theory must be further corrected by complicated supersonic-flow algebraic relations. *See* COMPRESSIBLE FLOW; GAS DYNAMICS; SHOCK WAVE.

A disadvantage of pitot and pitot-static tubes is that they have substantial dynamic resistance to changing conditions and thus cannot accurately measure unsteady, accelerating, or fluctuating flows. *See* ANEMOMETER; FLOW MEASUREMENT. Frank M. White

Pituitary gland The most structurally and functionally complex organ of the endocrine system. Through its hormones, the pituitary, also known as the hypophysis, affects every physiological process of the body. All vertebrates have a pituitary gland with a common basic structure and function. In addition to its endocrine functions, the pituitary may play a role in the immune response.

The hypophysis of all vertebrates has two major segments—the neurohypophysis (a neural component) and the adenohypophysis (an epithelial component)—each with a different embryological origin. The neurohypophysis develops from a downward process of the diencephalon (the base of the brain), whereas the adenohypophysis originates as an outpocketing of the primitive buccal epithelium, known as Rathke's pouch. The adenohypophysis has three distinct subdivisions: the pars tuberalis, the pars distalis, and the pars intermedia. The neurohypophysis comprises the pars nervosa and the infundibulum. The latter consists of the infundibular stalk and the median eminence of the tuber cinereum.

The structural intimacy of neurohypophysis and adenohypophysis that is established early during embryogenesis reflects the direct functional interaction between the central nervous system and endocrine system. The extent of this anatomical intimacy varies considerably among the vertebrate classes, from limited contact to intimate interdigitation. Vascular or neuronal pathways, or both, provide the means of exchanging chemical signals, thus enabling centers in the brain to exert control over the synthesis and release of adenohypophysial hormones.

Neurohormones, which are synthesized in specific regions of the brain, are conveyed to the neurohypophysis by way of axonal tracts, where they may be stored in distended axonal endings. Axons may also contact blood vessels and discharge their neurosecretory products into the systemic circulation or into a portal system leading to the adenohypophysis, or they may directly innervate pituitary gland cells. *See* NEUROSECRETION.

In most animals, the vascular link is the prime route of information transfer between brain and pituitary gland. This link begins in the tuber cinereum, the portion of the third ventricle floor that extends toward the infundibulum. The lower tuber cinereum, which is known as the median eminence, is well endowed with blood vessels that drain down into the pituitary stalk and ultimately empty into the anterior pituitary. The vascular link between the median eminence and the pituitary gland is known as the hypothalamo-hypophysial portal system. The median eminence in humans is vascularized by the paired superior hypophysial arteries. The pituitary gland is believed to have the highest blood flow rate of any organ in the body. However, its blood is received indirectly via the median eminence and the hypothalamo-hypophysial portal system. Most of the blood flow is from the brain to the pituitary gland, with retrograde flow from the adenohypophysis to the hypothalamus, suggesting a two-way communication between nervous and endocrine systems. Although the brain is protected from the chemical substances in the circulatory system by the blood–brain barrier, the median eminence lies outside that protective mechanism and is therefore permeable to intravascular substances. *See* BRAIN.

The hormones of the adenohypophysis may be grouped into three categories based on chemical and functional similarities. The first category consists of growth hormone (also known as somatotropin) and prolactin, both of which are large, single, polypeptide chains; the second category consists of the glycoprotein hormones; this family of hormones contains the gonadotropins and thyrotropin. The gonadotropins in many species, including humans, can be segregated into two distinct hormones, follicle-stimulating hormone and luteinizing hormone. The third group comprises adrenocortiotropic hormone and melanotropin (MSH; melanocyte-stimulating hormone). *See* ADENOHYPOPHYSIS HORMONE.

The regulation of the release of pituitary hormones is determined by precise monitoring of circulating hormone levels in the blood and by genetic and environmental factors that manifest their effect through the releasing and release-inhibiting factors of the hypothalamus. The hypothalamus is located at the base of the brain (the diencephalon) below the thalamus and above the pituitary gland, forming the walls and the lower portion of the third ventricle. It receives major neuronal inputs from the sense organs, hippocampus, thalamus, and lower brainstem structures, including the reticular formation and the spinal cord. Thus, the hypothalamus is designed and anatomically positioned to receive a diversity of messages from external and internal sources that can be transmitted by way of hypothalamic releasing factors to the pituitary gland, where they are translated into endocrine action. *See* NERVOUS SYSTEM (VERTEBRATE).

The neurohypophysis hormones, oxytocin and vasopressin, are synthesized in different neurons of the paraventricular and supraoptic nuclei of the hypothalamus and travel by axonal flow to the terminals in the neurohypophysis for storage and ultimate release into the vascular system. Oxytocin is important in stimulating milk release through its contractile action on muscle elements in the mammary gland. It also stimulates uterine smooth muscle contraction at parturition. Vasopressin affects water retention by its action on certain kidney tubules. Thus, it also affects blood pressure. *See* LACTATION; NEUROHYPOPHYSIS HORMONE.

The better-known neurotransmitters of the central nervous system include the catecholamines (dopamine, epinephrine, and norepinephrine), serotonin, acetylcholine, gamma-amino butyric acid (GABA), histamine, and the opioid peptides (enkephalins, endorphins, dynorphin, neoendorphin, rimorphin, and leumorphin). These substances are distributed widely in the central nervous system and, for most, also in the pituitary gland. If a particular amine or neurotransmitter is present in nerve fibers leading to the median eminence, it probably will influence pituitary gland activity via the portal system. Dopamine, serotonin, gamma-amino butyric acid, and acetylcholine are best known for such activity. These neurotransmitters play an important, but poorly understood, role in regulating pituitary function, either directly or by their action on neuropeptide-producing neurons. Understanding the pharmacology of neurotransmitters holds promise for the treatment of basic disorders of the hypothalamic-pituitary axis. *See* ACETYLCHOLINE; ENDOCRINE MECHANISMS; ENDOCRINE SYSTEM (VERTEBRATE); ENDORPHINS; HISTAMINE; HORMONE; NEUROBIOLOGY; NEUROIMMUNOLOGY; PITUITARY GLAND DISORDERS; SEROTONIN. Martin P. Schreibman

Pituitary gland disorders Inborn or acquired abnormalities in the structure or function of the human pituitary gland. Pituitary disorders can stem from any of five disease processes: tumors and other growths, intrinsic lesions of the anterior lobe, diseases affecting gland function, hypothalamic malfunction, and systemic disease that affects the adenohypophysis.

Tumors and other growths in and near the pituitary may cause failure of hormone secretion or impinge on nearby brain structures. The latter effect can give rise to neurological malfunctions, the most common of which is visual impairment such as narrowing of the visual fields. Intrinsic lesions of the anterior pituitary, usually benign, secrete excessive amounts of a single (rarely multiple) hormone, producing characteristic endocrine syndromes, the most dramatic of which are acromegaly and Cushing's disease. Infections, congenital anomalies, granulomas, vascular disorders, and, rarely, metastatic cancers may induce partial or total failure of one or more pituitary secretions, in which case growth hormone, follicle-stimulating hormone, and luteinizing hormone are the first to fail.

Diseases of the hypothalamus affect pituitary function through the mechanical effects of a mass or through disrupted secretion of the hypophysiotropic peptides. Typical manifestations include precocious or delayed puberty; diabetes insipidus; and derangements of sleep, eating, and temperature regulation. The dysfunction may be of congenital, traumatic, inflammatory, or neoplastic origin. However, the most important hypothalamic-pituitary diseases are tertiary hypothyroidism, precocious or delayed puberty, diabetes insipidus, and Kallmann's syndrome, which consists of a deficient sense of smell and lack of sexual development due to inborn failure of the hypothalamus to secrete gonadotropin-releasing hormone. Hypothalamic disorders can cause abnormalities of growth hormone secretion. Whereas most individuals with acromegaly (an excess of growth hormone) have intrinsic pituitary tumors that secrete growth hormone, some have excessive secretion of hypothalamic growth hormone-releasing hormone or insufficient secretion of somatostatin. Pituitary dwarfism, caused by the failure to secrete growth hormone in childhood, is usually due to primary disease of the adenohypophysis, but in some individuals the fundamental dysfunction (opposite to that in acromegaly) lies in the hypothalamus. *See* DIABETES; DWARFISM AND GIGANTISM; SOMATOSTATIN; THYROID GLAND DISORDERS.

The adenohypophysis can be affected subtly by systemic diseases, or more obviously by metastases from breast cancer. Prolonged treatment with large doses of adrenal corticosteroids, as in cases of lupus, asthma, or acute leukemia, can result in the failure of the adenohypophysis to secrete adrenocorticotropic hormone in times of physical stress. A deficiency in adrenocorticotropic hormone is a potentially fatal condition. *See* ADENOHYPOPHYSIS HORMONE; ADRENAL GLAND DISORDERS; ENDOCRINE MECHANISMS; PITUITARY GLAND.

Nicholas Christy

pK The logarithm (to the base 10) of the reciprocal of the equilibrium constant for a specified reaction under specified conditions (for example, solvent and temperature). The p*K* values are often more convenient to tabulate and use than the equilibrium constants themselves. The value of *K* for the dissociation of the HSO_4^- ion in aqueous solution at 25°C (77°F) is 0.0102 mole/liter. The logarithm is $0.008_6 - 2 = -1.991_4$ The p*K* is therefore $+ 1.991_4$. The choice of algebraic sign, although arbitrary, results in positive values for most dissociation constants applicable to aqueous solutions. The concept of p*K* is especially valuable in the study of solutions. *See* CHEMICAL EQUILIBRIUM; IONIC EQUILIBRIUM; PH.

Thomas F. Young

Placentation The intimate association or fusion of a tissue or organ of the embryonic stage of an animal to its parent for physiological exchange to promote the growth and development of the young. It enables the young, retained within the body or tissues of the mother, to respire, acquire nourishment, and eliminate wastes by bringing the bloodstreams of mother and young into close association but never into direct connection. Placentation characterizes the early development of all mammals except the egg-laying duckbill platypus and spiny anteater. It occurs in some species of all other orders of vertebrates except the birds. In fact, in certain sharks and reptiles it is almost as well developed as in mammals. A few examples are also known among invertebrates (*Peripatus*, certain tunicates, and insects). *See* FETAL MEMBRANE.

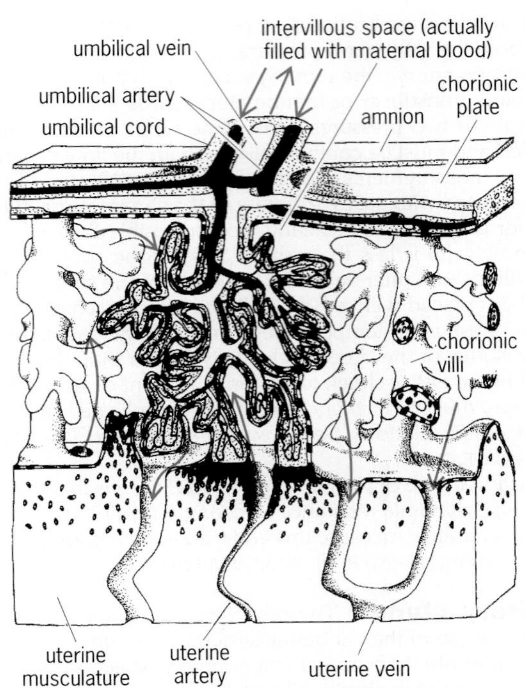

Block removed from center of human placenta.

Efficient interchange depends on close proximity of large areas of fetal tissues to maternal blood and glandular areas. This is provided in mammals by a remarkable regulatory cooperation between the developing outer layer (trophoblast) of the chorion, together with the vascular yolk sac or allantois or both, and the mother's uterine lining (endometrium). In the typical mammalian placenta, which is always formed by the chorion and the allantoic vessels, the fetal and maternal bloodstreams are as close as a few thousandths of a millimeter from each other (see illustration). The surface area of the fetal villi which contain the functional fetal capillaries is probably several times larger than the body surface of the female. In humans this ratio is known to be about 8:1.

Harland W. Mossman

Placer mining The exploitation of placer mineral deposits for their valuable heavy minerals. Placer mineral deposits consist of detrital natural material containing discrete mineral particles. They are formed by chemical and physical weathering of in-place heavy minerals, which are then concentrated through the action of wind or moving water. This concentration can be done through wave and current action in the ocean (beach and offshore placers), glacial action (moraine placers), wind action removing the lighter material (eolian placers), or the action of running water (stream placers). Stream placers are the most important of these deposits because of their common occurrence and their highly efficient concentration mechanisms. Marine placers, primarily beach placers, are the next most economically important, with the potential of offshore placers being the most recent to be recognized and developed. *See* MARINE MINING; ORE AND MINERAL DEPOSITS.

Minerals that are concentrated in placer deposits are a result of differences in specific gravity and, therefore, the economically important deposits are for minerals with high specific gravities [for example, gold (specific gravity 15–19), and platinum (14–19)].

Precious metals, primarily gold and platinum group metals, have been the most important product from placer mines. Their extremely high specific gravity coupled with their low chemical reactivity means that these minerals are efficiently concentrated in a placer environment and can be effectively recovered in a readily usable form. Although most modern gold is produced from lode, or "hard rock," deposits, the placer deposits of northern Canada, Alaska, and Siberia represent a virtually untapped source of the metal. *See* GOLD; PLATINUM.

Of more importance than gold are placer diamond deposits. Another important placer mineral is cassiterite, an ore of tin. Additionally, rutile and ilmenite, the principal ores of titanium, are found in commercial quantities only in beach placers. These same types of placers also yield monazite, a source of the rare earths yttrium, lanthanum, cerium, and thorium. *See* CASSITERITE; DIAMOND; ILMENITE; RUTILE.

Most placer mining operations involve surface mining methods, although underground methods are sometimes used. *See* SURFACE MINING; UNDERGROUND MINING.

The two major environmental problems associated with placer mining are water pollution and land disturbance. Both the mining and the processing of placer minerals require a great deal of water, and, once used, this water contains large amounts of suspended solids. If the water is allowed to run off into the rivers, these solids can have an adverse impact on the downstream environment. In suspension they can harm aquatic habitats, and when settled out can clog waterways and choke off irrigated crops. Since most placer mining is surface mining, surface disturbance is necessary, especially where dredging operations create vast piles of cobbles as mining progresses. One method of land reclamation is a mining plan which stockpiles top soil (where possible), recontours the spoil (waste) piles, and then returns the land to useful status. *See* ENVIRONMENTAL ENGINEERING; LAND RECLAMATION; WATER POLLUTION. Danny L. Taylor

Placodermi A class of fishes known from the Devonian Period, and a few that survived into the base of the Carboniferous. They were true fishes or gnathostomes, and can be distinguished from other fishes by the following characters: the gill chamber extends far under the cranium and is covered laterally by opercula; there is a neck joint between the cranium and the fused anterior vertebrae; often there is also a coaxial joint developed between the dermal bones of the cranial roof and shoulder girdle; the head and shoulder girdle are covered with dermal bones composed typically of cellular bone and superficially of semidentine instead of true dentine; the bones are commonly ornamented with tubercles or ridges; the endoskeleton is cartilage and may be calcified in a globular fashion or perichondrally ossified; the notochord is persistent, and the vertebrae consist only of neural and hemal arches; the tail is diphycercal or slightly heterocercal, and an anal fin is lacking.

Most placoderms were bottom-dwelling fishes, with the head and trunk dorsoventrally depressed; only the Stensioellida and some specialized arthrodires had laterally compressed and deepened bodies, suggesting a more nectic manner of life. Most of them were small or moderate-sized, but a few were, for their time, gigantic, reaching a length of as much as 20 ft (6 m). They were the dominant fishes of the Devonian, and are found in both marine and fresh-water deposits.

The Placodermi are subdivided into nine orders, each with its own distinct specializations: Stensioellida, Pseudopetalichthyida, Rhenanida, Ptyctodontida, Acathothoraci, Petalichthyida, Phyllolepida, Arthrodira, and Antiarcha. The evolution and interrelations of these orders are matters of disagreement.

Robert H. Denison; Everett C. Olson

Placodontia A small group of marine reptiles forming part of the Sauropterygia, known only from deposits of Triassic age in Europe, the Near East (Israel), North Africa (Tunisia), and China. Placodonts are reptiles with a grossly specialized dentition—flat-crowned teeth located in both the upper and lower jaws and on the palate—that functioned as a crushing device for hard-shelled invertebrate prey such as brachiopods, gastropods, clams, and cephalopods. *See* REPTILIA; SAUROPTERYGIA.

Placodonts are divided into two clades: the unarmored Placodontoidea (*Placodus, Paraplacodus*) and the armored ("turtle-like") Cyamodontoidea. The skeleton of unarmored placodonts shows few adaptations to life in nearshore habitats or shallow epicontinental seas. The thorax was box-shaped, and the digits of hand and foot were probably webbed; the joint surfaces and muscle attachment sites on the limb bones suggest that the animals did little if any walking on land. The most serious modifications to an aquatic environment occurred in the skull, and are correlated with the development of the durophagous (crushing) dentition. The entire skull became massive in relation to the body as a result of pachyostosis (thickening of the bone) and partial fusion of the skull elements. *See* DENTITION; SKULL.

The advanced cyamodontoids, such as *Placochelys*, resemble sea turtles in overall appearance and in the fact that their body was encased in an armor of dermal bones similar to that of the extinct dermochelyid sea turtle *Psephophorus*. Such similarities with turtles are, however, entirely due to convergent evolution. Cyamodontoids were much more diverse than placodontoids, evolving a variety of skull shapes that indicate different feeding strategies. Olivier C. Rieppel; Rainer Zangerl

Placozoa The most primitive metazoan animal phylum. At present, this phylum harbors a single named species, the enigmatic *Trichoplax adhaerens*. *See* ANIMAL KINGDOM; METAZOA.

In 1883, the German zoologist Franz Eilhard Schulze discovered this microscopic marine animal on the glass walls of a seawater aquarium at the University of Graz in Austria. The animal, measuring less than 5 mm (0.2 in.) in diameter and 10–15 micrometers (0.0004–0.0006 in.) in height, looked like an irregular and thin hairy plate sticking to the glass surface (see illustration). Schulze named this animal *Trichoplax adhaerens* (Greek for "sticky hairy plate"). In contrast to typical multicellular animals (metazoans), *Trichoplax* does not have a head or tail, nor does the animal possess any organs, nerve or muscle cells, basal lamina, or extracellular matrix. *Trichoplax* lacks an axis

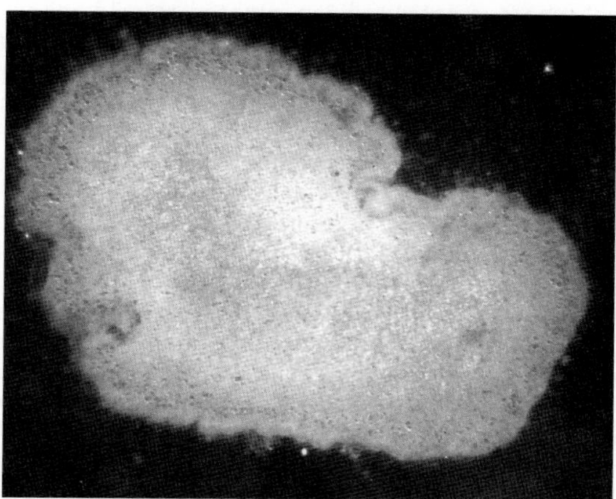

The placozoan *Trichoplax adhaerens*, the only described species of Placozoa. (*Photo by Wolgang Jakob and Bernd Schierwater*).

and any type of symmetry. It almost looks like a giant ameba (a protozoan) when it crawls over the substrate.

The immediate and defining characteristic between placozoas and protozoans is the number of somatic (non–germ line) cell types. In contrast to Protozoa, which consist of either a single cell or several cells of the same somatic cell type, Placozoa have at least four defined somatic cell types: lower epithelial cells, upper epithelial cells, gland cells, and fiber cells. These cells are arranged in a sandwichlike manner, with the lower epithelial and gland cells at the bottom, the upper epithelial cells at the top, and the fiber cells in between. Cells of the lower epithelium attach the animal to a solid substrate, enable the animal to crawl (with the aid of cilia), and allow feeding. In general, very little is known about the biology of Placozoa, and almost all current knowledge derives from laboratory observations. Field data are limited to records of finding *Trichoplax* on hard substrate surfaces from tropical and subtropical marine waters around the world. *See* FEEDING MECHANISMS (INVERTEBRATE); PROTOZOA.

Bernd Schierwater

Plague An infectious disease of humans and rodents caused by the bacterium *Yersinia pestis*. The sylvatic (wild-animal) form persists today in more than 200 species of rodents throughout the world. The explosive urban epidemics of the Middle Ages, known as the Black Death, resulted when the infection of dense populations of city rats living closely with humans introduced disease from the Near East. The disease then was spread both by rat fleas and by transmission between humans. During these outbreaks, as much as 50% of the European population died. At present, contact with wild rodents and their fleas, sometimes via domestic cats and dogs, leads to sporadic human disease. *See* INFECTIOUS DISEASE.

After infection by *Y. pestis*, fleas develop obstruction of the foregut, causing regurgitation of plague bacilli during the next blood meal. The rat flea, *Xenopsylla cheopsis*, is an especially efficient plague vector, both between rats and from rats to humans. Human (bubonic) plague is transmitted by the bite of an infected flea; after several days, a painful swelling (the bubo) of local lymph nodes occurs. Bacteria can then spread to other organ systems, especially the lung; fever, chills, prostration, and death may occur. Plague pneumonia develops in 10–20% of all bubonic infections. In some individuals, the skin may develop hemorrhages and necrosis (tissue death), probably the origin of the ancient name, the Black Death. The last primary pneumonic plague outbreak in the United States occurred in 1919, when 13 cases resulting in 12 deaths developed before the disease was recognized and halted by isolation of cases.

Bubonic plaque is suspected when the characteristic painful, swollen glands develop in the groin, armpit, or neck of an individual who has possibly been exposed to wild-animal fleas in an area where the disease is endemic. Immediate identification is possible by microscopic evaluation of bubo aspirate stained with fluorescent-tagged antibody. Antibiotics should be given if plague is suspected or confirmed. Such treatment is very effective if started early. The current overall death rate, approximately 15%, is reduced to less than 5% among patients treated at the onset of symptoms. *See* IMMUNOFLUORESCENCE; MEDICAL BACTERIOLOGY.

Darwin L. Palmer

Plains The relatively smooth sections of the continental surfaces, occupied largely by gentle rather than steep slopes and exhibiting only small local differences in elevation. Because of their smoothness, plains lands, if other conditions are favorable, are especially amenable to many human activities. Thus it is not surprising that the majority of the world's principal agricultural regions, close-meshed transportation networks, and concentrations of population are found on plains. Large parts of the Earth's plains, however, are hindered for human use by dryness, shortness of frost-free season, infertile soils, or poor drainage. Because of the absence of major differences in elevation or exposure or of obstacles to the free movement of air masses, extensive plains usually exhibit broad uniformity or gradual transition of climatic characteristics.

Somewhat more than one-third of the Earth's land area is occupied by plains. With the exception of ice-sheathed Antarctica, each continent contains at least one major expanse of smooth land in addition to numerous smaller areas. The largest plains of North America, South America, and Eurasia lie in the continental interiors, with broad extensions reaching to the Atlantic (and Arctic) Coast. The most extensive plains of Africa occupy much of the Sahara and reach south into the Congo and Kalahari basins. Much of Australia is smooth, with only the eastern margin lacking extensive plains. *See* TERRAIN AREAS.

Surfaces that approach true flatness, while not rare, constitute a minor portion of the world's\, plains. Most commonly they occur along low-lying coastal margins, the lower sections of major river systems, or the floors of inland basins. Nearly all are the products of extensive deposition by streams or in lakes or shallow seas. The majority of plains, however, are distinctly irregular in surface form, as a result of valley-cutting by streams or of irregular erosion and deposition by continental glaciers.

Edwin H. Hammond

Planck's constant A fundamental physical constant which represents the elementary quantum of action, action being defined as energy multiplied by time. Introduced by Max Planck in 1900, it has the value $h = 6.6261 \times 10^{-27}$ erg-second or 6.6261×10^{-34} joule-second. The symbol $\hbar$ sometimes called the Dirac h, is often used for convenience in physics to denote the quantity $h/2\pi$, where $\pi = 3.1416\ldots$

As used by Planck in deriving his radiation law, h multiplied by the frequency of radiation represented a bundle of energy, that is, a quantum of energy. Radiant energy at any wavelength can occur only as multiples of this energy; thus energy is quantized. *See* COMPTON EFFECT; FUNDAMENTAL CONSTANTS; HEAT RADIATION; QUANTUM MECHANICS.

Heinz G. Sell; Peter J. Walsh

Planck's radiation law A law of physics which gives the spectral energy distribution of the heat radiation emitted from a so-called blackbody at any temperature. Discovered by Max Planck, this law laid the foundation for the advent of the quantum theory because it was the first physical law to postulate that electromagnetic energy exists in discrete bundles, or quanta. *See* HEAT RADIATION; QUANTUM MECHANICS.

Heinz G. Sell; Peter J. Walsh

Plane curve The locus of points in the euclidean plane that satisfy some geometric or algebraic definition. Not all sets of points deserve to be called a curve, but the distinction is somewhat arbitrary. For most of this article, a curve is considered to be the locus of a set of points that satisfy an algebraic or transcendental equation in two variables.

The most interesting geometric properties are those preserved by linear transformations, especially translations, rotations, reflections, and magnifications. Useful geometric properties include the number of branches into which the curve is divided; the number and degree of nodes, cusps, isolated points, and flex points; the number of loops; symmetries; branches that go to infinity; and asymptotes.

These terms can be defined informally as follows. A branch is a maximal smooth continuous portion of the curve. A multiple point is a point in the plane that lies on two or more branches; its degree is the number of branches involved. A node is a multiple point where the branches cross. A cusp is a multiple point where the branches meet but do not pass; that is, each of the branches ends at that point. An isolated point is a point of the curve through which no branches pass. Multiple and isolated points are collectively termed singular points. Any point that is not singular is termed ordinary. A flex (or point of inflection) is

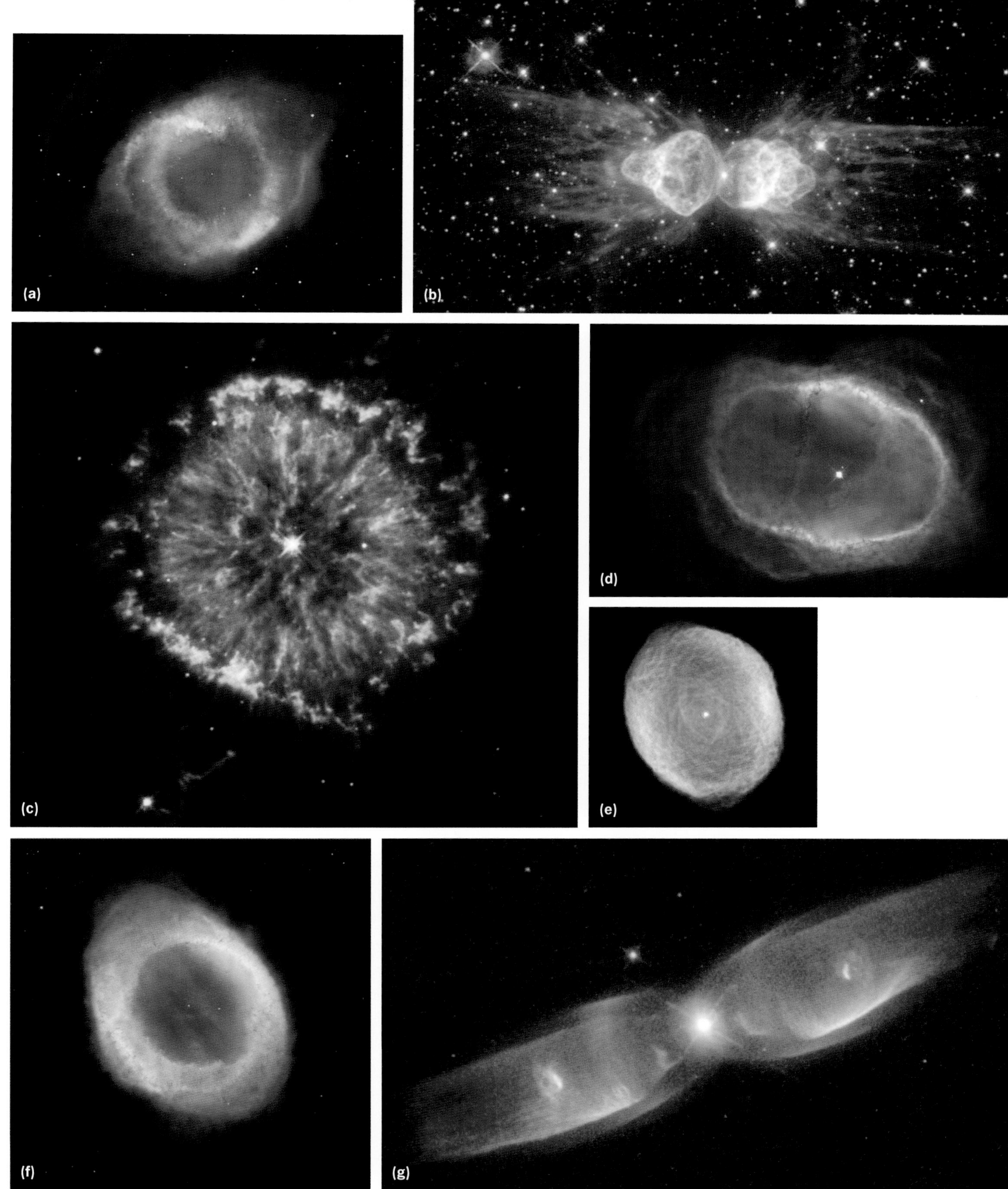

Hubble Space Telescope images of planetary nebulae. (*a*) Helix Nebula, NGC 7923, composite of Hubble Space Telescope images and wide view of the Mosaic Camera on the 0.9-m (35-in.) telescope at Kitt Peak National Observatory (*NASA, NOAO, ESA, Hubble Helix Nebula Team, M. Meixner, T. A. Rector*). (*b*) Ant Nebula, Menzel 3 (*NASA, ESA, Hubble Heritage Team*). (*c*) Puffball Nebula, NGC 6751 (*NASA, Hubble Heritage Team*). (*d*) Southern Ring, NGC 3132 (*NASA, Hubble Heritage Team*). (*e*) Spirograph Nebula, IC 418 (*NASA, Hubble Heritage Team*). (*f*) Ring Nebula, NGC 6720 (M57) (*NASA, Hubble Heritage Team*). (*g*) Butterfly Nebula, Minkowski 2–9 (*B. Balick, V. Icke, G. Mellema, NASA*).

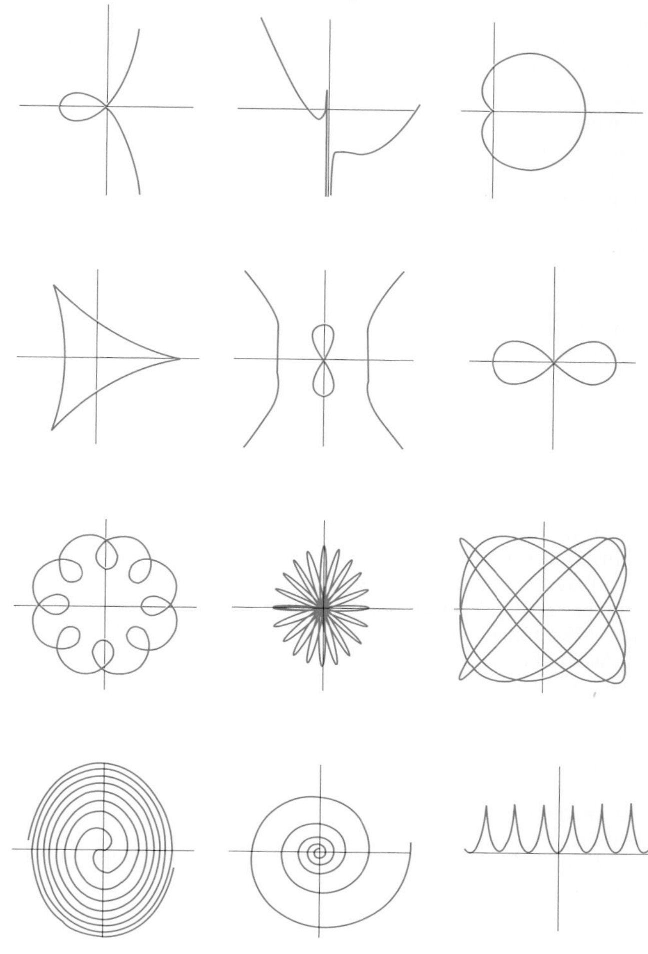

Plane curves. (*a*) **Right strophoid.** (*b*) **Trident of Newton.** (*c*) **Cardioid.** (*d*) **Deltoid.** (*e*) **Devil on two sticks.** (*f*) **Lemniscate of Bernoulli.** (*g*) **Epitrochoid.** (*h*) **Rhodona.** (*i*) **Bowditch curve.** (*j*) **Fermat's spiral.** (*k*) **Logarithmic spiral.** (*l*) **Cycloid.**

a point on the curve whose tangent cuts the curve. A smooth closed branch forms a loop. A curve is symmetric about a line L if every line perpendicular to L intersects the curve at equal distances from L on opposite sides of L; that is, portions of the curve form mirror images about L. The curve is symmetric about a point P if every line through P intersects the curve at equal distances from P in opposite directions. An asymptote is a line toward which a branch approaches as it moves to infinity from the origin; the curve and line are said to intersect at infinity.

In addition to these geometric properties, the form of the defining equation is of interest. This can be an algebraic (polynomial in x and y) or transcendental equation. In the former case, quadratic, cubic, and quartic equations are of special interest.

Once a coordinate system has been chosen, and the defining equation is known (in any of the forms, though the parametric form is usually the most useful), various properties of the curve can be defined in terms of the equation. These include the locations of x- and y-intercepts, local maxima and minima, flexes, nodes, and cusps. *See* ANALYTIC GEOMETRY.

The illustration shows some of the many plane curves of sufficient historical interest to have received names. *See* CARDIOID; CYCLOID; LEMNISCATE OF BERNOULLI; ROSE CURVE.

J. Dennis Lawrence

Plane geometry The branch of mathematics that deals with geometric figures, that is, collections of points that all lie in the same plane (coplanar). Although the words "point" and "plane" are undefined concepts, for elementary applications the intuitive meanings will serve: a point is a location, and a plane is a flat surface. For similar definitions, together with a discussion of the postulates and axioms (assumed truths) used in plane geometry, *See* EUCLIDEAN GEOMETRY.

Dimensions and measures. There are three spatial dimensions; geometric figures are classified as being zero, one, two, or three dimensional. Plane geometry deals only with geometric figures having fewer than three dimensions. Three-dimensional geometric figures, called solids, are dealt with in another branch of euclidean geometry. *See* SOLID (GEOMETRY).

A dimension is any measurement associated with a geometric figure that has units of length. The measure of a geometric figure is a number multiplied by a power of a length, with the result giving information about the size of the figure. The power of the length will be 1, 2, or 3, depending on whether the figure is one, two, or three dimensional. *See* UNITS OF MEASUREMENT.

Lines, line segments, and rays. Exactly one line passes through two given points. The part of a line between (and including) two points is called a line segment, with the two defining points being the end points of the segment. That part of a line that lies on one side of a point (together with that point) is called a ray.

Angles. An angle is the geometric figure formed by joining two rays having a common end point. Each ray is a side of the angle; the common end point is the vertex of the angle.

The concept of the measure of an angle may be understood by imagining that one ray of an angle is held fixed but the other ray is hinged at the vertex and allowed to rotate in the plane. A measure of the angle is a number, together with some unit of angular measure, that tells how much the hinged ray would need to be rotated so that it would overlie the fixed ray. If a ray were to be rotated exactly one revolution, it would return to its original position. The measure of an angle can be what fraction of one revolution would enable one side of the angle to become coincident with the other.

The revolution is a convenient unit of angular measure for many applications. However, a more commonly used unit is the degree, which is defined by $360^\circ = 1$ revolution. In the modern-day use of calculators and computers, fractions of degrees are most conveniently expressed by using decimal fractions; however, another subdivision of degrees is firmly entrenched in many applications: 1 degree = 60 minutes ($1^\circ = 60'$), and 1 minute = 60 seconds ($1' = 60''$).

An angle of measure 1/4 revolution, or 90°, is called a right angle. Two lines (or rays or segments) that intersect so as to form a right angle are said to be perpendicular. Two angles are complementary if their measures have a sum of 90°. An angle of measure 1/2 revolution, or 180°, is called a straight angle. Two angles are supplementary if their measures have a sum of 180°.

Polygons. A polygon is the geometric figure formed when line segments are joined end to end so as to enclose a region of the plane. The polygon having the least number of sides (three) is the triangle. An angle whose sides are two of those segments and whose arc lies inside the triangle is an interior angle of the triangle; an exterior angle of the triangle is any angle that is adjacent and supplementary to an interior angle. The interior angles of any triangle have measures that sum to 180°. This fact allows the determination of the measures of all angles formed by the intersections of three line if the measures of only two angles with different vertices are known. *See* POLYGON.

Congruent and similar geometric figures. Two geometric figures are congruent ($\simeq$) if they have exactly the same shape and size. If two geometric figures are congruent, one figure could be made to overlie the other by a combination of these types of motion: translation (sliding), rotation (twisting), and reflection about a line (flipping over). The parts of the geometric figures that would then coincide are called corresponding parts, where a part of a geometric figure is any set of points associated with that figure.

Two geometric figures are similar (~) if they have the same shape but (perhaps) have different sizes. If two geometric figures are similar, one figure could be made to overlie the other by a combination of translation, rotation, reflection, and either expanding or shrinking. The parts that would then coincide are corresponding.

Circles. A circle is a collection of points in the plane, all of which are the same distance from another point, called the center. The region bounded by a circle sometimes is called a disk. Circumference means either the circle that is the boundary curve of a disk or the distance around that circle. A radius of a circle is any line segment that joins the center and a point of a circle. A chord is any line segment whose end points lie on the circle. A diameter is any chord that contains the center. A secant is a line that intersects a circle in two points. A tangent is a line that intersects a circle in only one point, called the point of tangency.

Associated with a circle are several important dimensions, including the lengths of a radius and a diameter, usually denoted, respectively, by r and d. These symbols appear in almost all formulas involving the length of the circumference C or the area A enclosed by a circle. Relationships between these variables for any circle are given by the equations

$$d = 2r \qquad C = \pi d \qquad A = \pi r^2$$

In these formulas, π (the lowercase Greek letter pi) represents the irrational number (value 3.141592 . . .) that is usually defined as the ratio of the circumference to the diameter of any circle. *See* CIRCLE. Harry L. Baldwin, Jr.

Planer A machine for the shaping of long, flat, or flat contoured surfaces by reciprocating the workpiece under a stationary single-point tool or tools. Usually the workpiece is too large to be handled on a shaper.

Planers are built in two general types, open-side or double-housing. The former is constructed with one upright or housing to support the crossrail and tools. The double-housing type has an upright on either side of the reciprocating table connected by an arch at the top. *See* WOODWORKING. Alan H. Tuttle

Planet A relatively large celestial body moving in orbit around a star, in particular the Sun.

Solar system. Planets are by-products of the formation of the Sun, which condensed from a vast cloud of interstellar gas and dust. Initially the Sun was surrounded by a broad, flat disk of residual matter, from which the planets accumulated. *See* SOLAR SYSTEM.

The Sun is not the only star known to be encircled by planets. Astronomers now know of at least 180 other solar systems, more than 30 of which have multiple planets. *See* EXTRASOLAR PLANETS.

Prior to 2006, there was no formal definition for a planet. However, in that year members of the International Astronomical Union (IAU) defined a planet in our solar system as an object orbiting the Sun that is massive enough to have assumed a round shape and to have cleared the neighborhood around its orbit of smaller objects through gravitational scattering. Smaller, less massive objects that are round in shape but unable to clear their orbital neighborhood are considered dwarf planets.

The solar system's eight (major) planets fall into two basic groups: the small, dense, terrestrial planets—Mercury, Venus, Earth, and Mars—and the giant or Jovian planets Jupiter, Saturn, Uranus, and Neptune. The terrestrial planets are all located relatively close to the Sun (see illustration). The lower-density giant planets extend outward from Jupiter to great distances. This distribution is not accidental, but is related to the fractionation of rocky, icy, and gaseous materials present in the circumstellar disk during the earliest stages of the solar system's formation. *See* JUPITER; MARS; MERCURY (PLANET); NEPTUNE; SATURN; URANUS; VENUS.

Each of the planets from Earth to Neptune is accompanied by one or more secondary bodies called satellites. Many of the smallest satellites are not observable from Earth, but were discovered during spacecraft visits. *See* SATELLITE (ASTRONOMY).

The gas-and-dust disk that surrounded the infant Sun also gave rise to a multitude of smaller bodies, which fall into two main groups. The asteroids, of which nearly 200,000 have well-determined orbits, are rocky and largely confined between the orbits of Mars and Jupiter; objects in the Kuiper Belt, over 1000 of which are now known but which may number in the billions, are primarily icy and lie beyond the orbit of Neptune. *See* ASTEROID; KUIPER BELT.

Until recently, astronomers also considered Pluto to be a planet. However, the discovery of Kuiper Belt objects with orbits similar to Pluto's and in particular the discovery in 2003 of Eris, a very distant solar-system object larger than Pluto, triggered a protracted debate about whether Pluto truly qualified as a planet. In 2006, astronomers of the IAU approved a resolution that applies the term "planet" only to the solar system's eight largest worlds. In addition, Pluto, Eris, and the large asteroid Ceres became the first objects in a new class of "dwarf planets." By October 2008, two additional objects in the Kuiper Belt, Makemake and Haumea, had been accepted by the IAU as dwarf planets, and this designation may soon be applied to additional large asteroids and Kuiper Belt objects. *See* CERES; PLUTO.

Planetary orbits and motions. The motions of the planets in their orbits around the Sun are governed by three laws of motion discovered by Johannes Kepler at the beginning of the seventeenth century. *See* CELESTIAL MECHANICS; KEPLER'S LAWS.

First law: The orbit of a planet is an ellipse, with the Sun at one of its foci.

Second law (the law of areas): As a planet revolves in its orbit, the radius vector (the line from the Sun to the planet) sweeps out equal areas in equal intervals of time.

Third law (the harmonic law): The square of the period of revolution is proportional to the cube of the orbit's semimajor axis.

The mean distance from the Sun and sidereal periods of revolution of the planets and original three dwarf planets are given in Table 1.

Planetary configurations. In the course of their motions around the Sun, Earth and other planets occupy a variety of relative positions or configurations. The configurations for the inferior planets, Mercury and Venus, which are located inside Earth's orbit, differ from those for the superior planets, from Mars to Neptune, circulating outside Earth's orbit. Venus and Mercury are in conjunction with the Sun when closest to the Earth-Sun line, either between Earth and the Sun (inferior conjunction) or beyond the Sun (superior conjunction). On rare occasions when the planet is very close to the plane of Earth's orbit at the time of an inferior conjunction, a transit in front of the Sun is observed. Between conjunctions, a planet's elongation, its angular distance from to the Sun as measured from Earth's center, varies up to a maximum value; the greatest elongations of Mercury and Venus are 28° and 47°, respectively. The superior planets are not so limited, and their elongations can reach up to 180° when they are in opposition with the Sun. *See* TRANSIT (ASTRONOMY).

Because the orbits of the main planets are only slightly inclined to the plane of the orbit of Earth, the apparent paths of the planets are restricted to the zodiac, a belt 16° wide centered on the ecliptic. The ecliptic is the path in the sky traced out by the Sun in its apparent annual journey as Earth revolves around it. *See* ASTRONOMICAL COORDINATE SYSTEMS; ECLIPTIC.

For inferior planets, the apparent motions with respect to the celestial sphere, that is, to the fixed stars, appear as oscillations

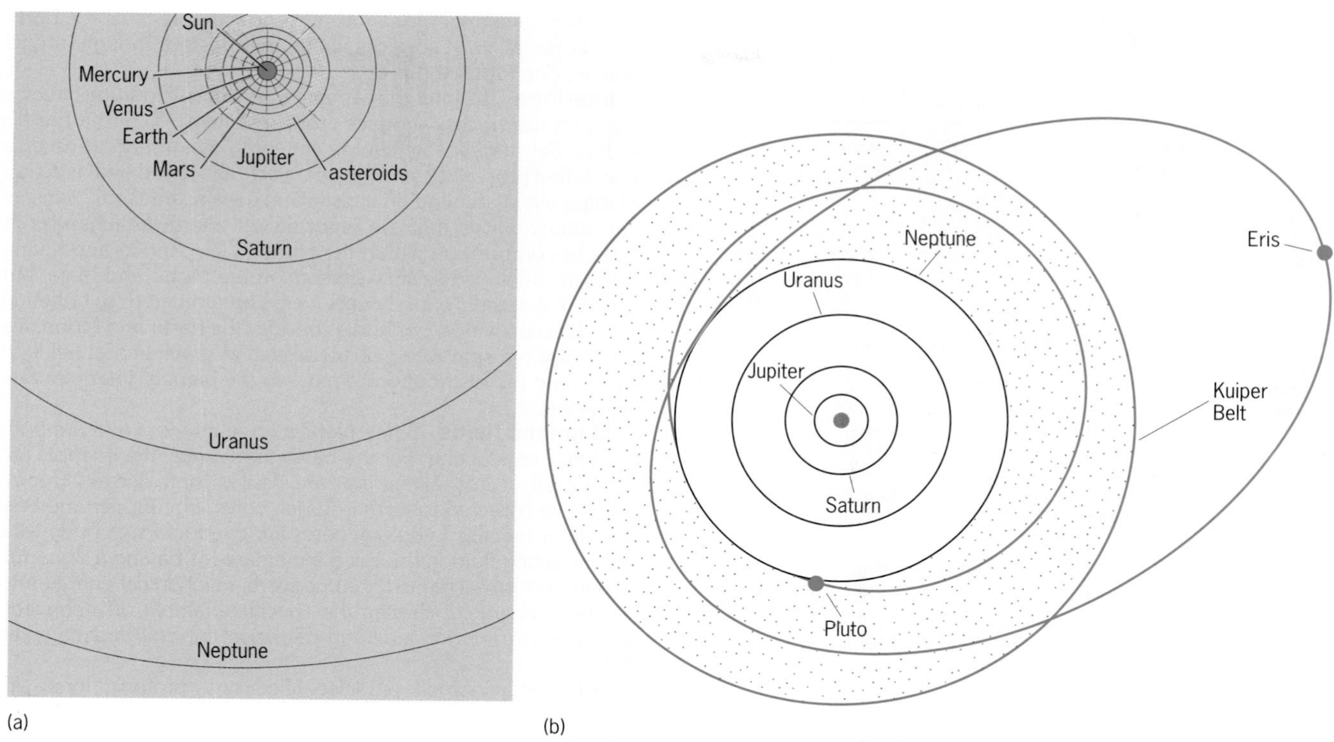

Plan of the solar system. (*a*) Solar system out to Neptune. (*b*) Outer solar system. Current (2006) positions of Pluto and Eris in their orbits are indicated.

back and forth about the position of the Sun steadily moving eastward among the stars. For a superior planet, the apparent motion is generally eastward or direct, but for short periods near the time of opposition it is westward or retrograde.

Planetary characteristics. The size, mass, density, and rotation period of each of the planets and original three dwarf planets are listed in Table 2.

Planetary radiations. The electromagnetic radiation received from a planet is made up of three main components: visible reflected sunlight, including some ultraviolet and near-infrared radiation; thermal radiation due to the planet's heat, including both infrared radiation and ultrashort radio waves; and nonthermal radio emission due to electrical phenomena, if any, in the planet's atmosphere or in its radiation belts.

The apparent brightness of a planet, as measured by visual, photographic, and electronic means, is usually expressed as its magnitude, based on the logarithmic scale for stellar magnitudes. A planet's apparent brightness varies in inverse proportion to the squares of the distances r from the Sun and Δ from Earth. Consequently the apparent brightnesses of large solar-system bodies are typically cited when the bodies are at opposition. For smaller bodies, the usual convention is an absolute magnitude H, indicating the body's apparent brightness at a distance from both the Sun and the observer of 1 AU and a phase angle of 0°. *See* MAGNITUDE (ASTRONOMY).

Planetary atmospheres. The principal constituents of the atmospheres of the terrestrial planets are carbon dioxide, nitrogen, water, and (on Earth only) oxygen; Mercury has a very tenuous envelope dominated by atoms of sodium and potassium. The atmospheres of the giant planets are composed primarily of hydrogen and helium, with lesser amounts of methane, ammonia, and water. Atmospheric motions are driven by temperature gradients—in general, those existing between the warm equatorial regions and the cooler polar areas. An atmosphere thus tends to redistribute heat over the planetary surface, lessening the temperature extremes found on airless bodies.

On planets having relatively dense atmospheres, heat from the Sun is trapped by the greenhouse effect. That is, visible radiation from the Sun passes readily through the atmosphere to heat the planetary surface, but infrared radiation reemitted from the surface is constrained from escaping back to space by the lower transparency of certain atmospheric gases (especially carbon dioxide, water vapor, and methane) to longer wavelengths. For example, although Venus absorbs approximately the same amount of energy from the Sun as does Earth, the greenhouse effect is responsible for heating the surface of Venus to a much higher temperature, approximately 750 K (900°F). *See* GREENHOUSE EFFECT.

J. Kelly Beatty

Origins of planets. The standard model of solar system formation assumes that the early solar nebula condensed from a diffuse interstellar cloud. The collapse of the cloud could have been initiated by cooling (such that thermal pressure in the cloud could no longer balance the cloud's self-gravity) or by perturbations that created a local high-density region. In any case, once self-gravity initiated condensation, the cloud's collapse would have been a runaway process. The cloud would have increased its rotational speed as it condensed (like a skater pulling in her arms), eventually forming a spinning disk with most of its mass in the center. This central mass became the basis for a protosun, and the molecules and atoms in the rest of the disk settled into the central plane of the disk. Some molecules would have condensed into solid particles, and the particles would have accreted into planetesimals, the building blocks of planets. *See* MOLECULAR CLOUD; PROTOSTAR.

Hydrogen and helium are thought to have made up 98% of the solar nebula, followed by hydrogenated forms of carbon, nitrogen, and oxygen (CH_4, H_2O, and NH_3, respectively), silicates (rocks), and metals (iron, nickel, and aluminum). These various constituents condensed into solid particles in the central plane of the disk, but because metals, rocks, and ices solidify at different temperatures, there was a compositional gradient as a function of distance from the protosun. Beyond the frost line (thought to

Table 1. Elements of planetary orbits

Planets	Symbol	Mean distance from Sun (semimajor axis of orbit)			Sidereal period of revolution	
		AU	10^6 mi	10^6 km	Years	Days
Mercury	☿	0.387	36.0	57.9	0.241	87.97
Venus	♀	0.723	67.2	108.2	0.615	224.70
Earth	⊕	1.000	93.0	149.6	1.000	365.26
Mars	♂	1.524	141.6	227.9	1.881	686.98
Jupiter	♃	5.203	483.6	778.4	11.86	4,332.7
Saturn	♄	9.555	886.7	1426.7	29.46	10,759.5
Uranus	⛢	19.19	1784.	2871.	84.00	30,685.0
Neptune	♆	30.07	2795.	4498.	164.79	60,190.0
Dwarf planets*						
Ceres	⚳	2.77	257.1	413.8	4.600	1,680.1
Pluto	♇	39.48	3670.	5906.	247.9	90,550.
Eris		67.76	6299	10137	557.7	203,720.

*By October 2008, two additional objects in the Kuiper Belt, Makemake and Haumea, had been designated dwarf planets, and several likely candidates exist.

be around 2–5 AU), volatiles like water, methane, and ammonia could condense into ices, which meant that availability of solid material was several times higher outside the frost line than inside. The planetary cores outside the frost line were therefore more massive, enough to attract and retain atmospheres of hydrogen and helium.

Orbital motion. The mass of the Sun is more than a thousand times greater than that of Jupiter, which in turn is about 2.5 more massive than the rest of the planets combined. As a result, the motion of the planets is mainly two-body Keplerian motion about a massive central body, which means that all of the planets execute elliptical orbits with the Sun at one focus. However, the presence of other planets, small objects, and even photons can have significant effects on the motion of solar system objects.

A resonance between two objects means that the same relative geometrical orientation occurs repeatedly. Even small interactions can build constructively over time, just as small pushes to a swing can produce large amplitudes if the pushes are applied at a consistent point in the swing cycle. Many of the objects in the solar system are constrained by gravitational resonances. One classic example is the 3:2 resonance between Neptune and Pluto (Neptune completes three orbits for every two Pluto orbits). *See* RESONANCE (ACOUSTICS AND MECHANICS).

Collisions. The most distinctive features on nearly all solid surfaces in the solar system are impact craters. Except for objects that have undergone recent resurfacing, such as the Earth or Europa, all solid surfaces reveal their history though impact craters. *See* METEORITE.

Interiors. The interiors of planets are difficult to study directly. Even on Earth, core samples probe less than 1% of the Earth's radius. Seismic sensors can detect compression and transverse waves that propagate through an object. The speed and direction of these waves depend on density and density gradients. Sensors can detect waves that are reflected off interfaces (changes in state or composition within the planet). The Apollo and Viking missions deployed seismic sensors on the Moon and Mars, but the Viking seismic experiments were contaminated by buffeting from Martian winds. For bodies besides the Earth and Moon, we must observe spin rates, oblateness, and gravitational fields to determine the distribution of mass in the planets' interiors. *See* SEISMOLOGY.

Magnetic fields. Some planets have interiors that are both fluid and conducting. For the Earth and Venus, the interiors are composed of iron alloys that are denser than silicates (rocks) and have collected to form iron-rich cores. For Jupiter and Saturn, their metallic hydrogen cores are conductors. A body with a conducting fluid interior is a candidate for having a dynamo. A dynamo is mechanism that converts mechanical energy into electrical energy, including the specific example of a convecting planetary interior generating electrical currents and magnetic fields.

Aurorae. Charged particles (electrons, protons, ions) are constrained to move along magnetic field lines. To first order, the planetary magnetic fields are all dominated by a dipole term (that is, they look similar to the magnetic fields produced by bar magnets), although Neptune's dipole is offset from the center of the planet. Charged particles inside a planet'smagnetosphere are guided along magnetic field lines until they intersect the planet's atmosphere, often near the north and south magnetic poles. Atoms in the atmosphere are then ionized and/or excited by collisions with the incoming charged particles. The excited atoms emit the auroral light when they return to their normal energy levels. Since the source of charged particles is typically larger than the planet (on Earth, the solar wind is the main source of charged particles), the auroras are produced in symmetric pairs, aurora borealis and aurora australis, at both magnetic poles. On Jupiter, the source is particles trapped in the Jovian magnetosphere as opposed to the solar wind. *See* AURORA; SOLAR WIND.

Atmospheric structure. In an isothermal atmosphere, density and pressure are both exponential functions of altitude. In real atmospheres, of course, there are changes in temperature,

Table 2. Physical characteristics of the Sun's planets

Planet	Equatorial radius (r_e) (Earth=1)	mi	km	Mass (Earth=1)	Density, g/cm^3	Rotation period
Mercury	0.38	1,515	2,440	0.055	5.43	58 d 15.5 h
Venus	0.95	3,761	6,052	0.815	5.20	243 d 0.5 h
Earth	1.00	3,963	6,378	1.000	5.52	23 h 56 m 23 s
Mars	0.53	2,110	3,396	0.107	3.93	24 h 37 m 23 s
Jupiter	11.21	44,423	71,492	317.832	1.33	9 h 55 m 30 s[1,2]
Saturn	9.45	37,449	60,268	95.159	0.69	10 h 39 m 22 s[1,3]
Uranus	4.01	15,882	25,559	14.536	1.32	17 h 22.2 m[1,4]
Neptune	3.88	15,389	24,764	17.147	1.64	16 h 6.6 m[1,5]
Dwarf planets[6]						
Ceres	0.08	303	487	0.0002	2.08	9 h 4.5 m
Pluto	0.18	716	1,153	0.002	2.1	6 d 9 h 17.6 m
Eris	0.19	745	1,200	?	?	>8 h

[1]Internal (System III) rotation period, the rotation period of the planet's core, as deduced from its magnetic field.
[2]Jupiter's equatorial (System I) rotation period is 9 h 50.5 m.
[3]Saturn's equatorial rotation period is 10 h 14.0 m.
[4]Uranus's equatorial rotation period is about 18.0 h.
[5]Neptune's equatorial rotation period is about 18.8 h.
[6]By October 2008, two additional objects in the Kuiper Belt, Makemake and Haumea, had been designated dwarf planets and several likely candidates exist.

composition, and phase changes with respect to altitude, but to first order, every density profile in the solar system is essentially an exponential function.

There is no sharp boundary between the atmosphere and space; the density and collision rates simply decrease to the point where individual molecules are on ballistic trajectories (in orbit around the planet) or they are escaping. A useful term is the exobase, the altitude above which molecules are more likely to escape than to collide with another molecule. Eliot Young

Planetarium An instrument that projects the stars, Sun, Moon, planets, and other celestial objects upon a large hemispherical dome, showing their motions as viewed from the Earth or space near the Earth. Days and years may be compressed into minutes. There are over 100 major planetariums around the world with domes 50 ft (15 m) or more in diameter; and there are also over 1000 smaller planetariums in communities, schools, and colleges.

The term planetarium originally applied to a mechanical model (also known as an orrery) that depicted the motions of the planets. Today the term refers to an optical projector. Most planetariums now have mechanical movements, but planetariums projecting computer-generated displays have also been developed. Additional optical devices and computer controls are common. The term planetarium also refers to the theater or building that houses the projector.

Many projectors are patterned after the basic design of the Carl Zeiss Company. Star spheres at each end of the projector show 8900 stars down to magnitude 6.5; 32 lenses (16 located in each globe) are used to project the stars. Cages between the two star spheres contain projectors for the Sun, Moon, and planets. The center part of the machine houses the driving motors. Additional projectors show such effects as variable stars, solar and lunar eclipses, the Milky Way, comets, and various circles and coordinates. Depending on the manufacturer, there may be constellation outlines, clouds, and built-in zoom effects for the planets.

Another design philosophy, begun in the late 1940s when A. Spitz designed a small planetarium for school classrooms and museums, was to manufacture a small and relatively inexpensive projector that would do for schools and small communities what the larger machines had done for the cities.

Some planetariums are installed in tipped domes which may include all-sky 70mm motion picture projection systems. Another innovation in planetarium design, based on computer graphics television projection, is the Evans & Sutherland Digistar 3 SP. The projector utilizes a high-resolution cathode-ray tube with a special 160° wide-angle lens for projection onto domes up to a diameter of 33 ft (10 m). Software and documentation include stellar, planetary, and constellation data files. The star and planet positions are fed into the high-intensity cathode-ray tube projector by a computer and thence onto the dome via a wide-angle lens. Evans & Sutherland also offers the Digistar 3 for domes up to 60 ft (18 m) in diameter. Six high-resolution/brightness projectors are used providing all-dome coverage. *See* CATHODE-RAY TUBE; COMPUTER GRAPHICS. Charles F. Hagar

Planetary gear train An assembly of meshed gears consisting of a central or sun gear, a coaxial internal or ring gear, and one or more intermediate pinions supported on a revolving carrier. Sometimes the term planetary gear train is used broadly as a synonym for epicyclic gear train, or narrowly to indicate that the ring gear is the fixed member. In a simple planetary gear train the pinions mesh simultaneously with the two coaxial gears. With the central gear fixed, a pinion rotates about it as a planet rotates about its sun, and the gears are named accordingly: the central gear is the sun, and the pinions are the planets.

In operation, input power drives one member of a planetary gear train, the second member is driven to provide the output, and the third member is fixed. If the third member is not fixed, no power is delivered. This characteristic provides a convenient clutch action. A clutch or brake band positioned about the intermediate member and fixed to the gearbox housing serves to lock or free the third member. The holding device itself does not enter into the power path.

Any one of these three elements can be fixed: the sun gear, the carrier, or the ring gear. Either of the two remaining elements can be driven and the other one used to deliver the output. There are six possible combinations, although three of these provide velocity ratios that are reciprocals of the other three. The ratios are entirely independent of the number of teeth on each planet.

Two simple planetary gear sets running on a common sun gear are known as a Simpson gear train. It is widely used in automotive automatic transmissions. In a compound planetary train, two planet gears are attached together on a common shaft. One planet meshes only with the central sun gear, the other only with the ring gear. As in simple planetary trains, there can be several of these planet pairs around the train to distribute the load and achieve balance. *See* AUTOMOTIVE TRANSMISSION; GEAR TRAIN; RECIPROCATING AIRCRAFT ENGINE. John R. Zimmerman; Donald L. Anglin

Planetary nebula An expanding nebula, or cloud, of gas thrown off by a dying star just before the star settles down to its ultimate endpoint as a white dwarf. Planetary nebulae are among the brightest and best-studied nebular objects in the sky, even though they are generally a few thousand light-years from Earth. The many shapes of planetary nebulae reflect poorly understood processes that occur inside most stars late in their lives. The Sun is likely to eject a planetary nebula in about 5 billion years. Many of the most prominent planetary nebulae seen today formed and emerged from their dusty cocoons within just the past 1000–2000 years. *See* SUN; WHITE DWARF STAR.

Planetary nebulae appear small and greenish in color, like the planet Uranus, when seen through small telescopes—hence the origin of the name by their discoverers during the nineteenth century. However, planetary nebulae are not planets but large (0.1–10 light-years), expanding (10–100 km/s; 6–60 mi/s), highly symmetrical gaseous clouds of stellar ejecta. The shapes of planetary nebulae vary, and are denoted as round, elliptical, or bipolar depending on their outlines. Delicate filaments, knots, and bubbles of material characterize their interiors. These features evolve as stellar winds of increasing speed (from 100 to 1000 km/s; 60 to 600 mi/s) plow into older, much slower gas ahead of them, ejected about 1000 years earlier by the geriatric, unstable star.

Many thousands of planetary nebulae have been cataloged. For every known planetary nebula, hunderds more are undoubtedly obscured by opaque dust clouds in the Milky Way. Planetary nebulae last for several thousand years before they expand and become too diffuse to be seen as discrete objects.

Complex patterns appear when the spectrum of light from a given nebula is analyzed. From these data the temperature, density, velocity, and chemical nature of the gas are measured. Such studies have shown that the gas ejected from the parent stars of planetary nebulae is enriched in carbon and nitrogen. Their masses, lifetimes, and large numbers suggest that planetary nebulae are the most prolific source enriching the interstellar medium with carbon and, to a lesser degree, nitrogen. Indeed, since no carbon could have survived the big bang, carbon-based life as we know it could not have occurred until generations of planetary nebuale seeded the clouds from which life formed with this element. *See* COSMOLOGY; INTERSTELLAR MATTER. Bruce Balick

Plant An organism that belongs to the Kingdom Plantae (plant kingdom) in biological classification. The study of plants is called botany. *See* BOTANY; CLASSIFICATION, BIOLOGICAL.

The Plantae share the characteristics of multicellularity, cellulose cell walls, and photosynthesis using chlorophylls *a* and *b* (except for a few plants that are secondarily heterotrophic).

Most plants are also structurally differentiated, usually having organs specialized for anchorage, support, and photosynthesis. Tissue specialization for photosynthetic, conducting, and covering functions is also characteristic. Plants have a sporic (rather than gametic or zygotic) life cycle that involves both sporophytic and gametophytic phases, although the latter is evolutionarily reduced in the majority of species. Reproduction is sexual, but diversification of breeding systems is a prominent feature of many plant groups. *See* PHOTOSYNTHESIS; PLANT REPRODUCTION.

A conservative estimate of the number of described species of plants is 250,000. There are possibly two or three times that many species as yet undiscovered, primarily in the Southern Hemisphere. Plants are categorized into nonvascular and vascular groups, and the latter into seedless vascular plants and seed plants. The nonvascular plants include the liverworts, hornworts, and mosses. The vascular plants without seeds are the ground pines, horsetails, ferns, and whisk ferns; seed plants include cycads, ginkgos, conifers, gnetophytes, and flowering plants. Each of these groups constitutes a division in botanical nomenclature, which is equivalent to a phylum in the zoological system. *See* PLANT TAXONOMY. Meredith Lane

Plant anatomy The area of plant science concerned with the internal structure of plants. It deals both with mature structures and with their origin and development.

The plant anatomist dissects the plant and studies it from different planes and at various levels of magnification. At the level of the cell, anatomy overlaps plant cytology, which deals exclusively with the cell and its contents. Sometimes the name plant histology is applied to the area of plant anatomy directed toward the study of cellular details of tissues. *See* PLANT CELL; PLANT ORGANS. Katherine Esau

Plant-animal interactions The examination of the ecology of interacting plants and animals by using an evolutionary, holistic perspective. For example, the chemistry of defensive compounds of a plant species may have been altered by natural-selection pressures resulting from the long-term impacts of herbivores. Also, the physiology of modern herbivores may be modified from that of thousands of years ago as adaptations for the detoxification or avoidance of plant defensive chemicals have arisen.

The application of the theories based on an understanding of plant-animal interactions provides an understanding of problems in modern agricultural ecosystems. In addition, plant-animal interactions have practical applications in medicine. For example, a number of plant chemicals, such as digitalin from the foxglove plant, that evolved as herbivore-defensive compounds have useful therapeutic effects on humans.

Effects of interaction types for each species*

Interaction	Effect on species A	Effect on species B
Mutualism	+	+
Commensalism	+	0
Antagonism	+	−
Competition	−	−
Amensalism	0	−
Neutralism	0	0

*+ = beneficial, − = harmful, 0 = neutral.

The evolutionary consequences of plant-animal interactions vary, depending on the effects on each participant. Interaction types range from mutualisms, that is, relationships which are beneficial to both participating species, to antagonisms, in which the interaction benefits only one of the participating species and negatively impacts the other. Interaction types are defined on the basis of whether the impacts of the interaction are beneficial, harmful, or neutral for each interacting species (see table). Warren G. Abrahamson

Fossil record. Plants and animals interact in a variety of ways within modern ecosystems. These interactions may range from simple examples of herbivory (animals eating plants) to more complex interactions such as pollination or seed and fruit dispersal. Animals also rely on plants for food and shelter. The complex interactions between these organisms over geologic time not only have resulted in an abundance and diversity of organisms in time and space but also have contributed to many of the evolutionary adaptations found in the biological world.

Paleobiologists have attempted to decipher some of the interrelationships that existed between plants and animals throughout geologic time. The ecological setting in which the organisms lived in the geologic past is being analyzed in association with the fossils. Thus, as paleobiologists have increased their understanding of certain fossil organisms, it has become possible to consider some aspects of the ecosystems in which they lived, and in turn, how various types of organisms interacted.

Herbivory. Perhaps the most widespread interaction between plants and animals is herbivory, in which plants are utilized as food. One method of determining the extent of herbivory in the fossil record is by analyzing the plant material that has passed through the digestive gut of the herbivore.

The stems of some fossil plants show tissue disruption similar to various types of wounds occurring in plant parts that have been pierced by animal feeding structures. As plants developed defense systems in the form of fibrous layers covering inner, succulent tissues, some animals evolved piercing mouthparts that allowed them to penetrate these thick-walled layers. In some fossil plants, it is also possible to see evidence of wound tissue that has grown over these penetration sites. *See* COAL BALLS; HERBIVORY.

Mimicry. Another example of the interactions between plants and animals that can be determined from the fossil record is mimicry. Certain fossil insects have wings that are morphologically identical to plant leaves, thus providing camouflage from predators as the insect rested on a seed fern frond.

Pollination. The transfer of pollen from the pollen sacs to the receptive stigma in angiosperms or to the seed in gymnosperms is an example of an ancient interaction between plants and animals. It has been suggested that pollination in some groups initially occurred as a result of indiscriminate foraging behavior by certain animals, and later evolved specifically as a method to effect pollination. The size, shape, and organization of fossil pollen grains provide insight into potential pollination vectors. *See* FOSSIL; PALEOBOTANY; POLLINATION. Thomas N. Taylor

Plant cell The basic unit of structure and function in nearly all plants. Although plant cells are variously modified in structure and function, they have many common features. The most distinctive feature of all plant cells is the rigid cell wall, which is absent in animal cells. The range of specialization and the character of association of plant cells is very wide. In the simplest plant forms a single cell constitutes a whole organism and carries out all the life functions. In just slightly more complex forms, cells are associated structurally, but each cell appears to carry out the fundamental life functions, although certain ones may be specialized for participation in reproductive processes. In the most advanced plants, cells are associated in functionally specialized tissues, and associated tissues make up organs such as the leaves, stem, and root. *See* CELL WALLS (PLANT).

Plant and animal cells are composed of the same fundamental constituents—nucleic acids, proteins, carbohydrates, lipids, and various inorganic substances—and are organized in the same fundamental manner. A characteristic of their organization is the presence of unit membranes composed of phospholipids and associated proteins and in some instances nucleic acids.

Perhaps the most conspicuous and certainly the most studied of the features peculiar to plant cells is the presence of plastids. The plastids are membrane-bound organelles with an inner membrane system. Chlorophylls and other pigments are associated with the inner membrane system. *See* CELL (BIOLOGY); CELL PLASTIDS; CHLOROPHYLL. W. Gordon Whaley

Plant communication Movement of signals or cues, presumably chemical, among individual plants or plant parts. These chemical cues are caused by damage to plant tissues or other stresses and stimulate physiological changes in the undamaged "receiving" plant or tissue. Communication among plants often involves induced plant defenses, which are chemicals produced actively by an individual plant in response to attack by insects or disease agents. *See* CHEMICAL ECOLOGY.

Plants produce a wealth of secondary metabolites that do not function in the main, or primary, metabolism of the plant, which includes photosynthesis, nutrient acquisition, and growth. Since many of these chemicals have very specific negative effects on animals or pathogens, ecologists speculate that they may be produced by plants as defenses. Plant chemical defenses either may be present all of the time (constitutive) or may be stimulated in response to attack (induced). Those produced in response to attack by pathogens are called phytoalexins. In order to demonstrate the presence of an induced defense, the chemistry of plant tissues or their suitability to some "enemy" (via a bioassay) must be compared before and after real or simulated attack. Changes found in the chemistry and suitability of the control or unattacked plants when nearby plants are damaged imply that some signal or cue has passed from damaged to undamaged plants. Responses in undamaged plants are related to the proximity of a damaged neighbor. *See* PHYTOALEXINS; PLANT GROWTH; PLANT METABOLISM. Jack C. Schultz

Plant evolution The process of biological and organic change within the plant kingdom by which the characteristics of plants differ from generation to generation. The main levels (grades) of evolution have long been clear from comparisons among living plants, but the fossil record has been critical in dating evolutionary events and revealing extinct intermediates between modern groups, which are separated from each other by great morphological gaps. Plant evolution has been clarified by cladistic methods for estimating relationships among both living and fossil groups. These methods attempt to reconstruct the branching of evolutionary lines (phylogeny) by using shared evolutionary innovations (for example, presence of a structure not found in other groups) as evidence that particular organisms are descendants of the same ancestral lineage (a monophyletic group, or clade). Many traditional groups are actually grades rather than clades; these are indicated below by names in quotes.

Most botanists restrict the term plants to land plants, which invaded the land after 90% of Earth history. There is abundant evidence of photosynthetic life extending back 3.5 billion years to the early Precambrian, in the form of microfossils resembling cyanobacteria (prokaryotic blue-green algae) and limestone reefs (stromatolites) made by these organisms. Larger cells representing eukaryotic "algae" appear in the late Precambrian, followed by macroscopic "algae" and animals just before the Cambrian. *See* ALGAE; EUKARYOTAE; FOSSIL; PROKARYOTAE.

Origin of land plants. Cellular, biochemical, and molecular data place the land plants among the "green algae," specifically the "charophytes," which resemble land plants in their mode of cell division and differentiated male and female gametes (oogamy). Land plants themselves are united by a series of innovations not seen in "charophytes," many of them key adaptations required for life on land. They have an alternation of generations, with a haploid, gamete-forming (gametophyte) and diploid, spore-forming (sporophyte) phase. Their reproductive organs (egg-producing archegonia, sperm-producing antheridia, and spore-producing sporangia) have a protective layer of sterile cells. The sporophyte, which develops from the zygote, begins its life inside the archegonium. The spores, produced in fours by meiosis, are air-dispersed, with a resistant outer wall that prevents desiccation. *See* CHAROPHYCEAE; PLANT REPRODUCTION.

Land plants have been traditionally divided into "bryophytes" and vascular plants (tracheophytes). These differ in the relative role of the sporophyte, which is subordinate and permanently attached to the gametophyte in "bryophytes" but dominant and independent in vascular plants. In vascular plants, tissues are differentiated into an epidermis with a waxy cuticle that retards water loss and stomates for gas exchange, parenchyma for photosynthesis and storage, and water- and nutrient-conducting cells (xylem, phloem). However, cladistic analyses imply that some "bryophytes" are closer to vascular plants than others. This implies that the land-plant life cycle originated before the full suite of vegetative adaptations to land life, and that the sporophyte began small and underwent a trend toward elaboration and tissue specialization. *See* EPIDERMIS (PLANT); PHOTOSYNTHESIS; PRIMARY VASCULAR SYSTEM (PLANT).

In the fossil record, the first recognizable macroscopic remains of land plants are Middle Silurian vascular forms with a branched sporophyte, known as "rhyniophytes." These differed from modern plants in having no leaves or roots, only dichotomously branching stems with terminal sporangia. However, spore tetrads formed by meiosis are known from older beds (Middle Ordovician); these may represent more primitive, bryophytic plants. *See* BRYOPHYTA; RHYNIOPHYTA.

In one of the most spectacular adaptive radiations in the history of life, vascular plants diversified through the Devonian. At the beginning of this period, vegetation was low and probably confined to wet areas, but by the Late Devonian, size had increased in many lines, resulting in large trees and forests with shaded understory habitats. Of the living groups of primitive vascular plants, the lycopsids (club mosses) branched off first, along with the extinct "zosterophyllopsids." A second line, the "trimerophytes," gave rise to sphenopsids (horsetails) and ferns (filicopsids). This radiation culminated in the coal swamp forests of the Late Carboniferous, with tree lycopsids (Lepidodendrales), sphenopsids (*Calamites*), and ferns (Marattiales). Remains of these plants make up much of the coal of Europe and eastern North America, which were then located on the Equator. *See* LYCOPODIALES; MARATTIALES; SPHENOPHYTA.

Seed plants. Perhaps the most significant event after the origin of land plants was evolution of the seed. Primitive seed plants ("gymnosperms") differ from earlier groups in their reproduction, which is heterosporous (producing two sizes of spores), with separate male and female gametophytes packaged inside the pollen grain (microspore), and the ovule (a sporangium with one functional megaspore, surrounded by an integument, which develops into the seed). The transfer of sperm (two per pollen grain) from one sporophyte to another through the air, rather than by swimming, represents a step toward independence from water for reproduction. This step must have helped plants invade drier areas than they had previously occupied. In addition, seed plants have new vegetative features, particularly secondary growth, which allows production of a thick trunk made up of secondary xylem (wood) surrounded by secondary phloem and periderm (bark). Together, these innovations have made seed plants the dominant organisms in most terrestrial ecosystems ever since the disappearance of the Carboniferous coal swamps. *See* ECOSYSTEM; PTERIDOSPERMS; SEED.

A major breakthrough in understanding the origin of seed plants was recognition of the "progymnosperms" in the Middle and Late Devonian. These plants, which were the first forest-forming trees, had secondary xylem, phloem, and periderm, but they still reproduced by spores, implying that the anatomical advances of seed plants arose before the seed. Like sphenopsids and ferns, they were apparently derived from "trimerophytes."

The earliest seed plants of the Late Devonian and Carboniferous, called "seed ferns" because of their frondlike leaves, show steps in origin of the seed. Origin of the typical mode of branching in seed plants, from buds in the axils of the leaves, occurred at about the same time. *See* PTERIDOSPERMS.

Seed plants became dominant in the Permian during a shift to drier climate and extinction of the coal swamp flora in the European-American tropical belt, and glaciation in the Southern Hemisphere Gondwana continents. Early conifers predominated in the tropics; extinct glossopterids inhabited Gondwana. Moderation of climate in the Triassic coincided with the appearance of new seed plant groups as well as more modern ferns. Many Mesozoic groups show adaptations for protection of seeds against animal predation, while flowers of the Bennettitales constitute the first evidence for attraction of insects for cross-pollination, rather than transport of pollen by wind.

Angiosperms. The last major event in plant evolution was the origin of angiosperms (flowering plants), the seed plant group that dominates the modern flora. The flower, typically made up of protective sepals, attractive petals, pollen-producing stamens, and ovule-producing carpels (all considered modified leaves), favors more efficient pollen transfer by insects. The ovules are enclosed in the carpel, so that pollen germinates on the sticky stigma of the carpel rather than in the pollen chamber of the ovule. The carpels (separate or fused) develop into fruits, which often show special adaptations for seed dispersal. Other advances include an extreme reduction of the gametophytes, and double fertilization whereby one sperm fuses with the egg, and the second sperm with two other gametophyte nuclei to produce a triploid, nourishing tissue called the endosperm. Angiosperms also developed improved vegetative features, such as more efficient water-conducting vessels in the wood and leaves with several orders of reticulate venation. These features may have contributed to their present dominance in tropical forests, previously occupied by conifers with scale leaves. *See* RAINFOREST.

Most botanists believed that the most primitive living angiosperms are "magnoliid dicots," based on their "gymnosperm"-like pollen, wood anatomy, and flower structure. Studies of Cretaceous fossil pollen, leaves, and flowers confirm this view by showing a rapid but orderly radiation beginning with "magnoliid"-like and monocotlike types, followed by primitive eudicots (with three pollen apertures), some related to sycamores and lotuses. *See* MAGNOLIOPHYTA.

Both morphological and molecular data imply that angiosperms are monophyletic and most closely related to Bennettitales and Gnetales, a seed plant group that also radiated in the Early Cretaceous but later declined to three living genera. Since all three groups have flowerlike structures, suggesting that the flower and insect pollination arose before the closed carpel, they have been called anthophytes. These relationships, plus problematical Triassic pollen grains and macrofossils with a mixture of angiospermlike and more primitive features, suggest that the angiosperm line goes back to the Triassic, although perhaps not as fully developed angiosperms. Within angiosperms, it is believed that "magnoliids" are relatively primitive, monocots and eudicots are derived clades, and wind-pollinated temperate trees such as oaks, birches, and walnuts (Amentiferae) are advanced eudicots. However, "magnoliids" include both woody plants, and herbs, and their flowers range from large, complex, and insect-pollinated to minute, simple, and wind-pollinated. These extremes are present among the earliest Cretaceous angiosperms, and cladistic analyses disagree on which is most primitive.

Although plant extinctions at the end of the Cretaceous have been linked with radiation of deciduous trees and proliferation of fruits dispersed by mammals and birds, they were less dramatic than extinctions in the animal kingdom. Mid-Tertiary cooling led to contraction of the tropical belt and expansion of seasonal temperate and arid zones. These changes led to the diversification of herbaceous angiosperms and the origin of open grassland vegetation, which stimulated the radiation of hoofed mammals, and ultimately the invention of human agriculture. *See* AGRICULTURE; FLOWER; PALEOBOTANY; PLANT KINGDOM.

James A. Doyle

Plant geography The study of the spatial distributions of plants and vegetation and of the environmental relationships which may influence these distributions. Plant geography (or certain aspects of it) is also known as phytogeography, phytochorology, geobotany, geographical botany, or vegetation science.

A flora is the collection of all plant species in an area, or in a period of time, independent of their relative abundances and relationships to one another. The species can be grouped and regrouped into various kinds of floral elements based on some common feature. For example, a genetic element is a group of species with a common evolutionary origin; a migration element has a common route of entry into the territory; a historical element is distinct in terms of some past event; and an ecological element is related to an environmental preference. An endemic species is restricted to a particular area, which is usually small and of some special interest. The collection of all interacting individuals of a given species, in an area, is called a population.

An area is the entire region of distribution or occurrence of any species, element, or even an entire flora. The description of areas is the subject of areography, while chorology studies their development. The local distribution within the area as a whole, as that of a swamp shrub, is the topography of that area. Areas are of interest in regard to their general size and shape, the nature of their margin, whether they are continuous or disjunct, and their relationships to other areas. Closely related plants that are mutually exclusive are said to be vicarious (areas containing such plants are also called vicarious). A relict area is one surviving from an earlier and more extensive occurrence. On the basis of areas and their floristic relationships, the Earth's surface is divided into floristic regions, each with a distinctive flora.

Floras and their distribution have been interpreted mainly in terms of their history and ecology. Historical factors, in addition to the evolution of the species themselves, include consideration of theories of shifting continental masses, changing sea levels, and orographic and climatic variations in geologic time, as well as theories of island biogeography, all of which have affected migration and perpetuation of floras. The main ecological factors include the immediate and contemporary roles played by climate, soil, animals, and humans. *See* ISLAND BIOGEOGRAPHY; PALEOBOTANY; PALEOECOLOGY.

Vegetation refers to the mosaic of plant life found on the landscape. The vegetation of a region has developed from the numerous elements of the local flora but is shaped also by nonfloristic physiological and environmental influences. Vegetation is an organized whole, at a higher level of integration than the separate species, composed of those species and their populations. Vegetation may possess emergent properties not necessarily found in the species themselves. Sometimes vegetation is very weakly integrated, as pioneer plants of an abandoned field. Sometimes it is highly integrated, as in an undisturbed tropical rainforest. Vegetation provides the main structural and functional framework of ecosystems. *See* ECOSYSTEM.

Plant communities are an important part of vegetation. No definition has gained universal acceptance, in part because of the high degree of independence of the species themselves. Thus, the community is often only a relative social continuity in nature, bounded by a relative discontinuity, as judged by competent botanists. *See* ECOLOGICAL COMMUNITIES.

In looking at vegetation patterns over larger areas, it is the basic physiognomic distinctions between grassland, forest, and desert, with such variants as woodland (open forest), savanna (scattered trees in grassland), and scrubland (dominantly shrubs), which are most often emphasized. These general classes of vegetation structure can be broken down further by reference to leaf types and seasonal habits (such as evergreen or deciduous). Geographic considerations may complete the names of the main vegetation formation types, also called biomes (such as tropical rainforest, boreal coniferous forest, or temperate grasslands). Such natural vegetation regions are most closely related to climatic patterns and secondarily to soil or other environmental factors. *See* ALTITUDINAL VEGETATION ZONES.

Vegetational plant geography has emphasized the mapping of such vegetation regions and the interpretation of these in terms of environmental (ecological) influences. Distinction has been made between potential and actual vegetation, the latter becoming more important due to human influence. *See* VEGETATION AND ECOSYSTEM MAPPING.

Some plant geographers point to the effects of ancient human populations, natural disturbances, and the large-herbivore extinctions and climatic shifts of the Pleistocene on the species composition and dynamics of so-called virgin vegetation. On the other hand, it has been shown that the site occurrence and geographic distributions of plant and vegetation types can be predicted surprisingly well from general climatic and other environmental patterns. Unlike floristic botany, where evolution provides a single unifying principle for taxonomic classification, vegetation structure and dynamics have no single dominant influence.

Basic plant growth forms (such as broad-leaved trees, stem-succulents, or forbs) have long represented convenient groups of species based on obvious similarities. When these forms are interpreted as ecologically significant adaptations to environmental factors, they are generally called life forms and may be interpreted as basic ecological types.

In general, basic plant types may be seen as groups of plant taxa with similar form and ecological requirements, resulting from similar morphological responses to similar environmental conditions. When similar morphological or physiognomic responses occur in unrelated taxa in similar but widely separated environments, they may be called convergent characteristics. *See* PLANTS, LIFE FORMS OF.

As human populations alter or destroy more and more of the world's natural vegetation, problems of species preservation, substitute vegetation, and succession have increased in importance. This is especially true in the tropics, where deforestation is proceeding rapidly. Probably over half the species in tropical rainforests have not yet even been identified. Because nutrients are quickly washed out of tropical rainforest soils, cleared areas can be used for only a few years before they must be abandoned to erosion and much degraded substitute vegetation. Perhaps the greatest current challenge in plant geography is to understand tropical vegetation and succession sufficiently well to design self-sustaining preserves of the great diversity of tropical vegetation. *See* BIOGEOGRAPHY; ECOLOGY; RAINFOREST. Elgene O. Box

Plant growth An irreversible increase in the size of the plant. As plants, like other organisms, are made up of cells, growth involves an increase in cell numbers by cell division and an increase in cell size. Cell division itself is not growth, as each new cell is exactly half the size of the cell from which it was formed. Only when it grows to the same size as its progenitor has growth been realized. Nonetheless, as each cell has a maximum size, cell division is considered as providing the potential for growth. *See* CELL (BIOLOGY); CELL DIVISION.

While growth in plants consists of an increase in both cell number and cell size, animal growth is almost wholly the result of an increase in cell numbers. Another important difference in growth between plants and animals is that animals are determinate in growth and reach a final size before they are mature and start to reproduce. Plants have indeterminate growth and, as long as they live, continue to add new organs and tissues. In a plant new cells are produced all the time, and some parts such as leaves and flowers may die, while the main body of the plant persists and continues to grow. The basic processes of cell division are similar in plants and animals, though the presence of a cell wall and vacuole in plant cells means that there are certain important differences. This is particularly true in plant cell enlargement, as plant cells, being restrained in size by a cellulose cell wall, cannot grow without an increase in the wall. Plant cell growth is thus largely a property of the cell wall. *See* CELL WALLS (PLANT).

Sites of cell division. Cell division in plants takes place in discrete zones called meristems. The stem and root apical meristems produce all the primary (or initial) tissues of the stem and root. The cylindrical vascular cambium produces more conducting cells at the time when secondary thickening (the acquisition of a woody nature) begins. The vascular cambium is a sheet of elongated cells which divide to produce xylem or water-conducting cells on the inside, and phloem or sugar-conducting cells on the outside. Unlike the apical meristems whose cell division eventually leads to an increase in length of the stem and root, divisions of the vascular cambium occur when that part of the plant has reached a fixed length, and lead only to an increase in girth, not in length. The final meristematic zone, the cork cambium, is another cylindrical sheet of cells on the outer edge of older stems and roots of woody plants. It produces new outer cells only, and these cells differentiate into the corky layers of the bark so that new protective layers are produced as the tree increases in circumference. *See* APICAL MERISTEM; BUD; LATERAL MERISTEM; PERIDERM; ROOT (BOTANY); STEM.

Controls. Plant growth is affected by internal and external factors. The internal controls are all the product of the genetic instructions carried in the plant. These influence the extent and timing of growth and are mediated by signals of various types transmitted within the cell, between cells, or all around the plant. Intercellular communication in plants may take place via hormones (or chemical messengers) or by other forms of communication not well understood. There are several hormones (or groups of hormones), each of which may be produced in a different location, that have a different target tissue and act in a different manner. *See* ABSCISIC ACID; AUXIN; CYTOKININS; GIBBERELLIN; PLANT HORMONES.

The external environments of the root and shoot place constraints on the extent to which the internal controls can permit the plant to grow and develop. Prime among these are the water and nutrient supplies available in the soil. Because cell expansion is controlled by cell turgor, which depends on water, any deficit in the water supply of the plant reduces cell turgor and limits cell elongation, resulting in a smaller plant. *See* PLANT-WATER RELATIONS.

Mineral nutrients are needed for the biochemical processes of the plant. When these are in insufficient supply, growth will be less vigorous, or in extreme cases it will cease altogether. *See* PLANT MINERAL NUTRITION.

An optimal temperature is needed for plant growth. The actual temperature range depends on the species. In general, metabolic reactions and growth increase with temperature, though high temperature becomes damaging. Most plants grow slowly at low temperatures, 32–50°F (0–10°C), and some tropical plants are damaged or even killed at low but above-freezing temperatures.

Light is important in the control of plant growth. It drives the process of photosynthesis which produces the carbohydrates that are needed to osmotically retain water in the cell for growth. *See* PHOTOSYNTHESIS.

Fruits and seeds. Fruits and seeds are rich sources of hormones. Initial hormone production starts upon pollination and is further promoted by ovule fertilization. These hormones promote the growth of both seed and fruit tissue. Fruits grow initially by cell division, then by cell enlargement, and finally sometimes by an increase in air spaces.

The growth of a seed starts at fertilization. A small undifferentiated cell mass is produced from the single-celled zygote. This proceeds to form a small embryo consisting of a stem tip bearing two or more leaf primordia at one end and a root primordium at the other. Either the endosperm or the cotyledons enlarge as a food store. *See* FRUIT; SEED.

Flowering. At a certain time a vegetative plant ceases producing leaves and instead produces flowers. This often occurs at a particular season of the year. The determining factor for this event is day length (or photoperiod). Different species of plants respond to different photoperiods. *See* FLOWER; PHOTOPERIODISM.

The light signal for flowering is received by the leaves, but it is the stem apex that responds. Exposing even a single leaf to the correct photoperiod can induce flowering. Clearly, then, a signal must travel from the leaf to the apex. Grafting a plant that has been photoinduced to flower to one not so induced can cause the noninduced plant to flower. It has been proposed that a flower-inducing hormone travels from the leaf to the stem apex and there induces changes in the development of the cells such that the floral morphology results.

Dormancy. At certain stages of the life cycle, most perennial plants cease growth and become dormant. Plants may cease growth at any time if the environmental conditions are unfavorable. When dormant, however, a plant will not grow even if the conditions are favorable. *See* DORMANCY.

Leaf abscission. As a perennial plant grows, new leaves are continuously or seasonally produced. At the same time the older leaves are shed because newer leaves are metabolically more efficient in the production of photosynthates. A total shedding of tender leaves may enable the plant to withstand a cold period or drought. In temperate deciduous trees, leaf abscission is brought about by declining photoperiods and temperatures. *See* ABSCISSION; PLANT MORPHOGENESIS. Peter J. Davies

Plant hormones A group of naturally occurring, organic substances that influence plant physiological processes at low concentrations. The processes consist mainly of growth, differentiation, and development, although other processes, such as stomatal movement, may also be affected. Typical effective concentrations [in the range of 1 μM externally, or about 1–100 nM internally (about one part per billion), depending on the compound] are far below those where either nutrients or vitamins would be effective.

Plant hormones have also been referred to as phytohormones or plant growth substances. The term "plant growth regulator" is mainly used by the agrichemical industry to denote synthetic growth regulators as distinct from endogenous growth regulators. Plant hormones include the auxins (represented principally by the compound indoleacetic acid), gibberellins, cytokinins, abscisic acid, ethylene, and brassinosteroids. Salicylic acid and jasmonic acid and some peptides are principally concerned with the defensive response to invasion by fungal diseases or insects. Polyamines can also be considered as plant hormones, although they operate at a higher concentration than those previously listed. The flowering hormone, florigen, is a phloem-transported gene product, probably a messenger ribonucleic acid (mRNA). *See* ABSCISIC ACID; AUXIN; CYTOKININS; GIBBERELLIN; ETHYLENE; HORMONE; PLANT GROWTH; PLANT MORPHOGENESIS; PLANT PHYSIOLOGY; SALICYLATE.

The individual details of hormone function vary from hormone to hormone. However, a general pattern is clear. First the hormone binds to a receptor, either on the cell membrane or soluble within the cell. This causes the initiation of a signal transduction cascade that culminates in the activation of the transcription of a suite of genes in response to the biosynthesis or activation of one or more DNA-binding transcription factors. In all cases to date, at least one member of the signal transduction cascade is a negative regulator; that is, it is present in the absence of the hormone and blocks the hormone-type response(s). Thus, the null response in mutants in which this component is defective is the hormone-type response, but in the absence of the hormone. When the hormone binds to the receptor, the transmission of the signal, often via phosphorylation of the signal transduction components, leads to the destruction of the negative regulator so that the hormone-type response can be expressed. The negative regulator is tagged for destruction by the protein ubiquitin, and it is then broken down in organelles called proteasomes. *See* PROTEASOME; PROTEIN. Peter J. Davies

Plant keys Artificial analytical constructs for identifying plants. The identification, nomenclature, and classification of plants are the domain of plant taxonomy, and one basic responsibility of taxonomists is determining if the plant at hand is identical to a known plant. The dichotomous key provides a shortcut for identifying plants that eliminates searching through numerous descriptions to find one that fits the unknown plant.

A key consists of series of pairs (couplets) of contradictory statements (leads). Each statement of a couplet must "lead" to another couplet or to a plant name. Each couplet provides an either-or proposition wherein the user must accept one lead as including the unknown plant in question and must simultaneously reject the opposing lead. The user then proceeds from acceptable lead to acceptable lead of successive couplets until a name for the unknown plant is obtained. For confirmation, the newly identified plant should then be compared with other known specimens or with detailed descriptions of that species.

Keys are included in monographic or revisionary treatments of groups of plants, most often for a genus or family. Books containing extended keys, coupled with detailed descriptions of each kind of plant (taxon), for a given geographic region are called manuals or floras, though the latter technically refers to a simple listing of names of plants for a given region. *See* PLANT KINGDOM; PLANT TAXONOMY. Donald J. Pinkava

Plant kingdom The worldwide array of plant life, including plants that have roots in the soil, plants that live on or within other plants and animals, plants that float on or swim in water, and plants that are carried in the air. Fungi used to be included in the plant kindom because they looked more like plants than animals and did not move about. It is now known that fungi are probably closer to animals in terms of their evolutionary relationships. Also once included in plants were the "blue-green algae," which are now clearly seen to be bacteria, although they are photosynthetic (and presumably the group of organisms from which the chloroplasts present in true plants were derived). The advent of modern methods of phylogenetic DNA analysis has allowed such distinctions, but even so, what remains of the plantlike organisms is still remarkably divergent and difficult to classify.

Plants range in size from unicellular algae to giant redwoods. Some plants complete their life cycles in a matter of hours, whereas the bristlecone pines are known to be over 4000 years old. Plants collectively are among the most poorly understood of all forms of life, with even their most basic functions still inadequately known, including how they sense gravity and protect themselves from infection by bacteria, viruses, and fungi. Furthermore, new species are being recorded every year.

Within the land plants, a great deal of progress has been made in sorting out phylogenetic (evolutionary) relationships of extant taxa based on DNA studies, and the system of classification listed below includes these changes. The angiosperms or flowering plants (Division Magnoliophyta) have recently been reclassified based on phylogenetic studies of DNA sequences. Within the angiosperms, several informal names are indicated in parentheses; these names may at some future point be formalized, but for the present they are indicated in lowercase letters because they have not been formally recognized under the Code of Botanical Nomenclature.

It is known that the bryophytes (Division Bryophyta) are not closely related to each other, but which of the three major groups is closest to the other land plants is not yet clear. Among the extant vascular plants, Lycophyta are the sister group to all the rest, with all of the fernlike groups forming a single monophyletic (natural) group, which is reflected here in the classification by putting them all under Polypodiophyta. This group is the sister to the extant seed plants, within which all gymnosperms form a group that is sister to the angiosperms. Therefore, if Division is taken as the highest category within Embryobionta (the embryo-forming plants), then the following scheme would reflect the present state of knowledge of relationships (an asterisk indicates that a group is known only from fossils). See separate articles on names marked by daggers.

Subkingdom Thallobionta (thallophytes)†
  Division Rhodophycota (red algae)
    Class Rhodophyceae†
  Division Chromophycota†
    Class: Chrysophyceae (golden or golden-brown algae)†
      Prymnesiophyceae†
      Xanthophyceae (yellow-green algae)†
      Eustigmatophyceae†
      Bacillariophyceae (diatoms)†
      Dinophyceae (dinoflagellates)†
      Phaeophyceae (brown algae)†
      Raphidophyceae (chloromonads)†
      Cryptophyceae (cryptomonads)†
  Division Euglenophycota (euglenoids)
    Class Euglenophyceae†
  Division Chlorophycota (green algae)†
    Class: Chlorophyceae†
      Charophyceae†
      Prasinophyceae
Subkingdom Embryobionta (embryophytes)†
  Division Rhyniophyta*†
    Class Rhyniopsida†
  Division Bryophyta†
    Class Hepaticopsida (liverworts)†
      Subclass Jungermanniidae†
        Order: Takakiales†
          Calobryales†
          Jungermanniales†
          Metzgeriales†
      Subclass Marchantiidae†
        Order: Sphaerocarpales†
          Monocleales†
          Marchantiales†
    Class: Anthocerotopsida (hornworts)†
      Sphagnopsida (peatmosses)†
      Andreaeopsida (granite mosses)†
      Bryopsida (mosses)†
      Subclass: Archidiidae†
        Bryidae†
        Order: Fissidentales†
          Bryoxiphiales†
          Schistostegales†
          Dicranales†
          Pottiales†
          Grimmiales†
          Seligeriales†
          Encalyptales†
          Funariales†
          Splachnales†
        Order: Bryales†
          Mitteniales†
          Orthotrichales†
        Order: Isobryales†
          Hookeriales†
          Hypnales†
      Subclass: Buxbaumiidae†
        Tetraphididae†
        Dawsoniidae†
        Polytrichidae†
  Division Lycophyta†
    Class Lycopsida†
      Order: Lycopodiales†
        Asteroxylales*†
        Protolepidodendrales*†
        Selaginellales*
        Lepidedendrales*†
        Isoetales†
    Class Zosterophyllopsida*†
  Division Polypodiophyta†
    Class Polypodopsida†
      Order: Equisetales†
        Marattiales†
        Sphenophyllales*
        Pseudoborniales*
        Psilotales
        Ophioglossales†
        Noeggerathiales*
        Protopteridales*
        Polypodiales†
    Class Progymnospermopsida*
  Division Pinopsida
    Class Ginkgoopsida†
      Order: Calamopityales*
        Callistophytales*
        Peltaspermales*
        Ginkgoales†
        Leptostrobales*
        Caytoniales†
        Arberiales*
        Pentoxylales*
    Class Cycadopsida†
      Order: Lagenostomales*
        Trigonocarpales*
        Cycadales†
        Bennettiales*
    Class Pinopsida†
      Order: Cordaitales*†
        Pinales†
        Podocarpales
        Gnetales†
  Division Magnoliophyta (angiosperms, flowering plants)†
    Class Magnoliopsida†
  unplaced groups: Amborellaceae, Ceratophyllaceae, Chloranthaceae, Nymphaeaceae, etc.
    eumagnoliids†
      Order: Magnoliales†
        Laurales†
        Piperales†
        Winterales
    monocotyledons†
      Order: Acorales†
        Alismatales†
        Asparagales†
        Dioscoreales†
        Liliales†
        Pandanales†
      commelinids
        Arecales†
        Commelinales†
        Poales†
        Zingiberales†

eudicotyledons†
(basal eudicots)
Order: Ranunculales†
Proteales†
Buxales
Trochodendrales†
(core eudicots)
Order: Berberidopsidales
Gunnerales
Dilleniales†
Santalales†
Caryophyllales†
Saxifragales†
Rosidae†
Order: Vitales
Myrtales†
Geraniales†
Crossosomatales
(eurosid I)
Order: Celastrales†
Cucurbitales
Fabales†
Fagales†
Malpighiales†
Oxalidales†
Rosales†
Zygophyllales†
(eurosid II)
Order: Brassicales†
Malvales†
Sapindales†
Asteridae†
Order: Cornales†
Ericales†
(euasterid I)
Order: Garryales
Gentianales†
Lamiales†
Solanales†
(euasterid II)
Order: Apiales†
Aquifoliales
Asterales†
Dipsacales†

See DEOXYRIBONUCLEIC ACID (DNA); PLANT EVOLUTION; PLANT PHYLOGENY; PLANT TAXONOMY. Mark W. Chase; Michael F. Fay

Plant metabolism The complex of physical and chemical events of photosynthesis, respiration, and the synthesis and degradation of organic compounds. Photosynthesis produces the substrates for respiration and the starting organic compounds used as building blocks for subsequent biosyntheses of nucleic acids, amino acids, and proteins, carbohydrates and organic acids, lipids, and natural products. *See* PHOTORESPIRATION; PHOTOSYNTHESIS; PLANT RESPIRATION. Irwin P. Ting

Plant mineral nutrition The relationship between plants and all chemical elements other than carbon, hydrogen, and oxygen in the environment. Plants obtain most of their mineral nutrients by extracting them from solution in the soil or the aquatic environment. Mineral nutrients are so called because most have been derived from the weathering of minerals of the Earth's crust. Nitrogen is exceptional in that little occurs in minerals: the primary source is gaseous nitrogen of the atmosphere.

Some of the mineral nutrients are essential for plant growth; others are toxic, and some absorbed by plants may play no role in metabolism. Many are also essential or toxic for the health and growth of animals using plants as food. Six basic facts have been established: (1) plants do not need any of the solid materials in the soil—they cannot even take them up; (2) plants do not need soil microorganisms; (3) plant roots must have a supply of oxygen; (4) all plants require at least 14 mineral nutrients; (5) all of the essential mineral nutrients may be supplied to plants as simple ions of inorganic salts in solution; and (6) all of the essential nutrients must be supplied in adequate but nontoxic quantities. These facts provide a conceptually simple definition of and test for an essential mineral nutrient. A mineral nutrient is regarded as essential if, in its absence, a plant cannot complete its life cycle.

Nutrients which plants require in relatively large amounts, that is, the essential macronutrients, are nitrogen, sulfur, phosphorus, calcium, potassium, and magnesium. Iron is not required in large amounts and hence is regarded as an essential micronutrient or trace element. With the progressive development of better techniques for purifying water and salts, the list of essential nutrients for all plants has expanded to include boron, manganese, zinc, copper, molybdenum, and chlorine. Evidence has accumulated in support of nickel being essential. In addition, sodium and silicon have been shown to be essential for some plants, beneficial to some, and possibly of no benefit to others. Cobalt has also been shown to be essential for the growth of legumes when relying upon atmospheric nitrogen. Claims that two other chemical elements (vanadium and selenium) may be essential micronutrients have still to be firmly established.

Mineral nutrients may be toxic to plants either because the specific nutrient interferes with plant metabolism or because its concentration in combination with others in solution is excessive and interferes with the plant's water relations. Other chemical elements in the environment may also be toxic. High concentrations of salts in soil solutions or aquatic environments may depress their water potential so that plants cannot obtain sufficient water to germinate or grow. Some desert plants growing in saline soils can accumulate salt concentrations of 20–50% dry weight in their leaves without damage, but salt concentrations of only 1–2% can damage the leaves of many species. *See* PLANT-WATER RELATIONS; PLANTS OF SALINE ENVIRONMENTS.

A number of elements interfere directly with other aspects of plant metabolism. Sodium is thought to become toxic when it reaches concentrations in the cytoplasm that depress enzyme activity or damage the structure of organelles, while the toxicity of selenium is probably due to its interference in metabolism of amino acids and proteins. The ions of the heavy metals, cobalt, nickel, chromium, manganese, copper, and zinc are particularly toxic in low concentrations, especially when the concentration of calcium in solution is low; increasing calcium increases the plant's tolerance. Aluminum is toxic only in acid soils. Boron may be toxic in soils over a wide pH range, and is a serious problem for sensitive crops in regions where irrigation waters contain excessive boron or where the soils contain unusually high levels of boron.

All plants grow poorly on very acid soils (pH $\leq$ 3.5); some plants may grow reasonably well on somewhat less acid soils. Several factors may be involved, and their interactions with plant species are complex. The harmful effects of soil acidity in some areas have been exacerbated by industrial emissions resulting in acid rain and in deposition of substances which increase the acidity on further reaction in the soil, with consequent damage to plants and animals in these ecosystems. *See* ACID RAIN.

The elemental composition of plants is important to the health and productivity of animals which graze them. With the exception of boron, all elements which are essential for plant growth are also essential for herbivorous mammals. Animals also require sodium, iodine, and selenium and, in the case of ruminant herbivores, cobalt. As a result, animals may suffer deficiencies of any one of this latter group of elements when ingesting plants which are quite healthy but contain low concentrations of these elements. In addition, nutrients in forage may be rendered unavailable to animals through a variety of factors that prevent their absorption from the gut. Plants and animals differ also in their tolerance of high levels of nutrients, sometimes with deleterious results for grazing animals. For example, the toxicity of high concentrations of selenium in plants to animals grazing them, known as selenosis, was recognized when the puzzling and long-known "alkali disease" and "blind staggers" in grazing livestock in parts of the Great Plains of North America were shown to be symptoms of chronic and acute selenium toxicity. *See* ABSORPTION (BIOLOGY); NITROGEN CYCLE; PLANT TRANSPORT OF SOLUTES; RHIZOSPHERE; ROOT (BOTANY); SOIL CHEMISTRY. J. F. Loneragan

Plant morphogenesis The origin and development of plant form and structure. Morphogenesis may be concerned with the whole plant, with a plant part, or with the subcomponents of a structure.

The establishment of differences at the two ends of a structure is called polarity. In plants, polar differences can be recognized very early in development. In the zygote, cytological differences at the two ends of the cell establish the position of the first cell division, and thus the fate of structures produced from the two newly formed cells. During the development of a plant, polarity is also exhibited in the plant axis (in the shoot and root tips). If a portion of a shoot or root is excised and allowed to regenerate, the end toward the shoot tip always regenerates shoots whereas the opposite end forms roots. Polarity is also evident on the two sides of a plant organ, such as the upper and lower surface of a leaf, sepal, or petal.

The diversity in plant form is produced mainly because different parts of the plant grow at different rates. Furthermore, the growth of an individual structure is different in various dimensions. Thus the rate of cell division and cell elongation as well as the orientation of the plane of division and of the axis of cell elongation ultimately establish the form of a structure. Such differential growth rates are very well orchestrated by genetic factors. Although the absolute growth rates of various parts of a plant may be different, their relative growth rates, or the ratio of their growth rates, are always constant. This phenomenon is called allometry (or heterogony), and it supports the concept that there is an interrelationship between the growth of various organs of a plant body. *See* PLANT GROWTH.

During development, either the removal of or changes in one part of the plant may drastically affect the morphogenesis of one or more other parts of the plant. This phenomenon is called correlation and is mediated primarily through chemical substances, such as nutrients and hormones. *See* APICAL DOMINANCE.

The ultimate factors controlling the form of a plant and its various organs are the genes. In general, several genes interact during the development of a structure, although each gene plays a significant role. Thus, a mutation in a single gene may affect the shape or size of a leaf, flower, or fruit, or the color of flower petals, or the type of hairs produced on stems and leaves. There are at least two classes of genes involved in plant morphogenesis: regulatory genes that control the activity of other genes, and effector genes that are directly involved in a developmental process. The effector genes may affect morphogenesis through a network of processes, including the synthesis and activity of proteins and enzymes, the metabolism of plant growth substances, changes in the cytoskeleton and the rates and planes of cell division, and cell enlargement. *See* PLANT HORMONES.

Plant form is also known to be affected by nutritional factors, such as sugars or nitrogen levels. For example, leaf shape can be affected by different concentrations of sucrose, and the sexuality of flowers is related to the nitrogen levels in the soil in some species. Inorganic ions (such as silver and cobalt) have also been known to affect the type of flower produced. *See* PLANT MINERAL NUTRITION.

Although genes are the ultimate controlling factors, they do not act alone, but interact with the existing environmental factors during plant development. Environmental factors, including light, temperature, moisture, and pressure, affect plant form. *See* PHYSIOLOGICAL ECOLOGY (PLANT); PLANT-WATER RELATIONS.

V. K. Sawhney

Plant movements The wide range of movements that allow plants to reorient themselves in relation to changed surroundings, to facilitate spore or seed dispersal, or, in the case of small free-floating aquatic plants, to migrate to regions optimal for their activities. There are two types of plant movement: abiogenic movements, which arise purely from the physical properties of the cells and therefore take place in nonliving tissues or organs; and biogenic movements, which occur in living cells or organs and require an energy input from metabolism.

Abiogenic movements. Drying or moistening of certain structures causes differential contractions or expansions on the two sides of cells and hence causes movements of curvature. Such movements are called hygroscopic and are usually associated with seed and spore liberation and dispersal. Examples of such movement occur in the "parachute" hairs of the fruit of dandelion (*Taraxacum officinale*), which are closed when damp but open when the air is dry to induce release from the heads and give buoyancy for wind dispersal.

Another type of abiogenic movement is due to changes in volume of dead water-containing cells. In the absence of a gas phase, water will adhere to lignocellulose cell walls. As water is lost by evaporation from the surface of these cells, considerable tensions can build up inside, causing them to decrease in volume while remaining full of water. The effect is most commonly seen in some grasses of dry habitats, such as sand dunes, where longitudinal rows of cells on one side of the leaf act as spring hinges, contracting in a dry atmosphere and causing the leaf to roll up into a tight cylinder, thus minimizing water loss by transpiration.

Biogenic movements. There are two types of biogenic movement. One of these is locomotion of the whole organism and is thus confined to small, simply organized units in an aqueous environment. The other involves the change in shape and orientation of whole organs of complex plants, usually in response to specific stimuli.

Locomotion. In most live plant cells the cytoplasm can move by a streaming process known as cyclosis. Energy for cyclosis is derived from the respiratory metabolism of the cell. The mechanism probably involves contractile proteins very similar to the actomyosin of animal muscles.

Cell locomotion is a characteristic of many simple plants and of the gametes of more highly organized ones. Motility in such cells is produced by cilia anchored in the peripheral layers of the cell and projecting into the surrounding medium. *See* CILIA AND FLAGELLA.

Cell locomotion is usually not random but is directed by some environmental gradient. Thus locomotion may be in response to specific chemicals, in which case it is called chemotaxis. Light gradients induce phototaxis; temperature gradients induce thermotaxis; and gravity induces geotaxis. One or more of these environmental factors may operate to control movement to optimal living conditions.

Movement of organs. In higher plants, organs may change shape and position in relation to the plant body. When bending or twisting of the organ is evoked spontaneously by some internal stimulus, it is termed autonomous movement. The most common movements, however, are those initiated by external stimuli such as light and the force of gravity. Of these there are two kinds. In nastic movements (nasties), the stimulus usually has no directional qualities (such as a change in temperature), and the movement is therefore not related to the direction from which the stimulus comes. In tropisms, the stimulus has a direction (for instance, gravitational pull), and the plant movement direction is related to it.

The most common autonomous movement is circumnutation, a slow, circular, sometimes waving movement of the tips of shoots, roots, and tendrils as they grow; one complete cycle usually takes from 1 to 3 h. These movements are due to differential growth, but some may be caused by turgor changes in the cells of special hinge organs and are thus reversible.

1. *Nastic movements*. There are two kinds of nastic movements, due either to differential growth or to differential changes in the turgidity of cells. They can be triggered by a wide variety of external stimuli.

Photonastic (light/dark trigger) movements are characteristic of many flowers and inflorescences, which usually open in the light and close in the dark. Thermonasty (temperature-change trigger) is seen in the tulip and crocus flowers, which open in a warm room and close again when cooled. The most striking nastic movements are seen in the sensitive plant (*Mimosa pudica*). Its multipinnate leaves are very sensitive to touch or slight injury. Leaflets fold together, pinnae collapse downward, and the whole leaf sinks to hang limply.

Epinasty and hyponasty occur in leaves as upward and downward curvatures respectively. They arise either spontaneously or as the result of an external stimulus, such as exposure to the gas ethylene in the case of epinasty; they are not induced by gravity.

2. *Tropisms*. Of these the most universal and important are geotropism (or more properly gravitropism) and phototropism; others include thigmotropism and chemotropism.

In geotropism, the stimulus is gravity. The main axes of most plants grow in the direction of the plumb line with shoots upward (negative geotropism) and roots downward (positive geotropism).

In phototropism the stimulus is a light gradient, and unilateral light induces similar curvatures; those toward the source are positively phototropic; those away from the source are negatively phototropic. Main axes of shoots are usually positively phototropic, while the vast majority of roots are insensitive.

In thigmotropism (sometimes called haptotropism), the stimulus is touch; it occurs in climbing organs and is responsible for tendrils curling around a support. In many tendrils the response may spread from the contact area, causing the tight coiling of the basal part of the tendril into an elaborate and elastic spring.

Chemotropism is induced by a chemical substance. Examples are the incurling of the stalked digestive glands of the insectivorous plant *Drosera* and incurling of the whole leaf of *Pinguicula* in response to the nitrogenous compounds in the insect prey. A special case of chemotropism concerns response to moisture gradients; for example, under artificial conditions in air, the primary roots of some plants will curve toward and grow along a moist surface. This is called hydrotropism and may be of importance under natural soil conditions in directing roots toward water sources. *See* PLANT HORMONES; PLANT PHYSIOLOGY.

Leslie J. Audus

Plant nomenclature The generally accepted system of naming plants. The study of plant variation and kinds leads to classification—the grouping of individual plants, the grouping of the groups, and the organizing of the groups (taxa) into ranks. Nomenclature deals with the naming of the taxa, the goal being to establish one correct scientific name for a given taxon with a given definition.

For most purposes, plants are considered in terms of three ranks: family, genus, and species. There are ranks above family (such as orders), ranks lower than species (such as subspecies), and even intercalated ranks (such as subgenera). Scientific family names are based on generic names and end in -aceae, such as Fagaceae (the oak family), which includes two major genera of trees, *Fagus* and *Quercus*.

The basic scientific plant name is a binomial, a combination of a Latin capitalized generic name and a Latin lowercase specific epithet, such as *Quercus alba*, the common American white oak. Scientific names may seem awkward, but to the extent that they are universally learned, they are also universally understood. The names also have meaning. *Quercus* is an ancient Roman word for the trees that provided the best and hardest woods, and *alba* means white, presumably a reference to the whitish undersides of the leaves. *See* PLANT KINGDOM.

Dan H. Nicolson

Plant organs Plant parts having rather distinct form, structure, and function. Organs, however, are interrelated through both evolution and development and are similar in many ways.

Roots, stems, and leaves are vegetative, or asexual, plant organs. They do not produce sex cells or play a direct role in sexual reproduction. In many species, nevertheless, these organs or parts of them (cuttings), may produce new plants asexually (vegetative reproduction). Sex organs are formed during the reproductive stage of plant development. In flowering plants, sex cells are produced in certain floral organs. The flower as a whole is sometimes called an organ, although it is more appropriate to consider it an assemblage of organs. *See* FLOWER; FRUIT; LEAF; PLANT REPRODUCTION; ROOT (BOTANY); STEM.

Katherine Esau

Plant pathology The study of disease in plants; it is an integration of many biological disciplines and bridges the basic and applied sciences. As a science, plant pathology encompasses the theory and general concepts of the nature and cause of disease, and yet it also involves disease control strategies, with the ultimate goal being reduction of damage to the quantity and quality of food and fiber essential for human existence.

Kinds of plant diseases. Diseases were first classified on the basis of symptoms. Three major categories of symptoms were recognized long before the causes of disease were known; necroses, destruction of cell protoplasts (rots, spots, wilts); hypoplases, failure in plant development (chlorosis, stunting); and hyperplases, overdevelopment in cell number and size (witches'-brooms, galls). This scheme remains useful for recognition and diagnosis.

When fungi, and then bacteria, nematodes, and viruses, were recognized as causes of disease, it became convenient to classify diseases according to the responsible agent. If the agents were infectious (biotic), the diseases were classified as being "caused by bacteria," "caused by nematodes," or "caused by viruses." To this list were added phanerogams and protozoans, and later mollicutes (mycoplasmas, spiroplasmas), rickettsias, and viroids. In a second group were those diseases caused by such noninfectious (abiotic) agents as air pollutants, inadequate oxygen, and nutrient excesses and deficiencies.

Other classifications of disease have been proposed, such as diseases of specific plant organs, diseases involving physiological processes, and diseases of specific crops or crop groups (for example, field crops, fruit crops, vegetable crops).

Symptoms of plant diseases. Symptoms are expressions of pathological activity in plants. They are visible manifestations of changes in color, form, and structure: leaves may become spotted, turn yellow, and die; fruits may rot on the plants or in storage; cankers may form on stems; and plants may blight and wilt. Diagnosticians learn how to associate certain symptoms with specific diseases, and they use this knowledge in the identification and control of pathogens responsible for the diseases.

Those symptoms that are external and readily visible are considered morphological. Others are internal and primarily histological, for example, vascular discoloration of the xylem of wilting plants. Microscopic examination of diseased plants may reveal additional symptoms at the cytological level, such as the formation of tyloses (extrusion of living parenchyma cells of the xylem of wilted tissues into vessel elements).

It is important to make a distinction between the visible expression of the diseased condition in the plant, the symptom, and the visible manifestation of the agent which is responsible for that condition, the sign. The sign is the structure of the pathogen, and when present it is most helpful in diagnosis of the disease.

All symptoms may be conveniently classified into three major types because of the manner in which pathogens affect plants. Most pathogens produce dead and dying tissues, and the symptoms expressed are categorized as necroses. Early stages of necrosis are evident in such conditions as hydrosis, wilting, and yellowing. As cells and tissues die, the appearance of the plant or plant part is changed, and is recognizable in such common conditions as blight, canker, rot, and spot.

Many pathogens do not cause necrosis, but interfere with cell growth or development. Plants thus affected may eventually become necrotic, but the activity of the pathogen is primarily inhibitory or stimulatory. If there is a decrease in cell number or size, the expressions of pathological activity are classified as hypoplases; if cell number or size is increased, the symptoms are grouped as hyperplases. These activities are very specific and most helpful in diagnosis. In the former group are such symptoms as mosaic, rosetting, and stunting, with obvious reduction in plant color, structure, and size. In the latter group are gall, scab, and witches'-broom, all visible evidence of stimulation of growth and development of plant tissues. *See* CROWN GALL.

Carl W. Boothroyd

The primary agents of plant disease are fungi, bacteria, viruses and viroids, nematodes, parasitic seed plants, and a variety of noninfectious agents.

Fungi. More plant diseases are caused by fungi than by any other agent. The fungi that cause plant disease derive their food from the plant (host) and are called parasites. Those that can live and grow only in association with living plant tissues are obligate parasites. Some fungi obtain their food from dead organic matter and are known as saprobes or saprophytes. Still others can utilize food from either dead organic matter or from living plant cells, and are referred to as either facultative parasites or facultative saprophytes. The classes with plant disease-causing fungi are Plasmodiophoromycetes, Chytridiomycetes, Zygomycetes, Oomycetes, Ascomycetes, Basidiomycetes, and their asexual stages. *See* FUNGI.

C. Wayne Ellett; J. W. Kimbrough

Bacteria. Eleven Gram-negative and seven Gram-positive genera of bacteria are associated with diseases of flowering plants. Species affected by these bacterial diseases include the major cereal, forage, vegetable, and fruit crops. The bacterial pathogens of plants are almost exclusively associated with plants, and do not cause serious diseases of humans or animals.

Each bacterial species produces distinctive symptoms on the host(s) that it attacks. The bacteria most commonly associated with plant diseases are Gram-negative, non-spore-forming, rod-shaped cells and include the genera *Agrobacterium, Acidovorax, Burkholderia, Erwinia, Ralstonia, Pantoea, Pseudomonas, Rhizomonas, Xanthomonas, Xylella*, and *Xylophilus*. Some of the bacterial species that are Gram-positive are in the genera of *Arthrobacter, Bacillus, Clavibacter, Clostridium, Cutrobacterium, Rhodococcus*, and *Streptomyces*. Many foliar bacterial pathogens are disseminated long distances by splashing rain and by wind-blown mist. Bacteria may also be moved from plant to plant by insects, in irrigation water, and by cultural activities used during crop production (such as pruning). Most bacterial plant pathogens survive adverse conditions in the host plant. Few bacterial species are able to survive for long periods in the soil free of the host plant.

Bacterial plant pathogens are not able to penetrate directly through the leaf or stem; instead they enter through natural openings or wounds. The mechanisms associated with pathogenesis are quite variable and are not completely understood. Bacterial cells produce many compounds that are involved in disease development. Some bacteria that are associated with vascular-wilt diseases produce gumlike substances or extracellular polysaccharides (EPS). The control of bacterial diseases relies heavily on the reduction of primary sources of inoculum. Therefore, control practices include the elimination of these sources of bacteria through seed certification and agronomic cultural practices to reduce or eliminate sources of primary inoculum. Plant varieties with resistance to bacterial diseases have been developed to control some of the foliar pathogens. Chemical control has been used quite extensively, but with limited success. Antibiotics and copper bactericides have provided some control, but numerous bacterial species have developed resistance to these compounds.

J. B. Jones

Viruses and viroids. Viruses and viroids are the simplest of the various causative agents of plant disease. The essential element of each of these two pathogens is an infective nucleic acid. The nucleic acid of viruses is covered by an exterior shell (coat) of protein, but that of viroids is not. *See* PLANT VIRUSES AND VIROIDS.

Approximately 400 plant viruses and about 10 viroids are known. The nucleic acid of most plant viruses is a single-stranded RNA; a number of isometric viruses have a double-stranded RNA. A few viruses contain double-stranded DNA, and several containing single-stranded DNA have been reported. The nucleic acid of viroids is a single-stranded RNA, but its molecular weight is much lower than that of viruses.

Some viruses, such as tobacco mosaic virus (TMV) and cucumber mosaic virus, are found in many plant species; others, such as wheat streak mosaic virus, occur only in a few grasses. Viruses are transmitted from plant to plant in several ways. The majority are transmitted by vectors such as insects, mites, nematodes, and fungi which acquire viruses during feeding upon infected plants. Some viruses are transmitted to succeeding generations by infected seed. Viroids are spread mainly by contact between healthy and diseased plants or by the use of contaminated cutting tools.

The control or prevention of virus diseases involves breeding for resistance, propagation of virus-free plants, use of virus-free seed, practices designed to reduce the spread by vectors, and, in some cases, the deliberate inoculation of plants with mild strains of a virus to protect them from the deleterious effects of severe strains. *See* PLANT VIRUSES AND VIROIDS.

Richard I. Hamilton

Nematodes. All soils that support plant life contain nematodes living in the water films that surround soil particles. Most nematodes feed primarily on microscopic plants, animals, and bacteria, but a few are parasites of animals; another relatively small group of nematodes parasitize plants. *See* NEMATA (NEMATODA).

Plant-parasitic nematodes are distinguished by their small size, about 1 mm average length, and mouthparts that are modified to form a hollow stylet which is inserted into plant cells. All of the plant parasites are placed into two orders, Tylenchida and Dorylaimida.

Plant injury is of three general types and is related to feeding habits. Migratory endoparasites destroy tissues as they feed, producing necrotic lesions in the root cortex. Other migratory endoparasites invade leaf tissues and produce extensive brown spots. Sedentary endoparasites do not kill host cells, but induce changes in host tissues, which lead to an elaborate feeding site or gall. The third general type of symptom is produced by certain migratory ectoparasites, where root tips are devitalized and cease to grow without any associated swelling or necrosis.

In addition to the plant injury that they cause directly, nematodes are important factors in disease complexes. Lesions and galls provide entrance courts for soil fungi and bacteria, and many diseases caused by soil-borne pathogens are more severe when nematodes are present. Important viruses are transmitted by nematodes of the order Dorylaimida.

Control of plant-parasitic nematodes often is based on selection of nonhosts for crop rotations or nematode-resistant

Many small plants of a dwarf mistletoe (*Arceuthobium vaginatum*) parasitizing a ponderosa pine branch. This is the most damaging disease of ponderosa pine in many parts of the West.

varieties. Some plant species release compounds into the soil that are toxic to nematodes. Animal manures, compost, and other organic amendments enhance the buildup of natural enemies of nematodes. R. A. Rohde

Seed plants. Many parasitic seed plants (estimated at nearly 3000) attack other higher plants. In some families (for example, the mistletoes Loranthaceae and Viscaceae) all members are parasitic; in others, only a single genus is parasitic in an otherwise autotrophic family.

Most parasitic plants are terrestrial; that is, the parasitic connection with the host plant is through the roots. Other parasitic plants grow on the above-ground parts of the host. Some plants are classed as semiparasites because they can live in the soil as independent plants for a time, but are not vigorous or may not flower if they do not become attached to a suitable host. The nutritional status of parasitic plants ranges from total parasites with no chlorophyll (for example, the broomrapes, Orobanchaceae) to plants that are well supplied with chlorophyll and obtain primarily water and minerals from their hosts (many mistletoes; see illustration). Frank G. Hawksworth

Noninfectious agents of disease. Plants with symptoms caused by noninfectious agents cannot serve as sources of further spread of the same disorder. Such noninfectious agents may be deficiencies or excesses of nutrients, anthropogenic pollutants, or biological effects by organisms external to the affected plants. On the farm, plant-damaging pollution may be caused by careless use of pesticides. Mishandled herbicides are by far the most damaging to plants. Off the farm, anthropogenic air pollutants are generated by industrial processes, and by any heating or transportation method that uses fossil fuels. The most common air pollutants that damage plants are sulfur oxides and ozone. Sulfur oxides are produced when sulfur-containing fossil fuels are burned or metallic sulfides are refined. Human-generated ozone is produced by sunlight acting on clouds of nitrogen oxides and hydrocarbons that come primarily from automobile exhausts. *See* AIR POLLUTION; WATER POLLUTION. Saul Rich

Epidemiology of plant disease. Epidemiology is the study of the intensification of disease over time and the spread of disease in space. The botanical epidemiologist is concerned with the interrelationships of the host plant (suscept), the pathogen, and the environment, which are the components of the disease triangle. With a thorough knowledge of these components, the outbreak of disease may be forecast in advance, the speed at which the epidemic will intensify may be determined, control measures can be applied at critical periods, and any yield loss to disease can be projected. The maximum amount of disease occurs when the host plant is susceptible, the pathogen is aggressive, and the environment is favorable.

Epidemiologically, there are two main types of diseases: monocyclic, those that have but a single infection cycle (with the rare possibility of a second or even third cycle) per crop season; and polycyclic, those that have many, overlapping, concatenated cycles of infection per crop season. For both epidemiological types, the increase of disease slows as the proportion of disease approaches saturation or 100%. R. D. Berger

Control of plant disease is defined as the maintenance of disease severity below a certain threshold, which is determined by economic losses. Diseases may be high in incidence but low in severity, or low in incidence but high in severity, and are kept in check by preventing the development of epidemics. The principles of plant disease control form the basis for preventing epidemics. However, the practicing agriculturist uses three approaches to the control of plant disease: cultural practices affecting the environmental requirement of the suscept-pathogen-environment triangle necessary for disease development, disease resistance, and chemical pesticides. Robert E. Stall

Plant phylogeny The evolutionary chronicle of plant life on the Earth. Understanding of this history is largely based on knowledge of extant plants, but the fossil record is playing an increasingly important role in refining and illuminating this picture. Study of deoxyribonucleic acid (DNA) sequences has also been revolutionizing this process in recent years. The molecular data (largely in the form of DNA sequences from several genes) have been demonstrated to be highly correlated with other information. *See* PHYLOGENY.

"Algae" was once a taxonomic designation uniting the lower photosynthetic organisms, but ultrastructural and molecular data have uncovered a bewildering diversity of species. Algae are now recognized as 10 divergent lineages on the tree of life that join organisms as distinct as bacteria and eukaryotic protozoans, ciliates, fungi, and embryophytes (including the land plants). In a biochemical context, the term "algae" defines species characterized by chlorophyll *a* photosynthesis (except Embryophyta); some of their descendants are heterotrophic (secondary chloroplast loss). Despite the variety of species it encompasses, the term "algae" also retains phylogenetic relevance. *See* ALGAE; CHLOROPHYLL; PHOTOSYNTHESIS. Gary W. Saunders

Embryobionta, or embryophytes, are largely composed of the land plants that appear to have emerged 475 million years ago. The evidence indicates that land plants have not evolved from different groups of green algae (Chlorophyta) as suggested in the past, but instead share a common ancestor, which was a green alga. Land plants all have adaptations to the terrestrial environment, including an alternation of generations (sporophyte or diploid and gametophyte or haploid) with the sporophyte generation producing haploid spores that are capable of resisting desiccation and dispersing widely, a cuticle covering their outside surfaces, and separate male and female reproductive organs in the gametophyte stage. The life history strategies of land plants fall into two categories that do not reflect their phylogenetic relationships. The mosses, hornworts, and liverworts represent the first type, and they have expanded the haploid generation, upon which the sporophyte is dependent. Several recent analyses of DNA data as well as evidence from mitochondrial DNA structure have demonstrated that the liverworts alone are the remnants of the earliest land plants and that the mosses and hornworts are closer to the vascular plants (tracheophytes). The tracheophytes include a large number of extinct and relatively simple taxa, such as the rhiniophytes and horneophytes known only as Silurian and Devonian fossils. All tracheophytes are of the second category, and they have expanded the sporophyte generation. Among extant tracheophytes, the earliest branching are the lycopods or club mosses (*Lycopodium* and *Selaginella*), and there are still a diversity of other forms, including sphenophytes (horsetails, *Equisetum*) and ferns (a large and diverse group in which the positions of several families still are not clear). *See* EMBRYOBIONTA.

All seeds plants take the reduction of the gametophyte generation a step further and make it dependent on the sporophyte, typically hiding it within reproductive structures, which are either cones or flowers. The first seed plants originated at least by the Devonian, and they are known to have a great diversity of extinct forms, including the seed ferns. There are two groups of extant seed-bearing plants, gymnosperms and angiosperms. In the gymnosperms, the seeds are not enclosed within tissue derived from the parent plant. There are four distinct groups of extant gymnosperms, often recognized as classes: Cycadopsida, Gnetopsida, Ginkgoopsida, and Pinopsida.

The angiosperms (also flowering plants or Magnoliopsida) are the dominant terrestrial plants, although the algae collectively must still be acknowledged as the most important in the maintenance of the Earth's ecological balance (fixation of carbon dioxide and production of oxygen). In angiosperms the seeds are covered by protective tissues derived from the parental plant. There are no generally accepted angiospermous fossils older than 120 million years, but the lineage is clearly much older based on DNA clocks and other circumstantial lines of evidence such as their current geographic distributions.

Traditionally the angiosperms have been divided into two groups, monocotyledons (monocots) and dicotyledons (dicots), based on the number of seed leaves. However, DNA sequence data have demonstrated that, although there are two groups, these are characterized by fundamentally different pollen organization, such that the monocots share with a group of dicots pollen with one pore whereas the rest of the dicots have pollen with three (or more) pores. Mark W. Chase

Plant physiology The branch of plant sciences that aims to understand how plants live, function, and grow. Its ultimate objective is to explain all life processes of plants by a minimal number of comprehensive principles founded in chemistry, physics, and mathematics. *See* BOTANY; PHOTOSYNTHESIS; PLANT GROWTH; PLANT METABOLISM; PLANT RESPIRATION.

Plant physiology seeks to understand all the aspects and manifestations of plant life. In agreement with the major characteristics of organisms, it is usually divided into three major parts: (1) the physiology of nutrition and metabolism, which deals with the uptake, transformation, and release of materials, and also their movement within and between the cells and organs of the plant, which may be subdivided into the processes involved in metabolism or transport; (2) the physiology of growth, development, and reproduction; and (3) environmental physiology, which seeks to understand the manifold responses of plants to the physical and biotic environment. The part of environmental physiology that deals with effects of and adaptations to adverse conditions is referred to as stress physiology.

Plant physiological research is carried out at various levels of organization and through the use of various methods. The main organizational levels are the genetic, molecular, subcellular, or organellar; the cellular; the organismal or whole-plant; and the population level. At the molecular level, work is carried out on biochemical processes, such as those involved in the biosynthesis of a particular compound, often without consideration of the cellular or tissue location. Work at the subcellular level is aimed at understanding metabolic processes and their regulation, and also the localization of molecules in particular structures of the cell. Work at the cellular level often deals with the same processes, but is concerned with their integration in the cell as a whole.

Research at the organismal level is concerned with the function of the plant as a whole and its different organs, and with the relationships between the organs. Research at the population level, which merges with experimental ecology, deals with physiological phenomena in plant associations which may consist either of one dominant species (like a field of corn), often in an agricultural context, or of numerous diverse species (such as a forest). Work at the organismal level, and to some extent the population level, may be carried out in facilities permitting maintenance of controlled environmental conditions (light intensity and quality, temperature, water, humidity, air composition, and nutrient supply), often in specially constructed spaces referred to as growth chambers. The purpose is to obtain results that can be quantitatively reproduced, given that natural conditions cannot be exactly duplicated. *See* PHYSIOLOGICAL ECOLOGY (PLANT); PLANT CELL. Peter J. Davies; Anton Lang

Plant pigment A substance in a plant that imparts coloration. The photosynthetic pigments are involved in light harvesting and energy transfer in photosynthesis. This group of pigments consists of the tetrapyrroles, which include chlorophylls (chl) and phycobilins, and the carotenoids. The light-absorbing groups of these molecules, the chromophores, contain conjugated double bonds (alternating single and double bonds) which make them effective photoreceptors. The sum of the absorption spectra of the chlorophylls and the carotenoids, evident in the absorption spectrum of a green leaf, is equivalent to the action spectrum of photosynthesis. *See* CAROTENOID; CHLOROPHYLL; LIGHT; PHOTOSYNTHESIS; PHYCOBILIN; PIGMENTATION.

The second major group of plant pigments comprises the anthocyanins, intensely colored plant pigments responsible for most scarlet, crimson, purple, mauve, and blue colors in higher plants. About 100 different anthocyanins are known. Unlike the chlorophylls and carotenoids, which are lipid-soluble chloroplast pigments, the anthocyanins are water-soluble and are located in the cell vacuole. Chemically, they are a class of flavonoids and are particularly closely related, both structurally and biosynthetically, to the flavonols. One of their major values to the plant lies in the contrasting colors they provide in flower and fruit against the green background of the leaf to attract insects and animals for pollination and seed dispersal. *See* FLAVONOID.

Shirley Raps; Jeffrey B. Harborne

Plant propagation The deliberate, directed reproduction of plants using plant cells, tissues, or organs. Asexual propagation, also called vegetative propagation, is accomplished by taking cuttings, by grafting or budding, by layering, by division of plants, or by separation of specialized structures such as tubers, rhizomes, or bulbs. This method of propagation is used in agriculture, in scientific research, and in professional and recreational gardening. It has a number of advantages over seed propagation: it retains the genetic constitution of the plant type almost completely; it is faster than seed propagation; it may allow elimination of the nonfruiting, juvenile phase of the plant's life; it preserves unique, especially productive, or esthetically desirable plant forms; and it allows plants with roots well adapted for growth on poor soils to be combined with tops that produce superior fruits, nuts, or other products. *See* BREEDING (PLANT); PLANT REPRODUCTION. C. E. LaMotte

Tissue cultures and protoplast cultures are among the techniques that have been investigated for plant propagation; the success of a specific technique depends on a number of factors. Practical applications of such methods include the clonal propagation of desirable phenotypes and the commercial production of virus-free plants.

Plant tissue cultures are initiated by excising tissue containing nucleated cells and placing it on an enriched sterile culture medium. The response of a plant tissue to a culture medium depends on a number of factors: plant species, source of tissue, chronological age and physiological state of the tissue, ingredients of the culture medium, and physical culturing conditions, such as temperature, photoperiod, and aeration.

Though technically more demanding, successful culture of plant protoplasts involves the same basic principles as plant tissue culture. Empirical methods are used to determine detailed techniques for individual species; such factors as plant species, tissue source, age, culture medium, and physical culture conditions have to be considered. *See* PLANT CELL; TISSUE CULTURE.

Karen Grady Ford

Plant reproduction The formation by a plant of offspring that are either exact copies or reasonable likenesses. When the process is accomplished by a single individual without fusion of cells, it is referred to as asexual; when fusion of cells is involved, whether from an individual or from different donors, the process is sexual.

Asexual reproduction. The higher green plant has remarkable powers of regeneration, as shown by the fact that almost all of its approximately 10 trillion cells are capable of reproducing the entire plant asexually. This can be demonstrated by the technique of tissue culture, in which individual cells of the plant are excised and grown in sealed containers on artificial media in the absence of contaminating microorganisms. When the experiment is properly done, a single isolated cell derived from root, stem, or leaf divides repeatedly to produce a mass of undifferentiated tissue. With proper chemical stimulation, such tissue can give rise to roots and buds, thus reconstituting a plant. Alternatively, the tissue may produce many embryos, each of which can yield a plant exactly like the original. Such plants regenerated from a single cell can usually flower and set seed normally when removed and placed in soil. This experiment shows that each cell of the plant body carries all the information required for formation of the entire organism. The culture of isolated cells or bits of tissue thus constitutes a means of vegetative propagation of the plant and can provide unlimited copies identical to the organism from which the cells were derived. *See* TISSUE CULTURE.

All other vegetative reproductive devices of higher plants are elaborations of this basic ability and tendency of plant cells to produce tissue masses that can organize into growing points (meristems) to yield the typical patterns of differentiated plant organs. For example, a stem severed at ground level may produce adventitious roots, a property used commercially in vegetative propagation. Similarly, the lateral buds formed along stems by action of the terminal meristem, can, if excised, give rise to entire plants. The "eyes" of the potato tuber, a specialized fleshy stem, are simply buds used in vegetative propagation of the crop. In many plants, cuttings made from fleshy roots can similarly form organized buds and reconstitute the plant by vegetative propagation. Thus, each of the vegetative organs of the plant (leaf, stem, and root) can give rise to new plants by asexual reproduction. *See* PLANT PROPAGATION.

Sexual reproduction. While in asexual reproduction, the genetic makeup of the progeny rarely differs greatly from that of the parent, the fusion of cells in sexual reproduction can give rise to new genetic combinations, resulting in new types of plants, some of which may be better able to cope with the daily environmental stresses. The life cycle of higher green plants consists of two distinct generations, based on the chromosomal complement of their cells. The sporophyte generation is independent and dominant in the flowering plants and ferns, but small, nongreen, and dependent in the mosses, and contains the $2n$ number of chromosomes. The diploidy results in each case from the fusion of sperm and egg to form the zygote, which then develops into an embryo and finally into the mature sporophyte, whatever its nature. The sporophyte generation ends with the formation of $1n$ spores by reduction division, or meiosis, in a spore mother cell. The spore then develops into the gametophyte generation, which in turn produces the sex cells, or gametes. The gametophyte generation ends when gametes fuse to form the zygote, restoring the $2n$ situation typical of sporophytes. *See* MEIOSIS.

In flowering plants, the gametophyte or $1n$ generation is reduced to just a few cells (generally three for the male and eight for the female). The male gametophyte is formed after meiosis occurs in the microspore mother cells of the anther, yielding a tetrad of $1n$ microspores. Each of these microspores then divides mitotically at least twice. The first division produces the tube nucleus and the generative nucleus. The generative nucleus then divides again to produce two sperms. These nuclei are generally not separated by cell walls, but at this stage the outer wall of the spore becomes thickened and distinctively patterned—a stage typical of the mature male gametophyte, the pollen grain. *See* FLOWER; MITOSIS; POLLEN; POLLINATION.

Each pollen grain has a weak pore in the sporopollenin wall, through which the pollen tube emerges at the time of germination. Pollen germinates preferentially in the viscous secretion on the surface of the stigma, and its progress down the style to the ovary is guided through specific cell-to-cell recognition processes. Throughout its growth, which occurs through the deposition of new cell wall material at the advancing tip, the pollen tube is controlled by the tube nucleus, usually found at or near the tip. When the pollen tube, responding to chemical signals, enters the micropyle of the ovule, its growth ceases and the tip bursts, discharging the two sperms into the embryo sac, the female gametophyte of the ovary.

The female gametophyte generation, like the male, arises through meiotic division of a $2n$ megaspore mother cell. This division forms four $1n$ megaspores, of which three usually disintegrate, the fourth developing into an eight-nucleate embryo sac by means of three successive mitotic divisions. The eight nuclei arrange themselves into two groups of four, one at each pole of the embryo sac. Then one nucleus from each pole moves to the center of the embryo sac. One of the three nuclei at the micropylar end of the embryo sac is the female gamete, the egg, which fuses with one of the sperm nuclei to form the zygote, the first cell of the sporophyte generation, which produces the embryo. The second sperm fuses with the two polar nuclei at the center of the embryo sac to form a $3n$ cell that gives rise to the endosperm of the seed, the tissue in which food is stored. The entire ovule ripens into the seed, with the integuments forming the protective seed coat. The entire ovary ripens into a fruit, whose color, odor, and taste are attractive to animals, leading to dispersal of the seeds. The life cycle is completed when the seed germinates and grows into a mature sporophyte with flowers, in which meiotic divisions will once again produce $1n$ microspores and megaspores.

Nonflowering higher plants such as the ferns and mosses also show a distinct alternation of generations. The familiar fern plant of the field is the sporophyte generation. Meiosis occurs in sporangia located in special places on the leaves, generally the undersides or margins. A spore mother cell produces a tetrad of $1n$ spores, each of which can germinate to produce a free-living, green gametophyte called a prothallus. On the prothallus are produced male and female sex organs called antheridia and archegonia, which give rise to sperms and eggs, respectively. Sperms, motile because of their whiplike flagella, swim to the archegonium, where they fertilize the egg to produce the zygote that gives rise to the sporophyte generation again.

In mosses, by contrast, the dominant green generation is the gametophyte. Antheridia or archegonia are borne at the tips of these gametophytes, where they produce sperms and eggs, respectively. When suitably wetted, sperms leave the antheridium, swim to a nearby archegonium, and fertilize the egg to produce a $2n$ zygote that gives rise to a nongreen, simple, dependent sporophyte. The moss sporophyte consists mainly of a sporangium at the end of a long stalk, at the base of which is a mass of tissue called the foot, which absorbs nutrients from the green, photosynthetic gametophyte. Meiosis occurs in the sporangium when a spore mother cell gives rise to four reduced spores. Each spore can germinate, giving rise to a filamentous structure from which leafy gametophytic branches arise, completing the life cycle.

Various members of the algae that reproduce sexually also display alternation of generations, producing sperms and eggs in antheridia and oogonia. Sporophyte and gametophyte generations may each be free-living and independent, or one may be partially or totally dependent on the other. *See* FRUIT; POPULATION DISPERSAL; SEED.

Arthur W. Galston; Vivian F. Irish

Plant respiration The metabolism of organic molecules using enzymes to generate usable energy in the form of adenosine triphosphate (ATP). The ATP made by respiration is used in plant cells to move things within the cell, to make new molecules needed for growth, and to power active transport (the pumping of ions or other substances across a cell membrane against an osmotic gradient). Most plant respiration is aerobic—that is, it requires the presence of oxygen—and involves the oxidation of sugars to carbon dioxide. When oxygen is not present, anaerobic respiration, or fermentation, occurs, allowing some ATP synthesis. Fermentation is much less efficient in harvesting energy from organic molecules than aerobic respiration is, and most plants do not tolerate long periods of anaerobiosis. Plant respiration encompasses several pathways, including glycolysis, the Krebs cycle, and oxidative electron transport. It occurs in all plant cells and is generally similar to animal respiration, with a few notable exceptions such as a cyanide-insensitive respiration mechanism lacking in animals. *See* BIOLOGICAL OXIDATION; CARBOHYDRATE METABOLISM; CITRIC ACID CYCLE; ENERGY METABOLISM; PLANT METABOLISM.

A major misconception about plant respiration is that it occurs only in plant cells lacking photosynthesis. Moreover, photosynthesis is often thought of as the *only* means for plants to generate energy. All plant cells respire. During photosynthesis, respiration may be reduced and cells may gain some ATP from photosynthetic reactions, but measurements indicate that there still is a significant amount of respiration in photosynthesizing cells. *See* PHOTOSYNTHESIS; PLANT CELL.

Glycolysis is a major component of both aerobic and anaerobic plant respiration. Glycolysis is the process by which glucose is broken down to form pyruvate, with energy captured in the form of ATP. It occurs in the cytosol of plant cells. Formally, glycolysis begins with glucose, but in many situations an intermediate of glycolysis is produced and the process begins with that product. Provided that oxygen is present, the pyruvate produced by glycolysis enters the mitochondria of plant cells. There, it loses one carbon atom and is converted to acetyl coenzyme A. The two carbons of the acetate component of acetyl CoA enter the Krebs cycle, also called the citric acid cycle or the tricarboxylic acid cycle. In one complete cycle, two carbon atoms are lost and significant reducing power is stored as the reduced form of nicotinamide adenine dinucleotide (NADH) and the similar molecule flavin adenine dinucleotide ($FADH_2$). The reducing power of NADH then can be used for oxidative electron transport on the inner membrane of the mitochondrion (the christae). *See* MITOCHONDRIA.

Electron transport is coupled to ATP synthesis by the movement of protons from the matrix to the intermembrane space. The energy gradient (proton motive force) resulting from the movement of hydrogen ions can be used to spin an ATP synthase complex on the inner mitochondrial membrane. This spinning forces adenosine diphosphate (ADP) to combine with inorganic phosphate (P_i) to make ATP.

Thomas D. Sharkey

Plant taxonomy The area of study focusing on the development of a classification system, or taxonomy, for plants based on their evolutionary relationships (phylogeny). The assumption is that if classification reflects phylogeny, reference to the classification will help researchers focus their work in a more accurate manner. The task is to make phylogeny reconstruction as accurate as possible. The basic unit of classification is generally accepted to be the species, but how a species should be recognized has been intensely debated. *See* PLANT KINGDOM; PLANT PHYLOGENY.

The earliest classifications of plants were those of the Greek philosophers such as Aristotle (384–322 B.C.) and Theophrastus (372–287 B.C.). The latter is often called the father of botany largely because he listed the names of over 500 species, some of which are still used as scientific names today. In the next 1600 years little progress occurred in plant taxonomy. It was not until the fifteenth century that there was renewed interest in botany, much of which was propelled by the medical use of plants. In 1753 Carolus Linnaeus, a Swedish botanist, published his *Species Plantarum*, a classification of all plants known to Europeans at that time. Linnaeus's system was based on the arrangement and numbers of parts in flowers, and was intended to be used strictly for identification (a system now referred to as an artificial classification as opposed to a natural classification, based on how closely related the species are).

In *Species Plantarum*, Linnaeus made popular a system of binomial nomenclature developed by the French botanist Gaspard Bauhin (1560–1624), which is still in use. Each species has a two-part name, the first being the genus and the second being the species epithet. For example, *Rosa alba* (italicized because it is Latin) is the scientific name of one species of rose; the genus is *Rosa* and the species epithet is *alba*, meaning white (it is not a requirement that scientific names be similar to common names or have real meaning, although such relevance is often the case). The genus name *Rosa* is shared by all species of roses, reflecting that they are thought to be more closely related to each other than to species in any other group.

Today, we understand that the best classification system is one that reflects the patterns of the evolutionary processes that produced these plants. The rules of botanical nomenclature (and those of zoology as well, although they are not identical) are part of an internationally accepted Code that is revised (minimally) at an international congress every 5 years. *See* PLANT EVOLUTION.

Use of common names in science and horticulture is not practical. Scientific names are internationally agreed upon so that a consistent taxonomic name is used everywhere for a given organism. In addition to genus and species, plants are classified by belonging to a family; related families are grouped into orders, and these are typically grouped into a number of yet higher and more encompassing categories. In general, higher categories are composed of many members of lower types—for example, a family may contain 350 genera, but some may be composed of a single genus with perhaps a single species if that species is distantly related to all others.

Many botanists use a number of intermediate categories between the level of genus and family, such as subfamilies, tribes, and subtribes, as well as some between species and genus, such as subgenera and sections, but none of these categories is formally mandated. They are useful nonetheless to reflect intermediate levels of relatedness, particularly in large families (composed of several hundreds or even thousands of species). Below the level of species, some botanists use the concept of subspecies (which is generally taken to mean a geographically distinct form of a species) and variety (which is often a genetic form or genotype, for example a white-flowered form of a typically blue-flowered species, or a form that is ecologically distinct).

The basic idea that plant classification should reflect evolutionary (genetic) relationships has been well accepted for some time, but the degree to which this could be assessed by the various means available differed. It has only recently become possible to assess genetic patterns of relatedness directly by analyzing DNA sequences. In the 1990s, DNA technology became much more efficient and less costly, resulting in a dramatic upsurge in the availability of DNA sequence data for various genes from each of the three genetic compartments present in plants (nuclear, mitochondrial, and plastid or chloroplast). In 1998 a number of botanists collectively proposed the first DNA-based classification of a major group of organisms, the angiosperms or flowering plants. For the first time, a classification was directly founded on assessments of the degree of relatedness made with objective, computerized methods of phylogeny reconstruction. Other data, such as chemistry and morphology, were also incorporated into these analyses, but by far the largest percentage of information came from DNA sequences—that is, relatedness was determined mostly on the basis of similarities in plants' genetic codes. The advantages of such a classification were immediately obvious: (1) it was not based on intuition about which category of information best reflected natural relationships; (2) it ended competition between systems based on differing emphases; (3) the analysis could be repeated by other researchers using either the same or different data (other genes or categories of information); and (4) it could be updated as new data emerged, particularly from studies of how chromosomes are organized and how morphology and other traits are determined by the genes that code for them. *See* DEOXYRIBONUCLEIC ACID (DNA); GENETIC MAPPING.

At the same time that DNA data became more widely available as the basis for establishing a classification, a more explicit methodology for turning the results of a phylogenetic analysis into a formal classification became popular. This methodology, called cladistics, allowed a large number of botanists to share ideas of how the various taxonomic categories could be better defined. Although there remain a number of dissenting opinions about some minor matters of classification, it is now impossible for scientists to propose alternative ideas based solely on opinion. *See* PHYLOGENY; TAXONOMY. Mark W. Chase; Michael F. Fay

Plant tissue systems The tissues of a plant are organized into three systems: the dermal tissue system, the ground tissue system, and the vascular tissue system. Most plants are composed of coherent masses of cells called tissues. Large units of tissues having some features in common are called tissue systems. In actual usage, however, the terms tissue and tissue system are not strictly separated.

Plant tissues are primary or secondary in origin. The primary tissues arise from apical meristems, the perennially embryonic tissues at the tips of roots and shoots. The primary tissues include the surface layer, or epidermis; the primary vascular tissues, xylem and phloem, which conduct water and food, respectively; and the ground tissues, comprising parenchyma, which is chiefly concerned with manufacture and storage of food, and collenchyma and sclerenchyma, which are the supporting tissues. Properties and functions of a tissue are usually related to the structure of the cells present. *See* APICAL MERISTEM; PLANT ANATOMY; PLANT CELL; PRIMARY VASCULAR SYSTEM (PLANT).

In the stem and root, the vascular tissues and some associated ground tissue are often treated as a unit, the stele. Ground tissue may be present in the center of the stele (pith) and on its periphery (pericycle). The ground tissue system enclosing the stele on the outside is called the cortex. *See* CORTEX (PLANT); PERICYCLE; PITH.

The secondary tissues arise from lateral meristems, and their formation is mainly responsible for the growth in thickness of stems and roots. They comprise secondary vascular tissues and the protective tissue called periderm. Secondary tissue development varies between the major plant groups and between species. Over time, this secondary growth may build up a massive core of wood (secondary xylem), but the outer tissue system—the bark (secondary phloem and other cells)—remains relatively thin because its outer or older part becomes compressed and, in many species, is continuously sloughed off. *See* BARK; LATERAL MERISTEM; PERIDERM; WOOD ANATOMY.

The production of flowers instead of vegetative shoots results from physiological and morphological changes in the apical meristem, which then becomes the flower meristem. The latter, however, produces tissue systems fundamentally similar to those in the vegetative body of the plant. *See* FLOWER.

Anna H. Lynch; Katherine Esau

Plant transport of solutes The movement of organic and inorganic compounds through plant vascular tissues. Transport can take place over considerable distances; in tree species, transport distances are often 100–300 ft (30–100 m).

This long-distance transport is necessary for survival in higher land plants, in which specialized organs of uptake or synthesis are separated by a considerable distance from the organs of utilization. Diffusion is not rapid enough to account for the amount of material moved over such long distances. Rather, transport depends on a flowing stream of liquid in vascular tissues (phloem and xylem) that are highly developed structurally.

The movement of organic solutes occurs mainly in the phloem, where it is also known as translocation. In the phloem the direction of transport is from places of production, such as mature leaves, to places of utilization or storage, such as the shoot apex or developing storage roots. Organic materials translocated in the phloem include the direct products of photosynthesis (sugars) as well as compounds derived from them (nitrogenous compounds and plant hormones, for example). The phloem also transmits signals between various plant organs in the form of regulatory molecules (such as RNAs), which move with the organic solutes.

Some movement of organic solutes does occur in the xylem of certain species. However, the xylem generally moves water and inorganic solutes or mineral elements from sites of uptake in the roots to sites where water is lost from the plant, primarily the leaves. Some redistribution of the ions throughout the plant may then occur in the phloem. *See* PLANT TISSUE SYSTEMS.

The phloem is the predominant tissue for the translocation of organic solutes, particularly sugars. In general, organic compounds are translocated in the phloem from areas called sources, which are regions of synthesis or mobilization, to areas of utilization called sinks, such as young leaves, root tips, shoot tips, and developing fruits. Mature photosynthesizing leaves are the major sources; materials can be transported from a given source leaf both acropetally and basipetally, that is, in the direction of the shoot tips and root tips, respectively. When multiple source leaves are present, the upper sources usually supply the shoot tips, while the lower sources supply the root tips. The distribution pattern from a leaf also depends on the pattern of vascular connections with other organs. *See* PHLOEM; PHOTOSYNTHESIS.

Long-term source-sink interactions are probably mediated by plant hormones, such as cytokinins, and by nutrients, such as potassium, phosphate, and sucrose. Molecules such as sucrose can change gene expression in source-sink systems. Small RNAs in the phloem play roles in defense against viral nucleic acids and in developmental processes. As they link sources and sinks in this role, they are a powerful means of communication and integration within the plant. The mechanism of translocation must account for the following observations: polarity, bidirectional movement, velocity, energy requirement, turgor pressure, and phloem structure.

Water and dissolved solutes move upward primarily in the tissues of the xylem.

Once in the xylem elements, inorganic solutes move with water to the transpiring surfaces. Water is lost in the leaves by evaporation from cell-wall surfaces; water vapor then diffuses into the atmosphere by way of small pores between two specialized cells (guard cells). The guard cells and the pore are collectively called a stomate. This loss of water from the leaf causes movement of water out of the xylem in the leaf to the surfaces where evaporation is occurring. Water has a high internal cohesive force, especially in small tubes with wettable walls. In

addition, the xylem elements and the cell walls provide a continuous water-filled system in the plant. Thus the loss of water from the xylem elements in the leaves causes a tension or negative pressure in the xylem sap. This tension is transmitted all the way down the stem to the roots, so that a flow of water occurs up the plant from the roots and eventually from the soil. The velocity of this sap flow in tree species ranges from 3 to approximately 165 ft/h (1 to 50 m/h), depending on the diameter of the xylem vessels. *See* PLANT MINERAL NUTRITION; PLANT-WATER RELATIONS; XYLEM.
S. Sovonick-Dunford

Plant viruses and viroids Plant viruses are pathogens which are composed mainly of a nucleic acid (genome) normally surrounded by a protein shell (coat); they replicate only in compatible cells, usually with the induction of symptoms in the affected plant. Viroids are among the smallest infection agents known. Their circular, single-stranded ribonucleic acid (RNA) molecule is less than one-tenth the size of the smallest viruses.

Viruses. Viruses can be seen only with an electron microscope (see illustration). Isometric (spherical) viruses range from 25 to 50 nanometers in diameter, whereas most anisometric (tubular) viruses are 12 to 25 nm in diameter and of various lengths (200–2000 nm), depending on the virus. The coat of a few viruses is covered by a membrane which is derived from its host.

Over 800 plant viruses have been recognized and characterized. The genomes of most of them, such as the tobacco mosaic virus (TMV), are infective single-stranded RNAs; some RNA viruses have double-stranded RNA genomes. Cauliflower mosaic virus and bean golden mosaic virus are examples of viruses having double-stranded and single-stranded deoxyribonucleic acid (DNA), respectively. The genome of many plant viruses is a single polynucleotide and is contained in a single particle, whereas the genomes of brome mosaic and some other viruses are segmented and distributed between several particles. There are also several low-molecular-weight RNAs (satellite RNAs) which depend on helper viruses for their replication. *See* DEOXYRIBONUCLEIC ACID (DNA); RIBONUCLEIC ACID (RNA).

The natural hosts of plant viruses are widely distributed throughout the higher-plant kingdom. Some viruses (TMV and cucumber mosaic virus) are capable of infecting over a hundred species in many families, whereas others, such as wheat streak mosaic virus, are restricted to a few species in the grass family. The replication of single-stranded RNA viruses involves release of the virus genome from the coat protein; the association of the RNA with the ribosomes of the cell; translation of the genetic information of the RNA into specific proteins, including subunits of the coat protein and possibly viral RNA-synthesizing enzymes (replicases); transmission by vectors and disease induction; synthesis of noninfective RNA using parental RNA as the template; and assembly of the protein subunits and viral RNA to form complete virus particles.

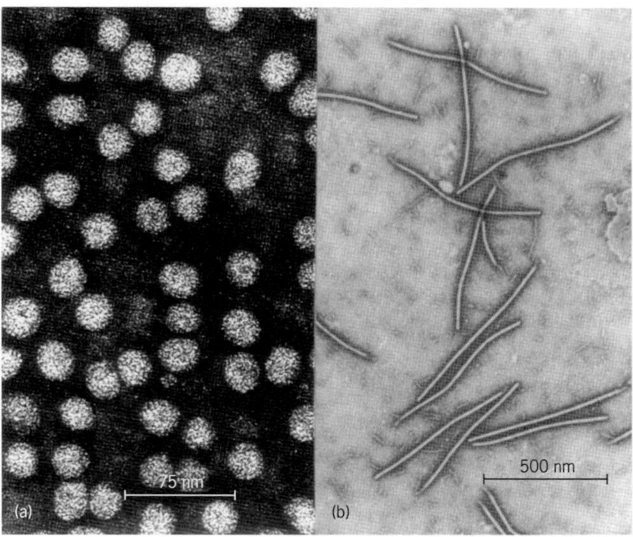

Representative plant viruses in purified virus preparation obtained from infected leaves. (*a*) Tobacco streak virus (isometric). (*b*) Pea seed-borne mosaic virus (anisometric).

In other RNA viruses, such as lettuce necrotic yellows virus, an enzyme which is contained in the virus must first make a complementary (infective) copy of the RNA; this is then translated into enzymes and coat protein subunits. The replication of double-stranded RNA viruses is similar to that of lettuce necrotic yellows virus.

With double-stranded DNA viruses, viral DNA is uncoated in a newly infected cell and transported to the nucleus, where it associates with histones to form a closed circular minichromosome. Two major RNA species (35S and 19S) are transcribed from the minichromosome by a host-encoded enzyme and are translated in the cytoplasm to produce virus-associated proteins. The 35S RNA serves as the template for a viral enzyme which transcribes it to viral DNA, which is then encapsidated to form virus particles.

Symptoms are the result of an alteration in cellular metabolism and are most obvious in newly developing tissues. In some plants, depending on the virus, the initial infection does not spread because cells surrounding the infected cells die, resulting in the formation of necrotic lesions. Such plants are termed hypersensitive. The size and shape of leaves and fruit may be adversely affected, and in some instances plants may be killed. Not all virus infections produce distinctive symptoms.

The most common mode of transmission for many viruses is by means of vectors, mainly insects (predominantly aphids and leafhoppers), and to a lesser extent mites, soil-inhabiting fungi, and nematodes which acquire viruses by feeding on infected plants. Viruses transmitted by one class of vector are rarely transmitted by another, and there is often considerable specificity between strains of a virus and their vectors.

Some viruses are transmitted to succeeding generations mainly by embryos in seeds produced by infected plants; over 200 viruses are transmitted in this way.

Viroids. Only about 30 viroids are known, but they cause very serious diseases in such diverse plants as chrysanthemum, citrus, coconut, and potato. They can also be isolated from plants that do not exhibit symptoms. Viroids are mainly transmitted by vegetative propagation, but some, such as potato spindle tuber viroid, are transmitted by seed or by contact between infected and healthy plants. Tomato planta macho viroid is efficiently transmitted by aphids. *See* PLANT PATHOLOGY; VIROIDS; VIRUS.
Richard I. Hamilton

Plant-water relations Water is the most abundant constituent of all physiologically active plant cells. Leaves, for example, have water contents which lie mostly within a range of 55–85% of their fresh weight. Other relatively succulent parts of plants contain approximately the same proportion of water, and even such largely nonliving tissues as wood may be 30–60% water on a fresh-weight basis. The smallest water contents in living parts of plants occur mostly in dormant structures, such as mature seeds and spores. The great bulk of the water in any plant constitutes a unit system. This water is not in a static condition. Rather it is part of a hydrodynamic system, which in terrestrial plants involves absorption of water from the soil, its translocation throughout the plant, and its loss to the environment, principally in the process known as transpiration.

Cellular water relations. The typical mature, vacuolate plant cell constitutes a tiny osmotic system, and this idea is central to any concept of cellular water dynamics. Although the cell walls of most living plant cells are quite freely permeable to water and solutes, the cytoplasmic layer that lines the cell wall is more permeable to some substances than to others.

If a plant cell in a flaccid condition—one in which the cell sap exerts no pressure against the encompassing cytoplasm and cell wall—is immersed in pure water, inward osmosis of water into the cell sap ensues. This gain of water results in the exertion of a turgor pressure against the protoplasm, which in turn is transmitted to the cell wall. This pressure also prevails throughout the mass of solution within the cell. If the cell wall is elastic, some expansion in the volume of the cell occurs as a result of this pressure, although in many kinds of cells this is relatively small.

If a turgid or partially turgid plant cell is immersed in a solution with a greater osmotic pressure than the cell sap, a gradual shrinkage in the volume of the cell ensues; the amount of shrinkage depends upon the kind of cell and its initial degree of turgidity. When the lower limit of cell wall elasticity is reached and there is continued loss of water from the cell sap, the protoplasmic layer begins to recede from the inner surface of the cell wall. Retreat of the protoplasm from the cell wall often continues until it has shrunk toward the center of the cell, the space between the protoplasm and the cell wall becoming occupied by the bathing solution. This phenomenon is called plasmolysis. *See* OSMOREGULATORY MECHANISMS.

In some kinds of plant cells movement of water occurs principally by the process of imbibition rather than osmosis. The swelling of dry seeds when immersed in water is a familiar example of this process.

Stomatal mechanism. Various gases diffuse into and out of physiologically active plants. Those gases of greatest physiological significance are carbon dioxide, oxygen, and water vapor. The great bulk of the gaseous exchanges between a plant and its environment occurs through tiny pores in the epidermis that are called stomates. Although stomates occur on many aerial parts of plants, they are most characteristic of, and occur in greatest abundance in, leaves. *See* EPIDERMIS (PLANT); LEAF.

Transpiration process. The term transpiration is used to designate the process whereby water vapor is lost from plants. Although basically an evaporation process, transpiration is complicated by other physical and physiological conditions prevailing in the plant. Whereas loss of water vapor can occur from any part of the plant which is exposed to the atmosphere, the great bulk of all transpiration occurs from the leaves. There are two kinds of foliar transpiration: (1) stomatal transpiration, in which water vapor loss occurs through the stomates, and (2) cuticular transpiration, which occurs directly from the outside surface of epidermal walls through the cuticle. In most species 90% or more of all foliar transpiration is of the stomatal type.

Transpiration is a necessary consequence of the relation of water to the anatomy of the plant, and especially to the anatomy of the leaves. Terrestrial green plants are dependent upon atmospheric carbon dioxide for their survival. In terrestrial vascular plants the principal carbon dioxide–absorbing surfaces are the moist mesophyll cells walls which bound the intercellular spaces in leaves. Ingress of carbon dioxide into these spaces occurs mostly by diffusion through open stomates. When the stomates are open, outward diffusion of water vapor unavoidably occurs, and such stomatal transpiration accounts for most of the water vapor loss from plants. Although transpiration is thus, in effect, an incidental phenomenon, it frequently has marked indirect effects on other physiological processes which occur in the plant because of its effects on the internal water relations of the plant.

Water translocation. In terrestrial rooted plants practically all of the water which enters a plant is absorbed from the soil by the roots. The water thus absorbed is translocated to all parts of the plant. The mechanism of the "ascent of sap" (all translocated water contains at least traces of solutes) in plants, especially tall trees, was one of the first processes to excite the interest of plant physiologists.

The upward movement of water in plants occurs in the xylem, which, in the larger roots, trunks, and branches of trees and shrubs, is identical with the wood. In the trunks or larger branches of most kinds of trees, however, sap movement is restricted to a few of the outermost annual layers of wood. *See* XYLEM.

Root pressure is generally considered to be one of the mechanisms of upward transport of water in plants. While it is undoubtedly true that root pressure does account for some upward movement of water in certain species of plants at some seasons, various considerations indicate that it can be only a secondary mechanism of water transport.

Upward translocation of water (actually a very dilute sap) is engendered by an increase in the negativity of water potential in the cells of apical organs of plants. Such increases in the negativity of water potentials occur most commonly in the mesophyll cells of leaves as a result of transpiration.

Water absorption. The successively smaller branches of the root system of any plant terminate ultimately in the root tips, of which there may be thousands and often millions on a single plant. Most absorption of water occurs in the root tip regions, and especially in the root hair zone. Older portions of most roots become covered with cutinized or suberized layers through which only very limited quantities of water can pass. *See* ROOT (BOTANY).

Whenever the water potential in the peripheral root cells is less than that of the soil water, movement of water from the soil into the root cells occurs. There is some evidence that, under conditions of marked internal water stress, the tension generated in the xylem ducts will be propagated across the root to the peripheral cells. If this occurs, water potentials of greater negativity could develop in peripheral root cells than would otherwise be possible. The absorption mechanism would operate in fundamentally the same way whether or not the water in the root cells passed into a state of tension. The process just described, often called passive absorption, accounts for most of the absorption of water by terrestrial plants.

The phenomenon of root pressure represents another mechanism of the absorption of water. This mechanism is localized in the roots and is often called active absorption. Water absorption of this type only occurs when the rate of transpiration is low and the soil is relatively moist. Although the xylem sap is a relatively dilute solution, its osmotic pressure is usually great enough to engender a more negative water potential than usually exists in the soil water when the soil is relatively moist. A gradient of water potentials can thus be established, increasing in negativity across the epidermis, cortex, and other root tissues, along which the water can move laterally from the soil to the xylem. *See* PLANT MINERAL NUTRITION.

Bernard S. Meyer

Plantaginales An order of flowering plants, division Magnoliophyta (Angiospermae), in the subclass Asteridae of the class Magnoliopsida (dicotyledons). The order consists of only the family Plantaginaceae, with about 250 species. Within its subclass the order is marked by its small, chiefly wind-pollinated flowers that have a persistent regular corolla. The perianth and stamens of the flowers are attached directly to the receptacle (hypogynous) and there are typically four petals. The plants are herbs or seldom half-shrubs with mostly basal, alternate leaves. The common plantain (*Plantago major*) is a lawn weed of this order. *See* ASTERIDAE; MAGNOLIOPSIDA.

Arthur Cronquist; T. M. Barkley

Plants, life forms of A term for the vegetative (morphological) form of the plant body. Life-form systems are based on differences in gross morphological features, and the categories bear no necessary relationship to reproductive structures, which form the basis for taxonomic classification. Features used in establishing life-form classes include deciduous versus evergreen leaves, broad versus needle leaves, size of leaves, degree of protection afforded the perennating tissue, succulence, and duration of life cycle (annual, biennial, or perennial).

There is a clear correlation between life forms and climates. For example, broad-leaved evergreen trees clearly dominate in the hot humid tropics, whereas broad-leaved deciduous trees prevail in temperature climates with cold winters and warm summers, and succulent cacti dominate American deserts. Although cacti are virtually absent from African deserts, members of the family Euphorbiaceae have evolved similar succulent life forms. Such adaptations are genetic, having arisen by natural selection.

Many life-form systems have been developed. The most successful and widely used system is that of C. Raunkiaer, proposed in 1905. Reasoning that it was the perennating buds (the tips of shoots which renew growth after a dormant season, either of cold or drought) which permit a plant to survive in a specific climate, Raunkiaer's classes were based on the degree of protection afforded the bud and the position of the bud relative to the soil surface. They applied to autotrophic, vascular, self-supporting plants. Raunkiaer's classificatory system is:

Phanerophytes: bud-bearing shoots in the air, predominantly woody trees and shrubs; subclasses based on height and on presence or absence of bud scales

Chamaephytes: bud within 10 in. (25 cm) of the surface, mostly prostrate or creeping shrubs

Hemicryptophytes: buds at the soil surface, protected by scales, snow, and litter

Cryptophytes: buds underneath the soil surface or under water

Therophytes: annuals, the seed representing the only perennating tissue

By determining the life forms of a sample of 1000 species from the world's floras, Raunkiaer showed a correlation between the percentage of species in each life-form class present in an area and the climate of the area. Raunkiaer concluded that there were four main phytoclimates: phanerophyte-dominated flora of the hot humid tropics, hemicryptophyte-dominated flora in moist to humid temperate areas, therophyte-dominated flora in arid areas, and a chamaephyte-dominated flora of high latitudes and altitudes.

Subsequent studies modified Raunkiaer's views. (1) Phanerophytes dominate, to the virtual exclusion of other life forms, in true tropical rainforest floras, whereas other life forms become proportionately more important in tropical climates with a dry season. (2) Therophytes are most abundant in arid climates and are prominent in temperate areas with an extended dry season, such as regions with Mediterranean climate. (3) Other temperate floras have a predominance of hemicryptophytes with the percentage of phanerophytes decreasing from summergreen deciduous forest to grassland. (4) Arctic and alpine tundra are characterized by a flora which is often more than three-quarters chamaephytes and hemicryptophytes, the percentage of chamaephytes increasing with latitude and altitude. *See* PLANT GEOGRAPHY.

There has been interest in developing systems which describe important morphologic features of plants and which permit mapping and diagramming vegetation. Descriptive systems incorporate essential structural features of plants, such as stem architecture and height; deciduousness; leaf texture, shape, and size; and mechanisms for dispersal. These systems are important in mapping vegetation because structural features generally provide the best criteria for recognition of major vegetation units. *See* ALTITUDINAL VEGETATION ZONES; VEGETATION AND ECOSYSTEM MAPPING.

Arthur W. Cooper

Plants of saline environments Plants that have evolved to flourish where the salt concentrations (notably those of sodium chloride) are 400 mol m^{-3} and beyond. These plants are known as halophytes. The majority of plant species are unable to tolerate one-quarter of this salt concentration and are called glycophytes. Apart from deserts and oceans, the other most extensive naturally saline areas are coastal marshes which are periodically inundated by seawater; this process leads to mangrove swamps in tropical climates. *See* BOG; SALT MARSH; SWAMP, MARSH, AND BOG.

There are many plant species adapted to saline environments. All those that normally grow in seawater are clearly halophytes. The flora of the oceans is dominated by algae, with few flowering plants adapted to this environment. As far as the angiosperms are concerned, there are no marine dicotyledonous plants and only one order of monocotyledons, the Helobiae, with some 12 genera and about 49 species. Most of these sea grasses grow in warmer waters, and since they cannot withstand long exposure to air, they occur at or about the level of the lowest tides. *See* ALGAE.

The distinction between terrestrial and aquatic plants is largely a matter of the degree to which the plant is predominantly in air or in water. For example, salt marsh plants are terrestrial but must withstand submergence, while algae of the upper shore are mostly submerged but must withstand periodic exposure to the atmosphere. The terrestrial halophytes are nearly all flowering plants, and there are few, if any, salt-tolerant mosses, liverworts, or even gymnosperms. Among the ferns, there are just two families with salt-tolerant genera, the Pteridaceae and the Ophioglossaceae. Among the flowering plants, salt tolerance occurs in many different families and appears to have been polyphyletic in origin. Salt-tolerant species are to be found in about one-third of the families of flowering plants. However, about half of the approximately 500 genera known to contain halophytic species belong to just 20 families. Among the monocotyledonous plants, grasses dominate as halophytes with 109 species, followed by salt-tolerant sedges with 83 species. Within the dicotyledonous plants, the Chenopodiaceae is the preeminent family for salt tolerance with 44 of its 100 genera containing salt-tolerant species (there are 312 such species); examples are *Atriplex*, *Salicornia*, and *Suaeda*. Other dicotyledonous families with significant numbers of halophytes are the Asteraceae (53 species) and the Aizoaceae (48 species).

Tropical salt marshes, or mangrove swamps, differ from those of temperate zones in being dominated by trees and shrubs rather than by herbs. There are at least 12 genera from eight different families, and besides their salt tolerance, these trees are remarkable for their root systems, which are adapted to provide stability in soft mud and facilitate gas exchange in an anaerobic substrate. Apart from coastal areas, saline environments are also frequent inland, especially in the hotter and drier regions of the world. In such environments, plants must be adapted not only to tolerate the high concentrations of salts in the soil but also to be drought-resistant. Many of the adaptations which have evolved to combat the former stress are advantageous in surviving the latter (and vice versa). *See* MANGROVE.

Many terrestrial halophytes, having small leaves which are often succulent, look somewhat different from glycophytes. In many species the leaves bear glands which secrete salt. These visible differences are the external manifestations of more fundamental differences in physiology. *See* PLANT-WATER RELATIONS.

Roberto A. Gaxiola; T. J. Flowers

Plasma (physics) The field of physics that studies highly ionized gases. Plasma is a gas of charged and neutral particles which exhibits collective behavior. All gases become ionized at sufficiently high temperatures, creating what has been called a fourth state of matter, together with solids, liquids, and gases. It has been estimated that more than 99% of the universe is in the plasma state. On the Earth, plasmas are much less common. Lightning is a familiar natural manifestation, and fluorescent lights are a practical application. Plasma applications and studies make use of an enormous range of plasma temperatures, densities, and neutral pressures. They extend from plasma processing applications at relatively low temperatures (such as plasma etching of semiconductor chips at low pressure, or plasma cutting torches at atmospheric pressure) to studies of controlled fusion at very high temperatures. *See* FLUORESCENT LAMP.

Plasma physics is a many-body problem that can be described by a combination of Newton's laws and Maxwell's equations. The charged particles in plasmas are usually ions, both positive and negative, and electrons. Plasmas are normally quasineutral; that is, the net positive ion charge density approximately equals the net negative charge density everywhere in the bulk of the plasma. Quasineutrality refers to charge density and does not imply equal densities of electrons and ions since ions can be multiply charged and can also have negative charge. In space and fusion plasmas, plasmas are normally magnetized, while in application plasmas on Earth, such as plasma processing, both magnetized and unmagnetized plasmas are employed. *See* MAXWELL'S EQUATIONS; NEWTON'S LAWS OF MOTION.

It is convenient to keep track of plasma properties in terms of characteristic lengths, frequencies, and velocities. Among these are the Debye length, the electron and ion plasma frequencies, the electron and ion gyrofrequencies and gyroradii, the electron and ion thermal velocities, the ion sound velocity, the Alfvén velocity, and various collision lengths. The definition of a plasma depends on several of these characteristic parameters, and the magnitude of ratios of these parameters to system size or applied frequencies determines most plasma behavior.

The simplest plasma is a collisionless, unmagnetized collection of ions and electrons with no significant currents. Such plasmas have quasineutral regions and nonneutral regions. The nonneutral regions are highly localized. They are usually located near boundaries (where they are known as sheaths), but are sometimes located within the plasma (where they are known as double layers).

Collective behavior refers to the plasma properties not present in single-particle motion. Collective behavior is a distinguishing characteristic of a plasma. It consists of flows, waves, instabilities, and so forth. Common examples are fluctuations in the aurora, generation of microwaves in devices such as magnetrons and klystrons, and reflection of electromagnetic waves from the ionosphere. *See* AURORA; KLYSTRON; MAGNETRON; RADIO-WAVE PROPAGATION.

Curiously, very high density collections of equal numbers of ions and electrons are not plasmas. Such systems are referred to as strongly coupled plasmas (even though, strictly speaking, they are not plasmas at all).

A collection of either electrons or ions can exhibit properties similar to those of an electrically neutral plasma if the charged-particle density is sufficiently large. For such so-called plasmas, the Debye length and the characteristic frequency of electrons or ions can still be defined, and collective behavior is still exhibited when the Debye length is less than the system's characteristic dimension. So-called pure electron plasmas or pure ion plasmas are unconfined in an unmagnetized system. However, particle traps consisting of a combination of electric and magnetic fields can be used to confine the charges. *See* PARTICLE TRAP.

The visual appearance of a plasma depends on the kind of ion present, the electron temperature, and the plasma density. Some plasmas are invisible. Curiously, if a plasma is present and not glowing, it is either very hot or very cold. For example, an H^+ plasma, or any other relatively hot plasma with fully stripped ions, contains atomic nuclei with no electrons, so there is no atomic physics and no optical emission or absorption. If plasma electrons and ions are very cold, there is insufficient energy to excite optical transitions. The glow often associated with plasmas indicates only where visible energy transitions are excited by energetic electrons or perhaps absorption of ultraviolet radiation, and may have little to do with the presence of bulk plasma. In fusion plasmas, the edges are often copious sources of emission associated with the dissociation and ionization of hydrogen and edge-generated impurities, while much of the hotter core plasma is fully ionized and invisible.

Direct-current glow-discharge plasmas originate from electrons created by secondary electron emission due to ion bombardment of a negatively biased cathode. The secondary electrons are accelerated through the cathode sheath potential (called the cathode fall) to energies the order of 1 keV, and partially ionize the neutral gas, releasing additional energetic electrons in a multiplicative process. The energetic electrons also undergo inelastic collisions with neutrals which result in optical emission that contributes to the so-called glow. *See* GLOW DISCHARGE; SECONDARY EMISSION.

The understanding of plasma physics begins with an understanding of the motion of single charged particles in a combination of electric and magnetic fields (E and B), produced by a combination of external fields and the motion of the charged particles themselves. The motion of a single particle, with mass m, charge q, and velocity $\mathbf{v}$, is governed by the Lorentz force, as given in Eq. (1). From the perpendicular component of Eq. (1),

$$m\frac{d\mathbf{v}}{dt} = q(\mathbf{E} + \mathbf{v} \times \mathbf{B}) \qquad (1)$$

it can be shown that the charged particles gyrate about magnetic field lines with a characteristic frequency (the cyclotron frequency). Ions rotate about the magnetic field in the clockwise direction, while electrons rotate counterclockwise with the magnetic field pointing outward. *See* ELECTRIC FIELD; PARTICLE ACCELERATOR.

In addition to the motion parallel to the magnetic field and the gyromotion about the magnetic field, there are drifts perpendicular to the magnetic field. For a general force, $\mathbf{F}$, in the presence of a magnetic field, the perpendicular drift velocity is given by Eq. (2).

$$\mathbf{v}_D = \frac{\mathbf{F} \times \mathbf{B}}{qB^2} \qquad (2)$$

Given a perpendicular electric field, particles can walk across a magnetic field. Forces associated with magnetic-field curvature give rise to a curvature drift in the direction orthogonal to the magnetic field, and to the radius of curvature of the magnetic field lines.

For gyro motion in a slowly changing magnetic field, which is approximately periodic, it can be shown that the ratio of the perpendicular energy to the magnetic field is approximately constant. This means that a charged particle moving parallel to a magnetic field and gyrating about the field will gyrate faster as the magnetic field increases. If the magnetic field changes in space and is constant in time, the total energy is conserved. For a sufficiently large magnetic field, a point is reached where the total energy equals the perpendicular energy, so that the parallel energy goes to zero and the particle reflects. This is known as magnetic mirroring.

Magnetic mirroring is the chief natural mechanism of charged-particle confinement. For example, this process confines charged particles in the ionosphere and magnetosphere. The magnetic field lines that connect the north and south magnetic poles of the Earth provide a mirror magnetic field which increases as either pole is approached. In the absence of collisions, a particle moving along and gyrating about such a magnetic field is magnetically confined, if it has a sufficiently large velocity perpendicular to the magnetic field. The Van Allen belts are composed of such mirror-trapped charged particles. The source of these particles is the solar wind, a stream of charged particles continuously emitted by the Sun.

For fully ionized plasmas, it is convenient to describe the plasma as a single fluid together with Maxwell's equations. This gives the magnetohydrodynamic (MHD) equations, which are used to describe plasma equilibria and plasma waves and instabilities. Their relative simplicity has made them ideal for solutions of fusion problems in complicated geometries, and they have been widely used to describe astrophysical plasmas and magnetohydrodynamic energy conversion. *See* MAGNETOHYDRODYNAMIC POWER GENERATOR; MAGNETOHYDRODYNAMICS.

Plasmas can support an impressive variety of electrostatic and electromagnetic waves not present in the absence of plasma. The waves are distinguished by their frequency, the presence or absence of dc magnetic fields, and the plasma temperature and density.

Ionization is the key to plasma production and can be accomplished in many different ways. The most common approach is to employ energetic electrons with energies greater than the ionization potential of the gas being ionized. In dc glow discharges, electrons produced by ion secondary electron emission are accelerated by the cathode sheath potential, as are electrons created by thermionic emission in hot-cathode plasmas. Electrons can also pick up energy by reflecting from oscillating radio-frequency sheath electric fields, or by cyclotron resonance in magnetic fields, or from collisions with other energetic electrons. *See* ELECTRICAL CONDUCTION IN GASES; GAS DISCHARGE; IONIZATION POTENTIAL; THERMIONIC EMISSION.

Several other approaches involving collisions, which do not require energetic electrons, also exist. These techniques include photoionization, ion-neutral charge exchange, surface ionization, and Penning ionization. Ions can also be produced in the dissociation of molecules. Yet another mechanism, called critical ionization velocity, is instability driven, and occurs when the kinetic energy of the neutral gas atoms streaming perpendicular to a magnetic field exceeds their ionization potential. *See* ION SOURCES; IONIZATION; PHOTOIONIZATION.

A vacuum chamber provides the simplest approach to confinement. In an unmagnetized plasma, electrons are lost more rapidly than ions, and the plasma acquires a net positive charge. The excess positive charge appears in a sheath at the plasma boundary with the bulk plasma potential more positive than the boundary potential. The decrease in potential at the boundary provides plasma electron confinement, reducing their loss rate to balance the ion loss rate.

Addition of a uniform magnetic field reduces the loss rate of ions and electrons transverse to the magnetic field, but has no effect on losses parallel to the magnetic field because the Lorentz force has no components along this field. Effective confinement by magnetic fields requires that the ion and electron gyroradii be small compared to device dimensions. Plasma transport across the magnetic field can still occur as a result of collisions or of perpendicular drifts.

In the absence of magnetic fields (both inside and outside the plasma), an equilibrium can be achieved by establishing a pressure balance between plasma and edge walls or edge gas. The existence of an equilibrium does not guarantee that a particular configuration is stable.

Plasma processing can be defined as the collection of techniques which make use of plasmas to create new materials or to modify properties of existing materials. It is used in a large variety of applications including semiconductor etching, preparing plastic surfaces to accept ink, depositing polymers, depositing diamond films, and hardening artificial hip joints. The technique has its foundations in plasma physics, chemistry, electrical and chemical engineering, and materials science.

Controlled fusion aims at taking advantage of nuclear fusion reactions to generate net power. Advances in fusion studies have been tied to the techniques developed for plasma confinement and heating. Fusion experiments employ either magnetic confinement or inertial confinement, in which fusion reactions take place before the plasma has a chance to expand to chamber boundaries. Magnetic mirrors are an example of open systems, while tokamaks, stellarators, and reversed-field pinches are examples of closed toroidal systems. Most magnetic confinement research experiments are done on tokamaks. *See* NUCLEAR FUSION.

Noah Hershkowitz

Naturally occurring plasmas exist throughout the solar system and beyond. Above the atmosphere, most matter is ionized. The lower-density ionized materials are considered to be plasmas, and they behave in manners very different from the behavior of nonplasmas. Some dense materials, such as stellar matter or electrolytic solutions, are often not considered to be plasmas even though they are ionized; they behave, for the most part, as do ordinary fluids.

Some of the major plasma-physics issues that are under study with naturally occurring plasmas are the energization of charged particles, the reconnection of magnetic fields (temporal changes in magnetic-field topology), the production of magnetic fields by dynamos, the production of electromagnetic waves, the interaction between waves and particles, and the transport of mass, momentum, and energy across magnetic fields.

Naturally occurring plasmas are in general difficult to measure. The solar-wind, ionospheric, and magnetospheric plasmas are diagnosed by single-point measurements by rockets and satellites; the solar atmosphere and all astrophysical plasmas are unreachable and must be diagnosed by the light and radio waves that they emit; and lightning is unpredictable and inhospitable to instruments and must be diagnosed primarily by the light that it emits. As a consequence of limited diagnostics, theoretical analysis and laboratory-plasma experiments play supporting roles in the investigations of naturally occurring plasmas.

Joseph E. Borovsky

Plasma diagnostics Techniques to measure the properties and parameters of ionized gases. Plasmas are gases in which a sufficient proportion of the atoms are ionized that the resulting mixture of free electrons and positively charged ions exhibits collective behavior resulting from electromagnetic interactions. *See* PLASMA (PHYSICS).

The simplest properties of a plasma that one might wish to measure are the density (n) of particles per unit volume, the fluid velocity ($\langle \mathbf{v} \rangle$), and the temperature ($T$). These gaseous parameters may be different for the different particle species in the plasma. There are normally at least two such species, electrons (subscript e) and ions (subscript i); but often multiple ion species and neutral particles are present. (The brackets around $\mathbf{v}$ indicate that it is the average value of the particle velocities of all the particles, or of the particles of a particular species.) Because plasma temperatures are generally high, from a few thousand degrees Celsius upward, they are usually expressed in energy units such as electronvolts ($1 \text{ eV} = 1.6 \times 10^{-19} \text{ J} \simeq 11{,}600^\circ\text{C}$).

Plasmas often possess nonthermal particle distributions. That is, the velocity distribution function of electrons or ions, f_s ($s = e, i$), which measures the numbers of electrons or ions that have various velocities, has not simply the Maxwell-Boltzmann form corresponding to thermal equilibrium at some temperature but has some more complicated form $f_s(\mathbf{v}_s)$. In that case an ideal that may sometimes be approached is to measure the entire velocity distribution $f_s(\mathbf{v}_s)$ and not just the parameters n, $\langle \mathbf{v} \rangle$, and T, which are moments of this distribution. *See* BOLTZMANN STATISTICS; KINETIC THEORY OF MATTER.

The electromagnetic fields $\mathbf{E}$ and $\mathbf{B}$ are also essential parameters that affect the plasma. They are affected in turn by the charge and current densities, ρ and $\mathbf{j}$, which are simple sums of the densities and velocities of the plasma species weighted by their charge per particle.

Techniques to measure the plasma parameters, and their space- and time-variation, are based on a wide variety of physical principles. Direct invasion of the plasma using solid probes is possible only for relatively cold and tenuous plasmas; otherwise the probe will be damaged by excessive heat flux. However, powerful noncontact diagnostics are based on measurements external to the plasma of electromagnetic fields, waves propagated through or scattered by the plasma, photons emitted by plasma electrons or atoms, and particle emission, transmission, or scattering.

By far the majority of probe measurements are based on measuring the electric current carried to the probe by the plasma particles. The simplest and most direct such measurement, called the Langmuir probe, consists of an electrode of some well-defined geometry (plane, cylinder, and sphere are all used). The potential of this probe is varied and the resulting probe current is measured.

The Langmuir probe cannot provide an independent measurement of the ion temperature because the ions are usually the attracted species and their current is almost independent of probe potential. More elaborate plasma flux measurements can be performed with devices called gridded energy analyzers, which can obtain energy distribution functions of both electrons and ions at the expense of far greater complexity, greater probe size and hence plasma perturbation, and often loss of absolute density calibration.

Langmuir probes can provide measurement of the electrostatic potential in the plasma, and hence the electric field, although not quite directly. The magnetic field can be measured by sensors based either on simple induction in a coil for pulsed fields, or typically on the Hall effect in a semiconductor for static fields. The former approach dominates fusion plasma research because of its linearity, noise insensitivity, and robustness, while the latter dominates space measurements in which sensitivity and dc response are essential. *See* ELECTRIC FIELD; HALL EFFECT; MAGNETIC INSTRUMENTS; MAGNETISM.

Propagating an electromagnetic wave through the plasma and measuring the plasma's effect on it is a powerful nonperturbing diagnostic technique. The plasma's presence modifies the refractive index. The index measurement is performed by measuring the phase difference caused by the plasma along an optical path, using some form of interferometer; the Michelson or Mach-Zehnder configurations are most popular. The phase shift is approximately proportional to the average plasma density along the optical path. *See* INTERFEROMETRY.

Electromagnetic wave emission from the free electrons in the plasma occurs predominantly from acceleration either by the ambient magnetic field (cyclotron or synchrotron emission) or by collisions with other plasma particles (bremsstrahlung). Bremsstrahlung from optically thin (hot and tenuous) plasmas has a photon energy spectrum that resembles the electron energy distribution; so, for example, its logarithmic slope gives the electron temperature. The spectrum for kiloelectronvolt-temperature plasmas is measured by x-ray techniques such as pulse-height analysis or approximately using foil filters. Large, dense, or cold plasmas are often optically thick to bremsstrahlung, in which case their spectrum is blackbody and provides electron temperature. Alternatively, by backlighting the plasma (for example with x-rays), their bremsstrahlung (collisional) absorption provides a measure of line-averaged density. This last technique is particularly important for laser-produced plasmas and inertial fusion research. *See* BREMSSTRAHLUNG.

Line radiation arises from electronic transitions between bound levels of atomic species in the plasma. The characteristic wavelengths serve to identify the species, and the intensities are proportional to the excited-state densities. Therefore line radiation is the primary method for determining the atomic and molecular composition of plasmas. Excitation of the atom from the ground state usually occurs by electron impact, requiring an electron energy greater than the excitation energy. Therefore the ratio of the intensities of lines from different excited states whose excitation energies differ by an amount comparable to the electron temperature is a measure of electron temperature. The rate of excitation is proportional to electron density; however, in order to eliminate from the measured intensity its dependence on the atomic density, a line ratio is generally chosen for measurement. If the population of one excited level of the line pair chosen is substantially affected by collisional deexcitation, because it is a metastable level for example, or is populated by recombination, then the ratio is sensitive to electron density. *See* ATOMIC STRUCTURE AND SPECTRA; SCATTERING EXPERIMENTS (ATOMS AND MOLECULES).

Neutral atoms are continually formed from plasma ions by charge exchange with other neutrals or recombination with electrons. The neutrals formed may escape promptly from the plasma, providing a means for detection of the ion distribution using a neutral-particle spectrometer. In fusion research plasmas, fusion reactions give rise to megaelectronvolt-energy neutrons and charged products. These too provide information on the ion temperature and density via the reaction rate and the emitted spectrum.

Energetic neutral or ion beams, generated specifically for the purpose, can be used for active plasma probing. Their attenuation is generally proportional to plasma density; but more important special-purpose applications include the internal measurement of potential (by local beam ionization in the plasma) and the provision of localized atomic species for other spectroscopic diagnostics.

I. H. Hutchinson

Plasma propulsion The imparting of thrust to a spacecraft through the acceleration of a plasma (ionized gas). A plasma can be accelerated by electrical means to exhaust velocities considerably higher than those attained by chemical rockets. The higher exhaust velocities (specific impulses) of plasma thrusters usually imply that, for a particular mission, the spacecraft would use less propellant than the amount required by conventional chemical rockets. This means that for the same amount of propellant a spacecraft propelled by a plasma rocket can increase in velocity over a set distance by an increment larger than that possible with a chemical propulsion system. An increasing number of near-Earth satellites have benefited from the propellant savings of plasma thrusters and rely on them for attitude control and station-keeping maneuvers. Generally, the larger the required velocity increment, the more beneficial is plasma propulsion compared to propulsion with rockets having lower exhaust velocities. Consequently, many ambitious missions—such as deep-space exploration including Moon tours and orbiters to outer planets, sample return missions, and heavy cargo and piloted missions—may be enabled by plasma propulsion. *See* PLASMA (PHYSICS); SPECIFIC IMPULSE.

Plasma propulsion is one of three major classes of electric propulsion, the others being electrothermal propulsion and ion (or electrostatic) propulsion. In electrothermal devices the working fluid is a hot nonionized or very weakly ionized gas. *See* ELECTROTHERMAL PROPULSION; ION PROPULSION; SPACECRAFT PROPULSION.

The most successful plasma thruster to date is the Hall thruster. It has been gaining acceptance in the West since the early 1990s, after three decades of improvement in the former Soviet Union. More than 200 Hall thrusters have been flown so far on about 30 spacecraft, mostly by the former Soviet Union and Russia. In a Hall thruster, a mostly radial magnetic field is applied inside the acceleration chamber. A hollow cathode, placed outside the chamber, is kept at a few hundred volts below the potential of an annular anode located upstream in the device, thus setting up a mostly axial electric field inside the chamber. As the cathode is heated, it emits electrons, which drift in the opposite direction of the electric field and enter the device, where they are subjected to the radial magnetic field and the axial electric field. This leads to an azimuthal (that is, about the device's axis of symmetry) drift for the electrons, which constitutes an electric current called the Hall current. The drifting electrons collide with propellant gas molecules (typically xenon) injected in the device through the anode. These impacts produce a plasma that finds itself acted on by an electromagnetic force, called the Lorentz force, which results from the interaction of the radial magnetic field and the azimuthal Hall current that are perpendicular to each other. The plasma is thus accelerated and exhausted out of the thruster at high exhaust velocities (from 10 km/s or 6 mi/s to above 50 km/s or 30 mi/s depending on the available power). The acceleration mechanism can also be described, equivalently, to be the result of the action of the applied electric field on the newly born ions (whose mass is too large to allow them to be magnetized) with the electrons acting as a neutralizing cloud whose momentum is coupled to the ions via interparticle electrostatic fields. Since the entire plasma is accelerated, and not only the positive ions, there is no space-charge buildup as in the ion thruster, and the device can have a thrust density many times that of the ion engine. *See* HALL EFFECT; MAGNETISM; MAGNETOHYDRODYNAMICS.

One of the most promising and thoroughly studied electromagnetic plasma accelerators is the magnetoplasmadynamic thruster. In this device the plasma is both created and accelerated by a high-current discharge between two coaxial electrodes. The discharge is due to the breakdown of the gas as it is injected in the interelectrode region. The acceleration process can be described as the result of the action of a body force acting on the plasma. This body force is the Lorentz force created by the interaction between the mostly radial current conducted through the plasma and the magnetic field. The latter can be either externally applied by a magnet or self-induced by the discharge, if the current is sufficiently high. The reliance on the radial component of the current (which is aligned with the electric field) instead of the Hall component (perpendicular to both electric and magnetic fields) allows operation at much higher plasma densities (and thrust densities) than with the Hall thruster.

Pulsed plasma thrusters rely on pulsing a high-current gas discharge on a time scale of a few microseconds. This is done by quickly discharging the energy stored in a capacitor into a gas that is either injected (gas-fed PPT) or vaporized from a solid propellant (ablative PPT), typically polytetrafluoroethylene. The quick rise in current through the gas forces the formation of a plasma sheet through which the current flows. The current sheet is subjected to the Lorentz force produced by the interaction of the current it carries and the induced magnetic field. This causes the current sheet to accelerate downstream, ionizing and entraining ambient gas as in a snowplow. The exhausted plasma and gas give the propulsive impulse. The pulsed operation of such electric propulsion devices allows power-limited spacecraft to take advantage of high-current plasma acceleration at the low average power available on current spacecraft (hundreds of watts). *See* SPACE POWER SYSTEMS.

Edgar Y. Choueiri

Plasmid A circular extrachromosomal genetic element that is ubiquitous in prokaryotes and has also been identified in a number of eukaryotes. In general, bacterial plasmids can be classified into two groups on the basis of the number of genes and functions they carry. The larger plasmids are deoxyribonucleic acid (DNA) molecules of around 100 kilobase (kb) pairs, which is sufficient to code for approximately 100 genes. There is usually a small number of copies of these plasmids per host chromosome, so that their replication must be precisely coordinated with the cell division cycle. The plasmids in the second group are smaller in size, about 6–10 kb. These plasmids may harbor 6–10 genes and are usually present in multiple copies (10–20 per chromosome). *See* GENE.

Plasmids have been identified in a large number of bacterial genera. Some bacterial species harbor plasmids with no known functions (cryptic plasmids) which have been identified as small circular molecules present in the bacterial DNA. The host range of a particular plasmid is usually limited to closely related genera. Some plasmids, however, are much more promiscuous and have a much broader host range.

The functions specified by different bacterial plasmids are usually quite specialized in nature. Moreover, they are not essential for cell growth since the host bacteria are viable without a plasmid when the cells are cultured under conditions that do not select for plasmid-specified gene products. Plasmids thus introduce specialized functions to host cells which provide versatility and adaptability for growth and survival. Plasmids which confer antibiotic resistance (R plasmids) have been extensively characterized because of their medical importance. Plasmids have played a seminal role in the spectacular advances in the area of genetic engineering. Individual genes can be inserted into specific sites on plasmids in cell cultures and the recombinant plasmid thus formed introduced into a living cell by the process of bacterial transformation. *See* GENETIC ENGINEERING.

Robert H. Rownd

Plasmin A proteolytic enzyme that can digest many proteins through the process of hydrolysis. Plasmin is found in plasma in the form of plasminogen, which is an inert precursor or zymogen; one site of synthesis is the liver. Plasma also contains inhibitors that limit either the transformation of plasminogen to plasmin or the action of plasmin.

Plasma itself has the potentiality of activating plasminogen, because it contains intrinsic activators. A tissue plasminogen activator is produced in blood vessel walls, from which it is released into the bloodstream following vascular injury. Further, the process of blood coagulation may foster the activation of plasminogen. The presence of a fibrin clot enhances activation of plasminogen by many agents.

Plasmin can act on many protein substrates. It liquefies coagulated blood by digesting fibrin, the insoluble meshwork of the clot; it also digests fibrinogen, the precursor of fibrin, rendering it incoagulable. The digestion products of fibrinogen and fibrin are anticoagulant substances that further interfere with the clotting process. Plasmin also inactivates several other protein procoagulant factors, particularly proaccelerin (factor V) and antihemophilic factor (factor VIII). Other substrates attacked by plasmin include gamma globulin and several protein or polypeptide hormones found in plasma. *See* BLOOD; ENZYME; FIBRINOGEN.

The action of plasmin may be related to certain body defenses against injury. Its capacity to digest fibrin is thought to bring about liquefaction of intravascular clots (thrombi) that might otherwise cause damage by reducing blood flow to tissues. It can convert the first component of complement, a group of plasma proteins important in immune reactions, to the proteolytic enzyme C1 esterase. It can also liberate polypeptide kinins (for example, bradykinin) from plasma precursors. The kinins can reproduce such elements of the inflammatory process as pain, dilatation and increased permeability of small blood vessels, and the migration of leukocytes. In addition to these physiologic actions, plasmin hydrolyzes casein, gelatin, denatured hemoglobin, and certain synthetic esters of arginine and lysine. *See* COMPLEMENT; IMMUNOLOGY; INFLAMMATION.

Oscar D. Ratnoff

Plasmodiophorida Protozoa composing an order of Mycetozoia. They are endoparasitic in plants, primarily causing, for example, club root of cabbage, powdery scab of potatoes, and galls in *Ruppia*. Underground portions of host plants are invaded by young parasites, often a flagellate which sometimes arises from a freshly excysted ameba. Becoming intracellular, the young parasite grows and develops into a plasmodium. At maturity, the plasmodium produces uninucleate cysts (spores) which are released upon degeneration of the damaged cell. Under favorable conditions, the released spores hatch into uninucleate stages which become the infective forms. In the invaded area, the host's tissue commonly undergoes hypertrophy to form a gall.

Richard P. Hall

Plasmodiophoromycota Obligate endoparasites (parasites restricted to living inside a host) of flowering plants, green and yellow-green algae, seaweeds, diatoms, water molds, and ciliates. The Plasmodiophoromycota (plasmodiophorids) are prevalent in both soil and aquatic habitats, where their biflagellate zoospores (independently motile spores) disperse the pathogen and infect hosts. A plasmodium develops within the host cell, often causing cell enlargement and proliferation, resulting in hypertrophy (abnormal enlargement) of host tissue. These organisms include devastating pathogens of food crops, such as *Plasmodiophora brassicae*, which causes clubroot (a plant disease involving malformed and abnormally enlarged roots) of crucifers (for example, mustard, cabbage, radishes, turnips, cauliflower, and broccoli). Plasmodiophoromycota were once thought to be kin to fungi, but recent molecular systematic studies have revealed that they are actually protozoa, members of the phylum Cercozoa and class Phytomyxea. *See* PARASITOLOGY; PROTOZOA.

The life history of plasmodiophorids is complex and includes (1) a flagellated motile stage, the zoospore; (2) an infectious stage, the encysted zoospore (cyst); (3) an unwalled assimilative and feeding stage, the multinucleate plasmodium; (4) a zoospore proliferation stage, the zoosporangium; and (5) a resistant stage, the resting spore. Zoosporangia and resting spores, cleaved from their respective plasmodia, cluster together with a characteristic organization in sori (spore balls).

Zoospores are uninucleate and microscopic, 3–5 micrometers in diameter. They bear two apically inserted smooth-surfaced flagella. Plasmodiophorid zoospores have a distinct method for penetration into the host. The zoospore swims to a suitable host, loses its flagella, crawls along the host surface, produces a wall around itself, and adheres to the host. Inside the cyst, a new structure shaped like a bullet (the stachel) forms within a channel (the rohr). A vacuole in the cyst expands and pushes the stachel into the host cell wall, piercing it. The parasite's protoplast is then injected into the host cytoplasm through the pore. *See* CILIA AND FLAGELLA.

There are approximately 12 currently recognized genera, but exploration of new habitats is leading to discovery of new taxa. Genera are differentiated based on the organization of their resting spores, especially as spore balls. Plasmodiophorids are economically important as pathogens of crop plants. In addition to being the causal agents of diseases, plasmodiophorids can serve as vectors (agents capable of transferring pathogens from one organism to another) for plant viruses. *See* PLANT VIRUSES AND VIROIDS.

Martha J. Powell

Plasmon The quanta of waves produced by collective effects of large numbers of electrons in matter when the electrons are disturbed from equilibrium. Plasmon excitations are easily detected in metals because they have a high density of electrons that are free to move. The result of plasmon stimulation by energetic electrons is seen in the illustration. The graph shows the probability of energy losses by fast electrons transmitted through a thin aluminum foil. The number of detected electrons in a beam is plotted against the energy loss during transit through the foil. Each energy-loss peak corresponds to excitation of one or more plasmons. Within experimental error, the peaks occur at integral multiples of a fundamental loss quantum.

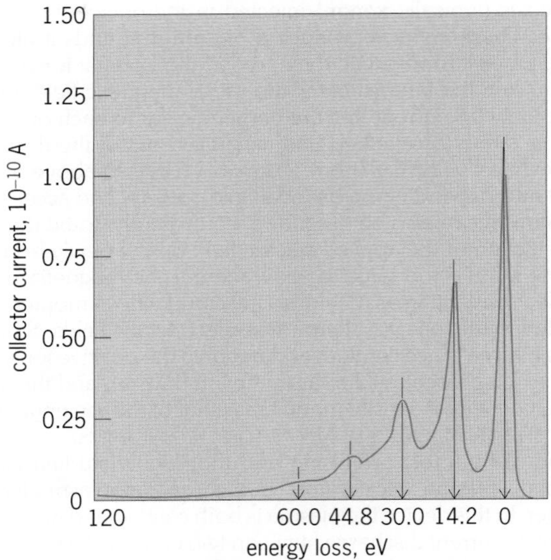

Number of detected electrons in a beam versus their energy loss during transit through a thin aluminum foil. (Number of electrons is expressed as a current; 10^{-10} A = 6.2 × 10^8 electrons per second.) Peaks at approximate multiples of 14.2 eV correspond to energy donated to plasmons in the aluminum. (*After T. L. Ferrell, T. A. Callcott, and R. J. Warmack, Plasmons and surfaces, Amer. Sci., 73:344–353, 1985*)

The name plasmon is derived from conventional plasma theory. A plasma is a state of matter in which the atoms are ionized and the electrons are free to move. At low particle densities this means an ionized gas, or classical plasma. In a metal the conduction electrons are also free to move, and there are many analogies to the classical low-density plasma. Similar to sound waves in a low-density gas of charged particles, the electron gas in a metal exhibits plasmon phenomena, that is, electron density waves. In both types of physical plasma, the frequency of plasma-wave oscillation is determined by the electronic density. Because of the high electron density in a metal (about 10^{23} electrons/cm^3), the frequencies are typically in the ultraviolet part of the electromagnetic spectrum. Quantum mechanics dictates that the energy of a plasmon is quantized, and one quantum of energy of a plasmon is given by its frequency multiplied by Planck's constant. *See* FREE-ELECTRON THEORY OF METALS; PLASMA (PHYSICS); QUANTUM MECHANICS.

Metallic objects support two kinds of plasmons: bulk and surface plasmons. Bulk plasmons are longitudinal density waves (like sound waves) that propagate through the volume of a metal. Electron density waves can also propagate at the interface between a metal substrate and a dielectric (surface plasmons). This fact can be easily derived from the electromagnetic theory developed by James Clerk Maxwell. Such electronic surface waves are strongly coupled to the electromagnetic field. They are officially termed surface plasmon-polaritons, but the term surface plasmon is more commonly used by many scientists. The charge fluctuations are extremely localized in the z direction (perpendicular to the interface) within the Thomas-Fermi screening length of about 0.1 nm. The corresponding electromagnetic wave is transverse magnetic (TM) in nature and has a beam (mode) size, D, that is not limited by diffraction in the direction normal to the metal surface. The field intensity, I, of such a wave has a maximum at the surface and decays exponentially away from the interface. *See* ELECTROMAGNETIC WAVE TRANSMISSION; MAXWELL'S EQUATIONS.

One of the most well-established applications of surface plasmons is biochemical sensors that can detect the presence of specific biochemical compounds. Selectivity can be obtained by the use of compounds or molecular impressions that bind only the specific target compound. By coating a thin metal film on a prism with antibodies, and tracking the shift in the surface plasmon excitation angle as antigens from a solution bind to the antibodies, the concentration of antigens can be determined. Similarly, since the surface plasmon excitation angle is sensitive to the index of refraction of the overcoating, it is possible to distinguish monomers from dimers just by exposing the thin film to the solution containing both, the monomer form having a different index from that of the dimer. The indices must be known from other measurements, and the thin film must be protected from deleterious effects of the solution. *See* ANTIBODY; ANTIGEN; BIOELECTRONICS; TRANSDUCER.

An important contribution of surface plasmons to analytical chemistry and biology is in Raman scattering. A portion of visible light incident on a molecule is frequency-shifted in accordance with the vibrational and rotational spectrum of the molecule. This constitutes the Raman signal that is characteristic of the scattering molecule. When a molecule is sitting on the surface of a metallic nanostructure, the fields can be substantially enhanced due to the resonant excitation of surface plasmons. This in turn leads to a surface-enhanced Raman spectroscopy signal. This signal can be many orders of magnitude stronger than the Raman signal obtained without the presence of the metal. This effect has enabled the successful implementation of Raman spectroscopy on trace amounts of chemical compounds and even single molecules. *See* RAMAN EFFECT. Mark L. Brongersma; Pieter G. Kik

Plaster A plastic mixture of solids and water which sets to a hard, coherent solid and which is used to line the interiors of buildings. A similar material of different composition, used to line the exteriors of buildings, is known as stucco. The term plaster is also used in the industry to designate plaster of paris.

Plaster is usually applied in one or more base (rough or scratch) coats up to $^3/_4$ in. (1.9 cm) thick, and also in a smooth, white, finish coat about $^1/_{16}$ in. (0.16 cm) thick. The solids in the base coats are hydrated (or slaked) lime, sand, fiber or hair (for bonding), and portland cement (the last may be omitted in some plasters). The finish coat consists of hydrated lime and gypsum plaster (in addition to the water). *See* LIME (INDUSTRY); MORTAR; PLASTER OF PARIS. J. F. McMahon

Plaster of paris The hemihydrate of calcium sulfate, composition $CaSO_4 \cdot {}^1/_2H_2O$, made by calcining the mineral gypsum, composition $CaSO_4 \cdot 2H_2O$, at temperatures up to 480°F (250°C). It is used for making plasters, molds, and models.

When the powdered hemihydrate is mixed with water to form a paste or slurry, the calcining reaction is reversed and a solid mass of interlocking gypsum crystals with moderate strength is formed. Upon setting there is very little (a slight contraction) dimensional change, making the material suitable for accurate molds and models.

Diverse types of plaster, varying in the time taken to set, the amount of water needed to make a pourable slip, and the final hardness, are made for different applications. These characteristics are controlled by the calcination conditions (temperature and pressure) and by additions to the plaster. *See* GYPSUM; PLASTER. J. F. McMahon

Plastic deformation of metals The permanent change in shape of a metal object as a result of applied or internal forces. This feature permits metals to be formed into pipe, wire, sheet, and so on, with an optimal combination of strength, ductility, and toughness. The onset of noticeable plastic deformation occurs when the applied tensile stress or applied effective stress reaches the tensile yield strength of the material. Continued plastic deformation usually requires a continued increase in the applied stress. The effects of plastic deformation may be removed by heat treatment. *See* MECHANICAL PROPERTIES OF METAL; PLASTICITY.

Mechanisms of plastic deformation in metals are understood in terms of crystal structure and incremental slip between atomic planes. These aspects are useful to understand: (1) regimes of stress and temperature for plastic deformation; (2) types of plastic stress-strain response; (3) reasons why metals exhibit plastic deformation; (4) mechanisms of plastic deformation in metal crystals and polycrystals; (5) strategies to strengthen metals against plastic deformation; (6) effect of strengthening, temperature, and deformation rate on ductility and toughness.

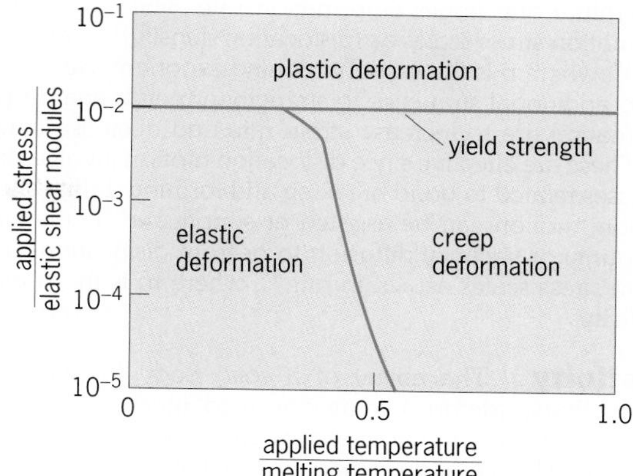

Deformation mechanism map. (*After M. F. Ashby and D. R. H. Jones, Engineering Materials I: An Introduction to Their Properties and Applications, 2d ed., Butterworth-Heimann, 1996*)

Plastic deformation in metals is observed macroscopically as a departure from linear-elastic stress-strain response. The illustration shows that plastic deformation predominates over other deformation mechanisms when the applied stress is greater than the yield strength. Unlike creep deformation, plastic deformation usually does not increase with time unless the applied stress is also increased. When the applied stress is less than the yield strength, other deformation mechanisms may dominate. *See* CREEP; ELASTICITY; MECHANICAL PROPERTIES OF METAL; STRESS AND STRAIN.

Metals are unique in that they tend to form crystals of densely packed atoms. This occurs in metals since the valence or outer electrons have a continuous succession of possible energy levels. These electrons create a "sea" of negative charge that pulls positively charged nuclei into densely packed crystal structures such as face-centered cubic (fcc), hexagonal close-packed (hcp), and body-centered cubic (bcc). Metallic glasses are an exception, since they are noncrystalline or amorphous. *See* CRYSTAL STRUCTURE; METALLIC GLASSES; METALS.

Under stress, metal crystals are prone to shear between densely packed planes. Plastic deformation occurs by shearing one plane relative to another. During the process, atomic bonds are broken and made. Densely packed slip planes are preferred in part because the depressions are shallow and allow for "smoother" slip. *See* SHEAR.

Experimental evidence confirms that slip produces plastic deformation in metals. Formation of slip traces during plastic deformation turns shiny metal surfaces to a dull state. In contrast, ceramic materials are brittle and lack plastic deformation at room temperature. This is due to the lack of closely packed planes and to the prohibitively large energy needed to break bonds during shearing. *See* BRITTLENESS; CERAMICS.

Four elementary strategies are used to make dislocation motion difficult and thereby increase the stress needed for plastic deformation. Grain-size strengthening involves the reduction of grain (crystal) size in polycrystals, so that grain boundaries provide more numerous trapping sites for dislocations. The contribution to deformation stress scales as (grain size)$^{-1/2}$. Solution strengthening involves adding impurity atoms to pure metals (alloying). Impurity atoms interact with dislocations via the local distortion or strain they induce. The contribution to deformation stress scales as (impurity concentration)$^{1/2}$. Precipitate strengthening is also the result of alloying to the extent that a new phase is produced inside the host phase. Precipitates induce strain and provide interfaces that can trap dislocations. The contribution to deformation stress scales as (precipitate spacing)$^{-1}$. Cold work increases dislocation density via deformation at lower temperature and larger deformation rate. The contribution to deformation stress scales as (dislocation density)$^{1/2}$ or as (plastic strain)^n, where n is the work hardening exponent. *See* ALLOY.

Two additional strategies to strengthen metals against plastic deformation are to increase strain rate and decrease temperature. These are effective since dislocation motion involves kinetic processes related to bond breaking and forming during shear. In addition, motion can be assisted or suppressed by the kinetics of impurity or vacancy diffusion to or from dislocations. Deformation stress scales as (strain rate)^m, where m is the strain-rate sensitivity. Peter M. Anderson

Plasticity The ability of a solid body to permanently change shape (deform) in response to mechanical loads or forces. Deformation characteristics are dependent on the material from which a body is made, as well as the magnitude and type of the imposed forces. In addition to plastic, other types of deformation are possible for solid materials.

One common test for measuring the plastic deformation characteristics of materials is the tensile test, in which a tensile (stretching) load is applied along the axis of a cylindrical specimen, with deformation corresponding to specimen elongation. The load is converted into stress; its units are megapascals (1 MPa = 10^6 newtons per square meter) or pounds per square inch (psi). Likewise, the amount of deformation is converted into strain, which is unitless. The test results are expressed as a plot of stress versus strain. *See* STRESS AND STRAIN.

Typical tensile stress-strain curves have been calculated for metal alloys and polymeric materials. For both materials, the initial regions of the curves are linear and relatively steep. Deformation that occurs within these regions is nonpermanent (nonplastic) or elastic. This means that the body springs back to its original dimensions once the stress is released, or that all of the deformation is recovered. In addition, stress is proportional to strain (Hooke's law), and the slope of this linear segment corresponds to the elastic (Young's) modulus. *See* ELASTICITY; HOOKE'S LAW; YOUNG'S MODULUS.

Plastic (permanent) deformation begins at the point where linearity ceases such that, upon removal of the load, not all deformation is recovered (the body does not assume its original or stress-free dimensions). The onset of plastic deformation is called yielding, and the corresponding stress value is called the yield strength. After yielding, all deformation is plastic and, until fracture, the curves are nonlinear. This behavior is characteristic of many metal alloys and polymeric materials. The concept of plasticity does not normally relate to ceramic materials such as glasses and metal oxides (for example, aluminum oxide). *See* PLASTIC DEFORMATION OF METAL. William D. Callister

Plastics processing The methods and techniques used to convert plastics materials in the form of pellets, granules, powders, sheets, fluids, or preforms into formed shapes or parts. Although the term plastics has been used loosely as a synonym for polymers and resins, plastics generally represent polymeric compounds that are formulated with plasticizers, stabilizers, fillers, and other additives for purposes of processability and performance. After forming, the part may be subjected to a variety of ancillary operations such as welding, adhesive bonding, machining, or surface decorating (painting, metallizing).

Injection molding. This process consists of heating and homogenizing plastics granules in an injection molding barrel until they are sufficiently fluid to allow for pressure injection into a relatively cold mold, where they solidify and take the shape of the mold cavity. For thermoplastics, no chemical changes occur within the plastic. Injection molding of thermosetting resins (thermosets) differs primarily in that the cylinder heating is designed to homogenize and preheat the reactive materials, and the mold is heated to complete the chemical cross-linking reaction to form an intractable solid.

Extrusion. In this typically continuous process, plastic pellets or granules are fluidized and homogenized through the rotating action of a screw (or screws) inside a barrel, and the melt is continuously pushed under pressure through a shaping die to form the final product. As material passes through the die, it initially acquires the shape of the die opening. The shape will change once the material exits from the die. Products made this way include extruded tubing, pipe, film, sheet, wire and substrate coatings, and profile shapes. The process is used to form very long shapes or many small shapes, which can be cut from the long shapes. The homogenizing capability of extruders is used for plastics blending and compounding. Pellets used for other processing methods, such as injection molding, are made by chopping long filaments of extruded plastic.

Blow molding. This process consists of melting plastics pellets or granules similar to that of extrusion or injection molding, forming a tube (called a parison or preform) and introducing air or other gas to cause the tube to expand into a free-blown hollow object or, most commonly, against the surface of a spilt mold with a hollow cavity for forming into a hollow object with a definite size and shape. The parison is traditionally made by extrusion (extrusion blow molding), although injection molded preforms (injection blow molding) have gained prominence. Bottles still represent a large share (80%) of the blow-molding market, with the remaining 20% considered as industrial blow molding (for example, fuel tanks).

Thermoforming. Plastics sheets are thermoformed into parts through the application of heat and pressure. Tooling for thermoforming is the least expensive of the plastics processes, accounting for the method's popularity. It can also accommodate very large parts as well as small parts, the latter being useful in low-cost prototype fabrication. Thermoformed products are typically categorized as permanent, or industrial products, and disposable products (for example, packaging).

Compression and transfer molding. Compression molding is one of the oldest molding techniques and consists of charging a plastics powder or preformed plug into a mold cavity, closing a mating mold half, and applying pressure to compress, heat, and cause the plastic to flow and conform to the cavity shape. The process is used primarily for thermosets, bulk molding compounds, and their composite counterparts, sheet molding compounds.

Foam processes. Foamed plastics materials have achieved a high degree of importance in the plastics industry. Foams range from soft and flexible to hard and rigid. There are three types of cellular plastics: blown (expanded matrix such as a natural sponge), syntactic (the encapsulation of hollow organic or inorganic microspheres into the matrix), and structural (dense outer skin surrounding a foamed core).

Blowing agents used for producing porous or cellular plastics can be roughly classified into two main categories: physical blowing agents and chemical blowing agents. Physical blowing agents are volatile liquids or compressed gases that change state during processing to form a cellular structure. Chemical blowing agents are solids that decompose above a certain temperature and release gaseous products. Depending on the blowing agents used and how they are being introduced, cellular plastics or plastics foams can be produced by a number of processes. Structural foams differ from other foams in that they are produced with a hard integral skin on the outer surface and a cellular core in the interior. They are made by injection molding liquefied resins containing chemical blowing agents. *See* FOAM.

Reinforced plastics/composites. These plastics have mechanical properties that are significantly improved due to the inclusion of fibrous reinforcements. The wide variety of resins and reinforcements that constitute this group of materials led to the more generalized use of the term composites.

For components using thermoset resins, the first step in the composite fabrication procedure is the impregnation of the reinforcement with the resin. The simplest method is to pass the reinforcement through a resin bath and use the wet impregnate directly. For easier handling and storage, the impregnated reinforcement can be subjected to heat to remove impregnating solvents or advance the resin cure to a slightly tacky or dry state. The composite in this form is called a prepreg (B stage). This B-stage condition allows the composite to be handled, yet the cross-linking reaction has not proceeded so far as to preclude final flow and conversion to a homogeneous part when further heat or pressure is applied. *See* COMPOSITE MATERIAL; POLYMERIC COMPOSITE.

Casting and encapsulation. Casting is a low-pressure process requiring nothing more than a container in the shape of the desired part. For thermoplastics, liquid monomer is poured into the mold and, with heat, allowed to polymerize in place to a solid mass. Thermosets, usually composed of liquid resins with appropriate curatives and property-modifying additives, are poured into a heated mold, wherein the cross-linking reaction completes the conversion to a solid. Encapsulation and potting are terms for casting processes in which a unit or assembly is encased or impregnated, respectively, with a liquid plastic, which is subsequently hardened by fusion or chemical reaction. These processes are predominant in the electrical and electronic industries for insulation and protection of components.

Calendering. In the calendering process, a plastic is masticated between two rolls that squeeze it out into a film that passes around one or more additional rolls before being stripped off as a continuous film. Fabric or paper may be fed through the latter rolls so that they become impregnated with the plastic. *See* POLYMER.

Lih-Sheng Turng; Sidney H. Goodman

Plate girder

A beam assembled from steel plates which are welded or bolted to one another, used to support the horizontal surface of a bridge. Plate girders generally have I-shaped cross sections (see illustration). Cross sections with two or more webs that form a U- or box-shape can also be considered plate girders, but are generally classified separately as box girders. Plate girders are commonly used in bridge structures where large transverse loads and spans are encountered. On average, the height (measured from the top of the top flange to the bottom of the bottom flange) of plate girders is one-tenth to one-twelfth the length of the span, which can vary based on the loading and architectural requirements. The dimensions of plate girders are tailored to specific applications by adjusting the width and thickness of the web and flange plates along the length of the beam. The depth is generally considered to be the distance between the centroids of the flanges which defines the effective moment arm in bending. *See* BRIDGE; LOADS, TRANSVERSE; STRUCTURAL STEEL.

Long-span plate girders are often fabricated in segments due to size or weight restrictions at the fabrication facility, during shipping, or at the construction site. These segments are spliced or joined together by either high-strength bolts or welding.

Plate girders commonly have stiffener plates which increase the strength of the web plate. Since plate girder webs are relatively slender, they are susceptible to buckling when subjected to shear and bending forces. The bending strength of webs can be enhanced using horizontal (longitudinal) stiffeners, while the shear strength of webs can be enhanced using vertical (transverse) stiffeners (*see* illustration).

Brian Chen

Plate tectonics

Plate tectonics theory provides an explanation for the present-day tectonic behavior of the Earth, particularly the global distribution of mountain building, earthquake activity, and volcanism in a series of linear belts. Numerous other geological phenomena such as lateral variations in surface heat flow, the physiography and geology of ocean basins, and various associations of igneous, metamorphic, and sedimentary rocks can also be logically related by plate tectonics theory.

The theory is based on a simple model of the Earth in which a rigid outer shell 30–90 mi (50–150 km) thick, the lithosphere, consisting of both oceanic and continental crust as well as the upper mantle, is considered to lie above a hotter, weaker semiplastic asthenosphere. The asthenosphere, or low-velocity zone, extends from the base of the lithosphere to a depth of about 400 mi (700 km). The brittle lithosphere is broken into a mosaic

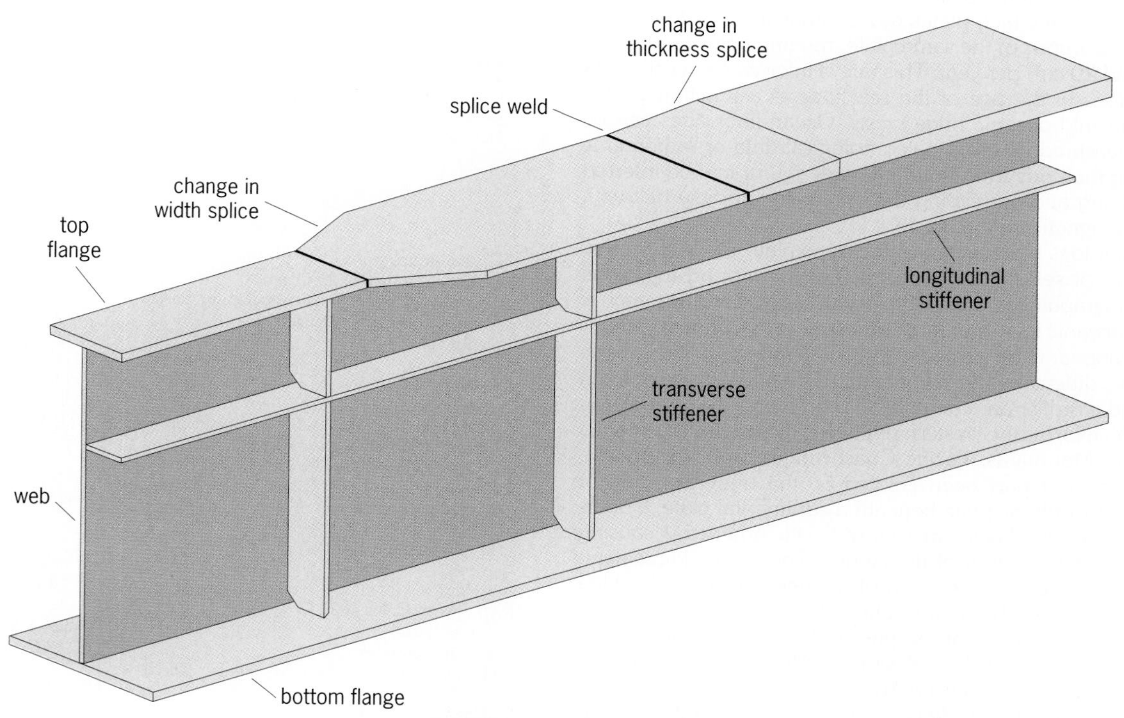

Plate girder with stiffeners and plate splices.

of internally rigid plates that move horizontally across the Earth's surface relative to one another. Only a small number of major lithospheric plates exist, which grind and scrape against each other as they move independently like rafts of ice on water. Most dynamic activity such as seismicity, deformation, and the generation of magma occurs only along plate boundaries, and it is on the basis of the global distribution of such tectonic phenomena that plates are delineated. *See* ASTHENOSPHERE; EARTH INTERIOR; LITHOSPHERE.

The plate tectonics model for the Earth is consistent with the occurrence of sea-floor spreading and continental drift. Convincing evidence exists that both these processes have been occurring for at least the last 600 million years (m.y.). This evidence includes the magnetic anomaly patterns of the sea floor, the paucity and youthful age of marine sediment in the ocean basins, the topographic features of the sea floor, and the indications of shifts in the position of continental blocks which can be inferred from paleomagnetic data on paleopole positions, paleontological and paleoclimatological observations, the match-up of continental margin and geological provinces across present-day oceans, and the structural style and rock types found in ancient mountain belts. *See* CONTINENTAL DRIFT; CONTINENTAL MARGIN; PALEOCLIMATOLOGY; PALEOMAGNETISM.

Geological observations, geophysical data, and theoretical considerations support the existence of three fundamentally distinct types of plate boundaries, named and classified on the basis of whether immediately adjacent plates move apart from one another (divergent plate margins), toward one another (convergent plate margins), or slip past one another in a direction parallel to their common boundary (transform plate margins). Plate margins are easily recognized because they coincide with zones of seismic and volcanic activity; little or no tectonic activity occurs away from plate margins. The boundaries of plates can, but need not, coincide with the contact between continental and oceanic crust. The nature of the crustal material capping a plate at its boundary may control the specific processes occurring there, particularly along convergent plate margins, but in general plate tectonics theory considers the continental crustal blocks as passive passengers riding on the upper surface of fragmenting, diverging, and colliding plates.

The velocity at which plates move varies from plate to plate and within portions of the same plate, ranging between 0.8 and 8 in. (2 and 20 cm) per year. This rate is inferred from estimates for variations in the age of the sea floor as a function of distance from mid-oceanic ridge crests. Ocean-floor ages can be directly measured by using paleontological data or radiometric age-dating methods from borehole material, or can be inferred by identifying and correlating the magnetic anomaly belt with the paleomagnetic time scale.

Not only does plate tectonics theory explain the present-day distribution of seismic and volcanic activity around the globe and physiographic features of the ocean basins such as trenches and mid-oceanic rises, but most Mesozoic and Cenozoic mountain belts appear to be related to the convergence of lithospheric plates. Two different varieties of modern mobile belts have been recognized: cordilleran type and collision type. The Cordilleran range, which forms the western rim of North and South America (the Rocky Mountains, Pacific Coast ranges, and the Andes), has for the most part been created by the underthrusting of an ocean lithospheric plate beneath a continental plate. Underthrusting along the Pacific margin of South America is causing the continued formation of the Andes. The Alpine-Himalayan belt, formed where the collision of continental blocks buckled intervening volcanic belts and sedimentary strata into tight folds and faults, is an analog of the present tectonic situation in the Mediterranean, where the collision of Africa and Europe has begun. *See* CORDILLERAN BELT; MOUNTAIN SYSTEMS; OROGENY.

Plate tectonics is considered to have been operative as far back as 2.5 b.y. Prior to that interval, evidence suggests that the plate tectonics may have occurred, although in a markedly different manner, with higher rates of global heat flow producing smaller convective cells or more densely distributed mantle plumes which fragmented the Earth's surface into numerous small, rapidly moving plates. Repetitious collision of plates may have accreted continental blocks by welding primitive "greenstone island arc terranes" to small granitic subcontinental masses. *See* CONTINENTS, EVOLUTION OF; GEODYNAMICS.

Walter C. Pitman III

Plateau Any elevated area of relatively smooth land. Usually the term is used more specifically to denote an upland of subdued relief that on at least one side drops off abruptly to adjacent lower lands. In most instances the upland is cut by deep but widely separated valleys or canyons. Small plateaus that stand above their surroundings on all sides are often called tables, tablelands, or mesas. The abrupt edge of a plateau is an escarpment or, especially in the western United States, a rim.

Edwin H. Hammond

Platinum A chemical element, Pt, atomic number 78, and atomic weight 195.08. Platinum is a soft, ductile, white noble metal. The platinum-group metals—platinum, palladium, iridium, rhodium, osmium, and ruthenium—are found widely distributed over the Earth. Their extreme dilution, however, precludes their recovery, except in special circumstances. For example, small amounts of the platinum metals, palladium in particular, are recovered during the electrolytic refining of copper. *See* IRIDIUM; OSMIUM; PALLADIUM; PERIODIC TABLE; RHODIUM; RUTHENIUM.

The platinum-group metals have wide chemical use because of their catalytic activity and chemical inertness. As a catalyst, platinum is used in hydrogenation, dehydrogenation, isomerization, cyclization, dehydration, dehalogenation, and oxidation reactions. *See* CATALYSIS; ELECTROCHEMICAL PROCESS.

Platinum is not affected by atmospheric exposure, even in sulfur-bearing industrial atmospheres. Platinum remains bright and does not visually exhibit an oxide film when heated, although a thin, adherent film forms below 450°C (840°F).

Physical properties of platinum

Properties	Value
Naturally occurring isotopes and % abundance	190, 0.0127% 192, 0.78% 194, 32.9% 195, 33.8% 196, 25.3% 198, 7.21%
Crystal structure	Face-centered cubic
Lattice constant a at 25°C, nm	0.39231
Thermal neutron capture cross section, barns	8.8
Common chemical valence	2, 4
Density at 25°C, g/cm^3	21.46
Melting point	1772°C (3222°F)
Boiling point	3800°C (6900°F)
Specific heat at 0°C, cal/g	0.0314
Thermal conductivity, 0–100°C, cal cm/cm^2 s°C	0.17
Linear coefficient of thermal expansion, 20–100°C, μin./in./°C	9.1
Electrical resistivity at 0°C, microhm-cm	9.85
Temperature coefficient of electrical resistance, 0–100°C/°C	0.003927
Tensile strength, 1000 lb/in.^{2}	
Soft	18–24
Hard	30–35
Young's modulus at 20°C	
lb/in.^{2}, static	24.8×10^6
lb/in.^{2}, dynamic	24.5×10^6
Hardness, Diamond Pyramid Number (DPN)	
Soft	37–42
Hard	90–95

Platinum may be worked to fine wire and thin sheet, and by special processes, to extremely fine wire. Important physical properties are given in the table.

Platinum can be made into a spongy form by thermally decomposing ammonium chloroplatinate or by reducing it from an aqueous solution. In this form it exhibits a high absorptive power for gases, especially oxygen, hydrogen, and carbon monoxide. The high catalytic activity of platinum is related directly to this property. *See* CRACKING; HYDROGENATION.

Platinum strongly tends to form coordination compounds. Platinum dioxide, PtO_2, is a dark-brown insoluble compound, commonly known as Adams catalyst. Platinum(II) chloride, $PtCl_2$, is an olive-green water-insoluble solid. Chloroplatinic acid, H_2PtCl_6, is the most important platinum compound. *See* COORDINATION CHEMISTRY.

In the glass industry, platinum is used at high temperatures to contain, stir, and convey molten glass. In the electrical industry, platinum is used in contacts and resistance wires because of its low contact resistance and high reliability in contaminated atmospheres. Platinum is clad over tungsten for use in electron tube grid wires. In the medical field, the simple coordination compounds cisplatin and carboplatin are two of the most active clinical anticancer agents. In combination with other agents, cisplatin is potentially curative for all stages of testicular cancer. Both agents are used for advanced gynecologic malignancies, especially ovarian tumors, and for head and neck and lung cancers. Carboplatin was developed in attempts to alleviate the severe toxic side effects of the parent cisplatin, with which is shares a very similar spectrum of anticancer efficacy. *See* CANCER (MEDICINE); CHEMOTHERAPY. Henry J. Albert; Nicholas Farrel

Platyasterida A Paleozoic (Ordovician and Devonian) order of sea stars of the class Asteroidea; only two genera have been recognized. Platyasteridans have relatively elongate flat arms; broad, transversely aligned ventral arm plates; and paxilliform aboral plates. Fundamental differences in ambulacral system construction have been found between modern asteroids and all of the well-known early Paleozoic asteroids, including the platyasteridans. Similarities between platyasteridans and modern asteroids are attributable to evolutionary convergence. *See* ASTEROIDEA; ECHINODERMATA. Daniel B. Blake

Platycopida An order of the Podocopa (Ostracoda) containing a single family, Cytherellidae, and two extinct superorders. Platycopids are small marine or brackish-water ostracods with an asymmetrical shell that is oblong, nearly rectangular, and laterally compressed; their six pairs of appendages are adapted for burrowing and filter feeding.

Species are dimorphic. The female carapace is wider near the rear and longer than that of the male, providing a brood chamber behind the body. All living species are benthic, and fossil forms probably were also. Some species live at great depths in the Atlantic Ocean, but others have been found in the Arabian, Caribbean, and Mediterranean seas in shallower water. *See* CRUSTACEA; OSTRACODA; PODOCOPA. Patsy A. McLaughlin

Platyctenida An order of the phylum Ctenophora comprising four families (Ctenoplanidae, Coeloplanidae, Tjalfiellidae, Savangiidae) and six genera. All species are highly modified from the planktonic ctenophores. The platyctenes are fairly small (1–6 cm or 0.4–2.4 in.) and brightly colored. They have adopted a variety of swimming, creeping, and sessile habits, with concomitant morphological changes and loss of typical ctenophoran characteristics. The body is compressed in the oral-aboral axis, and the oral part of the stomodeum is everted to form a creeping sole. Sexual reproduction involves internal fertilization in many species, with retention of the developing cydippid larvae in brood pouches. Some species also reproduce asexually by fission. Most platyctenids are found in tropical coastal waters, where many are ectocommensals on benthic organisms. *See* CTENOPHORA. Laurence P. Madin

Platyhelminthes A phylum of the invertebrates, commonly called the flatworms. They are bilaterally symmetrical, nonsegmented worms characterized by lack of coelom, anus, circulatory and respiratory systems, and exo- and endoskeletons. Many species are dorsoventrally flattened. They possess a protonephridial (osmoregulatory-excretory) system, a complicated hermaphroditic reproductive system, and a solid mesenchyme which fills the interior of the body (see illustration). Some parasitic species, that is, some trematodes, have secondarily acquired a lymphatic system resembling a true circulatory system. Some species of trematodes, the schistosomes, have separate sexes.

Traditionally, three classes were distinguished in the phylum: the Turbellaria, mainly free-living predacious worms; the Trematoda, or flukes, ecto- or endoparasites; and the Cestoda, or tapeworms, endoparasites found in the enteron (alimentary canal) of vertebrates, whose larvae are found in the tissues of invertebrates or vertebrates. However, recent cladistic analyses using morphology, including ultrastructure, as well as DNA analysis, have shown that the "Turbellaria" are an assemblage of taxa that are not monophyletic (that is, they are not a group containing all taxa with a common ancestor), and that the monogeneans, earlier included in the trematodes, do not belong to the trematodes. Most importantly, the Acoela do not belong to the Platyhelminthes, but are a very archaic group close to the base of the lower invertebrates; and all major groups of parasitic Platyhelminthes—that is, the Trematoda, Monogenea, and Cestoda—are monophyletic, constituting the Neodermata. *See* CESTODA; EUCESTODA; MONOGENEA; TREMATODA; TURBELLARIA.

Turbellaria are widespread in freshwater and the littoral zones (the biogeographic zones between the high- and low-water

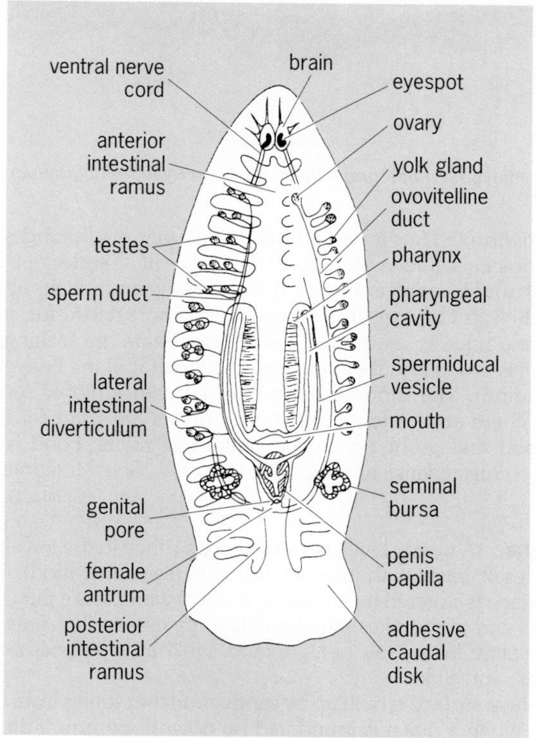

Bdelloura candida **(Tricladida), ectocommensal on the king crab, *Limulus*. Complete digestive and male systems are shown on the left, female systems on the right.**

marks) of the sea, while one group of triclads occurs on land in moist habitats. Adults of parasitic flatworms occur on or in many tissues and cavities of the vertebrates on which they feed. They are responsible for troublesome diseases in humans and animals. Monogeneans are of foremost importance as agents of fish disease, especially in aquaculture, whereas digeneans are important agents of disease in domestic animals and humans (schistosomiasis). Larval flukes are frequent in mollusks, mainly gastropods, and occasionally occur in pelecypods (bivalves). Vector hosts, such as insects and fish, are often interpolated between the mollusk and vertebrate. Adult tapeworms, living in the gut or the biliary ducts, compete with the host for food and accessory food factors such as vitamins. Larval tapeworms reside chiefly in arthropods, but larvae of the Cyclophyllidea develop in mammals, which may be severely impaired or even killed by the infection. *See* PARASITOLOGY; TRICLADIDA. Klaus Rohde; C. G. Goodchild

Platypus The single species *Ornithorhynchus anatinus* of the family Ornithorhynchidae in the order Monotremata, which occurs in eastern Australia and Tasmania. This mammal is a monotreme, which lays eggs and incubates them in a manner similar to birds. The platypus is also known as the duckbill, duckmole, and water mole. One of the most primitive mammals, it retains some reptilian characteristics. The female lacks a marsupium; however, marsupial or epipubic bones are well developed.

Platypus. (*Photograph by Paddy Ryan/Ryan Photographic*)

Ornithorhynchus is well adapted to aquatic life. It ranges from the tropical, sea-level streams to cold lakes at altitudes of 6000 ft (1800 m). The adult is about 2 ft (0.6 m) long and weighs approximately 4 lb (1.8 kg). It is covered with short dense fur, typical of many aquatic mammals; the external ears are lacking; and the tail is broad and flattened, resembling that of the beaver (see illustration). The rubbery bill is covered with a highly sensitive skin. There are no teeth, but horny ridges are used for crushing food and grubbing on the bottom of rivers. Food is principally crustaceans, worms, and tadpoles. *See* MONOTREMATA. Charles B. Curtin

Playa A nearly level, generally dry surface in the lowest part of a desert basin with internal drainage (see illustration). When its surface is covered by a shallow sheet of water, it is a playa lake. Playas and playa lakes are also called dry lakes, alkali flats, mud flats, saline lakes, salt pans, inland sabkhas, ephemeral lakes, salinas, and sinks.

A playa surface is built up by sandy mud that settles from floodwater when a playa is inundated by downslope runoff during a rainstorm. A smooth, hard playa occurs where ground-water discharge is small or lacking and the surface is flooded frequently. These mud surfaces are cut by extensive desiccation polygons caused by shrinkage of the drying clay. Puffy-ground playas form by crystallization of minerals as ground water evaporates in muds near the surface.

Light-colored playa in lowest part of Sarcobatus Flat in southern Nevada.

Subsurface brine is present beneath many playas. The type of brine depends on the original composition of the surface water and reflects the lithology of the rocks weathered in the surrounding mountains. *See* GROUND-WATER HYDROLOGY.

Numerous playas in the southwestern United States yield commercial quantities of evaporite minerals, commonly at shallow depths. Important are salt (NaCl) and the borates, particularly borax ($Na_2B_4O_7 \cdot 10H_2O$), kernite ($Na_2B_4O_7 \cdot 4H_2O$), ulexite ($NaCaB_5O_9 \cdot 8H_2O$), probertite ($NaCaB_5O_9 \cdot 5H_2O$), and colemanite ($Ca_2B_6O_{11} \cdot 5H_2O$). Soda ash (sodium carbonate; Na_2CO_3) is obtained from trona ($Na_3H(CO_3)_2 \cdot 2H_2O$) and gaylussite ($Na_2Ca(CO_3)_2 \cdot 5H_2O$). Lithium and bromine are produced from brine waters. *See* DESERT EROSION FEATURES; SALINE EVAPORITES. John F. Hubert

Plecoptera An order of primitive insects known as the stoneflies. The order comprises 16 families and more than 2000 species distributed on all continents except Antarctica. Stoneflies

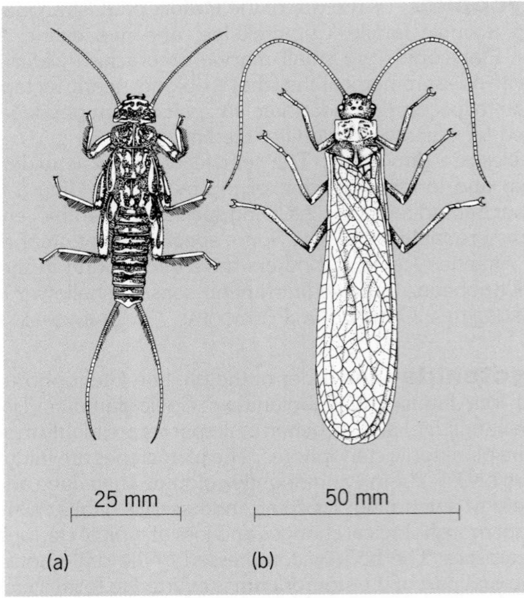

Plecoptera. (a) Nymph of *Paragnetina*. (b) Adult of *Pteronarcys*, one of the largest stoneflies. (*After A. H. Morgan, Field Book of Ponds and Streams, Putnam, 1930*)

spend the majority of their lives aquatically as nymphs (immature larval forms); they exit the water and live only briefly as adults on land. There are relatively slight differences between the aquatic nymphal stage and the terrestrial adult form, except for wings and tracheal gills. In stoneflies, the soft and somewhat flattened or cylindrical body ending in two tail filaments, the strong legs with paired tarsal claws, the chewing mouthparts, and the rusty blacks, dull yellows, and browns are characteristic of both the nymph and adult (see illustration). The name Plecoptera literally means pleated wing, referring to the hindwings that are folded and hidden under the forewings, which the adult holds close to the abdomen when at rest or walking. *See* INSECTA.

Stonefly nymphs live in the rapid stony parts of clean swift streams, although a few species occur along the rocky shores of large temperate lakes. Nymphs have diverse feeding habits, including species that are carnivores and others that are herbivores and/or detritivores. The adult life stage begins when full-grown nymphs climb onto rocks or plants, molt, and emerge as adults. In various parts of North America stonefly adults emerge during every month of the year. Adults live from one to a few weeks, and the time is mainly spent engaged in reproductive activities. *See* INSECT PHYSIOLOGY. Andrew K. Rasmussen; Lewis Berner

Pleiades A typical open cluster of stars (see illustration), and the best example in the sky of an open cluster for the casual observer. Open clusters are one of the two basic types of star clusters, distinguished from globular clusters by their smaller total populations and central densities, and by their younger age. *See* STAR CLUSTERS.

The Pleiades, located in the constellation Taurus, contains several hundred stars within a radius of about 1°, centered on its brightest star, Alcyone. (In comparison, the Moon is 0.5° across.) The cluster is permeated with a faint diffuse nebulosity. Though the Pleiades is always referred to as having seven bright stars, only six are seen today. *See* TAURUS; VARIABLE STAR.

The Pleiades. (*Lick Observatory photograph*)

The Pleiades is a young cluster, about 100 million years old, and still contains bright hot blue stars (type B) on the main sequence (hydrogen-burning stars) with abundances of elements similar to that of the Sun. It is compact, and close enough to see its smaller fainter members (including brown dwarfs). Although it has stars of widely varying masses and brightnesses, they are all approximately the same age and distance from Earth, about 130 parsecs (425 light-years). *See* BROWN DWARF; LIGHT-YEAR; PARSEC; STAR; STELLAR EVOLUTION. Nancy Remage Evans; Ronald Pitts

Pleistocene The older of the two epochs of the Quaternary Period. The Pleistocene Epoch represents the interval of geological time (and rocks accumulated during that time) extending from the end of the Pliocene Epoch (and the end of the Tertiary Period) to the start of the Holocene Epoch. Most recent time scales show the Pleistocene Epoch spanning the interval from 1.8 million years before present (m.y. B.P.) to 10,000 years B.P. The Pleistocene is commonly characterized as an epoch when the Earth entered its most recent phase of widespread glaciation. *See* HOLOCENE; PLIOCENE; QUATERNARY; TERTIARY.

In modern geological time scales, the Pleistocene is subdivided into a lower and an upper series. In Europe the lower series is considered equivalent to the Calabrian Stage, while the upper series is equated with the Sicilian and Tyrrehenian stages.

The onset of the Pleistocene brought glaciations that were more widespread than those in the Pliocene. Mountain glaciers expanded and continental ice fields covered large areas of the temperate latitudes. Sea ice also became more widespread. As evidence has accumulated in recent decades from both the land and the sea, it clearly shows at least 17 glacial events occurred during the Pleistocene. *See* CLIMATE HISTORY; GLACIAL EPOCH.

The expansion and decay of the ice sheets had a direct effect on the global sea level. Global sea-level fluctuations of 50–150 m (170–500 ft) have been estimated for various glacial-interglacial episodes during the Pleistocene. Since the last deglaciation, which began some 17,000 years ago, the sea level has risen by about 110 m (360 ft) worldwide, drowning all of the ancient lowstand shorelines. One important product of the sea-level drops was the migration of large river deltas to the edges of the continental shelves and to the deeper parts of the basins. Conversely, the last marine transgression that started in the late Pleistocene after rapid deglaciation and ended in the Holocene (6000 to 7000 years ago) resulted in new deltas that formed at the mouths of modern rivers. *See* DELTA; PALEOCEANOGRAPHY.

The onset of cooler climate and Pleistocene glaciation is also approximated with a wave of mammalian migration from the east to the west. A relatively modern-looking fauna that included the first true oxen, elephants, and the first one-toed horse appeared at the beginning of Pleistocene. The modern horse, *E. caballus*, made its first appearance some 250,000 years ago in the late Pleistocene in North America. From North America it migrated to Asia and then west to Europe. However, during the last glacial maximum, some 18,000 years ago, it became extinct in North America when it was unable to cross the deserts to migrate to South America. Oxen, deer and reindeer, large cats, mammoth, great elk, wolf, hyena, and woolly rhinos proliferated during the middle and late Pleistocene. Mammoths, which have been found preserved nearly intact in frozen soils in Siberia, ranged over much of Europe during the glacial times. *See* GEOLOGIC TIME SCALE; MAMMALIA.

The unique Pleistocene mammalian faunas of some of the isolated islands, such as Madagascar, the Philippines, Taiwan, and the Japanese Archipelago, indicate restriction in the dispersal of species during the Pleistocene. The Pliocene Epoch had given rise to the human precursor, *Homo habilis*, around 2 m.y. B.P. The appearance of *H. erectus* came on the scene almost at the Plio-Pleistocene boundary around 1.8 m.y. B.P. The first archaic *H. sapiens* are now considered to have arrived on the scene around about a million years ago. The appearance of

H. neanderthalensis or the Neanderthal Man, is now dated at least as far back as 250,000 years B.P. Recent datings have the appearance of the first true modern *H. sapiens* (the Cro-Magnon Man) to around 100,000 years B.P. *See* FOSSIL HUMANS; GEOLOGIC TIME SCALE. Bilal U. Haq

Pleochroic halos Spherical or elliptical regions up to 40 micrometers in diameter in which there is a change in color from the surrounding mineral when viewed with a petrographic microscope. Pleochroic halos are found around small inclusions of radioactive minerals—for example, zircon, monazite, allanite, xenotime, and apatite—and in rock-forming minerals, principally quartz, micas, amphiboles, and pyroxenes. Halos have also been identified in coalified wood preserved in deposits on the Colorado Plateau. *See* CORDIERITE; PETROFABRIC ANALYSIS; RADIOACTIVE MINERALS.

The change in color is a result of radiation damage caused by alpha particles emitted during the radioactive decay of nuclides in the decay chains of uranium-238, uranium-235, and thorium-232. The range of the alpha particle and ionization effects account for the size and color of the halos. The halos have a distinctive ring structure with varying degrees of discoloration between the rings: the coloration in the halos increases, saturates, and finally diminishes with increasing ion dose. *See* ALPHA PARTICLES; METAMICT STATE.

The early interest in pleochroic halos was in their use for geologic age dating. Careful attempts were made to correlate the halo color with the alpha-irradiation dose in order to estimate the age of the enclosing mineral. Additionally, the constant size of the rings of the uranium and thorium halos for minerals of different ages was taken as evidence that the decay constants for radionuclides used in age dating had remained constant throughout geologic time. Thermal annealing of the halos has been used to model the thermal histories of rock units. *See* FISSION TRACK DATING; GEOCHRONOMETRY. Rod Ewing

Pleochroism In some colored transparent crystals, the effect wherein the color is quite different in different directions through the crystals. In such a crystal the absorption of light is different for different polarization directions. In colored transparent tourmaline the effect may be so strong that one polarized component of a light beam is wholly absorbed, and the crystal can be used as a polarizer. *See* DICHROISM; TRICHROISM. Bruce H. Billings

Plesiosauria An order of extinct aquatic diapsid reptiles within the infraclass Sauropterygia, common throughout the world during the Jurassic and Cretaceous periods. These large carnivores are characterized by long, paddle-shaped limbs and short, dorsoventrally compressed bodies. Unlike the nothosaurs (more primitive members of the Sauropterygia), plesiosaurs have greatly expanded ventral portions of both pectoral and pelvic girdles to provide large areas for the attachment of muscles to move the limbs anteriorly and posteriorly. Lateral undulation of the trunk was severely restricted by elaboration of the ventral scales.

Approximately 40 genera of plesiosaurs are recognized, divided into two large groups: The plesiosauroids include the most primitive genera and others that have small heads and very long necks. Among the plesiosauroids, the elasmosaurids had as many as 76 cervical vertebrae. The pliosauroids had shorter necks but larger skulls. Both elasmosaurs and pliosaurs persisted until the end of the Mesozoic. *See* DIAPSIDA; NOTHOSAURIA; REPTILIA; SAUROPTERYGIA. Robert L. Carroll

Pleuronectiformes A very distinctive order of actinopterygian fishes, also known as the Heterosomata or commonly the flatfishes, which in turn are known by a variety of common names: halibut, turbots, plaice, flounders, soles, tonguefishes, and others. All adults are characterized by a flattened (laterally compressed) body with both eyes located on the same side of the head. Larval flatfishes are bilaterally symmetrical, but early in development one eye migrates across the top of the head to the other side. The asymmetry is reflected in complex modifications of the skull bones and associated muscles and nerves, as well as dentition, squamation (arrangement of scales), visceral anatomy, and paired fins. Flatfishes are further characterized by pigmented upper side; blind side usually lacking pigment; cycloid or ctenoid scales; tubercles in some species; long-based dorsal and anal fins, with the dorsal fin base extending to or beyond the eyes (except in the family Psettodidae); body cavity quite reduced in size; usually six or seven branchiostegal (gill bone) rays; and a swim bladder in adults of only a few species. *See* ACTINOPTERYGII; ANIMAL SYMMETRY; SWIM BLADDER.

Flatfishes are invariably benthic (bottom-dwelling) as postlarvae, spending their entire adult lives swimming on their side (except *Psettodes*, which occasionally swims upright), with the eyeless side down. They are carnivorous predators, keeping much to the bottom. The eyes are protrusible, allowing these fishes to see while buried in the substrate awaiting a wandering prey. In pursuit of prey, the flatfishes move far from the substrate. Worldwide many species of flatfish are valued as food and are important to the commercial fisheries industry. *See* MARINE FISHERIES.

The order Pleuronectiformes comprises about 678 extant species in about 134 genera, 14 families, and two suborders (Psettodoidei and Pleuronectoidei). About 10 species of soles are thought to be limited to freshwater habitats; a few more, primarily freshwater species enter estuaries or marine water, and another 20 that are normally marine species occasionally enter freshwater. Herbert Boschung

Pliocene The youngest of the five geological epochs of the Tertiary Period. The Pliocene represents the interval of geological time (and rocks deposited during that time) extending from the end of the Miocene Epoch to the beginning of the Pleistocene Epoch of the Quaternary Period. Modern time scales assign the duration of 5.0 to 1.8 million years ago (Ma) to the Pliocene Epoch. *See* MIOCENE; PLEISTOCENE; QUATERNARY; TERTIARY.

Pliocene marine sediments are commonly distributed along relatively restricted areas of the continental margins and in the deep-sea basins. Continental margin sediments are most often terrigenous and range from coarser-grained sandstone to finer-grained mudstone and clay. Major rivers of the world, such as the Amazon, Indus, and Ganges, contain thick piles of Pliocene terrigenous sediments in their offshore fans. The Pliocene deep-sea sediments are carbonate-rich (commonly biogenic oozes) and are often very thick (up to 5000 m or 16,400 ft). *See* BASIN; CONTINENTAL MARGIN.

Modern stratigraphic usage subdivides the Pliocene Epoch into two standard stages, the lower, Zanclean stage and the upper, Piacenzian stage.

The most notable tectonic events in the Pliocene include the beginning of the third and last phase of the Himalayan uplift, the Attican orogeny that began in the late Miocene and continued into the Pliocene, and the Rhodanian and Walachian orogenies that occurred during the later Pliocene. *See* OROGENY.

The latest Miocene is marked by a global cooling period that continued into the earliest Pliocene, and there is evidence that the East Antarctic ice sheet had reached the continental margins at this time. The global sea level had been falling through the late Miocene, and with the exception of a marked rise in the mid-Zanclean, the trend toward lowered sea levels continued through the Pliocene and Pleistocene. The mid-Zanclean sea-level rise (3.5–3 Ma) was also accompanied by a significant global warming event. The oxygen isotopic data, which record the prevailing sea surface temperatures and total ice volume on the ice caps, show little variations in the Equatorial Pacific during the middle Pliocene. By early Pliocene time, the major surface circulation patterns of the world ocean and the sources of supply of bottom waters were essentially similar to their modern counterparts. *See* GEOLOGIC THERMOMETRY.

By Pliocene time, much of the marine and terrestrial biota had essentially evolved its modern characteristics. The late Pliocene cooling led to the expansion of cooler-water marine assemblages of the higher latitudes into lower latitudes, particularly the foraminifers, bivalves, and gastropods. At the onset of cooling, the warm-water-preferring calcareous nannoplankton group of discoasters began waning in the late Pliocene and became extinct at the close of the epoch. *See* BIVALVIA; COCCOLITHOPHORIDA; FORAMINIFERIDA; GASTROPODA.

The widespread grasslands of the Pliocene were conducive to the proliferation of mammals and increase in their average size. The mid-Zanclean sea-level rise led to the geographic isolation of many groups of mammals and the increase in endemism. But the late Pliocene-Pleistocene lowering of sea level facilitated land connections and allowed extensive mammalian migration between continents with interchanges between North and South America. The arrival of the North American mammals led to increased competitive pressure and extinction of many typically South American groups. Horses evolved and spread widely in the Pliocene. *See* MAMMALIA.

The Pliocene Epoch also saw the appearance of several hominid species that are considered to be directly related to modern human ancestry. The earliest hominid bones have been discovered from Baringo, Kenya, in sediments that are dated to be of earliest Pliocene age. After this first occurrence, a whole suite of australopithicine species made their appearance in the Pliocene. *See* AUSTRALOPITHECINE; FOSSIL HUMANS; PALEONTOLOGY.

Bilal U. Haq

Plum Any of the smooth-skinned stone fruits grown on shrubs or small trees. Plums are widely distributed in all land areas of the North Temperate Zone, where many species and varieties are adapted to different climatic and soil conditions.

There are four principal groups: (1) Domestica (*Prunus domestica*) of European or Southwest Asian origin, (2) Japanese or Salicina (*P. salicina*) of Chinese origin, (3) Insititia or Damson (*P. insititia*) of Eurasian origin, and (4) American (*P. americana* and *P. hortulana*). The Domesticas are large, meaty, prune-type plums. A prune is a plum which dries without spoiling.

R. Paul Larsen

Plumbaginales An order of flowering plants, division Magnoliophyta (Angiospermae), in the subclass Caryophyllidae of the class Magnoliopsida (dicotyledons). The order consists of only the family Plumbaginaceae, with about 400 species. The plants are herbs or less often shrubs. The flowers are strictly pentamerous (that is, each floral whorl has five members); the petals are fused (sympetalous condition) and all alike in shape and size; the pollen is trinucleate; and there is a single basal ovule located in a compound ovary that has a single locule or cell. The Plumbaginaceae differ from most families of their subclass in their straight embryo, copious endosperm, absence of perisperm, and the presence of anthocyanin pigments instead of betalains. The family contains a few garden ornamentals, such as species of *Armeria*, known as thrift or sea-pink. *See* CARYOPHYLLIDAE; MAGNOLIOPHYTA; MAGNOLIOPSIDA.

Arthur Cronquist; T. M. Barkley

Pluto Formerly considered the outermost planet, tiny Pluto is now recognized to be one of the largest and closest members of a disk of icy planetesimals that surrounds the solar system beyond the orbit of Neptune. For this reason, the International Astronomical Union resolved to reclassify Pluto in a new category, dwarf planet, in 2006. Pluto was discovered on February 18, 1930.

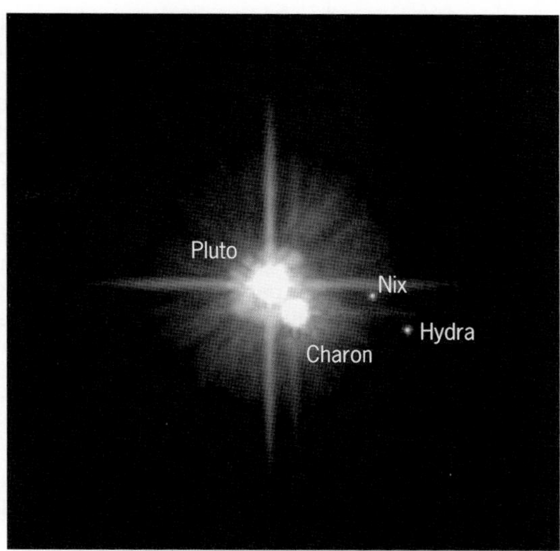

Pluto and its three known moons, imaged by the *Hubble Space Telescope* on February 15, 2006. (*NASA, ESA, H. Weaver (JHUAPL), A. Stern (SwRI), HST Pluto Companion Search Team*)

The near-infrared spectrum of Pluto reveals absorption features of solid nitrogen (N_2), carbon monoxide (CO), and methane (CH_4) ice. From these absorptions it is possible to conclude that nitrogen is the dominant ice on Pluto's surface (some 20–100 times more abundant than the other ices), which means it must also be the major constituent of the planet's tenuous atmosphere. The nitrogen absorption in the spectrum indicates a surface temperature near 38 K (−391°F), consistent with an atmospheric surface pressure roughly 10^{-5} times that on Earth.

Pluto has three small satellites: Charon, Nix, and Hydra, with Charon far larger than the other two (see illustration). Demonstrating a striking difference with Pluto, Charon apparently has little if any frozen methane on its surface, which instead exhibits the spectral signature of water ice. The orbit of Charon is unique in the solar system in that the satellite's period of revolution is identical to the rotational period of the planet. Thus an inhabitant of Pluto who lived on the appropriate hemisphere would see Charon hanging motionless in the sky.

The radius of Pluto is 1153 ± 10 km (716 ± 6 mi), making Pluto significantly smaller than the Earth's Moon (whose radius is 1738 km or 1080 mi), while Charon's radius is 603.5 ± 1.5 km (375.0 ± 0.9 mi). (The uncertainty in Pluto's radius includes the possibility that a lower atmosphere haze prevents the true surface of this dwarf planet from being measured by the occulation observation on which it is based.) Charon is thus about half the size of Pluto itself, making the most closely matched pair in the solar system. The density of Pluto is 2.0 g/cm^{-3}, while that of Charon is 1.65 g/cm^{-3}. These densities suggest compositions of ice and rock within the range exhibited by Saturn's regular satellites. Nix and Hydra are too small to permit determinations of size and mass.

Tobias C. Owen

Pluton A solid rock body that formed by cooling and crystallization of molten rock (magma) within the Earth. Most plutons, or plutonic bodies, are regarded as the product of crystallization of magma intruded into surrounding "country rocks" within the Earth (principally within the crust). Igneous rock bodies are referred to generally as either extrusive or volcanic on one hand, or as intrusive or plutonic on the other, although the term volcanic is sometimes also used to refer to small, shallow intrusive bodies associated with volcanoes. *See* IGNEOUS ROCKS; MAGMA.

Plutons occur in a nearly infinite variety of shapes and sizes, so that definition of types is arbitrary in many cases. In general, two modes of emplacement can be recognized with regard to the country rock. Concordant plutons are intruded between layers of stratified rock, whereas the more common discordant plutons are characterized by boundaries that cut across preexisting structures or layers in the country rock. The principal types of concordant plutons are sills, laccoliths, and lopoliths; the principal types of discordant plutons are dikes, volcanic necks or plugs, stocks, and batholiths.

Several mechanisms of magma intrusion are known or proposed. The most simple ones, pertaining to smaller plutonic bodies, are forceful injection or passive migration into fractures. Larger plutons may form by several processes. For example, less dense magma may migrate upward along a myriad of channelways to accumulate as a large molten body within the upper crust. Further migration could occur by forceful injection, by stoping (a process where the magma rises as blocks of the roof of the magma chamber break off and sink), and by diapiric rise, where country rocks flow around the upward-moving magma body. *See* PETROLOGY. W. Randall Van Schmus

Plutonium A chemical element, Pu, atomic number 94. Plutonium is a reactive, silvery metal in the actinide series of element elements. The principal isotope of chemical interest is ^{239}Pu, with a half-life of 24,131 years. It is formed in nuclear reactors by the process shown in the following reaction.

$$^{238}U + n \longrightarrow {}^{239}U \xrightarrow[23.5\ min]{\beta^-} {}^{239}Np \xrightarrow[2.33\ days]{\beta^-} {}^{239}Pu$$

Plutonium-239 is fissionable, but may also capture neutrons to form higher plutonium isotopes. *See* PERIODIC TABLE.

Plutonium-238, with a half-life of 87.7 years, is utilized in heat sources for space application, and has been used for heart pacemakers. Plutonium-239 is used as a nuclear fuel, in the production of radioactive isotopes for research, and as the fissile agent in nuclear weapons.

Plutonium exhibits a variety of valence states in solution and in the solid state. Plutonium metal is highly electropositive. Numerous alloys of plutonium have been prepared, and a large number of intermetallic compounds have been characterized.

Reaction of the metal with hydrogen yields two hydrides. The hydrides are formed at temperatures as low as 150°C (300°F). Their decomposition above 750°C (1400°F) may be used to prepare reactive plutonium powder. The most common oxide is PuO_2, which is formed by ignition of hydroxides, oxalates, peroxides, and nitrates of any oxidation state in air of 870–1200°C (1600–2200°F). A very important class of plutonium compounds are the halides and oxyhalides. Plutonium hexafluoride, the most volatile plutonium compound known, is a strong fluorinating agent. A number of other binary compounds are known. Among these are the carbides, silicides, sulfides, and selenides, which are of particular interest because of their refractory nature.

Because of its radiotoxicity, plutonium and its compounds require special handling techniques to prevent ingestion or inhalation. Therefore, all work with plutonium and its compounds must be carried out inside glove boxes. For work with plutonium and its alloys, which are attacked by moisture and by atmospheric gases, these boxes may be filled with helium or argon. *See* ACTINIDE ELEMENTS; NEPTUNIUM; NUCLEAR CHEMISTRY; TRANSURANIUM ELEMENTS; URANIUM. Fritz Weigel

Plywood A wood product in which thin sheets of wood are glued together, grains of adjacent sheets being at right angles to each other in the principal plane. Because of this cross-grained orientation, mechanical properties are less directional than those of natural lumber and more dimensionally stable. Tree farms are now cultivated specifically to yield logs suitable for processing into sheets for plywood.

The American Plywood Association identifies several grades of product. Plywood is designated group 1 when made from northern-grown Douglas-fir, western larch, and such southern pines as loblolly and longleaf, or other woods noted for their strength. Plywoods in groups 2, 3, and 4 are made from woods of successively lower strengths. Consequently, group 1 plywood offers the greatest stiffness, group 4 the least.

Plywood with waterproof glue is designated exterior type; it is also used interiorly where moisture is present. Plywood with nonmoisture-resistant glue is designated interior type; it can withstand an occasional soaking but neither repeated soakings nor continuous high humidity.

Veneer grades A through D extend from a smooth surface to a surface with occasional knotholes and limited splits. If the outer face of the plywood is cut from only heartwood or sapwood, free from open defects, the plywood is assigned veneer grade N, indicating that it will take a natural finish. *See* STEM; VENEER.

Most commonly used plywoods are $^1/_4$-in. (0.6-cm) sanded interior paneling or $^1/_2$-in. (1.3-cm) exterior grade plywood sheeting. Other standard thicknesses extend to 1 in. (2.5 cm) for interior types and to $1^1/_8$ in. (2.8 cm) for exterior types. The most common panel size is 4×8 ft (1.2×2.4 m); larger sizes are manufactured for such special purposes as boat hulls.

Finished plywood may be unsanded, sanded, or overlayed with several types of coatings for decorative and specialty uses. Plywood in appropriate grades is used in many different applications, such as furniture, wall facings, shelving, containers, crates, fences, forms, subflooring, and roof decking. *See* WOOD PRODUCTS. Frank H. Rockett; Roger M. Rowell

Pneumatolysis The alteration of rocks or crystallization of minerals by gases or supercritical fluids (generically termed magmatic fluids) derived from solidifying magma. At surface conditions, magmatic fluids contain steam with lesser amounts of carbon dioxide, sulfur dioxide, hydrogen sulfide, hydrogen chloride, and hydrogen fluoride, and trace amounts of many other volatile constituents. Magmatic fluids may contain relatively high concentrations of light and heavy elements, particularly metals, that do not crystallize readily in common rock-forming silicates constituting most of the solidifying magma; thus, valuable rare minerals and ores are sometimes deposited in rocks subjected to pneumatolysis. Magmatic fluids are acidic and may react extensively with rocks in the volcanic edifice or with wall rocks surrounding intrusions. Penetration of magmatic fluids into adjacent rocks is greatly aided by faults, fractures, and cracks developed during intrusion and eruption or created by earlier geologic events. *See* MAGMA; METAMORPHISM; METASOMATISM; ORE AND MINERAL DEPOSITS; VOLCANO.

Pneumatolysis describes specific mechanisms of mineral deposition, hydrothermal alteration, or metasomatism in which magmatic fluids play an extremely significant role. For example, lavas and ejecta at volcanoes may contain blocks (xenoliths) of wall rock that react with magmatic fluids to form pneumatolytic minerals such as vesuvianite (idocrase). Gases streaming from volcanic fumaroles deposit sublimates of sulfur, sulfates, chlorides, fluorides, and oxides of many metals. Wall rocks surrounding volcanic conduits may be thoroughly altered to mixtures of quartz, alunite, anhydrite, pyrite, diaspore, kaolin, as well as other minerals by acidic fluids degassed from magma. Rarely, gold, silver, base-metal sulfides, arsenides, and tellurides are deposited by the fluids, making valuable ores. *See* LAVA; METASOMATISM; PYROCLASTIC ROCKS; SUBLIMATION; VESUVIANITE; XENOLITH. Fraser Goff

Pneumonia An acute or chronic inflammatory disease of the lungs. More specifically when inflammation is caused by an infectious agent, the condition is called pneumonia; when the inflammatory process in the lung is not related to an infectious organism, it is called pneumonitis.

An estimated 45 million cases of infectious pneumonia occur annually in the United States, with up to 50,000 deaths directly attributable to it. Pneumonia is a common immediate cause of death in persons with a variety of underlying diseases. With the use of immunosuppressive and chemotherapeutic agents for treating transplant and cancer patients, pneumonia caused by infectious agents that usually do not cause infections in healthy persons (that is, pneumonia as an opportunistic infection) has become commonplace. Moreover, individuals with acquired immune deficiency syndrome (AIDS) usually die from an opportunistic infection, such as pneumocystis pneumonia or cytomegalovirus pneumonia. Concurrent with the variable and expanding etiology of pneumonia and the more frequent occurrence of opportunistic infections is the development of new antibiotics and other drugs used in the treatment of pneumonia. *See* ACQUIRED IMMUNE DEFICIENCY SYNDROME (AIDS); OPPORTUNISTIC INFECTIONS.

Bacteria, as a group, are the most common cause of infectious pneumonia, although influenza virus has replaced *Streptococcus pneumoniae* (*Diplococcus pneumoniae*) as the most common single agent. Some of the bacteria are normal inhabitants of the body and proliferate to cause disease only under certain conditions. Other bacteria are contaminants of food or water.

Most bacteria cause one of two main morphologic forms of inflammation in the lung. *Streptococcus pneumoniae* causes lobar pneumonia, in which an entire lobe of a lung or a large portion of a lobe becomes consolidated (firm, dense) and nonfunctional secondary to an influx of fluid and acute inflammatory cells that represent a reaction to the bacteria. This type of pneumonia is uncommon today, usually occurring in people who have poor hygiene and are debilitated. If lobar pneumonia is treated adequately, the inflammatory process may entirely disappear, although in some instances it undergoes a process called organization, in which the inflammatory tissue changes into fibrous tissue, usually rendering that portion of the lung nonfunctional.

The other morphologic form of pneumonia, which is caused by the majority of bacteria, is called bronchopneumonia. In this form there is patchy consolidation of lung tissue, usually around the small bronchi and bronchioles, again most frequently in the lower lobes. This type of pneumonia may also undergo complete resolution if there is adequate treatment, although rarely it organizes.

Viral pneumonia is usually a diffuse process throughout the lung and produces a different type of inflammatory reaction than is seen in bronchopneumonia or lobar pneumonia. Mycoplasma pneumonia, caused by *Mycoplasma pneumoniae*, is referred to as primary atypical pneumonia and causes an inflammatory reaction similar to that of viral pneumonia.

Pneumonia can be caused by a variety of other fungal organisms, especially in debilitated persons such as those with cancer or AIDS. *Mycobacterium tuberculosis*, the causative agent of pulmonary tuberculosis, produces an inflammatory reaction similar to fungal organisms. *See* MYCOBACTERIAL DISEASES; TUBERCULOSIS.

Legionella pneumonia, initially called Legionnaires' disease, is caused by bacteria of the genus *Legionella*. The condition is frequently referred to under the broader name of legionellosis. *See* LEGIONNAIRES' DISEASE.

The signs and symptoms of pneumonia and pneumonitis are usually nonspecific, consisting of fever, chills, shortness of breath, and chest pain. Fever and chills are more frequently associated with infectious pneumonias but may also be seen in pneumonitis. The physical examination of a person with pneumonia or pneumonitis may reveal abnormal lung sounds indicative of regions of consolidation of lung tissue. A chest x-ray also shows the consolidation, which appears as an area of increased opacity (white area). Cultures of sputum or bronchial secretions may identify an infectious organism capable of causing the pneumonia.

The treatment of pneumonia and pneumonitis depends on the cause. Bacterial pneumonias are treated with antimicrobial agents. If the organisms can be cultured, the sensitivity of the organism to a specific antibiotic can be determined. Viral pneumonia is difficult to treat, as most drugs only help control the symptoms. The treatment of pneumonitis depends on identifying its cause; many cases are treated with cortisone-type medicines.

Samuel P. Hammar

Poales An order of flowering plants (angiosperms) that includes the bromeliads, grasses, restios, rushes, and sedges. It comprises approximately 18 families and more than 18,000 species, over half of which are grasses (Poaceae). These families in general include plants without showy flowers (except for Bromeliaceae, Xyridaceae, and Rapateaceae) and with a grasslike form. These plants dominate vast areas of the Earth's surface and include some important grain crops, such as corn (*Zea mays*), rice (*Oryza sativa*), and wheat (*Triticum aestivum*), as well as rushes, sedges, grasses used for thatch, and bamboos used for construction. *See* DUNE VEGETATION; GRAIN CROPS; GRASS CROPS.

Mark W. Chase; Michael F. Fay

Podicipediformes A small order of aquatic birds that contains only a single living family, the Podicipedidae (grebes), with 20 species found throughout the world. Grebes are small to medium-sized birds that are specialized for aquatic life and can swim and dive skillfully. Their legs are placed far posteriorly with lobed—not webbed—toes that are used to propel the birds through the water, but their terrestrial locomotion is awkward. Their wings are small, and they rarely fly, preferring to dive to escape enemies; yet they migrate over long distances and have colonized many remote islands. Grebe bodies are compact with dense waterproof plumage. The neck is medium to long, the head is large, and the bill is short to medium and pointed. Grebes feed on fish and other aquatic animals that they pursue underwater. Grebes are monogamous, with an elaborate courtship and strong pair bond. They breed on freshwater, but some species winter along the coasts. *See* AVES.

Walter J. Bock

Podocopa One of two Recent subclasses of Ostracoda. The two may share many characteristics but differ significantly in others. Shared characteristics include a carapace (shell) with a straight-to-concave ventral margin and somewhat convex dorsal margin; adductor muscle-scar patterns varying from numerous to few; the absence of an anterior notch in the bivalve shell; and the absence of lateral eyes, heart, and frontal organ (Bellonci organ). Podocopid ostracods are small, rarely exceeding 0.25 in. (7 mm). The carapace valves are unequal, weakly or strongly calcified, and may be heavily ornamented. The hinge varies from simple to complex. Six or seven pairs of appendages are present; the seventh is adapted for walking, for cleaning, or as a clasping leg in males. The subclass is represented by the two extant orders, Platycopida and Podocopida.

Podocopid ostracods are found in fresh-water, marine, and brackish-water environments, and a few are terrestrial. The majority are burrowing or crawling inhabitants of the benthos and are filter or detritus feeders or herbivores. Some are capable of swimming, but none appear to ever lead a planktonic existence. *See* OSTRACODA.

Patsy A. McLaughlin

Podocopida An order of the subclass Podocopa, class Ostracoda, that consists of the extant suborder Podocopina and the extinct Metacopina. There is agreement on the assignment of the superfamilies Cypridoidea, Bairdioidea, Cytheroidea, and Darwinuloidea to the Podocopida; however, some researchers also include the Sigillioidea, and others consider the Terrestrichtheroidea a podocopid taxon of equivalent rank. In all podocopids the two valves fit firmly, hermetically sealing the animal inside when closed.

Appendages include antennules, antennae, mandibles, maxillules, and three pairs of thoracic legs. Antennules and antennae are both used in locomotion; swimming forms are provided with long, feathered setae, whereas crawling forms have only short, stout setae. Mandibles are strongly constructed and, with rare exceptions, used for mastication. The maxilla has setiferous endites and bears a large branchial plate to circulate water alongside the body. The furca is variously developed but never lamelliform. If eyes are present, the two lateral eyes and the median eye are joined into one central structure. No heart is developed. *See* CRUSTACEA; PODOCOPA. Patsy A. McLaughlin

Podostemales An order of flowering plants, division Magnoliophyta (Angiospermae), in the subclass Rosidae of the class Magnoliopsida (dicotyledons). The order consists of only the family Podostemaceae, with about 200 species, the greatest number occurring in tropical America. They are submerged aquatics with modified, branching, often thalluslike shoots, and small, perfect flowers with a much reduced perianth. *See* MAGNOLIOPSIDA; ROSIDAE. Arthur Cronquist

Poecilosclerida An order of sponges of the class Demospongiae in which the skeleton includes two or more types of megascleres, each localized in a particular part of the sponge colony. Frequently one type of megasclere is restricted to the dermis and another type occurs in the interior of the sponge. Sometimes one category is embedded in spongin fibers while a second category, usually spinose, protrudes from the fibers at right angles. Spongin is always present but varies in amount from species to species. Microscleres are usually present; often several types occur in one species. A wide variety of microsclere categories is found in the order but asters are never present. *See* DEMOSPONGIAE. Willard D. Hartman

Poecilostomatoida One of two major orders of parasitic Copepoda that were previously included in the Cyclopoida. The classification of parasitic copepods has been established on the basis of the structure of the mouth. In poecilostomatoids the mouth is represented by a transverse slit, partially covered by the overhanging labrum, which resembles an upper lip. Although there is variability in the form of the mandible among poecilostomatoids, it can be generalized as being falcate. Body segmentation is typically podoplean, having prosome-urosome articulation between the fifth and sixth thoracic somites; however, this segmentation is often lost with the molt to adulthood. The antennules frequently are reduced in size and the antennae modified to terminate in small hooks or claws that are used in attachment to host organisms.

Most poecilostomatoid copepods are ectoparasites of marine fishes or invertebrates, usually attaching to the external surface of the host or in the branchial cavity on the walls or gill surfaces. Representatives of one family, however, have successfully made the transition to fresh-water habitats, and a second family has evolved an endoparasitic mode of life. *See* COPEPODA; CRUSTACEA; CYCLOPOIDA. Patsy A. McLaughlin

Pogonophora (Siboglinidae) A family of sedentary marine worms (known as beard worms) that live in all the world's oceans, generally at depths between 330 and 13,200 ft (100 and 4000 m), shallower at higher latitudes and deeper in trenches. The first specimens of the Pogonophora, previously considered to be a phylum, but now known more accurately as Siboglinidae (a family of polychaete annelids), were dredged up late in the nineteenth century, but first investigated in the 1950s. Siboglinids construct a chitinous tube and have no mouth, gut, or anus in their postembryonic anatomy. Instead of actively feeding, they derive their nutrition from symbiotic bacteria hosted in specialized tissues or via absorption. With new technologies in deep-sea exploration, many new species of Siboglinidae have been discovered in unexpected habitats over the past three decades, notably around hydrothermal vents and cold-water hydrocarbon seeps, and on decaying whale bones. *See* DEEP-SEA FAUNA; HYDROTHERMAL VENT; MARINE ECOLOGY.

The body is generally divided into four regions. The anterior region (cephalic lobe) bears from one to thousands of tentacles or branchial filaments and is followed by a short collar region known as the forepart or vestimentum. The largest section is usually an extended trunk region, which can have regional specializations such as papillae, cilia, and chaetae. In the frenulates, vestimentiferans, and moniliferans, the terminal region (opisthosoma) has the internal septa and external chaetae typical of polychaete annelids. *See* ANNELIDA. Anja Schulze; Edward B. Cutler

Poison A substance which by chemical action and at low dosage can kill or injure living organisms. Broadly defined, poisons include chemicals toxic for any living form: microbes, plants, or animals. In common usage the word is limited to substances toxic for humans and mammals, particularly where toxicity is a substance's major property of medical interest. Because of their diversity in origin, chemistry, and toxic action, poisons defy any simple classification. Almost all chemicals with recognized physiological effects are toxic at sufficient dosage.

Origin and chemistry. Many poisons are of natural origin. Some bacteria secrete toxic proteins (for example, botulinus, diphtheria, and tetanus toxins) that are among the most poisonous compounds known. Lower plants notorious for poisonous properties are ergot (*Claviceps purpurea*) and a variety of toxic mushrooms. *See* ERGOT AND ERGOTISM; MUSHROOM; TOXIN.

Higher plants, which constitute the major natural source of drugs, contain a great variety of poisonous substances. Many of the plant alkaloids double as drugs or poisons, depending on dose. These include curare, quinine, atropine, mescaline, morphine, nicotine, cocaine, picrotoxin, strychnine, lysergic acid, and many others. *See* ATROPINE; COCAINE; MORPHINE ALKALOIDS; QUININE.

Poisons of animal origin (venoms) are similarly diverse. Toxic marine animals alone include examples of every phylum. Insects and snakes represent the best-known venomous land animals, but on land, too, all phyla include poison-producing species. Among mammalian examples are certain shrews with poison-producing salivary glands. *See* POISON GLAND.

Poisons of nonliving origin vary in chemical complexity from the toxic elements, for example, the heavy metals, to complex synthetic organic molecules. Most of the heavy metals (gold, silver, mercury, arsenic, and lead) are poisons of high potency in the form of their soluble salts. Strong acids or bases are toxic largely because of corrosive local tissue injury.

The chemically reactive gases hydrogen sulfide, hydrocyanic acid, chlorine, bromine, and ammonia are also toxic, even at low concentration, both because of their corrosiveness and because of more subtle chemical interaction with enzymes or other cell constituents.

Many organic substances of synthetic origin are highly toxic and represent a major source of industrial hazard. Most organic solvents are more or less toxic on ingestion or inhalation. Many alcohols, such as methanol, are much more toxic. Many solvents (for example, carbon tetrachloride, tetrachloroethane, dioxane, and ethylene glycol) produce severe chemical injury to the liver and other viscera, sometimes from rather low dosage.

Physiological actions. The action of poisons is generally described by the physiological or biochemical changes which they produce. For most poisons, a descriptive account can be given which indicates what organic system (for example, heart, kidney, liver, brain, and bone marrow) appears to be most critically involved and contributes most to seriously disordered body function or death. In many cases, however, organ effects are multiple, or functional derangements so generalized that a cause of death cannot be localized.

More precise understanding of the mechanism of poisons requires detailed knowledge of their action in chemical terms. Information of this kind is available for only a few compounds, and then in only fragmentary detail. Poisons that inhibit acetylcholinesterase have toxic actions traceable to a single blocked enzyme reaction, hydrolysis of normally secreted acetylcholine. Detailed understanding of the mechanism of chemical inhibition of cholinesterase is not complete, but allows some prediction of chemical structures likely to act as inhibitors. *See* ACETYLCHOLINE.

Carbon monoxide toxicity is also partly understood in chemical terms, since formation of carboxyhemoglobin, a form incapable of oxygen transport, is sufficient to explain the anoxic features of toxicity.

Heavy metal poisoning in many cases is thought to involve inhibition of enzymes by formation of metal mercaptides with enzyme sulfhydryl groups, the unsubstituted form of which is necessary for enzyme action. This is a general reaction that may occur with a variety of sulfhydryl-containing enzymes in the body. Specific susceptible enzymes whose inhibition explains toxicity have not yet been well documented.

Metabolic antagonists active as poisons function by competitive blocking of normal metabolic reactions. Some antagonists may act directly as enzyme inhibitors, others may be enzymatically altered to form derivatives which are even more potent inhibitors at a later metabolic step. *See* ENZYME INHIBITION.

Where poison mechanisms are relatively well understood, it has sometimes been possible to employ rationally selected antidotes.

Potency. The strength or potency of poisons is most frequently measured by the lethal dose, potency being inversely proportional to lethal dose. From statistically treated dose-response data, the dose killing 50% of the sample population can be determined, and is usually designated the MLD (median lethal dose) or LD_{50}. This is the commonest measure of toxic potency. *See* LETHAL DOSE 50; TOXICOLOGY. Elijah Adams

Poison gland The specialized gland of certain fishes, as well as the granular glands and some mucous glands of many aquatic and terrestrial Amphibia. The poison glands of fishes are simple or slightly branched acinous structures which use the holocrine method of secreting a mucuslike substance. The poison glands of snakes are modified oral or salivary glands. Amphibian glands are simple, acinous, holocrine, with granular secretion. In some cases these amphibian poison glands produce mucus by a merocrine method of secretion. These glands function as protective devices. *See* GLAND. Olin E. Nelsen

Poison ivy A general name applied to certain species of the genus *Toxicodendron*, previously known as *Rhus*, in the sumac family (Anacardiaceae). *Toxicodendron radicans* is the poison ivy of eastern North America; *T. diversiloba* is the poison oak of California. These plants are natives of North America. Both cause ivy poisoning, an annoying and often painful dermatitis. *Toxicodendron radicans*, the most widespread species, is extremely variable. It has a bushy or climbing habit and threefoliolate leaves which are smooth and glossy or hairy and are entire, toothed, or lobed. Poison ivy bears white fruits whereas the nonpoisonous sumacs bear red fruits. *See* SAPINDALES.

Perry D. Strausbaugh; Earl L. Core

Poison sumac The plant *Toxicodendron vernix* (previously in the genus *Rhus*), a member of the sumac family (Anacardiaceae). It is an inhabitant of swamps ranging from Quebec to Minnesota, and southward to Florida, Louisiana, and Texas. It is a tall bush or small tree bearing pinnately compound leaves with 7–13 entire (without marginal teeth) leaflets, and drooping, axillary clusters of persisting white fruits (see illustration). Like poison ivy, this plant is poisonous to touch, causing in many persons a severe inflammation of the skin, or dermatitis. The presence of white fruit separates this species from the nonpoisonous sumacs with their red fruits. *See* SAPINDALES.

Poison sumac fruits and leaf (*Toxicodendron vernix*).

Perry D. Strausbaugh; Earl L. Core

Poisonous plants More than 700 species of seed plants, ferns, horsetails, and fungi that cause toxic, though rarely fatal, reactions in humans and animals. Human allergic responses, including hay fever, asthma, and dermatitis, are widespread. Allergic responses are produced by many different plant species, but most common are poison ivy, poison sumac, and Pacific poison oak (all are species of *Toxicodendron*). Internal injury by toxic plants is less common but can be detrimental or lethal. *See* ASTHMA; POISON IVY; POISON SUMAC.

Glycoside-containing plants. Glycosides are common compounds in plants. They decompose to form one or more sugars, but sometimes the remaining compounds, aglycones, can be quite poisonous. Cyanogenic glycosides, which produce hydrocyanic acid, are found worldwide in many plant families; the best known are in the rose family (Rosaceae) and in the pea family (Fabaceae). Leaves, bark, and seeds of stone fruits such as cultivated and wild cherries, plums, peaches, bitter almonds, and apricots contain the glycoside amygdalin, which hydrolyzes to form hydrocyanic acid that can be fatally toxic to humans or animals. The same toxic substance is found in apple and pear seeds. Cardiac glycosides are found in many unrelated species of plants. Those of the foxglove (*Digitalis purpurea*) contain a number of these glycosides used medicinally to slow and strengthen the heartbeat. Oleander (*Nerium oleander*), which is cultivated in the warmer parts of the United States, contains a toxic glycoside that has an action similar to that of digitalis. *See* CYANIDE; DIGITALIS; GLYCOSIDE.

Alkaloid-containing plants. Alkaloids, compounds containing a nitrogen atom, have specific pharmacological effects on both humans and animals. Found in many different plant families, they have been used in drug therapy since ancient times, but misuse of these plants can produce poisonings. The potato family (Solanaceae) has many species that contain a number of alkaloids. Hyoscyamine and atropine are the alkaloids occurring in belladonna or deadly nightshade (*Atropa belladonna*), black henbane (*Hyoscyamus niger*), thornapples and jimsonweed (*Datura*), and tree daturas or angel's-trumpets (*Brugmansia*). The black nightshades (*Solanum*) contain glycoalkaloids. Plants of tobacco, *Nicotiana*, contain numerous alkaloids, principally the very toxic nicotine or its isomer anabasine. Plants of poison hemlock (*Conium macalatum*) have several alkaloids similar to nicotine, which affect the central nervous system. Plants of rattlebox (*Crotalaria*), groundsel (*Senecio*), and fiddleneck

(*Amsinckia*) have alkaloids of similar molecular structure. Anagyrine is a toxic alkaloid found in several species of lupine (*Lupinus*) in the western United States. Alkaloids present in species of monkshood (*Aconitum*) are extremely toxic. Larkspur plants (*Delphinium*) have similar toxic alkaloids affecting the central nervous system, causing excitability and muscular spasms. Plants of false hellebores (*Veratrum*) and death camas (*Zigadenus*) have complex alkaloids of similar structure and cause livestock deaths in the western United States. *See* ALKALOID; ATROPINE.

Heath plants. Toxic resins, andromedotoxins, formed by members of the heath family (Ericaceae) are derived from diterpenes. The most toxic species are mountain laurel, sheep laurel, and bog laurel, all in the genus *Kalmia*.

Pokeweed. Pokeweed (*Phytolacca americana*) is a garden weed throughout the United States. but it is native to the eastern and central areas. The entire plant, especially the seeds and the large root, is poisonous. Human poisonings have resulted from inadvertently including parts of the root along with the shoots.

Waterhemlock. The waterhemlocks (*Cicuta*) are widespread in North America. The large underground tubers, mistakenly considered edible, have caused human poisonings and death. Livestock usually do not eat waterhemlock, but have been fatally poisoned. The highly toxic principle is an unsaturated aliphatic alcohol which acts directly on the central nervous system.

Oxalate poisoning. Oxalic acid, as oxalate salts, accumulates in large amounts in some species of plants such as those of the genera *Halogeton*, *Bassia*, *Rumex*, and *Oxalis*.

Nitrate poisoning. Nitrate poisoning is widespread and results in many cattle deaths yearly. Any disruption of the normal synthesis of nitrates into amino acids and proteins causes large accumulations of nitrates in various species of plants, particularly in the goosefoot family (Chenopodiaceae).

Allelopathic toxins. Allelopathic phytotoxins are chemical compounds produced by vascular plants that inhibit the growth of other vascular plants. Residues of grain sorghum (*Sorghum bicolor*) can markedly reduce the following year's growth of wheat and some weedy grass seedlings. Locoweeds, belonging to the genera *Oxytropis* and *Astragalus*, produce an unknown toxin that causes loss of livestock that become addicted to eating these unpalatable plants. Other species of *Astragalus* produce toxic aliphatic nitro compounds. Still other species of *Astragalus* accumulate toxic quantities of selenium which usually causes chronic poisoning of livestock. *See* ALLELOPATHY.

Fungi. Every year human fatalities occur from ingestion of wild poisonous mushrooms. Those gathered in the wild require individual identification since toxic species may grow alongside edible ones. Ergot fungus, infecting many species of grasses, causes widespread poisoning of livestock. *See* ERGOT AND ERGOTISM; TOXICOLOGY.

Thomas C. Fuller

Polar meteorology

Polar meteorology The science of weather and climate in the high latitudes of the Earth. In the polar regions the Sun never rises far above the horizon and remains below it continuously for part of the year, so that snow and ice can persist for long periods even at low elevations. The meteorological processes that result have distinctive local and large-scale characteristics in both polar regions.

Uwe Radok

Polar molecule

Polar molecule A molecule possessing a permanent electric dipole moment. Molecules containing atoms of more than one element are polar except where forbidden by symmetry; molecules formed from atoms of a single element are nonpolar (except ozone). The dipole moments of polar molecules result in stronger intermolecular attraction, increased viscosities, higher melting and boiling points, and greater solubility in polar solvents than in nonpolar molecules.

Robert D. Waldron

Polar navigation

Polar navigation The complex of navigational techniques modified from those used in other areas to suit the distinctive regional character of polar areas. Although polar navigation has become routine to a rising number of navigators operating in and through such high-latitude parts of the word, their success continues to be based on a sound grasp of the regional differences and the developing adaptations of navigational principles and aids to suit these peculiar area needs. For example, in polar regions the meridians radiate outward from the poles, and parallels are concentric circles. Thus the rectangular coordinates familiar to the navigator accustomed to using the Mercator projection are replaced by polar coordinates.

Limitations. Piloting in polar regions is strongly affected by the absence of any great number of aids to navigation. Also, natural landmarks may not be shown on the chart, or may be difficult to identify. The appearance of some landmarks changes markedly under different ice conditions. When snow covers both the land and a wide ice foot attached to the shore and extending for miles seaward, even the shoreline is difficult to locate. *See* PILOTING.

Charts of polar regions are less reliable than those of other regions, because relatively little surveying has been done in the polar areas. Because relatively few soundings are shown on charts, ships entering harbors often send small boats ahead to determine the depth of water available. However, the reliability of charts of polar regions has steadily improved as additional information has become available.

Coverage of electronic navigation systems has also improved, but is still somewhat limited. Loran C sky waves are available throughout the Arctic, and ground waves extend to some parts of this area, but neither ground waves nor sky waves are available in the Antarctic. Radar is useful, but experience in interpretation of the scope in polar regions is essential for reliable results. This is particularly true in aircraft, where the relative appearance of water and land areas often reverse in winter and summer. A radio direction finder is useful, when radio signals are available. The use of electronics in polar regions is further restricted by magnetic storms, which are particularly severe in the auroral zones. *See* ELECTRONIC NAVIGATION SYSTEMS; RADAR.

Reliable dead reckoning depends upon the availability of accurate measurement of direction and distance (or speed and time). There are difficulties in meeting these requirements in polar regions. Direction is measured largely by a compass. The magnetic compass becomes unreliable in the vicinity of the magnetic poles of the Earth, and the north-seeking gyrocompass becomes unreliable in the vicinity of the geographical poles of the Earth. One solution to the directional problem in high latitudes is to use a directional gyro, a gyroscopic device that maintains its axis in a set direction but must be reset at frequent intervals, as every 15 min, because of gyroscopic drift. One of the devices discussed below for celestial direction determination can provide the directional data needed for setting or resetting a directional gyro. Directional gyros are not generally used aboard ship, and in aircraft they have been largely replaced by inertial navigators, which are also used by submarines and to a limited extent by surface ships. However, because inertial systems are self-contained, without reference to external signals, the insertion of incorrect data is always a possibility. *See* AIRCRAFT COMPASS SYSTEM; DEAD RECKONING; GYROSCOPE; INERTIAL NAVIGATION SYSTEM.

Several types of devices have been developed to facilitate the use of celestial bodies for determination of direction. The oldest is the sun compass, which utilizes the shadow of a shadow pin, or gnomon, and a suitable dial. A sky compass indicates direction of the Sun by means of polarized light in the sky when the Sun is near the horizon, even though it may be below the horizon or otherwise obscured. This device may offer the only means of determining direction during the brighter part of the long polar twilight. An astrocompass can be set for the coordinates of any celestial body and the latitude of the observer, and then gives an indication of azimuth, true north, and heading.

Distance or speed measurement in polar regions presents no problems in aircraft. When ships operate in ice, however, the sensing element in the water may be adversely affected or damaged by the ice. At best, dead reckoning is difficult aboard a ship operating in the ice, not only because of the difficulty encountered in measuring course and speed, but also because neither of these may be constant for very long.

Celestial navigation is of great importance in polar regions, sometimes providing the only means of determining position accurately, or establishing a directional reference. When operating in lower latitudes, navigators generally avoid observation of bodies near the horizon because of the uncertainty of the refraction correction there. In polar regions, even though refraction is more uncertain, navigators often have no choice. Near the equinoxes the Sun may be the only body available for several weeks, and it remains close to the horizon. During the polar summer the Sun is often the only celestial body available. When only one body is available, it is observed at frequent intervals, perhaps hourly, and a series of running fixes is plotted. *See* CELESTIAL NAVIGATION.

Modern technology. Advances in technology have contributed significantly to the safety and reliability of navigation in polar regions. The availability of modern echo sounders has made possible a continuous plot of the bottom profile beneath a vessel while under way. Sonar indicates the presence of an underwater obstruction. Reliable inertial navigators are available to make aircraft navigation in polar regions almost routine, providing both positional and directional data. Ship inertial navigation systems are available and have made practicable submarine operations under sea ice. *See* ECHO SOUNDER; SONAR.

The NAVSTAR Global Positioning System (GPS) of the U.S. Department of Defense has eliminated the limitations of the now defunct Navy Navigation Satellite System. Although all GPS satellites operate on the same frequencies (1227 and 1575 MHz), signals are modulated with two codes, one of which provides for identification of and lock-on to the desired signal. *See* NAVIGATION; SATELLITE NAVIGATION SYSTEMS.

Alton B Moody

Polarimetric analysis

Polarimetric analysis A method of chemical analysis based on the optical activity of the substance being determined. Optically active materials are asymmetric; that is, their molecules or crystals have no plane or center of symmetry. These asymmetric molecules can occur in either of two forms, *d*- and *l*-, called optical isomers. Asymmetric substances possess the power of rotating the plane of polarization of plane-polarized light. Measurement of the extent of this rotation, polarimetry, is performed by an instrument known as a polarimeter. Polarimetry is applied to both organic and inorganic materials. *See* OPTICAL ACTIVITY.

The extent of the rotation depends on the character of the substance, the length of the light path, the temperature of the solution, the wavelength of the light which is being used, the solvent (if there is one), and the concentration of the substance. In most work, the yellow light of the D line of the sodium spectrum (589.3 nanometers) is used to determine the specific rotation, according to the equation below. Here α is the measured angle

$$\text{Specific rotation} = [\alpha]_{\mathrm{D}}^{20} = \frac{\alpha}{l\rho}$$

of rotation, l is the length of the column of liquid in decimeters, and ρ the density of the solution. In other words, the specific rotation is the rotation in degrees which this plane-polarized light of the sodium D line undergoes in passing through a 10-cm-long (4-in.) sample tube containing a solution of 1 g/ml concentration at 20°C (68°F).

In the illustration, light from the sodium lamp is polarized by the polarizer (prism). It then passes through the cell containing the material being analyzed. After that, it passes through the analyzer (another prism) and then is detected (by eye or photocell). A comparison of the angular orientation of the analyzer as measured on the scale with the cell empty and with the cell filled with solution serves to measure the rotation of the polarized light by the sample. This rotation may be either clockwise (+) or counterclockwise (−).

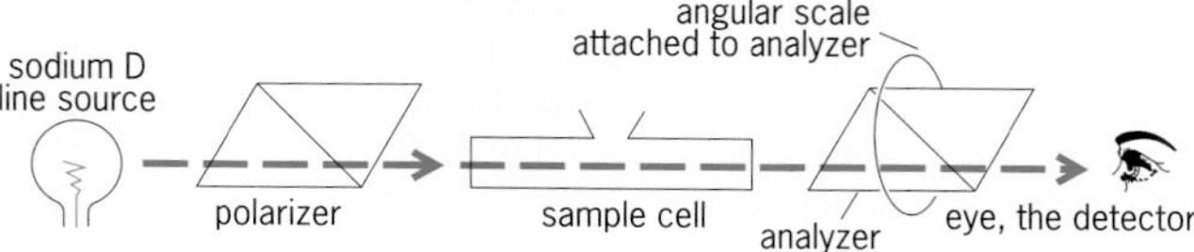

Simplified diagram of a polarimeter.

Polarimetry may be used for either qualitative or quantitative analytical work. In qualitative applications, the presence of an optically active material is shown, and then a calculation of specific rotation often leads to the identification of the unknown. In quantitative work, the concentration of a given optically active material is determined.

Polarimetry is used in carbohydrate chemistry, especially in the analysis of sugar solutions. Since there is great difference between the biological activities of the different optical forms of organic compounds, polarimetry is used in biochemical research to identify the molecular configurations.

Optical rotatory dispersion is the measurement of the specific rotation as a function of wavelength. The information obtained by this method has shown that minor changes in configuration of a molecule have a marked effect on its dispersion properties. By using the properties of compounds of known configuration, it has been possible to determine the absolute configurations of many other molecules and to identify various isomers. Most of the applications have been to steroids, sugars, and other natural products, including amino acids, proteins, and polypeptides. *See* COTTON EFFECT; OPTICAL ROTATORY DISPERSION; POLARIZED LIGHT.

Robert F. Goddu; James N. Little

Polarimetry

Polarimetry The science of determining the polarization state of electromagnetic radiation (x-rays, light or radio waves). Radiation is said to be linearly polarized when the electric vector oscillates in only one plane. It is circularly polarized when the x-plane component of the electric vector oscillates 90° out of phase with the *y*-plane component.

To completely specify the polarization state, it is necessary to make six intensity measurements of the light passed by a quarter-wave retarder and a rotatable linear polarizer, such as a Polaroid or a Nicol prism. The retarder converts circular light into linear light.

Most starlight is unpolarized. However, atoms in the presence of a magnetic field align themselves at fixed, quantized angles to the field direction. Then the spectral lines they emit are circularly polarized when the magnetic field is parallel to the line of sight, and linearly polarized when the field is perpendicular. The light from sunspots is polarized because the magnetic fields impose some direction in the emitting gas. Other phenomena also remove isotropy and produce polarization. *See* SOLAR MAGNETIC FIELD; ZEEMAN EFFECT.

Electrooptical devices are rapidly replacing rotating polarizers and fixed retarders. The magnetograph consists of a spectrograph to isolate the atomic spectral line for study; a Pockels cell, an electrooptic crystal whose retardance depends on an applied voltage; a polarizing prism to isolate the polarization state passed by the retarder; a pair of photocells to detect the transmitted light; and a scanning mechanism to sweep the solar image across the spectrograph entrance slit. Two photocells are needed to simultaneously measure left- and right-circular polarization. *See* SPECTROGRAPH.

A magnetograph can be made sensitive to linear polarization, but the signal levels are about 100 times weaker for the inferred transverse fields than for longitudinal fields of comparable strength. To improve signal-to-noise levels, the spectrograph can be replaced with an optical filter having a narrow passband, and the photocells can be replaced with an array of photosensitive picture elements (pixels). David M. Rust

Polaris The star α Ursae Minoris, also known as the North Star or Pole Star. It is perhaps the best-known star in the northern sky. Its location only 1 degree of arc from the north celestial pole, the point where the Earth's rotation axis intersects the celestial sphere, has made it a very useful reference point for navigation. It may easily be found by following the line joining the two bright stars at the end of the bowl of the Big Dipper. *See* URSA MAJOR; URSA MINOR.

Polaris (apparent magnitude 1.99) is a supergiant with an intrinsic brightness about 1500 times that of the Sun. It is accompanied by a 9th-magnitude main-sequence star, and its spectrum shows evidence of another, much closer companion in an eccentric orbit with a period of 30 years. *See* SUPERGIANT STAR.

Polaris is a variable star, displaying slight changes in brightness with a period close to 4 days. Polaris is a member of an important group of stars known as the Cepheid variables. However, it is atypical in that the amplitude of the variations is very small compared to other Cepheids and has decreased steadily over 100 years to the point where the pulsation of the star has virtually stopped. *See* CEPHEIDS; STAR; VARIABLE STAR. David W. Latham

Polarization of dielectrics A vector quantity representing the electric dipole moment per unit volume of a dielectric material. *See* DIELECTRIC MATERIALS.

Dielectric polarization arises from the electrical response of individual molecules of a medium and may be classified as electronic, atomic, orientation, and space-charge or interfacial polarization, according to the mechanism involved.

Electronic polarization represents the distortion of the electron distribution or motion about the nuclei in an electric field.

Atomic polarization arises from the change in dipole moment accompanying the stretching of chemical bonds between unlike atoms in molecules. *See* MOLECULAR STRUCTURE AND SPECTRA.

Orientation polarization is caused by the partial alignment of polar molecules, that is, molecules possessing permanent dipole moments, in an electric field. This mechanism leads to a temperature-dependent component of polarization at lower frequencies.

Space-charge or interfacial polarization occurs when charge carriers are present which can migrate an appreciable distance through a dielectric but which become trapped or cannot discharge at an electrode. This process always results in a distortion of the macroscopic field and is important only at low frequencies. *See* ELECTRIC FIELD; ELECTRIC SUSCEPTIBILITY. Robert D. Waldron

Polarization of waves The directional dependence of certain wave phenomena, at right angles to the propagation direction of the wave. In particular, ordinary light may be regarded as composed of two such asymmetrical components, referred to as its two states of linear polarization.

These two components are refracted differently by doubly refracting crystals, such as calcite, or Iceland spar. Each state of linear polarization is refracted according to its own separate refractive index. On a subsequent refraction by the same crystal, but now rotated through an angle θ about the direction of the beam, each component appears as a mixture of the original two polarization components, according to the proportions $\cos^2\theta : \sin^2\theta$. *See* BIREFRINGENCE; CRYSTAL OPTICS; REFRACTION OF WAVES.

In the early nineteenth century, T. Young suggested that light polarization arises from transverse oscillations. In J. C. Maxwell's theory of light as electromagnetic waves, visible light—and also other types of electromagnetic radiation such as radio waves, microwaves, and x-rays (distinguished from visible light only by wavelength)—consists of electric and magnetic fields, each oscillating in directions perpendicular to the propagation direction, the electric and magnetic field vectors being perpendicular to each other. The plane of polarization of the wave contains the electric vector (or magnetic vector; there is no general agreement which) and the propagation direction. *See* ELECTROMAGNETIC RADIATION; LIGHT; MAXWELL'S EQUATIONS.

If the plane of polarization remains constant along the wave (as in the case of each light component in a doubly refracting medium), the wave has linear (or plane) polarization. However, the plane of polarization can also rotate. If the rotation rate is constant, the intensity of the wave being also constant, a circularly polarized wave results. These are of two types: right-handed and left-handed.

Any electromagnetic wave can be considered to be composed of monochromatic components, and each monochromatic component can be decomposed into a left-handed and a right-handed circularly polarized part. The states of linear polarization are each made up of equal magnitudes of the two circularly polarized parts, with differing phase relations to provide the different possible directions of plane polarization. Monochromatic waves composed of unequal magnitudes of the two circularly polarized parts are called elliptically polarized. This refers to the fact that the electric and magnetic vectors trace out ellipses in the plane perpendicular to the direction of motion.

Photons have quantum-mechanical spin, which refers to the angular momentum of the photon, necessarily about its direction of motion. A photon's spin has magnitude 1, in fundamental units. This spin can point along the direction of motion (positive helicity, right-handed spin) or opposite to it (negative helicity, left-handed spin), and this corresponds (depending on conventions used) to a classical electromagnetic wave of right- or left-handed circular polarization. *See* HELICITY (QUANTUM MECHANICS); PHOTON.

Electromagnetic and gravitational waves both have the specific property that they are entirely transverse in character, which is a consequence of their speed of propagation being the absolute speed of relativity theory (the speed of light). This corresponds to the fact that their respective quanta, namely photons and gravitons, are massless particles. In the case of waves that travel at a smaller speed, as with fields whose quanta are massive rather than massless, there can be (unpolarized) longitudinal as well as transverse effects. Seismic waves traveling through the Earth's material, for example, can be transverse (polarized sideways oscillations) or longitudinal (unpolarized pressure waves). *See* SEISMOLOGY; SOUND; WAVE MOTION IN FLUIDS.

In most situations encountered in practice, light (or gravitational waves) consists of an incoherent mixture of different polarization states, and is referred to as unpolarized. However, light reflected off a refracting surface (for example, glass or water) is polarized to some extent; that is, there is a certain preponderence of one state of linear polarization over the orthogonal possibility. Complete polarization occurs for a particular angle of incidence, known as the Brewster angle. *See* POLARIZED LIGHT; REFLECTION OF ELECTROMAGNETIC RADIATION; WAVE MOTION. Roger Penrose

Polarized light Light which has its electric vector oriented in a predictable fashion with respect to the propagation direction. In unpolarized light, the vector is oriented in a random, unpredictable fashion. Even in short time intervals, it appears to be oriented in all directions with equal probability. Most light sources seem to be partially polarized so that some fraction of the light is polarized and the remainder unpolarized.

According to all available theoretical and experimental evidence, it is the electric vector rather than the magnetic vector of a light wave that is responsible for all the effects of polarization and other observed phenomena associated with light. Therefore, the electric vector of a light wave, for all practical purposes, can

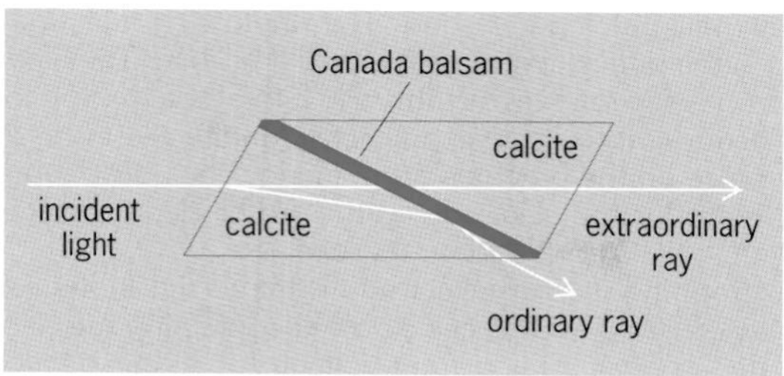

Fig. 1. Nicol prism. The ray for which Snell's law holds is called the ordinary ray.

be identified as the light vector. *See* CRYSTAL OPTICS; ELECTROMAGNETIC RADIATION; LIGHT; POLARIZATION OF WAVES.

Types. Polarized light is classified according to the orientation of the electric vector. In linearly polarized light, the electric vector remains in a plane containing the propagation direction. For monochromatic light, the amplitude of the vector changes sinusoidally with time. In circularly polarized light, the tip of the electric vector describes a circular helix about the propagation direction. The amplitude of the vector is constant. The frequency of rotation is equal to the frequency of the light. In elliptically polarized light, the vector also rotates about the propagation direction, but the amplitude of the vector changes so that the projection of the vector on a plane at right angles to the propagation direction describes an ellipse.

One of the simplest ways of producing linearly polarized light is by reflection from a dielectric surface. At a particular angle of incidence, known as Brewster's angle, the reflectivity for light whose electric vector is in the plane of incidence becomes zero. The reflected light is thus linearly polarized at right angles to the plane of incidence.

Linear polarizing devices. The first polarizers were glass plates inclined so that the incident light was at Brewster's angle. Such polarizers are quite inefficient since only a small percentage of the incident light is reflected as polarized light.

Certain natural materials absorb linearly polarized light of one vibration direction much more strongly than light vibrating at right angles. Such materials are termed dichroic. Tourmaline is one of the best-known dichroic crystals, and tourmaline plates were used as polarizers for many years. *See* DICHROISM.

Other natural materials exist in which the velocity of light depends on the vibration direction. These materials are called birefringent. One of the best-known of these birefringent crystals is transparent calcite (Iceland spar). The Nicol prism is made of two pieces of calcite cemented together (Fig. 1). The cement is Canada balsam, in which the wave velocity is intermediate between the velocity in calcite for the fast and the slow ray. The angle at which the light strikes the boundary is such that for one ray the angle of incidence is greater than the critical angle for total reflection. Thus the rhomb is transparent for only one polarization direction. *See* BIREFRINGENCE; CRYSTAL; OPTICS.

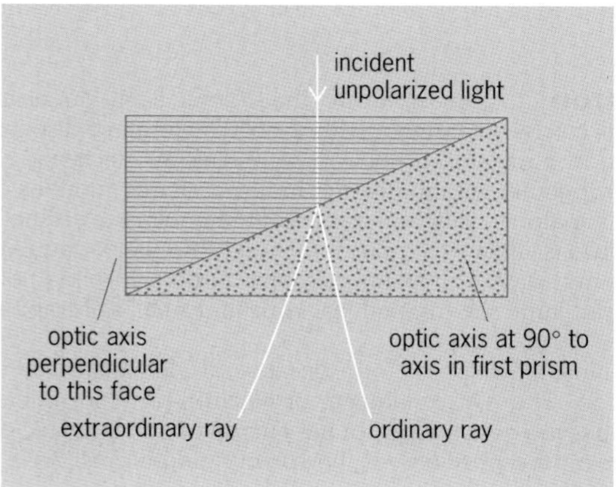

Fig. 2. Wollaston prism.

A different type of polarizer, made of quartz, is shown in Fig. 2. Here the vibration directions are different in the two pieces so that the two rays are deviated as they pass through the material. The incoming light beam is thus separated into two oppositely linearly polarized beams which have an angular separation between them, and it is possible to select either beam.

A third mechanism for obtaining polarized light is the Polaroid sheet polarizer, of which there are three types. The first is a microcrystalline polarizer in which small crystals of a dichroic material are oriented parallel to each other in a plastic medium. The second type depends for its dichroism on a property of an iodine-in-water solution. The iodine appears to form a linear high polymer. If the iodine is put on a transparent oriented sheet of material such as polyvinyl alcohol (PVA), the iodine chains apparently line themselves parallel to the PVA molecules and the resulting dyed sheet is strongly dichroic. A third type of sheet polarizer depends for its dichroism directly on the molecules of the plastic itself. This plastic consists of oriented polyvinylene.

Polarization by scattering. When an unpolarized light beam is scattered by molecules or small particles, the light observed at right angles to the original beam is polarized. The best-known example of polarization by scattering is the light of the north sky. *See* SCATTERING OF ELECTROMAGNETIC RADIATION.

Circular and elliptical polarizing devices. Circularly and elliptically polarized light are normally produced by combining a linear polarizer with a wave plate. A Fresnel rhomb can be used to produce circularly polarized light.

A plate of material (quartz, calcite, or other birefringent crystals) which is linearly birefringent is called a wave plate or retardation sheet. Wave plates have a pair of orthogonal axes which are designated fast and slow. Polarized light with its electric vector parallel to the fast axis travels faster than light polarized parallel to the slow axis. The thickness of the material can be chosen so that for light traversing the plate, there is a definite phase shift between the fast component and the slow component. A plate with a 90° phase shift is termed a quarter-wave plate.

If linearly polarized light is incident normally on a quarter-wave plate and oriented at 45° to the fast axis, the transmitted light will be circularly polarized. If the linearly polarized light is at an angle other than 45° to the fast axis, the transmitted radiation will be elliptically polarized.

Analyzing devices. Polarized light is one of the most useful tools for studying the characteristics of materials. The absorption constant and refractive index of a metal can be calculated by measuring the effect of the metal on polarized light reflected from its surface. *See* REFLECTION OF ELECTROMAGNETIC RADIATION.

The analysis of polarized light can be performed with a variety of different devices. If the light is linearly polarized, it can be extinguished by a linear polarizer and the direction of polarization of the light determined directly from the orientation of the polarizer. If the light is elliptically polarized, it can be analyzed with the combination of a quarter-wave plate and a linear polarizer. Any such combination of polarizer and analyzer is called a polariscope.

Bruce H. Billings; Hong Hua

Polarized light microscope A microscope that utilizes polarized light to form a highly magnified image of an object. Polarizing microscopes play an important role in crystallography, petrography, microchemistry, and biology. Although all light microscopes compare poorly with electron microscopes with respect to image resolution, polarized light microscopes have the unique ability to deliver information about the submicroscopic structure of the objects being examined. They also have the advantage of being relatively nondestructive, and may be used safely with living cells. Polarized light interactions with electromagnetically anisotropic structures, down to atomic dimensions, can be measured by polarized light microscopy. The sensitivity of polarized light microscopy as well as its importance to biology have been enhanced by the use of video technology. *See* POLARIZED LIGHT.

A polarizing microscope differs from a conventional light microscope in a number of ways. A polarizing microscope has a pair of polars (polarizing devices) in the optical train. The first polar (polarizer) defines the initial plane of polarization for light entering the microscope and is located between the illuminator and the condenser. The other polar (analyzer) is usually placed between the objective and the ocular tube and defines the plane of polarization of the light reaching the ocular. One or both must be accurately rotatable about the optical axis of the instrument. Usually the analyzer is also removable from the optical path.

The most frequently used type of polar is a dichroic sheet polarizing filter. For petrographic, crystallographic, and most microchemical applications, dichroic filter polars are the better choice. For critical biological applications, such as investigating the weak birefringence of cytoskeletal structures in living cells, the expense and complication attendant to the use of prism polars can be justified by the attainment of sensitivity unobtainable by other means.

In addition to the polars, all polarizing microscopes need rotatable specimen stages and one or more removable birefringence compensators. The compensators are birefringent devices used to measure magnitude and sign of retardation due to specimen birefringence, to enhance specimen image contrast, and to manipulate the state of polarization of light passing through any point in the specimen. *See* BIREFRINGENCE.

Polarizing microscopes for different applications have some differences in construction. For petrography and crystallography, the microscope should be able to accept a universal specimen stage capable of rotating the specimen about three axes. A polarizing microscope for biological use has less rigid requirements for angular orientation but needs rectified optics (to eliminate or greatly reduce depolarization at high numerical apertures), prism polars, and a sensitive elliptic compensator because the birefringence of the typical biological specimen is much smaller. An ocular telescope or a Bertrand lens (a built-in lens that optionally converts the ocular into a telescope) is essential for some applications and useful for all.

There are two traditional modes of use for the polarizing microscope, the orthoscopic mode and the conoscopic mode. In the orthoscopic mode, the ocular projects an image of the specimen, as in conventional microscopy. Rotation of the specimen stage reveals the location and orientation of any anisotropic features in the specimen. The conoscopic mode is used to characterize crystalline specimens. Here the ocular is not used to project an image of the specimen, but rather—with the aid of a special lens—the image of the objective exit pupil is examined to reveal the relative retardation experienced by polarized light as a function of its angle of incidence on the specimen. *See* CRYSTAL OPTICS; MICROSCOPE.

Gordon W. Ellis

Polarographic analysis An electrochemical technique used in analytical chemistry. Polarography involves measurements of current-voltage curves obtained when voltage is applied to electrodes (usually two) immersed in the solution being investigated. One of these electrodes is a reference electrode: its potential remains constant during the measurement. The second electrode is an indicator electrode. Its potential varies in the course of measurement of the current-voltage curve, because of the change of the applied voltage. In the simplest version, so-called dc polarography, the indicator electrode is a dropping-mercury electrode, consisting of a mercury drop hanging at the orifice of a fine-bore glass capillary. The capillary is connected to a mercury reservoir so that mercury flows through it at the rate of a few milligrams per second. The outflowing mercury forms a drop at the orifice, which grows until it falls off. The lifetime of each drop is several seconds (usually 2 to 5). Each drop forms a new electrode; its surface is practically unaffected by processes taking place on the previous drop. Hence each drop represents a well-reproducible electrode with a fresh, clean surface. *See* ELECTRODE; ELECTROCHEMICAL TECHNIQUES.

The dropping-mercury electrode is immersed in the solution to be investigated and placed in a cell containing the reference electrode. Polarographic current-voltage curves can be recorded with a simple instrument consisting of a potentiometer or another source of voltage and a current-measuring device. The voltage can be varied by manually changing the applied voltage in finite increments, measuring current at each, and plotting current as a function of the voltage. Alternatively, commercial instruments are available in which voltage is increased linearly with time (a voltage ramp), and current variations are recorded automatically.

Another polarographic technique is called pulse or differential pulse polarography. This technique is more sensitive by two orders of magnitude than dc polarography, and in inorganic trace analysis competes with atomic absorption and neutron activation analysis. The sensitivity of differential pulse polarography has found application in drug analysis. There is also a polarographic technique called ac polarography that is particularly useful for obtaining information on adsorption-desorption processes at the surface of the dropping-mercury electrode.

Polarographic studies can be applied to investigation of electrochemical problems, to elucidation of some fundamental problems of inorganic and organic chemistry, and to solution of practical problems. In electrochemistry, polarography allows measurement of potentials, and yields information about the rate of the electrode process, adsorption-desorption phenomena, and fast chemical reactions accompanying the electron transfer. In fundamental applications, polarography makes it possible to distinguish the form and charge of the species (for example, inorganic complex or organic ion) in the solution. Polarography also permits the study of equilibria (complex formation, acid-base, tautomeric), rates, and mechanisms.

Polarography can be used for investigation of the relationship between electrochemical data and structure. In inorganic analysis, polarography is used predominantly for trace-metal analysis (with increased sensitivity of differential pulse polarography and stripping analysis). In organic analysis, it is possible in principle to use polarography in elemental analysis and functional group analysis. The most important fields of application of inorganic determinations are in metallurgy, environmental analysis (air, water, and seawater contaminants), food analysis, toxicology, and clinical analysis. The possibility of being able to determine vitamins, alkaloids, hormones, terpenoid substances, and natural coloring substances has made polarography useful in analysis of biological systems, analysis of drugs and pharmaceutical preparations, and determination of pesticide or herbicide residues in foods.

Petr Zuman

Polaron A quasiparticle that forms from an electronic charge carrier interacting with a solid's vibrating displaceable atoms. In particular, a polaron comprises an electronic charge carrier together with the carrier-induced altered motions of surrounding atoms. Different classes of material have distinct types of polarons. Polarons may be divided into weak-coupling polarons and strong-coupling polarons. Strong-coupling polarons comprise large self-trapped polarons and small self-trapped polarons.

All types of polarons form because of the electron-lattice interaction. The electron-lattice, or electron-phonon, interaction describes the dependence of the energy of an electronic charge carrier on the positions of the surrounding atoms. As a result of this interaction neither electronic charge carriers nor atoms can move without affecting the other. Electronic carriers tend to follow atoms' motions. Polaron effects result as electronic charge carriers adjust to and modify atomic motion. *See* LATTICE VIBRATIONS; PHONON.

The electron-phonon interaction can be divided into two components. The long-range component of the electron-lattice interaction describes the dependence of the energy of an electronic charge carrier in an ionic (or polar) material on the positions of the surrounding ions. The short-range component of the electron-lattice interaction describes the dependence of the energy of an electronic charge carrier on the positions of nearby atoms with which the electronic carrier forms bonding or antibonding states. The short-range interaction predominates in covalent semiconductors. The term polaron was adopted in recognition of the electron-lattice interaction typically being especially strong in polar (ionic) materials. *See* COHESION (PHYSICS); COULOMB'S LAW; IONIC CRYSTALS; SEMICONDUCTOR.

Weak-coupling polaron. In some instances the electron-lattice interaction only slightly alters the motions of an electronic carrier and the atoms with which it interacts. Then the range over which an electronic carrier and the surrounding atoms affect one another is termed the weak-coupling polaron radius. The electron-lattice interaction simply acts to slightly lower the net energy and to slightly enhance the effective mass of an electronic charge carrier.

Strong coupling: self-trapping. More dramatic effects occur when the electron-lattice interaction is strong enough and the electronic carrier moves slowly enough for it to become self-trapped. The term self-trapped emphasizes that the electronic carrier is bound within a potential well that its very presence induces. The associated polaron comprises the self-trapped carrier and the associated pattern of displaced atomic equilibrium positions. Self-trapping may be regarded as a feedback phenomenon. In particular, confining a charge carrier induces atomic displacements that deepen the potential well that confines the carrier, thereby enhancing the charge carrier's confinement.

Large (self-trapped) polaron. The self-trapping polaron that is formed with only the long-range component of the electron-lattice interaction is termed a large polaron since its self-trapped carrier generally extends over multiple atomic sites.

Large polarons can move coherently through a solid. Since atomic motion is required for the self-trapped carrier to move, large polarons move slowly with very large effective masses, which cause them to move with exceptionally large momenta. As a result, large polarons are not easily scattered. Thus, it is the relatively small scattering rate, not the mobility, that best distinguishes large-polaron motion from that of a conventional charge carrier, a weak-coupling polaron. *See* BAND THEORY OF SOLIDS.

Small (self-trapped) polaron. The self-trapped carrier formed with the short-range component of the electron-lattice interaction is termed a small polaron. If the electron-lattice interaction is too weak to induce a collapse into a small polaron, the charge carrier does not self-trap at all. David Emin

Poliomyelitis An acute infectious viral disease which in its serious form affects the central nervous system and, by destruction of motor neurons in the spinal cord, produces flaccid paralysis. However, about 99% of infections are either inapparent or very mild. *See* ANIMAL VIRUS; CENTRAL NERVOUS SYSTEM.

The virus probably enters the body through the mouth; primary multiplication occurs in the throat and intestine. Transitory viremia occurs; the blood seems to be the most likely route to the central nervous system. The severity of the infection may range from a completely inapparent through minor influenzalike illness, or an aseptic meningitis syndrome (nonparalytic poliomyelitis) with stiff and painful back and neck, to the severe forms of paralytic and bulbar poliomyelitis. In all clinical types, virus is regularly present in the enteric tract. In paralytic poliomyelitis the usual course begins as a minor illness but progresses, sometimes with an intervening recession of symptoms (hence biphasic), to flaccid paralysis of varying degree and persistence. When the motor neurons affected are those of the diaphragm or of the intercostal muscles, respiratory paralysis occurs. Bulbar poliomyelitis results from viral attack on the medulla (bulb of the brain) or higher brain centers, with respiratory, vasomotor, facial, palatal, or pharyngeal disturbances.

Poliomyelitis occurs throughout the world. In temperate zones it appears chiefly in summer and fall, although winter outbreaks have been known. It occurs in all age groups, but less frequently in adults because of their acquired immunity. The virus is spread by human contact; the nature of the contact is not clear, but it appears to be associated with familial contact and with interfamily contact among young children. The virus may be present in flies.

Inactivated poliovirus vaccine (Salk; IPV), prepared from virus grown in monkey kidney cultures, was developed and first used in the United States, but oral poliovirus vaccine (Sabin; OPV) is now generally used throughout the world. The oral vaccine is a living, attenuated virus. Joseph L. Melnick

Pollen The small male reproductive bodies produced in the pollen sacs of seed plants (gymnosperms and angiosperms). On maturation in the pollen sac, a pollen grain may reach 0.00007 mg as in spruce, or less than 1/20 of this weight. A grain usually has two waxy, durable outer walls, the ectexine and the endexine, and an inner fragile wall, the intine, which surrounds the contents with its nuclei and reserves of starch and oil.

Pollen identification depends on interpretation of morphological features. Exine and aperture patterns are especially varied in the more highly evolved dicots, so that recognition at family, genus, or even species level may be possible despite the small surface area available on a grain (see illustration). Since the morphological characters are conservative in the extreme, usually changing very slowly through geologic time, studies of fine detail serve to establish the lineal descent of many plants living today. *See* PALYNOLOGY.

Extreme variations in size may occur within a family, but pollen grains range mainly from 15 to 50 micrometers, with the dicot range being from 2 μm in *Myosotis* to 250 μm in *Mirabilis*; the monocots range from about 15 to 300 μm or more, with eelgrass (*Zostera*) having pollen measuring 2550 $\times$ 3.7 μm in a class of its own; living gymnosperms range from 15 μm in *Gnetum* to about 180 μm in *Abies* (including sacs), while fossil types range from about 11 to 300 μm.

Most grains are free (monads) though often loosely grouped because of the spines, sticky oils, or viscin threads. Compound

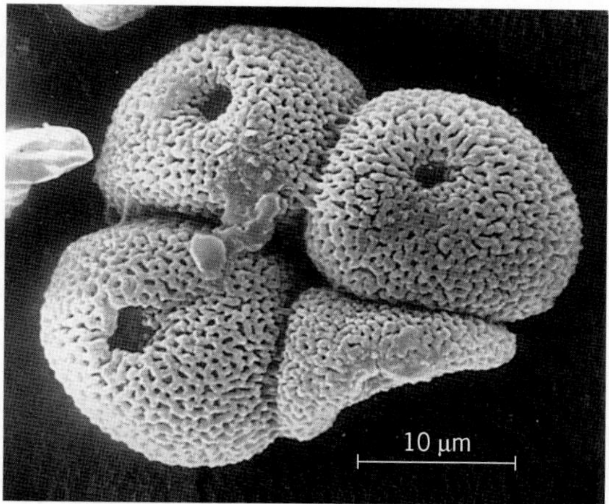

Tetrad of cattail (*Typha latifolia*), grains cohering, one-pored, exine reticulate. (*Scanning electron micrograph by C. M. Drew, U.S. Naval Weapons Center, China Lake, California*).

grains (polyads) are richly developed in some angiosperm families, commonly occurring in four (tetrads), or in multiples of four up to 64 or more.

In most pollen grains, the polar axis runs from the inner (proximal) face to the outer (distal) face, as oriented during tetrad formation. The equator crosses it at right angles. Bilateral grains dominate in the gymnosperms and monocots, the polar axis usually being the shorter one, with the single aperture on the distal side. On the other hand, almost all dicot pollen is symmetrical around the polar axis (usually the long axis), with shapes ranging mainly from spheroidal to ellipsoidal, with rounded equatorial outlines, and sometimes "waisted" as in the Umbelliferae. Three-pored grains often have strikingly triangular outlines in polar view. *See* FLOWER; PLANT REPRODUCTION; POLLINATION.

Lucy M. Cranwell

Pollination The transport of pollen grains from the plant parts that produce them to the ovule-bearing organs, or to the ovules (seed precursors) themselves. In gymnosperms, the pollen, usually dispersed by the wind, is simply caught by a drop of fluid excreted by each freely exposed ovule. In angiosperms, where the ovules are contained in the pistil, the pollen is deposited on the pistil's receptive end (the stigma), where it germinates. *See* FLOWER.

Without pollination, there would be no fertilization; it is thus of crucial importance for the production of fruit crops and seed crops. Pollination also plays an important part in plant breeding experiments aimed at increasing crop production through the creation of genetically superior types. *See* BREEDING (PLANT); PLANT REPRODUCTION.

Self- and cross-pollination. In most plants, self-pollination is difficult or impossible, and there are various mechanisms which are responsible. For example, in dichogamous flowers, the pistils and stamens reach maturity at different times; in protogyny, the pistils mature first, and in protandry, the stamens mature before the pistils. Selfing is also impossible in dioecious species, where some plants bear flowers that have only pistils (pistillate or female flowers), while other individuals have flowers that produce only pollen (staminate or male flowers). In monoecious species, where pistillate and staminate flowers are found in the same plant, self-breeding is at least reduced. Heterostyly is another device that promotes outbreeding. Here some flowers (pins) possess a long pistil and short stamens, while others (thrums) exhibit the reverse condition; each plant individual bears only pins or only thrums.

Flower attractants. As immobile organisms, plants normally need external agents for pollen transport. These can be insects, wind, birds, mammals, or water, roughly in that order of importance. In some plants the pollinators are simply trapped; in the large majority of cases, however, the flowers offer one or more rewards, such as sugary nectar, oil, solid food bodies, perfume, sex, an opportunity to breed, a place to sleep, or some of the pollen itself. For the attraction of pollinators, flowers provide either visual or olfactory signals. Color includes ultraviolet, which is perceived as a color by most insects and at least some hummingbird species. Fragrance is characteristic of flowers pollinated by bees, butterflies, or hawkmoths, while carrion or dung odors are produced by flowers catering to certain beetles and flies. A few orchids, using a combination of olfactory and visual signals, mimic the females of certain bees or wasps so successfully that the corresponding male insects will try to mate with them, thus achieving pollination (pseudocopulation).

While some flowers are "generalists," catering to a whole array of different animals, others are highly specialized, being pollinated by a single species of insect only. Extreme pollinator specificity is an important factor in maintaining the purity of plant species in the field, even in those cases where hybridization can easily be achieved artificially in a greenhouse or laboratory, as in most orchids. The almost incredible mutual adaptation between pollinating animal and flower which can frequently be observed exemplifies the idea of coevolution. *See* POLLEN.

Bastiaan J. D. Meeuse

Polonium A chemical element, Po, atomic number 84. Marie Curie discovered the radioisotope ^{210}Po in pitchblende. This isotope is the penultimate member of the radium decay series. All polonium isotopes are radioactive, and all are short-lived except the three α-emitters, artificially produced ^{208}Po (2.9 years) and ^{209}Po (100 years), and natural ^{210}Po (138.4 days). *See* PERIODIC TABLE.

Polonium (^{210}Po) is used mainly for the production of neutron sources. It can also be used in static eliminators and, when incorporated in the electrode alloy of spark plugs, is said to improve the cold-starting properties of internal combustion engines.

Most of the chemistry of polonium has been determined using ^{210}Po, 1 curie of which weighs 222.2 micrograms; work with weighable amounts is hazardous, requiring special techniques. Polonium is more metallic than its lower homolog, tellurium. The metal is chemically similar to tellurium, forming the bright red compounds $SPoO_3$ and $SePoO_3$. The metal is soft, and its physical properties resemble those of thallium, lead, and bismuth. Valences of 2 and 4 are well established; there is some evidence of hexavalency. Polonium is positioned between silver and tellurium in the electrochemical series.

Two forms of the dioxide are known: low-temperature, yellow, face-centered cubic (UO_2 type), and high-temperature, red, tetragonal. The halides are covalent, volatile compounds, resembling their tellurium analogs. *See* RADIOACTIVITY; TELLURIUM.

Kenneth W. Bagnall

Polyacetal A polyether derived from aldehydes (RCHO) or ketones (RR′CO) and containing —O—R—O—groups in the main chain. Of the many possible polyacetals, the most common is a polymer or copolymer of formaldehyde, polyoxymethylene $(\text{—O—CH}_2\text{—})_n$. While the substance paraformaldehyde contains oligomers or low-molecular-weight polyoxymethylenes (*n* very small), high-molecular-weight, crystalline polyoxymethylenes constitute an important class of engineering plastics that, in commerce, is often simply referred to as polyacetal. Cellulose and its derivatives also have a polyacetal structure. *See* ACETAL; CELLULOSE; FORMALDEHYDE; POLYETHER RESINS; POLYMER.

As shown below, formaldehyde can be readily polymerized by

nH—C(=O)—H (Formaldehyde) —[anionic or cationic initiator]→ $-[O-CH_2]_n-$ (Polyformaldehyde or polyoxymethylene)

$\left(\frac{n}{3}\right)$ Trioxane (ring: H_2C—O—CH_2—O—CH_2—O) —[cationic initiator]→ $-[O-CH_2]_n-$

using anionic initiators such as triphenylphosphine and, somewhat less readily, by using cationic initiators such as pro- tonic acids. Alternatively, a similar polymer can be obta- ined by the ring-opening polymerization of trioxane using, for example, a boron trifluoride complex as initiator. *See* POLYMERIZATION.

At temperatures above ~110°C (230°F; the ceiling temperature, above which depolymerization becomes favored over polymerization), the polymers degrade by an unzipping reaction to monomer. To prevent this, one of two approaches is commonly used: esterification of the hydroxyl end groups, or copolymerization with a small amount of a monomer such as ethylene oxide or 1,3-dioxolane.

Polyacetals are typically strong and tough, resistant to fatigue, creep, organic chemicals (but not strong acids or bases), and have low coefficients of friction. Electrical properties are also good. Improved properties for particular applications may be attained by reinforcement with fibers of glass or polytetrafluoroethylene, and by incorporation of an elastomeric toughening phase. The combination of properties has led to many uses such as plumbing fittings, pump and valve components, bearings and gears, computer hardware, automobile body parts, and appliance housings.

John A. Manson

Polyacrylate resins Polymers obtained from a variety of acrylic monomers, such as acrylic and methacrylic acids, their salts, esters, and amides, and the corresponding nitriles. The most important monomers with corresponding repeat units are shown here.

$$H_2C{=}C(CH_3){-}COOCH_3 \longrightarrow {+}CH_2{-}C(CH_3)(COOCH_3){+}_n$$

Methyl methacrylate → Poly(methyl methacrylate)

$$H_2C{=}C(H){-}COOC_2H_5 \longrightarrow {+}CH_2{-}CH(COOC_2H_5){+}_n$$

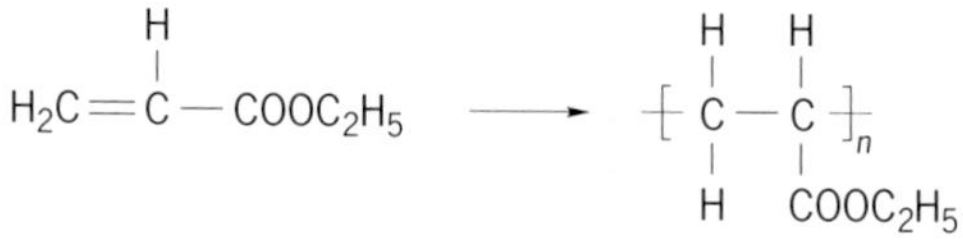

Ethyl acrylate → Poly(ethyl acrylate)

Poly(methyl methacrylate) is a hard, transparent polymer with high optical clarity, high refractive index, and good resistance to the effects of light and aging. It and its copolymers are useful for lenses, signs, indirect lighting fixtures, transparent domes and skylights, dentures, and protective coatings.

Solutions of poly(methyl methacrylate) and its copolymers are useful as lacquers. Aqueous latexes formed by the emulsion polymerization of methyl methacrylate with other monomers are useful as water-based paints and in the treating of textiles and leather.

Poly(ethyl acrylate) is a tough, somewhat rubbery product. The monomer is used mainly as a plasticizing or softening component of copolymers.

Methyl methacrylate is of interest as a polymerizable binder for sand or other aggregates, and as a polymerizable impregnant for concrete; usually a cross-linking acrylic monomer is also incorporated. The binder systems (polymer concrete) are used as overlays for bridge decks as well as for castings, while impregnation is used to restore concrete structures and protect bridge decks against corrosion by deicing salts. *See* PLASTICS PROCESSING; POLYMERIZATION.

John A. Manson

Polyacrylonitrile resins Hard, relatively insoluble, and high-melting materials produced by the polymerization of acrylonitrile, as shown in the reaction below. Polyacrylonitrile

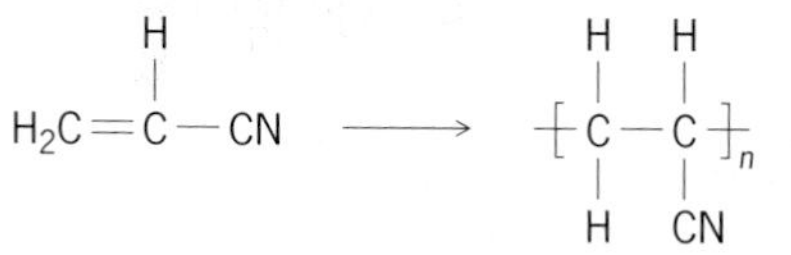

$$H_2C{=}C(H){-}CN \longrightarrow {+}CH_2{-}CH(CN){+}_n$$

Acrylonitrile → Polyacrylonitrile

is used almost entirely in copolymers. The copolymers fall into three groups: fibers, plastics, and rubbers. The presence of acrylonitrile in a polymericcomposition tends to increase its resistance to temperature, chemicals, impact, and flexing. The polymerization of acrylonitrile can be readily initiated by means of the conventional free-radical catalysts such as peroxides, by irradiation, or by the use of alkali metal catalysts. Although polymerization in bulk proceeds too rapidly to be commercially feasible, satisfactory control of a polymerization or copolymerization may be achieved in suspension and in emulsion, and in aqueous solutions from which the polymer precipitates. Copolymers containing acrylonitrile may be fabricated in the manner of thermoplastic resins.

The major use of acrylonitrile is in the form of fibers. By definition, an acrylic fiber must contain at least 85% acrylonitrile. The high strength; high softening temperature; resistance to aging, chemicals, water, and cleaning solvents; and the soft woollike feel or fabrics have made the product popular for many uses such as sails, cordage, blankets, and various types of clothing. *See* MANUFACTURED FIBER.

Copolymers of vinylidene chloride with small proportions of acrylonitrile are useful as tough, impermeable, and heat-sealable packaging films. Extensive use is made of copolymers of acrylonitrile with butadiene, often called NBR (formerly Buna N) rubbers, which contain 15–40% acrylonitrile. The NBR rubbers resist hydrocarbon solvents such as gasoline, abrasion, and in some cases show high flexibility at low temperatures. *See* RUBBER.

The development of blends and interpolymers of acrylonitrile-containing resins and rubbers represented a significant advance in polymer technology. The products, usually called ABS resins, typically are made by blending acrylonitrile-styrene copolymers with a butadiene-acrylonitrile rubber, or by interpolymerizing polybutadiene with styrene and acrylonitrile. The combination of low cost, good mechanical properties, and ease of fabrication by a variety of methods led to the rapid development of new uses for ABS resins. Applications include products requiring high impact strength, such as pipe, and sheets for structural uses, such as industrial duct work and components of automobile bodies. *See* ACRYLONITRILE; PLASTICS PROCESSING; STYRENE.

John A. Manson

Polyamide resins Products of polymerization of an amino acid or the condensation of a diamine with a dicarboxylic acid. They are used for fibers, bristles, bearings, gears, molded objects, coatings, and adhesives. The term nylon formerly referred specifically to synthetic polyamides as a class. Because of many applications in mechanical engineering, nylons are considered engineering plastics.

The most common commercial aliphatic polamides are nylons-6,6; -6; -6,10; -11; and -12. Nylon-6,6, nylon-6,10, nylon-6,12, and nylon-6 are the most commonly used polyamides for general applications as molded or extruded parts; nylon-6,6 and nylon-6 find general application as fibers.

As a group, nylons are strong and tough. Mechanical properties depend in detail on the degree and distribution of crystallinity, and may be varied by appropriate thermal treatment or by nucleation techniques. Because of their generally good mechanical properties and adaptability to both molding and extrusion, certain nylons are often used for gears, bearings, and electrical mountings. Nylon bearings and gears perform quietly and need little or no lubrication. Nylon resins are also used extensively as filaments, bristles, wire insulation, appliance parts, and film. Properties can also be modified by copolymerization. Reinforcement of nylons with glass fibers results in increased stiffness, lower creep and improved resistance to elevated temperatures. *See* HETEROCYCLIC POLYMER; PLASTICS PROCESSING; POLYETHER RESINS; POLYMERIZATION.

John A. Manson

Polychaeta The largest class of the phylum Annelida (multisegmented worms) withabout 13,000 described species.

Polychaetes range from less than 1 mm (0.04 in.) to up to 3 m (10 ft) in length. Represented worldwide, they are predominantly marine organisms and occur from the intertidal zone to the greatest ocean depths at all latitudes. Some notable exceptions occupy nonmarine habitats such as supralittoral zones (beyond the high-tide mark) and freshwater lakes. The only synapomorphy (a derived trait shared by two or more taxa that is believed to reflect their shared ancestry) that distinguishes them from other annelids is the pair of nuchal organs, chemosensory structures located on the prostomium (the portion of the head anterior to the mouth). *See* ANNELIDA; DEEP-SEA FAUNA.

As in other annelids, the polychaete body consists of two presegmental regions (the prostomium and the peristomium), a segmented trunk, and a postsegmental pygidium (tail region) [see illustration]. Often the prostomium and the peristomium are fused with the first anterior segments of the trunk, known as the cephalized segments, forming the head of the polychaete. The prostomium varies greatly in shape, from a simple conical structure to a simple T-shape, and may be fused to the peristomium. Head appendages, which are present or absent according to the species considered, include antennae, palps (various sensory, usually fleshy appendages near the oral aperture), peristomial cirri (slender hairlike appendages), and tentacular cirri. The trunk of a polychaete comprises serially repeated segments, also called metameres, numbering from few to many. Each segment basically bears a pair of lateral appendages, the parapodia. A parapodium is equipped with musculature, and bears chaetae (setae) in most taxa. Chaetae are epidermal chitin-sclerotized (hardened) proteinaceous rods (like bristles) sometimes impregnated with calcium carbonate. They act in anchoring the parapodium to the substrate during locomotion.

The digestive tract is essentially a simple tube supported by a dorsal mesentery and partial-to-complete septa. The anterior part of the digestive tract is often differentiated in a buccal (mouth) organ that is axial or ventral, well developed or simple. The muscular region of the buccal organ is called the pharynx; the eversible nonmuscular region is often referred to as the proboscis. Feeding methods in polychaetes are diverse. Polychaetes are raptorial (living on prey) feeders and use their buccal apparatus to seize food items; nonselective or selective deposit feeders; or filter feeders. The proboscis of raptorial feeders is well developed and often bears chitinous structures, the paragnaths (small, sharp jawlike structures) and jaws.

The great majority of polychaetes have a closed circulatory system made of medial dorsal and ventral longitudinal vessels linked by smaller capillaries and gut lacunae (cavities). Some taxa have more contractile hearts. Many species have elaborate branchiae (gills) developed as extensions of the body wall containing a loop of the vascular system and capillaries. Blood generally does not contain cells but respiratory pigments such as hemoglobins and chlorocruorins (green substances that are the probable cause of the color of the blood in some worms).

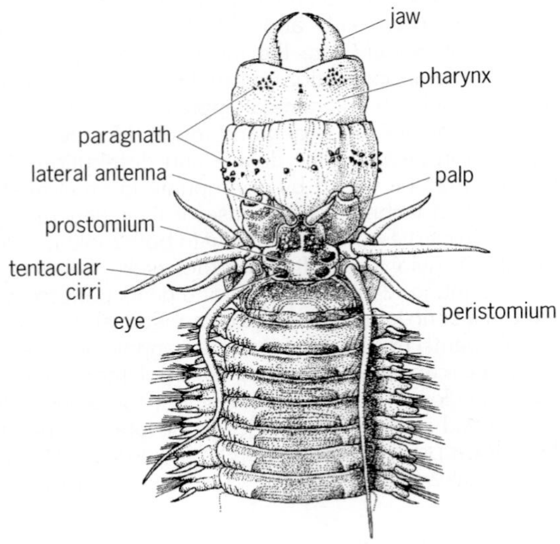

***Platynereis dumerilii antipoda*, showing the antero-dorsal end with the pharynx everted. (*After P. L. Beesley et al., 2000*)**

Polychaeta should be divided into two monophyletic groups, Scolecida and Palpata, with 81 accepted families. Palpata, which includes 72 families, are named for the presence of palps on the head. Palpata comprise two major clades, Canalipalpata and Aciculata. The Scolecida do not have palps but do have parapodia with similar-sized rami and two or more pairs of pygidial cirri; they include 9 families. Igor Eeckhaut; Olga Hartman

Polychlorinated biphenyls A generic term for a family of 209 chlorinated isomers of biphenyl. The biphenyl molecule is composed of two six-sided carbon rings connected at one carbon site on each ring. Ten sites remain for chlorine atoms to join the biphenyl molecule. The term polychlorinated biphenyl (PCB) has been used to refer to the biphenyl molecule with one to ten chlorine substitutions, as shown below. PCBs

3 2 2′ 3′ $Cl_nH_{(10-n)}$ 4 4′ 5 6 6′ 5′

Polychlorinated biphenyl (PCB)

were introduced into United States industry on a large scale in 1929. The qualities that made PCBs attractive were chemical stability, resistance to heat, low flammability, and high dielectric constant. The PCB mixture is a colorless, viscous fluid, is relatively insoluble in water, and can withstand high temperatures without degradation (higher-chlorinated isomers are not readily degraded in the environment).

The major use of PCBs has been as dielectric fluid in electrical equipment, particularly transformers capacitors, electromagnets, circuit breakers, voltage regulators, and switches. PCBs have also been used in heat transfer systems and hydraulic systems, and as plasticizers and additives in lubricating and cutting oils. *See* DIELECTRIC MATERIALS.

PCBs have been reported in animals, plants, soil, and water all over the world, even in animals living under 11,000 ft (3400 m) of water. These phenomena are the result of bioaccumulation and biomagnification in the food chain. In a few instances, poultry products, cattle, and hogs have been found to contain high concentrations of PCBs after the animals have eaten feed contaminated with PCBs. It is not known what quantities of PCBs have been released to the environment, but major sources are industrial and municipal waste disposal, spills and leaks from PCB-containing equipment, and manufacture and handling of PCB mixtures. *See* ATMOSPHERIC GENERAL CIRCULATION; BIOSPHERE; FOOD WEB; HUMAN ECOLOGY.

PCBs can enter the body through the lungs, gastrointestinal tract, and skin, circulate throughout the body, and be stored in adipose tissue. PCBs have been detected in human adipose tissues and in the milk of cows and humans. Some PCBs have the ability to alter reproductive processes in mammals. There is concern that PCBs may be carcinogenic in humans. Glenn Kuntz

Polycladida An order of marine Turbellaria which are several millimeters to several centimeters in length and whose leaflike bodies have a central intestine with radiating branches. Most species live in the littoral zone on the bottom, on seaweed or on other objects, or as commensals in the shells of mollusks and hermit crabs. None are parasitic. Except in warm waters, they are seldom brightly colored. *See* TURBELLARIA. E. Ruffin Jones

Polyester resins A large class of polymeric materials in which ester groups (see below) are in the main chains. Generally, polyesters are prepared by the condensation reactions of glycols (or dialcohols) with dicarboxylic acids, and they range from relatively soft aliphatics to aromatic derivatives which are considered engineering materials. Their properties may be modified by cross-linking, crystallization, plasticizers, or fillers. *See* ESTER.

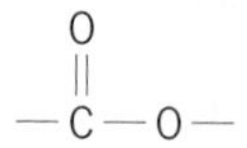

$$-\overset{\displaystyle O}{\overset{\|}{C}}-O-$$

Commercial products include alkyds, which are used in paints and enamels and in molding compounds; unsaturated polyesters or unsaturated alkyds, which are used extensively with fiberglass for boat hulls and panels; aliphatic saturated polyesters; aromatic polyesters such as poly(ethylene terephthalate), which is used in the form of fibers and films, in blow-molded bottles, and in injection-moldable resins such as poly(butylene terephthalate); and aromatic polycarbonates.

The alkyds are commonly used as coatings. Combinations of conventional vegetable drying oils and alkyd resins represent the basis of most of the oil-soluble paints. The drying oil–alkyd described above may be further modified by the inclusion of a vinyl monomer, such as styrene, in the original esterification process. Some of the styrene polymerizes, probably as a graft polymer, and the remainder polymerizes and copolymerizes in the final drying or curing of the paint. *See* COPOLYMER; DRYING OIL; PAINT AND COATINGS.

The unsaturated polyesters, in combination with glass fiber, have applications as panels, roofing, radar domes, boat hulls, and protective armor. The compositions are distinguished by ease of fabrication and high impact resistance. *See* POLYMERIC COMPOSITE.

Saturated aliphatic polyesters, made by the condensation of a diacid such as adipic acid with a diol such as diethylene glycol, the aliphatic polyesters have been frequently used as intermediates in the preparation of prepolymers for making segmented polyurethanes by reaction with diisocyanates. Lactone rings can also be opened to yield linear polyesters; for example, poly(ϵ-caprolactone) has been used as an intermediate in polyurethane technology, as a polymeric plasticizer for poly(vinyl chloride), and in other specialty applications. *See* RING-OPENING POLYMERIZATION.

A class of aliphatic polyesters that is growing in importance are those derived from bioresources. Poly(lactic acid), made by the polymerization of diolide, in turn derived from corn sugars, has become a commercial material, as in reaction (1). It is used in applications demanding biodegradability

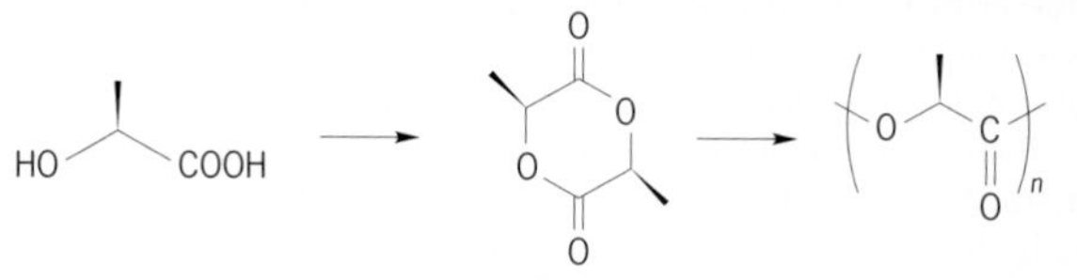

(1)

such as packaging, compost bags, and medical applications. The poly(hydroxyalkanoates), especially poly(hydroxybutyrate), are synthesized within the cells of bacteria in very high yield. Genetic modification of the bacteria has been used to prepare a family of these materials, which have extremely high molecular weights.

Several aromatic polyesters have achieved general importance as engineering materials, including poly(ethylene terephthalate), poly(butylene terephthalate), poly(ethylene naphthoate), and poly(trimethylene terephthalate). Polyethylene terephthalate is the most widely used material, in polyester fibers that are available in the United States and in Europe, in films, and in blow-molded bottles.

All aromatic polyesters include those commonly referred to as polyarylates, made from bisphenol A and mixtures of iso- and terephthalic acids (via their diacid chlorides). Although they may be crystalline, the commercial materials are amorphous, transparent polymers with glass transition temperatures in excess of 175°C (350°F).

An attractive class of aromatic polyesters are the liquid crystalline polyesters (LCPs), largely based on *p*-hydroxybenzoic acids in combination with other monomers such as terephthalic acid, hydroquinone, biphenol, or biphenol dicarboxylic acid. The LCP polyesters have excellent flow properties in the melt, but orient when spun to provide fibers with very high modulus (stiffness).

Polycarbonates (PC) are a strong, optically clear, tough group of thermoplastic polymers formed most frequently from bisphenol A. The products are noted for high glass-transition temperatures, usually greater than 145°C (284°F), high impact resistance, clarity, and resistance to creep.

Polycarbonates are prepared commercially by two methods: interfacial reaction of bisphenol A with phosgene, or transesterification with diphenyl carbonate in the melt as in reaction (2). The

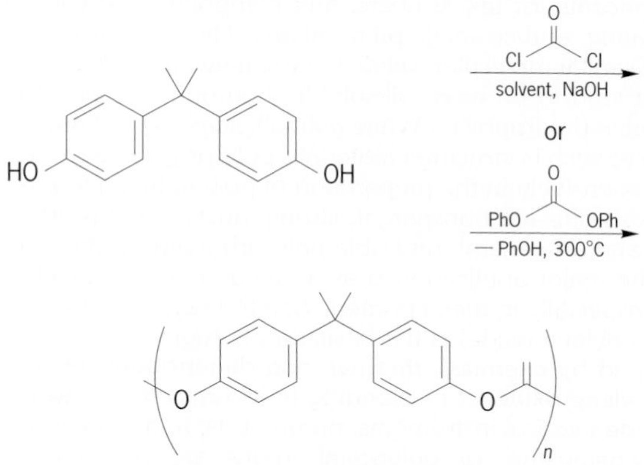

(2)

phosgene process is milder but requires removal of by-product NaCl, whereas the melt process produces resin pellets directly.

The polymer is usually available as an injection-molding resin. Because of its high strength, toughness, and softening point, the resin, both by itself and as a glass-reinforced material, has found many electrical and engineering applications. It is often used to replace glass and metals. Examples include bottles, unbreakable windows, appliance parts, and electrical housings. Films have excellent clarity and electrical characteristics, and have been used on capacitors and in solar collectors, as well as in electronic packaging. With excellent optical properties, polycarbonate has been the material of choice for optical media such as compact discs and DVDs.

Polydiallyl esters are polymers of diallyl esters, such as diallyl phthalate, diallyl carbonate, diallyl phenyl phosphonate, and diallyl succinate, in which cross-linked products are made by polymerization of the allyl groups. Thermosetting molding compounds may be produced by careful limitation of the initial polymerization to yield a product which is fusible.

Major applications are in electronic components, sealants, coatings, and glass-fiber composites. *See* PLASTICS PROCESSING; POLYMERIZATION.

Daniel J. Brunelle

Polyether resins Thermoplastic or thermosetting materials which contain ether-oxygen linkages, —C—O—C—, in the polymer chain. Depending upon the nature of the reactants and reaction conditions, a large number of polyethers with a wide range of properties may be prepared. The main groups of polyethers in use are epoxy resins, phenoxy resins, poly(ethylene oxide) and poly(propylene oxide) resins, polyoxymethylene, and poly(phenylene oxides).

The epoxy resins form an important and versatile class of cross-linked polyethers characterized by excellent chemical resistance, adhesion to glass and metals, electrical insulating properties, and ease and precision of fabrication. Various fillers such as calcium carbonate, metal fibers and powders, and glass fibers are commonly used in epoxy formulations in order to improve such properties as the strength and resistance to abrasion and high temperatures. Some reactive plasticizers act as curing agents, become permanently bound to the epoxy groups, and are usually called flexibilizers. Rubbery polymers are added to improve toughness and impact strength. Epoxies are commonly used in protective coatings. They are used as potting or encapsulating compositions for the protection of delicate electronic assemblies from the thermal and mechanical shock of rocket flight, and as dies for stamping metal forms.

Poly(ethylene oxide) and poly(propylene oxide) are thermoplastic products whose properties are greatly influenced by molecular weight. Low-to-moderate-molecular-weight poly(ethylene oxides) vary in form from oils to waxlike solids. They are relatively nonvolatile, are soluble in a variety of solvents, and have found many uses as thickening agents, plasticizers, lubricants for textile fibers, and components of various sizing, coating, and cosmetic preparations. The poly(propylene oxides) of similar molecular weight have somewhat similar properties, but tend to be more oil-soluble (hydrophobic) and less water-soluble (hydrophilic). While poly(alkylene oxides) are not of interest as such in structural materials, poly(propylene oxides) are used extensively in the preparation of polyurethane foams.

Phenoxy resins are transparent, strong, ductile, and resistant to creep, and, in general, resemble polycarbonates in their behavior. The major application is as a component in protective coatings, especially in metal primers. *See* POLYESTER RESINS.

Poly(phenylene oxide) is the basis for an engineering plastic characterized by chemical, thermal, and dimensional stability. Poly(phenylene oxide) is outstanding in its resistance to water. Uses include medical instruments, pump parts, and insulation.

Polyoxymethylene, or polyacetal, resins are polymers of formaldehyde. Having high molecular weights and high degrees of crystallinity, they are strong and tough and are established in the general class of engineering thermoplastics. Polyacetals are typically resistant to fatigue, creep, organic chemicals (but not strong acids or bases), and have low coefficients of friction. Electrical properties are also good. The combination of properties has led to many uses such as plumbing fittings, pump and valve components, bearings and gears, computer hardware, automobile body parts, and appliance housings. *See* PLASTICS PROCESSING; POLYMERIZATION.

John A. Manson

Poly(ethylene glycol) Any of a series of water-soluble polymers with the general formula $HO—(CH_2—CH_2—O)_n—H$. These colorless, odorless compounds range in appearance from viscous liquids to waxy solids. The low-molecular-weight members, diethylene glycol ($n = 2$) through tetraethylene glycol ($n = 4$), are produced as pure compounds and find use as humectants, dehydrating solvents for natural gas, textile lubricants, heat-transfer fluids, solvents for aromatic hydrocarbon extractions, and intermediates for polyester resins and plasticizers.

The intermediate members of the series with average molecular weights of 200 to 20,000 are used commercially in ceramic, metal-forming, and rubber-processing operations; as drug suppository bases and in cosmetic creams, lotions, and deodorants; as lubricants; as dispersants for casein, gelatins, and inks; and as antistatic agents. The highest members of the series have molecular weights from 100,000 to 10,000,000. They are of interest because of their ability at very low concentrations to reduce friction of flowing water. *See* ETHYLENE OXIDE; GLYCOLIPID; POLYMERIZATION.

Robert K. Barnes; Tacey X. Viegas

Poly(ethylene glycol) has a range of properties making it suitable for medical and biotechnical applications. Since poly(ethylene glycol) is soluble both in water and in most organic solvents, many applications are derived from this amphiphilicity. Other properties include lack of toxicity and immunogenicity, and a tendency to avoid other polymers and particles also present in aqueous solution.

Poly(ethylene glycol) is attached to drugs to enhance water and blood solubility. Similarly, poly(ethylene glycol) is attached to enzymes to impart solubility in organic solvents. These poly(ethylene glycol) enzymes are used as catalysts for industrial reactions in organic solvents. *See* CATALYSIS.

The tendency of poly(ethylene glycol) to avoid interaction with cellular and molecular components of the immune system results in the material being nonimmunogenic. This property leads to a greatly enhanced blood circulation lifetime of poly(ethylene glycol) proteins and to application as pharmaceuticals. Similarly, adsorption of proteins and cells to surfaces is greatly reduced by attaching poly(ethylene glycol) to the surface, and such coated materials find wide application as biomaterials. *See* BIOMEDICAL CHEMICAL ENGINEERING.

J. Milton Harris; Tacey X. Viegas

Polyfluoroolefin resins Resins distinguished by their resistance to heat and chemicals and by the ability to crystallize to a high degree. Several main products are based on tetrafluoroethylene, $F_2C{=}CF_2$ (TFE); hexafluoropropylene, $F_2C{=}CFCF_3$ (HFP); and monochlorotrifluoroethylene, $FClC{=}CF_2$ (CTFE).

Poly(tetrafluoroethylene) is the polymer of tetrafluoroethylene and is commonly known by the trade name Teflon®. It is insoluble, resistant to heat (up to 275°C or 527°F) and chemical attack, and has the lowest coefficient of friction of any solid. However, special surface treatments are required to ensure adhesion because poly(tetrafluoroethylene) does not adhere well to anything. Poly(tetrafluoroethylene) (TFE resin) is used for bearings, valve seats, packings, gaskets, coatings and tubing, and can withstand relatively severe conditions. Because of its excellent electrical properties, poly(tetrafluoroethylene) is useful when a dielectric material is required for service at a high temperature. The nonadhesive quality is often used to coat articles such as rolls and cookware to which materials might otherwise adhere.

The properties of poly(chlorotrifluoroethylene) (CTFE resin) are generally similar to those of poly(tetrafluoroethylene); however, the presence of the chlorine atoms in the former causes the polymer to be a little less resistant to heat and to chemicals. The applications of poly(chlorotrifluoroethylene) are in general similar to those for poly(tetrafluoroethylene). Because of its stability and inertness, the polymer is useful in the manufacture of gaskets, linings, and valve seats that must withstand hot and corrosive conditions. It is also used as a dielectric material, as a vapor and liquid barrier, and for microporous filters.

Poly(vinylidene fluoride) properties has generally similar to those of the other fluorinated resins: relative inertness, low dielectric constant, and thermal stability (up to about 150 or 300°F). The resins (PVF_2 resins) are, however, stronger and less susceptible to creep and abrasion than TFE and CTFE resins. Applications of poly(vinylidene fluoride) are mainly as electrical insulation, piping, process equipment, and as a protective coating in the form of a liquid dispersion.

Several types of fluorinated, noncrystallizing elastomers were developed in order to meet needs (usually military) for rubbers which possess good low-temperature behavior with a high degree of resistance to oils and to heat, radiation, and weathering. *See* HALOGENATED HYDROCARBON; PLASTICS PROCESSING; POLYMERIZATION.

John A. Manson

Polygalales An order of flowering plants, division Magnoliophyta (Angiospermae), in the subclass Rosidae of the class

Magnoliopsida (dicotyledons). The order consists of 7 families and nearly 2300 species, mostly of tropical and subtropical regions. The vast majority of the species belong to only 3 families, the Malpighiaceae (about 1200 species), Polygalaceae (about 750 species), and Vochysiaceae (about 200 species). Within its subclass the order is distinguished by its simple leaves and usually irregular flowers, which have the perianth and stamens attached directly to the receptacle (hypogynous), and often have the anthers opening by terminal pores instead of longitudinal slits. The Barbados cherry (*Malpighia glabra*), noted for the high vitamin-C content of its fruits, is a well-known member of the Polygalales. *See* MAGNOLIOPSIDA; PLANT KINGDOM; ROSIDAE.

Arthur Cronquist; T. M. Barkley

Polygon A geometric figure consisting of an ordered set of three or more (but a finite number of) points called vertices, each vertex connected by line segments called sides or edges to two other vertices. These two sides are said to be adjacent, and so are any two vertices that are end points of a side. The perimeter of the polygon is the sum of the lengths of the sides. The line segments that join two nonadjacent vertices of a polygon are called diagonals. A polygon is said to be directed, or oriented, if a preferred direction is assigned to each side so that at each vertex one of the adjacent sides is directed toward the vertex and the other away from it.

The angle between the two sides of a polygon at a vertex is called an angle of the polygon. Thus an n-sided polygon, called an n-gon, has n vertices and n angles. In particular, if n is 3, 4, 5, 6, 7, 8, 10, or 12, the polygon is called a triangle (3), quadrangle (4) [or quadrilateral, meaning four sides], pentagon (5), hexagon (6), heptagon (7), octagon (8), decagon (10), or dodecagon (12).

A plane polygon is one whose vertices all lie in the same plane. Other polygons are called skew polygons, except for the spherical polygons described below. Skew polygons can be constructed in any number of dimensions.

A plane polygon is called ordinary if no point belongs to more than two edges; it is proper if no two adjacent sides are collinear; it is simple if no two edges intersect each other except at vertices. A simple polygon divides the plane into two regions: an unbounded outside region and an inside region whose area is called the area of the polygon.

A simple polygon is called convex if its interior lies entirely on one side of each (infinite) line through any two adjacent vertices. A nonconvex polygon has at least one interior angle that exceeds 180°. The sum of the interior angles in a convex n-gon is $(n - 2)180°$. A plane polygon is called regular if all its sides are equal and all its angles are equal; it is semiregular if all its angles are equal. *See* REGULAR POLYTOPES.

A spherical polygon consists of points on a spherical surface called vertices, each connected to two adjacent vertices by great-circle arcs, called sides, that are measured by their central angles. The n angles of a convex spherical n-gon are the angles at each vertex between the tangent lines to the two sides that meet there, and their sum exceeds $(n - 2)180°$ by an amount, called the spherical excess, that is proportional to the spherical area enclosed by the polygon.

J. Sutherland Frame

Polygonales An order of flowering plants, division Magnoliophyta (Angiospermae), in the subclass Caryophyllidae of the class Magnoliopsida (dicotyledons). The order consists only of the family Polygonaceae, with about 800 species, most abundant in north temperate regions. Within its subclass the order is characterized by its well-developed endosperm; an ovary with but one chamber(unilocular), mostly tricarpellate with a single basal ovule which is usually straight and has the micropyle at the opposite end from the stalk (orthotropous); the flowers are often trimerous, that is, with floral parts in sets of three, and usually two or three sets of stamens. Rhubarb (*Rheum rhaponticum*) and buckwheat (*Fagopyrum esculentum*) are familiar members of the Polygonales. *See* BUCKWHEAT; CARYOPHYLLIDAE; MAGNOLIOPSIDA; RHUBARB.

Arthur Cronquist; T. M. Barkley

Polyhedron A solid whose boundary consists of a finite number of polygonal faces, that is, planar regions that are bounded by polygons. The sides of the faces are edges of the polyhedron; the vertices of the faces also are vertices of the polyhedron. *See* PLANE GEOMETRY; POLYGON.

Most polyhedra met in applied geometry are convex and simply connected. A polyhedron is convex if it passes this test: if any face is placed coincident with a plane, then all other points of the polyhedron lie on the same side of that plane. A more informal test is to imagine enclosing the polyhedron within a stretched elastic membrane; the polyhedron is convex if all points on the boundary are in contact with the membrane. A simply connected polyhedron has a boundary that is topologically equivalent to a sphere: if the boundary were made of some perfectly elastic material, then the boundary could be distorted into a sphere without tearing or piercing the surface. A simply connected polyhedron is said to be eulerian, because the number of faces F, the number of edges E, and the number of vertices V satisfy Euler's formula:

$$F + V = E + 2$$

Polyhedra exist having any number of faces greater than three. Some polyhedra have names that convey the number of faces (but not the shape) of the polyhedron: tetrahedron, 4 faces; pentahedron, 5 faces; hexahedron, 6 faces; octahedron, 8 faces; dodecahedron, 12 faces; and icosahedron, 20 faces.

A polyhedron is regular (or platonic) if all faces are congruent and all dihedral angles (the angles between adjacent faces) are equal. There are only five regular polyhedra: the tetrahedron, cube, octahedron, dodecahedron, and icosahedron.

Harry L. Baldwin, Jr.

Polymer Polymers, macromolecules, high polymers, and giant molecules are high-molecular-weight materials composed of repeating subunits. These materials may be organic, inorganic, or organometallic, and synthetic or natural in origin. Polymers are essential materials for almost every industry as adhesives, building materials, paper, cloths, fibers, coatings, plastics, ceramics, concretes, liquid crystals, photoresists, and coatings. They are also major components in soils and plant and animal life. They are important in nutrition, engineering, biology, medicine, computers, space exploration, health, and the environment. *See* CERAMICS; MANUFACTURED FIBER; NATURAL FIBER; SURFACE COATING.

Natural inorganic polymers include diamonds, graphite, sand, asbestos, agates, chert, feldspars, mica, quartz, and talc. Natural organic polymers include polysaccharides (or polycarbohydrates) such as starch and cellulose, nucleic acids, and proteins. Synthetic inorganic polymers include boron nitride, concrete, many high-temperature superconductors, and a number of glasses. Siloxanes or polysiloxanes represent synthetic organometallic polymers. *See* SILICONE RESINS.

Synthetic polymers used for structural components weigh considerably less than metals, helping to reduce the consumption of fuel in vehicles and aircraft. They even outperform most metals when measured on a strength-per-weight basis. Polymers have been developed which can also be used for engineering purposes such as gears, bearings, and structural members.

Nomenclature. Many polymers have both a common name and a structure-based name specified by the International Union of Pure and Applied Chemistry (IUPAC). Some polymers are commonly known by their acronyms. Some companies use trade names to identify the specific polymeric products they manufacture. For example, Fortrel® polyester is a poly(ethylene terephthalate) (PET) fiber. Polymers are often generically named, such as rayon, polyester, and nylon. *See* ORGANIC NOMENCLATURE; POLYAMIDE RESINS; POLYESTER RESINS; POLYACRYLATE RESIN.

Composition. Polymer structures can be represented by similar or identical repeat units. These are derived from smaller molecules, called monomers, which react to form the polymer. Propylene monomer and the repeat unit it forms in polypropylene are shown below. With the exception of its end groups,

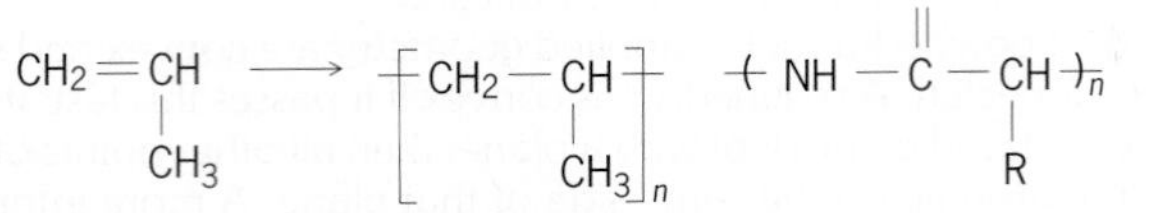

$$CH_2{=}CH(CH_3) \longrightarrow \left[CH_2{-}CH(CH_3) \right]_n \qquad {-}\!\left(NH{-}C({=}O){-}CH(R) \right)_n$$

Monomer — Polymer repeat unit — Protein repeat unit

polypropylene is composed entirely of this repeat unit. The number of units (n) in a polymer chain is called the degree of polymerization (DP). Other polymers, such as proteins, can be described in terms of the approximate repeat unit where the nature of R (a substituted atom or group of atoms) varies. *See* POLYVINYL RESINS; PROTEIN.

Primary structure. The sequence of repeat units within a polymer is called its primary structure. Unsymmetrical reactants, such as substituted vinyl monomers, react almost exclusively to give a "head-to-tail" product, in which the R substituents occur on alternate carbon atoms. A variety of head-to-head structures are also possible.

Each R-substituted carbon atom is a chiral center (an atom in a molecule attached to four different groups) with different geometries possible. Arrangements where the substitutes on the chiral carbon are random are referred to as atactic structures. Arrangements where the geometry about the chiral carbon alternates are said to be syndiotactic. Structures where the geometry about the chiral atom has the same geometry are said to be isotactic or stereoregular.

Stereoregular polymers are produced using special stereoregulating catalyst systems. A series of soluble catalysts have been developed that yield products with high stereoregularity and low chain-size disparity. As expected, polymers with regular structures—that is, isotactic and syndiotactic structures—tend to be more crystalline and stronger. *See* POLYMER STEREOCHEMISTRY AND PROPERTIES.

Polymers can be linear or branched with varying amounts and lengths of branching. Most polymers contain some branching.

Copolymers are derived from two different monomers, which may be represented as A and B. There exists a large variety of possible structures and, with each structure, specific properties. These varieties include alternating, random, block, and graft see (illustration). *See* COPOLYMER.

—A—B—A—B—A—B—A—B—
(a)

—A—B—B—A—B—A—A—A—B—A—B—B
(b)

—A—A—A—A—B—B—B—B—A—A—A
(c)

—A—A—A—A—A—A—A (with —B—B—B branches on the second and fifth A)
(d)

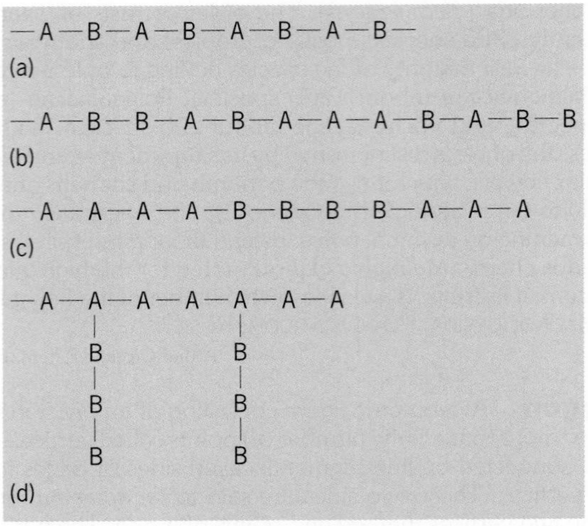

Copolymer structures: (a) alternating, (b) random, (c) block, (d) graft.

Secondary structure. This refers to the localized shape of the polymer, which is often the consequence of hydrogen bonding. Most flexible to semiflexible linear polymer chains tend toward two structures—helical and pleated sheet/skirtlike. The pleated skirt arrangement is most prevalent for polar materials where hydrogen bonding can occur. In nature, protein tissue is often of a pleated skirt arrangement. For both polar and nonpolar polymer chains, there is a tendency toward helical formation with the inner core having "like" secondary bonding forces. *See* HYDROGEN BOND.

Tertiary structure. This refers to the overall shape of a polymer, such as in polypeptide folding. Globular proteins approximate rough spheres because of a complex combination of environmental and molecular constraints, and bonding opportunities. Many natural and synthetic polymers have "superstructures," such as the globular proteins and aggregates of polymer chains, forming bundles and groupings.

Quaternary structure. This refers to the arrangement in space of two or more polymer subunits, often a grouping of tertiary structures. For example, hemoglobin (quaternary structure) is essentially the combination of four myoglobin (tertiary structure) units. Many crystalline synthetic polymers form spherulites. *See* HEMOGLOBIN.

Synthesis. For polymerization to occur, monomers must have at least two reaction points or functional groups. There are two main reaction routes to synthetic polymer formation—addition and condensation. In chain-type kinetics, initiation starts a series of monomer additions that result in the reaction mixture consisting mostly of unreacted monomer and polymer. Vinyl polymers, derived from vinyl monomers and containing only carbon in their backbone, are formed in this way. Examples of vinyl polymers include polystyrene, polyethylene, polybutadiene, polypropylene (see structure), and poly(vinyl chloride).

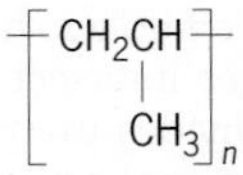

$$\left[CH_2CH(CH_3) \right]_n$$

Polypropylene, PP

The second main route is a step-wise polymerization. Polymerization occurs in a step-wise fashion so that the average chain size within the reaction mixture may have an overall degree of polymerization of 2, then 5, then 10, and so on, until the entire mixture contains largely polymer with little or no monomer left. Polymers typically produced using the step-wise process are called condensation polymers, and include polyamides, polycarbonates, polyesters, and polyurethanes (see structures). Condensation polymer chains are characterized as having a noncarbon atom in their backbone. For polyamides the noncarbon is nitrogen (N), while for polycarbonates it is oxygen (O). Condensation polymers are synthesized using melt (the reactants are heated causing them to melt), solution (the reactants are dissolved), and interfacial (the reactants are dissolved in immiscible solvents) techniques. *See* POLYMERIZATION; POLYOLEFIN RESINS; POLYURETHANE RESINS.

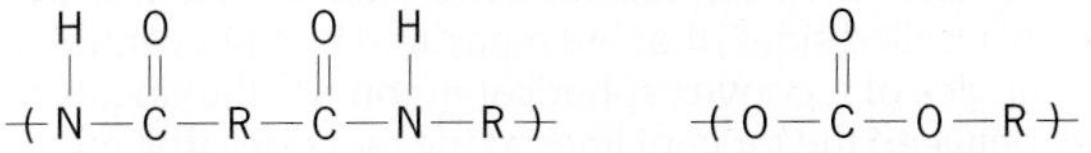

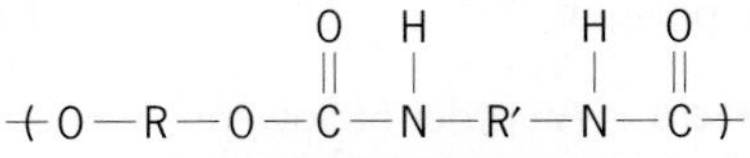

$$\left(N(H){-}C({=}O){-}R{-}C({=}O){-}N(H){-}R \right)$$

Polyamide, nylon

$$\left(O{-}C({=}O){-}O{-}R \right)$$

Polycarbonate

$$\left(O{-}R{-}O{-}C({=}O){-}N(H){-}R'{-}N(H){-}C({=}O) \right)$$

Polyurethane, PU

Molecular properties. These are used to help determine the structure and behavior of the polymer. The molecular weight of a particular polymer chain is the product of the number of units times the molecular weight of the repeating unit. Two statistical averages describe polymers, the number-average molecular weight and the weight-average molecular weight. *See* MOLECULAR WEIGHT.

Size is the most important property of polymers allowing for storage of information (nucleic acids and proteins). Polymeric materials remember any action that distorts or moves polymer chains or segments (such as bending, stretching, and melting). Size also accounts for an accumulation of the interchain and intrachain secondary attractive forces called van der Waals forces. For nonpolar polymers, such as polyethylene, the attractive forces for each repeating unit are less than that for polar polymers. Poly(vinyl chloride), a polar polymer, has attractive forces that include both dispersion and dipole-dipole forces so that the total attractive forces are proportionally larger than those for polyethylene. Polymers with hydrogen bonding (such as proteins, polysaccharides, nucleic acids, and nylons) have attractive forces that are even greater. Hydrogen bonding is so strong in cellulose that cellulose is not soluble in water until the inter- and intrachain hydrogen bonds are broken. *See* CELLULOSE.

Polymers often have a combination of ordered regions, called crystalline regions, and disordered or amorphous regions. Crystalline regions are more rigid, contributing to strength and resistance to external forces. The amorphous regions contribute to polymers' flexibility. Most commercial polymers have a balance between amorphous and crystalline regions, allowing a balance between flexibility and strength.

Polymers are viscoelastic materials. Ductile polymers, such as polyethylene and polypropylene, "give" or "yield," and at high elongations some strengthening and orientation occur. A brittle polymer, such as polystryene, does not give much and breaks at a low elongation. A fiber, a polymer material that is much longer than it is wide, exhibits high strength, high stiffness, and little elongation.

Materials. Fibers are polymer materials that are strong in one direction, and they are much longer (>100 times) than they are wide. Elastomers (or rubbers) are polymeric materials that can be distorted through the application of force, and when the force is removed, the material returns to its original shape. Plastics are materials that have properties between fibers and elastomers—they are hard and flexible. Coatings and adhesives are generally derived from polymers that are members of other groupings (for example, polysiloxanes are elastomers, but also are used as adhesives). Industrially important adhesives and coatings include laminates, sealants and caulks, composites, films, polyblends, liquid crystals, ceramics, cements, and smart materials. *See* ADHESIVE; CEMENT; COMPOSITE LAMINATES; LIQUID CRYSTALS; POLYMERIC COMPOSITE; RUBBER.

Additives. Processed polymeric materials are generally a combination of the polymer and the materials that are added to modify its properties, assist in processing, and introduce new properties. Additives can be solids, liquids, or gases. Typical additives are plasticizers, antioxidants, colorants, fillers, and reinforcements. *See* ANTIOXIDANT; INHIBITOR (CHEMISTRY).

Recycling. Many polymers are thermoplastics, that is, they can be reshaped through application of heat and pressure and used in the production of other thermoplastic materials. The recycling of thermosets, polymers that do not melt but degrade prior to softening, is more difficult. These materials are often ground into a fine powder, are blended with additives (often adhesives or binders), and then are reformed. *See* RECYCLING TECHNOLOGY.

Charles E. Carraher, Jr.

Polymer composites Polymer materials or matrices that are combined with fillers or fibers. Fillers are intentionally placed in polymers to make them stronger, lighter, electrically conductive, or cheaper. Any filler will affect the mechanical behavior of a polymeric material. For example, long fibers will make the material stiffer but usually denser, whereas foaming will make it more compliant but much lighter. A filler such as calcium carbonate will decrease the polymer's toughness, while making it considerably cheaper. Polymer composites are also found in nature, such as wood (cellulose fibers bonded with lignin) and bone (minerals bonded with collagen).

Mechanical properties of selected fibers

Fiber	Tensile strength, MPA	Tensile modulus, GPa	Elongation to break, %	Specific gravity
Polyethylene	3000	172	2.7	0.97
Aramid	2760	124	2.5	1.44
Graphite	2410	379	0.6	1.81
S-glass	4585	90	2.75	2.50

Polymer composites have become prevalent in automotive and aircraft design and manufacture. For example, newer commercial aircraft can have up to 80% of their structure made of polymer composites, including the whole fuselage. Many automobiles not only have body panels that are compression- or injection-molded out of fiber-reinforced polymers, but also polymer-composite structural components as well as many under-the-hood applications such as valve covers and oil pans. Other applications of polymer composites include filament-wound pressure vessels and pipes for the chemical industry, braided fiber–reinforced epoxy tubes used in the construction of composite bicycles, and resin transfer–molded boats, skies, and snowboards, to name a few. *See* COMPOSITE MATERIAL.

In a composite structure, the task of the polymer matrix is to hold the fibers in place or bond them together into one integral structure. Polymer matrices can be either thermoplastic or cross-linked polymers. Thermoplastic polymers are divided into amorphous and semicrystalline thermoplastics, and cross-linked polymers can be either elastomers or thermosets. *See* POLYMER.

In a well-designed composite material, the fiber is the load-bearing structure of a part. The purpose of introducing a fiber into a matrix is to transfer the load from the weaker material to the stronger one. This load transfer occurs over the length of the fiber. The length necessary to complete the load transfer from the matrix to the fiber, without fiber or matrix fracture, usually is referred to as the critical length. Traditionally, fibers were glass or carbon. However, fibers also can be made out of polymers by orienting or aligning the long molecular chains. The properties of polymeric fibers (polyethylene and aramid) are comparable to the properties of graphite and glass fibers, differing mostly in that polymers are less dense than inorganic fibers (see table). *See* CARBON; GLASS; GRAPHITE; MANUFACTURED FIBER.

Fiber-reinforced polymers are often anisotropic; that is, they show different properties in different directions. As a result, the usual relations that are used to represent and analyze homogeneous materials are not valid. The actual behavior of the composite is related not only to the individual properties of the matrix and the fibers but also to the volume fraction of fibers or fillers present in the composite, as well as the orientation of the loads relative to the fiber direction. For high volume-fraction fiber contents, only a slight misalignment of the fibers from the loading direction results in a drastic reduction of the properties. This is critical in composite structures that are constructed with aligned fibers, such as aircraft components. With help of commercially available computer-aided technology (CAD), one can design and predict properties of composites with reinforcing fibers that have orientation distributions that result from flow and deformation during manufacturing. Such systems are typical of compression-molded automotive body panels, as well as injection-molded fiber-reinforced components.

Tim A. Osswald

Polymer stereochemistry and properties

The properties of polymers are significantly influenced by their stereochemistry. Cellulose (structure **1**) and amylose (**2**), the

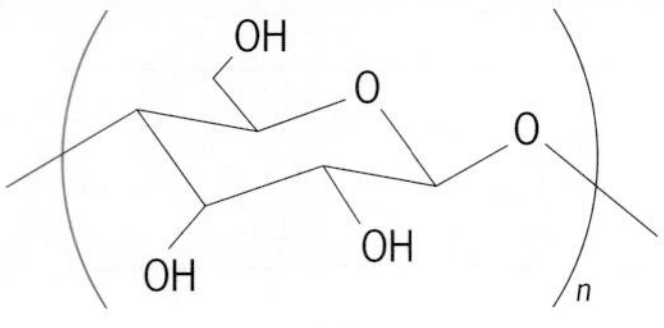

(1)

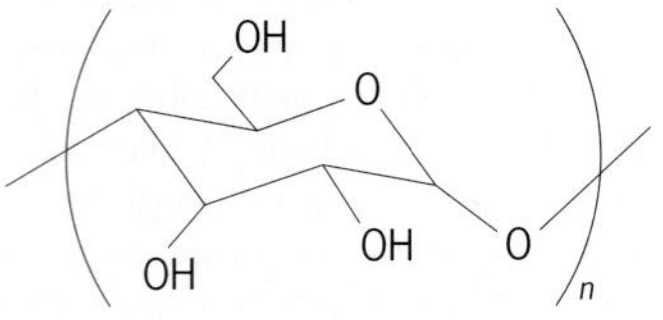

(2)

main components of wood and starch, respectively, are among the most abundant polymers on Earth. Both are produced from D-glucose and are stereoisomers of each other. These two have different linkages between 1- and 4-carbons; the former has a 1,4-β-structure and the latter a 1,4-α-structure. This difference in the linkage of these polysaccharides leads to different higher-order structures of the polymer chains; cellulose has a stretchable polymer chain, which facilitates the regular arrangement of the polymer chains through the interchain hydrogen bond, whereas amylose has a helical chain, in which about seven glucose residues construct one turn of a helix. Because of their structural differences, their properties are also very different; for instance, cellulose is not soluble in water because of the strong interchain hydrogen bonds, but amylose is. Human beings cannot digest cellulose to D-glucose to use as an energy source, but can digest amylose or starch to D-glucose.

When vinyl monomers ($CH_2{=}CH{-}X$) like propylene ($X = CH_3$), styrene ($X = C_6H_5$), vinyl acetate ($X = OCOCH_3$), and *N*-isopropylacrylamide ($X = CONHCH(CH_3)_2$) are polymerized, each step of the monomer addition to a propagating chain end yields an asymmetric *R* or *S* center. The continuing formation of the same center produces an isotactic polymer (**3**), and the alternating formation of the *R* and *S* centers produces a syndiotactic polymer (**4**). This kind of arrangement of side chains has

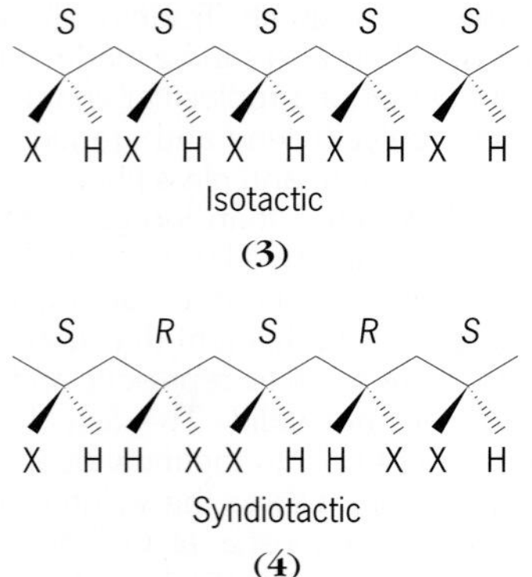

(3)

(4)

been called stereoregularity or tacticity. The control of tacticity was first realized by G. Natta in 1956 for the stereospecific coordination polymerization of propylene using the Ziegler-Natta catalyst consisting of $TiCl_3{-}Al(C_2H_5)_3$. The isotactic polymer had properties superior to those of the atactic polypropylene, which has randomly arranged side-chain methyl groups. Isotactic polypropylene is a hard, durable material with a high melting point (~165°C or 330°F) and high mechanical strength, whereas the atactic polypropylene is a greasy material. Isotactic polypropylene is one of the most widely used polymers. *See* POLYVINYL RESINS; POLYMER; STEREOCHEMISTRY. Yoshio Okamoto

Polymer-supported reaction

An organic chemical reaction where one of the species, such as the substrate, the reagent, or a catalyst, is bound to a cross-linked, and therefore insoluble, polymer support. A major attraction of polymer-supported reactions is that at the end of the reaction period the polymer-supported species can be separated cleanly and easily, usually by filtration, from the soluble species. This easy separation can greatly simplify product isolation procedures, and it may even allow the polymer-supported reactions to be automated. Because it is possible to reuse or recycle polymer-supported reactants and because they are insoluble, involatile, easily handled, and easily recovered, polymer-supported reactants are also attractive from an environmental point of view. *See* ASYMMETRIC SYNTHESIS; CATALYSIS.

Polymer-supported reactants are usually prepared in the form of beads of about 50–100 micrometers' diameter. Such beads can have a practically useful loading only when their interiors are functionalized, that is, carry functional groups. With a typical polymer-supported reactant, more than 99% of the reactive groups are inside the beads. An important consequence is that for soluble species to react with polymer-supported species the former must be able to diffuse freely into the polymer beads. *See* POLYMER.

In a typical application of reactions involving polymer-supported substrates, a substrate is first attached to an appropriately functionalized polymer support. The synthetic reactions of interest are then carried out on the supported species. Finally, the product is detached from the polymer and recovered. Often the polymer-supported species has served as a protecting group that is also a physical "handle" to facilitate separation. Combinatorial synthesis is carried out on polymer-supported substrates. This approach is used in the pharmaceutical industry to identify lead compounds, which can thus be identified in weeks rather than years.

Reactions involving polymer-supported reagents are generally more useful than those involving polymer-supported substrates, because no attachment or detachment reactions are needed, and it is not necessary for all the polymer-supported species to react in high yield. Indeed, polymer-supported reagents are often used in excess to drive reactions to high conversions. Polymer-supported catalysts are the most attractive type of polymer-supported reactants; in such reactions, the loadings of catalytic sites need not be high, not all sites need be active and, in most cases, the polymer-supported catalyst is recovered in a suitable form for reuse. Philip Hodge

Polymer thermodynamics

The relationships between properties of polymers, especially those affected by changes in temperature and the conversion of energy from one form to another. Like other materials, the bulk properties of polymers are determined by their chemical structure, but unlike many other classes of engineering materials, the range of properties that can be obtained is extremely wide. This is a consequence of the chemical and structural diversity of polymeric molecules. *See* POLYMER.

Polymeric molecules typically have negligible vapor pressure and exist only in solid or liquid forms. Some polymers form a regular crystalline solid with a well-defined melting temperature T_m. However, irregularities of molecular size and structure typically lead to solid phases that are amorphous or, at most, semi-crystalline. Thermodynamic studies on such materials reveal a gradual transition from an amorphous glassy phase to a viscoelastic liquid as the temperature is raised above the glass-transition temperature T_g (the temperature above which the polymers becomes flexible). Above either T_m or T_g, the viscosity drops very rapidly with increasing temperature.

Polymers are frequently blended in order to obtain materials with desired properties. In addition, polymers may be synthesized or processed in solution. In all these situations, the thermodynamics of mixing are of considerable importance. The most important issue is whether or not the components are miscible, and the key to understanding this is the behavior of the Gibbs free energy G of the system. For a system containing fixed amounts of the components under constant temperature T and pressure p, the fundamental thermodynamic criterion governing the number of phases present at equilibrium is that G should be minimized. Thus, a mixture held at constant T and p will split into different phases only if this leads to reduction in G. It follows that the phase behavior of a two-component (binary) mixture may be determined by examining the change of the molar Gibbs free energy upon mixing, Δg, as a function of the mole fraction x of one component.

Because of the huge number of possible blends and solutions of polymers, much of modern polymer thermodynamics is concerned with modeling the liquid-liquid and vapor-liquid equilibrium and other properties of such systems. The goal is to develop models that can predict the properties of mixtures containing polymers based on a minimum set of experimental data. Thermodynamic models fall into two broad categories: excess Gibbs free energy models and equation of state theories. The former seek to relate Δg to temperature, pressure, and composition and may be used to predict phase equilibrium, whereas the latter express the system pressure in terms of temperature, density, and composition. Equation-of-state theories also lead to Δg (and hence to predictions of phase behavior). They also permit other thermodynamic properties, such as compressibility and thermal expansivity, to be predicted. *See* COPOLYMER; ENTHALPY; FREE ENERGY; GLASS TRANSITION; POLYMER; POLYMERIZATION; THERMODYNAMIC PRINCIPLES. J. P. Martin Trusler

Polymerization The repetitive linking of small molecules (monomers) to make larger molecules (polymers). Polymerization requires that each small molecule have at least two reaction points or functional groups. The polymerization of a monomer usually involves many steps. In each step, new bonds are formed, resulting in the extension and growth of the respective chain. Synthetic polymers are usually polydisperse; that is, in any given sample, the many polymer chains present will have varying lengths, while being of the same chemical composition. Consequently, the value describing a molecular weight or size of a polymer sample always refers to an average, such as a weight average or number average. Size and architecture of polymer molecules are of greatest importance to the polymer chemist, as they determine the ultimate material properties, and hence the necessary polymerization process.

There are two distinct major types of polymerization processes: condensation polymerization, in which the chain growth is accompanied by elimination of small molecules such as water (H_2O) or methanol (CH_3OH), and addition polymerization, in which the polymer is formed without the loss of other materials.

The for mation of macromolecules by the condensation process, as in the production of polyamides, polyesters, and polysulfides, requires the elimination of small molecules; at the same time, strongly polar and strongly attracting groups are produced. Most condensations proceed in a stepwise fashion. Common types of condensation polymers are polyamide, polyester, polyurethane, polysiloxane, phenol-formaldehyde, urea-formaldehyde, melamine-formaldehyde, polysulfide, polyacetal, protein, wool, and silk. *See* PHENOLIC RESIN; POLYACETAL; POLYAMIDE RESINS; POLYESTER RESINS; POLYSULFIDE RESINS; POLYURETHANE RESINS; PROTEIN; SILICONE RESINS; SILK; WOOL.

An example of the condensation process is reaction (1) of

$$n\,H_2N-(CH_2)_5-\overset{\overset{\displaystyle O}{\|}}{C}-OH \xrightarrow[-(n-1)H_2O]{} H\!\left[\overset{\overset{\displaystyle H}{|}}{N}-(CH_2)_5-\overset{\overset{\displaystyle O}{\|}}{C}\right]_n\!OH \qquad (1)$$

ε-Aminocaproic acid — Eliminated by-product — Nylon-6 (polyamide)

ε-aminocaproic acid in the presence of a catalyst to form the polyamide, nylon-6. The repeating structural unit is equivalent to the starting material minus H and OH, the elements of water. A similar product would be obtained by reaction (2) of a diamine

$$n\,H_2N-(CH_2)_6-NH_2 + n\,HO-\overset{\overset{\displaystyle O}{\|}}{C}-(CH_2)_4-\overset{\overset{\displaystyle O}{\|}}{C}-OH \xrightarrow[-(2n-1)H_2O]{} H\!\left[\overset{\overset{\displaystyle H}{|}}{N}-(CH_2)_6-\overset{\overset{\displaystyle H}{|}}{N}-\overset{\overset{\displaystyle O}{\|}}{C}-(CH_2)_4-\overset{\overset{\displaystyle O}{\|}}{C}\right]_n\!OH \qquad (2)$$

Hexamethylene diamine — Adipic acid — Eliminated by-product — Nylon-6,6 (polyamide)

and a dicarboxylic acid. In both cases, the molecules formed are linear because the total functionality of the reaction system (functional groups per molecule) is always two.

Unsaturated compounds such as olefins and dienes polymerize without the elimination of other products. Certain ring structures, for example, lactams or alkylene oxides, may polymerize by an opening of the ring. The molecular weight and structure of the polymer are determined by the reaction conditions, that is, the nature of the catalyst or initiator, the temperature, and the concentration of reactants, monomer, initiator, and modifying agents. In general, a ceiling temperature exists above which polymerization cannot occur. Most addition polymerizations proceed by a chain reaction in which the average chain length of the polymer formed initially is high and may increase further through secondary branching reactions as the polymerization approaches completion.

The types of catalysis or initiation which are effective for addition (chain-growth) polymerization fall into four groups: (1) free-radical catalysis by peroxides, persulfates, azo compounds, oxygen, and ultraviolet and other radiation; (2) acid (cationic) catalysis by the Lewis acids, such as boron trifluoride, sulfuric acid, aluminum chloride, and other Friedel-Crafts agents; (3) basic (anionic) catalysis, by chromic oxide on silica-alumina, nickel, or cobalt on carbon black, molybdenum on alumina, and complexes of aluminum alkyls with titanium chloride. (4) The precise mechanisms of the fourth group are difficult to elucidate. However, evidence suggests that, at least in some cases, the mechanism involves an anionic process modified by coordination of the monomer and polymer with a surface.

Among the several kinds of polymerization catalysis, free-radical initiation has been most studied and is most widely employed. Atactic polymers are readily formed by free-radical polymerization, at moderate temperatures, of vinyl and diene monomers and some of their derivatives. *See* FREE RADICAL.

Some polymerizations can be initiated by materials, often called ionic catalysts, that contain highly polar reactive sites or complexes. A general mechanism is shown in reactions (3), in

$$\begin{array}{l}\text{Initiative complex:}\\ \quad \text{Complex catalyst} + M \rightarrow CM^*\\ \text{Initiation:}\\ \quad CM^* + M \rightarrow M^*CM\\ \text{Termination by decomposition of complex:}\\ \quad M_X^*CM \rightarrow M_X + CM\\ \text{Termination by transfer to monomer:}\\ \quad M_X^*CM + M \rightarrow M_X + M^*CM\end{array} \qquad (3)$$

which the growing chain is represented as an activated complex with the complex catalyst, without attempting to specify whether separate ions or free radicals are involved.

A distinguishing feature of complex catalysts is the ability of some representatives of each type to initiate stereoregular polymerization at ordinary temperatures or to cause the formation of polymers that can be crystallized. The polymerization process may often be visualized as the formation of an activated complex of the monomer with the complex catalyst. For stereoregular growth to take place, the entering monomer must collide with the complex and form a new, transient complex in which the new monomer molecule is held in a particular orientation. As reaction takes place, the new monomer assumes an activated condition within the complex catalyst and, at the same time, pushes the old monomer unit out. Chain growth is therefore similar to the growth of a hair from the skin. If conditions favor certain orientations of the old and new monomer units, a stereoregular polymer results. *See* CATALYSIS; SINGLE-SITE CATALYSTS (POLYMER).

Ralf M. Peetz; John A. Manson

Chain-growth polymerizations originate from an initiator fragment and involve the sequential addition of new monomer molecules to the growing chain end. The growth of any particular chain is stopped by transfer or termination reactions. In recent years, the toolbox of macromolecular engineering has grown to include a variety of polymerization chemistries that address the transfer and termination problem by either suppression or prevention, or by rendering them reversible. The result can be a polymer chain that stops growing when the monomer is depleted in the reaction mixture, but starts growing again when new monomer is added. Hence, the molecular weight scales linearly with conversion.

In such a living/controlled polymerization, a defined number of initiator fragments start a defined number of polymer chains, all of which ideally grow at the same rate, while no new chains are started. This results in a very uniform molecular weight distribution in the sample and a low polydispersity. This enables the synthesis of polymers with defined molecular weights and high uniformity. Because any impurity in the reaction mixture has the potential to act as a transfer or termination agent, or an initiator site, particular attention has to be paid to the purity of the monomers and solvents.

Even after isolation of the polymer, the chain ends often remain functional for either a next stable free-radical polymerization (SFRP), and reversible addition-fragmentation transfer (RAFT). *See* CONTROLLED/LIVING RADICAL POLYMERIZATION; MACROMOLECULAR ENGINEERING; POLYMER. Ralf M. Peetz

Polymicrobial diseases Contraction of multiple microbial infections. Such infections, which can occur in humans and animals, may be sequential or concurrent. For example, the initial infection may predispose the host to subsequent microbial infections, or the initial infecting microorganism may induce a state of immunosuppression. Concurrent infections are caused by various combinations of viruses, bacteria, fungi, and parasites. *See* BACTERIA; DISEASE; FUNGI; PARASITOLOGY; VIRUS.

Microorganisms can interact in a variety of ways, causing both acute and chronic diseases in humans and animals. One microorganism generates a niche for other pathogenic microorganisms to colonize; the presence of one microorganism predisposes the host to colonization by other microorganisms; two or more nonpathogenic microorganisms together cause pathology; or one microorganism interferes with infection or colonization of other microorganisms. Many of these polymicrobial interactions occur within biofilms that form on natural or artificial surfaces within the human host. Biofilms are sessile aggregate communities of bacteria attached to surfaces in hydrated polymeric matrices of exopolysaccharides. Organisms in biofilms are very hardy and resistant to antimicrobial agents as well as to host innate and acquired immune defenses. *See* BIOFILM; INFECTION; INFECTIOUS DISEASE; PATHOGEN.

Synergism results when one microorganism creates an environment that favors the infection and colonization of other, often pathogenic microorganisms. Recent examples are the concurrent infection of human metapneumovirus and coronavirus in severe acute respiratory syndrome (SARS) and concurrent infection of human metapneumovirus and respiratory syncytial virus in bronchiolitis and other respiratory infections. Other examples include measles virus, human T-lymphotropic virus type I (HTLV-I), HTLV-II, and human immunodeficiency virus (HIV), which induce immunosuppression favoring secondary infections with other viruses, bacteria, fungi, and protozoa.

The presence of one microorganism may predispose the host to colonization or infection by a second microorganism. Respiratory viral infections are notorious for predisposing humans and animals to secondary bacterial infections. In humans, influenza virus increases the incidence of bacterial pneumonia by *Staphylococcus aureus*, *Haemophilus influenzae*, and *Streptococcus pneumoniae*.

Polymicrobial diseases are serious diseases whose etiologic agents are sometimes difficult to diagnose and treat. Future work will undoubtedly focus on improving the identification of multiple microorganisms in polymicrobial diseases, defining microbe-to-microbe and microbe-to-host interactions, and improving prophylaxis and treatment of these diseases.

Kim A. Brogden; Janet M. Guthmiller

Polymixiiformes The only order of the superorder Polymixiomorpha, comprising one family (Polymixiidae) with one genus (*Polymixia*) and ten extant species. Several fossil genera are known from the Upper Cretaceous. Recent evidence suggests that the family may be the sister group to all other acanthomorphs (spiny-rayed fishes). The family is known commonly as beardfishes. *See* OSTEICHTHYES; TELEOSTEI.

Pacific beardfish (*Polymixia berndti*). (Photo by Jack E. Randall)

The beardfishes are so called for the pair of long hyoid (between the lower jaws) barbels (see illustration). They are further distinguished by a moderately elongate and compressed body; and two sets of intermuscular bones, which is a condition otherwise unknown among the acanthomorphs. Polymixiids are bathydemersal [living and feeding on the bottom below 200 m (660 ft)] fishes to depths of 700 m (2300 ft), occurring in warm temperate to tropical waters of the Atlantic, Indian, and Western Pacific oceans. They feed on other fishes, worms, and crustaceans and attain a maximum length of 38 cm (15 in.).

Herbert Boschung

Polymorphism (crystallography) The existence of different crystal structures with the same chemical composition. If only one chemical element is present, the forms are called allotropes. Graphite and diamond are allotropes of carbon, whereas quartz and cristobalite are polymorphs of silica (silicon dioxide, SiO_2). Although properties are different in these forms, reversible transformations, which involve small shifts in atom positions and no bulk transport of material, are common. The quartz transformation at 1063°F (573°C) is a reversible, atom-displacement transformation. *See* SILICA MINERALS.

In metals and ceramics, similar transformations are called martensitic. Advantage is taken of the localized nature of reversible transformation in steel by controlling the melting atmosphere, temperature, composition, mechanical working (alloying), and tempering and quenching operations. *See* HEAT TREATMENT (METALLURGY).

Control over transformations to achieve desirable properties as either devices or structural materials in extreme environments is a frequent objective. In the case of tin, reversibility on the atomic scale can have devastating consequences for bulk properties. Similar transformations may be beneficial in the right place and in the desired degree. Such transformation is attempted with metals and ceramics. *See* CERAMICS; CRYSTAL STRUCTURE.

Doris Evans

Polymorphism (genetics) A form of genetic variation, specifically a discontinuous variation, occurring within plant and animal species in which distinct forms exist together in the same population, even the rarest of them being too common to be maintained solely by mutation. Thus the human blood groups are examples of polymorphism, while geographical races are not; nor is the diversity of height among humans, because height is "continuous" and does not fall into distinct tall, medium, and short types. *See* MUTATION.

Distinct forms must be controlled by some switch which can produce one form or the other without intermediates such as those arising from environmental differences. This clear-cut control is provided by the recombination of the genes. Each gene may have numerous effects and, in consequence, all genes are nearly always of importance to the organism by possessing an overall advantage or disadvantage. They are very seldom of neutral survival value, as minor individual variations in appearance often are. Thus a minute extra spot on the hindwings of a tiger moth is in itself unlikely to be of importance to the survival of the insect, but the gene controlling this spot is far from negligible since it also affects fertility. *See* RECOMBINATION (GENETICS).

Genes having considerable and discontinuous effects tend to be eliminated if harmful, and each gene of this kind is therefore rare. On the other hand, those that are advantageous and retain their advantage spread through the population so that the population becomes uniform with respect to these genes. Evidently, neither of these types of genes can provide the switch mechanism necessary to maintain a polymorphism. That can be achieved only by a gene which has an advantage when rare, yet loses that advantage as it becomes commoner.

Occasionally there is an environmental need for diversity within a species, as in butterfly mimicry. Mimicry is the resemblance of different species to one another for protective purposes, chiefly to avoid predation by birds. Sexual dimorphism falls within the definition of genetic polymorphism. In any species, males and females are balanced at optimum proportions which are generally near equality. Any tendency for one sex to increase relative to the other would be opposed by selection.

In general, a gene having both advantageous and disadvantageous effects may gain some overall advantage and begin to spread because one of the features it controls becomes useful in a new environment. A balance is then struck between the advantages and disadvantages of such a gene, ensuring that a proportion of the species carry it, thus giving rise to permanent discontinuous variation, that is, to polymorphism. *See* PROTECTIVE COLORATION.

Polymorphism is increasingly known to be a very common situation. Its existence is apparent whenever a single gene having a distinct recognizable effect occurs in a population too frequently to be due merely to mutation. Even if recognized by some trivial effect on the phenotype, it must in addition have important other effects. About 30% of the people in western Europe cannot taste as bitter the substance phenylthiourea. This is truly an insignificant matter; indeed, no one even had the opportunity of tasting it until the twentieth century. Yet this variation is important since it is already known that it can affect disease of the thyroid gland. *See* GENETICS; POPULATION GENETICS.

E. B. Ford

Polynomial systems of equations Systems of mathematical equations which have the form of system (1). Each

$$\begin{aligned} f_1(x_1, x_2, \ldots, x_n) &= 0 \\ f_2(x_1, x_2, \ldots, x_n) &= 0 \\ \cdots\cdots\cdots\cdots\cdots\cdots \\ f_m(x_1, x_2, \ldots, x_n) &= 0 \end{aligned} \tag{1}$$

$f_i(x_1,x_2,\ldots, x_n)$, $i = 1, 2, \ldots, m$, is a sum of terms of the form shown as expression (2), where the coefficient $a_{i_1i_2\cdots i_n}$ is a constant, or fixed number, and the exponent i_j of the variable x_j is a nonnegative whole number. An example of such a system in

$$a_{i_1i_2\ldots i_n}x_1^{i_1}x_2^{i_2}\ldots x_n^{i_n} \tag{2}$$

two variables is system (3). The expressions $f_i(x_1,x_2,\ldots, x_n)$ are

$$\begin{aligned} x^2 - xy + y^2 - 1 &= 0 \\ x^2 + xy - 3y^2 - 2x + 2y + 1 &= 0 \end{aligned} \tag{3}$$

called polynomials in several variables. The problem posed by system (1) is to find necessary and sufficient conditions that there exist values of the variables $x_1 = a_1$, $x_2 = a_2, \ldots, x_n = a_n$ which simultaneously satisfy each equation of the system, and to find all such sets of values, which are called solutions of the system. In example (3), a complete set of solutions is given by $x = 1$, $y = 0$; $x = 0$, $y = 1$; $x = 1$; $y = 1$; and $x = -1$, $y = -1$. *See* EQUATIONS, THEORY OF; LINEAR SYSTEMS OF EQUATIONS.

Alexander P. Morgan

Polynuclear hydrocarbon One of a class of hydrocarbons possessing more than one ring. The aromatic polynuclear hydrocarbons may be divided into two groups. In the first, the rings are fused, which means that at least two carbon atoms are shared between adjacent rings. Examples are naphthalene (**1**), which has two six-membered rings, and acenaphthene (**2**), which has two six-membered rings and one five-membered ring.

In the second group of polynuclear hydrocarbons, the aromatic rings are joined either directly, as in the case of biphenyl (**3**), or through a chain of one or more carbon atoms, as in 1,2-diphenylethane (**4**).

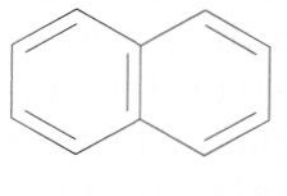

$CH_2 - CH_2$

(1) 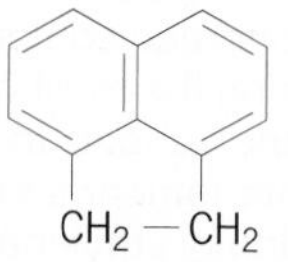 (2) 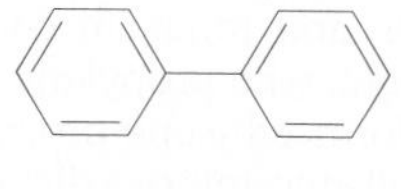 (3)

$C_6H_5CH_2CH_2C_6H_5$

(4)

The higher-boiling polynuclear hydrocarbons found in coal tar or in tars produced by the pyrolysis or incomplete combustion of carbon compounds are frequently fused-ring hydrocarbons, some of which may be carcinogenic. *See* AROMATIC HYDROCARBON; STEROID.

C. K. Bradsher

Polyol A compound containing more than one hydroxyl group (—OH). Each hydroxyl is attached to separate carbon atoms of an aliphatic skeleton. This group includes glycols, glycerol, and pentaerythritol and also such products as trimethylolethane, trimethylolpropane, 1,2,6-hexanetriol, sorbitol, inositol, and poly(vinyl alcohol). Polyols are obtained from many plant and animal sources and are synthesized by a variety of methods.

Polyols such as glycerol, pentaerythritol, trimethylolethane, and trimethylolpropane are used in making alkyd resins for decorative and protective coatings. Glycols, glycerol, 1,2,6-hexanetriol, and sorbitol find application as humectants and plasticizers for gelatin, glue, and cork.

The polymeric polyols used in manufacture of the urethane foams represent a series of synthetic polyols. These polyols are generally poly(oxyethylene) or poly(oxypropylene) adducts of di-to octahydric alcohols. *See* GLYCEROL; GLYCOL.

Philip C. Johnson

Polyolefin resins Polymers derived from hydrocarbon molecules that possess one or more alkenyl (or olefinic) groups. The term polyolefin typically is applied to polymers derived from ethylene, propylene, and other alpha-olefins, isobutylene, cyclic olefins, and butadiene, and other diolefins. *See* ALKENE.

Polyethylene. Polyethylene is any homopolymer or copolymer in which ethylene is the major component monomer. It is a semicrystalline polymer of low to moderate strength and high toughness; its stiffness, yield strength, and thermal and mechanical properties increase with crystallinity. Toughness and ultimate tensile strength increase with molecular weight. Polyethylene shows excellent toughness at low temperatures. Polyethylene is relatively inexpensive, extremely versatile, and adaptable to a large array of fabrication techniques. It is chemically inert; resistant to solvents, acids, and alkalis; and has good dielectric and barrier properties. It is used in many housewares, films, molded articles, and coatings.

Low-density polyethylene has densities ranging from 0.905 to 0.936 g/cm^3. High-pressure low-density polyethylene is referred to simply as LDPE; linear low-density polyethylene (LLDPE) is a copolymer of polyethylene that is produced by a low-pressure polymerization process. High-density polyethylene (HDPE) covers the density range from 0.941 to 0.967 g/cm^3. HDPE generally consists of a polymethylene $(CH_2)_n$ chain with no, or very few, side chains to disrupt crystallization, while LLDPE contains side chains whose length depends on the comonomer used. *See* COPOLYMER.

LLDPE finds wide application in plastic films such as garbage bags and stretch cling films. Sheathing and flexible pipe are applications that take advantage of the flexibility and low-temperature toughness of LLDPE. HDPE is used in food packaging, grocery bags, pickup truck bedliners, and large containers. Fibers have been produced that approach the strength of spider silk, and is used in fishing lines and in medical applications.

Rubbery ethylene copolymers are used in compounded mixtures and range in comonomer content from 25 to 60% by weight, with propylene being the most widely used comonomer to form ethylene-propylene rubber. In addition to propylene, small amounts of a diene are sometimes included, forming a terepolymer. Products containing ethylene-propylene rubber and terepolymer have many automotive uses such as in bumpers, facia, dashboard panels, steering wheels, and assorted interior trim. *See* RUBBER.

Ethylene copolymers are used to produce the polymer poly(ethylene-co-vinyl acetate) [EVA]. Applications include specialty film for heat sealing, adhesives, flexible hose and tubing, footwear components, bumper components, and gaskets. Foamed and cross-linked poly(ethylene-co-vinyl acetate) is used in energy-absorbing applications. Ionomers are ethylene copolymers that are produced from the copolymerization of ethylene with a comonomer containing a carboxylic group (COOH) such as methyl acrylic acid. Because of their toughness, ionomers are widely used in the covers for golf balls.

Bruce Bersted

Polypropylene. Commercial polypropylene (PP) homopolymers are isotactic, high-molecular-weight, semicrystalline solids having melting points around 160–165°C (320–329°F), low density (0.90–0.91 g/cm^3), and excellent stiffness and tensile strength. They have moderate impact strength (toughness), low density over a wide temperature range, excellent mechanical properties, and low electrical conductivity. Propylene, like ethylene, is produced in large quantities at low cost from the cracking of oil and other hydrocarbon feedstocks. Low-molecular-weight resins are used for melt spun and melt blown fibers and for injection-molding applications. Polypropylene resins are used in extrusion and blow-molding processes and to make cast, slit, and oriented films. Stabilizers are added to polypropylene to protect it from attack by oxygen, ultraviolet light, and thermal degradation; other additives improve resin clarity, flame retardancy, or radiation resistance.

Polypropylene homopolymers, random copolymers, and impact copolymers are used in such products as automotive parts, appliances, battery cases, carpeting, electrical insulation, fiber and fabrics, food packaging, and medical equipment. *See* PETROLEUM PRODUCTS.

Other poly(alpha-olefins). Poly(1-butene) is a tough and flexible resin that has been used in the manufacture of film and pipe. Poly(4-methyl-1-pentene) is used in the manufacture of chemical and medical equipment. High-molecular-weight polyisobutylenes are rubbery solids that are used as sealants, inner tubes, and tubeless tire liners. Low-molecular-weight polyisobutylenes are used in formulations for caulking, sealants, and lubricants. Butadiene and isoprene can be polymerized to give a number of polymer structures. The commercially important forms of polybutadiene and polyisoprene are similar in structure to natural rubber. *See* POLYACRYLONITRILE RESINS; POLYMER; POLYSTYRENE RESIN.

Steven A. Cohen

Polyoma virus A papovavirus (member of a DNA-containing group of animal viruses) that infects rodents. This virus can induce a wide variety of tumors when inoculated into newborn animals. The simple structure has made the virus a popular model for the study of lytic infection (a viral infection that results in the lysis of the host cell and release of viral progeny) by animal viruses, and viral-induced tumorigenesis (tumor formation). *See* TUMOR VIRUSES.

Polyoma virus is endemic in most wild populations of mice but causes few harmful effects. Tumors produced by this virus are unknown in the wild. Inoculation of large quantities of virus into newborn rodents, however, induces a variety of tumor types, particularly sarcomas and carcinomas. These tumors contain the viral genome but produce few infectious viral particles. Infected animals produce neutralizing antibodies directed against structural components of the virus; tumor-bearing animals produce antibodies to antigens (T antigens) which are present in tumor cells but not in the virus particle. *See* ANTIBODY; ANTIGEN.

The icosahedral viral particle consists of DNA and protein only. The genome is a small, double-stranded, closed circular DNA molecule approximately 5300 base pairs in length. It encodes the T antigens expressed early in the productive cycle and in transformed cells, and the viral capsid proteins, expressed late in the productive cycle. *See* DEOXYRIBONUCLEIC ACID (DNA).

The virus is easily propagated to high titer in mouse embryo tissue culture, resulting in cell lysis and the production of a hemagglutinin (a red blood cell-clumping antibody). This "productive infection" occurs only in mouse cells. Polyoma virus will infect cells of other rodents, but little viral proliferation occurs. Small numbers of "abortively infected" cells retain the viral genome and become transformed. Transformed cells are tumorigenic when injected into syngeneic animals, and contain the T antigens but produce no virus. The expression of the T antigens has been shown to be necessary and sufficient to induce cell transformation. *See* IMMUNOLOGY; ONCOLOGY; VIRUS.

S. M. Dilworth

Polyplacophora A class within the phylum Mollusca whose members are popularly called chitons, coat-of-mail shells, or sea cradles. All chitons are marine, and they typically live in the intertidal zone, although some live in deeper waters. They are found from subarctic to tropical latitudes but are most abundant in warmer waters. There are roughly 750 chiton species living today.

Polyplacophorans exhibit bilateral symmetry and are oval or elongate in outline. Chitons vary in length from a few millimeters ($^{1}/_{16}$ in.) to over 30 cm (12 in.), though most are a few centimeters (about 1 in.) long. They are flattened dorsoventrally and bear eight shell plates on their back. The plates, termed valves, are formed from crystals of the mineral aragonite, and the upper layer of the valves contains canals filled with sensory organs called esthetes. The larger of these structures serve as photoreceptors and hence act as eyes. *See* CHITON.

Chitons possess three valve types—head, intermediate, and tail. The valves are laterally embedded in the fleshy but tough girdle, which typically bears small aragonite spicules or scales of varying shapes. Ventrally, polyplacophorans possess a large foot which they use for moving about and for attaching to rocks in a manner similar to a limpet gastropod. The gills hang along both sides of the animal, in a deep groove between the foot and the shell. The anus is located near the posterior end of the foot. At the front of the foot is the head which bears the mouth with its associated rasping structure (the radula, which is characteristic of most mollusks). A unique feature of the chiton radula is that the largest teeth are hardened with magnetite, an iron mineral. Characters of the valves, the gills, and the radula typically have been used to differentiate major groups within the Polyplacophora. *See* LIMPET; MOLLUSCA.

Michael Vendrasco

Polyploidy The occurrence of related forms possessing chromosome numbers which are multiples of a basic number (n), the haploid number. Forms having $3n$ chromosomes are triploids; $4n$, tetraploids; $5n$, pentaploids, and so on. Autopolyploids are forms derived by the multiplication of chromosomes from a single diploid organism. As a result the homologous chromosomes come from the same source. These are distinguished from allopolyploids, which are forms derived from a hybrid between two diploid organisms. As a result, the homologous chromosomes come from different sources. About one-third of the species of vascular plants have originated at least partly by polyploidy, and as many more appear to have ancestries which involve ancient occurrences of polyploidy. The condition can be induced artificially with the drug colchicine and the production of polyploid individuals has become a valuable tool for plant breeding.

In animals, most examples of polyploidy occur in groups which are parthenogenetic, or in species which reproduce asexually by fission. *See* BREEDING (PLANT); CHROMOSOME ABERRATION; GENE; GENETICS; PLANT EVOLUTION; SPECIATION.

G. Ledyard Stebbins

In addition to polyploid organisms in which all of the body cells contain multiples of the basic chromosome number, most plants and animals contain particular tissues that are polyploid or polytene. Both polyploid and polytene cells contain extra copies of DNA, but they differ in the physical appearance of the chromosomes. In polytene cells the replicated copies of the DNA remain physically associated to produce giant chromosomes that are continuously visible and have a banded pattern. The term polyploid has been applied to several types of cells: multinucleate cells; cells in which the chromosomes cyclically condense but do not undergo nuclear or cellular division (this process is termed endomitosis); and cells in which the chromosomes appear to be continually in interphase, yet the replicated chromosomes are not associated in visible polytene chromosomes. *See* CHROMOSOME; MITOSIS.

Terry L. Orr-Weaver

Polypodiales The largest order of modern ferns, commonly called the true ferns, with approximately 250 genera and 9000 species; also known as Filicales. Although well represented in the temperate regions, they reach their greatest development in the moist tropics. They vary in habit from small filmy structures to large treelike plants. Many are epiphytic (live perched on other plants) and a number are climbing species. A few are aquatic. Perhaps the most striking species are the tropical tree ferns with their upright, unbranched stems and terminal clusters of large graceful leaves.

The Polypodiales differ from the other fern orders in being leptosporangiate—that is, their sporangium, or spore sac, arises from a single surface cell—and in having small sporangia with a definite number of spores. The wall of the sporangium is almost encircled with a ring of cells having unevenly thickened walls. This ring is called the annulus. When the sporangium is mature, the annulus, acting as a spring, causes the sporangium wall to rupture, thus discharging the spores. These plants are valued for their beauty and for the clues they give to the evolutionary history of the Polypodiales which extends back through the coal measures of the Paleozoic. *See* PALEOBOTANY.

The sporophyte is the conspicuous phase of the true ferns, and like other vascular plants it has true roots, stems, and leaves (Fig. 1). In most ferns, especially those of the temperate regions, the mature stem is usually a creeping rhizome (underground stem) without aerial branches. However, in several species the stems are branched, and in some they are erect. Whereas in the tropics the leaves are usually persistent and evergreen, in temperate regions the leaves of most species die back each year and are replaced by new ones the next growing season. Characteristic of this order is the apparent uncoiling of the leaves from the base toward the apex.

The internal structure of the blade of the leaf and of the root is very similar to that of these organs in the seed plants. The main difference is the presence of large intercellular spaces in the fern leaf and the frequent lack of apparent distinction between the

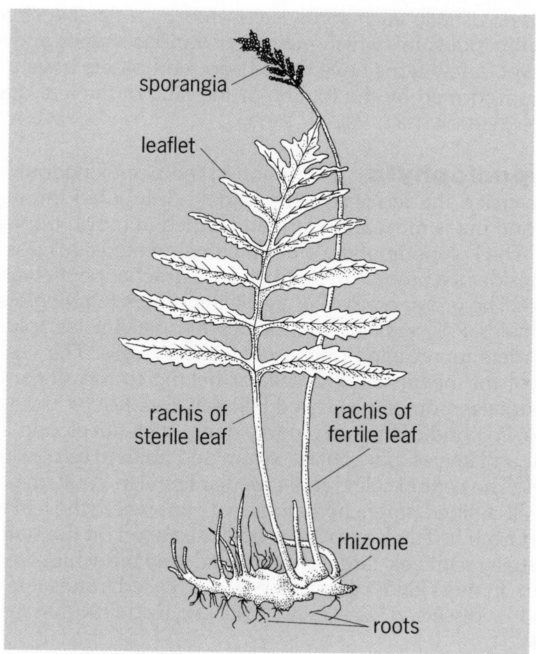

Fig. 1. The sensitive fern (*Onoclea sensibilis*), a representative of the Polypodiopsida. (*After W. W. Robbins, T. E. Weier, and C. R. Stocking, Botany: An Introduction to Plant Science, 3d ed., John Wiley and Sons, Inc., 1964*)

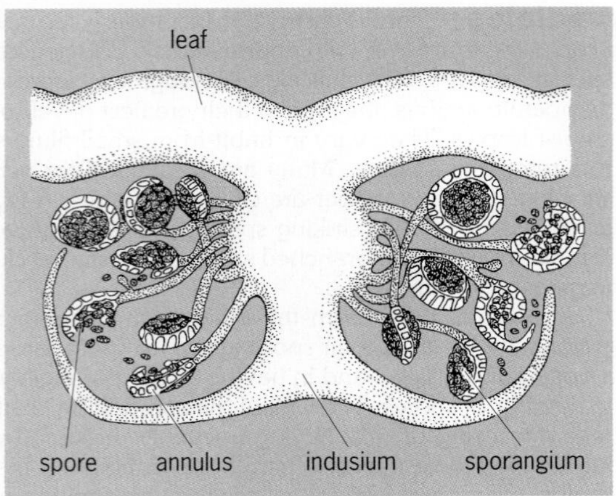

Fig. 2. Diagram of section through a fern leaf, showing details of a sorus. (*After W. W. Robbins, T. E. Weier, and C. R. Stocking, Botany: An Introduction to Plant Science, 3d ed., John Wiley and Sons, Inc., 1964*)

spongy and palisade cells of the mesophyll, possibly because most ferns grow in the shade.

The life cycle of the fern consists of two independent (self-sustaining) alternating generations. The common leafy fern plant is the sporophytic (spore-producing) generation. When the mature spores are discharged and reach a suitable substrate, they germinate and produce a small, flat, green, heart-shaped structure known as the prothallium or gametophytic (gamete-producing) generation. The gametophyte produces the sex organs antheridia (male) and archegonia (female). The gametes (sperm and egg) unite in fertilization and the resultant cell, or zygote, develops into the spore-bearing (sporophytic) fern plant.

In all ferns, the spores are produced in special multicellular organs known as sporangia. Except for a few genera, the sporangia are arranged in groups or clusters called sori (Fig. 2). These are on the lower surface of the leaves or fertile fronds, either along the midrib of the pinnae, near the leaf margins, or scattered. Usually each sorus is covered by a flaplike structure called the indusium, which may be of various shapes and sizes. However, a few ferns have naked sori, and others have a false indusium formed by the folding or inrolling of the leaf margin. *See* POLYPODIOPHYTA; PSILOTOPHYTA. Paul A. Vestal

Polypodiophyta A division of the plant kingdom, commonly called the ferns, which is widely distributed throughout the world but is most abundant and varied in moist, tropical regions. The Polypodiophyta are sometimes treated as a class Polypodiopsida of a broadly defined division Tracheophyta (vascular plants). The group consists of five living orders (Ophioglossales, Marattiales, Polypodiales, Marsileales, and Salviniales), plus several orders represented only by Paleozoic fossils. The vast majority of the nearly 10,000 species belong to the single order Polypodiales, sometimes called Filicales. *See* POLYPODIOPSIDA.

The Polypodiophyta ordinarily have well-developed roots, stems, and leaves that contain xylem and phloem as conducting tissues. The central cylinder of vascular tissue in the stem usually has well-defined parenchymatous leaf gaps where the leaf traces depart from it. The leaves are spirally arranged on the stem and are usually relatively large, with an evidently branching vascular system. In most kinds of ferns the leaves, called fronds, are compound or dissected. *See* LEAF; PHLOEM; ROOT (BOTANY); STEM; XYLEM.

The Polypodiophyta show a well-developed alternation of generations, both the sporophyte and the gametophyte generation being detached and physiologically independent of each other at maturity. The sporophyte is much the more conspicuous, and is the generally recognized fern plant. On some or all of its leaves it produces tiny sporangia which in turn contain spores. *See* MARATTIALES; POLYPODIALES. Arthur Cronquist

Polypodiopsida A class (also known as Filicineae) of the plant division Polypodiophyta containing a large group of plants commonly called ferns. They are widely distributed throughout the world with their greatest development in the moist tropics. Polypodiopsida is an old class with a good representation in the Paleozoic flora. Some of the plant fossils resemble contemporary living species. *See* MARATTIALES; POLYPODIALES; POLYPODIOPHYTA. Paul A. Vestal

Polypteriformes A distinctive and apparently ancient order of actinopterygian fishes comprising the subclass Cladistia, commonly known as bichirs and reedfishes. The following combination of characters distinguishes the order from all other actinopterygian fishes: elongate body covered with thick, rhombic, ganoid scales; slitlike spiracle behind each eye; well-ossified internal skeleton; maxillae firmly united with the skull; basically heterocercal tail, but with a symmetrical caudal fin, and with the upper part continuous with the dorsal fin; dorsal series of free sail-like finlets, each supported by a spine; lobed pectoral fin with its rays supported by ossified radials; pair of enlarged gular plates; very large intercalar bone; no branchiostegals; bilobed swim bladder (the left lobe less developed than the right) located ventral to and attached to the esophagus, serving as lungs; functional gills; young with a pair of feathery external gills, which are assimilated with age; spiral valve-type intestine; and four rather than five pairs of gill arches as is typical for actinopterygians (see illustration). *See* SCALE (ZOOLOGY); SWIM BLADDER.

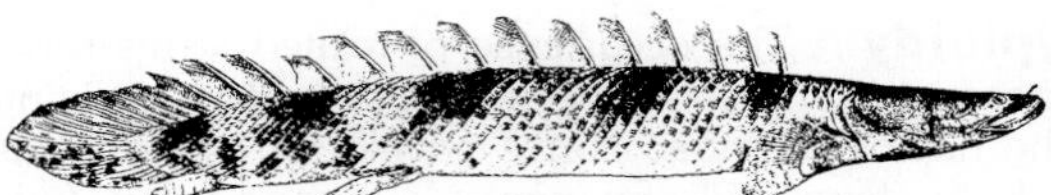

Bichir (*Polypterus endlicheri*), length to 3 ft (0.9 m). (*After G. A. Boulenger, Catalogue of the Fresh Water Fishes of Africa in the British Museum, vol. 1, 1909*)

The order is known from the Eocene and presently comprises a single extant family, Polypteridae, and two genera, *Polypterus*, with about 15 species, and *Erpetoichthys*, with one species. Both genera are confined to freshwaters of tropical Africa. They are common along lake shores and in swamps and are able to withstand anoxic (oxygen-deficient) conditions by breathing atmospheric oxygen. *See* ACTINOPTERYGII. Herbert Boschung; Reeve M. Bailey

Poly(p-xylylene) resins Linear, crystallizable resins based on an unusual polymerization of *p*-xylene and derivatives. The polymers are tough and chemically resistant, and may be deposited as adherent coatings by a vacuum process. The vapor deposition process makes it possible to coat small microelectronic parts with a thin layer of the polymer. John A. Manson

Polysaccharide A class of high-molecular-weight carbohydrates, colloidal complexes, which break down on hydrolysis to monosaccharides containing five or six carbon atoms. The polysaccharides are considered to be polymers in which monosaccharides have been glycosidically joined with the elimination of water. A polysaccharide consisting of hexose monosaccharide units may be represented by the reaction below.

$$nC_6H_{12}O_2 \rightarrow (C_6H_{10}O_5)_n + (n-1)H_2O$$

The term polysaccharide is limited to those polymers which contain 10 or more monosaccharide residues. Polysaccharides such as starch, glycogen, and dextran consist of several thousand D-glucose units. Polymers of relatively low molecular weight, consisting of two to nine monosaccharide residues, are referred to as oligosaccharides. *See* DEXTRAN; GLUCOSE; GLYCOGEN; MONOSACCHARIDE; STARCH.

Polysaccharides are often classified on the basis of the number of monosaccharide types present in the molecule. Polysaccharides, such as cellulose or starch, that produce only one monosaccharide type (D-glucose) on complete hydrolysis are termed homopolysaccharides. On the other hand, polysaccharides, such as hyaluronic acid, which produce on hydrolysis more than one monosaccharide type (*N*-acetylglucosamine and D-glucuronic acid) are named heteropolysaccharides. *See* CARBOHYDRATE.

William Z. Hassid

Polystyrene resin A hard, transparent, glasslike thermoplastic resin (see structure). Polystyrene is characterized by excellent electrical insulation properties, relatively high resistance to water, high refractive index, clarity, and low softening temperature.

$$(-CH-CH_2-)_n$$

Polystyrene

High-molecular-weight homopolymers, copolymers, and polyblends are used as extrusion and molding compounds for packaging, appliance and furniture components, toys, and insulating panels. Styrene-butadiene copolymers are still used for automobile tires and in various rubber articles. The effects of blending small amounts of a rubbery polymer, such as butadiene-styrene rubber, with a hard, brittle polymer are most dramatic when the latter is polystyrene. The polyblend may have impact strength greater than ten times that of polystyrene. Various combinations of complex polyblends and interpolymers of acrylonitrile, styrene, and butadiene (ABS) resins are important as molding resins. ABS resins are also used as toughening agents for polymers such as poly(vinyl chloride). Polystyrene is also used in combination with paints. The homopolymer and polyblends are used for panels or liners for refrigerator doors. Polystyrene may also be fabricated in the form of a rigid foam, which is used in packaging, food-service articles, and insulating panels. *See* ACRYLONITRILE; COPOLYMER; PLASTICS PROCESSING; POLYACRYLONITRILE RESINS; POLYMER; POLYMERIZATION; RUBBER; STYRENE.

John A. Manson

Polysulfide resins Resins that vary in properties from viscous liquids to rubberlike solids. Organic polysulfide resins are prepared by the condensation of organic dihalides with a polysulfide.

Compounding and fabrication of the rubbery polymers can be handled on conventional rubber machinery. The polysulfide rubbers are distinguished by their resistance to solvents such as gasoline, and to oxygen and ozone. The polymers are relatively impermeable to gases. The products are used to form coatings which are chemically resistant and special rubber articles, such as gasoline bags. The polysulfide rubbers were among the first polymers to be used in solid-fuel compositions for rockets. *See* ORGANOSULFUR COMPOUND; POLYMERIZATION; RUBBER.

John A. Manson

Polysulfone resins Polymers containing sulfone groups ($-SO_2-$) in the main chain, along with a variety of aromatic or aliphatic constituents. Polysulfones based on aromatic backbones constitute a useful class of engineering plastics, owing to their high strength, stiffness, and toughness together with high thermal and oxidative stability, low creep, transparency, and the ability to be processed by standard techniques for thermoplastics. The aromatic structural elements and the presence of sulfone groups are responsible for the resistance to heat and oxidation; ether and isopropylidene groups contribute some chain flexibility. Aromatic polysulfones can be used over wide temperature ranges. The high-temperature performance of poly(ethersulfones) to 200°C (390°F) is surpassed by few other polymers.

Because of the combination of properties discussed, polysulfone resins find many applications in electronic and automotive parts, medical instrumentation subject to sterilization, chemical and food processing equipment, and various plumbing and home appliance items. The inherent low flammability of polyphenylsulfone makes it one of the few materials that can pass the stringent regulations for aircraft interiors. Coating formulations are also available, as well as grades reinforced with glass or carbon fibers. *See* COPOLYMER; HETEROCYCLIC POLYMER; ORGANOSULFUR COMPOUND.

Theodore T. Moore; John A. Manson

Polytrichidae A subclass of the mosses (class Bryopsida), consisting of one family and 19 genera. The plants are acrocarpous and perennial, grow on soil, often in dry habitats, and are usually unbranched; annual increments often grow up through the male inflorescences, but do not give the appearance of forked stems. The plants are nearly always dioecious, with both male and female inflorescences terminal. The stems generally have well-developed vascular tissues. The long, narrow leaves are sheathed at the base; above the base the midrib often occupies most of the leaf. Stomata are usually present at or near the base of the capsule. *See* BRYOPHYTA; BRYOPSIDA; DAWSONIIDAE.

Howard Crum

Polytropic process A process which occurs with an interchange of both heat and work between the system and its surroundings. The nonadiabatic expansion or compression of a fluid is an example of a polytropic process. The interrelationships between the pressure (P) and volume (V) and pressure and temperature (T) for a gas undergoing a polytropic process are given by Eqs. (1) and (2), where a and b are the polytropic constants

$$PV^a = \text{constant} \qquad (1)$$

$$P^b/T = \text{constant} \qquad (2)$$

for the process of interest. These constants, which are usually determined from experiment, depend upon the equation of state of the gas, the amount of heat transferred, and the extent of irreversibility in the process. *See* GAS; ISENTROPIC PROCESS; ISOTHERMAL PROCESS; THERMODYNAMIC PROCESSES.

Stanley I. Sandler

Polyurethane resins Polymeric materials that are usually produced by reacting polyfunctional isocyanates [$R(N{=}C{=}O)_n$, where R is alkyl or aryl] with a macrodiol, or polyol, and a glycol extender in the presence of a catalyst. The macrodiols are based on polyethers or polyesters or a combination of both. Some of these have terminal hydroxyl groups, while those reacted with an excess of diisocyanate have terminal isocyanate groups, as in the prepolymer of propylene oxide [with $R(NCO)_2$]. In recent years, diamines have been used as extenders to achieve higher reaction rates in molding or spray applications. The polymerization reactions are done in a mold (casting, injection molding) or continuously in an extruder or on a conveyor (block foam). *See* GLYCOL; ISOCYANATE; POLYMER; POLYESTER RESINS; POLYETHER RESINS; POLYMERIZATION; POLYOL.

Polyurethanes are the most versatile class of reaction polymers, providing products ranging from soft thermoplastic elastomers or elastomeric (spandex) fibers to flexible or rigid foams. The elastomers are used mainly in automotive bumper applications produced by the reaction injection-molding process. Spandex fibers are used in sport clothing applications. Flexible foams are used in upholstery materials for furniture and bedding, automotive seating, and carpet underlay. Rigid foams are used as insulation materials in building and construction, refrigerated appliances, and in vehicles. In 2000, the total polyurethane global consumption was about 8000 kilotons. *See* PLASTICS PROCESSING.

The commodity aromatic isocyanates are toluene diisocyanate (a distilled 80:20 mixture of the 2,4- and 2,6-isomers; TDI; structure **1**), 4,4′-diphenylmethane diisocyanate (MDI; **2**),

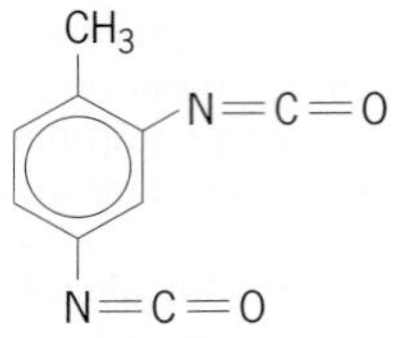

CH_3 N=C=O N=C=O

(1)

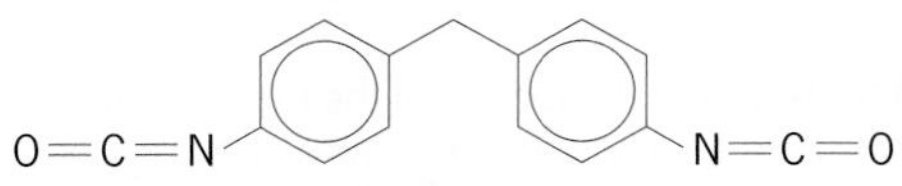

O=C=N N=C=O

(2)

and polymeric MDI (PMDI). When color stability is required in polyurethane resins, the more costly aliphatic diisocyanates are used. Examples are hexamethylene diisocyanate (HDI), isophorone diisocyanate (IPDI), and the reduced MDI (HMDI).

Henri Ulrich

Polyvinyl resins Polymeric materials generally considered to include polymers derived from monomers having the structure

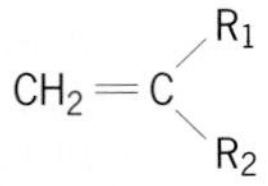

$$CH_2{=}C\begin{matrix}R_1\\R_2\end{matrix}$$

in which R_1 and R_2 represent hydrogen, alkyl, halogen, or other groups. This article refers to polymers whose names include the term vinyl. For discussions of other vinyl-type polymers *see* POLYACRYLATE RESIN; POLYACRYLONITRILE RESINS; POLYFLUOROOLEFIN RESINS; POLYOLEFIN RESINS; POLYSTYRENE RESIN.

Many of the monomers can be prepared by addition of the appropriate compound to acetylene. For example, vinyl chloride, vinyl fluoride, vinyl acetate, and vinyl methyl ether may be formed by the reactions of acetylene with HCl, HF, CH_3OOH, and CH_3OH, respectively. Processes based on ethylene as a raw material have also become common for the preparation of vinyl chloride and vinyl acetate.

The polyvinyl resins may be characterized as a group of thermoplastics which, in many cases, are inexpensive and capable of being handled by solution, dispersion, injection molding, and extrusion techniques. The properties vary with chemical structure, crystallinity, and molecular weight.

Poly(vinyl acetals) are relatively soft, water-insoluble thermoplastic products obtained by the reaction of poly(vinyl alcohol) with aldehydes. Properties depend on the extent to which alcohol groups are reacted. Poly(vinyl butyral) is rubbery and tough and is used primarily in plasticized form as the inner layer and binder for safety glass. Poly(vinyl formal) is the hardest of the group; it is used mainly in adhesive, primer, and wire-coating formulations, especially when blended with a phenolic resin.

Poly(vinyl acetate) is a leathery, colorless thermoplastic material which softens at relatively low temperatures and which is relatively stable to light and oxygen. The polymers are clear and noncrystalline. The chief applications are as adhesives and binders for water-based or emulsion paints. *See* PAINT.

Poly(vinyl alcohol) is a tough, whitish polymer which can be formed into strong films, tubes and fibers that are highly resistant to hydrocarbon solvents. Although poly(vinyl alcohol) is one of the few water-soluble polymers, it can be rendered insoluble in water by drawing or by the use of cross-linking agents. Two groups of products are available, those formed by the essentially complete hydrolysis of poly(vinyl acetate), and those formed by incomplete hydrolysis.

The former may be plasticized with water or glycols and molded or extruded into films, tubes, and filaments which are resistant to hydrocarbons. These products are used for liners in gasoline hoses, for grease-resistant coating and paper adhesives, for treating paper and textiles, and as emulsifiers and thickeners.

Poly(vinyl carbazole) is a tough, glassy thermoplastic with excellent electrical properties and a relatively high softening temperature. Uses of the product has been limited to small-scale electrical applications requiring resistance to high temperatures.

Poly(vinyl chloride) [PVC] is a tough, strong thermoplastic material which has an excellent combination of physical and electrical properties. The products are usually characterized as plasticized or rigid types. Poly(vinyl chloride)[and copolymers] is the second most commonly used polyvinyl resin and one of the most versatile plastics. The plasticized types are somewhat elastic materials which are familiar in the form of shower curtains, floor coverings, raincoats, dishpans, dolls, bottle-top sealers, prosthetic forms, wire insulation, and films. Rigid products, which may consist of the homopolymer, copolymer, or polyblends, are commonly used in the manufacture of phonograph records, pipe, chemically resistant liners for chemical reaction vessels, and siding and window sashes.

Poly(vinylidene chloride) is a tough, hornlike thermoplastic with properties generally similar to those of poly(vinyl chloride). Because of its relatively low solubility and decomposition temperature, the material is most widely used in the form of copolymers with other vinyl monomers, such as vinyl chloride. The copolymers are employed as packaging film, rigid pipe, and as filaments for upholstery and window screens.

Poly(vinyl ethers) exist in several forms varying from soft, balsamlike semisolids to tough, rubbery masses, all of which are readily soluble in organic solvents. Polymers of the alkyl vinyl ethers are used in adhesive formulations and as softening or flexibilizing agents for other polymers.

Poly(vinyl fluoride) is a tough, partially crystalline thermoplastic material which has a higher softening temperature than poly(vinyl chloride). Films and sheets are characterized by high resistance to impact and cracking caused by flexing and temperature and by resistance to weathering.

Poly(vinyl pyrrolidone) is a water-soluble polymer of basic nature which has film-forming properties, strong absorptive or complexing qualities for various reagents, and the ability to form water-solubles salts which are polyelectrolytes. The main uses are as a water-solubilizing agent for medicinal agents such as iodine, and as a semipermanent setting agent in hair sprays. Certain synthetic textile fibers containing small amounts of vinylpyrrolidone as a copolymer have improved affinity for dyes. *See* PLASTICS PROCESSING; POLYMER; POLYMERIZATION.

John A. Manson

Pomegranate A small deciduous tree, *Punica granatum*, belonging to the plant order Myrtales. Pomegranate is grown as an ornamental as well as for its fruit. The pomegranate is a native of Asia. It was originally known for its medicinal qualities, and cures for various ills were attributed to the fruit juice, the rind, and the bark of the roots. The fruit is a reddish, pomelike berry, containing numerous seeds imbedded in crimson pulp, from which an acid, reddish juice may be obtained. Limited quantities are grown in California and the Gulf states. *See* FRUIT, TREE; MYRTALES.

J. H. Clarke

Poplar Any tree of the genus *Populus*, family Salicaceae, marked by simple, alternate leaves which are usually broader than those of the willow, the other American representative of this family. Poplars have scaly buds, bitter bark, flowers and fruit in catkins, and a five-angled pith. See SALICALES; WILLOW.

Some species are commonly called cottonwood because of the cottony hairs attached to the seeds. Other species, called aspens, have weak, flattened leaf stalks which cause the leaves to flutter in the slightest breeze. One of the important species in the United States is the quaking, or trembling, aspen (*P. tremuloides*). The soft wood of this species is used for paper pulp. The European aspen (*P. nigra*), which is similar to the quaking aspen, is sometimes planted, and its variety, *italica*, the Lombardy poplar of erect columnar habit, is used in landscape planting. The black cottonwood (*P. trichocarpa*) is the largest American poplar and is also the largest broad-leaved tree in the forests of the Pacific Northwest. The cottonwood or necklace poplar (*P. deltoides*) is native in the eastern half of the United States. In the balsam or tacamahac poplar (*P. balsamifera*), the resin is used in medicine as an expectorant. The wood is used for veneer, boxes, crates, furniture, paper pulp, and excelsior. Arthur H. Graves; Kenneth P. Davis

Poppy A plant, *Papaver somniferum* (Papaveraceae), which is probably a native of Asia Minor. It is cultivated extensively in China, India, and elsewhere. This plant is the source of opium, obtained by cutting into the fruits (capsules) soon after the petals have fallen. The white latex (juice) flows from the cuts and hardens when exposed to the air. This solidified latex is collected, shaped into balls or wafers, and often wrapped in the flower petals. This is the crude opium, which contains at least 20 alkaloids, including morphine and codeine. See MORPHINE; PAPAVERALES. Perry D. Strausbaugh; Earl L. Core

Population dispersal The process by which groups of living organisms expand the space or range within which they live. Dispersal operates when individual organisms leave the space that they have occupied previously, or in which they were born, and settle in new areas. Natal dispersal is the first movement of an organism from its birth site to the site in which it first attempts to breed. Adult dispersal is a subsequent movement when an adult organism changes its location in space. As individuals move across space and settle into new locations, the population to which they belong expands or contracts its overall distribution. Thus, dispersal is the process by which populations change the area they occupy.

Migration is the regular movement of organisms during different seasons. Many species migrate between wintering and breeding ranges. Such migratory movement is marked by a regular return in future seasons to previously occupied regions, and so usually does not involve an expansion of population range. Some migratory species show astounding abilities to return to the exact locations used in previous seasons. Other species show no regular movements, but wander aimlessly without settling permanently into a new space. Wandering (called nomadism) is typical of species in regions where the availibility of food resources are unpredictable from year to year. Neither migration nor nomadism is considered an example of true dispersal. See MIGRATORY BEHAVIOR.

Virtually all forms of animals and plants disperse. In most higher vertebrates, the dispersal unit is an entire organism, often a juvenile or a member of another young age class. In other vertebrates and many plants, especially those that are sessile (permanently attached to a surface), the dispersal unit is a specialized structure (disseminule). Seeds, spores, and fruits are disseminules of plants and fungi; trochophores and planula larvae are disseminules of sea worms and corals, respectively. Many disseminules are highly evolved structures specialized for movement by specific dispersal agents such as wind, water, or other animals.

A special case of zoochory (dispersal using animal agents) involves transport by humans. The movement of people and cargo by cart, car, train, plane, and boat has increased the potential dispersal of weedy species worldwide. Many foreign aquatic species have been introduced to coastal areas by accidental dispersal of disseminules in ship ballast water. The zebra mussel is one exotic species that arrive in this manner and is now a major economic problem throughout the Great Lakes region of North America. Some organisms have been deliberately introduced by humans into new areas. Domestic animals and plants have been released throughout the world by farmers. A few pest species were deliberately released by humans; European starlings, for example.

Some of the most highly coevolved dispersal systems are those in which the disseminule must be eaten by an animal. Such systems have often evolved a complex series of signals and investments by both the plant and the animal to ensure that the seeds are dispersed at an appropriate time and that the animal is a dependable dispersal agent. Such highly evolved systems are common in fruiting plants and their dispersal agents, which are animals called frugivores. Fruiting plants cover their seeds with an attractive, edible package (the fruit) to get the frugivore to eat the seed. To ensure that fruits are not eaten until the seeds are mature, plants change the color of their fruits as a signal to show that the fruits are ready for eating.

Many plants in the tropical rainforests are coevolved to have their seeds dispersed by specific animal vectors, including birds, mammals, and ants. Many tropical trees, shrubs, and herbaceous plants are specialized to have their seeds dispersed by a single animal species. Temperate forest trees, in contrast, often depend on wind dispersal of both pollen and seeds.

Dispersal barriers are physical structures that prevent organisms from crossing into new space. Oceans, rivers, roads, and mountains are examples of barriers for species whose disseminules cannot cross such features. It is believed that the creation of physical barriers is the primary factor responsible for the evolution of new species. A widespread species can be broken into isolated fragments by the creation of a new physical barrier. With no dispersal linking the newly isolated populations, genetic differences that evolve in each population cannot be shared between populations. Eventually, the populations may become so different that no interbreeding occurs even if dispersal pathways are reconnected. The populations are then considered separate species. See SPECIATION.

Dispersal is of major concern for scientists who work with rare and endangered animals. Extinction is known to be more prevalent in small, isolated populations. Conservation biologists believe that many species exist as a metapopulation, that is, a group of populations interconnected by the dispersal of individuals or disseminules between subpopulations. The interruption of dispersal in this system of isolated populations can increase the possibility of extinction of the whole metapopulation. Conservation plans sometimes propose the creation of corridors to link isolated patches of habitat as a way of increasing the probability of successful dispersal. See EXTINCTION (BIOLOGY); POPULATION DISPERSION. John B. Dunning, Jr.

Population dispersion The spatial distribution at any particular moment of the individuals of a species of plant or animal. Under natural conditions organisms are distributed either by active movements, or migrations, or by passive transport by wind, water, or other organisms. The act or process of dissemination is usually termed dispersal, while the resulting pattern of distribution is best referred to as dispersion. Dispersion is a basic characteristic of populations, controlling various features of their structure and organization. It determines population density, that is, the number of individuals per unit of area, or volume, and its reciprocal relationship, mean area, or the average area per individual. It also determines the frequency, or chance of encountering one or more individuals of the population in a particular sample unit of area, or volume. The ecologist therefore studies not only the fluctuations in numbers of individuals in a population but also the changes in their distribution in space. *See* POPULATION DISPERSAL.

Principal types of dispersion. The dispersion pattern of individuals in a population may conform to any one of several broad types, such as random, uniform, or contagious (clumped). Any pattern is relative to the space being examined; a population may appear clumped when a large area is considered, but may prove to be distributed at random with respect to a much smaller area.

Random or haphazard implies that the individuals have been distributed by chance. In such a distribution, the probability of finding an individual at any point in the area is the same for all points. Hence a truly random pattern will develop only if each individual has had an equal and independent opportunity to establish itself at any given point. Examples of approximately random dispersions can be found in the patterns of settlement by free-floating marine larvae and of colonization of bare ground by airborne disseminules of plants. Nevertheless, true randomness appears to be relatively rare in nature.

Uniform distribution implies a regularity of distance between and among the individuals of a population. Perfect uniformity exists when the distance from one individual to its nearest neighbor is the same for all individuals. Patterns approaching uniformity are most obvious in the dispersion of orchard trees and in other artificial plantings, but the tendency to a regular distribution is also found in nature, as for example in the relatively even spacing of trees in forest canopies, the arrangement of shrubs in deserts, and the distribution of territorial animals.

The most frequent type of distribution encountered is contagious or clumped, indicating the existence of aggregations or groups in the population. Clusters and clones of plants, and families, flocks, and herds of animals are common phenomena. The formation of groups introduces a higher order of complexity in the dispersion pattern, since the several aggregations may themselves be distributed at random, evenly, or in clumps. An adequate description of dispersion, therefore, must include not only the determination of the type of distribution, but also an assessment of the extent of aggregation if the latter is present.

Factors affecting dispersion. The principal factors that determine patterns of population dispersion include (1) the action of environmental agencies of transport, (2) the distribution of soil types and other physical features of the habitat, (3) the influence of temporal changes in weather and climate, (4) the behavior pattern of the population in regard to reproductive processes and dispersal of the young, (5) the intensity of intra- and interspecific competition, and (6) the various social and antisocial forces that may develop among the members of the population. Although in certain cases the dispersion pattern may be due to the overriding effects of one factor, in general populations are subject to the collective and simultaneous action of numerous distributional forces and the dispersion pattern reflects their combined influence. When many small factors act together on the population, a more or less random distribution is to be expected, whereas the domination of a few major factors tends to produce departure from randomness.

Optimal population density. The degree of aggregation which promotes optimum population growth and survival varies according to the species and the circumstances. Groups or organisms often flourish best if neither too few nor too many individuals are present; they have an optimal population density at some intermediate level. The concept of an intermediate optimal population density is sometimes known as Allee's principle. *See* ECOLOGICAL COMMUNITIES; POPULATION ECOLOGY; POPULATION GENETICS.

Francis C. Evans

Population ecology The study of spatial and temporal patterns in the abundance and distribution of organisms and of the mechanisms that produce those patterns. Species differ dramatically in their average abundance and geographical distributions, and they display a remarkable range of dynamical patterns of abundance over time, including relative constancy, cycles, irregular fluctuations, violent outbreaks, and extinctions. The aims of population ecology are threefold: (1) to elucidate general principles explaining these dynamic patterns; (2) to integrate these principles with mechanistic models and evolutionary interpretations of individual life-history tactics, physiology, and behavior as well as with theories of community and ecosystem dynamics; and (3) to apply these principles to the management and conservation of natural populations.

In addition to its intrinsic conceptual appeal, population ecology has great practical utility. Control programs for agricultural pests or human diseases ideally attempt to reduce the intrinsic rate of increase of those organisms to very low values. Analyses of the population dynamics of infectious diseases have successfully guided the development of vaccination programs. In the exploitation of renewable resources, such as in forestry or fisheries biology, population models are required in order to devise sensible harvesting strategies that maximize the sustainable yield extracted from exploited populations. Conservation biology is increasingly concerned with the consequences of habitat fragmentation for species preservation. Population models can help characterize minimum viable population sizes below which a species is vulnerable to rapid extinction, and can help guide the development of interventionist policies to save endangered species. Finally, population ecology must be an integral part of any attempt to bring the world's burgeoning human population into harmonious balance with the environment. *See* ECOLOGY; MATHEMATICAL ECOLOGY; THEORETICAL ECOLOGY.

Robert Holt

Population genetics The study of both experimental and theoretical consequences of mendelian heredity on the population level, in contradistinction to classical genetics which deals with the offspring of specified parents on the familial level. The genetics of populations studies the frequencies of genes, genotypes, and phenotypes, and the mating systems. It also studies the forces that may alter the genetic composition of a population in time, such as recurrent mutation, migration, and intermixture between groups, selection resulting from genotypic differential fertility, and the random changes incurred by the sampling process in reproduction from generation to generation. This type of study contributes to an understanding of the elementary step in biological evolution. The principles of population genetics may be applied to plants and to other animals as well as humans. *See* GENETICS; MENDELISM.

C. C. Li

Population viability The ability of a population to persist and to avoid extinction. The viability of a population will increase or decrease in response to changes in the rates of birth, death, and growth of individuals. In natural populations, these rates are not stable, but undergo fluctuations due to external forces such as hurricanes and introduced species, and internal forces such as competition and genetic composition. Such factors can drive populations to extinction if they are severe or if several detrimental events occur before the population can recover. *See* ECOLOGY; POPULATION ECOLOGY.

One of the most important uses of population viability models comes from modern conservation biology, which uses these models to determine whether a population is in danger of extinction. This is called population viability analysis (PVA) and consists of demographic and genetic models that are used to make decisions on how to manage populations of threatened or endangered species. The National Research Council has called population viability analysis "the cornerstone, the obligatory tool by which recovery objectives and criteria [for endangered species] are identified." *See* ECOLOGICAL MODELING.

Gretchen LeBuhn; Thomas E. Miller

Porcelain A high-grade ceramic ware characterized by high strength, a white color (under the glaze), very low absorption, good translucency, and a hard glaze. Equivalent terms are European porcelain, hard porcelain, true porcelain, and hard paste porcelain. *See* GLAZING; POTTERY.

Porcelain is distinguished from other fine ceramic ware, such as china, by the fact that the firing of the unglazed ware (the bisque firing) is done at a lower temperature (1800–2200°F or 1000–1200°C) than the final or glost firing, which may be as high as 2700°F (1500°C). In other words, the ware reaches its final state of maturity at the maturing temperatures of the glaze.

The white color is obtained by using very pure white-firing kaolin or china clay and other pure materials, the low absorption results from the high firing temperature, and the translucency results from the glass phase. *See* CERAMICS. J. F. McMahon

Porcupine A mammal with spines or quills in addition to regular hair. The 23 species are classified in the order Rodentia and consist of two families: the Hystricidae, or Old World porcupines (11 species), and the Erethizontidae, or New World porcupines (12 species). Although traditionally considered to be related, the two families of porcupines may have evolved independently from different, unrelated ancestors. Porcupines are herbivorous, terrestrial, and generally nocturnal. *See* MAMMALIA; RODENTIA.

Quills are present on the back, sides, and tail of porcupines. Each quill consists of long sharp bristles of hair that are fused together. Porcupines defend themselves by striking attackers with their quilled tails. When a porcupine is disturbed, it rattles the quills on its tail. These produce a warning signal that resembles the noise created by a rattlesnake's tail. If the sound does not make the would-be aggressor retreat, the porcupine turns its back on it, erects its rapierlike quills, and charges backward. The quills come out easily and stick in the attacker's flesh. The porcupine grows new quills to replace the ones lost. In New World porcupines, the tip of each quill is covered with tiny backward-pointing projections called barbs. The barbs hook into the flesh, and the quills are almost impossible to remove. Porcupine victims may die from infections caused by germs on the quills or from damage to a vital organ.

Old World porcupines live in Africa, southeastern Asia, India, and southern Europe. Most species attain a length of about 3 ft (90 cm), including the tail. They are terrestrial or semifossorial (live both above and below ground) and make their homes in tunnels in the ground. Often six or seven porcupines share a communal burrow. They do not climb trees. New World porcupines are found in mixed coniferous forests, tropical forests, grasslands and deserts. Despite their generally chunky heavyset bodies, erethizontids can be arboreal unlike Old World porcupines. Donald W. Linzey

Porifera The sponges, a phylum of the animal kingdom which includes about 5000 described species. The body plan of sponges is unique among animals. Currents of water are drawn through small pores, or ostia, in the sponge body and leave by way of larger openings called oscula. The beating of flagella on collar cells or choanocytes, localized in chambers on the interior of the sponge, maintains the water current. Support for the sponge tissues is provided by calcareous or siliceous spicules, or by organic fibers, or by a combination of organic fibers and siliceous spicules. Some species have a compound skeleton of organic fibers, siliceous spicules, and a basal mass of aragonite or calcite. The skeletons of species with supporting networks of organic fibers have long been used for bathing and cleaning purposes. Because of their primitive organization, sponges are of interest to zoologists as an aid in understanding the origin of multicellular animals. *See* ANIMAL KINGDOM; PARAZOA.

The Porifera have a fossil record extending from the Precambrian to Recent times. More than 1000 genera of fossil sponges have been described from the Paleozoic, Mesozoic, and Cenozoic eras.

The living Porifera are divided into four classes on the basis of their skeletal structures. A taxonomic scheme of the Porifera follows.

Class Hexactinellida
  Subclass Amphidiscophora
    Order: Amphidiscosa
      Hemidiscosa
      Reticulosa
  Subclass Hexasterophora
    Order: Hexactinosa
      Lynchniscosa
      Lyssacinosa
Class Calcarea
  Subclass Calcinea
    Order: Clathrinida
      Leucettida
  Subclass Calcaronea
    Order: Leucosoleniida
      Sycettida
  Subclass Pharetronida
Class Demospongiae
  Subclass Tetractinomorpha
    Order: Homosclerophorida
      Choristida
      Spirophorida
      Hadromerida
      Axinellida
  Subclass Ceractinomorpha
    Order: Dendroceratida
      Dictyoceratida
      Halichondrida
      Haplosclerida
      Poecilosclerida
      (Permosphincta, extinct)
Class Sclerospongiae

Willard D. Hartman

Porocephalida One of two orders in the class Pentastomida of the phylum Arthropoda. In this order the larvae have four legs. The hooks on the adult articulate on a chitinous base or fulcrum and are arranged in a flattened trapezoidal pattern, or in a curved line or straight line. There are no podial or parapodial lobes. The two suborders are Porocephaloidea and Linguatuloidea.

The Porocephaloidea comprise five families—Porocephalidae, Sebekidae, Armilliferidae, Sambonidae, and Subtriquetridae—and include most of the pentastomid species. These animals are cylindrical, with rounded ends. The hooks are sessile; the mouth is anterior between the hooks. Adults occur in the respiratory passages of snakes, lizards, turtles, and crocodiles; the larvae are encysted in fishes, snakes, lizards, crocodiles, and mammals.

Linguatuloidea contains a single family, with a single genus, *Linguatula*. Adults live in the nasal cavities of carnivorous mammals; the larvae are encysted in herbivorous mammals. The body is elongate, flattened ventrally, annulate, and attenuated posteriorly. *See* PENTASTOMIDA. Horace W. Stunkard

Porphyrin One of a class of cyclic compounds in which the parent macrocycle consists of four pyrrole-type units linked together by single carbon bridges. Several porphyrins with selected peripheral substitution and metal coordination carry out vital biochemical processes in living organisms. Chlorins, bacteriochlorins, and corrins (see structures) are related tetrapyrrolic macrocycles that are also observed in biologically important compounds.

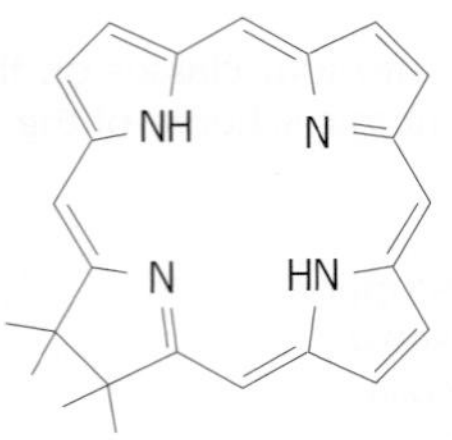

Chlorin

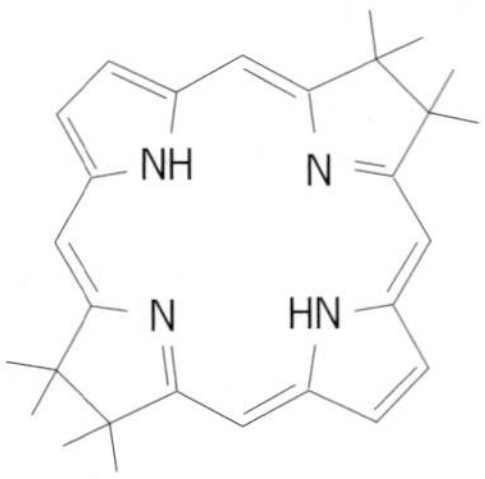

Bacteriochlorin

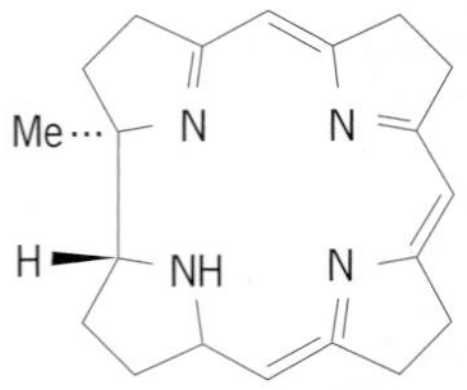

Corrin
Me = CH_3

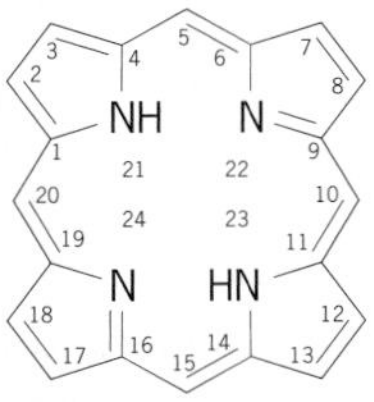

Porphyrin nucleus

The complexity of porphyrin nomenclature parallels the complex structures of the naturally occurring derivatives. Hans Fischer used a simple numbering system for the porphyrin nucleus and a set of common names to identify the different porphyrins and their isomers. A systematic naming based on the 1–24 numbering system for the porphyrin nucleus was later developed by the International Union of Pure and Applied Chemistry (IUPAC) and the International Union of Biochemistry (IUB), and this system has gained general acceptance. The need for common names is clear after examination of the systematic names; for example, protoporphyrin IX has the systematic name 2,7,12,18-tetramethyl-3,8-divinyl-13,17-dipropanoic acid.

The aromatic character (hence stability) of porphyrins has been confirmed by measurements of their heats of combustion. In addition, x-ray crystallographic studies have established planarity of the porphyrin macrocycle which is a basic requirement for aromatic character. *See* DELOCALIZATION; X-RAY CRYSTALLOGRAPHY.

Most metals and metalloids have been inserted into the central hole of the porphyrin macrocycle. The resulting metalloporphyrins are usually very stable and can bind a variety of small molecules (known as ligands) to the central metal atom. Heme, the iron complex of protoporphyrin IX, is the prosthetic group of a number of major proteins and enzymes that carry out diverse biological functions. These include binding, transport, and storage of oxygen (hemoglobin and myoglobin), electron-transfer processes (cytochromes), activation and transfer of oxygen to substrates (cytochromes P450), and managing and using hydrogen peroxide (peroxidases and catalases). *See* COORDINATION COMPLEXES; CYTOCHROME; HEMOGLOBIN.

Chlorophylls and bacteriochlorophylls are magnesium complexes of porphyrin derivatives known as chlorins and bacteriochlorins, respectively. They are the pigments responsible for photosynthesis. Several chlorophylls have been identified, the most common being chlorophyll *a*, which is found in all oxygen-evolving photosynthetic plants. Bacteriochlorophyll *a* is found in many photosynthetic bacteria. *See* CHLOROPHYLL; PHOTOSYNTHESIS.

Porphyrins and metalloporphyrins exhibit many potentially important medicinal and industrial properties. Metalloporphyrins are being examined as potential catalysts for a variety of processes, including catalytic oxidations. They are also being examined as possible blood substitutes and as electrocatalysts for fuel cells and for the electrochemical generation of hydrogen peroxide. The unique optical properties of porphyrins make them likely candidates for photovoltaic devices and in photocopying and other optical devices. A major area where porphyrins are showing significant potential is in the treatment of a wide range of diseases, including cancer, using photodynamic therapy. *See* CATALYSIS.

T. Wijesekera; D. Dolphin

Porphyroblast A relatively large crystal formed in a metamorphic rock. The presence of abundant porphyroblasts gives the rock a porphyroblastic texture. Minerals found commonly as porphyroblasts include biotite, garnet, chloritoid, staurolite, kyanite, sillimanite, andalusite, cordierite, and feldspar. Porphyroblasts are generally a few millimeters or centimeters across, but some attain a diameter of over 1 ft (30 cm). They may be bounded by well-defined crystal faces, or their outlines may be highly irregular or ragged. Very commonly they are crowded with tiny grains of other minerals that occur in the rock.

Most commonly, porphyroblasts develop in schist and gneiss during the late stages of recrystallization. As the rock becomes reconstituted, certain components migrate to favored sites and combine there to develop the large crystals. *See* GNEISS; METAMORPHIC ROCKS; SCHIST.

Carleton A. Chapman

Porphyry An igneous rock characterized by porphyritic texture, in which large crystals (phenocrysts) are enclosed in a matrix of very fine-grained to aphanitic (not visibly crystalline) material. Porphyries are generally distinguished from other porphyritic rocks by their abundance of phenocrysts and by their occurrence in small intrusive bodies (dikes and sills) formed at shallow depth within the earth. In this sense porphyries are hypabyssal rocks. *See* IGNEOUS ROCKS; PHENOCRYST.

Porphyries occur as marginal phases of medium-sized igneous bodies (stocks, laccoliths) or as apophyses (offshoots) projecting from such bodies into the surrounding rocks. They are also abundant as dikes cutting compositionally equivalent plutonic rock, or as dikes, sills, and laccoliths injected into the adjacent older rocks.

William Ingersoll Rose, Jr.

Positron The elementary antiparticle of an electron with mass equal to that of an electron but with positive rather than negative charge. Its existence was theoretically predicted by P. A. M. Dirac in 1928. It was first observed by C. D. Anderson in 1932. Positrons and electrons have the same spin and statistics. Positrons are naturally produced by radioactive decay of many heavier particles; this is referred to as beta-plus decay. Electron-positron pairs are also produced when high-energy photons or electrons interact with matter. *See* ANTIMATTER; ELECTRON; ELECTRON-POSITRON PAIR PRODUCTION; ELECTRON SPIN; ELEMENTARY PARTICLE; FERMI-DIRAC STATISTICS; PARTICLE ACCELERATOR; QUANTUM STATISTICS; RADIOACTIVITY; SPIN (QUANTUM MECHANICS).

In addition, quantum field theory predicts that positrons can be created when electrons are subjected to very strong, static electric fields. For this to occur, the field strength must exceed the energy equivalent of twice the rest mass of an electron, $2m_0c^2$, where m_0 is the electron rest mass and c is the speed of light. Experimentally, such strong fields can be achieved by using heavy-ion accelerators to collide two heavy atoms and temporarily form a superheavy atom with an atomic number $Z > 173$. However, to date there has been no definitive observation of positrons produced via this method. *See* NUCLEAR MOLECULE; QUASIATOM; SUPERCRITICAL FIELDS.

A positron is, in itself, stable, but cannot exist indefinitely in the presence of matter where it will ultimately collide with an electron. The two particles will be annihilated as a result of this collision, and photons will be created. However, a positron can become bound to an electron to form a short-lived "atom" termed positronium. *See* POSITRONIUM.

Positrons are of interest or have practical uses in many fields. In the field of physics, they are used to study differences between particles and antiparticles and how particles and antiparticles interact. In medicine, positron emission tomography (PET) is used to produce three-dimensional images of the inside of the body. In the field of material science, positrons are routinely used to study defects in the crystalline structure of solids or the properties of electrons within the solid. *See* CRYSTAL DEFECTS; MEDICAL IMAGING. Robert D. DuBois

Positronium The bound state of the electron and its antimatter counterpart, the positron. Positronium (abbreviated Ps) is perhaps the simplest bound state, since the constituents are structureless and the electromagnetic binding force is well understood. As in the hydrogen atom, the states of positronium are quantized. The spins of the two spin-$^1/_2$ constituents combine to yield total spin zero (singlet state) or one (triplet state). Singlet and triplet states are called parapositronium (p-Ps) and orthopositronium (o-Ps), respectively. States of positronium are labeled by principal quantum number n, orbital angular momentum, total spin, and total angular momentum (orbital plus spin). The energy levels are approximately -6.80 eV/n^2, with numerous small corrections due to relativity, virtual annihilation, and other electromagnetic effects. Additional corrections from strong and weak interactions are negligible. *See* ANGULAR MOMENTUM; ATOMIC STRUCTURE AND SPECTRA; ELECTRON; ELECTRON SPIN; POSITRON; SPIN (QUANTUM MECHANICS).

Positronium has a finite lifetime due to electron-positron annihilation into energy in the form of gamma-ray photons. The energetics of this reaction is governed by Albert Einstein's equation $E = mc^2$, where m is the total positronium mass (about twice the electron mass), c is the speed of light, and E is the total photon energy. A selection rule constrains the allowed decays, so that singlet positronium decays into an even number of photons (usually two), and triplet positronium into an odd number (usually three). Decay into a single photon is forbidden by energy-momentum conservation. The lifetimes are approximately 10^{-10} s for the singlet state and 10^{-7} s for the triplet state. *See* CONSERVATION LAWS (PHYSICS); GAMMA RAYS; REST MASS; SELECTION RULES (PHYSICS).

Positronium has been extensively studied since it was first produced by Martin Deutsch in 1951. High precision measurements have been done on the energy levels and decay rates of the lowest-energy states, and on energy levels of the first excited states. Because positronium is relatively free of strong and weak interaction complications, comparison between theory and experiment for these measured quantities provides a stringent check of quantum electrodynamics (QED). Positronium has been used to check for violations in fundamental symmetries and as a probe for physics beyond the standard model. Positronium is an analog for the more complicated quarkonium (quark-antiquark) bound states, particularly charmonium and bottomonium. *See* ELEMENTARY PARTICLE; MESON; QUANTUM ELECTRODYNAMICS; QUARKS; STANDARD MODEL; SYMMETRY LAWS (PHYSICS).

Annihilation radiation from positronium forms a component of the gamma-ray spectrum observed by astronomers, in particular from the galactic center. *See* GAMMA-RAY ASTRONOMY; MILKY WAY GALAXY. Gregory S. Adkins

Posttraumatic stress disorder An anxiety disorder in some individuals who have experienced an event that poses a direct threat to the individual's or another person's life. The characteristic features of anxiety disorders are fear, particularly in the ongoing absence of a real-life threat to safety, and avoidance behavior.

A diagnosis of posttraumatic stress disorder requires that four criteria be met. First, the individual musthave been exposed to an extremely stressful and traumatic event beyond the range of normal human experience (for example, combat, armed robbery, rape, or natural disasters such as earthquakes). Second, the individual must periodically and persistently reexperience the event. This reexperiencing can take different forms, such as recurrent dreams and nightmares, an inability to stop thinking about the event, flashbacks during which the individual relives the trauma, and auditory hallucinations. Third, there is persistent avoidance of events related to the trauma, and psychological numbing that was not present prior to the trauma. Fourth, enduring symptoms of anxiety and arousal are present. These symptoms can be manifested in different forms, including anger, irritability, a very sensitive startle response, an inability to sleep well, and physiological evidence of fear when the individual is reexposed to a traumatic event. Posttraumatic stress disorder can occur at any age, and the symptoms typically begin soon after the cessation of the traumatic event.

The development of posttraumatic stress disorder is likely influenced by a complex pattern of memory encoding, self-appraisal, coping strategies, cognitive styles, the nature of the trauma exposure, and biological responses that occur soon after the traumatic event and continue in a deteriorating cascade. Physiological arousal responses in individuals with posttraumatic stress disorder include increases in heart rate, respiration rate, and skin conductivity upon reexposure to traumatic stimuli. Posttraumatic stress disorder may also be associated with structural and physiological changes in the brain. Stressful events also affect the activity level of the pituitary and adrenal glands. All these physiological changes are probably complexly related to the persistence and waxing and waning of symptoms in posttraumatic stress disorder. In addition, extreme and prolonged stress is associated with a variety of physical ailments, including heart attacks, ulcers, colitis, and decreases in immunological functioning. *See* ADRENAL CORTEX; BRAIN; NEUROBIOLOGY; NORADRENERGIC SYSTEM; SEROTONIN.

Posttraumatic stress disorder is rarely the only psychological problem that an affected individual manifests. When an individual is diagnosed as having this disorder, particularly after it has been present for a number of years, it is common to also find significant depression, generalized anxiety, substance abuse and dependence, marital problems, and intense, almost debilitating anger. Although the primary symptoms of posttraumatic stress disorder (anxiety, fear, intrusion, and avoidance) are usually amenable to treatment efforts, these secondary problems commonly associated with the chronic disorder are more difficult to treat. *See* ADDICTIVE DISORDERS; AFFECTIVE DISORDERS; ANXIETY DISORDERS.

Posttraumatic stress disorder can be treated by pharmacological means and with psychotherapy. Most psychological treatments for the disorder involve reexposure to the traumatic event. This reexposure is typically imaginal and can range from simply talking about the trauma to having the person vividly imagine reliving the traumatic event. This latter behavioral procedure is called implosion therapy or flooding. While flooding is not appropriate for all posttraumatic stress disorder cases the procedure can dramatically decrease anxiety and arousal, intrusive thoughts, avoidance behavior, and emotional numbing. Along with specific behavior interventions, individuals with posttraumatic stress disorder should become involved in psychotherapeutic treatment for secondary problems. *See* PSYCHOPHARMACOLOGY; PSYCHOTHERAPY; STRESS (PSYCHOLOGY). Robert W. Butler

Postulate In a formal deductive system, a proposition accepted without proof, from which other propositions are deduced by the conventional methods of formal logic. There is a certain arbitrariness as to which propositions are to be treated as postulates, because when certain proved propositions are treated as such, other propositions which were originally postulates often become proved propositions. In strict usage, the term postulate is nearly equivalent to axiom, although axiom is often loosely used to denote a truth supposed to be self-evident. *See* LOGIC.

Percy W. Bridgman; Henry Margenau

Postural equilibrium A lifeless object is said to be in equilibrium, or in a state of balance, when all forces acting upon it cancel. The result is a state of rest. In an actively moving animal, internal as well as external forces have to be considered, and the maintenance of a balanced attitude in a body consisting of a number of parts that are loosely connected by movable joints is complex.

The maintenance of equilibrium is relatively easy in limbless animals. When the animal is turned on its back, the lack of contact pressure on the creeping surface and the tactile stimulation of the back initiate movements which return the animal to its normal position. This is known as the righting reflex. Free-swimming and flying animals are often in a precariously poised state of equilibrium, so that the normal attitude can be maintained only by the continuous operation of corrective equilibrating mechanisms. The same applies to a lesser degree to long-legged quadrupeds, such as many mammals, and to bipeds, such as birds and some primates (including humans).

For most species, there is a specific orientation of the whole body or of body segments (such as the head) with respect to gravity. This orientation is based on multisensory inputs and on a set of inborn reflexes acting on the musculature. These postural reflexes also stabilize the genetically defined body orientation against external disturbances. The sensory information relies on a number of sources: (1) static and dynamic information from the eyes; (2) static and dynamic mechanoreceptors incorporated in the various types of statocysts in invertebrate animals and in the vestibular organ or labyrinth of vertebrates; (3) proprioceptor organs such as muscle spindles, Golgi endings in tendons, Pacinian corpuscles and similar encapsulated endings associated with tendons and joints, and other pressure receptors in supporting surfaces (for example, the soles of feet); and (4) sensory endings in the viscera, capable of being differentially stimulated by changes in the direction of visceral pull on mesenteries, and on other structures. *See* SENSATION; SENSE ORGAN.

In higher terrestrial vertebrates, during stance the center of mass is usually situated high above the ground due to the support of the body by the limbs. A critical aspect of posture in quadrupedal and bipedal stance is equilibrium maintenance which is preserved only when under static conditions the projection of the center of mass remains inside the support base. This positioning of the center of mass is based on two main controls. A "bottom up" control is based on the afferent nerve impulses, cutaneous and proprioceptive, from the feet and the ankle joint muscles. These nerve impulse serve in building up posture from the feet to the head. A "top down" control starts from the head, and is predominant during dynamic activities such as locomotion. Due to labyrinthine afferent nerve impulses, the head axis orientation remains stable with respect to space. The movement-related visual afferents recorded by the retina monitor the head displacements with respect to space and adjust the body posture as a function of these inputs.

Two levels of control are involved in maintaining balance. A first level includes the spinal cord and the brainstem, where a set of inborn reflexes are organized for stance regulation and head orientation. Most postural reflexes rely on networks at that level. The cerebellum is involved in the adaptation of these reflexes to the external constraints. *See* REFLEX.

A second level of control includes cortical areas involved in multisensory integration and control as well as the basal ganglia. The postural body schema and the body orientation with respect to the external world are organized mainly at that level, with a predominant role in the right hemisphere. Coordination between balance control and locomotion or movements also depends on these higher levels.

Jean Massion

Potassium A chemical element, K, atomic number 19, and atomic weight 39.098. It stands in the middle of the alkali metal family, below sodium and above rubidium. This lightweight, soft, low-melting, reactive metal (see table) is very similar to sodium in its behavior in metallic forms. *See* ALKALI METALS; PERIODIC TABLE; RUBIDIUM; SODIUM.

Potassium chloride, KCl, finds its main use in fertilizer mixtures. It also serves as the raw material for the manufacture of other potassium compounds. Potassium hydroxide, KOH, is used in the manufacture of liquid soaps, and potassium carbonate in making soft soaps. Potassium carbonate, K_2CO_3, is also an important raw material for the glass industry. Potassium nitrate, KNO_3, is used in matches, in pyrotechnics, and in similar items which require an oxidizing agent. [M.Si.]

Potassium is a very abundant element, ranking seventh among all the elements in the Earth's crust, 2.59% of which is potassium in combined form. Seawater contains 380 parts per million, making potassium the sixth most plentiful element in solution.

Potassium is even more reactive than sodium. It reacts vigorously with the oxygen in air to form the monoxide, K_2O, and the peroxide, K_2O_2. In the presence of excess oxygen, it readily forms the superoxide, KO_2.

Physical properties of potassium metal

Property	Temperature °C	Temperature °F	SI units	Customary (engineering) units
Density	100	212	0.819 g/cm^3	51.1 lb/ft^3
	400	752	0.747 g/cm^3	46.7 lb/ft^3
	700	1292	0.676 g/cm^3	42.2 lb/ft^3
Melting point	63.7	147		
Boiling point	760	1400		
Heat of fusion	63.7	147	14.6 cal/g	26.3 Btu/lb
Heat of vaporization	760	1400	496 cal/g	893 Btu/lb
Viscosity	70	158	5.15 millipoises	6.5 kinetic units
	400	752	2.58 millipoises	3.5 kinetic units
	800	1472	1.36 millipoises	2 kinetic units
Vapor pressure	342	648	1 mm	0.019 lb/in.^{2}
	696	1285	400 mm	7.75 lb/in.^{2}
Thermal conductivity	200	392	0.017 cal/(s)(cm^2)(cm)(°C)	26.0 Btu(h)(ft^2)(°F)
	400	752	0.09 cal/(s)(cm^2)(cm)(°C)	21.7 Btu/(h)(ft^2)(°F)
Heat capacity	200	392	0.19 cal/(g)(°C)	0.19 Btu/(lb)(°F)
	800	1472	0.19 cal/(g)(°C)	0.19 Btu/(lb)(°F)
Electrical resistivity	150	302	18.7 microhm-cm	
	300	572	28.2 microhm-cm	
Surface tension	100–150	212–302	About 80 dynes/cm	

Potassium does not react with nitrogen to form a nitride, even at elevated temperatures. With hydrogen, potassium reacts slowly at 200°C (392°F) and rapidly at 350–400°C (662–752°F). It forms the least stable hydride of all the alkali metals.

The reaction between potassium and water or ice is violent, even at temperatures as low as −100°C (−148°F). The hydrogen evolved is usually ignited in reaction at room temperature. Reactions with aqueous acids are even more violent and verge on being explosive.

The potassium ion (K^+) is the most common intracellular cation and is essential for maintaining osmotic pressure and electrodynamic cellular properties in organisms. The intracellular potassium ion concentrations are typically high for most cells, whereas the potassium ion concentrations present in extracellular fluids are significantly lower. The hydrolysis of the coenzyme adenosine triphosphate (ATP) is mediated by the membrane-bound enzyme Na^+, K^+-ATPase. This enzyme is called the sodium pump and it is activated by both potassium and sodium ions; however, many enzymes are activated by potassium ions alone (for example, pyruvate kinase, aldehyde dehydrogenase, and phosphofructokinase).

Potassium deficiency may occur in several conditions, including malnutrition and excessive vomiting or diarrhea, and in patients undergoing dialysis; supplementation with potassium salts is sometimes required. *See* OSMOREGULATORY MECHANISMS.

Duarte Mota de Freitas

Potato, Irish A plant of the genus *Solanum* in the nightshade family, Solanaceae; it is related to tomatoes and peppers. There are more than 2000 species of *Solanum*, of which about 150 bear tubers. The potato of commerce, *S. tuberosum*, originated in South America, probably in the highlands of Peru and Bolivia, where it has been cultivated for several thousand years. Potatoes were introduced into Europe by Spanish explorers in the late sixteenth century and into the United States from Ireland in 1719. The crop became a staple in Europe, was a primary source of food in Ireland, and is known even today as the Irish potato. *See* SOLANALES.

The potato plant is an annual, herbaceous dicotyledon that is grown primarily for its edible tubers, which are short, thick underground stems that form on the ends of stolons (lateral stems). Lateral buds (eyes) on the mature tuber are the growing points for a new crop, provided that whole tubers or pieces with at least one eye are planted.

There are 95–100 varieties certified for seed production in the United States and Canada. In the United States, seven varieties account for more than 70% of the commercial acreage planted. These are Russet Burbank, Norchip, Atlantic, Russet Norkotah, Superior, Centennial Russet, and Kennebec.

Potatoes are grown in more countries than any crop except for corn (maize), and they are the fourth-most important food crop in supplying energy in the human diet, following rice, wheat, and corn. The leading potato producers are Russia, China, Poland, and the United States, in descending order.

Chemical constituents can be affected by variety, production area, cultural practices, maturity at harvest, and storage conditions. The average composition of potatoes is 78–80% water, 14–18% starch, 2% protein, 1% minerals, 0.4% fiber, and 0.1% fat, with some sugars, organic acids, amino acids, and vitamins. The potato is very nutritious, serving as a nearly complete food itself, and it produces more food per acre than any other crop. Starch constitutes about 65–80% of the dry weight of the potato and, calorically, is the most important component. Potatoes contain appreciable amounts of the B vitamins, and they are an excellent source of vitamin D. Potatoes contribute more vitamin C to the United States food supply than any other single food source. Inorganic constituents or minerals of potatoes are predominantly potassium, phosphorus, sulfur, chlorine, magnesium, and calcium. Potatoes contain sufficient quantities of iron to be a good nutritional source of this mineral as well.

J. Creighton Miller, Jr.

Potato, sweet The fleshy root of the plant *Ipomoea batatas*. The sweet potato was mentioned as being grown in Virginia as early as 1648. In 1930 the selection of outstanding strains of the Porto Rico variety, which was introduced into Florida in 1908, was begun in Louisiana, and the best strain, Unit I Porto Rico, was released in 1934. The Unit I Porto Rico is now being replaced by the Centennial, a yam type which has three times its vitamin A content. More than 70% of the commercial crop in the United States is of the Centennial variety. Louisiana is the leading state for commercial production of both the canned and fresh products.

There are two principal types of sweet potato, the kind erroneously called yam and the Jersey type. The chief difference between the two is that in cooking or baking the yam, much of the starch is broken down into simple sugars (glucose and fructose) and an intermediate product, dextrin. This gives it a moist, syrupy consistency somewhat sweeter than that of the dry (Jersey) type. On cooking, the sugar in the dry type remains as sucrose.

The yam is produced largely in the Southern states; however, because of the breeding of more widely adapted varieties, it is now being grown farther north. The Jersey sweet potato is grown largely along the eastern shore of Virginia. Maryland, Delaware, and New Jersey, and also in Iowa and Kansas. Julian C. Miller

Technically, the sweet potato is a perennial plant, but in commerical production the growth is terminated by harvest due to impending change of seasons (cold or rainy) or by achievement of optimum storage root size. Plantings are established from sprouts obtained from stored roots or from vine cuttings. Production of storage roots does not require pollination, so that adverse conditions only delay or reduce harvests; crop failures are rare. A variety of soil types may be used; however, light sandy soils are preferred for ease of harvest.

A number of factors contribute to the worldwide importance of sweet potatoes. Perhaps foremost is the high energy yield, which exceeds that of Irish potatoes and grains. Sweet potato roots are also high in nutritional quality.

The sweet potato is a versatile crop that is used as a food, an ingredient in other foods, an animal feed, and a feedstock for the production of starch and ethanol. *See* ANIMAL FEEDS; POTATO, IRISH; STARCH.

John C. Bouwkamp

Potential flow A fluid flow that is isentropic and that, if incompressible, can be mathematically described by Laplace's equation. For an ideal fluid, or a flow in which viscous effects are ignored, vorticity (defined as the curl of the velocity) cannot be produced, and any initial vorticity existing in the flow simply moves unchanged with the fluid. Ideal fluids, of course, do not exist since any actual fluid has some viscosity, and the effects of this viscosity will be important near a solid wall, in the region known as the boundary layer. Nevertheless, the study of potential flow is important in hydrodynamics, where the fluid is considered incompressible, and even in aerodynamics, where the fluid is considered compressible, as long as shock waves are not present. *See* BOUNDARY-LAYER FLOW; COMPRESSIBLE FLOW; ISENTROPIC FLOW.

In the absence of viscous effects, a flow starting from rest will be irrotational for all subsequent time. For an irrotational flow, the curl of the velocity is zero ($\nabla \times V = 0$). The curl of the gradient of any scalar function is zero ($\nabla \times \nabla\phi = 0$). It then follows mathematically that the condition of irrotationality can be satisfied identically by choosing the scalar function, ϕ, such that the velocity is the gradient of ϕ ($V = \nabla\phi$). For this reason, this scalar function ϕ has been traditionally referred to as the velocity potential, and the flow as a potential flow. *See* CALCULUS OF VECTORS; POTENTIALS.

By applying the continuity equation to the definition of the potential function, it becomes possible to represent the flow by the well-known Laplace equation ($\nabla^2\phi = 0$), instead of the coupled system of the continuity and nonlinear Euler equations. The linearity of the Laplace equation, which also governs other important physical phenomena such as electricity and magnetism, makes it possible to use the principle of superposition to combine elementary solutions in solving more complex problems. *See* EQUATION OF CONTINUITY; FLUID MECHANICS; LAPLACE'S DIFFERENTIAL EQUATION; LAPLACE'S IRROTATIONAL MOTION. Peter E. Raad

Potentials Functions or sets of functions from whose first derivatives a vector can be formed. A vector is a quantity which has a magnitude and a direction, such as force.

A single function, the scalar potential, is used in gravitation theory, electricity and magnetism, fluid mechanics, and other areas. The vectors obtained from it by partial differentiation are in these cases the gravitational, electric and magnetic field strengths, and the velocity, respectively. The vector potential is a set of three functions whose first derivatives give the magnetic induction. *See* ELECTRIC FIELD; GRAVITATION; LAPLACE'S IRROTATIONAL MOTION. F. Rohrlich

Potentiometer An instrument that precisely measures an electromotive force (emf) by balancing it against a known potential drop established by a three-terminal resistive divider (see illustration), across which a known voltage exists. A divider can adjust and control the voltage applied to some device or part of a circuit, for example, a volume control. The ratio of output to input voltage is equal to that of the tapped to total divider resistance when the current required by the controlled device is negligible. Potentiometers as voltage-measuring instruments have largely been replaced by the ratio instrument known as a digital voltmeter. The Weston standard cells which supplied a known standard voltage have also been superseded by solid-state voltage reference devices which are usually incorporated in the digital voltmeter.

Another type of potentiometer involves a current comparator which senses and corrects inequality of ampere-turns in two windings on a magnetic core. An advantage of this potentiometer over those whose long-term accuracy depends on the stability of a resistance ratio is that the ratio here is the turns-ratio of windings on a common magnetic core, dependent solely on conductor position and therefore not subject to drift with time. *See* CURRENT COMPARATOR; ELECTROMAGNETIC INDUCTION; MAGNETISM.

Potentiometer techniques may also be used for current measurement, the unknown current being sent through a known resistance and the *IR* drop opposed by balancing it at the voltage terminals of the potentiometer. Here, of course, internal heating and consequent resistance change of the current-carrying resistor (shunt) may be a critical factor in measurement accuracy; and the shunt design may require attention to dissipation of heat resulting from its I^2R power consumption. *See* CURRENT MEASUREMENT; JOULE'S LAW.

Potentiometer techniques have been extended to alternating-voltage measurements, but generally at a reduced accuracy level (usually 0.1% or so). Current is set on an ammeter which must have the same response on ac as on dc, where it may be calibrated with a potentiometer and shunt combination. Balance in opposing an unknown voltage is achieved in one of two ways: (1) a slide-wire and phase-adjustable supply; (2) separate in-phase and quadrature adjustments on slide wires supplied from sources that have a 90° phase difference. Such potentiometers have limited use in magnetic testing. *See* ALTERNATING CURRENT; ELECTRICAL MEASUREMENTS; VOLTAGE MEASUREMENT.

Forest K. Harris; Ronald F. Dziuba; Bryan P. Kibble

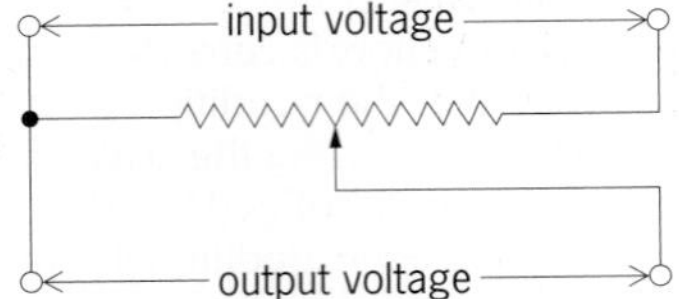

Three-terminal resistive voltage divider.

Pottery Vessels made entirely or partly of clay, and fired to a strong, hard product; occasionally, the term refers to just the lower grades of such ware. Pottery may be glazed or unglazed. *See* CERAMICS; GLAZING.

Grades of pottery (such as china, stoneware, earthenware, and other special types) are distinguished by their color, strength, absorption (the weight of water soaked up when the piece is submerged, expressed as a percentage of the original weight), and translucency (ability to transmit light). All these properties refer to the material or "body" under any glaze present. *See* PORCELAIN.

Absorption is due to the presence of open pores or voids in the fired material into which water can penetrate; in general, the higher the firing temperature, the lower the absorption. Body color is determined mainly by raw-material purity. Strength depends on the porosity and also on the amount and type of glass and crystals developed in the body on firing. Translucency is obtained in products in which there is low porosity and little differences in index of refraction between the glass and crystals in the body. *See* CLAY. J. F. McMahon

Pottiales An order of the true mosses (subclass Bryidae), consisting of three families and about 91 genera. The order is characterized by short, papillose leaf cells and deeply divided, commonly twisted peristome teeth. Members of the Pottiales typically grow erect, with stems simple or forked by subfloral innovation, and they often produce gemmae (asexual reproductive bodies) on leaves, stems, or rhizoids. The leaves have revolute margins, and are arranged in many rows. The single costa often ends in a hyaline hairpoint. *See* BRYIDAE; BRYOPHYTA; BRYOPSIDA.

Howard Crum

Poultry production Poultry production comprises two major categories, meat production and egg production. Most poultry produced in North America is grown under close control on highly specialized farms. The evolution from small flocks to large commercial units after World War II was facilitated by rapid advances in the knowledge of nutrition, breeding, housing, disease control, and processing of poultry and eggs, and by improvements in transportation and refrigeration which made possible distant marketing of fresh products.

Incubation. Artificial incubation was a major advance in poultry production because it became possible to hatch large numbers of chicks of the same age for farmers to raise for meat or egg production. Modern incubators are constructed of materials that can be effectively cleaned and disinfected and that provide good insulation of the chamber. Eggs are set in specially designed plastic flats which fit into channels in egg racks that move on wheels. The egg racks are equipped with mechanical systems to tilt the eggs 45°; turning is usually done hourly. Eggs are transferred from the setting trays into hatching trays 3 days before expected hatch. After the hatch is completed, the chicks are transferred to a conveyer belt for processing or directly into plastic boxes with absorbent paper pads or into disposable paper boxes with wood fiber pads. Chick servicing often involves sexing, vaccination, and beak trimming. Usually, chicks are held at the hatchery prior to shipment to farms, starting early on the day after hatching. Specially designed delivery trucks or buses are used to provide adequate ventilation for chicks during shipment.

Breeding. The genetic stock used for modern poultry production is produced by highly specialized breeding companies. Meat poultry is selected for good meat type, fast growth, disease resistance, and efficient conversion of feed to meat. Different strains of chickens are used for table egg production. These are selected for high egg production, large egg size, and small body weight for better conversion of feed to eggs and good livability. The body weights of meat and egg production strains are dramatically different. *See* BREEDING (ANIMAL).

Brooding and rearing. Day-old chicks require an ambient temperature of 85–87°F (29–30°C) for normal growth and health during the first week of life. As the chicks grow and feather, they can tolerate lower temperatures. Brooding heat is often provided by a radiant gas brooder stove. Most chicks are started on floors that are covered with 2–4 in. (5–10 cm) of a litter material such as pine shavings, rice hulls, or peanut hulls. Feeding is usually done is small troughs or on plastic trays until the chicks learn to eat and drink. Chicks are quickly trained to eat from mechanical feeders and drink from closed water delivery systems to reduce labor.

Feeding and nutrition. Poultry diets consist of common grains and protein sources with mineral and vitamin supplements. Animal or vegetable fats may be added to increase energy and reduce dustiness. Corn, grain sorghum, wheat, oats, and barley are often used for poultry feeding in the United States. Soybean meal is widely used as a protein supplement. Other important protein supplements are meat meal, fish meal, safflower meal, feather meal, and canola meal. *See* ANIMAL FEEDS.

Housing. The purpose of a poultry house is to confine the birds; to protect them from predators and environmental extremes which would cause mortality or reduce growth, feed efficiency, immunocompetence, fertility or egg production; to facilitate light control; and to facilitate bird management. Poultry houses can be constructed from locally available building materials. Smooth interior surfaces are preferred for effective sanitation. Houses are usually a maximum of 40 ft (12 m) wide to facilitate more uniform ventilation. House length is approximately 500 ft (152 m); most houses are constructed with a gable roof.

Production systems. Chickens for table egg production are often housed in cages to provide cleaner eggs and protect the birds from disease agents which are recycled to birds from the manure. Meat chickens and turkeys are usually grown in litter-floor houses because heavier poultry experience more lameness, breast blisters, and weaker bones and joints when grown in cages.

Ducks and geese. With some modification of husbandry, ducks and geese can be successfully raised in confinement. They do not require water for swimming and can be grown in litter-floor houses similar to meat chickens or turkeys.

Health maintenance. The production of commercial poultry in large flocks requires well-designed disease-control programs. The first requirement is for maintenance of biosecurity in production units. This means that entrance of contaminated workers and visitors, birds, feed, and equipment must be prevented. Some poultry diseases are more effectively or economically controlled by vaccination. Examples are Marek's disease, Newcastle disease, infectious bronchitis, avian pox, infectious bursal disease, and often several others, depending on the disease history of the farm and the area where the poultry are raised. *See* NEWCASTLE DISEASE.

Processing and marketing. Shell eggs are often processed in plants located on the farm. A mechanical system can detect and separate cracked eggs. Egg processing machines can process up to 300 cases of eggs (360 eggs per case) per hour. Machines are also available to separate the yolk and albumen from the egg shell. Poultry are usually processed at large central plants under inspection by the U.S. Department of Agriculture. The meat birds are loaded on trucks at farms during the night, and processing typically begins at midnight, and is followed by extensive cleaning and disinfection of the plant. Poultry are removed from transport racks or coops and hung on shackles. After stunning, bleeding, scalding, and feather removal, the carcasses must be transferred to a second line or table in a different room for evisceration, chilling, cutting, and packaging. Many poultry processing plants cut up poultry before packaging, and they may also separate meat from bone or skin. *See* AGRICULTURAL SCIENCE (ANIMAL); EGG (FOWL).

Ralph A. Ernst

Powder metallurgy

A metalworking process used to fabricate parts of simple or complex shape from a wide variety of metals and alloys in the form of powders. The process involves shaping of the powder and subsequent bonding of its individual particles by heating or mechanical working. Powder metallurgy is a highly flexible and automated process that is environmentally friendly, with a low relative energy consumption and a high level of materials utilization. Thus it is possible to fabricate high-quality parts to close tolerance at low cost. Powder metallurgy processing encompasses an extensive range of ferrous and nonferrous alloy powders, ceramic powders, and mixes of metallic and ceramic powders (composite powders). *See* METALLURGY.

Regardless of the processing route, all powder metallurgy methods of part fabrication start with the raw material in the form of a powder. A powder is a finely divided solid, smaller than about 1 mm in its maximum dimension. There are four major methods used to produce metal powders, involving mechanical comminution, chemical reactions, electrolytic deposition, and liquid-metal atomization. Metal powders exhibit a diversity of shapes ranging from spherical to acicular. Particle shape is an important property, since it influences the surface area of the powder, its permeability and flow, and its density after compaction. Chemical composition and purity also affect the compaction behavior of powders.

Powder metallurgy processes include pressing and sintering, powder injection molding, and full-density processing. *See* SINTERING.

Normally, parts made by pressing and sintering require no further treatment. However, properties, tolerances, and surface finish can be enhanced by secondary operations such as repressing, resintering, machining, heat treatment, and various surface treatments.

Powder injection molding is a process that builds on established injection molding technology used to fabricate plastics into complex shapes at low cost. It produces parts which have the shape and precision of injection-molded plastics but which exhibit superior mechanical properties such as strength, toughness, and ductility.

Parts fabricated by pressing and sintering are used in many applications. However, their performance is limited because of the presence of porosity. In order to increase properties and performance and to better compete with products manufactured by other metalworking methods (such as casting and forging), several powder metallurgy techniques have been developed that result in fully dense materials; that is, all porosity is eliminated. Examples of full-density processing are hot isostatic pressing, powder forging, and spray forming.

Powder metallurgy competes with several more conventional metalworking methods in the fabrication of parts, including casting, machining, and stamping. Characteristic advantages of powder metallurgy are close tolerances, low cost, net shaping, high production rates, and controlled properties. Other attractive features include compositional flexibility, low tooling costs, available shape complexity, and a relatively small number of steps in most powder metallurgy production operations.

Metal powders can be thermally unstable in the presence of oxygen. Very fine metal powders can burn in air (pyrophoricity) and are potentially explosive. Some respirable fine powders pose a health concern and can cause disease or lung dysfunction.

Control is exercised by the use of protective equipment and safe handling systems such as glove boxes. *See* INDUSTRIAL HEALTH AND SAFETY. Alan Lawley

Power The time rate of doing work. Like work, power is a scalar quantity, that is, a quantity which has magnitude but no direction. Some units often used for the measurement of power are the watt (1 joule of work per second) and the horsepower (550 foot-pounds of work per second). *See* WORK.

Power is a concept which can be used to describe the operation of any system or device in which a flow of energy occurs. In many problems of apparatus design, the power, rather than the total work to be done, determines the size of the component used. Any device can do a large amount of work by performing for a long time at a low rate of power, that is, by doing work slowly. However, if a large amount of work must be done rapidly, a high-power device is needed. High-power machines are usually larger, more complicated, and more expensive than equipment which need operate only at low power. A motor which must lift a certain weight will have to be larger and more powerful if it lifts the weight rapidly than if it raises it slowly. An electrical resistor must be large in size if it is to convert electrical energy into heat at a high rate without being damaged. Paul W. Schmidt

Power amplifier The final stage in multistage amplifiers, such as audio amplifiers and radio transmitters, designed to deliver appreciable power to the load. Power amplifiers may be called upon to supply power ranging from a few watts in an audio amplifier to many thousands of watts in a radio transmitter. In audio amplifiers the load is usually the dynamic impedance presented to the amplifier by a loudspeaker, and the problem is to maximize the power delivered to the load over a wide range of frequencies. The power amplifier in a radio transmitter operates over a relatively narrow band of frequencies with the load essentially a constant impedance. *See* AMPLIFIER. Harold F. Klock

Power factor In sinusoidal alternating-current theory, the cosine of the phase angle between the voltage across and the current through electrical circuitry. It equals the ratio of the mean power dissipated to the product of the root-mean-square voltage and current, and is unity for a circuit containing only pure resistances. It is less than unity if inductances and capacitances are involved. *See* ALTERNATING CURRENT. Bryan P. Kibble

Power-factor meter An instrument used to indicate whether load currents and voltages are in time-phase with one another. *See* POWER FACTOR.

The single-phase meter contains a fixed coil that carries the load current, and crossed coils that are connected to the load voltage. There is no spring to restrain the moving system, which takes a position to indicate the angle between the current and voltage. The scale can be marked in degrees or in power factor.

The angle between the currents in the crossed coils is a function of frequency, and consequently each power-factor meter is designed for a single frequency and will be in error at all other frequencies. Harry Sohon; Edward C. Stevenson

Power integrated circuits Integrated circuits that are capable of driving a power load. The key feature of a power integrated circuit that differentiates it from other semiconductor technologies is its ability to handle high voltage, high current, or a combination of both.

In its simplest form, a power integrated circuit may consist of a level-shifting and drive circuit that translates logic-level input signals from a microprocessor to a voltage and current level sufficient to energize a load. For example, such a chip may be used to operate electronic display, where the load is usually capacitive in nature but requires drive voltages above 100 V, which is much greater than the operating voltage of digital logic circuits (typically 5 V). At the other extreme, the power integrated circuit may be required to perform load monitoring, diagnostic functions, self-protection, and information feedback to the microprocessor, in addition to handling large amounts of power to actuate the load. An example of this is an automotive multiplexed bus system with distributed power integrated circuits for control of lights, motors, air conditioning, and so forth. *See* AUTOMOTIVE ELECTRICAL SYSTEM; ELECTRONIC DISPLAY; MICROPROCESSOR.

Power integrated circuits are expected to have an impact on all areas in which power semiconductor devices are presently being used. In addition, they are expected to open up new applications based upon their added features. The wide spectrum of voltages and currents over which power semiconductor devices are utilized are summarized in the table. *See* INTEGRATED CIRCUITS. B. Jayant Baliga

Power line communication A technology that uses electrical wiring as a medium to transmit digital signals from one device that has been rendered capable of employing this technology to another. Power line communication (PLC) enables the communication of data as well as multimedia messages by superimposing digital signals over the standard 50- or 60-Hz alternating current (AC). Low-speed power line communication has been used widely for some time in electrical utilities to provide voice communication, to control substations, and to protect high-voltage transmission lines. High-speed data transmission over power lines began in the late 1990s and was targeted to provide broadband Internet services and utility applications such as remote meter reading, as well as in home networking for data and multimedia communications. A short-range form of power line communication is also used for home automation and controls. *See* DATA COMMUNICATIONS; ELECTRIC POWER SYSTEMS; ELECTRIC POWER TRANSMISSION; INTERNET.

High-speed power line communication is of great interest in developing countries because of the lack of alternative broadband infrastructure such as cable or telephone lines. Using the existing electrical power distribution infrastructure, power line communication can provide digital communication without the need to lay new cabling. Thus, many countries are testing and deploying power line communication technology, despite concerns that power line communication can introduce electromagnetic interference (EMI). *See* ELECTRICAL INTERFERENCE.

In-home power line communication provides another networking option in addition to the popular wireless communication technology and other wired technologies like Ethernet, Home PNA (Phone Networking Alliance), and fiber optics. The major benefit of in-home power line communication is that people are able to use existing power plugs as access points for computer communications and other applications like surveillance video and home automation.

Background. Power line communication devices operate by injecting a carrier wave into the electrical wires at the transmitter. The carrier is modulated by digital signals. Each receiver in the system has a unique address and can receive and decode the signals transmitted over the wires. These devices may be either plugged into regular power outlets or permanently wired in place. Power line communication devices use advanced signal modulation, digital signal processing, and error detection and correction techniques.

Types of technology. Power line communication technologies may be distinguished by their voltage levels.

Low-voltage power line communication. Low-voltage power line communication devices usually are used in home networks. Power line communications do not require a specific access point since power outlets are the access points.

Low-voltage power line communication may also be used for temporary local-area network (LAN) setups for conferences, temporary medical centers, and other situations that require simple and fast setups.

Medium- and high-voltage power line communication. Medium- and high-voltage power line communications usually are used for broadband over power line (BPL) or home electricity usage metering. Medium- and high-voltage power line communication systems can often cover longer distances with less attenuation and higher data rates.

Haniph A. Latchman; Yu-Ju Lin; Sunguk Lee

Power plant A means for converting stored energy into work. Stationary power plants such as electric generating stations are located near sources of stored energy, such as coal fields or river dams, or are located near the places where the work is to be performed, as in cities or industrial sites. Mobile power plants for transportation service are located in vehicles, as the gasoline engines in automobiles and diesel locomotives for railroads. Power plants range in capacity from a fraction of a horsepower (hp) to over 10^6 kW in a single unit. Large power plants are assembled, erected, and constructed on location from equipment and systems made by different manufacturers. Smaller units are produced in manufacturing facilities.

Most power plants convert part of the stored raw energy of fossil fuels into kinetic energy of a spinning shaft. Some power plants harness nuclear energy. Elevated water supply or run-of-the-river energy is used in hydroelectric power plants. For transportation, the plant may produce a propulsive jet, as in some aircraft, instead of the rotary motion of a shaft. Other sources of energy, such as fuel cells, winds, tides, waves, geothermal, ocean thermal, nuclear fusion, photovoltaics, and solar thermal, have been of negligible commercial significance in the generation of power despite their magnitudes. *See* ENERGY SOURCES.

There is no practical way of storing the mechanical or electrical output of a power plant in the magnitudes encountered in power plant applications, although several small-scale concepts have been researched. As of now, however, the output must be generated at the instant of its use. This results in wide variations in the loads imposed upon a plant. The capacity, measured in kilowatts or horsepower, must be available when the load is imposed. Much of the capacity may be idle during extended periods when there is no demand for output. Hence much of the potential output, measured as kilowatt-hours or horsepower-hours, cannot be generated because there is no demand for output. Kilowatts cannot be traded for kilowatt-hours, and vice versa. *See* ENERGY STORAGE.

The efficiency of energy conversion is vital in most power plant installations. With thermal power plants the basic limitations of thermodynamics fix the efficiency of converting heat into work. The cyclic standards of Carnot, Rankine, Otto, Diesel, and Brayton are the usual criteria on which heat-power operations are variously judged. Performance of an assembled power plant, from fuel to net salable or usable output, may be expressed as thermal efficiency (%); fuel consumption (lb, pt, or gal per hp-h or per kWh); or heat rate (Btu supplied in fuel per hp-h or per kWh). American practice uses high or gross calorific value of the fuel for measuring heat rate or thermal efficiency and differs in this respect from European practice, which prefers the low or net calorific value.

In scrutinizing data on thermal performance, it should be recalled that the mechanical equivalent of heat (100% thermal efficiency) is 2545 Btu/hp-h and 3413 Btu/kWh (3.6 megajoules/kWh). Modern steam plants in large sizes (75,000–1,300,000 kW units) and internal combustion plants in modest sizes (1000–20,000 kW) have little difficulty in delivering a kilowatt-hour for less than 10,000 Btu (10.55 MJ) in fuel (34% thermal efficiency). For condensing steam plants, the lowest fuel consumptions per unit output (8200–9000 Btu/kWh or 8.7–9.5 MJ/kWh) are obtained in plants with the best vacuums, regenerative-reheat cycles using eight stages of extraction feed heating, two stages of reheat, primary pressures of 4500 lb/in.^{2} gage or 31 megapascals gage (supercritical), and temperatures of 1150°F (620°C). An industrial plant cogenerating electric power with process steam is capable of having a thermal efficiency of 5000 Btu/kWh (5.3 MJ/kWh).

Combustion turbines used in combined cycle configurations have taken a dominant role in new power generation capacity. The reason is the higher efficiency and lower emissions of the power plant in this arrangement. The rapid pace in advances in combustion turbine technology (such as higher firing temperatures that improve the Brayton cycle efficiency) has driven combined cycle efficiency to nearly 60% when using natural gas as fuel, while attaining low emission rates. Low fuel consumption (5700–6000 Btu/kWh or 6.0–6.3 MJ/kWh) is obtained by using higher firing temperatures, steam cooling on the combustor and gas turbine blades, a reheat steam cycle with a three-pressure heat recovery steam generator, and higher pressure and temperature of the steam cycle. These conditions are balanced with the need to keep the exhaust flue gas temperature as low as practical to achieve low emissions.

Gas turbines in simple cycle configuration are used mostly for peaking service due to their fast startup capabilities. The advances in the gas turbines have also increased the efficiency of simple cycle operations. Recuperation of the classic Brayton cycle gas turbine (simple cycle) is an accepted method of improving cycle efficiency that involves the addition of a heat exchanger to recover some portion of the exhaust heat that otherwise would be lost. *See* GAS TURBINE.

The nuclear power plant substitutes the heat of fission for the heat of combustion, and the consequent plant differs only in the method of preparing the thermodynamic fluid. It is otherwise similar to the usual thermal power plant. The pressure of a light-water reactor core is limited by material and safety considerations, while the temperature at which the steam is produced is determined by the core pressure. Because a nuclear reactor does not have the capability to superheat the steam above the core temperature, the steam temperature in a nuclear cycle is less than in a fossil cycle. *See* ELECTRIC POWER GENERATION; NUCLEAR REACTOR.

K. Keith Roe; Reginald S. Gagliardo

Power shovel A power-operated digging machine consisting of a lower frame and crawlers, a machinery frame, and a gantry supporting a boom which in turn supports a dipper handle and dipper. The machines are powered by on-board diesel engines or by electric motors. Diesel-powered machines utilize a series of clutches and brakes that allow the operator to control various motions. Electric motor machines generally have individual motors for each motion, but occasionally clutches and brakes are used allowing one motor to drive two motions. *See* BULK-HANDLING MACHINES; CONSTRUCTION EQUIPMENT; HOISTING MACHINES.

E. W. Sankey

Poynting's vector A vector, the outward normal component of which, when integrated over a closed surface in an electromagnetic field, represents the outward flow of energy through that surface. It is given by the equation below, where

$$\Pi = \mathbf{E} \times \mathbf{H} = \mu^{-1}\mathbf{E} \times \mathbf{B}$$

E is the electric field strength, **H** the magnetic field strength, **B** the magnetic flux density, and μ the permeability.

When an electromagnetic wave is incident on a conducting or absorbing surface, theory predicts that it should exert a force on the surface in the direction of the difference between the incident and the reflected Poynting's vector. *See* ELECTRIC FIELD; ELECTROMAGNETIC RADIATION; MAXWELL'S EQUATIONS; RADIATION PRESSURE.

William R. Smythe

Prairie dog A type of ground squirrel belonging to the family Sciuridae in the order Rodentia. These stout, short-tailed, short-legged ground squirrels inhabit open plains, short-grass prairies, and plateaus in the western part of North America from Canada to Mexico. The short coarse fur is grayish-brown. They have small beady eyes, pouched cheeks, and a short flat tail. Adults have a head-body length of 280–330 mm (11–13 in.), a tail length of 30–115 mm (1–4.5 in.), and weigh 0.7–1.4 kg (1.5–3 lb). *See* RODENTIA; SQUIRREL.

Prairie dogs are classified in the genus *Cynomys*. Obviously not dogs, they received this name because of their shrill doglike bark. Five species are currently recognized. Prairie dogs are diurnal, terrestrial, social mammals. They once lived predominantly in colonies known as towns, with miles of well-worn tunnels and dens extending in every direction beneath plateaus and upland prairies. A large town may have had millions of inhabitants. Today, few dog towns of any size remain outside of refuges, parks, or sanctuaries. *See* SOCIAL MAMMALS. Donald W. Linzey

Praseodymium A chemical element, Pr, atomic number 59, and atomic weight 140.91. Praseodymium is a metallic element of the rare-earth group. The stable isotope 140.907 makes up 100% of the naturally occurring element. The oxide is a black powder, the composition of which varies according to the method of preparation. If oxidized under a high pressure of oxygen it can approach the composition PrO_2. The black oxide dissolves in acid with the liberation of oxygen to give green solutions or green salts which have found application in the ceramic industry for coloring glass and for glazes. *See* PERIODIC TABLE; RARE-EARTH ELEMENTS. Frank H. Spedding

Prasinophyceae A small class of mostly motile, photosynthetic, unicellular algae in the chlorophyll *a-b* phyletic line (Chlorophycota). It is segregated from the Chlorophyceae primarily on the basis of ultrastructural characters, especially the possession of one or more layers of polysaccharide scales outside the plasmalemma. Prasinophytes are mainly members of the marine plankton, but they are also found in brackish- and freshwater habitats. A few are benthic, with both coccoid and colonial forms known, while others live symbiotically within dinoflagellates, radiolarians, and turbellarian worms. Approximately 180 species are known in 13 genera. *See* ALGAE; CHLOROPHYCOTA; PHYTOPLANKTON. Paul C. Silva; Richard L. Moe

Preamplifier A voltage amplifier suitable for operation with a low-level input signal. It is intended to be connected to another amplifier with a higher input level. Preamplifiers are necessary when an audio amplifier is to be used with low-output transducers such as magnetic phonograph pickups. A preamplifier may incorporate frequency-correcting networks to compensate for the frequency characteristics of a given input transducer and to make the frequency response of the preamplifier-amplifier combination uniform. *See* AMPLIFIER; VOLTAGE AMPLIFIER. Harold F. Klock

Prebiotic organic synthesis The plausible pathways by which the molecular precursors of life may have formed on the primitive Earth. Amino acids, the nitrogenous bases, and ribose phosphates can be prepared under conditions that might have prevailed on the primitive Earth. The linking together of amino acids to form polypeptides, and of nucleotides to form polynucleotides, has in principle been established.

Harold C. Urey's model of the primordial Earth postulates an atmosphere rich in methane, ammonia, water, and hydrogen. When this gas mixture is subjected to an electrical spark, analogous to the way that lightning may have initiated such syntheses 4 billion years ago, the identified products included several amino acids (glycine, alanine, and aspartic acid), the building blocks of proteins. This novel result lent credibility to a theory in which the origin of life was viewed as a cumulative, stepwise process, beginning with the gaseous synthesis of small molecules, which rained down into oceans, lagoons, and lakes. With water as a ubiquitous solvent, organic molecules could then react with one another to form larger molecules (biopolymers) and finally to assemble into primitive cells. This general scenario has guided the design of prebiotic simulations.

However, an accumulation of geophysical data and computational models has cast doubt on the relevance of the synthesis of amino acids to the primordial Earth. Hydrogen probably escaped rapidly as the Earth cooled, leaving an atmosphere in which methane and ammonia were virtually absent. As the input of hydrogen is diminished, the formation of biomolecules is inhibited. This problem has led some scientists to look for extraterrestrial sources of organic matter. For example, meteorites are known to contain a rich source of amino acids and other small biomolecules, and perhaps the infall of such cosmic bodies onto the young Earth gave life its start. Alternatively, there may have been localized environments on the Earth where methane and other hydrogen-rich precursors were abundant, such as deep-ocean hydrothermal vents, which would have been favorable for the formation of life. *See* AMINO ACIDS; HYDROTHERMAL VENT. William J. Hagan, Jr.

Precambrian A major interval of geologic time between about 540 million years (Ma) and 3.8 billion years (Ga) ago, comprising the Archean and Proterozoic eons and encompassing most of Earth history. The Earth probably formed around 4.6 Ga and was then subjected to a period of intense bombardment by meteorites so that there are few surviving rocks older than about 3.8 billion years. Ancient rocks are preserved exclusively in continental areas. All existing oceanic crust is younger than about 200 million years, for it is constantly being recycled by the processes of sea-floor spreading and subduction. Development of techniques for accurate determination of the ages of rocks and minerals that are billions of years old has revolutionized the understanding of the early history of the Earth. *See* DATING METHODS; GEOCHRONOMETRY; GEOLOGIC TIME SCALE; ROCK AGE DETERMINATION.

Detailed sedimentological and geochemical investigations of Precambrian sedimentary rocks and the study of organic remains have facilitated understanding of conditions on the ancient Earth. Microorganisms are known to have been abundant in the early part of Earth history. The metabolic activities of such organisms played a critical role in the evolution of the atmosphere and oceans. There have been attempts to apply the concepts of plate tectonics to Precambrian rocks. These diverse lines of investigation have led to a great leap in understanding the early history of the planet. *See* PLATE TECTONICS.

Rocks of the Archean Eon (2.5–3.8 Ga) are preserved as scattered small "nuclei" in shield areas on various continents. The Canadian shield contains perhaps the biggest region of Archean rocks in the world, comprising the Superior province. Much of the Archean crust is typified by greenstone belts, which are elongate masses of volcanic and sedimentary rocks that are separated and intruded by greater areas of granitic rocks. The greenstones are generally slightly metamorphosed volcanic rocks, commonly extruded under water, as indicated by their characteristic pillow structures. These structures develop when lava is extruded under water and small sac-like bodies form as the lava surface cools and they are expanded by pressure from lava within. Such structures are common in Archean greenstone assemblages in many parts of the world. *See* ARCHEAN; METAMORPHIC ROCKS.

The Proterozoic Eon extends from 2.5 Ga until 540 Ma, the beginning of the Cambrian Period and Phanerozoic Eon. Proterozoic successions include new kinds of sedimentary rocks, display proliferation of primitive life forms such as stromatolites, and contain the first remains of complex organisms, including metazoans (the Ediacaran fauna). Sedimentary rocks of the Proterozoic Eon contain evidence of gradual oxidation of the atmosphere. Abundant and widespread chemical deposits known as banded iron formations (BIF) make their appearance in Paleo-Proterozoic sedimentary basins. *See* BANDED IRON FORMATION; PROTEROZOIC. Grant M. Young

Precast concrete Concrete that has been cast into a form which is later incorporated into a structure. A concrete structure may be constructed by casting the concrete in place on the site, by building it of components cast elsewhere, or by a combination of the two. Concrete cast in other than its final position is called precast.

In contrast with cast-in-place concrete construction, in which columns, beams, girders, and slabs are cast integrally or bonded together by successive pours, precast concrete requires field connections to tie the structure together. These connections can be a major design problem.

Precast units can be standardized. Savings can then result from repeated reuse of forms and assembly-line production. Furthermore, high quality can be maintained because of the controls that can be kept on production under plant conditions. However, there is always the possibility that transportation, handling, and erection costs for the precast units will offset the savings. *See* CONCRETE; PRESTRESSED CONCRETE. Christine Beall

Precession The motion of an axis fixed in a body around a direction fixed in space. If the angle between the two is constant so that the axis sweeps out a circular cone, the motion is pure precession; oscillation of the angle is called nutation. An example of precession is the motion of the Earth's polar axis around the normal to the plane of the ecliptic; this is the precession of the equinoxes. A fast-spinning top, with nonvertical axis, which precesses slowly around the vertical direction, is another example. In both examples the precession is due to torque acting on the body. Another kind of precession, called free or fast precession, with a rate which is comparable to the rotation rate of the body, is seen, for instance, in a coin spun into the air. *See* NUTATION (ASTRONOMY AND MECHANICS); PRECESSION OF EQUINOXES.

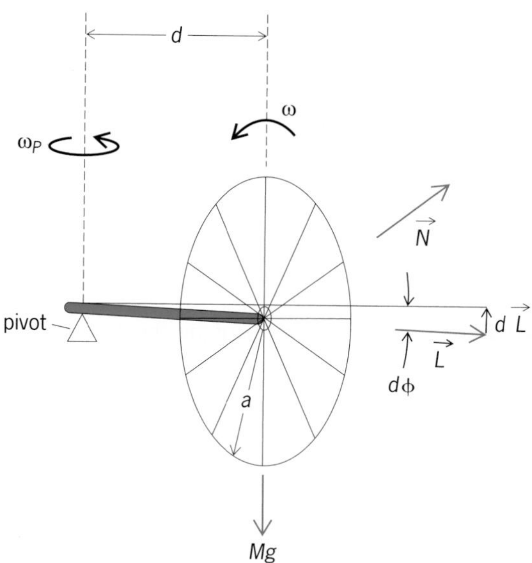

Simple precession of a rapidly spinning wheel with a horizontal axis supported by a pivot.

As a simple example of gyroscopic motion, consider a rapidly spinning wheel with a horizontal axis supported at a distance d from the plane of the wheel (see illustration). The angular momentum $\vec{L}$ is along the wheel symmetry axis and is approximately given by the angular momentum of the wheel about this axis; in the simple precession approximation to the motion, the angular momentum associated with precessional motion is neglected. The external torque $\vec{N}$ due to the gravitational force is perpendicular to the wheel axis in the horizontal plane. The change in the angular momentum $\vec{L}$ in an infinitesimal time interval dt is given by the rotational equation of motion in Eq. (1).

$$d\vec{L} = \vec{N}\,dt \qquad (1)$$

See RIGID-BODY DYNAMICS.

Since $d\vec{L}$ and $\vec{L}$ are perpendicular, the length $\vec{L}$ is unchanged to first order in dt. The direction of $\vec{L}$ is rotated counterclockwise in the horizontal plane. The angular velocity of precession ω_P about the vertical axis is then given by Eq. (2), where g is the

$$\omega_P = \frac{gd}{\omega a^2} \qquad (2)$$

gravitational acceleration, ω is the spin angular velocity, and a is the radius of the wheel. Thus ω_P is independent of the mass of the wheel and inversely proportional to ω. For ω very large, the precession rate ω_P is quite slow. *See* GYROSCOPE. Vernon D. Barger

Precession of equinoxes A slow change in the direction of the axis of rotation of the Earth and also a very small change in the orientation of the Earth's orbit around the Sun. Thus, both the celestial equator (the projection of the terrestrial equator onto the sky) and the ecliptic (the projection of the Earth's orbit onto the sky) move slowly in time and, consequently, so do their intersections, the vernal and autumnal equinoxes. *See* EARTH ROTATION AND ORBITAL MOTION; ECLIPTIC; EQUATOR; EQUINOX.

The change in direction of the Earth's axis of rotation is caused primarily by the Moon and secondarily by the Sun. The Earth is an oblate spheroid, bulging outward around the Equator by about 1 part in 300. Because the Earth's axis of rotation is tilted by about 23.4° to the plane of its orbit about the Sun, this bulge is also tilted. The extra gravitational pull of the Moon and Sun on this bulge produces a revolution, or precession, of the axis of rotation about the perpendicular to the orbit (see illustration); one complete revolution requires just under 26,000 years. This has been recognized for over 2000 years and was formerly called lunisolar precession. *See* PRECESSION.

In addition to uniform revolution, there are very small amplitude periodic oscillations; the primary one has a period of 18.6 years. These oscillations are collectively called nutation. They are caused by the same forces as precession, but are treated observationally and computationally as separate phenomena. *See* NUTATION (ASTRONOMY AND MECHANICS).

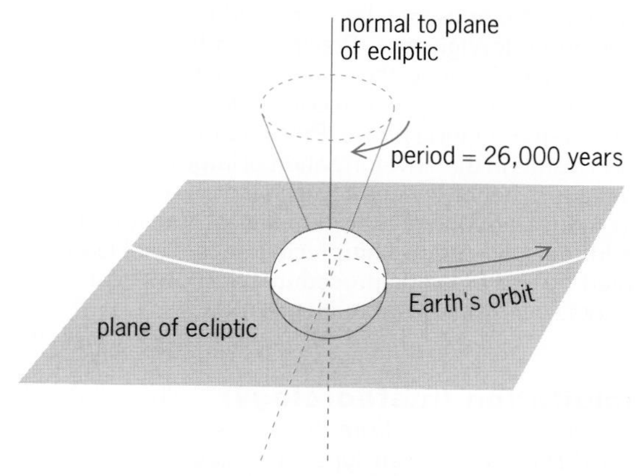

Motion of Earth's axis of rotation.

The slow change in the Earth's orbital plane (ecliptic) is caused by the gravitational perturbations of the other planets in the solar system. The principal effect is to produce an extremely small rotation of the axis of revolution of the Earth around an axis parallel to the total-angular-momentum axis of the solar system. This therefore produces a very small motion of the ecliptic itself and thus an additional small change in the apparent position of the equinox. This precession was formerly called planetary precession. E. Myles Standish; Robert S. Harrington

Precious stones The materials found in nature that are used frequently as gemstones, including amber, beryl (emerald and aquamarine), chrysoberyl (cat's-eye and alexandrite), coral, corundum (ruby and sapphire), diamond, feldspar (moonstone and amazonite), garnet (almandite, demantoid, and pyrope), jade (jadeite and nephrite), jet, lapis lazuli, malachite, opal, pearl, peridot, quartz (amethyst, citrine, and agate), spinel, spodumene (kunzite), topaz, tourmaline, turquois and zircon. *See* GEM.

The terms precious and semiprecious have been used to differentiate between gemstones on a basis of relative value. Because there is a continuous gradation of values from materials sold by the pound to those valued at many thousands of dollars per carat, and because the same mineral may furnish both, a division is essentially meaningless. Richard T. Liddicoat, Jr.

Precipitation (chemistry) The process of producing a separable solid phase within a liquid medium. In analytical chemistry, precipitation is widely used to effect the separation of a solid phase in an aqueous solution. For example, the addition of a water solution of silver nitrate to a water solution of sodium chloride results in the formation of insoluble silver chloride. Quite often, one of the components in the solution is thus virtually completely separated in a relatively pure form. It can then be isolated from the solution phase by filtration or centrifugation, and the substance determined by weighing. This procedure is known as gravimetric analysis. Precipitation may also be used merely to effect partial or complete separation of a substance for purposes other than that of gravimetric analysis. Such purposes might involve either the isolation of a relatively pure substance or the removal of undesirable components of the solution. *See* GRAVIMETRIC ANALYSIS.

The extent to which a component can be separated from solution can be determined from the solubility-product constant obtained by determining the quantity of dissolved substance present in a known amount of saturated solution. This value is known as the solubility. The solubility can be drastically altered merely by adding to the solution any of the ions that make up the precipitate, for example, by adding varying quantities of either silver nitrate or sodium chloride to a saturated solution of silver chloride. Although solubility can be altered over a wide range, the solubility product itself remains practically constant over this same range. *See* SOLUBILITY PRODUCT CONSTANT.

Various techniques may be employed in order to reduce contamination by foreign ions. Precipitation from dilute solution is often effective. Heating the reaction mixture speeds recrystallization processes by which incorporated foreign ions may be returned to the solution phase. Precipitation from homogeneous solution results in the slow formation of large crystals of small surface area and hence lessens coprecipitation. If all these methods fail to reduce adequately the quantity of foreign ions incorporated in the solid phase, the precipitate is dissolved and reprecipitated by the previous procedure. *See* CHEMICAL SEPARATION TECHNIQUES; CRYSTALLIZATION; NUCLEATION.

Louis Gordon; Royce W. Murray

Precipitation (meteorology) The fallout of water drops or frozen particles from the atmosphere. Liquid types are rain or drizzle, and frozen types are snow, hail, small hail, ice pellets (also called ice grains; in the United States, sleet), snow pellets (graupel, soft hail), snow grains, ice needles, and ice crystals. In England sleet is defined as a mixture of rain and snow, or melting snow. Deposits of dew, frost, or rime, and moisture collected from fog are occasionally also classed as precipitation. *See* HAIL; SNOW.

All precipitation types are called hydrometeors, of which additional forms are clouds, fog, wet haze, mist, blowing snow, and spray. Whenever rain or drizzle freezes on contact with the ground to form a solid coating of ice, it is called freezing rain, freezing drizzle, or glazed frost; it is also called an ice storm or a glaze storm, and sometimes is popularly known as silver thaw or erroneously as a sleet storm. *See* CLOUD; FOG.

Rain, snow, or ice pellets may fall steadily or in showers. Steady precipitation may be intermittent though lacking sudden bursts of intensity. Hail, small hail, and snow pellets occur only in showers; drizzle, snow grains, and ice crystals occur as steady precipitation. Showers originate from instability clouds of the cumulus family, whereas steady precipitation originates from stratiform clouds.

The amount of precipitation, often referred to as precipitation or simply as rainfall, is measured in a collection gage. It is the actual depth of liquid water which has fallen on the ground, after frozen forms have been melted, and is recorded in millimeters or inches and hundredths. A separate measurement is made of the depth of unmelted snow, hail, or other frozen forms. *See* SNOW GAGE. For discussions of other topics related to precipitation; CLOUD PHYSICS; DEW; DEW POINT; HUMIDITY; HYDROLOGY; HYDROMETEOROLOGY; RAIN SHADOW; VAPOR PRESSURE; WEATHER MODIFICATION. J. R. Fulks

Precipitation measurement Instruments used to measure the amount of rain or snow that falls on a level surface. Such measurements are made with instruments known as precipitation gages. A precipitation gage can be as simple as an open container on the ground to collect rain, snow, and hail; it is usually more complex, however, because of the need to avoid wind effects, enhance accuracy and resolution, and make a measurement representative of a large area. Precipitation is measured as the depth to which a flat horizontal surface would have been covered per unit time if no water were lost by runoff, evaporation, or percolation. Depth is expressed in inches or millimeters, typically per day. The unit of time is often understood and not stated explicitly. Snow and hail are converted to equivalent depth of liquid water. *See* METEOROLOGICAL INSTRUMENTATION; PRECIPITATION (METEOROLOGY); SNOW SURVEYING. Fred V. Brock

Accurate quantitative precipitation measurement is probably the most important weather radar application. It is extremely valuable for hydrological applications such as watershed management and flash flood warnings. Radar can make rapid and spatially contiguous measurements over vast areas of a watershed at relatively low cost. *See* METEOROLOGICAL RADAR; PRECIPITATION (METEOROLOGY); RADAR METEOROLOGY. Richard J. Doviak

Precipitin The visible result of the chemical interaction of antigen and antibody. Not all antibodies will result in precipitation, yet they may participate in agglutination reactions or add onto particulate antigens, and evidence for their occurrence together with precipitating antibody can be obtained for most sera. Precipitins may be noted qualitatively or be quantified by noting the end-point dilution (titer) of serum required to give a precipitate at the threshold of visibility, or the amount of antibody may be determined in milligrams or micrograms by analysis of the precipitate with correction for the antigen contained therein. *See* ANTIBODY; ANTIGEN. Donald Raum

Precipitin test A test to measure a specific reaction between antigen and antibody which results in a visible precipitate. Two types of qualitative tests are commonly used. The first is the ring test. A solution of antigen is laid over a solution of antibody or antiserum so that a sharp interface is formed. A precipitate at the interface is a positive reaction and occurs as antigen and antibody diffuse toward one another, creating a zone in which the ratio of their concentrations is optimal for precipitation. As little as 1.0 microgram of protein may be detected. This is an easy, rapid test. The second qualitative test involves mixing solutions of antigen and antibody. This is done by varying either the concentration of antibody or the concentration of antigen. When the amount of antigen is varied, three zones of precipitation are noted: antigen excess, equivalence, and antibody excess. *See* PRECIPITIN.

The quantitative determination of precipitating antibodies depends upon the two-stage nature of antigen-antibody interaction. The first stage of antigen-antibody combination reaches completion within a few minutes, while the second stage, in which aggregation of complexes occurs, may take days. To standardize these tests, it is necessary to consider time, temperature, and hydrogen ion and electrolyte concentration. Precipitates are digested and analyzed for nitrogen content, and the amount of antibody calculated. *See* IMMUNOASSAY; IMMUNOCHEMISTRY; IMMUNOLOGY.

Donald Raum

Precision agriculture The application of technologies and agronomic principles to manage spatial and temporal variability associated with all aspects of agricultural production for the purpose of improving crop performance and environmental quality. The intent of precision agriculture is to match agricultural inputs and practices to localized conditions within a field (site-specific management) and to improve the accuracy of their application. The finer-scale management of precision agriculture is in contrast to whole-field or whole-farm management strategies, where management decisions and practices are uniformly applied throughout a field or farmstead.

Successful implementation of precision agriculture requires three basic steps. First, farmers must obtain accurate maps of the spatial variability of factors (soils, plants, and pests) that determine crop yield and quality and/or factors that cause environmental degradation. Second, once known, variability can be managed using site-specific management recommendations and accurate input control technologies. Third, precision agriculture requires an evaluation component to understand the economic, environmental, and social impacts on the farm and adjacent ecosystems and to provide feedback on cropping system performance.

Precision agriculture is technology-enabled, information-based, and decision-focused, because it relies on an increasing level of detail in information acquired with technology to improve decision making in crop production. Consequently, precision agriculture will evolve as technology, information management, and decision tools emerge in this era of rapid technological advancement. *See* AGRICULTURAL SOIL AND CROP PRACTICES; AGRICULTURE; AGRONOMY; DECISION SUPPORT SYSTEM; INFORMATION SYSTEMS ENGINEERING.

F. J. Pierce; P. Nowak

Precision approach radar (PAR) A tracking system that provides a ground control approach (GCA) air-traffic controller with a precise display of an aircraft's position relative to a runway final-approach course. To ensure absolute safety, precise information is displayed on a plan position indicator (PPI). This display provides the controller with aircraft position information for control of heading and rate of descent. To accomplish this and maintain the required precision for a final-approach aid, the display shows the aircraft position in relation to range, azimuth, and elevation. The information presented on the precision approach radar display allows an air-traffic controller to direct a pilot down along a runway approach course to a precision landing. Precision radar approaches are accomplished in most weather conditions and do not require any on-board avionics equipment, such as an instrument landing system (ILS).

There are two ways to provide the required range, azimuth, and elevation information on the plan position indicator: use of two antennas, one scanning elevation and the other azimuth; and a single computer-controlled phased-array antenna that can provide pencil-beam tracking for both elevation and azimuth positions. *See* ANTENNA (ELECTROMAGNETISM); RADAR.

Mark A. DePlasco

Predator-prey interactions Predation occurs when one animal (the predator) eats another living animal (the prey) to utilize the energy and nutrients from the body of the prey for growth, maintenance, or reproduction. In the special case in which both predator and prey are from the same species, predation is called cannibalism. Sometimes the prey is actually consumed by the predator's offspring. This is particularly prevalent in the insect world. Insect predators that follow this type of lifestyle are called parasitoids, since the offspring grow parasitically on the prey provided by their mother.

Predation is often distinguished from herbivory by requiring that the prey be an animal rather than a plant or other type of organism (bacteria). To distinguish predation from decomposition, the prey animal must be killed by the predator. Some organisms occupy a gray area between predator and parasite. Finally, the requirement that both energy and nutrients be assimilated by the predator excludes carnivorous plants from being predators, since they assimilate only nutrients from the animals they consume. *See* FOOD WEB.

Population dynamics refers to changes in the sizes of populations of organisms through time, and predator-prey interactions may play an important role in explaining the population dynamics of many species. They are a type of antagonistic interaction, in which the population of one species (predators) has a negative effect on the population of a second (prey), while the second has a positive effect on the first. For population dynamics, predator-prey interactions are similar to other types of antagonistic interactions, such as pathogen-host and herbivore-plant interactions.

Community structure refers generally to how species within an ecological community interact. The simplest conception of a community is as a food chain, with plants or other photosynthetic organisms at the bottom, followed by herbivores, predators that eat herbivores, and predators that eat other predators. This simple conception works well for some communities. Nonetheless, the role of predator species in communities is often not clear. Many predators change their ecological roles over their lifetime. Many insect predators that share the same prey species are also quite likely to kill and devour each other. This is called intraguild predation, since it is predation within the guild of predators. Furthermore, many species are omnivores, feeding at different times as either predators or herbivores. Therefore, the role of particular predator species in a community is often complex.

Predator-prey interactions may have a large impact on the overall properties of a community. For example, most terrestrial communities are green, suggesting that predation on herbivores is great enough to stop them from consuming the majority of plant material. In contrast, the biomass of herbivorous zooplankton in many aquatic communities is greater than the biomass of the photosynthetic phytoplankton, suggesting that predation on zooplankton is not enough to keep these communities green. *See* POPULATION ECOLOGY.

Anthony R. Ives

Pregnancy The period during which a developing fetus is carried within the uterus. In humans, pregnancy averages 266 days (38 weeks) from conception to childbirth. Traditionally, pregnancy duration is counted from the woman's last menstrual period, which adds roughly 2 weeks to gestational age. This is how physicians arrive at a pregnancy length of 40 weeks (280 days).

The 9 months of pregnancy are typically divided into three periods (trimesters) of 3 months. The first sign of pregnancy is often the absence of an expected menstrual period. Common symptoms include nausea, breast tenderness, fatigue, and frequent urination. The diagnosis of pregnancy can be made as early as 10 days after fertilization by means of blood tests. By 6 weeks (from the last menstrual period), the uterus feels soft and is palpably enlarged. Pregnancy can be positively confirmed by observing cardiac motion of the fetus by ultrasound scanning (8 weeks) or by hearing fetal heart "tones" by using a Doppler detection instrument (10–12 weeks).

Early in the first trimester, the embryo's germ layers differentiate into organs and systems, a process that is nearly completed by the twelfth week. It is during this critical period of development that the fetus is most vulnerable to the adverse effects of drugs and other teratogenic influences. The second and third trimesters of pregnancy are characterized by increased fetal growth and gradual physiologic maturation of fetal organ systems. During this time, the maternal changes of pregnancy are greatest. The enlarging uterus encroaches on the abdominal region by the fourth month and at term nearly reaches the diaphragm. The breasts gradually enlarge in preparation for lactation. Striking cardiovascular changes, including nearly a 50% increase in cardiac output, provide the increased blood flow to accommodate the growing fetoplacental unit. Other changes in the renal, digestive, pulmonary, and endocrine systems reflect the numerous maternal adaptations that eventually must occur in a healthy pregnancy. *See* EMBRYOLOGY.

Early, regular prenatal care is associated with improved pregnancy outcome and seeks to identify risk factors in the pregnancy that may apply to mother or fetus. At 6–8 weeks, a complete physical examination, along with blood and urine analyses, should be performed. In addition to undergoing traditional tests, patients are now routinely screened for hepatitis B at the beginning of pregnancy, for fetal neural-tube defects such as open spine at 16 weeks, and for gestational diabetes at about 28 weeks. In addition to these blood tests, many physicians offer a sonogram at 16–18 weeks to establish gestational age, check for a multiple pregnancy, and screen for birth defects. During prenatal visits, a physician can evaluate nutrition, blood pressure, and fetal growth. *See* PRENATAL DIAGNOSIS.

Ideally, at the end of the third trimester, the process of labor begins. The muscles of the uterus contract, dilating the cervix and allowing the baby to begin moving into the vagina or birth canal. Continued contractions push the baby out of the mother's body. In the final stage of labor, the placenta detaches from the uterine walls and is expelled as the afterbirth. An alternative to vaginal delivery is the cesarean section, in which the baby is removed surgically through an abdominal incision.

The legal status of pregnancy termination (therapeutic abortion) varies from country to country, but about two-thirds of women in the world have access to legal abortion. Over 90% of abortions in the United States are performed in the first trimester by suction curettage, a technique that uses suctioning and removal of the uterine contents through the vagina with surgical instruments. Later pregnancies are terminated by a procedure called dilatation and evacuation (D&E) or by administration of drugs to stimulate uterine contractions. Medical and psychological sequelae to abortion are few, and are fewest for terminations in the first trimester. *See* PREGNANCY DISORDERS. Bruce D. Shephard

Pregnancy disorders Physical disorders that can arise as a consequence of pregnancy, ranging from mild to life-threatening. Extreme conditions can result in termination of the pregnancy or death of the mother.

The most common disorder of early pregnancy, persistent vomiting, is without known cause and usually subsides spontaneously within a few weeks. A more serious threat to pregnancy, vaginal bleeding (within the first 20 weeks), may be a sign of miscarriage (spontaneous abortion) or much less often, ectopic pregnancy (embryonic development outside the uterus, usually within the Fallopian tube). Over 50% of miscarriages are due to a spontaneous chromosomal abnormality in the sperm, egg, or developing embryo. Ectopics are nearly always removed surgically; the involved Fallopian tube usually can be preserved.

Some abnormal conditions in pregnancy are identified as a result of routine blood tests. Examples include rubella (German measles) and syphilis, both of which occur relatively infrequently and can cause birth defects. Rh disease is also uncommon because of prevention by prenatal blood tests and treatment with immunoglobulins in Rh-negative mothers. Routine screening is also recommended for hepatitis B. *See* HEPATITIS; PRENATAL DIAGNOSIS; RH INCOMPATIBILITY; RUBELLA; SYPHILIS.

Maternal conditions that may worsen during pregnancy include some forms of heart disease, seizure disorders, hypertensive disease, and acquired immune deficiency syndrome (AIDS). *See* ACQUIRED IMMUNE DEFICIENCY SYNDROME (AIDS).

Among the most important disorders during the second half of pregnancy are those associated with low birth weight due to either premature labor or fetal growth problems. Low birth weight is associated with twins, hypertensive disorders, smoking, and inadequate nutrition. The hypertensive disorders include pregnancy-induced hypertension (formerly known as toxemia) as well as chronic hypertension that exists before the pregnancy. Signs and symptoms of pregnancy-induced hypertension can include swelling of the hands and face, headaches, and a sudden weight gain of 5 lb (2 kg) or more in 1 week. Vaginal bleeding in late pregnancy, which is a potential emergency condition, often results from a placental problem; either the placenta is abnormally located near or over the cervix (placenta previa), or the placenta separates prematurely from the uterus (abruptio placenta). *See* HYPERTENSION.

Far less dramatic than these conditions, but equally important as a cause of fetal distress, is the "post dates" pregnancy, which occurs once pregnancy has extended 2 weeks beyond the date of expected delivery. When a pregnancy reaches 42 weeks, delivery is attempted as soon as possible as the aging placenta may lose its ability to provide adequate oxygen and nutrition.

Increased rest and proper nutrition are especially important in pregnancies complicated by high blood pressure, fetal growth problems, and twins. More immediate management often centers on the timing of delivery. *See* PREGNANCY. Bruce D. Shephard

Prehistoric technology The set of ideas that prescribe the manufacture and use of implements before written history. Technology is the principal means through which the human species has succeeded in occupying most of the world. Archeologists tend to use the term "artifact" for any material that was modified by ancient humans, whether this material was used or not, and the term "tool" for any material that was used by ancient humans, whether it was modified or not.

Archeological evidence has demonstrated that the human lineage has been making and using identifiable tools for at least 2.5–2.6 million years, beginning with stone implements and associated prehistoric remains. The advent of metalworking technologies, especially in the Old World, gradually brought about a decline in stone artifacts. By 2000 B.C., stone tools had been replaced by metal implements in much of Eurasia. Subsequent migrations of peoples and diffusion of ideas accelerated this replacement. *See* NEOLITHIC; PALEOLITHIC.

The archeological evidence for prehistoric technologies may be biased because of the nature of the raw materials that were used and the conditions of burial and preservation of these materials. In general, artifacts of stone, bone, pottery, and metal preserve fairly well in many areas, whereas artifacts of wood and other vegetable materials, skin, and horn tend to decay fairly rapidly and are normally found only in exceptional prehistoric contexts.

Primates

(*a*) Mouse lemur (*Microcebus murinus*) (*photo by David Haring/Duke Lemur Center*).

(*b*) Philippine tarsier (*Tarsius syrichta*) (*photo © 2006 William Flaxington*).

(*c*) Spider monkey (*Ateles geoffroyi*) from Panama (*photo by Roy Fontaine/WPRC AV Archives*).

(*d*) Long-tailed macaque (*Macaca fascicularis*) (*photo by Roy Fontaine/Monkey Jungle*).

(*e*) Vervet monkey, or green monkey (*Cercopithecus aethiops*) (*photo by Gary M. Stolz/U.S. Fish and Wildlife Service*).

(*f*) Guenon species, De Brazza's monkey (*Cercopithecus neglectus*) (*photo by Aaron Logan/LIGHTmatter Photography*).

In 1816, C. Thompsen, director of the Danish National Museum, began to chronologically order the museum's prehistoric collections into three major groups, based upon technology, and now called the Three Age System. The earliest was a Stone Age, followed by a Bronze Age, and finally an Iron Age. As time went on, the prehistory of Europe was further divided, based on regional sequences that could be documented through excavation. In other places, such as the Americas, Australia, Oceania, and sub-Saharan Africa, different nomenclature was often used, since the regional technological sequences differed from that of western Europe.

New technological traits can be introduced into a society in a number of ways: (1) innovation or invention, which is the development of new ideas, including new technological characteristics; (2) diffusion, which involves the spread of ideas, including technological knowledge, from one group to another; and (3) migration, which is the spread of peoples, often with new technologies, into new areas. One of the goals of a prehistorian is to try to ascertain, based on archeological evidence, which factors best explain the technological changes seen in the archeological record.

Nicholas Toth; Brooke Blades

Prehnite A mineral with the formula $Ca_2(Al,Fe^{3+})(OH)_2$ - $[Si_3AlO_{10}]$, with Al in parentheses in octahedral and Al in brackets in tetrahedral coordination by oxygens. The mineral usually occurs as stalactitic aggregates or as curved crystals, has a vitreous luster, and is yellowish green to pale green in color. Hardness is 6–$6^1/_2$ on Mohs scale; specific gravity 2.8–2.9. Common occurrences include vesicular basalts such as the Keweenaw basalts in the Upper Peninsula of Michigan, and the Watchung basalts in New Jersey. *See* SILICATE MINERALS.

Paul B. Moore

Prenatal diagnosis The identification of disease before the birth of a fetus. It often implies genetic diagnosis, but identification of anatomical defects as well as assessment of fetal functions and maturity are also considered. Some of the relatively common diseases that can be diagnosed prenatally are Tay-Sachs disease, cystic fibrosis, Duchenne's muscular dystrophy, hemophilia A, congenital adrenal hyperplasia, thalassemia, and sickle cell anemia.

Ultrasonic data have vastly improved the understanding of normal growth and development, thus permitting earlier and more accurate diagnosis of fetal disease. Fetal movements can also be observed, allowing assessment of functional well-being. *See* MEDICAL IMAGING; MEDICAL ULTRASONIC TOMOGRAPHY.

Obstetricians typically review the health history of the pregnant woman, the father-to-be, and their families in order to identify any possible heritable disorders in the families. Certain risks related to the patient's age, race, and geographic origin may be noted. Based on these assessments, genetic testing of the unborn child may be discussed or recommended. Using cells from the fetus, a prenatal genetic diagnosis can be made for at least 10% of those disorders that are known or assumed to result from a gene mutation. Such diagnosis is based on deoxyribonucleic acid (DNA) analysis or on detection of an abnormal enzyme or other protein produced by the defective gene. *See* HUMAN GENETICS.

Donald McNellis

Press fit A force fit that has negative allowance; that is, the bore in the fitted member is smaller than the shaft which is pressed into the bore. Tight fits have slight negative allowance so that light pressure is required to assemble the parts; they are used for gears, pulleys, cranks, and rocker arms. Medium force fits have somewhat greater negative allowance and require considerable pressure for assembly; they are used for fastening locomotive wheels, car wheels, and motor armatures. *See* ALLOWANCE; FORCE FIT; SHRINK FIT.

Paul H. Black

Pressure Pressure on a surface immersed in a fluid is defined as the normal force per unit area. Pressure always tends to compress the object on which it acts. This definition identifies the so-called static pressure p.

Another distinction is made between the absolute pressure, which is measured with respect to a perfect vacuum (that is, zero pressure), and the gauge pressure, which is measured with respect to the atmospheric value. In the International System of Units (SI), the pressure unit is 1 Pa (pascal), corresponding to a force of 1 N (1 newton = 1 kg m/s^2) acting on a surface area of 1 m^2. In the English system, the pressure unit is 1 psi (pound per square inch), corresponding to a force of 1 lbf acting on a surface area of 1 in.^{2}; 1 psi $\approx$ 6895 Pa. Other common units of pressure are 1 bar = 100,000 Pa, 1 atmosphere (atm) = 101,325 Pa, 1 in. $H_2O \approx$ 249 Pa, and 1 torr (1 mm Hg) $\approx$ 133 Pa. The standard atmospheric pressure at sea level is 101,325 Pa, whereas the pressure in interstellar space approaches zero. When considering mixtures of ideal gases (that is, such that each component does not influence the others), one may define the partial pressure of each component by assuming that it occupies the entire volume separately from the others; then the pressure of the mixture would be the sum of the partial pressures of all components. *See* DALTON'S LAW; FORCE; GAS; METRIC SYSTEM; PHYSICAL MEASUREMENT; UNITS OF MEASUREMENT.

In static fluids with uniform density, the pressure difference between two points at different elevations is called the hydrostatic pressure difference; it increases downward and is equal to $\rho g \Delta h$ (ρ is the density, $g \approx 9.81$ m/s$^2 \approx 32.2$ ft/s^2 is the gravitational acceleration, and Δh is the elevation difference). To define a "static" pressure within a moving fluid, one must average the normal stresses in three orthogonal directions. In addition to their static pressure, flowing fluids are characterized by their total pressure $p_o = p + {}^1/_2\, \rho V^2$ (V is the flow velocity). This is the pressure that would be achieved if the fluid were brought to rest, as would be the case if a fluid jet impacted on a wall. The term ${}^1/_2 \rho V^2$ is referred to as dynamic pressure. *See* PRESSURE MEASUREMENT.

Stavros Tavoularis

Pressure measurement The determination of the magnitude of a fluid force applied to a unit area. Pressure measurements are generally classified as gage pressure, absolute pressure, or differential pressure. *See* PRESSURE.

Pressure gages generally fall in one of three categories, based on the principle of operation: liquid columns, expansible-element gages, and electrical pressure transducers.

Liquid-column gages include barometers and manometers. They consist of a U-shaped tube partly filled with a nonvolatile liquid. Water and mercury are the two most common liquids used in this type of gage. *See* BAROMETER; MANOMETER.

There are three classes of expansible metallic-element gages: bourdon, diaphragm, and bellows. Bourdon-spring gages, in which pressure acts on a shaped, flattened, elastic tube, are by far the most widely used type of instrument. These gages are simple, rugged, and inexpensive. In diaphragm-element gages, pressure applied to one or more contoured diaphragm disks acts against a spring or against the spring rate of the diaphragms, producing a measurable motion. In bellows-element gages, pressure in or around the bellows moves the end plate of the bellows against a calibrated spring, producing a measurable motion.

Electrical pressure transducers convert a pressure to an electrical signal which may be used to indicate a pressure or to control a process. Such devices as strain gages and resistive, magnetic, crystal, and capacitive pressure transducers are commonly used to convert the measured pressure to an electrical signal. *See* PRESSURE TRANSDUCER; STRAIN GAGE.

John H. Zifcak

Pressure seal A seal used to make pressure-proof the interface (contacting surfaces) between two parts that have frequent or continual relative rotational or translational motion; such seals are known as dynamic seals, as compared with static seals. While the pressure in seals is lower than that in gaskets, the motion hinders their effectiveness so that there are more types of seals than gaskets, each type attempting to serve its environment. The materials are leather, rubber, cotton, and flax, and for piston rings, cast iron. The forms of nonmetallic seals are rectangular, V-ring, and O-ring. Cartridge seals are available for rolling-contact bearings. Special seals include carbon ring and labyrinth seals for turbines and mechanical seals for pumps. *See* GASKET. Paul H. Black

Pressure vessel A cylindrical or spherical metal container capable of withstanding pressures exerted by the material enclosed. Pressure vessels are important because many liquids and gases must be stored under high pressure. Special emphasis is placed upon the strength of the vessel to prevent explosions as a result of rupture. Codes for the safety of such vessels have been developed that specify the design of the container for specified conditions.

Most pressure vessels are required to carry only low pressures and thus are constructed of tubes and sheets rolled to form cylinders. Some pressure vessels must carry high pressures, however, and the thickness of the vessel walls must increase in order to provide adequate strength. Hydraulic and pneumatic cylinders are machine elements that are forms of pressure vessels. Y. S. Shin

Pressurized blast furnace A blast furnace operated under higher than normal pressure. The pressure is obtained by throttling the off-gas line, which permits a greater volume of air to be passed through the furnace at lower velocity and results in an increasing smelting rate. The process permits large increases in the weight of high-temperature air blown into the bottom of the furnace at lower gas velocities, thus increasing the rate of smelting and decreasing the rate of coke consumption, and also permitting smoother operation with less flue dust production through decreased pressure drop between bottom and top pressures. *See* FURNACE. Bruce S. Old

Prestressed concrete Concrete with stresses induced in it before use so as to counteract stresses that will be produced by loads. Prestress is most effective with concrete, which is weak in tension, when the stresses induced are compressive. One way to produce compressive prestress is to place a concrete member between two abutments, with jacks between its ends and the abutments, and to apply pressure with the jacks. The most common way is to stretch steel bars or wires, called tendons, and to anchor them to the concrete; when they try to regain their initial length, the concrete resists and is prestressed. The tendons may be stretched with jacks or by electrical heating.

Prestressed concrete is particularly advantageous for beams. It permits steel to be used at stresses several times larger than those permitted for reinforcing bars. It permits high-strength concrete to be used economically, for in designing a member with reinforced concrete, all concrete below the neutral axis is considered to be in tension and cracked, and therefore ineffective, whereas the full cross section of a prestressed concrete beam is effective in bending. *See* REINFORCED CONCRETE; STRESS AND STRAIN. Frederick S. Merritt

Priapulida One of the minor groups of wormlike marine animals, now regarded as a separate phylum of the animal kingdom with uncertain zoological affinities. The phylum is a small one with only two genera, *Priapulus* and *Halicryptus*.

Priapulida inhabit the colder waters of both hemispheres. They burrow in mud and sand of the sea floor, from the intertidal region to depths of 14,850 ft (4500 m).

Priapulids are small to medium-sized animals, the largest specimen attaining 6 in. (15 cm) in length. The body of *Priapulus* is made up of three distinct portions: proboscis, trunk, and caudal appendage (see illustration). Separated by a constriction from the trunk, the bulbous, introversible proboscis usually constitutes the anterior third of the body and is marked by 25 longitudinal ridges of papillae or spines. The mouth is located at the anterior end of the proboscis and is surrounded by concentric rows of teeth. The cylindrical trunk is annulated, but not segmented, and is often covered with irregularly dispersed spines and tubercles. At the posterior end of the trunk there are three openings: the anus and two urogenital apertures. Mary E. Rice

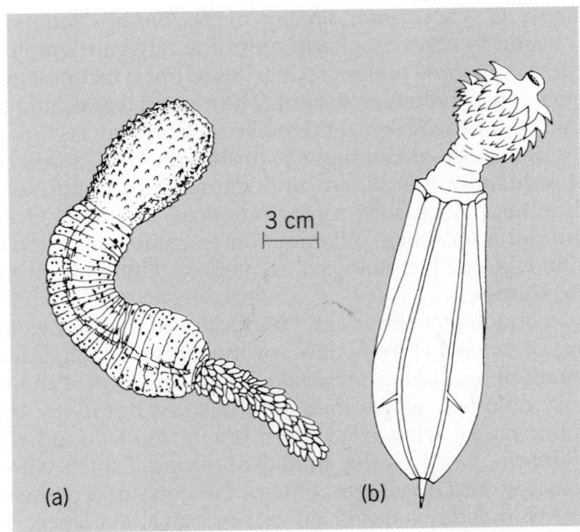

Priapulida. (*a*) *Priapulus* adult and (*b*) larva.

Prilling Solidification of droplets of molten material free-falling against an upward stream of air in a tower. It is a process used extensively in nitrogen fertilizer manufacturing. Melt is dispersed in the top of the tower at a temperature just above the solidification point of the material being processed. The dispersion arrangement, air flow, tower dimensions, and feed material are selected so the droplets approach a spherical shape and solidify before reaching the bottom of the tower. Prilling has long been a major way of agglomerating ammonium nitrate and urea because of its relative simplicity and economy. Prills of ammonium nitrate and urea are smooth, spherical, dust-free, and moderately rugged, but usually are smaller and more fragile than granulated fertilizers. *See* DRYING; FERTILIZER. E. O. Huffman

Primary vascular system (plant) The arrangement of conducting elements which serves for two-way transportation of substances between different parts of a plant. The conducting elements are of two principal kinds: xylem, which is mainly responsible for the conduction of water together with dissolved inorganic substances upward from the root to other plant organs; and phloem, which is mainly responsible for the conduction of food materials (assimilates), a flow which may take place in either direction. In the shoot region of the plant, xylem and phloem are usually associated into vascular bundles. In the root, however, they usually alternate with one another on different radii. *See* PHLOEM; XYLEM. William R. Philipson

Primates The mammalian order to which humans belong. Primates are generally arboreal mammals with a geographic distribution largely restricted to the Tropics. Unlike most other mammalian orders, the primates cannot be defined by a diagnostic suite of specializations, but are characterized by a combination of primitive features and progressive trends. These include:

1. Increased dominance of vision over olfaction, with eyes more frontally directed, development of stereoscopic vision, and reduction in the length of the snout.

2. Eye sockets of the skull completely encircled by bone.

3. Loss of an incisor and premolar from each half of the upper and lower jaws with respect to primitive placental mammals.

4. Increased size and complexity of the brain, especially those centers involving vision, memory, and learning.

5. Development of grasping hands and feet, with a tendency to use the hands rather than the snout as the primary exploratory and manipulative organ.

6. Progressive elaboration of the placenta in conjunction with longer gestation period, small litter size (only one or two infants), and precocial young.

7. Increased period of infant dependency and more intensive parenting.

8. A tendency to live in complex, long-lasting social groups.

It has been recognized for a long time that many of these features are adaptations for living in trees. However, it has been proposed more recently that primates may have developed their specializations as a consequence of being visually directed predators, living among the smaller branches of the forest canopy or undergrowth, that captured insects with their hands.

Classification of the primates is as follows:

Order Primates
- Suborder Strepsirhini
  - Infraorder Lorisiformes
    - Superfamily Lorisoidea
      - Family: Lorisidae (lorises)
      - Galagidae (bushbabies)
  - Infraorder Lemuriformes
    - Superfamily Lemuroidea
      - Family: Cheirogaleidae (dwarf lemurs)
      - Lepilemuridae (sportive lemur)
      - Lemuridae (true lemurs)
      - Indriidae (sifakas, indri, woolly lemur)
      - Daubentoniidae (aye-aye)
- Suborder Haplorhini
  - Hyporder Tarsiiformes
    - Superfamily Tarsioidea
      - Family Tarsiidae (tarsiers)
  - Hyporder Anthropoidea
    - Infraorder Platyrrhini
      - Superfamily Ceboidea
        - Family: Callitrichidae (marmosets, tamarins)
        - Cebidae (capuchins, squirrel monkeys, douroucoulis, titis)
        - Atelidae (sakis, uakaris, howler monkeys, spider monkeys, woolly monkeys)
    - Infraorder Catarrhini
      - Superfamily Cercopithecoidea
        - Family Cercopithecidae (Old World monkeys)
      - Superfamily Hominoidea
        - Family: Hylobatidae (gibbons, siamang)
        - Hominidae (orangutan, gorilla, chimpanzees, humans)

There are two major groups of primates: the strepsirhines or "lower" primates, and the haplorhines or "higher" primates. Strepsirhines have elongated and forwardly projecting lower front teeth that form a toothcomb, used for grooming the fur and for obtaining resins and gums from trees as a source of food. The digits of the hands and feet bear flattened nails, rather than claws, except for the second toe, which retains a sharp toilet claw for grooming. They also have a moist, naked rhinarium and cleft upper lip (similar to the wet noses of dogs). Most strepsirhines are nocturnal, with large eyes and a special reflective layer (the tapetum lucidum) behind the retina that intensifies images in low light. Compared with haplorhines, the brain size is relatively small and the snout tends to be longer.

The strepsirhines are subdivided into two major groups: the lorisoids, which are found throughout tropical Africa and Asia, and the lemuroids, which are restricted to Madagascar.

The lorisoids include the galagids or bushbabies (*Galago*, *Otolemur*, *Euoticus*, and *Galagoides*) and the lorisids or lorises (*Loris*, *Nycticebus*, *Perodicticus*, *Pseudopotto*, and *Arctocebus*). They are small nocturnal primates, in which the largest species, the greater bushbaby, weighs only about 1 kg (2 lb). Their diet consists mainly of a combination of insects, fruits, and gums. Lorisoids are semisolitary, living in small, dispersed social groups.

The greatest diversity of strepsirhines is found on Madagascar, where more than 30 species of lemuroids are represented, belonging to five different families.

Tarsiers, tiny primates (weighing only about 120 g) from the islands of Southeast Asia, all belong to a single genus, *Tarsius*. They are nocturnal with the largest eyes of any primate, and other adaptations for a specialized lifestyle as vertical clingers and leapers. In the past, tarsiers have been grouped together with the strepsirhines as prosimians, because they retain many primitive features lost in higher primates. However, tarsiers share a number of distinctive specializations with anthropoids that suggest that they are more closely related to each other than either is to the strepsirhines. For this reason, tarsiers and anthropoids are classified together as haplorhines.

The anthropoids include the platyrrhines or New World monkeys and the catarrhines or Old World monkeys, apes, and humans. Anthropoids are distinguished from strepsirhines and tarsiers in having a larger brain, relatively small eyes (all anthropoids are diurnal, active by day, except for the nocturnal douroucouli from South America), eye sockets almost completely enclosed by a bony septum, the two halves of the lower jaw fused in the midline rather than separated by a cartilage, small and immobile ears, the hands and feet bearing nails with no toilet claws (except for the callitrichids that have secondarily evolved claws on all fingers and toes), a single-chambered uterus rather than two-horned, and a more advanced placenta.

The platyrrhines from South and Central America are a diverse group of primates comprising more than 50 species and 16 genera. Primatologists have had a difficult time establishing a classification of platyrrhines that reflects their evolutionary interrelationships, and no consensus has been reached. There is agreement, however, that three distinct clusters can be defined: the callitrichids, the pitheciines, and the atelines. The last two groups appear to be closely related and are commonly included together in the family Atelidae. The relationships of the remaining platyrrhines are uncertain, and they are often placed together for convenience in the Cebidae.

All platyrrhines are arboreal, and they are widely distributed throughout tropical forests extending from Mexico to northern Argentina. They are small to medium-sized primates ranging from 100 g to 15 kg (0.2 to 33 lb). Platyrrhines exhibit a variety of quadrupedal locomotor types ranging from squirrellike scrambling to leaping and forelimb suspension. Atelines and capuchin monkeys are unique among primates in having a specialized prehensile tail that can grasp around branches for extra support.

The catarrhines include all anthropoid primates from Africa, Asia, and Europe. There are two main groups: the cercopithecids or Old World monkeys, and the hominoids or apes and humans. Catarrhines are distinguished from platyrrhines by a reduction in the number of premolars from three to two in each half of the upper and lower jaw, and the development of a tubelike (rather than ringlike) tympanic bone that supports the eardrum.

Old World monkeys are widely distributed throughout sub-Saharan Africa and tropical Asia, and also occur in the extreme southwestern tip of the Arabian Peninsula, northwest Africa,

Gibraltar (their only European record), and East Asia. They are a highly successful group comprising more than 80 species. They are distinguished from other anthropoids in having bilophodont molar teeth that bear a pair of transverse crests. They also have naked, roughened sitting pads on their rumps, called ischial callosities—a feature that they share with hylobatids. In addition, most Old World monkeys are highly sexually dimorphic, with males considerably larger than females.

Hominoidea is the superfamily to which apes and humans belong. Hominoids are distinguished from cercopithecoids in having primitive nonbilophodont molars, larger brains, longer arms than legs (except in humans), a broader chest, a shorter and less flexible lower back, and no tail. Many of these specializations relate to a more upright posture in apes, associated with a greater emphasis on vertical climbing and forelimb suspension.

Hominoids can be classified into two families: the Hylobatidae, which includes the gibbons and siamang, and the Hominidae, which includes the great apes and humans. The gibbons and siamang (*Hylobates*) are the smallest of the hominoids (4–11 kg or 9–24 lb), and for this reason they are sometimes referred to as the lesser apes. The nine or so species are common throughout the tropical forests of Asia. They are remarkable in having the longest arms of any primates, which are 30–50% longer than their legs. This is related to their highly specialized mode of locomotion, called brachiation, in which they swing below branches using only their forelimbs. Gibbons are fruit eaters, while the larger siamang incorporates a higher proportion of leaves in its diet. Hylobatids live in monogamous family groups in which males and females are similar in size.

The great apes include the orangutan (*Pongo*) from Asia and the gorilla (*Gorilla*) and chimpanzees (*Pan*) from Africa. These were formerly included together in their own family, the Pongidae, to distinguish them from humans, who were placed in the Hominidae. However, recent anatomical, molecular, and behavioral evidence has confirmed that humans are closely related to the great apes, especially to the African apes, and for this reason most scientists now classify them together in a single family, the Hominidae. The orangutan is restricted to the tropical rainforests of Borneo and northern Sumatra. They are large, arboreal primates that climb cautiously through the trees using all four limbs for support. Orangutans subsist mainly on fruits.

The gorilla, the largest of the hominoids, has a disjunct distribution in tropical Africa. Because of their great size, gorillas are almost entirely terrestrial, although females and young individuals frequently climb trees. Nests are often built on the ground. Gorillas move quadrupedally, and like chimpanzees, the hands are specialized for knuckle walking in which the weight of the animal is borne on the upper surface of the middle joints of the fingers. Mountain gorillas eat a variety of leaves, stems, and roots, while lowland gorillas eat a greater proportion of fruits. Groups consist of a dominant male, called a silverback, as well as several adult females, subadults, and infants.

There are two species of chimpanzees, the common chimpanzee (*Pan troglodytes*) and the bonobo or pygmy chimpanzee (*Pan panicus*). The common chimpanzee is widely distributed in the forests and woodlands stretching across equatorial Africa, while the pygmy chimpanzee is restricted to the tropical rainforests of the Congo. Both species nest and feed in trees, but they mostly travel on the ground. Common chimpanzees have eclectic diets, including meat, which they obtain by hunting small to medium-sized mammals. Tool-using behaviors are common, and more than a dozen simple tool types have been identified. Chimpanzees are gregarious and sociable, and they live in large multimale communities that divide into smaller subgroups for foraging. *See* APES; FOSSIL APES; FOSSIL HUMANS; FOSSIL PRIMATES; MAMMALIA; MONKEY. Terry Harrison

Prime mover The component of a power plant that transforms energy from the thermal or the pressure form to the mechanical form. Mechanical energy may be in the form of a rotating or a reciprocating shaft, or a jet for thrust or propulsion. The prime mover is frequently called an engine or turbine and is represented by such machines as waterwheels, hydraulic turbines, steam engines, steam turbines, windmills, gas turbines, internal combustion engines, and jet engines. These prime movers operate by either of two principles: (1) balanced expansion, positive displacement, intermittent flow of a working fluid into and out of a piston and cylinder mechanism so that by pressure difference on the opposite sides of the piston, or its equivalent, there is relative motion of the machine parts; or (2) free continuous flow through a nozzle where fluid acceleration in a jet (and vane) mechanism gives relative motion to the machine parts by impulse, reaction, or both. *See* GAS TURBINE; HYDRAULIC TURBINE; IMPULSE TURBINE; INTERNAL COMBUSTION ENGINE; POWER PLANT; REACTION TURBINE; STEAM ENGINE; STEAM TURBINE; TURBINE.

Theodore Baumeister

Primitive gut The tubular structure in embryos which differentiates into the alimentary canal. The method by which the primitive gut arises depends chiefly on the yolk content of the egg. Eggs with small or moderate amounts of yolk usually develop into spherical blastulae which invaginate at the vegetative pole to form double-walled gastrulae. The invaginated sac extends in length to become the primitive gut.

Animals such as fish, reptiles, and birds, having more yolk than can be cleaved, form flattened gastrulae consisting of three-layered blastoderms surmounting the yolk. Mammals also belong in this group, although the yolk has been lost secondarily in all except the monotremes. The head is formed by a folding of the blastoderm upon itself. The endodermal layer within the head fold becomes the pharynx. This foregut is extended by an anterior growth of the whole head and by the union of lateral endodermal folds at its posterior boundary. In most forms, the hindgut arises by a similar folding in the opposite direction, the tail fold, at the posterior end of the blastoderm. *See* CLEAVAGE (DEVELOPMENTAL BIOLOGY); GASTRULATION; OVUM. Howard L. Hamilton

Primulales An order of flowering plants, division Magnoliophyta (Angiospermae), in the subclass Dilleniidae of the class Magnoliopsida (dicotyledons). The order consists of three families: the Myrsinaceae, with about 1000 species; the Primulaceae, with about 1000 species; and the Theophrastaceae, with a little more than 100 species. These are plants with sympetalous flowers; that is, the petals are fused by their margins to form a corolla with a basal tube and terminal lobes. The functional stamens are opposite the corolla lobes, and there is a compound ovary that has a single style and two to numerous ovules which usually have two integuments and are on a free-central or basal placenta. The Myrsinaceae and Theophrastaceae are chiefly tropical and subtropical woody plants, but the Primulaceae are mostly herbaceous and are best developed in north temperate regions. Primrose (*Primula*) and cyclamen are familiar members of the Primulaceae. *See* DILLENIIDAE; MAGNOLIOPSIDA; PLANT KINGDOM.

Arthur Cronquist; T. M. Barkley

Printed circuit board The printed circuit board (PCB) has been in use for over a century, but it is still one of the most important parts of electronic devices, as the circuitry links all components together into an economical and efficiently integrated system. Once called printed wiring since they were first used to replace hand wiring in early telephone switchboards and radios, printed circuit boards are now ubiquitous. The printed circuit board consists of patterned arrays of conductors, such as copper, that are held onto and within dielectric materials, such as epoxy. But many other materials are used with dozens of processes to produce a myriad of circuit types that enable thousands of products.

The printed circuit board is at the center of modern electronics products. Every computer, Internet server, digital camera, cellular phone, television, pacemaker, and satellite has one or more printed circuit boards. Printed circuit boards have become specialized during a century of progress, and products range from tiny circuits for watches and hearing aids to extremely large ones for telecommunications and space stations.

Materials. The printed circuit board uses two basic materials: dielectrics and electrical conductors.

The dielectric must isolate the conductor array, while serving as the mechanical support platform. Although thousands of dielectrics exist, only a few are used by the printed circuit board industry. Dielectric materials can be divided into inorganic, including glass and ceramic, and organic, a much larger group of polymer-based compositions. *See* DIELECTRIC MATERIALS; POLYMER.

All printed circuits use electrical conductors to carry power, ground, and electronic signals. Copper is by far the most common conductor and the industry standard. But alloys, other metals, metal composites, and even nonmetals are used as conductors. *See* CONDUCTOR (ELECTRICITY); COPPER.

Types (material-based classification). Printed circuit boards can be divided into two categories, organic and inorganic, but three common types are recognized: organic rigid (rigid board), organic flexible (flex circuit or FPC), and ceramic.

Ceramic/inorganic are specialty circuit boards that are very robust, extremely reliable, and heat-tolerant, but they are more expensive than other types.

Circuits made from carbon-containing materials, especially plastics, were the earliest printed circuit board substrates and are the dominant materials today. They can be divided into rigid and flexible printed circuit boards.

Typified by epoxy-glass FR4, rigid circuits are the largest class of printed circuit boards (see illustration). Copper foil is applied to a lay-up of epoxy resin and glass weave, followed by heating in a press to polymerize the epoxy and bond the foil. Copper may be bonded to one side (single-sided) or both sides of the (double-sided) dielectric. The conductor patterns are generally formed by chemical etching.

Flexible circuits comprise a thin (25- to 125-μm) dielectric plastic film with copper on one or both sides. Processing may be similar to methods used for rigid printed circuit boards, but with special handling to accommodate the flexibility.

Manufacturing processes. Dozens of circuit processes have been developed, but the basic processes deal with conductor pattern formation, or image placement. The pattern-generating method is often used to classify the printed circuit board production and can determine board performance. Conductor patterns must be accurately manufactured, but vertical (z-axis) interconnects must also be fabricated for printed circuit boards with more than a one conductor layer. Ken Gilleo

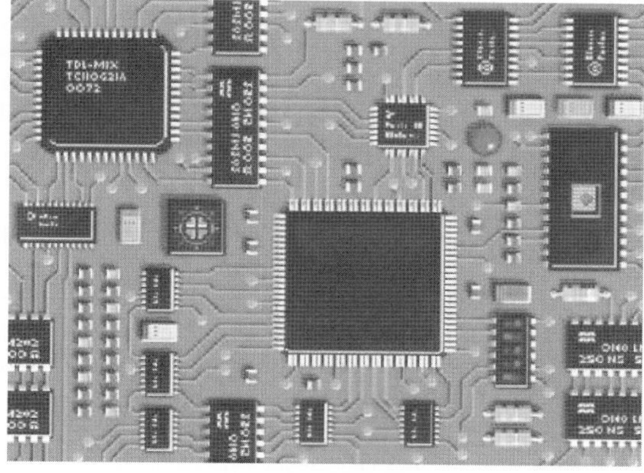

Common printed circuit board made from etched copper on epoxy-glass substrate, shown with assembled components.

Printing The processes for reproducing text and images using ink on paper. These include all the processes from typesetting through finishing of the printed product. Prepress operations for conventional plate processes consist of layout and design, typesetting, and platemaking. Traditionally, these were manual labor-intensive operations, but now most are digital, using desktop publishing systems.

There are five general printing categories: (1) relief printing, including letterpress and flexography; (2) planographic printing, including offset lithography, screenless lithography, collotype, and waterless printing; (3) intaglio, including gravure and engraving; (4) stencil and screen printing; and (5) digital printing, including electrostatic, inkjet, and laser printing.

Each printing process has specific requirements for its printing plates or image carriers. The printing press unit has a cylinder for mounting the plate and an inking system to feed ink to the plate, as well as a means for feeding paper or other substrates into the printing units and a delivery device for collecting the printed product. Digital printers differ, depending on the process; for example, electrophotographic printers are more like high-speed copiers.

After the sheets are printed, most are put through some postpress finishing operations to make a functional product. Postpress operations include folding, binding, embossing, die cutting, foil stamping, and so on.

The use of electronics and computers has rendered obsolete many techniques and led to the creation of new systems, equipment, and processes that have increased the speed precision of production, including scheduling and control; have simplified the operation of complicated systems in prepress, press, and postpress; and have improved the consistency and quality of the output of the printing processes.

Digital imaging and printing has resulted in the development of three specialized printing markets: on-demand printing, variable-data printing, and short-run color printing. On-demand printing produces documents at the time they are needed and usually within 24 hours. Variable-data printing allows for customized and personalized text or graphics. Short-run color printing allows for full-color printing in runs from 100 to 5000 impressions, which previously had been too expensive to print by conventional processes. *See* COMPUTER GRAPHICS; INK; INKJET PRINTING; PAPER; PHOTOCOPYING PROCESSES; TYPE (PRINTING). Harvey R. Levenson

Prion disease Transmissible spongiform encephalopathies in both humans and animals. Scrapie is the most common form in animals, while in humans the most prevalent form is Creutzfeldt-Jakob disease. This group of disorders is characterized at a neuropathological level by vacuolation of the brain's gray matter (spongiform change). They were initially considered to be examples of slow virus infections. Experimental work has consistently failed to demonstrate detectable nucleic acids—both ribonucleic acid (RNA) and deoxyribonucleic acid (DNA)—as constituting part of the infectious agent. Contemporary understanding suggests that the infectious particles are composed predominantly, or perhaps even solely, of protein, and from this concept was derived the acronym prion (proteinaceous infectious particles). Also of interest is the apparent paradox of how these disorders can be simultaneously infectious and yet inherited in an autosomal dominant fashion (from a gene on a chromosome other than a sex chromosome).

Disorders. Scrapie, which occurs naturally in sheep and goats, was the first of the spongiform encephalopathies to be described. An increasing range of animal species have been recognized as occasional natural hosts of this type of disease. Bovine spongioform encephalopathy, commonly known as mad cow disease, has been epidemic in British cattle. The first confirmed cases were reported in late 1986. By early 1995 it had been identified in almost 150,000 cattle and more than half of all British herds. Its exact origin is not known, but claims that it came from sheep are now discredited.

So far, animal models have indicated that only central nervous system tissue has been shown to transmit the disease after oral ingestion—a diverse range of other organs, including udder, skeletal muscle, lymph nodes, liver, and buffy coat of blood (white blood cells) proving noninfectious.

The currently recognized spectrum of human disorders encompasses kuru, Creutzfeldt-Jakob disease, Gerstmann-Straussler-Scheinker disease, and fatal familial insomnia. All, including familial cases, have been shown to be transmissible to animals and hence potentially infectious; all are invariably fatal with no effective treatments currently available.

Human-to-human transmission. A variety of mechanisms of human-to-human transmission have been described. Transmission is due in part to the ineffectiveness of conventional sterilization and disinfection procedures to control the infectivity of transmissible spongiform encephalopathies. Numerically, pituitary hormone–related Creutzfeldt-Jakob disease is the most important form of human-to-human transmission of disease. However, epidemiological evidence suggests that there is no increased risk of contracting Creutzfeldt-Jakob disease from exposure in the form of close personal contact during domestic and occupational activities. Incubation periods in cases involving human-to-human transmission appear to vary enormously, depending upon the mechanism of inoculation. Current evidence suggests that transmission of Creutzfeldt-Jakob disease from mother to child does not occur. Two important factors pertaining to transmissibility are the method of inoculation and the dose of infectious material administered. A high dose of infectious material administered by direct intracerebral inoculation is clearly the most effective method of transmissibility and generally provides the shortest incubation time. *See* BRAIN; MUTATION; NERVOUS SYSTEM DISORDERS; SCRAPIE; VIRUS INFECTION, LATENT, PERSISTENT, SLOW. Colin L. Masters; Steven J. Collins

Prism A polyhedron of which two faces are congruent polygons in parallel planes, and the other faces are parallelograms (see illustration). The bases B are the congruent polygons; the lateral faces are the parallelograms; the lateral edges are the edges not lying in the bases; and the perpendicular distance between the bases is the altitude h. Sections parallel to the bases are congruent to the bases. A prism is a right prism if its lateral edges are perpendicular to the bases; an oblique prism otherwise. A prism is called a triangular prism if its bases are triangles; a pentagonal prism if its bases are pentagons; and a parallelepiped if its bases are parallelograms. The volume of any prism is equal to the area of its base times its altitude ($V = Bh$). *See* POLYHEDRON. J. Sutherland Frame

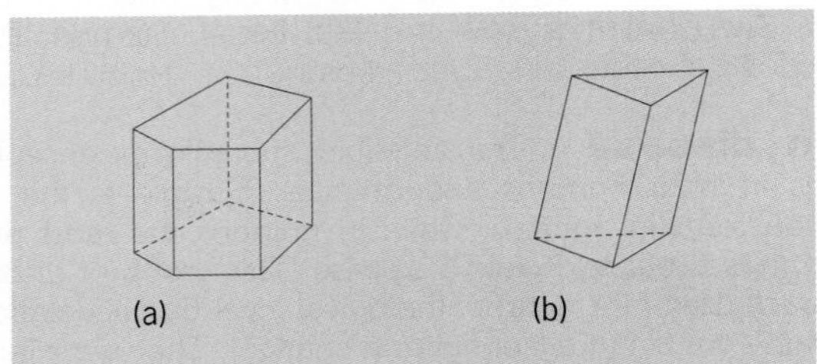

Prism configurations. (*a*) Right. (*b*) Oblique.

Prismatic astrolabe A surveying instrument used to make the celestial observations needed in establishing an astronomical position. The instrument (see illustration) consists of an accurate prism, a small pan of mercury to serve as an artificial horizon, an observing telescope with two eyepieces of different power, level bubbles and leveling screws, a magnetic compass and azimuth circle, adjusting screws, flashlight-battery power source, light, and a rheostat to control the intensity of illumination.

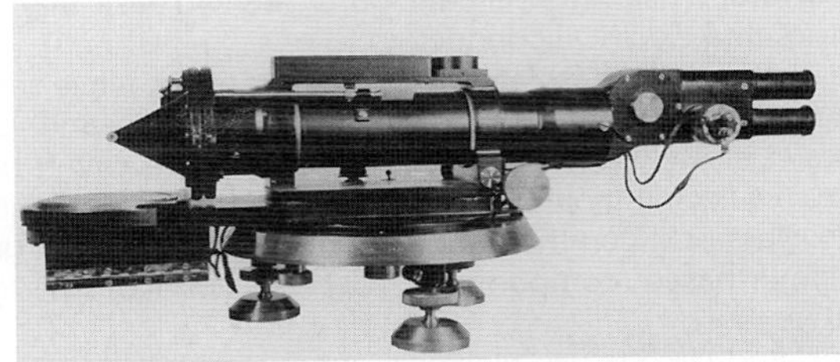
A prismatic astrolabe, used to make celestial observations. (*U.S. Naval Oceanographic Office*)

By using a fixed prism, the instrument measures a fixed altitude, usually 45°. As a rising star increases altitude past that for which the instrument was constructed, the direct image appears to move upward from the bottom of the field of vision to the top. The image reflected by the mercury horizon appears to move downward from top to bottom. At the established altitude the rays produce images at the center of the field of view. A fixed altitude is used to minimize error due to variations from standard atmospheric refraction. Each accurately timed observation provides one line of position. Alton B Moody

Pristiformes An order of batoid fishes comprising one family (Pristidae), two genera, and seven species and commonly known as sawfishes. Sawfishes are superficially similar to sawsharks (order Pristiophoriformes) in having an elongate sharklike body, sharklike locomotion, two distinct dorsal fins, a distinct caudal fin, and a long bladelike rostrum armed on both sides with large teeth but their fundamental structure is that of other batoids. *See* BATOIDEA; ELASMOBRANCHII; PRISTIOPHORIFORMES.

Sawfishes are found in tropical and subtropical shallow coastal waters and occasionally enter freshwater. The saw is used to grub in the sand and mud for bottom-dwelling invertebrates. Also, the toothed blade is used to slash to and fro in schools of small fish, stunning or killing them. The maximum length of the smallest species is 1.4 m (4.6 ft); the largest is 7.6 m (25 ft), possibly more, placing them among the largest elasmobranchs. Herbert Boschung

Pristiophoriformes An order of sharks comprising one family (Pristiophoridae) and two genera, commonly known as sawsharks. Sawsharks are superficially similar to sawfishes, which are rays, but are distinguishable from all other elasmobranchs by the following combination of characters: snout shaped as a long flat blade bearing alternate large and small teeth on each side that are weakly embedded and periodically replaced; one pair of long barbels on the underside of the blade; no visible dorsal fin spines; anal fin absent; large spiracles; and ovoviviparous, with embryos feeding solely on yolk. *See* ELASMOBRANCHII; PRISTIFORMES; SELACHII.

Sawsharks occur on continental and insular shelves and slopes in temperate and tropical waters of the western Atlantic (around the Bahamas, Florida, and Cuba), the southwestern Indian Ocean off South Africa, and the western Pacific from southern Australia to Japan, rarely entering estuaries. The genus *Pliotrema* (one species) has a maximum length of 170 cm (67 in.). *Pristiophorus* (four described species) has a maximum length of 80 to 137 cm (32 to 54 in.), depending on the species. Sawsharks cruise the bottom feeding on bony fishes, shrimps, and squids. Herbert Boschung

Private branch exchange A customer premises communications system that provides its station subscribers with a wide variety of telephony services and access to public switched telephone network (PSTN) trunk circuits provided by local exchange carrier (LEC) and interexchange carrier (IEC) services. Standard private branch exchange (PBX) system hardware elements include the common equipment (cabinets, shelf carriers, printed circuit boards) and telephone instruments. Peripheral servers may also be included in the system design configuration to support systems management and voice messaging requirements. A PBX system is privately owned and physically located at the user subscriber's premises. Telephony services provided by a PBX range from basic communications operations (provisioning of dial tone and switched connections between peripheral endpoints) to station user-controlled call processing features (such as call transfer, call forwarding, and call waiting) to advanced software applications (such as multisystem networking and call contact centers).

A defining attribute of a PBX system is that trunk circuits are segmented into customer-defined groups for incoming and outgoing call requirements, and allocated to station subscribers as pooled resources. Specific trunk routes and individual circuits are assigned on a call-by-call basis based on the PBX's automatic route selection (ARS) feature as programmed by the system administrator. PBX subscribers have no control over the trunk circuit they use.

Elements. There are several architecture design elements common to all PBXs, regardless of underlying system technology (analog or digital transmission, circuit-switched or packet-switched): common control elements, the switching network, service circuit boards, and port interface circuit cards.

Common control elements include, at minimum, a call processor and memory storage for generic feature software and the customer database.

The switching network used to connect PBX peripheral endpoints may be a circuit-switching system integrated into proprietary common equipment cabinets or an external Ethernet LAN (local-area network) packet-switched network. *See* DATA COMMUNICATIONS; LOCAL-AREA NETWORKS; PACKET SWITCHING.

Service circuits include tone senders (such as dial and busy), dual-tone multiple-frequency (DTMF) detectors for push-button analog telephones, registers to temporarily store dialed digits, and input/output (I/O) interfaces for peripheral devices such as printers. *See* COMPUTER PERIPHERAL DEVICES; TELEPHONE.

Port interface circuit cards are required to support station equipment [analog, digital, ISDN BRI (Integrated Services Digital Network Basic Rate Interface), IP (Internet Protocol)] and off-premises trunk circuits [GS/LS (Ground Start/Loop Start), DID (Direct Inward Dialing), E&M (Ear and Mouth), T1-interface, ISDN BRI/PRI (Primary Rate Interface), Auxiliary, FX (Foreign Exchange)]. *See* INTEGRATED SERVICES DIGITAL NETWORK (ISDN).

System design. PBX systems can be cost-effectively designed to support relatively small port capacity requirements (less than 20 station subscribers) or extremely large configuration needs (tens of thousands of station subscribers). PBX systems are the dominant system for enterprise customers with port size requirements above 80 stations.

Evolution of digital PBX. The two major PBX technology breakthroughs during the 1970s were computer-stored program control and digital circuit switching. Important PBX system design advances during the 1980s included stackable modular cabinets, dispersed processing, and digital telephone instruments with data terminals interfaces. Since the mid-1990s, the most important PBX technology innovation has been the integration of Internet Protocol (IP) communications standards into the core system design. *See* INTERNET; SWITCHING SYSTEMS (COMMUNICATIONS).

IP telephony PBX system designs. There are three basic IP telephony PBX system designs: IP-enabled, converged, and client-server (sometimes referred to as softswitch). IP-enabled PBXs are communications systems originally based on a circuit-switched network design that are upgraded to support IP telephony requirements using fully integrated media gateway port interface circuit cards to support IP station equipment and/or voice over IP (VoIP) trunk circuits. Converged IP-PBXs are communications systems based on either a traditional common control cabinet or a LAN-connected call telephony server. Client-server IP-PBXs are communications system based on a LAN-connected call telephony server with direct call control signaling to IP telephone instruments. *See* VOICE OVER IP. Allan Sulkin

Probability Although probability theory derives its notion and terminology from intuition, a vague statement such as "John will probably come" is as remote from it as the statement "John is forceful and energetic" is remote from mechanics. Probability theory constructs abstract models, mostly of a qualitative nature, and only experience can show whether these reasonably describe laws of nature or life. As always in mathematics, only logical relations and implications enter the theory, and the notion of probability is just as undefinable (and as intuitive) as are the notions of point, line, or mass.

The sample space. One speaks of probabilities only in connection with conceptual (not necessarily performable) experiments and must first define the possible outcomes. It is necessary to distinguish between elementary (indivisible) and compound outcomes or events. Each elementary outcome is called sample point; their aggregate is the sample space. The conceptual experiment is defined by the sample space, and it must be introduced and established at the outset.

Events. In examining a bridge hand, one may ask whether it contains an ace or satisfies some other condition. In principle each such event may be described by specifying the sample points which do satisfy the stipulated condition. Thus every compound event is represented by an aggregate of sample points, and in probability theory these terms are synonymous. The standard notations of set theory are used to describe relations among events. *See* SET THEORY.

Given an event A one may consider the case that A does not occur. This is the negation or complement of A, denoted by A'; it consists of those sample points that do not belong to A. Given two events A and B, the event C that either A or B or both occur is the union of A and B and denoted by $C = A \cup B$. In particular $A \cup A'$ is the whole sample space $\mathfrak{S}$ which therefore represents certainty. The event D, both A and B occur, is the intersection of A and B and written $D = A \cap B$. It consists of the points common to A and B.

Probabilities in finite spaces. If the sample space $\mathfrak{S}$ contains only N points $E_1, \ldots, E_N$ their probabilities may be any numbers such that $P\{E_j\} \geqq 0$ and $P\{E_1\} + \cdots + P\{E_N\} = 1$.

The probability $P\{A\}$ of an event A is the sum of the probabilities of all points contained in A; thus $P\{\mathfrak{S}\} = 1$.

Frequently considerations of symmetry lead one to consider all E_j as equally likely; that is, to set $P(E_j) = 1/N$. In this case $P(A) = n/N$ where n is the number of points in A; for a gambler betting on A, these represent the "favorable cases." For example, in throwing a pair of "perfect" dice, one naturally assumes that the 36 possible outcomes are equally likely. This model does not lose its justification or usefulness by the fact that actual dice do not live up to it. The assumption of perfect randomness in games, card shuffling, industrial quality control, or sampling is rarely realized, and the true usefulness of the model stems from the experience that noticeable departures from the ideal scheme lead to the detection of assignable causes and thus to theoretical or experimental improvements.

Conditional probability–independence. Suppose that a population of N people includes N_A color-blind persons and N_H females. To the event A "a randomly chosen person is color-blind" can be ascribed probability $P\{A\} = N_A/N$, and similarly for the event H that a person be female one has $P\{B\} = N_H/N$. If N_{AH} is the number of color-blind females, the ratio N_{AH}/N_H may be interpreted as probability that a randomly chosen female is color-blind; here the experiment "random choice in the population" is replaced by a selection from the female subpopulation. In the original experiment, N_{AH}/N is the probability of the simultaneous occurrence of both A and H, so that $N_{AH}/N_H = P\{A \cap H\}/P\{H\}$. Similar situations occur so frequently that it is convenient to define the conditional probability of the event A relative to H by Eq. (1). This concept is useful whenever it is desired to

$$P\{A \mid H\} = \frac{P\{A \cap H\}}{P\{H\}} \tag{1}$$

restrict the consideration to those cases where the event H occurs (or where the hypothesis H is fulfilled). Thus, in betting on an event A the knowledge that H occurred would induce one to replace $P\{A\}$ by $P\{A \mid H\}$.

Independent trials. The intuitive frequency interpretation of probability is based on the concept of experiments repeated under identical conditions; a theoretical model for this concept can be developed.

Consider an experiment described by a sample space $\mathfrak{S}$; for simplicity of language it can be assumed that $\mathfrak{S}$ consists of finitely many sample points $E_1, \ldots, E_N$. When the same experiment is performed twice in succession, the thinkable outcomes are the N^2 pairs of sample points (E_1, E_1), $(E_1, E_2), \ldots, (E_N, E_N)$, and these now constitute the new sample space. It is called the combinatorial product of $\mathfrak{S}$ by itself and denoted by $\mathfrak{S} \times \mathfrak{S}$.

Probabilities must be assigned to the events in $\mathfrak{S} \times \mathfrak{S}$. If the second trial is independent of the first, the probabilities in $\mathfrak{S} \times \mathfrak{S}$ follow the productive rule $P\{E_i, E_j\} = P\{E_i\}P\{E_j\}$.

In the case of n tossings of a coin, this rule leads to the probability 2^{-n} for each sample point in agreement with the requirement of equally likely cases. In the more general case of Bernoulli trials, each trial results in success S or failure F, and $P\{S\} = p$, $P\{F\} = q$ where $p + q = 1$. (This may be considered as the model of a skew coin.) A succession of n independent trials of this kind leads to the sample space of n-tuples $(SFFS \cdots FS)$, and the probability of such a point is the product $(pqqp \cdots qp)$ obtained on replacing each S by p and each F by q.

Markov chains. Markov chains represent an important scheme for dependent trials. Suppose that at each trial the possible outcomes are $E_1, \ldots, E_N$ and that whenever E_i occurs the conditional probability of E_j at the next trial is p_{ij}, independently of what happened at the preceding trials. Here, of course, $p_{ij} \geqq 0$ and $p_{i1} + p_{i2} + \cdots + p_{iN} = 1$ for each i. The p_{ij} are called transition probabilities. The whole process is now determined if the initial probabilities, π_i, at the first trial are known. For example, $P\{E_a\, E_b\, E_c\} = \pi_a p_{ab} p_{bc}$. The probability of the event "E_c at the third trial" is obtained by summation over all a and b, and so on. Markov chains, and their analog with continuous time, represent the simplest type of stochastic process. *See* STOCHASTIC PROCESS.

Random variables and their distributions. The theory of probability traces its origin to gambling, and the gambler's gain may still serve as the simplest example of a random variable. With every possible outcome (sample point) there is associated a number, namely, the corresponding gain. In other words, the gain is a function on the sample space, and such functions are called random variables. With the same experiment, one may associate many random variables.

Every random variable X has a distribution function $F(t) = P\{X \leqq t\}$. If X assumes only finitely many values, then $F(t)$ is a step function. The notion of independence carries over: Two random variables X and Y are independent if $p(X \leqq x, Y \leqq t\} = P(X \leqq s\} \cdot P\{Y \leqq t)$.

Expectations. Given a random variable X one may interpret its distribution function $F(t)$ as describing the distribution of a unit mass along the real axis such that the interval $a < x \leqq b$ carries mass $F(b) - F(a)$. In the case of a discrete variable assuming the values $x_1, x_2, \ldots$ with probabilities $p_1, p_2, \ldots$ the entire mass is concentrated at the points x_i; if $F'(x) = f(x)$ exists, it represents the ordinary mass density as defined in mechanics. The center of gravity of this mass distribution is called the expectation of X; the usual symbol for it is $E(X)$, but physicists and engineers use notations such as $\langle X\rangle$, $\langle X\rangle_{Av}$, or X. In the cases mentioned, $E(X)$ is given by Eqs.(2) and (3).

$$E(X) = \Sigma P_i x_j \tag{2}$$

$$E(X) = \int_{-\infty}^{+\infty} x f(x)\, dx \tag{3}$$

Before discussing the significance of the new concept, a few frequently used definitions are appropriate. Put $m = E(X)$. Then $(X - m)^2$ is, of course, a random variable. In mechanics, its expectation represents the moment of inertia of the mass distribution. In probability, it is called variance of X, given by Eq. (4). Its positive root is the standard deviation.

$$\mathrm{Var}(X) = E(X - m)^2 = E(X^2) - m^2 \tag{4}$$

The variance is a measure of spread: It is zero only if the entire mass is concentrated at the point m, and it increases as the mass is moved away from m. In the case of two variables X_1 and X_2 with expectations m_1 and m_2 it is necessary to consider not only the two variances $s_i^2 = E[(X_i - m_i)^2]$ but also the covariance $\mathrm{Cov}(X_1,X_2) = E[(X - m_1)(X_2 - m_2)] = E(X_1X_2) = m_1m_2$. The covariance divided by s_1s_2 is called the correlation coefficient of X_1, and X_2. If it vanishes, X_1 and X_2 are called uncorrelated. Every pair of independent variables is uncorrelated, but the converse is not true.

Laws of large numbers. To explain the meaning of the expectation and, at the same time, to justify the intuitive frequency interpretation of probability, consider a gambler who at each trial may gain the amounts $x_1, x_2, \ldots, x_n$ with probabilities $p_1, p_2, \ldots, p_n$. The gains at the first and second trials are independent random variables X_1, X_2 with the indicated distribution and the common expectation $m = \Sigma p_i x_i$. The event that an individual gain equals x, as probability p_i, and the frequency interpretation of probability leads one to expect that in a large number n of trials this event should happen approximately np_i times. If this is true, the total gain $S_n = X_1 + X_2 + \cdots + X_n$ should be approximately nm; that is, the average gain $(l/n)S_n$ should be close to m. The law of large numbers in its simplest form asserts this to be true. *See* GAME THEORY; STATISTICS.

William Feller

Problem solving (psychology)

Problems represent gaps between where one is and where one wishes to be, or between what one knows and what one wishes to know. Problem solving is the process of closing these gaps by finding missing information, reevaluating what is already known or, in some cases, redefining the problem.

Psychologists investigate the complexity of human behavior and seek to understand its nature and action. Pioneers realized that behavior was not an isolated response to objects and forces in the environment, but rather represented a systematic adaptation to specific environmental demands and opportunities. One kind of behavior that is created in response to the environment is problem solving. Research into this behavior has attempted to discover the information that is processed and the techniques for processing it.

People solve problems by first constructing an internal representation, called a problem space. This representation contains the information that must be processed in order for a problem to be solved. The problem space contains knowledge of the initial state and the goal state of the problem as well as possible intermediate states that must be searched in order to link up the beginning and the end of the task. Transitions between one state and another are achieved by selecting and applying operations that process information from memory or the task, in order to reduce the differences between what is given and what is to be achieved.

The structure of a given task helps determine the size and complexity of the problem space that is created to solve it. Characteristics of the human information-processing system, such as limited attention and a fallible memory, constrain the problem space so that the actual set of possibilities considered is usually smaller than the total set available. Human problem solvers attempt to limit and redefine a task so that the problem space to be attacked is a simpler version of the one that could be described. The fact that people tend to see new situations in terms of their own past experience enables them to cope with problem complexity. *See* INFORMATION PROCESSING (PSYCHOLOGY); LEARNING MECHANISMS; MEMORY. Paul E. Johnson

Proboscidea An order of placental mammals containing the largest and most powerful living terrestrial mammals—the elephants. Proboscideans once had a wide distribution, reaching every continent except Australia and Antarctica. Although early forms lacked both significant tusks and a proboscis, they are still formally considered proboscideans. *See* ELEPHANT; MAMMALIA.

The African elephant (*Loxodonta africana*) historically occurred throughout Africa from the Mediterranean Sea to the Cape of Good Hope, except in parts of the Sahara and some other desert regions. Most researchers agree that there are two kinds of African elephant: a smaller, darker forest elephant found in the tropical rainforests of West and Central Africa and a larger, paler bush or savannah elephant in the remainder of the current range. Some feel that these two forms should be regarded as separate subspecies—*L. a. cyclotis* for the forest elephant and *L. a. africana* for the bush elephant; others feel that the forest elephant should be a separate species.

The Asiatic or Indian elephant (*Elephas maximus*) historically occurred from Syria and Iraq east across Asia to Indochina, the Malay Peninsula, China, Sri Lanka, Sumatra, and possibly Java.

A third member of this order is the woolly mammoth (*Mammuthus primigenius*). It once occurred throughout most of Europe and the northern halves of North America and Asia. Mammoths survived up to about 3700 years ago on Wrangel Island in the Arctic Ocean off extreme northeastern Siberia. Mammoths had a dense covering of hair and small ears (both adaptations to prevent heat loss), and a trunk that ended in two fingerlike projections at its tip.

Proboscideans have a proboscis (trunk), enlarged incisors that form paired tusks, the largest ears of any mammal, and pillarlike legs whose structure prevents the elephant from jumping even a short distance. The nose and upper lip are drawn into the long trunk, which is used as a hand. The trunk measures about 6 ft (2 m) in length and weighs about 300 lb (140 kg). It is a boneless mass of flesh and muscles that bends easily. The elephant smells, drinks, and feeds itself with its trunk. It depends more on smell than on any other sense. Elephants feel with the sensitive tips of their trunks.

Wild elephants live in herds of 10 to 50 or more. The leader of the herd is usually a female in this matriarchal society. They may travel long distances in search of food or water or to escape pests. They are strictly herbivorous and feed on grasses, leaves, small branches, bark, berries, mangoes, coconuts, corn, and sugarcane. Donald W. Linzey

Procellariiformes A large order of strictly marine birds found far offshore except when breeding. The procellariiforms, or tube-nosed swimmers, are most closely related to their descendant group, the penguins. The Procellariiformes comprise four families: Diomedeidae (albatrosses; 13 species; southern and Pacific oceans); Procellariidae (shearwaters, petrels, and fulmars; 66 species; worldwide); Hydrobatidae (storm petrels; 21 species; worldwide); and Pelecanoididae (diving petrels; 4 species; subantarctic oceans).

The tube-nosed swimmers are characterized by having their nostrils enclosed in a tube, which is paired in albatrosses; dense plumage; webbed feet; and long wings. They range in size from the sparrow-sized storm petrels to large albatrosses, which have the greatest wingspan of all living birds, up to 12 ft (3.7 m). The procellariiforms are excellent fliers, as evidenced by the migratory—actually nomadic wandering—flights of many thousands of miles. Procellariiforms swim well, and the pelecanoidids and some shearwaters can dive under water, using their wings for propulsion. They are highly pelagic (oceanic) and feed on fish, squids, and crustaceans, which are concentrated into oils and waxy esters in their stomach for transport and feeding to their young, which can digest waxes. Procellariiforms have a well-developed olfactory sense to locate food and apparently to locate their nesting burrows at night. The larger species mature slowly; some albatrosses begin breeding only after reaching 6 to 8 years of age. Tight pair bonds are formed during courtship. *See* AVES. Walter J. Bock

Process control A field of engineering dealing with ways and means by which conditions of processes are brought to and maintained at desired values, and undesirable conditions are avoided as much as possible. In general, a process is understood to mean any system where material and energy streams are made to interact and to transform each other. Examples are the generation of steam in a boiler; the separation of crude oil by fractional distillation into gas, gasoline, kerosine, gas-oil, and residue; the sintering of iron ore particles into pellets; the polymerization of propylene molecules for the manufacture of polypropylene; the incineration of waste; and the roasting of coffee beans. In a wider sense, process control also encompasses determining the desired values.

Process control includes a number of basic functions, which can be arranged as follows:

Scheduling
Mode setting
Mode correction and quality control
Regulatory control Sequence control
Coping with faults

Computerized instrumentation has revolutionized the interaction with plant personnel, in particular the process operators. Traditionally, the central control room was provided with long panels or consoles, on which alarm lights, indicators, and recorders were mounted. Costs were rather high, and surveyability was poor. In computerized instrumentation, visual display units are the main components for information presentation in the human-machine interfaces. They can provide information in a concise and flexible way, adapted to human needs and capabilities. *See* AUTOMATION; CONTROL SYSTEMS. Gunnar Johannsen

Process engineering A branch of engineering in which a process effects chemical and mechanical transformations of matter, conducted continuously or repeatedly on a substantial scale. Process engineering constitutes the specification, optimization, realization, and adjustment of the process applied to manufacture of bulk products or discrete products. Bulk products are those which are homogeneous throughout and uniform in properties, are in gaseous, liquid, or solid form, and are made in separate batches or continuously. Examples of bulk product processes include petroleum refining, municipal water purification, the manufacture of penicillin by fermentation or synthesis, the forming of paper from wood pulp, the separation and crystallization of various salts from brine, the production of liquid oxygen and nitrogen from air, the electrolytic beneficiation of aluminum, and the manufacture of paint, whiskey, plastic resin, and so on. Discrete products are those which are separate and individual, although they may be identical or very nearly so. Examples of discrete product processes include the casting, molding, forging, shaping, forming, joining, and surface finishing of the component piece parts of end products or of the end products themselves. Processes are chemical when one or more essential steps involve chemical reaction. Almost no chemical process occurs without many accompanying mechanical steps such as pumping and conveying, size reduction of particles, classification of particles and their separation from fluid streams, evapor- ation and distillation with attendant boiling and condensation, absorption, extraction, membrane separations, and mixing. *See* DIALYSIS; DISTILLATION; EVAPORATION; EXTRACTION; ION-SELECTIVE MEMBRANES AND ELECTRODES; MECHANICAL CLASSIFICATION; MECHANICAL SEPARATION TECHNIQUES; MIXING; OPTIMIZATION; PRODUCTION ENGINEERING. Edward F. Leonard

Prochirality The property displayed by a prochiral molecule or a prochiral atom (prostereoisomerism). A molecule or atom is prochiral if it contains, or is bonded to, two constitutionally identical ligands (atoms or groups), replacement of one of which by a different ligand makes the molecule or atom chiral. Examples are shown below.

(1) H_1—Cα—H_2 with OH and CH_3

(2) CO_2H, H—C—OH, H—C—OH, CO_2H

(3) $H_2C{=}C{=}C(H)Cl$

(4)

(5) CO_2H, H_1—Cα—H_2, H—C—Br, CH_3

None of molecules **1–4** is chiral, but if one of the underlined pair of hydrogens is replaced, say, by deuterium, chirality results in all four cases. In compound **1**, ethanol, a prochiral atom or center can be discerned (Cα:CH_2); upon replacement of H by D, a chiral atom or center is generated, whose configuration depends on which of the two pertinent atoms (H_1 or H_2) is replaced. Molecule **5** is chiral to begin with, but separate replacement of H_1 and H_2 (say, by bromine) creates a new chiral atom at Cα and thus gives rise to a pair of chiral diastereomers. No specific prochiral atom can be discerned in molecules **2–4**, which are nevertheless prochiral (**3** has a prochiral axis).

Faces of double bonds may also be prochiral (and give rise to prochiral molecules), namely, when addition to one or other of the two faces of a double bond gives chiral products.

Although the term prochirality is widely used, especially by biochemists, a preferred term is prostereoisomerism. This is because replacement of one or other of the two corresponding ligands (called heterotopic ligands) or addition to the two heterotopic faces often gives rise to achiral diastereomers without generation of chirality. Thus not all compounds which display prostereoisomerism also display prochirality. *See* MOLECULAR ISOMERISM; STEREOCHEMISTRY. Ernest L. Eliel

Prochlorophyceae A class of prokaryotic organisms coextensive with the division Prochlorophycota in the kingdom Monera. Because prochlorophytes carry out oxygen-evolving photosynthesis, they may be considered algae. They are distinguished from Cyanophyceae, the only other prokaryotic algae, by the presence in their photosynthetic lamellae of chlorophyll *b* in addition to chlorophyll *a* and the absence of phycobilin pigments. Otherwise, they resemble Cyanophyceae biochemically and ultrastructurally. The class comprises a single genus, *Prochloron*, with one species. *See* ALGAE; CYANOPHYCEAE.

Paul C. Silva; Richard L. Moe

Procyon The brightest star in the constellation Canis Minor, apparent magnitude +0.36. Procyon (α Canis Minoris) is among the stars nearest to the Earth, at a distance of only 3.5 parsecs (1.08×10^{14} km or 6.7×10^{13} mi). Its spectral type is F5, but Procyon is slightly overluminous compared to a main-sequence star of the same spectral type, which indicates that Procyon has already begun to evolve off the main sequence. Its intrinsic luminosity is about seven times that of the Sun. *See* SPECTRAL TYPE; STELLAR EVOLUTION.

Procyon has a faint 11th-magnitude companion, Procyon B, a white dwarf in the final stages of its evolution, with a luminosity only 1/2000 that of the Sun. From the astrometric orbit the masses of the primary and its companion have been computed as 1.75 and 0.62 solar masses, respectively. The progenitor of the white dwarf was originally the more massive of the two stars and underwent the final stages of its stellar evolution sooner than the original secondary, which is now seen as Procyon A. *See* BINARY STAR; STAR; WHITE DWARF STAR. David W. Latham

Product design The determination and specification of the parts of a product and their interrelationship so that they become a unified whole. The design must satisfy a broad array of requirements in a condition of balanced effectiveness. A product is designed to perform a particular function or set of functions effectively and reliably, to be economically manufacturable, to be profitably salable, to suit the purposes and the attitudes of the consumer, and to be durable, safe, and economical to operate. For instance, the design must take into consideration the particular manufacturing facilities, available materials, know-how, and economic resources of the manufacturer. The product may need to be packaged; usually it will also need to be shipped so that it should be light in weight and sturdy of construction. The product should appear significant, effective, compatible with the culture, and appear to be worth more than the price. *See* PRODUCTION ENGINEERING; PRODUCTION PLANNING. Richard I. Felver

Product quality The collection of features and characteristics of a product that contribute to its ability to meet given requirements. Early work in controlling product quality was on creating standards for producing acceptable products. By the mid-1950s, mature methods had evolved for controlling quality, including statistical quality control and statistical process control, utilizing sequential sampling techniques for tracking the mean and variance in process performance. During the 1960s, these methods and techniques were extended to the service industry. During 1960–1980, there was a major shift in world markets, with the position of the United States declining while Japan and Europe experienced substantial growth in international markets. Consumers became more conscious of the cost and quality of products and services. Firms began to focus on total production systems for achieving quality at minimum cost. This trend has continued, and today the goals of quality control are largely driven by consumer concerns and preferences.

There are three views for describing the overall quality of a product. First is the view of the manufacturer, who is primarily concerned with the design, engineering, and manufacturing processes involved in fabricating the product. Quality is measured by the degree of conformance to predetermined specifications and standards, and deviations from these standards can lead to poor quality and low reliability. Efforts for quality improvement are aimed at eliminating defects (components and subsystems that are out of conformance), the need for scrap and rework, and hence overall reductions in production costs. Second is the view of the consumer or user. To consumers, a high-quality product is one that well satisfies their preferences and expectations. This consideration can include a number of characteristics, some of which contribute little or nothing to the functionality of the product but are significant in providing customer satisfaction. A third view relating to quality is to consider the product itself as a system and to incorporate those characteristics that pertain directly to the operation and functionality of the product. This approach should include overlap of the manufacturer and customer views. *See* MANUFACTURING ENGINEERING.

Quality control (QC) is the collection of methods and techniques for ensuring that a product or service is produced and delivered according to given requirements. This includes the development of specifications and standards, performance measures, and tracking procedures, and corrective actions to maintain control. The data collection and analysis functions for quality control involve statistical sampling, estimation of parameters, and construction of various control charts for monitoring the processes in making products. This area of quality control is formally known as statistical process control (SPC) and, along with acceptance sampling, represents the traditional perception of quality management. Statistical process control focuses primarily on the conformance element of quality, and to somewhat less extent on operating performance and durability. *See* PROCESS CONTROL; QUALITY CONTROL.

Concurrent engineering, quality function deployment, and total quality management (TQM) are modern management approaches for improving quality through effective planning and integration of design, manufacturing, and materials management functions throughout an organization. Quality improvement programs typically include goals for reducing warranty claims and associated costs because warranty data directly or indirectly impact most of the product quality dimensions. *See* ENGINEERING DESIGN.

Marlin U. Thomas

Product usability A concept in product design, sometimes referred to as ease of use or user-friendliness, that is related directly to the quality of the product and indirectly to the productivity of the work force. Customer surveys show that product quality is broken down into six components (in descending order of importance): reliability, durability, ease of maintenance, usability, trusted or brand name, and price. Ease of maintenance and usability both relate to product usability. Reliability also has a component of usability to it. If a product is too difficult to use and thus appears not to work properly, the customer may think that it has malfunctioned. Consequently, the customer may return the product to the store not because it is unreliable but because it does not work the way the customer thinks it should. *See* HUMAN-COMPUTER INTERACTION.

There are five criteria by which a product's usability can be measured, including time to perform a task, or the execution time; learnability; mental workload, or the mental effort required to perform a task; consistency in the design; and errors. The usability of a product usually cannot be optimized for all five criteria at the same time. Trade-offs will occur. As an example, a product that is highly usable in terms of fast execution times will often have poor usability in terms of the time needed to learn how to use the product. A product designer must be aware that it may not be possible for a product to be highly usable by all usability criteria, and so design according to the criteria that are most important to potential customers. Casual users of a product will have different demands on a product compared to expert users. *See* CONTROL SYSTEMS; HUMAN-FACTORS ENGINEERING; HUMAN-MACHINE SYSTEMS.

Many companies, especially computer or consumer electronics companies, have laboratories in which to test the usability of their products. The methods of usability testing are formal experimentation, informal experimentation, and task analyses. Although laboratory methods for improving usability can increase the cost of the product design, the benefits (market share, productivity) will outweigh the costs. *See* METHODS ENGINEERING; OPTIMIZATION.

Ray Eberts

Production engineering A branch of engineering that involves the design, control, and continuous improvement of integrated systems in order to provide customers with high-quality goods and services in a timely, cost-effective manner. It is an interdisciplinary area requiring the collaboration of individuals trained in industrial engineering, manufacturing engineering, product design, marketing, finance, and corporate planning. In many organizations, production engineering activities are carried out by teams of individuals with different skills rather than by a formal production engineering department.

In product design, the production engineering team works with the designers, helping them to develop a product that can be manufactured economically while preserving its functionality. Features of the product that will significantly increase its cost are identified, and alternative, cheaper means of obtaining the desired functionality are investigated and suggested to the designers. The process of concurrently developing the product design and the production process is referred to by several names such as design for manufacturability, design for assembly, and concurrent engineering. *See* ACTIVITY-BASED COSTING; DESIGN STANDARDS; PROCESS ENGINEERING; PRODUCT DESIGN; PRODUCTION PLANNING.

The specification of the production process should proceed concurrently with the development of the product design. This involves selecting the manufacturing processes and technology required to achieve the most economical and effective production. The technologies chosen will depend on many factors, such as the required production volume, the skills of the available work force, market trends, and economic considerations. In manufacturing industries, this requires activities such as the design of tools, dies, and fixtures; the specification of speeds and feeds for machine tools; and the specification of process recipes for chemical processes.

Actual production of physical products usually begins with a few prototype units being manufactured in research and development or design laboratories for evaluation by designers, the production engineering team, and sales and marketing personnel. The goal of this pilot phase is to give the production engineering team hands-on experience making the product, allowing problems to be identified and remedied before investing in additional production equipment or shipping defective products to the customer. The pilot production process involves changes to the product design and fine-tuning of unit manufacturing processes, work methods, production equipment, and materials to achieve an optimal trade-off between cost, functionality, and product quality and reliability. *See* PILOT PRODUCTION; PROTOTYPE.

The production facility itself can be designed around the sequence of operations required by the product, referred to as a product layout. General-purpose production machinery is used, and often must be set up for each individual job, incurring significant changeover times while this takes place. This type of production facility is usually organized in a process layout, where equipment with similar functions is grouped together. *See* HUMAN-MACHINE SYSTEMS; PRODUCTION METHODS.

The production engineering process does not stop once the product has been put into production. A major function of production engineering is continuous improvement—continually striving to eliminate inefficiencies in the system and to incorporate and advance the frontier of the best existing practice. The task of production engineering is to identify potential areas for improving the performance of the production system as a whole, and to develop the necessary solutions in these areas. *See* PRODUCT QUALITY.

Reha M. Uzsoy

Production lead time The total time required to convert raw materials into a finished product. More than the simple sum of machine operation and assembly times, it also includes time spent waiting for machines to be available, moving between machines, and performing quality inspections and other nonoperation functions. Short production lead times are ideal, as long times reduce the responsiveness of a company and result in high inventories.

While customer lead time is the interval of time between placing the order and receiving the order, production lead time typically focuses on the time required to produce the order once all the necessary materials are present. It does not take into consideration the time required to obtain the necessary materials or the time the materials wait in the plant before production begins.

Production lead time is very important in make-to-order industries (such as the aerospace industry). There, lead time can be a deciding factor in obtaining orders. In addition to low cost and superior quality, companies win orders by promising to deliver them quickly. Lead time is not as much of an issue in make-to-stock industries (such as small appliances and other retail products) in that these industries are building inventories against future demand, thus providing a buffer against longer than expected production lead times.

Production lead time is essential to the success of a company. In the global market place, customers demand low-cost, high-quality products in a short period of time. As a result, effective management of production lead time is as important as striving for low cost and high quality. Short production lead times allow a company to be responsive to its customers' needs. *See* INDUSTRIAL ENGINEERING; INVENTORY CONTROL; MANUFACTURING ENGINEERING; MATERIAL RESOURCE PLANNING; MATERIALS HANDLING; PRODUCTION PLANNING. Elin M. Wicks

Production planning The function of a manufacturing enterprise responsible for the efficient planning, scheduling, and coordination of all production activities. The planning phase involves forecasting demand and translating the demand forecast into a production plan that optimizes the company's objective, which is usually to maximize profit while in some way optimizing customer satisfaction. These twin objectives are not always synonymous. During the scheduling phase the production plan is translated into a detailed, usually day-by-day, schedule of products to be made. During the coordination phase actual product output is compared with scheduled product output, and this information is used to adjust production plans and production schedules. *See* OPTIMIZATION.

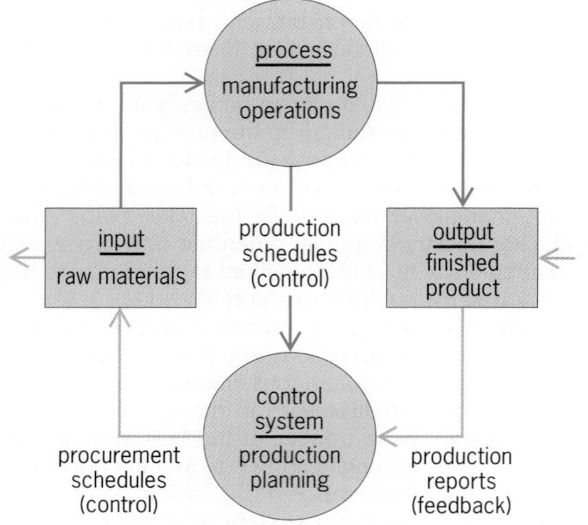

The production process as an input-output process.

If the production or manufacturing process is viewed as an input-output process, then the production planning function can be viewed as a control process with feedback (see illustration). The control is in the form of schedules and plans, while the feedback results from the comparison of the production reports with the production schedules. *See* CONTROL SYSTEMS; INVENTORY CONTROL; PRODUCTION ENGINEERING; PRODUCTION METHODS. John E. Biegel

Production systems A collection of workstations that perform operations such as manufacturing, assembly, inspection, finishing, and testing to create products. A workstation is a collection of resources that performs the same or similar operations for the same set of products. The resources at a workstation may be automated machines, machines operated by humans, or human operators performing manual operations. The storage and delivery of both raw materials and finished products are also part of the production system, but the focus here will be on the system that transforms materials and/or components into products.

The design or configuration of workstations in a production system is defined by the flow of products in the system. The flow of products is determined by how products move among the workstations during production, and is dictated by the volume and variety of products that the system will produce. In general, a simpler flow corresponds to lower product variety at higher volumes, and a more complex flow corresponds to higher product variety at lower volumes.

The specific production operations performed at workstations will vary greatly depending on the nature of the product and its requirements. For example, the production of automobile body components is performed through a series of metal-forming steps. The assembly of automobile bodies is generally accomplished in a set of automated welding process steps. The final assembly of the finished automobile is typically completed in a separate series of manual operations. Not all production systems are large and require the use of expensive machinery—for example, fast-food restaurants, office work, construction operations, and many others are production systems.

The design and operation of a production system can be quite complex and requires that many interrelated decisions be made. Examples of design decisions are the assignment of work to workstations (line balancing), the sizing of storage space for work-in-process between workstations, the methods used to control production, and the methods used for moving product from workstation to workstation. Examples of operational decisions are setting staff levels and shift lengths, production scheduling, and production system bottleneck isolation and improvement. *See* INDUSTRIAL ENGINEERING; MANUFACTURING ENGINEERING; SCHEDULING. David S. Kim

Productivity In a business or industrial context, the ratio of output production to input effort. The productivity ratio is an indicator of the efficiency with which an enterprise converts its resources (inputs) into finished goods or services (outputs). If the goal is to increase productivity, this can be done by producing more output with the same level of input. Productivity can also be increased by producing the same output with fewer inputs. One problem with trying to measure productivity is that a decision must be made in terms of identifying the inputs and outputs and how they will be measured. This is relatively easy when productivity of an individual is considered, but it becomes difficult when productivity involves a whole company or a nation.

Industry and government officials have adopted three common types of productivity measures. Partial productivity is the simplest type of productivity measure; a single type of input is selected for the productivity ratio. The company or organization selects an input factor that it monitors in daily activity. Direct labor hours is a factor that most companies monitor because they pay their employees based on hours worked.

Total factor productivity is a productivity measure combines that labor and capital, two of the most common input factors used in the partial productivity measure. This measure is often used at the national level, because many governments collect statistics on both labor and capital. In calculating at the national level, the gross national product (GNP) is used as the output.

Total productivity is a productivity measure that incorporates all the inputs required to make a product or provide a service. The inputs could be grouped in various categories as long as they determine the total inputs required to produce an output.

Many factors affect productivity. Some general categories for these factors are product, process, labor force, capacity, external influences, and quality.

There are many different plans that companies develop in an attempt to improve productivity. Wage incentive plans and changes in management structure are two ways that companies focus on the labor force. Investment in research and development allows companies to develop new products and processes that are more productive. Quality improvement programs can reduce waste and provide more competitive products at a lower cost. *See* METHODS ENGINEERING; OPERATIONS RESEARCH; PRODUCTION PLANNING. Gregory L. Tonkay

Progesterone A steroid hormone produced in the corpus luteum and placenta. The hormone has an important physiological role in the luteal phase of the menstrual cycle and in the maintenance of pregnancy. In addition, progesterone produced in the testis and adrenals has a key role as an intermediate in the biosynthesis of androgens, estrogens, and the corticoids (adrenal cortex steroids). *See* ANDROGEN; CHOLESTEROL; ESTROGEN; MENSTRUATION; PREGNANCY; STEROID; STEROL. Ralph I. Dorfman

Programmable controllers Electronic computers that are used for the control of machines and manufacturing processes through the implementation of specific functions such as logic, sequencing, timing, counting, and arithmetic. They are also known as programmable logic controllers (PLCs). Historically, process control of a single or a few related devices has been implemented through the use of banks of relays and relay logic for both the control of actuators and their sequencing. The advent of small, inexpensive microprocessors and single-chip computers, or microcontroller units, brought process control from the age of simple relay control to one of electronic digital control while neither losing traditional design methods such as relay ladder diagrams nor restricting their programming to that single paradigm. The computational power of programmable controllers and their integration into networks has led to capabilities approaching those of distributed control systems, and plantwide control is now a mixture of distributed control systems and programmable controllers. Applications for programmable controllers range from small-scale, local process applications in which as few as 10 simple feedback control loops are implemented, up to large-scale, remote supervisory process applications in which 50 or more process control loops spread across the facility are implemented. Typical applications include batch process control and materials handling in the chemical industry, machining and test-stand control and data acquisition in the manufacturing industry, wood cutting and chip handling in the lumber industry, filling and packaging in food industries, and furnace and rolling-mill controls in the metal industry. *See* DIGITAL COMPUTER; DISTRIBUTED SYSTEMS (CONTROL SYSTEMS); MICROPROCESSOR.

Although programmable controllers have been available since the mid-1970s, developments—such as the ready availability of local area networks (LANs) in the industrial environment, standardized hardware interfaces for manufacturer interchangability, and computer software to allow specification of the control process in both traditional (ladder logic) and more modern notations such as that of finite-state machines—have made them even more desirable for industrial process control. *See* LOCAL-AREA NETWORKS.

Programmable logic controllers are typically implemented by using commonly available microprocessors combined with standard and custom interface boards which provide level conversion, isolation, and signal conditioning and amplification. Microprocessors used in programmable controllers are similar or the same as those used in personal computers. The software of a programmable controller must respond to interrupts and be a real-time operating system, characteristics which the typical operating system of a personal computer does not possess. *See* MICROCOMPUTER; OPERATING SYSTEM; REAL-TIME SYSTEMS; SOFTWARE.

Perhaps the biggest benefit of programmable controllers is their small size, which allows computational power to be placed immediately adjacent to the machinery to be controlled, as well as their durability, which allows them to operate in harsh environments. This proximity of programmable controllers to the equipment that they control allows them to effect the sensing of the process and control of the machinery through a reduced number of wires, which reduces installation and maintenance costs. The proximity of programmable controllers to processes also improves the quality of the sensor data since it reduces line lengths, which can introduce noise and affect sensor calibration. Kenneth J. Hintz

Programming languages The different notations used to communicate algorithms to a computer. A computer executes a sequence of instructions (a program) in order to perform some task. In spite of much written about computers being electronic brains or having artificial intelligence, it is still necessary for humans to convey this sequence of instructions to the computer before the computer can perform the task. The set of instructions and the order in which they have to be performed is known as an algorithm. The result of expressing the algorithm in a programming language is called a program. The process of writing the algorithm using a programming language is called programming, and the person doing this is the programmer. *See* ALGORITHM.

In order for a computer to execute the instructions indicated by a program, the program needs to be stored in the primary memory of the computer. Each instruction of the program may occupy one or more memory locations. Instructions are stored as a sequence of binary numbers (sequences of zeros and ones), where each number may indicate the instruction to be executed (the operator) or the pieces of data (operands) on which the instruction is carried out. Instructions that the computer can understand directly are said to be written in machine language. Programmers who design computer algorithms have difficulty in expressing the individual instructions of the algorithm as a sequence of binary numbers. To alleviate this problem, people who develop algorithms may choose a programming language. Since the language used by the programmer and the language understood by the computer are different, another computer program called a compiler translates the program written in a programming language into an equivalent sequence of instructions that the computer is able to understand and carry out. *See* COMPUTER STORAGE TECHNOLOGY.

Machine language. For the first machines in the 1940s, programmers had no choice but to write in the sequences of digits that the computer executed. For example, assume we want to compute the absolute value of $A + B - C$, where A is the value at machine address 3012, B is the value at address 3013, and C is the value at address 3014, and then store this value at address 3015.

It should be clear that programming in this manner is difficult and fraught with errors. Explicit memory locations must be written, and it is not always obvious if simple errors are present. For example, at location 02347, writing 101... instead of 111... would compute $|A + B + C|$ rather than what was desired. This is not easy to detect.

Assembly language. Since each component of a program stands for an object that the programmer understands, using its name rather than numbers should make it easier to program. By naming all locations with easy-to-remember names, and by using symbolic names for machine instructions, some of the difficulties of machine programming can be eliminated. A relatively simple program called an assembler converts this symbolic notation into an equivalent machine language program.

The symbolic nature of assembly language greatly eased the programmer's burden, but programs were still very hard to write. Mistakes were still common. Programmers were forced to think in terms of the computer's architecture rather than in the domain of the problem being solved.

High-level language. The first programming languages were developed in the late 1950s. The concept was that if we want to compute $|A + B - C|$, and store the result in a memory location called D, all we had to do was write $D = |A + B - C|$ and let a computer program, the compiler, convert that into the sequences of numbers that the computer could execute. FORTRAN (an acronym for Formula Translation) was the first major language in this period.

FORTRAN statements were patterned after mathematical notation. In mathematics the = symbol implies that both sides of the equation have the same value. However, in FORTRAN and some other languages, the equal sign is known as the assignment operator. The action carried out by the computer when it encounters this operator is, "Make the variable named on the left of the equal sign have the same value as the expression on the right." Because of this, in some early languages the statement would have been written as $-D \rightarrow D$ to imply movement or change, but the use of $\rightarrow$ as an assignment operator has all but disappeared.

The compiler for FORTRAN converts that arithmetic statement into an equivalent machine language sequence. In this case, we did not care what addresses the compiler used for the instructions or data, as long as we could associate the names A, B, C, and D with the data values we were interested in.

Structure of programming languages. Programs written in a programming language contain three basic components: (1) a mechanism for declaring data objects to contain the information used by the program; (2) data operations that provide for transforming one data object into another; (3) an execution sequence that determines how execution proceeds from start to finish.

Data declarations. Data objects can be constants or variables. A constant always has a specific value. Thus the constant 42 always has the integer value of forty-two and can never have another value. On the other hand, we can define variables with symbolic names. The declaration of variable A as an integer informs the compiler that A should be given a memory location much like the way the variable A in example (2) was given the machine address 03012. The program is given the option of changing the value stored at this memory location as the program executes.

Each data object is defined to be of a specific type. The type of a data object is the set of values the object may have. Types can generally be scalar or aggregate. An object declared to be a scalar object is not divisible into smaller components, and generally it represents the basic data types executable on the physical computer. In a data declaration, each data object is given a name and a type. The compiler will choose what machine location to assign for the declared name.

Data operations. Data operations provide for setting the values into the locations allocated for each declared data variable. In general this is accomplished by a three-step process: a set of operators is defined for transforming the value of each data object, an expression is written for performing several such operations, and an assignment is made to change the value of some data object.

For each data type, languages define a set of operations on objects of that type. For the arithmetic types, there are the usual operations of addition, subtraction, multiplication, and division. Other operations may include exponentiation (raising to a power), as well as various simple functions such as modula or remainder (when dividing one integer by another). There may be other binary operations involving the internal format of the data, such as binary *and*, *or*, *exclusive or*, and *not* functions. Usually there are relational operations (for example, equal, not equal, greater than, less than) whose result is a boolean value of *true* or *false*. There is no limit to the number of operations allowed, except that the programming language designer has to decide between the simplicity and smallness of the language definition versus the ease of using the language.

Execution sequence. The purpose of a program is to manipulate some data in order to produce an answer. While the data operations provide for this manipulation, there must be a mechanism for deciding which expressions to execute in order to generate the desired answer. That is, an algorithm must trace a path through a series of expressions in order to arrive at an answer. Programming languages have developed three forms of execution sequencing: (1) control structures for determining execution sequencing within a procedure; (2) interprocedural communication between procedures; and (3) inheritance, or the automatic passing of information between two procedures.

Corrado Böhm and Giuseppi Jacopini showed in 1966 that a programming language needs only three basic statements for control structures: an assignment statement, an IF statement, and a looping construct. Anything else can simplify programming a solution, but is not necessary. If we add an input and an output statement, we have all that we need for a programming language. Languages execute statements sequentially with the following variations to this rule.

IF statement. Most languages include the IF statement. In the IF-THEN statement, the expression is evaluated, and if the value is *true*, then Statement_1 is executed next. If the value is *false*, then the statement after the IF statement is the next one to execute. The IF-THEN-ELSE statement is similar, except that specific true and false options are given to execute next. After executing either the THEN or ELSE part, the statement following the IF statement is the next one to execute.

The usual looping constructs are the WHILE statement and the REPEAT statement. Although only one is necessary, languages usually have both.

Inheritance is the third major form of execution sequencing. In this case, information is passed automatically between program segments. This is the basis for the models used in the object-oriented languages C++ and Java.

Inheritance involves the concept of a class object. There are integer class objects, string class objects, file class objects, and so forth. Data objects are instances of these class objects. Objects inherit the properties of the objects from which they were created. Thus, if an integer object were designed with the methods (that is, functions) of addition and subtraction, each instance of an integer object would inherit those same functions. One would only need to develop these operations once and then the functionality would pass on to the derived object.

All objects are derived from one master object called an Object. An Object is the parent class of objects such as magnitude, collection, and stream. Magnitude now is the parent of objects that have values, such as numbers, characters, and dates. Collections can be ordered collections such as an array or an unordered collection such as a set. Streams are the parent objects of files. From this structure an entire class hierarchy can be developed.

If we develop a method for one object (for example, *print* method for *object*), then this method gets inherited to all objects derived from that object. Therefore, there is not the necessity to always define new functionality. If we create a new class of integer that, for example, represents the number of days in a year (from 1 to 366), then this new integerlike object will inherit all of the properties of integers, including the methods to add, subtract, and print values. It is this concept that has been built into C++, Java, and current object-oriented languages.

Once we build concepts around a class definition, we have a separate package of functions that are self-contained. We are able to sell that package as a new functionality that users may be willing to pay for rather than develop themselves. This leads to an economic model where companies can build add-ons for existing software, each add-on consisting of a set of class definitions that becomes inherited by the parent class. *See* OBJECT-ORIENTED PROGRAMMING.

Current programming language models. C was developed by AT&T Bell Laboratories during the early 1970s. At the time, Ken Thompson was developing the UNIX operating system. Rather than using machine or assembly language as in (2) or (3) to write the system, he wanted a high-level language. *See* OPERATING SYSTEM.

C has a structure like FORTRAN. A C program consists of several procedures, each consisting of several statements, that include the IF, WHILE, and FOR statements. However, since the goal was to develop operating systems, a primary focus of C was to include operations that allow the programmer access to the underlying hardware of the computer. C includes a large number of operators to manipulate machine language data in the computer, and includes a strong dependence on reference variables so that C programs are able to manipulate the addressing hardware of the machine.

C++ was developed in the early 1980s as an extension to C by Bjarne Stroustrup at AT&T Bell Labs. Each C++ class would include a record declaration as well as a set of associated functions. In addition, an inheritance mechanism was included in order to provide for a class hierarchy for any program.

By the early 1990s, the World Wide Web was becoming a significant force in the computing community, and web browsers were becoming ubiquitous. However, for security reasons, the browser was designed with the limitation that it could not affect the disk storage of the machine it was running on. All computations that a web page performed were carried out on the web server accessed by web address (its Uniform Resource Locator, or URL). That was to prevent web pages from installing viruses on user machines or inadvertently (or intentionally) destroying the disk storage of the user.

Java bears a strong similarity to C++, but has eliminated many of the problems of C++. The three major features addressed by Java are:

1. There are no reference variables, thus no way to explicitly reference specific memory locations. Storage is still allocated by creating new class objects, but this is implicit in the language, not explicit.
2. There is no procedure call statement; however, one can invoke a procedure using the member of class operation. A call to *CreateAddress* for class *address* would be encoded as *address.CreateAddress*( ).
3. A large class library exists for creating web-based objects.

The Java bytecodes (called applets) are transmitted from the web server to the client web site and then execute. This saves transmission time as the executing applet is on the user's machine once it is downloaded, and it frees machine time on the server so it can process more web "hits" effectively. *See* CLIENT-SERVER SYSTEM.

Visual Basic, first released in 1991, grew out of Microsoft's GW Basic product of the 1980s. The language was organized around a series of events. Each time an event happened (for example, mouse click, pulling down a menu), the program would respond with a procedure associated with that event. Execution happens in an asynchronous manner.

Although Prolog development began in 1970, its use did not spread until the 1980s. Prolog represents a very different model of program execution, and depends on the resolution principle and satisfaction of Horn clauses of Robert A. Kowalski at the University of Edinburgh. That is, a Prolog statement is of the form *p*:- *q*, *r* which means *p* is true if both *q* is true or *r* is true.

A Prolog program consists of a series Horn clauses, each being a sequence of relations concerning data in a database. Execution proceeds sequentially through these clauses. Each relation can invoke another Horn clause to be satisfied. Evaluation of a relation is similar to returning a procedure value in imperative languages such as C or C++.

Unlike the other languages mentioned, Prolog is not a complete language. That means there are algorithms that cannot be programmed in Prolog. However, for problems that are amenable for searching large databases, Prolog is an efficient mechanism for describing those algorithms. *See* SOFTWARE; SOFTWARE ENGINEERING.

Marvin V. Zelkowitz

Progression (mathematics)

Progression (mathematics) Ordered, countable sets of numbers, $x_1, x_2, x_3, \ldots$, not necessarily all different. In general such sets are called sequences, whereas the term progression is usually confined to the special types: the arithmetic, in which the difference $x_k - x_{k-1}$ between successive terms is constant; the geometric, in which the ratio x_k/x_{k-1} is constant; and the harmonic, in which the reciprocals of the terms are in arithmetic progression.

If the first term of an arithmetic progression is a and the common difference b, then the terms of the progression are given by Eqs. (1). The sum of the first n terms S_n is given by Eq. (2).

$$x_1 = a, x_2 = a + b, x_3 = a + 2b, \ldots, x_n = a + (n-1)b, \ldots \tag{1}$$

$$S_n = n\frac{x_1 + x_n}{2} = n\left(a + \frac{n-1}{2}b\right) \tag{2}$$

If the first term of a geometric progression is a and the common ratio r, then the terms of the progression are given by Eqs. (3). Excluding the case $r = 1$ (when all terms are the same), the sum S_n of the first n terms is given by Eq. (4).

$$x_1 = a, x_2 = ar, x_3 = ar^2, \ldots, x_n = ar^{n-1}, \ldots \tag{3}$$

$$S_n = a\frac{1 - r^n}{1 - r} \tag{4}$$

The arithmetic mean A and the geometric mean G of n positive numbers are defined by Eqs. (5) and(6).

$$A = \frac{x_1 + x_2 + \cdots + x_n}{n} \tag{5}$$

$$G = \sqrt{x_1 x_2 \cdots x_n} \tag{6}$$

The reciprocals of sequence (1) form a harmonic progression. There is no compact expression for the sum of n terms.

If x_1, x_2, x_3 are in harmonic progression, then Eq. (7) is called

$$x_2 = \frac{2x_1 x_3}{x_1 + x_3} \tag{7}$$

their harmonic mean.

Louis Brand

Projective geometry

Projective geometry A geometry that investigates those properties of figures that are unchanged (invariant) when the figures are projected from a point to a line or plane.

Two features of plane projective geometry are (1) introduction of an ideal line that each ordinary line *g* intersects (the intersection being common to all lines parallel to *g*), and (2) the principle of duality, according to which any statement that is obtained from a valid one (theorem) by substituting for each concept involved, its dual, is also valid. ("Line" and "point" are dual, "connecting two points by a line" is dual to "intersecting two lines," and so on.) The subject has been developed both synthetically (as a logical consequence of a set of postulates) and analytically (by the introduction of coordinates and the application of algebraic processes). *See* CONFORMAL MAPPING; EUCLIDEAN GEOMETRY.

Leonard M. Blumenthal

Prokaryotae A taxonomic kingdom comprising microorganisms that lack a true nucleus or any other membrane-bound organelle. Instead of a nucleus, prokaryotic cells contain a nucleoid, which is a DNA-dense region of the cell where DNA replication and transcription occur. The chromosome of these cells is circular and, in contrast to eukaryotic cells, continuously codes for cellular proteins and other molecules. *See* CLASSIFICATION, BIOLOGICAL; EUKARYOTAE.

Prokaryotae include the Bacteria and the Archaea, which differ in their chemical composition and are unrelated. Bacteria typically have a peptidoglycan cell wall and a cytoplasmic membrane, while Archaea do not have peptidoglycan in their cell wall. Archaea can often be found in extreme environments, such as hot springs, in which most Bacteria cannot survive. *See* BACTERIA.

Prokaryotic cells undergo binary fission, where one cell divides into two equal daughter cells. Motile prokaryotic cells move by gliding over surfaces or by flagella. Different species of prokaryotes have different oxygen requirements. Some may be aerobic, requiring atmospheric oxygen; others are anaerobic, functioning in the absence of oxygen. A few prokaryotes are microaerophilic, requiring lowered oxygen and increased carbon dioxide concentrations to grow and reproduce. Prokaryotes respire to produce energy. *See* CELL MOTILITY; RESPIRATION.

Marcia Pierce

Prolacertiformes A group of diapsid reptiles classified within the Archosauromorpha and recorded from Upper Permian to Lower Jurassic deposits around the world. Prolacertiforms (also known as protorosaurs) share a suite of distinctive characters, including elongate neck vertebrae with long low neural spines and long slender ribs, reduction of the lower temporal bar (a bone forming a portion of the lateral aspect of the skull and part of the base of the cranium), and emargination of the quadrate (part of the upper jaw joint) to support a tympanum. *See* ARCHOSAURIA; DIAPSIDA; REPTILIA.

Protorosaurus from the Upper Permian of Britain and Germany was up to 6 ft (2 m) in length, but most specimens are smaller than this. It was one of the first fossil reptiles to be formally described (in 1710). In life, *Protorosaurus* may have resembled a large monitor lizard, but there is no relationship with true lizards. *Prolacerta* (Lower Triassic, South Africa) is smaller and more derived than *Protorosaurus*. The most distinctive prolacertiform was *Tanystropheus* from the Middle to Upper Triassic of Europe and the Middle East. Individuals of the largest species of *Tanystropheus* could reach 9 ft (3 m) in length, but much of this consisted of an extraordinary neck that was at least three times the length of the trunk region. Adults of *Tanystropheus* were aquatic, but juveniles may have been at least semiterrestrial. *Tanytrachelos* was closely similar to *Tanystropheus*, but smaller, from the Upper Triassic/Lower Jurassic of North America.

Susan E. Evans

Promethium A chemical element, Pm, atomic number 61. Promethium is the "missing" element of the lanthanide rare-earth series. The atomic weight of the most abundant separated radioisotope is 145. *See* PERIODIC TABLE.

Although a number of scientists have claimed to have discovered this element in nature as a result of observing certain spectral lines, no one has succeeded in isolating element 61 from naturally occurring materials. It is produced artificially in nuclear reactors, since it is one of the products that results from the fission of uranium, thorium, and plutonium.

All the known isotopes are radioactive. Its principal uses are for research involving tracers. Its main application is in the phosphor industry. It has also been used to manufacture thickness gages and as a nuclear-powered battery in space applications. *See* RARE-EARTH ELEMENTS.

Frank H. Spedding

Pronghorn An antelopelike animal, *Antilocapra americana*, the sole representative of the family Antilocapridae, and of uncertain taxonomic affinities. This animal is reputed to be the fastest ungulate in North America, sprinting as fast as 60 mi/h (96 km/h) and sustaining speeds of 30 mi/h (48 km/h) for about 15 mi (24 km). It is the only hollow-horned ungulate with branched horns present in both sexes. Like the deer, the pronghorn sheds its horns each fall; the new growth is complete by midsummer. The adult is about 3 ft (0.9 m) high, weighs about 100 lb (45 kg), and is covered with coarse brittle hair. *See* ANTELOPE.

The pronghorn lives in small herds in rather wild, rocky desert country. It blends well into the environment because of its cream-colored fur. Average lifespan is about 8 years. *See* ARTIODACTYLA.

Charles B. Curtin

Propagator (field theory) The probability amplitude for a particle to move or propagate to some new point of space and time when its amplitude at some point of origination is known. The propagator occurs as an important part of the probability in reactions and interactions in all branches of modern physics. Its properties are best described in the framework of quantum field theory for relativistic particles, where it is written in terms of energy and momentum. Concrete examples for electron-proton and proton-proton scattering are provided in the illustration. The amplitude for these processes contains the

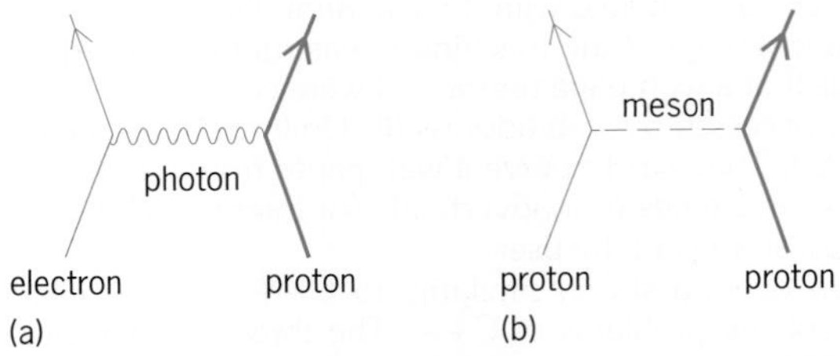

Feynman diagrams for scattering processes. (*a*) Electron-proton scattering via photon exchange. (*b*) Proton-proton scattering via meson exchange.

propagators for the exchanged proton and meson, which actually specify the dominant part of the probability of each process when the scattering occurs at small angles. In similar fashion, for any electromagnetic process, a propagator for each internal line of the Feynman diagram (each line not connected directly to the outside world) enters the probability amplitude. *See* FEYNMAN DIAGRAM; QUANTUM ELECTRODYNAMICS; QUANTUM FIELD THEORY; QUANTUM MECHANICS.

Kenneth E. Lassila

Propellant Usually, a combustible substance that produces heat and supplies ejection particles, as in a rocket engine. A propellant is both a source of energy and a working substance; a fuel is chiefly a source of energy, and a working substance is chiefly a means for expending energy. Because the distinction

(Top left) *Murex troscheli*. (Middle) *Conus gloriamaris*. (Top right) *Lambis crocata*. (Bottom) *Angaria vicdani*.

is more decisive in rocket engines, the term propellant is used primarily to describe chemicals carried by rockets for propulsive purposes. *See* AIRCRAFT FUEL; ROCKET PROPULSION; THERMODYNAMIC CYCLE.

Propellants are classified as liquid or as solid. Even if a propellant is burned as a gas, it may be carried under pressure as a cryogenic liquid to save space. For example, liquid oxygen and liquid hydrogen are important high-energy liquid bipropellants.

Liquid propellants. A liquid propellant releases energy by chemical action to supply motive power for jet propulsion. The three principal types of propellants are monopropellant, bipropellant, and hybrid propellant. Monopropellants are single liquids, either compounds or solutions. Bipropellants consist of fuel and oxidizer carried separately in the vehicle and brought together in the engine. Hybrid propellants use a combination of liquid and solid materials to provide propulsion energy and working substance. Typical liquid propellants are listed in the table. *See* METAL-BASE FUEL.

The availability of large quantities and their high performance led to selection of liquefied gases such as oxygen for early liquid-propellant rocket vehicles. Liquids of higher density with low vapor pressure (see table) are advantageous for the practical requirements of rocket operation under ordinary handling conditions. Such liquids can be retained in rockets for long periods ready for use and are convenient for vehicles that are to be used several times. The high impulse of the cryogenic systems is desirable for rocket flights demanding maximum capabilities, however, such as space exploration or transportation of great weights for long distances. Stanley Singer

Physical properties of liquid propellants

Propellant	Boiling point, °F (°C)	Freezing point, °F (°C)	Density g/ml	Specific impulse,* s
Monopropellants				
Acetylene	−119 (−84)	−115 (−82)	0.62	265
Hydrazine	236 (113)	35 (2)	1.01	194
Ethylene oxide	52 (11)	−168 (−111)	0.88	192
Hydrogen peroxide	288 (142)	13 (−11)	1.39	170
Bipropellants				
Hydrogen	−423 (−253)	−433 (−259)	0.07	
Hydrogen-fluorine	−306 (−188)	−360 (−218)	1.54	410
Hydrogen-oxygen	−297 (−183)	−362 (−219)	1.14	390
Nitrogen tetroxide	70 (21)	12 (−11)	1.49	—
Nitrogen-tetroxide-hydrazine	236 (113)	35 (2)	1.01	290
Red nitric acid	104 (40)	−80 (−62)	1.58	
Red fuming nitric acid–*uns*-dimethyl hydrazine	146 (63)	−71 (−57)	0.78	275

*Maximum theoretical specific impulse at 1000 psi (6.895 megapascals) chamber pressure expanded to atmospheric pressure.

Solid propellants. A solid propellant is a mixture of oxidizing and reducing materials that can coexist in the solid state at ordinary temperatures. When ignited, a propellant burns and generates hot gas. Although gun powders are sometimes called propellants, the term solid propellant ordinarily refers to materials used to furnish energy for rocket propulsion.

A solid propellant normally contains three essential components: oxidizer, fuel, and additives. Oxidizers commonly used in solid propellants are ammonium and potassium perchlorates, ammonium and potassium nitrates, and various organic nitrates, such as glyceryl trinitrate (nitroglycerin). Common fuels are hydrocarbons or hydrocarbon derivatives, such as synthetic rubbers, synthetic resins, and cellulose or cellulose derivatives. The additives, usually present in small amounts, are chosen from a wide variety of materials and serve a variety of purposes. Catalysts or suppressors are used to increase or decrease the rate of burning; ballistic modifiers may be used for a variety of reasons, as to provide less change in burning rate with pressure (platinizing agent); stabilizers may be used to slow down undesirable changes that may occur in tong-term storage.

Solid propellants are classified as composite or double base. The composite types consist of an oxidizer of inorganic salt in a matrix of organic fuels, such as ammonium perchlorate suspended in a synthetic rubber. The double-base types are usually high-strength, high-modulus gels of cellulose nitrate (guncotton) in glyceryl trinitrate or a similar solvent. H. W. Ritchey

Propeller (aircraft) A hub-and-multiblade device for changing rotational power of an aircraft engine into thrust power for the purpose of propelling an aircraft through the air (see illustration). An air propeller operates in a relatively thin medium compared to a marine propeller, and is therefore characterized by a relatively large diameter and a fairly high rotational speed. It is usually mounted directly on the engine drive shaft in front of or behind the engine housing. *See* PROPELLER (MARINE CRAFT).

Usually propellers have two, three, or four blades; for high-speed or high-powered airplanes, six or more blades are used. In some cases these propellers have an equal number of opposite rotating blades on the same shaft, and are known as dual-rotation propellers.

A propeller blade advances through the air along an approximate helical path which is the result of its forward and rotational velocity components. This action is similar to a screw being turned in a solid surface, except that in the case of the propeller a slippage occurs because air is a fluid. Because of the similarity to the action of a screw, a propeller is also known as an airscrew. To rotate the propeller blade, the engine exerts a torque force. This force is reacted on by the blade in terms of lift and drag force components produced by the blade sections in the opposite direction. As a result of the rational forces reacting on the air, a rotational velocity remains in the propeller wake with the same rotational direction as the propeller. This rotational velocity times the mass of the air is proportional to the power input. The sum of all the lift and drag components of the blade sections in the direction of flight are equal to the thrust produced. These forces react on the air, giving an axial velocity component opposite to the direction of flight. By the momentum theory, this velocity

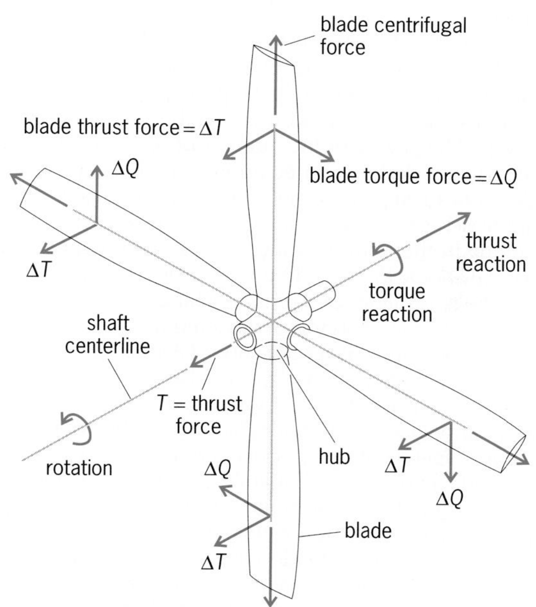

Typical four-bladed propeller system.

times the mass of the air going through the propeller is equal to the thrust.

A propeller blade must be designed to withstand very high centrifugal forces. The blade also must withstand the thrust force produced plus any vibratory forces generated, such as those due to uneven flow fields. To withstand the high stresses due to rotation, propeller blades have been made from a number of materials, including wood, aluminum, hollow steel, and plastic composites. The most common material used has been solid aluminum. However, the composite blade constructions are being used for new turboprop installations because of their very light weight and high strength characteristics.

For a small, low-power airplane, very simple, fixed-pitch, single-piece, two-blade propellers are used. The rotational speed of these propellers depends directly on the power input and forward speed of the airplane. Because of the fixed-blade angle of this type of propeller, it operates near peak efficiency only at one condition. To overcome the limitations of the simple fixed-pitch propeller, configurations that provide for variable blade angles are used. The blades of these propellers are retained in their hub so that they can be rotated about their centerline while the propeller rotates. For the normal range of operation, the blade angle varies from the low blade angle needed for takeoff to the high blade angle needed for the maximum speed of the airplane. *See* AIRCRAFT PROPULSION; AIRPLANE; HELICOPTER. Henry V. Borst

Propeller (marine craft) A component of a ship-propulsion power plant which converts engine torque into propulsive force or thrust, thus overcoming a ship's resistance to forward motion by creating a sternward accelerated column of water. Since 1860 the screw propeller has been the only propeller type used in ocean transport, mainly because of the evolution of the marine engine toward higher rotative speed.

The advantages of a screw propeller include light weight, flexibility of application, good efficiency at high rotative speed, and relative insensitivity to ship motion. The fundamental theory of screw propellers is applicable to all forms of marine propellers. In its present form a screw propeller consists of a streamlined hub attached outboard to a rotating engine shaft, on which are mounted two to seven blades. The blades are either solid with the hub, detachable, or movable. The screw propeller has the characteristic motion of a screw; it revolves about the axis along which it advances. The screw blades are approximately elliptical in outline.

One or more screw propellers are usually fitted as low as possible at the ship's stern to act as thrust-producing devices (see illustration). The low position of the propellers affords good protection and sufficient immersion during the pitching movements of the ship. The choice of the number of propellers to incorporate into a vessel design is based upon several factors. In general, a single-screw arrangement yields a higher propulsive efficiency than multiple screws, particularly when most of the propeller is operating in the boundary layer of the ship and can recover some of the energy loss. In addition, single-screw propulsion systems generally result in savings in machinery cost and weight in comparison to multiple-screw arrangements.

The formation and collapse of vapor-filled bubbles, or cavities, causes noise, vibration, and often rapid erosion of the propeller material, especially in fast, high-powered vessels. This phenomenon is known as cavitation. As long as the rotational and translational speeds of the propeller are not too high, the onset of cavitation can be delayed or limited to an acceptable amount by clever design of blade sections. *See* CAVITATION.

Supercavitating and superventilated propellers are designed to have fully developed blade cavities which spring from the leading edge of the blade, cover the entire back of the blade, and collapse well downstream of the blade trailing edge. The blade of such propellers has unique sections which usually are wedge-shaped with a sharp leading edge, blunt trailing edge, and concave face. Supercavitating propellers have cavities filled with water vapor and small amounts of gases dissolved in the fluid media. Superventilated propellers have cavities filled primarily with air from the water surface or gases other than water vapor from a gas supply system through the propeller shaft.

Stern view of *Great Land* in drydock, showing screw propeller. (*From E. Schorsch, R. T. Bicicchi, and J. W. Fu, Hull experiments on 24-knot RO/RO vessels directed toward fuel-saving application of copper-nickel, Soc. Nav. Archit. Mar. Eng. Trans., 86:254–276, 1978*)

For ships which normally operate at widely varying speeds and propeller loadings (towboats, rescue vessels, trawlers, and ferryboats), the application of controllable-pitch (rotatable-blade) propellers permits the use of full engine power at rated rpm under all operational conditions, ensuring maximum thrust production, utmost flexibility, and maneuverability. Since these propellers are also reversible, they permit the use of nonre-versible machinery (gas turbines). *See* MARINE ENGINE; MARINE MACHINERY.

Jacques B. Hadler

Proprioception The sense of position and movement of the limbs and the sense of muscular tension. The awareness of the orientation of the body in space and the direction, extent, and rate of movement of the limbs depend in part upon information derived from sensory receptors in the joints, tendons, and muscles. Information from these receptors, called proprioceptors, is normally integrated with that arising from vestibular receptors (which signal gravitational acceleration and changes in velocity of movements of the head), as well as from visual, auditory, and tactile receptors. Sensory information from certain proprioceptors, particularly those in muscles and tendons, need not reach consciousness, but can be used by the motor system as feedback to guide postural adjustments and control of well-practiced or semiautomatic movements such as those involved in walking.

Receptors for proprioception are the endings of peripheral nerve fibers within the capsule or ligaments of the joints or within muscle. These endings are associated with specialized end organs such as Pacinian corpuscles, Ruffini's cylinders, and Golgi organs, and muscle spindles. *See* CUTANEOUS SENSATION; SENSATION; SOMESTHESIS. Robert LaMotte

Propulsion The process of causing a body to move by exerting a force against it. Propulsion is based on the reaction principle, stated qualitatively in Newton's third law, that for every action there is an equal and opposite reaction. A quantitative description of the propulsive force exerted on a body is given by Newton's second law, which states that the force applied to any body is equal to the rate of change of momentum of that body, and is exerted in the same direction as the momentum change. *See* NEWTON'S LAWS OF MOTION.

In the case of a vehicle moving in a fluid medium, such as an airplane or a ship, the required change in momentum is generally produced by changing the velocity of the fluid (air or water) passing through the propulsive device or engine. In other cases, such as that of a rocket-propelled vehicle, the propulsion system must be capable of operating without the presence of a fluid medium; that is, it must be able to operate in the vacuum of space. The required momentum change is then produced by using up some of the propulsive device's own mass, which is called the propellant. *See* AERODYNAMIC FORCE; AIRFOIL; FLUID MECHANICS; PROPELLANT.

The two terms most generally used to describe propulsion efficiency are thrust specific fuel consumption for engines using the ambient fluid (air or water), and specific impulse for engines which carry all propulsive media on board. *See* SPECIFIC FUEL CONSUMPTION; SPECIFIC IMPULSE.

The energy source for most propulsion devices is the heat generated by the combustion of exothermic chemical mixtures composed of a fuel and an oxidizer. An air-breathing chemical propulsion system generally uses a hydrocarbon such as coal, oil, gasoline, or kerosine as the fuel, and atmospheric air as the oxidizer. A non-air-breathing engine, such as a rocket, almost always utilizes propellents that also provide the energy source by their own combustion.

Where nuclear energy is the source of propulsive power, the heat developed by nuclear fission in a reactor is transferred to a working fluid, which either passes through a turbine to drive the propulsive element such as a propeller, or serves as the propellant itself. Nuclear-powered ships and submarines are accepted forms of transportation. *See* TURBINE PROPULSION. Jerry Grey

Prosobranchia The largest and most diverse subclass of the molluscan class Gastropoda. The group includes mostly marine snails but with a few fresh-water and land genera, all retaining an anterior mantle cavity and internal evidence of torsion. Adult prosobranchs always retain the streptoneurous (twisted-commissure) condition of the central nervous system, with the commissures to the visceral ganglia in the characteristic figure-eight pattern. This pattern reflects the torsion through 180° during larval (or embryonic) development which has brought the mantle cavity to a position above the head and facing forward. This contrasts with conditions in the other two gastropod subclasses, Opisthobranchia and Pulmonata, in which the effects of torsion are reduced or obscured in adults by secondary processes of development and growth. *See* GASTROPODA; OPISTHOBRANCHIA; PULMONATA.

The diversity of functional morphology exhibited by the prosobranchs is not equaled by any comparable subclass in the entire animal kingdom. From two-gilled forms with symmetrical cardiac and renal structures (the "Diotocardia") which can be numbered among the most primitive of all living mollusks, evolution within prosobranch stocks has involved increasing asymmetry of pallial, cardiac, and renal systems and greater hydrodynamic efficiencies. Torsion and the anterior mantle cavity create locomotory, circulatory, sanitary, and hydraulic problems which have been solved in a variety of ways in different prosobranchs. Four orders are commonly recognized in subclass Prosobranchia: Archaeogastropoda, Neritacea, Mesogastropoda, and Neogastropoda. *See* NEOGASTROPODA. W. D. Russell-Hunter

Prospecting Exploration for mineral deposits. The result of prospecting is the discovery of potentially economic mineralization, that is, the prospect. Mineral exploration continues beyond prospecting to include the delineation and evaluation of the prospect to determine its minability as an orebody or economic mineral deposit. A successful prospect is developed into a mine. *See* MINING.

Prospecting generally pertains to the search for deposits of metallic ore minerals, but it also includes the search for nonmetallic or industrial minerals and rocks such as sulfur, potash, and limestone, and mineral fuels such as petroleum, coal, and oil shale.

With much of the Earth's readily accessible surface having been investigated for minerals, prospecting is increasingly directed toward the discovery of deeper mineralization in recognized mining districts; mineralization hidden beneath overlying rocks, sediments, and soils; and mineralization in the less-known jungle, arctic, and offshore parts of the world. *See* MARINE MINING; OIL AND GAS, OFFSHORE.

Prospecting is done on the basis of the guides to ore associated with a conceptual image of the anticipated orebody. The image is referred to as an exploration model, and it is drawn from the characteristics of known orebodies in similar terrain. The exploration model and its guides to ore are expressed in terms of the regional and local geologic pattern; it has a certain diagnostic mineralogical character, it will commonly have a halo or envelope of associated guide minerals, and it will be expected to have a recognizable geochemical and geophysical expression. *See* ORE AND MINERAL DEPOSITS.

The topography itself may give evidence of abrupt depressions related to the leaching and collapse of sulfide ore minerals, or it may show boldly exposed silicified zones associated with ore. Some of the latter expressions of ore mineralization represent outcrops of siliceous iron formation host rocks, quartz-filled breccia pipes, and the prospector's classic quartz reefs that indicate vein deposits. Aerial photography and satellite imagery are valuable in searching for the topographic expression of potential ore mineralization.

Outcrops of gossan (the residue of red, brown, and yellow iron oxides and silica that remains from the weathering and near-surface leaching of sulfide ore minerals) are examined in the field for evidence of underlying ore mineralization, and trains of float (fragments of ore and gossan) are traced toward their apparent topographic origin. In glaciated terrain, trains of ore boulders are mapped and traced systematically toward their apparent sources.

Placer gold and placer accumulations of other minerals such as platinum, cassiterite, rutile, and diamonds are sought as economic deposits in themselves and are used as guides to upstream deposits of associated minerals. In addition, resistant and relatively dense minerals in stream gravels and residual heavy minerals in soil are collected by the long-established prospector's method of panning the loose material, and these are traced to a source area.

Geochemical prospecting is based on two characteristics of orebodies: an association with anomalous concentrations of chemical elements within primary halos in the surrounding rock, and an association with secondary dispersal patterns of chemical elements in the surficial products of their weathering and erosion. Geochemical methods involve the field and laboratory analysis of sampled rock, soil, vegetation, and other natural materials for trace amounts (in parts per million or billion) of the principal indicator elements of an orebody and of the related pathfinder elements that provide more recognizable or farther-reaching anomalies. *See* GEOCHEMICAL PROSPECTING.

Imagery provided by remote sensing from aircraft and orbiting satellites is of fundamental importance in prospecting and in the patterns of exploration data associated with geographical information systems. Aerial photography in spectral bands of the near-infrared and near-ultraviolet frequencies is also used in photogeology for discriminating between types of exposed rock and soil and for emphasizing the appearance of bleached and stained areas as well as geobotanical anomalies. Airborne remote-sensing systems have provided radar imagery of terrain in the prospecting of cloud-covered jungle regions, and they have furnished thermal-band infrared surveys for recognizing anomalously warm areas that may be associated with mineralization. Airborne multispectral sensors with the capability of identifying some of the specific kinds of minerals in altered zones have been tested for use in prospecting. *See* AERIAL PHOTOGRAPH; GEOGRAPHIC INFORMATION SYSTEMS; REMOTE SENSING.

Geophysical exploration is based on the measurement of physical properties associated with geologic features. As a means of both airborne and ground prospecting for mineral deposits, it involves the recognition of contrasts in properties between the deposit and the adjacent rock, generally to depths on the order of 330–660 ft (100– 200 m), and the definition of deeper structural and lithologic features to be used as guides to ore mineralization. Magnetic, electrical, electromagnetic, and radioactive methods are the most widely used in prospecting for ore and industrial minerals deposits. Geophysical surveys are often made by several methods, so that more than one physical property can be taken into account. *See* GEOMAGNETISM; GEOPHYSICAL EXPLORATION; ROCK, ELECTRICAL PROPERTIES OF.

Drilling is the principal method of subsurface prospecting where evidence of ore mineralization and geophysical or geochemical anomalies indicates a target for prospecting at a depth of more than a few feet. Geophysical information is obtained by the probing or logging of drill holes. Electrical and electromagnetic logging is done in holes drilled in search of metallic orebodies; with these methods, the radius of search is extended considerably beyond that of the small-diameter cylinder of sampled rock. Gamma-ray methods of geophysical drill-hole logging have become standard practice in prospecting for uranium ore. *See* ENGINEERING GEOLOGY; WELL LOGGING. William C. Peters

Prostate gland A triangular body in men, the size and shape of a chestnut, that lies immediately in front of the bladder with its apex directed down and forward. It is found only in the male, having no female counterpart. The prostatic portion of the urethra extends through it, passing from the bladder to the penis. This organ contains 15–20 branched, tubular glands which form lobules. The gland ducts open into the urethra. Between the gland clusters, or alveoli, there is a dense, fibrous, connecting tissue, the stroma, which also forms a tough capsule around the gland, continuous with the bladder wall. Penetrating the prostate to empty into the urethra are the ejaculatory ducts from the seminal vesicles which are located above and behind the organ (see illustration). The prostatic gland secretes a viscid, alkaline fluid which aids in sperm motility and in neutralizing the acidity of the vagina, thus enhancing fertilization. After middle age, the prostate is sometimes subject to new tissue growth, usually benign, that may result in interference with urine flow through the compressed urethra. Walter J. Bock

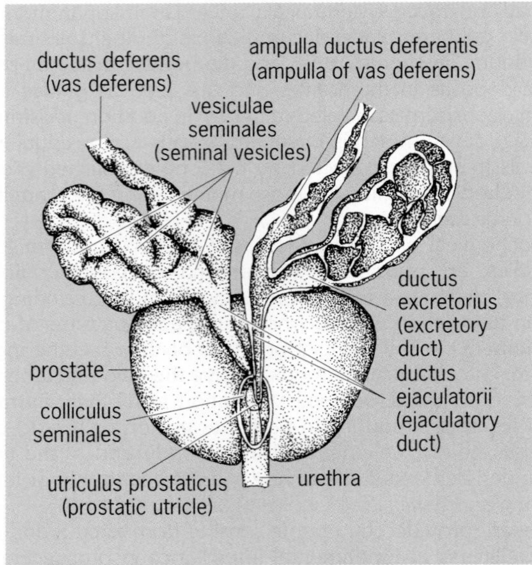

Prostate gland and seminal vesicles. (*After W. A. N. Darland, ed., American Illustrated Medical Dictionary, 19th ed., Saunders, 1942*)

Prostate gland disorders Disorders of the prostate gland that can cause voiding symptoms and sexual dysfunction. The three most common diseases are prostatitis, benign prostatic hyperplasia, and prostate cancer. Prostatitis is an inflammatory condition of the gland, causing a variety of voiding symptoms and pain in the area around the prostate. Benign prostatic hyperplasia is a benign growth of the inner portion of the gland that causes progressive obstruction of the urinary channel (urethra) at the base of the bladder. Prostate cancer is the most common nonskin malignancy in men and is a major health concern for the aging population. *See* CANCER (MEDICINE); CHEMOTHERAPY AND OTHER ANTINEOPLASTIC DRUGS; PROSTATE GLAND. Nelson N. Stone

Prosthesis An artificial replacement of a body part. It may be an internal replacement such as an artificial joint or an external replacement such as an artificial limb. Prostheses of all types are lighter and more functional than their predecessors; the broad field of prosthetics has benefited from advances in materials, miniaturization, and computer-generated fabrication.

Limb prosthetics. A standard nomenclature is used to refer to level of amputation and related prostheses. The term trans is used when an amputation goes across the axis of a long bone, such as transtibial (across the tibia of the leg) or transhumeral (across the humerus of the arm). When there are two bones together such as the tibia and fibula, the primary bone is identified. Amputations between long bones or through a joint are referred to as disarticulations and identified by the major body part, such as knee disarticulation. The term partial is used to refer to a part of the foot or hand distal to the ankle or wrist that may be amputated.

Socket design varies with the level of amputation and the configurations of the individual residual limb. The prosthetic socket must support body weight and hold the residual limb firmly and comfortably during all activities. Additionally, the socket needs to grip the residual limb firmly to reduce movement between the socket and the skin. Sockets are individually constructed for each client from a cast made of the residual limb. Sockets may be hard and rigid, or flexible and supported by a rigid frame.

There are several methods of suspending each type of prosthesis. Suction sockets allow suspension without belts, sleeves, or cuffs. In the transtibial prosthesis a rubber sleeve or a cuff that fits on the thigh may be used. In the transfemoral prosthesis a belt around the pelvis provides suspension.

Foot and ankle function is complex, and a variety of prosthetic feet are designed to respond dynamically to the pressure of walking and running. They store energy at the moment of heel contact, then return it at toe-off.

Of paramount importance is the prosthetic component, called a terminal device, that will substitute for the missing hand. There is no device that completely replaces the appearance or function of the anatomic hand. The two types of terminal device are the hand and the hook. Either is secured to a plastic socket encasing the forearm.

The hand may be active or passive: passive hands have no moving parts; active hands have a mechanism that permits the client to control finger position by appropriate action in the proximal part of the amputated limb. The most popular active hand is operated myoelectrically. The individual wears a socket with one or more skin electrodes that contact appropriate muscle groups.

A hook is made either of aluminum or steel, hooks have two fingers that the client can open and close. Myoelectrically controlled hooks are available; however, most individuals who wear hooks have cable-operated ones that are either voluntary-opening or voluntary-closing.

Joint replacements. Artificial replacements of joints, such as hip, knee, or shoulder, are another type of prosthesis. The knee and hip joints are the most frequently replaced. Lower limb joint replacements must have load-bearing capabilities and are fabricated in different sizes of metal. Joint replacements may sometimes be secured by polymer adhesive. A high degree of restoration of function is usually obtained with hip and knee joint replacements.

In the upper extremity, shoulder and metacarpal phalangeal joints are most frequently replaced. The joint replacement in the upper limb needs to be light and to allow a great range of movement.

Other prostheses. There are a great variety of other types of replacements of body parts. Women who lose a breast to cancer are fitted with a prosthesis that can be fabricated of a number of lightweight materials and molded to resemble in shape and texture the remaining breast. Prosthodontics are devices used to replace teeth. The cardiac pacemaker is a form of a prosthesis replacing the natural electrical stimulation of the heart with a battery-operated device inserted within the body. Damaged heart valves are replaced with artificial valves attached directly to the heart muscle and to the major blood vessels. Prosthetic eyes, fabricated to resemble the remaining eye in color and configuration, serve a cosmetic function only. Bella J. May

Biologic-prosthetic systems. Although mechanical or electromechanical prosthetic devices represent the main option in many situations, regeneration of destroyed or resected tissue is the preferred goal. The use of inert materials to provide a basis for regenerated tissue growth offers intermediate options for reconstruction. See BIOMEDICAL ENGINEERING; SURGERY. W. Rostoker

Protactinium A chemical element, Pa, atomic number 91 and atomic weight of 231.04. Isotopes of mass numbers 216, 217, and 222–238 are known, all of them radioactive. Only ^{231}Pa, the parent of actinium, ^{234}Pa, and ^{233}Pa occur in nature. The most important of these is ^{231}Pa, an α-emitter with a half-life of 32,500 years. The artificial isotope, ^{233}Pa, is important as an intermediary in the production of fissile ^{233}U. Both ^{231}Pa and ^{233}Pa can be synthesized by neutron irradiation of thorium. *See* ACTINIUM; PERIODIC TABLE; RADIOACTIVITY; URANIUM.

Protactinium is, formally, the third member of the actinide series of elements and the first in which a 5*f* electron appears, but its chemical behavior in aqueous solution resembles that of tantalum and niobium more closely than that of the other actinides. *See* NIOBIUM.

Metallic protactinium is silver in color, malleable, and ductile. The crystal structure is body-centered tetragonal. Samples exposed to air at room temperature show little or no tarnishing over a period of several months. The numerous compounds of protactinium that have been prepared and characterized include binary and polynary oxides, halides, oxyhalides, sulfates, oxysulfates, double sulfates, oxynitrates, selenates, carbides, organometallic compounds, and noble metal alloys. *See* ACTINIDE ELEMENTS.

H. W. Kirby

Protandry That condition in which an animal is first a male and then becomes a female. It occurs in many groups, including oysters and cyclostomes. The reverse condition is protogyny. *See* PROTOGYNY. Tracy I. Storer

Proteales An order of flowering plants, division Magnoliophyta, in the eudicots. Consisting of three families, the order is one of the most controversial in current classifications, with a lack of obvious morphological characters linking the family Nelumbonaceae (two species) to the other two families, Platanaceae (seven species) and Proteaceae (about 1350 species). However, DNA sequences indicate that the three families form a natural group. *See* MAGNOLIOPHYTA; MAGNOLIOPSIDA; WOOD ANATOMY.

Nelumbonaceae are aquatic, rhizomatous herbs with peltate leaves held above the water on long petioles, and have often been thought to be related to the true waterlilies (Nymphaeales). *Nelumbo* (sacred lotus and American lotus) is used as a source of food, as an ornamental, and as a sacred plant in several Asian countries.

Platanaceae, from the Northern Hemisphere, are deciduous, monoecious trees with simple, palmately lobed leaves, flaking bark, and branched hairs. Plane trees (*Platanus*) are common street trees, due to the regular loss of bark (which gives some resistance to pollution), and they also provide timber. *See* BARK; DECIDUOUS PLANTS; FRUIT; LEAF; NYMPHAEALES.

Proteaceae (predominantly tropical and subtropical, in the Southern Hemisphere) are evergreen shrubs and trees that often accumulate aluminum. Several genera (including *Banksia*, *Protea*, and *Leucospermum*) are widely cultivated for cut flowers; *Grevillea* and other genera are used for timber; and *Macadamia* yields edible nuts. *See* EVERGREEN PLANTS; ROOT (BOTANY).

Michael F. Fay

Proteasome A large proteolytic complex that degrades intracellular proteins. Intracellular proteins are continuously synthesized and degraded, and their levels in cells reflect the fine balance between these two processes. The rate of breakdown of individual proteins inside the cell varies widely and can be altered according to changes in the cellular environment. In eukaryotic cells, the site for degradation of most intracellular proteins is a large proteolytic particle termed the proteasome. Proteasomes are a major cell constituent, constituting up to 2% of cellular protein, and are essential for viability. They are found in the cytoplasm and nucleus of all eukaryotic cells. Simpler but homologous forms of the proteasome are also present in archaea and bacteria.

The form responsible for most intracellular protein breakdown is termed the 26S proteasome, which functions as a component of the ubiquitin-proteasome pathway (see illustration). This system catalyzes the rapid turnover of many critical regulatory proteins (for example, transcription factors, cell-cycle regulators, oncogenes); thus the proteasome plays a pivotal role in controlling a wide variety of cellular processes, ranging from cell division to circadian rhythms, gene transcription, and immune responses. Most normal long-lived proteins (which make up the bulk of proteins in cells) and damaged, misfolded, or mutated proteins (which could cause disease if they accumulated in cells) are also degraded by this pathway.

The proteasome differs from a typical proteolytic enzyme in many important respects. The typical protease is a single subunit enzyme of 20,000 to 40,000 daltons (Da). By contrast, the proteasome is up to 100-fold greater in size (2.5 million Da) and contains about 50 proteins with multiple enzymatic functions, some of which require adenosine triphosphate (ATP). Unlike typical protease pathways, the ubiquitin-proteasome pathway is ATP-dependent and uses the energy stored in ATP to mark, unfold, and transfer the protein substrate into the degradative chamber of the proteasome. Traditional proteases simply cleave a protein and release the partially digested fragments, whereas the proteasome binds and cuts the protein substrate into small peptides ranging from 3 to 25 residues in length, ensuring that partially digested proteins do not accumulate within cells. Peptides released by the proteasome are rapidly hydrolyzed to amino acids

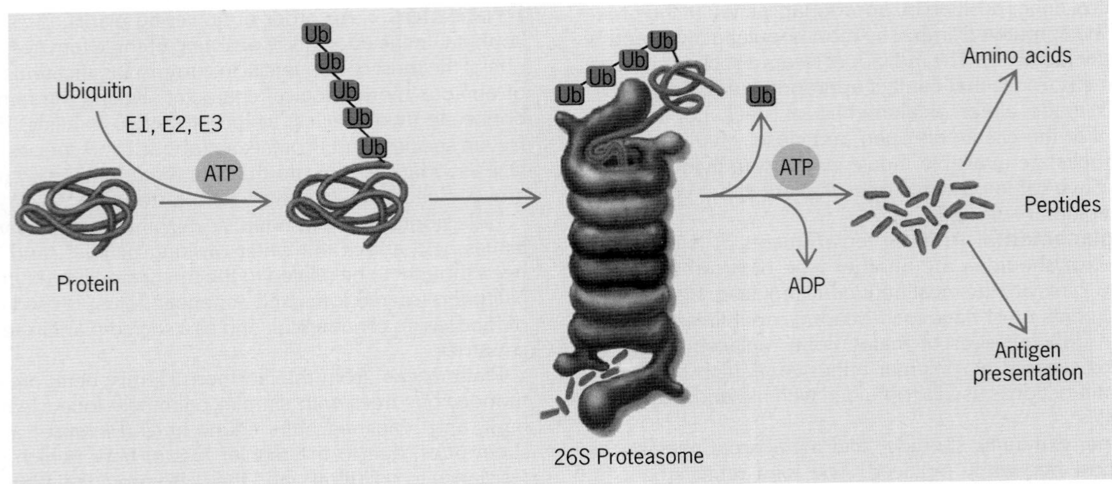

Most intracellular proteins destined for degradation by the 26S proteasome are bound to multiple ubiquitin molecules through the action of three enzymes: a ubiquitin-activating enzyme (E1), a ubiquitin-conjugating enzyme (E2), and a ubiquitin-protein ligase (E3). Degradation of ubiquitinated proteins occurs within the central chamber of the proteasome. Ubiquitin conjugation, unfolding of a substrate, and its translocation into the inner cavities of the proteasome require energy obtained by ATP hydrolysis to ADP. The great majority of peptides released by the proteasome are further digested into amino acids by other peptidases in the cytosol, but a small fraction of peptides escape destruction and serve in MHC class I antigen presentation.

by peptidases in the cytosol, which are then reutilized for new protein synthesis.

Ubiquitin is a small (76 amino acids) protein found in all eukaryotic cells but not in bacteria and archaea. Most proteins degraded by the 26S proteasome are first marked by the covalent linkage to multiple ubiquitin molecules. Ubiquitin conjugation represents a means of providing selectivity and specificity to the degradation process. Long chains of ubiquitin molecules are attached to proteins through the action of three enzymes called E1, E2, and E3. Proteins tagged with ubiquitin chains are then rapidly bound and destroyed by the 26S proteasome. Besides being involved in the rapid clearance of mutated and abnormal proteins which accumulate in various inherited diseases, the ubiquitin-proteasome pathway is implicated in regulated degradation of transcription factors, oncogenes, tumor-suppressor genes, and cell-cycle regulators. Therefore it is not surprising that several prevalent diseases are caused by the gain or loss of function of this proteolytic system. The ubiquitin-proteasome system is also very important in the onset of inflammation. *See* ADENOSINE TRIPHOSPHATE (ATP); AMINO ACIDS; CYTOLYSIS; ENZYME; NUCLEOPROTEIN; PROTEIN; TUMOR VIRUSES.

Tomo Saric; Stewart Lecker; Alfred L. Goldberg

Protective coloration A strategy that organisms use to avoid or deflect the attacks of predators by misleading the latter's visual senses.

Protective coloration can be classified according to whether the functioning or malfunctioning of the vertebrate visual system is exploited. Exploiting the malfunction of the system means simply "not being seen": the prey fails to attract the attention of the predator, usually because it is the same color as the general background or because it fails to cast a shadow. The organism avoids producing shadow by flattening itself against the substrate, or by countershading, in which the lower parts of a cylindrical prey such as a caterpillar are more lightly colored than the upper parts. As shadows normally form on the underside of cylinders, the shading cancels the shadow and makes the caterpillar optically flat. Animals that match their background often have an ability to select the appropriate background to rest on, or much less frequently can change their own color to match (as in the case of the chameleon).

Exploiting the functioning of the vertebrate perceptual system takes many forms. The vertebrate visual cortex decodes the image on the retina in a hierarchical process starting with the detection of edges. A moth may counter this by possessing strikingly contrasted patches of color on its wings, arranged in a random way. The outline of the moth is thus broken up, and the predator cannot decode it as a significant shape. The prey may also exploit the learning capacity of the predator. For example, insectivorous birds see leaves but do not attack them because they have learned (or perhaps know innately) that these are not edible. Resemblances to leaves, twigs, thorns, flowers, parts of flowers, and more bizarre objects like fresh turds (usually bird droppings) are very widespread. This type of camouflage is termed mimetic camouflage. Camouflage in general is often termed cryptic coloration.

Coloration may be considered mimetic if protection is achieved by a resemblance to some other existing object, which is recognized by the predator but not associated in its mind with feeding. Usually this negative or neutral association is learned, but in a minority of instances it is almost certainly innate. Small birds have an innate flight response to large eyes in close-up (which normally indicate that a cat or a predatory bird is dangerously close). This reaction is exploited by many moths and other insects, which have eyelike markings, sometimes very convincing in their shading and highlighting, on concealed parts of the wings. Attack by a bird causes such a moth to change its posture rapidly to reveal the fake eyes, thus frightening away the attacker. Motmots (birds which habitually prey on snakes) have a similar innate fear of the red, black, and yellow striping patterns of the deadly coral snakes. These patterns are mimicked by various nonvenomous snakes, and even some caterpillars.

Flash coloration describes the phenomenon in which the prey is cryptic when at rest, but reveals brilliantly colored parts while escaping. This behavior seems to function simply by startling the predator. Very small eye marks at the tips of the wings, or a false head at the wrong end of the body (shown by some coral reef fish, for example) may cause the predator to misdirect its attack.

Protection through the possession of a chemical or physical defense that is dangerous to one's potential predator, accompanied by a strikingly conspicuous pattern known as warning coloration (often black, red, yellow, and white), is widespread—the ensemble of defense and color is termed aposematic. The actual defense ranges from toxic venoms through stings (in wasps, for example), to the oozing of noxious foams or hemolymph (as in ladybirds), to the possession of toxic chemicals (cyanides, cardiac glycosides, alkaloids) that will poison the predator or simply produce a revolting taste. The function of the warning color is to remind the predator of its previous unpleasant experience.

Sometimes the term mimicry is restricted to resemblances between edible species and actively defended and warningly colored models (as opposed to inedible objects such as thorns). Much is known about the evolution of this kind of mimicry in butterflies (and to a lesser extent, in bees and flies). If the mimic is entirely edible, the relationship is parasitic; the mimic benefits from the resemblance, but as every encounter with a mimic reduces the predator's aversion, the model suffers some increase in the rate of attack. Such mimicry has traditionally been termed Batesian mimicry. Alternatively, the mimic may be almost or fully as defended as the model, leading to a mutualistic relationship known as Müllerian mimicry, in which both the model and the mimic species suffer a decreased rate of predation.

John R. G. Turner

Protein A polymeric compound made up of various monomeric units called amino acids. Amino acids are joined together in a chain by peptide (amide) bonds between the α-carboxyl groups and the α-amino groups of adjacent amino acids. The first amino acid in a protein usually contains a free α-amino group, as shown below. Proteins generally contain from

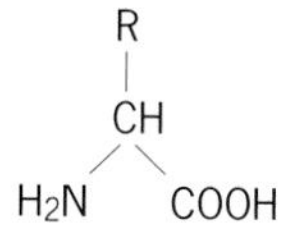

$$\begin{array}{c} R \\ | \\ CH \\ \diagup \quad \diagdown \\ H_2N \qquad COOH \end{array}$$

Amino acid

$$H_2N - \underset{R}{\overset{}{CH}} - CO - NH - \underset{R'}{CH} - CO - NH - \underset{R''}{CH} - CO$$

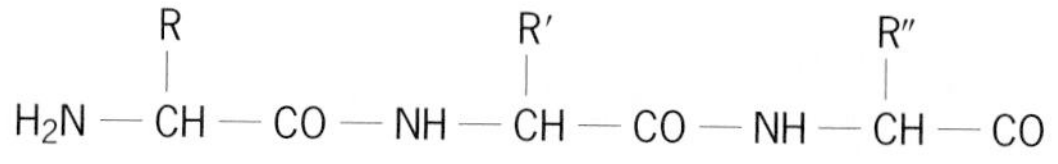

Peptide

50 to 1000 amino acids per chain. Small chains of up to 50 amino acids are usually referred to as peptides or polypeptides. Of the more than 200 amino acids that have been discovered either in the free state or in small peptides, only 20 amino acids, or derivatives of these 20, are present in mammalian proteins. *See* AMINO ACIDS; PEPTIDE.

Proteins are central to the processes of life. They are fundamental components of all biological systems, performing a wide variety of structural and functional roles. For example, proteins are primary constituents of structures such as hair, tendons, muscle, skin, and cartilage. Several hormones, such as insulin and growth hormone, are proteins. The substances responsible for oxygen and electron transport (hemoglobin and cytochromes, respectively) are conjugated proteins that contain a metalloporphyrin as the prosthetic group. Chromosomes are highly complex nucleoproteins, that is, proteins conjugated with nucleic acids. Viruses are also nucleoprotein in nature.

Protein enzymes are the catalysts of nearly all biochemical transformations. Pepsin and rennin are examples of digestive enzymes involved in breaking down food. DNA polymerases are enzymes that duplicate DNA for cell division, other enzymes are needed to repair damaged DNA, and gene expression is carried out by RNA polymerases. The chemical reactions used in metabolic pathways for carbohydrates (citric acid cycle), lipids, amino acids, and energy production (oxidative phosphorylation) are catalyzed by protein enzymes. All living things contain proteins because they serve as the molecular tools and machines of life. *See* ENZYME.

The linear arrangement of amino acids in a protein is termed its sequence (primary structure). The sequence is highly specific and characteristic for each particular protein. It is determined by the DNA sequence of each protein's gene that is expressed in the form of messenger RNA. Elucidation of the mechanism by which proteins are built up from free amino acids has been one of the key problems of molecular biology. *See* DEOXYRIBONUCLEIC ACID (DNA); MOLECULAR BIOLOGY; RIBONUCLEIC ACID (RNA).

Proteins are not stretched-out polymers; rather, each adopts a specific extended or compact and organized structure called its native structure. It is still not completely understood how proteins "fold" into their structures, nor can we accurately predict the complete structure from its amino acid sequence. In order to prevent the misfolding of certain proteins, chaperone proteins intervene to facilitate proper folding. Proteins that become misfolded or that have lost their native structure pose a major problem in cells. *See* MOLECULAR CHAPERONE; PROTEIN FOLDING.

The polypeptide backbone of protein can fold in several ways by means of hydrogen bonds between the carbonyl oxygen and the amide nitrogen. Structural elements created by backbone hydrogen bonding interactions in the polypeptide are called secondary structures and include features like α-helices, β-sheets, and turns. In helices, the backbone is coiled in a regular fashion that brings peptide bonds separated by several amino acids into close spatial approximation. The stability of a helix is attributed to hydrogen bonds between these peptide bonds. In addition to α-helices, polypeptides form β-sheet structures that are made of two or more segments that run parallel or antiparallel to each other and connect through backbone hydrogen bonds. *See* HYDROGEN BOND.

The third level of folding in a protein (tertiary structure) comes through interactions between different parts of the molecule. At this level of structure, various secondary structure elements are brought together and interact through many types of associations. Hydrogen bonds between different amino acids and peptide bonds, hydrophobic interactions between nonpolar side chains of amino acids such as phenylalanine and leucine, and salt bridges such as those between positively charged lysyl side chains and negatively charged aspartyl side chains all contribute to the tertiary structure specific to a given protein. Disulfide bridges formed between two cysteines at different linear locations in the molecule can stabilize parts of a three-dimensional structure by introducing a covalent bond as a cross-link. The result is a unique architecture that is predetermined by the particular sequence of amino acids in the protein. Finally, some proteins contain more than one polypeptide chain per molecule. This feature is referred to as the quarternary structure.

Karla L. Ewalt; Paul Schimmel; Gertrude E. Perlmann; James M. Manning

Protein degradation The enzymatic destruction of proteins. Like other macromolecular (large) components of an organism, proteins are in a dynamic state of synthesis (generation) and degradation (destruction, proteolysis). During proteolysis, the peptide bond that links the amino acids to one another is hydrolyzed, and free amino acids are released. The process is carried out by a group of enzymes called proteases. *See* ENZYME; PROTEIN.

Distinct proteolytic mechanisms serve different physiological requirements. It is necessary to distinguish between destruction of "foreign" and "self" proteins. Foreign proteins contained in the diet are degraded "outside" the body, in the digestive tract. This is necessary in order to prevent a response of the immune system to the intact proteins. The released amino acids are then absorbed and serve to build the organism's self proteins. Self proteins can be divided into extracellular and intracellular. The two groups are degraded by different mechanisms. Extracellular proteins in solution in the plasma (the blood fluid phase), such as blood coagulation factors, immunoglobulins, cargo carrier proteins (such as the low-density lipoprotein, LDL, that carries cholesterol; or transferrin that carries iron), and peptide hormones (such as insulin), are taken up by pinocytosis (engulfment of small droplets) or following binding to specific receptors. They are carried via a series of vesicles (endosomes) that fuse with lysosomes, where they are degraded by the proteases contained in these organelles. During this process, the extracellular proteins are never exposed to the intracellular environment (cytosol). Degradation of proteins in lysosomes is not specific, and all engulfed proteins are degraded at similar rates.

The proteolytic system involved in intracellular protein degradation must be distinct from lysosomal proteolysis. The discovery of the ubiquitin–proteasome proteolytic pathway has resolved this enigma. Degradation of a protein via the ubiquitin pathway involves two steps: tagging of the substrate by covalent attachment of multiple ubiquitin molecules, and degradation of the tagged protein by the 26S proteasome with release of free and reusable ubiquitin. In addition to serving as a proteolytic signal, ubiquitination can regulate other processes. *See* PROTEASOME.

The ubiquitin system is involved in regulating numerous basic cellular processes. Among them are cell division, differentiation and development, the cellular stress response, DNA repair, transcriptional regulation, and modulation of the immune and inflammatory responses. The list of cellular proteins that are targeted by ubiquitin is growing rapidly. The proteins include cell cycle regulators, tumor suppressors, transcriptional activators and their inhibitors, cell surface receptors, and endoplasmic reticulum proteins. Mutated/denatured/misfolded proteins are recognized specifically and removed efficiently. Thus, the system plays a key role in the cellular quality control and defense mechanisms.

Aaron Ciechanover; Kazuhiro Iwai

Protein engineering The design and construction of new proteins or enzymes with novel or desired functions, through the modification of amino acid sequences using recombinant DNA (deoxyribonucleic acid) technology. The sizes and three-dimensional conformations of protein molecules are also manipulated by protein engineering. The basic techniques of genetic engineering are used to alter the genes that encode proteins, generating proteins with novel activities or properties. Such manipulations are frequently used to discover structure-function relationships, as well as to alter the activity, stability, localization, and structure of proteins. *See* GENE; GENETIC ENGINEERING; PEPTIDE; PROTEIN.

Many subtle variations in a particular protein can be generated by making amino acid replacements at specific positions in the polypeptide sequence. Each protein is unique by virtue of the sequence of its amino acids. At any position in the sequence, an amino acid can be replaced by another to generate a mutant protein that may have different characteristics by virtue of the single replaced amino acid. Other mutations have very little effect on their proteins. This is particularly true when the amino acid being changed is substituted with other closely related amino acids and when the amino acid is not conserved in the same protein found in other organisms. Typically, site-directed protein engineering targets amino acids that are involved in a particular biological activity. *See* AMINO ACID; MUTATION.

In addition to the substitution of amino acids at specific positions, amino acids can be deleted from the sequence, either individually or in groups. These proteins are referred to as deletion mutants. Deletion mutants may or may not be missing one or more functions or properties of the full, naturally occurring protein. Protein sequences also can be joined or fused to that of another protein. The resulting protein is called a hybrid, fusion, or chimeric protein, which generally has characteristics that combine those of each of the joined partners. Protein fusions have been extensively used to study interactions between two or more proteins.

Point and deletion mutants and hybrid proteins are constructed to obtain polypeptides with new properties. These proteins are either created individually, by site-directed mutagenesis, or they are generated as a large pool or library of millions of variants. The library is then screened or subjected to a special selection procedure to obtain the protein or proteins with the desired characteristics. Protein engineering has been used to produce therapeutic proteins with improved properties such as increased solubility and stability. Proteins also can be engineered to acquire new biological activities.

Karla L. Ewalt; Paul Schimmel

Protein folding The process by which a long, flexible, unbranched chain of amino acids (the unfolded state of a protein) spontaneously changes shape to form a stable three-dimensional conformation (the native or folded state). This specific folding process is required for the protein's biological function. The sequential arrangement of amino acids is encoded by the genetic material, deoxyribonucleic acid (DNA) [see illustration]. The 20 most frequently used amino acids have a region in common that forms the covalent peptide bonds in the chain, termed the backbone, and a region that is chemically distinct, identified as the side chain. Amino acid side chains vary in size and shape, as well as the propensity to interact with water, oil, and ions. As a result, the side chains are differentially able to participate in noncovalent chemical bonds (hydrophobic interactions, electrostatic interactions, and hydrogen bonds) with each other and with the solvent. In addition, the amino acid cysteine can form covalent disulfide bonds with other cysteines. Together, this chemical interaction code determines the network of bonds that drives protein folding, stabilizes the native structure, and mediates protein function for any given sequence of amino acids.

Since a single protein chain can consist of thousands of amino acids, and 20 different amino acids are commonly used as assembly units, the number of possible protein sequences and configurations is immense. The function of a protein is determined

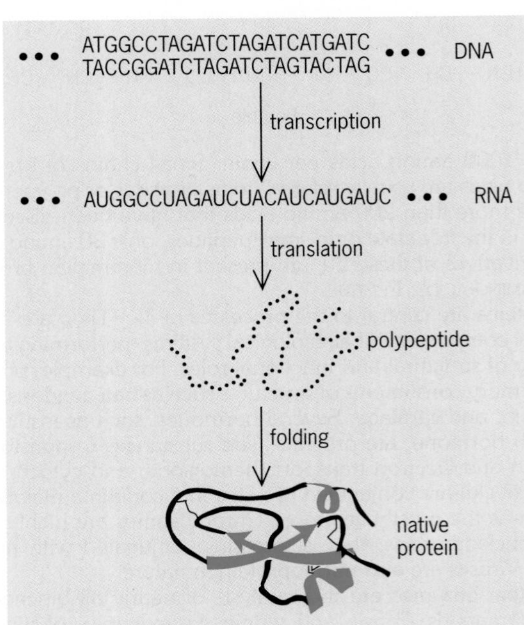

An integrated view of protein production with folding as the terminal step. The varying chemical properties of amino acid side chains drive protein folding.

by its three-dimensional structure, and hence protein functions are diverse: among many functions, these complex molecules act as efficient and specific biological machines to catalyze biochemical reactions for metabolism, provide structural support, regulate biological processes, and transport other molecules. The process of protein folding places the protein's atoms in the correct three-dimensional arrangement required for that protein's unique function. Loss of the protein's native structure through denaturation involves a simultaneous loss of the protein's function as well. Therefore, a comprehensive understanding of protein folding may explain protein malfunctions in vivo, as well as suggest a means to design artificial proteins with novel functions. *See* AMINO ACIDS; BIOORGANIC CHEMISTRY; PEPTIDE; PROTEIN.

Because protein folding is a balance between a large number of weak intraprotein and protein-solvent interactions, this process is strongly dependent on environmental conditions such as the solvent involved, salinity, temperature, and pressure. In vitro, partially folded or misfolded proteins form under nonoptimal conditions. These incorrect folding states can irreversibly coalesce to form various types of aggregates. Such aggregation is frequently an unwelcome side product during research and for production of protein-based pharmaceuticals. In vivo, misfolded proteins in a healthy cell are either refolded by chaperones or degraded, preventing a buildup of aggregates. However, nonoptimal conditions such as inflammation, overproduction of protein, aging/oxidation, and mechanical or chemical stresses may increase the amount of misfolded protein in vivo. These misfolded proteins may then form amyloid fibrils, which are associated with a large number of diseases, among them Alzheimer's disease, Down syndrome, Parkinson's disease, and cataracts. A comprehensive understanding of protein folding and structure will contribute toward the logical design of therapies for these diseases. Sarah Bondos; Kathleen Matthews

Protein kinase One of a family of enzymes that exert regulatory effects on a variety of cellular functions, as well as malignant transformation, by adding a phosphate group to proteins according to the equation

$$\text{ATP} + \text{protein—OH} \xrightarrow{\text{protein kinase}} \text{protein—OPO}_3^{2-} + \text{ADP}$$

(ATP represents adenosine triphosphate, ADP represents adenosine diphosphate, and —OH is a hydroxyl group attached to an amino acid residue.) Based upon the nature of the phosphorylated —OH group, these enzymes are classified as serine/threonine protein kinases (for example, protein kinase A) and tyrosine protein kinases, where serine, threonine, and tyrosine are amino acid residues found in proteins. Furthermore, there is a small group of dual-specificity kinases, which closely resemble serine/threonine kinases, that catalyze the phosphorylation of both threonine and tyrosine on target proteins.

Protein kinases are enzymes that play a role in nearly every aspect of cell biology. Work in the 1960s showed that protein kinases play a regulatory role in carbohydrate metabolism. Subsequent work indicates that protein kinases participate in nearly all cellular activities including regulation of gene expression, cell division, apoptosis (programmed cell death), differentiation, development, and antibody production. The brain, moreover, is an especially rich site of protein kinase activity. *See* CELL (BIOLOGY); NEUROBIOLOGY.

Protein phosphorylation represents the chief regulatory mechanism in animal cells. In some cases, phosphorylation leads to activation, and in other cases phosphorylation leads to inactivation of the phosphorylated protein. It is estimated that perhaps a quarter of all animal proteins can be phosphorylated by protein kinases. Moreover, many proteins can be phosphorylated at more than one residue by a given protein kinase, and many proteins can be phosphorylated by several protein kinases. *See* ENZYME; PROTEIN.

Protein phosphorylation is a reversible process. Protein kinases add phosphoryl groups to proteins, and protein phosphatases remove these groups during a hydrolysis reaction. To achieve coordinate regulation of cellular processes, both protein kinase activity and protein phosphatase activity are carefully regulated. Robert Roskoski, Jr.

Protein metabolism The transformation and fate of food proteins from their ingestion and assimilation to the elimination of their excretion products. Proteins are polymers of α-amino acids that are connected by peptide bonds. They are the chief structural and functional components of all living organisms. Proteins are the main building blocks of the cells, tissues, organs, and systems of the body. Proteins of one species differ from those of another species and, within a single animal, proteins of muscle differ from those of the brain, kidney, liver, and other organs. *See* AMINO ACIDS; METABOLISM; PEPTIDE; PROTEIN.

Body proteins are continually broken down and resynthesized, a process called turnover. There are two main pathways that participate in intracellular protein degradation: the ubiquitin-proteasome pathway and the lysosomal pathway. The proteasomal pathway is responsible for degrading proteins that are damaged because of oxidative stress or are no longer needed owing to changing metabolic needs. The lysosome is a membrane-enclosed vesicle inside the cell that contains a variety of proteolytic enzymes and operates under acidic conditions. It degrades proteins that combine with receptors found on the plasma membrane of the cell and that are subsequently taken up by the cell where they fuse with lysosomes. *See* CELL (BIOLOGY); LYSOSOME.

Besides the breakdown of endogenous proteins, ingested proteins contribute to the amino acid pool. Protein is digested to amino acids in the stomach and small intestine. These amino acids are absorbed and undergo a degree of metabolic interconversion prior to release into the portal vein, which takes them to the liver. Many of the amino acids are taken up by the liver, and the remainder circulates to the other tissues and organs in the body. The newly absorbed amino acids equilibrate with amino acids in cells that result from protein breakdown. The amino acid pool serves as the source of precursors for new protein synthesis and for the conversion to other metabolites. Twenty amino acids are required for protein synthesis. The amino acids of proteins fall into two nutritional categories: essential or indispensable and nonessential or dispensable. The essential amino acids cannot be synthesized by the body or cannot be synthesized in adequate quantities and therefore must be taken in the diet, while the nonessential amino acids can be synthesized by the body in sufficient quantities to sustain health.

Protein digestion is initiated in the stomach and completed in the duodenum of the small intestine. The main proteolytic enzyme of the stomach is pepsin, which is secreted by the chief cells in an inactive form called pepsinogen. The conversion of precursor pepsinogen to active pepsin is accelerated by stomach acid and by active pepsin. Pepsin preferentially hydrolyzes peptide bonds adjacent to an aromatic amino acid (phenylalanine, tyrosine, tryptophan), and pepsin requires an acid medium in which to function. Stomach acid also unfolds, or denatures, ingested proteins and increases their susceptibility to digestion. *See* DIGESTIVE SYSTEM; PEPSIN.

The acidic stomach contents, which contain partially degraded proteins, are discharged as chyme (semifluid, partially digested food mass) into the slightly alkaline fluid in the duodenum of the small intestine. The pancreatic juice released into the duodenum contains several digestive enzymes that are secreted as zymogens (inactive enzyme precursors) including trypsinogen, chymotrypsinogen, procarboxypeptidase, and proelastase. The activation of trypsinogen involves its proteolytic cleavage by yet another enzyme secreted by the pancreas, namely enteropeptidase, to produce active trypsin. Trypsin, in turn, catalyzes the formation of chymotrypsin, carboxypeptidase, and elastase from their inactive zymogen precursors. In each of these processes, certain peptide bonds are broken to yield the active enzymes.

Trypsin, chymotrypsin, and elastase are endopeptidases; that is, they cleave peptide bonds that occur in the interior of proteins. These three proteases catalyze the cleavage of proteins into small peptides. Aminopeptidases, which are exopeptidases that cleave amino acid residues from the amino-terminal ends of peptides, help to complete protein hydrolysis. Carboxypeptidases, which are also exopeptidases, catalyze the removal of amino acids at the C-terminus of substrate peptides or proteins. Amino acids, dipeptides, and tripeptides are absorbed by the intestinal cells. The hydrolysis of peptides to amino acids is completed intracellularly.

The amino acid digestion products are absorbed as rapidly as they are liberated. The absorption is confined chiefly to the small intestine and is a process that involves the metabolic participation of the cells of the intestinal lining. Amino acids are absorbed from the gut and liberated into the hepatic (liver) portal vein. Most amino acids are taken up and metabolized by the liver except for valine, leucine, and isoleucine—the branched-chain amino acids. The branched-chain amino acids are taken up chiefly by skeletal muscle. The liver amino acids are converted into other amino acids (the nonessential amino acids) and into cellular and plasma proteins.

The liver is the major organ for plasma protein synthesis. It synthesizes albumin, the predominant blood plasma protein. The liver also synthesizes fibrinogen, one of the main components of blood clots. *See* ALBUMIN; BLOOD; FIBRINOGEN; LIVER.

Robert Roskoski, Jr.; David M. Greenberg

Proteins, evolution of Proteins are large organic molecules that are involved in all aspects of cell structure and function. They are made up of polypeptide chains, each constructed from a basic set of 20 amino acids, covalently linked in specific sequences. Each amino acid is coded by three successive nucleotide residues in deoxyribonucleic acid (DNA); the sequence of amino acids in a polypeptide chain, which determines the structure and function of the protein molecule, is thus specified by a sequence of nucleotide residues in DNA. *See* GENE; PROTEIN.

Sequence analyses of polypeptides which are shared by diverse taxonomic groups have provided considerable information regarding the genetic events that have accompanied speciation. Interspecies comparison of the amino acid sequences of functionally similar proteins has been used to estimate the amount of genetic similarity between species; species that are genetically more similar to each other are considered to be evolutionarily more closely related than those that are genetically less similar.

The study of functionally related proteins from different animal species has suggested that single amino acid substitutions are the predominant type of change during evolution of such proteins. Insertions or deletions of one or more amino acids have also been reported. In proteins that serve the same function in dissimilar species small differences in the amino acid sequence will often not affect overall functioning of the protein molecule.

In taxonomic protein sequence analysis, the amino acid sequence of a protein from one species is compared with the amino acid sequence of the protein from another species, and the minimum number of nucleotide replacements (in DNA) required to shift from one amino acid to another is calculated. Peptide "genealogies" can be constructed from many such comparisons in a related group of organisms.

Classical versus protein-derived phylogenies. It has been recognized for a long time that the amino acid sequences of a protein are species-specific. Protein sequencing has been used widely since the mid-1960s to examine taxonomic relationships. Results indicate that, in general, genealogical relationships (phylogenies) based on sequence analyses correspond fairly well with the phylogeny of organisms as deduced from more classical methods involving morphological and paleontological data.

Evolutionary biologists are turning increasingly to the new nucleic acid sequencing technology as an alternative to determining the amino acid sequence of proteins. Knowing the actual nucleotide sequences of genes rather than having to infer them from protein sequence data allows more accurate data to be used in determining genealogical relationships of organisms. For example, silent nucleotide substitutions (that is, base changes in DNA codons that do not result in amino acid changes) can be detected.

Orthologous and paralogous sequences. The reconstruction of phylogenies from analysis of protein sequences is based on the assumption that the genes coding for the proteins are homologous, that is, descendants of a common ancestor. Those sequences whose evolutionary history reflects that of the species in which they are found are referred to as orthologous. The cytochrome *c* molecules (present in all eukaryotes) are an example of an orthologous gene family. Organisms as diverse as humans and yeast have a large proportion of the amino acids in these molecules in common; they derive from a single ancestral gene present in a species ancestral to both these organisms and to numerous others.

Sequences which are descendants of an ancestral gene that has duplicated are referred to as paralogous. Paralogous genes evolve independently within each species. The genes coding for the human α, β, γ, δ, ϵ, and ζ hemoglobin chains are paralogous. Their evolution reflects the changes that have accumulated since these genes duplicated. Analysis of paralogous genes in a species serves to construct gene phylogenies, that is, the evolutionary history of duplicated genes within a given lineage.

Rate of evolution. Sequence data from numerous proteins have shown that different proteins evolve at different rates. Some proteins show fewer amino acid substitutions, or more conservation, than others. Proteins such as immunoglobulins, snake venom toxins, and albumins have changed extensively. Their function apparently requires relatively less specificity of structure and therefore has relatively greater tolerance for variance. By contrast, certain proteins, such as various histones, have changed relatively little over long periods of time. Histone H4 shows extreme conservation; it has essentially the same sequence in all eukaryotes examined. Such extensive sequence conservation is generally interpreted to indicate that the functions of H4 are extremely dependent on its entire structure; thus little or no change is tolerated in its structure. The rates at which different proteins evolve are therefore thought to be due to different functional constraints on the structure of the proteins—the more stringent the conditions that determine the function of a protein molecule, the smaller the chance that a random change will be tolerated in its structure.

Each protein generally has a nearly constant evolutionary rate (the rate of acceptance of mutations) in each line of descent. Exceptions to this rule have been reported, however, and much effort has been spent on determining whether these anomalies are genuine. Some anomalies have been shown to be due to comparison of nonhomologous proteins, and others due to sequencing errors. Other deviations from constant rate of sequence evolution remain to be explained; once uncovered, these may provide useful information about the mechanisms of evolution at the molecular level.

Pamela K. Mulligan

Proteocephaloidea An order of tapeworms of the subclass Cestoda. With one exception, these worms are intestinal parasites of fresh-water fishes, amphibians, and reptiles. The holdfast organ bears four suckers and, frequently, an apical organ which may be suckerlike. The segmental anatomy is very similar to that of the Tetraphyllidea. Most authorities recognize two families, the Proteocephalidae and the Monticellidae. *See* CESTODA.

Clark P. Read

Proterozoic A major division of geologic time spanning from 2500 to 543 million years before present (Ma). The beginning of Proterozoic time is an arbitrary boundary that roughly coincides with the transition from a tectonic style dominated by extensive recycling of the Earth's continental crust to a style characterized by preservation of the crust as stable continental platforms. The end of the Proterozoic coincides with the Precambrian-Cambrian boundary, which is formally defined on the basis of the first appearance of diverse coelomate invertebrate animals. Proterozoic Earth history testifies to several remarkable biogeochemical events, including the formation and dispersal of the first supercontinent, the maturation of life and evolution of animals, the rise of atmospheric oxygen, and the decline of oceanic carbonate saturation. Tremendous iron and lead-zinc mineral deposits occur in Proterozoic rocks, as do the first preserved accumulations of oil and gas. *See* CAMBRIAN; PRECAMBRIAN.

Many of the Earth's Archean cratons are blanketed by little-deformed sequences of Proterozoic sedimentary rocks, which indicate that vigorous recycling of the Earth's crust, characteristic of Archean time, had slowed markedly by the beginning of Proterozoic time. This decrease in crustal recycling is attributed to the development of thick continental roots, which stabilized the cratons, and the decrease in heat that was escaping from the Earth's interior, believed to drive thermal convection in the Earth's mantle and recycling of the crust. Most of the Earth's Archean cratons appear to have participated in the formation of a supercontinent in Mesoproterozoic time, about 1200 Ma. This supercontinent, called Rodinia, seems to have assembled with the North American craton (Laurentia) at its center. Rodinia persisted until the latest part of the Neoproterozoic, about 600 Ma. *See* ARCHEAN; CONTINENTS, EVOLUTION OF; EARTH, CONVECTION IN; EARTH, HEAT FLOW IN; EARTH CRUST; EARTH INTERIOR; PLATE TECTONICS.

Giant iron oxide deposits were formed by precipitation from seawater about 2000 Ma, whereupon oxygen was free to accumulate in the atmosphere and shallow ocean. During most of Paleoproterozoic time the oceans and atmosphere were reducing and ferrous iron was abundant in seawater. *See* ATMOSPHERE, EVOLUTION OF.

The partial pressure of carbon dioxide on the early Earth was very high. During Proterozoic time, much of the mass of carbon shifted from the ocean and atmosphere to the solid Earth. Enormous volumes of limestone [$CaCO_3$] and dolostone [$CaMg(CO_3)_2$] were deposited and testify to this shift. *See* DOLOMITE ROCK; LIMESTONE; SEDIMENTARY ROCKS.

Glaciers covered significant parts of the Earth during two widely separated times in Proterozoic history. The first episode occurred about 2200 Ma, and glacial deposits of that age cover various parts of North America and Scandinavia. The second episode consisted of at least two different pulses spanning from 750 to 600 Ma during Neoproterozoic time. Glaciers formed at that time were of almost global extent, and were thought to have extended from the poles to the Equator, according to the snowball Earth hypothesis. *See* GLACIAL EPOCH.

A number of significant events in the evolution of life occurred during Proterozoic time. The record of biological activity is rich, consisting of actual body fossils, in addition to organism traces and impressions, and complex chemical biomarkers. Eukaryotic microbes appear to have evolved by about 1900 Ma, when they became major players in ecosystems present at that time. By the beginning of Neoproterozoic time, about 1000 Ma, multicellular eukaryotic algae are present in numerous sedimentary basins around the world. *See* EUKARYOTAE; RIBONUCLEIC ACID (RNA).

The evolution of animals did not take place until the close of Neoproterozoic time. Why these organisms evolved at this particular time in Earth history remains unanswered. General opinion proposes that it was likely the result of the confluence of a number of environmental factors, such as the rise in oxygen. Whatever the cause of their origin, these existed until at least 543 Ma, when another major evolutionary adaptive radiation began which marks the onset of Cambrian time and the end of the Proterozoic Eon. *See* ANIMAL EVOLUTION; EXTINCTION (BIOLOGY); GEOLOGIC TIME SCALE.

John P. Grotzinger

Protista The kingdom comprising all single-celled forms of living organisms in both the five-kingdom and six-kingdom systems of classification. Kingdom Protista encompasses both Protozoa and Protophyta, allowing considerable integration in the classification of both these animallike and plantlike organisms, all of whose living functions as individuals are carried out within a single cell membrane. Among the kingdoms of cellular organisms, this definition can be used to distinguish the Protista from the Metazoa (sometimes named Animalia) for many-celled animals, or from the Fungi and from the Metaphyta (or Plantae) for many-celled green plants. *See* METAZOA.

The most significant biological distinction is that which separates the bacteria and certain other simply organized organisms, including blue-green algae (collectively, often designated Kingdom Monera), from both Protista and all many-celled organisms. The bacteria are described as prokaryotic; both the Protista and the cells of higher plants and animals are eukaryotic. Structurally, a distinguishing feature is the presence of a membrane, closely similar to the bounding cell membrane, surrounding the nuclear material in eukaryotic cells, but not in prokaryotic ones. *See* EUKARYOTAE; PROKARYOTAE; PROTOZOA.

The definition that can separate the Protista from many-celled animals is that the protistan body never has any specialized parts of the cytoplasm under the sole control of a nucleus. In some protozoa, there can be two, a few, or even many nuclei, rather than one, but no single nucleus ever has separate control over any part of the protistan cytoplasm which is specialized for a particular function. In contrast, in metazoans there are always many cases of nuclei, each in control of cells of specialized function.

Most authorities would agree that the higher plants, the Metazoa, and the Parazoa (or sponges) almost certainly evolved (each independently) from certain flagellate stocks of protistans.

W. D. Russell-Hunter

Protobranchia A subclass of bivalve mollusks characterized by a foot with a sole that is divided sagittally and longitudinally and has papillate margins. The Protobranchia are divided into two orders, Solemyoida and Nuculoida. The Solemyoida have an edentulous shell, and if teeth are present, they are not chevron-shaped. The palps are small, triangular, and without palp proboscides. This is in contrast to the Nuculoida, which have well-developed, chevron-shaped hinge teeth and large palps with palp proboscides. The Nuculoida are subdivided into two superfamilies, Nuculacea and Nuculanacea.

The Protobranchia are found throughout the seas of the world, and they are particularly common in the deep sea, where they may form up to 10% of the invertebrate infauna. The subclass has one of the longest geological records within the animal kingdom, dating from the Early Ordovician if not the Late Cambrian. *See* BIVALVIA; MOLLUSCA.

John A. Allen

Protogyny A condition in hermaphroditic or dioecious animals in which the female reproductive structures mature before the male structures. It is of rare occurrence. Botanically, protogyny occurs in some plant species in which the stigma develops, withers, and dies before the anthers mature. *See* PROTANDRY.

Tracy I. Storer

Protolepidodendrales An extinct order of the class Lycopsida (clubmosses) of the Devonian Period. Most of these clubmosses are poorly preserved and not well understood; the notable exception is the widespread *Leclercqia complexa*, which has been largely reconstructed from fragmentary fossils.

Leclercqia (Early to Middle Devonian) was a medium-sized herb; its horizontal rhizomes generated vertical axes up to 0.4 in. (1 cm) in diameter. The vascular tissue formed a star-shaped actinostele, with inward maturation of the metaxylem tracheids. Additional structural support was provided in the cortical cylinder by thickened hypodermal fibers. The microphyllous leaves were not routinely shed. Most branched into five unequal, needle-like lobes, each supplied by a lateral branch of the median vascular trace, and arranged three-dimensionally. Zones of sterile leaves alternated with zones of fertile sporophylls; both sterile and fertile leaves bore a small ligule on the upper surface. The stomata-bearing sporangia were located on the same surface of the sporophyll, somewhat closer to the stem than the ligule. The sporangia were elliptical and released spores of the same size; thus, *Leclercqia* is primitively homosporous.

Complex leaf morphologies are important characteristics in assigning genera to the Protolepidodendrales, even though it may be that they represent multiple independent evolutionary events. The presence of a ligule has been demonstrated only in *Leclercqia*; yet it is the combination of homosporous reproduction and the presence of a ligule that delimits the order and testifies to its evolutionary intermediacy between the eligulate homosporous Lycopodiales and the ligulate heterosporous Selaginellales. *See* LYCOPODIALES; SELAGINELLALES.

Richard M. Bateman; William A. DiMichele

Proton A positively charged particle that is the nucleus of the lightest chemical element, hydrogen. The hydrogen atom consists of a proton as the nucleus, to which a single negatively charged electron is bound by an attractive electrical force (since opposite charges attract). The proton is about 1836 times heavier than the electron, so that the proton constitutes almost the entire mass of the hydrogen atom. Most of the interior of the atom is empty space, since the sizes of the proton and the electron are very small compared to the size of the atom. *See* ATOMIC STRUCTURE AND SPECTRA; ELECTRIC CHARGE; HYDROGEN.

For chemical elements heavier than hydrogen, the nucleus can be thought of as a tightly bound system of Z protons and N neutrons. An electrically neutral atom will then have Z electrons bound comparatively loosely in orbits outside the nucleus. *See* NEUTRON; NUCLEAR STRUCTURE.

The numerical values of some overall properties of the proton can be summarized as follows: charge, 1.602×10^{-19} coulomb; mass, 1.673×10^{-27} kg; spin, $(^1/_2)\hbar$ (where $\hbar$ is Planck's constant h divided by 2π); magnetic dipole moment, 1.411×10^{-26} joule/tesla; radius, about 10^{-15} m. *See* FUNDAMENTAL CONSTANTS; NUCLEAR MOMENTS; SPIN (QUANTUM MECHANICS).

It is instructive to contrast the proton's properties with those of the electron. All of the electron's properties have been found to be those expected of a spin-$^1/_2$ particle which is described by the Dirac equation of quantum mechanics. Such a Dirac particle has no internal size or structure. *See* ELECTRON; RELATIVISTIC QUANTUM THEORY.

By contrast, although it also has a spin of $^1/_2$, the proton's magnetic moment, which is different from that for a Dirac particle, and its binding with neutrons into nuclei strongly suggest that it has some kind of internal structure, rather than being a point particle. Two different kinds of high-energy physics experiments have been used to study the internal structure of the proton. An example of the first type of experiment is the scattering of high-energy electrons, above say 1 GeV, from a target of protons. The angular pattern and energy distribution of the scattered electrons give direct information about the size and structure of the proton. The second type of high-energy experiment involves the production and study of excited states of the proton, often called baryonic resonances. It has been found that the spectrum of higher-mass states which are produced in high-energy collisions follows a definite pattern. *See* BARYON.

In 1963, M. Gell-Mann and, independently, G. Zweig pointed out that this pattern is what would be expected if the proton were composed of three spin-$^1/_2$ particles, quarks, with two of the quarks (labeled u) each having a positive electric charge of magnitude equal to $^2/_3$ of the electron's charge (e), and the other quark (labeled d) having a negative charge of magnitude of $^1/_3 e$. Subsequently, the fractionally charged quark concept was developed much further, and has become central to understanding every aspect of the behavior and structure of the proton. *See* QUARKS.

An important class of fundamental theories, called grand unification theories (GUTs), makes the prediction that the proton will decay. The predicted lifetime of the proton is very long, about 10^{30} years or more—which is some 10^{20} times longer than the age of the universe—but this predicted rate of proton decay may be detectable in practical experiments. *See* GRAND UNIFICATION THEORIES.

If the proton is observed to decay, this new interaction will also have profound consequences for understanding of cosmology. The very early times of the big bang (about 10^{-30} s) are characterized by energies so high that the same grand unified interaction which would allow proton decay would also completely determine the subsequent evolution of the universe. This could then explain the remarkable astrophysical observation that the universe appears to contain only matter and not an equal amount of antimatter. *See* COSMOLOGY; ELEMENTARY PARTICLE.

Thomas H. Fields

Proton-induced x-ray emission (PIXE) A highly sensitive analytic technique for determining the composition of elements in small samples. Proton-induced x-ray emission (PIXE) is a nondestructive method capable of analyzing many elements simultaneously at concentrations of parts per million in samples as small as nanograms. PIXE is the preferred technique for surveying the environment for trace quantities of such toxic elements as lead and arsenic. There has also been a rapid development in the use of focused proton beams for PIXE studies in order to produce two-dimensional maps of the elements at spatial resolutions of micrometers.

The typical PIXE apparatus uses a small Van de Graaff machine to accelerate the protons which are then guided to the sample. Nominal proton energies are between 1 and 4 MeV; too low an energy gives too little signal while too high an energy produces too high a background. The energetic protons ionize some of the atoms in the sample, and the subsequent filling of empty inner orbits results in the characteristic x-rays. These monoenergetic x-rays emitted by the sample are then efficiently counted in a high-resolution silicon (lithium) detector which is sensitive to the x-rays of all elements heavier than sodium. *See* JUNCTION DETECTOR; X-RAYS.

The advantages of PIXE over electron-induced x-ray techniques derive from the heaviness of the proton which permits it to move through matter with little deflection. The absence of scattering results in negligible continuous radiation (bremsstrahlung). As a result, proton-induced x-ray techniques are two to three orders of magnitude more sensitive to trace elements than are techniques based on electron beams. *See* BREMSSTRAHLUNG.

Lee Grodzins

Protostar A hydrostatic sphere of gas in the process of becoming a star. A protostar forms by the gravitational collapse of a dense core within a giant cloud of dust and molecular gas (mostly H_2). As the core collapses from the inside out, it surrounds a central protostar with a cocoon of accreting dust and gas that hides it from view at optical wavelengths. Observations at longer wavelengths penetrate this material and reveal that the protostar is radiating due to the impact of infalling gas and dust. Most of the envelope material accretes first onto a circumstellar disk, from which it is then conveyed to the protostellar surface. As the envelope dissipates around a low-mass star such as the Sun, the protostar becomes visible as a pre-main-sequence T Tauri star that continues for a time to add to its mass from the accretion disk. It becomes a full-fledged star when the core temperature reaches the level required for nuclear fusion of hydrogen (temperature $T \sim 10^7$ K). For stars at least 10 times more massive than the Sun, nuclear burning begins before infalling material is dissipated, and the resulting high-intensity radiation quickly clears away the remaining envelope. *See* T TAURI STAR.

David Koerner

Prototheria One of the four subclasses of the class Mammalia. Prototheria contains a single order, the Monotremata. No ancestral genera of fossil monotremes are known, and the structure of the living monotremes is so specialized that the affinities of the Prototheria are largely conjectural. Most mammalogists believe that the prototheres arose from a different stock of therapsid reptiles than the one that gave rise to the Theria.

No fossils earlier than the Pleistocene are known, and these come from Australia. The duckbilled platypus and several species of the echidnas are living representatives of this group. Everything indicates that the Prototheria represent a very small and relatively unsuccessful group that has miraculously survived in an isolated corner of the Earth. *See* MAMMALIA; MONOTREMATA; THERAPSIDA.

D. Dwight Davis; Frederick S. Szalay

Prototype A first or original model of hardware or software. Prototyping involves the production of functionally useful and trustworthy systems through experimentation with evolving systems. Generally, this experimentation is conducted with much user involvement in the evaluation of the prototype.

A primary use for prototyping is the acquisition of information that affects early product development. For example, if requirements for human-computer interfaces are ambiguous or inadequate, prototyping is frequently used to define an acceptable functional solution. It is a method for increasing the utility of user knowledge for purposes of continuing development to a final product. Information obtained through prototyping is important to designers, managers, and users in identifying issues and problems. Prototyping conserves time and resources prior to the commitment of effort to construct a final product.

In many hardware and software development projects, the first prototype product built is barely usable. It is usually too slow, too big, too awkward in use. Hence, the term throwaway prototype is generally applied to describe this early use of prototyping. Usually this is due to lack of understanding of user requirements. There is no alternative but to start again and build a redesigned version in which these problems are solved.

A developmental prototyping approach for incremental design of subsystems is often used to reduce the risk involved in building a system-level prototype. In this prototyping environment an incremental approach to rapid prototyping of subsystems development is used. This provides for management oversight of the entire process to assure that resource usage is effective and efficient. Product assurance is implemented throughout the process to make certain that the prototype operation contains the necessary components to satisfy subsystem requirements. Requirements analysis is performed and reviewed, then incremental specifications are developed and reviewed, followed by design of the approved specifications, and completed by implementation of the product. *See* MODEL THEORY; SOFTWARE ENGINEERING; SYSTEMS ENGINEERING.

James D. Palmer

Protozoa A group of eukaryotic microorganisms traditionally classified in the animal kingdom. Although the name signifies primitive animals, some Protozoa (phytoflagellates and slime molds) show enough plantlike characteristics to justify claims that they are plants.

Protozoa are almost as widely distributed as bacteria. Free-living types occur in soil, wet sand, and in fresh, brackish, and salt waters. Protozoa of the soil and sand live in films of moisture on the particles. Habitats of endoparasites vary. Some are intracellular, such as malarial parasites in vertebrates, which are typical Coccidia in most of the cycle. Other parasites, such as *Entamoeba histolytica*, invade tissues but not individual cells. Most trypanosomes live in the blood plasma of vertebrate hosts. Many other parasites live in the lumen of the digestive tract or sometimes in coelomic cavities of invertebrates, as do certain gregarines. *See* COCCIDIA; GREGARINIA; TRYPANOSOMATIDAE.

Many Protozoa are uninucleate, others are binucleate or multinucleate, and the number of nuclei also may vary at different stages in a life cycle. Protozoa range in size from 1 to 10^6 micrometers. Colonies are known in flagellates, ciliates, and Sarcodina. Although marked differentiation of the reproductive and somatic zooids characterizes certain colonies, such as *Volvox*, Protozoa have not developed tissues and organs.

Morphology. A protozoan may be a plastic organism (ameboid type), but changes in form are often restricted by the pellicle. A protective layer is often secreted outside the pellicle, although the pellicle itself may be strengthened by incorporation of minerals. Secreted coverings may fit closely, for example, the cellulose-containing theca of Phytomonadida and Dinoflagellida, analogous to the cell wall in higher plants. The dinoflagellate theca (Fig. 1*a*) may be composed of plates arranged in a specific

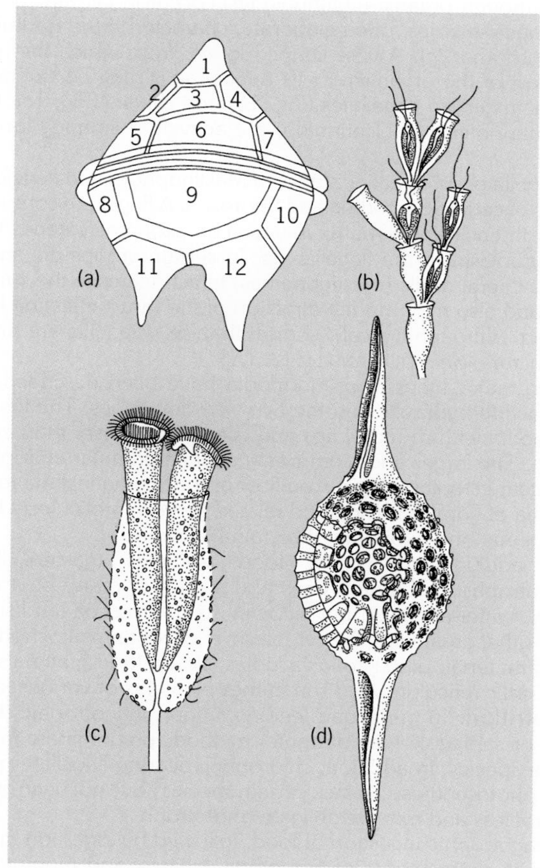

Fig. 1. External coverings of Protozoa. (*a*) Theca of dinoflagellate (*Peridinium*), showing separate plates. (*b*) Lorica of a colonial chrysomonad, *Dinobryon*. (*c*) Two zooids within a lorica of a peritrich, *Cothurnia*. (*d*) A radiolarian skeleton, siliceous type. (*After L. H. Hyman, The Invertebrates, vol. 1, McGraw-Hill, 1940*)

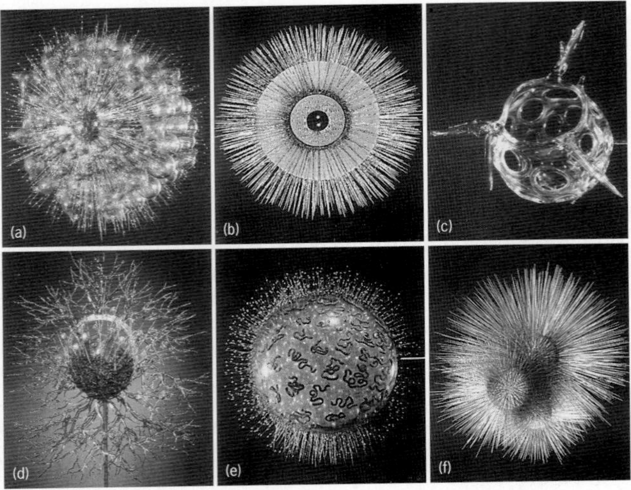

Fig. 2. Glass models of marine Protozoa. Radiolarian types: (a) *Trypanosphaera transformata* Haeckel (Indian Ocean); (b) *Actissa princeps* Haeckel (Indian and Pacific oceans); (c) *Peridium spinipes* Haeckel (Pacific Ocean); (d) *Lithocircus magnificus* Haeckel (Atlantic Ocean); (e) *Collozoum serpentinum* Haeckel (Atlantic Ocean). (f) Foraminiferan type: the pelagic *Globigerina bulloides* (which is found in all seas). (*American Museum of Natural History*)

pattern. Tests, as seen in Rhizopodea (Arcellinida, Gromiida, Foraminiferida), may be composed mostly of inorganic material, although organic (chitinous) tests occur in certain species. Siliceous skeletons, often elaborate, characterize the Radiolaria (Figs. 1*d* and 2*c*). A vase-shaped lorica, from which the anterior part of the organism or its appendages may be extended, occurs in certain flagellates (Fig. 1*b*) and ciliates (Fig. 1*c*). Certain marine ciliates (Tintinnida) are actively swimming loricate forms.

Flagella occur in active stages of Mastigophora and flagellated stages of certain Sarcodina and Sporozoa. A flagellum consists of a sheath enclosing a matrix in which an axoneme extends from the cytoplasm to the flagellar tip. In certain groups the sheath shows lateral fibrils (mastigonemes) which increase the surface area and also may modify direction of the thrust effecting locomotion. Although typically shorter than flagella, cilia are similar in structure. *See* CILIA AND FLAGELLA.

Two major types of pseudopodia have been described, the contraction-hydraulic and the two-way flow types. The first are lobopodia with rounded tips and ectoplasm denser than endoplasm. The larger ones commonly contain granular endoplasm and clear ectoplasm. Two-way flow pseudopodia include reticulopodia of Foraminiferida and related types, filoreticulopodia of Radiolaria, and axopodia of certain Heliozoia.

In addition to nuclei, food vacuoles (gastrioles) in phagotrophs, chromatophores and stigma in many phytoflagellates, water-elimination vesicles in many Protozoa, and sometimes other organelles, the cytoplasm may contain mitochondria, Golgi material, pinocytotic vacuoles, stored food materials, endoplasmic reticulum, and sometimes pigments of various kinds.

Nutrition. In protozoan feeding, either phagotrophic (holozoic) or saprozoic (osmotrophic) methods predominate in particular species. In addition, chlorophyll-bearing flagellates profit from photosynthesis; in fact, certain species have not been grown in darkness and may be obligate phototrophs.

Phagotrophic ingestion of food, followed by digestion in vacuoles, is characteristic of Sarcodina, ciliates, and many flagellates. Digestion follows synthesis of appropriate enzymes and their transportation to the food vacuole. Details of ingestion vary. Formation of food cups, or gulletlike invaginations to enclose prey, is common in more or less ameboid organisms, such as various Sarcodina, many flagellates, and at least a few Sporozoa. Entrapment in a sticky reticulopodial net occurs in Foraminiferida and certain other Sarcodina. A persistent cytostome and gullet are involved in phagotrophic ciliates and a few flagellates. Many ciliates have buccal organelles (membranes, membranelies, and closely set rows of cilia) arranged to drive particles to the cytostome. Particles pass through the cytostome into the cytopharynx (gullet), at the base of which food vacuoles (gastrioles) are formed. Digestion occurs in such vacuoles.

By definition saprozoic feeding involves passage of dissolved foods through the cortex. It is uncertain to what extent diffusion is responsible, but enzymatic activities presumably are involved in uptake of various simple sugars, acetate and butyrate. In addition, external factors, for example, the pH of the medium, may strongly influence uptake of fatty acids and phosphates.

Reproduction. Reproduction occurs after a period of growth which ranges, in different species, from less than half a day to several months (certain Foraminiferida). General methods include binary fission, budding, plasmotomy, and schizogony. Fission, involving nuclear division and replication of organelles, yields two organisms similar in size. Budding produces two organisms, one smaller than the other. In plasmotomy, a multinucleate organism divides into several, each containing a number of nuclei. Schizogony, characteristic of Sporozoa, follows repeated nuclear division, yielding many uninucleate buds.

Simple life cycles include a cyst and an active (trophic) stage undergoing growth and reproduction. In certain free-living and parasitic species, no cyst is developed. Dimorphic cycles show two active stages; polymorphic show several. The former include adult and larva (Suctoria); flagellate and ameba (certain Mastigophora and Sarcodina); flagellate and palmella (nonflagellated; certain Phytomonadida); and ameba and plasmodium (Mycetozoia especially).

Parasitic protozoa. Parasites occur in all major groups. Sporozoa are exclusively parasitic, as are some flagellate orders (Trichomonadida, Hypermastigida, and Oxymonadida), the Opalinata, Piroplasmea, and several ciliate orders (Apostomatida, Astomatida, and Entodiniomorphida). Various other groups contain both parasitic and free-living types. Protozoa also serve as hosts of other protozoa, certain bacteria, fungi, and algae.

Relatively few parasites are distinctly pathogenic, causing amebiasis, visceral leishmaniasis (kala azar), sleeping sickness, Chagas' disease, malaria, tick fever of cattle, dourine of horses, and other diseases. *See* CILIOPHORA; CNIDOSPORA; MALARIA; SARCOMASTIGOPHORA; SPOROZOA.

Richard P. Hall

Protura An order of ancestrally wingless insects (subclass Apterygota). The order is usually classified into four families in two suborders (Eosentomoidea and Acerentomoidea), and about 500 species are recognized.

The insects are under 0.08 in. (2 mm) in length and have an elongated body with a small head, relatively large abdomen, and three pairs of functional legs. The first pair of legs is normally held out in front of the head, antennalike. They serve to replace the sensory role of true antennae, which are absent. Paired pseudoculi (probably chemosensory) occupy eyelike positions on the head. The mouthparts include an elongate and styliform mandible, a maxilla, and a two-part labium. All parts are enclosed within a pocket in the head, so that only their tips are exposed. The thorax bears five-segmented legs, each with an apical claw. In the adult, the abdomen has 12 segments; the anus is terminal, and cerci (segmented sensory appendages on the last abdominal segment) are lacking. The sterna of the first, second, and third segments bear short styli, probably representing vestigial limbs. A large gland opens at the rear of the eighth dorsal plate, which sometimes has a comblike lid. The genitalia are enclosed within a ventral pouch between the eleventh and twelfth segments, and have paired protrusible stylets (slender, elongated appendages).

Protura live in forest soil, leaf litter, and similar places, and are thought to feed on mycorrhizal fungi. They are distributed globally but unevenly in temperate and tropical climates, and they sometimes occur in enormous numbers, as in some Oregon fir forests in the United States. *See* APTERYGOTA; INSECTA.

William L. Brown, Jr.

Proustite A mineral having composition Ag_3AsS_3. It occurs in prismatic crystals terminated by steep ditrigonal pyramids, but is more commonly massive or in disseminated grains. Hardness is 2–2.5 (Mohs scale) and specific gravity is 5.55. The luster is adamantine and the color ruby red. It is called light ruby silver in contrast to pyrargyrite, dark ruby silver. Proustite and pyrargyrite are found together in silver veins. Noted localities are at Chañarcillo, Chile; Freiberg, Germany; Guanajuato, Mexico; and Cobalt, Ontario, Canada. *See* PYRARGYRITE.

Cornelius S. Hurlbut, Jr

Provenance (geology) In sedimentary geology, all characteristics of the source area from which clastic (detrital) sediments and sedimentary rocks are derived, including relief, weathering, and source rocks. *See* WEATHERING PROCESSES.

The goal of most provenance studies of sedimentary rocks is the determination of source characteristics of the mountains or hills from which the constituent sediment was derived. Such determinations are difficult to make because sediment composition and texture are continually modified during erosion, transport, deposition, and diagenesis (postdepositional modification). It is most straightforward to determine provenance in situations in which these modifying effects are minimal; provenance may be indeterminate or ambiguous in situations involving extensive modification of sediment composition and texture. The former situation is most common in tectonically active areas, resulting in rapid uplift and erosion of mountains, rapid transport and deposition, and slight diagenetic modification after deposition. In contrast, stable continental areas (for example, cratons) provide ample opportunity for intense weathering so that chemical, mineralogical, and textural characteristics of sediment are intensely modified. *See* DIAGENESIS; EROSION; SEDIMENTARY ROCKS.

Clastic sediment is commonly recycled during multiple episodes of mountain building, erosion, sedimentation, lithification, and renewed mountain building. This process constitutes the rock cycle, within which igneous, sedimentary, and metamorphic rocks are created and modified. During this process, it is common for older sedimentary rocks to be uplifted and eroded, so that individual sedimentary particles (clasts) are recycled to form new sediment, which may be lithified to form new sedimentary rock. Provenance studies must determine the proportion of a sedimentary rock derived directly from indicated source rocks versus the proportion derived directly from another sedimentary rock (that is, rocks exposed during previous cycles of sedimentation). This determination is essential, but commonly difficult to accomplish because the multicyclic nature of sediment may be difficult to recognize. *See* METAMORPHIC ROCKS.

Fundamentally different methods of study are utilized, depending on what aspect of provenance is emphasized, what type of sediment is studied, and what scale of sampling and study is attempted. Grain size of detrital sediment is a dominant control over what methods may be employed.

Methods of determining provenance include direct determination of rock types (primarily used for coarse to medium grains); direct determination of mineralogy (used for all grain sizes); whole-rock geochemistry (used for medium to fine grains); geochemistry of individual mineral species (used for all grain sizes, but especially for medium grains); and radiometric dating of individual mineral species (primarily for medium grains). *See* GEOCHEMISTRY; MINERALOGY; ROCK AGE DETERMINATION.

Raymond V. Ingersoll

Prussian blue The common name for a material prepared by either the addition of trivalent iron (Fe^{III}) to ferrocyanide $[Fe(CN)_6]^{4-}$ or Fe^{II} to ferricyanide $[Fe(CN)_6]^{3-}$ with the composition of $Fe^{III}{}_4[Fe^{II}(CN)_6]_3 \cdot xH_2O$, where $x \sim 15$. Prussian blue is also known as Turnbull's blue or ferric hexacyanoferrate.

Uses of Prussian blue include artists' colors, carbon paper, typewriter ribbons, blueprinting, and printing inks. It is prescribed for the treatment of radioactive $^{137}Cs^+$ and thallium poisoning. Substitution of the Fe^{II} and Fe^{III} ions with other metal ions leads to Prussian blue–structured materials with different metal ions. These materials typically are magnets. Substitution of the Fe^{II} and Fe^{III} ions with Cr^{III} and V^{II} ions, respectively, results in a magnet with an ordering temperature around 100°C (212°F). *See* COORDINATION CHEMISTRY; COORDINATION COMPLEXES; CURIE TEMPERATURE; CYANIDE; FERRICYANIDE AND FERROCYANIDE; IRON.

Joel S. Miller

Prymnesiophyceae A class of algae (also known as Haptophyceae) in the chlorophyll *a-c* phyletic line (Chromophycota). In protozoological classification these organisms constitute an order, Prymnesiida or Haptomonadida, in the class Phytamastigophora. Most of the approximately 300 species of prymnesiophytes are biflagellate monads. *See* CHROMOPHYCOTA.

This class has been segregated from the Chrysophyceae, with which it shares many biochemical and ultrastructural characters. Prymnesiophytes differ from chrysophytes, however, in several significant characters: (1) the typical monad bears a filiform organelle, the haptonema, between the two flagella; (2) except in the order Pavlovales, the flagella are of equal length and smooth; (3) organic scales, which may be calcified, cover most motile cells.

Calcified scales (cellulosic scales impregnated with calcite) are characteristic of many prymnesiophytes. These scales, which are given the general term coccoliths, were discovered in marine sediment before they were observed on living cells. Coccoliths are classified into several morphological types (such as rhabdoliths, discoliths, zygoliths, and ceratoliths), which are of great diagnostic value in the taxonomy of the prymnesiophytes that bear them (coccolithophorids).

In most prymnesiophytes a nonmotile phase alternates with a motile phase. The nonmotile phase is a free-living unicell or a palmelloid or pseudofilamentous colony. In some cases the alternation is mediated by sexual reproduction. Usually, however, reproduction is effected by binary fission or the production of zoospores.

Prymnesiophytes are primary marine, with coccolithophorids constituting one of the three major components of phytoplankton (the others being diatoms and dinoflagellates). *See* ALGAE; COCCOLITHOPHORIDA; PHYTOPLANKTON; PROTOZOA.

Paul C. Silva; Richard L. Moe

Pseudoborniales An order of fossil plants found in Middle and Upper Devonian rocks. The group is related to Sphenophyllales and includes a single family and two monotypic genera. *Pseudobornia ursina* is known from Bear Island (north of Norway), Alaska, and Germany. *Prosseria grandis* is found in New York State. Sphenopsid characters are more firmly established in this order than in Hyeniales. *See* HYENIALES; SPHENOPHYLLALES.

Harlan P. Banks

Pseudomonas A genus of gram-negative, non-spore-forming, rod-shaped bacteria. Motile species possess polar flagella. They are strictly aerobic, but some members do respire anaerobically in the presence of nitrate. Some species produce acids oxidatively from carbohydrates; none is fermentative and none photosynthetic.

Members of the genus *Pseudomonas* cause a variety of infective diseases; some species cause disease of plants. One species,

P. mallei, is a mammalian parasite, and is the causative agent of glanders, an infectious disease of horses that occasionally is transmitted to humans by direct contact. *Pseudomonas aeruginosa* is the most significant cause of hospital-acquired infections, particularly in predisposed patients with metabolic, hematologic, and malignant diseases. The spectrum of clinical disease ranges from urinary tract infections to septicemia, pneumonia, meningitis, and infections of postsurgical and posttraumatic wounds. *See* GLANDERS; HOSPITAL INFECTIONS; MENINGITIS; PNEUMONIA.

Gerald L. Gilardi

Pseudophyllidea An order of tapeworms of the subclass Cestoda, parasitic in the intestine of all classes of vertebrates. Typically, the head is simple in structure with two groovelike attachment organs, the bothria. Most pseudophyllideans are segmented and polyzoic with replication of the reproductive systems, although there are a number which do not show such replication and are monozoic.

Dibothriocephalus latus, the broad or fish tapeworm of humans and certain piscivorous mammals, is a pseudophyllidean. In humans, this worm sometimes precipitates a pernicious anemia by competing with the host for vitamin B_{12}. Larval pseudophyllideans are occasionally found as parasites in the extraintestinal tissues of humans, producing a condition known as sparganosis. *See* CESTODA.

Clark P. Read

Pseudoscorpionida An order of terrestrial Arachnida having the general appearance of miniature scorpions without the postabdomen and sting. The body length is seldom greater than 0.2 in. (5.0 mm). Typically, each finger of the anterior appendages, or chelicerae, has a serrula composed of a row of ligulate plates. Ducts of silk glands open near the end of the movable finger, often in connection with a simple or branched spinneret. The second pair of appendages, or palpi, are large and conspicuous, usually with glands that discharge venom through a terminal tooth on one or both of the chelal fingers. The four pairs of legs are ambulatory.

Pseudoscorpions feed chiefly on small arthropods and, although frequently found on birds, mammals, and insects, are considered nonparasitic. Pseudoscorpions are common in the nests of mammals, birds, and social insects, in woody debris and forest litter, under stones, and in crevices in the bark of trees. About 2000 species have been described. *See* ARACHNIDA.

C. Clayton Hoff

Pseudosphaeriales (lichenized) An order of the class Ascolichenes. The order is also called the Pleosporales. They resemble the typical pyrenomycetous lichens except for the structure of the ascocarp, which is not a true perithecium. It is flask-shaped and lined with a layer of interwoven, branched pseudoparaphyses. The asci, with bitunicate walls, are located in scattered locules.

There are two major families: the larger one, Arthopyreniaceae, is a widespread family with at least five genera; the Mycoporaceae is a small family with two well-known genera, *Dermatina* and *Mycoporellum*. All of the species in this order are crustose and many lack a well-defined thallus.

Mason E. Hale

Psilophytales A group long recognized as an order of fossil plants (subdivision Psilopsida) collected in rocks of Late Silurian and Devonian age. It has been subdivided into three categories whose descriptions include the chief kinds of plants formerly included in Psilophytales. They are given in this Encyclopedia as three classes of the division Rhyniophyta: Rhyniopsida, Zosterophyllopsida, and Trimerophytopsida. *See* EMBRYOBIONTA; PALEOBOTANY; RHYNIOPHYTA; RHYNIOPSIDA; TRIMEROPHYTOPSIDA; ZOSTEROPHYLLOPSIDA.

Harlan P. Banks

Psilotophyta A division of the plant kingdom consisting of only two genera with three living species, *Psilotum nudum*, *P. complanatum*, and *Tmesipteris tannensis*. *Psilotum* is widespread in tropical and subtropical regions of both hemispheres, but *Tmesipteris* is confined to Australia and some of the Pacific islands. The Psilotophyta have the typical life cycle of vascular cryptograms, with an alteration of sporophyte and gametophyte generations, the sporophyte being much the larger and more complex. The Psilotophyta have no economic importance, but they are interesting as possible remnants of an ancient (Silurian and Devonian) group of plants, the Rhyniophyta, which is regarded as ancestral to all other vascular plants. *See* POLYPODIOPHYTA; PSILOPHYTALES; RHYNIOPHYTA.

Arthur Cronquist

Psittaciformes The parrots, a large order of land birds found worldwide but with most species concentrated on the landmasses of the Southern Hemisphere, particularly Australasia, the Neotropics, Africa, and southern Asia. Only a few species are found in the northern temperate regions. The parrots likely are closely allied to pigeons, from which they have probably evolved. *See* COLUMBIFORMES.

Although the parrots are generally placed in a single family, the Psittacidae, they are divided into a number of distinct subfamilies, which are listed below:

Order Psittaciformes
  Family Psittacidae
    Subfamily: Psittacinae (parrots; 247 species; worldwide, but few in the Palearctic)
      Cacatuinae (cockatoos; 18 species; Australasia to Indonesia)
      Micropsittinae (pygmy parrots; 6 species; New Guinea)
      Strigopinae (owl parrot; 1 species; New Zealand)
      Nestorinae (keas; 3 species; New Zealand)
      Loriinae (lories; 54 species; Australia to Indonesia)
      Psittrichadinae (Pesquet's parrot; 1 species; New Guinea)
      Loriculinae (hanging parrots; 10 species; southeast Asia)

Parrots have a characteristic strong hooked bill. The tongue is large and fleshy. A short neck connects the large head to the stocky body. The wings are of medium length, varying from pointed to rounded. The legs are short and stout with strong, clawed toes arranged in a zygodactyl pattern, two pointing forward and two backward. The tail varies from short to long. Parrots fly well and can attain high speeds, but even those few parrots that are migratory journey only short distances. One species, the owl parrot, is flightless. Parrots can walk well but not rapidly. Plumage of parrots is variable; most often it is green, but red, orange, yellow, blue, black, and white in bold bright patterns are also common. Parrots are mainly vegetarians, eating seeds, nuts, fruit, nectar, or pollen; a few consume animal food.

Parrots are social birds and usually live in flocks. Almost all species have a strong pair bond, remaining together year-round for life. Parrots are found mainly in the tropics and in the southern continents, and the center of parrot radiation is the Australasian-New Zealand region. Many parrots in captivity have an excellent ability to mimic words, and some can be taught a large vocabulary. However, parrots are not known to mimic the calls of other birds in the wild. Their vocalizing ability is probably important for the constant communication between mates.

Parrots are the most important birds in the avicultural and pet trade, with many species kept in captivity. Because the demand for parrots is so great, many countries have imposed exportation bans on native birds or importation restrictions on wild-trapped parrots. *See* AVES; ENDANGERED SPECIES. Walter J. Bock

Psocoptera An order of insects frequently referred to as the Corrodentia, or Copeognatha. Common names for members of this order are book lice, bark lice, and psocids, the latter a general term for all members of the order. They are usually less than 0.25 in. (0.6 cm) long, though rarely some may reach about 0.5 in. (1.2 cm). Wings may be absent, and when present are of differing distinctive venational types. Tarsi are two- or three-segmented, cerci are absent and metamorphosis is gradual. Chewing mouthparts usually have a much enlarged clypeus; the lacinia of the maxilla is usually elongate and chisellike, and the antennae have 13 or more segments.

Book lice are most common among old papers on dusty shelves, in cereals, or other domestic situations. They are usually pale, wingless types of insects. Many bark lice, the majority winged, occur on the bark or foliage of trees, and some are found under dead bark or beneath stones. Nymphs of a few species occur on tree trunks as clusters of gregarious individuals, but disperse when mature.

Psocoptera are worldwide, especially in warm countries, and some 1300 species are known. Current classification now lists about 27 families for this group. About 150 species, in 11 families, have been found in the United States. *See* INSECTA.

Ashley B. Gurney

Psychoacoustics All of the psychological interactions between humans (and animals) and the world of sound. It encompasses all studies of the perception of sound, as well as the production of speech. *See* HEARING (HUMAN); SPEECH.

Lawrence E. Marks

Psychoanalysis Psychoanalysis may be defined as (1) a psychological theory; (2) a form of psychotherapy, especially for the treatment of neurotic and character or personality disorders; and (3) a method for investigating psychological phenomena. Psychoanalysis was created and developed by Sigmund Freud, who presented his method, clinical observations, and theory in *Interpretation of Dreams* and other major works, including *The Psychopathology of Everyday Life* and *Three Essays on the Theory of Sexuality*, as well as in many of his case studies.

Psychoanalytic theory. Generally, psychoanalysis is concerned with the causal role of wishes and beliefs in human life. More specifically, it attempts to explain mental or behavioral phenomena that do not appear to make sense as the effects of unconscious wishes and beliefs. Such phenomena include dreams, disturbances in functioning such as slips of the tongue or pen and transient forgetting, and neurotic symptoms. Typically, unconscious wishes and beliefs are constituents of conflicts.

The term unconscious in psychoanalysis does not mean simply that mental contents are out of awareness. Its psychodynamic meaning is that the person does not want to be aware of these contents, and takes active steps to avoid being aware of them. A fundamental hypothesis of psychoanalysis is that because a mental entity is dynamically unconscious it has the causal power to produce the phenomena that are of interest to psychoanalysis.

At first, the dynamic unconscious was thought to consist of traumatic memories. Later, it was believed to consist of impulses or wishes—especially sexual (and aggressive) impulses or wishes. Psychoanalysis now emphasizes that the dynamic unconscious consists of fantasies, which have a history reaching back to childhood. These fantasies are internal scenarios in which sexual (and aggressive) wishes are imagined as fulfilled.

Psychoanalysis is distinct in attributing causal powers to unconscious sexual wishes. Such attribution depends on extending the meaning of sexual to encompass the quest for sensual pleasure in childhood (so-called infantile sexuality) and choices of objects and aims. One theme that is thought to have particular importance is the Oedipus complex, in which the child rivals one parent in seeking sensual gratifications of various kinds from the other parent.

When an unconscious fantasy is activated, it manifests itself in conscious mental states or in actions—importantly, in emotions; in interpretations of the significance of events or states of affairs; in attributions of motives to others; and in daydreams, dreams, and neurotic symptoms.

Unconscious fantasies, as distinct from both conscious reality-oriented imagining and conscious daydreaming, are constructed when imagination functions under very special conditions.

This emphasis on fantasy underscores the fact that psychoanalysis gives priority to the relation between wishes (including wishes a person knows could not conceivably be gratified in reality) and imagination (functioning under very special conditions).

Psychotherapy. Free association is the method of psychoanalysis. Patients are encouraged not to talk about some particular problem or aspect of their lives but rather to suspend any conscious purposive organization of what they say, speaking freely. Both psychoanalyst and patient follow the patient's productions: conscious purposes are replaced by unconscious purposes, which, under these conditions, can determine the direction of the patient's mental processes with less interference.

Interventions are predominantly interpretative; psychoanalysts do not seek primarily to tell their patients what to do, to educate them about the world, to influence their values, or to reassure them in one way or another that everything is or will be all right. Psychoanalysts look for patterns in what each patient says and for signs of feelings of which the patient is more or less unaware. They then engage their patients (who are increasingly aware of these patterns and able to experience and articulate these feelings) in an inquiry about the reasons for them or motives behind them. The focus is on what the patients do not know—and do not want to know—about themselves and their inner life, including strategies for avoiding such knowledge and the consequences of these strategies.

The goal of psychoanalytic psychotherapy is to extend the realm of what patients permit themselves to experience. It tries to mitigate the misery that patients with a neurotic, character, or personality disorder inflict on themselves.

The case-study method is characteristic of psychoanalytic research. The arguments that can be used in case studies are analogy (the use of familiar or homely models in which postulated causes and mechanisms can be shown to exist); consilience (the convergence of inferences from different kinds of information on a common cause); and abduction (inference to the best explanation). *See* PSYCHOTHERAPY. Marshall Edelson

Psycholinguistics An area of study which draws from linguistics and psychology and focuses upon the comprehension and production of language. Although psychologists have long been interested in language, and the field of linguistics is an older science than psychology, scientists in the two fields have had little contact until the work of Noam Chomsky was published in the late 1950s. Chomsky's writing had the effect of making psychologists acutely aware of their lack of knowledge about the structure of language, and the futility of focusing attention exclusively upon the surface structure of language. As a result, psycholinguists, who have a background of training in both linguistics and psychology, have been attempting since the early 1960s to gain a better understanding of how the abstract rules which determine human language are acquired and used to communicate appropriately created meaningful messages from one person to another via the vocal-auditory medium. Research has been directed to the evolutionary development of language, the biological bases of language, the nature of the sound system, the rules of syntax, the nature of meaning, and the process of language acquisition. David S. Palermo

Psychology The study of human behavior and mental processes. Psychology is sharply divided into applied and experimental areas. However, many fields are represented in both research and applied psychology.

Researchers in psychology study a wide range of areas. Cognitive research is often included as part of subdiscipline called cognitive science. This area examines central issues such as how mental process work, the relation between mind and brain, and the way in which biological transducing systems can convert physical regularities into perceptions of the world. Cognitive science is carved from the common ground shared by computer science, cognitive psychology, philosophy of mind, linguistics, neuropsychology, and cognitive anthropology. The study of human attention is a cognitive area that is central in the field. *See* COGNITION.

The study of consciousness involves such basic questions as the physiological basis of mental activity, the freedom of will, and the conscious and unconscious uses of memory. The latter topic can be classified under the rubric of implicit memory. *See* INSTINCTIVE BEHAVIOR; MEMORY; PSYCHOLINGUISTICS; SENSATION.

Social psychology includes the study of interactions between individuals and groups, as well as the effects of groups on the attitudes, opinions, and behavior of individuals. The field covers such topics as persuasion, conformity, obedience to authority, stereotyping, prejudice, and decision making in social contexts. *See* MOTIVATION; PERSONALITY THEORY.

Developmental psychology has three subfields: life-span development, child development, and aging. Most research in the area concentrates on child development, which examines the development of abilities, personality, social relations, and, essentially, every attribute and ability seen in adults. *See* AGING; INTELLIGENCE.

A clinical psychologist is usually known by the term psychologist, which in some states is a term that can be used only by a registered practitioner. A psychiatrist is a physician with a specialty in psychiatric treatment and, in most states, with certification as a psychiatrist by a board of medical examiners. A psychoanalyst is typically trained by a psychoanalytic institute in a version of the Freudian method of psychoanalysis. A large number of practitioners qualify both as psychoanalysts and psychiatrists. *See* PSYCHOANALYSIS.

Neuropsychologists are usually psychologists, who may come from an experimental or a clinical background but who must go through certification as psychologists. They treat individuals who have psychological disorders with a clear neurological etiology, such as stroke.

Clinical practice includes individual consultation with clients, group therapy, and work in clinics or with teams of health professionals. Psychological therapists work in many settings and on problems ranging from short-term crises and substance abuse, to psychosis and major disorders. While there are definite biases within each field, it is possible for a practitioner with any background to prefer behavior therapy, a humanistic approach, a Freudian (dynamic) approach, or an eclectic approach derived from these and other areas.

Nonclinical professional work in psychology includes the human-factors element, which traditionally is applied to the design of the interface between a machine and its human operator. Cognitive engineering is a branch of applied psychology that deals mainly with software and hardware computer design. Industrial psychology also includes personnel selection and management and organizational planning and consulting.

The use of psychology in forensic matters is a natural result of the fact that much of law is based on psychology. Psychologists have been involved in jury selection, organization of evidence, evaluation of eyewitness testimony, and presentation of material in court cases. Psychiatrists and psychologists are also called on to diagnose potential defendants for mental disorders and the ability to stand trial. William P. Banks

Psychoneuroimmunology The study of the interactions among behavioral, neural and endocrine, and immune functions. This convergence of disciplines has evolved to achieve a more complete understanding of adaptive processes. At one time, the immune system was considered an independent agency of defense that protected the organism against foreign material (that is, proteins that were not part of one's "self"). Indeed, the immune system is capable of considerable self-regulation. However, converging data from the behavioral and brain sciences indicate that the brain plays a critical role in the regulation or modulation of immunity. Thus, psychoneuroimmunology emphasizes the study of the functional significance of the relationship between these systems—not in place of, but in addition to, the more traditional analysis of the mechanisms governing the functions within a single system—and the significance of these interactions for health and disease. *See* NEUROIMMUNOLOGY.

Brain–immune system interactions. Evidence for nervous system–immune system interactions exists at several biological levels. Primary and secondary lymphoid organs are innervated by the sympathetic nervous system, and lymphoid cells bear receptors for many hormones and neurotransmitters. These substances, secreted by the pituitary gland, are thus able to influence lymphocyte function. Moreover, lymphocytes themselves can produce neuropeptide substances. Thus, there are anatomical and neurochemical channels of communication that provide a structural foundation for the several observations of functional relationships between the nervous and immune systems.

Stress and immunity. The link between behavior and immune function is suggested by experimental and clinical observations of a relationship between psychosocial factors, including stress, and susceptibility to or progression of disease processes that involve immunologic mechanisms. Abundant data document an association between stressful life experiences and changes in immunologic reactivity. The death of a family member and other, less severe, stressful experiences (such as taking examinations) result in transient impairments in several parameters of immune function. *See* DISEASE; STRESS (PSYCHOLOGY).

In animals, a variety of stressors can influence a variety of immune responses. Since immune responses are themselves capable of altering levels of circulating hormones and neurotransmitters, these interactions probably include complex feedback and feedforward mechanisms. *See* ENDOCRINOLOGY.

The direction, magnitude, and duration of stress-induced alterations of immunity are influenced by (1) the quality and quantity of stressful stimulation; (2) the capacity of the individual to cope effectively with stressful events; (3) the quality and quantity of immunogenic stimulation; (4) the temporal relationship between stressful stimulation and immunogenic stimulation; (5) the sampling times and the particular aspect of immune function chosen for measurement; (6) the experiential history of the individual and the existing social and environmental conditions upon which stressful and immunogenic stimulation are superimposed; (7) a variety of host factors such as species, strain, age, sex, and nutritional state; and (8) interactions among these variables.

Conditioning. Central nervous system involvement in the modulation of immunity is dramatically illustrated by the classical (Pavlovian) conditioning of the acquisition and extinction of suppressed and enhanced antibody- and cell-mediated immune responses. In a one-trial taste-aversion conditioning situation, a distinctively flavored drinking solution (the conditioned stimulus) was paired with an injection of the immunosuppressive drug cyclophosphamide (the unconditioned stimulus). When subsequently immunized with sheep red blood cells, conditioned animals reexposed to the conditioned stimulus showed a reduced antibody response compared to nonconditioned animals and conditioned animals that were not reexposed to the conditioned stimulus. *See* CONDITIONED REFLEX.

The acquisition and the extinction (elimination of the conditioned response by exposures to the conditioned stimulus without the unconditioned stimulus) of the conditioned enhancement and suppression of both antibody- and cell-mediated immune responses—and nonimmunologically specific host defense responses as well—have been demonstrated under a variety of experimental conditions.

Prospects. An elaboration of the integrative nature of neural, endocrine, and immune processes and the mechanisms underlying behaviorally induced alterations of immune function is likely to have clinical and therapeutic implications that will not be fully appreciated until more is known about the extent of these interrelationships in normal and pathophysiological states. *See* ENDOCRINE SYSTEM (VERTEBRATE); IMMUNOLOGY; NERVOUS SYSTEM (VERTEBRATE).

Robert Ader

Psychopharmacology A discipline that merges the subject matter of psychology, which studies cognition, emotion, and behavior, and pharmacology, which characterizes different drugs. Thus, psychopharmacology focuses on characterizing drugs that affect thinking, feeling, and action. In addition, psychopharmacology places particular emphasis on those drugs that affect abnormalities in thought, affect, and behavior, and thus has a relationship to psychiatry. Psychopharmacology is predominantly, but not exclusively, concerned with four major classes of drugs that are of clinical significance in controlling four major categories of psychiatric disorder: anxiety, depression, mania, and schizophrenia.

Anxiety is an emotional state that can range in intensity from mild apprehension and nervousness to intense fear and even terror. It has been estimated that 2–4% of the general population suffer from an anxiety disorder at some time. Although anxiety in some form is a common experience, it can become so intense and pervasive as to be debilitating; it may therefore require psychiatric attention and treatment with an anxiolytic drug. There are three major groups of anxiolytics. Members of the first group are called propanediols; meprobamate is the most widely used. The second group is the barbiturates, of which phenobarbital is the most generally prescribed. The third group, most frequently prescribed, is the benzodiazepines, the best known of which is diazepam.

A major advance in understanding the benzodiazepines was the identification of the cellular sites at which these drugs act (so-called benzodiazepine receptors). The distribution of these receptors in the brain has also been found to have a striking parallel to the distribution of the receptors for a naturally occurring substance called gamma-amino butyric acid (GABA). Furthermore, it is known that GABA has a ubiquitous inhibitory role in modulating brain function. Most importantly, it is now clear that benzodiazepines share a biochemical property in that all augment the activity of GABA. *See* ANXIETY DISORDERS; SEROTONIN; TRANQUILIZER.

The symptoms of depression can include a sense of sadness, hopelessness, despair, and irritability, as well as suicidal thoughts and attempts, which are sometimes successful. In addition, physical symptoms such as loss of appetite, sleep disturbances, and psychomotor agitation are often associated with depression. When depression becomes so pervasive and intense that normal functioning is impaired, antidepressant medication may be indicated. It has been estimated that as much as 6% of the population will require antidepressant medication at some time in their lives.

There are two major groups of antidepressant drugs. Members of the first group are called heterocyclics because of their characteristic chemical structures. Members of the second group, which are less often prescribed, are called monoamine oxidase inhibitors. *See* MONOAMINE OXIDASE.

The antidepressants typically require at least several weeks of chronic administration before they become effective in alleviating depression. This contrasts with the anxiolytics, which are effective in reducing anxiety in hours and even minutes. Another difference between these two classes of drugs is that the anxiolytics are more likely to be efficacious: anxiolytics are effective in the vast majority of nonphobic, anxious patients, whereas the antidepressants are effective in only about 65–70% of depressed patients. *See* AFFECTIVE DISORDERS.

Manic episodes are characterized by hyperactivity, grandiosity, flight of ideas, and belligerence; affected patients appear to be euphoric, have racing thoughts, delusions of grandeur, and poor if not self-destructive judgment. Periods of depression follow these episodes of mania in the majority of patients. The cycles of this bipolar disorder are typically interspersed among periods of normality that are, in most cases, relatively protracted.

Mania can usually be managed by chronic treatment with lithium salts and can be expected to be effective in 70–80% of the individuals treated. Furthermore, the period of depression that typically follows the manic episode can usually be prevented, or at least attenuated, if lithium treatment is maintained after the manic phase has subsided. Any periods of depression that do occur can be managed by antidepressant drugs. Lithium is no longer the only drug used in the management of mania. Carbamazepine, an anticonvulsant that is used in the treatment of epilepsy, is also useful in the treatment of periods of mania.

Schizophrenia is a form of psychosis; it incorporates a broad range of symptoms that can include bizarre delusions, hallucinations, incoherence of thought processes, inappropriate affect, and grossly disorganized movements. It affects 1–2% of the population. The symptoms of schizophrenia can be controlled, in varying degrees, by a large group of drugs called antipsychotics. Symptom management requires chronic medication and can be expected in about 80% or more of the schizophrenics treated. However, management is only partially successful in that normal functioning is not completely restored in most patients.

The antipsychotics have a broad range of side effects among which are disturbances of movement that fall into two general classes. The first class includes an array of symptoms very like those characteristic of Parkinson's disease. The second class of movement disorder is called tardive dyskinesia. Signs of this disturbance typically include involuntary movements that most often affect the tongue and facial and neck muscles but can also include the digits and trunk.

Although different antipsychotic drugs have different kinds and degrees of side effects, all share a single biochemical action: they all attenuate the activity of dopamine, a naturally occurring substance in the brain. The reduction in dopamine activity produced by the antipsychotics directly accounts for their effects on motor behavior. It is to be expected, therefore, that disrupted dopamine activity in this system would produce disturbances of movement. It is less clear, however, whether reduced dopamine function is also a factor in the process by which these drugs control psychotic (including schizophrenic) symptoms. *See* SCHIZOPHRENIA.

Peter L. Carlton

Psychophysical methods Methods for the quantitative study of the relations between physical stimulus magnitudes and the corresponding magnitudes of sensation, for example, between the physical intensity of a light and its perceived brightness or the concentration of a sugar solution and its observed sweetness. To establish these relations, measurement scales are needed, not only for physical magnitudes but also for subjective magnitudes. Subjective scales are not obtained directly from observation but are theoretical models which summarize observed relations between stimuli and responses. See SENSATION.

The term psychophysical methods is sometimes extended to include certain scaling techniques which are most often used with subjective dimensions to which there correspond no simple physical dimensions, for example, food preferences. See MULTIDIMENSIONAL SCALING.

In 1860, G. Fechner designed psychophysical methods to measure the absolute threshold, defined as the minimum stimulus energy that an organism can detect, and the differential threshold, defined as the minimum detectable change in a stimulus. Both quantities had to be defined as statistical averages. To obtain reliable measurements for these averages, Fechner devised the method of limits (also called the method of minimal changes) and the method of constant stimuli.

In the method of limits, the experimenter begins with a stimulus which is too weak for the subject to detect. In successive presentations, the stimulus intensity is increased in small, equal steps, the subject reporting after each presentation whether the stimulus was perceived until it has been detected. The descending series is then begun, the stimulus intensity beginning at an above-threshold value and decreasing in steps until the subject signals the disappearance of the stimulus. Many such series are given.

In measuring the difference threshold, essentially the same procedure is involved, except that the subject now signals the relation of a comparison stimulus to a standard stimulus. After a large number of such trials, the average of each of these four threshold values is computed.

To measure the absolute threshold by the method of constant stimuli, the experimenter selects a small number of stimulus values in the neighborhood of the absolute threshold (previously roughly located by informal use of the method of limits) and presents them to the subject a large number of times each, in an irregular order unknown to the subject. Each time a stimulus is presented, the subject reports the presence or absence of sensation.

The data provide the proportion of times that each stimulus resulted in a report of sensation by the subject. One can then estimate the stimulus value that has a probability of .50 of producing sensation, this value being defined as the absolute threshold. An analogous procedure is followed in obtaining difference thresholds.

Fechner proposed to use the results of threshold measurement in developing a subjective metric or scale. He defined the difference threshold, or just noticeable difference (jnd), as the subjective unit and the absolute threshold as the zero point of the subjective scale. Thus the subjective intensity of a particular brightness of light, for example, would be specified when it was given as 100 jnd's above threshold. The subjective scale so defined is not a linear function of the physical stimulus scale since jnd's, though defined as subjectively equal units, are not of physically equal magnitude throughout the intensity scale. The size of the jnd is approximately proportional to physical stimulus intensity. To the extent that this relation holds, Fechner deduced that subjective intensity should be proportional to the logarithm of the stimulus intensity.

Rather than requiring of the subject merely either yes-no or ordinal judgments, some methods require the subject to make direct-ratio discriminations. For instance, he or she may be presented with a moderately loud tone, and then required, by turning a knob, to adjust the loudness of a comparison tone until it is half as loud, or twice as loud, as the first. The first case illustrates the method of fractionation, the second the method of multiplication. In the method of magnitude estimation, the subject is given a stimulus, such as the brightness of a light, to serve as a modulus with a value assigned to it, for example, 10. The task, as other lights of different intensities are presented, is to assign them numbers which shall stand in the same ratio to 10 as their brightness stands to that of the modulus. One twice as bright is given the designation 20; one half as bright is 5. In these and other similar methods, whether the subject's task is to estimate or to produce the prescribed ratio or the prescribed fraction, there are certain common characteristics. Direct-ratio assessments are obtained from the subject; there can be experimental checks on internal consistency of the results, and since the individual judgments are not of high precision, repetition is required if stable averages are to be obtained.

The empirical results obtained by the various methods are in fairly good agreement. They agree in that, to at least a first approximation, subjective magnitudes on a variety of dimensions are found to be power functions of suprathreshold stimulus intensity; that is to say, subjective magnitude is proportional to the suprathreshold stimulus magnitude raised to a power. The powers have a range from 0.3 for auditory loudness to 3.5 for subjective intensity of alternating current that is applied to the skin.

In direct-matching methods the subject is not required to produce or assess the ratio of one subjective magnitude to another, but only to adjust a comparison stimulus until some attribute appears to match that of a standard stimulus. For example, the subject might be asked to adjust the physical intensities of tones of various frequencies until their loudness matched that of a 1000-Hz tone of fixed intensity. The result would be an equal-loudness contour, showing the intensities to which tones of various frequencies must be set to produce sensations of equal loudness. These data are of use in acoustics. See LOUDNESS.

The method of average error, the third of the three methods devised by Fechner, is a special application of direct-matching methods to cases in which the point of interest is in discrepancies between perception and stimulation. The subject adjusts a comparison stimulus to match a standard stimulus; the average of a number of such settings gives the point of subjective equality, and the difference between this point and the standard stimulus is the average error. Two illustrative uses of the method are the measurement of accuracy of distance perception and the measurement of the magnitude of so-called optical illusions. See HEARING (HUMAN); PSYCHOLOGY. John F. Hahn

Psychosis Any disorder of higher mental processes of such severity that judgments pertaining to the reality of external events are significantly impaired. A wide range of conditions can bring about a psychotic state. They include schizophrenia, mania, depression, ingestion of drugs, withdrawal from drugs, liver or kidney failure, endocrine disorders, metabolic disorders, and Alzheimer's disease, epilepsy, and other neurologic dysfunctions. The dreams of normal sleep are a form of psychosis.

Psychotic alterations of beliefs are called delusions. Psychotic alterations of perception are referred to as hallucinations. Psychotic states that are due to alcoholism, metabolic diseases, or other medical conditions are frequently accompanied by general mental confusion. On the other hand, psychiatric illnesses and drugs can produce hallucinations and delusions in the absence of general confusion. Few of those symptoms are unique to a particular illness, which can make proper diagnosis difficult and challenging. Correct diagnosis, however, is critical so that appropriate treatment can be provided. See ADDICTIVE DISORDERS; AFFECTIVE DISORDERS; ALZHEIMER'S DISEASE; NEUROTIC DISORDERS; PARANOIA; PSYCHOTOMIMETIC DRUGS; SCHIZOPHRENIA. Ralph E. Hoffman

Psychosomatic disorders Disorders characterized by physical symptoms that originate, at least in part, from emotional factors. The *Diagnostic and Statistical Manual of the American Psychiatric Association* (DSM) IV now assigns such conditions to two diagnostic categories: somatoform disorders and psychological factors affecting physical condition. Somatoform disorders are characterized by physicalsymptoms that are suggestive but not fully explained by a diagnosable medical disease. The second category, psychological factors affecting physical condition, refers to the presence of a general medical condition that is negatively affected (for example, prolonged or exacerbated) by psychological factors.

Psychological states influence body organs through a combination of three interrelated mechanisms: neural, hormonal, and immunologic. Voluntary movements (for example, clenching the teeth) are mediated through the motor neurons by the conscious command of the brain. In stress, clenching of the teeth, mediated by the same motor neurons, may also occur, but the act may not be voluntary and conscious. Stress usually causes an activation of the sympathetic nervous system and the hypothalamo-pituitary-adrenal axis followed by a decrease in immunocompetence. Immune mechanisms may be suppressed in part through corticosteroid activation, but a decrease in T-lymphocyte activity in stress (for example, following bereavement) may not be mediated by hormones. Individual specific, but inadvertent, conditioning of specific conflict or stress to specific bodily malfunction may be an important psychosomatic mechanism. *See* CONDITIONED REFLEX; NEUROIMMUNOLOGY; STRESS (PSYCHOLOGY).

There are six types of somatoform disorders: conversion disorder, somatization disorder, pain disorder, hypochondriasis, body dysmorphic disorder, and undifferentiated somatoform disorder. Conversion disorders are characterized by physical symptoms referrable to the somatosensory nervous system or special sensory organs that cannot be explained on the basis of a medical or neurologic disease and that are caused by psychological factors. Common symptoms include paralysis, blindness, ataxia (lack of muscular coordination, especially in walking), aphonia (loss of voice and power of speech), and numbness of the feet (stocking anesthesia). The symptoms may represent a psychological conflict or may be a form of body-language communication.

In somatization disorder (also known as Briquet's syndrome), the patient recurrently complains of multiple somatic symptoms that are referrable to practically every organ system in the body. Upon medical investigation, however, the symptoms turn out not to be a diagnosable physical disease. The chronicity and multiplicity of symptoms in somatization disorder distinguish it from conversion disorder. The symptoms do not usually symbolize psychological conflicts but may represent general dysphoria and distorted illness behavior.

When pain is the prominent clinical symptom and psychological factors are considered to play an important role in the onset, severity, exacerbation, or maintenance of the pain, pain disorder may be diagnosed.

Hypochondriasis is preoccupation with the idea of having a serious medical disease based on a misinterpretation of one or more bodily symptoms or signs. Such preoccupation continues in spite of a negative reasonable medical workup, and causes significant distress or impairment in functioning. Body dysmorphic disorder, otherwise known as dysmorphophobia, denotes a preoccupation with a defect in appearance that may be either imagined or greatly exaggerated, and that causes significant distress or impairment in function. *See* PAIN.

The management of psychological factors affecting physical condition should exemplify the biopsychosocial model, that is, effective treatment of the physical condition as well as psychotherapeutic and psychopharmacologic interventions as needed to ameliorate the psychological factors, including stress management and relaxation techniques. *See* PSYCHOPHARMACOLOGY; PSYCHOTHERAPY.

Hoyle Leigh

Psychotherapy Any treatment or therapy that is primarily psychological in nature. In recent years, counseling also has been included in this categorization.

Psychodynamic therapies. Historically, psychoanalysis—created by Sigmund Freud—has played an important role in the growth and development of psychotherapy. Central to Freud's theories was the importance of unconscious conflicts in producing the symptoms and defenses of the patient. The goal of therapy is to help the patient attain insight into the repressed conflicts which are the source of difficulty. Since patients resist these attempts bring to consciousness the painful repressed material, therapy must proceed slowly. Consequently, psychoanalysis is a long-term therapy requiring several years for completion and almost daily visits. Since Freud's time, there have been important modifications associated with former disciples such as Alfred Adler and Carl Jung. Self psychology and ego psychology are among more recent emphases. However, the popularity of psychoanalysis has waned. *See* PSYCHOANALYSIS.

Experiential therapies. A number of related therapies are included in this group. Probably best known was the patient-centered therapy of Carl Rogers appearing in the 1940s. In Rogers' therapy, a major emphasis is placed on the ability of the patient to change when the therapist is empathic and genuine and conveys nonpossessive warmth. The therapist is nondirective in the interaction with the patient and attempts to facilitate the growth potential of the patient. Other therapeutic approaches considered as experiential include Gestalt therapy, existential approaches, and transpersonal approaches. The facilitation of experiencing is emphasized as the basic therapeutic task, and the therapeutic relationship is viewed as a significant potentially curative factor.

Cognitive, behavioral, and interpersonal therapies. In behavioral therapies, therapists play a more directive role. The emphasis is on changing the patient's behavior, using positive reinforcement, and increasing self-efficacy. More recently, cognitive therapies such as those of A. T. Beck have tended to be combined with behavioral emphases. The cognitive-behavioral therapies have focused on changing dysfunctional attitudes into more realistic and positive ones and providing new information-processing skills. *See* COGNITION.

Most of the developments in interpersonal therapy have occurred in work with depressed patients. The goal of interpersonal therapy (a brief form of therapy) is centered on increasing the quality of the patient's interpersonal interactions. Emphasis is placed on enhancing the patient's ability to cope with stresses, improving interpersonal communications, increasing morale, and helping the patient deal with the effects of the depressive disorder. *See* PERSONALITY THEORY.

Eclectic and integrative therapies. The largest number of psychotherapists consider themselves to be eclectics. They do not adhere strictly to one theoretical orientation or school but use any procedures that they believe will be helpful for the individual patient. Eclecticism has been linked with the development of a movement for integration in psychotherapy. The emphasis in this new development is on openness to the views of other approaches, a less doctrinaire approach to psychotherapy, and an attempt to integrate two or more different theoretical views or systems of psychotherapy.

Group, family, and marital therapy. Most psychotherapy is conducted on a one-to-one basis—one therapist for one patient—and the confidentiality of these sessions is extremely important. However, there are other instances where more than one patient is involved because of particular goals. These include marital, family, and group therapy. Outpatient groups have been used for smoking cessation, weight loss, binge eating, and similar problems as well as for what were traditionally viewed as psychoneurotic problems. Inpatient group therapy was frequently employed in mental hospital settings.

There has been research on the combined use of medication and psychotherapy. In general, where two highly successful treatments are combined in cases with depressive or anxiety disorders, there appears to be little gain in effectiveness. However, in several studies of hospitalized patients with schizophrenia where individual psychotherapy has been ineffective, a combination of psychotherapy and medication has produced better results than medication alone. *See* AFFECTIVE DISORDERS; PSYCHOPHARMACOLOGY; SCHIZOPHRENIA. Sol L. Garfield

Psychotomimetic drugs A class of drugs, also called psychedelic or hallucinogenic drugs, that induce transient states of altered perception resembling or mimicking the symptoms of psychosis and are characterized by profound alterations in mood. The term psychotomimetic was originally coined to reflect the similarities between the mind-altering effects of certain psychoactive drugs and some of the symptoms of schizophrenia. *See* PSYCHOSIS; SCHIZOPHRENIA.

Psychotomimetic drugs produce a range of perceptual alterations and distortions. Sensory inputs are heightened or diminished (colors, textures, sounds, tastes, and touch become more or less intense). Sensations may assume compelling significance, to the point that an individual becomes completely immersed in an experience and disregards more salient stimuli. Individuals may experience illusions (distortions in the perceived nature or meaning of real objects). Typically hallucinations produced by psychotomimetic drugs are in the visual domain, and individuals realize that the perceptions are false. Synesthesias may occur, wherein individuals perceive stimuli from one sensory modality in another modality—for example, "seeing" sounds, or "hearing" colors. Time perception may be distorted, with time seeming to pass slower or faster. These effects on perceptions and thinking are invariably accompanied by changes in the individual's mood. Thus, individuals may experience intense and often rapidly shifting mood states ranging from mild apprehension to panic or severe depression to elation, or they may experience concurrent emotions (such as sadness and joy) which are not ordinarily experienced simultaneously.

Classical serotonergic hallucinogens include mescaline, lysergic acid diethylamide (LSD), and dimethyltryptamine (DMT), as well as the "magic mushroom" (*Psilocybe mexicana*) derivatives psilocybin and psilobin. Serotonergic hallucinogens produce their effects by partial stimulation of the 2_A subtype of serotonin receptors in the brain. While these drugs differ in the onset, duration, and intensity of effects, their acute behavioral effects are quite similar. *See* SEROTONIN.

Amphetamine and related drugs are stimulants, increasing the release of dopamine, norepinephrine, and serotonin in addition to blocking the reuptake of these neurotransmitters. They produce a "rush" (an orgasmlike state) followed by a state of euphoria, alertness, and a feeling of increased energy and self-confidence. Relative to classical serotonergic hallucinogens, perceptual alterations are not as prominent with amphetamine. *See* DOPAMINE; NORADRENERGIC SYSTEM.

One of the oldest and most widely used psychotomimetic drugs is derived from the flowers and leaves of the *Cannabis sativa* plant. Delta-9-tetrahydrocannabinol (THC) is the principal active ingredient of cannabis, but there are more than 80 other cannabinoids present in cannabis which may also contribute to its effects. Cannabinoids produce their psychotomimetic effects by actions at brain cannabinoid receptors. The effects of cannabis include changes in perception, cognition, and mood. *See* PSYCHOPHARMACOLOGY. Cyril D'Souza

Psychrometer An instrument consisting of two thermometers which is used in the measurement of the moisture content of air or other gases. The bulb or sensing area of one of the thermometers either is covered by a thin piece of clean muslin cloth wetted uniformly with distilled water or is otherwise coated with a film of distilled water. The temperatures of both the bulb and the air contacting the bulb are lowered by the evaporation which takes place when unsaturated air moves past the wetted bulb. An equilibrium temperature, termed the wet-bulb temperature (T_W), will be reached; it closely approaches the lowest temperature to which air can be cooled by the evaporation of water into that air. The water-vapor content of the air surrounding the wet bulb can be determined from this wet-bulb temperature and from the air temperature measured by the thermometer with the dry bulb (T_D by using an expression of the form $e = e_{SW} - aP\ (T_D - T_W)$. Here e is the water-vapor pressure of the air, e_{SW} is the saturation water-vapor pressure at the wet-bulb temperature, P is atmospheric pressure, and a is the psychrometric constant, which depends upon properties of air and water, as well as on speed of ventilation of air passing the wet bulb. *See* PSYCHROMETRICS. Richard M. Schotland

Psychrometrics A study of the physical and thermodynamic properties of the atmosphere. The properties of primary concern in air conditioning are (1) dry-bulb temperature, (2) wet- bulb temperature, (3) dew-point temperature, (4) absolute humidity, (5) percent humidity, (6) sensible heat, (7) latent heat, (8) total heat, (9) density, and (10) pressure.

The dry-bulb temperature is the ambient temperature of the air and water vapor as measured by a thermometer or other temperature-measuring device in which the thermal element is dry and shielded from radiation. *See* AIR TEMPERATURE; TEMPERATURE.

If the bulb of a dry-bulb thermometer is covered with a silk or cotton wick saturated with distilled water and the air is drawn over it at a velocity not less than 1000 ft/min (5 m/s), the resultant temperature will be the wet-bulb temperature. Where the dry-bulb and wet-bulb temperatures are the same, the atmosphere is saturated.

The dew-point temperature is the temperature at which the water vapor in the atmosphere begins to condense. This is also the temperature of saturation at which the dry-bulb, wet-bulb, and dew-point temperatures are all the same. *See* DEW POINT.

The actual quantity of water vapor in the atmosphere is designated as the absolute humidity. Percentage or relative humidity is the ratio of the actual water vapor in the atmosphere to the quantity of water vapor the atmosphere could hold if it were saturated at the same temperature. *See* HUMIDITY.

Sensible heat, or enthalpy of dry air, is heat which manifests itself as a change in temperature. *See* ENTHALPY.

Latent heat, or enthalpy of vaporization, is the heat required to change a liquid into a vapor without change in temperature. Latent heat is sometimes referred to as the latent heat of vaporization and varies inversely as the pressure.

The total heat, or enthalpy, of the atmosphere is the sum of the sensible heat, latent heat, and superheat of the vapor above the saturation or dew-point temperature. Total heat is relatively constant for a constant wet-bulb temperature, deviating only about 1.5–2% low at relative humidities below 30%.

The density of the atmosphere varies with both altitude and percentage humidity. The higher the altitude the lower the density, and the higher the moisture content the lower the density. *See* DENSITY.

Atmospheric pressure is usually referred to as barometric pressure. Pressure varies inversely as elevation, as temperature, and as percentage saturation. *See* MOISTURE-CONTENT MEASUREMENT; PSYCHROMETER. John Everetts, Jr.

Pteridosperms A group of extinct seed plants characterized by fernlike leaves that produced naked seeds. The discovery of the seed ferns was a major contribution to the study of plant evolution because it demonstrated the existence of a group of vascular plants that is today extinct. Although the seed ferns probably have their ultimate origin within the progymnosperm order Aneurophytales, the best evidence of the group comes from lower Carboniferous (Mississippian) and younger sediments. Some seed ferns are reconstructed as trees with stout stems, while perhaps the majority were vines or lianas that supported massive fernlike fronds. The seed ferns consist of six Paleozoic orders (Calamopityales, Buteoxylonales, Lyginopteridales,

Medullosales, Callistophytales, Glossopteridales) and three orders (Peltaspermales, Corystospermales, Caytoniales) found in Mesozoic rocks. *See* CAYTONIALES; PALEOBOTANY.

Thomas N. Taylor

Pteriomorphia An order containing seven superfamilies of marine and brackish bivalves with byssal attachment at some time in their geological history (the Ostreacea are exceptional, with the lower valve attached by byssal cement at settlement, and with no record of ancestral attachment by byssal threads). The Pteriomorphia possess some primitive features and some secondarily simple features: the hinge is edentulous or with inconspicuous teeth (except *Spondylus*); there are no siphons; minimal points of fusion occur between the left and right mantle lobes; the Mytilacea, Pteriacea, Pectinacea, and Anomiacea have filibranch ctenidia. Specialized features include simultaneous hermaphroditism in some Pectinacea, sex reversal and incubation of larvae in some Ostreacea, and protandrous hermaphroditism in others. Few features are common to all Pteriomorphia, and the systematic status is debatable. *See* BIVALVIA; MOLLUSCA.

Richard D. Purchon

Pterobranchia A group of small sessile hemichordates that may be colonial (*Rhabdopleura*), pseudocolonial (*Cephalodiscus*), or solitary (*Atubaria*). Each individual, or zooid, lives inside a nonchitinous tube secreted by the protosome, except for *Atubaria*; an aggregation of zooids is called a coenecium and can vary in shape. The protosome, or oral shield, is disciform, closes the mouth ventrally, and secretes the tube. The protocoel has symmetrical pores at the base of the first pair of arms. The mesosome, or collar, has an anteroventral mouth and one to nine pairs of dorsally ciliated tentaculated arms, which are used to collect small organisms for food. The mesocoels extend into both arms and tentacles.

The metasome is divided into a sacciform trunk and a slender ventral stalk that may be free at its end. The trunk contains the U-shaped digestive tract, with the pharynx having a single pair of gills, except in *Rhabdopleura*. The stomach is a sacciform expansion of the gut, and the tubular intestine curves dorsally and opens behind the arms by a middorsal anus.

The nervous system is very simple and lacks a neurochord. The buccal diverticulum is hollow, with an anteroventral heart vesicle and central sinus. The glomerulus is poorly developed, and the circulatory system is simpler that in enteropneusts. *See* HEMICHORDATA.

Jesús Benito

Pteropsida A large group of vascular plants characterized by having parenchymatous leaf gaps in the stele and by having leaves which are thought to have originated in the distant past as branched stem systems. Some botanists regard the Pteropsida as a natural group which they recognize as a class, subdivision, or division. Others regard it as an artificial assemblage of plants that have undergone certain similar changes from a rhyniophyte ancestry. The various components of the Pteropsida are here treated as three separate divisions under the names Magnoliophyta, Pinophyta, and Polypodiophyta. *See* MAGNOLIOPHYTA; PINOPHYTA; POLYPODIOPHYTA.

Arthur Cronquist

Pterosauria The "winged reptiles" of the Mesozoic Era, constituting the closest major group to Dinosauria and sharing many features with them. Their common ancestor was a small, bipedal, agile archosaur reptile probably similar to the small *Scleromochlus* of the Late Triassic. This ancestor probably possessed a large, lightly built skull, a short body, long hindlimbs, and digitigrade feet with four long metatarsals. Pterosaurs inherited all these features, and they further evolved the power of flight. Pterosaurs had a wing of skin that was internally supported by long, fine, possibly keratinous stiffening fibers and braced by the forelimb, including a greatly elongated fourth finger (the first three fingers remained small). Their brains were relatively large and somewhat birdlike; the canals of the ear region show that they were well suited for life in the air. *See* DINOSAUR; REPTILIA.

The earliest pterosaurs are known from the Late Triassic, about the time that the first dinosaurs appeared: they include *Eudimorphodon*, *Peteinosaurus*, and *Preondactylus*, all from marine rocks of Italy and Greenland. In recent decades, some of the most interesting finds have come from China and from a series of limestone nodules in the Santana region of Brazil. Pterodactyloids ("wing fingers") were a subgroup of Late Jurassic and Cretaceous pterosaurs that replaced the early pterosaurs. The best-known members of this group include the small *Pterodactylus* and the much larger *Pteranodon* with 7-m (23-ft) wing span (see illustration).

Late Cretaceous pterodactyloid *Pteranodon*, from western North America (reconstruction).

Experts agree that pterosaurs were true flying reptiles, powered by wings that flapped vigorously. The flight muscles had extensive attachments to the wing bones and the expanded breastbone, and the shoulder girdle was modified to brace the force of the flight stroke against the skeleton. The wing itself was made of thin skin attached to the bones of the forearm and wing finger. It was reinforced by thin, perhaps keratinous fibers only a few micrometers in diameter but several centimeters long. All but the largest pterosaurs were active flappers; the largest relied on soaring because muscle power cannot sustain flapping flight for extended periods in such large forms. *See* FLIGHT.

Kevin Padian

Pterygota A subclass of the class Insecta comprising the winged insects and their secondarily wingless descendants. This subclass, including the great majority of insects, is usually divided into two infraclasses: Paleoptera, or ancient-winged insects (mayflies, dragonflies, and several extinct orders, most of which could not fold the wings over the body), and Neoptera, or folding-winged insects (the rest of the winged insects and some secondarily wingless lineages), which ancestrally folded the wings over the body by a particular articulatory mechanism.

The acquisition of wings occurred once in the Insecta and was the major general adaptation in their evolution. The winged insects, both Paleoptera and Neoptera, first appear in the fossil record in the Early Pennsylvanian (upper Carboniferous), suggesting their origin in the Mississippian. *See* ENDOPTERYGOTA; EXOPTERYGOTA; INSECTA.

William L. Brown, Jr.

Public health An effort organized by society to protect, promote, and restore the people's health. It is the combination of sciences, skills, and beliefs that is directed to the maintenance and improvement of health through collective or social actions. The programs, services, and institutions of public health emphasize the prevention of disease and the health needs of the population as a whole. Additional goals include the reduction of the amount of disease, premature death, disability, and discomfort in the population.

The basic sciences of public health include epidemiology and vital statistics, which measure health status and assess health trends in the population. Epidemiology is also a powerful research method, used to identify causes and calculate risks of acquiring or dying of many conditions. Many sciences, including toxicology and microbiology, are applied to detect, monitor, and correct physical, chemical, and biological hazards in the environment. Such applications are being used to address concerns about a deteriorating global environment. The social and behavioral sciences have become more prominent in public health since the recognition that such factors as indolence, loneliness, personality type, and addiction to tobacco contribute to the risk of premature death and chronic disabling diseases. *See* EPIDEMIOLOGY.

In most industrial nations, public health services are organized nationally, regionally, and locally. National public health services are usually responsible for setting, monitoring, and maintaining health standards, for promoting good health, for collecting and compiling national health statistics, and for supporting and performing research on diseases important to public health. Regional (for example, state) public health services deal mainly with major health protection activities such as ensuring safe water and food supplies; they may also operate screening programs for early detection of disease and are responsible for health care of certain groups such as chronic mentally ill persons. Local public health services (in cities, large towns, and some rural communities) conduct a variety of personal public health services, such as immunization programs, health education, health surveillance and advice for mothers and newborn babies, and personal care of vulnerable groups such as the elderly and housebound long-term sick. Local health services also investigate and control epidemics and other communicable conditions such as sexually transmitted diseases.

National public health services communicate with each other in efforts to control diseases of international importance, and they collaborate worldwide under the auspices of the World Health Organization (WHO). While much of the work of WHO has been concentrated in the developing nations, it has also been involved in global efforts to control major epidemic diseases and to set standards for hazardous environmental and occupational exposures. John M. Last

Pulley A wheel with a flat, crowned, or grooved rim used with a flat belt, V-belt, or a rope to transmit motion and energy. Pulleys for use with V-belt and rope drives have grooved surfaces and are usually called sheaves. A combination of ropes, pulleys, and pulley blocks arranged to gain a mechanical advantage, as for hoisting a load, is referred to as block and tackle. *See* BELT DRIVE; BLOCK AND TACKLE.

Pulleys for flat belts are made of cast iron, fabricated steel, wood, and paper. A particular pulley design must be based on such considerations as the ability to resist shock, to conduct heat, and to resist corrosive environments. The face must be smooth enough to minimize belt wear; yet there must be adequate friction between belt and pulley face to carry the load.

The two common types of V-belt pulley are pressed-steel and cast-iron. The pressed-steel pulleys are suitable for single-belt drives. For multiple-belt drives, or in single-belt drives where pulley mass should be high to get a flywheel effect, cast-iron pulleys are used. John R. Zimmerman

Pulmonata A subclass of the molluscan class Gastropoda, containing about 23,500 species of snails that are grouped into three orders or superorders: Systellommatophora, Basommatophora, and Stylommatophora. Pulmonates include most of the common snails found on land or in freshwaters, although a few prosobranchs have invaded both habitats. Certain pulmonates are intertidal to subtidal on rocky shores of tropical and temperate regions. *See* BASOMMATOPHORA; STYLOMMATOPHORA; SYSTELLOMMATOPHORA.

Freshwater pulmonates belong to the superorder Basommatophora and are adapted to make use of accidental transport on birds or insects from one isolated body of water to another. Mostly they have short life-spans, and their numbers vary dramatically with the seasons.

Land pulmonates belong to the superorder Stylommatophora and are marginally terrestrial in that they can be active only when the humidity in their microhabitat is 90% or above. Thus many are cryptic or nocturnal in both feeding and reproduction. They can survive in deserts by estivating for up to 6 years. *See* GASTROPODA. G. Alan Solem

Pulsar A celestial radio source producing intense short bursts of radio emission. Since their discovery in 1968, over 1750 pulsars have been found (as of October 2006), and it has become clear that 100,000 pulsars must exist in the Milky Way Galaxy—most of them too distant to be detected with existing radio telescopes. *See* RADIO ASTRONOMY.

Pulsars are distinguished from most other types of celestial radio sources in that their emission, instead of being constant over time scales of years or longer, consists of periodic sequences of brief pulses. The interval between pulses, or pulse period, is nearly constant for a given pulsar, but for different sources ranges from 0.0014 to 8.5 s. The bursts of emission are generally confined to a window whose width is a few percent of the period. Individual pulses can vary widely in intensity; however, their periodic spacing is accurately maintained.

The association of pulsars with neutron stars, the collapsed cores left behind when moderate- to high-mass stars become unstable and collapse, is supported by many arguments. Prior to the discovery of pulsars, neutron stars were thought to be unobservable owing to their extremely small size (radius less than 15 km or 10 mi). The standard model for pulsars is a spinning neutron star with an intense dipole magnetic field (surface field of 10^8 teslas or 10^{12} gauss) misaligned with the rotation axis. The off-axis rotating dipole field develops a huge voltage difference between the neutron star surface and the surrounding matter. Charges accelerate in this voltage and generate an avalanche of electrons and positrons, a relativistic current leaving the polar zones of the star. Highly directive radio emission is formed in this current, which is observed as pulses, one per rotation, just like a rotating searchlight. *See* ELECTRON; POSITRON.

A fundamental observation that supports the rotating neutron star model is the remarkable stability of the basic pulsation periods, which typically remain constant to a few tens of nanoseconds over a year. This stability is natural to the free rotation of a compact, rigid object like a neutron star, but is extremely difficult to produce by any other known physical process. A small emitting body is also required by the observed rapid variations within pulses. A spinning, magnetized body will gradually slow down because it emits low-frequency electromagnetic radiation at harmonics of its rotation period. This slowing down is observed in all objects where observations of sufficient precision are available.

Further support of the neutron star model comes from the discovery of a 33-millisecond pulsar in the center of the Crab Nebula, the remnant of the supernova observed in A.D 1054. The lengthening of this pulsar's short period on a time scale of 1000 years matches the age of the remnant. The rotating, magnetized neutron star model is further supported by observations of diffuse radiation near the pulsar in the Crab Nebula by the *Hubble*, *XMM-Newton*, and *Chandra* space observatories. Currently, about 20 associations have been found between pulsars and supernova remnants which lend further support to these ideas. A 66-ms pulsar has been found in the supernova remnant 3C58. This is likely associated with the historical supernova of 1181, making this the youngest radio pulsar currently known. *See* CRAB NEBULA; NEUTRON STAR; SUPERNOVA.

Pulsars have provided a unique set of probes for the investigation of the diffuse gas and magnetic fields in interstellar space. Measurement of absorption at 1420 MHz, the frequency of the hyperfine transition in ground-state neutral hydrogen atoms, gives information on the structure of gas clouds, and in many cases provides an estimate of the pulsar distance. The index of refraction for radio waves in the ionized interstellar gas is strongly frequency-dependent, and low-frequency signals propagate more slowly than those at high frequencies. The broadband, pulsed nature of pulsar signals makes them ideal for measurements of this dispersion. Duncan R. Lorimer; Donald C. Backer

Pulse demodulator A device that extracts the analog information from a received pulse or from a set of pulses.

Constant-amplitude pulse-width-modulated (PWM) signals carry their information in the time interval that the pulse is "on." The source information is in the average value of the PWM signal; therefore, the source-information-recovery circuit can be a simple low-pass filter whose cutoff frequency must lie above the bandwidth of the source signal to preserve the information content, but below $^1/_2$ the frequency of the framing or timing signal to reject the timing-pulse noise. The source analog information behind ternary PWM signals of amplitudes $+V$, 0, and $-V$ is recovered in the same manner. *See* ELECTRIC FILTER; LOGIC CIRCUITS; PULSE MODULATOR.

Pulse-amplitude-modulated (PAM) signals are pulses of constant width, and usually the product of the source signal and a sampling-pulse train. Recovery again is achieved by simple low-pass filtering.

Pulse-density-modulated (PDM) signals consist of trains of constant-amplitude and constant-duration pulses whose pulse rates are proportional to the amplitudes of the source signals. The analog source information is again recovered by a simple low-pass filter.

Pulse-code-modulated (PCM) signals are digital representations of their source signals. A conventional digital-to-analog converter (DAC) recovers the analog source signal. *See* DEMODULATOR; DIGITAL-TO-ANALOG CONVERTER. Stanley A. White

Pulse generator An electronic circuit capable of producing a waveform that rises abruptly, maintains a relatively flat top for an extremely short interval, and then rapidly falls to zero. A relaxation oscillator, such as a multivibrator, may be adjusted to generate a rectangular waveform having an extremely short duration, and as such it is referred to as a pulse generator. However, there is a class of circuits whose exclusive function is generating short-duration, rectangular waveforms. These circuits are usually specifically identified as pulse generators. An example of such a pulse generator is the triggered blocking oscillator, which is a single relaxation oscillator having transformer-coupled feedback from output to input. *See* MULTIVIBRATOR.

Pulse generators sometimes include, but are usually distinguished from, trigger circuits. Trigger circuits generate a short-duration, fast-rising waveform for initiating or triggering an event or a series of events in other circuits. In the pulse generator, the pulse duration and shape are of equal importance to the rise and fall times. *See* TRIGGER CIRCUIT.

The term pulse generator is often applied not only to an electronic circuit generating prescribed pulse sequences but to an electronic instrument designed to generate sequences of pulses with variable delays, pulse widths, and pulse train combinations, programmable in a predetermined manner, often microprocessor-controlled.

A network, formed in such a way as to simulate the delay characteristics of a lossless transmission line, and appropriate switching elements to control the duration of a pulse form the basis for a variety of types of pulse generators. Some delay-line-controlled pulse generators are capable of generating pulses containing considerable amounts of power for such applications as modulators in radar transmitters. *See* DELAY LINE; WAVE-SHAPING CIRCUITS. Glenn M. Glasford

Pulse jet A type of jet engine characterized by periodic surges of thrust. The pulse jet engine was widely known for its use during World War II on the German V-l missile (see illustration). The basic engine cycle was invented in 1908. The inlet end of the engine is provided with a grid to which are attached flap valves. These valves are normally held by spring tension against the grid face and block the flow of air back out of the front of the engine. They can be sucked inward by a negative differential pressure to allow air to flow into the engine. Downstream from the flap valves is the combustion chamber. A fuel injection system is located at the entrance to the combustion chamber. The chamber is also fitted with a spark plug. Following the combustion chamber is a long exhaust duct which provides an inertial gas column.

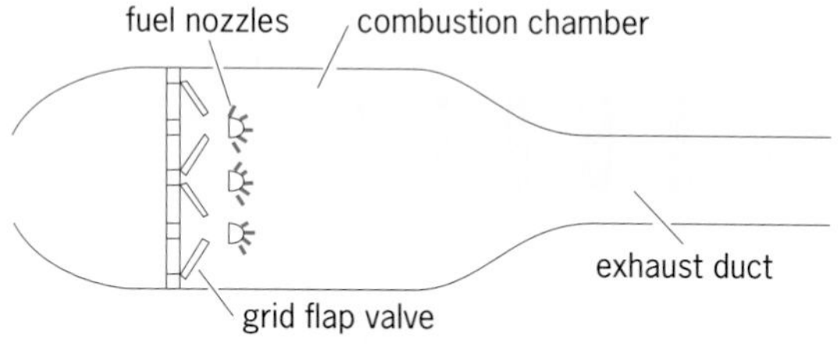

Diagram of a pulse jet.

Pulse jets have also been used to propel radio-controlled target drones and experimental helicopters. In the latter case, they were mounted on the blade tips for directly driving the rotor. The high fuel consumption, noise, and vibrations generated by the pulse jet limit its scope of applications. *See* PROPULSION. Benjamin Pinkel

Pulse modulation A set of techniques whereby a sequence of information-carrying quantities occurring at discrete instances of time is encoded into a corresponding regular sequence of electromagnetic carrier pulses. Varying the amplitude, polarity, presence or absence, duration, or occurrence in time of the pulses gives rise to the four basic forms of pulse modulation: pulse-amplitude modulation (PAM), pulse-code modulation (PCM), pulse-width modulation (PWM, also known as pulse-duration modulation, PDM), and pulse-position modulation (PPM).

Analog-to-digital conversion. An important concept in pulse modulation is analog-to-digital (A/D) conversion, in which an original analog (time- and amplitude-continuous) information signal $s(t)$ is changed at the transmitter into a series of regularly occurring discrete pulses whose amplitudes are restricted to a fixed and finite number of values. An inverse digital-to-analog (D/A) process is used at the receiver to reconstruct an approximation of the original form of $s(t)$. Conceptually, analog-to-digital conversion involves two steps. First, the range of amplitudes of $s(t)$ is divided or quantized into a finite number of predetermined levels, and each such level is represented by a pulse of fixed amplitude. Second, the amplitude of $s(t)$ is periodically measured or sampled and replaced by the pulse representing the level that corresponds to the measurement. *See* ANALOG-TO-DIGITAL CONVERTER; DIGITAL-TO-ANALOG CONVERTER.

According to the Nyquist sampling theorem, if sampling occurs at a rate at least twice that of the bandwidth of $s(t)$, the latter can be unambiguously reconstructed from its amplitude values at the sampling instants by applying them to an ideal low-pass filter whose bandwidth matches that of $s(t)$.

Quantization, however, introduces an irreversible error, the so-called quantization error, since the pulse representing a sample measurement determines only the quantization level in which the measurement falls and not its exact value. Consequently, the process of reconstructing $s(t)$ from the sequence of pulses yields only an approximate version of $s(t)$.

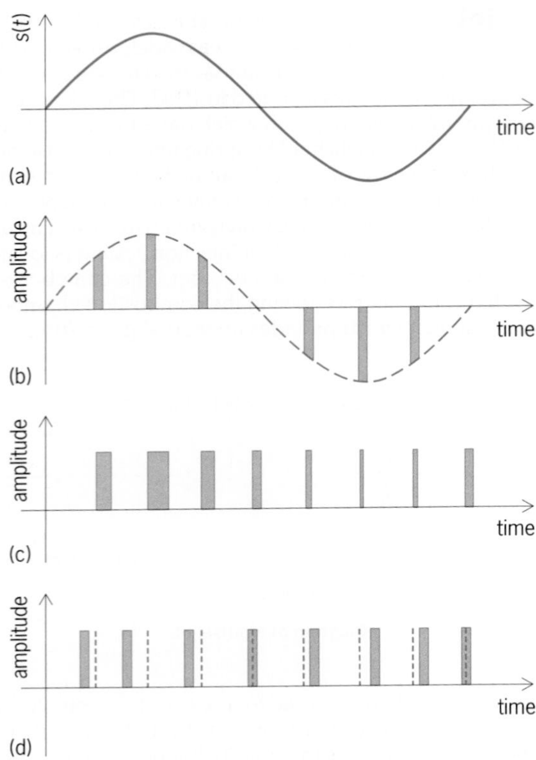

Fig. 1. Forms of pulse modulation for the case where the analog signal, *s(t)*, is a sine wave. (*a*) Analog signal, *s(t)*. (*b*) Pulse-amplitude modulation. (*c*) Pulse-width modulation. (*d*) Pulse-position modulation.

Pulse-amplitude modulation. In PAM the successive sample values of the analog signal $s(t)$ are used to effect the amplitudes of a corresponding sequence of pulses of constant duration occurring at the sampling rate. No quantization of the samples normally occurs (Fig. 1*a*, *b*). In principle the pulses may occupy the entire time between samples, but in most practical systems the pulse duration, known as the duty cycle, is limited to a fraction of the sampling interval. Such a restriction creates the possibility of interleaving during one sample interval one or more pulses derived from other PAM systems in a process known as time-division multiplexing (TDM). *See* MULTIPLEXING AND MULTIPLE ACCESS.

Pulse-width modulation. In PWM the pulses representing successive sample values of $s(t)$ have constant amplitudes but vary in time duration in direct proportion to the sample value. The pulse duration can be changed relative to fixed leading or trailing time edges or a fixed pulse center. To allow for time-division multiplexing, the maximum pulse duration may be limited to a fraction of the time between samples (Fig. 1*c*).

Pulse-position modulation. PPM encodes the sample values of $s(t)$ by varying the position of a pulse of constant duration relative to its nominal time of occurrence. As in PAM and PWM, the duration of the pulses is typically a fraction of the sampling interval. In addition, the maximum time excursion of the pulses may be limited (Fig. 1*d*).

Pulse-code modulation. Many modern communication systems are designed to transmit and receive only pulses of two distinct amplitudes. In these so-called binary digital systems, the analog-to-digital conversion process is extended by the additional step of coding, in which the amplitude of each pulse representing a quantized sample of $s(t)$ is converted into a unique sequence of one or more pulses with just two possible amplitudes. The complete conversion process is known as pulse-code modulation.

Figure 2*a* shows the example of three successive quantized samples of an analog signal $s(t)$, in which sampling occurs every T seconds and the pulse representing the sample is limited to $T/2$ seconds. Assuming that the number of quantization levels is limited to 8, each level can be represented by a unique sequence of three two-valued pulses. In Fig. 2*b* these pulses are of amplitude V or 0, whereas in Fig. 2*c* the amplitudes are V and $-V$.

PCM enjoys many important advantages over other forms of pulse modulation due to the fact that information is represented by a two-state variable. First, the design parameters of a PCM transmission system depend critically on the bandwidth of the original signal $s(t)$ and the degree of fidelity required at the point of reconstruction, but are otherwise largely independent of the information content of $s(t)$. This fact creates the possibility of deploying generic transmission systems suitable for many types of information. Second, the detection of the state of a two-state variable in a noisy environment is inherently simpler than the precise measurement of the amplitude, duration, or position of a pulse in which these quantities are not constrained. Third, the binary pulses propagating along a medium can be intercepted and decoded at a point where the accumulated distortion and attenuation are sufficiently low to assure high detection accuracy. New pulses can then be generated and transmitted to the next such decoding point. This so-called process of repeatering significantly reduces the propagation of distortion and leads to a quality of transmission that is largely independent of distance.

Time-division multiplexing. An advantage inherent in all pulse modulation systems is their ability to transmit signals from multiple sources over a common transmission system through the process of time-division multiplexing. By restricting the time duration of a pulse representing a sample value from a particular analog signal to a fraction of the time between successive samples, pulses derived from other sampled analog signals can be accommodated on the transmission system.

One important application of this principle occurs in the transmission of PCM telephone voice signals over a digital transmission system known as a T1 carrier. In standard T1 coding, an original analog voice signal is band-limited to 4000 hertz by

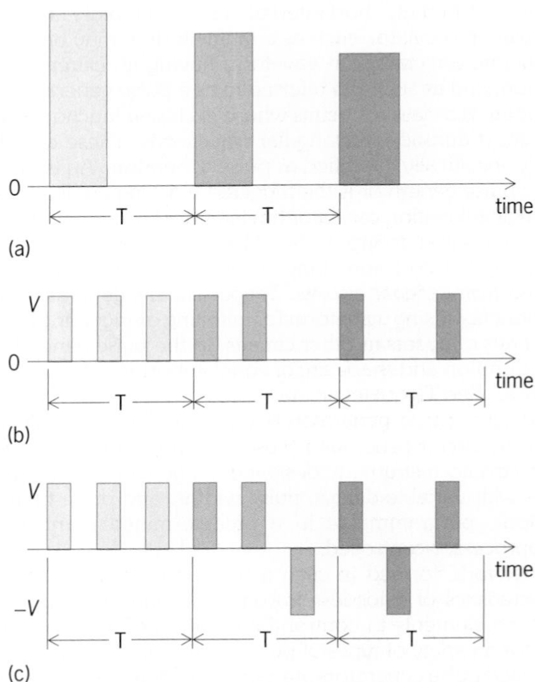

Fig. 2. Pulse-code modulation. (*a*) Three successive quantized samples of an analog signal. (*b*) With pulses of amplitude *V* or 0. (*c*) With pulses of amplitude *V* or −*V*.

passing it through a low-pass filter, and is then sampled at the Nyquist rate of 8000 samples per second, so that the time between successive samples is 125 microseconds. The samples are quantized to 256 levels, with each of them being represented by a sequence of 8 binary pulses. By limiting the duration of a single pulse to 0.65 microsecond, a total of 193 pulses can be accommodated in the time span of 125 microseconds between samples. One of these serves as a synchronization marker that indicates the beginning of such a sequence of 193 pulses, while the other 192 pulses are the composite of 8 pulses from each of 24 voice signals, with each 8-pulse sequence occupying a specified position. T1 carriers and similar types of digital carrier systems are in widespread use in the world's telephone networks.

Bandwidth requirements. Pulse modulation systems may incur a significant bandwidth penalty compared to the transmission of a signal in its analog form. An example is the standard PCM transmission of an analog voice signal band-limited to 4000 hertz over a T1 carrier. Since the sampling, quantizing, and coding process produces 8 binary pulses 8000 times per second for a total of 64,000 binary pulses per second, the pulses occur every 15.625 microseconds. Depending on the shape of the pulses and the amount of intersymbol interference, the required transmission bandwidth will fall in the range of 32,000 to 64,000 hertz. This compares to a bandwidth of only 4000 hertz for the transmission of the signal in analog mode.

Applications. PAM, PWM, and PPM found significant application early in the development of digital communications, largely in the domain of radio telemetry for remote monitoring and sensing. They have since fallen into disuse in favor of PCM.

Since the early 1960s, many of the world's telephone network providers have gradually, and by now almost completely, converted their transmission facilities to PCM technology. The bulk of these transmission systems use some form of time-division multiplexing, as exemplified by the 24-voice channel T1 carrier structure. These carrier systems are implemented over many types of transmission media, including twisted pairs of telephone wiring, coaxial cables, fiber-optic cables, and microwave. *See* COAXIAL CABLE; COMMUNICATIONS CABLE; MICROWAVE; OPTICAL COMMUNICATIONS; OPTICAL FIBERS; SWITCHING SYSTEMS (COMMUNICATIONS); TELEPHONE SERVICE.

The deployment of high-speed networks such as the Integrated Service Digital Network (ISDN) in many parts of the world has also relied heavily on PCM technology. PCM and various modified forms such as delta modulation (DM) and adaptive differential pulse-code modulation (ADPCM) have also found significant application in satellite transmission systems. *See* COMMUNICATIONS SATELLITE; DATA COMMUNICATIONS; ELECTRICAL COMMUNICATIONS; INTEGRATED SERVICES DIGITAL NETWORK (ISDN); MODULATION.

Hermann J. Helgert

Pulse modulator A device that converts an analog information signal to a pulse format. A pulse modulator encodes the amplitude information at a particular time by a set of short- or long-duration pulses using such methods as pulse-amplitude, pulse-width, pulse-position, pulse-density, or pulse-code modulation. A pulse modulator, therefore, accepts a continuous analog input signal and provides a pulse output that conveys the amplitude of the continuous analog signal as seen at an observation, or sampling, time t_s, often optimistically described as a sampling instant.

Pulse-amplitude modulation. A continuous analog signal, when viewed through equally spaced time windows, provides a pulse-amplitude modulation (PAM) signal. This is analogous to viewing the continuous signal through the vertical slats in a picket fence. The signal is sampled at equally spaced time intervals of T_s, the sampling period. The reciprocal of T_s is the sampling frequency f_s. If the highest frequency in the continuous analog signal is $f_{\max}$, then in order to avoid loss of information through sampling, the sampling theorem dictates that $f_s > 2 \cdot f_{\max}$. The duration of each sampling window, during which time we look at the analog signal, is the aperture time T_a (the spacing between the slats). The duty cycle, T_a/T_s, represents the fraction of the time that the PAM signal is present. The time that the PAM signal is absent (idle time) can be profitably used to multiplex $N = T_s/T_a$ different PAM signals together. Channel efficiency is improved by time-division-multiplexed (TDM) transmission, or the interleaving of N analog signals over a single channel. *See* INFORMATION THEORY; MULTIPLEXING AND MULTIPLE ACCESS.

A pulse-amplitude modulator is simply a transmission gate, or switch, that opens every T_s seconds for duration of T_a seconds to produce a PAM signal. On the other hand, a sample-and-hold circuit may be used. The value of the signal detected at the sampling instant is clamped or "held" until the next sample instant by a sample-and-hold (S&H) device.

The preceding discussion concerns baseband PAM. If the transmited signal is then multiplied by a sinusoidal carrier, it becomes passband PAM.

Phase modulator. A close relative of the pulse-amplitude modulator is the phase-shift-keyed (PSK) modulator. The amplitude of the signal is held constant, but the information is provided in the phase of a carrier. PSK and PAM can be combined to form QAM (quadrature-amplitude-modulated) signals, discussed below. *See* PHASE MODULATION.

Pulse-width modulation. Unlike a PAM signal, the amplitude of each pulse-width modulation (PWM) pulse is constant. The width τ of each output pulse carries the amplitude information of the continuous analog signal during the observation interval. The input signal $x(t)$ may be sampled and "held" or clamped to its value at the sampling instants, thereby generating a stair-step waveform; or it may simply be used in its continuous form.

PWM is especially useful in applications where power, as opposed to information or data, is being controlled because electrical power-switching devices are more power-efficient than proportional-control devices. Switched-device resistance is either very low or near-infinite. In dc-motor speed-control operation, PWM motor-drive signals virtually eliminate the static friction (stiction) of motor bearings and mechanical linkages because the torque is applied abruptly. PWM is also used for voltage regulation in switched power supplies, light dimmers, and class D amplifiers for audio applications. *See* ELECTRONIC POWER SUPPLY.

Pulse-density modulation. The pulse-density modulation (PDM) pulse rate during the sampling period is proportional to the amplitude of the analog input signal, as is its average value. A PDM signal may be generated by driving a voltage-controlled oscillator with the analog input signal, clipping the output to generate a square wave, then differentiating to produce a pulse train whose rate, or density, varies proportionally with the analog input voltage. *See* OSCILLATOR.

Pulse-code modulation. A pulse-code modulator produces numbers, rather than the analog values that are produced by all of the pulse-modulation methods discussed so far, which quantize the signal in time only. In those methods, the signal amplitude is specified at sampling instants t_s (quantized in time), but the signal amplitude can take on any values between the signal maximum and minimum values.

For additional robustness against noise, the digital format is often not straight binary, but reflected binary or Gray code, in which any two adjacent numerical values can differ by only one bit.

Corresponding to each sampling instant, the PCM produces one data word. A PCM signal is said to be digitized because its quantized amplitude values are in digital format. The PCM device itself is usually known as an analog-to-digital converter. A digitized signal is reasonably noise-resistant and can be modulated, demodulated, filtered, and otherwise processed more accurately than an analog signal. Many PCM data streams may be time-division-multiplexed together for transmission. *See* ANALOG-TO-DIGITAL CONVERTER. Stanley A. White

Pumice A rock froth, formed by the extreme puffing up of liquid lava by expanding gases liberated from solution in the lava prior to and during solidification. Some varieties will float in water for many weeks before becoming waterlogged. Typical pumice is siliceous (rhyolite or dacite) in composition, but the lightest and most vesicular pumice (known also as reticulite and thread-lace scoria) is of basaltic composition. *See* LAVA; VOLCANIC GLASS. Gordon A. MacDonald

Pump A machine that draws a fluid into itself through an entrance port and forces the fluid out through an exhaust port (see illustration). A pump may serve to move liquid, as in a cross-country pipeline; to lift liquid, as from a well or to the top of a tall building; or to put fluid under pressure, as in a hydraulic brake. These applications depend predominantly upon the discharge characteristic of the pump. A pump may also serve to empty a container, as in a vacuum pump or a sump pump, in which case the application depends primarily on its intake characteristic. *See* CENTRIFUGAL PUMP; COMPRESSOR; DISPLACEMENT PUMP; FAN; FUEL PUMP; PUMPING MACHINERY; VACUUM PUMP. Elliott F. Wright

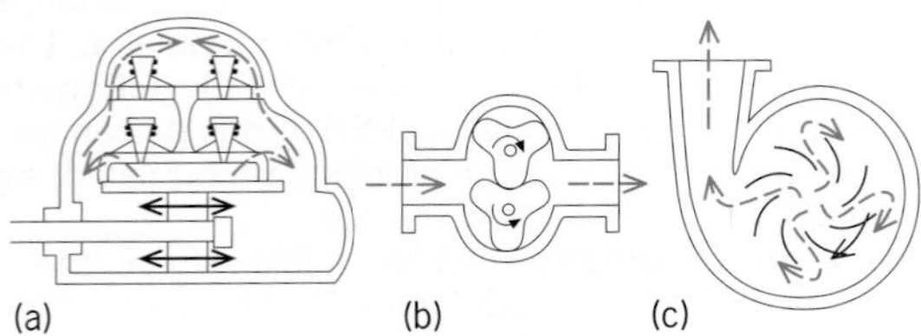

Pumps. (*a*) Reciprocating. (*b*) Rotary. (*c*) Centrifugal.

Pumped storage A process, also known as hydroelectric storage, for converting large quantities of electrical energy to potential energy by pumping water to a reservoir at a higher elevation, where it can be stored indefinitely and then released to pass through hydraulic turbines and generate electrical energy. An indirect process is necessary because electrical energy cannot be stored effectively in large quantities. Storage is desirable, as the consumption of electricity varies throughout the day, between weekday and weekend, as well as among seasons. Consequently, much of the generating equipment needed to meet the greatest daytime load is unused or lightly loaded at night or on weekends. During those times the excess capability can be used to generate energy for pumping, hence the necessity for storage.

A typical pumped-storage development consists of two reservoirs of essentially equal volume situated to maximize the difference between their levels. These reservoirs are connected by a system of waterways along which a pumping-generating station is located (see illustration). Under favorable geological conditions, the station will be located underground; otherwise it will be situated at the lower reservoir. The principal equipment of the station is the pumping-generating unit. In United States practice, the machinery is reversible and is used for both pumping and generating; it is designed to function as a motor and pump in one direction of rotation and as a turbine and generator in opposite rotation. *See* ELECTRIC POWER GENERATION; PUMPING MACHINERY; WATERPOWER. Dwight L. Glasscock

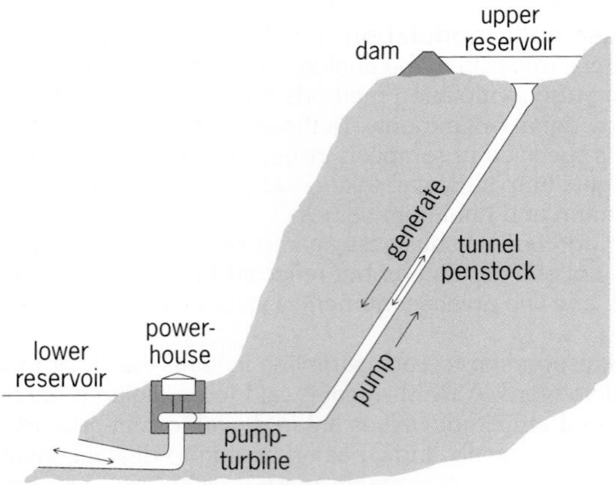

Schematic of a conventional pumped-storage development.

Pumping machinery Devices which convey fluids, chiefly liquids, from a lower to a higher elevation or from a region of lower pressure to one of higher pressure. Pumping machinery may be broadly classified as mechanical or as electromagnetic.

In mechanical pumps ghe fluid is conveyed by direct contact with a moving part of the pumping machinery. The two basic types are (1) velocity machines, centrifugal or turbine pumps, which impart energy to the fluid primarily by increasing its velocity, then converting part of this energy into pressure or head, and (2) displacement machines with plungers, pistons, cams, or other confining forms which act directly on the fluid, forcing it to flow against a higher pressure. *See* CENTRIFUGAL PUMP; DISPLACEMENT PUMP.

Where direct contact between the fluid and the pumping machinery is undersirable, as in atomic energy power plants for circulating liquid metals used as reactor coolants or as solvents for reactor fuels, electromagnetic pumps are used. There are no moving parts in these pumps; no shaft seals are required. The liquid metal passing through the pump becomes, in effect, the rotor circuit of an electric motor. Elliott F. Wright

Pumpkin The term commonly applied to the larger, orange-colored fruit of the *Cucurbita* species, used when ripe as a table vegetable, in pies, or for autumn decoration. Although some taxonomists would restrict the term pumpkin to the species *Cucurbita pepper* and *C. moschata*, it is also used in referring to *C. mixta*. New Jersey, Illinois, and California are important producing states. *See* SQUASH; VIOLALES. H. John Carew

Puna An alpine biological community in the central portion of the Andes Mountains of South America. Sparsely vegetated, treeless stretches cover high plateau country (altiplano) and slopes of central and southern Peru, Bolivia, northern Chile, and northwestern Argentina. The poor vegetative cover and the puna animals are limited by short seasonal precipitation as well as by the low temperatures of high altitudes.

Like the paramos of the Northern Andes, punas occur above timberline, and extend upward, in modified form, to perpetual snow. Due to greater heights of the Central Andean peaks, aeolian regions, that is, regions supplied with airborne nutrients above the upper limit of vascular plants, generally the snowline, are more extensive here than above the paramos. *See* PARAMO. Harriet G. Barclay

Purine A heterocyclic organic compound (**1**) containing

(**1**)

fused pyrimidine and imidazole rings. A number of substituted purine derivatives occur in nature; some, as components of nucleic acids and coenzymes, play vital roles in the genetic and metabolic processes of all living organisms. *See* COENZYME; NUCLEIC ACID.

Purines are generally white solids of amphoteric character. They can form salts with both acids and bases. Conjugated double bonds in purines results in aromatic chemical properties, that confers considerable stability, and accounts for their strong ultraviolet absorption spectra. With the exception of the parent compound, most substituted purines have low solubilities in water and organic solvents.

The purine bases, adenine (**2**) and guanine (**3**), together with pyrimidines, are fundamental components of all nucleic acids. Certain methylated derivatives of adenine and guanine are also present in some nucleic acids in low amounts. In biological systems, hypoxanthine (**4**), adenine, and guanine occur mainly as

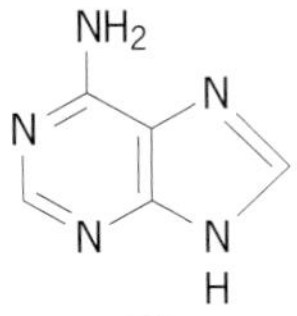

(2) 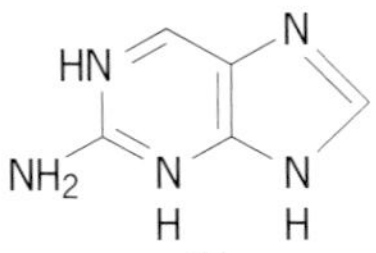 (3) 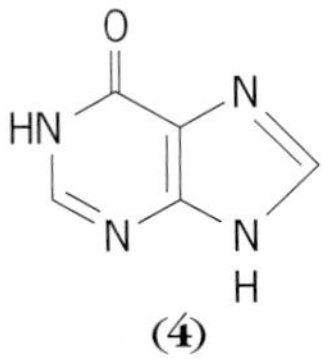 (4)

their 9-glycosides, the sugar being either ribose or 2-deoxyribose. Such compounds are termed nucleosides generically, and inosine (hypoxanthine nucleoside), adenosine, or guanosine specifically. The principal nucleotides contain 5′-phosphate groups, as in guanosine 5′-phosphate (GTP) and adenosine 5′-triphosphate (ATP).

Most living organisms are capable of synthesizing purine compounds. The sequence of enzymatic reactions by which the initial purine product, inosine 5′-phosphate, is formed utilizes glycine, carbon dioxide, formic acid, and amino groups derived from glutamine and aspartic acid. Adenosine 5′-phosphate and guanosine 5′-phosphate are formed from inosine 5′-phosphate.

Metabolic degradation of purine derivatives may also occur by hydrolysis of nucleotides and nucleosides to the related free bases. Deamination of adenine and guanine produces hypoxanthine and xanthine (**5**), both of which may be oxidized to uric acid (**6**).

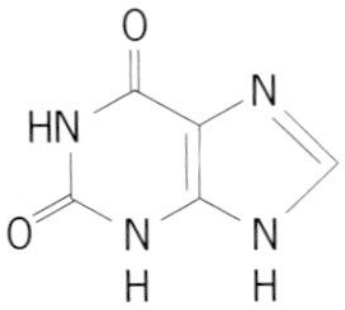

(5) 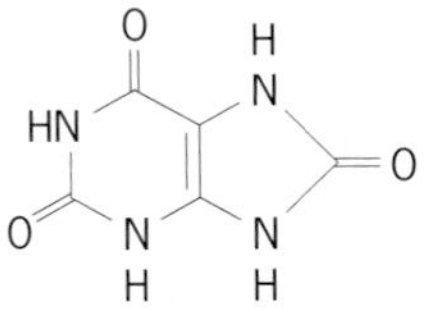 (6)

See URIC ACID.

Purine-related compounds have been investigated as potential chemotherapeutic agents. In particular, 6-mercaptopurine, in the form of its nucleoside phosphate, inhibits several enzymes required for synthesis of adenosine and guanosine nucleotides, and thus proves useful in selectively arresting the growth of tumors. The pyrazolopyrimidine has been used in gout therapy. As a purine analog, this agent serves to block the biosynthesis of inosine phosphate, as well as the oxidation of hypoxanthine and xanthine to uric acid. As a result of its use, overproduction of uric acid is prevented and the primary cause of gout is removed. *See* CHEMOTHERAPY AND OTHER NEOPLASTIC DRUGS; GOUT; PYRIMIDINE.

Standish C. Hartman

Push-pull amplifier An electronic circuit in which two transistors (or vacuum tubes) are used, one as a source of current and one as a sink, to amplify a signal. One device "pushes" current out into the load, while the other "pulls" current from it when necessary. A common example is the complementary-symmetry push-pull output stage widely used to drive loudspeakers (see illustration), where an *npn* transistor can source (push) current from a positive power supply into the load, or a *pnp* transistor can sink (pull) it into the negative power supply. The circuit functions as an amplifier in that the current levels at the output are larger than those at the input.

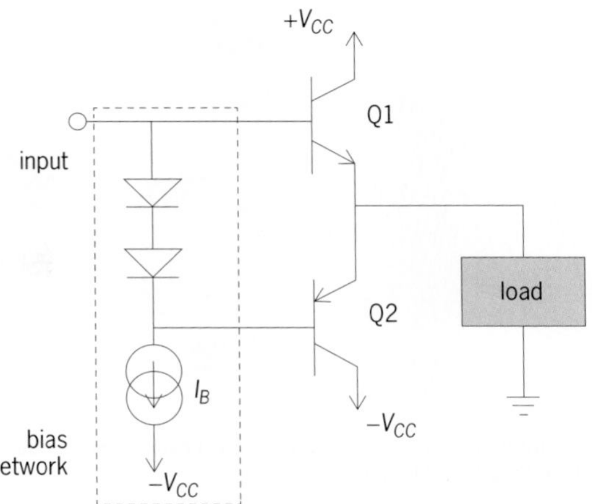

Complementary-symmetry push-pull output stage. Q1 is an *npn* transistor and Q2 is a *pnp* transistor; I_B is a bias current; positive ($+V_{CC}$) and negative ($-V_{CC}$) power supplies are shown.

A so-called bias network in a complementary-symmetry push-pull output stage (see illustration) functions to maintain a constant voltage difference between the bases of the two transistors. It can be designed either by setting a bias current, and diode sizes or by replacing it with a different network for class B, class A, or the common compromise, class AB mode of operation. In class B operation, where the bases of the transistors might simply be shorted together, only one transistor is "on" at a time and each is on average "on" for only 50% of the time; when the output current is zero, no current at all flows in the circuit. In class A operation a large voltage is maintained between the bases so that both devices stay "on" at all times, although their currents vary so that the difference flows into the load; and even when the output is zero, a large quiescent current flows from the power supplies. Class B operation is much more efficient than class A, which wastes a large amount of power when the signal is small. However, class B suffers from zero-crossing distortion as the output current passes through zero, because there is generally a delay involved as the input swings far enough to turn one transistor entirely off and then turn the other on. In class AB operation, some intermediate quiescent current is chosen to compromise between power and distortion.

Class AB amplifiers are conventionally used as loudspeaker drivers in audio systems because they are efficient enough to be able to drive the required maximum output power, often on the order of 100 W, without dissipating excessive heat, but can be biased to have acceptable distortion. Audio signals tend to be near zero most of the time, so good performance near zero output current is critical, and that is where class A amplifiers waste power and class B amplifiers suffer zero-crossing distortion. A class AB push-pull amplifier is also conventionally used as the output stage of a commercial operational amplifier. *See* AUDIO AMPLIFIER; OPERATIONAL AMPLIFIER; POWER AMPLIFIER; TRANSISTOR.

Martin Snelgrove

Pycnodontiformes An order of very specialized deep-bodied fishes near the holostean level of organization that are known only from the fossil record. They were a widespread group that first appeared in the Upper Triassic of Europe, flourished during the Jurassic and Cretaceous, and persisted to the Upper Eocene. Pycnodontiforms are a closely interrelated group most

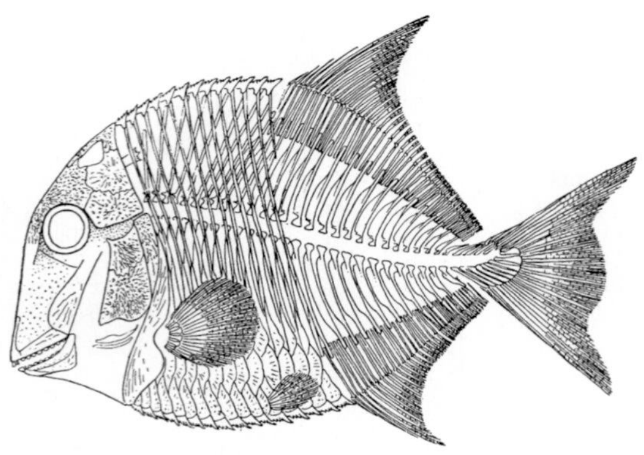

Coelodus costae, a pycnodont from Lower Cretaceous of Italy; length to 4 in. (10 cm).

commonly found preserved in marine limestone and associated with coraliferous facies.

They are characterized (see illustration) by a laterally compressed, disk-shaped body; long dorsal and anal fins, with each fin ray supported by its own endoskeletal element; an externally symmetrical tail; and an axial skeleton having greatly extended neural and hemal spines that are expanded in some forms, and neural and hemal arches which are well ossified—sometimes interdigitating around the notochord—but with no vertebral centra.

Ted M. Cavender

Pycnogonida A class of marine arthropods, consisting of about 600 Recent and 3 Devonian species. The Pycnogonida, or Pantopoda, are commonly called sea spiders.

The pycnogonids are characterized by reduction of the body to a series of cylindrical trunk somites supporting the appendages, a large specialized feeding apparatus called the proboscis, gonopores opening on the second joints of the legs, and a reduced abdomen. In many genera, there are seven pairs of appendages, of which the first four, namely the chelifores, palpi, ovigers, and first walking legs, are on the first or cephalic segment. This segment also bears a dorsal tubercle containing four simple eyes. Each of the remaining three trunk segments bears a single pair of legs.

Pycnogonids are found in all seas except the inner Baltic and Caspian, from intertidal regions to depths of 21,000 ft (6500 m), and one species is bathypelagic at about 3300 ft (1000 m). They are especially common in polar seas. Most of the intertidal species spend their lives in association with coelenterates as encysted, parasitic larval and juvenile stages, or are ectoparasitic as adults, being attached to anemones and hydroids by their claws and proboscides. Most of the deep-sea species are known only as adults, and their mode of life is a mystery.

The Pycnogonida are classified primarily on the presence or absence of various anterior appendages; about 60 genera are recognized, grouped in 8 families: Nymphonidae, Callipallenidae, Phoxichilidiidae, Endeidae, Ammotheidae, Austrodecidae, Colossendeidae, and Pycnogonidae. *See* ARTHROPODA; PALAEOISOPUS.

Joel W. Hedgpeth

Pygasteroida An order of Eognasthostomata which exhibits various stages in the backward migration of the anus out of the apical system. They have four genital pores, noncrenulate tubercles, and simple ambulacral plates. All members are referred to a single family, the Pygasteridae. They apparently arose from Triassic Pedinidae and occur in the Jurassic and Cretaceous of the Northern Hemisphere. They were formerly classified with other bilaterally symmetrical echinoids in the artificial assemblage Irregularia. *See* ECHINODERMATA; IRREGULARIA.

Howard B. Fell

Pyrenulales An order of the class Ascolichenes, also known as the Pyrenolichenes. The flask-shaped perithecia are uniformly immersed in the medulla of the thalli with a small ostiole opening at the surface. The asci and paraphyses arise from a blackened hypothecium and line the walls of the perithecium. The spores eventually burst the ascal walls and ooze out through the ostiole in a jelly matrix.

There are about 10 families, 50 genera, and more than 1500 species in the Pyrenulales. The major taxonomic criteria for separating genera and species are the septation and color of spores, since vegetative characters are so poorly developed.

Mason E. Hale

Pyridine An organic heterocyclic compound containing a triunsaturated six-membered ring of five carbon atoms and one nitrogen atom. Pyridine (**1**) and pyridine homologs are obtained

4 or γ; 5 or β′; 3 or β; 6 or α′; 2 or α; N 1

(1)

by extraction of coal tar or by synthesis. The pyridine system is found in natural products, for example, in nicotine (**2**) from tobacco, in ricinine (**3**) from castor bean, in pyridoxine or vitamin B_6 (**4**), in nicotinamide or niacinamide or vitamin P (**5**), and in several groups of alkaloids. *See* HETEROCYCLIC COMPOUNDS.

(2) **(3)**

(4) **(5)**

Pyridine (**1**) is a colorless, hygroscopic liquid with a pungent, unpleasant odor. When anhydrous it boils at 115.2–115.3°C (239.4–239.5°F). Pyridine is miscible with organic solvents as well as with water. The pyridine system is aromatic. It is stable to heat, to acid, and to alkali. Pyridine is used as a solvent for organic and inorganic compounds, as an acid binder, as a basic catalyst, and as a reaction intermediate.

Pyridine is an irritant to skin (eczema) and other tissues (conjunctivitis), and chronic exposure has been known to cause liver and kidney damage. Repeated exposure to atmospheric levels greater than 5 parts per million is considered hazardous.

Walter J. Gensler

Pyrimidine A heterocyclic organic compound (**1**) containing nitrogen atoms at positions 1 and 3. Naturally occurring

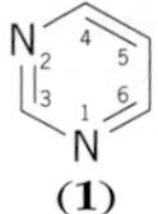

(1)

derivatives of the parent compound are of considerable biological importance as components of nucleic acids and coenzymes and, in addition, synthetic members of this group have found use as pharmaceuticals. *See* COENZYME; NUCLEIC ACID.

Pyrimidine compounds which are found universally in living organisms include uracil (**2**), cytosine (**3**), and thymine (**4**). Together with purines these substances make up the "bases" of nucleic acids, uracil and cytosine being found characteristically in ribonucleic acids, with thymine replacing uracil in deoxyribonucleic acids. A number of related pyrimidines also occur in lesser amounts in certain nucleic acids. Other pyrimidines of general natural occurrence are orotic acid and thiamine (vitamin B_1). *See* DEOXYRIBONUCLEIC ACID (DNA); PURINE; RIBONUCLEIC ACID (RNA).

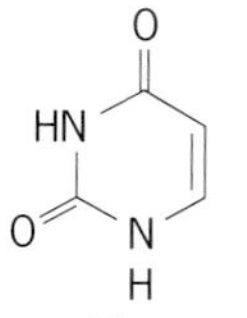

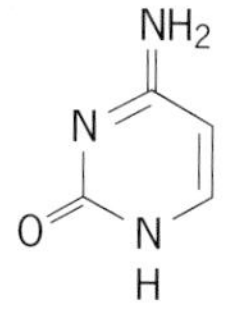

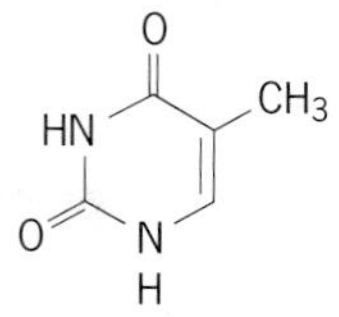

Among the sulfa drugs, the pyrimidine derivatives, sulfadiazine, sulfamerazine, and sulfamethazine, have general formula (**5**). These agents are inhibitors of folic acid biosynthesis in microorganisms. The barbiturates are pyrimidine derivatives which possess potent depressant action on the central nervous system. *See* BARBITURATES; SULFONAMIDE.

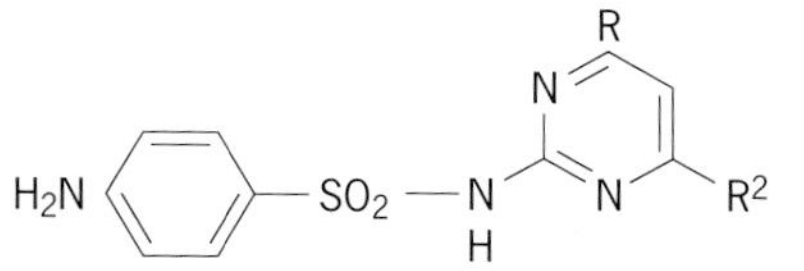

Standish C. Hartman

Pyroclastic rocks Rocks of extrusive (volcanic) origin, composed of rock fragments produced directly by explosive eruptions. Pyroclastic fragments may represent shattered and comminuted older rocks (volcanic, plutonic, sedimentary, or metamorphic) or solidified lava droplets formed by violent explosion. *See* TUFF; VOLCANO.

Robert I. Tilling

Pyroelectricity The property of certain crystals to produce a state of electric polarity by a change of temperature. Certain dielectric (electrically nonconducting) crystals develop an electric polarization (dipole moment per unit volume) when they are subjected to a uniform temperature change. This pyroelectric effect occurs only in crystals which lack a center of symmetry and also have polar directions (that is, a polar axis). These conditions are fulfilled for 10 of the 32 crystal classes. Typical examples of pyroelectric crystals are tourmaline, lithium sulfate monohydrate, cane sugar, and ferroelectric barium titanate.

Pyroelectric crystals can be regarded as having a built-in or permanent electric polarization. When the crystal is held at constant temperature, this polarization does not manifest itself because it is compensated by free charge carriers that have reached the surface of the crystal by conduction through the crystal and from the surroundings. However, when the temperature of the crystal is raised or lowered, the permanent polarization changes, and this change manifests itself as pyroelectricity.

The magnitude of the pyroelectric effect depends upon whether the thermal expansion of the crystal is prevented by clamping or whether the crystal is mechanically unconstrained. In the clamped crystal, the primary pyroelectric effect is observed, whereas in the free crystal, a secondary pyroelectric effect is superposed upon the primary effect. The secondary effect may be regarded as the piezoelectric polarization arising from thermal expansion, and is generally much larger than the primary effect. *See* PIEZOELECTRICITY.

H. Granicher

Pyroelectrics have a broad spectrum of potential scientific and technical applications. The most developed is the detection of infrared radiation. In addition, pyroelectric detectors can be used to measure the power generated by a radiation source (in radiometry), or the temperature of a remote hot body (in pyrometry, with corrections due to deviations from the blackbody emission). *See* PYROMETER; RADIOMETRY.

An infrared image can be projected on a pyroelectric plate and transformed into a relief of polarization on the surface. Other potential applications of pyroelectricity include solar energy conversion, refrigeration, information storage, and solid-state science.

A. Hadni

Pyroelectricity (biology) Electrical polarity in a biological material produced by a change in temperature. Pyroelectricity is probably a basic physical property of all living organisms. First discovered in 1966 in tendon and bone, it has since been shown to exist in most animal and plant tissues and in individual cells. Pyroelectricity appears to play a fundamental part in the growth processes (morphogenesis) and in physiological functions (such as sensory perception) of organisms. *See* PYROELECTRICITY.

The elementary components (for example, molecules) of biological (as well as of nonbiological) pyroelectric structures have a permanent electric dipole moment, and are arranged so that all positive dipole ends point in one direction and all negative dipole ends in the opposite direction. This parallel alignment of elementary dipoles is termed spontaneous polarization because it occurs spontaneously without the action of external fields or forces. In this state of molecular order, the structure concerned has a permanent electric dipole moment on a microscopic and macroscopic level. *See* DIPOLE MOMENT.

Spontaneous polarization is temperature-dependent; thus any change in temperature causes a change of the dipole moments, measurable as a change of electric charges at both ends of the polar axis. This is the pyroelectric effect. All pyroelectric structures are also piezoelectric, but the reverse is not true. *See* FERROELECTRICS; PIEZOELECTRICITY.

Prerequisites for the development of spontaneous polarization and pyroelectric activity in biological structures are (1) the presence of a permanent dipole moment in the molecules or molecular aggregates and (2) a molecular shape that favors a parallel alignment as much as possible (or at least does not impede it). Both these conditions are ideally fulfilled in bar- or board-shaped molecules with a permanent dipole moment along the longitudinal molecular axis. Several important organic substances have these molecular properties, and therefore behave pyroelectrically in biological structures. Examples include the epidermis of animals and plants, sensory receptors in animals, and tissues of the nervous and skeletal systems.

Living organisms are able to detect and discriminate between different stimuli in the environment, such as rapid changes of temperature, of illumination, and of hydrostatic and uniaxial pressure. These stimuli represent different forms of energy and are transduced, or converted, into the nearly uniform type of electrical signals whose voltage-time course frequently depends on dX/dt (X = external stimulus, t = time). Such electrical signals have been recorded on cutaneous sensory receptors, on external nerve endings, on epidermal structures, and even on the cell wall of single-cell organisms. The mechanisms of detection and transduction in these biological systems, still little understood, may lie in the pyroelectric behavior of the structures. Pyroelectric (and thus piezoelectric) behavior has been proved to exist in most biological systems, which means that these systems should in principle be able to function as pyroelectric detectors and transducers.

Herbert Athenstaedt

Pyrolysis A chemical process in which a compound is converted to one or more products by heat. By this definition, reactions that occur by heating in the presence of a catalyst, or in the presence of air when oxidation is usually a simultaneous reaction, are excluded. The terms thermolysis or thermal reaction have been used in essentially the same sense as pyrolysis. A simple example of pyrolysis is the classic experiment in which oxygen was first prepared by heating mercuric oxide [reaction (1)].

$$HgO \xrightarrow[(570°F)]{300°C} Hg + {}^1/_2 O_2 \quad (1)$$

Similar reactions occur with numerous other metallic oxides and salts. Thermal decomposition or calcining of limestone (calcium carbonate) is the basic step in the manufacture of lime [reaction (2)].

$$CaCo_3 \xrightarrow[(2200°F)]{1200°C} CaO + CO_2 \quad (2)$$

See LIME (INDUSTRY).

The term pyrolysis is most commonly associated with thermal reactions of organic compounds. Pyrolysis of material from plant and animal sources provided some of the first clues about constitution, as in the formation of isoprene from the thermal breakdown of rubber. A range of substances, including benzene, naphthalene, pyridine, and many other aromatic compounds, was obtained from coal tar, a pyrolysis product of coal. All of these pyrolysis processes lead to formation of volatile products characteristic of the source and also residues of char with high carbon content.

Pyrolysis reactions have been used as preparative methods and as means of generating transient intermediates that can be trapped or observed spectroscopically, or quenched by a further reaction. For preparative purposes, pyrolysis can generally be carried out by a flow process in which the reactant is vaporized with a stream of inert gas through a heated tube, sometimes at reduced pressure. In flash vacuum pyrolysis, the apparatus is placed under very low pressure, and the material to be pyrolyzed is vaporized by molecular distillation. *See* CHEMICAL DYNAMICS.

Types of reactions. At temperatures of 600–800°C (1100–1500°F), most organic compounds acquire sufficient vibrational energy to cause breaking of bonds with formation of free radicals. Alkanes undergo rupture of carbon-hydrogen (C-H) and carbon-carbon (C-C) bonds to two radicals that then react to give lower alkanes, alkenes, hydrogen, and also higher-molecular-weight compounds resulting from their recombination [reactions (3)].

$$CH_3CH_2CH_2CH_3 \xrightarrow[(1300°F)]{700°C} 2CH_3CH_2\cdot \rightarrow CH_3CH_3 + H_2C{=}CH_2 \quad (3a)$$

$$CH_3CH_2CH_2CH_3 \xrightarrow[(1300°F)]{700°C} CH_3CH_2\cdot CHCH_3 + H\cdot \rightarrow CH_3CH_2HC{=}CH_2 + H_2 \quad (3b)$$

These reactions are the basis of the thermal cracking processes used in petroleum refining. Pyrolysis of simple aromatic hydrocarbons such as benzene or naphthalene produces aryl radicals, which can attack other hydrocarbon molecules to give bi- and polyaryls, as shown in reaction (4) for the formation of biphenyl.

Benzene → Phenyl radical + H· —benzene→ Biphenyl (4)

See CRACKING; FREE RADICAL.

Pyrolytic eliminations can result in formation of a multiple bond by loss of HX from a compound H—C—C—X, where X = any leaving group. A typical example is the pyrolysis of an ester, which is one of the general methods for preparing alkenes. Pyrolytic elimination is particularly useful when acid-catalyzed dehydration of the parent alcohol leads to cationic rearrangement. Another useful application of this process is the production of ketenes from acid anhydrides [reaction (5)].

$$\underset{\text{Acid anhydride}}{RCH_2\overset{O}{\overset{\|}{C}}O\overset{O}{\overset{\|}{C}}CH_2R} \xrightarrow[(1300°F)]{700°C} \underset{\text{Ketene}}{RCH{=}C{=}O} + \underset{\text{Acid}}{RCH_2CO_2H} \quad (5)$$

See ACID ANHYDRIDE; ALKENE; ESTER.

Another type of thermal elimination occurs by loss of a small molecule such as nitrogen (N_2), carbon monoxide (CO), carbon dioxide (CO_2), or sulfur dioxide (SO_2), leading to reactive intermediates such as arynes, diradicals, carbenes, or nitrenes. The nitrene generated from aminobenzotriazole breaks down to benzyne at 0°C (32°F). Benzyne can be trapped by addition reaction or can dimerize to biphenylene. *See* REACTIVE INTERMEDIATES.

A number of pyrolytic reactions involve cleavage of specific C-C bonds in a carbon chain or ring. Fragmentation accompanied by transfer of hydrogen is a general reaction that occurs by a cyclic process. An example is decarboxylation of acids that contain a carbonyl group, which lose CO_2 on relatively mild heating. Acids with a double or triple carbon-to-carbon bond undergo decarboxylation at 300–400°C (570°–750°F). This type of reaction also occurs at higher temperatures with unsaturated alcohols and, by transfer of hydrogen from a C-H bond, with unsaturated ethers. Cleavage of a ring frequently occurs on pyrolysis. With alicyclic or heterocyclic four-membered rings, cleavage into two fragments is the reverse of 2 + 2 cycloaddition, as illustrated by the cracking of diketene [reaction (6)]. Pyrolysis is an important

$$\text{Diketene} \xrightarrow[(840°F)]{450°C} \underset{\text{Ketene}}{CH_2{=}C{=}O} \quad (6)$$

reaction in the chemistry of the pine terpenes, as in the conversion of β-pinene to myrcene. Benzocyclobutenes undergo ring opening to *o*-quinone dimethides. By combining this reaction in tandem with formation of the benzocyclobutene and a final Diels-Alder reaction, a versatile one-step synthetic method for the steroid ring system has been developed. *See* DIELS-ALDER REACTION; PINE TERPENE.

Many thermal reactions involve isomerization without elimination or fragmentation. These processes can occur by way of intermediates such as diradicals, as in the pyrolysis of pinene, or they may be concerted pericyclic reactions. An example of the latter is the Claisen-Cope rearrangement of phenyl or vinyl ethers and other 1,5-diene systems. These reactions can be carried out by relatively mild heating, and they are very useful in synthesis. *See* ORGANIC SYNTHESIS; PERICYCLIC REACTION.

Analytical applications. Thermal breakdown of complex structures leads to very complex mixtures of products arising from concurrent dissociation, elimination, and bond fission. Separation of these mixtures provides a characteristic pyrogram that is valuable as an analytical method, particularly for polymeric materials of both biological and synthetic origin. In this application, a small sample is heated on a hot filament or by laser. The pyrolysis products are then analyzed by gas chromatography, mass spectrometry, or a combination of both techniques. *See* GAS CHROMATOGRAPHY; MASS SPECTROMETRY.

Instrumentation with appropriate interfaces and data-handling systems has been developed to permit rapid and sensitive detection of pyrolysis products for a number of applications. One example is the optimization of conditions in petroleum cracking to produce a desired product from varied crude oils. The profile of pyrolysis fragments from a polymer can also be used to detect impurities.

James A. Moore

Pyrometallurgy The branch of extractive metallurgy in which processes employing chemical reactions at elevated temperatures are used to extract metals from raw materials, such as ores and concentrates, and to treat recycled scrap metal.

For metal production, the pyrometallurgical operation commences with either a raw material obtained by mining and subsequent mineral and ore processing steps to produce a concentrate, or a recycled material such as separated materials from scrapped automobiles, machinery, or computers.

Pyrometallurgical preparation processes convert raw materials to forms suitable for future processing. Reduction processes reduce metallic oxides and compounds to metal. Oxidizing processes oxidize the feed material to an intermediate or a semifinished metal product. Refining processes remove the last of the impurities from a crude metal. *See* ELECTROMETALLURGY; IRON METALLURGY; METALLURGY; PYROMETALLURGY, NONFERROUS.

Phillip J. Mackey

Pyrometallurgy, nonferrous The branch of extractive metallurgy in which processes employing chemical reactions at elevated temperatures are used to extract and refine nonferrous metals from ores, concentrates, and recycled materials. The entire process from feed to finished metal may be pyrometallurgical, or a pyrometallurgical step may be used in conjunction with other technologies. Increasingly, a mix of processes maximizes the efficiency and advantages of an overall operation.

The processes in pyrometallurgy in general can be classified as preparatory, reduction, oxidation, and metal refining. Treatment of a given raw material or metal may involve all these steps, or some of the steps may form a part of the total processing system, which may include nonpyrometallurgical operations.

In preparatory processes, the concentrate or upgraded ore or other feed is converted by chemical reaction to a form suitable for further processing. The most common subprocesses are drying and calcination, pyrolysis and hydrolysis, roasting, sintering, and chlorination. Even though a chemical reaction does not actually take place during drying, this subprocess is included since it is often part of a subsequent high-temperature operation such as smelting. In some cases, the preparatory process is carried out to provide a material that is amenable to treatment by hydrometallurgical processing, such as the roasting of zinc concentrates to produce a zinc calcine (essentially a zinc oxide intermediate product) which is leached with sulfuric acid solution for zinc production, or it is a step such as calcination in the preparation of alumina for aluminum smelting.

Reduction processes effect the high-temperature reaction of a metal compound to the metal and its separation from the residue, as represented by the reaction below, where MX is the metal

$$\mathrm{MX} + \mathrm{R} \longrightarrow \mathrm{M} + \mathrm{RX}$$

compound, R the reacting or reducing agent, and M the metal. The reducing agent and reaction conditions (for example, temperature and pressure) and the concentration of reactants and products are selected to achieve a rapid or spontaneous reaction. These reactions usually require energy input.

The amount of reducing agent used should be low and inexpensive, relative to the value of the metal produced, while the product RX should be readily separable from the metal. Reducing agents commonly used in nonferrous pyrometallurgy include carbon (usually as coke), carbon monoxide gas (from coke), natural gas, iron and ferrosilicon (for Mg production), aluminum (for Ca production), and magnesium (for Ti, Zr, and Hf production).

For thousands of years, pyrometallurgical smelting of sulfide materials has been the key production method for nonferrous metals, in particular for copper, nickel, tin, lead, and zinc. This still remains the case on account of lower costs associated with new intensive technology and lower overall energy consumption. Formerly, it was common to roast such feed materials prior to the actual smelting operation. Roasting is still a major processing step in zinc and tin production. However, the roasting step for copper and nickel production was gradually eliminated, and during the latter part of the twentieth century continuous smelting processes were developed to directly treat sulfide concentrates, producing (by oxidation of the sulfide material) a high-grade copper matte product (~70% Cu) in a single step. In the case of lead, the metal itself can be readily produced directly. Oxidation processes are normally exothermic, a characteristic that has generally led to the development of autogenous processes, requiring virtually no fossil fuel.

Two basic types of smelting processes are used for copper or nickel production: flash smelting and bath smelting. In flash smelting, a fine concentrate feed is introduced into the furnace chamber, along with oxygen-enriched air, and the reaction principally occurs in a gas-phase system between the oxygen-bearing gas and solid particles. In bath smelting, a concentrate feed is introduced into the furnace melt, which is blown and kept highly agitated by submerged tuyeres (injecting the oxygen-enriched air), such that the feed is enveloped and reacts within the turbulent bath.

The new continuous lead smelting process can produce lead directly, while on account of the thermodynamics of the copper smelting system with the immiscible $Cu\text{-}Cu_2S$ phases being present, copper production is normally carried out in two stages: (1) copper concentrate smelting to produce a high-grade matte (typically 60–75% Cu, 4–12% Fe, ~21% S), a slag (approximately 27–30% FeO, 15–20% Fe_3O_4, 25–30% SiO_2, 1–5% Cu), and a sulfur dioxide-rich gas (9–15% SO_2 at acid plant); and (2) copper matte converting, wherein the matte is oxidized or converted to metallic copper, producing a small amount of slag and sulfur dioxide gas.

A significant amount of copper is produced from recycled materials (such as from used automobiles, motors, old electrical appliances), and the pyrometallurgical processes are well able to handle this feed load on account of the flexibility as to feed type.

In metal refining processes, the starting material is generally an impure metal, usually produced in a primary production process. Impurities are removed to yield a final metal product, meeting a product specification. The processes are classified as (1) volatilization (separation of metal or metal compound as a gas from a liquid or solid); (2) drossing and precipitation (separation of the metal or impurities as a solid from the liquid melt); and (3) slag refining (separation of metal or impurities by their extraction from one liquid into a second immiscible liquid phase).

Philip J. Mackey

Pyrometer A temperature-measuring device, originally an instrument that measures temperatures beyond the range of thermometers, but now in addition a device that measures thermal radiation in any temperature range. This article discusses radiation pyrometers; for other temperature-measuring devices *see* BOLOMETER; THERMISTOR; THERMOCOUPLE.

The illustration shows a very simple type of radiation pyrometer. Part of the thermal radiation emitted by a hot object is intercepted by a lens and focused onto a thermopile. The resultant heating of the thermopile causes it to generate an electrical signal (proportional to the thermal radiation) which can be displayed on a recorder.

Unfortunately, the thermal radiation emitted by the object depends not only on its temperature but also on its surface characteristics. The radiation existing inside hot, opaque objects is so-called blackbody radiation, which is a unique function of temperature and wavelength and is the same for all opaque materials. However, such radiation, when it attempts to escape from the object, is partly reflected at the surface. In order to use the output of the pyrometer as a measure of target temperature, the effect of the surface characteristics must be eliminated. A cavity can be formed in an opaque material and the pyrometer sighted

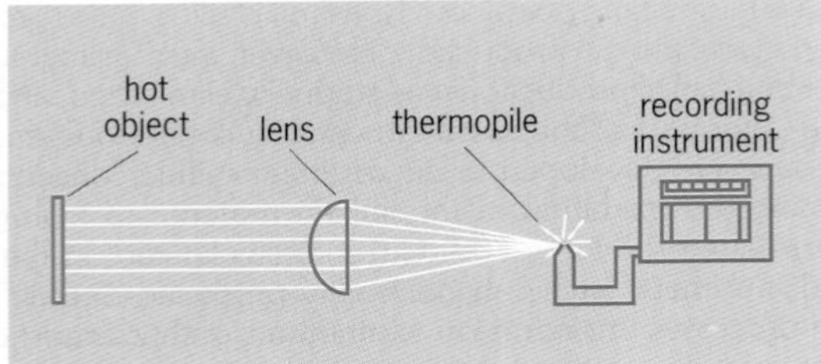

Elementary radiation pyrometer. (*After D. M. Considine and S. D. Ross, Process Instruments and Controls Handbook, 2d ed., McGraw-Hill, 1974*)

on a small opening extending from the cavity to the surface. The opening has no surface reflection, since the surface has been eliminated. Such a source is called a blackbody source, and is said to have an emittance of 1.00. By attaching thermocouples to the black-body source, a curve of pyrometer output voltage versus blackbody temperature can be constructed. *See* BLACKBODY; HEAT RADIATION.

Pyrometers can be classified generally into types requiring that the field of view be filled, such as narrow-band and total-radiation pyrometers; and types not requiring that the field of view be filled, such as optical and ratio pyrometers. The latter depend upon making some sort of comparison between two or more signals.

The optical pyrometer should more strictly be called the disappearing-filament pyrometer. In operation, an image of the target is focused in the plane of a wire that can be heated electrically. A rheostat is used to adjust the current through the wire until the wire blends into the image of the target (equal brightness condition), and the temperature is then read from a calibrated dial on the rheostat.

The ratio, or "two-color," pyrometer makes measurements in two wavelength regions and electronically takes the ratio of these measurements. If the emittance is the same for both wavelengths, the emittance cancels out of the result, and the true temperature of the target is obtained. This so-called gray-body assumption is sufficiently valid in some cases so that the "color temperature" measured by a ratio pyrometer is close to the true temperature. *See* TEMPERATURE MEASUREMENT; THERMOMETER. Thomas P. Murray

Pyromorphite A mineral series in the apatite group, or in the larger grouping of phosphate, arsenate, and vanadate-type minerals. In this series lead (Pb) substitutes for calcium (Ca) of the apatite formula $Ca_5(PO_4)_3(F,OH,Cl)$, and little fluorine (F) or hydroxide (OH) is present. *See* APATITE.

The pyromorphite series crystallizes in the hexagonal system. Crystals are prismatic. Other forms are granular, globular, and botryoidal. Pyromorphite colors range through green, yellow, and brown; vanadinite occurs in shades of yellow, brown, and red.

Pyromorphites are widely distributed as secondary minerals in oxidized lead deposits. Pyromorphite is a minor ore of lead; vanadinite is a source of vanadium and minor ore of lead. *See* LEAD; VANADIUM. Wayne R. Lowell

Pyrophyllite A hydrated aluminum silicate with composition $Al_2Si_4O_{10}(OH)_2$. The mineral is commonly white, grayish, greenish, or brownish, with a pearly to waxy appearance and greasy feel. It occurs as compact masses, as radiating aggregates (see illustration), and as foliated masses. Pyrophyllite belongs to the layer silicate (phyllosilicate) group of minerals. The mineral is soft (hardness $1\text{–}1^1/_2$ on the Mohs scale) and has easy cleavage parallel to the structural layers. The mineral is highly stable to acids.

Pyrophyllite is used principally for refractory materials and in other ceramic applications. The main sources for pyrophyllite

Specimen of pyrophyllite. (*Pennsylvania State University*)

in the United States are in North Carolina. An unusual form from the Transvaal is called African wonderstone. *See* SILICATE MINERALS. George W. Brindley

Pyrotechnics Mixtures of substances that produce noise, light, heat, smoke, or motion when ignited. They are used in matches, incendiaries, and other igniters; in fireworks and flares; in fuses and other initiators for primary explosives; in delay trains; for powering mechanical devices; and for dispersing materials such as insecticides.

Black powder, an intimate mixture of potassium nitrate, charcoal, and sulfur, was perhaps the earliest pyrotechnic and remains the most important one. It was discovered prior to A.D. 1000 in China during the Sung dynasty, where it was used in rockets and fireworks. Its introduction to Europe prior to 1242 eventually revolutionized warfare. Its use in blasting rock was also revolutionary, enabling rapid fragmentation of ore and other rock with a great reduction in manual labor.

A pyrotechnic mixture contains a fuel and an oxidizer, usually another ingredient to give a special effect, and often a binder. The oxidizer is usually a nitrate, perchlorate, chlorate, or peroxide of potassium, barium, or strontium. Fuels may be sulfur, charcoal, boron, magnesium, aluminum, titanium, or antimony sulfide. Examples of binders that are also fuels are dextrin and natural polysaccharides such as red gum. Salts of strontium, calcium, barium, copper, and sodium and powdered magnesium metal when combined with a fuel and oxidizer can give the special effects of scarlet, brick red, green, blue-green, yellow, and white flames, respectively.

Many pyrotechnic mixtures are easily ignited by impact, friction, flame, sparks, or static electricity. Even those that burn quietly in small quantities can explode violently when ignited under confinement or in larger quantities. D. Linn Coursen

Pyrotheria An extinct order of primitive, mastodonlike, herbivorous, hoofed mammals restricted to the Eocene and Oligocene deposits of South America. There is only one family (Pyrotheriidae) in the order and four genera in the family.

The characters of this group superficially resembling those in early proboscideans are nasal openings over orbits indicating the presence of a trunk, strong neck musculature, and six upper and four lower bilophodont cheek teeth. Pyrotheres are distantly related to the members of the superorder Paenungulata, including Proboscidea, Xenungulata, and others. *See* PROBOSCIDEA; XENUNGULATA. Gideon T. James

Pyroxene A large, geologically significant group of dark, rock-forming silicate minerals. Pyroxene is found in abundance in a wide variety of igneous and metamorphic rocks. Because of their structural complexity and their diversity of chemical composition and geologic occurrence, these minerals have been intensively studied by using a wide variety of modern analytical techniques. Knowledge of pyroxene compositions, crystal structures, phase relations, and detailed microstructures provides important information about the origin and thermal history of rocks in which they occur.

The general chemical formula for pyroxenes is $M2M1T_2O_6$, where T represents the tetrahedrally coordinated sites, occupied primarily by silicon cations (Si^{4+}). Names of specific end-member pyroxenes are assigned based on composition and structure type. Those pyroxenes containing primarily calcium (Ca^{2+}) or sodium (Na^+) cations in the M2 site are monoclinic. Pyroxenes containing primarily Mg^{2+} or iron(II) (Fe^{2+}) cations in the M2 site are orthorhombic at low temperatures, but they may transform to monoclinic at higher temperature. *See* CRYSTAL STRUCTURE; CRYSTALLOGRAPHY.

Common pyroxenes have specific gravity ranging from about 3.2 (enstatite, diopside) to 4.0 (ferrosilite). Hardnesses on the Mohs scale range from 5 to 6. Iron-free pyroxenes may be colorless (enstatite, diopside, jadeite); as iron content increases, colors range from light green or yellow through dark green or greenish brown, to brown, greenish black, or black (orthopyroxene, pigeonite, augite, hedenbergite, aegirine). Spodumene may be colorless, yellowish emerald green (hiddenite), or lilac pink (kunzite). *See* HARDNESS SCALES; SPODUMENE.

Pyroxenes in the rock-forming quadrilateral are essential constituents of ferromagnesian igneous rocks such as gabbros and their extrusive equivalents, basalts, as well as most peridotites. Pyroxenes may also be present as the dark constituents of more silicic diorites and andesites. *See* ANDESITE; BASALT; DIORITE; GABBRO; IGNEOUS ROCKS; PERIDOTITE.

Pyroxenes, especially those of the diopside-hedenbergite series, are found in medium- to high-grade metamorphic rocks of the amphibolite and granulite facies. *See* METAMORPHISM.

The peridotites found in the Earth's upper mantle contain Mg-rich, Ca-poor pyroxenes, in addition to olivine and other minor minerals. At successively greater depths in the mantle, these Mg-rich pyroxenes will transform sequentially to spinel (Mg_2SiO_4) plus stishovite (SiO_2), an ilmenite structure, or a garnet structure, depending on temperature; and finally at depths of around 360–420 mi (600–700 km) to an $MgSiO_3$ perovskite structure. *See* ASTHENOSPHERE; EARTH INTERIOR; GARNET; ILMENITE; OLIVINE; PEROVSKITE; SILICATE MINERALS; SPINEL; STISHOVITE.

Charles W. Burnham

Pyroxenite A heavy, dark-colored, phaneritic (visibly crystalline) igneous rock composed largely of pyroxene with smaller amounts of olivine or hornblende. Pyroxenite composed largely of orthopyroxene occurs with anorthosite and peridotite in large, banded gabbro bodies. Some of these pyroxenite masses are rich sources of chromium. Certain pyroxenites composed largely of clinopyroxene are also of magmatic origin, but many probably represent products of reaction between magma and limestone. Other pyroxene-rich rocks have formed through the processes of metamorphism and metasomatism. *See* GABBRO; IGNEOUS ROCKS; PERIDOTITE; PYROXENE. Carleton A. Chapman

Pyroxenoid A group of silicate minerals whose physical properties resemble those of pyroxenes. In contrast with the two-tetrahedra periodicity of pyroxene single silicate chains, the pyroxenoid crystal structures contain single chains of $(SiO_4)^{4-}$ silicate tetrahedra having repeat periodicities ranging from three to nine (see illustration). The tetrahedron is a widely used geometric representation for the basic building block of most silicate minerals, in which all silicon cations (Si^{4+}) are bonded to four oxygen anions arranged as if they were at the corners of a tetrahedron. In pyroxenoids, as in other single-chain silicates, two of the four oxygen anions in each tetrahedron are shared between two Si^{4+} cations to form the single chains, and the other two oxygen anions of each tetrahedron are bonded to divalent cations, such as calcium (Ca^{2+}), iron (Fe^{2+}), or manganese (Mn^{2+}). These divalent cations bond to six (or sometimes seven or eight) oxygen anions, forming octahedral (or irregular seven- or eight-cornered) coordination polyhedra. *See* PYROXENE; SILICATE MINERALS.

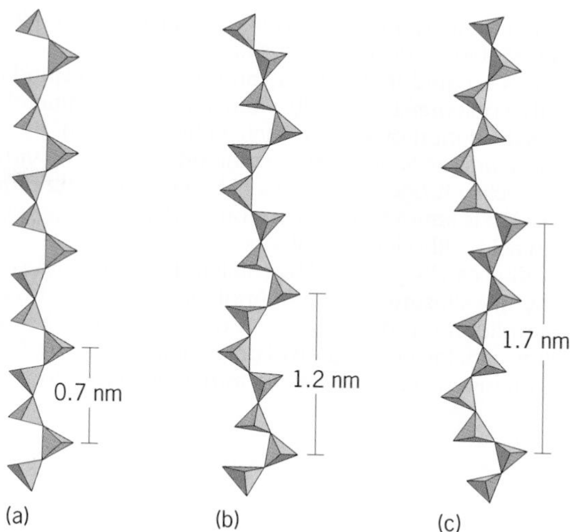

Tetrahedral silicate chains in three pyroxenoid structures: (*a*) wollastonite with three-tetrahedra periodicity, (*b*) rhodonite with five-tetrahedra periodicity, and (*c*) pyroxmangite-pyrox ferroite with seven-tetrahedra periodicity.

The pyroxenoid structures have composite structural units consisting of strips of octahedra (or larger polyhedra) two or more units wide formed by sharing of polyhedral edges, to which the silicate tetrahedral chains are attached on both top and bottom. The repeat periodicity of the octahedral strips is the same as that of the silicate chains to which they are attached. These composite units are cross-linked to form the three-dimensional crystal structures. The pyroxenoid minerals are triclinic, with either $C\bar{1}$ or $I\bar{1}$ space-group symmetry depending on the stacking of the composite units, and *c*-axis lengths ranging from about 0.71 nanometer for three-repeat silicate tetrahedral chains to about 2.3 nm for nine-repeat chains. *See* CRYSTAL STRUCTURE.

There are two series of pyroxenoid minerals, one anhydrous and one hydrous. The anhydrous pyroxenoids are significantly more abundant. A general formula for anhydrous pyroxenoids is $(Ca, Mn, Fe^{2+})SiO_3$. Silicate-chain repeat length is inversely proportional to mean divalent cation size. The hydrogen in hydrous pyroxenoids is hydrogen-bonded between two oxygen atoms; additional hydrogen in santaclaraite is bound as hydroxyl (OH) and as a water molecule (H_2O). Charles W. Burnham

Pyrrole One of a group of organic compounds containing a doubly unsaturated five-membered ring in which nitrogen occupies one of the ring positions. Pyrrole (**1**) is a representative compound. The pyrrole system is found in the green leaf pigment, chlorophyll, in the red blood pigment, hemoglobin, and in the blue dye, indigo. Interest in these colored bodies has been largely responsible for the intensive study of pyrroles. Tetrahydropyrrole, or pyrrolidine (**2**), is part of the structures of two

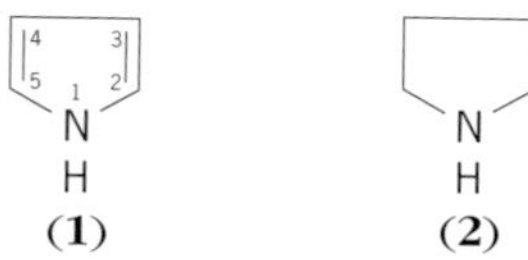

protein amino acids, proline and hydroxyproline. *See* HETEROCYCLIC COMPOUNDS; INDOLE; PORPHYRIN.

Pyrrole is a liquid that darkens and resinifies on standing in air, and that polymerizes quickly when treated with mineral acid. Familiar substitution processes, such as halogenation, nitration, sulfonation, and acylation, can be realized. Pyrrole, by virtue of its heterocyclic nitrogen, is very weakly basic. The hydrogen at the 1 position is removable as a proton, and accordingly, pyrrole is also an acid, although a weak one.

Pyrrolidine can be prepared by catalytic hydrogenation of pyrrole or by ring-closure reactions. Another derivative of pyrrole, 2-ketopyrrolidine, or pyrrolidone, is of considerable interest in connection with the preparation of polyvinylpyrrolidone. Pyrrolidone is combined with acetylene to form vinylpyrrolidone; polymerization of this material furnishes polyvinylpyrrolidone, which is suitable for maintaining osmotic pressure in blood and so acting as an extender for plasma or whole blood. *See* PYRIDINE.

Walter J. Gensler

Pythagorean theorem This theorem states that in any right triangle the square on the hypotenuse is equal to the sum of the squares on the other two sides: $r^2 = x^2 + y^2$. More than 100 different proofs have been given for this extremely important theorem of euclidean plane geometry.

The three-dimensional Pythagorean theorem may be phrased "the square of the diagonal of a rectangular box is equal to the sum of the squares of three adjacent edges that meet at a vertex: $r^2 = x^2 + y^2 + z^2$."

J. Sutherland Frame

Q (electricity) Often called the quality factor of a circuit, Q is defined in various ways, depending upon the particular application. In the simple RL and RC series circuits, Q is the ratio of reactance to resistance, as in Eqs. (1), where X_L is the inductive

$$Q = \frac{X_L}{R} \quad Q = \frac{X_C}{R} \quad \text{(a numerical value)} \tag{1}$$

reactance, X_C is the capacitive reactance, and R is the resistance. An important application lies in the dissipation factor or loss angle when the constants of a coil or capacitor are measured by means of the alternating-current bridge.

Q has greater practical significance with respect to the resonant circuit, and a basic definition is given by Eq. (2), where

$$Q_0 = 2\pi \frac{\text{max stored energy per cycle}}{\text{energy lost per cycle}} \tag{2}$$

Q_0 means evaluation at resonance. For certain circuits, such as cavity resonators, this is the only meaning Q can have.

For the RLC series resonant circuit with resonant frequency f_0, Eq. (3) holds, where R is the total circuit resistance, L is the

$$Q_0 = \frac{2\pi f_0 L}{R} = \frac{1}{2\pi f_0 CR} \tag{3}$$

inductance, and C is the capacitance. Q_0 is the Q of the coil if it contains practically the total resistance R. The greater the value of Q_0, the sharper will be the resonance peak.

The practical case of a coil of high Q_0 in parallel with a capacitor also leads to $Q_0 = 2\pi f_0 L/R$. R is the total series resistance of the loop, although the capacitor branch usually has negligible resistance.

In terms of the resonance curve, Eq. (4) holds, where f_0 is

$$Q_0 = \frac{f_0}{f_2 - f_1} \tag{4}$$

the frequency at resonance, and f_1 and f_2 are the frequencies at the half-power points. *See* RESONANCE (ALTERNATING-CURRENT CIRCUITS).

Burtis L. Robertson

Q meter A direct-reading instrument formerly widely used to measure the Q of an electrical component or passive network of components at radio frequencies. The instrument measures the voltage applied to a component or circuit together with the voltage across a key component and calculates the Q of the circuit from their ratio. An impedance-measuring instrument (bridge or analyzer) is more likely to be employed now. *See* ELECTRICAL MEASUREMENTS; Q (ELECTRICITY); RADIO-FREQUENCY IMPEDANCE MEASUREMENTS.

Bryan P. Kibble

Quadrature The condition in which the phase angle between two alternating quantities is 90°, corresponding to one-quarter of an electrical cycle. The electric and magnetic fields of electromagnetic radiation are in space quadrature, which means that they are at right angles in space. *See* ELECTROMAGNETIC RADIATION.

The current and voltage of a perfect coil are in quadrature because the coil current lags behind the coil voltage by exactly 90°. The current and voltage of an ideal capacitor are also in quadrature, but here the current through the capacitor leads the voltage across the capacitor by 90°. In these last two cases the current and voltage are in time quadrature.

John Markus

Quadric surface A surface defined analytically by an equation of the second degree in three variables. If these variables are x, y, z, such an equation has the form:

$$ax^2 + by^2 + cz^2 + 2exy + 2fxz + 2gyz + 2px + 2qy + 2rz + d = 0$$

Every plane section of such a surface is a conic. *See* CYLINDER; ELLIPSOID AND SPHEROID; HYPERBOLOID; PARABOLOID; SURFACE AND SOLID OF REVOLUTION.

J. Sutherland Frame

Qualitative chemical analysis The branch of chemistry concerned with identifying the elements and compounds present in a sample of matter. Inorganic qualitative analysis traditionally used classical "wet" methods to detect elements or groups of chemically similar elements, but instrumental methods have largely superseded the test-tube methods. Methods for the detection of organic compounds or classes of compounds have become increasingly available and important in organic, forensic, and clinical chemistry. Once it is known which elements and compounds are present, the role of quantitative analysis is to determine the composition of the sample. *See* ANALYTICAL CHEMISTRY; QUANTITATIVE CHEMICAL ANALYSIS.

Inorganic analysis. The operating principles of all systematic inorganic qualitative analysis schemes for the elements are similar: separation into groups by reagents producing a phase change; isolation of individual elements within a group by selective reactions; and confirmation of the presence of individual elements by specific tests.

Through usage and tradition, descriptive terms for sample sizes have been:

macro	0.1 gram or more
semimicro	0.01 to 0.1 gram
micro	1milligram (1 mg or 10^{-3} g)
ultramicro	1 microgram (1 μg or 10^{-6} g)
submicrogram	less than 1 microgram

For defining the smallest amount of a substance that can be detected by a given method, the term "limit of identification" is used. Under favorable conditions, an extremely sensitive method can detect as little as 10^{-15} g.

Spot tests are selective or specific single qualitative chemical tests carried out on a spot plate (a glass or porcelain plate with small depressions in which drop-size reactions can be carried out), on paper, or on a microscope slide. On paper or specially prepared adsorbent surfaces, spot tests become one- or two-dimensional through the use of solvent migration and differential adsorption (thin-layer chromatography). Containing selected indicator dyes, pH indicator paper strips are widely used for pH estimation. Solid reagent monitoring devices and indicator tubes are used for the detection and estimation of pollutant gases in air. *See* CHROMATOGRAPHY; PH.

By use of the microscope, crystal size and habit can be used for qualitative identification. With the addition of polarized light, chemical microscopy becomes a versatile method. *See* CHEMICAL MICROSCOPY.

Any instrumental method of quantitative analysis can be adapted to qualitative analysis. Some, such as electrochemical methods, are not often used in qualitative analysis (an exception would be the ubiquitous pH meter). Others, such as column chromatography (gas and liquid) and mass spectrometry, are costly and uncomplicated but capable of providing unique results. Emission spectroscopy is important in establishing the presence or absence of a suspected element in forensic analysis. The simple qualitative flame test developed into flame photometry and subsequently into atomic absorption spectrometry, *See* ATOMIC SPECTROMETRY; FLAME PHOTOMETRY; SPECTROCHEMICAL ANALYSIS.

Bombardment of surfaces by x-rays, electrons, and positive ions has given rise to a number of methods useful in analytical chemistry. X-ray diffraction is used to determine crystal structure and to identify crystalline substances by means of their diffraction patterns. X-ray fluorescence analysis employs x-rays to excite emission of characteristic x-rays by elements. The electron microprobe uses electron bombardment in a similar manner to excite x-ray emission. *See* SECONDARY ION MASS SPECTROMETRY (SIMS); X-RAY DIFFRACTION; X-RAY FLUORESCENCE ANALYSIS.

Neutron activation analysis, wherein neutron bombardment induces radioactivity in isotopes of elements not naturally radioactive, involves measurement of the characteristic gamma radiation and modes of decay produced in the target atoms. *See* ACTIVATION ANALYSIS. Jack L. Lambert

Organic analysis. Qualitative analysis of an organic compound is the process by which the characterization of its class and structure is determined. Due to the numerous classes of organic compounds and the complexity of their molecular structures, a systematic analytical procedure is often required.

A typical procedure entails an initial assignment of compound classification, followed by a complete identification of the molecular structure.

The initial step is an examination of the physical characteristics. Color, odor, and physical state can be valuable clues. The physical constants of an unknown compound provide pertinent data for the analyst. Constants such as melting point, boiling point, specific gravity, and refractive index are commonly measured.

The preliminary chemical tests which are applied are elemental analysis procedures. The elements generally associated with carbon, hydrogen, and oxygen are sulfur, nitrogen, and the halogens. The analyst is usually interested in the latter group. These elements are converted to water-soluble ionic compounds via sodium fusion. The resulting products are then detected with wet chemical tests. The solubility of a compound in various liquids provides information concerning the molecular weight and functional groups present in the compound. In order to indicate the presence or absence of a functional group, specific classification reactions are tested. The reactions are simple, are rapid, and require a small quantity of sample. No single test is conclusive evidence. A judicious choice of reactions can confirm or negate the presence of a functional group.

Instrumental methods are commonly applied for functional group determination and structure identification. Absorption spectroscopy and infrared absorption spectroscopy are among the most important techniques. Whenever a molecule is exposed to electromagnetic radiation, certain wavelengths cause vibrational, rotational, or electronic effects within the molecule. The radiation required to cause these effects is absorbed. The nature and configuration of the atoms determine which specific wavelengths are absorbed. Raman spectroscopy is slightly different from the other spectroscopic techniques. A sample is irradiated by a monochromatic source. Depending upon the vibrational and rotational energies of the functional groups in the sample, light is scattered from the sample in a way which is characteristic of the functional groups. Instrumental techniques are also valuable for structure identification. In addition to infrared spectroscopy, nuclear magnetic resonance and mass spectrometry are widely used. Other, less common techniques are electron spin resonance, x-ray diffraction, and nuclear quadrupole spectroscopy. *See* INFRARED SPECTROSCOPY; MASS SPECTROMETRY; NUCLEAR MAGNETIC RESONANCE (NMR); RAMAN EFFECT.

In addition to chemical structure, the physical structure of a sample can be important. Thermal analysis is a useful procedure for examining structural characteristics. Among the most popular methods are thermogravimetry, differential thermal analysis, differential scanning calorimetry, and thermal mechanical analysis. *See* CALORIMETRY.

Although the implementation of instrumental methods has greatly simplified qualitative analytical procedures, no individual instrument is capable of complete identification for all samples. A complement of instrumental and wet chemical techniques is generally required for adequate proof of identification. *See* SPECTROSCOPY. Sidney Siggia; Kenneth Longmoore

Quality control The operational techniques and the activities that sustain the quality of a product or service in order to satisfy given requirements. Quality control is a major component of total quality management and is applicable to all phases of the product life cycle: design, development, manufacturing, delivery and installation, and operation and maintenance.

The quality-control cycle consists of four steps: quality planning, data collection, data analysis, and implementation. Quality planning consists of defining measurable quality objectives. Quality objectives are specific to the product or service and to the phase in their life cycle, and they should reflect the customer's requirements.

The collection of data about product characteristics that are relevant to the quality objectives is a key element of quality control. These data include quantitative measurements (measurement by variables), as well as determination of compliance with given standards, specifications, and required product features (measurement by attributes). Measurements may be objective, that is, of physical characteristics, which are often used in the control of the quality of services. Since quality control was originally developed for mass manufacturing, which relied on division of labor, measurements were often done by a separate department. However, in the culture of Total Quality Management, inspection is often done by the same individual or team producing the item.

The data are analyzed in order to identify situations that may have an adverse effect on quality and may require corrective or preventive action. The implementation of those actions as indicated by the analysis of the data is undertaken, including modifications of the product design or the production process, to achieve continuous and sustainable improvement in the product and in customer satisfaction.

The methods and techniques for data analysis in quality control are generic and can be applied to a variety of situations. The techniques are divided into three main categories: diagnostic techniques; process control, which includes process capability assessment and control charts; and acceptance sampling.

Diagnostic techniques serve to identify and pinpoint problems or potential problems that affect the quality of processes and products, and include the use of flowcharts, cause-and-effect diagrams, histograms, Pareto diagrams, location diagrams, scatter plots, and boxplots.

Process-control methods are applicable to systems that produce a stream of product units, either goods or services. They serve to control the processes that affect those product characteristics that are relevant to quality as defined in the quality objectives. For example, in a system that produces metal parts, some of the processes that might need to be controlled are cutting, machining, deburring, bending, and coating. The relevant product characteristics are typically spelled out in the specifications in terms of physical dimensions, position of features, surface smoothness, material hardness, paint thickness, and so on. In a system that produces a service, such as a telephone help line, the relevant processes could be answering the call, identifying the problem, and solving the problem. The characteristics that are relevant to quality as perceived by the customer might include response time, number of referrals, frequency of repeat calls for the same problem, and elapsed time to closure.

Process control focuses on keeping the process operating at a level that can meet quality objectives, while accounting for random variations over which there is no control. There are two main aspects to process control: control charts and capability analysis. Control charts are designed to ascertain the statistical stability of the process and to detect changes in its level or variability that are due to assignable causes and can be corrected. Capability analysis considers the ability of the process to meet quality objectives as implied by the product specifications.

Process-control techniques were originally developed for manufactured goods, but they can be applied to a variety of situations as long as the statistical distribution of the characteristics of interest can be approximated by the normal distribution. In other cases, the principles still apply, but the formula may need to be modified to reflect the specific mathematical expression of the probability distribution functions. *See* PROCESS CONTROL.

Acceptance sampling refers to the procedures used to decide whether or not to accept product lots or batches based on the results of the inspection of samples drawn from the lots. Acceptance sampling techniques were originally developed for use by customers of manufactured products while inspecting lots delivered by their suppliers. These techniques are particularly well suited to situations where a decision on the quality level of product lots and their subsequent disposition needs to be made but it is not economic or feasible to inspect the entire production output.

Tzvi Raz

Quantitative chemical analysis The determination of the amount of an element on compound in a sample. Selection of a technique is based in part on the size of sample available, the quantity of analyte expected to be in the sample, the precision and accuracy of the technique, and the speed of analysis required. All techniques require calibration with respect to some standard of known composition. Caution is necessary to prevent other substances from giving signals falsely attributable to the sought-for substance, called the analyte. *See* CALIBRATION.

Direct measurement of a signal related to concentration or activity of the chemical species of interest is the most intuitive approach. Generally, a linear relationship between signal (or its logarithm) and concentration (or its logarithm) is sought. The relationship between signal and concentration is called a working curve. The slope of the line describing the relationship is known as sensitivity. The smallest quantity which is measurably different from the absence of analyte is the detection limit. Although a linear working curve is the simplest form to use, nonlinear curves may still be employed (either graphically or with the use of computers).

Titration is the process by which an unknown quantity of analyte (generally in solution) is determined by adding to it a standard reagent with which it reacts in a definite and known proportion. A chemical or instrumental means is provided to indicate when the standard reagent has consumed exactly the amount of analyte initially present. Each determination is performed with a reagent whose concentration is directly traceable to a primary standard, so that accuracy is frequently superior to that of other methods. *See* STOICHIOMETRY; TITRATION.

The sensitivity of a method is related to the method and analyte of interest, as well as to the presence of other species in the sample. The materials other than the analyte constitute a sample matrix. For example, seawater is a matrix quite different from distilled water because of the large amount of dissolved electrolyte present. If a signal can be derived from the analyte which is proportional to the amount of the analyte, but the sensitivity of the signal to concentration varies as a function of the sample matrix, the method of standard additions may be useful. The signal from the analyte is measured, after which small, known quantities of analyte are successively added to the sample, and the signal remeasured. The sensitivity of the method can thus be obtained, and the initial, preaddition signal interpreted to give the amount of analyte (C_1) initially present by using the relationship shown in the equation below, where S_1 is the initial

$$C_1 = \frac{\Delta C}{\Delta S} \cdot S_1$$

signal, C_1 the initial quantity of analyte, ΔC the standard addition of analyte, and ΔS the change in the signal caused by the addition. The sensitivity is $\Delta C/\Delta S$.

This approach is applicable only in the absence of reagent blanks, that is, signals caused by the presence of analyte in the reagent used for the determination.

In many methods, aliquots of samples are introduced into the measurement instrument. The signal from the analyte may vary with sample uptake rate or volume. To compensate for such effects, an internal standard, or species other than the analyte, may be added to the analyte in a known concentration prior to determination. The signal due to the internal standard is measured simultaneously with the analyte signal. Variation of the signal of the internal standard is interpreted to indicate the variation in sample uptake, which should be the same for both analyte and internal standard. The ratio of analyte signal to internal standard signal is independent of sample uptake. Thus the ratio of analyte to internal standard signal is used to establish the working curve, rather than using analyte signal alone.

If the relationship between signal and analyte concentration is nonlinear, quantitation may require the use of null comparison. A signal is observed from the analyte and from a standard whose concentration can be adjusted in a known way. When the signal from the adjustable standard equals the signal from the analyte, the two have identical concentrations. This condition is called a null, as there is no detectable difference between the sample and reference signals.

It is good laboratory practice to check every quantitative measurement for the influence of species other than the one being sought. For example, a glass electrode designed to sense hydrogen ion will also respond to high concentrations of sodium ion. The degree to which a given sensor responds to one species in preference to another is called the selectivity coefficient. If a general detector is desired, this coefficient ideally should be 1.0. If a species-specific detector is desired, the coefficient should be infinite. A signal of unknown or general origin which appears to underlie the analyte signal is known as background. This quantity, together with that for the reagent blank, may be subtracted from the raw signal and thus be compensated. However, variations in their level may prevent reliable compensation, particularly when small quantities of analyte are to be determined.

For further information on common quantitative techniques *see* ACTIVATION ANALYSIS; CALORIMETRY; CHEMICAL MICROSCOPY; CHROMATOGRAPHY; COMBUSTION; ELECTRON SPECTROSCOPY; ELECTROPHORESIS; GAS CHROMATOGRAPHY; GEL PERMEATION CHROMATOGRAPHY; IMMUNOASSAY; IMMUNOELECTROPHORESIS; ION EXCHANGE; ISOTOPE DILUTION TECHNIQUES; KINETIC METHODS OF ANALYSIS; MASS SPECTROMETRY; NUCLEAR MAGNETIC RESONANCE (NMR); POLARIMETRIC ANALYSIS; SPECTROSCOPY; X-RAY SPECTROMETRY.

Alexander Scheeline

Quantized electronic structure (QUEST) A material that confines electrons in such a small space that their wavelike behavior becomes important and their properties are strongly modified by quantum-mechanical effects. Such structures occur in nature, as in the case of atoms, but can be synthesized artificially with great flexibility of design and applications. They have been fabricated most frequently with layered semiconductor materials. Generally, the confinement regions for electrons in these structures are 1–100 nanometers in size. The allowable energy levels, motion, and optical properties of the electrons are strongly affected by the quantum-mechanical effects. The structures are referred to as quantum wells, wires, and dots, depending on whether electrons are confined with respect to motion in one, two, or three dimensions. Multiple closely spaced wells between which electrons can move by quantum-mechanical tunneling through intervening thin barrier-material layers are referred to as superlattices. *See* QUANTUM MECHANICS.

The most frequently used fabrication technique for quantized electronic structures is epitaxial growth of thin single-crystal semiconductor layers by molecular-beam epitaxy or by chemical vapor growth techniques. These artificially synthesized quantum structures find major application in high-performance transistors such as the microwave high-electron-mobility transistor (HEMT), and in high-performance solid-state lasers such as the semiconductor quantum-well laser. They also have important scientific applications for the study of fundamental two-dimensional, one-dimensional, and zero-dimensional physics problems in which particles are confined so that they have free motion in only two, one, or zero directions. Chemically formed nanocrystals, carbon nanotubes, zeolite cage compounds, and carbon buckyball C_{60} molecules are also important quantized electronic structures.

The optical applications are based on the interactions between light and electrons in the quantum structures. The absorption of a photon by an electron in a quantum well raises the electron from occupied quantum states to unoccupied quantum states. Electrons and holes in quantum wells may also recombine, with the resultant emission of photons from the quantized electronic structure as the electron drops from a higher state to a lower state. *See* ELECTRON-HOLE RECOMBINATION.

The photon emission is the basis for quantum-well semiconductor lasers, which have widespread applications in optical fiber communications and compact disk and laser disk optical recording. Quantum-well lasers operate by electrically injecting or pumping electrons into the lowest-conduction-band ($n = 1$) quantum-well state, where they recombine with holes in the highest-valence-band ($n = 1$) quantum-well state (that is, the electrons drop to an empty $n = 1$ valence-band state; illus. *a*), producing the emission of photons. These photons stimulate further photon emission and produce high-efficiency lasing. *See* COMPACT DISK; LASER; OPTICAL COMMUNICATIONS; OPTICAL RECORDING.

The photon absorption is the basis for quantum-well photodetectors and light modulators. In the quantum-well infrared photodetector an electron is promoted from lower (say, $n = 1$) to higher (say, $n = 2$) conduction band quantum-well states (illus. *b*) by absorption of an infrared photon. An electron in the higher state can travel more freely across the barriers, enabling it to escape from the well and be collected in a detector circuit. Changes in quantum-well shapes produced by externally applied electric fields can change the absorption wavelengths for light in a quantized electronic structure. The shift in optical absorption wavelength with electric field is known as the quantum-confined Stark effect. It forms the basis for semiconductor light modulators and semiconductor optical logic devices. *See* OPTICAL DETECTORS; OPTICAL MODULATORS; STARK EFFECT.

Modulation doping is a special way of introducing electrons into quantum wells for electrical applications. The electrons come from donor atoms lying in adjacent barrier layers (illus. *c*). Modulation doping is distinguished from conventional uniform doping in that it produces carriers in the quantum well without introducing impurity dopant atoms into the well. Since there are no impurity atoms to collide with in the well, electrons there are

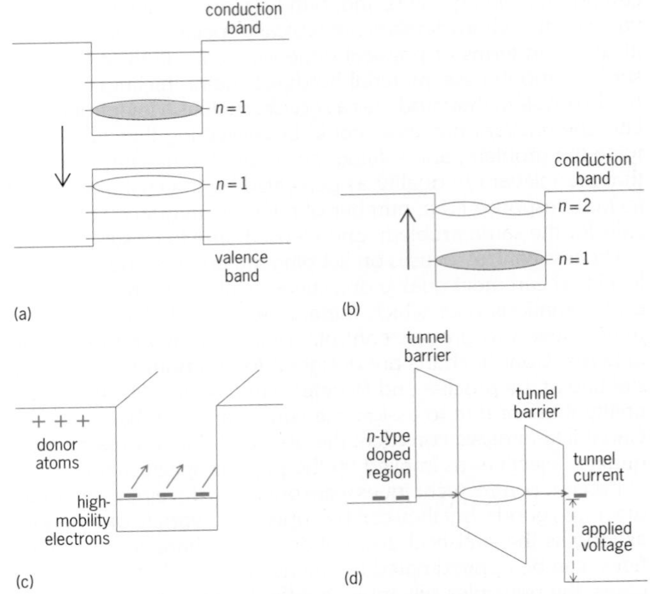

Principles of operation of quantum-well devices. (*a*) Quantum-well laser. (*b*) Quantum-well infrared detector. (*c*) High-electron-mobility transistor (HEMT or MODFET). Electrons in the quantum well that came from donor atoms in the barrier are free to move with high mobility in the direction perpendicular to the page. (*d*) Resonant tunneling device.

free to move with high mobility along the quantum-well layer. Resistance to electric current flow is thus much reduced relative to electrical resistance in conventional semiconductors. This enhances the low-noise and high-speed applications of quantum wells and is the basis of the high-electron-mobility transistor (HEMT), which is also known as the modulation-doped field-effect transistor (MODFET). HEMTs are widely used in microwave receivers for direct reception of satellite television broadcasts. *See* TRANSISTOR.

Electrical conductivity in carbon nanotubes occurs without doping and results from the absence of any energy gap in the electronic energy band structure of the nanotubes and the presence of allowed states at the Fermi energy. Individual nanotubes can be electrically contacted. Simple quantum wire transistors displaying quantized electron motion have been formed from single nanotubes.

Quantum-mechanical tunneling is another important property of quantized electronic structures. Tunneling of electrons through thin barrier layers between quantum wells is a purely quantum-mechanical effect without any real analog in classical physics or classical mechanics. It results from the fact that electrons have wavelike properties and that the particle waves can penetrate into the barrier layers. This produces a substantial probability that the particle wave can penetrate entirely through a barrier layer and emerge as a propagating particle on the opposite side of the barrier. The penetration probability has an exponential drop-off with barrier thickness. The tunneling is greatest for low barriers and thin barriers.

This effect finds application in resonant tunnel devices, which can show strong negative resistance in their electrical properties. In such a device (illus. *d*), electrons from an *n*-type doped region penetrate the barrier layers of a quantum well by tunneling. The tunneling current is greatest when the tunneling electrons are at the same energy as the quantum-well energy. The tunneling current actually drops at higher applied voltages, where the incident electrons are no longer at the same energy as the quantum-state energy, thus producing the negative resistance characteristic of the resonant tunneling diode. *See* ARTIFICIALLY

LAYERED STRUCTURES; NANOSTRUCTURE; NEGATIVE-RESISTANCE CIRCUITS; RESONANCE (QUANTUM MECHANICS); TUNNELING IN SOLIDS.

Arthur C. Gossard

Quantized vortices A type of flow pattern exhibited by superfluids, such as liquid ^{4}He below 2.17 K (−455.76°F). The term vortex designates the familiar whirlpool pattern where the fluid moves circularly around a central line and the velocity diminishes inversely proportionally to the distance from the center. The strength of a vortex is determined by the circulation, which is the line integral of the velocity around any path enclosing the central line. *See* VORTEX.

A superfluid is believed to be characterized by a macroscopic (that is, large-scale) quantum-mechanical wave function ψ . This wave function locks the superfluid into a coherent state. Since the velocity around the vortex increases without limit as the center is approached, the superfluid density and thus ψ must vanish at the center in order to avoid an infinite energy. Thus the central core of the vortex marks the zeros, or nodal lines, in the macroscopic wave function. *See* QUANTUM MECHANICS.

Quantized vortex lines are usually produced by rotating a vessel containing superfluid helium. At very low rotation speeds, no vortices exist: the superfluid remains at rest while the vessel rotates. At a certain speed the first vortex appears and corresponds to the first excited rotational state of the system. If the container continues to accelerate, additional quantized vortices will appear. At any given speed the vortices form a regular array which rotates with the vessel.

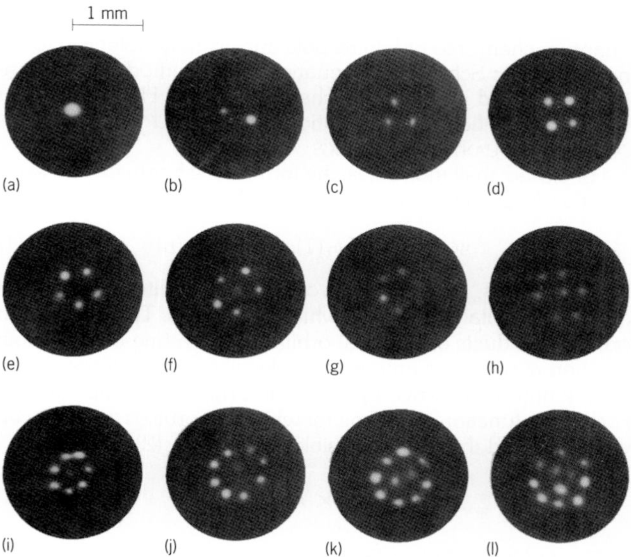

Stationary configurations of vortices which appear when a cylindrical container of superfluid ^{4}He is rotated about its axis. As the rotation speed increases from a to *f*, more vortices appear and the patterns become more complex. (*From E. J. Yarmchuk, M. J. V. Gordon, and R. E. Packard, Observation of stationary vortex arrays in rotating superfluid helium, Phys. Rev. Lett., 43:214–217, 1979*)

Quantized vortex lines were first detected in the mid-1950s by their influences on superfluid thermal waves traveling across the lines. In the late 1950s it was discovered that electrons in liquid helium form tiny charged bubbles which can become trapped on the vortex core but can move quite freely along the line. These electron bubbles (often referred to as ions) have been one of the most useful probes of quantized vortices. Researchers have been able to use ions to detect single quantized vortex lines. In one experiment the trapped ions are pulled out at the top of the vortex lines, accelerated, and focused onto a phosphor screen. The pattern of light thus produced on the phosphor is a map of the position of the vortices where they contact the liquid meniscus (see illustration). *See* LIQUID HELIUM; SUPERFLUIDITY.

Richard E. Packard

Quantum (physics) A term characterizing an excitation in a wave or field, connoting fundamental particlelike properties such as energy or mass, momentum, and angular momentum for this excitation. In general, any field or wave equation that is quantized, including systems already treated in quantum mechanics that are second-quantized, leads to a particle interpretation for the excitations which are called quanta of the field. This term historically was first applied to indivisible amounts of electromagnetic, or light, energy usually referred to as photons. The photon, or quantum of the electromagnetic field, is a massless particle, best interpreted as such by quantizing Maxwell's equations. Analogously, the electron can be said to be the quantum of the Dirac field through second quantization of the Dirac equation, which also leads to the prediction of the existence of the positron as another quantum of this field with the same mass but with a charge opposite to that of the electron. In similar fashion, quantization of the gravitational field equations suggests the existence of the graviton. The pi meson or pion was theoretically predicted as the quantum of the nuclear force field. Another quantum is the quantized lattice vibration, or phonon, which can be interpreted as a quantized sound wave since it travels through a quantum solid or fluid, or through nuclear matter, in the same manner as sound goes through air.

The use of quantum as an adjective (quantum mechanics, quantum electrodynamics) implies that the particular subject is to be treated according to the modern rules that have evolved for quantized systems. *See* ELEMENTARY PARTICLE; GRAVITATION; GRAVITON; MAXWELL'S EQUATIONS; MESON; PHONON; PHOTON; QUANTUM ELECTRODYNAMICS; QUANTUM FIELD THEORY; QUANTUM MECHANICS.

Kenneth E. Lassila

Quantum acoustics The investigation of the effects of the laws of quantum mechanics on the propagation and absorption of sound. At the present stage of development of physical science, quantum mechanics is the most fundamental theory of physical phenomena. However, for many applications in the everyday world, a sufficiently accurate description of nature is provided by classical mechanics. Quantum acoustics refers to acoustic experiments that are carried out under conditions such that the results can be understood only in terms of quantum theory. As a general tendency, quantum effects become more important in acoustic experiments that are performed with higher-frequency sound waves or that are carried out at lower temperatures. *See* QUANTUM MECHANICS.

Sound and phonons. To understand the quantum nature of sound, it is valuable to consider the origins of the quantum theory of light. Experiments in the latter part of the nineteenth century showed that the classical theory of electromagnetism combined with the laws of statistical mechanics could not explain the spectrum of light emitted by a heated surface. To reconcile theory and experiment, Max Planck in 1900 proposed that the energy of a light wave is quantized. This means that only certain values of the energy of the wave are allowed. These allowed values are given by the equation below, where $\hbar$ is Planck's constant divided

$$E = n\hbar\omega$$

by 2π, ω is the angular frequency of the light wave, and n is an integer. The modern interpretation of this formula describes this energy in terms of elementary quanta called photons. Each photon has an energy $\hbar\omega$, and the wave is made up of n photons. It was later realized that this same quantization of energy should apply to sound waves. These fundamental units of sound energy were given the name phonons. *See* HEAT RADIATION; LIGHT; PHONON; PHOTON.

Phonoatomic effect. One of the earliest experiments that confirmed the idea that light consists of photons was the photoelectric effect. In this effect the energy of light quanta is used to eject electrons from the surface of a metal. Only if the frequency of the light is sufficiently high do the photons have enough energy to knock the electrons out of the metal. The detection of this threshold frequency is thus strong evidence for quantization of energy. An analogous experiment has been performed with sound. In the experiment, sound is generated by a source in a vessel of liquid helium maintained at very low temperature, less than 0.1 K above absolute zero. In liquid helium, each atom is bound to the other atoms in the liquid by an energy that is unusually small compared to that of other liquids. For sound of sufficiently high frequency, the sound quanta when they arrive at the surface of the liquid have sufficient energy to knock helium atoms out of the liquid. These atoms can be detected by a suitable receiver placed above the surface of the liquid. The energy of an ejected atom, which determines its velocity, is equal to the energy of a phonon minus the binding energy. The time of arrival of the ejected atoms at the receiver is, in turn, determined by this velocity and hence is dependent on the frequency of the sound wave. Experiments have confirmed that the helium atoms arrive at the time expected based on this theory. *See* PHOTOEMISSION.

Phonon-phonon interactions. Sound waves in solids are attenuated during their propagation by a wide variety of physical processes. In many materials the most important mechanisms are related to impurities or defects in the solid, such as cracks, grain boundaries, or dislocations. Even when sound travels through a perfect crystal containing no defects, it is found that a measurable attenuation still occurs. In insulating crystals, where there are no free electrons, this attenuation is due to an interaction between the sound wave and the random thermal vibrations of the atoms in the solid. These random vibrations, which constitute the heat energy of the solid, are also quantized and are called thermal phonons. The attenuation of the sound wave can be attributed to collisions with the thermal phonons in which some of the sound quanta are scattered out of the sound beam. This mechanism is referred to as phonon-phonon scattering. *See* CRYSTAL DEFECTS; LATTICE VIBRATIONS; SOUND ABSORPTION.

In a linear elastic solid the elastic stress is exactly proportional to the strain, a phenomenon that is called Hooke's law. If Hooke's law holds, the presence of one wave does not affect the propagation of another, and so there are no interactions between phonons. Phonon-phonon scattering occurs because real solids always exhibit some deviations from linear elastic behavior. This nonlinearity is called anharmonicity. *See* HOOKE'S LAW; NONLINEAR ACOUSTICS.

Even when the temperature is very low and there are very few thermal phonons, anharmonicity can still give rise to an attenuation of a sound wave. This is because the sound phonons can spontaneously decay into phonons of lower frequencies. The rate at which this decay occurs is proportional to the frequency of the sound wave to the fifth power, and so the attenuation is important only for sound waves of very high frequency. If the number of phonons is sufficiently large, it is possible under some circumstances for there to be a large buildup in the population of some of the decay phonons. This process is called parametric amplification. *See* PARAMETRIC AMPLIFIER. Humphrey J. Maris

Quantum anomalies

Quantum anomalies Phenomena that arise when a quantity that vanishes according to the dynamical rules of classical physics acquires a finite value when quantum rules are used. For example, the classical Poisson bracket for some entities may vanish; yet the corresponding quantum commutator may be nonzero—this is a commutator anomaly. Alternatively, the flow of some material current may satisfy a continuity equation by virtue of the classical equations of motion, indicating conservative flow; but upon quantization the continuity equation may fail and the flow may no longer be conservative in the quantum theory—this is an anomalous divergence (of the current in question). Since the forms of Poisson brackets and quantum commutators as well as the occurrence of continuity equations for currents are related to symmetries and conservation laws of the theory, quantum anomalies serve to break some symmetries and destroy some conservation laws of classical models. This violation of symmetry is not driven by explicit symmetry-breaking terms in the dynamical equations—rather the quantization procedure itself violates the classical symmetry. The mathematical reason for this phenomenon is that classical dynamics, involving a finite number of degrees of freedom, usually leads to a quantum theory on an infinite-dimensional vector space (Hilbert space), and this "infinity" gives rise to novel effects. *See* CANONICAL TRANSFORMATIONS; CONSERVATION LAWS (PHYSICS); EQUATION OF CONTINUITY; HILBERT SPACE; NONRELATIVISTIC QUANTUM THEORY; SYMMETRY LAWS (PHYSICS).

The physically interesting setting for these phenomena is in quantum field theory, especially as applied to elementary particle physics, where the mechanism serves as an important source for symmetry breaking. Quantum anomalies also play a role in various other branches of physics, in which quantum field theory finds application, including condensed matter, supersymmetry, string theory, and motion in curved space-time. *See* ELEMENTARY PARTICLE; QUANTUM FIELD THEORY; SPACE-TIME; SUPERSTRING THEORY; SUPERSYMMETRY; SYMMETRY BREAKING. Roman Jackiw

Quantum chemistry

Quantum chemistry A branch of chemistry concerned with the application of quantum mechanics to chemical problems. More specifically, it is concerned with the electronic structure of molecules. Methods developed since 1960 permit the quantum chemist to obtain reliable approximate solutions to the nonrelativistic Schrödinger equation. The method which dominates the field of quantum chemistry is the Hartree-Fock or self-consistent-field approximation. *See* HAMILTON'S EQUATIONS OF MOTION; QUANTUM MECHANICS.

For closed-shell molecules, the form of the Hartree-Fock wave function is given by Eq. (1), in which $A(n)$, the antisymmetrizer

$$\Psi_{HF} = A(n)\phi_1(1)\phi_2(2)\ldots\phi_n(n) \tag{1}$$

for n electrons, has the effect of making a Slater determinant out of the orbital product on which it operates. The ϕ's are spin orbitals, products of a spatial orbital χ and a one-electron spin function α or β. For any given molecular system, there are an infinite number of wave functions of form (1), but the Hartree-Fock wave function is the one for which the orbitals ϕ have been varied to yield the lowest possible energy [Eq. (2)].

$$E = \int \Psi \times {}_{HF} H \Psi_{HF} d\tau \tag{2}$$

The resulting Hartree-Fock equations are relatively tractable due to the simple form of the energy E for single determinant wave functions [Eq. (3)].

$$E_{HF} = \sum_i I(i \mid i) + \sum_i \sum_{j>i} [(ij \mid ij) - (ij \mid ji)] \tag{3}$$

To solve the Hartree-Fock equations exactly, either the orbitals ϕ must be expanded in a complete set of analytic basis functions or strictly numerical (that is, tabulated) orbitals must be obtained. The former approach is impossible from a practical point of view for systems with more than two electrons, and the latter has been accomplished only for atoms and for a few diatomic molecules. Therefore, the exact solution of the Hartree-Fock equations is abandoned for polyatomic molecules. Instead an incomplete (but reasonable) set of analytic basis functions is adopted and solved for the best variational [that is, lowest energy given by Eq. (2)] wave function of form (1). Such a wave function is referred to as being of self-consistent-field (SCF)

Minimum-basis-set self-consistent-field geometry prediction compared with experiment for methylenecyclopropane

Parameter*	Theory	Experiment
$r\ (C_1{=}C_2)$	0.1298 nm	0.1332 nm
$r\ (C_2{-}C_3)$	0.1474 nm	0.1457 nm
$r\ (C_3{-}C_4)$	0.1522 nm	0.1542 nm
$r\ (C_1{-}H_1)$	0.1083 nm	0.1088 nm
$r\ (C_3H_3)$	0.1083 nm	0.109
$\theta(H_1C_1H_2)$	116.0°	114.3°
$\theta(H_3C_3H_4)$	113.6°	113.5°
$\theta(H_{34}C_3C_4)$	149.4°	150.8°

*Here r represents the carbon-carbon bond distance; θ represents the bond angle in degrees of H-C-H bonds; the numbers on C and H correspond to the numbered atoms in the displayed structure.

quality. For very-large-basis sets, then, it is reasonable to refer to the resulting SCF wave function as near-Hartree-Fock.

For large chemical systems, only minimum basis sets (MBS) can be used in ab initio theoretical studies. The term "large" includes molecular systems, with 100 or more electrons.

Ab initio theoretical methods have had the greatest impact on chemistry in the area of structural predictions. The most encouraging aspect of ab initio geometry predictions is their reliability. Essentially all molecular structures appear to be reliably predicted at the Hartree-Fock level of theory. Even more encouraging, many structures are accurately reproduced by using only minimum-basis-set self-consistent-field methods. A fairly typical example is methylenecyclopropane (see structure), with its

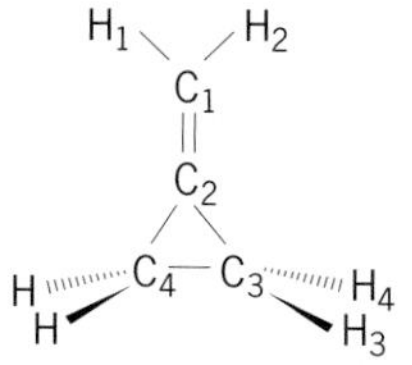

H_1 H_2 C_1 C_2 H H C_4 C_3 H_4 H_3

minimum-basis-set self-consistent-field structure compared with experiment in the table. Carbon-carbon bond distances differ typically by 0.002 nm from experiment, and angles are rarely in error by more than a few degrees. Thus, for many purposes, theory may be considered complementary to experiment in the area of structure prediction.

The most important post-Hartree-Fock methods for quantum chemistry are perturbation theory and the configuration interaction (CI) and coupled cluster (CC) methods. These three rigorous approaches may be labeled "convergent" quantum-mechanical methods, as each is ultimately capable of yielding exact solutions to Schrödinger's equation. The coupled cluster method treats excitations based on the number of electrons by which they differ from the Hartree-Fock reference function. Thus the CCSD method incorporates amplitudes differing by single (S) and double (D) excitations from Hartree-Fock. The CCSDT method adds all triple excitations to the CCSD treatment. As one goes to higher and higher excitations (for example, CCSDTQ includes all configurations differing by one, two, three, or four electrons from the Hartree-Fock reference configuration), one approaches the exact quantum-mechanical result.

In fact, coupled cluster theory beyond CCSD becomes impractical for large molecular systems. Thus, although the coupled cluster path to exact results is clear, it becomes a difficult road to follow. Triple excitations are sufficiently important that effective coupled cluster methods have been developed in which the effects of triples are approximated. The best of these methods, CCSD(T), is the closest thing to a panacea that exists today in quantum chemistry for very difficult problems involving smaller molecules. However, the range of applicability of the theoretically superior CCSD(T) method is much narrower than that of the popular hybrid Hartree-Fock/density functional methods. Thus, for most chemists the Hartree-Fock and density functional methods are likely to play central roles in molecular electronic structure theory for many years to come. *See* CHEMICAL BONDING; MOLECULAR ORBITAL THEORY; MOLECULAR STRUCTURE AND SPECTRA; RESONANCE (MOLECULAR STRUCTURE). Henry F. Schaefer III

Quantum chromodynamics A theory of the strong (nuclear) interactions among quarks, which are regarded as fundamental constituents of matter, structureless and indivisible at the current resolution of about 10^{-18} m. Quantum chromodynamics (QCD) explains why quarks combine in certain configurations to form the observed patterns of subnuclear particles, such as the proton and pi meson. According to this picture, the strong interactions among quarks are mediated by a set of force particles known as gluons. Quantum chromodynamics successfully accounts for many features of high-energy hard scattering of the strongly interacting particles. Interactions among gluons may lead to new structures that correspond to as-yet-undiscovered particles. The long-studied nuclear force that binds protons and neutrons together in atomic nuclei is regarded as a collective effect of the elementary interactions among constituents of the composite protons and neutrons. *See* NUCLEAR STRUCTURE.

By construction, quantum chromodynamics embodies many features abstracted from empirical observations, but the strength of the theory lies in the predictions it makes and the new understanding it brings. The property of asymptotic freedom implies a realm in which reliable calculations can be extracted using traditional Feynman-diagram techniques. These have been subjected to numerous quantitative tests with impressive success. The development of lattice gauge theory has made possible a growing range of computations of particle properties that emerge from the confinement of quarks. Part of the esthetic appeal of the theory is due to the fact that quantum chromodynamics is nearly identical in mathematical structure to quantum electrodynamics (QED) and to the unified theory of weak and electromagnetic interactions. This resemblance encourages the hope that a unified description of the strong, weak, and electromagnetic interactions may be at hand. *See* ELECTROWEAK INTERACTION; FEYNMAN DIAGRAM; QUANTUM ELECTRODYNAMICS; WEAK NUCLEAR INTERACTIONS.

Gauge theories. At the heart of current theories of the fundamental interactions is the idea of gauge invariance. Gauge theories constructed to embody various symmetry principles have emerged as the correct quantum-mechanical descriptions of the strong, weak, and electromagnetic interactions, at energies up to at least 1 TeV (10^{12} eV). *See* SYMMETRY LAWS (PHYSICS).

Color. Although the idea that the strongly interacting particles are built up of quarks brought new order to hadron spectroscopy and suggested new relations among mesons and baryons, the constituent description brought with it a number of puzzles. According to the Pauli exclusion principle, identical spin-$^1/_2$ particles cannot occupy the same quantum state. As a consequence, the observed baryons such as Δ^{++} (*uuu*) and Ω^- (*sss*), which would be composed of three identical quarks in the same state, would seem to be forbidden configurations. To comply with the Pauli principle, it is necessary to make the three otherwise identical quarks distinguishable by supposing that every flavor of quark exists in three varieties, fancifully labeled by the colors red, green, and blue. Color may be regarded as the strong-interaction analog of electric charge. Color cannot be created or destroyed by any of the known interactions. Like electric charge, it is said to be conserved. *See* COLOR (QUANTUM MECHANICS); EXCLUSION PRINCIPLE.

In light of evidence that color could be regarded as the conserved charge of the strong interactions, it was natural to seek a gauge symmetry that would have color conservation as its consequence. An obvious candidate for the gauge symmetry group is the unitary group SU(3), now to be applied to color rather than flavor. The theory of strong interactions among quarks that is prescribed by local color gauge symmetry is known as quantum chromodynamics. The mediators of the strong interactions are eight massless spin-1 bosons, one for each generator of the symmetry group. These strong-force particles are named gluons because they make up the "glue" that binds quarks together into hadrons. Gluons also carry color (in eight color-anticolor combinations) and hence have strong interactions among themselves.

Asymptotic freedom. The theoretical description of the strong interactions has historically been inhibited by the very strength of the interaction, which renders low-order perturbative calculations untrustworthy. However, in 1973 H. David Politzer, David J. Gross, and Frank Wilczek found that in many circumstances the effective strength of the interaction in Yang-Mills theories becomes increasingly feeble at short distances. For quantum chromodynamics, this remarkable observation implies that the interaction between quarks becomes weak at small separations. This discovery raises the hope that some aspects of the strong interactions might be treated by using familiar computational techniques that are predicated upon the smallness of the interaction strength.

Quarkonium. It was suggested in 1974 that the bound system of an extremely massive quark with its antiquark would be so small that the strong force would be extremely feeble. In this case, the binding between quark and antiquark is mediated by the exchange of a single massless gluon, and the spectrum of bound states resembles that of an exotic atom composed of an electron and an antielectron (positron) bound electromagnetically in a Coulomb potential generated by the exchange of a massless photon. Since the electron-positron atom is known as positronium, the heavy quark-antiquark atom has been called quarkonium. Two families of heavy quark-antiquark bound states, the ψ/J system composed of charmed quarks and the Υ system made up of *b* quarks, have been discovered. Both have level schemes characteristic of atomic spectra, which have been analyzed by using tools of nonrelativistic quantum mechanics developed for ordinary atoms. The atomic analogy has proved extremely fruitful for studying the strong interaction. *See* CHARM; J/PSI PARTICLE; POSITRONIUM.

Lattice models. To deal with the existence and properties of the hadrons themselves, it is necessary to devise a new computational approach that does not break down when the interaction becomes strong. The most promising method has been the crystal lattice formulation of the theory. By localizing quarks on individual lattice sites, it is possible to use many of the Monte Carlo simulation techniques developed in statistical physics for the study of spin systems such as magnetic substances. Modern calculations include the influence of dynamical quark-antiquark pairs. Lattice quantum chromodynamics analyses support the color-confinement paradigm. *See* ELEMENTARY PARTICLE; FUNDAMENTAL INTERACTIONS; GLUONS; ISING MODEL; QUANTUM FIELD THEORY; QUARKS; STANDARD MODEL; STATISTICAL MECHANICS; STRONG NUCLEAR INTERACTIONS.

C. Quigg

Quantum computation

Whereas classical computers perform operations on classical bits, which can be in one of two discrete states, 0 or 1, quantum computers perform operations on quantum bits, or qubits, which can be put into any superposition of two quantum states, $|0\rangle$ and $|1\rangle$. Quantum computers hold the promise to perform classically intractable tasks—computations that require astronomical time or hardware resources on the fastest classical computer—in minuscule time with minuscule resources. The building blocks of a quantum computer are atomic particles which, following the laws of quantum physics, behave like waves and exhibit interference phenomena as if they were in several locations at the same time. By equating different locations—for example, an electron in the lowest orbit or in an excited orbit of an atom—to binary digits 0 or 1, one may interpret the time-evolving state of the particles as executing several computations at the same time. One set of locations at a given time describes the result of one computation. Thus one atom can do two computations at once; two atoms can do four; three atoms can do eight. The challenge is to coerce the atoms to follow trajectories that amount to meaningful computations and to read out a definite result from the multitude of computations occurring in parallel. The control of trajectories is the hardware part of the challenge; the design of useful trajectories—algorithms that are superior to classical algorithms—is the software part of the challenge. *See* DIGITAL COMPUTER; SUPERCOMPUTER.

Principles. A classical computer manipulates strings of N classical bits, $(n_1, \ldots, n_N)$ with $n_j = 0$ or 1 $(j = 1, \ldots, N)$, in such a way that intermediate states of the computation are also strings of classical bits. A quantum computer manipulates states of N two-level atoms, nuclear spins, or other entities, $|n_1, \ldots, n_N\rangle$, with $n_j = 0$ if the jth atom is in the ground state and $n_j = 1$ if it is in the excited state, in such a way that intermediate states are superpositions of the states $|n_1, \ldots, n_N\rangle$. The 2^N states $|n_1, \ldots, n_N\rangle$ (the computational basis) are product states in which each atom is in either the ground state or excited state, and n_j is called the value of the jth qubit; they represent the strings of classical bits. The superpositions include states in which an atom no longer has a sharp value of n_j (it has indefinite bit value), and states in which an atom no longer exists in a state separate from the other atoms (entangled states); both have no classical counterpart. A quantum computation starts with a product state $|n_1, \ldots, n_N\rangle$; allows the state to evolve according to the Schrödinger equation (1), with initial condition $|\psi(0)\rangle =$

$$i\hbar\frac{d}{dt}|\psi(t)\rangle = H(t)|\psi(t)\rangle \qquad (1)$$

$|n_1, \ldots, n_N\rangle$ and time-dependent Hamiltonian $H(t)$ [energy operator] driving the coupled atoms; and ends with the measurement of the values of the qubits of the state $|\psi(t)\rangle$. The Hamiltonian generates the unitary time evolution operator $U(t)$ which takes the initial state into the final state according to $|\psi(t)\rangle = U(t)|\psi(0)\rangle$. Thus, to perform a specific computation, one must drive the atoms with a specific Hamiltonian; to read out the result, one must send the atoms through a series of state detectors. *See* BIT; SCHRÖDINGER'S WAVE EQUATION.

Power of quantum computation. A quantum computer is more powerful than a classical computer for two reasons:

1. The quantum state space is much larger than the classical state space: N qubits can be in an infinite number of different states; N classical bits can be in only 2^N different states. Thus a quantum computer can store and access an exponentially large number of states compared to a classical computer.

2. The quantum computer operates in a massively parallel way: If the initial state is the uniform superposition of all basis states, Eq. (2), the time evolution computes simultaneously $U(t)|n_1, \ldots, n_N\rangle$ for all 2^N possible inputs $|n_1, \ldots, n_N\rangle$ by linearity of $U(t)$.

$$|\psi(0)\rangle = 2^{-N/2} \sum_{n_1,\ldots,n_N=0,1} |n_1, \ldots, n_N\rangle \qquad (2)$$

The matrix element $\langle n'_1, \ldots, n'_N|U(t)|n_1, \ldots, n_N\rangle$ is the probability amplitude that the computation converts the input $|n_1, \ldots, n_N\rangle$ into the output $|n'_1, \ldots, n'_N\rangle$, along all possible classical computational paths in parallel (as given by Feynman's path integral). A classical computation can follow only a single path. *See* FEYNMAN INTEGRAL; SUPERPOSITION PRINCIPLE.

Outlook. A remarkable array of experimental realizations of quantum computing devices and implementations of algorithms has been achieved. Numerous other approaches have been proposed and are under investigation, for quantum computers with much longer decoherence times and control over many more qubits (a long-term goal is 200 qubits). These include solid-state NMR; solid-state CQED; electron spin resonance of donor atoms in semiconductors; electronic excitation in quantum dots; ion traps with hot ions; electrons trapped on superfluid helium films;

neutral atoms, including Bose-Einstein condensates, trapped in an optical lattice; linear optics (beam splitters, phase shifters, and single-photon sources); and topological quantum computing, using braided world lines of quasi-particles in the fractional quantum Hall effect as decoherence-free qubits. *See* BOSE-EINSTEIN CONDENSATION; ELECTRON PARAMAGNETIC RESONANCE (EPR) SPECTROSCOPY; HALL EFFECT; LIQUID HELIUM; QUANTUM MECHANICS.

Peter Pfeifer

Quantum electrodynamics The field of physics that studies the interaction of electromagnetic radiation with electrically charged matter within the framework of relativity and quantum mechanics. It is the fundamental theory underlying all disciplines of science concerned with electromagnetism, such as atomic physics, chemistry, biology, the theory of bulk matter, and electromagnetic radiation.

Efforts to formulate quantum electrodynamics (QED) were initiated by P. A. M. Dirac, W. Heisenberg, and W. Pauli soon after quantum mechanics was established. The first step was to remedy the obvious shortcoming of quantum mechanics: that it applies only to the case where particle speeds are small compared with that of light, *c*. This led to Dirac's discovery of a relativistic wave equation, in which the wave function has four components and is multiplied by certain 4×4 matrices. His equation incorporates in a natural manner the observed electron-spin angular momentum, which implies that the electron is a tiny magnet. The strength of this magnet (magnetic moment) was predicted by Dirac and agreed with observation. A detailed prediction of the hydrogen spectrum was also in good agreement with experiment. *See* ATOMIC STRUCTURE AND SPECTRA; ELECTRON SPIN; MATRIX THEORY.

In order to go beyond this initial success and calculate higher-order effects, however, the interaction of charge and electromagnetic field had to be treated dynamically. To begin with, a good theoretical framework had to be found for describing the wave-particle duality of light, that is, the experimentally well-established fact that light behaves like a particle (photon) in some cases but like a wave in others. Similarly, the electron manifests wave-particle duality, another observed fact. Once this problem was settled, the next question was how to deal with the interaction of charge and electromagnetic field. It is here that the theory ran into severe difficulties. Its predictions often diverged when attempts were made to calculate beyond lowest-order approximations. This inhibited the further development of the theory for nearly 20 years. Stimulated by spectroscopic experiments vastly refined by microwave technology developed during World War II, however, S. Tomonaga, R. P. Feynman, and J. Schwinger discovered that the difficulties disappear if all observable quantities are expressed in terms of the experimentally measured charge and mass of the electron. With the discovery of this procedure, called renormalization, quantum electrodynamics became a theory in which all higher-order corrections are finite and well defined. *See* PHOTON; QUANTUM MECHANICS; RELATIVISTIC QUANTUM THEORY; RELATIVITY; RENORMALIZATION; WAVE MECHANICS.

Quantum electrodynamics is the first physical theory ever developed that has no obvious intrinsic limitation and describes physical quantities from first principles. Nature accommodates forces other than the electromagnetic force, such as those responsible for radioactive disintegration of heavy nuclei (called the weak force) and the force that binds the nucleus together (called the strong force). A theory called the standard model has been developed which unifies the three forces and accounts for all experimental data from very low to extremely high energies. This does not mean, however, that quantum electrodynamics fails at high energies. It simply means that the real world has forces other than electromagnetism.

High-precision tests have provided excellent confirmation for the validity of the renormalization theory of quantum electrodynamics. In the high-energy regime, tests using electron-positron colliding-beam facilities at various high-energy physics laboratories have confirmed the predictions of quantum electrodynamics at center-of-mass energies up to 1.8×10^{11} electronvolts (180 GeV). The uncertainty principle implies that this is equivalent to saying that quantum electrodynamics is valid down to about 10^{-17} meter, a distance 100 times shorter than the radius of the proton.

High-precision tests of quantum electrodynamics have also been carried out at low energies by using various simple atomic systems. The most accurate is that of the measurement of the magnetic moment of the electron, or the gyromagnetic ratio *g*, the ratio of spin and rotation frequencies, which is correctly predicted by quantum electrodynamics to 12 significant digits. This constitutes the most precise confrontation of any experiment with a theoretical prediction in the history of science. *See* QUANTUM FIELD THEORY.

Toichiro Kinoshita

Quantum electronics Quantum electronics is concerned with the interaction of electromagnetic radiation and matter and with the use of radiation-matter interactions for applications in devices such as lasers and masers. The term "quantum electronics" for this field was coined in 1958 or 1959 by Charles H. Townes, who invented the maser, as this field in its early days brought together ideas from electrical engineering and quantum physics in the development of the maser and the laser. The field has grown immensely, and today quantum electronics is concerned not just with masers and lasers but with a wide range of topics related to linear as well as nonlinear radiation-matter interactions, quantum-mechanical properties of radiation, noise and photon statistics, physical properties of solid-state materials relevant to radiation interaction, and devices and systems that utilize these basic principles for practical applications. *See* LASER; MASER.

Many electronic devices, such as semiconductor resonant tunneling diodes or the superconductor Josephson junctions, may be considered quantum electronic devices, since their operation is dictated by the principles of quantum mechanics. However, quantum electronics is understood to be concerned with only those devices whose operation involves radiation-matter interactions, such as lasers, parametric oscillators, and optoelectronic devices, such as light modulators, photodetectors, and optical switches. *See* ELECTROOPTICS; OPTICAL BISTABILITY; OPTICAL DETECTORS; OPTICAL MODULATORS.

Radiation absorption and emission. Some of the most fundamental ideas in quantum electronics are related to the absorption and emission of radiation by matter. In 1901, Max Planck postulated that radiation can be absorbed or emitted by matter in only discrete bundles or packets that were later called photons. Soon after in 1917, Albert Einstein discovered that photon emission by matter can be a spontaneous process or a process stimulated by other photons. The number of photons can be increased substantially through stimulated emission, which is the basic principle behind the operation of lasers and masers. *See* PHOTOEMISSION; PHOTON.

Laser operation. In a laser, photons are confined inside a cavity along with atoms that can emit photons. When the rate at which photons are generated inside the cavity is larger than the rate at which photons are lost from the cavity, the number of photons inside the cavity increases dramatically and this is called lasing.

Coherent radiation. A feature of lasers that makes radiation from a laser source different from radiation from a black-body source is the coherence of the laser radiation. Lasers and masers can provide high-power coherent radiation all the way from microwave frequencies (10^9 Hz) to x-ray frequencies (10^{17} Hz). *See* COHERENCE.

Nonlinear optical effects. A wide assortment of interesting and useful nonlinear optical phenomena can be produced using light from lasers because of the unprecedented high power and coherence of laser radiation. Nonlinear optical effects can be understood in terms of the atomic or material polarization caused by the strong electric fields of laser radiation. Nonlinear optics has become a major part of the field of quantum electronics. *See* NONLINEAR OPTICS; POLARIZATION OF DIELECTRICS.

Ultrafast optics. High-intensity single-cycle optical pulses have been produced by mode-locking lasers. Ultrashort pulses from mode-locked lasers have enabled the study of dynamical processes in physics, chemistry, and biology with femtosecond time resolution. *See* OPTICAL PULSES; ULTRAFAST MOLECULAR PROCESSES.

Optics on microchips. It is now possible to generate, amplify, and control coherent radiation over a wide range of frequencies on microchips. Micrometer-scale lasers and optoelectronic devices are now possible. The development of inexpensive semiconductor lasers and integrated optoelectronic devices has been partly responsible for the wide deployment of fiber-optic communication links. *See* INTEGRATED OPTICS; OPTICAL COMMUNICATIONS.

Farhan Rana

Quantum field theory

Quantum field theory The quantum-mechanical theory of physical systems whose dynamical variables are local functions of space and time. As distinguished from the quantum mechanics of atoms, quantum field theories describe systems with an infinite number of degrees of freedom. Such theories provide the natural language for describing the interactions and properties of elementary particles, and have proved to be successful in providing the basis for the fundamental theories of the interactions of matter. The present understanding of the basic forces of nature is based on quantum field theories of the strong, weak, electromagnetic, and gravitational interactions. Quantum field theory is also useful in the study of many-body systems, especially in situations where the characteristic length of a system is large compared to its microscopic scale. *See* QUANTUM MECHANICS.

Quantum field theory originated in the attempt, in the late 1920s, to unify P. A. M. Dirac's relativistic electron theory and J. C. Maxwell's classical electrodynamics in a quantum theory of interacting photon and electron fields. This effort was completed in the 1950s and was extremely successful. At present the quantitative predictions of the theory are largely limited to perturbative expansions (believed to be asymptotic) in powers of the fine-structure constant. However, because of the extremely small value of this parameter, $\alpha = e^2/\hbar c \approx 1/137$ (where e is the electron charge, $\hbar$ is Planck's constant divided by 2π, and c is the speed of light), such an expansion is quite adequate for most purposes. The remarkable agreement of the predictions of quantum electrodynamics with high-precision experiments (sometimes to an accuracy of 1 part in 10^{12}) provides strong evidence for the validity of the basic tenets of relativistic quantum field theory. *See* CLASSICAL FIELD THEORY; ELECTROMAGNETIC RADIATION; MAXWELL'S EQUATIONS; PERTURBATION (QUANTUM MECHANICS); PHOTON; QUANTUM ELECTRODYNAMICS; RELATIVISTIC ELECTRODYNAMICS; RELATIVISTIC QUANTUM THEORY.

Quantum field theory also provides the natural framework for the treatment of the weak, strong, and gravitational interactions.

The first of such applications was Fermi's theory of the weak interactions, responsible for radioactivity, in which a hamiltonian was constructed to describe beta decay as a product of four fermion fields, one for each lepton or nucleon. This theory has been superseded by the modern electroweak theory that unifies the weak and the electromagnetic interactions into a common framework. This theory is a generalization of Maxwell's electrodynamics which was the first example of a gauge theory, based on a continuous local symmetry. In the case of electromagnetism the local gauge symmetry is the space-time-dependent change of the phase of a charged field. The existence of massless spin-1 particles, photons, is one of the consequences of the gauge symmetry. The electroweak theory is based on generalizing this symmetry to space-time-dependent transformations of the labels of the fields, based on the group SU(2) × U(1). However, unlike electromagnetism, part of this extended symmetry is not shared by the ground state of the system. This phenomenon of spontaneous symmetry breaking produces masses for all the elementary fermions and for the gauge bosons that are the carriers of the weak interactions, the $W^{\pm}$ and Z bosons. (This is known as the Higgs mechanism.) The electroweak theory has been confirmed by many precision tests, and almost all of its essential ingredients have been verified. *See* ELECTROWEAK INTERACTION; GAUGE THEORY; INTERMEDIATE VECTOR BOSON; SYMMETRY BREAKING; SYMMETRY LAWS (PHYSICS); WEAK NUCLEAR INTERACTIONS.

The application of quantum field theory to the strong or nuclear interactions dates from H. Yukawa's hypothesis that the short-range nuclear forces arise from the exchange of massive particles that are the quanta of local fields coupled to the nucleons, much as the electromagnetic interactions arise from the exchange of massless photons that are the quanta of the electromagnetic field. The modern theory of the strong interactions, quantum chromodynamics, completed in the early 1970s, is also based on a local gauge theory. This is a theory of spin-$^1/_2$ quarks invariant under an internal local SU(3) (color) gauge group. The observed hadrons (such as the proton and neutron) are SU(3) color-neutral bound states of the quarks whose interactions are dictated by the gauge fields (gluons). This theory exhibits almost-free-field behavior of quarks and gluons over distances and times short compared to the size of a hadron (asymptotic freedom), and a strong binding of quarks at large separations that results in the absence of colored states (confinement). *See* ELEMENTARY PARTICLE; GLUONS; MESON; QUANTUM CHROMODYNAMICS; QUARKS.

Quantum field theory has been tested down to distances of 10^{-20} m. There appears to be no reason why it should not continue to work down to Planck's length, $(G\hbar/c^3)^{1/2} \approx 10^{-35}$ m (where G is the gravitational constant), where the quantum effects of gravity become important. In the case of gravity, A. Einstein's theory of general relativity already provides a very successful classical field theory. However, the union of quantum mechanics and general relativity raises conceptual problems that seem to call for a radical reexamination of the foundations of quantum field theory. *See* FUNDAMENTAL INTERACTIONS; GRAVITATION; QUANTUM GRAVITATION; RELATIVITY.

David Gross

Quantum gravitation

Quantum gravitation The quantum theory of the gravitational field; also, the study of quantum fields in a curved space-time. In classical general relativity, the gravitational field is represented by the metric tensor $g_{\mu\nu}$ of space-time. This tensor satisfies Einstein's field equation, with the energy-momentum tensor of matter and radiation as a source. However, the equations of motion for the matter and radiation fields also depend on the metric.

Classical field theories such as Maxwell's electromagnetism or the classical description of particle dynamics are approximations valid only at the level of large-scale macroscopic observations. At a fundamental level, elementary interactions of particles and fields must be described by relativistic quantum mechanics, in terms of quantum fields. Because the geometry of space-time in general relativity is inextricably connected to the dynamics of matter and radiation, a consistent theory of the metric in interaction with quantum fields is possible only if the metric itself is quantized. *See* MAXWELL'S EQUATIONS; QUANTUM FIELD THEORY; RELATIVISTIC QUANTUM THEORY; RELATIVITY.

Under ordinary laboratory conditions the curvature of space-time is so extremely small that in most quantum experiments gravitational effects are completely negligible. Quantization in Minkowski space is then justified. Gravity is expected to play a significant role in quantum physics only at rather extreme conditions of strongly time-dependent fields, near or inside very dense matter. The scale of energies at which quantization of the metric itself becomes essential is given by $(\hbar c^5/G)^{1/2} \approx 10^{19}$ GeV, where

G is the gravitational constant, $\hbar$ is Planck's constant divided by 2π, and c is the velocity of light. Energies that can be reached in the laboratory or found in cosmic radiation are far below this order of magnitude. Only in the very early stages of the universe, within a proper time of the order of $(G\hbar/c^5)^{1/2} \approx 10^{-43}$ s after the big bang, would such energies have been produced. *See* BIG BANG THEORY.

In most physical systems the metric is quasistationary over macroscopic distances so that its fluctuations can be ignored. A quantum description of fields in a curved space-time can then be given by treating the metric as a classical external field in interaction with the quantum fields.

Quantum effects of black holes. The most striking quantum effect in curved space-time is the emission of radiation by black holes. A black hole is an object that has undergone gravitational collapse. Classically this means that it becomes confined to a space-time region in which the metric has a singularity (the curvature becomes infinite). This region is bounded by a surface, called the horizon, such that any matter or radiation falling inside becomes trapped. Therefore, classically the mass of a black hole can only increase. However, this is no longer the case if quantum effects are taken into account. When, because of fluctuations of the quantum field, particle-antiparticle or photon pairs are created near the horizon of a black hole, one of the particles carrying negative energy may move toward the hole, being absorbed by it, while the other moves out with positive energy. *See* BLACK HOLE; GRAVITATIONAL COLLAPSE.

It is found that the total rate of emission is inversely proportional to the square of the mass. For stellar black holes whose masses are of the order of a solar mass, the emission rate is negligibly small and unobservable. Only primordial black holes, of mass less than 10^{13} kg, formed very early in the quantum era of the universe, would have been small enough to produce quantum effects that could play any significant role in astrophysics or in cosmology. *See* ASTROPHYSICS.

Quantization of the metric. There are basically two approaches to the quantization of the metric, the canonical and the covariant quantization. A third method, which can be derived from the first and is now most widely used, is based on the Feynman path integral representation for the vacuum-to-vacuum amplitude, which is the generator of Green's functions for the quantum theory. One important feature of this method is that since the topology of the manifold is not specified at the outset it is possible to include a sum over paths in different topologies. The outcome of this idea is that the vacuum would, at the level of the Planck length, $(G\hbar/c^3)^{1/2} \approx 10^{-35}$ m, acquire a foamlike structure. *See* FEYNMAN INTEGRAL; GREEN'S FUNCTION.

At present a complete, consistent theory of quantum gravity is still lacking. The formal theory fails to satisfy the power-counting criterion for renormalizability. In every order of the perturbation expansion, new divergences appear which could only be canceled by counterterms that do not exist in the original lagrangian. This may not be just a technical problem but the reflection of a conceptual difficulty stemming from the dual role, geometric and dynamic, played by the metric. *See* RENORMALIZATION.

Supergravity and superstrings. Supergravity is a geometric extension of general relativity which incorporates the principle of supersymmetry. Supersymmetry is a kind of symmetry, discovered in the 1970s, that allows for the transformation of fermions and bosons into each other. (Fermions carry half-integer spin while bosons carry integer spin; they also obey different statistics.) Supergravity can be formulated in space-time manifolds with a total of $D = d + 1$ dimensions, where d, the number of space dimensions, can be as large as 10. They constitute truly unified theories of all interactions including gravity.

In the early 1980s, some encouraging results were found with a theory based on the idea that the basic objects of nature are not pointlike but actually one-dimensional objects like strings, which can be open or closed. Incorporating supersymmetry into the theory leads to a critical dimension $D = 10$.

In the approximation of neglecting string excitations, certain superstring models may be described in terms of local fields as a $D = 10$ supergravity theory. At present, these are the only theories that both include gravity and can be consistently quantized. Although a superstring theory may eventually become the ultimate theory of all the interactions, there is still a very long way to go in making the connection between its fundamental fields and the fields representing the particles and their interactions as observed at low energies. *See* FUNDAMENTAL INTERACTIONS; GRAVITATION; SUPERGRAVITY; SUPERSTRING THEORY; SUPERSYMMETRY.

Samuel W. MacDowell

Quantum mechanics The modern theory of matter, of electromagnetic radiation, and of the interaction between matter and radiation; also, the mechanics of phenomena to which this theory may be applied. Quantum mechanics, also termed wave mechanics, generalizes and supersedes the older classical mechanics and Maxwell's electromagnetic theory. Atomic and subatomic phenomena provide the most striking evidence for the correctness of quantum mechanics and best illustrate the differences between quantum mechanics and the older classical physical theories. Quantum mechanics is needed to explain many properties of bulk matter, for instance, the temperature dependence of the specific heats of solids.

The formalism of quantum mechanics is not the same in all domains of applicability. In approximate order of increasing conceptual difficulty, mathematical complexity, and likelihood of future fundamental revision, these domains are the following: (i) Nonrelativistic quantum mechanics, applicable to systems in which particles are neither created nor destroyed, and in which the particles are moving slowly compared to the velocity of light. Here a particle is defined as a material entity having mass, whose internal structure either does not change or is irrelevant to the description of the system, (ii) Relativistic quantum mechanics, applicable in practice to a single relativistic particle (one whose speed equals or nearly equals c); here the particle may have zero rest mass, in which event, its speed must equal c. (iii) Quantum field theory, applicable to systems in which particle creation and destruction can occur; the particles may have zero or nonzero rest mass. This article is concerned mainly with nonrelativistic quantum mechanics, which apparently applies to all atomic and molecular phenomena, with the exception of the finer details of atomic spectra. Nonrelativistic quantum mechanics also is well established in the realm of low-energy nuclear physics. *See* ATOMIC STRUCTURE AND SPECTRA; NUCLEAR PHYSICS; QUANTUM FIELD THEORY; RELATIVISTIC QUANTUM THEORY.

Planck's constant. The quantity 6.626×10^{-34} joule-second, first introduced into physical theory by Max Planck in 1901, is a basic ingredient of the formalism of quantum mechanics. Planck's constant commonly is denoted by the letter h; the notation $\hbar = h/2\pi$ also is standard.

Uncertainty principle. In classical physics the observables characterizing a given system are assumed to be simultaneously measurable (in principle) with arbitrarily small error. For instance, it is thought possible to observe the initial position and velocity of a particle and therewith, using Newton's laws, to predict exactly its future path in any assigned force field. According to the uncertainty principle, accurate measurement of an observable quantity necessarily produces uncertainties in one's knowledge of the values of other observables. In particular, for a single particle relation (1*a*) holds, where Δx represents the uncertainty

$$\Delta x \Delta p_x \gtrsim \hbar \qquad (1a)$$

$$\Delta t \ \Delta E \gtrsim \hbar \qquad (1b)$$

(error) in the location of the x coordinate of the particle at any instant, and Δp_x is the simultaneous uncertainty in the x component of the particle momentum. Relation (1a) asserts that under the best circumstances, the product $\Delta x \Delta p_x$ of the uncertainties cannot be less than about 10^{-34} joule second.

The uncertainty relation (1b) is derived and interpreted somewhat differently than relation (1a); it asserts that for any system, an energy measurement with error ΔE must be performed in a time not less than $\Delta t \sim \hbar/\Delta E$. If a system endures for only Δt seconds, any measurement of its energy must be uncertain by at least $\Delta E \sim \hbar \Delta t$. *See* UNCERTAINTY PRINCIPLE.

Wave-particle duality. It is natural to identify such fundamental constituents of matter as protons and electrons with the mass points or particles of classical mechanics. According to quantum mechanics, however, these particles, in fact all material systems, necessarily have wavelike properties. Conversely, the propagation of light, which, by Maxwell's electromagnetic theory, is understood to be a wave phenomenon, is associated in quantum mechanics with massless energetic and momentum-transporting particles called photons. The quantum-mechanical synthesis of wave and particle concepts is embodied in the de Broglie relations, given by Eqs. (2a) and (2b). These give the

$$\lambda = h/p \tag{2a}$$

$$f = E/h \tag{2b}$$

wavelength λ and wave frequency f associated with a free particle (a particle moving freely under no forces) whose momentum is p and energy is E; the same relations give the photon momentum p and energy E associated with an electromagnetic wave in free space (that is, in a vacuum) whose wavelength is λ and frequency is f. *See* PHOTON.

The wave properties of matter have been demonstrated conclusively for beams of electrons, neutrons, atoms (hydrogen, H, and helium, He), and molecules (H_2). When incident upon crystals, these beams are reflected into certain directions, forming diffraction patterns. Diffraction patterns are difficult to explain on a particle picture; they are readily understood on a wave picture, in which wavelets scattered from regularly spaced atoms in the crystal lattice interfere constructively along certain directions only. *See* ELECTRON DIFFRACTION; NEUTRON DIFFRACTION.

The particle properties of light waves are observed in the photoelectric effect and the Compton effect. *See* COMPTON EFFECT; PHOTOEMISSION.

Complementarity. Wave-particle duality and the uncertainty principle are thought to be examples of the more profound principle of complementarity, first enunciated by Niels Bohr (1928). According to the principle of complementarity, nature has "complementary" aspects; an experiment which illuminates one of these aspects necessarily simultaneously obscures the complementary aspect. To put it differently, each experiment or sequence of experiments yields only a limited amount of information about the system under investigation; as this information is gained, other equally interesting information (which could have been obtained from another sequence of experiments) is lost. Of course, the experimenter does not forget the results of previous experiments, but at any instant, only a limited amount of information is usable for predicting the future course of the system.

Quantization. In classical physics the possible numerical values of each observable, meaning the possible results of exact measurement of the observable, generally form a continuous set. For example, the x coordinate of the position of a particle may have any value between $-\infty$ and $+\infty$. In quantum mechanics the possible numerical values of an observable need not form a continuous set, however. For some observables, the possible results of exact measurement form a discrete set; for other observables, the possible numerical values are partly discrete, partly continuous; for example, the total energy of an electron in the field of a proton may have any positive value between 0 and $+\infty$, but may have only a discrete set of negative values, namely, -13.6, $-13.6/4$, $-13.6/9$, $-13.6/16$ eV,... Such observables are said to be quantized; often there are simple quantization rules determining the quantum numbers which specify the allowable discrete values. Spectroscopy, especially the study of atomic spectra, probably provides the most detailed quantitative confirmation of quantization.

Probability considerations. The uncertainty and complementarity principles, which limit the experimenter's ability to describe a physical system, must limit equally the experimenter's ability to predict the results of measurement on that system. Suppose, for instance, that a very careful measurement determines that the x coordinate of a particle is precisely $x = x_0$. This is permissible in nonrelativistic quantum mechanics. Then formally, the particle is known to be in the eigenstate corresponding to the eigenvalue $x = x_0$ of the x operator. Under these circumstances, an immediate repetition of the position measurement again will indicate that the particle lies at $x = x_0$. Knowing that the particle lies at $x = x_0$ makes the momentum p_x of the particle completely uncertain, however, according to relation (1a). A measurement of p_x immediately after the particle is located at $x = x_0$ could yield any value of p_x from $-\infty$ to $+\infty$.

More generally, suppose the system is known to be in the eigenstate corresponding to the eigenvalue α of the observable A. Then for any observable B, which is to some extent complementary to A, that is, for which an uncertainty relation of the form of relations (1) limits the accuracy with which A and B can simultaneously be measured, it is not possible to predict which of the many possible values $B = \beta$ will be observed. However, it is possible to predict the relative probabilities $P_\alpha(\beta)$ of immediately thereafter finding the observable B equal to β, that is, of finding the system in the eigenstate corresponding to the eigenvalue $B = \beta$.

To the eigenvalues correspond eigenfunctions, in terms of which $P_\alpha \lesssim \beta$ can be computed. In particular, when α is a discrete eigenvalue of A, and the operators depend only on x and p_x, the probability $P_\alpha(\beta)$ is postulated as in Eq. (3), where

$$P\alpha(\beta) = \left| \int_{-\infty}^{\infty} dx\, v^*(x, \beta) u(x, \alpha) \right|^2 \tag{3}$$

$u(x,\alpha)$ is the eigenfunction corresponding to $A = \alpha$; $v(x,\beta)$ is the eigenfunction corresponding to $B = \beta$; and the * denotes the complex conjugate. The integral in Eq. (3) is called the projection of $u(x,\alpha)$ on $u(x,\beta)$. The quantity $|u(x,\alpha)|^2\, dx$ is the probability that the system, known to be in the eigenstate $A = \alpha$, will be found in the interval x to $x + dx$. *See* EIGENVALUE (QUANTUM MECHANICS).

Wave function. When the system is known to be in the eigenstate corresponding to $A = \alpha$, the eigenfunction $u(x,\alpha)$ is the wave function; that is, it is the function whose projection on an eigenfunction $v(x,\beta)$ of any observable B gives the probability of measuring $B = \beta$. The wave function $\psi(x)$ may be known exactly; in other words, the state of the system may be known as exactly as possible (within the limitations of uncertainty and complementarity), even though $\psi(x)$ is not the eigenfunction of a known operator. This circumstance arises because the wave function obeys Schrödinger's wave equation. Knowing the value of $\psi(x)$ at time $t = 0$, the wave equation completely determines $\psi(x)$ at all future times. In general, however, if $\psi(x,0) = u(x,\alpha)$, that is, if $\psi(x,t)$ is an eigenfunction of A at $t = 0$, then $\psi(x,t)$ will not be an eigenfunction of A at later times $t > 0$.

A system described by a wave function is said to be in a pure state. Not all systems are described by wave functions, however. For example, a beam of hydrogen atoms streaming out of a small hole in a hydrogen discharge tube can be regarded as a statistical ensemble or mixture of pure states oriented with equal probability in all directions.

Schrödinger equation. Equation (4) describes a plane wave of frequency f, wavelength λ, and amplitude $A(\lambda)$, propa-

$$\psi(x,t) = A(\lambda)\exp\left[2\pi i\left(\frac{x}{\lambda} - ft\right)\right] \tag{4}$$

gating in the positive x direction. The previous discussion concerning wave-particle duality suggests that this is the form of the wave function for a beam of free particles moving in the x direction with momentum $p = p_x$, with Eq. (2) specifying the connections between f, λ, and E, p. Differentiating Eq. (4), it is seen that Eqs. (5) hold. Since for a free particle $E = p^2/2m$, it follows also that Eq. (6) is valid.

$$p_x\psi = \frac{h}{\lambda}\psi = \frac{\hbar}{i}\frac{\partial\psi}{\partial x} \tag{5a}$$

$$E\psi = hf\psi = -\frac{\hbar}{i}\frac{\partial\psi}{\partial t} \tag{5b}$$

$$\frac{-\hbar^2}{2m}\frac{\partial^2\psi}{\partial x^2} = -\frac{\hbar}{i}\frac{\partial\psi}{\partial t} \tag{6}$$

See WAVE MOTION.

Equation (6) holds for a plane wave of arbitrary λ, and therefore for any superposition of waves of arbitrary λ, that is, arbitrary p_x. Consequently, Eq. (6) should be the wave equation obeyed by the wave function of any particle moving under no forces, whatever the projections of the wave function on the eigenfunctions of p_x. Equations (5) and (6) further suggest that for a particle whose potential energy $V(x)$ changes, in other words, for a particle in a conservative force field, $\psi(x,t)$ obeys Eq. (7).

$$\frac{-\hbar^2}{2m}\frac{\partial^2\psi}{\partial x^2} + V(x)\psi = -\frac{\hbar}{i}\frac{\partial\psi}{\partial t} \tag{7}$$

Equation 7 is the time-dependent Schrödinger equation for a one-dimensional (along x), spinless particle. Noting Eq. (5b), and observing that Eq. (7) has a solution for the form of Eq. (8), it is inferred that $\psi(x)$ of Eq. (8) obeys the time-dependent Schrödinger equation, Eq. (9).

$$\psi(x,t) = \psi(x)\exp(-iEt/\hbar) \tag{8}$$

$$\frac{-\hbar^2}{2m}\frac{\partial^2\psi}{\partial x^2} + V(x)\psi = E\psi \tag{9}$$

See FORCE.

Equation (9) is solved subject to reasonable boundary conditions, for example, that ψ must be continuous and must not become infinite as x approaches $\pm\infty$. These boundary conditions restrict the values of E for which there exist acceptable solutions $\psi(x)$ to Eq. (9), the allowed values of E depending on $V(x)$. In this manner, the allowed energies of atomic hydrogen listed in the earlier discussion of quantization are obtained.

The forms of Eqs. (5a), (7), and (9) suggest that the classical observable p_x, must be replaced by the operator $(\hbar/i)\,(\partial/\partial x)$. With this replacement, Eq. (10) holds. In other words, whereas

$$(xp_x - p_x x)\psi = i\hbar\psi \tag{10}$$

the classical canonically conjugate variables x and p_x are numbers, obeying the commutative law in Eq. (11a), the quantum-mechanical quantities x and p_x are noncommuting operators, obeying Eq. (11b).

$$xp_x - p_x x = 0 \tag{11a}$$

$$xp_x - p_x x = i\hbar \tag{11b}$$

Correspondence principle. Since classical mechanics and Maxwell's electromagnetic theory accurately describe macroscopic phenomena, quantum mechanics must have a classical limit in which it is equivalent to the older classical theories. Although there is no rigorous proof of this principle for arbitrarily complicated quantum-mechanical systems, its validity is well established by numerous illustrations.

Edward Gerjuoy

Quantum numbers

The quantities, usually discrete with integer or half-integer values, which are needed to characterize a physical system of one or more atomic or subatomic particles. Specification of the set of quantum numbers serves to define such a system or, in other words, to label the possible states the system may have. In general, quantum numbers are obtained from conserved quantities determinable by performing symmetry transformations consisting of arbitrary variations of the system which leave the system unchanged. For example, since the behavior of a set of particles should be independent of the location of the origin in space and time (that is, the symmetry operation is translation in space-time), it follows that momentum and energy are rigorously conserved. *See* SYMMETRY LAWS (PHYSICS).

In general, each physical system must be studied individually to find the symmetry transformations, and thus the conserved quantities and possible quantum numbers. The quantum numbers themselves, that is, the actual state labels, are usually the eigenvalues of the physical operators corresponding to the conserved quantities for the system in question. *See* EIGENVALUE (QUANTUM MECHANICS); ELEMENTARY PARTICLE; PARITY (QUANTUM MECHANICS).

It is not necessary that the conserved quantity be "quantized" in order to be regarded as a quantum number; for example, a free particle possesses energy and momentum, both of which can have values from a continuum but which are used to specify the state of the particle.

Kenneth E. Lassila

Quantum solids

Solids in which the quantum-mechanical wave functions of individual atoms overlap the wave functions of neighboring atoms in the quantum ground state of the system. The spatial extension of the wave functions is called zero-point motion since the location of the atoms cannot be determined within the width of the wave function even at the absolute zero of temperature. The zero-point motion becomes large when the constituent atoms have small mass and the attractive force between them is small. The isotopes of helium have the weakest interaction of any atom or molecule and a small mass. The root-mean-square zero-point motion in those solids is approximately 25% of the mean distance between atoms. This zero-point motion results in some very unusual properties that are manifestations of the quantum statistical mechanics of many-particle systems. *See* HELIUM; INTERMOLECULAR FORCES; QUANTUM MECHANICS; QUANTUM STATISTICS.

Melting pressure of ^{3}He and ^{4}He. Unlike all other solids, helium cannot be solidified by simply cooling liquid helium to low temperatures. The liquid must be compressed to at least 25 bar for ^{4}He and 29 bar for ^{3}He in order to freeze (1 bar = 10^5 Pa $\approx$ 1 atm). In further contrast with other solids, the pressure required to freeze the isotopes of helium increases with decreasing temperature below a certain temperature.

Quantum statistics in solid phases. Quantum statistics becomes important when the wave functions of individual atoms or molecules overlap. ^{3}He atoms are "Fermi" particles (fermions) for which only one atom can occupy a given quantum state. In solid ^{3}He, the wave functions are much more localized than in liquid phase, so that the magnetization does not decrease until much lower temperatures. *See* FERMI-DIRAC STATISTICS.

^{4}He atoms are "Bose" particles (bosons) in which any number of atoms can occupy the same quantum state. At low temperatures, this leads to Bose-Einstein condensation to a highly ordered, low-entropy state in which a large fraction of the atoms are in the same quantum state below a critical temperature. Bose-Einstein condensation is responsible for the remarkable states of superfluidity and superconductivity. In the solid phase, the greater localization suppresses Bose-Einstein condensation. *See* BOSE-EINSTEIN CONDENSATION; BOSE-EINSTEIN STATISTICS; LIQUID HELIUM; SUPERCONDUCTIVITY; SUPERFLUIDITY.

Molecular quantum solids. The molecules H_2, HD, D_2, N_2, and a few others also have sufficient zero-point motion that some quantum solid properties might be expected. In the various isotopic combinations of solid hydrogen, these are manifested through the influence of quantum tunneling on nuclear magnetic resonance (NMR) relaxation times and on the orientational order of the molecules. *See* HYDROGEN; NUCLEAR MAGNETIC RESONANCE (NMR); TUNNELING IN SOLIDS. John M. Goodkind

Quantum statistics The statistical description of particles or systems of particles whose behavior must be described by quantum mechanics rather than by classical mechanics. As in classical, that is, Boltzmann statistics, the interest centers on the construction of appropriate distribution functions. However, whereas these distribution functions in classical statistical mechanics describe the number of particles in given (in fact, finite) momentum and positional ranges, in quantum statistics the distribution functions give the number of particles in a group of discrete energy levels. In an individual energy level there may be, according to quantum mechanics, either a single particle or any number of particles. This is determined by the symmetry character of the wave functions. For antisymmetric wave functions only one particle (without spin) may occupy a state; for symmetric wave functions, any number is possible. Based on this distinction, there are two separate distributions, the Fermi-Dirac distribution for systems described by antisymmetric wave functions and the Bose-Einstein distribution for systems described by symmetric wave functions. *See* BOLTZMANN STATISTICS; BOSE-EINSTEIN STATISTICS; EXCLUSION PRINCIPLE; FERMI-DIRAC STATISTICS; KINETIC THEORY OF MATTER; QUANTUM MECHANICS; STATISTICAL MECHANICS. Max Dresden

Quantum teleportation A way to transfer the state of a quantum system over large distances by employing entanglement. Entanglement is a nonclassical connection between objects that Albert Einstein called "spooky."

The concept of teleportation is frequently utilized in the literature of science fiction to overcome limitations imposed on space travel by the laws of physics. In the standard science fiction approach, the sender, Alice, scans the object to be teleported in order to read out all the information needed to describe it. She then sends that information to the receiver, Bob, who uses this information to reconstitute the object, not necessarily from the same material as that of the original. However, according to quantum mechanics, it is impossible to succeed in this way. If only one individual object is at hand, it is impossible to determine its quantum state by measurement. The quantum state represents all that can be known about the object, that is, all possible (in general, probabilistic) predictions that can be made about future observations of the object.

In fact, it is quantum mechanics that makes quantum teleportation possible using a feature of the theory, quantum entanglement. It is important to realize that there are significant differences between teleportation as portrayed in science fiction and quantum teleportation as realized in the laboratory. In the experiments, what is teleported is not the substance an object is made of but the information it represents.

Quantum entanglement. Entangled quantum states as used in teleportation were introduced into the discussion of the foundations of quantum mechanics by Einstein, Boris Podolsky, and Nathan Rosen in 1935. In the same year, Erwin Schrödinger introduced the notion of entanglement, which he called the essence of quantum mechanics.

The first teleportation experiments produced a pair of photons in a singlet polarization state. This means that neither photon enjoys a state of well-defined polarization; each one of the photons on its own is maximally unpolarized. Yet, when one of the two photons is subject to a polarization measurement, it assumes one specific polarization. That specific experimental outcome is completely random. As a consequence of the two photons being in the entangled singlet state, the other photon is instantly projected into a state orthogonal to that of the first photon. The fact that the measurement result on the second photon can be perfectly predicted on the basis of the measurement result of the first photon, even as neither one carries a well-defined quantum state, is known as the Einstein-Podolsky-Rosen paradox. In 1964 John Bell showed that these perfect correlations cannot be understood on the basis of properties that the entangled photons carry individually before the measurement. The resulting conflict between the philosophical position of local realism and the predictions of quantum mechanics is known as Bell's theorem. *See* HIDDEN VARIABLES; PHOTON; POLARIZATION OF WAVES.

Prospects. While the teleportation distance in the first experiments was of the order of 1 m (3 ft), experiments in 2004 extended the distance to the order of 600 m (2000 ft), and there are plans to perform such experiments over much larger distances and even from a satellite down to laboratories on the ground.

There is intense research in the development of both quantum communication networks and quantum computers. Future quantum computers would use individual quantum states, for example those of atoms, to represent information in so-called quantum bits. They are expected to allow some algorithms to be performed with significantly higher speed than any existing computers. Quantum teleportation would allow the transfer of the quantum output of one quantum computer to the quantum input of another quantum computer. *See* QUANTUM COMPUTATION; QUANTUM MECHANICS; QUANTUM THEORY OF MEASUREMENT. Anton Zeilinger

Quantum theory of matter The microscopic explanation of the properties of condensed matter, that is, solids and liquids, based on the fundamental laws of quantum mechanics. Without the quantum theory, some properties of matter such as magnetism and superconductivity have no explanation at all, while for others only a phenomenological description can be obtained. With the theory, it is at least possible to comprehend what is needed to approach a complete understanding.

The theoretical problem of condensed matter—large aggregates of elementary particles with mutual interactions—is the quantum-mechanical many-body problem: an enormous number, of order 10^{23}, of constituent particles in the presence of a heat bath and interacting with each other according to quantum-mechanical laws. What makes the quantum physics of matter different from the traditional quantum theory of elementary particles is that the fundamental constituents (electrons and ions) and their interactions (Coulomb interactions) are known but the solutions of the appropriate quantum-mechanical equations are not. This situation is not due to the lack of a sufficiently large computer, but is caused by the fact that totally new structures, such as crystals, magnets, ferroelectrics, superconductors, liquid crystals, and glasses, appear out of the complexity of the interactions among the many constituents. The consequence is that entirely new conceptual approaches are required to construct predictive theories of matter. The usual technique for approaching the quantum many-body problem for a condensed-matter system is to try to reduce the huge number of variables (degrees of freedom) to a number which is more manageable but still can describe the essential physics of the phenomena being studied. *See* CRYSTAL; FERROELECTRICS; GLASS; LIQUID CRYSTALS;

MAGNETIC MATERIALS; QUANTUM MECHANICS; SOLID-STATE PHYSICS; SUPERCONDUCTIVITY. Elihu Abrahams

Quantum theory of measurement The attempt to reconcile the counterintuitive features of quantum mechanics with the hypothesis that quantum mechanics is in principle a complete description of the physical world, even at the level of everyday objects. A paradox arises because, at the atomic level where the quantum formalism has been directly tested, the most natural interpretation implies that where two or more different outcomes are possible it is not necessarily true that one or the other is actually realized, whereas at the everyday level such a state of affairs seems to conflict with direct experience.

The resolution of this paradox that is probably most favored by practicing physicists proceeds in two stages. At stage 1, it is pointed out that, quite generically, whenever the quantum formalism appears to generate a superposition of macroscopically distinct states it is impossible to demonstrate the effects of interference between them. The reasons for this claim include the facts that the initial state of a macroscopic system is likely to be unknown in detail; the initial state has extreme sensitivity to random external noise; and most important, merely by virtue of its macroscopic nature any such system will rapidly have its quantum-mechanical state correlated (entangled) with that of its environment in such a way that no measurement on the system alone (without a simultaneous measurement of the complete state of the environment) can demonstrate any interference between the two states in question—a result often known as decoherence. Thus, it is argued, the outcome of any possible experiment on the ensemble of macroscopic systems prepared in this way will be indistinguishable from that expected if each system had actually realized one or the other of the two macroscopically distinct states in question. Stage 2 of the argument (often not stated explicitly) is to conclude that if this is indeed true, then it may be legitimately asserted that such realization of a definite macroscopic outcome has indeed taken place by this stage.

Most physicists agree with stage 1 of the argument. However, not all agree that the radical reinterpretation of the meaning of the quantum formalism which is implicit at stage 2 is legitimate; that is, an interpretation in terms of realization, by each individual system, of one alternative or the other, forbidden at the atomic level by the observed phenomenon of interference, is allowed once, on going to the macroscopic level, the phenomenon disappears. Consequently, various alternative interpretations have been developed. *See* QUANTUM MECHANICS. Anthony J. Leggett

Quark-gluon plasma A predicted state of nuclear matter containing deconfined quarks and gluons. According to the theory of strong interactions, called quantum chromodynamics, hadrons such as mesons and nucleons (the generic name for protons and neutrons) are bound states of more fundamental objects called quarks. The quarks are confined within the individual hadrons by the exchange of particles called gluons. However, calculations indicate that, at sufficiently high temperatures or densities, hadronic matter should evolve into a new phase of matter containing deconfined quarks and gluons, called a quark-gluon plasma when hot, or quark matter when cold and dense. Such a state of matter is thought to have existed briefly in the very early universe until about 10 microseconds after the big bang, and its cold form might exist today inside the dense cores of neutron stars. One important prediction is that the mechanism, called spontaneous chiral symmetry breaking, that is thought to be responsible for 98% of the mass of the nucleon under normal conditions is not operative in the quark-gluon plasma. *See* BIG BANG THEORY; HADRON; NEUTRON STAR; QUANTUM CHROMODYNAMICS.

The discovery and study of a new state of matter requires a means for producing it under controlled laboratory conditions. The extreme conditions of heat and compression required for the transition from the hadronic matter to the quark-gluon plasma phase can be achieved on Earth only by collisions of beams of heavy ions such as nuclei of gold or lead (although lighter nuclei can be used) with other heavy nuclei at extremely high energies. Calculations using the lattice gauge model of quantum chromodynamics indicate that energy densities of at least 1–2 GeV/fm^3 (1 fm = 1 femtometer = 10^{-15} m, 1 GeV = 1000 MeV), about 10 times that found in ordinary nuclear matter, must be produced in the collision for plasma formation to occur. *See* NUCLEAR REACTION; RELATIVISTIC HEAVY-ION COLLISIONS.

A systematic search for the quark-gluon plasma has been conducted at the Relativistic Heavy Ion Collider (RHIC) since 2000. Experiments at this accelerator facility colliding beams of gold nuclei with energies of up to 100 GeV per nucleon have found compelling evidence for the formation of a novel state of matter which exhibits many of the predicted properties of the quark-gluon plasma. The final state is found to be compatible with the assumption that particles are emitted from a thermally and chemically equilibrated source. The yield of energetic hadrons is strongly suppressed, indicating a high opacity of the medium to quarks and gluons. The intensity of the measured collective flow pattern among the emitted hadrons indicates that the produced matter is a nearly ideal fluid with an extremely low viscosity near the theoretical limit. Differences observed in the flow patterns of baryons and mesons suggest that the flow was generated in a phase in which quarks were not confined. Detailed studies of weakly interacting or rarely produced probes, such as lepton pairs and hadrons containing *c* quarks, are in progress. Experiments with nuclei of even higher energy are in preparation at the Large Hadron Collider at the European Center for Particle Physics (CERN). These experiments should provide critical tests of the theory of the strong interaction and illuminate the earliest moments of the universe. *See* ELEMENTARY PARTICLE; GLUONS; PARTICLE ACCELERATOR; QUARKS. Berndt Müller

Quarks The basic constituent particles of which elementary particles are understood to be composed. Theoretical models built on the quark concept have been very successful in understanding and predicting many phenomena in the physics of elementary particles.

The study of the elastic scattering of electrons on protons demonstrated that the proton has a finite form factor, that is, a finite radial extent of its electric charge and magnetic moment distributions. It was plausible that the charge cloud which constitutes the proton is a probability distribution of some smaller, perhaps pointlike constituents, just as the charge cloud of an atom was learned to be the probability distribution of electrons. Subsequent high-energy, deep inelastic scattering experiments of electrons on protons, leading to meson production, revealed form factors corresponding to pointlike constituents of the proton. These proton constituents, first referred to as partons, are now understood to include the constituent quarks of the proton.

These high-energy collisions also produced an abundance of resonance states, equivalent to short-lived particles. The spectroscopy of these hadronic states revealed an order and symmetry among the observed hadrons that could be interpreted in terms of representations of the SU(3) symmetry group. This in turn is interpreted as a consequence of the grouping of elementary constituents of fractional electric charge in pairs and triplets to form the observed particles. The general features of the quark model of hadrons have withstood the tests of time, and the static properties of hadrons are consistent with predictions of this model. *See* SYMMETRY LAWS (PHYSICS); UNITARY SYMMETRY.

Thus, the proton and neutron are not fundamental constituents of matter, but each is composed of three quarks, very much as the nuclei of ^{3}H and ^{3}He are made of protons and neutrons, and the molecules of NO_2 and N_2O are made of oxygen and nitrogen atoms.

Properties of quarks

Flavor	Mass[†], GeV/c^2	Electric charge[‡]	Baryon number	Spin[§]	Isotopic spin	Strangeness	Charm
u	0.0015–0.004	+2/3	+1/3	1/2	1/2	0	0
d	0.004–0.008	−1/3	+1/3	1/2	1/2	0	0
c	1.15–1.35	+2/3	+1/3	1/2	0	0	+1
s	0.080–0.130	−1/3	+1/3	1/2	0	−1	0
t	171.2±2.1[¶]	+2/3	+1/3	1/2	0	0	0
b	4.1–4.4	−1/3	+1/3	1/2	0	0	0

†As the mass of baryons composed of quarks is strongly influenced by the gluons binding the quarks, and as free quarks are not observed, the masses are theoretical estimates.
‡Charge is in units of the magnitude of the charge of an electron, 1.6×10^{-19} coulomb.
§Spin is in units of Planck's constant divided by 2π, written as $\hbar$.
¶The top quark mass is deduced from experimental measurements of its decay dynamics.

There are two kinds (or "flavors") of quarks of very low mass of which the proton, neutron, and pions are composed, and a third, more massive quark which is a constituent of "strange" particles, including the K mesons and hyperons such as the Λ^0. These are known as the up quark (u), the down quark (d), and the strange quark (s). Baryons are composed of three quarks, for example the proton (uud), neutron (udd), Λ^0 (uds), and Ξ^- (dss). Antiparticles such as antiprotons are formed by the antiquarks of those forming the particle, for example, the antiproton $\bar{p}(\bar{u}\bar{u}\bar{d})$. Mesons are composed of a quark-antiquark pair, such as the $\pi^+(u\bar{d})$, $\pi^-(\bar{u}d)$, $K^+(u\bar{s})$, and $K^-(\bar{u}s)$. *See* BARYON; HYPERON; MESON; STRANGE PARTICLES.

The quantum numbers of quarks are added to give the quantum numbers of the elementary particle which they form on combination. The unit of electrical charge of a quark is +2/3 or −1/3 of the charge on a proton (1.6×10^{-19} coulomb), and the baryon number of each quark is +1/3 (see table). The charge, baryon number, and so forth, of each antiquark are just the negative of that for each quark.

During the 1970s, experiments at electron-positron colliders and proton accelerators detected a relatively long-lived (that is, very narrow, in energy) resonant state of about 3.1 GeV total energy. This was interpreted as evidence for a new quark, the charm (c) quark, produced as a quark-antiquark resonance analogous to the ϕ. The discovery of this J/ψ resonance was followed by the observation and study of meson systems, now labeled D mesons, containing a single c or quark (paired with an antiquark of another flavor), as well as baryon states containing these quarks. *See* CHARM; J/PSI PARTICLE; PARTICLE ACCELERATOR.

A few years later, experiments with higher-energy proton beams, studying the spectra of muon-antimuon pairs at the Fermi National Accelerator Laboratory, discovered a more massive, narrow resonant state at about 9.4 GeV, which was labeled the Υ (upsilon). This was interpreted as evidence for a more massive quark, the b (bottom) quark. Subsequent experiments at proton and electron accelerators confirmed the existence of the b quark and also observed a corresponding family of meson resonant states, now referred to as B mesons.

During the 1990s, experiments observing collisions of protons and antiprotons at an energy of 1.8 TeV in the center of mass established the existence of the t (top) quark, primarily through analysis of its decay to a B meson and a W intermediate vector boson. The t mass of 171.2 ± 2.1 GeV/c^2 (about the mass of a tungsten atom) is so great that its weak decay through this channel is very fast, and mesonic states of the t and $\bar{t}$ quark (analogous to the Υ, the J/ψ, and the ϕ) are not observed, although the observed t's are from the production of pairs. *See* INTERMEDIATE VECTOR BOSON.

Quarks are understood to have a spin of 1/2; that is, their intrinsic angular momentum is $\hbar/2$ (where $\hbar$ is Planck's constant h divided by 2π), just as for the electron and muon. A problem arose when the structure of observed baryons required two or, in some cases, three quarks of the same flavor in the same quantum state, a situation forbidden for spin-1/2 particles by the Pauli exclusion principle. In order to accommodate this contradiction, a new quantum variable, arbitrarily labeled color, was introduced; the idea is that each quark is red, green, or blue (and the antiquarks, antired, and so forth). The color quantum number then breaks the degeneracy and allows up to three quarks of the same flavor to occupy a single quantum state. Confirmation of the color concept has been obtained from experiments with electron-positron storage rings, and the theory of quantum chromodynamics (QCD), based on this concept, has been developed. According to quantum chromodynamics, hadrons must be colorless; for example, baryons must consist of a red, a green, and a blue quark, and mesons of a quark-antiquark pair of the same color (for example, a red quark and an anti-red antiquark). *See* COLOR (QUANTUM MECHANICS); EXCLUSION PRINCIPLE; SPIN (QUANTUM MECHANICS).

The field quanta of quantum chromodynamics are gluons, massless, spin-1 quanta which interact with quarks. This is very analogous to the manner in which photons, the quanta of electromagnetic interaction, interact with particles containing electric charge and are responsible for electromagnetic forces. The QCD theory is part of the now widely accepted standard model of elementary particle interactions, together with the electroweak theory. Experiments have increasingly confirmed details of the standard model to the extent that most physicists are confident that it is fundamentally correct. *See* ELECTROWEAK INTERACTION; GLUONS; QUANTUM CHROMODYNAMICS; STANDARD MODEL.

There are three sets, or "generations," of quarks and leptons. Each generation contains a charged lepton (electron, muon, or tau lepton); a correspoding neutrino; a charge −1/3 quark color triad; and a charge +2/3 quark triad. *See* LEPTON; NEUTRINO.

Quarks and the theory of quantum chromodynamics are now firmly established as cornerstones of the standard model of elementary particles (together with the electroweak theory, charged leptons, neutrinos, and so forth). However, unanswered questions remain.

The advanced string and M theories have the property of supersymmetry, which demands that every spin-1/2 quark and lepton must have a partner with integral spin. As of 2008, no experimental evidence for any of these supersymmetric (SUSY) particles had been found. *See* SUPERSTRING THEORY; SUPERSYMMETRY.

Contemporary theories also predict that there exists one or more massive particles of integral spin, the Higgs paticles, responsible for the rest masses of the quarks and charged leptons. Again, the lack of evidence suggests that the Higgs particles must also have a rest mass of over 100 GeV/c^2, if they exist. *See* HIGGS BOSON.

Quarks may be permanently stable against decay via the weak interaction; however, it is also possible that quarks spontaneously decay to leptons. Intensive searches for the decay of the proton (into a neutral pion and a positron, for example) have been negative, setting a lower limit of over 10^{32} years for the proton lifetime. However, the apparent asymmetry of the universe between matter and antimatter (there is, at present, no evidence for primordial antimatter) suggests that antiprotons, for example, may spontaneously decay (or transform) more readily. *See* ANTIMATTER; PROTON.

Some theories have postulated that quarks are composed of smaller constituents, just as other objects that were originally believed to be fundamental subsequently were found to have internal structure. So far, all observations are compatible with the quarks as point objects, like the electron. Lawrence W. Jones

Quarrying The process of extracting rock material from the outer portions of the Earth's crust. The shape of the excavated space is dependent on the geologic condition and orientation of the rock mass being quarried. This excavated space is either an open pit or an underground cavity that is accessed by tunnel or shaft. An underground quarry is usually called a mine, even though the process for making the excavation is usually the same as on the surface. *See* OPEN-PIT MINING; ROCK; UNDERGROUND MINING.

The actual quarrying technique depends on the specific use of the excavated rock. Rock that is excavated for use as dimension stone—relatively flawless blocks or slabs of stone—is carefully extracted as large, 20-ton blocks. These blocks are sawed into thin (3-cm or 1.2-in.) slabs, and then cut to size and polished as finished pieces. Rock that is quarried for crushed stone is used either for construction aggregate or as an industrial mineral source. Crushed stone is broken in place by drilling and blasting into manageable-sized pieces for transport by truck or conveyor to a crushing plant, where it is crushed and screened into segregated, sized particles. *See* STONE AND STONE PRODUCTS.

Lance Mead

Quartz The most common oxide on the Earth's surface, constituting 12% of the crust by volume. Quartz is a crystalline form of silicon dioxide (SiO_2). Among the igneous rocks, quartz is especially common within granites, granodiorites, pegmatites, and rhyolites. In addition, quartz can be observed in low- to high-grade metamorphic rocks, including phyllites, quartzites, schists, granulites, and eclogites. Because hydrothermal fluids are enriched in dissolved silica, the passage of fluids through rock fractures results in the emplacement of quartz veins. *See* GRANITE; GRANODIORITE; IGNEOUS ROCKS; METAMORPHIC ROCKS; PEGMATITE; RHYOLITE.

Once quartz has formed, it persists through erosional reworking because of its low solubility in water (parts per million) and its high mechanical hardness (7 on Mohs scale). Consequently, quartz becomes increasingly concentrated in beach sands as they mature, and it is a major component of sandstone. In sedimentary environments, quartz also forms as the final crystallization product during silica diagenesis; amorphous silica on the sea floor that derives from the skeletons of diatoms, radiolarians, and sponges will transform to quartz upon prolonged exposure to increased temperatures ($\leq$300°C or 572°F) and pressures ($\leq$2 kilobars or 200 pascals) after burial. *See* DIAGENESIS; HARDNESS SCALES; SANDSTONE.

As with virtually all silicates, the atomic framework of the quartz structure consists of Si^{4+} cations that are tetrahedrally coordinated by oxygen anions (O^{2-}). Every oxygen anion is bonded to two silicon cations, so that the tetrahedral units are corner-linked to form continuous chains. In low-temperature quartz (or α-quartz), two distinct tetrahedral chains spiral about the crystallographic *c* axis.

Although the silica tetrahedra can be depicted as spirals about the *c* axis in a left-handed sense, right-handed quartz crystals are found in nature as abundantly as are left-handed crystals. These enantiomorphic varieties are known as the Brazil twins of quartz, and they may be distinguished by crystal shape (corresponding crystal faces occur in different orientations) and by opposite optical activities. *See* CRYSTAL OPTICS.

Impurity concentrations in natural α-quartz crystals usually fall below 1000 parts per million. The violet and yellow hues observed in amethyst and citrine are associated with Fe, and black smoky quartz contains Al. The white coloration of milky quartz reflects light scattering off minute fluid inclusions, and the pink tint in rose quartz is believed to arise from fine-scale intergrowths of a pegmatitic mineral called dumortierite [$Al_{27}B_4Si_{12}O_{69}(OH)_3$]. *See* AMETHYST; DUMORTIERITE.

Quartz is used predominantly by the construction industry as gravel and as aggregate in concrete. In addition, quartz is important in advanced technologies. Quartz is piezoelectric and has an extremely high quality factor. The high quality factor means that a bell made of quartz would resonate (ring) for a very long time. This property, combined with its piezoelectric behavior, makes quartz the perfect crystal for oscillators in watches.

Compression of α-quartz perpendicular to the *c* axis creates an electrostatic charge, and this property is exploited in oscillator plates in electronic components. Large flawless crystals of quartz are routinely synthesized for oscillators and for prisms in laser optic systems. Quartz also is employed in abrasives, fluxes, porcelains, and paints. *See* CONCRETE; OSCILLATOR; PIEZOELECTRICITY.

Peter J. Heaney

Quartz clock A clock that makes use of the piezoelectric property of a quartz crystal. When a quartz crystal vibrates, a difference of electric potential is produced between two of its faces. The actual displacement of the surface is generally only a few atom diameters. The crystal has a natural frequency of vibration that depends on its size and shape. If it is placed in an oscillating electric circuit having nearly the same frequency as the crystal, it is caused to vibrate at its natural frequency, and the frequency of the entire circuit becomes the same as the natural frequency of the crystal. *See* OSCILLATOR; PIEZOELECTRICITY.

In the quartz oscillator, this natural frequency may be used to produce such other frequencies as 1 or 5 MHz. A clock displaying the time of day can also be driven by using one of these frequencies.

The natural frequency of a quartz crystal is nearly constant if precautions are taken when it is cut and polished and it is maintained at nearly constant temperature and pressure. After a crystal has been placed in operation, its frequency usually varies slowly as a result of physical changes. If allowance is made for changes, laboratory quartz-crystal clocks may run for a year with accumulated errors of less than a few thousandths of a second. However, quartz crystals typically used in watches may ac- cumulate errors of several tens of seconds in one year. *See* WATCH.

The advantage of quartz clocks is that they are reliably relatively accurate and inexpensive, and easy to use in various applications such as computers and microprocessors. Thus, despite their inaccuracy relative to some other types of clocks, they enjoy wide popularity, particularly in applications requiring accurate timekeeping over a relatively short time span. In these applications, the rates and epochs of the quartz clocks may be readjusted periodically to account for possible accumulated errors. *See* CLOCK; COMPUTER; HOROLOGY; MICROPROCESSOR; TIME.

Dennis D. McCarthy

Quartzite A metamorphic rock consisting largely or entirely of quartz. Most quartzites are formed by metamorphism of sandstone; but some have developed by metasomatic introduction of quartz, SiO_2, often accompanied by other chemical elements, for example, metals and sulfur (ore quartzites). *See* METAMORPHIC ROCKS; METASOMATISM; SANDSTONE.

Pure sandstones yield pure quartzites. Impure sandstones yield a variety of quartzite types. The cement of the original sandstone is in quartzite recrystallized into characteristic silicate minerals, whose composition often reflects the mode of development.

Even the Precambrian quartzites correspond to types that are parallel to present-day deposits. *See* QUARTZ. T. F. W. Barth

Quasar An astronomical object that appears starlike on a photographic plate but possesses many other characteristics, such as a large redshift, that prove that it is not a star. The name quasar is a contraction of the term quasistellar object (QSO), which was originally applied to these objects for their photographic appearance. The objects appear starlike because their angular diameters are less than about 1 second of arc, which is the resolution limit of ground-based optical telescopes imposed by atmospheric effects.

Quasars were discovered in 1961 when it was noticed that very strong radio emission was coming from a localized direction in the sky that coincided with the position of a starlike object. When the positions of small-angular-diameter radio sources were accurately determined, the coincidence with starlike objects on optical photographs led to the discovery of a new, hitherto-unsuspected class of objects, the quasars. The full significance of the discovery was not appreciated until 1963, when it was noted that the hydrogen emission lines seen in the optical spectrum of the quasar 3C 273 were shifted by about 16% to the red from their normal laboratory wavelength. This redshift of the spectral lines is characteristic of galaxies whose spectra are redshifted because of the expansion of the universe and is not characteristic of stars in the Milky Way Galaxy.

The color of quasars is generally much bluer than that of most stars with the exception of white dwarf stars. The blueness of quasars as an identifying characteristic led to the discovery that many blue starlike objects have a large redshift and are therefore quasars. The quasistellar objects discovered this way turned out to emit little or no radio radiation and to be about 20 times more numerous than the radio-emitting quasistellar radio sources (QSSs). Why some should be strong radio emitters and most others not is unknown. Several orbiting x-ray satellites have found that most quasars also emit strongly at x-ray frequencies. Gamma rays have also been observed in many quasars. *See* GAMMA-RAY ASTRONOMY; X-RAY ASTRONOMY.

The emission from quasars varies with time. The shortest time scale of variability ranges from years to months at short radio wavelengths, to days at optical wavelengths, to hours at x-ray wavelengths. These different time scales suggest that the emissions from the different bands originate from different regions in the quasar. The rapid fluctuations indicate that there are some components in quasars that have diameters less than a light-hour or of the order of 10^9 km (10^9 mi), the size of the solar system. Very highly active quasars are sometimes referred to as optically violent variables (OVVs), blazars, or BL Lac's, after the prototype BL Lacertae, a well-known variable "star" that turned out to be a quasar. The optically violent variables have no or very weak emission lines in their optical spectrum.

The many similarities of the observed characteristics of quasars with radio galaxies, Seyfert galaxies, and BL Lacertae objects strongly suggest that quasars are active nuclei of galaxies. Quasars with large redshifts are spatially much more numerous than those with small redshifts. Because high-redshift objects are very distant and emitted their radiation at an earlier epoch, quasars must have been much more common in the universe about 10^{10} years ago. Observations with the *Hubble Space Telescope* have shown that this is the same epoch when galaxies are observed to be forming. Thus it is likely that quasars are associated with the birth of some galaxies.

More than 10^{53} J of energy are released in quasars over their approximately 10^6-year lifetime. Of the known energy sources, only gravitational potential energy associated with a mass about 10^9 times the mass of the Sun can provide this energy, but it is unknown how this gravitational energy produces jets of particles that are accelerated to very near the speed of light.

Several theories have been proposed for quasars. However, the most favored interpretation is that quasars are massive black holes surrounded by rapidly spinning disks of gas in the nuclei of some galaxies. The hot gas in the disk emits the x-ray and optical continuum, a heated halo around the disk produces the emission lines, and the relativistic radio jets are ejected along the rotation axis of the spinning disk. *See* ASTRONOMICAL SPECTROSCOPY; ASTROPHYSICS, HIGH-ENERGY; BLACK HOLE; INFRARED ASTRONOMY; NEUTRON STAR; RADIO ASTRONOMY. William Dent

Quasiatom A structure in which the nuclei of two atoms approach each other closely and their electrons are arranged in atomic orbitals characteristic of a single atom of atomic number equal to the sum of the nuclear charges. Quasiatoms can be formed for short times in atom-atom and ion-atom collisions when the nuclei are much closer than the mean orbital radius of the innermost K-shell electrons. The electrons are then bound in the electric field of both nuclear charges Z_1 and Z_2, which resembles the spherically symmetric $1/r^2$ Coulomb field of a single united atom having charge $Z_{ua} = Z_1 + Z_2$. *See* ATOMIC STRUCTURE AND SPECTRA.

An interesting effect is associated with quasiatoms with $Z >$ 173, in which the 1s binding energy is more than twice the electron rest mass, $E_{1s} > 2mc^2$. If a vacancy exists in this orbital, it is energetically favorable to create an electron-positron pair with the electron bound in this state. The positron would be repelled from the nucleus with kinetic energy equal to $E_{e^+} = |E_{1s}| - 2mc^2$. In the Dirac hole picture, in which the vacuum consists of a negative energy continuum ($E < -mc^2$) filled with electrons, the 1s level is said to fall into the negative-energy Dirac sea as Z increases above the critical value, $Z_{cr} = 173$. A 1s hole (vacancy) becomes embedded in the negative continuum as an unstable resonance state that decays in a time of $\sim 10^{-19}$ s to a bound electron and a spontaneously emitted monoenergetic positron.

The quantum electrodynamic vacuum in the presence of a bare supercritical nuclear charge is therefore unstable and decays to a fundamentally new charged vacuum, which consists of the nucleus with two 1s electrons (from the two spin orientations). At higher values of Z_{ua}, as additional quasiatomic levels enter the negative continuum, the charge of the quantum electrodynamic vacuum increases accordingly. If detected, spontaneous positron emission would represent the first observation of a phase transition in a gauge field theory. *See* ANTIMATTER; ELECTRON-POSITRON PAIR PRODUCTION; GAUGE THEORY; PHASE TRANSITIONS; POSITRON; QUANTUM ELECTRODYNAMICS; SUPERCRITICAL FIELDS. Thomas E. Cowan

Quasicrystal A solid with conventional crystalline properties but exhibiting a point-group symmetry inconsistent with translational periodicity. Like crystals, quasicrystals display discrete diffraction patterns, crystallize into polyhedral forms, and

Quasicrystals of an alloy of aluminum, copper, and iron, displaying an external form consistent with their icosahedral symmetry.

have long-range orientational order, all of which indicate that their structure is not random. But the unusual symmetry and the finding that the discrete diffraction pattern does not fall on a reciprocal periodic lattice suggest a solid that is quasiperiodic. Their discovery in 1982 contradicted a long-held belief that all crystals would be periodic arrangements of atoms or molecules.

It is easily shown that in two and three dimensions the possible rotations that superimpose an infinitely repeating periodic structure on itself are limited to angles that are $360°/n$, where n can be only 1, 2, 3, 4, or 6. Various combinations of these rotations lead to only 32 point groups in three dimensions, and 230 space groups which are combinations of the 14 Bravais lattices that describe the periodic translations with the allowed rotations. Until the 1980s, all known crystals could be classified according to this limited set of symmetries allowed by periodicity. Periodic structures diffract only at discrete angles (Bragg's law) that can be described by a reciprocal lattice, in which the diffraction intensities fall on lattice points that, like all lattices, are by definition periodic, and which has a symmetry closely related to that of the structure. *See* CRYSTAL; CRYSTALLOGRAPHY; X-RAY CRYSTALLOGRAPHY; X-RAY DIFFRACTION.

Icosahedral quasicrystals were discovered in 1982 during a study of rapid solidification of molten alloys of aluminum with one or more transition elements, such as manganese, iron, and chromium. Since then, many different alloys of two or more metallic elements have led to quasicrystals with a variety of symmetries and structures. The illustration shows the external polyhedral form of an icosahedral aluminum-copper-iron alloy.

The diffraction patterns of quasicrystals violate several predictions resulting from periodicity. Quasicrystals have been found in which the quantity n is 5, 8, 10, and 12. In addition, most quasicrystals exhibit icosahedral symmetry in which there are six intersecting fivefold rotation axes. Furthermore, in the electron diffraction pattern the diffraction spots do not fall on a (periodic) lattice but on what has been called a quasilattice. *See* ELECTRON DIFFRACTION.

John W. Cahn; Dan Shechtman

Quasielastic light scattering Small frequency shifts or broadening from the frequency of the incident radiation in the light scattered from a liquid, gas, or solid. The term quasielastic arises from the fact that the frequency changes are usually so small that, without instrumentation specifically designed for their detection, they would not be observed and the scattering process would appear to occur with no frequency changes at all, that is, elastically. The technique is used by chemists, biologists, and physicists to study the dynamics of molecules in fluids, mainly liquids and liquid solutions. It is often identified by a variety of other names, the most common of which is dynamic light scattering (DLS).

Several distinct experimental techniques are grouped under the heading of quasielastic light scattering (QELS). Photon correlation spectroscopy (PCS) is the technique most often used to study such systems as macromolecules in solution, colloids, and critical phenomena where the molecular motions to be studied are rather slow. This technique, also known as intensity fluctuation spectroscopy and, less frequently, optical mixing spectroscopy, is used to measure the dynamical constants of processes with relaxation time scales slower than about 10^{-6} s. For faster processes, dynamical constants are obtained by utilizing techniques known as filter methods, which obtain direct measurements of the frequency changes of the scattered light by utilizing a monochromator or filter much as in Raman spectroscopy. *See* RAMAN EFFECT; SCATTERING OF ELECTROMAGNETIC RADIATION.

Robert Pecora

Quaternary A period that encompasses at least the last 3,000,000 years of the Cenozoic Era, and is concerned with major worldwide glaciations and their effect on land and sea, on worldwide climate, and on the plants and animals that lived then. The Quaternary is divided into the Pleistocene Epoch and

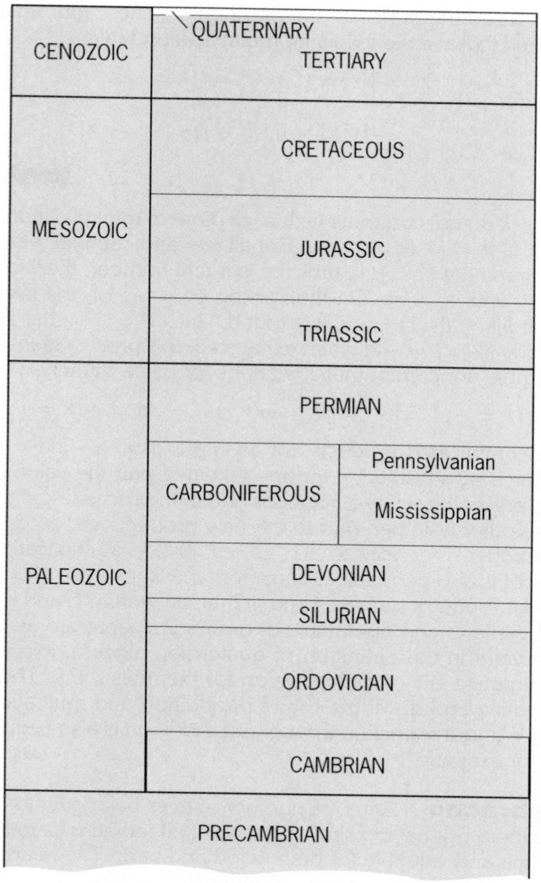

CENOZOIC	QUATERNARY	
	TERTIARY	
MESOZOIC	CRETACEOUS	
	JURASSIC	
	TRIASSIC	
PALEOZOIC	PERMIAN	
	CARBONIFEROUS	Pennsylvanian
		Mississippian
	DEVONIAN	
	SILURIAN	
	ORDOVICIAN	
	CAMBRIAN	
PRECAMBRIAN		

Holocene. The universal term Pleistocene is gradually replacing Quaternary; Holocene involves the last 7000 years since the Pleistocene. *See* CENOZOIC; GLACIAL EPOCH; HOLOCENE; PLEISTOCENE.

Sidney E. White

Quaternary ammonium salts Analogs of ammonium salts in which organic radicals have been substituted for all four hydrogens of the original ammonium cation. Substituents may be alkyl, aryl, or aralkyl, or the nitrogen may be part of a ring system. Such compounds are usually prepared by treatment of an amine with an alkylating reagent under suitable conditions. They are typically crystalline solids which are soluble in water and are strong electrolytes. Treatment of the salts with silver oxide, potassium hydroxide, or an ion-exchange resin converts them to quaternary ammonium hydroxides, which are very strong bases, as shown in the reaction below.

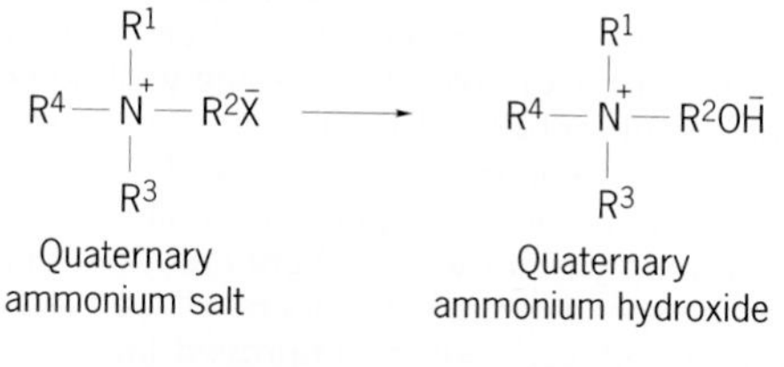

$$R^4\text{—}\overset{R^1}{\underset{R^3}{N^+}}\text{—}R^2\,X^- \longrightarrow R^4\text{—}\overset{R^1}{\underset{R^3}{N^+}}\text{—}R^2\,OH^-$$

Quaternary ammonium salt — Quaternary ammonium hydroxide

Some quaternary ammonium salts have found use as water repellents, fungicides, emulsifiers, paper softeners, antistatic agents, and corrosion inhibitors. *See* AMINE; AMMONIUM SALT; SURFACTANT.

Paul E. Fanta

Quaternions An associative, noncommutative algebra based on four linearly independent units or basal elements. Quaternions were originated in 1843, by W. R. Hamilton.

The four linearly independent units in quaternion algebra are commonly denoted by 1, i, j, k, where 1 commutes with i, j, k

and is called the principal unit or modulus. These four units are assumed to have the following multiplication table:

$$1^2 = 1i^2 = j^2 = k^2 = ijk = -1$$

$$i(jk) = (ij)k = ijk$$

$$1i = i1 \qquad 1j = j1 \qquad 1k = k1$$

The i, j, k do not commute with each other in multiplication, that is, $ij \neq ji$, $jk \neq kj$, $ik \neq ki$, etc. But all real and complex numbers do commute with i, j, k, thus if c is a real number, then $ic = ci$, $jc = ci$, and $kc = ck$. On multiplying $ijk = -1$ on the left by i, so that $iijk = i(-1) = -i$, it is found, since $i^2 = -1$, that $jk = i$. Similarly $jjk = ji = -k$; when exhausted, this process leads to all the simple noncommutative relations for i, j, k, namely,

$$ij = -ji = k \qquad jk = -kj = i \qquad ki = -ik = j$$

More complicated products, for example, $jikjk = -kki = i$, are evaluated by substituting for any adjoined pair the value given in the preceding series of relations and then proceeding similarly to any other adjoined pair in the new product, and so on until the product is reduced to ± 1, $\pm i$, $\pm j$, or $\pm k$. Multiplication on the right is also permissible; thus from $ij = k$, one has $ijj = kj$, or $-i = kj$. Products such as jj and jjj may be written j^2 and j^3.

All the laws and operations of ordinary algebra are assumed to be valid in the definition of quaternion algebra, except the commutative law of multiplication for the units i, j, k. Thus the associative and distributive laws of addition and multiplication apply Without restriction throughout. Addition is also commutative, for example, $i + j = j + i$. Don M. Yost

Quebracho Any of a number of trees belonging to different genera but having similar qualities, all indigenous to South America and valuable for both wood and bark. The heartwood of one South American tree, *Schinopsis lorentzii* (family Anacardiaceae), is called quebracho (meaning ax-breaker) in reference to the exceedingly hard wood, one of the hardest known. Quebracho is the world's most important source of tannin. *See* SAPINDALES; TANNIN. Perry D. Strausbaugh; Earl L. Core

Queueing theory The mathematical theory of the formation and behavior of queues or waiting lines. The name is also applied loosely to the mathematical study of a wide variety of problems connected with traffic congestion and storage systems. Uneven flow through a service point, with fluctuating arrivals and service times, constitutes a major topic of operations research. For the mathematician, queueing theory is particularly interesting because it is concerned with relatively simple stochastic processes, which are in general non-Markovian and possibly stationary. *See* OPERATIONS RESEARCH; STOCHASTIC PROCESS.

The principal pioneer of queueing theory, A. K. Erlang, began in 1908 to study problems of telephone congestion. It is of interest to study the waiting times of subscribers in a manually operated system—for example, the average waiting time and the chance that a subscriber will obtain service immediately without waiting—and to examine how much the waiting times will be affected if the number of operators is altered, or conditions are changed in any other way. If there are more operators or if service can be speeded up, subscribers will be pleased because waiting will be reduced, but the improved facility will be more expensive to maintain; therefore, a reasonable balance must be struck.

Related problems in the use of automatic telephone exchanges and of long-distance lines able to carry only a limited number of messages simultaneously have resulted in much mathematical study of telephone traffic problems. Similar problems arise in other contexts. In a factory a number of machines, such as looms, may be under the care of one or more repairer. If a machine breaks down, it must stand idle until a repairer is free from repairing other machines. Machines here correspond to telephone subscribers, breakdown corresponds to attempts to make a call, and repair corresponds to connection. Other examples of congestion situations are aircraft flying around in circles waiting to use an airport landing strip, automobiles lining up at a turnpike toll booth, and customers lining up at the counter of a retail shop, waiting for service.

What is most interesting to investigate varies with the circumstances. Sometimes it is the mean waiting time of customers, sometimes the frequency with which the queue length exceeds a given limit, sometimes the proportion of the servers' time that is idle, and sometimes the average duration of a period during which a server is continuously occupied. In the study of stocking a warehouse or retail shop, known generally as the theory of inventories, the frequency with which the stock will be exhausted is considered under various reordering policies. Similar considerations apply in the theory of dams and water storage. *See* LINEAR PROGRAMMING; SYSTEMS ENGINERRING. Francis J. Anscombe

Quince The deciduous tree *Cydonia oblonga*, originally from Asia, grown for its edible fruit. The fruit is a pear-shaped or apple-shaped pome, characteristically tomentose, aromatic, sour, astringent, and green, turning clear yellow at maturity. Used mostly for jam and jelly or as a stewed fruit, the fruit of the quince develops a pink color in cooking. *See* DECIDUOUS PLANTS; FRUIT; FRUIT, TREE; ROSALES. Harold B. Tukey

Quinine The chief alkaloid of the bark of the cinchona tree, which is indigenous to certain regions of South America. The structure of quinine is shown below.

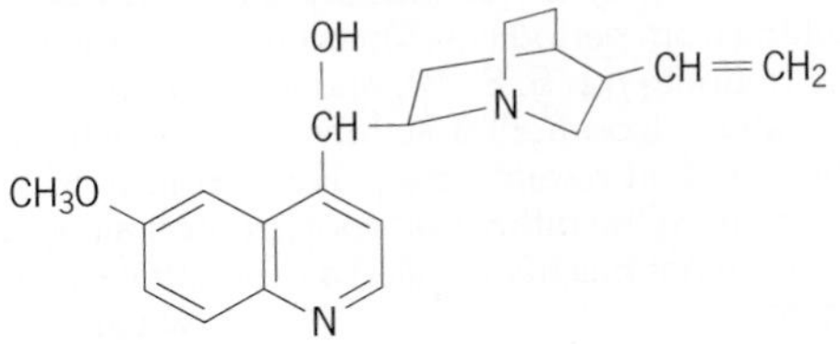

Until the 1920s quinine was the best chemotherapeutic agent for the treatment of malaria. However, clinical studies definitely established the superiority of the newer synthetic antimalarials such as primaquine, chloroquine, and chloroguanide. *See* ALKALOID; MALARIA. S. Morris Kupchan

Quinoa An annual herb, *Chenopodium quinoa* (family Chenopodiaceae), a native of Peru, and the staple food of many people in South America. These plants, grown at high altitudes, produce large quantities of highly nutritious seeds used whole in soups or ground into flour, which is made into bread or cakes. The seeds are also used as poultry feed, in medicine, and in making beer. In the United States the leaves are sometimes used as a substitute for spinach. *See* CARYOPHYLLALES. Perry D. Strausbaugh; Earl L. Core

Quinone One of a class of aromatic diketones in which the carbon atoms of the carbonyl groups are part of the ring structure. The name quinone is applied to the whole group, but it is often used specifically to refer to *p*-benzoquinone (**1**), *o*-Benzoquinone (**2**) is also known but the meta isomer does not exist.

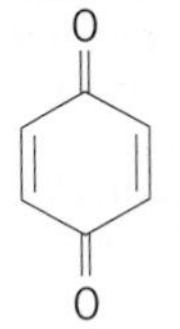

(1) 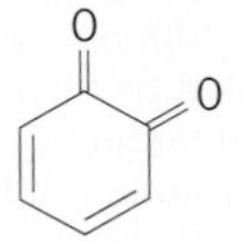 (2)

Quinones are prepared by oxidation of the corresponding aromatic ring systems containing amino ($—NH_2$) or hydroxyl ($—OH$) groups on one or both of the carbon atoms being converted to the carbonyl group.

Three of the several possible quinones derived from naphthalene are known: 1,4-naphthoquinone (**2**), 1,2-naphtho-quinone, and 2,6-naphthoquinone (**4**).

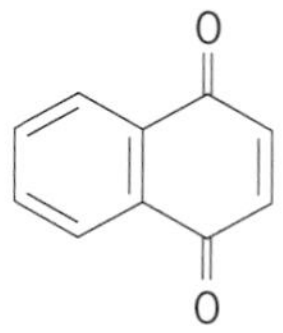

(3)

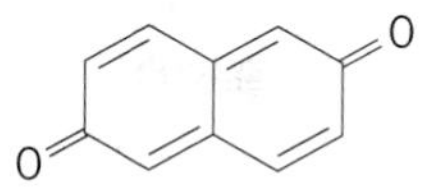

(4)

Important naturally occurring naphthoquinones are vitamins K_1 and K_2 which are found in blood and are responsible for proper blood clotting reaction. A number of quinone pigments have been isolated from plants and animals. Illustrative of these are juglone found in unripe walnut shells and spinulosin from the mold *Penicillium spinulosum*. 9,10-Anthraquinone derivatives form an important class of dyes of which alizarin is the parent type. *p*-Benzoquinone is manufactured for use as a photographic developer. *See* ANTHRAQUINONE PIGMENTS; AROMATIC HYDROCARBON; DYE; HYDROQUINONE; KETONE; OXDATION-REDUCTION; VITAMIN K.

David A. Shirley

Rabies An acute, encephalitic viral infection. Human beings are infected from the bite of a rabid animal, usually a dog. Canine rabies can infect all warm-blooded animals, and death usually results. *See* ANIMAL VIRUS.

The virus is believed to move from the saliva-infected wound through sensory nerves to the central nervous system, multiply there with destruction of brain cells, and thus produce encephalitis, with severe excitement, throat spasm upon swallowing (hence hydrophobia, or fear of water), convulsions, and death—with paralysis sometimes intervening before death.

All bites should immediately be cleaned thoroughly with soap and water, and a tetanus shot should be considered. The decision to administer rabies antibody, rabies vaccine, or both depends on four factors: the nature of the biting animal; the existence of rabies in the area; the manner of attack (provoked or unprovoked) and the severity of the bite and contamination by saliva of the animal; and recommendations by local public health officials. *See* PUBLIC HEALTH; TETANUS.

Diagnosis in the human is made by observation of Negri bodies (cytoplasmic inclusion bodies) in brains of animals inoculated with the person's saliva, or in the person's brain after death. A dog which has bitten a person is isolated and watched for 10 days for signs of rabies; if none occur, rabies was absent. If signs do appear, the animal is killed and the brain examined for Negri bodies, or for rabies antigen by testing with fluorescent antibodies. *See* FLUORESCENCE MICROSCOPE; VIRAL INCLUSION BODIES.

Individuals at high risk, such as veterinarians, must receive preventive immunization. If exposure is believed to have been dangerous, postexposure prophylaxis should be undertaken. If antibody or immunogenic vaccine is administered promptly, the virus can be prevented from invading the central nervous system. An inactivated rabies virus vaccine is available in the United States. It is made from virus grown in human or monkey cell cultures and is free from brain proteins that were present in earlier Pasteur-type vaccines. This material is sufficiently antigenic that only four to six doses of virus need be given to obtain a substantial antibody response. *See* VACCINATION.

Joseph L. Melnick

Raccoon A member of the mammalian family Procyonidae, along with ringtails, coatis, kinkajous, olingos, and lesser pandas. These carnivores inhabit areas from southern Canada to northern Argentina and Uruguay and have been introduced into Russia and Germany. Only one species, *Procyon lotor*, inhabits North America.

The raccoon is a medium-sized, heavily built, partially arboreal mammal with long, loose, gray to almost black fur. A white-bordered black mask extends across the eyes and down onto the cheeks. The foxlike head is broad with a pointed muzzle, and the ears are prominent, erect, and somewhat pointed. The moderately long, well-haired tail is banded with five to seven blackish rings alternating with wider, lighter bands. The plantigrade feet have naked smooth soles and five very long digits, each of which possesses a nonretractile claw. The "hands" are well adapted for grasping and manipulating objects. Raccoons have excellent senses of hearing, vision, smell, and touch.

Raccoons prefer moist areas and are found mainly in timbered swamps, on river bottoms, along the banks of streams and lakes, and in coastal salt marshes. These primarily nocturnal mammals are excellent climbers, and they swim well, but not far or often. Dens may be in hollow trees, hollow logs, old stumps, burrows, muskrat houses, or in outbuildings such as barns. Raccoons do not hibernate, although in northern areas they may remain in a den for much of the winter. Raccoons are opportunistic omnivores and feed on a wide variety of plant and animal foods including wild fruits, berries, acorns, shellfish, crustaceans, frogs, fish, small mammals, reptiles, and birds. *See* CARNIVORA; MAMMALIA.

Donald W. Linzey

Racemization The formation of a racemate from a pure enantiomer. Alternatively stated, racemization is the conversion of one enantiomer in a 50:50 mixture of the two enantiomers (+ and −, or R and S) of a substance. Racemization is normally associated with the loss of optical activity over a period of time since 50:50 mixtures of enantiomers are optically inactive. *See* OPTICAL ACTIVITY.

Racemization is an energetically favored process since it reflects a change from a more ordered to a more random state. But the rate at which enantiomers racemize is typically quite slow unless a suitable mechanistic pathway is available, since racemization usually, but not always, requires that a chemical bond at the chiral center of an enantiomer be broken. Racemization of enantiomers possessing more than one chiral center requires that all chiral centers of half of the molecules invert their configurations. *See* ENTROPY.

The observation and study of racemization have important implications for the understanding of the mechanisms of chemical reactions and for the synthesis and analysis of chiral natural products such as peptides. Moreover, racemization is of economic importance since it provides a way of converting an unwanted enantiomer into a useful one. Synthetic medicinal agents are often produced industrially as racemates. After resolution and isolation of the desired enantiomer, half of the product would have to be discarded were it not for the possibility of racemizing the unwanted isomer and of recycling the resultant racemate.

Samuel H. Wilen

Radar An acronym for radio detection and ranging, the original and basic function of radar. The name is applied to both the technique and the equipment used. Radar is a sensor; its purpose is to provide estimates of certain characteristics of its surroundings of interest to a user, commonly the presence, position, and motion of aircraft, ships, or other vehicles in its vicinity. In other uses, radars provide information about the Earth's surface (or that of other astronomical bodies) or about meteorological conditions. To provide the user with a full range of sensor capability, radars are often used in combinations or with other elements of the complete system.

Radar operates by transmitting electromagnetic energy into the surroundings and detecting energy reflected by objects. If a narrow beam of this energy is transmitted by the directive antenna, the direction from which reflections come and hence the bearing of the object may be estimated. The distance to the reflecting object is estimated by measuring the period between the transmission of the radar pulse and reception of the echo. In most radar applications, this period will be very short since electromagnetic energy travels with the velocity of light.

Kinds of radar. The physical nature of radars varies greatly. Familiar radars that are very small and compact include those used on small boats or in small aircraft for navigation and safety and, even smaller, the hand-held radars used for measurement of vehicle speed or even that of a baseball pitch. *See* MARINE NAVIGATION.

The largest radars cover acres of land, long arrays of antennas all operating together to monitor the flight of space vehicles and orbiting debris and also to contribute to radar astronomy. Other very large radars provide coverage at substantial distances well beyond the normal radar horizon; they monitor flight activity and sea-surface conditions, contributing to both air-traffic and meteorological interests. These radars are very large because they must use longer-than-usual radio wavelengths that result in the ionospheric containment of the signal for over-the-horizon operations. *See* RADAR ASTRONOMY; SPACECRAFT GROUND INSTRUMENTATION.

More common in size are those radars at airports, with rotating antennas 20–40 ft (6–12 m) wide. Radars intended for mobile use, particularly those in aircraft and in trucks, buses, and automobiles, are quite compact. *See* AIRBORNE RADAR.

Radars intended principally to determine the presence and position of reflecting targets in a region around the radar are called search radars. Other radars examine further the targets detected: examples are height finders with antennas that scan vertically in the direction of an assigned target, and tracking radars that are aimed continuously at an assigned target to obtain great accuracy in estimating target motion. In some modern radars, these search and track functions are combined, usually with some computer control. Surveillance radar connotes operation of this sort, somewhat more than just search alone. *See* SURVEILLANCE RADAR.

The development of quite complex and versatile radars, computer-controlled, generally using phased-array antennas and called multifunction radars, has been quite successful worldwide. Such antennas require no physical motion to position the radar beam, so various coverage routines and modes of operation can be scheduled with an efficiency and expediency unachievable in lesser radars.

Very accurate tracking radars intended for use at missile test sites or similar test ranges are called instrumentation radars. Radars designed to detect clouds and precipitation are called meteorological or weather radars. Improved techniques in Doppler-sensitive signal processing have enabled radars to examine more precisely the motion of reflecting objects (windborne dust, insects, and precipitants, for example) indicative of weather disturbances. Such radars are used in critical places such as airports where sudden wind shears (as from downdraft microbursts in the area) have threatened air safety. Many weather radars also use microwave polarimetry to examine the precipitants and characterize the disturbance more accurately. *See* DOPPLER EFFECT; DOPPLER RADAR; METEOROLOGICAL RADAR; STORM DETECTION.

Airborne and spaceborne radars have been developed to perform ground mapping with extraordinary resolution by special Doppler-sensitive processing while the radar is moved over a substantial distance. Such radars are called synthetic-aperture radars (SARs) because of the very large virtual antenna formed by the path covered while the processing is performed. Interferometry can provide topological information (3D SAR), and polarimetry and other signal analysis can provide more information on the nature of the surface (type of vegetation, for example). *See* REMOTE SENSING; SYNTHETIC APERTURE RADAR (SAR).

Some radars have separate transmit and receive antennas sometimes located miles apart. These are called bistatic radars, the more conventional single-antenna radar being monostatic. Some useful systems have no transmitter at all and are equipped to measure, for radarlike purposes, signals from the targets themselves. Such systems are often called passive radars, but the terms "radiometers" or "signal intercept systems" are generally more appropriate. Regular radars may, of course, occasionally be operated in passive modes. Doing so helps locate sources of interference, either deliberate (jamming) or unintentional.

The terms "primary" and "secondary" are used to describe, respectively, radars in which the signal received is reflected by the target and radars in which the transmission causes a transponder (transmitter-responder) carried aboard the target to transmit a signal back to the radar. The Identification Friend or Foe (IFF) system in both military and civil use is a secondary radar. *See* AIR-TRAFFIC CONTROL; ELECTRONIC NAVIGATION SYSTEMS.

Operation. It is convenient to consider radars as having four principal subsystems: the transmitter, the antenna, the receiver (and signal processor), and the control and interface apparatus (see illustation).

The transmitter provides the radio-frequency (RF) signal in sufficient strength (power) for the radar sensitivity desired and sends it to the antenna, which causes the signal to be radiated into space in a desired direction. The signal propagates (radiates) in space, and some of it is intercepted by reflecting (or "scattering") bodies. The part of the incident signal that is backscattered returns to the radar antenna, which collects it and routes it to the receiver (along with all incoming signals) for appropriate signal processing and detection. The presence of an echo of the transmitted signal in the received signal reveals the presence of a target. The echo is indicated by a sudden rise in the output of the detector, which produces a voltage (video) proportional to the sum of the RF signals being received and the RF noise inherent in the receiver itself. The range to the target is half the product of the speed of electromagnetic propagation (3×10^8 m/s), and the time between the transmission and the receipt of the echo. The direction or bearing of the target is disclosed by the direction to which the beam is pointed when an echo is received.

Most radars indeed use pulses in their operation. Those that do not are called continuous-wave (cw) radars. Some missile seekers, mostly interested only in the direction to a target, are

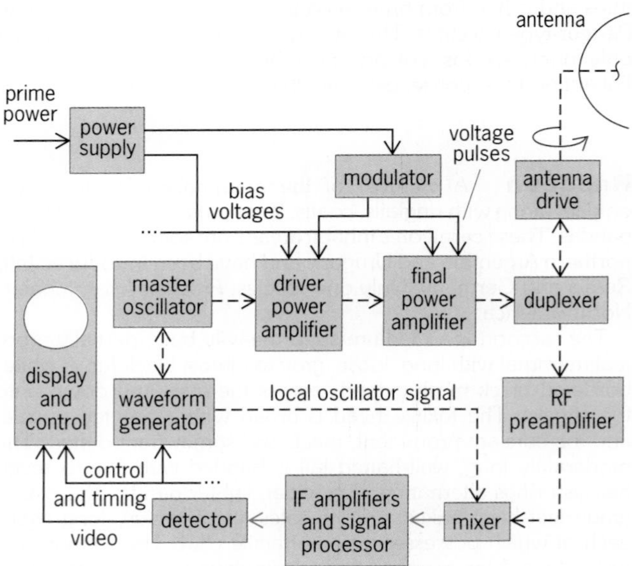

Block diagram of pulse radar.

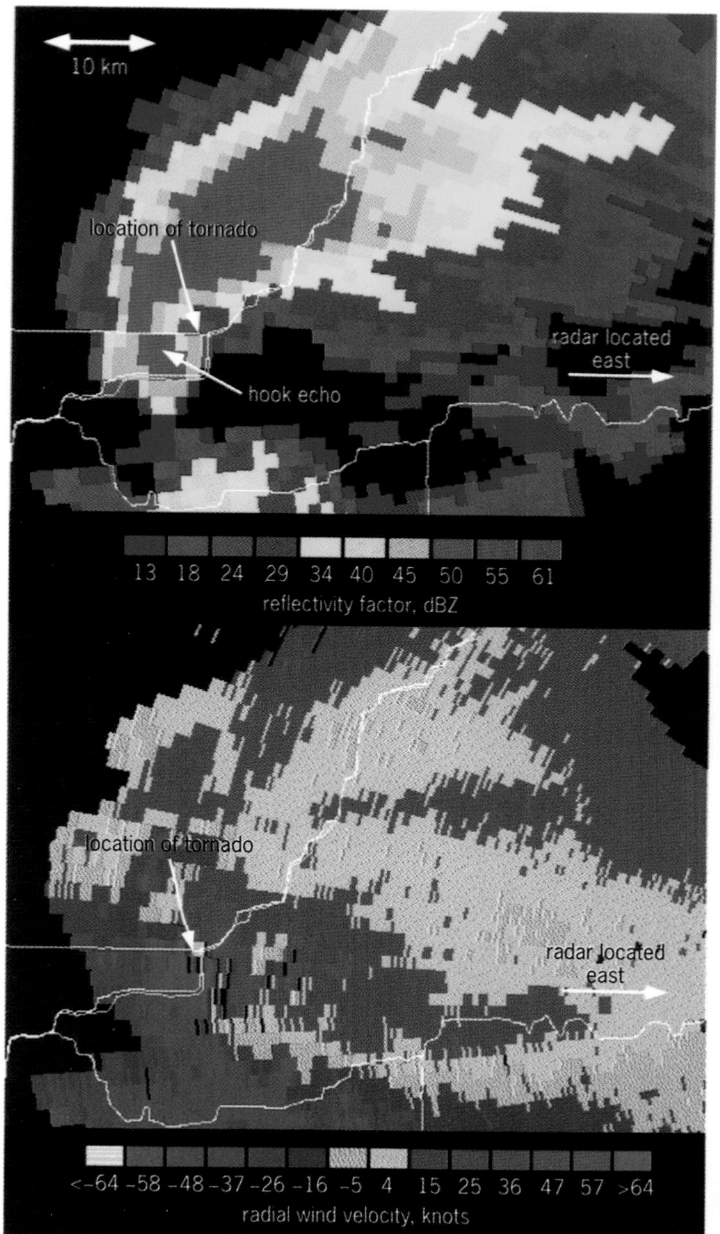

Colorplate 1. (Left) Depiction of the radar reflectivity factor (dBZ) and radial wind velocity (knots) for a supercell thunderstorm that produced a tornado northwest of Springfield, Illinois, on April 19, 1996. On the radial velocity display, reds represent air motion away from the radar.

Colorplate 2. (Below) Composite depiction of the radar reflectivity factor (dBZ) from radars on the United States east coast during the landfall of Hurricane Floyd on September 16, 1999.

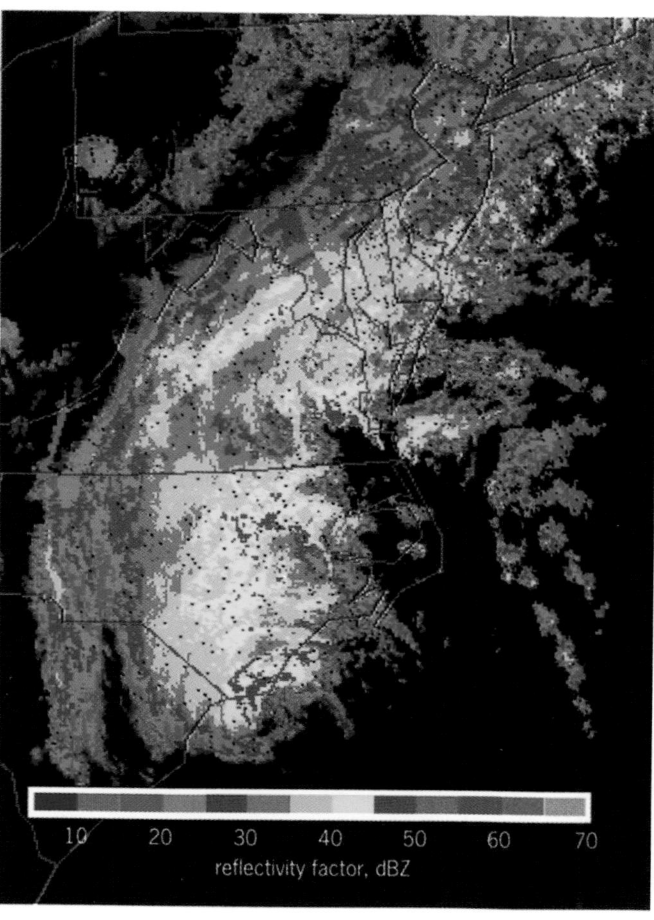

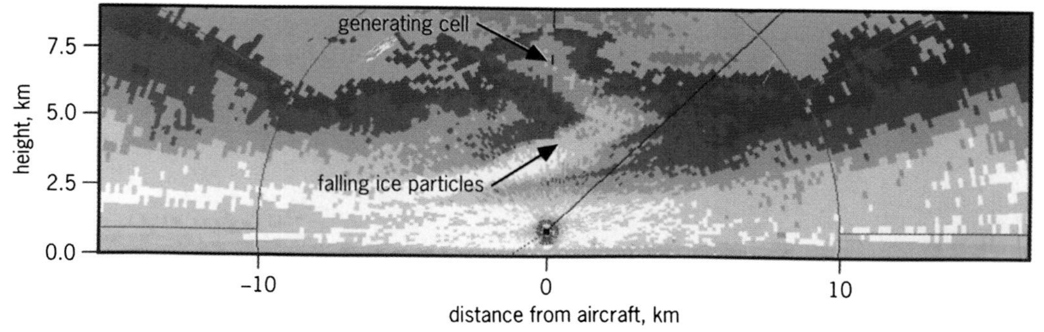

Colorplate 3. (Left) Vertical cross section of the radar reflectivity factor (dBZ) in a snowstorm measured by a radar mounted on an aircraft. A generating cell and stream of ice particles are indicated.

Radar carrier-frequency bands

Band designation	Nominal frequency range	Representative wavelength
HF	3–20 MHz	30 m at 10 MHz
VHF	30–300 MHz	3 m at 100 MHz
UHF	300–1000 MHz	1 m at 300 MHz
L	1.0–2.0 GHz	30 cm at 1 GHz
S	2.0–4.0 GHz	10 cm at 3 GHz
C	4.0–8.0 GHz	5 cm at 6 GHz
X	8.0–12.0 GHz	3 cm at 10 GHz
K_u	12.0–18 GHz	2 cm at 15 GHz
K	18–27 GHz	1.5 cm at 20 GHz
K_a	27–40 GHz	1 cm at 30 GHz
mm	40–300 GHz	0.3 cm at 100 GHz

of this type; police radars interested only in the Doppler shift of the echo for speed measurement are another example. *See* CONTINUOUS-WAVE RADAR.

A duplexer permits the same antenna to be used on both transmit and receive, and is equipped with protective devices to block the very strong transmit signal from going to the sensitive receiver and damaging it. The antenna forms a beam, usually quite directive, and, in the search example, rotates throughout the region to be searched, or otherwise causes the beam to be positioned as desired, perhaps without mechanical motion, as in radars using a computer-controlled phased-array antenna. In such phased arrays, the duplexing function is generally built into the feed system and phasing logic. *See* ANTENNA (ELECTROMAGNETISM).

Radars searching with mechanical antennas generally transmit several pulses (perhaps 20–40) as the antenna scans past a target, whereas phased-array radars may search with various "dwells," sometimes with just one transmission per beam position. The use of several pulses allows a buildup of the echo being received, a process called noncoherent or video integration. Most radars are equipped with low-noise RF preamplifiers to improve sensitivity. The signal is then "mixed" with (multiplied by) a local oscillator signal to produce a convenient intermediate-frequency (IF) signal, commonly at 30 or 60 MHz; the same principle is used in all heterodyne radio receivers. The local oscillator signal, kept offset from the transmit frequency by precisely this intermediate frequency, is supplied by the transmitter oscillators during reception. After other significant signal processing in the IF circuitry (of a digital nature in many newer radars), a detector produces a video signal, a voltage proportional to the strength of the processed IF signal. This video can be applied to a cathode-ray-tube (CRT) display so as to form a proportionately bright spot (a blip), which could be judged as resulting from a target echo. However, radars increasingly use artificial computerlike displays based on computer analysis of the video. Automatic detection and automatic tracking (based on a sequence of dwells) are typical of such data processing, reports being displayed for radar operator management and also made instantly available to the user system. *See* CATHODE-RAY TUBE; ELECTRONIC DISPLAY; HETERODYNE PRINCIPLE; MIXER; PREAMPLIFIER; RADIO RECEIVER.

Radar carrier frequencies are broadly identified by a widely accepted nomenclature. The spectrum is divided into bands (see table). The charged layers of the ionosphere present a highly refractive shell at radio frequencies well below the microwave frequencies of most radars. Consequently, over-the-horizon radars operate in the 10-MHz region to exploit this skip path.

Robert T. Hill

Radar-absorbing materials Materials that are designed to reduce the reflection of electromagnetic radiation from a conducting surface in the frequency range from approximately 100 MHz to 100 GHz. The level of reduction that is achieved varies from a few decibels to greater than 50 dB, reducing the reflected energy by as much as 99.999%.

Two methods have been widely adopted in order to produce such absorbers. The first is to avoid a discrete change of impedance at the material surface by gradually varying the impedance. The removal of the discontinuity at the surface allows the microwave energy to be transmitted into the absorbing medium without reflection. This transition from the impedance of free space to that of the bulk material is commonly achieved by a geometric profile. The carbon-loaded foam pyramids used as the lining of anechoic chambers are typical of this type of absorber. To produce such absorbers, it is necessary in practice to taper the material over distances which are large compared with the wavelength of the frequencies to be absorbed. Therefore, practical absorbers of this type giving greater than 20 dB absorption vary in thickness from about 0.8 in. (2 cm) at 10 GHz and above to 6 ft (2 m) at 100 MHz and above. The absorber performance improves with increasing thickness until the point is reached where all of the energy that enters the material is absorbed and only the front-face reflection is left. While this type of absorber is capable of producing a very high degree of absorption over a broad bandwidth, it is at the same time a relatively thick material. *See* ANECHOIC CHAMBER.

The second method of absorber design produces much thinner absorbing layers which are capable of producing good absorption ($\geq$25 dB) with restricted bandwidths. These materials achieve the absorption by a combination of attenuation within the material and destructive interference at the interface. The electromagnetic properties and the thickness of the layer are such that the initial reflected wave and the sum of the emergent rays resulting from the multiple reflections within the material are equal in magnitude and opposite in phase. The thickness of the layer is close to a quarter-wavelength at the frequency of operation, giving a 180° phase difference between the interface reflection and the emergent waves. *See* INTERFERENCE OF WAVES.

Microwave-absorbing materials are widely used both within the electronics industry and for defense purposes. Their uses can be classified into three major areas: (1) for test purposes so that accurate measurements can be made on microwave equipment unaffected by spurious reflected signals; (2) to improve the performance of any practical microwave system by removing unwanted reflections which can occur if there is any conducting material in the radiation path; and (3) to camouflage a military target by reducing the reflected radar signal. *See* ELECTRONIC WARFARE; MICROWAVE; MICROWAVE MEASUREMENTS; REFLECTION OF ELECTROMAGNETIC RADIATION.

S. B. Morris

Radar astronomy A powerful astronomical technique that uses radar echoes to furnish otherwise unavailable information about bodies in the solar system. By comparing a radar echo to the transmitted signal, information can be obtained about the target's size, shape, topography, surface bulk density, spin vector, and orbital elements, as well as the presence of satellites and in certain situations the target's mass and density. While other astronomical techniques rely on passive measurement of reflected sunlight or naturally emitted radiation, the illumination used in radar astronomy is a coherent signal whose polarization and time modulation or frequency modulation are tailored to meet specific scientific objectives. Through measurements of the distribution of echo power in time delay or Doppler frequency, radar achieves spatial resolution of a planetary target despite the fact that the radar beam is typically much larger than the angular extent of the target. This capability is particularly valuable for asteroids and planetary satellites, which appear as unresolved point sources through optical telescopes. Moreover, the centimeter-to-meter wavelengths used in radar astronomy readily penetrate cometary comas and the optically opaque clouds that conceal Venus and Titan, and also permit determination of near-surface roughness (abundance of wavelength-scale rocks), bulk density, and metal concentration in planetary regoliths. *See* ASTEROID; SATELLITE (ASTRONOMY); SATURN; VENUS.

A radar telescope is essentially a radio telescope equipped with a high-power transmitter (a klystron vacuum-tube amplifier) and specialized instrumentation that links the transmitter, low-noise receiver, high-speed data-acquisition computer, and antenna together in an integrated radar system. Planetary radars, which must detect echoes from targets at distances from about 10^6 km (10^6 mi) for closely approaching asteroids and comets to more than 10^9 km (nearly 10^9 mi) for Saturn's rings and satellites, are the largest and most sensitive radars on Earth. *See* KLYSTRON; RADAR; RADIO TELESCOPE. Steven J. Ostro

Radar meteorology The application of radar to the study of the atmosphere and to the observation and forecasting of weather. Meteorological radars transmit electromagnetic waves at microwave and radio-wave frequencies. Water and ice particles, inhomogeneities in the radio refractive index associated with atmospheric turbulence and humidity variations, insects, and birds scatter radar waves. The backscattered energy received at the radar constitutes the returned signal. Meteorologists use the amplitude, phase, and polarization state of the backscattered energy to deduce the location and intensity of precipitation, the wind speed along the direction of the radar beam, and precipitation type (for example, rain or hail). *See* METEOROLOGICAL RADAR; RADAR; WEATHER FORECASTING AND PREDICTION.

Much of the understanding of the structure of storms derives from measurements made with networks of Doppler radars. They are used to investigate the complete three-dimensional wind fields associated with storms, fronts, and other meteorological phenomena. *See* DOPPLER; PRECIPITATION (METEOROLOGY); STORM; STORM DETECTION; THUNDERSTORM. Robert M. Rauber

Radian measure A radian is the angle subtended at the center of a circle by an arc of the circle equal in length to its radius. It is proved in geometry that equal central angles of two circles subtend arcs proportional to their radii; and the converse is true. Hence the radian is independent of the length of the radius. The illustration represents two circles of radius r. Arc AB of length r

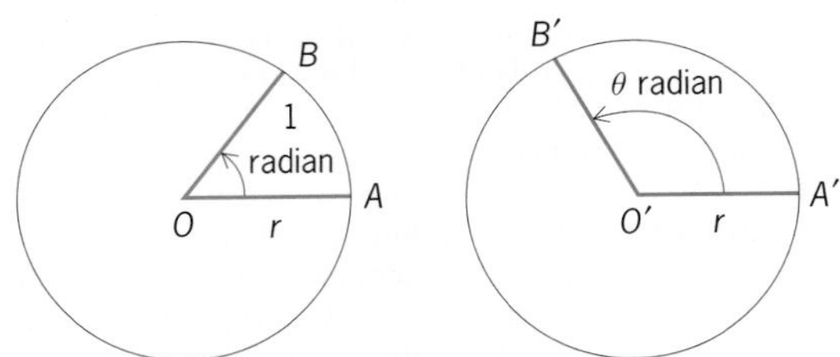

Diagrams showing radian measurement.

subtends 1 radian (rad) at the center O of the circle, and arc $A'B'$ of length s subtends θ rad at its center. Since arcs on equal circles are proportional to their subtended central angles, $s/r = \theta/1$, or the formula $s = r\theta$ holds. If $\theta = 2\pi$, $s = 2\pi r$, the circumference of the circle. Therefore 2π rad is the complete angle about a point or 360°, and 2π rad $= 360^\circ$, 1 rad $= 360^\circ/2\pi = 57.2958^\circ$.

The degree as a unit of angle has come down from antiquity. However, its use in various theories involves clumsy constants. The use of the radian avoids these constants. The radian is employed generally as a measure of angle in theoretical discussions; when no unit of angle is mentioned, the radian is understood. Lyman M. Kells

Radiance The physical quantity that corresponds closely to the visual brightness of a surface. A simple radiometer for measuring the (average) radiance of an incident beam of optical radiation (light, including invisible infrared and ultraviolet

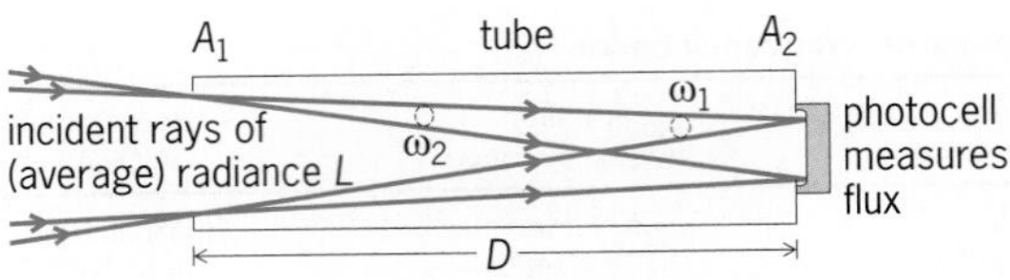

A simple radiometer.

radiation) consists of a cylindrical tube, with a hole in each end cap to define the beam cross section there, and with a photocell against one end to measure the total radiated power in the beam of all rays that reach it through both holes (see illustration). If A_1 and A_2 are the respective areas of the two holes, D is the length of the tube (distance between holes), and Φ is the radiant flux or power measured by the photocell, then the (average) radiance is approximately given by the equation $L = \Phi/(A_1 \cdot A_2/D^2)\ \mathrm{W \cdot m^{-2} \cdot sr^{-1}}$. Fred E. Nicodemus

Radiant heating Any system of space heating in which the heat-producing means is a surface that emits heat to the surroundings by radiation rather than by conduction or convection. The surfaces may be radiators such as baseboard radiators or convectors, or they may be the panel surfaces of the space to be heated. *See* PANEL HEATING AND COOLING.

The heat derived from the Sun is radiant energy. Radiant rays pass through gases without warming them appreciably, but they increase the sensible temperature of liquid or solid objects upon which they impinge. The same principle applies to all forms of radiant-heating systems, except that convection currents are established in enclosed spaces and a portion of the space heating is produced by convection. Any radiant-heating system using a fluid heat conveyor may be employed as a cooling system by substituting cold water or other cold fluid. However, the technique is not practical on the scale required for comfort control of an occupied space. Erwin L. Weber; Richard Koral

Radiation The emission and propagation of energy; also, the emitted energy itself. The etymology of the word implies that the energy propagates rectilinearly, and in a limited sense, this holds for the many different types of radiation encountered.

The major types of radiation may be described as electromagnetic, acoustic, and particle, and within these major divisions there are many subdivisions. Electromagnetic radiation is classified roughly in order of decreasing wavelength as radio, microwave, visible, ultraviolet, x-rays, and γ-rays. Acoustic or sound radiation may be classified by frequency as infrasonic, sonic, or ultrasonic in order of increasing frequency, with sonic being between about 16 and 20,000 Hz. The traditional examples of particle radiation are the α- and β-rays of radioactivity. *See* ELECTROMAGNETIC RADIATION; RADIOACTIVITY; SOUND. McAllister H. Hull, Jr.

Radiation biology The study of the action of ionizing and nonionizing radiation on biological systems. Ionizing radiation includes highly energetic electromagnetic radiation (x-rays, gamma rays, or cosmic rays) and particulate radiation (alpha particles, beta particles, neutrons, or heavy charged ions). Nonionizing radiation includes ultraviolet radiation, microwaves, and extralow-frequency (ELF) electromagnetic radiation. These two types of radiation have different modes of action on biological material: ionizing radiation is sufficiently energetic to cause ionizations, whereas nonionizing radiation causes molecular excitations. In both cases, the result is that chemical bonds of molecules may be altered, causing mutations, cell death, or other biological changes. *See* ELECTROMAGNETIC RADIATION; RADIATION.

Ionizing radiation originates from external sources (medical x-ray equipment, cathode-ray tubes in television sets or computer video displays) or from internal sources (ingested or inhaled radioisotopes, such as radon-222, strontium-90, and iodine-131), and is either anthropogenic or natural.

Nonionizing radiation originates from natural sources (sunlight, Earth's magnetic field, lightning, static electricity, endogenous body currents) and technological sources (computer video displays and television sets, microwave ovens, communications equipment, electric equipment and appliances, and high-voltage transmission lines).

Ionizing radiation. The action of ionizing radiation is best described by the three stages (physical, chemical, and biological) that occur as a result of energy release in the biological target material.

Physical stage. All ionizing radiation causes ionizations of atoms in the biological target material. The Compton effect, which predominates at the energies of electromagnetic radiation that are commonly encountered (for example, x-rays or gamma rays), strips orbital electrons from the atoms. These electrons (Compton electrons) travel through the target material, colliding with atoms and thereby releasing packets of energy. For low-energy x-rays, the photoelectric effect predominates, producing photoelectrons that transfer their energy in the same manner as Compton electrons. *See* COMPTON EFFECT; ELECTRON; GAMMA RAYS; X-RAYS.

The absorbed dose of ionizing radiation is measured as the gray (Gy, 1 joule of energy absorbed by 1 kilogram of material). Because of the very localized absorption of ionizing radiation, an amount of ionizing radiation energy equivalent to 1/100 the heat energy in a cup of coffee will result in a 50% chance that the person absorbing the radiation will die in 30 days.

Neutrons with energies between 10 keV and 10 MeV transfer energy mainly by elastic scattering, that is, billiard-ball-type collisions, of atomic nuclei in the target material. In this process the nucleus is torn free of some or all of the orbital electrons because its velocity is greater than that of the orbital electrons. The recoiling atomic nucleus behaves as a positively charged particle. Because the mass of the neutron is nearly the same as that of the hydrogen atom, hydrogenous materials are most effective for energy transfer. *See* NEUTRON.

Chemical stage. Chemical changes in biological molecules are caused by the direct transfer of radiation energy (direct radiation action) or by the production of chemically reactive products from radiolysis of water that diffuses to the biological molecule (indirect radiation action). More than half the biological action of low linear-energy-transfer (LET) ionizing radiation (for example, x-rays and gamma rays) results from indirect radiation action, about 90% of which is due to the action of the hydroxyl radical (OH·). For high linear-energy-transfer radiation, direct radiation action predominates. Chemicals that react with hydroxyl radicals, rendering them unreactive, provide protection against indirect radiation damage. *See* LINEAR ENERGY TRANSFER (BIOLOGY); RADIATION CHEMISTRY.

The most important biological targets for damage from ionizing radiation are probably the plasma membrane and DNA, because there is only one copy, or a few copies, in the cell; because they serve critical roles for the survival and propagation of cells; and because they are large. The last factor is important because ionizing radiation releases its energy in a random manner; thus the larger the target, the more likely that it will be damaged by radiation. Consequences of radiation damage in membranes are changes in ion permeability, with leakage of potassium ions; changes in active transport; and cell lysis. Lesions in DNA include single-strand breaks, double-strand breaks, base damage, interstrand cross-links, and DNA-protein cross-links. *See* CELL MEMBRANES; DEOXYRIBONUCLEIC ACID (DNA).

Biological stage. Various biological effects can result from the actions of ionizing radiation. Reproductive death is most pronounced in mammalian cells that are actively dividing and in nondifferentiated tissue. Thus, dividing tissues (bone marrow and the germinal cells of the ovary and testis) are radiosensitive, and nondividing tissues (liver, kidney, brain, muscle, cartilage, and connective tissue) are radioresistant. Developing embryos are quite radiosensitive. The radiosensitivity of organisms varies greatly, being related to their intrinsic sensitivity to radiobiological damage and to their ability to repair the damage. Radiation doses resulting in 10% survival range from 3 Gy (mouse and human cells), to greater than 1000 Gy (the bacterium *Deinococcus radiodurans*).

The three organ systems that generally contribute to the death of mammals following a single dose of whole-body irradiation are, in decreasing order of radiosensitivity, the hematopoietic system, the gastrointestinal system, and the cerebrovascular system. Late somatic effects may take years or decades to appear and include genetic mutations transmitted to subsequent generations, tumor development and carcinogenesis, and shortening of life span. *See* MUTAGENS AND CARCINOGENS; MUTATION; TUMOR.

Nonionizing radiation. Of all the nonionizing radiations, only ultraviolet radiation, microwaves, and high-voltage electromagnetic radiation are considered in the study of radiation biology.

Ultraviolet radiation. Since it can penetrate only several layers of cells, the effects of ultraviolet (UV) radiation on humans are restricted to the skin and the eyes. Ultraviolet radiation is divided into UV-C (wavelength of 200–280 nanometers), UV-B (280– 320 nm), and UV-A (320–400 nm). The most biologically damaging is UV-C, and the least damaging is UV-A. The solar spectrum at the Earth's surface contains only the UV-B and UV-A radiations.

Biological effects can arise only when absorption of ultraviolet radiation occurs. Absorption is dependent on the chemical bonds of the material, and it is highly specific. Sunburn is a form of erythema produced by overexposure to the UV-B portion of the solar spectrum. A rare but deadly form of skin cancer in humans, malignant melanoma, is induced by exposure to sunlight, with occurrences localized on those regions of the body that are most frequently exposed. Ultraviolet light can also cause photochemical damage. Cyclobutane pyrimidine dimers are the main photoproduct following exposure to UV-C and UV-B, and they can lead to cell death and precarcinogenic lesions. Other types of dimers are considered to be especially mutagenic. DNA-protein cross-links that are observed after ultraviolet radiation can be lethal. *See* PHOTOCHEMISTRY.

Survival from ultraviolet irradiation is reduced as the dose of radiation is increased. The shapes of survival curves are similar to those for lethality from ionizing radiation, they are dependent on the presence or absence of repair systems. The four repair systems that enhance biological survival include photoreactivation (splitting of cyclobutane dimers in the DNA of cells that have been irradiated by ultraviolet light); DNA excision repair; DNA recombination repair; and an inducible repair system of bacteria known as SOS repair. *See* ULTRAVIOLET RADIATION; ULTRAVIOLET RADIATION (BIOLOGY).

Microwaves. Microwaves are electromagnetic radiation in the region from 30 MHz to 300 GHz. They originate from devices such as telecommunications equipment and microwave ovens. Thermal effects of microwaves occur at exposure rates greater than 10 mW/cm^2 (70 mW/in.^{2}), while nonthermal effects are associated with exposure rates less than 10 mW/cm^2. Material with a high water content will have a higher absorption coefficient for microwaves, and thus a greater thermal response to microwave action. Microwave absorption is high in skin, muscle, and internal organs, and lower in bone and fat tissue. *See* MICROWAVE.

Cultured mammalian cells exposed to microwaves at a high power density show chromosome abnormalities after 15 min of exposure. Progression through the cell cycle is also temporarily interrupted, which interrupts DNA synthesis. Chromosome aberrations in peripheral blood lymphocytes are significantly greater for persons who are occupationally exposed to microwaves.

Microwaves can be lethal when the power intensity and exposure time are sufficient to cause a rise in temperature that exceeds an organism's homeostatic capabilities.

There are also some nonthermal effects associated with microwaves. A list of clinical symptoms includes increased fatigue, periodic or constant headaches, extreme irritability, decreased hearing acuity, and drowsiness during work. Laboratory studies involving exposure of animals to microwaves have produced changes in the electroencephalogram, blood-brain barrier, central nervous system, hematology, and behavior. Cell membrane permeability is also altered.

Extremely low frequency electromagnetic fields. This type of radiation is generated by the electric and magnetic fields associated with high-voltage current in power transmission lines, and also some household and industrial electrical equipment. Biological effects from ELF radiation are the least understood, and the potential consequences are the most controversial. The issue of potential biological damage from this type of radiation has arisen only since the introduction of very high voltage electric power transmission lines (440 kV and above) and the occurrence of widespread use of various electrical and electronic equipment. *See* ELECTROMAGNETIC PULSE (EMP).

Biological studies on ELF electromagnetic fields have been performed on cells and whole animals; and epidemiological studies have been carried out on populations exposed occupationally. The results share some common features: (1) There is not always a clear dose response; that is, increasing the exposure does not necessarily give rise to an increased biological effect. (2) Some biological effects are seen only at certain frequencies and dose rates. Some of the reported effects are subjective, and may be related to normal physiological adaptation to environmental changes. In humans, qualitative biological effects of low-frequency radiation (0 to 300 Hz) include headaches, lethargy, and decreased sex drive. Humans have been noted to perceive the presence of a 60-Hz electric field when the intensity is in the range of 2 to 12 kV/m (0.6 to 3.6 kV/ft), and animals were observed to avoid entering an area where the electric field was greater than 4 kV/m (1.2 kV/ft). *See* RADIATION INJURY (BIOLOGY).

Phillip M. Achey

Radiation chemistry The study of chemical changes resulting from the absorption of high-energy, ionizing radiation, including alpha particles, electrons, gamma rays, fission fragments, protons, deuterons, helium nuclei, and heavier charged projectiles. In absorbing materials of low and intermediate atomic weight such as aqueous systems and most biological systems, such radiation deposits energy in a largely indiscriminate manner, leaving behind a complex mixture of short-lived ions, free radicals, and electronically excited molecules. Radiation-induced chemical changes result from reaction with these intermediates. *See* PHOTOCHEMISTRY.

Sources of high-energy radiation include radioactive nuclides [for example, cobalt-60 (^{60}Co), strontium-90 (^{90}Sr), and hydrogen-3 (^{3}H)] and instruments such as x-ray tubes, Van de Graaff generators, the betatron, the cyclotron, and the synchrotron. An electron accelerator known as the Linac (linear electron accelerator) has proved particularly valuable for the study of transient species that have lifetimes as short as 16 picoseconds; and another electron accelerator, known as the Febetron, has been used for the study of the effects of single pulses of electrons with widths of several nanoseconds at very high currents.

The primary absorption processes for high-energy radiation are ionization and molecular excitation. The distribution of the absorbed energy, however, depends significantly upon the nature of the radiation and absorbing medium.

Evaluation of the yields of radiation-induced reactions requires knowledge of the energy imparted to the reacting system. The energy deposited in the system is termed the dose, and the measurement process is called dosimetry. Absorbed energy from ionizing radiation is described in terms of grays (Gy; joule/kg), in rads (100 ergs/g), or in electronvolts per gram or per cubic centimeter.

Because of its importance in both chemical and biological systems, the radiation chemistry of water has been extensively studied and serves as an example of radiation-induced chemical change. A primary radiation interaction process may be represented by the reaction below, where H_2O^* represents an electronically excited water molecule.

$$H_2O \rightarrow H_2O^+, e^-, H_2O^*$$

The secondary electron (e^-), if formed with sufficient energy, will form its own trail of ionization and excitation. Within 10^{-10} to 10^{-8} s, reactions within spurs form hydrogen (H) atoms, hydroxyl (OH) radicals, hydrated electrons and molecular products, molecular hydrogen (H_2), and hydrogen peroxide (H_2O_2). In pure water, radicals escaping the spurs undergo further radical-radical reactions and reactions with molecular products. Upon continuous irradiation, steady-state concentrations of H_2, H_2O_2, and smaller amounts of dioxygen (O_2) result and no further decomposition occurs.

In addition to basic kinetics and mechanistic studies, the principles of radiation chemistry find application in any process in which ionizing radiation is used to study, treat, or modify a biological or chemical system.

In radiation therapy, tumors are destroyed by the application of ionizing radiation from external or internally administered sources. Gamma rays are used for treatment of internal tumors; electron or charged-particle beams are applied to external or invasively accessible lesions.

The physiological concentration of iodine in the thyroid is the basis for the treatment of hyperthyroidism with ^{131}I. The beta radiation from this isotope is effective in localized tissue destruction.

A goal of any radiation therapy is maximum tumor cell destruction with minimum damage to healthy cells. *See* RADIATION THERAPY.

Radiation chemistry is used in food preservation by using ionizing radiation in doses that are lethal to microorganisms. The use of ionizing radiation for pathogen control has been approved by most governments for a wide range of foods. In general, limitations on dose have been specified for all products. Radiation is used to control pathogens in meat and meat products.

Processing of commercial quantities of food supplies requires a source of stable intensity and a radiation of sufficient penetrating power to deposit energy throughout the product in an economic, brief time period.

Francis J. Johnston

Radiation damage to materials Harmful changes in the properties of liquids, gases, and solids, caused by interaction with nuclear radiations. For a discussion of radiation damage in minerals *see* METAMICT STATE. For a description of damage caused to biological systems by radiation *see* RADIATION BIOLOGY.

Radiation damage is usually associated with materials of construction that must function in an environment of intense high-energy radiation from a nuclear reactor. Materials that are an integral part of the fuel element or cladding and nearby structural components are subject to such intense nuclear radiation that a decrease in the useful lifetime of these components can result. Radiation damage will also be a factor in thermonuclear reactors. *See* NUCLEAR FUSION; NUCLEAR REACTOR.

Electronic components are extremely sensitive to even moderate radiation fields. Transistors malfunction because of defect trapping of charge carriers. Ferroelectrics such as $BaTiO_3$ fail because of induced isotropy; quartz oscillators change frequency and ultimately become amorphous. High-permeability magnetic materials deteriorate because of hardening. Plastics used for electrical insulation rapidly deteriorate.

There are several mechanisms that function on an atomic and nuclear scale to produce radiation damage in a material if the radiation is sufficiently energetic, whether it be electrons, protons, neutrons, x-rays, fission fragments, or other charged particles.

Electronic excitation and ionization is most severe in liquids and organic compounds and appears in a variety of forms such as gassing, decomposition, viscosity changes, and polymerization in liquids. Rapid deterioration of the mechanical properties of plastics takes place either by softening or by embrittlement, while rubber suffers severe elasticity changes at low fluxes. Cross-linking, scission, free-radical formation, and polymerization are the most important reactions. *See* RADIATION CHEMISTRY.

In an environment of neutrons, transmutation effects may be important. Even materials that have a low cross section such as aluminum can show an appreciable accumulation of impurity atoms from transmutations. The elements boron and europium have very large cross sections and are used in control rods. Damage to the rods is severe in boron-containing materials. *See* TRANSMUTATION.

Displaced atoms are the most important source of radiation damage in nuclear reactors outside the fuel element. It is a consequence of the ability of the energetic neutrons bom in the fission process to knock atoms from their equilibrium position in their crystal lattice, displacing them many atomic distances away into interstitial positions and leaving behind vacant lattice sites.

Nuclear irradiations performed at low temperatures (4 K) result in the maximum retention of radiation-produced defects. As the temperature of irradiation is raised, many of the defects are mobile and some annihilation may take place. The increased mobility, particularly of vacancies and vacancy agglomerates, may lead to acceleration of solid-state reactions, such as precipitation, short- and long-range ordering, and phase changes. These reactions may lead to undesirable property changes.

The presence of small amounts of impurities may profoundly affect the behavior of engineering alloys in a radiation field. It has been observed that helium concentrations as low as 10^{-9} seriously reduce the high-temperature ductility of a stainless steel. Small amounts of copper, phosphorus, and nitrogen have a strong influence on the increase in the ductile-brittle transition temperature of pressure vessel steels under irradiation. Heat treatment prior to irradiation determines the retention of both major alloying components and impurities in solid solution in metastable alloys. It also affects the number and disposition of dislocations. Thus heat treatment is an important variable in determining subsequent radiation behavior. Douglas S. Billington

Radiation hardening The protection of semiconductor electronic devices and electronic systems from the effects of high-energy radiation. Applications for such devices are in three major areas: (1) satellites, which are exposed to natural space radiation from the Van Allen belts, solar flares, and cosmic rays; (2) electronics, especially sensor and control electronics for commercial nuclear power-generating plants; and (3) equipment designed to survive the radiation from nuclear explosions.

Although most radiation effects have been explained and are well understood, much remains to be done. Among the better-understood phenomena are displacement damage from neutrons and photocurrent transients produced by ionizing-radiation pulses. Basic electromagnetic-pulse interactions are also well understood, although their effects on complex electronic systems are extremely difficult to predict. *See* ELECTROMAGNETIC PULSE (EMP).

The quasipermanent effects of exposure to ionizing radiation are least understood, and understanding the response of semiconductor devices to this radiation is probably the single most important remaining radiation-hardening problem. Hardening of metal-oxide-semiconductor (MOS) devices has been accomplished by lower-temperature processing, which probably reduces physical or crystalline defects, and by developing extremely clean processes, which probably reduce chemical defects. The most important electrical manifestations of ionizing dose damage in MOS devices are an increase in leakage current; a shift in threshold voltage; and a decrease in speed, transconductance, and channel conductance. *See* CRYSTAL.

Dose-rate effects are well understood. The generation of electron-hole pairs in semiconductors is proportional to the dose. Carriers generated in or near a *pn* junction result in a transient photocurrent proportional to dose rate and the effective volume of the junction. High dose rates can damage semiconductor devices through logic upset, latch-up, and burnout. *See* PHOTOVOLTAIC EFFECT.

Displacement damage effects are caused by neutrons, protons, electrons, and other high-energy particles. The production of lattice defects is proportional to the nonionizing energy absorbed by the lattice. The dominant effect in bipolar silicon devices is a reduction in common-emitter current gain.

Single-event upsets are caused at a very low rate in logic and memory circuits by cosmic rays. Rates are low enough (less than 10^{-3} upset per bit per day) that error-detection-and-correction (EDAC) software can be effectively used. Single-event phenomena have led to the development of fault-tolerant architectures to mitigate the effects of random digital upsets, and to the design modification of integrated circuits to prevent an upset even when the active volume is struck by a cosmic ray. *See* FAULT-TOLERANT SYSTEMS.

Hardening of electronic systems is accomplished by a combination of selecting hardened components, designing circuits more tolerant to radiation-induced degradations, and shielding. Shielding is effective against x-rays and electrons, which have short ranges in materials, and relatively ineffective against gamma radiation, neutrons, and cosmic rays, which have long ranges in materials. *See* INTEGRATED CIRCUITS; RADIATION DAMAGE TO MATERIALS; RADIATION SHIELDING; SEMICONDUCTOR.

George C. Messenger

Radiation injury (biology) The harmful effects of ionizing radiation on plant and animal cells and tissues. The mechanisms of these changes are still poorly known. In multicellular organisms, effects on cells are complicated by the interaction of injured and intact cells. Consequently, proper understanding of radiation injury in such organisms calls for appreciation not only of the reaction of individual cells but also of groups of cells in organs and tissues. *See* GAMMA RAYS; X-RAYS.

Radiation is known to alter the genetic apparatus of the cell, to interfere with cell division, and to cause cell death. The most important action of radiation on the cell is the production of changes in the genetic or chromosomal apparatus. Once established, these changes are largely irreversible and, because of the strategic importance of each gene, may alter the fate of the cell and of its progeny as well. The changes include mutations and chromosome breaks. *See* MUTATION.

Cells differ markedly in their susceptibility to radiation-induced death, but any cell may be killed if it receives enough radiation. Although radiation death may occur during or immediately after irradiation, more often it ensues when the cell attempts to divide or after it has divided several times. The mechanism by which radiation kills cells is unknown, but the high radiosensitivity of dividing cells, with their tendency to die during cell division, suggests that the most critical type of injury involves the reproductive or chromosomal apparatus of the cell.

Tissues differ widely in radiosensitivity, but in general their susceptibility to radiation varies with the rate at which their component cells divide. Accordingly, the most radiosensitive tissues of the body, composed of cells that divide rapidly, are the blood-forming organs, gonads, skin, and intestine.

When the entire body is irradiated, killing of cells in the various radiosensitive organs causes a complex series of disturbances, the predominant signs and symptoms of which are referable to injury of the intestinal tract, blood-forming organs, and skin (acute radiation syndrome). The injury, if severe enough, is fatal within 30 days in most laboratory animals, but in humans death may be delayed until the second month after irradiation. Susceptibility to death varies from species to species, the average median lethal radiation dose for mammals being about 5 Gy.

It is paradoxical that ionizing radiation, which is effective in the treatment of cancer, should also be capable of causing cancer. As such, however, it is but one of many agents, including ultraviolet rays, viruses, and a variety of chemicals, that are known to have cancer-inducing potency. The earliest known example of radiation-induced cancer was reported in 1902, less than 10 years after the discovery of x-rays. Since then, numerous instances have been observed in humans, and the process has been studied extensively in experimental animals. *See* ONCOLOGY.

The essential change caused by radiation that leads ultimately to the development of cancer is unknown, as is the nature of the cancer change itself. There is growing evidence, however, that the transformation of a normal cell into a cancer cell entails a series of alterations and is not merely a one-step process. This is suggested by the relatively long induction period intervening between irradiation and the onset of malignant growth, a period averaging 5–20 years in humans, depending on the type of cancer induced. *See* CANCER (MEDICINE).

Plant cells have been used extensively for studying the genetic effects of radiation and for investigating the effects of radiation on chromosomes and on cell division. Relatively little work has been done in plants, however, on radiation injury at the tissue or organ level. From available evidence, it appears that radiation effects on plant and animal tissues are not qualitatively different if allowances are made for physiological discrepancies between the two types of organisms. *See* RADIATION BIOLOGY. Arthur C. Upton

Radiation pressure The force on an object exposed to electromagnetic radiation. Electromagnetic radiation (which includes visible light) carries both energy and momentum. If radiation impinges on a material body and becomes absorbed, the energy gives rise to heat and is readily detectable. When radiation interacts with an object and is absorbed or scattered, there is also a change in the momentum of the light. By conservation of momentum, this gives rise to a force on the object. This is called radiation pressure. The magnitude of this momentum for visible light is quite small and is difficult to detect. Only near or inside stars, where the intensity is enormous, do light forces have large effects. *See* ELECTROMAGNETIC RADIATION; LIGHT; MAXWELL'S EQUATION; POYNTING'S VECTOR.

This situation is different with lasers. Lasers generate coherent light beams and can be focused to spot sizes of about one wavelength, giving rise to very high intensities and intensity gradients, using modest total powers. Using Planck's law, the energy of a light photon is $E_{\text{photon}} = h\nu$, where h is Planck's constant and ν is the frequency. From Einstein's law, the photon's energy in terms of its effective mass is $E_{\text{photon}} = m_{\text{effective}}\, c^2$, where c is the velocity of light. The photon momentum then is $m_{\text{effective}}\, c = h\nu/c$. If, for example, a laser beam strikes a 100% reflecting mirror, it generates a radiation pressure force $F_{\text{rad}} = (P/h\nu)\ (2h\nu/c) = 2P/c$, where P is the power carried by the beam. For $P = 1$ watt, $F_{\text{rad}} \cong 10^{-8}\ \text{N} = 10^{-3}$ dynes. For a 1 micrometer-sized particle, with a density of about 1 g/cm^3 the acceleration due to the radiation is $A = F/m \cong 10^{-3}/10^{-12} = 10^9\ \text{cm/s}^2 \cong 10^6\ g$, where g is the acceleration of gravity, which is very large and should be readily observable. *See* LASER; PHOTON; PLANCK'S CONSTANT; QUANTUM MECHANICS; RELATIVITY.

Particle guidance and trapping. In 1970 Arthur Ashkin showed experimentally that a fraction of a watt of laser power striking a small transparent polystyrene particle several micrometers in diameter was sufficient to propel it many diameters per second through water in the direction of the light. This force in the direction of the light, called the scattering force, agreed with the above calculation. However, it was discovered that, when the particle was located off the beam axis in the region of a strong gradient of light intensity, there was an additional force component pushing the particle into the high-intensity region at the center of the beam. This component of force was termed the gradient component. It served to guide the particle and keep it on axis as it moved along the beam.

The understanding of the magnitude and properties of these two basic force components made it possible to devise the first stable three-dimensional (3D) optical trap for single neutral particles. At this time it was proposed that similar stable trapping of atoms was possible using the same two force components.

In 1973 Theo W. Hänsch and Arthur L. Shawlow proposed using the Doppler shift to provide optical damping or cooling of an atomic gas. If a pair of opposing laser beams in one dimension is tuned below an atomic resonance, then an atom moving in either direction feels a net damping because it is Doppler-shifted to a higher frequency, closer to resonance for the opposing beam, and to a lower frequency, away from resonance, for the trailing beam. With three pairs of beams in three dimensions, all the degrees of freedom can be damped. This creates what is known as an optical molasses. Residual fluctuations of the scattering force limit the cooling to a temperature known as the Doppler limit. For sodium atoms this is about 240 microkelvin. A single-beam gradient trap, later referred to as a tweezers, is the simplest of all traps, consisting of a single strongly focused Gaussian beam. *See* DOPPLER EFFECT; LASER COOLING; PARTICLE TRAP.

In 1978 experiments showing focusing and guiding of atoms in detuned light beams marked the beginning of what is known as atom optics, and also suggested that stable atom traps were feasible. By 1984 William D. Phillips and Wolfgang Ertmer succeeded in overcoming the strong Doppler shifts involved in slowing thermal atomic beams to a temperature of about 1 K, using the scattering force of an opposing light beam.

Overcoming the Doppler limit. In 1988 Phillips and his colleagues showed that by changing the molasses detuning, it was possible to reach a new cooling regime, where the atom's temperature was around 20 microkelvin, ten times lower than the Doppler limit. Applications of these ultracold atoms soon appeared, such as improved atomic clocks and atom interferometers. *See* ATOMIC CLOCK.

Biological applications. In experiments in 1987, Ashkin and J. M. Dziedzic discovered that infrared tweezers could trap living bacteria, viruses, and other biological organisms without causing optical damage. It was even possible to optically manipulate organelles inside larger cells and observe the mechanical properties of cytoplasm. This initiated the application of laser tweezers to biology.

Bose-Einstein condensation. Experiments have shown the coherence of atoms in Bose-Einstein condensates, and crude atom lasers have been demonstrated. Applications of Bose-Einstein condensates are possible: in Josephson devices, coherent atom lithography, quantum computers and cryptography, and even more accurate atomic clocks. *See* ATOM LASER; BOSE-EINSTEIN CONDENSATION. Anthur Ashkin

Radiation shielding Physical barriers designed to provide protection from the effects of ionizing radiation; also, the technology of providing such protection. Major sources of radiation are nuclear reactors and associated facilities, medical and industrial x-ray and radioisotope facilities, charged-particle accelerators, and cosmic rays. Types of radiation are directly ionizing (charged particles) and indirectly ionizing (neutrons, gamma rays, and x-rays). In most instances, protection of human life is the goal of radiation shielding. In other instances, protection may be required for structural materials which would otherwise be exposed to high-intensity radiation, or for radiation-sensitive materials such as photographic film and certain electronic components.

Charged particles lose energy and are thus attenuated and stopped primarily as a result of coulombic interactions with electrons of the stopping medium. Gamma-ray and x-ray photons lose energy principally by three types of interactions: photoemission, Compton scattering, and pair production. Neutrons lose energy in shields by elastic or inelastic scattering. Elastic scattering is more effective with shield materials of low atomic mass, notably hydrogenous materials, but both processes are important, and an efficient neutron shield is made of materials of both high and low atomic mass. *See* COMPTON EFFECT; ELECTRON-POSITRON PAIR PRODUCTION; PHOTOEMISSION.

The most common criteria for selecting shielding materials are radiation attenuation, ease of heat removal, resistance to radiation damage, economy, and structural strength.

For neutron attenuation, the lightest shields are usually hydrogenous, and the thinnest shields contain a high proportion of iron or other dense material. For gamma-ray attenuation, the high-atomic-number elements are generally the best. For heat removal, particularly from the inner layers of a shield, there may be a requirement for external cooling with the attendant requirement for shielding the coolant to provide protection from induced radioactivity.

Metals are resistant to radiation damage, although there is some change in their mechanical properties. Concretes, frequently used because of their relatively low cost, hold up well; however, if heated they lose water of crystallization, becoming somewhat weaker and less effective in neutron attenuation.

If shielding cost is important, cost of materials must be balanced against the effect of shield size on other parts of the facility, for example, building size and support structure. If conditions warrant, concrete can be loaded with locally available material such as natural minerals (magnetite or barytes), scrap steel, water, or even earth.

Radiation shields vary with application. The overall thickness of material is chosen to reduce radiation intensities outside the shield to levels well within prescribed limits for occupational exposure or for exposure of the general public. The reactor shield is usually considered to consist of two regions, the biological shield and the thermal shield. The thermal shield, located next to the reactor core, is designed to absorb most of the energy of the escaping radiation and thus to protect the steel reactor vessel from radiation damage. It is often made of steel and is cooled by the primary coolant. The biological shield is added outside to reduce the external dose rate to a tolerable level. *See* RADIATION BIOLOGY; RADIATION DAMAGE TO MATERIALS. Richard E. Faw

Radiation therapy The use of ionizing radiation to treat disease; the method is also known as radiotherapy and therapeutic radiology. Radiotherapy was widely used in the past to treat diseases of the skin, lymph nodes, and other organs. However, because radiation can cause cancer and because alternative treatments for these diseases have been discovered, radiation therapy is now mainly limited to treating malignant tumors: the medical specialty is called radiation oncology. *See* ONCOLOGY.

The exact mechanisms by which radiation kills cells remain uncertain. Most likely, electrons dislodged from water or biological molecules disrupt the bonds between atoms of the nuclear deoxyribonucleic acid (DNA), resulting in double-strand breaks. Although usually not immediately fatal, such damage may cause the death of cells when they attempt to divide. Complex enzymatic mechanisms can repair some of the damage if given sufficient time.

The fraction of cells surviving after irradiation depends on many factors. Radiation affects both normal and cancerous cells in a similar manner qualitatively, but different cell lines vary greatly in their quantitative sensitivity. Rapidly dividing tissues—such as the skin, bone marrow, and gastrointestinal mucosa—usually display the greatest sensitivity experimentally and the most immediate side effects clinically.

The goal of radiation therapy is either to cure the disease permanently (radical treatment) or to reduce or eliminate symptons (palliative treatment) by destroying tumors without causing unacceptable injuries to normal tissues. For many cancers—such as cancers of the reproductive organs, lymphomas, and small head and neck tumors—a cure usually is possible, and the chance of functionally significant complications is small. Some tumors, however, contain too many cells to be entirely destroyed by tolerable doses. Also, some neoplasms, such as sarcomas and glioblastomas, are relatively resistant to irradiation and are difficult to eradicate even when only small numbers of cells are present. Such situations are best handled by using surgery to remove all visible tumor. *See* CANCER (MEDICINE); RADIATION BIOLOGY. Abram Recht

Radiative transfer The study of the propagation of energy by radiative processes; it is also called radiation transport. Radiation is one of the three mechanisms by which energy moves from one place to another, the other two being conduction and convection. *See* ELECTROMAGNETIC RADIATION; HEAT TRANSFER.

The kinds of problems requiring an understanding of radiative transfer can be characterized by looking at meteorology, astronomy, and nuclear reactor design. In meteorology, the energy budget of the atmosphere is determined in large part by energy gained and lost by radiation. In astronomy, almost all that is known about the abundance of elements in space and the structure of stars comes from modeling radiative transfer processes. Since neutrons moving in a reactor obey the same laws as radiation being scattered by atmospheric particles, radiative transfer plays an important part in nuclear reactor design.

Each of these three fields—meteorology, astronomy, and nuclear engineering—concentrates on a different aspect of radiative transfer. In meteorology, situations are studied in which scattering dominates the interaction between radiation and matter; in astronomy, there is more interest in the ways in which radiation and the distribution of electrons in atoms affect each other; and in nuclear engineering, problems relate to complicated, three-dimensional geometry.

Radiative transfer is a complicated process because matter interacts with the radiation. This interaction occurs when the photons that make up radiation exchange energy with matter. These processes can be understood by considering the transfer of visible light through a gas made up of atoms. Similar processes occur when radiation interacts with solid dust particles or when it is transmitted through solids or liquids. *See* PHOTON.

If a gas is hot, collisions between atoms can convert the kinetic energy of motion to potential energy by raising atoms to an excited state. Emission is the process which releases this energy in the form of photons and cools the gas by converting the kinetic energy of atoms to energy in the form of radiation. The reverse process, absorption, occurs when a photon raises an atom to an excited state, and the energy is converted to kinetic energy in a collision with another atom. Absorption heats the gas by converting energy from radiation to kinetic energy. Occasionally an atom will absorb a photon and reemit another photon of the same energy in a random direction. If the photon is reradiated before the atom undergoes a collision, the photon is said to be scattered. Scattering has no net effect on the temperature of the gas. *See* ABSORPTION; ATOMIC STRUCTURE AND SPECTRA; SCATTERING OF ELECTROMAGNETIC RADIATION. Alan H. Karp

Radiator Any of numerous devices, units, or surfaces that emit heat, mainly by radiation, to objects in the space in which they are installed. Because their heating is usually radiant, radiators are of necessity exposed to view. They often also heat by conduction to the adjacent thermally circulated air.

Radiators are usually classified as cast-iron (or steel) or nonferrous. They may be directly fired by wood, coal, charcoal, oil, or gas (such as stoves, ranges, and unit space heaters). The heating medium may be steam, derived from a steam boiler, or hot water, derived from a water heater, circulated through the heat-emitting units.

Electric heating elements may be substituted for fluid heating elements in all types of radiators, convectors, and unit ventilators. *See* HOT-WATER HEATING SYSTEM; RADIANT HEATING; STEAM HEATING. Erwin L. Weber; Richard Koral

Radio Communication between two or more points, employing electromagnetic waves ("carriers") as the transmission medium. *See* ELECTROMAGNETIC RADIATION.

At the start of the twenty-first century, radio transmission services are in the midst of a fundamental transition from traditional (and increasingly antiquated) analog technology to highly sophisticated and advanced digital technology. The basic methods of analog transmission and reception of audio signals were developed principally in the twentieth century. For transmission, an audio pulse and a carrier pulse are blended into a modulated carrier wave, which is then amplified and fed into the antenna. The process of blending audio and carrier waves is called modulation. For reception, the receiving antenna and tuner catch the weak signal, amplify it, sort the audio pulse from the carrier, and play a now reamplified audio pulse through the speaker at home. Digital radio techniques differ from analog in the nature of the information being transmitted (audio represented digitally, as a sequence of 1's and 0's) and the methods of modulation employed. Digital techniques are significantly more flexible, robust (that is, less susceptible to interference and other deleterious effects), and more efficient than the analog techniques they replace. *See* ANTENNA (ELECTROMAGNETISM); ELECTRICAL COMMUNICATIONS; MODULATION; RADIO RECEIVER; RADIO TRANSMITTER; RADIO-WAVE PROPAGATION.

Radio waves transmitted continuously, with each cycle an exact duplicate of all others, indicate that only a carrier is present (also called an unmodulated carrier). The two principal characteristics of a carrier are its frequency (the number of cycles per second) and amplitude (which determines how much power is contained in the signal). The information desired to be transmitted must cause changes in the carrier which can be detected at a distant receiver. The method used for the transmission of the information is determined by the nature of the information which is to be transmitted as well as by the purpose of the communication system.

In code transmission, the carrier is keyed on and off to form dots and dashes. The technique, traditionally used in ship-to-shore and amateur communications, has been almost entirely superseded by more efficient methods.

In frequency-shift transmission, the frequency of the carrier is shifted a fixed amount to correspond with telegraphic dots and dashes or with combinations of pulse signals identified with the characters on a typewriter. This technique was widely used in handling the large volume of public message traffic on long circuits, principally by the use of teletypewriters. *See* TELETYPEWRITER.

In amplitude modulation, the amplitude of the carrier is varied in a way that conforms to the fluctuations of a sound wave, or, in the case of digital transmission, the specific sequence of 1's and 0's being transmitted. This technique is used in amplitude-modulation (AM) terrestrial radio broadcasting, analog television picture transmission, and many other services. One widely used digital method, called quadrature AM (QAM), is employed in some terrestrial digital radio systems. *See* AMPLITUDE MODULATION; AMPLITUDE-MODULATION RADIO.

In frequency modulation, the frequency of the carrier is varied in a way that conforms to the fluctuations of the modulating wave. This technique is used in frequency-modulation (FM) broadcasting, analog television sound transmission, and analog microwave relaying. *See* FREQUENCY MODULATION; FREQUENCY-MODULATION RADIO.

In phase modulation, the phase of the carrier is varied in a way that conforms to the fluctuations of the modulating wave. A family of popular digital methods, known as phase-shift keying (PSK), are used in digital satellite broadcasting. Specific forms include binary PSK (BPSK), quadrature PSK (QPSK), and octal PSK (8PSK), differentiated by the number of digital bits used to create each transmitted symbol. *See* PHASE MODULATION; PHASE-MODULATION-DETECTOR; PHASE MODULATOR.

In pulse transmission, the carrier is transmitted in short pulses, which change in repetition rate, width, or amplitude, or in complex groups of pulses which vary from group to succeeding group in accordance with the message information. These forms of pulse transmission are identified as pulse-code, pulse-time, pulse-position, pulse-amplitude, pulse-width, or pulse-frequency modulation. These complex techniques are employed principally in microwave relay and telemetry systems. *See* PULSE DEMODULATOR; PULSE MODULATION; PULSE MODULATOR.

In radar, the carrier is normally transmitted as short pulses in a narrow beam, similar to that of a searchlight. When a wave pulse strikes an object, such as an aircraft, energy is reflected to the radar station, which measures the round-trip time and converts it to distance. A radar receiver can display varying reflections in a maplike presentation on a video display. *See* RADAR.

Hundreds of thousands of radio transmitters exist, each requiring a carrier at some radio frequency. To prevent interference, different carrier frequencies are used for stations whose service areas overlap, and receivers are built which can select the carrier signal of the desired station. Resonant electronic circuits in the receiver are adjusted, or tuned, to accept the desired frequency and reject others. Digital technology, in particular the availability of high-speed, low-cost microprocessors and analog-to-digital converters, has made possible the development of software radios which select and demodulate radio signals according to a computer program which runs on a microprocessor. Software radios are more flexible and offer higher performance than do traditional radios.

All nations have a sovereign right to use freely any or all parts of the radio spectrum. But numerous international agreements and treaties divide the spectrum and specify sharing among nations for their mutual benefit and protection. Each nation designates its own regulatory agency. In the United States all nongovernmental radio communications are regulated by the Federal Communications Commission (FCC). *See* AMATEUR RADIO; RADIO BROADCASTING; RADIO SPECTRUM ALLOCATION.

David H. Layer; John D. Singleton; Michael C. Rau

Radio astronomy The study of celestial bodies by examination of the energy they emit at radio frequencies. Celestial radio noise originates by one of several processes. These include both broadband continuum radiation owing to (1) thermal radiation from solid bodies such as the planets, (2) thermal or bremsstrahlung radiation from hot gas in the interstellar medium, (3) synchrotron radiation from ultrarelativistic electrons moving in weak magnetic fields, (4) coherent processes as found in the Sun and on Jupiter, and (5) pulsed radiation from the rapidly rotating neutron stars, as well as (6) narrow "spectral line" radiation from atomic or molecular transitions that occur in the interstellar medium or in the gaseous envelopes around stars.

Radio astronomy observations cover the entire radio spectrum from less than 1 mm, where the cosmic signals become heavily attenuated by the atmosphere, to wavelengths of tens of meters, beyond which the incoming signals are attenuated by the ionosphere. However, radio waves between about 1-cm

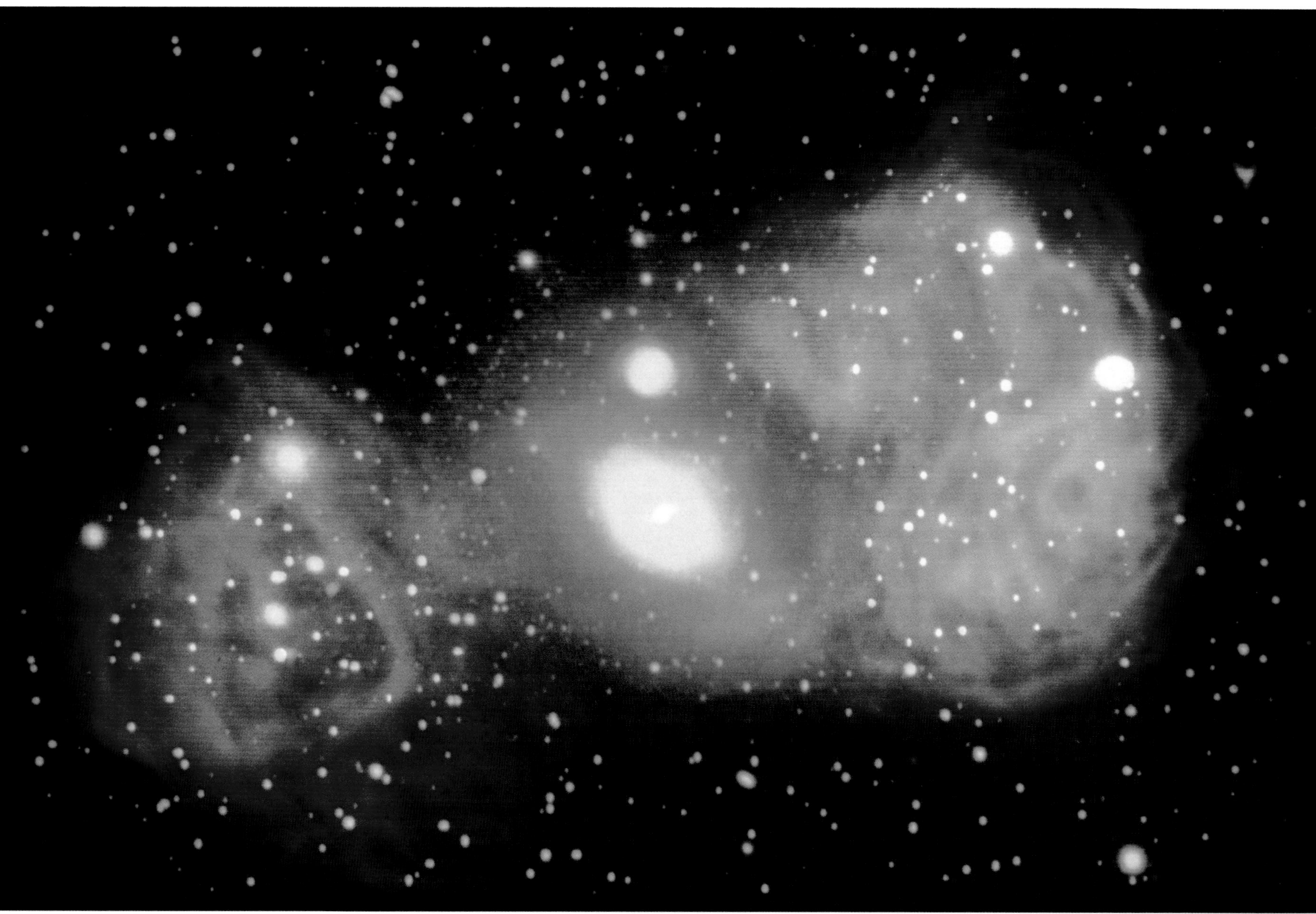

Image of the radio emission from the bright galaxy NGC 1316 shown in red with the visible emission superimposed in blue and white. The small galaxy above NGC 1316 is NGC 1317. The interaction between the two galaxies may help to fuel the black hole which energizes the radio source. (*NRAO/AUI/NSF*)

and 1-m wavelength not only get through the Earth's atmosphere and ionosphere with little distortion, they also penetrate much of the gas and dust in space as well as the clouds of planetary atmospheres. Radio astronomy can therefore sometimes give a much clearer picture of stars and galaxies than is possible by means of conventional observation using visible light. Sophisticated antennas equipped with very sensitive radio receivers that are used to detect cosmic radio emission are referred to as radio telescopes.

Spectral line emission occurs at specific wavelengths characteristic of the atomic and molecular species. However, due to the motion of gas clouds, the wavelength is shifted toward longer wavelengths if the source is moving away from the observer and toward shorter wavelengths if it is moving toward the observer. Precise measurements of this so-called Doppler shift are used to determine the radial velocities due to the random motion of gas clouds, the rotation of the Galaxy, and even the expansion of the universe. *See* DOPPLER EFFECT; REDSHIFT.

The construction of ever larger antenna systems and radio interferometers, improved radio receivers, and data processing methods have allowed radio astronomers to study fainter radio sources with increased angular resolution and improved image quality. Today more than 1 million sources of radio emission are known. The precise measurement of their position in the sky has usually allowed their identification with an optical, infrared, x-ray, or gamma-ray counterpart; however, some cosmic radio sources remain unidentified at other wavelengths even with the most sensitive instrumentation.

Solar system radio astronomy. Solar flares and sunspots are strong sources of radio emission. Their study has led to increased understanding of the complex phenomena near the surface of the Sun and provides advanced warning of dangerous solar flares which can interrupt radio communications on the Earth and endanger sensitive equipment in satellites and the health of astronauts. Even in the absence of activity, the quiet Sun is the strongest radio source in the sky, but only because it is so close to the Earth.

Radio telescopes are used to measure the surface temperatures of all the planets as well as some of the moons of Jupiter and Saturn. Radio studies of the planets have revealed the existence of a greenhouse effect on Venus, intense Van Allen belts surrounding Jupiter, powerful radio storms in the Jovian atmosphere, and an internal heating source deep within the interiors of Jupiter, Saturn, Uranus, and Neptune.

Astronomers also use radar observations to image features on the surface of Venus, which is completely obscured from visual scrutiny by the heavy cloud cover that permanently enshrouds the planet.

Radio emission from the Milky Way. Broadband continuum radio emission is observed throughout the radio frequency spectrum from a variety of stars including flare stars, binary stars, x-ray binaries, novae, and supernovae; from supernova remnants; and from magnetic fields and relativistic electrons in the interstellar medium thought to be produced in supernova explosions. One of the brightest observed radio sources in the Milky Way is the Crab Nebula, which is the remnant of a supernova observed by Chinese astronomers in the year 1054. High-resolution radio observations also show a weak, embedded, very compact radio source called Sgr A*, which is thought to lie at the precise center of the Galaxy (see illustration). Sgr A* is believed to be associated with a black hole containing the mass equivalent of 2 million suns within a region only 5 solar diameters across. *See* BLACK HOLE; CRAB NEBULA; MILKY WAY GALAXY; SUPERNOVA.

Atomic and molecular gas. The important spectral line of neutral atomic hydrogen (H I) is observed throughout the Milky Way Galaxy as well as in other spiral galaxies, although the galactic hydrogen is heavily concentrated toward the plane of the Milky Way. Measurements of the position and velocity of H I are used to trace out the spiral arms in the Milky Way system to study the rotation of the Galaxy.

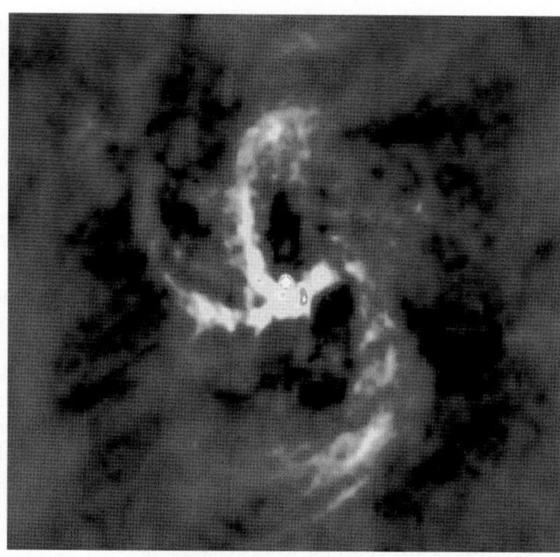

Radio image of the center of our Milky Way Galaxy known as Sagittarius A. The bright spot is the very compact radio source known as SgrA*, which is associated with the million-mass black hole at the very center of the Galaxy.

H I is also observed in nearby galaxies. The observed dependence of rotation velocity on distance from the center of the galaxy is referred to as a rotation curve and is used to study the distribution of mass in galaxies. Observations of rotation curves in a number of spiral galaxies show the presence of dark matter located far beyond the extent of visible stars and nebula. *See* DARK MATTER.

Molecular gas found in dense interstellar clouds radiates at radio wavelengths characteristic of each molecule. The wavelength of emission corresponds to transitions between different rotational and sometimes vibrational energy levels of the molecular species. More than 100 different molecular species have been observed, only a few of which can be observed at optical wavelengths. These include carbon monoxide, ammonia, water, ethyl and methyl alcohol, formaldehyde, and hydrogen cyanide as well as some heavy organic molecules. Most of the observed molecules are found only in dense giant molecular clouds, especially near the center of the Galaxy. Carbon monoxide has been detected in some of the most distant known galaxies and quasars. *See* INTERSTELLAR MATTER; MOLECULAR CLOUD; MOLECULAR STRUCTURE AND SPECTRA.

Pulsars. Pulsars (a term derived from pulsating radio stars) are neutron stars. Their electrons have all combined with protons in nuclei to form neutrons, and they have shrunk to a diameter of only a few kilometers following the explosion of the parent star in a supernova. Because they have retained the angular momentum of the much larger original star, neutron stars spin very rapidly, up to hundreds of times per second, and contain magnetic fields as strong as 10^8 tesla (10^{12} gauss) or more. The radio emission from pulsars is concentrated along a thin cone which produces a series of pulses corresponding to the rotation of the neutron star, much like a beacon from a rotating lighthouse lamp. The rotation rates of pulsars are extremely stable. They are the most precise clocks known in nature with pulse repetition rates accurate to better than 1 second in a million years. Pulsars are found mostly close to the plane of the Milky Way, especially toward the direction of the center of the Galaxy.

Galaxies, quasars, and active galactic nuclei (AGN). Normal galaxies, particularly spiral galaxies, are sources of weak radio emission comparable to the luminosity of our own Milky Way Galaxy. In some galaxies, regions of intense star formation release cosmic rays from supernovae causing intense radio emission which can be observed even at very great distances. Even more powerful are the radio galaxies, which can emit more than 100 million times more radio radiation than normal spiral galaxies do.

Quasars are found in the central regions of rare galaxies and may shine with the luminosity of up to 100 ordinary galaxies. While some quasars also contain distant radio lobes, more often the radio emission from quasars is confined to a very small region coincident with the optically visible quasar. Most quasars are only weak radio sources, and the reason why only some 10 to 15% of all quasars become strong radio sources is unclear. Quasars which are not powerful radio sources are referred to as radio-quiet quasars. *See* QUASAR.

The energy contained in the form of relativistic particles and magnetic fields surrounding radio galaxies and quasars is enormous, reaching in some cases as much as 10^{61} ergs. It is widely believed that the source of this energy is massive black holes at the center of the galaxy or quasar containing up to 10^9 times the mass of the Sun.

The strong compact radio sources associated with quasars and active galactic nuclei (AGN) are thought to be the result of relativistic boosting of synchrotron radiation from plasma clouds ejected toward the observer from galactic nuclei. But when the plasma clouds are ejected in other directions, astronomers see only the radiation from the large, stationary, more distant clouds characteristic of radio galaxies.

Sometimes the observed radio emission from a distant AGN, quasar, or galaxy passes very close to a massive galaxy or cluster of galaxies, which bends both the radio and optical waves so that astronomers may see more than one image arriving from different directions. This phenomenon is known as gravitational lensing. There may be multiple images or just a single ring image, known as an Einstein ring, surrounding the lens. *See* GRAVITATIONAL LENS.

The number or luminosity of radio galaxies increases with increasing distance from our Galaxy. Due to the finite signal travel time of radio waves, this means that in the early universe, conditions were very different from the local (current epoch) universe, and such studies provided the earliest evidence that we are living in an evolving or changing universe. *See* GALAXY, EXTERNAL.

Cosmic microwave background radiation. In 1965, Robert Wilson and Arno Penzias detected the faint comic microwave background (CMB) signal left over from the original big bang, which occurred some 14×10^9 years ago. As the universe expanded, it cooled to its present value of 2.7° above absolute zero. Subsequent observations with both ground- and space-based radio telescopes have detected the fine details in the CMB of less than one thousandth of a percent corresponding to the initial formation of structure in the early universe. These studies give new insight into the formation of galaxies and the rate of expansion of the universe, and show a surprising apparent acceleration of the expansion corresponding to an unknown form of dark energy. *See* ACCELERATING UNIVERSE; COSMIC BACKGROUND RADIATION; DARK ENERGY; RADIO TELESCOPE. Kenneth Kellermann

Radio broadcasting

Radio broadcasting The transmission, via radio-frequency electromagnetic waves, of audible program material for direct reception by the general public. Electromagnetic waves can be made to travel or propagate from a transmitting antenna to a receiving antenna. By modifying the amplitude, frequency, or relative phase of the wave in response to some message signal (a process known as modulation), it is possible to convey information from the transmitter to the receiver. In radio broadcasting, this information usually takes the form of voice or music. *See* ELECTRICAL COMMUNICATIONS; ELECTROMAGNETIC WAVE TRANSMISSION; MODULATION; RADIO.

Radio broadcasting occurs in seven frequency bands. Long-wave broadcasting is permitted by international agreement in a portion of the low-frequency band from 150 to 290 kHz in Europe. The most widely used broadcast band is in the medium-frequency (mf) range between 525 and 1700 kHz. It is commonly known as AM after amplitude modulation, the technique employed. Shortwave broadcasting is permitted worldwide in eight frequency bands between 5950 and 26,100 kHz. The very high frequency (VHF) band of 88 to 108 MHz is used for what is commonly called FM broadcasting, after frequency modulation that is used for transmissions. During the 1990s, a digital audio broadcasting (DAB) service in the 1452–1492- and 174–240-MHz frequency bands was put in place in Europe, Canada, and other countries. This band is unavailable in the United States for this service, so an alternate DAB system is being devised for use there. Radio broadcasting from satellites to listeners has been authorized in the 2310–2360-MHz frequency band. *See* AMPLITUDE MODULATION; AMPLITUDE-MODULATION RADIO; FREQUENCY MODULATION; FREQUENCY-MODULATION RADIO.

AM medium-frequency band. Broadcast stations in the medium-frequency band use amplitude modulation of a carrier wave to transmit information. The amplitude of the wave is modified in response to the changing amplitude of an audible voice or music signal. The AM receiver detects these amplitude changes and converts them back into audible signals, which can then be amplified and reproduced on acoustical transducers or speakers. *See* RADIO RECEIVER; RADIO TRANSMITTER.

The audible frequency range is generally considered to extend from 20 to 20,000 Hz. As a practical matter, AM broadcasting transmissions are limited to a range of 50 to 10,000 Hz. Because of transmission components and directional antennas, the fidelity of many stations is more severely restricted, resulting in voice transmission that is still acceptable but music transmission that is of relatively low fidelity.

An AM station may use a single tower for an antenna, resulting in an omnidirectional radiation pattern; or two or more transmitting towers to augment the radiation in certain directions while suppressing it in others, in order to comply with station allocation criteria. Since the allocation restrictions may be different during the daytime and nighttime hours, many stations employ two different directional antennas. The radiation from the antenna is expressed in millivolts per meter at 1 km (0.62 mi) from the antenna. *See* ANTENNA (ELECTROMAGNETISM).

AM broadcast signals propagate from the transmitter by three mechanisms: ground-wave, space-wave, and skywave. Ground waves travel along the ground surface (the boundary between the Earth and the atmosphere). Because they are surface waves, they penetrate into the ground, resulting in the energy being diminished because of losses in the ground. The current flowing in the antenna also produces space waves, which travel through the atmosphere from transmitter to receiver. Space-wave propagation is usually limited by intervening terrain obstacles or the curvature of the Earth. Sky-wave propagation occurs when space waves directed toward the ionosphere are reflected toward the Earth. This phenomenon can result in substantial signal strengths at distances of several hundred miles from the antenna. AM sky-wave propagation occurs primarily during nighttime hours by reflections from the E and F layers of the ionosphere at about 60 and 130 mi (100 and 220 km) altitude above the Earth's surface, respectively. *See* IONOSPHERE; RADIO-WAVE PROPAGATION.

The daytime ground-wave signal level protected by allocation criteria is usually 0.5 mV/m, although much higher signal strengths may be necessary to overcome noise from atmospheric and artificial sources, especially in highly urbanized areas. During nighttime hours, the service area for most AM stations (other than class A stations) is usually limited by interference from other cochannel stations. For low-power stations, this may be only 10 mi (16 km) or less from the transmitter. For the highest class of stations (class A), nighttime service is protected by the allocation criteria to the 0.5-mV/m, 50%-time sky-wave contour. Because such service is subject to the time variations and fading of sky-wave propagation, the 0.5-mV/m, 50%-time contour is considered to be a secondary service area.

FM VHF band. FM broadcasting has become the dominant broadcast service in the United States primarily because of its better fidelity and its superior reception, which is less subject to noise and interference than that of AM. Information is conveyed by frequency modulation or deviation of a carrier wave. In the United States, the carrier frequency may be deviated ±75 kHz around the assigned carrier frequency. The carrier frequencies, or channels, are spaced at 200-kHz intervals in the United States; a few other countries use slightly different channel spacings. Nearly all FM stations transmit in stereo.

The service area of an FM station depends on the propagation of space waves from the transmitter to the receiver. Space waves propagate through the atmosphere and are diffracted around, and reflected off, mountains, buildings, and other objects. Propagation within areas that have an unobstructed line-of-sight from transmitter to receiver is most reliable and predictable.

Shortwave broadcasting. For reaching audiences in foreign countries or other distant places, shortwave broadcasting is most often used. Nearly 600 million shortwave radio receivers are in use worldwide. Shortwave broadcasting is permitted worldwide in eight frequency bands from 5950 to 26,100 kHz. The assigned transmitting frequencies are spaced at 5-kHz intervals, resulting in a limited usable audio bandwidth. Voice transmissions are most effective, while music transmissions have limited fidelity.

Shortwave signals propagate via sky waves that are reflected one or more times from the E and F layers of the ionosphere. Multiple reflections are possible because a signal can also bounce off the Earth's surface after reflecting off the ionosphere in a "ping-pong" effect.

Digital audio broadcasting. To enhance audio quality, a digital audio broadcasting service has been put in place in Europe and elsewhere, operating primarily in the 1452– 1492-MHz band. As in VHF FM broadcasting, DAB uses transmitters located at elevated locations (mountaintops, building roofs) that provide the best line-of-sight paths to the intended service area. Digital broadcasting differs from VHF FM broadcasting in that the audio signal (voice or music) is first converted to a stream of binary digits (data bits) that represent the audio signal. These data bits are then used to modulate the radio-frequency carrier signal using one of several techniques. After transmission via the radio waves, the radio-frequency carrier is demodulated at the receiver to recover the stream of data bits, and the bits are then converted back to the audio signal. DAB offers improved reception quality and fidelity because error-correcting codes in the digital signal can be used to eliminate many flaws that may occur during transmission. *See* INFORMATION THEORY; MODULATION.

Satellite broadcasting. Satellite broadcasting also uses digital modulation techniques. However, in this case the transmitters are located on satellites high above the Earth. From this position, satellite broadcasting can achieve essentially universal coverage of an entire nation or most of a continent from one or two transmitters. Both geostationary and low-Earth-orbit satellites are used for these systems. Specialized receiving antennas placed in locations visible to the sky are usually needed, along with a special radio receiver specifically designed for satellite service. To extend coverage into areas that are not visible to the sky, some satellite broadcasting networks employ a network of ground-based transmitters (repeaters) to supplement the coverage of the satellite signal. Satellite broadcasting is capable of providing 100 or more channels of audio programming, usually on a subscription or fee basis. *See* DIRECT BROADCASTING SATELLITE SYSTEMS.

Harry R. Anderson

Radio broadcasting network

A group of radio broadcast stations interconnected by leased channels on wire, fiber, microwave, or satellite to one or more central feed points for the purpose of receiving and rebroadcasting program material of a timely nature to a geographically diverse audience; or alternatively, an organization which provides programming for, but does not actually own or operate any, broadcast stations. The terms "radio broadcasting network" and "radio network" are often used interchangeably.

The need for radio networks arises from the local nature of terrestrial radio broadcasting, which is due both to the physical nature of how radio waves propagate and to the way in which terrestrial radio services are regulated. At best, each AM and FM radio station will be receivable only on the order of 100 mi (160 km) from the transmitter site, with the exception of high-powered AM stations at night (so-called clear channel stations) which, because of a phenomenon known as skywave propagation, can be received over much greater distances. *See* RADIO BROADCASTING; RADIO-WAVE PROPAGATION.

In the United States, nongovernmental use of radio-frequency (RF) transmissions are regulated by the Federal Communications Commission (FCC), including the allocation of "channels" in the AM and FM broadcast bands which utilize RF frequencies 550–1705 kHz and 88–108 MHz, respectively. These channels, which listeners typically refer to by their broadcast frequency or call sign (a four-letter combination beginning with either W or K), are allocated by the FCC on a community-by-community basis. Before channels are allocated to a particular community, engineering analyses are performed to determine the impact on other nearby radio signals already allocated on either the same frequency or on the frequencies adjacent to the desired allocation channel. Based on these studies, the FCC determines the allowable distance a station can transmit without causing interference to its neighbors, which dictates the power level at which the broadcast station can operate. *See* RADIO SPECTRUM ALLOCATIONS.

The use of radio networks makes it possible to overcome these distance limitations and broadcast live and prerecorded programs to many communities, or even a national or global audience, through affiliated terrestrial radio stations (affiliates); the stations make national and regional markets available to advertisers and offer stations entertainment and public service programs.

Types of radio networks. Radio networks exist in many forms and are distinguished by whether they are government-owned (or sponsored) or are operated as a commercial enterprise (being either publicly or privately owned), whether they are free or subscription services, and the transmission medium used for reception.

Radio networks in the United States began as commercial enterprises and obtain revenue primarily through the sale of advertisements.

Public networks are nonprofit in nature and can be owned by either the government or not-for-profit corporations. International networks deliver programming outside of their own countries. Satellite Digital Audio Radio Service (SDARS) broadcasts directly from a satellite to the listener. Internet radio stations have the ability to reach a global audience since their transmission medium is the Internet. *See* INTERNET; SATELLITE RADIO.

Distribution technologies. Distribution of network programming to affiliates was accomplished using wired telephone circuits or wireless terrestrial microwave circuits until the 1980s, when the availability of affordable satellite uplink and downlink equipment and the ability for radio networks to lease satellite capacity made possible widespread use of geosynchronous satellite circuits as well. *See* COMMUNICATIONS SATELLITE; RADIO.

David H. Layer

Radio compass

A popular term for an automatic radio direction finder used for navigation purposes on ships and aircraft. It is not strictly a compass, because it indicates direction with respect to the radio station to which it is tuned, rather than to the north magnetic pole. The modern radio compass uses a nondirectional antenna in combination with a bidirectional loop antenna to provide a unidirectional bearing indication. Navigators thus know at all times whether they are traveling toward or away from the radio station used as a reference. The antennas are used with a special radio receiver that provides a visual indication of direction on a meter or cathode-ray indicator. *See* DIRECTION-FINDING EQUIPMENT.

John Markus

Radio-frequency amplifier A tuned amplifier that amplifies the signals commonly used in radio communications. Amplifier designs in the radio-frequency (RF) range differ significantly from conventional low-frequency circuit approaches; they consequently require special, distributed circuit considerations. The need for a distributed treatment is warranted whenever the size of the physical circuit is greater than about one-tenth of the signal wavelength.

The wavelength of a 3-gigahertz signal in free space, for example, is 10 cm (4 in.). Thus, lumped circuit theory will not accurately model the behavior of 3-GHz circuits whose dimensions are in the 1-cm range, and Kirchhoff's voltage and current laws must be abandoned in favor of voltage and current wave propagation phenomena. An immediate implication is that appropriate distributed matching networks are needed to reduce standing-wave effects and avoid undesirable oscillations. The designs of suitable matching networks in conjunction with stability and noise figure analyses are usually the first steps in the amplifier design process. Additional considerations for a radio-frequency amplifier include gain, gain flatness, gain compression, output power, bandwidth, harmonic generation, and bias conditions.

A lumped circuit design suffers serious drawbacks at high operational frequencies. Specifically, the lumped elements in the radio-frequency paths no longer act as ideal inductors and capacitors; resistive losses and parasitics cause undesired departures from the ideal electric circuit behaviors at high frequency. Fortunately, transmission lines—pairs or conductors of carefully chosen dimensions—can be made to function as capacitors or inductors at frequencies where discrete inductors and capacitors perform poorly. Whether a length of transmission line acts as a capacitor or an inductor depends on its length at its operating frequency and on whether its far end is open or shorted. *See* CAPACITOR; INDUCTOR; TRANSMISSION LINES.

Microstrip, a particularly useful form of transmission line for these applications, is implemented by laying down copper strips on the insulating surface of a circuit board whose opposite side is completely covered by a thin layer of copper. The critical parameters of the line are its copper thickness and width, the thickness and relative permittivity of the insulating (dielectric) layer between the strip and the copper substrate, and, to a lesser extent, the length of the line.

A practical matching network for a power amplifier uses a high-performance metal-oxide-semiconductor (MOS) field-effect transistor. This matching network is constructed from microstrip segments of varying width and length ratios. The widest section of microstrip adjacent to the device has very low characteristic impedance; this is necessary to transform the very low device input impedance to a reasonable level. *See* AMPLIFIER; ELECTRICAL IMPEDANCE; IMPEDANCE MATCHING; TRANSISTOR.

Reinhold Ludwig

Radio-frequency impedance measurements Measurements of electrical impedance at frequencies ranging from a few tens of kilohertz to about 1 gigahertz. In the electrical context, impedance is defined as the ratio of voltage to current (or electrical field strength to magnetic field strength), and it is measured in units of ohms (Ω). *See* ELECTRICAL IMPEDANCE.

At zero frequency, that is, when the current involved is a direct current, both voltage and current are expressible as real numbers. Their ratio, the resistance, is a scalar (real) number. However, at nonzero frequencies, the voltage is not necessarily in phase with the current, and both are represented by vectors, and therefore are conveniently described by using complex numbers. To distinguish between the scalar quantity of resistance at zero frequency and the vectorial quantity at nonzero frequencies, the word impedance is used for the complex ratio of voltage to current. *See* ALTERNATING CURRENT; DIRECT CURRENT; ELECTRICAL RESISTANCE.

The measurement of impedance at radio frequencies cannot always be performed directly by measuring an rf voltage and dividing it by the corresponding rf current, for the following reasons: (1) it may be difficult to measure rf voltages and currents without loading the circuit by the sensing probes; (2) the distributed parasitic reactances (stray capacitances to neighboring objects, and lead inductances) may be altered by the sensing probes; and (3) the spatial voltage and current distributions may prevent unambiguous measurements (in waveguides, for instance).

At low frequencies, impedance measurements are often carried out by measuring separately the resistive and reactive parts, using either Q-meter instruments (for resonance methods), or reconfigurable bridges, which are sometimes called universal LCR (inductance-capacitance-resistance) bridges. In one such bridge the resistive part of the impedance is measured at dc with a Wheatstone bridge. Capacitive reactance is measured with a series-resistance-capacitance bridge, and inductive reactance is measured with a Maxwell bridge, using alternating-current (ac) excitation and a standard capacitance. *See* WHEATSTONE BRIDGE.

Transformer bridges are capable of operating up to 100 MHz. The use of transformers offers the following advantages: (1) only two bridge arms are needed, the standard, and the unknown arms, and (2) both the detector and the source may be grounded at one of their terminals, minimizing ground-loop problems and leakage. *See* TRANSFORMER.

A coaxial line admittance bridge is usable from 20 MHz to 1.5 GHz. The currents flowing in three coaxial branch lines are driven from a common junction, and are sampled by three independently rotatable, electrostatically shielded loops, whose outputs are connected in parallel.

A quantity related to impedance is the complex (voltage) reflection coefficient, defined as the ratio of the reflected voltage to the incident voltage, when waves propagate along a uniform transmission line in both directions. Usually, uppercase gamma (Γ) or lowercase rho (ρ) is used to represent the reflection coefficient. When a transmission line of characteristic impedance Z_0 is terminated in impedance Z_T, the reflection coefficient at the load is given by Eq. (1), and the voltage standing-wave ratio (VSWR) is related to the magnitude of Γ by Eq. (2).

$$\Gamma = \frac{Z_T - Z_0}{Z_T + Z_0} \tag{1}$$

$$\text{VSWR} = \frac{1 + |\Gamma|}{1 - |\Gamma|} \tag{2}$$

See TRANSMISSION LINES.

When it is sufficient to measure only the voltage standing-wave ratio, resistive bridges may be used. Resistive bridges employed as reflectometers use a matched source and detector, and therefore differ from the Wheatstone bridge, which aims to use a zero-impedance voltage source and an infinite-impedance detector.

Some specialized electronic instruments make use of the basic definition of impedance, and effectively measure voltage and current. One such instrument is called an rf vector impedance meter. Instead of measuring both the voltage and the current, it drives a constant current into the unknown impedance, and the resultant voltage is measured.

Vector voltmeters (VVM) are instruments with two (high-impedance) voltmeter probes, which display the voltages at either probe (relative to ground) as well as the phase difference between them. One type operates from 1 MHz to 1 GHz, and linearly converts to a 20-kHz intermediate frequency by sampling.

When the magnitude of the reactive part of the impedance is much greater than the resistive part at a given frequency, resonance methods may be employed to measure impedance. The most commonly used instrument for this purpose is the Q meter. *See* Q METER.

At the upper end of the rf range, microwave methods of impedance measurement may also be used, employing slotted lines and six-port junctions. *See* MICROWAVE MEASUREMENTS.

Peter I. Somlo

Radio paging systems Systems, consisting of three basic elements—a personal paging receiver, radio transmitter, and an encoding device—whose primary purpose is to alert an individual, or group of individuals, and deliver a short message of a temporary or perishable nature. Characteristics that are used to define a specific paging system include distance covered, radio frequency, modulation type, paging code format, and message type.

On-site systems cover a single building or a small complex of buildings typically utilizing one low-power transmitter. Wide-area systems can cover an entire city or country and usually use multiple transmitters which simulcast the paging signals. Most paging systems now utilize the very high-frequency (VHF) or ultrahigh-frequency (UHF) radio spectrum using frequency modulation (FM). *See* FREQUENCY MODULATION; RADIO SPECTRUM ALLOCATIONS.

Paging receivers fall into four basic categories: tone alert, tone and voice, numeric, and alphanumeric. Tone pagers emit a "beep" when they are signaled. Some models silently alert the user with a vibration in place of a beep; other models use differing staccato beeps to provide the user with several alert messages. Tone and voice pagers allow the initiator of the page to transmit a simple voice message which will follow a pager's beep alert. Numeric pagers, sometimes called digital pagers, allow the initiator to convey numerical information. These messages are typically composed by using a tone telephone key pad. Alphanumeric pagers allow the initiator to send a complete textual message to the pager user. These messages are composed on word processors, personal computers, or dedicated terminals which can connect to a paging terminal. *See* MICROCOMPUTER; WORD PROCESSING.

James A. Wright

Radio receiver The part of the radio communications system that extracts information from radio-frequency (rf) energy intercepted by the antenna. Radio receivers are the most common electronic equipment worldwide and a vital part of all radio, television, and radar systems. Since the 1960s, radio receiver performance has improved greatly, while size, weight, and cost have fallen dramatically. In the past, radio receivers were built from analog circuits, but increasingly they are realized by digital signal processing. *See* RADIO; SIGNAL PROCESSING.

The antenna intercepts a band of energy in the radio frequencies containing many transmissions. These may have different modes of modulation; the two most common are amplitude modulation (AM) and frequency modulation (FM). Signals have a large size range, from a large fraction of a volt down to a small fraction of a microvolt. The receiver must be selective, responding to only one signal, must demodulate the signal, extracting the impressed information from the radio-frequency wave, and must raise it to an acceptable power level by amplification.

Single-sideband (SSB) transmissions are similar to amplitude modulation, but with one of the symmetrical pair of sidebands eliminated and the carrier suppressed. Single-sideband is significant because it conserves electromagnetic spectrum, introducing less spectrum pollution than any other modulation. Because of the growing prevalence of digital signals, digital modulation systems are increasingly important. *See* AMPLITUDE MODULATION; FREQUENCY MODULATION; MODULATION; SINGLE SIDEBAND.

Amplitude modulation. Figure 1 shows a simple receiver for amplitude-modulated signals. A band-pass filter selects the required signal, which, after optional amplification, is passed to the demodulator (obsolete term: detector), which in this version consists of a limiting amplifier and a multiplier. The high-gain limiting amplifier has an output of the same sign as the input but constant magnitude (a square wave). When multiplied by the amplitude-modulated waveform, this inverts negative half-cycles. After low-pass filtering to remove residual radio-frequency ripple, the modulating waveform is obtained, shown as an audio waveform, which passes to the loudspeaker. There are other amplitude demodulator circuits, all of which either suppress or invert

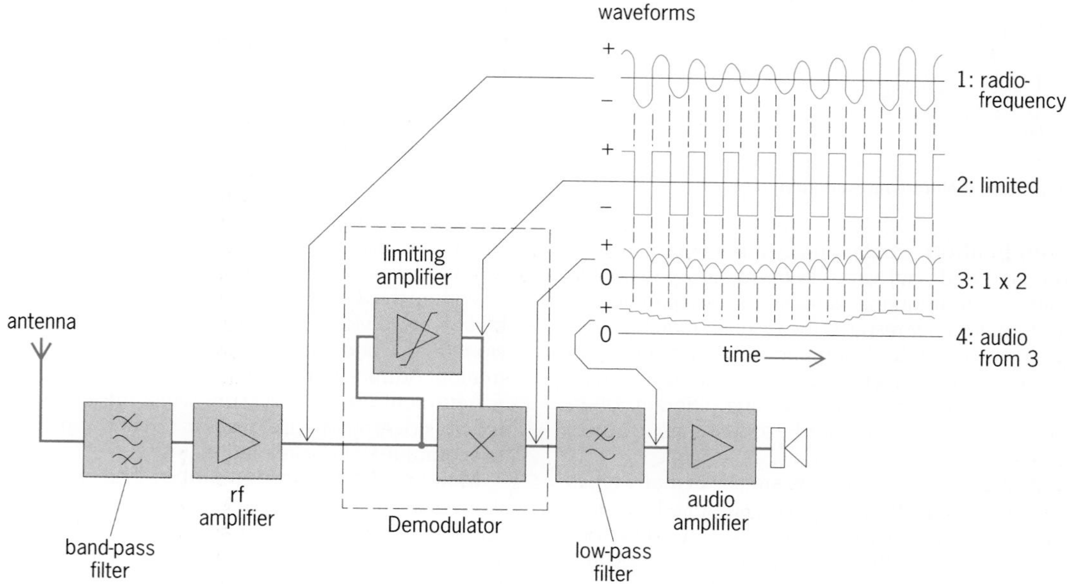

Fig. 1. Simple tuned-radio-frequency (TRF) receiver for amplitude-modulated (AM) signals.

alternate half-cycles of the amplitude-modulated waveform. *See* AMPLIFIER; AMPLITUDE-MODULATION DETECTOR; ELECTRIC FILTER.

Since the input signal may be on the order of microvolts whereas volts are required to operate a loudspeaker, the total receiver gain may range up to the order of a million or more. In this simple tuned-radio-frequency (TRF) receiver it is difficult to adjust the center frequency of a high-performance band-pass radio-frequency filter in order to receive other stations; also, the radio frequency may be too high for effective amplification. For this reason the superheterodyne (superhet) receiver (Fig. 2) superseded it.

In superheterodyne receivers, after a radio-frequency filter and amplifier, the signal passes to a mixer, from which the output is formed by the product of the signal and a locally generated wave. If the signal carrier is at f_c and the local oscillator is at f_o, the result is a wave at a frequency $\pm(f_o - f_c) = f_{\text{i-f}}$, known as the intermediate frequency (i-f), and this difference frequency can be kept constant, whatever the value of f_c, by a suitable choice of f_o. Subsequent to the mixer, the superheterodyne receiver becomes similar to the tuned-radio-frequency receiver except that it is now operating at a fixed frequency, so that, provided a low intermediate frequency is chosen, amplification is easy and high-performance filters can be used. Fixed-frequency intermediate-frequency filters, based on mechanical resonance in piezoelectric ceramics and crystals or on surface-acoustic waves, give extreme attenuation outside the passband. The superheterodyne architecture may be used with amplitude modulation, frequency modulation, and other modulation types. *See* HETERODYNE PRINCIPLE; MIXER; OSCILLATOR; PIEZOELECTRICITY; SURFACE-ACOUSTIC-WAVE DEVICES.

In amplitude-modulation systems, the amplitude of the signal carries the information, and so to keep the audio output from wide level changes it is desirable to adjust the strength of the signal at the demodulator to be roughly constant, despite variations at the antenna due to transmitter power or range. Thus, automatic gain control (AGC) is used in most amplitude-modulation receivers. *See* AUTOMATIC GAIN CONTROL (AGC).

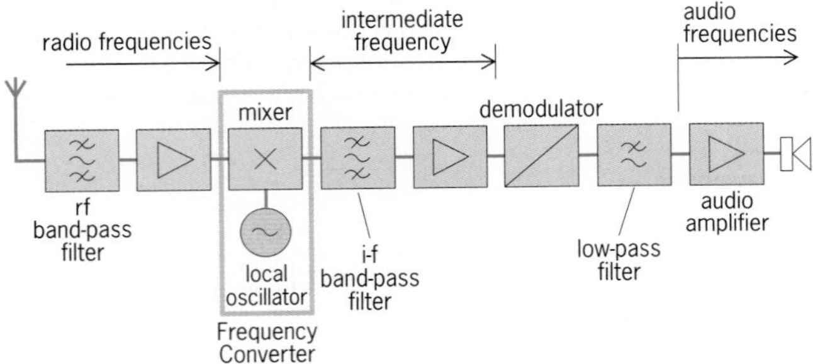

Fig. 2. Superheterodyne (superhet) receiver.

Frequency-modulation. For frequency modulation, the signal is of constant amplitude but varies in frequency, and so it is usual for the intermediate-frequency amplifier to be a limiter, giving a constant output. In other respects the receiver may be identical with one of those in Fig. 2. Automatic gain control is often omitted, but may be combined with limiting for the highest performance. The frequency demodulator is quite different, often utilizing a phase-sensitive detector. *See* FREQUENCY-MODULATION DETECTOR; LIMITER CIRCUIT.

Other receiver types. Radio recevers are use in many types of broadcast and communication systems, each of which uses some combination of amplifiers, filters, mixers, and demodulators to fulfill the purpose of that system. More sophisticated systems may combine the basic principles of superherodyne or homodyne architectures with digital modulation techniques to allow reception of complex signals, such as those containing large amounts of data. Many receiver systems now employ digital signal processing, which in many cases is required for their practical realization.

William Gosling; J. William Whikehart

Radio spectrum allocation The process of identifying specific frequency bands or blocks of the radio spectrum and determining the broad category of radio communication service or services that will be permitted there, and under what specified conditions. The radio spectrum is the part of the natural spectrum of electromagnetic radiation, lying between the frequency limits of 9 kilohertz and 300 gigahertz, that has been extensively used to provide communications services by means of electromagnetic waves. To make effective use of this resource, the radio spectrum is divided, in the frequency dimension, into a series of discrete bands or blocks, arranged in a Table of Frequency Allocations. Because different portions of the radio spectrum are marked by different propagation features and the nature and scope of radio communication services vary greatly, it has become vital that allocation decisions be made in a way that permits radio communication services to operate in an efficient and effective manner and that minimizes, to the extent feasible, the potential for interference between services. This process, generally, is referred to as frequency management and involves the work of governmental and international bodies. *See* ELECTROMAGNETIC RADIATION.

Need for spectrum allocation. Two broad principles underlie spectrum allocation. First, the radio spectrum is a finite resource that multiple users can easily render unusable if they cause harmful interference to each other by, for example, trying to simultaneously broadcast in the same band at the same power in the same geographic area. One means of avoiding this potential problem while simultaneously promoting robust radio use is for the government licensing bodies to adopt specific service and technical rules for the use of radio frequencies within a specific band, and to limit entry to specified parties (licensees). Such decisions are made within the framework of the allocation process, and the Table of Frequency Allocations remains the primary means of organizing the radio spectrum. *See* ELECTRICAL INTERFERENCE; ELECTROMAGNETIC COMPATIBILITY.

A second characteristic of radio waves is that radio communication use in one country can negatively affect the ability of citizens of another country to use the same radio spectrum. This is particularly significant in that the propagation characteristics in certain radio bands, such as those associated with shortwave radio and many of the amateur bands, can be heard on the other side of the world. Thus, radio spectrum allocations are necessarily a matter of worldwide concern. The International Telecommunication Union (ITU), an international organization within the United Nations System where governments and the private sector coordinate global telecommunications networks and services, is the primary entity in matters of international radio allocation. The ITU publishes the Radio Regulations, which includes an International Table of Frequency Allocations (International Table).

Within the United States, spectrum management responsibilities are divided by the Federal Communications Commission (FCC) and the National Telecommunications and Information Administration (NTIA). The NTIA governs all federal government use of spectrum, while the FCC governs all non–federal government use of spectrum and is responsible for other telecommunications-related issues.

Making allocation decisions. The characteristics of radio waves vary greatly throughout the radio spectrum, and the propagation characteristics of a particular band, along with the type of services expected to be deployed, will greatly influence allocation decisions. Efficient spectrum allocations take into account these characteristics. Because the wavelength of a broadcast signal can range from hundreds of meters to mere millimeters, the distance a radio signal can travel, its ability to penetrate buildings, and its susceptibility to atmospheric conditions can vary greatly. *See* BANDWIDTH REQUIREMENTS (COMMUNICATIONS); RADIO-WAVE PROPAGATION.

Allocated services. Specific allocated services include the broadcasting service, fixed and mobile services, unlicensed services, and amateur and personal radio services.

Broadcasting service. A broadcast allocation permits transmissions that are intended for direct reception by the general public, and includes sound, visual, and other types of transmissions. The commercial AM radio band consists of the frequency range 535 kHz–1.7 MHz, FM radio operates in the band 88–108 MHz, and television stations operate on several frequency bands—54–88 MHz for channels 2–6, 174–220 MHz for channels 7–13 (collectively, the very high frequency or VHF channels), and 470– 794 MHz for channels 14–69 (the ultrahigh-frequency or UHF channels). *See* AMPLITUDE-MODULATION RADIO; FREQUENCY-MODULATION RADIO; RADIO BROADCASTING; TELEVISION.

High-frequency (HF) broadcasting, also known as shortwave broadcasting, is an international service that operates between 5900 and 26,100 kHz. Its broadcast signals propagate over very long distances and thus can be received by the general public in foreign countries.

Fixed and mobile services. The fixed service is a radio communication service between specified fixed points, whereas a mobile station is intended to be used while in motion or during halts at unspecified points. From an allocation standpoint, it is most common to create a shared fixed and mobile band in order to promote the most flexible use of the radio spectrum. Because public safety services are provided under fixed and mobile allocations, a consistent theme in making these allocations has been to provide sufficient spectrum for public safety purposes. *See* MOBILE COMMUNICATIONS.

One development has been the identification of significant blocks of spectrum to support advanced commercial wireless needs. These voice and data services are commonly known as 3G (third-generation mobile). The 2000 World Administrative Radio Conference identified three bands that could be used for such purposes: 806–960 MHz, 1710–1885 MHz, and 2500–2690 MHz. While the bulk of this spectrum was intended for terrestrial-based mobile services, a small portion was also identified for a satellite-based component.

Spectrum in the 70-, 80- and 90-GHz bands, the socalled millimeter wave bands, has been made available for new fixed applications, including those that could provide increased broadband capacity.

Unlicensed services. An increasing number of services are provided in bands whose allocations permit unlicensed uses, in which an individual user has no rights to prevent interference caused by other users. Recent growth in the use of this spectrum has been in the area of wireless communication devices that use a common platform established by standards-setting bodies. Examples of these standards include Wi-Fi®, Bluetooth®, and WiMAX.

Wi-Fi is short for "wireless fidelity," and is the term used to describe wireless local-area networks operating on a common set of standards. Wi-Fi applications generally are used to connect personal computers and laptops to a network, and can be used both in private settings and via public wireless "hotspots." Wi-Fi is a relatively short-range service, with the best reception taking place within several hundred feet (or less) of a base station. Wi-Fi was initially deployed in the 2.4 GHz band, and it has expanded to the 5-GHz band. *See* WIRELESS FIDELITY TECHNOLOGY.

Bluetooth is similar to Wi-Fi in that it is based on a wireless standard and is designed to operate in the 2.4-GHz band. However, the Bluetooth standard operates over a much shorter distance, typically in the range of 30 ft (10 m), and was designed primarily to use short-range radio links to replace cables connecting fixed and portable electronic devices. Bluetooth has been incorporated in cellular phones, personal digital assistants (PDAs), laptop computers, and even automotive electronics.

WiMAX is a relatively new wireless networking standard that offers greater range and bandwidth than the Wi-Fi standard. For example, WiMAX can transfer around 70 megabits per second over a distance of 30 mi (50 km), and a single base station can serve thousands of users. WiMAX is based on standards that allow for the use of any frequency band between 2 and 11 GHz.

Amateur and personal radio services. Amateurs throughout the world communicate with each other directly or through ad hoc relay systems and amateur satellites, and use voice, teleprinting, telegraphy, facsimile, and television to communicate. The 2003 World Administrative Radio Conference realigned the allocations near 7 MHz in order to expand the worldwide 40-m amateur band by 100 kHz. Once international broadcasters operating in this band relocate to other frequencies in 2009, the band will be allocated exclusively to the amateur service. *See* AMATEUR RADIO.

The personal radio services provide short-range, low-power radio for personal communications, radio signaling, and business communications not provided for in other wireless services. The range of applications is wide, spanning from varied one- and two-way voice communications systems to non–voice data transmission devices used for monitoring patients or operating equipment by radio control. The Citizens Band (CB) Radio is included in the personal radio services.

Other allocations. Many other important radio services have distinct allocations; among the most notable are space- and satellite-based services, the maritime services, and aeronautical services. In addition, specific bands have been allocated for the radio astronomy and space research services. Radionavigation-satellite services, for example, include the United States' Global Positioning System (GPS) and Russia's *Glonass* satellite system. *See* RADIO ASTRONOMY; SATELLITE NAVIGATION SYSTEMS.

Trends and developments. A consistent theme in spectrum allocation is the quest to meet the demand for more spectrum. Because technological developments are only able to expand the range of usable radio spectrum to a small degree, another way to accommodate increased demand for radio spectrum is the promotion of increased sharing. Different services can use the same spectrum if the risk of interference is minimal and the uses are compatible or can be coordinated. A second way to make more use of spectrum is through efficient system design. A third mechanism is band clearing, where lightly used spectrum is made available for new services by moving existing operations to other bands.

Jamison Prime

Radio telescope

An instrument used in astronomical research to study naturally occurring radio emission from stars, galaxies, quasars, pulsars, interstellar clouds, and other astronomical bodies between wavelengths of about 1 mm (300 GHz) and 10 m (30 MHz). At the short wavelength end, the performance is limited by the opacity of the terrestrial atmosphere, and at the long wavelength end by the opacity of the ionosphere.

Radio telescopes vary widely in design and appearance, depending on the wavelength of operation and astronomical goals, but they are all composed of one or more antenna elements, each equipped with a radio receiver (illustration). The simplest type of radio telescope uses a parabolic antenna—the so-called single-dish or filled-aperture radio telescope—which operates in the same manner as a television satellite receiving antenna, to focus the incoming radiation onto a small antenna element called the feed, which is connected to the sensitive radio receiver. The larger the parabolic surface, the more energy is focused on the feed and the more sensitive the telescope.

The 110 by 100 m (361 by 328 ft) fully steerable radio telescope in Green Bank, West Virginia. (*NRAO/AUI/NSF*)

Because cosmic radio sources are extremely weak, radio telescopes are usually very large and equipped with the most sensitive radio receivers available. The weak cosmic signals can be easily masked by human-made radio interference, and great effort is taken to protect radio telescopes from locally generated interference by locating them in remote environments, often surrounded by mountains to provide shielding from terrestrial radio transmissions. With the increasing use of satellite radio transmissions to provide communications, entertainment, and navigation throughout the world, it is increasingly difficult to avoid the effects of radio interference.

In order to detect the faint broadband signals generated in interstellar or intergalactic space, radio telescopes are designed to receive over a very large bandwidth of hundreds of megahertz, and the data may be averaged for tens of hours or more to reduce the effect of random noise generated by the radio receiver itself. Also, many atoms and molecules found in interstellar space and around some stars radiate only at discrete frequencies, mostly at short, centimeter and millimeter wavelengths. These spectral lines are studied with radio telescopes able to simultaneously observe up to thousands of different narrow-frequency channels.

The angular resolution, or ability of a radio telescope to distinguish fine detail in the sky, depends on the wavelength of observations divided by the size of the instrument. But even the largest antennas when used at their shortest operating wavelength have an angular resolution only a little better than 1 arcminute, which is comparable to that of the unaided human eye at optical wavelengths. Because radio telescopes operate at much longer wavelengths than optical telescopes, they must be much larger than optical telescopes to achieve the same angular resolution. Fortunately at radio wavelengths, the distortions introduced by the atmosphere are less important than at optical wavelengths, and so the theoretical angular resolution of a radio telescope can be achieved in practice, even for the largest dimensions. The high angular resolution of radio telescopes is achieved by using the principles of interferometry to synthesize a very large effective aperture from a number of small elements.

Because radio signals can be divided and distributed over large distances without distortion, it is possible to build radio telescope systems of essentially unlimited dimensions. Although the laborious computational task to obtain images from the large amount of interferometer data was once formidable, these tasks are now easily accomplished by powerful PCs. The most powerful radio array of this type is the Very Large Array (VLA) located on the Plains of San Agustin near Socorro, in central New Mexico. The VLA consists of 27 parabolic antennas, each measuring 25 m (82 ft) in diameter. The VLA is used by nearly 1000 astronomers each year for a wide variety of research programs.

In cases in which antennas are spaced more than a few tens of kilometers apart, it becomes prohibitively expensive to employ real physical links to distribute the signals. The Multi-Element Radio-Linked Interferometer Network (MERLIN) at Jodrell Bank in the United Kingdom uses microwave radio links to connect seven antennas separated by up to 200 km (124 mi). It is used primarily to study compact radio sources associated with quasars, active galactic nuclei (AGN), and cosmic masers with a resolution of a few hundredths of an arcsecond.

Interferometer systems of essentially unlimited element separation can be formed by using the technique of very long baseline interferometry, or VLBI, in which the signals received at each element are recorded by broad-bandwidth tape recorders or large-capacity computer disks. The recorded tapes or disks are then transported to a common location where they are replayed and the signals combined to form interference fringes.

The Very Long Baseline Array (VLBA) consists of ten 25-m (82-ft) dishes spread across the United States from the Virgin Islands to Hawaii and is used to study quasars, galactic nuclei, cosmic masers, pulsars, and radio stars with a resolution as good as 0.0001 arcsecond, or more than 100 times better than that of the *Hubble Space Telescope*. *See* ASTRONOMICAL OBSERVATORY; RADIO ASTRONOMY.

Kenneth Kellermann

Radio transmitter A generator of radio-frequency (rf) signals for wireless communication over some distance, which can vary from the short ranges within a building to intercontinental distances. Most applications utilize signals from very low frequencies (VLF) to extremely high frequencies (EHF); some applications require frequencies as low as 45 Hz or as high as 100 GHz. The radio-frequency output power varies from a fraction of a watt in emergency beacons and portable equipment to several megawatts in long-range, low-frequency transmitters. *See* RADIO SPECTRUM ALLOCATIONS.

The architecture (organization) of a radio transmitter is determined by the type of signal it is intended to produce. The four basic architectures are those used for continuous-wave, frequency-modulation, amplitude-modulation, and single-sideband signals. Transmitters for some applications (for example, television) use a combination of these architectures (for example, frequency modulation for sound and single sideband for video), while others (for example, Loran C) use unique architectures. Alternative architectures such as envelope elimination and restoration or outphasing can be used to improve efficiency. *See* ELECTRICAL COMMUNICATIONS; MODULATION.

Continuous-wave (CW) transmitter. The most basic type of radio transmitter produces only a continuous-wave signal. Such transmitters are often switched on and off (keyed) to produce telegraph signals. The block diagram of a simple continuous-wave transmitter is shown in Fig. 1. The oscillator $G1$ produces a low-power signal, which is boosted to the final output power by a series of progressively larger power amplifiers. The optional inclusion of a frequency multiplier improves stability by allowing the frequencies of the oscillator and high-power amplifiers to be different. *See* FREQUENCY MULTIPLIER; OSCILLATOR; POWER AMPLIFIER.

The architecture includes both frequency translation and power splitting, which makes it more suitable for generating high-power signals at various frequencies. While at a relatively low level, the signal is translated by a mixer to the desired output

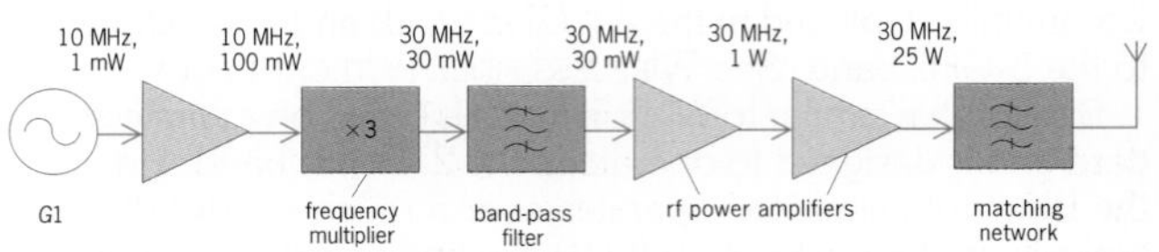

Fig. 1. Basic continuous-wave (CW) transmitter.

frequency. After amplification by a chain of power amplifiers, it is split into two parts to drive two final power amplifiers whose outputs are combined to produce the transmitter output. *See* MIXER; TELEGRAPHY.

Frequency-modulation (FM) transmitter. Analog frequency modulation is widely used for voice communication, high-quality audio broadcasting, and television audio. Frequency-shift keying (FSK) and phase-shift keying (PSK) are widely used for transmission of digital data via radio-frequency signals. *See* FREQUENCY MODULATION; FREQUENCY-MODULATION RADIO; MOBILE RADIO; RADIO BROADCASTING; TELEVISION.

Frequency-modulated and phase-modulated (PM) signals have constant amplitudes and are therefore produced by transmitters with architectures similar to those of the continuous-wave transmitter (Fig. 1). The principal change is the replacement of oscillator *G*1 by a frequency or phase modulator. In frequency-modulation transmitters, the frequency multiplier increases the frequency deviation as well as the carrier frequency of the frequency-modulated signal. *See* PHASE MODULATION.

In communication applications, the frequency modulator is typically a voltage-controlled crystal oscillator (VCXO) in which the capacitance of a varactor diode is used to vary slightly the frequency of a crystal oscillator. Other applications employ various types of modulators, including phase-shift, phase-locked-loop, comparator, and Armstrong. *See* FREQUENCY MODULATOR.

Amplitude-modulation transmitter. Full-carrier amplitude modulation is used in medium-frequency (MF) broadcasting, high-frequency (HF) international broadcasting, citizen-band communication, aircraft communication, and nondirectional navigation beacons. *See* AMPLITUDE MODULATION; ELECTRONIC NAVIGATION SYSTEMS.

Most modern full-carrier amplitude-modulation transmitters produce the output signal by amplitude modulation of the final radio-frequency power amplifier. Generally, the modulation is accomplished by varying the supply voltage of the radio-frequency power amplifier with a high-power radio-frequency amplifier. Since the radio-frequency carrier has constant amplitude until the final power amplifier, the architecture of the radio-frequency chain (Fig. 2) is similar to that of a continuous-wave or frequency-modulation transmitter. *See* AUDIO AMPLIFIER.

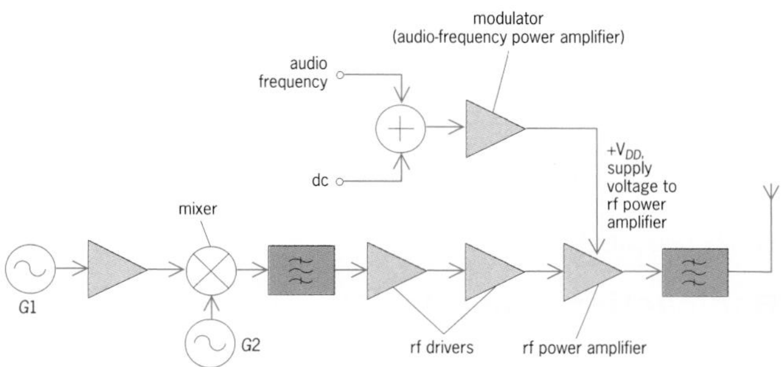

Fig. 2. Amplitude-modulation (AM) transmitter.

Single-sideband (SSB) transmitter. Single-sideband amplitude modulation is widely used for high-frequency voice communications, including military, marine, aeronautical, diplomatic, and amateur. It also finds use (as amplitude-compandored single-sideband, or ACSB) at very high frequencies (VHF) and ultrahigh frequencies (UHF).

Although single sideband is technically a form of amplitude modulation, the single-sideband signal itself has variations of both amplitude and phase. Signals such as multitone, independent sideband (ISB), and vestigial sideband (VSB), used for video, also possess such characteristics. Consequently, these signals are traditionally amplified by a chain of linear radio-frequency power amplifiers operating in class B. *See* SINGLE SIDEBAND.

The low-level output of the single-sideband modulator is first shifted by the local oscillator *G*2 to an intermediate frequency (i-f) that is at least twice the highest output frequency. The intermediate-frequency signal is then shifted downward to the desired output frequency by the variable-frequency oscillator (VFO) *G*3. The mixer output is low-pass-filtered and then amplified to the desired power. *See* ELECTRIC FILTER. Frederick H. Raab

Radio-wave propagation

Radio-wave propagation The means by which radio signals are transported through space from a transmitting antenna to a receiving antenna. *See* RADIO.

The frequencies around 20 kHz can be received reliably at distances of thousands of miles but are limited to telegraph-type signals and require very large transmitting antennas. Higher frequencies are needed for voice, and still higher frequencies for television transmission. As the frequency increases, the transmission range tends to decrease. Frequencies above 100 MHz can transmit wide-band signals, but they are limited to approximately line-of-sight distances with the usual type of equipment. However, distances of 200 mi (320 km) or more are possible by the use of high power and large antennas to provide narrow "searchlight" beams.

Reflections from the ionosphere (ionized layers 50–250 mi or 80–400 km above the Earth's surface) provide a useful but variable long-distance service at frequencies less than about 30 MHz. These reflections account for the long-range broadcast coverage at night and for the shortwave intercontinental communication. *See* RADIO BROADCASTING.

The principal components of the received radio signal are shown symbolically in the illustration. The vector sum of the direct, reflected, and surface waves has been called the space wave, ground wave, or tropospheric transmission to differentiate it from the ionospheric reflections. The ionospheric and surface waves are the principal components at frequencies below 10–30 MHz. The direct and reflected rays are the principal factors at frequencies above 30–50 MHz. Although the ionospheric, direct, and ground-reflected waves can be easily visualized as rays, the surface wave is more difficult to understand; it originates at the air-Earth boundary because the Earth is not a perfect reflector.

Variations in signal level with time are caused by changing atmospheric conditions. The severity of the fading usually increases as either the frequency or path length increases. Most fading is temporary diversion of energy to some direction other than the intended location, associated with refraction or interference, but absorption effects are important in the microwave region.

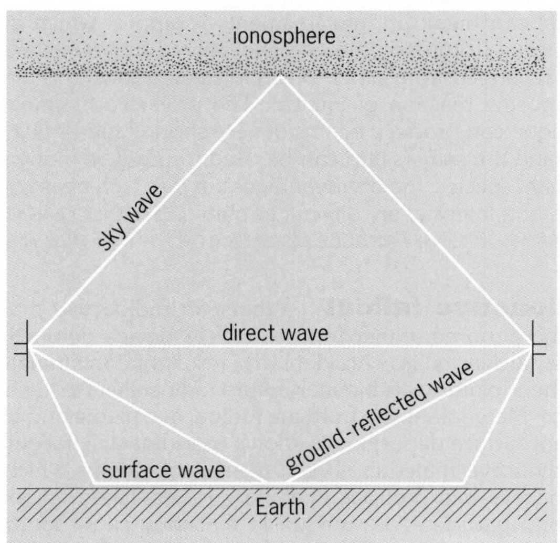

Possible transmission paths between antennas.

The dielectric constant of the atmosphere normally decreases gradually with increasing altitude. The result is that on the average the radio ray is bent or refracted toward the Earth so that the distance to the radio horizon is slightly greater than to the optical horizon. The amount of refraction is variable, and exceptionally long-range transmission may occur occasionally. Conversely, when the radio energy is bent away from the Earth (upward bending), the transmission loss is increased. *See* REFRACTION OF WAVES. Kenneth Bullington

Radioactive beams Beams of radioactive (unstable) nuclei. In several nuclear physics laboratories, a capability exists to produce such beams and, before these nuclei spontaneously decay, use them to gain insight into the reactions on and structure of nuclei never before accessible. Radioactive beams are particularly useful to study stellar explosions such as novae, supernovae, and x-ray bursts. These explosions are some of the most catastrophic events in the universe, generating enormous amounts of energy while synthesizing the elements that make up lifeforms and the world. These spectacular explosions involve, and in some cases are driven by, reactions where the atomic nuclei of hydrogen (protons) and helium (alpha particles) fuse with (are captured by) radioactive isotopes of heavier elements to form new elements. The capability to produce beams of radioactive nuclei allows direct measurements of these reactions, providing crucial information needed to theoretically model cataclysmic stellar events and to understand the origin of many chemical elements.

One approach to radioactive beam production is the isotope separator on-line (ISOL) technique. One accelerator bombards a target with a beam of stable nuclei, and a small number of the radioactive atoms of interest are produced through nuclear reactions. These atoms are transported, by various techniques, including thermal diffusion, to an ion source where they are ionized (removing or adding electrons to give atoms an electrical charge) and extracted. The radioactive ions are then mass-separated from other ions and accelerated to energies needed for nuclear physics experiments by a second accelerator. The ISOL technique can produce very high beam qualities, purities, and intensities; the disadvantages are that only a few radioactive beam species can be generated from each combination of production target and primary beam, and that beams with short lifetimes (less than 1 s) are difficult to produce. *See* ION SOURCES; MASS SPECTROSCOPE.

A complementary radioactive beam production technique is projectile fragmentation. When a high-energy beam of stable heavy ions passes through a thin target, the beam particles (projectiles) can break up into fragments—some of which are the radioactive isotope of interest. The desired fragments are then mass-separated from other ions and steered toward a target to undergo the reaction of interest. The projectile fragmentation technique can produce beams of very short lifetimes (10^{-6} s or less), and the same setup can be used to produce many different beam species; the disadvantages are that high beam quality, purity, and intensity are difficult to obtain. *See* NUCLEAR FUSION; NUCLEAR REACTION; PARTICLE ACCELERATOR. Michael S. Smith

Radioactive fallout Whenever radioactive materials become airborne, either from a nuclear device detonation or from a nuclear release accident, the resultant contaminated atmospheric plume will ultimately return radioactivity to the Earth's surface. Material settling from the radioactive plume and its subsequent surface deposition is known as radioactive fallout.

Radioactive materials consist of unstable atoms which emit gamma rays, beta particles, or alpha particles. These emissions—rays and particles—are unique in that they cause ionizations in neighboring atoms. The energy of the emitted radiation and subsequent ionizations can be a cause of concern if absorbed by living systems. *See* ALPHA PARTICLES; BETA PARTICLES; GAMMA RAYS; RADIOACTIVITY.

Fissioning of uranium and plutonium produces isotopes of about 70 different atoms; each atom may have several different isotopic forms. Examples of daughter products are the isotopes of elemental strontium, which are efficiently produced in nuclear fission. The isotope strontium-90 (^{90}Sr) has a 28-year half-life, while ^{89}Sr is produced in slightly higher concentrations but has a half-life of only 50 days. In addition to the differing half-lives, each isotope emits a unique radiation spectrum. The two strontium isotopes emit beta particles of different energies. Other isotopes such as cesium-137 (^{137}Cs, half-life 30 years) and iodine-131 (^{131}I, half-life 8 days) emit both beta particles and gamma rays. *See* CESIUM; IODINE; ISOTOPE; STRONTIUM.

Atmospheric fallout can be scavenged by rainfall. Wet deposition, involving washing out of atmospheric fallout, can increase local deposition patterns. This was the case following the Chernobyl accident, where local rainfall in Belarus, Ukraine, and Russia washed high concentrations of radioactive iodine and cesium out of the plume and onto the spring pasture. Radioactive iodine and cesium are relatively volatile and were more easily "boiled" out of Chernobyl's burning core. Cesium is a congener of potassium and therefore is fairly uniformly distributed throughout the body once inhaled or ingested. The result is a whole-body radiation dose. Furthermore, the energetic gamma-ray emission from ^{137}Cs adds a source of external radiation from surface deposits on the ground. These factors, in addition to its long half-life and relatively high concentration, make ^{137}Cs the major long-term contamination concern from fallout. For example, although the Chernobyl accident occurred in 1986, precautions must still be taken against potential intake doses of ^{137}Cs: inhalation doses can occur when burning wood from contaminated trees, and consuming mushrooms grown in contaminated forests delivers an ingestion dose.

Levels of ^{131}I in the plumes of radioactive fallout are of particular concern. With an 8 day half-life and a strong beta- and gamma-ray emission, this radionuclide concentrates almost exclusively in the thyroid gland. Recent dose reconstructions in the United States show that the 1950s and 1960s fallout radiation doses from ^{131}I were large enough to have increased the risk for thyroid cancer, especially in children. *See* THYROID GLAND.

In 1955, the United Nations established the Scientific Committee on the Effects of Atomic Radiation, due to concern over possible risks from fallout. It issues comprehensive reports about every 5 years and has collected and documented the world's literature on radioactive fallout and, more recently, on possible radiation consequences. *See* ATOMIC BOMB; NUCLEAR EXPLOSION; NUCLEAR FISSION; NUCLEAR FUELS; NUCLEAR REACTOR; RADIATION INJURY (BIOLOGY). Marvin Goldman

Radioactive minerals Minerals that contain uranium (U) or thorium (Th) as an essential component of their chemical composition. Examples are uraninite (UO_2) or thorite ($ThSiO_4$). There are radioactive minerals in which uranium and thorium substitute for ions of similar size and charge. There are approximately 200 minerals in which uranium or thorium are essential elements, although many of these phases are rare and poorly described. These minerals are important, as they are found in ores mined for uranium and thorium, most commonly uraninite and its fine-grained variety, pitchblende, for uranium. Thorite and thorogummite are the principal ore minerals of thorium. Minerals in which uranium and thorium occur in trace amounts, such as zircon ($ZrSiO_4$), are important because of their use in geologic age dating. The isotope uranium-238 (^{238}U) decays to lead-206 (^{206}Pb); ^{235}U decays to ^{207}Pb; ^{232}Th decays to ^{208}Pb; thus, the ratios of the isotopes of uranium, thorium, and lead can be used to determine the ages of minerals that contain these elements. *See* DATING METHODS; GEOCHRONOMETRY; LEAD ISOTOPES (GEOCHEMISTRY); THORITE; THORIUM; URANINITE; URANIUM. Rod Ewing

Radioactive tracer A radioactive isotope which, when injected into a chemically similar substance or artificially attached to a biological or physical system, can be traced by radiation detection devices. Many problems in biology and medicine not amenable to other approaches can be solved by the use of these tracers. *See* RADIOACTIVITY; RADIOACTIVITY AND RADIATION APPLICATIONS; RADIOISOTOPE; RADIOISOTOPE (BIOLOGY).

The simplest radioactive tracer studies consist of the tagging of a biological entity with a radioactive isotope (radioisotope). The entity is then tracked by following the radiation from the isotope. The operation becomes more complex when a large number of biological particles are labeled, for example, in the tagging of red blood cells or bacteria. When the labeled substance is injected into an animal, it is impossible to follow the individual labeled particles, but their average movement can be tracked by observations of the radiation. Finally, a radioisotope of a particular element can be used to tag that element. Phosphorus-32 can be introduced into the soil where a plant is growing, and the amount of phosphorus absorbed and its distribution throughout the plant can be studied.

In most biological tracer experiments, the radio-isotope is introduced into the system and its radiation subsequently measured with Geiger-Müller counters or scintillation detectors. Extremely soft (low-intensity) radiations can be detected by the use of photographic film. *See* GEIGER-MÜLLER COUNTER; SCINTILLATION COUNTER. Gordon L. Brownell

In medical applications, a radioactive atom can be attached to a molecule or more complex substance, which can then be used to examine a chemical reaction in a test tube, or it can be administered to a patient by ingestion or injection and subsequently be incorporated into a biochemical process. The radioactive emissions from the radioactive atom can be used to track (trace) the behavior of the labeled molecule or substance in biological processes by means of medical imaging, utilizing techniques such as positron emission tomography (PET) or single-photon-emission computed tomography (SPECT). *See* MEDICAL IMAGING.

The branch of medicine that uses radioactive tracers in the care of patients is called nuclear medicine. Radiotracers of practically every element can be produced in nuclear reactors or cyclotrons. Radioactive tracers are used as part of the diagnostic process. Three radionuclides—carbon-14, tritium (hydrogen-3), and phosphorus-32—remain the backbone of modern biomedical sciences. *See* NUCLEAR MEDICINE. Henry N. Wagner, Jr.

Radioactive waste management The treatment and containment of radioactive wastes. These wastes originate almost exclusively in the nuclear fuel cycle and in the nuclear weapons program. Their toxicity requires careful isolation from the biosphere. Their radioactivity is commonly measured in curies (Ci). Considering its toxicity, the curie is a rather large unit of activity. A more appropriate unit is the microcurie ($1\ \mu Ci = 10^{-6}$ Ci), but the nanocurie ($1\ nCi = 10^{-9}$ Ci) and picocurie ($1\ pCi = 10^{-12}$ Ci) are also frequently used. *See* UNITS OF MEASUREMENT.

Radioactive wastes are classified in four major categories: spent fuel elements and high-level waste (HLW), transuranic (TRU) waste, low-level waste (LLW), and uranium mill tailings. Examples of minor waste categories include radioactive gases produced during reactor operation, radioactive emissions resulting from the burning of uranium-containing coal, or contaminated uranium mine water.

Spent fuel elements arise when uranium is fissioned in a reactor to generate energy. Most of the existing radioactivity is contained in spent nuclear fuel and high-level waste. For the first 100 years, the toxicity is dominated by the beta- and gamma-emitting fission products [such as strontium-90 (^{90}Sr) and cesium-137 (^{137}137), with half-lives of approximately 30 years]; thereafter, the long-lived, alpha-emitting transuranium elements [for example, plutonium-239 (^{239}Pu), with a half-life of 24,000 years] and their radioactive decay daughters [for example, americium-241 (^{241}Am), with a half-life of 432 years, a daughter of plutonium-241 (^{241}Pu), with a half-life of 13 years] are important. Burial in geologic formations at a depth of 500–1000 m (1600–3200 ft) appears at present the most practical and attractive disposal method. *See* NUCLEAR FISSION; NUCLEAR FUEL CYCLE; NUCLEAR FUELS REPROCESSING; TRANSURANIUM ELEMENTS.

However, geology as a predictive science is still in its infancy, and many of the parameters entering into model calculations of the long-term retention of the waste in geologic media are questionable. The major single problem is the heating of the waste and its surrounding rock by the radioactive decay heat. This heating can accelerate the penetration of groundwater into the repository, the dissolution of the waste, and its transport to the biosphere. Much effort has been devoted to the development of canisters to encapsulate the spent fuel elements or the glass blocks containing high-level waste, and of improved waste forms and overpacks that promise better resistance to attack by groundwater.

Although the radioactivity of the transuranic wastes is considerably smaller than that of high-level waste or spent fuel, the high radiotoxicity and long lifetime of these wastes also require disposal in a geologic repository. Waste with less than 100 nCi/g (3.78 Bq/kg) of transuranic elements will be treated as low-level waste.

Uranium is naturally radioactive, decaying in a series of steps to stable lead. It is currently a rare element, averaging between 0.1 and 0.2% in the mined ore. At the mill, the rock is crushed to fine sand, and the uranium is chemically extracted. The residues are discharged to the tailings pile. The tailings contain the radioactive daughters of the uranium. The long-lived isotope thorium-230 (^{230}Th, half-life 80,000 years) decays into radium-226 (^{226}Ra, half-life 1600 years), which in turn decays to radon-222 (^{222}Rn, half-life 3.8 days). Radium and radon are known to cause cancer, the former by ingestion, the latter by inhalation. Radon is an inert gas and thus can diffuse out of the mill tailings pile and into the air. Ground-water pollution by radium that has leached from the pile has also been observed around tailings piles, but its health effects are more difficult to estimate, since the migration in the ground water is difficult to assess and also highly site-specific. *See* RADIOACTIVITY; RADIUM; RADON; URANIUM.

Although the radioactivity contained in the mill tailings is very small relative to that of the high-level waste and spent fuel, it is comparable to that of the transuranic waste. It is mainly the dilution of the thorium and its daughters in the large volume of the mill tailings that reduces the health risks to individuals relative to those posed by the transuranium elements in the transuranic wastes. However, this advantage is offset by the great mobility of the chemically inert radon gas, which emanates into the atmosphere from the unprotected tailings. New mill tailings piles will be built with liners to protect the ground water, and will be covered with earth and rock to reduce atmospheric release of the radon gas.

By definition, practically everything that does not belong to one of the three categories discussed above is considered low-level waste. This name is misleading because some wastes, though low in transuranic content, may contain very high beta and gamma activity. The current method of low-level waste disposal is shallow-land burial, which is relatively inexpensive but provides less protection than a geologic repository.

At the end of their lifetime, nuclear facilities have to be dismantled (decommissioned) and the accumulated radioactivity disposed of. Nuclear power plants represent the most important category of nuclear facilities, containing the largest amounts of radioactive wastes, which can be grouped in three classes: neutron-activated wastes, surface-contaminated wastes, and miscellaneous wastes. *See* NUCLEAR POWER; NUCLEAR REACTOR.

The neutron-activated wastes are mainly confined to the reactor pressure vessel and its internal components, which have been exposed to large neutron fluences during reactor operation. These components contain significant amounts of long-lived nontransuranic radioactive isotopes such as niobium-94 (^{94}Nb, an impurity in the stainless steel), which emits highly penetrating gamma rays and has a half-life of 20,000 years. These wastes are unacceptable for shallow-land disposal as low-level wastes. Disposal in a geologic repository is envisioned. *See* DECONTAMINATION OF RADIOACTIVE MATERIALS. Robert O. Pohl

Radioactivity A phenomenon resulting from an instability of the atomic nucleus in certain atoms whereby the nucleus experiences a spontaneous but measurably delayed nuclear transition or transformation with the resulting emission of radiation. The discovery of radioactivity by H. Becquerel in 1896 marked the birth of nuclear physics.

All chemical elements may be rendered radioactive by adding or by subtracting (except for hydrogen and helium) neutrons from the nucleus of the stable ones. Studies of the radioactive decays of new isotopes far from the stable ones in nature continue as a major frontier in nuclear research. The availability of this wide variety of radioactive isotopes has stimulated their use in many different fields, including chemistry, biology, medicine, industry, artifact dating, agriculture, and space exploration. *See* ALPHA PARTICLES; BETA PARTICLES; GAMMA RAYS; ISOTOPE; RADIOACTIVITY AND RADIATION APPLICATIONS.

A particular radioactive transition may be delayed by less than a microsecond or by more than a billion years, but the existence of a measurable delay or lifetime distinguishes a radioactive nuclear transition from a so-called prompt nuclear transition, such as is involved in the emission of most gamma rays. The delay is expressed quantitatively by the radioactive decay constant, or by the mean life, or by the half-period for each type of radioactive atom, discussed below.

The most commonly found types of radioactivity are alpha, beta negatron, beta positron, electron capture, and isomeric transition. Each is characterized by the particular type of nuclear radiation which is emitted by the transforming parent nucleus. In addition, there are several other decay modes that are observed more rarely in specific regions of the periodic table.

Transition rates and decay laws. The rate of radioactive transformation, or the activity, of a source equals the number A of identical radioactive atoms present in the source, multiplied by their characteristic radioactive decay constant λ. Thus Eq. (1) holds, where the decay constant λ has dimensions of s^{-1}.

$$\text{Activity} = A\lambda \text{ disintegrations per second} \tag{1}$$

The numerical value of λ expresses the statistical probability of decay of each radioactive atom in a group of identical atoms, per unit time. For example, if $\lambda = 0.01\ s^{-1}$ for a particular radioactive species, then each atom has a chance of 0.01 (1%) of decaying in 1 s, and a chance of 0.99 (99%) of not decaying in any given 1-s interval. The constant λ is one of the most important characteristics of each radioactive nuclide: λ is essentially independent of all physical and chemical conditions such as temperature, pressure, concentration, chemical combination, or age of the radioactive atoms.

Many radioactive nuclides have two or more independent and alternative modes of decay. For example, ^{238}U can decay either by alpha-particle emission or by spontaneous fission. When two or more independent modes of decay are possible, the nuclide is said to exhibit dual decay. The competing modes of decay of any nuclide have independent partial decay constants given by the probabilities $\lambda_1, \lambda_2, \lambda_3, \ldots$, per second, and the total probability of decay is represented by the total decay constant λ, defined by Eq. (2).

$$\lambda = \lambda_1 + \lambda_2 + \lambda_3 + \cdots \tag{2}$$

The actual life of any particular atom can have any value between zero and infinity. The average or mean life of a large number of identical radioactive atoms is, however, a definite and important quantity. The total L of the life-spans of all the A_0 atoms initially present is given by Eq. (3). Then the average lifetime L/A_0, which is called the mean life τ, is given by Eq. (4).

$$L = \frac{A_0}{\lambda} \tag{3}$$

$$\tau = 1/\lambda \tag{4}$$

The time interval over which the chance of survival of a particular radioactive atom is exactly one-half is called half-period T (also called the half-life, written $T_{1/2}$). The half-period T is related to the total radioactive decay constant λ, and to the mean life τ, by Eq. (5). For mnemonic reasons, the half-period T is much

$$T = 0.693/\lambda = 0.693\tau \tag{5}$$

more frequently employed than the total decay constant λ or the mean life τ.

Radioactive series decay. In a number of cases a radioactive nuclide A decays into a nuclide B which is also radioactive; the nuclide B decays into C which is also radioactive, and so on. For example, $^{232}_{90}$Th decays into a series of 10 successive radioactive nuclides. Substantially all the primary products of nuclear fission are negatron beta-particle emitters which decay through a chain or series of two to six successive beta-particle emitters before a stable nuclide is reached as an end product.

Alpha-particle decay. Alpha-particle decay is that type of radioactivity in which the parent nucleus expels an alpha particle (a helium nucleus), which contains two protons and two neutrons. Thus, the atomic number, or nuclear charge, of the decay product is 2 units less than that of the parent, and the nuclear mass of the product is 4 atomic mass units less than that of the parent, because the emitted alpha particle carries away this amount of nuclear charge and mass. This decrease of 2 units of atomic number or nuclear charge between parent and product means that the decay product will be a different chemical element, displaced by 2 units to the left in a periodic table of the elements.

In the simplest case of alpha decay, every alpha particle would be emitted with exactly the same velocity and hence the same kinetic energy. However, in most cases there are two or more discrete energy groups called lines. For example, in the alpha decay of a large group of ^{238}U atoms, 77% of the alpha decays will be by emission of alpha particles whose kinetic energy is 4.20 MeV, while 23% will be by emission of 4.15-MeV alpha particles. When the 4.20-MeV alpha particle is emitted, the decay product nucleus is formed in its ground (lowest energy) level. When a 4.15-MeV alpha particle is emitted, the decay product is produced in an excited level, 0.05 MeV above the ground level. This nucleus promptly transforms to its ground level by the emission of a 0.05-MeV gamma ray or alternatively by the emission of the same amount of energy in the form of a conversion electron and the associated spectrum of characteristic x-rays. Thus in all alpha-particle spectra, the alpha particles are emitted in one or more discrete and homogeneous energy groups, and alpha-particle spectra are accompanied by gamma-ray and conversion electron spectra whenever there are two or more alpha-particle groups in the spectrum.

Among all the known alpha-particle emitters, most alpha-particle energy spectra lie in the domain of 4–6 MeV, although a few extend as low as 2 MeV and as high as 10 MeV. There is a systematic relationship between the kinetic energy of the emitted alpha particles and the half-period of the alpha emitter. The highest-energy alpha particles are emitted by short- lived nuclides, and the lowest-energy alpha particles are emitted by the very-long-lived alpha-particle emitters. H. Geiger and J. M. Nuttall showed that there is a linear relationship between log λ and the energy of the alpha particle.

The Geiger-Nuttall rule is inexplicable by classical physics, but emerges clearly from quantum, or wave, mechanics. In 1928 the hypothesis of transmission through nuclear potential barriers was shown to give a satisfactory account of the alpha-decay data, and it has been altered subsequently only in details.

Beta-particle decay. Beta-particle decay is a type of radioactivity in which the parent nucleus emits a beta particle. There are two types of beta decay established: in negatron beta decay (β^-) the emitted beta particle is a negatively charged electron (negatron); in positron beta decay (β^+) the emitted beta particle is a positively charged electron (positron). In beta decay the atomic number shifts by one unit of charge, while the mass number remains unchanged. In contrast to alpha decay, when beta decay takes place between two nuclei which have a definite energy difference, the beta particles from a large number of atoms will have a continuous distribution of energy.

For each beta-particle emitter, there is a definite maximum or upper limit to the energy spectrum of beta particles. This maximum energy, E_{max}, corresponds to the change in nuclear energy in the beta decay. As in the case of alpha decay, most beta-particle spectra include additional continuous spectra which have less maximum energy and which leave the product nucleus in an excited level from which gamma rays are then emitted.

For nuclei very far from stability, the energies of these excited states populated in beta decay are so large that the excited states may decay by proton, two-proton, neutron, two-neutron, three-neutron, or alpha emission, or spontaneous fission.

The continuous spectrum of beta-particle energies implies the simultaneous emission of a second particle besides the beta particle, in order to conserve energy and angular momentum for each decaying nucleus. This particle is the neutrino. The neutrino has zero charge and nearly zero rest mass, travels at essentially the same speed as light (3×10^{10} cm/s), and is emitted as a companion particle with each beta particle. By postulating the simultaneous emission of a beta particle and a neutrino, E. Fermi developed in 1934 a quantum-mechanical theory which satisfactorily gives the shape of beta-particle spectra, and the relative half-periods of beta-particle emitters for allowed beta decays. *See* NEUTRINO.

When the ground state of a nucleus differing by two units of charge from nucleus A has lower energy than A, then it is theoretically possible for A to emit two beta particles, either $\beta^+\beta^+$ or $\beta^-\beta^-$ as the case may be, and two neutrinos or antineutrinos, and go from Z to $Z \pm 2$. Here two protons decay into two neutrons, or vice versa. This is a second-order process and so should go much slower than beta decay. There are a number of cases where such decays should occur, but their half-lives are of the order of 10^{20} years or greater. Such decay processes are obviously very difficult to detect. The first direct evidence for two-neutrino double-beta-minus decay of selenium-82, was found only in 1987.

Whenever it is energetically allowed by the mass difference between neighboring isobars, a nucleus Z may capture one of its own atomic electrons and transform to the isobar of atomic number $Z - 1$. Usually the electron-capture (EC) transition involves an electron from the K shell of atomic electrons, because these innermost electrons have the greatest probability density of being in or near the nucleus. *See* ELECTRON CAPTURE.

Gamma-ray decay. Gamma-ray decay involves a transition between two excited levels of a nucleus, or between an excited level and the ground level. A nucleus in its ground level cannot emit any gamma radiation. Therefore gamma-ray decay occurs only as a sequel of another radioactive decay process or of some other process whereby the product nucleus is left in an excited state. Such additional processes include the fusion of two nuclei, Coulomb excitation, and induced nuclear fission. *See* COULOMB EXCITATION; NUCLEAR FISSION; NUCLEAR FUSION.

A gamma ray is high-frequency electromagnetic radiation (a photon) in the same family with radio waves, visible light, and x-rays. The energy of a gamma ray is given by $h\nu$, where h is Planck's constant and ν is the frequency of oscillation of the wave in hertz. The gamma-ray or photon energy $h\nu$ lies between 0.05 and 3 MeV for the majority of known nuclear transitions. Higher-energy gamma rays are seen in neutron capture and some reactions. *See* ELECTROMAGNETIC RADIATION.

Gamma rays carry away energy, linear momentum, and angular momentum, and account for changes of angular momentum, parity, and energy between excited levels in a given nucleus. This leads to a set of gamma-ray selection rules for nuclear decay and a classification of gamma-ray transitions as "electric" or as "magnetic" multipole radiation of multipole order 2^l, where $l = 1$ is called dipole radiation, $l = 2$ is quadrupole radiation, and $l = 3$ is octupole, l being the vector change in nuclear angular momentum. The most common type of gamma-ray transition in nuclei is the electric quadrupole (E2). There are cases where several hundred gamma rays with different energies are emitted in the decays of atoms of only one isotope. *See* MULTIPOLE RADIATION.

An alternative type of deexcitation which always competes with gamma-ray emission is known as internal conversion. Instead of the emission of a gamma ray, the nuclear excitation energy can be transferred directly to a bound electron of the same atom. Then the nuclear energy difference is converted to energy of an atomic electron, which is ejected from the atom.

When the energy between two states in the same nucleus exceeds 1.022 MeV, twice the rest mass energy of an electron, it is also possible for the nucleus to give up its excess energy to an electron-positron pair—a pair creation process. *See* ELECTRON-POSITRON PAIR PRODUCTION.

Spontaneous fission. This involves the spontaneous breakup of a nucleus into two heavy fragments and neutrons. Spontaneous fission can occur when the sum of the masses of the two heavy fragments and the neutrons is less than the mass of the parent undergoing decay. After the discovery of fission in 1939, it was subsequently discovered that isotopes like ^{238}U had very weak decay branches for spontaneous fission, with branching ratios on the order of 10^{-6}. Some isotopes with relatively long half-lives such as ^{252}Cf have large (3.1%) spontaneous fission branching.
Joseph H. Hamilton

Heavy cluster decays. Alpha-particle decay and spontaneous fission are two natural phenomena in which an atomic nucleus spontaneously breaks into two fragments, but the fragments are of very different size in one case and almost equal size in the other. On the basis of fragmentation theory and the two-center shell model, new kinds of radioactivities that are intermediate between alpha particle decay and fission were predicted in 1980. Subsequently, it was shown theoretically that the new processes should occur throughout a very broad region of the nuclear chart, including elements with atomic numbers higher than 40. However, experimentally observable emission rates could be expected only for nuclei heavier than lead, in a breakup leading to a very stable heavy fragment with proton and neutron numbers equal or very close to $Z = 82$, $N = 126$ ($^{208}_{82}$Pb or its neighborhood). The main competitor is always alpha-particle decay. In 1984, a series of experimental confirmations began with the discovery of $^{14}_{6}$C radioactivity of $^{223}_{88}$Ra. A very promising technique uses solid-state track-recording detectors with special plastic films and glasses that are sensitive to heavier clusters but not to alpha particles.
Walter Greiner; Joseph H. Hamilton

Proton radioactivity. Proton radioactivity is a mode of radioactive decay that is generally expected to arise in proton-rich nuclei far from the stable isotopes, in which the parent nucleus changes its chemical identity by emission of a proton in a single-step process. Its physical interpretation parallels almost exactly the quantum-mechanical treatment of alpha-particle decay. For many years only a few examples of this decay mode were observed, because of the narrow range of half-lives and decay energies where this mode can compete with other modes. However, in the late 1990s, experimental techniques using new recoil mass spectrometers, which can separate rare reaction products, and new double-sided silicon strip detectors became available and opened up the discovery of many new proton radioactivities. Two-proton radioactivities from ground states of nuclei are now observed. The first example was the two-proton decay of ^{45}Fe to ^{43}Cr. Two-proton decay of an excited state in ^{18}Ne to ^{16}O was observed earlier.

Delayed particle emissions. Thirteen types of beta-delayed particle emissions have been observed. Over 100 beta-delayed particle radioactivities are now known. Theoretically, the number of isotopes which can undergo beta-delayed particle emission could exceed 1000. Thus, this mode, which was observed in only a few cases prior to 1965, is among the important ones in nuclei very far from the stable ones in nature. Studies of these decays can provide insights into the nucleus which can be gained in no other way. Joseph H. Hamilton

Radioactivity and radiation applications

The field in which the subatomic fragments emitted in radioactive decay (alpha-, beta-, gamma-rays) or produced by high-voltage accelerators (electrons, protons, x-rays) are applied to the problems of science, engineering, industry, and medicine. The techniques are extraordinarily versatile and sensitive and are basically inexpensive. A disadvantage that limits the range and extent of these applications is the health hazard that may be involved. *See* RADIOACTIVITY; RADIOISOTOPE.

Tracer applications are based on two principles. First is the chemical similarity of radioactive atoms and other atoms of the same element. Periodically a few of the radioactive atoms decay, emitting some penetrating subatomic fragments that can be detected one by one, usually through their ability to cause ionization. Thus the movement of a particular element can be followed through various chemical, physical, and biological steps. The second principle involves the characteristic half-life and nature of the emitted fragments. This makes a radioactive species unique and thereby detectable above a background of radioactive emitters associated with elements. For discussions of radioisotope techniques relating to tracer methodology *see* ACTIVATION ANALYSIS; ISOTOPE DILUTION TECHNIQUES; RADIOACTIVE TRACER; RADIOECOLOGY.

Penetration and scattering applications arise from the fact that subatomic fragments can penetrate a thick section of a material, and yet a small fraction of the incident particles can be backscattered by a relatively thin section. The oldest application of the penetrating properties of energetic ionizing photons is radiography. An extension of this technique is autoradiography. Since World War II the penetration and scattering properties of beta- and gamma-rays have been applied in industry in the form of thickness gages. *See* AUTORADIOGRAPHY; RADIOGRAPHY.

The absorption of small amounts of energy from ionizing particles and ionizing photons has chemical effects that have been the basis of several practical applications. The oldest application of this principle is radiation therapy. For example, in cancer therapy the local affected areas are irradiated by external beams of gammas from cobalt-60 or of radiation from accelerators. Radioactive sources have also been administered internally to induce beneficial biochemical reactions in patients afflicted with various ailments. *See* ISOTOPIC IRRADIATION; RADIOLOGY.

A related area is the radiation sterilization of biomedical supplies. The advantages to this method of biochemical destruction of microscopic life are that (1) unlike steam sterilization, it can be performed at low temperatures on plastics and other thermally unstable materials, and (2) unlike germicidal gases, ionizing radiation can reach every point in the treated product. Radiation-sterilized objects are not radioactive.

The radiation preservation of food is an area of considerable promise. Small doses can inhibit sprouting in potatoes, kill insects in wheat, and sterilize pork products but practical applications have been sharply limited due to a cautious role by regulatory authorities in approving such procedures.

Kinetic energy of emissions in radioactive decay can be converted to useful forms of light, heat, and electricity. *See* LUMINOUS PAINT; NUCLEAR BATTERY. Joseph Silverman

Radioactivity standards

Calibrated standard sources of radioactive substances used to determine, by comparison, the strength or activity of samples of the same substances in terms of the number of radioactive atoms they contain or in terms of some figure proportional to this number. The calibration of the standard source in terms of the number of radioactive atoms is usually an elaborate procedure but need only be carried out once, and the calibration may be made at a standardizing laboratory, such as the National Institute of Standards and Technology, having special equipment for the work. Comparison between a sample and the standard is usually made by finding the ratio of the response of an ionization chamber, or other detector of radiation, to the radiation from a sample and from the standard. In each case the intensity of the radiation, and therefore the response of the detector under identical conditions, is proportional to the number of radioactive atoms in the source, because this number is also proportional to the activity or disintegration rate of a source. *See* HALF-LIFE; RADIOACTIVITY. Leon F. Curtiss; Karl Z. Morgan

Radiocarbon dating

A method of obtaining age estimates on organic materials which has been used to date samples as old as 75,000 years. The method has provided age determinations in archeology, geology, geophysics, and other branches of science.

Radiocarbon (^{14}C) determinations can be obtained on wood; charcoal; marine and fresh-water shell; bone and antler; peat and organic-bearing sediments; carbonate deposits such as tufa, caliche, and marl; and dissolved carbon dioxide (CO_2) and carbonates in ocean, lake, and ground-water sources. Each sample type has specific problems associated with its use for dating purposes, including contamination and special environmental effects. While the impact of ^{14}C dating has been most profound in archeological research and particularly in prehistoric studies, extremely significant contributions have also been made in hydrology and oceanography. In addition, beginning in the 1950s the testing of thermonuclear weapons injected large amounts of artificial ^{14}C ("bomb ^{14}C") into the atmosphere, permitting it to be used as a geochemical tracer.

Carbon (C) has three naturally occurring isotopes. Both ^{12}C and ^{13}C are stable, but ^{14}C decays by very weak beta decay (electron emission) to nitrogen-14 (^{14}N) with a half-life of approximately 5700 years. Naturally occurring ^{14}C is produced as a secondary effect of cosmic-ray bombardment of the upper atmosphere. As $^{14}CO_2$, it is distributed on a worldwide basis into various atmospheric, biospheric, and hydrospheric reservoirs on a time scale much shorter than its half-life. Metabolic processes in living organisms and relatively rapid turnover of carbonates in surface ocean waters maintain ^{14}C levels at approximately constant levels in most of the biosphere. The natural ^{14}C activity in the geologically recent contemporary "prebomb" biosphere was approximately 13.5 disintegrations per minute per gram of carbon. *See* COSMOGENIC NUCLIDE; ISOTOPE.

To the degree that ^{14}C production has proceeded long enough without significant variation to produce an equilibrium or steady-state condition, ^{14}C levels observed in contemporary materials may be used to characterize the original ^{14}C activity in the corresponding carbon reservoirs. Once a sample has been removed from exchange with its reservoir, as at the death of an organism, the amount of ^{14}C begins to decrease as a function of its half-life. A ^{14}C age determination is based on a measurement of the residual ^{14}C activity in a sample compared to the activity of a sample of assumed zero age (a contemporary standard) from the same reservoir. The relationship between the ^{14}C age and the ^{14}C activity of a sample is given by the equation below, where

$$t = \frac{1}{\lambda} \ln \frac{A_o}{A_s}$$

t is radiocarbon years B.P. (before the present), λ is the decay constant of ^{14}C (related to the half-life $t_{1/2}$ by the expression $t_{1/2} = 0.693/\lambda$), A_o is the activity of the contemporary standards, and A_s is the activity of the unknown age samples. Conventional radiocarbon dates are calculated by using this formula, an internationally agreed half-life value of 5568 ± 30 years, and a specific contemporary standard.

The naturally occurring isotopes of carbon occur in the proportion of approximately 98.9% ^{12}C, 1.1% ^{13}C, and 10^{-10}% ^{14}C. The extremely small amount of radiocarbon in natural materials was one reason why ^{14}C was one of the isotopes which had been produced artificially in the laboratory before being detected in natural concentrations. A measurement of the ^{14}C content of an organic sample will provide an accurate determination of the sample's age if it is assumed that (1) the production of ^{14}C by cosmic rays has remained essentially constant long enough to establish a steady state in the $^{14}C/^{12}C$ ratio in the atmosphere, (2) there has been a complete and rapid mixing of ^{14}C throughout the various carbon reservoirs, (3) the carbon isotope ratio in the sample has not been altered except by ^{14}C decay, and (4) the total amount of carbon in any reservoir has not been altered. In addition, the half-life of ^{14}C must be known with sufficient accuracy, and it must be possible to measure natural levels of ^{14}C to appropriate levels of accuracy and precision. R. E. Taylor

Radiochemical laboratory A laboratory or facility used for investigation and handling of radioactive chemicals that provides a safe environment for the worker and the public. Features vary depending on the type of radioactive emissions to be handled, the quantity, the half-life, and the physical form (solid, liquid, gas, or powder). Special measures to minimize spread of contaminated material and to dispose of radioactive waste are required. Working surfaces should be smooth and easily washable to permit effective decontamination if necessary. Good ventilation and detectors for monitoring radiation and contamination on surfaces or people are also typical features. *See* VENTILATION.

Investigations utilizing only very small amounts (a few microcuries) of beta or gamma emitters which are not readily dispersed (no powders or volatile liquids) may sometimes be performed without special facilities on the bench top. In this case, precautions such as working on plastic-backed absorbent paper and wearing protective gloves and lab coat may be sufficient. A special bag or can for disposing of the paper and gloves as radioactive waste is required. If the radioactive isotopes are solely alpha-particle emitters, containment and isolation from direct contact are more serious concerns. Due to the limited penetration but high biological toxicity of alpha particles, it is essential to avoid ingestion or inhalation. For very small quantities, work may take place with double rubber gloves in a fume hood with appropriate filter. Generally, an enclosed glove box is used, situated inside a hood and maintained at negative pressure with respect to the face of the hood and the room. Sensors to monitor proper differential pressure and adequate airflow are usually used to assure containment and to generate an alarm if conditions degrade.

For work with pure beta-emitting isotopes, long-handled tongs or other tools are used for higher levels of radioactivity in order to shield the hands. Generally, since most beta emission is also accompanied by penetrating gamma emission, the entire work area must be enclosed in heavily shielded enclosures. *See* ALPHA PARTICLES; BETA PARTICLES; RADIATION CHEMISTRY; RADIOACTIVITY; RADIOCHEMISTRY; RADIOISOTOPE.

Hot laboratories contain walled enclosures for remotely handling larger quantities of gamma-emitting isotopes. A small enclosure is usually referred to as a cave, while large ones are called hot cells. Hot cells are usually equipped with remote manipulators and thick windows made from high-density lead glass. Leonard F. Mausner

Radiochemistry A subject which embraces all applications of radioactive isotopes to chemistry. It is not precisely defined and is closely linked to nuclear chemistry. The widespread use of isotopes in chemistry is based on two fundamental properties exhibited by all radioactive substances. The first property is that the disintegration rate of an isotopic sample is directly proportional to the number of radioactive atoms in the sample. Thus, measurement of its disintegration rate (with a Geiger counter, for example) serves to analyze a radioactive compound. With nearly all chemical elements (notable exceptions being nitrogen and oxygen, which have no suitable radioactive isotopes), an isotope may be incorporated in a chemical compound, and thereafter, masses of this compound as small as 10^{-6} to 10^{-10} g may be measured with a high precision. The second property is that the disintegration rate is completely unaffected by the chemical form of the isotope, and conversely, the property of radioactivity does not affect the chemical properties of the isotope. By substituting or labeling a particular atom within a molecule, isotopes can be used to trace the fate of that atom during a chemical reaction. Radiochemistry has been used to study the efficiency of chemical separations, rates of chemical reaction and diffusion, isotopic exchange reactions, and chemical reaction mechanisms. *See* NUCLEAR CHEMISTRY; RADIATION CHEMISTRY; RADIOACTIVITY; RADIOCHEMICAL LABORATORY. Donald R. Stranks

Radioecology The study of the fate and effects of radioactive materials in the environment. It derives its principles from basic ecology and radiation biology.

Responses to radiation stress have consequences for both the individual organism and for the population, community, or ecosystem of which it is a part. When populations or individuals of different species differ in their sensitivities to radiation stress, for example, the species composition of the entire biotic community may be altered as the more radiation-sensitive species are removed or reduced in abundance and are replaced in turn by more resistant species. Such changes have been documented by studies in which natural ecological systems, including grasslands, deserts, and forests, were exposed to varying levels of controlled gamma radiation stress. *See* POPULATION ECOLOGY.

Techniques of laboratory toxicology are also available for assessing the responses of free-living animals to exposure to low levels of radioactive contamination in natural environments. This approach uses sentinel animals, which are either tamed, imprinted on the investigator, or equipped with miniature radio transmitters, to permit their periodic relocation and recapture as they forage freely in the food chains of contaminated habitats. When the animals are brought back to the laboratory, their level of radioisotope uptake can be determined and blood or tissue samples taken for analysis. In this way, even subtle changes in deoxyribonucleic acid (DNA) structure can be evaluated over time. These changes may be suggestive of genetic damage by radiation exposure. In some cases, damage caused by a radioactive contaminant may be worsened by the synergestic effects of other forms of environmental contaminants such as heavy metals.

Because of the ease with which they can be detected and quantified in living organisms and their tissues, radioactive materials are often used as tracers. Radioactive tracers can be used to trace food chain pathways or determine the rates at which various processes take place in natural ecological systems. Although most tracer experiments were performed in the past by deliberately introducing a small amount of radioactive tracer into the organism or ecological system to be studied, they now take advantage of naturally tagged environments where trace amounts of various radioactive contaminants were inadvertently released from operating nuclear facilities such as power or production reactors or waste burial grounds. *See* RADIOACTIVE TRACER.

An important component of radioecology, and one that is closely related to the study of radioactive tracers, is concerned with the assessment and prediction of the movement and concentration of radioactive contaminants in the environment in general, and particularly in food chains that may lead to humans. *See* ECOLOGY; ENVIRONMENTAL RADIOACTIVITY; ENVIRONMENTAL TOXICOLOGY; FOOD WEB; RADIATION BIOLOGY. I. Lehr Brisbin, Jr.

Radiography The technique of producing a photographic image of an opaque specimen by the penetration of radiation such as gamma rays, x-rays, neutrons, or charged particles. When a beam of radiation is transmitted through any heterogeneous object, it is differentially absorbed, depending upon the varying thickness, density, and chemical composition of this object. The image registered by the emergent rays on a photographic film adjacent to the specimen under examination constitutes a shadowgraph or radiograph of its interior. Radiography is the general term applied to this nondestructive film technique of testing the gross internal structure of any object, whether it be of the chest of a patient for evidence of tuberculosis, silicosis, heart pathology, or embedded foreign objects; of bones in case of fractures or of arthritis or other bone diseases; or of a weld in a pipe to observe cracks, inclusions, or voids. Radiography with x-rays is commonly used in both medical and industrial applications. Industrial work also involves gamma and neutron radiography. Radiography with charged particles is under development. Most of this discussion will be concerned with radiography with x-rays and gamma rays. *See* CHARGED PARTICLE BEAMS; GAMMA RAYS; NEUTRON; X-RAYS.

Industrial radiography enables detection of internal physical imperfections such as voids, cracks, flaws, segregations, porosities, and inclusions. It is frequently used for visualization of inaccessible internal parts in order to check their location or condition. It is extensively applied wherever internally sound metallic components are required such as (1) in the foundry industry to guarantee the soundness of castings; (2) in the welding of pressure vessels, pipelines, ships, and reactor components to guarantee the soundness of welds; (3) in the manufacture of fuel elements for reactors to guarantee their size and soundness; (4) in the solid-propellant and high-explosives industry to guarantee the soundness and physical purity of the material; and (5) in the automotive, aircraft, nuclear, space, oceanic, and guided-missile industries, whenever internal soundness is required.

The general term applied to radiation imaging and inspection is radiology. This includes film and similar photographic image methods, such as radiographic paper, under the term radiography. In medical circles, the term roentgenography, derived from the name of the discoverer of x-rays, W. C. Roentgen, is used. The older technique of registering an image on a fluorescent screen is called fluoroscopy. The fluoroscopic prompt-response imaging of radiation has largely been replaced by electronic detection with image intensifiers or sensitive television cameras. This technique, called radioscopy, is now widely used in both medical and industrial applications. *See* RADIOLOGY.

Other variations of radiation imaging include xeroradiography, microradiography, flash radiography, and computerized tomography. Xeroradiography is a dry-plate, electrostatic image method similar to that used in photocopy machines. Microradiography involves a magnified image to improve spatial resolution and the detection of small detail. Flash radiography is the production of an x-ray image in a very short time, of the order of nanoseconds, in order to stop fast motion. The computerized tomography (CT) method recreates an image that is essentially a slice through the object. Computerized tomography has made a strong impact on medical diagnosis and industrial inspection. *See* COMPUTERIZED TOMOGRAPHY; MICRORADIOGRAPHY.

Neutron beams, obtained from nuclear reactors, accelerators, or radioactive sources, can penetrate matter with relative ease since they are not electrically charged. The attenuation of neutrons by most materials is relatively small because the neutron carries no electric charge and consequently is neither attracted nor repelled by the charged particles in the nucleus, nor by the electron clouds associated with the atoms of the material through which the neutron passes. On the other hand, the neutron absorption coefficients of some elements with low atomic numbers are high; hydrogen, lithium, and boron are particularly attenuating. This reversal of attenuation properties between neutrons and x-rays leads to complementary properties for the two radiographic methods. With neutrons, it is possible to visualize materials such as liquids, adhesives, rubber, plastic, or explosives even when they are in metal assemblies.

Proton radiography employs beams of protons. A rapidly moving, high-energy proton or other charged particle moves through material with little attenuation until it slows sufficiently for the charges on the particle and the material to interact. A monoenergetic charged particle travels a well-defined distance, called the range, in a given material before it is stopped. Since most of the attenuation of the charged particles occurs near the end of the range, a very small change in material thickness will result in a large change in radiation transmission. Therefore, the sensitivity of this method to small changes in object thickness is very great, if the total path for the radiation approximates the range. This is a major advantage of proton radiography. Changes in object thickness as small as 0.05% have been imaged with one-step film. Gerold H. Tenney; George L. Clark; Harold Berger

Radioimmunoassay A general method employing the reaction of antigen with specific antibody, permitting measurement of the concentration of virtually any substance of biologic interest, often with unparalleled sensitivity. The basis of the method is summarized in the competing reactions shown in the illustration. The unknown concentration of the antigenic substance in a sample is obtained by comparing its inhibitory effect on the binding of radioactively labeled antigen to a limited amount of specific antibody with the inhibitory effect of known standards.

A typical radioimmunoassay is performed by the simultaneous preparation of a series of standard and unknown mixtures in test tubes, each containing identical concentrations of labeled antigen and specific antibody, as well as variable amounts of standards or the unknown sample. After an appropriate reaction time, which may be hours or days depending on the association constant for reaction with the particular antiserum, the antibody-bound (B) and free (F) fractions of the labeled antigen are separated by one of a variety of techniques. The B/F ratios in the standards are plotted as a function of the concentration of unlabeled antigen (standard curve), and the unknown concentration of antigen is determined by comparing the observed B/F ratio with the standard curve.

The radioimmunoassay principle has found wide application in the measurement of a large and diverse group of substances in a variety of problems of clinical and biological interest. It is therefore not unexpected that there are differences in the specific methods employed for the assay of a particular substance. It seems that virtually any substance of biologic interest can be measured using radioimmunoassay, modified according to the

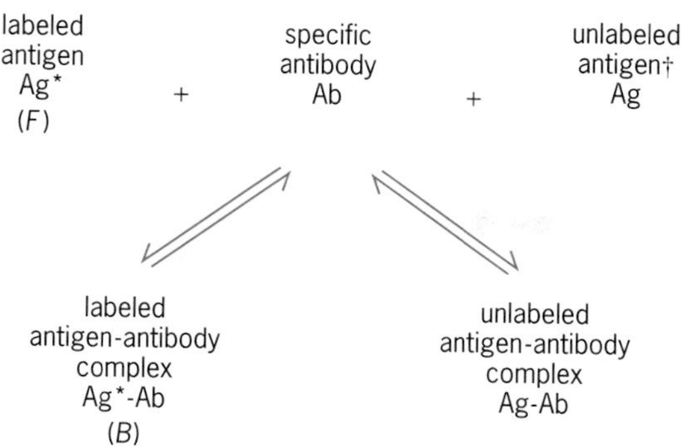

Competing reactions that form the basis of radioimmunoassay. Dagger indicates antigen in known standard solutions or unknown samples. B indicates bound antigen; F, free antigen.

characteristics of the particular substance. *See* ANTIBODY; ANTIGEN; IMMUNOASSAY. Rosalyn S. Yalow; J. John Cohen

Radioisotope A radioactive isotope (as distinguished from a stable isotope) of an element. Atomic nuclei are of two types, unstable and stable. Those in the former category are said to be radioactive and eventually are transformed, by radioactive decay, into the latter. One of the three types of particles or radiation (alpha particles, beta particles, and gamma rays) is emitted during each stage of the decay. *See* ISOTOPE; RADIOACTIVITY.

The term radioisotope is also loosely used to refer to any radioactive atomic species. Whereas approximately a dozen radioisotopes are found in nature in appreciable amounts, hundreds of different radioisotopes have been artificially produced by bombarding stable nuclei with various atomic projectiles.

A very wide variety of radioisotopes are produced in particle accelerators, such as the cyclotron. Charged particles, such as deuterons (D^+) and protons (H^+), are accelerated to great speeds by high-voltage electrical fields and allowed to strike targets in which nuclear reactions take place; for example, proton in, neutron out (p,n), increasing the target-atom atomic number by one without changing the atomic mass; and deuteron in, proton out (d,p), increasing the atomic mass by one without changing the atomic number. The target elements become radioactive because the nuclei of the atoms are unbalanced, having an excess or deficit of neutrons or protons. Although the particle-accelerating machines are most versatile in producing radioisotopes, the amount of radioactive material that can be produced is relatively smaller than that made in a nuclear reactor [less than curie amounts; a curie (abbreviated Ci) is that quantity of a radioisotope required to supply 3.7×10^{10} disintegrations per second or 3.7×10^{10} becquerels (Bq)]. For large-scale production, nuclear reactors with neutron fluxes of 1×10^{10} to 5×10^{15} neutrons per square centimeter per second are required. *See* NUCLEAR REACTION; NUCLEAR REACTOR; PARTICLE ACCELERATOR; REACTOR PHYSICS; UNITS OF MEASUREMENT. Arthur F. Rupp

Radioisotope (biology) A radioactive isotope used in studying living systems, such as in the investigation of metabolic processes. The usefulness of radioisotopes as tracers arises chiefly from three properties: (1) At the molecular level the physical and chemical behavior of a radioisotope is practically identical with that of the stable isotopes of the same element. (2) Radioisotopes are detectable in extremely minute concentrations. (3) Analysis for radioisotopic content often can be achieved without alteration of the sample or system. In some applications, principally those in which reaction rates and transfer rates are studied, isotopes, particularly radioisotopes, have unique advantages as tracers. *See* ISOTOPE; RADIOISOTOPE.

The amount of isotope to be used and the path by which it is introduced into the system are governed by many factors. Sufficient tracer to be detectable must be used, but the amounts of material which are introduced must be small enough not to disturb the system by their mass, pharmacological effects, or the effects of radiation. The mass of 1 curie, the unit of disintegration rate, depends inversely upon the half-life and directly upon the atomic weight of the particular radioisotope; it is 1 gram for ^{226}Ra (half-life 1620 years), but only 8 micrograms for ^{131}I (half-life 8.0 days). In tracer experiments with small animals, microcurie quantities are usually adequate.

There are many methods for detecting the presence of radioactive material. The Geiger counter has largely been displaced by thallium-activated sodium iodide scintillation crystals for counting gamma rays, but Geiger counters and proportional counters are still useful for counting alpha and beta particles. In histological and cytological studies the method of autoradiography, in which photographic film is exposed through contact with the specimen, is very useful. The autoradiographic method is also used extensively in conjunction with paper or column chromatography, particularly in studies of metabolic pathways. *See* AUTORADIOGRAPHY; CHROMATOGRAPHY; GEIGER-MÜLLER COUNTER; PARTICLE DETECTOR; SPECTROSCOPY.

One of the outstanding achievements in which radioisotopes have played a role has been the use of carbon-14 in the elucidation of the metabolic path of carbon in photosynthesis. The products produced in the first few seconds following exposure to light have been identified by combinations of paper chromatography and autoradiography. The extrathyroidal metabolism of iodine, the path of iodine in the thyroid gland, and other problems of intermediary metabolism have been studied intensively. The concept of the dynamic state of cell constituents is largely attributable to discoveries made with isotopic tracers. At one time it was thought that concentration gradients across cell membranes depended upon their being impermeable, but the use of isotopes has refuted this hypothesis by proving that in many such cases the substances involved are normally transported in both directions across the membrane. In physiology, radioisotopes have been used in a wide variety of permeability, absorption, and distribution studies. *See* ABSORPTION (BIOLOGY); CELL MEMBRANE; PHOTOSYNTHESIS.

The kinetics of cellular proliferation has provided a rich vein for application of radioisotopic methods. For example, the lifetime of human red blood cells (about 120 days) was established with the use of ^{59}Fe-labeled cells. Some applications, such as the intake of ^{131}I by the thyroid, the measurement of the red-cell mass with ^{51}Cr-labeled red cells, and the absorption of ^{60}Co-labeled vitamin B_{12}, are of practical clinical importance in the diagnosis and treatment of disease, and knowledge of the rates of distribution and disposal of a wide variety of radioactive substances is basic to the problem of evaluating the hazard from fallout radiation. James S. Robertson

Radioisotope geochemistry A branch of environmental geochemistry and isotope geology concerned with the occurrence of radioactive nuclides in sediment, water, air, biological tissues, and rocks. The nuclides have relatively short half-lives ranging from a few days to about 10^6 years, and occur only because they are being produced by natural or anthropogenic nuclear reactions or because they are the intermediate unstable daughters of long-lived naturally occurring radioactive isotopes of uranium and thorium. The nuclear radiation, consisting of alpha particles, beta particles, and gamma rays, emitted by these nuclides constitutes a potential health hazard to humans. However, their presence also provides opportunities for measurements of the rates of natural processes in the atmosphere and on the surface of the Earth. *See* ALPHA PARTICLES; BETA PARTICLES; GAMMA RAYS.

The unstable daughters of uranium and thorium consist of a group of 43 radioactive isotopes of 13 chemical elements, including all of the naturally occurring isotopes of the chemical elements radium, radon, polonium, and several others. A second group of radionuclides is produced by the interaction of cosmic rays with the chemical elements of the Earth's surface and atmosphere. This group includes hydrogen-3 (tritium), beryllium-10, carbon-14, aluminum-26, silicon-32, chlorine-36, iron-55, and others. A third group of radionuclides is produced artificially by the explosion of nuclear devices, by the operation of nuclear reactors, and by various particle accelerators used for research in nuclear physics. Some of the radionuclides produced in nuclear reactors decay sufficiently slowly to be useful for geochemical research, including strontium-90, cesium-137, iodine-129, and isotopes of plutonium. The explosion of nuclear devices in the atmosphere has also contributed to the abundances of certain radionuclides that are produced by cosmic rays such as tritium and carbon-14. *See* COSMOGENIC NUCLIDE; DATING METHODS; ENVIRONMENTAL RADIOACTIVITY; RADIOACTIVE FALLOUT; RADIOISOTOPE; TRANSURANIUM ELEMENTS. Gunter Faure

Radiolaria A group of marine protists, regarded as a subclass of Actinopodea in older classifications, but not recognized as a natural group in some modern systems owing to its heterogeneity. Radiolarians occur almost exclusively in the open ocean as part of the plankton community, and are widely recognized for their ornate siliceous skeletons produced by most of the groups (illustration *a–c*). Their skeletons occur abundantly in ocean sediments and are used in analyzing the layers of the sedimentary record (biostratigraphy).

In some classification systems, the Radiolaria are subdivided into two classes, Polycystinea and Phaeodarea. However, clarification of the evolutionary relationships (phylogeny) of the Radiolaria using modern molecular genetic analyses indicates that Polycystinea and Phaeodarea are undoubtedly polyphyletic: that is, they are not derived from closely related phylogenetic ancestors. Therefore, current taxonomic systems separate them.

A characteristic feature of the group is the capsule, a central mass of cytoplasm bearing one or more nuclei, food reserves, and metabolic organelles. This is surrounded by a perforated wall and a frothy layer of cytoplasm known as the extracapsulum, where food digestion generally occurs and numerous axopodia (stiffened strands of cytoplasm) and rhizopodia radiate toward the surrounding environment. Radiolarians feed on microplankton captured by the sticky axopodia. Algal symbionts, including dinoflagellates, green algae, and golden-brown pigmented algae, occur profusely in the extracapsulum. The algal symbionts living within the protection of the extracapsulum provide photosynthetically derived food for the radiolarian host.

In the order Spumellarida (class Polycystinea) the central capsule is perforated by numerous pores distributed evenly on the surface of the capsular wall. These pores, containing strands of cytoplasm, provide continuity between the cytoplasm in the central capsule and the surrounding extracapsulum. The skeletons of the Spumellarida are characteristically developed on a spherical organizational plan, but some are spiral-shaped (resembling snail shells) or produce elongate skeletons composed of numerous chambers built one upon another. In some genera, such as *Thalassicolla* (illus. *d*), there is no skeleton; in others there are rods or spicules, or often a single or multiple concentric latticework skeleton (illus. *c*).

Multicellular aggregates (colonies of spumellaridans), measuring several centimeters in diameter (or even several meters in some rare elongate forms), consist of numerous radiolarian central capsules enclosed within a gelatinous envelope and interconnected by a web of rhizopodia that bears abundant algal symbionts. A thin halo of feeding rhizopodia protrudes from the surface of the colony and is used to capture prey. Reproduction is poorly understood. In some colonial forms, daughter colonies are produced by asexual reproduction (fission). Flagellated swarmers (illus. *e*) released from mature central capsules of some species are possibly gametes and have a large crystal inclusion of strontium sulfate.

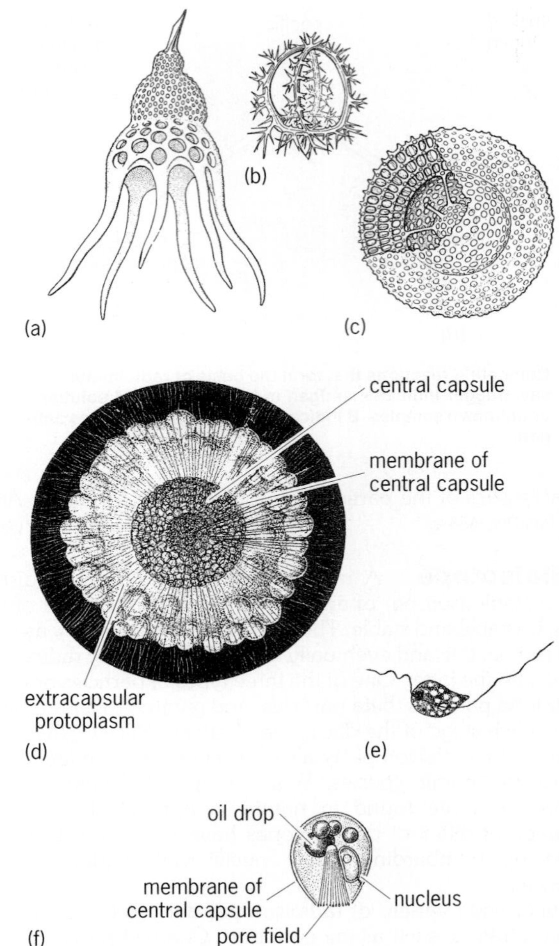

Radiolaria. Skeletons representing (*a*, *b*) certain Nasselarida and (*c*) certain Spumellarida. (*d*) *Thalassicolla*, one of the Spumellarida without skeletal elements. (*e*) Biflagellate gametes. (*f*) Central capsule, one of the Nasselarida showing one group of pores. (*After L. H. Hyman, The Invertebrates, vol. 1, McGraw-Hill, 1940*)

In the Nasselarida (Polycystinea), the central capsule is often ovate and the pores are localized at one pole (illus. *f*). The axopodia and rhizopodia emerge from this pore field and are supplied by a conelike array of microtubules within the central capsule. The skeleton, when present, is often shaped like a dome or helmet.

Radiolarians in the class Phaeodarea possess a central capsule with two types of pore areas, a larger one (astropyle) that serves as a kind of cytopharynx where food is carried into the central capsule, and two accessory openings (parapylae) where smaller strands of cytoplasm emerge. The phaeodarian skeleton exhibits a wide range of shapes, including geodesic-like lattice spheres and small porous clam-shaped shells. No symbiotic algae have been reported in the Phaeodarea.

Radiolarians have a fossil record that extends back to early Paleozoic time, about a half billion years ago. Compared with other groups of skeleton-bearing marine microplankton, they are highly diverse, several hundred species having inhabited the oceans at any given time. Because they are planktonic and have undergone continuous evolutionary change, radiolarians are particularly useful for determining time equivalence (and geological ages) of marine sedimentary deposits at widely separated localities. The Cenozoic record of radiolarians in sediments, particularly on the deep-sea floor, is sufficiently complete to show the course of evolutionary change in considerable detail. Assemblages of fossil radiolarians also provide clues to oceanic conditions during the geological past. Each of the major oceanic water masses has its characteristic radiolarian fauna, and so changes in the distribution or composition of these assemblages can be interpreted in terms of changes in the pattern of water masses, or in their oceanographic properties. *See* ACTINOPODEA; PROTOZOA. O. Roger Anderson; William R. Riedl

Radiology The medical science concerned with x-rays, radioactive materials, and other ionizing radiations, and the application of the principles of this science to diagnosis and treatment of disease. Nonionizing radiations of infrared and ultrasound are also used for diagnosis.

Diagnostic radiology uses radiation, usually x-rays, to study the configuration of anatomical structures or the function of body organ systems. *See* RADIOGRAPHY.

Radioactive isotopes are used to obtain images of organ systems and functions. The accumulation of isotope in a tumor or an organ such as the thyroid is recorded by a suitable γ-ray detector attached to an electronic amplifier and recording equipment. The image of the radioactivity concentrated in an organ is viewed on a television-type screen and recorded on a photographic print. *See* RADIOACTIVE TRACER.

Sound waves of 1–10 MHz are transmitted from a crystal transducer, and after amplification are displayed on an oscilloscope and recorded on a photographic print. The ultrasound pulses demonstrate organ structures such as the heart, liver, and spleen. Although the resolution is less fine than that obtained with x-ray, there is an advantage in that the ultrasound is nonionizing radiation. Ultrasound is particularly useful, therefore, in determining the size and degree of development of the human fetus. *See* ULTRASONICS.

Infrared radiation from the human body is used to detect tumors such as breast tumors, which are near the body surface. The technique, thermography, is based on the idea that tumors are warmer than the surrounding normal tissue. This increase in temperature is detected by an infrared device, and the "hot spot" scan is displayed on a television-type screen, with permanent records kept on photographic prints.

Radiation therapy deals with the treatment of disease with ionizing radiation. The diseases most commonly treated are cancer and allied diseases. Radiation therapy has been found useful in the management of some diseases such as ringworm of the scalp and bursitis, but because of possible serious complications occurring many years later, the use of ionizing radiation is generally avoided if alternative methods of treatment are available.

In cancer therapy the objective is to destroy a tumor without causing irreparable radiation damage in normal body tissues that must of necessity be irradiated in the process of delivering a lethal dose to the tumor. This applies particularly to important normal structures in the vicinity of the tumor. The relative radiosensitivity of the tumor with respect to these normal structures is the chief factor determining the success of the treatment. Lee B. Lusted

Radiometry A branch of science that deals with the measurement or detection of radiant electromagnetic energy. Radiometry is divided according to regions of the spectrum in which the same experimental techniques can be used. Thus, vacuum ultraviolet radiometry, intermediate-infrared radiometry, far-infrared radiometry, and microwave radiometry are considered separate fields, and all of these are to be distinguished from radiometry in the visible spectral region. Curiously, radiometry in the visible is called radiometry, optical radiation measurement science, or photometry, but it is not called visible radiometry. *See* ELECTROMAGNETIC RADIATION; INFRARED RADIATION; LIGHT; MICROWAVE; ULTRAVIOLET RADIATION.

Any radiation detector (such as a thermometer) that responds to an increase in temperature caused by the absorption of radiant energy is known as a thermal detector. Similarly, any detector (such as a photochemical reaction) that responds to the excitation of a bound electron is called a photon or quantum detector.

Liquid-in-glass thermometers are sluggish and relatively insensitive. The key to developing thermal detectors with better performance than liquid-in-glass thermometers has been to secure a large and rapid rise in temperature associated with a high sensitivity to temperature changes.

Thermal detectors have been based upon a number of different principles. Radiation thermocouples produce a voltage, bolometers undergo a change in resistance, pyroelectric detectors undergo a change in spontaneous electric polarization, and the gas in pneumatic detectors (Golay cells) and photoacoustic detectors expands in response to incident radiation. The periodic expansion and contraction of the gas in response to high-frequency modulated radiation is detected by a sensitive microphone in the case of the photoacoustic detector. The Golay cell, on the other hand, uses a sensitive photomultiplier and a reference beam of light to detect distortion of a flexible membrane mirror caused by the expansion and contraction of the gas. *See* BOLOMETER; PYROELECTRICITY; THERMOCOUPLE.

The main problem with thermal detectors is that they respond not only to electromagnetic radiation but to any source of heat. This makes their design, construction, and use rather difficult, because they must be made sensitive to the radiation of interest while remaining insensitive to all other sources of heat, such as conduction, convection, and background radiation, that are of no interest in the particular measurement.

Photon detectors respond only to photons of electromagnetic radiation that have energies greater than some minimum value determined by the quantum-mechanical properties of the detector material. Since heat radiation from the environment at room temperature consists of infrared photons, photon detectors for use in the visible can be built so that they do not respond to any source of heat except the radiation of interest.

Following the introduction of planar silicon technology for microelectronics, the same technology was quickly exploited to make planar photodiodes based on the internal photoelectric effect in silicon. In these devices, the separation of a photogenerated electron-hole pair by the built-in field surrounding the p^+n junction induces the flow of one electron in an external short circuit (such as the inputs to an operational amplifier) across the electrodes. The number of electrons flowing in an external short circuit per absorbed photon is called the quantum efficiency. The use of these diodes has grown to the point where they are the most widely used detector for the visible and nearby spectral regions. Their behavior as a radiation detector in the visible is so nearly ideal that they can be used as a standard, their cost is so low that they can be used for the most mundane of applications, and their sensitivity is so high that they can be used to measure all but the weakest radiation (which requires the most sensitive photomultipliers). *See* JUNCTION DIODE; PHOTODIODE; SEMICONDUCTOR DIODE.

Research efforts have been directed at producing photon detectors based on more exotic semiconductors, and more complicated structures to extend the sensitivity, time response, and spectral coverage. *See* OPTICAL DETECTORS. Jon Geist

Radish A cool-season annual or biennial crucifer, *Raphanus sativus*, of Chinese origin belonging to the plant order Capparales. The radish is grown for its thickened hypocotyl, which is eaten uncooked as a salad vegetable. Colors include red, yellow, white, black, pink, and red-white combinations. *See* CAPPARALES. H. John Carew

Radium A chemical element, Ra, with atomic number 88. The atomic weight of the most abundant naturally occurring isotope is 226. Radium is a rare radioactive element found in uranium minerals to the extent of 1 part for about every 3×10^6 parts of uranium. Chemically, radium is an alkaline-earth metal having properties quite similar to those of barium. Radium is important because of its radioactive properties and is used primarily in medicine for the treatment of cancer, in atomic energy technology for the preparation of standard sources of radiation, as a source for actinium and protactinium by neutron bombardment, and in certain metallurgical and mining industries for preparing gamma-ray radiographs. *See* PERIODIC TABLE.

Thirteen isotopes of radium are known; all are radioactive; four occur naturally; the rest are produced synthetically. Only ^{226}Ra is technologically important. It is distributed widely in nature, usually in exceedingly small quantities. The most concentrated source is pitchblende, a uranium mineral containing about 0.014 oz (0.4 g) of radium per ton of uranium.

Biologically, radium behaves as a typical alkaline-earth element, concentrates in bones by replacing calcium and, as a result of prolonged irradiation, causes anemia and cancerous growths. The tolerance dose for the average human being has been estimated at a total of 1 μg of radium fixed within the body. However, because radiations from radium and its decay products preferentially destroy malignant tissue, radium and radon, the gaseous decay product of radium, have been used to check the growth of cancer.

When first prepared, nearly all radium compounds are white, but they discolor on standing because of intense radiation. Radiation causes a purple or brown coloration in glass on long contact with radium compounds. Eventually the glass crystallizes and becomes crazed. Radium salts ionize the surrounding atmosphere, thereby appearing to emit a blue glow, the spectrum of which consists of the band spectrum of nitrogen. Radium compounds will discharge an electroscope, fog a light-shielded photographic plate, and produce phosphorescence and fluorescence in certain inorganic compounds such as zinc sulfide. The emission spectrum of radium compounds is similar to those of the other alkaline earths; radium halide imparts a carmine color to a flame.

Physical properties of radium

Property	Value
Atomic number	88
Atomic weight	226.05
Valence states	0, 2+
Specific gravity	6.0 at 20°C
Melting point	700°C (1290°F)
Boiling point	~1140°C (2080°F)
Ionic radius, Ra^{2+}	0.245 nm (estimated)
Atomic parachor	~140
Decomposition potential	1.718 volt
Heat of formation of oxide	130 kcal/mole
Magnetic susceptibility	Feebly paramagnetic

When freshly prepared, radium metal has a brilliant white metallic luster. Some of its physical properties are shown in the table. Chemically, the metal is highly reactive. It blackens rapidly on exposure to air because of the formation of a nitride. Radium reacts readily with water, evolving hydrogen and forming a soluble hydroxide. *See* ALKALINE-EARTH METALS; NUCLEAR REACTION; RADIOACTIVE MINERALS; RADIOACTIVITY; RADON. Murrell L. Satutsky

Radius of gyration The radial distance from a rotation axis at which the mass of an object could be concentrated without altering the moment of inertia of the body about that axis. If the mass M of the body were actually concentrated at a distance k from the axis, the moment of inertia about that axis would be Mk^2. If this quantity is equal to the actual moment of inertia I about the rotation axis, then k is the radius of gyration given by $k = \sqrt{I/M}$. The quantity k has dimensions of length and is measured in appropriate units of length such as meters. *See* MOMENT OF INERTIA. Edwin R. Jones

Radome A protective housing for radar, satellite communications, or similar antennas, often spherical but sometimes shaped to satisfy aerodynamic or other requirements of the installation. Radomes are designed to satisfy both electromagnetic and mechanical requirements. *See* RADAR.

Radomes protect antennas from harmful environmental conditions. Shielding mechanically scanning antennas from winds, for example, reduces the mechanical power needed to ensure proper motion. Properly designed radomes can lessen the detrimental effects of ice, snow, and rain accumulation. *See* ANTENNA (ELECTROMAGNETISM).

Radomes on aircraft must be shaped for proper flight characteristics. Top-mounted larger antennas are housed in rotating radomes or "rotodomes," turning with the antenna inside; smaller antennas operate behind shaped stationary radomes at the nose of the aircraft or beneath it.

Spherical radomes for surface installations must be large enough to allow the motion of the antenna inside. They may be constructed of thin dielectric sheets or made of a thin flexible material. *See* DIELECTRIC MATERIALS.

In general terms, the radome must be simultaneously transparent to the radar or communication signal and structurally sound. The designer has available many materials of varying characteristics to help in meeting these joint requirements. The dielectric nature of all such materials, causing both absorptive loss in the material itself and surface reflections, is of prime interest to the designer. *See* ABSORPTION OF ELECTROMAGNETIC RADIATION; RADAR-ABSORBING MATERIALS; REFLECTION OF ELECTROMAGNETIC RADIATION. Robert T. Hill

Radon A chemical element, Rn, atomic number 86. Radon is produced as a gaseous emanation from the radioactive decay of radium. The element is highly radioactive and decays by the emission of energetic alpha particles. Radon is the heaviest of the noble, or inert, gas group and thus is characterized by chemical inertness. More than 25 isotopes of radon have been identified. All isotopes are radioactive with short half-lives. *See* PERIODIC TABLE.

Radon is found in natural sources only because of its continuous replenishment from the radioactive decay of longer-lived precursors in minerals containing uranium, thorium, or actinium. ^{222}Rn (half-life 3.82 days), ^{220}Rn (thoron; half-life 55 s), and ^{219}Rn (actinon; half-life 4.0 s), occur in nature as members of the uranium (U), thorium (Th), and actinium (Ac) series, respectively. All three decay by the emission of energetic alpha particles. *See* ACTINIUM; ALPHA PARTICLES; RADIOACTIVITY; RADIUM; THORIUM; URANIUM.

Any surface exposed to ^{222}Rn becomes coated with an active deposit which consists of a group of short-lived daughter products. The radiations of this active deposit include energetic alpha particles, beta particles, and gamma rays. The ultimate decay products of radon following the rapid decay of the active deposit to lead-210 include bismuth-210, polonium-210, and finally, stable lead-206. Radon possesses a particularly stable electronic configuration, which gives it the chemical properties characteristic of noble gas elements. It has a boiling point of −62°C (−80°F) and a melting point of −71°C (−96°F). The spectrum of radon has been extensively studied, and resembles that of the other inert gases. Radon is readily adsorbed on charcoal, silica gel, and other adsorbents, and this property can be used to separate the element from gaseous impurities. Earl K. Hyde

The rocks and soils of the Earth's crust contain approximately 3 parts per million of ^{238}U, the long-lived head of the uranium series; 11 ppm of ^{232}Th, the head of the thorium series; but only about 0.02 ppm of ^{235}U, the long-lived member of the actinium group. The radon isotopes ^{222}Rn and ^{220}Rn are produced in proportion to the amount of the parent present. Some of the newly formed radon atoms which originate in or on the surface of mineral grains escape into the soil gas, where they are free to diffuse within the soil capillaries. Some of the radon atoms eventually find their way to the surface, where they become a part of the atmosphere. Even though thorium (^{232}Th) is generally more abundant than uranium in the Earth's crust, the probability for decay is smaller; hence, the production rate of ^{222}Rn and ^{220}Rn in the soil is roughly the same. Much of the ^{220}Rn decays before reaching the Earth's surface due to its short half-life.

When radon (^{222}Rn or ^{220}Rn) passes from soil to air, it is mixed throughout the lower atmosphere by eddy diffusion and the prevailing winds. Mean radon levels are found to be higher during those times of year when atmospheric stability is the greatest such as may occur during the fall months. Radon and its daughters play an important role in atmospheric electricity. Near the Earth's surface almost half of the ionization of the air is due to ^{220}Rn and ^{222}Rn and their daughter products. The alpha emitters from these chains typically produce about 10^7 ion pairs per second per cubic meter. *See* ATMOSPHERIC CHEMISTRY; ATMOSPHERIC ELECTRICITY.

Radon is readily soluble in water. Since ground and surface waters are in close contact with soil and rocks containing small quantities of radium, it is not surprising to find radon in public water supplies.

The radon isotopes ^{220}Rn and ^{222}Rn are used widely in the study of gaseous transport processes both in the underground environment and in the atmosphere. Radon accumulates to high levels of the order of 4000 becquerels/m^3 or more in caves unless natural or artificial ventilation occurs. Changes in ^{222}Rn concentrations in spring and well water and in soil and rocks have been suggested as a means of predicting earthquakes.

The tendency of the decay products of radon to attach to aerosols means that these nuclides will be inhaled and deposited in the bronchial epithelium and lungs. The daughter products, therefore, make up the major part of the internal radiation dose from radon. Ways of reducing radon levels within homes or workplaces include increased ventilation and sealing of major sources of entry from soil and building materials. Workers in uranium mines may encounter radon and decay product levels of the order of 50,000 Bq/m^3 or more. Ventilation procedures and special filters for the miners must be used. *See* BIOSPHERE; RADIATION INJURY (BIOLOGY); RADIOECOLOGY; RADIOISOTOPE (GEOCHEMISTRY).

M. Wilkening

Rafflesiales An order of flowering plants, division Magnoliophyta (Angiospermae), in the subclass Rosidae of the class Magnoliopsida (dicotyledons). The order consists of three families and fewer than a hundred species, all tropical or subtropical. The plants are highly specialized, nongreen, rootless parasites which grow from the roots of the host. They have few or solitary, rather large to very large flowers with numerous ovules and a single set of tepals that are commonly united into a conspicuous, corolloid calyx. *See* MAGNOLIOPSIDA; ROSIDAE. Arthur Cronquist

Railroad control systems Those devices and systems used to direct or restrain the movement of trains, cars, or locomotives on railroads, rapid-transit lines, and similar guided ground-transportation networks. Such control varies from the use of simple solenoid valves to fully automatic electronic-electromechanical systems.

A primary function of railroad control systems is to ensure the safe movement of trains. This is generally accomplished by providing train operators and track-side operators with visual indications of equipment status. The simplest form of control consists of track-switch-position indicators combined with track-side manually operated "stop" or "proceed" signals, which the train operator follows. Advanced systems incorporate fully automated train control, subject to human supervisory control and potential intervention when faults occur in automated systems. *See* CONTROL SYSTEMS; RAILROAD ENGINEERING; TRAFFIC-CONTROL SYSTEMS.

Block signaling significantly improves the safety of railroad operations. Automatic block signaling is accomplished by sectionalizing the track into electrical circuits to detect the presence of other trains, engines, or cars. Logic circuits in the control system detect the locations of the trains and the positions of switches, and then set the necessary signals to inform the train operators when to stop, run slowly, or proceed at posted speeds. The control system automatically detects the presence of a leading train, selects the signal to be given, and then sets the signal indications for the following train operators to read so that they may perform accordingly. In conjunction with automatic block signals, many subway rapid-transit lines incorporate automatic trip stops along the tracks to ensure that train operators obey the stop signals.

Automatic cab signaling systems display signaling information (traditionally, permitted speeds) on board the train. Coded information is transmitted to the train, generally via the running rails. Antennas and receivers aboard the train pick up, amplify, decode, and distribute the intelligence, which then causes the proper signal aspects to be displayed in the cab. Automatic cab signaling reduces or eliminates the need for wayside signals and improves the all-weather capability of trains and the train-handling capacity of the track.

Automatic train control (ATC) subsystems, located wholly on board the train, sense whether or not the train is operating within safe speed limits. If it is not, automatic train control sets the brake to bring the train to a stop or to a speed below the allowed speed. Automatic cab signaling with automatic train control is used on many railroads and several rapid-transit lines in the United Sates and on systems in Europe and Japan. Automatic train operation (ATO) subsystems perform nonvital operating functions such as starting, running at the prescribed speeds, slowing down, and stopping, and on some rapid-transit installations include passenger-door controls. Automatic train operation builds upon the information transmitted to the train as part of automatic cab signaling, and is a logical next step in automating train operations.

Station stopping presents a special set of requirements for rapid transit, commuter railroads, and mainline railroad passenger operations. Accurate positioning of car doors at the station platform and smooth deceleration at relatively high rates are desirable for passenger comfort and efficient operating performance. A special subsystem of control referred to as programmed train-stop systems (or station-stop systems) are a combination of on-board and wayside electronic and electromechanical equipment that can bring a train to rest within inches of its stopping-point target.

Car identification systems is an example of central line supervision. This system scans and decodes a series of colored and patterned lines placed on the side of each car to identify an individual car. This information is transmitted to the operations area, where a computer system is used to establish routing, determine maintenance schedules, and so on. Dispatchers and central operators can also use computer workstations to obtain information on system status.

Railroad terminals (points of origin and destination of trains) are critical to the efficient, cost-effective operation of railroads, so they represent a major focus for automation. A terminal generally contains three types of yards: a receiving yard, where incoming trains from the main line are temporarily stored; the hump yard, where cars are classified and resorted into new trains; and a departure yard, where trains are assembled and stored for dispatch onto the main line. James Costantino; Donna C. Nelson

Railroad engineering A branch of engineering concerned with the design, construction, maintenance, and operation of railways. Railway engineering includes elements of civil, mechanical, industrial, and electrical engineering. It is unique in being concerned with the interaction between moving vehicles (mechanical engineering) and infrastructure (civil engineering). The employment of both a load-supporting guideway and groups or strings of connected vehicles on flanged wheels for the transport of goods and people sets railways apart from other modes of transport. *See* CIVIL ENGINEERING; MECHANICAL ENGINEERING.

The plan view of a railroad track is known as the horizontal alignment. It is made up of a series of curves (arcs of simple circles), tangents (straight tracks), and spirals joining the curves and tangents. Deviations from any of the three are flaws. These imperfections are corrected periodically by a technique known as lining the track.

The side or elevation view of track, composed of a series of straight portions and the vertical curves joining them, is known as the vertical alignment. The vertical change in elevation, in feet, over a horizontal distance of 100 ft is the percent grade. Because the friction coefficient of steel wheels on steel rails is low, railroad grades must also be low, with values from zero to 1.5% fairly common. Two-percent grades are severe, usually requiring helper locomotives. Grades that are more severe, up to about 4%, can be surmounted only with considerable extra operating care and at significant additional expense.

Track gage is the distance between rails. The standard gage throughout the world is 4 ft 8.5 in. Narrow gages of 3 ft and broad gages of 5 and 6 ft have all been tried at various times and in different places.

The function of rail is to guide wheels and distribute their vertical and lateral loads over a wider area. Neither cast nor wrought iron was ideally suited to this task. The development of steel that was three times harder than wrought iron at reasonable cost made it possible for the weight of vehicles and therefore the productivity of railways to increase. Rails are joined end to end by butt welding, whereby continuous rails of over 1000 ft (300 m) in length can be produced. When laid in track, the rails are heavily anchored to restrain movement due to temperature changes. *See* CAST IRON; STEEL; STEEL MANUFACTURE; WROUGHT IRON.

Crossties play important roles in the distribution of wheel loads vertically, longitudinally, and laterally. Each tie must withstand loads up to one-half that imposed on the rail by a wheel. The crosstie must then distribute that load to the ballast surrounding it. Timber crossties vary in section from 6 in. × 6 in. (15 cm × 15 cm) for the lightest applications to 7 in. × 9 in. (18 cm × 23 cm) for heavy-duty track and in length from 8 ft 6 in. (2.6 m) to 9 ft (2.7 m) in length. Well-treated hardwood ties in well-maintained track may be expected to last 30 years or more. Timber crossties become unserviceable after time because of splitting, decay, insect attack, center cracking, mechanical wear, and crushing. Prestressed concrete monoblock crossties are standard in the United Kingdom and parts of continental Europe.

The granular material that supports crossties vertically and restrains them laterally is known as ballast. Ideal ballast is made up of hard, sharp, angular interlocking pieces that drain well and yet permit adjustments to vertical and horizontal alignment. Materials that crush and abrade, creating fines that block drainage or that cement, should not be used. Soft limestones and gravel, including rounded stones, are examples of poor ballast, while crushed granite, trap rock, and hard slags are superior.

The earliest diesel-electric switching locomotives developed about 600 horsepower and road freight units 1350 hp. Single locomotive units of 4000 hp are common. Common practice since dieselization began has been to employ a number of locomotive units coupled together to form a single, more powerful power source.

Locomotive development includes designs using liquefied natural gas as a fuel to reduce environmental pollutants. Locomotives equipped with the necessary power conditioning equipment and squirrel-cage asynchronous motors, made possible by the advent of high-capacity solid-state electronics, exhibit superior adhesion and have no troublesome commutators. They are adept at hauling heavy-tonnage mineral freight, fast passenger trains, or high-speed merchandise trains (freight trains that haul primarily high-value merchandise, as opposed to low-cost raw materials such as coal or grain). *See* LOCOMOTIVE.

Rail passenger systems such as the Shinkansen in Japan (1964); TGV (Très Grande Vitesse; 1981) in France; ICE (Inter City Eisenbahn; 1991) in Germany; X-2000 in Sweden (1990); or Britain's several High-Speed Intercity Trains are notable for speed and convenience. These systems have developed in context of an awareness of deteriorating highway infrastructures, serious concern with the air pollution generated by automobiles and trucks, increasing traffic congestion in urban areas, and worries over petroleum usage and supply. *See* AIR POLLUTION; MAGNETIC LEVITATION.

High-speed passenger trains require significantly different track configurations (for example, curve superelevation and turnout designs) than do slower freight trains, even those high-valued merchandise or intermodal highway trailers and containers, which frequently travel at speeds of 70 mi/h (110 km/h). These engineering differences are impractical on lines primarily moving heavy mineral freight, where axle loadings and speeds differ even more radically from high-speed passenger trains.

George H. Way

Rain shadow An area of diminished precipitation on the lee side of mountains. There are marked rain shadows, for example, east of the coastal ranges of Washington, Oregon, and California, and over a larger region, much of it arid, east of the Cascade Range and Sierra Nevadas. All mountains decrease precipitation on their lee; but rain shadows are sometimes not marked if moist air often comes from different directions, as in the Appalachian region.

The causes of rain shadow are (1) precipitation of much of the moisture when air is forced upward on the windward side of the mountains, (2) deflection or damming of moist air flow, and (3) downward flow on the lee slopes, which warms the air and lowers its relative humidity. J. R. Fulks

Rainbow An optical effect of the sky formed by sunlight falling on the spherical droplets of water associated with a rain shower. The circular arc of colors in the rainbow is seen on the side of the sky away from the Sun. The bright, primary rainbow shows the spectrum of colors running from red, on the outside of the bow, to blue on the inside. Sometimes a fainter, secondary bow is seen outside the primary bow with the colors reversed from their order in the primary bow. The shape of each bow is that of a circle, centered on the antisolar point, a point in the direction exactly opposite to that of the Sun, which is marked by the shadow of the observer's head.

As a light ray from the Sun strikes the surface of a water drop, some light is reflected and some passes through the surface into the drop. The primary bow results from light that enters the drop, reflects once inside the drop, and then leaves the drop headed toward the observer's eye. Light that is reflected twice inside the drop produces the secondary bow. The change of direction that occurs when a light ray enters or leaves the waterdrop (refraction) is different for the different colors that make up white sunlight. As a result, the size of the circle is different for each color, thereby separating the colors into the rainbow sequence. *See* METEOROLOGICAL OPTICS. Robert Greenler

Rainforest Forests that occur in continually wet climates with no dry season. There are relatively small areas of temperate rainforests in the Americas and Australasia, but most occur in the tropics and subtropics.

The most extensive tropical rainforests are in the Americas. These were originally 1.54×10^6 mi^2 (4×10^6 km^2) in extent, about half the global total, and mainly in the Amazon basin. A narrow belt also occurs along the Atlantic coast of Brazil, and a third block lies on the Pacific coast of South America, extending from northern Peru to southern Mexico.

Tropical rainforests have a continuous canopy (commonly 100–120 ft or 30–36 m tall) above which stand huge emergent trees, reaching 200 ft (60 m) or taller. Within the rainforest canopy are trees of many different sizes, including pygmies, that reach only a few feet. Trees are the main life form and are often, for purposes of description and analysis, divided into strata or layers. Trees form the framework of the forest and support an abundance of climbers, orchids, and other epiphytes, adapted to the microclimatic conditions of the different zones of the canopy, from shade lovers in the gloomy, humid lower levels, to sun lovers in the brightly lit, hotter, and drier upper levels. Most trees have evergreen leaves, many of which are pinnate or palmate. These features of forest structure and appearance are found throughout the world's lowland tropical rainforests. There are other equally distinctive kinds of rainforest in the lower and upper parts of perhumid tropical mountains, and additional types on wetlands.

Rainforests occur where the monthly rainfall exceeds 4 in. (100 mm) for 9–12 months. They merge into other seasonal or monsoon forests where there is a stronger dry season (3 months or more with 2.5 in. or 60 mm of rainfall). The annual mean temperature in the lowlands is approximately 64°F (18°C). There is no season unfavorable for growth.

Primary rainforests are exceedingly rich in species of both plants and animals. There are usually over 100 species of trees 2.5 in. (10 cm) in diameter or bigger per 2.4 acres (1 ha). There are also numerous species of climbers and epiphytes. Flowering and fruiting occur year-round, but commonly there is a peak season; animal breeding may be linked to this. Secondary rainforests are much simpler. There are fewer tree species, less variety from location to location, and fewer epiphytes and climbers; the animals are also somewhat different. *See* ECOLOGICAL SUCCESSION.

Tropical rainforests are a source of resins, dyes, drugs, latex, wild meat, honey, rattan canes, and innumerable other products essential to rural life and trade. Modern technology for extraction and for processing has given timber of numerous species monetary value, and timber has come to eclipse other forest products in importance. The industrial nations use much tropical hardwood for furniture, construction, and plywood. Rainforest timbers, however, represent only 11% of world annual industrial wood usage, a proportion that has doubled since 1950. West Africa was the first main modern source, but by the 1960s was eclipsed by Asia, where Indonesia and Malaysia are the main producers of internationally traded tropical hardwoods. Substantial logging has also developed in the neotropics. *See* FOREST ECOSYSTEM; FOREST TIMBER RESOURCES. T. C. Whitmore

Rajiformes One of four orders making up the Batoidei. Aside from a few exceptions, the combination of characters that distinguish the order from other batoids are a single unbranched rostral process on front of the cranium, extending to the tip of the snout; antorbital (in front of the eye socket) cartilage that does not extend forward to help support the anterior part of the disc; two dorsal fins on the tail; lack of spines on the tail; lack of electric organs; a snout that is not a long flat blade; a head lacking anterior finlike expansions of pectoral fins; and ovoviviparous reproduction. Their food consists of various crustaceans, mollusks, and small fishes. Classification is as follows:

Order Rajiformes
  Rhinidae (genus *Rhina*)
  Rhynchobatidae (*Rhynchobatus*)
  Rhinobatidae (*Aptychotrema, Rhinobatos, Trygonorrhina, Zapteryx*)
  Rajidae (skates; 18 genera)

The family Rajidae (skates) consists of 18 genera and about 200 species. It is represented in all seas of the world, from the tropics to subpolar waters, from shallow depths to at least 1000 m (3300 ft). Like other rajiforms, skates are bottom feeders, relying primarily on their pectoral fins for locomotion. In body form, skates are more raylike than sharklike. The head and body are strongly depressed; the snout is moderately slender; and the tail sector is sharply distinct from the body sector. *Raja* is the typical genus and has the largest number of species. *See* BATOIDEA.

Herbert Boschung

Raman effect A phenomenon observed in the scattering of light as it interacts with a material medium in which the incident light suffers a change in frequency due to internal energy change of the molecular scatterers. Raman scattering differs in this respect from Rayleigh scattering in which the incident and scattered light have the same frequency. Shifts in frequency are determined by the type of molecules in the scattering medium, and spectral analysis of the scattered light can provide a "fingerprint" of the chemical structure of the scatterers. Both incoherent and coherent forms of the Raman effect exist. Spontaneous Raman scattering, the usual (incoherent) form, is very weak, with Raman signals 4–5 orders of magnitude smaller than Rayleigh scattering and about 14 orders of magnitude smaller than fluorescence. In stimulated Raman scattering, a coherent form, the signals may be quite large. In addition, there are nonlinear forms of Raman scattering, including hyper-Raman scattering and coherent anti-Stokes Raman scattering (CARS). *See* SCATTERING OF ELECTROMAGNETIC RADIATION.

Discovery. Raman scattering was experimentally observed in 1928 by the Indian physicists C. V. Raman and K. S. Krishnan, who examined sunlight scattered by a number of liquids. With the help of complementary filters, they found frequencies in the scattered light that were lower than the frequencies in the filtered sunlight. By using light of a single frequency from a mercury arc, they showed that the new frequencies were characteristic of the scattering medium.

Physical principles. The mechanism of the Raman effect can be understood either by the particle picture of light or by wave theory. Both features emerge in the quantum theory of radiation.

Photon picture. The particle model envisages light quanta or photons traversing a molecular medium undergoing scattering. If the interaction is elastic, the photons scatter with unchanged energy E and momentum, and hence with unchanged frequency ν, using Planck's relationship, $E = h\nu$, where h is Planck's constant. Such a process gives rise to Rayleigh scattering. If, however, the collision is inelastic, the photons can gain or lose energy from the molecules. A change ΔE in the internal energy of the scatterer causes a change $-\Delta E$ in photon energy, hence a corresponding frequency change $\Delta\nu = -\Delta E/h$. Frequency-shifted scattered photons can occur at lower or higher energy relative to the incident photons, depending on whether the scattering molecule is in the ground or excited vibrational state. In the former case, photons lose energy by exciting a vibration and the scattered light appears at a lower frequency, $\nu_s = \nu - \Delta E/h$, called the Stokes frequency. In the latter case, by interacting with a molecule in an excited vibrational state, the photons gain energy from the molecular vibrations and the scattered signal appears at higher frequency $\nu_{aS} = \nu + \Delta E/h$, the anti-Stokes frequency. See LIGHT; QUANTUM MECHANICS; SCATTERING EXPERIMENTS (ATOMS AND MOLECULES).

Wave picture. In the wave picture, light as an electromagnetic wave interacts with the molecules in the medium, inducing an oscillating dipole moment by separating the positively charged nuclei and negatively charged electrons. In Rayleigh scattering (elastic), the oscillation frequency is ν, the same as that of the incident light. However, in the case of Raman scattering (inelastic), scattering from the molecules, vibrating at frequency $\Delta E/h$, gives rise to two frequency components at $\nu \pm \Delta E/h$, with ΔE the internal energy of the scatterer. This induced dipole moment interacts with the light wave, producing emission not only at the frequency ν of the incoming field (elastic component), but also at sidebands at the sum and difference frequencies between the incoming light and the molecular vibrations (Stokes and anti-Stokes components ν_S and ν_{aS}). See DIPOLE MOMENT.

In general, any scatterer with internal energy structure can undergo the Raman effect by exchanging internal energy with the radiation field. In molecules, rotations, as well as vibrations, can produce Raman effects. Similarly, atoms can give rise to Raman scattering.

Raman spectroscopy. Raman scattering is analyzed by spectroscopic means. When a medium is excited by monochromatic incident light, the collection of new frequencies in the spectrum of scattered light is called its Raman spectrum. As the Raman effect probes vibrations of the molecule, which depend on the kinds of constituent atoms and their bond strengths and arrangements, a Raman spectrum provides a high degree of chemical structural information. The kind of information provided by this spectrum is similar to that provided by infrared spectroscopy. See INFRARED SPECTROSCOPY.

The advent of lasers as intense sources of monochromatic light was a milestone in the history of Raman spectroscopy and resulted in dramatically improved scattering signals. Today, intense sources of laser light over a wide range of frequencies from the near-ultraviolet to the near-infrared region are used in Raman scattering studies, enabling selection of optimum excitation conditions for each sample. See LASER; LASER SPECTROSCOPY.

Incoherent Raman scattering. The most common form of incoherent Raman spectroscopy, is known as the spontaneous Raman effect. Several other variants of incoherent Raman scattering such as resonance Raman scattering (RSS), surface-enhanced Raman scattering (SERS), and hyper-Raman scattering (HRS) are also important.

Coherent Raman scattering. In addition to the incoherent Raman processes discussed above, the availability of lasers has also opened the field of study of coherent or stimulated Raman processes. Types of coherent scattering processes include stimulated Raman scattering (SRS), Raman lasers, inverse Raman scattering, and coherent anti-Stokes Raman scattering (CARS).

Applications. Raman spectroscopy is of considerable value in determining molecular structure and in chemical analysis. It can be used to characterize the chemical and structural composition of nearly all kinds of molecules in liquid, gas phase, and solid samples for both scientific study and industrial monitoring, such as control of products and processes. With developments in SERS, Raman spectroscopy not only is a tool for structural analysis, but also allows ultrasensitive and trace detection of substances down to the single-molecule level. Raman spectroscopy has become an important tool for characterizing nanostructures such as carbon nanotubes and nanowires. Raman spectroscopy can also be used to characterize the chemical and structural composition of complex systems, such as biological materials. Materials can be studied remotely, either with line-of-sight techniques or via optical fiber probes (with appropriate filters). See FIBER-OPTIC SENSOR; MOLECULAR STRUCTURE AND SPECTRA.

Katrin Kneipp; Michael S. Feld

Ramie

Ramie The plant *Boehmeria nivea*, a stingless member of the nettle family; the only member of the family used commercially for fiber. This herbaceous perennial is erect, usually nonbranching, and 3 to 6 ft (1 to 2 m) tall at maturity. The fiber, which comes from the inner bark, is exceptionally strong and has uses similar to those for fiber flax. Ramie is grown most in the tropics and subtropics of the Far East and Brazil. See NATURAL FIBER.

Elton G. Nelson

Random matrices

Random matrices Collections of large matrices, chosen at random from some ensemble. Random-matrix theory is a branch of mathematics which emerged from the study of complex physical problems, for which a statistical analysis is often more enlightening than a hopeless attempt to control every degree of freedom, or every detail of the dynamics. Although the connections to various parts of mathematics are very rich, the relevance of this approach to physics is also significant.

Random matrices were introduced by Eugene Wigner in nuclear physics in 1950. In quantum mechanics the discrete energy levels of a system of particles, bound together, are given by the eigenvalues of a hamiltonian operator, which embodies the interactions between the constituents. This leads to the Schrödinger equation which, in most cases of interest in the physics of nuclei, cannot be solved exactly, even with the most advanced computers. For a complex nucleus, instead of finding the location of the nuclear energy levels through untrustworthy approximate solutions, Wigner proposed to study the statistics of eigenvalues of large matrices, drawn at random from some ensemble. The only constraint is to choose an ensemble which respects the symmetries that are present in the forces between the nucleons in the original problem, and to select a sequence of levels corresponding to the quantum numbers that are conserved as a consequence of these symmetries, such as angular momentum and parity. The statistical theory does not attempt to predict the detailed sequence of energy levels of a given nucleus, but only the general properties of those sequences and, for instance, the presence of hidden symmetries. In many cases this is more important than knowing the exact location of a particular energy level. This program became the starting point of a new field, which is now widely used in mathematics and physics for the understanding of quantum chaos, disordered systems, fluctuations in mesoscopic systems, random surfaces, zeros of analytic functions, and so forth. See CONSERVATION LAWS (PHYSICS); EIGENVALUE (QUANTUM MECHANICS); MATRIX THEORY; QUANTUM MECHANICS.

The mathematical theory underlying the properties of random matrices overlaps with several active fields of contemporary mathematics, such as the asymptotic behavior of orthogonal polynomials at large-order, integrable hierachies, tau functions, semiclassical expansions, combinatorics, and group theory; and it is the subject of active research and collaboration between physics and mathematics.

Edouard Brézin

Rangefinder (optics)

Rangefinder (optics) An optical instrument for measuring distance, usually from its position to a target point. Light from the target enters the optical system through two windows spaced apart, the distance between the windows being termed

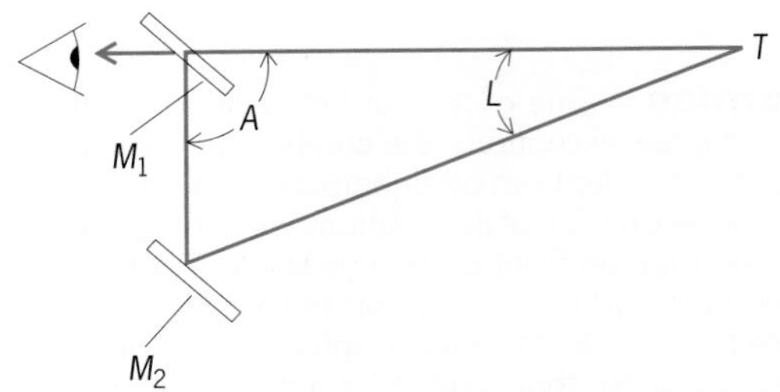

Simple coincidence rangefinder. *A* is a right angle; *L* = convergence angle at target *T*.

the base length of the rangefinder. The rangefinder operates as an angle-measuring device for solving the triangle comprising the rangefinder base length and the line from each window to the target point. Rangefinders can be classified in general as being of either the coincidence or the stereoscopic type.

In coincidence rangefinders, one-eyed viewing through a single eyepiece provides the basis for manipulation of the rangefinder adjustment to cause two images or parts of each to match or coincide. This type of device is used, in its simpler forms, in photographic cameras. The basic optical arrangement is shown in the illustration, where M_1 and M_2 are a semitransparent mirror and a reflecting mirror, respectively. When coincidence is obtained, that is, when the target T is seen in the same apparent position along either path, the rangefinder equation for the range D is satisfied:

$$D = B \cot L$$

Stereoscopic rangefinders are entirely different, although externally they resemble coincidence rangefinders except for the fact that they possess two eyepieces. It is essentially a large stereobinocular fitted with special reticles which allow a skilled user to superimpose the stereo image formed by the pair of reticles over the images of the target seen in the eyepieces, so that the reticle marks appear to be suspended over the target and at the same apparent distance. Edward K. Kaprelian

Rankine cycle A thermodynamic cycle used as an ideal standard for the comparative performance of heat-engine and heat-pump installations operating with a condensable vapor as the working fluid. Applied typically to a steam power plant, as shown in the illustration, the cycle has four phases: (1) heat addition *bcde* in a boiler at constant pressure p_1 changing water at b to superheated steam at e, (2) isentropic expansion *ef* in a prime mover from initial pressure P_1 to backpressure P_2, (3) heat rejection *fa* in a condenser at constant pressure p_2 with wet steam at f converted to saturated liquid at a, and (4) isentropic compression *ab* of water in a feed pump from pressure p_2 to pressure p_1.

This cycle more closely approximates the operations in a real steam power plant than does the Carnot cycle. Between given temperature limits it offers a lower ideal thermal efficiency for the conversion of heat into work than does the Carnot standard. Losses from irreversibility, in turn, make the conversion efficiency in an actual plant less than the Rankine cycle standard. *See* CARNOT CYCLE; REFRIGERATION CYCLE; THERMODYNAMIC CYCLE; VAPOR CYCLE. Theodore Baumeister

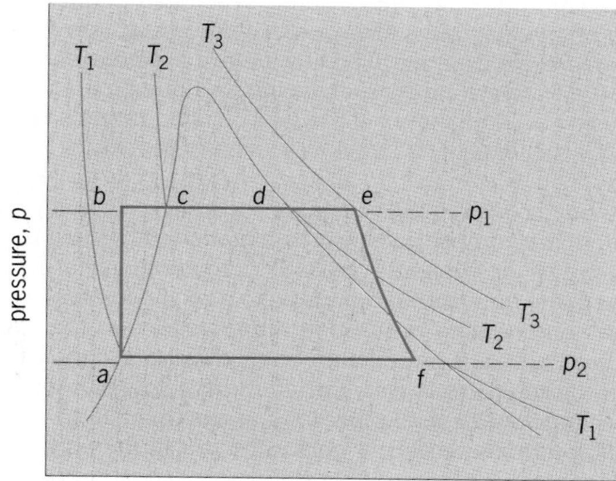

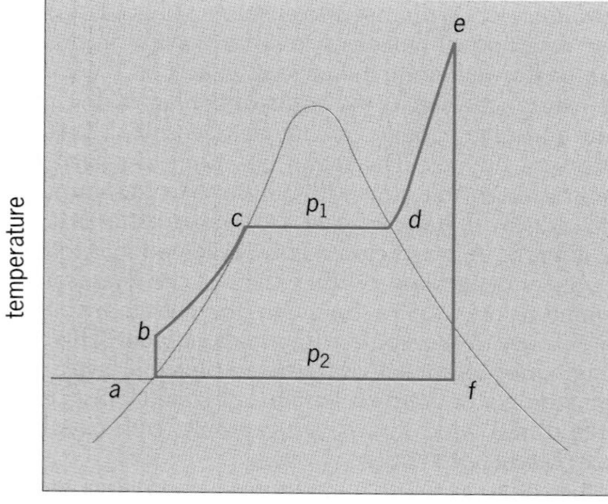

Rankine-cycle diagrams (pressure-volume and temperature-entropy) for a steam power plant using superheated steam. Pressure-volume diagrams shows curves for constant temperatures T_1, T_2, and T_3 (isothermals).

Ranunculales An order of flowering plants, division Magnoliophyta (Angiospermae), in the subclass Magnoliidae of the class Magnoliopsida (dicotyledons). The order consists of 8 families and about 3200 species. The vast majority belong to only 3 families, the Ranunculaceae (2000), Berberidaceae (650), and Menispermaceae (425). Within its subclass, the order is characterized by its mostly separate carpels, triaperturate pollen, herbaceous or only secondarily woody habit, frequently numerous stamens, generally more than two sepals, and lack of ethereal oil cells. Many members of the order contain isoquinoline alkaloids. The barberry (*Berberis*), in the family Berberidaceae, and the buttercup (*Ranunculus*), columbine (*Aquilegia*), larkspur (*Delphinium*), and wind flower (*Anemone*), in the family Ranunculaceae, are familiar genera. *See* MAGNOLIIDAE; MAGNOLIOPSIDA. Arthur Cronquist; T. M. Barkley

Rape Rape (*Brassica napus*) and turnip rape (*B. campestris*) plants are members of the Cruciferae family. The name is derived from the Latin *rapum*, meaning "turnip," to which these plants are closely related. The aerial portions of rape plants have been bred to produce oilseeds, fodder, and vegetable crops. Rape seed is small, round, and usually black, although varieties with yellow seed coats are also grown. The seeds contain over 40% oil. *See* CAPPARALES.

Brassica oilseed crops are the world's third most important source of edible vegetable oil, with production centered in Canada, northern Europe, China, and the Indian subcontinent. Seed, oil, and meal from rapeseed plants that contain superior nutritional qualities are known worldwide as canola. Canola oil is a highly nutritious salad and cooking oil and is used in the manufacture of margarine and shortening. It is also used in Europe as a diesel fuel, while rape oil is used industrially in the manufacture of plastics and as a lubricant. Canola meal, the by-product of oilseed extraction, is a high-quality protein feed supplement for livestock and poultry. R. K. Downey

Raphidophyceae A small class of poorly known biflagellate unicellular algae (raphidomonads) in the chlorophyll *a-c* phyletic line (Chromophycota). It is sometimes called Chloromonadophyceae, and its members called chloromonads, but this nomenclature is confusing because it calls to mind *Chloromonas*, a totally unrelated genus of green algae. The alternative nomenclature used here is derived from *Raphidomonas*, a generic name within the class. *See* CHROMOPHYCOTA.

Two families are recognized. All photosynthetic raphidomonads are placed in the Vacuolariaceae. The Thaumatomastigaceae comprises a few colorless forms that bear pseudopodia. They are osmotrophic or phagotrophic or both. Most genera in both families occur in fresh water (acidic to neutral), but there are brackish-water and marine forms that produce conspicuous blooms. *Heterosigma*, for example, is a frequent cause of red tide in the inland seas of Japan. *See* ALGAE.

Paul C. Silva; Richard L. Moe

Rare-earth elements The group of 17 chemical elements with atomic numbers 21, 39, and 57–71; the name lanthanides is reserved for the elements 58–71. The name rare earths is a misnomer, because they are neither rare nor earths. *See* ACTINIDE ELEMENTS; LANTHANIDE CONTRACTION; PERIODIC TABLE.

Most of the early uses of the rare earths took advantage of their common properties and were centered principally in the glass, ceramic, lighting, and metallurgical industries. Today these applications use a very substantial amount of the mixed rare earths just as they are obtained from the minerals, although sometimes these mixtures are supplemented by the addition of extra cerium or have some of their lanthanum and cerium fractions removed.

The elements exhibit very complex spectra, and the mixed oxides, when heated, give off an intense white light which resembles sunlight, a property finding application in cored carbon arcs, such as those employed in the movie industry.

The rare-earth metals have a great affinity for the nonmetallic elements, as, for example, hydrogen, carbon, nitrogen, oxygen, sulfur, phosphorus, and the halides. Considerable amounts of the mixed rare earths are reduced to metals, such as misch metal, and these alloys are used in the metallurgical industry. Alloys made of cerium and the mixed rare earths are used in the manufacture of lighter flints. Rare earths are also used in the petroleum industry as catalysts. Yttrium aluminum garnets (YAG) are used in the jewelry trade as artificial diamonds.

Although the rare earths are widely distributed in nature, they generally occur in low concentrations. They are found in high concentrations as mixtures in a number of minerals. The relative abundance of the different rare earths in various rocks, geological formations, and the stars is of great interest to the geophysicist, astrophysicist, and cosmologist.

The rare-earth elements are metals possessing distinct individual properties. Many of the properties of the rare-earth metals and alloys are quite sensitive to temperature and pressure. They are also different when measured along different crystal axes of the metal; for example, electrical conductivity, elastic constants, and so on. The rare earths form organic salts with certain organic chelate compounds. These chelates, which have replaced some of the water around the ions, enhance the differences in properties among the individual rare earths. Advantage is taken of this technique in the modern ion- exchange methods of separation. *See* CHELATION; ION EXCHANGE; TRANSITION ELEMENTS.

Frank H. Spedding

Rare-earth minerals Naturally occurring solids, formed by geological processes, that contain the rare-earth elements—the lanthanides (atomic numbers 57–71) and yttrium (atomic number 39)—as essential constituents. In a rare-earth mineral, at least one crystallographic site contains a total atomic ratio of lanthanides and yttrium that is greater than that of any other element. The mineral name generally has a suffix, called a Levinson modifier, indicating the dominant rare-earth element; for example, monazite-(La) [$LaPO_4$] contains predominantly lanthanum, and monazite-(Ce) [$CePO_4$] contains predominantly cerium. *See* MINERAL; MONAZITE; PERIODIC TABLE; RARE-EARTH ELEMENTS.

So far, about 170 distinct species of rare-earth minerals have been described. A large number of carbonates, phosphates, silicates, niobates, and fluorides are known as rare-earth minerals. It is necessary to obtain structural as well as chemical information about a mineral to judge the essentiality of its rare-earth elements (that is, whether the rare-earth element is part of the mineral's ideal formula or is an impurity). Sometimes, minerals with significant rare-earth content are treated as rare-earth minerals, even if the rare-earth element content appears unessential to the mineral. More than 60 mineral species, including the apatite group minerals, garnet group minerals, and fluorite, are in this category. *See* APATITE; FLUORITE; GARNET.

Rare-earth minerals can be observed as accessory minerals in igneous rocks, such as monazite-(Ce) in granite. Carbonatite is the typical host rock of rare-earth minerals such as bastnäsite-(Ce) [$CeCO_3F$] and monazite-(Ce). Rare-earth minerals also often occur in pegmatite. In both carbonatite and pegmatite, rare-earth elements are concentrated by primary crystallization from melt and by hydrothermal reactions. Carbonatite deposits containing rare-earth elements are found throughout the world. Chemically stable, rare-earth minerals are not weathered easily. As a result, they have been deposited as heavy minerals in beach sand. Such deposits are found in Southeast Asia and Western Australia. *See* CARBONATITE; PEGMATITE.

Among the rare-earth minerals, bastnäsite-(Ce) is the most important source of rare-earth elements. Monazite-(Ce), synchysite-(Ce) [$CaCe(CO_3)_2F$], xenotime-(Y), britholite-(Ce) [$(Ce_3Ca_2)(SiO_4)_3(OH)$], and allanite-(Ce) [$CaCeAl_2Fe(Si_2O_7)(SiO_4)O(OH)$] are also sources. Rare-earth elements have been leached with acid from the surface of clay minerals. Rare-earth minerals containing radioactive nuclear species, such as thorium and uranium, are not used as source materials. *See* CLAY MINERALS.

Ritsuro Miyawaki

Rarefied gas flow Flow of gases below standard atmospheric pressure, sometimes called low-pressure gas flow. The flow may be confined to pipes between a chamber or vessel to be evacuated and a pump, or it may be the beam of molecules issuing from an orifice into a large evacuated chamber or the plume of exhaust gases from a rocket launched into the upper atmosphere, for example. The flow velocity is measured with respect to a fixed boundary such as the wall of a pipe, the surface of a rocket or jet plane, or a model in a wind tunnel. *See* FLUID FLOW; GAS; MOLECULAR BEAMS; PIPE FLOW; VACUUM PUMP; WIND TUNNEL.

For flow through ducts, the gases concerned are initially those of the original atmosphere inside a chamber that must be evacuated. However, even after the bulk of the original gas has been removed, the pumps must continue to remove gas evolved from surfaces and leaking in through imperfections in the walls. In some cases, gas is introduced through valves at a controlled rate as part of the process being carried out at a low pressure.

Since the flow through pipes involves an interaction or drag at the walls, a pressure drop is generated across the entrance and exit of the pipe. Also, gaseous impurities from the pump may flow toward the chamber when the pressure is very low. Proper design of the duct system therefore involves selecting pipes and valves of adequate internal diameter to ensure a minimal pressure drop and the insertion of baffles or traps to prevent impurities from the pumps from entering the process chamber.

The resistance due to the walls depends on the mass flow velocity, and may depend on the gas viscosity and the pressure or density of the gas. The mean free path of molecules is the distance between collisions with other molecules in the gas. *See* KINETIC THEORY OF MATTER; MEAN FREE PATH; VISCOSITY.

The analysis of low-pressure flow is divided into three or four flow regimes depending on the value of the Knudsen number Kn defined as the ratio of the mean free path to a characteristic length, and the dimensionless Reynolds number. The characteristic length may be chosen as the mean pipe diameter in the case of confined flow or as some length associated with a test model suspended in a wind tunnel, for example. *See* GAS DYNAMICS; KNUDSEN NUMBER; REYNOLDS NUMBER.

Another dimensionless number used in gas flow dynamics is the Mach number (Ma), defined as the ratio of the mass flow velocity to the local velocity of sound in the gas. *See* MACH NUMBER.

When the mean free path is much smaller than the pipe diameter ($Kn < 0.01$), the gas flows as a continuous viscous fluid with velocity near the axis of the pipe at locations well beyond the pipe entrance much higher than the velocity in gas layers near the wall. The velocity profile as a function of radial distance from the axis depends on the distance from the entrance and the viscosity of the gas. When the Reynolds number is less than 2000, the profile is a simple curved surface so that the flow is laminar (laminar flow regime). When the mean free path becomes greater than about 0.01 times the diameter, the profile is distorted by boundary-layer effects, and the velocity near the wall does not approach zero (sometimes referred to as slip flow). *See* LAMINAR FLOW.

For Reynolds numbers above the critical value (approximately 2100), the flow is subject to instabilities depending on the geometry of the boundary and at high Reynolds numbers becomes turbulent (turbulent flow regime). *See* TURBULENT FLOW.

When the mean free path is about equal to or greater than the pipe diameter ($Kn \geq 1$), the gas molecules seldom collide with each other, but can either pass through the pipe without striking the wall or scatter randomly back and forth between various points on the wall and eventually escape through the exit or pass back through the entrance. This type of gas flow is known as free-molecule flow (molecular flow regime). The transition region ($0.01 < Kn < 1$) between the laminar flow regime and the molecular flow regime is referred to as the Knudsen or transition flow regime.

The flow may also be classified by the boundary conditions or by the Mach number. For example, Couette flow involves the flow of rarefied gas between two surfaces that are moving with respect to each other with different parallel tangential velocities. For hypersonic flow, $Ma \geq 5$. Benjamin B. Dayton

Raspberry The horticultural name for certain species of the genus *Rubus*, plant order Rosales. In these species the fruit, when ripe (unlike the blackberry), separates thimblelike from the receptacle. Raspberry plants are upright shrubs with perennial roots and prickly, biennial canes (stems). There are several species, both American and European, from which the cultivated raspberries have been developed. Varieties are grouped as to color of fruit—black, red, and purple, the last being hybrids between the red and black types. Leading states in commercial production are Michigan, Oregon, New York, Washington, Ohio, Pennsylvania, New Jersey, and Minnesota. The fruit is sold fresh for dessert purposes, is canned, and is made into jelly or jam, but quick freezing is the most important processing method. *See* FRUIT; ROSALES. J. Harold Clarke

Ratchet A wheel, usually toothed, operating with a catch or a pawl so as to rotate in a single direction (see illustration). A ratchet and pawl mechanism locks a machine such as a hoisting winch so that it does not slip. The locking action may serve to produce rotation in a desired direction and to disengage in the undesired direction as in a drill brace. The catch or pawl may be of various shapes such as an eccentrically mounted disk or ball bearing. Gravity, a spring, or centrifugal force (with the catch mounted internal to the ratchet) are commonly used to hold the pawl against the ratchet. A ratchet and pawl provides an arresting action. *See* BRAKE; ESCAPEMENT; PAWL. Frank H. Rockett

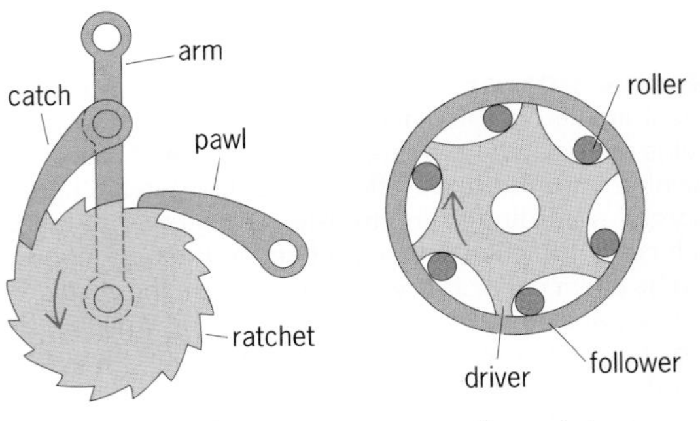

Ratchets. Toothed ratchet is driven by catch when arm moves to left; pawl holds ratchet during return stroke of catch. In roller ratchet, rollers become wedged between driver and follower when driver turns faster than follower in direction of arrow.

Rate-of-climb indicator An aircraft instrument that provides an indication of the vertical change of the aircraft position within the air mass. It is more commonly known as the vertical-velocity or vertical-speed indicator. Contained within a sealed case, it is connected to the aircraft static pressure source, the so-called at-rest air pressure outside the aircraft, through a calibrated leak. Although the instrument operates from a static pressure source, it is a differential pressure indicator. The differential pressure is established between the static pressure in the diaphragm or pressure capsule and the trapped static pressure within the case. When the aircraft changes vertical position, the static pressure in the diaphragm changes immediately but, because of the metering action of the calibrated leak, the case pressure will remain at its prior value and cause the needle to show a change in vertical speed. The needle is usually calibrated in feet per minute but may be calibrated in any appropriate unit of length over time.

In many modern aircraft with flight computers, the rate of climb or descent is electronically calculated by differentiating the altitude from the pitot-static source. *See* AIRCRAFT INSTRUMENTATION; PITOT TUBE. Grady W. Wilson

Ratites A group of flightless, mostly large, running birds characterized by their flat, keelless sternum, which were formerly segregated as a superorder of birds, the Palaeognathae, but whose interrelationships have been a long-standing controversy. Species include the ostriches, emus, cassowaries, and rheas. The ratites apparently represent two or three phyletic lineages [the African ostriches and elephant birds; the Australasian emus, cassowaries, moas and kiwis; and the Neotropical rheas and tinamous] which evolved from a common ancestral stock—the Holarctic fossil Lithornithidae of the Paleocene-Eocene periods, which were possibly much like the still volant (capable of flight) tinamous of Central and South America. Each phyletic line probably reached its present center of distribution before becoming gigantic in size and flightless. *See* AVES; NEOGNATHAE; PHYLOGENY; STRUTHIONIFORMES. Walter J. Bock

Rauwolfia A genus of mostly poisonous, tropical trees and shrubs of the dogbane family (Apocynaceae). Certain species are the source of valuable emetics and cathartics. The species *Rauwolfia serpentina* has received special attention as the source of tranquilizing drugs. Among the purified alkaloids obtained from *R. serpentina*, reserpine is perhaps the one most used as a tranquilizing agent. *See* GENTIANALES; TRANQUILIZER. Perry D. Strausbaugh; Earl L. Core

Raw water Water obtained from natural sources such as streams, reservoirs, and wells. Natural water always contains impurities in the form of suspended or dissolved mineral or organic matter and as dissolved gases acquired from contact with earth and atmosphere. Industrial or municipal wastes may also contaminate raw water. *See* WATER POLLUTION.

If admitted to a steam-generating unit, such contaminations may corrode metals or form insulating deposits of sediments or scale on heat-transfer surfaces, with resultant overheating and possible failure of pressure parts.

Raw water can be treated to remove objectionable impurities or to convert them to forms that can be tolerated. For steam generation, suspended solids are removed by settling or filtration. Scale-forming hardness is diminished by chemical treatment to produce insoluble precipitates that are removable by filtration, or soluble compounds that do not form scale. Essentially complete purification is achieved by demineralizing treatment or evaporation. *See* WATER SOFTENING; WATER TREATMENT. F. G. Ely

Reactance The imaginary part of the impedance of an alternating-current circuit.

The impedance Z of an alternating current circuit is a complex number given by Eq. (1). The imaginary part X is the reactance.

$$Z = R + jX \quad (1)$$

The units of reactance, like those of impedance, are ohms. Reactance may be positive or negative. For example, the impedance of an inductor L at frequency ω is given by Eq. (2), so X is

$$Z = jL\omega = jX \quad (2)$$

positive, and X is referred to as the inductive reactance. The impedance of a capacitor C is given by Eq. (3), so X is negative, and X is referred to as the capacitive reactance.

$$Z = -\frac{j}{C\omega} = jX \quad (3)$$

The reactance of a circuit may depend on both the resistors and the inductors or capacitors in the circuit. For example, the circuit in the illustration has admittance [Eq. (4)]

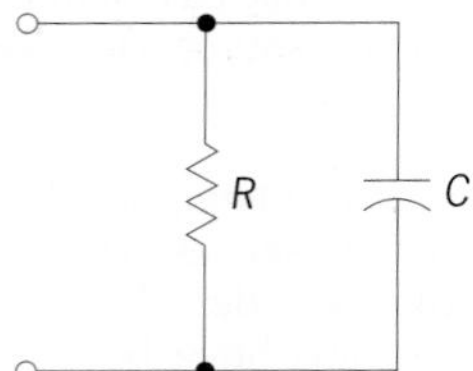

Circuit with a resistor and capacitor in parallel.

and impedance [Eq. (5)], so that the value of the reactance [Eq. (6)], depends on both the capacitor C and the resistor R.

$$Y = \frac{1}{R} + jC\omega \quad (4)$$

$$Z = \frac{R}{1 + jRC\omega} \quad (5)$$

$$X = -\frac{R^2C\omega}{1 + R^2C^2\omega^2} \quad (6)$$

See ADMITTANCE; ALTERNATING CURRENT; ELECTRICAL IMPEDANCE. J. O. Scanlan

Reaction turbine A power-generation prime mover utilizing the steady-flow principle of fluid acceleration, where nozzles are mounted on the moving element. The rotor is turned by the reaction of the issuing fluid jet and is utilized in varying degrees in steam, gas, and hydraulic turbines. All turbines contain nozzles; the distinction between the impulse and reaction principles rests in the fact that impulse turbines use only stationary nozzles, while reaction turbines must incorporate moving nozzles. *See* IMPULSE TURBINE; PRIME MOVER. Theodore Baumeister

Reactive intermediates Unstable compounds which are formed as necessary intermediate stages during a chemical reaction. Thus, if a reaction in which A is converted to B requires that A first be converted to C, then C is an intermediate in the reaction (A $\rightarrow$ C $\rightarrow$ B). The term reactive further implies a certain degree of instability of the intermediate; reactive intermediates are typically isolable only under special conditions, and most of the information regarding the structure and properties of reactive intermediates comes from indirect experimental evidence.

In organic reactions the most common types of reactive intermediates are those arising from dissociative reactions, in which carbon has a decreased valence. Associative reactions can also give rise to some of the same intermediates, and to others in which carbon has an increased valence. *See* VALENCE.

Carbocations are compounds in which carbon bears a positive charge. Classical carbocations (also called carbenium ions) are trivalent, and have only six valence electrons. Nonclassical carbocations (also called carbonium ions) are tetra- or pentavalent, and have eight valence electrons. Examples are the methyl cation (classical) CH_3^+, and the methonium ion (nonclassical) CH_5^+.

Carbanions are compounds in which carbon bears a negative charge. A carbanion will always have a positive counterion in association with it; depending upon the particular cation and the stability of the carbanion, the association may be ionic, covalent, or some intermediate combination of ionic and covalent bonding, as shown below (M = metal). Carbanions are trivalent, with eight valence electrons.

—C—M ⟷ —C:⁻ ⁺M

Covalent   Ionic

Free radicals are neutral compounds having an odd number of electrons and therefore one unpaired electron. Carbon free radicals are trivalent, with seven valence electrons, and typically assume a planar structure. Free radicals are primarily electron-deficient species and are stabilized by structural features which donate electron density or delocalize the odd electron by resonance. *See* FREE RADICAL; RESONANCE (MOLECULAR STRUCTURE).

Radical ions are charged compounds with an unpaired electron, and are either radical cations (positively charged) or radical anions (negatively charged). In many cases a radical ion is derived from a stable neutral molecule by addition of one electron (radical anion) or removal of one electron (radical cation).

Carbenes are compounds which have a divalent carbon. The divalent carbon also has two nonbonded electrons, for a total of six valence electrons. The two nonbonded electrons may have either the same spin quantum number, which is a triplet state, or an opposite spin quantum number, which is a singlet state. Generation of carbenes is most commonly by photolysis or thermolysis of diazo compounds or ketenes, or by alpha-elimination reactions.

There are many other kinds of reactive intermediates which do not fit into the previous classifications. Some are simply compounds which are unstable for a variety of possible reasons, such as structural strain or an unusual oxidation state. *See* CHEMICAL DYNAMICS; ELECTROPHILIC AND NUCLEOPHILIC REAGENTS; MOLECULAR ORBITAL THEORY. Carl C. Wamser

Reactive power The concept of reactive power arises in electrical circuits operated from alternating-current sources. Electric fields (associated with capacitive effects) and magnetic fields (associated with inductive effects) are alternately charged and discharged according to the transfer of energy that takes place in each cycle. Reactive power is related to the peak of power absorbed by either capacitance or inductance in the circuit. By convention, a capacitor produces reactive power, while an inductor absorbs reactive power. All power system equipment generates or absorbs reactive power, including overhead lines and underground cables, transformers, and consumer loads. Real power is used to describe the average value of the power waveform. Real power controls the frequency of the system, while reactive power controls the voltage. Both are required for successful operation of the power system. *See* CAPACITOR; INDUCTOR.

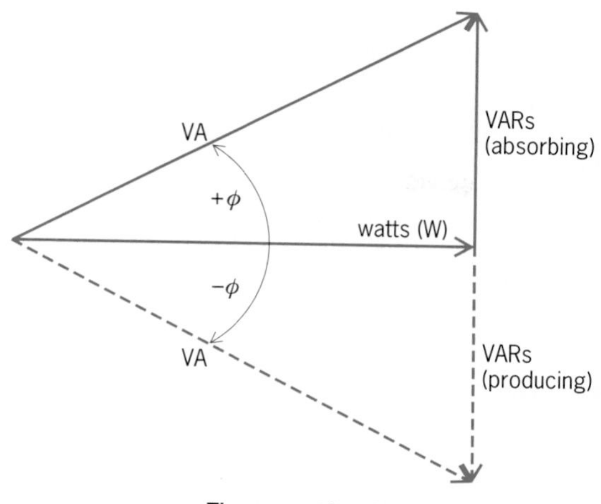

The power triangle.

Power triangle. The real power consumed by a single-phase device will be denoted as P, the reactive power as Q, and the complex power as S. The reactive power is measured by a unit called a VAR, for volt-ampere reactive, while the real power is measured in watts (W). As shown in the illustration, the vector (complex) sum of these two components defines the complex power, S, whose magnitude $|S|$, termed the apparent power, is measured in units of volt-amperes (VA). The graphical relation between S, P, and Q is known as the power triangle. *See* POWER; VOLT-AMPERE.

D. McGillis; K. El-Arroudi; Francisco D. Galiana

Reactor (electricity) A device for introducing an inductive reactance into a circuit. Inductive reactance x is a function of the product of frequency f and inductance L; thus, $x = 2\pi fL$. For this reason, a reactor is also called an inductor. Since a voltage drop across a reactor increases with frequency of applied currents, a reactor is sometimes called a choke. All three terms describe a coil of insulated wire. *See* CHOKE (ELECTRICITY); INDUCTOR.

According to their construction, reactors can be divided into those that employ iron cores and those where no magnetic material is used within the windings. The first type consists of a coil encircling a circuit of iron which usually contains an air gap or a series of air gaps. The air gaps are used to attenuate the effects of saturation of the iron core. The second type, called an air-core reactor, is a simple circular coil, wound around a cylinder constructed of nonmagnetic material for greater mechanical strength. This strength is necessary for the coil to withstand the electromagnetic forces acting on each conductor. These forces become very large with heavy current flow, and their direction tends to compress the coil into less space: radial forces tend to elongate internal conductors in the coil and to compress the external ones while the axial forces press the end sections toward the center of the coil.

Both iron-core and air-core reactors may be of the air-cooled dry type or immersed in oil or a similar cooling fluid. Both types of reactors are normally wound with stranded wire in order to reduce losses due to eddy currents and skin effect. In addition, it is important to avoid formation of short-circuited metal loops when building supporting structures for air-core reactors since these reactors usually produce large magnetic fields external to the coil. If these fields penetrate through closed-loop metal structures, induced currents will flow, causing both losses and heating of the structures. Which of these two reactor types should be used depends on the particular application. *See* EDDY CURRENT; SKIN EFFECT (ELECTRICITY).

Victor R. Stefanovic

Reactor physics The science of the interaction of the elementary particles and radiations characteristic of nuclear reactors with matter in bulk. These particles and radiations include neutrons, beta rays, and gamma rays of energies between zero and about 10^7 electronvolts. *See* BETA PARTICLES; GAMMA RAYS.

The study of the interaction beta and gamma radiations with matter is, within the field of reactor physics, undertaken primarily to understand the absorption and penetration of energy through reactor structures and shields. *See* RADIATION SHIELDING.

With this exception, reactor physics is the study of those processes pertinent to the chain reaction involving neutron-induced nuclear fission with consequent neutron generation. Reactor physics is differentiated from nuclear physics, which is concerned primarily with nuclear structure. Reactor physics makes direct use of the phenomenology of nuclear reactions. Neutron physics is concerned primarily with interactions between neutrons and individual nuclei or with the use of neutron beams as analytical devices, whereas reactor physics considers neutrons primarily as fission-producing agents. In the hierarchy of professional classification, neutron physics and reactor physics are both ranked as subfields of the more generalized area of nuclear physics. *See* CHAIN REACTION (PHYSICS); NEUTRON; NUCLEAR FISSION; NUCLEAR PHYSICS.

Concepts. Reactor physics borrows most of its basic concepts from other fields. From nuclear physics comes the concept of the nuclear cross section for neutron interaction, defined as the effective target area of a nucleus for interaction with a neutron beam. The total interaction is the sum of interactions by a number of potential processes, and the probability of each of them multiplied by the total cross section is designated as a partial cross section. An outgrowth of this is the definition of macroscopic cross section, which is the product of cross section (termed microscopic, for specificity) with atomic density of the nuclear species involved.

Cross sections vary with energy according to the laws of nuclear structure. In reactor physics this variation is accepted as input data to be assimilated into a description of neutron behavior. Common aspects of cross section dependence, such as variation of absorption cross section inversely as the square root of neutron energy, or the approximate regularity of resonance structure, form the basis of most simplified descriptions of reactor processes in terms of mathematical or logical models.

The concept of neutron flux is related to that of macroscopic cross section. This may be defined as the product of neutron density and neutron speed, or as the rate at which neutrons will traverse the outer surface of a sphere embedded in the medium, per unit of spherical cross-sectional area. The product of flux and macroscopic cross section yields the reaction rate per unit volume and time.

Criticality. The critical condition is what occurs when the arrangement of materials in a reactor allows, on the average, exactly one neutron of those liberated in one nuclear fission to cause one additional nuclear fission. If a reactor is critical, it will have fissions occurring in it at a steady rate. This desirable condition is achieved by balancing the probability of occurrence of three competing events: fission, neutron capture which does not cause fission, and leakage of neutrons from the system. If υ is the average number of neutrons liberated per fission, then criticality is the condition under which the probability of a neutron causing fission is $1/\upsilon$. Generally, the degree of approach to criticality is evaluated by computing k_{eff}, the ratio of fissions in successive links of the chain, as a product of probabilities of successive processes.

Reactivity. Reactivity is a measure of the deviation of a reactor from the critical state at any frozen instant of time. The term reactivity is qualitative, because several sets of units are in current use to describe it.

Reflectors. Reflectors are bodies of material placed beyond the chain-reacting zone of a reactor, whose function is to return to the active zone (or core) neutrons which might otherwise leak. Reflector worth can be crudely measured in terms of the albedo, or probability that a neutron passing from core to reflector will return again to the core.

Good reflectors are materials with high scattering cross sections and low absorption cross sections. The first requirement ensures that neutrons will not easily diffuse through the reflector, and the second, that they will not easily be captured in diffusing back to the core.

Beryllium is the outstanding reflector material in terms of neutronic performance. Water, graphite, D_2O, iron, lead, and ^{238}U are also good reflectors.

Reactor dynamics. Reactor dynamics is concerned with the temporal sequence of events when neutron flux, power, or reactivity varies. The inclusive term takes into account sequential events, not necessarily concerned with nuclear processes, which may affect these parameters. There are basically three ways in which a reactor may be affected so as to change reactivity. A control element, absorbing rod, or piece of fuel may be externally actuated to start up, shut down, or change reactivity or power level; depletion of fuel and poison, buildup of neutron-absorbing fission fragments, and production of new fissionable material from the fertile isotopes ^{232}Th, ^{234}U, ^{238}U, and ^{240}Pu make reactivity depend upon the irradiation history of the system; and changes in power level may produce temperature changes in the system, leading to thermal expansion, changes in neutron cross sections, and mechanical changes with consequent change of reactivity.

Reactor control physics. Reactor control physics is the study of the effect of control devices on reactivity and power level. As such, it includes a number of problems in reactor statics, because the primary question is to determine the absorption of the control elements in competition with the other neutronic processes. It is, however, a problem in dynamics, given the above information, to determine what motions of the control devices will lead to stable changes in reactor output.

Reactivity changes. Long-term reactivity changes may represent a limiting factor in the burning of nuclear fuel without costly reprocessing and refabrication. As the chain reaction proceeds, the original fissionable material is depleted, and the system would become subcritical if some form of slow addition of reactivity were not available. This is the function of shim rods in a typical reactor. The reactor is originally loaded with enough fuel to be critical with the rods completely inserted. As the fuel bums out, the rods are withdrawn to compensate. *See* NUCLEAR FUELS.

Reactor kinetics. This is the study of the short-term aspects of reactor dynamics with respect to stability, safety against power excursion, and design of the control system. Control is possible because increases in reactor power often reduce reactivity to zero (the critical value) and also because there is a time lapse between successive fissions in a chain resulting from the finite velocity of the neutrons and the number of scattering and moderating events intervening, and because a fraction of the neutrons is delayed. *See* DELAYED NEUTRON; NUCLEAR REACTOR.

Bernard I. Spinrad

Reagent chemicals

High-purity chemicals used for analytical reactions, for the testing of new reactions where the effects of impurities are unknown, and in general for chemical work where impurities must either be absent or at known concentrations.

Chemicals are purified by a variety of methods, the most common being recrystallization from solution. If the desired chemical is volatile and the impurities are not volatile, sublimation is an effective method of purification. For liquid chemicals, distillation is an effective procedure. Finally, the simplest procedure may be to synthesize the desired reagent from pure materials. *See* CHEMICAL SEPARATION TECHNIQUES; CRYSTALLIZATION; DISTILLATION.

Commercial chemicals are available at several levels of purity. Chemicals labeled "technical" or "commercial" are usually quite impure. The grade "USP" indicates only that the chemical meets the requirements of the United States Pharmacopeia. The term "CP" means only that the chemical is purer than "technical." Chemicals designated "reagent grade" or "analyzed reagent" are specially purified materials which usually have been analyzed to establish the levels of impurities. The American Chemical Society has established specifications and tests for purity for some chemicals. Materials which meet these specifications are labeled "Meets ACS Specifications."

Kenneth G. Stone; Charles Rulfs

Real-time systems

Computer systems in which the computer is required to perform its tasks within the time restraints of some process or simultaneously with the system it is assisting. Usually the computer must operate faster than the system assisted in order to be ready to intervene appropriately.

Real-time computer systems and applications span a number of different types.

In real-time control and real-time process control the computer is required to process systems data (inputs) from sensors for the purpose of monitoring and computing system control parameters (outputs) required for the correct operation of a system or process. The type of monitoring and control functions provided by the computer for subsystem units ranges over a wide variety of tasks, such as turn-on and turn-off signals to switches; feedback signals to controllers (such as motors, servos, and potentiometers) to provide adjustments or corrections; steering signals; alarms; monitoring, evaluation, supervision, and management calculations; error detection, and out-of-tolerance and critical parameter detection operations; and processing of displays and outputs.

In real-time assistance the computer is required to do its work fast enough to keep up with a person interacting with it (usually at a computer terminal device of some sort, for example, a screen and keyboard). The computer supports the person or persons interacting with it and provides access, retrieval, and storage functions, usually through some sort of database management system, as well as data processing and computational power. System access allows the individual to intervene in the system's operation. The real-time computer also often provides monitoring or display information, or both. *See* MULTIACCESS COMPUTER.

In real-time robotics the computer is a part of a robotic or self-contained machine. Often the computer is embedded in the machine, which then becomes a smart machine. If the smart machine also has access to, or has embedded within it, artificial intelligence functions (for example, a knowledge base and knowledge processing in an expert system fashion), it becomes an intelligent machine. *See* COMPUTER; DIGITAL COMPUTER; EMBEDDED SYSTEMS; EXPERT SYSTEMS; ROBOTICS.

Earl C. Joseph

Real variable

A variable whose range is a subset of the real numbers. By extension the term is also used to refer to the theory of functions of one or more real variables. This theory has to do with properties of broad classes of functions, such as continuity, types of discontinuities, differentiability of functions, oscillation and variation of functions, and the various kinds of integrals. *See* CALCULUS; INTEGRATION.

Real numbers are those commonly used in the geometric theory of measurement. The integers and fractions, also called rational numbers, are included among the real numbers. In practice an irrational number x is specified by telling which rational numbers are less than x and which are greater than x. Such a division of the rational numbers into two classes was used by J. W. R. Dedekind as the formal definition of a real number and is called a Dedekind cut.

Lawrence Murray Graves

Rearrangement reaction A reaction in which an atom or bond moves or migrates, having been initially located at one site in a reactant molecule and ultimately located at a different site in a product molecule. A rearrangement reaction may involve several steps, but the key feature defining it as a rearrangement is that a bond shifts from one site of attachment to another. The simplest examples of rearrangement reactions are intra- molecular, that is, reactions in which the product is simply a struc- tural isomer of the reactant [reaction (1)].

(1)

See MOLECULAR ISOMERISM.

More complex rearrangement reactions occur when the rearrangement is accompanied by another reaction, for example, a substitution reaction (2).

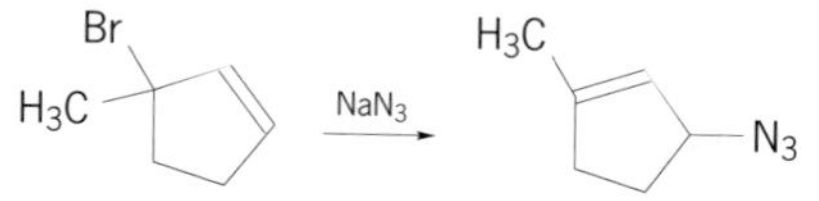

Br, H_3C —NaN_3→ H_3C, N_3 (2)

Rearrangement reactions are classified and named on the basis of the group that migrates and the initial and final location of the migrating bond. The initial bond location is designated as position 1, and the final location as position i, where the number of atoms is simply counted along the connection from 1 to i. Such a migration is called a [1,i] rearrangement or [1,i] shift. If the migrating group also reattaches itself at a different site from the one to which it had originally been attached, then both shifts are indicated, as in [i,j] shift. Reaction (1) is an example of a [3,3] rearrangement, because the initial carbon-oxygen (C-O) bond that breaks designates the 1 position for each component, and both components then rearrange and form a new C-C bond by reattaching at position 3 for each component. Reaction (2) is an example of a [1,3] rearrangement with substitution, commonly called an S_N2' reaction.

In addition, classification can be based on how many electrons move with the migrating group. Of the two electrons in the initial bond that breaks, the migrating group may bring with it both electrons (nucleophilic or anionotropic), one electron (radical), or no electrons (electrophilic or cationotropic). If the rearrangement is a concerted reaction in which there is a cyclic delocalized transition state that results in shifts of pi bonds as well as sigma bonds, the reaction is called a sigmatropic rearrangement [for example, reaction (1)]. *See* DELOCALIZATION; PERICYCLIC REACTION.

The great majority of rearrangement reactions occur when a molecule develops a severely electron-deficient site. Shift of a nearby atom or group, with its pair of electrons, can serve to satisfy the electron deficiency at the original site, although it typically leaves behind another site with electron deficiency. As long as the final site can bear the electron deficiency better than the original site, the rearrangement will be favorable. *See* CHEMICAL BONDING.

Carl W. Wamser

Reciprocating aircraft engine A fuel-burning internal combustion piston engine specially designed and built for minimum fuel consumption and light weight in proportion to developed shaft power. The rotating output shaft of the engine may be connected to a propeller, ducted fan, or helicopter rotor that, in turn, accelerates the surrounding air, imparting an equal and opposite thrust force for propulsion of the aircraft.

Reciprocating aircraft engines are used in about 75% of all powered aircraft flying in the United States. Most of the aircraft powered by these engines belong to the general aviation segment of the domestic aviation fleet, which includes all aircraft except those flown by scheduled airlines and the military. There are over 200,000 active aircraft in the United States general aviation fleet. The reciprocating aircraft engine is used to power single-engine and multiengine airplanes, helicopters, and airships. It is the principal engine used in aircraft for air taxi, pilot training, business, personal, and sport flying as well as aerial application of seed, fertilizer, herbicides, and pesticides for farming. *See* AGRICULTURAL AIRCRAFT; GENERAL AVIATION.

Predominantly, reciprocating aircraft engines operate on a four-stroke cycle, where each piston travels from one end of its stroke to the other four times in two crankshaft revolutions to complete one cycle. The cycle is composed of four distinguishable events called intake, compression, expansion (or power), and exhaust, with ignition taking place late in the compression stroke and combustion of the fuel-air charge occurring early in the expansion stroke. These spark-ignition engines burn specially formulated aviation gasolines and produce shaft power by the force of combustion gas pressure on pistons acting on connecting rods turning a crankshaft. *See* INTERNAL COMBUSTION ENGINE.

Most modern aircraft using engines with up to 336 kW (450 hp) output are powered by air-cooled, horizontally opposed, reciprocating engines. The trend in modern reciprocating engine development is toward lower engine weight and improved fuel economy rather than increased power.

Kenneth J. Stuckas

Reciprocity principle In the scientific sense, a theory that expresses various reciprocal relations for the behavior of some physical systems. Reciprocity applies to a physical system whose input and output can be interchanged without altering the response of the system to a given excitation. Optical, acoustical, electrical, and mechanical devices that operate equally well in either direction are reciprocal systems, whereas unidirectional devices violate reciprocity. The theory of reciprocity facilitates the evaluation of the performance of a physical system. If a system must operate equally well in two directions, there is no need to consider any nonreciprocal components when designing it.

Some systems that obey the reciprocity principle are any electrical network composed of resistances, inductances, capacitances, and ideal transformers; systems of antennas, which obey certain restrictions; mechanical gear systems; and light sources, lenses, and reflectors.

Devices that violate the theory of reciprocity are transistors, vacuum tubes, gyrators, and gyroscopic couplers. Any system that contains the above devices as components must also violate the reciprocity theory. The gyrator differs from the transistor and vacuum tube in that it is linear and passive, as opposed to the active and nonlinear character of the other two devices. *See* GYRATOR; TRANSISTOR.

Hugh S. Landes

Recombination (genetics) The formation of new genetic sequences by piecing together segments of previously existing ones. Recombination often follows deoxyribonucleic acid (DNA) transfer in bacteria and, in higher organisms, is a regular feature of sexual reproduction. *See* ANIMAL REPRODUCTION; DEOXYRIBONUCLEIC ACID (DNA); REPRODUCTION (ANIMAL); PLANT REPRODUCTION.

The fact that recombinants occur in sexual reproduction is due to reciprocal exchanges between chromosomes (crossing over) that take place in the first meiotic division. *See* CROSSING-OVER (GENETICS).

Crossing-over between homologous chromosome pairs can also occur during the prophase of mitotic nuclear division. The frequency is very much lower than in meiosis, presumably because the mitotic cell does not form the synaptic apparatus for efficient pairing of homologs. *See* MITOSIS.

Recombination was once thought to occur only between genes, never within them. Indeed, the supposed indivisibility of the gene was regarded as one of its defining features, the other being that it was a single unit of function. However, examination of very large progenies shows that, in all organisms studied, nearly all functionally allelic mutations of independent origin can recombine with each other to give nonmutant products, generally at frequencies ranging from a few percent (the exceptionally high frequency found in *Saccharomyces*) down to 0.001% or less. Recombination within genes is most frequently nonreciprocal.

Bacteria have no sexual reproduction in the true sense, but many or most of them are capable of transferring fragments of DNA from cell to cell by one of three mechanisms. (1) Fragments of the bacterial genome can become joined to plasmid DNA and transferred by cell conjugation. (2) Genomic fragments can be carried from cell to cell in the infective coats of bacterial viruses (phages), a process called transduction. (3) Many bacteria have the capacity to assimilate fragments of DNA from solution and so may acquire genes from disrupted cells. Fragments of DNA acquired by any of these methods can be integrated into the DNA of the genome in place of homologous sequences previously present. Homologous integration in bacteria is similar in its nonreciprocal nature to recombination within genes of eukaryotic organisms. *See* BACTERIAL GENETICS; BACTERIOPHAGE; TRANSDUCTION (BACTERIA).

Bacteriophages, plasmids, bacteria, and unicellular eukaryotes provide many examples of differentiation through controlled and site-specific recombination of DNA segments. In vertebrates, a controlled series of deletions leads to the generation of the great diversity of gene sequences encoding the antibodies and T-cell receptors necessary for immune defense against pathogens. All these processes depend upon interaction and recombination between specific DNA sequences, catalyzed by site-specific recombinase enzymes. The molecular mechanisms may have some similarities with those responsible for general meiotic recombination, except that the latter does not depend on any specific sequence, only on similarity (homology) of the sequence recombined.

Techniques have been devised for the artificial transfer of DNA fragments from any source into cells of many different species, thus conferring new properties upon them (transformation). In bacteria and the yeast *S. cerevisiae*, integration of such DNA into the genome requires substantial sequence similarity between incoming DNA and the recipient site. However, cells of other fungi, higher plants, and animals are able to integrate foreign DNA into their chromosomes with little or no sequence similarity. These organisms appear to have some system that recombines the free ends of DNA fragments into chromosomes regardless of their sequences. It may have something in common with the mechanism, equally obscure, whereby broken ends of chromosomes can heal by nonspecific mutual joining. *See* TRANSFORMATION (BACTERIA).

The science of genetics has been revolutionized by the development of techniques using isolated cells for specific cleaving and rejoining of DNA segments and the introduction of the reconstructed molecules into living cells. This artificial recombination depends on the use of site-specific endonucleases (restriction enzymes) and DNA ligase. *See* GENE; GENETIC ENGINEERING; GENETICS; RESTRICTION ENZYME.

J. R. S. Fincham

Rectifier A nonlinear circuit component that allows more current to flow in one direction than in the other. An ideal rectifier is one that allows current to flow in one (forward) direction unimpeded but allows no current to flow in the other (reverse) direction. Thus, ideal rectification might be thought of as a switching action, with the switch closed for current in one direction and open for current in the other direction. Rectifiers are used primarily for the conversion of alternating current (ac) to direct current (dc). *See* ELECTRONIC POWER SUPPLY.

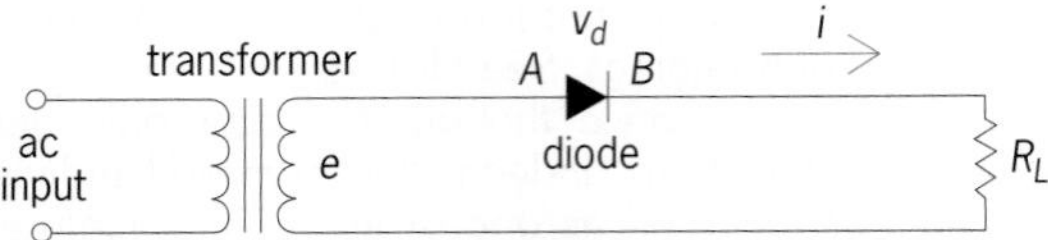

Fig. 1. Half-wave diode rectifier. V_d = voltage across diode. Ideal diode allows current *i* to flow only in forward direction from *A* to *B*.

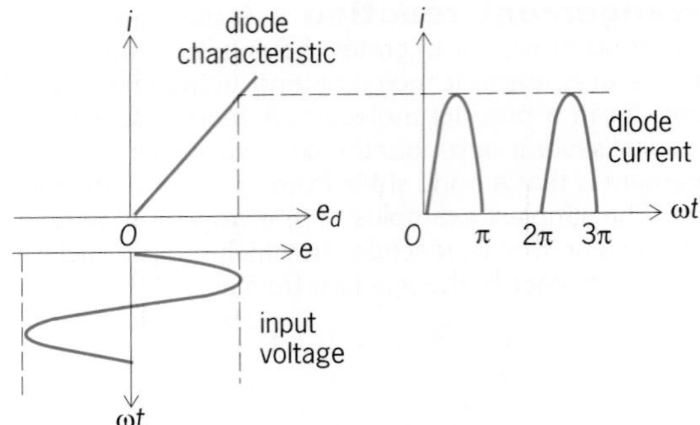

Fig. 2. Rectifying action of half-wave diode rectifier. t = time; ω = angular frequency of input voltage.

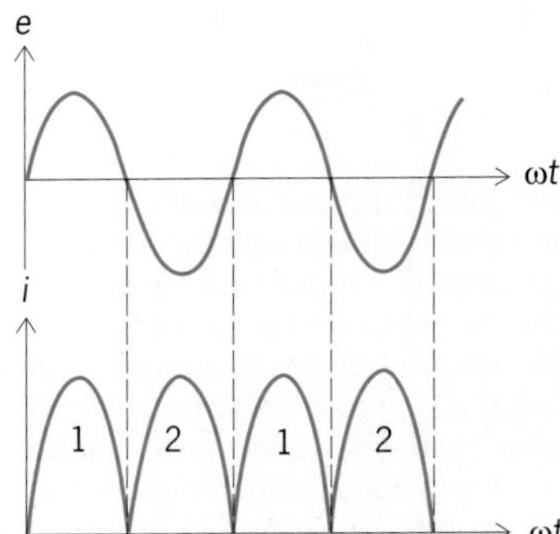

Fig. 3. Applied voltage and output current of full-wave rectifier.

A variety of rectifier elements are in use. The vacuum-tube rectifier can efficiently provide moderate power. Its resistance to current flow in the reverse direction is essentially infinite because the tube does not conduct when its plate is negative with respect to its cathode. In the forward direction, its resistance is small and almost constant. Gas tubes, used primarily for higher power requirements, also have a high resistance in the reverse direction. The semiconductor rectifier has the advantage of not requiring a filament or heater supply. This type of rectifier has approximately constant forward and reverse resistances, with the forward resistance being much smaller. Mechanical rectifiers can also be used. The most common is the vibrator, but other devices are also used. *See* GAS TUBE; MECHANICAL RECTIFIER; SEMICONDUCTOR RECTIFIER.

If the average current is subtracted from the current flowing in the rectifier, an alternating current results. This ripple current flowing through a load produces a ripple voltage which is often undesirable. Filter and regulator circuits are used to reduce it to as low a value as is required. *See* ELECTRIC FILTER; VOLTAGE REGULATOR.

A half-wave rectifier circuit is shown in Fig. 1. The rectifier, a diode, is practically ideal. The ac input is applied to the primary of the transformer; secondary voltage *e* supplies the rectifier and load resistor R_L. The rectifying action of the diode is shown in Fig. 2, in which the current *i* of the rectifier is plotted against the voltage e_d across the diode. The applied sinusoidal voltage from the transformer secondary is shown under the voltage axis; the resulting current *i* flowing through the diode is shown at the right to be half-sine loops.

A full-wave rectifier circuit uses two separate diodes. The resulting current wave shape is shown in Fig. 3. A more continuous flow of direct current is produced because the first diode conducts for the positive half-cycle and the second diode conducts for the negative half-cycle.

When high dc power is required by an electronic circuit, a polyphase rectifier circuit may be used. It is also desirable when expensive filters must be used. This is particularly true of power supplies for the final radio-frequency and audiofrequency stages of large radio and television transmitters. Donald L. Waidelich

Rectilinear motion Motion is defined as continuous change of position of a body. If the body moves so that every particle of the body follows a straight-line path, then the motion of the body is said to be rectilinear. *See* MOTION.

When a body moves from one position to another, the effect may be described in terms of motion of the center of mass of the body from a point *A* to a point *B* (see illustration). If the center of mass of the body moves along a straight line connecting the points *A* and *B*, then the motion of the center of mass of the body is rectilinear. If the body as a whole does not rotate while it is moving, then the path of every particle of which the body is composed is a straight line parallel to or coinciding with the path of the center of mass, and the body as a whole executes rectilinear motion. This is shown by the straight line connecting points P_1 and P_2 in the illustration. *See* CENTER OF MASS.

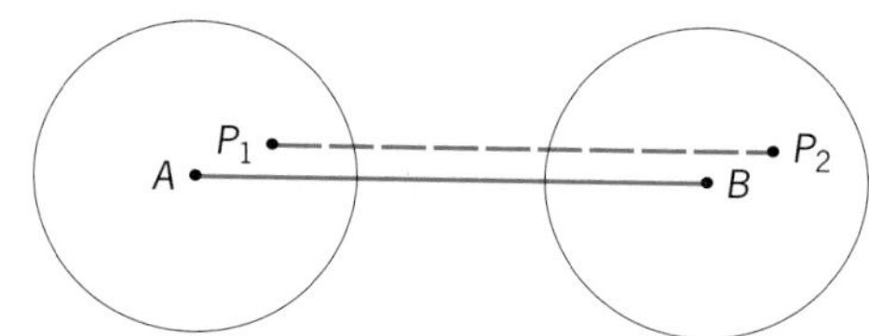

Rectilinear motion. All points move parallel to the center of mass.

Rectilinear motion is an idealized form of motion which rarely, if ever, occurs in actual experience, but it is the simplest imaginable type of motion and thus forms the basis for the analysis of more complicated motions. However, many actual motions are approximately rectilinear and may be treated as such without appreciable error. For example, a ball thrown directly upward may follow, for all practical purposes, a straight-line path. The motion of a high-speed rifle bullet fired horizontally may be essentially rectilinear for a short length of path, even though in its larger aspects the ideal path is a parabola. The motion of an automobile traveling over a straight section of roadway is essentially rectilinear if minor variations of path are neglected. The motion of a single wheel of the car is not rectilinear, although the motion of the center of mass of the wheel may be essentially so. *See* BALLISTICS. Rogers D. Rusk

Recursive function A function that maps natural numbers to natural numbers and is special in that it must be calculable by using a precisely specified algorithm. The mathematical definitions of partial recursive functions and recursive functions were developed to give a precise mathematical characterization of those functions or operations on the natural numbers which are computable by using effective procedures.

An effective procedure is a procedure or process determined by a finite list of precise instructions. There is no upper bound on the number of instructions in this list. These instructions can be carried out in a discrete one-step-at-a-time fashion. The only equipment needed is pencil and paper, and an unlimited supply is assumed to be available. No creativity is involved in applying these instructions, and no random devices such as flipping a coin are allowed in carrying them out.

An effective procedure may be used to calculate a function as follows: The calculator inputs to the effective procedure a natural number (0, 1, 2, 3, . . .) and allows the procedure to be carried out on this input. If an ouput is obtained, the function is said to be defined at this input and its value at this input is the output so obtained. Any function which can be computed in this way is called effectively computable. For example, addition and multiplication of natural numbers are effectively computable functions.

The notion of effective procedure is intuitive and vague. In the setting of an idealized computing device called a Turing machine, a more precise definition of certain effectively computable functions can be given. *See* AUTOMATA THEORY; GÖDEL'S THEOREM; LOGIC. George C. Nelson

Recycling technology Methods for reducing solid waste by reusing discarded materials to make new products. The three integral phases of recycling are the collection of recyclable materials, manufacture or reprocessing of these materials into new products, and purchase of these products. Various techniques have been developed to recycle plastics, glass, metals, paper, and wood.

Plastics. Plastic discards represent an estimated 10% by weight and up to 26% by volume of the municipal solid waste in the United States discarded after materials recovery. About 2% by weight of discarded plastics is recovered. Approximately half of plastic waste consists of single-use convenience packaging and containers. Many manufacturers prefer plastic for packaging because it is lightweight, resists breakage and environmental deterioration, and can be processed to suit specific needs. Once plastics are discarded, these attractive physical properties become detriments.

The collected plastic waste is usually separated manually from the waste stream, and often it is cleaned to remove adhesives or other contaminants. It is sorted further, based on different resins. Mechanical separation techniques can be used to sort plastics based on unique physical or chemical properties.

A significant problem is the presence of contaminants such as dirt, glass, metals, chemicals from previous usage, toxicants from metallic-based pigments, and other materials that are part of or have adhered to the plastic products. Other constraints involve inconsistencies in the amount of different plastic resins in commingled plastic wastes used for recycling, and engineering aspects of recycled plastic products, such as lessened chemical and impact resistance, strength, and stiffness, and the need for additional chemicals to counteract other types of degradation for reprocessing. There may be limitations to the number of times that a particular plastic product can be effectively recycled as compared to steel, glass, or aluminum, which can be recycled many times with no loss of their properties and virtually no contamination. R. L. Swanson; Vincent T. Breslin; Marci L. Boutman

Glass. Glass containers are a usual ingredient in community recycling programs; they are 100% recyclable and can be recycled indefinitely. In 1993 in the United States, glass containers, which constitute 6% of the solid-waste stream by weight, were recycled at a rate of 35%. Nearly one-third of the glass containers available for consumption in the United States were cycled back into glass containers and other useful items such as glasphalt, or were returned as refillable bottles.

The process of recycling glass is straightforward. Cullet (scrap glass) in the form of used glass bottles and jars is mixed with silica sand, soda ash, and limestone in a melting furnace at temperatures up to 2800°F (1540°C). The molten glass is poured into a forming machine, where it is blown or pressed into shape. The new containers are gradually cooled, inspected, and shipped to the customer. Before glass can be recycled, however, it must be furnace ready, that is, sorted by color and free of contaminants.

Cullet must meet a standard of quality similar to that of the raw material it replaces. Contamination from foreign material will result in the cullet being rejected by the plants, as it poses a serious threat to the integrity and purity of the glass packaging being produced. Contaminants include metal caps, lids, stones, dirt, and ceramics. Paper labels do not need to be removed for recycling, as they burn off at high furnace temperatures. Nathan Tyler; Natalie U. Roy

Metals. Metals must be recycled to alleviate the need to mine more ore, to reduce energy consumption, to limit the dissemination of metals into the environment, and to reduce the cost of metals. In the United States a substantial portion of these needs are met by recycling metals.

The extensive recycling is important for three reasons. (1) The energy required to recycle a metal is considerably less in comparison to producing it from ore. (2) Extracting the metal from ore produces a tremendous amount of waste material. (3) Metals that are not recycled become dissipated throughout the environment; since many metals are toxic, this can result in the pollution of water and soil. *See* HAZARDOUS WASTE.

While it is beneficial to recycle, two important problems hinder recycling: collection and impurity buildup. When a metal becomes scrap and is a candidate for recycling, it must be collected at a cost that makes it attractive to recyclers. It is useful to divide scrap into three categories—home scrap, new or prompt scrap, and old or obsolete scrap—whose methods for collection differ significantly. Home scrap is waste produced during fabrication, and includes casting waste (for example, risers), shearings and trimmings, and rejected material. This scrap is usually recycled within the plant, and therefore it is not recognized as recycled material in recycling statistics. New or prompt scrap is waste generated by the user of semifinished material, that is, scrap from machining operations (such as turnings or borings), trimmings, and rejected material. This material is collected and sold to recyclers and, if properly labeled and segregated, it is easy to recycle and is valuable. Old or obsolete scrap is waste derived from products that have completed their life cycle, such as used beverage cans, old automobiles, and defunct batteries. The collection and the impurity buildup problems are most severe when considering old or obsolete scrap. Dale F. Stein

Paper. Paper and paperboard for recycling come from a variety of sources, including offices, retail businesses, coverters, printers, and households. Paper products that have been distributed, have been purchased, and have served their intended purposes are considered postconsumer waste. Other sources, such as scrap paper generated in the papermaking process (mill broke) or converting operations (such as trimmings from envelopes and boxes), are considered preconsumer waste.

Recycled paper fibers are used in the manufacture of many recycled-content paper products such as paperboard, corrugated containers, tissue products, newspapers, and printing and writing paper. They can also be used in other products such as insulation, packing materials, and molded egg cartons and flowerpots. *See* PRINTING.

Collection is the crucial first step in recycling. It occurs in curbside programs, in drop-off centers, in paper drives, and increasingly in commercial collection systems run side by side with waste collection for landfill or incineration.

Reprocessing begins by sorting waste papers by grade and level of cleanliness. Next, the waste paper (usually in bales) is mixed with water in a slusher or pulper to produce a fiber-and-water slurry. In this pulping stage the paper is agitated until broken down into fibers, and large-size contaminants (greater than about 5 mm or 0.2 in.) are removed when the pulper is emptied through the screen plate. Depending on the intended product, chemicals such as surfactants are added to the pulper to help remove undesirable materials from the fibers for separation in later operations.

The pulp is then pumped through several different-size slotted or perforated screens to separate medium-size contaminants (usually 5–0.2 mm or 0.2–0.05 in.) from the pulp. Screening is generally followed by centrifugal cleaning, where the pulp is subjected to a vortex in a tapered cone. Using specific designs, cleaners separate high-specific-gravity materials, such as dirt and sand, and low-specific-gravity materials, such as styrofoam and some plastics, from the pulp. *See* PAPER. Jeffrey S. Shaw

Wood. Waste is generated at every stage of the process by which a forest tree is turned into consumer and industrial products. Additional waste is generated in the disposal of those products. Wood waste is also produced by the homeowner and by large and small businesses and is generated from landscaping and agricultural operations such as pruning and tree removal. While these processes do not strictly return wood to the economy in its original form, they have the effect of diverting wood residues from the landfill, and thus they may be included under a broadly interpreted definition of recycling.

Wood recycling begins with wood separation from the waste stream. Recovered materials can be processed into various products, including fuel, raw material for particleboard or other wood-composite panel products, compost, landscaping mulch, animal bedding, landfill cover, amendments for municipal solid waste and sludge compost, artificial firewood, wood-plastic composite lumber and other composite products, charcoal, industrial oil absorbents, insulation, and specialty concrete.

Most of these products require that the wood be ground into small particles. A typical grinder is a hammermill, although a variety of grinders are used. The size of the particles is determined by the end use of the wood; sizes smaller than about 20 mesh are called wood flour (the particles passing the 20-mesh screen are usually less than 0.8 mm in size). Wood may then be passed over an electromagnet which removes items made of ferrous metals such as nails or staples. If additional processing is performed, it is typically to separate wood particles by size. This is accomplished in two ways: the particles can be passed through a series of screens of different mesh size and the various-sized particles can be collected from the screens; or the particles may be separated in a tower with air blown in the bottom and out the top; the particles distribute themselves in the tower, with the small, light particles on top and the large, dense ones at the bottom. *See* WOOD PRODUCTS. John Simonsen

Red dwarf star

Red dwarf star A low-mass main-sequence star of spectral classes M and L. Red dwarf stars range from about 0.6 solar mass at class M0 down to 0.08 solar mass in cool M and warm L, below which the proton-proton chain cannot run. Lower-mass bodies are termed brown dwarfs. Effective temperatures range from 3800 K (6400°F) at class M0 down to 2000 K (3100°F) at class L0, and absolute visual magnitudes from +9 to +20. Downward along the main sequence, red dwarfs produce progressively more radiation in the infrared. Luminosities range from about 0.05 down to 2×10^{-4} the solar luminosity. Radii range from about 0.5 to down 0.1 solar. Spectra become increasingly complex. *See* BROWN DWARF; MAGNITUDE (ASTRONOMY); PROTON-PROTON CHAIN.

Red dwarfs constitute over 70% of all stars, and as a result constitute about half the visible mass of the Milky Way Galaxy. Yet their intrinsic faintness renders them all invisible to the unaided eye. Their lifetimes are so long that none has ever evolved in the Milky Way Galaxy's lifetime. *See* DWARF STAR; MILKY WAY GALAXY; SPECTRAL TYPE; STAR; STELLAR EVOLUTION. James B. Kaler

Red Sea

Red Sea A body of water that separates northeastern Africa from the Arabian Peninsula. The Red Sea forms part of the African Rift System, which also includes the Gulf of Aden and a complex series of continental rifts in East Africa extending as far south as Malawi. The Red Sea extends for 1920 km (1190 mi) from Ras (Cape) Muhammed at the southern tip of the Sinai Peninsula to the Straits of Bab el Mandab at the entrance to the Gulf of Aden. At Sinai the Red Sea splits into the Gulf of Suez, which extends for an additional 300 km (180 mi) along the northwest trend of the Red Sea and the nearly northward-trending Gulf of Aqaba. The 175-km-long (109-mi) Gulf of Aqaba forms the southern end of the Levant transform, a primarily strike-slip fault system extending north into southern Turkey. The Levant transform also includes the Dead Sea and Sea of Galilee and forms the northwestern boundary of the Arabian plate. *See* ESCARPMENT; FAULT AND FAULT STRUCTURES.

The Red Sea consists of narrow marginal shelves and coastal plains and a broad main trough with depths ranging from about 400 to 1200 m (1300 to 3900 ft). The main trough is bisected by a narrow (<60 km or 37 mi wide) axial trough with a very rough bottom morphology and depths of greater than 2000 m (6600 ft). The maximum recorded depth is 2920 m (9580 ft). *See* REEF.

Water circulation in the Red Sea is driven by monsoonal wind patterns and changes in water density due to evaporation. Evaporation in the Red Sea is sufficient to lower the sea level by over 2 m (6.6 ft) per year. No permanent rivers flow into the sea, and there is very little rainfall. As a result, there must be a net inflow of water from the Gulf of Aden to compensate for evaporative losses. During the winter monsoon, prevailing winds in the Red Sea are from the south, and there is a surface current from the Gulf of Aden into the Red Sea. During the summer monsoon, the wind in the Red Sea blows strongly from the north, causing a surface current out of the Red Sea. *See* MONSOON METEOROLOGY; SEAWATER. James R. Cochran

Redbeds Clastic sediments and sedimentary rocks that are pigmented by red ferric oxide which coats grains, fills pores as cement, or is dispersed as a muddy matrix. These conspicuously colored rocks commonly constitute thick sequences of nonmarine, paralic (marginal marine), and less commonly shallow marine deposits. Clastic redbeds accumulated in many parts of the globe during the past 10^9 years of Earth history. Ferric oxides also pigment marine chert, limestone, and cherty iron formations and ooidal ironstones, but these chemical deposits are not usually included among redbeds.

Some redbeds contain abundant grains of sedimentary and low-grade metamorphic rocks and relatively few grains of iron-bearing minerals. Most of them, however, contain feldspar and relatively abundant grains of opaque black oxides derived from igneous and high-grade metamorphic source rocks. Clay minerals in older redbeds, as in most other ancient clastic deposits, are predominantly illite and chlorite, thus providing no specific clue to the climate in the source area or at the place of deposition.

In many of the younger redbeds the pigmenting ferric oxide mineral cannot be identified specifically because of its poor crystallinity. In most of the older ones, however, hematite is the pigment. As seen under the scanning electron microscope, the hematite is in the form of hexagonal crystals scattered over the surface of grains and clay mineral platelets. In red mudstones most of the pigment is associated with the clay fraction.

Redbeds do not contain significantly more total iron than nonred sedimentary rocks. Normally, iron increases with decreasing grain size of redbeds. Moreover, the amount of iron in the grain-coating pigments is small compared with that in opaque oxides, dark silicates, and clay minerals.

On a global scale, paleomagnetic evidence of the distribution of redbeds relative to their pole position corroborates paleogeographic data, suggesting that most redbeds, evaporites, and eolian sandstones accumulated less than 30° north and south of a paleo equator where hot, dry climate generally prevailed. But diagenetic development of red hematite may be acquired long after deposition. Moreover, continental drift reconstructions reveal that the most widespread redbeds in the geologic record developed near the Equator in late Paleozoic and early Mesozoic time when the continents were assembled in a great landmass, Pangaea. *See* CONTINENTS, EVOLUTION OF; PALEOGEOGRAPHY; PALEOMAGNETISM. Franklyn B. Van Houten

Redshift A systematic displacement toward longer wavelengths of lines in the spectra of distant galaxies, and also of the continuous part of the spectrum. First studied systematically by E. Hubble, redshift is central to observational cosmology, in which it provides the basis for the modern picture of an expanding universe.

There are two fundamental properties of redshifts. First, the fractional redshift $\Delta\lambda/\lambda$ is independent of wavelength. This rule has been verified from 21 cm (radio radiation from neutral hydrogen atoms) to about 6×10^{-5} cm (the visible region of the electromagnetic spectrum) and leads to the interpretation of redshift as resulting from a recession of distant galaxies. Though this interpretation has been questioned, no other mechanism is known that would explain the observed effect.

Second, redshift is correlated with apparent magnitude in such a way that when redshift is translated into recession speed and apparent magnitude into distance, the recession speed is found to be nearly proportional to the distance. This rule was formulated by Hubble in 1929, and the constant of proportionality bears his name. *See* HUBBLE CONSTANT; MAGNITUDE (ASTRONOMY). David Layzer; Anthony Aguirre

Redtop grass One of the bent grasses, *Agrostis alba* and its relatives, which occur in cooler, more humid regions of the United States on a wide variety of soils. Redtop tolerates both wet and dry lands and acid and infertile soils, and it is used where other species of grasses do not thrive. Redtop is a perennial and makes a coarse, loose turf. The inflorescence is a reddish open panicle. Redtop is used for pasture and hay and is fairly nutritious if harvested promptly when heading occurs. It is effective in preventing erosion by holding banks of drainage ditches, waterways, and terrace channels. *See* CYPERALES. Howard B. Sprague

Redwood A member of the pine family, *Sequoia sempervirens*, is the tallest tree in the Americas, attaining a height of 350 ft (107 m) and a diameter of 27 ft (8.2 m). Its present range is limited to a strip along the Pacific Coast, extending from southwest Oregon to south of San Francisco. The leaves are evergreen, sharply pointed, small, disposed in two vertical rows on short branches, and scalelike on the main stem. The cones are egg-shaped. The bark is a dull red-brown, on old trees sometimes 1 ft (0.3 m) thick, densely fibrous, and highly resistant to fire. The tree gets its common name from the color of the bark as well as that of the heartwood.

The wood holds paint well and is used for bridge timbers, tanks, flumes, silos, posts, shingles, paneling, doors, caskets, furniture, siding, and many other building purposes. *See* PINALES; PINE. Arthur H. Graves; Kenneth P. Davis

Reef A rigid and wave-resistant marine structure that stands above its surroundings. Biologists and geologists specify that reefs are constructed by organisms that secrete calcium carbonate skeletons. To navigators, a reef is any rocky structure that poses a threat to navigation.

Modern reef-builders include algae, mollusks, bryozoans, worms, and sponges, but the most important are corals. Coral reefs are most common in warm water (typically ≥20°C) in which calcium carbonate precipitates more easily. Most shallow-water corals are hermatypic, and they host symbiotic zooxanthellae (algal-like dinoflagellates) that provide metabolites through photosynthesis. This provides energy for the coral while removing carbon dioxide that inhibits calcification. While coral reefs are more common in the shallow tropics, they can be found in both deeper and colder water. *See* SCLERACTINIA.

Generally, coral reefs can be divided into three categories: fringing, barrier, and atoll. Coral reefs initially form along the steep slopes of a new volcano in shallow waters close to shore. As the volcano subsides (moves away and sinks), reefs build vertically to stay close to sea level and the light needed for photosynthesis. Fringing reefs, which are closest to shore, occur around newly formed and thus high volcanic islands. As the volcanic island slowly subsides, reefs transition to barriers, and eventually to atolls, once the volcano has sunk from sight beneath the surface waters. The species and shapes of corals on a reef are controlled primarily by light availability, wave energy, and sedimentation. *See* ATOLL; OCEANIC ISLANDS; VOLCANO.

Since the mid-1970s, coral abundance has plummeted on many reefs, especially those closer to population centers. Arguments abound over the specific causes of recent declines, but the scientific community is unanimous in its assessment of likely anthropogenic ties for the global degradation of coral reefs.

Dennis Hubbard

Reengineering The application of technology and management science to the modification of existing systems, organizations, processes, and products in order to make them more effective, efficient, and responsive. Responsiveness is a critical need for organizations in industry and elsewhere. It involves providing products and services of demonstrable value to customers, and thereby to those individuals who have a stake in the success of the organization. Reengineering can be carried out at the level of the organization, at the level of organizational processes, or at the level of the products and services that support an organization's activities. The entity to be reengineered can be systems management, process, product, or some combination. In each case, reengineering involves a basic three-phase systems-engineering life cycle comprising definition, development, and deployment of the entity to be reengineered.

Systems-management reengineering. At the level of systems management, reengineering is directed at potential change in all business or organizational processes, including the systems acquisition process life cycle itself. Systems-management reengineering may be defined as the examination, study, capture, and modification of the internal mechanisms or functionality of existing system-management processes and practices in an organization in order to reconstitute them in a new form and with new features, often to take advantage of newly emerged organizational competitiveness requirements, but without changing the inherent purpose of the organization itself.

Process reengineering. Reengineering can also be considered at the levels of an organizational process. Process reengineering is the examination, study, capture, and modification of the internal mechanisms or functionality of an existing process or systems-engineering life cycle, in order to reconstitute it in a new form and with new functional and nonfunctional features, often to take advantage of newly emerged or desired organizational or technological capabilities, but without changing the inherent purpose of the process that is being reengineered.

Product reengineering. The term "reengineering" could mean some sort of reworking or retrofit of an already engineered product, and could be interpreted as maintenance or refurbishment. Reengineering could also be interpreted as reverse engineering, in which the characteristics of an already engineered product are identified, such that the product can perhaps be modified or reused. Inherent in these notions are two major facets of reengineering: it improves the product or system delivered to the user for enhanced reliability or maintainability, or to meet a newly evolving need of the system users; and it increases understanding of the system or product itself. This interpretation of reengineering is almost totally product-focused.

Thus, product reengineering may be redefined as the examination, study, capture, and modification of the internal mechanisms or functionality of an existing system or product in order to reconstitute it in a new form and with new features, often to take advantage of newly emerged technologies, but without major change to the inherent functionality and purpose of the system. This definition indicates that product reengineering is basically structural reengineering with, at most, minor changes in purpose and functionality of the product. This reengineered product could be integrated with other products having rather different functionality than was the case in the initial deployment. Thus, reengineered products could be used, together with this augmentation, to provide new functionality and serve new purposes. There are a number of synonyms for product reengineering, including renewal, refurbishing, rework, repair, maintenance, modernization, reuse, redevelopment, and retrofit.

Much of product reengineering is very closely associated with reverse engineering to recover either design specifications or user requirements. Then follows refinement of these requirements or specifications and forward engineering to achieve an improved product. Forward engineering is the original process of defining, developing, and deploying a product, or realizing a system concept as a product; whereas reverse engineering, sometimes called inverse engineering, is the process though which a given system or product is examined in order to identify or specify the definition of the product either at the level of technological design specifications or at system- or user-level requirements.

Andrew P. Sage

Reference electrode An electrode with an invariant potential. In electrochemical methods, where it is necessary to observe, measure, or control the potential of another electrode (denoted indicator, test, or working electrode), it is necessary to use a reference electrode, which maintains a potential that remains practically unchanged during the course of an electrochemical measurement. Potentials of indicator or working electrodes are measured or expressed relative to reference electrodes. *See* POTENTIALS.

One such electrode, the normal hydrogen electrode, has been chosen as a reference standard, relative to which potentials of other electrodes and those of oxidation-reduction couples are often expressed. By maintaining a constant pressure of hydrogen gas the potential of a hydrogen electrode can be used for determination of the activity of hydrogen ions in the tested solution. However, in practice the determination of the hydrogen-ion activity (pH) is performed by using a glass electrode. The hydrogen electrode itself is used only in fundamental studies and some nonaqueous solutions. The hydrogen electrode, however, remains important for providing a reference standard. *See* ACTIVITY (THERMODYNAMICS); PH.

In practice, potentials are measured against reference electrodes that are easier to work with than the normal hydrogen electrode. Such electrodes are known as secondary reference electrodes; the most common are the calomel and silver–silver chloride electrodes. *See* ELECTRODE; SOLVENT.

Petr Zuman

Reflecting microscope A microscope whose objective is composed of two mirrors, one convex and the other concave (see illustration). The imaging properties are independent of the wavelength of light, and this freedom from chromatic aberration allows the objective to be used even for infrared and ultraviolet radiation. Although the reflecting microscope is simple in appearance, the construction tolerances are so small and so difficult to achieve that the system is used only when refracting objectives are unsuitable. The distance from the objective to the specimen can be made very large; this large working distance is useful in special applications, such as examining objects situated

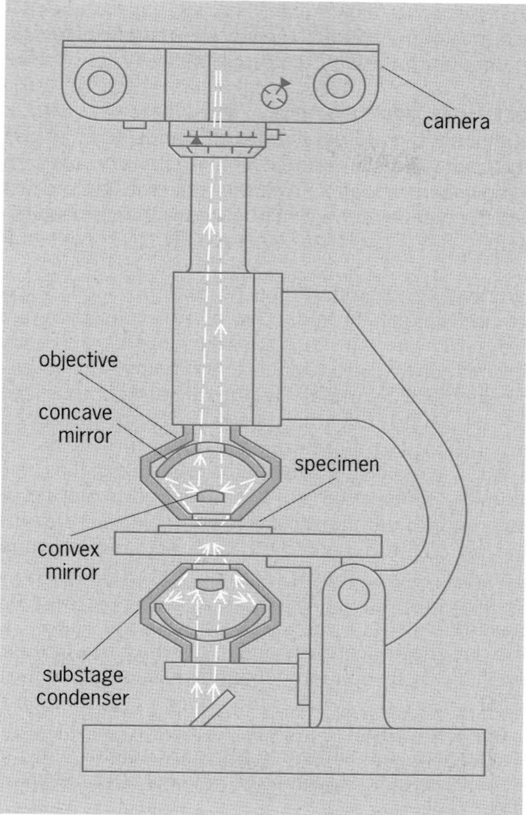

Reflecting microscope arranged for photomicrography.

within metallurgical furnaces. Reflecting microscopes have been mainly used for microspectrometry in the infrared and the ultraviolet, and for ultraviolet microphotography. *See* MICROSCOPE; OPTICAL MICROSCOPE.

David S. Grey

Reflection and transmission coefficients

When an electromagnetic wave passes from a medium of permeability μ_1 and dielectric constant ϵ_1 to one with values μ_2 and ϵ_2, part of the wave is reflected at the boundary and part transmitted. The ratios of the amplitudes in the reflected wave and the transmitted wave to that in the incident wave are called the reflection and transmission coefficients, respectively. For oblique incidence, the reflection and refraction formulas of optics are most convenient, but for normal incidence of plane waves on plane boundaries, such as occur with transmission lines, waveguides, and some free waves, the concept of wave impedance and characteristic impedance is useful. *See* ELECTROMAGNETIC RADIATION; TRANSMISSION LINES; WAVEGUIDE.

William R. Smythe

Reflection of electromagnetic radiation

The returning or throwing back of electromagnetic radiation such as light, ultraviolet rays, radio waves, or microwaves by a surface upon which the radiation is incident. In general, a reflecting surface is the boundary between two materials of different electromagnetic properties, such as the boundary between air and glass, air and water, or air and metal. Devices designed to reflect radiation are called reflectors or mirrors.

The simplest reflection laws are those that govern plane waves of radiation. The law of reflection concerns the incident and reflected rays (as in the case of a beam from a flashlight striking a mirror) or, more precisely, the wave normals of the incident and reflected waves. The law states that the incident and reflected rays and the normal to the reflecting surface all lie in one plane, called the plane of incidence, and that the reflection angle

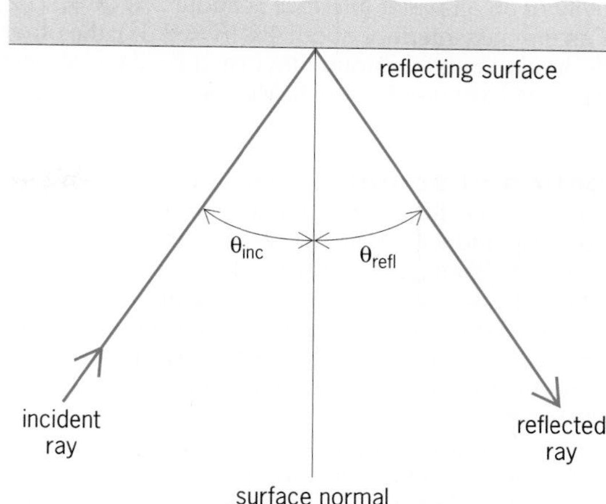

Reflection of electromagnetic radiation from smooth surface.

equals the angle of incidence as in Eq. (1) [see illustration]. The

$$\theta_{\text{refl}} = \theta_{\text{inc}} \tag{1}$$

angles θ_{inc} and θ_{refl} are measured between the surface normal and the incident and reflected rays, respectively. The surface (in the above example, that of the mirror) is assumed to be smooth, with surface irregularities small compared to the wavelength of the radiation. This results in so-called specular reflection. In contrast, when the surface is rough, the reflection is diffuse. An example of this is the diffuse scattering of light from a screen or from a white wall where light is returned through a whole range of different angles.

The reflectivity of a surface is a measure of the amount of reflected radiation. It is defined as the ratio of the intensities of the reflected and incident radiation. The reflectivity depends on the angle of incidence, the polarization of the radiation, and the electromagnetic properties of the materials forming the boundary surface. These properties usually change with the wavelength of the radiation. Reflecting materials are divided into two groups: transparent materials also called dielectrics, and opaque conducting materials, usually metals.

The reflectivity of polished metal surfaces is usually quite high. Silver and aluminum, for example, reflect more than 90% of visible light. In ordinary mirrors the reflecting surface is the interface between metal and glass, which is thus protected from oxidation, dirt, and other forms of deterioration. When it is not permissible to use this protection for technical reasons, one uses "front-surface" mirrors, which are usually coated with evaporated aluminum.

The material property that determines the amount of radiation reflected from an interface between two dielectric media is the phase velocity v of the electromagnetic radiation in the two materials. In optics one uses as a measure for this velocity the refractive index n of the material, which is defined by Eq. (2) as

$$n = c/v \tag{2}$$

the ratio of the velocity of light c in vacuum and the phase velocity in the material. For visible light, for example, the refractive index of air is about $n = 1$, the index of water is about $n = 1.33$, and the index of glass is about $n = 1.5$. *See* PHASE VELOCITY; REFRACTION OF WAVES.

For normal incidence ($\theta_{\text{inc}} = 0$) the reflectivity R of the interface is given by Eq. (3), in which the material constants are

$$R = \left(\frac{v_1 - v_2}{v_1 + v_2}\right)^2 = \left(\frac{n_2 - n_1}{n_2 + n_1}\right)^2 \tag{3}$$

labeled 1 and 2, where the radiation is incident in material 1. The reflectivity of an air-water interface is about 2% ($R = 0.02$) and that of an air-glass interface about 4% ($R = 0.04$); the other 98% or 96% are transmitted through the water or glass, respectively. *See* ALBEDO; GEOMETRICAL OPTICS; MIRROR OPTICS; REFLECTION OF SOUND. Herwig Kogelnik

Reflection of sound The return of sound waves from surfaces on which they are incident. The geometrical laws for reflection of sound waves are the same as those for light waves. The apparent differences involve only questions of scale, because the average wavelength of sound is about 100,000 times that of light. For example, a mirror or lens used to produce a beam of sound waves must be enormously large compared to mirrors and lenses used in optical systems. *See* REFLECTION OF ELECTROMAGNETIC RADIATION.

A concave surface tends to concentrate the reflected sound waves. Convex reflectors tend to spread the reflected waves. Therefore, when placed at the boundaries of a room, they tend to diffuse the sound throughout the room. For this reason, some radio-broadcasting studios employ cylindrical convex panels as part of their wall construction to promote diffusion. *See* ARCHITECTURAL ACOUSTICS; ECHO; SOUND. Cyril M. Harris

Reflex A simple, unlearned, yet specific behavioral response to a specific stimulus. Reflexes are exhibited by virtually all animals from protozoa to primates. Along with other, more complex stimulus-bound responses such as fixed action patterns, they constitute much of the behavioral repertoire of invertebrates. In higher animals, such as primates, where learned behavior dominates, reflexes nevertheless persist as an important component of total behavior.

The simplest known reflexes require only one neuron or, in the strictest sense, none. For example, ciliated protozoa, which are single cells and have no neurons, nevertheless exhibit apparently reflexive behaviors. However, most reflexes require activity in a large circuit of neurons. The neurons involved in most reflexes are connected by specific synapses to form functional units in the nervous system. Such a circuit begins with sensory neurons and ends with effector cells such as skeletal muscles, smooth muscles, and glands, which are controlled by motor neurons. The central neurons which are often interposed between the sensory and motor neurons are called interneurons. The sensory side of the reflex arc conveys specificity as to which reflex will be activated. The remainder of the reflex response is governed by the specific synaptic connections that lead to the effector neurons. A familiar reflex is the knee-jerk or stretch reflex. It involves the patellar (kneecap) tendon and a group of upper leg muscles. Other muscle groups show similar reflexes. J. L. Larimer

Reforestation The reestablishment of forest cover either naturally or artificially. Given enough time, natural regeneration will usually occur in areas where temperatures and rainfall are adequate and when grazing and wildfires are not too frequent.

Reforestation occurs on land where trees have been recently removed due to harvesting or to natural disasters such as a fire, landslide, flooding, or volcanic eruption. When abandoned cropland, pastureland, or grasslands are converted to tree cover, the practice is termed afforestation (where no forest has existed in recent memory). Afforestation is common in countries such as Australia, South Africa, Brazil, India, and New Zealand. Although natural regeneration can occur on abandoned cropland, planting trees will decrease the length of time required until the first harvest of wood. Planting also has an advantage in that both tree spacing and tree species can be prescribed. The selection of tree species can be very important since it affects both wood quality and growth rates. Direct seeding is also used for both afforestation and reforestation, although it often is less successful and requires more seed than tree planting. Unprotected seed are often eaten by birds and rodents, and weeds can suppress growth of newly germinated seed. For these reasons, direct seeding accounts for only about 5% and 1% of artificial reforestation in Canada and the United States, respectively. David B. South

Reforming processes Those processes used to convert, with limited cracking, petroleum liquids into higher-octane gasoline. Due to the demand for higher-octane gasoline, thermal reforming was developed (from thermal cracking processes) to improve the octane number of fractions within the boiling range of gasoline. *See* CRACKING; GASOLINE; OCTANE NUMBER; PETROLEUM.

Upgrading by reforming may be accomplished, in part, by an increase in volatility (reduction of molecular size) or by the conversion of *n*-paraffins to isoparaffins, olefins, and aromatics, and of naphthenes (cycloalkanes) to aromatics. The nature of the final product is influenced by the structure and composition of the straight-run (virgin) naphtha (hydrocarbon mixture) feedstock. In thermal reforming, the reactions resemble those in the cracking of gas oils. The molecular size is reduced, while olefins and some aromatics are synthesized. For example, hydrocracking of high-molecular-weight paraffins yields lower-molecular-weight paraffins and an olefin; dehydrocyclization of paraffin compounds yields aromatic compounds; isomerization of *n*-paraffins yields isoparaffins; and isomerization of methylcyclopentane yields cyclohexane.

In the presence of catalysts and in the presence of the hydrogen available from dehydrogenation reactions, hydrocracking of paraffins to yield two lower-molecular-weight paraffins takes place, and olefins that do not undergo dehydrocyclization are dehydrogenated so that the end product contains only traces of olefins. *See* DEHYDROGENATION; HYDROCRACKING; ISOMERIZATION; PARAFFIN.

Thermal reforming was a natural development from thermal cracking, since reforming is also a thermal decomposition reaction. Cracking converts heavier oils into gasoline constituents, whereas reforming converts these gasoline constituents into higher-octane molecules. The equipment for thermal reforming is essentially the same as for thermal cracking, but higher temperatures are used. The higher octane number of the product (reformate) is due primarily to the cracking of longer-chain paraffins into higher-octane olefins. *See* DISTILLATION.

The products of thermal reforming are gases, gasoline, and residual oil. The amount and quality of the reformate are very dependent on the temperature. As a rule, the higher the reforming temperature, the higher the octane number of the product but the lower the reformate yield. Adding catalysts increases the yield for higher-octane gasolines at a given temperature.

Thermal reforming is less effective and less economical than catalytic processes and has been largely supplanted. The octane number was changed by the severity of the cracking, and the product had increased volatility, compared to the volatility of the feedstock.

Modifications of the thermal reforming process due to the inclusion of hydrocarbon gases with the feedstock are known as gas reversion and polyforming. These are essentially the same but differ in the manner in which the gases and naphtha are passed through the heating furnace. In gas reversion, the naphtha and gases flow through separate lines in the furnace and are heated independently of one another. In naphtha reforming, the C_3 and C_4 gases are premixed with the naphtha and pass together through the furnace.

Like thermal reforming, catalytic reforming converts low-octane gasoline into high-octane gasoline (reformate). Although thermal reforming can produce reformate with a research octane number of 65–80 depending on the yield, catalytic reforming produces reformate with octane numbers on the order of 90–95. Catalytic reforming is conducted in the presence of hydrogen over hydrogenation-dehydrogenation catalysts. Depending on the catalyst, a definite sequence of reactions takes place, involving structural changes in the feedstock. *See* CATALYSIS; PETROLEUM PROCESSING AND REFINING. James G. Speight

Refraction of waves The change of direction of propagation of any wave phenomenon which occurs when the wave velocity changes. The term is most frequently applied to visible light, but it also applies to all other electromagnetic waves, as well as to sound and water waves.

The physical basis for refraction can be readily understood with the aid of the illustration. Consider a succession of equally spaced wavefronts approaching a boundary surface obliquely. The direction of propagation is in ordinary cases perpendicular to the wavefronts. In the case shown, the velocity of propagation is less in medium 2 than in medium 1, so that the waves are slowed down as they enter the second medium. Thus, the direction of travel is bent toward the perpendicular to the boundary surface (that is, $\theta_2 < \theta_1$). If the waves enter a medium in which the velocity of propagation is faster than in their original medium, they are refracted away from the normal.

Snell's law. The simple mathematical relation governing refraction is known as Snell's law. If waves traveling through a medium at speed v_1 are incident on a boundary surface at angle θ_1 (with respect to the normal), and after refraction enter the second medium at angle θ_2 (with the normal) while traveling at speed v_2, then Eq. (1) holds. The index of refraction n of a

$$\frac{v_1}{v_2} = \frac{\sin\theta_1}{\sin\theta_2} \tag{1}$$

medium is defined as the ratio of the speed of waves in vacuum c to their speed in the medium. Thus $c = n_1 v_1 = n_2 v_2$, and therefore Eq. (2) holds. The refracted ray, the normal to the surface,

$$n_1 \sin\theta_1 = n_2 \sin\theta_2 \tag{2}$$

and the incident ray always lie in the same place.

The relative index of refraction of medium 2 with respect to that of medium 1 may be defined as $n = n_1/n_2$. Snell's law then becomes Eq. (3). For sound and other elastic waves which

$$\sin\theta_1 = n \sin\theta_2 \tag{3}$$

require a medium in which to propagate, only this last form has meaning. Equation (3) is frequently used for light when one medium is air, whose index of refraction is very nearly unity.

When the wave travels from a region of low velocity (high index) to one of high velocity (low index), refraction occurs only if (n_1/n_2) sin $\theta_1 \leq 1$. If θ_1 is too large for this relation to hold, then sin $\theta_2 > 1$, which is meaningless. In this case the waves are totally reflected from the surface back into the first medium. The largest value that θ_1 can have without total internal reflection taking place is known as the critical angle θ_c. Thus sin $\theta_c = n_2/n_1$.

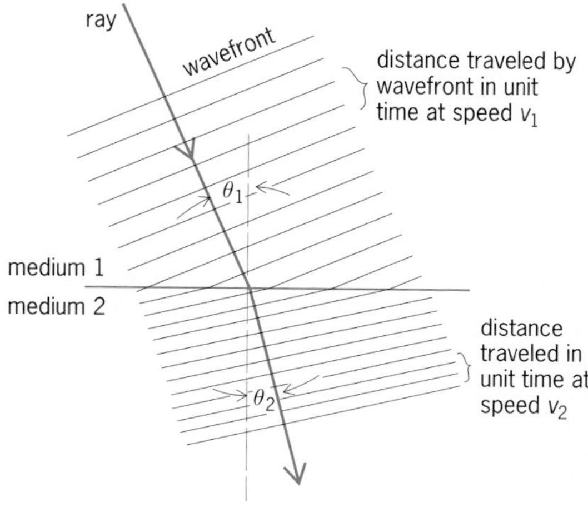

Physical basis for Snell's law.

Atmospheric refraction. The index of refraction of the Earth's atmosphere increases continuously from 1.000000 at the edge of space to 1.000293 (yellow light) at 0°C and 760 mmHg (101.3 kilopascals) pressure. Thus celestial bodies as seen in the sky are actually nearer to the horizon than they appear to be. The effect decreases from a maximum of about 35 minutes of arc for an object on the horizon to zero at the zenith, where the light enters the atmosphere at perpendicular incidence.

Other manifestations of atmospheric refraction are the mirages and "looming" of distant objects which occur over oceans or deserts, where the vertical density gradient of the air is quite uniform over a large area. *See* MIRAGE.

Sound waves. The velocity of sound in a gas is proportional to the square root of the absolute temperature. Because of the vertical temperature gradients in the atmosphere, refraction of sound can be quite pronounced. As in mirage formation, to allow large-scale refraction the temperature at a given height must be uniform over a rather large horizontal area. *See* ATMOSPHERIC ACOUSTICS; SOUND.

Seismic waves. The velocity of elastic waves in a solid depends upon the modulus of elasticity and upon the density of the material. Waves propagating through solid earth are refracted by changes of material or changes of density. Worldwide observations of earthquake waves enable scientists to draw conclusions on the distribution of density within the Earth. *See* SEISMOLOGY.

Water waves. As the waves enter shallower water they travel more slowly. As a train of waves approaches a coastline obliquely, its direction of travel becomes more nearly perpendicular to the shore because of refraction. *See* WAVE MOTION. John W. Stewart

Refractometric analysis A method of chemical analysis based on the measurement of the index of refraction of a substance. The most common type of refractometer is the Abbe refractometer. It is simple to use, requiring but a drop or two of sample and allowing a measurement of refractive index to be made in 1–2 min, with a precision of 0.0001. More precise measurements of refractive indices may be made by using a dipping or immersion refractometer, the prism of which is completely immersed in the sample. The most precise measurements of the refractive indices of gases or solutions containing small traces of impurities are made with an interferometer. For measurements in flowing systems, differential refractometers are used.

The measurement of refractive index is used to identify compounds whose other physical constants are quite similar. Because minute amounts of impurities often cause a measurable change in the refractive index of a pure material, refractive index is often used as a criterion for purity. A measurement of refractive index gives information as to the gross amount of impurity; it does not serve to identify the impurity. *See* REFRACTION OF WAVES. Robert F. Goddu; James N. Little

Refractory One of a number of ceramic materials for use in high-temperature structures or equipment. The term high temperatures is somewhat indefinite but usually means above about 1830°F (1000°C), or temperatures at which, because of melting or oxidation, the common metals cannot be used. In some special high-temperature applications, the so-called refractory metals such as tungsten, molybdenum, and tantalum are used. *See* CERAMICS.

The greatest use of refractories is in the steel industry, where they are used for construction of linings of equipment such as blast furnaces, hot stoves, and open-hearth furnaces. Other important uses of refractories are for cement kilns, glass tanks, nonferrous metallurgical furnaces, ceramic kilns, steam boilers, and paper plants. Special types of refractories are used in rockets, jets, and nuclear power plants. Many refractory materials, such as aluminum oxide and silicon carbide, are also very hard and are used as abrasives; some applications, for example, aircraft brake linings, make use of both characteristics.

Refractory materials are commonly grouped into (1) those containing mainly aluminosilicates; (2) those made predominantly of silica; (3) those made of magnesite, dolomite, or chrome ore, termed basic refractories (because of their chemical behavior); and (4) a miscellaneous category usually referred to as special refractories. John F. McMahon

Refrigeration

Refrigeration The cooling of a space or substance below the environmental temperature. Mechanical refrigeration is primarily an application of thermodynamics wherein the cooling medium, or refrigerant, goes through a cycle so that it can be recovered for reuse. The commonly used basic cycles, in order of importance, are vapor-compression, absorption, steam-jet or steam-ejector, and air. Each cycle operates between two pressure levels, and all except the air cycle use a two-phase working medium which alternates cyclically between the liquid and vapor phases.

The term "refrigeration" is used to signify cooling below the environmental temperature to lower than about 150 K (−190°F; −123°C). The term "cryogenics" is used to signify cooling to temperatures lower than 150 K. *See* CRYOGENICS.

Vapor-compression cycle. The vapor-compression cycle consists of an evaporator in which the liquid refrigerant boils at low temperature to produce cooling, a compressor to raise the pressure and temperature of the gaseous refrigerant, a condenser in which the refrigerant discharges its heat to the environment, usually a receiver for storing the liquid condensed in the condenser, and an expansion valve through which the liquid expands from the high-pressure level in the condenser to the low-pressure level in the evaporator. This cycle may also be used for heating if the useful energy is taken off at the condenser level instead of at the evaporator level. *See* HEAT PUMP.

Absorption cycle. The absorption cycle accomplishes compression by using a secondary fluid to absorb the refrigerant gas, which leaves the evaporator at low temperature and pressure. Heat is applied, by means such as steam or gas flame, to distill the refrigerant at high temperature and pressure. The most-used refrigerant in the basic cycle is ammonia; the secondary fluid is then water. This system is used for the lower temperatures. Another system is lithium bromide-water, where the water is used as the refrigerant. This is used for higher temperatures. Due to corrosion, special inhibitors must be used in the lithium bromide-water system. The condenser, receiver, expansion valve, and evaporator are essentially the same as in any vapor-compression cycle. The compressor is replaced by an absorber, generator, pump, heat exchanger, and controlling-pressure reducing valve.

Steam-jet cycle. The steam-jet cycle uses water as the refrigerant. High-velocity steam jets provide a high vacuum in the evaporator, causing the water to boil at low temperature and at the same time compressing the flashed vapor up to the condenser pressure level. Its use is limited to air conditioning and other applications for temperatures above 32°F (0°C).

Air cycle. The air cycle, used primarily in airplane air conditioning, differs from the other cycles in that the working fluid, air, remains as a gas throughout the cycle. Air coolers replace the condenser, and the useful cooling effect is obtained by a refrigerator instead of by an evaporator. A compressor is used, but the expansion valve is replaced by an expansion engine or turbine which recovers the work of expansion. Systems may be open or closed. In the closed system, the refrigerant air is completely contained within the piping and components, and is continuously reused. In the open system, the refrigerator is replaced by the space to be cooled, the refrigerant air being expanded directly into the space rather than through a cooling coil.

Refrigerants. The working fluid in a two-phase refrigeration cycle is called a refrigerant. A useful way to classify refrigerants is to divide them into primary and secondary. Primary refrigerants are those fluids (pure substances, azeotropic mixtures which behave physically as a single pure compound, and zeotropes which have temperature glides in the condenser and evaporator) used to directly achieve the cooling effect in cycles where they alternately absorb and reject heat. Secondary refrigerants are heat transfer or heat carrier fluids. *See* AIR CONDITIONING; AIR COOLING; AUTOMOTIVE CLIMATE CONTROL; COLD STORAGE; COOLING TOWER; MARINE REFRIGERATION; REFRIGERATOR. Peter E. Liley; Carl F. Kayan

Refrigeration cycle

Refrigeration cycle A sequence of thermodynamic processes whereby heat is withdrawn from a cold body and expelled to a hot body. Theoretical thermodynamic cycles consist of nondissipative and frictionless processes. For this reason, a thermodynamic cycle can be operated in the forward direction to produce mechanical power from heat energy, or it can be operated in the reverse direction to produce heat energy from mechanical power. The reversed cycle is used primarily for the cooling effect that it produces during a portion of the cycle and so is called a refrigeration cycle. It may also be used for the heating effect, as in the comfort warming of space during the cold season of the year. *See* HEAT PUMP; THERMODYNAMIC PROCESSES.

In the refrigeration cycle a substance, called the refrigerant, is compressed, cooled, and then expanded. In expanding, the refrigerant absorbs heat from its surroundings to provide refrigeration. After the refrigerant absorbs heat from such a source, the cycle is repeated. Compression raises the temperature of the refrigerant above that of its natural surroundings so that it can give up its heat in a heat exchanger to a heat sink such as air or water. Expansion lowers the refrigerant temperature below the temperature that is to be produced inside the cold compartment or refrigerator. The sequence of processes performed by the refrigerant constitutes the refrigeration cycle. When the refrigerant is compressed mechanically, the refrigerative action is called mechanical refrigeration.

There are many methods by which cooling can be produced. The methods include the noncyclic melting of ice, or the evaporation of volatile liquids, as in local anesthetics; the Joule-Thomson effect, which is used to liquefy gases; the reverse Peltier effect, which produces heat flow from the cold to the hot junction of a bimetallic thermocouple when an external emf is imposed; and the paramagnetic effect, which is used to reach extremely low temperatures. However, large-scale refrigeration or cooling, in general, calls for mechanical refrigeration acting in a closed system. *See* PARAMAGNETISM; REFRIGERATION.

The purpose of a refrigerator is to extract as much heat from the cold body as possible with the expenditure of as little work as possible. The yardstick in measuring the performance of a refrigeration cycle is the coefficient of performance, defined as the ratio of the heat removed to the work expended. The coefficient of performance of the reverse Carnot cycle is the maximum obtainable for stated temperatures of source and sink. *See* CARNOT CYCLE.

The reverse Brayton cycle it was one of the first cycles used for mechanical refrigeration. Before Freon and other condensable fluids were developed for the vapor-compression cycle, refrigerators operated on the Brayton cycle, using air as their working substance. Air undergoes isentropic compression, followed by reversible constant-pressure cooling. The high-pressure air next expands reversibly in the engine and exhausts at low temperature. The cooled air passes through the cold storage chamber, picks up heat at constant pressure, and finally returns to the suction side of the compressor. *See* BRAYTON CYCLE.

Theodore Baumeister; Peter E. Liley

Refrigerator An insulated, cooled compartment. If it is large enough for the entry of a person, it is termed a walk-in box; otherwise it is called a reach-in refrigerator. Cooling may be by mechanical or gas refrigeration, by water or dry ice, or by brine circulation. Temperatures maintained depend upon the requirements of the product stored, generally varying from 55°F (13°C) down to 0°F (−18°C), and sometimes lower.

A household or domestic refrigerator is a factory-built, self-contained cabinet. The range of storage capacities is wide and varies among manufacturers. Modem designs have a main compartment for holding food above freezing, a second compartment for storage below freezing, and trays for the freezing of ice cubes. Low-temperature household refrigerators, or home freezers, for the storage of frozen foods are manufactured in both the chest and the upright, or vertical, types.

A commercial refrigerator is any factory-built refrigerated fixture, cabinet, or room that can be assembled and disassembled readily. Commercial or built-in refrigerators are used in restaurants, markets, hospitals, hotels, and schools for the storage of food and other perishables. *See* REFRIGERATION. Carl F. Kayan

Regeneration (engineering) Any of the various processes for restoring a system to its original state by restoring some property of a system to its original value, or for using the properties of a system at one point in a cycle to modify the properties of the same system at another point in that cycle. The term has a wide variety of uses in different fields of engineering. William L. Beasley

Regenerative biology The field within developmental biology that focuses on regeneration, the process by which an animal restores a lost part of its body. Broadly defined, the term regeneration can include many kinds of restorative activities shown by a wide variety of organisms. However, regeneration in this context connotes reproduction of a more perfect or complete replacement of a missing tissue or organ than usually results from processes of tissue repair or wound healing. Within the field of developmental biology, most research in regeneration involves systems in which removing a complex structure or major part of an organism initiates a chain of events that produces a structure that duplicates the missing part both functionally and anatomically. *See* DEVELOPMENTAL BIOLOGY.

The best-known and most widely studied examples of regeneration are those involving epimorphosis, in which the lost structure is reproduced directly by a combination of cell proliferation and redifferentiation of new tissue. Examples can be found throughout the animal kingdom, from the extensive regenerative capacities of *Hydra* and *Planaria* to the regrowth of antlers in deer. Research on regenerating systems such as in annelid worms, hydra, and amphibian tails and limbs has yielded information regarding basic mechanisms of animal development. Regeneration can involve growth of an existing population of unspecialized cells or cells arising by dedifferentiation of tissues at the site of injury. *See* CELL DIFFERENTIATION; STEM CELLS.

Mammals, birds, and reptiles have a much more poorly developed ability than amphibians and fish to regenerate complete organs, but nevertheless can reform missing tissue and restore function after partial removal of certain organs. If part of the liver is cut away, the remaining portion increases in size to compensate for the missing tissue and to restore the normal functional capacity of the organ. The process of liver regeneration involves the triggering of active growth in the remaining liver cells, in cells of bile ductules, and in unspecialized stem cells, all of which are usually quiescent in the normal liver. In the musculoskeletal system, different populations of quiescent stem cells allow efficient replacement of damaged or partially removed bones and muscles. *See* BONE; LIVER; MUSCLE.

Of all vertebrates, amphibians have the most highly developed capacity for regeneration. Certain species have the ability to regenerate not only limbs and tails but also parts of the eye, lower jaw, intestine, and heart. Complete regeneration of amputated limbs can occur throughout the lifetime of most urodeles (salamanders and newts) but usually occurs much more rapidly in larval animals. In anurans (frogs and toads) the ability to regenerate limbs is lost during metamorphosis to the adult form. *See* METAMORPHOSIS.

A well-developed capacity for tail regeneration is found in larval amphibians, as well as in certain species of adult salamanders. Many lizards can also regenerate tails lost through injury or by autotomy during escape from a predator. Autotomy is produced by a strong contraction of the segmental muscles of the tail, causing one of the vertebrae to come apart along a preformed breakage plane and separating the muscles and other tissues so that the distal portion of the tail is detached from the body. Whether lost by autotomy or injury, the tail is regenerated by a process superficially similar to that of amphibian limbs.

Protozoa and simple multicellular animals, including sponges, cnidaria (for example, *Hydra*), and flatworms (for example, *Planaria*), display remarkable capacities for regeneration following various experimental manipulations. Regenerative ability in such organisms correlates closely with their capacity to reproduce asexually, most commonly by fission or by budding, and the mechanisms of growth involved in regeneration are often very similar to those of asexual propagation.

Most annelids, such as the earthworm, can readily regenerate segments after their removal: some species can regenerate whole organisms from any fragment. Like more primitive invertebrates, certain annelids can reproduce asexually by transverse fission. The capacities for fission and for reconstitution from fragments in annelids are remarkable, considering the anatomical complexity of animals in this phylum. *See* ANNELIDA.

The ability of certain echinoderms, such as starfish, to regenerate missing arms is well known. Cutting such an animal into several pieces results in each piece forming a new organism, a phenomenon that usually requires the presence of at least some of the central portion of the body. Some starfish species, however, can actually regenerate an entire body from the cut end of an isolated arm. *See* ECHINODERMATA.

The capacity for appendage regeneration is widespread among the many diverse members of the phylum *Arthropoda*. In these complex animals with well-developed exoskeletons and no asexual mode of reproduction, regeneration shows a close correlation with the molting process. *See* ARTHROPODA. Anthony Mescher

Regolith The mantle or blanket of unconsolidated or loose rock material that overlies the intact bedrock and nearly everywhere forms the land surface. The regolith may be residual (weathered in place), or it may have been transported to its present site. The undisturbed residual regolith may grade from agricultural soil at the surface, through fresher and coarser weathering products, to solid bedrock several feet or more beneath the surface. The transported regolith includes the alluvium of rivers, sand dunes, glacial deposits, volcanic ash, coastal deposits, and the various mass-wasting deposits that occur on hillslopes. The lunar surface also has a regolith. This layer of fragmental debris is believed to derive from prolonged meteoritic and secondary fragment impact. *See* SOIL; WEATHERING PROCESSES. Victor R. Baker

Regular polytopes The n-dimensional analogs of the regular polygons ($n = 2$) and platonic solids ($n = 3$). They are conveniently denoted by their Schläfli symbols $\{p, q \ldots\}$; for instance, the pentagon, hexagon, octagon, tetrahedron, octahedron are denoted by $\{5\}$, $\{6\}$, $\{8\}$, $\{3, 3\}$, $\{3, 4\}$. The cube is $\{4, 3\}$ because its faces are squares $\{4\}$ and there are three of them at each vertex. The five platonic solids $\{p, q\}$, are determined by the inequality below. The numbers of vertices, edges, faces

$$(p-2)(q-2) < 4$$

(V, E, F) can be deduced from the obvious relations $pF = 2E = qV$ with the help of Euler's formula $V - E + F = 2$.

The general polytope (sometimes loosely called a "polyhedron" [see illustration] regardless of the number of dimensions) is a finite region of n-dimensional space enclosed by a finite number of hyperplanes. When any redundant hyperplanes have been discarded, those that remain contain $(n-1)$-dimensional polytopes called cells or "facets." For instance, the cells of a polygon are its sides, those of a polyhedron are its faces, and those of a four-dimensional polytope are solids. *See* ANALYTIC GEOMETRY; EUCLIDEAN GEOMETRY. H. S. M. Coxeter

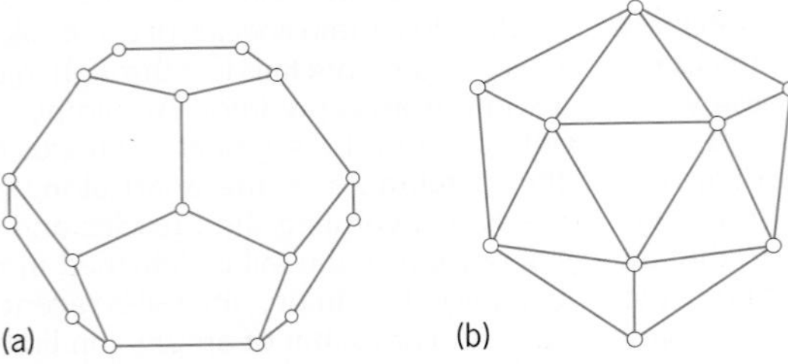

Two regular polyhedrons. (*a*) Dodecahedron $\{5, 3\}$. (*b*) Icosahedron $\{3, 5\}$.

Regulator

A control device designed to maintain the value of some quantity substantially constant. The value to be maintained can usually be established at any value within the range of the regulator by making an appropriate setting. A regulated system is a feedback control system employing a regulator to maintain some quantity of the system at a constant value. *See* CONTROL SYSTEMS. John A. Hrones

Reheating

The addition of heat to steam of reduced pressure after the steam has given up some of its energy by expansion through the high-pressure stages of a turbine. The reheater tube banks are arranged within the setting of the steam-generating unit in such relation to the gas flow that the steam is restored to a high temperature. Under suitable conditions of initially high steam pressure and superheat, one or two stages of reheat can be advantageously employed to improve thermodynamic efficiency of the cycle. *See* STEAM-GENERATING UNIT; STEAM TURBINE; SUPERHEATER; VAPOR CYCLE. R. A. Miller

Reinforced concrete

Portland cement concrete containing higher-strength, solid materials to improve its structural properties. Generally, steel wires or bars are used for such reinforcement, but for some purposes glass fibers or chopped wires have provided desired results.

Unreinforced concrete cracks under relatively small loads or temperature changes because of low tensile strength. The cracks are unsightly and can cause structural failures. To prevent cracking or to control the size of crack openings, reinforcement is incorporated in the concrete. Reinforcement may also be used to help resist compressive forces or to improve dynamic properties.

Steel usually is used in concrete. It is elastic, yet has considerable reserve strength beyond its elastic limit. Under a specific axial load, it changes in length only about one-tenth as much as concrete. In compression, steel is more than 10 times stronger than concrete, and in tension, more than 100 times stronger. *See* STEEL.

During construction, the bars are placed in a form and then concrete from a mixer is cast to embed them. After the concrete has hardened, deformation is resisted and stresses are transferred from concrete to reinforcement by friction and adhesion along the surface of the reinforcement. Individual wires or bars resist stretching and tensile stress in the concrete only in the direction in which such reinforcement extends. Tensile stresses and deformations, however, may occur simultaneously in other directions. Therefore reinforcement must usually be placed in more than one direction. For this purpose, reinforcement sometimes is assembled as a rectangular grid. Bars, grids, and fabric have the disadvantage that the principal effect of reinforcement occurs primarily in the plane of the layer in which they are placed. Consequently, the reinforcement often must be set in several layers or formed into cages. Under some conditions, fiber-reinforced concrete is an alternative to such arrangements. *See* COMPOSITE BEAM; CONCRETE; CONCRETE BEAM; CONCRETE COLUMN; CONCRETE SLAB; PRESTRESSED CONCRETE. Frederick S. Merritt

Relapsing fever

An acute infectious disease characterized by recurring fever. It is caused by spirochetes of the genus *Borrelia* and transmitted by the body louse (*Pediculus humanus humanus*) and by ticks of the genus *Ornithodoros*.

Louse-borne relapsing fever, caused by *Borrelia recurrentis*, is typically epidemic. Epidemics, once widespread on all continents, are rare but still occur in certain parts of South America, Africa, and Asia. Tick-borne relapsing fevers are endemic. They are more widely distributed throughout the Eastern and Western hemispheres. At least 15 species of *Borrelia* have been recognized as causative agents.

After incubation of 2–10 days, the initial attack begins abruptly with chills, high fever, headache, and pains in muscles and joints, and lasts 2–8 days, ending by crisis. A remission period of 3–10 days is followed by a relapse similar to the initial attack but milder. There may be 4–5 relapses, although occasionally 10 or more have been recorded. Mortality varies from 2 to 5% but may be considerably higher during epidemics.

Chlortetracycline is the most effective antibiotic drug, but penicillin, oxytetracycline, and streptomycin also have therapeutic value. *See* ANTIBIOTIC.

The best way to prevent relapsing fever is to control louse and tick populations with effective insecticides and acaricides. *See* MEDICAL BACTERIOLOGY; PEDICULOSIS. Willy Burgdorfer

Relative atomic mass

The ratio of the average mass per atom of the natural nuclidic composition of an element to $^1/_{12}$ of the mass of an atom of nuclide ^{12}C. For example, $\mu(\mathrm{Cl}) = 35.453$. Relative atomic mass replaces the concept of atomic weight. It is also known as relative nuclidic mass. *See* NUCLIDE. Thomas C. Waddington

Relative molecular mass

The ratio of the average mass per formula unit of the natural nuclidic composition of a substance to $^1/_{12}$ of the mass of an atom of nuclide ^{12}C. For example, $\mu(\mathrm{KCl}) = 74.555$. Relative molecular mass replaces the concept of molecular weight. *See* NUCLIDE. Thomas C. Waddington

Relative motion

All motion is relative to some frame of reference. The simplest laboratory frame of reference is three mutually perpendicular axes at rest with respect to an observer. In terms of the frame of reference of an observer some distance from Earth, the laboratory frame of reference would be moving with Earth as it rotates on its axis and as it revolves about the Sun. What would be a simple form of motion in the laboratory frame of reference would appear to be a much more complicated motion in the frame of reference of the distant observer. *See* FRAME OF REFERENCE.

Motion means continuous change of position of an object with respect to an observer. To another observer in a different frame of reference the object may not be moving at all, or it may be moving in an entirely different manner. The motions of the planets were found in ancient times to appear quite complicated in the laboratory frame of reference of an observer on Earth. By transferring to the frame of reference of an imaginary observer on the Sun, Johannes Kepler showed that the relative motion of the planets could be simply described in terms of elliptical orbits. The validity of one description is no greater than the other, but the latter description is far more convenient. Rogers D. Rusk

Relativistic electrodynamics The study of the interaction between electrically charged particles and electromagnetic fields when the velocities involved are comparable to that of light.

A group of charged particles in motion can be represented by a distribution in charge and distribution in current. During the latter part of the eighteenth century and the early part of the nineteenth century, experiments by C. A. Coulomb, M. Faraday, A. M. Ampère, and others showed that electric and magnetic fields are produced by charge and current distributions. These fields, in turn, act on other charges and currents. The interaction between charges and currents on the one hand and electric and magnetic fields on the other is the topic of study of electrodynamics. This field of study was established as a quantitative and self-contained subject in 1864 when J. C. Maxwell formulated his equations for the electromagnetic field. Maxwell conjectured that a time-varying electric field is equivalent to an electric current in its effect of producing a magnetic field, and named it the displacement current. The inclusion of this displacement current enabled Maxwell to combine all the previously established laws of electromagnetism into a coherent whole in his equations. *See* CLASSICAL FIELD THEORY; DISPLACEMENT CURRENT; MAXWELL'S EQUATIONS.

With the inclusion of the displacement current, the Maxwell equations are relativistically covariant, meaning that they are valid for all velocities, even those approaching the velocity of light. However, the implications of the covariance of the equations were not fully appreciated until A. Einstein formulated the special theory of relativity in 1905. Relativistic electrodynamics was then rapidly developed into a powerful and precise field of physics. It describes and predicts all macroscopic electrodynamic phenomena to the minutest detail and with perfect accuracy, and now forms the foundation on which the entire electrical industry is based. However, its limitations soon became evident when attempts were made to apply it to atomic phenomena: Straightforwardly applied, relativistic electrodynamics failed to explain many of these phenomena, and its predictions frequently disagreed with experimental observations. For these microscopic phenomena, quantum electrodynamics (QED) was developed in the 1930s to replace classical relativistic electrodynamics. In 1967 quantum electrodynamics was further unified by S. Weinberg and A. Salam with the theory of weak interactions to form the electroweak theory. *See* QUANTUM ELECTRODYNAMICS; RELATIVISTIC MECHANICS; RELATIVITY; WEAK NUCLEAR INTERACTIONS.

Electrodynamic problems generally fall into one of two categories:

1. Finding the electromagnetic field produced by prescribed charge and current distributions. For example, one may want to determine the electromagnetic field produced or radiated by a given oscillatory electric current in a transmitting antenna, or the field radiated by an accelerating electron.

2. Finding the effect of a predetermined electromagnetic field on the motion of charges and currents. This is the inverse problem corresponding to that of the receiving antenna or of the motion of charged particles in an accelerator.

All other electrodynamic problems are combinations or iterations of these two basic types. For instance, the scattering of light (electromagnetic radiation) by a charged particle is composed of, first, the incident light shaking the charge and, second, the subsequent emission of the scattered light by the shaken charge. *See* SCATTERING OF ELECTROMAGNETIC RADIATION. Lee C. Teng

Relativistic heavy-ion collisions Collisions between heavy atomic nuclei at relative velocities close to the speed of light. These high-energy nuclear collisions are usually divided into two different domains, relativistic and ultrarelativistic collisions, depending on whether the kinetic energy per nucleon (the generic name for protons and neutrons) is either close to the rest mass of the nucleon (relativistic collisions) or much larger than the nucleon rest mass (ultrarelativistic collisions).

By utilizing high-energy nuclear collisions, it is possible to study nuclear matter under conditions of very high temperatures and densities. The most common form of nuclear matter, at least under terrestrial conditions, is found in the atomic nucleus, which consists of protons and neutrons bound together by the strong nuclear force. If nuclear matter is heated up to temperatures comparable to the rest mass of the pion, it becomes a mixture of nucleons, pions, and various other particles, collectively denoted hadrons. Under these circumstances, nuclear matter is referred to as hadronic matter. *See* HADRON; NEUTRON; NEUTRON STAR; NUCLEAR STRUCTURE; PROTON.

According to the quantum chromodynamics (QCD) theory, all hadrons are bound states of a set of more fundamental entities called quarks. The quarks are confined within the hadrons by the exchange of gluons. Quantum chromodynamics calculations using the most powerful computers available show that if hadronic matter is further heated or compressed to very high densities it will undergo a phase transition into a new phase of matter, called the quark-gluon plasma. In this phase the hadrons will lose their identity, and the quarks and gluons will be deconfined within volumes much larger than the typical hadron volume of 0.1–0.5 cubic femtometer. Quantum chromodynamics calculations indicate that the phase transition will occur at a critical density around 5–10 times the normal nuclear matter density of approximately 0.2 nucleon/fm^3, or at a critical temperature around 150 MeV. *See* GLUONS; QUANTUM CHROMODYNAMICS; QUARK-GLUON PLASMA; QUARKS.

When two nuclei collide at high energies, some of the nucleons in each nucleus, called spectators, will continue their motion unaffected, while other nucleons, called participants, will strike one or several nucleons in the other nucleus. In the overlap volume a hot and dense fireball will develop. If the temperature or density of the fireball becomes larger than the critical values, a quark-gluon plasma will be created with an estimated lifetime of $1–5 \times 10^{-23}$ s. The fireball will start to expand and cool, and the quarks in the plasma will eventually be reconfined into a large number of hadrons (hadronization). After further expansion the hadrons will cease interacting with each other (freeze out) and leave the collision zone without further mutual interactions.

In the search for the quark-gluon plasma, a fundamental problem is that even if the plasma is created in the early phases of the collisions, the subsequent hadronization and scattering of the hadrons before freeze-out might mask any traces of the plasma. In order to circumvent this problem, many plasma signatures have been proposed. Soren P. Sorensen

Relativistic quantum theory The quantum theory of particles which is consistent with the special theory of relativity, and thus can describe particles moving arbitrarily close to the speed of light. It is now realized that the only satisfactory rela-tivistic quantum theory is quantum field theory; the attempt to relativize the Schrödinger equation for the wave function of a single particle fails [Eq. (1)]. However, with a change of interpretation, relativistic wave equations do correctly describe some aspects of the motions of particles in an electromagnetic field. *See* QUANTUM FIELD THEORY; QUANTUM MECHANICS; RELATIVITY.

The Schrödinger equation for the wave function $\psi(\mathbf{r},t)$ of a particle is Eq. (1), where E is the energy operator $i\hbar(\partial/\partial t)$, $\mathbf{p}$ is the

$$E\psi = H(\mathbf{p},\mathbf{r})\psi \tag{1}$$

momentum operator $-i\hbar\Delta$, $H(\mathbf{p},\mathbf{r})$ is the classical hamiltonian, and $\hbar$ is Planck's constant divided by 2π. For a nonrelativistic free particle, $H = \mathbf{p}^2/2m$. The naive way to relativize Eq. (1) would be to use the relativistic hamiltonian, Eq. (2). However, this equation is not relativistically invariant.

$$H = \sqrt{(mc^2)^2 + \mathbf{p}^2c^2} \tag{2}$$

The so-called Klein-Gordon equation, Eq. (3), is relativistically

$$E^2\varphi = [(mc^2)^2 + \mathbf{p}^2c^2]\varphi \tag{3}$$

invariant. However, the only possible density of a conserved quantity formed from [φ is of the form shown in (4). But this

$$\rho \propto \varphi^*E\varphi - \varphi E\varphi^* \tag{4}$$

cannot be a probability density, because it is not positive definite (it changes sign when φ is replaced by φ^*).

But ρ, in relation (4), can be interpreted as a charge density (when multiplied by a unit charge e); φ is then to be interpreted as a matrix element of a field operator Φ of a quantized field whose quanta are particles with mass m and charge e or $-e$ and zero spin.

P. A. M. Dirac found a relativized form of Eq. (1), Eq. (5), which is both linear in E and has a positive definite density form ρ, where β and α are constants which obey Eqs. (6). Obviously

$$E\psi = [\beta mc^2 + \alpha\cdot\mathbf{p}c] \quad \psi(\rho \propto \psi^*\psi) \tag{5}$$

$$\begin{gathered} \alpha_i\alpha_j + \alpha_j\alpha_i = 0, \quad i \neq j \\ \alpha\beta + \beta\alpha = 0 \quad \alpha_i^2 = 1 \quad \beta^2 = 1 \quad i,j = 1,2,3 \end{gathered} \tag{6}$$

the four constants β and α_i cannot be numbers; however, they can be 4×4 matrices, and Ψ is then a four-component object called a Dirac spinor.

If plane wave solutions of Dirac's equation (5) are considered, then $\mathbf{p}$ is now a number. Taking Eq. (5) as an eigenequation for E, four eigenstates are found (because H is a 4×4 matrix), two with

$$E = +\sqrt{(mc^2)^2 + p^2c^2}$$

and two with

$$E = -\sqrt{(mc^2)^2 + p^2c^2}$$

The interpretation of the two positive energy states is that they are the two spin states of a particle with spin $^1/_2\hbar$. But the two negative energy states are an embarrassment; even a particle that was initially in a positive energy state would quickly make radiative transitions down through the negative energy states. Dirac's solution was to observe that if the particle described by ψ obeyed the Pauli principle, then one can suppose that all the negative energy states are already filled with particles, thus excluding any more. There are still four single-particle states for a given momentum $\mathbf{p}$: the two spin states of a particle with positive energy, and the two states obtained by removing a negative energy particle (of momentum $-\mathbf{p}$). These last states ("hole states") have positive energy and a charge opposite the charge of the particle. The hole is in fact the antiparticle; if the particle is an electron, the hole is a positron. *See* ANTIMATTER; POSITRON.

With the filling up of the negative energy states, one no longer has a single-particle system, and ψ, just as in the Klein-Gordon case, no longer can be interpreted as a wave function but must be interpreted as a matrix element of a field operator Ψ.

Charles J. Goebel

Relativity A general theory of physics, primarily conceived by Albert Einstein, which involves a profound analysis of time and space, leading to a generalization of physical laws, with far-reaching implications in important branches of physics and in cosmology. Historically, the theory developed in two stages. Einstein's initial formulation in 1905 (now known as the special, or restricted, theory of relativity) does not treat gravitation; and one of the two principles on which it is based, the principle of relativity (the other being the principle of the constancy of the speed of light), stipulates the form invariance of physical laws only for inertial reference systems. Both restrictions were removed by Einstein in his general theory of relativity developed in 1915, which exploits a deep-seated equivalence between inertial and gravitational effects, and leads to a successful "relativistic" generalization of Isaac Newton's theory of gravitation.

Special theory. The key feature of the theory of special relativity is the elimination of an absolute notion of simultaneity in favor of the notion that all observers always measure light to have the same velocity, in vacuum, c, independently of their own motion. The impetus for the development of the theory arose from the theory of electricity and magnetism developed by J. C. Maxwell. This theory accounted for all observed phenomena involving electric and magnetic fields and also predicted that disturbances in these fields would propagate as waves with a definite speed, c, in vacuum. These electromagnetic waves predicted in Maxwell's theory successfully accounted for the existence of light and other forms of electromagnetic radiation. However, the presence of a definite speed, c, posed a difficulty, since if one inertial observer measures light to have velocity c, it would be expected that another inertial observer, moving toward the light ray with velocity v with respect to the first, would measure the light to have velocity $c + v$. Hence, it initially was taken for granted that there must be a preferred rest frame (often referred to as the ether) in which Maxwell's equations would be valid, and only in that frame would light be seen to travel with velocity c. However, this viewpoint was greatly shaken by the 1887 experiment of A. A. Michelson and E. W. Morley, which failed to detect any motion of the Earth through the ether. By radically altering some previously held beliefs concerning the structure of space and time, the theory of special relativity allows Maxwell's equations to hold, and light to propagate with velocity c, in all frames of reference, thereby making Maxwell's theory consistent with the null result of Michelson and Morley. *See* ELECTROMAGNETIC RADIATION; LIGHT; MAXWELL'S EQUATIONS.

Simultaneity in prerelativity physics. The most dramatic aspect of the theory of special relativity is its overthrowing of the notion that there is a well-defined, observer-independent meaning to the notion of simultaneity. To explain more precisely what is meant by absolute simultaneity, the following terminology will be introduced: An event is a point of space at an instant of time. Since it takes four numbers to specify an event—one for the time at which the event occurred and three for its spatial position—it follows that the set of all events constitutes a four-dimensional continuum, which is referred to as space-time.

A space-time diagram (Fig. 1) is a plot of events in space-time, with time, t, represented by the vertical axis and two spatial directions (x, y) represented by the horizontal axes. (The z direction is not shown.) For any event A shown in the diagram, there are many other events in this diagram—say, an event B—having the property that an observer or material body starting at event B can, in principle, be present at event A. The collection of all such events constitutes the past of event A. Similarly, there are many events—say, an event C—having the property that an observer or material body starting at event A can, in principle, be present at event C. These events constitute the future of A. Finally, there remain some events in space-time which lie neither to past nor future of A. In prerelativity physics, these events are assumed to

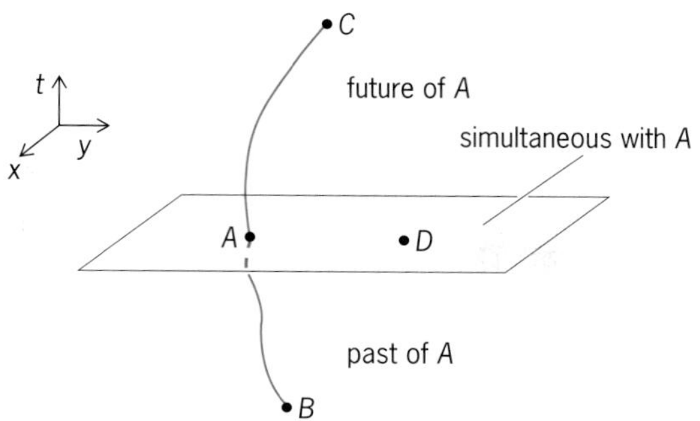

Fig. 1. Space-time diagram illustrating the causal relationships with respect to an event, *A*, in prerelativity physics. Event *B* lies to the past of *A*, event *C* lies to the future of *A*, and event *D* is simultaneous with *A*.

make a three-dimensional set, and are referred to as the events which are simultaneous with event A.

In both prerelativity physics and special relativity, an inertial observer is one who is not acted upon by any external forces. In both theories it is assumed that any inertial observer, $\mathcal{O}$ can build a rigid cartesian grid of meter sticks, all of which intersect each other at right angles. Observer $\mathcal{O}$ may then label the points on this cartesian grid by the coordinates (x, y, z) representing the distance of the point from $\mathcal{O}$ along the three orthogonal directions of the grid. A clock may then be placed at each grid point. In prerelativity physics, these clocks may be synchronized by requiring that they start simultaneously with each other. Any event in space-time may then be labeled by the four numbers t, x, y, z as follows: The numbers x, y, z assigned to the event are the spatial coordinates of the grid point at which the event occurred, whereas t is the time of the event as determined by the synchronized clock situated at that grid point. *See* FRAME OF REFERENCE.

It is of interest to compare the coordinate labelings given to events in space-time by two inertial observers, $\mathcal{O}$ and $\mathcal{O}'$, who are in relative motion. The relationship occurring in prerelativity physics is called a galilean transformation. In the simple case where $\mathcal{O}'$ moves with velocity v in the x direction with respect to $\mathcal{O}$, and these observers meet at the event A labeled by $(t, x, y, z) = (t', x', y', z') = (0, 0, 0, 0)$, with the axes of the grid of meter sticks carried by $\mathcal{O}'$ aligned (that is, not rotated) with respect to those of $\mathcal{O}$, the transformation is given by Eqs. (1). The galilean

$$t' = t \qquad (1a)$$

$$x' = x - vt \qquad (1b)$$

$$y' = y \qquad (1c)$$

$$z' = z \qquad (1d)$$

transformation displays in an explicit manner that the two inertial observers, $\mathcal{O}$ and $\mathcal{O}'$, agree upon the time labeling of events and, in particular, agree upon which events are simultaneous with a given event.

Causal structure in special relativity. In special relativity there is a different causal relationship between an arbitrary event A and other events in space-time (Fig. 2). As in prerelativity physics, there are many events, B, which lie to the past of A. There also are many events which lie to the future of A. However, there is now a much larger class of events which lie neither to the past nor to the future of A. These events are referred to as being spacelike-related to A.

The most striking feature of this causal structure (Fig. 2) is the absence of any three-dimensional surface of simultaneity. Indeed, the closest analog to the surface of simultaneity in prerelativity physics is the double-cone-shaped surface that marks the boundaries of the past and future of event A. This surface comprises the paths in space-time of all light rays which pass through event A, and for this reason it is referred to as the light cone of A. Thus, the statement that the events lying to the future of A are contained within the light cone of A is equivalent to the statement that a material body present at event A can never overtake a light ray emitted at event A. In special relativity, the light cone of an event A replaces the surface of simultaneity with event A as the absolute, observer-independent structure of space-time related to causality.

As in prerelativity physics, it is assumed in special relativity that an inertial observer, $\mathcal{O}$, can build a rigid grid of meter sticks, place clocks at the grid points, and label events in space-time by global inertial coordinates (t, x, y, z). The only difference from the procedure used for the construction of the similar coordinates in prerelativity physics is that the synchronization of clocks is now nontrivial, since the causal structure of space-time no longer defines an absolute notion of simultaneity. Nevertheless, any pair of clocks in $\mathcal{O}$'s system can be synchronized—and thereby all of $\mathcal{O}$'s clocks can be synchronized—by having an assistant stationed half-way between the clocks send a signal to the two clocks in a symmetrical manner. This synchronization of clocks allows $\mathcal{O}$ to define a notion of simultaneity; that is, $\mathcal{O}$ may declare events A_1 and A_2 to be simultaneous if time readings, t_1 and t_2, of the synchronized clocks at events A_1 and A_2 satisfy $t_1 = t_2$. However, events judged by $\mathcal{O}$ to be simultaneous will, in general, be judged by $\mathcal{O}'$ to be nonsimultaneous. *See* LORENTZ TRANSFORMATIONS.

The key assumptions of special relativity are encapsulated by the following two postulates.

Postulate 1: The laws of physics do not distinguish between inertial observers; in particular, no inertial observer can be said to be at rest in an absolute sense. Thus, if observer $\mathcal{O}$ writes down equations describing laws of physics obeyed by physically measurable quantities in her global inertial coordinate system (t, x, y, z), then the form of these equations must be identical when written down by observer $\mathcal{O}'$ in his global inertial coordinates (t', x', y', z').

Postulate 2: All inertial observers (independent of their relative motion) must always obtain the same value, c, when they measure the velocity of light in vacuum. In particular, the path of a light ray in space-time must be independent of the motion of the emitter of the light ray. Furthermore, no material body can have a velocity greater than c.

The precise relationship between the labeling of events in space-time by the global inertial coordinate systems of two inertial observers, $\mathcal{O}$ and $\mathcal{O}'$, in special relativity is given by the Lorentz transformation formulas. In the simple case where $\mathcal{O}'$

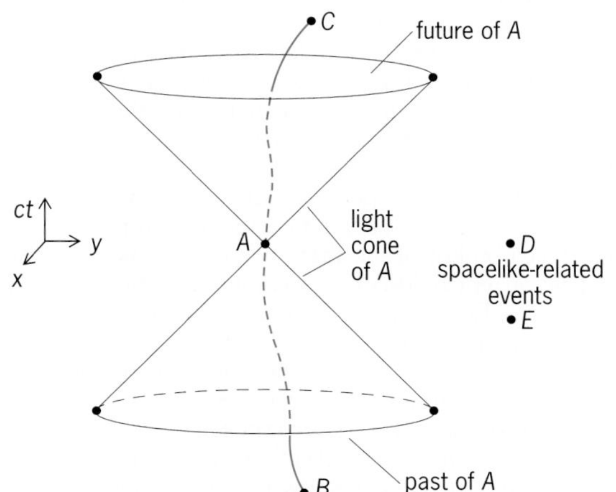

Fig. 2. Space-time diagram illustrating the causal relationships with respect to an event, *A*, in special relativity. Event *B* lies to the past of *A*, event *C* lies to the future of *A*, and events *D* and *E* are spacelike-related to *A*.

moves with velocity v in the x direction with respect to $\mathcal{O}$ and crosses $\mathcal{O}$'s world line at the event labeled by $(t, x, y, z) = (t', x', y', z') = (0, 0, 0, 0)$ with spatial axes aligned, the Lorentz transformation is given by Eqs. (2).

$$t' = \frac{t - \dfrac{xv}{c^2}}{\sqrt{1 - \dfrac{v^2}{c^2}}} \tag{2a}$$

$$x' = \frac{x - vt}{\sqrt{1 - \dfrac{v^2}{c^2}}} \tag{2b}$$

$$y' = y \tag{2c}$$

$$z' = z \tag{2d}$$

Space-time geometry. A key question both in prerelativity physics and in special relativity is what quantities, describing the space-time relationships between events, are observer independent. Such quantities having observer-independent status may be viewed as describing the fundamental, intrinsic structure of space-time.

It has already been seen that in special relativity the time interval, Δt, between two events is no longer observer independent. Furthermore, since different inertial observers disagree over simultaneity, the spatial interval between two simultaneous events is not even a well-defined concept, and cannot be observer independent. Remarkably, however, in special relativity, all inertial observers agree upon the value of the space-time interval, I, between any two events, where I is defined by Eq. (3). What is most

$$I = (\Delta x)^2 + (\Delta y)^2 + (\Delta z)^2 - c^2(\Delta t)^2 \tag{3}$$

remarkable about this formula for I is that it is very closely analogous to the formula for squared distance in euclidean geometry. The minus sign occurring in the last term in Eq. (3) is of considerable importance, since it distinguishes between the notions of time and space in special relativity. Nevertheless, this minus sign turns out not to be a serious obstacle to the mathematical development of the theory of lorentzian geometry based upon the space-time interval, I, in a manner which parallels closely the development of euclidean geometry. In particular, notions such as geodesics (straightest possible lines) can be introduced in lorentzian geometry in complete analogy with euclidean geometry. The Lorentz transformation between the two inertial observers is seen from this perspective to be the mathematical analog of a rotation between two cartesian frames in euclidean geometry.

The formulation of special relativity as a theory of the lorentzian geometry of space-time is of great importance for the further development of the theory, since it makes possible the generalization which describes gravitation. The lorentzian geometry defined by Eq. (3) is a flat geometry, wherein initially parallel geodesics remain parallel forever. The theory of general relativity accounts for the effects of gravitation by allowing the lorentzian geometry of space-time to be curved. *See* SPACE-TIME.

Consequences. The theory of special relativity makes many important predictions, the most striking of which concern properties of time. One such effect, known as time dilation, is predicted directly by the Lorentz transformation formula (2*a*). If observer $\mathcal{O}$ carries a clock, then the event at which $\mathcal{O}$'s clock reads time τ would be labeled by her as $(\tau, 0, 0, 0)$. According to Eq. (2*a*), the observer $\mathcal{O}'$ would label the event as $(t', x', 0, 0)$ where t' is given by Eq. (4). Thus, $\mathcal{O}'$ could say that a clock carried by $\mathcal{O}$

$$t' = \frac{\tau}{\sqrt{1 - \dfrac{v^2}{c^2}}} > \tau \tag{4}$$

slows down on account of $\mathcal{O}$'s motion relative to $\mathcal{O}'$. Similarly, $\mathcal{O}$ would find that a clock carried by $\mathcal{O}'$ slows down with respect to hers. This apparent disagreement between $\mathcal{O}$ and $\mathcal{O}'$ as to whose clock runs slower is resolved by noting that $\mathcal{O}$ and $\mathcal{O}'$ use different notions of simultaneity in comparing the readings of their clocks.

The decay of unstable elementary particles provides an important direct application of the time dilation effect. If a particle is observed to have a decay lifetime T when it is at rest, special relativity predicts that its observed lifetime will increase according to Eq. (4) when it is moving. Exactly such an increase is routinely observed in experiments using particle accelerators, where particle velocities can be made to be extremely close to c. *See* PARTICLE ACCELERATOR.

An even more striking prediction of special relativity is the clock paradox: Two identical clocks which start together at an event A, undergo different motions, and then rejoin at event B will, in general, register different total elapsed time in going from A to B. This effect is the lorentzian geometry analog of the mundane fact in euclidean geometry that different paths between two points can have different total lengths. *See* CLOCK PARADOX.

Robert H. Wald

General theory. General relativity is the geometric theory of gravitation developed by Einstein in the decade before 1915. It is a generalization of special relativity, and includes the classical gravitational theory of Newton as its limit for weak gravitational fields and low velocities. The most important applications of the theory are to cosmology and to the structure of neutron stars and black holes. Due to astronomical observations since the 1990s, general relativity is now in the forefront of fundamental physics research.

Need for a relativistic theory of gravity. One of the basic results of special relativity is that no physical effect can propagate with a velocity greater than that of light c; nature has a universal speed limit. However, classical gravitational theory describes the gravitational field of a body everywhere in space as a function of its instantaneous position, which is equivalent to the assumption that gravitational effects propagate with an infinite velocity. Thus, special relativity and classical gravitational theory are inconsistent, and a relativistic theory of gravity is necessary.

Principle of equivalence. Dating to Galileo, it is a fundamental question as to why bodies of different mass fall with the same acceleration in a gravitational field. This is now called the universality of free fall or the weak equivalence principle, and has been well tested experimentally. The universality of free fall follows from classical Newtonian gravitational theory since both the gravitational force on a body and its inertial resistance to acceleration are proportional to its mass. Thus the mass cancels out of the equations for the motion. However, this is not really an explanation but only an ad hoc description.

A deeper and more natural explanation occurred to Einstein. If an observer in the gravitational field of the Earth and another in an accelerating elevator or rocket in free space both drop a test body, they will both observe it to accelerate relative to the floor. According to classical theory, the Earth-based observer would attribute this to a gravitational force, while the elevator-based observer would attribute it to the accelerated floor overtaking the uniformly moving body. For both observers the motion is identical, and in particular the acceleration is independent of the mass of the test body. Einstein elevated this fact to a general principle, called the Einstein principle of equivalence. This states that on a local scale all physical effects of a gravitational field are indistinguishable from the physical effects of an accelerated reference frame. This profound principle is a cornerstone of the theory of general relativity. From the point of view of the principle of equivalence, it is obvious why the motion of a test body in a gravitational field is independent of its mass. But the principle applies not only to gravitational effects but to all physical phenomena, and thereby has profound consequences.

Tensors and Einstein's field equations. The observational fact that the motion of a body in a gravitational field is independent of its mass suggested to Einstein that gravity is fundamentally a geometric phenomenon, and that bodies should follow a special path determined only by geometry (see below). Because of this, gravity is naturally described by the mathematics of four-dimensional geometry. This geometry involves tensors, which are a simple generalization of vectors, and systems of tensor equations that are manifestly independent of the coordinate system. *See* TENSOR ANALYSIS.

The space-time of relativity contains one rank-2 tensor of particular importance, called the metric tensor $g_{\mu\nu}$, which is a generalization of the Lorentz metric of special relativity, introduced in Eq. (3). The metric tensor completely describes the geometry of space-time. Nearby points in space-time, also called events, which are separated by coordinate distances dx^μ, have an invariant physical separation whose square, called the line element, is defined by Eq. (5). (In this and subsequent equations, the in-

$$ds^2 = g_{\mu\nu} dx^\mu dx^\nu \tag{5}$$

dices that appear twice in an expression are to be summed over; this is the Einstein summation convention.) This quantity is a generalization of the space-time interval in special relativity.

In a tensor equation one tensor of a given type is set equal to another of the same type. The field equations of general relativity are tensor equations for the metric tensor. The Riemann tensor (or curvature tensor), $R^\alpha_{\mu\beta\nu}$ plays a central role in the geometric structure of space-time; if it is zero, the space-time is termed flat and has no gravitational field; if nonzero, the space-time is termed curved and a gravitational field is present. In terms of the Ricci tensor, which is the Riemann tensor summed over $\alpha = \beta$, the Einstein field equations for empty space-time are given in Eq. (6):

$$R^\alpha_{\mu\alpha\nu} \equiv R_{\mu\nu} = 0 \qquad \text{(empty space)} \tag{6}$$

The field equations are a set of 10 second-order partial differential equations, since $R^{\mu\nu}$ is a four-by-four symmetric tensor and has 10 independent components; the equations are to be solved for the metric tensor in a coordinate system that we are free to choose. Since the curvature of this space-time corresponds to the intrinsic presence of a gravitational field, the idea of a force field in classical gravitational theory is replaced in general relativity by the geometric concept of curved space-time. *See* DIFFERENTIAL EQUATION.

In a nonempty region of space-time the field equations (6) must be modified to include a tensor that represents the matter or energy which is the source of gravity, called the energy-momentum tensor $T_{\mu\nu}$. The modified equations are given in Eq. (7), where G is the gravitational constant, equal to 6.67×10^{-11} N · m^2 · kg^{-2}: [For empty space, $G_{\mu\nu} = 0$, which can be shown to imply $R_{\mu\nu} = 0$ also, so this is consistent with Eq. (6).]

$$\begin{aligned} G_{\mu\nu} &\equiv R_{\mu\nu} - {}^1\!/_2\, g_{\mu\nu} R^\alpha_\alpha \\ &= -\frac{8\pi G}{c^4} T_{\mu\nu} \qquad \text{(nonempty space)} \end{aligned} \tag{7}$$

The tensor $G_{\mu\nu}$, called the Einstein tensor, represents the space-time geometry, and the tensor $T_{\mu\nu}$ represents the mass or energy that produces gravity. These equations automatically imply the conservation of energy and momentum, which is an extremely important result. Moreover, in the limiting case when the fields are weak and velocities small compared to c, the equations reduce to the equations of classical Newtonian gravity.

Cosmological term. The field equations were given in the form of Eq. (7) by Einstein in 1916. However, they can be consistently generalized by the addition of another term on the left side, which he called the cosmological term, $\Lambda g_{\mu\nu}$. The more general equations are given in Eq. (8).

$$G_{\mu\nu} + \Lambda g_{\mu\nu} = -\frac{8\pi G}{c^4} T_{\mu\nu} \tag{8}$$

The constant Λ is called the cosmological constant. The extra term corresponds to a repulsive force between bodies that increases with distance.

Einstein introduced the cosmological term in 1917 in order to balance the usual attractive force of gravity and obtain static models of the universe, since it was then believed that the universe was static. When it was discovered by E. P. Hubble in 1929 that the universe is expanding, as evidenced by the Doppler shifts of distant galaxies, Einstein abandoned the cosmological term. However, beginning in 1998 measurements of the distance and recession velocity of very distant supernovae indicated that the cosmological term, or something that behaves just like it, is present and is causing the expansion of the universe to accelerate. The observed value of the constant is about (10^{10} light-years)$^{-2}$, so the cosmological term is only relevant on the cosmological scale. *See* COSMOLOGICAL CONSTANT; SUPERNOVA.

In the more modern view the cosmological term may be taken to the right side of the field equations and interpreted as the energy-momentum tensor of the vacuum; the energy density of the vacuum is $\rho_v = \Lambda c^4/8\pi G$. Just such a vacuum energy is predicted by quantum field theory, but its magnitude is, very roughly, an astounding 120 orders of magnitude greater than allowed by observation. This nonsensical result (with related questions) is called the cosmological constant problem.

The energy of the vacuum, in whatever guise, is now generally referred to as dark energy, and it comprises about 73% of the cosmic energy density; although it constitutes the dominant stuff of the universe, its nature is unknown, which is one of the biggest unresolved questions in physics today. *See* DARK ENERGY.

Motion of test bodies. Test bodies are objects with very small mass and size and no internal structure. In general relativity the path of a test body in space-time is a generalized straight line called a geodesic; a geodesic may be thought of as a curve that is "everywhere parallel to itself," or as an "extremal curve," either the shortest or longest curve between two given points. In relativity a geodesic is the longest curve between two points in space-time, with the distance being defined as the integral of ds along the curve. General relativity theory possesses an extraordinary property regarding geodesics: because the field equations are nonlinear (unlike those of classical theory), the motion of a test body in a gravitational field cannot be specified independently of the field equations, since the body has mass and contributes a small amount to the field. Indeed the field equations are so restrictive that the equations of motion may be derived from them and need not be treated as a postulate.

Schwarzschild's solution. A very important solution of the field equations was obtained by K. Schwarzschild in 1916, surprisingly soon after the inception of general relativity. This solution represents the field in empty space around a spherically symmetric body such as the Sun. It is the basis for a relativistic description of the solar system and most of the experimental tests of general relativity to date.

Gravitational redshift. Electromagnetic radiation of a given frequency emitted in a gravitational field will appear to an outside observer to have a lower frequency; that is, it will be redshifted. The formula for the redshift can be derived from the principle of equivalence, so redshift experiments do not test the field equations. The most accurate test of the redshift to date was performed using a hydrogen maser on a rocket. Comparison of the maser frequency with Earth-based masers gave a measured redshift in agreement with theory to about 1 part in 10^4. *See* GRAVITATIONAL REDSHIFT.

Perihelion shift of Mercury. The geodesic equations of motion can be solved for the orbit of a planet, considered as a test body in the Schwarzschild field of the Sun. As should be expected, the planetary orbit obtained is very similar to the ellipse of classical theory. However, according to relativity the ellipse rotates very slowly in the plane of the orbit so that the perihelion, the point of closest approach of the planet to the Sun, is at a slightly different angular position for each orbit. This shift is extremely small for all the planets of the solar system, but it is greatest for Mercury, whose perihelion advance is predicted to be 43 seconds of arc in a century. This agrees with the discrepancy between classical theory and observation that was well known for many years before the discovery of general relativity. *See* CELESTIAL MECHANICS.

Deflection of light. The principle of equivalence suggests an extraordinary phenomenon of gravity. Light crossing the Einstein elevator horizontally will appear to be deflected downward in a parabolic arc because of the upward acceleration of the elevator. The same phenomenon must therefore occur for light in the gravitational field of the Sun; that is, it must be deflected toward the Sun. A calculation of this deflection gives 1.75″ for the net deflection of starlight grazing the edge of the Sun. Modern measurements, made by tracking quasars with radio telescopes as they pass near or behind the Sun, find the deflection to be within 0.1% of the value predicted by general relativity.

In 1936 Einstein observed that if two stars were exactly lined up with the Earth, the more distant star would appear as a ring of light, distorted from its point appearance by the lens effect of the gravitational field of the nearer star. It was soon pointed out that a very similar phenomenon was much more likely to occur for entire galaxies instead of individual stars. Many examples of such gravitational lens systems have now been found. *See* GRAVITATIONAL LENS.

Radar time delay. In the curved space around the Sun the distance between points in space, for example between two planets, is not the same as it would be in flat space. In particular, the round-trip travel time of a radar signal sent between the Earth and another planet will be increased by the curvature effect when the Earth, the Sun, and the planet are approximately lined up. Using radar ranging to planets and spacecraft, the time delay has been found to agree with the predictions of general relativity to an accuracy of better than 0.1%

Precession of a gyroscope. Relativity theory predicts that a spinning gyroscope in orbit around the Earth will precess by a small amount. Most of this relativistic precession is due to the curvature of space-time described by the Schwarzschild metric; it is called the geodetic precession, and amounts to about 6.6 arcseconds per year for a low Earth orbit.

However, there is an additional smaller precession of great interest. For weak gravitational fields the Einstein field equations resemble Maxwell's equations of electromagnetism, and contain two types of fields: one resembles the electric field and is called the gravito-electric field, since it is due simply to the presence of matter; the other resembles the magnetic field and is called the gravito-magnetic field, since it is due to matter in motion, that is, a mass current. Effects of the gravito-magnetic field are generally referred to as Lense-Thirring effects (following the work of J. Lense and H. Thirring in 1918) or as frame dragging effects. For an appropriately oriented low polar orbit and gyroscope spin direction the Lense-Thirring precession is about 0.042 arcseconds per year, and is orthogonal to the geodetic precession, so the two precessions may be separately measured. The Gravity Probe B (GP-B) spacecraft was launched in April 2004 and has made a high-precision measurement of the geodetic precession and the first direct measurement of the Lense-Thirring precession—and thus the gravito-magnetic field of the Earth. *See* GYROSCOPE; MAXWELL'S EQUATIONS.

For applications of general relativity to neutron stars, the binary pulsar, gravitational radiation, black holes, and cosmology, and the relation of general relativity to quantum theory, *see* BLACK HOLE; COSMOLOGY; GRAVIATIONAL RADIATION; NEUTRON STAR; PULSAR; QUANTUM GRAVITATION.

Ronald J. Adler

Relaxation time of electrons The characteristic time for a distribution of electrons in a solid to approach or "relax" to equilibrium after a disturbance is removed. A familiar example is the property of electrical conductivity, in which an applied electric field generates an electron current which relaxes to an equilibrium zero current after the field is turned off. The conductivity of a material is directly proportional to this relaxation time; highly conductive materials have relatively long relaxation times. The closely related concept of a lifetime is the mean time that an electron will reside in a given quantum state before changing state as a result of collision with another particle or intrinsic excitation. This lifetime is related to equilibrium properties of the material, whereas the relaxation time relates to the thermal and electrical transport properties. The average distance that an electron travels before a collision is called the mean free path. Although typical collision times in metals are quite short (on the order of 10^{-14} s at room temperature), mean free paths range from about 100 atomic distances at room temperature to 10^6 atomic distances in pure metals near absolute zero temperature. Considering the very dense packing of atoms in a solid, these surprisingly long electron path lengths are analogous to the unlikely event that a rifle bullet might travel for miles through a dense forest without hitting a tree. The detailed explanation of the electron mean free path in metals is a major success of the modern theory of solids.

A relaxation time appears in the simplest expression for the transport property of electrical conductivity, which states that the electrical conductivity equals the product of the relaxation time, the density of conduction electrons, and the square of the electron charge, divided by the electron effective mass in the solid. *See* BAND THEORY OF SOLIDS; ELECTRICAL CONDUCTIVITY OF METALS.

The conduction process is a steady-state balance between the accelerating force of an electric field and the decelerating friction of electron collisions which occur on the time scale of the relaxation time. This process may be described in terms of the probability distribution function for the electrons, which depends on the electron momentum (proportional to the wave vector, **k**, which labels the quantum state of the electrons), the position, and the time. Viewed in **k**-space, the entire distribution will shift from an equilibrium state under the influence of a perturbation such as an electric field (see illustration). For example, in the ground state, the collection of occupied electron states in **k**-space is bounded by the Fermi surface centered at the origin, while in an electric field this region is shifted. Because of electron collisions with impurities, lattice imperfections, and vibrations (also called phonons), the displaced surface may be maintained in a steady state in an electric field. These collisions also restore the equilibrium distribution after the field is turned off, and the relaxation time is determined by the rate at which the shifted distribution returns to equilibrium. Specifically, the contribution of collisions to the rate of change of the shifted distribution after the field is turned off equals the difference between the shifted and equilibrium distributions divided by the relaxation time. This statement is referred to as the relaxation-time approximation, and is a simple way of expressing the role of collisions in the maintenance of thermodynamic equilibrium. The details of the various collision mechanisms are lumped into the parameter of the relaxation time. For example, in the case of mixed scattering by impurities and phonons, the inverse of the relaxation time can

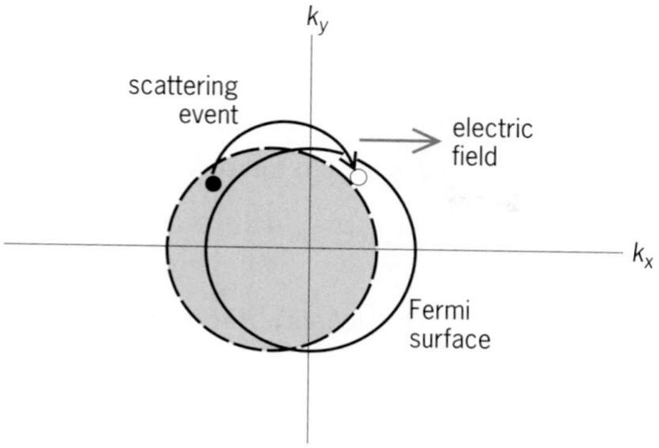

Effect of the electric field on electron distribution in a solid, viewed in k-space, where k is the electron wave vector. The shaded area indicates occupied states in distribution which result when the field is applied.

be determined from the sum of collision rates as the sum of the inverses of the electron-impurity and electron-phonon scattering times. *See* CRYSTAL DEFECTS; FERMI SURFACE; LATTICE VIBRATIONS; PHONON.

In pure metals at low temperatures, the long mean free path of conduction electrons results from their large velocity (on the order of 10^6 m/s near the Fermi surface) and relatively long relaxation time, on the order of 10^{-9} s. From a practical standpoint, this is what makes metals useful as electrical conductors even at room temperature, where a relaxation time on the order of 10^{-14} s and a mean free path (equal to the product of the velocity and the relaxation time) on the order of 10^{-8} m (or about 100 atomic distances) is typical.

As compared to poorly conducting solids or insulators, an excess of so-called free electrons in a metal contributes to the long mean free paths. From a quantum mechanical standpoint, the free-electron wave function readjusts in a perfectly periodic atomic lattice to avoid the atomic ion cores and spend most of the time in the spaces between. In the analogy of the rifle fired into a forest, the bullet will not travel far, but the sound of the gunshot can, because the sound waves bend around the trees in a constructive manner. In a perfectly periodic lattice, the electron waves scatter constructively from the atomic ion cores, resulting in screening of the cores and coherent transmission of the waves over large distances. Any disturbance to the lattice periodicity tends to destroy this wave phenomenon, resulting in lower transmission or energy loss. At room temperature, the conductivity is usually limited by scattering from lattice vibrations (phonons). At lower temperatures, vibrations are greatly reduced, but the conductivity is still limited by scattering from impurities and imperfections. *See* ELECTRICAL RESISTIVITY; FREE-ELECTRON THEORY OF METALS; QUANTUM MECHANICS.

In the superconductive state, observed in many metals and certain complex compounds at sufficiently low temperatures, impurities, imperfections, and lattice vibrations become completely ineffective in retarding current flow, leading to persistent electric currents even after the driving electric field is removed. This state of resistanceless conduction is described as an ordered quantum state of pairs of electrons resulting from lattice vibrations which deform the ion core potential in such a way as to provide an attractive electron-electron interaction. The relaxation time thus becomes infinite (as nearly as can be measured), a rare macroscopic manifestation of a quantum effect. *See* SUPERCONDUCTIVITY.

Gary L. Eesley

Relay An electromechanical, solid-state, or digital device operated by variations in the input that, in turn, operate or control other devices connected to the output. Relays are used in a wide variety of applications throughout industry, such as in telephone exchanges, digital computers, motor and sequencing controls, and automation systems. Highly sophisticated relays are utilized to protect electric power systems against trouble and power blackouts as well as to regulate and control the generation and distribution of power. In the home, relays are used in refrigerators, automatic washing machines and dishwashers, and heating and air-conditioning controls. Although relays are generally associated with electrical circuitry, there are many other types, such as pneumatic and hydraulic. Input may be electrical and output directly mechanical, or vice versa. *See* ELECTRIC PROTECTIVE DEVICES; SWITCHING SYSTEMS (COMMUNICATIONS).

Relays using discrete solid-state components, operational amplifiers, or microprocessors can provide more sophisticated designs. For industrial, commercial, or residential control, they are used particularly in applications where the relay and associated equipment are packaged together. For industrial or utility power systems they are used separately in conjunction with appropriate switching devices. *See* AMPLIFIER; MICROPROCESSOR; OPERATIONAL AMPLIFIER.

Classifying electrical-type relays by function and somewhat in the order of increasing complexity, there are (1) auxiliary relays, (2) monitoring relays, (3) regulating relays, (4) reclosing, synchronism check, and synchronizing relays, and (5) protective relays. The relay units may be electromechanical, solid-state, or based on digital microprocessors. Electromechanical relays are still used in many applications because of their relative simplicity, long life, and proven high reliability. However, relays based on digital technology are now preferred, particularly for protective relays, and are replacing the electromechanical or solid-state relays as existing applications or facilities are modified or new ones are installed.

J. Lewis Blackburn; Stanley H. Horowitz

Reliability, availability, and maintainability Reliability is the probability that an engineering system will perform its intended function satisfactorily (from the viewpoint of the customer) for its intended life under specified environmental and operating conditions. Maintainability is the probability that maintenance of the system will retain the system in, or restore it to, a specified condition within a given time period. Availability is the probability that the system is operating satisfactorily at any time, and it depends on the reliability and the maintainability. Hence the study of probability theory is essential for understanding the reliability, maintainability, and availability of the system. *See* PROBABILITY.

Reliability is basically a design parameter and must be incorporated into the system at the design stage. It is an inherent characteristic of the system, just as is capacity, power rating, or performance. A great deal of emphasis is placed on quality of products and services, and reliability is a time-oriented quality characteristic. There is a relationship between quality or customer satisfaction and measures of system effectiveness, including reliability and maintainability. Customers are concerned with the performance of the product over time.

To analyze and measure the reliability and maintainability characteristics of a system, there must be a mathematical model of the system that shows the functional relationships among all the components, the subsystems, and the overall system. The reliability of the system is a function of the reliabilities of its components. A system reliability model consists of some combination of a reliability block diagram or a cause-consequence chart, a definition of all equipment failure and repair distributions, and a statement of spare and repair strategies. All reliability analyses and optimizations are made on these conceptual mathematical models of the system.

Maintainability is a measure of the ease and rapidity with which a system or equipment can be restored to operational status following a failure. It is a characteristic of equipment design and installation, personnel availability in the required skill levels, adequacy of maintenance procedures and test equipment, and the physical environment under which maintenance is performed. Maintainability is expressed as the probability that an item will be retained in or restored to a specific condition within a given period of time, when the maintenance is performed in accordance with prescribed procedures and resources. Kailash C. Kapur

Reluctance A property of a magnetic circuit analogous to resistance in an electric circuit.

Every line of magnetic flux is a closed path. Whenever the flux is largely confined to a well-defined closed path, there is a magnetic circuit. That part of the flux that departs from the path is called flux leakage.

For any closed path of length l in a magnetic field H, the line integral of $H \cos \alpha \, dl$ around the path is the magnetomotive force (mmf) of the path, as in Eq. (1), where α is the angle between

$$\text{mmf} = \oint H \cos \alpha \, dl \qquad (1)$$

H and the path. If the path encloses N conductors, each with current I, Eq. (2) holds.

$$\text{mmf} = \oint H \cos \alpha \, dl = NI \qquad (2)$$

Consider the closely wound toroid shown in the illustration. For this arrangement of currents, the magnetic field is almost

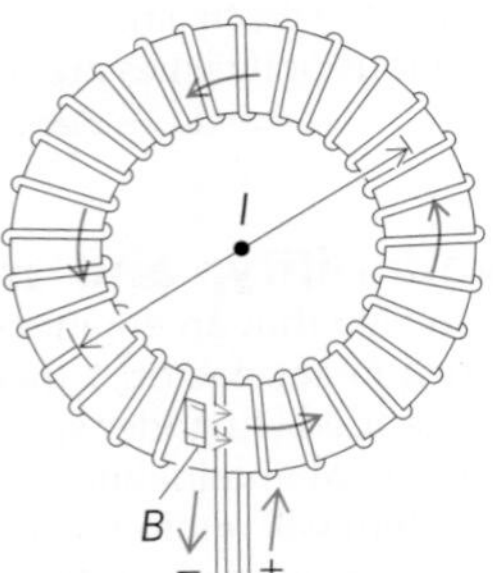

A toroidal coil.

entirely within the toroidal coil, and there the flux density or magnetic induction B is given by Eq. (3), where l is the mean

$$B = \mu \frac{NI}{l} \qquad (3)$$

circumference of the toroid and μ is the permeability. The flux Φ within the toroid of cross-sectional area A is given by either form of Eqs. (4), which is similar in form to the equation for the

$$\Phi = BA = \frac{\mu A}{l} NI$$
$$\Phi = \frac{NI}{l/\mu A} = \frac{\text{mmf}}{l/\mu A} = \frac{\text{mmf}}{\mathcal{R}} \qquad (4)$$

electric circuit, although nothing actually flows in the magnetic circuit. The factor $l/\mu A$ is called the reluctance $\mathcal{R}$ of the magnetic circuit. The reluctance is not constant because the permeability μ varies with changing flux density. Kenneth V. Manning

Reluctance motor An alternating-current motor with a stator winding like that of an induction motor, and a rotor that has projecting or salient poles of ferromagnetic material. When connected to an alternating-current source, the stator winding produces a rotating magnetic field, with a speed of $4\pi f/p$ radians per second ($120f/p$ revolutions per minute), where f is the frequency of the source and p the number of magnetic poles produced by the winding. When the rotor is running at the same speed as the stator field, its iron poles tend to align themselves with the poles of that field, producing torque. If a mechanical load is applied to the shaft of the motor, the rotor poles lag farther behind the stator-field poles, and increased torque is developed to match that of the mechanical load. This torque is given by the equation below, where ϕ is the flux per pole, determined largely

$$\tau = \frac{p}{2} \phi^2 \frac{dR}{d\delta}$$

by the applied voltage. The quantity $dR/d\delta$ is the rate of change of magnetic reluctance per pole with respect to δ, the angle of lag in mechanical radians. This quantity typically varies as sin $p\delta$. Here, $p\delta = 2\delta_e$, where δ_e is the lag angle in electrical radians. Therefore, at constant torque load the rotor runs in synchronism with the stator field, with the rotor poles lagging the field poles by a constant angle. *See* ELECTRICAL DEGREE.

This motor develops torque only at synchronous speed, and thus no starting torque is produced. For that reason, induction-motor rotor bars are usually built into the pole faces, and the motor starts as an induction motor. When the rotor speed approaches that of the magnetic field, the pole pieces lock in step with the magnetic poles of the field, and the rotor runs at synchronous speed.

Single-phase reluctance motors may be started by the methods used for single-phase induction motors, such as capacitor, split-phase, or shaded-pole starting. *See* ALTERNATING-CURRENT MOTOR; ELECTRIC ROTATING MACHINERY; INDUCTION MOTOR; MOTOR; SYNCHRONOUS MOTOR. George McPherson, Jr.

Remipedia A unique class of the Crustacea whose members, at first glance, resemble polychaete worms in overall body form. This impression comes from their multiple serial segments and lack of obvious body divisions. The lack of body organization suggests that remipeds are the most primitive crustaceans yet recognized. When compared with the other presumably primitive crustacean class, the Cephalocarida, remipeds are relatively large, with mature individuals ranging from 0.6 to 1.7 in. (15 to 42 mm) in body length. Although the trunk appendages are biramous and paddle-shaped in both classes, endites and epipods that are present in cephalocarids are lacking in remipeds. *See* CEPHALOCARIDA; CRUSTACEA.

As now constituted, the Remipedia comprise two orders, three families, and four genera; however, as more marine caves are explored additional species will be discovered. Patsy A. McLaughlin

Remote-control system A control system in which the issuing of the control command and its execution are separated by a relatively significant distance. The system normally includes a command device where the control command is entered, and an actuator that executes it. These are connected by a transmission medium that transmits the command, usually in a coded format.

The transmission medium may be a mechanical link, where the command is transmitted as force; a pneumatic or hydraulic line, where pressure represents the command; an electrical line with a voltage or current signal; or radio or infrared waves that are modulated according to the command. *See* MODULATION.

The simplest remote-control systems are limited to switching-type functions. These systems operate basically in an open loop, that is, without relying on feedback. Some typical examples are a ceiling lamp turned on and off by a light switch via an electrical wire; the on/off function of a television receiver with an infrared remote controller; and railway switches operated from a remote-control room.

The most characteristic remote-control systems involve feedback that is provided by the human operator. The person issuing the control command senses the result of the control action and guides the system accordingly. This kind of operation can be found, for example, in remote control of toy cars and airplanes by wire or radio, remote operation of large construction cranes, and cockpit control of an airplane's engines and control surfaces. *See* FLIGHT CONTROLS.

Teleoperation represents an important class within remote-control systems with human feedback. Teleoperators (or remote manipulators) act as extensions of the human hand. They are employed in situations where access is difficult or impossible or where the environment is hazardous for humans, such as in underwater and space operations, or in the presence of radiation, chemical, or biological contamination. *See* REMOTE MANIPULATORS.

Many automatic control systems may also be considered as remote controllers. This is the case whenever the sensing of the controlled variable and the automatic formation of the control command are removed from the actuator. A typical example is the heating and air-conditioning system of a building, where room thermostats operate remotely located furnaces, compressors, and fans. *See* CONTROL SYSTEMS. Janos Gertler

Remote manipulators Mechanical, electromechanical, or hydromechanical devices which enable a person to perform manual operations while separated from the site of the work. Remote manipulators are designed for situations where direct contact would be dangerous to the human (working with radioactive material), where direct human contact is ill-advised or impossible (certain medical procedures), and where human force-producing capabilities are absent (the disabled) or need to be amplified to complete some task (industrial assembly or construction).

Basic defining elements are common to almost all remote manipulators. An input device or control handle allows the operator to command the remote manipulator. The movement of the input device is received by a control station that translates the inputs into a form that can be transmitted over the distance separating the human and remote manipulator. This translation can be mechanical, using cables and linkages, or electrical/electromagnetic, using the movement of the input device to generate an electrical or electromagnetic signal that is easily transmitted to the remote manipulator. Since vision is an important cue that humans use in direct manipulation, visual feedback of a remote manipulator's actions typically must be provided. In some remote manipulation systems, tactile feedback to the human operator is provided; that is, forces proportional to those being exerted by the remote manipulator on the object are fed back to the human through the input device. Such force feedback is important in certain tasks where the possibility of damage to the manipulator or object can occur.

The growth in the number and variety of remote manipulator applications has been aided by enabling technologies such as digital computers, lightweight materials, and video communication links. Space applications include the space shuttle remote manipulator arm, which has been used to retrieve and launch large satellites. The arm is operated by astronauts in the shuttle orbiter cabin, and employs graphic displays of the forces and torques being applied by the manipulator arm to the satellite. Other applications include use in crewless underwater vehicles and surgical procedures. *See* SPACE SHUTTLE; UNDERWATER VEHICLE.

Aiding the disabled is an important use of remote manipulators. One example is devices that allow individuals with little or no control of their upper extremities to feed themselves. Such devices have been referred to as teletheses (alluding to the extrasensory perception of distant objects).

Devices that augment the strength of the human have been proposed for industrial applications. For example, a load-sharing manipulator has been envisioned in which human arm-manipulator coordination effectively allows the human to work with a "partner" that has considerably more strength. *See* CONTROL SYSTEMS; HUMAN-MACHINE SYSTEMS; ROBOTICS. Ronald A. Hess

Remote sensing The gathering and recording of information about terrain and ocean surfaces without actual contact with the object or area being investigated. Remote sensing uses the visual, infrared, and microwave portions of the electromagnetic spectrum. Remote sensing is generally conducted by means of remote sensors installed in aircraft and satellites.

Photography. Photography is probably the most useful remote sensing system. Much of the experience gained over the years from photographs of the terrain taken from aircraft is being drawn upon for use in space.

Multispectral photography isolates the reflected energy from a surface in a number of given wavelength bands and records each spectral band separately on film. This technique allows selection of the significant bandwidths in which a given area of terrain displays maximum tonal contrast and, hence, increases the effective spectral resolution of the system over conventional black-and-white or color systems. Because of its spectral selectivity capabilities, the multispectral approach provides a means of collecting a great amount of specific information.

Multispectral imagery. Multispectral scanning systems record the spectral reflectance by photoelectric means (rather than by photochemical means as in multispectral photography) simultaneously in several individual wavelengths within the visual and near-infrared portions of the electromagnetic spectrum.

In satellite applications, optical energy is sensed by an array of detectors simultaneously in four spectral bands from 0.47 to 1.1 micrometers. As the optical sensors for the various frequency bands sweep across the underlying terrain in a plane perpendicular to the flight direction of the satellite, they record energy from individual areas on the ground. The smallest individual area distinguished by the scanner is called a picture element or pixel, and a separate spectral reflectance is recorded in analog or digital form for each pixel. The spectral reflectance values for each pixel can be transmitted electronically to ground receiving stations in near-real time, or stored on magnetic tape in the satellite until it is over a receiving station. When the signal intensities are received on the ground, they can be reconstructed almost instantaneously into the virtual equivalent of conventional aerial photographs.

Infrared. Thermal infrared radiation is mapped by means of infrared scanners similar to multispectral scanners, but in this case radiated energy is recorded generally in the 8–14-μm portion of the electromagnetic spectrum. The imagery provided by an infrared scanning system gives information that is not available from ordinary photography or from multispectral scanners operating in the visual portion of the electromagnetic spectrum. *See* EMISSIVITY; INFRARED RADIATION.

In the past, thermal infrared images were generally recorded on photographic film. Videotape records are replacing film as the primary recording medium and permit better imagery to be produced and greater versatility in interpretation of data.

Thermal infrared mapping (thermography) from satellite altitudes is proving to be useful for a number of purposes, one of which is the mapping of thermal currents in the ocean. Thermal infrared mapping from aircraft and satellite altitudes has many other uses also, including the mapping of volcanic activity and geothermal sites, location of groundwater discharge into surface and marine waters, and regional pollution monitoring.

Microwave radar. This type of remote sensing utilizes both active and passive sensors. The active sensors such as radar supply their own illumination and record the reflected energy. The passive microwave sensors record the natural radiation. A variety of sensor types are involved. These include imaging radars, radar scatterometers and altimeters, and over-the-horizon radars using large ground-based antenna arrays, as well as passive microwave radiometers and imagers. One of the most significant advantages of these instruments is their all-weather capability, both day and night. *See* MICROWAVE; RADAR.

High-frequency (hf) radar. Such radars utilize frequencies in the 3–30-MHz portion of the electromagnetic spectrum (median wavelength of about 20 m) and are thus not within the microwave part of the spectrum. The energy is transmitted by ground-based antennas in either a sky-wave or surface-wave mode. In the sky-wave mode, the energy is refracted by the various ionospheric layers back down to the Earth's surface some 480–1800 mi or 800–3000 km (on a single-hop basis) away from the hf radar antenna site. The incident waves are reflected from such surface features as sea waves. Peter C. Badgley

Renewable resources Agricultural materials used as feedstocks for industrial processes. For many centuries agricultural products were the main sources of raw material for the manufacturing of soap, paint, ink, lubricants, grease, paper, cloth, drugs, and a host of other nonfood products. During the early 1900s, the advances in organic synthesis in western Europe and the United States led to the use of coal as an alternative resource; in the 1940s, oil and natural gas were added as starting materials as a result of great advances in catalysis and polymer sciences. Since then the petrochemical industry has grown rapidly as the result of the abundance and low price of the starting materials as well as the development of new products. However, with the rapidly increasing economies of the nations of the world, these developments did not ever result in reduction in the utilization of agricultural products as industrial materials.

Animal fats, marine and vegetable oils, and their fatty acid derivatives have always played a major role in the manufacturing of many industrial products. Some of these commodities are produced solely for industrial end uses; examples are linseed, tung, castor (not counting minor amounts used for medicinal purposes), and sperm whale oils. Others, such as tallow and soybean oil, are used for both edible and industrial products. *See* DETERGENT; FAT AND OIL; LUBRICANT; SOAP.

Starch, cellulosics, and gums also have been used for many centuries as industrial materials, whereas sugar crops, such as sugarcane and sugarbeet, have mainly satisfied world food requirements. *See* CELLULOSE; GUM; STARCH.

Natural rubber and turpentine are excellent examples of plant-derived hydrocarbons. The development of synthetic rubbers during and after World War II has never threatened the demand for natural rubber; there is generally a world shortage. Turpentine is a product of the wood and paper pulp industry and is used as a solvent and thinner in paints and varnishes. *See* RUBBER.

The threat that industrial nations might be separated from part or all of their traditional sources of raw materials through political and economic upheavals or natural calamities has resulted in a renewed effort to develop additional crops for local agriculture. In the United States, research has provided a number of candidate species that either are now in commercial development or are ready for the time when circumstances warrant such development. Examples are jojoba (liquid wax ester to replace sperm whale oil), guayule (alternate source of natural rubber), kenaf (paper fiber with annual yields much higher than available from trees), and crambe and meadow-foam (long-chain fatty acids, since erucic acid is no longer available from rapeseed oil). There is also active research involving *Cuphea* species (alternate source of lauric and other medium-chain fatty acids, to augment coconut oil), *Vernonia* (source of epoxy oil), and several other promising plants. For example, the Chinese tallow tree has the potential of producing 2.2 tons per acre (5 metric tons per hectare) of seed oil that could be used for manufacturing fuel and other chemicals. L. H. Princen

Renner-Teller effect The splitting, into two, of the potential function along the bending coordinate in degenerate electronic states of linear triatomic or polyatomic molecules. Most of the areas and methods of molecular physics and spectroscopy assume the validity of the Born-Oppenheimer approximation. The nuclei generally move much more slowly than the electrons, the frequencies associated with electronic transitions are much higher than vibrational frequencies, and one can consider separately the three types of molecular motion: electronic, vibrational, and rotational. These statements are no longer necessarily valid for electronic states which are degenerate or at least close to degeneracy, and the Born-Oppenheimer approximation breaks down.

Degenerate electronic states usually occur in molecules having a high degree of symmetry. The symmetric equilibrium geometry which causes the electronic degeneracy is, in general, lowered in the course of molecular vibrations, and this may lead to splitting of the potential. The molecular potential is usually expressed in terms of a polynomial expansion in displacements r, and, in nonlinear molecules, the linear terms may lead to coupling of the electronic and vibrational degrees of freedom. The resulting breakdown of the Born-Oppenheimer approximation is in this case known as the Jahn-Teller effect. In linear molecules the symmetry is lowered during bending vibrations. In the bending potential the linear (and other odd) terms are zero by symmetry. The first nonvanishing terms which can couple the degenerate electronic states are quadratic in the bending coordinate. The results of this coupling in linear molecules are referred to as the Renner-Teller effect, or simply the Renner effect. *See* JAHN-TELLER EFFECT; MOLECULAR STRUCTURE AND SPECTRA.

Vladimir E. Bondybey; Terry A. Miller

Rennin The common name for chymosin, a proteolytic enzyme that is used to coagulate milk in cheese making. Rennin participates in the cheese-ripening process through its proteolytic activity.

The traditional source of rennin is the fourth stomach (abomasum) of calves that have been fed only milk. The stomachs are dried or salted, cut into small pieces, and soaked in 10% salt brine to extract enzyme components from the stomach lining. Rennin also is produced commercially by using genetically engineered microorganisms.

Approximately 6% of the rennin used to coagulate milk is retained in active form in cheese curd. During cheese ripening, rennin modifies the curd protein structure through its proteolytic action on α-casein, leading to textural changes described as a loss of curdiness. Casein peptides resulting from rennin action become precursors for flavor compounds in some cheeses such as Cheddar. *See* CASEIN; CHEESE; ENZYME; MILK. Paul Kindstedt

Renormalization A program in quantum field theory consisting of a set of rules for calculating S-matrix amplitudes which are free of ultraviolet (or short-distance) divergences, order by order in perturbative calculations in an expansion with respect to coupling constants. *See* SCATTERING MATRIX.

So far the only field theories known to be renormalizable in four dimensions are those which include spin-0, spin-$^1/_2$, and spin-1 fields such that no term in the lagrangian exceeds operator dimension 4. The operator dimension of any term is calculated by assigning dimension 1 to bosons and derivatives ∂_μ, and dimension $^3/_2$ to fermions. Spin-1 fields are allowed only if they correspond to the massless gauge potentials of a locally gauge-invariant Yang-Mills-type theory associated with any compact Lie group. The gauge invariance can remain exact or can be allowed to break via spontaneous breakdown without spoiling the renormalizability of the theory. In the latter case the spin-1 field develops a mass. The successful quantum chromodynamics theory describing the strong forces and the SU(2) $\times$ U(1) Weinberg-Salam-Glashow gauge model of unified electroweak particle interactions are such renormalizable gauge models containing spin 0, $^1/_2$, and 1 fields. *See* ELECTROWEAK INTERACTION;

Reptilia

(*a*) American crocodile (*Crocodylus acutus*) from Ding Darling National Wildlife Refuge, Florida (*photo by Gerald and Buff Corsi; © California Academy of Sciences*).

(*b*) Five-lined skink (*Eumeces fasciatus*) (*photo by George W. Robinson; © California Academy of Sciences*).

(*c*) Desert tortoise (*Gopherus agassizii*) (*photo by Joshua Tree National Park/NPS*).

(*d*) Collared lizard (*Crotaphytus collaris*) (*photo by Petrified Forest National Park/NPS*).

(*e*) Short-horned lizard (*Phrynosoma hernandesi*) (*photo by Lake Mead National Recreation Area/NPS*).

(*f*) Ground snake (*Sonora semiannulata*) (*photo by NPS/Tonto National Monument, Arizona*).

(*g*) Longnose snake (*Rhinocheilus lecontei*) (*photo by NPS/Tonto National Monument, Arizona*).

FUNDAMENTAL INTERACTIONS; GAUGE THEORY; LIE GROUP; QUANTUM CHROMODYNAMICS; QUANTUM ELECTRODYNAMICS; WEAK NUCLEAR INTERACTIONS. Itzhak Bars

Effective field theory is a general and powerful method for analyzing quantum field theories over a wide range of length scales. Together with a closely related idea, the Wilson renormalization group, it places renormalization theory on a more general, physical, and rigorous basis. This method is most naturally developed in the Feynman path integral formulation of quantum field theory, where amplitudes are given by an integral over all histories. Each history is weighted by a phase equal to the classical action divided by Planck's constant. *See* ACTION; FEYNMAN INTEGRAL.

Joseph Polchinski

Reproductive behavior Behavior related to the production of offspring; it includes such patterns as the establishment of mating systems, courtship, sexual behavior, parturition, and the care of young. Successful reproductive efforts require the establishment of a situation favorable for reproduction, often require behavior leading to the union of male and female gametes, and often require behavior that facilitates or ensures the survival and development of the young; the mere union of gametes is not generally sufficient for successful reproduction. For each species, there is a complex set of behavioral adaptations that coordinate the timing and patterning of reproductive activity. Typically, this entails integration of both overt behavioral and internal physiological events in both male and female, all of which are intricately enmeshed in manners adapted to the environment in which the animals live. The behavioral patterns related to reproduction tend to be relatively stereotyped within a species, but diverse among different species—especially distantly related species. The end products of cycles of reproductive activity are viable, fertile offspring which, in turn, will reproduce and thus perpetuate the species.

The relationships between individual males and females and the degree of exclusivity in mating are part of the mating system of a species or population. There are three basic mating system types: monogamy, polygamy, and promiscuity. In monogamy the reproductive unit is generally a single male and a single female, the partners copulate only with each other, there may be shared parental care, and there is some kind of prolonged pair bond. In polygamous mating systems, there is again a prolonged association, but more than two individuals are involved in the relationship. In the promiscuous mating system, there is no prolonged bond formed, and there are multiple matings by members of at least one of the sexes.

Territoriality or dominance occurs in many kinds of mating systems. A territory is an area that is defended against conspecific animals (those of the same species). It may be occupied by a single individual, a bonded male-female pair, or a larger group. The resident of a territory generally has privileged access to the resources on that territory. Where the territory is relatively large, as in many diurnal songbirds, it may include sufficient resources to support a bonded pair and their offspring. By contrast, in many colonially nesting marine birds the territories may encompass little more than a nest site, while food and other resources are collected at a distance. In a special form of territoriality, a lek, males of some species, such as sage grouse, defend small territories that are used only for breeding. *See* TERRITORIALITY.

Whereas in a territorial system the outcome of a contest for resources is generally predictable given only the location of the encounter, in a dominance relationship an individual wins regardless of location. In troops of various species of primates, for example, there may be a single dominant male and a hierarchy of males ranking below him. There are many varieties of dominance relationships, with male hierarchies, female hierarchies, mixed-sex hierarchies, and triangular relationships that are departures from linearity. Dominance-related contests may occur seasonally, generally peaking in intensity during the breeding season, and can be of great importance in determining which individuals reproduce.

Mate choice and then courtship are both essential to successful reproduction in animals. The importance of female choice remains controversial. Ultimately, females can benefit by mating with males that are exceptional either in their ability to accrue resources or in the possession of "good genes," or both. Males forming prolonged pair bonds invest much, and it is not surprising that they may exercise mate choice as well. In a variety of species of invertebrates, for example, males prefer large females to small ones.

Courtship entails a sequence of behavioral patterns that eventually may lead to the completed mating. Patterns of courtship are quite diverse among different species but generally entail reciprocal signaling between male and female. Mate choice is an important function of courtship. Many bouts of courtship break off without going to completed matings, often as a result of choice on the part of one or both partners. Another function of courtship relates to synchronization. The gametes must be shed at a time when sperm are viable, eggs are ripe, male and female are in the appropriate state of readiness, and the environment is supportive of reproductive effort. The progressive interactive sequence of the courtship episode allows for coordinated events to occur at times appropriate for successful reproduction.

Prior to birth or hatching of offspring, parents may engage in behavioral patterns that will aid the young when they arrive. This may entail preparation of a burrow or nest, provision of stored food, or acquisition of other resources. In some species the parent's aid ends with such preparations, but in others parental care may be extensive and prolonged. Parental care, especially maternal care, is highly developed in all species of mammals. By definition, the females of all mammalian species possess mammary glands for the nourishment of young. It is probably for this physiological reason that role reversal is less common among mammals than in other taxa.

Parental investment entails any investment by the parent that increases the ability of the young to survive and reproduce at some cost to the parent. Much parental investment, like milk, cannot be shared; that which is given to one offspring cannot be given to another. Other kinds of parental investment, like defense against predators, is shareable. In addition, the parent making any investment is prevented from engaging in other activities, such as searching for his or her own food or seeking additional mates.

Because they contribute larger gametes and often engage in more extensive parental behavior, the females of most species display a higher level of parental investment than males. Members of the sex investing more (typically females) thus become a limiting resource for the sex investing less (typically males). It is generally agreed that this gives rise to the disparity between female and male reproductive strategies—with males more often competing for access to females and females more choosy than males. *See* ANIMAL COMMUNICATION; ANIMAL REPRODUCTION; BEHAVIORAL ECOLOGY; ETHOLOGY. Donald A. Dewsbury

Reproductive system The structures concerned with the production of sex cells (gametes) and perpetuation of the species. The reproductive function constitutes the only vertebrate physiological function that necessitates the existence of two morphologically different kinds of individuals in each animal species, the males and the females (sexual dimorphism). The purpose of the reproductive function is fertilization, that is, the fusion of a male and a female sex cell produced by two distinct individuals.

Anatomy. Egg cells, or ova, and sperm cells, or spermatozoa, are formed in the primary reproductive organs, which are collectively known as gonads. Those of the male are called testes; those of the female are ovaries. The gonads are paired structures, although in some forms what appears to be an unpaired gonad is the result either of fusion of paired structures or of unilateral degeneration.

The reproductive elements formed in the gonads must be transported to the outside of the body. In most vertebrates, ducts are utilized for this purpose. These ducts, together with the structures that serve to bring the gametes of both sexes together, are known as accessory sex organs. The structures used to transport the reproductive cells in the male are known as deferent ducts and those of the female as oviducts. In a few forms no ducts are present in either sex, and eggs and sperm escape from the body cavity through genital or abdominal pores.

Oviducts, except in teleosts and a few other fishes, are modifications of Möllerian ducts formed during early embryonic development. In all mammals, each differentiates into an anterior, nondistensible Fallopian tube and a posterior, expanded uterus. In all mammals except monotremes the uterus leads to a terminal vagina which serves for the reception of the penis of the male during copulation. The lower part, or neck, of the uterus is usually telescoped into the vagina to a slight degree. This portion is referred to as the cervix.

In most vertebrates the reproductive ducts in both sexes open posteriorly into the cloaca. In some, modifications of the cloacal region occur and the ducts open separately to the outside or, in the male, join the excretory ducts to emerge by a common orifice. *See* ANIMAL REPRODUCTION; COPULATORY ORGAN; OVARY; PENIS; TESTIS.

Charles K. Weichert

Physiology. The physiological process by which a living being gives rise to another of its kind is considered one of the outstanding characteristics of plants and animals. It is one of the two great drives of all animals: self-preservation and racial perpetuation.

Estrous and menstrual cycles. The cyclic changes of reproductive activities in mammalian females are known as estrous or menstrual cycles. Most mammalian females accept males only at estrus (heat). Estrus in mammals can occur several times in one breeding season; the mare, ewe, and rat come to estrus every 21, 16, and 5 days respectively if breeding does not take place. This condition is called polyestrus. The bitch is monestrous; she has only one heat, or estrus, to the breeding season and if not served then, she does not come into heat again for a prolonged interval, 4–6 months according to different breeds. In monestrous and seasonally polyestrous species the period of sexual quiescence between seasons is called anestrus. *See* ESTRUS.

The reproductive cycle of the female in the primate and human is well marked by menstruation, the period of vaginal blood flow. Menstruation does not correspond to estrus but occurs between the periods of ovulation at the time the corpus luteum declines precipitously. *See* MENSTRUATION.

Mating. Mating, also called copulation or coitus, is the synchronized bodily activity of the two sexes which enables them to deposit their gametes in close contact. It is essential for successful fertilization because sperm and ovum have a very limited life span.

The logistics of sperm transport to the site of fertilization in the oviduct present many interesting features in mammals, but it is important to distinguish between passive transport of sperm cells in the female genital tract, and sperm migration, which clearly attributes significance to the intrinsic motility of the cell. Viable spermatozoa are actively motile, and although myometrial contractions play a major role in sperm transport through the uterus, progressive motility does contribute to migration into and within the oviducts. Even though a specific attractant substance for spermatozoa has not yet been demonstrated to be released from mammalian eggs or their investments, some form of chemotaxis may contribute to the final phase of sperm transport and orientation toward the egg surface.

Although in most mammalian species the oocyte is shed from the Graafian follicle in a condition suitable for fertilization, ejaculated spermatozoa must undergo some form of physiological change in the female reproductive tract before they can penetrate the egg membranes. The interval required for this change varies according to species, and the process is referred to as capacitation. The precise changes that constitute capacitation remain unknown, although there is strong evidence that they are—at least in part—membrane-associated phenomena, particularly in the region of the sperm head, that permit release of the lytic acrosomal enzymes with which the spermatozoon gains access to the vitelline surface of the egg.

Fertilization takes place in the oviducts of mammals and the fertilized eggs or embryos do not descend to the uterus for some 3 to 4 days in most species. During this interval, the embryo undergoes a series of mitotic divisions until it comprises a sphere of 8 or 16 cells and is termed a morula. Formation of a blastocyst occurs when the cells of the morula rearrange themselves around a central, fluid-filled cavity, the blastocoele. As the blactocyst develops within the uterine environment, it sheds its protective coat and undergoes further differentiation before developing an intimate association with the endometrium, which represents the commencement of implantation or nidation.

Association of the embryo with the uterine epithelium, either by superficial attachment or specific embedding in or beneath the endometrium, leads in due course to the formation of a placenta and complete dependence of the differentiating embryo upon metabolic support from the mother. Implantation and placentation exhibit a variety of forms, but in all instances the hormonal status of the mother is of great importance in determining whether or not implantation can proceed. *See* PREGNANCY.

Endocrine function. The endocrine glands secrete certain substances (hormones) which are necessary for growth, metabolism, reproduction, response to stress, and various other physiological processes. The endocrine glands most concerned with the process of reproduction are the pituitary and the gonads.

The formation of gametes (spermatogenesis and oogenesis) is controlled by anterior pituitary hormones. The differentiation of male and female reproductive tracts is influenced, and mating behavior and estrous cycles are controlled, by male or female hormones. The occurrence of the breeding season is mainly dependent upon the activity of the anterior lobe of the pituitary, which is influenced through the nervous system by external factors, such as light and temperature. The transportation of ova from the ovary to the Fallopian tube and their subsequent transportation, development, and implantation in the uterus are controlled by a balanced ratio between estrogen and progesterone. Furthermore, it is known that estrogens, androgens, and progesterone can all have the effect of inhibiting the production or the secretion, or both, of gonadotrophic hormones, permitting the cyclic changes of reproductive activity among different animals. *See* ESTROGEN; PITUITARY GLAND; PROGESTERONE.

Mammary glands are essential for the nursing of young. Their growth, differentiation, and secretion of milk, and in fact the whole process of lactation, are controlled by pituitary hormones as well as by estrogen and progesterone. Other glands and physiological activities also influence lactation, although this is largely via the trophic support of other pituitary hormones.

M. C. Chang; Michael J. K. Harper; R. H. F. Hunter

Reproductive system disorders

Reproductive system disorders Those disorders which involve the structures of the human female and male reproductive systems.

Female system. Disorders of the female reproductive system may involve the ovaries, Fallopian tubes, uterus, cervix, vagina, or vulva.

Failure of the ovaries to form normally results in short stature, sterility, and lack of development of female secondary sex characteristics, such as breast growth, fat deposition in buttocks and thighs and mons pubis, and female escutcheon. Destruction of the ovaries after puberty results in loss of fertility, cessation of menses, loss of secondary sex characteristics, and osteoporosis. *See* OSTEOPOROSIS.

Neoplastic enlargement of the ovary can be cystic or solid, benign or malignant. Malignant ovarian tumors are commonly asymptomatic in their early stages, and may be quite widely spread in the pelvic and abdominal cavity before they are discovered. *See* OVARIAN DISORDERS; OVARY.

Endometriosis, which is a condition involving the presence of ectopic endometrial tissue, can affect the ovaries. Endometriosis can be found as large "chocolate" cysts of the ovary, called endometriomas, or small "blue domed" cysts; both are filled with old, dark brown blood. Tiny implants of endometriosis commonly called powder burns also occur on the surface of the ovaries, on other pelvic peritoneum, and over the Fallopian tubes and uterus as well. Endometriomas of the ovary may be painful, or may rupture and cause diffuse pelvic pain, while the smaller endometrial implants may cause severe pain with menstrual periods, generalized pelvic pain, pain with intercourse, and infertility.

Inflammation is the most common disorder of the Fallopian tubes, and if it is repeated or severe, destruction of the tubal lining with closure of the outer ends of the tubes can occur. This inflammation can be caused by several organisms. Sterility commonly results because the tube is permanently closed to passage of egg and sperm. The second most common problem involving the tube is pregnancy. The egg is fertilized in the outer portion of the tube and descends to implant in the uterus, but in some cases the passage is delayed and the conceptus attaches to the wall of the tube. The tube has a small lumen and thin wall, and the growing pregnancy quickly enlarges and grows through the tube, leading usually to rupture and hemorrhage into the peritoneal cavity. *See* PREGNANCY DISORDERS.

Developmental abnormalities can occur during formation of the uterus. The Fallopian tubes might not join at all, or might join partially from the cervical end upward. Septae or walls in the vagina and uterus can also occur. These abnormalities are more common in females who are exposed to diethylstilbesterol (DES) during the mother'spregnancy. The muscle (myometrium) and lining (endometrium) of the uterus are susceptible to various problems, including tumors, infections, and hormonal derangements. Benign tumors of the myometrium (fibroids) are a common disorder, producing irregular enlargement of the uterus and sometimes causing pain, obstruction of the urinary tract, and heavy vaginal bleeding. The uterus can be one site of a significant infection which produces fever and pain and usually involves other organs, such as ovaries and tubes. Hormonal abnormalities resulting in anovul can lead to overgrowth and irregular shedding of the endometrium. Pregnancy usually proceeds uneventfully, but certain accidents of pregnancy, such as threatened, incomplete, or missed abortion, can produce irregular bleeding and some uterine discomfort. *See* PREGNANCY; UTERINE DISORDERS; UTERUS.

Infection of the endocervical glands with gonococcus or chlamydia trachomatis agent can occur. This may be asymptomatic, except for producing a mucopurulent discharge, or it may cause pain when the cervix is manipulated, particularly during intercourse. The infection can ascend from this area into the internal genital organs and adjacent structures. The cervix can be affected by malignant tumors which can be adenocarcinomas, or tumors of the glandular cells, or more commonly squamous tumors. Cervical cancer is associated with several factors, including early age at first coitus, multiple sex partners, especially at an early age, smoking, and infection with certain subtypes of human papilloma virus. There is no single factor that "causes" this cancer, but rather a combination of these factors is critical. Papanicolaou smears are used to detect early changes suggestive of this cancer. Further evaluation is by low-power magnification, called colposcopy, and biopsy. *See* GONORRHEA.

Developmental abnormalities of the vagina may include imperforate hymen, septae, both vertical and horizontal, and complete failure of development. Inflammation of the lining of the vagina can be due to several common organisms. Specific chemical treatment is available for each of these entities. Postmenopausal women can experience atrophy or shrinkage of the vagina and mucosa secondary to estrogen deprivation. This produces itching, bleeding, and pain with intercourse. Estrogen therapy can relieve these symptoms. The vaginal support can be weakened through childbirth, or may simply be naturally poor, allowing the bladder and rectum to bulge inward, and the cervix and uterus to protrude from the introitus. This condition can cause a variety of symptoms, including involuntary loss of urine with a cough or laugh, inability to move the bowels without mechanically pushing the stool out, and a sensation of pelvic heaviness. Intravaginal devices called pessaries may improve support, but surgical correction may be necessary. Vaginal cancer can occur, but it is rare, and cervical cancer can also involve the vagina. Treatment is usually surgical. Vaginal trauma can lead to significant bleeding. Penetrating straddle injuries can occur, and trauma with intercourse or childbirth can produce lacerations. Surgical repair is usually undertaken. Vaginismus, or painful spasm of the muscular sidewalls of the vagina, can occur, making intercourse painful. *See* VAGINAL DISORDERS.

Infection of the female external genitalia can be diffuse or localized, involving bacteria, viruses, or fungal agents. Inflammation can be caused by allergic reactions to soap, powders, semen, lubricant, or even clothing.

Male system. The principal organs of the male reproductive system are the testicles, epididymis, vas deferens and seminal vesicles, prostate gland, urethra, and penis.

During fetal life the testicles form in the abdominal cavity near the kidneys and migrate down into the scrotum. Failure of descent permanently damages the sperm-producing cells, but allows the interstitial cells which produce hormones to survive. Other causes of male infertility are organic problems which block passage of seminal fluid or interfere with sperm production. The testes are sites of malignant tumors which carry a high mortality rate. Undescended testicles are more susceptible to malignant transformation, and for that reason should be removed when discovered, and appropriate hormonal replacement instituted. The membranes covering the testicle can become filled with fluid, a condition known as hydrocele. *See* TESTIS.

The major disorder of the epididymis is infection and inflammation, which can lead to scarring and permanent blockage of the ducts.

The vas deferens and seminal vesicles are rarely afflicted by disease, although the network of veins surrounding the vas deferens can become engorged and tortuous, and is called a variocoele.

Inflammation and development of small calculi are relatively minor ailments of the prostate. Benign overgrowth of this gland is the most common and troublesome complaint. This enlargement results in constriction of the urethra, which obstructs urinary outflow and leads to an increasing residual of urine in the bladder. Cancer of the prostate is another relatively frequent problem in older males. This tumor spreads primarily to bones, where it is markedly painful. *See* PROSTATE GLAND DISORDERS.

Inflammation is the chief disorder of the urethra, causing dysuria and a discharge.

Inflammation (balanitis) or narrowing (phimosis) of the foreskin, which covers the glans penis, can occur and may interfere with urination. Viral infections can produce warty growths (condyloma acuminata) or painful ulcers (herpetic lesions) of the glans or shaft of the penis, and a syphilitic infection can result in the firm, painless ulcer known as a chancre. *See* HERPES; SYPHILIS.

Disorders of erectile capability range from tumescence unaccompanied by sexual desire (Peyronie's disease, priapism) to impotence, both psychologic and secondary to old age. Carcinoma can arise from the surface epithelium of the penis and spread both locally and via the lymphatic system to the nodes of the groin. This disease is virtually unknown in populations where males are circumcised at birth. *See* INFERTILITY; ONCOLOGY; REPRODUCTIVE SYSTEM; REPRODUCTIVE TECHNOLOGY. Gay M. Guzinski

Reproductive technology Any procedure undertaken to aid in conception, intrauterine development, and birth when natural processes do not function normally. The most common are in vitro fertilization and gamete intrafallopian transfer.

Infertility, which is the inability to conceive during at least 12 months of unprotected intercourse, is an increasingly common problem. Hormonal therapy and microsurgery are used to overcome many hormonal and mechanical forms of infertility, but many types of infertility do not respond to such treatment. *See* INFERTILITY.

In vitro fertilization bypasses the Fallopian tubes. Immediately prior to ovulation the mature oocyte is removed from the ovary and placed, together with prepared sperm, in a petri dish for 2–3 days. Fertilization takes place during this time and the fertilized oocyte develops into a two- to eight-cell embryo, which is then transferred into the uterus. In vitro fertilization is useful for females with absent or severely damaged Fallopian tubes; couples in which the female has endometriosis; when the male has severely reduced sperm counts; or when the couple has immunologic or unexplained infertility for a period of 2 or more years. The four principal steps of in vitro fertilization are induction and timing of ovulation, oocyte retrieval, fertilization, and embryo transfer.

Gamete intrafallopian transfer, or GIFT, is similar to in vitro fertilization with a few important distinctions. In gamete intrafallopian tube transfer, the spermatozoa and oocytes are placed into the fimbriated end of the Fallopian tube during the laparoscopy. It is unusual for more than two oocytes to be placed into either Fallopian tube. Indications for gamete intrafallopian transfer include unexplained infertility of two or more years' duration, cervical stenosis, immunologic infertility, oligospermia, and endometriosis. At least one Fallopian tube must appear normal; where there has been severe pelvic adhesions or distorted tubal anatomy from any cause, gamete intrafallopian transfer should not be considered. *See* PREGNANCY; REPRODUCTIVE SYSTEM DISORDERS.

Machelle M. Seibel

Reptilia As traditionally defined, a class of vertebrates composed of four living orders—Testudines (turtles), Rhyncocephalia (tuataras), Squamata (lizards and snakes), and Crocodylia (alligators and crocodiles)—as well as numerous extinct orders. A classical taxonomy for the class is given below, followed by the classification scheme for extant reptiles currently accepted by most professional herpetologists. All orders other than the Testudines, Rhyncocephalia, Squamata, and Crocodylia are extinct.

Classical taxonomy:

Class Reptilia
- Subclass Anapsida
  - Order Captothinida
    - Mesosauria
    - Testudines (turtles)
- Subclass Diapsida
  - Infraclass Sauropterygia
    - Order Nothosauria
      - Plesiosauria
      - Placodontia
  - Infraclass Lepidosauria
    - Order Araeoscelida
      - Eosuchia
      - Rhyncocephalia (tuataras)
      - Squamata (lizards and snakes)
  - Infraclass Archosauria
    - Order Protosauria
      - Rhynchosauria
      - Thecodontia
      - Crocodylia (alligators and crocodiles)
    - Order Pterosauria
      - Saurischia
      - Ornithischia
- Subclass Ichthyopterygia
- Subclass Synapsida
  - Order Pelycosauria
    - Therapsida

Current classification scheme for extant reptiles:

- Subclass Anapsida
  - Order Testudines (turtles)
- Subclass Lepidosauria
  - Order Rhyncocephalia (tuataras)
    - Squamata
      - Suborder Sauria (lizards and skinks)
        - Amphisbaenia (amphisbaenids)
        - Ophidia (snakes)
- Subclass Archosauria
  - Order Crocodylia (alligators and crocodiles)

Globally, there are approximately 8200 extant species of reptiles described, including 300 species of turtles, 2980 species of snakes, and 4760 species of lizards and skinks. Reptiles are found on all continents excluding Antarctica, with their greatest diversity centered in tropical and subtropical ecosystems.

Reptiles are considered to have certain fundamental characteristics. Most basic is an integument (skin) with well-developed scales. These scales cover most of the body and serve to reduce water loss in the terrestrial environment. Another fundamental characteristic of reptiles is that they are ectothermic, meaning that the predominant amount of their internal body temperature comes from external sources such as sunlight. Reptiles do generate metabolic heat, but most rely on thermoregulatory behaviors to maintain their optimal temperature, such as can be seen in the behavior of the basking turtle. Reptile reproduction is as varied as their diversity and forms. Most species are oviparous, laying eggs in the sand, soil, or other materials such as decaying wood, but some species of lizards and snakes are viviparous, giving birth to live young.

The evolution of reptiles originates with their diversification from early amphibians. Adaptations of reptiles are rooted in a more general early vertebrate move from a strong reliance on aquatic environments. One of the central events in this process was the evolution of the amniotic egg, hence the Amniota, which includes reptiles, birds, and mammals.

The earliest reptiles are best discussed within the context of the earliest amniotes, which date from the mid-Carboniferous period beginning approximately 330 million years before present (mybp). Based on the fossil record to date, there are two important time frames of reptile radiation. The first, from the late Carboniferous (290 mybp) through the Permian period (250 mybp), was an important phase of evolution and radiation of the ancestral groups of species we are familiar with today (turtles, snakes, and lizards), and the mammal-like reptiles (the reptile group that is the ancestral group to modern-day mammals). The second is often referred to as the Age of Reptiles. This period begins in the mid-Triassic (~225 mybp), with the increasing diversity of groups like the Ichthyosaurs, Pterosaurs, Plesiosaurs, Synapsids, Ornithischia, and Saurischia (the latter two orders are commonly referred to as dinosaurs). The success of these groups, along with other groups such as the squamates and testudines, continued until a very abrupt decline (particularly among the dinosaurs) at the end of the Cretaceous Period (65 mybp). While the outcome was catastrophic for dinosaurs, other groups began to be more successful, among them the evolutionary lines that ultimately led to modern mammals, birds, turtles, snakes, and lizards.

W. Ben Cash

Repulsion motor An alternating-current (ac) commutator motor designed for single-phase operation. The chief distinction between the repulsion motor and the single-phase series motors is the way in which the armature receives its power. In the series motor the armature power is supplied by conduction from the line power supply. In the repulsion motor, however, armature power is supplied by induction (transformer action) from the field of the stator winding. For discussion of the ac series motor *see* UNIVERSAL MOTOR.

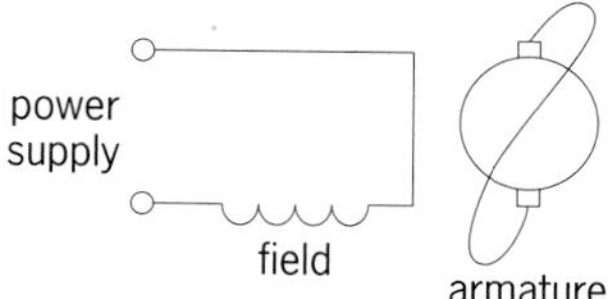

Schematic of a repulsion motor.

The repulsion motor primary or stationary field winding is connected to the power supply. The secondary or armature winding is mounted on the motor shaft and rotates with it. The terminals of the armature winding are short-circuited through a commutator and brushes. There is no electrical contact between the stationary field and rotating armature (see illustration). *See* ALTERNATING-CURRENT MOTOR; WINDINGS IN ELECTRIC MACHINERY.

Irving L. Kosow

Reservoir A place or containment area where water is stored. Where large volumes of water are to be stored, reservoirs usually are created by the construction of a dam across a flowing stream. When water occurs naturally in streams, it is sometimes not available when needed. Reservoirs solve this problem by capturing water and making it available at later times. *See* DAM.

In addition to large reservoirs, many small reservoirs are in service. These include varieties of farm ponds, regulating lakes, and small industrial or recreational facilities. In some regions, small ponds are called tanks. Small reservoirs can have important cumulative effects in rural regions

Reservoirs can be developed for single or multiple purposes, such as to supply water for people and cities, to provide irrigation water, to lift water levels to make navigation possible on streams, and to generate electricity.

Another purpose of reservoirs is to control floods by providing empty spaces for flood waters to fill, thereby diminishing the rate of flow and water depth downstream of the reservoir.

Reservoirs also provide for environmental uses of water by providing water to sustain fisheries and meet other fish and wildlife needs, or to improve water quality by providing dilution water when it is needed in downstream sections of rivers. Reservoirs may also have esthetic and recreational value, providing boating, swimming, fishing, rafting, hiking, viewing, photography, and general enjoyment of nature. *See* PUMPED STORAGE; RIVER ENGINEERING; WATER SUPPLY ENGINEERING.

Neil S. Grigg

Resin Originally a category of vegetable substances soluble in ethanol but insoluble in water, but generally in modern technology an organic polymer of indeterminate molecular weight. The class of flammable, amorphous secretions of conifers or legumes are considered true resins. Water- swellable secretions of various plants, especially the Burseraceae, are called gum resins. The natural vegetable resins are largely polyterpenes and their acid derivatives, which find application in the manufacture of lacquers, adhesives, varnishes, and inks.

The synthetic resins, originally viewed as substitutes for certain natural resins, have a large place of their own in industry and commerce. Phenol-formaldehyde, phenol-urea, and phenol-melamine resins are important commercially. Any unplasticized organic polymer is considered a resin, thus nearly any one of the common plastics may be viewed as a synthetic resin. Water-soluble resins are marketed chiefly as substitutes for vegetable gums and in their own right for highly specialized applications. Carboxymethy[cellulose, hydroxyalkylated cellulose derivatives, modified starches, polyvinyl alcohol, polyvinylpyrrolindone, and polyacrylamides are widely used as thickening agents for foods, paints, and drilling muds, as fiber sizings, in various kinds of protective coatings, and as encapsulating substances. *See* POLYMER.

Frank Wagner

Resistance heating The generation of heat by electric conductors carrying current. The degree of heating for a given current is proportional to the electrical resistance of the conductor. If the resistance is high, a large amount of heat is generated, and the material is used as a resistor rather than as a conductor.

In addition to having high resistivity, heating elements must be able to withstand high temperatures without deteriorating or sagging. Other desirable characteristics are low temperature coefficient of resistance, low cost, formability, and availability of materials. Most commercial resistance alloys contain chromium or aluminum or both, since a protective coating of chrome oxide or aluminum oxide forms on the surface upon heating and inhibits or retards further oxidation.

Since heat is transmitted by radiation, convection, or conduction or combinations of these, the form of element is designed for the major mode of transmission. The simplest form is the helix, using a round wire resistor, with the pitch of the helix approximately three wire diameters. This form is adapted to radiation and convection and is generally used for room or air heating. It is also used in industrial furnaces, utilizing forced convection up to about 1200°F (650°C). Such helixes are stretched over grooved high-alumina refractory insulators and are otherwise open and unrestricted.

The electrical resistance of molten salts between immersed electrodes can be used to generate heat. Limiting temperatures are dependent on decomposition or evaporization temperatures of the salt, Parts to be heated are immersed in the salt. Heating is rapid and, since there is no exposure to air, oxidation is largely prevented. Disadvantages are the personnel hazards and discomfort of working close to molten salts.

A major application of resistance heating is in electric home appliances, including electric ranges, clothes dryers, water heaters, coffee percolators, portable radiant heaters, and hair dryers. Resistance heating also has application in home or space heating.

If the resistor is located in a thermally insulated chamber, most of the heat generated is conserved and can be applied to a wide variety of heating processes. Such insulated chambers are called ovens or furnaces, depending on the temperature range and use. The term oven is generally applied to units which operate up to approximately 800°F (430°C). Typical uses are for baking or roasting foods, drying paints and organic enamels, baking foundry cores, and low-temperature treatments of metals. The term furnace generally applies to units operating above 1200°F (650°C). Typical uses of furnaces are for heat treatment or melting of metals, for vitrification and glazing of ceramic wares, for annealing of glass, and for roasting and calcining of ores. *See* ELECTRIC HEATING; FURNACE.

Willard Roth

Resistance measurement The quantitative determination of that property of an electrically conductive material, component, or circuit called electrical resistance. The ohm, which is the International System (SI) unit of resistance, is defined through the application of Ohm's law as the electric resistance between two points of a conductor when a constant potential difference of 1 volt applied to these points produces in the conductor a current of 1 ampere. Ohm's law can thus be taken to define resistance R as the ratio of dc voltage V to current

I, Eq. (1). For bulk metallic conductors, for example, bars, sheets,

$$R = \frac{V}{I} \qquad (1)$$

wires, and foils, this ratio is constant. For most other substances, such as semiconductors, ceramics, and composite materials, it may vary with voltage, and many electronic devices depend on this fact. The resistance of any conductor is given by the integral of expression (2), where l is the length, A the cross-sectional

$$\int_0^l \frac{\rho\, dl}{A} \qquad (2)$$

area, and ρ the resistivity. *See* ELECTRICAL CONDUCTIVITY OF METALS; ELECTRICAL RESISTANCE; ELECTRICAL RESISTIVITY; OHM'S LAW; SEMICONDUCTOR.

Since January 1, 1990, all resistance measurements worldwide have been referred to the quantized Hall resistance standard, which is used to maintain the ohm in all national standards laboratories. Conventional wire-wound working standards are measured in terms of the quantized Hall resistance and then used to disseminate the ohm through the normal calibration chain. These working standards can be measured in terms of the quantized Hall resistance with a one-standard-deviation uncertainty of about 1 part in 10^8. *See* HALL EFFECT.

The value of an unknown resistance is determined by comparison with a standard resistor. The Wheatstone bridge is perhaps the most basic and widely used resistance- or impedance-comparing device. Its principal advantage is that its operation and balance are independent of variations in the supply. The greatest sensitivity is obtained when all resistances are similar in value, and the comparison of standard resistors can then be made with a repeatability of about 3 parts in 10^8, the limit arising from thermal noise in the resistors. In use, the direction of supply is reversed periodically to eliminate effects of thermal or contact emf's.

The bridge is normally arranged for two-terminal measurements, and so is not suitable for the most accurate measurement at values below about 100 Ω, although still very convenient for lower resistances if the loss of accuracy does not matter. However, a Wheatstone bridge has also been developed for the measurement of four-terminal resistors. This involves the use of auxiliary balances, and resistors of the same value can be compared with uncertainties of a few parts in 10^8.

Typically a bridge will have two decade-ratio arms, for example, of 1, 10, 100, 1000, and 10,000 Ω, and a variable switched decade arm of 1–100,000 Ω, although many variations are encountered. For the measurement of resistors of values close to the decade values, a considerable increase in accuracy can be obtained by substitution measurement, in which the bridge is used only as an indicating instrument. The resistors being compared can be brought to the same value by connecting a much higher variable resistance across the larger of them, and the accuracy of this high-resistance shunt can be much less than that of the resistance being compared. *See* WHEATSTONE BRIDGE.

The Kelvin double bridge is a double bridge for four-terminal measurements, and so can be used for very low resistances. The addition to its use for accurate laboratory measurement of resistances below 100 Ω, it is very valuable for finding the resistance of conducting rods or bars, or for the calibration in the field of air-cooled resistors used for measurement of large currents. *See* KELVIN BRIDGE.

Measurements of resistances from 10 megohms to 1 terohm (10^{12} Ω) or even higher with a Wheatstone bridge present additional problems. The resistance to be measured will usually be voltage-dependent, and so the measurement voltage must be specified. The resistors in the ratio arms must be sufficiently high in value that they are not overloaded. If a guard electrode is fitted, it is necessary to eliminate any current flowing to the guard from the measurement circuit. The power dissipated in the 1-MΩ resistor is then 10 mW, and the bridge ratio is 10^6. The guard is connected to a subsidiary divider of the same ratio, so that any current flowing to it does not pass through the detector. Automated measurements can be made by replacing the ratio arms of the Wheatstone bridge by programmable voltage sources. An alternative method that can also be automated is to measure the RC time constant of the unknown resistor R combined with a capacitor of known value C. *See* INSULATION RESISTANCE TESTING.

An obvious and direct way of measuring resistance is by the simultaneous measurement of voltage and current, and this is usual in very many indicating ohmmeters and multirange meters. In most digital instruments, which are usually also digital voltage meters, the resistor is supplied from a constant-current circuit and the voltage across it is measured by the digital voltage meter. This is a convenient arrangement for a four-terminal measurement, so that long leads can be used from the instrument to the resistor without introducing errors. The simplest systems, used in passive pointer instruments, measure directly the current through the meter which is adjusted to give full-scale deflection by an additional resistor in series with the battery. This gives a nonlinear scale of limited accuracy, but sufficient for many practical applications. *See* CURRENT MEASUREMENT; VOLTAGE MEASUREMENT.

Cyril H. Dix; R. Gareth Jones

Resistance welding A process in which the heat for producing the weld is generated by the resistance to the flow of current through the parts to be joined. The application of external force is required; however, no fluxes, filler metals, or external heat sources are necessary. Most metals and their alloys can be successfully joined by resistance welding processes. Several methods are classified as resistance welding processes: spot, roll-spot, seam, projection, upset, flash, and percussion.

In resistance spot welding, coalescence at the faying surfaces is produced in one spot by the heat obtained from the resistance to electric current through the work parts held together under pressure by electrodes. The size and shape of the individually formed welds are limited primarily by the size and contour of the electrodes. *See* SPOT WELDING.

In roll resistance spot welding, separated resistance spot welds are made with one or more rotating circular electrodes. The rotation of the electrodes may or may not be stopped during the making of a weld.

In resistance seam welding, coalescence at the faying surfaces is produced by the heat obtained from resistance to electric current through the work parts held together under pressure by electrodes. The resulting weld is a series of overlapping resistance spot welds made progressively along a joint by rotating the electrodes.

In projection welding, coalescence is produced by the heat obtained from resistance to electric current through the work parts held together under pressure by electrodes. The resulting welds are localized at predetermined points by projections, embossments, or intersections.

In upset welding, coalescence is produced simultaneously over the entire area of abutting surfaces or progressively along a joint, by the heat obtained from resistance to electric current through the area of contact of those surfaces. Pressure is applied before heating is started and is maintained throughout the heating period.

In flash welding, coalescence is produced simultaneously over the entire area of abutting surfaces by the heat obtained from resistance to electric current between the two surfaces and by the application of pressure after heating is substantially completed. Flash and upsetting are accompanied by expulsion of the metal from the joint. *See* FLASH WELDING.

In percussion welding, coalescence is produced simultaneously over the entire abutting surfaces by the heat obtained from an arc produced by a rapid discharge of electrical energy with pressure percussively applied during or immediately following the electrical discharge.

Most metals and alloys can be resistance-welded to themselves and to each other. The weld properties are determined by the metal and by the resultant alloys which form during the welding process. Stronger metals and alloys require higher electrode forces, and poor electrical conductors require less current. Copper, silver, and gold, which are excellent electrical conductors, are very difficult to weld because they require high current densities to compensate for their low resistance. Medium- and high-carbon steels, which are hardened and embrittled during the normal welding process, must be tempered by multiple impulses. Ernest F. Nippes

Resistor One of the three basic passive components of an electric circuit that displays a voltage drop across its terminals and produces heat when an electric current passes through it. The electrical resistance, measured in ohms, is equal to the ratio of the voltage drop across the resistor terminals measured in volts divided by the current measured in amperes. *See* OHM'S LAW.

Resistors are described by stating their total resistance in ohms along with their safe power-dissipating ability in watts. The tolerance and temperature coefficient of the resistance value may also be given. *See* ELECTRICAL RESISTANCE; ELECTRICAL RESISTIVITY.

All resistors possess a finite shunt capacitance across their terminals, leading to a reduced impedance at high frequencies. Resistors also possess inductance, the magnitude of which depends greatly on the construction and is largest for wire-wound types. *See* CAPACITANCE; ELECTRICAL IMPEDANCE; INDUCTANCE.

Resistors may be classified according to the general field of engineering in which they are used. Power resistors range in size from about 5 W to many kilowatts and may be cooled by air convection, air blast, or water. The smaller sizes, up to several hundred watts, are used in both the power and electronics fields of engineering.

Direct-current (dc) ammeters employ resistors as meter shunts to bypass the major portion of the current around the low-current elements. These high-accuracy, four-terminal resistors are commonly designed to provide a voltage drop of 50–100 mV when a stated current passes through the shunt. *See* AMMETER.

Voltmeters of both the dc and the ac types employ scale-multiplying resistors designed for accuracy and stability. The arc-over voltage rating of these resistors is of importance in the case of high-voltage voltmeters. *See* VOLTMETER.

Standard resistors are used for calibration purposes in resistance measurements and are made to be as stable as possible, in value, with time, temperature, and other influences. Resistors with values from 1 ohm to 10 megohms are wound by using wire made from special alloys. The best performance is obtained from quaternary alloys, which contain four metals. The proportions are chosen to give a shallow parabolic variation of resistance with temperature, with a peak, and therefore the slowest rate of change, near room temperature. *See* ELECTRICAL UNITS AND STANDARDS.

By far the greatest number of resistors manufactured are intended for use in the electronics field. The major application of these resistors is in transistor analog and digital circuits which operate at voltage levels between 0.1 and 200 V, currents between 1 μA and 100 mA, and frequencies from dc to 100 MHz. Their power-dissipating ability is small, as is their physical size.

Since their exact value is rarely important, resistors are supplied in decade values (0.1, 1, 10, 100 ohms, and so forth) with the interval between these divided into a geometric series, thus having a constant percentage increase. For noncritical applications, values from a series with intervals of 20% (12 per decade) are appropriate. A series with 10% intervals (24 per decade) is often used for resistors having a tolerance of 1%. Where the precise value of a resistor is important, a series with 2.5% intervals (96 per decade) may be used.

Resistors are also classified according to their construction, which may be composition, film-type, wire-wound, or integrated circuit.

The composition resistor is in wide use because of its low cost, high reliability, and small size. Basically it is a mixture of resistive materials, usually carbon, and a suitable binder molded into a cylinder. Copper wire leads are attached to the ends of the cylinder, and the entire resistor is molded into a plastic or ceramic jacket. Composition resistors are commonly used in the range from several ohms to 10–20 MΩ, and are available with tolerances of 20, 10, or 5%.

The film-type resistor is now the preferred type for most electronic applications because its performance has surpassed that of composition resistors and mass-production techniques have reduced the cost to a comparable level. Basically this resistor consists of a thin conducting film of carbon, metal, or metal oxide deposited on a cylindrical ceramic or glass former. The resistance is controlled by cutting a helical groove through the conducting film. This helical groove increases the length and decreases the width of the conducting path, thereby determining its ohmic value. By controlling the conductivity, thickness of the film, and pitch of the helix, resistors over a wide range of values can be manufactured. Film construction is used for very high value resistors, up to and even beyond 1 TΩ (10^{12} ohms).

Wire remains the most stable form of resistance material available; therefore, all high-precision instruments rely upon wire-wound resistors. Wire also will tolerate operation at high temperatures, and so compact high-power resistors use this construction. Power resistors are available in resistance values from a fraction of an ohm to several hundred thousand ohms, at power ratings from one to several thousand watts, and at tolerances from 10 to 0.1%. The usual design of a power resistor is a helical winding of wire on a cylindrical ceramic former. After winding, the entire resistor is coated in vitreous enamel. Alternatively, the wound element may be fitted inside a ceramic or metal package, which will assist in heat dissipation. The helical winding results in the resistor having significant inductance, which may become objectionable at the higher audio frequencies and all radio frequencies. Precision wire-wound resistors are usually wound in several sections on ceramic or plastic bobbins and are available in the range from 0.1 Ω to 10 MΩ.

Integrated circuit resistors must be capable of fabrication on a silicon integrated circuit chip along with transistors and capacitors. There are two major types: thin-film resistors and diffused resistors. Thin-film resistors are formed by vacuum deposition or sputtering of nichrome, tantalum, or Cermet (Cr-SiO). Such resistors are stable, and the resistance may be adjusted to close tolerances by trimming the film by using a laser. Typical resistor values lie in the range from 100 Ω to 10 kΩ with a matching tolerance of ±0.2% and a temperature coefficient of resistance of ±10 to ±200 ppm/°C.

Diffused resistors are based upon the same fabrication geometry and techniques used to produce the active transistors on the silicon chip or die. A diffused base, emitter, or epitaxial layer may be formed as a bar with contacts at its extremities. The resistance of such a semiconductor resistor depends upon the impurity doping and the length and cross section of the resistor region. In the case of the base-diffused resistor, the emitter and collector regions may be formed so as to pinch the base region to a very small cross-sectional area, thereby appreciably increasing the resistance. The relatively large impurity carrier concentration in *n*- and *p*-type regions limits the resistance value. Resistor values between 100 Ω and 10 kΩ are common. *See* INTEGRATED CIRCUITS.

The deposited-film and wire-wound resistors lend themselves to the design of adjustable resistors or rheostats and potentiometers. Adjustable-slider power resistors are constructed in the same manner as any wire-wound resistor on a cylindrical form except that when the vitreous outer coating is applied an uncovered strip is provided. The resistance wire is exposed along this strip, and a suitable slider contact can be used to adjust the overall resistance, or the slider can be used as the tap on a potentiometer. *See* POTENTIOMETER; RHEOSTAT. R. B. D. Knight

Resolving power (optics) A quantitative measure of the ability of an optical instrument to produce separable images. The images to be resolved may differ in position because they represent (1) different points on the object, as in telescopes and microscopes, or (2) images of the same object in light of two different wavelengths, as in prism and grating spectroscopes. For the former class of instruments, the resolving limit is usually quoted as the smallest angular or linear separation of two object points, and for the latter class, as the smallest difference in wavelength or wave number that will produce separate images. Since these quantities are inversely proportional to the power of the instrument to resolve, the term resolving power has generally fallen into disfavor. It is still commonly applied to spectroscopes, however, for which the term chromatic resolving power is used, signifying the ratio of the wavelength itself to the smallest wavelength interval resolved. The figure quoted as the resolving power or resolving limit of an instrument may be the theoretical value that would be obtained if all optical parts were perfect, or it may be the actual value found experimentally. Aberrations of lenses or defects in the ruling of gratings usually cause the actual resolution to fall below the theoretical value, which therefore represents the maximum that could be obtained with the given dimensions of the instrument in question. This maximum is fixed by the wave nature of light and may be calculated for given conditions by diffraction theory. *See* DIFFRACTION; OPTICAL IMAGE. Francis A. Jenkins; George R. Harrison

Resonance (acoustics and mechanics) When a mechanical or acoustical system is acted upon by an external periodic driving force whose frequency equals a natural free oscillation frequency of the system, the amplitude of oscillation becomes large and the system is said to be in a state of resonance.

A knowledge of both the resonance frequency and the sharpness of resonance is essential to any discussion of driven vibrating systems. When a vibrating system is sharply resonant, careful tuning is required to obtain the resonance condition. Mechanical standards of frequency must be sharply resonant so that their peak response can easily be determined. In other circumstances, resonance is undesirable. For example, in the faithful recording and reproduction of musical sounds, it is necessary either to have all vibrational resonances of the system outside the band of frequencies being reproduced or to employ heavily damped systems. *See* ACOUSTIC RESONATOR; SYMPATHETIC VIBRATION; VIBRATION. Lawrence E. Kinsler

Resonance (alternating-current circuits) A condition in a circuit characterized by relatively unimpeded oscillation of energy from a potential to a kinetic form. In an electrical network there is oscillation between the potential energy of charge on capacitance and the kinetic energy of current in inductance. This is analogous to the mechanical resonance seen in a pendulum.

Three kinds of resonant frequency in circuits are officially defined. Phase resonance is the frequency at which the phase angle between sinusoidal current entering a circuit and sinusoidal voltage applied to the terminals of the circuit is zero. Amplitude resonance is the frequency at which a given sinusoidal excitation (voltage or current) produces the maximum oscillation of electric charge in the resonant circuit. Natural resonance is the natural frequency of oscillation of the resonant circuit in the absence of any forcing excitation. These three frequencies are so nearly equal in low-loss circuits that they do not often have to be distinguished.

Resonance is of great importance in communications, permitting certain frequencies to be passed and others to be rejected. Thus a pair of telephone wires can carry many messages at the same time, each modulating a different carrier frequency, and each being separated from the others at the receiving end of the line by an appropriate arrangement of resonant filters. A radio or television receiver uses much the same principle to accept a desired signal and to reject all the undesired signals that arrive concurrently at its antenna; tuning a receiver means adjusting a circuit to be resonant at a desired frequency. Hugh Hildreth Skilling

Resonance (molecular structure) A feature of the valence-bond method, which is a mathematical procedure to obtain approximate solutions to the Schrödinger equation for molecules. The valence-bond method is based on the theorem that if two or more solutions to the Schrödinger equation are available, certain linear combinations of them will also be solutions. It has this basis in common with its rival, the molecular orbital method. The valence-bond and molecular orbital approaches are both approximations and, if carried out to their logical and exact extremes, must yield identical results; nevertheless, both are often described as theories. In the valence-bond theory, combinations of solutions represent hypothetical structures of the molecule in question. These structures are said to be resonance (or contributing) structures, and the real molecule is said to be the resonance hybrid (or just simply the hybrid) of these structures. *See* MOLECULAR ORBITAL THEORY; SCHRÖDINGER'S WAVE EQUATION.

The resonance theory provided a solution for a molecule which had baffled and preoccupied chemists for a century—benzene. The principal use of resonance still lies in the qualitative description of molecules whose properties would otherwise be difficult to understand. *See* BENZENE.

Until the beginning of the 20th century, benzene posed a baffling challenge to organic chemists. In spite of its relatively simple formula, C_6H_6, they were unable to conceive of a suitable structure for it. While a great many structures were proposed, the properties of benzene corresponded to none of them.

In the early 1870s F. A. Kekulé proposed a revolutionary idea; benzene must be represented by two structures, (**1**) and (**2**), rather than one, and all compounds containing the benzene skeleton must be subject to a rapid equilibration (oscillation) between the two. Kekulé's description of benzene was not completely satisfactory. While it accounted for the number of substituted benzene isomers, it did not explain why the compound failed to exhibit reactivity indicating the presence of multiple bonds. The problem was resolved with the advent of quantum mechanics in the early part of this century. In a sense, this solution is an expansion of Kekulé's oscillating pair; the so-called activation energy (the energy which must be imparted to a molecule in order to make it overcome the barrier that keeps it from being converted into another molecule) is negative in the case of benzene with respect to the oscillation, and this molecule therefore exists neither as (**1**) nor as (**2**) at any time, but it is an

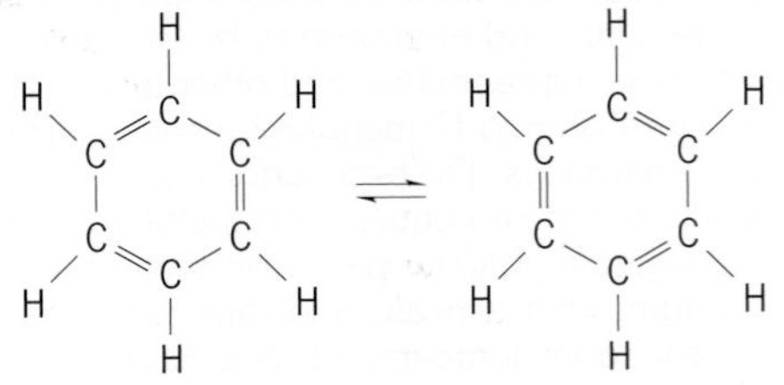

(1) ⇌ (2)

intermediate form (**3**) all the time. This intermediate structure of

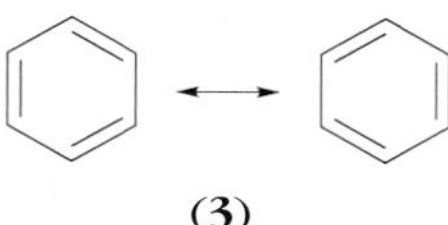

benzene is described in terms of Kekulé's structures with the symbol ↔ between them; this is intended to signify that benzene has neither structure, but in fact is a hybrid of the two. The properties of benzene are thereby indicated to be those of neither (**1**) nor (**2**), but to be intermediate between the two.

The only property of the hybrid which is not intermediate between those of the hypothetical contributing structures is the energy: the energy of a resonance hybrid is by definition always at a minimum. This fact is responsible for the abnormal reluctance of benzene to undergo addition reactions; such reactions would lead to products that no longer have the resonance energy.

Although benzene is the classical example of resonance, the phenomenon is certainly not limited to it. Furthermore, the properties of all compounds are affected by resonance to some degree.

Although the molecular orbital approach has largely supplanted the valence-bond method, the resonance language remains so convenient that it is still used. *See* CHEMICAL BONDING; MOLECULAR STRUCTURE AND SPECTRA; QUANTUM CHEMISTRY.

William J. le Noble

Resonance (quantum mechanics) An enhanced coupling between quantum states with the same energy. The concept of resonance in quantum mechanics is closely related to resonances in classical physics. *See* RESONANCE (ACOUSTICS AND MECHANICS); RESONANCE (ALTERNATING-CURRENT CIRCUITS).

The matching of frequencies is central to the concept of resonance. An example is provided by waves, acoustic or electromagnetic, of a spectrum of frequencies propagating down a tube or waveguide. If a closed side tube is attached, its characteristic natural frequencies will couple and resonate with waves of those same frequencies propagating down the main tube. This simple illustration provides a description of all resonances, including those in quantum mechanics. The propagation of all quantum entities, whether electrons, nucleons, or other elementary particles, is represented through wave functions and thus is subject to resonant effects. *See* ACOUSTIC RESONATOR; CAVITY RESONATOR; HARMONIC (PERIODIC PHENOMENA); WAVEGUIDE.

An important allied element of quantum mechanics lies in its correspondence between frequency and energy. Instead of frequencies, differences between allowed energy levels of a system are considered. In the presence of degeneracy, that is, of different states of the system with the same energy, even the slightest influence results in the system resonating back and forth between the degenerate states. These states may differ in their internal motions or in divisions of the system into subsystems. The above example of wave flow suggests the terminology of channels, each channel being a family of energy levels similar in other respects. These energies are discretely distributed for a closed channel, whereas a continuum of energy levels occurs in open channels whose subsystems can separate to infinity. If all channels are closed, that is, within the realm of bound states, resonance between degenerate states leads to a theme of central importance to quantum chemistry, namely, stabilization by resonance and the resulting formation of resonant bonds. *See* DEGENERACY (QUANTUM MECHANICS); ENERGY LEVEL (QUANTUM MECHANICS); RESONANCE (MOLECULAR STRUCTURE).

Resonances occur in scattering when at least one channel is closed and one open. Typically, a system is divided into two parts: projectile + target, such as electron + atom or nucleon + nucleus. One channel consists of continuum states with their two parts separated to infinity. The other, closed channel consists of bound states. In the atomic example, a bound state of the full system would be a state of the negative ion and, in the nuclear example, a state of the larger nucleus formed by incorporating one extra nucleon in the target nucleus. *See* QUANTUM MECHANICS; SCATTERING EXPERIMENTS (ATOMS AND MOLECULES); SCATTERING EXPERIMENTS (NUCLEI).

A. R. P. Rau

Resonance ionization spectroscopy A form of atomic and molecular spectroscopy in which wavelength-tunable lasers are used to remove electrons from (ionize) a given kind of atom or molecule. Laser-based resonance ionization spectroscopy (RIS) methods have been developed and used with ionization detectors, such as proportional counters, to detect single atoms. Resonance ionization spectroscopy can be combined with mass spectrometers to provide analytical systems for a wide range of applications, including physics, chemistry, materials sciences, medicine, and the environmental sciences.

When an atom or molecule is irradiated with a light source of frequency ν, photons at this selected frequency are absorbed only when the energy $h\nu$ (h is Planck's constant) is almost exactly the same as the difference in energy between some excited state (or intermediate state) and the ground state of the atom or molecule. If a laser source is tuned to a very narrow bandwidth (energy width or frequency width) at a frequency that excites a given kind of atom (see illustration), it is highly unlikely that any other kind of atom will be excited. An atom in an excited state can be ionized by photons of the specified frequency, ν, provided that $2h\nu$ is greater than the ionization potential of the atom. Even though the final ionization step can occur with any energy above a threshold, the entire process of ionization is a resonance one, because the intermediate state must first be excited in a resonance photon absorption. Resonance ionization spectroscopy is a selective process in which only those atoms that are in resonance with the light source are ionized. Modern pulsed lasers can provide a sufficient number of photons in a single pulse to remove one electron from each atom of any selected type. The first transition from the ground state to the intermediate state is easily saturated as these states come into equilibrium with a moderate photon flux. To saturate the final transition from the intermediate state to the ionization continuum (to remove one electron from each atom) requires laser pulses of considerable energy per pulse and is generally difficult. These frequency-tunable (or wavelength-tunable) lasers have made resonance ionization spectroscopy a potential method for the sensitive (and highly selective) detection of nearly every type of atom in the periodic table. *See* ATOMIC STRUCTURE AND SPECTRA; IONIZATION

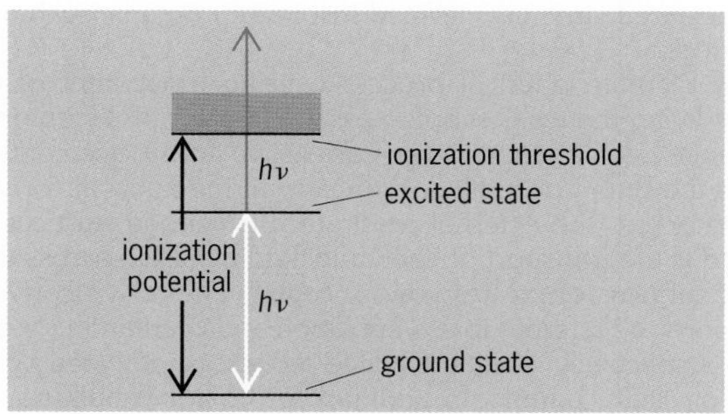

Basic laser scheme for resonance ionization spectroscopy (RIS). The atom or molecule is irradiated by a light source with frequency ν, or photons of energy $h\nu$, where h is Planck's constant. The shading shows the beginning of the ionization continuum, which starts at several electronvolts above the ground state, depending on the type of atom. An electron from the ground state of the atom or molecule is being excited by two photons, thus resonantly ionizing the atom or molecule.

POTENTIAL; LASER; LASER SPECTROSCOPY; PHOTOIONIZATION; RESONANCE (QUANTUM MECHANICS).

Resonance ionization spectroscopy is used in a wide variety of applications such as the trace analysis of low levels of elements in extremely pure materials, for example, semiconductors in the electronics industry. These analyses employ sputter-initiated resonance ionization spectroscopy (SIRIS), in which an argon ion beam sputters a tiny cloud of atoms from a sample placed in a high-vacuum system and a pulsed laser is tuned to detect the specified impurity atom. Ions are further selected for atomic mass by using a mass spectrometer. *See* MASS SPECTROMETRY; SPUTTERING.

The sputter-initiated resonance ionization spectroscopy method is also used for chemical and materials research, geophysical research and explorations, medical diagnostics, biological research, and environment analysis. *See* CHEMICAL DYNAMICS; DIFFUSION.

Resonance ionization spectroscopy is used in sophisticated nuclear physics studies involving high-energy accelerators. It is used as an on-line detector to record the hyperfine structure of nuclei with short lifetimes and hence to determine several nuclear properties such as nuclear spin and the shape of nuclei. *See* HYPERFINE STRUCTURE; NUCLEAR STRUCTURE.

Resonance ionization spectroscopy is used for measurements of krypton-81 in the natural environment to determine the ages of polar ice caps and old ground-water deposits. Oceanic circulation and the mixing of oceans could also be studied by measuring the concentrations of noble-gas isotopes by resonance ionization spectroscopy. *See* DATING METHODS; RADIOISOTOPE (GEOCHEMISTRY). G. Samuel Hurst

Respiration Commonly, the processes by which living organisms take up oxygen and eliminate carbon dioxide in order to provide energy. A more formal, comprehensive definition is the various processes associated with the biochemical transformation of the energy available in the organic substrates derived from foodstuffs to energy usable for synthetic and transport processes, external work, and, eventually, heat. This transformation, generally identified as metabolism, most commonly requires the presence of oxygen and involves the complete oxidation of organic substrates to carbon dioxide and water (aerobic respiration). If the oxidation is incomplete, resulting in organic compounds as end products, oxygen is typically not involved, and the process is then identified as anaerobic respiration.

The term external respiration is more appropriate for describing the exchange of oxygen and carbon dioxide between the organism and its environment. In most multicellular organisms, and nearly all vertebrates (with the exception of a few salamanders lacking both lungs and gills), external respiration takes place in specialized structures termed respiratory organs, such as gills and lungs.

The ultimate physical process causing movement of gases across living tissues is simple passive diffusion. Respiratory gas exchange also depends on two convective fluid movements. The first is the bulk transport of the external medium, air or water, to and across the external respiratory exchange surfaces. The second is the transport of coelomic fluid or blood across the internal surfaces of the respiratory organ. These two convective transports are referred to as ventilation and circulation (or perfusion), respectively. They are active processes, powered by muscular or ciliary pumps. In addition to providing bulk transport of the respiratory gases, the convection pumps are important in maximizing the concentration gradients for diffusion across the gas-exchanging tissue surfaces.

In almost all vertebrates and many invertebrates, the circulating internal medium (blood, hemolymph, or coelomic fluid) contains a respiratory pigment, for example, hemoglobin or hemocyanin, which binds reversibly with oxygen, carbon dioxide, and protons. Respiratory pigments augment respiratory gas exchange, both by increasing the capacity for bulk transport of the gases and by influencing gas partial pressure (concentration) gradients across tissue exchange surfaces. *See* BLOOD; HEMOGLOBIN; RESPIRATORY PIGMENTS (INVERTEBRATE).

The physiological adjustment of organisms to variations in their need for aerobic energy production involves regulated changes in the exchange and transport of respiratory gases. The adjustments are effected by rapid alterations in the ventilatory and circulatory pumps and by longer-term modifications in the respiratory properties of blood. *See* LUNG; METABOLISM; RESPIRATORY SYSTEM. Kjell Johansen; John B. West

Respirator A device designed to protect the wearer from noxious gases, vapors, and aerosols or to supply oxygen or doses of medication to the wearer. Respirators are used widely in industry to protect workers against harmful atmospheres, and in the military to protect personnel against chemical, biological, or radioactive warfare agents. Respirators are classified according to whether they are atmosphere-supplying or air-purifying.

Atmosphere-supplying respirators are used in atmospheres deficient in oxygen or extremely hazardous to the health of the wearer. Such atmospheres can occur in unventilated cellars, wells, mines, burning buildings, and enclosures containing inert gas. The self-contained breathing apparatus (SCBA) is a completely self-contained unit with the air supply or the oxygen-generating material being carried by the wearer. Air-supplied respirators are equipped with the same variety of facepieces as the SCBA, however these respirators can have the air supplied to the facepiece by means of a hose and a blower—the hose mask—or from a compressed-air source equipped with proper airflow and pressure-regulating equipment—the air-line mask.

Negative-pressure air-purifying respirator.

In an air-purifying respirator, ambient air is passed through a purifying medium to remove the contaminants. However, these devices do not provide oxygen or protect against oxygen-deficient atmospheres. A widely used air-purifying respirator is the nonpowered, or negative-pressure, respirator (see illustration). Ambient air is inhaled through the purifying medium in the replaceable cartridges and exhaled through an exhaust valve. In the case of the powered air-purifying respirator, an external blower, usually powered by a belt or helmet-mounted battery pack, forces air through the purifying medium and supplies it to the wearer under positive pressure, thus minimizing the problem of face-seal leakage. Benjamin Y. H. Liu; Daniel A. Japuntich

Respiratory pigments (invertebrate) Colored, metal-containing proteins that combine reversibly with oxygen, found in the body fluids or tissues of multicellular invertebrate animals and microorganisms. The role of these pigments is primarily to aid in the transport of molecular oxygen. Thus they are distinguished from respiratory enzymes, which are concerned with the metabolic consumption of oxygen. Four distinctly colored groups of respiratory pigments exist among invertebrates: hemoglobins (purple, become orange-red with oxygen), chlorocruorins (green, become red with oxygen), hemocyanins (colorless, become blue with oxygen), and hemerythrins (colorless, become red with oxygen). Those hemoglobins confined to muscle cells are called myoglobins. *See* HEMOGLOBIN.

Each pigment is composed of two parts, a large protein molecule to which is bound one or more small moieties called prosthetic groups, each of which is or contains a metal. The metal binds the oxygen, and this binding imparts the characteristic color to the pigment. In hemoglobins the prosthetic group is an iron porphyrin compound called heme. Chlorocruorin contains a similar iron porphyrin which differs from heme only in that a vinyl group in the molecule is replaced by formyl. The prosthetic group of hemerythrin consists of two adjacent iron atoms which bind an oxygen molecule between them. The prosthetic group of hemocyanin is analogous and consists of two adjacent copper atoms. Pigments containing vanadium have been found in tunicates, but these substances do not combine reversibly with oxygen and so cannot be considered respiratory pigments.

The protein part of the pigment confers reversibility upon the combination of the metal with oxygen. In the absence of protein, the prosthetic groups lose their capacity to combine with oxygen reversibly. Instead, the metals are irreversibly oxidized: Electrons are transferred from metal to oxygen. The bonds between metal and protein so alter the electronic energy levels of the metal that this transfer, if it occurs, is reversible. For this reason, the combination of hemoglobin with oxygen is described as oxygenation rather than oxidation. The protein is also responsible for certain physiological adaptations of the pigment to the environment. Thus the affinity of the pigment for oxygen is often highest in animals that inhabit environments with the lowest oxygen content. *See* RESPIRATION.

Austen F. Riggs

Respiratory syncytial virus A virus belonging to the Paramyxoviridae, genus *Pneumovirus*. This virus, although unrelated to any other known respiratory disease agent and differing from the parainfluenza viruses in a number of important characteristics, has been associated with a large proportion of respiratory illnesses in very young children, particularly bronchiolitis and pneumonia. It appears to be one of the major causes of these serious illnesses of infants. It is the only respiratory virus that occurs with its greatest frequency in infants in their first 6 months of life. In older infants and children, a milder illness is produced.

The clinical disease in young infants may be the result of an antigen-antibody reaction that occurs when the infecting virus meets antibody transmitted from the mother. For this reason respiratory syncytial vaccines that stimulate production of antibodies in the serum, but not in the nasal secretions, may do more harm than good. *See* ANIMAL VIRUS; VIRUS CLASSIFICATION.

Joseph L. Melnick; M. E. Reichmann

Respiratory system The system of organs involved in the acquisition of oxygen and the elimination of carbon dioxide by an organism. The lungs and gills are the two most important structures of vertebrates involved in the phase known as external respiration, or gaseous exchanges, between the blood and environment. Internal respiration refers to the gaseous exchanges which occur between the blood and cells. Certain other structures in some species of vertebrates serve as respiratory organs; among these are the integument or skin of fishes and amphibians. The moist, highly vascular skin of anuran amphibians is important in respiration. Certain species of fish have a vascular rectum which is utilized as a respiratory structure, water being taken in and ejected regularly by the animal. Saclike cloacal structures occur in some aquatic species of turtles. These are vascular and are intermittently filled with, and emptied of, water. It is thought that they may function in respiration. During embryonic life the yolk sac and allantois are important respiratory organs in certain vertebrates. *See* ALLANTOIS; YOLK SAC.

Structurally, respiratory organs usually present a vascular surface that is sufficiently extensive to provide an adequate area of absorption for gaseous exchange. This surface is moist and thin enough to allow for the passage of gases.

Robert S. McEwen; Thomas S. Parsons

The shape and volume of the lung, because of its pliability, conforms almost completely to that of its cavity. The lungs are conical; each has an apex and a base, two surfaces, two borders, and a hilum. The apex extends into the superior limit of the thoracic cavity. The base is the diaphragmatic surface. The costal surface may show bulgings into the intercostal spaces. The medial surface has a part lying in the space beside the vertebral column and a part imprinted by the form of structures bulging outward beneath the mediastinal pleura. The cardiac impression is deeper on the left lung because of the position of the heart.

For convenience the lung may be divided into anatomical areas. The bronchial tree branches mainly by dichotomy. The ultimate generations, that is, the respiratory bronchioles, alveolar ducts, and alveoli constitute all of the respiratory portion of the lung. The trachea and extrapulmonary bronchi are kept open by C-shaped bars of hyaline cartilage. When in their branching the bronchi and bronchioles are reduced to a diameter of 1 mm or less, they are then free of cartilage and are called terminal bronchioles. One of the terminal bronchioles enters the apex of a secondary lobule of the lung. These secondary lobules are anatomic units of the lung, whose hexagonal bases rest on the pleura or next to a bronchiole or blood vessel. Finer lines divide the bases of the secondary lobules into smaller areas. These are the bases of primary lobules, each served by a respiratory bronchiole. *See* LUNG; RESPIRATION.

The blood supply to the lung is provided by the pulmonary and the bronchial arteries. The nerves which supply the lung are branches of the vagus and of the thoracic sympathetic ganglia 2, 3, and 4. Efferent vagal fibers are bronchoconstrictor and secretory, whereas the afferents are part of the arc for the breathing reflex. Efferent sympathetic fibers are bronchodilators; hence, the use of adrenalin for relief of bronchial spasm resulting from asthma. *See* ASTHMA; NERVOUS SYSTEM (VERTEBRATE).

Leo P. Clements

Respiratory system disorders Dysfunction of the respiratory system, which supplies the body with the oxygen needed for metabolic activities in the cells and removes carbon dioxide, a product of cellular metabolism. The respiratory system includes the nose, mouth, throat, larynx, trachea, bronchi, lungs, and the muscles of respiration such as the intercostal muscles and the diaphragm. *See* RESPIRATION.

The lung has a great reserve capacity, and therefore a significant amount of disease usually must be present to produce clinical signs and symptoms. Shortness of breath (dyspnea) on exertion is the most common symptom of a respiratory disorder. Shortness of breath while at rest is indicative of severe respiratory disease and usually implies a severe abnormality of the lung tissue. If the respiratory system is so diseased that normal oxygenation of the blood cannot occur, blood remains dark, and a bluish color can be seen in the lips or under the fingernails; this condition is referred to as cyanosis. Other signs and symptoms of respiratory disorder can include fever, chest pain, coughing, excess sputum production, and hemoptysis (coughing up blood). Most of these signs and symptoms are nonspecific. *See* HYPOXIA.

Most diseases of the airways increase the resistance against which air is sucked in and pushed out of the lungs. Diseases of the nose usually have little influence since collateral respiration through the mouth compensates easily. Diseases of the throat, larynx, and trachea can significantly inhibit the flow of air into the lungs. Infections in the back of the throat, such as in diphtheria, can cause marked swelling of mucous membranes, resulting in air obstruction. Edema (swelling) of the mucosal lining of the larynx can also cause a reduction in air flow. Likewise, air flow can be inhibited in asthma, in which the smooth muscle in the trachea and bronchi episodically constricts. Chronic bronchitis results in inflammation of and excess mucus production by the bronchi and this also can lead to a reduction in air flow. Bronchiolitis, a condition that usually occurs in children and is often caused by a respiratory virus, results in narrowing and inflammation of small airways and a decrease in air flow.

Pneumonia, cancer, and emphysema are the most common lung diseases and are a major cause of morbidity and mortality in the United States. Of the four major types of lung cancer, approximately 90% can be attributed to the carcinogens present in cigarette smoke. Lung cancer may be detected in asymptomatic persons with a routine chest x-ray, or it may be discovered because of pain, excess coughing, or hemoptysis. *See* CANCER (MEDICINE); EMPHYSEMA; PNEUMONIA.

Among the diseases of pulmonary circulation, congenital malformations of the heart and pulmonary artery account for many cases of respiratory insufficiency in newborn and younger children. In adults, acquired heart diseases can cause backup of blood into the lung and an increased pressure in the pulmonary circulatory system. Also, blood clots, which usually develop first in the deep veins of the legs, can break free and flow to the heart. There they enter the pulmonary arteries and wedge in their small branches, where the clots are referred to as pulmonary thromboemboli, and can cause areas of death in the lung tissue (pulmonary infarcts). Persons who develop thromboemboli usually have chest pain and shortness of breath, and some have hemoptysis. *See* CARDIOVASCULAR SYSTEM; CIRCULATION DISORDERS.

Some neuromuscular diseases, such as poliomyelitis and amyotrophic lateral sclerosis (Lou Gehrig's disease), can cause dysfunction of the muscles of respiration. The resulting inability of these muscles to move air into the lungs can cause severe shortness of breath and predispose the patient to pneumonia. *See* MUSCULAR SYSTEM DISORDERS; POLIOMYELITIS; RESPIRATORY SYSTEM.

Samuel P. Hammar

Response A quantitative expression of the manner in which a microphone, amplifier, loudspeaker, or other component or system performs its intended function. A linear response means that the output signal is exactly proportional to the input signal for the entire range of frequencies over which the device is intended to operate. A logarithmic response means that the output signal is a logarithmic function of the input signal. The frequency response of a device, often presented as a curve on a graph, is the deviation over the frequency range from the response at some selected frequency, such as 1000 Hz. *See* AMPLIFIER; CHARACTERISTIC CURVE; LOUDSPEAKER; MICROPHONE.

John Markus

Rest mass A constant intrinsic to a body which determines its inertial and energy-momentum properties. It is a fundamental concept of special relativity, and in particular it determines the internal energy content of a body. It is the same as the inertial mass of classical mechanics. According to the principle of equivalence, the basic physical principle of general relativity, the inertial mass of a body is also equal to its gravitational mass. *See* CLASSICAL MECHANICS; GRAVITATION; RELATIVISTIC MECHANICS; RELATIVITY.

The rest mass or inertial mass of a body, m, is a measure of its resistance to being accelerated at a by **a** force $\mathbf{F}$; in classical mechanics the relation between inertial mass, acceleration, and force is given by Newton's law, Eq. (1). In special relativity Newton's law holds exactly only in the body's rest frame, that is, the frame in which the body is instantaneously at rest. *See* NEWTON'S LAWS OF MOTION.

$$\mathbf{F} = m\mathbf{a} \qquad (1)$$

Associated with the rest mass of a body, there is an internal or rest energy. In the system where the body is at rest, the energy of the body is given by Eq. (2).

$$E = mc^2 \qquad \text{(body at rest)} \qquad (2)$$

The experimental realization of the interconversion of mass and energy is accomplished in the reactions of nuclei and elementary particles. In particular, the energy source of nuclear bombs and nuclear fission reactors is a small decrease in the total mass of the interacting nuclei, which gives rise to a large energy release because of the large numerical value of c^2. *See* ELEMENTARY PARTICLE; NUCLEAR FISSION.

Ronald J. Adler

Restionales An order of flowering plants, division Magnoliophyta (Angiospermae), in the subclass Commelinidae of the class Liliopsida (monocotyledons). The order consists of 4 families and about 450 species, some 400 of them belonging to the Restionaceae. The vast majority of the species grow in temperate regions of the Southern Hemisphere. The Restionales are wind- or self-pollinated, with reduced flowers and a single, pendulous, orthotropous ovule in each of the 1–3 locules of the ovary. *See* COMMELINIDAE; CYPERALES; LILIOPSIDA; MAGNOLIOPHYTA.

Arthur Cronquist; T. M. Barkley

Restoration ecology A field in the science of conservation that is concerned with the application of ecological principles to restoring degraded, derelict, or fragmented ecosystems. The primary goal of restoration ecology (also known as ecological restoration) is to return a community or ecosystem to a condition similar in ecological structure, function, or both, to that existing prior to site disturbance or degradation.

A reference framework is needed to guide any restoration attempt—that is, to form the basis of the design (for example, desired species composition and density) and monitoring plan (for example, setting milestones and success criteria for restoration projects). Such a reference system is derived from ecological data collected from a suite of similar ecosystems in similar geomorphic settings within an appropriate biogeographic region. Typically, many sites representing a range of conditions (for example, pristine to highly degraded) are sampled, and statistical analyses of these data reveal what is possible given the initial conditions at the restoration site. *See* ECOLOGY, APPLIED; ECOSYSTEM.

Peggy L. Fiedler

Restriction enzyme An enzyme, specifically an endodeoxyribonuclease, that recognizes a short specific sequence within a deoxyribonucleic acid (DNA) molecule and then catalyzes double-strand cleavage of that molecule. Restriction enzymes have been found only in bacteria, where they serve to protect the bacterium from the deleterious effects of foreign DNA. *See* DEOXYRIBONUCLEIC ACID (DNA).

There are three known types of restriction enzymes. Type I enzymes recognize a specific sequence on DNA, but cleave the DNA chain at random locations with respect to this sequence. They have an absolute requirement for the cofactors adenosine triphosphate (ATP) and *S*-adenosylmethionine. Because of the random nature of the cleavage, the products are a heterogeneous array of DNA fragments. Type II enzymes also recognize a specific nucleotide sequence but differ from the type I enzymes in that they do not require cofactors and they cleave specifically within or close to the recognition sequence, thus generating a specific set of fragments. It is this exquisite specificity which has made these enzymes of great importance in DNA research, especially in the production of recombinant DNAs. Type III enzymes have properties intermediate between those of the type I and type II enzymes. They recognize a specific sequence and cleave specifically a short distance away from the recognition sequence. They have an absolute requirement for the ATP cofactor, but they do not hydrolyze it.

A key feature of the fragments produced by restriction enzymes is that when mixed in the presence of the enzyme DNA ligase, the fragments can be rejoined. Should the new fragment carry genetic information that can be interpreted by the bacterial cell containing the recombinant molecule, then the information will be expressed as a protein and the bacterial cell will serve as an ideal source from which to obtain that protein. For instance, if the DNA fragment carries the genetic information encoding the hormone insulin, the bacterial cell carrying that fragment will produce insulin. By using this method, the human gene for insulin has been cloned into bacterial cells and used for the commercial production of human insulin. The potential impact of this technology forms the basis of the genetic engineering industry. *See* ENZYME; GENETIC ENGINEERING. Richard Roberts

Resultant of forces A system of at most a single force and a single couple whose external effects on a rigid body are identical with the effects of the several actual forces that act on the body. For analytic purposes, forces are grouped and replaced by their resultant. Forces can be added graphically (see illustration) or analytically. The sum of more than two vector forces can be found by extending the method of illustration *c* to a three-dimensional vector polygon in which one force is drawn from the tip of the previous one until all are laid out.

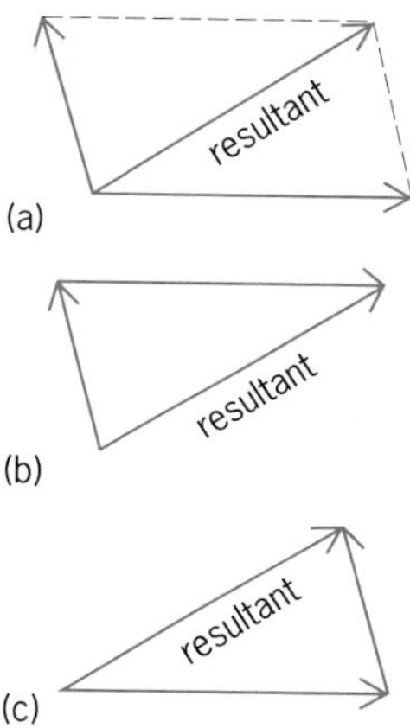

Resultant of two forces acting through a common center. (*a*) Diagonal of parallelogram. (*b, c*) Hypotenuse of triangle.

The resultant force is the force vector required to close the polygon directed from the tail of the first force vector to the tip of the last. A force system has a zero force resultant if its vector polygon closes. *See* CALCULUS OF VECTORS.

Two force systems are equivalent if their resultant forces, as described above, are equal and if their total vector moments about the same point are also equal. Vector moments are combined in the same manner as forces, that is, by parallelograms, triangles, or polygons. A resultant is the equivalent force system having the fewest possible forces and couples. *See* COUPLE; FORCE; STATICS. Nelson S. Fisk

Retaining wall A generic structure that is employed to restrain a vertical-faced or near-vertical-faced mass of earth. The earth behind the wall may be either the natural embankment or the backfill material placed adjacent to the retaining wall. Retaining walls must resist the lateral pressure of the earth, which tends to cause the structure to slide or overturn.

There are several types of retaining walls. A gravity wall is typically made of concrete and relies on its weight for stability (illus. *a*). The mass of the structure must be sufficient to develop enough frictional resistance to sliding, and the base or footing of the structure must be wide enough to develop sufficient moment

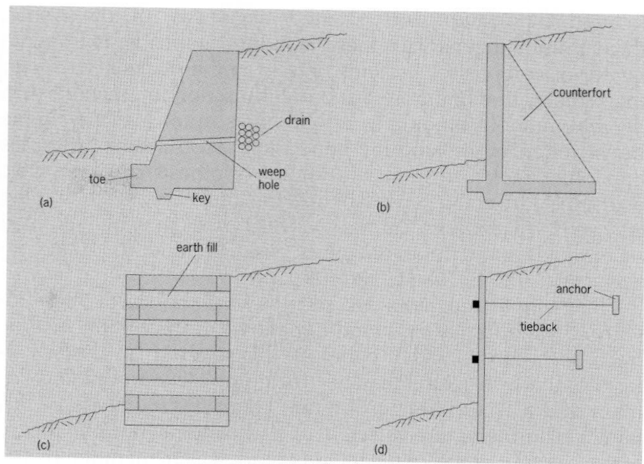

Common types of retaining walls. (*a*) Gravity wall. (*b*) Cantilever wall. (*c*) Crib wall. (*d*) Bulkhead.

to resist overturning earth forces. A cantilever retaining wall (illus. *b*) gains a larger effective mass by virtue of the soil placed on the horizontal cantilevered section of the wall. Reinforced counterforts are spaced along the wall to increase its strength. A variation of the gravity retaining wall is the crib wall (illus. *c*) is usually constructed of prefabricated interlocking concrete units. The crib is then filled with soil before the backfill adjacent to the crib is placed. Bulkhead retaining walls (illus. *d*) consist of vertical sheet piling that extends down into the soil and is stabilized by one or more tiebacks and anchors periodically spaced along the structure. The sheet piling may be made of reinforced concrete, steel, or aluminum. *See* CANTILEVER.

Retaining walls are often used in the marine environment, where they separate the retained soil from the water. Gravity walls (known as seawalls) can be constructed where strong wave and current forces are exerted on the wall. Bulkheads are more commonly found in sheltered areas such as harbors and navigation channels. *See* HARBORS AND PORTS. R. M. Sorensen

Reticular formation Characteristic clusters of nerve cell bodies (gray matter) and their meshwork, or reticulum, of fibers which are found in the brainstem and the diencephalon. The reticular formation is thought to be a complex, highly integrated mechanism which exerts both inhibition and facilitation on almost every type of activity of the central nervous system. *See* NERVOUS SYSTEM (VERTEBRATE). Douglas B. Webster

Reticulosa An order of the subclass Hexasterophora in the class Hexactinellida. This is a group of Paleozoic hexactinellids with a branching form. Each branch is provided with dermal, parenchymal, and gastral spicule reticulations. *See* HEXASTEROPHORA. Willard D. Hartman

Retinoid receptor A protein in the cell nucleus that mediates the actions of retinoids (the natural and synthetic analogs of vitamin A) by regulating the rates of transcription of retinoid-responsive genes. Retinoid receptors facilitate the normal biological actions of vitamin A within the body. Thus, they play an important role in the maintenance of normal growth and development; in the immune response; in male and female reproduction; in blood cell development; in the maintenance of healthy skin and bones; and ultimately in the general good health of the organism. *See* VITAMIN A.

The biochemical details of how retinoid receptors regulate gene transcription are well understood. The retinoid receptors recognize specific deoxyribonucleic acid (DNA) sequences, called retinoid response elements, within the regulatory (promoter) regions of genes. The transcription of a retinoid-responsive gene is either increased or diminished through interactions with these response elements, with coactivator or corepressor proteins present in the nucleus, and with the proteins that compose the basal transcription machinery. In general, retinoid binding to a retinoid receptor facilitates gene transcription by making the gene more accessible to enzymes and other factors responsible for the synthesis of ribonucleic acid (RNA) from the gene, whereas the absence of such binding lessens RNA synthesis. As many as 500 genes may be regulated by retinoids and retinoid receptors. *See* CELL NUCLEUS; DEOXYRIBONUCLEIC ACID (DNA); GENE; PROTEIN.

The retinoid receptors are members of the steroid/thyroid/retinoid superfamily of ligand-dependent transcription factors. Each of the proteins that compose this superfamily shares certain common structural features, including a ligand-binding domain and a DNA-binding domain. Often the members of this family of proteins are referred to as hormone nuclear receptors or, more simply, nuclear receptors. Each member of the steroid/thyroid/retinoid superfamily recognizes a specific natural ligand (or ligands), and in response to the availability of this ligand acts to influence either positively or negatively the expression of genes that are responsive to the ligand. *See* NUCLEAR HORMONE RECEPTORS.

Six different retinoid nuclear receptors have been identified, and each is the product of its own individual gene. Based on similarities in their protein structure, three of these are classified as retinoic acid receptors (RARs) and three are classified as retinoid X receptors (RXRs). The three distinct RAR subtypes are termed RAR-α, RAR-β, and RAR-γ; and the RXR subtypes are termed RXR-α, RXR-β, and RXR-γ. Different combinations of RAR and RXR species are found in tissues and cells. Thus, not all tissues and cells have the same complement of RARs or RXRs. It is not fully understood what significance these differences in RAR and RXR distribution have for mediating retinoid actions in the body. Solveig Halldorsdottir; William S. Blaner

Retortamonadida An order of parasitic flagellate protozoa belonging to the class Zoomastigophorea. All retortamonads are medium to large in size and have a complicated blepharoplast-centrosome-axo-style apparatus. Retortamonadida have two or four flagella, one turned ventrally into a cytostomal depression. The nucleus, containing a distinct endosome, is located at the anterior tip. The body is twisted. These organisms are actually symbionts, ingesting bacteria in the digestive tracts of their hosts. *Retortamonas* has several species, some of which infest insects and vertebrates. *Chilomastix* also has a number of species, found in vertebrates and invertebrates (see illustration). *See* PROTOZOA; SARCOMASTIGOPHORA; ZOOMASTIGOPHOREA. James B. Lackey

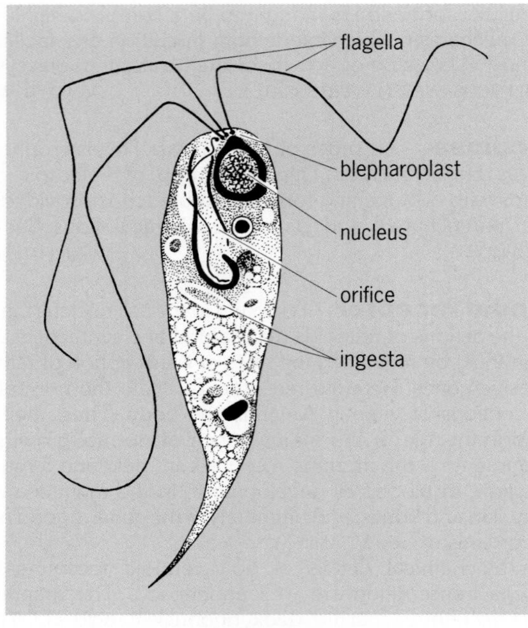

A retortamonad, *Chilomastix aulastomi*.

Retrograde motion (astronomy) In astronomy, either an apparent east-to-west motion of a planet or comet with respect to the background stars or a real east-to-west orbital motion of a comet about the Sun or of a satellite about its primary. The majority of the objects in the solar system revolve from west to east about their primaries. However, near the time of closest approach of Earth and a superior planet, such as Jupiter, because of their relative motion, the superior planet appears to move from east to west with respect to the background stars. The same apparent motion occurs for an inferior planet, such as Venus, near the time of closest approach to Earth.

Actual, rather than apparent, retrograde motion occurs among the satellites and comets; the eighth and ninth satellites of Jupiter and the ninth satellite of Saturn are examples. *See* ORBITAL MOTION. Raynor L. Duncombe

Retrovirus A family of viruses distinguished by four characteristics: (1) genetic information is in the form of ribonucleic acid (RNA); (2) each single virus particle or virion possesses the enzyme reverse transcriptase; (3) virion morphology consists of two proteinaceous structures, a dense core and an envelope that surrounds the core; and (4) the viral reproductive cycle requires the synthesis of a deoxyribonucleic acid (DNA) copy of the viral genome and insertion of that DNA into the genome of the host cell. *See* ANIMAL VIRUS; RIBONUCLEIC ACID (RNA).

The genome is composed of two identical molecules of single-stranded RNA. This genomic RNA is similar in function to messenger RNA (positive polarity).The core contains the RNA genome of the virus. DNA is not present in the virions of retroviruses. The reverse transcriptase in each virion copies the RNA genome into DNA shortly after entry of the virus into the animal or human host cell. The discovery of this enzyme changed thinking in biology. Previously, the only known direction for the flow of genetic information was from DNA to RNA, yet retroviruses make DNA copies of their genome by using an RNA molecule. This reversal of genetic information was "backwards," and hence the family name retrovirus, meaning backward virus. *See* REVERSE TRANSCRIPTASE.

The DNA copy of the RNA genome is inserted directly into one of the chromosomes of the host cell. This results in new genetic information being acquired by the host cell. One such class of new genes carried by retroviruses is oncogenes (tumor genes). Oncogenes are carried either in place of or in addition to one of the normal genes of retroviruses. Retroviral oncogenes appear to be involved in the induction of tumors in animals. *See* ONCOGENES.

The human T cell leukemia virus (HTLV) is one of two retroviruses known to cause disease in humans. The other is human immunodeficiency virus (HIV), the cause of acquired immune deficiency sydrome (AIDS). HTLV is associated with human leukemias of T cell origin and neurologic diseases. Preston A. Marx

Reverberation After sound has been produced in, or enters, an enclosed space, it is reflected repeatedly by the boundaries of the enclosure, even after the source ceases to emit sound. This prolongation of sound after the original source has stopped is called reverberation. A certain amount of reverberation adds a pleasing characteristic to the acoustical qualities of a room. However, excessive reverberation can ruin the acoustical properties

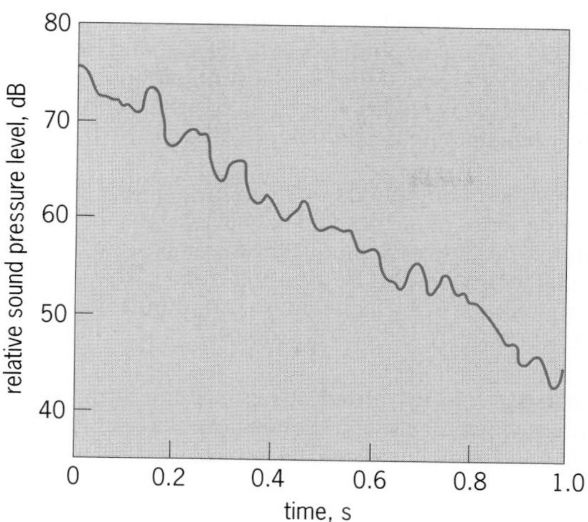

Typical decay curve illustrating reverberation.

of an otherwise well-designed room. A typical record representing the sound-pressure level at a given point in a room plotted against time, after a sound source has been turned off, is given in the decay curve shown in the illustration. The rate of sound decay is not uniform noindent but fluctuates about an average slope. *See* ARCHITECTURAL ACOUSTICS; SOUND. Cyril M. Harris

Reverse transcriptase Any of the deoxyribonucleic acid (DNA) polymerases present in particles of retroviruses which are able to carry out DNA synthesis using an RNA template. This reaction is called reverse transcription since it is the opposite of the usual transcription reaction, which involves RNA synthesis using a DNA template. *See* RETROVIRUS.

The transfer of genetic information from RNA to DNA in retrovirus replication was proposed in 1964 by H. M. Temin in the DNA provirus hypothesis for the replication of Rous sarcoma virus, an avian retrovirus which causes tumors in chickens and transformation of cells in culture, and reverse transcriptase has since been purified from virions of many retroviruses. The avian, murine, and human retrovirus DNA polymerases have been extensively studied.

Studies indicate that reverse transcriptase is widely distributed in living organisms and that all reverse transcriptases are evolutionarily related. For example, the organization of the nucleotide sequence of integrated retroviral DNA has a remarkable resemblance to the structure of bacterial transposable elements, in particular, transposons.

Reverse transcriptase genes are present in the eukaryotic organisms in retrotransposons and in retroposons or long interspersed (LINE) elements. Both of these types of elements can transpose in cells. *See* DEOXYRIBONUCLEIC ACID (DNA); RIBONUCLEIC ACID (RNA); TRANSPOSONS. Howard M. Temin

Revetment A facing or veneer of stone, concrete, or other materials constructed on a sloping embankment, dike, or beach face to protect it against erosion caused by waves or currents. The revetment may be a rigid cast-in-place concrete structure; but more commonly it is a flexible structure constructed of stone riprap or interlocking concrete blocks. It is sometimes an articulated block structure where the armor blocks are set in a form known as a flexible carpet; that is, the blocks interlock for stability, but the interlocking makes them flexible enough to respond to settlement of the underlying soil. A flexible revetment provides protection from exterior hydraulic forces, and it also can tolerate some settlement or consolidation of the underlying soil.

A typical revetment might employ stone riprap as the armor material (see illustration). A revetment typically has three major components: (1) the armor layer, which resists the wave or current-induced hydraulic forces; (2) a filter layer under the armor layer to allow water seepage out of the underlying soil without the removal of fine soil particles; and (3) a mechanism to stabilize the structure toe. Toe stabilization is particularly important where waves break on the structure, but may not be necessary if the revetment extends to sufficient depths where hydraulic forces will not erode the toe of the slope. The design water level (see illus.) for the structure may be higher than the normal water level during nonstorm conditions. If the revetment is exposed to waves that will break and run up the face of the revetment, the upper extent of the revetment must be sufficiently high to counter the force exerted by the waves.

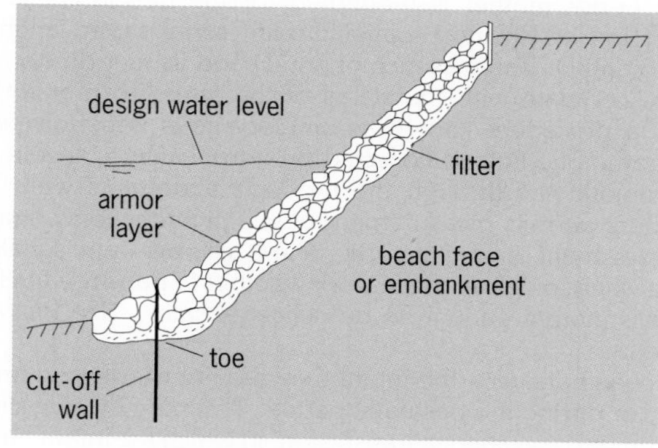

Cross-sectional profile of a typical stone revetment.

Although stone riprap is the most commonly used material for revetment armor layers, a wide variety of other materials have been used, including cast-in-place concrete and poured asphalt, wire bags filled with stone (gabions), interlocking concrete blocks, soil cement, cement-filled bags, interlocked tires, woven wooden mattresses, and vegetation (only used for surfaces exposed to very low waves or slow-moving currents). *See* RETAINING WALL; RIVER ENGINEERING. R. M. Sorensen

Reynolds number In fluid mechanics, the ratio $\rho v d/\mu$, where ρ is fluid density, v is velocity, d is a characteristic length, and μ is fluid viscosity. The Reynolds number is significant in the design of a model of any system in which the effect of viscosity is important in controlling the velocities or the flow pattern. In the evaluation of drag on a body submerged in a fluid and moving with respect to the fluids, the Reynolds number is important.

The Reynolds number also serves as a criterion of type of fluid motion. In a pipe, for example, laminar flow normally exists at Reynolds numbers less than 2000, and turbulent flow at Reynolds numbers above about 3000. *See* DYNAMIC SIMILARITY; FLUID MECHANICS; LAMINAR FLOW; TURBULENT FLOW. Glenn Murphy

Rh incompatibility A condition in which red blood cells of the fetus become coated with an immunoglobulin (IgG) antibody [anti-Rh (rhesus) D antibody] of maternal origin which is directed against an antigen (Rh D antigen) of paternal origin that is present on fetal cells. *See* ANTIBODY; ANTIGEN; IMMUNOGLOBULIN.

A pregnant woman may develop an antibody to a red blood cell antigen that the fetus has but that she does not possess. This occurs because the fetus has inherited the antigen from the father; fetal red blood cells pass into the maternal circulation and are recognized as foreign by the mother. The mother develops antibodies to the foreign protein; the antibodies pass across the placenta and bind to the fetal red blood cells. The fetal mononuclear phagocytic system recognizes the antibody-sensitized red blood cells as foreign and removes them from the circulation producing a hemolytic anemia. The maternal immune response is known as red blood cell alloimmunization. Alloimmunization of the mother results in hemolytic disease of the fetus and newborn (also known as erythroblastosis fetalis). *See* ANEMIA; AUTOIMMUNITY; PREGNANCY.

The development of Rh incompatibility is determined by the Rh types of the parents. Some Rh-incompatible pregnancies do not result in sensitization. Factors influencing the occurrence of sensitization include the timing and extent of the transplacental hemorrhage, the degree and strength of antibody development in the mother, and the ABO status of mother and fetus. Most immunizations result from transplacental hemorrhage during placental separation at delivery.

The at-risk mother is monitored for the development and progression of Rh incompatibility with serial serum antibody titers to anti-D. The presence of anti-D and its titer do not necessarily correlate with the status of the fetus. To monitor the status, a procedure known as amniocentesis is performed in which amniotic fluid is collected by introducing a needle into the amniotic sac through the mother's abdominal wall. This procedure permits measurement of bile pigment concentration and assessment of lung maturity. These tests are done serially to decide when to deliver the fetus or whether to transfuse the fetus until it is mature enough to be delivered safely. *See* PRENATAL DIAGNOSIS.

Following delivery, the infant may require transfusion immediately or during the postnatal period. This may involve simple replacement or exchange transfusion, in which the infant's blood volume is replaced with compatible blood. The aim of exchange is to remove bilirubin and antibody-coated red blood cells and to correct the anemia while replacing the infant's blood with blood that is antigen negative for the offending antibody. Other medical treatment for jaundice may also be required. *See* TRANSFUSION.

Naomi L. C. Luban; Chrysanthe Gaitatzes; Aimee M. Barton

Rhabditida An order of nematodes in which the number of labia varies from a full complement of six to three or two or none. The tubular stoma may be composed of five or more sections called rhabdions. The three-part esophagus always ends in a muscular bulb that is invariably valved. The excretory tube is cuticularly lined, and paired lateral collecting tubes generally run posteriorly from the excretory cell; some taxa have anterior tubules also. Females have one or two ovaries; when only one is present, the vulva shifts posteriorly. The cells of the intestine may be uninucleate, binucleate, or tetranucleate, and the hypodermal cells may also be multinucleate.

There are eight superfamilies in the order: Rhabditoidea, Alloionematoidea, Bunonematoidea, Cephaloboidea, Panagrolaimoidea, Robertioidea, Chambersiellolidea, and Elaphonematoidea. The Rhabditoidea are one of the largest nematode superfamilies and contain many important parasites of humans and domestic animals. This superfamily is distinguished by the well-developed cylindrical stoma and three-part esophagus that ends in a valved terminal bulb. In parasitic species, adult stages and some larval stages lack the valved terminal bulb. Though most species of Rhabditoidea are free-living feeders on terrestrial bacteria, others are important in the biological control of insects or as parasites of mammals. *See* MEDICAL PARASITOLOGY.

Armand R. Maggenti

Rhenium A chemical element, Re, with atomic number 75 and atomic weight 186.21. Rhenium is a transition element. It is a dense metal (21.04) with the very high melting point of 3440°C (6220°F). *See* PERIODIC TABLE.

Rhenium is similar to its homolog technetium in that it may be oxidized at elevated temperatures by oxygen to form the volatile heptoxide, Re_2O_7; this in turn may be reduced to a lower oxide, ReO_2. The compounds ReO_3, Re_2O_3, and Re_2O, are well known. Perrhenic acid, $HReO_4$, is a strong monobasic acid and is only a very weak oxidizing agent. Complex perrhenates, such as cobalt hexammine perrhenate, $[Co(NH_3)_6(ReO_4)_3]$, are also known.

The halogen compounds of rhenium are very complicated, and a large series of halides and oxyhalides have been reported. Rhenium forms two well-characterized sulfides, Re_2S_7 and ReS_2, as well as two selenides, Re_2Se_7 and $ReSe_2$. The sulfides have their counterparts in the technetium compounds, Tc_2S_7 and TcS_2. *See* TECHNETIUM; TRANSITION ELEMENTS.

Sherman Fried

Rheology In the broadest sense of the term, that part of mechanics which deals with the relation between force and deformation in material bodies. The nature of this relation depends on the material of which the body is constituted. It is customary to represent the deformation behavior of metals and other solids by a model called the linear or hookean elastic solid (displaying the property known as elasticity) and that of fluids by the model of the linear viscous or newtonian fluid (displaying the property known as viscosity). These classical models are, however, inadequate to depict certain nonlinear and time-dependent deformation behavior that is sometimes observed. It is these nonclassical behaviors which are the chief interest of rheologists and hence referred to as rheological behavior. *See* ELASTICITY; STRESS AND STRAIN; VISCOSITY.

Rheological behavior is particularly readily observed in materials containing polymer molecules which typically contain thousands of atoms per molecule, although such properties are also exhibited in some experiments on metals, glasses, and gases. Thus rheology is of interest not only to mathematicians and physicists, who consider it to be a part of continuum mechanics, but also to chemists and engineers who have to deal with these materials. It is of special importance in the plastics, rubber, film, and coatings industries. *See* FLUID MECHANICS; PAINT; PLASTICS PROCESSING; POLYMER; RUBBER; SURFACE COATING.

Models and properties. Consider a block of material of height h deformed in the manner indicated in Fig. 1; the bottom surface is fixed and the top moves a distance w parallel to itself. A measure of the deformation is the shear strain γ given by Eq. (1).

$$\gamma = \frac{w}{h} \tag{1}$$

To achieve such a deformation if the block is a linear elastic material, it is necessary to apply uniformly distributed tangential forces on the top and bottom of the block as shown in Fig. 1*b*. The intensity of these forces, that is, the magnitude of the net force per unit area, is called the shear stress S. For a linear elastic material, γ is much less than unity and is related

$$S = G\gamma \tag{2}$$

to S by Eq. (2), where the proportionality constant G is a property of the material known as the shear modulus.

If the material in the block is a newtonian fluid and a similar set of forces is imposed, the result is a simple shearing flow, a

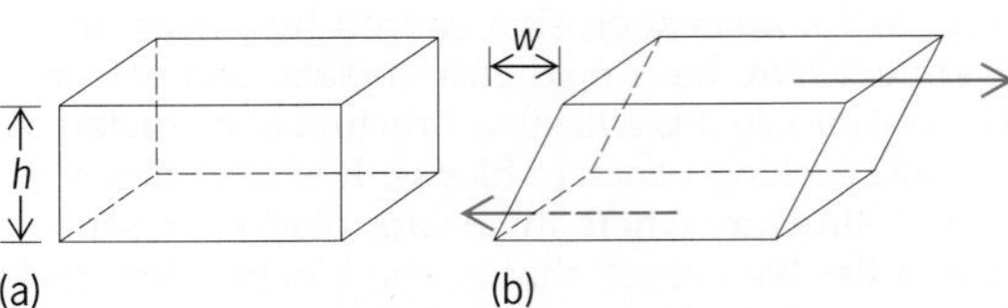

Fig. 1. Simple shear. (*a*) Undeformed block of height *h*. (*b*) Deformed block after top has moved a distance *w* parallel to itself. The arrows indicate the net forces acting on the top and bottom faces. The forces which must be applied to left and right faces to maintain a steady state are not indicated.

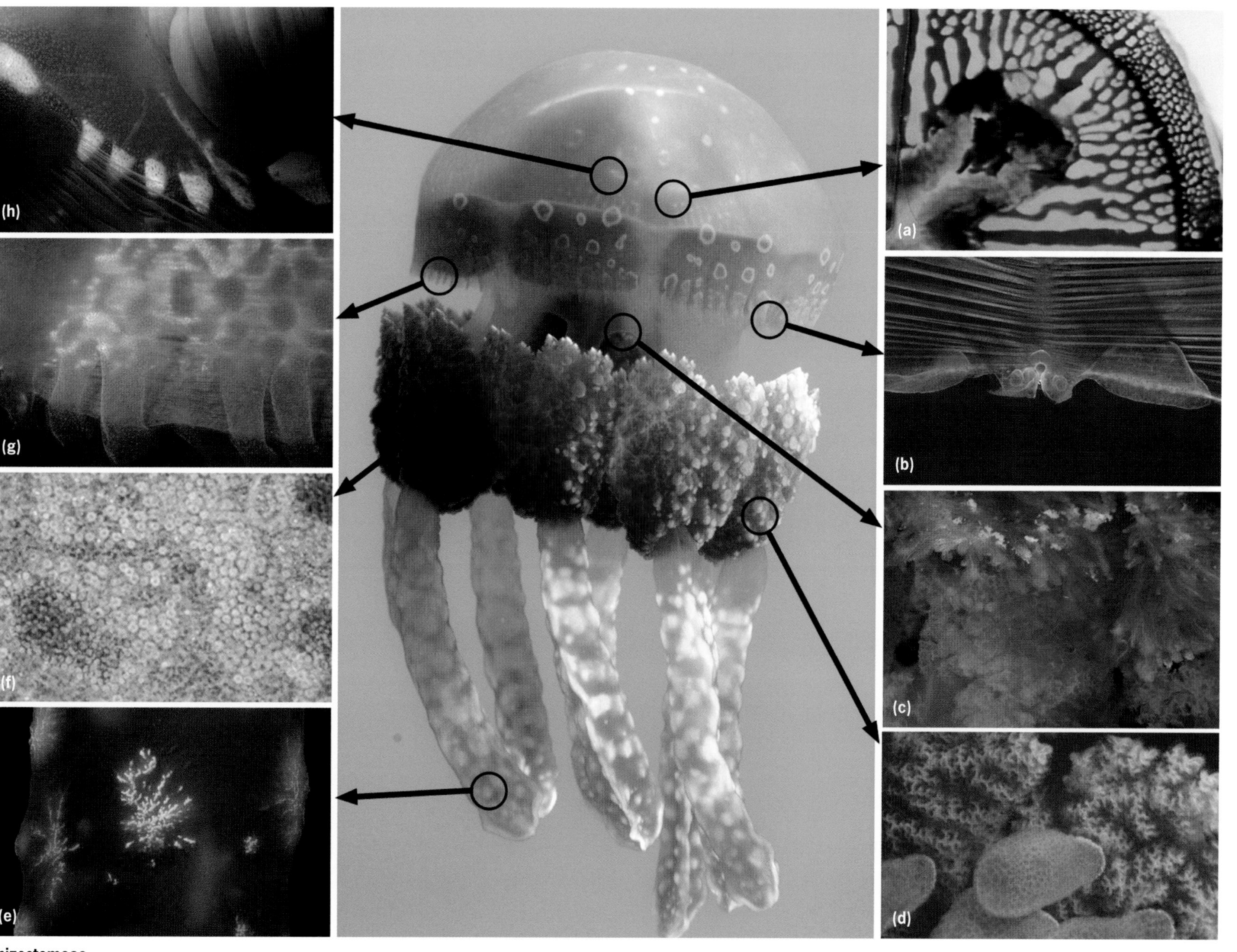

Rhizostomeae

The morphology and anatomy of a rhizostome medusa, *Mastigias*. The enlarged images show details of (*a*) the radial canal system and a segment of the circular canal in the umbrella, (*b*) muscle bands and a rhopalium (as seen looking at the subumbrella), (*c*) brood filaments (tubelike protrusions) with planulae (white dots) overlying suctorial mouths (bottom half of the image), (*d*) the crenulated pattern formed by the numerous minute suctorial mouths which are overlain in places by lily-pad-like filaments on the oral arms, (*e*) a portion of a terminal club, (*f*) zooxanthellae (each brown dot is an individual zooxanthella), (*g*) the bell margin showing several lappets, and (*h*) the subumbrellar region where radial canals originate from the gastrovascular cavity (each origin is, in this case, marked by a white fleck; a gonad can be seen in the upper right of the image).

deformation as pictured in Fig. 1*b* with the top surface moving with a velocity dw/dt. This type of motion is characterized by a rate of shear $\dot{\gamma} = (dw/dt)/h$, which is proportional to the shear stress S as given by Eq. (3), where η is a property of the material called the viscosity.

$$S = \eta\dot{\gamma} \quad (3)$$

Linear viscoelasticity. If the imposed forces are small enough, time-dependent deformation behavior can often be described by the model of linear viscoelasticity. The material properties in this model are most easily specified in terms of simple experiments.

In a creep experiment a stress is suddenly applied and then held constant; the deformation is then followed as a function of time. This stress history is indicated in the solid line of Fig. 2*a* for the case of an applied constant shear stress S_0. If such an experiment is performed on a linear elastic solid, the resultant deformation is indicated by the full line in Fig. 2*b* and for the linear viscous fluid in Fig. 2*c*. In the case of elasticity, the result is an instantly achieved constant strain; in the case of the fluid, an instantly achieved constant rate of strain. In the case of viscoelastic materials, there are some which eventually attain a constant equilibrium strain (Fig. 2*d*) and hence are called viscoelastic solids. Others eventually achieve constant rate of strain (Fig. 2*e*) and are called viscoelastic fluids. If the material is linear viscoelastic, the deformation $\gamma(S_0, t)$ is a function of the time t since the stress was applied and also a linear function of S_0; that is, Eq. (4) is satisfied, where $J(t)$ is independent of S_0. The

$$\gamma(S_0, t) = S_0 J(t) \quad (4)$$

function $J(t)$ is a property of the material known as the shear creep compliance. *See* CREEP (MATERIALS).

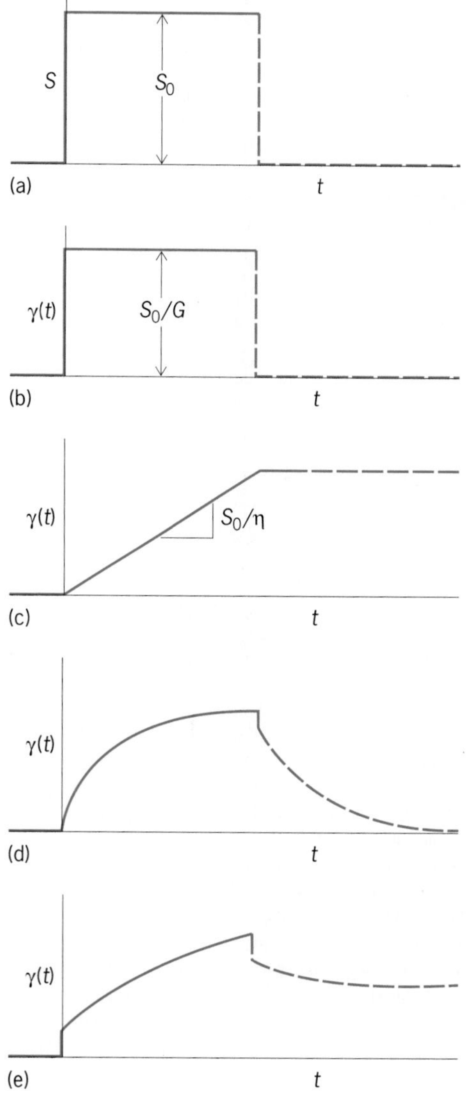

Fig. 2. Creep and recovery; solid lines indicate creep; broken lines indicate recovery. (*a*) Applied stress history. (*b*) Corresponding strain history for linear elastic solid, (*c*) linear viscous fluid, (*d*) viscoelastic solid, and (*e*) viscoelastic fluid.

Nonlinear viscoelasticity. If stresses become too high, linear viscoelasticity is no longer an adequate model for materials which exhibit time-dependent behavior. In a creep experiment, for example, the ratio of the strain to stress, $\gamma(t, S_0)/S_0$, is no longer independent of S_0; this ratio generally decreases with increasing S_0. Two examples of nonlinear viscoelasticity are shear thinning and thixotropy.

For polymer melts, solutions, and suspensions, generally speaking, the viscosity decreases as the shear rate increases. This type of behavior, called shear thinning, is of considerable industrial significance. For example, paints are formulated to be shear-thinning. A high viscosity at low flow rates keeps the paint from dripping from the brush or roller and prevents sagging of the paint film newly applied to a vertical wall. The lower viscosity at the high deformation rates while brushing or rolling means that less energy is required, and hence the painter's arm does not become overly tired.

Thixotropy is a property of suspensions (for example, bentonite clay in water) which, after remaining at rest for a long time, act as solids; for example, they cannot be poured. However, if it is stirred, such a suspension can be poured quite freely. If the suspension is then allowed to rest, the viscosity increases with time and finally sets again. This whole process is reversible; it can be repeated again and again. *See* GEL; NON-NEWTONIAN FLUID.

Hershel Markovitz

Rheostat A variable resistor constructed so that its resistance value may be changed without interrupting the circuit to which it is connected. It is used to vary the current in a circuit. The resistive element of a rheostat may be a metal wire or ribbon, carbon disks, or a conducting liquid. *See* RESISTOR.

Frank H. Rockett

Rheumatic fever An illness that follows an upper respiratory infection with the group A streptococcus (*Streptococcus pyogenes*) and is characterized by inflammation of the joints (arthritis) and the heart (carditis). Arthritis typically involves multiple joints and may migrate from one joint to another. The carditis may involve the outer lining of the heart, the heart muscle itself, or the inner lining of the heart. A minority of affected individuals also develop a rash (erythema marginatum), nodules under the skin, or Sydenham's chorea (a neurologic disorder characterized by involuntary, uncoordinated movements of the legs, arms, and face). Damage to heart valves may be permanent and progressive, leading to severe disability or death from rheumatic heart disease years after the initial attack. The disease occurs an average of 19 days after the infection and is thought to be the result of an abnormal immunologic reaction to the group A streptococcus. Initial attacks of rheumatic fever generally occur among individuals aged 5 to 15. Those who have had one attack are highly susceptible to recurrences after future streptococcal infections.

Initial attacks of rheumatic fever can be prevented by treatment of strep throat with penicillin for at least 10 days. Patients who have had an episode of rheumatic fever should continue taking antibiotics for many years to prevent group A streptococcal infections that may trigger a recurrence of rheumatic fever. *See* HEART DISORDERS; STREPTOCOCCUS.

A. L. Bisno; James B. Dale

Rheumatism Any combination of muscle or joint pain, stiffness, or discomfort arising from nonspecific disorders. It is generally used as a lay expression to indicate a chronic or recurrent condition affecting a certain area and precipitated by cold, dampness, or emotional stress.

Lumbago, wryneck, charleyhorse and shinsplint are commonly used expressions included under the catchall category of rheumatism.

Robert Searles

Rhinoceros An odd-toed ungulate (order Perissodactyla) belonging to the family Rhinoceratidae. The bodies and limbs of these mammals are massive and thick-skinned. Also characteristic of the family are the horns (epidermal derivatives), of which there may be one or two, composed of a solid mass of hairs attached to a bony prominence on the skull. The feet are tridactyl, with three hooves; the middle one is the most developed. Five species are found in Asia and Africa.

Sight is poorly developed in these animals; however, hearing is quite good and the sense of smell is excellent. Most species are solitary and aggressive. All species are vegetarians. The great Indian rhinoceros (*Rhinoceros unicornis*) is about 14 ft (4.3 m) long and nearly 6 ft (1.8 m) high, with a single horn that may be 2 ft (0.6 m) long in both sexes. Some species show a strict territoriality, having definite tracks through the grass, definite sites for dung deposition, and personal mud wallows. The gradual disappearance of the rhinoceros is attributed to habitat destruction and poaching, especially the unrestrained hunting for the horns (which were used as a medicine, although of no medicinal value), the meat, and the hide.

Donald W. Linzey

Rhinovirus A genus of the family Picornaviridae. Members of the human rhinovirus group include at least 113 antigenically distinct types. Like the enteroviruses, the rhinoviruses are small (17–30 nanometers), contain ribonucleic acid (RNA), and are not inactivated by ether. Unlike the enteroviruses, they are isolated from the nose and throat rather than from the enteric tract, and are unstable if kept under acid conditions (pH 3–5) for 1–3 h. Rhinoviruses have been recovered chiefly from adults with colds and only rarely from patients with more severe respiratory diseases. *See* COMMON COLD; ENTEROVIRUS.

In a single community, different rhinovirus types predominate during different seasons and during different outbreaks in a single season, but more than one type may be present at the same time.

Although efforts have been made to develop vaccines, none is available. Problems that hinder development of a useful rhinovirus vaccine include the short duration of natural immunity even to the specific infecting type, the large number of different antigenic types of rhinovirus, and the variation of types present in a community from one year to the next. *See* ANIMAL VIRUS; PICORNAVIRIDAE; VIRUS CLASSIFICATION.

Joseph L. Melnick; M. E. Reichmann

Rhizocephala An order of crustacean parasites related to the barnacles. Worldwide in distribution, they prey on other crustaceans, principally Decapoda, such as crabs, shrimp, and their allies. Rhizocephala produce modifications affecting the abdomen of the crab, making males resemble females and causing immature females to acquire precociously the adult form. These parasites have become so modified by their mode of life that, as adults, they are no longer recognizable as barnacles, or even as crustaceans.

An adult rhizocephalan is a thin-walled sac enclosing a visceral mass, composed chiefly of ovaries and testes. It shows no trace of segmentation, appendages, or sense organs. Even an alimentary tract is missing. Instead, it possesses a threadlike root system which penetrates the interior of the host in all directions and absorbs nourishment from the body fluids of the crab.

Placidus G. Reischman

Rhizophorales An order of flowering plants, division Magnoliophyta (Angiospermae), in the subclass Rosidae of the class Magnoliopsida (dicotyledons). The order contains a single family, Rhizophoraceae, with about 100 species widely distributed in the tropics. The plants are mostly tanniferous trees and shrubs with the leaves opposite, simple, and entire. The flowers are regular, mostly perfect, and variously perigynous or epigynous. The sepals are four or five and commonly fleshy or leathery; the petals are the same number as the sepals and likewise small and fleshy. The stamens are twice as many as the petals or sometimes more. The fruit is berrylike or rarely a capsule. Most members of the family are inland species, but the most conspicuous group are some 17 species of shoreline shrubs, the mangroves. *See* MAGNOLIOPHYTA; MAGNOLIOPSIDA; PLANT KINGDOM.

T. M. Barkley

Rhizopodea A class of Sarcodina including both parasitic and free-living species found in fresh and salt water and the soil. No species forms true axopodia; instead, pseudopodia may be filopodia, lobopodia, or reticulopodia; or there may be no pseudopodia in some cases. Rhizopodea include five subclasses: Lobosia, Filosia, Granuloreticulosia, Mycetozoia, and Labyrinthulia. *See* FILOSIA; GRANULORETICULOSIA; LABYRINTHULIA; LOBOSIA; SARCODINA.

Richard P. Hall

Rhizosphere The volume of soil around living plant roots that is influenced by root activity, either directly or via the activities of the microbiota that are stimulated by rhizodeposits (materials released by roots). Root–soil–microbiota interactions ultimately alter the biological, biochemical, chemical, and physical properties of the surrounding soil. The rhizosphere is thus a complex, dynamic, and heterogeneous microenvironment that plays a key role in plant nutrition and health, diversity of plant and soil microbial communities, soil functioning and genesis, and biogeochemical cycling of elements. *See* ROOT (BOTANY); SOIL.

The prime rhizosphere effect is a consequence of rhizodeposition. On average, 20% of the carbon assimilated by plants via photosynthesis is released by roots in the soil as a range of rhizodeposits: sloughed-off cells and decaying root hairs, respired CO_2, and organic compounds that are released via active secretion (for example, root mucilage) or passive diffusion (for example, root exudates such as sugars, amino acids, or organic acids). Rhizodeposition is a significant component of carbon cycling in ecosystems. Although sometimes viewed as a net loss of carbon for the plant, it is a major source of carbon for soil microbiota, making the rhizosphere a hot spot for trophic relationships. Consequently, considerable stimulation of microorganisms is known to occur around roots. This rhizosphere effect is evidenced by an increased microbial biomass in the rhizosphere relative to bulk soil. *See* SOIL MICROBIOLOGY.

The outer boundary of the rhizosphere is loosely defined. Depending on which measure of root activity is considered, the rhizosphere volume can thus represent from much less than 1% up to 100% of the soil volume colonized by roots. Thus, one cannot draw a single borderline between the rhizosphere and the bulk soil, and most rhizosphere processes do not show a sharp boundary but a more or less steep gradient. *See* MICROBIAL ECOLOGY.

Philippe Hinsinger

Rhizostomeae An order of the class Scyphozoa with the most highly organized features of this class. The umbrella is generally higher than it is wide. The margin of the umbrella is divided into many lappets but is not provided with tentacles. Many radial canals, which are connected with each other to form

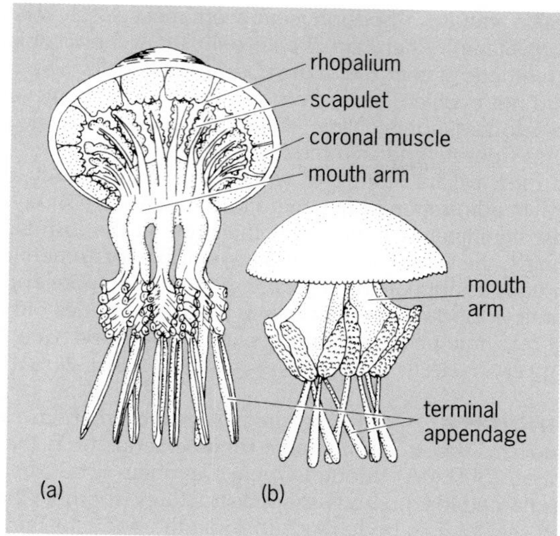

Rhizostomeae. (*a*) *Rhizostoma*. (*b*) *Mastigias*. (After L. Hyman, *The Invertebrates*, vol. 1, McGraw-Hill, 1940)

a complicated network, issue from the cruciform stomach. The oral part is eight-sided, with many suctorial mouths surrounded by numerous small tentacles (see illustration). Usually there is no central mouth. No species of this group is injurious to the human skin. Some large forms, such as *Rhopilema*, are used as food in China and Japan.

A fair number of fossils of this order were found in the strata of the Jurassic Period. *See* SCYPHOZOA. Tohru Uchida

Rho-theta system A navigation system in which one or more signals are emitted from a facility (or colocated facilities) to produce simultaneous indication of bearing and distance. Since a bearing is a radial line of position and a distance is a circular line of position, the rho-theta system always ensures a position fix produced by the intersection of two lines of position which are at right angles to each other. This produces a minimum geometric dilution of position, a figure of merit for all radio navigation systems, and is one of the chief advantages of a rho-theta system. Another major advantage is that it is a single-site system and can thus be installed on a ship or an island. This has also made it attractive politically, enabling small countries to have their own navigation systems. *See* DOPPLER VOR; ELECTRONIC NAVIGATION SYSTEMS; TACAN; VOR (VHF OMNIDIRECTIONAL RANGE). Sven H. Dodington

Rhodium A chemical element, Rh, atomic number 45, relative atomic weight 102.905. Rhodium is a transition metal and one of the group of platinum metals (ruthenium, osmium, rhodium, iridium, palladium, and platinum) that share similar chemical and physical properties. *See* METAL; PERIODIC TABLE; PLATINUM.

The terrestrial abundance of rhodium is exceedingly low; it is estimated to be 0.0004 part per million in the Earth's crust. It is found as a single isotope, ^{103}Rh, with a nuclear spin of 12. Since the platinum metals share common reactivities and are mined from a common source, there is an involved chemical process that is used to separate the individual elements, including rhodium. *See* ISOTOPE.

Metallic rhodium is the whitest of the platinum metals and does not tarnish under atmospheric conditions. Its surface is normally covered by a thin, firmly bound layer of rhodium(IV) oxide (RhO_2). Rhodium is insoluble in all acids, including aqua regia. It dissolves in molten potassium bisulfate ($KHSO_4$), a useful property for its extraction from platinum ores, since iridium, ruthenium, and osmium are insoluble in this melt. Important physical properties of metallic rhodium are given in the table. *See* ACID AND BASE; AQUA REGIA; HALOGEN ELEMENTS.

Physical properties of rhodium metal

Property	Value
Crystal structure	Face-centered cubic
Lattice constant *a*, at 25°C (77°F), nm	0.38031
Thermal neutron capture cross section, barns (10^{-28} m^2)	149
Density at 25°C (77°F), g/cm^3	12.43
Melting point	1963°C (3565°F)
Boiling point	3700°C (6700°F)
Specific heat at 0°C, cal/g (J/kg)	0.0589 (246)
Thermal conductivity, 0–100°C, cal cm/cm^{2}s°C (J · m/m^2 · s · °C)	0.36 (151)
Linear coefficient of thermal expansion, 20–100°C, μin./(in./°C) or m/(m · °C)	8.3
Electrical resistivity at 0°C, microhm-cm	4.33
Temperature coefficient of electrical resistance, 0–100°C/°C	0.00463
Tensile strength, 10^3 lb/in.^{2} (6.895 MPa)	
Soft	120–130
Hard	200–230
Young's modulus at 20°C (68°F), lb/in.^{2} (Gpa)	
Static	46.2 × 10^6 (319)
Dynamic	54.8 × 10^6 (378)
Hardness, diamond pyramid number	
Soft	120–140
Hard	300
ΔH_{fusion}, kJ/mol	21.6
$\Delta H_{vaporization}$, kJ/mol	494
ΔH_f monoatomic gas, kJ/mol	556
Electronegativity	2.2

Metallic rhodium is available as powder, sponge, wire, and sheets. It is ductile when hot and retains its ductility when cold. However, it work-hardens rapidly. Molten rhodium dissolves oxygen. Upon cooling, the oxygen gas is liberated, and this can lead to ruptures in the external surface of the crust of the metal. As a result, molten rhodium is best handled under an inert atmosphere of argon, which does not dissolve in rhodium.

Complexes of Rh(III), including $RhCl_3(pyridine)_3$, $Rh(CO)Cl_3[P(C_6H_5)_3]_2$, and $RhCl_6^{3-}$, are diamagnetic six-coordinate with octahedral geometry. The most common chemical form of rhodium is $RhCl_3 \cdot 3H_2O$, a red-brown, deliquescent material that is a useful starting material for the preparation of other rhodium compounds. In contrast to the hydrated material, red anhydrous rhodium(III) chloride ($RhCl_3$) is a polymeric, paramagnetic compound that does not dissolve in water. *See* DIAMAGNETISM.

The low natural abundance and high cost of rhodium limit its uses to specialty applications. The major use is in catalysis, which accounts for over 60% of its production. Rhodium is a component of catalytic converters used in the control of exhaust emissions from automobiles. *See* CATALYTIC CONVERTER.

Rhodium is also used in the hydrogenation of olefins to alkanes. For hydrogenation, both heterogeneous catalysis and homogeneous catalysis are used. Heterogeneous conditions are achieved with rhodium metal finely dispersed on an inert support (activated carbon, charcoal, or alumina).

Rhodium complexes have been developed as catalysts for the synthesis of one optical isomer of L-dopa (used in treatment of Parkinson's disease). Greater selectivity makes rhodium catalysts more useful in hydroformylation or oxo reactions than the less expensive cobalt catalysts. A platinum-rhodium alloy is an efficient commercial catalyst for the formation of nitric acid through ammonia oxidation. *See* CATALYSIS; HETEROGENEOUS CATALYSIS; HOMOGENEOUS CATALYSIS; HYDROFORMYLATION; HYDROGENATION.

Rhodium-platinum alloys are favored for high-temperature applications. The International Temperature Scale over the range 630.5–1063°C (1134.9–1945.4°F) is defined by a thermocouple using a 10% rhodium-platinum alloy. Electroplated rhodium retains its bright surface under atmospheric conditions and finds use as electrical contacts and reflective surfaces. The reflectivity of rhodium surfaces is high (80%) and does not tarnish. About 6% of the rhodium production goes into jewelry manufacturing. *See* ELECTROPLATING OF METALS; TRANSITION ELEMENTS.

Alan A. Balch

Rhodochrosite The mineral form of manganese carbonate. Calcium, iron, magnesium, and zinc have all been reported to replace some of the manganese. The equilibrium replacement of manganese by calcium increases with the temperature of crystallization.

Rhodochrosite occurs more often in massive or columnar form than in distinct crystals. The color ranges from pale pink to brownish pink. Hardness is 3.5–4 on Mohs scale, and specific gravity is 3.70.

Well-known occurrences of rhodochrosite are in Europe, Asia, and South America. In the United States large quantities occur at Butte, Montana. As a source of manganese, rhodochrosite is also important at Chamberlain, South Dakota, and in Aroostook County, Maine. *See* CARBONATE MINERALS; MANGANESE.

Robert I. Harker

Rhodonite A mineral inosilicate with composition $MnSiO_3$. Hardness is 5.5–6 on Mohs scale, and specific gravity is 3.4–3.7. The luster is vitreous and the color is rose red, pink, or brown. Rhodonite is similar in color to rhodochrosite, manganese carbonate, but it may be distinguished by its greater hardness and insolubility in hydrochloric acid. It has been found at Langban, Sweden; near Sverdlovsk in the Ural Mountains; and at Broken Hill, Australia. Fine crystals of a zinc-bearing variety, fowlerite, are found at Franklin, New Jersey. *See* SILICATE MINERALS.

Cornelius S. Hurlbut, Jr.

Rhodophyceae A large class of plants, commonly called red algae, coextensive with the division Rhodophycota. Most red algae are found in the ocean, growing on rocks, wood, other plants, or animals in the intertidal zone and to depths limited by the availability of light. A few genera and species occur in fresh water, and these are usually found in rapidly flowing, well-aerated, cold streams. Some, however, grow in quiet warm water, while a few are subaerial. Most red algae are photosynthetic, but some grow on other algae with varying degrees of parasitism. Approximately 675 genera and 4100 species are recognized. *See* ALGAE.

Rhodophyceae are characterized by a unique combination of biochemical, reproductive, and ultrastructural features. The primary photosynthetic pigment is chlorophyll *a*. Water-soluble tetrapyrrolic compounds called phycobilins serve as accessory photosynthetic pigments. The chief food reserve, floridean starch, is a branched polymer of glucose similar to amylopectin of green plants. It occurs as granules in the cytoplasm. Multinucleate cells are common. Rhodophyceae are distinctive among eukaryotic algae in their lack of flagella, a feature shared among major groups only by the chlorophycean order Zygnematales. Some unicellular forms and many spores and male gametes of multicellular forms are capable of gliding or feeble ameboid motion. Unicellular red algae, which may form mucilaginous colonies, are considered primitive. Most red algae have multicellular thalli of microscopic or macroscopic size, including individual filaments, blades, and complex plants of distinctive form produced by the interplay of filamentous systems. *See* CORALLINALES; ZYGNEMATALES.

Two subclasses of Rhodophyceae are traditionally recognized: the Bangiophycidae and the Florideophycidae. Classification within the Bangiophycidae is based largely on vegetative and asexual reproductive features, while that of the Florideophycidae is based primarily on details of the development of the female reproductive system and carposporophyte, secondarily on vegetative features. Rhodophyceae seem most closely related to Cyanophyceae in their use of chlorophyll *a* and phycobilins as photosynthetic pigments and the absence of flagella. They probably did not evolve directly from Cyanophyceae, however, but from a colorless, nonflagellate, eukaryotic ancestor that acquired pigments from an endosymbiotic blue-green alga.

The most salient feature of red algae is their beauty, which has drawn admiration from generations of seaside visitors. Their greatest significance, however, is their role in the formation of coral reefs, the Corallinales being responsible for cementing together various animal and algal components. Of more apparent economic importance is their use as food, a centuries-old tradition of maritime peoples in many parts of the world. *See* AGAR; CARRAGEENAN; REEF.

Paul C. Silva; Richard L. Moe

Rhombifera A diverse class of extinct, marine, stalked echinoderms whose stratigraphic range extends from the Late Cambrian (505 MA) through Upper Devonian (360 MA). The class is named for rhombus-shaped structures (rhombs) that circulated seawater or body fluids through the body for the respiratory extraction of dissolved oxygen. The main body, or theca, held the visceral mass and was attached to the seafloor by a short to long stem. Atop the theca were variously developed small appendages that were used for suspension feeding in slowly moving currents. This class is composed of three distantly related groups: Glyptocystitida, Hemicosmititida, and Caryocystitida. A fourth group, Polycosmititida, is too poorly known to assign to Rhombifera with certainty.

The glyptocystitid *Cheirocystis fulotonensis*.

Glyptocystitids (see illustration) comprise the bulk of rhombiferans. They include most of the variation with sessile and mobile species, and thecal shapes including barrel-shaped, spherical, and flattened triangular. *See* ECHINODERMATA.

Colin D. Sumrall

Rhubarb A herbaceous perennial, *Rheum rhaponticum*, of Mediterranean origin, belonging to the plant order Polygonales. Rhubarb is grown for its thick petioles which are used mainly as a cooked dessert; it is frequently called the pieplant. The leaves, which are high in oxalic acid content, are not commonly considered edible. Outdoor rhubarb is a common garden vegetable in most areas of the United States except the South. Michigan and Washington are important centers for forced or hothouse rhubarb. *See* POLYGONALES.

H. John Carew

Rhynchobdellae An order of the class Hirudinea. These leeches possess an eversible proboscis and lack hemoglobin in the blood. They may be divided into two families, the Glossiphoniidae and the Ichthyobdellidae. Glossiphoniidae are flattened, mostly small leeches occurring chiefly in freshwater. Ichthyobdellidae typically have cylindrical bodies with conspicuous, powerful suckers used to attach themselves to passing fish. They frequently have lateral appendages which aid in respiration. *See* HIRUDINEA.

Kenneth H. Mann

Rhynchocephalia One of the two surviving clades of lepidosaurian reptiles, represented today by only two species, *Sphenodon punctatus* and *Sphenodon guntheri*, commonly known as tuatara. *Sphenodon* (a name meaning wedge-tooth) has a fully diapsid skull that was secondarily acquired from an ancestral state represented by *Gephyrosaurus*, the most primitive known member of the clade Rhynchocephalia. Typically, the teeth of rhynchocephalians are fused to the edges of the jaws (a condition known as acrodont). *See* DIAPSIDA; LEPIDOSAURIA.

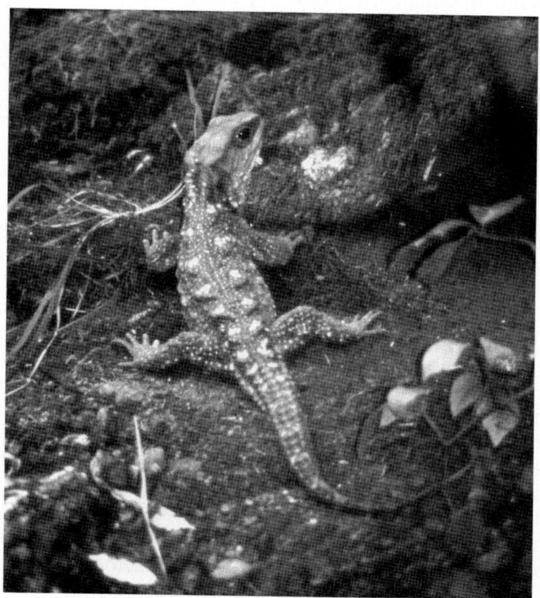

Young *Sphenodon* (tuatara). (*Photo by Susan Evans*)

The living New Zealand tuatara, or *Sphenodon*, is a medium-sized (2–2$^1/_2$ ft, or 60–80 m) reptile, rather like a heavy-bodied iguana in appearance but very different in its lifestyle (see illustration). Tuatara is a Maori word meaning "peaks on the back," a reference to the jagged dorsal crest. Once widespread on the main islands of New Zealand, they are now restricted to about 30 small offshore islets, threatened mainly by introduced animals such as pack-rats and cats. *Sphenodon* is unique among lepidosaurs in having only very rudimentary hemipenes (the paired male copulatory organs that characterize lizards and snakes) and an extended, tortoiselike lifespan. It also possesses a well-developed parietal eye, a light-sensitive organ on the top of the head that helps in the control of seasonal cycles. By day, tuatara live in shallow burrows excavated by seabirds; at night they emerge to feed. The typical diet consists of worms, snails, and insects, most notably large local crickets called wetas. Fertilization is internal. The eggs are buried and take 12–14 months to develop. *See* REPTILIA.

Susan E. Evans

Rhynchonellida An extant order of brachiopods that has been an important component of marine benthic communities since the Ordovician. Rhynchonellids possess unequally biconvex valves typically with a fold and sulcus; many species contain strong radial ribs that produce deflections in the commissure, the line of junction between the two valves. Their shells are generally impunctate and also typically lack a hinge line parallel to the hinge axis, resulting in a pointed beak or umbo when viewed in lateral profile. Internally, rhynchonellids possess calcareous processes (crura) that in extant species provide support for the lophophore. Rhynchonellids are sessile, attached, epifaunal suspension feeders. They have a functional pedicle that they use to attach to the substrate. Although rhynchonellids were never diverse compared to other brachiopod orders, they were commonly important members of local communities. They achieved a diversity peak in the Devonian and again in the Jurassic.

Rhynchonellids have shifted their habitat preference in the oceans since their origin in the Middle Ordovician when they originated in shallow low-latitude seas; however, presently they are more common in deep-water habitats from middle and high latitudes and are rare members of benthic communities in low-latitude shallow seas. *See* ARTICULATA (ECHINODERMATA); BRACHIOPODA.

Mark E. Patzkowsky

Rhynchonelliformea One of three subphyla currently recognized in the phylum Brachiopoda, a clade of lophophore-bearing protostome metazoan animals. The name is derived from the stratigraphically oldest subclade with extant representatives, the order Rhynchonellida. Rhynchonelliforms constitute over 90% of the more than 4500 named brachiopod genera, most of which (over 95%) are extinct. Rhynchonelliformea includes all taxa formerly referred to the class Articulata, in addition to several groups newly discovered as fossils, formerly placed in class Inarticulata, or of uncertain taxonomic affiliation.

All rhynchonelliforms have an organic-rich, bivalved calcite shell and can be distinguished from other brachiopods by the following shared, derived characters: the presence of a fibrous secondary shell layer, a pedicle without a coelomic core, a diductor (valve opening) muscle system, and the presence of articulatory structures between the two valves.

Rhynchonelliformea includes 19 orders organized in five classes.

Phylum Brachiopoda
- Subphylum Linguliformea
  - Craniiformea
  - Rhynchonelliformea
    - Class Chileata
      - Order: Chileida
        - Dictyonellida
    - Class Obolellata
      - Order: Obolellida
        - Naukatida
    - Class Kutorginata
      - Order: Kutorginida
    - Class Strophomenata
      - Order: Strophomenida
        - Productida
        - Orthotetida
        - Billingsellida
    - Class Rhynchonellata
      - Order: Protorthida
        - Orthida
        - Pentamerida
        - Rhynchonellida
        - Atrypida
        - Spiriferida
        - Spiriferinida
        - Thecideida
        - Athyridida
        - Terebratulida

Living rhynchonelliforms are distributed globally, and enjoy their greatest abundance and diversity in temperate latitudes. They live from rather deep intertidal to abyssal depths, reaching greatest abundance and diversity in the shallow to relatively deep (few hundred meters) subtidal. Extinct rhynchonelliforms, particularly in the Paleozoic Era, were far more abundant and diverse than today in virtually all geographic realms, and lived at nearly all depths. *See* BRACHIOPODA; RHYNCHONELLIDA Sandra J. Carlson

Rhynchosauria An order of herbivorous diapsid reptiles in the infraclass Archosauria, limited to the Triassic System but with a worldwide distribution. Rhynchosaurs were pig- or sheep-sized quadrupedal reptiles that were common in the Middle and Upper Triassic Series of India, South America, Europe, and eastern Africa. Except for the earliest South African genus, *Mesosuchus*, they are characterized by multiple rows of teeth on both the upper and lower jaws that were fused into deep sockets. Teeth were not replaced as they were worn but were added posteriorly as the jaw grew. In most genera, the premaxillae are devoid of teeth and overhang the front of the lower jaw like a beak. The external nostril is median rather than paired.

Rhynchosaurs were elements of the early archosauromorph radiation that included the protorosaurs and primitive archosaurs. The structure of the ankle is nearly identical in the early members of these groups, but later rhynchosaurs enlarged the centrals which contributed to a simple hinge joint between the lower leg and the tarsals.

Rhynchosaurs were locally common in the Late Triassic but are unknown in the Jurassic. Their rapid extinction may have resulted from changes in the vegetation on which they fed, or from predation from the expanding community of predaceous archosaurs, including early dinosaurs. *See* ARCHOSAURIA; DIAPSIDA; PROTOROSAURIA; REPTILIA. Robert L. Carroll

Rhyniophyta A division of the subkingdom Embryobionta. The bryophytes and vascular plants are included in this subkingdom. The category Rhyniophyta was devised for the relatively simple Silurian-Devonian vascular plants long held to be ancestral to other groups of vascular plants and usually referred to as Psilophytales. These plants have leafless stems and lack roots; their general morphological structure is not complex. The three classes of Rhyniophyta currently recognized are Rhyniopsida, Zosterophyllopsida, and Trimerophytopsida. *See* EMBRYOBIONTA; PSILOPHYTALES; RHYNIOPSIDA; TRIMEROPHYTOPSIDA; ZOSTEROPHYLLOPSIDA. Harlan P. Banks

Rhyniopsida The earliest demonstrable vascular land plants, appearing in Silurian (mid-Ludlovian) time. Their small, leafless axes were usually branched dichotomously in three planes; adventitious and perhaps pseudomonopodial branching also occurred. Sporangia were usually terminal on the main axes. Some terminated in lateral branches, and some were subtended by adventitious branches. *See* RHYNIOPHYTA. Harlan P. Banks

Rhyolite A very light-colored, aphanitic (not visibly crystalline), volcanic rock that is rich in silica and broadly equivalent to granite in composition. Migration of rhyolitic magma through the Earth's crust, which causes much of the Earth's explosive and hazardous volcanic activity, represents a major process of chemical fractionation by which continental crust grows and evolves. *See* GRANITE.

Rhyolites are formed by the process of molten silica-rich magma flowing toward the Earth's surface. Small differences in this process, notably those related to the release of gas from the magma at shallow depth, produce extremely diverse structural features. The high silica content gives rhyolitic lava a correspondingly high viscosity; this hinders crystallization and often causes young rhyolite to be a mixture of microcrystalline aggregates and glassy material. Because of the glassy nature of most rhyolites, they are best characterized by chemical analysis. They typically have 70–75 wt % silicon dioxide (SiO_2) and more potassium oxide (K_2O) than sodium oxide (Na_2O). *See* LAVA; MAGMA; VOLCANIC GLASS; VOLCANO.

Rhyolite is one of the most common volcanic rocks in continental regions; it is virtually absent in the ocean basins. The rock often occurs in large quantities associated with andesite and basalt. It is common in environments ranging from accretionary prisms at continental margins to magmatic arcs related to subduction zones. Rhyolite is also prevalent in extensional regions and hot spots in continental interiors. *See* ANDESITE; BASALT.

Leland W. Younker

Riboflavin Also known as vitamin B_2, riboflavin is widely distributed in nature, and is found mostly in milk, egg white, liver, and leafy vegetables. It is a water-soluble yellow-orange fluorescent pigment.

Riboflavin deficiency results in poor growth and other pathologic changes in the skin, eyes, liver, and nerves. Riboflavin deficiency in humans is usually associated with a cracking at the corners of the mouth called cheilosis; inflammation of the tongue, which appears red and glistening (glossitis); corneal vascularization accompanied by itching; and a scaly, greasy dermatitis about the corners of the nose, eyes, and ears. *See* VITAMIN.

Stanley N. Gershoff

Ribonuclease A group of enzymes, widely distributed in nature, which catalyze hydrolysis of the phosphodiester bonds in ribonucleic acid (RNA). The sites of hydrolysis may vary considerably, depending upon the specificity of the particular enzyme. Differences in specificity for the site of cleavage have led to the use of these various ribonucleases as tools in determining the structure and chemistry of RNA. The most thoroughly studied of the ribonucleases has been that isolated from bovine pancreas.

Research on ribonuclease has played a prime role in advancing the understanding of protein structure and function; also, it was the first protein to be totally synthesized from its component amino acids. The small size and great stability of ribonuclease, coupled with the ease of isolation of pure enzyme in large quantities, have contributed notably to its successful characterization. Since the elucidation of the amino acid sequence of ribonuclease, much information has been compiled with regard to the three-dimensional structure of the enzyme and to specific regions of the molecule that are catalytically important. *See* ENZYME; NUCLEIC ACID. Robert L. Heinrikson; M. Todd Washington

Ribonucleic acid (RNA) One of the two major classes of nucleic acid, mainly involved in translating the genetic information carried in deoxyribonucleic acid (DNA) into proteins. Various types of ribonucleic acids function in protein synthesis: transfer RNAs (tRNAs) and ribosomal RNAs (rRNAs) function in the synthesis of all proteins, while messenger RNAs (mRNAs) are a diverse set, each member of which acts specifically in the synthesis of one protein. Messenger RNA is the intermediate in the usual biological pathway DNA → RNA → protein. Other types of RNA serve other important functions for cells and viruses, such as the involvement of small nuclear RNAs (snRNAs) in mRNA splicing. In some cases, RNA performs functions typically considered DNA-like, such as serving as the genetic material for certain viruses, or roles typically carried out by proteins, such as RNA enzymes or ribozymes. *See* DEOXYRIBONUCLEIC ACID (DNA).

Structure and synthesis. RNA is a linear polymer of four different nucleotides. Each nucleotide is composed of three parts: a five-carbon sugar known as ribose, a phosphate group, and one of four bases attached to each ribose, adenine (A), cytosine (C), guanine (G), or uracil (U). The structure of RNA is basically a repeating chain of ribose and phosphate moieties, with one of the four bases attached to each ribose. The structure and function of the RNA vary depending on its sequence and length. *See* NUCLEOTIDE; RIBOSE.

In its basic structure, RNA is quite similar to DNA. It differs by a single change in the sugar group (ribose instead of deoxyribose) and by the substitution of uracil for the base thymine (T). Typically, RNA does not exist as long double-stranded chains as does DNA, but as short single chains with higher-order structure due to base pairing and tertiary interactions within the RNA molecule. Within the cell, RNA usually exists in association with specific proteins in a ribonucleoprotein (RNP) complex.

The nucleotide sequence of RNA is encoded in genes in the DNA, and it is transcribed from the DNA by a complementary templating mechanism that is catalyzed by one of the RNA polymerase enzymes. In this templating scheme, the DNA base T specifies A in the RNA, A specifies U, C specifies G, and G specifies C.

Transfer RNA. These small RNAs (70–90 nucleotides) act as adapters that translate the nucleotide sequence of mRNA into a protein sequence. They do this by carrying the appropriate amino acids to ribosomes during the process of protein synthesis. Each cell contains at least one type of tRNA specific for each of the 20 amino acids, and usually several types. The base sequence in the mRNA directs the appropriate amino acid–carrying tRNAs to the ribosome to ensure that the correct protein sequence is made. *See* PROTEIN; RIBOSOMES.

Ribosomal RNA. Ribosomes are complex ribonucleoprotein particles that act as workbenches for the process of protein synthesis, that is, the process of linking amino acids to form proteins. Each ribosome is made up of several structural rRNA molecules and more than 50 different proteins, and it is divided into two subunits, termed large and small. The RNA components of the ribosome account for more than half of its weight. Like tRNAs, rRNAs are stable molecules and exist in complex folded structures. Each of these rRNAs is closely associated with ribosomal proteins and is essential in determining the exact structure of the ribosome. In addition, the rRNAs, rather than the ribosomal proteins, are likely the basic functional elements of the ribosome.

Messenger RNA. Whereas most types of RNA are the final products of genes, mRNA is an intermediate in information transfer. It carries information from DNA to the ribosome in a genetic code that the protein-synthesizing machinery translates into protein. Specifically, mRNA sequence is recognized in a sequential fashion as a series of nucleotide triplets by tRNAs via base pairing to the three-nucleotide anticodons in the tRNAs. There are specific triplet codons that specify the beginning and end of the protein-coding sequence. Thus, the function of mRNA involves the reading of its primary nucleotide sequence rather than the activity of its overall structure. Messenger RNAs are typically shorter-lived than the more stable structural RNAs, such as tRNA and rRNA. *See* GENETIC CODE.

Small nuclear RNA. Small RNAs, generally less than 300 nucleotides long and rich in uridine (U), are localized in the nucleoplasm (snRNAs) and nucleolus (snoRNAs) of eukaryotic cells, where they take part in RNA processing. These RNAs have characteristic folded structures and function as ribonucleoproteins (snRNPs and snoRNPs), in association with specific proteins. Five snRNPs participate in intron removal during eukaryotic mRNA splicing, whereas many others participate in the extensive posttranscriptional modification that occurs during production of mature rRNA. *See* CELL NUCLEUS; INTRON.

Catalytic RNA. RNA enzymes, or ribozymes, are able to catalyze specific cleavage or joining reactions either in themselves or in other molecules of nucleic acid.

Viral RNA. While most organisms carry their genetic information in the form of DNA, certain viruses, such as polio and influenza viruses, have RNA as their genetic material. The viral RNAs occur in different forms in different viruses. For example, some are single-stranded and some are double-stranded; some occur as a single RNA chromosome while others are multiple. In any case, the RNA is replicated as the genetic material, and either its sequence, or a complementary copy of itself, serves as mRNA to encode viral proteins. The RNA viruses known as retroviruses contain an enzyme that promotes synthesis of complementary DNA in the host cell, thus reversing the typical flow of information in biological systems. *See* ANIMAL VIRUS; RETROVIRUS; VIRUS.

Other types of RNA. RNA molecules serve other important and diverse cellular functions. For example, a ribonucleoprotein enzyme (telomerase) is responsible for replication of chromosome ends, with the RNA component functioning as a template for the reverse transcriptase activity of telomerase. An essential RNA component is also present in the signal recognition particle (SRP), a ribonucleoprotein complex that directs ribosomes synthesizing membrane and secreted proteins to the endoplasmic reticulum, where synthesis and posttranslational processing are completed.

Ann L. Beyer; Michael W. Gray

Ribosomes Small particles, present in large numbers in every living cell, whose function is to convert stored genetic information into protein molecules. In this synthesis process, a molecule of messenger ribonucleic acid (mRNA) is fed through the ribosome, and each successive trinucleotide codon on the messenger is recognized by complementary base-pairing to the anticodon of an appropriate transfer RNA (tRNA) molecule, which is in turn covalently bound to a specific amino acid. The successive amino acids become linked together on the ribosome, forming a polypeptide chain whose amino acid sequence has thus been determined by the nucleic acid sequence of the mRNA. The polypeptide is subsequently folded into an active protein molecule. Ribosomes are themselves complex arrays of protein and RNA molecules, and their fundamental importance in molecular biology has prompted a vast amount of research, with a view to finding out how these particles function at the molecular level. *See* GENETIC CODE; PROTEIN; RIBONUCLEIC ACID (RNA).

Ribosomes are composed of two subunits, one approximately twice the size of the other. In the bacterium *Escherichia coli*, whose ribosomes have been the most extensively studied, the smaller subunit (30S) contains 21 proteins and a single 16S RNA molecule. The larger (50S) subunit contains 32 proteins, and two RNA molecules (23S and 5S). The overall mass ratio of RNA to protein is about 2:1. Cations, in particular magnesium and polyamines, play an important role in maintaining the integrity of the ribosomal structures. The ribosomes are considerably larger in the cytoplasm of higher organisms (eukaryotes). Nevertheless, all ribosomal RNA molecules have a central core of conserved structure, which presumably reflects the universality of the ribosomal function. *See* CELL (BIOLOGY); ORGANIC EVOLUTION.

The process of protein biosynthesis is essentially very similar in both prokaryotes and eukaryotes; what follows is a brief summary of what happens in *E. coli*. The first step is that an initiator tRNA molecule attached to the amino acid *N*-formyl methionine recognizes its appropriate codon on a mRNA molecule, and binds with the mRNA to the 30S subunit. A 50S subunit then joins the complex, forming a complete 70S ribosome. A number of proteins (initiation factors, which are not ribosomal proteins) are also involved in the process. At this stage, the initiator aminoacyl tRNA occupies one binding site, the P-site (peptidyl site), on the ribosome, while a second tRNA binding site (the

A-site, or aminoacyl site) is free to accept the next aminoacyl tRNA molecule. In the subsequent steps, the elongation process aminoacyl tRNA molecules are brought to the A-site as ternary complexes together with guanosine triphosphate (GTP) and a protein factor (elongation factor Tu). Once an aminoacyl tRNA is in the A-site, the initiator amino acid (or at later stages the growing polypeptide chain) is transferred from the P-site tRNA to the A-site aminoacyl tRNA. It is not clear whether this peptidyl transferase activity requires the active participation of a ribosomal component. After peptide transfer has taken place, the peptide is attached to the A-site tRNA, and an "empty" tRNA molecule is at the P-site. The peptidyl tRNA complex must now be translocated to the P-site, in order to free the A-site for the next incoming tRNA molecule. Here again protein factors and GTP are involved, and it is probable that the empty tRNA occupies a third ribosomal site (exit or E-site) before finally leaving the complex. Studies show that elongation factor G hydrolyzes GTP to bring about translocation of the tRNA molecules. The protein chain is completed by the appearance of a "stop" codon in the mRNA. This is recognized by yet another set of protein factors, which causes the completed polypeptide chain to be released from the ribosome. At any one time a number of ribosomes are engaged in the reading of a single mRNA molecule, which leads to the appearance of polyribosomes (polysomes). *See* MOLECULAR BIOLOGY.

R. Brimacombe; H. G. Wittmann; S. Joseph

Ribozyme A ribonucleic acid (RNA) molecule that, like a protein, can catalyze specific biochemical reactions. Examples include self-splicing rRNA and RNase P, both involved in catalyzing RNA processing reactions (that is, the biochemical reactions that convert a newly synthesized RNA molecule to its mature form). Different ribozyme structures catalyze quite distinct RNA processing reactions, just as protein enzyme families that are composed of different structures catalyze different types of biochemical reactions.

Ribozymes share many similarities with protein enzymes, as assessed by two parameters that are used to describe a biological catalyst. The Michaelis-Menten constant K_m relates to the affinity that the catalyst has for its substrate, and ribozymes possess K_m values which are comparable to K_m values of protein enzymes. The catalytic rate constant describes how efficiently a catalyst converts substrate into product. The values of this constant for ribozymes are markedly lower than those values observed for protein enzymes. Nevertheless, ribozymes accelerate the rate of chemical reaction with specific substrates by 10^{11} compared with the rate observed for the corresponding uncatalyzed, spontaneous reaction. Therefore, ribozymes and protein enzymes are capable of lowering to similar extents the activation energy for chemical reaction. *See* ENZYME; PROTEIN; RIBONUCLEIC ACID (RNA).

Daniel W. Celander

Rice The plant *Oryza sativa* is the major source of food for nearly one-half of the world's population. The most important rice-producing countries are mainland China, India, and Indonesia, but in many smaller countries rice is the leading food crop. In the United States, rice production is largely concentrated in selected areas of Arkansas, California, Louisiana, and Texas. *See* CARBOHYDRATE; CYPERALES; WHEAT.

Over 95% of the world rice crop is used for human food. Although most rice is boiled, a considerable amount is consumed as breakfast cereals. Rice starch also has many uses. Broken rice is used as a livestock feed and for the production of alcoholic beverages. The bran from polished rice is used for livestock feed; the hulls are used for fuel and cellulose. The straw is used for thatching roofs in the Orient and for making paper, mats, hats, and baskets. Rice straw is also woven into rope and used as cordage for bags. This crop serves a multitude of purposes in countries where agriculture is dependent largely upon rice.

Rice is unlike many other cereal grains in that all cultivated varieties belong to the same species and have 12 pairs of chromosomes, as do most wild types. The extent of variation in morphological and physiological characteristics within this single species is greater than for any other cereal crop. *See* GENETICS.

Rice is an annual grass plant varying in height from 2 to 6 ft (0.6 to 1.8 m). Plants tiller, that is, develop new shoots freely, the number depending upon spacing and soil fertility. The inflorescence is an open panicle. Flowers are perfect and normally self-pollinated, with natural crossing seldom exceeding 3–4%. A distinct characteristic of the flower is the six anthers rather than the customary three of other grasses. Spikelets have a single floret, lemma and palea completely enclosing the caryopsis or fruit, which may be yellow, red, brown, or black. Lemmas may be awnless, partly awned, or fully awned. Threshed rice, which retains its lemma and palea, is called rough rice or paddy. *See* FLOWER; FRUIT; GRASS CROPS; INFLORESCENCE; PLANT REPRODUCTION.

In the United States, only about 25 varieties are in commercial production. Cultivated rices are classified as upland and lowland. Upland types, which can be grown in high-rainfall areas without irrigation, produce relatively low yields. The lowland types, which are grown submerged in water for the greater part of the season, produce higher yields. In contrast to most plants, rice can thrive when submerged because oxygen is transported from the leaves to the roots. All rice in the United States is produced under lowland or flooded conditions. Rice varieties are also classified as long- or short-grain. Most long-grain rices have high amylose content and are dry or fluffy when cooked, while most short-grain rices have lower amylose content and are sticky when cooked. In the United States a third grain length is recognized: medium-grain. The medium-grain rices have cooking qualities similar to those of short-grain varieties. *See* AGRICULTURAL SCIENCE (PLANT); GRAIN CROPS.

J. N. Rutger

The rice kernel has four primary components: the hull or husk, the seedcoat or bran, the embryo or germ, and the endosperm. The main objective of milling rice is to remove the indigestible hull and additional portions of bran to yield whole unbroken endosperm. Rice milling involves relatively uncomplicated abrasive and separatory procedures which provide a variety of products dependent on the degree of bran removal or the extent of endosperm breakage.

Instant rice is made from whole grain rice by pretreating under controlled cooking, cooling, and drying conditions to impart the quick-cooking characteristic. Ready-to-eat breakfast rice cereals are prepared from milled rice as flakes or puffs. Rice bran oil was developed as a result of increased extraction of lipids from rice bran. It is utilized as an edible-grade oil in a variety of applications as well as an industrial feedstock for soap and resin manufacture. *See* CEREAL; FAT AND OIL (FOOD); FOOD MANUFACTURING; SOLVENT EXTRACTION.

Mark A. Uebersax

Ricinulei An order of extremely rare arachnids, also known as the Podogona, with a body less than 1 in. (2.5 cm) in length. Superficially, they resemble ticks in general appearance and movement, and are found only in tropical Africa and in the Americas, from the Amazon to Texas. The two anterior pairs of appendages are chelate. Less than 25 modern species are known. The occurrence of several fossils from Carboniferous time suggests that the group was formerly more common. *See* ARACHNIDA.

C. Clayton Hoff

Rickettsioses Often severe infectious diseases caused by several diverse and specialized bacteria, the rickettsiae and rickettsia-like organisms. The best-known rickettsial diseases infect humans and are usually transmitted by parasitic arthropod vectors.

Rickettsiae and rickettsia-like organisms are some of the smallest microorganisms visible under a light microscope. Although originally confused with viruses, in part because of their small size and requirements for intracellular replication, rickettsiae and rickettsia-like organisms are characterized by basic bacterial (gram-negative) morphologic features. Their key metabolic enzymes are variations of typical bacterial enzymes. The genetic material of rickettsiae and rickettsia-like organisms likewise seems to conform to basic bacterial patterns. The genome of all rickettsia-like organisms consists of double-stranded deoxyribonucleic acid (DNA).

Rickettsiae enter host cells by phagocytosis and reproduce by simple binary fission. The site of growth and reproduction varies among the various genera.

Clinically, the rickettsial diseases of humans are most commonly characterized by fever, headache, and some form of cutaneous eruption, often including diffuse rash, as in epidemic and murine typhus and Rocky Mountain spotted fever, or a primary ulcer or eschar at the site of vector attachment, as in Mediterranean spotted fever and scrub typhus. Signs of disease may vary significantly between individual cases of rickettsial disease. Q fever is clinically exceptional in several respects, including the frequent absence of skin lesions.

All of the human rickettsial diseases, if diagnosed early enough in the infection, can usually be effectively treated with the appropriate antibiotics. Tetracycline and chloramphenicol are among the most effective antibiotics used; they halt the progression of the disease activity, but do so without actually killing the rickettsial organisms. Presumably, the immune system is ultimately responsible for ridding the body of infectious organisms. Penicillin and related compounds are not considered effective. *See* ANTIBIOTIC.

Most rickettsial diseases are maintained in nature as diseases of nonhuman vertebrate animals and their parasites. Human infection may usually be regarded as peripheral to the normal natural infection cycles, and human-to-human transmission is not the rule. However, the organism responsible for epidemic typhus (*Rickettsia prowazekii*) and the agent responsible for trench fever (*Rochalimaea quintana*) have the potential to spread rapidly within louse-ridden human populations. *See* ZOONOSES.

All known spotted fever group organisms are transmitted by ticks. Despite a global distribution in the form of various diseases, nearly all spotted fever group organisms share close genetic, antigenic, and certain pathologic features. Examples of human diseases include Rocky Mountain spotted fever (in North and South America), fièvre boutonneuse or Mediterranean spotted fever (southern Europe), South African tick-bite fever (Africa), Indian tick typhus (Indian subcontinent), and Siberian tick typhus (northeastern Europe and northern Asia). If appropriate antibiotics are not administered, Rocky Mountain spotted fever, for example, is a life-threatening disease. *See* INFECTIOUS DISEASE.

Russell L. Regnery

Riemann surface A generalization of the complex plane that was originally conceived to make sense of mathematical expressions such as $\sqrt{z}$ or $\log z$. These expressions cannot be made single-valued and analytic in the punctured plane $\mathbf{C}\backslash\{0\}$ (that is, the complex plane with the point 0 removed). The difficulty is that for some closed paths the value of the expression when reaching the end of the path is not the same as it is at the beginning. For example, the closed path can be chosen to be the unit circle centered at $z = 0$ and followed counterclockwise from $z = 1$. If $\sqrt{z}$ is assigned the value $+1$ at $z = 1$, its value at the end of the circuit is -1. Similarly, if $\log z$ is assigned the value 0 at $z = 1$, at the end of the circuit, allowing the values to change continuously, the value is $2\pi i$. *See* COMPLEX NUMBERS.

The construction of an abstract surface for $\sqrt{z}$ on which $\sqrt{z}$ has a single value at each point and on which the values of $\sqrt{z}$ vary continuously around each closed path, always ending with the starting value, proceeds as follows. From the complex plane $\mathbf{C}$ the nonnegative half of the real axis $\{z = x + iy : y = 0, 0 \leq x\}$ is removed. This slit is thought of as having two edges: an upper or + edge and a lower or − edge ($z = 0$ is not counted). Another copy of this slit plane is then placed over, and parallel to, the original, as the first and second floors in a building. Next, the two sheets are connected by attaching the − edge of the bottom sheet to the + edge of the top sheet, and the + edge of the bottom sheet to the − edge of the top sheet. The abstract surface thus constructed is not embedded in euclidean 3-space because the attachments cannot be done in 3-space without creating self-intersections. The expression $\sqrt{z}$ is defined at the point over z on the upper sheet to have positive imaginary part and on the lower to have negative part.

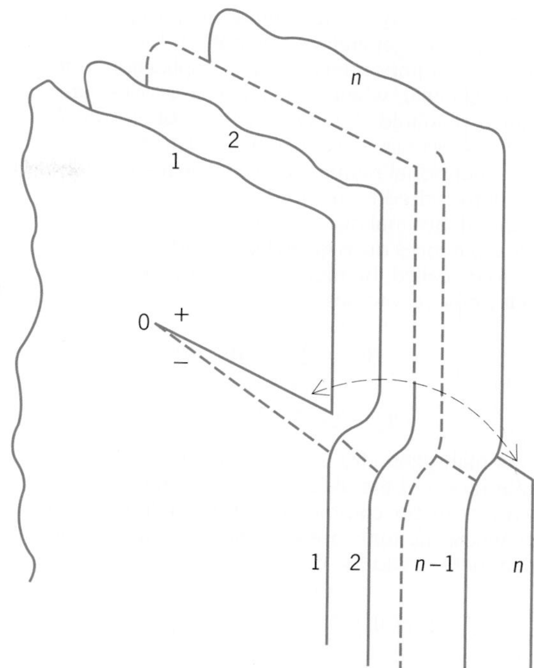

Riemann surface for $\sqrt[n]{z}$. ***(After A. D. Sveshnikob and A. N. Tikhonov, The Theory of Functions of a Complex Variable, Mir Publishers, 1973)***

The abstract surface constructed is called the Riemann surface for $\sqrt{z}$. It is two-sheeted over the z plane, except that only one point lies over $z = 0$, which is therefore called a branch point. The Riemann surface for $\sqrt[n]{z}$, which has n branches instead of 2, is shown in the illustration.

In the early 1920s, H. Weyl defined a Riemann surface in the context of the newly developing field of topology. This definition does not depend on having a particular function in hand, and does not require the use of such seemingly artificial devices as slits. A Riemann surface is a two-dimensional connected topological manifold (surface) with a complex structure. *See* MANIFOLD (MATHEMATICS); TOPOLOGY.

Albert Marden

Riemannian geometry The geometry of riemannian manifolds. Riemannian geometry was initiated by B. Riemann in 1854, following the pioneering work of C. F. Gauss on surface theory in 1827. Riemann introduced a coordinate space in which the infinitesimal distance between two neighboring points is specified by a quadratic differential form, given below. Such a space is a natural generalization of euclidean geometry and Gauss's geometry of surfaces in three-dimensional euclidean space, as well as the noneuclidean geometries: hyperbolic geometry (previously discovered by J. Bolyai and N. I. Lobachevsky) and elliptic geometry. A riemannian manifold is a topological space that further generalizes this notion. Riemannian geometry derives great importance from its application in the general theory of relativity, where the universe is considered to be a riemannian manifold. *See* DIFFERENTIAL GEOMETRY; EUCLIDEAN GEOMETRY; NONEUCLIDEAN GEOMETRY; RELATIVITY.

An n-dimensional riemannian space is a space whose points can be characterized by n coordinates u^i ($i = 1, 2, \ldots, n$), and where an infinitesimal distance ds between two points with coordinate differences du^i is given by a quadratic differential form, Eq. (1), to be called the metric. In general, the quantities g_{ij} are functions of the coordinates u^i.

$$ds^2 = \sum_{i,j} g_{ij}\, du^i\, du^j \qquad (1)$$

$$g_{ij} = g_{ji} \qquad 1 \le i, j \le n$$

Riemannian geometry is local in the sense that it is valid in a neighborhood of the u^i coordinates. It studies properties that are invariant under coordinate transformations of the form of Eq. (2), where the functions are smooth with the functional determinant det $(\partial u'^i/\partial u^j) \neq 0$.

$$u'^i = u'^i(u^1, \ldots, u^n) \qquad 1 \le i \le n \qquad (2)$$

At the foundation of riemannian geometry is the notion of a riemannian manifold. A differentiable manifold is a topological space with a covering by coordinate neighborhoods, such that in the domain where the u^i coordinates and the u'^i coordinates are both valid, they are related by the transformation of Eq. (2) with the conditions mentioned there. The manifold is called riemannian if a riemannian metric is defined in each neighborhood that gives the same value of ds^2 in the intersection of any two neighborhoods. Riemannian geometry is divided between local and global: the former dealing with properties in a neighborhood, the latter with properties of the manifold as a whole. *See* MANIFOLD (MATHEMATICS); TOPOLOGY. S. S. Chern

Rift valley One of the geomorphological expressions between two tectonic plates that are opening relative to each other or sliding past each other. The term originally was used to describe the central graben structures of such classic continental rift zones as the Rhinegraben and the East African Rift, but the definition now encompasses mid-oceanic ridge systems with central valleys such as the Mid-Atlantic Ridge. *See* MID-OCEANIC RIDGE; PLATE TECTONICS.

Continental and oceanic rift valleys are end members in what many consider to be an evolutionary continuum. In the case of continental rift valleys, plate separation is incomplete, and the orientation of the stress field relative to the rift valley can range from nearly orthogonal to subparallel. Strongly oblique relationships are probably the norm. In contrast, oceanic rift valleys mark the place where the trailing edges of two distinctly different plates are separating. The separation is complete, and the spreading is organized and focused, resulting in rift valleys that tend to be oriented orthogonal or suborthogonal to the spreading directions.

The basic cross-sectional form of rift valleys consists of a central graben surrounded by elevated flanks. It is almost universally accepted that the central grabens of continental rift valleys are subsidence features. The crystalline basement floors of some parts of the Tanganyika and Malawi (Nyasa) rift valleys in East Africa lie more than 5 mi (9 km) below elevated flanks. *See* GRABEN.

In continental rift valleys the true cross-sectional form is typically asymmetric, with the rift floors tilted toward the most elevated flank. Most of the subsidence is controlled by one border fault system, and most of the internal faults parallel the dip of the border faults. *See* FAULT AND FAULT STRUCTURES.

Oceanic rift valleys are also distinctly separated into segments by structures known as transform faults. The cross-sectional form of oceanic rift valleys can be markedly asymmetric. It is unlikely that the cross-sectional form of oceanic rift valleys is related genetically to that of continental rift valleys, except in the broadest possible terms. *See* TRANSFORM FAULT. Bruce R. Rosendahl

Rift Valley fever An arthropod-borne (primarily mosquito), acute, febrile, viral disease of humans and numerous species of animals, particularly ruminants. Rift Valley fever is caused by a ribonucleic acid (RNA) virus in the genus *Phlebovirus* of the family Bunyaviridae. In sheep and cattle, it is also known as infectious enzootic hepatitis. First described in the Rift Valley of Africa, the disease presently occurs in west, east, and south Africa and has extended as far north as Egypt. Outside of Africa, Rift Valley fever has spread to the island of Madagascar and the Arabian Peninsula (Saudi Arabia and Yemen). Historically, outbreaks of Rift Valley fever have occurred at 10–15-year intervals in normally dry areas of Africa subsequent to a period of heavy rainfall.

In humans, clinical signs of Rift Valley fever are influenzalike, and include fever, headache, muscular pain, weakness, nausea, epigastric pain, and photophobia. Most people recover within 4–7 days, but some individuals may have impaired vision or blindness in one or both eyes; a small percentage of infected individuals develop encephalitis or a hemorrhagic syndrome and die.

Rift Valley fever should be suspected when the following observations are made in a disease outbreak: (1) high abortion rates (possibly approaching 100%) in ewes, cows, and female dogs, with lower rates in goats and in other ruminants; (2) high mortality (possibly approaching 100%) in lambs and calves less than 7 days old, and lower rates of disease and mortality in older animals; (3) extensive liver lesions in aborted fetuses and newborn animals; (4) an influenzalike disease in people, particularly in individuals associated with livestock; (5) occurrence of the disease during a period of high insect activity following heavy seasonal rainfall with persistent flooding; (6) rapid spread. The diagnosis is confirmed by isolating the virus from tissues of an infected animal or human.

Animals should not be moved from an infected area to a Rift Valley–free area if there is any indication of active disease. Control of the disease is best accomplished by widespread vaccination of susceptible animals to prevent amplification of the virus and, thus, infection of vectors. Any individual that works with infected animals or live virus in a laboratory should be vaccinated. *See* ANIMAL VIRUS; VACCINATION. Charles A. Mebus

Rigel The bright star in the southwest corner of the constellation Orion (apparent magnitude +0.12), also referred to as β Orionis. It is a blue-white supergiant of spectral type B8, one of the most luminous stars known. Its intrinsic brightness is estimated to be more than 50,000 times that of the Sun. Rigel is a very young star by stellar standards, with such an enormous rate of energy output that its life span is only a few million years. By comparison, the Sun is approximately 5×10^9 years old, and is still only half way through its main-sequence evolution. Rigel is the brightest member of the Orion OB1 Association, a large molecular cloud complex in which active star formation is currently taking place. In addition to its high luminosity, Rigel has a photosphere which is so large that if the star were placed where the Sun is, it would almost fill the orbit of Mercury. *See* MOLECULAR CLOUD; ORION; SPECTRAL TYPE; STAR; STELLAR ROTATION; SUPERGIANT STAR. David W. Latham

Rigid body An idealized extended solid whose size and shape are definitely fixed and remain unaltered when forces are applied. Treatment of the motion of a rigid body in terms of Newton's laws of motion leads to an understanding of certain important aspects of the translational and rotational motion of real bodies without the necessity of considering the complications involved when changes in size and shape occur. Many of the principles used to treat the motion of rigid bodies apply in good approximation to the motion of real elastic solids. *See* RIGID-BODY DYNAMICS.

Dudley Williams

Rigid-body dynamics The study of the motion of a rigid body under the influence of forces. A rigid body is a system of particles whose distances from one another are fixed. The general motion of a rigid body consists of a combination of translations (parallel motion of all particles in the body) and rotations (circular motion of all particles in the body about an axis). Its equations of motion can be derived from the equations of motion of its constituent particles. *See* RECTILINEAR MOTION; ROTATIONAL MOTION.

The location of a mass point m_i can be specified relative to a fixed-coordinate system by a position vector $\vec{r}_i$ with cartesian components (x_i,y_i,z_i). The vector force $\vec{F}_i$ which acts on the mass points has corresponding components (F_{ix},F_{iy},F_{iz}). Newton's second law for the motion of m_i is stated in Eq. (1). Here $\ddot{\vec{r}}_i \equiv$

$$\vec{F}_i = m_i\ddot{\vec{r}}_i \tag{1}$$

$d^2\vec{r}_i/dt^2 = \vec{a}$ is the acceleration of m_i. *See* ACCELERATION; FORCE; NEWTON'S LAWS OF MOTION.

Translational motion. If Eq. (1) is summed over all particles in the rigid body, the left-hand side becomes the total force acting

$$\vec{F} = \sum_i \vec{F}_i$$

on the rigid body. If the internal forces satisfy Newton's third law (to each action there is an equal but opposite reaction), the contributions of the internal forces cancel in pairs and $\vec{F}$ is the total external force on the rigid body, $\vec{F}^{\text{ext}}$. The right-hand side can be expressed in terms of the center-of-mass (CM) position vector defined by Eq. (2), where M is the total mass of the body. Then the sum of the equations of motion, Eq. (1), takes the form of Eq. (3). This equation of motion for the center of mass

$$\vec{R} = \frac{1}{M}\sum_i m_i\vec{r}_i \qquad M = \sum_i m_i \tag{2}$$

$$\vec{F}^{\text{ext}} = M\ddot{\vec{R}} \tag{3}$$

of the rigid body has exactly the form of the equation of motion for a particle of mass M and position $\vec{R}$, under the influence of an external force $\vec{F}^{\text{ext}}$. Consequently, the second law of motion holds, not just for a particle, but for an arbitrary rigid body, if the position of the body is interpreted to mean the position of its center of mass. *See* CENTER OF MASS.

The momentum of a mass point is given by the product of the mass and the velocity, $\vec{p}_i = m_i\dot{\vec{r}}_i$, where $\dot{\vec{r}}_i \equiv d\vec{r}_i/dt$. The total momentum of the center of mass of the rigid body, obtained by summing over the momenta $\vec{p}_i$ of its constituent masses, is given by Eq. (4). In terms of the center-of-mass momentum $\vec{P}$, the equation of motion for the center of mass is expressed by Eq. (5). For an isolated rigid body, the external force is zero and

$$\vec{P} = M\dot{\vec{R}} \tag{4}$$

$$\dot{\vec{P}} = \vec{F}^{\text{ext}} \tag{5}$$

therefore $\vec{P}$ is constant. According to Eq. (4), this implies that the center of mass moves with constant velocity $\vec{V} = \vec{P}/M$. *See* CONSERVATION OF MOMENTUM; MOMENTUM.

In fact, the preceding equations for translational motion hold for any body, rigid or nonrigid.

Rotational motion. The total angular momentum of a rigid body about a point O with coordinate $\vec{r}_O$ is the sum of the angular momenta of its constituent masses, and is given by Eq. (6). Here

$$\vec{L}_O = \sum_i (\vec{r}_i - \vec{r}_O) \times m_i(\dot{\vec{r}}_i - \dot{\vec{r}}_O) \tag{6}$$

$\times$ denotes the cross-product of the coordinate vector $(\vec{r}_i - \vec{r}_O)$ with the momentum vector $m_i(\dot{\vec{r}}_i - \dot{\vec{r}}_O)$. The time derivative of $\vec{L}$ is given in Eq. (7). Hereafter the point O is taken to be either a

$$\dot{\vec{L}}_O = \sum_i (\vec{r}_i - \vec{r}_O) \times m_i(\ddot{\vec{r}}_i - \ddot{\vec{r}}_O) \tag{7}$$

fixed point (in which case $\dot{\vec{r}}_O = \ddot{\vec{r}}_O = 0$) or the center-of-mass point. Using the equation of motion (1), $m_i\ddot{\vec{r}}_i$ can be replaced by $\vec{F}_i$. Thus the rotational equation of motion (8) is obtained.

$$\dot{\vec{L}}_O = \sum_i (\vec{r}_i - \vec{r}_O) \times \vec{F}_i \tag{8}$$

The right-hand side of Eq. (8) is known as the torque, $\vec{N}$. The contribution of the internal forces to the torque vanishes if the "extended third law" holds; namely, action equals reaction and is directed along a line between the particles. In this circumstance the rotational equation of motion is given by Eqs. (9), where $\vec{N}^{\text{ext}}$

$$\dot{\vec{L}}_0 = \vec{N}_0^{\text{ext}} \tag{9a}$$

$$\vec{N}_0^{\text{ext}} = \sum_i (\vec{r}_i - \vec{r}_O) \times \vec{F}_i^{\text{ext}} \tag{9b}$$

is the total torque associated with external forces that act on the rigid body. *See* ANGULAR MOMENTUM; TORQUE.

It is straightforward to show from Eq. (6) that the angular momentum about an arbitrary point O is related to the angular momentum about the center of mass by Eq. (10). The torque about an arbitrary point O can also be easily related to the torque about the center of mass, by Eq. (11).

$$\vec{L}_O = \vec{L}_{\text{CM}} + (\vec{R} - \vec{r}_O) \times \vec{P} \tag{10}$$

$$\vec{N}_O^{\text{ext}} = \vec{N}_{\text{CM}}^{\text{ext}} + (\vec{R} - \vec{r}_O) \times \vec{F}^{\text{ext}} \tag{11}$$

Six coordinates determine the positions of all particles in a rigid body, and the motion of a rigid body is described by six equations of motion. The translational motion of the center of mass is determined by Eq. (5), and the rotational motion about the center of mass, or a fixed point, is determined by Eq. (9). These six equations, which hold for any system of particles, completely describe the motion of a rigid body.

Motion of an isolated system. The equation $\dot{\vec{L}} = \vec{N}^{\text{ext}}$ has the same form as $\dot{\vec{P}} = \vec{F}^{\text{ext}}$. Both $\vec{L}$ and $\vec{P}$ are constants for an isolated system since $\vec{N}^{\text{ext}} = 0$ and $\vec{F}^{\text{ext}} = 0$. Even though the two conditions $\vec{N}^{\text{ext}} = 0$ and $\vec{F}^{\text{ext}} = 0$ appear similar, there are some important differences for systems in which internal motion is possible. If $\vec{F}^{\text{ext}} = 0$, a center of mass which is at rest will remain so, regardless of internal forces or internal motion. If $\vec{N}^{\text{ext}} = 0$, the total angular momentum is constant, and if initially zero, will remain zero. However, $\vec{L} = 0$ does not exclude changes in orientation of the system by the use of merely internal forces. There is no rotational analog to the equation $\vec{r}(t) = (\vec{P}/M)t + \vec{r}(0)$ for linear motion of the center of mass.

Static equilibrium. In the design of permanent structures, the conditions under which a rigid body remains in steady motion under the action of a set of forces are of great importance. The six conditions for complete equilibrium of a rigid body are given in Eqs. (12). However, in many circumstances equilibrium

$$\vec{F}^{\text{ext}} = \textstyle\sum_i \vec{F}_i^{\text{ext}} = 0 \qquad \vec{N}_{\text{CM}}^{\text{ext}} = \textstyle\sum_i (\vec{r}_i - \vec{R}) \times \vec{F}_i^{\text{ext}} = 0 \tag{12}$$

is desired only for a subset of the six independent directions of motion. To illustrate, the external force in the direction of motion of an accelerating automobile is nonzero, but equilibrium is maintained in all other directions. *See* STATICS. Vernon D. Barger

Rinderpest An acute or subacute, contagious viral disease of ruminants and swine, manifested by high fever, lachrymal discharge, profuse diarrhea, erosion of the epithelium of the mouth and of the digestive tract, and high mortality. Rinderpest (also known as cattle plague) is caused by a ribonucleic acid (RNA) virus classified in the genus *Morbillivirus* within the Paramyxoviridae family. This virus is closely associated with the viruses of human measles, peste des petits ruminants of sheep and goats, canine distemper, phocine distemper, and the dolphin morbillivirus. Although only one serotype of rinderpest virus is known, the various field strains, grouped by molecular characterization into three lineages known as the Asian lineage and African lineages 1 and 2, vary significantly in virulence. The rinderpest virus is easily inactivated by heat and survives outside the host for a short time. Therefore, transmission of rinderpest is by direct contact between animals. *See* ANIMAL VIRUS; MEASLES; PARAMYXOVIRUS.

Rinderpest is one of the oldest diseases known to affect cattle. It is characterized by the development of high fever that lasts for several days until just prior to death. Cattle suffering from rinderpest develop a very severe, profuse diarrhea and a drop in the number of white blood cells. The morbidity in susceptible cattle and buffalo is greater than 90%, with death of almost all clinically affected animals. *See* DIARRHEA.

Rinderpest affects virtually all cloven-hoofed animals, both domesticated and wild. Cattle, Asian buffaloes, yaks, African buffaloes, lesser kudus, elands, warthogs, and giraffes are particularly susceptible. All swine are susceptible and may suffer severe disease. Modified (attenuated), live rinderpest vaccines generate a strong, protective immune response that lasts for life. One such vaccine developed in East Africa has been widely used in Africa and Asia and has proved safe and efficacious in large-scale vaccination campaigns and outbreak control. *See* VACCINATION. Alfonso Torres; Peter Roeder

Ring A tie member or chain link. Tension or compression applied through the center of a ring produces bending moment, shear, and normal force on radial sections. Because shear stress is zero at the boundaries of the section where bending stress is maximum, it is usually neglected. W. J. Krefeld; W. G. Bowman

Ring-opening polymerization The formation of macromolecules from cyclic monomers such as cyclic hydrocarbons, ethers, esters, amides, siloxanes, and sulfur (eight-membered ring). Thus, ring-opening polymerization is of particular interest, since macromolecules of almost any chemical structure can be prepared.

Polymerization is initiated with the breaking of a single (sigma) bond in the cyclic monomer. The driving force of ring-opening polymerization comes from the strain of the rings. The major source of the ring strain is the angular strain, with the release of the strain being energetically favorable.

As the ring size increases, ring strain decreases, although this change is not absolute. Some cyclic monomers, such as six-membered cyclic ethers, are virtually strain-free, and their polymerization to a high-molecular-weight polymer is not possible. By contrast, six-membered cyclic esters and siloxanes (for example, hexamethylcyclotrisiloxane) polymerize easily, due to their particular conformation. Conformational strain (related to the opposition of hydrogen atoms across the rings) is responsible for the strain for some of the larger cyclic monomers.

The majority of heterocyclic monomers polymerize by ionic mechanisms. That is, either anions or cations located at the end of a growing macromolecule attack the monomer molecule, breaking a bond between the heteroatom (for example, nitrogen or oxygen) and the adjacent atom so that both bonding electrons remain with one of the atoms (heterolytic cleavage). *See* HETEROCYCLIC COMPOUNDS; POLYMER; POLYMERIZATION. S. Penczek

Ring theory The mathematical term ring is used to designate a type of algebraic system with two compositions satisfying most but not all the properties of addition and multiplication in the system of integers, $0, \pm 1, \pm 2, \ldots$. In precise terms a ring is a set R with two binary compositions called addition and multiplication whose results on an ordered pair (a,b), a,b in R, are denoted by $a + b$ and ab, respectively.

These compositions must satisfy the following conditions:

C. $a + b$ and ab belong to R (closure).

Al. $a + b = b + a$ (commutative law).

A2. $(a + b) + c = a + (b + c)$ (associative law).

A3. There exists an element 0 (called zero) in R satisfying $a + 0 = a$ for every a in R.

A4. For each a in R there exists an element $-a$ (called the negative of a) in R such that $a + (-a) = 0$.

M1. $(ab)c = a(bc)$.

D. $a(b + c) = ab + ac$; $(b + c)a = ba + ca$ (distributive laws).

In the ring I of integers (addition and multiplication as usual) there are further conditions, for example, the commutative law of multiplication ($ab = ba$) and the cancellation law that if $a \neq 0$ and $ab = ac$, then $b = c$. *See* SET THEORY.

The importance of the concept of a ring stems from the fact that it embraces many special cases which are fundamental in all branches of mathematics. Thus it includes the ring I of integers, the ring R_0 of rational numbers, the ring $R^{\#}$ of real numbers, the ring C of complex numbers, various rings of functions, rings of matrices, and so on.

The conditions A1–A4 on the addition composition are exactly equivalent to the statement that any ring is a commutative group relative to its addition composition. This group is called the additive group of the ring. The algebraic system consisting of the set of elements of a ring together with its multiplication composition is called the multiplicative semigroup of the ring. *See* GROUP THEORY.

Various classes of rings are singled out by imposing conditions on the multiplicative semigroup. Thus integral domains are rings in which the product of nonzero elements is nonzero. Division rings are rings whose sets of nonzero elements are groups relative to the multiplication composition, and fields are division rings satisfying the commutative law of multiplication. Nathan Jacobson

Ripple voltage The time-varying part of a voltage that is ideally time-invariant. Most electronic systems require a direct-current voltage for at least part of their operation. An ideal direct-current voltage is available from a battery, but batteries are impractical for many applications. To obtain a direct-current voltage from the alternating-current power mains requires using some type of power supply.

A typical linear power supply system configuration (see illustration) consists of a transformer to change the voltage at the mains to the desired level, a rectifier to convert the alternating-current input voltage v_1 to a pulsating direct-current voltage v_2, followed by a low-pass filter. The output voltage v_{out} of the filter consists of a large direct-current voltage with a superimposed alternating-current voltage. This remaining superimposed alternating-current voltage is called the ripple voltage.

Practical linear power supplies often include a voltage regulator between the low-pass filter and the load. The voltage regulator is usually an electronic circuit that is specifically designed to provide a very stable dc output voltage even if large variations occur in the input. Nonlinear power supplies, which are often termed switching power supplies or switched-mode power

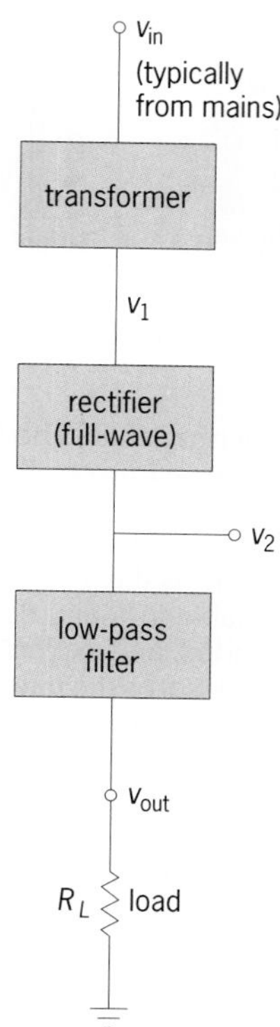

System configuration of a power supply system.

supplies, are becoming increasingly popular as a practical alternative for producing a low-ripple dc output. *See* ELECTRIC FILTER; ELECTRONIC POWER SUPPLY; RECTIFIER; VOLTAGE REGULATOR.

Stanley G. Burns

Risk assessment and management The scientific study of risk, the potential realization of undesirable consequences from hazards arising from a possible event, the assessment of the acceptability of the risks, and the management of unacceptable risks. For example, the probability of contracting lung cancer (unwanted consequence) is a risk caused by carcinogens (hazards) contained in second-hand tobacco smoke (event). The risk is estimated using scientific methods, and then the acceptability of that risk is assessed by public health officials. Risk management is the term for the systematic analysis and control of risk, such as prohibiting smoking in public places. Risks are caused by exposure to hazards. Sudden hazards are referred to as acute (for example, a flash flood caused by heavy rains); prolonged hazards are referred to as chronic (for example, carcinogens in second-hand tobacco smoke and polluted air).

The definition of risk contains two components: the probability of an undesirable consequence of an event and the seriousness of that consequence. In the example of a flash flood, risk can be defined as the probability of having a flood of a given magnitude. Sometimes the probability is expressed as a return period, which means, for instance, that a flood of a specified magnitude is expected to occur once every 100 years. The scope of a flood can be expressed as the level or stage of a river, or the dollar amount of property damage.

Most human activities involve risk. The risk of driving, for example, can be subdivided according to property damage, human injuries, fatalities, and harm to the environment. Even the stress and lack of exercise due to driving create health risks. Although risk pervades modern society and is widely acknowledged, it continues to cause unending controversy and debate.

Risk estimates are seldom accurate to even two orders of magnitude, and widely varying perceptions of risk by different interest groups can add confusion and conflict to the risk management process. Environmental risk assessment is laden with uncertainty, particularly with respect to the quantification of chemical emissions; the nature of contaminant transport (such as the region over which a chemical may spread and the velocity of movement) in the water, air, and soil; the type of exposure pathway (such as inhalation, ingestion, and dermal contact); the effects on people based on dose-response studies (which are extrapolated from animals); ecological impacts; and so forth.

Thousands of natural and other hazards are subjected to the statistical analysis of mortality and morbidity data. Society selects a small number of risks to manage, but often some high risks (such as radon in houses) may not be managed, while some low risks (such as movement of dangerous goods) may be selected for management. Management alternatives include banning of the hazard (drugs), regulating the hazard (drivers' tests and licensing), controlling the release and exposure of hazardous materials, treatment after exposure, and penalties for damages. Each management alternative may be analyzed to estimate the impact on risk.

Risk estimates are uncertain, are described in technical language, and are outside the general understanding or experience of most people. Perception plays a crucial role, tending to exaggerate the significance, for example, of risks that are involuntary, catastrophic, or newsworthy. Effective risk management therefore requires effective risk communication.

Risk assessment is the evaluation of the relative importance of an estimated risk with respect to other risks faced by the population, the benefits of the activity source of the risk, and the costs of managing the risk. For risks due to long-term exposure to chemicals, the risk assessment activity generally incorporates the estimation of the response of people to the exposure (that is, risk analysis is a part of risk assessment). The methods used include studies on animals, exposure of tissues, and epidemiology. *See* ENVIRONMENTAL ENGINEERING; ENVIRONMENTAL TOXICOLOGY; HAZARDOUS WASTE.

Keith W. Hipel; John Shortreed

Ritz's combination principle The empirical rule, for mulated by W. Ritz in 1905, that sums and differences of the frequencies of spectral lines often equal other observed frequencies. The rule is an immediate consequence of the quantum-mechanical formula $hf = E_{i\cdot} - E_f$ relating the energy hf of an emitted photon to the initial energy $E_{i\cdot}$ and final energy E_f, of the radiating system; h is Planck's constant and f is the frequency of the emitted light. *See* ATOMIC STRUCTURE AND SPECTRA; ENERGY LEVEL (QUANTUM MECHANICS); QUANTUM MECHANICS.

Edward Gerjuoy

River A natural, fresh-water surface stream that has considerable volume compared with its smaller tributaries. The tributaries are known as brooks, creeks, branches, or forks. Rivers are usually the main stems and larger tributaries of the drainage systems that convey surface runoff from the land. Rivers flow from headwater areas of small tributaries to their mouths, where they may discharge into the ocean, a major lake, or a desert basin.

Rivers flowing to the ocean drain about 68% of the Earth's land surface. Regions draining to the sea are termed exoreic, while those draining to interior closed basins are endoreic. Areic regions are those which lack surface streams because of low rainfall or lithoogic conditions.

Sixteen of the largest rivers account for nearly half of the total world river flow of water. The Amazon River alone carries nearly 20% of all the water annually discharged by the world's rivers. Rivers also carry large loads of sediment. The total sediment load for all the world's rivers averages about 22×10^9 tons (20×10^9 metric tons) brought to the sea each year. Sediment loads for individual rivers vary considerably. The Yellow River of northern China is the most prolific transporter of sediment. Draining an agricultural region of easily eroded loess, this river averages about 2×10^9 tons (1.8×10^9 metric tons) of sediment per year, one-tenth of the world average. *See* DEPOSITIONAL SYSTEMS AND ENVIRONMENTS; LOESS.

River discharge varies over a broad range, depending on many climatic and geologic factors. The low flows of the river influence water supply and navigation. The high flows are a concern as threats to life and property. However, floods are also beneficial. The ancient Egyptian civilization was dependent upon the Nile River floods to provide new soil and moisture for crops. Floods are but one attribute of rivers that affect human society. Means of counteracting the vagaries of river flow have concerned engineers for centuries. In modern times many of the world's rivers are managed to conserve the natural flow for release at times required by human activity, to confine flood flows to the channel and to planned areas of floodwater storage, and to maintain water quality at optimum levels. *See* FLOODPLAIN; RIVER ENGINEERING. Victor R. Baker

River engineering A branch of civil engineering that involves the control and utilization of rivers for the benefit of humankind. Its scope includes river training, channel design, flood control, water supply, navigation improvement, hydraulic structure design, hazard mitigation, and environmental enhancement. River engineering is also necessary to provide protection against floods and other river disasters. The emphasis is often on river responses, long-term and short-term, to changes in nature, and stabilization and utilization, such as damming, channelization, diversion, bridge construction, and sand or gravel mining. Evaluation of river responses is essential at the conceptual, planning, and design phases of a project and requires the use of fundamental principles of river and sedimentation engineering. *See* CANAL; DAM; RIVER; STREAM TRANSPORT AND DEPOSITION. Howard H. Chang

River tides Tides that occur in rivers with an open connection to a tidal estuary or ocean. These tides are highly modified from coastal ocean tides as they propagate landward into shallow water, often through a brackish estuary and then into a fresh-water river. Ocean tides form in response to the gravitational forces exerted by the Sun and Moon acting on the water masses on the surface of the rotating Earth. Within estuaries or rivers, water masses are small enough that tidal motions are not generated directly from lunar and solar forces; rather, the motions occur because of tide-generating forces at the estuary entrance or river mouth. However, variations in the tide-generating forces in a coastal ocean produce corresponding variations in tides within the upland estuary/river system. *See* ESTUARINE OCEANOGRAPHY; RIVER; TIDE.

Typically, tidal range increases as the width of the estuary or river narrows. Both high- and low-water levels rise with bed level, and tidal range begins to decrease. River tides propagate upstream until friction dissipates the tidal energy or some other physical impediment is encountered.

Tidal currents in rivers are the horizontal motions of water produced by the river tides. In the lower reaches of an estuary or river, tidal currents are usually bidirectional but become unidirectional (but still oscillatory) further upstream. In a river, tidal current reversals may occur far upstream under conditions of large tidal amplitudes and low river flows. Tidal currents are highly modified by geometry and bathymetry, with stronger currents occurring in deep, narrow channels and weaker currents occurring in wide, shallow areas. Upstream-directed currents that occur on rising tides are referred to as flood currents and downstream-directed currents that occur on falling tides are referred to as ebb currents. Slack water occurs at times when the flow changes direction and there is little or no water motion. Jeffrey W. Gartner

Rivet A short rod with a head formed on one end. A rivet is inserted through aligned holes in two or more parts to be joined; then by pressing the protruding end, a second head is formed to hold the parts together permanently. The first head is called the manufactured head and the second one the point. In forming the point, a hold-on or dolly bar is used to back up the manufactured head and the rivet is driven, preferably by a machine riveter.

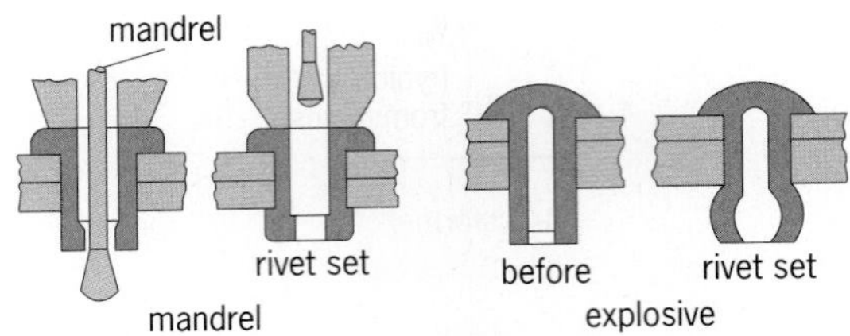

Two types of blind rivet.

Blind rivets are special rivets that can be set without access to the point. They are available in many designs but are of three general types: screw, mandrel, and explosive (see illustration). In the mandrel type the rivet is set as the mandrel is pulled through. In the explosive type an explosive charge in the point is set off by a special hot iron; the explosion expands the point and sets the rivet. Paul H. Black

Robotics A field of engineering concerned with the development and application of robots, and computer systems for their control, sensory feedback, and information processing. The many types of robotic systems include robotic manipulators, robotic hands, mobile robots, walking robots, aids for disabled persons, telerobots, and microelectromechanical systems.

The term "robotics" has been broadly interpreted. It includes research and engineering activities involving the design and development of robotic systems. Planning for the use of industrial robots in manufacturing or evaluation of the economic impact of robotic automation can also be viewed as robotics. This breadth of usage arises from the interdisciplinary nature of robotics, a field involving mechanisms, computers, control systems, actuators, and software. *See* BIOMECHANICS; COMPUTER; CONTROL SYSTEMS; CYBERNETICS; ELECTRICAL ENGINEERING; INDUSTRIAL ENGINEERING; MECHANICAL ENGINEERING; SOFTWARE ENGINEERING.

Robots produce mechanical motion that, in most cases, results in manipulation or locomotion. Mechanical characteristics for robotic mechanisms include degrees of freedom of movement, size and shape of the operating space, stiffness and strength of the structure, lifting capacity, velocity, and acceleration under load. Performance measures include repeatability and accuracy of positioning, speed, and freedom from vibration.

A robot control system directs the motion and sensory processing of a robot or system of cooperating robots. The controller may consist of only a sequencing device for simple robots, although most multiaxis industrial robots today employ servo-controlled positioning of their joints by a microprocessor-based system.

The robot sensory system gathers specific information needed by the control system and, in more advanced systems, maintains an internal model of the environment to enable prediction and decision making. The joint position transducers on industrial robots provide a minimal sensory system for many industrial applications, but other sensors are needed to gather data about the external environment. Sensors may detect position, velocity, acceleration, visual, proximity, acoustic, force-torque, tactile, thermal, and radiation data.

As information moves up from the sensory device, the amount of information increases and the speed of data acquisition decreases. These control architectures form the basis for computer integrated manufacturing (CIM), a hierarchical approach to organizing automated factories. A new paradigm has emerged, based on the interconnection of intelligent system elements that can learn, reason, and modify their configuration to satisfy overall system requirements. One of the most important of these approaches is based on holonic systems. *See* AUTOMATION; COMPUTER-INTEGRATED MANUFACTURING; INTELLIGENT MACHINE.

A telerobotic system augments humans by allowing them to extend their ability to perform complex tasks in remote locations. It is a technology that couples the human operator's visual, tactile, and other sensory perception functions with a remote manipulator or mobile robot. These systems are useful for performing tasks in environments that are dangerous or not easily accessible for humans. Telerobotic systems are used in nuclear handling, maintenance in space, undersea exploration, and servicing electric transmission lines. Perhaps the most important sensory data needed for telepresence are feedback of visual information, robot position, body motion and forces, as well as tactile information. Master-slave systems have been developed in which, for example, a hand controller provides control inputs to an articulated robotic manipulator. These systems are capable of feeding back forces felt by the robot to actuators on the exoskeletal master controller so that the operator can "feel" the remote environment. *See* HUMAN-MACHINE SYSTEMS; REMOTE MANIPULATORS.

Graphical simulation is used to design and evaluate a workcell layout before it is built. The robot motion can be programmed on the simulation and downloaded to the robot controller. Companies market software systems that include libraries of commercially available robots and postprocessors for off-line robot programming. *See* SIMULATION. William A. Gruver

Roche limit The closest distance which a satellite, revolving around a parent body, can approach the parent without being pulled apart tidally. The simplest formal definition is that the Roche limit is the minimum distance at which a satellite can be in equilibrium under the influence of its own gravitation and that of the central mass about which it is describing a circular orbit. If the satellite is in a circular orbit and has negligible mass, the same density as the primary, and zero tensile strength, the Roche limit is 2.46 times the radius of the primary.

When a star has exhausted the supply of hydrogen in its core, its radius will increase by a factor of 10 to 100. A star in a binary system may then exceed its Roche limit, material will thus escape from that star, and its companion will receive the excess material. *See* BINARY STAR. P. K. Seidelmann

Rock A relatively common aggregate of mineral grains. Some rocks consist essentially of but one mineral species (monomineralic, such as quartzite, composed of quartz); others consist of two or more minerals (polymineralic, such as granite, composed of quartz, feldspar, and biotite). Rock names are not given for those rare combinations of minerals that constitute ore deposits, such as quartz, pyrite, and gold. In the popular sense rock is considered also to denote a compact substance, one with some coherence; but geologically, friable volcanic ash also is a rock. A genetic classification of rocks is shown below.

- Igneous
  - Intrusive
    - Plutonic (deep)
    - Hypabyssal (shallow)
  - Extrusive
    - Flow
    - Pyroclastic (explosive)
- Sedimentary
  - Clastic (mechanical or detrital)
  - Chemical (crystalline or precipitated)
  - Organic (biogenic)
- Metamorphic
  - Cataclastic
  - Contact metamorphic and pyrometasomatic
  - Regional metamorphic (dynamothermal)
- Hybrid
  - Metasomatic
  - Migmatitic

Exceptions to the requirement that rocks consist of minerals are obsidian, a volcanic rock consisting of glass; and coal, a sedimentary rock which is a mixture of organic compounds. *See* COAL; IGNEOUS ROCKS; METAMORPHIC ROCKS; OBSIDIAN; ROCK MECHANICS; SEDIMENTARY ROCKS; VOLCANIC GLASS. E. William Heinrich

Rock, electrical properties of The effect of changes in pressure and temperature on electrical properties of rocks. There has been increasing interest in the electrical properties of rocks at depth within the Earth and the Moon. The reason for this interest has been consideration of the use of electrical properties in studying the interior of the Earth and its satellite, particularly to depths of tens or hundreds of kilometers. At such depths pressures and temperatures are very great, and laboratory studies in which these pressures and temperatures are duplicated have been used to predict what the electrical properties at depth actually are. More direct measurements of the electrical properties deep within the Earth have been made by using surface-based electrical surveys of various sorts. An important side aspect of the study of electrical properties has been the observation that, when pressures near the crushing strength are applied to a rock, marked changes in electrical properties occur, probably caused by the development of incipient fractures. Such changes in resistivity might be used in predicting earthquakes, if they can be measured in the ground. *See* GEOELECTRICITY; GEOPHYSICAL EXPLORATION.

Attempts to measure the electrical properties of rocks to depths of tens or hundreds of kilometers in the Earth indicate that the Earth's crust is zoned electrically. The surface zone, with which scientists are most familiar, consists of a sequence of sedimentary rocks, along with fractured crystalline and metamorphic rocks, all of which are moderately good conductors of electricity because they contain relatively large amounts of water in pore spaces and other voids. This zone, which may range in thickness from a kilometer to several tens of kilometers, has conductivities varying from about $\frac{1}{2}$ ohm-m in recent sediments to 1000 ohm-m or more in weathered crystalline rock.

The basement rocks beneath this surface zone are crystalline, igneous, or metamorphic rocks which are much more dense, having little pore space in which water may collect. Since most rock-forming minerals are good insulators at normal temperatures, conduction of electricity in such rocks is determined almost entirely by the water in them. As a result, this part of the Earth's crust is electrically resistant (the resistivity lies in the range 10,000–1,000,000 ohm-m).

At rather moderate depths beneath the surface of the second zone, resistivity begins to decrease with depth. This decrease is considered to be the result of higher temperatures, which almost certainly are present at great depths. High temperatures lead to partial ionization of the molecular structure of minerals composing a rock, and the ions render even the insulating minerals conductive. George V. Keller

Rock age determination Finding the age of rocks based on the presence of naturally occurring long-lived radioactive isotopes of several elements in certain minerals and rocks. Measurements of rock ages have enabled geologists to reconstruct the geologic history of the Earth from the time of its formation 4.6×10^9 years ago to the present. Age determinations of rocks from the Moon have also contributed to knowledge of the history of the Moon, and may someday be used to study the history of Mars and of other bodies within the solar system. *See* EARTH, AGE OF; RADIOACTIVITY.

Many rocks and minerals contain radioactive atoms that decay spontaneously to form stable atoms of other elements. Under certain conditions these radiogenic daughter atoms accumulate within the mineral crystals so that the ratio of the daughter atoms divided by the parent atoms increases with time. This ratio can be measured very accurately with a mass spectrometer, and is then used to calculate the age of the rock by means of an equation

Parent-daughter pairs used for dating rocks and minerals

Parent	Daughter	Half-life, 10^9years
Potassium-40	Argon-40	11.8
Potassium-40	Calcium-40	1.47
Rubidium-87	Strontium-87	48.8
Samarium-147	Neodymium-143	107
Rhenium-187	Osmium-187	43
Thorium-232	Lead-208	14.008
Uranium-235	Lead-207	0.7038
Uranium-238	Lead-206	4.468

based on the law of radioactivity. The radioactive atoms used for dating rocks and minerals have very long half-lives, measured in billions of years. They occur in nature only because they decay very slowly. The pairs of parents and daughters used for dating are listed in the table. *See* DATING METHODS.

The rubidium-strontium method is based on rubidium-87, which decays to stable strontium-87 (^{87}Sr) by emitting a beta particle from its nucleus. The abundance of the radiogenic strontium-87 therefore increases with time at a rate that is proportional to the Rb/Sr ratio of the rock or mineral. The method is particularly well suited to the dating of very old rocks such as the ancient gneisses near Godthaab in Greenland, which are almost 3.8×10^9 years old. This method has also been used to date rocks from the Moon and to determine the age of the Earth by analyses of stony meteorites.

The potassium-argon method is based on the assumption that all of the atoms of radiogenic argon-40 that form within a potassium-bearing mineral accumulate within it. This assumption is satisfied only by a few kinds of minerals and rocks, because argon is an inert gas that does not readily form bonds with other atoms. The K-Ar method of dating has been used to establish a chronology of mountain building events in North America beginning about 2.8×10^9 years ago and continuing to the present. In addition, the method has been used to date reversals of the polarity of the Earth's magnetic field during the past 1.3×10^7 years. *See* OROGENY; PALEOMAGNETISM.

The uranium, thorium-lead method is based on uranium and thorium atoms which are radioactive and decay through a series of radioactive daughters to stable atoms of lead (Pb). Minerals that contain both elements can be dated by three separate methods based on the decay of uranium-238 to lead-206, uranium-235 to lead-207, and thorium-232 to lead-208. The three dates agree with each other only when no atoms of uranium, thorium, lead, and of the intermediate daughters have escaped. Only a few minerals satisfy this condition. The most commonly used mineral is zircon ($ZrSiO_4$), in which atoms of uranium and thorium occur by replacing zirconium. *See* LEAD ISOTOPES (GEOCHEMISTRY); RADIOACTIVE MINERALS.

The common-lead method is based on the common ore mineral galena (PbS) which consists of primordial lead that dates from the time of formation of the Earth and varying amounts of radiogenic lead that formed by decay of uranium and thorium in the Earth. The theoretical models required for the interpretation of common lead have provided insight into the early history of the solar system and into the relationship between meteorites and the Earth.

The fission-track method is based on uranium-238 which can decay both by emitting an alpha particle from its nucleus and by spontaneous fission. The number of spontaneous fission tracks per square centimeter is proportional to the concentration of uranium and to the age of the sample. When the uranium content is known, the age of the sample can be calculated. This method is suitable for dating a variety of minerals and both natural and manufactured glass. Its range extends from less than 100 years to hundreds of millions of years. *See* FISSION TRACK DATING.

The samarium-neodymium method of dating separated minerals or whole-rock specimens is similar to the Rb-Sr method. The Sm-Nd method is even more reliable than the Rb-Sr method of dating rocks and minerals, because samarium and neodymium are less mobile than rubidium and strontium. The isotopic evolution of neodymium in the Earth is described by comparison with stony meteorites. *See* METEORITE.

The rhenium-osmium method is based on the beta decay of naturally occurring rhenium-187 to stable osmium-187. It has been used to date iron meteorites and sulfide ore deposits containing molybdenite. Gunter Faure

Rock burst A seismic event that is caused by the mining of underground or surface openings in a high-stress environment. The Mine Safety and Health Administration (MSHA) defines a rock burst as "a sudden and violent failure of a large volume of overstressed rock, resulting in the instantaneous release of large amounts of accumulated energy." Underground, a rock burst sounds like a loud blast. On the surface, a rock burst sounds like a sonic boom, while the ground and structures on the surface shake as if a small earthquake had occurred.

Rock bursts mainly occur in deep underground hard-rock mines. They also occur in some soft-rock mines, such as coal, potash, and trona ($NaHCO_3 \cdot Na_2CO_3$). A rock burst is a very serious hazard to underground mining personnel. Worldwide, there are over 100 fatalities per year from rock bursts. Because of the violent nature of a rock burst, its occurrence often causes significant damage to nearby underground openings. *See* ORE AND MINERAL DEPOSITS; ROCK MECHANICS; UNDERGROUND MINING. Wilson Blake

Rock cleavage A secondary, planar structure of deformed rocks. A cleavage is penetrative and systematic, as opposed to fractures and shear zones which may occur alone or in widely spaced sets. It is generally better developed in fine grained rocks than in coarse ones. Application of the term derives from the ability to split rocks along the structure.

Simple cleavages are generally parallel to the axial surfaces of folds. This is true for folds formed by uniform flow of the rock mass where primary layering behaves as a passive marker. When rock layers are buckled, the primary layering behaving as mechanical discontinuities, cleavage that develops early may be fanned by subsequent growth of the fold. Cleavage is also associated with faults. *See* FAULT AND FAULT STRUCTURES; FOLD AND FOLD SYSTEMS.

Continuous (microscopically penetrative) cleavages, as in slates, are the earliest tectonic fabric elements that can be recognized in rocks. Cleavage also develops in rocks of lower deformational and metamorphic grade than slates. In these rocks, the cleavage occurs as discrete surfaces or seams, often coated with a film of clay or carbonaceous material; the cleavage surfaces are separated by zones of undeformed sedimentary rock. The surfaces may be smooth, anastomosing, or stylolitic. *See* METAMORPHISM; ROCK MECHANICS; SLATE. David B. Bieler

Rock magnetism The permanent and induced magnetism of rocks and minerals on scales ranging from the atomic to the global, including applications to magnetic field anomalies and paleomagnetism. Natural compasses, concentrations of magnetite (Fe_3O_4) called lodestones, are one of humankind's oldest devices. W. Gilbert in 1600 discovered that the Earth itself is a giant magnet, and speculated that its magnetism might be due to subterranean lodestone deposits. Observations by B. Brunhes in 1906 that some rocks are magnetized reversely to the present Earth's magnetic field, and by M. Matuyama in 1929 that reversely and normally magnetized rocks correspond to different geological time periods, made it clear that geomagnetism is dynamic, with frequent reversals of north and south poles.

Nevertheless, permanent magnetism of rocks remains important because it alone provides a memory of the intensity, direction, and polarity of the Earth's magnetic field in the geological past. From this magnetic record comes much of the evidence for continental drift, sea-floor spreading, and plate tectonics. *See* CONTINENTAL DRIFT; GEOMAGNETISM; MAGNET; MAGNETISM; PALEOMAGNETISM; PLATE TECTONICS.

The magnetism of rocks arises from the ferromagnetism or ferrimagnetism of a few percent or less of minerals such as magnetite. The magnetic moments of neighboring atoms in such minerals are coupled parallel or antiparallel, creating a spontaneous magnetization M_S. All magnetic memory, including that of computers, permanent magnets, and rocks, is due to the spontaneous and permanent nature of this magnetism. Spontaneous magnetization requires no magnetic field to create it, and cannot be demagnetized.

The magnetism can be randomized on the scale of magnetic mineral grains because different regions of a crystal tend to have their M_S vectors in different directions. These regions are called magnetic domains. Grains so small that they contain only one domain (single-domain grains) are the most powerful and stable paleomagnetic recorders. Larger, multidomain grains can also preserve a paleomagnetic memory, through imbalance in the numbers, sizes, or directions of domains, but this memory is more easily altered by time and changing geological conditions. *See* FERRIMAGNETISM; FERROMAGNETISM; MAGNETISM; MAGNETITE; MAGNETIZATION; MAGNETOMETER. David J. Dunlop

Rock mechanics The scientific discipline that deals with understanding how rocks deform and fail due to natural and human-induced forces at scales ranging from micrometers to kilometers. Rock mechanics is also the engineering discipline for designing and stabilizing underground and surface excavations and rock foundations.

The deformation and failure of rocks influence our lives in many ways. Structures that are built with or into rocks include highways, dams, bridges, tunnels, mines, and water and petroleum wells, to name some. These structures must be designed to remain stable over their expected lifespan. Natural occurrences of rock failure include earthquakes, volcanic eruptions, and landslides. These geologic hazards must be understood and methods for predicting rock failure events must be developed. *See* ENGINEERING GEOLOGY; GEODYNAMICS.

Discontinuities cause the strength of a rock mass to be much weaker than the intact rock. Before an engineering structure is made in a rock mass, it is very important to characterize the properties of the discontinuities, a process called rock-mass characterization. Rock-mass characterization involves the careful measurement of discontinuity type, orientation, spacing, length, roughness, fill, and other properties. This information can be collected in the field from existing rock outcrops or from borehole drill core.

Rock-mass classification is another aspect of rock-mass characterization. Rock-mass classification uses subsets of the rock characterization data to determine numerical or descriptive ratings. These ratings can then be used to estimate underground support requirements, rock-mass strength and deformation, and other engineering design criteria.

Rocks deform and fail due to forces, including gravity, tectonic, seismic, water, and thermal. Force is not a useful measure of rock strength because it depends on the size of the rock specimen subjected to the force; that is, it takes a small force to break a small rock sample and a larger force to break a larger sample of the same rock type. Instead, we use stress as a measure of rock strength. Rock failure will occur at a point in a rock when the stress at that point exceeds the strength of the rock.

Strain is a measure of the deformation of a rock. Strain is defined as the change in length divided by the original length, and is unitless.ε Under conditions in the shallow crust, rock is brittle and cannot strain very much before failure. A strain of 0.05 (0.5%) is usually sufficient to cause most rocks to fail. *See* STRESS AND STRAIN.

In many circumstances, a rock engineering design will be unstable without artificial rock support. Rock support technologies include rock bolts, cable bolts, shotcrete, concrete, steel sets, wire mesh, and other methods.

New technologies and research are essential elements of the field of rock mechanics. for modeling rock-mass deformation and failure. The numerical programs that are used in rock mechanics include finite-element models, boundary-element models, distinct-element models, finite difference models, micromechanical models, and many others. Technologies being applied to rock-mass characterization for increasing accuracy, safety, access, and human bias include digital image processing, ground-based lidar, Geographical Information Systems (GIS), Geographical Positioning Systems (GPS), remote sensing, and geophysics. *See* FINITE ELEMENT METHOD; GEOGRAPHIC INFORMATION SYSTEMS; GEOPHYSICS; IMAGE PROCESSING; LIDAR; REMOTE SENSING. John M. Kemeny

Rock varnish A dark coating on rock surfaces exposed to the atmosphere. Rock varnish is probably the slowest-accumulating sedimentary deposit, growing at only a few micrometers to tens of micrometers per thousand years. Its thickness ranges from less than 5 μm to over 600 μm, and is typically 100 μm or so. Although found in all terrestrial environments, varnish is mostly developed and well preserved in arid to semiarid deserts; thus, another common name is desert varnish. Rock varnish is composed of about 30% manganese and iron oxides, up to 70% clay minerals, and over a dozen trace and rare-earth elements. The building blocks of rock varnish are mostly blown in as airborne dust. Although the mechanism responsible for the formation of rock varnish remains unclear, two hypotheses have been proposed to explain the great enrichment in manganese within varnish (typically over 50 times compared with the adjacent environment such as soils, underlying rock, or dust). The abiotic hypothesis assumes that small changes in pH can concentrate manganese by geochemical processes. The biotic hypothesis suggests that bacteria, and perhaps other microorganisms, concentrate manganese; this is supported by culturing experiments and direct observations of bacterial enhancement of manganese. *See* SEDIMENTARY ROCKS.

Since rock varnish records environmental, especially climatic, events that are regionally or even globally synchronous, varnish microstratigraphy can be used as a tool for age dating. Without radiometric calibration, varnish microstratigraphy itself may be used to estimate relative ages of varnished geomorphic or archeological features in deserts. Once calibrated, varnish microstratigraphy can provide numerical age estimates for geomorphic and archeological features. Specifically, for petroglyphs and geoglyphs, the layering patterns of rock varnish hold the greatest potential for assigning ages. *See* ARCHEOLOGICAL CHRONOLOGY; DATING METHODS; ROCK AGE DETERMINATION. Tanzhuo Liu

Rocket Either a propulsion system or a complete vehicle driven by such a propulsive engine. A rocket engine provides the means whereby chemical matter is burned to release the energy stored in it and the energy is expended, by ejection at high velocity of the products of combustion (the working fluid). The ejection imparts motion to the vehicle in a direction opposite to that of the ejected matter. A rocket vehicle is propelled by rocket reaction and includes all components necessary for such propulsion, and a payload such as an explosive charge, scientific instruments, or a crew. A rocket vehicle also includes guidance and control equipment mounted in a structural air-frame or spaceframe. *See* ROCKET PROPULSION; ROCKET STAGING; SATELLITE (SPACECRAFT); SPACE FLIGHT; SPACE PROBE. George P. Sutton

Rocket astronomy The discipline that makes use of sounding rockets that fly near-vertical paths carrying scientific instruments to altitudes ranging from 25 to more than 900 mi (40 to 1500 km). Altitudes up to 30 mi (48 km) can be reached by balloons, so sounding rockets are typically used for higher altitudes in order to measure emissions from the Sun or other celestial sources that do not penetrate the Earth's atmosphere. Sounding rockets do not achieve orbital velocity; after completion of the launch phase, the payload follows a ballistic trajectory that permits 5–15 min of data taking before reentry. *See* ROCKET.

In comparison with experiments launched on satellites, rockets offer the advantages of simplicity, relatively frequent access to launch opportunities, a shorter time scale from conception to reality, lower cost, and recoverability of the payload and the possibility of postflight instrument calibration, refurbishment, and reflight. Major disadvantages are short observing time, localized coverage, and size and weight restrictions on payloads. *See* SATELLITE ASTRONOMY.

Scientific rocket programs focus on the disciplines of aeronomy, magnetospheric physics, meteorology, and material sciences, as well as astronomy and astrophysics. The ability to carry out vertical profile measurements of relevant atmospheric parameters at heights of 25–125 mi (40–200 km) is essential in many of these scientific disciplines. To the extent that the Sun influences or controls conditions in the upper atmosphere and magnetosphere of the Earth, there is a strong connection between solar astronomy and the more local research areas. *See* AERONOMY; IONOSPHERE; MAGNETOSPHERE.

The emission from the solar corona is dominated by ultraviolet and x-ray photons. During the late 1950s and early 1960s, techniques were developed for focusing x-rays and thereby providing direct imaging of the corona. These early studies revealed the highly structured nature of the atmosphere, with approximately semicircular loops of hot plasma outlining the shape of the underlying magnetic field, which confines the hot gas. Among the notable advances in coronal studies from rockets has been the development of a new technique for x-ray imaging, with conconitant improvements in imaging spectroscope methods: the use of multilayer coatings for enhanced x-ray reflectivity. *See* SUN; X-RAY ASTRONOMY; X-RAY TELESCOPE.

Observations of nonsolar sources from sounding rockets are hindered by the low intensity of the emission and the problem of pointing the payload at the source during the flight. Leon Golub

Rocket propulsion The process of imparting a force to a flying vehicle, such as a missile or a spacecraft, by the momentum of ejected matter. This matter, called propellant, is stored in the vehicle and ejected at high velocity. In chemical rockets the propellents are chemical compounds that undergo a chemical combustion reaction, releasing the energy for thermodynamically accelerating and ejecting the gaseous reaction products at high velocities. Chemical rocket propulsion is thus differentiated from other types of rocket propulsion, which use nuclear, solar, or electrical energy as their power source and which may use mechanisms other than the adiabatic expansion of a gas for achieving a high ejection velocity. Propulsion systems using liquid propellants (such as kerosine and liquid oxygen) have traditionally been called rocket engines, and those that use propellants in solid form have been called rocket motors. *See* ELECTROTHERMAL PROPULSION; INTERPLANETARY PROPULSION; ION PROPULSION; PLASMA PROPULSION; PROPULSION; SPACECRAFT PROPULSION.

Performance. The performance of a missile or space vehicle propelled by a rocket propulsion system is usually expressed in terms of such parameters as range, maximum velocity increase of flight, payload, maximum altitude, or time to reach a given target. Propulsion performance parameters (such as rocket exhaust velocity, specific impulse, thrust, or propulsion system weight) are used in computing these vehicle performance criteria. The table gives typical performance values. *See* SPECIFIC IMPULSE; THRUST.

Typical performance values of rocket propulsion systems*

Propulsion system parameter	Typical range of values
Specific impulse at sea level	180–390 s
Specific impulse at altitude	215–470 s
Exhaust velocity at sea level	5800–15,000 ft/s (1800–4500 m/s)
Combustion temperature	4000–7200°F (2200–4000°C)
Chamber pressures	100–3000 lb/in.^{2} (0.7–20 MPa)
Ratio of thrust to propulsion system weight	20–150
Thrust	0.01–6.6 × 10^6 lb (0.05–2.9 × 10^7 n)†
Flight speeds	0–50,000 ft/s (0–15,000 m/s)

*Exact values depend on application, propulsion system design, and propellant selection.

†Maximum value applies to a cluster; for a single rocket motor it is 3.3 × 10^6 lb (14,700 kN).

Applications. Rocket propulsion is used for different military missiles or space-flight missions. Each requires different thrust levels, operating durations, and other capabilities. In addition, rocket propulsion systems are used for rocket sleds, jet-assisted takeoff, principal power plants for experimental aircraft, or weather sounding rockets. For some space-flight applications, systems other than chemical rockets are used or are being investigated for possible future use. *See* GUIDED MISSILE; ROCKET-SLED TESTING; SATELLITE (SPACECRAFT); SPACE FLIGHT; SPACE PROBE.

Liquid-propellant rocket engines. These use liquid propellants stored in the vehicle for their chemical combustion energy. The principal hardware subsystems are one or more thrust chambers, a propellant feed system, which includes the propellant tanks in the vehicle, and a control system.

Bipropellants have a separate oxidizer liquid (such as liquefield oxygen or nitrogen tetroxide) and a separate fuel liquid (such as liquefied hydrogen or hydrazine). Monopropellants consist of a single liquid that contains both oxidizer and fuel ingredients. A catalyst is required to decompose the monopropellant into gaseous combustion products. Bipropellant combinations allow higher performance (higher specific impulse) than monopropellants. *See* PROPELLANT.

The three principal components of a thrust chamber are the combustion chamber, where rapid, high-temperature combustion takes place; the converging-diverging nozzle, where the hot reaction-product gases are accelerated to supersonic velocities; and an injector, which meters the flow of propellants in the desired mixture of fuel and oxidizer, introduces the propellants into the combustion chamber, and causes them to be atomized or broken up into small droplets. Some thrust chambers (such as the space shuttle's main engines and orbital maneuvering engines) are gimbaled or swiveled to allow a change in the direction of the thrust vector for vehicle flight motion control.

Solid-propellant rocket motors. In rocket motors the propellant is a solid material that feels like a soft plastic or soap. The solid propellant cake or body is known as the grain. It can have a complex internal geometry and is fully contained inside the solid motor case, to which a supersonic nozzle is attached.

The propellant contains all the chemicals necessary to maintain combustion. Once ignited, a grain will burn on all exposed surfaces until all the usable propellant is consumed; small unburned residual propellant slivers often remain in the chamber. As the grain surface recedes, a chemical reaction converts the solid propellant into hot gas. The hot gas then flows through internal passages within the grain to the nozzle, where it is accelerated to supersonic velocities. A pyrotechnic igniter provides the energy for starting the combustion.

The nozzle must be protected from excessive heat transfer, from high-velocity hot gases, from erosion by small solid or liquid particles in the gas (such as aluminum oxide), and from chemical reactions with aggressive rocket exhaust products. The highest heat transfer and the most severe erosion occur at the nozzle throat and immediately upstream from there. Special composite materials, called ablative materials, are used for heat protection, such as various types of graphite or reinforced plastics with fibers made of carbon or silica. The development of a new composite material, namely, woven carbon fibers in a carbon matrix, has allowed higher wall temperatures and higher strength at elevated temperatures; it is now used in nozzle throats, nozzle inlets, and exit cones. It is made by carbonizing (heating in a nonoxidizing atmosphere) organic materials, such as rayon or phenolics. Multiple layers of different heat-resistant and heat-insulating materials are often particularly effective. A three-dimensional pattern of fibers created by a process similar to weaving gives the nozzle extra strength. *See* NOZZLE.

Nozzles can have sophisticated thrust-vector control mechanisms. In one such system the nozzle forces are absorbed by a doughnut-shaped, confined, liquid-filled bag, in which the liquid moves as the nozzle is canted. The space shuttle solid rocket boosters have gimbaled nozzles for thrust-vector control, with actuators driven by auxiliary power units and hydraulic pumps.

Hybrid rocket propulsion. A hybrid uses a liquid propellant together with a solid propellant in the same rocket engine. The arrangement of the solid fuel is similar to that of the grain of a solid-propellant rocket; however, no burning takes place directly on the surface of the grain because it contains little or no oxidizer. Instead, the fuel on the grain surface is heated, decomposed, and vaporized, and the vapors burn with the oxidizer some distance away from the surface. The combustion is therefore inefficient.

Testing. Because flights of rocket-propelled vehicles are usually fairly expensive and because it is sometimes difficult to obtain sufficient and accurate data from fast-moving flight vehicles, it is accepted practice to test rocket propulsion systems and components extensively on the ground under simulated flight conditions. Components such as an igniter or a turbine are tested separately. Complete engines are tested in static engine test stands; the complete vehicle stage is also tested statically. In the latter two tests the engine and vehicle are adequately secured by suitable structures. Only in flight tests are they allowed to leave the ground.

George P. Sutton

Rocket staging

The use of successive rocket sections, each having its own engine or engines. One way to minimize the weight of large missiles, or space vehicles, is to use multiple stages. The first or initial stage is usually the heaviest and biggest and often called the booster; the next few stages are successively smaller and are generally called sustainers. Each stage is a complete vehicle in itself and carries its own propellant (either solid or liquid; both fuel and oxidizer), its own propulsion system, and has its own tankage and control system.

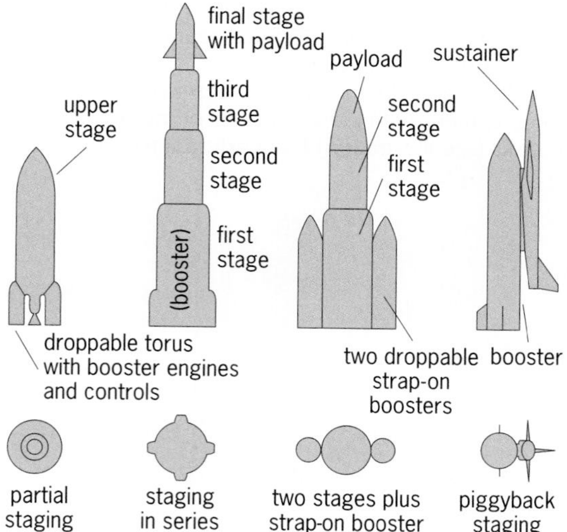

Typical schemes for staging missiles.

Once the propellant of a given stage is expended, the dead weight of that stage including empty tanks, rocket engine, and controls is no longer useful in contributing additional kinetic energy to the succeeding stages. By dropping off this useless weight, the mass that remains to be accelerated is made smaller; therefore it is possible to accelerate the payload to higher velocity than would be attainable if multiple staging were not used.

It is quite possible to employ different types of power plants, different types of propellants, and entirely different configurations in successive stages of any one multistage vehicle (see illustration). Because staging adds complications, it is impractical to have more than four to seven stages in any one vehicle. *See* ROCKET PROPULSION.

George P. Sutton

Rodentia

The mammalian order consisting of the rodents, often known as the gnawing mammals. This is the most diverse group of mammals in the world, consisting of over 2000 species, more than 40% of the known species of mammals on Earth today. Rodents range in size from mice, weighing only a few grams, to the Central American capybara, which is up to 130 cm (4 ft) in length and weighs up to 79 kg (170 lb). Rodents have been found in virtually every habitat and on every continent except Antarctica. Rodents have adapted to nearly every mode of life, including semiaquatic swimming (beavers and muskrats), gliding ("flying" squirrels), burrowing (gophers and African mole rats), arboreal (dormice and tree squirrels), and hopping (kangaroo rats and jerboas). Nearly all rodents are herbivorous, with a few exceptions that are partially insectivorous to totally omnivorous, such as the domestic rat. The great adaptability and rapid evolution and diversity of rodents are mainly due to their short gestation periods (only 3 weeks in some mice) and rapid turnover of generations.

The most diagnostic feature of the Rodentia is the presence of two pair of ever-growing incisors (one pair above and one below) at the front of the jaws. These teeth have enamel only on the front surface, which allows them to wear into a chisellike shape, giving rodents the ability to gnaw. Associated with these unique teeth are a number of other anatomical features that enhance this ability. Behind the incisors is a gap in the jaws where no teeth grow, called a diastema. The diastema of the upper jaw is longer than that of the lower jaw, which allows rodents to engage their gnawing incisors while their chewing teeth (molars and premolars) are not being used. The reverse is also true; rodents can use their chewing teeth (also called cheek teeth) while their incisors are disengaged.

The entire skull structure of rodents is designed to accommodate this task of separating the use of the different types of teeth. Rodent skulls have long snouts; the articulation of the lower jaw with the skull is oriented front to back rather than sideways as in other mammals; the jaw muscles (masseter complex) are extended well forward into the snout; and the number of cheek teeth is less than in most other mammals—all features unique to rodents.

The classification of rodents has always been difficult because of the great diversity of both Recent and fossil species. Traditionally, there are two ways that rodents have been divided: into three major groups based on the structure of the attachment of the jaw muscle on the skull (Sciuromorpha, Hystricomorpha, Myomorpha); or into two groups based on the structure of the lower jaw (Sciurognathi, Hystricognathi). The difficulty in using these groups (usually considered suborders or infraorders) is that the distinctive adaptations of one group of rodents are also present in others, derived in completely separate ways.

William W. Korth

Rodenticide A toxic chemical that is used to kill pest rodents and sometimes other pest mammals, including moles, rabbits, and hares. Most rodenticides are used to control rats and house mice.

Rodenticides are generally combined with some rodent-preferred food item such as grain (corn, wheat, oats) or a combination of grains in low yet effective amounts. Bait formulations may be in pellet forms or incorporated in paraffin blocks of varying sizes. As a safeguard against accidental ingestion by nontarget species, baits are placed either where they are inaccessible to children, domestic animals, or wildlife, or within tamper-resistant bait boxes designed to exclude all but rodent-size animals.

As a group, anticoagulant rodenticides dominate the market and are sold under a wide variety of trade names. In order of their development, they are warfarin, pindone (Pival), diphacinone, and chlorophacinone. When small amounts of these anticoagulants are consumed over several days, death results from internal bleeding. The newer, second-generation anticoagulant rodenticides, such as brodifacoum, bromadiolone, and difethialone, were developed to counteract the growing genetic resistance in rats and house mice to the earlier anticoagulants, especially warfarin. The second-generation compounds are more potent and capable of being lethal following a single night's feeding, although death is generally delayed by several days. Rodenticides that do not belong to the anticoagulant group include zinc phosphide, bromethalin, and cholecalciferol (vitamin D_3). The feeding and lethal characteristics differ among them. Acutely toxic strychnine baits are also available but are restricted to underground application, primarily for pocket gophers and moles. Several lethal fumigants or materials that produce poisonous gases are used to kill rodents in burrows and within other confined areas such as unoccupied railway cars or buildings. Lethal fumigants include aluminum phosphide, carbon dioxide, chloropicrin, and smoke or gas cartridges, which are ignited to produce carbon monoxide and other asphyxiating gases.

Because of their high toxicity, rodenticides are inherently hazardous to people, domestic animals, and wildlife. They are highly regulated, as are certain other types of pesticides. Some rodenticides can be purchased and used only by trained certified or licensed pest control operators, while others with a greater safety margin can be used by the general public. All rodenticides must be used in accordance with the label directions and may be prohibited where they may jeopardize certain endangered species. *See* PESTICIDE; RODENTIA.

Rex E. Marsh

Roentgenium The nineteenth of the synthetic transuranium elements, symbol Rg, atomic number 111. Discovered in late 1994, roentgenium should be a homolog of the elements gold, silver, and copper. It is the ninth element in the 6*d* shell. *See* COPPER; GOLD; HALF-LIFE; RADIOISOTOPE; SILVER.

The element was discovered on December 17, 1994, at GSI (Gesellschaft für Schwerionenforschung), Darmstadt, Germany, by detection of the isotope ^{272}Rg (the isotope of roentgenium with mass number 272), which was produced by fusion of a nickel-64 projectile and a bismuth-209 target nucleus after the fused system was cooled by emission of one neutron. The optimum bombarding energy for producing ^{272}Rg corresponds to an excitation energy of 15 MeV for the fused system. Sequential alpha decays to meitnerium-268, bohrium-264, dubnium-260, and lawrencium-256 allowed identification from the known decay properties of ^{260}Db and ^{256}Lr. In the decay chain (see illustration), the first three members are new isotopes. The isotope ^{272}Rg has a half-life of 1.5 ms, and is produced with a cross section of 3.5×10^{-36} cm^2. Altogether, three chains were observed during the 17 days of irradiation. The new isotopes meitnerium-268 and bohrium-264 with their half-lives of 70 ms and 0.4 s, respectively, show a trend toward longer half-lives than those

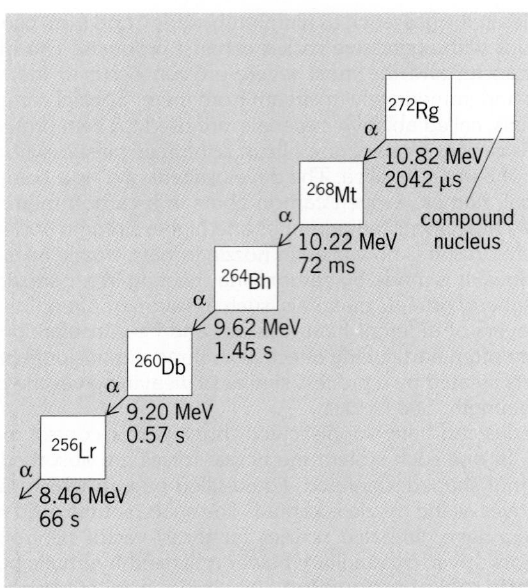

Sequence of decay chains that document the discovery of roentgenium. Numbers below boxes are alpha energies and correlation times. Roentgenium is produced in the reaction $^{64}Ni + ^{209}Bi \rightarrow ^{272}Rg + 1n$.

of the previously known isotopes of these elements. *See* ALPHA PARTICLES; NEUTRON; NUCLEAR REACTION.

Peter Armbruster

Roll mill A series of rolls operating at different speeds and used to grind paint or to mill flour. In paint grinding, a paste is fed between two low-speed rolls running toward each other at different speeds. Because the next roll in the mill is turning faster, it develops shear in the paste and draws the paste through the mill. The film is scraped from the last high-speed roll. For grinding flour, rolls are operated in pairs, rolls in each pair running toward each other at different speeds. Grooved rolls crush the grain; smooth rolls mill the flour to the desired fineness. *See* GRINDING MILL.

Ralph M. Hardgrove

Rolling contact Contact between bodies such that the relative velocity of the two contacting surfaces at the point of contact is zero. Common applications of rolling contact are the friction gearing of phonograph turntables, speed changers, and wheels on roadways. Rolling contact mechanisms are, generally speaking, a special variety of cam mechanisms. The concepts of rolling contact are used in the study of antifriction bearings and in the study of the behavior of toothed gearing. *See* ANTIFRICTION BEARING; CAM MECHANISM; GEAR.

Pure rolling contact can exist between two cylinders rotating about their centers, with either external or internal contact. Two friction disks (see illustration) have external rolling contact if no

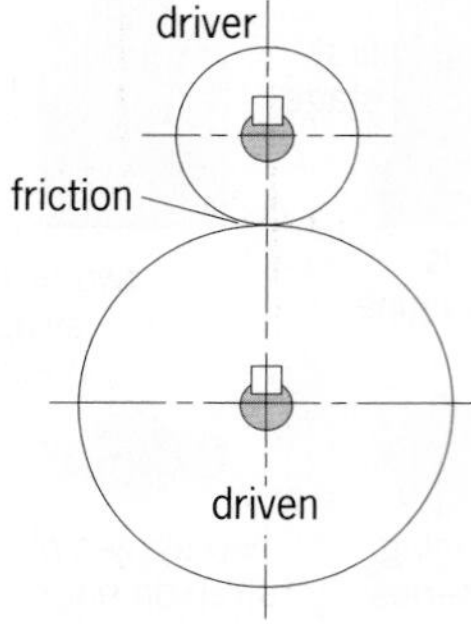

Rolling friction disks.

slipping occurs between them. The rotational speeds of the disks are then inversely proportional to their radii. John R. Zimmerman

Roof construction An assemblage to provide cover for homes, buildings, and commercial, industrial, and recreational areas. Roofs are constructed in different forms and shapes with various materials. A properly designed and constructed roof protects the structure beneath it from exterior weather conditions, provides structural support for superimposed loads, provides diaphragm strength to maintain the shape of the structure below, suppresses fire spread, and meets desired esthetic criteria.

Modern roof construction usually consists of an outer roofing assembly that is attached atop a deck or sheathing surface, which in turn is supported by a primary framework such as a series of beams, trusses, or arches. The shape of the roof and type of roof construction are usually determined by, and consistent with, the materials and deck of the primary structure underneath. *See* ARCH; BEAM; TRUSS.

Roof shapes include flat; hipped, where two sloping deck surfaces intersect in a line, the ridge or hip; pyramidal, which involves three of more sloping planes; domed, or other three-dimensional-surface, such as spherical, parabolic, or hyperbolic, shells; and tentlike, which are suspended fabric or membrane surfaces.

A roof assembly is a series of layers of different materials placed on and attached to the roof deck. Each type of roof assembly—related to protection against water entry from rain, snow, or ice; and insulation for temperature change, fire propagation, wind uplift, and moisture migration—has its own design requirements and methods of construction and attachment.

A roof is built upward from the structure below. The framework, or primary structural components, rest on the walls and columns of the structure, and these support the roof deck or the sheathing, which in turn carries the roofing assembly. The walls and columns may have girders framing into them. Beams rest on or are connected to girders. The roof deck or sheathing, the components that provide the basic support for the roofing assembly, span between and are anchored to the primary structural framing.

Long-span roofs are use space trusses, usually, of steel; but reinforced-concrete domes or other shell shapes, including folded plates, may be employed. In cable-supported roofs, the primary framework is composed of cables in tension that are slung between separate posts or from the top of the surrounding building perimeter. Tentlike or membrane roofs are a special application of a cable-supported roof. Air-supported roofs utilize a waterproof coated fabric that is inflated to its rigid shape by developing and maintaining a positive air pressure inside the structure, which keeps the roof surface under tension. Tennis-court "bubbles" utilize this design. *See* BUILDINGS; REINFORCED CONCRETE; STRUCTURAL STEEL. Milton Alpern

Root (botany) The absorbing and anchoring organ of vascular plants. Roots are simple axial organs that produce lateral roots, and sometimes buds, but bear neither leaves nor flowers. Elongation occurs in the root tip. The older portion of the root, behind the root tip, may thicken through cambial activity. Some roots, grass for example, scarcely thicken, but tree roots can become 4 in. (10 cm) or more in diameter near the stem. Roots may be very long. The longest maple (*Acer*) roots are usually as long as the tree is tall, but the majority of roots are only a few inches long. The longest roots may live for many years, while small roots may live for only a few weeks or months.

Root tips and the root hairs on their surface take up water and minerals from the soil. They also synthesize amino acids and growth regulators (gibberellins and cytokinins). These materials move up through the woody, basal portion of the root to the stem. The thickened, basal portion of the root anchors the plant in the soil. Thickened roots, such as carrots, can store food that is later used in stem growth. *See* CYTOKININ; GIBBERELLIN.

Roots usually grow in soil where: it is not too dense to stop root tip elongation; there is enough water and oxygen for root growth; and temperatures are high enough (above 39°F or 4°C) to permit root growth, but not so high that the roots are killed (above 104°F or 40°C). In temperate zones most roots are in the uppermost 4 in. (10 cm) of the soil; root numbers decrease so rapidly with increasing depth that few roots are found more than 6 ft (2 m) below the surface. Roots grow deeper in areas where the soil is hot and dry; roots from desert shrubs have been found in mines more than 230 ft (70 m) below the surface. In swamps with high water tables the lack of oxygen restricts roots to the uppermost soil layers. Roots may also grow in the air. Poison ivy vines form many small aerial roots that anchor them to bark or other surfaces.

The primary root originates in the seed as part of the embryo, normally being the first organ to grow. It grows downward into the soil and produces lateral second-order roots that emerge at right angles behind the root tip. Sometimes it persists and thickens to form a taproot. The second-order laterals produce third-order laterals and so on until there are millions of roots in a mature tree root system. In contrast to the primary root, most lateral roots grow horizontally or even upward. In many plants a few horizontal lateral roots thicken more than the primary, so no taproot is present in the mature root system.

Adventitious roots originate from stems or leaves rather than the embryo or other roots. They may form at the base of cut stems, as seen in the horticultural practice of rooting cuttings. B. F. Wilson

Root (mathematics) If a function $f(x)$ has the value 0 for $x = a$, a is a root of the equation $f(x) = 0$. The fundamental theorem of algebra states that any algebraic equation of the form $a_0x^n + a_1x^{n-1} + \cdots a_{n-1}x + a_n = 0$, where the a_k's are real numbers, has at least one root. From this it follows readily that such an equation has roots, real or complex, in number equal to the index (here n) of the highest power of x.

Furthermore, if $a + ib$ (where $i = \sqrt{-1}$) is a complex root of the given equation, so is $a - ib$, the conjugate of $a + ib$. Equations of degrees up to four may be solved algebraically. This statement means that the roots may be expressed as functions of the coefficients, the functions involving the elementary arithmetical processes of addition, multiplication, raising a number to a power, or extracting the root of a certain order of a given number. It was proved by H. Abel and by E. Galois that it is not possible to solve algebraically the *general* algebraic equation of degree higher than four. However, it is possible to determine the real roots of an algebraic equation to any desired degree of approximation. *See* CALCULUS; EQUATIONS, THEORY OF; NUMERICAL ANALYSIS. Arnold N. Lowan; Salomon Bochner

Root-mean-square The square root of the arithmetic mean of the squares of a set of numbers is called their root-mean-square. If the numbers are $x_1, x_2, x_3, \ldots, x_n$, the root-mean-square is equal to

$$\sqrt{\frac{x_1{}^2 + x_2{}^2 + x_3{}^2 \cdots + x_n{}^2}{n}}$$

It is valuable as an average of the magnitudes of quantities, and it is not affected by the signs of the quantities.

Among applications of root-mean-square the most important is the standard deviation from the arithmetic mean. If the arithmetic mean $\bar{x}$ is equal to

$$\frac{x_1 + x_2 + x_3 \cdots + x_n}{n}$$

then the standard deviation s is equal to

$$\sqrt{\frac{(x_1-\bar{x})^2+(x_2-\bar{x})^2+(x_3-\bar{x})^2+\cdots(x_n-\bar{x})^2}{n}}$$

Thus standard deviation from the mean is the root-mean-square of the deviations from the mean. *See* STATISTICS.

Hollis R. Cooley

Rope A long flexible structure consisting of many strands of wire, plastic, or vegetable fiber such as manila. Rope is classified as a flexible connector and is used generally for hoisting, conveying, or transporting loads; transmitting motion; and occasionally transmitting power. For flexibility and to reduce stresses as the rope bends over the sheave (pulley), a rope is made of many small strands. *See* PULLEY.

Paul H. Black

Rosales An order of flowering plants in the eurosid I group of the rosid eudicots. Recently on the basis of DNA sequence studies the number of families was greatly reduced and changed. The order as now recognized contains only 11 families, many of which are small. Many families in this order are wind-pollinated (the former Urticales) and exhibit the typical syndrome of small petalless flowers with dangling anthers, whereas the other families are insect-pollinated with large, showy flowers in which the carpels are sometimes free. *See* FLOWER; MAGNOLIOPHYTA; MAGNOLIOPSIDA; POLLINATION.

The largest family is Rosaceae with nearly 3000 species. The great number of economically important trees, shrubs, and herbs in this family include apples (*Malus*), pears (*Pyrus*), almonds, cherries, plums, and prunes (*Prunus*), strawberries (*Fragaria*), blackberries, raspberries, and their relatives (*Rubus*), as well as many minor fruits. Many are grown as ornamental plants, including roses (*Rosa*), avens (*Geum*), cinquefoil (*Potentilla*), firethorn (*Pyracantha*), redbush (*Photinia*), and spirea (*Spiraea*). *See* ALMOND; APPLE; BLACKBERRY; CHERRY; PEAR; PLUM; RASPBERRY; STRAWBERRY.

The second-largest family, Rhamnaceae (900 species), includes mostly woody species (shrubs and trees) that differ from Rosaceae in having fused carpels and only five stamens (rather than many) that are placed in the same position as the petals. A number of species in Rhamnaceae are of minor economic importance as timbers, some are of medicinal use or as dyes, and one is a minor fruit crop, the jujube (*Zizyphus*). A few are ornamentals, such as *Ceanothus* and *Colletia*.

Of the families formerly placed in Urticales, the largest are Moraceae (950 species) and Urticaceae (700 species). Many of these are large forest trees, but some are forest herbs and weeds. In Moraceae, many species are sources of timber and fruit, including figs (*Ficus*), mulberries (*Morus*), and breadfruits (*Artocarpus*). In Urticaceae, fiber-producing species are common, including hemp (*Cannabis*), ramie or China grass (*Boehmeria*), and bast-fiber (*Urtica*). Ulmaceae and Celtidaceae are small families of north temperate forest trees, many of which provide useful timbers, including elm (*Ulmus*), hackberries (*Celtis*), and *Zelkova*. *See* ELM; FIBER CROPS; FIG; HACKBERRY; HEMP; MULBERRY; RAMIE.

Mark W. Chase

Rose curve A type of plane curve that consists of loops (leaves, petals) emanating from a common point and that has a roselike appearance. Taking the common point O as the pole of a polar coordinate system (see illustration), these curves have equations of the form $\rho = a \cdot \sin n\theta$, where $a > 0$ and n is a positive integer (also $\rho = a \cdot \cos n\theta$, with a different choice of the initial line of the coordinate system). The curve is a circle of diameter a for $n = 1$. It has n or $2n$ leaves, according as n is an odd or even integer, respectively. The lemniscate is sometimes called a two-leaved rose, though its equation $\rho^2 = a^2 \cos 2\theta$ is not of the form given above. *See* LEMNISCATE OF BERNOULLI.

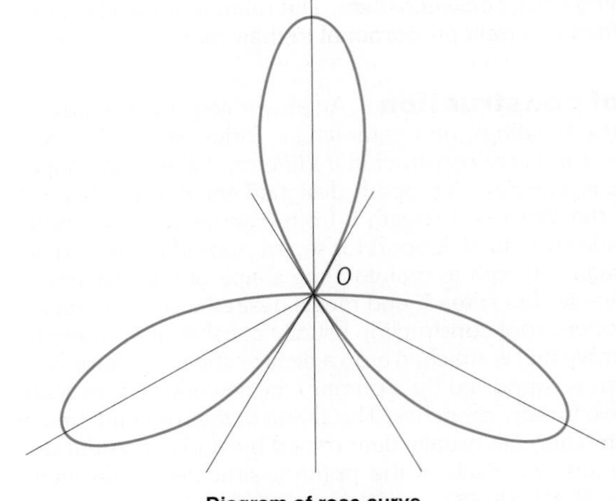

Diagram of rose curve.

Leonard M. Blumenthal

Rosemary *Rosmarinus officinalis*, a member of the mint family, grown for its highly aromatic leaves and as an ornamental. Rosemary is an evergreen and perennial which can live as long as 20 years under favorable conditions. Although many varieties exist, *R. officinalis* is the only species. Most varieties are suitable for culinary use, although some (such as Pine Scented) contain high levels of terpenes and have a turpentine-like scent. *See* LAMIALES.

Rosemary is native to the Mediterranean area, and much of the world production is harvested from wild plants growing there. Though it is widely cultivated mainly in France, Spain, and California, it will grow in most temperate areas not subject to hard frosts.

Rosemary oil has been found to contain chemicals such as rosmanol which inhibit oxidation and bacterial growth. Rosemary is used in poultry seasoning and for flavoring soups and vegetables. *See* SPICE AND FLAVORING.

Seth Kirby

Rosidae A former widely recognized subclass of Magnoliophyta (angiosperms or flowering plants). Deoxyribonucleic acid (DNA) and morphological studies have demonstrated that this group includes many more families than previously thought. Approximately 136 families (roughly 30% of the total angiosperms) in 11 orders are included in the rosids. With the asterids (often recognized as Asteridae), the rosids represent the two most advanced lines of dicots.

The primary distinguishing features of rosids are their diplostemenous flowers (anthers in two whorls, each whorl typically numbering the same as the perianth parts, although in some groups the number of stamens has increased) with unfused parts, except for the carpels (asterids have haplostemenous flowers with fused parts, typically with the stamens fused at least to the bases of the petals). Rosids have ovules with two integuments (that is, they are bitegmic, as opposed to the unitegmic asterids) and a several-layered nucellus (that is, they are crassinucellate, as opposed to the tenuinucellate asterids, which have a single-layered nucellus). There is nuclear endosperm formation (cellular in asterids), a reticulate pollen layer, mucilaginous leaf epidermis, and generally simple perforation vessel end-walls in the wood. The largest rosid orders are Malpighiales (16,100 species), Fabales (17,500), and Myrtales (10,600). *See* APIALES; CELASTRALES; CORNALES; EUPHORBIALES; FABALES; GERANIALES; HALORAGALES; LINALES; MAGNOLIOPSIDA; MYRTALES; PODOSTEMALES; POLYGALALES; PROTEALES; RAFFLESIALES; RHIZOPHORALES; ROSALES; SANTALALES; SAPINDALES.

Mark W. Chase

Rosin A brittle resin ranging in color from dark brown to pale lemon yellow and derived from the oleoresin of pine trees. Rosin is insoluble in water, but soluble in most organic solvents. It softens at about 180–190°F (80–90°C). Rosin consists of about 90% resin acids and about 10% neutral materials such as anhydrides, sterols, and diterpene aldehydes and alcohols.

Rosin is obtained by wounding living trees and collecting the exudate (gum rosin), by extraction of pine stumps (wood rosin), and as a by-product from the kraft pulping process (sulfate or tall oil rosin).

The largest single use of rosin is in sizing paper to control water absorption, an application in which fortified rosin is important. Rosin soaps are used as emulsifying and tackifying agents in synthetic rubber manufacture. Other rosin uses include adhesives, printing inks, and chewing gum. Irving S. Goldstein

Rotary engine Internal combustion engine that duplicates in some fashion the intermittent cycle of the piston engine, consisting of the intake-compression-power-exhaust cycle, wherein the form of the power output is directly rotational.

Four general categories of rotary engines can be considered: (1) cat-and-mouse (or scissor) engines, which are analogs of the reciprocating piston engine, except that the pistons travel in a circular path; (2) eccentric-rotor engines, wherein motion is imparted to a shaft by a principal rotating part, or rotor, that is eccentric to the shaft; (3) multiple-rotor engines, which are based on simple rotary motion of two or more rotors; and (4) revolving-block engines, which combine reciprocating piston and rotary motion. *See* AUTOMOBILE; COMBUSTION CHAMBER; DIESEL CYCLE; DIESEL ENGINE; GAS TURBINE; INTERNAL COMBUSTION ENGINE; OTTO CYCLE. Wallace Chinitz

Rotational motion The motion of a rigid body which takes place in such a way that all of its particles move in circles about an axis with a common angular velocity; also, the rotation of a particle about a fixed point in space. Rotational motion is illustrated by (1) the fixed speed of rotation of the Earth about its axis; (2) the varying speed of rotation of the flywheel of a sewing machine; (3) the rotation of a satellite about a planet; (4) the motion of an ion in a cyclotron; and (5) the motion of a pendulum. Circular motion is a rotational motion in which each particle of the rotating body moves in a circular path about an axis. Such motion is exhibited by the first and second examples. For information concerning the other examples *see* HARMONIC MOTION; PARTICLE ACCELERATOR; PENDULUM.

The speed of rotation, or angular velocity, remains constant in uniform circular motion. In this case, the angular displacement θ experienced by the particle or rotating body in a time t is $\theta = \omega t$, where ω is the constant angular velocity.

A special case of circular motion occurs when the rotating body moves with constant angular acceleration. If a body is moving in a circle with an angular acceleration of α radians/s^2, and if at a certain instant it has an angular velocity ω_0, then at a time t seconds later, the angular velocity may be expressed as $\omega = \omega_0 + \alpha t$, and the angular displacement as $\theta = \omega_0 t + {}^1\!/_2 \alpha t^2$. *See* ACCELERATION; VELOCITY.

A rotating body possesses kinetic energy of rotation which may be expressed as $T_{\text{rot}} = {}^1\!/_2 I\omega^2$, where ω is the magnitude of the angular velocity of the rotating body and I is the moment of inertia, which is a measure of the opposition of the body to angular acceleration. The moment of inertia of a body depends on the mass of a body and the distribution of the mass relative to the axis of rotation. For example, the moment of inertia of a solid cylinder of mass M and radius R about its axis of symmetry is ${}^1\!/_2 MR^2$.

The action of a torque L is to produce an angular acceleration α according to the equation below, where $I\omega$, the product of

$$L = I\alpha = I\frac{d\omega}{dt} = \frac{d}{dt}(I\omega)$$

moment of inertia and angular velocity, is called the angular momentum of the rotating body. This equation points out that the angular momentum $I\omega$ of a rotating body, and hence its angular velocity ω, remains constant unless the rotating body is acted upon by a torque. Both L and $I\omega$ may be represented by vectors.

It is readily shown that the work done by the torque L acting through an angle θ on a rotating body originally at rest is exactly equal to the kinetic energy of rotation. *See* ANGULAR MOMENTUM; MOMENT OF INERTIA; RIGID-BODY DYNAMICS; TORQUE; WORK. Carl E. Howe; R. J. Stephenson

Rotifera A phylum of microscopic, mainly free-living aquatic animals, characterized primarily by an anterior ciliary apparatus, the corona. When the cilia of the corona are in motion the animals appear to have a pair of rapidly rotating wheels on their heads, thus the name "Rotifera," Latin for "wheel bearers."

Rotifers show considerable diversity in form and structure. They are bilaterally symmetrical animals with complete digestive, excretory, nervous, and reproductive systems but lack separate respiratory and circulatory systems. Most have about 1000 nuclei and are 0.1–1.0 mm in length. They possess three characteristic morphological features: the corona; a syncytial epidermis with an intracytoplasmic lamina; and the mastax, a gizzardlike structure derived from a modified pharynx.

Older classifications placed rotifers in the phylum Aschelminthes, an incongruous assemblage including also the Nematoda, Priapulida, Kinorhyncha, and Gastrotricha, and divided rotifers into three groups: Monogononts, Bdelloids, and Seisonids. Because of new molecular and ultrastructural information, Aschelminthes is no longer considered a valid group and Rotifera is generally considered to be a phylum, with the three groups now given class status. Only Bdelloidea and Monogononta (but not Seisonacea) possess all the classical rotifer features. The obligately parasitic Acanthocephala is a closely related phylum or may be a fourth group of Rotifera.

The body of a rotifer is usually divided into three parts: the head, trunk, and foot. The head carries the corona, the mouth, and mastax of the digestive system, as well as the central ganglion (brain) of the nervous system. Most organs are located in the trunk, including the stomach, intestine, cloaca, anus, and gastric glands of the digestive system, the simple excretory system, and the reproductive organs. The foot contains cement glands whose secretions enable many species to attach themselves to the substratum. The three major groups of rotifers each have a characteristic form of reproduction. Seisonids reproduce only sexually. The monogonont life cycle is a complex combination of asexual and sexual forms of reproduction. Bdelloid rotifers reproduce parthogenetically. *See* ACANTHOCEPHALA; BDELLOIDEA; MONOGONONTA; SEISONACEA. J. B. Jennings; David B. Mark Welch

Rous sarcoma The first filterable agent (virus) known to cause a solid tumor in chickens. It was discovered in 1911 by P. F. Rous, who won the Nobel prize in 1967 for his discovery. It is a ribonucleic acid virus and belongs to the avian leukosis group. Certain strains of the virus cause tumors in hamsters, rabbits, monkeys, and other species. The Rous virus is known as a "defective" virus in that it is incapable of producing tumors by itself but requires another closely related virus of the avian leukosis group to act as a "helper" for the production of the foci. *See* ANIMAL VIRUS; TUMOR VIRUSES; VIRUS, DEFECTIVE. Alice E. Moore

Rubber Originally, a natural or tree rubber, which is a hydrocarbon polymer of isoprene units. With the development of synthetic rubbers having some rubbery characteristics but differing in chemical structure as well as properties, a more general designation was needed to cover both natural and synthetic rubbers. The term elastomer, a contraction of the words elastic and polymer, was introduced, and defined as a substance that can be stretched at room temperature to at least twice its original length and, after having been stretched and the stress removed, returns with force to approximately its original length in a short time.

Three requirements must be met for rubbery properties to be present in both natural and synthetic rubbers: long thread like molecules, flexibility in the molecular chain to allow flexing and coiling, and some mechanical or chemical bonds between molecules.

Natural rubber and most synthetic rubbers are also commercially available in the form of latex, a colloidal suspension of polymers in an aqueous medium. Natural rubber comes from trees in this form; many synthetic rubbers are polymerized in this form; some other solid polymers can be dispersed in water. *See* POLYMER.

Latexes are the basis for a technology and production methods completely different from the conventional methods used with solid rubbers.

In the crude state, natural and synthetic rubbers possess certain physical properties which must be modified to obtain useful end products. The raw or unmodified forms are weak and adhesive. They lose their elasticity with use, change markedly in physical properties with temperature, and are degraded by air and sunlight. Consequently, it is necessary to transform the crude rubbers by compounding and vulcanization procedures into products which can better fulfill a specific function.

Although natural rubber may be obtained from hundreds of different plant species, the most important source is the rubber tree (*Hevea brasiliensis*). Natural rubber is *cis*-1,4-polyisoprene, containing approximately 5000 isoprene units in the average polymer chain. *See* RUBBER TREE.

Styrene-butadiene rubber (SBR) is the most important synthetic rubber and the most widely used rubber in the entire world. Formerly designated GR-S, SBRs are obtained by the emulsion polymerization of butadiene and styrene in varying ratios. However, in the most commonly used type, the ratio of butadiene to styrene is approximately 78:22. Unlike natural rubber, SBR does not crystallize on stretching and thus has low tensile strength unless reinforced. The major use for SBR is in tires and tire products. Other uses include belting, hose, wire and cable coatings, flooring, shoe products, sponge, insulation, and molded goods.

Butyl rubber is essentially a polyisobutylene except for the presence of diolefin, usually isoprene, to provide the unsaturation necessary for vulcanization. Butyl rubbers have excellent resistance to oxygen, ozone, and weathering. In addition, these rubbers exhibit good electrical properties and high impermeability to gases. The high impermeability to gases results in use of butyl as an inner liner in tubeless tires. Other widespread uses are for wire and cable products, injection-molded and extruded products, hose, gaskets, and sealants, and where good damping characteristics are needed.

Ethylene-propylene polymers are produced by the copolymerization of ethylene and propylene. These copolymers exhibit outstanding resistance to heat, oxygen, ozone, and other aging and degrading agents. Abrasion resistance in tire treads is excellent. The mechanical properties of their vulcanizates are generally approximately equivalent to those of SBR.

One of the first synthetic rubbers used commercially in the rubber industry is neoprene, a polymer of chloroprene, 2-chlorobutadiene-1,3. The neoprenes have exceptional resistance to weather, sun, ozone, and abrasion. They are good in resilience, gas impermeability, and resistance to heat, oil, and flame. They are fairly good in low temperature and electrical properties. This versatility makes them useful in many applications requiring oil, weather, abrasion, or electrical resistance or combinations of these properties, such as wire and cable, hose, belts, molded and extruded goods, soles and heels, and adhesives.

Nitrile-type rubbers are copolymers of acrylonitrile and a diene, usually butadiene. The nitrile rubbers can be blended with natural rubber, polysulfide rubbers, and various resins to provide characteristics such as increased tensile strength, better solvent resistance, and improved weathering resistance.

Fluoroelastomers are basically copolymers of vinylidene fluoride and hexafluoropropylene. Because of their fluorine content, they are the most chemically resistant of the elastomers and also have good properties under extremes of temperature conditions. They are useful in the aircraft, automotive, and industrial areas.

Polyurethane elastomers are of interest because of their versatility and variety of properties and uses. They can be used as liquids or solids in a number of manufacturing methods. The largest use has been for making foam for upholstery and bedding. *See* POLYURETHANE RESINS.

Polysulfide rubbers have a large amount of surlfur in the main polymer chain and are therefore very chemically resistant, particularly to oils and solvents. They are used in such applications as putties, caulks, and hose for paint spray, gasoline, and fuel. Polyacrylate rubbers are useful because of their resistance to oils at high temperatures, including sulfur-bearing extreme-pressure lubricants. *See* POLYSULFIDE RESINS.

Proper choice of catalyst and order of procedure in polymerization have led to development of thermoplastic elastomers. The leading commercial types are styrene block copolymers having a structure which consists of polystyrene segments or blocks connected by rubbery polymers such as polybutadiene, polyisoprene, or ethylene-butylene polymer. Thermoplastic elastomers are very useful in providing a fast and economical method of producing a variety of products. One of the disadvantages for many applications is the low softening point of the thermoplastic elastomers.

Edward G. Partridge

Rubber tree *Hevea brasiliensis*, a member of the spurge family (Euphorbiaceae) and a native of the Amazon valley. It is the natural source of commercial rubber.

It has been introduced into all the tropical countries supporting the rainforest type of vegetation, and is grown extensively in established plantations, especially in Malaysia. The latex from the trees is collected and coagulated. The coagulated latex is treated in different ways to produce the kind of rubber desired. Rubber is made from the latex of a number of other plants, but *Hevea* is the rubber plant of major importance. *See* EUPHORBIALES; RUBBER.

Perry D. Strausbaugh; Earl L. Core

Rubella A benign, infectious virus disease of humans characterized by coldlike symptoms and transient, generalized rash. This disease, also known as German measles, is primarily a disease of childhood. However, maternal infection during early pregnancy may result in infection of the fetus, giving rise to serious abnormalities and malformations. The congenital infection persists in the infant, who harbors and sheds virus for many months after birth.

In rubella infection acquired by ordinary person-to-person contact, the virus is believed to enter the body through respiratory pathways. Antibodies against the virus develop as the rash fades, increase rapidly over a 2–3-week period, and then fall during the following months to levels that are maintained for life. One attack confers life-long immunity, since only one antigenic type of the virus exists. Immune mothers transfer antibodies to their offspring, who are then protected for approximately 4–6 months after birth. *See* IMMUNITY.

Live attenuated rubella vaccines have been available since 1969. The vaccine induces high antibody titers and an enduring and solid immunity. It may also induce secretory immunoglobulin (IgA) antibody in the respiratory tract and thus interfere with establishment of infection by wild virus. This vaccine is available as a single antigen or combined with measles and mumps vaccines (MMR vaccine). The vaccine induces immunity in at least 95% of recipients, and that immunity endures for at least 10 years. *See* BIOLOGICALS; VACCINATION. Joseph L. Melnick

Rubellite The red to red-violet variety of the gem mineral tourmaline. Perhaps the most sought-for of the many colors in which tourmaline occurs, it was named for its resemblance to ruby. The color is thought to be caused by the presence of lithium. It has a hardness of 7–7.5 on Mohs scale, a specific gravity near 3.04, and refractive indices of 1.624 and 1.644. Fine gem-quality material is found in Brazil, Madagascar, Maine, southern California, the Ural Mountains, and elsewhere. *See* GEM; TOURMALINE. Richard T. Liddicoat, Jr.

Rubiales An order of flowering plants, division Magnoliophyta (Angiospermae), in the subclass Asteridae of the class Magnoliopsida (dicotyledons). The order consists of the large family Rubiaceae, with about 6500 species and the family Theligoniaceae with only 3 species. The Rubiales are marked by their inferior ovary; regular or nearly regular corolla with the petals grown together by their margins; stamens (equal in number to the petals) which are attached to the corolla tube alternate with the lobes; and opposite leaves with interpetiolar stipules or whorled leaves without stipules. The most familiar species of temperate regions are herbs with whorled leaves, such as madder (*Rubia tinctorium*, the traditional source of red dye), but opposite-leaved tropical shrubs such as *Coffea* (the source of coffee) are more typical of the group. *See* COFFEE; IPECAC; MAGNOLIOPHYTA; MAGNOLIOPSIDA. Arthur Cronquist; T. M. Barkley

Rubidium A chemical element, Rb, atomic number 37, and atomic weight 85.47. Rubidium is an alkali metal. It is a light, low-melting, reactive metal.

Most uses of rubidium metal and rubidium compounds are the same as those of cesium and its compounds. The metal is used in the manufacture of electron tubes, and the salts in glass and ceramic production.

Rubidium is a fairly abundant element in the Earth's crust, being present to the extent of 310 parts per million (ppm). This places it just below carbon and chlorine and just above fluorine and strontium in abundance. Sea water contains 0.2 ppm of rubidium, which (although low) is twice the concentration of lithium. Rubidium is like lithium and cesium in that it is tied up in complex minerals; it is not available in nature as simple halide salts as are sodium and potassium.

Rubidium has a density of 1.53 g/cm^3 (95.5 lb/ft^3, a melting point of 39°C (102°F), and a boiling point of 688°C (1270°F).

Rubidium is so reactive with oxygen that it will ignite spontaneously in pure oxygen. The metal tarnishes very rapidly in air to form an oxide coating, and it may ignite. The oxides formed are a mixture of Rb_2O, Rb_2O_2, and RbO_2. The molten metal is spontaneously flammable in air.

Rubidium reacts violently with water or ice at temperatures down to −100°C (−148°F). It reacts with hydrogen to form a hydride which is one of the least stable of the alkali hydrides. Rubidium does not react with nitrogen. With bromine or chlorine, rubidium reacts vigorously with flame formation. Organorubidium compounds can be prepared by techniques similar to those used for sodium and potassium. *See* ALKALI METALS; CESIUM. Marshall Sittig

Ruby The red variety of the mineral corundum, in its finest quality the most valuable of gemstones. Only medium to dark tones of red to slightly violet-red or very slightly orange-red are called ruby; light reds, purples, and other colors are properly called sapphires. In its pure form the mineral corundum, with composition Al_2O_3, is colorless. The rich red of fine-quality ruby is the result of the presence of a minute amount of chromic oxide. The chromium presence permits rubies to be used for lasers producing red light. *See* CORUNDUM; LASER; SAPPHIRE.

The finest ruby is the transparent type with a medium tone and a high intensity of slightly violet-red, which has been likened to the color of pigeon's blood. Star rubies do not command comparable prices, but they, too, are in great demand. The ruby was among the first of the gemstones to be duplicated synthetically and the first to be used extensively in jewelry. *See* GEM. Richard T. Liddicoat, Jr.

Rugosa One of the two principal orders of Paleozoic corals, the other being the Tabulata. Rugose corals first appeared in the Middle Ordovician. The Rugosans diversified steadily during the Early and Middle Paleozoic to reach their greatest diversity in the Middle Devonian. A series of extinction events in the Late Devonian almost resulted in the disappearance of the group, but a handful of survivors rediversified to almost match peak diversity by the Late Mississippian. Thereafter, the group declined and became extinct at the end of the Permian. *See* PALEOZOIC; PERMIAN.

The rugose coral skeleton (or corallum) was secreted by a polyp (coral animal) assumed to have been similar in many respects to the polyps of living corals. Rugose corals existed in either solitary or colonial form, though solitary forms were more common. The skeleton of a solitary rugose coral (corallite) consists of a cup, cylinder, or (more rarely) a plate that is almost always defined by a solid outer wall (epitheca), in or on which radial vertical blades or spines (septa) are disposed. The upper surface, to which the soft parts were attached, is the calice. In about one-third of rugose corals, the initial corallite (protocorallite) produces offsets to form a colony. There was no specialization of individuals in rugose coral colonies and all the corallites are essentially similar in appearance. *See* CNIDARIA; TABULATA. Colin Scrutton

Runge vector The Runge vector describes certain unchanging features of a nonrelativistic two-body interaction for which the potential energy is inversely proportional to the distance r between the bodies or, alternatively, in which each body exerts a force on the other that is directed along the line between them and proportional to r^{-2}. Two basic interactions in nature are of this type: the gravitational interaction between two masses (called the classical Kepler problem), and the Coulomb interaction between like or unlike charges (as in the hydrogen atom). Both at the classical level and the quantum-mechanical level, the existence of a Runge vector is a reflection of the symmetry inherent in the interaction. *See* COULOMB'S LAW; QUANTUM MECHANICS; SYMMETRY LAWS (PHYSICS). D. M. Fradkin

Running fit The intentional difference in dimensions of mating mechanical parts that permits them to move relative to each other. A free running fit has liberal allowance; it is used on high-speed rotating journals or shafts. A medium fit has less allowance; it is used on low-speed rotating shafts and for sliding parts. Running fits are affected markedly by their surface finish and the effectiveness of lubrication. *See* ALLOWANCE. Paul H. Black

Rutabaga The plant *Brassica napobrassica*, a cool-season, hardy biennial crucifer of European origin, belonging to the order Capparales and probably resulting from the natural crossing of cabbage and turnip. The fleshy roots are cooked and usually eaten mashed as a vegetable. Rutabagas have been widely grown as a livestock feed in northern Europe and eastern Canada. Commercial production is limited to Canada and the northern part of the United States. *See* CAPPARALES; TURNIP.

H. John Carew

Ruthenium A chemical element, Ru, atomic number 44. The element is a brittle gray-white metal of low natural abundance, usually found alloyed with other platinum metals in nature. Ruthenium occurs as seven stable isotopes, and more than ten radioactive (unstable) isotopes are known. Four allotropes of the metal are known. Ruthenium is a hard white metal, workable only at elevated temperatures. It can be melted with an electric arc or an electron beam. *See* ISOTOPE; METAL; PERIODIC TABLE; RADIOISOTOPE; TRANSITION ELEMENTS.

The metal is not oxidized by air at room temperature, but it does oxidize to give a surface layer of ruthenium dioxide (RuO_2) at about 900°C (1650°F); at about 1000°C (1830°F) the volatile compounds ruthenium tetraoxide (RuO_4) and ruthenium monoxide (RuO) form, which can result in loss of the metal. Metallic ruthenium is insoluble in common acids and aqua regia up to 100°C (212°F). The principal properties of ruthenium ore given in the table. *See* AQUA REGIA.

Principal properties of ruthenium

Property	Value
Atomic number	44
Atomic weight	101.07
Crystal structure	Hexagonal close-packed
Lattice constant *a* at 25°C (77°F),	0.27056
nm *c*/*a* at 25°C (77°F)	1.5820
Density at 25°C (77°F), g/cm^3	12.37
Thermal neutron capture cross section, barns (10^{-28} m^2)	2.50
Melting point	2310°C (4190°F)
Boiling point	4080°C (7380°F)
Specific heat at 0°C, cal/g (J/kg)	0.0551 (231)
Thermal conductivity, 0–100°C, cal cm/cm^2°C	0.25
Linear coefficient of thermal expansion, 20–100°C, μin./(in.)(°C) or μm/(m)(°C)	9.1
Electrical resistivity at 0°C, microhm-cm	6.80
Temperature coefficient of electrical resistance, 0–100°C/°C	0.0042
Tensile strength (annealed), kN · m^{-2}	4.96×10^5
Young's modulus at 20°C (68°F), lb/in.^{2} (Pa)	
Static	60×10^6 (4.1×10^{11})
Dynamic	69×10^6 (4.75×10^{11})
Vickers hardness number (diamond pyramid hardness)	200–350

Ruthenium is relatively rare, having a natural abundance in the Earth's crust of about 0.0004 part per million. It is always found in the presence of other platinum metals. The major commercial sources of the element are the native alloys osmiridium and iridiosmium and the sulfide ore laurite. The element is also separated from other platinum metals by an intricate process, involving treatment with aqua regia (in which ruthenium, osmium, rhodium, and iridium are insoluble), to yield the pure metal.

Ruthenium is used commercially to harden alloys of palladium and platinum. The alloys are used in electrical contacts, jewelry, and fountain-pen tips. Application of ruthenium to industrial catalysis (hydrogenation of alkenes and ketones) and to automobile emission control (catalytic reduction of nitric oxide) and detection have been active areas of research. In medicine, ruthenium complexes have attracted some attention as potential antitumor reagents and imaging reagents. Ruthenium tetraoxide is finding increasing use as an oxidant for organic compounds. *See* CATALYSIS; CATALYTIC CONVERTER; HYDROGENATION; OSMIUM; PALLADIUM; PLATINUM; TECHNETIUM.

Carol Creutz

Rutherfordium A chemical element, symbol Rf, atomic number 104. Rutherfordium is the first element beyond the actinide series. In 1964 G. N. Flerov and coworkers at the Dubna Laboratories in Russia claimed the first identification of rutherfordium. A. Ghiorso and coworkers made a definitive identification at the Lawrence Radiation Laboratory, Berkeley, University of California, in 1969.

The Dubna group claimed the preparation of rutherfordium, mass number 260, by irradiating plutonium-242 with neon-22 ions in the heavy-ion cyclotron. The postulated nuclear reaction was $^{242}Pu + ^{22}Ne \rightarrow ^{260}Rf + 4$ neutrons. By 1969 the Berkeley group had succeeded in discovering two alpha-emitting isotopes of rutherfordium with mass numbers 257 and 259 by bombarding ^{249}Cf with ^{12}C and ^{13}C projectiles from the Berkeley heavy-ion linear accelerator (HILAC).

A number of years after the discovery at Berkeley, a team at Oak Ridge National Laboratory confirmed discovery of the isotope ^{257}Rf by detecting the characteristic nobelium x-rays following alpha decay. *See* ACTINIDE ELEMENTS; NOBELIUM; TRANSURANIUM ELEMENTS.

Albert Ghiorso

Rydberg atom An atom which possesses one valence electron orbiting about an atomic nucleus within an electron shell well outside all the other electrons in the atom. Such an atom approximates the hydrogen atom in that a single electron is interacting with a positively charged core. Early observations of atomic electrons in such Rydberg quantum states involved studies of the Rydberg series in optical spectra. Electrons jumping between Rydberg states with adjacent principal quantum numbers, n and $n - 1$, with n near 80 produce microwave radiation. Microwave spectral lines due to such electronic transitions in Rydberg atoms have been observed both in laboratory experiments and in the emissions originating from certain low-density partially ionized portions of the universe called HII regions. *See* ELECTRON CONFIGURATION; INTERSTELLAR MATTER; MICROWAVE.

The advent of the laser has made possible the production of sizable numbers of Rydberg atoms within a bulb containing gas at low pressures, 10^{-2} torr (1.3 pascals) or less. The rapid energy-resonance absorption of several laser light photons by an atom in its normal or ground state results in a Rydberg atom in a state with a selected principal quantum number. Aggregates of Rydberg atoms have been used as sensitive detectors of infrared radiation, including thermal radiation. They have also been observed to collectively participate in spontaneous photon emission, called superradiance. Such aggregates form the active medium for infrared lasers that operate through the usual laser mechanism of collective stimulated photon emission. All these developments are based upon the great sensitivity of Rydberg atoms to external electromagnetic radiation fields. Atoms with n near 40 can absorb almost instantaneously over a hundred microwave photons and become ionized at easily achievable microwave power levels. Isotope separation techniques have been developed that combine the selectivity of laser excitation of Rydberg states with the ready ionizability of Rydberg atoms. Such applications have been pursued for atoms ranging from deuterium through uranium. *See* INFRARED RADIATION; ISOTOPE SEPARATION; LASER.

James E. Bayfield

Rydberg constant The most accurately measured of the fundamental constants; it is a universal scaling factor for any spectroscopic transition and an important cornerstone in the determination of other constants.

This constant was introduced empirically. J. Balmer's formula described the visible spectral lines of atomic hydrogen, while J. Rydberg's formula applied to the spectra of many elements. Their results may be summarized by Eq. (1), where λ is the wavelength

$$\frac{1}{\lambda} = R\left(\frac{1}{n_1^2} - \frac{1}{n_2^2}\right) \tag{1}$$

of the spectral line and R is a constant. In Balmer's account of the visible hydrogen spectrum, n_1 was equal to 2, while n_2 took on the integer values 3, 4, 5, and so forth. In Rydberg's more general work, n_1 and n_2 differed slightly from integer values. A remarkable result of Rydberg's work was that the constant R was the same for all spectral series he studied, regardless of the element. This constant R has come to be known as the Rydberg constant.

Applied to hydrogen, Niels Bohr's atomic model leads to Balmer's formula with a predicted value for the Rydberg constant given by Eq. (2), where m_e is the electron mass, e is the

$$R_\infty = \frac{m_e e^4}{8h^3\epsilon_0^2 c} \tag{2}$$

electron charge, h is Planck's constant, ϵ_0 is the permittivity of vacuum, and c is the speed of light. The equation expresses the Rydberg constant in SI units. To express it in cgs units, the right-hand side must be multiplied by $(4\pi\epsilon_0)^2$. The subscript ∞ means that this is the Rydberg constant corresponding to an infinitely massive nucleus.

E. Schrödinger's wave mechanics predicts the same energy levels as the simple Bohr model, but the relativistic quantum theory of P. A. M. Dirac introduces small corrections or fine-structure splittings. The modern theory of quantum electrodynamics predicts further corrections. Additional small hyperfine-structure corrections account for the interaction of the electron and nuclear magnetic moments. *See* FINE STRUCTURE (SPECTRAL LINES); HYPERFINE STRUCTURE; QUANTUM ELECTRODYNAMICS; QUANTUM MECHANICS; RELATIVISTIC QUANTUM THEORY.

The Rydberg constant is determined by measuring the wavelength or frequency of a spectral line of a hydrogenlike atom or ion. The highest resolution and accuracy has been achieved by the method of Doppler-free two-photon spectroscopy, which permits the observation of very sharp resonance transitions between long-living states. The 2006 adjustment of the fundamental constants, taking into account different measurements, adopted the value $R_\infty = 10{,}973{,}731.568{,}527 \pm 0.000{,}073\ \text{m}^{-1}$ for the Rydberg constant. The measurements provide an important cornerstone for fundamental tests of basic laws of physics. *See* ATOMIC STRUCTURE AND SPECTRA; FUNDAMENTAL CONSTANTS; LASER; LASER SPECTROSCOPY. Theodor W. Hänsch; Martin Weitz

Rye A winter-hardy and drought-resistant cereal plant, *Secale cereale*, in the grass family (Graminae). It resembles wheat, with which it intercrosses to a limited extent. Rye is propagated almost completely by cross-pollination. The inflorescence is a spike or ear (see illustration). Spikelets are arranged flatwise against a zigzag rachis; they usually have two flowers, enclosed by a lemma and palea with two adjacent glumes. The young florets contain three stamens and a pistil. The fertilized pistil develops into a naked grain, or kernel, that is easily threshed. There are several recognized species of *Secale*, most of which have shattering spikes and small kernels. There are both perennial and winter-annual species of rye, with winter forms being favored over spring types for production. The only commercially cultivated species is the nonshattering *S. cereale*. *See* CYPERALES; GRASS CROPS; WHEAT.

Rye spikes or ears.

Rye is more important in Europe and Asia than in the Western Hemisphere. Russia is the leading world producer, followed by Poland and Germany. Canada and Argentina produce significant amounts, and Switzerland and northwest Europe have high yields. Rye production in the United States is mostly in South Dakota, North Dakota, Minnesota, and Georgia.

Rye grain is used for animal feed, human food, and production of spirits. Ground rye is mixed with other feeds for livestock. It is often fall-sown to provide soil cover and pasturage for livestock. Egg yolks of chickens and butter from cows fed on rye have a rich yellow color. *See* DISTILLED SPIRITS.

Compared to other small grains, rye has a fewer number of cultivars (agricultural varieties). Short-strawed types are gaining favor. Plant and kernel characteristics of rye are variable, partly because of cross-pollination. Height may range from 4 to 6 ft (120 to 180 cm) under moderately fertile conditions. Kernel color may be amber, gray, green, blue, brown, or black.

Tetraploid forms, whose chromosome number has been doubled, are available. Tetraploid wheat and rye have been hybridized and chromosomes doubled to form Triticales, which is increasing in usage. *See* BREEDING (PLANT); GRAIN CROPS. H. L. Shands

Rye grain is milled into flour in a manner similar to that used for wheat flour. Variations are made based on the compositional and structural differences between these grains. Rye bread production requires blending of rye flours with wheat flours to provide sufficient dough strength. Specialty varieties of rye breads are classified according to ethnic origins or as sweet or sour doughs. Sour rye breads may be developed from natural lactic fermentations or through the incorporation of cultured milk. Swedish rye crisp breads are generally prepared from whole ground meal. *See* FOOD ENGINEERING. Mark A. Uebersax

S

Sable A carnivore classified in the family Mustelidae along with martens and fishers. The sable (*Martes zibellina*) has a very slender, long, and supple body. Coloration varies but is usually dark brown on the back and slightly lighter on the flanks, belly, and head. The winter pelage is long, silky, and luxurious with a particularly strong, thick undercoat; the summer pelage is shorter, coarser, duller, and darker. The limbs are short and terminate in five toes with semiretractile claws. Adult sables have a head and body length of 35–55 cm (14–22 in.), including a tail length of 12–19 cm (5–7 in.). They weigh 0.5–2 kg (1–4 lb).

Sables are generally solitary and inhabit both coniferous and deciduous forests, preferably near streams. They move by leaping and are very agile and quick. Sables may be active at any time during the day or night. Food consists principally of rodents, but pikas, birds, fish, honey, nuts, and berries may also be consumed.

The sable originally ranged throughout the entire taiga (boreal coniferous forest) zone from Scandinavia (northern Finland) to eastern Siberia and North Korea including Sakhalin and Hokkaido islands. Because of excessive trapping for its valuable fur, it disappeared from much of its range by the early 1900s. However, programs of protection and reintroduction have allowed the species to increase its numbers and distribution, so that exceptionally dense populations are once again found in various areas, chiefly in the mountains of the northernmost parts of its former range. *See* CARNIVORA. Donald W. Linzey

Saccharin The sodium salt of *o*-sulfobenzimide, manufactured by processes that start with toluene or phthalic anhydride. The free imide, called insoluble saccharin because it is insoluble in water, has limited use as a flavoring agent in pharmaceuticals. The sodium and calcium salts are very soluble in water and are widely used as sweetening agents.

Sodium saccharin is 300–500 times sweeter than cane sugar (sucrose). The saccharin salts are used to improve the taste of pharmaceuticals and toothpaste and other toiletries, and as nonnutritive sweeteners in special dietary foods and beverages. Using noncaloric saccharin in place of sugar permits the formulation of low-calorie products for people on calorie-restricted diets and of low-sugar products for diabetics. *See* ASPARTAME. Karl M. Beck

Saccopharyngiformes Teleost fishes in the subdivision Elopomorpha whose member taxa have a leptocephalous larval stage. They have lost a number of structures common to most fishes, such as the symplectic bones, opercular bones, branchiostegal bones, scales, pelvic fins, ribs, pyloric caeca, and swim bladder. These fishes live at great depths in total darkness. The enormous jaws and distensible gut allow them to capture and swallow prey much larger than themselves. Twenty-six species in five genera and four families are known from the bathypelagic areas of the Atlantic, Indian, and Pacific oceans. *See* OSTEICHTHYES; TELEOSTEI. Herbert Boschung

Sacoglossa An order of the gastropod subclass Opisthobranchia containing a thousand living species of herbivorous sea slugs; sometimes called Ascoglossa. They occur in all the oceans, at shallow depths, reaching their greatest size and diversity in tropical seas.

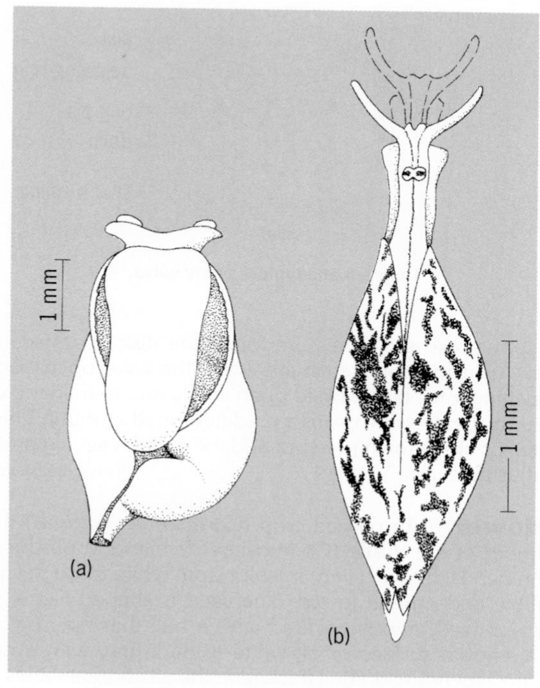

Representative sacoglossans: (*a*) *Volvatella*; (*b*) *Berthelinia*.

Members of this abundant order have two common features: the herbivorous habit (except for *Olea* and a species of *Stiliger* which eat the eggs of other mollusks) and the possession of a uniseriate radula in which the oldest, often broken, teeth are usually retained within the body, not discarded.

Sacoglossans possess varied shells; there may be a single shell, capacious and coiled (*Volvatella*, illustration *a*; *Cylindrobulla*), a flattened open dorsal shell (*Lobiger*), or two lateral shells (the bivalved gastropods, *Berthelinia*, illustration *b*). In the highest sacoglossans (*Elysia*, *Limapontia*) the true shell is completely lost after larval metamorphosis. *See* OPISTHOBRANCHIA. T. E. Thompson

Safety glass A unitary structure formed of two or more sheets of glass between each of which is interposed a sheet of plastic, usually polyvinyl butyral. In usual manufacture, two clean and dry sheets of plate glass and a sheet of plastic are preliminarily assembled as a sandwich under slight pressure to produce a void-free bond. The laminate is then pressed under heat long enough to unite. For use in surface vehicles the finished laminated glass is approximately $^{1}/_{4}$ in. (6 mm) thick; for aircraft it is thicker. *See* GLASS. Frank H. Rockett

Safety valve A relief valve set to open at a pressure safely below the bursting pressure of a container, such as a boiler or compressed air receiver. Typically, a disk is held against a seat by a spring; excessive pressure forces the disk open (see illustration). Construction is such that when the valve opens slightly, the opening force builds up to open it fully and to hold the valve open until the pressure drops a predetermined amount. This differential or blow-down pressure and the initial relieving pressure are adjustable. *See* VALVE. Theodore Baumeister

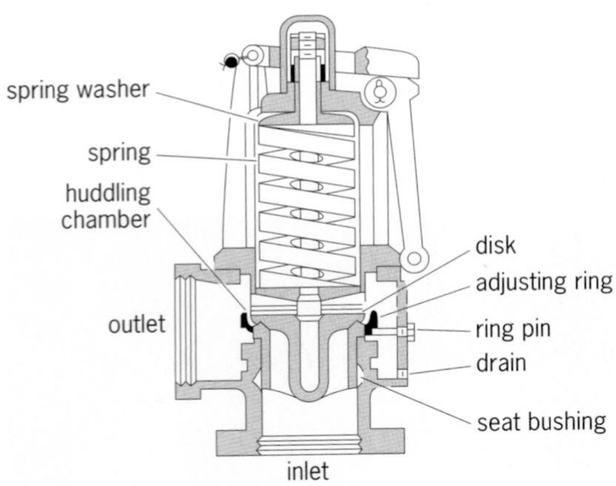

Diagram of a typical safety valve.

Safflower An oilseed crop (*Carthamus tinctorius*) that is a member of the thistle (Compositae) family and produces its seed in heads. Flowers vary in color from white through shades of yellow and orange to red. The seed is shaped like a small sunflower seed, and is covered with a hull that may be white with a smooth surface or off-white to dark gray with a ridged surface. Depending on hull thickness, the oil content varies from 25 to 45%. Safflower is grown commercially in Mexico, Australia, and California. *See* ASTERALES.

There are two types of safflower oil, both with 6–8% palmitic acid and 1–2% stearic acid. One type, the standard or polyunsaturated type, has 76–79% linoleic acid and 11–17% oleic acid. This high-linoleic type has been used in soft margarines, in salad oils, and in the manufacture of paints and varnishes. It has had limited use for frying foods, because heat causes it to polymerize and form a tough film on the cooking vessel. The second type of oil, called the high–oleic-acid or monounsaturated type, has 76–79% oleic acid and 11–17% linoleic acid. Its fatty-acid composition is similar to that of olive oil, but the flavor is bland. High-oleic safflower oil is a premium frying oil. The meal left after the extraction of the oil may contain 20–45% protein, depending on the amount of hull removed from the seed before processing. The meal is used as a poultry and livestock feed. *See* FAT AND OIL (FOOD). P. F. Knowles

Saffron The plant *Crocus sativus*, a member of the iris family (Iridaceae). A native of Greece and Asia Minor, it is now cultivated in various parts of Europe, India, and China. This crocus is the source of a potent yellow dye used for coloring foods and medicine. The dye is extracted from the styles and stigmas of the flowers. *See* LILIALES. Perry D. Strausbaugh; Earl L. Core

Sage A shrubby perennial plant in the genus *Salvia* of the mint family (Limiaceae). There are several species, including garden, or true, sage (*S. officinalis*), the sage most commonly used in foods. Many varieties of garden sage are known, but the Dalmatian type possesses the finest aroma. Garden sage is native to southern and eastern Europe, and is still cultivated extensively there and in the United States and Russia. It is a plant of low stature (2 ft or 60 cm), with hairy, oblong grayish-green leaves about $1\frac{1}{2}$–2 in. (4–5 cm) long. Sage does best in warm, dry regions, with full sun.

To preserve the essential oil content and leaf color, sage is dried, as are most other herbs. Once dried, sage is separated from the stems and made available to consumers as whole, rubbed (crushed), or ground leaves. The dried leaves are among the most popular spices in western foods. Sage is highly aromatic and fragrant, with a pungent, slightly bitter and astringent taste. Both the dried leaves and essential oil of sage are used in flavoring and for antioxidant properties in cheeses, pickles, processed foods, vermouth, and bitters. *See* LAMIALES; SPICE AND FLAVORING. Seth Kirby

Sagittarius The Archer, a large, zodiacal, southern summer constellation (see illustration). Sagittarius contains the center of the Milky Way Galaxy, so it is especially rich in star clusters and star clouds. However, it never rises high in the sky in the Northern Hemisphere. Visitors to the Southern Hemisphere marvel at Sagittarius's overall brightness and richness. *See* CENTAURUS; MILKY WAY GALAXY; STAR CLOUDS; STAR CLUSTERS; ZODIAC.

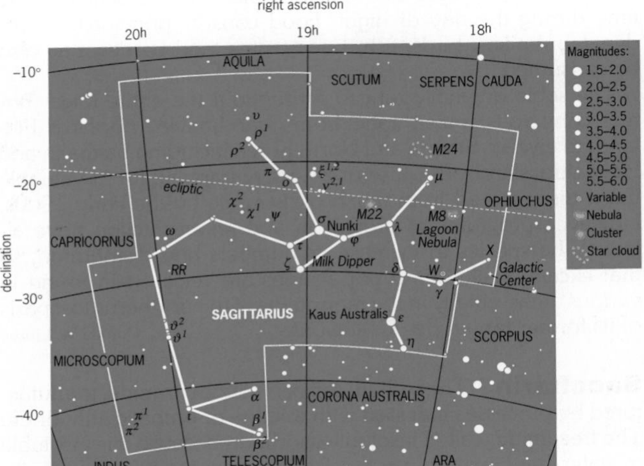

Modern boundaries of the constellation Sagittarius, the Archer. The celestial equator is 0° of declination, which corresponds to celestial latitude. Right ascension corresponds to celestial longitude, with each hour of right ascension representing 15° of arc. Apparent brightness of stars is shown with dot sizes to illustrate the magnitude scale, where the brightest stars in the sky are 0th magnitude or brighter and the faintest stars that can be seen with the unaided eye at a dark site are 6th magnitude. (*Wil Tirion*)

At the center of this large constellation, there is an asterism (star pattern) known as the Milk Dipper, since the Milky Way can be seen through its dipper shape. Several Messier objects (noted in Charles Messier's eighteenth-century catalog of nonstellar objects) include the Lagoon Nebula (M8), the globular star cluster M22, and the open star cluster M24. *See* CONSTELLATION; MESSIER CATALOG. Jay M. Pasachoff

Sahel A semiarid bioclimatic zone bordering the southern margins of the Sahara Desert in Africa. It extends between 13 and 18°N and stretches across the width of the African continent from the Atlantic Ocean to the Red Sea, an area of about 5 million square kilometers (1.9 million square miles). Within this area are large portions of the countries of Mauritania, Senegal, Gambia, Mali, Niger, Chad, Sudan, and Eritrea, as well as the northern parts of Burkina Faso, Nigeria,

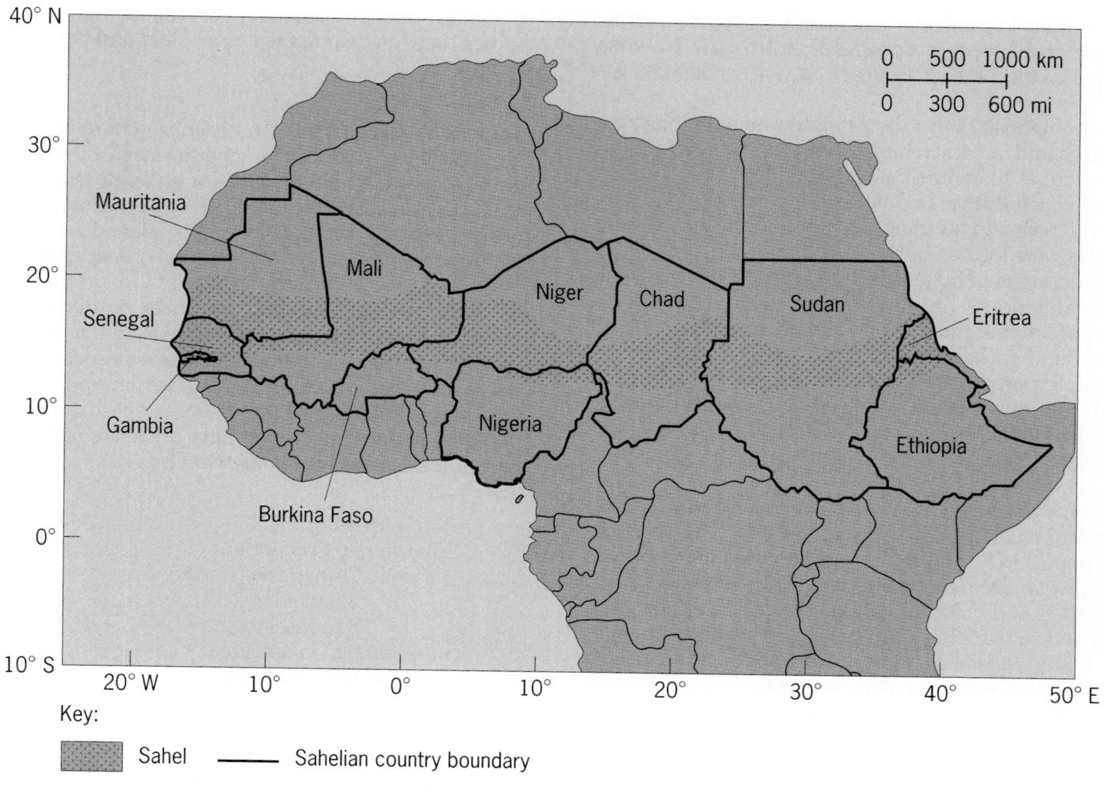

Location of the Sahel. The Sahelian countries are named.

and Ethiopia (see illustration). The actual boundary of the Sahel is neither as precise nor as rigid as suggested by lines of latitude. For practical purposes, the Sahel is usually defined in terms of its climate and vegetation. *See* AFRICA.

Two major but contrasting air masses dominate the climate of the Sahel: the dry, tropical continental air from the northeast and the humid, tropical maritime air from the southwest. These air masses converge in a zone of humidity and temperature contrast known as the Inter-Tropical Discontinuity (ITD). The mean annual rainfall also shows pronounced south–north gradients, decreasing from about 600–750 mm (24–30 in.) at the boundary of the Sahel and the Sudan savanna in the south to 100–200 mm (4–8 in.) in the north. *See* AIR MASS; CLIMATOLOGY; TROPICAL METEOROLOGY.

Temperature is high throughout the year, but tends to be highest between March and June with mean monthly temperatures between 30 and 35°C (86 and 95°F), although daily maximum temperatures frequently exceed 40°C (104°F).

Sahel vegetation is transitional between the Sudan savanna woody grassland to the south and the open desert to the north. The vegetation consists mostly of open grassland and thorny woody species. Annual grasslands dominate the northern Sahel, while wooded grasslands occur on sandy soils in the southern part. Among the most important woody plant species are various species of acacia, combretum, and baobab (*Adansonia digitata*). *See* DESERT; SAVANNA. Aondover Tarhule

Saint Elmo's fire A type of corona discharge observed on ships under conditions approaching those of an electrical storm. The charge in the atmosphere induces a charge on the masts and other elevated structures. The result of this is a corona discharge which causes a spectacular glow around these points. *See* CORONA DISCHARGE. Glenn H. Miller

Salicales An order of flowering plants, division Magnoliophyta (Angiospermae), in the subclass Dilleniidae of the class Magnoliopsida (dicotyledons). The order consists of the single family Salicaceae, with about 350 species. There are only two genera, *Salix* (willow) and *Populus* (poplar and cottonwood), the former by far the larger. The Salicales are dioecious, woody plants, with alternate, simple, stipulate leaves and much reduced flowers that are aggregated into catkins. The mature seeds are plumose-hairy and are distributed by the wind. *See* DILLENIIDAE; MAGNOLIOPSIDA; POPLAR; WILLOW. Arthur Cronquist; T. M. Barkley

Salicylate A salt or ester of salicylic acid having the general formula shown below and formed by replacing the carboxylic hydrogen of the acid by a metal (M) to give a salt

$$(o)\text{-}C_6H_4(OH)C(=O){-}O{-}M \quad \text{or} \quad {-}R$$

or by an organic radical (R) to give an ester. Alkali-metal salts are water-soluble; the others, insoluble. Sodium salicylate is used in medicines as an antirheumatic and antiseptic, in the manufacture of dyes, and as a preservative (illegal in foods). Salicylic acid is used in the preparation of aspirin. The methyl ester is the chief component of oil of wintergreen. This ester is used in pharmaceuticals as a component of rubbing liniment. It is also used as a flavoring agent and an odorant. *See* ASPIRIN. Elbert H. Hadley

Saline evaporites Deposits of bedded sedimentary rocks composed of salts precipitated during solar evaporation of surface or near-surface brines derived from seawater or continental waters. Dominant minerals in ancient evaporite beds are anhydrite (along with varying amounts of gypsum) and halite, which make up more than 85% of the total sedimentary evaporite salts. Many other salts make up the remaining 15%; their varying proportions in particular beds can be diagnostic of the original source of the mother brine. *See* ANHYDRITE; EVAPORATION; GYPSUM; HALITE; SALT (CHEMISTRY); SEAWATER; SEDIMENTARY ROCKS.

Today, brines deposit their salts within continental playas or coastal salt lakes and lay down beds a few meters thick and tens of kilometers across. In contrast, ancient, now-buried evaporite beds are often much thicker and wider; they can be up to hundreds of meters thick and hundreds of kilometers wide. Most ancient evaporites were formed by the evaporation of saline waters within hyperarid areas of huge seaways typically located within arid continental interiors. The inflow brines in such seaways were combinations of varying proportions of marine and continental ground waters and surface waters. There are few modern depositional counterparts to these ancient evaporites, and none to those beds laid down when whole oceanic basins dried up, for example, the Mediterranean some 5.5 million years ago. *See* BASIN; DEPOSITIONAL SYSTEMS AND ENVIRONMENTS; MEDITERRANEAN SEA; PLAYA.

Evaporite salts precipitate by the solar concentration of seawater, continental water, or hybrids of the two. The chemical makeup, salinity (35‰), and the proportions of the major ions in modern seawater are near-constant in all the world's oceans, with sodium (Na) and chloride (Cl) as the dominant ions and calcium (Ca) and sulfate (SO_4) ions present in smaller quantities [a $Na(Ca)SO_4Cl$ brine]. Halite and gypsum anhydrites have been the major products of seawater evaporation for at least the past 2 billion years, but the proportions of the more saline minerals, such as sylvite/magnesium sulfate ($MgSO_4$) salts, appear to have been more variable. *See* CALCIUM; CHLORINE; ION; MAGNESIUM; SODIUM; SULFUR. John Warren

Salmonelloses Diseases caused by *Salmonella*. These include enteritis and septicemia with or without enteritis. *Salmonella typhi*, *S. paratyphi A*, *B*, and *C*, and occasionally *S. cholerae suis* cause particular types of septicemia called typhoid and paratyphoid fever, respectively; while all other types may cause enteritis or septicemia, or both together.

Typhoid fever has an incubation period of 5–14 days. It is typified by a slow onset with initial bronchitis, diarrhea or constipation, a characteristic fever pattern (increase for 1 week, plateau for 2 weeks, and decrease for 2–3 weeks), a slow pulse rate, development of rose spots, swelling of the spleen, and often an altered consciousness; complications include perforation of the bowel and osteomyelitis. Typhoid fever leaves the individual with a high degree of immunity. Vaccination with an oral vaccine gives an individual considerable protection for about 3 years. The only effective antibiotic is chloramphenicol. *See* ANTIBIOTIC; IMMUNITY; VACCINATION.

Paratyphoid fever has a shorter course and is generally less severe than typhoid fever. Vaccination is an ineffective protective measure.

Enteric fevers, that is, septicemias due to types of *Salmonella* other than those previously mentioned, are more frequent in the United States than typhoid and paratyphoid fever but much less frequent than *Salmonella* enteritis. In children and in previously healthy adults, enteric fevers are most often combined with enteritis and have a favorable outlook. The organisms involved are the same as those causing *Salmonella* enteritis. Chloramphenicol or ampicillin are used in treatment. However, resistant strains have been observed. *See* DRUG RESISTANCE.

Inflammation of the small bowel due to *Salmonella* is one of the most important bacterial zoonoses. The most frequent agents are *S. typhimurium*, *S. enteritidis*, *S. newport*, *S. heidelberg*, *S. infantis*, and *S. derby*. The incubation period varies from 6 h to several days. Diarrhea and fever are the main symptoms; the intestinal epithelium is invaded, and early bacteremia is probable. Predisposed are persons with certain preexisting diseases (the same as for enteric fevers), very old and very young individuals, and postoperative patients. Antimicrobial treatment serves only to prolong the carrier state and has no effect on the disease. Alexander von Graevenitz

Salmoniformes An order of generalized fishes, including the salmons and trouts, characterized by soft or articulated fin rays; the presence (usually) of an adipose dorsal fin; cycloid scales (scales absent in few species); pelvic fins in the abdominal region and each with more than six rays and an associated axillary process (absent in Osmeridea); a pelvic girdle free from the pectoral girdle; a pectoral fin placed low on the side and more or less horizontal; an upper jaw usually bordered by premaxillae and maxillae; nonprotractile premaxillae; the presence (usually) of an orbitosphenoid bone as well as a mesocoracoid arch in the pectoral girdle; the absence of Weberian apparatus connecting the swim bladder with the inner ear; a physostomous (connected to the esophagus) air bladder; the larvae are not leptocephalous-like; and there are no luminescent organs.

Order Salmoniformes
- Superfamily Osmeridea
  - Family Osmeridae (smelts)
    - Salangidae
    - Sundasalangidae
- Superfamily Galaxioidea
  - Family Retropinnidae
    - Lepidogalaxiidae
    - Galaxiidae
- Superfamily Salmonoidea
  - Family Salmonidae (trouts and salmons)
    - Subfamily Coregoninae
      - Thymallinae
      - Salmoninae

The salmonids are distinguished by having fewer than 16 dorsal fin rays, small scales numbering 110 or more (usually many more), and maxillary teeth. Generally the large seagoing anadromous species that return to their natal stream to spawn and die are called salmons, whereas the smaller species that live to spawn several seasons are called trouts. All species spawn in freshwater. Salmons and trouts are Holarctic in native distribution; however, species have been introduced throughout the world. Worldwide salmons and trouts are probably more studied, more fished for recreation purposes, more exploited commercially, more genetically manipulated, and more introduced in foreign countries than any other group of fishes. These fishes are very important economically. There are seven genera and about 30 species. *See* AQUACULTURE. Herbert Boschung

Salt (chemistry) A compound formed when one or more of the hydrogen atoms of an acid are replaced by one or more cations of the base. The common example is sodium chloride in which the hydrogen ions of hydrochloric acid are replaced by the sodium ions (cations) of sodium hydroxide. There is a great variety of salts because of the large number of acids and bases which has become known.

Salts are classified in several ways. One method—normal, acid, and basic salts—depends upon whether all the hydrogen ions of the acid or all the hydroxide ions of the base have been replaced:

Class	*Examples*
Normal salts	$NaCl$, NH_4Cl, Na_2SO_4, Na_2CO_3, Na_3PO_4, $Ca_3(P0_4)_2$
Acid salts	$NaHCO_3$, NaH_2PO_4, Na_2HPO_4, $NaHSO_4$
Basic salts	$Pb(OH)Cl$, $Sn(OH)Cl$

The other method—simple salts, double salts (including alums), and complex salts—depends upon the character of completeness of the ionization:

Class	Examples
Simple salts	NaCl, $NaHCO_3$, Pb(OH)Cl
Double salts	KCl·$MgCl_2$
Alums	$KAl(SO_4)_2$, $NaFe(SO_4)_2$, $NH_4Cr(SO_4)_2$
Complex salts	$K_3Fe(CN)_6$, $Cu(NH_3)_4Cl_2$, $K_2Cr_2O_7$

See ACID AND BASE; CHEMICAL BONDING. Alfred B. Garrett

Salt (food) The chemical compound sodium chloride. It is used extensively in the food industry as a preservative and flavoring, as well as in the chemical industry to make chlorine and sodium. *See* CHLORINE; FOOD PRESERVATION; SALT (CHEMISTRY); SODIUM.

Salt was originally made by evaporating sea water (solar salt). This method is still in common usage; however, impurities in solar salt make it unsatisfactory for most commercial uses, and these impurities also lead to clumping. Salt, freshly produced from sea-water evaporation ponds, may contain large numbers of halophilic (salt-loving) microorganisms. In the United States refined salt is obtained from underground mines located in Michigan and Louisiana. Salt is usually handled during the refining processes as brine.

Salt is liable to clumping during periods of high humidity, so preventives are added. Materials used include magnesium carbonate and certain silicates. Iodides are added in those areas where iodine deficiencies exist. Roy E. Morse

Salt dome An upwelling of crystalline rock salt and its aureole of deformed sediments. A salt pillow is an immature salt dome comprising a broad salt swell draped by concordant strata. A salt stock is a more mature, pluglike diapir of salt that has pierced, or appears to have pierced, overlying strata. Most salt stocks are 0.6–6 mi (1–10 km) wide and high. Salt domes are closely related to other salt upwellings, some of which are much larger. Salt canopies, which form by coalescence of salt domes and tongues, can be more than 200 mi (300 km) wide. *See* DIAPIR.

Exploration for oil and gas has revealed salt domes in more than 100 sedimentary basins that contain rock salt layers several hundred meters or more thick. The salt was precipitated from evaporating lakes in rift valleys, intermontaine basins, and especially along divergent continental margins. Salt domes are known in every ocean and continent. *See* BASIN.

Salt domes consist largely of halite (NaCl, common table salt). Other evaporites, such as anhydrite ($CaSO_4$) and gypsum ($CaSO_4 \cdot 2H_2O$), form thinner layers within the rock salt. *See* HALITE; SALINE EVAPORITES.

Salt domes supply industrial commodities, including fuel, minerals, chemical feedstock, and storage caverns. Giant oil or gas fields are associated with salt domes in many basins around the world, especially in the Middle East, North Sea, and South Atlantic regions. Salt domes are also used to store crude oil, natural gas (methane), liquefied petroleum gas, and radioactive or toxic wastes. *See* OIL AND GAS STORAGE. M. P. A. Jackson

Salt-effect distillation A process of extractive distillation in which a salt that is soluble in the liquid phase of the system being separated is used in place of the normal liquid additive introduced to the extractive distillation column in order to effect the separation.

Extractive distillation is a process used to separate azeotrope-containing systems or systems in which relative volatility is excessively low. An additive, or separating agent, that is capable of raising relative volatility and eliminating azeotropes in the system being distilled is supplied to the column, where it mixes with the feed components and exerts its effect. The agent is subsequently recovered from one or both product streams by a separate process and recycled for reuse. *See* AZEOTROPIC MIXTURE.

In salt-effect distillation, the process is essentially the same as for a liquid agent, although the subsequent process used to recover the agent for recycling is different; that is, evaporation is used rather than distillation. The salt is added to the system by being dissolved in the reentering reflux stream at the top of the column. Being nonvolatile, it will reside in the liquid phase, flowing down the column and out in the bottom product stream.

The major commercial use of salt-effect distillation is in the concentration of aqueous nitric acid, using the salt magnesium nitrate as the separating agent. Other commercial applications include acetone-methanol separation using calcium chloride and isopropanol-water separation using the same salt. *See* AZEOTROPIC DISTILLATION; DISTILLATION. William F. Furter

Salt gland A specialized gland located around the eyes and nasal passages in marine turtles, snakes, and lizards, and in birds such as the petrels, gulls, and albatrosses, which spend much time at sea. In the marine turtle it is an accessory lacrimal gland which opens into the conjunctival sac. In seagoing birds and in marine lizards it opens into the nasal passageway. Salt glands copiously secrete a watery fluid containing a high percentage of salt, higher than the salt content of urine in these species. As a consequence, these animals are able to drink salt-laden sea water without experiencing the dehydration necessary to eliminate the excess salt via the kidney route. *See* GLAND. Olin E. Nelsen

Salt marsh A maritime habitat characterized by grasses, sedges, and other plants that have adapted to continual, periodic flooding. Salt marshes are found primarily throughout the temperate and subarctic regions.

The tide is the dominating characteristic of a salt marsh. The salinity of the tide defines the plants and animals that can survive in the marsh area. The vertical range of the tide determines flooding depths and thus the height of the vegetation, and the tidal cycle controls how often and how long vegetation is submerged. Two areas are delineated by the tide: the low marsh and the high marsh. The low marsh generally floods and drains twice daily with the rise and fall of the tide; the high marsh, which is at a slightly higher elevation, floods less frequently. *See* MANGROVE.

Salt marshes usually are developed on a sinking coastline, originating as mud flats in the shallow water of sheltered bays, lagoons, and estuaries, or behind sandbars. They are formed where salinity is high, ranging from 20 to 30 parts per thousand of sodium chloride. Proceeding up the estuary, there is a transitional zone where salinity ranges from 20 to less than 5 ppt. In the upper estuary, where river input dominates, the water has only a trace of salt. This varying salinity produces changes in the marsh—in the kinds of species and also in their number. Typically, the fewest species are found in the salt marsh and the greatest number in the fresh-water tidal marsh. *See* ESTUARINE OCEANOGRAPHY.

The salt marsh is one of the most productive ecosystems in nature. In addition to the solar energy that drives the photosynthetic process of higher rooted plants and the algae growing on the surface muds, tidal energy repeatedly spreads nutrient-enriched waters over the marsh surface. Some of this enormous supply of live plant material may be consumed by marsh animals, but the most significant values are realized when the vegetation dies and is decomposed by microorganisms to form detritus. Dissolved organic materials are released, providing an essential energy source for bacteria that mediate wetland biogeochemical cycles (carbon, nitrogen, and sulfur cycles). *See* BIOGEOCHEMISTRY; BIOLOGICAL PRODUCTIVITY.

The salt marsh serves as a sediment sink, a nursery habitat for fishes and crustaceans, a feeding and nesting site for waterfowl and shorebirds, a habitat for numerous unique plants and animals, a nutrient source, a reservoir for storm water, an erosion control mechanism, and a site for esthetic pleasures. Appreciation for the importance of salt marshes has led to federal and state legislation aimed at their protection. Franklin C. Daiber

Salviniales A small order of heterosporous, leptosporangiate ferns (division Polypodiophyta) which float on the surface of the water. The delicate, branching stem is provided with small, simple to bifid or more or less dissected leaves. The sporangia are enclosed in specialized appendages of the leaves, called sporocarps. The order contains only a single family, with two widely distributed genera, *Salvinia* and *Azolla*, with only about 20 species in all. *See* POLYPODIOPHYTA. Arthur Cronquist

Samarium A chemical element, Sm, atomic number 62, belonging to the rare-earth group. Its atomic weight is 150.35, and there are 7 naturally occurring isotopes; ^{147}Sm, ^{148}Sm, and ^{149}Sm are radioactive and emit α particles.

Samarium oxide is pale yellow, is readily soluble in most acids, and gives topaz-yellow salts in solutions. Samarium has found rather limited use in the ceramic industry, and it is used as a catalyst for certain organic reactions. One of its isotopes has a very high cross section for the capture of neutrons, and therefore there has been some interest in samarium in the atomic industry for use as control rods and nuclear poisons. *See* LANTHANUM; RARE-EARTH ELEMENTS. Frank H. Spedding

Sampled-data control system A type of digital control system in which one or more of the input or output signals is a continuous, or analog, signal that has been sampled. There are two aspects of a sampled signal: sampling in time and quantization in amplitude. Sampling refers to the process of converting an analog signal from a continuously valued range of amplitude values to one of a finite set of possible numerical values. This sampling typically occurs at a regular sampling rate, but for some applications the sampling may be aperiodic or random.

While the device to be controlled is usually referred to as the plant, sampled-data control systems are also used to control processes. The term plant refers to machines or mechanical devices which can usually be mathematically modeled by an analysis of their kinematics, such as a robotic arm or an engine. A process refers to a system of operations such as a batch reactor for the production of a particular chemical, or the operation of a nation's economy. The output of the plant which is to be controlled is called the controlled variable. A regulator is one type of sampled-data control system, and its purpose is to maintain the controlled variable at a preset value (for example, the robotic arm at a particular position, or an airplane turboprop engine at a constant speed) or the process at a constant value (for example, the concentration of an acid, or the inflation rate of an economy). This input is called the reference or setpoint. The second type of sampled-data control system is a servomechanism, whose purpose is to make the controlled variable follow an input variable. Examples of servomechanisms are a robotic arm used to paint automobiles which may be required to move through a predefined path in three-dimensional space while holding the sprayer at varying angles, an automobile engine which is expected to follow the input commands of the driver, a chemical process that may require the pH of a batch process to change at a specified rate, and an economy's growth rate which is to be changed by altering the money supply. *See* ANALOG-TO-DIGITAL CONVERTER; PROCESS CONTROL; REGULATOR; SERVOMECHANISM.

The analog-to-digital converter changes the sampled signal into a binary number so that it can be used in calculations by the digital compensator. Since a digital controller computes the control signal used to drive the plant, a digital-to-analog converter must be used to change this binary number to an analog voltage. The digital compensator in the typical sampled-data control system takes the digitized values of the analog feedback signals and combines them with the setpoint or desired trajectory signals to compute a digital control signal, to actuate the plant through the digital-to-analog converter. A compensator is used to modify the feedback signals in such a way that the dynamic performance of the plant is improved relative to some performance index. *See* CONTROL SYSTEMS; DIGITAL COMPUTER; DIGITAL CONTROL; DIGITAL-TO-ANALOG CONVERTER. Kenneth J. Hintz

Sand Unconsolidated granular material consisting of mineral, rock, or biological fragments between 63 micrometers and 2 mm in diameter. Finer material is referred to as silt and clay; coarser material is known as gravel. Sand is usually produced primarily by the chemical or mechanical breakdown of older source rocks, but may also be formed by the direct chemical precipitation of mineral grains or by biological processes. Accumulations of sand result from hydrodynamic sorting of sediment during transport and deposition. *See* CLAY MINERALS; DEPOSITIONAL SYSTEMS AND ENVIRONMENTS; GRAVEL; MINERAL; ROCK; SEDIMENTARY ROCKS.

Most sand originates from the chemical and mechanical breakdown, or weathering, of bedrock. Since chemical weathering is most efficient in soils, most sand grains originate within soils. Rocks may also be broken into sand-size fragments by mechanical processes, including diurnal temperature changes, freeze-thaw cycles, wedging by salt crystals or plant roots, and ice gouging beneath glaciers. *See* WEATHERING PROCESSES.

Because sand is largely a residual product left behind by incomplete chemical and mechanical weathering, it is usually enriched in minerals that are resistant to these processes. Quartz not only is extremely resistant to chemical and mechanical weathering but is also one of the most abundant minerals in the Earth's crust. Many sands dominantly consist of quartz. Other common constituents include feldspar, and fragments of igneous or metamorphic rock. Direct chemical precipitation or hydrodynamic processes can result in sand that consists almost entirely of calcite, glauconite, or dense dark-colored minerals such as magnetite and ilmenite. *See* FELDSPAR; QUARTZ.

Although sand and gravel has one of the lowest average per ton values of all mineral commodities, the vast demand makes it among the most economically important of all mineral resources. Sand and gravel is used primarily for construction purposes, mostly as concrete aggregate. Pure quartz sand is used in the production of glass, and some sand is enriched in rare commodities such as ilmenite (a source of titanium) and in gold. *See* CONCRETE. Mark J. Johnsson

Sandalwood The name applied to any species of the genus *Santalum* of the sandalwood family (Santalaceae). However, the true sandalwood is the hard, close-grained, aromatic heartwood of a parasitic tree, *S. album*, of the Indo-Malayan region. This fragrant wood is used in ornamental carving, cabinet work, and as a source of certain perfumes. The odor of the wood is an insect repellent, and on this account the wood is much used in making boxes and chests. The fragrant wood of a number of species in other families bears the same name, but none of these is the real sandalwood. *See* SANTALALES. Perry D. Strausbaugh; Earl L. Core

Sandstone A clastic sedimentary rock comprising an aggregate of sand-sized (0.06–2.0-mm) fragments of minerals, rocks, or fossils held together by a mineral cement. Sandstone forms when sand is buried under successive layers of sediment. During burial the sand is compacted, and a binding agent such as quartz, calcite, or iron oxide is precipitated from ground water which moves through passageways between grains. Sandstones grade upward in grain size into conglomerates and breccias; they grade downward in size into siltstones and shales. When the proportion of fossil fragments or carbonate grains is greater than 50%, sandstones grade into clastic limestones. *See* BRECCIA; CONGLOMERATE; LIMESTONE; SAND; SHALE.

The basic components of a sandstone are framework grains (sand particles), which supply the rock's strength; matrix or mud-sized particles, which fill some of the space between grains; and crystalline cement. The composition of the framework grains reveals much about the history of the derivation of the sand grains, including the parent rock type and weathering history of the parent rock. Textural attributes of sandstone are the same as those for sand, and they have the same genetic significance. *See* SAND.

Lee J. Suttner

Sandstones are classified according to the relative proportion of quartz to other grain types, and according to the ratio of feldspar grains to finely crystalline lithic fragments. Quartz-rich sandstones are commonly called quartz-arenite. Sandstones poor in quartz are commonly called arkose, when feldspar grains are more abundant than lithic fragments, and litharenite (or graywacke) when the reverse is true. Subarkose and sublitharenite (or subgraywacke) refer to analogous sandstones of intermediate quartz content. Sandstones composed dominantly of calcareous grains are called calcarenite, and represent a special variety of limestone. Other sandstones composed exclusively of volcanic debris are called volcanic sandstone, and are gradational, through the interplay of eruptive and erosional processes, to tuff, the fragmental volcanic rocks produced by the disintegration of magma during explosive volcanic eruptions. *See* ARENACEOUS ROCKS; ARKOSE; FELDSPAR; GRAYWACKE; QUARTZ; TUFF.

William R. Dickinson

Because sandstone can possess up to 35% connected pore space, it is the most important reservoir rock in the Earth's crust. In the future sandstone may serve as a reservoir into which hazardous fluids, such as nuclear wastes, are injected for storage. *See* HAZARDOUS WASTE.

Sandstone which is easily split (flagstone) and has an attractive color is used as a building stone. Sandstone is also an important source of sand for the glass industry and the construction industry, where it is used as a filler in cement and plaster. Crushed sandstone is used as road fill and railroad ballast. Silica-cemented sandstone is used as firebrick in industrial furnaces. Some of the most extensive deposits of uranium are found in sandstones deposited in ancient stream channels. *See* GLASS; SEDIMENTARY ROCKS; STONE AND STONE PRODUCTS; URANIUM.

Lee J. Suttner

Santalales An order of flowering plants, division Magnoliophyta (Angiospermae), in the subclass Rosidae of the class Magnoliopsida (dicotyledons). The order consists of 10 families and about 2000 species. The largest families are the Loranthaceae (about 900 species), Santalaceae (about 400 species), Viscaceae (about 350 species), and Olacaceae (about 250 species). A few of the more primitive members of the Santalales are autotrophic, but otherwise the order is characterized by progressive adaptation to parasitism, accompanied by progressive simplification of the ovules. Some members of the Santalales, such as sandalwood (*Santalum album*, a small tree of southern Asia), are rooted in the ground and produce small branch roots which invade and parasitize the roots of other plants. Others, such as mistletoe (*Viscum* and other genera of the Viscaceae), grow on trees, well above the ground. *See* FLOWER; MAGNOLIOPHYTA; MAGNOLIOPSIDA; MISTLETOE; SANDALWOOD.

Arthur Cronquist; T. M. Barkley

Sapindales An order of flowering plants, division Magnoliophyta (Angiospermae), in the subclass Rosidae of the class Magnoliopsida (dicotyledons). The order consists of 15 families and about 5400 species. The largest families are the Rutaceae (about 1500 species), Sapindaceae (about 1500 species), Meliaceae (about 550 species), Anacardiaceae (about 600 species), and Burseraceae (about 600 species). Most of the Sapindales are woody plants, with compound or lobed leaves and polypetalous, hypogynous to perigynous flowers with one or two sets of stamens and only one or two ovules in each locule of the ovary. A large proportion of the species have glandular cavities in the leaves, or resin ducts in the bark and wood, or other sorts of secretory structures.

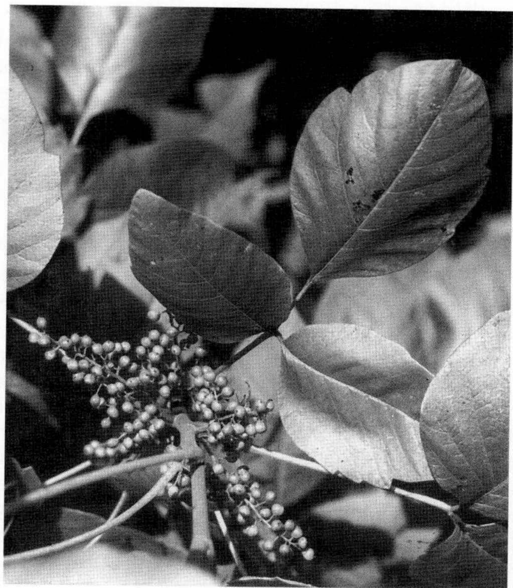

Poison ivy (*Toxicodendron radicans*). (*John Gerard, National Audubon Society*)

Some of the Anacardiaceae, including poison ivy (*Toxicodendron radicans*; see illustration) and the lacquer tree (*T. vernicifluum*), are notoriously allergenic to humans. Some other well-known members of the order are orange, lemon, lime, and grapefruit (all species of *Citrus* in the Rutaceae); the various kinds of maple (*Acer*, in the Aceraceae); and the horse chestnut (*Aesculus hippocastanum*, in the Hippocastanaceae). *See* BUCKEYE; CASHEW; CITRON; GRAPEFRUIT; KUMQUAT; LEMON; LIGNUMVITAE; LIME (BOTANY); MAGNOLIOPHYTA; MAGNOLIOPSIDA; MAHOGANY; MANDARIN; MAPLE; ORANGE; PISTACHIO; POISON IVY; QUEBRACHO; TANGERINE; VARNISH TREE.

Arthur Cronquist; T. M. Barkley

Sapphire The name given to all gem varieties of the mineral corundum, except those that have medium to dark tones of red that characterize ruby. Although the name sapphire is most commonly associated with the blue variety, there are many other colors of gem corundum to which sapphire is applied correctly; these include yellow, brown, green, pink, orange, purple, colorless, and black. Sapphire has a hardness of 9, a specific gravity near 4.00, and refractive indices of 1.76–1.77. Asterism, the star effect, is the result of reflections from tiny, lustrous, needlelike inclusions of the mineral rutile, plus a domed form of cutting. *See* CORUNDUM; GEM; RUBY; RUTILE.

Richard T. Liddicoat, Jr.

Sapropel A term originally defined as an aquatic sediment rich in organic matter that formed under reducing conditions (lack of dissolved oxygen in the water column) in a stagnant water body. Such inferences about water-column dissolved-oxygen contents are not always easy to make for ancient environments. Therefore, the term sapropel or sapropelic mud has been used loosely to describe any discrete black or dark-colored sedimentary layers (>1 cm or 0.4 in. thick) that contain greater than 2 wt % organic carbon. Sapropels may be finely laminated (varved) or homogeneous, and may less commonly exhibit structures indicating reworking or deposition of the sediment by currents. Sapropels largely contain amorphous organic matter derived from planktonic organisms (such as planktonic or benthic algae in lakes or plankton in marine settings). Such organic matter possesses a large hydrogen-to-carbon ratio; therefore, sapropelic sequences are potential petroleum-forming deposits. The enhanced preservation of amorphous organic matter in sapropels may indicate conditions of exceptionally great surface-water productivity, extremely low bottom-water dissolved-oxygen contents, or both. Some sapropels may, however, contain substantial amounts of organic matter derived from land plants. *See* ANOXIC ZONES; MARINE SEDIMENTS; ORGANIC GEOCHEMISTRY; PETROLEUM; VARVE.

Michael A. Arthur

Sarcodina A superclass of Protozoa (in the subphylum Sarcomastigophora) in which movement involves protoplasmic flow, often with recognizable pseudopodia. Gametes may be flagellated, as in certain Foraminiferida. Most species are floating or creeping; a few are sessile. The pellicle is relatively thin, and the body is apt to be plastic unless restrained by skeletal structures. Sarcodina live in fresh, brackish, or salt waters; soil or sand; and as endoparasites in animals and plants. A group may be limited to a specific habitat, but many have a rather wide range. Sarcodina include two major classes: Rhizopodea and Actinopodea. *See* ACTINOPODEA; RHIZOPODEA; SARCOMASTIGOPHORA.

Richard P. Hall

Sarcomastigophora A subphylum of Protozoa, including those forms that possess flagella or pseudopodia or both. Organisms have a single type of nucleus, except the developmental stages of some Foraminiferida. Sexuality, if present, is syngamy, the fusion of two gametes. Spores typically are not formed. Flagella may be permanent or transient or confined to a certain stage in the life history; this is true also of pseudopodia. Both flagella and pseudopodia may be present at the same time.

Three superclasses are included: (1) Mastigophora, commonly flagellates, contains 19 orders. (2) Opalinata includes 1 order; these organisms were once considered as ciliates, but further research has indicated flagellate kinships. (3) Sarcodina comprises organisms which normally possess pseudopodia and are flagellated only in the developmental stages; 13 orders possess irregularly distributed lobose or filose and branching pseudopodia, while 7 orders have radially distributed axopodia, often with axial filaments. *See* MASTIGOPHORA; SARCODINA.

Most of the plant flagellates will live in either fresh water or in both fresh and salt water. The zooflagellates are small and are not sufficiently abundant to enter markedly into the food chain. But the parasites and symbionts are of considerable interest economically and theoretically, for example, the trypanosomes and the peculiar xylophagous (wood-eating) symbionts of termites. In the termites these parasites actually digest the wood eaten by the host. Conspicuous in the ecology of marine waters are the dinoflagellates, radiolarians, and acantharians, especially in tropical waters of otherwise low productivity. *See* PROTOZOA.

James B. Lackey

Sarcopterygii A class of vertebrates that contains the orders Coelacanthiformes (lobefin fishes) and Ceratodontiformes (lungfishes), as well as the Tetrapoda. Following is a recent classification of the extant sarcopterygians:

Class Sarcopterygii
- Subclass Coelacanthimorpha
  - Order Coelacanthiformes (lobefin fishes)
    - Family Coelacanthidae
- Subclass Dipnotetrapodomorpha
  - Order Ceratodontiformes (lungfishes)
    - Family Ceratodontidae
    - Lepidoserinidae
    - Protopteridae
- Infraclass Tetrapoda

Debates concerning the relationships of the sarcopterygians and the ancestry of tetrapods—that is, whether the origin of the tetrapods occurred within the lungfishes, coelacanths, or rhipidistians—have persisted for many years, and will probably continue into the future. *See* COELACANTHIFORMES; DIPNOI; OSTEICHTHYES; TETRAPODA.

Herbert Boschung

Sarcosporida An order of Protozoa of the class Haplosporea which comprises parasites in skeletal and cardiac muscle of vertebrates. The organisms have a very wide distribution both geographically and in host species, infecting reptiles, birds, and mammals (including marsupials). Humans are occasionally infected by the parasite referred to as *Sarcocystis lindemanni*. It is doubtful whether Sarcosporida found in different hosts are themselves always different, though this has often been assumed. Host specificity is known not to be strict; however, it is unlikely that there is only one species—*S. miescheriana*—as proposed by some investigators.

The criteria of species are very hard to define for *Sarcocystis*. Species descriptions have usually been based on an assumed host specificity, size of spores, and cyst characteristics. But host specificity has seldom been proved, spore size is variable (degree of variation often in part dependent on the host species), and cyst morphology is frequently inconstant. Nevertheless, cyst morphology is probably more stable than anything else about the parasite.

In general, the cyst wall is said to have one or two layers, the outer one being either smooth or provided with spines or villi. The genesis of the wall is in some dispute; some think it is formed by the host as a reaction to the parasite, but the majority opinion is that the parasite itself forms the cyst. The cyst is divided internally into compartments, and the outer wall is smooth or rough, depending on the species, with numerous, minute villuslike projections on the outer surface. The spores are crescent-shaped and have a minute projection, termed a conoid, at the more pointed or anterior end, from which fibrils pass anteriorly. These fibrils are called toxonemes at the point of origin, and sarconemes (in *Sarcocystis*) farther back. The nucleus and several mitochondria are in the posterior half.

The life cycle of *Sarcocystis* is quite typical of the Sporozoa, since sexual stages are lacking and schizogony (multiple fission), though sometimes claimed, apparently does not occur. Instead, reproduction is by binary fission, with the eventual development of cysts (Miescher's tubules) in the muscles. These cysts are relatively large structures, easily visible with the unaided eye, and contain myriads of the minute crescentic spores. Infection of a new host is believed to be by the ingestion of these spores with food and water contaminated by feces from an infected animal. Most natural infections appear to be mild and to do the host little harm. *See* HAPLOSPOREA; PROTOZOA; SPOROZOA.

Reginald D. Manwell

Sarsaparilla A flavoring material obtained from the roots of at least four species of the genus *Smilax* (Liliaceae). These are *S. medica* of Mexico, *S. officinalis* of Honduras, *S. papyracea* of Brazil, and *S. ornata* of Jamaica, all tropical American vines found in the dense, moist jungles. The flavoring is used mostly in combination with other aromatics such as wintergreen. *See* LILIALES; SPICE AND FLAVORING.

Perry D. Strausbaugh; Earl L. Core

Sassafras A medium-sized tree, *Sassafras albidum*, of the eastern United States, extending north as far as southern Maine. Sometimes it is only a shrub in the north, but from Pennsylvania southward heights of 90 ft (27 m) or more with diameters of 4–7 ft (1.2–2.1 m) have been reported for this plant. Sassafras is said to live from 700 to 1000 years. It can be recognized by the bright-green color and aromatic odor of the twigs and leaves. The leaves are simple or mitten-shaped (hence a common name

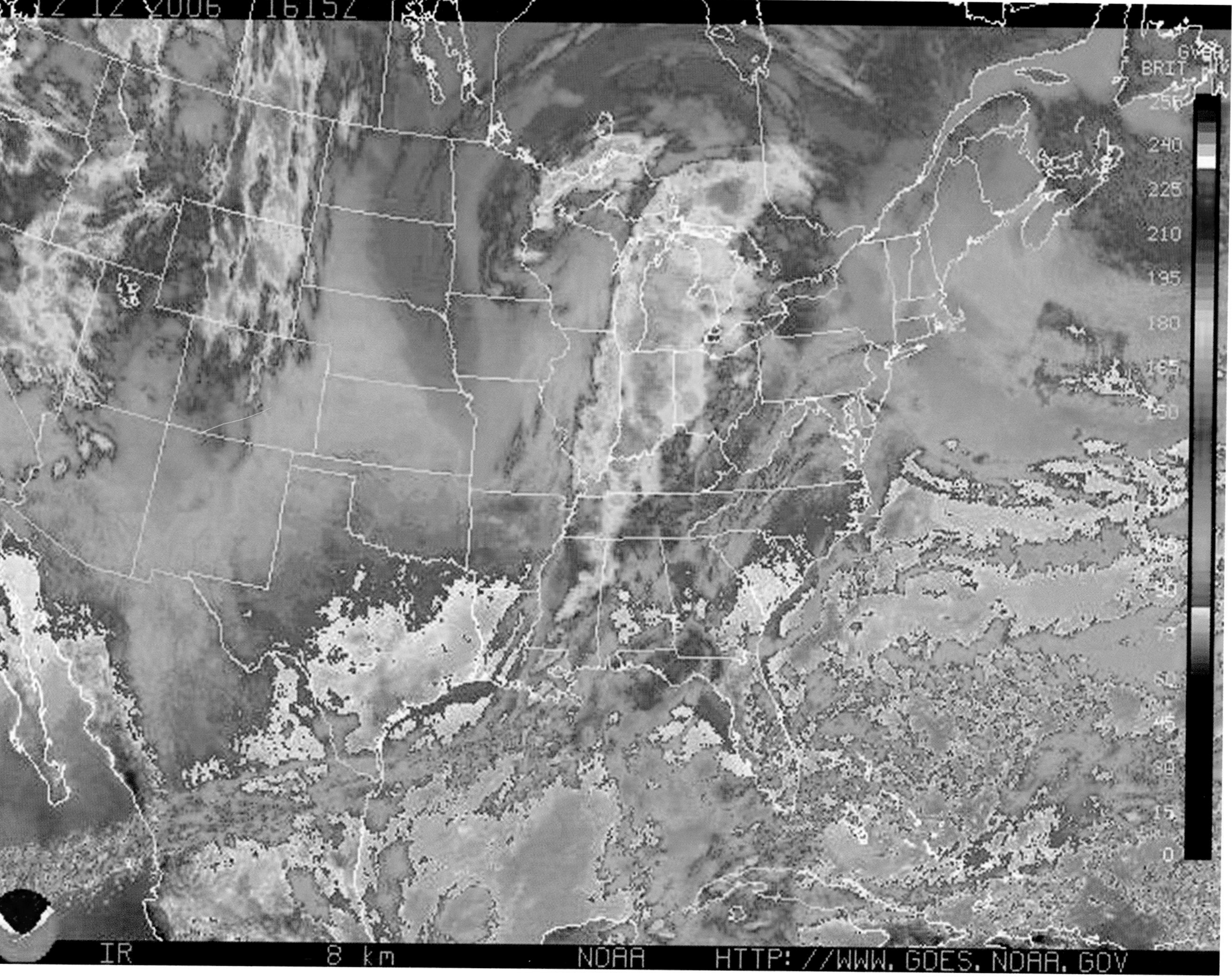

Infrared image from the *GOES 12* satellite, showing the eastern United States on December 12, 2006, at 1903 UTC (2:03 p.m. EST). Cloud tops are enhanced with colors to help establish temperature in Kelvin per the color bar at the right. In an infrared image, cold clouds are high clouds, so cold cloud tops and cloud distribution can easily be identified. (*NOAA*)

"mitten-tree"), or they may have lobes on both sides of the leaf blade. *See* MAGNOLIALES. Arthur H. Graves; Kenneth P. Davis

Satellite (astronomy) A relatively small body orbiting a larger one that in turn orbits a star. In the solar system, all of the planets except Mercury and Venus have satellites. Over 150 were known by September 25, 2008, with 168 in well-established orbits and distributed as follows: Earth, 1; Mars, 2; Jupiter, 62; Saturn, 60; Uranus, 27; Neptune, 13; Pluto, 3. More satellites are being discovered each year as a result of deliberate searches by increasingly powerful Earth-based techniques and by the *Cassini-Huygens* mission to the Saturn system. The close flyby of the *Galileo* spacecraft (enroute to Jupiter) of the asteroid 243 Ida in 1993 revealed the presence of a 0.9-mi-diameter (1.5-km) satellite, now known as Dactyl. This unexpected discovery has been followed by the detection of several other asteroid satellites by Earth-based observers. Several Kuiper Belt objects (distant comet nuclei) have also been observed to be binaries. *See* ASTEROID; BINARY ASTEROID; KUIPER BELT.

For planets, it is customary to distinguish between regular satellites that have nearly circular orbits lying essentially in the plane of a planet's equator and irregular satellites whose orbits are highly inclined, very elliptical, or both. The former almost certainly originated with the parent planet, while the latter must be captured objects. The Earth's Moon is a special case. The most widely favored hypothesis for its origin invokes an impact with Earth by a Mars-sized planetesimal, and ejection of material that first formed a ring around the Earth and then coalesced to form the Moon. Pluto's Charon may have formed through a similar collision. *See* MOON; PLUTO; SOLAR SYSTEM. Tobias C. Owen

Satellite (spacecraft) A spacecraft that is in orbit about a planet (usually the Earth or a moon). Spacecraft are devices intended for observation, research, or communication in space. Even those spacecraft that are on the way to probe the outer reaches of the solar system usually complete at least a partial revolution around Earth before being accelerated into an interplanetary trajectory. Devices such as sounding rockets follow ballistic (approximately parabolic) paths after fuel exhaustion, but they are not satellites because they do not achieve velocities great enough to avoid falling back to Earth before completing even one revolution. *See* ROCKET ASTRONOMY; SPACE PROBE.

The space shuttle, the *International Space Station* (*ISS*), and many automated (crewless, robotic) satellites travel in low Earth orbits (LEO) about 100 mi (160 km) or more above Earth's surface. They have typical orbital periods of about 90 min. These satellites have lifetimes of days, weeks, months, or years, depending on their altitudes, their mass-to-drag ratios, and atmospheric drag variations caused by solar activity. The *International Space Station* would have an orbital lifetime of only a few years without the periodic orbital boosts provided by the space shuttle or other rocket-powered space vehicles such as the Russian *Progress*. Most LEO satellites spend up to nearly half of their time in Earth's shadow. The space shuttle provides its electric power from fuel cells, but almost every other LEO spacecraft depends on solar cells for its power and batteries to operate through the sixteen 35-min "nights" which occur during each 24-h terrestrial day. *See* SPACE SHUTTLE; SPACE STATION.

Earth is not a perfect sphere. Its rotation causes its equatorial diameter to be 26 mi (42 km) greater than its corresponding polar dimension. For LEO satellites at altitudes below 3700 mi (6000 km), a retrograde inclination slightly greater than 90° may be selected which will cause the orbital plane to rotate eastward at exactly one revolution per year. This equatorial bulge phenomenon has the desirable result of permitting the plane of such Sun-synchronous orbits, as viewed from the Sun, to remain in the same apparent orientation throughout the year. In more practical terms, if such a Sun-synchronous satellite crosses the Equator in Brazil at 10:00 a.m. on January 1, it will also do so on June 30 or on any other day of the year. Since the orbital plane remains fixed relative to the Earth-Sun axis, the equatorial crossing time also occurs at the same local time at any longitude. This is ideal for weather, Earth resources monitoring, and reconnaissance purposes, because shadows will fall with the same relative length in the same direction and daily weather buildups will be imaged at essentially the same stage from each orbit to the next. *See* METEOROLOGICAL SATELLITES; MILITARY SATELLITES; REMOTE SENSING.

Earth has a period of rotation relative to the fixed stars of 23 h 56 min 4 s, which is one sidereal day. A satellite orbit of this period is said to be a geosynchronous orbit (GEO). If this orbit is also circular and equatorial, the spacecraft is said to be geostationary, because it remains in a fixed position relative to any observer on the approximately one-third of Earth from which the satellite is visible. Its principal advantage for communications is that, once pointed at the GEO spacecraft, an antenna on Earth never needs to be repointed. *See* COMMUNICATIONS SATELLITE.

In addition to communications, the GEO arc is used for weather observation spacecraft. Three such spacecraft evenly spaced along the Equator can monitor continuously severe weather around the entire globe, with the exception of regions within about 10° of the North and South poles, where hurricanes and tornadoes are absent.

Many *Explorer*-class spacecraft [special-purpose smaller satellites, typically 150–500 lb (70–230 kg) mass] have been devoted to studying phenomena whose investigation requires direct sampling of the local environment, such as magnetic fields and associated ionized plasma particles in Earth's magnetosphere and radiation (Van Allen) belts. Their orbits have been quite varied. Most have traveled in highly eccentric Earth orbits, characterized by perigees (lowest altitudes) of 100–200 mi (160–320 km) and apogees (highest altitudes) out to lunar distances. *See* SCIENTIFIC AND APPLICATIONS SATELLITES.

During the 1990s, radio navigation satellite systems assumed global importance. The two leading systems are the U.S. Global Positioning Satellite (GPS) constellation and the Russian GLONASS system. *See* SATELLITE NAVIGATION SYSTEMS.

The size and shape of a spacecraft is almost always dictated primarily by its mission requirements. The principal constraints are usually imposed by the dimensions and shape of the satellite payload provisions of the launch vehicle. An important requirement of virtually all powered automated spacecraft is sufficient solar cell mounting area both to power the payload in sunlight and to charge its batteries to continue payload operations during solar eclipse periods. Another requirement is to provide spacecraft attitude stabilization and control so that sensors and antennas can be pointed in the required directions. *See* SPACE FLIGHT; SPACECRAFT STRUCTURE. John F. Clark

Satellite meteorology The branch of meteorological science that uses meteorological sensing elements on satellites to define the past and present state of the atmosphere. Meteorological satellites can measure a wide spectrum of electromagnetic radiation in real time, providing the meteorologist with a supplemental source of data.

Modern satellites are sent aloft with multichannel high-resolution radiometers covering an extensive range of infrared and microwave wavelengths. Radiometers sense cloudy and clear-air atmospheric radiation at various vertical levels, atmospheric moisture content, ground and sea surface temperatures, and ocean winds, and provide visual imagery as well. *See* METEOROLOGICAL SATELLITES.

There are two satellite platforms used for satellite meteorology: geostationary and polar. Geostationary (geo) satellites orbit the Earth at a distance that allows them to make one orbit every 24 hours. By establishing the orbit over the Equator, the satellite appears to remain stationary in the sky. This is important for continuous scanning of a region on the Earth for mesoscale (approximately 10–1000 km horizontal) forecasting.

Polar satellites orbit the Earth in any range of orbital distances with a high inclination angle that causes part of the orbit to fly over polar regions. The orbital distance of 100–200 mi (160–320 km) is selected for meteorological applications, enabling the satellite to fly over a part of the Earth at about the same time every day. With orbital distances of a few hundred miles, the easiest way to visualize the Earth-satellite relationship is to think of a satellite orbiting the Earth pole-to-pole while the Earth rotates independently beneath the orbiting satellite. The advantage of polar platforms is that they eventually fly over most of the Earth. This is important for climate studies since one set of instruments with known properties will view the entire world.

The enormous aerial coverage by satellite sensors bridges many of the observational gaps over the Earth's surface. Satellite data instantaneously give meteorologists up-to-the minute views of current weather phenomena.

Images derived from the visual channels are presented as black and white photographs. The brightness is solely due to the reflected solar light illuminating the Earth. Visible images are useful for determining general cloud patterns and detailed cloud structure. In addition to clouds, visible imagery shows snowcover, which is useful for diagnosing snow amount by observing how fast the snow melts following a storm. Cloud patterns defined by visual imagery can give the meteorologist detailed information about the strength and location of weather systems, which is important for determining storm motion and provides a first guess or forecast as to when a storm will move into a region. *See* CLOUD.

More quantitative information is available from infrared sensors, which measure radiation at longer wavelengths (from infrared to microwave). By analyzing the infrared data, the ground surface, cloud top, and even intermediate clear air temperatures can be determined 24 hours a day. By relating the cloud top temperature in the infrared radiation to an atmospheric temperature profile from balloon data, cloud top height can be estimated. This is a very useful indicator of convective storm intensity since more vigorous convection will generally extend higher in the atmosphere and appear colder. *See* BLACKBODY; MICROWAVE; PLANCK'S RADIATION LAW.

The advent of geosynchronous satellites allowed the position of cloud elements to be traced over time. These cloud movements can be converted to winds, which can provide an additional source of data in an otherwise unobserved region. These techniques are most valuable for determination of mid- and high-level winds, particularly over tropical ocean areas. Other applications have shown that low-level winds can be determined in more spatially limited environments, such as those near thunderstorms, but those winds become more uncertain when the cloud elements grow vertically into air with a different speed and direction (a sheared environment). *See* WIND.

By using a wide variety of sensors, satellite data provide measurements of phenomena from the largest-scale global heat and energy budgets down to details of individual thunderstorms. Having both polar orbiting and geosynchronous satellites allows coverage over most Earth locations at time intervals from 3 minutes to 3 hours.

The greatest gain with the introduction of weather satellites was in early detection, positioning, and monitoring of the strength of tropical storms (hurricanes, typhoons). Lack of conventional meteorological data over the tropics (particularly the oceanic areas) makes satellite data indispensable for this task. The hurricane is one of the most spectacular satellite images. The exact position, estimates of winds, and qualitative determination of strength are possible with continuous monitoring of satellite imagery in the visible channels. In addition, infrared sensors provide information on cloud top height, important for locating rain bands. Microwave sensors can penetrate the storm to provide an indication of the interior core's relative warmth, closely related to the strength of the hurricane, and sea surface temperature to assess its development potential. *See* HURRICANE; TROPICAL METEOROLOGY.

Most significant weather events experienced by society—heavy rain or snow, severe thunderstorms, or high winds—are organized by systems that have horizontal dimensions of about 60 mi (100 km). These weather systems, known as mesoscale convective systems, often fall between stations of conventional observing networks. Hence, meteorologists might miss them were it not for satellite sensing. *See* HAIL; MESOMETEOROLOGY; METEOROLOGY; PRECIPITATION (METEOROLOGY); STORM; THUNDERSTORM; TORNADO; WEATHER FORECASTING AND PREDICTION.

Daniel L. Birkenheuer; John A. McGinley

Satellite navigation systems Radionavigation systems using artificial satellites as sources of radio signals and position references. Development of satellite-based systems for global positioning and navigation began almost immediately after the launch of *Sputnik I* by the Soviet Union in 1957. The U.S. Navy Navigation Satellite System, better known as *Transit*, became operational in 1964. The Soviet Union responded with its own satellite navigation systems, *Parus* and *Tsikada*. All three systems were based on measurement of Doppler shift in signals received from a satellite, and were aimed mainly at offering navigational guidance to ships and submarines. *See* DOPPLER EFFECT.

Starting in the mid-1990s, the term satellite navigation system became synonymous with the U.S. Navstar Global Positioning System (GPS), but this would change. GPS is used every day by millions around the world for accurate estimates of position, velocity, and time. These estimates can be obtained instantaneously and continuously anywhere and by anyone equipped with an inexpensive, pocket-sized GPS receiver. The only requirement is that the receiver must be able to "see" at least four GPS satellites in the sky.

The success of GPS has led to efforts by other countries to develop systems modeled after it. The development of GLONASS began in the Soviet Union a few steps behind GPS. The system was acquired by Russia following the breakup of the Soviet Union, but it is not yet ready for operational use. GPS and GLONASS both started out as military systems. By contrast, Galileo, a European system on the drawing boards in 2008, is being designed primarily as a civil system. GPS, GLONASS, and Galileo are based on the same principles and exploit the same technologies. The generic name for each is Global Navigation Satellite System (GNSS).

Principles of satellite navigation. The basic idea behind position estimation with GPS (as well as GLONASS and Galileo) is very simple: One's position can be determined if one knows, or can measure, distances to objects whose position coordinates are known. Estimation of a position based on measurement of distances is referred to as trilateration.

While the idea of trilateration is not new, a global navigation satellite system based on this idea would not have been feasible without several technologies which were developed or came to maturity in the second half of the twentieth century. The key technologies are stable space-based platforms in predictable orbits, extremely stable atomic clocks, spread spectrum signals, global coordinate frames, and microelectronics.

Trilateration requires measurement of ranges to three position references (that is, objects at known positions). A satellite navigation system uses satellites as the position references.

Measurement of ranges in terms of transit times of signals from the satellites to a receiver must be accurate. Therefore, measurement of transit times must also be accurate to the nanosecond (10^{-9} s) level. Such precision is obtained by marking the times of transmission on the signals in accordance with ultrastable cesium and rubidium atomic clocks carried aboard the satellites. *See* ATOMIC CLOCK; SPREAD SPECTRUM COMMUNICATION.

A global positioning system requires a global coordinate frame in which to express the positions of its satellites and the users. Previously, the coordinate frames were defined at most on a country or regional basis. GPS provided the impetus for development by the U.S. Department of Defense of the World Geodetic System 1984 (WGS 84), which has now become a de facto international standard. *See* GEODESY.

GPS. GPS was developed in the 1970s and 1980s by the U.S. Department of Defense, and was declared ready for operational use in 1995. The system was intended primarily for the U.S. military; civil use was a secondary objective. Through its many civil applications in industry, commerce, and outdoor recreational activities, GPS is essential to daily life as a new utility.

GPS can be thought of in terms of its three elements, generally referred to as segments: space segment, control segment, and user segment. The space segment comprises the satellite constellation. The baseline constellation consists of 24 satellites in circular orbits with a radius of 26,560 km (16,504 mi) and orbital period of about 12 h. The satellites are arranged in six planes containing four satellites each and inclined at 55° relative to the equator. With this satellite constellation, more than 99% of the GPS users worldwide with a clear view of the sky have at least four satellites in view at all times.

The control segment encompasses the facilities and functions required for operating GPS. The facilities include crewless monitoring stations and uplink radar sites distributed around the globe, and the Master Control Station located at Schriever Air Force Base in Colorado. The main functions include satellite health monitoring, and prediction and upload of satellite orbits and clock parameters.

The user segment encompasses all aspects of design, development, production, and utilization of user equipment for the different applications.

Systems under development. Besides GPS, other satellite navigation systems are under development.

GLONASS. While GPS was under development, the Soviet Union undertook to develop a similar system called GLONASS. Like GPS, GLONASS was designed primarily for the military, and a subset of its signals was offered for civil use in the late 1980s. Since the dissolution of the Soviet Union, the Russian Federation has assumed responsibility for GLONASS.

GLONASS, like GPS, fielded a 24-satellite constellation in medium earth orbits but, unlike GPS, placed the satellites in three planes with an inclination of 65°. Each satellite transmits in two frequency bands in the L-band and offers GPS-like services for the civil and military users.

Galileo. A more promising GNSS is the European system, Galileo, planned as "an open, global system, fully compatible with GPS, but independent from it." Galileo was conceived as a joint public-private enterprise under civilian control, to be financed and managed by the European Commission, the European Space Agency, and industry.

Galileo would be a second-generation GNSS, following the path of GPS and benefiting from technological advances since the mid-1970s when GPS was designed. Galileo would offer an Open Service without any user fees.

Compatibility and interoperability of systems. GPS, GLONASS, and Galileo are autonomous systems. From a user's perspective, at a minimum, these systems should be compatible or noninterfering: None should degrade the performance of the others in a significant way. Ideally, the systems would be interoperable, allowing a user to combine measurements from the three systems so that the performance is as least as good as the best of the three at that place and time.

Applications. While its positioning capability receives most attention, GPS is also a precise timing service. For example, before GPS, it would have been a challenge to synchronize clocks at multiple sites to conduct simultaneous observations of events (perhaps celestial) at distant locations. An inexpensive GPS receiver can now serve as a precise clock, keeping time with an accuracy of better than 0.1 microseconds relative to Coordinated Universal Time (UTC). Timing is critical in today's telecommunication systems, and GPS is being used increasingly to synchronize the elements of networks. *See* ATOMIC TIME; TIME.

The applications of GPS for navigation and fleet tracking in land transportation, aviation, and maritime commerce continue to grow. The convergence of GPS and personal computer technology has made it possible to collect vast amounts of positional data and organize them into geographic information systems (GIS), showing, for example, a map showing concentrations of toxic or radioactive wastes at a site. The centimeter-level positioning capability of GPS has revolutionized surveying and mapping, and has created novel applications in construction, mining, agriculture, and the earth sciences. *See* GEOGRAPHIC INFORMATION SYSTEMS.

Millions of bikers, runners, hikers, sailors, and fishermen have come to rely on GPS for their position and velocity. GPS receivers are now routine in cell phones and personal digital assistants.

Pratap Misra

Satellite radio Direct audio broadcasting from satellites to mobile and stationary receivers. This concept originated in the late 1980s as an outgrowth of direct television broadcasting from satellites to home receivers. There are currently two satellite radio systems (also known as satellite digital audio radio service) operating in the continental United States: one is providing service to users in Japan and Korea, and another is in advanced planning to serve Europe. *See* SATELLITE TELEVISION BROADCASTING.

The useful range of frequencies for radio broadcast transmission from satellites is 1–3 GHz. The Federal Communications Commission (FCC) has allocated a frequency band of 2320–2345 MHz for satellite radio, which has been coordinated internationally. This frequency band cannot be used in the United States for any other service because the sensitivity, mobility, and antenna omnidirectionality of the ground receivers require an interference-free environment. Of the available 25 MHz, a satellite radio system requires between 10 and 15 MHz bandwidth to support a reasonable number of audio programs. As a result, the FCC auctioned two 12.5-MHz-wide bands for satellite radio service, and granted licenses for two such systems in October 1997 to Sirius Satellite Radio Inc. and to XM Satellite Radio, Inc. *See* RADIO; RADIO BROADCASTING; RADIO SPECTRUM ALLOCATION; RADIO-WAVE PROPAGATION.

The technical challenge of satellite radio is in keeping service outages very infrequent and short. Service outages are caused by blockages, such as a building, structure, or mountain that obstructs the line-of-sight between a satellite and receiving antenna; by multipath, where received reflected signal components have sufficient amplitude and phase offset to corrupt (distort) the unreflected received signal; by foliage attenuation; and by interference, particularly the out-of-band emissions from transmitters in adjacent frequency bands.

The elevation angle from the satellite to the radio is extremely important in mitigating blockage, multipath, and foliage attenuation. For geostationary orbits, the elevation angle is reduced as the location of the radio moves further north or south of the Equator. An alternative is to use a constellation of satellites in highly inclined, elliptical geosynchronous orbits so that high elevation angles can be provided to radios located at latitudes above 30°N. *See* COMMUNICATIONS SATELLITE.

Robert D. Briskman

Satellite television broadcasting The transmission of television and other program material via satellite directly to individual homes and businesses. Direct broadcasting satellite (DBS) systems, or direct-to-home (DTH) satellite systems, are operational in many nations and regions around the world. The highly successful "small dish" DTH systems are characterized by all-digital transmission with the links to the customer dishes at frequencies above 10 GHz using radio spectrum sometimes referred to as Ku-band. *See* SATELLITE RADIO.

DBS systems use a satellite in geostationary orbit to receive television signals sent up from a broadcasting center, amplify them, and transmit them down to the customer receiving antennas called dishes. The satellite also shifts the signal frequency, so that, for example, a signal sent up to the satellite in the 17.3–17.8-GHz uplink band is transmitted back down in the 12.2–12.7-GHz downlink band. The downlink signal is picked up by a receive antenna that is permanently pointed at the satellite. *See* ANTENNA (ELECTROMAGNETISM).

The signal from the dish antenna is first passed to a downconverter, mounted outdoors on the antenna, which shifts it to (typically) the 0.95–1.45-GHz band. This signal is then conducted by cable to the receiver near the television set. The receiver contains the channel selector or tuner, a demodulator, and a decryptor (descrambler) as well as a video-audio decoder to permit the user to view authorized channels. *See* TELEVISION RECEIVER.

Television broadcasting by satellite may use radio-frequency spectrum licensed as either the broadcasting satellite service (BSS) or the fixed satellite service (FSS). The two major DBS systems in the United States, DIRECTV and Dish, have used BSS spectrum with a downlink frequency of 12.2–12.7 GHz. Other major systems in the Western Hemisphere use FSS spectrum with a downlink frequency around 11 GHz. The term DBS is generally used to denote systems using the BSS band. The term DTH is more general and applies to television delivery systems employing any band. *See* RADIO SPECTRUM ALLOCATION.

A direct broadcasting satellite typically contains 32 or more transponders, each with a radio-frequency power output in the range 120–240 W. Each transponder acts as a separate amplifier, frequency translator, and transmitter chain for a signal uplinked from the broadcasting facility. The transmitter of each transponder drives one of the satellite's downlink antennas. Each downlink antenna creates either a broad national beam or spot beams.

All recently deployed DBS and DTH systems use digital signals. Typically in these systems, a single 24-MHz-wide satellite transponder can carry an error-corrected digital signal of 30 megabits per second or greater. A wide variety of communications services can be converted to digital form and carried as part of this digital signal, including conventional television, high-definition television, multichannel audio, and other forms of digital data. *See* COMMUNICATIONS SATELLITE; TELEVISION.

John P. Godwin

Saturable reactor An iron-core inductor in which the effective inductance is changed by varying the permeability of the core. Saturable-core reactors are used to control large alternating currents where rheostats are impractical. Theater light dimmers often employ saturable reactors.

In the illustration of two types of saturable-core reactors, illustration *a* shows two separate cores, while in illustration. *b* a three-legged core is formed by placing two two-legged cores together. The load winding, connected in series with the load, carries the alternating current and acts as an inductive element. The control winding carries a direct current of adjustable magnitude, which can saturate the magnetic core.

Reducing the magnitude of the control current reduces the intensity of saturation. This increases the reactance of the load winding. As the reactance increases, the voltage drop in the load winding increases and causes a reduction in the magnitude of

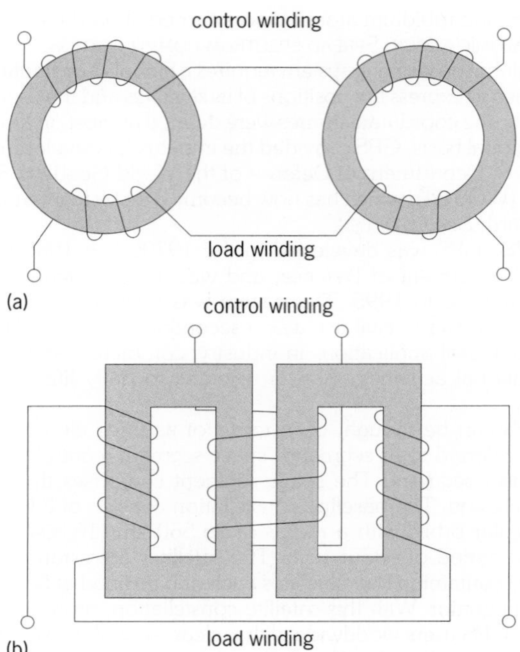

Typical construction of saturable-core reactors. (*a*) Two separate cores. (*b*) Three-legged cores.

the voltage applied to the load. *See* INDUCTANCE; MAGNETIC PERMEABILITY; MAGNETIZATION.

Wilson S. Pritchett

Saturation The condition in which, after a sufficient increase in a causal force, further increase in the force produces no additional increase in the resultant effect. Many natural phenomena display saturation. For example, after a magnetizing force becomes sufficiently strong, further increase in the force produces no additional magnetization in a magnetic circuit; all the magnetic domains have been aligned, and the magnetic material is saturated. *See* MAGNETIC MATERIALS.

After a sponge has absorbed all the liquid it can hold, it is saturated. In thermionic vacuum tubes thermal saturation is reached when further increase in cathode temperature produces no (or negligible) increase in cathode current; anode saturation is reached when further increase in plate voltage produces substantially no increase in anode current. *See* DISTORTION (ELECTRONIC CIRCUITS); SATURATION CURRENT; VACUUM TUBE.

In colorimetry the purer a color is, the higher its saturation. Radiation from a color of low saturation contains frequencies throughout much of the visible spectrum. *See* COLORIMETRY.

Frank H. Rockett

Saturation current A term having a variety of specific applications but generally meaning the maximum current which can be obtained under certain conditions.

In a simple two-element vacuum tube, it refers to either the space-charge-limited current on one hand or the temperature-limited current on the other. In the first case, further increase in filament temperature produces no significant increase in anode current, whereas in the latter a further increase in voltage produces only a relatively small increase in current. *See* VACUUM TUBE.

In a gaseous-discharge device, the saturation current is the maximum current which can be obtained for a given mode of discharge. Attempts to increase the current result in a different type of discharge. *See* ELECTRICAL CONDUCTION IN GASES.

A third case is that of a semiconductor. Here again the saturation current is that maximum current which just precedes a change in conduction mode. *See* SEMICONDUCTOR.

Glenn H. Miller

Saturn The second largest planet in the solar system and the sixth in order of distance to the Sun. The outermost planet known prior to 1781, Saturn is surrounded by a beautiful system of rings, more than 30 moons, and a complex magnetosphere. Despite the planet's huge size, its mean density is so low it could float in water. Saturn is also the only planet that has a satellite (Titan) with a dense atmosphere. This distant planetary system has been visited by four spacecraft: *Pioneer 11* in 1979, *Voyagers 1* and *2* in 1980–1981, and *Cassini-Huygens* since 2004. The first three were flybys; the last is a joint United States–European mission consisting of a Saturn orbiter and a probe into the atmosphere of Titan.

Saturn makes one revolution about the Sun in 29.42 years. The equatorial diameter of Saturn is about 75,000 mi (120,540 km), and the polar diameter about 67,600 mi (108,700 km). The volume is 769 (Earth = 1) with a few percent uncertainty. The mass is about 95.2 (Earth = 1) or 1/3500 (Sun = 1). The mean density is 0.70 g/cm^3, the lowest mean density of all the planets. The rotation axis of both the planet and the rings is inclined 27° to the perpendicular to the orbital plane. The visible cloud layers of Saturn are much more homogeneous in appearance than those of Jupiter. There is no feature comparable to the Great Red Spot, and the contrast of the features (huge storm systems in the cloud decks) that are visible is very low (Fig. 1).

The optical spectrum of Saturn is characterized by strong absorption bands of methane (CH_4) and by much weaker bands of ammonia (NH_3). Absorption lines of molecular hydrogen (H_2) have also been detected.

Fig. 1. Saturn viewed from the *Cassini Orbiter*. The soft velvety appearance of the low-contrast banded structure is due to scattering by the haze layer above the planet's cloud deck. The image has been rotated so that north is up. The Sun illuminates Saturn from below and Saturn's tilt throws shadows of the rings onto the northern hemisphere during the current season. The image was taken on January 23, 2005 at a distance of 1.7 × 10^6 mi (2.8 × 10^6 km) from Saturn. (*NASA/JPL/Space Science Institute*)

The temperature the planet should assume in response to solar heating is calculated to be about 76 K (−323°F), somewhat lower than the measured value of 92 K (−294°F). This suggests that Saturn has an internal heat source of roughly the same magnitude as that on Jupiter. As in the case of Jupiter, a thermal inversion exists in the upper atmosphere. The inversion region is well above the main cloud layer, which should consist primarily of frozen ammonia crystals, with an admixture of some other substances such as sulfur to provide the yellowish color sometimes observed in the equatorial zone.

Theoretical models for the internal structure of Saturn are similar to those for Jupiter, that is, a dense core surrounded by hydrogen compressed to a liquid metallic state that gradually merges into an extremely deep atmosphere. As on Jupiter, the gradual precipitation of helium from solution in the metallic hydrogen surrounding the core liberates heat from Saturn's interior and produces the subsolar abundance of helium (relative to hydrogen) found in Saturn's atmosphere. The existence of a magnetic field and belts of trapped electrons was initially deduced from observations of nonthermal radiation at dekameter wavelengths and was mapped out in detail by in situ measurements from instruments on the spacecraft that have visited the planet. The *Cassini Orbiter* is continuing these investigations. *See* MAGNETOSPHERE.

Jupiter and Saturn are relatively similar bodies. Both seem to have bulk compositions close to that of the Sun and the other stars, and both are rich in hydrogen and helium. In that sense, a large fraction of the mass they contain consists of the same primitive material from which the entire solar system was formed, whereas the inner planets such as Earth and Mars never accumulated a solar proportion of the light gases. This conclusion has been strengthened by measurements of the heavy isotope of hydrogen known as deuterium. They demonstrate that the hydrogen that makes up most of the mass of both Jupiter and Saturn was captured directly from the solar nebula when these planets formed 4.6 billion years ago. However, both Jupiter and Saturn show an enhancement of the carbon/hydrogen ratio (as determined from methane and hydrogen) compared with the Sun. This suggests that both planets formed in a two-stage process that led initially to formation of a large core of approximately the same size that generated an outgassed, secondary atmosphere, followed by the attraction of an envelope of gases from the surrounding nebula. *See* SOLAR SYSTEM.

The most remarkable feature associated with Saturn is the complex ring system that surrounds the planet (Fig. 2; colorplate 1). The system is divided into four main regions, designated A through D. The narrow F ring is located just beyond the edge of ring A, and there are G and E rings still farther out. Each of the four main regions is subdivided into many individual "ringlets," so that Saturn is actually surrounded by thousands of rings. The

Fig. 2. Visible-light image of Saturn's rings from the *Cassini Orbiter*. The Sun is shining through the rings so that they take on the appearance of a photonegative; the dense B ring at the center blocks much of the incoming light, while the less dense regions scatter and transmit light. The image was taken on April 15, 2005 at a distance of 350,000 mi (570,000 km) from Saturn. (*NASA/JPL/Space Science Institute*)

ring system is made up of myriad separate particles that move independently in flat, mostly circular orbits in Saturn's equatorial plane. Periodic perturbations by the major satellites are responsible, in part, for the main divisions of Saturn's rings.

Satellites. As of September 25, 2008, Saturn had 60 confirmed satellites. Still more are expected to be discovered. The largest and brightest, Titan, is visible with small telescopes; the other satellites are much fainter. Titan's mean apparent diameter corresponds to a linear diameter of approximately 3440 mi (5550 km). But this diameter refers to the satellite's atmosphere, which contains several layers of smog consisting of aerosols produced photochemically by incident sunlight. The solid surface of Titan has a diameter of 3200 mi (5150 km), making this satellite larger than Mercury but slightly smaller than Jupiter's giant Ganymede. This object contains a large fraction of icy material and is thus quite different from the Moon or the inner planets in composition. Furthermore, it is large and cold enough to retain a thick, nitrogen (N_2)-dominated atmosphere that contains a few percent of methane (CH_4) and exerts a surface pressure of 1.5 bars (1.5×10^5 Pa), or 1.5 times the sea-level pressure on Earth. The atmosphere is sufficiently thick to have allowed the safe entry and 2 h, 20 min descent by parachute of the *Huygens* probe on January 14, 2005. After landing, the probe continued to transmit data for another hour (colorplate 2). The surface of Titan is so cold (94 ± 2 K or -290 ± 4°F) that rivers and lakes of liquid methane must be present. Branching channel systems resembling rivers on Earth were seen by the probe, which also detected liquid methane in the surface material at the landing site. The cameras and the radar on the orbiter have found very few craters, indicating that Titan has some active processes at work that are burying and erasing them. A search for lakes, seas, and geysers of methane is continuing. The atmospheric chemistry and the solid surface on which liquids and aerosols can accumulate make Titan an attractive natural laboratory in which future investigators will be able to test ideas about chemical evolution on the primitive Earth.

The other inner satellites encompass a variety of characteristics. Most have densities near unity, indicating a predominantly icy composition. The surfaces of these objects are covered with impact craters, with the exception of Enceladus. Large regions of the surface of Enceladus are free of craters, indicating reworking of the surface in relatively recent times. Coupled with the unusually high reflectivity of this satellite (close to 100%), the modified surface suggests internal activity leading to partial melting or pulverization of the icy surface, and the production and expulsion of the tiny ice grains that populate the E ring. The *Cassini* spacecraft discovered the emission of water vapor from fissures at the satellite's south pole.

Saturn also has a number of more distant satellites that are classified as "irregular" because of the large eccentricities or inclinations (or both) of their orbits. These satellites must have been captured by Saturn early in its history rather than forming together with it from the solar nebula. The largest of these is Phoebe, which moves in a retrograde direction. *See* RETROGRADE MOTION (ASTRONOMY); SATELLITE (ASTRONOMY). Tobias C. Owen

Saurischia One of two major monophyletic clades of Dinosauria, the other being Ornithischia. Saurischian dinosaurs take their name (meaning "lizard hipped") from the structure of their pelvis. In saurischians, the pubis bone points downward and forward from the hip joint, and the ischium extends downward and backward, so that the two bones are separated by a wide angle in side view. This feature allows the majority of saurischians to be easily distinguished from ornithischian dinosaurs (in which the pubis has rotated so that it points downward and backward and thus lies alongside the ischium). However, the saurischian condition is actually a retention of the primitive reptilian condition (which is also present in crocodilians, squamates, as well as many extinct groups) and, therefore, cannot be regarded as a unique feature of saurischians. To add to the confusion, some saurischian dinosaurs (various advanced theropod groups and their direct descendants, the birds) have independently evolved an ornithischian-type pelvis. *See* DINOSAURIA; ORNITHISCHIA.

The earliest saurischians are known from deposits of Late Triassic age (approximately 225 million years ago) in Brazil and Argentina. Early in its evolutionary history, Saurischia split into two major monophyletic clades: the Sauropodomorpha and Theropoda. Both groups diversified rapidly in the Late Triassic–Early Jurassic interval, each increasing in species richness and abundance to become the dominant vertebrates in terrestrial ecosystems. They quickly achieved a global distribution, and the remains of early saurischians are known from all continents.

Traditionally, sauropodomorphs have been divided into two groups: Sauropoda and Prosauropoda. In general, all sauropodomorphs share the same body plan: a small head, elongate neck, barrel-shaped body, and long, counterbalancing tail. The majority of prosauropods were facultatively bipedal, spending most of their time on all fours, but occasionally walking bipedally: very large prosauropods (such as *Riojasaurus*) were obligate quadrupeds. Prosauropods were confined to the Late Triassic–Early Jurassic, an interval in which they were the dominant terrestrial herbivores. Sauropods also appeared in the Late Triassic. In general, sauropod evolution followed several trends, including increased body size [culminating in animals such as *Argentinosaurus*, which may have reached a weight of 70 metric tons (77 tons)] and increased neck length. All sauropods were obligate quadrupeds and were more advanced herbivores than prosauropods. Sauropods reached their acme in the Late Jurassic, after which they declined in diversity and abundance, though a second peak in diversity also occurred in the middle to Late Cretaceous. The group survived until the end of the Cretaceous Period. *See* CRETACEOUS; JURASSIC; TRIASSIC.

Theropoda is a monophyletic group that also includes birds. All nonavian theropods were bipedal, and the majority were carnivorous. Theropods ranged in size from animals less than a meter in length (for example, *Microraptor*), which were probably largely insectivorous, to gigantic superpredators (such as *Tyrannosaurus rex*) over 12 m (39 ft) long that weighed up to 6.5 metric tons (7.2 tons) and preyed upon other dinosaurs. The dominant theme in theropod evolution relates to their transformation from scaly, fast-running, terrestrial bipeds to flying birds, with the subsequent transformation of forelimbs into wings and many other modifications associated with flight, including the origin of feathers. Nonavian theropods survived until the end of the Cretaceous Period; avian theropods survived the Cretaceous/Tertiary (K/T) extinction event and, therefore, represent the only group of living dinosaurs. *See* AVES; FEATHER.

Paul M. Barrett

Sauropterygia An infraclass of Mesozoic reptiles that are, without exception, adapted to the marine environment. The infraclass includes the nothosaurs, plesiosaurs, and placodonts. These reptiles, along with the ichthyosaurs, played a significant role as predators within the marine animal community of the Mesozoic Era.

The placodonts are a distinctive but highly varied assemblage of aquatic reptiles. They had short bodies, paddlelike limbs, and flat cheek teeth designed for crushing hard-shelled prey. The genus *Helodus* was covered by a dorsal bony armor and a roofing of dermal scutes, but most other genera lacked armor. Placodonts have come only from rocks of the Middle and Upper Triassic of Europe, North Africa, and the Middle East. *See* PLACODONTIA.

The Nothosauria are the relatively generalized stem group from which the plesiosaurs evolved. With the exception of a single New World species and a record from Japan, nothosaurs are known primarily from Europe and the Near East (Israel) in rocks of Triassic age. The nothosaurs are notably diverse in the mode and degree of secondary aquatic modification. The directions of aquatic specialization involve shortening, or more often lengthening, of the neck; enlargement of the orbits or the temporal fenestrae; and reduction, or more commonly increase, in the number of phalanges in manus (hand), pes (foot), or both. A feature of considerable evolutionary significance in the light of plesiosaurian differentiation is the great individual variability in the number of presacral vertebrae (32–42) in *Pachypleurosaurus edwardsi*.

The Plesiosauria are the successful, compact, and highly specialized offshoot of the nothosaurs that attained worldwide distribution. The early steps of aquatic adaptation, initiated by the nothosaurs, have led to extensive anatomical modifications: The region comprising chest and abdomen became short, stout, and inflexible; the ventral bones of shoulder girdle and pelvis increased in area enormously; and the limbs, transformed into large flippers, became the principal organs of propulsion. Two major trends of plesiosaur evolution may be discerned since the Early Jurassic: In the one group there was a tendency toward a shortening of the neck from 27 to 13 vertebrae and an increase in skull size: in the other group the opposite trend led to forms of bizarre body proportions, for example, *Elasmosaurus*, with a neck containing 76 vertebrae. The plesiosaurs were carnivorous.

The precise affinities of placodonts, nothosaurs, and plesiosaurs remain uncertain. No earlier group of reptiles from which they might have come is known, and they left no descendants. *See* NOTHOSAURIA; PLACODONTIA; PLESIOSAURIA; REPTILIA.

Rainer Zangerl; Everett C. Olson

Savanna The term savanna was originally used to describe a tropical grassland with more or less scattered dense tree areas. This vegetation type is very abundant in tropical and subtropical areas, primarily because of climatic factors. The modern definition of savanna includes a variety of physiognomically or environmentally similar vegetation types in tropical and extratropical regions. The physiognomically savannalike extratropical vegetation types (forest tundra, forest steppe, and everglades) differ greatly in environment and species composition.

In the widest sense savanna includes a range of vegetation zones from tropical savannas with vegetation types such as the savanna woodlands to tropical grassland and thornbush. In the extratropical regions it includes the "temperate" and "cold savanna" vegetation types known under such names as taiga, forest tundra, or glades. *See* GRASSLAND ECOSYSTEM; TAIGA; TUNDRA.

Helmut Lieth

Savory A herb of the mint family in the genus *Satureja*. There are more than 100 species, but only *S. hortensis* (summer savory) and *S. montana* (winter savory) are grown for flavoring purposes. *See* LAMIALES.

Summer savory, an annual herb, is characterized by long thin wiry stems with long internodes between small leaves. Winter savory is a perennial in most climates and, unlike summer savory, will tolerate some freezing weather. It can become woody after one or two growing seasons.

Savory is indigenous to areas surrounding the Mediterranean Sea. *Satureja montana* occurs wild from North Africa to as far north as Russia but is little cultivated. *Satureja hortensis*, which is widely cultivated, is native to Europe. Both types of savory are harvested or cut two or three times a year, after which the leaves are dehydrated and separated from the stems to be used as a spice in various foods, including poultry seasoning and beans. *See* SPICE AND FLAVORING.

Seth Kirby

Saxifragales An order of flowering plants near the base of the eudicots. The order's exact relationships to other orders are still under study, but the Saxifragales appear to be related to the rosid eudicots. They include 13 small to moderately sized families and perhaps around 2300 species. There are two categories: woody species that are sometimes wind-pollinated, and insect-pollinated herbs. The members are difficult to characterize morphologically but fall into several sets of families that have been considered closely related by most authors. *See* MAGNOLIOPHYTA; MAGNOLIOPSIDA; POLLINATION; ROSIDAE.

None of the families are of particular economic importance except as ornamentals, although *Cercidiphyllum* (Cercidiphyllaceae), *Liquidambar* (sweet gums), *Altingia* (both Altingiaceae), and a few others produce timbers. The largest family is Crassulaceae (1500 species), which are plants of arid zones, including the stonecrops (*Sedum*) and kalanchoes (*Kalanchoe*). A number of species are commonly cultivated ornamentals, including peony (*Paeonia*, Paeoniaceae), saxifrages (*Saxifraga*, Saxifagaceae), and witch hazel (*Hamamelis*, Hamamelidaceae); a few, the currants and gooseberries (*Ribes*, Grossulariaceae), are cultivated as fruits. *See* HAMAMELIDALES.

Mark W. Chase

Scalar A term synonymous in mathematics with "real" in real number or real function. The magnitude of a vector two units in length is the real number of scalar 2. The dot or scalar product of two vectors is the product of three real numbers associated with them (the magnitude of the first vector times the magnitude of the second, times the cosine of the angle between them) and is therefore a scalar. If in the functional relationships $S = S(x, y, z)$, $\mathbf{F} = \mathbf{F}(x, y, z)$, S is a real number and $\mathbf{F}$ is a vector, then $S(x, y, z)$ is a scalar function but $\mathbf{F}(x, y, z)$ is a vector function. *See* CALCULUS OF VECTORS.

Homer V. Craig

Scale (music) As a piece of music progresses, it typically outlines a set of pitches by repeatedly sounding a subset of all the possible notes. When these notes are rearranged into ascending or descending order, they are called a musical scale.

There is often a range of pitches that will be heard as the same. Perhaps the trumpet plays middle C a bit flat, while the guitar plays a bit sharp, in accordance with the artistic requirements of the musical context. The mind hears both pitches as the same note, C, and the limits of acceptability are far cruder than the ear's powers of resolution. This suggests a kind of categorical perception, where a continuum of possible stimuli (in this case, pitch) is perceived as consisting of a small number of disjoint classes. Thus scales partition pitch space into disjoint chunks. *See* MUSICAL INSTRUMENTS; PITCH; SENSATION.

In the modern Western tradition, scales are standardized subsets of the 12-tone equal temperament (abbreviated 12-tet, and also called the chromatic scale) in which each octave is divided into 12 (approximately) equal sounding divisions. These are further classified into major and minor depending on the exact ordering of the intervals, and are classified into modes depending on the starting point. Historically, however, scales based on 12-tet are fairly recent. In addition, other cultures use musical scales that are quite different. *See* TUNING.

William A. Sethares

Scale (zoology) The fundamental unit of the primary scaled integument of vertebrates, of which the epidermal or dermal component may be the more conspicuous or elaborated. Although there is great diversity in detailed structure among vertebrate groups, scaled integuments may be interpreted as the most effective way of producing a physically strong external body surface, with maximum flexibility necessitated by the fundamental pattern of vertebrate locomotion relying on lateral sinusoidal movements of the body.

In the majority of extant fish the epidermis is extremely thin. The term "scale" when used in an ichthyological context usually refers to the relatively large, prominent, dermal ossification, on the outer surface of which lie so-called dental tissues. The latter—enamel or enamellike mineralizations and dentine—are therefore present at the dermoepidermal boundary. The scaled integument of fish shows a definite pattern which is governed by the orientation of the underlying myomeres and by changes in body depth and length. The general evolutionary tendency toward thinning, or size reduction, of integumentary sclerifications seen in various fish lineages is probably associated with weight reduction and increased locomotor efficiency.

Scaled integuments are absent from modern amphibians, although nonoverlapping dermal ossifications are seen in some apodans. It seems probable that the absence of scales in other modern species is associated with the secondary utilization of the integument as a respiratory surface. In most reptilian lineages, there is a ubiquity of scaled integuments, with or without dermal ossifications, but always with elaboration of epidermal tissues. The distribution of keratinaceous protein types varies in different groups of reptiles, and in this respect modern crocodilian scales exactly resemble the leg scales of birds. Scaled integuments prompted the subclass name for lepidosaurs—the tuatara, lizards, and snakes, the last two constituting the order Squamata. Lepidosaurian scales may or may not possess dermal ossifications. Among extant reptiles the extraordinary development of the dermal skeleton in turtles forms the characteristic carapace.

P. F. Maderson

Scandium A chemical element, Sc. The only naturally occurring isotope is ^{45}Sc. The electronic configuration of the ground-state, gaseous atom consists of the argon rare-gas core plus three more electrons in the $3d14s2$ levels. It has an unfilled inner shell (only one $3d$ electron) and is the first transition metal. It is one of the elements of the rare-earth group. *See* PERIODIC TABLE; RARE-EARTH ELEMENTS; TRANSITION ELEMENTS.

The principal raw materials for the commercial production of scandium are uranium and tungsten tailings and slags from tin smelters or blast furnaces used in cast iron production. Wolframite (WO_3) concentrates contain 500–800 ppm scandium. *See* WOLFRAMITE.

Scandium is the least understood of the $3d$ metals. The major reason has been the unavailability of high-purity scandium metal, especially with respect to iron impurities. Many of the physical properties reported in the literature vary considerably, but the availability of electrotransport-purified scandium has allowed measurement of the intrinsic properties of scandium (see table).

Scandium increases the strength of aluminum. It also strengthens magnesium alloys when added to magnesium together with silver, cadmium, or yttrium. Scandium inhibits the oxidation of the light rare earths and, if added along with molybdenum, inhibits the corrosion of zirconium alloys in high-pressure steam. The addition of ScC to TiC has been reported to form the second-hardest material known. Sc_2O_3 can be used in many other oxides to improve electrical conductivity, resistance to thermal shock, stability, and density. Scandium is used in the preparation of the laser material $Gd_3ScGa_4O_{12}$, gadolinium scandium gallium garnet (GSGG). This garnet when doped with both Cr^{3+} and Nd^{3+} ions is said to be $3^1/_2$ times as efficient as the widely used Nd^{3+}-doped yttrium aluminum garnet (YAG:Nd^{3+}) laser. Ferrites and garnets containing scandium are used in switches in computers; in magnetically controlled switches that modulate light passing through the garnet; and in microwave equipment. Scandium is used in high-intensity lights. Scandium iodide is added because of its broad emission spectrum. Bulbs with mercury, NaI, and ScI_3 produce a highly efficient light output of a color close to sunlight. This is especially important when televising presentations indoors or at night. When used with night displays, the bulbs give a natural daylight appearance. Scandium metal has been used as a neutron filter. It allows 2-keV neutrons to pass through, but stops other neutrons that have higher or lower energies. *See* YTTRIUM.

Jennings Capellin; Karl A. Gschneidner

Room-temperature properties of scandium metal (unless otherwise specified)

Property	Value
Atomic number	21
Atomic weight	44.956
Lattice constant (hcp, α–Sc), a_0	0.33088 nm
c_0	0.52680 nm
Density	2.989 g/cm^3
Metallic radius	0.16406 nm
Atomic volume	15.041 cm^3/mol
Transformation point	1337°C (2439°F)
Melting point	1541°C (2806°F)
Boiling point	2836°C (5137°F)
Heat capacity	25.51 J/mol K
Standard entropy, $S^\circ_{298.15}$	34.78 J/mol K
Heat of transformation	4.01 kJ/mol
Heat of fusion	14.10 kJ/mol
Heat of sublimation (at 298 K)	377.8 kJ/mol
Debye temperature (at 0 K)	345.3 K
Electronic specific heat constant	10.334 mJ/mol K^2
Magnetic susceptibility, χ_A^{298} (a)	297.6×10^{-6} emu/mol
χ_A^{298} (c)	288.6×10^{-6} emu/mol
Electrical resistivity, ρ_a^{300}	70.90 μohm-cm
ρ_c^{300}	26.88 μohm-cm
Thermal expansion, $\alpha_{a,i}$	7.55×10^{-6}
$\alpha_{c,i}$	15.68×10^{-6}
Isothermal compressibility	17.8×10^{-12} m^2/N
Bulk modulus	5.67×10^{10} N/m^2
Young's modulus	7.52×10^{10} N/m^2
Shear modulus	2.94×10^{10} N/m^2
Poisson's ratio	0.279

Scanning electron microscope An electron microscope that builds up its image as a time sequence of points in a manner similar to that employed in television.

The imaging method of the scanning electron microscope (SEM) allows separation of the two functions of a microscope, localization and information transfer. The SEM utilizes a very fine probing beam of electrons which sweeps over the specimen to emit a variety of radiations. The signal, which is proportional to the amount of radiation leaving an individual point of the specimen at any instant, can be used to modulate the brightness of the beam of the display cathode-ray tube as it rests on the corresponding point of the image. In practice, the points follow one another with great rapidity so that the image of each point becomes an image of a line, and the line in turn can move down the screen so rapidly that the human eye sees a complete image as in television. The image can also be recorded in its entirety by

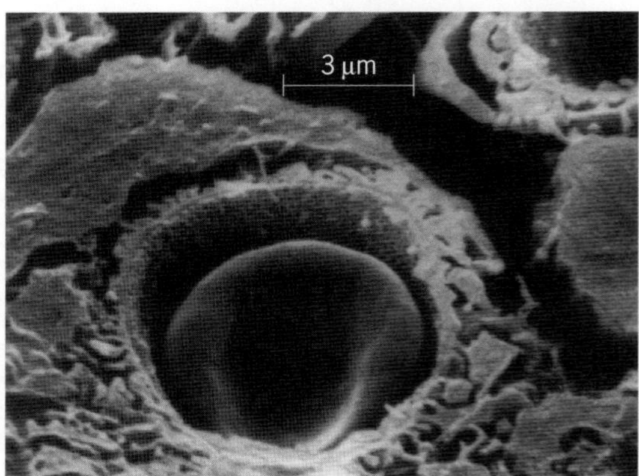

Red blood cell in a capillary of the kidney. Ethanol cryofracture technique was followed by critical-point drying. (*From W. J. Humphreys, B. O. Spurdock, and J. S. Johnson, Critical point drying of ethanol-infiltrated, cryofractured biological specimens for scanning electron microscopy, Proceedings of the Scanning Electron Microscopy Symposium, pp. 275–282, 1974*)

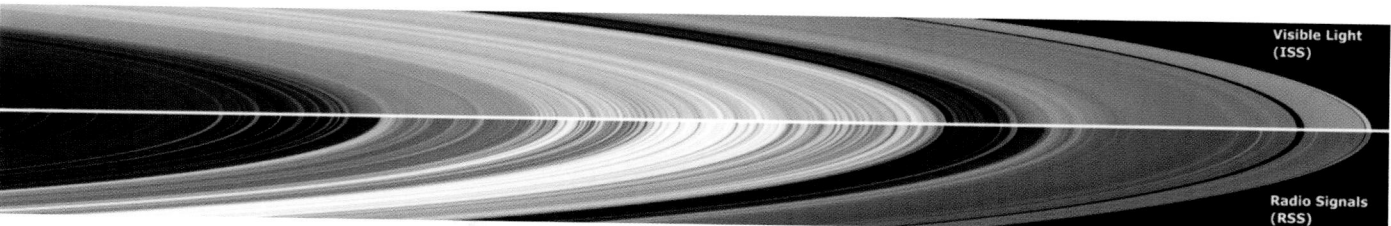

Colorplate 1. (Top) Structure of Saturn's rings observed by two different approaches. The upper half of the image is a natural-color mosaic of images by the *Cassini* narrow-angle camera. Such images have information about how ring structure differs both with distance from the planet and with position around the equatorial circle, but resolution is limited to a few kilometers at best. The bottom simulated image is constructed from a radio occultation observation on May 3, 2005. Color in this image is used to represent information about ring particle sizes. The radial resolution from such an image can be as fine as 50 m (164 ft), but the image provides information only along a one-dimensional track through the rings and is constructed by assuming circular symmetry over the ring region of interest. (*NASA/JPL/Space Science Institute*)

6 in (15 cm)

Colorplate 2. (Left) Image from the *Huygens* probe on the surface of Titan. This colored view, following processing to add reflection spectra data, gives an indication of the actual color of the surface. The rocklike object indicated as being 6 in. (15 cm) across is about 2.8 ft (85 cm) from *Huygens*. The surface is darker than expected, consisting of a mixture of water and hydrocarbon ice. There is evidence of erosion at the base of the objects on the surface, indicating possible fluvial activity. (*ESA/NASA/JPL/University of Arizona*)

allowing the point-by-point information to build up in sequence on a photographic film (see illustration).

As with all microscopy research, it is very often the preparative methodology that determines the success or failure of the research. A specific requirement of scanning electron microscope preparation is that the material be dried. This preparative step must be done very carefully to minimize surface tension effects. The two methods most often used—critical-point drying and freeze drying—have yielded excellent results even for very fragile biological tissue. Alternatively, the drying step can be avoided completely by observing the specimen while it is still frozen. Interior structure can be revealed by using techniques in which the specimen is broken at a low temperature. *See* ELECTRON MICROSCOPE; MICROTECHNIQUE. Thomas L. Hayes

Scanning tunneling microscope An instrument for producing surface images with atomic-scale lateral resolution, in which a fine probe tip is scanned over the surface at a distance of 0.5–1 nanometer, and the resulting tunneling current, or the position of the tip required to maintain a constant tunneling current, is monitored.

Scanning tunneling microscopes have pointed electrodes that are scanned over the surface of a conducting specimen, with help from a piezoelectric crystal whose dimensions can be altered electronically. They normally generate images by holding the current between the tip of the electrode and the specimen at some constant (set-point) value by using a piezoelectric crystal to adjust the distance between the tip and the specimen surface, while the tip is piezoelectrically scanned in a raster pattern over the region of specimen surface being imaged. By holding the force, rather than the electric current, between tip and specimen at a set-point value, atomic force microscopes similarly allow the exploration of nonconducting specimens. In either case, when the height of the tip is plotted as a function of its lateral position over the specimen, an image that looks very much like the surface topography results.

It is becoming increasingly possible to record other signals (such as lateral force, capacitance, scan-related tip displacement, temperature, light intensity, or magnetic resonance) as the tip scans. For example, modern atomic force microscopes can map lateral force and conductivity along with height, while image pairs from scanning tunneling microscopes scanning to and fro can provide information about friction as well as topography.

Scanning tunneling microscopes make it possible not just to view atoms but to push them and even to rearrange them in unlikely combinations (sometimes whether or not these rearrangements are desirable). A few considerations of scale are important in understanding this process. Atoms comprise a positive nucleus and a surrounding cloud of negative electrons. These charges rearrange when another atom approaches, with unlike charges shifting to give rise to the van der Waals force of attraction between neutral atoms. This force makes gravity (and most accelerations) ignorable when contact between solid objects in the micrometer size range and smaller is involved, since surface-to-volume ratios are inversely proportional to object size.

The electric field in the scanning tunneling microscope allows plucking as well, in which adsorbed or substrate atoms are removed and transferred to the electrode tip with a suitable voltage pulse. Because the electric field from the tip falls off less rapidly with separation than do van der Waals forces, the most weakly attached nearby atom rather than the nearest may end up being removed. One solution to this problem is a hybrid approach. By invoking the tip electric field for bond breaking only when the tip is sufficiently close to the target atom that the van der Waals forces contribute as well, atoms on silicon could be singly removed and redeposited at will.

A third kind of selective bond breaking was also demonstrated. It involved the selective breaking of silicon-hydrogen bonds using electron energies (that is, pulse voltages) below those necessary to break bonds directly. Since the desorption probability was observed to vary exponentially with the tip-specimen current, it is believed that vibrational heating from inelastic electron tunneling mediated the chemical transition in this work. This work involves bond alteration at the level of signal atoms, the ultimate frontier for lithographic miniaturization. Philip P. Fraundorf

Scaphopoda A class in the phylum Mollusca, comprising two orders, Gadilida and Dentaliida. The class is wholly marine; it is common and probably most diverse in the deep sea, and poorly represented in brackish, littoral, or estuarine habitats. Scaphopods are characterized by a tapering tubular shell with two apertures: the large ventral aperture that the foot extends out of and a small dorsal aperture. The tapering shells found in the Dentaliida give the class its common names tusk or tooth shells. Adult scaphopods range in length from about 3 to 150 mm. The shell consists of aragonite arranged in three or four layers.

Systematically the class is related to bivalves, and probably arose from a member of the extinct molluscan class, the Rostroconchia. The two orders can be separated on the basis of shell structure. Typically gadilids have the widest part of the shell more toward the middle of the animal, while in the dentaliids the widest part of the shell is the ventral aperture. Gadilid shells are typically highly polished, while dentaliid shells lack this luster. In addition, there are significant internal morphological differences between the two orders.

Scaphopods are bilaterally symmetrical and surrounded by the mantle, which forms a tube. The body is suspended in the tube from the dorsal-anterior part of the shell. The largest part of the mantle cavity is adjacent to the ventral aperture and contains the foot, head, and feeding tentacles, called captacula. Each captaculum consist of a long extendable stalk and a terminal bulb. The captacular bulb is covered with cilia, and the stalk may have a ciliated band or tufts of cilia along it.

Scaphopods eat foraminiferans, and other small shelled organisms such as bivalves or shelled eggs. Dentaliids can also ingest sediment. Prey are captured by the use of the captacula, which have a chemical adhesive system to attach to prey. The foot is used to burrow and construct a feeding cavity.

Scaphopods live in unconsolidated sediments, and most members of any population are typically completely buried at any one time. However, scaphopods back up to the surface and extend the dorsal shell apex to release eggs or sperm, and presumably to flush the mantle cavity with fresh water. *See* MOLLUSCA. Ronald Shimek

Scapolite An aluminosilicate mineral. It is commonly found as light-colored, translucent tetragonal prisms. Scapolite is normally white, but many other colors are known, including some used as semiprecious gems resembling amethyst and citrine. The mineral has a Mohs hardness of 5–6. The formula of scapolite is $(Na,Ca)_4(Al,Si)_6SI_6O_{24}(Cl,CO_3,SO_4)$. In nature significant amounts of K and SO_4 substitute for Na and CO_3.

Scapolite is a common mineral in metamorphic rocks, particularly in those which contain calcite. It is found in marbles, gneisses, skarns, and schists. Scapolite probably forms about 0.1% of the Earth's upper crust. It is commonly found as inclusions in igneous rocks derived from deep within the Earth's crust, and probably makes up several percent of the lower crust. *See* SILICATE MINERALS. David E. Ellis

Scarlet fever An acute contagious disease that results from infection with *Streptococcus pyogenes* (group A streptococci). It most often accompanies pharyngeal (throat) infections with this organism but is occasionally associated with wound infection or septicemia. Scarlet fever is characterized by the appearance, about 2 days after development of pharyngitis, of a red rash that blanches under pressure and has a sandpaper texture.

Usually the rash appears first on the trunk and neck and spreads to the extremities. The rash fades after a week, with desquamation, or peeling, generally occurring during convalescence. The disease is usually self-limiting, although severe forms are occasionally seen with high fever and systemic toxicity. Appropriate antibiotic therapy is recommended to prevent the onset in susceptible individuals of rheumatic fever and rheumatic heart disease. *See* MEDICAL BACTERIOLOGY; RHEUMATIC FEVER; STREPTOCOCCUS.

Ernest D. Gray

Scattering experiments (atoms and molecules) Experiments in which a beam of incident electrons, atoms, or molecules is deflected by collisions with an atom or molecule. Such experiments provide tests of the theory of scattering as well as information about atomic and molecular forces. Scattering experiments can be designed to simulate conditions in planetary atmospheres, electrical discharges, gas lasers, fusion reactors, stars, and planetary nebulae. *See* ELECTRICAL CONDUCTION IN GASES; GAS DISCHARGE; LASER; NUCLEAR FUSION; PLANETARY NEBULA; PLANET; STAR.

In general, in any type of collision, scattering occurs, which causes the direction of relative motion of the two systems to be rotated to a new direction after the collision. More than two systems may also result from such an impact. A complete description of a collision event requires measurement of the directions, speeds, and internal states of all the products. *See* COLLISION (PHYSICS).

There are two basic types of scattering experiments. The simpler involves passing a collimated beam of particles (electrons, atoms, molecules, or ions) through a dilute target gas (in a cell or a jet) and measuring the fraction of incident particles that are deflected into a certain angle relative to the incident beam direction. In the second method, a collimated beam of particles intersects a second beam. The scattering events are usually registered by measuring the deflection or internal-state change of the beam particles. *See* MOLECULAR BEAMS.

Scattering in a particular type of collision is specified in terms of a differential cross section. The probability that, in a particular type of collision, the direction of motion of the electron is turned through a specified scattering angle into a specified solid angle is proportional to the corresponding differential scattering cross section. Collision cross sections can be measured with appropriately designed experimental apparatus. Depending on the type of collision process, that apparatus may measure the scattering angle, energy, charge, or mass of the scattered systems.

For the simplest case, the scattering of a beam of structureless particles of specified mass and speed by a structureless scattering center, the differential cross section may be calculated exactly by using the quantum theory. In the special case where the Coulomb force fully describes the interaction, both the quantum and classical theory give the same exact value for the differential cross section at all values of the scattering angle. *See* QUANTUM MECHANICS.

For scattering of systems with internal structure (for example, molecules, and their ions), no exact theoretical calculation of the cross section is possible. Methods of approximation specific to different types of collisions have been developed. The power of modern high-speed computers has greatly increased their scope and effectiveness, with scattering experiments serving as benchmarks. *See* ATOMIC STRUCTURE AND SPECTRA; SUPERCOMPUTER.

Ronald A. Phaneuf

Scattering experiments (nuclei) Experiments in which beams of particles such as electrons, nucleons, alpha particles and other atomic nuclei, and mesons are deflected by elastic collisions with atomic nuclei. Much is learned from such experiments about the nature of the scattered particle, the scattering center, and the forces acting between them. Scattering experiments, made possible by the construction of high-energy particle accelerators and the development of specialized techniques for detecting the scattered particles, are one of the main sources of information regarding the structure of matter. *See* NUCLEAR STRUCTURE; PARTICLE ACCELERATOR; PARTICLE DETECTOR; SCATTERING MATRIX.

K. A. Erb

Scattering layer A layer of organisms in the sea which causes sound to scatter and returns echoes. Recordings by sonic devices of echoes from sound scatterers indicate that the scattering organisms are arranged in approximately horizontal layers in the water, usually well above the bottom. The layers are found in both shallow and deep water.

John B. Hersey

Scattering matrix An infinite-dimensional matrix or operator that expresses the state of a scattering system consisting of waves or particles or both in the far future in terms of its state in the remote past; also called the S matrix. In the case of electromagnetic (or acoustic) waves, it connects the intensity, phase, and polarization of the outgoing waves in the far field at various angles to the direction and polarization of the beam pointed toward an obstacle. It is used most prominently in the quantum-mechanical description of particle scattering, in which context it was invented in 1937 by J. A. Wheeler to describe nuclear reactions. Because an analog of the Schrödinger equation for the description of particle dynamics is lacking in the relativistic domain, W. Heisenberg proposed in 1943 that the S matrix rather than the hamiltonian or the lagrangian be regarded as the fundamental dynamical entity of quantum mechanics. This program played an important role in high-energy physics during the 1960s but is now largely abandoned. The physics of fundamental particles is now described primarily in terms of quantum gauge fields, and these are used to determine the S matrix and its elements for the collision and reaction processes observed in the laboratory. *See* ELEMENTARY PARTICLE; GAUGE THEORY; NUCLEAR REACTION; QUANTUM MECHANICS; RELATIVISTIC QUANTUM THEORY; SCATTERING EXPERIMENTS (ATOMS AND MOLECULES); SCATTERING EXPERIMENTS (NUCLEI).

The mathematical properties of the S matrix in nonrelativistic quantum mechanics have been thoroughly studied and are, for the most part, well understood. If the potential energy in the Schrödinger equation, or the scattering obstacle, is spherically symmetric, the eigenfunctions of the S matrix are spherical harmonics and its eigenvalues are of the form exp $(2i\delta_l)$, where the real number δ_l is the phase shift of angular momentum l. In the nonspherically symmetric case, analogous quantities are called the eigenphase shifts, and the eigenfunctions depend on both the energy and the dynamics. In the relativistic regime, without an underlying Schrödinger equation for the particles, the mathematical properties are not as well known. Causality arguments (no signal should propagate faster than light) lead to dispersion relations, which constitute experimentally verifiable consequences of very general assumptions on the properties of nature that are independent of the detailed dynamics. *See* ANGULAR MOMENTUM; CAUSALITY; DISPERSION RELATIONS; EIGENFUNCTION; SPHERICAL HARMONICS.

Roger G. Newton

Scattering of electromagnetic radiation The process in which energy is removed from a beam of electromagnetic radiation and reemitted with a change in direction, phase, or wavelength. All electromagnetic radiation is subject to scattering by the medium (gas, liquid, or solid) through which it passes.

It has been known since the work of J. Maxwell in the nineteenth century that accelerating electric charges radiate energy and, conversely, that electromagnetic radiation consists of fields which accelerate charged particles. Light in the visible, infrared, or ultraviolet region interacts primarily with the electrons in gases, liquids, and solids—not the nuclei. The scattering process in these wavelength regions consists of acceleration of the electrons by the incident beam, followed by reradiation from the accelerating charges. *See* ELECTROMAGNETIC RADIATION.

Scattering processes may be divided according to the time between the absorption of energy from the incident beam and the subsequent reradiation. True "scattering" refers only to those processes which are essentially instantaneous. Mechanisms in which there is a measurable delay between absorption and reemission are usually termed luminescence. *See* LUMINESCENCE.

Instantaneous scattering processes may be further categorized according to the wavelength shifts involved. Some scattering is "elastic"; there is no wavelength change, only a phase shift. In 1928 C. V. Raman discovered the process in which light was inelastically scattered and its energy was shifted by an amount equal to the vibrational energy of a molecule or crystal.

In liquids or gases two distinct processes generate inelastic scattering with small wavelength shifts. The first is Brillouin scattering from pressure waves. When a sound wave propagates through a medium, it produces alternate regions of high compression (high density) and low compression (or rarefaction). Brillouin scattering of light to higher (or lower) frequencies occurs because the medium is moving toward (or away from) the light source. This is an optical Doppler effect. *See* DOPPLER EFFECT.

The second kind of inelastic scattering studied in fluids is due to entropy and temperature fluctuations, and is known as Rayleigh scattering. These entropy fluctuations produce a broadening in the scattered radiation centered about the exciting wavelength, rather than sharp, well-defined wavelength shifts. Under the assumption that the scattering in fluids is from particles much smaller than the wavelength of the exciting light, Lord Rayleigh derived in 1871 an equation for such scattering. The dependence of scattering intensity upon the inverse fourth power of the wavelength given in Rayleigh's equation is responsible for the fact that daytime sky looks blue and sunsets red: blue light is scattered out of the sunlight by the air molecules more strongly than red; at sunset, more red light passes directly to the eyes without being scattered. *See* ENTROPY.

Rayleigh's derivation of his scattering equation relies on the assumption of small, independent particles. Under some circumstances of interest, both of these assumptions fail. Colloidal suspensions provide systems in which the scattering particles are comparable to or larger than the exciting wavelengths. Such scattering is called the Tyndall effect and results in a nearly wavelength-independent (that is, white) scattering spectrum. The Tyndall effect is the reason clouds are white (the water-droplets become larger than the wavelengths of visible light). *See* TYNDALL EFFECT.

The breakdown of Rayleigh's second assumption—that of independent particles—occurs in all liquids. There is strong correlation between the motion of neighboring particles. This leads to fixed phase relations and destructive interference for most of the scattered light. The remaining scattering arises from fluctuations in particle density discussed above. J. F. Scott

Scent gland A specialized skin gland of the tubuloalveolar or acinous variety found in many mammals. These glands produce substances having peculiar odors. In some instances they are large, in others small. Examples of large glands are the civet gland in the civet cat, the musk gland in the musk deer, and the castoreum gland in the beaver. The civet gland is an anal gland, whereas the musk and castoreum are preputial. Examples of small scent glands are the preputial or Tyson's glands in the human male which secrete the smegma, and the vulval glands in the female. The secretions in all of the above glands are sebaceous. *See* GLAND. Olin E. Nelsen

Scheduling A decision-making function that plays an important role in most manufacturing and service industries. Scheduling is applied in procurement and production, in transportation and distribution, and in information processing and communication. A scheduling function typically uses mathematical optimization techniques or heuristic methods to allocate limited resources to the processing of tasks.

Project scheduling is concerned with a set of activities that are subject to precedence constraints, specifying which jobs have to be completed before a given job is allowed to start its processing. All activities belong to a single (and typically large) project that has to be completed in a minimum time; for example, a large real estate development or the construction of an aircraft carrier.

Production or job shop scheduling is important in manufacturing settings, for example, semiconductor manufacturing. Customer orders have to be executed. Each order entails a number of operations that have to be processed on the resources or the machines available. Each order has a committed shipping date that plays the role of a due date. Production scheduling often also includes lot sizing and batching.

Timetabling occurs often in class room scheduling, scheduling of meetings, and reservation systems. In many organizations, especially in the service industries, meetings must be scheduled in such a way that all necessary participants are present; often other constraints have to be satisfied as well (in the form of space and equipment needed). Such problems occur in schools with classroom and examination scheduling as well as in the renting of hotel rooms and automobiles.

Work-force scheduling (crew scheduling, and so on) is increasingly important, especially in the service industries. For example, large call centers in many types of enterprises (airlines, financial institutions, and others) require the development of complicated personnel scheduling techniques.

In order to determine satisfactory or optimal schedules, it is helpful to formulate the scheduling problem as a mathematical model. Such a model typically describes a number of important characteristics. One characteristic specifies the number of machines or resources as well as their interrelationships with regard to the configuration, for example, machines set up in series, machines set up in parallel. A second characteristic of a mathematical model concerns the processing requirements and constraints. These include setup costs and setup times, and precedence constraints between various activities. A third characteristic has to do with the objective that has to be optimized, which may be a single objective or a composite of different objectives. For example, the objective may be a combination of maximizing throughput (which is often equivalent to minimizing setup times) and maximizing the number of orders that are shipped on time.

The scheduling function is often incorporated in a system that is embedded in the information infrastructure of the organization. This infrastructure may be an enterprise-wide information system that is connected to the main databases of the company. Many other decision support systems may be plugged into such an enterprise-wide information system—for example, forecasting, order promising and due date setting, and material requirements planning (MRP).

The database that the scheduling system relies on usually has some special characteristics. It has static data as well as dynamic data. The static data—for example, processing requirements, product characteristics, and routing specifications—are fixed and do not depend on the schedules developed. The dynamic data are schedule-dependent; they include the start times and completion times of all the operations on all the different machines, and the length of the setup times (since these may also be schedule-dependent).

The economic impact of scheduling is significant. In certain industries the viability of a company may depend on the effectiveness of its scheduling systems, for example, airlines and semiconductor manufacturing. Good scheduling often allows an organization to conduct its operations with a minimum of resources. *See* MATERIAL RESOURCE PLANNING; PRODUCTION PLANNING. Michael Pinedo; Sridhar Seshadri

Scheelite A mineral consisting of calcium tungstate, $CaWO_4$. Scheelite occurs in colorless to white, tetragonal crystals; it may also be massive and granular. Its fracture is uneven, and its luster is vitreous to adamantine. Scheelite has a hardness of 4.5–5 on Mohs scale and a specific gravity of 6.1. Its streak is white. The mineral is transparent and fluoresces bright bluish-white under ultraviolet light.

Scheelite is an important tungsten mineral and occurs in small amounts in vein deposits. The most important scheelite deposit in the United States is near Mill City, Nevada. *See* TUNGSTEN.

Edward C. T. Chao

Schematic drawing Concise, graphical symbolism whereby the engineer communicates to others the functional relationship of the parts in a component and, in turn, of the components in a system. The symbols do not attempt to describe in complete detail the characteristics or physical form of the elements, but they do suggest the functional form which the ensemble of elements will take in satisfying the functional requirements of the component. They are different from a block diagram in that schematics describe more specifically the physical process by which the functional specifications of a block diagram are satisfied.

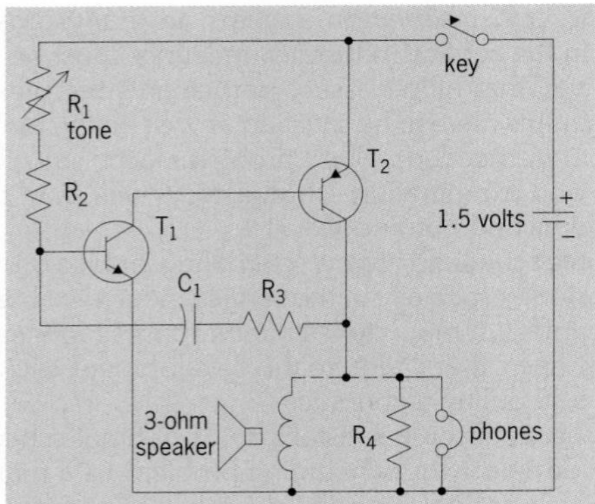

Fig. 1. Simple transistorized code practice oscillator, using standard symbols. (*Adapted from J. Markus, Sourcebook of Electronic Circuits, McGraw-Hill, 1968*)

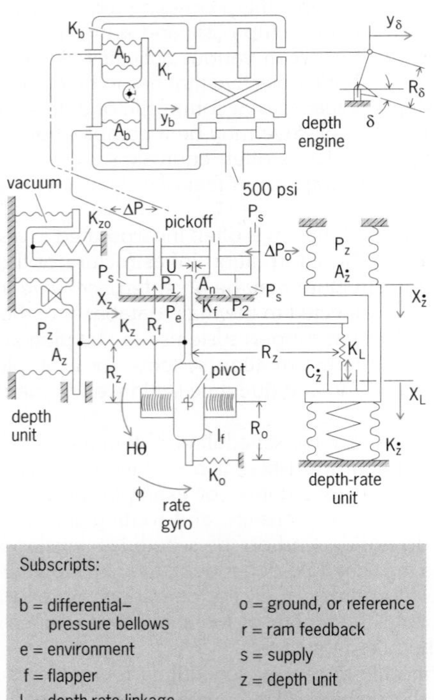

Fig. 2. Mechanical schematic of the depth-control mechanism of a torpedo.

An electrical schematic is a functional schematic which defines the interrelationship of the electrical elements in a circuit, equipment, or system. The symbols describing the electrical elements are stylized, simplified, and standardized to the point of universal acceptance (Fig. 1).

In a mechanical schematic, the graphical descriptions of elements of a mechanical system are more complex and more intimately interrelated than the symbolism of an electrical system and so the graphical characterizations are not nearly as well standardized or simplified (Fig. 2). However, a mechanical schematic illustrates such features as components, acceleration, velocity, position force sensing, and viscous damping devices. *See* DRAFTING; ENGINEERING DRAWING.

Robert W. Mann

Schist Medium- to coarse-grained, mica-bearing metamorphic rock with well-developed foliation (layered structure) termed schistosity. Schist is derived primarily from fine-grained, mica-bearing rocks such as shales and slates. The schistosity is formed by rotation, recrystallization, and new growth of mica; it is deformational in origin. The planar to wavy foliation is defined by the strong preferred orientation of platy minerals, primarily muscovite, biotite, and chlorite. The relatively large grain size of these minerals (up to centimeters) produces the characteristic strong reflection when light shines on the rock. *See* BIOTITE; CHLORITE; MUSCOVITE.

Schists are named by the assemblage of minerals that is most characteristic in the field; for example, a garnet-biotite schist contains porphyroblasts of garnet and a schistosity dominated by biotite. Schists can provide important information on the relationship between metamorphism and deformation. *See* METAMORPHIC ROCKS; METAMORPHISM; PETROFABRIC ANALYSIS.

Ben A. van der Pluijm

Schistosomiasis A disease in which humans are parasitized by any of three species of blood flukes: *Schistosoma mansoni*, *S. haematobium*, and *S. japonicum*. Adult *S. mansoni* prefer the veins of the hemorrhoidal plexus, *S. haematobium* those of the vesical plexus, and *S. japonicum* those of the small intestine. The disease is also known as bilharziasis. *See* DIGENEA.

An embryonated egg passed in feces or urine hatches in fresh water, liberating a miracidium larva which penetrates into specific gastropod snails. The larval cycle in the snail lasts for about 1 month. The cercaria emerges from the mollusk, swims in the water, and penetrates the skin of the final host upon coming in contact with it.

Schistosomiasis is an agricultural hazard for all ages in irrigated lands or swamps. Elsewhere fluvial waters are the main source of infection, in which case incidence is marked in human beings who are less than 15 years old and is higher among boys than among girls.

José F. Maldonado-Moll

Schistostegales An order of the true mosses (subclass Bryidae), consisting of a single species, *Schistostega pennata*, the cave moss, which is especially characterized by its leaf arrangement and the form of its protonema. The plants grow in dimly lit, cavelike places, such as the undersides of upturned tree roots. The cave moss is widely distributed in north temperate regions. An unrelated moss found in Australia and New Zealand, *Mittenia plumula*, has an identical protonema and habitat.

The leafy plants are erect, unbranched, and dimorphous: sterile shoots have leaves wide-spreading in two rows and confluent

Computer-graphic images of the surface of a high-temperature superconductor using data from a scanning tunneling electron microscope. The superconductor is a film of sputtered $YBa_2Cu_3O_7$ on magnesium oxide (MgO), showing spiral grains with step heights of about one unit cell (1.2 nanometers). Colors represent increasing altitude from blue to red-violet. *(From L. Prueitt, C-6, Los Alamos National Laboratory; data supplied by M. Hawley, I. Raistrick, and R. Houlton, Los Alamos National Laboratory)*

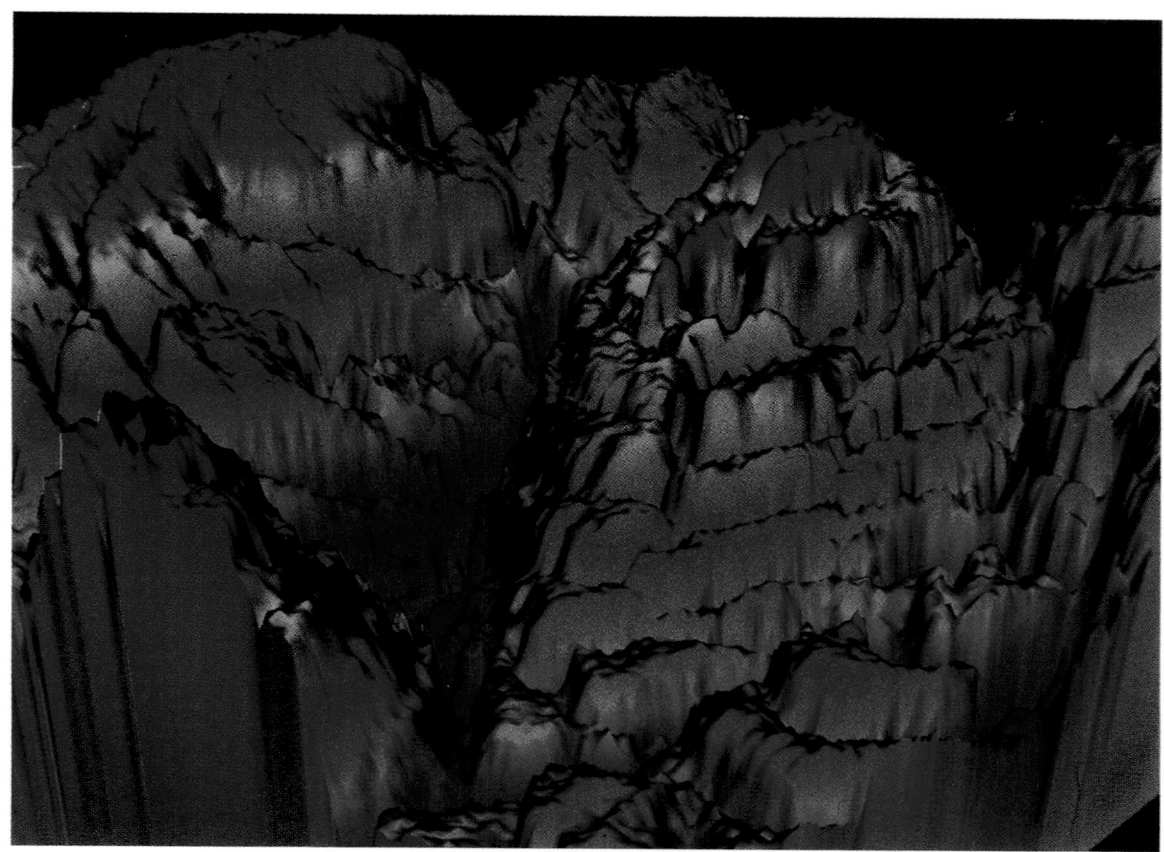

because of broad basal decurrencies. Fertile plants have leaves in five rows and erect-spreading in a terminal tuft. All leaves are ecostate, entire, and unbordered. The inflorescences are terminal, the setae elongate, and the capsules subglobose, with a flat operculum but no peristome. *See* BRYIDAE; BRYOPHYTA; BRYOPSIDA.

Howard Crum

Schizomida An order of Arachnida with two families, Protoschizomidae (*Agastoschizomus, Protoschizomus*) and Schizomida (*Schizomus, Trithyreus, Megaschizomus*), including about 130 described species, mainly in the genus *Schizomus*. Schizomids occur in most tropical and warm temperate regions, but each species has a restricted distribution.

Many undescribed species probably occur in tropical leaf litter and caves. Schizomids are most closely related to Uropygi, with which they share several evolutionary novelties. *See* UROPYGI.

Schizomids are 0.12–0.44 in. (3–11 mm) in size, are white to tan, and lack eyes, although some retain vestigial "eye-spots." The carapace is distinctively tripartite and the abdomen has a short, terminal flagellum.

Because schizomids are vulnerable to desiccation, they inhabit only moist places such as leaf litter, soil, caves, or beneath stones and logs. They are strict carnivores and fast runners. *See* ARACHNIDA.

Jonathan A. Coddington

Schizophrenia A brain disorder that is characterized by unusual mental experiences, such as hallucinations, changes in emotional experience, and severe decrements in social, cognitive, and occupational functioning. The popular belief that schizophrenia is characterized by multiple personalities is a misconception. Patients with schizophrenia demonstrate a series of biological differences when compared as a group with healthy individuals without schizophrenia. At present, however, there is no single biological marker available to indicate the presence of schizophrenia. A diagnosis is made on the basis of a cluster of symptoms reported by the patient and signs identified by the clinician.

People with schizophrenia may report perceptual experiences in the absence of a real environmental stimulus. The most common of these is auditory hallucinations, most often reported in the form of words spoken to the person with schizophrenia. The voices are often derogatory in nature, and can be tremendously frightening.

People with schizophrenia often maintain beliefs that are not held by the overwhelming majority of the general population. To be considered delusions, the beliefs must be unshakable. In many cases, these beliefs may be "bizarre." A bizarre delusion is defined as an idea that is completely impossible (for example, "a computer has been inserted into my skull"), in contrast to nonbizarre delusions that are plausible ("The FBI is watching me"), but still not true. In some instances, the delusions have an element of suspiciousness to them, such as the incorrect belief that others are planning to cause the person with schizophrenia harm. The delusions may or may not be related to hallucinatory experiences.

People with schizophrenia may manifest an odd outward appearance due to disorganization. This presentation may include speech that does not follow logically or sensibly, at times to the point of being incoherent. Facial expression may be odd or inappropriate, such as laughing for no reason. In some cases, people with schizophrenia may move in a strange and awkward manner. The extreme of this behavior, referred to as catatonia, has become very rare since pharmacological treatments became available.

Although less striking than delusions and hallucinations, perhaps the most devastating feature of schizophrenia is the cognitive impairment found in most people with the disorder. On average, people with schizophrenia perform in the lowest 2–10% of the general population on tests of attention, memory, reasoning and problem solving, motor skills, and language abilities. These cognitive deficits are perhaps the most important explanation for the difficulties that people with schizophrenia have functioning in everyday society. *See* COGNITION.

The onset of schizophrenia is generally in the late teens to early twenties; however, onset is possible throughout the life span. While the onset of symptoms is abrupt in some people, others experience a more insidious process, including extreme social withdrawal, reduced motivation, mood changes, and cognitive and functional decline prior to the onset of full-blown schizophrenia symptoms.

It is likely that there are various forms of schizophrenia, perhaps all with different causes. Although schizophrenia appears to be inherited, at least in some cases, the influence of genes is far from complete. Although many arguments have been put forth regarding environmental factors that could cause schizophrenia, very few of these theories are consistently supported.

One of the new developments in the study of schizophrenia is associated with imaging technologies. Magnetic resonance imaging (MRI) allows for the visualization of the brain at high levels of resolution, allowing the identification of structures as small as 1 mm. People with schizophrenia often have changes in the structure of their brain, such as enlargement of the cerebral ventricles (fluid-filled spaces in the brain close to the midline). Various brain regions have been found to be smaller in patients with schizophrenia, including the frontal cortex, temporal lobes, thalamus, and the hippocampi. Studies of patients with schizophrenia also have found patterns of abnormal activation of the brain while performing tests of memory and problem solving. *See* BRAIN; MEDICAL IMAGING.

Both pharmacological and behavioral approaches are used to treat schizophrenia. Various medications, referred to as neuroleptics or antipsychotics, have been developed and act primarily to block dopamine receptors in the brain. These medications have an impact on delusions and hallucinations in some patients, but treatment is often accompanied by a variety of side effects. There are several targets for behavioral treatments in schizophrenia. Patients with schizophrenia have difficulty acquiring skills in social, occupational, and independent living domains. Structured training programs have attempted to teach patients with schizophrenia how to function more effectively in these areas. Family interventions have also been designed to provide a supportive environment for patients with schizophrenia. Teaching patients how to cope with their hallucinations and delusions is another target of current behavioral treatment. Cognitive-behavioral treatments have been employed to help patients realize the nature of their symptoms and to develop plans for coping with them. *See* PSYCHOPHARMACOLOGY; PSYCHOTHERAPY.

Richard S. E. Keefe; Philip D. Harvey

Schlieren photography Any technique for the photographic recording of schlieren, which are regions or stria in a medium that is surrounded by a medium of different refractive index. Refractive index gradients in transparent media cause light rays to bend (refract) in the direction of increasing refractive index. This is a result of the reduced light velocity in a higher-refractive-index material. This phenomenon is exploited in viewing the schlieren, with schlieren photographs as the result. Electronic video recorders, scanning diode array cameras, and holography are widely used as supplements. *See* HOLOGRAPHY; OPTICAL DETECTORS; PHOTOGRAPHY; REFRACTION OF WAVES; SHADOWGRAPH.

There are many techniques for optically enhancing the appearance of the schlieren in an image of the field of interest. In the oldest of these, called the knife-edge method (see illustration), a point or slit source of light is collimated by a mirror and passed through a field of interest, after which a second mirror focuses the light, reimaging the point or slit where it is intercepted

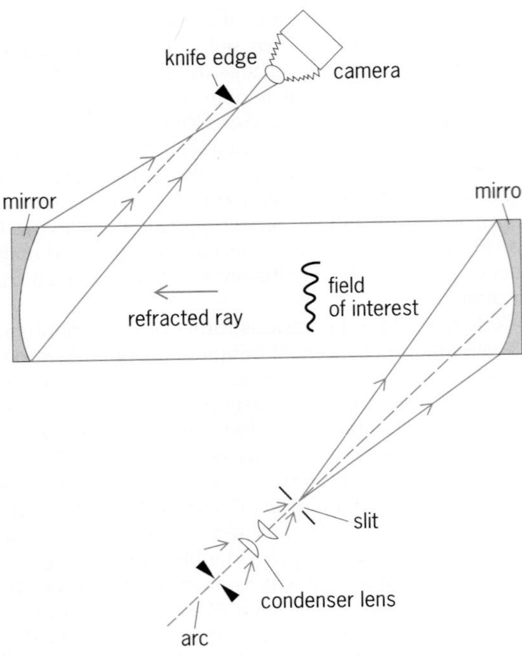

Knife-edge method of viewing schlieren, employing the "z" configuration.

by an adjustable knife edge (commonly a razor blade). The illustration shows the "z" configuration which minimizes the coma aberration in the focus. Mirrors are most often used because of the absence of chromatic aberration. *See* ABERRATION (OPTICS).

Rays of light that are bent by the schlieren in the direction of the knife edge are intercepted and removed from the final image of the region of interest, causing those regions to appear dark. Consequently, the system is most sensitive to the density gradients that are perpendicular to the knife edge. The knife edge is commonly mounted on a rotatable mount so that it can be adjusted during a measurement to optimally observe different gradients in the same field of interest. The intensity in the processed image is proportional to the refractive index gradient. A gradient in the same direction as the knife edge appears dark. Gradients in the opposite direction appear bright. This method, employed with arc light sources, is still one of the simplest ways to view refractive index changes in transparent solids, liquids, and gases. A well-designed schlieren system can easily detect the presence of a refractive index gradient that causes 1 arc-second deviation of a light ray.

Except for locating and identifying schlieren-causing events such as turbulent eddies, shock waves, and density gradients, schlieren systems are usually considered to be qualitative instruments. Quantitative techniques for determining density are possible but are much more difficult to employ. The most common of these is color schlieren. The knife edge is replaced with a multicolored filter. Rays of light refracted through different angles appear in different colors in the final image.

The availability of lasers and new optical components has expanded the method considerably. When a coherent light source such as a laser is used, the knife edge can be replaced by a variety of phase-, amplitude-, or polarization-modulating filters to produce useful transformations in the image intensity. *See* INTERFEROMETRY; LASER; POLARIZED LIGHT. James D. Trolinger

Schmidt camera A fast, compact, wide-field optical system that uses a thin aspheric front lens at the center of curvature of a larger concave spherical mirror. The image is focused onto a curved focal plane between the two elements (see illustration). Schmidt cameras are sometimes used in microscopes, astronomical spectrographs, and projection televisions. The largest such devices can image several degrees of the sky and are known as Schmidt telescopes. The Estonian optician Bernhard Schmidt devised the scheme in 1930. The field of best focus is located midway between the lens and the mirror and is curved toward the mirror, with a radius of curvature equal to the focal length. Historically film or photographic plates were bent to match this curved focus, and used to make the first all-sky survey pictures. Field-flattening lenses permit the use of charge-coupled devices (CCDs) in newer applications. Large mosaics of CCDs have replaced photographic emulsions in some applications, and large numbers of optical fibers have also been used in multiobject spectrograph configurations. *See* CHARGE-COUPLED DEVICES; GEOMETRICAL OPTICS; LENS (OPTICS); MIRROR OPTICS.

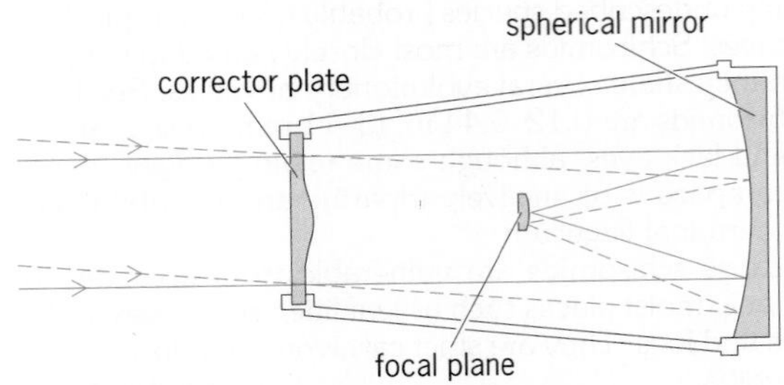

Cross section of Schmidt camera with aspherical corrector plate. (*After J. M. Pasachoff, Astronomy: From the Earth to the Universe, 6th ed., Brooks/Cole Publishing, 2002*)

Schmidt telescopes are very fast, some having focal ratios in the vicinity of $f/1$. Their distinguishing feature is the very wide angle over which good images are obtained. Thus, under a dark sky they are sensitive to extended objects of low surface brightness. Schmidt telescopes have no coma; because the only lens element is so thin, they suffer only slightly from chromatic aberration and astigmatism. The front element, smaller than the main mirror, is sometimes known as a corrector plate. Often the outer surface of the corrector plate is plane, with the inner surface bearing the figure; however, a slight curvature is sometimes introduced to prevent misleading ghost images. This corrector plate almost eliminates the spherical aberration, giving extremely sharp images. For the most critical work, over a wide range of wavelengths, it must be made achromatic. *See* ABERRATION (OPTICS); OPTICAL SURFACES.

The largest Schmidt telescopes are 39–53 in. (1.0–1.3 m) in diameter. Larger corrector plates would sag too much, as they can only be supported around their circumference.

The success of the Schmidt design has led to many other types of catadioptric systems, with a combination of lenses and mirrors. The desire for a wide field has led to most modern large telescopes being built to the Ritchey-Chrétien design, which uses two hyperboloid mirrors instead of the traditional paraboloids to achieve coma-free fields of up to 1°, with a small amount of focal plane curvature. These fields are much smaller than those of Schmidts, but the telescope aperture can be larger, because no large corrector lens is needed. *See* TELESCOPE.

Jay M. Pasachoff; Andrew Pickles

Schottky anomaly A contribution to the heat capacity of a solid arising from the thermal population of discrete energy levels as the temperature is raised. The effect is particularly prominent at low temperatures, where other contributions to the heat capacity are generally small. *See* SPECIFIC HEAT.

Discrete energy levels may arise from a variety of causes, including the removal of orbital or spin degeneracy by magnetic fields, crystalline electric fields, and spin orbit coupling, or from the magnetic hyperfine interaction. Such effects commonly occur in paramagnetic ions. *See* LOW-TEMPERATURE THERMOMETRY.

Corresponding to the Schottky heat capacity, there is a contribution to the entropy. This can act as a barrier to the attainment of low temperatures if the substance is to be cooled either by adiabatic demagnetization or by contact with another cooled substance. Conversely, a substance with a Schottky anomaly can be used as a heat sink in experiments at low temperatures (generally below 1 K or −457.9°F) to reduce temperature changes resulting from the influx or generation of heat. *See* ADIABATIC DEMAGNETIZATION; LOW-TEMPERATURE PHYSICS. W. P. Wolf

Schottky barrier diode A metal-semiconductor diode (two terminal electrical device) that exhibits a very nonlinear relation between voltage across it and current through it; formally known as a metallic disk rectifier. Original metallic disk rectifiers used selenium of copper oxide as the semiconductor coated on a metal disk. Today, the semiconductor is usually single-crystal silicon with two separate thin metal layers deposited on it to form electrical contacts. One of the two layers is made of a metal which forms a Schottky barrier to the silicon. The other forms a very low resistance, so-called ohmic, contact. The Schottky barrier is an electron or hole barrier caused by an electric dipole charge distribution associated with the contact potential difference which forms between a metal and a semiconductor under equilibrium conditions. The barrier is very abrupt at the surface of the metal because the charge is primarily on the surface. However, in the semiconductor, the charge is distributed over a small distance, and the potential gradually varies across this distance. *See* CONTACT POTENTIAL DIFFERENCE.

A basic useful feature of the Schottky diode is the fact that it can rectify an alternating current. Substantial current can pass through the diode in one direction but not in the other. If the semiconductor is *n*-type, electrons can easily pass from the semiconductor to the metal for one polarity of applied voltage, but are blocked from moving into the semiconductor from the metal by a potential barrier when the applied voltage is reversed. If the semiconductor is *p*-type, holes experience the same type of potential barrier but, since holes are positively charged, the polarities are reversed from the case of the *n*-type semiconductor. In both cases the applied voltage of one polarity (called forward bias) can reduce the potential barrier for charge carriers leaving the semiconductor, but for the other polarity (called reverse bias) it has no such effect. *See* DIODE; SEMICONDUCTOR; SEMICONDUCTOR RECTIFIER. James E. Nordman

Schottky effect The enhancement of the thermionic emission of a conductor resulting from an electric field at the conductor surface. Since the thermionic emission current is given by the Richardson formula, an increase in the current at a given temperature implies a reduction in the work function of the emitter.

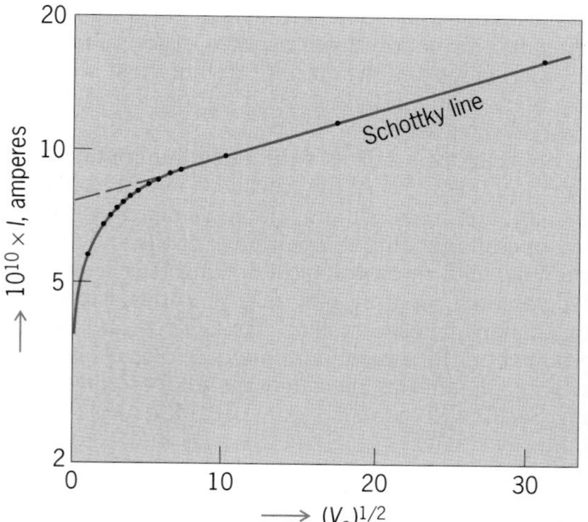

Logarithm of thermionic emission current *I* of tungsten as function of square root of anode voltage V_a. (*After W. B. Nottingham, Phys. Rev., 58:927–928, 1940*)

The reduction in work function can be calculated by considering the effect of a constant externally applied field on the potential energy of an electron near the conductor surface, and is found to be proportional to the square root of the field. *See* THERMIONIC EMISSION; WORK FUNCTION (ELECTRONICS).

A plot of the logarithm of the current versus the square root of the anode voltage should yield a straight line. An example is given in the illustration for tungsten; the deviation from the straight line for low anode voltages is due to space-charge effects. *See* SPACE CHARGE. Adrianus J. Dekker

Schrödinger's wave equation A linear, homogeneous partial differential equation that determines the evolution with time of a quantum-mechanical wave function.

Quantum mechanics was developed in the 1920s along two different lines, by W. Heisenberg and by E. Schrödinger. Schrödinger's approach can be traced to the notion of wave-particle duality that flowed from A. Einstein's association of particlelike energy bundles (photons, as they were later called) with electromagnetic radiation, which, classically, is a wavelike phenomenon. For radiation of definite frequency f, each bundle carries energy hf. The proportionality factor, $h = 6.626 \times 10^{-34}$ joule-second, is a fundamental constant of nature, introduced by M. Planck in his empirical fit to the spectrum of blackbody radiation. This notion of wave-particle duality was extended in 1923 by L. de Broglie, who postulated the existence of wavelike phenomena associated with material particles such as electrons. *See* PHOTON; WAVE MECHANICS.

There are certain purely mathematical similarities between classical particle dynamics and the so-called geometric optics approximation to propagation of electromagnetic signals in material media. For the case of a single (nonrelativistic) particle moving in a potential $V(\mathbf{r})$, this analogy leads to the association with the system of a wave function, $\Psi(\mathbf{r})$, which obeys Eq. (1).

$$-\frac{\hbar^2}{2m}\nabla^2\Psi + V\Psi = E\Psi \qquad (1)$$

Here m is the mass of the particle, E its energy, $\hbar = h/(2\pi)$, and ∇^2 is the laplacian operator. *See* CALCULUS OF VECTORS; GEOMETRICAL OPTICS; LAPLACIAN.

It is possible to ask what more general equation a time-, as well as space-dependent wave function, $\Psi(\mathbf{r}, t)$, might obey. What suggests itself is Eq. (2), which is now called the Schrödinger equation.

$$i\hbar\frac{\partial\Psi}{\partial t} = -\frac{\hbar^2}{2m}\nabla^2\Psi + V\Psi \qquad (2)$$

The wave function can be generalized to a system of more than one particle, say N of them. A separate wave function is not assigned to each particle. Instead, there is a single wave function, $\Psi(\mathbf{r}_1, \mathbf{r}_2, \ldots, \mathbf{r}_N, t)$, which depends at once on all the position coordinates as well as time. This space of position variables is the so-called configuration space. The generalized Schrödinger equation is Eq. (3),

$$ih\frac{\partial\Psi}{\partial t} = -\sum_{i=1}^{N}\frac{\hbar^2}{2m_i}\nabla_i^2\Psi + V\Psi \qquad (3)$$

where the potential V may now depend on all the position variables. Three striking features of this equation are to be noted:

1. The complex number i (the square root of minus one) appears in the equation. Thus Ψ is in general complex. *See* COMPLEX NUMBERS AND COMPLEX VARIABLES.

2. The time derivative is of first order. Thus, if the wave function is known as a function of the position variables at any one instant, it is fully determined for all later times.

3. The Schrödinger equation is linear and homogeneous in Ψ, which means that if Ψ is a solution so is $c\Psi$, where c is an arbitrary complex constant. More generally, if Ψ_1 and Ψ_2 are solutions, so too is the linear combination $c_1\Psi_1 + c_2\Psi_2$, where c_1

and c_2 are arbitrary complex constants. This is the superposition principle of quantum mechanics. *See* SUPERPOSITION PRINCIPLE.

The Schrödinger equation suggests an interpretation in terms of probabilities. Provided that the wave function is square integrable over configuration space, it follows from Eq. (3) that the norm, $\langle\Psi|\Psi\rangle$, is independent of time, where the norm is defined by Eq. (4). It is possible to normalize Ψ (multiply it by a suitable

$$\langle\Psi \mid \Psi\rangle = \int d^3x_1 d^3x_2 \ldots d^3x_N \Psi^* \Psi \quad (4)$$

constant) to arrange that this norm is equal to unity. With that done, the Schrödinger equation itself suggests that expression (5) is the joint probability distribution at time t for finding parti-

$$\Psi^* \Psi d^3x_1 d^3x_2 \ldots d^3x_N \quad (5)$$

cle 1 in the volume element d^3x_1, particle 2 in d^3x_2, and so forth.

Sam Treiman

Schuler pendulum Any apparatus which swings, because of gravity, with a natural period of 84.4 min, that is, with the same period as a hypothetical simple pendulum whose length is the Earth's radius. In 1923 Max Schuler showed that such an apparatus has the unique property that the pendulum arm will remain vertical despite any motions of its pivot. It is therefore useful as a base for navigational instruments. Schuler also showed how gyroscopes can be used to increase the period of a physical pendulum to the desired 84.4 min.

Gyrocompasses employ the Schuler principle to avoid errors due to ship accelerations. The principle has become the foundation of the science of inertial navigation. *See* GYROCOMPASS; INERTIAL GUIDANCE SYSTEM.

Robert H. Cannon

Sciatica A specific pattern of leg pain that follows the course of the sciatic nerves. It is usually caused by injury to one or more of the sciatic root nerves within the lower spinal canal. Sciatic pain is radicular pain, that is, pain that follows the distribution of a nerve. The typical pain locations are in the back of the thigh, behind the knee, over the calf, and into one or both sides of the foot as far as the toes. The precise location of the pain varies, but it is always constant and dominant in the leg as opposed to the back. Sciatica may be associated with diminished sensation or decreased power in the muscles supplied by the affected nerve roots. More often the pain occurs with signs of root irritation only but with otherwise normal function.

Sciatica results from the combination of mechanical pressure and chemical insult acting on a nerve root. Because the physical deformation of a nerve root caused by a herniated disc is always associated with an inflammatory reaction, an acute disc rupture is by far the most frequent cause of sciatica. Other but rare possibilities include pyogenic infections with abscess formation, synovial cysts (cysts occurring in proximity to facet joints in the spine), diabetic neuropathy, vasculitis (inflammation of blood or lymph vessels), injected irritating agents, and a range of metabolic and autoimmune disorders producing neuritis (nerve inflammation). *See* PAIN; SPINE; VERTEBRA.

Hamilton Hall

Science In common usage the word science is applied to a variety of disciplines or intellectual activities which have certain features in common. Usually a science is characterized by the possibility of making precise statements which are susceptible of some sort of check or proof. This often implies that the situations with which the special science is concerned can be made to recur in order to submit themselves to check, although this is by no means always the case. There are observational sciences such as astronomy or geology in which repetition of a situation at will is intrinsically impossible, and the possible precision is limited to precision of description.

A common method of classifying sciences is to refer to them as either exact sciences or descriptive sciences. Examples of the former are physics and, to a lesser degree, chemistry; and of the latter, taxonomical botany or zoology. The exact sciences are in general characterized by the possibility of exact measurement. One of the most important tasks of a descriptive science is to develop a method of description or classification that will permit precision of reference to the subject matter. *See* PHYSICAL SCIENCE.

Percy W. Bridgman; Gerald Holton

Scientific and applications satellites Satellites used to gain basic knowledge. They may be satellites of Earth or of other planets. Satellites used to apply space knowledge and techniques to practical purposes are called applications satellites. Many spacecraft are used for multiple purposes, combining space exploration, science, and applications in various ways. Scientific satellites are often called research satellites, and applications satellites are commonly designated by the application field, for example, navigation satellites. *See* SATELLITE (SPACECRAFT).

The space environment was one of the first areas investigated by scientific satellites. In this environment are cosmic rays, dust, magnetic fields, and various radiations from the Sun and galaxies. The Van Allen radiation belts were discovered by the first United States satellites. Soon thereafter satellites and space probes confirmed the existence of the solar wind. The continued investigation of these phenomena by satellites and space probes has led to the knowledge that a planet with a strong magnetic field, like Earth, is surrounded by a complex region, called a magnetosphere.

Numerous investigations are made of the effect of the space environment on materials and processes, including effects of radiation damage, meteoric erosion processes, and extremely high vacuum for very long periods. The influence of prolonged weightlessness on welding, alloying metals, growing crystals, and biological processes constitutes a promising field for future practical applications.

Scientific satellites, sounding rockets, and space probes are having a profound effect on earth science. The Earth and its atmosphere can now be studied in comparison with the other planets and their atmospheres, providing greater insight into the formation and evolution of the solar system. Satellites also make possible precision measurements of the size and shape of Earth and its gravitational field, and even measurements of the slow drifting of continents relative to each other.

Satellites are also having a great influence on astronomy by making it possible to observe celestial objects in all the wavelengths that reach the vicinity of Earth, whereas the ionosphere and atmosphere prevent most of these radiations from reaching telescopes on the ground. Moreover, even in the visible wavelengths, the small-scale turbulence and continuously varying refraction in the lower atmosphere distorts the image of a stellar object viewed at the ground, so that there is a limit to the improvement that can be achieved by increasing the size of a ground-based telescope.

The applications satellites emphasize the continuing day-to-day practical utilization of the satellite. They are operational in nature, although some of them, either directly or as a by-product of their operational output, contribute significantly to research, for example, with meteorological or Earth survey satellites. An applied research program generally precedes establishment of an applications satellite system. Thus, Tiros and Nimbus laid the groundwork for operational meteorological satellites. Likewise, Syncom satellites preceded the Intelsat communications system. *See* COMMUNICATIONS SATELLITE; METEOROLOGICAL SATELLITES; MILITARY SATELLITES; SATELLITE NAVIGATION SYSTEMS; SPACE FLIGHT.

Homer E. Newell; John F. Clark

Scientific methods Strategies or uniform rules of procedure used in some scientific research with a measure of success. Scientific methods differ in generality, precision, and the extent to which they are scientifically justified. Thus, whereas the experimental method can in principle be used in all the sciences dealing with ascertainable facts, the various methods for measuring the electron charge are specific. The search for increasing quantitative precision involves the improvement or invention of special methods of measurement, also called techniques. All scientific methods are required to be compatible with confirmed scientific theories capable of explaining how the methods work. The most general of all the methods employed in science is called the scientific method.

The scientific method may be summarized as the following sequence of steps: identification of a knowledge problem; precise formulation or reformulation of the problem; examination of the background knowledge in a search for items that might help solve the problem; choice or invention of a tentative hypothesis that looks promising; conceptual test of the hypothesis, that is, checking whether it is compatible with the bulk of the existing knowledge on the matter; drawing some testable consequences of the hypothesis; design of an empirical (observational or experimental) test of the hypothesis or a consequence of it; actual empirical test of the hypothesis, involving a search for both favorable and unfavorable evidence (examples and counterexamples); critical examination and statistical processing of the data (for example, calculation of average error and elimination of outlying data); evaluation of the hypothesis in the light of its compatibility with both the background knowledge and the fresh empirical evidence; if the test results are inconclusive, design and performance of new tests, possibly using different special methods; if the test results are conclusive, acceptance, modification, or rejection of the hypothesis; if the hypothesis is acceptable, checking whether its acceptance forces some change (enrichment or correction) in the background knowledge; identifying and tackling new problems raised by the confirmed hypothesis; and repetition of the test and reexamination of its possible impact on existing knowledge.

The scientific method is not a recipe for making original discoveries or inventions; it does not prescribe the pathway that scientists must follow to attain success. The goal of the scientific method is to ascertain whether a hypothesis is true to some degree. Indeed, the nucleus of the scientific method is the confrontation of an idea (hypothesis) with the facts it refers to, regardless of the source of the idea in question. In sum, the scientific method is a means for checking hypotheses for truth rather than for finding facts or inventing ideas. *See* SCIENCE. Mario Bunge

Scintillation counter A particle or radiation detector which operates through emission of light flashes that are detected by a photosensitive device, usually a photomultiplier or a silicon PIN diode. The scintillation counter not only can detect the presence of a particle, gamma ray, or x-ray, but can measure the energy, or the energy loss, of the particle or radiation in the scintillating medium. The sensitive medium may be solid, liquid, or gaseous, but is usually one of the first two. The scintillation counter is one of the most versatile particle detectors, and is widely used in industry, scientific research, medical diagnosis, and radiation monitoring, as well as in exploration for petroleum and radioactive minerals that emit gamma rays. Many low-level radioactivity measurements are made with scintillation counters. *See* LOW-LEVEL COUNTING; PARTICLE DETECTOR; PHOTOMULTIPLIER.

Scintillation counters are made of transparent crystalline materials, liquids, plastics, or glasses. In order to be an efficient detector, the bulk scintillating medium must be transparent to its own luminescent radiation, and since some detectors are quite extensive, covering meters in length, the transparency must be of a high order. One face of the scintillator is placed in optical contact with the photosensitive surface of the photomultiplier or PIN diode (see illustration). In order to direct as much as possible of the light flash to the photosensitive surface, reflecting material is placed between the scintillator and the inside surface of the container.

In many cases it is necessary to collect the light from a large area and transmit it to the small surface of a photomultiplier. In this case, a "light pipe" leads the light signal from the scintillator surface to the photomultiplier with only small loss. The best light guides and light fibers are made of glass, plastic, or quartz. It is also possible to use lenses and mirrors in con- junction with scintillators and photomultipliers. *See* OPTICAL FIBERS.

A charged particle, moving through the scintillator, loses energy and leaves a trail of ions and excited atoms and molecules. Rapid interatomic or intermolecular transfer of electronic excitation energy follows, leading eventually to a burst of luminescence characteristic of the scintillator material. When a particle stops in the scintillator, the integral of the resulting light output, called the scintillation response, provides a measure of the particle energy, and can be calibrated by reference to particle sources of known energy. Photomultipliers or PIN diodes may be operated so as to generate an output pulse of amplitude proportional to the scintillation response.

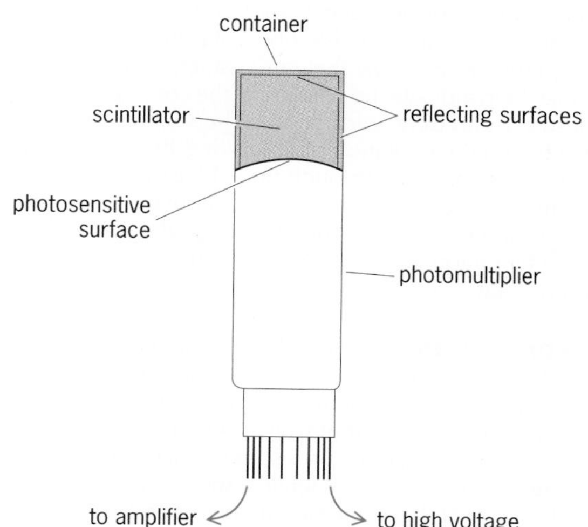

Diagram of a scintillation counter.

When a particle passes completely through a scintillator, the energy loss of the particle is measured. When a gamma ray converts to charged particles in a scintillator, its energy may also be determined. When the scintillator is made of dense material and of very large dimensions, the entire energy of a very energetic particle or gamma ray may be contained within the scintillator, and again the original energy may be measured. Such is the case for energetic electrons, positrons, or gamma rays which produce electromagnetic showers in the scintillator. Energy spectra can be determined in these various cases by using electronic equipment to convert amplitudes of the output pulses from the photomultiplier or PIN diode to digital form, for further processing by computers or pulse-height analyzers. Frank D. Brooks; Robert Hofstadter

Scleractinia An order of the subclass Zoantharia which comprises the true or stony corals (see illustration). These are solitary or colonial anthozoans which attach to a firm substrate. They are profuse in tropical and subtropical waters and contribute to the formation of coral reefs or islands. Some species are free and unattached.

Most of the polyp is impregnated with a hard calcareous skeleton secreted from ectodermal calcioblasts. The solitary corals form cylindrical, discoidal, or cuneiform skeletons, whereas colonial skeletons are multifarious. The polyps increase rapidly by intra- or extratentacular budding, and the skeletons of polyps

Solitary coral polyps, *Oulangia* sp.

which settle in groups may fuse to form a colony. The pyriform, ciliated planula swims with its aboral extremity, which is composed of an ectodermal sensory layer, directed anteriorly. Planulation occurs periodically in conformity with lunar phases in many tropical species. *See* ZOANTHARIA. Kenji Atoda

Scleractinian corals possess robust skeletons, so they have a rich fossil record. Because they are restricted mainly to tropical belts, they help indicate the position of the continents throughout the Mesozoic and Cenozoic periods. They are also important for understanding the evolution of corals and the origin and maintenance of reef diversity through time. Pleistocene corals shows persistent reef coral communities throughout the last several hundred thousand years. Environmental degradation has led to the dramatic alteration of living coral communities during the past several decades. John M. Pandolfi

Sclerenchyma Single cells or aggregates of cells whose principal function is thought to be mechanical support of plants or plant parts. Sclerenchyma cells have thick secondary walls and may or may not remain alive when mature. They vary greatly in form and are of widespread occurrence in vascular plants. Two general types, sclereids and fibers, are widely recognized, but since these intergrade, the distinction is sometimes arbitrary. Norman H. Boke

Sclerosponge A class of sponges that lay down a compound skeleton including an external, basal mass of calcium carbonate, either aragonite or calcite, and internal siliceous spicules and protein fibers. The living tissue forms a thin layer over the basal calcareous skeleton and extends into its surface depressions; the organization of the tissue is similar to that of encrusting demosponges. Sclerosponges are common inhabitants of cryptic habitats on coral reefs in both the Caribbean and Indo-Pacific biogeographic regions. *See* DEMOSPONGIAE. Willard D. Hartman

Scorpaeniformes An order of fishes, called the mail-cheeked fishes, distinguished by a unique character, the suborbital stay. The stay is an extension of the hypertrophied third infraorbital bone, which crosses the cheek obliquely and usually is firmly attached to the preoperculum. The head is usually spiny, and the body may be spiny or have bony plates, or both, or be smooth and appear naked. Almost all scorpaenids have prominent pectoral fins, either broad and fanlike or winglike, with the lower rays deeply incised. The order consists of about 1271 species and 266 genera in 25 families.

Order Scorpaeniformes
Suborder Dactylopteroidei (flying gurnards)
Scorpaenoidei (scorpionfishes)
Platycephaloidei (searobins and flatheads)
Anoplopomatoidei (sablefishes)
Suborder Hexagrammoidei (greenlings)
Normanichthyiodei (no common name)
Cottoidei
Superfamily Cottoidea (sculpins and poachers)
Cyclopteroidea (lumpfishes and snailfishes)

Thirty-five percent of all scorpaeniform fishes are in the suborder Scorpaenoidei. Of its six families, the principal one is Scorpaenidae (scorpionfishes), which accounts for 80% of the 444 species in the suborder. These fishes are represented in all of the tropical and temperate seas of the world and are the most generalized scorpaeniforms. The family Scorpaenidae contains the world's most venomous fishes. Scorpaenids have venomous glands at the base of the dorsal, anal, and pelvic spines. Most scorpionfishes are bottom dwellers that sit motionless among coral, rock, or grass, and, depending on their color patterns for concealment, wait for a venturesome prey to come within striking distance. Among the venomous fishes is the beautiful redlionfish (*Pterois volitans*). Its sting causes a number of debilitating symptoms in humans which could lead to death if not properly and promptly treated. However, far more dangerous are the stonefishes (for example, *Synanceia*). Their spines are like hypodermic needles, ready to deliver the most venomous neurotoxin (ichthyocanthotoxin = toxin from fish spines) known in the fish kingdom.

The cottoidei are the largest group of scorpaeniform fishes, accounting for 49% of the species. The physiognomy of cottoids is quite variable, with overall shapes from elongate to globose. Cottidae (sculpins) are the largest family, with 70 genera and 300 species. Sculpins are essentially Holarctic in distribution. The marine species reach their greatest diversity in the coastal waters of the North Pacific. *See* ACTINOPTERYGII; OSTEICHTHYES; PERCIFORMES. Herbert Boschung

Scorpiones An order of the Arachnida characterized by chelate pedipalps and chelicerae (pincerlike appendages), pectines (feathery chemo- and mechanoreceptors used to survey the texture of the ground surface and detect pheromones), and a narrow, flexible postabdomen bearing a venomous segment (telson) with a terminal sting (aculeus). The scorpion body is divided into a cephalothorax (prosoma), covered by an unsegmented carapace, and a segmented abdomen (opisthosoma) (see illustration). The opisthosoma is differentiated into an anterior preabdomen (mesosoma) and a postabdomen (metasoma), which, together with the telson, constitutes the "tail" or cauda. The cephalothorax bears the chelicerae, pedipalps, and four pairs of walking legs. The preabdomen contains seven segments, the postabdomen five. One pair of small, simple median eyes (ocelli) and, depending on the species, 2–5 (usually 3) pairs of anterolateral eyes are situated on the carapace. Troglobitic (cave-dwelling) scorpions lack all or some eyes.

Scorpions are derived from amphibious ancestors that lived in the Silurian, more than 400 million years ago. Approximately 1490 extant species of scorpions, placed in 170 genera, and between 13 and 20 families, have been described worldwide. They vary in color from translucent through tan or brown to black, or combinations thereof, and in size from 10 mm to 21 cm (0.4–8.3 in.). Scorpions occur on all continents except Antarctica, but are most abundant and diverse in tropical and subtropical regions, especially in desert and semidesert habitats.

Scorpion venoms contain multiple low-molecular-weight proteinaceous neurotoxins that block sodium and potassium channels, preventing the transmission of nerve impulses across synapses. Scorpion envenomation represents a significant cause of morbidity and mortality in some regions (such as Mexico, North Africa, and the Middle East). However, most scorpions are harmless. The sting may be painful, but not dangerous. *Centruroides exilicauda*, from Arizona, California, and New Mexico,

Scorpion, *Hadrurus arizonensis* (Ewing, 1928), adult male. (*Photo by Randy Mercurio; copyright © American Museum of Natural History*)

is the only species known to be lethal in the United States. The venom of these scorpions has proved fatal to healthy children up to 16 years of age and to adults suffering from hypertension and general debility. *See* ARACHNIDA.

Lorenzo Prendini

Scorpius The Scorpion, a large zodiacal constellation (see illustration). The constellation loops through the sky like a giant fish-hook. The bright supergiant star Antares is the 15th brightest star in the sky. Its name means Rival of Mars, since both are bright and reddish in color, and Mars sometimes passes close to Antares. *See* ANTARES; ZODIAC.

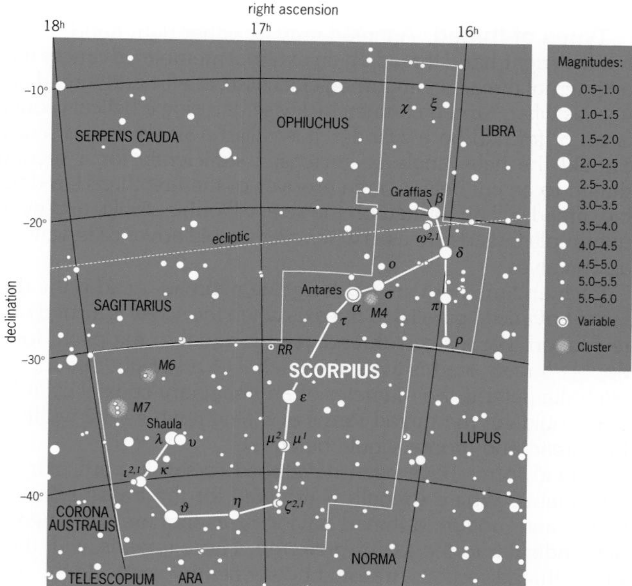

Modern boundaries of the constellation Scorpius, the Scorpion. The celestial equator is 0° of declination, which corresponds to celestial latitude. Right ascension corresponds to celestial longitude, with each hour of right ascension representing 15° of arc. Apparent brightness of stars is shown with dot sizes to illustrate the magnitude scale, where the brightest stars in the sky are 0th magnitude or brighter and the faintest stars that can be seen with the unaided eye at a dark site are 6th magnitude. (*Wil Tirion*)

Scorpius lies along the Milky Way and contains many beautiful star clusters and star clouds. The globular cluster M4 and the open clusters M6 and M7 are especially noticeable. *See* CONSTELLATION; MESSIER CATALOG; STAR CLOUDS; STAR CLUSTERS.

Jay M. Pasachoff

Scramjet An air-breathing jet engine that relies for propulsion on the compressing or ramming effect on air taken into the engine inlet at supersonic speeds, normally when the aircraft is traveling at speeds above Mach 4. The term is derived from "supersonic combustion ramjet." *See* MACH NUMBER.

Scramjet operation. A ramjet is a jet engine that relies on the compressing or ramming effect on the air taken into the inlet while the aircraft is in motion. The geometry of the engine compresses the air rather than a piston, as in an internal combustion engine, or a compressor, as in a gas turbine. In a ramjet, the compressed air is slowed to subsonic speeds (less than Mach 1) as it is routed into the combustion chamber, where it is mixed with fuel, burned, and expanded through a nozzle to generate thrust.

To get an aircraft to Mach 4, the ignition condition for a scramjet, a rocket or jet engine is typically used for the initial acceleration. As the vehicle is traveling through the atmosphere on its way to Mach 4, air is being compressed at hypersonic (greater than Mach 5) and supersonic (Mach 1–5) speeds on the lower surface of the aircraft and enters the engine at high pressure. From the scramjet inlet, the air flows at supersonic speed into the combustor, where fuel is mixed with the air, burned, and expanded to make thrust. *See* HYPERSONIC FLIGHT.

Advantages. Scramjets offer several advantages over conventional jet and rocket engines. A scramjet has the ability to operate at much faster speeds. While burning hydrocarbon fuel, a scramjet engine can operate at speeds up to Mach 8. Using hydrogen as the fuel, a scramjet engine can operate at speeds in excess of Mach 12. The performance (specific impulse) of a scramjet is significantly higher than rocket engines. A scramjet engine does not require the use of high-pressure pumps, compressors, or turbines as do gas turbine or rocket engines. While reducing moving parts increases engine reliability and durability, benefits such as high speed, relative fuel efficiency, and range are the reasons that scramjets will revolutionize the aerospace industry. *See* ROCKET PROPULSION; TURBINE PROPULSION.

Scramjet-powered access-to-space systems that will take off and land horizontally like commercial airplanes are being conceptualized today. These systems will be designed to achieve orbit while employing relatively low acceleration profiles, potentially opening access to orbital flight to a much larger section of the population.

Joaquin Castro

Scrapie A transmissible, invariably fatal disease of adult sheep and goats characterized by degeneration of the central nervous system. It is one of the transmissible spongiform encephalopathies, a disease family which includes bovine spongiform encephalopathy (BSE) in cattle (commonly known as mad cow disease), chronic wasting disease in deer, and Kuru and variant Creutzfeldt-Jakob disease in humans. Scrapie is now present in many sheep populations worldwide, excluding Australia and New Zealand. *See* NERVOUS SYSTEM DISORDERS.

Scrapie has a long incubation period. It is widely believed that most animals are infected at or shortly after birth, with about 90% of sheep developing signs of disease at 2–5 years of age. Scrapie affects both sexes and is insidious and progressive in its onset, with a clinical course that lasts weeks or months. The clinical signs often begin with hyperexcitability or dullness, and proceed to pruritus (itching), ataxia (incoordination), trembling, and loss of condition. Pruritus is accompanied by scratching or scraping (hence the name), leading to wool loss. There is no treatment for scrapie, and affected animals will, if not euthanized, become recumbent and die.

Scrapie is caused by a transmissible agent, which can be found in cell-free filtrates of the brain and other organs from scrapie-affected sheep. The prion (proteinaceous infectious particle) hypothesis proposes that the etiologic agent of scrapie and other transmissible spongiform encephalopathies is the host protein, PrP, which can exist in a nonpathological form and a pathological form. *See* PRION DISEASE; VETERINARY MEDICINE.

M. Baylis

Screening A mechanical method of separating a mixture of solid particles into fractions by size. The mixture to be separated, called the feed, is passed over a screen surface containing openings of definite size. Particles smaller than the openings fall through the screen and are collected as undersize. Particles larger than the openings slide off the screen and are caught as oversize. A single screen separates the feed into only two fractions. Two or more screens may be operated in series to give additional fractions. Screening occasionally is done wet, but most commonly it is done dry.

Industrial screens may be constructed of metal bars, perforated or slotted metal plates, woven wire cloth, or bolting cloth. The openings are usually square but may be circular or rectangular. *See* MECHANICAL CLASSIFICATION; MECHANICAL SEPARATION TECHNIQUES; SEDIMENTATION (INDUSTRY). Warren L. McCabe

Screw A cylindrical body with a helical groove cut into its surface. For practical purposes a screw may be considered to be a wedge wound in the form of a helix so that the input motion is a rotation while the output remains translation. The screw is to the wedge much the same as the wheel and axle is to the lever in that it permits the exertion of force through a greatly increased distance.

The screw is by far the most useful form of inclined plane or wedge and finds application in the bolts and nuts used to fasten parts together; in lead and feed screws used to advance cutting tools or parts in machine tools; in screw jacks used to lift such objects as automobiles, houses, and heavy machinery; in screw-type conveyors used to move bulk materials; and in propellers for airplanes and ships. *See* PROPELLER (AIRCRAFT); PROPELLER (MARINE CRAFT); SCREW FASTENER; SCREW JACK; SCREW THREADS; SIMPLE MACHINE. Richard M. Phelan

Screw fastener A threaded machine part used to join parts of a machine or structure. Screw fasteners are used when a connection that can be disassembled and reconnected and that must resist tension and shear is required. A nut and bolt is a common screw fastener. Bolt material is chosen to have an extended stress-strain characteristic free from a pronounced yield point. Nut material is chosen for slight plastic flow.

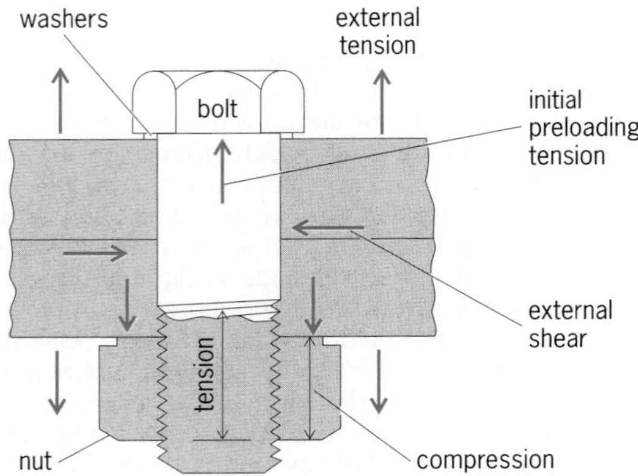

Forces on a bolt fastener under preloading.

The nut is tightened on the bolt to produce a preload tension in the bolt (see illustration). This preload has several advantageous effects. It places the bolt under sufficient tension so that during vibration the relative stress change is slight with consequent improved fatigue resistance and locking of the nut. Preloading also increases the friction between bearing surfaces of the joined members so that shear loads are carried by the friction forces rather than by the bolt. *See* BOLT; JOINT (STRUCTURES); NUT (ENGINEERING); SCREW. Frank H. Rockett

Screw jack A mechanism for lifting and supporting loads, usually of large size. A screw jack mechanism consists of a thrust collar and a nut which rides on a bolt; the threads between the nut and bolt normally have a square shape. A standard form of screw jack has a heavy metal base with a central threaded hole into which fits a bolt capable of rotation under a collar thrusting against the load. *See* SCREW; SIMPLE MACHINE. James J. Ryan

Screw threads Continuous helical ribs on a cylindrical shank. Screw threads are used principally for fastening, adjusting, and transmitting power. To perform these specific functions, various thread forms have been developed. A thread on the outside of a cylinder or cone is an external (male) thread; a thread on the inside of a member is an internal (female) thread (Fig. 1). *See* SCREW.

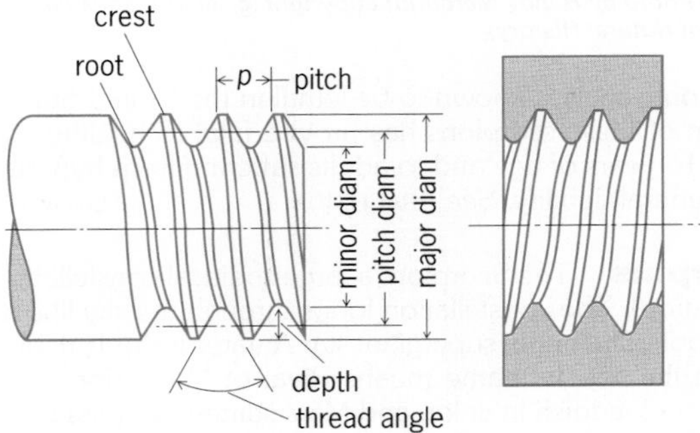

Fig. 1. Screw thread nomenclature.

Types of thread. A thread may be either right-hand or left-hand. A right-hand thread on an external member advances into an internal thread when turned clockwise; a left-hand thread advances when turned counterclockwise. If a single helical groove is cut or formed on a cylinder, it is called a single-thread screw. Should the helix angle be increased sufficiently for a second thread to be cut between the grooves of the first thread, a double thread will be formed on the screw. Double, triple, and even quadruple threads are used whenever a rapid advance is desired, as on valves.

Pitch and major diameter designate a thread. Lead is the distance advanced parallel to the axis when the screw is turned one revolution. For a single thread, lead is equal to the pitch; for a double thread, lead is twice the pitch. For a straight thread, the pitch diameter is the diameter of an imaginary coaxial cylinder that would cut the thread forms at a height where the width of the thread and groove would be equal.

Thread forms have been developed to satisfy particular requirements. Where strength is required for the transmission of power and motion, a thread having faces that are more nearly perpendicular to the axis is preferred. These threads, with their strong thread sections, transmit power nearly parallel to the axis of the screw.

Thread fastener. Most threaded fasteners are a threaded cylindrical rod with some form of head on one end. Of the many forms available, five types meet most requirements for threaded fasteners and are used for the bulk of production work: bolt, stud, cap screw, machine screw, and set screw (Fig. 2). Bolts and screws can be obtained with varied heads and points.

A bolt is generally used for drawing two parts together. A stud is a rod threaded on both ends. Studs are used for parts that must be removed frequently and for applications where bolts would be impractical.

Cap screws (plated or unplated) are widely used in machine tools and for assembling parts in automotive and aeronautical equipment. They are available in four standard heads: hexagon,

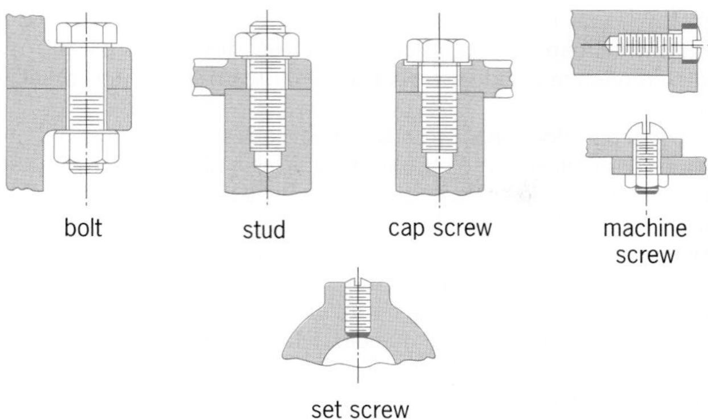

Fig. 2. Common types of fasteners. ***(After T. E. French and C. J. Vierck, Engineering Drawing and Graphic Technology, 12th ed., McGraw-Hill, 1978).***

flat, round, and fillister. Flathead and roundhead screws have slotted heads (Fig. 3). Machine screws are similar to cap screws

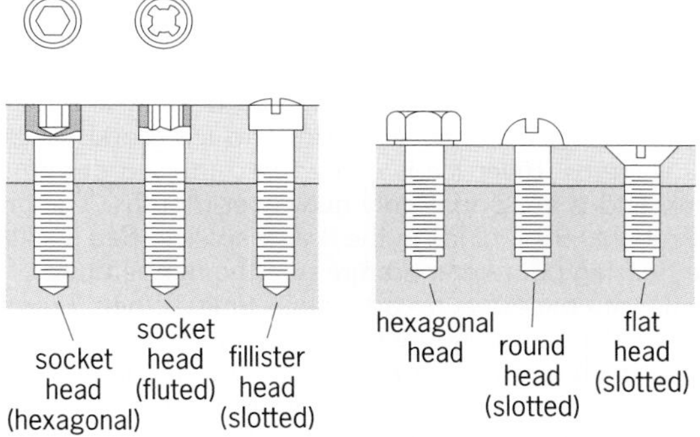

Fig. 3. Cap screws. ***(After W. J. Luzadder, Fundamentals of Engineering Drawing, 8th ed., Prentice-Hall, 1981).***

and fulfill the same purpose, being employed principally on small work. Set screws made of hardened steel are used to hold parts in a position relative to one another. Wood screws, lag screws, and hanger bolts are used in wood. Warren J. Luzadder

Scrophulariales An order of flowering plants, division Magnoliophyta (Angiospermae), in the subclass Asteridae of the class Magnoliopsida (dicotyledons). The order consists of 12 families and more than 11,000 species. The largest families are the Scrophulariaceae (about 4000 species), Acanthaceae (about 2500 species), Gesneriaceae (about 2500 species), Bignoniaceae (about 800 species), and Oleaceae (about 600 species). The Scrophulariales are Asteridae, with a usually superior ovary and generally either with an irregular corolla or with fewer stamens than corolla lobes, or commonly both. They uniformly lack stipules.

The Scrophulariaceae are characterized by their usually herbaceous habit; irregular flowers with usually only two or four functional stamens; axile placentation; and dry, dehiscent fruits with more or less numerous seeds that have endosperm. The Acanthaceae are distinguished by their chiefly herbaceous habit, mostly simple and opposite leaves, lack of endosperm, and especially the enlarged and specialized funiculus, which is commonly developed into a jaculator that expels the seeds at maturity.

Some well-known members of the Scrophulariales are the African violet (*Saintpaulia ionantha*, in the Gesneriaceae), catalpa (Bignoniaceae), and lilac (*Syringa vulgaris*, in the Oleaceae). *See* ASTERIDAE; FLOWER; MAGNOLIOPHYTA; MAGNOLIOPSIDA; OLIVE; PLANT KINGDOM. Arthur Cronquist; T. M. Barkley

Scyphozoa A class of the phylum Cnidaria traditionally containing four major groups—Coronatae, Rhizostomeae, Semaeostomeae, and Stauromedusae—all marine or estuarine. Most occur in two forms, the polyp, or scyphistoma, and the medusa, or scyphomedusa. However, some are polyplike and sessile throughout their lives, while others are always pelagic and lack the sessile scyphistoma stage. Among the Cnidaria, the Scyphozoa are characterized by having well-developed medusae of large size and fairly well organized scyphistomae of small size. The Scyphozoa are most closely related to the other medusae-containing classes—Cubozoa and Hydrozoa—within the Cnidaria.

Scyphistomae are usually solitary and small, measuring much less than an inch (only a few millimeters) in height. They are sessile. Each solitary scyphistoma is composed of three major parts—pedal disk, stalk, and calyx—and is shaped, overall, something like a wine glass or vase. The calyx is crowned by the oral disk with a circlet of tentacles around its perimeter and a four-sided mouth at its center.

Scyphomedusae, when fully grown, are generally large, sometimes attaining a diameter of 3 ft (1 m) and weighing over 300 lb (140 kg). Most scyphomedusae are natatory (adapted for swimming). The scyphomedusa is composed of the umbrella (or bell), which is the major propulsive structure, generally appended with oral arms. Differences in the shape, structure, ornamentation, and color of the bell and oral arms are the main features used to distinguish major groups and species of Scyphozoa.

Important differences in morphology, DNA, and proteins used to identify jellyfish indicate that there are several hundreds of species of Scyphozoa. Most are typically found near the coast because scyphistomae are benthic. Exceptions are *Pelagia* and other wholly pelagic forms, including many of the mid- and deep-water scyphomedusae. Order Rhizostomeae contains the most species; about 95 are currently recognized. The rhizostomes are predominantly found in tropical regions, especially the Indo-West Pacific center of marine biodiversity. Order Semaeostomeae contains fewer species (approximately 75) which are more widely distributed, with representatives in polar, deep-sea, open-ocean, tropical, and temperate coastal regions. Coronatae, with only approximately 50 species recognized, is typically regarded as a deep-sea group although it does have a number of shallow-water representatives in both the tropics and temperate regions. There are only approximately 35 known species of Stauromedusae, which are distributed predominantly in cold temperate and polar regions of northern and southern hemispheres. Some Semaeostomeae and Rhizostomeae are injurious to humans because of their cnidocytes. *See* CNIDARIA. Michael N. Dawson

Sea anemone Any polyp of the nearly 1000 anthozoan coelenterates belonging to the order Actiniaria. They occur intertidally and subtidally in marine and estuarine habitats, attached to solid substrates or burrowing into soft sediments. No freshwater or truly planktonic species are known. Anemones may be very small, a fraction of an inch long, to large individuals more than 3 ft (90 cm) in length or diameter. *See* ACTINIARIA.

Sea anemones enter into a number of interesting symbiotic partnerships. Some are host to single-celled marine algae, which grow within the cells of the anemone. The algae provide organic materials which aid in the nutrition of the anemone. Other anemones live on gastropod shells inhabited by hermit crabs. *See* ANTHOZOA; CNIDARIA. Cadet Hand

Sea breeze A diurnal, thermally driven circulation in which a surface convergence zone often exists between airstreams having over-water versus over-land histories. The sea breeze is one of the most frequently occurring small-scale (mesoscale) weather systems. It results from the unequal sensible heat flux of the lower atmosphere over adjacent solar-heated land and water masses. Because of the large thermal inertia of a water body, during daytime the air temperature changes little over the water while over land the air mass warms. Occurring during periods of fair skies and generally weak large-scale winds, the sea breeze is recognizable by a wind shift to onshore, generally several hours after sunrise. On many tropical coastlines the sea breeze is an almost daily occurrence. It also occurs with regularity during the warm season along mid-latitude coastlines and even occasionally on Arctic shores. Especially during periods of very light winds, similar though sometimes weaker wind systems occur over the shores of large lakes and even wide rivers and estuaries (lake breezes, river breezes). At night, colder air from the land often will move offshore as a land breeze. Typically the land breeze circulation is much weaker and shallower than its daytime counterpart. *See* ATMOSPHERIC GENERAL CIRCULATION; MESOMETEOROLOGY; METEOROLOGY.

The occurrence and strength of the sea breeze is controlled by a variety of factors, including land-sea surface temperature differences; latitude and day of the year; the synoptic wind and its orientation with respect to the shoreline; the thermal stability of the lower atmosphere; surface solar radiation as affected by haze, smoke, and stratiform and convective cloudiness; and the geometry of the shoreline and the complexity of the surrounding terrain. *See* WIND.

Walter A. Lyons

Sea-floor imaging The process whereby mapping technologies are used to produce highly detailed images of the sea floor. High-resolution images of the sea floor are used to locate and manage marine resources such as fisheries and oil and gas reserves, identify offshore faults and the potential for coastal damage due to earthquakes, and map out and monitor marine pollution, in addition to providing information on what processes are affecting the sea floor, where these processes occur, and how they interact. *See* MARINE GEOLOGY.

Side-scan sonar provides a high-resolution view of the sea floor. In general, a side-scan sonar consists of two sonar units attached to the sides of a sled tethered to the back of a ship. Each sonar emits a burst of sound that insonifies a long, narrow corridor of the sea floor extending away from the sled. Sound reflections from the corridor that echo back to the sled are then recorded by the sonar in their arrival sequence, with echoes from points farther away arriving successively later. The sonars repeat this sequence of "talking" and listening every few seconds as the sled is pulled through the water so that consecutive recordings build up a continuous swath of sea-floor reflections, which provide information about the texture of the sea floor. *See* ECHO SOUNDER; HYDROPHONE; SONAR; SONOBUOY; UNDERWATER SOUND.

The best technology for mapping sea-floor depths or bathymetry is multibeam sonar. These systems employ a series of sound sources and listening devices that are mounted on the hull of a survey ship. As with side-scan sonar, every few seconds the sound sources emit a burst that insonifies a long, slim strip of the sea floor aligned perpendicular to the ship's direction. The listening devices then begin recording sounds from within a fan of narrow sea-floor corridors that are aligned parallel to the ship and that cross the insonified strip. By running the survey the same way that one mows a lawn, adjacent swaths are collected parallel to one another to produce a complete sea-floor map of an area.

The most accurate and detailed view of the sea floor is provided by direct visual imaging through bottom cameras, submersibles, remotely operated vehicles, or if the waters are not too deep, scuba diving. Because light is scattered and absorbed in waters greater than about 33 ft (10 m) deep, the sea-floor area that bottom cameras can image is no more than a few meters. This limitation has been partly overcome by deep-sea submersibles and remotely operated vehicles, which provide researchers with the opportunity to explore the sea floor close-up for hours to weeks at a time. But even the sea-floor coverage that can be achieved with these devices is greatly restricted relative to side-scan sonar, multibeam sonar, and satellite altimetry.

The technology that provides the broadest perspective but the lowest resolution is satellite altimetry. A laser altimeter is mounted on a satellite and, in combination with land-based radars that track the satellite's altitude, is used to measure variations in sea-surface elevation to within 2 in. (5 cm). Removing elevation changes due to waves and currents, sea-surface height can vary up to 660 ft (200 m). These variations are caused by minute differences in the Earth's gravity field, which in turn result from heterogeneities in the Earth's mass. These heterogeneities are often associated with sea-floor topography. By using a mathematical function that equates sea-surface height to bottom elevations, global areas of the sea floor can be mapped within a matter of weeks. However, this approach has limitations. Sea-floor features less than 6–9 mi (10–15 km) in length are generally not massive enough to deflect the ocean surface, and thus go undetected. Furthermore, sea-floor density also affects the gravity field; and where different-density rocks are found, such as along the margins of continents, the correlation between Earth's gravity field and sea-floor topography breaks down. *See* ALTIMETER.

Lincoln F. Pratson

Sea ice Ice formed by the freezing of seawater. Ice in the sea includes sea ice, river ice, and land ice. Land ice is principally icebergs. River ice is carried into the sea during spring breakup and is important only near river mouths. The greatest part, probably 99% of ice in the sea, is sea ice. *See* ICEBERG.

The freezing point temperature and the temperature of maximum density of seawater vary with salinity. When freezing occurs, small flat plates of pure ice freeze out of solution to form a network which entraps brine in layers of cells. As the temperature decreases more water freezes out of the brine cells, further concentrating the remaining brine so that the freezing point of the brine equals the temperature of the surrounding pure ice structure. The brine is a complex solution of many ions.

The brine cells migrate and change size with changes in temperature and pressure. The general downward migration of brine cells through the ice sheet leads to freshening of the top layers to near zero salinity by late summer. During winter the top surface temperature closely follows the air temperature, whereas the temperature of the underside remains at freezing point, corresponding to the salinity of water in contact.

The sea ice in any locality is commonly a mixture of recently formed ice and old ice which has survived one or more summers. Except in sheltered bays, sea ice is continually in motion because of wind and current.

Waldo Lyon

Sea of Okhotsk A semienclosed basin adjacent to the North Pacific Ocean, bounded on the north, east, and west by continental Russia, the Kamchatka Peninsula, and northern Japan. On its southeast side the Sea of Okhotsk is connected to the North Pacific via a number of straits and passages through the Kuril Islands. The sea covers a surface area of approximately 590,000 mi^2 (1.5 million km^2), or about 1% of the total area of the Pacific, and has a maximum depth of over 9000 ft (3000 m). Its mean depth is about 2500 ft (830 m). Owing to the cold, wintertime Arctic winds that blow to the southeast, from Russia toward the North Pacific Ocean, the Sea of Okhotsk is partially covered with ice during the winter months, from November through April. *See* BASIN; PACIFIC OCEAN.

The amount of water exchanged between the Sea of Okhotsk and the North Pacific Ocean is not well known, but it is thought that the waters of the Sea of Okhotsk that do enter the North Pacific may play an important role in the Pacific's large-scale circulation. The reason is the extreme winter conditions over the Sea of Okhotsk: its waters are generally colder and have a lower salinity than the waters at the same density in the North Pacific. *See* OCEAN CIRCULATION; SEAWATER. Stephen C. Riser

Seaborgium A chemical element, symbol Sg, atomic number 106. Seaborgium has chemical properties similar to tungsten. It was synthesized and identified in 1974. This discovery of seaborgium took place nearly simultaneously in two nuclear laboratories, the Lawrence Berkeley Laboratory at the University of California and the Joint Institute for Nuclear Research at Dubna in Russia.

The Berkeley group, under the leadership of A. Ghiorso, used as its source of heavy ions the Super-Heavy Ion Linear Accelerator (SuperHILAC). The production and positive identification of the isotope of seaborgium with the mass number 263 decays with a half-life of 0.9 + 0.2 s by the emission of alpha particles of principal energy 9.06 + 0.04 MeV. This isotope is produced in the reaction in which four neutrons are emitted: $^{249}Cf(^{18}O,4n)$.

The Dubna group, under the leadership of G. N. Flerov and Y. T. Oganessian, produced its heavy ions with a heavy-ion cyclotron. They found a product that decays by the spontaneous fission mechanism with the very short half-life of 7 ms. They assigned it to the isotope ^{259}Sg, suggesting reactions in which two or three neutrons are emitted: $^{207}Pb(^{54}Cr,2n)$ and $^{208}Pb(Cr^{54},3n)$. *See* NOBELIUM; NUCLEAR CHEMISTRY; PERIODIC TABLE; TRANSURANIUM ELEMENTS. Glenn T. Seaborg

Seamount and guyot A seamount is a mountain that rises from the ocean floor; a submerged flat-topped seamount is termed a guyot. By arbitrary definition, seamounts must be at least 3000 ft (about 900 m) high, but in fact there is a continuum of smaller undersea mounts, down to heights of only about 300 ft (100 m). Some seamounts are high enough temporarily to form oceanic islands, which ultimately subside beneath sea level. There are on the order of 10,000 seamounts in the world ocean, arranged in chains (for example, the Hawaiian chain in the North Pacific) or as isolated features. In some chains, seamounts are packed closely to form ridges. Very large oceanic volcanic constructions, hundreds of kilometers across, are called oceanic plateaus. *See* MARINE GEOLOGY; OCEANIC ISLANDS; VOLCANO.

Almost all seamounts are the result of submarine volcanism, and most are built within less than about 1 million years. Seamounts are made by extrusion of lavas piped upward in stages from sources within the Earth's mantle to vents on the seafloor. Seamounts provide data on movements of tectonic plates on which they ride, and on the rheology of the underlying lithosphere. The trend of a seamount chain traces the direction of motion of the lithospheric plate over a more or less fixed heat source in the underlying asthenosphere part of the Earth's mantle. *See* LITHOSPHERE; PLATE TECTONICS. Edward L. Winterer

Seaplane An airplane capable of navigating on, taking off from, and alighting upon the surface of water. Seaplanes are grouped into two main types: flying boats and float planes. In the flying boat, the hull, which provides buoyancy and planing area, is an integral part of the airframe, a specially designed fuselage which supports the wings and tail surfaces and houses the crew, equipment, and cargo. Although multihull flying boats have been built, modern use is confined to single-hull boats with lateral stability on the water provided by small floats or pontoons attached to the wings. The float plane is a standard landplane made capable of water operation by the addition of floats which are attached to the airframe by struts. In practice the twin float is used exclusively, lateral stability on the water being provided by the separation of the two identical floats (see illustration). A seaplane with retracting wheels which permit either land or water operation is known as an amphibian. *See* AIRPLANE. Richard A. Hoffman

A single-engine float plane.

Seasons The four divisions of the year based upon variations of sunlight intensity (solar energy per unit area at the Earth's surface) at local solar noon (noontime) and daylight period. The variations in noontime intensity and daylight period are the result of the Earth's rotational axis being tilted 23.5° from the perpendicular to the plane of the Earth's orbit around the Sun. The direction of the Earth's axis with respect to the stars remains fixed as the Earth orbits the Sun. If the Earth's axis were not tilted from the perpendicular, there would be no variation in noontime sunlight intensity or daylight period and no seasons. The Earth-Sun distance does not influence the seasons because it varies only slightly, and it is overwhelmed by the effects of variations in sunlight intensity and daylight period due to the alignment of the Earth's axis.

Sunlight intensity at a location depends upon the angle from the horizon to the Sun at local solar noon; this angle in turn depends upon the location's latitude and the position of the Earth in its orbit. At increased angles, a given amount of sunlight is spread over smaller surface areas, resulting in a greater concentration of solar energy, which produces increased surface heating. Intensity is a more important factor than the number of daylight hours in determining the heating effect at the Earth's surface. *See* EARTH ROTATION AND ORBITAL MOTION. Harold P. Coyle

Seawater An aqueous solution of salts of a rather constant composition of elements whose presence determines the climate and makes life possible on the Earth and which constitutes the oceans, the mediterranean seas, and their embayments. The physical, chemical, biological, and geological events therein are the studies that are grouped as oceanography. Water is most often found in nature as seawater (about 98%). The rest is ice, water vapor, and fresh water. The basic properties of seawater, their distribution, the interchange of properties between sea and atmosphere or land, the transmission of energy within the sea, and the geochemical laws governing the composition of seawater and sediments are the fundamentals of oceanography. *See* HYDROSPHERE; OCEANOGRAPHY.

Major constituents of seawater (salinity 35 psu)*

Positive ions	Amount, g/kg	Negative ions	Amount, g/kg
Sodium (Na^C)	10.752	Chloride (Cl^-)	19.345
Magnesium (Mg^{2+})	1.295	Bromide (Br^-)	0.066
Potassium (K^C)	0.390	Fluoride (F^-)	0.0013
Calcium (Ca^{2+})	0.416	Sulfate (SO_4^-)	2.701
Strontium (Sr^{2C})	0.013	Bicarbonate (HCO_3^-)	0.145
		Boron hydroxide ($B(OH)_3^-$)	0.027

*Water, 965 psu; dissolved materials, 35 psu.

The major chemical constituents of seawater are cations (positive ions) and anions (negative ions) [see Table]. In addition, seawater contains the suspended solids, organic substances, and dissolved gases found in all natural waters. A standard salinity of 35 practical salinity units (psu; formerly parts per thousand, or ‰) has been assumed. While salinity does vary appreciably in oceanic waters, the fractional composition of salts is remarkably constant throughout the world's oceans. In addition to the dissolved salts, natural seawater contains particulates in the form of plankton and their detritus, sediments, and dissolved organic matter, all of which lend additional coloration beyond the blue coming from Rayleigh scattering by the water molecules. Almost every known natural substance is found in the ocean, mostly in minute concentrations. *See* SCATTERING OF ELECTROMAGNETIC RADIATION. John R. Apel

Seawater fertility A measure of the potential ability of seawater to support life. Fertility is distinguished from productivity, which is the actual production of living material by various trophic levels of the food web. Fertility is a broader and more general description of the biological activity of a region of the sea, while primary production, secondary production, and so on, is a quantitative description of the biological growth at a specified time and place by a certain trophic level. Primary production that uses recently recycled nutrients such as ammonium, urea, or amino acids is called regenerated production to distinguish it from the new production that is dependent on nitrate being transported by mixing or circulation into the upper layer where primary production occurs. New production is organic matter, in the form of fish or sinking organic matter, that can be exported from the ecosystem without damaging the productive capacity of the system. *See* BIOLOGICAL PRODUCTIVITY.

The potential of the sea to support growth of living organisms is determined by the fertilizer elements that marine plants need for growth. Fertilizers, or inorganic nutrients as they are called in oceanography, are required only by the first trophic level in the food web, the primary producers; but the supply of inorganic nutrients is a fertility-regulating process whose effect reaches throughout the food web. When there is an abundant supply to the surface layer of the ocean that is taken up by marine plants and converted into organic matter through photosynthesis, the entire food web is enriched, including zooplankton, fish, birds, whales, benthic invertebrates, protozoa, and bacteria. *See* DEEP-SEA FAUNA; FOOD WEB; MARINE FISHERIES.

The elements needed by marine plants for growth are divided into two categories depending on the quantities required: The major nutrient elements that appear to determine variations in ocean fertility are nitrogen, phosphorus, and silicon. The micronutrients are elements required in extremely small, or trace, quantities including essential metals such as iron, manganese, zinc, cobalt, magnesium, and copper, as well as vitamins and specific organic growth factors such as chelators. Knowledge of the fertility consequences of variations in the distribution of micronutrients is incomplete, but consensus among oceanographers is that the overall pattern of ocean fertility is set by the major fertilizer elements—nitrogen, phosphorus, and silicon—and not by micronutrients.

Two types of marine plants carry out primary production in the ocean: microscopic planktonic algae collectively called phytoplankton, and benthic algae and sea grasses attached to hard and soft substrates in shallow coastal waters.

The benthic and planktonic primary producers are a diverse assemblage of plants adapted to exploit a wide variety of marine niches; however, they have in common two basic requirements for the photosynthetic production of new organic matter: light energy and the essential elements of carbon, hydrogen, nitrogen, oxygen, phosphorus, sulfur, and silicon for the synthesis of new organic molecules. These two requirements are the first-order determinants of photosynthetic growth for all marine plants and, hence, for primary productivity everywhere in the ocean.

The regions of the world's oceans differ dramatically in overall fertility. In the richest areas, the water is brown with diatom blooms, fish schools are abundant, birds darken the horizon, and the sediments are fine-grained black mud with a high organic content. In areas of low fertility, the water is blue and clear, fish are rare, and the bottom sediments are well-oxidized carbonate or clay. These extremes exist because the overall pattern of fertility is determined by the processes that transport nutrients to the sunlit upper layer of the ocean where there is energy for photosynthesis. *See* SEAWATER. Richard T. Barber

Sebaceous gland A gland which produces and liberates sebum, a mixture composed of fat, cellular debris, and keratin. When the gland arises in association with a hair follicle, it forms a thickened outpushing from the side of the developing follicle near the epidermis. Central cells in these sebaceous glands form oil droplets within the cytoplasm. These cells disintegrate to liberate the sebaceous substance and are therefore of the holocrine type. The Meibomian or tarsal glands, within the tarsus or supporting plate at the edge of the eyelids, are sebaceous and complex tubuloacinous structures. The numerous separate glands open along the entire edge of the upper and lower lids. Retained secretions of the tarsal glands produce a chalozion or Meibomian cyst. *See* GLAND. Olin E. Nelsen

Second sound A type of wave propagated in the superfluid phase of liquid helium (helium II) and in certain other substances under special conditions. The name is misleading since second sound is not in any sense a sound wave, but a temperature or entropy wave. In ordinary or first sound, pressure and density variations propagate with very small accompanying variations in temperature; in second sound, temperature variations propagate with no appreciable variation in density or pressure. *See* LIQUID HELIUM; SUPERFLUIDITY.

The two-fluid model of helium II provides further insight into the nature of second sound. In this model the liquid can be described as consisting of superfluid and normal components of densities ρ_s and ρ_n, respectively, such that the total density $\rho = \rho_s + \rho_n$. The superfluid component is frictionless and devoid of entropy; the normal component has a normal viscosity and contains the entropy and thermal energy of the system. In a temperature or second-sound wave, the normal and superfluid flows are oppositely directed so that $\rho_s\mathbf{V}_s + \rho_n\mathbf{V}_n = 0$, where $\mathbf{V}_s$ and $\mathbf{V}_n$ are the superfluid and normal flow velocities. Thus a variation in relative densities of the two components, and hence a temperature fluctuation, propagates with no change in total density or pressure. In a first-sound wave, the two components move in phase, that is, $\mathbf{V}_n \cong \mathbf{V}_s$.

Theoretical predictions that second sound should exist in certain solid dielectric crystals under suitable conditions have been confirmed experimentally for solid helium single crystals at temperatures between 0.4 and 1.0 K (−459.0 and −457.9°F). *See* DIELECTRIC MATERIALS.

Another quite different class of materials can exhibit second sound. In smectic A liquid crystals, when the wave vector is oblique with respect to the layers of these ordered structures, a modulation of the interlayer spacing can propagate at nearly constant density. *See* LIQUID CRYSTALS. Henry A. Fairbank

Secondary emission The emission of electrons from the surface of a solid into vacuum caused by bombardment with charged particles, in particular with electrons. The mechanism of secondary emission under ion bombardment is quite different from that under electron bombardment; it is only in the latter case that the term secondary emission is generally used.

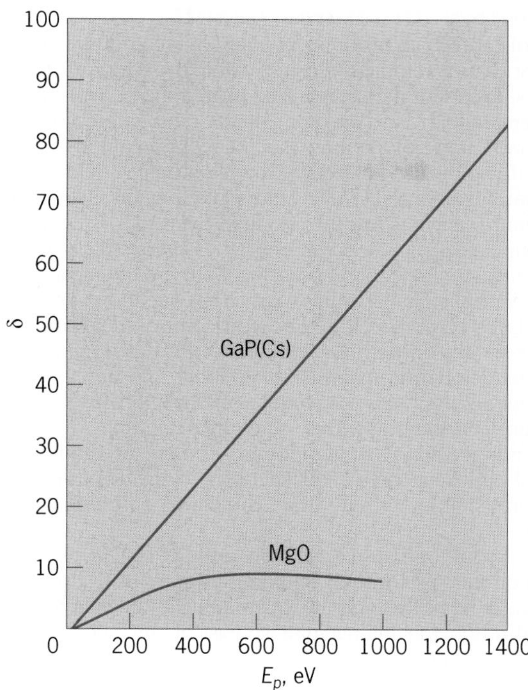

Secondary emission yield versus primary energy for GaP(Cs). The curve for MgO is shown for comparison.

The bombarding electrons and the emitted electrons are referred to, respectively, as primaries and secondaries. Secondary emission has important practical applications because the secondary yield, that is, the number of secondaries emitted per incident primary, may exceed unity. Thus, secondary emitters are used in electron multipliers, especially in photomultipliers, and in other electronic devices such as television pickup tubes, storage tubes for electronic computers, and so on.

The emission of secondary electrons can be described as the result of three processes: (1) excitation of electrons in the solid into high-energy states by the impact of high-energy primary electrons, (2) transport of these secondary electrons to the solid-vacuum interface, and (3) escape of the electrons over the surface barrier into the vacuum. The efficiency of each of these three processes, and hence the magnitude of the secondary emission yield δ, varies greatly for different materials.

Most of the materials used in practical devices are semiconductors or insulators whose band-gap energies are much larger than their electron affinities. Examples are magnesium oxide (MgO), beryllium oxide (BeO), cesium antimonide (Cs_3Sb), and potassium chloride (KCl). Maximum δ values in the 8–15 range are typically obtained at primary energies of several hundred volts.

In certain semiconductors the bands are bent downward to such an extent that the vacuum level lies below the bottom of the conduction band in the bulk. A material with this characteristic is said to have negative effective electron affinity. The most important material in this category is cesium-activated gallium phosphide, GaP(Cs). The illustration shows the curve of yield δ versus primary energy E_p for GaP(Cs) by comparison with MgO. Values of δ exceeding 100 are readily obtained, with maximum yields at energies in the 5–10-kV region. *See* BAND THEORY OF SOLIDS; SEMICONDUCTOR.

Alfred H. Sommer

Secondary ion mass spectrometry (SIMS)

An instrumental technique that measures the elemental and molecular composition of solid materials. SIMS provides methods of visualizing the two- and three-dimensional composition of solids at lateral resolutions approaching several hundred nanometers and depth resolutions of 1–10 nm. This technique employs an energetic ion beam to remove or sputter the atomic and molecular constituents from a surface in a very controlled manner. The sputtered products include atoms, molecules, and molecular fragments that are characteristic of the surface composition within each volume element sputtered by the ion beam. A small fraction of the sputtered atoms and molecules are ionized as either positive or negative ions, and a measurement by SIMS determines the mass and intensity of these secondary ions by using various mass analysis or mass spectrometry techniques. In this technique, the sputtering ions are referred to as the primary ions or the primary ion beam, while the ions produced in sputtering the solid are the secondary ions. Most elements in the periodic table produce secondary ions, and SIMS can quantitatively detect elemental concentrations in the part-per-million to part-per-billion range.

Secondary ions are formed by kinetic and chemical ionization processes in which sputtering is achieved by energy transfer from the primary ions to the solid surface. Typically, SIMS primary ions impact the surface with kinetic energies of 5–20 keV, and this energy is ultimately transferred to the sample atoms and molecules. The energy transfer initiates a collision cascade within the solid that ejects atoms and molecules at the solid surface-vacuum interface.

SIMS analyses are divided into two broad categories known as dynamic and static. In the dynamic type, the most common SIMS method, a relatively intense primary ion beam sputters the sample surface at high sputter rates, providing a very useful way to determine the in-depth concentration of different elements in a solid. The most common secondary ions detected in a dynamic secondary ion mass spectrometry analysis are elemental ions or clusters of elemental ions.

Static or molecular SIMS utilizes a very low intensity primary ion beam, and static analyses are typically completed before a single monolayer has been removed from the surface. Most static analyses are stopped before 1% of the top surface layer has been chemically damaged or eroded; under these conditions, molecular and molecular fragment ions characteristic of the chemical structure of the surface are often detected. Thus, static SIMS is best suited for near-surface analysis of molecular composition or chemical structure information, while dynamic SIMS provides the best technique for bulk and in-depth elemental analysis. *See* SPUTTERING.

Instrument designs for SIMS require a primary ion gun or column to generate and transport the primary ion beam to the sample surface, a sample chamber with sample mounting facilities, a mass spectrometer which performs mass-to-charge separation of the different secondary ions, and an ion-detection system. The complete instrument is typically housed in an ultrahigh vacuum chamber. *See* ION SOURCES; MASS SPECTROSCOPE.

Dynamic SIMS has been successfully utilized in diverse applications. The analytical issues are the detection and localization of specific elements on the surface or in the bulk of the materials. The most important and common application area of dynamic SIMS is the bulk and in-depth analysis of semiconductor materials. Another important contribution of SIMS to materials science is the identification and localization of trace elements in metal grain boundaries, providing detailed insight into the chemistry of welds and alloys. Dynamic SIMS also has the unique ability to detect specific catalyst poisons on the surface and in the bulk of used or spent catalysts. In geological sciences, SIMS has been used to detect isotopic anomalies in the composition of various geological and meteoritic samples to help determine the age of the universe. Dynamic SIMS has found extensive applications to the characterization of both hard and soft biological tissue. *See* SEMICONDUCTOR.

Applications of static secondry ion mass spectrometry to analyses of the near-surface region of solids has become increasingly important as ever more sophisticated materials are developed. The near-surface region is generally defined as the top five monolayers (2–3 nm) of a solid. Examples of technology areas in which the chemistry of the near-surface region plays a critical role include high-performance glass coatings, liquid-crystal displays, manufacturing, semiconductor processing, biopolymer and biocompatible materials development, and polymer adhesives and coatings. *See* MASS SPECTROMETRY; MATERIALS SCIENCE AND ENGINEERING; SEMICONDUCTOR DIODE.

Robert W. Odom

Secretion The export of proteins by cells. With few exceptions, in eukaryotic cells proteins are exported via the secretory pathway, which includes the endoplasmic reticulum and the Golgi apparatus. Secreted proteins are important in many physiological processes, from the transport of lipids and nutrients in the blood, to the digestion of food in the intestine, to the regulation of metabolic processes by hormones. *See* CELL (BIOLOGY); CELL ORGANIZATION; ENDOCRINE MECHANISMS.

Proteins destined for export are synthesized on ribosomes attached to the outside of the rough endoplasmic reticulum, a portion of the endoplasmic reticulum that is specialized for the synthesis of secretory proteins and most of the cell's membrane proteins. After they are folded, the proteins enter small vesicles in which they are transported to the Golgi apparatus. When the proteins reach the last cisterna of the Golgi, a highly tubulated region known as the *trans*-Golgi network, they are sorted and packaged again into transport vesicles, some of which are in the form of elongated tubules. From here, there are two pathways that proteins can take to the cell surface, depending on the cell type. Proteins can be transported directly to the plasma membrane (constitutive secretion) or to secretory granules (regulated secretion). *See* ENDOPLASMIC RETICULUM; GOLGI APPARATUS.

In all cells, there exists a constitutive secretion pathway whereby vesicles and tubules emerging from the *trans*-Golgi network fuse rapidly with the plasma membrane. The emerging vesicles and tubules attach to microtubules, cytoskeletal elements emanating from the Golgi region, that accelerate their transport to the plasma membrane. *See* ABSORPTION (BIOLOGY); CELL MEMBRANES.

In cells that secrete large amounts of hormones or digestive enzymes, most secretory and membrane proteins emerging from the *trans*-Golgi network are not immediately secreted, but are stored in membrane-bounded secretory granules. Secretory granules release their contents into the extracellular space in a process known as exocytosis, when their membranes fuse with the plasma membrane. Exocytosis occurs only after the cell receives a signal, usually initiated by the binding of a hormone or neurotransmitter to a receptor on the cell surface. The receptor triggers a signal transduction cascade that results in increased concentrations of second messengers such as cyclic adenosine 3′,5′-monophosphate and phosphatidylinositol triphosphate. In most secretory cells, the second messengers or the hormone receptors themselves trigger the opening of calcium channels through which calcium ions stream into the cytoplasm. Calcium initiates the docking of the secretory granules with the plasma membrane and the activation of the fusion apparatus. *See* ENZYME; HORMONE; SIGNAL TRANSDUCTION.

In exceptional cases, proteins can be exported directly from the cytoplasm without using the secretory pathway. One such protein is fibroblast growth factor, a hormone involved in the growth and development of tissues such as bone and endothelium. Several interleukins, proteins that regulate the immune response, are also released via an unconventional route that may involve transport across the plasma membrane through channel proteins. These channels have adenosine 5′-triphosphatase (ATPase) enzyme activity and use the energy derived from the hydrolysis of ATP to catalyze transport. *See* CELLULAR IMMUNOLOGY.

Michael Rindler

Secretory structures (plant) Cells or organizations of cells which produce a variety of secretions. The secreted substance may remain deposited within the secretory cell itself or may be excreted, that is, released from the cell. Substances may be excreted to the surface of the plant or into intercellular cavities or canals. Some of the many substances contained in the secretions are not further utilized by the plant (resins, rubber, tannins, and various crystals), while others take part in the functions of the plant (enzymes and hormones). Secretory structures range from single cells scattered among other kinds of cells to complex structures involving many cells; the latter are often called glands.

Epidermal hairs of many plants are secretory or glandular. Such hairs commonly have a head composed of one or more secretory cells borne on a stalk. The hair of a stinging needle is bulbous below and extends into a long, fine process above. If one touches the hair, its tip breaks off, the sharp edge penetrates the skin, and the poisonous secretion is released.

Glands secreting a sugary liquid—the nectar—in flowers pollinated by insects are called nectaries. Nectaries may occur on the floral stalk or on any floral organ: sepal, petal, stamen, or ovary.

The hydathode structures discharge water—a phenomenon called guttation—through openings in margins or tips of leaves. The water flows through the xylem to its endings in the leaf and then through the intercellular spaces of the hydathode tissue toward the openings in the epidermis. Strictly speaking, such hydathodes are not glands because they are passive with regard to the flow of water.

Some carnivorous plants have glands that produce secretions capable of digesting insects and small animals. These glands occur on leaf parts modified as insect-trapping structures. In the sundews (*Drosera*) the traps bear stalked glands, called tentacles. When an insect lights on the leaf, the tentacles bend down and cover the victim with a mucilaginous secretion, the enzymes of which digest the insect. *See* INSECTIVOROUS PLANTS; VENUS' FLYTRAP.

Resin ducts are canals lined with secretory cells that release resins into the canal. Resin ducts are common in gymnosperms and occur in various tissues of roots, stems, leaves, and reproductive structures.

Gum ducts are similar to resin ducts and may contain resins, oils, and gums. Usually, the term gum duct is used with reference to the dicotyledons, although gum ducts also may occur in the gymnosperms.

Oil ducts are intercellular canals whose secretory cells produce oils or similar substances. Such ducts may be seen, for example, in various parts of the plant of the carrot family (Umbelliferae).

Laticifers are cells or systems of cells containing latex, a milky or clear, colored or colorless liquid. Latex occurs under pressure and exudes from the plant when the latter is cut.

Katherine Esau

Sedative A medication capable of producing a mild state of inhibition of the central nervous system (CNS) associated with reduced awareness of external stimuli. Numerous pharmacologic agents can induce different degrees of sedation, depending on the following variables: dosage; route of administration; absorption, metabolism, and excretion rates of the compound; specific receptor sites in the central nervous system that are affected by the agent; environmental setting; and state of the patient. *See* CENTRAL NERVOUS SYSTEM.

Ethanol was probably the first sedative compound and was widely used for its analgesic and hypnotic properties, as well as for its ability to decrease inhibitory anxiety with resultant relaxation and occasional euphoria. Subsequent sedative compounds include the barbiturates (for example, phenobarbital and secobarbital) and the benzodiazepines (for example, diazepam and alprazolam). Although these sedative compounds possess properties of tolerance and habituation, they vary in their addictive potential according to specific receptor sites and the particular type of patient. *See* ADDICTIVE DISORDERS; BARBITURATES.

Other classes of chemical agents that are used as sedatives include the antihistamines as well as some antidepressant drugs which possess sedative side effects in addition to their primary pharmacologic properties. Since these classes of compounds are not addicting, they can be safely used as hypnotics. *See* ANXIETY DISORDERS; TRANQUILIZER.

Donald M. Gallant

Sedentaria A group of 28 or more families of polychaete annelids in which the anterior, or cephalic, region is more or less completely concealed by overhanging peristomial structures, or the body is divided into an anterior thoracic and a posterior abdominal region; the pharynx or proboscis is usually soft and epithelial, lacking hard jaws or paragnaths. *See* POLYCHAETA.

Olga Hartman

Sedimentary rocks Rocks that accumulate at the surface of the Earth, under ambient temperatures. Together with extruded hot lavas, sedimentary rocks form a thin cover of stratified material (the stratisphere) over the deep-seated igneous and metamorphic rocks that constitute the bulk of the Earth's crust. Sediments cover about three-quarters of the land and of the ocean floor. The thickness of the stratisphere is generally measured in kilometers, and locally reaches about 15 km (50,000 ft). *See* EARTH CRUST; IGNEOUS ROCKS; METAMORPHIC ROCKS; ROCK.

Most sediments accumulate as sand and dust or mud. Being deposited from fluids (air, water) under the influence of gravity, they tend to assume level surfaces (though locally steep slopes may be developed, as in dunes and reefs). Changes in supply of sediment and in depositing agencies change the nature of the deposits from day to day and from millennium to millennium, and commonly interrupt the process altogether. As a result, the accumulated mantle of sediment has a layered structure, divided into beds or strata. Sediments become compacted as waters are squeezed out of them during burial and tectonism, and become cemented as remaining pore space becomes filled by newly growing minerals, mainly calcite or quartz. Bacterial degradation of organic matter, invasion by other fluids, and changes in temperature continue to alter the chemical environment, and lead to alteration of unstable mineral phases. Such processes are included in the term diagenesis. Soft sediment thus becomes converted to rock, but the geologist includes both in the concept of sedimentary rocks. *See* CALCITE; DIAGENESIS; QUARTZ.

When sediments are carried to greater depths or are otherwise subjected to high heat or pressure, growth of new minerals and plastic deformation destroy sedimentary structures and metamorphose the rock. Alternatively, the sediment melts in transition to igneous rock. Thus, sedimentary rocks are recycled through geologic time. Most of the crust under the continents, consisting of igneous and metamorphic rocks, has probably passed through the sedimentary state at some point. Despite such losses, sedimentary rocks have locally survived from very early (Archean) times, nearly 4 billion years ago. *See* ARCHEAN.

Sediments are almost entirely derived from transfer of materials within the Earth's crust. First in importance is gradation, the wearing away of the highlands and the deposition of the products in the low spots: subsiding basins and the oceans. Second is crustal volcanism, which produces large ash falls from explosive volcanoes, and recycles ions to the surface in hot springs. Small amounts are contributed from the mantle underlying the crust: mainly pumice produced when mantle-derived oceanic basalts interact with water. A small fraction of sediment consists of organic matter created by organisms from carbon dioxide and water. Water frozen in the atmosphere transiently covers parts of the stratisphere with ice, while traces of extraterrestrial matter continue to be added from meteorites. *See* COSMIC SPHERULES; METEORITE; WEATHERING PROCESSES.

Though sediments contain such a large range of diverse constituents occurring in a wide variety of mixtures, such mixtures are generally dominated by one or two constituents, and thus may be grouped into a number of classes, each of which can be divided into families.

Detrital sediments are alternately transported and deposited, reeroded, and redeposited on their way to a more permanent resting place, so that their constituents may carry the imprints of a complex history, while the structure of the deposit testifies to the last depositional episode. *See* DEPOSITIONAL SYSTEMS AND ENVIRONMENTS.

Pyroclastic sediments originate from volcanic vents. Submarine eruptions form pumice, or frothy glass, much of which floats widely. The important contributions are great eruptions of glass droplets are ejected into atmosphere and stratosphere to fall as a rain of pumice, sand, and silt, in some cases mixed with crystals. Pyroclastic rocks, largely composed of glass, are readily altered to clay minerals (montmorillonite) in weathering. They produce excellent soils. Beds of montmorillonite (bentonites) are mined for preparation of artificial muds such as those used in well drilling. *See* MONTMORILLONITE; PUMICE; VOLCANO; VOLCANOLOGY.

Chemical sediments represent the precipitation of materials carried in solution, either by simple chemical precipitation or by the activity of organisms.

Carbonate rocks form about 20% of all sediments. In natural waters, calcium and magnesium are mainly held in solution by virtue of carbon dioxide. In many fresh waters and in the surficial ocean, withdrawal of carbon dioxide—by warming of the water or by the consumption of carbon dioxide in green-plant photosynthesis—leads to supersaturation and to the deposition of calcium carbonate. This normally yields a lime mud of microscopic crystals. Even more important is the secretion of calcium carbonate skeletons, ultimately deposited on ocean floors, by some algae and by a large variety of animals, ranging from microscopic foraminifera to corals and molluscan shells. Carbonate rocks are a major ingredient of portland cement. They are crushed in large quantities for use in road building, agriculture, and smelting, and in the chemical industry. They also furnish building and ornamental stone. Carbonate rocks contain a large share of the world's petroleum resources. *See* ARAGONITE; CALCITE; CEMENT; DOLOMITE; LIMESTONE; OOLITE; STYLOLITES.

Evaporites are formed in bays, estuaries, and lakes of arid regions. On progressive evaporation, seawater first forms deposits of calcium sulfate as gypsum or anhydrite, followed by halite (NaCl) and ultimately potash and magnesium salts. Evaporation of lake water may yield different precipitates such as trona, borax, and silicates. Much of what is sold as table salt is mined from evaporite deposits, as is potash fertilizer. Plaster of paris is produced from gypsum or anhydrite, and the chemical industry relies on evaporite deposits of various types. *See* FERTILIZER; GYPSUM; HALITE; PLASTER OF PARIS; SALINE EVAPORITES; SALT (FOOD).

Nondetrital siliceous rocks such as silicon dioxide (silica) is second only to carbonate in the dissolved load of most streams. Organisms take up nearly all silica supplied, covering much of the deep-sea floor with radiolarian and diatomaceous ooze. Over geological time spans, diagenetic alteration converts these into dull white opal-ct or quartz porcellanites, or into the solid, waxy-looking mosaics of fine quartz grains known as chert or flint. Diatom ooze is mined for abrasives and filters, as well as for insulation. *See* CHERT; SILICA MINERALS.

Carbonaceous sediments are the result of organic activity, and are of two sorts: the peat-coal series and the kerogens. Peat is used for local fuel in boggy parts of the world. Lignite and bituminous coals continue to be important fuels. *See* COAL; LIGNITE; PEAT.

A. G. Fischer

Sedimentation (industry) The separation of a dilute suspension of solid particles into a supernatant liquid and a concentrated slurry. If the purpose of the process is to concentrate the solids, it is termed thickening; and if the goal is the removal of the solid particles to produce clear liquid, it is called clarification. Thickening is the common operation for separating fine solids from slurries. Examples are magnesia, alumina red mud, copper middlings and concentrates, china clay (kaolin), coal tailings, phosphate slimes, and pulp-mill and other industrial wastes. Clarification is prominent in the treatment of municipal water supplies.

The driving force for separation is the difference in density between the solid and the liquid. Ordinarily, sedimentation is effected by the force of gravity, and the liquid is water or an aqueous solution. For a given density difference, the solid settling process proceeds more rapidly for larger-sized particles. For fine particles or small density differences, gravity settling may be too slow to be practical; then centrifugal force rather than gravity can be used. Further, when centrifugal force is inadequate, the more positive method of filtration may be employed. All those methods of separating solids and liquids belong to the generic group of mechanical separations. *See* CENTRIFUGATION; CLARIFICATION; FILTRATION; THICKENING.

Particles too minute to settle at practical rates may form flocs by the addition of agents such as sodium silicate, alum, lime, and alumina. Because the agglomerated particles act like a single large particle, they settle at a feasible rate and leave a clear liquid behind. Vincent W. Uhl

Sedimentology The study of natural sediments, both lithified (sedimentary rocks) and unlithified, and of the processes by which they are formed. Sedimentology includes all those processes that give rise to sediment or modify it after deposition: weathering, which breaks up or dissolves preexisting rocks so that sediment may form from them; mechanical transportation; deposition; and diagenesis, which modifies sediment after deposition and burial within a sedimentary basin and converts it into sedimentary rock. Sediments deposited by mechanical processes (gravels, sands, muds) are known as clastic sediments, and those deposited predominantly by chemical or biological processes (limestones, dolomites, rock salt, chert) are known as chemical sediments. *See* SEDIMENTARY ROCKS; WEATHERING PROCESSES.

The raw materials of sedimentation are the products of weathering of previously formed igneous, metamorphic, or sedimentary rocks. In the present geological era, 66% of the continents and almost all of the ocean basins are covered by sedimentary rocks. Therefore, most of the sediment now forming has been derived by recycling previously formed sediment. Identification of the oldest rocks in the Earth's crust, formed more than 3×10^9 years ago, has shown that this process has been going on at least since then. Old sedimentary rocks tend to be eroded away or converted into metamorphic rocks, so that very ancient sedimentary rocks are seen at only a few places on Earth. *See* EARTH CRUST; IGNEOUS ROCKS; METAMORPHIC ROCKS.

Major controls. The major controls on the sedimentary cycle are tectonics, climate, worldwide (eustatic) changes in sea level, the evolution of environments with geological time, and the effect of rare events.

Tectonics are the large-scale motions (both horizontal and vertical) of the Earth's crust. Tectonics are driven by forces within the interior of the Earth but have a large effect on sedimentation. These crustal movements largely determine which areas of the Earth's crust undergo uplift and erosion, thus acting as sources of sediment, and which areas undergo prolonged subsidence, thus acting as sedimentary basins. Rates of uplift may be very high (over 10 m or 33 ft per 1000 years) locally, but probably such rates prevail only for short periods of time. Over millions of years, uplift even in mountainous regions is about 1 m (3.3 ft) per 1000 years, and it is closely balanced by rates of erosion. Rates of erosion, estimated from measured rates of sediment transport in rivers and from various other techniques, range from a few meters per 1000 years in mountainous areas to a few millimeters per 1000 years averaged over entire continents. *See* BASIN; PLATE TECTONICS.

Climate plays a secondary but important role in controlling the rate of weathering and sediment production. The more humid the climate, the higher these rates are. A combination of hot, humid climate and low relief permits extensive chemical weathering, so that a larger percentage of source rocks goes into solution, and the clastic sediment produced consists mainly of those minerals that are chemically inert (such as quartz) or that are produced by weathering itself (clays). Cold climates and high relief favor physical over chemical processes. *See* CLIMATOLOGY.

Tectonics and climate together control the relative level of the sea. In cold periods, water is stored as ice at the poles, which can produce a worldwide (eustatic) lowering of sea level by more than 100 m (330 ft). Changes in sea level, whether local or worldwide, strongly influence sedimentation in shallow seas and along coastlines; sea-level changes also affect sedimentation in rivers by changing the base level below which a stream cannot erode its bed.

One of the major conclusions from the study of ancient sediments has been that the general nature and rates of sedimentation have been essentially unchanged during the last billion years of geological history. However, this conclusion, uniformitarianism, must be qualified to take into account progressive changes in the Earth's environment through geological time, and the operation of rare but locally or even globally important catastrophic events. The most important progressive changes have been in tectonics and atmospheric chemistry early in Precambrian times, and in the nature of life on the Earth, particularly since the beginning of the Cambrian. *See* CAMBRIAN.

Throughout geological time, events that are rare by human standards but common on a geological time scale, such as earthquakes, volcanic eruptions, and storms, produced widespread sediment deposits. There is increasing evidence for a few truly rare but significant events, such as the rapid drying up of large seas (parts of the Mediterranean) and collisions between the Earth and large meteoric or cometary bodies (bolides).

Sediment is moved either by gravity acting on the sediment particles or by the motions of fluids (air, water, flowing ice), which are themselves produced by gravity. Deposition takes place when the rate of sediment movement decreases in the direction of sediment movement; deposition may be so abrupt that an entire moving mass of sediment and fluid comes to a halt (mass deposition, for example, by a debris flow), or so slow that the moving fluid (which may contain only a few parts per thousand of sediment) leaves only a few grains of sediment behind. The settling velocity depends on the density and viscosity of the fluid, as well as on the size, shape, and density of the grains. *See* STREAM TRANSPORT AND DEPOSITION.

Chemical sedimentation. Chemical weathering dissolves rock materials and delivers ions in solution to lakes and the ocean. The concentrations of ions in river and ocean water are quite different, showing that some ions must be removed by sedimentation. Comparison of the modern rate of delivery of ions to the ocean, with their concentration in the oceans, shows that some are removed very rapidly (residence times of only a few thousand years) whereas others, such as chlorine and sodium, are removed very slowly (residence times of hundreds of millions of years).

Biological effects. Many so-called chemical sediments are actually produced by biochemical action. Much is then reworked by waves and currents, so that the chemical sediment shows clastic textures and consists of grains rounded and sorted by transport. Depositional and diagenetic processes, however, are often strongly affected by organic action, no matter what the origin of the sediment. Plants in both terrestrial and marine environments tend to trap sediment, enhancing deposition and slowing erosion.

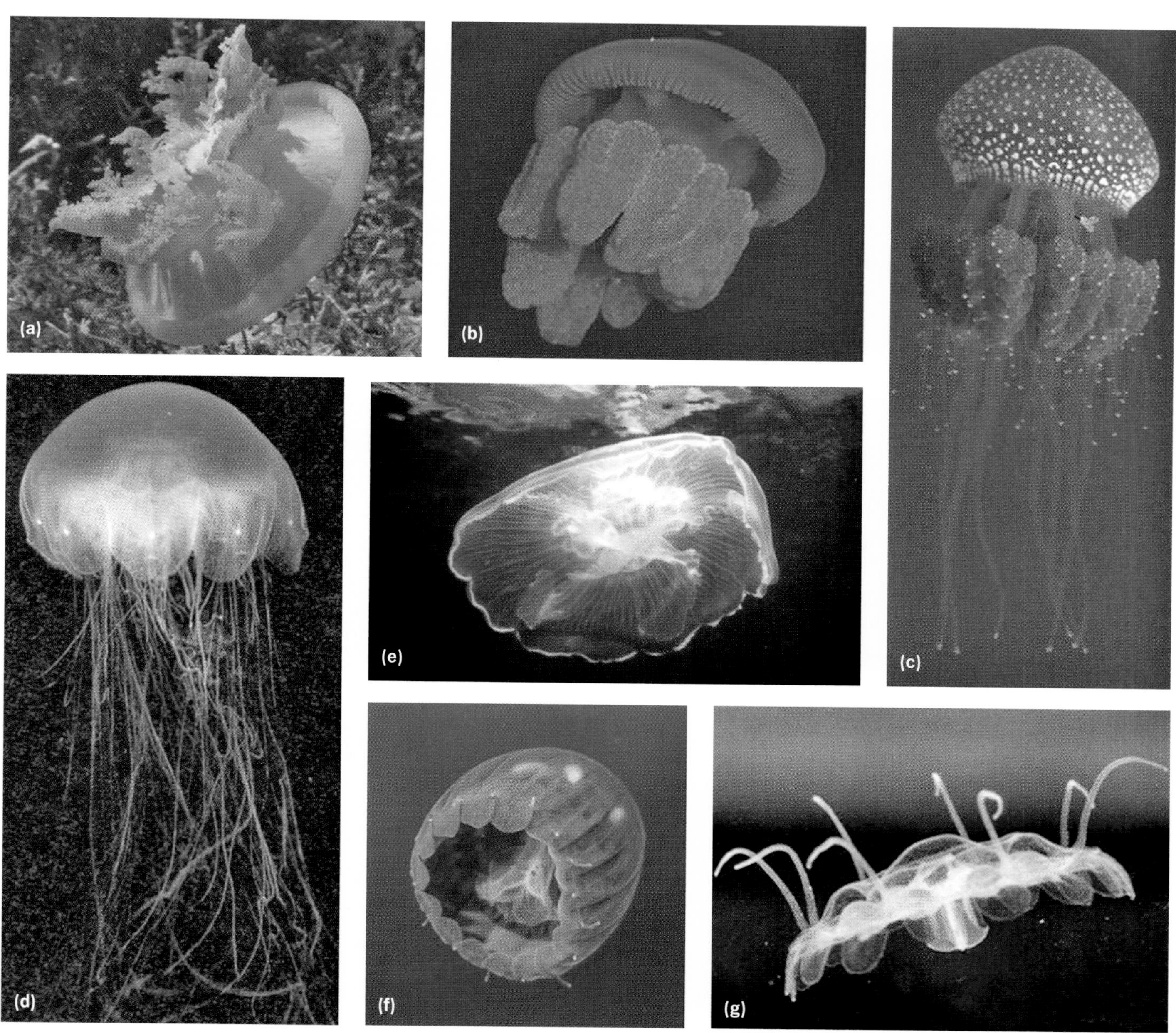

Some common genera of Scyphomedusae.

Order Rhizostomeae: (*a*) *Cassiopeia* (*photo by L.E. Martin*), (*b*) *Catostylus* (*photo by K. Pitt*), (*c*) *Phyllorhiza* (*photo by K. Pitt*).

Order Semaeostomeae: (*d*) *Cyanea* (*photo by G. Edgar, reproduced with permission from Edgar, 2000*), (*e*) *Aurelia*.

Order Coronatae: (*f*) *Linuche* (*photo by L.E. Martin*), (*g*) *Nausithoe* (*photo by A. Migotto*).

Sedimentary environments and facies. Sedimentary rocks preserve the main direct evidence about the nature of the surface environments of the ancient Earth and the way they have changed through geological time. Thus, besides trying to understand the basic principles of sedimentation, sedimentologists have studied modern and ancient sediments as records of ancient environments. For this purpose, fossils and primary sedimentary structures are the best guide. These structures are those formed at the time of deposition, as opposed to those formed after deposition by diagenesis, or by deformation. In describing sequences of sedimentary rocks in the field (stratigraphic sections), sedimentologists recognize compositional, structural, and organic aspects of rocks that can be used to distinguish one unit of rocks from another. Such units are known as sedimentary facies, and they can generally be interpreted as having formed in different environments of deposition. Though there are a large number of different sedimentary environments, they can be classified in a number of general classes, and their characteristic facies are known from studies of modern environments. *See* FACIES (GEOLOGY); STRATIGRAPHY; TRACE FOSSILS. G. V. Middleton

Seebeck effect

The generation of a temperature-dependent electromotive force (emf) at the junction of two dissimilar metals. This phenomenon provides the physical basis for the thermocouple. In 1821, T. J. Seebeck discovered that near a closed circuit composed of two linear conductors of two different metals a magnetic needle would be deflected if, and only if, the two junctions were at different temperatures, and that if the temperatures of the two junctions were reversed the direction of deflection would also be reversed. He investigated 35 different metals and arranged them in a series such that at a hot junction, current flows from a metal earlier in the series to a later one. *See* ELECTROMOTIVE FORCE (EMF).

A thermocouple consists of a pair of wires of dissimilar metals, joined at the ends. One junction is kept at an accurately known cold temperature, usually that of melting ice, and the other is used for the measurement of an unknown temperature, by measuring the emf generated as a result of the Seebeck effect. *See* THERMOCOUPLE; THERMOELECTRICITY. A. Earle Bailey

Seed

A fertilized ovule containing an embryo which forms a new plant upon germination. Seed-bearing characterizes the higher plants—the gymnosperms (conifers and allies) and the angiosperms (flowering plants). Gymnosperm (naked) seeds arise on the surface of a structure, as on a seed scale of a pine cone. Angiosperm (covered) seeds develop within a fruit, as the peas in a pod. *See* FLOWER; FRUIT.

Structure. One or two tissue envelopes, or integuments, form the seed coat which encloses the seed except for a tiny pore, the micropyle (see illustration). The micropyle is near the funiculus (seed stalk) in angiosperm seeds. The hilum is the scar left when the seed is detached from the funiculus. Some seeds have a raphe, a ridge near the hilum opposite the micropyle, and a bulbous strophiole. Others such as nutmeg possess arils, outgrowths of the funiculus, or a fleshy caruncle developed from the seed coat near the hilum, as in the castor bean. The embryo consists of an axis and attached cotyledons (seed leaves). The part of the axis above the cotyledons is the epicotyl (plumule); that below, the hypocotyl, the lower end of which bears a more or less developed primordium of the root (radicle). The epicotyl, essentially a terminal bud, possesses an apical meristem (growing point) and, sometimes, leaf primordia. The seedling stem develops from the epicotyl. An apical meristem of the radicle produces the primary root of the seedling, and transition between root and stem occurs in the hypocotyl. *See* APICAL MERISTEM; ROOT (BOTANY); STEM.

Two to many cotyledons occur in different gymnosperms. The angiosperms are divided into two major groups according to number of cotyledons: the monocotyledons and the dicotyledons. Mature gymnosperm seeds contain an endosperm (albumen or nutritive tissue) which surrounds the embryo. In some mature dicotyledon seeds the endosperm persists, the cotyledons are flat and leaflike, and the epicotyl is simply an apical meristem. In other seeds, such as the bean, the growing embryo absorbs the endosperm, and food reserve for germination is stored in fleshy cotyledons. The endosperm persists in common monocotyledons, for example, corn and wheat; and the cotyledon, known as the scutellum, functions as an absorbing organ during germination. Grain embryos also possess a coleoptile and a coleorhiza sheathing the epicotyl and the radicle, respectively. The apical meristems of lateral seed roots also may be differentiated in the embryonic axis near the scutellum of some grains.

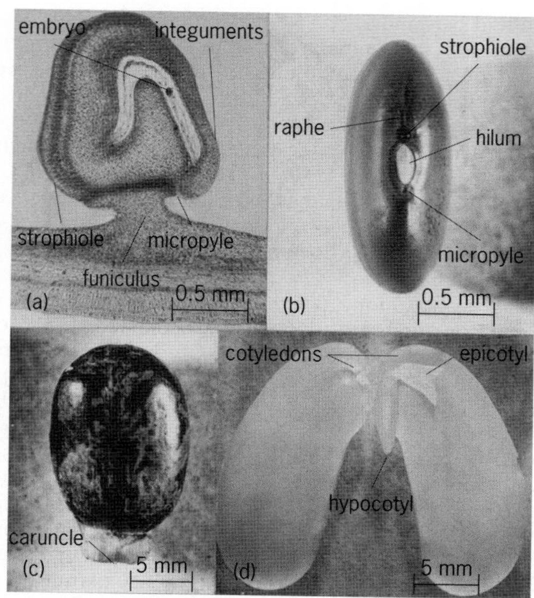

Seed structures. (*a*) Median longitudinal section of pea ovule shortly after fertilization, showing attachment to pod tissues. (*b*) Mature kidney bean. (*c*) Mature castor bean. (*d*) Opened embryo of mature kidney bean.

Many so-called seeds consist of hardened parts of the fruit enclosing the true seed which has a thin, papery seed coat. Among these are the achenes, as in the sunflower, dandelion, and strawberry, and the pits of stone fruits such as the cherry, peach, and raspberry. Many common nuts also have this structure. Mechanisms for seed dispersal include parts of both fruit and seed. *See* POPULATION DISPERSAL.

Economic importance. Propagation of plants by seed and technological use of seed and seed products are among the most important activities of modern society. Specializations of seed structure and composition provide rich sources for industrial exploitation apart from direct use as food. Common products include starches and glutens from grains, hemicelluloses from guar and locust beans, and proteins and oils from soybeans and cotton seed. Drugs, enzymes, vitamins, spices, and condiments are obtained from embryos, endosperms, and entire seeds, often including the fruit coat. Most of the oils of palm, olive, and pine seeds are in the endosperm. Safflower seed oil is obtained mainly from the embryo, whereas both the seed coat and embryo of cotton seed are rich in oils. *See* FOOD; PLANT ANATOMY; PLANT REPRODUCTION. Roger M. Reeve

Physiology. Physical and biochemical processes of seed growth and germination are controlled by genetic and environmental factors. Conditions of light, temperature, moisture, and oxygen affect the timing and ability of a seed to mature and germinate. Seed development (embryogenesis) is concerned with the synthesis and storage of carbohydrate, protein, and oil to supply nutrients to the germinating seedling prior to soil emergence. Seed development occurs in several stages: rapid cell division, seed fill, and desiccation. The timing of each stage is species-specific and environmentally influenced.

Dormancy. Seed dormancy is the inability of a living seed to germinate under favorable conditions of temperature, moisture, and oxygen. Dormancy does not occur in all seeds, but typically occurs in plant species from temperate and colder habitats. This process allows for a delay in seed germination until environmental conditions are adequate for seedling survival. At least three types of seed dormancy are recognized: primary, secondary (induced), and enforced. Primary dormancy occurs during seed maturation, and the seed does not germinate readily upon being shed. Secondary and enforced dormancy occur after the seed is shed and may be caused by adverse environmental factors such as high or low temperature, absence of oxygen or light, low soil moisture, and presence of chemical inhibitors. Seeds with secondary dormancy will not germinate spontaneously when environmental conditions improve, and need additional environmental stimuli. Seeds with enforced dormancy germinate readily upon removal of the environmental limitation. Regulation of dormancy may be partly controlled by hormones. *See* ABSCISIC ACID; DORMANCY.

Dormancy is terminated in a large number of species when an imbibed seed is illuminated with white light. Biochemical control of this process is related to the functioning of a single pigment, phytochrome, frequently located in the seed coat or embryonic axis. Phytochrome imparts to the seed the ability to interpret light quality, such as that under an existing vegetative canopy, and to distinguish light from dark with respect to its position in the soil. Phytochrome also is affected by temperature and is involved in the seasonal control of the ending of dormancy. Hormones that promote germination of dormant seeds include gibberellins, cytokinins, ethylene, and auxins. *See* AUXIN; CYTOKININS; ETHYLENE; GIBBERELLIN.

Germination. Germination is the process whereby a viable seed takes up water and the radicle (primary root) or hypocotyl emerges from the seed under species-specific conditions of moisture, oxygen, and temperature. Dormant seeds must undergo additional environmental stimuli to germinate. The germinating seed undergoes cell expansion, as well as increases in respiration, protein synthesis, and other metabolic activities prior to emergence of the growing seedling. Cadance A. Lowell

Seiche A short-period oscillation in an enclosed or semienclosed body of water, analogous to the free oscillation of water in a dish. The initial displacement of water from a level surface can arise from a variety of causes, and the restoring force is gravity, which always tends to maintain a level surface. Once formed, the oscillations are characteristic only of the geometry of the basin itself and may persist for many cycles before decaying under the influence of friction. The term "seiche" appears to have been first used to describe the rhythmic oscillation of the water surface in Lake Geneva, which occasionally exposed large areas of the lake bed that are normally submerged. *See* WAVE MOTION IN LIQUIDS.

Seiches can be generated when the water is subject to changes in wind or atmospheric pressure gradients or, in the case of semi-enclosed basins, by the oscillation of adjacent connected water bodies having a periodicity close to that of the seiche or of one of its harmonics. Other, less frequent causes of seiches include heavy precipitation over a portion of the lake, flood discharge from rivers, seismic disturbances, submarine mudslides or slumps, and tides. The most dramatic seiches have been observed after earthquakes. Alfred Wüest; David M. Farmer

Seismic exploration for oil and gas Prospecting for oil and gas using exploration seismology, a geophysical method of determining geologic structure by means of prospector-induced elastic waves. By studying body waves such as compressional and shear waves propagating through the Earth's interior, the constituent and elastic properties of its solid and liquid core, its solid mantle, and its thin crust are defined. The major differences between earthquake seismology and petroleum exploration seismology are scales and knowledge of the location of seismic disturbances. Earthquake seismology studies naturally generated seismic waves, which have periods in minutes and resolution in kilometers. In exploration seismology, artificial sources are used that have periods of tenths of a second and tens of meters of resolution. Production seismology requires higher-frequency seismic waves and better resolution, often resolution in the order of a few meters. *See* EARTH; EARTH INTERIOR.

Computer technology allows resolution of some of the theoretical complexities of elastic wave propagation so that deeper insight into the wave field phenomena can be obtained. The availability of a large number of channels in the recording instrument facilitates three-dimensional and three-component acquisitions. Powerful supercomputers allow manipulation of larger and larger data sets, and they have facilitated display and interpretation of them as a single data unit through the use of advanced computer visualization techniques. *See* COMPUTER; SUPERCOMPUTER.

The availability of powerful workstations led to the wide use of interactive processing and interpretation. When such is coupled with technically advanced algorithms, the amount of information that the interpreter can obtain from the data increases significantly. Better quality control is provided, fine-tuning analysis is achieved more easily, and the data can be enhanced to meet specific objectives. If there are discrepancies between the model and the real data, a hypothesis can be proposed based on information derived from the data. This process can be iterated until the Earth model derived is consistent with all available surface and subsurface geophysical, petrophysical, geological, and engineering data sets. *See* ALGORITHM; MODEL THEORY; SIMULATION.

The seismic method as applied to exploration of oil and gas involves field acquisition, data processing, and geologic interpretation. Seismic field acquisition requires placement of acoustic receivers (geophones) on the surface in the case of land exploration, or strings of hydrophones in the water in the case of marine exploration. Seismic data processing is usually done in large computing centers with digital mainframe computers or a large number of processors in parallel configurations. The end result of seismic data processing is the production of a subsurface profile similar to a geologic cross section. It is commonly plotted in a time scale, but it is also possible to plot it in depth. These time or depth profiles are used for geologic interpretation. Geologic interpretation of seismic data has two key components, structural and stratigraphic. Structural interpretation of seismic data involves mapping of the geologic relief of different subsurface strata by using seismic data as well as information from boreholes and outcrops. Stratigraphic interpretation looks at attributes within a common stratum and interprets changes to infer varying reservoir conditions such as lithology, porosity, and fluid content.

Historically, surface seismic acquisition is done by placing sources and receivers along a straight line so that it can be assumed that all the reflection points fall in a two-dimensional plane formed between the line of traverse and the vertical. This is known as two-dimensional seismic. Three-dimensional seismic is a method of acquiring surface seismic data by placing sources and receivers in an areal pattern. One example of a simple three-dimensional layout is to place the receivers along a line and shoot into these receivers along a path perpendicular to this line. *See* COMPUTER GRAPHICS.

A three-dimensional seismic survey provides a more accurate and detailed image of the subsurface. It offers significantly higher signal quality than the two-dimensional data commonly acquired. It also improves both spatial and temporal resolutions. The three-dimensional seismic technique is being applied to exploration and production of oil and gas, accounting for more than half of the seismic activity in the Gulf of Mexico and North Sea.

Production seismology is the application of seismic techniques to problems related to the production and exploitation of petroleum reservoirs. Since production geophysics is the only effective method available that can image the reservoirs under in-place conditions, it has become an active field of applied research aimed at improving descriptions and understanding of reservoirs and their fluid flow behaviors. *See* GEOPHYSICAL EXPLORATION; PETROLEUM ENHANCED RECOVERY; PETROLEUM GEOLOGY; PETROLEUM RESERVOIR ENGINEERING; SEISMOLOGY. Tai. P. Ng

Seismic risk The probability that social or economic consequences of earthquakes will equal or exceed specified values at a site, at several sites, or in an area, during a specified exposure time.

Although the term seismic risk is occasionally used in a general sense to mean the potential for both the occurrence of natural phenomena and the economic and life loss associated with earthquakes, it is useful to differentiate between the concepts of seismic hazard and seismic risk. Seismic hazard may be defined as any physical phenomena that result either from surface faulting during shallow earthquakes or from the ground shaking resulting from an earthquake and that may produce adverse effects on human activities.

The exposure time is the time period of interest for seismic hazard or risk calculations. In practical applications, the exposure time may be considered to be the design lifetime of a building or the length of time over which the numbers of casualties will be estimated. S. T. Algermissen

Seismic stratigraphy Determination of the nature of sedimentary rocks and their fluid content from analysis of seismic data. Seismic stratigraphy is divided into seismic-sequence (facies) analysis and reflection-character analysis.

In seismic-sequence analysis the first step is to separate seismic-sequence units, also called seismic-facies units. This is usually done by mapping unconformities where they are shown by angularity. Angularity below an unconformity may be produced by erosion at an angle across the former bedding surfaces or by toplap (offlap), and angularity above an unconformity may be produced by onlap or downlap, the latter distinction being based on geometry. The unconformities are then followed along reflections from the points where they cannot be so identified, advantage being taken of the fact that the unconformity reflection is often relatively strong. The procedure often followed is to mark angularities in reflections by small arrows before drawing in the boundaries. *See* UNCONFORMITY.

Seismic-facies units are three-dimensional, and many of the conclusions from them are based on their three-dimensional shape. The appearance on seismic lines in the dip and strike directions is often very different. For example, a fan-shaped unit might show a progradational pattern in the dip direction and discontinuous, overlapping arcuate reflections in the strike direction. *See* FACIES (GEOLOGY).

Reflection-character analysis may be based on information from boreholes which suggests that a particular interval may change nearby in a manner which increases its likelihood to contain hydrocarbon accumulations. Lateral changes in the wave shape of individual reflection events may suggest where the stratigraphic changes or hydrocarbon accumulations may be located. *See* DRILLING AND BORING, GEOTECHNICAL; SEISMIC EXPLORATION FOR OIL AND GAS.

Where sufficient information is available to develop a reliable model, expected changes are postulated and their effects are calculated and compared to observed seismic data. The procedure is called synthetic seismogram manufacture; it usually involves calculating seismic data based on sonic and density logs from boreholes, sometimes based on a model derived in some other way. The sonic and density data are then changed in the manner expected for a postulated stratigraphic change, and if the synthetic seismogram matches the actual seismic data sufficiently well, it implies that the changes in earth layering are similar to those in the model. *See* GEOPHYSICAL EXPLORATION; SEISMOGRAPHIC INSTRUMENTATION; SEISMOLOGY; STRATIGRAPHY. Robert E. Sheriff

Seismographic instrumentation Various devices or systems of devices for measuring movement in the Earth. Ground motion is generally the result of passing seismic waves, gravitational tides, atmospheric processes, and tectonic processes. Seismographic instrumentation typically consists of a sensing element (seismometer), a signal-conditioning element or elements (galvanometer, mechanical or electronic amplifier, filters, analog-to-digital conversion circuitry, telemetry, and so on), and a recording element (analog visible or direct, frequency modulation, or digital magnetic tape or disk). Seismographs are used for earthquake studies, investigations of the Earth's gravity field, nuclear explosion monitoring, petroleum exploration, and industrial vibration measurement.

Seismographic instruments may be required to measure ground motions accurately over a range approaching 12 orders of magnitude, from as small as 10^{-11} m to as large as several meters (a very large earthquake). The instruments may be required to measure frequencies as low as $\sim 10^{-5}$ Hz (the semidiurnal gravitational tides) to as high as $\sim 10^{4}$ Hz (as observed from acoustic emissions from rock failures in mines). Seismic waves from earthquakes are observed in the bandwidth of $\sim 3 \times 10^{-4}$ Hz (the gravest free oscillations of the Earth) to ~ 200 Hz (a local earthquake). In exploration seismology the frequency range of interest is typically 10–1000 Hz.

The seismometer is the basic sensing element in seismographic instruments, and there are two fundamentally different types: inertial and strain. The inertial seismometer generates an output signal that is proportional to the relative motion between its frame (usually attached to the ground or a point of interest) and an internal inertial reference mass. The strain seismometer (or linear extensometer) generates an output that is proportional to the distance between two points.

A seismoscope is a device that indicates only the occurrence of relatively strong ground shaking and not its time of occurrence or duration. A typical seismoscope inscribes a hodograph of horizontal strong ground motion on a smoked watch glass.

A dilatometer continuously and precisely measures volumetric strain. The quantity measured is the change ΔV in the reference volume V, and the ratio $\Delta V/V$ gives the volumetric strain. Dilatometers are typically installed in a borehole in competent rock (preferably granite) at a depth of 100–300 m (330–1000 ft).

A tiltmeter monitors the relative change in the elevation between two points, usually with respect to a liquid-level surface. The horizontal distance between the reference points may be as little as a few millimeters or as large as several hundred meters.

The gravity meter is just a vertical-component accelerometer, that is, a pendulum sensing ground motion and equipped with a displacement transducer, analogous to the inertial tiltmeter. Gravimeters are widely used in geophysical exploration, in the study of earth tides, and in the recording of very low frequency (0.0003–0.01 Hz) seismic waves from earthquakes. *See* ACCELEROMETER.

The complete seismograph produces a record of the properly conditioned signal from the seismometer, along with appropriate timing information. The recording system may be as simple as a mechanical stylus scratching a line on a smoke-covered drum in a portable microearthquake seismograph, or as complex as a multichannel computer-controlled system handling 25,000 24-bit digital words per second in a modern seismic reflection survey for petroleum exploration. The range between these extremes includes many special-purpose seismographs, all designed to record ground motion in a particular application. *See* EARTH TIDES; EARTHQUAKE; GEOPHYSICAL EXPLORATION; SEISMIC EXPLORATION FOR OIL AND GAS; SEISMOLOGY.

Thomas V. McEvilly; Robert A. Uhrhammer

Seismology The study of the shaking of the Earth's interior caused by natural or artificial sources. Throughout the period in which plate tectonics was advanced and its basic tenets tested and confirmed in the early 1960s, and into the latest phase of inquiry into basic processes, seismology (and particularly seismic imaging) has provided critical observational evidence upon which discoveries have been made and theory has been advanced regarding the structure of the Earth's crust, mantle, and core. *See* PLATE TECTONICS.

Theoretical seismology. A seismic source is an energy conversion process that over a short time (generally less than a minute and usually less than 1–10 s) transforms stored potential energy into elastic kinetic energy. This energy then propagates in the form of seismic waves through the Earth until it is converted into heat by internal (molecular) friction. Large sources, that is, sources that release large amounts of potential energy, can be detected worldwide. Earthquakes above Richter magnitude 5 and explosions above 50 kilotons or so are large enough to be observed globally before the seismic waves dissipate below modern levels of detection. Small charges of dynamite or small earthquakes are detectable at a distance of a few tens to a few hundreds of kilometers, depending on the type of rock between the explosion and the detector. *See* EARTHQUAKE.

Seismic vibrations are recorded by instruments known as seismometers that sense the change in the position of the ground (or water pressure) as seismic waves pass underneath. The record of ground motion as a function of time is a seismogram, which may be in either analog or digital form. Advances in computer technology have made analog recording virtually obsolete: most seismograms are recorded digitally, which makes quantitative analysis much more feasible.

The response of the Earth to a seismic disturbance can be approximated by the equation of motion for a disturbance in a perfectly elastic body. This equation holds regardless of the type of source, and is closely related to the acoustic-wave equation governing the propagation of sound in a fluid. The equation of motion for an isotropic perfectly elastic solid separates into two equations describing the propagation of purely dilatational (volume changing, curl-free) and purely rotational (no volume changing, divergence-free) disturbances. These propagate with wave speeds α and β, respectively. These velocities are also known as the compressional or primary (P) and shear or secondary (S) velocities, and the corresponding waves are called P and S waves. The compressional velocity is always faster than the shear velocity. In the Earth, α can range from a few hundred meters per second in unconsolidated sediments to more than 13.7 km/s (8.2 mi/s) just above the core–mantle boundary. Wave speed β ranges from zero in fluids (ocean, fluid outer core) to about 7.3 km/s (4.4 mi/s) at the core-mantle boundary. *See* HOOKE'S LAW; SPECIAL FUNCTIONS.

A P wave has no curl and thus only causes the material to undergo a volume change with no other distortion. An S wave has no divergence, thus causing no volume change, but right angles embedded in the material are distorted. Explosions are relatively efficient generators of compressional disturbances, but earthquakes generate both compressional and shear waves. Compressional waves, by virtue of the mechanical stability condition, always arrive before shear waves.

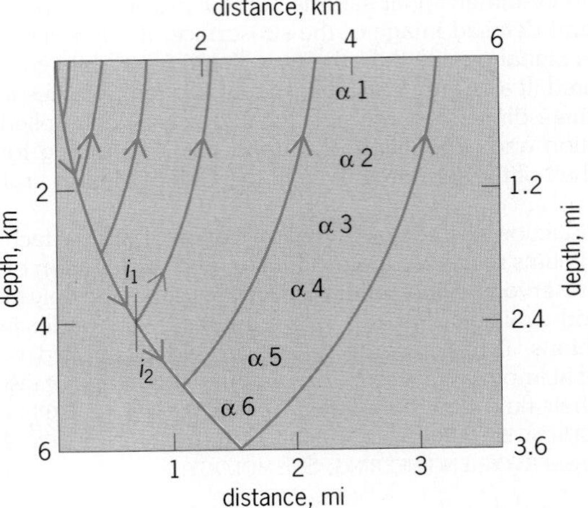

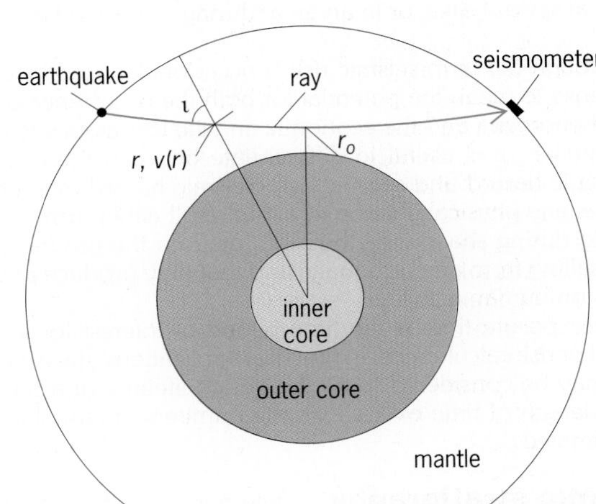

Seismic ray paths. (*a*) A single ray passing through a multilayered Earth comprising a stack of uniform velocity layers will be reflected from each layer and also be refracted as it passes from one layer into the layer below in a manner that obeys Snell's law. Each ray therefore is considered to give rise to a new system of rays. (*b*) Ray diagram for a cross section of the spherical Earth. At the point labeled $v(r)$, r = radial distance and v = velocity. r_0 = radial distance to the turning point. α is the angle of incidence.

Compressional and shear waves can exist in an elastic body irrespective of its boundaries. For this reason, seismic waves traveling with speed α or β are known as body waves. A third type of wave motion is produced if the elastic material is bounded by a free surface. The free-surface boundary conditions help trap energy near the surface, resulting in a boundary or surface wave. This in turn can be of two types. A Rayleigh wave combines both compressional and shear motion and requires only the presence of a boundary to exist. A Love wave is a pure-shear disturbance that can propagate only in the presence of a change in the elastic properties with depth from the free surface. Both are slower than body waves.

Solutions of the elastic-wave equation in which a wave function of a particular shape propagates with a particular speed are known as traveling waves. An important property of traveling waves is their causality; that is, the wave function has no amplitude before the first predicted arrival of energy. The complete seismic wavefield can be constructed by summing up every possible traveling wave.

Traveling-wave or full-wave theory provides the basis for a very useful theoretical abstraction of elastic-wave propagation in terms of the more common notions of wavefronts and their outwardly directed normals, called rays. Ray theory makes the prediction of certain kinematic quantities such as ray path, travel time, and distance by a simple geometric exercise. Ray theory can be developed in the context of an Earth comprising flat-lying layers of uniform velocities; this is a very useful approximation for most problems in crustal seismology and can be extended to spherical geometry for global studies.

Kinematic equations have been developed to describe what happens to rays as they impinge on the boundaries between layers. The illustration shows a single ray propagating in the stack of horizontal layers that define the model Earth. At each interface, part of the ray's energy is reflected, but a portion also passes through into the layer below. The transmitted portion of the ray is refracted; that is, it changes the angle at which it is propagating. The relationship between the incident angle and the refracted angle is exactly the same as that describing the refraction of light between two media of differing refractive index. *See* REFRACTION OF WAVES.

These simple geometric equations can be extended to the computation of amplitudes provided that there are no sharp discontinuities in the velocity as a function of depth. More exact representations of the amplitudes and wave shapes that solve the full-wave equation to varying extents can be constructed with the aid of powerful computers; these methods are collectively known as seismogram synthesis, and the seismograms thus computed are known as synthetics. Synthetics can be computed for elastic or dissipative media that vary in one, two, or three dimensions.

In a typical experiment for crustal imaging, a source of seismic energy is discharged on the surface, and instruments record the disturbance at numerous locations. Many different types of sources have been devised, from simple explosives to mechanical vibrators and devices known as airguns that discharge a "shot" of compressed air. The details of the source-receiver geometry vary with the type of experiment and its objective, but the work always involves collecting a large number of recordings at increasing distance from the source. This seismogram is complex, exhibiting a number of distinct arrivals with a variety of shapes and having amplitudes that change with distance. Although this seismogram clearly does not resemble the structure of the Earth in any sensible way and is therefore not what would normally be thought of as an image, it can be analyzed to recover estimates of those physical properties of the Earth that govern seismic-wave propagation.

Two- and three-dimensional imaging. A volume of the crust can be directly imaged by seismic tomography. In crustal tomography, active sources are used (explosives on land, airguns at sea) so that the source location and shape are already known. Experiments can be constructed in which sources and receivers are distributed in such a way that many rays pass through a particular volume and the tomographic inversion can produce relatively high-resolution images of velocity pertubations in the crust. Crustal tomography uses transmitted rays like those that pass from a surface source through the crust to receivers that are also on the same surface. *See* GEOPHYSICAL EXPLORATION; GROUP VELOCITY; PHASE VELOCITY; SEISMIC EXPLORATION FOR OIL AND GAS; WAVE EQUATION.

Seismic source imaging. Another imaging problem in global seismology is constructing models of the seismic source. So-called first-motion representations of seismic sources (earthquakes) are the result of measurements made on the very first P waves or S waves arriving at an instrument; therefore they represent the very beginning of the rupture on the fault plane. This is not a problem if the rupture is approximately a point source, but this is true in practice only if the earthquake is quite small or exceptionally simple. An alternative is to examine only longer-period seismic phases, including surface waves, to obtain an estimate of the average point source that smooths over the space and time complexities of a large rupture. This so-called centroid-moment-tensor representation is routinely computed for events with magnitudes greater than about 5.5. Because an estimate for a centroid moment tensor is derived from much more of the seismogram than the first arrivals, it gives a better estimate of the energy content of the earthquake. This estimate, known as the seismic moment, represents the total stress reduction resulting from the earthquake; it is the basis for a new magnitude number M_W. This value is equivalent to the Richter body wave (m_b) or surface-wave magnitude (M_S) at low magnitudes, but it is much more accurate for magnitudes above about 7.5.

Some large events comprise smaller subevents distributed in space and time and contributing to the total rupture and seismic moment. The position and individual rupture characteristics of these subevents can be mapped with remarkable precision, given data of exceptional bandwidth and good geographical distribution. An outstanding problem is whether the location of these subevents is related to stress heterogeneities within the fault zone. These stress heterogeneities are known as barriers or asperities, depending on whether they stop or initiate rupture. *See* SEISMOGRAPHIC INSTRUMENTATION.

John Mutter; Art Lerner-Lam

Seisonacea A class of the phylum Rotifera which comprises a group of little-known marine animals. They form a single family with about seven species and are found only in Europe. The Seisonacea are epizoic or possibly ectoparasitic on crustacea. They have a very elongated jointed body with a small

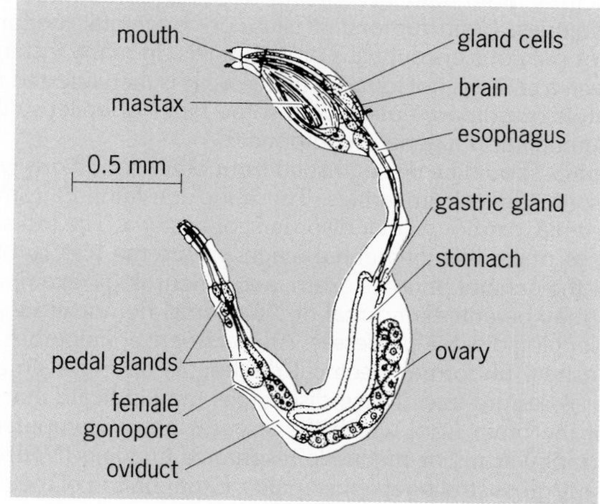

***Seison*, a rotifer.**

head; a long, slender neck region; a thicker, fusiform trunk; and an elongated foot, terminating in a perforated disk (see illustration). The Seisonacea are larger than other rotifers, attaining sizes up to 0.12 in. (3 mm) in length. *See* ROTIFERA.

Elbert H. Ahlstrom

Seizure disorders Conditions in which there are recurrent seizures. Such conditions are also known as epilepsy; the isolated occurrence of a seizure, however, is not designated as epilepsy. A seizure (ictus) is an event in which there is a sudden alteration in function of nerve cells, most commonly involving excessive electrical activity of the cells. This sudden change in nerve cell function is usually relatively brief, lasting seconds to minutes. Soon after a seizure, the brain may function quite normally. The manifestation of a seizure varies depending on which area of the brain is involved. *See* BRAIN; NERVOUS SYSTEM (VERTEBRATE).

The following are examples of the four common seizure types.

Focal motor epilepsy: This type of seizure, also known as a simple partial seizure with motor symptoms, is manifested by uncontrolled rhythmic jerking of the face, arm, or leg, caused by excessive abnormal discharges of nerve cells within the area of the brain, which under usual circumstances controls movement in that part of the body.

Temporal lobe seizures: These seizures are known as simple partial seizures and may be manifested by a myriad of symptoms depending upon which part of the lobe is involved. Psychomotor, or complex partial, seizures are the most common type. Clinically, they are characterized by an alteration in the state of consciousness and performance of repetitive, patterned, non-goal-directed activity. These types of seizures are characterized electrically by abnormal discharges occurring within the temporal lobe for the duration of the seizure.

Grand mal: These seizures, also referred to as generalized tonic-clonic convulsive seizures, major motor seizures, or convulsions, occur if the abnormal discharges involve the entire brain all at once. In this condition there are forceful, generalized, symmetrical musculature contractions accompanied by loss of consciousness, and, at times, by urinary incontinence and tongue biting.

Generalized nonconvulsive seizures: These may be atonic seizures, which are characterized by a sudden loss of muscle tone, or they may be absences (petit mal), which consist of brief periods of loss of consciousness and immediate recovery. This type of seizure is more frequently seen in children than in adults.

Epilepsy is not a disease in itself. It is a symptom of an underlying disease process. That disease process may be metabolic, such as uremia, decreased brain oxygen, or low calcium levels; or structural brain damage, such as from head trauma at birth, brain injuries, brain tumors, strokes, or congenital malformations, or previous encephalitis or meningitis. In many instances, however, a cause is not found. The disorder is then referred to as idiopathic or primary epilepsy. In some types of epilepsy there is apparently also a genetic component.

Seizures should be distinguished from other conditions which have some clinical similarities. These include fainting from hypoglycemia, cardiovascular disorders, or hysteria. The tools necessary to make this differential diagnosis are the history of the illness, the general physical exam, and a neurologic exam, all of which may be entirely normal or which may demonstrate signs of underlying disease processes. An electroencephalogram may demonstrate abnormal electrical discharges during a time the patient is seizure-free. These discharges may indicate that portion of the brain from which the seizures arise. Computerized tomography scans or magnetic resonance imaging (MRI) may show any gross structural abnormality. Examination of the blood may demonstrate abnormal circulating chemicals. *See* COMPUTERIZED TOMOGRAPHY; ELECTROENCEPHALOGRAPHY; MEDICAL IMAGING.

The ideal treatment of epilepsy is removal of the cause, such as excision of a tumor, or correction of a metabolic disorder, such as uremia or hypocalcemia. In many instances, however, the cause cannot be established or may not be amenable to direct treatment. When the cause cannot be removed, the symptoms (seizures) are treated. The initial method of obtaining seizure control is antiepileptic drugs. About 80% of the people with epilepsy obtain good control or elimination of seizures with medication. These medications are chemicals of varying structures and may be effective by a number of different brain mechanisms. Brain surgery is a therapeutic potential for some of the 20% of the people with epilepsy who do not achieve control on medication. Surgery may be considered if the abnormally discharging nerve cells which cause the epilepsy are in a dispensable area of brain, that is, one of the frontal or temporal lobes.

Linda Moretti Ojemann; Mark D. Holmes

Selachii One of two subdivisions of the subclass Elasmobranchii, consisting of nine extant orders and 34 extant families, collectively known as sharks. The other subdivision, Batoidea, includes the sawfishes, guitarfishes, skates, and rays. In the classification system below, daggers indicate extinct orders.

Class Chondrichthyes
 Subclass Elasmobranchii
  Subdivision Selachii
   †Order Cladoselachiformes
   †Order Symmoriiformes
   †Order Ctenacanthiformes
   †Order Hybodontiformes
   Order Heterodontiformes
     (bullhead sharks)
    Orectolobiformes
     (carpet sharks)
    Lamniformes
     (mackerel sharks)
    Carcharhiniformes
     (ground sharks)
    Hexanchiformes
     (frill shark and cow sharks)
    Echinorhiniformes
     (bramble sharks)
    Squaliformes
     (sleeper sharks and dogfish sharks)
    Squatiniformes
     (angel sharks)
    Pristiophoriformes
     (saw sharks)

There are several characters common to all sharks that distinguish them unequivocally from all other elasmobranchs: the gill openings are at least partly lateral (that is, they can be seen in lateral view); the edges of the pectoral fins are not attached to the sides of the head anterior to the gill openings; and the upper margin of the orbit is free from the eyeball, that is, there is a free lid. *See* BATOIDEA; ELASMOBRANCHII.

Sharks vary considerably in body form. On one end of the physiognomic spectrum are the mackerel and ground sharks with sleek fusiform bodies built for speed and power, while on the other end are angel sharks with a batoid-shaped body adapted for cruising on or near the sea floor. Sharks vary greatly in size, ranging from only 0.4 m (1.3 ft) in adult smooth dogfishes to, reputedly, 18 m (59 ft) in whale sharks.

The great majority of sharks are found in relatively shallow waters around the world, primarily tropical and subtropical latitudes of both hemispheres. Their numbers decrease in temperate waters and drop precipitously in north temperate zones.

Herbert Boschung

Selaginellales An order of the class Lycopsida (club mosses), regarded as more advanced than the extinct Protolepidodendrales because they produce spores of two sizes, but less advanced than the Lepidodendrales and Isoetales because they lack wood and finite growth. All are perennial herbs, varying in growth habit from prostrate to climbing. *See* ISOETALES; LEPIDODENDRALES; PROTOLEPIDODENDRALES.

Although the approximately 700 extant species are traditionally assigned to a single genus, *Selaginella*, it is possible that this genus should include only the few species that possess undivided steles and microphyllous leaves of a single morphology. The bulk of the selaginellalean species are included in the genus *Stachygynandrum*. Both genera are geographically widespread and occupy a wide range of habitats.

Both selaginellalean genera show several remarkable biological features (see illustration). *Stachygynandrum* has leaves that

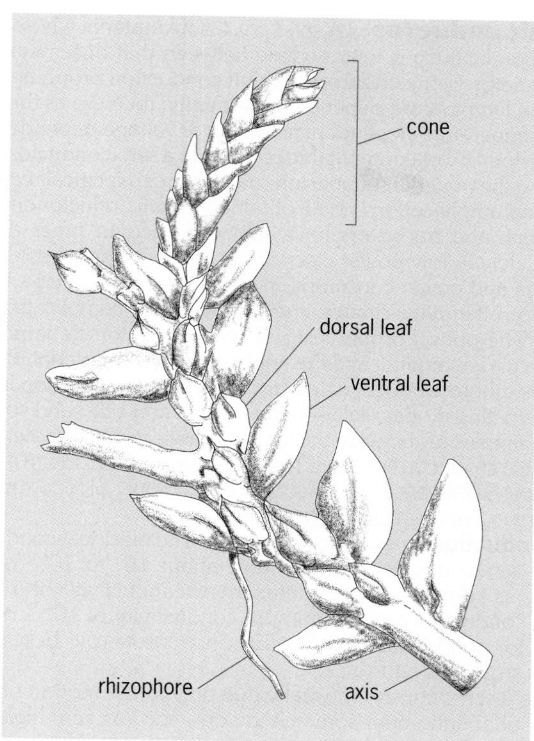

Characteristic features of a typical heterophyllous selaginellalean of the *Stachygynandrum* type. (*After T. R. Webster, Developmental problems in Selaginella in an evolutionary context, Ann. Mo. Bot. Gard., 79:633–647, 1992*)

are positioned both ventrally and dorsally. This character is regarded as an adaptation to efficient light capture. Moreover, in many species the lower surface of the dorsal leaf and upper surface of the ventral leaf consist of more or less cubic cells that contain only one large chloroplast, whose orientation within the cell can be altered to maintain optimal levels of incident light for photosynthesis.

The origin of the rhizophores (specialized aerial rooting structures) is unclear.

Selaginellaleans have a free-sporing heterosporous life history. Sporangia are borne on sporophylls that are aggregated into cones. Many species possess active mechanisms for ejecting either microspores or entire microsporangia. The megaspores and microspores germinate to produce independent, unisexual gametophytes. Both genders remain largely within the spore wall. The megagametophytes produce several archegonia that yield eggs, which chemically attract the swimming biflagellate spermatozoids released by the microspores. *See* LYCOPHYTA; LYCOPODIALES; LYCOPSIDA. Richard M. Bateman; William A. DiMichele

Selection rules (physics) General rules concerning the transitions which may occur between the states of a quantum-mechanical physical system. They derive in almost all cases from the symmetry properties of the states and of the interaction which gives rise to the transitions. The system may have a classical (nonquantum) counterpart, and in this case the selection rules may often be related to the classical conserved quantities. A first use of selection rules is in determining the symmetry classes of the states; but in a great variety of ways they may yield other information about the system and the conservation laws. *See* QUANTUM MECHANICS; SYMMETRY LAWS (PHYSICS).

For an isolated system the total angular momentum is a conserved quantity; this fact derives from a fundamental fact of nature, namely, that space is isotropic. Each state is then classifiable by angular momentum J and its z component M $(= -J, -J + 1, \ldots, +J)$. Angular momenta combine in a vectorial fashion. Thus, if the system makes a particle-emitting transition $J_1,M_1 \rightarrow J_2,M_2$, the emitted particles must carry away angular momentum (j, μ), where $\mathbf{j} = \mathrm{j}_1 - \mathrm{j}_2$. This implies that $\mu = M_1 - M_2$ and that j takes on values $J_1 - J_2, J_1 - J_2 + 1, \ldots, (J_1 + J_2)$. Thus in transitions $(J = 4 \leftrightarrow J = 2)$ the possible j values comprise only 2, 3, 4, 5, 6, and, if it is also specified that $M_1 - M_2 = \pm 4$, only 4, 5, 6. Observe that J_z is additive. *See* ANGULAR MOMENTUM; QUANTUM NUMBERS.

Another fundamental symmetry, the parity, which determines the behavior of a system (or of its description) under inversion of the coordinate axes, is conserved by the strong and electromagnetic interactions, and gives a classification of systems as even $(\pi = +1)$ or odd $(\pi = -1)$. Under combination the parity combines multiplicatively. Thus, if the transition above is $4^{\pm} \rightarrow 2^{\pm}$, it follows that $j^{\pi} = 2^-, \ldots, 6^-$, while $4^{\pm} \rightarrow 2^{\pm}$ would give $j^{\pi} = 2^+, \ldots, 6^+$. The angular momentum $\mathbf{j}$ may be a combination of intrinsic spin $\mathbf{s}$ and orbital angular momentum $\mathbf{l}$. Scalar, pseudoscalar, vector, and pseudovector particles are respectively characterized by $s^{\pi_s} = 0^+, 0^-, 1^-, 1^+$, where π_s is the "intrinsic" parity, while l always carries $\pi_l = (-1)^l$. *See* PARITY (QUANTUM MECHANICS); SPIN (QUANTUM MECHANICS).

The isospin symmetry of the elementary particles is almost conserved, being broken by electromagnetic and weak interactions. It is described by the group SU(2), of unimodular unitary transformations in two dimensions. Since the SU(2) algebra is identical with that of the angular momentum SO(3), isospin behaves like angular momentum with its three generators $\mathbf{T}$ replacing $\mathbf{J}$.

The isospin group is a subgroup of SU(3) which defines a more complex fundamental symmetry of the elementary particles. Two of its eight generators commute, giving two additive quantum numbers, T_z and strangeness S' (or, equivalently, charge and hypercharge). The strangeness is conserved $(\Delta S' = 0)$ for strong and electromagnetic, but not for weak, interactions. The selection rules and combination laws for SU(3) and its many extensions, and the quark-structure underlying them, correlate an enormous amount of information and make many predictions about the elementary particles. *See* BARYON; ELEMENTARY PARTICLE; MESON; QUARKS; UNITARY SYMMETRY.

A great variety of other groups have been introduced to define relevant symmetries for atoms, molecules, nuclei, and elementary particles. They all have their own selection rules, representing one aspect of the symmetries of nature. J. B. French

Selectivity The ability of a radio receiver to separate a desired signal frequency from other signal frequencies, some of which may differ only slightly from the desired value. Selectivity is achieved by using tuned circuits that are sharply peaked and by increasing the number of tuned circuits. With a sharply peaked circuit, the output voltage falls off rapidly for frequencies increasingly lower or higher than that to which the circuit is tuned. *See* Q (ELECTRICITY); RADIO RECEIVER; RESONANCE (ALTERNATING-CURRENT CIRCUITS). John Markus

Selenium A chemical element, Se, atomic number 34, atomic weight 78.96. The properties of this element are similar to those of tellurium. *See* PERIODIC TABLE; TELLURIUM.

Selenium burns in air with a blue flame to give selenium dioxide, SeO_2. The element also reacts directly with a variety of metals and nonmetals, including hydrogen and the halogens. Nonoxidizing acids fail to react with selenium, but nitric acid, concentrated sulfuric acid, and strong alkali hydroxides dissolve the element. The only important compound of selenium with hydrogen is hydrogen selenide, H_2Se, a colorless flammable gas possessing a distinctly unpleasant odor, and a toxicity greater and a thermal stability less than that of hydrogen sulfide. Selenium oxyhalide, $SeOCl_2$, is a colorless liquid widely used as a nonaqueous solvent. The oxybromide, $SeOBr_2$, is an orange solid having chemical properties similar to those of $SeOCl_2$. The oxyflouride, $SeOF_2$, a colorless liquid with a pungent smell, reacts with water, glass, and silicon, and also forms additional compounds. Compounds in which C-Se bonds appear are numerous and vary from the simple selenols, RSeH, to molecules exhibiting biological activity such as selenoamino acids and selenopeptides. *See* ORGANOSELENIUM COMPOUND.

The abundance of this widely distributed element in the Earth's crust is estimated to be about 7×10^{-5} % by weight, occurring as the selenides of heavy elements and to a limited extent as the free element in association with elementary sulfur. Examples of the variety of selenide minerals are berzelianite (Cu_2Se), eucairite (AgCuSe), and jermoite [$As(S,Se)_2$]. Selenium minerals do not occur in sufficient quantity to be useful as commercial sources of the element.

Major uses of selenium include the photocopying process of xerography, which depends on the light sensitivity of thin films of amorphous selenium, the decolorization of glasses tinted by the presence of iron compounds, and use as a pigment in plastics, paints, enamels, glass, ceramics, and inks. Selenium is also employed in photographic exposure meters and as a metallurgical additive to improve the machinability of certain steels. Minor uses include application as a nutritional additive for numerous animal species, use in photographic toning, metal-finishing operations, metal plating, high-temperature lubricants, and as catalytic agents, particularly in the isomerization of certain petroleum products. John W. George

The biological importance of selenium is well established, as all classes of organisms metabolize selenium. In humans and other mammals, serious diseases arise from either excessive or insufficient dietary selenium. The toxic effects of selenium have long been known, particularly for grazing animals. In soils with high selenium content, some plants accumulate large amounts of selenium. Animals that ingest these selenium-accumulating plants develop severe toxic reactions. *See* SOIL CHEMISTRY.

Although toxic at high levels, selenium is an essential micronutrient for mammalian species. The accepted minimum daily requirement of selenium for adult humans is 70 micrograms. Many types of food provide selenium, particularly seafood, meats, grains, and the onion family. Mammals and birds require selenium for production of the enzyme glutathione peroxidase, which protects against oxidation-induced cancers. Other seleno-proteins of unknown function are found in mammalian blood, various tissues, and spermatozoa. *See* AMINO ACIDS.

Milton J. Axley; Thressa G. Stadtman

Seligeriales An order of the true mosses (Bryidae), consisting of one family and five or six genera. The plants grow on rocks, and may be exceedingly small and gregarious to moderate in size and tufted. The stems are erect and simple or forked. The linear to lance-subulate leaves have a single costa which often fills the subula. The operculate capsules are terminal, and the peristome is usually present. The calyptra is generally cucullate. *See* BRYIDAE; BRYOPHYTA; BRYOPSIDA. Howard Crum

Semaeostomeae An order of the class Scyphozoa including most of the common medusae. The umbrella of these medusae is more flat than high and is usually domelike. The margin of the umbrella is divided into many lappets. Sensory organs are situated between the lappets. The tentacles are generally well developed and very long except in a few forms.

The life history of this group shows the typical alternation of generations, with forms passing through several larval stages. The Semaeostomeae are distributed mostly in temperate zones and are generally coastal forms with a few exceptions. Some of them are known to have violent poison on their tentacles. Several fossils of this group were found in the strata of the Jurassic period. *See* SCYPHOZOA. Tohru Uchida

Semiconductor A solid crystalline material whose electrical conductivity is intermediate between that of a metal and an insulator. Semiconductors exhibit conduction properties that may be temperature-dependent, permitting their use as thermistors (temperature-dependent resistors), or voltage-dependent, as in varistors. By making suitable contacts to a semiconductor or by making the material suitably inhomogeneous, electrical rectification and amplification can be obtained. Semiconductor devices, rectifiers, and transistors have replaced vacuum tubes almost completely in low-power electronics, making it possible to save volume and power consumption by orders of magnitude. In the form of integrated circuits, they are vital for complicated systems. The optical properties of a semiconductor are important for the understanding and the application of the material. Photodiodes, photoconductive detectors of radiation, injection lasers, light-emitting diodes, solar-energy conversion cells, and so forth are examples of the wide variety of optoelectronic devices. *See* INTEGRATED CIRCUITS; LASER; LIGHT-EMITTING DIODE; PHOTODIODE; PHOTOELECTRIC DEVICES; SEMICONDUCTOR DIODE; SEMICONDUCTOR RECTIFIER; THERMISTOR; TRANSISTOR; VARISTOR.

Conduction in semiconductors. The electrical conductivity of semiconductors ranges from about 10^3 to 10^{-9} ohm^{-1} cm^{-1}, as compared with a maximum conductivity of 10^7 for good conductors and a minimum conductivity of 10^{-17} ohm^{-1} cm^{-1} for good insulators. *See* ELECTRIC INSULATOR; ELECTRICAL CONDUCTIVITY OF METALS.

The electric current is usually due only to the motion of electrons, although under some conditions, such as very high temperatures, the motion of ions may be important. The basic distinction between conduction in metals and in semiconductors is made by considering the energy bands occupied by the conduction electrons. *See* BAND THEORY OF SOLIDS; IONIC CRYSTALS.

At absolute zero temperature, the electrons occupy the lowest possible energy levels, with the restriction that at most two electrons with opposite spin may be in the same energy level. In semiconductors and insulators, there are just enough electrons to fill completely a number of energy bands, leaving the rest of the energy bands empty. The highest filled energy band is called the valence band. The next higher band, which is empty at absolute zero temperature, is called the conduction band. The conduction band is separated from the valence band by an energy gap, which is an important characteristic of the semiconductor. In metals, the highest energy band that is occupied by the electrons is only partially filled. This condition exists either because the number of electrons is not just right to fill an integral number of energy bands or because the highest occupied energy band overlaps the next higher band without an intervening energy gap. The electrons in a partially filled band may acquire a small amount of energy from an applied electric field by going to the higher levels in the same band. The electrons are accelerated in a direction opposite to the field and thereby constitute an electric current. In semiconductors and insulators, the electrons are found only in completely filled bands, at low temperatures. In order to increase the energy of the electrons, it is necessary to raise electrons from the valence band to the conduction band across the energy gap. The electric fields normally encountered are not large enough to accomplish this with appreciable probability. At sufficiently high temperatures, depending on the magnitude of the energy gap, a significant number of valence electrons gain enough energy thermally to be raised to the conduction band. These electrons in an unfilled band can easily participate in conduction. Furthermore, there is now a corresponding number of vacancies in the electron population of the valence band. These vacancies, or holes as they are called, have the effect of carriers of positive charge, by means of which the valence band makes a contribution to the conduction of the crystal. *See* HOLE STATES IN SOLIDS.

The type of charge carrier, electron or hole, that is in largest concentration in a material is sometimes called the majority carrier and the type in smallest concentration the minority carrier.

The majority carriers are primarily responsible for the conduction properties of the material. Although the minority carriers play a minor role in electrical conductivity, they can be important in rectification and transistor actions in a semiconductor.

Intrinsic semiconductors. A semiconductor in which the concentration of charge carriers is characteristic of the material itself rather than of the content of impurities and structural defects of the crystal is called an intrinsic semiconductor. Electrons in the conduction band and holes in the valence band are created by thermal excitation of electrons from the valence to the conduction band. Thus an intrinsic semiconductor has equal concentrations of electrons and holes. The carrier concentration, and hence the conductivity, is very sensitive to temperature and depends strongly on the energy gap. The energy gap ranges from a fraction of 1 eV to several electronvolts. A material must have a large energy gap to be an insulator.

Extrinsic semiconductors. Typical semiconductor crystals such as germanium and silicon are formed by an ordered bonding of the individual atoms to form the crystal structure. The bonding is attributed to the valence electrons which pair up with valence electrons of adjacent atoms to form so-called shared pair or covalent bonds. These materials are all of the quadrivalent type; that is, each atom contains four valence electrons, all of which are used in forming the crystal bonds. *See* CRYSTAL STRUCTURE.

Atoms having a valence of +3 or +5 can be added to a pure or intrinsic semiconductor material with the result that the +3 atoms will give rise to an unsatisfied bond with one of the valence electrons of the semiconductor atoms, and +5 atoms will result in an extra or free electron that is not required in the bond structure. Electrically, the +3 impurities add holes and the +5 impurities add electrons. They are called acceptor and donor impurities, respectively. Typical valence +3 impurities used are boron, aluminum, indium, and gallium. Valence +5 impurities used are arsenic, antimony, and phosphorus.

Semiconductor material "doped" or "poisoned" by valence +3 acceptor impurities is termed *p*-type, whereas material doped by valence +5 donor material is termed *n*-type. The names are derived from the fact that the holes introduced are considered to carry positive charges and the electrons negative charges. The number of electrons in the energy bands of the crystal is increased by the presence of donor impurities and decreased by the presence of acceptor impurities. *See* ACCEPTOR ATOM; DONOR ATOM.

At sufficiently high temperatures, the intrinsic carrier concentration becomes so large that the effect of a fixed amount of impurity atoms in the crystal is comparatively small and the semiconductor becomes intrinsic. When the carrier concentration is predominantly determined by the impurity content, the conduction of the material is said to be extrinsic. Physical defects in the crystal structure may have similar effects as donor or acceptor impurities. They can also give rise to extrinsic conductivity.

Materials. The group of chemical elements which are semiconductors includes germanium, silicon, gray (crystalline) tin, selenium, tellurium, and boron. Germanium, silicon, and gray tin belong to group 14 of the periodic table and have crystal structures similar to that of diamond. Germanium and silicon are two of the best-known semiconductors. They are used extensively in devices such as rectifiers and transistors.

A large number of compounds are known to be semiconductors. A group of semiconducting compounds of the simple type AB consists of elements from columns symmetrically placed with respect to column 14 of the periodic table. Indium antimonide (InSb), cadmium telluride (CdTe), and silver iodide (AgI) are examples of III–V, II–IV, and I–VI compounds, respectively. The various III–V compounds are being studied extensively, and many practical applications have been found for these materials. Some of these compounds have the highest carrier mobilities known for semiconductors. The compounds have zincblende crystal structure which is geometrically similar to the diamond structure possessed by the elemental semiconductors, germanium and silicon, of column 14, except that the four nearest neighbors of each atom are atoms of the other kind. The II–VI compounds, zinc sulfide (ZnS) and cadmium sulfide (CdS), are used in photoconductive devices. Zinc sulfide is also used as a luminescent material. *See* LUMINESCENCE; PHOTOCONDUCTIVITY.

The properties of semiconductors are extremely sensitive to the presence of impurities. It is therefore desirable to start with the purest available materials and to introduce a controlled amount of the desired impurity. The zone-refining method is often used for further purification of obtainable materials. The floating zone technique can be used, if feasible, to prevent any contamination of molten material by contact with the crucible. *See* ZONE REFINING.

For basic studies as well as for many practical applications, it is desirable to use single crystals. Various methods are used for growing crystals of different materials. For many semiconductors, including germanium, silicon, and the III–V compounds, the Czochralski method is commonly used. The method of condensation from the vapor phase is used to grow crystals of a number of semiconductors, for instance, selenium and zinc sulfide. *See* CRYSTAL GROWTH.

The introduction of impurities, or doping, can be accomplished by simply adding the desired quantity to the melt from which the crystal is grown. When the amount to be added is very small, a preliminary ingot is often made with a larger content of the doping agent; a small slice of the ingot is then used to dope the next melt accurately. Impurities which have large diffusion constants in the material can be introduced directly by holding the solid material at an elevated temperature while this material is in contact with the doping agent in the solid or the vapor phase.

A doping technique, ion implantation, has been developed and used extensively. The impurity is introduced into a layer of semiconductor by causing a controlled dose of highly accelerated impurity ions to impinge on the semiconductor. *See* ION IMPLANTATION.

An important subject of scientific and technological interest is amorphous semiconductors. In an amorphous substance the atomic arrangement has some short-range but no long-range order. The representative amorphous semiconductors are selenium, germanium, and silicon in their amorphous states, and arsenic and germanium chalcogenides, including such ternary systems as Ge-As-Te. Some amorphous semiconductors can be prepared by a suitable quenching procedure from the melt. Amorphous films can be obtained by vapor deposition.

Rectification in semiconductors. In semiconductors, narrow layers can be produced which have abnormally high resistances. The resistance of such a layer is nonohmic; it may depend on the direction of current, thus giving rise to rectification. Rectification can also be obtained by putting a thin layer of semiconductor or insulator material between two conductors of different material.

A narrow region in a semiconductor which has an abnormally high resistance is called a barrier layer. A barrier may exist at the contact of the semiconductor with another material, at a crystal boundary in the semiconductor, or at a free surface of the semiconductor. In the bulk of a semiconductor, even in a single crystal, barriers may be found as the result of a nonuniform distribution of impurities. The thickness of a barrier layer is small, usually 10^{-3} to 10^{-5} cm.

A barrier is usually associated with the existence of a space charge. In an intrinsic semiconductor, a region is electrically neutral if the concentration n of conduction electrons is equal to the concentration p of holes. Any deviation in the balance gives a space charge equal to $e(p - n)$, where e is the charge on an electron. In an extrinsic semiconductor, ionized donor atoms give a positive space charge and ionized acceptor atoms give a negative space charge.

Surface electronics. The surface of a semiconductor plays an important role technologically, for example, in field-effect transistors and charge-coupled devices. Also, it presents an interesting case of two-dimensional systems where the electric field in the surface layer is strong enough to produce a potential wall which is narrower than the wavelengths of charge carriers. In such a case, the electronic energy levels are grouped into subbands, each of which corresponds to a quantized motion normal to the surface, with a continuum for motion parallel to the surface. Consequently, various properties cannot be trivially deduced from those of the bulk semiconductor. *See* CHARGE-COUPLED DEVICES; SURFACE PHYSICS. H. Y. Fan

Semiconductor diode A two-terminal electronic device that utilizes the properties of the semiconductor from which it is constructed. In a semiconductor diode without a *pn* junction, the bulk properties of the semiconductor itself are used to make a device whose characteristics may be sensitive to light, temperature, or electric field. In a diode with a *pn* junction, the properties of the *pn* junction are used. The most important property of a *pn* junction is that, under ordinary conditions, it will allow electric current to flow in only one direction. Under the proper circumstances, however, a *pn* junction may also be used as a voltage-variable capacitance, a switch, a light source, a voltage regulator, or a means to convert light into electrical power. *See* SEMICONDUCTOR.

The conductivity of a semiconductor is proportional to the number of electrical carriers (electrons and holes) it contains. In a temperature-compensating diode, or thermistor, the number of carriers changes with temperature. *See* THERMISTOR.

In a photoconductor the semiconductor is packaged so that it may be exposed to light. Light photons whose energies are greater than the band gap can excite electrons from the valence band to the conduction band, increasing the number of electrical carriers in the semiconductor. *See* PHOTOCONDUCTIVITY.

In some semiconductors the conduction band has more than one minimum. This results in a region of negative differential conductivity, and a device operated in this region is unstable. The current pulsates at microwave frequencies, and the device, a Gunn diode, may be used as a microwave power source. *See* MICROWAVE SOLID-STATE DEVICES.

A rectifying junction is formed whenever two materials of different conductivity types are brought into contact. Most commonly, the two materials are an *n*-type and a *p*-type semiconductor, and the device is called a junction diode. However, rectifying action also occurs at a boundary between a metal and a semiconductor of either type. If the metal contacts a large area of semiconductor, the device is known as a Schottky barrier diode; if the contact is a metal point, a point-contact diode is formed. *See* SCHOTTKY EFFECT.

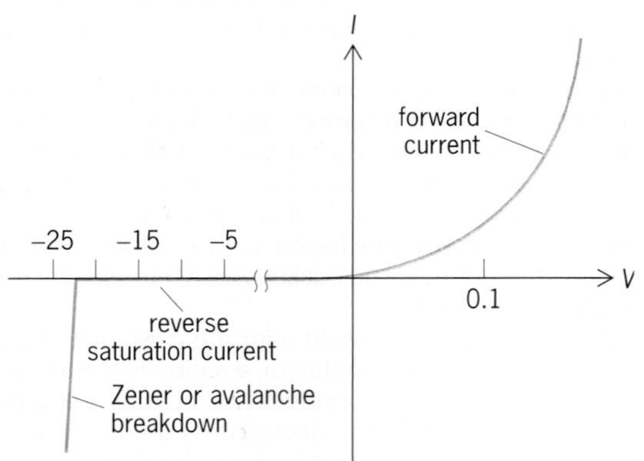

Current-voltage characteristic of a *pn* junction.

The contact potential between the two materials in a diode creates a potential barrier which tends to keep electrons on the *n* side of the junction and holes on the *p* side. When the *p* side is made positive with respect to the *n* side by an applied field, the barrier height is lowered and the diode is forward biased. Majority electrons from the *n* side may flow easily to the *p* side, and majority holes from the *p* side may flow easily to the *n* side. When the *p* side is made negative, the barrier height is increased and the diode is reverse-biased. Then, only a small leakage current flows: Minority electrons from the *p* side flow into the *n* side, and minority holes from the *n* side flow into the *p* side. The current-voltage characteristic of a typical diode is shown in the illustration. Rectifying diodes can be made in a variety of sizes, and much practical use can be made of the fact that such a diode allows current to flow in essentially one direction only. *See* JUNCTION DIODE; SEMICONDUCTOR RECTIFIER; TUNNEL DIODE. Stephen Nygren

Semiconductor heterostructures Structures consisting of two different semiconductor materials in junction contact, with unique electrical or electrooptical characteristics. A heterojunction is a junction in a single crystal between two dissimilar semiconductors. The most important differences between the two semiconductors are generally in the energy gap and the refractive index. In semiconductor heterostructures, differences in energy gap permit spatial confinement of injected electrons and holes, while the differences in refractive index can be used to form optical waveguides. Semiconductor heterostructures have been used for diode lasers, light-emitting diodes, optical detector diodes, and solar cells. In fact, heterostructures must be used to obtain continuous operation of diode lasers at room temperature. Heterostructures also exhibit other interesting properties such as the quantization of confined carrier motion in ultrathin heterostructures and enhanced carrier mobility in modulation-doped heterostructures. Structures of current interest utilize III–V and IV–VI compounds having similar crystal structures and closely matched lattice constants. *See* BAND THEORY OF SOLIDS; LASER; LIGHT-EMITTING DIODE; OPTICAL DETECTORS; REFRACTION OF WAVES; SOLAR CELL.

The most intensively studied and thoroughly documented materials for heterostructures are GaAs and $Al_xGa_{1-x}As$. Several other III–V and IV–VI systems also are used for semiconductor heterostructures. A close lattice match is necessary in heterostructures in order to obtain high-quality crystal layers by epitaxial growth and thereby to prevent excessive carrier recombination at the heterojunction interface.

When the narrow energy gap layer in heterostructures becomes a few tens of nanometers or less in thickness, new effects that are associated with the quantization of confined carriers are observed. These ultrathin heterostructures are referred to as superlattices or quantum well structures, and they consist of alternating layers of GaAs and $Al_xGa_{1-x}As$. These structures are generally prepared by molecular-beam epitaxy. Each layer is 5 to 40 nanometers thick.

In the GaAs layers, the motion of the carriers is restricted in the direction perpendicular to the heterojunction interfaces, while they are free to move in the other two directions. The carriers can therefore be considered as a two-dimensional gas. The Schrödinger wave equation shows that the carriers moving in the confining direction can have only discrete bound states. *See* QUANTUM MECHANICS.

Another property of semiconductor heterostructures is illustrated by a modulation doping technique that spatially separates conduction electrons in the GaAs layer and their parent donor impurity atoms in the $Al_xGa_{1-x}As$ layer. Since the carrier mobility in semiconductors is decreased by the presence of ionized and neutral impurities, the carrier mobility in the modulation-doped GaAs is larger than for a GaAs layer doped with impurities to give the same free electron concentration. Higher carrier mobilities should permit preparation of devices that operate at higher frequencies than are possible with doped layers. *See* SEMICONDUCTOR. H. C. Casey, Jr.

Semiconductor memories Devices for storing digital information that are fabricated by using integrated circuit technology. Semiconductor memories are widely used to store programs and data in almost every digital system, and have replaced core memory as the main active computer memory.

Many different types of semiconductor memories are used in digital systems to perform various functions—bulk data storage, program storage, temporary storage, and cache (or intermediate) storage. Almost all of the memories are a form of random-access memory (RAM), in which any storage location can be accessed in the same amount of time.

Even though most semiconductor memories can be randomly accessed, they are not all referred to as RAMs. RAMs are memory chips that cannot retain data without power but permit data to be both read from or written into the memory chip's storage locations.

Within the category of read/write RAMs, many subdivisions have been created to satisfy the performance and system architecture requirements of the various applications. Basically there are two types of read/write RAMs—dynamic and static (DRAMs and SRAMs). The terms "dynamic" or "static" refer to the structure of the actual storage circuit (the cell structure) used to hold each data bit within the memory chip. A dynamic memory uses a storage cell based on a transistor and capacitor combination, in which the digital information is represented by a charge stored on each of the capacitors in the memory array. The memory gets the name "dynamic" from the fact that the capacitors are imperfect and will lose their charge unless the charge is repeatedly replenished (refreshed) on a regular basis (every few milliseconds) by externally supplied signals. Static memories, in contrast, do not use a charge-storage technique; instead, they use either four transistors and two resistors to form a passive-load flip-flop, or six transistors to form a flip-flop with dynamic loads for each cell in the array. Once data are loaded into the flip-flop storage elements, the flip-flops will indefinitely remain in that state until the information is intentionally changed or the power to the memory circuit is shut off.

In addition to static and dynamic RAMs, there is an attempt to combine both technologies, thus merging the high storage density of dynamic memory cells with the simplicity of use of static RAMs. Referred to as pseudostatic or pseudodynamic RAMs, these memories include circuits on the chip to automatically provide the refresh signals needed by the dynamic cells in the memory array. Since the signals do not have to be supplied by the external system, the memory appears to function like a static RAM.

There are many other forms of semiconductor memories in use—mask-programmable read-only memories (ROMs), fuse-programmable read-only memories (PROMs), ultraviolet-erasable programmable read-only memories (UV EPROMs), electrically alterable read-only memories (EAROMs), electrically erasable programmable read-only memories (EEPROMs), flash EPROMs, nonvolatile static RAMs (NV RAMs), and ferroelectric memories. Most of these memory types are randomly accessible, but their main distinguishing feature is that once information has been loaded into the storage cells, the information stays there even if the power is shut off.

The ROM is programmed by the memory manufacturer during the actual device fabrication. Here, though, there are two types of ROMs: one is called late-mask or contact-mask programmable, and the other is often referred to as a ground-up design.

As an alternative to the mask-programmable memories, all the other nonvolatile memory types permit the users to program the memories themselves. The fuse PROM is a one-time programmable memory—once the information is programmed in, it cannot be altered.

The birth of the microprocessor in the early 1970s brought with it nonvolatile memory types that offered reusability. Information stored in the memory can be erased—in the case of the UV EPROM, by an ultraviolet light, and in the case of the EAROM, EEPROM, flash EPROM (often referred to as just a flash memory device), nonvolatile (NV) RAM, or ferroelectric memory, by an electrical signal. Then the circuit can be reprogrammed with new information that can be retained indefinitely. All of these memory types are starting to approach the ideal memory element for the computer, an element that combines the flexibility of the RAM with the permanence of the ROM when power is removed. *See* COMPUTER STORAGE TECHNOLOGY; INTEGRATED CIRCUITS; LOGIC CIRCUITS; MICROPROCESSOR. David Bursky

Semiconductor rectifier A semiconductor diode that is used in rectification and power control. The semiconductor diode conducts current preferentially in one direction and inhibits the flow of current in the opposite direction by utilizing the properties of a junction formed from two differently doped semiconductor materials. Doped silicon is by far the most widely used semiconductor. Semiconductor diodes are intrinsic to integrated circuits and discrete device technology and are used to perform a wide variety of isolation, switching, signal processing, level shifting, biasing, control, and alternating-current (ac) to direct-current (dc) conversion (rectification) functions. *See* CONTROLLED RECTIFIER; RECTIFIER; SEMICONDUCTOR; SEMICONDUCTOR DIODE.

Either as a key element of an integrated circuit or as a discrete packaged part, the silicon rectifier diode is used in a plethora of applications from small power supplies for consumer electronics to very large power-rectification industrial installations. Many semiconductor diodes are used in non-power-conversion applications in signal processing and communications. These include avalanche or Zener diodes; diodes used for amplitude-modulation radio detection, mixing, and frequency translation; IMPATT, PIN, and step-recovery diodes, used at microwave frequencies; diodes fabricated from gallium arsenide and related compounds, used in optoelectronics; and light-emitting diodes (LEDs) and solid-state lasers. *See* AMPLITUDE-MODULATION DETECTOR; LASER; LIGHT-EMITTING DIODE; MICROWAVE SOLID-STATE DEVICES; MIXER; ZENER DIODE.

Silicon rectifier diodes. The electrical heart of the semiconductor diode is the junction between *p*-type and *n*-type doped silicon regions. Discrete silicon diodes are commercially available with forward-current specifications from under 1 A to several thousands of amperes. Diodes may be connected in parallel for greater current capability as long as the design provides for the current being uniformly distributed between the parallel diodes. This is usually done with a ballast resistor in series with each diode. *See* BALLAST RESISTOR.

Ideally, the current through a reverse-biased diode, called the saturation current (I_S) or reverse current (I_R), approaches zero. Practically speaking, this current is several orders smaller than the forward current (I_F). The maximum value of the reverse blocking voltage is limited primarily by the structure and doping of the semiconductor layers. This maximum voltage is referred to as the avalanche breakdown voltage, or the peak reverse voltage (PRV) or peak inverse voltage (PIV). It is a very important parameter for power supply and power conversion designs. Exceeding the peak inverse voltage is usually destructive unless the circuit design provides for limiting the avalanche current and resultant heating. In summary, at positive voltages and currents (quadrant I of the voltage-current characteristic), the silicon rectifier diode shows the on-state conducting characteristic, with high current and low forward voltage drop; at negative voltages and currents (quadrant III), it shows the reverse-blocking or reverse-bias, off-state characteristic, with high blocking voltage and low (ideally zero) reverse blocking current. See ELECTRICAL BREAKDOWN.

Integrated-circuit diode-junction avalanche breakdown voltages are of the order of several tens of volts. Single silicon rectifier diodes designed for power conversion applications are available with ratings from a few hundred to a few thousand volts. Several diodes can be connected in series for greater voltage capability. Prepackaged series diode strings can be rated to tens of thousands of volts at several amperes. This series connection must ensure equal voltage division across each diode to guard against catastrophic failure of the entire series. Typically this is done by including a high-value equal-value resistor in parallel with each diode to obtain equal voltages, and a parallel capacitor to provide a low-impedance path for high-voltage transients that are often present in industrial environments. See JUNCTION DIODE.

Schottky diodes. Unlike a silicon diode formed from a *pn* junction, the Schottky diode makes use of the rectification effect of a metal-to-silicon interface and the resultant barrier potential. The Schottky diode, sometimes called the Schottky-barrier diode, overcomes the major limitation of the *pn* junction diode; being a majority carrier device, it has a lower forward voltage drop (0.2–0.3 V, compared to 0.7–1.0 V) and faster switching speed than its minority-carrier *pn* junction counterpart. However, other factors confine its use to low-voltage power applications, chiefly the relatively small breakdown voltage, typically 45 V. Secondary shortcomings include a high reverse current and restricted temperature of operation, with commercial devices providing a maximum of 175°C (347°F) compared with 200°C (392°F) for *pn* junction diodes.

Integrated circuits used in computer and instrument systems commonly require voltages less than 15 V and as low as 3.3 V. Thus the advantage of low forward-voltage drop and faster switching favors the Schottky diode. This is particularly true for high-frequency switching voltage regulator power supply applications where voltages at 20–50 kHz must be rectified. The higher reverse current can be tolerated. However, cooling or heat sinking is more critical because of the higher reverse-current temperature coefficient and lower maximum operating temperature. See SCHOTTKY BARRIER DIODE.

Rectifier circuits. The greatest usage of rectifier diodes is the conversion of ac to dc. The single diode of a half-wave rectifier for a single-phase ac voltage conducts only on the positive half-cycle. Because of this, the output voltage across the load resistance is unidirectional and has a nonzero average value. This output waveform is called a pulsating dc. Therefore the input ac voltage has been rectified to a dc voltage. For most applications, a filter, usually consisting of large electrolytic capacitors, must be employed at the output to smooth the ripple present on the pulsating dc voltage to come close to a constant dc voltage value. See CAPACITOR; ELECTRIC FILTER; ELECTRONIC POWER SUPPLY; RIPPLE VOLTAGE.

In lower-power applications from a few watts to a few hundred watts, such as used in computers, television receivers, and laboratory instruments, a switching voltage regulator is commonly used to generate a 10-kHz–50-kHz ac signal from the high-ripple ac power supply voltage. The advantage is the ease and lower cost in filtering the ripple resulting from rectifying high-frequency ac as opposed to filtering low-frequency ac. See VOLTAGE REGULATOR.

Thyristors. Whereas the basic semiconductor rectifier has two terminals, an anode and cathode, a silicon controlled rectifier (SCR) has three terminals: an anode, cathode, and control electrode called the gate. The silicon controlled rectifier is a four-layer device modeled as two interconnected *pnp* and *npn* transistors.

Normally, there is no current flow from the anode to cathode. Both transistors are off; that is, they are blocking any current flow. By applying a relatively small trigger pulse control signal to the gate electrode, the *npn* transistor is switched on. When the *npn* transistor is switched on, the *pnp* transistor is also switched on. Consequently the silicon controlled rectifier is turned on and a current flows through the silicon controlled rectifier and external circuit. The resultant internal voltages keep both the *npn* and *pnp* transistors on even when the gate voltage is removed. The device is said to exhibit regenerative, positive-feedback, or latching-type switching action. There is a voltage drop of about 1 V across the on-state silicon controlled rectifier. The power dissipation rating required in specifying a silicon controlled rectifier is given by this 1-V drop multiplied by the peak current flowing through the device. See TRANSISTOR.

Current continues to flow even when the gate signal is reduced to zero. To reset the silicon controlled rectifier, the external current must be reduced below a certain value. Thus, the thyristor can be switched into the on state (conducting condition) by applying a signal to the gate, but must be restored to the off state by circuit action. If the anode current momentarily drops below some holding current or if the anode voltage is reversed, the silicon controlled rectifier reverts to its blocking state and the gate terminal regains control. Typical silicon controlled rectifiers turn on in 1–5 microseconds and require 10–100 μs of momentary reverse voltage on the anode to regain their forward-blocking ability.

Other semiconductor diode topologies are also used for power control. A generic term for these power-control devices is the thyristor.

Thyristor applications fall into two general categories. The devices can be used from an ac supply, much like silicon rectifier diodes. However, unlike the rectifier diode, which conducts load current as soon as the anode voltage exceeds about 0.7 V, the thyristor will not conduct load current until it is triggered into conduction. Therefore, the power delivered to the load can be controlled. This mode of operation is called ac phase control. It is extensively used in applications requiring conversion from ac to variable-voltage dc output, such as adjustable-speed dc motor drives, and in lighting and heating control. See DIRECT-CURRENT MOTOR.

The other category of applications is operation in dc circuits. This allows power conversion from a battery or rectified ac line to a load requiring either an alternating supply (dc-to-ac conversion) or a variable-voltage dc supply (dc-to-dc conversion). Since the rate of switching the thyristors in dc circuits can be varied by the control circuit, a thyristor inverter circuit can supply an ac load with a variable frequency. The fundamental approach in both cases is to convert a dc voltage to a chopped voltage of controllable duty cycle. Changing the duty cycle either at a variable rate (frequency power modulation) or by varying the pulse width at a fixed frequency (pulse-width power modulation) effectively controls the power delivered to the load. See CHOPPING; PULSE MODULATION.

Important applications for dc-to-dc conversion, dc-to-ac power conversion at variable frequency, and dc-to-ac power conversion at fixed frequency are, respectively, control of battery-powered industrial vehicles such as forklift trucks and mining locomotives, adjustable-speed operation of ac synchronous and induction motors in industrial processing, and power transmission conversion. See ALTERNATING-CURRENT MOTOR; CONVERTER; INDUSTRIAL TRUCKS.

Stanley G. Burns

Semionotiformes An order of actinopterygian fishes which appeared first in the upper Permian, reached maximum development in the Triassic and Jurassic, and persists in the Recent fauna as the gars. In the Semionotiformes the body is encased in a heavy armor of interlocking ganoid scales, which are thick, are more or less rhomboidal, and have an enamellike surface. Modern forms have an elongate body and bony jaws provided with enlarged conical teeth (see illustration).

The single Recent family, Lepisosteidae, contains one genus, *Lepisosteus*, with seven species restricted to lowland fresh and

Spotted gar (*Lepisosteus oculatus*). (*After G. B. Goode, Fishery Industries of the United States, 1884*)

brackish waters of North and Central America. *See* ACTINOPTERYGII.

Reeve M. Bailey

Sendai virus A member of the viruses in the type species Parainfluenza 1, genus *Paramyxovirus*, family Paramyxoviridae; it is also called hemagglutinating virus of Japan (HVJ). Sendai virus was originally recovered in Sendai, Japan, from mice inoculated with autopsy specimens from newborns who died of fatal pneumonitis in an epidemic in 1952. Subsequent attempts to isolate this virus from humans were, however, mostly unsuccessful, although mice are commonly infected with Sendai virus along with rats, guinea pigs, hamsters, and pigs. It is believed that the natural host of Sendai virus is the mouse and that the virus is usually nonpathogenic for humans. *See* ANIMAL VIRUS; PARAINFLUENZA VIRUS.

Nakao Ishida

Sensation The obtaining of information from the environment through the agency of the sense organs. Sensation commonly includes the interpretation of that information as a particular attribute of the world or self. *See* SENSE ORGAN.

Sensation is the intermediary between reception and perception, but the boundaries are vague. Reception refers to the absorption of some form of energy (a stimulus) from the external or internal environment through a process called transduction. Perception includes the interpretation of that stimulus to generate a representation of the world. Within the nervous system, the interpretation is generated by multiple feedback loops, so there is no distinct point at which sensation ends and perception begins. *See* BRAIN; COGNITION; PERCEPTION.

It is common to speak of the five senses (sight, hearing, taste, smell, and touch), but the senses also include hot and cold, pain, limb position, acceleration, and internal states such as hunger, thirst, and fatigue. Some senses have dedicated organs, such as the eyes, ears, tongue, or nasal mucosa, while others are more diffuse, with receptors throughout the skin or within other organs. Each sense has a specific pathway to the brain and primary areas within the brain where the signals are analyzed.

Sensation is initiated by the stimulation of receptors. A receptor is a specialized cell that can absorb energy from its environment and signal its presence to other neurons. This is accomplished by a change in electrical potential of the receptor, resulting in a change in the amount of chemical transmitter released to secondary neurons. *See* NEUROBIOLOGY; NEURON.

Most sensory systems adapt to the general level of stimulation. By doing so, they concentrate on what is different from the general environment. The process of adaptation is slower than most responses, so one does not adapt to, and thus ignore, the stimuli of interest. *See* CUTANEOUS SENSATION; MECHANORECEPTORS; PAIN; PROPRIOCEPTION.

Michael W. Levine

Sense amplifier An electronic amplifier circuit used to sense and refresh the value of a bit stored in a memory cell of a dynamic random access memory (DRAM) integrated circuit.

In DRAM, bits are represented in memory by the presence or absence of an electric charge stored on tiny capacitors. Cell capacitance is directly proportional to cell area, and since the chip area devoted to each bit-cell must decrease with an increase in the memory capacity of a given size of chip, the resulting cell capacitance is tiny, typically on the order of femtofarads (1 femtofarad = 10^{-15} farad). Since the voltage used to charge these capacitors is usually limited to 5 V or less, the charge stored in each memory cell also is quite small, on the order of femtocoulombs. *See* CAPACITANCE; CAPACITOR.

In order to read out the value of a given bit of a word in this type of memory, the bit-cell voltage, or equivalently the magnitude of its charge, needs to be sensed, and the results of this sense operation must be delivered to the rest of the circuit. Both the sense amplifier used to read the bit-cell voltage and the interconnect wiring (the bit line) between the bit cell and the amplifier have inherent parasitic capacitances which, in total, are much larger than the bit-cell capacitance. If the sense amplifier is simply switched across the bit cell, the charge on the bit-cell capacitor will redistribute across the parasitic capacitor and the bit-cell capacitor in accordance with Kirchhoff's law, which requires that the two capacitors have the same voltage, where the total charge must come from the original charge stored on the memory-cell capacitor. Thus the voltage seen by the amplifier is equal to the original bit-cell voltage multiplied by the ratio of the bit-cell capacitance to the sum of the two capacitances, and the charge remaining on the memory cell capacitor after sensing would be reduced from the original charge in the same proportion. If the parasitic capacitance is much larger than the cell capacitance, it follows that the voltage available at the sense amplifier input will be small and the voltage and charge remaining on the memory cell will be considerably reduced. More sophisticated approaches need to be taken to sense this tiny voltage and to refresh the memory cell during sense, so that the memory cell capacitor is left fully charged (or fully discharged) when sensing is complete. *See* KIRCHHOFF'S LAWS OF ELECTRIC CIRCUITS.

The sense amplifier action is based on metal-oxide-semiconductor field-effect transistors (MOSFETs). These can be thought of as simple switches that are opened and closed in a predetermined sequence to carry out the read/refresh memory cycle. To assist the sense, a memory reference cell, containing a capacitor of approximately one-half the capacitance of the bit-cell capacitor, is added to the circuit. *See* SEMICONDUCTOR MEMORIES; TRANSISTOR.

Philip V. Lopresti

Sense organ A structure which is a receptor for external or internal stimulation. A sense organ is often referred to as a receptor organ. External stimuli affect the sensory structures which make up the general cutaneous surface of the body, the exteroceptive area, and the tissues of the body wall or the proprioceptive area. These somatic area receptors are known under the general term of exteroceptors. Internal stimuli which originate in various visceral organs such as the intestinal tract or heart affect the visceral sense organs or interoceptors. A receptor structure is not necessarily an organ; in many unicellular animals it is a specialized structure within the organism. Receptors are named on the basis of the stimulus which affects them, permitting the organism to be sensitive to changes in its environment.

Photoreceptors are structures which are sensitive to light and in some instances are also capable of perceiving form, that is, of forming images. Light-sensitive structures include the stigma of phytomonads, photoreceptor cells of some annelids, pigment cup ocelli and retinal cells in certain asteroids, the eye-spot in many turbellarians, and the ocelli of arthropods. The compound eye of arthropods, mollusks, and chordates is capable of image formation and is also photosensitive. *See* PHOTORECEPTION.

Phonoreceptors are structures which are capable of detecting vibratory motion or sound waves in the environment. The most common phonoreceptor is the ear, which in the vertebrates has other functions in addition to sound perception. *See* EAR (VERTEBRATE); PHONORECEPTION.

Statoreceptors are structures concerned primarily with equilibration, such as the statocysts found throughout the various phyla of invertebrates and the inner ear or membranous labyrinth filled with fluid.

The sense of smell is dependent upon the presence of olfactory neurons, called olfactoreceptors, in the olfactory epithelium of the nasal passages among the vertebrates. *See* OLFACTION.

The sense of taste is mediated by the taste buds, or gustatoreceptors. In most vertebrates these taste buds occur in the oral cavity, on the tongue, pharynx, and lining of the mouth; however, among certain species of fish, the body surface is supplied with taste buds as are the barbels of the catfish. *See* TASTE.

The surface skin of vertebrates contains numerous varied receptors associated with sensations of touch, pain, heat, and cold. *See* CHEMICAL SENSES; CUTANEOUS SENSATION; SENSATION.

Charles B. Curtin

Sensitivity (engineering) A property of a system, or part of a system, that indicates how the system reacts to stimuli. The stimuli can be external (that is, an input signal) or a change in an element in the system. Thus, sensitivity can be interpreted as a measure of the variation in some behavior characteristic of the system that is caused by some change in the original value of one or more of the elements of the system.

Sensitivity is commonly used as a figure of merit for characterizing system performance. As a figure of merit, the sensitivity is a numerical indicator of system performance that is useful for predicting system performance in the presence of elemental variations or comparing the relative performance of two or more systems that ideally have the same performance. In the latter case, the performance of the systems relative to some parameter of interest is rank-ordered by the numerical value of the corresponding sensitivity functions. If T is the performance characteristic and X is the element or a specified input level, then mathematically sensitivity is expressed as a normalized derivative of T with respect to X.

A limiting factor in using the sensitivity of a system to characterize performance at low signal levels is the noise. Noise is a statistical description of a random process inherent in all elements in a physical system. The noise is related to the minimum signal that can be processed in a system as a function of physical variables such as pressure, visual brightness, audible tones, and temperature. *See* ELECTRICAL NOISE.

There exist many situations where the sensitivity measure indicates the ability of a system to meet certain design specifications. For example, in an electronic system the sensitivity of the output current with respect to the variation of the power-supply voltage can be very critical. In that case, a system with a minimum sensitivity of the output current with respect to the power-supply voltage must be designed. Another example is a high-fidelity audio amplifier whose sensitivity can be interpreted as the capacity of the amplifier to detect the minimum amplifiable signal.

Edgar Sánchez-Sinencio

Sensory cell regeneration The replacement of receptor cells within sensory end organs (retina, cochlea, taste buds, olfactory epithelium), most commonly via addition of newly differentiated cells to the systems. Humans and other organisms use a variety of sensory receptors to detect and interpret information from the surrounding environment. These receptors may be damaged by environmental toxins, injury, or overstimulation, thereby reducing the sensory input received by the organism and the animal's ability to respond appropriately to stimuli. In some systems, such as taste, new sensory receptors are produced on a regular basis. In others, including hearing in humans, the loss of sensory cells is permanent. *See* CHEMICAL SENSES; SENSATION.

Regeneration of sensory receptor cells occurs in three manners: repair of damaged receptors, direct conversion (transdifferentiation) of a local nonreceptor cell into a receptor, or proliferation of new receptor cells from a pool of progenitor cells. While all three regenerative processes may occur in sensory systems, proliferation of new cells is arguably the most important.

In most adult vertebrates, chemical receptor cells possess the greatest regenerative potential. Aging taste receptor cells and olfactory receptor neurons undergo programmed cell death (apoptosis) and are replaced by newly proliferated receptor cells that arise within the sensory epithelia. In the mouse, the average taste receptor cell lives for only nine days, demonstrating that continuous receptor turnover is normal in this system. There is a close link between apoptosis and proliferation in gustatory and olfactory systems such that a constant number of mature receptor cells exist in the system at any given time. Taste buds and olfactory epithelia can also increase production of receptor cells following end organ damage, leading to the restoration of epithelial structure and function. *See* APOPTOSIS; EPITHELIUM; OLFACTION; TASTE.

In contrast, regeneration of photoreceptors and sensory hair cells is dependent upon the species and age of the animal. In teleost fish and amphibians, the retina and inner ear epithelia grow throughout the life of the organism and new receptor cells are added to the growing sensory structure. Fishes and amphibians also possess increased regenerative capabilities in the eye and ear following sensory overstimulation or drug-induced damage to sensory receptors. The situation in amniotic vertebrates (reptiles, birds, and mammals) is very different. It was originally thought that these vertebrates lost all regenerative capacity in the eye and ear. However, landmark studies in the past few decades show that birds and perhaps mammals do retain limited regenerative potential. *See* CELL DIFFERENTIATION; EAR (VERTEBRATE); EYE (VERTEBRATE).

External intervention may be necessary to achieve regeneration in the mammalian retina and cochlea. Stem cell therapy provides one possible strategy. However, functional recovery following stem cell transplantation has not yet been demonstrated in either sensory system. *See* STEM CELLS.

Pamela Lanford; Allison Coffin

Sepioidea An order of the class Cephalopoda (subclass Coleoidea) including the cuttlefishes (*Sepia*), the bobtail squids (*Sepiola*), and the ram's-horn squid (*Spirula*). The group is characterized by an internal shell that is calcareous and broad with closely packed laminate chambers (the cuttlebones of cuttlefishes). The mouth is surrounded by ten appendages (eight arms and two longer tentacles) that bear suckers with chitinous rings. The tentacles are contractile and retractile into pockets at their bases (see illustration).

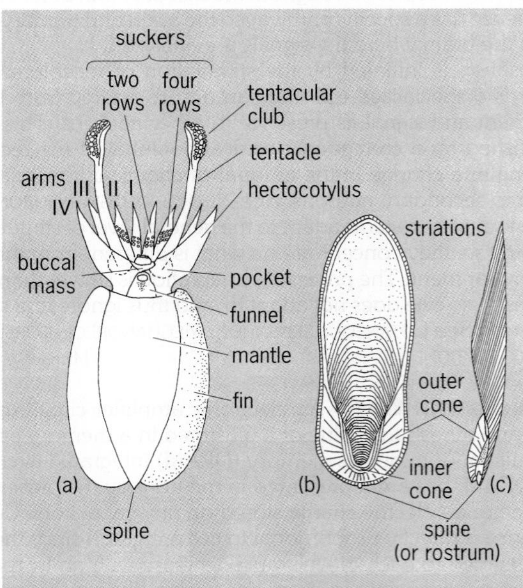

Diagram of some features of (*a*) a cuttlefish and of (*b*) a cuttlebone shown in ventral view and (*c*) in cross section.

Cuttlefishes are common benthic or epibenthic (living on or just above the bottom, respectively) animals that occur in the warm and temperate waters of the nearshore and continental shelf zones of the Old World, but they are excluded from the Western Hemisphere (North and South America). They prey on shrimps, crabs, small fishes, and other cuttlefishes. The sexes are separate, and during mating, which follows a colorful ritualistic courting behavior, sperm is transferred to the female in cylindrical packets (spermatophores) by a modified arm (the hectocotylus) of the male.

The cuttlebones are used to control the buoyancy of the cuttlefishes as fluid is pumped into and out of the laminar chambers. Dried cuttlebones are a source of calcium for cage birds and are used for fine jewelry molds, dentifrices, and cosmetics. The cuttlefishes eject an attention-getting blob of brownish-black ink when threatened by predators, then change color, become transparent, and jet-swim away, leaving the predator to attack the false body (pseudomorph) of ink. Artists have used the ink, called sepia, for centuries. Cuttlefishes are important in world fisheries; about 200,000 metric tons are caught each year for human food. *See* CEPHALOPODA; COLEOIDEA. Clyde F. E. Roper

Sepiolite A complex hydrated magnesium silicate mineral named for its resemblance to cuttlefish bone, alternately named meerschaum (sea foam). The ideal composition, $Mg_8(H_2O)_4(OH)_4Si_{12}O_{30}$, is modified by some additional water of hydration, but is otherwise quite representative. Interlaced disoriented fibers aggregate into a massive stone so porous that it floats on water. These stones are easily carved, take a high polish with wax, and harden when warmed. *See* CLAY MINERALS; SILICATE MINERALS. William F. Bradley

Septibranchia A subclass of bivalve mollusks (class Bivalvia) that are unique in their possession of a muscular septum instead of a filamentous gill. The Septibranchia equate in great part to the superfamily Poromyacea, which includes the septibranch families Cuspidariidae and Poromyidae, and the Verticordiidae. The Verticordiidae have gills that are greatly reduced in size. Although there are a few cuspidariid species in shallow seas, the great majority of septibranchs are found at lower slope and abyssal depths which are deficient in food for filter-feeding bivalves. They live close to the surface in soft sediments. Most species are less than 20 mm maximum length. *See* BIVALVIA; MOLLUSCA.

The term septibranch remains extremely useful because it describes mollusks having a septum and other morphological specializations that relate to the septibranch's unique carnivorous habits. The septum is a muscular, horizontal partition dividing the mantle cavity. It is derived from the enormous gill found in the mollusks of the subclass Lamellibranchia: The filaments are reduced in size and modified. Other modifications include a muscular stomach for crushing the prey and high proteolytic activity of the gastric juice. J. A. Allen

Septic tank A single-story, watertight, on-site treatment system for domestic sewage, consisting of one or more compartments, in which the sanitary flow is detained to permit concurrent sedimentation and sludge digestion. The septic tank is constructed of materials not subject to decay, corrosion, or decomposition, such as precast concrete, reinforced concrete, concrete block, or reinforced resin and fiberglass. The tank must be structurally capable of supporting imposed soil and liquid loads. Septic tanks are used primarily for individual residences, isolated institutions, and commercial complexes such as schools, prisons, malls, fairgrounds, summer theaters, parks, or recreational facilities. Septic tanks have limited use in urban areas where sewers and municipal treatment plants exist. *See* CONCRETE; REINFORCED CONCRETE; STRUCTURAL MATERIALS.

Septic tanks do not treat sewage; they merely remove some solids and condition the sanitary flow so that it can be safely disposed of to a subsurface facility such as a tile field, leaching pools, or buried sand filter. The organic solids retained in the tank undergo a process of liquefaction and anaerobic decomposition by bacterial organisms. The clarified septic tank effluent is highly odorous, contains finely divided solids, and may contain enteric pathogenic organisms. The small amounts of gases produced by the anaerobic bacterial action are usually vented and dispersed to the atmosphere without noticeable odor or ill effects. *See* RURAL SANITATION; SEWAGE; SEWAGE TREATMENT. Gerald Palevsky

Sequence stratigraphy The study of stratigraphic sequences, defined as stratigraphic units bounded by unconformities. With improvements in the acquisition and processing of reflection-seismic data by petroleum exploration companies in the 1970s came the recognition that unconformity-bounded sequences could be recognized in most sedimentary basins. This was the beginning of an important development, seismic stratigraphy, which also included the use of seismic reflection character to make interpretations about large-scale depositional facies and architecture. *See* SEISMIC STRATIGRAPHY; UNCONFORMITY.

Underpinning sequence-stratigraphic methods are the following interrelated principles: (1) The volume of sediment accumulating in any part of a sedimentary basin is dependent on the space made available for sediment by changes in sea level or basin-floor elevation. This space is referred to as accommodation. (2) Changes in accommodation tend to be cyclic, and they are accompanied by corresponding changes in sedimentary environment and depositional facies. Thus, a rise in base level typically leads to an increase in accommodation, deepening of the water in the basin, with corresponding changes in facies, and a transgression, with a consequent landward shift in depositional environments and in depositional facies. A fall in base level may lead to exposure and erosion (negative accommodation), with the development of a widespread unconformity. (3) These predictable changes provide the basis for a model of the shape and internal arrangement or architecture of a sequence, including the organization and distribution of sedimentary facies and the internal bedding surfaces that link these facies together. *See* BASIN; DEPOSITIONAL SYSTEMS AND ENVIRONMENTS; FACIES (GEOLOGY).

Clastic-dominated sequences are bounded by unconformities. These surfaces (sequence boundaries) are typically well developed within coastal and shelf sediments, where they form as a result of subaerial exposure and erosion during falling sea level. In deeper-water settings, including the continental slope and base of slope, there may be no corresponding sedimentary break; and sequences may be mapped into such settings only if the unconformity can be correlated to the equivalent conformable surface (the correlative conformity). In some instances, the surface of marine transgression, which develops during the initial rise in sea level from a lowstand, forms a distinctive surface that is close in age to the subaerial unconformity and may be used as the sequence boundary. *See* MARINE GEOLOGY; MARINE SEDIMENTS.

The cycle of rise and fall of sea level may be divided into four segments: lowstand, transgressive, highstand, and falling stage. The deposits that form at each stage are distinctive, and are assigned to systems tracts named for each of these stages.

Carbonate-dominated sequences are derived from carbonate sedimentation which is most active in warm, clear, shallow, shelf seas. During the sea-level cycle, these conditions tend to be met during the highstand phase. Sediment production may be so active, including that of reef development at the platform margin, that it outpaces accommodation generation, leading to deposition on the continental slope. Oversteepened sediment slopes there may be remobilized, triggering sediment gravity flows and transportation into the deep ocean. This process is called highstand shedding.

There are several processes of sequence generation that range from a few tens of thousands of years to hundreds of millions of years for the completion of a cycle of rise and fall of sea level. More than one such process may be in progress at any one time within a basin, with the production of a range of sequence styles nested within or overlayering each other.

High-frequency sequence generation is driven by orbital forcing of climate (the so-called Milankovitch effects), of which glacial eustasy is the best-known outcome. The effects of glacioeustasy have dominated continental-margin sedimentation since the freeze-up of Antarctica in the Oligocene. Regional tectonism—such as the process of thermal subsidence following rifting, and flexural loading in convergent plate settings—develops changes in basement elevation that drive changes in relative sea level. These cycles have durations of a few millions to a few tens of millions of years, and they are confined to individual basins or the flanks of major orogens or plate boundaries. *See* PALEOGEOGRAPHY.

Sequence concepts enable petroleum exploration and development geologists to construct predictive sequence models for stratigraphic units of interest from the limited information typically available from basins undergoing petroleum exploration. These models can guide regional exploration, and can also assist in the construction of production models that reflect the expected partitioning of reservoir-quality facies within individual stratigraphic units. *See* CLIMATE HISTORY; GEOLOGIC TIME SCALE; GEOPHYSICAL EXPLORATION; GLACIOLOGY; PALEOCLIMATOLOGY; STRATIGRAPHY.

Andrew D. Miall

Sequoia The giant sequoia or big tree (*Sequoia gigantea*) occupies a limited area in California and is said to be the most massive of all living things. The leaves are evergreen, scalelike, and overlapping on the branches. In height sequoia is a close second to the redwood (300–330 ft or 90–100 m) but the trunk is more massive. Sequoia trees may be 27–30 ft (8–9 m) in diameter at 10 ft (3 m) from the ground. The stump of one tree showed 3400 annual rings. The red-brown bark is 1–2 ft (0.3–0.6 m) thick and spongy. Vertical grooves in the trunk give it a fluted appearance. The heartwood is dull purplish-brown and lighter and more brittle than that of the redwood. The wood and bark contain much tannin, which is probably the cause of the great resistance to insect and fungus attack. The most magnificent trees are within the Kings Canyon and Sequoia National Parks. *See* PINALES; REDWOOD.

Arthur H. Graves; Kenneth P. Davis

Series The indicated sum of a succession of numbers or terms. Series are used to obtain approximate values of infinite repeating decimals, to solve transcendental equations, to obtain values of logarithms or trigonometric functions, to evaluate integrals, and to solve boundary value problems.

For a finite series, with only a limited number of terms, the sum is found by addition. For an infinite series, with an unlimited number of terms, a sum or value can be assigned only by some limiting process. When the simplest such process yields a value, the infinite series is convergent. Many tests for convergence enable one to learn whether a sum can be found without actually finding it.

If each term of an infinite series involves a variable x and the series converges for each value of x in a certain range, the sum will be a function of x. Often the sum is a given function of x, $f(x)$, for which a series having terms of some given form is desired. Thus the Taylor's series expansion

$$f(x) = \sum_{n=0}^{\infty} f^{(n)}(a)\frac{(x-a)^n}{n!}$$

can be found for a large class of functions, the analytic functions, and represent such functions for sufficiently small values of $|x-a|$. For a much less restricted type of function on the interval $-\pi < x < \pi$, a Fourier series expansion of the form

$$\tfrac{1}{2}A_0 + \sum_{n=1}^{\infty}(A_n \cos nx + B_n \sin nx)$$

can be found.

Finite series. Here the problem of interest is to determine the sum of the first n terms,

$$S_n = u_0 + u_1 + u_2 + \cdots + u_{n-1}$$

when u_n is a given function of n. Examples are the arithmetic series, with $u_n = a + nd$ and $S_n = (n/2)[2a+(n-1)d]$, and the geometric series, with $u_n = ar^n$ and $S_n = a(1-r^n)/(1-r)$. *See* PROGRESSION (MATHEMATICS).

Convergence and divergence. An infinite series is the indicated sum of an unlimited number of terms

$$u_0 + u_1 + u_2 + \cdots u_n \cdots$$

or more briefly

$$\sum_{n=0}^{\infty} u_k$$

or simply Σu_n, read "sigma of u_n." The sum S_n of the first n terms is known as the nth partial sum. Thus S_n is the finite sum

$$\sum_{k=0}^{n-1} u_k$$

If, as n increases indefinitely or becomes infinite, the partial sum S_n approaches a limit S, then the infinite series Σu_n is convergent. S denotes the sum or value of the series. For example, if $|r| < 1$,

$$S = \sum ar^n = \frac{a}{1-r} \quad \text{since } S_n = a\frac{1-r^n}{1-r}$$

If, as n becomes infinite, the partial sum S_n does not approach a finite limit, then the infinite series Σu_n is divergent. For example, $\Sigma 1$ diverges, since here $S_n = n$ becomes infinite with n. Also $\Sigma(-1)^n$ diverges, since here $S_n = \tfrac{1}{2}[1-(-1)^{n+1}]$ which is alternately one and zero.

Positive series are series each of whose terms is a positive number or zero. For such series, the partial sum S_n increases as n increases. If for some fixed number A no sum S_n ever exceeds A, the sums are bounded and admit A as an upper bound. In this case, S_n must approach a limit, and the series is convergent. If every fixed number is exceeded by some S_n, the sums are unbounded. In this case, S_n must become positively infinite and the series is divergent. The tests for convergence of positive series are tests for boundedness, and this is shown by a comparison of S_n with the partial sums of another series or with an integral.

For any series Σu_n, which may have both positive and negative terms, the series of absolute values, $\Sigma|u_n|$, is a positive series whose convergence may be proved by one of the tests for positive series. If $\Sigma|u_n|$ converges, then Σu_n necessarily converges and is said to converge absolutely. The sum of an absolutely convergent series is independent of the order of the terms.

A series which converges but which does not converge absolutely is said to be conditionally convergent. For such a series, a change in the order of the terms may change the sum or cause divergence.

Power series. There are series with $u_n = a_n x^n$. For such a series, it may happen that

$$\lim_{n\to\infty}\left|\frac{a_{n+1}}{a_n}\right| = A$$

If $A = 0$, the series converges for all values of x. If $A \neq 0$, the series converges for all x of the interval $-1/A < x < 1/A$. It will diverge for all x with $|x|$ $1/A$. For any power series, the interval of convergence is related in this way to a number A, which, however, in the general case has to be given by the superior limit of

$$\sqrt{|a_n|}$$

Similar remarks apply to the series with $u_n = a_n(x - c)^n$. Here the interval of convergence is $|x - c| < 1/A$.

One of the most important power series is the binomial series:

$$1 + mx + \frac{m(m-1)}{1.2}x^2 + \cdots + \frac{m(m-1)(m-2)\cdots(m-n+1)}{n!}x^n + \cdots$$

When m is a positive integer, this is a finite sum of $m + 1$ terms which equals $(1 + x)^m$ by the binomial theorem. When m is not a positive integer, the interval of convergence is $-1 < x < 1$, and for x in this interval, the sum of the series is $(1 + x)^m$. *See* BINOMIAL THEOREM.

Let the power series $\Sigma a_n(x - c)^n$ have the sum function $f(x)$. Then $a_0 = f(c)$, $a_n = f^{(n)}(c)/n!$, and the series is the Taylor series of $f(x)$ at $x = c$. Thus, every power series whose interval of convergence has positive length can be put in the form

$$f(x) = f(c) + f'(c)\frac{x-c}{1!} + \cdots + f^{(n)}(c)\frac{(x-c)^n}{n!} + \cdots$$

where $f(x)$ is the sum function.

The Maclaurin series is the special case of Taylor series with $c = 0$:

$$f(x) = f(0) + f'(0)\frac{x}{1!} + \cdots + f^{(n)}(0)\frac{x^n}{n!} + \cdots$$

Uniform convergence. Let each term of a series be a function of z, $u_n = g_n(z)$. Let S_n be the sum of the first n terms, and S the sum to which the series converges for a particular value of z. Then $R_n = S - S_n$ is the remainder after n terms, and for the particular value of z, lim R_n must equal zero. If, for a given range of z, it is possible to make $R_n(z)$ arbitrarily small for sufficiently large n without specifying which z in the range is under consideration, the series converges uniformly. *See* FOURIER SERIES AND TRANSFORMS.

Applications. Series sometimes appear in disguised form in arithmetic. Thus the approximation of a rational number by an infinite repeating decimal is really a geometric series.

Philip Franklin; Salomon Bochner

Series circuit An electric circuit in which the principal circuit elements have their terminals joined in sequence so that a common current flows through all the elements. The circuit may consist of any number of passive and active elements, such as resistors, inductors, capacitors, electron tubes, and transistors. *See* CIRCUIT (ELECTRICITY).

Robert Lee Ramey

Serology The study of antibodies in the serum or other body fluids as well as the study of the antigens present on cells of the blood. It is the oldest branch of immunology. Blood consists of red and white cells, platelets, and plasma (the fluid medium). After blood coagulates, serum (consisting of plasma minus the clotting proteins) separates from the clot. It is generally easier to prepare and more convenient to study in a laboratory than plasma. Even before antibodies were identified, it was noted that the serum of a person who had recovered from a bacterial infection could agglutinate, or clump, those bacteria, so a simple test became available to help identify the organisms with which a person was, or had been, infected. It is easier and more reliable to measure antibodies in serum than the microorganisms themselves, which often are present at low or undetectable numbers. *See* AGGLUTINATION REACTION; ANTIBODY; ANTIGEN; ANTIGEN-ANTIBODY REACTION; IMMUNOLOGY; SERUM.

In modern laboratories there are many techniques for identifying antibodies against microorganisms. Agglutination is still used, as well as complement fixation, enzyme-linked immunosorbent assays (ELISA), virus neutralization assays, and immunofluorescence. Most serologic tests can be set up quickly in the face of an emerging infectious disease epidemic. *See* COMPLEMENT-FIXATION TEST; IMMUNOASSAY; IMMUNOFLUORESCENCE.

J. John Cohen

Serotonin A neurotransmitter, also known as 5-hydroxytryptamine (5-HT), that is released throughout most of the brain and spinal cord by the terminals of neurons originating in the raphe nuclei of the brain stem. Serotonin is also released from platelets to promote blood clotting and from enterochromaffin cells in the intestines as a hormone; moreover, in the pineal gland serotonin is an intermediate in the synthesis of the hormone melatonin (5-methoxy *N*-acetyltryptamine).

Serotonin is made from tryptophan, an essential amino acid, and the concentration of tryptophan in the central nervous system (CNS) normally controls the rate at which serotonin is synthesized. Thus, eating carbohydrate foods that elevate brain tryptophan levels (by increasing the amino acid's transport across the blood-brain barrier) can increase serotonin's synthesis and its release into synapses. This unusual ability of serotonin-releasing neurons to couple the release of their neurotransmitter to the composition of each meal allows these neurons to participate in maintaining nutritional homeostasis. However, this ability also underlies the "carbohydrate craving" and weight gain associated with a number of mood disturbances, in which patients learn to use dietary carbohydrates as though they were antidepressant drugs. A large number of protein molecules—including enzymes, receptors, and binding proteins—are involved in serotonin's actions, allowing an impressive array of relatively specific drugs to be developed for treating serotonin-related disorders. Serotonin is metabolized by the enzyme monoamine oxidase (MAO), which is present in virtually all cells, including the neurons that produce serotonin. *See* NEUROSECRETION.

Serotonin in the CNS has numerous known physiological and behavioral roles. Drugs that increase its levels within brain synapses, or interact with its receptors, are used to relieve depression, anxiety, excessively aggressive behavior, and bulimia; to enhance satiety and suppress carbohydrate craving; to treat premenstrual syndrome; to suppress vomiting (for example, in patients receiving cancer chemotherapy); and to treat migraines. Serotonin release within the brain also promotes sleep, and within the spinal cord it lowers blood pressure and acts as a "gate," limiting pain-generated signals ascending to the brain. *See* BRAIN; NEUROBIOLOGY; PSYCHOPHARMACOLOGY; PSYCHOTOMIMETIC DRUGS; SYNAPTIC TRANSMISSION.

Richard J. Wurtman

Serpentine The name traditionally applied to three hydrated magnesium silicate minerals, antigorite, chrysotile, and lizardite. All have similar chemical compositions but with three different but closely related layered crystal structures. Serpentine also has been used as a group name for minerals with the same layered structures but with a variety of compositions. The general formula is $M_3T_2O_5(OH)_4$, where M may be magnesium (Mg), ferrous iron (Fe^{2+}), ferric iron (Fe^{3+}), aluminum (Al), nickel (Ni), manganese (Mn), cobalt (Co), chromium (Cr), zinc (Zn), or lithium (Li); and T may be silicon (Si), Al, Fe^{3+}, or boron (B).

Lizardite has a planar structure, with the misfit accommodated by slight adjustments of the atomic positions within the layers. Chrysotile has a cylindrical structure in which the layers are either concentrically or spirally rolled to produce fiber commonly ranging from 15 to 30 nanometers in diameter, and micrometers to centimeters in length. These fibers have great strength and flexibility and are the most abundant and commonly used form of asbestos. Antigorite has a modulated wave structure, with wavelengths generally varying between 3 and 5 nm. *See* ASBESTOS.

Frederick J. Wicks

Serpentinite A common rock composed of serpentine minerals; usually formed through the hydration of ultramafic rocks, dunites, and peridotites in a process known as serpentinization. The result is the formation of hydrated magnesium-rich minerals, such as antigorite, chrysotile, or lizardite, commonly with magnetite or, less frequently, brucite. *See* ASBESTOS; DUNITE; PERIDOTITE.

Serpentinites can be distinguished by, and are named for, the dominant serpentine mineral in the rock, that is, antigorite-serpentinite, chrysotile-serpentinite, and lizardite-serpentinite. Lizardite-serpentinites are the most abundant. They have been formed in retrograde terrains and are characterized by the pseudomorphic replacement of the original olivine, pyroxenes, amphiboles, and talc by lizardite with or without magnetite or brucite. Antigorite-serpentinites can form directly from minerals such as olivine, pyroxene, and so forth in retrograde terrains similar to lizardite, but at a high temperature. Chrysotile-serpentinites usually occur only in chrysotile asbestos deposits. The occurrence of serpentinites is widespread, particularly in greenstone belts, mountain chains, and mid-ocean ridges, where they have formed through the serpentinization of ultramafic rocks. *See* MID-OCEANIC RIDGE; SERPENTINE. Frederick J. Wicks

Serum The liquid portion that remains when blood is allowed to clot spontaneously and is then centrifuged to remove the blood cells and clotting elements. It has approximately the same volume (55%) as plasma and differs from it only by the absence of fibrinogen. *See* FIBRINOGEN.

Blood serum contains 6–8% solids, including macromolecules such as albumin, antibodies and other globulins, and enzymes; peptide and lipid-based hormones; and cytokines; as well as certain nutritive organic materials in small amounts, such as amino acids, glucose, and fats. Somewhat less than 1% of the serum consists of inorganic substances. Small amounts of respiratory gases are dissolved in the serum, as is the gas nitric oxide, which serves as a chemical messenger and vasodilator. Small amounts of waste material are also present. These substances, along with other small molecules which are not bound to blood proteins, are filtered out as blood flows through the kidney. *See* BLOOD; CLINICAL PATHOLOGY; KIDNEY.

Certain types of sera, both human and animals, are used in clinical medicine. Immune serum and hyperimmune serum either are developed by naturally occurring disease or are deliberately prepared by repeated injection of antigens to increase antibody titer for either diagnostic tests or the treatment of active disease. These sera are referred to as antisera, since they have a specific antagonistic action against specific antigens. *See* ANTIBODY; ANTIGEN; BIOLOGICALS; IMMUNITY.

By custom, the clear portion of any liquid material of animal origin separated from its solid or cellular elements is also referred to as sera. These fluids are more properly referred to as effusions. *See* SEROLOGY. Reuben Straus; Bruce A. Stanley

Servomechanism A system for the automatic control of motion by means of feedback. The term servomechanism, or servo for short, is sometimes used interchangeably with feedback control system (servosystem). In a narrower sense, servomechanism refers to the feedback control of a single variable (feedback loop or servo loop). In the strictest sense, the term servomechanism is restricted to a feedback loop in which the controlled quantity or output is mechanical position or one of its derivatives (velocity and acceleration). *See* CONTROL SYSTEMS.

The purpose of a servomechanism is to provide one or more of the following objectives: (1) accurate control of motion without the need for human attendants (automatic control); (2) maintenance of accuracy with mechanical load variations, changes in the environment, power supply fluctuations, and aging and deterioration of components (regulation and self-calibration); (3) control of a high-power load from a low-power command signal (power amplification); (4) control of an output from a remotely located input, without the use of mechanical linkages (remote control, shaft repeater).

The illustration shows the basic elements of a servomechanism and their interconnections; in this type of block diagram the connection between elements is such that only a unidirectional cause-and-effect action takes place in the direction shown by the arrows. The arrows form a closed path or loop; hence this is a single-loop servomechanism or, simply, a servo loop. More complex servomechanisms may have two or more loops (multiloop servo), and a complete control system may contain many servomechanisms. *See* BLOCK DIAGRAM.

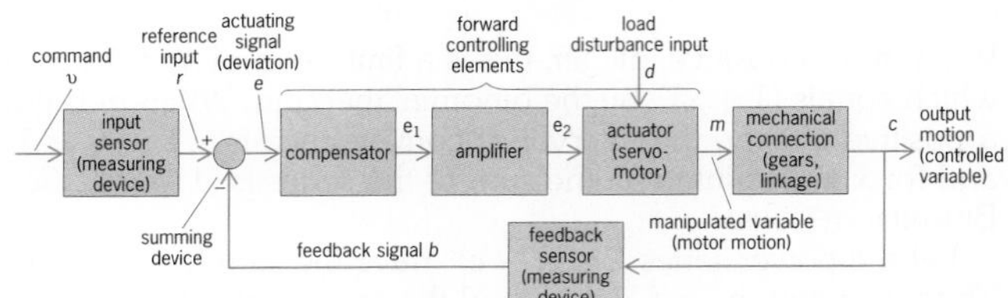

Servo loop elements and their interconnections. Cause-and-effect action takes place in the directions of arrows. (*After American National Standards Institute, Terminology for Automatic Control, ANSI C85.1*)

Servomechanisms were first used in speed governing of engines, automatic steering of ships, automatic control of guns, and electromechanical analog computers. Today, servomechanisms are employed in almost every industrial field. Among the applications are cutting tools for discrete parts manufacturing, rollers in sheet and web processes, elevators, automobile and aircraft engines, robots, remote manipulators and teleoperators, telescopes, antennas, space vehicles, mechanical knee and arm prostheses, and tape, disk, and film drives. *See* COMPUTER STORAGE TECHNOLOGY; FLIGHT CONTROLS; GOVERNOR; MAGNETIC RECORDING; REMOTE MANIPULATORS; ROBOTICS. Gerald Weiss

Set theory A mathematical term referring to the study of collections or sets. Consider a collection of objects (such as points, dishes, equations, chemicals, numbers, or curves). This set may be denoted by some symbol, such as X. It is useful to know properties that the set X has, irrespective of what the elements of X are. The cardinality of X is such a property.

Two sets A and B are said to have the same cardinal written $C(A) = C(B)$, provided there is a one-to-one correspondence between the elements of A and the elements of B. For finite sets this notion coincides with the phrase "A has the same number of elements as B." However, for infinite sets the above definition yields some interesting consequences. For example, let A denote the set of integers and B the set of odd integers. The function $f(n) = 2n-1$ shows that $C(A) = C(B)$. Hence, an infinite set may have the same cardinal as a part or subset of itself.

A is called a subset of B if each element of A is an element of B, and it is expressed as $A \subset B$. The collection of odd integers is a subset of itself.

One approved method of forming a set is to consider a property P possessed by certain elements of a given set X. The set of elements of X having property P may be considered as a set Y. The expression $p \in X$ is used to denote the fact that p is an element of X. Then $Y = \{p \mid p \in X \text{ and } p \text{ has property } P\}$. Another approved method is to consider the set Z of all subsets of a given set X. Paradoxically, it is not permissible to regard the collection of all sets as a set.

In set theory, one is interested not only in the properties of sets but also in operations involving sets: addition, subtraction, multiplication, and mapping. The sum of A and B ($A + B$ or $A \cup B$) is the set of all elements in either A or B; that is, $A + B = \{p \mid p \in A \text{ or } p \in B)$. The intersection of A and B ($A \cdot B$, $A \cap B$, or AB) is the set of all elements in both A and B; that is, $A \cdot B = \{p \mid p \in A \text{ and } p \in B)$. If there is no element which is in both

A and B, one says that A does not intersect B and writes $A \cdot B = 0$. The expression $A - B$ is used to denote the collection of elements of A that do not belong to B; that is $A - B = \{p \mid p \in A \text{ and } p \notin B\}$.

R. H. Bing

Sewage Water-carried wastes, in either solution or suspension, that flow away from a community. Also known as wastewater flows, sewage is the used water supply of the community. It is more than 99.9% pure water and is characterized by its volume or rate of flow, its physical condition, its chemical constituents, and the bacteriological organisms that it contains. Depending on their origin, wastewaters can be classed as sanitary, commercial, industrial, or surface runoff.

The spent water from residences and institutions, carrying body wastes, ablution water, food preparation wastes, laundry wastes, and other waste products of normal living, are classed as domestic or sanitary sewage. Liquid-carried wastes from stores and service establishments serving the immediate community, termed commercial wastes, are included in the sanitary or domestic sewage category if their characteristics are similar to household flows. Wastes that result from an industrial process or the production or manufacture of goods are classed as industrial wastes. Their flows and strengths are usually more varied, intense, and concentrated than those of sanitary sewage. Surface runoff, also known as storm flow or overland flow, is that portion of precipitation that runs rapidly over the ground surface to a defined channel. Precipitation absorbs gases and particulates from the atmosphere, dissolves and leaches materials from vegetation and soil, suspends matter from the land, washes spills and debris from urban streets and highways, and carries all these pollutants as wastes in its flow to a collection point. Discharges are classified as point-source when they emanate from a pipe outfall, or non-point-source when they are diffused and come from agriculture or unchanneled urban land drainage runoff. *See* HYDROLOGY; PRECIPITATION (METEOROLOGY).

Wastewaters from all of these sources may carry pathogenic organisms that can transmit disease to humans and other animals; contain organic matter that can cause odor and nuisance problems; hold nutrients that may cause eutrophication of receiving water bodies; and may contain hazardous or toxic materials. Proper collection and safe, nuisance-free disposal of the liquid wastes of a community are legally recognized as a necessity in an urbanized, industrialized society. *See* ANALYTICAL CHEMISTRY; PUBLIC HEALTH; SEWAGE SOLIDS; SEWAGE TREATMENT; TOXICOLOGY.

Gerald Palevsky

Sewage collection systems Configurations of inlets, catch basins, manholes, pipes, drains, mains, holding basins, pump stations, outfalls, controls, and special devices to move wastewaters from points of collection to discharge. The system of pipes and appurtenances is also known as the sewerage system. Wastewaters may be sanitary sewage, industrial wastes, storm runoff, or combined flows.

A sewer is a constructed ditch or channel designed to carry away liquid-conveyed wastes discharged by houses and towns. Modern sewer systems typically are gravity-flow pipelines installed below the ground surface in streets and following the ground slope. The depth of cover over pipelines is controlled by factors such as the location of rock and ground water, the ability to receive flows from all buildings by gravity, depth to frost line, economics of maintaining gravity flow as compared with pumping, and location and elevation of other existing utilities and infrastructures.

Sewerage systems are designed to carry the liquid wastes smoothly, without deposition, with a minimum of wasted hydraulic energy, and at minimum costs for excavation and construction; they should provide maximum capacity for future populations and flows. Engineered construction, controlled by availability of time, material, personnel, and finances, affects the choice and use of individual components within sewerage systems. *See* INDUSTRIAL HEALTH AND SAFETY; SEWAGE; SEWAGE DISPOSAL.

Gerald Palevsky

Sewage solids The accumulated, semiliquid material consisting of suspended, colloidal, and dissolved organic and inorganic matter separated from wastewater during treatment. Sludges are developed as contained pollutants, and contaminants are separated by mechanical, hydraulic, biological, or chemical processes. The various classes of solids that are removed and collected must be disposed of in a safe, nuisance-free manner without hazard to health or the environment. Collection, handling, transporting, and disposal of removed solids are difficult and costly, since they are offensive and putrescible, with 92–99.5% water content. Sewage solids must be treated by thickening, chemical conditioning, mechanical dewatering, thermal action, biological stabilization, or digestion to convert putrescible organic matter to relatively inert end products, remove water, and reduce weight and volume.

Sewage solids are classified as screenings, scum, grit, septage, or sewage sludges. Screenings are large solids, carried by incoming wastewater, that are captured mechanically on screens or racks with openings of various sizes. These protective units remove floating debris, including wood, clothing, cans, rags, paper, rubber and plastic goods, and stringy material that could damage equipment or create problems in plant maintenance and operation.

Scum is defined as the floating fraction of sewage solids, with specific gravity under 1.0, that, under quiescent conditions, rises to the surface of the wastewater. Primary tank skimmings contain oils, fats, soaps, rubber and plastic hygienic products, cigarette filter tips, paper, and similar materials.

Heavy suspended solids consisting of sand, cinders, coffee grounds, seeds, small metal objects, and other generally inorganic particles carried in wastewater inflow are collectively known as grit. The amount of grit varies with type of sewer, season, weather, intensity of runoff, condition of streets and sewers, and use of household garbage disposal units.

Septage consists of partially digested material pumped from on-site sanitary waste-water disposal systems. It contains a mixture of grit, scum, and suspended solids, adding to treatment plant sludge. *See* LEACHING; SEPTIC TANK.

Sludge derives its name from the unit process from which it settles out. Primary sludge, or raw sludge, develops as solids in incoming wastewater settle hydraulically. Raw sludge, containing up to 5% solids by weight, is gray, greasy, viscous, unsightly, contains visible fecal solids and scraps of household wastes, and has a disagreeable odor. Sludge thickening is a process that is used to remove water, increase the concentration of solids, reduce weight and volume, and prepare sludges for further treatment and handling. *See* PRECIPITATION (CHEMISTRY); WATER TREATMENT.

Solids are generally disposed of in landfills, buried, composted, or recycled as soil amendments. *See* AIR POLLUTION; HAZARDOUS WASTE; SEWAGE; SEWAGE DISPOSAL; SEWAGE TREATMENT.

Gerald Palevsky

Sewage treatment Unit processes used to separate, modify, remove, and destroy objectionable, hazardous, and pathogenic substances carried by wastewater in solution or suspension in order to render the water fit and safe for intended uses. Treatment removes unwanted constituents without affecting or altering the water molecules themselves, so that wastewater containing contaminants can be converted to safe drinking water. Stringent water quality and effluent standards have been developed that require reduction of suspended solids (turbidity), biochemical oxygen demand (related to degradable organics), and coliform organisms (indicators of fecal pollution); control of pH as well as the concentration of certain organic chemicals and heavy metals; and use of bioassays to guarantee safety of treated discharges to the environment.

In all cases, the impurities, contaminants, and solids removed from all wastewater treatment processes must ultimately be collected, handled, and disposed of safely, without damage to humans or the environment. *See* SEWAGE SOLIDS.

Treatment processes are chosen on the basis of composition, characteristics, and concentration of materials present in solution or suspension. The processes are classified as pretreatment, preliminary, primary, secondary, or tertiary treatment, depending on type, sequence, and method of removal of the harmful and unacceptable constituents. Pretreatment processes equalize flows and loadings, and precondition wastewaters to neutralize or remove toxics and industrial wastes that could adversely affect sewers or inhibit operations of publicly owned treatment works. Preliminary treatment processes protect plant mechanical equipment; remove extraneous matter such as grit, trash, and debris; reduce odors; and render incoming sewage more amenable to subsequent treatment and handling. Primary treatment employs mechanical and physical unit processes to separate and remove floatables and suspended solids and to prepare wastewater for biological treatment. Secondary treatment utilizes aerobic microorganisms in biological reactors to feed on dissolved and colloidal organic matter. As these microorganisms reduce biochemical oxygen demand and turbidity (suspended solids), they grow, multiply, and form an organic floc, which must be captured and removed in final settling tanks. Tertiary treatment, or advanced treatment, removes specific residual substances, trace organic materials, nutrients, and other constituents that are not removed by biological processes. Most advanced wastewater treatment systems include denitrification and ammonia stripping, carbon adsorption of trace organics, and chemical precipitation. Evaporation, distillation, electrodialysis, ultrafiltration, reverse osmosis, freeze drying, freeze-thaw, floatation, and land application, with particular emphasis on the increased use of natural and constructed wetlands, are being studied and utilized as methods for advanced wastewater treatment to improve the quality of the treated discharge to reduce unwanted effects on the receiving environment. *See* ABSORPTION; DISTILLATION; EVAPORATION; SEWAGE; SEWAGE DISPOSAL; ULTRAFILTRATION; WETLANDS.

On-site sewage treatment for individual homes or small institutions uses septic tanks, which provide separation of solids in a closed, buried unit. Effluent is discharged to subsurface absorption systems. *See* SEPTIC TANK; STREAM POLLUTION; UNIT PROCESSES; WATER TREATMENT.

Gerald Palevsky

Sewing machine A mechanism that stitches cloth, leather, book pages, and other material by means of a double-pointed needle or any eye-pointed needle. In ordinary two-threaded machines, a lock stitch is formed (see illustration). A presser foot held against the material with a yielding spring adjusts itself automatically to variations in thickness of material and allows the operator to turn the material as it feeds through the machine. A cluster of cams, any one of which can be selected to guide the needle arm, makes possible a variety of stitch patterns. *See* CAM MECHANISM.

Frank H. Rockett

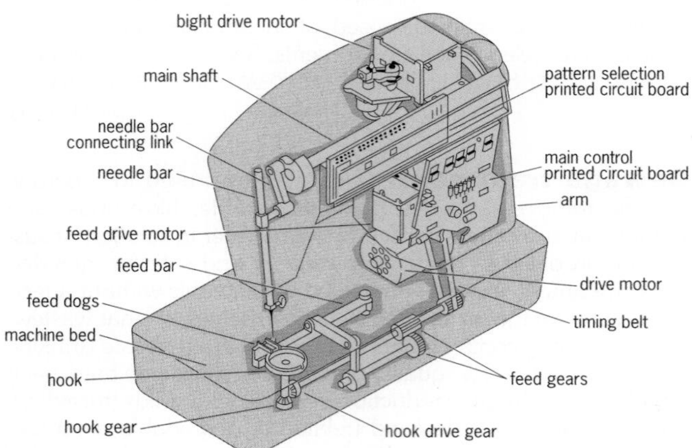

Lock stitch; upper or needle thread is dark color, and under or bobbin thread is light color.

Sex determination The genetic mechanisms by which sex is determined in all living organisms. The nature of the genetic basis of sex determination varies a great deal, however, among the various forms of life.

There are two aspects of sexuality: the primary form involves the gametes, and the secondary aspect is gender. In its broadest usage the term "sex" refers to the processes that enable species to exchange materials between homologous chromosomes, that is, to effect recombination. Generally, recombination is essential to their mechanism for reproduction. For most organisms this involves, either exclusively or as one stage in the life cycle, the formation of special cells, known as gametes, by meiosis. *See* GAMETOGENESIS.

Most sexually reproducing species produce two different kinds of gametes. The relatively large and sessile form, an ovum or egg, usually accumulates nutriments in its cytoplasm for the early development of the offspring. The relatively mobile form, a sperm (or pollen grain in many plants), contributes little beyond a haploid chromosome set. Thus the primary form of sex differentiation determines which kind of gamete will be produced. The formation of gametes usually involves the concomitant differentiation of specialized organs, the gonads, to produce each kind of gamete. The ova-producing gonad is usually known as an archegonium or ovary (in flowering plants it is part of a larger organ, the pistil or carpel); the gonad producing the more mobile gametes is usually known as a testis in animals and an antheridium or stamen in plants. *See* OVARY; OVUM; SPERM CELL; TESTIS.

In most animals and many plants, individuals become specialized to produce only one kind of gamete. These individuals usually differ not only in which kind of gonad they possess but also in a number of other morphological and physiological differences, or secondary sex characteristics. The latter may define a phenotypic sex when present, even if the typical gonad for that sex is absent or nonfunctional. The form that usually produces ova is known as female; the one that usually produces sperm or pollen is known as male. Since some sexual processes do not involve gametes, the more universal application of the term "gender" refers to any donor of genetic material as male and the recipient as female.

Sexual reproduction in plants is not always accompanied by the kinds of differentiation described above. The majority of plant species are monoecious, with both kinds of gonads on the same plant. Plants such as the date palm, willow, poplar, and papaya that bear male and female gonads on separate plants are dioecious. They occur in about 60 of the 400 or so families of flowering plants, 20 of which are thought to contain exclusively dioecious species. *See* PLANT REPRODUCTION.

Although the sexes are distinct in most animals, many species are hermaphroditic; that is, the same individual is capable of producing both eggs and sperm. This condition is particularly common among sessile or sluggish, slowly moving forms, including most flatworms, a large percentage of the worms that are parasitic in humans and other mammals, many members of the annelid phylum, many mollusks, and some fishes.

Sex differentiations are often accompanied by consistent chromosomal dimorphisms, leading to the presumption that the chromosomal differences are related to, and possibly responsible for, the sex differences. Indeed, the chromosomes that are not alike in the two sexes were given the name sex chromosomes very soon after their discovery. Some workers use the term "heterosomes" to distinguish them from the autosomes, which are the chromosome pairs that are morphologically identical to each other in both sexes.

In most species, one of the sex chromosomes, the X chromosome, normally occurs as a pair in one gender but only singly in the other. The gender with two X chromosomes is known as the homogametic sex, because each gamete normally receives an X chromosome after meiosis. The gender with only one X chromosome generally also has a morphologically different sex chromosome, the Y chromosome. The X and Y chromosomes usually pair to some extent at meiosis, with the result that the XY is the heterogametic sex, with half its gametes containing an X and half containing a Y. Geneticists soon noted that the fundamental dimorphism of X and Y chromosomes lay in their genic contents: X chromosomes of the species share homologous loci, just as do pairs of autosomes, whereas the Y chromosome usually has few, if any, loci that are also represented on the X. Thus X and Y chromosomes are sometimes very similar in shape or size but are almost always very different in genetic materials. *See* CHROMOSOME.

The major factor in sex differentiation in humans is a locus on the short arm of the Y chromosome designated SRY or SrY (for sex-determining region of the Y). This comparatively small gene contains no introns and encodes for a protein with only 204 amino acids. The protein appears to be a deoxyribonucleic acid (DNA)-binding type that causes somatic cells of the developing gonad to become Sertoli cells that secrete a hormone, Müllerian inhibiting substance (MIS), that eliminates the Müllerian duct system (the part that would produce major female reproductive organs). The gonad is now a testis, and certain cells in it become the Leydig cells that produce testosterone, which causes the primordial Wolffian duct system of the embryo to develop the major male reproductive organs. If no MIS is produced, further development of the Müllerian duct structures occurs, and in the absence of testosterone the Wolffian ducts disappear, producing the normal female structures. Embryos lacking SRY or having mutated forms of it normally become females even if they are XY. This system of sex determination is called Y-dominant. It appears to be characteristic of almost all mammals, even marsupials, among others. While SRY is the primary gene, many other genes, both autosomal and X-chromosomal, are involved in the course of developing the two sexes in mammals, as manifested by disorders in gender development or sex reversal when they mutate. *See* MUTATION.

In some organisms, sex determination seems to depend on the environment rather than on genes or chromosomes. In the horsetail plant, *Equisetum*, sex depends on growth conditions: if conditions are good, the plants become female, and if conditions are poor, they are male. Similarly, in many fishes and reptiles the temperature at which the eggs develop determines the resultant sex ratio. In alligators, for example, warm temperatures result in an excess of males, whereas cool temperatures produce an excess of females. In a number of turtle species the opposite is true, with females predominating in the warmer environment.

M. Levitan

Sex-linked inheritance The inheritance of a trait (phenotype) that is determined by a gene located on one of the sex chromosomes. These traits, including many diseases, have characteristic patterns of familial transmission; for a century these patterns were used to localize the genes for these traits to a specific chromosome. As the genomes of many organisms have now been completely sequenced, reliance on the specific pattern of inheritance to map genes on sex chromosomes has waned.

The expectations of sex-linked inheritance in any species depend on how the chromosomes determine sex. For example, in humans, males are heterogametic, having one X chromosome and one Y chromosome, whereas females are homogametic, having two X chromosomes. In human males, the entire X chromosome is active (although not all genes are active in every cell), whereas one of a female's X chromosomes is largely inactive. Random inactivation of one X chromosome occurs during the early stages of female embryogenesis, and every cell that descends from a particular embryonic cell has the same X chromosome inactivated. The result is dosage compensation for X-linked genes between the sexes. A specific gene on the long arm of the X chromosome, called *XIST* at band q13, is a major controller of X inactivation. This pattern of sex determination occurs in most vertebrates, but in birds and many insects and fish the male is the homogametic sex. *See* SEX DETERMINATION.

In general terms, traits determined by genes on sex chromosomes are not different from traits determined by autosomal genes. Enzymes, structural proteins, receptor proteins, and so forth are encoded by genes on both autosomes and sex chromosomes. Sex-linked traits are distinguishable by their mode of transmission through successive generations of a family. Sex-linked inheritance has been studied extensively in humans, for whom it is preferable to speak in terms of X-linked or Y-linked inheritance.

Red-green color blindness was the first human trait proven to be due to a gene on a specific chromosome. The characteristics of this pattern of inheritance are readily evident. Males are more noticeably or severely affected than females; in the case of red-green color blindness, women who have one copy of the mutant gene (that is, are heterozygous or carriers) are not at all affected. Among offspring of carrier mothers, on average one-half of their sons are affected, whereas one-half of their daughters are carriers. Affected fathers cannot pass their mutant X chromosome to their sons, but do pass it to all of their daughters, who thereby are carriers. A number of other well-known human conditions behave in this manner, including the two forms of hemophilia, Duchenne muscular dystrophy, the fragile X syndrome that is a common cause of mental retardation, and glucose-6-phosphate dehydrogenase deficiency, which predisposes to hemolytic anemia. *See* ANEMIA; COLOR VISION; HEMOPHILIA; MUSCULAR DYSTROPHY.

Refined cytogenetic and molecular techniques have supplemented family studies as a method for characterizing sex-linked inheritance and for mapping genes to sex chromosomes in many species. Over 500 human traits and diseases seem to be encoded by genes on the X chromosome, and nearly 500 genes have been mapped. Among mammals, genes on the X chromosome are highly conserved. Thus, identifying an X-linked trait in mice is strong evidence that a similar trait, and underlying gene, exists on the human X chromosome. The human Y chromosome harbors about 50 genes. *See* GENETICS; HUMAN GENETICS.

Reed E. Pyeritz

Sextant A navigation instrument used for measuring angles, primarily altitudes of celestial bodies. Originally, the sextant

A marine sextant.

had an arc of 60°, or 1/6 of a circle, from which the instrument derived its name. Because of the double-reflecting principle used, such an instrument could measure angles as large as 120°. In modem practice, the name sextant is commonly applied to all instruments of this type regardless of the length of the arc, which is seldom exactly 60°. The optical principles of the sextant are similar to those of the prismatic astrolabe. *See* PRISMATIC ASTROLABE.

Modern sextants may be grouped into two classes, marine and air. The marine sextant (see illustration) is designed for use by mariners. It utilizes the visible sea horizon as the horizontal reference. An instrument designed for use in aircraft is called an air sextant. Such sextants have built-in artificial horizons. Most modern air sextants are periscopic to permit observation of celestial bodies without need of an astrodome in the aircraft. *See* CELESTIAL NAVIGATION. Alton B. Moody

Sexual dimorphism Any difference, morphological or behavioral, between males and females of the same species. In many animals, the sex of an individual can be determined at a glance. For example, roosters have bright plumage, a large comb, and an elaborate tail, all of which are lacking in hens. Sexual dimorphism arises as a result of the different reproductive functions of the two sexes and is a consequence of both natural selection and sexual selection. Primary differences, such as the structure of the reproductive organs, are driven by natural selection and are key to the individual's function as a mother or father. Other differences such as the peacock's enormous tail are driven by sexual selection and increase the individual's success in acquiring mates. *See* ORGANIC EVOLUTION.

A less obvious sexual dimorphism is the difference in size of male and female gametes. In nearly all cases, the sperm (or pollen) are substantially smaller and more numerous than the ova. Eggs are large because they contain nutrients essential for development of the embryo. However, the sole purpose of sperm is to fertilize the egg. Sperm do not contain any nutrients and can therefore be small. *See* GAMETOGENESIS.

In nearly all animal groups (apart from mammals and birds), females are larger than males because bigger females tend to produce more eggs. Therefore large female size is favored. In contrast, among mammals and birds males are generally the larger sex. Differences in body size and shape can be caused by factors other than reproductive success. Sexual dimorphism can arise as a consequence of competition between the sexes over resources, or because the sexes use different resources. For example, in many species of snake, males and females use different habitats and eat different food, which has led to differences in their head shape and feeding structures.

Plants also differ in showiness. Many plants bear both male and female flowers (simultaneous hermaphrodites), but male flowers are sometimes larger and more conspicuous. For example, the female catkins of willow are dull gray compared with the bright yellow male catkins, because male flowers compete with each other to attract pollinators. In plant species with separate sexes (dioecious), males tend to produce more flowers than females in order to increase their chances of pollen transfer to females. *See* FLOWER; POLLINATION.

Animals and plants show marked sexual dimorphism in other traits. Calling, singing, pheromones, and scent marking can all be explained by competition between males and by female mate choice. *See* ANIMAL COMMUNICATION.

Associated with morphological sexual dimorphism are several behavioral differences between males and females. Many of these are related to locating a mate, competition between males, and female choosiness. Animals also show sexual dimorphism relating to their roles as parents. Many parents continue to provide for their young after birth, with the female performing the bulk of the care in most species. Female mammals suckle their young, whereas males cannot because they lack mammary glands. However, some mammals and many birds share parental duties, with both males and females feeding and protecting the young. *See* MATERNAL BEHAVIOR. Alison M. Dunn; Nina Wedell

Sexually transmitted diseases Infections that are acquired and transmitted by sexual contact. Although virtually any infection may be transmitted during intimate contact, the term sexually transmitted disease is restricted to conditions that are largely dependent on sexual contact for their transmission and propagation in a population. The term venereal disease is literally synonymous with sexually transmitted disease but traditionally is associated with only five long-recognized diseases (syphilis, gonorrhea, chancroid, lymphogranuloma venereum, and donovanosis). Sexually transmitted diseases occasionally are acquired nonsexually (for example, by newborn infants from their mothers, or by clinical or laboratory personnel handling pathogenic organisms or infected secretions), but in adults they are virtually never acquired by contact with contaminated intermediaries such as towels, toilet seats, or bathing facilities. However, some sexually transmitted infections (such as human immunodeficiency virus infection, viral hepatitis, and cytomegalovirus infection) are transmitted primarily by sexual contact in some settings and by nonsexual means in others. *See* GONORRHEA; SYPHILIS.

The sexually transmitted diseases may be classified in the traditional fashion, according to the causative pathogenic organisms, as follows:

Bacteria
- *Chlamydia trachomatis*
- *Neisseria gonorrhoeae*
- *Treponema pallidum*
- *Mycoplasma genitalium*
- *Mycoplasma hominis*
- *Ureaplasma urealyticum*
- *Haemophilis ducreyi*
- *Calymmatobacterium granulomatis*
- *Salmonella* species
- *Shigella* species
- *Campylobacter* species

Viruses
- Human immunodeficiency viruses (types 1 and 2)
- Herpes simplex viruses (types 1 and 2)
- Hepatitis viruses B, C, D
- Cytomegalovirus
- Human papillomaviruses
- Molluscum contagiosum virus
- Kaposi sarcoma virus

Protozoa
- *Trichomonas vaginalis*
- *Entamoeba histolytica*
- *Giardia lamblia*
- *Cryptosporidium* and related species

Ectoparasites
- *Phthirus pubis* (pubic louse)
- *Sarcoptes scabiei* (scabies mite)

Sexually transmitted diseases may also be classified according to clinical syndromes and complications that are caused by one or more pathogens as follows:

1. Acquired immunodeficiency syndrome (AIDS) and related conditions
2. Pelvic inflammatory disease
3. Female infertility
4. Ectopic pregnancy
5. Fetal and neonatal infections
6. Complications of pregnancy

7. Neoplasia
8. Human papillomavirus and genital warts
9. Genital ulcer-inguinal lymphadenopathy syndromes
10. Lower genital tract infection in women
11. Viral hepatitis and cirrhosis
12. Urethritis in men
13. Late syphilis
14. Epididymitis
15. Gastrointestinal infections
16. Acute arthritis
17. Mononucleosis syndromes
18. Molluscum contagiosum
19. Ectoparasite infestation

See ACQUIRED IMMUNE DEFICIENCY SYNDROME (AIDS); CANCER (MEDICINE); DRUG RESISTANCE; GASTROINTESTINAL TRACT DISORDERS; HEPATITIS; PUBLIC HEALTH; URINARY TRACT DISORDERS.

Most of these syndromes may be caused by more than one organism, often in conjunction with nonsexually transmitted pathogens. They are listed in the approximate order of their public health impact. H. Hunter Handsfield

Sferics Electromagnetic radiations produced primarily by lightning strokes from thunderstorms. It is estimated that globally there occur about 2000 thunderstorms at any one time, and that these give rise to about 100 lightning strokes every second. The radiations are short impulses that usually last a few milliseconds, with a frequency content ranging from the low audio well into the gigahertz range. Sferics (short for atmospherics) are easily detected with an ordinary amplitude-modulation (AM) radio tuned to a region between radio stations, especially if there are thunderstorms within a few hundred miles. These sounds or noises have been identified and characterized with specific names, for example, hiss, pop, click, whistler, and dawn chorus. They fall into what is generally known as radio noise. *See* ATMOSPHERIC ELECTRICITY; DUST STORM; ELECTROMAGNETIC RADIATION; LIGHTNING; THUNDERSTORM.

The various types of sferics include terrestrial, magnetospheric, and Earth-ionospheric. Terrestrial sferics includes anthropogenic noise from sources such as automobile ignition, motor brushes, coronas from high-voltage transmission lines, and various high-current switching devices. Dust storms and dust devils have also been observed to produce sferics. *See* ELECTRICAL NOISE.

Lightning-generated sferics are sometimes coupled into the magnetosphere, where they are trapped and guided by the Earth's magnetic field. In this mode, the impulse travels in an ionized region. As a result, the frequencies present in the original impulse are separated by dispersion (the higher frequencies travel faster than the lower) and produce the phenomena known as whistlers. *See* MAGNETOSPHERE.

By far the dominant and most readily observed sferics are the lightning-produced impulses that travel in the spherical cavity formed by the ionosphere and the Earth's surface. Lightning currents produce strong radiation in the very low-frequency band, 3–30 kHz, and in the extremely low-frequency band, 6 Hz–3 kHz. *See* IONOSPHERE; RADIO-WAVE PROPAGATION. Marx Brook

Shadow A region of darkness caused by the presence of an opaque object interposed between such a region and a source of light. A shadow can be totally dark only in that part called the umbra, in which all parts of the source are screened off. With a point source, the entire shadow consists of an umbra, since there can be no region in which only part of the source is eclipsed. If the source has an appreciable extent, however, there exists a transition surrounding the umbra, called the penumbra, which is illuminated by only part of the source. Depending on what fraction of the source is exposed, the illumination in the penumbra varies from zero at the edge of the full shadow to the maximum where the entire source is exposed. The edge of the umbra is not perfectly sharp, even with an ideal point source, because of the wave character of light. *See* DIFFRACTION; ECLIPSE. Francis A. Jenkins; William W. Watson

Shadowgraph An optical method of rendering fluid flow patterns visible by using index-of-refraction differences in the flow. The method relies on the fact that rays of light bend toward regions of higher refractive index while passing through a transparent material. The fluid is usually illuminated by a parallel beam of light. The illustration depicts the method as it might be applied to a fluid sample undergoing thermal convection between two parallel plates, with the lower plate being kept warmer than the upper one. As illustrated, the rays bend toward the cooler down-flowing regions, where the refractive index is higher, and away from the warmer up-flowing ones. After they have passed through the fluid layer, the rays tend to focus above the cooler regions and defocus above the warmer regions. If an image of the light beam is recorded not too far from the sample, brighter areas of the image will lie above regions of down flow, where the rays have been concentrated, and darker areas will lie above regions of up flow. Because the light passes completely through the sample, the bending effect for each ray is averaged over the sample thickness. *See* CONVECTION (HEAT); REFRACTION OF WAVES.

In convection experiments the refractive index varies because of thermal expansion of the fluid, but the method is not restricted regarding the mechanism responsible for disturbing the refractive index. Thus the same method may be used to visualize denser and less dense regions in a gas flowing in a wind tunnel, including Mach waves and shock waves, where the denser regions have a higher-than-average refractive index. *See* SHOCK WAVE; SUPERSONIC FLOW; WIND TUNNEL.

Images are usually recorded by means of a charge-coupled-device (CCD) camera, digitized, and stored in a computer. Such a digitized image consists of an array of numbers, each number being proportional to the brightness at a particular point in the image. The image points (pixels) form a closely spaced rectangular grid. A reference image may be taken in the absence of any fluid flow, and the reference image may be divided point by point

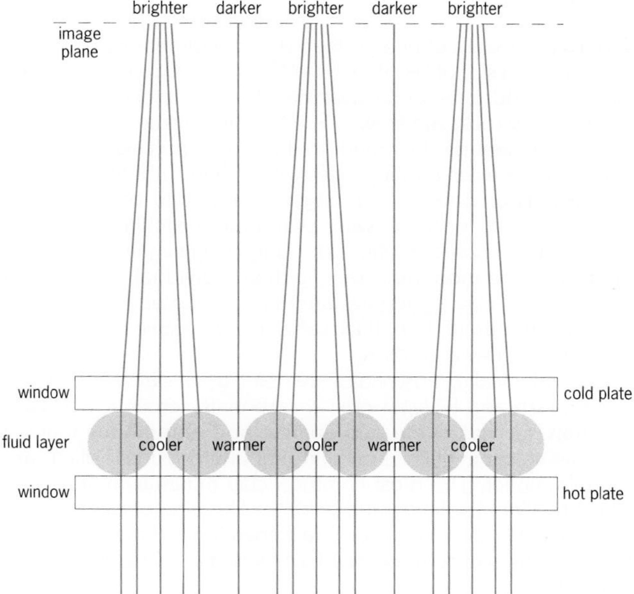

Schematic of the shadowgraph method applied to reveal convection patterns in a fluid layer. A cross section of the apparatus perpendicular to the convection rolls is shown.

into images taken with the fluid moving. *See* CHARGE-COUPLED DEVICES. David S. Cannell

Shaft balancing The process (often referred to as rotor balancing) of redistributing the mass attached to a rotating body in order to reduce vibrations arising from centrifugal force.

A rotating shaft supported by coaxial bearings (for example, ball bearings together with any attached mass, such as a turbine disk or motor armature) is called a rotor. If the center of mass of a rotor is not located exactly on the bearing axis, a centrifugal force will be transmitted via the bearings to the foundation. The horizontal and vertical components of this force are periodic shaking forces that can travel through the foundation to create serious vibration problems in neighboring components.

Any rigid shaft may be dynamically balanced by adding or subtracting a definite amount of mass at any convenient radius in each of two arbitrary transverse cross sections of the rotor. The so-called balancing planes selected for this purpose are usually located near the ends of the rotor, where suitable shoulders or balancing rings have been machined to permit the convenient addition of mass (lead weights, calibrated bolts, and so on) or the removal of mass (by drilling or grinding). Long rotors, running at high speeds, may undergo appreciable elastic deformations. For such flexible rotors it is necessary to utilize more than two balancing planes. *See* MECHANICAL VIBRATION. Burton Paul

Shafting The machine element that supports a roller and wheel so that they can perform their basic functions of rotation. Shafting, made from round metal bars of various lengths and machined to dimension the surface, is used in a great variety of shapes and applications. Because shafts carry loads and transmit power, they are subject to the stresses and strains of operating machine parts. Most shafting is rigid and carries bending loads without appreciable deflection. Some shafting is highly flexible; it is used to transmit motion around comers.

Shafts used in special ways are given specific names, although fundamentally all applications involve transmission of torque. The primary shafting connection between a wheel and a housing is an axle. A short shaft is a spindle. A short stub shaft mounted as part of a motor or engine or extending directly therefrom is a head shaft. A secondary shaft that is driven by a main shaft and from which power is supplied to a machine part is called a countershaft. *See* BELT DRIVE; PULLEY. James J. Ryan

Shale A class of fine-grained clastic sedimentary rocks with a mean grain size of less than 0.0625 mm (0.0025 in.), including siltstone, mudstone, and claystone. One-half to two-thirds of all sedimentary rocks are shales. *See* SEDIMENTARY ROCKS.

Shale is deposited as mud, that is, small particles of silt and clay. The particles are deposited when fluid turbulence caused by currents or waves is no longer adequate to counteract the force of gravity, or if the water evaporates or infiltrates into the ground. Clay particles often form larger aggregates which settle from suspension more rapidly than individual particles. Silt particles and clay aggregates are often deposited as thin layers less than 10 mm (0.4 in.) thick called laminae. *See* DEPOSITIONAL SYSTEMS AND ENVIRONMENTS.

Mineralogically, most shales are made up of clay minerals, silt-sized quartz and feldspar grains, carbonate cements, accessory minerals such as pyrite and apatite, and amorphous material such as volcanic glass, iron and aluminum oxides, silica, and organic matter. The most common clay minerals in shales are smectite, illite, kaolinite, and chlorite. The type of clay particles depo- sited is dependent on the mineralogy, climate, and tectonics of the source area. *See* CLAY MINERALS; CHLORITE; ILLITE; KAOLINITE.

Shales are usually classified or described according to the amount of silt, the presence and type of lamination, mineralogy, chemical composition, and color. Variations in these properties are related to the type of environment in which the shale was deposited and to postdepositional changes caused by diagenesis and compaction. *See* DIAGENESIS.

The small size of pores in shale relative to those in sandstone causes shale permeability to be much lower than sand permeability. Although fracturing due to compaction stresses or to tectonic movements can create deviations from this general trend, shales often form permeability barriers to fluid movement; this has important bearing on the occurrence of subsurface water and hydrocarbons. Ground-water aquifers are commonly confined by an underlying low-permeability shale bed or aquiclude, which prevents further downward movement of the water. Hydrocarbon reservoirs are often capped by low-permeability shale which forms an effective seal to prevent hydrocarbons from escaping. *See* AQUIFER. Joseph R. Davis

Shape memory alloys A group of metallic materials that can return to some previously defined shape or size when subjected to the appropriate thermal procedure. That is, shape memory alloys can be plastically deformed at some relatively low temperature and, upon exposure to some higher temperature, will return to their original shape. Materials that exhibit shape memory only upon heating are said to have a one-way shape memory, while those which also undergo a change in shape upon recooling have a two-way memory. Typical materials that exhibit the shape memory effect include a number of copper alloy systems and the alloys of gold-cadmium, nickel-aluminum, and iron-platinum. *See* ALLOY; CRYSTALLOGRAPHY; HEAT TREATMENT (METALLURGY); METAL, MECHANICAL PROPERTIES OF. Mel Schwartz

Shear A straining action wherein applied forces produce a sliding or skewing type of deformation. A shearing force acts parallel to a plane as distinguished from tensile or compressive forces, which act normal to a plane. Examples of force systems producing shearing action are forces transmitted from one plate to another by a rivet that tend to shear the rivet, forces in a beam that tend to displace adjacent segments by transverse shear, and forces acting on the cross section of a bar that tend to twist it by torsional shear (see illustration). Shear forces are usually accompanied by normal forces produced by tension, thrust, or bending. Shearing stress is the intensity of distributed force expressed as force per unit area. *See* STRESS AND STRAIN. John B. Scalzi

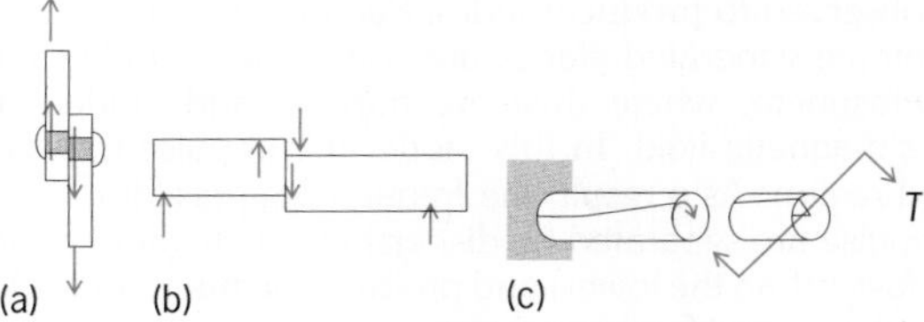

Shearing actions. (*a*) Single shear on rivet. (*b*) Transverse shear in beam. (*c*) Torsion.

Shear center A point on a line parallel to the axis of a beam through which any transverse force must be applied to avoid twisting of the section. A beam section will rotate when the resultant of the internal shearing forces is not collinear with the externally applied force. The shear center may be determined by locating the line of action of the resultant of the internal shear forces. A rolled wide flange beam section has two axes of symmetry, and therefore the shear center coincides with the geometric center or centroid of the section. When such a beam member is loaded transversely in the plane of the axes, it will bend without twisting. *See* LOADS, TRANSVERSE. John B. Scalzi

Sheep Sheep are members of the family Bovidae in the order Artiodactyla, the even-toed, hoofed mammals. Sheep were possibly the first animals to be domesticated about 11,000 years ago in southwestern Asia. *See* ARTIODACTYLA.

Wild sheep range over the mountainous areas of the Northern Hemisphere. They have long, curved, massive horns and mixed coats with a hairy outercoat and a woolly undercoat. Although wild sheep vary in size, they are usually larger than domestic sheep and have shorter tails.

Sheep are called lambs until about 12 months of age, from which time to 24 months they are referred to as yearlings, and thereafter as two-year-olds, and so on. The female sheep is called a ewe and the male, a ram or buck; the castrated male is called a wether. Sheep meat is called lamb or mutton depending on the age at slaughter.

A precise definition is difficult since sheep are so variable. Horns, if present, tend to curl in a spiral. If horns occur in both sexes, they are usually larger for the ram. In some breeds only the rams have horns but many breeds are hornless. The hairy outercoat common to wild sheep was eliminated by breeding in most domestic sheep. Wool may cover the entire sheep, or the face, head, legs, and part of the underside may be bare; or wool may be absent as in some breeds (bred for meat) with a short hairy coat or a coat which constantly sheds. Although domestic sheep are mostly white, shades of brown, gray, and black occur, sometimes with spotting or patterns of color. *See* WOOL.

Hundreds of breeds of sheep of all types, sizes, and colors are found over the world. Wool-type breeds, mostly of Merino origin, are important in the Southern Hemisphere, but both fine- and long-wool types are distributed all over the world. Sheep with fat tails or fat rumps are common in the desert areas of Africa and Asia. These usually produce carpet wool. Milk breeds are found mostly in central and southern Europe. Meat breeds from the British Isles are common over the world. Clair E. Terrill

Sheet-metal forming The shaping of thin sheets of metal (usually less than 1/4 in. or 6 mm) by applying pressure through male or female dies or both. Parts formed of sheet metal have such diverse geometries that it is difficult to classify them. Sheet forming is accomplished basically by processes such as stretching, bending, deep drawing, embossing, bulging, flanging, roll forming, and spinning. In most of these operations there are no intentional major changes in the thickness of the sheet metal. *See* METAL FORMING.

Stretch forming is a process in which the sheet metal is clamped between jaws and stretched over a form block. The process is used primarily in the aerospace industry to form large panels with varying curvatures.

Bending is one of the most common processes in sheet forming. The part may be bent not only along a straight line, but also along a curved path (stretching, flanging). In addition to male and female dies used in most bending operations, the female die can be replaced by a rubber pad (Fig. 1). The roll-forming process replaces the vertical motion of the dies by the rotary motion of rolls with various profiles. Each successive roll bends the strip a little further than the preceding roll.

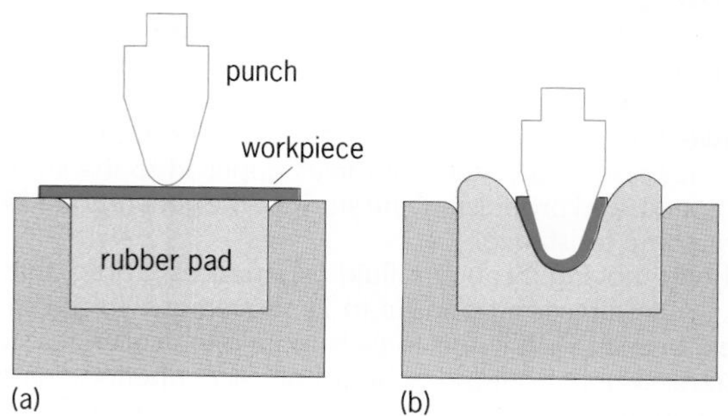

Fig. 1. Bending process with a rubber pad. (*a*) Before forming. (*b*) After forming.

While many sheet-forming processes are carried out in a press with male and female dies usually made of metal, there are some processes which utilize rubber to replace one of the dies. The simplest of these processes is the Guerin process (Fig. 2).

A great variety of parts are formed by the deep-drawing process (Fig. 3), the successful operation of which requires a careful control of factors such as blank-holder pressure, lubrication, clearance, material properties, and die geometry.

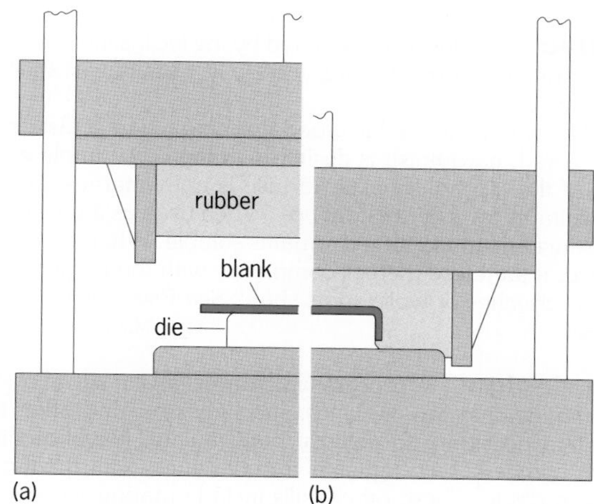

Fig. 2. Guerin process, the simplest rubber forming process. (*a*) Before forming. (*b*) After forming.

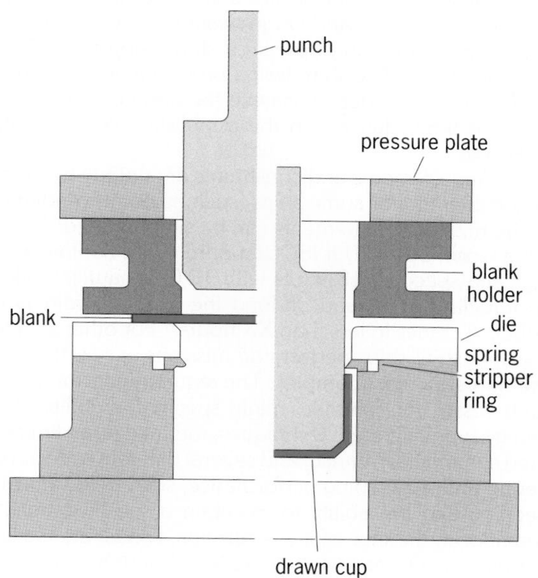

Fig. 3. The deep-drawing process.

Many parts require one or more additional processes. Embossing consists of forming a pattern on the sheet by shallow drawing. Coining consists of putting impressions on the surface by a process that is essentially forging, the best example being the two faces of a coin. Shearing is separation of the material by the cutting action of a pair of sharp tools, similar to a pair of scissors. *See* COINING.

The spinning process forms parts with rotational symmetry over a mandrel with the use of a tool or roller. There are two basic types of spinning: conventional or manual spinning, and shear spinning. The conventional spinning process forms the material over a rotating mandrel with little or no change in the thickness of the original blank. In shear spinning (hydrospinning, floturning) the deformation is carried out with a roller in such a manner that the diameter of the original blank does not change but the thickness of the part decreases by an amount dependent on the mandrel angle. Shear spinning produces parts with various shapes (conical, curvilinear, and also tubular by tube spinning on a cylindrical mandrel) with good surface finish, close tolerances, and improved mechanical properties. *See* METAL COATINGS; SPINNING (METALS). Serope Kalpakjian

Shellac The lac resin (secreted by the lac insect) when used in flake (or shell) form. Shellac varnish is a solution of shellac in denatured alcohol.

Shellac varnish is used in wood finishing where a fast-drying, light-colored, hard finish is desired. Drying is by simple evaporation of the alcohol. Shellac varnish is not water-resistant and is not suitable for exterior coatings. When used as a finish, it has the distinct advantage that it remains soluble. When touch-up is required, it therefore merges completely with the original finish, and no scratches or worn spots show. *See* PAINT AND COATINGS; VARNISH. C. R. Martinson; C. W. Sisler

Ship design A process which translates a set of owner's requirements into the drawings, specifications, and other technical data necessary to actually build a ship. Naval architects lead the process, but engineers and designers with many other skills contribute. These other skills include marine engineering, structural design, and production engineering. The ship design process is iterative, and is subdivided into several phases during which the design is developed in increasing degrees of detail. Typically, the owner's requirements specify the mission that the new ship must perform and define such parameters as required speed, fuel endurance, and cargo weight and capacity. Generally, the cost to build and operate a ship is constrained by the prospective owner. The ship design process involves numerous trade-off studies in order to achieve the desired capability and, at the same time, stay within the established cost. *See* NAVAL ARCHITECTURE.

Mission requirements and constraints are unique to each ship being considered. For some ships, such as point-to-point cargo ships, the mission requirements can be simply stated; for example, "Transport 5000 20 ft ISO standard cargo containers at an average sea speed of 18 knots with 10,000 nautical miles between refuelings. On- and off-load the 5000 containers using shore-based cranes in less than XX hours." For other ship types, such as industrial ships performing missions at sea, the mission requirements are more complex. The requirements for a fisheries research vessel, for example, might specify the ability to catch fish using several different techniques, radiated noise limitations, required sonar performance, and several different aspects of maneuvering and seakeeping performance, such as low-speed stationkeeping and the ability to maintain a specified track over the sea floor in the face of cross currents, winds, and seas. *See* MERCHANT SHIP; OCEANOGRAPHIC VESSELS; SHIP POWERING, MANEUVERING, AND SEAKEEPING.

Cost, both to design and build the ship and to operate it, is usually constrained. The two primary elements of operating cost are crew and fuel, so there is nearly always pressure on the designer to reduce crew size and fuel consumption. Physical constraints may also be imposed on the design related to construction, operational, or maintenance requirements. Weight or dimensional constraints may be imposed if the ship is to be built or maintained in a specific dry dock. Pier or harbor limitations may also impose dimensional constraints. Ship length may be limited by the requirement to tie up to a certain pier. Ship air draft (vertical distance from the water surface to the highest point on the ship) may be limited by the need to pass under bridges of a certain height. Ship navigational draft (vertical distance from the water surface to the lowest point on the ship) may be limited by the depth of a dredged channel in a particular harbor.

In addition to unique mission requirements and constraints, every ship must satisfy certain physical principles. The fundamental principles are that (1) the ship hull and superstructure must have adequate storage space, and (2) the ship must float at an acceptable waterline (draft neither too great nor too small) when it is fully loaded. Another principle is that the ship must be statically stable; that is, when it is displaced from its equilibrium condition, it must tend to return to that condition. For example, when the ship is heeled to one side by a disturbing force such as a wind gust, it must tend to return to the vertical rather than continuing to roll and capsizing. The ship's hull must have sufficient strength to withstand the forces that will act upon it over a range of loading and sea conditions. The ship must possess sufficient propulsive power to achieve the desired speed even with a fouled bottom and in adverse sea conditions. In addition, it must generate sufficient electric power to satisfy the requirements of mission systems; ship machinery; heating, ventilation, and air conditioning (HVAC) systems; hotel; and other ship services. *See* BUOYANCY; HYDROSTATICS; MARINE ENGINEERING; MARINE MACHINERY. Peter A. Gale

Ship nuclear propulsion Nuclear reactors for shipboard propulsion can be of any type used for the production of useful heat. Nuclear power is particularly suitable for vessels which need to be at sea for long periods without refueling or for powerful submarine propulsion. Only the pressurized water reactor and the liquid metal reactor have actually been applied to operating vessels. The pressurized water reactor has been most widely applied since it uses a readily available coolant and has a relatively simple cycle and control system and a large industrial and technical base. The supposed advantages of a liquid metal reactor (compactness, fast response, and higher propulsion plant efficiency) have not been proven in application, and liquid metal reactors are not now in marine service. *See* NUCLEAR POWER; NUCLEAR REACTOR; REACTOR PHYSICS.

In all the shipboard nuclear power plants that have been built, energy conversion is based on the steam turbine cycle, and that portion of the plant is more or less conventional. There are two types in use: a steam turbine geared to a fixed-pitch propeller (called a geared turbine), and a steam turbine generator whose output drives an electric motor connected to a propeller (called a turboelectric unit). Any energy conversion process that converts heat into mechanical energy could be used to propel a ship. For example, a closed-cycle helium gas turbine has been studied, but none has been built for ship propulsion. *See* MARINE ENGINE; MARINE MACHINERY; PROPELLER (MARINE CRAFT); SHIP POWERING, MANEUVERING, AND SEAKEEPING. Alan R. Newhouse

Ship powering, maneuvering, and seakeeping The three central areas of ship hydrodynamics. Basic concepts of powering, maneuvering, and seakeeping are critical to an understanding of high-speed craft. *See* NAVAL ARCHITECTURE; NAVAL SURFACE SHIP; SHIP DESIGN; SHIPBUILDING.

Powering. The field of powering is divided into two related issues: resistance, the study of forces opposed to the ship's forward speed, and propulsion, the study of the generation of forces to overcome resistance.

A body moving through a fluid experiences a drag, that is, a force in the direction opposite to its movement. In the specific context of a ship's hull, this force is more often called resistance. Resistance arises from a number of physical phenomena, all of which vary with speed, but in different ways. These phenomena are influenced by the size, shape, and condition of the hull, and other parts of the ship. They include frictional resistance and form

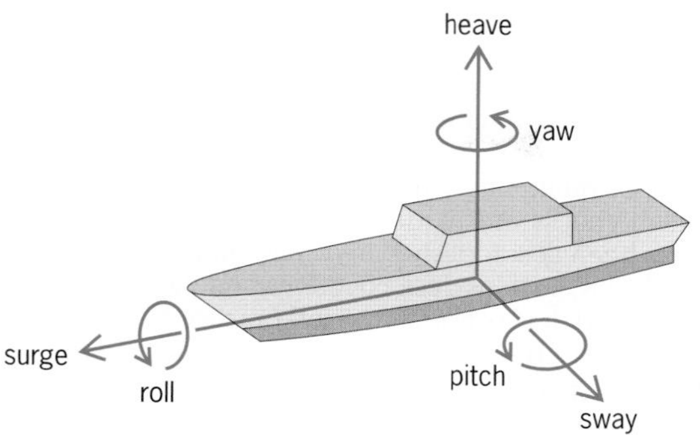

Ship motion degrees of freedom.

drag (often grouped together as viscous resistance), wavemaking resistance, and air resistance. *See* STREAMLINING; VISCOSITY.

Many devices have been used to propel a ship. In approximate historical order they include paddles, oars, sails, draft animals (working on a canal towpath), paddlewheels, marine screw propellers, vertical-axis propellers, airscrews, and waterjets. A key distinction is whether or not propulsive forces are generated in the same body of fluid that accounts for the main sources of the ship's resistance, resulting in hull-propulsor interaction. *See* PROPELLER (MARINE CRAFT).

Any propulsor can be understood as a power conversion device. Delivered power for a rotating propulsor is the product of torque times rotational speed. The useful power output from the system is the product of ship resistance times ship speed, termed effective power. The efficiency of this power conversion is often termed propulsive efficiency. *See* BOAT PROPULSION.

Maneuvering. Maneuvering (more generally, ship controllability) includes consideration of turning, course-keeping, acceleration, deceleration, and backing performance. The field of maneuvering has also come to include more specialized problems of ship handling, for example, the production of sideways motion for docking or undocking, turning in place, and position-keeping using auxiliary thrusters or steerable propulsion units. In the case of submarines, maneuvering also includes depth-change maneuvers, either independently or in combination with turning.

Seakeeping. The modern term "seakeeping" is used to describe all aspects of a ship's performance in waves, affected primarily by its motions in six degrees of freedom (see illustration). Seakeeping issues are diverse, including the motions, accelerations, and structural loads caused by waves. Some are related to the comfort of passengers and crew, some to the operation of ship systems, and others to ship and personnel safety. Typical issues include the incidence of motion sickness, cargo shifting, loss of deck cargo, hull bending moments due to waves, slamming (water impact loads on sections of the hull), added powering in waves, and the frequency and severity of water on deck.

In the past, initial powering, maneuvering, and seakeeping predictions for new designs depended almost entirely on design rules of thumb or, at best, applicable series or regression data from previous model tests, subsequently refined by additional model tests. With the increase in computing power available to the naval architect, computational fluid dynamic methods are now applied to some of these problems in various stages of the ship design process. *See* COMPUTATIONAL FLUID DYNAMICS.

Robert M. Scher

Ship routing The selection of the most favorable route for a ship's transit in the open ocean. Safety, economics, and environmental protection play a key role in determining this selection. In the late 1890s, the practice of following predetermined routes for shipping was adopted by shipping companies operating in the North Atlantic. As ships grew both in size and number, the ability to plan open-ocean transits based upon the environmental conditions expected during a voyage and the effect of these conditions upon the ship's performance became important factors in planning a ship's route. Ship routing is also referred to as weather routing, optimum track ship routing, and meteorological navigation.

Routes can be computed to minimize transit time, avoid waves over a specific height, or, more typically, provide a compromise between travel time and the roughness of the seaway to be encountered. Routing offices compute the initial route based on the long-range weather forecast, but maintain a daily plot of both the ship and the storm positions. If conditions warrant, an amended route is sent to the ship master based on the latest weather changes. *See* OCEAN WAVES; WEATHER FORECASTING AND PREDICTION.

Although the effect of waves is the predominant factor considered, the ship router also utilizes knowledge of ocean currents, wind, and hazards to navigation such as ice fields or fog banks. On U.S. east coast runs, the currents are more important than the waves. Most major oil companies have their tankers follow the Gulf Stream on northbound runs and avoid it on the return trip. *See* GULF STREAM; OCEAN CIRCULATION.

Richard W. James; Douglas Taggart

Ship salvage Voluntary response to a maritime peril by other than the ship's own crew. The property in danger can be any type or size of vessel or maritime cargo. Salvage is encouraged by compensating the salvor based on risk to the salvor and the salvor's equipment, conditions under which the salvage service was performed, the value of the vessel and cargo saved and, more recently, minimizing environmental damage.

Salvage services can be performed using either dedicated salvage vessels or vessels hired for the specific operation. Salvage ships are typically large seagoing tugs, usually capable of 5000–10,000 hp (3750–7500 kW), designed and outfitted to work at remote sites in all weather. Salvage ships carry a variety of portable salvage equipment such as firefighting gear, pumps and patching material, electrical generators, compressors, diving equipment, wire and chain, beach gear, pollution control equipment, and the material to make whatever might be needed on-site.

True salvage is time-critical, involving assets that are immediately available. Assistance is provided in situations involving collisions, firefighting, flooding and damage control, damage from hostile action, and breakdowns of propulsion or steering systems. The salvage team may fight fire from off-ship or board the casualty to fight fires. Salvors also may board the ship to control flooding and stabilize the casualty if required. Rescue towing often results from one or more of the above and the need to reach a safe haven or avoid going aground.

Stranding results when a ship drives, or is driven, aground through engine or steering casualty, error, problem with navigation or navigation aid, or bad weather, and cannot extricate itself. Time-criticality may be due to coming bad weather, tidal range, ongoing damage to the ship from the grounding, or other requirement such as clearing a blocked shipping channel. The salvor must assess the condition of the casualty and what is required to refloat it, the potential for further damage to the ship, and the potential for loss of cargo and damage to the environment.

Environmental salvage is the protection of the environment from damage by pollutants that may result from a maritime casualty. International agreements such as the International Convention for the Prevention of Pollution from Ships (MARPOL 73/78) and national laws such as the Oil Pollution Act of 1990 highlight the need for protecting the environment. This is most effectively accomplished by keeping pollutants in a ship during a casualty, minimizing dispersion, and rapidly recovering any spilled pollutants, usually cargo crude oil or petroleum products, or ship's fuel stored in bunkers. Salvage actions such as pumping air into a damaged tank to restore buoyancy may blow oil out through the damaged hull below the waterline. Salvors must be prepared to deal with pollutants when this occurs, and plan salvage operations to minimize oil release.

Wreck removal involves recovery of a stranded or sunken vessel, and is usually not time-critical. Many salvage techniques can be brought to bear. If the main deck of the stranded or sunken vessel is above water, a combination of patching and pumping may be sufficient to refloat the casualty. If the main deck is not submerged too deeply, a watertight wall, known as a cofferdam, may be built around the main deck to above the water surface. Patching and pumping may then refloat the casualty.

A ship's cargo may be of more value than the ship itself, and the salvor may recover the cargo without recovering the ship. The best-known type of cargo salvage is the recovery of jewels or precious metals. But even something as mundane as copper ingots may be considered worth recovering by both the salvor and the owner or underwriter. Richard P. Fiske

Shipbuilding The construction of large vessels which travel over seas, lakes, or rivers. Many different approaches have been used in the construction of ships. Sometimes a ship must be custom-built to suit the particular requirements of a low-volume trade route with unique cargo characteristics. On the other hand, there are many instances where a significant number of similar ships are constructed, providing an opportunity to employ procedures which take advantage of repetitive processes.

The building of a ship can be divided into seven phases: design, construction planning, work prior to keel laying, ship erection, launching, final outfitting, and sea trials. *See* SHIP DESIGN.

The construction planning process establishes the construction techniques to be used and the schedules which all of the shipbuilding activities must follow. Construction planners generally start with an erection diagram on which the ship is shown broken down into erection zones and units. To facilitate the fabrication of steel, insofar as possible, the erection units are designed to be identical. The size (or weight) of the erection units selected is usually limited by the amount of crane capacity available. Once the construction planners have established the manner in which the ship is to be erected and the sequence of construction, the schedules for construction can be developed. Working backward from the time an erection unit is required in the dock, with allowances made for the many processes involved, a schedule of working plans and for procurement of purchased equipment is prepared.

Before the keel of a ship is laid (or when the first erection unit is placed in position) a great deal of work must have been accomplished for work to proceed efficiently. The working drawings prepared by ship designers completely define a ship, but often not in a manner that can be used by the construction trades people. Structural drawings prescribe the geometry of the steel plates used in construction, but they cannot be used, in the form prepared, to cut steel plates. Instead, the detailed structural drawings must be translated into cutting sketches, or numerical-control cutting tapes, which are used to fabricate steel. Several organizations have developed sophisticated computer programs which readily translate detailed structural drawings into machine-sensible tapes which can be used to drive cutting torches.

If all of the preceding work has been accomplished properly and on schedule, the erection of a ship can proceed rapidly; however, problem areas invariably arise. When erecting a ship one plate at a time, there are no serious fitting problems; but when 900-metric-ton erection units do not fit (or align) properly, there are serious problems which tend to offset some of the advantages for this practice.

A ship is launched as soon as the hull structure is sufficiently complete to withstand the strain. Ships may be launched endwise, sidewise, or by in-place flotation (for example, graving docks). The use of a graving dock requires a greater investment in facilities than either of the other two methods, but in some cases there may be an overall advantage due to the improved access to the ship and the simplified launch procedure. *See* DRY-DOCKING.

The final outfitting of a ship is the construction phase during which checks are made to ensure that all of the previous work has been accomplished in a satisfactory manner; and last-minute details, such as deck coverings and the top coat of paint, are completed. It is considered good practice to subject as much of the ship as possible to an intensive series of tests while at the dock, where corrections and final adjustments are more easily made than when at sea. As a part of this test program, the main propulsion machinery is subjected to a dock trial, during which the ship is secured to the dock and the main propulsion machinery is operated up to the highest power level permissible.

When a comprehensive program of dockside tests have been completed, the only capabilities which have not been demonstrated are the operation of the steering gear during rated-power conditions and the operation of the main propulsion machinery at rated power; these capabilities must be demonstrated during trials at sea. R. L. Harrington

Shipping fever A severe inflammation of the lungs (pneumonia) commonly seen in North American cattle after experiencing the stress of transport. This disease occurs mainly in 6–9-month-old beef calves transported to feedlots. The characteristic shipping fever pneumonia is caused primarily by the bacteria *Mannheimia* (formerly *Pasteurella*) *haemolytica* serotype A1 and, to a lesser extent, *Pasteurella multocida*; thus a synonym for shipping fever is bovine pneumonic pasteurellosis. With the stress of shipment and infection, these pathogenic bacteria replicate rapidly in the upper respiratory tract and are inhaled into the lungs, where pneumonia develops in the deepest region of the lower respiratory tract (pulmonary alveoli). Viruses can also concurrently damage pulmonary alveoli and enhance the bacterial pneumonia. *See* ANIMAL VIRUS; INFLAMMATION; PASTEURELLA.

Symptoms of shipping fever occur 1–2 weeks after transport. Disease can occur in 50% or more of a group of cattle. Initially, cattle have reduced appetite, high fever, rapid and shallow respiration, depression, and a moist cough. During the later stages of disease, cattle lose weight and have labored breathing. The lesion causing these symptoms is a severe pneumonia accompanied by inflammation of the lining of the chest cavity, and is called a fibrinous pleuropneumonia. Without vigorous treatment, shipping fever can cause death in 5–30% of affected cattle.

Treatment is aimed at eliminating bacteria, limiting the inflammatory reaction in the lung, and providing supportive care for cattle. To eliminate bacteria, diseased cattle are injected with antibacterial drugs, usually antibiotics. To reduce inflammation, nonsteroidal antiinflammatory drugs are given. Supportive care includes separating affected cattle from the stress of crowding, supplying adequate water, and feeding a ration high in roughage. Vaccines can stimulate immunity to the viruses and bacteria associated with shipping fever. Although vaccination of cattle does not prevent shipping fever, it may reduce the number of cattle affected and the severity of the disease in individual animals. *See* PNEUMONIA; RESPIRATORY SYSTEM DISORDERS; VACCINATION. Anthony W. Confer

Shock absorber Effectively a spring, a dashpot, or a combination of the two, arranged to minimize the acceleration of the mass of a mechanism or portion thereof with respect to its frame or support.

The spring type of shock absorber is generally used to protect delicate mechanisms, such as instruments, from direct impact or instantaneously applied loads. Such springs are often made of rubber or similar elastic material. *See* SHOCK ISOLATION.

An example of the dashpot type of shock absorber is the direct-acting shock absorber in an automotive spring suspension system (see illustration). Here the device is used to dampen and control a spring movement. The energy of the mass in motion is converted to heat by forcing a fluid through a restriction, and the heat is dissipated by radiation and conduction from the shock absorber. *See* VIBRATION DAMPING.

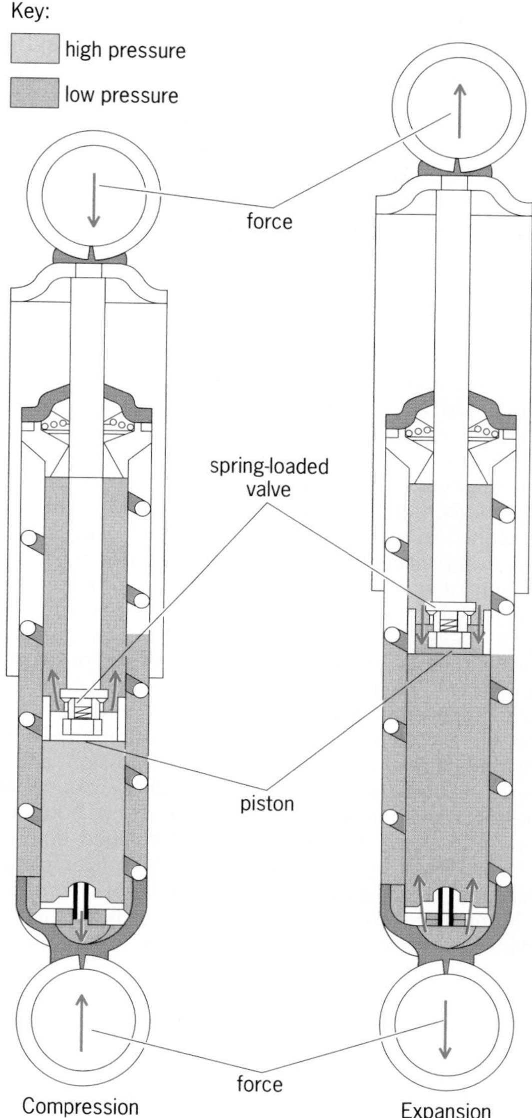

A dashpot-type shock absorber. (*Plymouth Division, Chrysler Corp.*)

There are also devices available which combine springs and viscous damping (dashpots) in the same unit. They use elastic solids (such as rubber or metal), compressed gas (usually nitrogen), or both for the spring. A flat-viscosity hydraulic fluid is used for the viscous damping. L. Sigfred Linderoth, Jr.

Shock isolation

The application of isolators to alleviate the effects of shock on a mechanical device or system. The term shock generally denotes suddenness, either in the application of a force or in the inception of a motion. *See* SHOCK WAVE.

Shock isolation is accomplished by storing energy in a resilient medium (isolator, cushion, and so on) and releasing it at a slower rate. The effectiveness of an isolator depends upon the duration of the shock impact.

Rubber is the most common material used in commercial shock isolators. Rubber isolators are generally used where the shock forces are created through small displacements. For larger displacement shock forces, such as those experienced by shipping containers in rough handling conditions, thick cushions of felt, rubberized hair, sponge rubber, cork, or foam plastics are used.

The shock load must be divided between the case, the shock cushion, and the equipment. The case, since it must withstand effects of rough handling such as sliding and dropping, is by necessity rigid. The more rigid the case the closer to a 1:1 ratio will be the transfer of the shock from outside to inside. The absorption of the shock is primarily between the cushion and the equipment. *See* DAMPING; SHOCK ABSORBER; SPRING (MACHINES); VIBRATION DAMPING. K. W. Johnson

Shock syndrome

A clinical condition caused by widespread impairment of blood flow resulting in an inadequate supply of oxygen to the body's tissues. Shock is a sudden and progressive condition that, if not reversed, can lead to death. Shock affects the entire body, as opposed to other conditions where blood flow is reduced to only a portion of a single organ, such as stroke or heart attack. It is typically, but not always, associated with a low blood pressure that can be felt as a weak pulse.

Shock can be categorized according to the initiating event: hypovolemic (loss of circulating blood volume), cardiogenic (loss of cardiac pumping activity), distributive (shunting of blood flow away from the tissues), or obstructive (a blockage of blood flow through the heart or major vessels). Regardless of the initiating event, shock will incite responses from the body. These responses will often maintain blood flow to vital organs for a period of time but at the expense of metabolic stress and exhaustion in the tissues. The nervous system reacts to shock by sending signals to the blood vessels in nonvital organs (such as the skin and bowel) that constrict these blood vessels, reduce blood flow to those organs, and allow blood to go preferentially to the vital organs (heart and brain). The nervous system also increases the heart rate in an attempt to pump more blood.

As shock progresses, the metabolically active tissues are starved for oxygen (a condition termed ischemia) and change their metabolic processes from those that use oxygen (aerobic) to those that do not require it (anaerobic). Anaerobic metabolism is a very inefficient process. Once cells are unable to maintain basal metabolic activities, the cell dies and releases harmful intracellular substances.

Advanced shock is generally associated with impaired cardiac contractility (even if this was not the initial cause) and slugging of blood flow in the microcapillaries. By this time, shock has progressed to the degree that, despite reversal of the inciting process, the impaired tissues do not recover and death results. *See* BLOOD; CARDIOVASCULAR SYSTEM; RESPIRATION. J. Stephan Stapczynski

Shock tube

A laboratory device for rapidly raising confined samples of fluids (primarily gases) to preselected high temperatures and densities. This is accomplished by a shock wave, generated when a partition (diaphragm) that separates a low-pressure (driven) section from a high-pressure (driver) section is rapidly removed. Shock tubes can be circular, square, or rectangular in cross section. In the driven section, gaseous samples can be heated to temperatures as high as 27,500°F (15,000 K) under strictly homogeneous conditions. At the shock front the transition from the unshocked to the high-temperature condition is of short but of finite duration. The incident shock wave generates a slower-moving wave that additionally heats the compressed gas. *See* SHOCK WAVE.

With the diaphragm in place the driven section is filled to a modest pressure with an inert gas plus the gas of interest. The driver section is filled to a high pressure with a low-molecular-weight gas: helium or hydrogen. When the diaphragm is rapidly ruptured, expansion of the high-pressure gas acts as a low-mass piston that generates a steepening pressure front, which moves ahead of the boundary between the driver and the test gases.

Shock tubes are used to investigate the gas dynamics of shocks and for preparing test samples for equilibrium or kinetic studies. Many gas dynamic problems can be investigated in shock tubes, such as thermal boundary-layer growth, shock bifurcation, and shock-wave focusing and reflection. Shock tubes are used to prepare gases for study at very low or very high temperatures without thermal contact with extraneous surfaces. Significant applications of shock tubes to chemical kinetics includes the determination of diatom dissociation rates and the study of polyatomic molecules. Simon H. Bauer

Shock wave A mechanical wave of large amplitude, propagating at supersonic velocity, across which pressure or stress, density, particle velocity, temperature, and related properties change in a nearly discontinuous manner. Unlike acoustic waves, shock waves are characterized by an amplitude-dependent wave velocity. Shock waves arise from sharp and violent disturbances generated from a lightning stroke, bomb blast, or other form of intense explosion, and from various types of supersonic flow over stationary objects. Extraterrestrial examples include supernovae expanding against neighboring intergalactic gas clouds and solar wind flowing over magnetized planets and satellites.

The abrupt nature of a shock wave in a gas can best be visualized from a schlieren photograph or shadow graph of supersonic flow over objects. Such photographs show well-defined surfaces in the flow field across which the density changes rapidly, in contrast to waves within the range of linear dynamic behavior of the fluid. Measurements of fluid density, pressure, and temperature across the surfaces show that these quantities always increase along the direction of flow, and that the rates of change are usually so rapid as to be beyond the spatial resolution of most instruments except when the shock wave is very weak or at very low gas densities. These surfaces of abrupt change in fluid properties are called shock waves or shock fronts. *See* SCHLIEREN PHOTOGRAPHY; SHADOWGRAPH; WAVE MOTION IN FLUIDS.

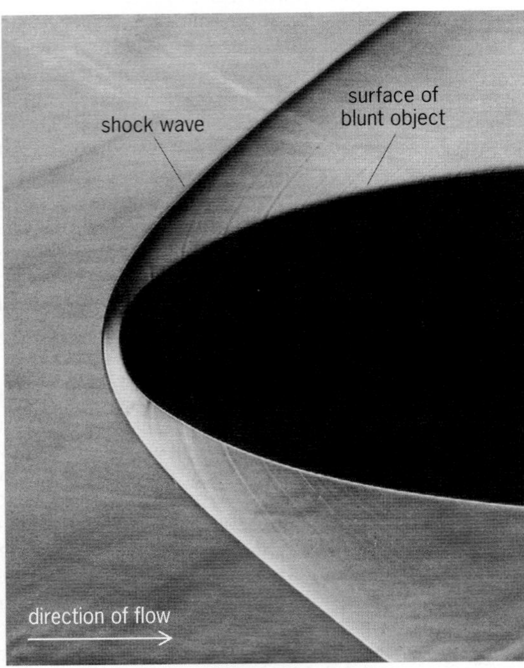

Schlieren photograph of supersonic flow over blunt object. Shock wave is approximately parabolic, and detached from object. (*Avco Everett Research Laboratory, Inc.*)

Shock waves in supersonic flow may be classified as normal or oblique according to whether the orientation of the surface of abrupt change is perpendicular or at an angle to the direction of flow. A schlieren photograph of a supersonic flow over a blunt object is shown in the illustration. Although this photograph was obtained from a supersonic flow over a stationary model in a shock tube, the general shape of the shock wave around the object is quite typical of those observed in a supersonic wind tunnel, or of similar objects (or projectiles) flying at supersonic speeds in a stationary atmosphere. The shock wave in this case assumes an approximately parabolic shape and is clearly detached from the blunt object. The central part of the wave, just in front of the object, may be considered an approximate model of the normal shock; the outer part of the wave is an oblique shock wave of gradually changing obliqueness and strength. *See* WIND TUNNEL

Shao-chi Lin

Some common examples of shock waves in condensed materials are encountered in the study of underground or underwater explosions, meteorite impacts, and ballistics problems. The field of shock waves in condensed materials (solids and liquids) has grown into an important interdisciplinary area of research involving condensed matter physics, geophysics, materials science, applied mechanics, and chemical physics. The nonlinear aspect of shock waves is an important area of applied mathematics.

Experimentally, shock waves are produced by rapidly imparting momentum over a large flat surface. This can be accomplished in many different ways: rapid deposition of radiation using electron or photon beams (lasers or x-rays), detonation of a high explosive in contact with the material, or high-speed impact of a plate on the sample surface. The impacting plate itself can be accelerated by using explosives, electrical discharge, underground nuclear explosions, and compressed gases. The use of compressed gas to accelerate projectiles with appropriate flyer plates provides the highest precision and control as well as convenience in laboratory experiments.

Large-amplitude one-dimensional compression and shear waves have been studied in solids. In these experiments, a macroscopic volume element is subjected to both a compression and shear deformation. The combined deformation state is produced by impacting two parallel flyer plates that are inclined at an angle to the direction of the plate motion. Momentum conservation coupled with different wave velocities for compression and shear waves leads to a separation of these waves in the sample interior. These experiments provide direct information about the shear response of shocked solids, and subject samples to more general loading states than the uniaxial strain state. *See* STRESS AND STRAIN.

Shock waves subject matter to unusual conditions and therefore provide a good test of understanding of fundamental processes. The majority of the studies on condensed materials have concentrated on mechanical and thermodynamic properties. These are obtained from measurements of shock velocity, stress, and particle velocity in well-controlled experiments. Advanced techniques using electromagnetic gages, laser interferometry, piezoelectric gages, and piezoresistance gages have given continuous, time-resolved measurements at different sample thicknesses.

The study of residual effects, that is, the postshock examination of samples subjected to a known pulse amplitude and duration, is of considerable importance to materials science and metallurgy. The conversion of graphite to diamonds is noteworthy. Other effects that have been observed are microstructural changes, enhanced chemical activity, changes in material hardness and strength, and changes in electrical and magnetic properties. The generation of shock-induced lattice defects is thought to be important for explaining these changes in material properties. There has been growing interest in using shock methods for material synthesis and powder compaction. *See* DIAMOND; HIGH-PRESSURE CHEMISTRY; HIGH-PRESSURE MINERAL SYNTHESIS; HIGH-PRESSURE PHYSICS. Y. M. Gupta

Short circuit An abnormal condition (including an arc) of relatively low impedance, whether made accidentally or intentionally, between two points of different potential in an electric network or system. *See* CIRCUIT (ELECTRICITY); ELECTRICAL IMPEDANCE.

Common usage of the term implies an undesirable condition arising from failure of electrical insulation, from natural causes (lightning, wind, and so forth), or from human causes (accidents, intrusion, and so forth). From an analytical viewpoint, however, short circuits represent a severe condition that the circuit designer must consider in designing an electric system that must withstand all possible operating conditions. *See* ELECTRIC PROTECTIVE DEVICES; ELECTRICAL INSULATION; LIGHTNING AND SURGE PROTECTION.

In circuit theory the short-circuit condition represents a basic condition that is used analytically to derive important information concerning the network behavior and operating capability. Thus, along with the open-circuit voltage, the short-circuit current provides important basic information about the network at a given point. *See* DIRECT-CURRENT CIRCUIT THEORY; NETWORK THEORY; THÉVENIN'S THEOREM (ELECTRIC NETWORKS).

The short-circuit condition is also used in network theory to describe a general condition of zero voltage. Thus the term short-circuit admittance (or impedance) is used to describe a network condition in which certain terminals have had their voltage reduced to zero, for the purpose of analysis. This leads to the terms short-circuit driving point admittance, short-circuit transfer admittance, and similar references to the zero voltage condition. *See* ADMITTANCE.

Short-circuit protection is a separate discipline dedicated to the study, analysis, application, and design of protective apparatus that are intended to minimize the effect of unintentional short circuits in power supply systems. For these analyses the short circuit is an important limiting (worst) case, and is used to compute the coordination of fuses, circuit reclosers, circuit breakers, and other devices designed to recognize and isolate short circuits. The short circuit is also an important parameter in the specification of these protective devices, which must have adequate capability for interrupting the high short-circuit current. *See* CIRCUIT BREAKER; FUSE (ELECTRICITY).

Short circuits are also important on high-frequency transmission lines where shorted stub lines, one-quarter wavelength long and shorted at the remote end, are used to design matching sections of the transmission lines which also act as tuning elements. *See* TRANSMISSION LINES. Paul M. Anderson

Short takeoff and landing (STOL) The term applied to heavier-than-air craft that cannot take off and land vertically, but can operate within areas substantially more confined than those normally required by aircraft of the same size. A pure STOL aircraft is a fixed-wing vehicle that derives lift primarily from free-stream airflow over the wing and its high lift system, sometimes with significant augmentation from the propulsion system. Although all vertical takeoff and landing (VTOL) machines, including helicopters, can lift greater loads by developing forward speed on the ground before liftoff, they are still regarded as VTOL (or V/STOL craft), operating in the STOL mode. *See* VERTICAL TAKEOFF AND LANDING (VTOL).

It has been customary to define STOL capability in terms of the runway length required to take off or land over a 50-ft (15-m) obstacle, the concept of "short" length being variously defined as from 500 to 2000 ft (150 to 600 m), depending on the high-lift concept employed and on the mission of the aircraft. In addition to being able to operate from short runways, STOL aircraft are usually expected to be able to maneuver in confined airspace so as to minimize the required size of the terminal area. Such aircraft must therefore have unusually good slow-flight stability and control characteristics, especially in turbulence and under instrument flight conditions. *See* AIRPLANE. W. P. Nelms

Shrew A mammal of the family Soricidae. Shrews are found in Asia, Africa, Europe, North America, and northern South America. The family includes 376 species in 26 genera grouped into three subfamilies: the red-toothed shrews (Soricinae), the white-toothed shrews (Crocidurinae), and the African white-toothed shrews (Myosoricini).

Shrews are small secretive animals that constantly run about searching for food. Shrews have the first upper and lower incisors enlarged, thereby effectively forming a set of tweezers with which they can capture prey and extract them from small places. The eyes are minute, the snout is sharp-pointed, and the ears are very small. Musk glands on the flank region exude an odor, possibly having a protective function.

Because of their small size and the fact that they lose heat rapidly, these animals have a very high metabolism. They alternate between short (3 h) stretches of feeding and sleeping throughout the 24 h, and eat approximately their own weight in food per day. Major foods of some of the larger shrews are earthworms, snails, and centipedes, whereas small ones often feed on small insect larvae. *See* MAMMALIA. John O. Whitaker, Jr.

Shrink fit A fit that has considerable negative allowance so that the diameter of a hole is less than the diameter of a shaft that is to pass through the hole, also called a heavy force fit. Shrink fits are used for permanent assembly of steel external members, as on locomotive wheels. The difference between a shrink fit and a force fit is in method of assembly. In shrink fits, the outer member is heated, or the inner part is cooled, or both, as required. The parts are then assembled and returned to the same temperature. *See* ALLOWANCE; FORCE FIT. Paul H. Black

Shunting The act of connecting an electrical element in parallel with (across) another element. The shunting connection is shown in illus. *a*.

An example of shunting involves a measuring instrument whose movement coil is designed to carry only a small current for a full-scale deflection of the meter. To protect this coil from

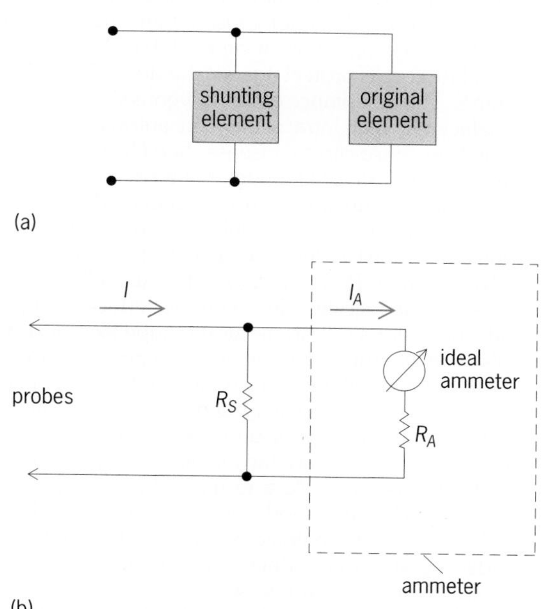

Shunting. (*a*) **Shunting connection.** (*b*) **Ammeter shunted by resistor** R_S.

an excessive current that would destroy it when measuring currents that exceed its rating, a shunt resistor carries the excess current.

Illustration *b* shows an ammeter (a current-measuring instrument) with internal resistance R_A. It is shunted by a resistor R_S. The current through the movement coil is a fraction of the measured current, and is given by the equation below. With different

$$I_A = \frac{R_S}{R_A + R_S} I$$

choices of R_S, the measuring range for the current I can be changed. *See* AMMETER; CURRENT MEASUREMENT.

Similar connections and calculations are used in a shunt ohmmeter to measure electrical resistance. Shunt capacitors are often used for voltage correction in power transmission lines. A shunt capacitor may be used for the correction of the power factor of a load. In direct current shunt motors, the excitation (field) winding is connected in parallel with the armature. *See* DIRECT-CURRENT MOTOR; OHMMETER; POWER FACTOR; RESISTANCE MEASUREMENT.

In electronic applications, a shunt regulator is used to divert an excessive current around a particular circuit. In broadband electronic amplifiers, several techniques may be used to extend the bandwidth. For high-frequency extension, a shunt compensation is used where, typically, a capacitor is shunted across an appropriate part of the circuit. Shunt capacitors (or more complicated circuits) are often used to stabilize and prevent undesired oscillations in amplifier and feedback circuits. *See* AMPLIFIER; DIRECT-CURRENT CIRCUIT THEORY; FEEDBACK CIRCUIT. Shlomo Karni

Sickle cell disease An inherited disorder of red blood cells characterized by lifelong anemia and recurrent painful episodes. The sickle cell mutation is caused by a single nucleotide effecting a change in the β-globin gene, resulting in the substitution of valine for glutamic acid as the sixth amino acid of β-globin. The short circulatory survival of red blood cells that contain sickle cell hemoglobin S results in anemia, and their abnormal rigidity contributes to painful obstruction of small blood vessels. *See* ANEMIA; GENETIC CODE; HEMOGLOBIN.

The sickle cell gene is found most commonly among individuals of African ancestry, but also has a significant incidence in Mediterranean, Middle Eastern, and Asian Indian populations. Inheritance of one sickle gene and one normal β-globin allele results in a simple heterozygous condition known as sickle cell trait. This benign carrier condition is associated with a normal life expectancy, and it does not cause either anemia or recurrent pain. The large amounts of hemoglobin A within sickle-cell-trait red blood cells protect against the deleterious effects of hemoglobin S. The inheritance of homozygous sickle cell anemia results in sufficiently high intracellular concentration of sickle cell hemoglobin S to cause clinical disease. *See* HUMAN GENETICS.

The property of sickle cell hemoglobin S responsible for clinical disease is its insolubility when deoxygenated. Oxygenated sickle cell hemoglobin S is as soluble as oxygenated normal hemoglobin, but when it is deoxygenated it aggregates and forms an insoluble polymer. Polymerization of sickle cell hemoglobin within deoxygenated sickle cells reversibly reduces cellular deformability and distorts cells to the sickle shape (see illustration). Sickle cells usually return from the venous circulation to the arterial, where the hemoglobin is reoxygenated and the cells unsickle. Persistent cycles of sickling and unsickling result in the generation of dehydrated, very dense sickle cells; these are irreversibly sickled cells that are incapable of resuming a normal shape when reoxygenated. As a result of the poor deformability of individual sickle red blood cells, sickle cell blood has high viscosity. The impaired rheologic properties of sickle blood are compounded by abnormal adherence of sickle red cells to endothelial cells lining the blood vessels.

The short-lived nature of sickle red blood cells results in lifelong chronic hemolytic anemia with which accelerated red blood

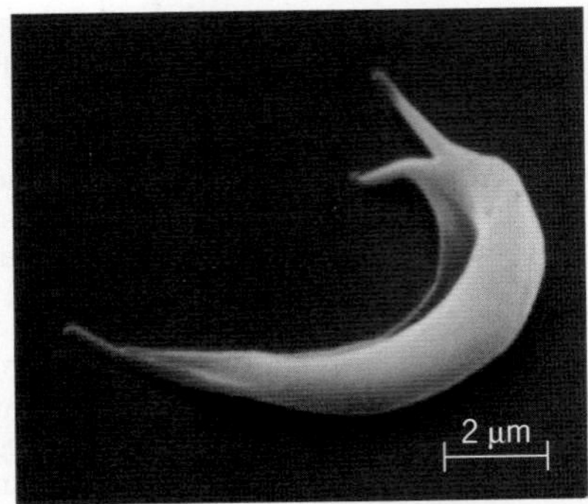

Scanning electron micrograph of a deoxygenated sickle red blood cell. (*Electron micrograph by Dr. James White*)

cell production cannot keep pace. The increased turnover of red blood cells results in elevated levels of hemoglobin degradation and bilirubin production by the liver and in very frequent formation of gallstones. Vasoocclusive complications of sickle cell disease include the episodic painful crises and both chronic and acute organ dysfunction. Average life expectancy is in the fifth decade. One disease manifestation that is particularly problematic in young children is susceptibility to infections.

The standard method of diagnosing sickle cell syndromes is hemoglobin electrophoresis. There are a variety of diagnostic tests based on deoxyribonucleic acid (DNA). These DNA-based diagnostic methods are particularly useful for prenatal diagnosis of sickle cell disease. *See* DEOXYRIBONUCLEIC ACID (DNA); ELECTROPHORESIS; PRENATAL DIAGNOSIS.

Despite the profound understanding of sickle cell disease, treatment of painful episodes often consists of only symptomatic therapy, including analgesics for pain, antibiotics for infections, and transfusions for episodes of severe anemia. Genetic counseling and prenatal diagnosis remain important therapeutic approaches. *See* BLOOD; GENETIC ENGINEERING. Stephen H. Embury

Sideband The range of the electromagnetic spectrum located either above (the upper sideband) or below (the lower sideband) the frequency of a sinusoidal carrier signal $c(t)$. The sidebands are produced by modulating the carrier signal in amplitude, frequency, or phase in accordance with a modulating signal $m(t)$ to produce the modulated signal $s(t)$. The resulting distribution of power in the sidebands of the modulated signal depends on the modulating signal and the particular form of modulation employed. *See* AMPLITUDE MODULATION; FREQUENCY MODULATION; MODULATION; PHASE MODULATION. Hermann J. Helgert

Sidereal time One of several kinds of time scales used in astronomy, whose primary application is as part of the coordinate system to locate objects in the sky. It is also the basis for determining the solar time used in everyday living.

The common measurements of time are based on the motions of the Earth that most affect everyday life: Earth's rotation on its axis, and revolution in orbit around the Sun. Objects in the sky reflect these motions and appear to move westward, crossing the meridian each day. A particular object or point is chosen as a marker, and the interval between its successive crossings of the local meridian is defined to be a day, divided into 24 equal parts called hours. The actual length of the day for comparison between systems depends on the reference object chosen. The time of day is reckoned by the angular distance around the sky that the reference object has moved westward since it last crossed the meridian. In fact, the angular distance west of the meridian is called the hour angle. *See* MERIDIAN.

The reference point for marking sidereal time is the vernal equinox, one of the two points where the planes of the Earth's Equator and orbit appear to intersect on the celestial sphere. The sidereal day is the interval of time required for the hour angle of the equinox to increase by 360°. One rotation of the Earth with respect to the Sun is a little longer, because the Earth has moved in its orbit as it rotates and hence must turn approximately 361° to complete a solar day. A sidereal year is the time required for the mean longitude of the Sun to increase 360°, or for the Sun to make one circuit around the sky with respect to a fixed reference point. *See* EARTH ROTATION AND ORBITAL MOTION; TIME.

Alan D. Fiala

Siderite A mineral ($FeCO_3$) with the same space group and hexagonal crystal system as calcite ($CaCO_3$). Siderite has a gray, tan, brown, dark brown, or red color, has rhombohedral cleavages, and occasionally may show rhombohedral crystal terminations. It may display curved crystal faces like dolomite ($CaMg[CO_3]_2$), but more commonly is found as massive, compact, or earthy masses. It has a high specific gravity of 3.94, a medium hardness of 3.5–4, and a high index of refraction, 1.88. *See* CARBONATE MINERALS; CRYSTAL STRUCTURE; DOLOMITE.

Siderite, a widespread mineral in near-surface sediments and ore deposits, occurs in hydrothermal veins, lead-silver ore deposits, sedimentary concretions formed in limestones and sandstones, and Precambrian banded iron formations that precipitated under acidic conditions. Famous localities for siderite are found in Styria (Austria), Westphalia (Germany), Cornwall (Britain), Wawa (Northern Ontario, Canada), Minas Gerais (Brazil), and Llallagua and Potosi (Bolivia). These and other occurrences have provided locally significant quantities of siderite as iron ore. *See* IRON; MAGNETITE; METAMORPHISM; ORE AND MINERAL DEPOSITS.

Eric J. Essene

Siderophores Low-molecular-mass molecules that have a high specificity for chelating or binding iron. Siderophores are produced by many microorganisms, including bacteria, yeast, and fungi, to obtain iron from the environment. More than 500 different siderophores have been identified from microorganisms. Some bacteria produce more than one type of siderophore. *See* BACTERIA; BIOINORGANIC CHEMISTRY; CHELATION; FUNGI; IRON; YEAST.

Iron is required by aerobic bacteria and other living organisms for a variety of biochemical reactions in the cell. Although iron is the fourth most abundant element in the Earth's crust, it is not readily available to bacteria. Iron is found in nature mostly as insoluble precipitates that are part of hydroxide polymers. Bacteria living in the soil or water must have a mechanism to solubilize iron from these precipitates in order to assimilate iron from the environment. Iron is also not freely available in humans and other mammals. Most iron is found intracellularly in heme proteins and ferritin, an iron storage compound. Iron outside cells is tightly bound to proteins. Therefore, bacteria that grow in humans or other animals and cause infections must have a mechanism to remove iron from these proteins and use it for their own energy and growth needs. Siderophores have a very high affinity for iron and are able to solubilize and transport ferric iron (Fe^{3+}) in the environment and also to compete for iron with mammalian proteins such as transferrin and lactoferrin. The majority of bacteria and fungi use siderophores to solubilize and transport iron. Microorganisms can use either siderophores produced by themselves or siderophores produced by other microorganisms. *See* IRON METABOLISM.

The many different types of siderophores can generally be classified into two structural groups, hydroxamates and catecholate compounds. Despite their structural differences, all form an octahedral complex with six binding coordinates for Fe^{3+}.

Siderophores have potential applications in the treatment of some human diseases and infections. Some siderophores are used therapeutically to treat chronic or acute iron overload conditions in order to prevent iron toxicity in humans. Individuals who have defects in blood cell production or who receive multiple transfusions can sometimes have too much free iron in the body. However, in order to prevent infection during treatment for iron overload, it is important to use siderophores that cannot be used by bacterial pathogens.

A second clinical application of siderophores is in antibiotic delivery to bacteria. Some gram-negative bacteria are resistant to antibiotics because they are too big to diffuse through the outer-membrane porins. However, siderophore-antibiotic combination compounds have been synthesized that can be transported into the cell using the siderophore receptor. *See* ANTIBIOTIC; DRUG RESISTANCE.

Pamela A. Sokol

Sigma-delta converter A class of electronic systems containing both analog and digital subsystems whose most common application is to the conversion of analog signals to digital form and vice versa. The device is also known as a delta-sigma converter. The main advantage of the sigma-delta approach to signal conversion is its minimal reliance on the quality of the analog components required. To achieve this end, the system uses pulse density modulation to create a high-rate stream of single-amplitude pulses. For analog-to-digital conversion, the rate at which the pulses are generated depends on the amplitude of the analog voltage being sensed. For digital-to-analog conversion, the pulse density depends on the numeric digital quantity applied at the converter input. *See* PULSE MODULATION.

The simplest implementation of a sigma-delta analog-to-digital converter uses an analog circuit to generate the single-valued pulse stream from an analog source, and a digital system to repeatedly sum the number of these pulses over a fixed number of pulse intervals. The summing operation converts the pulses to a numeric value, achieving analog-to-digital conversion. Conversely, in a digital-to-analog converter, a digital circuit is used to convert numeric values from a digital processor to a pulse stream, and these pulses then are low-pass-filtered by a relatively simple analog system to produce an analog waveform. This low-pass filtering effectively sums the uniform analog pulse amplitudes over a fixed interval. The circuit—analog or digital—used to generate the pulse stream is called a sigma-delta (or delta-sigma) modulator. *See* ELECTRIC FILTER.

In each case, analog information is contained not in the pulse amplitude but in the number of pulses that occur during the conversion interval. This distribution of the analog information makes the conversion process essentially independent of the amplitude of the pulses and greatly simplifies the design and fabrication of the analog portion of the converter. It does, however, require that the sampling process be rapid, since the resolution of the conversion depends on the number of pulses that can exist in the conversion interval.

Philip V. Lopresti

Signal detection theory A theory in psychology which characterizes not only the acuity of an individual's discrimination but also the psychological factors that bias the individual's judgments. Failure to separate these two aspects of discrimination had tempered the success of theories based upon the classical concept of a sensory threshold. The theory provides a modern and more complete account of the process whereby an individual makes fine discriminations.

The theory of signal detection has two parts of quite different origins. The first comes from mathematical statistics and is a translation of the theory of statistical decisions. The major contribution of this part of the theory is that it permits a determination of the individual's discriminative capacity, or sensitivity, that is independent of the judgmental bias or decision criterion the individual may have had when the discrimination was made. The second part of the theory comes from the study of electronic communications. It provides a means of calculating for simple signals, such as tones and lights, the best discrimination that can be attained. The prediction is based upon physical measurements of the signals and their interfering noise.

This opportunity to compare the sensitivity of human observers with the sensitivity of an "ideal observer" for a variety of signals is of considerable usefulness, and of growing interest, in sensory psychology. Signal detection theory has been applied to several topics in experimental psychology in which separation of intrinsic discriminability from decision factors is desirable. Included are attention, imagery, learning, conceptual judgment, personality, reaction time, manual control, and speech.

The analytical apparatus of the theory has been of value in the evaluation of the performance of systems that make decisions based on uncertain information. Such systems may involve only people, or people and machines together, or only machines. Examples come from medical diagnosis, where clinicians may base diagnostic decisions on a physical examination, or on an x-ray image, or where machines make diagnoses, perhaps by counting blood cells of various types. John A. Swets

Signal generator A piece of electronic test equipment that delivers a sinusoidal output of accurately calibrated frequency. The frequency may be anywhere from audio to microwave, depending upon the intended use of the instrument. The frequency and the amplitude are adjustable over a wide range. The oscillator must have excellent frequency stability, and its amplitude must remain constant over the tuning range.

The Wien-bridge oscillator is commonly used for frequencies up to about 200 kHz. For a radio-frequency signal generator up to about 200 MHz, a resonant circuit oscillator is used (such as a tuned-plate tuned-grid, Hartley, or Colpitts). Beyond this range VHF and microwave oscillators are used. *See* OSCILLATOR.

Jacob Millman

Signal processing Operations that modify or create signals.

For a signal to be useful, it must have one or more characteristics that vary, usually over time or distance, to represent information. Amplitude, frequency, and phase are common signal characteristics that can carry information. *See* AMPLITUDE (WAVE MOTION); FREQUENCY (WAVE MOTION); PHASE (PERIODIC PHENOMENA).

A simple signal processing operation is scaling, which multiplies the input signal by a constant value to produce the output signal. Another simple operation is the addition of a constant value to the input to produce the output.

A signal processing operation may have more than one input. For example, two input signals may be added together to produce an output signal. Addition is a linear mathematical operation. An operation may also be nonlinear, as is the case where an input signal is squared or when two input signals are multiplied together.

Logical operations may also be employed in signal processing to make decisions and route signals. For example, a switch may be used to select between two signals based on the value of a third signal. *See* LOGIC CIRCUITS.

An important signal processing operation is time delay. The output signal is a delayed copy of the input signal.

Simple mathematical, logical, and time-delay operations are combined into signal processing systems that perform more complex processing operations. The simple operations, or blocks, are connected together in an appropriate manner to implement the desired complex operation.

Signal processing is performed in analog and digital circuits. Analog signal processing has been used since the beginning of audio and radio electronics. Over the past few decades, digital signal processing has become an increasingly common method, especially for more complex systems.

Because computers use numbers and perform calculations on numbers, they can be programmed to perform digital signal processing. By approximating analog signals and operations in digital form, analog signal processing can also be performed on computers using simulation programs. *See* COMPUTER; DIGITAL COMPUTER; SIMULATION.

Characteristics of operations and systems. Signal processing operations and systems can be described in different ways. The choice depends on the physical process being described, the approach to designing the processing system, the desired outcomes of the design, the modeling of the design, and the capabilities of the designer and design tools.

Linear and nonlinear signal processing. One characteristic already discussed is whether an operation is linear or nonlinear. An operation with input x and output y is linear if the equation below holds true, where m and b are constants. A nonlinear

$$y(t) = m \cdot x(t) + b$$

operation does not meet this requirement.

If an operation does not vary over time, as shown in the equation, it is said to be time-invariant.

Linear time-invariant systems are by far the signal processing systems most commonly designed, analyzed, and implemented. They can be represented by simple mathematical equations called transfer functions, which describe completely the input-output relationship of the system for any input signal. *See* LINEAR SYSTEM ANALYSIS; LINEARITY.

Feedback systems. A system may contain feedback, which means at least one signal within the system returns to a point that indirectly feeds into the same signal, creating a loop. The output signal $B(t)$ is fed back through operations that include time delay, with the delayed result added to the processed input to create the output. *See* CONTROL SYSTEMS.

Analog signal processing. Analog processing is performed in traditional analog circuits, which are built from analog components. The signals in these circuits carry information according to the time-varying values of their voltage and current. Analog signals and operations are continuous in value and in time.

Analog processing components. Two kinds of analog components are used: passive and active. Resistors, capacitors, inductors, and transformers are passive components. Active components are based on semiconductors or in older or high-power circuits on electron tubes. Transistors and diodes are common types of semiconductor components. *See* AMPLIFIER; DIODE; ELECTRON TUBE; TRANSISTOR; CAPACITOR; INDUCTOR; RESISTOR; TRANSFORMER.

Integrated analog processing. A very useful analog device is the operational amplifier, or op amp. It is a circuit consisting of active and passive components that have been packaged, or integrated, into a single, higher-level active component. Very complex and high-performance analog signal processing systems can be integrated into a single component, or integrated circuit. This allows, for example, most of the signal processing for a radio or television to be contained in a package the size of a quarter, helping to reduce the cost to manufacture these products. *See* INTEGRATED CIRCUITS; OPERATIONAL AMPLIFIER.

Digital signal processing. Two important aspects of all-digital signal processing systems separate them from analog systems: sampling and quantization.

Sampling. In digital signal processing (DSP) systems, signals exist only at discrete points, usually in time or space. Such signals are called sampled signals and have a value at each of these points. Thus, a sampled signal A can also be written as $A(n)$, where n is an index of the values, or samples.

How a signal varies over time or space can be described by how fast it changes, that is, by its frequency content. A sampled signal can contain frequencies no higher than one-half its sample rate. And, for a signal with bandwidth B to be accurately represented as a sampled signal, it must be sampled at a rate $2 \cdot B$ or higher. This minimum rate is called the Nyquist rate, and half this rate is the Nyquist frequency. In practice, the sample rate must be somewhat higher than the Nyquist rate. *See* INFORMATION THEORY.

Quantization and signal representations. In digital signal processing systems, signals are represented by binary numbers, and as such are quantized, meaning they can take on only a limited set of values. For example, an 8-bit form of a signal can take on values from 0 to 255, or −128 to 127, depending on what the signal represents or how it is processed. The quantization of a signal means that for each sample, its value must be approximated to one of the values, or quantization steps, allowed by the binary form representing the signal.

Analog-to-digital and digital-to-analog converters. Many applications of digital signal processors involve processing analog signals, such as the signal from a microphone. Analog-to-digital (A/D) and digital-to-analog (D/A) converters provide the interfaces between analog signals and the digital signals inside processors. Converters must sample at the Nyquist rate of the signal being converted. *See* ANALOG-TO-DIGITAL CONVERTER; DIGITAL-TO-ANALOG CONVERTER.

Types of signal processing blocks. Signal processing systems are built from various types of signal processing blocks.

The blocks that will be described are examples of the many types of operations used in signal processing systems.

Linear processing blocks and filters. Addition, scaling, and delay blocks are the most basic linear blocks, and the simplest to implement, especially using digital signal processing. From these, other linear blocks, like filters, can be created.

Nonlinear processing blocks. Nonlinear blocks for multiplication, division, squares and higher powers, square roots, and logarithms can all be performed in signal processing. Analog implementations require the use of active components. The digital forms of some of these operations, like multiplication and division, are usually executed in the same way long-hand arithmetic is performed.

Modulator blocks process audio, image, and data signals so they can be transmitted from one point to another, as in a radio broadcast, or stored into physical media, like a compact disk. An amplitude modulator of a radio transmitter varies the amplitude of a high-frequency carrier in proportion to the level of the relatively low-frequency audio signal to be transmitted. A demodulator performs the opposite operation of its corresponding modulator to recover the original modulated signal. *See* AMPLITUDE MODULATOR; DEMODULATOR; MODULATION; MODULATOR.

Logical blocks. Logic signals can be created in signal processing systems and used to control other signals. For example, in a comparator block an input signal is compared to a threshold to produce a logical 0 or 1 output. A switch block outputs one of its two inputs according to a logic selection input to the block. Powerful signal processing systems can be created from combining logic-based blocks with other types of blocks. An example of this is a neural network, which is capable of complex tasks like pattern recognition in images. *See* NEURAL NETWORK.

Examples of signal processing systems. Radio signal processing is a key element of many electronic systems, including cellular telephones, radar, point-to-point radio communication, and transmitters and receivers for AM, FM, television, and more recently satellite broadcast systems. These systems usually employ a combination of analog and digital processing, and consist of many processing blocks. *See* RADIO.

Signal processing on space probes allows large amounts of telemetry data, like images of other planets, to be transmitted to radio receivers on Earth, where more processing converts the data back into pictures and other usable forms. *See* TELEMETERING.

Devices for producing and reproducing music make heavy use of signal processing, especially digital processing. Electronic keyboards contain processing to create multiple sound sources. An example of sound manipulation is reverberation, which uses delay and gain blocks arranged into feedback loops to create echoes. Portable MP3 music players employ signal processing to convert highly compressed stored sound data into music output. *See* MUSICAL INSTRUMENTS; SOUND-REINFORCEMENT SYSTEM.

A few more examples of the many systems that employ signal processing are medical-imaging devices, sonar, DVD recorders, engine controllers in cars, multichannel sound systems, night-vision goggles, and guidance systems for rockets and missiles.

J. William Whikehart

Signal-to-noise ratio

The quantity that measures the relationship between the strength of an information-carrying signal in an electrical communications system and the random fluctuations in amplitude, phase, and frequency superimposed on that signal and collectively referred to as noise. For analog signals, the ratio, denoted *S/N*, is usually stated in terms of the relative amounts of electrical power contained in the signal and noise. For digital signals the ratio is defined as the amount of energy in the signal per bit of information carried by the signal, relative to the amount of noise power per hertz of signal bandwidth (the noise power spectral density), and is denoted E_b/N_0. Since both signal and noise fluctuate randomly with time, *S/N* and E_b/N_0 are specified in terms of statistical or time averages of these quantities.

The magnitude of the signal-to-noise ratio in a communications systems is an important factor in how well a receiver can recover the information-carrying signal from its corrupted version and hence how reliably information can be communicated. Generally speaking, for a given value of *S/N* the performance depends on how the information quantities are encoded into the signal parameters and on the method of recovering them from the received signal. The more complex encoding methods such as phase-shift keying or quadrature amplitude-shift keying usually result in better performance than simpler schemes such as amplitude- or frequency-shift keying. As an example, a digital communication system operating at a bit error rate of 10^{-5} requires as much as 7 dB less for E_b/N_0 when employing binary phase-shift keying as when using binary amplitude-shift keying. *See* ELECTRICAL COMMUNICATIONS; ELECTRICAL NOISE; INFORMATION THEORY; MODULATION.

Hermann J. Helgert

Signal transduction

The transmission of molecular signals from a cell's exterior to its interior. Molecular signals are transmitted between cells by the secretion of hormones and other chemical factors, which are then picked up by different cells. Sensory signals are also received from the environment, in the form of light, taste, sound, smell, and touch. The ability of an organism to function normally is dependent on all the cells of its different organs communicating effectively with their surroundings. Once a cell picks up a hormonal or sensory signal, it must transmit this information from the surface to the interior parts of the cell—for example, to the nucleus. This occurs via signal transduction pathways that are very specific, both in their activation and in their downstream actions. Thus, the various organs in the body respond in an appropriate manner and only to relevant signals. *See* CELL (BIOLOGY).

All signals received by cells first interact with specialized proteins in the cells called receptors, which are very specific to the signals they receive. These signals can be in various forms. The most common are chemical signals, which include all the hormones and neurotransmitters secreted within the body as well as the sensory (external) signals of taste and smell. The internal hormonal signals include steroid and peptide hormones, neurotransmitters, and biogenic amines, all of which are released from specialized cells within the various organs. The external signals of smell, which enter the nasal compartment as gaseous chemicals, are dissolved in liquid and then picked up by specialized receptors. Other external stimuli are first received by specialized receptors (for example, light receptors in the eye and touch receptors in the skin), which then convert the environmental signals into chemical ones, which are then passed on to the brain in the form of electrical impulses.

Once a receptor has received a signal, it must transmit this information effectively into the cell. This is accomplished either by a series of biochemical changes within the cell or by modifying the membrane potential by the movement of ions into or out of the cell. Receptors that initiate biochemical changes can do so either directly via intrinsic enzymatic activities within the receptor or by activating intracellular messenger molecules. Receptors may be broadly classified in four groups that differ in their mode of action and in the molecules that activate them.

The largest family of receptors are the G-protein-coupled receptors (GPCRs), which depend on guanosine triphosphate (GTP) for their function. Many neurotransmitters, hormones, and small molecules bind to and activate specific G-protein-coupled receptors.

A second family of membrane-bound receptors are the receptor tyrosine kinases (RTKs). They function by phosphorylating themselves and recruiting downstream signaling components.

Ion channels are proteins open upon activation, thereby allowing the passage of ions across the membrane. Ion channels are responsive to either ligands or to voltage changes across the membrane, depending on the type of channel. The movement of ions changes the membrane potential, which in turn changes cellular function. *See* BIOPOTENTIALS AND IONIC CURRENTS.

Steroid receptors are located within the cell. They bind cell-permeable molecules such as steroids, thyroid hormone, and vitamin D. Once these receptors are activated by ligand, they translocate to the nucleus, where they bind specific DNA sequences to modulate gene expression. *See* STEROID.

The intracellular component of signal propagation, also known as signal transduction, is receptor-specific. A given receptor will activate only very specific sets of downstream signaling components, thereby maintaining the specificity of the incoming signal inside the cell. In addition, signal transduction pathways amplify the incoming signal by a signaling cascade (molecule A activates several molecule B's, which in turn activate several molecule C's) resulting in an appropriate physiological response by the cell.

Several small molecules within the cell act as intracellular messengers. These include cAMP, cyclic guanosine monophosphate (cGMP), nitric oxide (NO), and Ca^{2+} ions. Increased levels of Ca^{2+} in the cell can trigger several changes, including activation of signaling pathways, changes in cell contraction and motility, or secretion of hormones or other factors, depending on the cell type. Increased levels of nitric oxide cause relaxation of smooth muscle cells and vasodilation by increasing cGMP levels within the cell. Increasing cAMP levels can modulate signaling pathways by activating the enzyme protein kinase A (PKA).

One of the most important functions of cell signaling is to control and maintain normal physiological balance within the body. Activation of different signaling pathways leads to diverse physiological responses, such as cell proliferation, death, differentiation, and metabolism. Signaling pathways in cells may also interact with each other and serve as signal integrators. For example, negative and positive feedback loops in pathways can modulate signals within a pathway; positive interactions between two signaling pathways can increase duration of signals; and negative interactions between pathways can block signals. *See* CELL NUCLEUS; CELL ORGANIZATION; ENDOCRINE SYSTEM (VERTEBRATE); NORADRENERGIC SYSTEM.

Prahlad T. Ram; Ravi Iyengar

Significant figures Digits that show the number of units in a measurement expressed in decimal notation.

Scientific notation is useful in showing which digits are significant. In scientific notation, a number is expressed as the product of a number 1 to 10 and a power of 10, or the product of 1 and a power of 10. Thus, the number 123,000 is 1.23×10^5.

The precision of a measurement is based on the size of the unit of measurement. The smaller the unit, the more precise is the measurement.

Computations cannot improve the precision of the measurement. To add measures, they should all be rounded to the unit of the least precise measurement. The sum 8.6 cm + 0.14 cm + 2.75 cm is found by rounding each to tenths: 8.6 cm + 0.1 cm + 2.8 cm = 11.5 cm. Even by doing this, the absolute error might be as large as 0.05 + 0.005 + 0.005 or 0.06, and affect the result by as much as 0.1.

In multiplying and dividing approximate numbers, the product or quotient is rounded to the number of significant digits in the number with the fewest significant digits. For example, 6.2 m × 8.75 m by computation is 54.25. The product needs to be rounded to 54 m^2 so as to show two significant digits. *See* NUMERICAL ANALYSIS; STATISTICS.

Joseph N. Payne

Silica minerals Silica (SiO_2) occurs naturally in at least nine different varieties (polymorphs), which include tridymite, cristobalite, coesite, and stishovite, in addition to high (β) and low (α) quartz. These forms are characterized by distinctive crystallography, optical characteristics, physical properties, pressure-temperature stability ranges, and occurrences.

The crystal structures of all silica polymorphs except stishovite contain silicon atoms surrounded by four oxygens, thus producing tetrahedral coordination polyhedra. Each oxygen is bonded to two silicons, creating an electrically neutral framework. Stishovite differs from the other silica minerals in having silicon atoms surrounded by six oxygens (octahedral coordination.) Ideal high tridymite is composed of sheets of SiO_4 tetrahedra oriented perpendicular to the *c* crystallographic axis (Fig. 1) with adjacent tetrahedra in these sheets pointing in opposite directions. High cristobalite, like tridymite, is composed

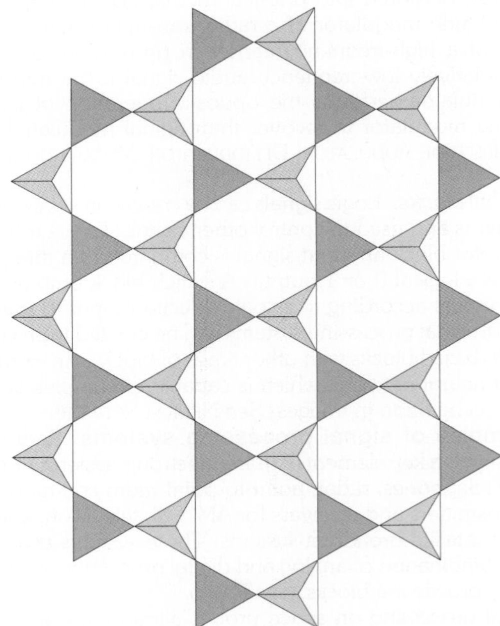

Fig. 1. Portion of an idealized sheet of tetrahedrally coordinated silicon atoms similar to that found in tridymite and cristobalite. Sharing of apical oxygens (which point in alternate directions) between silicons in adjacent sheets generates a continuous framework. (*After J. J. Papike and M. Cameron, Crystal chemistry of silicate minerals of geophysical interest, Rev. Geophys. Space Phys., 14:37–80; copyright © 1976 by American Geophysical Union*)

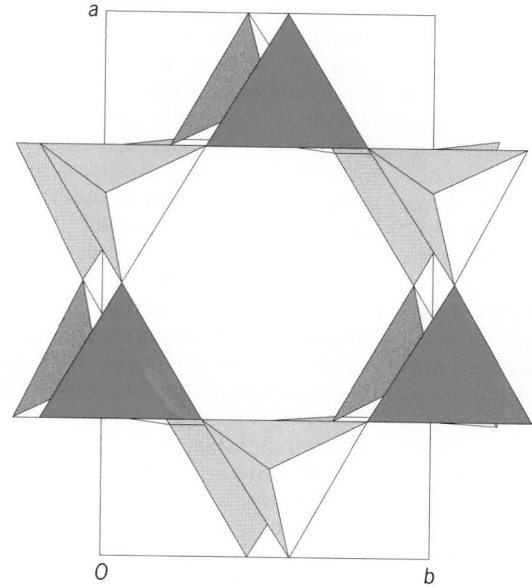

Fig. 2. Portion of cubic high cristobalite illustrating the distortion of tetrahedral sheets, with are oriented parallel to (111), and the 60° rotation of adjacent sheets. (*After J. J. Papike and M. Cameron, Crystal chemistry of silicate minerals of geophysical interest, Rev. Geophys. Space Phys., 14:37–80*; copyright © *1976 by American Geophysical Union*)

of parallel sheets of SiO_4 tetrahedra with neighboring tetrahedra pointing in opposite directions. However, the hexagonal rings are distorted and adjacent sheets are rotated 60° with respect to one another, resulting in the geometry shown in Fig. 2. Coesite also contains silicon atoms tetrahedrally coordinated by oxygen. These polyhedra share corners to form chains composed of four-membered rings. Silicon in stishovite is octahedrally coordinated by oxygen. These coordination polyhedra share edges and corners to form chains of octahedra parallel to the *c* crystallographic axis.

Chemically, all silica polymorphs are ideally 100% SiO_2. However, unlike quartz which commonly contains few impurities, the compositions of tridymite and cristobalite generally deviate significantly from pure silica. This usually occurs because of a coupled substitution in which a trivalent ion such as Al^{3+} or Fe^{3+} substitutes for Si^{4+}, with electrical neutrality being maintained by monovalent or divalent cations occupying interstices In the relatively open structures of these two minerals. *See* COESITE; QUARTZ.

John C. Drake

Silicate minerals All silicates are built of a fundamental structural unit, the so-called SiO_4 tetrahedron. The crystal structure may be based on isolated SiO_4 groups or, since each of the four oxygen ions can bond to either one or two silicon (Si) ions, on SiO_4 groups shared in such a way as to form complex isolated groups or indefinitely extending chains, sheets, or three-dimensional networks. Mixed structures in which more than one type of shared tetrahedra are present also are known. *See* SILICON.

Silicates are classified according to the nature of the sharing mechanism, as revealed by x-ray diffraction study. The sharing mechanism gives rise to a characteristic ratio of Si to O, but it is possible for oxygen ions that are not bonded to Si to be present in the structure, and sometimes some or all of any aluminum present must be counted as equivalent to Si.

The detailed crystallographic and physical properties of the various silicates are broadly related to the type of silicate framework that they possess. Thus, the phyllosilicates as a group typically have a platy crystal habit, with a cleavage parallel to the plane of layering of the structure, and are optically negative with rather high birefringence. The inosilicates, based on an extended one-dimensional rather than two-dimensional linkage of the SiO_4 tetrahedra, generally form crystals of prismatic habit; if cleavage is present, it will be parallel to the direction of elongation. The tectosilicates commonly are equant in habit, without marked preference for cleavage direction, and tend to have a relatively low birefringence.

Silicate minerals make up the bulk of the outer crust of the Earth and form in a wide range of geologic environments. Many silicates are of economic importance. For discussions of certain silicate mineral groups *see* AMPHIBOLE; ANDALUSITE; CHLORITE; CHLORITOID; EPIDOTE; FELDSPAR; FELDSPATHOID; GARNET; HUMITE; MICA; OLIVINE; PYROXENE; SCAPOLITE; SERPENTINE; ZEOLITE.

Clifford Frondel

Silicate phase equilibria Silicate phase equilibria studies define the conditions of temperature, composition, and pressure at which silicates can stably coexist. Silicate phase equilibria relations are used by geologists, ceramists, and cement manufacturers to explain the variation of composition of silica-bearing minerals, as well as their number and order of appearance in rocks, slags, glasses, and cements. They are also useful to interpret the chemistry of refractories, boiler scale deposits, and welding fluxes.

Silica itself makes up nearly 60% by weight of the Earth's crust. The next most abundant oxides, in decreasing order, are Al_2O_3, CaO, Na_2O, FeO, MgO, K_2O, and Fe_2O_3; all of these occur principally combined with silica as silicates. Free silica and the hundreds of silicate minerals make up nearly 97% of the Earth's crust. The study of silicate phase equilibria was initiated by geologists seeking to apply the phase rule of J. Willard Gibbs to these abundant natural substances. *See* PHASE EQUILIBRIUM; SILICATE MINERALS.

Dynamic and static methods are used to determine equilibrium in silicate systems. Dynamic methods require large samples of silicates that are difficult to prepare, and require also that equilibrium be reached quickly. Silicates in general are slow to react, and supercooling or superheating of hundreds of degrees before reaction occurs is common. Many silicates react sluggishly at temperatures of 1000°C (1830°F) or higher.

Most silicate phase equilibria are determined by the static method of holding a sample under controlled conditions until equilibrium is attained, then quenching the sample for examination.

Equilibrium is established when the products obtained by heating a sample to a given temperature are identical with the products of cooling a sample to that temperature; no requirement is made regarding the texture, shape, or grain size of the products. Another criterion used in recognizing equilibrium is that no change of the sample can be observed after holding the charge at a given temperature for very long periods of time.

The use of diagrams to present silicate phase equilibria is customary, since such diagrams express quantitatively the amount and composition of each phase present at any bulk composition in the system at any temperature.

David B. Stewart

Siliceous sediment Fine-grained sediment and sedimentary rock dominantly composed of the microscopic remains of the unicellular, silica-secreting plankton diatoms and radiolarians. Minor constituents include extremely small shards of sponge spicules and other microorganisms such as silicoflagellates. Siliceous sedimentary rock sequences are often highly porous and can form excellent petroleum source and reservoir rocks. *See* SEDIMENTARY ROCKS.

Given their biologic composition, siliceous sediments provide some of the best geologic records of the ancient oceans. Diatoms did not evolve until the late Mesozoic; thus the majority of siliceous rocks older than approximately 150 million years are formed by radiolarians. Geologists map the distribution of ancient siliceous sediments now pushed up onto land by plate tectonic processes, and can thus determine which portions of the ancient seas were biologically productive; this knowledge in turn can give great insight into regions of the Earth's crust that may be economically productive (for example oil-containing regions). The vast oil reserves of coastal California are predominantly found in the Monterey Formation, a highly porous diatomaceous siliceous sedimentary sequence distributed along the western seaboard of the United States. *See* CHALK; CHERT; LIMESTONE.

R. W. Murray

Siliceous sinter A porous silica deposit formed around hot springs. It is white to light gray and sometimes friable. Geyserite is a variety of siliceous sinter formed around geysers. The siliceous sinters are deposited as the hot subterranean waters cool after issuing at the surface and become supersaturated with silica that was picked up at depth. The sinters are frequently deposited on algae that live in the pools around the hot springs. *See* GEYSER.

Raymond Siever

Silicoflagellata A class of marine planktonic chromists. Their skeletons are composed of hollow, siliceous rods and resemble those of the Radiolaria, with which they have been grouped. They are usually subpyramidal or hemispherical in shape, delicately filigreed, and range in size from 10 to 150 micrometers (see illustration). Two families and 11 genera of silicoflagellates have been described from siliceous sedimentary rocks ranging in age from the early Cretaceous to Recent, in association with abundant diatoms and siliceous sponge spicules. One genus, *Dictyocha*, lives in the ocean today. *See* MICROPALEONTOLOGY; PHYTOPLANKTON.

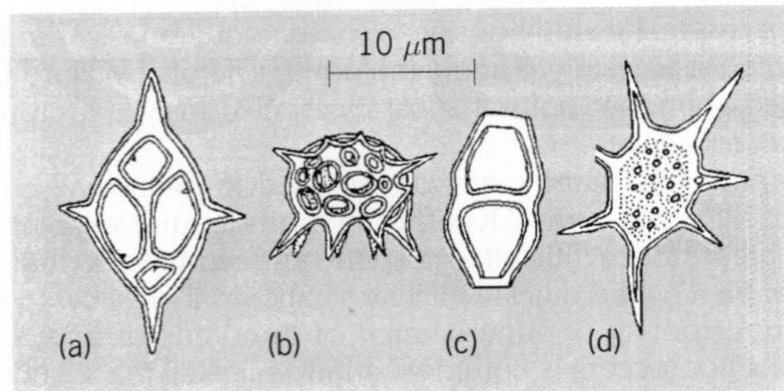

Examples of fossil and modern Silicoflagellata. (*a*) *Dictyocha*, Cretaceous to Recent; (*b*) *Cannopilus*, Miocene; (*c*) *Naviculopsis*, Eocene to Miocene; and (*d*) *Vallacerta*, Upper Cretaceous.

Daniel J. Jones; Sally Walker

Silicoflagellida An order of the phylum Protozoa, class Phytamastigophorea. These organisms are marine flagellates which have an internal, siliceous, tubular skeleton; numerous small, discoid, yellow chromatophores; and a single flagellum. At times the organisms also put forth, from the ends of their skeletal tubes, long, rather threadlike pseudopodia. The skeleton forms a basket within which the moiety of protoplasm lies, but skeletal elements always have at least a thin covering. There is a single genus, *Dictyocha*, with four species. *See* PHYTAMASTIGOPHOREA.

James B. Lackey

Silicon A chemical element, Si, atomic number 14, and atomic weight 28.086. Silicon is the most abundant electropositive element in the Earth's crust. The element is a metalloid with a decided metallic luster; it is quite brittle. It has a specific gravity of 2.42 at 20°C (68°F), melts at 1420°C (2588°F), and boils at 3280°C (5936°F). The element is usually tetravalent in its compounds, although sometimes divalent, and is decidedly electropositive in its chemical behavior. In addition, pentacoordinate and hexacoordinate compounds of silicon are known. *See* METALLOID; PERIODIC TABLE.

Crude elementary silicon and its intermetallic compounds are used in alloying constituents to strengthen aluminum, magnesium, copper, and other metals. Metallurgical silicon of 98–99% purity is used as the starting material for manufacturing organosilicon compounds and silicone resins, elastomers, and oils. Silicon chips are used in integrated circuits. Photovoltaic cells for direct conversion of solar energy to electricity use wafers sliced from single crystals of electronic-grade silicon. Silicon dioxide is used as the raw material for making elementary silicon and for silicon carbide. Sizable crystals of it are used for piezoelectric crystals. Fused quartz sand becomes silica glass, used in chemical laboratories and plants as well as an electrical insulator. A colloidal dispersion of silica in water is used as a coating agent and as an ingredient in certain polishes.

Naturally occurring silicon contains 92.2% of the isotope of mass number 28, 4.7% of silicon-29, and 3.1% of silicon-30. In addition to these stable, natural isotopes, several artificially radioactive isotopes are known. Elementary silicon has the physical properties of a metalloid, resembling germanium below it in group 14 of the periodic table. In very pure form silicon is an intrinsic semiconductor, although the extent of its semiconduction is greatly increased by the introduction of minute amounts of impurities. Silicon resembles the metals in its chemical behavior. It is about as electropositive as tin, and decidedly more positive than germanium or lead. In keeping with this rather metallic character, silicon forms tetrapositive ions and a variety of covalent compounds; it appears as a negative ion in only a few silicides and as a positive constituent of oxy acid or complex anions.

Several series of hydrides are formed, a variety of halides (some of which contain silicon-to-silicon bonds), and also many series of oxygen-containing compounds which may be either ironic or covalent in their properties.

Silicon occurs in many forms of the dioxide and as almost numberless variations of the natural silicates. For a discussion of the structures and compositions of the representative classes, *see* SILICATE MINERALS.

In abundance, silicon exceeds by far every other element except oxygen. It constitutes 27.72% of the solid crust of the Earth, whereas oxygen constitutes 46.6%, and the next element after silicon, aluminum, accounts for 8.13%.

Silicon is reported to form compounds with 64 of the 96 stable elements, and it probably forms silicides with 18 other elements. Besides the metal silicides, used in large quantities in metallurgy, silicon forms useful and important compounds with hydrogen, carbon, the halogen elements, nitrogen, oxygen, and sulfur. In addition, useful organosilicon derivatives have been prepared.

Edwin P. Plueddemann

Silicone resins Soluble or meltable materials based on highly branched polymers composed of alternating silicon and oxygen atoms and having organic (R) substituents attached to the silicon atoms. On curing, these polymers are converted into highly cross-linked, insoluble materials. They may contain four types of units, usually symbolized by the letters M, D, T, Q, as follows:

```
     R              R              R              |
     |              |              |             -O-
                                                  |
R — Si — O —   — O — Si — O —   — O — Si — O —   — O — Si — O —
     |              |              |              |
     R              R             -O-            -O-
                                   |              |
     M              D              T              Q
```

Each of the oxygen atoms is shared by neighboring units or by the unit and a functional OR group (R = H or alkyl). Therefore, the formulas of units are M—$R_3SiO_{1/2}$, D—R_2SiO, T—$RSiO_{3/2}$, Q—SiO_2.

Silicone resins are combinations of these units, the proportions of which are varied over a broad range. They are characterized by the R/Si ratio, which usually is in the range 1–1.7. Silicone resins constitute one of the three main classes of silicone products. The other two are silicone rubbers (R/Si = 2) and silicone fluids (R/Si > 2). The structure and properties of silicone resins are strongly dependent on the R/Si value. The lower the R/Si value, the less miscible the resin with organic polymers, the easier it undergoes gelation, and the harder it is after curing. Conversely, resins with higher R/Si values (containing more organic groups) are more flexible and compatible with organic resins. The basic component of silicone resins is usually the T unit. Polymers composed exclusively of T units, having the $(RSiO_{3/2})_n$ structure, are called polysilsesquioxanes and may contain fragments of a ladder or a cage structure. T resins having R/Si = 1 are known, but more common are DT resins. The D units give the resin softness and flexibility, which is particularly important in paints and varnishes. M and Q units also are often used. Resins composed exclusively of M and Q units, called MQ resins, are very important and have recently found numerous applications, such as pressure-sensitive adhesives and flexible silicone coatings, as well as high-performance defoamers, release agents, and additives for improving the mechanical properties of silicone rubber. *See* INORGANIC AND ORGANOMETALLIC POLYMERS; ORGANOSILICON COMPOUND; POLYMER; RUBBER.

The organic group is most often methyl or phenyl. Commonly used phenylmethyl silicone resins contain both phenyl and methyl in various proportions. Phenyl groups give the material superior thermal resistance. Cured resins containing a large proportion of phenyl groups are less brittle and more readily miscible with organic polymers. Other organic groups are introduced to give silicone resins special properties. For example, enhanced water repellency is attained by introducing fluoroalkyl groups or long-chain hydrocarbon groups, such as dodecyl ($C_{12}H_{25}$). Silicone resins contain a considerable number of hydroxy (HO—) or alkoxy (for example, CH_3O—) groups which, in the curing process, cross-link the polymer by condensation. These reactive groups may also form hybrids with organic resins. Vinyl groups or SiH groups are sometimes introduced in silicone resins to give them the ability to cure or form hybrids by the addition to olefinic bonds in organic polymers.

The presence of reactive silanol (SiOH) groups in the resins may make them unstable, particularly if the R/Si ratio is low and the molecular mass is relatively high. Their stability may be increased by the partial replacement of hydroxy groups by alkoxy groups. The stability of T resins (R/Si = 1) used for masonry is attained by conversion of the ≡SiOH to ≡SiONa or ≡SiOK groups. [Here ≡ means a triply bonded Si atom; that is, in ≡SiOH only SiOH group reacts and the other three (undefined) do not.]

Silicone resins are usually sold as solutions of the highly branched polymer with a molecular mass of 1000–60,000 daltons in an inert solvent such as an aromatic hydrocarbon, chlorinated hydrocarbon, or alkyl ester. Commercial solvent-free resins are either viscous liquids or powders which melt at 50–80°C (120–180°F). Some silicone resins are available as emulsions or dispersions in water.

Silicone resins have outstanding thermal stability and weathering resistance. They change little on exposure to humidity, oxygen, heat, and sunlight. They are also physiologically inert, retain their gloss for a long time, and have good electrical resistance. Their mechanical and electrical properties are not very dependent on temperature. Silicone resins have very interesting surface properties, including very low surface energy. They are water-repellent and foreign bodies do not stick to them. They also show high permeability for gases. Silicone resins are particularly useful in applications where the material is exposed to extreme temperatures and adverse weather conditions, with many types of silicone resins produced for specific applications.

The range of use of silicone resins is expanding, with many new applications such as antifouling coatings, sensor coatings, resins for microelectronics, materials for photopattering, adhesives for porous ceramic materials, abrasion-resistant coatings, materials for optical communication, and biomedical materials. Particularly important are various silicone-organic hybrid resin materials.

Julian Chojnowski

Silk The lustrous fiber produced by the larvae of silkworms; also the thread or cloth made from such fiber.

The cocoons of the silkworm are delivered to a factory, called a filature, where the silk is unwound from the cocoons and the strands are collected into skeins. The process of unwinding the filament from the cocoon is called reeling. As the filament of a single cocoon is too fine for commercial use, 3–10 strands are usually reeled at a time to produce the desired diameter of raw silk thread. The usable length of the reeled filament is from 1000 to 2000 ft (300 to 600 m). The remaining part of the filament is valuable raw material for the manufacture of spun silk.

The term reeled silk is applied to the raw silk strand formed by combining several filaments from separate cocoons. It is reeled into skeins, which are packed in small bundles called books. From the filature, the books of reeled silk go to the throwster where they are transformed into silk yarn, also called silk thread, by a process known as throwing. Silk throwing is analogous to the spinning process that changes cotton, linen, or wool fibers into yarn. The manufacture of silk yarn, however, does not include carding, combing, and drawing out, the usual processes for producing a continuous yarn.

M. David Potter

Sillimanite A nesosilicate mineral of composition $Al_2O[SiO_4]$, crystallizing in the orthorhombic system. Sillimanite commonly occurs in slender crystals or parallel groups, and is frequently fibrous, hence the synonym fibrolite. There is one perfect cleavage, luster is vitreous, color is brown, pale green, or white, hardness is 6–7 on Mohs scale, and the specific gravity is 3.23. *See* HARDNESS SCALE.

Sillimanite, andalusite, and kyanite are polymorphs of $Al_2O[SiO_4]$. The three $Al_2O[SiO_4]$ polymorphs are important in assessing the metamorphic grade of the rocks in which they crystallized. *See* ANDALUSITE; KYANITE; SILICATE MINERALS.

Paul B. Moore

Silurian The third oldest period of the Paleozoic Era, spanning an interval from about 412 to 438 million years before the present. The Silurian system includes all sedimentary rocks deposited and all igneous and metamorphic rocks formed in the Silurian Period. Both the base and top of the Silurian have been designated by international agreement at the first appearances of certain graptolite species in rock sequences at easily examined and well-studied outcrops. *See* GEOLOGIC TIME SCALE.

The most prominent feature of Silurian paleogeography was the immense Gondwana plate. It included much of present-day South America, Africa, the Middle East, Antarctica, Australia, and the Indian subcontinent. During the Silurian, many plates continued the relative northward motion that had commenced during the mid-Ordovician. Plate positions and plate motions as well as topographic features of the plates controlled depositional environments and lithofacies. These, in turn, significantly influenced organismal development and distributions. Silurian Northern Hemisphere plates, other than a portion of Siberia, are not known north of the Northern Hemisphere tropics. Presumably, nearly all of the Northern Hemisphere north of the tropics was ocean throughout the Silurian. *See* CONTINENTS, EVOLUTION OF; PALEOGEOGRAPHY.

Absence of plates bearing continental or shallow shelf marine environments north of about 45° north latitude indicates that ocean circulation in most of the Silurian Northern Hemisphere was zonal. Ocean surface currents in the tropics would have been influenced strongly by the prevailing westerlies. The large size of the Gondwana plate and the presence of land over much of it would have led to development of seasonal monsoon conditions. Surface circulation south of 30° south would have hit the western side of Laurentia and flowed generally northward. *See* PALEOCEANOGRAPHY.

Collision of the Avalonian and Laurentian plates in the latest Ordovician coincides with development of the Southern Hemisphere continental glaciation. Erosion of the land area formed at the Avalon-Laurentian plate collision generated a large volume of coarse to fine-grained siliclastic materials. That part of South America (modern eastern South America) near the South Pole for the early part of the Silurian was the site of as many as four brief glacial episodes. *See* PALEOCLIMATOLOGY.

Both nonvascular and vascular plants continued to develop in land environments following their originations in the early mid-Ordovician. Many of these Silurian plants were mosslike and bryophytelike. Psilophytes assigned to the genus *Cooksonia* were relatively widespread in Late Silurian terrestrial environments. The probable lycopod (club moss) *Baragwanathia* apparently lived in nearshore settings in modern Australia during the latter part of the Silurian. Silurian land life also included probable arthopods and annelid worms. Fecal pellets of wormlike activity have been found as well as remains of centipede-, millepede-, and spiderlike arthropods. *See* PALEOBOTANY; PALEOECOLOGY.

Shallow marine environments in the tropics were scenes of rich growths of algae, mat-forming cyanobacteria, spongelike organisms, sponges, brachiopods, bryozoans, corals, crinoids, and ostracodes. Nearshore marine siliclastic strata bear ostracodes, small clams, and snails and trilobites. Certain nearshore strata bear the remains of horseshoe-crab-like eurypterids. *See* ALGAE; BRACHIOPODA; BRYOZOA; CRINOIDEA; CYANOBACTERIA; OSTRACODA; TRILOBITA.

Fish are prominent in a number of Silurian nearshore and some offshore marine environments. Jawless armored fish include many species of thelodonts that had bodies covered with minute bony scales, heterostracans, and galeaspids that had relatively heavily armored head shields, and anaspids that possessed body armor consisting of scales and small plates. Jawed fish were relatively rare in the Silurian. They were primarily spiny sharks or acanthodians. As well, there are remains of true sharklike fish and fish with interior bony skeletons (osteichthyans) in Late Silurian rocks. *See* ANASPIDA; HETEROSTRACI; OSTEICHTHYES; THELODONTIDA.

William B. N. Berry

Siluriformes The catfishes, a large and highly diverse order of the Ostariophysi. Another ordinal name commonly used is Nematognathi, from the Greek *nematos*, thread, and *gnathos*, jaw, a descriptive name referring to the long barbels around the mouth of most catfishes. Catfishes differ from other members of the Ostariophysi in having a more complex Weberian apparatus and in lacking intermuscular bones, as well as lacking subopercle, symplectic, or intercalar bones in the head. They are further characterized by either lacking scales (in which case they are covered with tough skin) or being covered with overlapping bony plates; having vomer, pterygoid, and palatine bones that are usually toothed; having maxillae that are rudimentary and toothless; lacking a parietal bone; usually having one to four pairs of barbels; having (usually) an adipose fin; and having pectoral and dorsal fins that are each preceded by a strong spine in most species.

The order includes about 2800 species in 35 families and is represented on all continents, albeit only by fossils in Antarctica. Catfishes are chiefly freshwater species. Twelve families of catfishes are known only in Central and South America; three families are restricted to Africa; five are endemic to eastern Asia; three are shared by Africa and Asia; two are Indonesian, one Eurasian; one ranges from India to southeast Asia; and one, the Ictaluridae (see illustration), is endemic to North America. Not

Blue catfish (*Ictalurus furcatus*). (*Drawing © Joseph R. Tomelleri*)

only are catfishes diverse in morphology, biology, and life histories, but also they vary tremendously in size. Adults of the little madtoms (*Noturus*) of North America may weigh only a few grams, while the Mekong giant catfish (*Pangasianodon gigas*, family Pangasidae) has been reported to weigh up to 293 kg (646 lb). *See* ACTINOPTERYGII; CYPRINIFORMES; OSTEICHTHYES.

Herbert Boschung

Silver A chemical element, Ag, atomic number 47, atomic mass 107.868. It is a gray-white, lustrous metal. Chemically it is one of the heavy metals and one of the noble metals; commercially it is a precious metal. There are 25 isotopes of silver with atomic masses ranging from 102 to 117. Ordinary silver is made up of the isotopes of masses 107 (52% of natural silver) and 109 (48%). *See* PERIODIC TABLE.

Although silver is the most active chemically of the noble metals, it is not very active in comparison with most other elements. It does not oxidize as iron does when it rusts, but it reacts with sulfur or hydrogen sulfide to form the familiar silver tarnish. Electroplating silver with rhodium prevents this discoloration. Silver itself does not react with dilute nonoxidizing acids (hydrochloric or sulfuric acids) or strong bases (sodium hydroxide). However, oxidizing acids (nitric or concentrated sulfuric acids) dissolve it by reaction to form the unipositive silver ion, Ag^+. The Ag^+ ion is colorless, but a number of silver compounds are colored because of the influence of their other constituents.

Silver is almost always monovalent in its compounds, but an oxide, fluoride, and sulfide of divalent silver are known. Some coordination compounds of silver, also called silver complexes, contain divalent and trivalent silver. Although silver does not oxidize when heated, it can be oxidized chemically or electrolytically to form silver oxide or peroxide, a strong oxidizing agent. Because of this activity, silver finds considerable use as an oxidation catalyst in the production of certain organic materials.

Soluble silver salts, especially $AgNO_3$, have proved lethal in doses as small as 0.07 oz (2 g). Silver compounds may be slowly absorbed by the body tissues, with a resulting bluish or blackish pigmentation of the skin (argyria).

Silver is a rather rare element, ranking 63rd in order of abundance. Sometimes it occurs in nature as the free element (native silver) or alloyed with other metals. For the most part, however, silver is found in ores containing silver compounds. The principal silver ores are argentite, Ag_2S, cerargyrite or horn silver, AgCl, and several minerals in which silver sulfide is combined with sulfides of other metals; stephanite, $5Ag_2S \cdot Sb_2S_5$; polybasite, $9(Cu_2S, Ag_2S) \cdot (Sb_2S_3, As_2S_3)$; proustite, $3Ag_2S \cdot As_2S_3$; and pyragyrite, $3Ag_2S \cdot Sb_2S_3$. About three-fourths of the silver produced is a by-product of the extraction of other metals, copper and lead in particular. *See* SILVER METALLURGY.

Pure silver is a white, moderately soft metal (2.5–3 on Mohs hardness scale), somewhat harder than gold. When polished, it has a brilliant luster and reflects 95% of the light falling on it.

Silver is second to gold in malleability and ductility. Its density is 10.5 times that of water. The quality of silver, its fineness, is expressed as parts of pure silver per 1000 parts of total metal. Commercial silver is usually 999 fine. Silver is available commercially as sterling silver (7.5% copper) and in ingots, plate, moss, sheets, wire, castings, tubes, and powder.

Silver, with the highest thermal and electrical conductivities of all the metals, is used for electrical and electronic contact points and sometimes for special wiring. Silver has well-known uses in jewelry and silverware. Silver compounds are used in many photographic materials. In most of its uses, silver is alloyed with one or more other metals. Alloys in which silver is an ingredient include dental amalgam and metals for engine pistons and bearings. *See* PHOTOGRAPHIC MATERIALS; SILVER ALLOYS.

William E. Cooley

Silver alloys Combinations of silver with one or more other metals. Pure silver is very soft and ductile but can be hardened by alloying. Copper is the favorite hardener and normally is employed in the production of sterling silver, which must contain a minimum of 92.5% silver, and also in the production of coin silver.

Silver-copper eutectic and modifications containing other elements such as zinc, tin, cadmium, phosphorus, or lithium are widely used for brazing purposes, where strong joints having relatively good corrosion resistance are required. Where higher strengths at elevated temperature are required, silver-copper-palladium alloys and other silver-palladium alloys are suitable. The addition of a small amount of silver to copper raises the recrystallizing temperature without adverse effect upon the electrical conductivity.

Silver may be alloyed with gold or palladium in any ratio, producing soft and ductile alloys; certain of these intermediate alloys are useful for electrical contacts, where resistance to sulfide formation must be achieved.

Silver has proved to be a useful component for high-duty bearings in aircraft engines, where it may be overlaid with a thin layer of lead and finally with a minute coating of indium. Specially developed alloys of silver with tin, plus small percentages of copper and zinc in the form of moderately fine powder, can be mixed with mercury to yield a mass which is plastic for a time and then hardens, developing relatively high strength despite the fact that it contains about 50% mercury. This material was developed specifically for dental use and is generally known as amalgam, although the term amalgam actually includes all the alloys of mercury with other metals. *See* AMALGAM; SILVER.

Edmund M. Wise

Silver metallurgy The art and science of extracting silver metal economically from various ores, and the reclamation of silver from the myriad types of industrial processes or scrap produced therefrom. It includes all processes of separating silver from its ores, alloys, and solutions, as well as the smelting, refining, and working of the metal and its alloys and compounds. It deals with the technical application of the chemical and physical properties of silver to its concentration, extraction, purification, alloying, working, and compounding to meet the requirements of technical needs.

Chloridization is an extractive process in which silver is precipitated from aqueous solutions. In chloridization methods the silver contained in ores or metallurgical products is first converted by means of a chloridizing roast into a compound which is soluble in water or in certain aqueous solutions. The silver is then precipitated as an insoluble compound by suitable reagents and the precipitate worked for the metal.

The cyanidation process differs only in minor details, such as strength of the solution and time of treatment, from the cyanide process used in the recovery of gold. No preliminary metallurgical treatment other than fine grinding is required. *See* GOLD METALLURGY.

The Parkes process is based on the greater affinity of silver for zinc than for lead. Slab zinc is added to an argentiferous lead bath, the temperature of which has been raised higher than the melting point of the zinc. When the zinc has been melted and thoroughly mixed into the lead bath, the bath temperature is lowered and a silver-zinc alloy separates and floats on the top of the kettle. This zinc crust, as it is now called, is pressed off to remove the excess lead. The resultant high silver-lead retort bullion is then cupeled to recover the silver and gold.

Cupellation is the oldest and most widely known method of separating and recovering gold and silver from lead. The silver-lead bullion is charged to a reverberatory-type cupellation furnace. After the charge has melted, air is blown across the top of the molten bath, causing oxidation of the lead and other impurities which separate as in impure litharge slag, leaving behind the silver and gold as a doré alloy.

By far the most important method of silver production results from the smelting and subsequent treatment of the silver-rich slimes resulting from the electrolytic refining of copper. Slimes which have been leached with sulfuric acid or air are filtered and charged to a reverberatory-type furnace with appropriate fluxes. The resulting melt slag contains most of the base-metal impurities. Upon completion of this step, the material remaining in the furnace consists of gold, silver, selenium, tellurium, and residual base metals, and is called matte. Treatment of the matte eventually yields dore which is then cast into anodes for subsequent electrolytic refining. *See* SILVER.

R. D. Mushlitz

Silviculture The theory and practice of controlling the establishment, composition, and growth of stands of trees for any of the goods (including timber, pulp, energy, fruits, and fodder) and benefits (water, wildlife habitat, microclimate amelioration, and carbon sequestration) that they may be called upon to produce. In practicing silviculture, the forester draws upon knowledge of all natural factors that affect trees growing upon a particular site, and guides the development of the vegetation, which is either essentially natural or only slightly domesticated, to best meet the demands of society in general and ownership in particular. Based on the principles of forest ecology and ecosystem management, silviculture is more the imitation of natural processes of forest growth and development than a substitution for them.

The spatial patterns in which old trees are removed and the species that replace them determine the structure and developmental processes of the new stands. If all the trees are replaced at once with a single species, the result is so-called pure stand or monoculture in which all of the trees form a single canopy of foliage that is lifted ever higher as the stand develops. If several species start together from seeds, from small trees already present, or from sprouts, as a single cohort, the different species usually tend to grow at different rates in height. Some are adapted to develop in the shade of their sun-loving neighbors. The result is a stratified mixture. Such stands grow best on soils or in climates, such as in tropical moist forests, where water is not a limiting factor and the vegetation collectively uses most of the photosynthetically active light. If trees are replaced in patches or strips, the result is an uneven-aged stand which may be of one species or many.

These different spatial and temporal patterns of stand structure are created by different methods of reproduction. The simplest is the clear-cutting method, in which virtually all of the vegetation is removed. Although it is sometimes possible to rely on adjacent uncut stands as sources of seed, it is usually necessary to reestablish the new stand by artificial seeding or planting after clear-cutting. The seed tree method differs only in that a limited number of trees are temporarily left on the area to provide seed.

In the shelterwood method, enough trees are left on the cutting area to reduce the degree of exposure significantly and to provide a substantial source of seed. In this method the growth of a major portion of the preexisting crop continues, and the old trees are not entirely removed until the new stand is well established. The three methods just described lead to the creation of essentially even-aged stands.

The choice of methods of regeneration cutting depends on the ecological status of the species and stands desired. Species that characterize the early stages of succession, the so-called pioneers, will endure, and usually require the kind of exposure to sunlight resulting from heavy cutting or, in nature, severe fire, catastrophic windstorms, floods, and landslides. These pioneer species usually grow rapidly in youth but are short-lived and seldom attain large size. The species that attain greatest age and largest size are ordinarily those which are intermediate in successional position and tolerance of shade. Some of them will become established after the severe exposure of clear-cutting, but they often start best with light initial shade such as that created by shelterwood cutting. Their longevity and large size result from the fact that they are naturally adapted to reproduce after disturbances occurring at relatively long intervals. The shade-tolerant species representing late or climax stages in the succession are adapted to reestablish themselves in their own shade. These species represent natural adaptations to the kinds of fatal disturbance caused by insects, disease, and atmospheric agencies rather than to the more complete disturbance caused by fire. *See* ECOLOGICAL SUCCESSION; FOREST ECOSYSTEM.

Much silvicultural practice is aimed at the creation and maintenance of pure, even-aged stands of single species. This approach is analogous to that of agriculture and simplifies administration, harvesting, and other operations. The analogy is often carried to the extent of clear-cutting and planting of large tracts with intensive site preparation, especially with species representative of the early or intermediate stages of succession. Mixed stands are more difficult to handle from the operational standpoint but are more resistant to injury from insects, fungi, and other damaging agencies which usually tend to attack single species. They are also more attractive than pure stands and make more complete use of the light, water, and nutrients available on the site. They usually do not develop unless these site factors are comparatively favorable; where soil moisture or some other factor is limiting, it may be possible for only a single well-adapted species to grow. If the site is highly favorable, as in tropical rainforest, river floodplains, or moist ravines, it is so difficult to maintain pure stands that mixed stands are inevitable.

The application of silviculture involves a number of accessory practices other than cutting. In localities of high fire risks, it may be desirable to burn the slash (logging debris) after cutting. Not only does this reduce the potential fuel, but it may also help the establishment of seedlings by baring the mineral soil or reducing the physical barrier represented by the slash. Slash disposal is most often necessary where the cutting has been very heavy or where the climate is so cold or dry that decay is slow. Deliberate prescribed burning of the litter beneath existing stands of fire-resistant species is sometimes carried out even in the absence of cutting to reduce the fuel for wild fires, to kill undesirable understory species, to enhance the production of forage for wild and domestic animals, and to improve seedbed conditions. *See* FOREST FIRE.

Integrated schedules of treatment for stands are called silvicultural systems. They cover both intermediate and reproduction treatments but are classified and named in terms of the general method of reproduction cutting contemplated. Such programs are evolved for particular situations and kinds of stands with due regard for all the significant biological and economic considerations involved. These considerations include the desired uses of the land, kinds of products and services sought, prospective costs and returns of the enterprise presented by management of the stand, funds available for long-term investment in stand treatments, harvesting techniques and equipment employed, reduction of losses from damaging agencies, and the natural requirements that must be met in reproducing the stand and fostering its growth. *See* FOREST AND FORESTRY; FOREST MANAGEMENT.

David M. Smith; Mark S. Ashton

Simple machine Any of several elementary machines, one or more of which is found in practically every machine. The group of simple machines usually includes only the lever, wheel and axle, pulley (or block and tackle), inclined plane, wedge, and screw. However, the gear drive and hydraulic press may also be considered as simple machines. The principles of operation and typical applications of simple machines depend on several closely related concepts. *See* EFFICIENCY; FRICTION; MECHANICAL ADVANTAGE; POWER; WORK.

Two conditions for static equilibrium are used in analyzing the action of a simple machine. The first condition is that the sum of forces in any direction through their common point of action is zero. The second condition is that the summation of torques about a common axis of rotation is zero. Corresponding to these two conditions are two ways of measuring work. In machines with translation, work is the product of force and distance. In machines with rotation, work is the product of torque and angle of rotation. *See* BLOCK AND TACKLE; GEAR DRIVE; HYDRAULIC PRESS; INCLINED PLANE; LEVER; SCREW; TORQUE; WEDGE; WHEEL AND AXLE.

Richard M. Phelan

Simulation A broad collection of methods used to study and analyze the behavior and performance of actual or theoretical systems. Simulation studies are performed, not on the real-world system, but on a (usually computer-based) model of the system created for the purpose of studying certain system dynamics and characteristics. The purpose of any model is to enable its users to draw conclusions about the real system by studying and analyzing the model. The major reasons for developing a model, as opposed to analyzing the real system, include economics, unavailability of a "real" system, and the goal of achieving a deeper understanding of the relationships between the elements of the system.

Simulation can be used in task or situational training areas in order to allow humans to anticipate certain situations and be able to react properly; decision-making environments to test and select alternatives based on some criteria; scientific research contexts to analyze and interpret data; and understanding and behavior prediction of natural systems, such as in studies of stellar evolution or atmospheric conditions.

With simulation a decision maker can try out new designs, layouts, software programs, and systems before committing resources to their acquisition or implementation; test why certain phenomena occur in the operations of the system under consideration; compress and expand time; gain insight about which variables are most important to performance and how these variables interact; identify bottlenecks in material, information, and product flow; better understand how the system really operates (as opposed to how everyone thinks it operates); and compare alternatives and reduce the risks of decisions.

The word "system" refers to a set of elements (objects) interconnected so as to aid in driving toward a desired goal. This definition has two connotations: First, a system is made of parts (elements) that have relationships between them (or processes that link them together). These relationships or processes can range from relatively simple to extremely complex. One of the necessary requirements for creating a "valid" model of a system is to capture, in as much detail as possible, the nature of these interrelationships. Second, a system constantly seeks to be improved. Feedback (output) from the system must be used to measure the performance of the system against its desired goal. Both of these elements are important in simulation. *See* SYSTEMS ENGINEERING.

Systems can be classified in three major ways. They may be deterministic or stochastic (depending on the types of elements that exist in the system), discrete-event or continuous (depending on the nature of time and how the system state changes in relation to time), and static or dynamic (depending on whether or not the system changes over time at all). This categorization affects the type of modeling that is done and the types of simulation tools that are used.

Models, like the systems they represent, can be static or dynamic, discrete or continuous, and deterministic or stochastic. Simulation models are composed of mathematical and logical relations that are analyzed by numerical methods rather than analytical methods. Numerical methods employ computational procedures to run the model and generate an artificial history of the system. Observations from the model runs are collected, analyzed, and used to estimate the true system performance measures. *See* MODEL THEORY; STOCHASTIC PROCESS.

There is no single prescribed methodology in which simulation studies are conducted. Most simulation stuides proceed around four major areas: formulating the problem, developing the model, running the model, and analyzing the output. Statistical inference methods allow the comparison of various competing system designs or alternatives. For example, estimation and hypothesis testing make it possible to discuss the outputs of the simulation and compare the system metrics.

Many of the applications of simulation are in the area of manufacturing and material handling systems. Simulation is taught in many engineering and business curricula with the focus of the applications also being on manufacturing systems. The characteristics of these systems, such as physical layout, labor and resource utilization, equipment usage, products, and supplies, are extremely amenable to simulation modeling methods. *See* COMPUTER-INTEGRATED MANUFACTURING; FLEXIBLE MANUFACTURING SYSTEM.

James Pomykalski

Sine wave A wave having a form which, if plotted, would be the same as that of a trigonometric sine or cosine function. The sine wave may be thought of as the projection on a plane of the path of a point moving around a circle at uniform speed. It is characteristic of one-dimensional vibrations and one-dimensional waves having no dissipation. *See* HARMONIC MOTION.

The sine wave is the basic function employed in harmonic analysis. It can be shown that any complex motion in a one-dimensional system can be described as the superposition of sine waves having certain amplitude and phase relationships. The technique for determining these relationships is known as Fourier analysis. *See* FOURIER SERIES; WAVE EQUATION; WAVE MOTION; WAVEFORM.

William J. Galloway

Single crystal In crystalline solids the atoms or molecules are stacked in a regular manner, forming a three-dimensional pattern which may be obtained by a three-dimensional repetition of a certain pattern unit called a unit cell. When the periodicity of the pattern extends throughout a certain piece of material, one speaks of a single crystal. A single crystal is formed by the growth of a crystal nucleus without secondary nucleation or impingement on other crystals. *See* CRYSTAL STRUCTURE; CRYSTALLOGRAPHY.

When grown from a melt, single crystals usually take the form of their container. Crystals grown from solution (gas, liquid, or solid) often have a well-defined form which reflects the symmetry of the unit cell. *See* CRYSTAL GROWTH; CRYSTALLIZATION; ZONE REFINING.

Ideally, single crystals are free from internal boundaries. They give rise to a characteristic x-ray diffraction pattern.

Many types of single crystal exhibit anisotropy, that is, a variation of some of their physical properties according to the direction along which they are measured. For example, the electrical resistivity of a randomly oriented aggregate of graphite crystallites is the same in all directions. This anisotropy exists both for structure-sensitive properties, which are strongly affected by crystal imperfections (such as cleavage and crystal growth rate), and for structure-insensitive properties, which are not affected by imperfections (such as elastic coefficients).

The structure-sensitive properties of crystals (for example, strength and diffusion coefficients) seem governed by internal defects, often on an atomic scale. *See* CRYSTAL DEFECTS.

David Turnbull

Single sideband An electronic signal-processing technique in which a spectrum of intelligence is translated from a zero reference frequency to a higher frequency without a change of frequency relationships within the translated spectrum. Single-sideband (SSB) signals have no appreciable carrier.

Amplitude-modulated (AM) signals have identical upper and lower sidebands symmetrically located on each side of the translation frequency, which is often called the carrier. The SSB spectrum differs from the AM spectrum in having little or no carrier and only one sideband. *See* AMPLITUDE MODULATION.

In the SSB signal-processing action, the intelligence spectrum to be translated is applied to the signal input port of a balanced modulator. A higher-frequency sinusoidal signal, often called a carrier, is applied to the other input port of this circuit. Its function is to translate the zero reference spectrum to the carrier frequency and to produce the upper and lower sidebands, which are symmetrically located on each side of the carrier. The carrier frequency power is suppressed to a negligible value by the balanced operation of the modulator and does not appear at the output. Generally, the balanced modulator operates at an intermediate frequency which is lower than the frequency of transmission. Following the balanced modulator is a sideband filter which is designed to remove the unwanted sideband signal power and to allow only the desired intelligence spectrum to pass. *See* AMPLITUDE MODULATOR; MODULATOR.

There are many advantages in the use of SSB techniques for communication systems. The two primary advantages are the reduction of transmission bandwidth and transmission power. The bandwidth required is not greater than the intelligence bandwidth and is one-half that used by amplitude modulation. The output power required to give equal energy in the intelligence bandwidth is one-sixth that of amplitude modulation.

Propagation of radio energy via ionospheric refraction provides the possibility for multiple paths of differing path length which can cause a selective cancellation of frequency components at regular frequency spacings. This produces in amplitude modulation a severe distortion of the intelligence because of the critically dependent carrier-to-sideband amplitude and phase relationships. SSB is much less affected under these conditions.

David M. Hodgin

Single-site catalysts (polymer science) Metallocenes and related compounds which have a single catalytic site and are used to produce uniform polymers and copolymers with tailored structures and physical properties. Metallocenes are "sandwich compounds" in which a π-bonded metal atom is situated between two aromatic ring systems (see illustration). *See* CATALYSIS; METALLOCENES; POLYMER.

In 1953, Karl Ziegler succeeded in polymerizing ethylene into high-density polyethylene at standard pressure and room temperature, based on the use of the catalyst titanium tetrachloride combined with diethylaluminum chloride. Some months later, Giulio Natta demonstrated a catalyst system that was capable of polymerizing propylene into semicrystalline polypropylene. For their work, Ziegler and Natta shared the Nobel Prize for Chemistry in 1963. Modern Ziegler-Natta catalysts are mixtures of solid and liquid compounds, often containing $MgCl_2/TiCl_4/Al(C_2H_5)_3$ and, for propylene polymerization, different internal and external

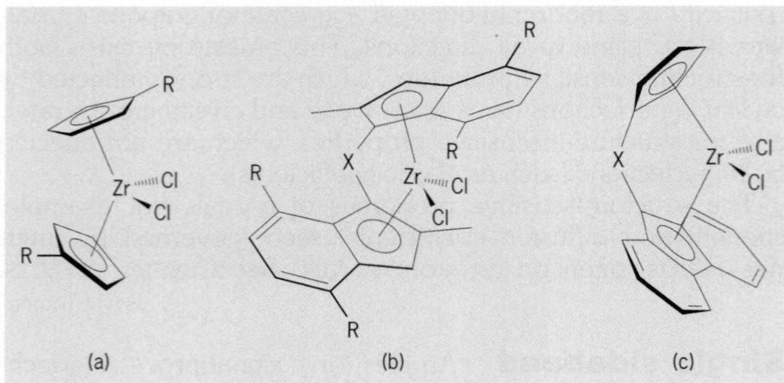

Structures of metallocenes (in this case, zirconocenes). (*a*) Nonchiral structure produces atactic polypropylene. (*b*) Racemic mixture of this chiral structure produces isotactic polypropylene. (*c*) C_s-symmetric structure produces syndiotactic polypropylene. R = alkyl substitution [H, CH_3, C_2H_5, C_6H_5]; X = bridge [—CH_2—CH_2—, $(CH_3)_2Si$, $(CH_3)_2C$].

donors such as ethylbenzoate, silanes, or ethers to increase stereoregularity (tacticity). An important catalyst for ethylene polymerization is the Phillips catalyst, that is, chromium trioxide on a silica support. This catalyst is used in the gas-phase polymerization of ethylene, but is unable to produce polypropylene with methyl groups always on the same side of the chain (isotactic). As Ziegler-Natta catalysts are heterogeneous and complex systems with different active sites, the resulting polymer structure can be controlled only to a limited degree. *See* HETEROGENEOUS CATALYSIS; POLYMERIZATION; POLYMER STEREOCHEMISTRY AND PROPERTIES; POLYOLEFIN RESINS.

In contrast to Ziegler systems, metallocene catalysts have only one active site (single-site catalysts) and produce polymers with a narrow molecular weight distribution $M_w/M_n = 2$ (M_w and M_n are the weight average and number average molecular weights). Their catalytic activity is 10–100 times higher than that of the classical Ziegler-Natta systems. In addition, metallocene catalysts are soluble in hydrocarbons or liquid propylene, and their structure can be easily changed. These properties allow one to predict accurately the properties of the resulting polyolefins by knowing the structure of the catalyst used during their manufacture, as well as to control the resulting molecular weight and distribution, comonomer content, and tacticity by careful selection of the appropriate reactor conditions. *See* MOLECULAR WEIGHT.

Metallocenes, in combination with the conventional aluminumalkyl cocatalysts used in Ziegler systems, are capable of polymerizing ethylene, but only at a very low activity. In 1977 in Hamburg, it was discovered that methylalumoxane (MAO) greatly improved catalysis with metallocenes, enhancing their activity, surprisingly, by a factor of 10,000. Methylaluminoxane is a com-

$$\begin{array}{ccccccccccccc} & & CH_3 & & & & CH_3 & & & & CH_3 & & & & CH_3 & & \\ & & | & & & & | & & & & | & & & & | & & \\ CH_3 & — & Al & — & O & — & Al & — & O & — & Al & — & O & — & Al & — & H_3C \end{array}$$

pound in which aluminum and oxygen atoms are arranged alternately and free valences are saturated by methyl substituents. Metallocene catalysts, especially zirconocenes, treated with MAO can polymerize up to 100 tons of ethylene per gram of zirconium (see illustration). Such highly active catalysts can remain in the product. The insertion time (for the insertion of one molecule of ethylene into the growing chain) is only 10^{-5} s. A comparison to enzymes is not far-fetched. It is generally assumed that the function of MAO is to first undergo a fast ligand exchange reaction with the metallocene dichloride, thus producing the metallocene methyl and dimethylaluminum compounds. In the next step, either Cl^- or $CH_3{}^-$ is abstracted from the metallocene compound by an aluminum center in MAO, thus forming a metallocene cation and an MAO anion. The alkylated metallocene cation represents the active center. Meanwhile, other weakly coordinating cocatalysts, such as tetra(perfluorophenyl)borate anions, $[(C_6F_5)_4B]^-$, have been successfully used to activate metallocenes and other single-site catalysts. Polyolefins with different microstructures and characteristics can be custom-made by varying the ligands of the metallocene (see illustration). By combining different olefins and cycloolefins with one another, the range of characteristics can be further broadened. Conventional heterogeneous catalysts have not yet produced polyolefins with tailored microstructures or chemically uniform copolymers. Using metallocene catalysts, it was possible for the first time to produce polyethylenes, polypropylenes, and copolymers with narrow molecular-weight distribution, syndiotactic (alternating stereoregular structure) polypropylene, syndiotactic polystyrene, cyclopolymers of 1,5-hexadiene, and cycloolefin copolymers (COCs) with high catalytic activity. *See* COPOLYMER; POLYSTYRENE RESIN. W. Kaminsky

Sintering The welding together and growth of contact area between two or more initially distinct particles at temperatures below the melting point, but above one-half of the melting point in kelvins. Since the sintering rate is greater with smaller than with larger particles, the process is most important with powders, as in powder metallurgy and in firing of ceramic oxides.

Although sintering does occur in loose powders, it is greatly enhanced by compacting the powder, and most commercial sintering is done on compacts. Compacting is generally done at room temperature, and the resulting compact is subsequently sintered at elevated temperature without application of pressure. For special applications, the powders may be compacted at elevated temperatures and therefore simultaneously pressed and sintered. This is called hot pressing or sintering under pressure.

Certain compacts from a mixture of different component powders may be sintered under conditions where a limited amount of liquid, generally less than 25 vol%, is formed at the sintering temperature. This is called liquid-phase sintering, important in certain powder-metallurgy and ceramic applications. *See* CERAMICS; POWDER METALLURGY. F. V. Lenel

Sinus Any space in an organ, tissue, or bone, but usually referring to the paranasal sinuses of the face. In humans, four such sinuses, lined with ciliated, mucus-producing epithelium, communicate with each nasal passage through small apertures. The ethmoid and sphenoid sinuses are located centrally between and behind the eyes. The frontal sinuses lie above the nasal bridge, and the maxillary sinuses are contained in the upper jaw beneath the orbits. The mastoid portion of the temporal bone contains air cells lined with similar epithelium. Thomas S. Parsons

Siphonaptera An order of insects commonly known as fleas. These insects are bloodsucking ectoparasites of animals and humans and may transmit serious diseases. Adult fleas are easily recognized. They are small, wingless, about $^1/_8$ in. (3 mm) in length, dark brown or black, and laterally flattened. They have three pairs of legs, with the third pair modified for jumping. The body of a flea is characteristically oval-shaped and armed with combs, setae, and spines, which prevent the flea from being dislodged from the host. The head bears mouthparts adapted for bloodsucking (see illustration). *See* INSECTA.

Fleas undergo complete metamorphosis (holometabolous); that is, their life history involves four distinct phases: the egg, larva, pupa, and adult. The only exception is the neosomic *Tunga monositus* (sand flea), which has only two stages. Female adult fleas may lay large numbers of eggs (400 or more). They are laid in the bedding or nest of the host, where they eventually hatch into the larva. Under favorable conditions, the entire life cycle may be completed in as little as 2 weeks.

Adult fleas feed exclusively on the blood of the host. Adult fleas spend considerable time off the host, and may be found in

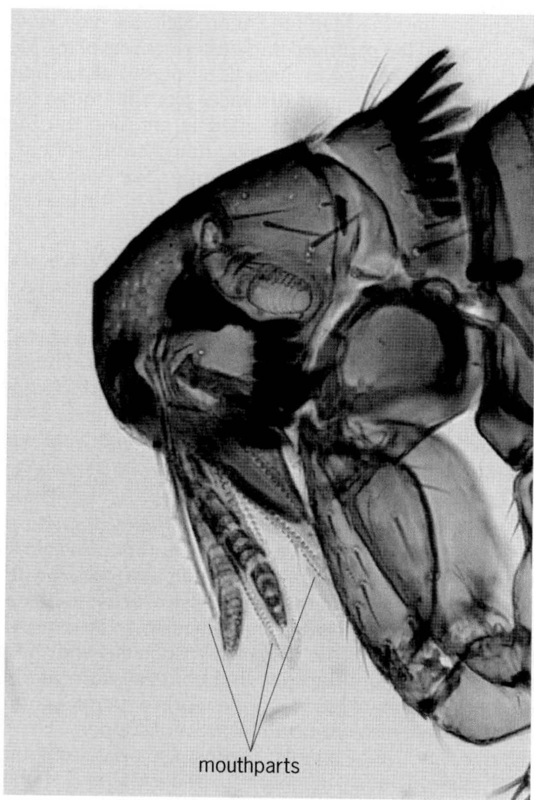

Head of *Spilopsyllus cuniculi* (rabbit flea), with extended mouthparts.

or around the bedding or nest of the host. They only visit the host to feed, and may feed multiple times during one blood meal.

Specialists now recognize 2575 species, belonging to 244 genera. The great majority of fleas occur on mammals (95%); the remainder live on birds. The host specificity of fleas, as well as their seasonal and geographical distribution, is largely dependent on specific requirements of the larvae. Adult fleas may occasionally attack humans if given the opportunity of contact. Since many fleas are carriers and transmitters (vectors) of disease, humans can contract infections through their bite. Plague, which is responsible for millions of deaths throughout the history of humankind, is a disease of rats. It is transmitted between rats and from rats to humans by fleas, particularly the oriental rat flea (*Xenopsylla cheopis*). Murine or endemic typhus fever is another disease transmitted from domestic rats to humans by rat fleas. Flea allergy is the term applied by physicians to the severe reactions which sometimes result from flea bites. *See* PLAGUE.

Katharina Dittmar; Irving Fox

Siphonophora An order of the class Hydrozoa of the phylum Coelenterata, characterized by an extremely complex organization of components of several different types, some having the basic structure of a jellyfish, others of a polyp. The components may be connected by a stemlike region or may be more closely united into a compact organism.

Most siphonophores possess a float and are animals of the open seas. Best known is the Portuguese man-of-war, *Physalia*, with a float as much as 16 in. (40 cm) long and tentacles which extend downward for many feet. These animals may be swept shoreward and may make swimming not only unpleasant but dangerous. *See* COELENTERATA; HYDROZOA.

Sears Crowell

Siphonostomatoida An order of Copepoda; all members are parasites of marine and fresh-water fishes or a variety of invertebrate hosts. In fact, it is estimated that 67% of copepod parasites of fishes belong to the Siphonostomatoida. Their obvious success in the parasitic mode of life appears at least in part to be a result of two morphologic adaptations. The first is the modification of the buccal apparatus (mouth and appendages) resulting in a mouth cone that has a small opening near the base through which the mandibles are free to enter. The maxilliae are subchelate or brachiform and serve as the appendage for attachment to the host. The second adaptation is the development of a frontal filament, which is a larval organ of attachment. In those siphonostomatoids possessing a frontal filament, the brachiform second maxilla may be used to manipulate the frontal organ, and in some species fusion of those two structures forms the attachment structure. *See* COPEPODA; CRUSTACEA; POECILOSTOMATOIDA.

Patsy A. McLaughlin

Sipuncula A phylum of sedentary marine vermiform coelomates that are unsegmented, but possibly distantly related to the annelids; they are commonly called peanut worms. There are 17 genera and approximately 150 species living in a wide variety of oceanic habitats within the sediment or inside any protective shelter such as a discarded mollusk shell, foraminiferan test, or crevice in rock or coral.

Adult sipunculans range in trunk length from 2 to over 500 mm (0.08 to over 20 in.). The shape of the body ranges from almost spherical to a slender cylinder. Sipunculans have a variety of epidermal structures (papillae, hooks, or shields). Many species lack color, but shades of yellow or brown may be present. Internal anatomy is relatively simple. The digestive tract has a straight esophagus and a double-coiled intestine extending toward the posterior end of the body and back terminating in a rectum, sometimes bearing a small cecum. A ventral nerve cord with lateral nerves and circumenteric connectives to the pair of cerebral ganglia are present. Two or four pigmented eyespots may be present on the cerebral ganglia, and a chemoreceptor (nuchal organ) is usually present. Most species of sipunculans are dioecious (have two sexes) that are similar, that is, they lack any sexual dimorphism.

Most sipunculans are deposit feeders. Therefore, these worms play a part in the recycling of detritus and probably consume smaller invertebrates in the process. They are in turn preyed on by fishes and probably other predators (including humans in China). *See* FEEDING MECHANISMS (INVERTEBRATE).

Edward B. Cutler; A. Schulze

Siren (acoustics) A sound source that is based on the regular interruption of a stream of fluid (usually air) by a perforated rotating disk or cylinder. The components of a siren are a source of air, a rotor containing a number of ports which interrupt the airflow at the desired frequency, and ports in a stator through which the air escapes. The air is supplied by a compressor, and a motor drives the rotor. The frequency of the sound wave produced by the siren is the product of the speed of rotation and the number of ports in the rotor. The shape of the rotor and stator ports determines the wave shape at the entrance of the stator port. The stator ports feed into a horn in order to improve radiation. Siren performance parameters are sound power output, acoustic pressure, and efficiency, that is, the ratio of acoustic power output to compressor power. *See* SOUND PRESSURE.

Applications of sirens include acoustic levitation (the use of radiation pressure to levitate small objects), broadband underwater sound projectors, and sonic fatigue (fatigue life and failure of structures subjected to fluctuating pressures generated by acoustic waves). *See* ACOUSTIC LEVITATION.

Electromechanical sirens use an electric motor instead of a compressed air supply to generate the acoustic signal. A second motor spins the rotor. The stator and horn increase the sound power output and efficiency. Electromechanical sirens are widely used as warning devices. *See* SOUND. Bart Lipkens

Sirenia An order of herbivorous aquatic placental mammals, commonly known as sea cows, that includes the living manatees and dugongs and the recently exterminated Steller's sea cow. The order has an extensive fossil record dating from the early Eocene Epoch, some 50 million years ago.

The earliest known sirenians were quadrupedal and capable of locomotion on land. Fossils clearly document the evolutionary transition from these amphibious forms to the modern, fully aquatic species, which have lost the hindlimbs and transformed the forelimbs into paddlelike flippers. The living species have streamlined, fusiform bodies with short necks and horizontal tail fins like those of cetaceans, but no dorsal fins. The skin is thick and nearly hairless. The nostrils are separate, and the ears lack external pinnae.

Sirenians typically feed on aquatic angiosperms, especially seagrasses, but in ecologically marginal situations they also eat algae and even some animal material. They are normally found in tropical or subtropical marine waters, but some have become adapted to fresh water or colder latitudes. Body sizes have ranged from less than 3 m up to 9–10 m (30–33 ft). Sirenians mate and give birth in the water, bearing a single calf (occasionally twins) after about 13–14 months of gestation and then nursing it from one pair of axillary mammae. The closest relatives of sirenians among living mammals are the Proboscidea (elephants). *See* PROBOSCIDEA.

A classification scheme is given below.

Order Sirenia
- Family: Prorastomidae
  - Protosirenidae
  - Trichechidae
    - Subfamily: Miosireninae
      - Trichechinae
- Family Dugongidae
  - Subfamily: Halitheriinae
    - Hydrodamalinae
    - Dugonginae

Daryl P. Domning

Sirius The star α Canis Majoris, also referred to as the Dog Star, the brightest of all the stars in the night sky (apparent magnitude -1.47). Sirius owes its apparent brightness both to its close distance to Earth, only 2.64 parsecs (8.14×10^{13} km or 5.06×10^{13} mi), and to its intrinsic luminosity, which is more than 20 times that of the Sun. It is a main-sequence star of spectral type A1 with an effective temperature of about 9400 K (16,500°F). *See* SPECTRAL TYPE.

Sirius is a very interesting binary system in which the companion is a degenerate star (α CMa B). It is the brightest among the known white dwarfs (apparent magnitude 8.4) and the closest to the Sun. It orbits around the bright star once every 50 years. The luminosity of the white dwarf is 400 times smaller than that of the Sun, which makes it extremely difficult to see because of the overpowering brightness of the nearby primary. Its temperature is estimated to be 24,790 K (44,654°F), much hotter than the primary. The white dwarf's mass is comparable to the Sun, yet is concentrated in a body the size of the Earth. The mean density of the material in α CMa B is about 2×10^6 times that of water, or 2 metric tons/cm^3 (nearly 40 tons/in.^{3}). *See* BINARY STAR; STAR; STELLAR EVOLUTION; WHITE DWARF STAR. David W. Latham

Sirocco A southerly or southeasterly wind current from the Sahara or from the deserts of Saudi Arabia which occurs in advance of cyclones moving eastward through the Mediterranean Sea. The sirocco is most pronounced in the spring, when the deserts are hot and the Mediterranean cyclones are vigorous. It is observed along the southern and eastern coasts of the Mediterranean Sea from Morocco to Syria as a hot, dry wind capable of carrying sand and dust great distances from the desert source. The sirocco is cooled and moistened in crossing the Mediterranean and produces an oppressive, muggy atmosphere when it extends to the southern coast of Europe. *See* AIR MASS; WIND.

Frederick Sanders

Sisal A fiber obtained from the leaves of *Agave sisalana*, produced in Brazil, Haiti, and several African countries, including Tanzania, Kenya, Angola, and Mozambique.

Sisal is used mainly for twine and rope, but some of the lower grades are used for upholstery padding and paper. The greatest quantity goes into farm twines, followed by industrial tying twine and rope. Most sisal-fiber ropes made in the United States are small to medium in size, intended for light duty. Sisal is sometimes used for marine cordage in Europe. *See* NATURAL FIBER.

Elton G. Nelson

Skarn A broad range of rock types made up of calc-silicate minerals such as garnet, regardless of their association with ores, that originate by replacement of precursor rocks. It was a term originally coined by miners in reference to rock consisting of coarse-grained, calc-silicate minerals associated with iron ores in central Sweden. Ore deposits that contain skarn are termed skarn deposits; such deposits are the world's premier sources of tungsten. They are also important sources of copper, iron, molybdenum, zinc, and other metals. Skarns also serve as sources of industrial minerals such as graphite, asbestos, and magnesite. *See* ORE AND MINERAL DEPOSITS; SILICATE MINERALS.

Based on mineralogy, three idealized types of skarn are recognized: calcic skarn characterized by calcium- and iron-rich silicates (andradite, hedenbergite, wollastonite); magnesian skarn characterized by calcium- and magnesium-rich silicates (forsterite, diopside, serpentine); and aluminous skarn characterized by aluminum- and magnesium-rich calc-silicates (grossularite, vesuvianite, epidote). M. T. Einaudi

Skeletal system The supporting tissues of animals which often serve to protect the body, or parts of it, and play an important role in the animal's physiology.

Skeletons can be divided into two main types based on the relative position of the skeletal tissues. When these tissues are located external to the soft parts, the animal is said to have an exoskeleton. If they occur deep within the body, they form an endoskeleton. All vertebrate animals possess an endoskeleton, but most also have components that are exoskeletal in origin. Invertebrate skeletons, however, show far more variation in position, morphology, and materials used to construct them.

The vertebrate endoskeleton is usually constructed of bone and cartilage; only certain fishes have skeletons that lack bone. In addition to an endoskeleton, many species possess distinct exoskeletal structures made of bone or horny materials. This dermal skeleton provides support and protection at the body surface.

Various structural components make up the human skeleton, including collagen, three different types of cartilage (hyaline, fibrocartilage, and elastic), and a variety of bone types (woven, lamellar, trabecular, and plexiform). *See* BONE; COLLAGEN; CONNECTIVE TISSUE.

The vertebrate skeleton consists of the axial skeleton (skull, vertebral column, and associated structures) and the appendicular skeleton (limbs or appendages). The basic plan for vertebrates is similar, although large variations occur in relation to functional demands placed on the skeleton.

Axial skeleton. The axial skeleton supports and protects the organs of the head, neck, and torso, and in humans it comprises the skull, ear ossicles, hyoid bone, vertebral column, and rib cage.

Skull. The adult human skull consists of eight bones which form the cranium, or braincase, and 13 facial bones that support the eyes, nose, and jaws. There are also three small, paired ear ossicles—the malleus, incus, and stapes—within a cavity in the temporal bone. The total of 27 bones represents a large reduction in skull elements during the course of vertebrate evolution. The three components of the skull are the neurocranium, dermatocranium, and visceral cranium. *See* EAR (VERTEBRATE).

The brain and certain sense organs are protected by the neurocranium. All vertebrate neurocrania develop similarly, starting as ethmoid and basal cartilages beneath the brain, and as capsules partially enclosing the tissues that eventually form the olfactory, otic, and optic sense organs. Further development produces cartilaginous walls around the brain. Passages (foramina) through the cartilages are left open for cranial nerves and blood vessels. Endochondral ossification from four major centers follows in all vertebrates, except the cartilaginous fishes. *See* TETRAPODA.

The visceral skeleton, the skeleton of the pharyngeal arches, is demonstrated in a general form by the elasmobranch fishes, where all the elements are cartilaginous and support the jaws and the gills. The mandibular (first) arch consists of two elements on each side of the body: the palatoquadrates dorsally, which form the upper jaw, and Meckel's cartilages, which join ventrally to form the lower jaw. The hyoid (second) arch has paired dorsal hyomandibular cartilages and lateral, gill-bearing ceratohyals. This jaw mechanism attaches to the neurocranium for support. In all jawed vertebrates except mammals, an articulation between the posterior ends of the palatoquadrate and Meckel's cartilages occurs between the upper and lower jaws. The bony fishes have elaborated on the primitive condition, where the upper jaw was fused to the skull and the lower jaw or mandible could move only in the manner of a simple hinge. Teleosts are able to protrude the upper and lower jaws. In the course of mammalian evolution, the dentary of the lower jaw enlarged and a ramus expanded upward in the temporal fossa. This eventually formed an articulation with the squamosal of the skull. With the freeing of the articular bone and the quadrate from their function in jaw articulation, they became ear ossicles in conjunction with the columella, that is, a skeletal rod that formed the first ear ossicle. The remaining visceral skeleton has evolved from jaw and gill structures in the fishes to become an attachment site for tongue muscles and to support the vocal cords in tetrapods. *See* MAMMALIA.

Vertebral column. The vertebral column is an endoskeletal segmented rod of mesodermal origin. It provides protection to the spinal cord, sites for muscle attachment, flexibility, and support, particularly in land-based tetrapods where it has to support the weight of the body (see illustration). Hard, spool-shaped bony vertebrae alternate with tough but pliable intervertebral discs. Each typical vertebral body (centrum) has a bony neural arch extending dorsally. The spinal cord runs through these arches, and spinal nerves emerge through spaces. Bony processes and spines project from the vertebrae for the attachment of muscles and ligaments. Synovial articulations between adjacent vertebrae effectively limit and define the range of vertebral motion.

Vertebral morphology differs along the length of the column. There are two recognized regions in fishes (trunk and caudal) and five in mammals (cervical, thoracic, lumbar, sacral, and caudal), reflecting regional specializations linked to function. Humans have seven cervical, twelve thoracic, five lumbar, five (fused) sacral, and four coccygeal vertebrae. Most amphibians, reptiles, and mammals have seven cervical vertebrae regardless of neck length, whereas the number is variable in birds. Specific modification to the first two cervical vertebrae in most reptiles, birds, and mammals gives the head extra mobility. The presence of large ribs in the thoracic region often limits spinal flexibility. In typical tetrapods, the sacral region is usually modified for support of the pelvic girdle, while the number of caudal vertebrae varies greatly (from 0 to 50) between and within animal groups. *See* SPINE; VERTEBRA.

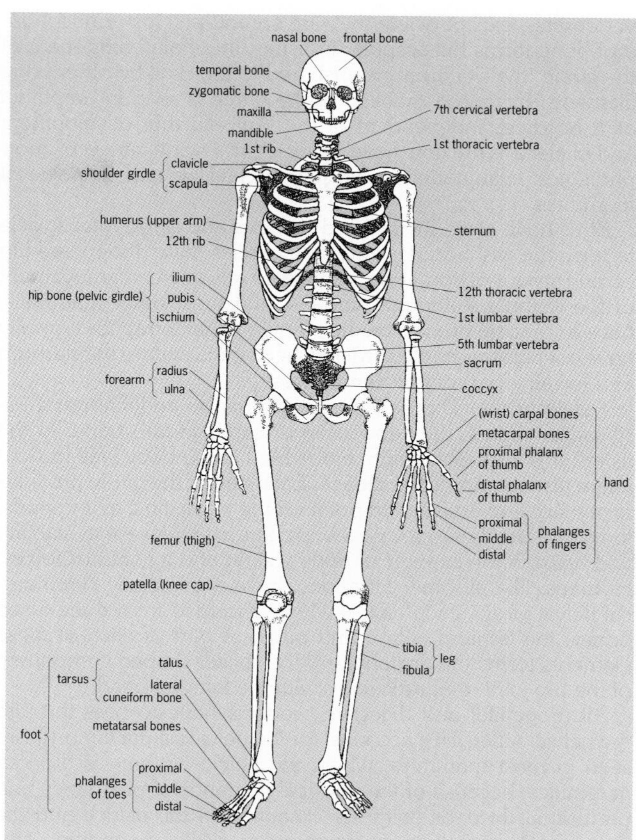

Human skeleton (anterior view). (*After G. J. Romanes, ed., Cunningham's Textbook of Anatomy, 10th ed., Oxford University Press, 1964*)

Sternum and ribs. Jawed fishes have ribs that help maintain the rigidity and support of the coelomic cavity. These ribs typically follow the connective tissue septa that divide successive muscle groups. In the caudal region, they are often small paired ventral ribs, fused on the midline to form the haemal arches. Ancestral tetrapods had ribs on all vertebrae, and their lengths varied between the vertebral regions. Modern amphibia (frogs and toads) have few thoracic ribs, and these are much reduced and never meet ventrally. Reptiles have varied rib arrangements, ranging from snakes with ribs on each vertebra (important for locomotor requirements) to turtles with only eight ribs which are fused to the inside of the carapace. Flying birds and penguins have a greatly enlarged sternum that links the ribs ventrally. In humans there are twelve pairs of ribs which form a strong but movable cage encompassing the heart and lungs.

Appendicular skeleton. This section of the skeletal system comprises the pectoral and pelvic limb girdles and bones of the free appendages. The girdles provide a supporting base onto which the usually mobile limbs attach.

Pectoral girdle. The pectoral girdle has both endoskeletal and dermal components. The dermal components are derived from postopercular dermal armor of primitive fishes, and are represented by the clavicles and interclavicles in modern vertebrates, except where they are secondarily lost. Endochondral bone forms the scapula. In fishes, the main component of the girdle (the cleithrum) is anchored to the skull by other bony elements. Increased mobility of the girdle is seen in amphibia as it becomes independent of the skull. Further development and skeletal reduction have resulted in a wide range of morphologies, culminating in the paired clavicles and scapulae of mammals.

Birds have fused their paired clavicles and single interclavicle to form the wishbone or furcula. Clavicles have disappeared in certain groups of bounding mammals to allow greater movement of the scapula. Although humans, and most other mammals, have a coracoid process on the scapula, other tetrapods typically have a separate coracoid bracing the scapula against the sternum and forming part of the glenoid fossa.

Pelvic girdle. The pelvic girdle forms by endochondral ossification, that is, the conversion of cartilage into bone. In the fishes, it is a small structure embedded in the body wall musculature just anterior to the cloaca. Each half of the girdle provides an anchor and articulation point for the pelvic fins. In tetrapods, the girdle attaches to the vertebral column to increase its stability and assist in the support of body weight and locomotor forces. Humans, like all other tetrapods, have a bilaterally symmetrical pelvic girdle, each half of which is formed from three fused bones: the ischium, ilium, and pubis. A part of each of these elements forms the acetabulum, the socket-shaped component of the hip joint, that articulates with the femoral head.

All urogenital and digestive products have to pass through the pelvic outlet. This accounts for the pelvic sexual dimorphism seen in most mammals, where the pelvic opening is broader in females, because of the physical demands of pregnancy and parturition. In birds (with the exception of the ostrich and the rhea), both sexes have an open pelvic girdle, a condition also found in female megachiropteran bats (flying foxes), gophers, and mole-rats.

Paired fins and tetrapod limbs. Paired fins in fishes come in different forms, but all are involved in locomotion. In the simplest form they are fairly rigid and extend from the body, functioning as stabilizers, but they are also capable of acting like a wing to produce lift as in sharks. In many fishes, the pectoral fins have narrow bases and are highly maneuverable as steering fins for low-speed locomotion. In addition, some fishes use their pectoral and pelvic fins to walk on the river bed, while others have greatly enlarged pectoral fins that take over as the main propulsive structures.

The basic mammalian pectoral limb consists of the humerus, radius, ulna, carpals, five metacarpals, and fourteen phalanges; and the pelvic limb consists of the femur, tibia, fibula, tarsal, five metatarsals, and fourteen phalanges. A typical bird pelvic limb consists of a femur, tibiotarsus (formed by fusion of the tibia with the proximal row of tarsal bones), fibula, and tarsometatarsus (formed by fusion of metatarsals II–IV), metatarsal I, and four digits (each consisting of two to five phalanges). Mike Bennett

Skin The entire outer surface of the body and the principal boundary between the external environment and the body's internal environment of cells and fluids. Skin serves as the primary barrier against the intrusion of foreign elements and organisms into the body, and also as a large and complex sense organ through which animals explore and learn about the external world. In addition, skin functions to maintain the homeostasis of the body's constituents, acting as a barrier to the loss of various ions and nutrients by diffusion. For terrestrial animals, it also serves as an effective barrier to water loss, without which most land animals would rapidly become desiccated and die.

The skin of humans and other mammals can be divided into two distinct regions, the epidermis and the dermis.

The epidermis is the outermost layer of the skin. It varies in thickness from 0.1 mm in most of the protected areas of the skin to approximately 1 mm in those regions exposed to considerable friction, such as the soles of the feet and palms of the hands. The epidermis consists of a great many horizontally oriented layers of cells. The outermost layer, the stratum corneum, consists of many layers of this packed cellular debris, forming an effective barrier to water loss from lower layers of the skin. The lowest levels of stratum germinativum constitute the portion of the skin that contains melanocytes, cells that produce the dark pigment melanin. Different levels of melanin secretion are responsible for the large range of pigmentation observed among humans.

The dermis plays a supportive and nutritive role for the epidermis. The epidermis has no blood supply of its own. However, nutrients and oxygen are apparently provided by diffusion from the blood supply of the underlying dermis. The average thickness of the dermis is 1–3 mm. It is in this layer that the sebaceous and sweat glands are located and in which the hair follicles originate. The products of all these sets of glands are derived from the rich blood supply of the dermis. Hair, sweat glands, and mammary glands (which are modified sweat glands) are skin inclusions unique to mammals. *See* HAIR; INTEGUMENTARY PATTERNS; SWEAT GLAND; THERMOREGULATION. Albert F. Bennett

Skin disorders The skin is subject to localized and generalized disorders, as well as those of primary occurrence in the skin and those secondary to involvement of other tissue. Diseases and disorders may affect any of the structures of the skin. They may be caused by external agents, either infectious or noninfectious, or by the abnormal accumulation of normal or abnormal skin elements, either inborn or acquired.

Infectious diseases. These are classified by the type of infectious agent—bacterial, parasitic, fungal, or viral.

Bacterial infections are distinguished clinically by the skin layer or appendage affected, but treatment is based upon the organism causing the infection. Impetigo and cellulitis are the most common infections of skin. Both infections may be caused by streptococcus, including group B hemolytic streptococcus, and staphylococcus, often species resistant to penicillin. Skin appendages such as hair follicles may be similarly infected. *See* ANTIBIOTIC; DRUG RESISTANCE; STAPHYLOCOCCUS; STREPTOCOCCUS.

Parasitic skin conditions are most often seen in epidemics among individuals who are in close contact, or where hygiene is poor. Head, body, and pubic lice, and scabies are the most common. Pubic lice and scabies are also often transmitted by sexual contact. *See* MEDICAL PARASITOLOGY.

Fungal infections are extremely common. In the United States, most fungal infections in humans are with species incapable of infecting tissue other than keratinized epidermis. These organisms are known as dermatophytes; they cause tinea pedis (athlete's foot), tinea cruris (some forms of jock itch), and tinea capitis (a scalp condition responsible for some forms of hair loss). *See* MEDICAL MYCOLOGY.

Viral diseases often involve the skin; however, viral warts (verruca vulgaris) and molluscum contagiosum are the primary examples of viral diseases that affect only the skin. Both are characterized by single or multiple, somewhat contagious, skin tumors that usually are small but can in rare instances exceed 0.4 in. (1 cm) in diameter. *See* ANIMAL VIRUS; CHICKENPOX AND SHINGLES; HERPES; MEASLES; SMALLPOX.

Inflammatory disease. Most itchy rashes are due to inflammation of the skin; they are usually known as eczema or dermatitis. In the acute stage, eczematous dermatitis is characterized by a vesicular, oozing condition. Seborrheic dermatitis is a common eczematous condition affecting primarily the areas of skin that bear sebaceous glands, that is, scalp, central face, chest, axilla, and groin.

Hereditary disease. Atopic dermatitis is the skin manifestation of atopy, a clinically apparent hypersensitivity. This condition may also be associated with asthma and pollen allergies. Psoriasis is a common disease of unknown etiology. The typical psoriasis consists of well-defined patches and plaques of red skin with a silvery scale that often results in pinpoint bleeding when removed. The most common and persistent sites are the elbows, knees, and scalp, but any area of skin may be involved; there may be significant morbidity and disability.

Other conditions. Acne vulgaris is an extremely common skin disorder, affecting 80–90% of young adults, usually during adolescence. The pilosebaceous organ of the skin is the primary target, particularly the sebaceous gland, its duct, and the infundibulum of the hair follicle. In addition to increased production of sebum, bacteria contribute to the development of acne lesions. These lesions consist of open and closed comedones (blackheads), papules, pustules, and cysts.

Reactions to ingested materials, such as food or medications, often appear in the skin, usually either as a red, itchy, measleslike rash or as urticaria (hives). These reactions indicate allergy to the material, which must then be avoided. *See* ALLERGY.

Neoplasms. Skin neoplasms may be benign or malignant, congenital or acquired, and they may arise from any component of the skin. Almost all skin neoplasms are benign and acquired. The common mole (melanocytic nevus) is a neoplasm of benign melanocytes; usually it is acquired, but it may be present at birth, when it is often known as a birthmark. Other common congenital nevi or birthmarks are of vascular origin, including strawberry and cavernous hemangiomas and port wine stains.

Malignant neoplasms may arise from cellular elements of the epidermis or dermis, or by infiltration of the skin by malignant cells arising from other tissues. By far the most common are basal cell and squamous cell cancers, which arise from basal and squamous keratinocytes of the epidermis, respectively. They usually are characterized by a nonhealing sore, persistent red scaling or crusting patch, or a slowly growing pearly nodule on skin that has been exposed to the sun; they occur mostly on the head, neck, hands, and arms.

Malignant melanoma arises from the pigment-forming melanocyte, and thus it is usually pigmented. It is a metastasizing cancer that is often fatal if not removed surgically in the early stage. It can be recognized as a pigmented lesion, often thought to be a benign nevus initially, that increases in size, changes color, particularly with admixtures of black, blue, red, or white along with the usual tans or browns; and it becomes irregular in size and shape. *See* CANCER (MEDICINE).

Two unusual multicentric primary skin malignancies are mycosis fungoides and Kaposi's sarcoma. Mycosis fungoides is a lymphoma of the skin, that may remain confined to the skin for 10 or more years before eventually spreading to internal organs and causing death. It may be extremely difficult to diagnose, both clinically, when it can appear only as eczematous patches, and histologically, for months or years. Kaposi's sarcoma occurs in two forms, the classic form seen on the legs of elderly Mediterranean men, and a form associated with HIV-1 infection and AIDS that may occur on any part of the body. It is derived from skin blood vessels, is multicentric, and usually appears as red to violet patches, plaques, or nodules. It is usually not fatal, although it may eventually spread to internal organs and may cause significant morbidity. There are numerous other primary skin cancers, but they are rare. *See* ACQUIRED IMMUNE DEFICIENCY SYNDROME (AIDS); SKIN.

Alan N. Moshell

Skin effect (electricity)

The tendency for an alternating current to concentrate near the outer part or "skin" of a conductor. For a steady unidirectional current through a homogeneous conductor, the current distribution is uniform over the cross section; that is, the current density is the same at all points in the cross section. With an alternating current, the current is displaced more and more to the surface as the frequency increases. The conductor's effective cross section is therefore reduced so the resistance and energy dissipation are increased compared with the values for a uniformly distributed current. The effective resistance of a wire rises significantly with frequency; for example, for a copper wire of 1-mm (0.04-in.) diameter, the resistance at a frequency of 1 MHz is almost four times the dc value. *See* ALTERNATING CURRENT; ELECTRICAL RESISTANCE.

A skin depth or penetration depth δ is frequently used in assessing the results of skin effect; it is the depth below the conductor surface at which the current density has decreased to $1/e$ (approximately 37%) of its value at the surface. This concept applies strictly only to plane solids, but can be extended to other shapes provided the radius of curvature of the conductor surface is appreciably greater than δ.

At a frequency of 60 Hz the penetration depth in copper is 8.5 mm (0.33 in.); at 10 GHz it is only 6.6×10^{-7} m. Waveguide and resonant cavity internal surfaces for use at microwave frequencies are therefore frequently plated with a high-conductivity material, such as silver, to reduce the energy losses since nearly all the current is concentrated at the surface. Provided the plating material is thick compared to δ, the conductor is as good as a solid conductor of the coating material. *See* CAVITY RESONATOR; MICROWAVE; WAVEGUIDE.

Frank A. Benson

Skink

A widely dispersed reptile belonging to the family Scincidae. The family contains approximately 1200 species and is represented on every continent except Antarctica. Characteristics include elongate bodies, moderate to long tails, and proportionately short, reduced, or absent limbs. Some scales on the head and body have cores of bone. In many species the eyelids are transparent. Skinks exhibit pleurodont dentition; that is, the teeth are attached on the side of the jaw. *See* DENTITION.

Five-lined skink (*Eumeces fasciatus*). (*Photo by George W. Robinson; © California Academy of Sciences*)

Skinks vary widely in size and longevity (some have been known to live to 20 years). They are carnivorous, eating invertebrates and, in the case of larger species, even small mammals, reptiles, and birds. Three reproductive modes can be found among skink species: oviparity (egg-laying), viviparity (bearing live young), and ovoviviparity (producing eggs that develop internally and hatch before or soon after extrusion). Three genera, *Neoseps*, *Scincella*, and *Eumeces* (see illustration), occur in the United States. A number of species of the genus *Tiliqua*, represented by the largest skinks in terms of body size, inhabit Malaysia and Australia. *Tiliqua gigas* reaches a length of 2 ft (0.6 m). *See* REPTILIA; SQUAMATA.

W. Ben Cash

Skunk A carnivore long classified in the subfamily Mephitinae in the family Mustelidae (otters, weasels, mink, badgers, and their relatives), but since 1997 placed in the independent family Mephitidae along with the Oriental stink badger *Mydaus*. This new placement was based on convincing DNA evidence linking skunks (genera *Mephitis*, *Spilogale*, and *Conepatus*) and *Mydaus* genetically. Skunks are found only in the New World, ranging from Canada to Chile. These mammals have short limbs with five clawed toes on each foot. The claws are curved and nonretractile. A pair of well-developed anal scent glands, used for defense against predators, are present. A short rostrum is present on the elongate and flattened skull. Carnassial teeth (used to shear flesh and bone) are well developed.

Striped and hooded skunks (*Mephitis*) comprise two species that range from southern Canada to Costa Rica. Adults are about the size of a house cat, standing approximately 18 cm (7 in.) at the shoulder, with adults usually attaining a head and body length between 550 and 750 mm (21 and 29 in.), including a tail 175–300 mm (7–12 in.) long. Adults usually weigh between 1.3 and 4.5 kg (3 and 10 lb). The relatively small head possesses small, short, rounded ears, small eyes, and a pointed muzzle. The pelage (the coat, including fur and hairs) is long, coarse, and oily. The body is black with a narrow white stripe running up the middle of the forehead and a broad white area on the nape of the neck that usually divides into a V at about the shoulders. The resulting two white stripes may continue to the base of the bushy tail. There is, however, much variation in the length and width of the stripes; some skunks have very broad, well-defined stripes, whereas in others the stripes are almost or completely lacking. Striped and hooded skunks are omnivorous and feed on insects, crayfish, fish, frogs, turtle eggs, snakes, small birds, and small mammals. During fall and winter, fruit, carrion, grains and nuts, grasses, leaves, and buds are important. Donald W. Linzey

Slate Any of the deformed fine-grained, mica-rich rocks that are derived primarily from mudstones and shales, containing a well-developed, penetrative foliation that is called slaty cleavage. Slaty cleavage is a secondary fabric element that forms under low-temperature conditions (less than 540°F or 300°C), and imparts to the rock a tendency to split along planes. It is a type of penetrative fabric; that is, the rock can be split into smaller and smaller pieces, down to the size of the individual grains. If there is an obvious spacing between fabric elements (practically, greater than 1 mm), the fabric is called spaced. Slates typically contain clay minerals (for example, smectite), muscovite/illite, chlorite, quartz, and a variety of accessory phases (such as epidote or iron oxides). Under increasing temperature conditions, slate grades into phyllite and schist. See ARGILLACEOUS ROCKS; CLAY MINERALS; PHYLLITE; SCHIST; SHALE.

Slaty cleavage is defined by a strong dimensional preferred orientation of clay in a very clay rich, low-grade metamorphic rock, and the resulting rock is a slate. Slaty cleavage tends to be smooth and planar. Coupled with the penetrative nature of slaty cleavage, these characteristics enable slates to split into very thin sheets. This and the durability of the rock are reasons why slates are used in the roofing industry, in the tile industry, and in the construction of pool tables. Ben A. van der Pluijm

Sleep and dreaming Sleep is generally defined as an easily reversible, temporary, periodic state of suspended behavioral activity, unresponsiveness, and perceptual disengagement from the environment. Compared with other states of temporary unresponsiveness such as syncope or coma, sleep is easily reversible with strong or meaningful sensory stimuli (for example, the roar of a nearby tiger, or a voice speaking the sleeper's name). Sleep should not be considered a state of general unconsciousness. The sleeper is normally unconscious (but not always) of the nature of events in the surrounding environment; this is the meaning of perceptual disengagement. However, the sleeper's attention may be fully engaged in experiencing a dream. And if reportability is accepted as a sufficient condition for conscious mental processes, any dream that can be recalled must be considered conscious. Dreaming, then, can be simply defined as the world-modeling constructive process through which people have experiences during sleep, and a dream is just whatever the dreamer experienced while sleeping. See CONSCIOUSNESS.

Nature of sleep. In general, biological organisms do not remain long in states of either rest or activity. For example, if a cat's blood sugar level drops below a certain point, the cat is motivated by hunger to venture from its den in search of a meal. After satisfying the urge to eat, the cat is no longer motivated to expend energy tracking down uncooperative prey; now its biochemical state motivates a return to its den, to digest in peace, conserve energy, and generally engage in restful, regenerative activities, including sleep. This example tracks a cat through one cycle of its basic rest-activity cycle (BRAC). Such cyclic processes are ubiquitous among living systems.

Sleep and wakefulness are complementary phases of the most salient aspects of the brain's endogenous circadian rhythm, or biological clock. Temporal isolation studies have determined the biological clock in humans to be slightly longer than 24 hours. Several features of sleep are regulated by the circadian system, including sleep onset and offset, depth of sleep, and rapid eye movement (REM) sleep intensity and propensity. In the presence of adequate temporal cues (for example, sunlight, noise, social interactions, and alarm clocks), the internal clock keeps good time, regulating a host of physiological and behavioral processes. See BIOLOGICAL CLOCKS; NORADRENERGIC SYSTEM.

Sleep is not a uniform state of passive withdrawal from the world. In fact, a version of the BRAC continues during sleep, showing a periodicity of approximately 90 minutes. There are two distinct kinds of sleep: a quiet phase (also known as quiet sleep or QS, slow-wave sleep) and an active phase (also known as active sleep or AS, REM sleep, paradoxical sleep), which are distinguished by many differences in biochemistry, physiology, psychology, and behavior. Recordings of electrical activity changes of the brain (electroencephalogram or EEG), eye movements (electrooculogram or EOG), and chin muscle tone (electromyogram or EMG) are used to define the various stages and substages of sleep. See ELECTROENCEPHALOGRAPHY.

Sleep cycle. If sleepy enough, most people can fall asleep under almost any condition. After lying in bed for a few minutes in a quiet, dark room, drowsiness usually sets in. The subjective sensation of drowsiness can be objectively indexed by a corresponding change in brain waves (EEG activity): formerly continuous alpha rhythms (see illustration) gradually break up into progressively shorter trains of regular alpha waves and are replaced by low-voltage mixed-frequency EEG activity. When less than half of an epoch [usually the staging epoch is the 20–30 seconds it takes to fill one page of polygraph (sleep recording) paper] is occupied by continuous alpha rhythm, sleep onset is considered to have occurred and stage 1 sleep is scored. At this stage, the EOG usually reveals slowly drifting eye movements (SEMs) and muscle tone might or might not decrease. Awakenings at this point frequently yield reports of hypnagogic (leading into sleep) imagery, which can often be extremely vivid and bizarre.

Stage 1 is a very light stage of sleep described by most subjects as "drowsing" or "drifting off to sleep." Normally, it lasts only a few minutes before further EEG changes occur, defining another sleep stage. It is at this point that startlelike muscle jerks known as hypnic myoclonias or hypnic jerks occasionally briefly interrupt sleep. As the subject descends deeper into sleep, the EEG of stage 2 sleep is marked by the appearance of relatively high-amplitude slow waves called K-complexes as well as 12–14-Hz rhythms called sleep spindles. The EOG would generally indicate little

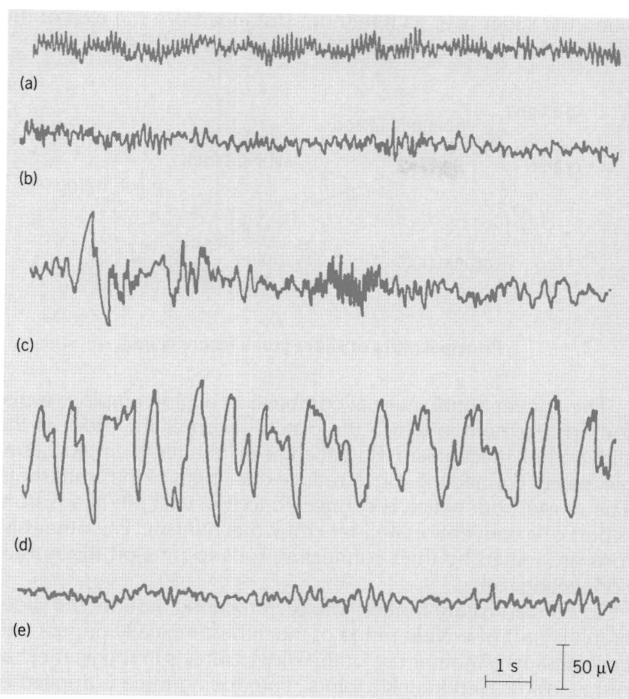

Human EEG associated with different stages of sleep and wakefulness. (*a*) Relaxed wakefulness (eyes shut) shows rhythmic 8-12-Hz alpha waves. (*b*) Stage 1 non-REM sleep shows mixed frequencies, especially 3–7-Hz theta waves. (*c*) Stage 2 non-REM sleep shows 12–14-Hz sleep spindles and K-complexes. (*d*) Delta sleep shows large-amplitude (>75 μV) 0.5–2-Hz delta waves. (*e*) REM sleep shows low-amplitude, mixed frequencies with sawtooth waves.

eye movement activity, and the EMG would show somewhat decreased muscle tone. Reports of mental activity from this stage of sleep are likely to be less bizarre and more realistic than those from stage 1.

After several minutes in stage 2, high-amplitude slow waves (delta waves) gradually begin to appear in the EEG. When at least 20% of an epoch is occupied by these (1–2-Hz) delta waves, stage 3 is defined. Usually this slow-wave activity continues to increase until it completely dominates the appearance of the EEG. When the proportion of delta EEG activity exceeds 50% of an epoch, the criterion for the deepest stage of sleep, stage 4, is met. During stages 3 and 4, often collectively referred to as delta sleep, the EOG shows few genuine eye movements but is obscured by the high-amplitude delta waves. Muscle tone is normally low, although it can be remarkably high, as when sleepwalking or sleep-talking occurs. Recall of mental activity on arousal from delta sleep is generally very poor and fragmentary and is more thoughtlike than dreamlike. It should be noted that cognitive functioning immediately after abrupt wakening from sleep is likely to carry over some of the characteristics of the preceding sleep state. This phenomenon, known as sleep inertia, can be used as an experimental tool for studying (by inference) cognition during different stages of sleep.

After about 90 minutes, the progression of sleep stages is reversed, back through stage 3 and stage 2 to stage 1 again. But now the EMG shows virtually no activity at all, indicating that muscle tone has reached its lowest possible level, and the EOG discloses the occurrence of rapid eye movements—at first only a few at a time, but later in dramatic profusion. This is REM (or active) sleep. Breathing rate and heart rate become more rapid and irregular, and both males and females show signs of sexual arousal. Brain metabolic rate increases to levels that typically exceed the waking state average. This state of intense brain activation is normally accompanied by experiences that seem vividly real while they last, but often vanish within seconds of waking. When people are abruptly awakened from REM sleep, 80–90% of the time they recall vivid and sometimes extremely detailed dreams.

While all this activity is happening in the brain, the body remains almost completely still (except for small twitches), because it is temporarily paralyzed during REM sleep to prevent dreams from being acted out. The brainstem system that causes the paralysis of REM sleep does not always inactivate immediately upon awakening. The resulting experience, known as sleep paralysis, can be terrifying, but it is quite harmless and normal if it occurs only occasionally. However, frequent sleep paralysis can be a symptom of a disorder of REM sleep called narcolepsy. *See* SLEEP DISORDERS.

After a REM period lasting perhaps 5 to 15 minutes, a young adult will typically go back through the preceding cycle stages, dreaming vividly three or four more times during the remainder of the night, with two major modifications. First, decreasing amounts of slow-wave EEG activity (stages 3 and 4 or delta sleep) occur in each successive cycle. Later in the night, after perhaps the second or third REM period, no delta sleep appears on the EEG at all, only non-REM, stage-2, and REM sleep. Second, as the night proceeds, successive REM periods tend to increase in length, up to a point. The fact that for humans most REM occurs in the last portion of the sleep cycle as dawn approaches suggests that REM serves a function related to preparation for waking activity.

Finally, after four or five periods of dreaming sleep, the sleeper wakes up (for perhaps the tenth time during the night) and gets up for the day. It may be difficult to believe that brief awakenings occur this frequently during an average night; however, they are promptly forgotten. This retrograde amnesia is a normal feature of sleep: information in short-term memory at sleep onset is usually not transferred into more permanent storage.

Evolution and function of sleep. Most organisms are adapted to either the dark or light phase of the day-night cycle. Therefore, a biological process that limits activity to the phase of the cycle to which the organism is adapted will enhance survival. Most likely sleep developed out of the rest phase of the BRAC, allowing organisms to minimize interactions with the world during the less favorable phase, while engaging in a variety of "off-line" internal maintenance operations, including growth, repair, synthesis of biochemicals consumed during waking activity, and energy conservation, as well as memory consolidation. The fact that different species have many differences in sleep structure, process, and function fits with the idea that sleep serves the specific adaptive needs of each species. Quiet sleep appears to be an older form of sleep with simpler and more universal functions related to energy conservation, growth, and restoration. Active sleep is a mammalian invention with functions that appear to be related to specifically mammalian needs such as live birth. The portion of total sleep composed of REM is at its highest level perinatally: newborn humans spend 8 hours per day in REM sleep, with even more time during the last 6 weeks before birth. The time of maximal REM corresponds to the time of maximal growth of the brain. This is not a coincidence but points to the main evolutionary function of REM: to provide a source of endogenous stimulation supporting the unfolding of genetic programming and self-organization of the brain. The fact that REM time does not decrease to zero after the full development of the nervous system suggests secondary adaptive advantages afforded by REM during adulthood, which may include facilitation of difficult learning and preparation of the brain for optimal functioning on arousal.

Dreams. From the biological perspective, the basic task of the brain is to predict and control the organism's actions and regulate those actions to achieve optimal outcomes (in terms of survival and reproduction). To accomplish this task, the brain in some sense internally "models" the world. The waking brain bases the features of its world model primarily on the current information received from the senses and secondarily on expectations derived from past experience. In contrast, the sleeping brain acquires little information from the senses. Therefore in sleep, the primary sources of information available to the brain are the current motivational state of the organism (for example, current concerns, fears, hunger, and desires) and past experience (for example, both species-specific instincts and personal memories of recent and associated past experiences of potential relevance). According to this theory, dreams result from brains using internal information to create a simulation of the external world, in a manner directly parallel to the process of waking perception, minus most sensory input. *See* PERCEPTION. Stephen LaBerge

Sleep disorders Disorders of the reversible state during which the individual's voluntary functions are suspended but the involuntary functions, such as circulation and respiration, are uninterrupted; the sleeping subject assumes a characteristic posture with relative immobility and decreased responses to external stimuli. The latest International Classification of Sleep Disorders (ICSD) divides sleep disorders into eight broad categories: (1) insomnias; (2) sleep-related breathing disorders, including obstructive sleep apnea syndrome; (3) hypersomnias of central origin (including narcolepsy) that are not secondary to sleep-related breathing disorders, circadian rhythm sleep disorders, or other causes of disturbed night sleep; (4) circadian rhythm sleep disorders; (5) parasomnias (abnormal movements or behavior intruding into sleep, such as sleepwalking, sleep terrors, confusional arousals, etc.); (6) sleep-related movement disorders, such as restless legs syndrome; (7) isolated symptoms (for example, long sleeper, short sleeper, snoring, sleep talking, etc.), including apparently normal variants and unresolved issues due to insufficient information; and (8) other sleep disorders, which do not conform to the listing outlined above.

Excessive daytime sleepiness, insomnia (sleeplessness), and abnormal movements and behavior during sleep are the major sleep complaints. Individuals with hypersomnia complain of excessive daytime sleepiness. An individual with insomnia may have difficulty initiating or maintaining sleep; repeated awakenings or early morning awakenings; or daytime fatigue and impairment of performance. The commonest cause of insomnia is psychiatric or psychophysiologic disorder (for example, depression, anxiety, or stress), but other causes include medical disorders or pain. Chronic insomnia can best be treated by sleep hygiene and behavioral modification.

There are two forms of sleep apnea. In central apnea, both the airflow at the mouth and nose and the effort by the diaphragm decrease. In obstructive apnea, the airflow stops but the effort by the diaphragm continues. Obstructive sleep apnea syndrome is common in middle-aged and elderly obese men.

Narcoleptic sleep attacks usually begin in individuals between the ages of 15 and 25. Narcolepsy is characterized by an irresistible desire to sleep, and the attacks may last 15–30 min. Most individuals also have cataplexy, during which there is transient loss of muscle tone that causes the individual to fall to the ground or slump the head forward for a few seconds. Narcolepsy cannot be cured. *See* BIOLOGICAL CLOCKS; SLEEP AND DREAMING. Sudhansu Chokroverty

Slider-crank mechanism A four-bar linkage with output crank and ground member of infinite length. A slider crank (see illustration) is most widely used to convert reciprocating to rotary motion (as in an engine) or to convert rotary to reciprocating motion (as in pumps), but it has numerous other applications. Positions at which slider motion reverses are called dead centers. When crank and connecting rod are extended in a straight line and the slider is at its maximum distance from the axis of the crankshaft, the position is top dead center (TDC); when the slider is at its minimum distance from the axis of the crankshaft, the position is bottom dead center (BDC).

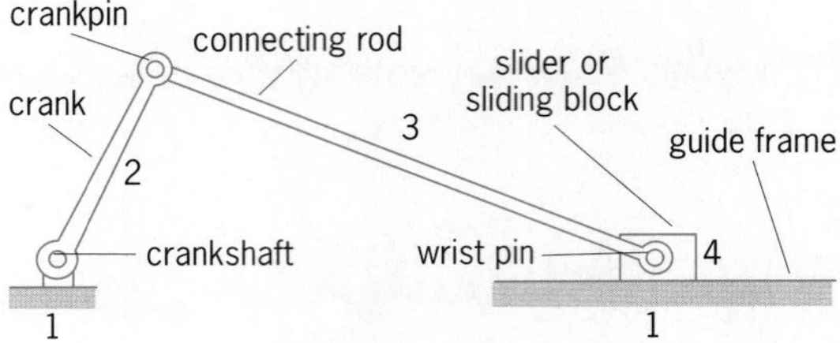

Principal parts of slider-crank mechanism.

The conventional internal combustion engine employs a piston arrangement in which the piston becomes the slider of the slider-crank mechanism. Radial engines for aircraft employ a single master connecting rod to reduce the length of the crankshaft. The master rod, which is connected to the wrist pin in a piston, is part of a conventional slider-crank mechanism. The other pistons are joined by their connecting rods to pins on the master connecting rod.

To convert rotary motion into reciprocating motion, the slider crank is part of a wide range of machines, typically pumps and compressors. Another use of the slider crank is in toggle mechanisms, also called knuckle joints. The driving force is applied at the crankpin so that, at TDC, a much larger force is developed at the slider. *See* FOUR-BAR LINKAGE. Douglas P. Adams

Slip (electricity) A numerical value used in describing the performance of electrical couplings and induction machines. In an electrical coupling, slip is defined simply as the difference between the speeds of the two rotating members. In an induction motor, slip is a measure of the difference between synchronous speed and shaft speed.

When the stator windings of an induction motor are connected to a suitable alternating voltage supply, they set up a rotating magnetic field within the motor. The speed of rotation of this field is called synchronous speed, and is given by Eq. (1) or Eq. (2),

$$\omega_s = \frac{4\pi f}{p} \quad \text{rad/s} \tag{1}$$

$$n_s = 120\frac{f}{p} \quad \text{rev/min} \tag{2}$$

where f is the line frequency and p is the number of magnetic poles of the field. The number of poles is determined by the design of the windings. In accord with Faraday's voltage law, a magnetic field can induce voltage in a coil only when the flux linking the coil varies with time. If the rotor were to turn at the same speed as the stator field, the flux linkage with the rotor would be constant. No voltages would be induced in the rotor windings, no rotor current would flow, and no torque would be developed. For motor action it is necessary that the rotor windings move backward relative to the magnetic field so that Faraday's law voltages may be induced in them. That is, there must be slip between the rotor and the field. *See* ELECTROMAGNETIC INDUCTION; INDUCTION MOTOR.

The amount of slip may be expressed as the difference between the field and rotor speeds in revolutions per minute or radians per second. However, the slip of an induction motor is most commonly defined as a decimal fraction of synchronous speed, as in Eq. (3) or Eq. (4). Here n is the motor speed in

$$s = \frac{n_s - n}{n_s} \tag{3}$$

$$s = \frac{\omega_s - \omega}{\omega_s} \tag{4}$$

revolutions per minute, ω is its speed in radians per second, and s is the slip, or more properly the per unit slip. Typical full-load values of slip for an induction motor range from 0.02 to 0.15, depending on rotor design. Slip is sometimes expressed in percent of synchronous speed, rather than per unit. If an induction machine is driven faster than synchronous speed, the slip becomes negative, and the machine acts as a generator, forcing energy back into the electrical supply line. *See* ELECTRIC ROTATING MACHINERY.

George McPherson, Jr.

Slip rings Electromechanical components which, in combination with brushes, provide a continuous electrical connection between rotating and stationary conductors. Typical applications of slip rings are in electric rotating machinery, synchros, and gyroscopes. Slip rings are also employed in large assemblies where a number of circuits must be established between a rotating device, such as a radar antenna, and stationary equipment.

Slip rings are usually constructed of steel with the cylindrical outer surface concentric with the axis of rotation. Insulated mountings insulate the rings from the shaft and from each other. Conducting brushes are arranged about the circumference of the slip rings and held in contact with the surface of the rings by spring tension. *See* ELECTRIC ROTATING MACHINERY; MOTOR.

Arthur R. Eckels

Sloan Digital Sky Survey A large-scale imaging and redshift survey which, when complete, will map in detail over one quarter of the sky, determining the positions and brightness of over 100 million celestial objects and the redshifts of over a million galaxies and quasars. A primary objective of the survey is to obtain a map of the three-dimensional distribution of galaxies in the universe.

Many decades of observational and theoretical research have converged on a unified picture of the nature of the universe. Instrumental in these advances have been detailed measurements of the expansion rate of the universe by scientists using the *Hubble Space Telescope*, of the radiation from the cosmic microwave background radiation by the Wilkinson Microwave Anisotropy Probe, and of the distribution of galaxies by several redshift surveys, of which the largest to date have been the Two Degree Field Survey and the Sloan Digital Sky Survey (SDSS). *See* HUBBLE SPACE TELESCOPE; UNIVERSE; WILKINSON MICROWAVE ANISOTROPY PROBE.

Redshifts are caused by the expansion of the universe. Because of this expansion, the farther away from us a galaxy lies the more rapidly it is receding from us. This relation can be turned around, for once the proportionality between distance and redshift (the Hubble "constant" or parameter) has been measured, it can be used together with a galaxy's redshift to give its distance. Thus observations of the positions on the sky and of the redshifts of a sample of galaxies can be used to measure the three-dimensional distribution of galaxies. *See* HUBBLE CONSTANT; REDSHIFT.

Observing strategy. The Sloan Digital Sky Survey is unique in several important ways: It is the largest redshift survey, with more than 400,000 galaxy redshifts measured as of June 2004 and plans to measure a million; and it acquires the imaging data from which the positions of the galaxies can be measured during the same time period as it measures the redshifts. *See* ASTRONOMICAL IMAGING.

The Sloan Digital Sky Survey follows a dual observing strategy: (1) imaging is done in the best weather; (2) objects are selected from the imaging according to carefully defined criteria by powerful computer software; and their spectra are observed during the less-than-perfect weather. The spectra are observed with devices that can obtain 640 spectra simultaneously; 7 or 8 sets, or several thousand spectra, can be observed in a single night. To do this, the Sloan Digital Sky Survey uses a dedicated 2.5-m (98-in.) telescope at the Apache Point Observatory, New Mexico. The telescope is equipped with a multi-charge-coupled-device camera and two fiber-fed spectrographs, and observes every possible clear, dark night year-round except for a short maintenance period in the summer. *See* SPECTROGRAPH; TELESCOPE.

Results. The Sloan Digital Sky Survey has obtained several fundamental results. A "Great Wall" of galaxies, 1.4×10^9 light-years long and 80% longer than the "Great Wall" found by Margaret Geller, John Huchra, and Valerie de Lapparent, is the largest structure in the universe observed so far. The distribution of galaxies is in excellent agreement with the current cosmological model. The Sloan Digital Sky Survey has been instrumental in discovering that the Milky Way Galaxy is surrounded by streams of stars from the tidal destruction of small satellite galaxies that used to orbit the Galaxy. It has discovered many gravitational lenses, in which light from distant quasars is bent by the warping of space by lumps of mass (the lenses) between the Earth and the quasar. It has discovered the lowest-luminosity galaxy ever found, a satellite of the nearby Andromeda Galaxy. It has discovered large numbers of brown dwarfs, objects intermediate in mass between planets and stars. *See* BROWN DWARF; GALAXY, EXTERNAL; GRAVITATIONAL LENS; MILKY WAY GALAXY; QUASAR.

Gillian R. Knapp

Slope The trigonometric tangent of the angle α that a line makes with the x axis. In the illustration the slope of a plane curve C at a point P of C is the slope of the line that is tangent to C at P. If $y = f(x)$ is an equation in rectangular coordinates of curve C, the slope of C at $P(x_0, y_0)$ is the value of the derivative $dy/dx = f'(x)$ at P, denoted by $f'(x_0)$, and hence an equation of the nonvertical tangent to C at P is $y - y_0 = f'(x_0)(x - x_0)$. *See* ANALYTIC GEOMETRY; CALCULUS.

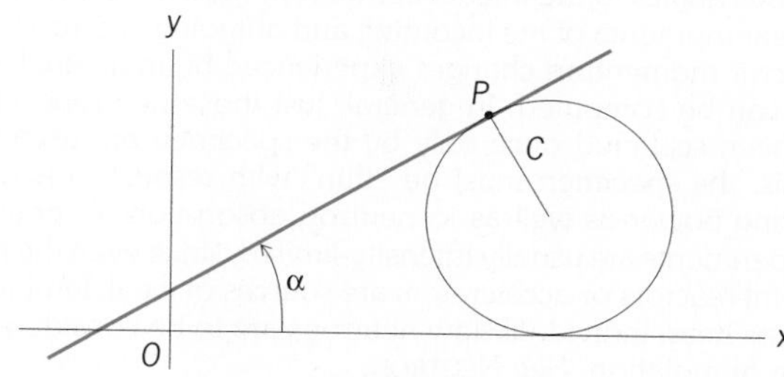

Slope of a curve.

Leonard M. Blumenthal

Sloth A mammal classified in the order Xenarthra along with anteaters, tamanduas, and armadillos. The sloth differs from all other mammals by having additional articulations (xenarthrales) between their lumbar vertebrae (called xenarthrous vertebrae). They also have a double rather than a single inferior vena cava, the vein that returns blood from the posterior portion of the body to the heart. Females have a primitive, divided uterus and a common urinary and genital tract; males have internal testes.

Two extant families of tree sloths (Bradypodidae and Megalonychidae) inhabit the lowland and upland tropical forests of Central and South America. The family Bradypodidae includes a single genus (with three species), the three-toed tree sloth (*Bradypus*). This is a slender-bodied animal with a head and body length of 413–700 mm (16–27 in.) and a tail length of 20–90 mm (0.75–3.5 in.). The three-toed sloth weighs 2.25–6.20 kg (5–13.5 lb). It has three toes on each front and rear foot. The animal can neither hear nor see well but depends on its senses of smell and touch. These sloths are drab brown or gray with patches of white. The family Megalonychidae (two-toed tree sloths) contains a single genus (*Choloepus*) with two species. Two-toed tree sloths, or unau, with two toes on their front feet and three on their hind feet, are somewhat larger and more active than their three-toed cousins.

Tree sloths are solitary and spend most of their lives hanging upside down from the upper branches of a tree. They hang by the long, hooked claws at the ends of their toes and slowly move along, hand over hand in the inverted position. Sloths sleep and give birth while hanging upside down. They spend their waking hours slowly picking and eating leaves, especially the leaves of the Cecropia tree. Sloths are so lethargic that two blue-green algae, which look like moss, actually grow in the grooves of their coarse hair. In the rainy season, their fur often has a distinct greenish tinge due to the luxuriant growth of algae; in times of drought, the fur turns yellowish. The greenish color serves to camouflage the animals in the tree canopy. *See* MAMMALIA.

Donald W. Linzey

Slow neutron spectroscopy The use of beams of slow neutrons, from nuclear reactors or nuclear accelerators, in studies of the structure or structural dynamics of solid, liquid, or gaseous matter. Studies of the chemical or magnetic structure of substances are usually referred to under the term neutron diffraction, while studies of atomic and magnetic dynamics go under the terms slow neutron spectroscopy, inelastic neutron scattering, or simply neutron spectroscopy. *See* NEUTRON DIFFRACTION; NUCLEAR REACTOR; PARTICLE ACCELERATOR; SPECTROSCOPY; X-RAY DIFFRACTION.

In a neutron spectroscopy experiment, a beam of neutrons is scattered by a specimen and the scattered neutrons are detected at various angles to the initial beam. From these measurements, the linear momenta of the incoming and outgoing neutrons (and the vector momentum changes experienced by individual neutrons) can be computed. In general, just those neutrons which have been scattered once only by the specimen are useful for analysis; the specimen must be "thin" with respect to neutron scattering power as well as to neutron absorption. In practice, the experiments are usually intensity-limited, since even the most powerful reactors or accelerators are sources of weak luminosity when, as here, individual slow neutrons are to be considered as quanta of radiation. *See* NEUTRON.

Neutron spectroscopy requires slow neutrons, with energies of the order of neutrons in equilibrium with matter at room temperature, or approximately 0.025 eV. The corresponding de Broglie wavelengths are approximately 0.2 nanometer, of the order of interatomic spacings in solids or liquids. The fast neutrons emitted in nuclear or slow fission reactions can be slowed down to thermal velocities in matter which is transparent to neutrons and which contains light elements, such as hydrogen, carbon, and beryllium, by a process of diffusion and elastic (billiard-ball) scattering known as neutron moderation. By selection of those diffusing neutrons which travel in a certain restricted range of directions (collimation), a beam of thermal and near-thermal neutrons can be obtained. *See* NEUTRON; QUANTUM MECHANICS; THERMAL NEUTRONS.

The bulk of the observations can be accounted for in terms of scattering of semiclassical neutron waves by massive, moving-point scatterers in the forms of atomic nuclei and their bound electron clouds. The spatial structure of the scatterers, time-averaged over the duration of the experiment, gives rise to the elastic scattering from the specimen that is studied in neutron diffraction; the spatial motions of the scatterers give rise to the Doppler-shifted inelastic scattering involved in slow neutron spectroscopy.

Just as (slow) neutron diffraction is the most powerful available scientific tool for study of the magnetic structure of matter on an atomic scale, so slow neutron spectroscopy is the most powerful tool for study of the atomic magnetic and nuclear dynamics of matter in all its phases. The direct nature of the analysis has in some cases added considerable support to the conceptual structure of solid-state and liquid-state physics and thus to the confidence with which the physics is applied. For example, neutron spectroscopy has confirmed the existence of phonons, magnons, and the quasiparticles (rotons) of liquid helium II. Detailed information has been obtained on the lattice vibrations of most of the crystalline elements and numerous simple compounds, on the atomic dynamics of many simple liquids, on the dynamics of liquid helium in different phases, and on the atomic magnetic dynamics of a great variety of ferromagnetic, ferrimagnetic, antiferromagnetic, and modulated magnetic substances.

Bertram N. Brockhouse

Smallpox An acute infectious viral disease characterized by severe systemic involvement and a single crop of skin lesions that proceeds through macular, papular, vesicular, and pustular stages. Smallpox is caused by variola virus, a brick-shaped, deoxyribonucleic acid-containing member of the Poxviridae family. Strains of variola virus are indistinguishable antigenically, but have differed in the clinical severity of the disease caused. Following a 13-year worldwide campaign coordinated by the World Health Organization (WHO), smallpox was declared eradicated by the World Health Assembly in May 1980. Smallpox is the first human disease to be eradicated.

Humans were the only reservoir and vector of smallpox. The disease was spread by transfer of the virus in respiratory droplets during face-to-face contact. Before vaccination, persons of all ages were susceptible. It was a winter-spring disease; there was a peak incidence in the drier spring months in the Southern Hemisphere and in the winter months in temperate climates. The spread of smallpox was relatively slow. The incubation period was an average of 10–12 days, with a range of 7–17 days. Fifteen to forty percent of susceptible persons in close contact with an infected individual developed the disease.

There were two main clinically distinct forms of smallpox, variola major and variola minor. Variola major, prominent in Asia and west Africa, was the more severe form, with widespread lesions and case fatality rates of 15–25% in unvaccinated persons, exceeding 40% in children under 1 year. From the early 1960s to 1977, variola minor was prevalent in South America and south and east Africa; manifestations were milder, with a case fatality rate of less than 1%.

There is no specific treatment for the diseases caused by poxviruses. Supportive care for smallpox often included the systemic use of penicillins to minimize secondary bacterial infection of the skin. When lesions occurred on the cornea, an antiviral agent (idoxuridine) was advised.

Edward Jenner, a British general medical practitioner who used cowpox to prevent smallpox in 1796, is credited with the discovery of smallpox vaccine (vaccinia virus). However, the global smallpox eradication program did not rely only on vaccination. Although the strategy for eradication first followed a mass vaccination approach, experience showed that intensive efforts to identify areas of epidemiologic importance, to detect outbreaks and cases, and to contain them would have the greatest effect on interrupting transmission. In 1978, WHO established an International Commission to confirm the absence of smallpox worldwide. The recommendations made by the commission included abandoning routine vaccination except for laboratory workers at special risk. *See* ANIMAL VIRUS; VACCINATION.

Joel Breman

Smart card A plastic card, the size of a credit card, that contains an embedded silicon computer chip. Its primary purpose is the portable storage and retrieval of data used to authorize various types of electronic transactions (see illustration). Compared to their predecessors, magnetic stripe cards (for example, credit cards), smart cards can store data relating to more than one institution, facilitate more than one type of use, carry substantially larger volumes of data, process transactions at higher rates, and provide logical and physical security to the card's data.

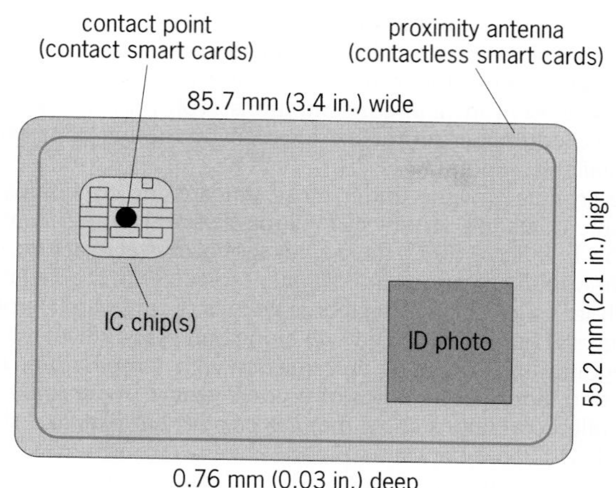

Smart card showing both the contact point and the proximity antenna.

Smart cards work by interacting with a card reader—an interface between the smart card and an institution.

Smart cards support a wide range of transactions based on their ability to identify and store information about the cardholder, card issuer, authorized accounts and their activities, authorized merchants, and authorized transactions. The validation of a cardholder's identity may be knowledge-based, where the individual knows something, such as a personal identification number (PIN). It may also be possession-based, where the presence of a "token" (for example, an ID card or a bank card) validates the person's identity. In addition, smart cards can carry biometric information (for example, fingerprints or iris scans), providing greater security for the identification process.

Some common smart card applications include credit transactions, debit transactions, stored-value transactions, information management applications, loyalty/affinity transactions, and multiple applications. Katherine M. Shelfer; Kevin Meadows

Smith chart A graphical procedure for converting impedances into reflection coefficients and vice versa. At high-frequency circuit operation, voltages and currents behave like traveling waves propagating over finite-length components. Copper traces on printed circuit boards, coaxial cables, and even simple wires become transmission lines. Changes in the length or operating frequency of these transmission lines result in periodic impedance behaviors not encountered in low-frequency circuits. Frequently, the impedance is replaced by the reflection coefficient, a more convenient way to quantify the transmitted and reflected voltage-current waves. To show how the impedance can be converted into a reflection coefficient and vice versa, P. H. Smith developed an ingenious graphical procedure based on conformal mapping principles. His approach permits an easy and intuitive display of the reflection coefficient as well as the complex impedance in a single graph. Although such a graphical procedure, nowadays known as the Smith chart, was developed in the 1930s prior to the computer age, it has retained its popularity and can be found in every data book describing passive and active high-frequency components and systems. Almost all computer-aided design programs utilize the Smith chart for the analysis of circuit impedances, design of matching networks, and computations of noise figures, gain, and stability circles. The Smith chart also finds widespread use for a number of RF applications. Even instruments such as the ubiquitous network analyzer have the option to represent certain measurement data in a Smith chart format. *See* CONFORMAL MAPPING; ELECTRICAL IMPEDANCE; RADIO-FREQUENCY AMPLIFIER; REFLECTION AND TRANSMISSION COEFFICIENTS; TRANSMISSION LINES. Reinhold Ludwig

Smithsonite A naturally occurring rhombohedral zinc carbonate ($ZnCO_3$), with a crystal structure similar to that of calcite ($CaCO_3$). Smithsonite has a hardness on the Mohs scale of $4^1/_2$, has a specific gravity of 4.30–4.45, and exhibits perfect rhombohedral cleavage. *See* CALCITE; CRYSTAL STRUCTURE; HARDNESS SCALES.

Smithsonite most commonly forms as an alteration product of the mineral sphalerite (ZnS) during supergene enrichment of zinc ores in arid or semiarid environments. It is seldom pure zinc carbonate, commonly containing other divalent metal ions such as manganese (Mn^{2+}), ferrous iron (Fe^{2+}), magnesium (Mg^{2+}), calcium (Ca^{27+}), cadmium (Cd^{2+}), cobalt (Co^{2+}), or lead (Pb^{2+}) substituting for zinc ion. Substitution of other elements for zinc may result in different colors such as blue-green (copper), yellow (cadmium), and pink (cobalt). Smithsonite has been mined as an ore of zinc and has also been used as an ornamental stone. *See* CARBONATE MINERALS; ZINC. John C. Drake

Smog The noxious mixture of gases and particles commonly associated with air pollution in urban areas. Harold Antoine des Voeux is credited with coining the term in 1905 to describe the air pollution in British towns. *See* AIR POLLUTION.

The constituents of smog affect the human cardio-respiratory system and pose a health threat. Individuals exposed to smog can experience acute symptoms ranging from eye irritation and shortness of breath to serious asthmatic attacks. Under extreme conditions, smog can cause mortality, especially in the case of the infirm and elderly. Smog can also harm vegetation and likely leads to significant losses in the yields from forests and agricultural crops in affected areas.

The only characteristic of smog that is readily apparent to the unaided observer is the low visibility or haziness that it produces, due to tiny particles suspended within the smog. Observation of the more insidious properties of smog—the concentrations of toxic constituents—requires sensitive analytical instrumentation. Technological advances in these types of instruments, along with the advent of high-speed computers to simulate smog formation, have led to an increasing understanding of smog and its causes.

Smog is an episodic phenomenon because specific meteorological conditions are required for it to accumulate near the ground. These conditions include calm or stagnant winds which limit the horizontal transport of the pollutants from their sources, and a temperature inversion which prevents vertical mixing of the pollutants from the boundary layer into the free troposphere. *See* METEOROLOGY; STRATOSPHERE; TROPOSPHERE.

Smog can be classified into three types: London smog, photochemical smog, and smog from biomass burning.

London smog arises from the by-products of coal burning. These by-products include soot particles and sulfur oxides. During cool damp periods (often in the winter), the soot and sulfur oxides can combine with fog droplets to form a dark acidic fog. As nations switch from coal to cleaner-burning fossil fuels such as oil and gas as well as alternate energy sources such as hydroelectric and nuclear, London smogs cease. *See* ACID RAIN; COAL.

Photochemical smog is a more of a haze than a fog and is produced by chemical reactions in the atmosphere that are triggered by sunlight. A. J. Hagen-Smit first unraveled the chemical mechanism that produces photochemical smog. He irradiated mixtures of volatile organic compounds (VOC) and nitrogen oxides (NO_x) in a reaction chamber. After a few hours, Hagen-Smit observed the appearance of cracks in rubber bands stretched across the chamber. Knowing that ozone (O_3) can harden and crack rubber, Hagen-Smit correctly reasoned that photochemical smog was caused by photochemical reactions involving VOC and NO_x, and that one of the major oxidants produced in this smog was O_3. *See* ATMOSPHERIC CHEMISTRY; OZONE; PHOTOCHEMISTRY.

While generally not as dangerous as London smog, photochemical smog contains a number of noxious constituents. Ozone, a strong oxidant that can react with living tissue, is one of these noxious compounds. Another is peroxyacetyl nitrate (PAN), an eye irritant that is produced by reactions between NO_2 and the breakdown products of carbonyls. Particulate matter having diameters of about 10 micrometers or less are of concern because they can penetrate into the human respiratory tract during breathing and have been implicated in a variety of respiratory ailments.

Probably the oldest type of smog known to humankind is produced from the burning of biomass or wood. It combines aspects of both London smog and photochemical smog since the burning of biomass can produce copious quantities of smoke as well as VOC and NO_x. William L. Chameides

Smoke Fumes and smoke are dispersions of finely divided solids or liquids in a gaseous medium. The particle-size range is 0.01–5.0 micrometers. Typical dispersions are smokes from incomplete combustion of organic matter such as tobacco, wood, and coal; soot or carbon black; oil-vapor mists; chemical fumes such as sulfur trioxide (SO_3) and phosphorus pentoxide (P_2O_5) mists, ammonium chloride (NH_4Cl), and metal oxides; and the products of hydrolysis of metal chlorides by moist air. Oil-vapor and P_2O_5 mists (formed by burning phosphorus in moist air) have been extensively used in military operations to produce screening smokes. *See* AIR POLLUTION. Richard W. Bukowski

Snap ring A form of spring used principally as a fastener. Piston rings are a form of snap ring used as seals. The ring is elastically deformed, put in place, and allowed to snap back toward its unstressed position into a groove or recess. The snap ring may be used externally to provide a shoulder that retains a wheel or a bearing race on a shaft, or it may be used internally to provide a construction that confines a bearing race in the bore of a machine frame, as illustrated. L. Sigfred Linderoth, Jr.

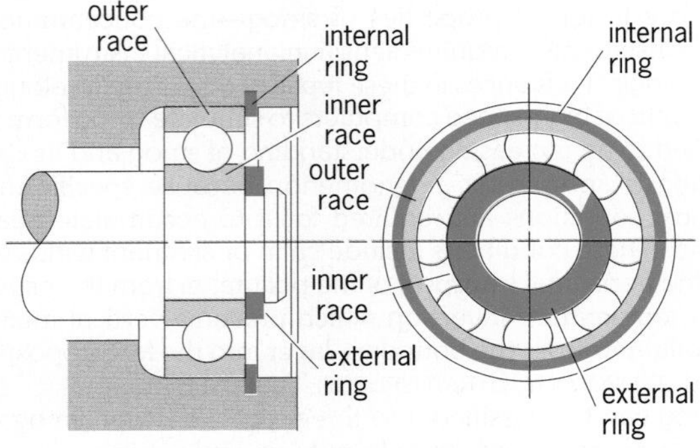

Snap rings hold ball-bearing race in place. Internal ring supports axial thrust of axle. External ring aligns Inner race against shoulder of shaft.

Snow Frozen precipitation resulting from the growth of ice crystals from water vapor in the Earth's atmosphere.

As ice particles fall out in the atmosphere, they melt to raindrops when the air temperature is a few degrees above 32°F (0°C), or accumulate on the ground at colder temperatures. At temperatures above −40°F (−40°C), individual crystals begin growth on icelike aerosols (often clay particles 0.1 micrometer in diameter), or grow from cloud droplets (10 μm in diameter) frozen by similar particles. At lower temperatures, snow crystals grow on cloud droplets frozen by random molecular motion. At temperatures near 25°F (−4°C), crystals sometimes grow on ice fragments produced during soft hail (graupel) growth. Snow crystals often grow in the supersaturated environment provided by a cloud of supercooled droplets; this is known as the Bergeron-Findeisen process for formation of precipitation. When crystals are present in high concentrations (100 particles per liter) they grow in supersaturations lowered by mutual competition for available vapor.

Ice crystals growing under most atmospheric conditions (air pressure down to 0.2 atm or 20 kilopascals and temperatures 32 to −58°F or 0 to −50°C) have a hexagonal crystal structure, consistent with the arrangement of water molecules in the ice lattice, which leads to striking hexagonal shapes during vapor growth. The crystal habit (ratio of growth along and perpendicular to the hexagonal axis) changes dramatically with temperature. Both field and laboratory studies of crystals grown under known or controlled conditions show that the crystals are platelike above 27°F (−3°C) and between 18 and −13°F (−8 and −25°C), and columnlike between 27 and 18°F (−3 and −8°C) and below −13°F (−25°C).

Individual crystals fall in the atmosphere at velocity up to 0.5 m s^{-1} (1.6 ft s^{-1}). As crystals grow, they fall at higher velocity, which leads, in combination with the high moisture availability in a supercooled droplet cloud, to sprouting of the corners to form needle or dendrite skeletal crystals.

Under some conditions crystals aggregate to give snowflakes. This happens for the dendritic crystals that grow near 5°F (−15°C), which readily interlock if they collide with each other, and for all crystals near 32°F (0°C). Snowflakes typically contain several hundred individual crystals.

When snow reaches the ground, changes take place in the crystals. At temperatures near 32°F (0°C) the crystals rapidly lose the delicate structure acquired during growth, sharp edges evaporate, and the crystals take on a rounded shape, some 1–2 mm (0.04–0.08 in.) in diameter. These grains sinter together at their contact points to give snow some structural rigidity. The specific gravity varies from ~0.05 for freshly fallen "powder" snow to ~0.4 for an old snowpack. *See* CRYSTAL GROWTH; PRECIPITATION (METEOROLOGY). John Hallett

Snow line A term generally used to refer to the elevation of the lower edge of a snow field. In mountainous areas, it is not truly a line but rather an irregular, commonly patchy border zone, the position of which in any one sector has been determined by the amount of snowfall and ablation. These factors may vary considerably from one part to another. On glacier surfaces the snow line is sometimes referred to as the glacier snow line or névé line (the outer limit of retained winter snow cover on a glacier).

Year-to-year variation in the position of the orographical snow line is great. The mean position over many decades, however, is important as a factor in the development of nivation hollows and protalus ramparts in deglaciated cirque beds. *See* GLACIOLOGY; SNOWFIELD AND NÉVÉ. Maynard M. Miller

Snow surveying A technique for providing an inventory of the total amount of snow covering a drainage basin or a given region. Most of the usable water in western North America originates as mountain snowfall that accumulates during the winter and spring and appears several months later as streamflow. Snow surveys were established to provide an estimate of the snow water equivalent (that is, the depth of water produced from melting the snow) for use in predicting the volume of spring runoff. They are also extremely useful for flood forecasting, reservoir regulation, determining hydropower requirements, municipal and irrigation water supplies, agricultural productivity, wildlife survival, and building design, and for assessing transportation and recreation conditions.

Conventional snow surveys are made at designated sites, known as snow courses, at regular intervals each year throughout the winter period. A snow sampler is used to measure the snow depth and water equivalent at a series of points along a snow course. Average depth and water equivalent are calculated for each snow course. Satellite remote sensing and data relay are technologies used to obtain information on snow cover in more remote regions. *See* REMOTE SENSING; SNOW GAGE; SURFACE WATER.

Barry E. Goodison

Snowfield and névé The term snowfield is usually applied to mountain and glacial regions to refer to an area of snow-covered terrain with definable geographic margins. Where the connotation is very general and without regard to geographical limits, the term snow cover is more appropriate; but glaciology requires more precise terms with respect to snowfield areas. These terms differentiate according to the physical character and age of the snow cover. Technically, a snowfield can embrace only new or old snow (material from the current accumulation year). Anything older is categorized as firn or ice. The word firn is a derivative of the German adjective *fern*, meaning "of last year," and hence refers to hardened snow not yet metamorphosed to ice which has been retained from the preceding year or years. Thus, by definition, a snowfield composed of firn can be called a firn field. Another term familiar to glaciologists is névé, from the French word for a mass of hardened snow in a mountain or glacier environment. In English, rather than a specific word for the material itself, a descriptive phrase is used such as: "consolidated granular snow not yet changed to glacier ice." Because of the need for simple terms, however, it is acceptable to use the French term *névé* when specifically referring to a geographical area of snowfields on mountain slopes or glaciers (that is, an area covered with perennial "snow" and embracing the entire zone of annually retained accumulation). For reference to the compacted remnant of the snowpack itself, that is, the material which is retained at the end of an annual melting period on a snowfield or névé, it is appropriate to use the derivative German term firn. *See* GLACIOLOGY.

Maynard M. Miller

Soap A cleansing agent described chemically as an alkali metal salt of a long-carbon-chain monocarboxylic acid, such as sodium myristate ($NaCOOC_{13}H_{27}$), as represented below.

$$\left(\begin{matrix} CH_2 & & CH_2 & & CH_2 & & CH_2 & & CH_2 \\ & CH_2 & & CH_2 & & CH_2 & & CH_2 & \end{matrix}\right)_n \quad \vdots \quad COO^- \quad Na^+$$

Hydrocarbon chain | Carboxylate

The hydrocarbon portion is hydrophobic and the carboxylate portion is hydrophilic. This duality enables soap to physically remove dirt and oils from surfaces and disperse or emulsify them in water. For detergency purposes, the most useful hydrophobic portion contains 12–18 carbon atoms. When the chain length exceeds 18 carbons, they become insoluble in water.

A soap-making process that was formerly used was based on an alkaline hydrolysis reaction, saponification, according to the reaction below, where R represents the hydrocarbon chain.

$$\underset{\text{Fat}}{C_3H_5(OOCR)_3} + \underset{\substack{\text{Caustic}\\ \text{soda}}}{3NaOH} \rightarrow \underset{\text{Soap}}{3NaOOCR} + \underset{\text{Glycerin}}{C_3H_5(OH)_3}$$

Most soap is now made by a continuous process. Fats and oils may be converted directly to soap by the reaction with caustic, and the neat soap separated by a series of centrifuges or by countercurrent washing. The saponification may be carried out with heat, pressure, and recirculation. Other important modern processes hydrolyze the fats directly with water and catalysts at high temperatures. This permits fractionation of the fatty acids, which are neutralized to soap in a continuous process. Advantages for this process are close control of the soap concentration, the preparation of soaps of certain chain lengths for specific purposes, and easy recovery of the by-product glycerin. Tallow and coconut oil are the most common fatty materials used for soapmaking.

Because of the marked imbalance of polarity within the molecules of synthetic detergents and soaps, they have unusual surface and solubility characteristics. The feature of their molecular structure which causes these properties is the location of the hydrophilic function at or near the end of a long hydrocarbon (hydrophobic) chain. One part of the molecule is water-seeking while the other portion is oil-seeking. Thus these molecules concentrate and orient themselves at interfaces, such as an oil-solution interface where the hydrophobic portion enters the oil and the hydrophilic portion stays in the water. Consequently, the interfacial tension is lowered and emulsification can result. At an agitated air-solution interface, the excess detergent concentration leads to sudsing.

Considerable research has also been done on the aqueous solutions of soap. Many of the phase characteristics now known to occur in most surfactant-water systems were first discovered in soap systems. These include micelle formation where, at a critical concentration, soap molecules in solution form clusters (micelles). The hydrocarbon chains associate with each other in the interior, and the polar groups are on the outside. *See* MICELLE; SURFACTANT.

There is one serious disadvantage to the use of soap which has largely caused its replacement by synthetic detergents. The problem is that carboxylate ions of the soap react with the calcium and magnesium ions in natural hard water to form insoluble materials which manifest as a floating curd. Bar soap for personal bathing is the greatest remaining market for soap. Some commercial laundries having soft water continue to use soap powders. Metallic soaps are alkaline-earth or heavy-metal long-chain carboxylates which are insoluble in water but soluble in nonaqueous solvents. They are used as additives to lubricating oils, in greases, as rust inhibitors, and in jellied fuels. *See* DETERGENT.

Roy C. Mast

Soapstone A soft talc-rich rock. Soapstones are rocks composed of serpentine, talc, and carbonates (magnesite, dolomite, or calcite). They represent original periodites which were altered at low temperatures by hydrothermal solutions containing silicon dioxide, carbon dioxide, and other dissolved materials (products of low-grade metasomatism). The whole group of rocks may loosely be referred to as soapstones because of their soft, soapy consistency.

T. F. W. Barth

Social hierarchy A fundamental aspect of social organization that is established by fighting or display behavior and results in a ranking of the animals in a group. Social, or dominance, hierarchies are observed in many different animals, including insects, crustaceans, mammals, and birds. In many species, size, age, or sex determines dominance rank. Dominance hierarchies often determine first or best access to food, social interactions, or mating within animal groups.

When two animals fight, several different behavioral patterns can be observed. Aggressive acts and submissive acts are both parts of a fight. Aggression and submission, together, are known as agonistic behavior. An agonistic relationship in which one animal is dominant and the other is submissive is the simplest type of dominance hierarchy. In nature, most hierarchies involve more than two animals and are composed of paired dominant-subordinate relationships. The simplest dominance hierarchies are linear and are known as pecking orders. In such a hierarchy the top individual (alpha) dominates all others. The second-ranked individual (beta) is submissive to the dominant alpha but dominates the remaining animals. The third animal (gamma) is submissive to alpha and beta but dominates all others. This pattern is repeated down to the lowest animal in the hierarchy, which cannot dominate any other group member.

Other types of hierarchies result from variations in these patterns. If alpha dominates beta, beta dominates gamma, but gamma dominates alpha, a dominance loop is formed. In some species a single individual dominates all members of the social group, but no consistent relationships are formed among the other animals. In newly formed hierarchies, loops or other nonlinear relationships are common, but these are often resolved over time so that a stable linear hierarchy is eventually observed.

Males often fight over access to females and to mating with them. Male dominance hierarchies are seen in many hooved mammals (ungulates). Herds of females use dominance hierarchies to determine access to food. Agonistic interactions among females are often not as overtly aggressive as those among males, but the effects of the dominance hierarchy can easily be observed. In female dairy cattle, the order of entry into the milking barn is determined by dominance hierarchy, with the alpha female entering first. *See* REPRODUCTIVE BEHAVIOR.

Because dominant animals may have advantages in activities such as feeding and mating, they will have more offspring than subordinate animals. If this is the case, then natural selection will favor genes for enhanced fighting ability. Heightened aggressive behavior may be counterselected by the necessity for amicable social interactions in certain circumstances. Many higher primates live in large groups of mixed sex and exhibit complex social hierarchies. In these groups, intra- and intersexual dominance relationships determine many aspects of group life, including feeding, grooming, sleeping sites, and mating. Macaque, baboon, and chimpanzee societies are characterized by cooperative alliances among individuals that are more important than individual fighting ability in maintaining rank. *See* PRIMATES.

Social hierarchies provide a means by which animals can live in groups and exploit resources in an orderly manner. In particular, food can be distributed among group members with little ongoing conflict. Another motivation for group living is mutual defense. Even though subordinates receive less food or have fewer opportunities to mate, they may have greatly increased chances of escaping predation. *See* BEHAVIORAL ECOLOGY; POPULATION ECOLOGY; SOCIAL MAMMALS; TERRITORIALITY. Michael D. Breed

Social insects Insects that share resources and reproduce cooperatively. The shared resources are shelter, defense, and food (collection or production). After a period of population growth, the insects reproduce in several ways. As social insect groups grow, they evolve more differentiation between members but reintegrate into a more closely organized system known as eusocial. These are the most advanced societies with individual polymorphism, and they contain insects of various ages, sizes, and shapes. All the eusocial insects are included in the orders Isoptera (termites) and Hymenoptera (wasps, bees, and ants). *See* HYMENOPTERA; INSECTA; ISOPTERA; POLYMORPHISM (GENETICS).

The social insects have evolved in various patterns. In the Hymenoptera, the society is composed of only females; males are produced periodically for their sperm. They usually congregate and attract females, or they visit colonies with virgin females and copulate there. In the Hymenoptera, sex is determined largely by whether the individual has one or two sets of chromosomes. Thus the queen has the power to determine the sex of her offspring: if she lets any of her stored sperm reach the egg, a female is produced; if not, a male results. In the more primitive bees and wasps, social role (caste) is influenced by interaction with like but not necessarily related individuals. The female that can dominate the others assumes the role of queen, even if only temporarily. Domination is achieved by aggression, real or feigned, or merely by a ritual that is followed by some form of salutation by the subordinates. This inhibits the yolk-stimulating glands and prevents the subordinates from contributing to egg production; if it fails to work, the queen tries to destroy any eggs that are laid. Subordinate females take on more and more of the work of the group for as long as the queen is present and well. At first, all the eggs are fertilized and females develop, with the result that virgin females inhabit the nest for the first batches. They are often undernourished, and this, together with their infertility, reduces their urge to leave the nest and start another one. Such workers are said to be produced by maternal manipulation.

Reproductive ants, like termites, engage in a massive nuptial flight, after which the females, replete with sperm, go off to start a new nest. At some stage after the nuptials, the reproductives break off their wings, which have no further use. Workers, however, never have wings because they develop quickly and pass right through the wing-forming stages; their ovaries and genitalia are also reduced. Ant queens can prevent the formation of more queens; as with the honeybee, they do this behaviorally by using pheromones. They also force the workers to feed all larvae the same diet. To this trophogenic caste control is added a blastogenic control; eggs that are laid have a developmental bias toward one caste or another. This is not genetic; bias is affected by the age of the queen and the season: more worker-biased eggs are laid by young queens and by queens in spring. In some ants, workers mature in various sizes. Since they have disproportionately large heads, the biggest workers are used mainly for defense; they also help with jobs that call for strength, like cutting vegetation or cracking nuts.

Social insects make remarkable nests that protect the brood as well as regulate the microclimate. The simplest nests are cavities dug in soil or soft wood, with walls smoothed and plastered with feces that set hard. Chambers at different levels in the soil are frequently connected by vertical shafts so that the inhabitants can choose the chamber with the best microclimate. Termites and ants also make many different types of arboreal nests. These nests are usually made of fecal material, but one species of ant (*Oecophylla*) binds leaves together with silk produced in the salivary glands of their larvae that the workers hold in their jaws and spin across leaves. A whole group of ants (for example, *Pseudomyrmex*) inhabit the pith of plants.

Social bees use wax secreted by their cuticular glands and frequently blended with gums from tree exudates for their nest construction. Cells are made cooperatively by a curtain of young bees that scrape wax from their abdomen, chew it with saliva, and mold it into the correct shape; later it is planed and polished. With honeybees the hexagonal comb reaches perfection as a set of back-to-back cells, each sloping slightly upward to prevent honey from running out. The same cells are used repeatedly for brood and for storage; or they may be made a size larger for rearing males. Only the queen cell is pendant, with a circular cross section and an opening below.

The ubiquity and ecological power of social insects depend as much on their ability to evolve mutualistic relations with other organisms as on the coherence of their social organization. Wood and the cellulose it contains is normally available as a source of energy only to bacteria and fungi. However, it is used as a basic resource by both termites and ants that have evolved a technique of culturing these organisms. Though lower termites have unusual protozoa as intestinal symbionts, higher termites have bacteria in pouches in their hind gut. Many have a fungus that they culture in special chambers in their nests. The termites feed on woody debris, leaves, and grass cuttings; the fungus digests these materials with the aid of termite feces and produces soft protein-rich bodies that the termites share with their juveniles and reproductives, neither of which are able to feed themselves. Protected from the weather, the fungus can remain active throughout the dry seasons—an inestimable advantage in the subtropics.

Many ants collect and store seeds that they mash and feed directly to their larvae. Provided they are collected when dry and stored in well-ventilated chambers, these seeds can remain viable and edible for an entire season. The plants benefit because not all the seeds are eaten; some that start to germinate are thrown out with the rubbish of the colony—in effect a way of planting them. Others are left behind by the ants when they change nests. In this way, grass seeds can extend their range into dry areas that they could not reach alone.

The dispersal of plant pollen by bees is a well-known symbiosis, and it has led to the evolution of many strange shapes, colors, and scents in flowers. Quite specific flower-bee relationships may exist in which one plant may use very few species of bees for the transfer of pollen. *See* POLLINATION; POPULATION ECOLOGY. M. V. Brian

Social mammals Mammals that exhibit social behavior. This may be defined as any behavior stimulated by or acting upon another animal of the same species. In this broad sense, almost any animal which is capable of behavior is to some degree social. Even those animals which are completely sedentary, such as adult sponges and sea squirts, have a tendency to live in colonies and are social to that extent. Social reactions are occasionally given by species other than the animals' own; an example would be the relations between domestic animals and humans.

The postnatal development of each species is closely related to the social organization typical of the adults. Every highly social animal has a short period early in life when it readily forms attachments to any animal with which it has prolonged contact. The process of socialization begins almost immediately after birth in ungulates like the sheep, and the primary relationship is formed with the mother. In dogs and wolves the process does not begin until about 3 weeks of age, at a time when the mother is beginning to leave the pups. Consequently the strongest relationships are formed with litter mates, thus forming the foundation of a pack. Many rodents stay in the nest long after birth; primary relationships are therefore formed with nest mates. Young primates are typically surrounded by a group of their own kind, but because they are carried for long periods the first strong relationship tends to be with the mother.

Mammals may develop all types of social behavior to a high degree, but not necessarily in every species. Mammals have great capacities for learning and adaptation, which means that social relationships are often highly developed on the basis of learning and habit formation as well as on the basis of heredity and biological differences. The resulting societies tend to be malleable and variable within the same species and to show considerable evidence of cultural inheritance from one generation to the next. Mammalian societies have been completely described in relatively few forms, and new discoveries will probably reveal the existence of a greater variety of social organization.

Basic human social organization and behavior obviously differs from that of all other primates, although it is related to them. At the same time, the range of variability of human societies as seen in the nuclear family does not approach that in mammals as a whole. Human societies are characterized by the presence of all fundamental types of social behavior and social relationships rather than by extreme specialization. *See* ETHOLOGY; HUMAN ECOLOGY; REPRODUCTIVE BEHAVIOR; SOCIOBIOLOGY. John P. Scott

Sociobiology A scientific discipline that applies principles of evolutionary biology to the study of animal and human social behavior. It is a synthesis of ethology, ecology, and evolutionary theory in which social behavior is viewed as a product of natural selection and other biological processes. Although most of the research in sociobiology has focused on understanding the behavior of nonhumans, sociobiological explanations have been used to interpret patterns of human behavior as well. *See* ECOLOGY; ETHOLOGY; ORGANIC EVOLUTION.

Sociobiology predicts that individuals will behave in ways that maximize their fitness (their success at projecting copies of their genetic material into succeeding generations) and argues that such behaviors can arise through the same evolutionary processes that operate on other trait systems. The central principle underlying sociobiology is that an individual's behavior is shaped, in part, by its genes, and thus is heritable and subject to natural selection. Natural selection is simply the result of the differential survival and reproduction of individuals who show heritable variation in a trait. The variants of a trait that convey greater fitness will increase in frequency in a population over time. The distinctive sociobiological perspective on behavior views behaviors as strategies that have evolved through natural selection to maximize an individual's fitness. Note that under this view there is absolutely no need to assume that maximizing fitness is in any way a conscious goal underlying an animal's behavior—successful behavior is simply an emergent result of the evolutionary process and not the force driving that process. *See* BEHAVIORAL ECOLOGY.

Altruism. Animals sometimes behave in ways that seem to reduce their own personal fitness while increasing that of other individuals. For example, in many species of social mammals and in numerous species of birds, individuals commonly give alarm vocalizations that alert others to the presence of a predator while seemingly rendering themselves more conspicuous to attack. Evolutionary biologists since Charles Darwin have noted that these seemingly altruistic acts pose a challenge to the notion of an animal's behavior being a strategy to maximize individual fitness. *See* SOCIAL INSECTS; SOCIAL MAMMALS.

One resolution to this paradox comes from taking a "gene's eye view" of evolution, which recognizes that Darwinian fitness (an individual's success in producing offspring) is simply a specific case of the more general concept of inclusive fitness. Because an individual shares genes with its relatives, there are actually two different routes by which it can pass on copies of those genes to the next generation: first, through personal or direct reproduction; and second, by helping relatives to reproduce. Inclusive fitness is a composite measure of an individual's genetic contribution to the next generation that considers both of these routes.

Animals who behave altruistically toward relatives may, in fact, be behaving selfishly in the genetic sense if their behavior sufficiently enhances their inclusive fitness through its effects on the survival and reproduction of relatives. This phenomenon is referred to as kin selection. Because close relatives share more genes, on average, with an individual than do distant relatives, such seemingly altruistic behavior tends to be directed toward closer kin.

Another solution to the altruism paradox—one that can even operate among nonrelatives under certain conditions—is reciprocity. If the recipient of some altruistic behavior is likely to repay a donor in the future, it may be beneficial for the donor to perform the behavior even at some immediate cost to its own fitness. This is especially true if the cost is low and the anticipated benefit is high. Examples include food sharing among unrelated vampire bats and alliance behavior in some primates.

Under some conditions, it is possible for altruism to result from natural selection operating at the level of groups, rather than at the level of individuals or genes (for example, if groups containing altruists were at a sufficient selective advantage compared with groups containing only selfish individuals). However, within such groups, selection acting on individuals would still tend to favor selfishness over altruism. In any event, much of the cogency of sociobiology has resulted from the recognition that selection seems to be most powerful when it acts at lower levels—notably, genes and individuals rather than groups or species.

Other social strategies. Over the past three decades, the sociobiological view of behavior as an evolved strategy for maximizing inclusive fitness has proven to be a powerful explanatory paradigm for investigating many other aspects of animal social behavior. The phenomenon of sociality, for example, has been extensively explored from this perspective. Each species—and individuals within each species—can be investigated as to the relative costs and benefits associated with the decision of whether or not to be social. Similar considerations apply to many other strategic (though typically unconscious) decisions that animals make during their lives, such as whether to reproduce sexually, when in life to begin reproducing (with regard to season and age at maturation), how many partners to mate with and who those partners should be, whether to compete with other individuals over access to partners, and whether to bestow parental care and the level of care to give. *See* MATERNAL BEHAVIOR; REPRODUCTIVE BEHAVIOR; SEXUAL DIMORPHISM. David P. Barash; Anthony Di Fiore

Soda niter A nitrate mineral having chemical composition $NaNO_3$ (sodium nitrate); also known as nitratite, it is by far the most abundant of the nitrate minerals. It sometimes occurs as simple rhombohedral crystals but is usually massive granular. The mineral has a perfect rhombohedral cleavage, conchoidal fracture, and is rather sectile. Its hardness is 1.5 to 2 on Mohs scale, and its specific gravity is 2.266. It has a vitreous luster and is transparent. It is colorless to white, but when tinted by impurities, it is reddish brown, gray, or lemon yellow. *See* HARDNESS SCALES.

Soda niter is a water-soluble salt found principally as a surface efflorescence in arid regions, or in sheltered places in wetter climates. It is usually associated with niter, nitrocalcite, gypsum, epsomite, mirabilite, and halite. The only large-scale commercial deposits of soda niter in the world occur in a belt roughly 450 mi (725 km) long and 10–50 mi (16–80 km) wide along the eastern slope of the coast ranges in the Atacama, Tarapaca, and Antofagasta Deserts of northern Chile. Chilean nitrate had a monopoly of the world's fertilizer market for many years, but now occupies a subordinate position owing to the development of synthetic processes for nitrogen fixation which permit the production of nitrogen from the air. *See* CALICHE; FERTILIZER; NITER; NITRATE MINERALS; NITROGEN. George Switzer

Sodalite A mineral of the feldspathoid group with chemical composition $Na_4Al_3Si_3O_{12}Cl$. Sodalite is usually massive or granular with poor cleavage. The Mohs hardness is 5.5–6.0, and the density 2.3. The luster is vitreous, and the color is usually blue but may also be white, gray, or green. Notable occurrences are at Mount Vesuvius; Bancroft, Ontario; and on the Kola Peninsula of Russia. *See* FELDSPATHOID; SILICATE MINERALS. Lawrence Grossman

Sodium A chemical element, Na, atomic number 11, and atomic weight 22.9898. Sodium is between lithium and potassium in the periodic table. The element is a soft, reactive, low-melting metal with a specific gravity of 0.97 at 20°C (68°F). Sodium is commercially the most important alkali metal. The physical properties of metallic sodium are summarized in the table below. *See* PERIODIC TABLE.

Physical properties of sodium metal

Property	Temperature C	Temperature F	Metric (scientific) units	British (engineering) units
Density	0	32	0.972 g/cm^3	60.8 lb/ft^3
	100	212	0.928 g/cm^3	58.0 lb/ft^3
	800	1472	0.757 g/cm^3	47.3 lb/ft^3
Melting point	97.5	207.5		
Boiling point	883	1621		
Heat of fusion	97.5	207.5	27.2 cal/g	48.96 Btu/lb
Heat of vaporization	883	1621	1005 cal/g	1809 Btu/lb
Viscosity	250	482	3.81 millipoises	4.3 kinetic units
	400	752	2.69 millipoises	3.1 kinetic units
Vapor pressure	440	824	1 mm	0.019 lb/in.^{2}
	815	1499	400 mm	7.75 lb/in.^{2}
Thermal conductivity	21.2	70.2	0.317 cal/(s)(cm)(°C)	76 Btu/(h)(ft)(°F)
	200	392	0.193 cal/(s)(cm)(°C)	46.7 Btu(h)(ft)(°F)
Heat capacity	20	68	0.30 cal/(g)(°C)	0.30 Btu/(lb)(°F)
	200	392	0.32 cal/(g)(°C)	0.32 Btu/(lb)(°F)
Electrical resistivity	100	212	965 microhm-cm	
Surface tension	100	212	206.4 dynes/cm	
	250	482	199.5 dynes/cm	

Sodium ranks sixth in abundance among all the elements in the Earth's crust, which contains 2.83% sodium in combined form. Only oxygen, silicon, aluminum, iron, and calcium are more abundant. Sodium is, after chlorine, the second most abundant element in solution in seawater. The important sodium salts found in nature include sodium chloride (rock salt), sodium carbonate (soda and trona), sodium borate (borax), sodium nitrate (Chile saltpeter), and sodium sulfate. Sodium salts are found in seawater, salt lakes, alkaline lakes, and mineral springs. *See* ALKALI METALS.

Sodium reacts rapidly with water, and even with snow and ice, to give sodium hydroxide and hydrogen. The reaction liberates sufficient heat to melt the sodium and ignite the hydrogen. When exposed to air, freshly cut sodium metal loses its silvery appearance and becomes dull gray because of the formation of a coating of sodium oxide.

Sodium does not react with nitrogen. Sodium and hydrogen react above about 200°C (390°F) to form sodium hydride. Sodium reacts with ammonia, forming sodium amide. Sodium also reacts with ammonia in the presence of coke to form sodium cyanide.

Sodium does not react with paraffin hydrocarbons but does form addition compounds with naphthalene and other polycyclic aromatic compounds and with arylated alkenes. The reaction of sodium with alcohols is similar to, but less rapid than, the reaction of sodium with water. Sodium reacts with organic halides in two general ways. One of these involves condensation of two organic, halogen-bearing compounds by removal of the halogen, allowing the two organic radicals to join directly. The second type of reaction involves replacement of the halogen by sodium, giving an organosodium compound. *See* ORGANOMETALLIC COMPOUND.

Sodium chloride, or common salt, NaCl, is not only the form in which sodium is found in nature but (in purified form) is the most important sodium compound in commerce as well. Sodium hydroxide, NaOH, is also commonly known as caustic soda. Sodium carbonate, Na_2CO_3, is best known under the name soda ash.

The largest single use for sodium metal, accounting for about 60% of total production, is in the synthesis of tetraethyllead, an antiknock agent for automotive gasolines. A second major use is in the reduction of animal and vegetable oils to long-chain fatty alcohols; these alcohols are raw materials for detergent manufacture. Sodium is used to reduce titanium and zirconium halides to their respective metals. Sodium chloride is used in curing fish, meat packing, curing hides, making freezing mixtures, and food preparation (including canning and preserving). Sodium hydroxide is used in the manufacture of chemicals, cellulose film, rayon soap pulp, and paper. Sodium carbonate is used in the glass industry and in the manufacture of soap, detergents, various cleansers, paper and textiles, nonferrous metals, and petroleum products. Sodium sulfate (salt cake) is used in the pulp industry and in the manufacture of flat glass. *See* GLASS; HALITE; PAPER. Marshall Settig

The sodium ion (Na^+) is the main positive ion present in extracellular fluids and is essential for maintenance of the osmotic pressure and of the water and electrolyte balances of body fluids. Hydrolysis of adenosine triphosphate (ATP) is mediated by the membrane-bound enzyme Na^+,K^+-ATPase (this enzyme is also called sodium pump). The potential difference associated with the transmembrane sodium and potassium ion gradients is important for nerve transmission and muscle contraction. Sodium ion gradients are also responsible for various sodium ion–dependent transport processes, including sodium-proton exchange in the heart, sugar transport in the intestine, and sodium-lithium exchange and amino acid transport in red blood cells. *See* POTASSIUM. Duarte Mota de Freitas

Sodium-vapor lamp An arc discharge lamp with sodium vapor as the emitting species. Low-pressure and high-pressure types are used commercially.

The low-pressure sodium lamp has remarkably high luminous efficiency, or efficacy, producing as much as 200 lumens per watt of input power. The radiation is nearly monochromatic yellow in color, and is used where color rendition is unimportant. The very high luminous efficacy is due to very stringent control of heat losses from the arc. The key to the heat conservation is the indium oxide film used to reflect infrared energy back to the arc tube and the evacuated outer jacket that minimizes the conducted thermal losses. Lamps at 18 and 10 W of power have been developed for very low-energy-cost uses. The 10-W lamp produces light at 100 lm/W. *See* ILLUMINATION; LUMINOUS EFFICACY.

When the sodium pressure is increased by a higher liquid-sodium-pool temperature, the yellow sodium D-line resonance radiation reaches a condition where pressure broadening and self-absorption occur, producing a lamp no longer monochromatic yellow but golden-white in color. The high-pressure sodium lamp has a maximum arc tube temperature typically above 2000°F (1100°C) with a sodium amalgam reservoir temperature about 1300°F (700°C). The term high-pressure is used merely to distinguish this light source from the low-pressure sodium lamp. The plasma arc column of the high-pressure sodium lamp has a total pressure of sodium, mercury, and inert gas of typically slightly less than 1 atm (10^5 Pa).

Because of the more compact source size, the light from a high-pressure sodium lamp can be more effectively directed by reflectors and refractors than that from the low-pressure sodium lamp. Therefore the higher efficacy of the low-pressure sodium lamp is offset by its reduced luminous efficiency. Sodium-vapor lamps are replacing high-pressure mercury lamps in many floodlight and roadway applications. More sodium-vapor lamps are manufactured annually than the high-pressure mercury and metal halide lamps combined. *See* LAMP; VAPOR LAMP. Charles I. McVey

Sofar The acronym for sound fixing and ranging. The concept is based on finding the geographic location of sound that is created by an explosive or other impulsive event, such as a missile impact, by using the time of arrival of the signal at three or more receiving stations. This is known as hyperbolic fixing. This concept was used to locate aircraft that crashed at sea. Although other techniques have replaced this application, the term sofar channel has remained in use, referring to what is also known as the deep sound channel. *See* HYPERBOLIC NAVIGATION SYSTEM; UNDERWATER SOUND. Ralph R. Goodman

Software A set of instructions that cause a computer to perform one or more tasks. The set of instructions is often called a program or, if the set is particularly large and complex, a system. Computers cannot do any useful work without instructions from software; thus a combination of software and hardware (the computer) is necessary to do any computerized work. A program must tell the computer each of a set of minuscule tasks to perform, in a framework of logic, such that the computer knows exactly what to do and when to do it. *See* COMPUTER PROGRAMMING.

Programs are written in programming languages, especially designed to facilitate the creation of software. In the 1950s, programming languages were numerical languages easily understood by computer hardware; often, programmers said they were writing such programs in machine language.

Machine language was cumbersome, error-prone, and hard to change. In the latter 1950s, assembler (or assembly) language was invented. Assembler language was nearly the same as machine language, except that symbolic (instead of numerical) operations and symbolic addresses were used, making the code considerably easier to change.

The programmable aspects of computer hardware have not changed much since the 1950s. Computers still have numerical operations, and numerical addresses by which data may be accessed. However, programmers now use high-level languages, which look much more like English than a string of numbers or operation codes. *See* NUMBERING SYSTEMS; NUMERICAL REPRESENTATION (COMPUTERS); PROGRAMMING LANGUAGES.

Well-known programming languages include Basic, Java, and C. Basic has been modified into Visual Basic, a language useful for writing the portion of a program that the user "talks to" (i.e., the user interface or graphical user interface or GUI). Java is especially useful for creating software that runs on a network of computers. C and C++ are powerful but complex languages for writing such software as systems software and games. *See* HUMAN-COMPUTER INTERACTION; LOCAL-AREA NETWORKS; WIDE-AREA NETWORKS.

Packaged software such as word processors, spreadsheets, graphics and drawing tools, email systems, and games are widely available and used. Some software packages are enormous; for example, enterprise resource planning (ERP) software can be used by companies to perform almost all of their so-called back-office software work. *See* COMPUTER GRAPHICS; ELECTRONIC MAIL; VIDEO GAMES; WORD PROCESSING.

Systems software is necessary to support the running of an application program. Operating systems are needed to link the machine-dependent needs of a program with the capabilities of the machine on which it runs. Compilers translate programs from high-level languages into machine languages. Database programs keep track of where and how data are stored on the various storage facilities of a typical computer, and simplify the task of entering data into those facilities or retrieving the data. Networking software provides the support necessary for computers to interact with each other, and with data storage facilities, in a situation where multiple computers are necessary to perform a task, or when software is running on a network of computers (such as the Internet or the World Wide Web). *See* DATABASE MANAGEMENT SYSTEM; INTERNET; OPERATING SYSTEM; TELEPROCESSING; WORLD WIDE WEB.

Business applications software processes transactions, produces paychecks, and does the myriad of other tasks that are essential to running any business. Roughly two-thirds of software applications are in the business area.

Scientific and engineering software satisfies the needs of a scientific or engineering user to perform enterprise-specific tasks. Because scientific and engineering tasks tend to be very enterprise-specific, there has been no generalization of this application area analogous to the that of the ERP for backoffice business systems. The scientific-engineering application usually is considered to be in second place only to business software in terms of software products built.

Edutainment software instructs (educates) or plays games with (entertains) the user. Such software often employs elaborate graphics and complex logic. This is one of the most rapidly growing software application areas, and includes software to produce special effects for movies and television programs.

Real-time software operates in a time-compressed, real-world environment. Although most software is in some sense real-time, since the users of modern software are usually interacting with it via a GUI, real-time software typically has much shorter time constraints. For example, software that controls a nuclear reactor must make decisions and react to its environment in minuscule fractions of a second.

With the advent of multiple program portions, software development has become considerably more complicated. Whereas it was formerly considered sensible to develop all of a software system in the same programming language, now the different portions are often developed in entirely different languages. The relatively complex GUI, for example, can most conveniently be developed in one of the so-called visual languages, since those languages contain powerful facilities for creating it. The server software, on the other hand, will likely be built using a database package and the database language SQL (a Structured Query Language, for inquiring into the contents of a database). If the server software is also responsible for interacting with a network such as the Internet, it may also be coded in a network-support language such as Java. An object-oriented approach may be adopted in its development, since the software will need to manipulate objects on the Internet. *See* COMPUTER PROGRAMMING; OBJECT-ORIENTED PROGRAMMING; SOFTWARE ENGINEERING.

Robert L. Glass

Software engineering The process of manufacturing software systems. A software system consists of executable computer code and the supporting documents needed to manufacture, use, and maintain the code. For example, a word processing system consists of an executable program (the word processor), user manuals, and the documents, such as requirements and designs, needed to produce the executable program and manuals. *See* SOFTWARE.

Software engineering is ever more important as larger, more complex, and life-critical software systems proliferate. The rapid decline in the costs of computer hardware means that the software in a typical system often costs more than the hardware it runs on. Large software systems may be the most complex things ever built. This places great demands on the software engineering process, which must be disciplined and controlled.

To meet this challenge, software engineers have adapted many techniques from older engineering fields, as well as developing new ones. For example, divide and conquer, a well-known technique for handling complex problems, is used in many ways in software engineering. The software engineering process itself, for example, is usually divided into phases. The definition of these phases, their ordering, and the interactions between the phases specify a software life-cycle model. The best-known life-cycle model is the waterfall model consisting of a requirements definition phase, a design phase, a coding phase, a testing phase, and a maintenance phase. The output of each phase serves as the input to the next. *See* SYSTEMS ENGINEERING.

The purpose of the requirements phase is to define what a system should do and the constraints under which it must operate. This information is recorded in a requirements document. A typical requirements document might include a product overview; a specification of the development, operating, and maintenance environment for the product; a high-level conceptual model of the system; a specification of the user interface; specification of functional requirements; specification of nonfunctional requirements; specification of interfaces to systems outside the system under development; specification of how errors will be handled; and a listing of possible changes and enhancements to the system. Each requirement, usually numbered for reference, must be testable.

In the design phase, a plan is developed for how the system will implement the requirements. The plan is expressed using a design method and notation. Many methods and notations for software design have been developed. Each method focuses on certain aspects of a system and ignores or minimizes others. This is similar to viewing a building with an architectural drawing, a plumbing diagram, an electrical wiring diagram, and so forth.

The coding phase of the software life-cycle is concerned with the development of code that will implement the design. This code is written is a formal language called a programming language. Programming languages have evolved over time from sequences of ones and zeros directly interpretable by a computer, through symbolic machine code, assembly languages, and finally to higher-level languages that are more understandable to humans. *See* PROGRAMMING LANGUAGES.

Most coding today is done in one of the higher-level languages. When code is written in a higher-level language, it is translated into assembly code, and eventually machine code, by a compiler. Many higher-level languages have been developed, and they can be categoriged as functional languages, declarative languages, and imperative languages.

Following the principle of modularity, code on large systems is separated into modules, and the modules are assigned to individual programmers. A programmer typically writes the code using a text editor. Sometimes a syntax-directed editor that "knows" about a given programming language and can provide programming templates and check code for syntax errors is used. Various other tools may be used by a programmer, including a debugger that helps find errors in the code, a profiler that shows which parts of a module spend most time executing, and optimizers that make the code run faster.

Testing is the process of examining a software product to find errors. This is necessary not just for code but for all life-cycle products and all documents in support of the software such as user manuals.

The software testing process is often divided into phases. The first phase is unit testing of software developed by a single programmer. The second phase is integration testing where units are combined and tested as a group. System testing is done on the entire system, usually with test cases developed from the system requirements. Acceptance testing of the system is done by its intended users.

The basic unit of testing is the test case. A test case consists of a test case type, which is the aspect of the system that the test case is supposed to exercise; test conditions, which consist of the input values for the test; the environmental state of the system to be used in the test; and the expected behavior of the system given the inputs and environmental factors.

When software is changed to fix a bug or add an enhancement, a serious error is often introduced. To ensure that this does not happen, all test cases must be rerun after each change. The process of rerunning test cases to ensure that no error has been introduced is called regression testing. *See* SOFTWARE TESTING AND INSPECTION.

Walkthroughs and inspections are used to improve the quality of the software development process. Consequently, the software products created by the process are improved. A quality system is a collection of techniques whose application results in continuous improvement in the quality of the development process. Elements of the quality system include reviews, inspections, and process audits.

Large software systems are not static; rather, they change frequently both during development and after deployment. Maintenance is the phase of the software life-cycle after deployment. The maintenance phase may cost more than all of the others combined and is thus of primary concern to software organizations. The Y2K problem was, for example, a maintenance problem.

Maintenance consists of three activities: adaptation, correction, and enhancement. Enhancement is the process of adding new functionality to a system. This is usually done at the request of system users. This activity requires a full life-cycle of its own. That is, enhancements demand requirements, design, implementation, and test. Studies have shown that about half of maintenance effort is spent on enhancements.

Adaptive maintenance is the process of changing a system to adapt it to a new operating environment, for example, moving a system from the Windows operating system to the Linux operating system. Adaptive maintenance has been found to account for about a quarter of total maintenance effort. Corrective maintenance is the process of fixing errors in a system after release. Corrective maintenance takes about 20% of maintenance effort.

Since software systems change frequently over time, an important activity is software configuration management. This consists of tracking versions of life-cycle objects, controlling changes to them, and monitoring relationships among them. Configuration management activities include version control, which involves keeping track of versions of life-cycle objects; change control, an orderly process of handling change requests to a system; and build control, the tracking of which versions of work products go together to form a given version of a software product.

William B. Frakes

Software metric A rule for quantifying some characteristic or attribute of a computer software entity. For example, a simple one is the FileSize metric, which is the total number of characters in the source files of a program. The FileSize metric can be used to determine the measure of a particular program, such as 3K bytes. It provides a concrete measure of the abstract attribute of program size. Other metrics can be used for software entities such as requirements documents, design object models, or database structure models. Metrics for requirements and design documents can be used to guide decisions about development and as a basis for predictions, such as for cost and effort. Metrics for programs can be used to support decisions about testing and maintenance and as a basis for comparing different versions of programs. Ideally, metrics for the development cost of software and for the quality of the resultant program are desirable. *See* COMPUTER PROGRAMMING; SOFTWARE ENGINEERING.

David A. Gustafson; William Hankley

Software testing and inspection Procedures for the detection of software faults. When software does not operate as it is intended to do, a software failure is said to occur. Software failures are caused by one or more sections of the software program being incorrect. Each of these incorrect sections is called a software fault. The fault could be as simple as a wrong value. A fault could also be complete omission of a decision in the program. Faults have many causes, including misunderstanding of requirements, overlooking special cases, using the wrong variable, misunderstanding of the algorithm, and even typing mistakes. Software that can cause serious problems if it fails is called safety-critical software. Many applications in aircraft, medicine, nuclear power plants, and transportation involve such software.

Software testing is the execution of the software with the purpose of detecting faults. Software inspection is a manual process of analyzing the source code to detect faults. Many of the same techniques are used in both procedures. Other techniques can also be used to minimize the possibility of faults in the software. These techniques include the use of formal specifications, formal proofs of correctness, and model checking. However, even with the use of these techniques, it is still important to execute software with test cases to detect possible faults.

Software testing involves selecting test cases, determining the correct output of the software, executing the software with each test case, and comparing the actual output with the expected output. More testing is better, but costs time and effort. The value of additional testing must be balanced against the additional cost of effort and the delay in delivering the software. Another consideration is the potential cost of failure of the software. Safety-critical software is usually tested much more thoroughly than any other software.

The number of potential test cases is huge. For example, in the case of a simple program that multiplies two integer numbers, if each integer is internally represented as a 32-bit number (a common size for the internal representation), then there are 2^{32} possible values for each number. Thus, the total number of possible input combinations is 2^{64}, which is more than 10^{19}. If a test case can be done each microsecond (10^{-6} second), then it will take hundreds of thousands of years to try all of the possible test cases. Trying all possible test cases is called exhaustive testing and is usually not a reasonable approach because of the size of the task.

One approach to software testing is to find test cases so that all statements in a program are executed. A more extensive criterion for test selection is "every branch coverage." This means that each branch coming out of every decision is tested. Instead of just requiring the whole decision to be true or false, the "multiple condition coverage" criterion requires all combinations of truth values for each simple comparison in a decision to be covered. Another approach is called dataflow coverage. The basis for coverage is the execution paths between the statement where a variable is assigned a value (a def or definition) and a statement where that value is used. These paths must be free of other definitions of the variable of interest. *See* DATAFLOW SYSTEMS.

Functional testing compares the actual behavior of the software with the expected behavior. That expected behavior is usually described in a specification. More involved functional test case selection involves analyzing the conditions inherent in the task.

Another approach to test selection concentrates on the boundaries between the subdomains. This approach recognizes that many faults are related to the boundary conditions. Test cases are chosen to check whether the boundary is correct. Test cases on the boundary and test cases just off the boundary are chosen. *See* SOFTWARE; SOFTWARE ENGINEERING.

David A. Gustafson

Soil Finely divided rock-derived material containing an admixture of organic matter and capable of supporting vegetation. Soils are independent natural bodies, each with a unique morphology resulting from a particular combination of climate, living plants and animals, parent rock materials, relief, the ground waters, and age. Soils support plants, occupy large portions of the Earth's surface, and have shape, area, breadth, width, and depth. Soil, as used here, differs in meaning from the term as used by engineers, where the meaning is unconsolidated rock material. *See* PEDOLOGY.

Origin and classification. Soil covers most of the land surface as a continuum. Each soil grades into the rock material below and into other soils at its margins, where changes occur in relief, groundwater, vegetation, kinds of rock, or other factors which influence the development of soils. Soils have horizons, or layers, more or less parallel to the surface and differing from those above and below in one or more properties, such as color, texture, structure, consistency, porosity, and reaction (see illustration). The succession of horizons is called the soil profile.

Soil formation proceeds in stages, but these stages may grade indistinctly from one into another. The first stage is the accumulation of unconsolidated rock fragments, the parent material. Parent material may be accumulated by deposition of rock fragments moved by glaciers, wind, gravity, or water, or it may accumulate more or less in place from physical and chemical weathering of hard rocks. The second stage is the formation of horizons. This stage may follow or go on simultaneously with the accumulation of parent material. Soil horizons are a result of dominance of one or more processes over others, producing a layer which differs from the layers above and below. *See* WEATHERING PROCESSES.

Systems of soil classification are influenced by concepts prevalent at the time a system is developed. The earliest classifications were based on relative suitability for different crops, such as rice soils, wheat soils, and vineyard soils. Over the years, many systems of classification have been attempted but none has been found markedly superior. Two bases for classification have been tried. One basis has been the presumed genesis of the soil; climate and native vegetation were given major emphasis. The

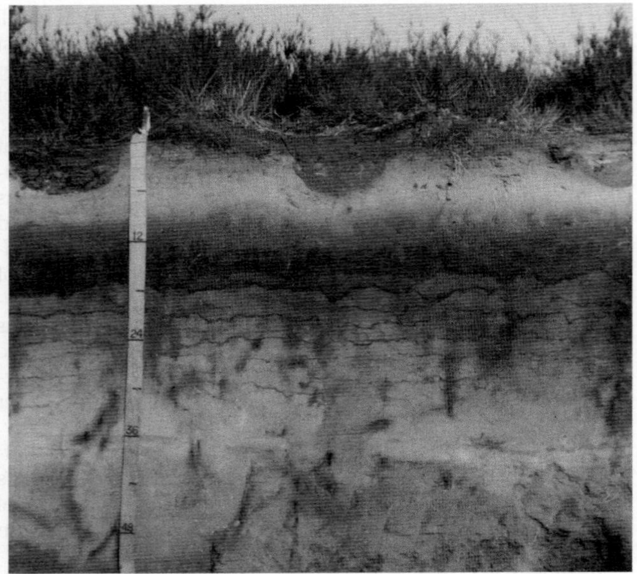

Photograph of a soil profile showing horizons. The dark crescent-shaped spots at the soil surface are the result of plowing. The dark horizon is the principal horizon of accumulation of organic matter that has been washed down from the surface. The thin wavy lines were formed in the same manner. 1 in. = 2.5 cm.

other basis has been the observable or measurable properties of the soil.

The Soil Survey staff of the U.S. Department of Agriculture and the land-grant colleges adopted the current classification scheme in 1965. This system differs from earlier systems in that it may be applied to either cultivated or virgin soils. Previous systems have been based on virgin profiles, and cultivated soils were classified on the presumed characteristics or genesis of the virgin soils. The new system has six categories, based on both physical and chemical properties. These categories are the order, suborder, great group, subgroup, family, and series, in decreasing rank. The orders and the general nature of the included soils are given in the table. The suborder narrows the ranges in soil moisture and temperature regimes, kinds of horizons, and composition, according to which of these is most important. The taxa (classes) in the great group category soils have the same kinds of horizons in the same sequence and have similar moisture and temperature regimes. The great groups are divided into subgroups that show the central properties of the great group, intergrade subgroups that show properties of more than one great group, and other subgroups for soils with atypical properties that are not characteristic of any great group. The families are defined largely on the basis of physical and mineralogic properties of importance to plant growth. The soil series is a group of soils having horizons similar in differentiating characteristics and arrangement in the soil profile, except for texture of the surface portion, and developed in a particular type of parent material.

Soil orders

Order	Formative element in name	General nature of soils
Alfisols	alf	Gray to brown surface horizons, medium to high base supply, with horizons of clay accumulation; usually moist, but may be dry during summer
Aridisols	id	Pedogenic horizons, low in organic matter, and usually dry
Entisols	ent	Pedogenic horizons lacking
Histosols	ist	Organic (peats and mucks)
Inceptisols	ept	Usually moist, with pedogenic horizons of alteration of parent materials but not of illuviation
Mollisols	oil	Nearly black organic-rich surface horizons and high base supply
Oxisols	ox	Residual accumulations of inactive clays, free oxides, kaolin, and quartz; mostly tropical
Spodosols	od	Accumulations of amorphous materials in subsurface horizons
Ultisols	ult	Usually moist, with horizons of clay accumulation and a low supply of bases
Vertisols	ert	High content of swelling clays and wide deep cracks during some season

Surveys. Soil surveys include those researches necessary (1) to determine the important characteristics of soils, (2) to classify them into defined series and other units, (3) to establish and map the boundaries between kinds of soil, and (4) to correlate and predict adaptability of soils to various crops, grasses, and trees; behavior and productivity of soils under different management systems; and yields of adapted crops on soils under defined sets of management practices. Although the primary purpose of soil surveys has been to aid in agricultural interpretations, many other purposes have become important, ranging from suburban planning, rural zoning, and highway location, to tax assessment and location of pipelines and radio transmitters. This has happened because the soil properties important to the growth of plants are also important to its engineering uses.

Two kinds of soil maps are made. The common map is a detailed soil map, on which soil boundaries are plotted from direct observations throughout the surveyed area. Reconnaissance soil maps are made by plotting soil boundaries from observations made at intervals. The maps show soil and other differences that are of significance for present or foreseeable uses. Guy D. Smith

Physical properties. Physical properties of soil have critical importance to growth of plants and to the stability of cultural structures such as roads and buildings. Such properties commonly are considered to be: size and size distribution of primary particles and of secondary particles, or aggregates, and the consequent size, distribution, quantity, and continuity of pores; the relative stability of the soil matrix against disruptive forces, both natural and cultural; color and textural properties, which affect absorption and radiation of energy; and the conductivity of the soil for water, gases, and heat. These usually would be considered as fixed properties of the soil matrix, but actually some are not fixed because of influence of water content. The additional property, water content—and its inverse, gas content—ordinarily is transient and is not thought of as a property in the same way as the others. However, water is an important constituent, despite its transient nature, and the degree to which it occupies the pore space generally dominates the dynamic properties of soil. Additionally, the properties listed above suggest a macroscopic homogeneity for soil which it may not necessarily have. In a broad sense, a soil may consist of layers or horizons of roughly homogeneous soil materials of various types that impart dynamic properties which are highly dependent upon the nature of the layering. Thus, a discussion of dynamic soil properties must include a description of the intrinsic properties of small increments as well as properties it imparts to the system.

From a physical point of view it is primarily the dynamic properties of soil which affect plant growth and the strength of soil beneath roads and buildings. While these depend upon the chemical and mineralogical properties of particles, particle coatings, and other factors discussed above, water content usually is the dominant factor. Water content depends upon flow and retention properties, so that the relationship between water content and retentive forces associated with the matrix becomes a key physical property of a soil. *See* EROSION; GROUND-WATER HYDROLOGY; SOIL MECHANICS. Walter H. Gardner

Soil chemistry The study of the composition and chemical properties of soil. Soil chemistry involves the detailed investigation of the nature of the solid matter from which soil is constituted and of the chemical processes that occur as a result of the action of hydrological, geological, and biological agents on the solid matter. Because of the broad diversity among soil components and the complexity of soil chemical processes, the application of a wide variety of concepts and methods employed in the chemistry of aqueous solutions, of amorphous and crystalline solids, and of solid surfaces is required.

Elemental composition. The elemental composition of soil varies over a wide range, permitting only a few general statements to be made. Those soils that contain less than 12–20% organic carbon are termed mineral. All other soils are termed organic. Carbon, oxygen, hydrogen, nitrogen, phosphorus, and sulfur are the most important constituents of organic soils and of soil organic matter in general. Carbon, oxygen, and hydrogen are most abundant; the content of nitrogen is often about one-tenth that of carbon, while the content of phosphorus or sulfur is usually less than one-fifth that of nitrogen (Table 1).

Table 1. Average percentages of total carbon, total nitrogen, and organic phosphorus in selected soils

Soil	% C	% N	% P
Sand	2.5	.23	.04
Fine sandy loam	3.3	.23	.06
Medium loam	2.3	.22	.05
Clay loam, well drained	4.6	.36	.10
Clay loam, poorly drained	8.0	.43	.05
Peat	46.1	1.32	.03

Besides oxygen, the most abundant elements found in mineral soils are silicon, aluminum, and iron. The distribution of chemical elements will vary considerably from soil to soil and, in general, will be different in a specific soil from the distribution of elements in the crustal rocks of the Earth. The most important micro or trace elements in soil are boron, copper, manganese, molybdenum, and zinc, since these elements are essential in the nutrition of green plants. Also important are cobalt, selenium, cadmium, and nickel. The average distribution of trace elements in soil is not greatly different from that in crustal rocks (Table 2).

The elemental composition of soil varies with depth below the surface because of pedochemical weathering. The principal processes of this type that result in the removal of chemical elements from a given soil horizon are: (1) soluviation (ordinary dissolution in water), (2) cheluviation (complexation by organic or inorganic ligands), (3) reduction, and (4) suspension. The principal effect of these four processes is the appearance of alluvial horizons in which compounds such as aluminum and iron oxides, aluminosilicates, or calcium carbonate have been precipitated from solution or deposited from suspension. *See* WEATHERING PROCESSES.

Table 2. Average amounts of trace elements commonly found in soils and crustal rocks

Trace element	Soil, ppm*	Crustal rocks, ppm
As	6	1.8
B	10	10
Cd	.06	.2
Co	8	25
Cr	100	100
Cu	20	55
Mo	2	1.5
Ni	40	75
Pb	10	13
Se	.2	.05
V	100	135
Zn	50	70

*ppm = parts per million.

Minerals. The minerals in soils are the products of physical, geochemical, and pedochemical weathering. Soil minerals may be either amorphous or crystalline. They may be classified further, approximately, as primary or secondary minerals, depending on whether they are inherited from parent rock or are produced by chemical weathering, respectively.

The bulk of the primary minerals that occur in soil are found in the silicate minerals. Chemical weathering of the silicate minerals is responsible for producing the most important secondary minerals in soil. These are found in the clay fraction and include aluminum and iron hydrous oxides (usually in the form of coatings on other minerals), carbonates, and aluminosilicates. *See* CLAY MINERALS; SILICATE MINERALS.

Ion exchange. A portion of the chemical elements in soil is in the form of cations that are not components of inorganic salts but that can be replaced reversibly by the cations of leaching salt solutions or acids. These cations are said to be exchangeable, and their total quantity is termed the cation exchange capacity (CEC) of the soil. The CEC of a soil generally will vary directly with the amounts of clay and organic matter present and with the distribution of clay minerals.

The stoichiometric exchange of the anions in soil for those in a leaching salt solution is a phenomenon of relatively small importance in the general scheme of anion reactions with soils. Under acid conditions ($pH < 5$) the exposed hydroxyl groups at the edges of the structural sheets or on the surfaces of clay-sized particles become protonated and thereby acquire a positive charge. The degree of protonation is a sensitive function of pH, the ionic strength of the leaching solution, and the nature of the clay-sized particle.

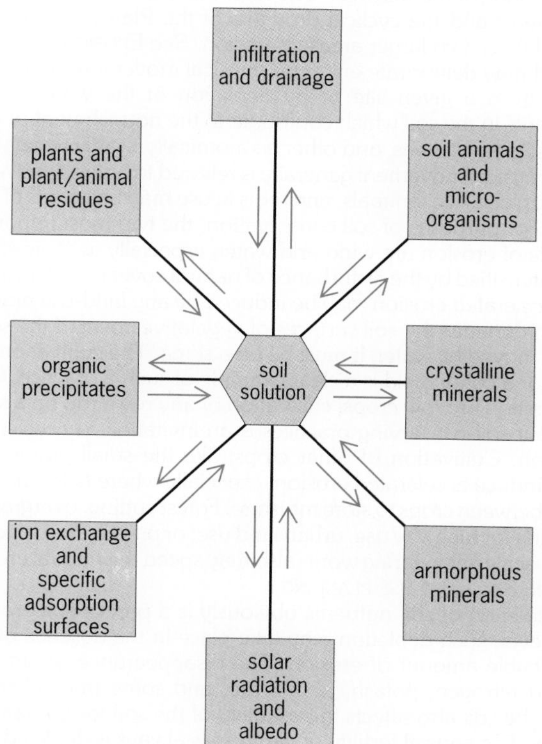

Factors influencing the chemistry of soil solution. (*Modified from J. F. Hodgson, Chemistry of the micronutrients in soil, Adv. Agron., 15:141, 1963*)

Soil solution. The solution in the pore space of soil acquires its chemical properties through time-varying inputs and outputs of matter and energy that are mediated by the several parts of the hydrologic cycle and by processes originating in the biosphere (see illustration). The soil solution thus is a dynamic and open natural water system whose composition reflects the many reactions that can occur simultaneously between an aqueous solution and an assembly of mineral and organic solid phases that varies with both time and space. *See* SOIL. Garrison Sposito

Soil conservation

The practice of arresting or minimizing artificially accelerated soil deterioration. Its importance has grown because cultivation of soils for agricultural production, deforestation and forest cutting, grazing of natural range, and other disturbances of the natural cover and position of the soil have increased greatly since the middle of the nineteenth century in response to the growth in world population and human technical capacity. Accelerated soil deterioration has been the unfortunate consequence.

Erosion. Accelerated erosion has been known throughout history wherever people have tilled or grazed slopes or semiarid soils. The exact extent of accelerated soil erosion in the world today is not known, particularly as far as the rate of soil movement is concerned. However, it may be safely said that nearly every semiarid area with cultivation or long-continued grazing, every hill land with moderate to dense settlement in humid middle latitude and subtropical climates, and all cultivated or grazed hill lands in the Mediterranean climate areas suffer to some degree from such erosion. These recognized and often growing problems of erosion occur in such culturally diverse areas as southern China, the Indian plateau, South Africa, Russia and Ukraine, and Spain.

Within the United States the most critical areas have been the hill lands of the interior Southeast, the Great Plains, the Palouse area hills of the Pacific Northwest, southern California hills, and slope lands of the Midwest. The high-intensity rainstorms of the Southeast and the cyclical droughts of the Plains have predisposed these two larger areas to erosion. *See* EROSION.

Soil may deteriorate either by physical movement of soil particles from a given site or by depletion of the water-soluble elements in the soil which contribute to the nourishment of crop plants, grasses, trees, and other economically usable vegetation. The physical movement generally is referred to as erosion. Wind, water, glacial ice, animals, and tools in use may be agents of erosion. For purposes of soil conservation, the two most important agents of erosion are wind and water, especially as their effects are intensified by the disturbance of natural cover or soil position.

Accelerated erosion may be induced by any land-use practice which denudes the soil surfaces of vegetative cover. If the soil is to be moved by water, it must be on a slope. The cultivation of a corn or a cotton field is a clear example of such a practice. Corn and cotton are row crops; cultivation of any row crop on a slope without soil-conserving practices is an invitation to accelerated erosion. Cultivation of other crops, like the small grains, also may induce accelerated erosion, especially where fields are kept bare between crops to store moisture. Forest cutting, overgrazing, grading for highway use, urban land use, or preparation for other large-scale engineering works also may speed the natural erosion of soil. *See* LAND-USE PLANNING.

Depletion of soil nutrients obviously is a part of soil erosion. However, such depletion may take place in the absence of any noticeable amount of erosion. The disappearance of naturally stored nitrogen, potash, phosphate, and some trace elements from the soil also affects the usability of the soil for human purposes. The natural fertility of virgin soils always is depleted over time as cultivation continues, but the rate of depletion is highly dependent on management practices. *See* PLANT MINERAL NUTRITION; SOIL; SOIL FERTILITY.

Soil mismanagement. One of the chief causes of erosion-inducing agricultural practices in the United States has been ignorance of their consequences. The cultivation methods of the settlers of western European stock who set the pattern of land use in the United States came from physical environments which, because of the mild nature of their rainstorms and their prevailing soil textures, were in general far less susceptible to erosion than were the environments into which they migrated in North America. The principal European-sown grains, moreover, were wheat, barley, oats, and rye which covered the soil surface completely as they grew. Corn, cotton, and tobacco, however, were crops unfamiliar to European agriculture. In eastern North America the combination of European cultivation methods and American intertilled crops resulted in generations of soil mismanagement. In later years the Plains environment, with its alternation of drought and plentiful moisture, was also an unfamiliar one to settlers from western Europe.

On the Plains and in other susceptible western areas, small-grain monoculture, particularly of wheat, encouraged the exposure of the uncovered soil surface so much of the time that water and wind inevitably took their toll. On rangelands, the high percentage of public range (for whose management little individual responsibility could be felt), lack of knowledge as to the precipitation cycle and range capacity, and the desire to maximize profits every year contributed to a slower but equally sure denudation of cover. *See* AGRICULTURAL SOIL AND CROP PRACTICES.

Finally, the United States has experienced extensive erosion in mountain areas because of forest mismanagement. Clearcutting of steep slopes, forest burning for grazing purposes, inadequate fire protection, and shifting cultivation of forest lands have allowed vast quantities of soil to wash out of the slope sites where they could have produced timber and other forest values indefinitely. *See* FOREST MANAGEMENT.

Conservation measures. Measures of soil management designed to reduce the effects of accelerated erosion have been known in both the Western world and in the Far East since long before the beginning of the Christian Era. The value of forests for watershed protection was known in China at least 10 centuries ago. The most important of the ancient measures on agricultural lands was terrace construction, although actual physical restoration of soil to original sites also has been practiced. Certain land management practices that were soil-conserving have been a part of western European agriculture for centuries, principally those centering on livestock husbandry and crop rotation. However, it is principally since 1920 that techniques for soil conservation have been developed for many types of environment in terms of an integrated approach. The measures include farm, range, and forest management practices, and the building of engineered structures on land and in stream channels. A first and most important step in conservational management is the determination of land capability—the type of land use and economic production to which a plot is suited by slope, soil type, drainage, precipitation, wind exposure, and other natural attributes. The objective of such determination is to achieve permanent productive use as nearly as possible. Donald J. Patton

Soil degradation

Loss in the quality or productivity of soil that is often the result of human activities, such as agriculture practices, deforestation, mining, waste disposal, and chemical spills. Degradation is attributed to changes in soil nutrient status, biota, loss of organic matter, deterioration of soil structure, and toxicity due to accumulations of naturally occurring or anthropogenic (human-made) materials. The effects of soil degradation include loss of agricultural productivity, negative impacts on the environment and economic stability, and exploitation of marginally suitable or virgin land. *See* SOIL.

Healthy soil is rich with life, particularly microorganisms. The soil microbial community is essential to the cycling of nutrients and the decomposition of wastes; thus hindrance of microbial activities may have serious ecological results. Soil microbial activity is generally closely related to the amount of soil organic matter; thus degradative forces such as erosion that deplete organic matter also result in long-term losses of microbial activity. *See* SOIL ECOLOGY; SOIL MICROBIOLOGY.

Failing to replace nutrients removed with crop harvests depletes the soil over the long term. Excessive leaching of the basic cations that buffer soil pH may result in soil acidification, changing the solubility, and thus availability, of certain nutrients to plants. Land management practices, such as clear-cutting of forests, may result in long-term changes in nutrient cycling. *See* SOIL CHEMISTRY; SOIL FERTILITY.

Historically the mining of coal and various metal ores has been among the most devastating forms of land use. Strip (surface) mining can expose large amounts of reduced (decreased oxidation state) minerals in the form of mine tailings and rubble. The acidic mine tailings are extremely difficult to revegetate, and without intervention remain exposed, continuing to produce acidic drainage until oxidizable material is depleted. Left untreated, strip mines may require 50 to 150 years for recovery. *See* AUTOXIDATION; LAND RECLAMATION; SURFACE MINING.

In some cases, such as petroleum spills, compounds are toxic to soil organisms when present at high enough concentrations, and thus produce large shifts in microbial community structure. Spillage of other toxic chemicals can cause similar long-term effects on microbiology. Eventually, microorganisms present in the soil degrade most organic contaminants, thus alleviating toxic effects. Unlike organic contaminants, which are usually degraded to nontoxic forms, toxic metals become essentially permanent features of soils once introduced. Fortunately, adsorption of metals to soil colloids decreases the availability of the metals for movement in the environment or uptake by plants, animals, and microorganisms. *See* BIODEGRADATION; SOIL CONSERVATION.

Gerald K. Sims

Soil ecology The study of the interactions among soil organisms, and between biotic and abiotic aspects of the soil environment. Soil is made up of a multitude of physical, chemical, and biological entities, with many interactions occurring among them. Soil is a variable mixture of broken and weathered minerals and decaying organic matter. Together with the proper amounts of air and water, it supplies, in part, sustenance for plants as well as mechanical support. *See* SOIL.

Abiotic and biotic factors lead to certain chemical changes in the top few decimeters (8–10 in.) of soil. The work of the soil ecologist is made easier by the fact that the surface 10–15 cm (4–6 in.) of the A horizon has the majority of plant roots, microorganisms, and fauna. A majority of the biological-chemical activities occur in this surface layer.

The biological aspects of soil range from major organic inputs, decomposition by primary decomposers (bacteria, fungi, and actinomycetes), and interactions between microorganisms and fauna (secondary decomposers) which feed on them. The detritus decomposition pathway occurs on or within the soil after plant materials (litter, roots, sloughed cells, and soluble compounds) become available through death or senescence. Plant products are used by microorganisms (primary decomposers). These are eaten by the fauna which thus affect flows of nutrients, particularly nitrogen, phosphorus, and sulfur. The immobilization of nutrients into plants or microorganisms and their subsequent mineralization are critical pathways. The labile inorganic pool is the principal one that permits subsequent microorganism and plant existence. Scarcity of some nutrient often limits production. Most importantly, it is the rates of flux into and out of these labile inorganic pools which enable ecosystems to successfully function. *See* ECOLOGY; ECOSYSTEM; GUILD; SOIL; SYSTEMS ECOLOGY.

David C. Coleman

Soil fertility The ability of a soil to supply plant nutrients. Sixteen chemical elements are required for the growth of all plants: carbon, oxygen, and hydrogen (these three are obtained from carbon dioxide and water), plus the elements nitrogen, phosphorus, potassium, calcium, magnesium, sulfur, iron, manganese, zinc, copper, boron, molybdenum, and chlorine. Some plant species also require one or more of the elements cobalt, sodium, vanadium, and silicon. *See* PLANT MINERAL NUTRITION.

While carbon and oxygen are supplied to plants from carbon dioxide in the air, the other essential elements are supplied primarily by the soil. Of the latter, all except hydrogen from water are called mineral nutrients. Only part of the 13 essential mineral nutrients in soil are in a chemical form that can be immediately used by plants. The unusable (unavailable) parts, which eventually do become available to plants, are of two kind: they may be in organic combination (such as nitrogen in soil humus) or in solid inorganic soil particles (such as potassium in soil clays). The time for complete decomposition and dissolution of these compounds varies widely, from days to hundreds of years.

Soils exhibit a variable ability to supply the mineral nutrients needed by plants. This characteristic allows soils to be classified according to their level of fertility. This can vary from a deficiency to a sufficiency, or even toxicity (too much), of one or more nutrients. A serious deficiency of only one essential nutrient can still greatly reduce crop yields. Several soil properties are important in determining a soil's inherent fertility. One property is the adsorption and storage of nutrients on the surfaces of soil particles. Such adsorption of a number of nutrients is caused by an attraction of positively charged nutrients to negatively charged soil particles. This adsorption is called cation exchange (adsorbed cations can be exchanged with other cations in solution), and the quantity of nutrient cations a soil can adsorb is called its cation-exchange capacity. *See* ADSORPTION; ION EXCHANGE; SOIL CHEMISTRY.

The negative charge in soils is associated with clay particles, but some of the soil's cation-exchange capacity may arise from organic matter (humus) in the soil. Negative charges of organic matter arise largely from carboxylic and phenolic acid functional groups. Since these functional groups are weak acids, the negative charge from organic matter increases as the soil pH increases. The negative charges of soil organic matter can adsorb the same cations as described for the soil clays. The proportion of cation-exchange capacity arising from mineral clays and from organic matter depends on the proportions of each in the soil and on the kinds of clays. In most mineral soils, the soil clays comprise the greater proportion of cation-exchange capacity. Within the class of mineral soils, those soils with more clay and less sand and silt have the greatest cation-exchange capacity.

The amount and kind of acids on the cation exchange sites can have a substantial influence on a soil's perceived fertility. Two factors cause soils to become acid: when crops are harvested, exchangeable bases are removed as part of the crop; and exchangeable bases move with drainage water below the crop's root zone (leaching). Since much of the nitrogen fertilizer supplied to crops contains ammonium nitrogen (this is true of both manure and chemical fertilizers), the addition of high rates of nitrogen also enhances soil acidification. *See* FERTILIZER; NITROGEN CYCLE; SOIL.

David E. Kissel

Soil mechanics The study of the response of masses composed of soil, water, and air to imposed loads. Because both water and air are able to move through the soil pores, the discipline also involves the prediction of these transport processes. Soil mechanics provides the analytical tools required for foundation engineering, retaining wall design, highway and railway subbase design, tunneling, earth dam design, mine excavation design, and so on. Because the discipline relates to rock as well as soils, it is also known as geotechnical engineering. *See* ENGINEERING GEOLOGY.

Soil consists of a multiphase aggregation of solid particles, water, and air. This fundamental composition gives rise to unique engineering properties, and the description of the mechanical behavior of soils requires some of the most sophisticated principles of engineering mechanics. The terms multiphase and aggregation both imply unique properties. As a multiphase material, soil exhibits mechanical properties that show the combined attributes of solids, liquids, and gases. Individual soil particles behave as solids, and show relatively little deformation when subjected to either normal or shearing stresses. Water behaves as a liquid, exhibiting little deformation under normal stresses, but deforming greatly when subjected to shear. Being a viscous liquid, however, water exhibits a shear strain rate that is proportional to the shearing stress. Air in the soil behaves as a gas, showing appreciable deformation under both normal and shear stresses. When the three phases are combined to form a soil mass, characteristics that are an outgrowth of the interaction of the phases are manifest. Moreover, the particulate nature of the solid particles contributes other unique attributes. *See* VISCOSITY.

When dry soil is subjected to a compressive normal stress, the volume decreases nonlinearly; that is, the more the soil is compressed, the less compressible the mass becomes. Thus, the more tightly packed the particulate mass becomes, the more it resists compression. The process, however, is only partially reversible, and when the compressive stress is removed the soil does not expand back to its initial state.

When this dry particulate mass is subjected to shear stress, an especially interesting behavior owing to the particulate nature of the soil solids results. If the soil is initially dense (tightly packed), the mass will expand because the particles must roll up and over each other in order for shear deformation to occur. Conversely, if the mass is initially loose, it will compress when subjected to a shear stress. Clearly, there must also exist a specific initial density (the critical density) at which the material will display zero volume change when subjected to shear stress. The term dilatancy has been applied to the relationship between shear stress and volume change in particulate materials. Soil is capable of resisting shear stress up to a certain maximum value. Beyond this value, however, the material undergoes large, uncontrolled shear deformation. *See* SHEAR.

The other limiting case is saturated soil, that is, a soil whose voids are entirely filled with water. When such a mass is initially loose and is subjected to compressive normal stress, it tends to decrease in volume; however, in order for this volume decrease to occur, water must be squeezed from the soil pores. Because water exhibits a viscous resistance to flow in the microscopic pores of fine-grained soils, this process can require considerable time, during which the pore water is under increased pressure. This excess pore pressure is at a minimum near the drainage face of the soil mass and at a maximum near the center of the soil sample. It is this gradient (or change in pore water pressure with change in position within the soil mass) that causes the outflow of water and the corresponding decrease in volume of the soil mass.

Conversely, if an initially dense soil mass is subjected to shear stress, it tends to expand. The expansion, however, may be time-dependent because of the viscous resistance to water being drawn into the soil pores. During this time the pore water will be under decreased pressure. Thus, in saturated soil masses, changes in pore water pressure and time-dependent volume change can be induced by either changes in normal stress or by changes in shear stress. *See* ROCK MECHANICS; SOIL.

Francisco Silva-Tulla

Soil microbiology The study of biota that inhabit the soil and the processes that they mediate. The soil is a complex environment colonized by an immense diversity of microorganisms. Soil microbiology focuses on the soil viruses, bacteria, actinomycetes, fungi, and protozoa, but it has traditionally also included investigations of the soil animals such as the nematodes, mites, and other microarthropods. These organisms, collectively referred to as the soil biota, function in a belowground ecosystem based on plant roots and litter as food sources. Modern soil microbiology represents an integration of microbiology with the concepts of soil science, chemistry, and ecology to understand the functions of microorganisms in the soil environment.

The surface layers of soil contain the highest numbers and variety of microorganisms, because these layers receive the largest amounts of potential food sources from plants and animals. The soil biota form a belowground system based on the energy and nutrients that they receive from the decomposition of plant and animal tissues. The primary decomposers are the bacteria and fungi.

Microorganisms, especially algae and lichen, are pioneering colonizers of barren rock surfaces. Colonization by these organisms begins the process of soil formation necessary for the growth of higher plants. After plants have been established, decomposition by microorganisms recycles the energy, carbon, and nutrients in dead plant and animal tissues into forms usable by plants. Therefore, microorganisms have a key role in the processing of materials that maintain life on the Earth. The transformations of elements between forms are described conceptually as the elemental cycles.

In the carbon cycle, microorganisms transform plant and animal residues into carbon dioxide and the soil organic matter known as humus. Humus improves the water-holding capacity of soil, supplies plant nutrients, and contributes to soil aggregation. Microorganisms may also directly affect soil aggregation. The extent of soil aggregation determines the workability or tilth of the soil. A soil with good tilth is suitable for plant growth because it is permeable to water, air, and roots. *See* HUMUS.

Soil microorganisms play key roles in the nitrogen cycle. The atmosphere is approximately 80% nitrogen gas (N_2), a form of nitrogen that is available to plants only when it is transformed to ammonia (NH_3) by either soil bacteria (N_2 fixation) or by humans (manufacture of fertilizers). Soil bacteria also mediate denitrification, which returns nitrogen to the atmosphere by transforming NO_3^- to N_2 or nitrous oxide (N_2O) gas. Microorganisms are crucial to the cycling of sulfur, phosphorus, iron, and many micronutrient trace elements.

In addition to the elemental cycles, there are several interactions between plants and microbes which are detrimental or beneficial to plant growth. Some soil microorganisms are pathogenic to plants and cause plant diseases such as root rots and wilts. Many plants form symbiotic relationships with fungi called mycorrhizae (literally fungus-root). Mycorrhizae increase the ability of plants to take up nutrients and water. The region of soil surrounding plant roots, the rhizosphere, may contain beneficial microorganisms which protect the plant root from pathogens or supply stimulating growth factors. The interactions between plant roots and soil microorganisms is an area of active research in soil microbiology. *See* BIOGEOCHEMISTRY; MYCORRHIZAE; NITROGEN CYCLE; NITROGEN FIXATION; NITROGEN OXIDES; RHIZOSPHERE.

The incredible diversity of soil microorganisms is a vast reserve of potentially useful organisms. Many of the medically important antibiotics are produced by filamentous bacteria known as actinomycetes. The soil is the largest reservoir of these medically important microorganisms. *See* ANTIBIOTIC.

The numerous natural substances that are used by microorganisms indicate that soil microorganisms have diverse mechanisms for degrading a variety of compounds. Human activity has polluted the environment with a wide variety of synthetic or processed compounds. Many of these hazardous or toxic substances can be degraded by soil microorganisms. This is the basis for the treatment of contaminated soils by bioremediation, the use of microorganisms or microbial processes to detoxify and degrade environmental contaminants. Soil microbiologists study the microorganisms, the metabolic pathways, and the controlling environmental conditions that can be used to eliminate pollutants from the soil environment. *See* HAZARDOUS WASTE.

Microbiologists traditionally isolate pure strains of microorganisms by using culture methods. Methods that do not rely on culturing microorganisms include microscopic observation and biochemical or genetic analysis of specific cell constituents. The rates or controlling factors for microbial processes are studied by using methods from chemistry, biology, and ecology. Typically, these studies involve measuring the rate of production and consumption of a compound of interest. The results of these studies are commonly analyzed by using mathematical models. Models allow the information from one system to be generalized for different environmental conditions. *See* FLUORESCENCE MICROSCOPE; IMMUNOFLUORESCENCE; MICROBIOLOGY; MOLECULAR BIOLOGY; SOIL; SOIL CHEMISTRY; SOIL ECOLOGY. Jeanette M. Norton

Soil sterilization A chemical or physical process that results in the death of soil organisms. This control method affects many organisms, even though the elimination of only specific weeds, fungi, bacteria, viruses, nematodes, or pests is desirable. Even if complete sterilization is achieved, it is short lived since organisms will recolonize this biological vacuum quite rapidly. Soil sterilization can be achieved through both physical and chemical means. Physical control measures include steam and solar energy. Chemical control methods include herbicides and fumigants. Dielectric heating and gamma irradiation are used less frequently as soil sterilization methods. Composting can be used to sterilize organic materials mixed with soil, but it is not used for the sterilization of soil alone. Soil sterilization is used in greenhouse operations, the production of high-value or specialty crops, and the control of weeds. Carl A. Strausbaugh

Sol-gel process A chemical synthesis technique for preparing gels, glasses, and ceramic powders. The synthesis of materials by the sol-gel process generally involves the use of metal alkoxides, which undergo hydrolysis and condensation polymerization reactions to give gels.

The production of glasses by the sol-gel method is an area that has important scientific and technological implications. For example, the sol-gel approach permits preparation of glasses at far lower temperatures than is possible by using conventional melting. It also makes possible synthesis of compositions that are difficult to obtain by conventional means because of problems associated with volatilization, high melting temperatures, or crystallization. In addition, the sol-gel approach is a high-purity process that leads to excellent homogeneity. The sol-gel approach is adaptable to producing films and fibers as well as bulk pieces, that is, monoliths (solid materials of macroscopic dimensions, at least a few millimeters on a side). *See* GLASS.

The sol-gel process can ordinarily be divided into the following steps: forming a solution, gelation, drying, and densification. The preparation of a silica glass begins with an appropriate alkoxide such as tetraethylorthosilicate [$Si(OC_2H_5)_4$; TEOS], which is mixed with water (H_2O) and a mutual solvent, such as ethanol (C_2H_5OH), to form a solution. Hydrolysis leads to the formation of silanol groups (Si—OH). These species are only intermediates. Subsequent condensation reactions produce siloxane bonds (Si—O—Si) plus the by-product alcohol (ROH) or water.

The silica gel formed by this process leads to a rigid, interconnected three-dimensional network consisting of submicrometer pores and polymeric chains. During the drying process (at ambient pressure), the solvent liquid is removed and substantial shrinkage occurs. The resulting material is known as a xerogel. When solvent removal occurs under hypercritical (supercritical) conditions, the network does not shrink and a highly porous, low-density material known as an aerogel is produced. Heat treatment of a xerogel at elevated temperature produces viscous sintering (shrinkage of the xerogel due to a small amount of viscous flow) and effectively transforms the porous gel into a dense glass.

The sol-gel process is widely applicable to other inorganic oxides. By using various metal alkoxides $M(OR)_n$, where M is Al, Ti, V, Cr, Mo, W, Zr, and so on and OR is an alkoxy functionality as described previously, different oxides can be prepared. The resulting xerogels are either amorphous or nanocrystalline. Heat treatment of these materials leads to crystallization rather than the formation of a dense glass as occurs in the case of silica.

The sol-gel process offers advantages for a broad spectrum of materials applications. The types of materials go well beyond silica and include inorganic compositions that possess specific properties such as ferroelectricity, electrochromism, or superconductivity. The most successful applications utilize the composition control, microstructure control, purity, and uniformity of the method combined with the ability to form various shapes at low temperatures. Films and coatings were the first commercial applications of the sol-gel process. The development of sol-gel-based optical materials has also been quite successful, and applications include monoliths (lenses, prisms, lasers), fibers (waveguides), and a wide variety of optical films. One application of sol-gel materials that has emerged in recent years is that of bioactive glasses for biosensors.

Aerogels are a class of sol-gel materials that possess very low density and high surface area. Aerogels are mesoporous solids in which nanometer-scale solid domains are networked with a high volume of continuous pores. Aerogel materials have been widely used as catalysts, catalyst supports, and thermal insulation materials. In recent times, the applications of aerogels have expanded to electrochemical devices, low dielectric constant materials for integrated circuits, and the capture of cosmic dust particles. Bruce Dunn; Esther H. Lan

Solanales An order of flowering plants, division Magnoliophyta, in the euasterid I group of the asterid eudicotyledons. The order consists of two large and three small families, of approximately 4275 species. Solanaceae and Convolvulaceae account for all but 25 of the species. Solanales are generally characterized by sympetalous flowers with a superior ovary, and alternate leaves. *See* ASTERIDAE; FLOWER; FRUIT; LEAF; MAGNOLIOPHYTA; MAGNOLIOPSIDA.

Solanaceae, approximately 2600 species, are cosmopolitan herbs, shrubs, lianas, and trees, with branched hairs and often spines and alkaloids. The family is of great economic significance, yielding potatoes, tomatoes, and eggplant (*Solanum*), peppers (*Capsicum*), tobacco (*Nicotiana*), and many ornamentals. Deadly nightshade (*Atropa*), jimson weed (*Datura*), and henbane (*Hyoscyamus*) are well-known poisonous members of the family. *See* BELLADONNA; EGGPLANT; PEPPER; POTATO, IRISH; TOBACCO.

Convolvulaceae, approximately 1650 species, include herbaceous and woody members that are often climbers. The main economic crop is sweet potato (*Ipomoea*), but the family also includes ornamentals (morning glory, *Ipomoea*; and *Convolvulus*), noxious weeds (*Calystegia* and *Convolvulus*), and parasites (*Cuscuta*). *See* POTATO, SWEET; WEEDS. Michael F. Fay

Solar cell A semiconductor electrical junction device which directly and efficiently absorbs the radiant energy of sunlight and converts it into electrical energy. Solar cells may be used individually as light detectors, for example in cameras, or connected in series and parallel to obtain the required values of current and voltage for electric power generation.

Most solar cells are made from single-crystal silicon and have been very expensive for generating electricity, but have found application in space satellites and remote areas where low-cost conventional power sources have been unavailable.

The conversion of sunlight into electrical energy in a solar cell involves three major processes: absorption of the sunlight in the semiconductor material; generation and separation of free positive and negative charges to different regions of the solar cell, creating a voltage in the solar cell; and transfer of these separated charges through electrical terminals to the outside application in the form of electric current.

When light is absorbed in the semiconductor, a negatively charged electron and positively charged hole are created. The heart of the solar cell is the electrical junction which separates these electrons and holes from one another after they are created by the light. An electrical junction may be formed by the contact of: a metal to a semiconductor (this junction is called a Schottky barrier); a liquid to a semiconductor to form a photoelectrochemical cell; or two semiconductor regions (called a *pn* junction).

The fundamental principles of the electrical junction can be illustrated with the silicon *pn* junction. Pure silicon to which a trace amount of a group V element (in the periodic table) such as phosphorus has been added is an *n*-type semiconductor, where electric current is carried by free electrons. Each phosphorus atom contributes one free electron, leaving behind the phosphorus atom bound to the crystal structure with a unit positive charge. Similarly, pure silicon to which a trace amount of a group III element such as boron has been added is a *p*-type semiconductor, where the electric current is carried by free holes. The interface between the *p*- and *n*-type silicon is called the *pn* junction. The fixed charges at the interface due to the bound boron and phosphorus atoms create a permanent dipole charge layer with a high electric field. When photons of light energy from the Sun produce electron-hole pairs near the junction, the built-in electric field forces the holes to the *p* side and the electrons to the *n* side. This displacement of free charges results in a voltage difference between the two regions of the crystal. When a load is connected at the terminals, electron current flows and electrical power is available at the load. *See* SEMICONDUCTOR; SOLAR-ELECTRIC POWER GENERATION; SOLAR ENERGY. K. W. Mitchell

Solar constant The total solar radiant energy flux incident upon the top of the Earth's atmosphere at a standard distance (1 astronomical unit, 1.496×10^8 km or 9.3×10^7 mi) from the Sun. In 1980 it was discovered that the so-called solar constant actually varies with time, though only by small amounts, around a value of about 1366 $W \cdot m^{-2}$ (1.96 $cal \cdot cm^{-2} \cdot min^{-1}$).

Both expected and unexpected items contribute to the variability of the solar constant. Sunspots can produce deficits of up to a few tenths of 1% of the solar constant, on typical time scales of 1 week. Other surface manifestations of solar magnetic activity, faculae, contribute excesses rather than deficits. Convective motions and global oscillations of the solar interior, analogous to seismic waves on the Earth, produce variations of a few parts per million on time scales of a few hundred seconds. Finally, and unexpectedly, there is an apparent 11-year sunspot cycle variation amounting to an approximately 0.1% increase of the solar constant during the sunspot maxima. This long-term effect has the opposite dependence from that found for individual sunspots, which block the solar radiant energy and cause decreases rather than increases. *See* HELIOSEISMOLOGY; SUN. Hugh S. Hudson

Solar corona The outer atmosphere of the Sun. The corona is characterized by extremely high temperatures of several million kelvins, which cause it to extend far above the denser surface regions of the Sun. Coronal gas is constrained to follow the magnetic field of the Sun, forming it into the shapes seen during a solar eclipse, with a coronagraph, or in x-rays. These shapes include long streamers that penetrate interplanetary space, looplike tubes over the strongest fields, and vast regions of very low density called coronal holes. *See* CORONAGRAPH; ECLIPSE; SOLAR MAGNETIC FIELD; X-RAY ASTRONOMY.

Coronal holes are the source of solar matter streaming into interplanetary space. These streams often interact with the Earth's magnetic field to cause auroras and other geophysical phenomena. The interaction between the hot coronal plasma and the evolving magnetic fields in the corona sometimes leads to instabilities which eject large amounts of matter into space. These coronal mass ejections (CMEs) can have geophysical effects similar to those of the streams and ejecta. *See* AURORA; MAGNETOSPHERE; SOLAR WIND; SUN. Leon Golub

Solar energy The energy transmitted from the Sun. The upper atmosphere of Earth receives about 1.5×10^{21} watt-hours (thermal) of solar radiation annually. This vast amount of energy is more than 23,000 times that used by the human population of this planet, but it is only about one two-billionth of the Sun's massive outpouring—about 3.9×10^{20} MW. *See* SUN.

The power density of solar radiation measured just outside Earth's atmosphere and over the entire solar spectrum is called the solar constant. According to the World Meteorological Organization, the most reliable (1981) value for the solar constant is 1370 ± 6 W/m^2. *See* SOLAR CONSTANT.

Solar radiation is attenuated before reaching Earth's surface by an atmosphere that removes or alters part of the incident energy by reflection, scattering, and absorption. In particular, nearly all ultraviolet radiation and certain wavelengths in the infrared region are removed. However, the solar radiation striking Earth's surface each year is still more than 10,000 times the world's energy use. Radiation scattered by striking gas molecules, water vapor, or dust particles is known as diffuse radiation. Clouds are a particularly important scattering and reflecting agent, capable of reducing direct radiation by as much as 80 to 90%. The radiation arriving at the ground directly from the Sun is called direct or beam radiation. Global radiation is all solar radiation incident on the surface, including direct and diffuse. *See* SOLAR RADIATION.

Solar research and technology development aim at finding the most efficient ways of capturing low-density solar energy and developing systems to convert captured energy to useful purposes. Also of significant potential as power sources are the indirect forms of solar energy: wind, biomass, hydropower, and the tropical ocean surfaces. With the exception of hydropower, these energy resources remain largely untapped. *See* ENERGY SOURCES.

Five major technologies using solar energy are being developed. (1) The heat content of solar radiation is used to provide moderate-temperature heat for space comfort conditioning of buildings, moderate- and high-temperature heat for industrial processes, and high-temperature heat for generating electricity. (2) Photovoltaics convert solar energy directly into electricity. (3) Biomass technologies exploit the chemical energy produced through photosynthesis (a reaction energized by solar radiation) to produce energy-rich fuels and chemicals and to provide direct heat for many uses. (4) Wind energy systems generate mechanical energy, primarily for conversion to electric power. (5) Finally, a number of ocean energy applications are being pursued; the most advanced is ocean thermal energy conversion, which uses temperature differences between warm ocean surface water and cooler deep water to produce electricity. *See* BIOMASS; PHOTOVOLTAIC CELL; SOLAR HEATING AND COOLING; WIND.

Solar energy can be converted to useful work or heat by using a collector to absorb solar radiation, allowing much of the Sun's radiant energy to be converted to heat. This heat can be used directly in residential, industrial, and agricultural operations; converted to mechanical or electrical power; or applied in chemical reactions for production of fuels and chemicals.

A solar energy system is normally designed to be able to deliver useful heat for 6 to 10 h a day, depending on the season and weather. Storage capacity in the solar thermal system is one way to increase a plant's operating capacity.

There are four primary ways to store solar thermal energy: (1) sensible-heat-storage systems, which store thermal energy in materials with good heat-retention qualities; (2) latent-heat-storage systems, which store solar thermal energy in the latent heat of fusion or vaporization of certain materials undergoing a change of phase; (3) chemical energy storage, which uses reversible reactions (for example, the dissociation-association reaction of sulfuric acid and water); and (4) electrical or mechanical storage, particularly through the use of storage batteries (electrical) or compressed air (mechanical). *See* ENERGY STORAGE.

Photovoltaic systems convert light energy directly to electrical energy. Using one of the most versatile solar technologies, photovoltaic systems can, because of their modularity, be designed for power needs ranging from milliwatts to megawatts. They can be used to provide power for applications as small as a wristwatch to as large as an entire community. They can be used in centralized systems, such as a generator in a power plant, or in dispersed applications, such as in remote areas not readily accessible to utility grid lines.

Biomass energy is solar energy stored in plant and animal matter. Through photosynthesis in plants, energy from the Sun transforms simple elements from air, water, and soil into complex carbohydrates. These carbohydrates can be used directly as fuel (for example, burning wood) or processed into liquids and gases (for example, ethanol or methane). Biomass is a renewable energy resource because it can be harvested periodically and converted to fuel. *See* CARBOHYDRATE; PHOTOSYNTHESIS.

Wind is a source of energy derived primarily from unequal heating of Earth's surface by the Sun. Energy from the wind has been used for centuries to propel ships, to grind grain, and to lift water. Wind turbines extract energy from the wind to perform mechanical work or to generate electricity.

Ocean thermal energy conversion uses the temperature difference between surface water heated by the Sun and deep cold water pumped from depths of 2000 to 3000 ft (600 to 900 m). This temperature difference makes it possible to produce electricity from the heat engine concept. Since the ocean acts as an enormous solar energy storage facility with little fluctuation of temperature over time, ocean thermal energy conversion, unlike most other renewable energy technologies, can provide electricity 24 h a day.

Robert L. San Martin

Solar heating and cooling The use of solar energy to produce heating or cooling for technological purposes. Beneficial uses include distillation of sea water to produce salt or potable water; heating of swimming pools; space heating; heating of water for domestic, commercial, and industrial purposes; cooling by absorption or compression refrigeration; and cooking. *See* SOLAR ENERGY.

Distillation. Production of potable water from sea water by solar distillation is accomplished in several parts of the world by use of glass-roofed solar stills (see illustration). Production of salt from the sea has been accomplished for hundreds of years by trapping ocean water in shallow ponds at high tide and simply allowing the water to evaporate under the influence of the Sun. *See* DISTILLATION.

Swimming pool heating. Swimming pool heating is a moderate-temperature application which, under suitable weather conditions, can be accomplished with a simple unglazed and uninsulated collector. For applications where a significant temperature difference exists between the fluid within the collector passages and the ambient air, both glazing and insulation are essential.

Space heating. Space heating can be carried out by active systems which use separate collection, distribution, and storage subsystems, or by passive designs which use components of a building to admit, store, and distribute the heat resulting from absorbing the incoming solar radiation within the building itself.

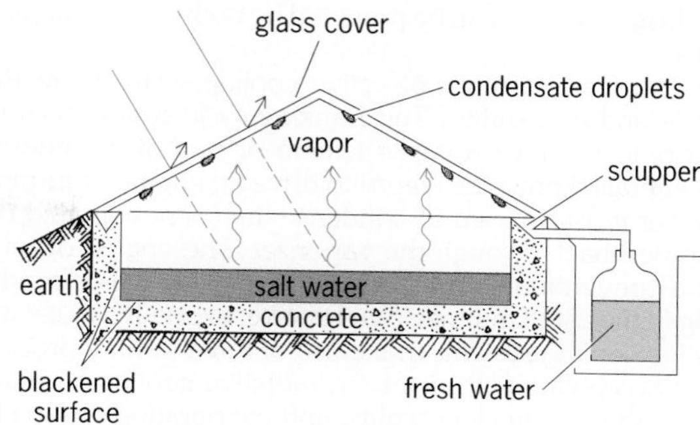

Roof-type solar still.

Passive systems can be classified as direct-gain when they admit solar radiant energy directly into the structure through large south-facing windows, or as indirect-gain when a wall or a roof absorbs the solar radiation, stores the resulting heat, and then transfers it into the building. Passive systems are generally effective where the number of hours of sunshine during the winter months is relatively high, where moderate indoor temperature fluctuations can be tolerated, and where the need for summer cooling and dehumidification is moderate or non- existent.

Active systems may use either water or air to transport heat from roof-mounted south-facing collectors to storage in rock beds or water tanks. The stored heat may be withdrawn and used directly when air is the transfer fluid. When the heat is collected and stored as hot water, fan-coil units are generally used to transfer the heat to air which is then circulated through the warmed space. Standby energy sources are included in designs for active systems, since some method of providing warmth must be included for use when the Sun's radiant energy is inadequate for long periods of time. The standby heater may be something as simple as a wood-burning stove or fireplace, or as complex as an electrically powered heat pump. *See* COMFORT HEATING; HEAT PUMP.

Service water heating. Solar water heating for domestic, commercial, or industrial purposes is an old and successful application of solar-thermal technology. The most widely used water heater, and one that is suitable for use in relatively warm climates where freezing is a minor problem, is the thermosiphon type. A flat-plate collector is generally used with a storage tank which is mounted above the collector. A source of water is connected near the bottom of the tank, and the hot water outlet is connected to its top. A downcomer pipe leads from the bottom of the tank to the inlet of the collector, and an insulated return line runs from the top of the collector to the upper part of the storage tank which is also insulated.

The system is filled with water, and when the Sun shines on the collector, the water in the tubes is heated. It then becomes less dense than the water in the downcomer, and the heated water rises by thermosiphon action into the storage tank. It is replaced by cool water from the bottom of the tank, and this action continues as long as the Sun shines on the collector with adequate intensity.

For applications where the elevated storage tank is undesirable or where very large quantities of hot water are needed, the tank is placed at ground level. A small pump circulates the water in response to a signal from a controller which senses the temperatures of the collector and the water near the bottom of the tank. Heat exchangers may also be used with water at operating pressure within the tubes of the exchanger and the collector water outside to eliminate the necessity of using high-pressure collectors. *See* HOT-WATER HEATING SYSTEM.

Cooling. Cooling can be provided by both active and passive systems.

The two feasible types of active cooling systems are Rankine cycle and absorption. The Rankine cycle system uses solar collectors to produce a vapor (steam or one of the fluorocarbons generally known as Freon) to drive an engine or turbine. A condenser must be used to condense the spent vapor so it can be pumped back through the vaporizer. The engine or turbine drives a conventional refrigeration compressor which produces cooling in the usual manner. *See* RANKINE CYCLE; REFRIGERATION.

Passive cooling systems make use of three natural processes: convection cooling with night air; radiative cooling by heat rejection to the sky on clear nights; and evaporative cooling from water surfaces exposed to the atmosphere. The effectiveness of each of these processes depends upon local climatic conditions. *See* ENERGY STORAGE; SOLAR CELL. John I. Yellott

Solar magnetic field The magnetic field rooted in the Sun and extending out past the planets into the solar system. The field at the Sun's surface is detected remotely by its effect (the Zeeman effect) on atoms whose radiation is observed from Earth. This technique was first applied in 1908 by G. E. Hale to detect the fields in sunspots. In 1952, H. D. Babcock and his son H. W. Babcock used a scanning technique to make the first magnetic maps of the entire visible disk of the Sun. Their daily "magnetograms" soon revealed a variety of magnetic features, including bipolar regions associated with sunspot groups, and unipolar regions whose fields extend far from the Sun and are responsible for recurrent geomagnetic activity at Earth. The field strengths range from a few gauss (a few hundred microtesla) in quiet areas to 3500 gauss (0.35 tesla) in sunspots. In 1962, the *Mariner 2* spacecraft, en route to Venus, made the first on-site sampling of the extended solar field in space. The average field strength was only 50 microgauss (5 nanotesla), reflecting the rapid (inverse square) fall-off of field strength with distance from the Sun. *See* ZEEMAN EFFECT.

Solar magnetic fields are related to the 11-year variation in the occurrence of sunspots. As the new sunspot cycle begins, concentrations of bipolar flux break through the Sun's surface in each hemisphere, beginning at about 40° latitude and gradually progressing toward the solar equator over the next few years. The bipolar regions are oriented approximately east-west with their leading polarities (in the sense of solar rotation) all positive in the northern hemisphere and negative in the southern hemisphere during a given sunspot cycle (Hale's law). They are tilted slightly so that their trailing polarities are closer to the Sun's poles and the leading polarities are closer to the equator. This small but systematic effect ultimately leads to the formation of unipolar regions at the Sun's poles, positive in one hemisphere and negative in the other.

It is presently believed that convection and differential rotation are responsible for the eruption of bipolar magnetic regions. *See* MAGNETISM; SUN. Neil R. Sheeley, Jr.

Solar neutrinos Neutrinos produced in nuclear reactions inside the Sun. Neutrinos are produced as well in laboratory nuclear reactions. The first direct tests of how the Sun produces its luminosity (observed most conspicuously on Earth as sunlight) have been carried out by observing solar neutrinos. The results of these experiments confirm the theory of how the Sun shines and stars evolve. Moreover, the results show that neutrinos behave differently than predicted by the standard model of particle physics.

Many explanations have been advanced for the discrepancy between the observed and the predicted event rates in the solar neutrino experiments. These explanations can be divided into three general classes: (1) the standard solar model must be significantly modified; (2) something is seriously wrong with the experiments; (3) the standard model of how neutrinos behave must be significantly modified.

Precise measurements of the thousands of frequencies with which the Sun pulsates on its surface (with characteristic periods of the order of 5 minutes) have confirmed to an accuracy of 0.1% the predictions of the standard solar model for these pulsation frequencies. This agreement is convincing evidence that the standard solar model is an accurate description of the Sun. *See* HELIOSEISMOLOGY.

All of the solar neutrino experiments have been examined carefully by many different searchers. A variety of checks have been made to test whether there was a significant error or a large uncertainty in one of the experiments that might explain the difference between prediction and observation. No significant previously unknown errors or uncertainties have been found. Moreover, intense laboratory sources of neutrinos have been placed near the gallium neutrino detectors, and the expected number of events have been observed from these artificial sources. The consensus view among scientists in the field is that the solar neutrino experiments are yielding a valid but surprising result.

The only remaining possibility is that the theory of how the neutrino behaves must be changed. Indeed, in 2000, the results of a decisive experiment showed unequivocally that solar neutrinos change their type on their way from the center of the Sun. All of the results from solar neutrino experiments are consistent with the conclusion that the standard solar model predicts accurately the number of neutrinos of different energies that are emitted by the Sun but that some of the neutrinos change their type on the way from the center of the Sun to the detectors on Earth. *See* NEUTRINO; STANDARD MODEL; SUN. John N. Bahcall

Solar radiation The electromagnetic radiation and particles (electrons, protons, alpha particles, and rarer heavy atomic nuclei) emitted by the Sun. The electromagnetic radiation covers a wavelength range from x-rays to radio waves, that is, from about 0.01 nanometer to 30 km. The annual mean irradiance at Earth, integrated over the whole spectrum, amounts to 1365 $W \cdot m^{-2}$, and 99% of its energy is carried by radiation with wavelengths between 278 and 4600 nm, with the maximum at 472 nm. *See* ELECTROMAGNETIC RADIATION; SOLAR CONSTANT.

The Sun also emits a continuous stream of particles, the solar wind, which originates in coronal holes and the upper corona. Explosive events on the Sun, the solar flares and coronal mass ejections, emit particles that are much more energetic and numerous than those of the solar wind. Solar flares are produced by the most powerful explosions, releasing energies of up to 10^{25} joules in 100–1000 s and high-speed electrons that emit intense radiation at radio and x-ray wavelengths. They also produce nuclear reactions in the solar atmosphere with the emission of gamma rays and of neutrons that move nearly at the speed of light. Coronal mass ejections expand away from the Sun at speeds of hundreds of kilometers per second, becoming larger than the Sun and removing up to 5×10^{13} kg of coronal material. Both events are believed to be ignited by the reconnection of magnetic fields. If the emitted particles reach the Earth, they give rise to the aurora at high latitudes, and they can damage satellites, endanger humans in space, and on the Earth disturb telecommunications and even disrupt power systems. *See* AURORA; COSMIC RAYS; IONOSPHERE; MAGNETOSPHERE; SOLAR WIND; SUN. Claus Fröhlich

Solar system The Sun and the bodies moving in orbit around it. The most massive body in the solar system is the Sun, a typical single star that is itself in orbit about the center of the Milky Way Galaxy. Nearly all of the other bodies in the solar system—the terrestrial planets, outer planets, asteroids, and comets—revolve on orbits about the Sun. Various types of satellites revolve around the planets; in addition, the giant planets all have orbiting rings. The orbits for the planets appear to be fairly stable over long time periods and hence have undergone little change since the formation of the solar system. It is thought that some 4.56×10^9 years ago a rotating cloud of gas and dust collapsed to form a flattened disk (the solar nebula) in which the Sun and other bodies formed. The bulk of the gas in the solar nebula moved inward to form the Sun, while the remaining gas and dust are thought to have formed all the other solar system bodies by accumulation proceeding through collisions of intermediate-sized bodies called planetesimals. Planetary systems are believed to exist around many other stars in the Milky Way Galaxy. Solid evidence for the existence of Jupiter-mass planets around nearby solarlike stars now exists. *See* EXTRASOLAR PLANETS; PLANET.

Composition. The Sun is a gaseous sphere with a radius of about 7×10^5 km (4×10^5 mi), composed primarily of hydrogen and helium and small amounts of the other elements. The terrestrial planets (Mercury, Venus, Earth, and Mars) are the closest to the Sun. They are composed primarily of silicate rock (mantles) and iron (cores). The Earth is the largest terrestrial planet; Mercury is the smallest, with a mass of 0.053 times that of Earth. *See* EARTH; MARS; MERCURY (PLANET); PLANETARY PHYSICS; SUN; VENUS.

The outer planets are subdivided into the gas giant or Jovian planets (Jupiter and Saturn), the ice giant planets (Uranus and Neptune), and Pluto. By far the largest planet is Jupiter, with a mass 318 times that of the Earth, while the other giant planets are more massive by a factor of 15 or more than Earth. Jupiter and Saturn are composed primarily of hydrogen and helium gas, like the Sun, but with rock and ices, such as frozen water, methane, and ammonia, concentrated in their cores. Uranus and Neptune also have rock and ice cores surrounded by envelopes with smaller amounts of hydrogen and helium. Pluto, slightly smaller than the Earth's Moon, is probably composed primarily of rock and ice; it has now been reclassified as a dwarf planet. *See* JUPITER; NEPTUNE; PLUTO; SATURN; URANUS.

The region between Mars and Jupiter is populated by a large number of rocky bodies called asteroids. The asteroids are smaller than the terrestrial planets. with most known asteroids being about 1 km (0.6 mi) in radius, though a few have radii of hundreds of kilometers. Some asteroids have orbits that take them within the orbits of Earth and the other terrestrial planets. Small fragments of asteroids (or comets) that impact the Earth first appear as meteors in the sky; any meteoric material that survives the passage through the Earth's atmosphere and reaches the surface is called a meteorite. *See* ASTEROID; METEOR; METEORITE.

Comets are icy bodies (so-called dirty snowballs) with diameters on the order of 10 km (6 mi). In contrast to the orbits of most planets, cometary orbits often are highly elliptical and have large inclinations that take them far from the plane where the planets orbit. The region well beyond Pluto's orbit is populated with a very large number (perhaps 10^{12}) of comets, out to a limiting distance of about 10^5 AU. The distribution of comets within this huge volume, the Oort Cloud, is uncertain. Comets have been detected orbiting in the plane of the solar system at distances of 30 to 55 AU; this flattened distribution is called the Edgeworth-Kuiper Belt. Pluto itself is now thought to be one of the largest members of this population. *See* COMET; KUIPER BELT.

Origin. The nebular hypothesis, advanced in 1796 by P. S. de Laplace, holds that the Sun and the rest of the bodies in the solar system formed from the same rotating, flattened cloud of gas and dust, now called the solar nebula. The nebular hypothesis explains the gross orbital properties of the solar system: all planets orbit (and most rotate) in the same sense as the Sun rotates, with their nearly circular orbits being confined largely to a single plane almost perpendicular to the Sun's rotation axis.

Observations of present-day regions of star formation in the Milky Way Galaxy confirm the stellar implications of Laplace's nebular hypothesis: very young stars (protostars) are indeed found embedded in dense clouds of gas and dust that often show evidence for flattening and rotation.

The solar nebula was produced by the collapse of a dense interstellar cloud. Radio telescopes have shown that such clouds exist with masses comparable to that of the Sun. Eventually they enter the collapse phase, where supersonic inward motions develop that lead to the formation of a stellar-sized core at the center of the cloud in about 10^5–10^6 years. *See* INTERSTELLAR MATTER; MOLECULAR CLOUD; RADIO ASTRONOMY; STELLAR EVOLUTION.

In a rotating cloud, not all of the in-falling gas and dust falls directly onto the central protostar, because of the conservation of angular momentum. Instead, a disklike solar nebula forms. The disk must evolve in such a way as to transfer mass inward to feed the growing Sun, while transporting outward the excess angular momentum undesired by the Sun but required for the planets. While this sort of evolution may appear to be contrived if not miraculous, it is actually to be expected on very general grounds for any viscous disk that is undergoing a loss of energy, as the solar nebula will, through radiation to space.

The portion of the nebula that is to form the planets must decouple from the gaseous nebula to avoid being swallowed by the Sun. This occurs by the process of coagulation of dust grains through mutual collisions; when solid bodies become large enough (roughly kilometer-sized), they will no longer be tied to the nebula through brownian motion (as is the case with dust grains) or gas drag (as happens with smaller bodies).

About 10^{12} kilometer-sized planetesimals are needed to form just the terrestrial planets; significantly greater numbers of similarly sized bodies would be needed to form the giant planets. These planetesimals are already roughly the size of many asteroids and comets, suggesting that many of these bodies are simply leftovers from intermediate phases of the planet formation process.

The subsequent growth of the planetesimals through gravitational accumulation is in two distinct phases. In the first phase, planetesimals grew by accumulation of other planetesimals at essentially the same distance from the Sun. Once the nearby planetesimals were all swept up, this phase ended.

In the second phase, accumulation requires bodies at significantly different distances from the Sun to collide. This phase may have involved violent collisions between planetary-sized bodies. A glancing collision between a Mars-sized and an Earth-sized body appears to be the best explanation for the formation of the Earth-Moon system; debris from the giant impact would end up in orbit around the Earth and later form the Moon.

Forming gas giant planets by the two-step process requires about 10^7 years for a 10-Earth-mass core to form and then accrete a massive gaseous envelope. The alternative means for forming the gas giant planets is much more rapid, requiring only about 10^3 years for a gravitational instability of the gaseous nebula to produce a massive clump of gas and dust. The formation of ice giant planets could occur in the disk instability mechanism if the solar nebula was born in a region of high-mass stars. Their radiation would photoevaporate the disk gas beyond Saturn's orbit as well as the gaseous envelopes of the protoplanets orbiting there.

Alan P. Boss

Solar wind A flow of completely ionized material away from the Sun. As this plasma moves through space, it creates a bubble known as the heliosphere. At heights of a few solar radii above the Sun's visible surface, the flow becomes supersonic, reaching speeds of 350–750 km/s (220–470 mi/s) in interplanetary space. The solar wind finally terminates at a distance of perhaps 150 astronomical units (1 AU is the distance from the Sun to Earth), where it interacts with the local interstellar medium.

This flow can be conveniently categorized as either ambient or time-dependent. The combination of solar rotation together with the ejection of material of different speeds from the Sun leads to a dynamical interaction between neighboring parcels of plasma and the formation of interaction regions and shocks that bound them. This forms the basis of the ambient solar wind. These structures appear stationary in a frame of reference rotating with the Sun and are thus known as corotating interaction regions (CIRs). Coronal mass ejections (CMEs) represent the most spectacular form of time-dependent phenomena.

During the course of the 11-year solar cycle, the structure of the solar wind changes at a fundamental level. At the minimum of the solar activity cycle a simple flow pattern exists consisting of (1) a slow wind about the magnetic equator, and (2) a fast wind emanating from large polar coronal holes. In contrast, at solar maximum the flow is considerably more complex and time-dependent. The rate, launch site, and properties of coronal mass ejections also vary with the phase of the solar cycle. The interaction of the solar wind with the Earth's geomagnetic environment represents an important and practical application of space physics known as space weather, which seeks to describe the conditions in space that affect both the Earth and its technological systems.

The first clue that the Sun was emitting a continuous wind can be traced back to observations of comet tails, which were found to always point away from the Sun no matter whether the comet was moving toward or away from the Sun. In 1957 L. Biermann proposed the presence of a continuous stream of particles propagating away from the Sun that pushed the ions making up the ion tail of a comet. The following year, E. N. Parker derived a theory for the solar corona that required there to be a flow of plasma away from it. He named this flow the solar wind. In 1959, the Soviet spacecraft *Lunik 2* made the first direct measurements of the particles making up the solar wind. *See* COMET; SPACE PROBE.

Coronal heating and solar wind acceleration. The visible surface of the Sun is known as the photosphere. Beyond this lies the chromosphere, and above this lies the corona.

The temperature of the photosphere is a modest 10,000 K, while the solar corona is more than 10^6 K. How can the corona be hotter than the photosphere? An explanation for coronal heating comes from a process known as magnetic reconnection. Coronal heating is strongest in regions of intense magnetic fields that are constantly merging with one another and liberating large amounts of energy. Fountains of plasma lead to the formation of giant loops hundreds of thousands of kilometers above the surface of the Sun. The plasma on these loops can reach temperatures of several million kelvin and may be the energy source that heats the solar corona. The corona extends many solar radii away from the Sun, eventually becoming the solar wind. *See* SOLAR CORONA.

Ambient solar wind. Two types of ambient solar wind can be distinguished: slow and fast. The slow flow, which is associated with coronal streamers, travels away from the Sun at approximately 350 km/s (220 mi/s). It is denser and colder than the fast wind. The fast wind travels at more than twice the speed of the slow wind. It is also considerably steadier and emanates from coronal holes. The helium-to-proton abundance acts as a unique tracer of the origin of the solar wind in which the spacecraft is immersed. Instruments on board spacecraft now make it possible to identify the source regions of the solar wind in the corona.

Interplanetary magnetic field. The electrical conductivity of the solar wind plasma is sufficiently high that the magnetic field is effectively frozen to it. Thus, as the solar wind propagates away from the Sun, it drags the solar field with it. Because the field lines remain attached to the Sun at one end, solar rotation causes them to become wrapped into spirals. *See* MAGNETOHYDRODYNAMICS; PLASMA (PHYSICS).

Dynamic processes. As material flows away from the Sun, the combined effects of solar rotation together with different intrinsic speeds of the flow lead to slower material being caught by faster material, and faster material outrunning slower material. In the former case, the fast material plows into the slower material compressing and heating it. This compressed region is known as an interaction region (IR), and when the flow pattern back at the Sun does not change much with time, it is further known as a corotating interaction region.

Time-dependent effects. Coronal mass ejections are spectacular events that involve the expulsion of massive amounts of solar material into the heliosphere. To initiate such an eruption requires a large amount of energy. It is thought that this energy is supplied by the magnetic field in the corona. Coronal mass ejections contribute a small, but significant amount to the overall solar wind flow at low latitudes. Fast coronal mass ejections have been identified as the leading cause of strong geomagnetic storms.

Space weather. The field of space weather is a relatively new discipline that aims to understand the conditions on the Sun, solar wind, magnetosphere, ionosphere, and thermosphere that can (1) affect the performance and reliability of space-borne and ground-based technological systems, and (2) endanger human life or health. An important component of space weather is the interaction of the solar wind with the Earth's magnetosphere. The boundary between these two regions is known as the magnetopause, which provides a barrier to the flow of the solar wind. When solar activity increases, such as in a fast coronal mass ejection, the system is sufficiently perturbed, leading to a geomagnetic storm. *See* GEOMAGNETIC VARIATIONS; MAGNETOSPHERE; SPACE BIOLOGY; SUN.

Pete Riley

Soldering A low-temperature metallurgical joining method in which the solder (joining material) has a much lower melting point than the surfaces to be joined (substrates). Because of its lower melting point, solder can be melted and brought into contact with the substrates without melting them. During the soldering process, molten solder wets the substrate surfaces (spreads over them) and solidifies on cooling to form a solid joint.

The most important technological applications of solders are in the assembly of electronic devices, where they are used to make metallic joints between conducting wires, films, or contacts. They are also used for the routine low-temperature joining of copper plumbing fixtures and other devices. In addition, solder is used in the fusible joints of fire safety devices and other high-temperature detectors; the solder joint liquefies if the ambient temperature exceeds the solder's melting point, releasing a sprinkler head or triggering some other protective operation.

Tin or indium content is included in solder to facilitate bonding to the metals that are most commonly soldered, such as copper (Cu), nickel (Ni), and gold (Au). Tin and indium form stable intermetallic compounds with copper and nickel, and indium also forms intermetallics with gold. The intermetallic reaction at the solder-substrate interface creates a strong, stable bond. *See* ALLOY; INTERMETALLIC COMPOUNDS.

J. W. Morris, Jr.

Solenodon A rare insectivorous mammal classified in the family Solenodontidae. Solenodons are among the largest living insectivores. There are only two extant species. The Hispaniolan, or Haitian, solenodon (*Solenodon paradoxus*) is restricted to the remote, wet, densely vegetated central highland regions on the island of Hispaniola (Dominican Republic and Haiti). The coarse pelage (coat) varies from blackish to reddish brown. Most individuals possess a small, square, whitish area on the nape of the neck. Adults are 28–33 cm (11–13 in.) in body length, have a 17.5–26-cm-long (7–10.4-in.) tail, and weigh 600–1000 g (1.3–2.2 lb). The smaller Cuban solenodon, or almique (*S. cubanus*), has longer and finer fur which is blackish brown with white or buff. It is now restricted to Oriente Province in Cuba. Adults are 20–30 cm (10–12 in.) in body length, have a 26–30-cm-long (10.5–12 in.) tail, and weigh 600–700 g (1.3–1.5 lb). Two additional species are known only from skeletal remains in Haiti and Cuba.

Solenodons resemble shrews in general appearance except that they are much larger and more stoutly built. They have relatively large heads, tiny eyes, and large, partially naked ears. The head tapers to a long, flexible, and mobile proboscis that extends well beyond the lower jaw. The senses of touch, smell, and hearing are highly developed. All limbs possess five toes, with the toes on the forelimbs possessing long, stout, sharp claws. Solenodons are nocturnal and inhabit forests and brushy edges. Food consists of a variety of invertebrates, including millipedes, ground beetles and other insects, insect larvae, earthworms, and snails, as well as amphibians and small reptiles and birds. *See* INSECTIVORA. Donald W. Linzey

Solenoid (electricity) An electrically energized coil of insulated wire which produces a magnetic field within the coil. If the magnetic field produced by the coil is used to magnetize and thus attract a plunger or armature to a position within the coil, the device may be considered to be a special form of electromagnet and in this sense the words solenoid and electromagnet are synonymous. In a wider scientific sense the solenoid may be used to produce a uniform magnetic field for various investigations. So long as the length of the coil is much greater than its diameter (20 or more times), the magnetic field at the center of the coil is sensibly uniform, and the field intensity is almost exactly that given by the equation for a solenoid of infinite length.

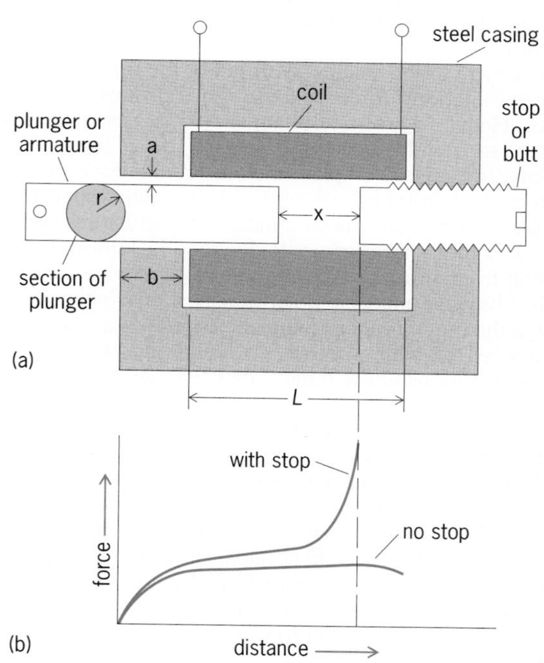

Steel-clad solenoid. (*a*) Cross-sectional view. (*b*) Relation of the force acting on the armature to the displacement of the armature.

When used as an electromagnet of the plunger type, the solenoid usually has an iron or steel casing. The casing increases the mechanical force on the plunger and also serves to constrain the magnetic field. The addition of a butt or stop at one end of the solenoid greatly increases the force on the plunger when the distance between the plunger and the stop is small. The illustration shows a steel-clad solenoid with plunger and plunger stop. The relation of force versus distance with and without the stop is also shown. *See* ELECTROMAGNET. Jerome Meisel

Solenoid (meteorology) In meteorological usage, solenoids are hypothetical tubes formed in space by the intersection of a set of surfaces of constant pressure (isobaric surfaces) and a set of surfaces of constant specific volume of air (isosteric surfaces) or density (isopycnic surfaces). The isobaric and isosteric surfaces are such that the values of pressure and specific volume, respectively, change by one unit from one surface to the next. The state of the atmosphere is said to be barotropic when there are no solenoids, that is, when isobaric and isosteric surfaces coincide. The number of solenoids cut by any plane surface element of unit area is a measure of the torque exerted by the pressure gradient force, tending to accelerate the circulation of air around the boundary of the area. *See* BAROCLINIC FIELD; BAROTROPIC FIELD; ISOPYCNIC. Frederick Sanders; Howard B. Bluestein

Solenopora A genus of extinct calcareous red algae (Rhodophyta). *Solenopora* is characterized by encrusting, rounded, nodular growth forms ranging in size from several millimeters to a few centimeters. Internally, this alga consists of calcified rows or filaments of cells, commonly polygonal in shape, ranging from 20 to 100 micrometers in diameter. Solenoporids probably provided an ancestral stock to the true coralline algae.

Solenopora appeared first in Cambrian time and became extinct in the early part of the Cenozoic. The genus flourished during the Ordovician, Silurian, and Jurassic periods. The environmental distribution of *Solenopora* was comparable to some modern coralline algae. *See* ALGAE. John L. Wray

Solid (geometry) A three-dimensional geometric figure consisting of points continuously connected, and separated from the rest of space by a surface called the boundary of the solid. Points on the boundary are usually considered part of the solid. A solid is bounded if there exists a sphere having finite radius that could enclose the solid. A solid is convex if all points of any line segment having end points on the boundary also are points of the solid. Harry L. Baldwin, Jr.

Solid solution Compositional variation of a crystalline substance due to substitution or omission of various atomic constituents within a crystal structure. Solid solutions can be classified as substitutional, interstitial, or omissional. They may also be categorized by the nature of their thermodynamic properties, such as enthalpy, entropy, and free energy (for example ideal, nonideal, and regular solid solutions).

The concept of solid solution can be understood by considering a specific mineral group such as olivine. Although olivine-group minerals may exhibit a range of compositions they all have a similar crystal structure. Thus all of the minerals in the olivine group are isostructural (having a similar crystal structure). Yet within this structural framework, compositions vary considerably. Such chemical variation can be described by defining the nature and extent of the atomic substitutions involved or by describing intermediate compositions in terms of limiting end-member compositions. In addition to substitution within the crystal structure, compositional variation may take place by interstitial substitution or omission solid solution. Interstitial solid solution occurs when ions or atoms occupy a position in a crystal structure that is usually vacant. Omission solid solution occurs when an atom or ion is missing from a specific crystallographic position. *See* OLIVINE.

Solid solution is widespread among minerals. In fact, very few naturally occurring minerals exist as pure end-member substances but exhibit trace to extensive solid solution. The extent of solid solution depends upon the relative sizes of the atoms or ions involved, the charges of the ions, the coexistence and composition of other minerals or liquid (for example, magma), and the temperature and pressure conditions of formation (with temperature having a more pronounced effect). John C. Drake

Solid-state battery A battery in which both electrodes and the electrolyte are solids (see illustration). Solid electrolytes are a class of materials also known as superionic conductors and fast ion conductors, and their study belongs to an area of science known as solid-state ionics. As a group, these materials are very good conductors of ions but are essentially insulating toward electrons, properties that are prerequisites for any electrolyte. The high ionic conductivity minimizes the internal resistance of the battery, thus permitting high power densities, while the high electronic resistance minimizes its self-discharge rate, thus enhancing its shelf life. Examples of such materials include Ag_4RbI_5 for Ag^+ conduction, LiI/Al_2O_3 mixtures for Li^+ conduction, and the clay and β-alumina group of compounds ($NaAl_{11}O_{17}$) for Na^+ and other mono- and divalent ions. At room temperature the ionic conductivity of a single crystal of sodium β-alumina is 0.035 S/cm, comparable to the conductivity of a 0.1 *M* HCl solution. This conductivity, however, is reduced in a battery by a factor of 2–5, because of the use of powdered or ceramic material rather than single crystals. Of much interest are glassy and polymeric materials that can be readily made in thin-film form, thus enhancing the rate capability of the overall system. *See* ELECTROLYTE; ELECTROLYTIC CONDUCTANCE; IONIC CRYSTALS.

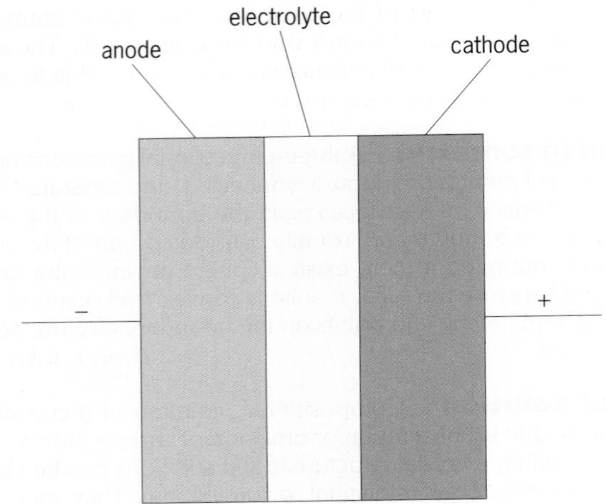

Schematic diagram of solid-state battery.

Solid-state batteries generally fall into the low-power-density and high-energy-density category. The former limitation arises because of the difficulty of getting high currents across solid–solid interfaces. However, these batteries do have certain advantages that outweigh this disadvantage: They are easy to miniaturize (for example, they can be constructed in thin-film form), and there is no problem with electrolyte leakage. They tend to have very long shelf lives, and usually do not have any abrupt changes in performance with temperature, such as might be associated with electrolyte freezing or boiling. Being low-power devices, they are also inherently safer. The major applications of these batteries are in electronic devices such as cardiac pacemakers, cameras, electrochromic displays, watches, and calculators. *See* BATTERY.

M. Stanley Whittingham

Solid-state chemistry The science of the elementary, atomic compositions of solids and the transformations that occur in and between solids and between solids and other phases to produce solids. Solid-state chemistry deals primarily with those microscopic features which are uniquely characteristic of solids and which are the causes for the macroscopic chemical properties and the chemical reactions of solids. As with other branches of the physical sciences, solid-state chemistry also includes related areas that furnish concepts and explanations of those phenomena which are more characteristic of the subject itself.

The overlap of solid-state chemistry and solid-state physics is extensive. However, the perspectives of the two are different. In general, solid-state physics treats properties, such as energy and entropy, which are continuously variable in the solid, whereas solid-state chemistry concerns those properties which are discontinuous because of chemical reactions. Also, solid-state chemistry tends to be based on structure in configuration space, whereas solid-state physics tends to be based on momentum space. *See* INORGANIC CHEMISTRY; SOLID-STATE PHYSICS.

Solid-state chemistry has no single unifying theoretical base and tends to be largely an experimental science supported by several theoretical bases. Consequently, its separation into topics is not well established. Aspects of solid-state chemistry include chemical bonding, crystal defects, crystal structures, crystal field theory, diffusion in solids, ionic crystals, lattice vibrations, and nonstoichiometry. *See* NONSTOICHIOMETRIC COMPOUNDS.

Studies of structures provide a basis for understanding the chemical bonding in solids, their properties, reactions among them, and their variabilities of composition. Numerous binary compounds or phases have layered structures. The layers are displaced relative to each other in such a way that the structures can be classified in the space groups, but when other elements or compounds are incorporated between the layers, the structure is highly distorted. The most widely recognized material to possess the layered structure and the associated solid-state chemistry is graphite. Structural information is represented by the location of atoms on the lattice network. The structures are determined by diffraction of x-rays, neutrons, or electrons. Two other features of the structure of solids are the local structure about a given atom and the extended structure on a more global scale. *See* GRAPHITE.

The structure of a solid is the result of the operation of interatomic or interionic forces and the size and shape of the atoms or ions. Hence, logically, bonding should be described first and structure second. However, the detailed role of the electrons in interionic forces is so complex, and the quantitative aspects of the problem of minimizing the potential energy with respect to all the possible configurations is so difficult, that structures cannot be derived. Rather, it is necessary to derive some information about bonding from structures, cohesive energies, refractive indices, electron binding energies, polarizabilities, and other properties through the use of models. Simple, classically based models can be classified generally as ionic, covalent, and metallic bonding and combinations of the three. *See* CHEMICAL BONDING; STRUCTURAL CHEMISTRY.

The mechanisms of chemical reactions within and between solids are through lattice vibrations, lattice defects, and changes in valence states. These are the structural features through which migration of mass, charge, and energy occur. Consequently, diffusion and conductivity are integral, basic parts of solid-state chemistry. *See* DIFFUSION; ELECTRICAL CONDUCTIVITY OF METALS.

A feature of the electronic structure of solids, which is particularly essential to solid-state chemistry, is the valence state of the ion and the energy required to change the valency in the solid. *See* VALENCE.

Many of the photo-induced processes which occur in solids can be imagined to be solid-state chemical reactions. The classical ones, of course, are those involved in photographic plates and films. *See* PHOTOCHEMISTRY.

R. J. Thorn

Solid-state physics The study of the physical properties of solids, such as electrical, dielectric, elastic, and thermal properties, and their understanding in terms of fundamental physical laws. Most problems in solid-state physics would be called solid-state chemistry if studied by scientists with chemical training, and vice versa. Solid-state physics emphasizes the properties common to large classes of compounds rather than the dependence of properties upon compositions, the latter receiving greater emphasis in solid-state chemistry. In addition, solid-state chemistry tends to be more descriptive, while solid-state physics focuses upon quantitative relationships between properties and the underlying electronic structure. *See* SOLID-STATE CHEMISTRY.

Many of the scientists who study the physics of liquids identify with solid-state physics, and the term "condensed-matter physics" has been used by some researchers to replace "solid-state physics" as a division of physics. It includes noncrystalline solids such as glass as well as crystalline solids. *See* AMORPHOUS SOLID; GLASS.

In solid-state physics it is generally assumed that the electronic states can be described as wavelike. The individual electronic states, called Bloch states, have energies which depend upon the wave number (a vector equal to the momentum divided by $\hbar$, which is Planck's constant divided by 2π), and the wave number is restricted to a domain called the Brillouin zone. This energy given as a function of the wave number is called the band structure. There are several curves, called bands, for each line in the Brillouin zone. *See* BRILLOUIN ZONE.

The total energy of a solid includes a sum of the energies of the occupied electronic states. Since the energy bands depend upon the positions of the atoms, so does the total energy, and the stable crystal structure is that which minimizes this energy. The theory has not proved adequate to really predict the crystal structure of various solids, but it is possible to predict the changes in energy under various distortions of the lattice. There are in fact three times as many independent distortions, called normal modes, as there are atoms in the solid. Each has a wave number, and the frequencies of the normal vibrational modes, as a function of wave number in the Brillouin zone, form vibrational bands in direct analogy with the electronic energy bands. These can be directly calculated from quantum theory or measured by using neutron or x-ray diffraction. *See* CRYSTAL; LATTICE VIBRATIONS; NEUTRON DIFFRACTION; X-RAY DIFFRACTION. Walter A. Harrison

Solids pump A device used to move solids upward through a chamber or conduit. It is able to overcome the large dynamic forces at the base of a solids bed and cause the entire bed to move upward.

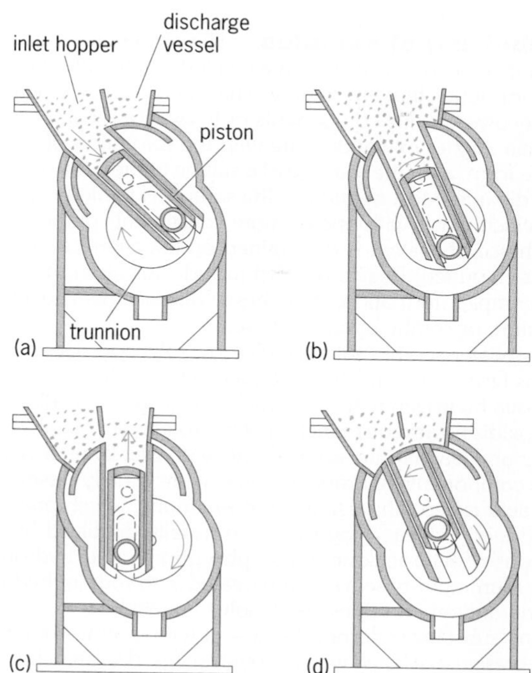

Operation cycle of mechanically driven solids pump. (*a*) Filling with solids from inlet hopper. (*b*) Piston rotating on a trunnion toward its discharge position. (*c*) The discharge position, with piston pushing charge of solids upward. (*d*) Piston rotating back toward original filling position.

Solids pumps are used to cause motion of solids in process-type equipment in which treatment of solids under special conditions of temperature, oxidation, and reduction can be combined with upward motion and discharge of the spent solids overhead from the reacting vessel. The solids pump has found its principal application in the operation of oil-shale retorts.

Solids pumps are inherently of the positive displacement type. One practical method uses a reciprocating piston mounted on a trunnion permitting it to swing into an inclined position for filling and then to swing back into vertical position for discharge. The illustration shows a mechanically driven solids pump in four positions through its cycle of operation. *See* BULK-HANDLING MACHINES. Clyde Berg

Solifugae An order of nonvenomous, spiderlike, predatory arachnids found chiefly in arid and semiarid, tropical, and warm-temperate regions worldwide. Solifugids can range in size from a few centimeters to almost 3 in. (7.5 cm). Most of the 1100 known species are nocturnal. They are commonly known as sun spiders, camel spiders, or wind scorpions. The last name stems from the fact that they exhibit the most rapid sprint speeds for any known group of terrestrial arthropods.

Solifugids possess relatively large anterior appendages, known as chelicerae, that are used for grasping and crushing prey. Ounce for ounce, these jaws produce one of the most powerful bites in the animal kingdom. Depending on their size, solifugids feed on a variety of arthropods and small vertebrates. These arachnids are most frequently found within burrows or rock crevices, or under rocks, logs, or surface debris. *See* ARACHNIDA; ARTHROPODA. Fred Punzo; C. Clayton Hoff

Soliton An isolated wave that propagates without dispersing its energy over larger and larger regions of space. In most of the scientific literature, the requirement that two solitons emerge unchanged from a collision is also added to the definition; otherwise the disturbance is termed a solitary wave.

There are many equations of mathematical physics which have solutions of the soliton type. Correspondingly, the phenomena which they describe, be it the motion of waves in shallow water or in an ionized plasma, exhibit solitons. The first observation of this kind of wave was made in 1834 by John Scott Russell, who followed on horseback a soliton propagating in the windings of a channel. In 1895, D. J. Korteweg and H. de Vries proposed an equation for the motion of waves in shallow waters which possesses soliton solutions, and thus established a mathematical basis for the study of the phenomenon. Interest in the subject, however, lay dormant for many years, and the major body of investigations began only in the 1950s. Researches done by analytical methods and by numerical methods made possible with the advent of computers gradually led to a complete understanding of solitons. *See* DIFFERENTIAL EQUATION; SURFACE WAVES; WAVE MOTION.

Eventually, the fact that solitons exhibit particlelike properties, because the energy is at any instant confined to a limited region of space, received attention, and solitons were proposed as models for elementary particles. However, it is difficult to account for all of the properties of known particles in terms of solitons. More recently it has been realized that some of the quantum fields which are used to describe particles and their interactions also have solutions of the soliton type. The solitons would then appear as additional particles, and may have escaped experimental detection because their masses are much larger than those of known particles. In this context the requirement that solitons emerge unchanged from a collision has been found too restrictive, and particle theorists have used the term soliton where traditionally the term solitary wave would be used. *See* ELEMENTARY PARTICLE; QUANTUM FIELD THEORY. Claudio Rebbi

A hydrodynamic soliton is simply described by the equation of Korteweg and de Vries, which includes a dispersive term and a term to represent nonlinear effects. Easily observed in a wave tank, a bell-shaped solution of this equation balances the effects of dispersion and nonlinearity, and it is this balance that is the essential feature of the soliton phenomenon. Tidal waves in the Firth of Forth were found by Scott Russell to be solitons, as are internal ocean waves and tsunamis. At an even greater level of energy, it has been suggested that the Great Red Spot of the planet Jupiter is a hydrodynamic soliton. *See* JUPITER; OCEAN WAVES; TSUNAMI.

The most significant technical application of the soliton is as a carrier of digital information along an optical fiber. The optical soliton is governed by the nonlinear Schrödinger equation, and again expresses a balance between the effects of optical dispersion and nonlinearity that is due to electric field dependence of the refractive index in the fiber core. If the power is too low, nonlinear effects become negligible, and the information spreads (or disperses) over an ever increasing length of the fiber. At a pulse power level of about 5 milliwatts, however, a robustly stable soliton appears and maintains its size and shape in the presence of disturbing influences. Present designs for data transmission systems based on the optical soliton have a data rate of 4×10^9 bits per second. *See* OPTICAL COMMUNICATIONS.

A carefully studied soliton system is the transverse electromagnetic (TEM) wave that travels between two strips of superconducting metal separated by an insulating layer thin enough (about 2.5 nanometers) to permit transverse Josephson tunneling. Since each soliton carries one quantum of magnetic flux, it is also called a fluxon if the magnetic flux points in one direction, and an antifluxon if the flux points in the opposite direction. Oscillators based on this system reach into the submillimeter wave region of the electromagnetic spectrum (frequencies greater than 10^{11} Hz). *See* JOSEPHSON EFFECT; WAVEGUIDE.

The all-or-nothing action potential or nerve impulse that carries a bit of biological information along the axon of a nerve cell shares many properties with the soliton. Both are solutions of nonlinear equations that travel with fixed shape at constant speed, but the soliton conserves energy, while the nerve impulse balances the rate at which electrostatic energy is released from the nerve membrane to the rate at which it is consumed by the dissipative effects of circulating ionic currents. The nerve process is much like the flame of a candle. *See* BIOPOTENTIALS AND IONIC CURRENTS.

Alwyn Scott

Solstice The two days during the year when the Earth is so located in its orbit that the inclination (about $23^1/_2^\circ$, or 23.45°) of the polar axis is toward the Sun. This occurs on June 21, called the summer solstice, when the North Pole is tilted toward the Sun; and on December 22, called the winter solstice, when the South Pole is tilted toward the Sun (see illustration). The adjectives summer and winter, used above, refer to the Northern Hemisphere; seasons are reversed in the Southern Hemisphere.

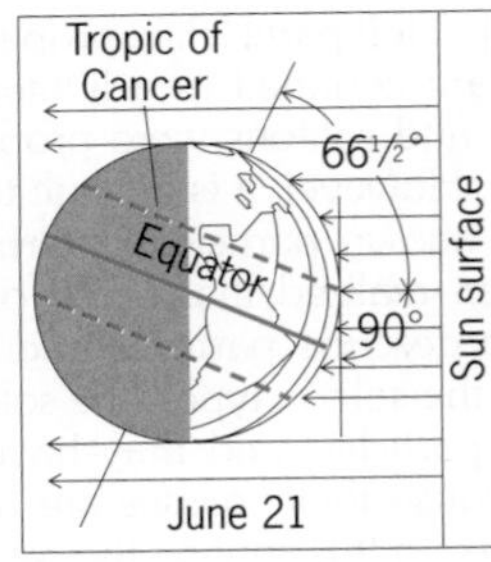

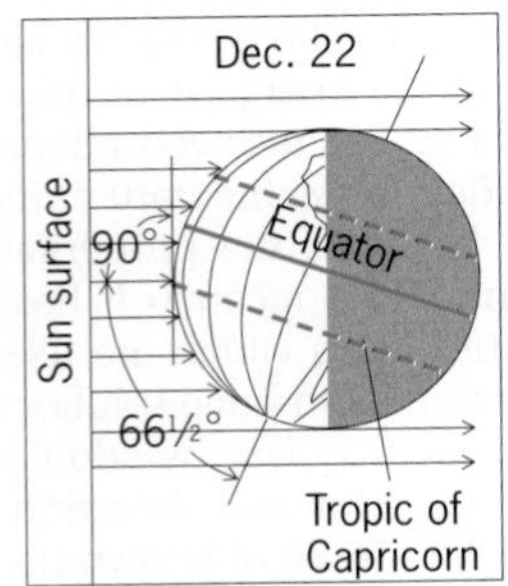

The Earth at the time of the summer and winter solstices. The dates may vary because of the extra one-fourth day in the year.

At the time of the summer solstice every place north of the Arctic Circle will have 24 h of sunlight and the length of day at all places north of the Equator will be more than 12 h, increasing in length with increasing latitude. Identical conditions are found in the Southern Hemisphere at the time of the Northern Hemisphere's winter solstice. *See* MATHEMATICAL GEOGRAPHY.

Van H. English

Solubility product constant A special type of simplified equilibrium constant (symbol K_{sp}) defined for, and useful for, equilibria between solids (s) and their respective ions in solution, for example, reaction (1). For this relatively simple equilibrium, Eqs. (2) and (3) apply.

$$AgCl(s) \leftrightharpoons Ag^+ + Cl^- \quad (1)$$

$$[Ag^+][Cl^-] \cong K_{sp} \quad (2)$$

$$(Ag^+)(Cl^-)/(AgCl) = K_{sp} \quad (3)$$

It can be demonstrated experimentally that a small increase in the molar concentration of chloride ion $[Cl^-]$ causes a reduction in the concentration of silver present as Ag^+. Similarly, an increase in $[Ag^+]$ reduces $[Cl^-]$. The product of the two concentrations is approximately constant as indicated by Eq. (2) and equal to the K_{sp} of Eq. (3). Equation (3) is exact since the variables are activities instead of concentrations. In accordance with the choice of standard state usually made for a solid, the activity of solid AgCl is unity, hence Eq. (4) holds.

$$(Ag^+)(Cl^-) = K_{sp} = 1.8 \times 10^{-10} \text{ mole}^2 \text{ liter}^{-2} \quad (4)$$

In practice, various complications arise: addition of too much of either ion produces more complicated ions and hence actually increases the apparent concentration of the other ion. Addition of a salt without a common ion (that is, a salt supplying neither Ag^+ nor Cl^-) either may react with Ag^+ or Cl^- or may merely increase the concentration of both ions by a lowering of the mean ionic activity coefficient. *See* IONIC EQUILIBRIUM; PRECIPITATION (CHEMISTRY).

Thomas F. Young

Solubilizing of samples The process by which samples that do not dissolve easily are converted into different chemical compounds that are soluble. The sample may be heated in air to evolve volatile components or to oxidize a component to a volatile higher oxidation state with the formation of an acid-soluble form, as in the roasting of a sulfide to form the oxide and sulfur dioxide. Most frequently, the sample is treated with a solvent which reacts with one or more constituents of the sample. The choice of solvent is determined by the chemical reactions that are required. Reactions used include solvation, neutralization, complex formation, metathesis, displacement, oxidation-reduction, or combinations of these.

Most water-soluble salts dissolve by solvation. Basic oxides such as ferric oxide dissolve in aqueous hydrochloric acid. Metals above hydrogen in the electrochemical series will dissolve in a nonoxidizing acid by reduction of hydrogen ion. Metals such as copper and lead require an oxidizing acid, usually nitric acid. All of the components of brass and bronze are usually dissolved by nitric acid except tin, arsenic, and antimony, which precipitate as hydrated oxides. Alloy steels are usually dissolved by combinations of hydrochloric, nitric, phosphoric, and hydrofluoric acids. Aluminum-base alloys are treated with sodium hydroxide solution and any residues are dissolved in acid.

Many substances do not dissolve at temperatures obtainable in the presence of liquid water. However, fused salt reactions employing temperatures of 400–1100°C (750–2000°F) are necessary for the attack and decomposition of many types of samples. The material used as the solvent is called a flux, and the process of melting the mixture of dry, solid flux with the sample is called a fusion.

Charles L. Rulfs

Solution A homogeneous mixture of two or more components whose properties vary continuously with varying proportions of the components. A liquid solution can be distinguished experimentally from a pure liquid by the fact that during transfers into other single phases at equilibrium (freezing and vaporizing at constant pressure) the temperature and other properties vary continuously, whereas those of a pure liquid remain constant. For an apparent exception *see* AZEOTROPIC MIXTURE. *See also* SOLVENT.

Gases, unless highly compressed, are mutually soluble in all proportions.

A solid solution is, similarly, a single phase whose composition and other properties vary continuously with changing composition of the liquid phase with which it is in equilibrium. *See* SOLID SOLUTION.

The extent to which substances can form solutions depends upon the kind and strength of the attractive forces between the several molecular species involved. It is necessary to consider the attractive forces exerted by molecules of the following types: (1) nonpolar molecules; (2) polar molecules, that is, those containing electric dipoles; (3) ions; and (4) metallic atoms.

Actual solutions may be considered in terms of their departure from a simple idealized model—a mixture of components having the same attractive fields, which mix without change in volume or heat content. This is analogous to an ideal gas mixture, which is formed with no heat of mixing and in which the total pressure is the sum of the partial pressures. In such a solution the escaping tendency of the individual molecules is the same, whether they are surrounded by similar or by different molecules.

Joel H. Hildebrand

Solution mining The extraction of the valuable components from a mineral deposit using an aqueous leaching solution. In its original sense, solution mining refers to evaporite mining, the dissolution of soluble rock material such as salt by using borehole wells to pump water into the deposit and remove the resulting saturated brine. In its current usage, solution mining also includes ore leaching, the in-place (in-situ) leaching of valuable metal components from an orebody, and the mine-site procedures of heap leaching and dump leaching. Often included is the Frasch process of using superheated water to melt sulfur in its deep deposits and recover the molten sulfur in borehole wells. *See* ORE AND MINERAL DEPOSITS.

Evaporites represent a broad class of water-soluble minerals (salts). Commercially important evaporites include halite (sodium chloride), sylvite and silvinite (potash), bischofite (magnesium chloride), nahcolite (sodium bicarbonate), trona (raw soda ash), and langbeinite and carnallite [both sources of potash and magnesia (magnesium oxide)]. *See* SALINE EVAPORITES.

Solution mining involves injecting a solvent into the pay zone of the deposit through a cased borehole. For evaporites, the solvent is hot water, which forms brine as the soluble minerals dissolve. The brine is brought to the surface via the casing system in the same or another borehole and sent to a processing facility for recovery by the controlled crystallization of the desired product, followed by dewatering and drying. Some minerals may require additional purification steps, such as the flotation of potash crystals. The depleted brine is chemically reconditioned and injected back into the deposit. Thus, solution mining creates minimal surface disturbance and little waste, compared to conventional mining. *See* WELL.

Although solution mining is simple in principle, there are several key issues for successful operation. One is to maintain close control over the solvent parameters, such as pH, oxidation potential, temperature, and pressure. Another issue is isolation of the solution mining zone from the surrounding geologic structures. This requires effective well completion techniques that are compatible with the brine. Once the well is drilled and cased, the casing must be cemented into the formation. This prevents brine from migrating along the annular space outside the casing and contaminating adjacent aquifers. *See* ELECTROCHEMICAL SERIES; OIL AND GAS WELL COMPLETION.

W. Joseph Schlitt

Leaching large stockpiles and mine waste dumps (ore heaps) annually accounts for about one-third of both new copper and gold production in the United States, the world's second largest source of both commodities. Ore leaching is an important contributor to the total world production of these commodities, along with silver. Ore leaching's very low unit processing cost, combined with low-cost earthmoving, allows profitable treatment of huge tonnages of low-grade material that could not otherwise be exploited.

Ore containing rock must first be fragmented to allow chemical leaching solutions to percolate through it. The metal-bearing minerals are liberated using an aqueous wetting fluid that penetrates and saturates the microfractures within the ore particles by capillary action. The ore minerals, originally deposited geochemically from hydrothermal solutions, are typically located in these micropores. Once dissolved, the metals are removed from the ore particles by diffusion through the solution-filled micropores and swept out of the ore heap by the flowing leachate. In effect, nature's ore deposition process has been reversed at a quicker pace. Nevertheless, commercial leaching times are measured in weeks, months, and even years. Leachates, or "pregnant liquors," are processed in surface plants to recover the metal and regenerate the leachant for reuse in the ore heap in a closed-loop system. *See* HYDROMETALLURGY; LEACHING.

Robert W. Bartlett

Modern sulfur mining dates from the invention of the Frasch process in the late nineteenth century. H. Frasch accomplished in-place sulfur mining by using superheated water. Frasch mining begins by drilling wells into the sulfur deposit. Steel tubing (casing) is run into the drill hole to case off the barren overlying formation, and it is cemented in at the top of the sulfur deposit. Within this casing, three concentric strings of pipe are set within the sulfur deposit. Superheated water at 325°F (163°C) can be pumped down the annulus between the two strings of pipe to leave the casing string through the perforations above the packer, and to circulate in the sulfur deposit. Native sulfur melts at 275°F (135°C) in a cone-shaped area of influence around the well, and because liquid sulfur is denser than water, it settles to the bottom of the well.

The pressure from the heated water forces the molten sulfur into the lower set of perforations and into the inner string of casing. The molten sulfur rises to its hydrostatic head. Compressed air pumped through this inner tubing expands when it leaves the tubing, thus "jetting" the sulfur to the well collar, where it is collected in a tank (sulfur pan).

Alan F. Edwards

Solvation The association or combination of a solute unit (ionic, molecular, or particulate) with solvent molecules. This association may involve chemical or physical forces, or both, and may vary in degree from a loose, indefinite complex to the formation of a distinct chemical compound. Such a compound contains a definite number of solvent molecules per solute molecule. Solvation occurring in aqueous solutions is referred to as hydration. In certain colloidal suspensions, solvation is, to a large extent, responsible for the stability of the sol. *See* COLLOID; HYDRATION; SOLUTION; SOLVENT.

Francis J. Johnston

Solvent By convention, the component present in the greatest proportion in a homogeneous mixture of pure substances (solutions). Components of mixtures present in minor proportions are called solutes. Thus, technically, homogeneous mixtures are possible with liquids, solids, or gases dissolved in liquids; solids in solids; and gases in gases. In common practice this terminology is applied mostly to liquid mixtures for which the solvent is a liquid and the solute can be a liquid, solid, or gas. See SOLUTION.

Three broad classes of solvents are recognized—aqueous, nonaqueous, and organic. Formalistically, the nonaqueous and organic classifications are both not aqueous, but the term organic solvents is generally applied to a large body of carbon-based compounds that find use industrially and as media for chemical synthesis. Organic solvents are generally classified by the functional groups that are present in the molecule, for example, alcohols, halogenated hydrocarbons, or hydrocarbons; such groups give an indication of the types of physical or chemical interactions that can occur between solute and solvent. Nonaqueous solvents are generally taken to be inorganic substances and a few of the lower-molecular-weight, carbon-containing substances such as acetic acid, methanol, and dimethylsulfoxide. Nonaqueous solvents can be solids (for example, fused LiI), liquids (H_2SO_4), or gases (NH_3) at ambient conditions; the solvent properties of fused sodium iodide (NaI) are manifested in the molten state, whereas hydrogen sulfate (H_2SO_4) and ammonia (NH_3) must be liquefied to act as solvents. J. J. Lagowski

Solvent extraction A technique, also called liquid extraction, for separating the components of a liquid solution. This technique depends upon the selective dissolving of one or more constituents of the solution into a suitable immiscible liquid solvent. It is particularly useful industrially for separation of the constituents of a mixture according to chemical type, especially when methods that depend upon different physical properties, such as the separation by distillation of substances of different vapor pressures, either fail entirely or become too expensive.

Industrial plants using solvent extraction require equipment for carrying out the extraction itself (extractor) and for essentially complete recovery of the solvent for reuse, usually by distillation. See DISTILLATION.

The petroleum refining industry is the largest user of extraction. In refining virtually all automobile lubricating oil, the undesirable constituents such as aromatic hydrocarbons are extracted from the more desirable paraffinic and naphthenic hydrocarbons. By suitable catalytic treatment of lower boiling distillates, naphthas rich in aromatic hydrocarbons such as benzene, toluene, and the xylenes may be produced. The latter are separated from paraffinic hydrocarbons with suitable solvents to produce high-purity aromatic hydrocarbons and high-octane gasoline. Other industrial applications include so-called sweetening of gasoline by extraction of sulfur-containing compounds; separation of vegetable oils into relatively saturated and unsaturated glyceride esters; recovery of valuable chemicals in by-product coke oven plants; pharmaceutical refining processes; and purifying of uranium.

Solvent extraction is carried out regularly in the laboratory by the chemist as a commonplace purification procedure in organic synthesis, and in analytical separations in which the extraordinary ability of certain solvents preferentially to remove one or more constituents from a solution quantitatively is exploited. Batch extractions of this sort, on a small scale, are usually done in separatory funnels, where the mechanical agitation is supplied by handshaking of the funnel. See EXTRACTION. Robert E. Treybal

Somatic cell genetics The study of mechanisms of inheritance in animals and plants by using cells in culture. In such cells, chromosomes and genes can be reshuffled by parasexual means rather than having to depend upon the genetic recombination and chromosome segregation that occur during the meiotic cell divisions preceding gamete formation and sexual reproduction. Genetics is concerned with the role of genes and chromosomes and their environmental interactions in the development and function of individuals and the evolution of species. Genetic analysis of complex multicellular organisms classically requires multigeneration families. Because fairly large numbers of progeny of defined matings have to be scored, genetic analysis of animals and plants with long generation times, small families, or lack of controlled matings is difficult and slow. Somatic cell genetics circumvents many of these limitations. It has enhanced the scope and speed of genetic analysis in higher plants and animals, especially when combined with the powerful techniques of molecular biology. Cells in culture can be used to generate fertile plants and animals from single cells, including stem cells. With these methods, virtually every gene in any species of interest can be identified, its functions clarified, and the structural and functional changes it has undergone in particular phylogenetic lineages determined. See CHROMOSOME; GENE; GENETICS.

Plant meristem cells can be grown readily in culture. They can be totipotent (capable of differentiating into all cell types) and can even grow into a mature plant. In mammals, the only totipotent cells, in addition to fertilized eggs, are the embryonic stem cells of early preimplantation embryos and some embryonal carcinoma cells. Even adults contain some stem cells, but these appear to be only pluripotent or even lineage-specific, able to differentiate into only a limited number of cell types. Totipotent cells appear to be immortal in culture. Other diploid cells senesce, dying after a limited number of cell divisions. This is sufficient for chromosome studies, and hundreds of thousands of these are performed each year on normal and cancerous human cells. In addition, cells in early passage cultures of many species can be frozen and maintained in a viable state for years (cryopreservation). The cells can be thawed at any time and grown in culture. Thus, a single biopsy can provide cultured diploid cells for many years, including cells from endangered species. See STEM CELLS.

When a foreign gene (transgene) is introduced into a cell nucleus, it inserts at random into some chromosomal site, sometimes disrupting an important gene and causing deleterious effects. Fortunately, methods have been developed that target the transgene to the gene's normal chromosomal position. This has made it possible to replace one of the two copies of a gene in diploid cells with a normal or engineered copy of the gene from an external source. Transgenic methods have tremendous potential for the development of plants resistant to insects, microorganisms, or environmental stress. Transgenic animals are generated in ever-increasing numbers. Increasing numbers of transgenic mice are being produced to study gene function, cell differentiation, and morphogenesis and to create animal models of human diseases whose study may speed the development of treatments. See GENETIC ENGINEERING; MOLECULAR BIOLOGY.

Orlando J. Miller

Somatostatin A naturally occurring regulatory peptide that carries out numerous functions in the human body, including the inhibition of growth hormone secretion from the anterior pituitary gland. Somatostatin consists of 14 amino acids; two cysteine residues are joined by a disulfide bond so that the peptide forms a ring structure. A larger variant of this peptide, called somatostatin-28, is produced in some cells and has an additional 14 amino acids attached at the amino-terminal end of normal somatostatin (somatostatin-14).

Somatostatin acts primarily as a negative regulator of a variety of different cell types, blocking processes such as cell secretion, cell growth, and smooth muscle contraction. It is secreted from the hypothalamus into the portal circulation and travels to the anterior pituitary gland, where it inhibits the production and release of both growth hormone and thyroid-stimulating hormone. Many tissues other than the hypothalamus contain somatostatin, suggesting that this peptide has numerous roles.

Each of the functions of somatostatin is initiated by the binding of the peptide to one or more of five different cell-surface receptor proteins, thereby activating one or more intracellular G-proteins and initiating biochemical signaling pathways within the cell. *See* SIGNAL TRANSDUCTION.

Analogs of somatostatin have been synthesized that are smaller, more potent, longer-lasting, and more specific in their biological effects than natural somatostatin. Some of these analogs have become useful as drugs. *See* ENDOCRINE SYSTEM (VERTEBRATE); HORMONE; NEUROSECRETION; PITUITARY GLAND.

William H. Simmons

Somesthesis A general term for the somatic sensibilities aroused by stimulation of bodily tissues such as the skin, muscles, tendons, joints, and the viscera. Six primary qualities of somatic sensation are commonly recognized: touch-pressure (including temporal variations such as vibration), warmth, coolness, pain, itch, and the position and movement of the joints. These basic sensory qualities exist because each is served by a different set of sensory receptors (the sensory endings of certain peripheral nerve fibers) which differ not only in their sensitivities to different types of stimuli, but also in their connections to structures within the central nervous system.

The somatic sensory pathways are dual in nature. One major part, the lemniscal system, receives input from large-diameter myelinated peripheral nerve fibers (for example, those serving the sense of touch-pressure). The second major somatic pathway is called the anterolateral system. It receives input from small-diameter myelinated and unmyelinated peripheral nerve fibers carrying pain and temperature information. *See* CUTANEOUS SENSATION; PAIN; PROPRIOCEPTION.

Robert LaMotte

Sonar A remote sensing technique or device that uses sound waves to detect, locate, and sometimes identify objects in water. The term is an acronym for sound navigation and ranging. There are many applications, using a wide variety of equipment. Naval uses include detection of submarines, sea mines, torpedoes, and swimmers; torpedo guidance; acoustic mines; and navigation. Civilian uses include determining water depth; finding fish; mapping the ocean floor; locating various objects in the ocean, such as pipelines, wellheads, wrecks, and obstacles to navigation; measuring water current profiles; and determining characteristics of ocean bottom sediments. Sound waves rather than electromagnetic waves (for example, radar and light) are used in these applications because their attenuation in seawater is much less. Some marine mammals use sound waves to find food and to navigate. *See* ACOUSTIC TORPEDO; ANTISUBMARINE WARFARE; ECHOLOCATION; MARINE GEOLOGY; MARINE NAVIGATION; UNDERWATER NAVIGATION; UNDERWATER SOUND.

There are two generic types of sonar: active (echolocation) and passive. An active sonar projects a signal (typically a short pulse of sound) into the water in a narrow beam, which propagates at a speed of about 1500 m/s (5000 ft/s). If there is an object (target) in the beam, it reflects a fraction of the sound energy to the sonar, which detects the echo. By measuring the elapsed time between projection and reception, the range to the target can be computed (range = sound speed × travel time ÷ 2).

Direction to the target is determined from the orientation of the sound beam at the time of reception. Passive sonar does not radiate sound but depends on detecting sounds radiated by targets such as submarines and ships. Passive sonar determines direction to a target in the same manner as active sonar, but range determination is more difficult.

In an elementary active pulse sonar, a pulse signal of certain frequency and duration is generated, amplified, and sent to an electroacoustic transducer, which converts the electrical signal into a sound signal, which then radiates into the water. If the transducer is reciprocal in character (typically the case), it also can be used to sense (detect) the returning echoes. The receiver amplifies the weak echoes and measures the range to each target, as well as the orientation of the receiving beam at the time of reception. This information is displayed in some form of range-direction plot.

Most active sonar transducers are mounted on the hulls of submarines or near the keels of surface ships. Sometimes, transducers are towed at a water depth that provides better operation. There are three basic transducer orientations. In the conventional depth sounder, the sound beam is directed downward. Echoes are reflected from the ocean bottom (and from fish that may be in the beam), and the depth of the ocean beneath the sonar can be determined. In the side-scan sonar configuration, the beam is oriented to the side of the ship (normal to the direction of travel) and (usually) slightly downward. As the ship moves forward, a volume of water to the side of the ship is searched. Generally, two sonars are used, one searching to the right and one to the left. Side-scan sonars are well suited to search at a constant speed and along straight lines, such as in mapping the ocean bottom and in general searches of an area. The third, and most popular, sonar configuration involves rotating the sound beam about the vertical axis to search (scan) a sector of the water centered on the sonar platform. *See* ECHO SOUNDER.

The range at which a target can be detected depends on the strength of the projected signal (source level), propagation losses to the target and return, reflection characteristics of the target (target strength), and sensitivity of the receiver. Also, the target echo must be stronger than various masking signals (noise and reverberation), which also are received by the sonar. An important sonar performance characteristic is its ability to locate a target accurately and determine whether an echo is from a single target or from several targets close together. The uncertainty in the direction to the target is approximately the width of the receiving beam.

Passive sonars are used primarily to detect submarines and, to a lesser extent, surface ships. Because passive sonar does not radiate any sound that would reveal its location, it is the primary sensor used by submarines. The major weakness is that it cannot directly measure range to a target. To determine target location, the sonar must take bearings on a target from different locations. Passive sonars depend on detecting noise radiated by targets, a mixture of sounds generated by propellers and hull vibrations (caused by motors, engines, pumps, and hydrodynamic forces). The noise has a continuous spectrum and discrete tones related to rotational speeds of propellers, engines, and so forth. By analysis of the received signals the sonar often can identify the type of target. Most of the radiated energy is in the audible frequency band and decreases in intensity with increasing frequency.

Most passive sonars use large receiving transducer arrays in order to achieve high sensitivity, discriminate against ambient noise, and determine precisely the direction to a target. In submarines, these arrays may be recessed into the structure or mounted on the hull. Submarines also may tow long slender line arrays. A number of very large fixed receiving arrays, placed on the ocean floor with cables running to shore stations, constantly observe strategically important ocean areas. Detection ranges for large passive sonars vary from hundreds of kilometers against noisy targets under good conditions to a very few kilometers against quiet targets.

The sonobuoy consists of a small surface buoy with a hydrophone array suspended beneath the water. Sounds received by the array are telemetered by radio link to an aircraft overhead. (There are also active sonobuoys.) Sonobuoys have relatively short detection ranges and are used primarily in tactical situations. *See* PARAMETRIC ARRAY; SONOBUOY; ULTRASONICS.

Chester M. McKinney

Sonic boom An audible sound wave generated by an object that moves faster than the speed of sound (supersonic object). The sonic boom forms because the air is pushed away faster than the air molecules can move. The displaced air becomes highly compressed and creates a very strong sound wave, referred to as a compressional head shock or bow shock. At the back of the supersonic object the air has to fill the void left as the object moves forward; in this case, the gas becomes rarefied and a rarefractional tail shock develops. These shock waves are the main components of a sonic boom, and they are generated the entire time that an object flies faster than the speed of sound, not just when it breaks the sonic barrier. *See* SHOCK WAVE.

Sonic booms may be natural or generated by human activity. A natural sonic boom is thunder, created when lightning ionizes air, which expands supersonically. Meteors can create sonic booms if they enter the atmosphere at supersonic speeds. Human sources of sonic booms include aircraft, rockets, the space shuttle during reentry, and bullets. *See* METEOR; THUNDER.

Sonic booms are commonly associated with supersonic aircraft. The shock waves associated with sonic booms propagate away from the aircraft in a unique fashion. These waves form a cone, called the Mach cone, that is dragged behind the aircraft. The illustration shows the outline of the Mach cone generated by an F-18 fighter aircraft flying at Mach 1.4. The schlieren photographic technique was used to display the sonic boom, which is normally invisible. The half-angle of the cone is determined solely by the Mach number of the aircraft, $\theta = \arctan (1/M)$, 36° for $M = 1.4$. The shock waves travel along rays that are perpendicular to the Mach cone (see illustration). As the Mach number increases, θ becomes smaller and the sound travels almost directly downward. *See* MACH NUMBER; SCHLIEREN PHOTOGRAPHY; SUPERSONIC FLIGHT.

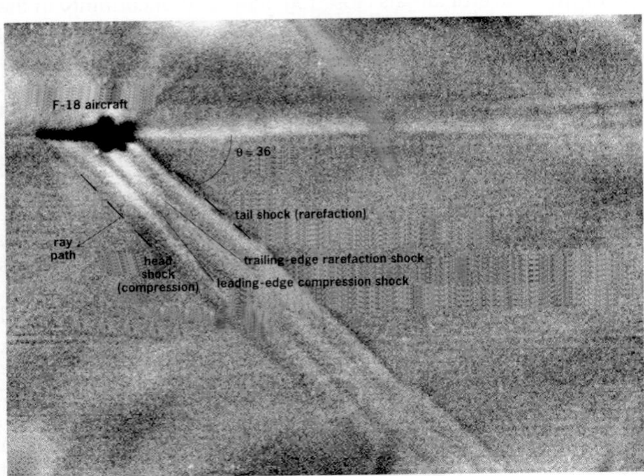

Schlieren image of the shock waves generated by an F-18 aircraft flying at Mach 1.4. The Mach cone generated by the head and tail shocks can be seen as well as the shocks generated by the leading and trailing edges of the wings. (*NASA Dryden Flight Research Center*)

The typical peak pressure amplitude (or overpressure) of a sonic boom on the ground is about 50–100 pascals. A sonic boom with 50 Pa (1 lbf/ft^2 or 0.007 psi) overpressure will produce no damage to buildings. Booms in the range of 75–100 Pa are considered disturbing by some people. Occasionally there is minor damage to buildings from sonic booms in the range of 100–250 Pa; however, buildings in good condition will be undamaged by overpressures up to 550 Pa. Very low flying aircraft (30 m or 100 ft) can produce sonic booms of 1000–7000 Pa. These pressures are still about five times less than that needed to injure the human ear, but can lead to damage to buildings, such as the breaking of glass windows and the cracking of plaster. Although sonic booms are not dangerous, they can evoke a strong startle response in people and animals. Robin Cleveland

Sonobuoy An expendable device that enables aircraft to detect underwater objects, such as submarines, acoustically. Acoustics is the preferred energy form for use in salt water, because it tends to be the least attenuated by the medium.

A sonobuoy consists of an electronic radio link and antenna connected to a miniature sonar system. It contains the means for its launch from the aircraft, its entry into the water, separation of a floating antenna from the underwater transducer and sonar, and activation of a seawater battery upon entry, as well as a scuttling means for final sinking upon completion of its intended useful life. The short life requires that each component be highly reliable and effective at low cost. The package must also be small and lightweight, since large numbers of packages are to be carried on an aircraft. *See* SONAR.

The simplest sonobuoy is a passive sonar that senses sound with its hydrophone, amplifies and converts it to a radio signal, and transmits the signal from its antenna for analysis, evaluation, and storage in the aircraft. The sonobuoy system may be independently activated by an underwater sound source, usually an explosive device dropped from the aircraft. Two or more simple sonobuoys may be deployed to permit processing of directional information, passively or actively. *See* HYDROPHONE.

Buoys of various types are used as sensors for oceanographic data such as sound speed, geophysical data such as earthquakes, bioacoustic data such as snapping shrimp, and other signal and noise sources. Some have been designed for long-term deployment on station, with provision for storing data until interrogated from an aircraft or a ship. *See* BUOY. Stanley L. Ehrlich

Sonochemistry The study of the chemical changes that occur in the presence of sound or ultrasound. Industrial applications of ultrasound include many physical and chemical effects, for example, cleaning, soldering, welding, dispersion, emulsification, disinfection, pasteurization, extraction, flotation of minerals, degassing of liquids, defoaming, and production of gas-liquid sols. *See* SOUND; ULTRASONICS.

When liquids are exposed to intense ultrasound, high-energy chemical reactions occur, often accompanied by the emission of light. There are three classes of such reactions: homogeneous sonochemistry of liquids, heterogeneous sonochemistry of liquid-liquid or liquid-solid systems, and sonocatalysis (which overlaps the first two). In some cases, ultrasonic irradiation can increase reactivity by nearly a millionfold. Especially for liquid-solid reactions, the rate enhancements via ultrasound have proved extremely useful for the synthesis of organic and organometallic compounds. Because cavitation occurs only in liquids, chemical reactions are not generally seen in the ultrasonic irradiation of solids or solid-gas systems.

Ultrasound spans the frequencies of roughly 20 kHz to 10 MHz (human hearing has an upper limit of less than 18 kHz). Ultrasound has acoustic wavelengths of roughly 7.5–0.015 cm which are much larger than molecular dimensions. As a result, the chemical effects of ultrasound are not from direct interaction, but are derived from several different physical mechanisms, depending on the nature of the system. For both sonochemistry and sonoluminescence, the most important of these mechanisms is acoustic cavitation: the formation, growth, and implosive collapse of bubbles in liquids irradiated with high-intensity sound. During the final stages of cavitation, compression of the gas inside the bubbles produces enormous local heating and high pressures. *See* CAVITATION.

When a liquid-solid interface is subjected to ultrasound, cavitation occurs, but a markedly asymmetric bubble collapse occurs, which generates a jet of liquid directed at the surface with velocities greater than 330 ft/s (100 m/s). The impingement of this jet can create a localized erosion (and even melting), responsible for surface pitting and ultrasonic cleaning. Enhanced chemical reactivity of solid surfaces is associated with these processes.

Ultrasonic irradiation of liquid-powder suspensions produces another effect: high-velocity interparticle collisions. Cavitation and the shock waves that it creates in a slurry can accelerate solid particles to high velocities. The resultant collisions are capable of inducing dramatic changes in surface morphology, composition, and reactivity.

The predominant reactions of homogeneous sonochemistry are bond breaking and radical formation. In addition to the initiation or enhancement of chemical reactions, irradiation of liquids with high-intensity ultrasound generates the emission of visible light. The production of such luminescence is a consequence of the localized hot spot created by the implosive collapse of gas- and vapor-filled bubbles during acoustic cavitation. In general, sonoluminescence may be considered a special case of homogeneous sonochemistry. Under conditions where an isolated, single bubble undergoes cavitation, recent studies on the duration of the sonoluminescence flash suggest that a shock wave may be created within the collapsing bubble. *See* CHEMILUMINESCENCE; HOMOGENEOUS CATALYSIS.

A major industrial application of ultrasound is emulsification. The first reported and most studied liquid-liquid heterogeneous systems have involved ultrasonically dispersed mercury. The effect of the ultrasound in this system appears to be due to the large surface area of mercury generated in the emulsion. *See* EMULSION.

The effects of ultrasound on liquid-solid heterogeneous organometallic reactions have been a matter of intense investigation. Various research groups have dealt with extremely reactive metals, such as lithium (Li), magnesium (Mg), or zinc (Zn), as stoichiometric reagents for a variety of common transformations.

Sonochemistry can be used as a synthetic tool for the creation of unusual inorganic materials. Ultrasound has proved extremely useful in the synthesis of a wide range of nanostructured materials, including high-surface-area transition metals, alloys, carbides, oxides, and colloids. Sonochemistry is also proving to have important applications with polymeric materials. Substantial work has been accomplished in the sonochemical initiation of polymerization and in the modification of polymers after synthesis. Sonochemistry has found another recent application in the preparation of unusual biomaterials, notably protein microspheres. The mechanism responsible for microsphere formation is a combination of two acoustic phenomena: emulsification and cavitation. These protein microspheres have a wide range of biomedical applications, including their use as echo contrast agents for sonography, magnetic resonance imaging contrast enhancement, and drug delivery. *See* NANOCHEMISTRY; NANOSTRUCTURE; POLYMER; PROTEIN.

Kenneth S. Suslick

Sonoluminescence The transformation of acoustic energy into light through collapsing bubbles. In particular, single-bubble sonoluminescence (SBSL) is the periodic light emission of a single acoustically trapped and periodically driven gas bubble. It occurs when the bubble collapses so violently that the energy focusing during collapse is intense enough for partial ionization of the gas inside the bubble to occur, leading to subsequent light emission. The phenomenon of single-bubble sonoluminescence was discovered in 1989 by Felipe Gaitan, then a graduate student, working with Lawrence A. Crum. A micrometer-sized bubble is acoustically trapped in a water-filled flask at resonance. The air saturation in the water is typically 20–40%. Once per cycle, at bubble collapse, the bubble emits a short pulse of light, that typically lasts 100–300 picoseconds. The origin of the light is thermal bremsstrahlung: During the adiabatic collapse, the gas inside the bubble is heated, presumably to about 15,000 K. Consequently, the gas partly ionizes. The ions and electrons interact, decelerate, and finally recombine, and the deceleration is connected with light emission. *See* BREMSSTRAHLUNG.

Bubble dynamics. The backbone of the theoretical understanding of single-bubble sonoluminescence is the Rayleigh-Plesset equation, given below. It describes the dynamics of a

$$R\ddot{R} + \frac{3}{2}\dot{R}^2 = \frac{1}{\rho}\left(p_g - P_0 - P(t) - 4\eta\frac{\dot{R}}{R} - \frac{2\sigma}{R}\right)$$

spherical bubble, with radius $R(t)$, in a liquid. Here σ denotes the surface tension, η the dynamic viscosity, P_0 the ambient pressure (usually 1 atm), $p_g(R(t))$ the gas pressure inside the bubble, and $P(t)$ the time-dependent external driving pressure, which in the case of single-bubble sonoluminescence is simply $P(t) = P_a \sin(2\pi f t)$, with the driving amplitude P_a and the driving frequency f.

On pressure reduction, the bubble strongly expands to typically 10 times its equilibrium radius R_0. Once the pressure increases, the bubble violently collapses, followed by a "ringing" of the bubble at its eigenfrequency.

Chemical activity. The chemical activity inside the bubble is a consequence of the high temperatures achieved. The sonoluminescing bubble can be viewed as a high-temperature, high-pressure microlaboratory or reaction chamber, which can be controlled through the external parameters such as forcing pressure, frequency, or dissolved gas concentration. When the bubble is expanding, gas dissolved in the liquid and liquid vapor enter the bubble. During the adiabatic collapse, these gases are partly trapped inside the hot bubble and react. *See* ADIABATIC PROCESS; SONOCHEMISTRY.

Detlef Lohse

Sorghum Sorghum includes many widely cultivated grasses having a variety of names in various countries. Cultivated sorghums in the United States are classified as a single species, *Sorghum bicolor*, although there are many varieties and hybrids. The two major types of sorghum are the grain, or nonsaccharine, type, cultivated for grain production and to a lesser extent for forage, and the sweet, or saccharine, type, used for forage production and for making syrup and sugar.

Grain sorghum is grown in the United States chiefly in the Southwest and the Great Plains. It is a warm-season crop which withstands heat and moisture stress better than most other crops, but extremely high temperatures and extended drought may reduce yields. It is extensively grown in Texas, Kansas, Nebraska, Oklahoma, Missouri, Colorado, and South Dakota. This grain production is fed to cattle, poultry, swine, and sheep primarily. Sorghum is considered nearly equal to corn in feed value.

Sorghums originated in the northeastern quadrant of Africa. Until recent years, practically all grain sorghums of importance introduced into the United States were tall, late maturing, and generally unadapted. Since its introduction into the United States in colonial times, the crop has been altered in many ways, these changes coming as a result of naturally occurring genetic mutations combined with hybridization and selection work of plant breeders. The fact that hybrid grain sorghums with high yield potential could be produced with stems that are short enough for harvesting mechanically made the crop appealing to many farmers. *See* BREEDING (PLANT).

Grain sorghum is difficult to distinguish from corn in its early growth stages, but at later stages it becomes strikingly different. Sorghum plants may tiller (put out new shoots), producing several head-bearing culms from the basal nodes. Secondary culms may also develop from nodal buds along the main stem. The inflorescence (head) varies from a dense to a lax panicle, and the spikelets produce perfect flowers that are subject to both self- and cross-fertilization. Mature grain in different varieties varies in size and color from white to cream, red, and brown. Grain sorghum seeds are small and should not be planted too deep since sorghum lacks the soil-penetrating ability of corn. The seeds are planted either in rows wide enough for tractor cultivation or in narrower rows if cultivation is not intended.

Frederick R. Miller

Commonly known as sorgo, sweet sorghum was introduced into North America from China in 1850, although its ancestry traces back to Egypt. It is an annual, rather drought-resistant crop. The culms are from 2 to 15 ft (0.6 to 4.6 m) tall, and the hard cortical layer, or shell, encloses a sweet, juicy pith that is interspersed with vascular bundles. At each node both a leaf and a lateral bud alternate on opposite sides; the internodes are alternately grooved on one side. Leaves are smooth with glossy or waxy surfaces and have margins with small, sharp, curved teeth. The leaves fold and roll up during drought. The inflorescence is a panicle of varying size having many primary branches with paired ellipsoidal spikelets containing two florets in each fertile sessile spikelet. The plant is self-pollinated. Seed is planted in cultivated rows and fertilized similarly to corn. The main sorghum-syrup-producing area is in the south-central and southeastern United States. *See* CORTEX (PLANT); PITH. Leonard D. Baver

Sound The mechanical excitation of an elastic medium. Originally, sound was considered to be only that which is heard. This admitted questions such as whether or not sound was generated by trees falling where no one could hear. A more mechanistic approach avoids these questions and also allows acoustic disturbances too high in frequency (ultrasonic) to be heard or too low (infrasonic) to be classed as extensions of those events that can be heard.

A source of sound undergoes rapid changes of shape, size, or position that disturb adjacent elements of the surrounding medium, causing them to move about their equilibrium positions. These disturbances in turn are transmitted elastically to neighboring elements. This chain of events propagates to larger and larger distances, constituting a wave traveling through the medium. If the wave contains the appropriate range of frequencies and impinges on the ear, it generates the nerve impulses that are perceived as hearing. *See* HEARING (HUMAN).

Acoustic pressure. A sound wave compresses and dilates the material elements it passes through, generating associated pressure fluctuations. An appropriate sensor (a microphone, for example) placed in the sound field will record a time-varying deviation from the equilibrium pressure found at that point within the fluid. The changing total pressure P measured will vary about the equilibrium pressure P_0 by a small amount called the acoustic pressure, $p = P - P_0$. The SI unit of pressure is the pascal (Pa), equal to 1 newton per square meter (N/m^2). Standard atmospheric pressure (14.7 lb/in.^{2}) is approximately 1 bar $= 10^6$ dyne/cm^2 $= 10^5$ Pa. For a typical sound in air, the amplitude of the acoustic pressure may be about 0.1 Pa (one-millionth of an atmosphere); most sounds cause relatively slight perturbations of the total pressure. *See* MICROPHONE; PRESSURE; PRESSURE MEASUREMENT; PRESSURE TRANSDUCER; SOUND PRESSURE.

Plane waves. One of the more basic sound waves is the traveling plane wave. This is a pressure wave progressing through the medium in one direction, say the $+x$ direction, with infinite extent in the y and z directions. A two-dimensional analog is ocean surf advancing toward a very long, straight, and even beach. *See* WAVE (PHYSICS); WAVE EQUATION; WAVE MOTION.

A most important plane wave, called a harmonic wave, is the smoothly oscillating monofrequency plane wave described by Eq. (1). The amplitude of this wave is P. The phase (argument

$$p = P\cos\left[2\pi f\left(t - \frac{x}{c}\right)\right] \quad (1)$$

of the cosine) increases with time, and at a point in space the cosine will pass through one full cycle for each increase in phase of 2π. The period T required for each cycle must therefore be such that $2\pi fT = 2\pi$, or $T = 1/f$, so that $f = 1/T$ can be identified as the frequency of oscillation of the pressure wave. During this period T, each portion of the waveform has advanced through a distance $\lambda = cT$, and this distance λ must be the wavelength. This gives the fundamental relation (2) between the frequency, wave-

$$\lambda f = c \quad (2)$$

length, and speed of sound c in any medium. For example, in air at room temperature the speed of sound is 343 m/s (1125 ft/s). A sound of frequency 1 kHz (1000 cycles per second) will have a wavelength of $\lambda = c/f = 343/1000$ m $= 0.34$ m (1.1 ft). Lower frequencies will have longer wavelengths: a sound of 100 Hz in air has a wavelength of 3.4 m (11 ft). For comparison, in fresh water at room temperature the speed of sound is 1480 m/s (4856 ft/s), and the wavelength of 1-kHz sound is nearly 1.5 m (5 ft), almost five times greater than the wavelength for the same frequency in air.

Description of sound. The characterization of a sound is based primarily on human psychological responses to it. Because of the nature of human perceptions, the correlations between basically subjective evaluations such as loudness, pitch, and timbre and more physical qualities such as energy, frequency, and frequency spectrum are subtle and not necessarily universal.

The strength of a sound wave is described by its intensity. From basic physical principles, the instantaneous rate at which energy is transmitted by a sound wave through unit area is given by the product of acoustic pressure and the component of particle velocity perpendicular to the area. The time average of this quantity is the acoustic intensity. If all quantities are expressed in SI units (pressure amplitude or effective pressure amplitude in Pa, speed of sound in m/s, and density in kg/m^3), then the intensity will be in watts per square meter (W/m^2). *See* SOUND INTENSITY.

Because of the way the strength of a sound is perceived, it has become conventional to specify the intensity of sound in terms of a logarithmic scale with the (dimensionless) unit of the decibel (dB). An individual with unimpaired hearing has a threshold of perception near 10^{-12} W/m^2 between about 2 and 4 kHz, the frequency range of greatest sensitivity. As the intensity of a sound of fixed frequency is increased, the subjective evaluation of loudness also increases, but not proportionally. Rather, the listener tends to judge that every successive doubling of the acoustic intensity corresponds to the same increase in loudness. For sounds lying higher than 4 kHz or lower than 500 Hz, the sensitivity of the ear is appreciably lessened. Sounds at these frequency extremes must have higher threshold intensity levels before they can be perceived, and doubling of the loudness requires smaller changes in the intensity with the result that at higher levels sounds of equal intensities tend to have more similar loudnesses. It is because of this characteristic that reducing the volume of recorded music causes it to sound thin or tinny, lacking both highs and lows of frequency. Since most sound-measuring equipment detects acoustic pressure rather than intensity, it is convenient to define an equivalent scale in terms of the sound pressure level. The intensity level and sound-pressure level are usually taken as identical, but this is not always true. *See* DECIBEL; LOUDNESS.

How "high" sound of a particular frequency appears to be is described by the sense of pitch. A few minutes with a frequency generator and a loudspeaker show that pitch is closely related to the frequency. Higher pitch corresponds to higher frequency, with small influences depending on loudness, duration, and the complexity of the waveform. For the pure tones (monofrequency sounds) encountered mainly in the laboratory, pitch and frequency are not found to be proportional. Doubling the frequency less than doubles the pitch. For the more complex waveforms usually encountered, however, the presence of harmonics favors a proportional relationship between pitch and frequency. *See* PITCH.

Propagation of sound. Plane waves are a considerable simplification of an actual sound field. The sound radiated from a source (such as a loudspeaker, a hand clap, or a voice) must spread outward much like the widening circles from a pebble thrown into a lake. A simple model of this more realistic case is a spherical source vibrating uniformly in all directions with a single frequency of motion. The sound field must be spherically symmetric with an amplitude that decreases with increasing distance from the source, and the fluid elements must have particle velocities that are directed radially.

Not all sources radiate their sound uniformly in all directions. When someone is speaking in an unconfined space, for example an open field, a listener circling the speaker hears the voice most well defined when the speaker is facing the listener. The voice loses definition when the speaker is facing away from the listener. Higher frequencies tend to be more pronounced in front of the speaker, whereas lower frequencies are perceived more or less uniformly around the speaker.

Diffraction. It is possible to hear but not see around the corner of a tall building. However, higher-frequency sound (with shorter wavelength) tends to bend or "spill" less around edges and corners than does sound of lower frequency. The ability of a wave to spread out after traveling through an opening and to bend around obstacles is termed diffraction. This is why it is often difficult to shield a listener from an undesired source of noise, like blocking aircraft or traffic noise from nearby residences. Simply erecting a brick or concrete wall between source and receiver is often an insufficient remedy, because the sounds may diffract around the top of the wall and reach the listeners with sufficient intensity to be distracting or bothersome. *See* ACOUSTIC NOISE; DIFFRACTION.

Rays. Since the speed of sound varies with the local temperature (and pressure, in other than perfect gases), the speed of a sound wave can be a function of position. Different portions of a sound wave may travel with different speeds of sound.

Each small element of a surface of constant phase traces a line in space, defining a ray along which acoustic energy travels. The sound beam can then be viewed as a ray bundle, like a sheaf of wheat, with the rays distributed over the cross-sectional area of the surface of constant phase. As the major lobe spreads with distance, this area increases and the rays are less densely concentrated. The number of rays per unit area transverse to the propagation path measures the energy density of the sound at that point.

It is possible to use the concept of rays to study the propagation of a sound field. The ray paths define the trajectories over which acoustic energy is transported by the traveling wave, and the flux density of the rays measures the intensity to be found at each point in space. This approach, an alternative way to study the propagation of sound, is approximate in nature but has the advantage of being very easy to visualize.

Reflection and transmission. If a sound wave traveling in one fluid strikes a boundary between the first fluid and a second, then there may be reflection and transmission of sound. For most cases, it is sufficient to consider the waves to be planar. The first fluid contains the incident wave of intensity I_i and reflected wave of intensity I_r; the second fluid, from which the sound is reflected, contains the transmitted wave of intensity I_t. The directions of the incident, reflected, and transmitted plane sound waves may be specified by the grazing angles θ_i, θ_r, and θ_t (measured between the respective directions of propagation and the plane of the reflecting surface). *See* REFLECTION OF SOUND.

Absorption. When sound propagates through a medium, there are a number of mechanisms by which the acoustic energy is converted to heat and the sound wave weakened until it is entirely dissipated. This absorption of acoustic energy is characterized by a spatial absorption coefficient for traveling waves. *See* SOUND ABSORPTION.

Alan B. Coppens

Sound absorption The process by which the intensity of sound is diminished by the conversion of the energy of the sound wave into heat. The absorption of sound is an important case of sound attenuation. Regardless of the material through which sound passes, its intensity, measured by the average flow of energy in the wave per unit time per unit area perpendicular to the direction of propagation, decreases with distance from the source. This decrease is called attenuation. In the simple case of a point source of sound radiating into an ideal medium (having no boundaries, turbulent fluctuations, and the like), the intensity decreases inversely as the square of the distance from the source. This relationship exists because the spherical area through which the energy propagates per unit time increases as the square of the propagation distance. This attenuation or loss may be called geometrical attenuation.

In addition to this attenuation due to spreading, there is effective attenuation caused by scattering within the medium. Sound can be reflected and refracted when incident on media of different physical properties, and can be diffracted and scattered as it bends around obstacles. These processes lead to effective attenuation, for example, in a turbulent atmosphere; this is easily observed in practice and can be measured, but is difficult to calculate theoretically with precision. *See* DIFFRACTION; REFLECTION OF SOUND; REFRACTION OF WAVES.

In actual material media, geometrical attenuation and effective attenuation are supplemented by absorption due to the interaction between the sound wave and the physical properties of the propagation medium itself. This interaction dissipates the sound energy by transforming it into heat and hence decreases the intensity of the wave. In all practical cases, the attenuation due to such absorption is exponential in character. *See* SOUND INTENSITY.

The four classical mechanisms of sound absorption in material media are shear viscosity, heat conduction, heat radiation, and diffusion. These attenuation mechanisms are generally grouped together and referred to as classical attenuation or thermoviscous attenuation. *See* CONDUCTION (HEAT); DIFFUSION; HEAT RADIATION; VISCOSITY.

Sound absorption in fluids can be measured in a variety of ways, referred to as mechanical, optical, electrical, and thermal methods. All these methods reduce essentially to a measurement of sound intensity as a function of distance from the source.

The amount of sound that air absorbs increases with audio frequency and decreases with air density, but also depends on temperature and humidity. Sound absorption in air depends heavily on relative humidity. The reason for the strong dependence on relative humidity is molecular relaxation. One can note the presence of two transition regimes in most of the actual absorption curves, representing the relaxation effects of nitrogen and oxygen, the dominant constituents of the atmosphere. *See* ATMOSPHERIC ACOUSTICS.

Sound absorption in water is generally much less than in air. It also rises with frequency, and it strongly depends on the amount of dissolved materials (in particular, salts in seawater), due to chemical relaxation. *See* UNDERWATER SOUND.

The theory of sound attenuation in solids is complicated because of the presence of many mechanisms responsible for it. These include heat conductivity, scattering due to anisotropic material properties, scattering due to grain boundaries, magnetic domain losses in ferromagnetic materials, interstitial diffusion of atoms, and dislocation relaxation processes in metals. In addition, in metals at very low temperature the interaction between the lattice vibrations (phonons) due to sound propagation and the valence of electrons plays an important role, particularly in the superconducting domain. *See* CRYSTAL DEFECTS; DIFFUSION; FERROMAGNETISM; SUPERCONDUCTIVITY.

Henry E. Bass; Angelo J. Campanella; James P. Chambers; R. Bruce Lindsay

Sound field enhancement A system for enhancing the acoustical properties of both indoor and outdoor spaces, particularly for unamplified speech, song, and music. Systems are subdivided into those that primarily change the natural reverberation in the room, increasing its level and decay time (reverberation enhancement systems); and those that essentially replace the natural reverberation (sound field synthesis systems). Both systems may use amplifiers, electroacoustic elements, and signal processing to add sound field components to change the natural acoustics. Sound field enhancement is used to produce variable acoustics, to produce a particular acoustics which is not attainable by passive means, or because a venue has one or all of the following deficiencies: (1) unsuitable ratio between direct, early reflected, and reverberant sound; (2) unsatisfactory early reflection pattern; and (3) short reverberation times. *See* AMPLIFIER.

The purpose of sound field enhancement systems is not to provide higher speech intelligibility or clarity—in contrast to sound reinforcement systems—but to adjust venue characteristics to best suit the program material and venue, and in such a way optimize the subjectively perceived sound quality and enjoyment. *See* SOUND-REINFORCEMENT SYSTEM.

The use of sound field enhancement ranges from systems for homes (surround sound systems and home theater sound systems) to systems for rooms seating thousands of listeners (concert halls, performing arts centers, opera houses, and churches). Outdoor spaces may have elaborate designs to approach the sound quality of indoor spaces (advanced concert pavilions, outdoor concert venues). Systems may be used to improve conditions for both performers and listeners. *See* ARCHITECTURAL ACOUSTICS; REVERBERATION.

Basic elements. A sound field enhancement system may consist of one or more microphones, systems for amplification and electronic signal processing, and one or more loudspeakers. The signal processing equipment may include level controls, equalizers, delays and reverberators, systems for prevention of howl and ringing, as well as frequency dividing networks and crossover networks. *See* EQUALIZER; LOUDSPEAKER; MICROPHONE.

Prerequisites. Many indoor venues have acoustical characteristics unsuitable for communication such as speech, song, or music. Outdoor venues seldom have satisfactory acoustical conditions for unamplified acoustic communication. Satisfactory acoustic conditions include for speech, good speech intelligibility and sound quality; and for music, appropriate loudness, clarity, reverberation envelopment, spaciousness, and other sound quality characteristics for the music to be performed and enjoyed. Rooms with insufficient reverberation are often perceived to be acoustically “dry” or “dead.” Insufficient reverberation usually is due to lack of room volume for the existing sound absorption areas in the room. The reverberation time is essentially proportional to room volume and inversely proportional to sound absorption, as indicated by Sabine's formula. In these cases, sound field enhancement systems may be used to improve acoustical conditions, by increasing the sound level and reverberation time. Absence of noise is also necessary for good acoustic conditions. *See* ACOUSTIC NOISE; SOUND ABSORPTION.

System design. A simple reverberation enhancement system uses a microphone to pick up the sound signal to be reverberated, amplified, and reradiated in the auditorium. In most cases, reverberation is added to the original signal by electronic digital signal processing. An advanced sound field enhancement system uses several, possibly directional microphones to pick up sound selectively from the stage area. Signal processing and many loudspeakers are used to obtain the desired sound field for the audience and performers. Most performing art centers need variable acoustic conditions that can be changed instantaneously using electronic controls.

Categories. Sound field enhancement systems can be classified according to different criteria. One possible scheme is shown in the illustration. Sound field synthesis systems are designed to provide the necessary sound field components, such as early reflections from side walls and ceiling (the latter are usually called envelopmental sound), and reverberation by using digital filters, which can be designed to imitate a specific desired room

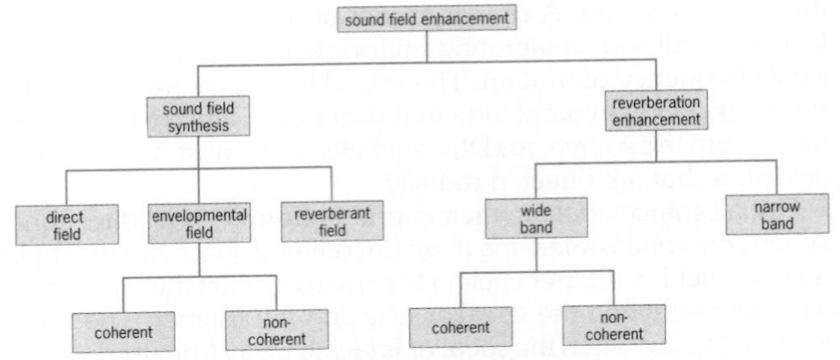

Classification scheme for sound field enhancement systems.

response. Reverberation enhancement systems are designed to primarily increase the reverberation time and sound level of the reverberant field, while having negligible influence on the direct sound and the early reflected sound. Reverberation enhancement systems can be further subdivided into regenerative systems, based on the acceptance of unavoidable positive feedback between loudspeakers and microphones to increase the reverberation time; and nonregenerative systems, based on the use of electronic reverberators. Regenerative systems can be designed so as to work over a wide or narrow frequency range; nonregenerative systems are usually wide-frequency-range systems. *See* REFLECTION OF SOUND; SOUND. Mendel Kleiner

Sound intensity

Sound intensity A fundamental acoustic quantity which describes the rate of flow of acoustic energy through a unit of area perpendicular to the flow direction. The unit of sound intensity is watt per square meter. The intensity is calculated at a field point (x) as a product of acoustic pressure p and particle velocity u. Generally, both p and u are functions of time, and therefore an instantaneous intensity vector is defined by the equation below.

$$\vec{I}_i(x,t) = p(x,t) \cdot \vec{u}(x,t)$$

The time-variable instantaneous intensity, $\vec{I}_i(x,t)$, which has the same direction as $\vec{u}(x,t)$, is a nonmoving static vector representing the instantaneous power flow through a point (x). *See* POWER; SOUND PRESSURE.

Many acoustic sources are stable at least over some time interval so that both the sound pressure and the particle velocity in the field of such a source can be represented in terms of their frequency spectra.

The applications of sound intensity were fully developed after a reliable technique for intensity measurement was perfected. Sound intensity measurement requires measuring both the sound pressure and the particle velocity. Very precise microphones for sound-pressure measurements are available.

An application of the intensity technique is the measurement of sound power radiated from sources. The knowledge of the radiated power makes it possible to classify, label, and compare the noise emissions from various pieces of equipment and products and to provide a reliable input into environmental design. *See* SOUND. Jiri Tichy

Sound pressure

Sound pressure The incremental variation in the static pressure of a medium when a sound wave is propagated through it. Sound refers to small-amplitude, propagating pressure perturbations in a compressible medium. These pressure disturbances are related to the corresponding density perturbation via the material equation of state, and the manner in which these disturbances propagate is governed by a wave equation. Since a pressure variation with time is easily observed, the science of sound is concerned with small fluctuating pressures and their spectral characteristics. The unit of pressure commonly used in acoustics is the micropascal ($1\ \mu\text{Pa} = 1\ \mu\text{N/m}^2 = 10^{-5}\ \text{dyne/cm}^2 = 10^{-5}\ \mu\text{bar}$). One micropascal is approximately 10^{-11} times the normal atmospheric pressure. *See* PRESSURE; PRESSURE MEASUREMENT; WAVE MOTION.

The instantaneous sound pressure at a point can be harmonic, transient, or a random collection of waves. This pressure is usually measured with an instrument that is sensitive to a particular band of frequencies. A concept widely used in acoustics is "level," which refers to the logarithm of the ratio of any two field quantities. When the ratio is proportional to a power ratio, the unit for measuring the logarithm of the ratio is called a bel, and the unit for measuring this logarithm multiplied by 10 is called a decibel (dB). The sound intensity, which describes the rate of flow of acoustic energy (acoustic power flow) per unit area, is given by the mean square pressure divided by the acoustic impedance, defined as the product of the medium density and compressional wave speed. *See* DECIBEL; NOISE MEASUREMENT; SOUND; SOUND INTENSITY.

William M. Carey

Sound recording The technique of entering sound, especially music, on a storage medium for playback at a subsequent time. The storage medium most widely used is magnetic tape. *See* MAGNETIC RECORDING.

Monophonic. In the simplest form of sound recording, a single microphone picks up the composite sound of a musical ensemble, and the microphone's output is recorded on a reel of $^{1}/_{2}$-in. (6.4-mm) magnetic tape. This single-track, or monophonic, recording suffers from a lack of dimension, since in playback the entire ensemble will be heard from a single point source: the loudspeaker.

Stereophonic. An improved recording, with left-to-right perspective and an illusion of depth, may be realized by using two microphones which are spaced or angled appropriately and whose outputs are routed to separate tracks on the tape recorder. These two tracks are played back over two loudspeakers, one each on the listener's left and right. Under ideal conditions this stereophonic system will produce an impressive simulation of the actual ensemble. *See* STEREOPHONIC SOUND.

Binaural. In this system, two microphones are placed on either side of a small acoustic baffle in an effort to duplicate the human listening condition. The recording is played back over headphones, so that the microphones are, in effect, an extension of the listener's hearing mechanism.

Multitrack. To give the recording engineer more technical control over the recording medium, many recordings are now made using a multiple-microphone technique. In place of the stereo microphone, one or more microphones are located close to each instrument or group of instruments. *See* MICROPHONE.

In the control room the engineer mixes the outputs of all microphones to achieve the desired musical balance. As a logical extension of this technique, the microphone outputs may not be mixed at the time of the recording, but may be routed to 16 or more tracks on a tape recorder, for mixing at a later date.

When many microphones are so used, each instrument in the ensemble is recorded—for all practical purposes—simultaneously. The complex time delay and acoustic relationships within the room are lost in the recording process, and the listener hears the entire ensemble from one perspective only, as though he or she were as close to each instrument as is each microphone. Electronic signal processing devices may not be entirely successful in restoring the missing information, and the listener hears a recording that may be technically well executed, yet lacking the apparent depth and musical cohesiveness of the original. However, the multitrack technique becomes advantageous when it is impractical or impossible to record the entire ensemble at once stereophonically. *See* SOUND-REPRODUCING SYSTEMS.

John M. Woram

Sound-recording studio An enclosure treated acoustically to provide controlled recording of music or speech. Professional sound-recording studios range from small home studios (such as a converted garage or basement) to large facilities using state-of-the-art equipment. They consist of two rooms, the main studio or stage (performance area) and the audio control room. The stage includes one or more microphones, is acoustically isolated to ensure that external sounds do not interfere with the recording, and can be broken up into small booths to separate individual performers. The audio control room contains the mixing console and recording devices. *See* ACOUSTIC NOISE; ARCHITECTURAL ACOUSTICS; SOUND; SOUND RECORDING.

Michael Rettinger; Angelo Campanella

Sound-reinforcement system A system that amplifies or enhances the sound of a speaking person, singer, or musical instrument. Sound-reinforcement systems are used to increase the clarity of the original sound as well as to add loudness and reverberation. Those that improve loudness and clarity are often called public address systems, and those that improve reverberation characteristics are called reverberation enhancement systems. Systems for increased loudness are used when listeners are located far from the sound source, as in a large audiorium or outdoors. Systems for increased clarity are used to overcome the effects of unwanted noise or excessive reverberation in a room (such as concert halls and churches with reverberation characteristics designed for music). Reverberation enhancement systems are used to add sound field components, such as reverberation to rooms having unsuitable levels of direct, early reflected sound and reverberant sound or having short reverberation times. *See* ARCHITECTURAL ACOUSTICS; REVERBERATION; SOUND; SOUND FIELD ENHANCEMENT.

The basic elements of a sound-reinforcement system are (1) one or more microphones, (2) amplification and electronic signal processing systems, and (3) one or more loudspeakers. The signal processing equipment may include level controls, equalizers or tone controls, systems for prevention of howl and ringing, delays and reverberators, as well as frequency dividing and crossover networks.

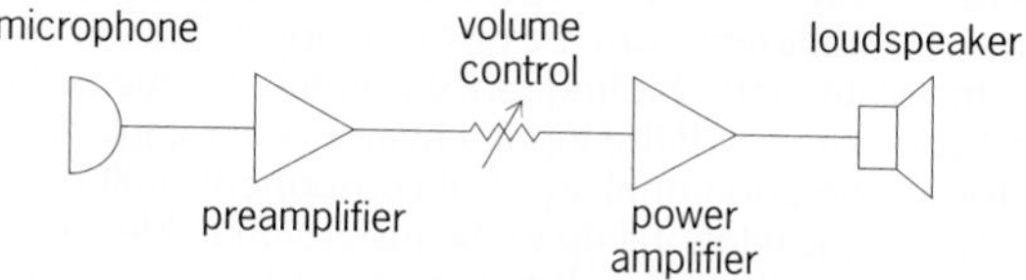

Basic sound-reinforcement system.

The illustration shows the signal flow diagram of a simple system. In the simplest systems, all elements are joined as one physical unit. An example of a simple public address system is the hand-held battery-driven electronic megaphone, consisting of a microphone, a frequency-response-shaping filter, an amplifier, and a horn loudspeaker. Most sound-reinforcement systems, however, are assembled from individual elements to optimize and adapt the system performance for the needs of a particular installation.

David Lloyd Klepper; Mendel Kleiner

Sound-reproducing systems Systems that attempt to reconstruct some or all of the audible dimensions of an acoustic event that occurred elsewhere. A sound-reproducing system includes the functions of capturing sounds with microphones, manipulating those sounds using elaborate electronic mixing consoles and signal processors, and then storing the sounds for reproduction at later times and different places. *See* MICROPHONE.

Certain technical variables are relevant to the reliable reconstruction of audio events. These include frequency range, dynamic range, and linear and nonlinear distortions. Traditionally the audible frequency range has been considered to be 20 Hz to 20 kHz. However, lower frequencies can generate interesting tactile impressions, and it is now argued by some that higher frequencies (even 50 kHz and beyond) add other perceptual nuances. *See* HEARING (HUMAN).

In humans, dynamic range is the range of sound level from the smallest audible sound to the largest sound that can be tolerated. In devices, it is the range from background noise to unacceptable distortion. From the threshold of audibility to the onset of discomfort is about 120 decibels at middle and high frequencies. Microphones are not a limiting factor since the best have dynamic ranges in excess of 130 dB. In the recording studio, dynamics are manipulated by electronic gain compensation, so the ultimate limitation is the storage medium. For example, a 16-bit compact disk (CD) can exceed 100 dB signal-to-noise ratio, while 24-bit digital media can exceed any reasonable need for dynamic range. Background acoustical noise in studios and concert halls sets the lower limit for recordings, just as background noise in homes and cars sets a lower limit for playback. Quiet concert halls and homes are around 25 dB sound pressure level (SPL) at middle and high frequencies. Crescendoes in music and movies approximate 105 dB. *See* ACOUSTIC NOISE; COMPACT DISK; DECIBEL; DISTORTION (ELECTRONIC CIRCUITS); GAIN; LOUDNESS; SIGNAL-TO-NOISE RATIO; SOUND; SOUND PRESSURE.

Variations in amplitude and phase as functions of frequency are known as linear distortions, since (ideally) they do not vary with signal level. In practice they often do, and in the process generate nonlinear distortions. In terms of their importance in preserving the timbre of voices and musical instruments, the amplitude versus frequency characteristic, known commonly as the frequency response, is the dominant factor. Humans are very sensitive to variations in frequency response, the amount depending on the bandwidth of the variation. It is convenient to discuss this in terms of the quality factor (Q) of resonances, since most such variations are the result of resonances in loudspeaker transducers or enclosures. *See* DISTORTION (ELECTRONIC CIRCUITS); Q (ELECTRICITY); RESONANCE (ACOUSTICS AND MECHANICS); RESPONSE; REVERBERATION.

Nonlinear distortions occur when a device behaves differently at different signal levels. The waveform coming out of a distorting device will be different from the one entering it. The difference in shape, translated into the frequency domain, is revealed as new spectral components. If the input waveform is a single frequency (pure tone), the additional spectral components will be seen to have a harmonic relationship to the input signal. Hence they are called harmonic distortion. If two or more tones are applied to the device, nonlinearities will create harmonics of all of the tones (harmonic distortion) and, in addition, more new spectral components that are sum-and-difference multiples of combinations of the tones. These additional components are called intermodulation distortion.

Perceptually, listeners are aided by a phenomenon called masking, in which loud sounds prevent some less loud sounds from being heard. This means that the music causing the distortion inhibits the ability to hear it. Rough guidelines suggest that, in music, much of the time we can be unaware of distortions measuring in whole percentages, but that occasionally small fractions of a percent can be heard. *See* MASKING OF SOUND.

Loudspeakers radiate sound in all directions, so that measurements made at a single point represent only a tiny fraction of the total sound output. In rooms, most of the sound we hear from loudspeakers reaches our ears after reflections from room boundaries and furnishings, meaning that our perceptions may be more influenced by measures of reflected and total sound than by a single measurement, say, on the principal axis. *See* ARCHITECTURAL ACOUSTICS; SOUND FIELD ENHANCEMENT.

To be useful, technical measurements must allow us to anticipate how these loudspeakers might sound in rooms. Consequently, it is necessary to measure sounds radiated in many different directions at points distributed on the surface of an imaginary sphere surrounding the loudspeaker. From these data, it is possible to calculate the direct sound from the loudspeaker to the listener, estimates of strong early reflected sounds from room boundaries, and estimates of later reflected or reverberant sounds. It is also possible to calculate measures of total sound output, regardless of the direction of radiation (sound power) and of the uniformity of directivity as a function of frequency. All of these measured quantities are needed in order to fully evaluate the potential for good sound in rooms. *See* DIRECTIVITY.

Binaural system configurations are sound-reproduction techniques in which a dummy head, equipped with microphones in the ear locations, captures the original performance. Listeners then audition the reproduced performance through headphones, with the left and right ears hearing, ideally, what the dummy head "heard." The system is good, but flawed in that sounds that should be out in front (usually the most important sounds) tend to be perceived to be inside, or close to, the head. Addressing this limitation, systems have been developed that use two loudspeakers, combined with signal processing to cancel the acoustical crosstalk from the right loudspeaker to the left ear, and vice versa. The geometry of the loudspeakers and listener is fixed. In this mode of listening, sounds that should be behind are sometimes displaced forward, but the front hemisphere can be very satisfactorily reproduced. *See* BINAURAL SOUND SYSTEM; EARPHONES; VIRTUAL ACOUSTICS.

Multichannel audio systems began with stereophonic systems using two channels, because in the 1950s the technology was limited to one modulation for each groove wall of a record. Good stereo imaging is possible only for listeners on the axis of symmetry, equidistant from both loudspeakers. *See* MODULATION; STEREOPHONIC SOUND.

Quadraphonic systems appeared in the 1970s, and added two more loudspeakers behind the listeners, mirroring the ones in front. The most common systems exhibited generous interchannel crosstalk or leakage. To hear the proper spatial perspective, the preferred listening location was restricted to front/back as well as left/right symmetry. The failure to agree on a single standard resulted in the system's demise.

For film sound applications, Dolby Surround modified the matrix technology underlying one of the quadraphonic systems, rearranging the four matrixed channels into a left, center, right, and surround configuration. In cinemas, the single, limited-bandwidth, surround channel information is sent to several loudspeakers arranged along the side walls and across the back of the audience area. For home reproduction, the surround channel is split between two loudspeakers placed above and to the sides of the listeners.

Even with the best active (electronically steered) matrix systems, the channels are not truly discrete. Sounds leak into unintended channels where they dilute and distort directional and spatial impressions. Digital recording systems now provide five discrete full-bandwidth channels, plus a low-frequency special effects channel for movies with truly big bass sounds. For homes, 5.1 channels refers to five satellite loudspeakers, operating with a common low-frequency subwoofer. The subwoofer channel is driven through a bass-management system that combines the low frequencies of all channels, usually crossing over at 80–100 Hz so that, with reasonable care in placement, it cannot be localized.

Floyd Toole

Sourwood A deciduous tree, *Oxydendrum arboreum*, of the heath family, indigenous to the southeastern section of the United States, and found from Pennsylvania to Florida and west to Indiana and Louisiana. It is usually a small or medium-sized tree. The wood is not used commercially. Sourwood is also known as sorrel tree, and it is widely planted as an ornamental. *See* ERICALES.

Arthur H. Graves; Kenneth P. Davis

South America The southernmost of the New World or Western Hemisphere continents, with three-fourths of it lying within the tropics. South America is approximately 4500 mi

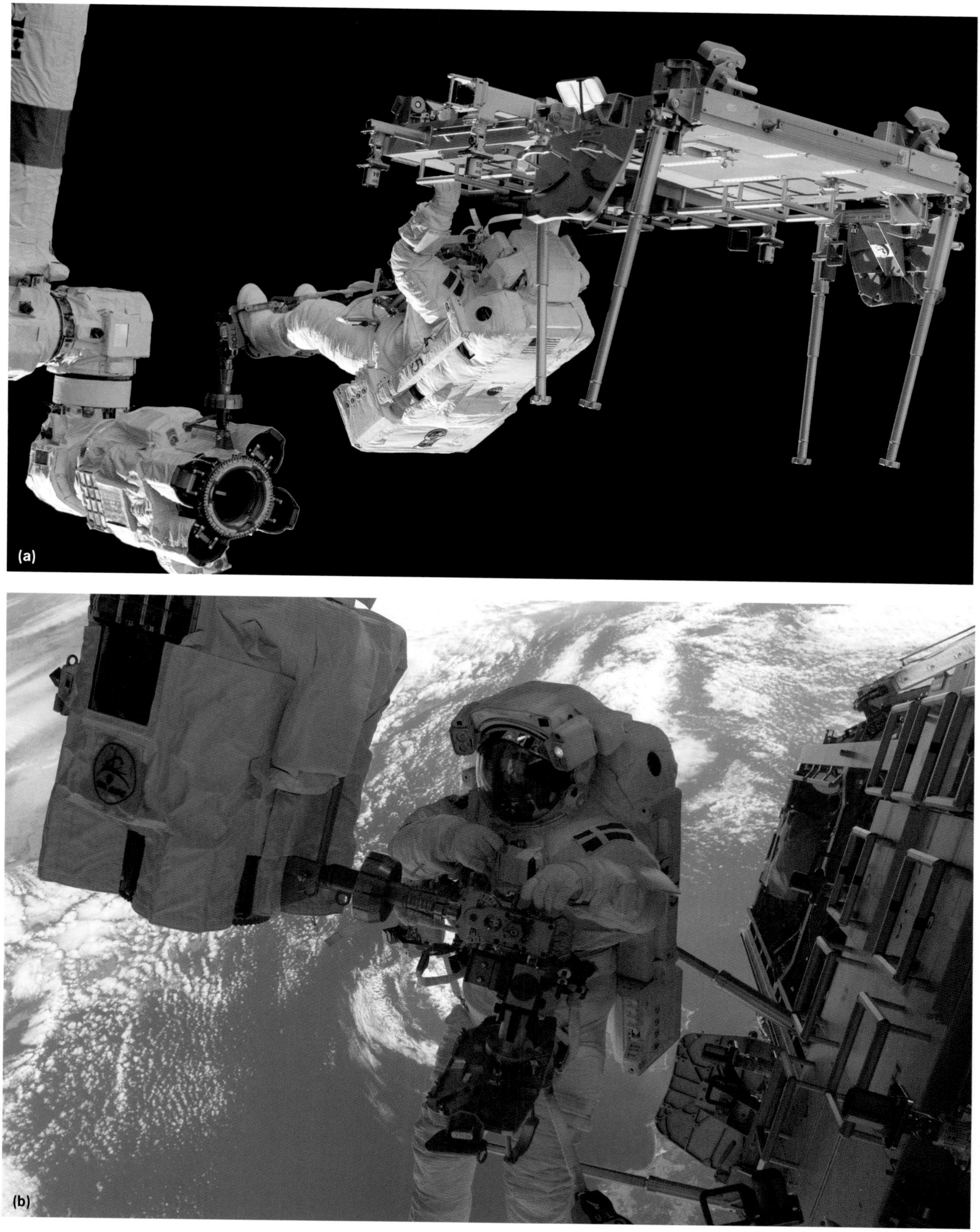

(a)

(b)

European Space Agency (ESA) astronaut Christer Fuglesang on December 14, 2006, during the second of four sessions of extravehicular activity (EVA) of mission *STS 116* of the space shuttle *Discovery* to the *International Space Station*. Astronaut Robert L. Curbeam, Jr. (not shown) also participated in this spacewalk. (*a*) Fuglesang working to relocate one of the two Crew Equipment Translation Aid (CETA) rail carts, with his feet secured on the Canadarm2 Space Station Remote Manipulator System. (*b*) Fuglesang using the Canadarm2 end effector, at left.

(7200 km) long and at its greatest width 3000 mi (4800 km). Its area is estimated to be about 7,000,000 mi^2 (18,000,000 km^2). South America has many unique physical features, such as the Earth's longest north-south mountain range (the Andes), highest waterfall (Angel Falls), highest navigable freshwater lake (Lake Titicaca), and largest expanse of tropical rainforest (Amazonia). The western side of the continent has a deep subduction trench offshore, whereas the eastern continental shelf is more gently sloping and relatively shallow. *See* CONTINENT.

South America has three distinct regions: the relatively young Andes Mountains located parallel to the western coastline, the older Guiana and Brazilian Highlands located near the eastern margins of the continent, and an extensive lowland plains, which occupies the central portion of the continent. The regions have distinct physiographic and biotic features.

The Andes altitudes often exceed 20,000 ft (6000 m) and perpetual snow tops many of the peaks, even along the Equator. So high are the Andes in the northern half of the continent that few passes lie below 12,000 ft (3600 m). Because of the vast extent of the Andes, a greater proportion of South America than of any other continent lies above 10,000 ft (3000 m). The young, rugged, folded Andean peaks stand in sharp contrast to the old, worn-down mountains of the eastern highlands. Although the Andes appear to be continuous, most geologists believe that they consist of several structural units, more or less joined. They are a single range in southern Chile, two ranges in Bolivia, and dominantly three ranges in Peru, Ecuador, and Colombia.

Except in Bolivia, where they attain their maximum width of 400 mi (640 km), the Andes are seldom more than 200 mi (320 km) wide. The average height of the Andes is estimated to be 13,000 ft (3900 m). However, it is only north of latitude 35°S that the mountains exceed elevations of 10,000 ft (3000 m).

From the southern tip of Cape Horn north to 41°S latitude, the western coastal zone consists of a broad chain of islands where a mountainous strip subsided and the ocean invaded its valleys. This is one of the world's finest examples of a fiorded coast. Nowhere along the Pacific coast is there a true coastal plain. South of Arica, Chile, the bold, precipitous coast is broken by only a few deep streams, the majority of which carry no water for years at a time. Between Arica and Caldera, Chile, there are no natural harbors and almost no protected anchorages. In fact, South America's coastline is the least indented of all the continents except Africa's. *See* FIORD.

The Caribbean coast of Colombia is a lowland formed largely of alluvium, deposited by the Magdalena and Cauca rivers, and bounded by mountains on three sides. In Venezuela, the Central Highlands rise abruptly from the Caribbean, with lowlands around Lake Maracaibo, west of Puerto Cabello, and around the mouth of the Río Tuy of the Port of Guanta. The coastal region of Guyana, Suriname, and French Guiana is a low, swampy alluvial plain 10–30 mi (16–48 km) wide, and as much as 60 mi (96 km) wide along the larger rivers. This coastal plain is being built up by sediments carried by the Amazon to the Atlantic and then deflected westward by the equatorial current and cast upon the shore by the trade winds.

There is no broad coastal plain south of the Amazon and east of the Brazilian Highlands to afford easy access to the interior. The rise from the coastal strip to the interior is quite gradual in northeastern Brazil; but southward, between Bahia and Río Grande do Sul, the steep Serra do Mar is a formidable obstacle to transportation.

Along coastal Uruguay there is a transition between the hilly uplands and plateaus of Brazil and the flat Pampas of Argentina, whereas coastal Argentina as far south as the Río Colorado, in Patagonia, is an almost featureless plain. In Patagonia, steep cliffs rise from the water's edge. Behind these cliffs lies a succession of dry, flat-topped plateaus, surmounted occasionally by hilly land composed of resistant crystalline rocks. Separating southern Patagonia from Tierra del Fuego is the Strait of Magellan, which is 350 mi (560 km) long and 2–20 mi (3–32 km) wide. Threading through numerous islands, the strait is lined on each side with fiords and mountains.

There are three great river systems in South America and a number of important rivers that are not a part of these systems. The largest river system is the Amazon which, with its many tributaries, drains a basin covering 2,700,000 mi^2 (7,000,000 km^2), or about 40% of the continent. The next largest is the system composed of the Paraguay, Paraná, and La Plata rivers, the last being a huge estuary. The third largest river system, located in southern Venezuela, is the Orinoco, which drains 365,000 mi^2 (945,000 km^2) of land, emptying into the Atlantic Ocean along the northeast edge of the continent.

The plants and animals of the South American tropics are classified as Neotropical, defined by the separation of the South American and African continents during the Middle Cretaceous (95 million years ago). The Paraná basalt flow, which caps the Brazilian shield in southern Brazil and adjacent parts of Uruguay and Argentina, as well as western Africa, indicates the previous linkage between the South American and African continents. South America has many biotic environments, including the constantly moist tropical rainforest, seasonally dry deciduous forests and savannas, and high-altitude tundra and glaciated environments.

Amazonia contains the largest extent of tropical rainforest on Earth. It is estimated to encompass up to 20% of the Earth's higher plant species and is a critically important source of fresh water and oxygen. Structurally complex, the rainforest is composed of up to four distinct vertical layers of plants and their associated fauna. The layers often cluster at 10, 20, 98, and 164 ft (3, 6, 30, and 50 m) in height. The lower canopy and forest floor are usually open spaces because of the low intensity of light (around 1%) that reaches the forest floor. Over 75% of Amazonian soils are classified as infertile, acidic, or poorly drained, making them undesirable for agriculture because of nutrient deficiencies. Most of the nutrients in the tropical rainforest are quickly absorbed and stored in plant biomass because the high annual rainfall and associated leaching make it impossible to maintain nutrients in the soils. In addition to the high structural complexity of the tropical rainforest, there is considerable horizontal diversity or patchiness. As many as 300 separate species of trees can be found in a square mile (2.6 km^2) sample tract of rainforest in Brazil. The high complexity and species diversity of the rainforest are the result of long periods of relative stability in these regions. *See* RAINFOREST.

Deciduous forest are found in areas where there is seasonal drought and the trees lose their leaves in order to slow transpiration. The lower slopes of the Andes, central Venezuela, and central Brazil are areas where these formations are found. Conifer forests occur in the higher elevations of the Andes and the higher latitudes of Chile and Argentina. *See* DECIDUOUS PLANTS.

Tropical savannas occupy an extensive range in northern South America through southeastern Venezuela and eastern Colombia. Temperate savannas are found in Paraguay, Uruguay, the Pampas of Argentina, and to the south, Patagonia. Savannas are composed of a combination of grass and tree species. The climate in these areas is often quite hot with high rates of evapotranspiration and a pronounced dry season. Most of the plants and animals of these zones are drought-adapted and fire-adapted. Tall grasses up to 12 ft (3.5 m) are common as are thorny trees of the Acacia (Fabaceae) family. Many birds and mammals are found in these zones, including anteater, armadillo, capybara (the largest rodent on Earth), deer, jaguar, and numerous species of venomous snake, including rattlesnake and bushmaster (*mapanare*). *See* SAVANNA.

South America is unique in having a west-coast desert (the Atacama) that extends almost to the Equator, probably receiving less rain than any on Earth, and an east coast desert located poleward from latitude 40°S (the Patagonian). *See* DESERT.

In Bolivia and Peru the zone from 10,000 to 13,000 ft (3000 to 3900 m), though occasionally to 15,000–16,000 ft (4500 to 4800 m), is known as the *puna*. Here the hot days contrast sharply with the cold nights. Above the *puna*, from timberline to snowline, is the *paramo*, a region of broadleaf herbs and grasses found in the highest elevations of Venezuela, Colombia, and Ecuador. Many of the plant species in these environments are similar to those found at lower elevations; however, they grow closer to the ground in order to conserve heat and moisture. *See* PARAMO; PUNA. Deborah A. Salazar; C. Langdon White

South Pole That end of the Earth's rotational axis opposite the North Pole. It is the southernmost point on the Earth and one of the two points through which all meridians pass (the North Pole being the other point). This is the geographic pole and is not the same as the south magnetic pole. The South Pole lies inland from the Ross Sea, within the land mass of Antarctica, at an elevation of about 9200 ft (2800 m).

There is no natural way to determine local Sun time because there is no noon position of the Sun, and shadows point north at all times, there being no other direction from the South Pole. *See* MATHEMATICAL GEOGRAPHY; NORTH POLE. Van H. English

Southeast Asian waters All the seas between Asia and Australia and the Pacific and the Indian oceans. They form a geographical and oceanographical unit because of their special structure and position, and make up an area of 3,450,000 mi^2 (8,940,000 km^2), or about 2.5% of the surface of all oceans.

The surface circulation is completely reversed twice a year by the changing monsoon winds. The subsurface circulation carries chiefly the outrunners of the intermediate waters of the Pacific Ocean into these seas. The tides are mostly of the mixed type. Diurnal tides are found in the Java Sea, in the Gulf of Tonkin, and in the Gulf of Thailand. Semidiurnal tides with high amplitudes occur in the Malacca Straits. *See* INDIAN OCEAN; PACIFIC OCEAN. Klaus Wyrtki; Gary Meyers

The Southeast Asian seas are characterized by the presence of numerous major plate boundaries. In Southeast Asia the plate boundaries are identified, respectively, by young, small ocean basins with their spreading systems and associated high heat flow; deep-sea trenches and their associated earthquake zones and volcanic chains; and major strike-slip faults such as the Philippine Fault (similar to the San Andreas Fault in California). The Southeast Asian seas are thus composed of a mosaic of about 10 small ocean basins whose boundaries are defined mainly by trenches and volcanic arcs. The dimensions of these basins are much smaller than the basins of the major oceans. The major topographic features of the region are believed to represent the surface expression of plate interactions, the scars left behind on the sea floor. *See* PLATE TECTONICS. Dennis E. Hayes

Soybean *Glycine max*, a legume native to China that has become a major source of vegetable protein and oil for human and animal consumption and for industrial usage. The valued portion of the plant is the seed, which contains about 40% protein and 20% oil. Illinois, Iowa, Arkansas, Missouri, Indiana, Mississippi, Minnesota, Ohio, Louisiana, and Tennessee are the major soybean producers in the United States. *See* FAT AND OIL (FOOD). Walter R. Fehr

Space Physically, space is that property of the universe associated with extension in three mutually perpendicular directions. Space, from a newtonian point of view, may contain matter, but space exists apart from matter. Through usage, the term space has come to mean generally outer space or the region beyond Earth. Geophysically, space is that portion of the universe beyond the immediate influence of Earth and its atmosphere. From the point of view of flight, space is that region in which a vehicle cannot obtain oxygen for its engines or rely upon an atmospheric gas for support (either by buoyancy or by aerodynamic effects). Astronomically, space is a part of the space-time continuum by which all events are uniquely located. Relativistically, the space and time variables of uniformly moving (inertial) reference systems are connected by the Lorentz transformations. Gravitationally, one characteristic of space is that all bodies undergo the same acceleration in a gravitational field and therefore that inertial forces are equivalent to gravitational forces. Perceptually, space is sensed indirectly by the objects and events within it. Thus, a survey of space is more a survey of its contents. *See* EUCLIDEAN GEOMETRY; LORENTZ TRANSFORMATIONS; RELATIVITY; SPACE-TIME. S. F. Singer

Space biology Collectively, the various biological sciences concerned with the study of organisms and their components when exposed to the environment of space (or of a spacecraft beyond the immediate influence of Earth and its atmosphere). Space biology studies include all the biological science disciplines, such as molecular, cellular, developmental, environmental, behavioral, and radiation biology; physiology; medicine; biophysics; and biochemistry. Space biology encompasses both basic and applied sciences, and focuses on using space research to understand and resolve biological problems on Earth, as well as alleviating or preventing the deleterious physiological changes associated with space flight or habitation in non-Earth environments. *See* SPACE.

As crewed space flight developed, medical research focused on the health and well-being of astronauts, which was paramount. This medical research has historically been included in the area of aerospace medicine. From these research efforts, directed at survival of living systems in the space environment, has evolved the broader discipline of space biology, which uses the space environment to gain an understanding of basic biological processes. *See* AEROSPACE MEDICINE.

Space biology considers all of the inherent physical factors of the space environment and their effects on living organisms: (1) factors due to the environment of space, such as radiation and reduced pressure; (2) factors due to the flight dynamics of spacecraft, including acceleration, weightlessness, and vibration; (3) factors due to the environment of the spacecraft, such as air composition, atmospheric pressure, toxic materials, noise, and confinement; and (4) factors due to the environment on nonterrestrial bodies, such as the level of gravity, composition of soil, and temperature extremes. Of these factors, only weightlessness and space radiation are unique to the space environment, but the effects of the other factors must be considered and mimicked in ground-based or flight controls when conducting space biological experiments. In addition, space experiments must take into consideration the physical effects of weightlessness on the environment surrounding and supporting the organisms. For example, neither gases nor liquids behave in a space environment as they do under the conditions found on Earth, as sedimentation, buoyancy, and convection do not occur in a weightless environment. *See* ACCELERATION; WEIGHTLESSNESS. Charles Edwin Wade

Space charge The net electric charge within a given volume. If both positive and negative charges are present, the space charge represents the excess of the total positive charge diffused through the volume in question over the total negative charge. Edward G. Ramberg

Space communications Communications between a vehicle in outer space and Earth, using high-frequency electromagnetic radiation (radio waves). Provision for such communication is an essential requirement of any space mission. The total communication system ordinarily includes (1) command, the transmission of instructions to the spacecraft; (2) telemetry, the transmission of scientific and applications data from the spacecraft to Earth; and (3) tracking, the determination of the distance (range) from Earth to the spacecraft and its radial velocity (range-rate) toward or away from Earth by the measurement of the round-trip radio transmission time and Doppler frequency shift (magnitude and direction). A specialized but commercially important application, which is excluded from consideration here, is the communications satellite system in which the spacecraft serves solely as a relay station between remote points on Earth. See COMMUNICATIONS SATELLITE; DOPPLER EFFECT; MILITARY SATELLITES; SATELLITE (SPACECRAFT); SCIENTIFIC AND APPLICATIONS SATELLITES; SPACE FLIGHT; SPACE NAVIGATION AND GUIDANCE; SPACE PROBE; SPACECRAFT GROUND INSTRUMENTATION; TELEMETERING.

Certain characteristic constraints distinguish space communication systems from their terrestrial counterparts. Although only line-of-sight propagation is required, both the transmitter and the receiver are usually in motion. The movement of satellites relative to the rotating Earth, for example, requires geographically dispersed Earth stations to achieve adequate communication with the spacecraft on each orbit.

Because enormous distances are involved (over a billion miles to the planets beyond Jupiter), the signal received on Earth from deep-space probes is so small that local interference, both artificial and natural, has to be drastically reduced. For this purpose, the transmitted frequency has to be sufficiently high, in the gigahertz range, to reduce noise originating in the Milky Way Galaxy (galactic noise background). The receiver site must be remote from technologically advanced population centers to reduce artificial noise, and at a dry location to avoid precipitation attenuation of the radio signal as well as the higher antenna thermal noise associated with higher atmospheric absolute humidity and relatively warm cloud droplets. The receiving antennas must be steerable and large, typically 85 ft (26 m) or at times 210 ft (64 m) in diameter, to enhance the received signal strength relative to the galactic noise background. Special low-noise preamplifiers such as cooled masers are mounted on the Earth receiver antenna feed to reduce the receiver input thermal noise background. Sophisticated digital data processing is required, and the ground-receiver complex includes large high-speed computers and associated processing equipment. See MASER; PREAMPLIFIER; RADIO RECEIVER; RADIO TELESCOPE.

The spacecraft communications equipment is constrained by severe power, weight, and space limitations. Typical communications equipment mass ranges from 25 to 220 lb (12 to 100 kg). Another major challenge is reliability, since the equipment must operate for years, sometimes for decades, unattended, in the difficult radiation, vacuum, and thermal environment of space. Highly reliable components and equipment have been developed, and redundancy is employed to eliminate almost all single-point failures. For example, it is not unusual to have as many as three redundant command receivers operating continuously, because without at least one such receiver in operation no command can get through, including a command to switch from a failed command receiver to a backup radio. Power can be saved by putting some or all of the redundant radios on timers, and to switch to a backup receiver if no commands have been received through the primary receiver within a predetermined interval; but the saved power may come at the cost of a possible delay in emergency response initiation.

Spacecraft power is always at a premium, and other techniques must also be used to minimize its consumption by the communication system. The transmitter is a major power consumer, so its efficiency must be maximized. All aspects of data transmission must contribute to error-free (very low bit error rate) reproduction of the telemetry data using no more power or bandwidth than is absolutely essential. Pulse-code modulation is a common technique which helps meet this goal. In general terms, space communication systems are far less forgiving than terrestrial systems and must be designed, constructed, and tested to much higher standards. See PULSE MODULATION; SPACE POWER SYSTEMS; SPACE TECHNOLOGY.

The Tracking and Data Relay Satellite System (TDRSS) consists of a series of geostationary spacecraft and an Earth terminal located at White Sands, New Mexico. The purpose of TDRSS is to provide telecommunication services between low-Earth-orbiting (LEO) user spacecraft and user control centers. A principal advantage of the system is the elimination of the need for many of the worldwide ground stations for tracking such spacecraft. The *Tracking and Data Relay Satellite* (*TDRS*) provides no processing of data; rather, it translates received signals in frequency and retransmits them. User orbits are calculated from range and range-rate data obtained through the *TDRS* by using transponders on the user spacecraft. John F. Clark; Donald Platt

Space flight The penetration by humans into the reaches of the universe above the terrestrial atmosphere, and investigation of these regions by automated, remote-controlled and crewed vehicles.

The purpose of space flight is to provide significant contributions to the physical and mental needs of humanity on a national and global basis. Such contributions fall specifically in the areas of (1) Earth resources of food, forestry, atmospheric environment, energy, minerals, water, and marine life; (2) Earth and space sciences for research; (3) commercial materials processing, manufacturing in space, and public services. More general goals of space flight include expansion of knowledge; exploration of the unknown, providing a driving force for technology advancement and, hence, improved Earth-based productivity; development and occupation of new frontiers with access to extraterrestrial resources and unlimited energy; strengthening of national prestige, self-esteem, and security; and providing opportunity for international cooperation and understanding.

This article focuses on crewed space flight. For discussion of other missions see COMMUNICATIONS SATELLITE; METEOROLOGICAL SATELLITES; MILITARY SATELLITES; SATELLITE (SPACECRAFT); SATELLITE NAVIGATION SYSTEMS; SCIENTIFIC AND APPLICATIONS SATELLITES; SPACE PROBE.

To conduct crewed space flight, the two leading space-faring nations, the United States and Russia, formerly the Soviet Union, have developed spacecraft systems and the necessary ground facilities, research and development base, operational know-how, planning experience, and management skills. In the United States, crewed space programs are conducted by the National Aeronautics and Space Administration (NASA), a federal agency established in 1958 for the peaceful exploration of space. In Russia, crewed space flights have been under the auspices of the U.S.S.R. Academy of Sciences; they are now the responsibility of the Russian Space Agency (Rosaviakosmos, or RKA). The first spacecraft with a human on board, the *Vostok 1*, piloted by Yuri A. Gagarin, was launched on April 12, 1961, from the Baikonur Cosmodrome in Kazakhstan and returned after completing one revolution of the Earth. The first American space flight of a human took place 3 weeks later, when NASA launched Alan B. Shepard on May 5 on the *Mercury-Redstone 3* for a 15-min suborbital test flight.

The early spacecraft of both nations were built for only one space flight. The first multiply reusable space vehicle, the space shuttle, was launched by the United States on April 12, 1981.

In November 1987, the 13 member nations of the European Space Agency (ESA) agreed to become the third major power engaging in human space flight. European efforts today focus on a two-pronged program: participation as a partner in the *International Space Station*, primarily with the Columbus Orbital Facility (COF), and extension of the Ariane family of expendable launch vehicles with the heavy-lift carrier Ariane 5. By joining the original partnership behind the space station and assigning astronauts to United States space shuttle missions (as did the European Space Agency and Russia), Canada and Japan have also entered the ranks of space-faring nations, while other countries, such as Brazil, are preparing to join. On October 15, 2003, the People's Republic of China launched Yang Liwei on a 21-hour orbital flight aboard the spacecraft *Shenzhou 5*, and this achievement was followed on October 12, 2005 by the 115-hour orbital flight of two taikonauts aboard *Shenhzou 6*, on September 25, 2008, by the 3-day flight of *Shenzhou 7* with three crewmembers, which included extravehicular activity.

Crewed spacecraft. A crewed spacecraft is a vehicle capable of sustaining humans above the terrestrial atmosphere. In a more limited sense, the term "crewed spacecraft" is usually understood to apply to vehicles for transporting and sustaining human crews in space for time periods limited by prestored on-board supplies, as distinct from orbital space stations which support theoretically unlimited habitation of humans in space by autonomous systems, crewed maintenance, and periodic resupply.

The basic requirements of crewed spacecraft are quite different from those of crewless space probes and satellites. The presence of humans on board necessitates a life-sustaining environment and the means to return safely to Earth. The major common feature of all crewed spacecraft, therefore, is the atmospheric return element or reentry module. It consists basically of a pressure-tight cabin for the protection, comfort, and assistance of the crew, similar to the cockpits of fighter aircraft, but shaped externally in accordance with the desired airflow and its forces, and surrounded by heat-resistant material, the thermal protection system (TPS), to cope with the high frictional and radiative heating accompanying the energy dissipation phase of the atmospheric return flight to Earth.

The main system that distinguishes crewed spacecraft from other spacecraft is the environmental control and life support system. It provides the crew with a suitably controlled atmosphere. Because all atmospheric supplies must be carried into space, it is essential to recirculate and purify the spacecraft atmosphere to keep the total weight of the vehicle within reasonable limits.

Space suits or pressure suits are mobile spacecraft or chambers that house the astronauts and protect them from the hostile environment of space. They provide atmosphere for breathing, pressurization, and thermal control; protect astronauts from heat, cold, glare, radiation, and micrometeorites; contain a communication link and hygiene equipment; and must have adequate mobility. The suit is worn during launch, docking, and other critical flight phases.

Because of the potential hazards of space flight to personnel, crewed spacecraft must meet stringent requirements of safety and hardware reliability. Reliability is associated with the probability that the spacecraft systems will operate properly for the required length of time and under the specified conditions. To assure survival and return in all foreseeable emergencies, the design of a crew-rated spacecraft includes standby systems (double or triple redundancy) and allows for launch escape, alternative and degraded modes of operation, contingency plans, emergency procedures, and abort trajectories to provide maximum probability of mission success. The priority order of this reliability requirement is (1) crew safety, (2) minimum achievable mission fulfillment, (3) mission data return, and (4) minimal degradation. *See* RELIABILITY, AVAILABILITY, AND MAINTAINABILITY.

Soviet/Russian programs. After the first *Sputnik* launch on October 4, 1957, developments of crewed space flight capability followed in quick succession, leading from the first-generation *Vostok* (East) to the second-generation *Voskhod* (Ascent) and to the third-generation *Soyuz* (Union) spacecraft. Originally engaged in an aggressive program to land a cosmonaut on the Moon before the United States lunar orbit mission of *Apollo 8* in December 1968, the Soviet Union redirected its aims, after four test failures of its N1 heavy-lift launcher, toward the development of permanent human presence in Earth orbit.

The *Vostok*, a single-seater for short-duration missions and ballistic reentry from Earth orbits, consisted of a near-spherical cabin covered entirely with thermal protection system material, having three small viewports and external radio antennas. The 7-ft (2-m) sphere of the cabin was attached to a service module. The second-generation *Voskhod*, essentially a greatly modified *Vostok*, was a short-duration multiperson craft and was designed to permit extravehicular activity or spacewalking by one of the crew.

The *Soyuz* design is much heavier, larger, and more advanced in its orbital systems, permitting extended orbital stay times. It consists of three main sections or modules: a descent vehicle, an orbital module, and an instrument-assembly module. The three elements are joined together, with the descent vehicle in the middle. Shortly before atmospheric reentry, the two outer modules are jettisoned.

Soyuz spacecraft carried originally up to three cosmonauts. However, when an accidental explosive decompression of the descent cabin during reentry caused the death of the *Soyuz 11* crew in 1971, the third seat was removed to make room for the necessary additional life-support equipment, and the *Soyuz* thereafter carried only two cosmonauts. An improved version, the *Soyuz T* (for "Transport") was introduced in 1980. A new version, the *Soyuz-TM*, with extended mission duration and new subsystems, replaced this vehicle in February 1987, after a crewless test flight in May 1986. (The third seat was reintroduced with *Soyuz TM-24* in August 1996.) In October 2002, the *Soyzuz TM* was, in turn, replaced by the *Soyuz TMA*, with modifications financed by NASA for improved safety and widened crewmember size range.

Between 1971 and 1981, five *Salyut* space stations operated in low Earth orbit successively. The most successful of them, *Salyut 6*, was the first of a "second generation" of *Salyut* space stations. Unlike its predecessors, it had two docking ports, instead of only one. This enabled it to receive visiting crews and resupply ships. Together with other new on-board systems, this feature was the key to missions of considerably extended duration. Its successor, *Salyut 7*, was launched in 1982.

On February 19, 1986, the Soviet Union launched the core vehicle in its new *Mir* (Peace) space station complex series. This complex represented a new-generation space station which evolved from the *Salyut* and *Kosmos* series vehicles. The core, an advanced version of *Salyut*, had six docking ports and consisted of four sections. Connected to various docking ports were four laboratory modules, which were launched separately between 1989 and 1996. These modules were dedicated to different scientific and technical disciplines or functions, including technological production with a shop, astrophysics, biological research, and medical research.

In its later years, *Mir* required increasing maintenance and repairs by its crews, particularly after two serious emergencies in 1997. When U.S. support of *Mir* ended at the conclusion of the joint Shuttle/Mir program (ISS Phase 1), further crewed operations of the space station were increasingly difficult and impractical for the Russian Space Agency. The last crew returned to Earth on August 27, 1999, and the station was deorbited on March 23, 2001. *See* SPACE STATION.

With the successful first test flight, from Baikonur, of the powerful expendable heavy-lift launcher *Energia* on May 15, 1987, the Soviet Union gained a tremendous new launch capability. With its second, and last, flight on November 15, 1988, the *Energia* launched the Soviet space shuttle *Buran* on its only (crewless) orbital test flight. After the collapse of the Soviet Union, the Energia/Buran programs were terminated.

United States programs. During the first decade after its inception in 1961, the United States crewed space program was conducted in three major phases—Mercury, Gemini, and Apollo.

Mercury was in its basic characteristics similar to the Soviet *Vostok*, but it weighed only about a third as much, as necessitated by the smaller missiles of the United States at that time (the Redstone and Atlas). The one-person *Mercury* capsules used ballistic reentry and were designed to answer the basic questions about humans in space: how they were affected by weightlessness, how they withstood the gravitational forces of boost and entry, how well they could perform in space. *See* WEIGHTLESSNESS.

The second United States step into space was the Gemini Program. With the two-person *Gemini* capsule, for the first time a crewed spacecraft had been given operational maneuverability in space. Its reentry module flew a lifting reentry trajectory for precise landing point control. In addition, its design permitted extravehicular activity by one of the crew.

The third-generation spacecraft, *Apollo*, was 82 ft (25 m) tall and had five distinct parts: the command module (CM), the service module (SM), the lunar module (LM), the launch escape system (LES), and the spacecraft/lunar module adapter (SLA). The three modules made up the basic spacecraft; the LES and SLA were jettisoned early in the mission after they had outlived their function. The command module served as the control center for the spacecraft and provided living and working quarters for the three-member crew for the entire flight, except for the period when two persons entered the lunar module for the descent to the Moon and return. The command module was the only part of the spacecraft that returned to Earth, flying a lifting trajectory with computer-steered maneuvers. The lunar module carried two astronauts from the orbiting command/service module (CSM) down to the surface of the Moon, provided a base of operations there, and returned the two astronauts to a rendezvous with the command service module in orbit. On the last three lunar landings, *Apollo 15–17*, lunar exploration was supported by the lunar roving vehicle.

A more powerful booster was required to lift the *Apollo* spacecraft to Earth orbit and thence to the Moon. At the Army Ballistic Missile Agency (ABMA) in 1958, a team of engineers under Wernher von Braun set out to prove that vastly more powerful space rockets could be built from existing hardware by clustering engines and tanks. The project evolved into the Saturn Program of NASA. The Saturn 5, which became the Apollo lunar launch vehicle, had the capability to lift 250,000 lb (113 metric tons) into low Earth orbit and to send 100,000 lb (45 metric tons) to the Moon.

In addition to its lunar mission in the Apollo program, the Saturn 5, in a two-stage version, served also—in its last and thirteenth flight—to launch the first United States space station, *Skylab*.

Skylab. The experimental space station *Skylab* was the largest object ever placed in space until that time, and the first crewed project in the U.S. Space Program with the specific purpose of developing the utility of space flight in order to expand and enhance humanity's well-being on Earth. To that end, *Skylab's* equipment included Earth resources remote sensing instrument and the first crewed solar telescopes in space. A total of three crews of three astronauts each carried out experiments and observations on *Skylab*. *Skylab* was launched crewless by Saturn 5 in 1973. The space station underwent repair in orbit, a first for the space program.

Apollo-Soyuz test project. The last flight of an *Apollo* spacecraft occurred as the United States part in the world's first international crewed space flight, the Apollo-Soyuz Test Project on July 15–24, 1975. During that time, three American astronauts and two Soviet cosmonauts linked up in space.

Space transportation system. After the end of the joint American-Soviet space mission, attention began to focus on the routine application of newly acquired know-how, systems, and experience, specifically in the form of the emerging Space Transportation System (STS), with the ability to transport inexpensively a variety of useful payloads to orbit, as the mainstay and "work horse" of the United States space program. The two major components of the Space Transportation System were the space shuttle and the *Spacelab*.

The space shuttle is a reusable system. It has three major elements: the orbiter, an external tank (ET) containing the liquid propellants to be used by the orbiter main engines for ascent, and two solid-propellant rocket boosters (SRBs) "strapped on" to the external tank. The orbiter and solid rocket booster casings are reusable; the external tank is expended on each launch. *See* SPACE SHUTTLE.

The orbiter *Columbia* made its first orbital flight on April 12–14, 1981. Three additional shuttles were added by the end of 1985. With the launch of the twenty-fifth shuttle mission, *51-L*, on January 28, 1986, tragedy struck the American space program. At approximately 11:40 a.m. EST, 73 s after liftoff, the flight of *Challenger*, on its tenth mission, abruptly ended in an explosion triggered by a leak in the right solid rocket booster, killing the seven-member crew. Continuation of the United States crewed space program was suspended pending a thorough reassessment of flight safety issues and implementation of necessary improvements. Shuttle flight operations resumed on September 29, 1988.

Highlights of post-*Challenger* missions included the launch of the interplanetary research probes *Magellan* to Venus and *Galileo* to Jupiter, the retrieval of the 10.5-metric-ton (11.5-ton) research satellite Long Duration Exposure Facility (LDEF), the placement in orbit of the *Hubble Space Telescope*, the launch of the European solar probe *Ulysses*, and the intricate repair and maintenance missions to the *Hubble* telescope.

On February 1, 2003, after completing a successful research mission, *Columbia* was lost with its crew during reentry. The Columbia Accident Investigation Board concluded that one of the left wing's leading-edge reinforced carbon-carbon (RCC) elements had been punctured during ascent to orbit by a "suitcase-sized" chunk of foam insulation, blown off the external tank by the supersonic airstream, hitting the leading edge and rendering the wing unable to withstand reentry longer than about 8 min after entry interface. Shuttle operations were halted as NASA and its contractors labored on intensive return-to-flight (RTF) efforts.

The space shuttle returned to the skies on July 26, 2005, with the liftoff of *Discovery*. The flight was successful but engineers were concerned about the shedding of some insulating foam material off the external tank. The second RTF test flight was delayed for nearly a year. However, with test flights in July and September 2006 and an additional flight in December, the shuttle fleet resumed its operations.

The *Spacelab* was a major adjunct of the shuttle. Developed and funded by member nations of the European Space Agency, the large pressurized *Spacelab* module with an external equipment pallet was designed to be the most important payload carrier during the space shuttle era, and flew on 16 shuttle missions between 1983 and 1997. With its large transport capacity (in weight, volume, and power supply), the *Spacelab* was intended to support a wide spectrum of missions in science, applications, and technology by providing versatile and economical laboratory and observation facilities in space for many users. Another objective was to reduce significantly the time from experiment concept to experiment result, compared with previous space practice, and also to reduce the cost of space experimentation. It allowed the direct participation of qualified men and women scientists and engineers to operate their own equipment in orbit.

International Space Station. Interest in the development of a permanent crewed platform in Earth orbit dates back to the very beginnings of human space flight. While the practical realization of the concept was accomplished and proven by the Soviet Salyut/Mir and the American Skylab programs, the real breakthrough happened on January 29, 1998, when representatives of 16 nations signed a partnership agreement for the joint development and operation of an *International Space Station* (*ISS*). The goal of the *ISS* program is to establish a permanent platform for humans to live and work in space in support of science and technology research, business, education, and exploration. Its objectives are to provide a world-class laboratory complex uniquely located in a microgravity and vacuum environment where long-term scientific research can be carried out to help fight diseases on Earth, unmask fundamental processes leading to new manufacturing processes and products to benefit life on Earth, and observe and understand the Earth's environment and the universe. When completed, the *International Space Station* will have a mass of about 1,040,000 lb (470 metric tons). It will be 356 ft (109 m) across and 290 ft (88 m) long, with almost an acre (0.4 hectare) of solar panels to provide up to 110 kilowatts of power to six state-of-the-art laboratories.

The on-orbit assembly of the *ISS* began with the launch of the U.S.-owned, Russian-built Functional Cargo Block (FGB) *Zarya* ("Dawn") on November 20, 1998. On October 31, 2000, a Soyuz-U carrier lifted off from Baikonur and placed in orbit *Soyuz TM-31*, carrying the first resident crew for the space station. Since then, each crew has remained on the station for about six months before rotating with the subsequent crew.

Early in 2003, further progress in *ISS* assembly was brought to a halt by the standdown of the space shuttles after the *Columbia* loss. Crew rotation flights could be accomplished only by Russian Soyuz vehicles. As a consequence of the reduction in resupply missions to the station, which now could be supported only by Russian crewless automated Progress cargo ships, station crew size was reduced from three to a two-person "caretaker" crew per expedition. With the return to flight of the shuttle, *ISS* assembly resumed in 2006.

Vision for space exploration. On January 14, 2004, U.S. President Bush announced the Vision for Space Exploration, a long-term plan for returning astronauts to the Moon to prepare for voyages to Mars and other planets in the solar system. The plan called for completion of the *International Space Station* and retirement of the space shuttle by 2010, and the development of new spacecraft and launch systems. To support this initiative, NASA is developing a family of spacecraft, launchers, and associated hardware, collectively known as Project Constellation.

The Orion spacecraft will be shaped like an Apollo capsule but will be significantly larger. Plans call for it to be launched into low-earth orbit, either for solo flights or for missions to the *International Space Station*, using the proposed Ares I rocket, which would consist of a single solid rocket booster derived from those used in the space shuttle system, and a liquid-fueled second stage.

Lunar flights would involve the use of a larger launch vehicle, the proposed Ares V, which would use five RS-68 engines (more powerful and less expensive than the space shuttle main engines) and a pair of five-segment solid rocket boosters. The Ares V would carry a combination of the Earth Departure Stage (EDS), an enlarged version of the S-IVB upper stage of the Saturn V rocket, and the Lunar Surface Access Module (LSAM). Like the Apollo lunar module, the LSAM would consist of an ascent stage, which would house the four-person crew, and a descent stage, which would have the landing legs, most of the crew's oxygen and water, and scientific equipment.

In 2006, NASA announced plans to eventually construct a solar-powered lunar base located near one of the poles of the Moon, rather than landing at several sites, as was done in the Apollo missions. An incremental buildup is envisioned.

Jesco von Puttkamer; Jonathan Weil

Space navigation and guidance The determination of the position and velocity of a space probe with respect to a target body or a known reference body such as the Earth (navigation) and, based upon this determination, the application of propulsive maneuvers to alter the subsequent path of the probe (guidance).

Space navigation can be viewed as determining the current position and velocity of the probe and then using that determination as the basis of predicting future motion. This determination is made by taking a series of measurements relating to the probe's location and motion and combining these measurements in such a manner as to make the most accurate estimate of the probe's current position and velocity, taking into account possible small errors or inaccuracies in the measurements themselves. One of the most powerful (and accurate) measurements which can be made is the relative velocity between an Earth tracking station and the space probe itself. This is accomplished by broadcasting to the probe an electromagnetic signal which consists of a single tone having a stable frequency. The probe will receive this signal shifted in frequency in exact proportion to the relative station-probe velocity and then immediately will rebroadcast it back to the Earth. This frequency shift is known as the Doppler effect. It is also possible to measure the station-probe distance (or range) using electromagnetic signals. An additional, very powerful measurement, known as delta-differential-one-way-range (denoted Δ-DOR or delta-DOR), is performed by simultaneously observing both a space probe and a nearby extragalactic radio source (quasar) from two tracking stations. These four simultaneous measurements are differenced to form an extremely accurate estimate of the angular distance between the probe and quasar as seen in the sky. *See* DOPPLER EFFECT.

For missions to the satellites of the outer planets (Jupiter and beyond) or Pluto, or for a mission to a body such as a comet or asteroid, the position of the target may be sufficiently uncertain as to make a strictly Earth-relative navigation scheme inadequate. Here it is necessary to make measurements that directly involve the target. Typical of these is to obtain optical measurements of the locations of the target relative to a star as seen from the probe itself using an onboard camera.

In its simplest form, guidance consists of comparing the predicted future motion of the probe against that which is desired, and if these are sufficiently different, executing a propulsive maneuver to modify that future position. Such maneuvers are called trajectory correction maneuvers (TCMs). Typically, the probe will contain a number of small rocket motors which can be fired in any desired direction by first rotating the spacecraft away from its cruise orientation and then holding the new attitude fixed while the rocket motor is firing.

Modern missions generally employ sequences of these basic navigation and guidance functions. An entirely different method of making very large changes to the orbit of a space probe utilizes the gravity of intermediate bodies rather than the engines on the probe. These multiplanet missions require very accurate navigation to the intermediate targets. Such missions may gather scientific information at both the intermediate bodies and the ultimate target or may simply use the intermediate bodies as an extremely efficient means of getting to the final target. At its simplest, the method of "gravity-assist" uses a close flyby of a massive body to exchange energy and momentum between the space probe and the body.

For missions designed to orbit another body, a large terminal guidance maneuver, also called an orbit insertion maneuver is applied at the encounter with the target body. These maneuvers are quite large and may use the large majority of the propellant carried by the probe. For orbiter missions, navigation and guidance takes place much as for interplanetary cruise with both orbit determination and correction maneuvers being performed on a regular basis. *See* GUIDANCE SYSTEMS; SPACE PROBE.

Dennis V. Byrnes; David W. Curkendall

Space power systems On-board assemblages of equipment to generate, store, and distribute electrical energy on satellites and spacecraft. A reliable source of electrical power is required for supplying energy to the spacecraft and its payloads during launch and through several years of operational lifetime in a space environment. Present-generation spacecraft fly power systems from tens of watts to several kilowatts. Each of the three fuel cells on the space shuttle delivers 12 kW continuous power and 16 kW peak power. The International Space Station's solar arrays will generate 110 kW total power, with approximately 46 kW available for research activities. *See* SATELLITE (SPACECRAFT); SPACE FLIGHT; SPACE PROBE; SPACE SHUTTLE; SPACE STATION.

With few exceptions, the power systems for United States satellites have used photovoltaic generation for power, batteries for energy storage, and a host of electrical equipment for appropriate regulation, conversion, and distribution. Fuel cells have been limited primarily to use in the crewed space program, supplying power for *Gemini, Apollo*, and the space shuttle. Radioisotope thermoelectric generators (RTGs) have powered, or augmented solar power on, many planetary missions and probes, powered lunar instruments left on the Moon by the *Apollo* missions, and augmented the solar-array battery power system on at least one Earth-orbiting spacecraft. *See* ENERGY STORAGE; FUEL CELL; SOLAR CELL; THERMOELECTRICITY.

Many factors influence the final configuration of a power system. Basic to the initial consideration are the nature of the mission (Earth-orbiting or planetary) and mission lifetime. Other relevant factors include (1) spacecraft and payload requirements with consideration to average and peak power loads; (2) effect of environment such as orbit period, length of time the spacecraft is in sunlight and shadow, radiation particle damage, and space charging; and (3) constraints imposed by the spacecraft such as weight, volume, spacecraft shape and appendages, satellite attitude control, electromagnetic radiation limitations, characteristics of payloads, and thermal dissipation. For spacecraft that are shuttle-launched, additional considerations include compatibility with shuttle payload safety requirements and, in some cases, the retrievability by the shuttle for return to Earth. The weight of the power system ranges from 15 to 25% of the spacecraft weight.

Jeffrey C. Mitchell

Space probe An automated crewless vehicle, the payload of a rocket-launching system, designed for flight missions to other planets, to the Moon, and into interplanetary space, as distinguished from Earth-orbiting satellites (see table).

Types of missions. The space probe is used primarily for scientific purposes, which are stated as the mission objectives. Missions have been designed to explore many diverse targets, such as the rocky bodies of the inner solar system, including Mercury, Venus, Mars, and the asteroids; the giant gaseous planets in the outer solar system, Jupiter, Saturn, Uranus, and Neptune; and icy bodies originating in the outer solar system, such as comets, Kuiper Belt objects, and the dwarf planet Pluto. Other missions are designed to study solar physics, the properties of interplanetary space, or targets far beyond the solar system. Most spacecraft launched to a planet or other body also study the environment of charged particles and electromagnetic fields in interplanetary space during their cruise phase en route to the destination.

Missions may also be categorized by complexity. The simplest are flyby spacecraft, which study their target body during a relatively brief encounter period from a distance of hundreds to thousands of miles as they continue past. Next are orbiters, which circle a planet or other body for extended study; some may carry atmospheric descent probes. Even more complex are lander missions, which touch down on a planet or other body for the collection of on-site data; some may bear exploration rovers designed to range beyond the immediate landing site. Finally, the most complex space probes envisaged are sample-return missions, which would collect specimen material from a target body and return it to Earth for detailed study.

Spacecraft subsystems. In the broadest terms, a space probe may be considered a vehicle that transports a payload of sensing instruments to the vicinity of a target body. Thus, the spacecraft must include a number of subsystems to provide power, to communicate with Earth, to maintain and modify attitude and perform maneuvers, to maintain acceptable on-board temperature, and to manage the spacecraft overall. *See* SPACE TECHNOLOGY.

Scientific instruments. The scientific payload may be divided into remote-sensing instruments such as cameras, and direct-sensing instruments such as magnetometers or dust detectors. They may be classified as passive instruments, which detect radiance given off by a target body, or active, which emit energy such as radar pulses to characterize a target body. *See* ASTRONOMICAL IMAGING; REMOTE SENSING.

Power subsystem. Electrical power is required for all spacecraft functions. The total required power ranges from about 300 to 2500 W for current missions, depending on the complexity of the spacecraft. The power subsystem must generate, store, and distribute electrical power. All space probes launched so far have generated power either via solar panels or via radioisotope power systems. *See* NUCLEAR BATTERY; SOLAR CELL; SPACE POWER SYSTEMS.

Telecommunications subsystem. In order to accomplish its mission, the spacecraft must maintain communications with Earth. This includes receiving commands sent from ground controllers, as well as transmitting scientific data and routine engineering "housekeeping" data. All of these transmissions are made in various segments of the microwave spectrum. The design of the telecommunications subsystem takes into account the volume of data to be transmitted and the distance from Earth at which the spacecraft will operate, dictating such considerations as the size of antennas and the power of the on-board transmitter. *See* MICROWAVE; TELEMETERING.

Advanced planetary probes have carried a dish-shaped high-gain antenna which is the chief antenna used to both transmit and receive. These antennas typically consist of a large parabolic reflector, with a subreflector mounted at the main reflector's focus in a Cassegrain-type configuration. In the interest of redundancy and in the event that Earth pointing is lost, spacecraft virtually always carry other on-board antennas. These may be low-gain antennas, which typically offer nearly omnidirectional coverage except for blind spots shadowed by the spacecraft body, or medium-gain antennas, which provide a beam width of perhaps 20–30°. *See* ANTENNA (ELECTROMAGNETISM).

Attitude-control subsystem. It would be impossible to navigate the spacecraft successfully or point its scientific instruments or antennas without closely controlling its orientation in space, or attitude. Some spacecraft, particularly earlier ones, have been spin-stabilized; during or shortly after launch, the spacecraft is set spinning at a rate on the order of a few revolutions per minute. Much like a rotating toy top, the spacecraft's orientation is stabilized by the gyroscopic action of its spinning mass. Most planetary spacecraft, however, are three-axis-stabilized, meaning that their attitude is fixed in relation to space. The spacecraft's attitude is maintained and changed via on-board thruster jets or reaction wheels, or a combination of both.

Propulsion subsystem. Most spacecraft are outfitted with a series of thruster jets, each of which produces approximately 0.2–2 pounds-force (1–10 newtons) of thrust. Thrusters are usually fueled with a monopropellant, hydrazine, which decomposes explosively when it contacts an electrically heated metallic catalyst within the thruster. In addition to maintaining the spacecraft's attitude, on-board thrusters are used for trajectory-correction maneuvers. Spacecraft designed to orbit a planet or similar target body must carry a larger propulsion element capable of decelerating the spacecraft into orbit upon arrival. *See* SPACECRAFT PROPULSION.

Thermal control subsystem. In order to minimize the impact of temperature variations on the electronics onboard, spacecraft nearly always incorporate some form of thermal control. Mechanical louvers, controlled by bimetallic strips similar to those in terrestrial thermostats, are often used to selectively

Important space probes

Name	Launch date	Comments
Luna 1	Jan. 2, 1959	Lunar flyby; passed within 3278 mi (5257 km) of the Moon
Pioneer 4	Mar. 3, 1959	Cosmic rays; passed 37,300 mi (60,000 km) from Moon
Luna 2	Sept. 2, 1959	Impacted Moon
Luna 3	Oct. 4, 1959	Photographed far side of Moon
Pioneer 5	Mar. 11, 1960	Study of space between Earth and Venus; magnetic fields and cosmic rays
Mariner 2	Aug. 26, 1962	First planetary flyby; Venus probe
Ranger 7	July 28, 1964	Lunar impact and approach photography
Mariner 4	Nov. 28, 1964	Mars flyby; photography, magnetic fields, cosmic rays
Ranger 8	Feb. 17, 1965	Lunar impact and approach photographs
Ranger 9	Mar. 21, 1965	Lunar impact in Alphonsus; approach photography
Zond 3	July 18, 1965	Flyby of Moon; photographs, other data
Pioneer 6	Dec.16, 1965	Solar orbiter
Luna 9	Jan. 31, 1966	First soft landing on Moon
Luna 10	Mar. 31, 1966	First probe to orbit Moon; collected data, but no camera instrument
Surveyor 1	May 30, 1966	Soft landing on Moon; environmental data and photography
Lunar Orbiter 1	Aug. 10, 1966	Lunar photographs
Pioneer 7	Aug. 17, 1966	Solar orbiter
Luna 11	Aug. 24, 1966	Lunar orbiter; TV camera failed
Luna 12	Oct. 22, 1966	Lunar orbiter; successful return of photographs and other data
Lunar Orbiter 2	Nov. 6, 1966	Lunar orbital photography
Luna 13	Dec, 21, 1966	Lunar lander; surface photography and soil information
Lunar Orbiter 3	Feb. 5, 1967	Lunar orbital photography
Venera 4	June 12, 1967	Analysis of Venus atmosphere; lander failed during descent
Mariner 5	June 14, 1967	Venus flyby; atmospheric and magnetospheric data
Surveyor 3	Apr. 17, 1967	Lunar surface photography and surface properties
Lunar Orbiter 4	May 4, 1967	Lunar orbital photography
Lunar Orbiter 5	Aug. 1, 1967	Lunar orbital photography
Surveyor 5	Sept. 8, 1967	Lunar surface photography and surface properties, including elemental analysis of surface
Surveyor 6	Nov. 7, 1967	Same as *Surveyor 5*; landing in Sinus Medii
Pioneer 8	Dec. 13, 1967	Solar orbiter
Surveyor 7	Jan. 7, 1968	Same as *Surveyor 5*
Luna 14	Apr. 7, 1968	Lunar orbiter
Pioneer 9	Nov. 8, 1968	Solar orbiter
Zond 6	Nov. 10, 1968	Lunar flyby; crashed on return to Earth
Venera 5	Jan. 5, 1969	Same as *Venera 4*
Venera 6	Jan. 10, 1969	Same as *Venera 4*
Mariner 6	Feb. 25, 1969	Photography and analysis of surface and atmosphere of Mars
Mariner 7	Mar. 27, 1969	Same as *Mariner 6*
Luna 15	July 14, 1969	Lunar orbiter (crashed during attempted lunar landing)
Zond 7	Aug. 8, 1969	Reentered Aug. 14, 1969; third crewless circumlunar flight; recovered in the Soviet Union
Venera 7	Aug. 17, 1970	Lander capsule transmitted 23 min from surface of Venus, Dec. 15, 1970
Luna 16	Sept. 12, 1970	Reentered Sept. 24, 1970; crewless Moon lander touched down on Sea of Fertility Sept. 20, 1970; returned lunar soil samples
Zond 8	Oct. 20, 1970	Circled Moon; recovered Oct. 27, 1970
Luna 17	Nov. 10, 1970	Landed on Moon Nov. 17, 1970; crewless Moon rover
Mars 2	May 19, 1971	First Mars landing
Mars 3	May 28, 1971	Mars probe
Mariner 9	May 30, 1971	Mars probe
Luna 18	Sept. 2, 1971	Lunar lander; failed during descent
Luna 19	Sept. 28, 1971	Lunar photography mission
Luna 20	Feb. 14, 1972	Lunar sample return
Pioneer 10	Mar. 2, 1972	Jupiter encounter; transjovian interplanetary probe
Venera 8	Mar. 27, 1972	Venus landing July 22, 1972
Luna 21	Jan. 8, 1972	Lunar lander with rover
Pioneer 11	Apr. 5, 1973	Jupiter encounter and trasjovian interplanetary probe; also Saturn encounter
Mars 4	July 21, 1973	Mars orbiter
Mars 5	July 25, 1973	Mars orbiter
Mars 6	Aug. 5, 1973	Mars lander
Mars 7	Aug. 9, 1973	Mars lander
Mariner 10	Nov. 3, 1973	Venus and Mercury encounter
Luna 22	May 29, 1974	Lunar orbiter
Helios 1	Dec. 10, 1974	Inner solar system, solar wind exploration
Venera 9	June 8, 1975	Venus probe
Venera 10	June 14, 1975	Venus probe
Viking 1	Aug. 20, 1975	Mars lander and orbiter
Viking 2	Sept. 9, 1975	Mars lander and orbiter
Helios 2	Jan. 15, 1976	Interplanetary; similar objectives to those of *Helios 1*
Luna 24	Aug. 9, 1976	Lunar sample return
Voyager 2	Aug. 20, 1977	Jupiter, Saturn, Uranus, and Neptune encounters; also satellites and ring systems
Voyager 1	Sept. 5, 1977	Same objectives as *Voyager 2* with some orbital differences giving differing encounter trajectories
Pioneer Venus Orbiter	May 20, 1978	Returning atmospheric, surface, and particle and field information
Pioneer Venus Multi-Probe Bus	Aug. 8, 1978	Penetration of Venus atmosphere by four probes; returned atmospheric data
Venera 11	Sept. 8, 1978	Venus lander; returned information on surface properties; detection of lightning and thunderlike sounds
Venera 12	Sept. 14, 1978	Similar mission to *Venera 11*

(conti.)

Important space probes (*cont.*)

Name	Launch date	Comments
Venera 13	Oct. 30, 1981	Venus lander
Venera 14	Nov. 4, 1981	Venus lander
Venera 15	June 2, 1983	Venus lander; surface topography
Venera 16	June 7, 1983	Similar mission to *Venera 15*
International Cometary Explorer (*ICE*)	—	Originally *International Sun-Earth Explorer 3* (*ISEE 3*) Earth satellite, redirected using a lunar swingby on Dec. 22, 1983, to encounter with Comet Giacobini-Zinner; plasma and magnetic field measurements
Vega 1	Dec. 15, 1984	Venus probe–Halley intercept
Vega 2	Dec. 21, 1984	Venus probe–Halley intercept
Sakigake	Jan. 8, 1985	Halley intercept; precursor to *Suisei*, upgraded to full mission
Giotto	July 2, 1985	Halley intercept
Suisei	Aug. 19, 1985	Halley intercept; plasma and magnetic field measurements
Phobos 1	July 7, 1988	Mars/Phobos probe, lost by command error
Phobos 2	July 12, 1988	Mars/Phobos probe; some data, but communications lost
Magellan	May 4, 1989	Venus radar mapper
Galileo	Oct. 18, 1989	Jupiter orbiter and atmospheric probe
Hiten/Hagoromo	Jan. 24, 1990	Moon orbiter and relay probe; orbiter transmitter malfunctioned
Ulysses	Oct. 6, 1990	Solar polar orbiter
Mars Observer	Sept. 25, 1992	Contact lost 3 days before Mars arrival
Clementine	Jan. 25, 1994	Orbited Moon; thruster malfunction prevented asteroid flyby
Solar and Heliospheric Observatory (SOHO)	Dec. 2, 1995	Orbits L1 libration point to study the Sun
Near Earth Asteroid Rendezvous	Feb. 17, 1996	Asteroid orbiter
Mars Global Surveyor	Nov. 7, 1996	Mars orbiter
Mars 96	Nov. 16, 1996	Mars orbiter and landers; launch vehicle failed
Mars Pathfinder	Dec. 4, 1996	Mars lander and rover
Advanced Composition Explorer	Aug. 25, 1997	Orbits L1 libration point to study charged particles
Cassini	Oct. 15, 1997	Saturn orbiter/Titan descent probe
Lunar Prospector	Jan. 6, 1998	Lunar orbiter
Nozomi	July 4, 1998	Mars orbiter; failed to reach Mars orbit
Deep Space 1	Oct. 24, 1998	Test of ion engine and 11 other advanced technologies; asteroid and comet flybys
Mars Climate Orbiter	Dec. 11, 1998	Lost during Mars arrival
Mars Polar Lander	Jan. 3, 1999	Lost during Mars arrival
Stardust	Feb. 7, 1999	Comet flyby, dust sample return
Mars Odyssey	April 7, 2001	Mars orbiter
Wilkinson Micowave Anisotropy Probe	June 30, 2001	Solar orbiter studying cosmic background radiation
Genesis	Aug. 8, 2001	Solar wind sample return; hard impact upon landing in Utah
Comet Nucleus Tour	July 3, 2002	Intended for two comet flybys; failed shortly after launch
Hayabusa	May 9, 2003	Comet sample return mission; sample collection uncertain
Mars Express	June 2, 2003	Mars orbiter
Beagle 2	June 2, 2003	Lander carried on Mars Express; failed during descent
Spirit (*Mars Exploration Rover A*)	June 10, 2003	Mars rover
Opportunity (*Mars Exploration Rover B*)	July 7, 2003	Mars rover
Spitzer Space Telescope	Aug. 25, 2003	Infrared telescope in Earth-trailing heliocentric orbit
Smart 1	Sept. 27, 2003	Lunar orbiter
Rosetta	March 2, 2004	Comet orbiter-lander
Messenger	Aug. 3, 2004	Mercury orbiter
Deep Impact	Jan. 12, 2005	Released impactor to excavate crater in comet nucleus
Mars Reconnaissance Orbiter	Aug. 12, 2005	Mars orbiter
Venus Express	Nov. 9, 2005	Venus orbiter
New Horizons	Jan. 19, 2006	Pluto flyby
Solar Terrestrial Relations Observatory (STEREO)	Oct. 26, 2006	Twin spacecraft in solar orbit
Phoenix	Aug. 4, 2007	Mars lander
Kayuga	Sept. 14, 2007	Lunar orbiter
Dawn	Sept. 27, 2007	To orbit asteroid Vesta (2011) and dwarf planet Ceres (2015)
Chang'e-1	Oct. 24, 2007	Lunar orbiter
Chandrayaan-1	Oct. 22, 2008	Lunar orbiter and impactor

radiate heat from the interior of the spacecraft into space. Other thermal strategies include painting exterior surfaces. In some cases, spacecraft may also carry one or more active forms of heating to maintain temperature at required minimums.

Command and data subsystem. This designation is given to the main computer that oversees management of spacecraft functions and handling of collected data. Blocks of commands transmitted from Earth are stored in memory in the command and data subsystem and are executed at prescribed times. This subsystem also contains the spacecraft clock in order to accurately pace its activities, as well as all the activities of the spacecraft.

Structure subsystem. The spacecraft's physical structure is considered a subsystem itself for the purposes of planning and design. Usually the heart of this structure is a spacecraft bus, often consisting of a number of bays, which houses the spacecraft's main subsystems. *See* SPACECRAFT STRUCTURE. Franklin O'Donnell

Space processing Experiments conducted in space in order to take advantage of the reduced gravity in studies of the growth, behavior, and properties of materials. Spacelab, developed by the European Space Agency (ESA), is a laboratory module that flies in the space shuttle payload bay. First launched in 1983 on the *Columbia*, the module has become the workhorse for United States and international science missions emphasizing low gravity. The experiments in fluids, combustion, materials science, and biotechnology conducted on these missions, together with their related ground-based research, have resulted in more than 2200 scientific publications.

Since biotechnology experiments generally have modest space and power requirements, they were able to be accommodated by middeck lockers in other shuttle missions as well as on the *Spacelab* microgravity emphasis missions, which provided them with many more flight opportunities. Even though the microgravity environment on some of these missions was not ideal, biomolecular crystal growth experiments for structural analysis as well as cell and tissue culturing experiments produced many interesting results, some of which could have commercial applications.

There are two compelling reasons for the study of combustion in microgravity. One is the issue of fire safety in the design and operation procedures of orbiting laboratories; the other is to take advantage of the weightless state to study certain combustion phenomena in more detail and to test various models in which convection has been ignored in order to be mathematically tractable. Examples of the latter are a number of droplet combustion experiments in which the burning of free-floating or tethered droplets was studied. The absence of gravity allowed the droplet size to be increased to as much as 5 mm (0.2 in.) so that more detailed observation could be made. The objective is to test theories of droplet combustion and soot formation that are of importance to improving the efficiency of internal combustion engines, gas turbine engines, and home and industrial oil-burning heating systems. *See* COMBUSTION; GAS TURBINE; INTERNAL COMBUSTION ENGINE.

In materials science, the processing of metallic alloys and composites has been carried out in space in order to study their microstructure, thermal properties, and crystal growth.

Robert J. Naumann

Space shuttle A reusable crewed orbital transportation system. The space shuttle, along with crewless (robotic) expendable launch vehicles such as Atlas and Delta, make up the United States Space Transportation System (STS). The shuttle provides the unique capability for in-flight rendezvous and retrieval of faulty or obsolescent satellites, followed by satellite repair, update, and return to orbit or return to Earth for repair and relaunch. The space shuttle also plays an essential continuing role in the construction and provisioning of the *International Space Station* (*ISS*) by transporting major components, such as the giant solar-cell arrays and the Canadian computer-driven double-ended robotic arm, from Earth to the *ISS* and installing them using extended extravehicular activity (EVA, or space walks) by trained *ISS* resident and shuttle-visiting astronauts. *See* SATELLITE (SPACECRAFT); SPACE STATION.

Early in its history the space shuttle was expected to be a low-cost replacement for expendable launch vehicles. Following the *Challenger* shuttle accident in 1986, it became clear that crewless vehicles would have a continuing place in the United States launch vehicle fleet, that they have advantages over the shuttle such as lower cost and shorter lead time for tasks within their capability, and that the shuttle fleet would be fully occupied for much of the first decade of the twenty-first century doing complex jobs requiring human presence, in particular associated with the *ISS*, which no other launch system can do. For such reasons the STS was expanded to include the expendable launch vehicle families in use by the Department of Defense (DOD) and the National Aeronautics and Space Administration (NASA).

The space shuttle orbiter accommodates a crew of four to seven for orbital mission durations up to about 10 days. The space shuttle flight system (see illustration) consists of the orbiter, which includes the three liquid-fueled shuttle main engines; an external fuel tank; and two solid-fuel strap-on booster rockets. The external tank is discarded during each launch. The solid-fuel rocket casings are recovered and reused. The orbiter lands horizontally as an unpowered ("dead stick") aircraft on a long runway at the NASA Kennedy Space Center in Florida or, when conditions for landing these are unacceptable, at the Edwards Air Force Base in California. *See* ROCKET PROPULSION.

The shuttle is launched with all three main engines and both strap-on solid-fuel booster rockets burning. The booster rockets are separated 2 min after liftoff at an altitude of 30 mi (48 km), 28 mi (44 km) downrange. Main engine cutoff occurs about 8 min after liftoff, nearly 70 mi (110 km) above the Atlantic Ocean, 890 mi (1430 km) downrange from Kennedy. The external tank (ET) is separated from the orbiter shortly after the main engine cutoff, before orbital velocity (speed) is reached, so that the relatively light empty tank can burn up harmlessly during its reentry into the atmosphere above the ocean. Main propulsion for the rest of the mission is provided by the orbital maneuvering system (OMS) engines, whose first task is to complete insertion of the shuttle into its final, nearly circular path during its first orbit (flight once around Earth) after liftoff.

After the shuttle orbiter completes the orbital phase of its mission and is ready to return to Kennedy, its pilot rotates the spacecraft 180° (tail-first relative to the orbiter's direction of motion) and fires the orbital maneuvering system engines to decrease the space vehicle's speed. This maneuver reduces the orbiter's speed just enough to allow the orbiter to fall into the tenuous outer atmosphere, where atmospheric drag causes the orbit to decay (altitude and velocity decrease) along a spiral path. This reentry slowdown occurs in a precise, computer-controlled manner so that the landing may be made manually, halfway around the Earth, on the desired runway at Kennedy or Edwards. At orbital altitudes, small reaction-control engines (jets) maintain the desired orientation of the orbiter. As the spacecraft becoming-airplane descends into the denser lower air, its aerodynamic control surfaces gradually take over their normal maintenance of heading, pitch, and roll, and the small jets are turned off. Landing is made at a speed of 220 mi/h (100 m/s). The brakes are helped to bring the vehicle to a stop by a parachute (drag chute) which is deployed from the tail after touchdown of the nose wheel and released well before the vehicle comes to a complete stop.

Major orbiter systems are the environmental control and life support, electric power, hydraulic, avionics and flight control, and the space shuttle main engines. Each system is designed with enough redundancy to permit either continuation of mission operations or a safe return to Earth after any single-element failure. For example, three fuel-cell power plants generate the in-flight electric power. Each fuel cell feeds one of three independent electrical distribution buses. Similarly, three independent hydraulic systems, each powered by an independent auxiliary power unit, operate the aerosurfaces. *See* FUEL CELL; SPACE POWER SYSTEMS.

The avionics system uses five general-purpose computers, four of which operate in a redundant set while the fifth operates independently. These computers perform guidance, navigation, and control calculations, and operate the attitude (orientation) controls and other vehicle systems. They also monitor system performance and can automatically reconfigure redundant systems in the event of a failure during flight-critical mission phases. Each of the three space shuttle main engines has an independent computer for engine control during ascent. *See* DIGITAL COMPUTER; MULTIPROCESSING.

The thermal protection system, which is not redundant, presented a major technical challenge to the orbiter development schedule. This system, unlike those of previous single-use spacecraft, has a design requirement of reuse for 100 missions. Performance requirements also dictate that the system withstand temperatures as high as 3000°F (1650°C) while maintaining the

Space shuttle lifting off from the Kennedy Space Center.

vehicle's structure at no more than 350°F (177°C). *See* ATMOSPHERIC ENTRY.

The key to meeting this challenge was to develop a material possessing an extremely low specific heat capacity and thermal conductivity, together with adequate mechanical strength to withstand the launch and reentry vibration and acceleration. These thermal characteristics must remain stable up to temperatures of at least 3000°F (1650°C). The solution was found in silica ceramic tiles, some 24,000 of which cover most of the orbiter's surface. During the highest thermal load portion of the reentry trajectory (path), the orbiter's wings are level and the nose is elevated well above the flight path. This causes the temperature of the undersurface to be substantially higher than that of the upper surface. For this reason, the undersurface is covered with black borosilicate glass-coated high-temperature tiles. Most of the upper surface of the shuttle is covered with a lower-temperature silica blanket made of two outer layers of woven fabric and an insulating buntinglike center layer stitched together in a quiltlike pattern. Finally, the nose cap and wing leading edges are subject to the highest (stagnation) temperatures, which requires the use of molded reinforced carbon-carbon material.

The first operational shuttle accomplished the first commercial satellite deployments in 1982. In 1984 the first in-orbit satellite repair was accomplished on the *Solar Maximum Mission*'s control system and a primary experiment sensor. Later that same year *Palapa* and *Westar*, two communications satellites in useless orbits, were recovered and returned to Earth on the same space shuttle mission. These repair and recovery missions demonstrated the usefulness of humans in these new space tasks.

Shortly before noon on January 28, 1986, *Challenger* lifted off from Kennedy Space Center on the twenty-fifth shuttle mission. At 72 s into the flight, with no apparent warning from real-time data, the external tank exploded and *Challenger* was destroyed. All seven crew members perished, the first to die in NASA's human space-flight program spanning 25 years and 56 crewed launches. Further flight operations were suspended while a Presidential (Blue-ribbon) Commission reviewed the circumstances surrounding the accident, determined its probable cause, and developed recommendations for corrective actions. The cause of the accident was determined to be inadequate design of the solid rocket motor field joint. Deflection of the joint with deformation of the seals at cold temperature allowed hot combustion products to bypass both O-rings, resulting in erosion and subsequent failure of both primary and secondary seals.

In September 1988, a *Tracking and Data Relay Satellite* (*TDRS*) was launched by the first space shuttle to fly after the *Challenger* accident. The initial *TDRS* network was completed 6 months later by the second shuttle launch of a *TDRS*. *See* SPACE COMMUNICATIONS; SPACECRAFT GROUND INSTRUMENTATION.

In the realm of solar system exploration, shuttle launches of *Magellan* to Venus and *Galileo* to Jupiter in 1989 were followed in 1990 by the shuttle launch of *Ulysses* to investigate the magnetic field configuration and variations around the Sun's poles, after the craft's shuttle launch. STS 32 launched the long-awaited *Hubble Space Telescope* (*HST*) in 1990; however, this telescope's most important work was not accomplished until later that decade after two shuttle EVA visits—the first in 1993 to correct a bad case of astigmatism produced by the *HST* mirror manufacturer, and the second in 1997 to update the sensors and replace major spacecraft subsystem components that were failing. Additional servicing missions were carried out in 1999 and 2002.

On February 3, 2003, while the shuttle *Columbia* was reentering the atmosphere, it disintegrated. All seven crew members perished, and pieces of the orbiter fell in a long swath over East Texas. As much of the debris as possible was collected and taken to Kennedy Space Center, where pieces were placed in their appropriate position on a hanger floor. The Columbia Accident Investigation Board (CAIB) was formed to determine the cause of the accident and to present recommendations. As during the *Challenger* investigation, all flights were suspended.

Seven months after the disaster, the CAIB issued its report. The cause of this accident was a breach in the thermal protection sytem on the leading edge of the left wing, caused by a piece of insulating foam that separated during launch from the left bipod ramp section on the external tank. This breach in the reinforced carbon-carbon panel allowed superheated air to penetrate through the leading edge of the left wing, which melted the aluminum structure of the wing. Aerodynamic forces caused loss of control of the shuttle and its subsequent breakup.

The CAIB made 15 recommendations. The independent NASA Return to Flight Task Group agreed that NASA had met the intent of 12 of the recommendations. NASA could not meet the major recommendation, to eliminate foam from leaving the external tank, due to engineering problems.

When *STS 114* launched in July 2005, the new cameras on the external tank, installed for high resolution of the orbiter, recorded a large piece of foam coming off the tank. Six in-flight anomalies were taken due to the unexpected foam loss. The next mission, *STS 121*, did not launch until July 4, 2006, almost exactly one year after *STS 114*. The most significant change in the tank was the elimination of the liquid hydrogen and liquid oxygen protuberance air load (PAL) ramps.

In 2004, President Bush announced the Vision for Space Exploration. Part of the Vision required the shuttle to be retired no later than 2010, only 5 years after the first return-to-flight mission. In 2006, after much debate, NASA decided to perform one more *Hubble* servicing mission, enhancing its capabilities and extending its life by replacing failed components and updating some sensors and data processors. John F. Clark; Jeffrey C. Mitchell

Space station A complex physical structure specifically designed to serve as a multipurpose platform in low Earth orbit. Functioning independently and often without a crew actively involved in onboard operations, a space station contains the structures and mechanisms to operate and maintain such support systems as command and data processing, communications and tracking, motion control, electrical power, and thermal and environmental control. Evolving together with technology and increasing in scope and complexity, the space station has a history in which each program was based upon the developments and achievements of its predecessor.

Salyut. The Soviet Union began construction of the world's first space station, called *Salyut*, in 1970. It was the primary Soviet space endeavor for the next 15 years. The design was retained not only through a series of Salyuts that were launched within that decade and later, but also through the development of *Mir*, the most famous, long-lived Russian achievement in space. The Salyuts were cylindrical in shape and contained compartments with specialized functions for operating a space station. The docking compartment was designed to accept the Soyuz spacecraft that transported the cosmonauts. The transfer compartment gave the cosmonauts access to the various work compartments. The work compartments contained the mechanisms that operated and controlled the station, as well as the laboratories in which the cosmonauts performed experiments while they were onboard. *Salyut 1* initially carried a crew of three, but after three cosmonauts died when a valve in the crew compartment of their descent module burst and the air leaked out, the crews were reduced to two cosmonauts, and both crew members were outfitted with pressurized space suits.

Skylab. This program was developed by the National Aeronautics and Space Administration (NASA) building on the success of its heavy-lift rocket, the Saturn, which had boosted the Apollo rockets and helped place the first human being on the Moon. *Skylab* weighed just less than 100 tons (90 metric tons). It was launched on May 14, 1973, from the Kennedy Space Center aboard a Saturn 5 rocket, which enjoyed a reputation for never having failed. Although the launch was flawless, a shield designed to shade *Skylab's* workshop deployed about a minute after liftoff and was torn away by atmospheric drag. That began a series of problems, most involving overheating, that had to be overcome before *Skylab* was safe for human habitation. Eventually three crews served aboard *Skylab* throughout 1973 for periods of 28, 59, and 84 days, respectively. The single greatest contribution made by each crew was the extravehicular activities that restored *Skylab's* ability to serve as a space station.

Apollo-Soyuz test program. In 1975, during the period of dètente, plans were made for a joint United States–Soviet space venture, known as the Apollo-Soyuz Test Program (ASTP). For the first time, United States astronauts and Soviet cosmonauts became members of each other's flight teams, training together, touring their launch sites, and working in their mission control rooms. For 15 months, they trained and prepared for their historical joint mission in space. In July 1975, the astronauts of *Apollo 18* linked with the cosmonauts of *Soyuz 22*; their docked configuration functioned as a miniature international space station.

Mir. *Mir* (which in Russian means either "peace" or "world") was the name of the vehicle for the world's first multipurpose, permanently operating crewed space station. Based upon the Salyut design and configuration, *Mir* essentially was an extension and expansion of the shell of the basic Soviet space vehicle. It would incorporate the standard Salyut/Soyuz/Progress profile. External ports were added for docking of Soyuz vehicles which would carry crew members, and Progress (resupply) vehicles which would bring foodstuffs, drinking water, extra equipment and apparatus, sanitary requisites, medical apparatus, and propellant. These same Progress vehicles would return space "junk" to Earth. Construction was based upon a modular design, permitting the Soviets to replace modules whenever significant improvements in technology made the earlier modules obsolete. Two cylindrical modules formed the basic shape of *Mir* and served as the living and central control compartments for the crews. Additional modules were used for scientific experiments.

The *Mir* core station was a 49-ft (15-m) module. Its launch aboard a Soviet Proton rocket on February 20, 1986, was televised internationally for the first time in Soviet space history. *Mir* was assembled in space and was composed of six modules. In addition to the core, the Kvant-1 module (launched in April 1987) was a 19-ft (6-m) pressurized lab used for astrophysics research. The Kvant-2 module was launched in December 1989, and was used for biological research and for Earth observation. The Kristall module was launched in August 1990, and provided the docking port for the United States space shuttles that visited *Mir*. Between 1994 and 2001, there were nine dockings between space shuttle vehicles and *Mir*. The final two modules, Spektr and Priroda (launched in June 1995 and April 1996, respectively) were remote sensing modules used to study the Earth.

International Space Station. The Soviet successes with *Mir* prompted the United States to respond with what became known as *Space Station Freedom*. Moved from the NASA drawing board in 1993, components of the space station were tested by shuttle crews during a variety of missions. Supporters of this space station advocated such unique opportunities as manufacturing drugs, scientific materials research in microgravity, and studying the health and status of the Earth's environment from outer space. At the direction of President Clinton, the United States transformed the single-nation concept for *Space Station Freedom* into a multinational partnership with the European Space Agency and the Russian Space Agency to create what is now known as the *International Space Station* (*ISS*). *See* SPACE PROCESSING.

Five space agencies and sixteen nations have united to build the *International Space Station*. When completed, the *ISS* will be the largest human-made object ever to orbit the Earth (see illustration). It will include six laboratories built by a consortium of nations. United States space shuttle, Russian Soyuz, and partner-nation heavy transport vehicle flights will transport the structures and mechanism necessary to construct the station over 12 years. The United States space shuttle must deliver most *ISS* modules and major components. The *ISS* will weigh approximately 1 million pounds (almost 450,000 kilograms); it will orbit the Earth at an inclination of 51.6° to either side of the Equator; and it will fly at an altitude between 230 and 285 mi (370–460 km).

The first permanent crew of the *International Space Station* launched from the Baikonur Cosmodrome on October 31, 2000. Astronauts and cosmonauts assigned to flights to the *ISS* compose an Expedition crew. As of November 2008, there had been 18 Expedition crews posted to the *ISS*. Expedition crews are composed of either two or three members: a Commander and either one or two Flight Engineers. The primary missions of each Expedition crew are to assemble the station and to learn how to live and work in space.

In addition to serving as a test bed and springboard for future space exploration, the *ISS* will serve as a research laboratory for innovation in science and technology. Research has focused on such topics as biomedical research and countermeasures to understand and control the effects of space and zero gravity on crew members; biological study of gravity's influence on the evolution, development, and internal processes of plants and animals; biotechnology to develop superior protein crystals for drug development; and fluid physics. Advanced research will be oriented toward topics such as human support technology, materials science, combustion science, and fundamental physics.

Construction of the *International Space Station* is projected to be completed in 2010. It will then host an international crew of

Artist's rendering of how the *International Space Station* may look when its assembly sequence is completed in 2010. (Changes to the final station configuration will likely occur before then.) The 1-million-pound (450,000-kg) station will have a pressurized volume equal to two jumbo jets, and an acre of solar panels. (*NASA*)

six members. *See* SPACE FLIGHT; SPACE TECHNOLOGY; SPACECRAFT STRUCTURE.

James J. Pelosi

Space technology The systematic application of engineering and scientific disciplines to the exploration and utilization of outer space. Space technology developed so that spacecraft and humans could function in this environment that is so different from the Earth's surface. Conditions that humans take for granted do not exist in outer space. Objects do not fall. There is no atmosphere to breathe, to keep people warm in the shade, to transport heat by convection, or to enable the burning of fuels. Stars do not twinkle. Liquids evaporate very quickly and are deposited on nearby surfaces. The solar wind sends electrons to charge the spacecraft, with lightninglike discharges that may damage the craft. Cosmic rays and solar protons damage electronic circuits and human flesh. The vast distances require reliable structures, electronics, mechanisms, and software to enable the craft to perform when it gets to its goal—and all of this with the design requirement that the spacecraft be the smallest and lightest it can be while still operating as reliably as possible.

All spacecraft designs have some common features: structure and materials, electrical power and storage, tracking and guidance, thermal control, and propulsion. The spacecraft structure is designed to survive the forces of launching and ground handling. The structure is made of metals (aluminum, beryllium, magnesium, titanium) or a composite (boron/epoxy, graphite/epoxy). It must also fit the envelope of the launcher. *See* SPACECRAFT STRUCTURE.

To maintain temperatures at acceptable limits, various active and passive devices are used: coatings or surfaces with special absorptivities and emissivities, numerous types of thermal insulation, such as multilayer insulation and aerogel, mechanical louvers to vary the heat radiated to space, heat pipes, electrical resistive heaters, or radioisotope heating units.

The location of a spacecraft can be measured by determining its distance from the transit time of radio signals or by measuring the direction of received radio signals, or by both. The direction of a spacecraft can be determined by turning the Earth station antenna to obtain the maximum signal, or by other equivalent and more accurate methods. *See* SPACE NAVIGATION AND GUIDANCE.

The velocity of a spacecraft is changed by firing thrusters. Solid propellant thrusters are rarely used. Liquid propellant thrusters are either monopropellant or bipropellant. Electric thrusters, such as mercury or cesium ion thrusters, have also been used. Electric thrusters have the highest efficiency (specific impulse) but the lowest thrust. *See* INTERPLANETARY PROPULSION; ION PROPULSION; PROPELLANT; SPACECRAFT PROPULSION.

Most spacecraft are spin-stabilized or are three-axis body-stabilized. The former uses the principles of a gyroscope; the latter uses sensors and thrusters to maintain orientation. Some body-stabilized spacecraft (such as astronomical observatories) are fixed in inertial space, while others (such as Earth observatories) have an axis pointed at the Earth and rotate once per orbit. A body-stabilized spacecraft is simpler than a spinner but requires more hardware. The orientation of a spacecraft is measured with Sun sensors (the simplest method), star trackers (the most accurate), and horizon (Earth or other body) or radio-frequency (rf) sensors (usually to determine the direction toward the Earth). Attitude corrections are made by small thrusters or by reaction or momentum wheels; as the motor applies a torque to accelerate or decelerate the rotation, an equal and opposite torque is imparted to the spacecraft.

Primary electrical power is most often provided by solar cells made from a thin section of crystalline silicon protected by a thin glass cover. Excess power from the solar cells is stored in rechargeable batteries so that when power is interrupted during an eclipse, it can be drawn from the batteries. Other sources of power generation include fuel cells, radioisotope thermoelectric generators (RTGs), tethers, and solar dynamic power. Fuel cells have been used on the Apollo and space shuttle programs and produce a considerable amount of power, with drinkable water as a by-product. *See* BATTERY; FUEL CELL; SOLAR CELL; SPACE POWER SYSTEMS.

The status and condition of a spacecraft are determined by telemetry. Temperatures, voltages, switch status, pressures, sensor data, and many other measurements are transformed into voltages, encoded into pulses, and transmitted to Earth. This information is received and decoded at the spacecraft control center. Desired commands are encoded and transmitted from the control center, received by the satellite, and distributed to the appropriate subsystem. Commands are often used to turn equipment on or off, switch to redundant equipment, make necessary adjustments, and fire thrusters and pyrotechnic devices. *See* SPACE COMMUNICATIONS; SPACECRAFT GROUND INSTRUMENTATION; TELEMETERING.

Many spacecraft missions have special requirements and hence necessitate special equipment. Satellites that leave the Earth's gravitational field to travel around the Sun and visit other planets have special requirements due to the greater distances, longer mission times, and variable solar radiation involved.

Spacecraft that return to Earth require special protection for reentry into Earth's atmosphere. In some missions one spacecraft must find, approach, and make contact with another spacecraft. *See* SATELLITE (SPACECRAFT); SPACE PROBE.

Space is distant not only in kilometers but also in difficulty of approach. Large velocity changes are needed to place objects in space, which are then difficult to repair and expensive to replace. Therefore spacecraft must function when they are launched, and continue to function for days, months, or years. The task is similar to that of building a car that will go 125,000 mi (200,000 km) without requiring mechanical repair or refueling. Not only must space technology build a variety of parts for many missions, but it must achieve a reliability far greater than the average. This is accomplished by building inherent reliability into components and adding redundant subsystems, supported by a rigorous test schedule before launch. Efforts are made to reduce the number of single points of failure, that is, components that are essential to mission success and cannot be bypassed or made redundant. Jeffrey C. Mitchell; Gary D. Gordon

Space-time A term used to denote the geometry of the physical universe as suggested by the theory of relativity. It is also called space-time continuum. Whereas in Newtonian physics space and time had been considered quite separate entities, A. Einstein and H. Minkowski showed that they are actually intimately intertwined.

Einstein showed that in general two observers, each using the same techniques of observation but being in motion relative to each other, will disagree concerning the simultaneity of distant events. But if they do disagree, they are also unable to compare unequivocally the rates of clocks moving in different ways, or the lengths of scales and measuring rods. Instead, clock rates and scale lengths of different observers and different frames of reference must be established so as to assure the principal observed fact. Each observer, using his or her own clocks and scales, must measure the same speed of propagation of light. This requirement leads to a set of relationships known as the Lorentz transformations. *See* LORENTZ TRANSFORMATIONS.

In accordance with the Lorentz transformations, both the time interval and the spatial distance between two events are relative quantities, depending on the state of motion of the observer who carries out the measurements. There is, however, a new absolute quantity that takes the place of the two former quantities. It is known as the invariant, or proper, space-time interval τ and is defined by Eq. (1), where T is the ordinary time interval, R

$$\tau^2 = T^2 - \frac{1}{c^2}R^2 \qquad (1)$$

the distance between the two events, and c the speed of light in empty space. Whereas T and R are different for different observers, τ has the same value. In the event that Eq. (1) would render τ imaginary, its place may be taken by σ, defined by Eq. (2). If both τ and σ are zero, then a light signal leaving the loca-

$$\sigma^2 = R^2 - c^2T^2 \qquad (2)$$

tion of one event while it is taking place will reach the location of the other event precisely at the instant the signal from the latter is coming forth.

The existence of a single invariant interval led the mathematician Minkowski to conceive of the totality of space and time as a single four-dimensional continuum, which is often referred to as the Minkowski universe. In this universe, the history of a single space point in the course of time must be considered as a curve (or line), whereas an event, limited both in space and time, represents a point. So that these geometric concepts in the Minkowski universe may be distinguished from their analogs in ordinary three-dimensional space, they are referred to as world curves (world lines) and world points, respectively. *See* GRAVITATION; RELATIVITY. Peter G. Bergmann

Spacecraft ground instrumentation Instrumentation located on the Earth for monitoring, tracking, and communicating with satellites, space probes, and crewed spacecraft. Radars, communication antennas, and optical instruments are classified as ground instrumentation. They are deployed in networks and, to a lesser extent, in ranges. Ranges are relatively narrow chains of ground instruments used to follow the flights of missiles, sounding rockets, and spacecraft ascending to orbit. Some ranges are a few miles long; others, such as the U.S. Air Force's Eastern Test Range, stretch for thousands of miles. Networks, in contrast, are dispersed over wide geographical areas so that their instruments can follow satellites in orbit as the Earth rotates under them at 15° per hour, or space probes on their flights through deep space.

Networks are of two basic kinds: networks supporting satellites in Earth orbit, and networks supporting spacecraft in deep space far from Earth. A third concept was added in the 1980s with the Tracking and Data Relay Satellite System (TDRSS), also called the Space Network (SN). TDRSS replaced most of the ground stations used for Earth orbital support. The *Tracking and Data Relay Satellites* (*TDRS*) are placed in geosynchronous orbits to relay signals to and from other orbiting spacecraft during more than 85% of each orbit, to and from a single ground station.

Ranges and networks have various technical functions:

1. Tracking: determination of the positions and velocities of space probes and satellites through radio and optical means.

2. Telemetry: reception of telemetered signals from scientific instruments and spacecraft housekeeping functions.

3. Voice reception and transmission: provision for communication with the crew of a spacecraft, such as the space shuttle.

4. Command: transmission of coded commands to spacecraft equipment, including scientific instruments.

5. Television reception and transmission: provision for observation of the crew, spacecraft environment, and so on.

6. Ground communications: telemetry, voice, television, command, tracking data, and spacecraft acquisition data

transmission between network sites and the central mission control center, and payload information to user facilities.

7. Computing: calculation of orbital elements and radar acquisition data prior to transmission to users; also, computation of the signals that drive visual displays at a mission control center.
See SPACE COMMUNICATIONS; SPACE NAVIGATION AND GUIDANCE; TELEMETERING.

NASA operates two ground-based networks and the TDRSS. The ground-based networks are the Spaceflight Tracking and Data Network (STDN), which tracks, commands, and receives telemetry from United States and foreign satellites in Earth orbit; and the Deep Space Network (DSN), which performs the same functions for all types of spacecraft sent to explore deep space, the Moon, and solar system planets. The TDRSS provides the same support to Earth orbital spacecraft as the STDN. The U.S. Department of Defense operates two classified networks: the Satellite Control Facility (SCF); and the National Range Division Stations, which include those of all United States military ranges. Russia and the European Space Agency (ESA) also maintain similar networks.

The Laser Tracking Network consists of a series of both fixed and mobile laser systems used for ranging to retroreflector-equipped satellites in highly stable orbits. Laser stations obtain ranging data for these satellites by bouncing a highly concentrated pulse of laser light off the retroreflector corner cube installed on the spacecraft exterior. The exact position of the spacecraft in orbit can then be mathematically determined for a given point in time. By comparing several ranging operations, orbital predictions can be interpolated and extrapolated. The resultant data have a variety of applications, such as precise prediction of satellite orbit and measurement of the Earth's gravitational field, polar motion, earth tides, Earth rotation, tectonic plate motion, and crustal motion to accuracies within the centimeter range. The Laser Tracking Network is a multinational cooperative effort with over 30 laser sites located in North and South America, Europe, China, Japan, and Australia. H. William Wood

Spacecraft propulsion A system that provides control of location and attitude of spacecraft by using rocket engines to generate motion. Spacecraft propulsion systems come in various forms depending on the specific mission requirements. Each exhibits considerable variation in such parameters as thrust, specific impulse, propellant mass and type, pressurization schemes, cost, and materials. All of these variables must be considered in deciding which propulsion system is best suited to a given mission. Typical spacecraft applications include communications satellites, science and technology spacecraft, and Earth-monitoring missions such as weather satellites. Orbital environments range from low-Earth to geosynchronous to interplanetary. *See* ASTRONAUTICAL ENGINEERING; ROCKET PROPULSION; SATELLITE (SPACECRAFT); SPACE FLIGHT; SPACE PROBE; SPECIFIC IMPULSE; THRUST.

The two fundamental variables that define the design of spacecraft propulsion systems are the total velocity change to be imparted to the spacecraft for translational purposes, and the impulse necessary to counteract the various external torques imposed on the spacecraft body. From these, the required quantity of a given propellant combination can be specified. Propellant accounts for almost 60% of the lift-off mass of a communications satellite.

The specific impulse has a significant effect on the total propellant load that a spacecraft must carry to perform its assigned mission. Since a massive satellite must be boosted into space by the use of expensive launch vehicles, such as the space shuttle and Ariane, significant cost savings may be gained if smaller, less expensive launch vehicles may be used. The size of the required launch vehicle is directly proportional to the mass of the payload. Since most of the other components that make up spacecraft are relatively fixed in weight, it is critical to utilize propellant combinations that maximize specific impulse.

For modern spacecraft the choices are either bipropellants, which utilize a liquid oxidizer and a separate liquid fuel; solid propellants, which consist of oxidizer and fuel mixed together; or monopropellants, which are liquid fuels that are easily dissociated by a catalyst into hot, gaseous reaction products. High specific impulse is offered by bipropellants, followed by solid propellants and monopropellants.

Spacecraft attitude control schemes play an important role in defining the detailed characteristics of spacecraft propulsion systems. Essentially, there are three methods for stabilizing a spacecraft: three-axis control, spin control, and gravity gradient. In three-axis systems the body axes are inertially stabilized with reference to the Sun and stars, and utilize rocket engines for control in all six degrees of freedom. Spin-stabilized spacecraft use the inertial properties of a gyroscope to permanently align one of the axes by rotating a major portion of the spacecraft body about this axis. This approach significantly reduces the number of thrusters needed for control. Gravity gradient control is a nonactive technique that relies on the Earth's tidal forces to permanently point a preferred body axis toward the Earth's center. *See* GYROSCOPE; INERTIAL GUIDANCE SYSTEM; SPACECRAFT STRUCTURE.

Translation of a spacecraft, independent of its control technique, requires thrusters aligned parallel to the desired translational axis. Usually, all three axes require translational capability. Combining the two requirements for attitude control and translation results in the minimum number of rocket engines required to perform the mission. These are supplemented with additional thrusters to allow for failures without degrading the performance of the propulsion system. Simplistically, it would be reasonable to assume that the propellant-engine combination with the highest specific impulse would be the preferable choice. However, the ultimate requirement is the lowest possible mass for the entire propulsion system. The complexity of the system is greatly influenced by, and is roughly proportional to, the specific impulse, since bipropellants require more tanks, valves, and so forth than either solid systems or monopropellant systems. This is primarily due to the differences in density between liquids and solids, and the fact that bipropellants require high-pressure gas sources to expel the fluid from the tanks and into the rocket engine chamber. For communications satellites in the lift-off weight range of 3000 lbm (1360 kg), the trade-off between specific impulse and system mass dictates the use of a solid rocket motor for the main-orbit circularizing burn and a monopropellant propulsion system for on-orbit attitude control and translation. Spacecraft launch masses above about 5000 lbm (2268 kg) require the use of all-bipropellant systems. Keith Davies

Spacecraft structure The supporting structure for systems capable of leaving the Earth and its atmosphere, performing a useful mission in space, sometimes returning to the Earth and sometimes landing on other bodies. Among the principal technologies that enter into the design of spacecraft structures are aerodynamics, aerothermodynamics, heat transfer, structural mechanics, structural dynamics, materials technology, and systems analysis. In applying these technologies to the structural design of a spacecraft, trade studies are made to arrive at a design which fulfills system requirements at a minimum weight with acceptable reliability and which is capable of being realized in a reasonable period of time.

The structural aspects of space flight can be divided into six broad regions or phases: (1) transportation, handling, and storage; (2) testing; (3) boosting; (4) Earth-orbiting flight; (5) reentry, landing, and recovery on the Earth; (6) interplanetary flight with orbiting of or landing on other planets. Each phase has its own structural design criteria requiring detailed consideration of heat, static loads, dynamic loads, rigidity, vacuum effects, radiation, meteoroids, acoustical loads, atmospheric pressure loads, foreign atmospheric composition, solar pressure, fabrication techniques, magnetic forces, sterilization requirements, accessibility for repair, and interrelation of one effect with the others. Heavy reliance is placed on computer-generated mathematical models and ground testing.

The basic spacecraft structural design considerations apply equally well to both crewed and crewless spacecraft. The degree of reliability of the design required is, however, much greater for crewed missions. Also, the spacecraft structures in the case of crewed missions must include life-support systems and reentry and recovery provisions. In the case of lunar or planetary missions where landing on and leaving the foreign body are required, additional provisions for propulsion, guidance, control, spacecraft sterilization, and life-support systems must be realized, and the structure must be designed to accommodate them.

Testing. To ensure that spacecraft structures will meet mission requirements criteria in general requires testing to levels above the expected environmental conditions by a specific value. The test level must be set to provide for variations in materials, manufacture, and anticipated loads. In cases where structures are required to perform dynamic functions repeatedly, life testing is required to ensure proper operation over a given number of cycles of operation. *See* INSPECTION AND TESTING.

Boost. The purpose of the boost phase is to lift the vehicle above the sensible atmosphere, to accelerate the vehicle to the velocity required, and to place the spacecraft at a point in space, heading in the direction required for the accomplishment of its mission. For space missions, the required velocities range from 26,000 ft/s (8 km/s) for nearly circular Earth orbits to 36,000 ft/s (11 km/s) for interplanetary missions. Achievement of these velocities requires boosters many times the size of the spacecraft itself. Generally, this boosting is accomplished by a chemically powered rocket propulsion system using liquid or solid propellants. Multiple stages are required to reach the velocities for space missions. Vertical takeoff requires a thrust or propulsive force that exceeds the weight of the complete flight system by approximately 30%. An example of a multiple-stage booster is the Delta launch vehicle used for the crewless missions (see illustration). The Delta II 7925 vehicle has the capability to place 4000 lb (1800 kg) into a geosynchronous transfer orbit. *See* INTERPLANETARY PROPULSION; PROPELLANT; ROCKET PROPULSION.

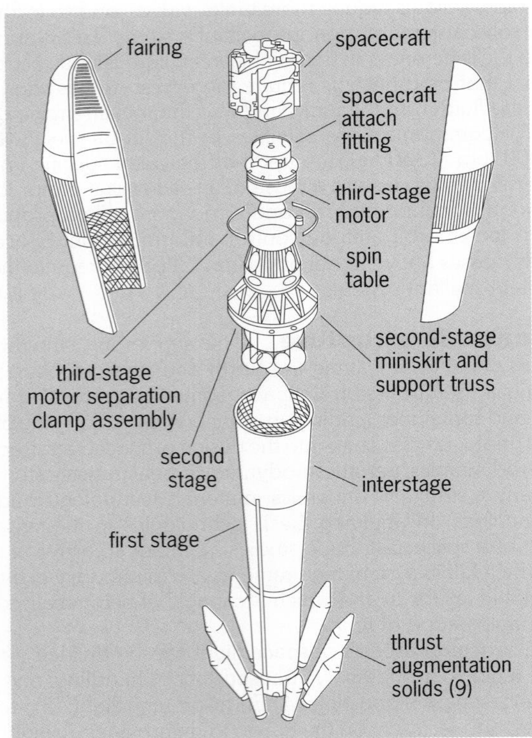

Exploded view of typical Delta II 7925 three-stage structure. (*After Commercial Delta II Payload Planners Guide, McDonnell Douglas Commercial Delta Inc., 1990*)

Space phase and design considerations. The space phase begins after the boost phase and continues until reentry. In this phase, the structures that were stowed for launch are deployed. Important design considerations include control system interaction, thermally induced stress, and minimization of jitter and creaks.

The spacecraft control system imparts inertial loads throughout the structure. In the zero gravity environment, every change in loading or orientation must be reacted through the structure.

Spacecraft structural design usually requires that part of the principal structure be a pressure vessel. Efficient pressure vessel design is therefore imperative. An important material property, especially in pressure vessel design, is notch sensitivity. Notch sensitivity refers to the material's brittleness under biaxial strain. This apparent brittleness contributed to premature failure of some early boosters. *See* PRESSURE VESSEL.

Meteoric particles may have extremely high velocities relative to the spacecraft (up to 225,000 ft/s or 68 km/s). Orbital debris also include residual particles resulting from human space-flight activities. Collisions involving these bodies and a space station and other long-duration orbiting spacecraft are inevitable. The worst-case effects of such collisions include the degradation of performance and the penetration of pressure vessels, including high-pressure storage tanks and habitable crew modules. An essential parameter in the design of these structures is the mitigation of these effects. *See* METEOR.

Radiation shielding is required for some vehicles, particularly those operating for extended times within the Earth's magnetically trapped radiation belts or during times of high sunspot activity. The shielding may be an integral part of the structure. Computer memories are particularly susceptible to radiation and cosmic-ray activity and must be shielded to survive. Effects of radiation on most metallic structures over periods of 10–20 years is not severe. The durability of composite structures in space is a major concern for long life. Based on available data, the synergistic effects of vacuum, heat, ultraviolet, and proton and electron radiation degrade the mechanical, physical, and optical properties of polymers.

Temperature extremes in the structure and the enclosed environment are controlled by several techniques. Passive control is accomplished by surface coatings and multilayer thermal blankets which control the radiation transfer from the spacecraft to space and vice versa. Because incident solar radiation varies inversely with the square of distance from the Sun, means of adjusting surface conditions are required for interplanetary missions. Heat generated by internal equipment or other sources must be considered in the heat balance design. Other techniques used to actively control spacecraft temperatures are thermal louvers and heat pipes. *See* HEAT PIPE.

Thermal gradients must be considered in spacecraft design, especially when the spacecraft has one surface facing the Sun continuously. In some cases it is desirable to slowly rotate the spacecraft to eliminate such gradients.

Spacecraft structures usually are required to be lightweight and rigid, which results in the selection of high-modulus materials. Titanium and beryllium have low densities and relatively high modulus-density ratios. Alloys of these metals are relatively difficult to fabricate, and therefore their application is quite limited. The more common aluminum, magnesium, and stainless steel alloys are basic spacecraft structural materials. They are easy to fabricate, relatively inexpensive, and in general quite suitable

Backdropped by a blue and white part of Earth, the *International Space Station* is seen from space shuttle *Discovery* as the two spacecraft begin their relative separation on June 11, 2008. Earlier, the STS-124 and Expedition 17 crews concluded almost 9 days of cooperative work on board the shuttle and station. (*NASA*)

for use in the space environment. Plastics are used in spacecraft structures when radio-frequency or magnetic isolation is required. They are also used in situations where some structural damping is desired.

The modern requirements for low weight, high strength, high stiffness, and low thermal expansion (for precision optical pointing) have prompted the use of composite materials for spacecraft structure. These materials consist of high-strength reinforcement fibers which are supported by a binder material referred to as the matrix. The fibers are typically made of glass, graphite, or carbon, and the matrix is an epoxy resin. *See* COMPOSITE MATERIAL.

Reentry phase. Although the atmospheric layer of the Earth is relatively thin, it is responsible for the reduction of vehicle velocity and the resulting deceleration loads, as well as for the severe heating experienced by reentering vehicles. A body entering the Earth's atmosphere possesses a large amount of energy. This energy must be dissipated in a manner which allows the reentering vehicle to survive. Most of the vehicle's original energy can be transformed into thermal energy in the air surrounding the vehicle, and only part of the original energy is retained in the vehicle as heat. The fraction that appears as heat in the vehicle depends upon the characteristics of the flow around the vehicle. In turn, the flow around the vehicle is a function of its geometry, attitude, velocity, and altitude. *See* ATMOSPHERIC ENTRY.

Spacecraft are seldom designed to reenter the Earth's atmosphere (the space shuttle being an exception), but may be designed to enter extraterrestrial atmospheres. In either case, the structural design is similar.

High-speed reentry causes extreme friction and heat buildup on spacecraft that must be dissipated by using high-temperature ceramic or ablative materials. The space shuttle is covered with special thermal insulating tiles that allow the structural elements to remain cool when the surface reaches 1200°F (650°C), and its leading edges are protected by a carbon-carbon reinforced material that can withstand temperatures as high as 2300°F (1260°C).

Satellites whose orbits decay into the Earth's upper atmosphere become flaming objects as they rapidly descend. Generally, most or all of the satellite is consumed before it reaches the surface, but there are exceptions such as the March 22, 2001, reentry of the Russian space station, *Mir.*

In crewed applications, vehicles employing aerodynamic lift during reentry have several advantages over zero-lift ballistic bodies: (1) The use of lift allows a more gradual descent, thus reducing the deceleration forces on both vehicle and occupants. (2) The vehicle's ability to glide and maneuver within the atmosphere gives it greater accuracy in either hitting a target or landing at a predetermined spot. (3) It can accommodate greater errors of guidance systems because for a given deceleration it can tolerate a greater range of entry angles. (4) Greater temperature control is afforded because aerodynamic lift may be varied to control altitude with velocity.

Structures. Erectable structures take many and varied shapes. They are sometimes relatively simple hinged booms, while on other occasions they become quite large and massive. Many more spacecraft structures are rigid than erectable or inflatable.

The space shuttle or Space Transportation System (STS) can carry 65,000 lb (30,000 kg) of cargo to and from low Earth orbit. *See* SPACE SHUTTLE.

The *International Space Station* (*ISS*) is a cooperative, 16-nation effort. It will include six laboratories and weigh a million pounds when assembled. *See* COMMUNICATIONS SATELLITE; METEOROLOGICAL SATELLITES; MILITARY SATELLITES; SATELLITE NAVIGATION SYSTEMS; SCIENTIFIC SATELLITES; SPACE STATION.

William Haile; Peter S. Weinberger

Spallation reaction A nuclear reaction that can take place when two nuclei collide at very high energy (typically 500 MeV per nucleon and up), in which the involved nuclei are either disintegrated into their constituents (protons and neutrons), light nuclei, and elementary particles, or a large number of nucleons are expelled from the colliding system resulting in a nucleus with a smaller atomic number. This mechanism is clearly different from fusion reactions induced by heavy or light ions with modest kinetic energy (typically 5 MeV per nucleon) where, after formation of a compound nucleus, only a few nucleons are evaporated. A spallation reaction can be compared to a glass that shatters in many pieces when it falls on the ground. The way that the kinetic energy is distributed over the different particles involved in a spallation reaction and the process whereby this results in residues and fluxes of outgoing particles are not well understood. *See* NUCLEAR FUSION.

Spallation reactions take place in interstellar space when energetic cosmic rays (such as high-energy protons) collide with interstellar gas, which contains atoms such as carbon, nitrogen, and oxygen. This leads to the synthesis of light isotopes, such as ^{6}Li, ^{9}Be, ^{10}Be, and ^{11}B, that cannot be produced abundantly in nucleosynthesis scenarios in the big bang or stellar interiors. *See* BIG BANG THEORY; NUCLEOSYNTHESIS.

In terrestrial laboratories spallation reactions are initiated by bombarding targets with accelerated light- or heavy-ion beams, and they are used extensively in basic and applied research, such as the study of the equation of state of nuclear matter, production of energetic neutron beams, and radioactive isotope research. *See* NEUTRON DIFFRACTION; RELATIVISTIC HEAVY-ION COLLISIONS; SLOW NEUTRON SPECTROSCOPY.

Piet Van Duppen

Sparganosis An infection by the plerocercoid, or sparganum, of certain species of the genus *Spirometra*. The adult normally occurs in the intestine of dogs and cats, the procercoid in copepods, and the plerocercoid in the musculature of frogs, snakes, or aquatic mammals, but not in fish. Humans may become infected accidentally by drinking water with infected copepods or by eating hosts containing plerocercoids. Since humans are not suitable hosts for the adult stage, the plerocercoids leave the gut and enter the musculature, body cavities, or other sites and remain as plerocercoids. Human sparganosis is rare in North America. *See* PSEUDOPHYLLIDEA.

Reino S. Freeman

Spark gap The region between two electrodes in which a disruptive electrical spark may take place. The gap should be taken to mean the electrodes as well as the intervening space. Such devices may have many uses. The ignition system in a gasoline engine furnishes a very important example. Another important case is the use of a spark gap as a protective device in electrical equipment. Here, surges in potential may be made to break down such a gap so that expensive equipment will not be damaged. *See* BREAKDOWN POTENTIAL; ELECTRIC SPARK.

Glenn H. Miller

Spark knock The sound and related effects produced in spark-ignited internal combustion engines by instantaneous ignition and combustion (autoignition) of the gaseous fuel-air mixture ahead of the advancing flame front.

After spark ignition, a flame travels outward from the spark plug and, under normal combustion, will progressively burn the entire fuel-air charge. The burned gas liberates heat and expands, leading to increased pressure and temperature in the unburned gas ahead of the flame front. The unburned gas may be raised above its autoignition temperature. If the flame front velocity is too small the unburned gas spontaneously ignites and burns instantaneously. The instantaneous combustion results in a very intense pressure wave that produces the audible, high-frequency (pinging) sound known as spark knock. *See* COMBUSTION CHAMBER; EXPLOSIVE.

Besides sound, spark knock can result in pitting, or erosion, of the combustion chamber, damage to spark plug electrodes, and possible structural damage to the engine. Spark knock also leads to loss of engine efficiency by inducing spark plug preignition, resulting in overly advanced spark timing. Spark knock also causes intense turbulence within the cylinder, aggravating heat loss from the burned gas to the colder cylinder and head surfaces and reducing efficiency. *See* AUTOMOTIVE ENGINE; COMBUSTION; INTERNAL COMBUSTION ENGINE. Joseph R. Asik; Donald L. Anglin

Spark plug A device that screws into the combustion chamber of an internal combustion engine to provide a pair of electrodes between which an electrical discharge is passed to ignite the combustible mixture. The spark plug consists of an outer steel shell that is electrically grounded to the engine and a ceramic insulator, sealed into the shell, through which a center electrode passes (see illustration). The high-voltage current jumps the gap between the center electrode and the ground electrode fixed to the outer shell.

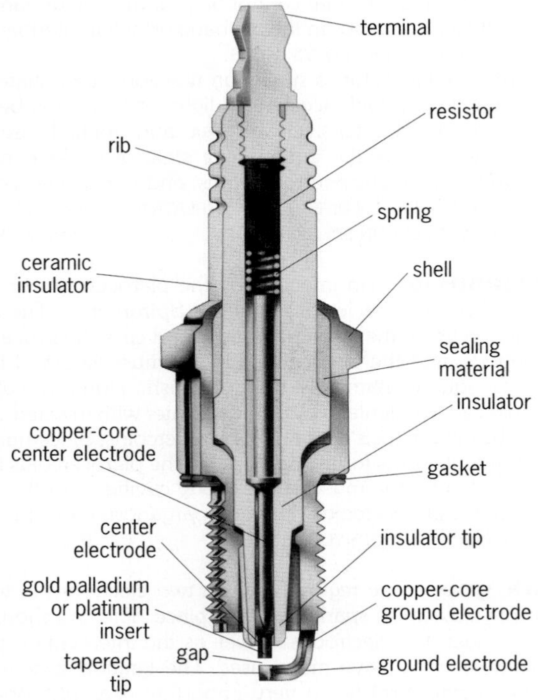

Cross section of a typical spark plug. (*Champion Spark Plug Co.*)

The electrodes are made of nickel and chrome alloys that resist electrical and chemical corrosion. Some center electrodes have a copper core, while others have a platinum tip. Many spark plugs have a resistor in the center electrode to help prevent radio-frequency interference. The parts exposed to the combustion gases are designed to operate at temperatures hot enough to prevent electrically conducting deposits but cool enough to avoid ignition of the mixture before the spark occurs. *See* IGNITION SYSTEM. Donald L. Anglin

Spatangoida An order of exocyclic Euechinoidea in which the posterior ambulacral plates form a shield-shaped area behind the mouth. The plates are arranged in two similar parallel longitudinal series. The apical system is compact. The families are defined mainly by reference to the fascicles, which are ribbonlike bands of minute, close-set uniform ciliated spinules on various parts of the test. The order reached its maximum in the mid-Tertiary, but is still richly represented today in most seas. *See* EUECHINOIDEA. Howard B. Fell

Spearmint Either of two vegetatively propagated, clonal cultivar species (*Mentha spicata* and *M. longifolia*) of mints of the family Lamiaceae (Labiatae). They are grown primarily in Idaho, Indiana, Michigan, Washington, and Wisconsin as a source of essential oil of spearmint.

Principal uses of the oil are in flavoring gum, toothpaste, and candy. Chopped fresh leaves of *M. spicata* preserved in vinegar are used as a condiment served with lamb, especially in England, and dried or freeze-dried leaves of several strains are used in flavoring soups, stews, tea, or sauces. Sprigs of the decorative curly mint *M. crispa* (or *M. spicata* var. *crispata*) are often used in mixed drinks such as mint juleps. *See* LAMIALES. Merritt J. Murray

Special functions Functions which occur often enough to acquire a name. Some of these, such as the exponential, logarithmic, and the various trigonometric functions, are extensively taught in school and occur so frequently that routines for calculating them are built into many pocket calculators. *See* DIFFERENTIATION; LOGARITHM; TRIGONOMETRY.

The more complicated special functions, or higher transcendental functions as they are often called, have been extensively studied by many mathematicians because they arose in the problems which were being studied. Among the more useful functions are the following: the gamma function defined by Eq. (1), which generalizes the factorial; the related beta function defined by Eq. (2), which generalizes the binomial coefficient; and ellip-

$$\Gamma(x) = \int_0^\infty t^{x-1} e^{-t} dt \tag{1}$$

$$B(x, y) = \int_0^1 t^{x-1}(1-t)^{y-1} dt \tag{2}$$

tic integrals, which arose when mathematicians tried to determine the arc length of an ellipse, and their inverses, the elliptic functions. The hypergeometric function and its generalizations includes many of the special functions which occur in mathematical physics, such as Bessel functions, Legendre functions, error functions, and the classical orthogonal polynomials of Jacobi, Laguerre, and Hermite. The zeta function defined by Eq. (3)

$$\zeta(s) = \sum_{n=1}^{\infty} n^{-s} \tag{3}$$

has many applications in number theory, and it also arises in M. Planck's work on radiation. *See* BESSEL FUNCTIONS; ELLIPTIC FUNCTION AND INTEGRAL; GAMMA FUNCTION; HEAT RADIATION; HYPERGEOMETRIC FUNCTIONS; LEGENDRE FUNCTIONS; NUMBER THEORY; ORTHOGONAL POLYNOMIALS. Richard Askey

Speciation The process by which new species of organisms evolve from preexisting species. It is part of the whole process of organic evolution. The modern period of its study began with the publication of Charles Darwin's and Alfred Russell Wallace's theory of evolution by natural selection in 1858, and Darwin's *On the Origin of Species* in 1859.

Belief in the fixity of species was almost universal before the middle of the nineteenth century. Then it was gradually realized that all species continuously change, or evolve; however, the causative mechanism remained to be discovered. Darwin proposed a mechanism. He argued that (1) within any species population there is always some heritable variation; the individuals differ among themselves in structure, physiology, and behavior; and (2) natural selection acts upon this variation by eliminating the less fit. Thus if two members of an animal population differ from each other in their ability to find a mate, obtain food, escape from predators, resist the ravages of parasites and pathogens, or survive the rigors of the climate, the more successful will be more likely than the less successful to leave descendants. The more successful is said to have greater fitness, to be better adapted, or to be selectively favored. Likewise among plants: one plant individual is fitter than another if its heritable characteristics make it more successful than the other in obtaining light, water, and nutrients, in protecting itself from herbivores and disease organisms, or in surviving adverse climatic conditions. Over the course of time, as the fitter members of a population leave more descendants than the less fit, their characteristics become more common.

This is the process of natural selection, which tends to preserve the well adapted at the expense of the ill adapted in a variable population. The genetic variability that must exist if natural selection is to act is generated by genetic mutations in the broad sense, including chromosomal rearrangements together with point mutations. *See* GENETICS; MUTATION.

If two separate populations of a species live in separate regions, exposed to different environments, natural selection will cause each population to accumulate characters adapting it to its own environment. The two populations will thus diverge from each other and, given time, will become so different that they are no longer interfertile. At this point, speciation has occurred: two species have come into existence in the place of one. This mode of speciation, speciation by splitting, is probably the most common mode. Two other modes are hybrid speciation and phyletic speciation; many biologists do not regard the latter as true speciation.

Many students of evolution are of the opinion that most groups of organisms evolve in accordance with the punctuated equilibrium model rather than by phyletic gradualism. There are two chief arguments for this view. First, it is clear from the fossil record that many species persist without perceptible change over long stretches of time and then suddenly make large quantum jumps to radically new forms. Second, phyletic gradualism seems to be too slow a process to account for the tremendous proliferation of species needed to supply the vast array of living forms that have come into existence since life first appeared on Earth. *See* ANIMAL EVOLUTION; POPULATION GENETICS; SPECIES CONCEPT.

E. C. Pielou

Species concept The idea that the diversity of nature is divisible into a finite number of definable species. In general, species concepts grow out of attempts to understand the very nature of biological organization above the level of the individual organism. There are two basic questions: (1) What does it mean to be a species in general? Do all species have certain characteristics, such as forming genealogical lineages, just as all atoms have certain characteristics, such as the ability to undergo chemical reactions? (2) What does it mean to be a particular species? The first question addresses species concepts. The second question addresses how to apply a species concept to living organisms of the world. Does the name *Homo sapiens* apply to a group of organisms existing in nature? If so, does it belong, as a member, to a natural kind that can be characterized by some set of properties? The difference between the questions of what it is to be a species in general versus what does it mean to be a particular species represents the dividing line between what it is to be a natural kind and what it is to be a natural individual. *See* SPECIATION.

The ideas that organisms could be grouped into more or less discrete units, and the idea that some of these units were more similar to each other than to other units, long predated ideas about evolution. That there was a pattern to life's diversity gave rise to the first species concept: members of one species are different in their physical characteristics from members of other species, and their offspring look like them. This concept was the basis for both folk taxonomies (names of different groups of individual organisms in common language) and scientific taxonomies (formal names meant to have scientific import). As the practice of scientific taxonomy developed, species were gathered into larger groups based on similarities and differences. These groupings were hierarchical and seemed to reflect patterns of similarities and differences seen in nature. Current evolutionary theories seek to explain why life's diversity seems to be arranged in the hierarchical manner that we see in nature. *See* TAXONOMIC CATEGORIES; TAXONOMY; ZOOLOGICAL NOMENCLATURE.

Species concepts, in one form or another, have been around for several hundred years. Each concept (and there are over 20 current ones) is an attempt to capture the properties of the largest (or smallest) kind and thus permit a search for particulars (*Homo sapiens*, etc.) that are members of the kind. Of the plethora of species concepts, evolutionary biologists (including systematists and taxonomists) seem to be converging on the concept of species-as-lineages, the Evolutionary Species Concept—species are those things that exist between speciation events and that originate through speciation events. Other concepts capture part of this more general concept.

E. O. Wiley

Specific charge The ratio of charge to mass expressed as e/m, of a particle. The acceleration of a particle in electromagnetic fields is proportional to its specific charge. Specific charge can be determined by measuring the velocity which the particle acquires in falling through an electric potential; by measuring the frequency of revolution in a magnetic field; or by observing the orbit of the particles in combined electric and magnetic fields. *See* ELEMENTARY PARTICLE.

Charles J. Goebel

Specific fuel consumption The ratio of the fuel mass flow of an aircraft engine to its output power, in specified units. Specific fuel consumption (abbreviated sfc or SFC) is a widely used measure of atmospheric engine performance. For reciprocating engines it is usually given in U.S. Customary units of pound-mass per hour per horsepower [(lbm/h)/hp or lbm/(hp·h)], and International System (SI) units of kilograms per hour per kilowatt [(kg/h)/kW]. *See* RECIPROCATING AIRCRAFT ENGINE.

For the gas turbine family of atmospheric aircraft engines, and for ramjets, performance is usually given in terms of thrust specific fuel consumption (abbreviated tsfc or TSFC) expressed as fuel mass flow per unit thrust output with Customary units of pound-mass per hour per pound-force [(lbm/h)/lbf] or SI units of kilograms per hour per newton [(kg/h)/N; 1 N equals approximately 0.225 lbf]. For high-supersonic and hypersonic ramjets, specific fuel consumption is sometimes given in pound-mass per second per pound-force [(lbm/s)/lbf] or kilograms per second per newton [(kg/s)/N]. *See* AIRCRAFT PROPULSION; PROPULSION; RAMJET; TURBINE PROPULSION; TURBOJET.

J. Preston Layton

Specific heat A measure of the heat required to raise the temperature of a substance. When the heat ΔQ is added to a body of mass m, raising its temperature by ΔT, the ratio C given in Eq. (1) is defined as the heat capacity of the body. The quantity c defined in Eq. (2) is called the specific heat capacity or specific

$$C = \frac{\Delta Q}{\Delta T} \qquad (1)$$

$$c = \frac{C}{m} = \frac{1}{m}\frac{\Delta Q}{\Delta T} \qquad (2)$$

heat. A commonly used unit for heat capacity is joule · kelvin^{-1} (J · K^{-1}); for specific heat capacity, the unit joule · gram^{-1} · K^{-1} (J · g^{-1} · K^{-1}) is often used. Joule should be preferred over the unit calorie = 4.18 J. As a unit of specific heat capacity, Btu · lb^{-1} · °F^{-1} = 4.21 J · g^{-1} · K^{-1} is also still in use in English-language engineering literature. If the heat capacity is referred to the amount of substance in the body, the molar heat capacity c_m results, with the unit J · mol^{-1} · K^{-1}.

If the volume of the body is kept constant as the energy ΔQ is added, the entire energy will go into raising its temperature. If, however, the body is kept at a constant pressure, it will change its volume, usually expanding as it is heated, thus converting some of the heat ΔQ into mechanical energy. Consequently, its temperature increase will be less than if the volume is kept constant. It is therefore necessary to distinguish between these two processes, which are identified with the subscripts V (constant volume) and p (constant pressure): C_V, c_V, and C_p, c_p. For gases at low pressures, which obey the ideal gas law, the molar heat capacities differ by R, the molar gas constant, as given in Eq. (3), where $R = 8.31\ \text{J} \cdot \text{mol}^{-1} \cdot \text{K}^{-1}$; that is, the expanding gas heats up less.

$$c_p - c_V = R \tag{3}$$

For solids, the difference between c_p and c_V is of the order of 1% of the specific heat capacities at room temperature. This small difference can often be ignored. *See* CALORIMETRY; CHEMICAL THERMODYNAMICS; HEAT CAPACITY; THERMODYNAMIC PROCESSES.

Robert O. Pohl

Specific impulse The impulse produced by a rocket divided by the mass m_p of propellant consumed. Specific impulse I_{sp} is a widely used measure of performance for chemical, nuclear, and electric rockets. It is usually given in seconds for both U.S. Customary and International System (SI) units.

The impulse produced by a rocket is the thrust force F times its duration t in seconds. The specific impulse is given by the equation below. Its equivalent, specific thrust F_{sp}, that is sometimes

$$I_{sp} = \frac{Ft}{m_p} \tag{1}$$

used alternatively, is the rocket thrust divided by the propellant mass flow rate F/m_p. *See* IMPULSE (MECHANICS); THRUST.

Calculation of specific impulse for the various forms of electric rockets involves electrothermal, resistance or arc heating of the propellant or its ionization and acceleration to high jet velocity by electrostatic or electromagnetic body forces. Ions in the exhaust jets of these devices must be neutralized so the spacecraft will not suffer from space charging or other effects from the plumes of the devices' operation. *See* ELECTROTHERMAL PROPULSION; ION PROPULSION; PLASMA PROPULSION; ROCKET PROPULSION; SPACECRAFT PROPULSION.

J. Preston Layton

Speckle The generation of a random intensity distribution, called a speckle pattern, when light from a highly coherent source, such as a laser, is scattered by a rough surface or inhomogeneous medium. *See* LASER.

The surfaces of most materials are extremely rough on the scale of an optical wavelength (approximately 5×10^{-7} m). When nearly monochromatic light is reflected from such a surface, the optical wave resulting at any moderately distant point consists of many coherent wavelets, each arising from a different microscopic element of the surface. Since the distances traveled by these various wavelets may differ by several wavelengths if the surface is truly rough, the interference of the wavelets of various phases results in the granular pattern of intensity called speckle. If a surface is imaged with a perfectly corrected optical system, diffraction causes a spread of the light at an image point, so that the intensity at a given image point results from the coherent addition of contributions from many independent surface areas. As long as the diffraction-limited point-spread function of the imaging system is broad by comparison with the microscopic surface variations, many dephased coherent contributions add at each image point to give a speckle pattern.

The basic random interference phenomenon underlying laser speckle exists for sources other than lasers. For example, it explains radar "clutter," results for scattering of x-rays by liquids, and electron scattering by amorphous carbon films. Speckle theory also explains why twinkling may be observed for stars, but not for planets. *See* COHERENCE; DIFFRACTION; INTERFERENCE OF WAVES; TWINKLING STARS.

In metrology, the most obvious application of speckle is to the measurement of surface roughness. If a speckle pattern is produced by coherent light incident on a rough surface, then surely the speckle pattern, or at least the statistics of the speckle pattern, must depend upon the detailed surface properties. An application of growing importance in engineering is the use of speckle patterns in the study of object displacements, vibration, and distortion that arise in nondestructive testing of mechanical components.

James C. Wyant

Astronomical speckle interferometry is a technique for obtaining spatial information on astronomical objects at the diffraction-limited resolution of a telescope, despite the presence of atmospheric turbulence. Speckle interferometry techniques have proven to be an invaluable tool for astronomical research, allowing studies of a wide range of scientifically interesting problems. They have been widely used to determine the separation and position angle of binary stars, and for accurate diameter measurements of a large number of stars, planets, and asteroids. Speckle imaging techniques have successfully uncovered details in the morphology of a range of astronomical objects, including the Sun, planets, asteroids, cool giants and supergiants, young stellar objects, the supernova SN1987A in the Large Magellanic Cloud, Seyfert galaxies, and quasars. *See* BINARY STAR; INTERFEROMETRY.

Margarita Karovska

Spectral type An indicator of the physical and chemical characteristics of a star, based on study of the star's spectrum. Stars possess a remarkable variety of spectra, some simple, others complex. To understand the natures of the stars, it was first necessary to bring order to the subject and to classify the spectra.

The modern system of classification was initiated about 1890. The spectra were ordered by letter, A through O, largely on the basis of the strengths of the hydrogen lines. Several letters were found to be unnecessary or redundant, and on the basis of continuity of lines other than hydrogen, it was found that B preceded A and O preceded B. The result is the classical spectral sequence, OBAFGKM. The classes were decimalized, setting up the sequence O5, . . . , O9, B0, . . . , B9, A0, and so forth. (Not all the numbers are used.) The modern Harvard sequence, after the observatory where it was formulated, runs from O2 to M9. Classes L and T were added to the sequence in 1999.

Class A has the strongest hydrogen lines, B is characterized principally by neutral helium (with weaker hydrogen), and O by ionized helium. Hydrogen weakens notably through F and G, but the metal lines, particularly those of ionized calcium, strengthen. In K, hydrogen becomes quite weak, while the neutral metals grow stronger. The M stars effectively exhibit no hydrogen lines at all but are dominated by molecules, particularly titanium oxide (TiO). L stars are dominated by metallic hydrides and neutral alkali metals, while T is defined by methane, ammonia, and water. At G, the sequence branches downward into R and N, whose stars are rich in carbon molecules. In class S, the titanium oxide molecular bands of class M are replaced by zirconium oxide (ZrO). Classes R and N are commonly combined into class C.

At first appearance, the different spectral types seem to reflect differences in stellar composition. However, within the sequence OBAFGKMLT the elemental abundances are roughly similar. The dramatic variations in spectra are strictly the result of changes in temperature. The different spectra of the R, N, and S stars, however, are caused by true and dramatic variations in the chemical composition, the result of internal thermonuclear processing and convection. *See* STELLAR EVOLUTION.

In the 1940s, W. W. Morgan, P. C. Keenan, and E. Kellman expanded the Harvard sequence to include luminosity. A system of roman numerals is appended to the Harvard class to indicate position on the Hertzsprung-Russell diagram: I for supergiant, II

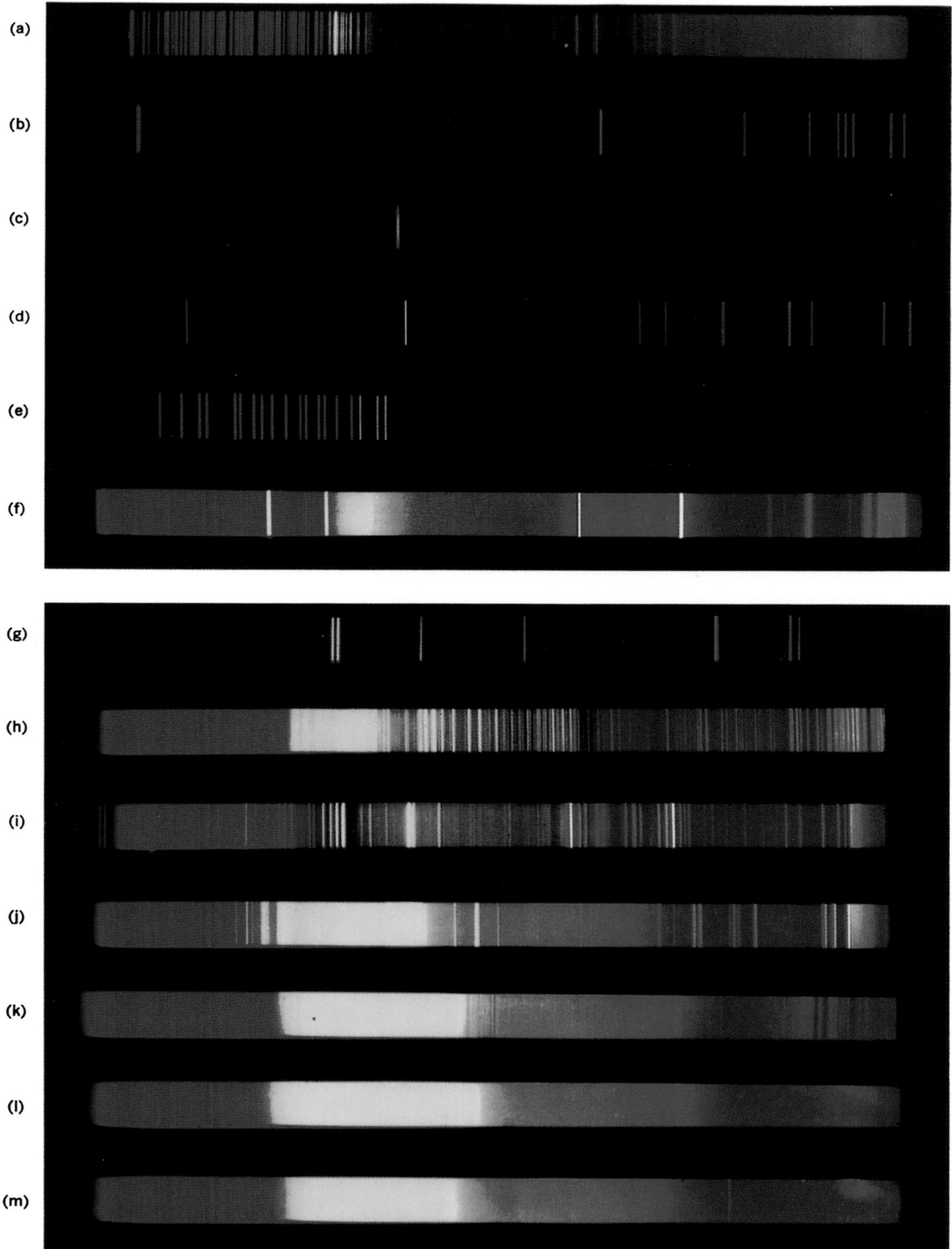

Typical spectra obtained in the visible region. (*a*) Molecular hydrogen. (*b*) Atomic hydrogen. (*c*) Sodium-vapor lamp (D lines). (*d*) Helium. (*e*) Neon. (*f*) Lithium. (*g*) Mercury. (*h*) Iron. (*i*) Barium. (*j*) Calcium. (*k*) Fraunhofer absorption lines. (*l*) Tungsten filament lamp. (*m*) Fluorescent lamp. (*Bausch & Lomb*)

for bright giant, III for giant, IV for subgiant, and V for dwarf or main sequence. *See* ASTRONOMICAL CATALOGS; HERTZSPRUNG-RUSSELL DIAGRAM.

James B. Kaler

Spectrograph An optical instrument that consists of an entrance slit, collimator, disperser, camera, and detector and that produces and records a spectrum. A spectrograph is used to extract a variety of information about the conditions that exist where light originates and along the paths of light. It reveals the details that are stored in the light's spectral distribution, whether this light is from a source in the laboratory or a quasistellar object a billion light-years away.

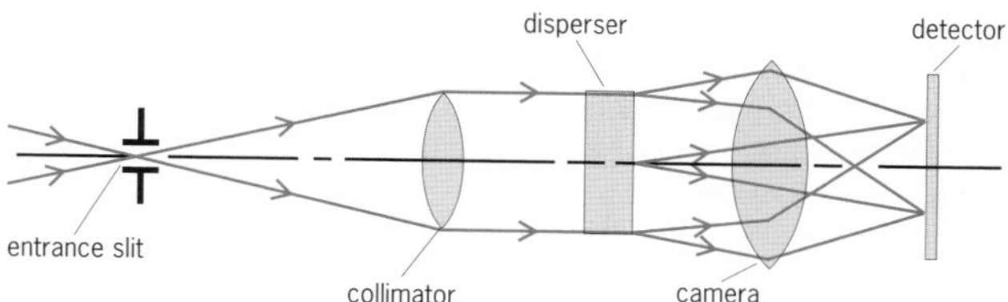

Basic optical components of a spectrograph.

Spectrograph design takes into account the type of light source to be measured, and the circumstances under which these measurements will be made. Since observational astronomy presents unusual problems in these areas, the design of astronomical spectrographs may also be unique.

Astronomical spectrographs have the same general features as laboratory spectrographs (see illustration). The width of the entrance slit influences both spectral resolution and the amount of light entering the spectrograph, two of the most important variables in spectroscopy. The collimator makes this light parallel so that the disperser (a grating or prism) may properly disperse it. The camera then focuses the dispersed spectrum onto a detector, which records it for further study.

Laboratory spectrographs usually function properly only in a fixed orientation under controlled environmental conditions. By contrast, most astronomical spectrographs are used on a moving telescope operating at local temperature. Thus, their structures must be mechanically and optically insensitive to orientation and temperature.

The brightness, spectral characteristics, and geometry of laboratory sources may be tailored to experimental requirements and to the capabilities of a spectrograph. Astronomical sources, in the form of images at the focus of a telescope, cannot be manipulated, and their faintness and spectral diversity make unusual and difficult demands on spectrograph performance.

Typical laboratory spectrographs use either concave gratings, which effectively combine the functions of collimator, grating, and camera in one optical element, or plane reflection gratings with spherical reflectors for collimators and cameras. *See* ASTRONOMICAL SPECTROSCOPY.

Ron Hilliard

Spectroheliograph A spectrographic instrument that produces monochromatic images of the Sun. In a simple form of the instrument, an image of the Sun from a solar telescope is focused on a plane containing the entrance slit of the spectroheliograph (see illustration). The light passing through the slit is collimated by a concave mirror that is tilted such that the light is incident on a plane diffraction grating. Part of the dispersed light from the grating is focused by a second concave mirror, identical to the first mirror, at an exit slit identical to the entrance slit. By symmetry of the optical system, the portion of the solar disk imaged on the entrance slit is reimaged in the plane of the exit slit with the same image scale but in dispersed wavelength. The light imaged along the exit slit then corresponds to the portion of the solar image falling on the entrance slit, but in the light of only a narrow region of the spectrum, as determined by the spectrographic dispersion. The particular wavelength sampled is set by the grating angle. By uniform transverse motion of the instrument such that the entrance slit is scanned across the solar image, the light passing through the exit slit maps out a corresponding monochromatic image of the Sun, which can be recorded photographically with a stationary camera. Alternatively, the solar image can be scanned across the entrance slit of a stationary spectroheliograph, the camera then synchronously moved in step with the image. *See* DIFFRACTION GRATING.

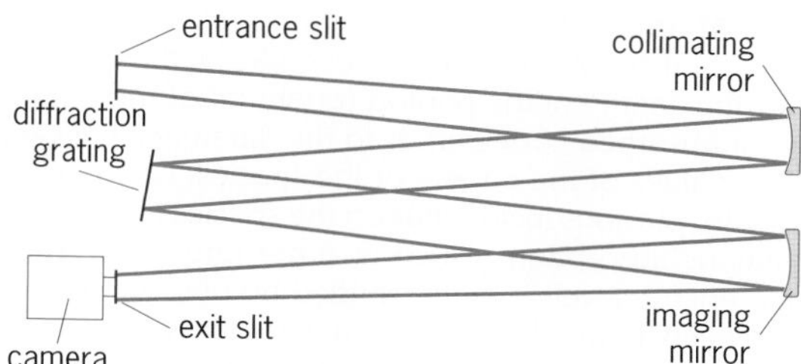

Optical configuration of a simple spectroheliograph.

Digital recording has greatly expanded the performance capabilities, versatility, and range of applications of spectoheliographs, reducing the observing time and allowing flexibility in observing programs. Sequential readout of a linear array that is aligned with the exit slit, and that moves with the slit as the solar image is scanned at the entrance slit, provides a fast and convenient observing system. A charge-coupled-device (CCD) array, used directly at the focal plane without a slit, allows simultaneous recording of an image at the wavelength of a spectral line as well as the adjacent spectral region. Computer processing of the digital data can then provide velocity and magnetic field information, and extensive solar fields can be recorded in a few seconds. *See* ASTRONOMICAL IMAGING; CHARGE-COUPLED DEVICES; SPECTROGRAPH.

Raymond N. Smartt

Spectroscopy An analytic technique concerned with the measurement of the interaction (usually the absorption or the emission) of radiant energy with matter, with the instruments necessary to make such measurements, and with the interpretation of the interaction both at the fundamental level and for practical analysis.

A display of such data is called a spectrum, that is, a plot of the intensity of emitted or transmitted radiant energy (or some function of the intensity) versus the energy of that light. Spectra due to the emission of radiant energy are produced as energy is emitted from matter, after some form of excitation, then collimated by passage through a slit, then separated into components of different energy by transmission through a prism (refraction) or by reflection from a ruled grating or a crystalline solid (diffraction), and finally detected. Spectra due to the absorption of radiant energy are produced when radiant energy from a stable source, collimated and separated into its components in a monochromator, passes through the sample whose absorption spectrum is to be measured, and is detected. Instruments which produce spectra are variously called spectroscopes, spectrometers, spectrographs, and spectrophotometers. *See* SPECTRUM.

Interpretation of spectra provides fundamental information on atomic and molecular energy levels, the distribution of species within those levels, the nature of processes involving change from one level to another, molecular geometries, chemical bonding, and interaction of molecules in solution. At the practical level, comparisons of spectra provide a basis for the determination of qualitative chemical composition and chemical structure, and for quantitative chemical analysis.

Origin of spectra. Atoms, ions, and molecules emit or absorb characteristically; only certain energies of these species are possible; the energy of the photon (quantum of radiant energy) emitted or absorbed corresponds to the difference between two permitted values of the energy of the species, or energy levels. (If the flux of photons incident upon the species is great enough, simultaneous absorption of two or more photons may occur.) Thus the energy levels may be studied by observing the differences between them. The absorption of radiant energy is accompanied by the promotion of the species from a lower to a higher energy level; the emission of radiant energy is accompanied by falling from a higher to a lower state; and if both processes occur together, the condition is called resonance.

Instruments. Spectroscopic methods involve a number of instruments designed for specialized applications.

An optical instrument consisting of a slit, collimator lens, prism or grating, and a telescope or objective lens which produces a spectrum for visual observation is called a spectroscope.

If a spectroscope is provided with a photographic camera or other device for recording the spectrum, the instrument is called a spectrograph.

A spectroscope that is provided with a calibrated scale either for measurement of wavelength or for measurement of refractive indices of transparent prism materials is called a spectrometer.

A spectrophotometer consists basically of a radiant-energy source, monochromator, sample holder, and detector. It is used for measurement of radiant flux as a function of wavelength and for measurement of absorption spectra.

An interferometer is an optical device that measures differences of geometric path when two beams travel in the same medium, or the difference of refractive index when the geometric paths are equal. Interferometers are employed for high-resolution measurements and for precise determination of relative wavelengths. *See* INTERFEROMETRY.

Methods and applications. Since the early methods of spectroscopy there has been a proliferation of techniques, often incorporating sophisticated technology.

Acoustic spectroscopy uses modulated radiant energy that is absorbed by a sample. The loss of that excess produces a temperature increase that can be monitored around the sample by using a microphone transducer. This is the optoacoustic effect. *See* PHOTOACOUSTIC SPECTROSCOPY.

In astronomical spectroscopy, the radiant energy emitted by celestial objects is studied by combined spectroscopic and telescopic techniques to obtain information about their chemical composition, temperature, pressure, density, magnetic fields, electric forces, and radial velocity. *See* ASTRONOMICAL SPECTROSCOPY; SPECTROHELIOGRAPH.

Atomic absorption and fluorescence spectroscopy is a branch of electronic spectroscopy that uses line spectra from atomized samples to give quantitative analysis for selected elements at levels down to parts per million, on the average.

Attenuated total reflectance spectroscopy is the study of spectra of substances in thin films or on surfaces obtained by the technique of attenuated total reflectance or by a closely related technique called frustrated multiple internal reflection. In either method the radiant-energy beam penetrates only a few micrometers of the sample. The technique is employed primarily in infrared spectroscopy for qualitative analysis of coatings and of opaque liquids.

Electron spectroscopy includes a number of subdivisions, all of which are associated with electronic energy levels. The outermost or valence levels are studied in photoelectron spectroscopy. Electron impact spectroscopy uses low-energy electrons (0–100 eV).

X-ray photoelectron spectroscopy (XPS), also called electron spectroscopy for chemical analysis (ESCA), and Auger spectroscopy use x-ray photons to remove inner-shell electrons. Ion neutralization spectroscopy uses protons or other charged particles instead of photons. *See* AUGER EFFECT; ELECTRON SPECTROSCOPY; SURFACE AND INTERFACIAL CHEMISTRY; SURFACE PHYSICS.

Fourier transform spectroscopy is a technique that has been applied to infrared spectrometry and nuclear magnetic resonance spectrometry to allow the acquisition of spectra from smaller samples in less time, with high resolution and wavelength accuracy. *See* FOURIER SERIES AND TRANSFORMS.

Gamma-ray spectroscopy employs the techniques of activation analysis and Mössbauer spectroscopy. *See* ACTIVATION ANALYSIS; MÖSSBAUER EFFECT; NEUTRON SPECTROMETRY.

Information on processes which occur on a picosecond time scale can be obtained by making use of the coherent properties of laser radiation, as in coherent anti-Stokes-Raman spectroscopy. Laser fluorescence spectroscopy provides the lowest detection limits for many materials of interest in biochemistry and biotechnology. Ultrafast laser spectroscopy may be used to study some aspects of chemical reactions, such as transition states of elementary reactions and orientations in bimolecular reactions. *See* LASER SPECTROSCOPY.

In mass spectrometry, the source of the spectrometer produces ions, often from a gas, but also in some instruments from a liquid, a solid, or a material absorbed on a surface. The dispersive unit provides either temporal or spatial dispersion of ions according to their mass-to-charge ratio. *See* MASS SPECTROMETRY; SECONDARY ION MASS SPECTROMETRY (SIMS); TIME-OF-FLIGHT SPECTROMETERS.

In multiplex or frequency-modulated spectroscopy, each optical wavelength exiting the spectrometer output is encoded or modulated with an audio frequency that contains the optical wavelength information. Use of a wavelength analyzer then allows recovery of the original optical spectrum.

When a beam of light passes through a sample, a small fraction of the light exits the sample at a different angle. If the wavelength of the scattered light is different than the original wavelength, it is called Raman scattering. Raman spectroscopy is used in structural chemistry and is a valuable tool for surface analysis. A related process, resonance Raman spectroscopy, makes use of the fact that Raman probabilities are greatly increased when the exciting radiation has an energy which approaches the energy of an allowed electronic absorption. *See* RAMAN EFFECT.

In x-ray spectroscopy, the excitation of inner electrons in atoms is manifested as x-ray absorption; emission of a photon as an electron falls from a higher level into the vacancy thus created is x-ray fluorescence. The techniques are used for chemical analysis. *See* X-RAY FLUORESCENCE ANALYSIS; X-RAY SPECTROMETRY.

Maurice M. Bursey

Spectrum The term spectrum is applied to any class of similar entities or properties strictly arrayed in order of increasing or decreasing magnitude. In general, a spectrum is a display or plot of intensity of radiation (particles, photons, or acoustic radiation) as a function of mass, momentum, wavelength, frequency, or some other related quantity. For example, a β-ray spectrum represents the distribution in energy or momentum of negative electrons emitted spontaneously by certain radioactive nuclides, and when radionuclides emit α-particles, they produce an α-particle spectrum of one or more characteristic energies. A mass spectrum is produced when charged particles (ionized atoms or molecules) are passed through a mass spectrograph in which electric and magnetic fields deflect the particles according to their charge-to-mass ratios. The distribution of sound-wave energy over a given range of frequencies is also called a spectrum. *See* MASS SPECTROSCOPE; SOUND.

In the domain of electromagnetic radiation, a spectrum is a series of radiant energies arranged in order of wavelength or of frequency. The entire range of frequencies is subdivided into wide intervals in which the waves have some common characteristic of generation or detection, such as the radio-frequency spectrum, infrared spectrum, visible spectrum, ultraviolet spectrum, and x-ray spectrum.

Spectra are also classified according to their origin or mechanism of excitation, as emission, absorption, continuous, line, and band spectra. An emission spectrum is produced whenever the radiations from an excited light source are dispersed. An absorption spectrum is produced against a background of continuous radiation by interposing matter that reduces the intensity of radiation at certain wavelengths or spectral regions. The energies removed from the continuous spectrum by the interposed absorbing medium are precisely those that would be emitted by the medium if properly excited. A continuous spectrum contains an unbroken sequence of waves or frequencies over a long range. Line spectra are discontinuous spectra characteristic of excited atoms and ions, whereas band spectra are characteristic of molecular gases or chemical compounds. *See* ATOMIC STRUCTURE AND SPECTRA; ELECTROMAGNETIC RADIATION; LINE SPECTRUM; MOLECULAR STRUCTURE AND SPECTRA; SPECTROSCOPY. W. F. Meggers; W. W. Watson

Spectrum analyzer An instrument for the analysis and measurement of signals throughout the electromagnetic spectrum. Spectrum analyzers are available for subaudio, audio, and radio-frequency measurements, as well as for microwave and optical signal measurements.

Generally, a spectrum analyzer separates the signal into two components: amplitude (displayed vertically) and frequency (displayed horizontally). On some low-frequency analyzers, phase information can also be displayed. Low-frequency analyzers are sometimes grouped under the heading "harmonic analyzers," although this term is becoming less common.

On a conventional spectrum analyzer, a screen with a calibrated graticule displays the components of the input signal. The vertical scale displays the amplitude of each component, and the chosen frequency band is displayed horizontally. Components of the signal being analyzed are displayed as vertical lines whose height is proportional to amplitude and whose horizontal displacement equates to frequency. Originally, cathode-ray tubes were used for the display; solid-state displays such as liquid-crystal displays now are used. *See* CATHODE-RAY TUBE; ELECTRONIC DISPLAY.

Early radio-frequency and microwave analyzers were developed to measure the performance of microwave radar transmitters and to analyze signals from single-sideband transmitters. *See* RADAR; SINGLE SIDEBAND.

A typical use for radio-frequency and microwave spectrum analyzers is the measurement of spurious radiation (noise) from electrical machinery and circuits, known as radio-frequency interference (RFI). Other uses include monitoring and surveillance to detect unauthorized or unintended transmissions, such as the civil monitoring of broadcast and communication channels and the detection of electronic warfare signals. Another application is the analysis of radio communication transmitters and receivers, including those used in radio and television broadcasting, satellite systems, and mobile radio and cellular telephone communications. *See* ELECTRICAL INTERFERENCE; ELECTRONIC WARFARE.

Low-frequency spectrum analyzers are used in a variety of applications. The most obvious use is the measurement of distortion and unwanted signals in all types of audio equipment, from recording and broadcast studios to amplifiers used in the home. *See* SOUND RECORDING; SOUND-REPRODUCING SYSTEMS.

Further uses include the analysis of speech waveforms, measurement of vibration and resonances in mechanical equipment and structures, determination of echo delays in seismic signals, investigation of noise such as from aircraft engines or from machinery in factories, analysis of sonar signals used to detect objects underwater, and study of ultrasonic waves to determine the internal structure of objects such as human tissue and metal castings. *See* BIOMEDICAL ULTRASONICS; MECHANICAL VIBRATION; NOISE MEASUREMENT; NONDESTRUCTIVE EVALUATION; SEISMOLOGY; SONAR; ULTRASONICS.

Optical spectrum analyzers use techniques such as a collimating mirror and a diffraction grating or a Michelson interferometer to separate out the light-wave components. They are used for a variety of applications, including measurements on lasers and light-emitting diodes, and for the analysis of optical-fiber equipment used to carry multichannel, digital telephony. *See* DIFFRACTION GRATING; INTERFEROMETRY; LASER; LIGHT-EMITTING DIODE; OPTICAL COMMUNICATIONS; OPTICAL FIBERS. S. J. Gledhill

Speech A set of audible sounds produced by disturbing the air through the integrated movements of certain groups of anatomical structures. Humans attach symbolic values to these sounds for communication. There are many approaches to the study of speech.

Speech production. The physiology of speech production may be described in terms of respiration, phonation, and articulation. These interacting processes are activated, coordinated, and monitored by acoustical and kinesthetic feedback through the nervous system.

Most of the speech sounds of the major languages of the world are formed during exhalation. Consequently, during speech the period of exhalation is generally much longer than that of inhalation. The aerodynamics of the breath stream influence the rate and mode of the vibration of the vocal folds. This involves interactions between the pressures initiated by thoracic movements and the position and tension of the vocal folds. *See* RESPIRATION.

The phonatory and articulatory mechanisms of speech may be regarded as an acoustical system whose properties are comparable to those of a tube of varying cross-sectional dimensions. At the lower end of the tube, or the vocal tract, is the larynx. It is situated directly above the trachea and is composed of a group of cartilages, tissues, and muscles. The upper end of the vocal tract may terminate at the lips, at the nose, or both. The length of the vocal tract averages 6.5 in. (16 cm) in men and may be increased by either pursing the lips or lowering the larynx.

The larynx is the primary mechanism for phonation, that is, the generation of the glottal tone. The vocal folds consist of connective tissue and muscular fibers which attach anteriorly to the thyroid cartilage and posteriorly to the vocal processes of the arytenoid cartilages. The vibrating edge of the vocal folds measures about 0.92– 1.08 in. (23–27 mm) in men and considerably less in women. The aperture between the vocal folds is known as the glottis. The tension and position of the vocal folds are adjusted by the intrinsic laryngeal muscles, primarily through movement of the two arytenoid cartilages. *See* LARYNX.

When the vocal folds are brought together and there is a balanced air pressure to drive them, they vibrate laterally in opposite directions. During phonation, the vocal folds do not transmit the major portion of the energy to the air. They control the energy by regulating the frequency and amount of air passing through the glottis. Their rate and mode of opening and closing are dependent upon the position and tension of the folds and the pressure and velocity of airflow. The tones are produced by the recurrent puffs of air passing through the glottis and striking into the supralaryngeal cavities.

Speech sounds produced during phonation are called voiced. Almost all of the vowel sounds of the major languages and some of the consonants are voiced. In English, voiced consonants may be illustrated by the initial and final sounds in the following words: "bathe," "dog," "man," "jail." The speech sounds produced when the vocal folds are apart and are not vibrating are called unvoiced; examples are the consonants in the words "hat," "cap," "sash," "faith." During whispering all the sounds are unvoiced.

The rate of vibration of the vocal folds is the fundamental frequency of the voice (F0). It correlates well with the perception of pitch. The frequency increases when the vocal folds are made taut. Relative differences in the fundamental frequency of the voice are utilized in all languages to signal some aspects of linguistic information.

Many languages of the world are known as tone languages, because they use the fundamental frequency of the voice to distinguish between words. Chinese is a classic example of a tone language. There are four distinct tones in Chinese speech. Said with a falling fundamental frequency of the voice, *ma* means "to scold." Said with a rising fundamental frequency, it means "hemp." With a level fundamental frequency it means "mother," and with a dipping fundamental frequency it means "horse." In Chinese, changing a tone has the same kind of effect on the meaning of a word as changing a vowel or consonant in a language such as English.

The activity of the structures above and including the larynx in forming speech sound is known as articulation. It involves some muscles of the pharynx, palate, tongue, and face and of mastication.

The primary types of speech sounds of the major languages may be classified as vowels, nasals, plosives, and fricatives. They may be described in terms of degree and place of constriction along the vocal tract. *See* PHONETICS.

The only source of excitation for vowels is at the glottis. During vowel production the vocal tract is relatively open and the air flows over the center of the tongue, causing a minimum of turbulence. The phonetic value of the vowel is determined by the resonances of the vocal tract, which are in turn determined by the shape and position of the tongue and lips.

The nasal cavities can be coupled onto the resonance system of the vocal tract by lowering the velum and permitting airflow through the nose. Vowels produced with the addition of nasal resonances are known as nasalized vowels. Nasalization may be used to distinguish meanings of words made up of otherwise identical sounds, such as *bas* and *banc* in French. If the oral passage is completely constricted and air flows only through the nose, the resulting sounds are nasal consonants. The three nasal consonants in "meaning" are formed with the constriction successively at the lips, the hard palate, and the soft palate.

Plosives are characterized by the complete interception of airflow at one or more places along the vocal tract. The places of constriction and the manner of the release are the primary determinants of the phonetic properties of the plosives. The words "par," "bar," "tar," and "car" begin with plosives. When the interception is brief and the constriction is not necessarily complete, the sound is classified as a flap. By tensing the articulatory mechanism in proper relation to the airflow, it is possible to set the mechanism into vibrations which quasiperiodically intercept the airflow. These sounds are called trills.

These are produced by a partial constriction along the vocal tract which results in turbulence. Their properties are determined by the place or places of constriction and the shape of the modifying cavities. The fricatives in English may be illustrated by the initial and final consonants in the words "vase," "this," "faith," "hash."

The ability to produce meaningful speech is dependent in part upon the association areas of the brain. It is through them that the stimuli which enter the brain are interrelated. These areas are connected to motor areas of the brain which send fibers to the motor nuclei of the cranial nerves and hence to the muscles. Three neural pathways are directly concerned with speech production, the pyramidal tract, the extrapyramidal, and the cerebellar motor paths. It is the combined control of these pathways upon nerves arising in the medulla and ending in the muscles of the tongue, lips, and larynx which permits the production of speech. *See* NERVOUS SYSTEM (VERTEBRATE).

Six of the 12 cranial nerves send motor fibers to the muscles that are involved in the production of speech. These nerves are the trigeminal, facial, glossopharyngeal, vagus, spinal accessory, and the hypoglossal. *See* PSYCHOACOUSTICS; PSYCHOLINGUISTICS.

Development. In the early stages of speech development the child's vocalizations are quite random. The control and voluntary production of speech are dependent upon physical maturation and learning.

It is possible to describe the development of speech in five stages. In the first stage the child makes cries in response to stimuli. These responses are not voluntary but are part of the total bodily expression. The second stage begins between the sixth and seventh week. The child is now aware of the sounds he or she is making and appears to enjoy this activity. During the third stage the child begins to repeat sounds heard coming from himself or herself. This is the first time that the child begins to link speech production to hearing. During the ninth or tenth month the child enters the fourth stage and begins to imitate without comprehension the sounds that others make. The last stage begins between the twelfth and eighteenth month, with the child intentionally employing conventional sound patterns in a meaningful way. The exact time at which each stage may occur varies greatly from child to child.

Speech technology. Speech technology has been developing within three areas. One has to do with identifying a speaker by analyzing a speech sample. Since the idea is analogous to that of identifying an individual by fingerprint analysis, the technique has been called voice print. However, fingerprints have two important advantages over voice prints: (1) they are based on extensive data that have accumulated over several decades of use internationally, whereas no comparable reference exists for voice prints; and (2) it is much easier to alter the characteristics of speech than of fingerprints. Consequently, this area has remained largely dormant. Most courts in the United States, for instance, do not admit voice prints as legal evidence.

The two other areas of speech technology, synthesis and recognition, have seen explosive growth. In many applications where a limited repertoire of speech is required, computer-synthesized speech is used instead of human speakers. A common technology currently used in speech synthesis involves an inventory of pitch-synchronized, prestored human speech. These prestored patterns are selected according to the particular requirements of the application and recombined with some overlap into the desired sentence by computer, almost in real time. The quality of synthesized speech for English is remarkably good, though it is limited at present to neutral, emotionless speech. Many other languages are being synthesized with varying degrees of success.

The recognition of speech by computer is much more difficult than synthesis. Instead of just reproducing the acoustic wave, the computer must understand something of the semantic message that the speech wave contains, in order to recognize pieces of the wave as words in the language. Humans do this easily because they have a great deal of background knowledge about the world, because they are helped by contextual clues not in the speech wave, and because they are extensively trained in the use of speech. Nonetheless, given various constraints, some of the existing systems do remarkably well. These constraints include (1) stable acoustic conditions in which speech is produced, (2) a speaker trained by the system, (3) limited inventory of utterances, and (4) short utterances. The research here is strongly driven by the marketplace, since all sorts of applications can be imagined where spoken commands are required or highly useful. *See* SPEECH DISORDERS.

William S.-Y. Wang

Speech disorders

Typically, problems of speech production, articulation, or voice, or disruptions of the ability to use language. Another way to categorize these disorders is on the basis of the cause, organic or nonorganic. Organic speech and language disorders typically result from disease or injury to the structures, muscles, and nerves necessary to move and control the mechanisms for speech and voice production. Damage to the brain, particularly the left hemisphere, as the result of stroke, dementia, or traumatic head injury can result in disruptions of normal language function. In children, failure of the brain or systems subserving language and speech to develop adequately can result in significant developmental delays in the acquisition of appropriate language and speech functions. Disorders of language and speech for which there is no demonstrable organic etiology are often the result of poor learning or inadequate speech and language models. In some instances, there may be a psychogenic basis for the disorder.

Articulation for speech production involves the appropriate sequential movement of structures known as the articulators: mandible, tongue, lips, velum, and so forth. There is a wide range of acceptable productions of a given sound (phoneme) in the language. However, when the production of one or more sounds in the language is imprecise and cannot be understood, then an articulation disorder is present. Disorders can range from mild problems such as the distortions associated with a lisp, or can be so severe as to render the speaker unintelligible. Most articulation problems can be categorized in terms of how the sounds are produced. Speech sounds can be misarticulated by distortion, omission, substitution, or addition. These problems can arise from organic causes, such as impairment to motor control of the speech musculature following a stroke or cerebral palsy (dysarthria), cleft palate or lip, or hearing loss. Many articulation problems have no discernible organic basis; these are labeled as functional, since the etiology of the problem is unclear. Many of these problems begin in childhood and can persist into adult life if not treated appropriately. *See* CLEFT LIP AND CLEFT PALATE; HEARING IMPAIRMENT; PHONETICS.

Voice disorders occur when the phonatory mechanism is not functioning properly. Some impairments of function can be the result of vocal abuse or misuse. These voice disorders are said to be functional, although prolonged abuse or misuse can result in temporary or permanent damage to the phonatory mechanism. There are, however, voice disorders that have a true organic etiology. These disorders result from disease or some other condition that affects the laryngeal, oral, or respiratory structures, which are involved in the production of voice. Aphonia is a complete absence of voice. More commonly, individuals with voice disorder are dysphonic, retaining some voice quality. Whether functional or organic in origin, voice disorders are described in terms of pitch, quality (resonance), or loudness. *See* LOUDNESS; PITCH.

Among the most difficult disorders of speech to treat are those of fluency, which are characterized by a disruption of the normal rate of speech production. There are three major types of fluency disorders: stuttering, cluttering, and those related to motor speech disorders of neurological origin (dysarthria and apraxia). Of these, stuttering is the most commonly observed.

Language disorders involve impairment of the ability to voluntarily use or comprehend the language of the individual's linguistic community. Such disorders need not involve a concomitant impairment of speech or voice production. These disorders can be divided into two groups. In delayed language acquisition, normal language development fails to occur because of neurologic or intellectual problems. Aphasia typically occurs in the adult following a stroke or traumatic injury to the brain after the individual had acquired and was using language appropriate to age, gender, and linguistic community. In addition, there has been increasing recognition of language disorders associated with the dementias. *See* APHASIA; SPEECH.

R. S. Tikofsky

Speech perception A term broadly used to refer to how an individual understands what others are saying. More narrowly, speech perception is viewed as the way a listener can interpret the sound that a speaker produces as a sequence of discrete linguistic categories such as phonemes, syllables, or words. *See* PHONETICS; PSYCHOLINGUISTICS.

Classical work in the 1950s and 1960s concentrated on uncovering the basic acoustic cues that listeners use to hear the different consonants and vowels of a language. It revealed a surprisingly complex relationship between sound and percept. The same physical sound (such as a noise burst at a particular frequency) can be heard as different speech categories depending on its context (as "k" before "ah," but as "p" before "ee" or "oo"), and the same category can be cued by different sounds in different contexts. Spoken language is thus quite unlike typed or written language, where there is a relatively invariant relationship between the physical stimulus and the perceived category.

The reasons for the complex relationship lie in the way that speech is produced: the sound produced by the mouth is influenced by a number of continuously moving and largely independent articulators. This complex relationship has caused great difficulties in programming computers to recognize speech, and it raises a paradox. Computers readily recognize the printed word but have great difficulty recognizing speech. Human listeners, on the other hand, find speech naturally easy to understand but have to be taught to read (often with difficulty). It is possible that humans are genetically predisposed to acquire the ability to understand speech, using special perceptual mechanisms usually located in the left cerebral hemisphere. *See* HEMISPHERIC LATERALITY.

Building on the classical research, the more recent work has drawn attention to the important contribution that vision makes to normal speech perception; has explored the changing ability of infants to perceive speech and contrasted it with that of animals; and has studied the way that speech sounds are coded by the auditory system and how speech perception breaks down in those with hearing impairment.There has also been substantial research on the perception of words in continuous speech.

Adult listeners are exquisitely sensitive to the differences between sounds that are distinctive in their language. The voicing distinction in English (between "b" and "p") is cued by the relative timing of two different events (stop release and voice onset). At a difference of around 30 milliseconds, listeners hear an abrupt change from one category to another, so that a shift of only 5 ms can change the percept. On the other hand, a similar change around a different absolute value, where both sounds are heard as the same category, would be imperceptible. The term categorical perception refers to this inability to discriminate two sounds that are heard as the same speech category.

Categorical perception can arise for two reasons: it can have a cause that is independent of the listener's language—for instance, the auditory system may be more sensitive to some changes than to others; or it can be acquired as part of the process of learning a particular language. The example described above appears to be language-independent, since similar results have been found in animals such as chinchillas whose auditory systems resemble those of humans. But other examples have a language-specific component. The ability to hear a difference between "r" and "l" is trivially easy for English listeners, but Japanese perform almost at chance unless they are given extensive training. How such language-specific skills are developed has become clearer following intensive research on speech perception in infants.

Newborn infants are able to distinguish many of the sounds that are contrasted by the world's languages. Their pattern of sucking on a blind nipple signals a perceived change in a repeated sound. They are also able to hear the similarities between sounds such as those that are the same vowel but have different pitches. The ability to respond to such a wide range of distinctions changes dramatically in the first year of life. By 12 months, infants no longer respond to some of the distinctions that are outside their native language, while infants from language communities that do make those same distinctions retain the ability. Future experience could reinstate the ability, so it is unlikely that low-level auditory changes have taken place; the distinctions, although still coded by the sensory system, do not readily control the infant's behavior.

Although conductive hearing losses can generally be treated adequately by appropriate amplification of sound, sensorineural hearing loss involves a failure of the frequency-analyzing mechanism in the inner ear that humans cannot yet compensate for. Not only do sounds need to be louder before they can be heard, but they are not so well separated by the ear into different frequencies. Also, the sensorineurally deaf patient tolerates only a limited range of intensities of sound; amplified sounds soon become unbearable (loudness recruitment).

These three consequences of sensorineural hearing loss lead to severe problems in perceiving a complex signal such as speech. Speech consists of many rapidly changing frequency components that normally can be perceptually resolved. The lack of frequency resolution in the sensorineural patient makes it harder for the listener to identify the peaks in the spectrum that distinguish the simplest speech sounds from each other; and the use of frequency-selective automatic gain controls to alleviate the recruitment problem reduces the distinctiveness of different sounds further. These patients may also be less sensitive than people with normal hearing to sounds that change over time, a disability that further impairs speech perception.

Some profoundly deaf patients can identify some isolated words by using multichannel cochlear implants. Sound is filtered into different frequency channels, or different parameters of the speech are automatically extracted, and electrical pulses are then conveyed to different locations in the cochlea by implanted electrodes. The electrical pulses stimulate the auditory nerve directly, bypassing the inactive hair cells of the damaged ear. Such devices cannot reconstruct the rich information that the normal cochlear feeds to the auditory nerve. *See* HEARING (HUMAN); HEARING AID; HEARING IMPAIRMENT; PERCEPTION; PSYCHOACOUSTICS; SPEECH.

C. J. Darwin

Speech recognition In a strict sense, the process of electronically converting a speech waveform (as the acoustic realization of a linguistic expression) into words (as a best-decoded sequence of linguistic units). At times it can be generalized to the process of extracting a linguistic notion from a sequence of sounds, that is, an acoustic event, which may encompass linguistically relevant components, such as words or phrases, as well as irrelevant components, such as ambient noise, extraneous or partial words in an utterance, and so on. Applications of speech recognition include an automatic typewriter that responds to voice, voice-controlled access to information services (such as news and messages), and automated commercial transactions (for example, price inquiry or merchandise order by telephone), to name a few. Sometimes, the concept of speech recognition may include "speech understanding," because the use of a speech recognizer often involves understanding the intended message expressed in the spoken words. Currently, such an understanding process can be performed only in an extremely limited sense, often for the purpose of initiating a particular service action among a few choices. For example, a caller's input utterance "I'd like to borrow money to buy a car" to an automatic call-routing system of a bank would connect the caller to the bank's loan department.

Converting a speech waveform into a sequence of words involves several essential steps. First, a microphone picks up the acoustic signal of the speech to be recognized and converts it into an electrical signal. A modern speech recognition system also requires that the electrical signal be represented digitally by means of an analog-to-digital (A/D) conversion process, so that it can be processed with a digital computer or a microprocessor. This speech signal is then analyzed (in the analysis block) to produce a representation consisting of salient features of the speech. The most prevalent feature of speech is derived from its short-time spectrum, measured successively over short-time windows of length 20–30 milliseconds overlapping at intervals of 10–20 ms. Each short-time spectrum is transformed into a feature vector, and the temporal sequence of such feature vectors thus forms a speech pattern.

The speech pattern is then compared to a store of phoneme patterns or models through a dynamic programming process in order to generate a hypothesis (or a number of hypotheses) of the phonemic unit sequence. (A phoneme is a basic unit of speech and a phoneme model is a succint representation of the signal that corresponds to a phoneme, usually embedded in an utterance.) A speech signal inherently has substantial variations along many dimensions. First is the speaking rate variation—a speaker cannot produce a word of identical duration at will. Second, articulation variation is also abundant, in terms both of talker-specific characteristics and of the manner in which a phoneme is produced. Third, pronunciation variation occurs among different speakers and in various speaking contexts (for example, some phonemes may be dropped in casual conversation). Dynamic programming is performed to generate the best match while taking these variations into consideration by compressing or stretching the temporal pattern and by probabilistically conjecturing how a phoneme may have been produced. The latter includes the probability that a phoneme may have been omitted or inserted in the utterance. The knowledge of probability (often called a probabilistic model of speech) is obtained via "training," which computes the statistics of the speech features from a large collection of spoken utterances (of known identity) according to a mathematical formalism.

The hypothesized phoneme sequence is then matched to a stored lexicon to reach a tentative decision on the word identity. The decoded word sequence is further subject to verification according to syntactic constraints and grammatical rules, which in turn define the range of word hypotheses for lexical matching. This process of forming hypotheses about words and matching them to the observed speech pattern, and vice versa, in order to reach a decision according to a certain criterion is generally referred to as the "search." In limited domain applications in which the number of legitimate expressions is manageably finite, these constraints can be embedded in an integrated dynamic programming process to reduce search errors.

The degree of sophistication of a speech recognition task is largely a function of the size of the vocabulary it has to deal with. A task involving, say, less than 100 words is called small vocabulary recognition and is mostly for command-and-control applications with isolated word utterances as input. There are usually very few grammatical constraints associated with these types of limited tasks. When the vocabulary size grows to 1000 words, it is possible to construct meaningful sentences, although the associated grammatical rules are usually fairly rigid. For dictation, report writing, or other tasks such as newspaper transcription, a speech recognition system with a large vocabulary (on the order of tens of thousands of word entries) is needed. *See* LINGUISTICS; PSYCHOACOUSTICS; SPEECH.

The technology of speech recognition often finds applications in speaker recognition tasks as well. Speaker recognition applications can be classified into two essential modes, speaker identification and speaker verification. The goal of speaker identification is to use a machine to find the identity of a talker, in a known population of talkers, using the speech input. Speaker verification aims to authenticate a claimed identity from the voice signal.

Biing Hwang Juang

Speed The time rate of change of position of a body without regard to direction. It is the numerical magnitude only of a velocity and hence is a scalar quantity. Linear speed is commonly measured in such units as meters per second, miles per hour, or feet per second.

Average linear speed is the ratio of the length of the path traversed by a body to the elapsed time during which the body moved through that path. Instantaneous speed is the limiting value of the foregoing ratio as the elapsed time approaches zero. *See* VELOCITY.

Rogers D. Rusk

Speed regulation The change in steady-state speed of a machine, expressed in percent of rated speed, when the load on the machine is reduced from rated value to zero. The definition of regulation usually means the net change in a steady-state characteristic, and does not include any transient deviation or oscillation that may occur prior to reaching the new operation point. This same definition is used for stating the speed regulation of electric motors and for certain prime movers, such as steam turbines.

Paul M. Anderson

Speedometer A device for indicating the speed of a vehicle. There are three types of speedometers in general use: mechanical analog, quartz electric analog, and digital microprocessor.

The mechanical analog speedometer is driven by a cable housed in a casing and connected to a gear at the transmission. This gear is designed for the particular vehicle model, considering the vehicle's tire size and rear axle ratio. In most cases, the speedometer is designed to convert 1001 revolutions of the drive cable into registering 1 mi on the odometer, which records distance traveled by the vehicle. The speed-indicating portion of the speedometer operates on the magnetic principle. In the speedometer head, the drive cable attaches to a revolving permanent magnet that rotates at the same speed as the cable. Floating on bearings between the upper frame and the revolving permanent magnet is a nonmagnetic movable speed cup. The magnet revolves within the speed cup, producing a rotating magnetic field. The magnetic field is constant, and the amount of speed cup movement is at all times in proportion to the speed of the magnet rotation. A pointer, attached to the speed cup spindle, indicates the speed on the speedometer dial. *See* MAGNETIC FIELD.

The quartz speedometer utilizes an accurate clock signal supplied by a quartz crystal, along with integrated electronic circuitry to process an electrical speed signal. This signal is generated by a permanent-magnet generator mounted in the transmission. This permanent-magnet generator, designed to be used with both quartz and digital speedometers, provides a sinusoidal speed signal that is proportional to vehicle speed at the rate of 4004 pulses per mile (2503 per kilometer).

In the digital microprocessor speedometer, the vehicle speed is monitored by the permanent speed sensor mounted in the transmission. The signal is transmitted to the microprocessor where the counter converts the speed signal to a digital signal and stores it in memory. The timing circuit has the capacity to handle the counter and memory storage in less than 0.25 s. Memory circuit signals are sent to the electronic display circuit, which selects the display numerals representing the vehicle's speed, according to the number of pulses received from the speed sensor. *See* AUTOMOTIVE TRANSMISSION; ELECTRONIC DISPLAY.

Robert A. Grimm

Spelaeogriphacea A crustacean order within the class Malacostraca, superorder Peracarida. Few spelaeogriphacean species are known: four extant species and two supposed fossil species. *Spelaeogriphus lepidops* is from a stream in a cave in Table Mountain, South Africa. *Potiicoara brasiliensis* inhabits a pool in a cave in the Mato Grosso, Brazil. *Mangkurtu mityula* and *M. kutjarra* have been reported from aquifers in the Pilbara region of Western Australia. *Acadiocaris novascotica* is a fossil from Carboniferous marine sediments in Canada, and *Liaoningogriphus quadripartitus* is a fossil from lacustrine (lake-associated) deposits of Jurassic age in China.

Spelaeogriphus lepidops is the largest of the living species, reaching 9 mm (0.36 in.) in length; the others attain half that length at most. All have a slender flexible body of similar-sized segments. The head, incorporating the first thoracic segment, supports a short carapace that extends backward over the top and sides of the second thoracic segment. The carapace encloses a branchial space in which sit a pair of maxillipedal gills. The seven pairs of walking legs on the remaining thoracic segments are thin and biramous. The animals creep over the bottom with their walking legs and swim by undulations of the whole body. All extant species are inhabitants of freshwater in subterranean environments, caves, or aquifers. Because all species are found in such small areas and are dependent on the persistence of ground water, it could be advocated that they are critically endangered. *See* PERACARIDA.

Gary C. B. Poore

Sperm cell The male gamete. The typical sperm of most animals has a head containing the nucleus and acrosome, a middle piece with the mitochondria, and a tail with the 9 + 2 microtubule pattern (see illustration). Sperm, as well as the acrosome shape, varies with the species. The nucleus consists of condensed chromatin (deoxyribonucleic acid, DNA) and histone proteins. The acrosome, which is derived from the Golgi complex, contains hydrolytic enzymes, that is, hyaluronidase capable of lysing the egg coats at fertilization. Actin molecules which aid in the interaction between sperm and egg are found in the area between the acrosome and nucleus. The mitochondria in the middle piece apparently provide the energy necessary for the motility created by the tail. The tail has a central core, or axial filament, made up of nine double tubules and two central tubules. *See* CILIA AND FLAGELLA.

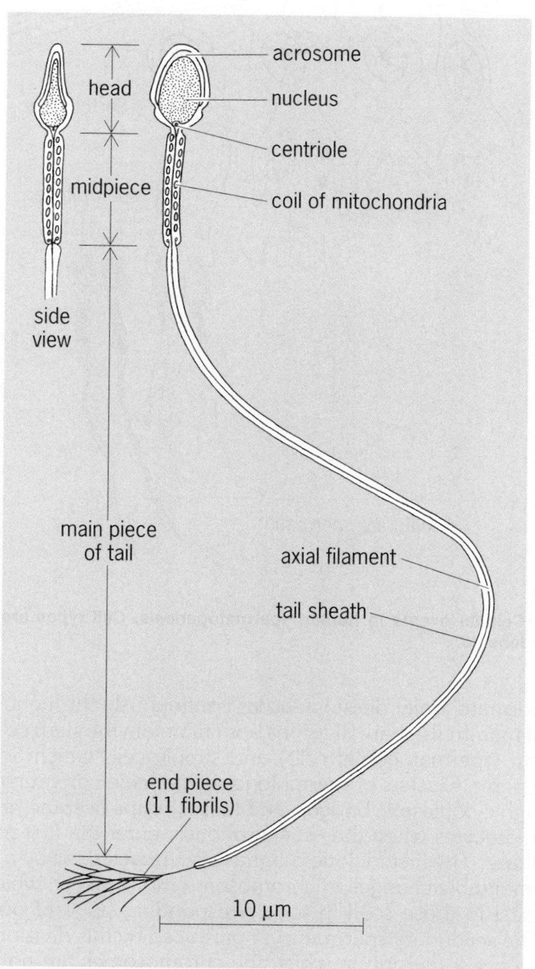

Diagram of a human spermatozoon, based on electron micrographs.

Many groups, including nematodes, myriapods, and crustaceans, have atypical sperm which lack a flagellum and are presumably nonmotile. The sperm of *Ascaris* is round and moves by ameboid means. The crustacean sperm have a large acrosome of several components. In the anomurans a middle with mitochondria and arms filled with microtubules precedes the nucleus. In the true crabs the nucleus forms arms which possess microtubules and also surrounds the many component acrosomes. The nucleus of crustacean sperm does not condense with maturation as it does in typical sperm mentioned above. *See* SPERMATOGENESIS. Gertrude W. Hinsch

Spermatogenesis

Spermatogenesis The differentiation of spermatogonial cells (primordial germ cells in the testes) into spermatozoa (see illustration).

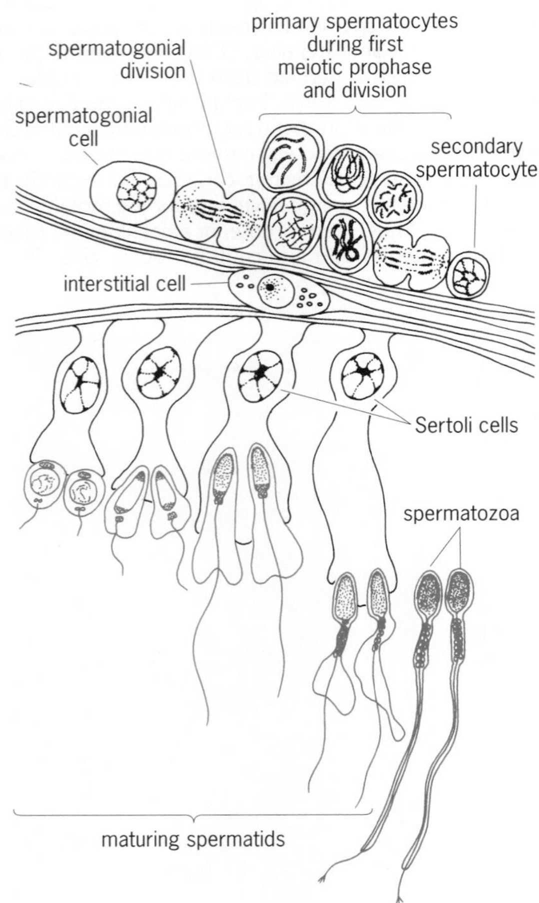

Cellular events in human spermatogenesis. Cell types are labeled.

Spermatogonial divisions occur continuously throughout the life of mammals; these divisions both maintain the stem cell population (spermatogonial cells) and supply cells which develop into sperm. Clusters of spermatogonia maintain communication through cytoplasmic bridges, and these groups become primary spermatocytes when they synchronously enter the first meiotic prophase. The first meiotic prophase is characterized by a series of remarkable changes in chromosome morphology, which are identical to those seen in the corresponding stage of oogenesis. The secondary spermatocyte produced by this division then undergoes a division in which the chromosomes are not replicated; the resulting spermatids contain half the somatic number of chromosomes. *See* MEIOSIS.

The spermatids become embedded in the cytoplasm of Sertoli cells, and there undergo the distinctive changes which result in formation of spermatozoa. These morphological transformations include the conversion of the Golgi apparatus into the acrosome and progressive condensation of the chromatin in the nucleus. A centriole migrates to a position distal to the nucleus and begins organizing the axial filament which will form the motile tail of the sperm. Mitochondria may fuse to form a nebenkern as is the case for many vertebrates, or there may be less extensive fusion as in mammals. In all cases the resulting structures become located around the axial filament in the midpiece. The cytoplasm of the spermatid is reflected distally away from the nucleus during spermatid maturation; eventually, most of the cytoplasm is sloughed off and discarded.

The Sertoli cells are thought to provide nutrition for the developing sperm, because their cytoplasm contains large stores of glycogen which diminish as spermatids mature. There is no direct evidence for this nutritive function, but some forms of male sterility are associated with the failure to produce normal Sertoli cells. Electron microscopy has revealed distinct plasma membranes surrounding the two cell types at the points of contact, and thus the Sertoli cell-spermatid relationship is not syncytial as once thought.

Spermatogenesis is cyclical to a varying extent depending on the species, and under endocrine control. Spermatogenesis is maintained and regulated by male steroid hormones such as testosterone, which is produced by the interstitial or Leydig cells found in the connective tissue of the testis. Interstitial cells, in turn, are stimulated by luteinizing hormone (LH) which is produced by the pituitary gland. The male testis-regulating hormone was formerly known as interstital cell-stimulating hormone (ICSH), but it is now known to be identical to LH. *See* ENDOCRINE MECHANISMS; GAMETOGENESIS; TESTIS. Spencer J. Berry

Sphaeractinoidea

Sphaeractinoidea An extinct group of fossil organisms generally referred to as the Mesozoic stromatoporoids. These stromatoporoids appear in the late Permian with the genus *Circopora*, but did not reach abundance until the Jurassic and early Cretaceous. They declined rapidly in late Cretaceous time, and their presence in younger beds is in doubt. They grew large, calcareous, internally laminate skeletons in encrusting, dome-shaped, bulbous, fingerlike, and branched forms. Internally, the skeleton is composed of an irregularly repetitive three-dimensional network of longitudinal pillars or walls, tangential laminae with beams and cyst plates much like that of their Paleozoic look-alikes. In some genera, longitudinal canals crossed by tabulae are prominent. The microstructure of the structural elements is fibrous in well-preserved fossils and has suggested to some that they secreted aragonite rather than calcite. However, some living sponges secrete calcite in a similar microstructure, and it may be that the Mesozoic stromatoporoids did the same. The Mesozoic stromatoporoids are important reef-building organisms in the Middle East, particularly in areas that surrounded the Mesozoic Tethys seaway. *See* REEF; SPONGIOMORPHIDA; STROMATOPOROIDEA. C. W. Stearn

Sphaerocarpales

Sphaerocarpales An order of liverworts in the subclass Marchantiidae, consisting of two families, the terrestrial Sphaerocarpaceae (*Sphaerocarpos* and *Geothallus*) and the aquatic Riellaceae (*Riella*). The plants are characterized by envelopes surrounding each antheridium and archegonium, absence of elaters, poor development of seta, and absence of thickenings in the unilayered wall of an indehiscent capsule.

The rhizoids are smooth, and air chambers and pores are lacking. Oil bodies may be present. The antheridia are separately contained within receptacles on the dorsal surface or at the margins, while the archegonia are separately produced in receptacles dorsally or behind the growing point. The sporophyte consists of a bulbous foot, short or obsolescent seta, and globose indehiscent capsule with a unilayered wall. *See* BRYOPHYTA.

Howard Crum

Sphagnopsida A class of the plant division Bryophyta containing plants commonly called peatmosses. The spongelike plants grow as perennials in soft cushions or lawns in wet habitats (rarely they grow submerged). The class consists of a single genus, *Sphagnum*, of some 200 species. The thallose protonema, fascicled branches, dimorphous leaf cells, and spores developed from amphithecial tissue are characters unique to the class. The plants are ecologically important, owing to the spongelike construction of leaves and outer cells of stems and branches, and the ability, whether living or dead, to create acid conditions by exchanging hydrogen ions for cations in solution. The dried plant parts are used as a mulch in horticultural practice, and as fuel where the plant is abundant. *See* BRYOPHYTA. Howard Crum

Sphalerite A mineral, β-ZnS, also called blende. It is the low-temperature form and more common polymorph of ZnS, Pure β-ZnS on heating inverts to wurtzite, α-ZnS, at 1868°F (1020°C).

The mineral is most commonly in coarse to fine, granular, cleavable masses. The luster is resinous to submetallic; the color is white when pure, but is commonly yellow, brown, or black, darkening with increased percentage of iron. There is perfect dodecahedral cleavage; the hardness is $3^1/_2$ on Mohs scale; specific gravity is 4.1 for pure sphalerite.

Sphalerite is a common and widely distributed mineral. It occurs both in veins and in replacement deposits in limestones. As the chief ore mineral of zinc, sphalerite is mined on every continent. The United States is the largest producer, followed by Canada, Mexico, Russia, Australia, Peru, the Congo River area, and Poland. *See* WURTZITE; ZINC

Cornelius S. Hurlbut, Jr.; Paul B. Moore

Sphenisciformes The penguins, a small monotypic order of flightless marine swimming birds found in the colder southern oceans. They have most likely evolved from the order Procellariiformes, perhaps from a diving petrel-like ancestor. Classification schemes that hypothesize a link between penguins and loons have no factual support. *See* GAVIIFORMES; PROCELLARIIFORMES.

Penguins are medium-sized to large birds. They are completely flightless, their wings having been modified into stiff flattened flippers. They stand upright on legs that are placed far posterior and that terminate in four toes, the anterior three of which are webbed. Penguins swim and dive well, using only their wings for propulsion: their feet are used only for steering during swimming. Terrestrial locomotion is by walking, hopping, or sliding on the belly while pushing with the wings. The plumage consists of dense scalelike feathers that are black dorsally and white ventrally. A distinctive pattern or crest, often yellow, occurs on the head. Penguins are gregarious, breeding in large colonies along the coast. The males and females form strong pair bonds and share in the incubation and care of the downy nestlings. The older young of some species are kept in large groups, or creches. The emperor penguin (*Aptenodytes forsteri*) breeds on the ice pack along the Antarctic coasts during the fall. Incubation of the one egg is the responsibility of the male, which remains on the nest for over 2 months in the winter without eating.

Penguins are found only in the cold southern oceans, on the Antarctic continent and its surrounding islands and northward to Australia, New Zealand, South America, and Africa. One species, the Galápagos penguins (*Spheniscus mendiculus*), is found in the Galápagos Islands, which are on the Equator but are surrounded by the cold Humboldt Current. *See* AVES. Walter J. Bock

Sphenophyllales An extinct group of articulate land plants, common during Late Pennsylvanian and Early Permian times. They are typified by *Sphenophyllum*, a small, branching plant, probably of trailing habit. The long, jointed stems had superposed, longitudinal, surficial ribs between nodes. The vascular system contained a solid xylem core with triangular primary wood. The leaves were wedge-shaped and had toothed, notched, or rounded distal margins. Long, terminal cones, when found detached, contained sporangia and spores. Most species were homosporous (produced spores of a single type). *See* PALEOBOTANY. Sergius H. Mamay

Sphenophyta One of the major divisions (formerly known as Equisetophyta) of vascular plants that includes both living and fossil representatives. The three principal orders are Pseudoborniales (Devonian), Sphenophyllales (Devonian-Triassic), and Equisetales (Devonian-Recent); the Hyeniales (Devonian) may also be sphenophytes. *See* EQUISETALES; HYENIALES; PSEUDOBORNIALES; SPHENOPHYLLALES.

All sphenophytes are characterized by axes with distinct nodes that produce whorls of small leaves or branches; branches often contain longitudinal ribs and furrows. Internally the stems of sphenophytes are characterized by longitudinally oriented canals, some of which functioned in gaseous exchange. Secondary tissues were produced in a few fossil forms. The reproductive organs of this group are loosely arranged strobili or cones consisting of a central axis bearing whorls of modified branches, each terminating in a recurved, thick-walled sporangium. Most sphenophytes produced one type of spore (homospory), although a few fossil forms were heterosporous.

Today all that remains of the sphenophytes is the genus *Equisetum*, commonly called the horsetail or scouring rush. Except in Australia and New Zealand, the members are worldwide in distribution and typically grow in damp habitats along the edges of streams, although some species are adapted to mesic conditions. A few species attain considerable size, but none of these produce secondary tissues. *See* EMBRYOBIONTA. Thomas N. Taylor

Sphere Both in euclidean solid geometry and in common usage the word sphere denotes a solid of revolution obtained by revolving a semicircle of radius r about its diameter. Its total volume is $V = {}^4/_3\pi r^3$.

However, in analytic geometry, and more generally in modern mathematics, the word sphere denotes a spherical surface that bounds a solid sphere. In this sense a sphere is the locus of all points P in three-dimensional space whose distance from a fixed point O (called the center) is equal to a given number. The word radius may refer either to one of the segments OP, or to their common length r. A plane that intersects a sphere in just one point is called a tangent plane and is perpendicular to the radius drawn from the center of the sphere to that point. A plane that intersects a sphere in more than one point intersects it in a circle. The circle is called a great circle or a small circle of the sphere according to whether the plane does or does not pass through the center of the sphere. If two parallel planes intersect a sphere, the spherical surface between them is called a zone.

Any great circle of a sphere divides it into two hemispheres. A second great circle cuts a hemisphere into two lunes. A third great circle cuts each lune into two spherical triangles. *See* SURFACE AND SOLID OF REVOLUTION. J. Sutherland Frame

Spherical harmonics A spherical harmonic or solid spherical harmonic of degree n is a homogeneous function, $R_n(x, y, z)$, of degree n which satisfies Laplace's equation below.

$$\Delta R \equiv \frac{\partial^2 R}{\partial y^2} + \frac{\partial^2 R}{\partial y^2} + \frac{\partial^2 R}{\partial z^2} \equiv 0$$

Here n is any number, and $(x^2 + y^2 + z^2)^{(-n-1)/2}$. $R_n(x, y, z)$ is a spherical harmonic of degree $-n - 1$. There are analogous definitions for spaces of any number of dimensions. In the present article, n is a nonnegative integer and R_n, a polynomial in x, y, z (polynomial spherical harmonic). In terms of spherical coordinates r, θ, ϕ, $R_n(x, y, z) = r^n S_n(\theta, \phi)$, where S_n, a polynomial in $\cos\theta$, $\sin\theta$, $\cos\phi$, $\sin\phi$, is a spherical surface harmonic of degree n. There are $2n + 1$ linearly independent spherical surface harmonics of degree n; any spherical surface harmonic of degree n is a linear combination of these, and conversely any linear combination of spherical surface harmonics of degree n is again a spherical surface harmonic of degree n.

Spherical harmonics occur in potential theory. They occur in connection with Laplace's equation not only in spherical coordinates but also in spheroidal coordinates (spheroidal harmonics) and confocal coordinates (ellipsoidal surface harmonics). In spherical coordinates, spherical surface harmonics occur in connection with Laplace's and Poisson's equations, the wave equation, and the Schrödinger equation. In mathematical physics, spherical harmonics appear in the theories of gravitation, electricity and magnetism, hydrodynamics, and in other fields.

A. Erdélyi

Sphingolipid Any lipid containing the long-chain amino alcohol sphingosine (structure **1**) or a variation of it, such as dihydrosphingosine, phytosphingosine (**2**), or dehydrophytosphingosine.

$$\underset{\textbf{(1)}}{CH_3(CH_2)_{12}CH{=}CH{-}\underset{OH}{CH}{-}\underset{NH_2}{CH}{-}CH_2OH}$$

$$\underset{\textbf{(2)}}{CH_3(CH_2)_{13}\underset{OH}{CH}{-}\underset{OH}{CH}{-}\underset{NH_2}{CH}{-}CH_2OH}$$

Sphingosine itself is synthesized by condensing a long-chain fatty acid with the amino acid serine.

Sphingosine is converted into a variety of derivatives to form the family of sphingolipids. The simplest form is a ceramide which contains a sphingosine and a fatty acid residue joined by an amide linkage. Ceramide is the basic building block of practically all of the naturally occurring sphingolipids. It can be further modified by the addition of a phosphorylcholine at the primary alcohol group to form sphingomyelin, a ubiquitous phospholipid in the plasma membranes of virtually all cells. Modification of a ceramide by addition of one or more sugars at the primary alcohol group converts it to a glycosphingolipid, which occurs widely in both the plant and animal kingdoms. *See* GLYCOSIDE; LIPID.

Sphingolipids participate in diverse cellular functions. A number of inheritable diseases that can cause severe mental retardation and early death occur as the result of a deficiency in one or more of the degradative enzymes, resulting in the accumulation of a particular sphingolipid in tissues. These diseases are collectively called sphingolipidoses and include Niemann-Pick disease, Gaucher disease, Krabbe disease, metachromatic leukodystrophy, and several forms of gangliosidoses, such as Tay-Sachs disease. Functionally, glycosphingolipids are known to serve as important cell-surface molecules for mediating cell-to-cell recognition, interaction, and adhesion. They also serve as receptors for a variety of bacterial and viral toxins. Many glycosphingolipids can modulate immune responses as well as the function of hormones and growth factors by transmitting signals from the exterior to the interior of the cell. A number of glycolipids are also found to participate in a variety of immunological disorders by serving as autoantigens. Other sphingolipids and their metabolites may serve as second messengers in several signaling pathways that are important to cell survival or programmed cell death (apoptosis). *See* AUTOIMMUNITY; METABOLIC DISORDERS.

Robert K. Yu

Spica A 1st-magnitude star in the constellation Virgo. Spica (α Virginis) is a hot main-sequence star of spectral type B1, effective temperature about 25,000 K (45,000°F), at a distance of 80 parsecs from the Sun (2.59×10^{15} km or 1.61×10^{15} mi). *See* SPECTRAL TYPE.

It is a double-line spectroscopic binary with an orbital period of almost exactly 4 days. The system constitutes an example of an ellipsoidal variable, where the separation between the two stars is so small that powerful tidal forces distort the stars. The changing aspect of the primary as seen from Earth causes light variations with a period equal to the orbital period. The separation between the two stars is only about 3.5 times the radius of the primary. The primary itself is an intrinsic variable of the β Cephei type and is pulsating with a period of about 4 h. *See* BINARY STAR; STAR; VARIABLE STAR.

David W. Latham

Spice and flavoring Ingredients added to food to provide all or a part of the flavor. Spices are pungent or aromatic substances of vegetable origin used in foods at levels that yield no significant nutritive value. Flavor is the perception of those characteristics of a substance taken orally that affect the senses of taste and olfaction. The term flavoring refers to a substance which may be a single chemical species or a blend of natural or synthetic chemicals whose primary purpose is to provide all or part of the particular flavor effect to any food or other product taken orally. Flavorings are categorized by source: animal, vegetable, mineral, and synthetic. *See* OLFACTION; TASTE.

Other than by source, flavorings can be divided into two groups; one group affects primarily the sense of taste, and the other affects primarily the sense of olfaction. The members of the first group are called seasonings, the members of the second group are called flavors. The same terms can be used to divide flavorings in another way. The term flavors can be applied to those products which provide a characterizing flavor to a food or beverage. The term seasoning can then be applied to those products which modify or enhance the flavor of a food or beverage—spices added to meats, blends added to potato chips, lemon added to apple pie.

In the United States most flavorings are processed. Therefore another classification is by method of manufacture. Chemicals are produced by chemical synthesis or physical isolation and include such substances as vanillin, salt, monosodium glutamate, citric acid, and menthol. Concentrates are powders manufactured by dehydrating vegetables, such as onions and garlic, or concentrated fruit juices. Condiments are single ingredients or blends of flavorful foods, spices, and seasonings, some of which may have been derived by fermentation, enzyme action, roasting, or heating. They are usually designed to be added to prepared food at the table (for example, chutney, vinegar, soy sauce, prepared mustard). Other manufactured flavorings include essential oils, hydrolyzed plant proteins, process flavors, and compounded or blended flavorings. *See* ESSENTIAL OILS.

Spices are natural substances that have been dehydrated and consist of whole or ground aromatic and pungent parts of plants, for example, anise, cinnamon, dill, nutmeg, and pepper. Such spices are also classified as herbs if grown in temperate climates.

Extracts are natural substances produced by extraction from solutions of the sapid constituents of spices and other botanicals in food-grade solvents. They are also available as synthetics.

Earl J. Merwin

Spider silk Spider silks are fine fibers produced from fibrous-protein solutions for use by spiders in the various functions required for their survival. Spiders are the only animals that use silk for a variety of functions in their daily lives, including reproduction, food capture, and construction. The exceptional toughness, stiffness, and extensibility of spider-silk fibers outperform even the best synthetic materials. With the recent and expected advances in molecular genetics and proteomics, as well as in the understanding of material properties in the nanoscale size range, orb-web-weaving spiders are ideal organisms for studying the relationship among the physical properties of protein-based materials in terms of their function and evolution. *See* ARANEAE; FIBROUS PROTEIN; NATURAL FIBER; SILK.

All types of silk have the same basic building blocks—protein polymers made of amino acids. The silk glands in an orb-weaving spider, such as *N. clavipes*, differ significantly from one another in morphology and composition, and produce different kinds of silk. The best-characterized silk protein and associated fiber is the dragline silk of *N. clavipes*. Dragline-silk fibers are composed primarily of two protein polymers: spidroin I and spidroin II. Both spidroin I and II consist of repeated alternating protein sequences rich in alanine (Ala) and glycine (Gly) amino acids. These proteins are stored in the glands as concentrated solutions (30–50 wt %). A combination of physicochemical and rheological processes (including ion exchange and elongational flow) cause the protein molecules to self-assemble and transform from an unaggregated and soluble state in the gland, through an intermediate liquid-crystalline state, into an insoluble fiber, while passing through the spinning canal.

After spinning, the dry silk fibers consist of stiff nanometer-scale crystals embedded in a softer protein matrix. The nanocrystals are formed by the aggregation of hydrophobic polyalanine regions into a highly ordered crystalline β-sheet via self-assembly, while the glycine-rich protein regions form the softer rubbery matrix. This combination of hard crystallites embedded in a soft matrix is considered the primary reason for the exceptional mechanical strength and extensibility of dragline fibers. *See* AMINO ACIDS; POLYMER; PROTEIN.

Although spidroin I and spidroin II have been sequenced, a complete understanding of the various physicochemical processes during the spinning of silk fibers is still lacking. Before we can completely mimic spider silk, we must fully understand the biochemistry, processing strategy, and molecular basis of the mechanical properties of silk. Nitin Kumar; Gareth H. McKinley

Spilite An aphanitic (microscopically crystalline) to very-fine-grained igneous rock, with more or less altered appearance, resembling basalt but composed of albite or oligoclase, chlorite, epidote, calcite, and actinolite.

In spite of the highly sodic plagioclase, spilites are generally classed with basalts because of the low silica content (about 50%). They also retain many textural and structural features characteristic of basalt.

Spilites are found most frequently as lava flows and more rarely as small dikes and sills. Spilitic lavas typically show pillow structure, in which the rock appears composed of closely packed, elongated, pillow-shaped masses up to a few feet across. Pillows are typical of subaqueous lava flows. Vesicles, commonly filled with various minerals, may give the rock an amygdaloidal structure. *See* AMYGDULE; BASALT; IGNEOUS ROCKS

William Ingersoll Rose, Jr.

Spin (quantum mechanics) The intrinsic angular momentum of a particle. It is that part of the angular momentum of a particle which exists even when the particle is at rest, as distinguished from the orbital angular momentum. The total angular momentum of a particle is the sum of its spin and its orbital angular momentum resulting from its translational motion. The general properties of angular momentum in quantum mechanics imply that spin is quantized in half integral multiples of $\hbar$ ($=h/2\pi$, where h is Planck's constant); orbital angular momentum is restricted to half even integral multiples of $\hbar$. A particle is said to have spin $^3/_2$, meaning that its spin angular momentum is $^3/_2$. *See* ANGULAR MOMENTUM.

A nucleus, atom, or molecule in a particular energy level, or a particular elementary particle, has a definite spin. The spin is an intrinsic or internal characteristic of a particle, along with its mass, charge, and isotopic spin. *See* QUANTUM MECHANICS; SYMMETRY LAWS (PHYSICS). Charles J. Goebel

Spin-density wave The ground state of a metal in which the conduction-electron-spin density has a sinusoidal variation in space, with a wavelength usually incommensurate with the crystal structure. This antiferromagnetic state normally occurs in metals, alloys, and compounds with a transition-metal component. It occurs also, however, in quasi-one-dimensional organic conductors. *See* ANTIFERROMAGNETISM; CRYSTAL STRUCTURE; ELECTRON SPIN; ORGANIC CONDUCTOR.

There are well over 100 materials which, over a temperature range, support a spin-density wave. These include some of the rare-earth elements of the lanthanide series and the 3*d* transition metals, manganese and chromium, the latter being the prototypical itinerant electron antiferromagnet. The occurrence of inelastic neutron-scattering peaks at incommensurate points indicates the existence of spin-density-wave fluctuations in some metals thought to be nonmagnetic (for example, copper and yttrium) when doped with magnetic impurities (manganese and gadolinium, respectively). This behavior suggests that the spin-density-wave instability may be common, even in nontransition metals. *See* RARE-EARTH ELEMENTS; TRANSITION ELEMENTS.

Eric Fawcett

Spin glass One of a wide variety of materials which contain interacting atomic magnetic moments and also possess some form of disorder, in which the temperature variation of the magnetic susceptibility undergoes an abrupt change in slope, that is, a cusp, at a temperature generally referred to as the freezing temperature. At lower temperatures the spins have no long-range magnetic order, but instead are found to have static or quasistatic orientations which vary randomly over macroscopic distances. The latter state is referred to as spin-glass magnetic order. Spin-glass ordering is usually detected by means of magnetic susceptibility measurements, although additional data are required to demonstrate the absence of long-range order. Closely related susceptibility cusps can also be observed by using neutron diffraction. It is not generally agreed whether spin glasses undergo a phase transition or not. *See* MAGNETIC SUSCEPTIBILITY; NEUTRON DIFFRACTION; PHASE TRANSITIONS. R. E. Walstedt

Spin label A molecule which contains an unpaired electron spin which can be detected with electron spin resonance (ESR) spectroscopy. Molecules are labeled when an atom or group of atoms which exhibits some unique physical property is chemically bonded to a molecule of interest. Groups containing unpaired electrons include organic free radicals and a variety of types of transition-metal complexes (such as vanadium, copper, iron, and manganese). Through analysis of ESR spectra, rates of molecular motion whose motion is restrained by surrounding molecules can be determined.

Analysis of the rate and type of motion of a spin label is important for a wide variety of biological problems. The type of label used in these studies is generally a nitroxide free radical. Spin-labeling studies provide a powerful technique for the study of the geometry and dimensions of receptors in enzymes. Spin labels have been used extensively to study the structure of membranes, and can provide important information about the organization and rates of motion in membranes. Spin labels have also been used to study the structure and organization of synthetic polymers and to study phase transitions. *See* CELL MEMBRANES;

ELECTRON PARAMAGNETIC RESONANCE (EPR) SPECTROSCOPY; ENZYME. Robert Kreilik

Spinach A cool-season annual of Asiatic origin, *Spinacia oleracea*, belonging to the plant order Caryophyllales. It is grown for its foliage and served as a cooked vegetable or as a salad. New Zealand spinach (*Tetrogonia expansa*) and Mountain spinach (*Atriplex hortense*) are also called spinach but are less commonly grown. Spinach plants are usually dioecious. *See* CARYOPHYLLALES. H. John Carew

Spinal cord The portion of the central nervous system within the spinal canal of the vertebral column, that is, the entire central nervous system except the brain. The spinal cord extends from the foramen magnum at the base of the skull to a variable level of the spinal canal; it terminates at the lumbar level in humans and extends well into the caudal region in fishes.

The outer portion of the spinal cord is made up of nerve fibers most of which are oriented longitudinally and carry information between parts of the spinal cord, between spinal cord and brain, and between brain and spinal cord. The outer white matter is divided into dorsal, lateral, and ventral columns. The interior of the spinal cord consists of gray matter and is divided into a dorsal sensory horn (or column) and a ventral motor horn (or column). In the thoracic and lumbar regions of the cord there is also a small lateral horn (or column) which contains preganglionic sympathetic neurons. In the very center of the bilaterally symmetrical spinal cord is a small central canal, containing cerebrospinal fluid. *See* SYMPATHETIC NERVOUS SYSTEM.

Paired spinal nerves enter the spinal canal between each pair of vertebrae and connect with the spinal cord. The number of spinal nerves varies widely in vertebrates; in humans there are 31 pairs (8 cervical, 12 thoracic, 5 lumbar, 5 sacral, and 1 coccygeal). Each spinal nerve divides into a dorsal sensory root and a ventral motor root before entering the spinal cord. The motor neurons of the ventral horn, in addition to receiving synapses from dorsal root axons, also receive synaptic endings from neurons in other parts of the spinal cord and from long axons coming from the brain. The axons of the ventral horn neurons leave the cord through the ventral root of the spinal nerve and run with peripheral nerves to innervate the muscles of the body. With this complex synaptic and fiber organization, the spinal cord can act as the integrating center for spinal reflexes (such as the knee jerk reflex), send sensory information from the brain, and receive information from the brain to initiate or inhibit muscular activity. *See* MOTOR SYSTEMS; NERVOUS SYSTEM (VERTEBRATE); SENSATION. Douglas B. Webster

Spinal cord disorders In addition to those disorders common to the brain, the spinal cord is subject to certain lesions because of its position or structure. A few of the more important are mentioned.

Spinal cord injury results from dislocations, fracture, or compression in many cases, but a special form, called spinal shock, may result from a severe blow without actual distortion of adjacent tissue. In this case, there is a temporary paralysis which gradually clears. In direct damage, the cord may be slightly, partially, or completely damaged at one or more levels. Typical motor and sensory losses follow, with a poor prognosis for recovery if the nerve tissue is severely injured.

A fairly common type of potential cord injury is seen in a number of cases of slipped disks, in which the inner, soft part of the vertebral column extrudes into the spinal canal. If this compresses the cord, functional loss of temporary or permanent degree can follow; more often, pressure is exerted on spinal roots so that pain, numbness, and some type of muscle weakness intervene.

Spinal cord tumors are not infrequent and most of these are of two types, the metastatic, from a primary source elsewhere in the body, and the tumors of the meninges or connective tissue related to the cord. The latter include neurofibromas, meningiomas, and gliomas, which occur most often. The signs and symptoms and the extent of damage relate largely to the physical compression of the cord at a particular level. *See* TUMOR.

A few of the more common congenital defects involving the cord include an unclosed neural canal, or spina bifida, and reduplicated or otherwise malformed cords, such as those caught in an external sac of other tissues, the meningomyelocoele.

Inflammations may result from known or unknown agents and in meningitis may involve primarily the coverings; in myelitis, the cord itself. The meningococcus, pneumococcus, streptococcus, tubercle bacillus, and other microorganisms frequently cause meningitis. The most widely known cause of myelitis, of course, is the poliomyelitis virus group.

An ill-defined group of disorders characterized by degeneration of nerve tracts or myelin sheaths of the cord is found more often than one would suspect. In these, some unknown or poorly understood mechanism causes the deterioration of cells and fibers so that function is altered, then lost, and the nervous tissue is either replaced by a scar or a softening of cystlike area remains. Multiple sclerosis, combined degeneration associated with pernicious anemia, Parkinson's disease, postinfectious encephalomyelitis, and syringomyelia are examples. *See* MULTIPLE SCLEROSIS; PARKINSON'S DISEASE; SPINAL CORD. Edward G. Stuart; N. Karle Mottet

Spinicaudata An order of fresh-water branchiopod crustaceans formerly included in the order Conchostraca. The entire body and its appendages are covered by a bivalve carapace up to about 17 mm (0.7 in.) long, hence the name clam shrimp. The carapace is often bilaterally compressed and usually displays several lines of growth. The head cannot be protruded. Reproduction is usually bisexual. All species frequent temporary pools in many parts of the world, but *Cyclestheria* also occurs in permanent waters. *See* BRANCHIOPODA. Geoffrey Fryer

Spinning (metals) A production technique for shaping and finishing metal. In the spinning of metal, a sheet is rotated and worked by a round-ended tool. The sheet is formed over a mandrel. Spinning may serve to smooth wrinkles in drawn parts, provide a fine finish, or complete a forming operation as in curling an edge of a deep-drawn part. Spun products range from precision reflectors and nose cones to kitchen utensils. *See* SHEET-METAL FORMING. Ralph L. Freeman

Spinning (textiles) The fabrication of yarn (thread) from either discontinuous natural fibers or bulk synthetic polymeric material. In a textile context the term spinning is applied to two different processes leading to the yarns used to make threads, cords, ropes, or woven or knitted textile products.

Natural fibers, such as wool, cotton, or linen, are generally found as short, entangled filaments. Their conversion into yarn is referred to as spinning. After a carding operation on the raw material to disentangle the short filaments, the filaments are drawn (drafted) to promote alignment in an overlapping pattern and then twisted to form, by mechanical interlocking of the discontinuous filaments, a resistant continuous yarn. *See* COTTON; LINEN; WOOL.

The term spinning is also used for the production of monofilaments from synthetic polymers—for example, polyamides or nylons, polyesters, and acrylics—or modified natural polymers, such as cellulose-rayon. Generally the monofilaments are stretched (drawn) to increase their strength by promoting molecular orientation and are wound as yarn which can be used directly for threads, cords, or ropes. Such yarn, however, is often cut into relatively short lengths (staple) and reformed by a process similar to that used for natural fibers into a yarn more suitable, in terms of appearance and feel, for making certain textile products. *See* MANUFACTURED FIBER; NATURAL FIBER; TEXTILE. J. M. Charrier

Spintronics The term "spintronics" usually refers to the branch of physics concerned with the manipulation, storage, and transfer of information by means of electron spins in addition to or in place of the electron charge as in conventional electronics. Introduced in 1996, spintronics was originally the name for a Defense Advanced Research Projects Agency (DARPA) program. In conventional electronics, only the charge of the electrons is of consequence for device operation, but using the electron's other fundamental property, its spin, has opened up the new field of spintronics. Major advances in electron spin transport started in 1979–1980 with the discovery of large low-temperature magnetoresistance in metallic superlattices. Later demonstrations of the "giant" effect at room temperature evolved toward application in practical devices.

Spintronics promises the possibility of integrating memory and logic into a single device. In certain cases, switching times approaching a picosecond are possible, which can greatly increase the efficiency of optical devices such as light-emitting diodes (LEDs) and lasers. The control of spin is central as well to efforts to create entirely new ways of computing, such as quantum computing, or analog computing that uses the phases of signals for computations. *See* QUANTUM COMPUTATION; SPIN (QUANTUM MECHANICS).

Spin is a fundamental quantum-mechanical property. It is the intrinsic angular momentum of an elementary particle, such as the electron. Of course, any charged object possessing spin also possesses an intrinsic magnetic moment. In ferromagnetism, the spins of electrons are preferentially aligned in one direction. In 1988, it was demonstrated that currents flowing from a ferromagnet into an ordinary metal retain their spin alignment for distances longer than interatomic spaces, so that spin and its associated magnetic moment can be transported just as charge. This means that magnetization as well can be transferred from one place to another. *See* DOMAIN (ELECTRICITY AND MAGNETISM); FERROMAGNETISM; MAGNETISM; MAGNETIZATION.

Giant magnetoresistive effect. The first practical application of this phenomenon is in the giant magnetoresistive effect (GMR). The GMR is observed in thin-film materials composed of alternate ferromagnetic and nonmagnetic layers. The resistance of the material is lowest when the magnetic moments in ferromagnetic layers are aligned in the same direction, and highest when they are antialigned. The first GMR-based magnetic field sensor was created in 1994, and high-performance disk drives utilizing GMR-based read heads to detect magnetic fields were realized in 1997 and now are ubiquitous. *See* COMPUTER STORAGE TECHNOLOGY; MAGNETIC RECORDING; MAGNETORESISTANCE.

Spin-dependent tunneling device. A spin-dependent tunneling device (SDT) is similar to a GMR cell but replaces the metal between the two ferromagnetic layers with a very thin insulator through which a current can tunnel preferentially when the two magnetic orientations are aligned. The difference in resistance between the spin-aligned and nonaligned cases is much greater than for GMR devices and large enough that the low-resistance state can encode, say, a "1" and the high-resistance state, a "0." Recently, an SDT device was used in the first commercial magnetoresistive random access memory (MRAM), a fast RAM that is nonvolatile, meaning it does not require power to retain information.

Spin momentum transfer effect. Significant developments ensure that MRAM will be able to scale down to 60 nm and below. The most notable of these was the discovery of the spin momentum transfer effect (SMT), in which the angular momentum carried by a spin-polarized current can exert a torque on the magnetization of a magnetic film that is magnetized in any nonparallel direction. This effect is also known as spin torque. SMT potentially offers orders-of-magnitude lower switching currents and concomitantly much lower energy per bit to write. Apparently SMT switching can significantly improve the performance of MRAM and make it a truly universal memory.

Spin-polarized field-effect transistors. A key bottleneck to increasing computer capabilities is moving information between memories and logic circuits. Ideally, if individual devices could both process and store information, transfer delays could be eliminated, at least for data in immediate use. A spin-based device that could accomplish this dual task is a spin-polarized field-effect transistor (spin FET).

Spin resonant tunneling diodes. The addition of spin sensitivity can potentially produce devices that switch faster than any transistor. One such extremely fast switch is a spin resonant tunneling diode (RTD). This consists of a quantum well sandwiched between two insulating barriers. Current can flow only when the applied voltage reaches a precise value that allows a quantum-mechanically resonant state to exist within the quantum well. Such switches can turn on and off within less than 1 ps. A spin RTD can act like a transistor. *See* TRANSISTOR.

Quantum computing. Another avenue for using the spins of elementary particles comes from the rapidly developing field of quantum computing. The states of spin of electrons or other spin-$^1/_2$ particles can be used as an implementation of a qubit (quantum bit, the unit of quantum information). Information can be encoded using the polarization of the spin, manipulation (computation) can be done using external magnetic fields or laser pulses, and readout can be done by measuring spin-dependent transport. One viable candidate for quantum information is electron spins in coupled quantum dots. Daryl Treger; Stuart A. Wolf

Spiral A term used generically to describe any geometrical entity that winds about a central point or axis while also receding from it. Spiral staircases, helices, and nonplanar loxodromes (curves that intersect those of a given class at a constant angle, for example, rhumb lines, in case the curves are on a sphere whose meridians form the given class) are examples of spirals whose windings do not lie in a plane. *See* HELIX. Leonard M. Blumenthal

Spiriferida An extinct order of brachiopods, in the subphylum Rhynchonelliformea, that inhabited shallow seas of the Paleozoic and early Mesozoic. It was the most diverse group of spire-bearing brachiopods (those that contain a spirally coiled calcareous structure called the brachidium used to support the lophophore). Spiriferids also possess unequally biconvex valves that are externally smooth or radially ribbed. The valves are generally strophic (straight hinge) with a well-developed interarea commonly limited to the ventral valve. Shells may be punctate (possessing small perforations filled by outer epithelial tissue) or impunctate. Punctate shells arose in stratigraphically younger spiriferids; the presence or absence of punctae serves to distinguish two major taxonomic groups. Spiriferids were sessile, attached, epifaunal suspension feeders. Most had a functional pedicle, used for attachment to the substrate. *See* ARTICULATA (ECHINODERMATA); BRACHIOPODA. Mark E. Patzkowsky

Spirometry The measurement, by a form of gas meter, of volumes of gas that can be moved in or out of the lungs. The classical spirometer is a hollow cylinder (bell) closed at its top. With its open end immersed in a larger cylinder filled with water, it is suspended by a chain running over a pulley and attached to a counterweight. The magnitude of a gas volume entering or leaving is proportional to the vertical excursion of the bell. Volume changes can also be determined from measurements of flow, or rate of volume change, that can be sensed and recorded continuously by a transducer that generates an electrical signal. The flow signal can be continuously integrated to yield a volume trace.

The volume of gas moved in or out with each breath is the tidal volume; the maximal possible value is the vital capacity. Even after the most complete expiration, a volume of gas that cannot be measured by the above methods, that is, the residual volume, remains in the lungs. It is usually measured by a gas dilution method or by an instrument that measures blood flow in the lungs. Lung volumes can also be estimated by radiological or optical methods.

At the end of an expiration during normal resting breathing, the muscles of breathing are minimally active. Passive (elastic and gravitational) forces of the lungs balance those of the chest wall. In this state the volume of gas in the lungs is the functional residual capacity or relaxation volume. Displacement from this volume requires energy from natural (breathing muscles) or artificial (mechanical) sources. *See* RESPIRATION. Arthur B. Otis

Spirophorida An order of sponges of the class Demospongiae, subclass Tetractinomorpha, with a globular shape and a skeleton of oxeas and triaenes as megascleres, and microspined, contorted, sigmalike microscleres. Members of this order range down to depths of at least 5900 ft (1800 m). *See* DEMOSPONGIAE. Willard D. Hartman

Spirurida An order of nematodes in which the labial region is usually provided with two lateral labia or pseudolabia; in some taxa there are four or more lips; rarely lips are absent. Because of the variability in lip number, there is variation in the shape of the oral opening, which may be surrounded by teeth. The amphids are most often laterally located; however, in some taxa they may be located immediately posterior to the labia or pseudolabia. The stoma may be cylindrical and elongate or rudimentary. The esophagus is generally divisible into an anterior muscular portion and an elongate swollen posterior glandular region, where the multinucleate glands are located. Eclosion larvae are usually provided with a cephalic spine or hook and a porelike phasmid on the tail.

All known spirurid nematodes utilize an invertebrate in their life cycle; the definitive hosts are mammals, birds, reptiles, and rarely amphibians. The order contains four superfamilies: Spiruroidea, Physalopteroidea, Filarioidea, and Drilonematoidea.

Spiruroidea. The Spiruroidea comprise parasitic nematodes whose life cycle always requires an intermediate host for larvae to the third stage. The definitive hosts are mammals, birds, fishes, reptiles, and rarely amphibians; and spiruroids may be located in the host's digestive tract, eye, or nasal cavity or in the female reproductive system. Morphologically, the lip region is variable in Spiruroidea, ranging from four lips to none. When lips are present, the lateral lips are well developed and are referred to as pseudolabia. The cephalic and cervical region may be ornamented with cordons, collarettes, or cuticular rings. The stoma is always well developed and is often provided with teeth just inside the oral opening. In birds the nematodes are often associated with the gizzard, and the damage caused results in death, generally by starvation. When the muscles of the gizzard are destroyed, seeds pass intact and cannot be digested.

This superfamily contains the largest of all known nematodes, *Placentonema gigantissima*, parasitic in the placenta of sperm whales. Mature females attain a length of 26 ft (8 m) and a diameter of 1 in. (2.5 cm). The adult female has 32 ovaries, which produce great numbers of eggs. Armand Maggenti

Filarioidea. The Filarioidea contain highly specialized parasites of most groups of vertebrates. They are particularly common in amphibians, birds, and mammals. While they cannot be classified as completely harmless, most of the many hundreds of known species are not associated with any recognized disease. A limited number of species produce serious diseases in humans, and a few others produce serious diseases in domestic or wild animals. The filarial parasites of humans are found almost exclusively in the tropics, with some extension into the subtropics.

There are no conspicuous divisions into distinct body regions. Sexual dimorphism is the rule; in common with other nematodes, the female filarioid is at least twice as long as the male, and often the difference is much greater. The adult worms are found in a wide variety of places in the body of the vertebrate host, but each species has its preferred host and preferred location within that host.

All the known filariae require a bloodsucking arthropod intermediate host, usually an insect and commonly a dipteron, in which to complete embryonation. The microfilariae are ingested as the arthropod feeds. After embryonation is completed, the resulting infective larvae gain entrance into the definitive vertebrate in association with the next feeding of the arthropod. *See* NEMATA (NEMATODA).

Filariasis. Filariasis is a disease caused by Filarioidea in humans or lower animals. The term is loosely used to indicate mere infection by such organisms. In human medicine, filariasis commonly refers to the disease caused by, or to infection with, one of the mosquito-borne, elephantoid-producing filarioids—most frequently *Wuchereria bancrofti*, less frequently *Brugia malayi*, and more recently *B. timori*. The only specific laboratory aid to diagnosis is the detection and identification of the microfilariae.

Onchocerciasis. This disease is caused by *Onchocerca volvulus* in the subcutaneous lymphatics. It is characterized by subcutaneous nodules which are most conspicuous where the skin lies close over bony structures, via cranium, pelvic girdle, joints, and shoulder blades. When they are on the head, the microfilariae reach the eyes. Ocular disturbances vary from mild transient bleary vision to total and permanent blindness.

Loa loa. The African eye worm, *Loa loa*, is the filarioidean worm most commonly acquired by Caucasian immigrants, including missionaries, in Africa. Transmission is by daytime-feeding sylvan deer-flies, genus *Chrysops*. The only preventive measures are protective clothing, including head nets. Repellents have some value. Fortunately, serious damage is rare even when the worm gets into the eye. The areas of pitting edema known as calabar swellings are painful and diagnostic. They commonly occur on the wrists, hands, arms, or orbital tissues. Gilbert F. Otto

Spitzer Space Telescope A high-performance infrared telescope that is one of the four Great Observatories launched by the National Aeronautics and Space Administration (NASA) between 1990 and 2003. It utilizes modern detector arrays in space, where they are limited only by the faint glow of the zodiacal dust cloud. Ground-based infrared telescopes can operate only at the wavelengths where the atmosphere is transparent, lying between 1 and 25 micrometers. Even within these windows, the thermal emission of the atmosphere is more than a million times greater than the dilute emission of the zodiacal cloud; there is additional foreground thermal emission from the telescope itself. High-sensitivity detectors are blinded by these bright foreground signals. Operating in space eliminates the atmospheric absorption and emission; also, a telescope in the vacuum of space can be cooled sufficiently to virtually eliminate its emission. *See* ZODIACAL LIGHT.

Spitzer was known for most of its life as *SIRTF* (*Space Infrared Telescope Facility*). *SIRTF* was launched on August 25, 2003, and was renamed after Lyman Spitzer, Jr., the first prominent astronomer to advocate putting telescopes into space.

Design. The approved concept introduces a number of innovations. It uses an Earth-trailing orbit to get far from the thermal radiation of the Earth. The instruments are cooled inside a Dewar containing liquid helium, as with its predecessors, *IRAS* and *ISO*. However, the telescope is mounted on the outside of this Dewar and was launched warm. The very cold environment away from the Earth allows the outer shell of the satellite to cool passively, greatly reducing the heat load on the telescope. *See* DEWAR FLASK.

Spitzer has three instruments. The Infrared Array Camera (IRAC) images in bands at wavelengths of 3.6, 4.5, 5.8, and

8 μm. The Infrared Spectrograph (IRS) provides spectra from 5 to 40 μm at low resolution (1–2%) and from 10 to 38 μm at moderate resolution (about 0.15%). The Multiband Imaging Photometer for *Spitzer* (MIPS) provides imaging at 24, 70, and 160 μm. The instruments use advanced infrared detector arrays that operate at or near the fundamental natural background limit in space.

Objectives. *Spitzer* is operated as a general-access observatory from the Jet Propulsion Laboratory and a science center at the California Institute of Technology. Eighty percent of the observing time is assigned by competitive application from the astronomical community. The science program includes investigations of star formation, planetary systems, and the assembly and evolution of galaxies. *See* EXTRASOLAR PLANETS; GALAXY, EXTERNAL; GALAXY FORMATION AND EVOLUTION; STELLAR EVOLUTION.

George Rieke

Splachnales An order of the true mosses (subclass Bryidae), whose members are remarkable perennial plants that grow mainly on nitrogenous substrates, such as dung, and show considerable differentiation of neck tissue below the spore-bearing part of the capsule. The order consists of two families with about eight genera.

The plants are gregarious or dense-tufted, erect, and often forked. The leaves are soft, lanceolate or obovate, sometimes bordered at the margins, and commonly toothed. The single costa may be excurrent or terminate at or below the apex. The setae are elongate and the capsules erect with a noticeably differentiated neck. Stomata are very numerous in the neck, and have two guard cells. A single peristome is usually present, entire or forked with 16 teeth sometimes paired or joined in twos and fours. *See* BRYIDAE; BRYOPHYTA; BRYOPSIDA.

Howard Crum

Spleen An organ of the circulatory system present in most vertebrates, lying in the abdominal cavity usually in close proximity to the left border of the stomach.

In humans the spleen normally measures about 1 by 3 by 5 in. (2.5 × 7.5 × 12.5 cm) and weighs less than 1/2 lb (230 g). It is a firm organ with an oval shape and is indented on its inner surface to form the hilum, or stalk of attachment to the peritoneum. This mesentery fold also carries the splenic artery and vein to the organ.

The spleen is an important part of the blood-forming, or hematopoietic, system; it is also one of the largest lymphoid organs in the body and as such is involved in the defenses against disease attributed to the reticuloendothelial system. Although the chief functions of the spleen appear to be the production of lymphocytes, the probable formation of antibodies, and the destruction of worn-out red blood cells, other less well-understood activities are known. For example, in some animals it may act as a reservoir for red blood cells, contracting from time to time to return these cells to the bloodstream as they are needed. In the fetus and sometimes in later life, the spleen may be a primary center for the formation of red blood cells. Another function of the spleen is its role in biligenesis. Because the spleen destroys erythrocytes, it is one of the sites where extrahepatic bilirubin is formed. *See* BILIRUBIN; SPLEEN DISORDERS.

Walter Bock

Spleen disorders The spleen is rarely the site of primary disorders except those of vascular origin, but it is frequently involved in systemic inflammations, metabolic diseases, and generalized blood disorders.

Among vascular disturbances, acute and chronic congestion are prominent, particularly chronic congestion caused by cardiac failure, cirrhosis of the liver, and obstruction of the blood flow from the spleen by thrombi, scarring, or tumor tissue. Obstruction of the splenic artery or its branches by thrombi may result in an infarct caused by either cardiac or blood disease.

Inflammations include acute and chronic forms. The characteristic engorgement of blood often causes a marked enlargement of the organ. Bacteremias frequently produce this enlargement, or splenomegaly, and inflammation, but any severe infectious disease such as diphtheria or pneumonia may do so.

The leukemias, especially when of the lymphocytic or neutrophilic varieties, cause some of the most prominent cases of splenomegaly as well as other changes.

Tumors originating in the spleen are rare and usually limited to such benign growths as hemangiomas, lymphangiomas, and fibromas, but malignant lymphomas and lymphosarcomas also occur. Secondary tumors, which originate elsewhere and metastasize to the spleen, are not uncommon, particularly the lymphoma group. *See* SPLEEN.

Edward G. Stuart; N. Karle Mottet

Splines A series of projection and slots used instead of a key to prevent relative rotation of cylindrically fitted machine parts. Splines are several projections machined on the shaft; the shaft fits into a mating bore called a spline fitting. Splines are made in two forms, square and involute, as illustrated. Since there are several projections (integral keys) to share the force in transmitting power, the splines can be shallow, thereby not weakening the shaft as much as a standard key.

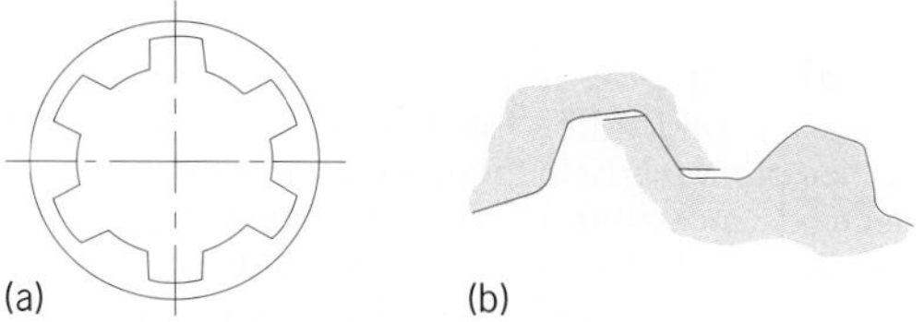

Diagrams of (*a*) square spline and (*b*) involute spline profile.

Three classes of fits are used for square splines: sliding (as for gear shifting) under load, sliding when not loaded, and permanent fit. Square splines have been used extensively for machine parts. In the automotive industry, square splines have been replaced generally by involute splines.

Involute splines are used to prevent relative rotation of cylindrically fitted machine parts and have the same functional characteristics as square splines. The involute spline, however, is like an involute gear, and the spline fitting (internal part) is like a mating internal gear. *See* MACHINE KEY.

Paul H. Black

Spodumene The name given to the monoclinic lithium pyroxene $LiAl(SiO_3)_2$. Spodumene commonly occurs as white to yellowish prismatic crystals, often with a "woody" appearance.

Spodumene is usually found as a constituent in certain granitic pegmatites. The emerald-green variety, hiddenite, and a lilac variety, kunzite, are used as precious stones. Spodumene from pegmatites is used as an ore for lithium. *See* LITHIUM; PYROXENE.

George W. DeVore

Spongiomorphida An extinct group of genera of Mesozoic fossils. Informally known as the spongiomorphs, they were established as a family of scleractinian corals by Fritz Frech in 1890 on the basis of specimens from the Zlambach beds of late Triassic age in Austria. He grouped the fossils into four genera (*Spongiomorpha, Heptastylopsis, Heptastylis*, and *Stromatomorpha*). The fossils consist of closely spaced longitudinal rods of calcite or aragonite that are perpendicular to the growth surface and that give off lateral expansions which join the rods and in some forms unite to form laminae parallel to the growth surface. The term spongiomorph should be used only in an informal sense for the group of genera in which the septa are reduced to closely spaced rods and the radial structure and individuality of the corallites is obscure. *See* SCLERACTINIA; SCLEROSPONGE; SPHAERACTINOIDEA; STROMATOPOROIDEA.

C. W. Stearn

Sporozoa A subphylum of Protozoa, typically with spores. The spores are simple and have no polar filaments. There is a single type of nucleus. There are no cilia or flagella except for flagellated microgametes in some groups. In most Sporozoa there is an alternation of sexual and asexual stages in the life cycle. In the sexual stage, fertilization is by syngamy, that is, the union of male and female gametes. All Sporozoa are parasitic. The subphylum is divided into three classes—Telosporea, Toxoplasmea, and Haplosporea. *See* HAPLOSPOREA; PROTOZOA; TELOSPOREA; TOXOPLASMEA. Norman D. Levine

Sports medicine A branch of medicine concerned with the effects of exercise and sports on the human body, including treatment of injuries. Sports medicine can be divided into three general areas: clinical sports medicine, sports surgery, and the physiology of exercise. Clinical sports medicine includes the prevention and treatment of athletic injuries and the design of exercise and nutrition programs for maintaining peak physical performance. Sports surgery is also concerned with the treatment of injuries from contact (human or object) sports. Exercise physiology, a growing field of sports medicine, involves the study of the body's response to physical stress. It comprises the science of fitness, the preservation of fitness, and the role of fitness in the prevention and treatment of disease. Otto Appenzeller

Spot welding A resistance-welding process in which coalescence is produced by the flow of electric current through the resistance of metals held together under pressure. Usually the upper electrode moves and applies the clamping force. Pressure must be maintained at all times during the heating cycle to prevent flashing at the electrode faces. Electrodes are water-cooled and are made of copper alloys because pure copper is soft and deforms under pressure. The electric current flows through at least seven resistances connected in series for any one weld (see illustration). After the metals have been fused together, the

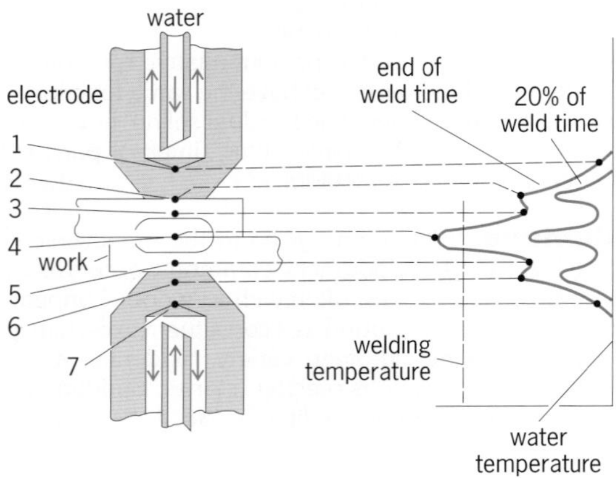

Distribution of temperature in local (numbered) elements of a spot-welding operation.

electrodes usually remain in place sufficiently long to cool the weld. *See* RESISTANCE WELDING; WELDING AND CUTTING OF METALS. Eugene J. Limpel

Spray flow A special case of a two-phase (gas and liquid) flow in which the liquid phase is the dispersed phase and exists in the form of many droplets. The gas phase is the continuous phase, so abstract continuous lines (or surfaces) can be constructed through the gas at any instant without intersection of the droplets. The droplets and the gas have velocities that can be different, so both phases can move through some fixed volume or chamber and the droplets can move relative to the surrounding gas. *See* GAS; LIQUID; PARTICULATES; TWO-PHASE FLOW.

Spray flows have many applications. Sprays are used to introduce liquid fuel into the combustion chambers of diesel engines, turbojet engines, liquid-propellant rocket engines, and oil-burning furnaces. They are used in agricultural and household applications of insecticides and pesticides, for materials and chemicals processing, for fire extinguishing, for cooling in heat exchangers, for application of medicines, and for application of coatings (including paint and various other types of layered coatings). Common liquids (such as water, fuels, and paints) are used in sprays. It is sometimes useful to spray uncommon liquids such as molten metals. In the various applications, the approximately spherical droplets typically have submillimeter diameters that can be as small as a few micrometers. *See* METAL COATINGS.

Sprays are formed for industrial, commercial, agricultural, and power generation purposes by injection of a liquid stream into a gaseous environment. In addition, sprays can form naturally in a falling or splashing liquid. Injected streams of liquid tend to become unstable when the dynamic pressure (one-half of the gas density times the square of the liquid velocity) is much larger than the coefficient of surface tension divided by the transverse dimension. Typically, the liquid stream disintegrates into ligaments (coarse droplets) and then into many smaller spherical droplets. The breakup (or atomization) process is faster at higher stream velocity, and the final droplet sizes are smaller for higher stream velocities. Spray droplet sizes vary and typically are represented statistically by a distribution function. The number of droplets in a spray can be as high as a few million in a volume smaller than a liter. *See* ATOMIZATION; JET FLOW; SURFACE TENSION. William A. Sirignano

Spread spectrum communication A means of communicating by purposely spreading the spectrum (frequency extent or bandwidth) of the communication signal well beyond the required bandwidth of the data modulation signal. Spread spectrum signals are typically transmitted by electromagnetic waves in free space, with usage in both nonmilitary and military systems.

Motivation for using spread spectrum signals is based on the following:

1. Spread spectrum systems have the ability to reject hostile as well as unintentional jamming by interfering signals.

2. Spread spectrum signals have a low probability of being intercepted or detected since the power in the transmitted wave is "spread" over a large bandwidth or frequency extent.

3. Since these signals cannot be readily demodulated without knowing the code and its precise timing, a level of message privacy is obtained.

4. The wide bandwidth of the spread spectrum signals provides tolerance to multipath (reflected waves that typically take longer to arrive at the receiver than the direct desired signal so that the two can be distinguished).

5. A high degree of precision in ranging (distance measuring) can be obtained by using one type of spread spectrum signaling called direct sequence, with applications to navigation.

6. Multiple access, or the ability to send many independent signals over the same frequency band, is possible in spread spectrum signaling.

See COMMUNICATIONS SCRAMBLING; ELECTRICAL INTERFERENCE; ELECTRONIC WARFARE; RADIO-WAVE PROPAGATION.

There are four generic types of spread spectrum signals: direct sequence (DS) or pseudonoise (PN), frequency hopping (FH), linear frequency modulation (chirp), and time hopping (TH). The first two methods are much more commonly used today than the other two.

Direct sequence modulation is characterized by phase-modulating a sine wave by an unending string of pseudonoise

Squamata

(*a*) Western banded gecko (*Coleonyx variegatus*).

(*b*) Tree lizard (*Urosaurus ornatus*).

(*c*) Gila monster (*Heloderma suspectum*).

(*d*) Western coral snake (*Micruroides euryxanthus*).

(*e*) Arizona black rattlesnake (*Crotalus viridis cerberus*).

(*f*) Eastern ribbon snake (*Thamnophis sauritus*).

(*Photos a–e by NPS/Tonto National Monument, Arizona. Photo f by NPS/Little River Canyon National Preserve, Alabama.*)

code chips (symbols of much smaller time duration than a bit). This unending string is typically based on a pseudonoise code that generates an apparently random sequence of code chips that repeats only after the pseudonoise code period. Digital data representing the information to be transmitted are binary phase-shift keyed onto the carrier. Then the pseudonoise code generator also binary phase-shift keys the carrier, and the composite signal is transmitted. *See* PHASE MODULATION.

In a direct-sequence system, the phase of the carrier changes pseudorandomly with the pseudonoise code. In a frequency-hopping system, the frequency of the carrier changes according to a pseudonoise code with a consecutive group of pseudonoise code chips defining a particular frequency. Typically either multiple frequency-shift keying (MFSK) or differential phase-shift keying (DPSK) is used as the data modulation. Multiple frequency-shift keying is a modulation scheme in which one of a number of tones (2, 4, 8, and so forth) is transmitted at a given time according to a group of consecutive data bits (n bits produce 2^n tones). At each hop frequency one of the 2^n tones is selected according to the n bits, and one of 2^n corresponding frequencies, centered about the hop frequency, is transmitted. In conjunction with frequency hopping, multiple frequency-shift keying would imply, at each instant of time, a given carrier frequency that depends on the hop-pseudonoise code sequence and the consecutive group of the most recent n data bits. Differential phase-shift keying is similar to phase-shift keying except that only the differences of the phases (not the actual phases) are encoded and noncoherent techniques (not requiring a carrier loop) can be employed at the receiver.

A device called a frequency synthesizer achieves the actual frequency selection. For example, a 12-bit segment of the pseudonoise code may correspond to one of 2^{12} different frequencies, so that one of approximately 4000 (2^{12}) frequencies is selected each hop time. Frequency synthesizers are used in both the transmitter and the receiver. The transmitter modulates the data by typically using either multiple frequency-shift keying or differential phase-shift keying modulation, which in turn is frequency-hopped by the frequency synthesizer. At the receiver, an acquisition process is utilized to synchronize the receiver frequency synthesizer with the received hopping signal, and then a tracking system maintains synchronism. Finally a bit synchronizer provides timing for the data demodulator which demodulates the original, transmitted data bits.

An important aspect of spread spectrum communications is multiple access. Code-division multiple access (CDMA) is a method by which spread spectrum signals are utilized to allow the use of multiple signals over the same frequency band. The two common types are direct-sequence CDMA (DS/CDMA) systems and frequency-hopped CDMA (FH/CDMA) systems. *See* MULTIPLEXING AND MULTIPLE ACCESS.

Although the early evolution of spread spectrum systems was motivated primarily by military interests, nonmilitary applications have enjoyed considerable development. One important example that has both military and nonmilitary users is the Global Positioning Systems (GPS), which is a direct-sequence, CDMA, spread spectrum system for transmitting the satellite ranging codes. *See* SATELLITE NAVIGATION SYSTEMS.

The space shuttle utilizes a direct-sequence spread spectrum communication system on its forward link. It relays data through the geostationary *Tracking and Data Relay Satellites* (*TDRS*). Another example in the military arena is the Milstar system, which utilizes frequency-hopping spread spectrum communication over a very large bandwidth to achieve considerable immunity from unfriendly jamming signals. *See* MILITARY SATELLITES; SPACE COMMUNICATIONS; SPACE SHUTTLE.

Globalstar is a commercial satellite system that utilizes a CDMA signal structure along with a bent-pipe transponder to provide communications much like a cell phone, except that Globalstar is satellite-based. Currently the handsets are considerably more expensive than cell phones. However, these phones have the advantage that they can be reached beyond the range of cell phones. *See* COMMUNICATIONS SATELLITE.

The main reception source of news stories for radio and newspapers is small spread spectrum ground stations. Another application of code-division multiple access is transmission directly by satellite from a bank's automatic teller machine to that bank's computer facility. A home security system using spread spectrum techniques imposed on the ac power line has been used in Japan. Many cordless telephones utilize direct-sequence spread spectrum techniques.

Jack K. Holmes

Spring (hydrology) A place where groundwater discharges upon the land surface because the natural flow of groundwater to the place exceeds the flow from it. Springs are ephemeral, discharging intermittently, or permanent, discharging constantly. Springs are usually at mean annual air temperatures. The less the discharge, the more the temperature reflects seasonal temperatures. Spring water usually originates as rain or snow (meteoric water).

Hot-spring water may differ in composition from meteoric water through exchange between the water and rocks. Common minerals consist of component oxides. Oxygen of minerals has more ^{18}O than meteoric water. Upon exchange, the water is enriched in ^{18}O. Most minerals contain little deuterium, so that slight deuterium changes occur. Some hot-spring waters are acid from the oxidation of hydrogen sulfide to sulfate.

Mineral spring waters have high concentrations of solutes and wide ranges in chemistry and temperatures; hot mineral springs may be classified as hot springs as well as mineral springs. Most mineral springs are high either in sodium chloride or sodium bicarbonate (soda springs) or both; other compositions are found, such as a high percentage of calcium sulfate from the solution of gypsum.

The chemical compositions of spring waters are seldom in chemical equilibrium with the air. Groundwaters whose recharge is through grasslands may contain a thousand times as much CO_2 as would be in equilibrium with air, and those whose recharge is through forests may contain a hundred times as much as would be in equilibrium with air. Sulfate in ground water may be reduced in the presence of organic matter to H_2S, giving some springs the odor of rotten eggs. *See* GEYSER; GROUND-WATER HYDROLOGY.

Ivan Barnes

Spring (machines) A machine element for storing energy as a function of displacement. Force applied to a spring member causes it to deflect through a certain displacement, thus absorbing energy.

A spring may have any shape and may be made from any elastic material. Even fluids can behave as compression springs and do so in fluid pressure systems. Most mechanical springs take on specific and familiar shapes such as helix, flat, or leaf springs. All mechanical elements behave to some extent as springs because of the elastic properties of engineering materials.

The most frequent use of springs is to supply motive power in a mechanism. Common examples are clock and watch springs, toy motors, and valve springs in auto engines. A special case of the spring as a source of motive power is its use for returning displaced mechanisms to their original positions, as in the door-closing device, the spring on the cam follower for an open cam, and the spring as a counterbalance. Frequently a spring in the form of a block of very elastic material such as rubber absorbs shock in a mechanism. Springs also serve an important function in vibration control. *See* SHOCK ABSORBER; SHOCK ISOLATION.

Springs may be classified into six major types according to their shape. These are flat or leaf, helical, spiral, torsion bar, disk, and constant force springs. A leaf spring is a beam of cantilever design with a deliberately large deflection under a load. The helical spring consists essentially of a bar or wire or uniform

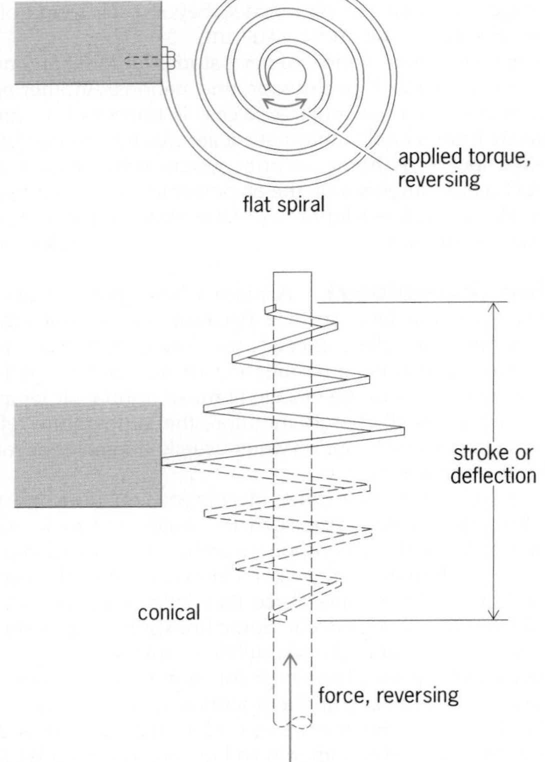

Spiral spring is unique in responding to torsional or translation forces.

cross section wound into a helix. In a spiral spring, the spring bar or wire is wound in an Archimedes spiral in a plane. A spiral spring is unique in that it may be deflected in one of two ways or a combination of both of them (see illustration). A torsion bar spring consists essentially of a shaft or bar of uniform section. The disk spring consists essentially of a disk or washer supported at the outer periphery by one force and an opposing force on the center or hub of the disk. A constant force spring is used when a constant force must be applied regardless of displacement.

L. Sigfred Linderoth, Jr.

Spruce Evergreen tree belonging to the genus *Picea* of the pine family. The needles are single, usually four-sided, and borne on little peglike projections; the cones are pendulous. Resin ducts in the wood may be seen with a magnifying lens, but they are fewer than in *Pinus*.

The white spruce (*P. glauca*), ranging from northern New England to the Lake States and Montana and northward into Alaska, is distinguished by the somewhat bluish cast of its needles, small cylindrical cones, and gray or pale-brown twigs without pubescence (hairs). Red spruce (*P. rubens*) is a similar tree but with greener foliage; smaller, more oval cones; and more or less pubescent twigs. Occurring naturally with white spruce in the northeastern United States and adjacent Canada, red spruce extends southward along the Appalachians into North Carolina. Black spruce (*P. mariana*) ranges from northern New England and Newfoundland to Alaska. However, it occurs sparingly in the Appalachians to West Virginia. The cones are smaller than in the white and red species and are egg-shaped or nearly spherical and persistent. The twigs are pubescent.

Blue spruce (*P. pungens*), also known as Colorado blue spruce, is probably the best known of the western species because of its wide use as an ornamental tree. The twigs are glabrous (without pubescence). Engelmann spruce (*P. engelmanni*) has needles usually of a deep blue-green color, sometimes much like those of the blue spruce but the young twigs are slightly hairy. The cones, although cylindrical, are smaller than in blue spruce. This species is also a Rocky Mountain tree like the blue spruce, but it is more widely distributed from British Columbia to Arizona and also in the mountains of Oregon and Washington. Sitka spruce (*P. sitchensis*) is the largest spruce in the Northern Hemisphere. The leaves have a pungent odor, are considerably flattened, and stand out from the twig in all directions. It ranges from Alaska to northern California. The Norway spruce (*P. abies*), the common spruce of Europe, is much planted in the United States for timber, as well as for ornamental purposes. It can be recognized by the dark-green color of the leaves; glabrous, pendent, short branchlets; and large cones, usually near the top of the tree. *See* PINALES.

Arthur H. Graves; Kenneth P. Davis

Sputtering The ejection of material from a solid or liquid surface following the impact of energetic ions, atoms, or molecules. Sputtering is the basis of a large variety of methods for the synthesis and analysis of materials.

Sputtering can be classified according to the mode of energy loss of the incident (primary) particle. Nuclear stopping involves billiard ball-like atomic collisions in which a significant momentum transfer occurs; it dominates for incident ion energies below about 1–2 keV per nucleon. Electronic stopping involves collisions in which little momentum is transferred, but significant electronic excitation is caused in the target; it dominates for energies above about 10 keV per nucleon.

Sputtering has also been classified into physical and chemical sputtering. Physical sputtering involves a transfer of kinetic energy from the incident particle to the surface atoms leading to ejection, while chemical sputtering occurs when the incident species react chemically with the target surface leading to the formation of a volatile reaction product which evaporates thermally from the surface.

Sputtering of complex materials—metal alloys, inorganic and organic compounds and polymers, and minerals—can produce complex results. The relative efficiencies with which different elemental species are ejected following ion impact can differ, giving rise to preferential sputtering. When preferential sputtering occurs, the species sputtered with the lower efficiency accumulates to a higher concentration at the surface. Subsurface collisions of the incident ion cause atomic motion leading to atomic mixing of surface and subsurface layers over the ion penetration depth. Chemical bonds can be broken, and sometimes new bonds can be formed. Sputtering of solids which have multiple phases, or which are polycrystalline, leads to the development of surface roughness due to the differences in sputtering yields between different regions. *See* ION BEAM MIXING.

Sputtering is widely used in the manufacture of semiconductor devices; sputter deposition is used to deposit thin films with a high degree of control by sputtering material from a target onto a substrate; sputter etching is used to remove unwanted films in a reversal of this process. Reactive ion etching is a chemical sputtering process in which chemically active sputtering species form volatile compounds with the target material leading to significantly higher etch rates and great selectivity. For example, fluorine-containing compounds etch silicon rapidly by forming volatile silicon tetrafluoride but do not etch aluminum or other metals used to make electrical interconnections between devices on a semiconductor chip because the metal fluorides are involatile. Sputter etching and reactive ion etching have the useful advantage of being anisotropic—that is, they etch only in one direction so that very fine surface features can be delineated. *See* CRYSTAL GROWTH; GLOW DISCHARGE; INTEGRATED CIRCUITS.

In materials characterization, sputtering is used to remove surface material controllably, allowing in-depth concentration profiles of chemical composition to be determined with a surface-sensitive sampling technique. *See* SURFACE PHYSICS.

Peter Williams

Squalene A C_{30} triterpenoid hydrocarbon. Squalene is made up of six (*trans*-1,4)-isoprene units linked as two farnesyl (head-to-tail) groups that are joined tail to tail in the center (see illustration).

Structure of squalene; the tail-to-tail joining is indicated by T.

Squalene can be isolated in large quantities from the liver oils of the shark and other elasmobranch fishes, and is a relatively inexpensive compound. Complete hydrogenation of the liver oil gives the saturated hydrocarbon squalane, which is used in lotions and skin lubricants.

The major significance of squalene is its role as a central intermediate metabolite in the biogenesis of all steroids and triterpenoids. *See* STEROID; TERPENE.

James A. Moore

Squaliformes An order of sharks comprising six families, 24 genera, and 97 species, commonly known as dogfish sharks. The order differs from all other sharks by the following combination of characters: two dorsal fins, with or without spines; no anal fin; spiracle behind eye; and lower nictitating eye membrane absent. For the most part, squaliform sharks inhabit continental and insular shelves and slopes, and sea mounts. Some species live at great depths. So far as known, squaliforms are ovoviviparous and produce from one to 22 pups per litter, depending on the species. The dwarf lantern shark (*Etmopterus perryi*), at 17 cm (6.7 in.) in length, probably is the world's smallest shark; the largest squaliform is the Pacific sleeper shark (*Somniosus pacificus*), which is up to 7 m (23 ft) in length. The species *Squalus acanthias* (spiny dogfish) (see illustration) is probably the world's most abundant shark, attaining a length of 160 cm (63 in.), although most are much smaller. The diet consists of bony fishes, cephalopods, crustaceans and other invertebrates, mammals, and other sharks. *See* ELASMOBRANCHII; SELACHII.

Spiny dogfish (*Squalus acanthias*). (*Photo courtesy of Alaska Fisheries Science Center, National Marine Fisheries Service, National Oceanic and Atmospheric Administration, U.S. Department of Commerce*)

Herbert Boschung

Squall A strong wind with sudden onset and more gradual decline, lasting for several minutes. Wind speeds in squalls commonly reach 30–60 mi/h (13–27 m/s), with a succession of brief gusts of 80–100 mi/h (36–45 m/s) in the more violent squalls. Squalls may be local in nature, as with isolated thunderstorms, or may occur over a wide area in the vicinity of a well-developed cyclone, where the squalls locally reinforce already strong winds. Because of their sudden violent onset, and the heavy rain, snow, or hail showers which often accompany them, squalls cause heavy damage to structures and crops and present severe hazards to transportation.

The most common type of squall is the thundersquall or rain squall associated with heavy convective clouds, frequently of the cumulonimbus type. Such a squall usually sets in shortly before onset of the thunderstorm rain, blowing outward from the storm and generally lasting for only a short time. It is formed when cold air, descending in the core of the thunderstorm rain area, reaches the Earth's surface and spreads out. Particularly in desert areas, the thunderstorm rain may largely or wholly evaporate before reaching the ground, and the squall may be dry, often associated with dust storms. *See* DUST STORM; SQUALL LINE; THUNDERSTORM.

Squalls of a different type result from cold air drainage down steep slopes. The force of the squall is derived from gravity and depends on the descending air which is colder and more dense than the air it replaces. So-called fall winds of this kind are common on mountainous coasts of high latitudes, where cold air forms on elevated plateaus and drains down fiords or deep valleys. Violent squalls also characterize the warm foehn winds of the Alps and the similar chinook winds on the eastern slopes of the Rocky Mountains. *See* CHINOOK; WIND.

Chester W. Newton; Edwin Kessler

Squall line A line of thunderstorms, near whose advancing edge squalls occur along an extensive front. The thundery region, 12–30 mi (20–50 km) wide and up to 1200 mi (2000 km) long, moves at a typical speed of 30 knots (15 m/s) for 6–12 h or more and sweeps a broad area. In the United States, severe squall lines are most common in spring and early summer when northward incursions of maritime tropical air east of the Rockies interact with polar front cyclones. Ranking next to hurricanes in casualities and damage caused, squall lines also supply most of the beneficial rainfall in some regions. *See* FRONT; SQUALL; THUNDERSTORM.

Chester W. Newton

Squamata The dominant order of living reptiles, composed of the lizards and snakes. The group first appeared in Jurassic times and today is found in all but the coldest regions. Various forms are adapted for arboreal, burrowing, or aquatic lives, but most squamates are fundamentally terrestrial. There are over 7700 extant species: 4800+ lizards and 2900+ snakes. *See* REPTILIA.

The order is readily distinguished from all known reptiles by its highly modified skull. The immediate ancestors of primitive lizards are members of the diapsid order Eosuchia, in which a pair of temporal foramina are present on both sides of the head. In the Squamata, however, in the case of the quadratojugal and jugal bones, which border the lower margin of the inferior temporal opening, the former has been lost completely and the latter greatly reduced. In consequence, there is only a single temporal opening present, and the quadrate has become enlarged and movable. Even this temporal opening is lost or reduced in many forms. No other reptiles show these modifications, which allow for great kinesis in the lower jaw since it articulates with the quadrate. In addition, the order is distinct from other living reptile groups because its members have no shells or secondary palates, and the males possess paired penes (hemipenes). Another distinguishing characteristic of the squamates is the synchronous shedding of the outer layer of the integument, called ecdysis. *See* EOSUCHIA; LEPIDOSAURIA; SPHENODONATA.

Traditionally, the order Squamata has been divided into two major subgroups: the lizards, suborder Sauria (= Lacertilia), and the snakes, suborder Serpentes. The latter group is basically a series of limbless lizards, and it is certain that snakes are derived from some saurian ancestor. There are many different legless lizards, and it has been suggested that more than one line has evolved to produce those species currently grouped together as snakes.

Sauria. The majority of lizards are insectivorous, but a few species feed on plants. Others, notably the Varanidae and allies, feed on larger prey, including birds and mammals. The largest living lizard is the Komodo dragon (*Varanus komodoensis*), which attains a length of 10 ft (3 m) and a weight of 250 lb (113 kg).

The majority of lizards are quadrupedal in locomotion and are usually ambulatory scamperers or scansorial (adapted for climbing). Some forms are bipedal, at least when in haste. The coloration of each species of lizard is characteristic. Most forms exhibit marked differences in coloration between the sexes, at least during the breeding season, and frequently the young are markedly different from the parents. Color changes occur in rapid fashion among some species, and all are capable of metachrosis or changing color to a certain extent. *See* CHROMATOPHORE; SEXUAL DIMORPHISM.

There are only two species of venomous lizards, both members of the genus *Heloderma*, in the family Helodermatidae. The Gila monster (*H. suspectum*) is found in western New Mexico, Arizona, extreme southern Nevada, southwestern Utah, eastern California, and northwestern Mexico. The beaded lizard (*H. horridum*) is a Mexican species. Venom is produced in glands along the lower jaw and penetrates wounds inflicted by the anterior recurved teeth.

The following list indicates the major evolutionary lines, families, and distribution of lizards. Families indicated by an asterisk contain limbless, snakelike species.

Iguania line
Family Iguanidae: the Americas, Madagascar, Fiji
Agamidae: Africa, southern and central Asia, Australia
Chamaeleonidae: Africa, Madagascar, southern India, Near East, southern Spain
Gekkota line
Family Gekkonidae: circumtropical, all continents and most continental and oceanic islands
Pygopodidae: Australia, New Guinea
Scincomorpha line
Family Xantusiidae: North and Middle America, Cuba
Teiidae: the Americas
Lacertidae: Africa, Eurasia
*Gerrhosauridae: Africa, Madagascar
*Scincidae: cosmopolitan, except frigid areas
Gymnophthalmidae: Central and South America
Cordylidae: southern Africa
Annulata line
*Family Amphisbaenidae: tropics of Africa and America, Iberian Peninsula
*Bipedidae: Mexico
Trogonophidae: northern Africa and Middle East
Rhineuridae: Florida (U.S.)
Anguimorpha line
*Family Anguidae: the Americas, Eurasia
*Anniellidae: the Californias
Xenosauridae: southeastern China, Central America
Helodermatidae: North America
Varanidae: Africa, southern Asia, Australia
Lanthonotidae: North Borneo

Serpentes. Snakes are basically specialized, limbless lizards which probably evolved from burrowing forms but have now returned from subterranean habitats to occupy terrestrial, arboreal, and aquatic situations. The following characteristics are typical of all serpents. There is no temporal arch so that the lower jaw and quadrate are very loosely attached to the skull. This gives the jaw even greater motility than is the case in lizards. The body is elongate with 100–200 or more vertebrae, and the internal organs are elongate and reduced. A spectacle covers the eye, and there is no tail autotomy.

The largest living snake is the Indian python (*Python reticulatus*), which reaches 30 ft (9 m) in length and a weight of 250 lb (113 kg). The largest venomous snake is the king cobra (*Ophiophagus hannah*) of southern Asia, which is known to attain a length of 18 ft (5.5 m).

The senses of snakes are fundamentally similar to those of lizards. Great dependence is placed upon olfaction and the Jacobson's organs (specialized derivatives of the olfactory apparatus). The tongue of all snakes is elongate and deeply bifurcated. It is protrusible and is constantly being projected to pick up samples for the organs of Jacobson from the surrounding environment. Snakes are deaf to airborne sounds and receive auditory stimuli only through the substratum via the bones of the head. The eyes are greatly modified from those in lizards, and there is no color vision. Some groups are totally blind and have vestigial eyes covered by scales or skin.

Four basic patterns of locomotion are found in snakes, and several may be used by a particular individual at different times. The most familiar type is serpentine curvilinear (lateral undulatory). In this pattern, the snake moves forward by throwing out lateral undulations of the body and pushing them against any irregularity in the surface. Snakes using rectilinear locomotion move forward in a straight line, without any lateral undulations, by producing wavelike movements in the belly plates. Laterolinear locomotion, or sidewinding, is used primarily on smooth or yielding surfaces and is very complex. Concertina locomotion movement resembles the expansion and contraction of that musical instrument.

The vast majority of living snakes are harmless to humans, although a number are capable of inflicting serious injury with their venomous bites. The venom apparatus has evolved principally as a method of obtaining food, but it is also advantageous as a defense against attackers. Fangs are teeth modified for the injection of venom into the victim, and the venom glands are modified salivary glands connected to the grooved fangs by a duct. The venom itself is a complex substance containing a number of enzymes. Certain of these enzymes attack the blood, others the nervous system, and some are spreaders.

The following list indicates the major groups of living snakes and their distribution.

Family Typhlopidae: circumtropical
Leptotyphlopidae: circumtropical
Anomalepididae: southern Central America and northern South America
Aniliidae: southwestern Asia and Central America
Anomochilidae: Malaysia, Sumatra, Borneo
Boidae: circumtropical, western United States
Bolyeriidae: Round Island (Indian Ocean)
Tropidophiidae: West Indies, Central America, South America
Acrochordidae: India, Solomon Islands, Australia
Colubridae: cosmopolitan, except frigid areas
Elapidae: Africa, Asia, Australia, the Americas
Viperidae: Eurasia, Africa

W. Ben Cash; Jay M. Savage

Squash The common name for edible fruits of several species of the genus *Cucurbita: C. pepo, C. moschata, C. maxima*, and *C. mixta*. Those species originated in the Americas but are now grown in most countries around the world. Within squash there is tremendous variation in size, shape, color, and usage.

The most clearly defined group is summer squash, fruit of any species of *Cucurbita* eaten as a vegetable when immature. It is most commonly *C. pepo*. Fruit color may be white, yellow, or light or dark green, and the green may be solid or striped. Shapes may be flattened disks as in Pattypan, cylindrical as in Zucchini and Cocozelle, or with necks as in the straightneck and crookneck types. Summer squash has mild flavor, high water content, and relatively low nutritional value.

Winter squash is fruit of *Cucurbita* eaten when mature and derives its name from its ability to be stored for several weeks or months before consumption. Varieties of winter squash are found in all four species. The Table Queen group, synonymous with Acorn, is *C. pepo*, Butternut belongs to *C. moschata*, Green-striped Cushaw is *C. mixta*, while *C. maxima* has the widest range of types, including Buttercup, Hubbards, and Delicious of various colors, Banana, and Boston Marrow. Flesh color varies from light yellow to dark orange, and the edible portion ranges from thin to very thick.

H. M. Munger

Squatiniformes An order of bizarre sharks, commonly known as angel sharks, comprising one family, one genus (*Squatina*), and 15 species that superficially resemble rays (see illustration). The following combination of characters identify the angel sharks: body flattened dorsoventrally; pectoral fins winglike and spread forward alongside head but not attached to it; mouth almost terminal; nostrils terminal, with barbels on anterior margin; eyes and large spiracles on top of head; five gill slits partly on sides of head and partly on ventral surface; two small spineless dorsal fins on tail; anal fin absent; and lower lobe of caudal fin longer than upper lobe. Maximum lengths of three species vary from 63 to 89 cm (25 to 35 in.), and 12 species vary from 120 to 200 cm (47 to 79 in.). Angel sharks are found primarily in cool temperate to tropical zones of the eastern and western Pacific, eastern and western Atlantic, and extreme southwestern Indian Ocean. They feed on a variety of small bony fishes, crustaceans, cephalopods, gastropods, and bivalves by using their highly protrusible jaws to snap up prey at high speed. *See* ELASMOBRANCHII; SELACHII.

Pacific angel shark (*Squatina californica*). (*Photo by Tony Chess/NOAA Fisheries, Southwest Fisheries Science Center*)

Herbert Boschung

Squeezed quantum states Quantum states for which certain variables can be measured more accurately than is normally possible.

All matter and radiation fluctuate. Much random fluctuation derives from environmental influence, but even if all these influences are removed, there remains the intrinsic uncertainty prescribed by the laws of quantum physics. The position and momentum of a particle, or the electric and magnetic components of an electromagnetic field, are conjugate variables that cannot simultaneously possess definite values (Heisenberg uncertainty principle). It is possible, however, to have the position of a particle more and more accurately specified at the expense of increasing momentum uncertainty; the same applies to electromagnetic field amplitudes. This freedom underlies the phenomenon of squeezing or the possibility of having squeezed quantum states. With squeezed states, the inherent quantum fluctuation may be partly circumvented by focusing on the less noisy variable, thus permitting more precise measurement or information transfer than is otherwise possible. *See* UNCERTAINTY PRINCIPLE.

According to quantum electrodynamics, the vacuum is filled with a free electromagnetic field in its ground state that consists of fluctuating field components with significant noise energy. If ϕ is a phase angle and $a(\phi)$ and $a[\phi + (\pi/2)]$ are two quadrature components of the field (for example, the electric and magnetic field amplitudes), the vacuum mean-square field fluctuation is given by Eq. (1), independently of the phase angle.

$$\langle \Delta a^2(\phi) \rangle = \frac{1}{4} \tag{1}$$

Equation (1) is normalized to a photon; the corresponding equivalent noise temperature at optical frequencies is thousands of kelvins. Equation (1) also gives the general fluctuation of an arbitrary coherent state, which is the quantum state of ordinary lasers. Further environment-induced randomness is introduced in addition to Eq. (1) for other conventional light sources, including light-emitting diodes. *See* COHERENCE; ELECTRICAL NOISE; LASER; QUANTUM ELECTRODYNAMICS.

In a squeezed state, the quadrature fluctuation is reduced below Eq. (1) for some ϕ, as given in Eq. (2). At that point,

$$\langle \Delta a^2(\phi) \rangle < \frac{1}{4} \tag{2}$$

squeezing, that is, reduction of field fluctuation below the coherent state level, occurs. The fluctuation of the conjugate quadrature is correspondingly increased to preserve the uncertainty relation, Eq. (3). In a two-photon coherent state, or squeezed state

$$\langle \Delta a^2(\phi) \rangle \left\langle \Delta a^2 \left(\phi + \frac{\pi}{2}\right) \right\rangle \geq \frac{1}{16} \tag{3}$$

in the narrow sense, Eq. (3) is satisfied with equality. As seen in the illustration, the designation "squeezed state" is partly derived

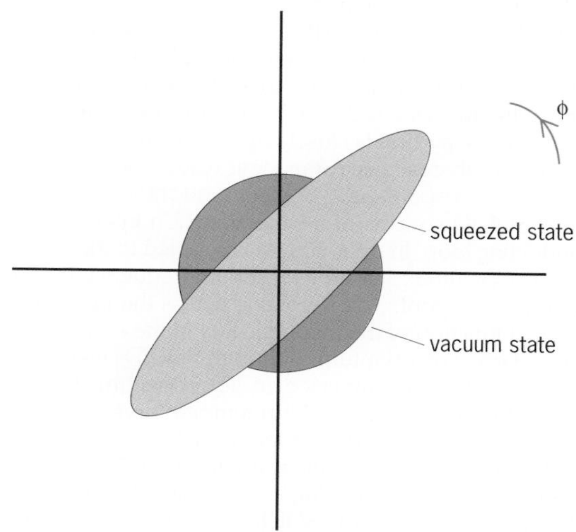

Field-amplitude fluctuation $\langle \Delta a^3(\phi) \rangle$ as a function of phase angle ϕ. The noise circle of the vacuum state, Eq. (1) in the text, is squeezed to an ellipse according to Eq. (3).

from the fact that the noise circle of Eq. (1) is squeezed to an ellipse when Eq. (3) is satisfied with equality.

Squeezed light can be generated by a variety of processes, especially nonlinear optical processes. The first successful experimental demonstration of squeezing, in 1985, involved a four-wave mixing process in an atomic beam of sodium atoms.

Squeezing was first studied in connection with optical communication, although it is evident that reduced quantum fluctuation might find applications in precision measurements. *See* NONLINEAR OPTICS; OPTICAL COMMUNICATIONS. Horace P. Yuen

Squid The common name applied to cephalopods of the order Teuthoidea. They are marine mollusks that inhabit the oceans of the world. Squids are characterized by having eight arms and two longer tentacles around the mouth; an elongated, tapered, usually streamlined body; an internal rod- or blade-like shield (gladius); and fins on the body (mantle). The arms have two (infrequently four or six) rows of suckers and occasionally clawlike hooks, and the tentacles have terminal clubs with suckers, hooks, or both. The muscular, elastic tentacles are contractile, not retractile into pockets like those of cuttlefishes (Sepioidea). *See* SEPIOIDEA.

Squids have an exceptionally well-developed brain and organs of the central nervous system that approach in complexity and function those of fishes and even some birds and mammals. Squids are active, powerful swimmers, driven by jet propulsion as water taken into the mantle cavity is forcefully expelled through the funnel. Prey, normally shrimps, fishes, or other squids, are captured with the two tentacles and held with the arms while the beaks cut off bites that the radula and tongue shove down the throat. *See* NERVOUS SYSTEM (INVERTEBRATE).

Two groups (suborders) of squids are recognized: Myopsida and Oegopsida. *See* CEPHALOPODA; COLEOIDEA; TEUTHOIDEA. Clyde F. E. Roper

SQUID An acronym for superconducting quantum interference device, which actually refers to two different types of device, the dc SQUID and the rf SQUID.

The dc SQUID consists of two Josephson tunnel junctions connected in parallel on a superconducting loop (see illustration). A small applied current flows through the junctions as a supercurrent, without developing a voltage, by means of Cooper pairs of electrons tunneling through the barriers. However, when the applied current exceeds a certain critical value, a voltage is generated. When a magnetic field is applied so that a magnetic flux threads the loop, the critical value oscillates as the magnetic flux is changed, with a period of one flux quantum, $\Phi_0 = h/(2e) \approx 2.07 \times 10^{-15}$ weber, where h is Planck's constant and e is the electron charge. The oscillations arise from the interference of the two waves describing the Cooper pairs at the two junctions, in a way that is closely analogous to the interference between two coherent electromagnetic waves. *See* INTERFERENCE OF WAVES; JOSEPHSON EFFECT; SUPERCONDUCTIVITY.

The rf SQUID consists of a single junction interrupting a superconducting loop. In operation, it is coupled to the inductor of an *LC*-tank circuit excited at its resonant frequency by a radio-frequency (rf) current. The rf voltage across the tank circuit oscillates as a function of the magnetic flux in the loop, again with a period of one flux quantum. Although SQUIDs were for many years operated while immersed in liquid helium, ceramic superconductors with high transition temperatures make possible devices operating in liquid nitrogen at 77 K.

SQUIDs have important device applications. Usually with the addition of a superconducting input circuit known as a flux transformer, both dc and rf SQUIDs are used as magnetometers to detect tiny changes in magnetic field. The output of the SQUID is amplified by electronic circuitry at room temperature and fed back to the SQUID so as to cancel any applied

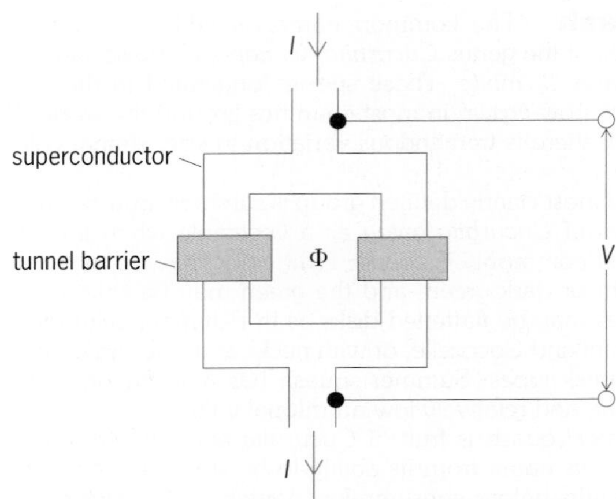

Direct-current (dc) SQUID with enclosed magnetic flux Φ. I = applied current; V = generated voltage.

flux. This makes it possible to detect changes in flux as small as 10^{-6} of one flux quantum with SQUIDs based on low-transition-temperature superconductors, corresponding to magnetic field changes of the order of 1 femtotesla in a 1-hertz bandwidth. Suitable modifications to the input circuit enable the SQUID to measure other physical quantities, including voltages, displacement, or magnetic susceptibility. SQUIDs are also used for logic and switching elements in experimental digital circuits and high-speed analog-to-digital converters. *See* ANALOG-TO-DIGITAL CONVERTER; INTEGRATED CIRCUITS; MAGNETOMETER; SUPERCONDUCTING DEVICES. John Clarke

Squirrel A rodent in the family Sciuridae. This family includes all tree squirrels and flying squirrels, as well as squirrels that live on the ground such as the woodchuck, marmot, chipmunk, and prairie dog. There are 51 living genera and 272 species living in a wide variety of habitats ranging through tropical rainforests, tundra, alpine meadows, and semiarid deserts in Europe, Asia, Africa, North America, and most of South Africa. They are absent from Australia, Madagascar, Polynesia, and the Sahara Desert. Locomotion is plantigrade (walking with the whole sole of the foot touching the ground). Sensitive vibrissae are present on the head, feet, and outsides of the legs. *See* CHIPMUNK; MARMOT; PRAIRIE DOG; RODENTIA.

Tree and flying squirrels are mostly small and slender-bodied with bushy tails often as long as their bodies. The tail is primarily used to maintain and correct the balance of the animal in leaping from branch to branch, but it may also be used as a rudder when leaping, as a flag to communicate social signals, and as a wrap-around blanket when the animal sleeps. Soft pads are present on the soles of the feet, and sharp claws are present on all toes. Tree squirrels are agile and are active during the day, unlike flying squirrels, which are nocturnal and have large eyes for seeing in dim light. Flying squirrels are not really fliers but gliders. Their "flying" equipment consists of a thin, loose furry membrane (patagium) extending along each side of the body between the fore and hind limbs. As the animal launches into the air from some lofty bough, it spreads its legs, drawing the membranes taut. This wide expanse gives the squirrel enough support to permit it to glide in a downward course for a considerable distance.

Ground-dwelling squirrels are usually larger and more heavy-bodied than tree squirrels. They have powerful forelimbs and large claws for digging. They live in subterranean burrows or have their underground dens among rocks. Most ground squirrels hibernate. *See* HIBERNATION AND ESTIVATION. Donald W. Linzey

SS 433 A binary star system in the Milky Way Galaxy consisting of a neutron star or a black hole accreting matter in the form of a disk from an early-type companion star. The most remarkable feature of the system is the presence of bipolar narrow outflows of matter (jets), with a velocity of 26% the speed of light, whose ejection axes change with time. *See* BLACK HOLE; NEUTRON STAR.

SS 433 owes its name to a catalog published in 1977 by C. B. Stephenson and N. Sanduleak, listing sources with a strong hydrogen emission line. Further optical observations of SS 433 in 1978 showed, besides the very strong hydrogen emission line in the spectrum, a few other lines whose wavelengths did not correspond to any known element. The extraordinary discovery, at first incomprehensible, was that the wavelengths of these "strange" lines also moved with time. A well-known effect that can cause shifts in the wavelength of a line is the Doppler effect. If the gas that emits a line is moving toward the observer, the line in the spectrum appears at a shorter wavelength than the wavelength at which it has been emitted at rest with the gas (that is, blueshift of the line), while if the gas is moving away from the observer, the line appears at a larger wavelength (that is, redshift of the line). What was difficult to understand was the fact that these moving lines appeared always in pairs, and moved smoothly in a periodic pattern. *See* ATOMIC STRUCTURE AND SPECTRA; DOPPLER EFFECT.

All the above mentioned unusual spectral features of SS 433 can be well fit with a relatively simple geometrical model, the kinematic model. The system is ejecting matter in form of two antiparallel narrow (that is, with an opening angle less than 5°) jets. These jets precess in a 21° half-opening angle cone with a period of about 162 days. The components of the jets (probably discrete blobs of matter) move in a straight line away from the system. Since the ejection axes rotate on a cone (like a spinning top) and the angle between the precession axes and the line of sight is about 78°, the projection of the jets onto the plane of the sky results in a twisted trace (see illustration). The blobs of matter have a constant velocity of about 0.26*c*, where *c* is the speed of light. For the first time, relativistic jets (that is, moving at a significant fraction of the speed of light, in this case 26%), which had already been discovered in distant galaxies, were observed relatively close-by, from a stellar object in our own galaxy.

Most jets are thought to originate in the innermost regions of an accretion disk, and the outflowing axis to lie in a direction approximately perpendicular to the disk plane. Emission lines in the spectrum of SS 433 are observed not only in the optical band but also in x-ray spectra, revealing the presence in the jets of heavy elements (such as iron) and highly ionized atoms. The gas that emits lines is thermal (that is, the state of the matter can be described by a macroscopic quantity like the temperature). The process that emits the light observed in the radio band from the jets of SS 433 is, instead, nonthermal and is due to synchrotron emission: Charged particles wrapping around a magnetic field, accelerate and thus emit light. *See* SYNCHROTRON RADIATION.

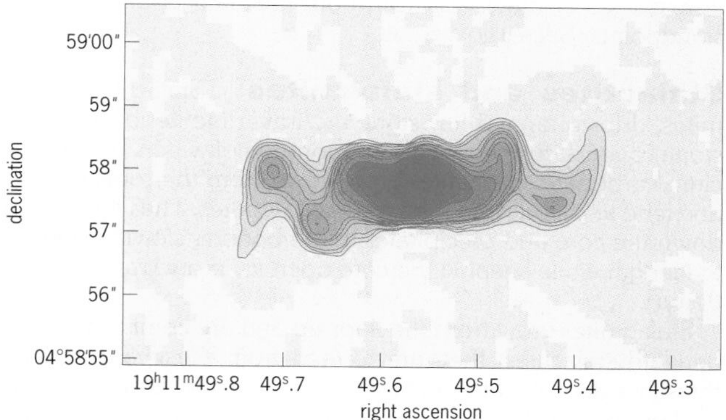

Radio image of SS 433 obtained in 2003 with the Very Large Array. A contour plot of the intensity of radio emission is superimposed on a pixel image. The binary core is the brightest spot in the center, from which matter is outflowing in two antiparallel jets. On each side of the core is a twisted trace that is the projection onto the plane of the sky of the precessing jets.

SS 433 is also an eclipsing binary system with an orbital period of about 13 days. The mass of the companion star should exceed 5 times the mass of the Sun. The binary system is imbued with a dense envelope of gas probably coming from a strong stellar wind from the high-mass companion star. *See* BINARY STAR; ECLIPSING VARIABLE STARS.

Simone Migliari

Stability augmentation The alteration of the inherent behavior of a system. As an example, ships tend to exhibit significant rolling motions at sea. To dampen these rolling motions, a roll stabilization (feedback) system can be used. Such a system consists of a set of vanes (that is, small wings) extending outward from the hull, below the waterline. By varying the vane incidence angle relative to the hull, a hydrodynamic lift is generated on the vane. The vanes are driven by a feedback system so that the rolling motions are opposed by creating positive lift on one side of the hull and negative lift on the other side.

As a second example, nearly all satellites require some form of stability augmentation to help in keeping the antennas or sensors aligned with receiving equipment on Earth. The stability augmentation is effected by thrusters which receive their commands from a feedback system. *See* SPACECRAFT PROPULSION.

As a third example, stability augmentation systems are used on aircraft. This is usually achieved by a system which controls one or more flight-control surfaces (or engines) automatically without inputs from the pilot. The inherent stability and response behavior of many modern airplanes tends toward low damping or even instability. The physical reasons have to do with the configuration of the airplane and the combination of flight speed and altitude at which the airplane is operated. Several modern fighters and even some transports are intentionally designed with no or little inherent stability. There are a number of reasons for such a design condition. In the case of fighters, excellent maneuverability in combat is essential. By making a fighter intentionally inherently unstable, it is easy to design the control system so that load factors in pull-ups or in turns can be built up rapidly. In the case of transports, the motivation to design for little or no inherent stability is to lower the size of the tail and thereby achieve a reduction in drag and weight. *See* AIRPLANE; MILITARY AIRCRAFT.

The control exercised by the stability augmentation system contrasts with that exercised by the pilot. The pilot may be connected with the flight-control surface via a direct mechanical link. Alternatively, in many modern airplanes the pilot cockpit control movement is sensed by a position transducer. The output of the position transducer in turn is sent, via a computer-amplifier combination, to a hydraulic actuator, referred to as a servo, which drives the flight-control surface. Command signals which come from the pilot or from the stability augmentation system are sent by wire (fly-by-wire) or by optical conduit (fly-by-light) to the electromagnetic valve. A valve distributes high-pressure hydraulic fluid either to the left or to the right of the piston so that the piston is forced to move. The piston in turn moves the flight-control surface. *See* FLIGHT CONTROLS.

With the introduction of fast in-flight digital computers, it has become possible to equip airplanes with so-called full flight envelop protection systems. Such systems are designed to refuse any pilot input which might get the airplane into a flight condition from which recovery is no longer possible. Such systems can easily be arranged to prevent a pilot from rolling a commercial airplane too much or to prevent the pilot from stalling the airplane. Such systems can also be arranged so that loads acting on the wing or tail do not approach dangerously high levels. In that case the system is referred to as a load-alleviation system. Jan Roskam

Stabilizer (aircraft) The horizontal or vertical aerodynamic wing surfaces that provide aircraft stability and longitudinal balance in flight. Horizontal and vertical stabilizers (fins) are similar to the aircraft wing in structural design and function

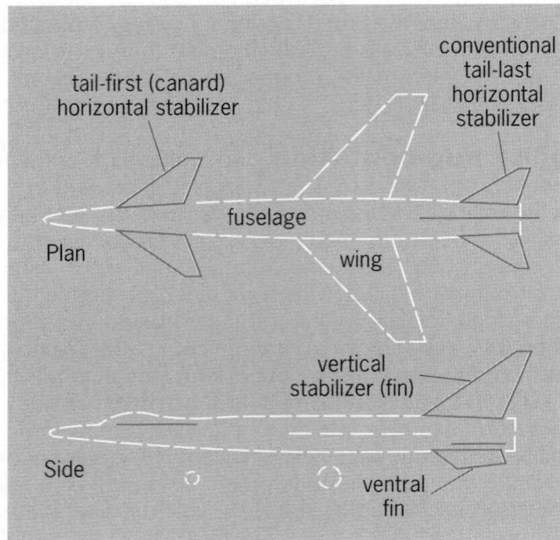

Various stabilizer arrangements.

of providing lift at angle of attack to the wind. However, stabilizers are not required to supply lift to overcome aircraft weight during flight, and when the wing-fuselage center of pressure is behind the aircraft center of gravity, the aerodynamic load on the horizontal stabilizer may be downward. Various stabilizer arrangements are shown in the illustration. *See* FLIGHT CONTROLS. Malcolm J. Abzug

Stabilizer (chemistry) Any substance that tends to maintain the physical and chemical properties of a material. Degradation, that is, irreversible changes in chemical composition or structure, is responsible for the premature failure of materials. Stabilizers are used to extend the useful life of materials as well as to maintain their critical properties above the design specifications. Oxygen and water are the principal degradants, but ultraviolet radiation also can have a significant effect (photodegradation).

A wide variety of additives has been developed to stabilize polymers against degradation. Stabilizers are available that inhibit thermal oxidation, burning, photodegradation, and ozone deterioration of elastomers. Research in the chemistry of the low-temperature oxidation of natural rubber has revealed that hydrocarbon polymers oxidize by a free-radical chain mechanism. In pure, low-molecular-weight hydrocarbons, an added initiator is required to produce the first radicals. In contrast, initiation of polymer oxidation occurs in the complex molecules of elastomers through impurities already present, for example, hydroperoxides. *See* CHAIN REACTION (CHEMISTRY); FREE RADICAL.

Stabilization of hydrocarbon polymers can be accomplished with preventative or chain-breaking antioxidants. Preventative antioxidants stabilize by reducing the number of radicals formed in the initiation stage. Where hydroperoxides are responsible for initiation, the induced decomposition of these reactive intermediates into nonradical products provides effective stabilization. Suppressing the catalytic effects of metallic impurities that increase the rate of radical formation can also provide stabilization. Chain-breaking antioxidants interrupt the oxidative chain by providing labile hydrogens to compete with the polymer in reaction with the propagating radicals. The by-product of this reaction is a radical which is not capable of continuing the oxidative chain.

Stabilization of polymers against photooxidation, the principal component of outdoor weathering, is accomplished by addition of ultraviolet absorbers, and radical scavengers. Ultraviolet absorbers absorb and harmlessly dissipate damaging radiation. Another class of additives, known as hindered-amine light stabilizers, function by scavenging destructive radicals. *See* ANTIOXIDANT; CARBON BLACK; INHIBITOR (CHEMISTRY); PHOTODEGRADATION. W. Lincoln Hawkins

Stain (microbiology) Any colored, organic compound, usually called dye, used to stain tissues, cells, cell components, or cell contents. The dye may be natural or synthetic. The object stained is called the substrate. The small size and transparency of microorganisms make them difficult to see even with the aid of a high-power microscope. Staining facilitates the observation of a substrate by introducing differences in optical density or in light absorption between the substrate and its surround or between different parts of the same substrate. In electron microscopy, and sometimes in light microscopy (as in the silver impregnation technique of staining flagella or capsules), staining is accomplished by depositing on the substrate ultraphotoscopic particles of a metal such as chromium or gold (the so-called shadowing process); or staining is done by treating the substrate with solutions of metallic compounds such as uranyl acetate or phosphotungstic acid. Stains may be classified according to their molecular structure. They may also be classified according to their chemical behavior into acid, basic, neutral, and indifferent. This classification is of more practical value to the biologist. *See* ELECTRON MICROSCOPE; MEDICAL BACTERIOLOGY. Georges Knaysi

Stainless steel The generic name commonly used for that entire group of iron-base alloys which exhibit phenomenal resistance to rusting and corrosion because of chromium (Cr) content. Contents of Cr exceeding 10%, with carbon (C) held suitably low, make iron effectively rustproof.

Other alloy elements, notably nickel (Ni) and molybdenum (Mo), can also be added to the basic stainless composition to produce both variety and improvement of properties. Over 100 different stainless steels are produced commercially, about half as standardized grades. Some are more properly classed as stainless irons since they do not harden as steel; others are true steels to which corrosion resistance becomes an added feature. Still others that are neither properly steels nor irons introduce totally new classes of materials, from both mechanical and chemical standpoints. *See* ALLOY; STEEL. Carl A. Zapffe

Stalactites and stalagmites Stalactites, stalagmites, dripstone, and flowstone are travertine deposits in limestone caverns, formed by the evaporation of waters bearing calcium carbonate. Stalactites grow down from the roofs of caves and tend to be long and thin, with hollow cores. The water moves down the core and precipitates at the bottom, slowly extending the length while keeping the core open for more water to move down.

Stalagmites grow from the floor up and are commonly found beneath stalactites; they are formed from the evaporation of the same drip of water that forms the stalactite. Stalagmites are thicker and shorter than stalactites and have no central hollow core. *See* CAVE; LIMESTONE. Raymond Siever

Stall-warning indicator A device that determines the critical angle of attack for a given aircraft, at which point the drag coefficient overpowers the lift coefficient and the aircraft will no longer sustain itself in steady-state condition (level flight

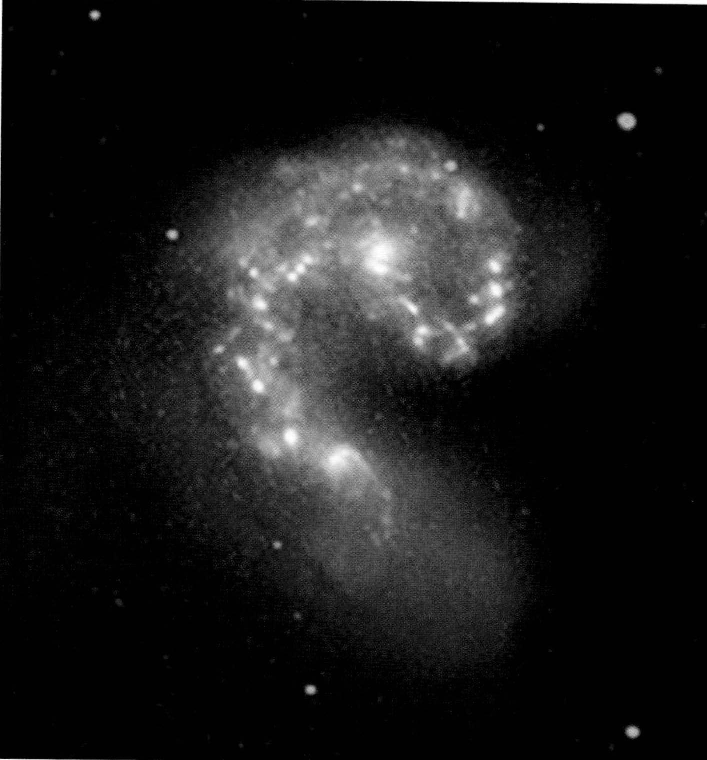

Colorplate 1. (Above) Composite multi-wavelength, false-color image of nearest starburst galaxy, M82, from NASA's *Spitzer*, *Hubble*, and *Chandra* space observatories. The galaxy is about 11.7 million light-years away. X-ray data recorded by *Chandra* appears in blue, infrared light recorded by *Spitzer* in red, *Hubble* observations of hydrogen emission in orange, and the bluest visible light in yellow-green. *(NASA/JPL-Caltech/Space Telescope Science Institute/Chandra X-ray Center/University of Arizona/ESA/AURA/Johns Hopkins University)*

Colorplate 2. (Left) Image of the colliding Antennae Galaxies, NGC 4038 and 4039, around 68 million light-years away. The image is a false-color composite of infrared data from the *Spitzer Space Telescope* and visible-light data from the Kitt Peak National Observatory. Visible light from stars in the galaxies (blue and green) is shown together with infrared light from warm dust clouds heated by newborn stars (red). The galaxies have been merging together for about the last 800 million years, triggering a tremendous burst of star formation, particularly at the site where the two galaxies overlap. The two nuclei or centers of the merging galaxies show up as yellow-white areas, one above the other. The brightest clouds of forming stars lie in the overlap region between and left of the nuclei. *(NASA/JPL-Caltech/Z. Wang, Harvard-Smithsonian Center for Astrophysics/M. Rushing/NOAO)*

or climb/descent). The indicator usually operates from vane sensors, airflow pressure sensors, tabs on the leading edge of the wings, and computing devices which include accelerometers, airspeed detectors, and vertical gyros. James W. Angus

Standard (physical measurement) An accepted reference sample which is used for establishing a unit for the measurement of physical quantities. A physical quantity is specified by a numerical factor and a unit; for example, a mass might be expressed as 8 g, a length as 6 cm, and a time interval as 2 min. Here the gram is a mass unit defined in terms of the international kilogram, which serves as the primary standard of mass. The centimeter is defined in terms of the international meter, which is the primary standard of length and is defined as the length of path traveled by light in a vacuum during a time interval of 1/299,792,458 of a second. In similar fashion, the minute is a time interval defined as 60 s, where the second is the international standard of time and is defined as the duration of 9 192 631 770 periods of the radiation corresponding to the transition between the two hyperfine energy levels of the ground state of the cesium-133 atom.

The National Institute of Standards and Technology in the United States and comparable laboratories in other countries are responsible for maintaining accurate secondary standards for various physical quantities. *See* ELECTRICAL UNITS AND STANDARDS; LIGHT; METRIC SYSTEM; PHYSICAL MEASUREMENT; TIME. Dudley Williams

Standard model The theory that explains the three major interactions of elementary particle physics—the strong interaction responsible for nuclear forces, the weak interaction responsible for radioactive decay, and the electromagnetic interaction—in terms of a common physical picture. The model for this picture is quantum electrodynamics, the fundamental theory underlying electromagnetism. In that theory, electrons, viewed as structureless elementary constituents of matter, interact with photons, structureless elementary particles of light. The standard model extends quantum electrodynamics to explain all three interactions of subnuclear physics in terms of similar basic constituents. *See* ELECTRON; ELECTROWEAK INTERACTION; ELEMENTARY PARTICLE; LIGHT; PHOTON; QUANTUM CHROMODYNAMICS; QUANTUM ELECTRODYNAMICS; STRONG NUCLEAR INTERACTIONS; WEAK NUCLEAR INTERACTIONS. Michael E. Peskin

Staphylococcus A genus of bacteria containing at least 28 species that are collectively referred to as staphylococci. Their usual habitat is animal skin and mucosal surfaces. Although the genus is known for the ability of some species to cause infectious diseases, many species rarely cause infections. Pathogenic staphylococci are usually opportunists and cause illness only in compromised hosts. *Staphylococcus aureus*, the most pathogenic species, is usually identified by its ability to produce coagulase (proteins that affect fibrinogen of the blood-clotting cascade). Since most other species of staphylococci do not produce coagulase, it is useful to divide staphylococci into coagulase-positive and coagulase-negative species. Coagulase-negative staphylococci are not highly virulent but are an important cause of infections in certain high-risk groups. Although *Staphylococcus* infections were once readily treatable with antibiotics, some strains have acquired genes making them resistant to multiple antimicrobial agents. *See* BACTERIA; DRUG RESISTANCE; MEDICAL BACTERIOLOGY.

Staphylococcus cells are spherical with a diameter of 0.5–1.5 micrometers. Clumps of staphylococci resemble bunches of grapes when viewed with a microscope, owing to cell division in multiple planes. The staphylococci have a gram-positive cell composition, with a unique peptidoglycan structure that is highly cross-linked with bridges of amino acids. *See* STAIN (MICROBIOLOGY).

Most species are facultative anaerobes. Within a single species, there is a high degree of strain variation in nutritional requirements. Staphylococci are quite resistant to desiccation and high-osmotic conditions. These properties facilitate their survival in the environment, growth in food, and communicability.

In addition to genetic information on the chromosome, pathogenic staphylococci often contain accessory elements such as plasmids, bacteriophages, pathogenicity islands (DNA clusters containing genes associated with pathogenesis), and transposons. These elements harbor genes that encode toxins or resistance to antimicrobial agents and may be transferred to other strains. Genes involved in virulence, especially those coding for exotoxins and surface-binding proteins, are coordinately or simultaneously regulated by loci on the chromosome. *See* BACTERIAL GENETICS; BACTERIOPHAGE; PLASMID; TRANSPOSONS.

Most *Staphylococcus aureus* infections develop into a pyogenic (pus-forming) lesion caused by acute inflammation. Inflammation helps eliminate the bacteria but also damages tissue at the site of infection. Typical pyogenic lesions are abscesses with purulent centers containing leukocytes, fluid, and bacteria. Pyogenic infections can occur anywhere in the body. Blood infections (septicemia) can disseminate the organism throughout the body and abscesses can form internally.

Certain strains of *S. aureus* produce exotoxins that mediate two illnesses, toxic shock syndrome and staphylococcal scalded skin syndrome. In both diseases, exotoxins are produced during an infection, diffuse from the site of infection, and are carried by the blood (toxemia) to other sites of the body, causing symptoms to develop at sites distant from the infection. Toxic shock syndrome is an acute life-threatening illness mediated by staphylococcal superantigen exotoxins. Staphylococcal scalded skin syndrome, also known as Ritter's disease, refers to several staphylococcal toxigenic infections. It is characterized by dermatologic abnormalities caused by two related exotoxins, the type A and B exfoliative (epidermolytic) toxins. *See* CELLULAR IMMUNOLOGY; TOXIC SHOCK SYNDROME.

Staphylococcal food poisoning is not an infection, but an intoxication that results from ingestion of staphylococcal enterotoxins in food. The enterotoxins are produced when food contaminated with *S. aureus* is improperly stored under conditions that allow the bacteria to grow. Although contamination can originate from animals or the environment, food preparers with poor hygiene are the usual source. Effective methods for preventing staphylococcal food poisoning are aimed at eliminating contamination through common hygiene practices, such as wearing gloves, and proper food storage to minimize toxin production. *See* FOOD POISONING.

Coagulase-positive staphylococci are the most important *Staphylococcus* pathogens for animals. Certain diseases of pets and farm animals are very prominent. *Staphylococcus aureus* is the leading cause of infectious mastitis in dairy animals. Greg Bohach

Star A self-luminous body that during its life generates (or will generate) energy and support by thermonuclear fusion.

Names. Over 6000 stars can be seen with the unaided eye. The brightest carry proper names from ancient times, most of Arabic origin. A more general system names stars within constellations by Greek letters roughly in accord with apparent brightness, followed by the Latin genitive of the constellation name (for example, Vega, in Lyra, is also Alpha Lyrae). More generally yet, brighter stars carry numbers in easterly order within a constellation (Vega is also 3 Lyrae). All naked-eye stars also have HR (Harvard Revised) numbers assigned in order east of the vernal equinox. A variety of catalogs list millions of telescopic stars. *See* ASTRONOMICAL CATALOGS; CONSTELLATION.

Magnitudes and colors. About 130 B.C., Hipparchus assigned naked-eye stars to six brightness groups or "apparent magnitudes" (m), with first magnitude the brightest. This scheme is now quantified as a logarithmic system such that five magnitudes correspond to a factor of 100 in brightness, rendering first magnitude 2.512... times brighter than second, and so on; calibration extends the brightest stars into negative numbers. The faintest stars now observed are around magnitude 30, 10^{12} times fainter than magnitude zero.

Stars assume subtle colors from red to blue-white, reflecting different spectral energy distributions that result from temperatures ranging from under 2000 K (3100°F) to over 100,000 K (180,000°F). The magnitude of a star therefore depends on the detector's color sensitivity. Numerous magnitude systems range from the ultraviolet into the infrared, though the apparent visual magnitude ($m_v = V$) is still standard. The differences among the systems allow measures of stellar color and temperature. *See* COLOR INDEX; MAGNITUDE (ASTRONOMY).

Distances. The fundamental means of finding stellar distances is parallax. As the Earth moves in orbit around the Sun, a nearby star will appear to shift its location against the background. The parallax (p in arcseconds) is defined as one half the total shift, and is the angle subtended by the Earth's orbital radius as seen from the star. Distance in parsecs (pc) is 1/pc, where 1 pc = 206,265 AU = 3.26 light-years. (A light-year is the distance that a ray of light will travel in a year.) The nearest star, a telescopic companion to Alpha Centauri, is 1.34 pc = 4.39 light-years away. *See* LIGHT-YEAR; PARALLAX (ASTRONOMY); PARSEC.

Distribution and motions. All the unaided-eye stars and 200 billion more are collected into the Milky Way Galaxy (or simply, the Galaxy), 98% concentrated into a thin disk over 100,000 light-years across. From the Sun, inside the disk and 27,000 light-years from the Galaxy's center in Sagittarius, the disk appears as the Milky Way. Surrounding the disk is a vast but sparsely populated halo.

Angular proper motions across the line of sight depend on velocities across the line of sight and distances. Statistical analysis of these motions shows the Sun to be moving through the local stars at a speed of 15–20 km/s (9–12 mi/s), roughly toward Vega. From radial velocities of sources outside the Galaxy, it is found that the Sun moves in a roughly circular orbit at 220 km/s (137 mi/s), which when combined with the space motions of other stars allows their galactic orbits to be determined. Stars in the disk have closely circular orbits; those in the halo have elliptical ones. *See* DOPPLER EFFECT; MILKY WAY GALAXY.

Absolute magnitudes. The apparent visual magnitude of a star depends on its intrinsic visual luminosity and on the inverse square of the distance. Knowledge of the distance allows the determination of the true visual luminosity, expressed as the absolute visual magnitude, M_v. This quantity is defined as the apparent visual magnitude that the star would have at a distance of 10 pc. Absolute visual magnitudes range from $M_v = -10$ to +23 (a factor of 10^{13}). The Sun's absolute visual magnitude is in the middle of this range, +4.83.

Spectral classes. Stars exhibit a variety of absorption-line spectra. The absorptions, narrow cuts in the spectra, are produced by atoms, ions, and molecules in the stars' thin, semitransparent outer layers, or atmospheres. Over a century ago, Edward C. Pickering lettered them according to the strengths of their hydrogen lines. After he and his assistants dropped some letters and rearranged others for greater continuity, they arrived at the standard spectral sequence, OBAFGKM, which was later decimalized (the Sun is in class G2). Classes L and T were added in 1999. Since the sequence correlates with color, it must also correlate with temperature, which ranges from near 50,000 K (90,000°F) for hot class O, through about 6000 K (10,000°F) for solar-type class G, to under 2000 K (3100°F) for class L, and below 1000 K (1300°F) for class T (all of which are brown dwarf substars).

The majority of stars in the galactic disk have chemical compositions like that of the Sun. Differences in spectra result primarily from changes in molecular and ionic composition and in the efficiencies of absorption, all of which correlate with temperature. *See* ASTRONOMICAL SPECTROSCOPY; CARBON STAR; SPECTRAL TYPE.

Hertzsprung-Russell diagram. Shortly after the invention of the spectral sequence, H. N. Russell and E. Hertzsprung showed that luminosity correlates with spectral class. A graph of absolute visual (or bolometric) magnitude (luminosity increasing upward) plotted against spectral class (or temperature, decreasing toward the right) is called the Hertzsprung-Russell (HR) diagram (see illustration). The majority of stars lie in a band in which luminosity climbs up and to the left with temperature (with the Sun in the middle). But another band begins near the location of the Sun and proceeds up and to the right (to lower temperature), luminosities increasing to thousands solar as temperature drops to class M. To be bright and cool, such stars must have large radiating areas and radii. To distinguish between the bands, the larger stars are called giants, while those of the main band are termed dwarfs (or main-sequence stars). Yet brighter stars to the cool side of the main sequence, with luminosities approaching 10^6 solar, are called supergiants. In between the giants and the dwarfs lie a few subgiants. At the top, superior to the supergiants, are the very rare hypergiants. *See* DWARF STAR; GIANT STAR; SUBGIANT STAR; SUPERGIANT STAR.

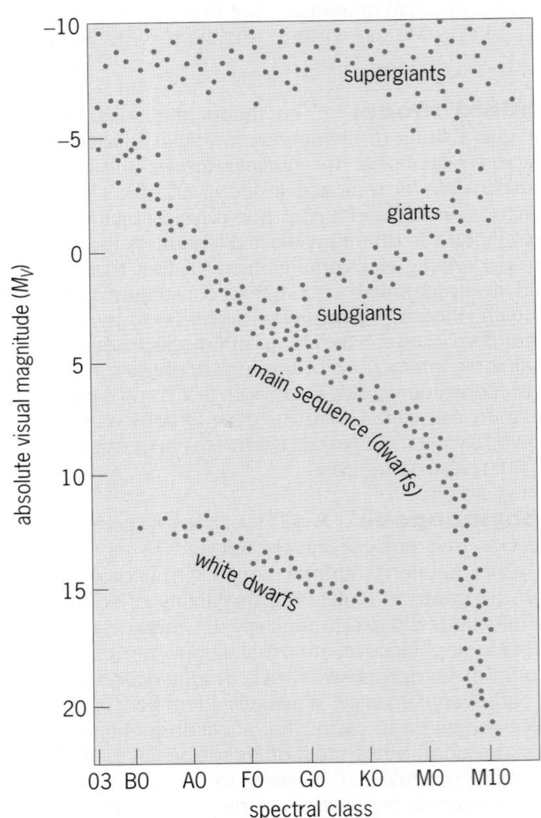

Hertzsprung-Russell (HR) diagram, showing the positions of the major kinds of stars.

In the lower left corner of the diagram, beneath the main sequence, are stars so dim that they must be very small. Since the first ones found were hot and white, they became known as white dwarfs in spite of their actual temperatures or colors. White dwarfs must be positioned on the HR diagram according to their temperatures (rather than their spectral classes). *See* WHITE DWARF STAR.

Stars in the galactic halo are deficient in heavy elements. Low metal content makes halo dwarfs bluer than those of the standard main sequence, shifting them to the left and seemingly downward on the HR diagram, where they are known as subdwarfs. *See* HERTZSPRUNG-RUSSELL DIAGRAM; STELLAR POPULATION.

Double and multiple stars. Most stars are members of some sort of community, from doubles through multiples (double-doubles, and so forth) to clusters, which themselves contain doubles. Separations between components of double stars range from thousands of astronomical units (with orbital periods of a million years) to stars that touch each other (and orbit in hours). *See* BINARY STAR.

Masses and main-sequence properties. Observations of hundreds of binary stars show that mass (M) increases upward along the main sequence from about 7.5% solar at cool class M to over 20 solar in the cooler end of class O. Extrapolation by theory, as well as through observations of clusters, suggests masses of 120 solar at the extreme hot end (class O3) of the main sequence. Fainter than the Sun, luminosity is proportional to M^3; from the Sun and brighter, luminosity goes as M^4 and then back to M^3. This mass-luminosity relation is the result of higher internal temperatures and pressures in more massive stars caused by gravitational compression. *See* MASS-LUMINOSITY RELATION.

As internal temperature climbs above the 8×10^6 K limit, hydrogen fuses to helium via the proton-proton cycle at an ever-increasing rate. Above about 15×10^6 K (27×10^{6}°F), so does fusion by the carbon cycle, in which carbon acts as a nuclear catalyst. *See* CARBON-NITROGEN-OXYGEN CYCLES; PROTON-PROTON CHAIN.

The onset of carbon-cycle dominance coincides with a change in stellar structure. The Sun has a radiatively stable core surrounded by an envelope whose outer parts are in a state of convection that helps produce a magnetic field and magnetic sunspots. Hotter dwarfs have shallower convection layers, and above class F, envelope convection disappears, the cores becoming convective. The convective layers of cooler dwarfs deepen, until below about 0.3 solar mass (class M3), convection takes over completely and the stars are thoroughly mixed. *See* CONVECTION (HEAT).

Below a lower mass limit of 7.5% solar, internal temperatures and densities are not great enough to allow the proton-proton chain to operate. Such stars, called brown dwarfs, glow dimly and redly from gravitational contraction and from the fusion of natural deuterium (^{2}H) into helium. *See* BROWN DWARF.

Clusters. Doubles and multiples are highly structured. Clusters are not, the member stars orbiting a common center of mass. Open clusters are fairly small collections in which a few hundred or a thousand stars are scattered across a few tens of light-years. About 150 globular clusters occupy the Galaxy's halo, the poorest about as good as a rich open cluster, the best containing over a million stars within a volume 100 light-years across.

The HR diagrams of clusters are radically different from the HR diagrams of the general stellar field. Those of open clusters differ among themselves in having various portions of the upper main sequence removed. The effect is related to the cluster's age, since high-mass stars die first. Globular clusters, which lack an upper main sequence and are therefore all old (up to about 12×10^9 years), contain a distinctive "horizontal branch" composed of modest giants.

All clusters disintegrate with time as their stars escape through a form of evaporation aided by tides raised by the Galaxy. Old open clusters (more than 10^9 years old) are rare, while the compact globulars can survive for the age of the Galaxy. *See* STAR CLUSTERS.

Variable stars. Dwarfs are generally stable. Giants and supergiants, however, can have structures that allow them to pulsate. Cepheids (named after the star Delta Cephei) are F and G supergiants and bright giants that occupy a somewhat vertical instability strip in the middle of the HR diagram. They vary regularly by about a magnitude over periods ranging from 1 to 50 days. The pulsation is driven by a deep layer of gas in which helium is becoming ionized. This layer alternately traps and releases heat from the outgoing radiation. Larger and more luminous Cepheids take longer to pulsate. Once this period-luminosity relationship is calibrated through parallax and main-sequence cluster fitting, the period of a Cepheid gives its absolute magnitude, which in turn gives its distance. Cepheids are vital in finding distances to other galaxies. *See* CEPHEIDS; HUBBLE CONSTANT.

Miras (after the star Mira, Omicron Ceti), or long-period variables, are luminous M, C, or S giants that can vary visually by more than 10 magnitudes over periods that range from 50 to 1000 days, the pulsations again driven by deep ionization layers. *See* MIRA.

Duplicity produces its own set of intrinsic variations. If the members of a binary system are close enough together, one of them can transfer mass to the other, and instabilities in the transfer process cause the binary to flicker. If one companion is a white dwarf, infalling compressed hydrogen can erupt in a thermonuclear runaway, producing a sudden nova that can reach absolute magnitude −10. If the white dwarf gains enough matter such that it exceeds its allowed limit (the Chandrasekhar limit) of 1.4 solar masses, it can even explode as a supernova. *See* CATACLYSMIC VARIABLE; NOVA; VARIABLE STAR.

Off the HR diagram. Various kinds of stars are not placeable on the classical HR diagram. The most common examples are the central stars of planetary nebulae, which are complex shells and rings of ionized gas that surround hot blue stars. *See* PLANETARY NEBULA.

Even hotter neutron stars (10^6 K, 1.8×10^{6}°F) are found associated with the exploded remains of supernovae (supernova remnants). Only 25 or so kilometers (15 mi) across, the visible ones spin rapidly, and are highly magnetic, with fields 10^{12} times the strength of Earth's field. Radiation beamed from tilted, wobbling magnetic axes can strike the Earth to produce seeming pulses of radiation. *See* NEUTRON STAR; PULSAR.

Evolution. The different kinds of stars can be linked through theories of stellar evolution. Stars are born by the gravitational collapse of dense knots within cold dusty molecular clouds found only in the Galaxy's disk. When the new stars are hot enough inside to initiate fusion, their contraction halts and they settle onto the main sequence. The higher the stellar mass, the greater the internal compression and temperature, and the more luminous the star. But the higher the internal temperature, the greater the rate at which hydrogen is fused, and the shorter the star's life. The rate of fusion is proportional to the luminosity, L. Lifetime t is proportional to (fuel supply)/(rate of use) and thus to M/L. But L is on average proportional to $M^{3.5}$, so t is proportional to $1/M^{2.5}$. For example, a 100-solar-mass star will exhaust its core hydrogen in only 2.5 million years. *See* MOLECULAR CLOUD; PROTOSTAR; STELLAR EVOLUTION.

James B. Kaler

Star clouds

Star clouds Large groupings of stars with dimensions of 1000 to several thousand light-years. Star clouds are not stable clusterings of stars or even groups of a common origin, such as stellar associations. Instead, they are large areas where the stellar density is higher than average in a galaxy. They are nevertheless physical entities in that they result from large-scale star formation events or series of events that are to some extent limited in size by the characteristics of the galactic environment.

Large star clouds are found commonly in the spiral arms of galaxies such as the Milky Way Galaxy and the Andromeda Galaxy (M31), but are not found in smaller, dwarf galaxies or in very old galaxies. Conspicuous star clouds in the Milky Way Galaxy are found in the constellations Carina, Cygnus, Sagittarius, and Scutum. *See* ANDROMEDA GALAXY; MILKY WAY GALAXY; STAR.

Paul Hodge

Star clusters Groups of stars held together by mutual gravitational attraction. There are two basic morphological types: open clusters and globular clusters. Typical densities of field stars near the Sun are 0.1 star per cubic parsec. (One parsec equals 1.9×10^{13} mi or 3.1×10^{13} km.) Open clusters, with dozens to thousands of stars, have central densities of 0.3 to 10 stars per cubic parsec, and are often elongated or amorphous in shape. Globular clusters, with a thousand to several million stars, and central densities of a few hundred to over 100,000 stars per cubic parsec, are generally spherical (see illustration). Associations are even looser assemblages, with a few hundred stars and central densities lower than the field. They are often recognized because of unusually large numbers of special types of stars. OB associations are dominated by hot, luminous, young O and B stars. T associations are dominated by young T Tauri variable stars. *See* SPECTRAL TYPE; T TAURI STAR.

Star clusters are important because, within each cluster, member stars probably are at the same distance from the Sun, and have the same age and the same initial chemical composition. Stars with larger masses begin life hotter and brighter than lower-mass stars and end their lives earlier. Most stars within clusters lie along a band in the luminosity-temperature plane (where temperature is measured by spectral type or color). In older clusters, this main-sequence band terminates at fainter and cooler levels. Ages derived from such observations provide a chronology of which clusters formed first. Differences observed among stars away from the main sequence, such as in luminosity and temperature or in chemical composition, reflect changes to the stars during the ends of their lives and provide laboratories for the study of stellar evolution. *See* HERTZSPRUNG-RUSSELL DIAGRAM; MILKY WAY GALAXY; PARALLAX (ASTRONOMY); STELLAR EVOLUTION.

Open clusters, once called galactic clusters, are found mostly along the band of Milky Way, and are rarely found more than 1000 parsecs above or below the plane. A dozen are visible to the naked eye, such as Ursa Major, the Hyades (the horns of Taurus), the Pleiades (also in Taurus), and Praesepe ("the Beehive," in Cancer). Over a 1100 are cataloged, and the Milky Way probably contains tens of thousands, most hidden by interstellar obscuration. *See* HYADES; PLEIADES.

The youngest clusters may be hidden within dark dust clouds opaque to optical light. The oldest open clusters may reach 10^{10} years. Few old clusters are known, partly because of the dissolution of clusters with time.

Great globular cluster in Hercules, Messier 13, photographed with 200-in. (500-cm) telescope. (*California Institute of Technology, Palomar Observatory*)

Generally more distant, more massive, older, and more deficient in heavy elements than open clusters, globular clusters are dispersed all over the sky, but with a strong concentration in the direction of the galactic center (Sagittarius). In 1917, H. Shapley estimated distances to globulars to infer the distance to the galactic center. A total of 153 globular clusters are known, some of which are visible to the unaided eye. A significant fraction of globular clusters may have been formed in dwarf galaxies that have been absorbed by the Milky Way Galaxy. The Sagittarius dwarf galaxy is in the process of being absorbed, and several globular clusters are or have been part of it. The globular cluster M54 may be the Sagittarius dwarf galaxy's nucleus. *See* GALAXY, EXTERNAL.

The nearest globular lies about 2000 parsecs away, while the most distant ones in the Milky Way Galaxy lie over 100,000 parsecs from the Sun. There seem to be two chemically distinct classes of globular clusters: metal-deficient clusters, typically 30 times poorer in heavy elements than the Sun; and metal-rich clusters, typically about 3 times more deficient. Those associated with the Sagittarius dwarf galaxy have different element-to-iron abundance patterns than most other globular clusters. The metal-deficient globulars have very high velocities relative to the Sun, and so do not belong to the disk of the Milky Way; they are found in a spherical distribution around the galactic center. The more metal-rich globulars have motions characteristic of disk stars and clusters, and are found within a few kiloparsecs of the plane. All globular clusters in the Milky Way are very old, typically about 1.2×10^{10} years. *See* STAR.

Bruce W. Carney

Star tracker A device that automatically measures the angular separation of stellar observations with respect to a reference platform. It is also referred to as an astrotracker.

By using the star tracker in conjunction with a precise time reference (chronometer) and a dead-reckoning device consisting of gyroscopes and accelerometers (inertial navigator), a digital computer can correct many of the inertial navigator errors so that precise, autonomous (free from any radio position aids), terrestrial navigation can be achieved. The major errors corrected by the star tracker are introduced by the inertial navigator's gyroscopes that result in attitude deviations. In this configuration, called a stellar inertial system, very high precision aircraft autonomous navigation is achieved. Since the availability of radio position aids poses no problem for commercial aircraft, stellar inertial navigation technology is applied only on military vehicles. These navigation devices are used when radio position aids, such as Loran and the Global Positioning System (GPS), may be unavailable. *See* CHRONOMETER; ELECTRONIC WARFARE; INERTIAL NAVIGATION SYSTEM.

Star trackers are also used for both military and nonmilitary applications on space probes, space-based interceptors, and satellites. In these applications the precise attitude capabilities of these devices provide the fiducial precision reference for pointing of the vehicle and Earth or planet sensors. On space missions, star trackers are the only sensor presently available that can provide arc-second attitude accuracy. *See* SPACE NAVIGATION AND GUIDANCE.

A gimbaled stellar inertial system is mounted on the stable element of an inertial navigator. The tracking system measures the telescope azimuth rotation angle and elevation angle to the acquired star with respect to a stable inertial reference. The tracker is programmed to observe different, widely angular separated stars in order to achieve accurate operation. The deviation of the measured stellar observations from their ideal stellar positions is utilized to enhance the performance of the inertial navigator.

The preponderance of modern systems do not contain gimbals. These systems are completely strapped down (that is, with no moving parts) and observe stars at random as the stars come into the rigid, fixed star tracker's field of view. Modern star trackers can operate in two modes. The first (lost in space) uses sophisticated star pattern matching algorithms to determine initial position of the vehicle. The second uses its knowledge of the telescope's location and direction, know which stars should be in the field of view and where their images are located on the star tracker's optical detector. In order to make a three-axis attitude correction, there should be a least two stellar observations ideally separated by 90°. Smaller angular separations lead to a dilution of attitude correction precision. The multiple measurements need not be performed simultaneously since the star tracker corrects the attitude of each axis based on the stellar observations available. Thus, these strapdown startracker systems consist of either multiple telescopes with moderate fields of view or a single telescope with a very wide field of view for proper operation. *See* CELESTIAL NAVIGATION.

The preponderance of star-tracker photodetectors used to measure the stellar irradiance are solid-state, silicon semiconductor photosensors. These devices have their peak responsivity in the near-infrared region (0.6–1.0 micrometer). Imaging focal-plane arrays can have several million photosensors on a single semiconductor silicon chip arranged in a rectangular grid matrix. Each of the individual photosensors on the focal-plane array is called a pixel. The devices are similar to the photosensors found in home video-camera recorders. *See* CHARGE-COUPLED DEVICES.

Seymour Levine

Starburst galaxy A galaxy that is observed to be undergoing an unusually high rate of formation of stars. It is often defined as a galaxy that, if it continues to form stars at the currently observed rate, will exhaust its entire supply of star-forming material, the interstellar gas and dust, in a time period that is very short compared to the age of the universe. For a typical starburst galaxy, this gas and dust exhaustion time scale is less than 10^8 years, that is, less than 1% of the age of the universe. Such galaxies must be undergoing a passing burst of star formation.

The term "starburst," coined in the 1981 by Daniel W. Weedman and his collaborators, usually implies that the burst of star formation is occurring in the nuclear regions of the galaxy, the term first being used to describe a sample of luminous spiral galaxies with bright pointlike nuclei. Indeed, starbursts are very frequently confined to the inner approximately 1000-light-year-diameter regions of massive galaxies. However, there exist galaxies that meet the definition of short gas-exhaustion time scale but exhibit more galaxy-wide star formation. These include the nearest, best-studied starburst galaxy, M82. The related term "starburst region" is also now often used to refer to large regions within galaxies where exceptionally energetic star formation is occurring.

Starburst galaxies often appear blue because of the dominance of their energy output by hot massive stars. Many blue starburst galaxies were discovered in the extensive ultraviolet-sensitive objective prism surveys of B. E. Markarian. Most starburst galaxies are also strong emitters of mid- and far-infrared radiation, because the newly formed stars illuminate their natal clouds of dust and gas with ultraviolet and optical radiation, and as the dust warms, it reradiates this energy at far-infrared wavelengths. *The Infrared Astronomical Satellite* (*IRAS*) discovered over 100,000 galaxies that emit strongly at 60 micrometers, the majority of which are likely to be starburst galaxies. The most luminous of these galaxies are commonly termed ultraluminous infrared galaxies (ULIRGs). The *Spitzer Space Telescope* has been able to image starburst galaxies in the mid- and far-infrared with unprecedented sensitivity and detail. One of the most spectacular results from *Spitzer* is regions with intense emission from large molecules, probably polycyclic aromatic hydrocarbons (PAHs). *See* INFRARED ASTRONOMY; INTERSTELLAR MATTER; MOLECULAR CLOUD; PROTOSTAR; SPITZER SPACE TELESCOPE.

There are several theories as to why starbursts occur. One likely cause is the interaction between two galaxies as they pass close to or collide with one another. The tidal forces generated result in shock-wave compression of the interstellar material, loss of angular momentum and infall of material into the central regions of the galaxy, and star formation in the compressed clouds. Many starburst galaxies show evidence of interactions, including distorted appearance and long wispy tails of material. Not all starbursts can be due to interactions, however, since some display no evidence for any recent disturbance. Other mechanisms that are thought responsible for high star-formation rates in galaxies are very strong spiral density waves and central bar instabilities. *See* GALAXY FORMATION AND EVOLUTION.

Dramatic observational breakthroughs have demonstrated that star formation rates in galaxies were probably much higher at very early times in the universe than they are now. In particular, evidence has been found that ultraluminous infrared galaxies were much more common and much more important at earlier times in the history of the universe than now. It is not known whether starbursts are responsible for the tremendous energy of these distant, young, highly infrared-luminous galaxies; if they are, then, starburst events may even have dominated the processes of galaxy building long ago, and much of the total energy radiated by early galaxies may be detectable only at infrared or submillimeter wavelengths because ultraviolet and optical light would have been very strongly obscured by the dense clouds of gas and dust involved in the starburst. *See* GALAXY, EXTERNAL; SUBMILLIMETER ASTRONOMY; TELESCOPE; UNIVERSE.

Carol J. Lonsdale

Starch A carbohydrate that occurs as discrete, partially crystalline granules in the seeds, roots (tubers), stems (pith), leaves, fruits, and pollen grains of higher plants. Starch functions as the main storage or reserve form of carbohydrate; it is second in abundance only to cellulose, a major structural component of plants. Cereal grains, tuber and root crops, and legumes (seeds) have long been used as major sources of carbohydrate in human diets. *See* CELLULOSE.

Starch is isolated commercially from the following sources: cereal grain seeds [maize (corn), wheat, rice, sorghum], roots and tubers [potato, sweet potato, tapioca (cassava), arrowroot], and stems and pith (sago). Cereal grains are often steeped first to loosen the starch granules in the endosperm matrix, followed by wet grinding or milling. Roots and tubers are ground to give a suspension containing starch granules. Then follows screening or sieving, washing, centrifuging, dewatering, and drying.

Starch, a polymer of glucose, is an alpha-glucan, predominantly containing alpha-1,4-glucosidic linkages with a relatively small amount of alpha-1,6-glucosidic linkages forming branch points. Two major polymeric components are present: amylose and amylopectin. Normally a white powder of 98–99.5% purity, starch is insoluble in cold water, ethanol, and most common solvents. *See* GLUCOSE.

Starches are involved in important roles in foods, either naturally occurring in an ingredient or added to achieve a desired functional characteristic. Often the desired functional characteristic (thickening; gelling; adhesive; binding; improving acid, heat, and shear stability) cannot be achieved by using a native starch. Starches may be altered physically, chemically, or enzymatically to produce modified starches with improved functional properties.

The paper, textile, adhesive, chemical, pharmaceutical, and polymer industries use starch and starch derivatives. Organic acids and organic solvents for use as chemical intermediates, enzymes, hormones, antibiotics, and vaccines are industrially produced from starch. *See* CARBOHYDRATE.

David R. Lineback

Stark effect The effect of an electric field on spectrum lines. The electric field may be externally applied; but in many cases it is an internal field caused by the presence of neighboring ions or atoms in a gas, liquid, or solid. Discovered in 1913 by J. Stark, the effect is most easily studied in the spectra of hydrogen and helium, by observing the light from the cathode dark space of an electric discharge. Because of the large potential drop across this region, the lines are split into several components. For observation perpendicular to the field, the light of these components is linearly polarized.

The linear Stark effect exhibits large, nearly symmetrical patterns. The interpretation of the linear Stark effect was one of the first successes of the quantum theory. According to this theory, the effect of the electric field on the electron orbit is to split each energy level of the principal quantum number n into $2n - 1$ equidistant levels, of separation proportional to the field strength. *See* ATOMIC STRUCTURE AND SPECTRA.

The quadratic Stark effect occurs in lines resulting from the lower energy states of many-electron atoms. The quadratic Stark effect is basic to the explanation of the formation of molecules from atoms, of dielectric constants, and of the broadening of spectral lines.

The intermolecular Stark effect is produced by the action of the electric field from surrounding atoms or ions on the emitting atom. The intermolecular effect causes a shifting and broadening of spectrum lines. The molecules being in motion, these fields are inhomogeneous in space and also in time. Hence the line is not split into resolved components but is merely widened.
F. A Jenkins; W. W. Watson

The quantum-confined Stark effect is the Stark effect observed in structures in which the hydrogenic system is confined in a layer of thickness much less than its normal diameter. This is not practical with atoms, but the effect is observed with excitons in semiconductor quantum-well heterostructures. It is important that quantum-confined Stark shifts can be much larger than the binding energy of the hydrogenic system. The resulting shifts of the exciton optical absorption lines can be used to make optical beam modulators and self-electrooptic-effect optical switching devices. *See* ARTIFICIALLY LAYERED STRUCTURES; ELECTROOPTICS; EXCITON; OPTICAL MODULATORS; SEMICONDUCTOR HETEROSTRUCTURES.
David A. B. Miller

Static A hissing, crackling, or other sudden sharp noise that tends to interfere with the reception, utilization, or enjoyment of desired signals or sounds. Perhaps the commonest form of static is that heard in ordinary broadcast receivers during electrical storms. Interference in radio receivers caused by improperly operating electric devices in the vicinity is sometimes also called static. *See* SFERICS.

The crackling sounds heard when long-playing plastic phonograph records are played are also called static. These sounds are caused by sudden deflection of the phonograph needle by dust particles, which are attracted by the grooves of the record by surface electric charges caused by friction on dry days. Static appears as momentary white specks in a television picture. *See* ELECTRICAL INTERFERENCE; ELECTRICAL NOISE.
John Markus

Static electricity Electric charge at rest, generally produced by friction or electrostatic induction. Triboelectrification is the process whereby charge transfer between dissimilar materials, at least one of which must have a high electrical resistivity, occurs due to rubbing or mere contact. *See* ELECTRIC CHARGE; ELECTRICAL RESISTIVITY.

In modern industry, highly insulating synthetic materials, such as plastic powders and insulating liquids, are used in large quantities in an ever-increasing number of applications. Such materials charge up readily, and large quantities of electrical energy may develop with an attendant risk of incendiary discharges. When, for example, powder is pneumatically transported along pipes, charge levels of up to about 100 microcoulombs per kilogram can develop and potentials of thousands of volts are generated within powder layers and the powder cloud. Energetic sparking from charged powder may initiate an explosion of the powder cloud. Similar problems occur when insulating liquids, such as certain fuels, are pumped along pipes, and it is essential that strict grounding procedures are followed during the refueling of aircraft, ships, and other large vehicles.

The capacity of a person for retaining charge depends upon stature, but is typically about 150 picofarads. Even the simple operations of removing items of clothing or sliding off a chair can lead to body discharges to ground of about 0.1 μC, which are energetic enough to ignite a mixture of natural gas and air. Human body capacitance is sufficiently high that, if poorly conducting shoes are worn, body potential may rise to 15,000 V or so above ground during industrial operations such as emptying bags of powder. Sparking may then occur with energy exceeding the minimum ignition energy of powder or fumes, so initiating a fire or explosion. Conducting footware should be used to prevent charge accumulation on personnel in industrial situations where triboelectrification may occur. *See* CAPACITANCE.

In the microelectronics industry, extremely low-energy discharges, arising from body potentials of only a few tens of volts, can damage microelectronics systems or corrupt computer data. During the handling of some sensitive semiconductor devices, it is imperative that operators work on metallic grounded surfaces and are themselves permanently attached to ground by conducting wrist straps. *See* ELECTROSTATICS.
A. G. Bailey

Static var compensator A thyristor-controlled (hence static) generator of reactive power, either lagging or leading, or both. The word var stands for volt ampere reactive, or reactive power. The device is also called a static reactive compensator.

Need for reactive compensation. Reactive power is the product of voltage times current where the voltage and current are 90° out of phase with one another. Thus, reactive power flows one way for one-quarter of a cycle, the other way for the next quarter of a cycle, and so on (in contrast to the real power, or active power, which flows in one direction only). This back-and-forth flow results in no net power being delivered by the generator to the load. However, current associated with reactive power does flow through the conductor and creates extra losses. *See* ALTERNATING CURRENT; ELECTRIC POWER MEASUREMENT; VOLT-AMPERE.

Most loads draw lagging reactive power, which causes electric power system voltage to sag. On the other hand, under light loads, the capacitance of high-voltage lines can create excessive leading reactive power, causing the voltage at some locations to rise above the nominal value. Finally, it is prudent to keep reactive power flows to a minimum in order to allow the lines to carry more active power.

Mechanical versus static compensation. Utilities frequently install capacitors connected from line to ground to compensate for lagging reactive power and reactors connected from line to ground to compensate for leading reactive power. These reactors and capacitors are switched in and out with mechanical switches based on the level of line loading as it varies throughout the day. However, frequent operation of these mechanical switches may reduce their reliability. *See* CAPACITOR; REACTOR (ELECTRICITY).

It is desirable to have a controllable source of reactive power (leading or lagging); and the static var compensator, controlled with static switches, called thyristors, for higher reliability, fulfills this function. It is more expensive than mechanically switched capacitors and reactors (due to the cost of thyristor valves and associated equipment), and hence its use is based on an economic trade-off of benefits versus cost. *See* SEMICONDUCTOR RECTIFIER.
Narain G. Hingorani

Statics The branch of mechanics that describes bodies which are acted upon by balanced forces and torques so that they remain at rest or in uniform motion. This includes point particles, rigid bodies, fluids, and deformable solids in general. Static point particles, however, are not very interesting, and special branches of mechanics are devoted to fluids and deformable solids. For example, hydrostatics is the study of static fluids, and elasticity and plasticity are two branches devoted to deformable bodies. Therefore this article will be limited to the discussion of the statics of rigid bodies in two- and three-space dimensions. *See* BUOYANCY; ELASTICITY; HYDROSTATICS; MECHANICS.

In statics the bodies being studied are in equilibrium. The equilibrium conditions are very similar in the planar, or two-dimensional, and the three-dimensional rigid body statics. These are that the vector sum of all forces acting upon the body must be zero; and the resultant of all torques about any point must be zero. Thus it is necessary to understand the vector sums of forces and torques.

In studying statics problems, two principles, superposition and transmissibility, are used repeatedly on force vectors. They are applicable to all vectors, but specifically to forces and torques (first moments of forces). The principle of superposition of vectors is that the sum of any two vectors is another vector. The principle of transmissibility of a force applied to a rigid body is that the same mechanical effect is produced by any shift of the application of the force along its line of action. To use the superposition principle to add two vectors, the principle of transmissibility is used to move some vectors along their line of action in order to add to their components.

The moment of a force about a directed line is a signed number whose value can be obtained by applying these two rules: (1) The moment of a force about a line parallel to the force is zero. (2) The moment of a force about a line normal to a plane containing the force is the product of the magnitude of the force and the least distance from the line to the line of the force. *See* EQUILIBRIUM OF FORCES; FORCE; TORQUE. Brian De Facio

Statistical mechanics That branch of physics which endeavors to explain the macroscopic properties of a system on the basis of the properties of the microscopic constituents of the system. Usually the number of constituents is very large. All the characteristics of the constituents and their interactions are presumed known; it is the task of statistical mechanics (often called statistical physics) to deduce from this information the behavior of the system as a whole.

Scope. Elements of statistical mechanical methods are present in many widely separated areas in physics. For instance, the classical Boltzmann problem is an attempt to explain the thermodynamic behavior of gases on the basis of classical mechanics applied to the system of molecules.

Statistical mechanics gives more than an explanation of already known phenomena. By using statistical methods, it often becomes possible to obtain expressions for empirically observed parameters, such as viscosity coefficients, heat conduction coefficients, and virial coefficients, in terms of the forces between molecules. Statistical considerations also play a significant role in the description of the electric and magnetic properties of materials. *See* BOLTZMANN STATISTICS; INTERMOLECULAR FORCES; KINETIC THEORY OF MATTER.

If the problem of molecular structure is attacked by statistical methods, the contributions of internal rotation and vibration to thermodynamic properties, such as heat capacity and entropy, can be calculated for models of various proposed structures. Comparison with the known properties often permits the selection of the correct molecular structure.

Perhaps the most dramatic examples of phenomena requiring statistical treatment are the cooperative phenomena or phase transitions. In these processes, such as the condensation of a gas, the transition from a paramagnetic to a ferromagnetic state, or the change from one crystallographic form to another, a sudden and marked change of the whole system takes place. *See* PHASE TRANSITIONS.

Statistical considerations of quite a different kind occur in the discussion of problems such as the diffusion of neutrons through matter. In this case, the probability of the various events which affect the neutron are known, such as the capture probability and scattering cross section. The problem here is to describe the physical situation after a large number of these individual events. The procedures used in the solution of these problems are very similar to, and in some instances taken over from, kinetic considerations. Similar problems occur in the theory of cosmic-ray showers.

It happens in both low-energy and high-energy nuclear physics that a considerable amount of energy is suddenly liberated. An incident particle may be captured by a nucleus, or a high-energy proton may collide with another proton. In either case, there is a large number of ways (a large number of degrees of freedom) in which this energy may be utilized. To survey the resulting processes, one can again invoke statistical considerations. *See* SCATTERING EXPERIMENTS (NUCLEI).

Of considerable importance in statistical physics are the random processes, also called stochastic processes or sometimes fluctuation phenomena. The brownian motion, the motion of a particle moving in an irregular manner under the influence of molecular bombardment, affords a typical example. The stochastic processes are in a sense intermediate between purely statistical processes, where the existence of fluctuations may safely be neglected, and the purely atomistic phenomena, where each particle requires its individual description. *See* BROWNIAN MOVEMENT; STOCHASTIC PROCESS.

All statistical considerations involve, directly or indirectly, ideas from the theory of probability of widely different levels of sophistication. The use of probability notions is, in fact, the distinguishing feature of all statistical considerations. *See* PROBABILITY; STATISTICS.

Methods. For a system of N particles, each of the mass m, contained in a volume V, the positions of the particles may be labeled $x_1, y_1, z_1, \ldots, x_N, y_N, z_N$, their cartesian velocities $v_{x1}, \ldots, v_{zN}$, and their momenta $P_{x1}, \ldots, P_{zN}$. This simplest statistical description concentrates on a discussion of the distribution function $f(x,y,z;v_x,v_y,v_z;t)$. The quantity $f(x,y,z;v_x,v_y,v_z;t) \cdot (dx dy dz dv_x dv_y dv_z)$ gives the (probable) number of particles of the system in those positional and velocity ranges where x lies between x and $x + dx$; v_x between v_x and $v_x + dv_x$, and so on. These ranges are finite.

Observations made on a system always require a finite time; during this time the microscopic details of the system will generally change considerably as the phase point moves. The result of a measurement of a quantity Q will therefore yield the time average, as in Eq. (1). The integral is along the trajectory in phase

$$\overline{Q}_t = \frac{1}{t}\int_0^t Q\,dt \qquad (1)$$

space; Q depends on the variables $x_1, \ldots, P_{zN}$, and t. To evaluate the integral, the trajectory must be known, which requires the solution of the complete mechanical problem.

Ensembles. J. Willard Gibbs first suggested that instead of calculating a time average for a single dynamical system, a collection of systems, all similar to the original one, should instead be considered. Such an ensemble of systems is to be constructed in harmony with the available knowledge of the single system, and may be represented by an assembly of points in the phase space, each point representing a single system. If, for example, the energy of a system is precisely known, but nothing else, the appropriate representative example would be a uniform distribution of ensemble points over the energy surface, and no ensemble points elsewhere. An ensemble is characterized by a density function $\rho(x_1, \ldots, z_N; p_{x1}, \ldots, p_{zN}; t) \equiv p(x,p,t)$. The significance of this function is that the number of ensemble systems dN_e contained in the volume element $dx_1 \ldots dz_N; dp_x \ldots dp_{zN}$ of the phase space (this volume element will be called $d\Gamma$) at time t is as given in Eq. (2).

$$\rho(x,p,t)\, d\Gamma = dN_e \tag{2}$$

The ensemble average of any quantity Q is given by Eq. (3).

$$\overline{Q}_{\text{ens}} = \frac{\int Q\rho\, d\Gamma}{\int \rho\, d\Gamma} \tag{3}$$

The basic idea now is to replace the time average of an individual system by the ensemble average, at a fixed time, of the representative ensemble. Stated formally, the quantity $\overline{Q}_t$ defined by Eq. (1), in which no statistics is involved, is identified with $\overline{Q}_{\text{ens}}$ defined by Eq. (3), in which probability assumptions are explicitly made.

Relation to thermodynamics. It is certainly reasonable to assume that the appropriate ensemble for a thermodynamic equilibrium state must be described by a density function which is independent of the time, since all the macroscopic averages which are to be computed as ensemble averages are time-independent.

The so-called microcanonical ensemble is defined by Eq. (4*a*), where c is a constant, for the energy E between E_0 and $E_0 + \Delta E$; for other energies Eq. (4*b*) holds. By using Eq. (3), any micro-

$$\rho(p,x) = c \tag{4a}$$

$$\rho(p,x) = 0 \tag{4b}$$

canonical average may be calculated. The calculations, which involve integrations over volumes bounded by two energy surfaces, are not trivial. Still, many of the results of classical Boltzmann statistics may be obtained in this way. For applications and for the interpretation of thermodynamics, the canonical ensembles is much more preferable. This ensemble describes a system which is not isolated but which is in thermal contact with a heat reservoir.

There is yet another ensemble which is extremely useful and which is particularly suitable for quantum-mechanical applications. Much work in statistical mechanics is based on the use of this so-called grand canonical ensemble. The grand ensemble describes a collection of systems; the number of particles in each system is no longer the same, but varies from system to system. The density function $p(N,p,x)\, d\Gamma_N$ gives the probability that there will be in the ensemble a system having N particles, and that this system, in its $6N$-dimensional phase space Γ_N, will be in the region of phase space $d\Gamma_N$. Max Dresden

Statistical process control A collection of strategies and statistical methods for ensuring high-quality results in manufacturing and service processes. The origins of statistical process control coincide with a shift in focus by manufacturing companies from quantity to quality. The pioneers in this transformation include W. Shewhart, W. E. Deming, J. Juran, K. Ishakawa, G. Taguchi, and others. At the heart of this transformation were two simple ideas: understanding the customer's needs and meeting the customer's needs consistently. Accomplishing both of these tasks while minimizing wasted time, effort, and resources insured a competitive advantage. *See* PROCESS CONTROL; QUALITY CONTROL; STATISTICS.

Total quality management refers to both the philosophy and practices associated with continuously improving processes within a company. An all-encompassing view of processes included not only manufacturing operations but also service areas such as sales and accounting functions. Companies adopted various strategies for continuous improvement. An important commonality was reliance upon actual process data to monitor systems, inform decision makers, and guide the continuous improvement activities. International standards were eventually established to ensure that total quality management programs followed certain basic principles and practices. The standards known as ISO 9000:2000 are a product of the International Standards Organization (ISO), the American Society for Quality (ASQ), and the American National Standards Institute (ANSI).

The ideas surrounding total quality management grew into the Six Sigma quality movement, which continues to evolve. Among

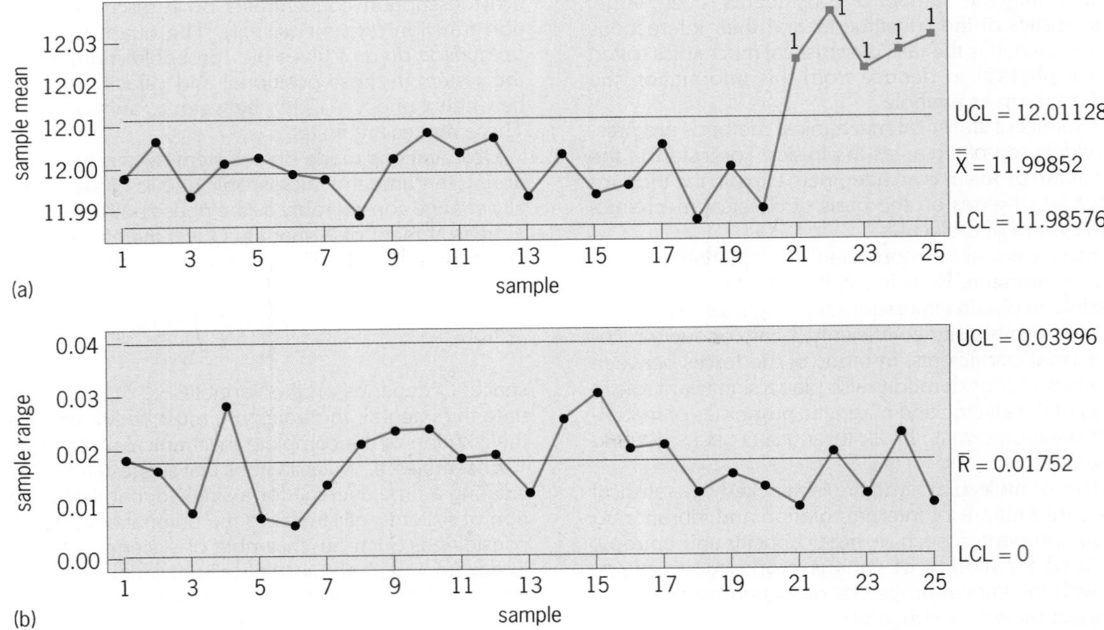

Control charts for length: (*a*) mean and (*b*) range. UCL = upper central limit. LCL = lower central limit. $\bar{\bar{X}}$ = average of all $\bar{X}$ points. $\bar{R}$ = average of all R points.

the unique features of Six Sigma programs are the certification levels for managers trained in statistical problem solving and the use of the DMAIC process. The acronym DMAIC refers to five steps: define/deploy, measure, analyze, improve, and control.

All measurable processes exhibit variation that can be characterized as belonging to one of two classes: common-cause variation and special-cause variation. Common-cause variation is often unexplainable and unavoidable without fundamental changes to the underlying process. Special-cause variation is attributable to a specific cause that may be identified and corrected. The quality control chart is used to detect and address special-cause variation. Additionally, the control chart is used to quantify and avoid overreaction to common-cause variation. Overadjustment of equipment by operators can result in a deterioration of overall quality. *See* CONTROL CHART.

The control chart is displayed graphically to allow easy use by operators. For example, the illustration shows the control charts for the mean length of parts and the range in the length of parts. For each sample, the average length for four randomly selected parts is calculated and plotted on the mean chart, and the sample range for the lengths is plotted on the range chart. Observations falling beyond the control limits of a control chart are called signals. The mean chart signal at time = 21 suggests that something unusual has occurred that requires the operator's attention. No values are beyond the control limits of the range chart, which indicates that the problem affects only the part length, not the process consistency (that is, the process is consistently making bad parts).

In addition to common control charts, there are many specialized monitoring techniques available to practitioners. Multivariate control charts are used when several important process variables are related to one another. Forecast-based or regression-based charts are useful when data values change systematically over time. Process change due to tool wear is an example. Practitioners are advised to select an appropriate technique for each process.

The primary goals of control charting are to identify and eliminate special causes of variation, ensure that the process remains in control, and quantify common-cause variation. Once a process is behaving in a stable and predictable manner, one may evaluate the capability of the process and use designed experiments to gain further improvements. *See* EXPERIMENT; INDUSTRIAL ENGINEERING; MANUFACTURING ENGINEERING. Benjamin M. Adams

Statistics The field of knowledge concerned with collecting, analyzing, and presenting data. Not only workers in the physical, biological, and social sciences, but also engineers, business managers, government officials, market analysts, and many others regularly use statistical methods in their work. The methods range from simple counting to complex mathematical systems designed to extract the maximum amount of information from very extensive data.

In an important sense statistics may be regarded as a field of application of probability theory. The common problem faced by any worker who must collect, analyze, and present data is that of random variation which prevents repetition of exactly the same result when a measurement is repeated. Statistical methods are employed to assess the magnitude of random variation, to minimize it, to balance it out, to remove it by calculation procedures, and to analyze it by suitably arranged patterns of observation. The theory of probability is concerned with the properties of random variables and hence furnishes the basis for developing techniques for controlling them. *See* PROBABILITY.

Viewing statistics from another direction, it is the science of deriving information about populations by observing only samples of those populations. A population is any well-specified collection of elements. Thus, one may refer to the population of adults in the continental United States viewing television screens at 8:14 P.M. on August 6, 1970. Populations may be finite or

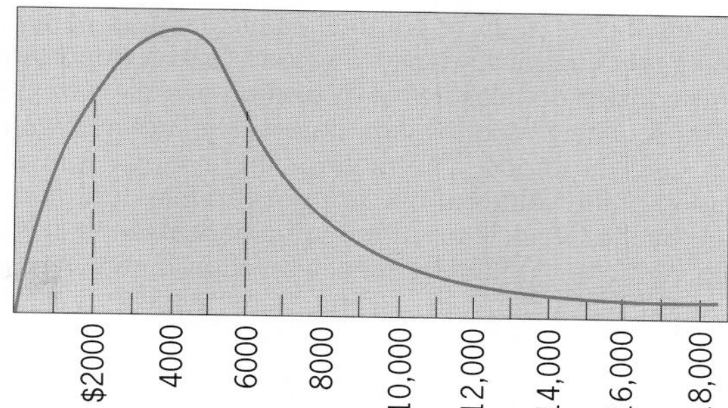

Fig. 2. Distribution of incomes.

infinite. An element of a univariate population is characterized by the value of a random variable which measures some single attribute of interest in the population. Thus, one may be interested in whether or not individuals of the television audience were or were not viewing program *A*; with each individual one may associate a random variable; let it be *X*, which takes on the value of 1 if the individual is watching *A* and 0 if he or she is not. If one were interested in a second characteristic of the elements of the television audience (such as age), one would be dealing with a bivariate population; a third characteristic (such as economic status) would make it a trivariate or, less specifically, a multivariate population.

Distributions. In a univariate population, the population distribution is a curve (function of the random variable which characterizes the elements of the population) from which one can determine the proportion of the population which has elements in a certain range of the random variable. For example, the curve of Fig. 1 provides the distribution of annual incomes of family units in the United States in 1954. The total area under the curve is 1. The area under the curve between any two vertical lines gives the proportion of the families having annual incomes between the two values marked on the horizontal scale by the two vertical lines.

The distribution is also referred to as the distribution function, the density function, the frequency function, or the probability density.

The total area under the distribution curve to the left of each point can also be plotted to give a curve which starts at zero and reaches unity as the variable becomes large; the resulting curve is sometimes called the cumulative distribution function, the probability distribution, or simply the distribution. The cumulative form of the curve of Fig. 1 is shown in Fig. 2; the height of the curve at any point on the horizontal scale equals the area to the left of that point under the curve of Fig. 1 and is the proportion of the population having incomes less than the value at that point. The distribution (in either frequency or cumulative form) gives complete information about the way the characterizing variable is spread through the population.

Population parameters. Populations (or population distributions) are often specified incompletely by certain population

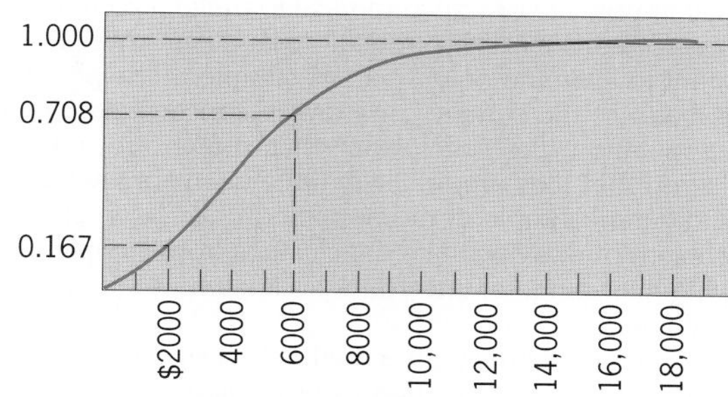

Fig. 2. Cumulative distribution of incomes.

parameters. Some of these parameters are location parameters or measures of central tendency; a second class of important parameters consists of measures of dispersion or scale parameters.

The most widely used location parameters are the mean, the median, and the mode. The mean is the average over all the population of the values of the random variable. It is often represented by the Greek letter μ. In mathematical terms, if x is the random variable, $f(x)$ the frequency function for a given population, and $F(x)$ its cumulative form, then the mean is as shown in Eq. (1).

$$\mu = \int_{-\infty}^{\infty} xf(x)dx = \int_{-\infty}^{\infty} xdF(x) \quad (1)$$

The median, often designated by Med, M, or X_{50}, is a number such that, at most, one-half the values of the variable associated with the elements of the population fall above or below it. The mode is the most frequent value of the random variable; if the frequency function has a unique maximum value, the mode is the value of the random variable at which the frequency function reaches its maximum. Location parameters are numbers near the center of the range over which the random variable of the population varies.

The extent to which a population is scattered on either side of its center is roughly indicated by measures of dispersion such as the standard deviation, the mean deviation, the interquartile range, the range, and sometimes others. The standard deviation is the square root of the mean square of the deviations from the mean; it is usually denoted by the Greek letter σ; σ^2 is called the variance and is expressed by Eq. (2). The mean deviation,

$$\sigma^2 = \int_{-\infty}^{\infty} (x-\mu)^2 f(x)dx = \int_{-\infty}^{\infty} (x-\mu)^2 dF(x) \quad (2)$$

shown in Eq. (3), is the average over the population of the deviations from the mean, all taken to be positive.

$$\text{Mean deviation} = \int_{-\infty}^{\infty} |x-\mu|\, f(x)dx = \int_{-\infty}^{\infty} |x-\mu| dF(x) \quad (3)$$

The interquartile range (often denoted by Q) is the difference $X_{.75} - X_{.25}$, where $X_{.75}$ is the value of the random variable such that one-quarter of the population has values larger than $X_{.75}$, and $X_{.25}$ is the number such that one-quarter of the population has values smaller than $X_{.25}$. The three numbers, $X_{.25}$, $X_{.50}$, $X_{.75}$, are called quartiles; these divide the population into quarters. The range is the difference between the largest and the smallest of the population elements.

Sampling. If one examines every element of a population and records the value of the random variable for each, complete information is obtained about the distribution of the random variable in the population, and there is no statistical problem. It is usually impossible or uneconomical to make a complete enumeration (or census) of a population, and one must therefore be content to examine only a part or sample of the population. On the basis of the sample, one draws conclusions about the entire population; the conclusions thus drawn are not certain in the sense that they would likely have been somewhat different if a different sample of the population had been examined. The problem of drawing valid conclusions from samples and of specifying their range of uncertainty is known as the problem of statistical inference.

Simple random sampling is a method of selecting a sample of n elements out of a population of N elements so that all such samples have an equal probability of being drawn. This may be done by selecting a first element at random from the population, then a second element at random from the remaining population, and so on until the n elements are selected.

The observations of a sample, besides providing estimates of population parameters, can also be used to obtain an estimate of the population's frequency function. This estimate is determined by dividing the range of the sample observations into several intervals of equal length L and counting the number of observations occurring in each interval; these numbers are then divided by nL to determine fractions giving the relative density of the sample occurring in each interval; then on a sheet of graph paper one lays out the intervals on a horizontal axis and plots horizontal lines above each interval at a height equal to the fraction corresponding to the interval; finally the successive plotted horizontal lines are connected by vertical lines to form a broken line curve known as a histogram (Fig. 3).

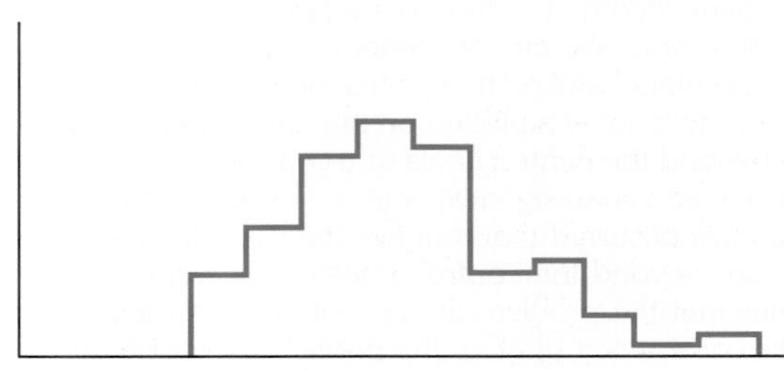

Fig. 3. Histogram.

When a population can be regarded as being made up of several nonoverlapping subpopulations, one may draw a sample from it by drawing a simple random sample from each subpopulation (or stratum); this procedure is called stratified random sampling. Systematic sampling may be regarded as a sampling of units at regular intervals in the population, for example, every tenth unit. When the elements of a population are in mutually exclusive groups called the primary units of the population, then a sample of these primary units might be made, and then, from those selected, a sample of the individual elements would be made. This procedure is called two-stage sampling or subsampling. Multistage sampling can involve more than two stages of sampling.

Many important sampling distributions are derived for random samples drawn from a normal or Gaussian distribution, which is a bell-shaped symmetrical distribution centered at its mean μ.

Estimation. In making an estimate of the value of a parameter of a population from a sample, a function (called the estimator) of the observations is used. For example, for estimating the mean of a normal population the mean of the sample observations is usually taken as the estimator. Another estimator is the average of the two most extreme observations. In fact, there is an infinitude of estimators. The problem of estimation is to find a "good" estimator.

A good estimator may be regarded as one which results in a distribution of estimates concentrated near the true value of the parameter and which can be applied without excessive effort.

Tests of hypotheses. Besides estimation of parameters, another major area of statistical inference is the testing of hypotheses. A hypothesis is merely an assertion that a population has a specific property. The test consists of drawing a sample from the population and determining whether or not it is consistent with the assertion. Very often the hypothesis is a statement about the mean of a population; that it has a given value, that it is the same as that of another population, that it exceeds that of another population by at least 10 units, and the like. Thus, one may be comparing a new blend of gasoline with a current blend, a new drug with a standard one, a new manufacturing process with an existing one.

Experimental design. An experiment is performed to obtain information about the relations between several variables. For example, one may study the effect of storage temperature and duration of storage on the flavor of a frozen food. Three variables (flavor, temperature, and duration) are involved; one (flavor) is called the subject of the experiment; the other two are called factors which influence the subject. Sometimes the factors have intrinsic value in themselves; sometimes they are merely nuisance variables which must be taken into account because it is impossible to perform the experiment without them.

There exist in the statistical literature great numbers of specific experimental designs. These are patterns for making experimental observations; the actual construction of the designs requires quite advanced mathematics based on group theory, finite geometries, and combinatorial analysis. The mathematical problem is to find a pattern from which it is possible to extract the desired information and yet minimize the number of observations.

Regression and correlation. The regression problem is that of estimating certain unknown constants or parameters occurring in a function which relates several variables; the variables may be random or not. By far the most easily handled cases are those in which the function is linear in the unknown parameters, and it is worth considerable effort to transform the function to that form if at all possible. *See* ANALYSIS OF VARIANCE; DISTRIBUTION (PROBABILITY).

Alexander M. Mood

Statoblasts

Chitin-encapsulated bodies, resistant to freezing and a limited amount of desiccation. They serve as a special means of asexual reproduction in the Phylactolaemata, a class of fresh-water Bryozoa. They are 0.01–0.06 in. (0.26–1.5 mm) long, and their shape and structure are important to the taxonomy of the group.

Statoblasts are classified as follows: sessoblasts, those which attach to zooecial tubes (structures of the outer layer or zooecial of an individual in the colony) or the substratum; floatoblasts and spinoblasts, both of which have a float of "air" cells and therefore are free-floating; and piptoblasts, which are free but have no float (see illustration).

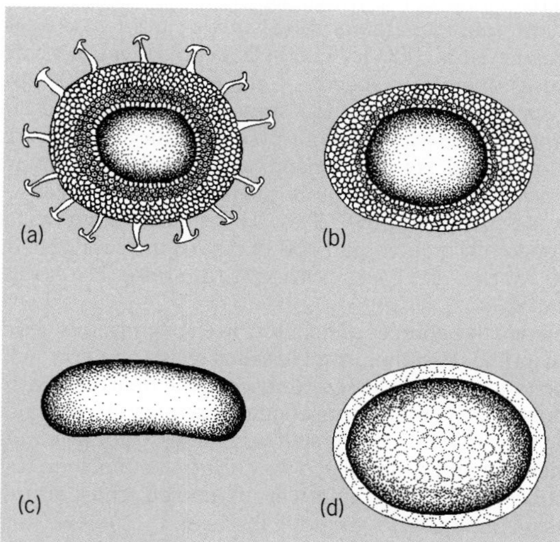

Some types of statoblasts. (*a*) Spinoblast of *Pectinatella magnifica*. (*b*) Floatoblast of *Plumatella repens*. (*c*) Piptoblast or sessoblast of *Fredericella sultana*. (*d*) Sessoblast of *Stolella indica*.

These bodies are produced in enormous quantities from spring to autumn. They develop by organization of masses of peritoneal cells and epidermal cells that bulge into the coelom. Each mass then secretes protective upper and lower chitinous valves, the rims of which often project peripherally. Statoblasts are therefore somewhat disk-shaped.

These bodies remain dormant for variable periods of time and serve to tide species over adverse ecological conditions, such as freezing or drying, which kill the colony. During this time they may be dispersed over considerable distances, being carried by animals, floating vegetation, or the action of water currents. When environmental conditions become favorable, which is usually in the spring, statoblasts germinate and a zooid develops from the mass of cells lying between the two valves. Statoblasts of *Lophopodella* have germinated after 50 months of drying. *See* BRYOZOA.

Sybil P. Parker

Staurolite

A nesosilicate mineral occurring in metamorphic rocks. The chemical formula of staurolite may be written as $A_4B_4C_{18}D_4T_8O_{40}X_8$, where A = Fe^{2+}, Mg; B = Fe^{2+}, Zn, Co, Mg, Li, Al, Fe^{3+}, Mn^{2+}; C = Al, Fe^{3+}, Cr, V, Ti; D = Al, Mg; T = Si, Al; X = OH, F, O. Staurolite occurs as well-formed, often-twinned, prismatic crystals. It is brown-black, reddish brown, or light brown in color and has a vitreous to dull luster. Light color and dull luster can result from abundant quartz inclusions. There is no cleavage, specific gravity is 3.65–3.75, and hardness is 7–7.5 (Mohs scale). *See* HARDNESS SCALES.

Typical minerals occurring with staurolite are quartz, micas (muscovite and biotite), garnet (almandine), tourmaline, and kyanite, sillimanite, or andalusite. Staurolite is common where pelitic schists reach medium-grade metamorphism. Examples are the Swiss and Italian Alps (notable at Saint Gotthard, Switzerland), and all the New England states, Virginia, the Carolinas, Georgia, New Mexico, Nevada, and Idaho. *See* METAMORPHISM; SILICATE MINERALS.

Frank C. Hawthorne

Stauromedusae

An order of the cnidarian class Scyphozoa, usually found in circumpolar regions. The egg develops into a planula which can only creep since it lacks cilia. The planula changes into a polyp that metamorphoses directly into a combined polyp and medusa form. The medusa is composed of a

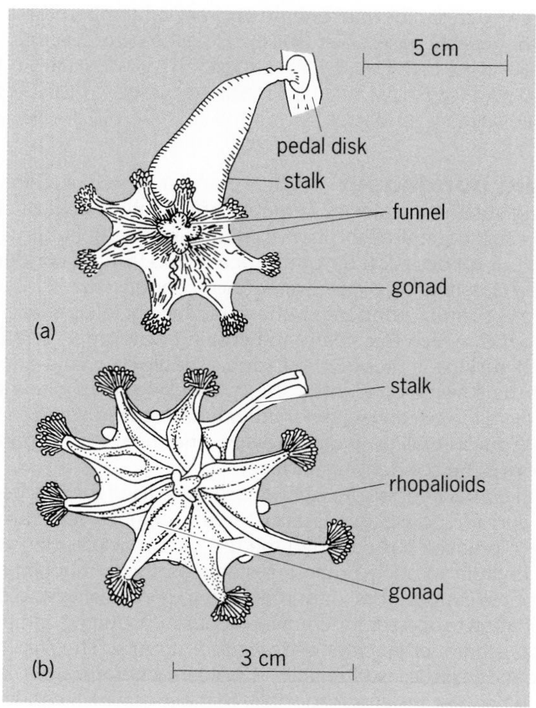

Stauromedusae. (*a*) *Lucernaria*. (*b*) *Haliclystus*. (After L. H. Hyman, *The Invertebrates*, vol. 1, McGraw-Hill, 1940)

cuplike bell called a calyx (medusan part) and a stem (polyp part) which terminates in a pedal disk (see illustration). The calyx is eight-sided and has eight groups of short, capped tentacles and eight sensory bodies, called anchors (rhopalioids), on its margin. The mouth, situated at the center of the calyx, has four thin lips and leads to the stomach in which gastral filaments are arranged in a row on either side of each interradius. Though sessile, the medusa can move in a leechlike fashion by alternate attachment and release of the pedal disk, using the substratum as an anchor. *See* SCYPHOZOA. Tohru Uchida

Steam Water vapor, or water in its gaseous state. Steam is the most widely used working fluid in external combustion engine cycles, where it will utilize practically any source of heat, that is, coal, oil, gas, nuclear fuel (uranium and thorium), waste fuel, and waste heat. It is also extensively used as a thermal transport fluid in the process industries and in the comfort heating and cooling of space. The universality of its availability and its highly acceptable, well-defined physical and chemical properties also contribute to the usefulness of steam.

The temperature at which steam forms depends on the pressure in the boiler. The steam formed in the boiler (and conversely steam condensed in a condenser) is in temperature equilibrium with the water. Under these conditions, with steam and water in contact and at the same temperature, the steam is termed saturated. Steam can be entirely vapor when it is 100% dry, or it can carry entrained moisture and be wet. After the steam is removed from contact with the liquid phase, the steam can be further heated without changing its pressure. If initially wet, the additional heat will first dry it and then raise it above its saturation temperature. This is a sensible heat addition, and the steam is said to be superheated. Superheated steam at temperatures well above the boiling temperature for the existing steam pressure follows closely the laws of a perfect gas. Chiefly because of its availability, but also because of its nontoxicity, steam is widely used as the working medium in thermodynamic processes. It has a uniquely high latent heat of vaporization. Steam has a specific heat about twice that of air and comparable to that of ammonia. The specific heat of steam is relatively high so that it can carry more thermal energy at practical temperatures than can other usable gases. *See* BOILER; BOYLE'S LAW; CHARLES' LAW; DALTON'S LAW; ENTROPY; STEAM ENGINE; STEAM-GENERATING UNIT; STEAM HEATING; STEAM TURBINE; THERMODYNAMIC CYCLE; THERMODYNAMIC PRINCIPLES; WATER. Theodore Baumeister

Steam condenser A heat-transfer device used for condensing steam to water by removal of the latent heat of steam and its subsequent absorption in a heat-receiving fluid, usually water, but on occasion air or a process fluid. Steam condensers may be classified as contact or surface condensers.

In the contact condenser, the condensing takes place in a chamber in which the steam and cooling water mix. The direct contact surface is provided by sprays, baffles, or void-effecting fill. In the surface condenser, the condensing takes place separated from the cooling water or other heat-receiving fluid (or heat sink). A metal wall, or walls, provides the means for separation and forms the condensing surface.

Both contact and surface condensers are used for process systems and for power generation serving engines and turbines. Modern practice has confined the use of contact condensers almost entirely to such process systems as those involving vacuum pans, evaporators, or dryers, and to condensing and dehumidification processes inherent in vacuum-producing equipment such as steam jet ejectors and vacuum pumps. The steam surface condenser is used chiefly in power generation but is also used in process systems, especially in those in which condensate recovery is important. Air-cooled surface condensers are used in process systems and in power generation when the availability of cooling water is limited. *See* STEAM; STEAM TURBINE; VAPOR CONDENSER. Joseph F. Sebald

Steam electric generator An alternating-current (ac) synchronous generator driven by a steam turbine for 50- or 60-Hz electrical generating systems operation. *See* STEAM TURBINE.

The synchronous generator is made of two parts: a stator (stationary) and a rotor (rotating). The stator is a cylindrical steel frame inside of which a cylindrical iron core made of thin insulated laminations is mounted on a support system. The iron core has equally spaced axial slots on its inside circumference, and wound within the core slots is a stator winding. The stator winding copper is electrically insulated from the core. The rotor consists of a forged solid steel shaft. Wound into axial slots on the outside diameter of the shaft is a copper rotor winding that is held in the slots with wedges. Retaining rings support the winding at the rotor body ends. The rotor winding, referred to as the field, is electrically insulated from the shaft and is arranged in pole pairs (always an even number) to form the magnetic field which produces the flux. The rotor shaft (supported by bearings) is coupled to the steam turbine, and rotates inside the stator core. *See* ELECTRIC ROTATING MACHINERY; WINDINGS IN ELECTRIC MACHINERY.

The stator winding (armature) is connected to the ac electrical transmission system through the bushings and output terminals. The rotor winding (field) is connected to the generator's excitation system. The excitation system provides the direct-current (dc) field power to the rotor winding via carbon brushes riding on a rotating collector ring mounted on the generator rotor. The synchronous generator's output voltage amplitude and frequency must remain constant for proper operation of electrical load devices. During operation, the excitation system's voltage regulator monitors the generator's output voltage and current. The voltage regulator controls the rotor winding dc voltage to maintain a constant generator stator output ac voltage, while allowing the stator current to vary with changes in load. Field windings typically operate at voltages between 125 and 575 V dc. The synchronous generator's output frequency is directly proportional to the speed of the rotor, and the speed of the generator rotor is held constant by a speed governor system associated with the steam turbine. *See* ALTERNATING-CURRENT GENERATOR; GENERATOR.

Synchronous generators range in size from a few kilovolt-amperes to 1,650,000 kVA. 60-Hz steam-driven synchronous generators operate at speeds of either 3600 or 1800 rpm; for 50-Hz synchronous generators these speeds would be 3000 or 1500 rpm. These two- and four-pole generators are called cylindrical rotor units. For comparison, water (hydro)-driven and air-driven synchronous generators operate at lower speeds, some as low as 62 rpm (116 poles). The stator output voltage of large (generally greater than 100,000 kVA) units ranges 13,800–27,000 V. *See* ELECTRIC POWER GENERATION; HYDROELECTRIC GENERATOR.

There are five sources of heat loss in a synchronous generator: stator winding resistance, rotor winding resistance, core, windage and friction, and stray losses. Removing the heat associated with these losses is the major challenge to the machine designer. The cooling requirements for the stator windings, rotor windings, and core increase proportionally to the cube of the machine size. The early synchronous generators were air-cooled. Later, air-to-water coolers were required to remove the heat. James R. Michalec

Steam engine A machine for converting the heat energy in steam to mechanical energy of a moving mechanism, for example, a shaft. The steam engine can utilize any source of heat in the form of steam from a boiler. Most modern machine elements had their origin in the steam engine: cylinders, pistons, piston rings, valves and valve gear crossheads, wrist pins, connecting rods, crankshafts, governors, and reversing gears. *See* BOILER; STEAM.

The twentieth century saw the practical end of the steam engine. The steam turbine replaced the steam engine as the major prime mover for electric generating stations. The internal combustion engine, especially the high-speed automotive types which burn volatile (gasoline) or nonvolatile (diesel) liquid fuel, has completely displaced the steam locomotive with the diesel locomotive and marine steam engines with the motorship and motorboat. Because of the steam engine's weight and speed limitations, it was also excluded from the aviation field. *See* DIESEL ENGINE; GAS TURBINE; INTERNAL COMBUSTION ENGINE; STEAM TURBINE.

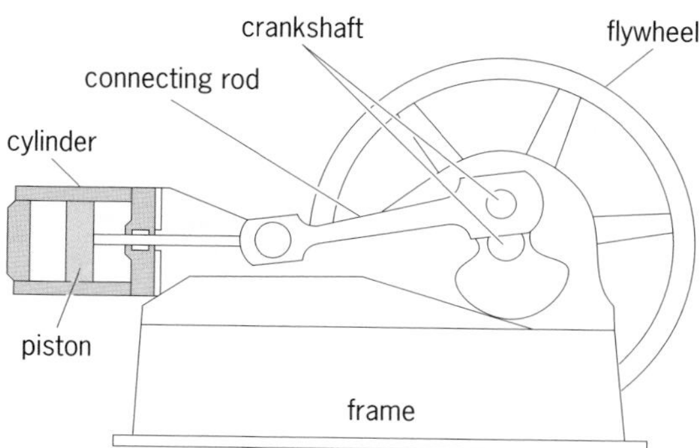

Fig. 1. Principal parts of horizontal steam engine.

A typical steam reciprocating engine consists of a cylinder fitted with a piston (Fig. 1). A connecting rod and crankshaft convert the piston's to-and-fro motion into rotary motion. A flywheel tends to maintain a constant-output angular velocity in the presence of the cyclically changing steam pressure on the piston face. A D slide valve admits high-pressure steam to the cylinder and allows the spent steam to escape (Fig. 2). The power developed by the engine depends upon the pressure and quantity of steam admitted per unit time to the cylinder.

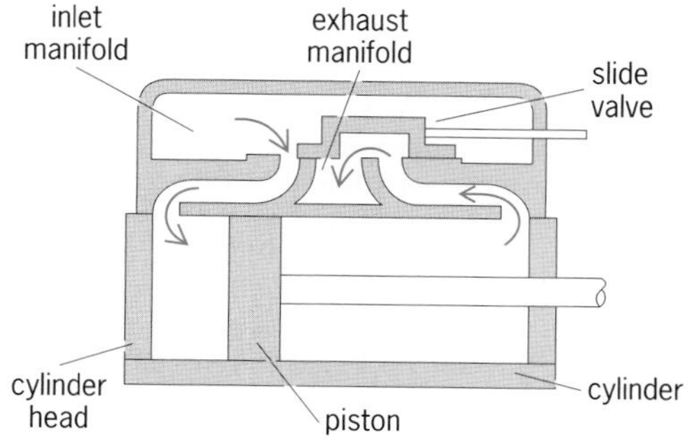

Fig. 2. Single-ported slide valve on counterflow double-acting cylinder.

Engines are classified as single- or double-acting, and as horizontal (Fig. 1) or vertical depending on the direction of piston motion. If the steam does not fully expand in one cylinder, it can be exhausted into a second, larger cylinder to expand further and give up a greater part of its initial energy. Thus, an engine can be compounded for double or triple expansion.

Steam engines can also be classed by functions, and are built to optimize the characteristics most desired in each application. Stationary engines drive electric generators, in which constant speed is important, or pumps and compressors, in which constant torque is important. Theodore Baumeister

Steam-generating furnace An enclosed space provided for the combustion of fuel to generate steam. The closure confines the products of combustion and is capable of withstanding the high temperatures developed and the pressures used. Its dimensions and geometry are adapted to the rate of heat release, to the type of fuel, and to the method of firing so as to promote complete burning of the combustible and suitable disposal of ash. In water-cooled furnaces the heat absorbed materially affects the temperature of gases at the furnace outlet and contributes directly to the generation of steam. *See* FURNACE CONSTRUCTION; STEAM; STEAM-GENERATING UNIT. George W. Kessler

Steam-generating unit The wide diversity of parts, appurtenances, and functions needed to release and utilize a source of heat for the practical production of steam at pressures to 5000 lb/in.^{2} (34 megapascals) and temperatures to 1100°F (600°C), often referred to as a steam boiler for brevity. *See* STEAM.

The essential steps of the steam-generating process include (1) a furnace for the combustion of fuel, or a nuclear reactor for the release of heat by fission, or a waste heat system; (2) a pressure vessel in which feedwater is raised to the boiling temperature, evaporated into steam, and generally superheated beyond the saturation temperature; and (3) in many modern central station units, a reheat section or sections for resuperheating steam after it has been partially expanded in a turbine. This aggregation of functions requires a wide assortment of components, which may be variously employed in the interests, primarily, of capacity and efficiency in the steam-production process. The selection, design, operation, and maintenance of these components constitute a complex process. *See* BOILER; REHEATING; SUPERHEATER. Theodore Baumeister

Steam heating A heating system that uses steam generated from a boiler. The steam heating system conveys steam through pipes to heat exchangers, such as radiators, convectors, baseboard units, radiant panels, or fan-driven heaters, and returns the resulting condensed water to the boiler. Such systems normally operate at pressure not exceeding 15 lb/in.^{2} gage or 103 kilopascals gage, and in many designs the condensed steam returns to the boiler by gravity because of the static head of water in the return piping. With utilization of available operating and safety control devices, these systems can be designed to operate automatically with minimal maintenance and attention.

In a one-pipe steam heating system, a single main serves the dual purpose of supplying steam to the heat exchanger and conveying condensate from it. Ordinarily, there is but one connection to the radiator or heat exchanger, and this connection serves as both the supply and return. A two-pipe system is provided with two connections from each heat exchanger, and in this system steam and condensate flow in separate mains and branches.

Another source for steam for heating is from a high-temperature water source (350–450°F or 180–230°C) using a high-pressure water to low-pressure steam heat exchanger. *See* BOILER; COMFORT HEATING; OIL BURNER; STEAM-GENERATING FURNACE. John W. James

Steam jet ejector A steam-actuated device for pumping compressible fluids, usually from subatmospheric suction pressure to atmospheric discharge pressure. A steam jet ejector is most frequently used for maintaining vacuum in process equipment in which evaporation or condensation takes place. Because of its simplicity, compactness, reliability, and generally low first cost, it is often preferred to a mechanical vacuum pump for removing air from condensers serving steam turbines, especially for marine service.

Two or more stages may be arranged in series, depending upon the total compression ratio required. Two or more sets of series stages may be arranged in parallel to accommodate variations in capacity. Vapor condensers are usually interposed between the compression stages of multistage steam jet ejectors to condense and remove a significant portion of the motive steam and other condensable vapors. Joseph F. Sebald

Steam separator A device for separating a mixture of the liquid and vapor phases of water. Steam separators are used in most boilers and may also be used in saturated steam lines to separate and remove the moisture formed because of heat loss.

Steam separators have many forms and may be as fundamental as a simple baffle that utilizes inertia caused by a change of direction. Most modern, high-capacity boilers use a combination of cyclone separators and steam dryers. Cyclone separators slice the steam-water mixture into thin streams so that the steam bubbles need travel only short distances through the mixture to become disengaged, and then they whirl the mixture in a circular path, creating a centrifugal force many times greater than the force of gravity. Steam dryers remove small droplets of water from the steam by providing a series of changes in direction and a large surface area to intercept the droplets. *See* BOILER FEEDWATER; STEAM; STEAM TURBINE. Earl E. Coulter

Steam temperature control Means for regulating the operation of a steam-generating unit to produce steam at the required temperature. The temperature of steam is affected by the change in the relative heat absorption as load varies, by changes in ash or slag deposits on the heat-absorbing surface, by changes in fuel, by changes in the proportioning of fuel and combustion air, or by changes in feedwater temperature. Low steam temperature lowers the efficiency of the thermal cycle. However, high steam temperature, which increases thermal efficiency, is restricted by the strength and durability of materials used in superheaters. Control of steam temperature is, therefore, a matter of primary concern in the design of modern steam-generating units.

Steam temperature can be controlled by one or more of several methods. These include (1) the damper control of gases to the superheater, to the reheater, or to both, thus changing the heat input; (2) the recirculation of low-temperature flue gas to the furnace, thus changing the relative amounts of heat absorbed in the furnace and in the superheater, reheater, or both; (3) the selective use of burners at different elevations in the furnace or the use of tilting burners, thus changing the location of the combustion zone with respect to the furnace heat-absorbing surface; (4) the attemperation, or controlled cooling, of the steam by the injection of spray water or by the passage of a portion of the steam through a heat exchanger submerged in the boiler water; (5) the control of the firing rate in divided furnaces; and (6) the control of the firing rate relative to the pumping rate of the feedwater to forced-flow once-through boilers. Generally, these various controls are adjusted automatically. *See* STEAM-GENERATING UNIT. George W. Kessler

Steam turbine A machine for generating mechanical power in rotary motion from the energy of steam at temperature and pressure above that of an available sink. By far the most widely used and most powerful turbines are those driven by steam. Until the 1960s essentially all steam used in turbine cycles was produced in boilers burning fossil fuels (coal, oil, and gas) or, in minor quantities, certain waste products. However, modern turbine technology includes nuclear steam plants as well as production of steam supplies from other sources. *See* NUCLEAR REACTOR.

The illustration shows a small, simple mechanical-drive turbine of a few horsepower. It illustrates the essential parts for all steam turbines regardless of rating or complexity: (1) a casing, or shell, usually divided at the horizontal center line, with the halves bolted together for ease of assembly and disassembly; it contains the stationary blade system; (2) a rotor carrying the moving buckets (blades or vanes) either on wheels or drums, with bearing journals on the ends of the rotor; (3) a set of bearings

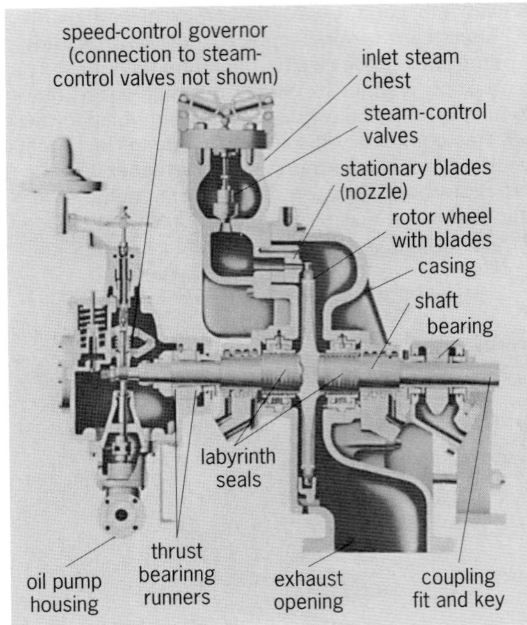

Cutaway of small, single-stage steam turbine. (*General Electric Co.*)

attached to the casing to support the shaft; (4) a governor and valve system for regulating the speed and power of the turbine by controlling the steam flow, and an oil system for lubrication of the bearings and, on all but the smallest machines, for operating the control valves by a relay system connected with the governor; (5) a coupling to connect with the driven machine; and (6) pipe connections to the steam supply at the inlet and to an exhaust system at the outlet of the casing or shell.

Steam turbines are ideal prime movers for applications requiring rotational mechanical input power. They can deliver constant or variable speed and are capable of close speed control. Drive applications include centrifugal pumps, compressors, ship propellers, and, most importantly, electric generators.

Steam turbines are classified (1) by mechanical arrangement, as single-casing, cross-compound (more than one shaft side by side), or tandem-compound (more than one casing with a single shaft); (2) by steam flow direction (axial for most, but radial for a few); (3) by steam cycle, whether condensing, noncon-densing, automatic extraction, reheat, fossil fuel, or nuclear; and (4) by number of exhaust flows of a condensing unit, as single, double, triple flow, and so on. Units with as many as eight exhaust flows are in use. *See* TURBINE. Frederick G. Baily

Steel Any of a great number of alloys that contain the element iron as the major component and small amounts of carbon as the major alloying element. These alloys are more properly referred to as carbon steels. Small amounts, generally on the order of a few percent, of other elements such as manganese, silicon, chromium, molybdenum, and nickel may also be present in carbon steels. However, when large amounts of alloying elements are added to iron to achieve special properties, other designations are used to describe the alloys. For example, large amounts of chromium, over 12%, are added to produce the important groups of alloys known as stainless steels. *See* STAINLESS STEEL.

Low-carbon steels, sometimes referred to as mild steels, usually contain less than 0.25% carbon. These steels are easily hot-worked and are produced in large tonnages for beams and other structural applications. The relatively low strength and high ductility of the low-carbon steels make it possible also to cold-work these steels. Cold-rolled low-carbon steels are extensively used for sheet applications in the appliance and automotive industries. Cold-rolled steels have excellent surface finishes, and both hot- and cold-worked mild steels are readily welded.

Medium-carbon steels contain between 0.25 and 0.70% carbon, and are most frequently used in the heat-treated condition for machine components that require high strength and good fatigue resistance.

Steels containing more than 0.7% carbon are in a special category because of their high hardness and low toughness. This combination of properties makes the high-carbon steels ideal for bearing applications where wear resistance is important and the compressive loading minimizes brittle fracture that might develop on tensile loading. George Krauss

Steel manufacture A sequence of operations in which pig iron and scrap steel are processed to remove impurities and are separated into the refined metal and slag.

Reduction of iron ores by carbonaceous fuel directly to a steel composition was practiced in ancient times, but liquid processing was unknown until development of the crucible process, in which iron ore, coal, and flux materials were melted in a crucible to produce small quantities of liquid steel. Modern steelmaking processes began with the invention of the airblown converter by H. Bessemer in 1856. The Thomas process was developed in 1878; it modified the Bessemer process to permit treatment of high-phosphorus pig iron. The Siemens-Martin process, also known as the open-hearth process, was developed at about the same time. The open-hearth process utilizes regenerative heat transfer to preheat air used with a burner; it can generate sufficient heat to refine solid steel scrap and pig iron in a reverberatory furnace. After World War II, various oxygen steelmaking processes were developed.

Steelmaking can be divided into acid and basic processes depending upon whether the slag is high in silica (acid) or high in lime (basic). The furnace lining in contact with the slag should be a compatible material. A silica or siliceous material is used in acid processes, and a basic material such as burned dolomite or magnesite is used in basic processes. Carbon, manganese, and silicon, the principal impurities in pig iron, are easily oxidized and separated; the manganese and silicon oxides go into the slag, and the carbon is removed as carbon monoxide and carbon dioxide in the off-gases. Phosphorus is also oxidized but does not separate from the metal unless the slag is basic. Removal of sulfur occurs to some extent by absorption in a basic slag. Thus, the basic steelmaking processes are more versatile in terms of the raw materials they can handle, and have become the predominant steelmaking processes.

A typical pig iron charged to the steelmaking process might contain roughly 4% carbon, 1% manganese, and 1% silicon. The phosphorus and sulfur levels in the pig iron vary. The composition of the steel tapped from the steelmaking furnace generally ranges from 0.04 to 0.80% carbon, 0.06 to 0.30% manganese, 0.01 to 0.05% phosphorus, and 0.01 to 0.05% sulfur, with negligible amounts of silicon.

Electric arc furnace technology began late in the nineteenth century with the original design of P. L. T. Heroult. The three-graphite electrode furnace with a swinging roof for top charging and a rocker base for tilting to tap the finished molten steel has been continuously improved and developed further. *See* ELECTRIC FURNACE.

The rapid development of steelmaking technology using the electric arc furnace, not only for alloy and stainless steels but especially for carbon steel production, has increased its share of production capacity to about 20% of the steel industry.

Remelting and refining of special alloys are carried out in duplex or secondary processes; the principal ones are argon-oxygen decarburization, electroslag refining, vacuum arc remelting, and vacuum induction melting. *See* ELECTROMETALLURGY; VACUUM METALLURGY.

Ladle metallurgy was used first to produce high-quality steels, but has been extended to producing many grades of steel because of the economic advantages of higher productivity. The purpose of these ladle treatments is to produce clean steel; introduce reactive additions, such as calcium or rare earths; add alloying additions, as for microalloyed steels, with high recovery; and increase furnace utilization, allowing higher-productivity smelting operations of the blast furnace, and melting and refining operations in steelmaking. Ladle treatments in steel production generally are classified as synthetic slag systems; gas stirring or purging; direct immersion of reactants, such as rare earths; lance injection of reactants; and wire feeding of reactants. These are often used in combination to produce synergistic effects, for example, synthetic slag and gas stirring for desulfurization followed by direct immersion, injection, or wire feeding for inclusion shape control. *See* METAL CASTING; PYROMETALLURGY; REFRACTORY; STAINLESS STEEL; STEEL. Robert D. Pehlke

Stellar evolution The large-scale, systematic, and irreversible changes with time of the structure and composition of a star.

Star formation. Stars are born from compact knots within dark molecular clouds that are refrigerated by dust that blocks heating starlight. If the random knots, compressed by supernovae or other means, are dense enough, they can contract under their own gravity. Conservation of angular momentum demands that as they collapse they must spin faster. *See* ANGULAR MOMENTUM; MOLECULAR CLOUD.

High-speed particles (cosmic rays) from exploding stars partially ionize the dusty knots. The ions grab onto the weak magnetic field of the Galaxy and, as a result of their physical interaction with neutral atoms and molecules, provide the initial means to slow the rotation. If the rotation is still too fast, the contracting body may split into a double (or more complex) star, though the origins of doubles are not clearly solved. A contracting protostar still indeed rotates progressively faster until the part of its mass not accreted by the star itself is spun out into a dusty disk, from which planets might later accumulate. From the disk shoot powerful molecular flows that slow the star still more. *See* PROTOSTAR.

When the protostar's interior reaches about 10^6 K (1.8×10^6 °F), it can fuse its internal deuterium. That and convection, which brings in fresh deuterium from outside the nuclear-burning zone, bring some stability, and a star can now be said to be born. Stars like the Sun shrink at constant temperature until deuterium fusion dies down. Then they heat at roughly constant luminosity until the full proton-proton chain begins, which provides the stars' luminosity and stops the contraction. The stars settle onto the zero-age main sequence. The whole process takes only 10 or so million years.

Early evolution. The main sequence is that zone on the Hertzsprung-Russell (HR) diagram in which stars are stabilized against gravitational contraction by fusion. The higher the stellar mass, the greater the internal compression and temperature, and the more luminous the star. Hydrogen fusion is highly sensitive to temperature, a small increase in stellar mass meaning a much higher fusion rate. Although greater mass means a greater nuclear-burning core mass and therefore a larger fuel supply, the increased fusion rate is more than offsetting and thereby shortens stellar life. While the new Sun was destined to survive on the main sequence for 10^{10} years, a 0.1-solar-mass star will live there for 10^{13} years, while a 100-solar-mass star will exhaust its core hydrogen in only 2.5×10^6 years.

The main sequence is divided into three parts. Below 0.8 solar mass (roughly class G8), no star has ever had time to evolve.

Between 0.8 solar mass and around 10 solar masses (classes G8 to B1), the stars die as white dwarfs. Above it (classes O and B0), they explode. Binary stars contribute further to the richness of stellar phenomena.

Intermediate-mass evolution. Main-sequence life lasts until the core hydrogen is almost gone, at which time hydrogen fusion rather suddenly shuts down. With no support, the now-quiet helium core can contract more rapidly under gravity's force. It heats, causing hydrogen fusion to spread into a thick enclosing shell that runs on the carbon cycle. With a new (though temporary) energy source, the star first dims some while it expands and cools at the surface, changing its spectrum to class K. The transition takes only a few hundred million years or less, leaving few stars in the middle of the HR diagram, the lower masses appearing as F, G, and K subgiants.

As core contraction proceeds beyond the rightward transition in the HR diagram, stars from about 1 to 5 solar masses (still fusing hydrogen in a shell) suddenly and dramatically increase their luminosities. The future Sun will eventually grow 1000 times brighter than it is today, and a 5-solar-mass star (which begins at about 600 solar luminosities) will reach nearly 3000 solar. At the same time, the stars swell to become red giants. While the core (roughly half a solar mass) of the future Sun shrinks to the size of Earth, the radius will expand to that of Mercury's orbit, or even beyond. When the core temperature climbs to 10^8 K, helium nuclei (alpha particles) begin fusing to unstable beryllium (^{8}Be), which quickly decays back to alpha particles, setting up an equilibrium. The tiny amount of ^{8}Be present reacts with additional alpha particles to create carbon via the triple alpha process:

$$^4\text{He} + {}^4\text{He} \leftrightarrow {}^8\text{Be}$$

$$^8\text{Be} + {}^4\text{He} \rightarrow {}^{12}\text{C} + \text{gamma ray}$$

Fusion with additional helium nuclei creates oxygen and even neon. *See* NUCLEOSYNTHESIS.

The star, now stabilized by a helium-burning core that is surrounded by a hydrogen-fusing shell, retreats about halfway down the red giant branch. The numerous lower-mass stars reside in the class K "red giant clump" (see illustration). Low-mass metal-deficient globular cluster stars that have suffered different rates of mass loss spread out from the clump toward higher temperatures to create the distinctive horizontal branch. Energy-generating fusion reactions will try to proceed toward iron (the most stable of all nuclei). Some 80% of the energy is generated in hydrogen burning, so (discounting further burning modes) the helium-burning stage lasts only around 20% of the main-sequence lifetime. From 5 solar masses up, evolution proceeds similarly, but instead of settling into a distinct location on the HR diagram, the stars loop to the blue (higher temperatures) where they fuse helium as class G, F, and A giants. *See* GIANT STAR.

Asymptotic giant branch (AGB). When the helium has fused to carbon and oxygen, the core again contracts. Helium fusion spreads outward into a shell, and for the second time the star climbs the HR diagram's giant branch. Since the second climb is roughly asymptotic to the first, the second climb creates the asymptotic giant branch. The shrinking carbon-oxygen core is now surrounded by a shell of fusing helium, while the old hydrogen-fusing shell expands, cools, and shuts down. Eventually, however, the helium shell runs out of fuel, and the hydrogen shell reignites. Hydrogen burning feeds fresh helium into the space between it and the carbon core, and when there is enough of it, helium burning reignites explosively in a helium flash (or thermal pulse) that can affect the star's surface. The flash squelches hydrogen fusion, and the whole process starts again, helium flashes coming at progressively shorter intervals. AGB stars become larger and brighter than before, passing into the cool end of class M, where they eventually become unstable

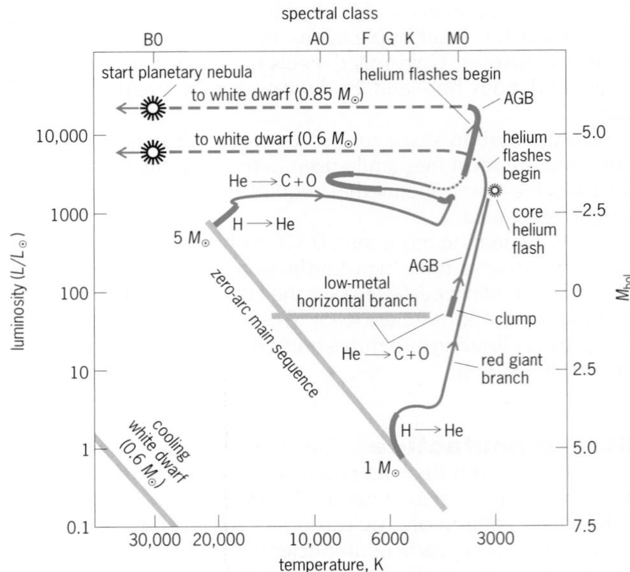

Hertzsprung-Russell diagram of evolution of intermediate-mass stars. These stars evolve from the main sequence to become giants that stabilize during helium fusion in a "clump." When helium fusion is done, they brighten again as AGB stars, lose their envelopes, and evolve to the left to produce planetary nebulae, finally at lower left becoming white dwarfs. (*After J. B. Kaler, Stars, W. H. Freeman, 1992, 1993, from work of I. Iben, Jr.*)

enough to pulsate as long-period variables (Miras). The Sun will become 5000 times brighter than now and will reach out to the Earth's orbit. *See* MIRA; VARIABLE STAR.

Mass loss and planetary nebulae. During the giant stages, stellar winds greatly increase. Mira pulsations cause shock waves that help drive mass from the stellar surfaces, where the cooled gas becomes ever richer in molecules, some even condensing into dust grains. High luminosity pushes the dust outward, and the dust couples with the gas, resulting in slow (10 km/s or 6 mi/s) thick winds tens of millions of times stronger than the solar wind (up to 10^{-5} solar mass per year). Because stars are in this state for hundreds of thousands of years, they will lose much of themselves back into space—the Sun about half of its mass, a 10-solar-mass star over 80%.

So much mass is lost that an evolving star becomes stripped nearly to its fusion zone, which is protected from the outside by a low-mass hydrogen envelope. As the inner region becomes exposed, the wind diminishes in mass but increases in speed and temperature. Hammering at the surrounding dusty, molecule-filled shroud of lost mass, the high-speed wind compresses the inner edge into a thick ring. Eaten away from the top by the wind and from below by fusion, the stellar envelope shrinks, slowly exposing the hot shell-core structure beneath. When the stripped star's surface reaches 25,000 K (45,000°F), the dense ring that its high-speed wind had previously created is ionized. Subsequent recapture of electrons by ions, along with collisional excitation of heavy atoms, causes the shell to glow, and a planetary nebula is born. *See* PLANETARY NEBULA.

The star inside first heats at constant luminosity to over 100,000 K (180,000°F), the luminosity and final temperature depending on the old core's mass (which ranges from around 0.5 solar masses to nearly 1.4 solar masses, the Chandrasekhar limit, above which white dwarfs cannot exist). As residual nuclear fusion shuts down, the star cools and dims at constant radius to become a white dwarf. *See* WHITE DWARF STAR.

High-mass evolution. As the mass of a star increases, so does the mass of the core. The Sun will turn into a white dwarf

Mosaic image taken by the *Hubble Space Telescope* of the Crab Nebula, the remnant of a supernova explosion recorded by Chinese and Japanese astronomers in 1054. The image was assembled from 24 Wide Field and Planetary Camera 2 exposures in 1999 and 2000. The blue interior glow is synchrotron radiation from high-energy electrons accelerated by the Crab's central pulsar. Colors in the intricate filaments trace the light from atoms of hydrogen, oxygen, and sulfur expelled during the explosion. (*NASA; ESA; J. Hester and A. Loll, Arizona State University*)

Mosaic of the Orion Nebula assembled from the *Hubble Space Telescope*'s Advanced Camera for Surveys in one of the most detailed astronomical images ever produced. The nebula is a gas cloud excited to incandescence by hot young stars in its interior, at the edge of an immense molecular cloud. At 1500 light-years, it is the nearest large star-forming region, and its energetic stars have blown away obscuring gas and dust. A range of star-formation stages is observed, from massive, young stars that are shaping the nebula to pillars of dense gas that may be the homes of budding stars. (*NASA; ESA; M. Robberto, Space Telescope Science Institute/ESA; Hubble Space Telescope Orion Treasury Project Team*)

of around 0.6 solar mass. At 10 solar masses, the core reaches the Chandrasekhar limit, and the star cannot become a white dwarf. At first, high-mass evolution proceeds similarly to that of stars of lower mass. As high-mass stars use their core hydrogen, they too migrate to the right on the HR diagram, becoming not so much brighter but larger, cooling at their surfaces and turning into supergiants. Though almost all these supergiants vary to some extent, the most massive become unstable and undergo huge eruptions. *See* SUPERGIANT STAR.

While intermediate-star fusion stops at carbon and oxygen, supergiants continue onward. The carbon-oxygen core shrinks and heats to the point that carbon fusion can begin, and then carbon and oxygen convert to a more complex mix dominated by oxygen, neon, and magnesium. Helium fusion now continues in a surrounding shell that is nested in one that is fusing hydrogen. Once carbon burning has run its course, the unsupported oxygen-neon-magnesium core shrinks and heats, and now it is carbon burning's turn to move outward into a shell. When hot enough, the oxygen-neon-magnesium mix fires up to burn to one dominated by silicon and sulfur. Continuing the process, the developed silicon and sulfur core finally becomes hot enough to fuse to iron, the silicon-burning core wrapped in oxygen-neon-magnesium, helium, and hydrogen-burning shells.

Supernova. Each nuclear fusion stage generates less energy, and since each takes on the role of supporting the star, each lasts a shorter period of time. While hydrogen fusion takes millions of years, the iron core is created from silicon fusion in a matter of weeks. Iron cannot fuse and produce energy. The core, about 1.5 solar masses and the size of Earth, suddenly collapses at a speed a good fraction that of light. The iron atoms are broken back to neutrons and protons. Under crushing densities, free electrons merge with protons to make yet more neutrons. When the collapsing neutron core hits nuclear densities of 10^{14} g/cm^3 (10^{14} times the density of water), it violently bounces, and the rebound tears away all the outer layers. From the outside, the observer sees the explosion as a type II supernova that can reach an absolute magnitude of -18. *See* SUPERNOVA.

The debris of the explosion, the supernova remnant, highly enriched in iron and other heavy elements, expands for centuries into space. Exposed by the explosion is the compact hot neutron star. Only 25 km (15 mi) across, it is stabilized by the pressure of neutron degeneracy. Conservation of angular momentum makes the little star spin dozens of times per second, its magnetic field compacted to 10^{12} or more times stronger than Earth's. Radiation beams out along a tilted, wobbling magnetic axis, and if Earth is in its path, a "pulse" of radiation is observed, the neutron star now a "pulsar." The most highly magnetized neutron stars, the magnetars, can have magnetic field strengths greater than 10^{14} times that of Earth, and are related to anomalous x-ray pulsars (AXPs) and to soft gamma-ray repeaters (SGRs), the latter occasionally releasing bursts so powerful that they can affect the Earth. *See* NEUTRON STAR; PULSAR.

Beyond the Chandrasekhar limit of solar masses, electron degeneracy pressure cannot stabilize a white dwarf, and it must collapse. Neutron stars are similarly limited to around 3 solar masses. Stars at the upper end of the main sequence are expected to create iron cores that exceed even this limiting mass. The dense remains cannot stabilize, and they too must therefore collapse. When such a collapsing star passes a critical radius at which the escape velocity is about that of light, it become invisible, and a black hole is born. About half a dozen black hole candidates are recognized in binary systems in which the black hole affects the companion. *See* BLACK HOLE.

James B. Kaler

Stellar magnetic field A magnetic field, far stronger than the Earth's magnetic field, which is possessed by many stars. Magnetic fields are important throughout the life cycle of a star. Initially, magnetic fields regulate how quickly interstellar clouds collapse into protostars. Later in the star formation process, circumstellar disk material flows along magnetic field lines, either accreting onto the star or flowing rapidly out along the rotation axis. Outflowing material (stellar winds) carries away angular momentum, slowing rotation at a rate that depends on stellar magnetic field strength. On the Sun, dark sunspots, prominences, flares, and other forms of surface activity are seen in regions where there are strong magnetic fields. There is some evidence that long-term variations in solar activity may affect the Earth's climate. Turbulence in the solar atmosphere drives magnetic waves which heat a tenuous corona (seen during eclipses) to millions of degrees. About 10% of hotter stars (with temperatures of about 10,000 K) with stable atmospheres are Ap stars, which have stronger magnetic fields that control the surface distribution of exotic elements. Even after stars end their internal fusion cycle and become compact remnants, magnetic fields channel accreting material from binary companions, occasionally producing spectacular novae. Despite the enormous gravity around pulsars, magnetic forces far exceed gravitational forces, creating intense electromagnetic beams that spin down (slow the rotation of) the pulsar. *See* BINARY STAR; CATACLYSMIC VARIABLE; NOVA; PROTOSTAR; PULSAR; SOLAR MAGNETIC FIELD; STAR; STELLAR EVOLUTION.

Jeff A. Valenti

Stellar population One of the categories into which stars may be classified, based on their place in the evolution of the galaxy containing them. The stellar component of the Milky Way Galaxy consists of three populations: the thin disk, the thick disk, and the halo.

The thin disk, originally referred to as population I, is the youngest component of the galactic stellar population. Still actively forming massive stars from molecular clouds, it is confined to within about 0.35 kiloparsec of the plane. (1 kpc equals 3300 light-years or 3.1×10^{16} km or 1.9×10^{16} mi.) All of the stars have metallicities lying between about one-fifth and twice the solar value, and star formation appears to have remained constant in this population for about the past 8×10^9 years. One reason for the relatively small thickness of the disk is the low velocity dispersion of the component stars; their motion is completely dominated by the differential rotation of the disk. These stars are found associated with H II regions and OB associations as well as open clusters. *See* INTERSTELLAR MATTER; MOLECULAR CLOUD; SUPERNOVA.

The thick disk is an older population, approximately 9–10 $\times$ 10^9 years, roughly corresponding to the range between what was once called population II and population I. Its metallicity lies between about one-tenth and one-third of the solar value. The stars in this population are distributed over greater distances from the plane, up to 1.5 kpc, and have correspondingly larger velocity dispersion. This population also includes globular clusters and subdwarfs that overlap at the lowest end of the abundances with the properties of the halo globulars, although the system of old disk globulars is distributed differently than those of the halo.

Lying around the disk and the nuclear spheroidal bulge, there is a halo, roughly corresponding to the original population II, that extends to considerable distances from the plane, some as distant as 30 kpc. This population has an age of order 10–15 $\times$ 10^9 years. The stars in this region have very large velocity dispersions and do not appear to participate in the differential rotation as much as other stars. Their metallicities are all lower than about one-twentieth that of the Sun and may extend down to 10^{-3} of the solar value. The most metal-poor globular clusters belong to this population. This stellar halo is not the same as the dark matter halo, but is probably embedded within it. *See* MILKY WAY GALAXY; STAR.

Steven N. Shore

Stellar rotation The spinning of stars, due to their angular momentum. Stars do not necessarily rotate as solid bodies, and their angular momentum may be distributed nonuniformly, depending on radius or latitude. However, it is nearly impossible to resolve the surfaces of stars, or see their interiors, and the limited ability to observe them means that their rotation is generally expressed as a single number, $v_{eq} \sin i$. In this measured quantity, v_{eq} is the star's equatorial rotational velocity (in kilometers per second), and i is the angle between the star's rotation axis and the line of sight to the star. In other words, $v_{eq} \sin i$ is the component of a star's rotation that is projected onto the line of sight between the star and the observer. Measurements of $v_{eq} \sin i$ in stars range from as little as 1 km s^{-1} (0.6 mi s^{-1}) up to 400 km s^{-1} (250 mi s^{-1}) or more. A more physically useful measure of rotation is Ω, a star's angular velocity, or P_{rot}, the rotation period (the inverse of Ω). In some cases it is possible to measure P_{rot} directly. *See* ANGULAR MOMENTUM; STAR.

A $v_{eq} \sin i$ value is determined from the breadth of absorption lines in the star's spectrum. The Doppler effect causes the lines to be broadened because one limb of the star is receding and the other approaching. *See* DOPPLER EFFECT.

Late-type stars (cooler than about 6500 K or 11,200 °F) exhibit spots on their surfaces analogous to sunspots, and these can be large enough to produce observable changes in the light of the star. In such cases it is possible to measure P_{rot} directly and, in a few instances, changes in P_{rot} have led to estimates of differential rotation (the dependence of Ω on latitude).

There is a broad trend in which typical rotation rates decline with the mass of the star in going down the main sequence. There is an especially prominent break in rotation such that stars with masses below about 1.2 times the solar mass have very low rotation rates, due to their intrinsic structure.

The early-type stars (types O, B, A, and early-F) typically have $v_{eq} \sin i$ values of 50 km s^{-1} (30 mi s^{-1}) or more. The high $v_{eq} \sin i$ values (200 km s^{-1} or 125 mi s^{-1} and more) are seen only in the main-sequence O stars.

Late-type stars are cooler than about 6500 K (11,200 °F) and have spectral types of late-F, G, K, or M. The key property of late-type stars that determines much of their phenomenology is the presence of a convective envelope. In such stars, rotation (especially differential rotation) interacts with the convection to produce complex motions in the electrically conductive plasma of the star. These circulatory patterns enable the star to regenerate a seed magnetic field, the so-called dynamo mechanism. The magnetic field can grip an ionized wind beyond the star's surface, leading to gradual angular momentum loss, a process that is seen occurring on the Sun. More rapidly rotating stars produce stronger magnetic fields. *See* MAGNETISM; STELLAR MAGNETIC FIELD.

In this scenario, late-type main-sequence stars spin down over their lifetimes, and that is observed to occur because steadily declining rotation rates can be seen among stars in clusters of increasing age. Moreover, stars that are born together but with different rotation rates will tend to end up with the same P_{rot} because the faster-rotating star will lose angular momentum more quickly. Young stars tend to rotate rapidly because they have not had time to lose the angular momentum with which they formed, while old low-mass stars rotate slowly.

Significant numbers of very low mass stars and brown dwarfs have been found and studied at high spectroscopic resolution. These objects are fully convective from their surfaces all the way to their cores, in contrast to the Sun and similar stars which have a convective envelope. This change in structure corresponds to a change in the observed rotation in that very low mass stars and brown dwarfs appear not to lose angular momentum and so continue to spin rapidly. These objects are very small in size, so the rotation rates seen—15-60 km s^{-1} (10-40 mi s^{-1})—imply high angular velocities of as much as 100 times that of the Sun. *See* BROWN DWARF.

The Sun is a very slowly rotating star ($v_{eq} \sin i = 1.8$ km s^{-1} or 1.1 mi s^{-1}). Over its 4.5-billion-year main-sequence lifetime, it has gradually lost angular momentum to get to the rate that is seen today. Helioseismological studies of the Sun's interior show it to rotate as a solid body, so if it ever had a rapidly rotating core, it does no longer. *See* HELIOSEISMOLOGY; SUN.

David Soderblom

Stem The organ of vascular plants that usually develops branches and bears leaves and flowers. On woody stems a branch that is the current season's growth from a bud is called a twig. The stems of some species produce adventitious roots. *See* ROOT (BOTANY).

General characteristics. While most stems are erect, aerial structures, some remain underground, others creep over or lie prostrate on the surface of the ground, and still others are so short and inconspicuous that the plants are said to be stemless, or acaulescent. When stems lie flattened immediately above but not on the ground, with tips curved upward, they are said to be decumbent, as in juniper. If stems lie flat on the ground but do not root at the nodes (joints), the stem is called procumbent or prostrate, as in purslane. If a stem creeps along the ground, rooting at the nodes, it is said to be repent or creeping, as in ground ivy. Most stems are cylindrical and tapering, appearing circular in cross section; others may be quadrangular or triangular.

Herbaceous stems (annuals and herbaceous perennials) die to the ground after blooming or at the end of the growing season. They usually contain little woody tissue. Woody stems (perennials) have considerable woody supporting tissue and live from year to year. A woody plant with no main stem or trunk, but usually with several stems developed from a common base at or near the ground, is known as a shrub. Nelle Ammons

External features. A shoot or branch usually consists of a stem, or axis, and leafy appendages. Stems have several distinguishing features. They arise either from the epicotyl of the embryo in a seed or from buds. The stem bears both leaves and buds at nodes, which are separated by leafless regions or internodes, and sometimes roots and flowers (see illustration).

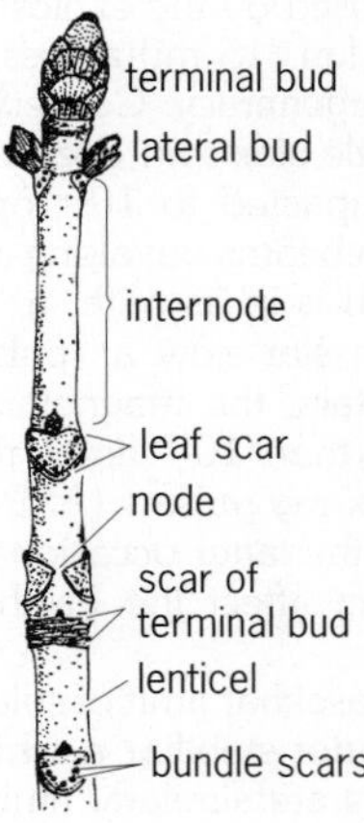

Winter woody twig (horse chestnut) showing apical dominance. (*After E. W. Sinnott and K. S. Wilson, Botany: Principles and Problems, 5th ed., McGraw-Hill, 1955*)

The nodes are the regions of the primary stem where leaves and buds arise. The number of leaves at a node is usually specific for each plant species. In deciduous plants which are leafless during winter, the place of former attachment of a leaf is marked by the leaf scar. The scar is formed in part by the abscission zone formed at the base of the leaf petiole. The stem regions between nodes are called internodes. Internode length varies greatly among species, in different parts of the same stem, and under different growing conditions.

Lenticels are small, slightly raised or ridged regions of the stem surface that are composed of loosely arranged masses of cells in the bark. Their intercellular spaces are continuous with those in the interior of the stem, therefore permitting gas exchange similar to the stomata that are present before bark initiation.

There are three major types of stem branching: dichotomous, monopodial, and sympodial. Dichotomy occurs by a division of the apical meristem to form two new axes. If the terminal bud of an axis continues to grow and lateral buds grow out as branches, the branching is called monopodial. If the apical bud terminates growth in a flower or dies back and one or more axillary buds grow out, the branching is called sympodial. Often only one bud develops so that what appears to be single axis is in fact composed of a series of lateral branches arranged in linear sequence.

The large and conspicuous stems of trees and shrubs assume a wide variety of distinctive forms. Columnar stems are basically unbranched and form a terminal leaf cluster, as in palms, or lack obvious leaves, as in cacti. Branching stems have been classified as excurrent, when there is a central trunk and a conical leaf crown, as in firs and other conifers, or as decurrent (or deliquescent), when the trunk quickly divides up into many separate axes so that the crown lacks a central trunk, as in elm. *See* TREE.

Jack B. Fisher

Internal features. The stem is composed of the three fundamental tissue systems that are found also in all other plant organs: the dermal (skin) system, consisting of epidermis in young stems and periderm in older stems of many species; the vascular (conducting) system, consisting of xylem (water conduction) and phloem (food conduction); and the fundamental or ground tissue system, consisting of parenchyma and sclerenchyma tissues in which the vascular tissues are embedded. The arrangement of the vascular tissues varies in stems of different groups of plants, but frequently these tissues form a hollow cylinder enclosing a region of ground tissue called pith and separated from the dermal tissue by another region of ground tissue called cortex. *See* CORTEX (PLANT); EPIDERMIS (PLANT); PARENCHYMA; PHLOEM; PITH; SCLERENCHYMA; XYLEM.

Part of the growth of the stem results from the activity of the apical meristem located at the tip of the shoot. The derivatives of this meristem are the primary tissues: epidermis, primary vascular tissues, and the ground tissues of the cortex and pith. In many species, especially those having woody stems, secondary tissues are added to the primary. These tissues are derived from the lateral meristems, oriented parallel with the sides of the stem: cork cambium (phellogen), which gives rise to the secondary protective tissue periderm, which consists of phellum (cork), phellogen (cork cambium), and phelloderm (secondary cortex) and which replaces the epidermis; and vascular cambium, which is inserted between the primary xylem and phloem and forms secondary xylem (wood) and phloem. *See* APICAL MERISTEM; LATERAL MERISTEM.

The vascular tissues and the closely associated ground tissues—pericycle (on the outer boundary of vascular region), interfascicular regions (medullary or pith rays), and frequently also the pith—may be treated as a unit called the stele. The variations in the arrangement of the vascular tissues serve as a basis for distinguishing the stelar types. The word stele means column and thus characterizes the system of vascular and associated ground tissues as a column. This column is enclosed within the cortex, which is not part of the stele. *See* PERICYCLE.

James E. Gunckel

Stem cells

Stem cells Cells that have the ability to self-replicate and to give rise to mature cells. The concept of stem cells was originally based on renewing tissues. Self-renewal, together with the capacity for differentiation, defines the properties of stem cells.

Stem cells can be found at different stages of fetal development and are present in a wide range of adult tissues. Many of the terms used to distinguish stem cells are based on their origins and the cell types of their progeny. There are three basic types of stem cells. Totipotent stem cells, meaning that their potential is total, have the capacity to give rise to every cell type of the body and to form an entire organism. Pluripotent stem cells, such as embryonic stem cells, are capable of generating virtually all cell types of the body but are unable to form a functioning organism. Multipotent stem cells can give rise only to a limited number of cell types. For example, adult stem cells, also called organ- or tissue-specific stem cells, are multipotent stem cells found in specialized organs and tissues after birth. Their primary function is to replenish cells lost from normal turnover or disease in the specific organs and tissues in which they are found. *See* CELL DIFFERENTIATION; HISTOGENESIS.

Totipotent stem cells occur at the earliest stage of embryonic development. The union of sperm and egg creates a single totipotent cell. This cell divides into identical cells in the first hours after fertilization. All these cells have the potential to develop into a fetus when they are placed into the uterus. The first differentiation of totipotent cells forms a sphere of cells called the blastocyst, which has an outer layer of cells and an inner cell mass. The outer layer of cells will form the placenta and other supporting tissues during fetal development, whereas cells of the inner cell mass go on to form all three primary germ layers: ectoderm, mesoderm, and endoderm. The three germ layers are the embryonic source of all types of cells and tissues of the body. Embryonic stem cells are derived from the inner cell mass of the blastocyst. They retain the capacity to give rise to cells of all three germ layers. However, embryonic stem cells cannot form a complete organism because they are unable to generate the entire spectrum of cells and structures required for fetal development. Thus, embryonic stem cells are pluripotent, not totipotent, stem cells. *See* EMBRYOLOGY; EMBRYONIC DIFFERENTIATION; GERM LAYERS.

Embryonic germ cells differ from embryonic stem cells in the tissue sources from which they are derived, but appear to be similar to embryonic stem cells in their pluripotency. Human embryonic germ cell lines are established from the cultures of the primordial germ cells obtained from the gonadal ridge of late-stage embryos, a specific part that normally develops into the testes or the ovaries. Embryonic germ cells in culture, like cultured embryonic stem cells, form embryoid bodies, which are dense, multilayered cell aggregates consisting of partially differentiated cells. The embryoid body–derived cells have high growth potential. The cell lines generated from cultures of the embryoid body cells can give rise to cells of all three embryonic germ layers, indicating that embryonic germ cells may represent another source of pluripotent stem cells. Much of the knowledge about embryonic development and stem cells has been accumulated from basic research on mouse embryonic stem cells.

Since 1998, research teams have succeeded in growing human embryonic stem cells in culture. Human embryonic stem cell lines have been established from the inner cell mass of human blastocysts that were produced through in vitro fertilization procedures. The techniques for growing human embryonic stem cells are similar to those used for growth of mouse embryonic stem cells. However, human embryonic stem cells must be grown on a mouse embryonic fibroblast feeder layer or in media conditioned by mouse embryonic fibroblasts. Human embryonic stem cell lines can be maintained in culture to generate indefinite numbers of identical stem cells for research. As with mouse embryonic stem cells, culture conditions have been designed to direct differentiation into specific cell types (for example, neural and hematopoietic cells).

Adult stem cells, also referred to as somatic stem cells, occur in a wide variety of mature tissues in adults as well as in children. Like all stem cells, adult stem cells can self-replicate. Their ability to self-renew can last throughout the lifetime of individual organisms. Unlike embryonic stem cells, though, it is usually difficult to expand adult stem cells in culture. Adult stem cells reside in specific organs and tissues but account for a very small number of the cells in tissues. They are responsible for maintaining a stable state of the specialized tissues. To replace lost cells, stem cells typically generate intermediate cells called precursor or progenitor cells, which are no longer capable of self-renewal. However, they continue undergoing cell division, coupled with maturation, to yield fully specialized cells. Such stem cells have been identified in many types of adult tissues, including bone marrow, blood, skin, gastrointestinal tract, dental pulp, retina of the eye, skeletal muscle, liver, pancreas, and brain. Adult stem cells are usually designated according to their source and their potential. Adult stem cells are multipotent because their potential is normally limited to one or more lineages of specialized cells. However, a special multipotent stem cell that can be found in bone marrow, called the mesenchymal stem cell, can produce all cell types of bone, cartilage, fat, blood, and connective tissues. *See* REGENERATIVE BIOLOGY.

Blood stem cells, or hematopoietic stem cells, are the most studied type of adult stem cells. The concept of hematopoietic stem cells is not new, as it has been long realized that mature blood cells are constantly lost and destroyed. Billions of new blood cells are produced each day to make up the loss. This process of blood cell generation, called hematopoiesis, occurs largely in the bone marrow. An alternative source of blood stem cells is human umbilical cord blood, a small amount of blood remaining in the placenta and blood vessels of the umbilical cord. It is traditionally treated as a waste material after delivery of the newborn. However, since the recognition of the presence of blood stem cells in umbilical cord blood in the late 1980s, its collection and banking has grown quickly. Similar to bone marrow, umbilical cord blood can be used as a source material of stem cells for transplant therapy. However, because of the limited number of stem cells in umbilical cord blood, most of the procedures are performed on young children of relatively low body weight. A current focus of study is to promote the growth of umbilical cord blood stem cells in culture in order to generate sufficient numbers of stem cells for adult recipients. *See* HEMATOPOIESIS; TRANSPLANTATION BIOLOGY.

Neural stem cells, the multipotent stem cells that generate nerve cells, are a new focus in stem cell research. Active cellular turnover does not occur in the adult nervous system as it does in renewing tissues such as blood or skin. Because of this observation, it had been dogma that the adult brain and spinal cord were unable to regenerate new nerve cells. However, since the early 1990s, neural stem cells have been isolated from the adult brain as well as from fetal brain tissues. Stem cells in the adult brain are found in the areas called the subventricular zone and the ventricle zone. Another location of brain stem cells occurs in the hippocampus, a special structure of the cerebral cortex related to memory function. Stem cells isolated from these areas are able to divide and to give rise to nerve cells (neurons) and neuron-supporting cell types in culture.

Stem cell plasticity refers to the phenomenon of adult stem cells from one tissue generating the specialized cells of another tissue. The long-standing concept of adult organ-specific stem cells is that they are restricted to producing the cell types of their specific tissues. However, a series of recent studies have challenged the concept of tissue restriction of adult stem cells. Although the stem cells appear able to cross their tissue-specific boundaries, crossing occurs generally at a low frequency and mostly only under conditions of host organ damage. The finding of stem cell plasticity is unorthodox and unexpected (since adult stem cells are considered to be organ/tissue-specific), but it carries significant implications for potential cell therapy. For example, if differentiation can be redirected, stem cells of abundant source and easy access, such as blood stem cells in bone marrow or umbilical cord blood, could be used to substitute stem cells in tissues that are difficult to isolate, such as heart and nervous system tissue.

Chen Wang

Stenolaemata A class of Bryozoa confined to fully marine water. The Stenolaemata include several thousand species distributed among four orders: Cystoporata, Trepostomata, Cryptostomata, and Cyclostomata (=Tubuliporata). The commonly recognized order Fenestrata likely is part of the Cryptostomata. First appearing late in the Early Ordovician, stenolaemates expanded quickly to dominate bryozoan assemblages through the Early Cretaceous. Stenolaemates are now a minor part of the bryozoan fauna in most marine environments. *See* BRYOZOA; CRYPTOSTOMATA; CYCLOSTOMATA (BRYOZOA); CYSTOPORATA; TREPOSTOMATA.

Stenolaemate colonies vary greatly in size and shape. Many Paleozoic through mid-Mesozoic forms were large and massive, but more recent representatives are commonly small and delicate. Colonies are encrusting, erect, or free-living. Some erect forms have single or regularly spaced, flexible cuticular joints; segments between joints and all other entire colonies are rigidly calcified and enclosed within a thin cuticular membrane.

Living stenolaemate colonies are made up of individual feeding units called zooids with lophophores (rings of tentacles) that are circular in basal outline. Zooidal body cavities are enclosed within tubular or prismatic zooecia (skeletons of individual zooids), gradually expanding from their proximal ends. Zooecia typically are relatively thin-walled in colony interiors (endozone) and relatively thick-walled in outer regions (exozone). Individual colonies are hermaphroditic. Colonies grow by asexual budding.

Frank K. McKinney

Stephanoberyciformes An order of fishes containing the pricklefishes and whalefishes. Stephanoberyciformes is recognized as the sister group of all remaining acanthomorphs. It is defined by the following characters: pelvic girdle attached to cleithrum or coracoid of pectoral girdle; skull bone usually exceptionally thin (fragile bones often characteristic of pelagic fishes); subocular shelf absent; supramaxilla absent or reduced; and body shape variable, from elongate to short and rounded. The biology is poorly known, and many species are known from only a few specimens. These are small deep-water fishes, ranging in total length from 2.1 to 43 cm (0.8 to 17 in.), but most are less than 10 cm (4 in.), and occupying depths to 5300 m (17,400 ft or 3.3 mi). Several species are known to be oviparous (the assumed mode of reproduction of the entire order), with planktonic eggs and larvae. The order comprises nine families, 28 genera, and about 75 described species, plus many others known but yet undescribed. *See* DEEP-SEA FAUNA; OSTEICHTHYES; TELEOSTEI.

Herbert Boschung

Stepping motor An electromagnetic incremental-motion actuator which converts digital pulse inputs to analog output motion. The device is also termed a step motor. When energized in a programmed manner by a voltage and current input, usually dc, a step motor can index in angular or linear increments. With proper control, the output steps are always equal in number to the input command pulses. Each pulse advances the rotor shaft one step increment and latches it magnetically at the precise point to which it is stepped. Advances in digital computers and particularly microcomputers revolutionized the controls of step motors. These motors are found in many industrial control systems, and large numbers are used in computer peripheral equipment, such as printers, tape drives, capstan drives, and memory-access mechanisms. Step motors are also used in numerical control systems, machine-tool controls, process control, and many systems in the aerospace industry. *See* COMPUTER GRAPHICS; COMPUTER NUMERICAL CONTROL; COMPUTER PERIPHERAL DEVICES; CONTROL SYSTEMS; DIGITAL-TO-ANALOG CONVERTER; PROCESS CONTROL.

There are many types of step motors. Most of the widely used ones can be classified as variable-reluctance, permanent-magnet, or hybrid permanent-magnet types. A variable-reluctance step motor is simple to construct and has low efficiency. The permanent-magnet types are more complex to construct and have a higher efficiency. Benjamin C. Kuo

Stereochemistry The study of the three-dimensional ar- rangement of atoms or groups within molecules and the properties which follow from such arrangement. Molecules that have identical molecular structures but differ in the relative spatial arrangement of component parts are stereoisomers. Inorganic and organic compounds exhibit stereoisomerism. Examples are structures (**1**)–(**8**).

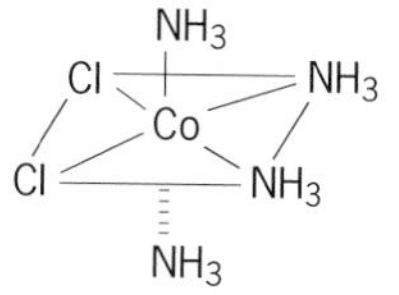

(**1**) 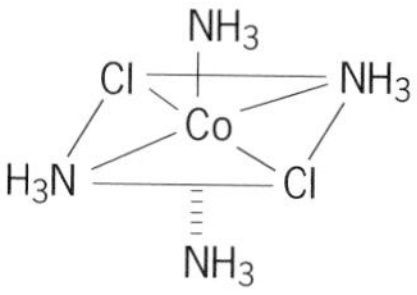 (**2**)

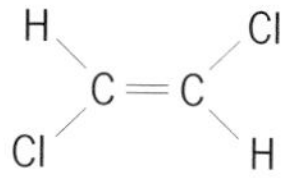

(**3**) 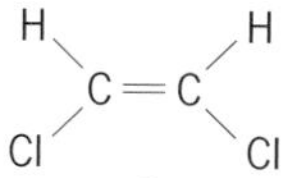 (**4**)

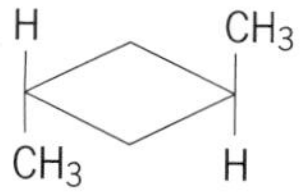

(**5**) 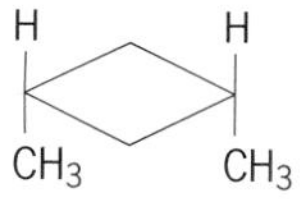 (**6**)

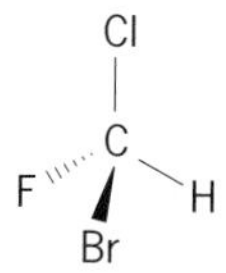

(**7**) 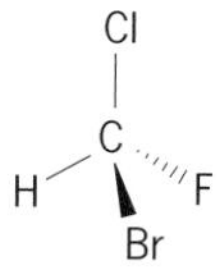 (**8**)

The nature of the stereochemistry of a molecule is determined by its symmetry. The symmetry elements to be considered are: planes of symmetry, axes of symmetry, centers of symmetry, and reflection or mirror symmetry. Two types of stereoisomers are known. Those such as (**7**) and (**8**), which are devoid of reflection symmetry—which cannot be superimposed on their image in a mirror—are called enantiomers. All other stereoisomers, such as the pairs (**1**)–(**2**), (**3**)–(**4**), and (**5**)–(**6**), are called diastereomers. The configuration of a stereoisomer designates the relative position of the atoms associated with a specific structure. The structures of stereoisomers (**1**) and (**2**) differ only in configuration. The same is true for (**3**) and (**4**), (**5**) and (**6**), and (**7**) and (**8**). *See* ENANTIOMER. Samuel H. Wilen

Stereophonic radio transmission The transmission of stereophonic audio signals over broadcast radio systems, including terrestrial analog amplitude-modulation (AM), frequency-modulation (FM) systems, and digital radio systems (terrestrial and satellite). *See* STEREOPHONIC SOUND.

Two-channel stereo broadcasting of audio was first implemented for FM radio signals in the 88–108-MHz frequency band (allocated for FM radio broadcasting) in the 1960s, and became the dominant mode of transmission, in particular for music programming, in the 1970s. A multiplexing method was used to accomplish this and allowed the signal transmitted by a stereo broadcast station to be decipherable by either a stereo receiver or a monophonic receiver. For the monophonic receiver to receive a stereo broadcast, some component of the broadcast must be a single signal that includes information from both the left and the right channels. This is called the $L + R$ signal. A monophonic receiver simply demodulates the $L + R$ signal and delivers it to the listener.

The stereo receiver must use the $L + R$ signal and other transmitted information to demodulate separate L and R signals. For this, stereo broadcasts include an $L - R$ signal. In the stereo receiver, after detection of the $L + R$ and $L - R$ signals, the L and R signals are separated as shown in Eqs. (1) and (2).

$$(L + R) + (L - R) = 2L \quad (1)$$
$$(L + R) - (L - R) = 2R \quad (2)$$

The subtraction used in Eq. (2) is realized through phase shifting and addition. In FM stereo transmissions, the $L + R$ signal is directly frequency-modulated onto the radio-frequency carrier, assuring compatibility with monophonic systems (see illustration). The $L - R$ signal is used to amplitude-modulate a 38-kHz hypersonic subcarrier. The 38-kHz carrier is suppressed, creating a double-sideband (DSB), suppressed-carrier AM signal at 38 kHz. This hypersonic signal is then frequency-modulated onto the radio-frequency carrier. A 19-kHz pilot signal is also transmitted to indicate that the broadcast is stereo and to provide an accurate frequency reference for demodulation of the 38-kHz $L - R$ signal. *See* AMPLITUDE MODULATION; FREQUENCY MODULATION.

In the FM stereo receiver, the entire baseband signal that has been frequency-modulated onto a radio-frequency carrier ($L + R$, pilot, and double-sideband $L - R$) is first detected with an FM discriminator. The $L + R$ signal is isolated by a low-pass filter. The pilot tone is doubled and mixed with the double-sideband suppressed-carrier $L - R$ signal, so that the $L - R$ signal can be demodulated by an AM envelope detector. The $L + R$ and $L - R$ audio signals are then passed through a network that implements the functions expressed in Eqs. (1) and (2). The resulting left- and right-channel audio signals are amplified and presented to the listener. *See* AMPLITUDE-MODULATION DETECTOR; ELECTRIC FILTER; FREQUENCY-MODULATION DETECTOR; FREQUENCY-MODULATION RADIO.

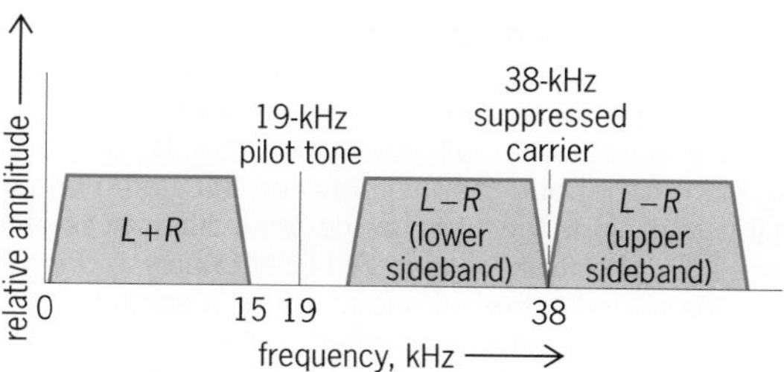

Baseband spectrum of an FM stereo signal. This audio and supersonic spectrum is frequency-modulated onto a radio-frequency carrier.

Stereo broadcasting for the AM band (550–1701 kHz) was first authorized by the FCC in 1982, when five different competing (and incompatible) systems were authorized. Due in part to the confusion created by so many systems being available, AM stereo never became a popular service. In 2002 the FCC authorized a new digital radio service for the AM band, called in-band/on-channel (IBOC) digital radio, which supports stereo audio, effectively ending the analog AM stereo saga in the United States.

A number of digital radio systems are being implemented around the world, and in most cases these systems support the transmission of not only two-channel stereo but also "surround sound" stereo signals with five or more channels. In 2005, satellite digital audio radio service (SDARS) provider XM Radio began broadcasting some of its music channels in 5.1-channel surround using technology developed by Neural Audio, and a number of terrestrial FM radio broadcasters were also experimenting with surround sound using the IBOC digital radio broadcast system authorized for the FM band by the FCC in 2002.

David H. Layer; Jeffrey Ward

Stereophonic sound A system of sound recording or transmission in which signals are captured, mixed, or synthesized using two or more audio channels in such a way as to deliver a spatial or three-dimensional auditory impression to a listener when these audio channels are connected to loudspeakers in a listening room. Binaural or "head related" stereo is a term used to describe the capture or synthesis of a pair of signals that closely resemble those present at a listener's ears in a natural spatial sound field. Such signals are intended for reproduction using a pair of headphones worn by the listener, although they can be reproduced using loudspeakers if suitable signal processing is employed. From the late 1990s, "surround sound" stereo systems involving five or more channels began to achieve commercial success, primarily as an adjunct to home television systems. Wavefield synthesis systems, which reached the market in the early years of the twenty-first century, use very large numbers of loudspeakers in an attempt to recreate acoustical wavefronts in a physically accurate fashion.

Auditory cues in stereophony. The aim of stereophonic recording and reproduction is to deliver consistent auditory spatial cues to the listener to give rise to an intended spatial impression. It is nonetheless common for stereophony to rely on a degree of auditory illusion for this purpose, as it is not usually possible to reconstruct acoustical sound fields in a physically accurate manner over a wide listening area when using a limited number of loudspeakers.

Stereophonic signals are created either by capturing the relevant cues using two or more microphones, or by artificially introducing the necessary interchannel differences using audio signal processing. The perceived source images that result from these cues during reproduction are known as phantom images because they are not real sources. In binaural stereo or wavefield synthesis, they are sometimes known as virtual sources. *See* VIRTUAL ACOUSTICS.

Stereophonic sound pickup. Since the late 1960s, it has become common for much stereo program material to be recorded using a large number of microphones, each located close to one source or a small group of sources. By proportionally mixing the microphone outputs between the stereo channels, it is possible to provide the interchannel level differences necessary for perceiving phantom images. Artificial effects and reverberation may be added subsequently to create a spatial impression, or signals from microphones located in the reverberant field of a studio may be mixed with the dry signals. Surround sound recording extends these concepts to more than two channels to create an immersive sound field.

Stereophonic reproduction. The ability to create convincing stereophonic reproduction is highly dependent upon the performance of the loudspeakers, the acoustics of the listening room, and the hearing process of the listener. For two-channel stereo, the ideal arrangement is one where the loudspeakers subtend an angle of $\pm 30^\circ$ at the listening position. This arrangement is maintained in the configuration for 3-2 Stereo (5.1 surround), in which a center loudspeaker is added between the left and right front loudspeakers, as well as two rear-side loudspeakers at $\pm 110^\circ$. The rear-side locations were chosen as a compromise between the requirements for movie applications and those for music reproduction. *See* HEARING (HUMAN); SOUND RECORDING; SOUND-REPRODUCING SYSTEMS.

Rollins Brook; Francis Rumsey

Stereoscopy The phenomenon of simultaneous vision with two eyes, producing a visual experience of the third dimension, that is, a vivid perception of the relative distances of objects in space. In this experience the observer seems to see the space between the objects located at different distances from the eyes.

Stereopsis, or stereoscopic vision, is believed to have an innate origin in the anatomic and physiologic structures of the retinas of the eyes and the visual cortex. It is present in normal binocular vision because the two eyes view objects in space from two points, so that the retinal image patterns of the same object points in space are slightly different in the two eyes. The stereoscope, with which different pictures can be presented to each eye, demonstrates the fundamental difference between stereoscopic perception of depth and the conception of depth and distance from the monocular view. *See* VISION.

Kenneth N. Ogle

Steric effect (chemistry) The influence of the spatial configuration of reacting substances upon the rate, nature, and extent of reaction. The sizes and shapes of atoms and molecules, the electrical charge distribution, and the geometry of bond angles influence the courses of chemical reactions. The steric course of organochemical reactions is greatly dependent on the mode of bond cleavage and formation, the environment of the reaction site, and the nature of the reaction conditions (reagents, reaction time, and temperature). The effect of steric factors is best understood in ionic reactions in solution. *See* STEREOCHEMISTRY.

Ernest Wenkert

Sterilization An act of destroying all forms of life on and in an object. A substance is sterile, from a microbiological point of view, when it is free of all living microorganisms. Sterilization is used principally to prevent spoilage of food and other substances and to prevent the transmission of diseases by destroying microbes that may cause them in humans and animals. Microorganisms can be killed either by physical agents, such as heat and irradiation, or by chemical substances.

Heat sterilization is the most common method of sterilizing bacteriological media, foods, hospital supplies, and many other substances. Either moist heat (hot water or steam) or dry heat can be employed, depending upon the nature of the substance to be sterilized. Moist heat is also used in pasteurization, which is not considered a true sterilization technique because all microorganisms are not killed; only certain pathogenic organisms and other undesirable bacteria are destroyed. *See* PASTEURIZATION.

Many kinds of radiations are lethal, not only to microorganisms but to other forms of life. These radiations include both high-energy particles as well as portions of the electromagnetic spectrum. *See* RADIATION BIOLOGY.

Filtration sterilization is the physical removal of microorganisms from liquids by filtering through materials having relatively small pores. Sterilization by filtration is employed with liquid that may be destroyed by heat, such as blood serum, enzyme solutions, antibiotics, and some bacteriological media and medium constituents. Examples of such filters are the Berkefeld filter (diatomaceous earth), Pasteur-Chamberland filter (porcelain), Seitz filter (asbestos pad), and the sintered glass filter.

Chemicals are used to sterilize solutions, air, or the surfaces of solids. Such chemicals are called bactericidal substances. In lower concentrations they become bacteriostatic rather than bactericidal; that is, they prevent the growth of bacteria but may not kill them. Other terms having similar meanings are employed. A disinfectant is a chemical that kills the vegetative cells of pathogenic microorganisms but not necessarily the endospores of spore-forming pathogens. An antiseptic is a chemical applied to living tissue that prevents or retards the growth of microorganisms, especially pathogenic bacteria, but which does not necessarily kill them.

The desirable features sought in a chemical sterilizer are toxicity to microorganisms but nontoxicity to humans and animals, stability, solubility, inability to react with extraneous organic materials, penetrative capacity, detergent capacity, noncorrosiveness, and minimal undesirable staining effects. Rarely does one chemical combine all these desirable features. Among chemicals that have been found useful as sterilizing agents are the phenols, alcohols, chlorine compounds, iodine, heavy metals and metal complexes, dyes, and synthetic detergents, including the quaternary ammonium compounds. Charles F. Niven, Jr.

Steroid Any of a subset of biological molecules known as lipids that share a specific, fused four-ring structure comprising three rings of six carbon atoms and one ring of five carbon atoms. Different types of steroids have different chemical modifications to this common ring structure. Steroids are found in animals, plants, fungi, and protozoa. In mammals, steroids are components of cell membranes, aid in digestion, and act as hormones to control a wide range of physiological responses. Because of their role in governing physiological processes, naturally occurring and synthetic steroids are used extensively in medicine to treat a wide range of conditions. *See* LIPID; STEROL.

Cholesterol is the most abundant steroid in animals. It is found primarily in the plasma membrane of cells, where it is a major component of the lipid bilayer. It is also found, usually to a lesser extent, in the membranes of subcellular organelles such as mitochondria. Because cholesterol possesses a fused ring structure, it has greater rigidity than the other lipid components of these membranes and therefore tends to reduce the fluidity of membranes by restricting the movements of the more flexible lipid components. Thus, cholesterol plays an important role in determining the physical characteristics of biological membranes. While plants have very little cholesterol, a similar steroid, stigmasterol, plays an analogous role in the membranes of their cells. *See* CELL (BIOLOGY); CELL MEMBRANES; CHOLESTEROL.

In addition to being an important component of cell membranes, cholesterol is a precursor of a variety of critical steroid hormones. Among these are glucocorticoids, androgens, estrogens, progestins, and mineralocorticoids. Each family of steroid hormones elicits a different physiological response. Cortisol (also known as hydrocortisone) is an example of a glucocorticoid. This hormone is produced by the adrenal gland and participates in the regulation of carbohydrate metabolism. Testosterone is the primary androgen hormone. It is produced in the gonads and is a male sex hormone. Estradiol is the primary estrogen hormone. It is produced by the ovaries and placenta and is a female sex hormone. Progesterone is the primary progestin. This hormone is produced by the adrenal gland and the corpus luteum (an ovarian follicle that has ruptured after releasing an egg). Progesterone is a female sex hormone. Aldosterone is an example of a mineralocorticoid hormone. It is produced by the adrenal gland and regulates electrolyte balance. *See* ADRENAL GLAND; ALDOSTERONE; ESTROGEN; PROGESTERONE. M. Todd Washington

Sterol Any of a group of naturally occurring or synthetic organic compounds with a steroid ring structure, having a hydroxyl (—OH) group, usually attached to carbon-3. This hydroxyl group is often esterified with a fatty acid (for example, cholesterol ester). The hydrocarbon chain of the fatty-acid substituent varies in length, usually from 16 to 20 carbon atoms, and can be saturated or unsaturated. Sterols commonly contain one or more double bonds in the ring structure and also a variety of substituents attached to the rings. Sterols and their fatty-acid esters are essentially water insoluble. For transport in an aqueous milieu (for example, the bloodstream of mammals), sterols and other lipids are bound to specific proteins, forming lipoprotein particles. These particles are classified based on their composition and density. One lipoprotein class is abnormally high in the blood of humans prone to heart attacks.

Sterols are widely distributed in nature. Modifications of the steroid ring structure are made by specific enzyme systems, producing the sterol characteristic for each species, such as ergosterol in yeast. The major regulatory step in the sterol biosynthetic pathway occurs early in the process. Drugs that lower blood cholesterol levels in humans are designed to inhibit this regulatory enzyme. In addition to their conversion to sterols, several intermediates in the pathway are precursors of other important biological compounds, including chlorophyll in plants, vitamins A, D, E, and K, and regulators of membrane functions and metabolic pathways.

A universal role of sterols is to function as part of membrane structures. In addition, some insects require sterols in their diets. Cholesterol also serves as a precursor of steroid hormones (estrogens, androgens, glucocorticoids, and mineralocorticoids) and bile acids. *See* CHOLESTEROL; STEROID. Mary E. Dempsey

Stichosomida An order of nematodes (roundworms) formerly constituting at least two important groups of animal parasites: the mermithids, which are parasites of invertebrates, and the trichinellids, which are parasites of vertebrates. These nematodes are characterized by a pharynx that is narrow and thin-walled anteriorly and is surrounded posteriorly by unicellular, glandular stichocytes, each with a duct into the pharyngeal lumen. The pharynx extends one-fourth to nine-tenths of the body length in various taxa and is almost devoid of musculature. The region of the pharynx surrounded by stichocytes is known as the stichosome. Recent phylogenetic analysis based on a synthesis of molecular and morphological data suggests that the stichosome may be an example of parallel evolution and that the trichinellids and mermithids are more appropriately separated as two orders: Trichinellida, with at least six families, and Mermithida, with two families. *See* MEDICAL PARASITOLOGY; NEMATA (NEMATODA). Howard Ferris

Stickleback Any member of the fish family Gasterosteidae, order Gasterosteiformes. The body may be elongate, naked, or with bony scutes along the sides. Three to 16 well-developed isolated dorsal spines precede a normal dorsal fin having 6 to 14 rays. Additional identifying features are as follows: 12 or 13 caudal fin rays; a single spine and 1 or 2 soft rays in the pelvic fin; 3 branchiostegal rays; circumorbital ring incomplete posteriorly; epineurals present; and 28 to 42 vertebrae. A maximum length of about 18 cm (7 in.) is reached in *Spinachia spinachia*. *See* GASTEROSTEIFORMES; OSTEICHTHYES; TELEOSTEI.

Sticklebacks are small freshwater and marine fishes of the cold and temperate waters of the Northern Hemisphere. All species are of some economic importance since their diet includes mosquito larvae, and the freshwater species are especially popular aquarium fishes. Also, sticklebacks are famous as subjects of numerous studies regarding evolution, genetics, behavior, and physiology. *See* INSTINCTIVE BEHAVIOR; SEXUAL DIMORPHISM. Herbert Boschung; Charles B. Curtin

Stilbite A mineral belonging to the zeolite family of silicates. It crystallizes in sheaflike aggregates of thin tabular crystals. There is perfect cleavage parallel to the side pinacoid, and here the mineral has a pearly luster; elsewhere the luster is vitreous. The color is usually white but may be brown, red, or yellow. Hardness is $3^1/_2$–4 on Mohs scale; specific gravity is 2.1–2.2. *See* ZEOLITE.

Stilbite is found in Iceland, India, Scotland, Nova Scotia, and in the United States at Bergen Hill, New Jersey, and the Lake Superior copper district in Michigan. Cornelius S. Hurlbut, Jr.

Stirling engine An engine in which work is performed by the expansion of a gas at high temperature to which heat is supplied through a wall. Like the internal combustion engine, a Stirling engine provides work by means of a cycle in which a piston compresses gas at a low temperature and allows it to

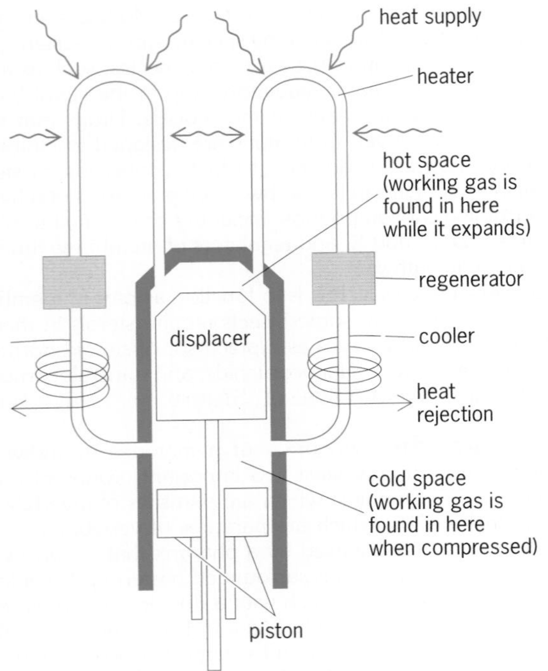

Principle of Stirling engine, displacer type.

expand at a high temperature. In the former case the heat is provided by the internal combustion of fuel in the cylinder, but in the Stirling engine the heat (obtained from externally burning fuel) is supplied to the gas through the wall of the cylinder (see illustration).

The rapid changes desired in the gas temperature are achieved by means of a second piston in the cylinder, called a displacer, which in moving up and down transfers the gas back and forth between two spaces, one at a fixed high temperature and the other at a fixed low temperature. When the displacer is raised, the gas will flow from the hot space via the heater and cooler tubes into the cold space. When it is moved downward, the gas will return to the hot space along the same path. During the first transfer stroke the gas has to yield up a large amount of heat to the cooler; an equal quantity of heat has to be taken up from the heater during the second stroke. *See* INTERNAL COMBUSTION ENGINE. Roelof J. Meijer

Stishovite Naturally occurring stishovite, SiO_2, is a mineral formed under very high pressure with the silicon atom in sixfold, or octahedral, coordination instead of the usual fourfold, or tetrahedral, coordination. The presence of stishovite indicates formation pressures in excess of 10^6 lb/in.^{2} (7.5 gigapascals). The possibility of the existence of stishovite at great depths strongly influences the interpretations of geophysicists and solid-state physicists regarding the phase transitions of mineral matter, as well as the interpretation of seismic data in the study of such regions of the interior of the Earth. *See* SILICA MINERALS.

Stishovite occurs in submicrometer size in very small amounts (less than 1% of the rock) in samples of Coconino sandstones from the Meteor Crater of Arizona, which contains up to 10% of coesite, the other high-pressure polymorph of silica. Because of its extremely fine grain size and because of the sparsity of this mineral in the rock, positive identification of the mineral is possible only by the x-ray diffraction method after chemical concentration. *See* COESITE; METEORITE.

The specific gravity of stishovite, calculated from the x-ray data, is 4.28, compared with the value of 4.35 for the synthetic material. It is 46% denser than coesite and much denser than other modifications of silica. *See* RUTILE. Edward C. T. Chao

Stochastic control theory A branch of control theory which aims at predicting and minimizing the magnitudes and limits of the random deviations of a control system through optimizing the design of the controller. Such deviations occur when random noise and disturbance processes are present in a control system, so that the system does not follow its prescribed course but deviates from the latter by a randomly varying amount.

In contrast to deterministic signals, random signals cannot be described as given functions of time such as a step, a ramp, or a sine wave. The exact function is unknown to the system designer; only some of its average properties are known.

A random signal may be generated by one of nature's processes, for instance, radar noise and wind- or wave-induced forces and moments on a radar antenna or a ship. Alternatively, it may be generated by human intelligence, for instance, the bearing of a zigzagging aircraft, or the contour to be followed by a duplicating machine.

One outstanding experimental fact about nature's random processes is that these signals are very closely Gaussian. The word "Gaussian" is a mathematical concept which describes one or more signals, $i_1, i_2, \ldots, i_n$ having the following properties:

1. The amplitude of each signal is normally distributed.
2. The joint distribution function of any number of signals at the same or different times taken from the set is a multivariate normal distribution. This experimental fact is not surprising in view of the fact that a random process of nature is usually the sum total of the effects of a large number of independent contributing factors. For instance, thermal noise is due to the thermal motions of billions of electrons and atoms. An ocean-wave height at any particular time and place is the sum of wind-generated waves at previous times over a large area. *See* DISTRIBUTION (PROBABILITY); ELECTRICAL NOISE; STOCHASTIC PROCESS.

The underlying mechanism that generates a random process can usually be described in physical or mathematical terms. For instance, the underlying mechanism that generates shot-effect noise is thermionic emission. If the generating mechanism does not change with time, any measured average property of the random process is independent of the time of measurement aside from statistical fluctuations, and the random process is called stationary. If the generating mechanism does change, the random process is called nonstationary. *See* CONTROL SYSTEMS. S. S. L. Chang; Gerald Cook

Stochastic process A physical stochastic process is any process governed by probabilistic laws. Examples are (1) development of a population as controlled by Mendelian genetics; (2) Brownian motion of microscopic particles subjected to molecular impacts or, on a different scale, the motion of stars in space; (3) succession of plays in a gambling house; and (4) passage of cars by a specified highway point.

In each case, a probabilistic system is evolving; that is, its state is changing with time. Thus the state at time t depends on chance: It is a random variable $x(t)$. The parameter set of values of t involved is usually (and will always be in this article) either an interval (continuous parameter stochastic process) or a set of integers (discrete parameter stochastic process). Some authors, however, apply the term stochastic process only to the continuous parameter case.

If the state of the system is described by a single number, $x(t)$ is numerical-valued. In other cases, $x(t)$ may be vector-valued or even more complicated. For the numerical case, as the state changes, its values determine a function of time, the sample function, and the probability laws governing the process determine the probabilities assigned to the various possible properties of sample functions.

A mathematical stochastic process is a mathematical structure inspired by the concept of a physical stochastic process, and studied because it is a mathematical model of a physical stochastic process or because of its intrinsic mathematical interest and its applications both in and outside the field of probability. The mathematical stochastic process is defined simply as a family of random variables. That is, a parameter set is specified, and to each parameter point t a random variable $x(t)$ is specified. If one recalls that a random variable is itself a function, if one denotes a point of the domain of the random variable $x(t)$ by ω, and if one denotes the value of this random variable at ω by $x(t, \omega)$, it results that the stochastic process is completely specified by the function of the pair (t, ω) just defined, together with the assignment of probabilities. If t is fixed, this function of two variables defines a function of ω, namely, the random variable denoted by $x(t)$. If ω is fixed, this function of two variables defines a function of t, a sample function of the process.

Probabilities are ordinarily assigned to a stochastic process by assigning joint probability distributions to its random variables. These joint distributions, together with the probabilities derived from them, can be interpreted as probabilities of properties of sample functions. For example, if t_0 is a parameter value, the probability that a sample function is positive at time t_0 is the probability that the random variable $x(t_0)$ has a positive value. The fundamental theorem at this level is that, to any self-consistent assignment of joint probability distributions, there corresponds a stochastic process.

Stationary processes are the stochastic processes for which the joint distribution of any finite number of the random variables is unaffected by translations of the parameter; that is, the distribution of $x(t_1 + h), \ldots, x(t_n + h)$ does not depend on h. *See* PROBABILITY.

A Markov process is a process for which, if the present is given, the future and past are independent of each other. More precisely, if $t_1 < \cdots < t_n$ are parameter values, and $1 < j < n$, then the sets of random variables $[x(t_1), \ldots, x(t_{j-1})]$ and $[x(t_{j+1}), \ldots, x(t_n)]$ are mutually independent for given $x(t_j)$. Equivalently, the conditioned probability distribution of $x(t_n)$ *for* given $x(t_{n1}), \ldots, x(t_{n-1})$ depends only on the specified value of $x(t_{n-1})$ and is in fact the conditional probability distribution of $x(t_n)$, given $x(t_{n-1})$. An important and simple example is the Markov chain, in which the number of states is finite or denumerably infinite.

A martingale is a stochastic process with the property that, if $t_1 < \cdots < t_n$ are parameter values, the expected value of $x(t_n)$ for given $x(t_1), \ldots, x(t_{n-1})$ is equal to $x(t_{n-1})$. That is, the expected future value, given present and past values, is equal to the present value. The interpretation that a martingale can be thought of as the fortune of a player after the successive plays of a fair gambling game is obvious.

A process with independent increments is a continuous-parameter process with the property that, if $t_1, < \cdots < t_n$ are parameter values, the successive increments in $x(t_2) - x(t_1), \ldots, x(t_n) - x(t_{n-1})$ are mutually independent. If $y(t) = x(t) - x(t_0)$, where t_0 is fixed, the $y(t)$ process is then a Markov process. *See* GAME THEORY; INFORMATION THEORY; LINEAR PROGRAMMING; OPERATIONS RESEARCH.

Joseph L. Doob

Stoichiometry All chemical measurements, such as the measurements of atomic and molecular weights and sizes, gas volumes, vapor densities, deviation from the gas laws, and the structure of molecules. In determining the relative weights of the atoms, scientists have relied upon combining ratios, specific heats, and measurements of gas volumes. All such measurements, and the calculations that relate them to each other, constitute the field of stoichiometry. Since measurements are expressed in mathematical terms, stoichiometry can be considered to be the mathematics of general chemistry.

In a general usage, the term stoichiometry refers to the relationships between the measured quantitites of substances or of energy involved in a chemical reaction; the calculations of these quantities include the assumption of the validity of the laws of definite proportions and of conservation of matter and energy. *See* CONSERVATION OF ENERGY; CONSERVATION OF MASS.

A typical stoichiometric problem involves predicting the weight of reactant needed to produce a desired amount of a product in a chemical reaction. For example, phosphorus can be extracted from calcium phosphate, $Ca_3(PO_4)_2$, by a certain process with a 90% yield (some calcium phosphate fails to react or some phosphorus is lost). In a specific problem, it might be necessary to determine the mass of calcium phosphate required to prepare 16.12 lb of phosphorus by this process. The balanced equation for the preparation is shown in reaction (1). In this reaction,

$$2Ca_3(PO_4)_2 + 10C + 6SiO_2 \rightarrow 6CaSiO_3 + 10CO + P_4 \quad (1)$$

2 moles of calcium phosphate are required to produce 1 mole of phosphorus. Two moles of calcium phosphate have a mass of 620 lb, and 1 mole of phosphorus as P_4 has a mass of 124 lb. Using these relationships, calculation (2) is made. Since

$$\begin{aligned} \frac{2 \text{ moles } Ca_3(PO_4)}{1 \text{ mole } P_4} &= \frac{620 \text{ lb } Ca_3(PO_4)_2}{124 \text{ lb } P_4} \\ &= \frac{X \text{ lb } Ca_3(PO_4)_2}{16.12 \text{ lb } P_4} \\ X &= 80.6 \text{ lb} \end{aligned} \quad (2)$$

the yield of phosphorus is only 90%, extra $Ca_3(PO_4)_2$ must be used: 88.1 lb is the mass of calcium phosphate required to yield 16.12 lb of phosphorus by this process.

Calculations of this sort are important in chemical engineering processes, in which amounts and yields of products must be known. The same reasoning is used in calculations of energy generated or required. In this case, the energy involved in the reaction of a known weight of the material in question must be known or determined.

The calculations discussed involve compounds in which the ratio of atoms is generally simple. For a discussion of compounds in which the relative number of atoms cannot be expressed as ratios of small whole numbers, *see* NONSTOICHIOMETRIC COMPOUNDS

John C. Bailar, Jr.

Stoker A mechanical means for feeding coal into, and for burning coal in, a furnace. There are three basic types of stokers. Chain or traveling-grate stokers have a moving grate on which the coal burns; they carry the coal from a hopper into the furnace and move the ash out (see illustration). Spreader stokers mechanically or pneumatically distribute the coal from a hopper at the furnace front wall and move it onto the grate which usually moves continuously to dispose of the ash after the coal is burned. Underfeed stokers are arranged to force fresh coal from the hopper to the bottom of the burning coal bed, usually

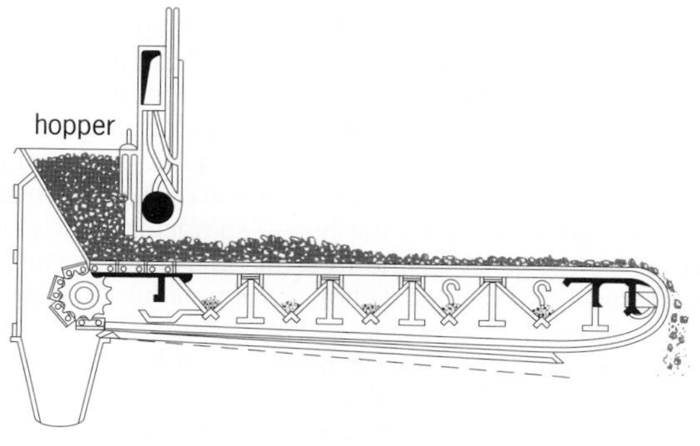

Chain grate stoker.

by means of a screw conveyor. The ash is forced off the edges of the retort peripherally to the ashpit or is removed by hand.
George W. Kessler

Stokes' theorem The assertion that under certain light restrictions the surface integral of $(\Delta \mathfrak{F} \times \mathfrak{F}) \cdot \nu$ over a surface patch S is equal to the line integral of $\mathfrak{F} \cdot \tau$ taken around C, the boundary curve of S, provided the sense of transcription of C is right-handed relative to ν. This can be expressed as

$$\iint (\nabla \times \mathfrak{F}) \cdot \nu \, dS = \int \mathfrak{F} \cdot \tau \, ds$$

Here $\mathfrak{F}$ is a vector function, ν is one of the two unit normals to the two-sided surface S, s is arc length measured positively in the sense which is right-handed relative to ν, and τ is the unit tangent vector to C in the sense of increasing s. *See* CALCULUS OF VECTORS.
Homer V. Crais

Stolonifera An order of the cnidarian subclass Alcyonaria (Octocorallia) which lacks coenenchyme. They form either simple or rather complex colonies. The polyp has a cylindrical body with a retractile oral portion which can withdraw into a solid anthostele or calyx protected by many calcareous spicules. The base of the mature polyp is attached to a creeping stolon which is a ribbonlike network or thin flat mat from which daughter polyps arise. Daughter polyps never bud from the wall of the primary polyp. Each polyp is connected by solenial tubes of the stolons, or by transverse platforms. *See* OCTOCORALLIA (ALCYONARIA).
Kenji Atoda

Stomach The tubular or saccular abdominal organ of the digestive system adapted for temporary food storage and preliminary stages of food breakdown.

In some primitive vertebrates the stomach may be little more than a simple tube quite similar to other portions of the gastrointestinal tract. In other forms the stomach is a distinct, and frequently large, saclike structure of variable shape. Carnivorous forms typically have a better-developed stomach than herbivores, probably reflecting the larger but less numerous feedings characteristic of the former, but exceptions are numerous.

In birds the stomach consists of a proventriculus and a gizzard. The former is well supplied with glands which secrete softening and digestive enzymes; the latter is a strong, muscular grinding organ whose action is often enhanced by the ingestion of small stones.

Mammals have stomachs which vary considerably in structure. Although a single chamber is most common, some mammals, such as cows and their relatives (ruminants), have as many as four. These chambers may have developed either from modifications of the posterior portions of the esophagus or from alterations of the stomach itself.

The human stomach is located beneath the diaphragm, through which the posterior, terminal end of the esophagus passes. The stomach appears as a dilated tube continuous with the distal end of the esophagus. The upper curvature of the stomach is usually above and to the left of the esophageal orifice. This expanded anterior portion is the fundus and is commonly filled with air or gas. The body (corpus) of the stomach is directed toward the attenuated right extremity or pyloric region and is subject to variations in size and shape, depending upon functional activities, habits, disease, and volume of the contents. The pyloric walls are marked by the heavy sphincter muscle which controls the passage of chyme (a semiliquid fluid produced by the mechanical and chemical changes of preliminary digestion) into the duodenum.

The stomach of vertebrates is lined by a mucous membrane that is usually thrown into longitudinal folds called rugae. Most of the surface is covered with mucus-secreting epithelial cells, but scattered throughout the lining are many small glandular pits which are lined with one or more types of secretory cells. *See* DIGESTIVE SYSTEM.
Thomas S. Parsons

Stomatopoda The only extant order of the subclass Hoplocarida belonging to the class Malacostraca of the Crustacea. This order of the mantis shrimps contains 17 families. Stomatopoda are among the larger Crustacea; the size of an adult ranges from about 0.4 in. (10 mm) to over 12 in. (300 mm).

The bodies of stomatopods are narrow and elongate, almost eruciform. Only part of the cephalic and thoracic somites are fused and covered by the dorsal shield, the carapace. Those cephalic somites which bear the antennulae and the eyes are free and visible anterior to the carapace, and the last four thoracic somites are similarly exposed. The tail fan consists of a well-developed, sometimes peculiarly sculptured or deformed, median plate, the telson, and the two uropods. Anteriorly, the carapace bears a flattened movable plate, the rostrum. The eyes are large, stalked, and movable. Of eight pairs of thoracic appendages the first is narrow, slender, and hairy, probably being used for cleaning purposes. The second thoracic leg is very strong and heavy. It has become a large raptorial claw that shows a great resemblance to that of the praying mantis, and for this reason the Stomatopoda are often given the name mantis shrimps.

The Stomatopoda are marine animals rarely found in brackish water. Most species are confined to tropical and subtropical areas, though some occur in the boreal and antiboreal regions. The majority of stomatopods live in the littoral and sublittoral zones, but a few species have been found in greater depths, down to 2500 ft (760 m). *See* MALACOSTRACA.
Lipke B. Holthuis

Stomiiformes An order of neoteleost fishes which is probably the sister group to all other neoteleosts. The following characters serve to distinguish the Stomiiformes: body form long and slender to very deep, and moderately to extremely compressed; luminescent organs (photophores) present below axis of body; chin barbel present in some; premaxilla and maxilla toothed and in gape of mouth; gape extending well past eye in most species; scales, if present, cycloid and easily lost; pectoral, dorsal, or adipose fins absent in some; ventral adipose fin (in front of anal fin) present in some; 4 to 9 pelvic fin rays; 5 to 24 branchiostegal rays; and color usually dark brown or black, silvery in some.

Stomiiforms are small fishes, ranging in maximum total length from about 2 to 58 cm (0.8 to 23 in.). Species for which the mode of reproduction is known are oviparous, and their eggs and larvae are planktonic. Food items are known to be copepods, ostracodes, amphipods, euphausiids, chaetognaths, and small fishes. Most species occur in the tropical to temperate zones of the Atlantic, Indian, and Pacific oceans. Commonly known as dragonfishes, these fishes are mostly mesopelagic, occupying the water column to depths of about 1500 m (4900 ft) during the day and migrating to near the surface at night. Some species live at greater depths, but probably none at depths greater than 3880 m (12,730 ft or 2.4 mi) as occupied by the spotlight loosejaw dragonfish (*Malacosteus niger*). Stomiiformes comprises five families, 53 genera, and 391 species. *See* BIOLUMINESCENCE; DEEP-SEA FAUNA; OSTEICHTHYES; PHOTOPHORE GLAND; TELEOSTEI.
Herbert Boschung

Stone and stone products The term stone is applied to rock that is cut, shaped, broken, crushed, and otherwise physically modified for commercial use. The two main divisions are dimension stone and crushed stone. Other descriptive terms may be used, for example, building stone, roofing stone, or precious stone. *See* GEM; ROCK.

The term dimension stone is applied to blocks that are cut and milled to specified sizes, shapes, and surface finishes. The principal uses are for building and ornamental applications. Granites, limestones, sandstones, and marbles are widely used; basalts, diabases, and other dark igneous rocks are used less extensively. Soapstone is used to some extent. Rock suitable for use as dimension stone must be obtainable in large, sound blocks, free from incipient cracks, seams, and blemishes, and must be without mineral grains that might cause stains as a result of weathering. It must have an attractive color, and generally a uniform texture.

Slate differs from other dimension stone because it can be split into thin sheets of any desired thickness. Commercial slate must be uniform in quality and texture and reasonably free from knots, streaks, or other imperfections, and have good splitting properties. Roofing slates are important products of most slate quarries. However, the roofing-slate industry has declined considerably because of competition from other types of roofing. Slate is also used for milled products such as blackboards, electrical panels, window and door sills and caps, baseboards, stair treads, and floor tile. *See* SLATE.

Nearly all the principal types of stone—granite, diabase, basalt, limestone, dolomite, sandstone, and marble—may be used as sources of commercial crushed stone; limestone is by far the most important. Crushed stone is made from sound, hard stone, free from surface alteration by weathering. Stone that breaks in chunky, more or less cubical fragments is preferred. Commercial stone should be free from certain deleterious impurities, such as opalescent quartz, and free from clay or silt. Crushed stone is used principally as concrete aggregate, as road stone, or as railway ballast. Other uses for limestone are as a fluxing material to remove impurities from ores smelted in metallurgical furnaces, in the manufacture of alkali chemicals, calcium carbide, glass, paper, paint, and sugar, and for filter beds and for making mineral wool. Robert L. Bates

Storage tank A container designed to store liquids or gases at or near atmospheric pressures. When the container is used to store a solid or granular material, it is called a storage bin or silo.

Aboveground large petroleum storage tanks.

Tanks can be constructed underground, partially underground, on the ground, or supported up to 100 ft (300 m) above the ground. In the context of industrial operations and regulations, the term tank applies to liquid containers having a storage volume exceeding 660 gallons (2.5 m^3). The largest tanks are found in the petroleum industry for storage of crude oil, gasoline, and various liquid hydrocarbons. These cylindrical containers can measure over 330 ft (100 m) in diameter and up to 65 ft (20 m) in height, with capacities over 40×10^6 gal (150,000 m^3). Municipal water supply companies often build tanks on hilltops so that there is always a reliable supply of pressurized water.

Most tanks are constructed of mild carbon steels, because this is usually the most economic alternative, considering utility, lifespan, construction schedule, and costs. However, a substantial number of tanks are constructed of various materials such as alloy steels, austenitic stainless steels, aluminum alloys, reinforced concrete, glass-reinforced epoxy plastic, and various plastics. There are still some wood tanks in use.

The storage of gases depends on the ability to liquefy gases. Storage tanks for petroleum liquids than can be liquefied at atmospheric storage temperatures are simply pressure vessels. When the required storage capacity is large and the vapor pressure of the liquefied gas is relatively high, refrigeration is used to liquefy the gas and store it at reasonably low tank pressures. An example of this technique is storage of liquefied natural gas. *See* LIQUEFIED NATURAL GAS (LNG). Philip E. Myers

Storm An atmospheric disturbance involving perturbations of the prevailing pressure and wind fields on scales ranging from tornadoes (0.6 mi or 1 km across) to extratropical cyclones (1.2–1900 mi or 2–3000 km across); also, the associated weather (rain storm, blizzard, and the like). Storms influence human activity in such matters as agriculture, transportation, building construction, water impoundment and flood control, and the generation, transmission, and consumption of electric energy. *See* WIND.

The form assumed by a storm depends on the nature of its environment, especially the large-scale flow patterns and the horizontal and vertical variation of temperature; thus the storms most characteristic of a given region vary according to latitude, physiographic features, and season. Extratropical cyclones and anticyclones are the chief disturbances over roughly half the Earth's surface. Their circulations control the embedded smaller-scale storms. Large-scale disturbances of the tropics differ fundamentally from those of extratropical latitudes. *See* HURRICANE; SQUALL; THUNDERSTORM; TORNADO; TROPICAL METEOROLOGY.

Cyclones form mainly in close proximity to the jet stream, that is, in strongly baroclinic regions where there is a large increase of wind with height. Weather patterns in cyclones are highly variable, depending on moisture content and thermodynamic stability of air masses drawn into their circulations. Warm and occluded fronts, east of and extending into the cyclone center, are regions of gradual upgliding motions, with widespread cloud and precipitation but usually no pronounced concentration of stormy conditions. Extensive cloudiness also is often present in the warm sector. Passage of the cold front is marked by a sudden wind shift, often with the onset of gusty conditions, with a pronounced tendency for clearing because of general subsidence behind the front. Showers may be present in the cold air if it is moist and unstable because of heating from the surface. Thunderstorms, with accompanying squalls and heavy rain, are often set off by sudden lifting of warm, moist air at or near the cold front, and these frequently move eastward into the warm sector. *See* CYCLONE; JET STREAM; WEATHER.

Extratropical cyclones alternate with high-pressure systems or anticyclones, whose circulation is generally opposite to that of the cyclone. The circulations of highs are not so intense as in well-developed cyclones, and winds are weak near their centers. In low levels the air spirals outward from a high; descent in upper levels results in warming and drying aloft. Anticyclones fall into two main categories, the warm "subtropical" and the cold "polar" highs.

Between the scales of ordinary air turbulence and of cyclones, there exist a variety of circulations over a middle-scale or mesoscale range, loosely defined as from about one-half up to a few hundred miles. Alternatively, these are sometimes referred to as subsynoptic-scale disturbances because their dimensions are so small that they elude adequate description by the ordinary synoptic network of surface weather stations. Thus their detection often depends upon observation by indirect sensing systems. *See* METEOROLOGICAL SATELLITES; RADAR METEOROLOGY; STORM DETECTION. Chester W. Newton

Storm detection Identifying storm formation, monitoring subsequent storm evolution, and assessing the potential for destruction of life and property through application of various methods and techniques. Doppler radars, satellite-borne instruments, lightning detection networks, and surface observing networks are used to detect the genesis of storms, to diagnose their nature, and to issue warnings when a threat to life and property exists. *See* STORM.

Radar surveillance. Radars emit pulses of electromagnetic radiation that are broadcast in a beam, whose angular resolution is about 1° with a range resolution of about 0.5 km (0.3 mi). The radar beam may intercept precipitation particles in a storm that reflect a fraction of the transmitted energy to the transmitter site (generally called reflectivity or the scatter cross section per unit volume). As the transmitter sweeps out a volume by rotating and tilting the transmitting antenna, the reflectivity pattern of the precipitation particles embodied in the storm is defined. Doppler radars also can measure the velocities of precipitation particles along the beam (radial velocity). Reflectivity and velocity patterns of the storm hydrometeors then make it possible to diagnose horizontal and vertical circulations that may arise within the storm, and to estimate the type and severity of weather elements attending the storm, such as rainfall, hail, damaging winds, and tornadoes. *See* DOPPLER RADAR; PRECIPITATION (METEOROLOGY); RADAR; RADAR METEOROLOGY; WIND.

Satellite surveillance. Since the early 1960s, meteorological data from satellites have had an increasing impact on storm detection and monitoring. In December 1966 the first geostationary *Applications Technology Satellite (ATS 1)* allowed forecasters in the United States to observe storms in animation. A Geostationary Operational Environmental Satellite (GOES) program was initiated within the National Oceanic and Atmospheric Administration (NOAA) with the launch of *GOES 1* in October 1975. The visible and infrared spin scan radiometer (VISSR) provided imagery, which significantly advanced the ability of meteorologists to detect and observe storms by providing frequent-interval visible and infrared imagery of the Earth surface, cloud cover, and atmospheric moisture patterns.

The first of NOAA's next generation of geostationary satellites, *GOES 8* was launched in the spring of 1994. *GOES 8* introduced improved capabilities to detect and observe storms. The *GOES 8* system includes no conflict between imaging and sounding operation, multispectral imaging with improved resolution and better signal-to-noise in the infrared bands, and more accurate temperature and moisture soundings of the storm environment. The Earth's atmosphere is observed nearly continuously.

Derived-product images showing fog and stratus areas from *GOES 8* are created by combining direct satellite measurements, such as by subtracting brightness temperatures at two different wavelengths. *GOES 8* shows the fog and stratus much more clearly because of its improved resolution. This capability enables forecasters to detect boundaries between rain-cooled areas having fog or low clouds, and clear areas. Such boundaries are frequently associated with future thunderstorm development. The sounder on *GOES 8* is capable of fully supporting routine forecasting operations. This advanced sounding capability consists of better vertical resolution in both temperature and moisture, and improved coverage of soundings in and around cloudy weather systems. *See* CLOUD; FOG; METEOROLOGICAL SATELLITES; SATELLITE METEOROLOGY.

Surface observing systems. Larger convective storm systems such as squall lines and mesoscale convective systems can be detected (but not fully described) by the temperature, moisture, wind, and pressure patterns observed by appropriate surface instrumentation. Automatic observing systems provide frequent data on pressure, temperature, humidity, wind, cloud base, and most precipitation types, intensity, and accumulation. Analyses of these data, combined with improved conceptual models of convective storm systems, enable forecasters to detect and monitor the intense mesoscale fluctuations in pressure and winds that often accompany the passage of convective weather systems such as bow echoes, derechos (strong, straight-line winds), and squall lines. A bow echo is a specific radar reflectivity pattern associated with a line of thunderstorms. The middle portion of the thunderstorm line is observed to move faster than the adjacent portions, causing the line of storms to assume a bowed-out configuration. Other analyses of these mesoscale data fields aid the forecaster in detecting favorable areas for thunderstorm cell regeneration, which may produce slowly moving mesoscale convective storms attended by heavy rains and flash floods. *See* SQUALL LINE; WEATHER OBSERVATIONS.

Cloud-to-ground lightning detectors. Lightning location stations provide forecasters with the location, polarity, peak current, and number of strokes in a flash to ground within seconds of the flash occurrence. Useful applications have emerged with regard to the detection and tracking of thunderstorms, squall lines, other mesoscale convective systems, and the weather activity that accompany these phenomena, such as tornadoes and hail. *See* LIGHTNING; MESOMETEOROLOGY; SFERICS; THUNDERSTORM; WEATHER FORECASTING AND PREDICTION. Charles F. Chappell

Storm electricity Processes responsible for the separation of positive and negative electric charges in the atmosphere during storms, including the spectacular manifestation of this charge separation: lightning discharges. Cloud electrification is almost invariably associated with convective activity and with the formation of precipitation in the form of liquid water (rain) and ice particles (graupel and hail). The most vigorous convection and active lightning occurs in the summertime, when the energy source for convection, water vapor, is most prevalent. Winter snowstorms can also be strongly electrified, but they produce far less lightning than summer storms. Electrified storm clouds occasionally occur in complete isolation; more commonly they are found in convective clusters or in lines that may extend horizontally for hundreds of kilometers. *See* PRECIPITATION (METEOROLOGY).

Measurements of electric field at the ground and from instrumented balloons within thunderclouds have disclosed an electrostatic structure that appears to be fairly systematic throughout the world. The measurements show that the principal variations in charge occur in the vertical and are affected by the temperature of the cloud. The charge structure within a thundercloud is tripolar, with a region of dominant negative charge sandwiched between an upper region of positive charge and a subsidiary lower region of positive charge. In addition to the charge accumulations described within the cloud, electrical measurements disclose the existence of charge-screening layers at the upper cloud boundary and a layer of positive charge near the Earth's surface beneath the cloud. These secondary charge accumulations arising from charge migration outside the cloud are caused by electrostatic forces of attraction set up by the charges within the cloud.

Large differences of electric potential are associated with the distribution of charge maintained by active thunderclouds. These large differences in potential are maintained by charging currents that result from the motions of air and particles. The charging currents range from milliamperes in small clouds that are not producing lightning to several amperes for large storms with high rates of lightning. *See* CLOUD PHYSICS.

In response to charge separation within a thundercloud, the electric field increases to a value of approximately 10^6 V/m (300,000 V/ft) at which point dielectric breakdown occurs and lightning is initiated. Most lightning extends through the cloud at speeds of 10,000–100,000 m/s (22,000–220,000 mi/h). The peak temperature of lightning, which is a highly ionized plasma, may exceed 30,000 K (54,000°F). The acoustic disturbance caused by the sudden heating of the atmosphere by lightning is thunder. *See* LIGHTNING; THUNDER.

Meteorologists have shown a growing interest in the large-scale display of real-time lightning activity, since lightning is one

(*a*) Adult male *Hemisquilla californiensis*, 10 in. (25.4 cm) in length. The red patches on the carapace are strongly polarized and occur only in males. (*b*) Adult *Raoulserenea komaii*, a 3.5-in. (8.9-cm) spearing stomatopod from the South Pacific. (*c*) An *Odontodactylus scyllarus* female (5.5 in.; 14 cm) carrying her pink eggs using her maxillipeds. (*d*) A 3-in. (7.6-cm) adult *Neogonodactylus curacaoensis*, a common stomatopod found in cavities in coral throughout the Caribbean. (*Photos by Roy L. Caldwell*)

(a)

(b)

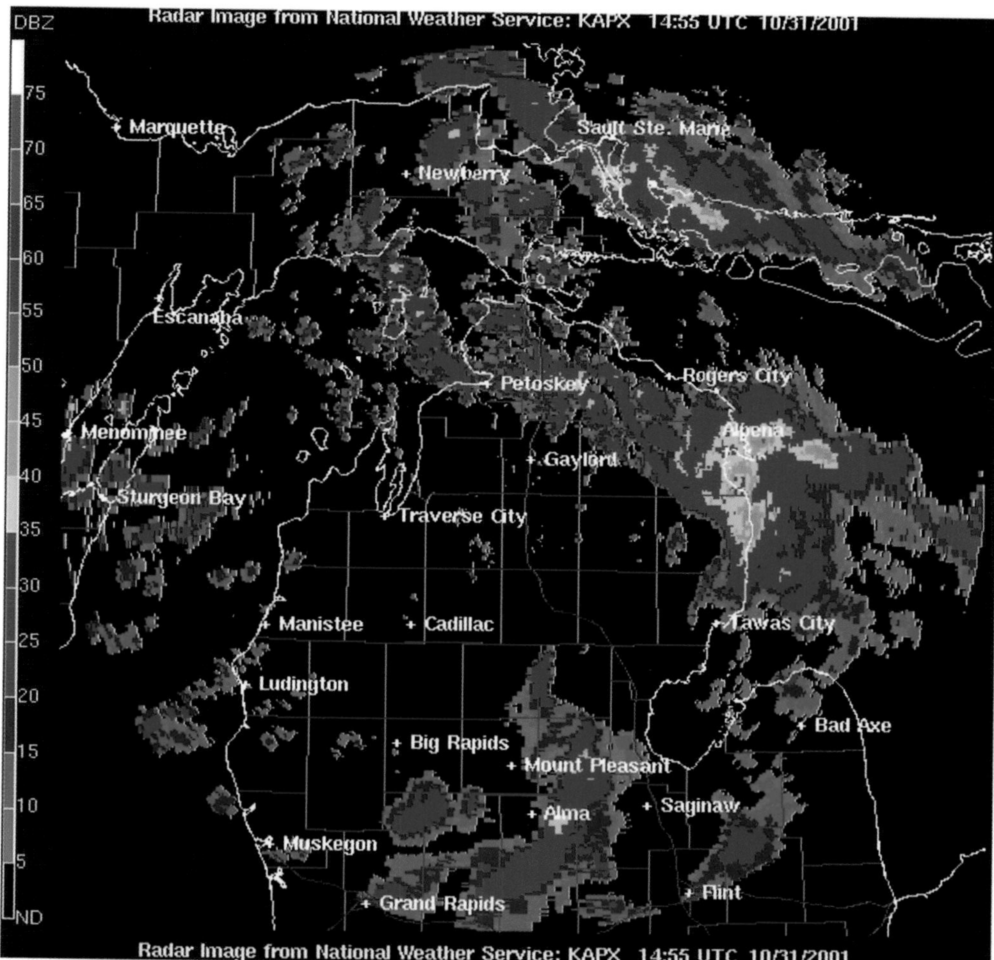

(c)

Images from *GOES 8*. (*a*) In an infrared image, cold clouds are high clouds, so the colors typically highlight the colder regions (*NOAA*). (*b*) In a water vapor image, white areas indicate moisture and dark areas indicate little or no moisture, so the colors typically highlight areas with large amounts of moisture (*NOAA*). (*c*) Short range base reflectivity patterns from the NEXRAD of Gaylord, Michigan, on October 31, 2001. Base reflectivity images display echo intensity (reflectivity) measured in dBZ (decibels of Z; Z represents the energy reflected back to the radar). Maximum range of the short range base reflectivity product is 124 nautical miles (about 143 mi) from the radar location. The colors are the different echo intensities (the amount of transmitted power returned to the radar receiver). Typically, light rain is occurring when the dBZ value reaches 20. (*National Weather Service, NOAA*)

of the most sensitive indicators of convective activity. Research has expanded into relationships between lightning characteristics and the meteorological evolution of different types of storms. The discovery of the sensitive dependence of local lightning activity on the temperature of surface air has led to research focused on the use of the global electrical circuit as a diagnostic for global temperature change. *See* ATMOSPHERIC ELECTRICITY; STORM DETECTION; THUNDERSTORM. Earle Williams

Storm surge An anomalous rise in water elevations caused by severe storms approaching the coast. A storm surge can be succinctly described as a large wave that moves with the storm that caused it. The surge is intensified in the nearshore, shallower regions where the surface stress caused by the strong onshore winds pile up water against the coast, generating an opposing pressure head in the offshore direction. However, there are so many other forces at play in the dynamics of the storm surge phenomenon, such as bottom friction, Earth's rotation, inertia, and interaction with the coastal geometry, that a simple static model cannot explain all the complexities involved. Scientists and engineers have dedicated many years in the development and application of sophisticated computer models to accurately predict the effects of storm surges.

The intensity and dimension of the storm causing a surge, and thus the severity of the ensuing surge elevations, depend on the origin and atmospheric characteristics of the storm itself. Hurricanes and severe extratropical storms are the cause of most significant surges. In general, hurricanes are more frequent in low to middle latitudes, and extratropical storms are more frequent in middle to high latitudes. *See* HURRICANE; STORM. Sergio R. Signorini

Straight-line mechanism A mechanism that produces a straight-line (or nearly so) output motion from an input element that rotates, oscillates, or moves in a straight line. Common machine elements, such as linkages, gears, and cams, are often used in ingenious ways to produce the required controlled motion. The more elegant designs use the properties of special points on one of the links of a four-bar linkage. *See* MECHANISM.

Four-bar linkages that generate approximate straight lines are not new. In 1784 James Watt applied the concept to the vertical-cylinder beam engine. By selecting the appropriate link lengths, the designer can easily develop a mechanism with a high-quality approximate straight line. Contemporary kinematicians have contributed to more comprehensive studies of the properties of the mechanisms that generate approximate straight lines. The work not only describes the various classical mechanisms, but also provides design information on the quality (the amount of deviation from a straight line) and the length of the straight-line output. *See* FOUR-BAR LINKAGE; LINKAGE (MECHANISM).

Gears can also be used to generate straight-line motions. The most common combination would be a rack-and-pinion gear. *See* GEAR.

Cam mechanisms are generally not classified as straight-line motion generators, but translating followers easily fall into the classical definition. *See* CAM MECHANISM. John A. Smith

Strain gage A device which measures mechanical deformation (strain). Normally it is attached to a structural element, and uses the change of electrical resistance of a wire or semiconductor under tension. Capacity, inductance, and reluctance are also used.

The strain gage converts a small mechanical motion to an electrical signal by virtue of the fact that when a metal (wire or foil) or semiconductor is stretched, its resistance is increased. The change in resistance is a measure of the mechanical motion. In addition to their use in strain measurement, these gages are used in sensors for measuring the load on a mechanical member, forces due to acceleration on a mass, or stress on a diaphragm or bellows. *See* STRAIN. John H. Zifcak

Strake The slender forward extension of the inboard region of the wing of a combat aircraft used to provide increased lift in the high angle-of-attack maneuvering condition. In contrast to the normal attached-flow design principles, these strakes are built to allow the flow to separate along the leading edge in the high angle-of-attack range and to roll up into strong leading-edge vortices. The illustration shows a typical strake installation

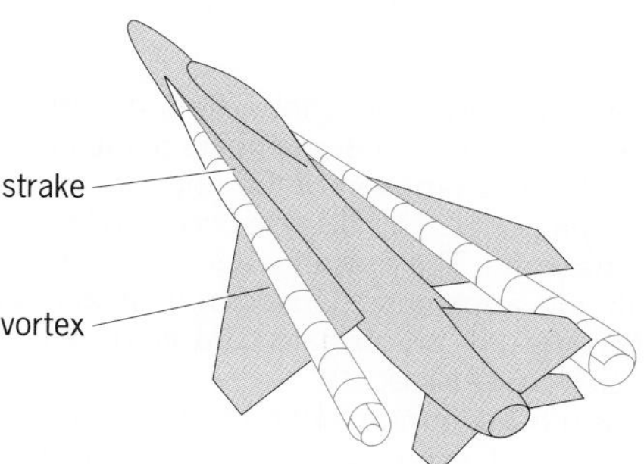

Strake and resulting vortex flow.

and the vortices generated. These are highly stable, and their strong swirling motion creates a lower pressure area on the strake upper surface, resulting in a large incremental increase in lift force known as vortex lift. This vortex lift increases rapidly in the high angle-of-attack range and is less susceptible to the normal stalling characteristics encountered with conventional lifting surfaces. *See* AIRFOIL; SUBSONIC FLIGHT; VORTEX; WING. Edward C. Polhamus

Strand line The line that marks the separation of land and water along the margin of a pond, lake, sea, or ocean; also called the shoreline. The strand line is very dynamic. It changes with the tides, storms, and seasons, and as long-term sea-level changes take place. The sediments on the beach respond to these changes, as do the organisms that live in this dynamic environment. On a beach organisms move with the tides, and on a rocky coast they tend to be organized relative to the strand line because of special limitations or adaptations to exposure.

Geologists who study ancient coastal environments commonly try to establish where the strand line might be in the rock strata. This can sometimes be determined by a combination of the nature and geometry of individual laminations in the rock, by identifying sedimentary structures that occur at or near the strand line. The most indicative of these structures are swash marks, which are very thin accumulations of sand grains that mark the landward uprush of a wave on the beach. *See* FACIES (GEOLOGY); PALEOGEOGRAPHY; STRATIGRAPHY. Richard A. Davis, Jr.

Strange particles Bound states of quarks, in which at least one of these constituents is of the strange (s) type. Strange quarks are heavier than the up (u) and down (d) quarks, which form the neutrons and protons in the atomic nucleus. Neutrons (udd) and protons (uud) are the lightest examples of a family of particles composed of three quarks, known as baryons. These and other composite particles which interact dominantly through the strong (nuclear) force are known as hadrons. The first strange hadron discovered (in cosmic rays in 1947) was named the lambda baryon, Λ; it is made of the three-quark combination uds. A baryon containing a strange quark is also called a hyperon. Although strange particles interact through the strong (nuclear) force, the strange quark itself can decay only by conversion to a quark of different type (such as u or d) through the weak interaction. For this reason, strange particles have very long lifetimes, of the order of 10^{-10} s, compared to the lifetimes of the order of 10^{-23} s for particles which decay directly through the strong interaction. This long lifetime was the origin of the term strange particles. *See* BARYON; HADRON; NEUTRON; PROTON; STRONG NUCLEAR INTERACTIONS.

In addition to strange baryons, strange mesons occur. The lightest of these are the kaons ($K^+ = u\,\bar{s}$ and $K^0 = d\,\bar{s}$) and the antikaons ($\bar{K}^0 = s\,\bar{d}$ and $K^- = s\,\bar{u}$). Kaons and their antiparticles have been very important in the study of the weak interaction and in the detection of the very weak *CP* violation, which causes a slow transition between neutral kaons and neutral antikaons. *See* ELEMENTARY PARTICLE; MESON; QUARKS. Ted Barnes

Strangles A highly contagious disease of the upper respiratory tract of horses and other members of the family Equidae, characterized by inflammation of the pharynx and abscess formation in lymph nodes. This disease occurs in horses of all ages throughout the world. The causative agent is *Streptococcus equi*, a clonal pathogen apparently derived from an ancestral strain of *S. zooepidemicus*. It is an obligate parasite of horses, donkeys, and mules. *See* STREPTOCOCCUS.

Strangles is most common and most severe in young horses, and is very prevalent on breeding farms. The causative agent has been reported to survive for 7 weeks in pus but dies in a week or two on pasture. Transmission is either direct by nose or mouth contact or aerosol, or indirect by flies, drinking buckets, pasture, and feed. The disease is highly contagious under conditions of crowding, exposure to severe climatic conditions such as rain and cold, and prolonged transportation. Carrier animals, although of rare occurrence, are critical in maintenance of the streptococcus and in initiation of outbreaks.

The mean incubation period is about 10 days, with a range of 3–14 days. The animal becomes quieter, has fever of 39–40.5°C (102–105°F), nasal discharge, loss of appetite, and swelling of one or more lymph nodes of the mouth. Pressure of a lymph node on the airway may cause respiratory difficulty. Abscesses in affected lymph nodes rupture in 7–14 days, and rapid clinical improvement and recovery then ensues. Recovery is associated with formation of protective antibodies in the nasopharynx and in the serum. *See* ANTIBODY.

Streptococcus equi is easily demonstrated in smears of pus from abscesses and in culture of pus or nasal swabs on colistin–nalidixic acid blood agar. Acutely affected animals also show elevated white blood cell counts and elevated fibrinogen.

Commercially available vaccines are injected in a schedule of two or three primary inoculations followed by annual boosters. However, the clinical attack rate may be reduced by only 50%, a level of protection much lower than that following the naturally occurring disease. *See* IMMUNITY.

Procaine penicillin G is the antibiotic of choice and quickly brings about reduction of fever and lymph node enlargement. *See* BIOLOGICALS. John F. Timoney

Stratigraphy A discipline involving the description and interpretation of layered sediments and rocks, and especially their correlation and dating. Correlation is a procedure for determining the relative age of one deposit with respect to another. The term "dating" refers to any technique employed to obtain a numerical age, for example, by making use of the decay of radioactive isotopes found in some minerals in sedimentary rocks or, more commonly, in associated igneous rocks. To a large extent, layered rocks are ones that accumulated through sedimentary processes beneath the sea, within lakes, or by the action of rivers, the wind, or glaciers; but in places such deposits contain significant amounts of volcanic material emplaced as lava flows or as ash ejected from volcanoes during explosive eruptions. *See* DATING METHODS; IGNEOUS ROCKS; ROCK AGE DETERMINATION; SEDIMENTARY ROCKS.

Sedimentary successions are locally many thousands of meters thick owing to subsidence of the Earth's crust over millions of years. Sedimentary basins therefore provide the best available record of Earth history over nearly 4 billion years. That record includes information about surficial processes and the varying environment at the Earth's surface, and about climate, changing sea level, the history of life, variations in ocean chemistry, and reversals of the Earth's magnetic field. Sediments also provide a record of crustal deformation (folding and faulting) and of large-scale horizontal motions of the Earth's lithospheric plates (continental drift). Stratigraphy applies not only to strata that have remained flat-lying and little altered since their time of deposition, but also to rocks that may have been strongly deformed or recrystallized (metamorphosed) at great depths within the Earth's crust, and subsequently exposed at the Earth's surface as a result of uplift and erosion. As long as original depositional layers can be identified, some form of stratigraphy can be undertaken. *See* BASIN; CONTINENTAL DRIFT; FAULT AND FAULT STRUCTURES; SEDIMENTOLOGY.

An important idea first articulated by the Danish naturalist Nicolaus Steno in 1669 is that in any succession of strata the oldest layer must have accumulated at the bottom, and successively younger layers above. It is not necessary to rely on the present orientation of layers to determine their relative ages because most sediments and sedimentary rocks contain numerous features, such as current-deposited ripples, minor erosion surfaces, or fossils of organisms in growth position, that have a well-defined polarity with respect to the up direction at the time of deposition (so-called geopetal indicators). This principle of superposition therefore applies equally well to tilted and even overturned strata. Only where a succession is cut by a fault is a simple interpretation of stratigraphic relations not necessarily possible, and in some cases older rocks may overlie younger rocks structurally. *See* DEPOSITIONAL SYSTEMS AND ENVIRONMENTS.

The very existence of layers with well-defined boundaries implies that the sedimentary record is fundamentally discontinuous. Discontinuities are present in the stratigraphic record at a broad range of scales, from that of a single layer or bed to physical surfaces that can be traced laterally for many hundreds of kilometers. Large-scale surfaces of erosion or nondeposition are known as unconformities, and they can be identified on the basis of both physical and paleontological criteria. *See* PALEONTOLOGY; UNCONFORMITY.

Most stratal discontinuities possess time-stratigraphic significance because strata below a discontinuity tend to be everywhere older than strata above. To the extent that unconformities can be recognized and traced widely within a sedimentary basin, it is possible to analyze sedimentary rocks in a genetic framework, that is, with reference to the way they accumulated. This is the basis for the modern discipline of sequence stratigraphy, so named because intervals bounded by unconformities have come to be called sequences.

Traditional stratigraphic analysis has focused on variations in the intrinsic character or properties of sediments and rocks—properties such as composition, texture, and included fossils (lithostratigraphy and biostratigraphy)—and on the lateral tracing of distinctive marker beds such as those composed of ash from a single volcanic eruption (tephrostratigraphy). The techniques of magnetostratigraphy and chemostratigraphy are also based on intrinsic characteristics, although these techniques require sophisticated laboratory analysis. Sequence stratigraphy attempts to integrate these approaches in the context of stratal geometry, thereby providing a unifying framework in which to investigate the time relations between sediment and rock bodies as well as to measure their numerical ages (chronostratigraphy and geochronology). Seismic stratigraphy is a variant of the technique of sequence stratigraphy in which unconformities are identified and traced in seismic reflection profiles on the basis of reflection geometry. *See* GEOCHRONOMETRY; SEISMIC STRATIGRAPHY; SEISMOLOGY.

Conventional stratigraphy currently recognizes two kinds of stratigraphic unit: material units, distinguished on the basis of some specified property or properties or physical limits; and temporal or time-related units. A common example of a material unit is the formation, a lithostratigraphic unit defined on the basis of lithic characteristics and position within a stratigraphic succession. Each formation is referred to a section or locality where it is well developed (a type section), and assigned an appropriate geographic name combined with the word formation or a descriptive lithic term such as limestone, sandstone, or shale (for example, Tapeats Sandstone). Some formations are divisible into two or more smaller-scale units called members and beds. In other cases, formations of similar lithic character or related genesis are combined into composite units called groups and supergroups.

Sequence stratigraphy differs from conventional stratigraphy in two important respects. The first is that basic units (sequences) are defined on the basis of bounding unconformities and correlative conformities rather than material characteristics or age. The second is that sequence stratigraphy is fundamentally not a system for stratigraphic classification, but a procedure for determining how sediments accumulate. *See* SEQUENCE STRATIGRAPHY.

Nicholas Christie-Blick; Bilal U. Haq

Stratosphere The atmospheric layer that is immediately above the troposphere and contains most of the Earth's ozone. Here temperature increases upward because of absorption of solar ultraviolet light by ozone. Since ozone is created in sunlight from oxygen, a by-product of photosynthesis, the stratosphere exists because of life on Earth. In turn, the ozone layer allows life to thrive by absorbing harmful solar ultraviolet radiation. The mixing ratio of ozone is largest (10 parts per million by volume) near an altitude of 30 km (18 mi) over the Equator. The distribution of ozone is controlled by solar radiation, temperature, wind, reactive trace chemicals, and volcanic aerosols. *See* ATMOSPHERE; TROPOSPHERE.

The heating that results from absorption of ultraviolet radiation by ozone causes temperatures generally to increase from the bottom of the stratosphere (tropopause) to the top (stratopause) near 50 km (30 mi), reaching 280 K (45°F) over the summer pole. This temperature inversion limits vertical mixing, so that air typically spends months to years in the stratosphere. *See* TEMPERATURE INVERSION; TROPOPAUSE.

The lower stratosphere contains a layer of small liquid droplets. Typically less than 1 micrometer in diameter, they are made primarily of sulfuric acid and water. Occasional large volcanic eruptions maintain this aerosol layer by injecting sulfur dioxide into the stratosphere, which is converted to sulfuric acid and incorporated into droplets. Enhanced aerosol amounts from an eruption can last several years. By reflecting sunlight, the aerosol layer can alter the climate at the Earth's surface. By absorbing upwelling infrared radiation from the Earth's surface, the aerosol layer can warm the stratosphere. The aerosols also provide surfaces for a special set of chemical reactions that affect the ozone layer. Liquid droplets and frozen particles generally convert chlorine-bearing compounds to forms that can destroy ozone. They also tend to take up nitric acid and water and to fall slowly, thereby removing nitrogen and water from the stratosphere. The eruption of Mount Pinatubo (Philippines) in June 1991 is believed to have disturbed the Earth system for several years, raising stratospheric temperatures by more than 1 K (1.8°F) and reducing global surface temperatures by about 0.5 K (0.9°F). *See* AEROSOL.

Ozone production is balanced by losses due to reactions with chemicals in the nitrogen, chlorine, hydrogen, and bromine families. Reaction rates are governed by temperature, which depends on amounts of radiatively important species such as carbon dioxide. Human activities are increasing the amounts of these molecules and are thereby affecting the ozone layer. Evidence for anthropogenic ozone loss has been found in the Antarctic lower stratosphere. Near polar stratospheric clouds, chlorine and bromine compounds are converted to species that, when the Sun comes up in the southern spring, are broken apart by ultraviolet radiation and rapidly destroy ozone. This sudden loss of ozone is known as the anthropogenic Antarctic ozone hole. *See* STRATOSPHERIC OZONE.

Matthew H. Hitchman

Stratospheric ozone While ozone is found in trace quantities throughout the atmosphere, the largest concentrations are located in the lower stratosphere in a layer between 9 and 18 mi (15 and 30 km). Atmospheric ozone plays a critical role for the biosphere by absorbing the ultraviolet radiation with wavelength (λ) 240–320 nanometers. This radiation is lethal to simple unicellular organisms (algae, bacteria, protozoa) and to the surface cells of higher plants and animals. It also damages the genetic material of cells and is responsible for sunburn in human skin. The incidence of skin cancer has been statistically correlated with the observed surface intensity of ultraviolet wavelength 290– 320 nm, which is not totally absorbed by the ozone layer. *See* OZONE; STRATOSPHERE; ULTRAVIOLET RADIATION (BIOLOGY).

Ozone also plays an important role in photochemical smog and in the purging of trace species from the lower atmosphere. Furthermore, it heats the upper atmosphere by absorbing solar ultraviolet and visible radiation ($\lambda < 710$ nm) and thermal infrared radiation ($\lambda \simeq 9.6$ micrometers). As a consequence, the temperature increases steadily from about −60°F (220 K) at the tropopause (5–10 mi or 8–16 km altitude) to about 45°F (280 K) at the stratopause (30 mi or 50 km altitude). This ozone heating provides the major energy source for driving the circulation of the upper stratosphere and mesosphere. *See* ATMOSPHERIC GENERAL CIRCULATION; TROPOPAUSE.

Above about 19 mi (30 km), molecular oxygen (O_2) is dissociated to free oxygen atoms (O) during the daytime by ultraviolet photons, ($h\nu$), as shown in reaction (1). The oxygen atoms produced then form ozone (O_3) by reaction (2), where M is an

$$O_2 + h\nu \rightarrow O + O \qquad \lambda < 242 \text{ nm} \tag{1}$$

$$O + O_2 + M \rightarrow O_3 + M \tag{2}$$

arbitrary molecule required to conserve energy and momentum in the reaction. Ozone has a short lifetime during the day because of photodissociation, as shown in reaction (3). However,

$$O_3 + h\nu \rightarrow O_2 + O \qquad \lambda < 710 \text{ nm} \tag{3}$$

except above 54 mi (90 km), where O_2 begins to become a minor component of the atmosphere, reaction (3) does not lead to a net destruction of ozone. Instead the O is almost exclusively converted back to O_3 by reaction (2). If the odd oxygen concentration is defined as the sum of the O_3 and O concentrations, then odd oxygen is produced by reaction (1). It can be seen that reactions (2) and (3) do not affect the odd oxygen concentrations but merely define the ratio of O to O_3. Because the rate of reaction (2) decreases with altitude while that for reaction (3) increases, most of the odd oxygen below 36 mi (60 km) is in the form of O_3 while above 36 mi (60 km) it is in the form of O. The reaction that is responsible for a small fraction of the odd oxygen removal rate is shown as reaction (4). A significant fraction

$$O + O_3 \rightarrow O_2 + O_2 \tag{4}$$

of the removal is caused by the presence of chemical radicals [such as nitric oxide (NO), chlorine (Cl), bromine (Br), hydrogen (H), or hydroxyl (OH)], which serve to catalyze reaction (4) (see illustration).

The discovery in the mid-1980s of an ozone hole over Antarctica, which could not be explained by the classic theory of ozone and had not been predicted by earlier chemical models,

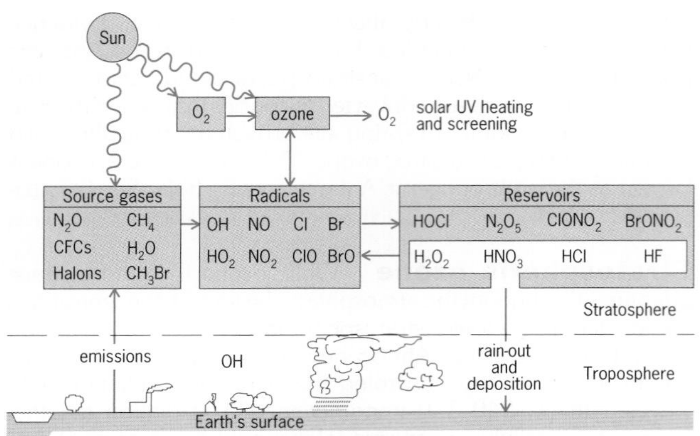

Principal chemical cycles in the stratosphere. The destruction of ozone is affected by the presence of radicals which are produced by photolysis or oxidation of source gases. Chemical reservoirs are relatively stable but are removed from the stratosphere by transport toward the troposphere and rain-out.

led to many speculations concerning the causes of this event, which can be observed each year in September and October. As suggested by experimental and observational evidence, heterogeneous reactions on the surface of liquid or solid particles that produce Cl_2, HOCl, and $ClNO_2$ gas, and the subsequent rapid photolysis of these molecules, produces chlorine radicals (Cl, ClO) which in turn lead to the destruction of ozone in the lower stratosphere by a catalytic cycle [reactions (5)–(7)].

$$Cl + O_3 \rightarrow ClO + O_2 \tag{5}$$

$$ClO + ClO \rightarrow Cl_2O_2 \tag{6}$$

$$Cl_2O_2 + h\nu \rightarrow 2Cl + O_2 \tag{7}$$

Solar radiation is needed for these processes to occur.

Sites on which the reactions producing Cl_2, HOCl, and $ClNO_2$ can occur are provided by the surface of ice crystals in polar stratospheric clouds (PSCs). These clouds are formed between 8 and 14 mi (12 and 22 km) when the temperature drops below approximately −123°F (187 K). Other types of particles are observed at temperatures above the frost point of −123°F (187 K). These particles provide additional surface area for these reactions to occur. Clouds are observed at high latitudes in winter. Because the winter temperatures are typically 20–30°F (10–15 K) colder in the Antarctic than in the Arctic, their frequency of occurrence is highest in the Southern Hemisphere. Thus, the formation of the springtime ozone hole over Antarctica is explained by the activation of chlorine and the catalytic destruction of O_3 which takes place during September, when the polar regions are sunlit but the air is still cold and isolated from midlatitude air by a strong polar vortex. Satellite observations made since the 1970s suggest that total ozone in the Arctic has been abnormally low during the 1990s, probably in relation to the exceptionally cold winter temprature in the Arctic lower stratosphere recorded during that decade. Guy P. Brasseur; Ronald G. Prinn

Strawberry Low-growing perennials, spreading by stolons, with fruit consisting of a fleshy receptacle, and "seeds" in pits or nearly superficial on the receptacle. The strawberry in the United States is derived from two species: *Fragaria chiloensis*, which grows along the Pacific Coast of North and South America, and *F. virginiana*, the eastern meadow strawberry, both members of the order Rosales. *See* ROSALES.

The strawberry is the most universally grown of the small fruits, both in the home garden and in commercial plantings. Home garden production is possible in nearly all of the states, provided water can be supplied where rainfall is insufficient. Commercial production is important in probably three-fourths of the states. The following states are large producers: Oregon, California, Tennessee, Michigan, Louisiana, Washington, Arkansas, Kentucky, and New York. *See* FRUIT. J. Harold Clarke

Stream function In fluid mechanics, a mathematical idea which satisfies identically, and therefore eliminates completely, the equation of mass conservation. If the flow field consists of only two space coordinates, for example, x and y, a single and very useful stream function $\psi(x, y)$ will arise. If there are three space coordinates, such as (x, y, z), multiple stream functions are needed, and the idea becomes much less useful and is much less widely employed.

The stream function not only is mathematically useful but also has a vivid physical meaning. Lines of constant ψ are streamlines of the flow; that is, they are everywhere parallel to the local velocity vector. No flow can exist normal to a streamline; thus, selected ψ lines can be interpreted as solid boundaries of the flow.

Further, ψ is also quantitatively useful. In plane flow, for any two points in the flow field, the difference in their stream function values represents the volume flow between the points. *See* CREEPING FLOW; FLUID FLOW. Frank M. White

Stream gaging The measurement of water discharge in streams. Discharge is the rate of movement of the stream's water volume. It is the product of water velocity times cross-sectional area of the stream channel. Several techniques have been developed for measuring stream discharge; selection of the gaging method usually depends on the size of the stream. The most accurate methods for measuring stream discharge make use of in-stream structures through which the water can be routed, such as flumes and weirs.

A flume is a constructed channel that constricts the flow through a control section, the exact dimensions of which are known. Through careful hydraulic design and calibration by laboratory experiments, stream discharge through a flume can be determined by simply measuring the water depth (stage) in the inlet or constricted sections. Appropriate formulas relate stage to discharge for the type of flume used.

A weir is used in conjunction with a dam in the streambed. The weir itself is usually a steel plate attached to the dam that has a triangular, rectangular, or trapezoidal notch over which the water flows. Hydraulic design and experimentation has led to calibration curves and appropriate formulas for many different weir designs. To calculate stream discharge through a weir, only the water stage in the reservoir created by the dam needs to be measured. Stream discharge can be calculated by using the appropriate formula that relates stage to discharge for the type of weir used. *See* HYDROLOGY; SURFACE WATER. Thomas C. Winter

Stream pollution Biological, or bacteriological, pollution in a stream indicated by the presence of the coliform group of organisms. While nonpathogenic itself, this group is a measure of the potential presence of contaminating organisms. Because of temperature, food supply, and predators, the environment provided by natural bodies of water is not favorable to the growth of pathogenic and coliform organisms. Physical factors, such as flocculation and sedimentation, also help remove bacteria. Any combination of these factors provides the basis for the biological self-purification capacity of natural water bodies.

Nonpolluted natural waters are usually saturated with dissolved oxygen. They may even be supersaturated because of the oxygen released by green water plants under the influence of sunlight. When an organic waste is discharged into a stream, the dissolved oxygen is utilized by the bacteria in their metabolic processes to oxidize the organic matter. The oxygen is replaced by reaeration through the water surface exposed to the atmosphere.

This replenishment permits the bacteria to continue the oxidative process in an aerobic environment. In this state, reasonably clean appearance, freedom from odors, and normal animal and plant life are maintained.

An increase in the concentration of organic matter stimulates the growth of bacteria and increases the rates of oxidation and oxygen utilization. If the concentration of the organic pollutant is so great that the bacteria use oxygen more rapidly than it can be replaced, only anaerobic bacteria can survive and the stabilization of organic matter is accomplished in the absence of oxygen. Under these conditions, the water becomes unsightly and malodorous, and the normal flora and fauna are destroyed. Furthermore, anaerobic decomposition proceeds at a slower rate than aerobic. For maintenance of satisfactory conditions, minimal dissolved oxygen concentrations in receiving streams are of primary importance.

Municipal sewage and industrial wastes affect the oxygen content of a stream. Cooling water, used in some industrial processes, is characterized by high temperatures, which reduce the capacity of water to hold oxygen in solution. Municipal sewage requires oxygen for its stabilization by bacteria. Oxygen is utilized more rapidly than it is replaced by reaeration, resulting in the death of the normal aquatic life. Further downstream, as the oxygen demands are satisfied, reaeration replenishes the oxygen supply.

Polluted waters are deprived of oxygen by the exertion of the biochemical oxygen demand, which is defined as the quantity of oxygen required by the bacteria to oxidize the organic matter. Factors such as the turbulence of the stream flow, biological growths on the stream bed, insufficient nutrients, and inadequate bacteria in the river water influence the rate of oxidation in the stream as well as the removal of organic matter.

When a significant portion of the waste is in the suspended state, settling of the solids in a slow-moving stream is probable. The organic fraction of the sludge deposits decomposes anaerobically, except for the thin surface layer which is subjected to aerobic decomposition due to the dissolved oxygen in the overlying waters. In warm weather, when the anaerobic decomposition proceeds at a more rapid rate, gaseous end products, usually carbon dioxide and methane, rise through the supernatant waters. The evolution of the gas bubbles may raise sludge particles to the water surface. Although this phenomenon may occur while the water contains some dissolved oxygen, the more intense action during the summer usually results in depletion of dissolved oxygen.

Water may absorb oxygen from the atmosphere when the oxygen in solution falls below saturation. Dissolved oxygen for receiving waters is also derived from two other sources: that in the receiving water and the waste flow at the point of discharge, and that given off by green plants.

Unpolluted water maintains in solution the maximum quantity of dissolved oxygen. The saturation value is a function of temperature and the concentration of dissolved substances, such as chlorides. When oxygen is removed from solution, the deficiency is made up by the atmospheric oxygen, which is absorbed at the water surface and passes into solution. The oxygen balance in a stream is determined by the concentration of organic matter and its rate of oxidation, and by the dissolved oxygen concentration and the rate of reaeration. *See* ESTUARINE OCEANOGRAPHY; FRESHWATER ECOSYSTEM; SEWAGE DISPOSAL; WATER POLLUTION.

Donald J. O'Connor

Stream transport and deposition The sediment debris load of streams is a natural corollary to the degradation of the landscape by weathering and erosion. Eroded material reaches stream channels through rills and minor tributaries, being carried by the transporting power of running water and by mass movement, that is, by slippage, slides, or creep. The size represented may vary from clay to boulders. At any place in the stream system the material furnished from places upstream either is carried away or, if there is insufficient transporting ability, is accumulated as a depositional feature. The accumulation of deposited debris tends toward increased ease of movement, and this tends eventually to bring into balance the transporting ability of the stream and the debris load to be transported.

Luna B. Leopold

Streaming potential The potential which is produced when a liquid is forced to flow through a capillary or a porous solid. The streaming potential is one of four related electrokinetic phenomena which depend upon the presence of an electrical double layer at a solid-liquid interface. This electrical double layer is made up of ions of one charge type which are fixed to the surface of the solid and an equal number of mobile ions of the opposite charge which are distributed through the neighboring region of the liquid phase. In such a system the movement of liquid over the surface of the solid produces an electric current, because the flow of liquid causes a displacement of the mobile counterions with respect to the fixed charges on the solid surface. The applied potential necessary to reduce the net flow of electricity to zero is the streaming potential.

Quentin Van Winkle

Streamlining The contouring of a body to reduce its resistance (drag) to motion through a fluid.

For fluids with relatively low viscosity such as water and air, effects of viscous friction are confined to a thin layer of fluid on the surface termed the boundary layer. Under the influence of an increasing pressure, the flow within the boundary layer tends to reverse and flow in an upstream direction. Viscosity tends to cause the flow to separate from the body surface with consequent formation of a region of swirling or eddy flow (termed the body wake; illus. *a*). This eddy formation leads to a reduction in the downstream pressure on the body and hence gives rise to a force opposite to the body motion, known as pressure drag. *See* WAKE FLOW.

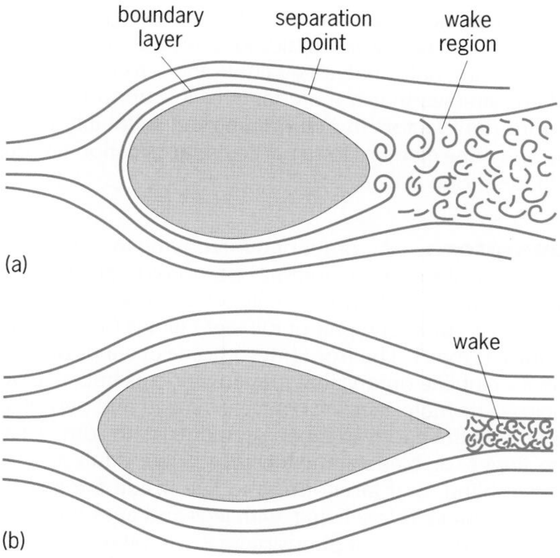

Flow about bodies in uniform subsonic flow. (*a*) Blunt body. (*b*) Streamlined body.

In general, streamlining in subsonic flow involves the contouring of the body in such a manner that the wake is reduced and hence the pressure drag is reduced. The contouring should provide for gradual deceleration to avoid flow separation, that is, reduced adverse pressure gradients. These considerations lead to the following general rules for subsonic streamlining: The forward portion of the body should be well rounded, and the body should curve back gradually from the forward section to a tapering aftersection with the avoidance of sharp corners along the body surface. These conditions are well illustrated by teardrop shapes (illus. *b*).

At supersonic speeds the airflow can accommodate sudden changes in direction by being compressed or expanded. Where this change in direction occurs at the nose of the body, a compression wave is created, the strength of which depends upon the magnitude of the change in flow direction. Lowering the body-induced flow angle weakens this compression shock wave. When the flow changes direction again at the midpoint of the body, the air will expand to follow the shape of the body. This change in direction creates expansion waves. At the tail of the body the direction changes again, creating another compression or shock wave. At each of these shock waves, changes in pressure, density, and velocity occur, and in this process energy is lost. This loss results in a retarding force known as wave drag. *See* SHOCK WAVE.

Bodies which are streamlined for supersonic speeds are characterized by a sharp nose and small flow deflection angles. Because the intensity of the shock wave and the drag level is dependent upon the magnitude of the change in flow direction, the width or thickness of the body should be minimal. *See* BOUNDARY-LAYER FLOW. Arthur G. Hansen; Dennis M. Bushnell

Strength of materials A branch of applied mechanics concerned with the behavior of materials under load, relationships between externally applied loads and internal resisting forces, and associated deformations. Knowledge of the properties of materials and analysis of the forces involved are fundamental to the investigation and design of structures and machine elements. *See* MACHINE DESIGN; STRUCTURAL MATERIALS.

Investigation of the resistance of a member, dealing with internal forces, is called free-body analysis. Determination of the distribution and intensity of the internal forces and the associated deformations is called stress analysis. *See* STRESS AND STRAIN.

A material offers resistance to external load only insofar as the component elements can furnish cohesive strength, resistance to compaction, and resistance to sliding. The relations developed in strength of materials analysis evaluate the tensile, compressive, and shear stresses that a material is called upon to resist. The most important factors in determining the suitability of a structural or machine element for a particular application are strength and stiffness. *See* SHEAR. John B. Scalzi

Strepsiptera An order of twisted-wing insects that spend most of their life cycle as internal parasitoids of other insects. The adult male is only a few millimeters in wing span, is free living, and has only one pair of full flight wings (the posterior, or metathoracic, pair). The front (mesothoracic) wings are reduced to narrow, clublike organs that may function as halteres, or flight balancers. The male eyes are coarsely faceted and berrylike, and the antennae have four to seven segments, with some segments having finger- or bladelike extensions. Most adult females are immobile, blind, and larviform, and live inside the insect host. Rarely, the female is free living, with legs and eyes but no wings. More than 600 species of Strepsiptera are known, many of them not yet formally described and named. The order's relationship to other insect orders remains uncertain.

Both male and female begin larval life as mobile first-stage organisms with eyes and three pairs of functional legs. These triunguloids eventually attack the immature or adult forms of the host and enter their bodies, where they molt to an apodous instar. Two subsequent molts result in the divergence of the sexes and the differentiation of a hardened forebody that eventually protrudes through the host integument; the molted larval integuments are not shed but are incorporated in the general puparial wall arid the neotenous adult female capsule. Males (and some adult females) emerge from the puparium and begin the search for a mate, probably through the mediation of airborne pheromones.

Strepsiptera are found worldwide in temperate and tropical habitats. They parasitize insects, mostly of orders Hemiptera and Hymenoptera, but also some cockroaches, mantids, orthopterans, flies, and silverfish. Their effect on the host is variable, ranging from reproductive failure to death. *See* ENDOPTERYGOTA; PHEROMONE. William L. Brown, Jr.

Streptococcus A large genus of spherical or ovoid bacteria that are characteristically arranged in pairs or in chains resembling strings of beads. Many of the streptococci that constitute part of the normal flora of the mouth, throat, intestine, and skin are harmless commensal forms; other streptococci are highly pathogenic. The cells are gram-positive and can grow either anaerobically or aerobically, although they cannot utilize oxygen for metabolic reactions. Glucose and other carbohydrates serve as sources of carbon and energy for growth. All members of the genus lack the enzyme catalase. Streptococci can be isolated from humans and other animals.

Streptococcus pyogenes is well known for its participation in many serious infections. It is a common cause of throat infection, which may be followed by more serious complications such as rheumatic fever, glomerulonephritis, and scarlet fever. Other beta-hemolytic streptococci participate in similar types of infection, but they are usually not associated with rheumatic fever and glomerulonephritis. Group B streptococci, which are usually beta-hemolytic, cause serious infections in newborns (such as meningitis) as well as in adults. Among the alpha-hemolytic and nonhemolytic streptococci, *S. pneumoniae* is an important cause of pneumonia and other respiratory infections. Vaccines that protect against infection by the most prevalent capsular serotypes are available. The viridans streptococci comprise a number of species commonly isolated from the mouth and throat. Although normally of low virulence, these streptococci are capable of causing serious infections (endocarditis, abscesses). *See* PNEUMONIA. Kathryn Ruoff

Streptococcus pneumoniae (pneumococcus) The major causative microorganism of lobar pneumonia. Pneumococci occur singly, as pairs, or in short chains of oval or lancet-shaped cocci, 0.05–1.25 micrometers each, flattened at proximal sides and pointed at distal ends. A capsule of polysaccharide envelops each cell or pair of cells. The microorganism is nonmotile and stains gram-positive unless degenerating. *See* PNEUMONIA; STREPTOCOCCUS.

The pneumococcus resembles many species of alpha-hemolytic streptococci. Pneumococci have been isolated from the upper respiratory tract of healthy humans, monkeys, calves, horses, and dogs. Epizootics (outbreaks) of pneumococcal infection have been described in monkeys, guinea pigs, and rats, but are not the source of human infection. In humans, pneumococci may be found in the upper respiratory tract of nearly all individuals at one time or another. Following damage to the epithelium lining the respiratory tract, most commonly of viral origin, pneumococci may invade the lungs. They are the principal cause of lobar pneumonia in humans and may cause pleural empyema, pericarditis, endocarditis, meningitis, peritonitis, arthritis, and infection of the middle ear (otitis media) or paranasal sinuses. Approximately one in four cases of pneumococcal pneumonia is accompanied by invasion of the bloodstream by pneumococci, producing bacteremia (presence of bacteria in the blood).

Although the high mortality of untreated pneumococcal infection has been reduced significantly by treatment with antibiotics, one of every six patients with bacterial lobar pneumonia still succumbs despite optimal therapy. In addition, the number of pneumococcal strains resistant to one or more antimicrobial drugs has been gradually but steadily increasing. For these reasons, prophylactic vaccination is recommended, especially for those segments of the population that are at high risk of fatal infection. A polyvalent vaccine containing the purified capsular polysaccharides of the 23 types that are responsible for 80% of bacterial pneumococcal infection, and with an aggregate efficacy of 65–70% in preventing infection with any of the types represented in it, is available. *See* IMMUNOLOGY; OPPORTUNISTIC INFECTIONS; VACCINATION.

Robert Austrian

Streptothricosis An acute or chronic infection of the epidermis, caused by the bacterium *Dermatophilus congolensis*. which results in an oozing dermatitis with scab formation. Streptothricosis includes dermatophilosis, mycotic dermatitis, lumpy wool, strawberry foot-rot, and cutaneous streptothricosis—diseases having a worldwide distribution and affecting a wide variety of species, including humans.

The infectious form of the bacterium is a coccoid, motile zoospore that is released when the skin becomes wet. Thus, the disease is closely associated with rainy seasons and wet summers. Zoospores lodge on the skin of susceptible animals and germinate by producing filaments which penetrate to the living epidermis, where the organism proliferates by branching mycelial growth.

Early cutaneous lesion of dermatophilosis in cattle reveals small vesicles, papules, and pus formation under hair plaques. An oozing dermatitis then appears as the disease progresses and exudates coalesce to form scabs, which change to hard crusts firmly adherent to the skin. The crusts enlarge and harden, and are often devoid of hair. Lesions occur on most areas of sheep, but the characteristic lesions in the wooled areas occur as numerous hard masses of crust or scab scattered irregularly over the back, flanks, and upper surface of the neck. Lesion resolution has been found to correlate with the presence of immunoglobulin A–containing plasma cells in the dermis and with the antibody levels to *D. congolensis* at the skin surface of infected sheep and cattle.

No single treatment is considered specific for dermatophilosis. Some observers claim that topical agents are successful for sheep and cattle. A number of systemic antibiotics are effective in treating the disease. A combination of streptomycin and penicillin has given good therapeutic results in both bovine and equine infections.

Shelley S. Sutherland

Stress (psychology) Generally, environmental events of a challenging sort as well as the body's response to such events. Of particular interest has been the relationship between stress and the body's adaptation to it on the one hand and the body's susceptibility to disease on the other. Both outcomes involve behavioral and brain changes as well as psychosomatic events, that is, changes in body function arising from the ability of the brain to control such function through neural output as well as hormones. One problem is that both environmental events and bodily responses have been referred to interchangeably as stress. It is preferable to refer to the former as the stressor and the latter as the stress response. The stress response consists of a cascade of neural and hormonal events that have short- and long-lasting consequences for brain and body alike. A more serious issue is how an event is determined to be a stressor. One view is to define a stressor as an environmental event causing a negative outcome, such as a disease. Another approach is to view stressors as virtually any challenge to homeostasis and to regard disease processes as a failure of the normal operation of adapative mechanisms, which are part of the stress response. With either view, it is necessary to include psychological stressors, such as fear, that contain implied threats to homeostasis and that evoke psychosomatic reactions. These are reactions that involve changes in neural and hormonal output caused by psychological stress. Psychosomatic reactions may lead to adaptive responses, or they may exacerbate disease processes. Whether the emphasis is on adaptation or disease, it is essential to understand the processes in the brain that are activated by stressors and that influence functions in the body. *See* HOMEOSTASIS; PSYCHOSOMATIC DISORDERS.

Among the many neurotransmitter systems activated by stress is noradrenaline, produced by neurons with cell bodies in the brainstem that have vast projections up to the forebrain and down the spinal cord. Stressful experiences activate the noradrenergic system and promote release of noradrenaline; severe stress leads to depletion of noradrenaline in brain areas such as the hypothalamus. This release and depletion of noradrenaline stores results in changes at two levels of neuronal function: phosphorylation is triggered by the second-messenger cyclic AMP and occurs in the presynaptic and postsynaptic sites where noradrenaline is released and where it also acts; synthesis of new protein is induced via actions on the genome. Both processes enhance the ability of the brain to form noradrenaline when the organism is once again confronted with a stressful situation. Other neurotransmitter systems may also show similar adaptive changes in response to stressors. *See* NORADRENERGIC SYSTEM.

Stress also activates the neurally mediated discharge of adrenaline from the adrenal medulla and of hypothalamic hormones that initiate the neuroendocrine cascade, culminating in glucocorticoid release from the adrenal cortex. Thus, the activity of neurons triggered by stressful experiences, physical trauma, fear, or anger leads to hormone secretion that has effects throughout the body. Virtually every organ of the body is affected by stress hormones. The hypothalamic hormone (corticotrophin-releasing hormone) that triggers the neuroendocrine cascade directly stimulates the pituitary to secrete ACTH. In response to certain stressors, the hypothalamus also secretes vasopressin and oxytocin, which act synergistically with corticotrophin-releasing hormone on the pituitary to potentiate the secretion of ACTH. Various stressors differ in their ability to promote output of vasopressin and oxytocin, but all stressors stimulate release of corticotrophin-releasing hormone. Other hormones involved in the stress response include prolactin and thyroid hormone; the metabolic hormones insulin, epinephrine, and glucagon; and the endogenous opiates endorphin and enkephalin. *See* ENDORPHINS.

Of all the hormones in the endocrine cascade initiated by stress, the glucocorticoids are the most important because of their widespread effects throughout the body and in the brain. The brain contains target cells for adrenal glucocorticoids secreted in stress, and receptors in these cells are proteins that interact with the genome to affect expression of genetic information. Thus, the impact of stress-induced activation of the endocrine cascade that culminates in glucocorticoid release is the feedback of glucocorticoids on target brain cells. The effect is to alter the structure and function of the brain cells over a period of time ranging from hours to days.

In the case of noradrenaline, glucocorticoids have several types of feedback effects that modify how the noradrenergic system responds to stress. Glucocorticoids inhibit noradrenaline release, and they reduce the second-messenger response of brain structures such as the cerebral cortex to noradrenaline. Glucocorticoid feedback also affects the serotonin system, facilitating serotonin formation during stress but at the same time altering the levels of several types of serotonin receptors in different brain regions, which has the net effect of shifting the balance within the serotonergic system. Taken together, evidence points to a role of glucocorticoid secretion in leading to restoration of homeostatic balance by counteracting the acute neural events such as increased activity of noradrenaline and serotonin, which are turned on by stressful experiences. Other neurotransmitter systems may also respond to glucocorticoid action. Moreover, the other hormones activated by stress have effects on the brain and body that must be considered. *See* SEROTONIN.

In general, stress hormones are protective and adaptive in the immediate aftermath of stress, and the organism is more vulnerable to many conditions without them. However, the same hormones can promote damage and accelerate pathophysiological changes, such as bone mineral loss, obesity, and cognitive impairment, when they are overproduced or not turned off. This wear-and-tear on the body has been called allostatic load. It is based upon the notion that allostasis is the active process of maintaining stability, or homeostasis, through change, and allostatic load is the almost inevitable cost to the body of doing so.

Stress hormone actions have important effects outside the brain on such systems as the immune response. Glucocorticoids and catecholamines from sympathetic nerves and the adrenal medulla participate in the mobilization and enhancement of immune function in the aftermath of acute stress. These effects improve the body's defense against pathogens but can exacerbate autoimmune reactions. When they are secreted chronically, the stress-related hormones are generally immunosuppressive; such effects can be beneficial in the case of an autoimmune disease but may compromise defense against a virus or bacterial infections. At the same time, glucocorticoids are important agents for containing the acute-phase response to an infection or autoimmune disturbance. In the absence of such containment, the organism may die because of the excessive inflammatory response. *See* IMMUNOLOGY.

Besides affecting the immune response, stressors are believed to exacerbate endogenous depressive illness in susceptible individuals. Major depressive illness frequently results in elevated levels of cortisol in the blood. It is not clear whether the elevated cortisol is a cause or strictly a result of the brain changes involved in depressive illness. *See* AFFECTIVE DISORDERS. Bruce McEwen

Stress and strain Related terms defining the intensity of internal reactive forces in a deformed body and associated unit changes of dimension, shape, or volume caused by externally applied forces. Stress is a measure of the internal reaction between elementary particles of a material in resisting separation, compaction, or sliding that tend to be induced by external forces. Total internal resisting forces are resultants of continuously distributed normal and parallel forces that are of varying magnitude and direction and are acting on elementary areas throughout the material. These forces may be distributed uniformly or nonuniformly. Stresses are identified as tensile, compressive, or shearing, according to the straining action.

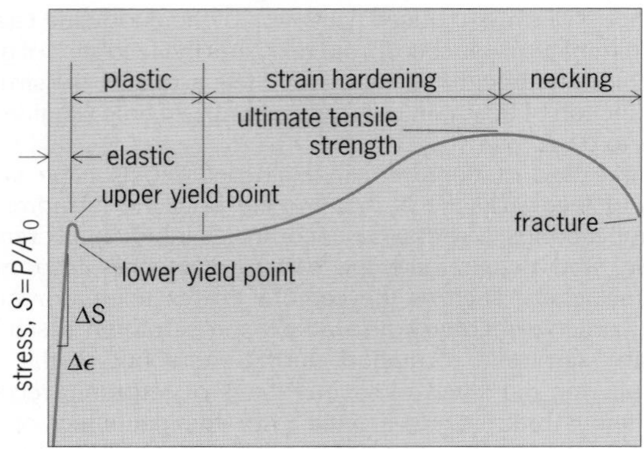

Stress-strain diagram for a low-carbon steel. ΔS = change in stress; $\Delta \epsilon$ = change in strain; P = force; A_0 = area of cross section.

Strain is a measure of deformation such as (1) linear strain, the change of length per unit of linear dimensions; (2) shear strain, the angular rotation in radians of an element undergoing change of shape by shearing forces; or (3) volumetric strain, the change of volume per unit of volume. The strains associated with stress are characteristic of the material. Strains completely recoverable on removal of stress are called elastic strains. Above a critical stress, both elastic and plastic strains exist, and that part remaining after unloading represents plastic deformation called inelastic strain. Inelastic strain reflects internal changes in the crystalline structure of the metal. Increase of resistance to continued plastic deformation due to more favorable rearrangement of the atomic structure is strain hardening.

A stress-strain diagram is a graphical representation of simultaneous values of stress and strain observed in tests and indicates material properties associated with both elastic and inelastic behavior (see illustration). It indicates significant values of stress-accompanying changes produced in the internal structure. *See* ELASTICITY; STRENGTH OF MATERIALS. John B. Scalzi

Strigiformes The owls, an order of nocturnal predacious birds found worldwide that are probably related most closely to the goatsuckers. The strigiforms are arranged in four families: Ogygoptyngidae (fossil; Paleocene of North America), Protostrigidae (fossil; Eocene of North America), Tytonidae (barn owls; 11 species; worldwide), and Strigidae (typical owls; 135 species; worldwide). The bay owls (*Phodilus*), occurring in southern Asia and Africa, are somewhat intermediate between the Tytonidae and the Strigidae and are sometimes placed in a separate family; here they are treated as a subfamily, Phodilinae, of the Tytonidae. *See* CAPRIMULGIFORMES.

Owls are predators of small to medium size with a strong hooked bill and strong claws, and with soft plumage of somber colors. The large head has a facial disk that covers the feathered parabolic reflectors of the bird's acute hearing system. The eyes are large and capable of sight in very dim light. The wings are long and rounded, and the flight feathers are fringed for silent flight. The ulna has a unique bony arch on the shaft that supports a series of sensory receptors which may be associated with their silent flight. Their strong legs are short to medium in length and terminate in strong feet. Owls are excellent fliers, but walk poorly. Of the four toes on each foot, two point forward and two backward; all bear strong claws. Prey is detected by acute night vision or by directional hearing; some species of owls can locate and catch their prey in total darkness.

Owls are generally nonmigratory and solitary, but some species live in small flocks. A few northern species migrate. Courtship takes place at night with a male hooting to a female, which answers. A strong pair bond exists between the monogamous male and female, which usually build their nest in a tree cavity or take over the abandoned nest of some other large bird. The clutch of up to seven eggs is incubated by both sexes; after hatching, the young stay in the nest and are cared for by both parents. *See* AVES. Walter J. Bock

Stripping (chemical engineering) The removal of one or more components from a mixed system. Usually the components being removed are impurities or are considered less valuable than the material from which they are stripped. Examples of stripping operations are (1) the removal of overburden from a deposit of coal or ore, (2) the removal of color from a dyed fabric using a reactive agent, (3) the removal of an organic or a metallic coating from a solid surface, (4) the removal of a dissolved component from a liquid by extraction using a second liquid which is insoluble in the first, and (5) the removal of a volatile component from a liquid by vaporization. *See* DISTILLATION; DYEING; ELECTROPLATING OF METALS; SOLVENT EXTRACTION.

The stripping operation is an important step in many industrial processes that use absorption to purify gases or recover valuable components from the gas phase. In such processes, the solvent leaving the absorber (the rich solution) must be stripped of absorbed materials in order to regenerate the solvent for recycle to

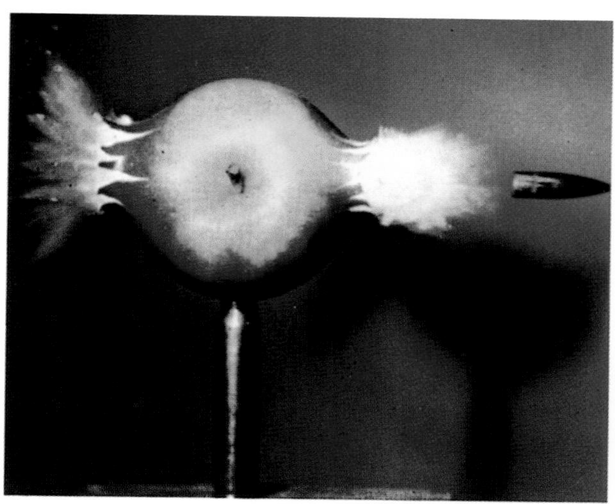

Stroboscopic photographs of bullets in flight and of a milk drop. (*EG&G, Inc.*)

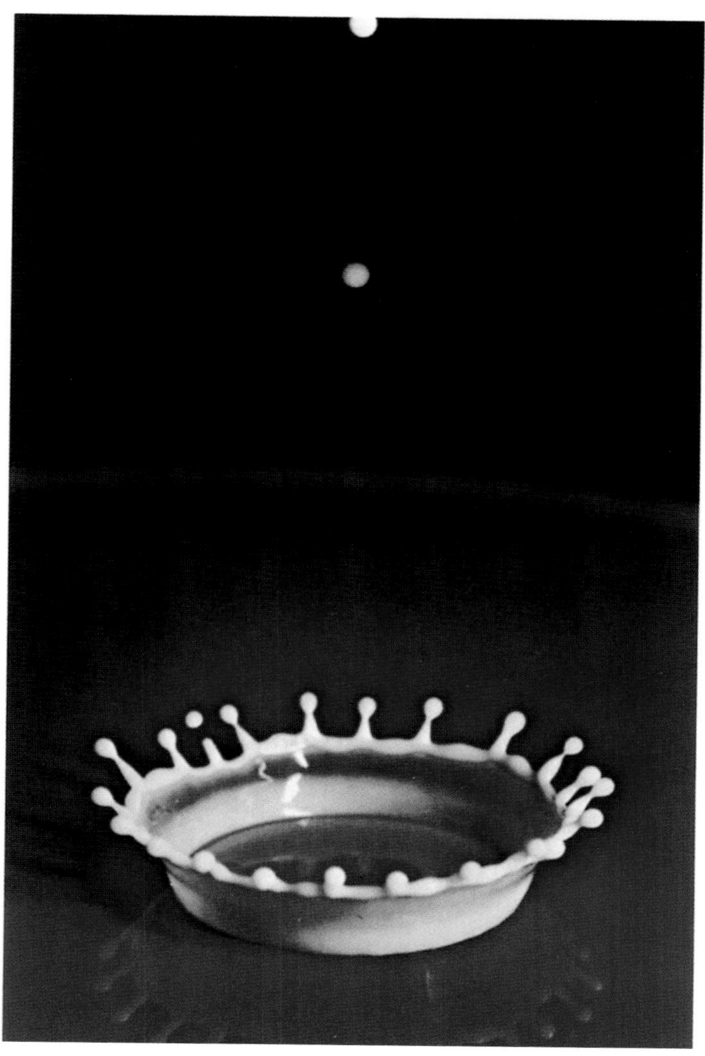

the absorber (as lean solution). The stripping operation also provides a concentrated stream of solute components for possible recovery.

The stripping of a volatile component from a liquid may be accomplished by pressure reduction, the application of heat, or the use of an inert gas (stripping vapor). Many processes employ a combination of all three; that is, after absorption at elevated pressure, the solvent is flashed to atmospheric pressure, heated, and admitted into a stripping column which is provided with a bottom heater (reboiler). Solvent vapor generated in the reboiler or inert gas injected at the bottom of the column serves as stripping vapor which rises countercurrently to the downflowing solvent. When steam is used as stripping vapor for a solvent not miscible with water, the process is called steam stripping. Arthur L. Kohl

Stroboscope An instrument for observing moving bodies by making them visible intermittently and thereby giving them the optical illusion of being stationary. A stroboscope (sometimes called strobe) may operate by illuminating the object with brilliant flashes of light or by imposing an intermittent shutter between the viewer and the object.

Stroboscopes are used to measure the speed of rotation or frequency of vibration of a mechanical part or system. They have the advantage over other instruments of not loading or disturbing the equipment under test. Mechanical equipment may be observed under actual operating conditions with the aid of stroboscopes. Parasitic oscillations, flaws, and unwanted distortion at high speeds are readily detected.

The flashing-light stroboscopes employ gas discharge tubes to provide a brilliant light source of very short duration. Tubes may vary from neon glow lamps, when very little light output is required, to special stroboscope tubes capable of producing flashes of several hundred thousand candlepower with duration of only a few millionths of a second. *See* NEON GLOW LAMP; VAPOR LAMP. Arthur R. Eckels

Stroboscopic photography Stroboscopic or "strobe" photography generally refers to pictures of both single and multiple exposure taken by flashes of light from electrical discharges. Originally the term referred to multiple-exposed photographs made with a stroboscopic disk as a shutter. One essential feature of modern stroboscopic photography is a short exposure time, usually much shorter than can be obtained by a mechanical shutter.

High-speed photography with Stroboscopic light has proved to be one of the most powerful research tools for observing fast motions in engineering and in science. Likewise, the electrical system of producing flashes of light in xenon-filled flash lamps is of great utility for studio, candid, and press photography. *See* PHOTOGRAPHY; STROBOSCOPE. Harold E. Edgerton

Stromatolite A laminated, microbial structure in carbonate rocks (limestone and dolomite). Stromatolites are the oldest macroscopic evidence of life on Earth, at least 2.5 billion years old, and they are still forming in the seas. During the 1.5 billion years of Earth history before marine invertebrates appeared, stromatolites were the most obvious evidence of life, and they occur sporadically throughout the remainder of the geologic record. In Missouri and Africa, stromatolite reefs have major accumulations of lead, zinc, or copper; and in Montana, New Mexico, and Oman, stromatolites occur within oil and gas reservoirs. For geologists, the shapes of stromatolites are useful indications of their environmental conditions, and variations in form and microstructure of the laminations may be age-diagnostic in those most ancient sedimentary rocks that lack invertebrate fossils. *See* DOLOMITE; LIMESTONE; REEF.

Stromatolites are readily recognizable in outcrops by their characteristic convex-upward laminated structure. Individual, crescent-shaped laminations, which are generally about a millimeter thick, are grouped together to produce an enormous range of shapes and sizes.

The tiny, filamentous cyanobacteria (blue-green algae) that make present-day stromatolites, and similar filaments associated with the oldest stromatolites known, are considered one of the most successful organisms on Earth. Living stromatolites in the Bahamas and Western Australia possess laminations that record the episodic trapping and binding of sediment particles by the microbial mat. In the modern oceans, stromatolites develop almost exclusively in extreme marine conditions that exclude or deter browsing invertebrates and fish from destroying the microbial mats and inhibit colonization by competing algae. *See* ALGAE; CYANOBACTERIA. Robert Ginsburg

Stromatoporoidea A common group of fossil organisms that lived during the Ordovician, Silurian, and Devonian periods. They are preserved in rocks of these systems as large (cabbage-sized) calcium carbonate fossils in the shape of crusts, plates, domes, fingers, bulbs, cylinders, and bushes consisting internally of a three-dimensional network of regularly repeating structural elements such as pillars, laminae, cyst plates, and walls. Although recognized as a class, or subclass, of sponges, these fossils, unlike most sponges, are lacking in siliceous or calcareous spicules (spikelike supporting structures). The term stromatoporoid has been used for a very similar group of Mesozoic fossils that are discussed here as sphaeractinids. Unlike the Paleozoic stromatoporoids, some of the sphaeractinids show the remains of siliceous spicules and can be assigned to several subclasses of the sponge class Demospongiae. Stromatoporoid has also been used as a descriptive term to apply to a grade of evolution of a wide range of different groups of sponges in which a calcareous basal skeleton of stromatoporoid-like architecture was secreted. The implication of such usage is that the Paleozoic fossils are not a natural group but a collection of disparate lineages of sponges that are at a similar stage of evolution. The term stromatoporoid is best restricted to the fossils of Ordovician, Silurian, and Devonian age defined by their lack of spicules and characteristic internal structures that can be grouped as the class Stromatoporoidea of the Porifera. *See* DEMOSPONGIAE; PORIFERA; SPHAERACTINOIDEA.

Paleozoic stromatoporoids secreted large carbonate sheets probably originally of the mineral calcite, although some of the Ordovician representatives could have been aragonite. In these growth forms, the stromatoporoids closely resemble several groups of colonial fossils such as the bryozoans, corals, and hydrozoans. However, unlike most of these fossils, they are not formed of the union of a series of tubes or cups that housed the individuals of a colony, but of a more or less continuous network of skeletal elements like those of a sponge. The living tissue of the stromatoporoid sponge must have occupied only the surface and uppermost layers of the skeleton; the remainder was filled with seawater in life.

Paleozoic stromatoporoids were among the most important reef-building organisms of the middle Paleozoic seas. They lived in shallow tropical and subtropical marine waters. They reached their greatest diversity and abundance in the worldwide reefs of middle Devonian time. The Stromatoporoidea declined in late Devonian time, suffered a crisis in the middle of the epoch, and became extinct at the end of the period. *See* PALEOECOLOGY; REEF. C. W. Stearn

Strong nuclear interactions One of the fundamental physical interactions, which acts between a pair of hadrons. Hadrons include the nucleons, that is, neutrons and protons; the strange baryons, such as lambda (Λ) and sigma (Σ); the mesons, such as pion (π) and rho (ρ); and the strange meson, kaon (K). The nature of the interaction is determined principally through observations of the collision of a hadron pair. From this it is found that the interaction has a short range of about 10^{-15} m (10^{-13} in.) and is by far the dominant force within this range, being much larger than the electromagnetic interaction, which is next in magnitude. The strong interaction conserves parity and is time-reversal-invariant. *See* BARYON; HADRON; MESON; NUCLEON; PARITY (QUANTUM MECHANICS); STRANGE PARTICLES; SYMMETRY LAWS (PHYSICS); TIME REVERSAL INVARIANCE.

Interaction between nucleons (*a*) from exchange of single pion, (*b*) from exchange of ρ-meson, a two-pion system, (*c*) from exchange of two separate pions with formation of excited state of nucleon, and (*d*) without formation of excited state.

The interaction between baryons for distances greater than 10^{-15} m arises from the exchange of mesons. At relatively large distances, single-pion exchange dominates (illus. *a*). At shorter separation distances, the two-pion systems such as the ρ become important (illus. *b*). The interaction between the strange baryons, and between the strange baryons and the nucleons, is moderated by kaon exchange. To summarize this description, the interaction between baryons in the SU(3) multiplet is the consequence of the exchange of SU(3) spin 0 and spin 1 bosons. In a second approximation, the exchange of two pions (illus. *c, d*), or more generally, the exchange of two members of the SU(3) spin 0 and spin 1 multiplet, is responsible for a component of the strong nuclear interaction. *See* UNITARY SYMMETRY.

The range of the interaction generated by these exchanges can be calculated by using the formula below, where m is the mass

$$\text{Range} = \frac{\hbar}{mc}$$

of the exchanged particles, $\hbar$ is Planck's constant divided by 2π, and c is the speed of light. According to the above equation, the range of the interaction developed when a single pion is exchanged (illus. 1*a*) is equal to 1.4×10^{-15} m (5.5×10^{-14} in.), while that due to two-pion exchange (illus. 1*d*) is 0.7×10^{-15} m (2.8×10^{-14} in.).

At short separation distances the quark-gluon structure of the baryons must be taken into account. The interaction must be considered as a property of the six-quark-plus-gluon system. The decisive elements are the Pauli principle obeyed by the quarks, and the mismatch between the six-quark wave function and the two-baryon wave function. Thus, at short distances the interaction is effectively repulsive or more generally independent of the kinetic energy of the baryons at infinite separations. *See* ELEMENTARY PARTICLE; EXCLUSION PRINCIPLE; FUNDAMENTAL INTERACTIONS; GLUONS; QUARKS.
Herman Feshbach

Strongylida An order of nematodes in which the cephalic region may be adorned with three or six labia or the labia may be replaced by a corona radiata. All strongylid nematodes are parasitic. The order embraces eight superfamilies: Strongyloidea, Diaphanocephaloidea, Ancylostomatoidea, Trichostrongyloidea, Metastrongyloidea, Cosmocercoidea, Oxyuroidea, and Heterakoidea.

The Strongyloidea contain important parasites of reptiles, birds, and mammals. The early larval stages may be free-living microbivores, but the adults are always parasitic. Three species are important parasites of horses, *Strongylus vulgaris, S. equinus*, and *S. edentatus*. All three undergo direct life cycles; that is, infestations are acquired by ingestion of contaminated food.

The Trichostrongyloidea comprise obligate parasites of all vertebrates but fishes. Normally they are intestinal parasites, but some are found in the lungs. The species are important parasites of sheep, cattle, and goats. The adult females lay eggs in the intestinal tract, which are passed out with the feces. In the presence of oxygen the eggs hatch in a few days. When the larvae are ingested by an appropriate host, their protective sheath is lost, and they proceed through the fourth larval stage to adulthood in the intestinal tract, where they may enter the mucosa. No migration takes place outside the gastrointestinal tract.

The Metastrongyloidea comprise obligate parasites of terrestrial and marine mammals, found commonly in the respiratory tract. In their life cycle they utilize both paratenic and intermediate hosts, among them a variety of invertebrates, including earthworms and mollusks. Two important species are *Metastrongylus apri* (swine lungworm) and *Angiostrongylus cantonensis* (rodent lungworm).

The Heterakoidea are capable of parasitizing almost any warm-blooded vertebrate as well as reptiles and amphibians. The species *Ascaridia galli* is the largest known nematode parasite of poultry; males are 2–3 in. (50–76 mm) long, and females 3–4.5 in. (75–116 mm).
Armand R. Maggenti

The Oxyuroidea constitute a large group of the phylum Nemata. Hosts include terrestrial mammals, birds, reptiles, amphibians, fishes, insects, and other arthropods. The species are small to medium sized and thin bodied. With one exception, known life cycles are direct. Typically the eggs pass out of the host's alimentary tract onto the ground, where they become fully embryonated and infective. Normally the infective egg does not hatch until a susceptible animal ingests it. The cecum and colon of the host are the typical locations of these parasites. Larvae in all stages of development and adults occur in the gut.
John T. Lucker

The human pinworm, *Enterobius vermicularis*, is probably the most contagious of all helminthic diseases. It is estimated that 10% of the world's population suffer from this parasite, the majority being children. Indeed, incidence among schoolchildren in the cooler regions of the world often approaches 100%. Infection occurs when eggs are inhaled or ingested. The most common method of transmission is from anus to mouth. Because of the aerial transmission, this disease is highly contagious. Though the infection is seldom serious, the behavioral symptoms are disturbing: nail biting, teeth grinding, anal scratching, insomnia, nightmares, and even convulsions. Several medical treatments are available, but there is often the danger of reinfestation from contaminated objects within the household or institution. *See* NEMATA (NEMATODA).
Armand R. Maggenti

Strontianite The mineral form of strontium carbonate, usually with some calcium replacing strontium. It characteristically occurs in veins with barite or celestite or as masses in certain sedimentary rocks. Strontianite is normally prismatic, but it may also be massive. It may be colorless or gray with yellow, green, or brownish tints. The hardness is $3\frac{1}{2}$ on Mohs scale, and the specific gravity of 3.76. It occurs at Strontian, Scotland, and in Germany, Austria, Mexico, and India and, in the United States, in the Strontium Hills of Calfornia. *See* CARBONATE MINERALS; STRONTIUM.
Robert I. Harker

Strontium A chemical element, Sr. Strontium is the least abundant of the alkaline-earth metals. The crust of the Earth is 0.042% strontium, making this element as abundant as chlorine and sulfur. The main ores are celestite, $SrSO_4$, and strontianite, $SrCO_3$. *See* ALKALINE-EARTH METALS; PERIODIC TABLE; STRONTIANITE.

Strontium nitrate is used in pyrotechnics, railroad flares, and tracer bullet formulations. Strontium hydroxide forms soaps and

Properties of strontium

Property	Value
Atomic number	38
Atomic weight	87.62
Isotopes (stable)	84, 86, 87, 88, 90
Boiling point, °C	1638(?)
Melting point, °C	704(?)
Density, g/cm3 at 20°C	2.6

greases with a number of organic acids which are structurally stable, resistant to oxidation and breakdown over a wide temperature range.

Strontium is divalent in all its compounds which are, aside from the hydroxide, fluoride, and sulfate, quite soluble. Strontium is a weaker complex former than calcium, giving a few weak oxy complexes with tartrates, citrates, and so on. Some physical properties of the element are given in the table. Reed F. Riley

Strophanthus A genus of woody climbers of the dogbane family (Apocynaceae), natives of tropical Asia and Africa. They are the source of arrow poisons. The dried, greenish, ripe seeds of *Strophanthus hispidus* and *S. kombe* contain the glucoside strophanthin, which is much used in treating heart ailments. Strophanthin acts directly on heart muscle, increasing muscular force. It causes the heart to beat more regularly and decreases the pulse rate. Strophanthin is a precursor of cortisone, which is used in the treatment of arthritis. *See* GENTIANALES. Perry D. Strausbaugh; Earl L. Core

Strophomenida An extinct order of brachiopods, in the subphylum Rhynchonelliformea, that inhabited shallow shelf seas of the Paleozoic Era. It contains the greatest taxonomic diversity of all rhynchonelliform brachiopod orders, and its members exhibit a diverse range of morphologies and modes of life.

Strophomenida contains four suborders: Strophomenidina, Chonetidina, Productidina, and Strophalosiidina. A fifth suborder, Orthotetidina, shares characteristics with both strophomenids and primitive orthids. Strophomenids are distinguished from all other articulate brachiopods by their concavo-convex or plano-convex lateral profiles and the lack of a functional pedicle in adults. They also possess strophic (straight) hinge lines and variable ornamentation from fine-ribbed early Paleozoic forms to more elaborate spine-bearing late Paleozoic forms. Adult strophomenids were mostly free-living, sessile, epifaunal suspension feeders. The concavo-convex lateral profile and elaborate spines were adaptations for living on soft substrates. *See* BRACHIOPODA; RHYNCHONELLIFORMEA. Mark E. Patzkowsky

Structural analysis A detailed evaluation intended to assure that, for any structure, the deformations will be sufficiently below allowable values that structural failure will not occur. The deformations may be elastic (fully recoverable) or inelastic (permanent). They may be small, with an associated structural failure that is cosmetic; for example, the deflection of a beam supporting a ceiling may cause cracking of the plaster. They may be large, with an associated structural failure that is catastrophic; for example, the buckling of a column or the fracture of a tension member causes complete collapse of the structure.

Structural analysis may be performed by tests on the actual structure, on a physical model of the structure to some scale, or through the use of a mathematical model. Tests on an actual structure are performed in those cases where many similar structures will be produced, for example, automobile frames, or where the cost of a test is justified by the importance and difficulty of the project, for example, a lunar lander. Physical models are sometimes used where subassemblages of major structures are to be investigated. The vast majority of analyses, however, are on mathematical models, particularly in the field of structural engineering which is concerned with large structures such as bridges, buildings, and dams. *See* BRIDGE; BUILDINGS; DAM; STRUCTURE (ENGINEERING).

The advent of the digital computer made it possible to create mathematical models of great sophistication, and almost all complex structures are now so analyzed. Programs of such generality have been written as to permit the analysis of any structure. These programs permit the model of the structure to be two- or three-dimensional, elastic or inelastic, and determine the response to forces that are static or dynamic. Most of the programs utilize the stiffness method, in which the stiffnesses of the individual elements are assembled into a stiffness matrix for the entire structure, and analysis is performed in which all behavior is assumed to be linearly elastic. *See* DIGITAL COMPUTER; ELASTICITY.

The structural engineer's function continues to require training and experience in conceptualizing the structure, choosing the appropriate model, estimating the loads that will be of importance, coding the information for the program, and interpreting the results. The analyst usually enters the process after the conceptualization. Most structures consist of assemblies of members connected at joints. While all real members transmit axial, torsional, and bending actions, the majority of buildings and bridges are analyzed as trusses, beams, and frames with either axial or bending forces predominant. *See* BEAM; ENGINEERING DESIGN; STRESS AND STRAIN; STRUCTURAL DESIGN; TRUSS.

Whether the model selected is detailed or simplified, one extremely important part of the analysis consists of the estimate of the loads to be resisted. For bridges and buildings, the primary vertical loads are gravity loads. These include the weight of the structure itself, and such appurtenances as will be permanent in nature. These are referred to as dead loads. The loads to be carried, the live loads, may consist of concentrated loads (heavy objects occupying little space, for example, a printing press), or loads distributed over relatively large areas (such as floor and deck coverings). Horizontal loads on buildings are produced by wind and by the inertia forces created during earthquakes. In seismic analysis, computers are used to simulate the dynamic characteristics of the structure. The accelerations actually measured during earthquakes are then used to determine the response of the structure. *See* LOADS, DYNAMIC; LOADS, TRANSVERSE; SEISMIC RISK. G. Donald Brandt

Structural chemistry Much of chemistry is explainable in terms of the structures of chemical compounds. The understanding of these structures hinges very strongly on understanding the electronic configurations of the elements. The union of atoms, and therefore the formation of compounds from the elements, is associated with interactions among the extranuclear electrons of the individual atoms. Electronic interactions among atoms may occur in two ways: Electrons may be transferred from one atom to another, or they may be shared by two (or more) atoms. The first type of interaction is called electrovalence and results in the formation of electrically charged monatomic ions. The second, covalence, leads to the formation of molecules and complex ions. *See* CHEMICAL BONDING. Daryle H. Busch

In considering structures more complex than those derived from simple monoatomic ions, the logical step is to consider single polyhedral aggregates of atoms. In its most precise sense, structure is used to denote a knowledge of the bonding distances and angles between atoms in chemical compounds and, in turn, the geometrical arrangements which they form. These atomic arrangements and the associated distances and angles serve uniquely as "fingerprints" of these atom spatial configurations, and depend very much on the electronic configurations around atoms. The chemical combination of neutral atoms to produce uncharged species results in molecule formation, whereas the similar combination of atoms or ions possessing a net charge results in the formation of complex ions. A basic understanding of the species formed involves the concept of the coordination polyhedron, which allows a simple classification of the structures of many polyatomic molecules and ions. This type of classification is particularly useful because it conveniently explains the packing together of simple chemical molecules or ions in terms of highly symmetrical polyhedra. There is an obvious connection between polyhedra and the structures found in crystalline solids formed from them. Crystal formation often involves the linking of convex polyhedra by the sharing of corners, edges, or faces, ultimately forming space-filling assemblies in which all faces of each polyhedron are in contact with faces of other polyhedra. The most important simple polyhedrons are the tetrahedron, the trigonal bipyramid, the octahedron, the pentagonal bipyramid, and the cube. The most commonly observed of these polyhedral configurations are the tetrahedron (four faces) and the octahedron (six faces).

The simplest correlative device which accurately summarizes a very large number of structures and enables the chemist to predict, with a good chance of success, the geometric array of the atoms in a compound of known composition, is based on an extreme electrostatic model. This model, or theory, represents the bonds in a purely formal way. The central atom is considered to be a positive ion having a charge equal to its oxidation state. The groups attached to the central atom (the ligands) are then treated either as negative ions or as neutral dipolar molecules. The principal justification for this approach lies in its successful correlation of a vast amount of information.

A number of significant observations can be made with regard to these formulations. There are series of ions, or ions and molecules, having the same type of composition, differing only in the nature of the central ion and the net charge on the aggregate. Examples are found in the series: NO_3^-, CO_3^{2-}, BO_3^{3-}, ClO_3^-, SO_3^{2-}, PO_3^{3-}; ClO_4^-, SO_4^{2-}, PO_4^{3-}, SiO_4^{4-}; AlF_6^{3-}, SiF_6^{2-}, PF_6^-. The numbers of atomic nuclei and of electrons are the same for all the members of each series; consequently, these are called isoelectronic series. Not only are the several chemical entities in such series isoelectronic, but they are usually identical in geometrical structure (isostructural).

It may also be observed that corresponding ions from a given vertical family of the periodic table commonly vary in coordination number. A useful example is found in N^{5+} and P^{5+} which form NO_3^- and PO_4^{3-}, respectively. In addition, some neutral molecules expand their coordination numbers to form stable anionic halo complexes, whereas others do not. Thus, SiF_4 reacts with fluoride ion to form SiF_6^{2-}, whereas CF_4 does not form a similar complex ion. The most satisfactory explanation of these and many related observations is conveniently formulated in terms of the electrostatic model chosen here.

The necessary condition for stability of the coordination polyhedron MA_n requires that the anions A are each in contact with the central atom M. As a consequence of this condition, the limit of stability of the structure arises in those cases where the anions are also mutually in contact. Larger ligands, or anions, would not be in contact with the central ion. This relationship is usually summarized in terms of the limiting ratio of the radius of the cation, r_M, to that of the anion, r_A, below which the anions would no longer be in contact with the cation.

According to the valence-bond theory, the principal requirements for the formation of a covalent bond are a pair of electrons and suitably oriented electron orbitals on each of the atoms being bonded. The geometry of the atoms in the resulting coordination polyhedron is correlated with the orientation of the orbitals on the central atom. The orbitals used depend on the energies of the electrons in them. In general, the order of increasing energy of the electron orbitals is $(n - 1)d < ns < np < nd$. It is concluded that a nontransition atom having one valence electron will form a covalent bond utilizing an *s* orbital. In those cases where an unshared pair of electrons may be assigned to the *ns* orbital, as many as three equivalent bonds may be formed by utilizing the three *np* orbitals of the central atom. Because of the orientation of these *p* orbitals with respect to each other, the three resulting bonds should be at 90° to each other. This expectation is nearly realized in PH_3. In order to account for four or six equivalent bonds, or for that matter in order to account for all the remaining polyhedral and polygonal structures, except the angular structure for a coordination number of 2 (with two unshared pairs of electrons on the central atom), an additional assumption is necessary. It is assumed that *s* and *p*, *s* and *d*, or *s*, *p*, and *d* orbitals, are replaced by new orbitals, called hybridized orbitals. These hybridized orbitals are derived from the original orbitals (mathematically) in such a way that the required number of equivalent bonds may be formed. In the simplest case, it is shown that *s* and *p* may be combined to form two equivalent *sp* hybridized orbitals directed at 180° to each other. Other sets of hybridized orbitals have been shown to be appropriate to describe the bonding in other structures. *See* LIGAND FIELD THEORY.

Among inert-gas ions of the first row of eight elements in the periodic table, there are four orbitals available for covalent bond formation, one 2*s* and three 2*p*. Consequently, a maximum of four bonds may be formed. This is in general agreement with the existence of the tetrahedron as the limiting coordination polyhedron among these elements, for example, BeF_4^{2-}, BF_4^-, CCl_4, NH_4^+. Although only Li^+ deviates from this pattern, having a coordination number of 6 in its crystalline halides, these compounds are best treated as simple electrovalent salts. In keeping with the limitation of only four orbitals, the formation of double or triple bonds between atoms of these elements reduces the coordination number of the central atom. Thus, the highest coordination number of a first-row element forming one double bond is 3. This is illustrated by the structures below.

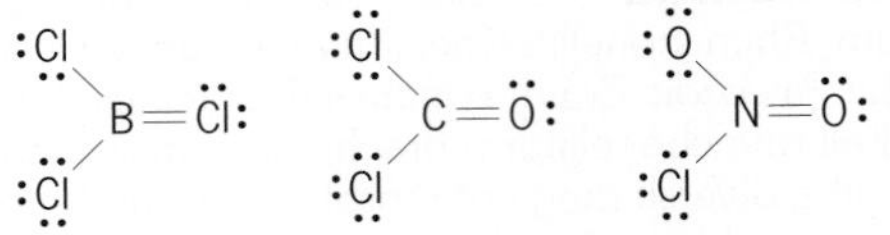

```
:Cl                 :Cl                 :O
    \                   \                   \
     B=Cl:               C=O:                N=O:
    /                   /                   /
:Cl                 :Cl                 :Cl
```

In these and similar examples, the geometric array is determined by the formation of three single bonds utilizing sp^2 hybridized orbitals on the central atom and *p* orbitals on the ligand. In general, the bonds determining the geometry of a molecule or ion (in this way) are called σ bonds. The double bond results from the superposition of a second bond, a π bond, between two atoms. In this example the formation of a π bond reduces the number of σ bonds from four to three, thus changing the geometries of the corresponding molecules or ions from tetrahedral to trigonal planar.

The formation of a second π bond (a triple bond or two double bonds) reduces the coordination number of the atom in question still further, resulting in the linear *sp* set of hybridized orbitals being utilized in σ-bond formation. *See* VALENCE.

With regard to the nature of doubly bonded compounds, another problem arises when such structures are viewed from the standpoint of valence-bond theory. In the species BCl_3, $COCl_2$, NO_2Cl, and many similar substances, nonequivalent bonds are predicted. The doubly bonded oxygen should be closer to the central carbon atom than the singly bonded ones. This is not found to be true experimentally so long as the similar atoms are otherwise equivalent. There is only one observable C—O distance in carbonate, one N—O distance in nitrate, and so on. To account for such facts as these, the concept of resonance must be introduced. If the π bond exists, it must exist equally between the central atom and all the equivalent oxygen atoms. The resonance method of describing this situation is to say that one of the pictorial structures is inadequate to describe the substance properly, but that enough pictorial structures (resonance structures) should be considered to permute the double bond about all the equivalent bonds. The true structure is assumed to be something intermediate to all the resonance structures and more stable than any of them because it exists in preference to any one of them. The resonance structures for CO_3^{2-} are the following:

See CONJUGATION AND HYPERCONJUGATION; RESONANCE (MOLECULAR STRUCTURE).

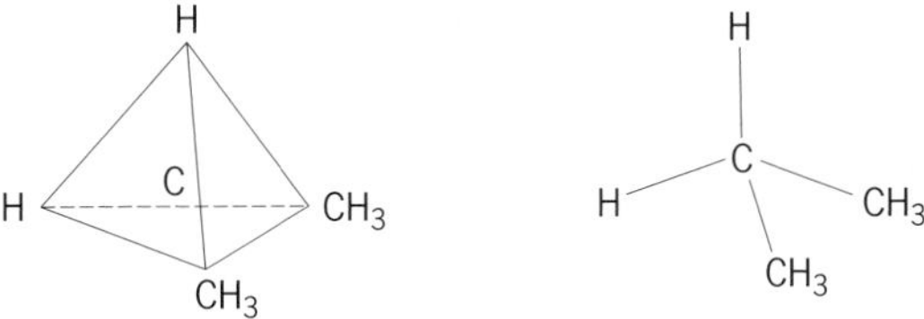

Fig. 1. Geometric structure of propane.

The classic homologous series of compounds in organic chemistry provide useful examples involving the condensation of polyhedrons containing the same central element in the individual units. The general formula, $C_nH_{2n} + 2$, represents a large number of compounds extending from the lowest member, methane, CH_4, to polyethylene, a plastic of economic importance in which n is a very large number. Two ways exist for the linking together of these tetrahedrons. This gives rise to two molecular forms, both of which are stable, well-known compounds. It is an essential part of these structures that each —C link is linear (because it is merely a σ-bond); however, when two carbon atoms are linked to a third, the C—C—C angle is essentially determined by the bond angle of the central carbon atom (that is, the other carbons may be treated as ligands to the first (Fig. 1). The other familiar homologous series of organic chemistry differ from the saturated hydrocarbons in having at least one unique coordination polyhedron of a different type. The olefins contain two doubly bonded, or unsaturated, carbon atoms, whose polyhedral structures are trigonal planar, and the remainder of the carbons are tetrahedral. As in the case of the aliphatic hydrocarbons, the olefins exhibit an isomerism which is associated with the branching of the chain structure. In addition, the presence of two linked trigonal planar carbon atoms and the fact that the polyhedrons cannot rotate about the double bond give rise to a different kind of isomerism, called cis-trans isomerism (Fig. 2). See BOND ANGLE AND DISTANCE.

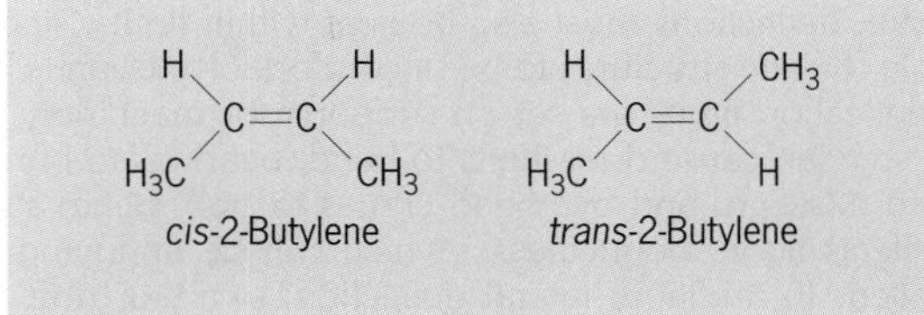

Fig. 2. Cis-trans isomerism among olefins.

The existence of a predicted isomerism provides one of the most important confirmations of the theories of chemical structure. In general, the polyhedral view of molecular structure, as described here, has been thoroughly verified by the discovery of the many types of predicted isomerism. The first really convincing proof of the tetrahedral structures of saturated carbon atoms involved optical isomerism. See OPTICAL ACTIVITY.

The aromatic hydrocarbons are characterized by cyclic arrangements of trigonal planar carbon atoms (Fig. 3*a*). The highly symmetrical nature of the benzene molecule is not fully represented by such a structure. The figure indicates the presence of three double and three single bonds in the ring. It has been shown that the C—C bonds are all the same and, consequently, the true structure of the substance must be represented by two resonance structures which interchange the single and double bonds (Fig. 3*b*). See BENZENE.

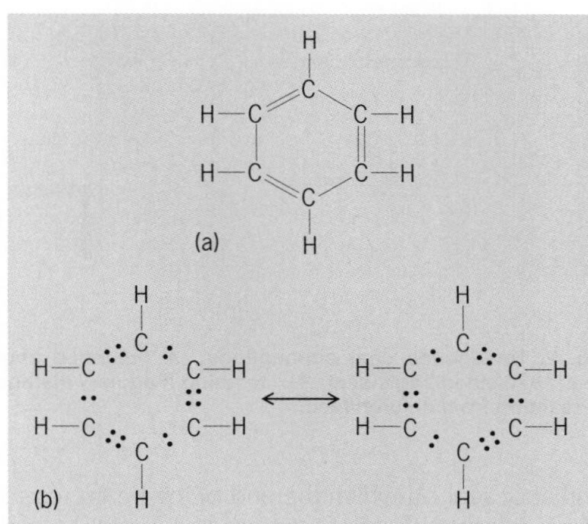

Fig. 3. Benzene molecule. (*a*) Structural formula. (*b*) Two forms in resonance.

Jack M. Williams

Structural connections Methods of joining the individual members of a structure to form a complete assembly. The connections furnish supporting reactions and transfer loads from one member to another. Loads are transferred by fasteners (rivets, bolts) or welding supplemented by suitable arrangements of plates, angles, or other structural shapes. When the end of a member must be free to rotate, a pinned connection is used.

The suitability of a connection depends on its deformational characteristics as well as its strength. Rotational flexibility or complete rigidity must be provided according to the degree of end restraint assumed in the design. A rigid connection maintains the original angles between connected members virtually unchanged after loading. Flexible or nonrestraining connections permit rotation approximately equal to that at the ends of a simply supported beam. Intermediate degrees of restraint are called semirigid.

A commonly used form of connection for rolled-beam sections, called a web connection, consists of two angles which are attached to opposite sides of a member and which are in turn connected to the web of a supporting beam, girder, column, or framing at right angles. A shelf angle may be added to facilitate erection (Fig. 1).

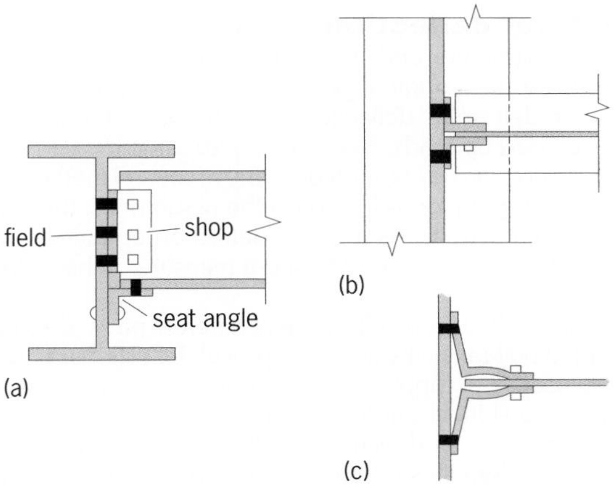

Fig. 1. Riveted or bolted web connections. (*a*) Beam-to-ginder, elevation view. (*b*) Plan view. (*c*) Bending of angle legs.

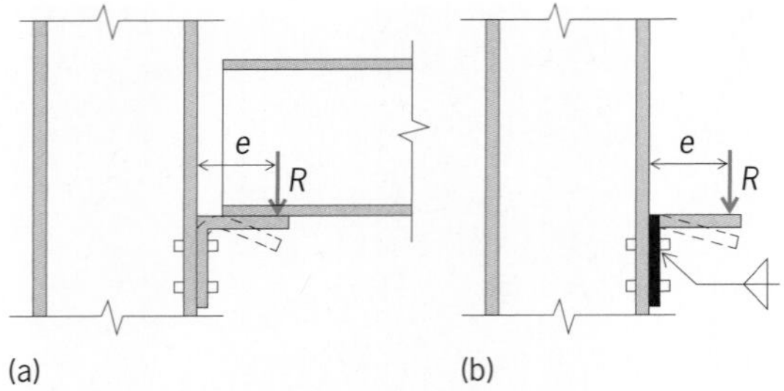

Fig. 2. Unstiffened seat connections. (*a*) Fastened angle seat. (*b*) Welded angle seat. R = reaction load; e = distance to reaction from column face.

A bracket or seat on which the end of the beam rests is a seat connection; it is intended to furnish the end reaction of the supported beam. Two general types are used: The unstiffened seat provides bearing for the beam by a projecting plate or angle leg which offers resistance only by its own flexural strength (Fig. 2); the stiffened seat is supported by a vertical plate or angle which transfers the reaction force to the supporting member without flexural distortion of the outstanding seat.

When the action line of a transferred force does not pass through the centroid of the connecting fastener group or welds, the connection is subjected to rotational moment which produces additional shearing stresses in the connectors. The load transmitted by diagonal bracing to a supporting column flange through a gusset plate is eccentric with reference to the connecting fastener group.

In beam-to-column connections and stiffened seat connections or when members transfer loads to columns by a gusset plate or a bracket, the fasteners are subjected to tension forces caused by the eccentric connection. Although there are initial tensions in the fasteners, the final tension is not appreciably greater than the initial tension.

Rigidity and moment resistance are necessary at the ends of beams forming part of a continuous framework which must resist lateral and vertical loads. Wind pressures tend to distort a building frame, producing bending in the beams and columns which must be suitably connected to transfer moment and shear. The resisting moment can be furnished by various forms of angle T for fasteners or welded or bracket connections.

Where appreciably angular change between members is expected, and in special cases where a hinge support without moment resistance is desired, connections are pinned. Many bridge trusses and large girder spans have pin supports. *See* BOLT; JOINT (STRUCTURES); RIVET.

John B. Scalzi

Structural deflections The deformations or movements of a structure and its components, such as beams and trusses, from their original positions. It is as important for the designer to determine deflections and strains as it is to know the stresses caused by loads. *See* STRESS AND STRAIN.

Deflections may be computed by any of several methods. Generally the computation is based on the assumption that stress is proportional to strain. As a result, deflection equations involve the modulus of elasticity E, which is a measure of the stiffness of a material.

The relation between deflections at different parts of a structure is indicated by Maxwell's law of reciprocal deflections. This states that if a load P is applied at any point A in any direction a and causes a shift of another point B in direction b, the same load applied at B in direction b will cause an equal shift of A in direction a (see illustration). The law is used in a number of ways such as in simplifying deflection calculations, checking the accuracy of computations, and producing influence lines. *See* STRUCTURAL ANALYSIS.

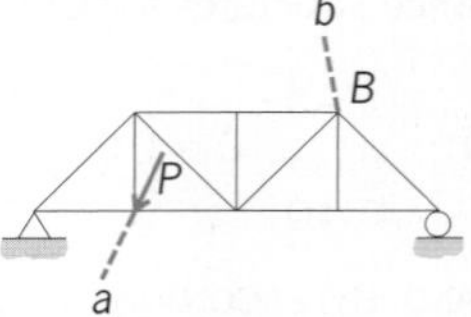

Example of Maxwell's law of reciprocal deflections.

Beam and truss deflections usually are computed by similar methods, except that integration is used for equations and summation for trusses. Beam deflection equations involve bending moments and moments of inertia. Truss deflection equations are based on the stresses and cross-sectional areas of chords and web members. Deflections may also be determined graphically. *See* BEAM; TRUSS.

John B. Scalzi

Structural design The selection of materials and member type, size, and configuration to carry loads in a safe and serviceable fashion. In general, structural design implies the engineering of stationary objects such as buildings and bridges, or objects that may be mobile but have a rigid shape such as ship hulls and aircraft frames. Devices with parts planned to move with relation to each other (linkages) are generally assigned to the area of mechanical design.

Structural design involves at least five distinct phases of work: project requirements, materials, structural scheme, analysis, and design. For unusual structures or materials a sixth phase, testing, should be included. These phases do not proceed in a rigid progression, since different materials can be most effective in different schemes, testing can result in changes to a design, and a final design is often reached by starting with a rough estimated design, then looping through several cycles of analysis and redesign. Often, several alternative designs will prove quite close in cost, strength, and serviceability. The structural engineer, owner, or end user would then make a selection based on other considerations.

Before starting design, the structural engineer must determine the criteria for acceptable performance. The loads or forces to be resisted must be provided. For specialized structures this may be given directly, as when supporting a known piece of machinery, or a crane of known capacity. For conventional buildings, building codes adopted on a municipal, county, or state level provide minimum design requirements for live loads (occupants and furnishings, snow on roofs, and so on). The engineer will calculate dead loads (structure and known, permanent intallations) during the design process. For the structure to be serviceable or useful, deflections must also be kept within limits, since it is possible for safe structures to be uncomfortably "bouncy." Very tight deflection limits are set on supports for machinery, since beam sag can cause driveshafts to bend, bearings to burn out, parts to misalign, and overhead cranes to stall. Beam stiffness also affects floor "bounciness," which can be annoying if not controlled. In addition, lateral deflection, sway, or drift of tall buildings is often held within approximately height/500 (1/500 of the building height) to minimize the likelihood of motion discomfort in occupants of upper floors on windy days. *See* LOADS, DYNAMIC; LOADS, TRANSVERSE.

Technological advances have created many novel materials such as carbon fiber- and boron fiber-reinforced composites, which have excellent strength, stiffness, and strenth-to-weight properties. However, because of the high cost and difficult or unusual fabrication techniques required, glass-reinforced composites such as fiberglass are more common, but are limited to lightly loaded applications. The main materials used in structural design are more prosaic and include steel, aluminum, reinforced concrete, wood, and masonry. *See* COMPOSITE MATERIAL; MASONRY; PRECAST CONCRETE; PRESTRESSED CONCRETE; REINFORCED CONCRETE; STRUCTURAL MATERIALS.

In an actual structure, various forces are experienced by structural members, including tension, compression, flexure (bending), shear, and torsion (twist). However, the structural scheme selected will influence which of these forces occurs most frequently, and this will influence the process of material selection. *See* SHEAR; TORSION.

Analysis of structures is required to ensure stability (static equilibrium), find the member forces to be resisted, and determine deflections. It requires that member configuration, approximate member sizes, and material properties be known or assumed. Aspects of analysis include: equilibrium; stress, strain, and elastic modulus; linearity; plasticity; and curvature and plane sections. Various methods are used to complete the analysis.

Once a structure has been analyzed (by using geometry alone if the analysis is determinate, or geometry plus assumed member sizes and materials if indeterminate), final design can proceed. Deflections and allowable stresses or ultimate strength must be checked against criteria provided either by the owner or by the governing building codes. Safety at working loads must be calculated. Several methods are available, and the choice depends on the types of materials that will be used. Once a satisfactory scheme has been analyzed and designed to be within project criteria, the information must be presented for fabrication and construction. This is commonly done through drawings, which indicate all basic dimensions, materials, member sizes, the anticipated loads used in design, and anticipated forces to be carried through connections. Richard L. Tomasetti; Leonard M. Joseph

Structural geology The branch of geology that deals with study and interpretation of deformation of the Earth's crust. Deformation brings about changes in size (dilation), shape (distortion), position (translation), or orientation (rotation). Evidence for the changes caused by deformation are commonly implanted into geologic bodies in the form of recognizable structures, such as faults and joints, folds and cleavage, and foliation and lineation. The geologic record of structures and structural relations is best developed and most complicated in mountain belts, the most intensely deformed parts of the Earth's crust. *See* MOUNTAIN SYSTEMS.

The discipline of structural geology harnesses three interrelated strategies of analysis: descriptive analysis, kinematic analysis, and dynamic analysis. Descriptive analysis is concerned with recognizing and describing structures and measuring their orientations. Kinematic analysis focuses on interpreting the deformational movements responsible for the structures. Dynamic analysis interprets deformational movements in terms of forces, stresses, and mechanics. The ultimate goal of these interdependent approaches is to interpret the physical evolution of crustal structures, that is, tectonic analysis. A major emphasis in modern structural geology is strain analysis, the quantitative analysis of changes in size and shape of geologic bodies, regardless of scale.

There are many significant practical applications of structural geology. An understanding of the descriptive and geometric properties of folds and faults, as well as mechanisms of folding and faulting, is of vital interest to exploration geologists in the petroleum industry. Ore deposits commonly are structurally controlled, or structurally disturbed, and for these reasons detailed structural geologic mapping is an essential component of mining exploration. Other applications of structural geology include the evaluation of proposals for the disposal of radioactive waste in the subsurface, and the targeting of safe sites for dams, hospitals, and the like in regions marked by active faulting. *See* FAULT AND FAULT STRUCTURES; FOLD AND FOLD SYSTEMS; ORE AND MINERAL DEPOSITS; PETROLEUM GEOLOGY. George H. Davis

Structural materials Construction materials which, because of their ability to withstand external forces, are considered in the design of a structural framework.

Brick is the oldest of all artificial building materials. It is classified as face brick, common brick, and glazed brick. Face brick is used on the exterior of a wall and varies in color, texture, and mechanical perfection. Common brick consists of the kiln run of brick and is used behind whatever facing material is employed providing necessary wall thickness and additional structural strength. Glazed brick is employed largely for interiors where beauty, ease of cleaning, and sanitation are primary considerations. *See* BRICK.

Structural clay tiles are burned-clay masonry units having interior hollow spaces termed cells. Such tile is widely used because of its strength, light weight, and insulating and fire-protection qualities. *See* TILE.

Architectural terra-cotta is a burned-clay material used for decorative purposes. The shapes are molded either by hand in plaster-of-paris molds or by machine, using the stiff-mud process.

Building stones generally used are limestone, sandstone, granite, and marble. Until the advent of steel and concrete, stone was the most important building material. Its principal use now is as a decorative material because of its beauty, dignity, and durability. *See* GRANITE; LIMESTONE; MARBLE; SANDSTONE; STONE AND STONE PRODUCTS.

Concrete is a mixture of cement, mineral aggregate, and water, which, if combined in proper proportions, form a plastic mixture capable of being placed in forms and of hardening through the hydration of the cement. *See* CONCRETE; PRESTRESSED CONCRETE; REINFORCED CONCRETE.

The cellular structure of wood is largely responsible for its basic characteristics, unique among the common structural materials. When cut into lumber, a tree provides a wide range of material which is classified according to use as yard lumber, factory or shop lumber, and structural lumber. Laminated lumber is used for beams, columns, arch ribs, chord members, and other structural members. Plywood is generally used as a replacement for sheathing, or as form lumber for reinforced concrete structures. *See* LUMBER; PLYWOOD; WOOD ANATOMY; WOOD PRODUCTS.

Important structural metals are the structural steels, steel castings, aluminum alloys, magnesium alloys, and cast and wrought iron. Steel castings are used for rocker bearings under the ends of large bridges. Shoes and bearing plates are usually cast in carbon steel, but rollers are often cast in stainless steel. Aluminum alloys are strong, lightweight, and resistant to corrosion. The alloys most frequently used are comparable with the structural steels in strength. Magnesium alloys are produced as extruded shapes, rolled plate, and forgings. The principal structural applications are in aircraft, truck bodies, and portable scaffolding. Gray cast iron is used as a structural material for columns and column bases, bearing plates, stair treads, and railings. Malleable cast iron has few structural applications. Wrought iron is used extensively because of its ability to resist corrosion. It is used for blast plates to protect bridges, for solid decks to support ballasted roadways, and for trash racks for dams. *See* ALUMINUM; CAST IRON; MAGNESIUM ALLOYS; STEEL; STRUCTURAL STEEL; WROUGHT IRON. Charles M. Antoni

Composite materials are engineered materials that contain a load-bearing material housed in a relatively weak protective matrix. A composite material results when two or more materials, each having its own, usually different characteristics, are combined, producing a material with properties superior to its components. The matrix material (metallic, ceramic, or polymeric) bonds together the reinforcing materials (whiskers, laminated fibers, or woven fabric) and distributes the loading between them. *See* COMPOSITE MATERIAL; CRYSTAL WHISKERS.

Fiber-reinforced polymers (FRP) are a broad group of composite materials made of fibers embedded in a polymeric matrix. Compared to metals, they generally have relatively high strength-to-weight ratios and excellent corrosion resistance. They can be formed into virtually any shape and size. Glass is by far the most used fiber in FRP (glass-FRP), although carbon fiber (carbon-FRP) is finding greater application. Although complete FRP shapes and structures are possible, the most promising application of FRP in civil engineering is for repairing structures or infrastructure. FRP can be used to repair beams, walls, slabs, and columns. *See* POLYMERIC COMPOSITE. Mel Schwartz

Structural mechanics A branch of applied mechanics that deals with the behavior of solid bodies subjected to various types of loading. The major concepts covered by structural mechanics include the fundamental ones, such as stresses, strains, material laws, and equations of equilibrium, and the more advanced ones, such as energy principles, elasticity, dynamics, stability, and load-carrying capacities. These concepts underlie the design and analysis of a huge variety of mechanical and structural systems and components. A good knowledge of structural mechanics is essential for engineers, typically in the professions of civil, mechanical, aeronautical, and aerospace engineering, whose major duty is to design structures or solid components. *See* STRENGTH OF MATERIALS; STRESS AND STRAIN.

Three rules must be obeyed in the analysis of a structure or its components. The first is equilibrium, which means that all the forces acting on any part of a structure or its entity must be balanced out under any loading condition. The second is compatibility or displacement continuity, which means that any two connected parts of a structure must remain connected in the same way after deformation. The third is material law, which describes the deformation of a member in response to loading, based on empirical or theoretical formulas.

Various methods can be used for the analysis of a structure under given loads, based on the consideration of equilibrium, compatibility (that is, displacement continuity), and material laws, or their variants in energy forms. One powerful method is the finite element method, by which a structure is conceptually decomposed into a number of discrete elements, and the structural equations are assembled from those of the elements making up the structure. *See* FINITE ELEMENT METHOD; STRUCTURAL ANALYSIS.

The design of structures may be regarded as a mixture of science and art. A goal of structural engineers is to design structures not only to meet their functional needs but also to make them esthetically appealing, so that they can become an ingredient of the environment or a cultural element of the society. Yeong-Bin Yang

Structural petrology The study of the structural aspects of rocks, as distinct from the purely chemical and mineralogical studies that are generally emphasized in other branches of petrology. The term was originally used synonymously with petrofabric analysis, but is sometimes restricted to denote the analysis of only microscopic structural and textural features. *See* PETROFABRIC ANALYSIS; PETROGRAPHY. John M. Christie

Structural plate A simple rolled steel section used as an isolated structural element, as a support of other structural elements, or as part of other structural elements. When isolated plates are extremely thick, their design is controlled by shear; plates of moderate thickness are controlled by bending (with some torsion), and very thin plates carry their loads principally by tensile membrane action. Although stresses of all types exist in all plates, it is usually sufficient to deal with only the most significant. Bending is the most common design criterion.

Plates are commonly used as cover plates on wide-flange beams, as the flanges and webs of plate girders, and as the sides of tube-shaped beams and columns. In all these cases, serious consideration must be given to the fact that the plate may buckle when compressed. Fortunately, the plates have edge supports in the direction of the stress, so they function as panels rather than as beams. Their ratios of length to width are large enough that the resistance to local buckling of the plate element depends upon its width-thickness ratio, practically independent of its length. (The length of the overall section is still significant in determining the member's capacity.) *See* BEAM; COLUMN; JOINT (STRUCTURES); LOADS, TRANSVERSE; STRUCTURAL STEEL. G. Donald Brandt

Structural stability A structure's ability to resist compressive stress without compromising its load-carrying capacity. When a structure is subjected to a sufficiently high compressive force (or stress), it has a tendency to lose its stiffness, experience a noticeable change in geometry, and become unstable. Examples of structural instability include the buckling of a column under a compressive axial force, the lateral torsional buckling of a beam under a transverse load, the sideways buckling of an unbraced frame under a set of concentric column forces, and the buckling of a plate under a set of in-plane forces. *See* BEAM; COLUMN; LOADS, TRANSVERSE; STRESS AND STRAIN; STRUCTURAL DEFLECTIONS; TORSION.

Systems that are geometrically imperfect experience deformations resembling their buckled configurations as soon as the loads are applied. Their buckling behavior is often one of limit point instability. If the system is elastic and the amount of geometrical imperfections is not appreciable, the limit load often approaches that of the critical load. There are exceptions such as the finite disturbance buckling of axially loaded cylindrical shells and snap-through buckling of shallow arches and spherical caps.

Material inelasticity occurs when the stress in the system reaches its yield value. To account for yielding in a stability analysis, one can use reduced values for the system's elastic material constants such as the modulus of elasticity E and shear modulus G in the analysis. *See* SHEAR; YOUNG'S MODULUS.

Stability design equations for structural members in a number of design specifications are often arrived at by the use of a semiempirical approach in which experimental data are curve-fitted in a form that is congruent with fundamental mechanics and stability principles. *See* STRUCTURAL DESIGN. Eric M. Lui

Structural steel Steel used in engineering structures, usually manufactured by either the open-hearth or the electric-furnace process. The exception is carbon-steel plates and shapes whose thickness is $^{7}/_{16}$ in. (11 mm) or less and which are used in structures subject to static loads only. These products may be made from acid-Bessemer steel. The physical properties and chemical composition are governed by standard specifications of the American Society for Testing and Materials (ASTM). Structural steel can be fabricated into numerous shapes for various construction purposes. *See* STEEL. Waldo G. Bowman

Structure (engineering) An arrangement of designed components that provides strength and stiffness to a built artifact such as a building, bridge, dam, automobile, airplane, or missile. The artifact itself is often referred to as a structure, even though its primary function is not to support but, for example, to house people, contain water, or transport goods. *See* AIRPLANE; AUTOMOBILE; BRIDGE; BUILDINGS; DAM.

The primary requirements for structures are safety, strength, economy, stiffness, durability, robustness, esthetics, and ductility. The safety of the structure is paramount, and it is achieved by adhering to rules of design contained in standards and codes, as well as in exercising strict quality control over all phases of planning, design, and construction. The structure is designed to be strong enough to support loads due to its own weight, to human activity, and to the environment (such as wind, snow, earthquakes, ice, or floods). The ability to support loads during its intended lifetime ensures that the rate of failure is insignificant for practical purposes. The design should provide an economical structure within the constraints of all other requirements. The structure is designed to be stiff so that under everyday conditions of loading and usage it will not deflect or vibrate to an extent that is annoying to the occupants or detrimental to its function. The materials and details of construction have durability, such that the structure will not corrode, deteriorate, or break under the effects of weathering and normal usage during its lifetime. A structure should be robust enough to withstand intentional or unintentional misuse (for example, fire, gas explosion, or collision with a vehicle) without totally collapsing. A structural design takes into consideration the community's esthetic sensibilities. Ductility is necessary to absorb the energy imparted to the structure from dynamic loads such as earthquakes and blasts. *See* Construction engineering; Engineering design.

Common structural materials are wood, masonry, steel, reinforced concrete, aluminum, and fiber-reinforced composites. Structures are classified into the categories of frames, plates, and shells, frequently incorporating combinations of these. Frames consist of "stick" members arranged to form the skeleton on which the remainder of the structure is placed. Plated structures include roof and floor slabs, vertical shear walls in a multistory building, or girders in a bridge. Shells are often used as water or gas containers, in roofs of arenas, or in vehicles that transport gases and liquids. The connections between the various elements of a structure are made by bolting, welding or riveting. *See* Composite material; Concrete; Steel; Structural materials.

Theodore V. Galambos

Struthioniformes A small order of weak-flying, partridge-like birds and giant, flightless ratite birds found in the southern continents. Their relationship to other birds is unknown. The struthioniforms are characterized by a palaeognathous palate, a break in the postnasal strut, close approximation of the zygomatic process to the quadrate, and the structure of the rhamphotheca (the horny sheath covering a bird's beak). They are frequently placed in a separate superorder, the Palaeognathae. However, that distinction places too much emphasis on their separation from other birds. Contrary to common opinion, no evidence supports the concept that the palaeognathous birds are primitive among living birds. *See* Ratites.

The order Struthioniformes can be divided into three suborders, each with both fossil and extant representatives, as follows:

Order Struthioniformes
  Suborder Lithornithi
    Family: Lithornithidae (fossil; Paleocene and Eocene of North America and Europe)
      Palaeotididae (fossil; Eocene of Europe)
  Suborder Tinami
   Superfamily Tinamoidea
    Family: Tinamidae (tinamous; 47 species; Neotropics)
      Diogenomithidae (fossil; Paleocene of South America)
      Rheidae (rheas; 2 species; South America)
   Superfamily Apterygoidea
    Family: Dromornithidae (fossil; Miocene to Pleistocene of Australia)
      Dromaiidae (emus; 2 species; Australia)
      Casuariidae (cassowaries; 3 species; Australia and New Guinea)
      Dinomithidae (moas; subfossil, about 15 species; New Zealand)
      Apterygidae (kiwis; 3 species; New Zealand)
  Suborder Struthioni
    Family: Eleutheromithidae (fossil; Eocene of Europe)
      Struthionidae (ostriches; 1 species; Africa and formally Arabia)
      Aepyomithidae (elephant birds; fossil, about 8 species; Quaternary of Madagascar)

The struthioniforms are medium-sized to giant birds. The ostriches are the largest extant birds, but some of the moas, elephant birds, and dromornithids were even larger. The heads of the struthioniforms are small; a medium, flattened bill is a common feature except in the kiwis, which have a long bill used for probing into the ground for worms. The plumage is soft, and the wings are reduced in all forms except the lithionithids and tinamous, which can still fly. The legs are strong, and all forms run well. The struthioniforms eat a variety of foods, especially large fruits and other large food items. Breeding is polygamous, with two or more females laying eggs in a single nest. The males are responsible for incubation, and they assume the major or sole role in caring for the downy young, which leave the nest after hatching. *See* Aves; Endangered species.

Walter J. Bock

Strychnine alkaloids Alkaloid substances derived from the seeds and bark of plants of the genus *Strychnos* (family Loganiaceae). This genus serves as the source of poisonous, nitrogen-containing plant materials, such as strychnine (see structure; R = H). The seeds of the Asian species of *Strychnos* contain 2–3% alkaloids, of which about half is strychnine and the rest is closely related materials; for example, brucine (see structure; R = OCH_3) is a more highly oxygenated relative. Strychnine and brucine are isolated by extraction of basified plant residue with chloroform and then, from the chloroform solution, by dilute sulfuric acid. Precipitation from the dilute acid is accomplished with ammonium hydroxide. Strychnine is separated from brucine by fractional crystallization from ethanol. *See* Crystallization; Strychnos.

At one time strychnine was used as a tonic and a central nervous system stimulant, but because of its high toxicity (5 mg/kg is a lethal dose in the rat) and the availability of more effective substances, it no longer has a place in human medicine.

David Dalton

Strychnos A genus of tropical trees and shrubs belonging to the Logania family (Loganiaceae). *Strychnos nux-vomica*, a native of India and Sri Lanka, is the source of strychnine. The alkaloid, strychnine, has been used medicinally in the treatment of certain nervous disorders and paralysis. Curare, used by the Indians to poison arrows, is obtained from *S. toxifera* and *S. castelnaei* in Guiana and Amazonas and from *S. tieute* in the Sunda Islands. Curare paralyzes the motor nerve endings in striated muscles and is used in medical practice in cases in which a state of extreme muscular relaxation or even immobility is desirable. It has become an important drug in the field of anesthesiology. *See* Gentianales; Strychnine alkaloids.

Perry D. Strausbaugh; Earl L. Core

Sturgeon Any of the ~25 species of large fishes that comprise the family Acipenseridae in the order Acipenseriformes. These fish are found in North Temperate Zone waters, where they are almost exclusively bottom-living and feed on organisms such as mollusks, worms, and larvae. The body has five rows of bony plates, of which one is situated dorsally, two laterally, and two ventrally. The snout is elongate, and there are four barbels on its lower surface; the mouth is ventrally located, and in the adult the jaws lack teeth (see illustration). The skeleton is mainly cartilaginous. *See* ACIPENSERIFORMES; OSTEICHTHYES.

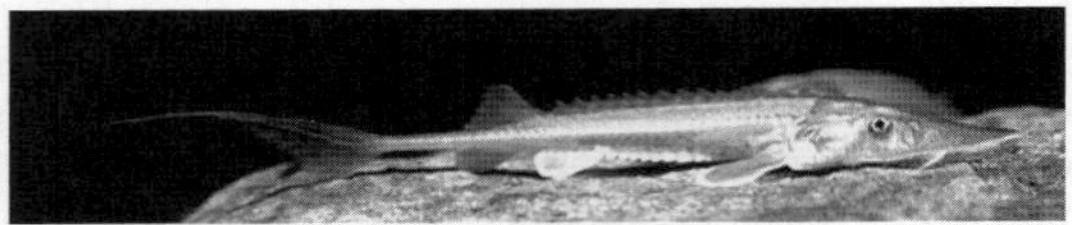

Shovel-nose sturgeon (*Scaphirhynchus platorynchus*). (*Photo by Konrad Schmidt*)

Sturgeons are known primarily for the roe, which is processed as caviar. A single female can produce millions of eggs, which may be removed from dead fish or stripped from living fish. The smoked flesh of the sturgeon is a delicacy. Unfortunately, sturgeons have been seriously depleted in North America and most of Europe, although substantial fisheries remain in Western and Central Asia. *See* AQUACULTURE; MARINE FISHERIES.

Charles B. Curtin

Stylasterina An order of the class Hydrozoa of the phylum Cnidaria, including several brightly colored branching or encrusting "corals" of warm seas. (True corals belong to a different class, the Anthozoa.) The calcareous skeleton is covered by living tissue and is penetrated by ramifying tubes. Nutritive polyps, the gastrozooids, lie in cups on one surface or along certain edges of the skeletal substance. A spine, or style, at the base of each cup gives the order its name. Some authorities combine the Stylasterina with the Milleporina in a single order, the Hydrocorallina. *See* ANTHOZOA; HYDROZOA.

Sears Crowell

Stylolites Irregular surfaces occurring in certain rocks, mostly parallel to bedding planes, in which small toothlike projections on one side of the surface fit into cavities of like shape on the other side (see illustration). Stylolites are most common in limestones and dolomites but are also present in many other kinds of rock, including sandstones, gypsum beds, and cherts. *See* DOLOMITE; LIMESTONE; SEDIMENTARY ROCKS.

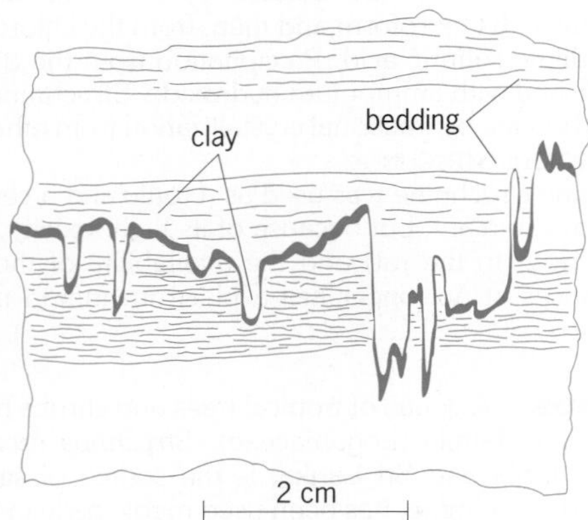

Stylolite in limestone.

Raymond Siever

Stylommatophora An order or suborder of the gastropod molluscan subclass Pulmonata containing about 20,500 species that are grouped into 56 families. Nearly all land snails without an operculum are stylommatophorans. They have eyespots on the tips of a pair of retractile tentacles, hermaphroditic reproduction with partial fusion of the male and female system, and, normally, a second pair of retractile tentacles that function as chemoreceptors. All are air-breathing and most are truly terrestrial. Three orders, based on excretory structures, are recognized. The more primitive Orthurethra and Mesurethra have less efficient water conservation devices than do the more specialized members of the order Sigmurethra, in which a closed ureter functions in water conservation. The latter specialization apparently is a prerequisite for evolution toward a sluglike structure, since all 16 families with slugs or sluglike taxa belong to the Sigmurethra. *See* PULMONATA.

G. Alan Solem

Styrene A colorless, liquid hydrocarbon with the formula $C_6H_5CH{=}CH_2$. It boils at 145.2°C (293.4°F) and freezes at −30.6°C (−23.1°F). The ethylenic linkage of styrene readily undergoes addition reactions and under the influence of light, heat, or catalysts undergoes self-addition or polymerization to yield polystyrene.

The majority of the styrene used is converted into polystyrene, but other thermoplastic or even thermosetting resins are prepared from styrene by copolymerization with suitable comonomers. A smaller quantity of styrene goes into the manufacture of elastomers or synthetic rubbers.

Styrene is a skin irritant. Prolonged breathing of air containing more than 400 ppm of styrene vapor may be injurious to health. *See* POLYMERIZATION; POLYSTYRENE RESIN.

Charles K. Bradsher

Subduction zones Regions where portions of the Earth's tectonic plates are diving beneath other plates, into the Earth's interior. Subduction zones are defined by deep oceanic trenches, lines of volcanoes parallel to the trenches, and zones of large earthquakes that extend from the trenches landward.

Plate tectonic theory recognizes that the Earth's surface is composed of a mosaic of interacting lithospheric plates, with the lithosphere consisting of the crust (continental or oceanic) and associated underlying mantle, for a total thickness of about 100 km (60 mi). Oceanic lithosphere is created by sea-floor spreading at mid-ocean ridges (divergent, or accretionary, plate boundaries) and destroyed at subduction zones (at convergent, or destructive, plate boundaries). At subduction zones, the oceanic lithosphere dives beneath another plate, which may be either oceanic or continental. Part of the material on the subducted plate is recycled back to the surface (by being scraped off the subducting plate and accreted to the overriding plate, or by melting and rising as magma), and the remainder is mixed back into the Earth's deeper mantle. This process balances the creation of lithosphere that occurs at the mid-ocean ridge system. The convergence of two plates occurs at rates of 1–10 cm/yr (0.4–4 in./yr) or 10–100 km (6–60 mi) per million years (see illustration).

During subduction, stress and phase changes in the upper part of the cold descending plate produce large earthquakes in the upper portion of the plate, in a narrow band called the Wadati-Benioff zone that can extend as deep as 700 km (420 mi). The plate is heated as it descends, and the resulting release of water leads to melting of the overlying mantle. This melt rises to produce the linear volcanic chains that are one of the most striking features of subduction zones. *See* LITHOSPHERE.

Subduction zones can be divided in two ways, based either on the nature of the crust in the overriding plate or on the age of the subducting plate. The first classification yields two broad categories: those beneath an oceanic plate, as in the Mariana or Tonga trenches, and those beneath a continental plate, as along the west coast of South America (see illustration). The first type is known as an intraoceanic convergent margin, and the second is known as an Andean-type convergent margin. *See* MID-OCEANIC RIDGE; PLATE TECTONICS.

Active volcanoes are highly visible features of subduction zones. The volcanoes that have developed above subduction

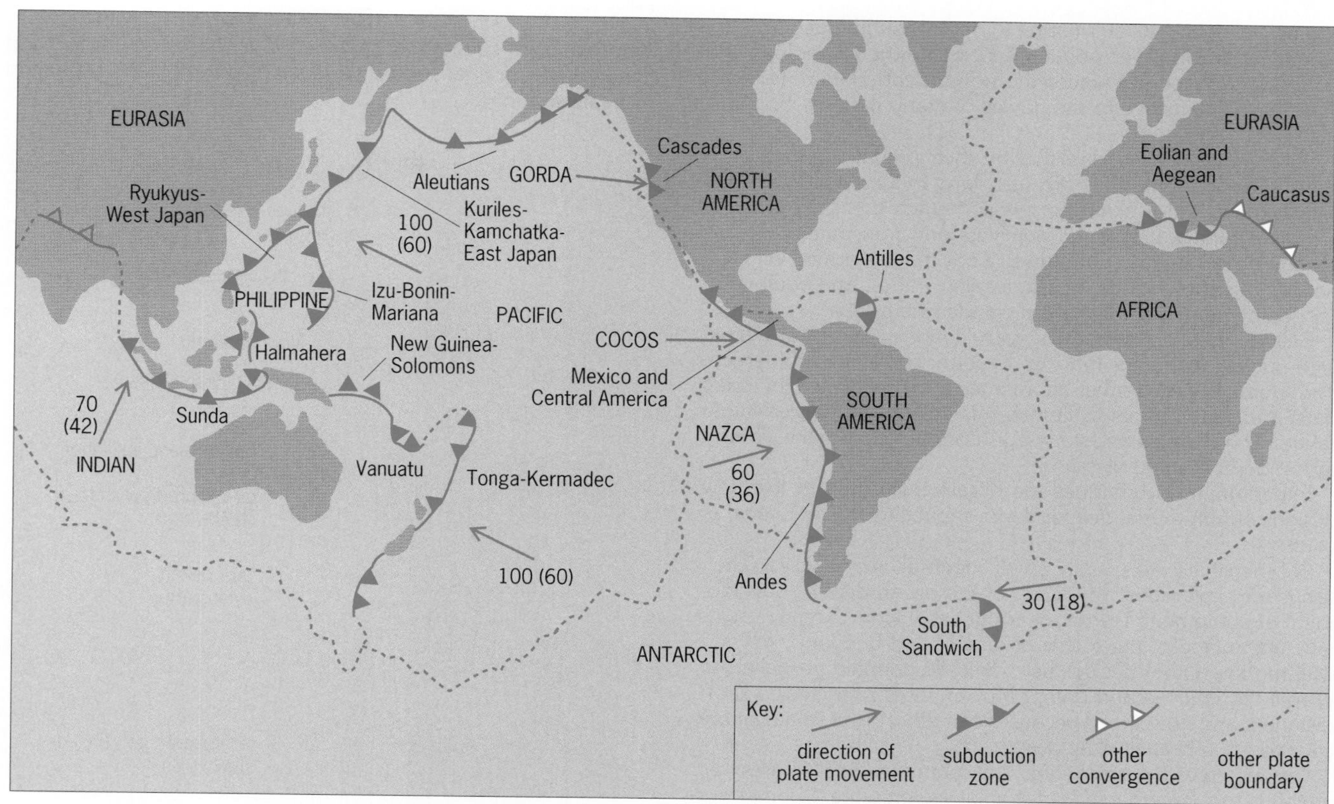

Principal subduction zones. Plate names are in all-capitals. Numbers at arrows indicate velocity of plate movement in kilometers (miles) per million years. (*After J. Gill, Orogenic Andesites and Plate Tectonics, Springer-Verlag, 1981*)

zones in East Asia, Australasia, and the western Americas surround the Pacific Ocean in a so-called Ring of Fire. At intraoceanic convergent margins, volcanoes may be the only component above sea level, leading to the name "island arc." The more general term "volcanic arc" refers to volcanoes built on either oceanic or continental crust. *See* VOLCANO.

An eventual consequence of subduction is orogeny, or mountain building. Subduction zones are constantly building new crust by the production of volcanic material or the accretion of oceanic sediments. However, the development of the greatest mountain ranges—the Alps or the Himalayas—occurs not during "normal" subduction but during the death of a subduction zone, when it becomes clogged with a large continent or volcanic arc. *See* OROGENY.

R. J. Stern; S. H. Bloomer

Subgiant star An evolving star of luminosity class IV. Such a star is brighter than the main-sequence dwarfs and fainter than the true giants in its spectral class, lying between the two on the Hertzsprung-Russell diagram. The classic subgiants fall in a small region from class F to K (with effective temperatures ranging from 7000 to 4000 K or 12,000 to 7000°F). In class G they lie at absolute visual magnitude +3 with luminosities about five times the solar luminosity. Classic subgiants have masses similar to or a bit greater than around 1.3 times that of the Sun and violate the mass-luminosity relation as too bright for their masses. The concept is extended to the hot part of the Hertzsprung-Russell diagram from class F to O, in which the distinctions between subgiants and neighboring dwarfs and giants are much smaller, only a magnitude or less.

Main-sequence dwarfs run on the fusion of hydrogen to helium in their cores. The subgiant stage begins when the core hydrogen mass fraction drops to a few hundredths, and then continues as the fraction goes to zero and the star evolves toward, but not onto, the red-giant branch with a contracting helium core. *See* DWARF STAR; GIANT STAR; HERTZSPRUNG-RUSSELL DIAGRAM; SPECTRAL TYPE; STAR; STELLAR EVOLUTION.

James B. Kaler

Sublimation The process by which solids are transformed directly to the vapor state without passing through the liquid phase. Sublimation is of considerable importance in the purification of certain substances such as iodine, naphthalene, and sulfur.

Sublimation is a universal phenomenon exhibited by all solids at temperatures below their triple points. For example, it is a common experience to observe the disappearance of snow from the ground even though the temperature is below the freezing point and liquid water is never present. The rate of disappearance is low, of course, because the vapor pressure of ice is low below its triple point. Sublimation is a scientifically and technically useful phenomenon, therefore, only when the vapor pressure of the solid phase is high enough for the rate of vaporization to be rapid. *See* PHASE EQUILIBRIUM; TRIPLE POINT; VAPOR PRESSURE.

Norman H. Nachtrieb

Submarine A ship which can operate completely submerged in the water. The term formerly applied to any ship capable of operating completely underwater, but now usually describes a ship built for military purposes. The term "submersible" usually is applied to small, underwater vehicles that are built for research, rescue, commercial work, or pleasure.

By the end of World War II, antisubmarine warfare had progressed significantly by exploiting the limited underwater endurance and speed of the diesel-electric designs of that era. The application of nuclear power to submarines after World War II reestablished the near-invulnerability of the submarine to antisubmarine warfare from surface ships and aircraft. Nuclear power depends on nuclear fission rather than the oxidation of fossil fuels and thus requires no oxygen source as do diesel engines, allowing the submarine to operate submerged for very long periods. However, advances in submarine technology and nonnuclear propulsion cause the nonnuclear submarine to remain highly attractive to the navies of many nations. *See* SHIP NUCLEAR PROPULSION.

Submarines can be classified by their primary military missions. Attack submarines are fast, long-range ships equipped with torpedo tubes or cruise missile launch tubes. They carry sensitive underwater sound receivers and transmitters (sonar) used to detect enemy submarines. They may be armed with torpedoes of various kinds, cruise missiles, mines, and equipment for deployment of small units of clandestine troops.

Ballistic-missile submarines carry long-range missiles fitted with nuclear warheads that can be launched while submerged. The submarine can remain submerged and undetected for many days and, on command, launch missiles on any target within range. The missiles are stowed in and launched from vertical tubes. *See* BALLISTIC MISSILE.

Experimental submarines are occasionally built to test new designs of hull shape, deeper depth capability, power plants, or controls.

Submersibles are usually small, deep-diving vehicles. Their use is for exploration and study of the ocean depths, development of equipment, rescue, or commercial work. Some designs take advantage of the forces of gravity and buoyancy for vertical motion. Other designs use vertically oriented propellers to propel the craft up and down. Movement is restricted to short distances and slow speed because of small size and small battery capacity. *See* UNDERWATER VEHICLE.

Compared with surface ships, the submarine has features that enable it to submerge and resist great sea pressure. Submarines have a pressure hull and a nonpressure hull. The pressure hull is the watertight, pressure-proof envelope in which equipment operates and the officers and crew live. In certain areas of the submarine there is a nonpressure hull of lighter structure, forming the main ballast tanks. A nonwatertight superstructure provides a smooth, fair envelope to cover pipes, valves, and fittings on top of the hull. Above the superstructure the fairwater similarly encloses the bridge, the periscope, and multiple mast supports. *See* PERISCOPE; SHIP DESIGN.

The principal means of detecting the presence of a submerged submarine is to listen for sounds which may have been generated on board or by its movement through the water. Very small amounts of acoustic energy can be detected by sophisticated sonars. Therefore, modern submarines are designed with multiple features to greatly reduce the amount of noise they generate.

Millard S. Firebaugh; James H. Webber

Submarine cable A cable, primarily for communications, laid on the ocean floor to provide international links. This term is sometimes applied to power cables in water, but these are not usually of great length.

Coaxial cables with 1840 circuits (each having 3-kHz bandwidth) are used extensively for shorter cables. Cables using digital transmission over optical fibers (lightguide transmission) offer a significant cost-per-channel advantage over coaxial systems. Voice signals are transmitted by laser-generated digital light pulses over single-mode fiber pairs. *See* COAXIAL CABLE; COMMUNICATIONS CABLE; MULTIPLEXING AND MULTIPLE ACCESS; TRANSMISSION LINES.

SL280 digital lightguide transatlantic submarine cables utilize semiconductor laser diodes operating at a wavelength of 1.3-micrometers and carries 280 megabits per second on six hair-thin fibers, of which two pairs are active and the third is available as standby. Each fiber is a glass strand 125 μm in diameter, plastic-coated to 250 μm, and embedded in an elastomer that is part of the unit fiber structure. The fibers are helically wrapped around a central kingwire for strength during manufacture and an outer sheath of nylon to ensure dimensional stability. This structure is then helically wrapped with steel armor wires for strength, ensheathed in a copper conductor, and coated with a final protective layer of polyethylene insulation (see illustration).

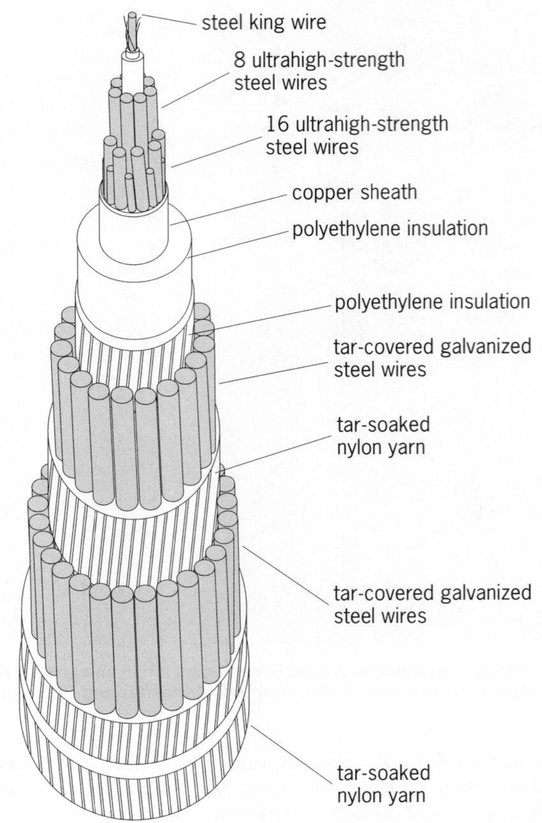

Cutaway view of double-armored SL cable.

The fiber-optic digital repeaters regenerate the signal in its original form, unlike the earlier analog repeaters which were signal amplifiers. The repeaters detect the optical signal entering the repeater, transform it into an electrical signal and amplify it, regenerate it by putting it through a high-speed clocked decision circuit, and then convert the regenerated electrical signal into an optical one by using a 1.3-μm laser transmitter. Transmission rates of 560 megabits per second are provided in the subsequent SL560 design, utilizing lasers operating at 1.5 μm.

Virtually all analog (coaxial) submarine cable systems were configured as a single line from one landing point to another. Traffic requirements, however, are not identical for all countries, and it was realized to be advantageous to implement a tree configuration, where the various branches go to different countries. Digital fiber-optic technology makes branching possible by dedicating certain fiber pairs to certain subsystems.

With the increased volume and capacity of fiber-optic systems, customers and system owners cannot afford or support the potential loss of business and revenue associated with a cable failure. To reduce this risk, many systems employ a ring-network design architecture. These systems use a self-healing loop network with built-in backup and redundancy, which provide fiber-on-fiber restoration and connectivity. In the unlikely event of a cable failure of any nature, communications traffic can be shifted from one fiber cable to another so that digital voice, data, or video communications can continue without disruption. *See* LASER; OPTICAL COMMUNICATIONS; OPTICAL FIBERS; PULSE MODULATION.

Gary J. Stout

Submarine canyon A steep-sided valley developed on the sea floor of the continental slope. Submarine canyons serve as major conduits for sediment transport from land and the continental shelf to the deep sea. Modern canyons are relatively narrow, deeply incised, steeply walled, often sinuous valleys with predominantly V-shaped cross sections. Most canyons originate near the continental-shelf break, and generally extend to the base of the continental slope. Canyons often occur off the mouths of large rivers, such as the Hudson River or the Mississippi River, although many other canyons, such as the Bering Canyon in the southern Bering Sea, are developed along structural trends. *See* CONTINENTAL MARGIN.

Modern submarine canyons vary considerably in their dimensions. The average lengths of canyons has been estimated to be about 34 mi (55 km); although the Bering Canyon is more than 930 mi (1495 km) long and is the world's longest submarine canyon. The shortest canyons are those off the Hawaiian Islands, and average about 6 mi (10 km) in length. Submarine canyons are characterized by relatively steep gradients. The average slope of canyon floors is 309 ft/mi (58 m/km). In general, shorter canyons tend to have higher gradients. For example, shorter canyons of the Hawaiian group have an average gradient of 766 ft/mi (144 m/km), whereas the Bering Canyon has a slope of only 42 ft/mi (7.9 m/km).

In comparison to modern canyons, dimensions of ancient canyons are considerably smaller. Deposits of ancient canyons are good hydrocarbon reservoirs. This is because submarine canyons and channels are often filled with sand that has the potential to hold oil and gas. Examples of hydrocarbon-bearing canyon-channel reservoirs are present in the subsurface in California, Louisiana, and Texas.

Physical and biological processes that are common in submarine canyons are mass wasting, turbidity currents, bottom currents, and bioerosion. Mass wasting is a general term used for the failure, dislodgement, and downslope movement of sediment under the influence of gravity. Common examples of mass wasting are slides, slumps, and debris flows. Major slumping events can lead to formation of submarine canyons. The Mississippi Canyon in the Gulf of Mexico is believed to have been formed by retrogressive slumping during the late Pleistocene fall in sea level and has been partially infilled during the Holocene rise in sea level. *See* DEPOSITIONAL SYSTEMS AND ENVIRONMENTS; MARINE GEOLOGY; MARINE SEDIMENTS; MASS WASTING; TURBIDITY CURRENT.

G Shanmugam

Submillimeter astronomy Investigation of the universe by probing the electromagnetic spectrum at wavelengths from approximately 0.3 to 1.0 millimeter: the submillimeter waveband. This waveband is bounded at longer wavelengths by the millimeter waveband (1–10 mm), and at shorter wavelengths by the far-infrared waveband (20–300 micrometers). Obtaining an accurate view of the processes going on in stars and galaxies requires observations at a range of different wavelengths, with each available window providing complementary information. Astronomical objects with temperatures between about 10 kelvins and several hundred kelvins, typically in the interstellar medium of galaxies, emit radiation strongly in the submillimeter waveband. *See* ELECTROMAGNETIC RADIATION.

The submillimeter waveband is one of the last parts of the electromagnetic spectrum to be investigated. This results from both the technical challenge of making sensitive detectors, which must be held at temperatures close to absolute zero, and the effects of the Earth's atmosphere. Atmospheric molecules, primarily water vapor, both absorb (dim) the signal from astronomical sources and emit their own radiation that acts to mask the astronomical signals. The effects are most severe at the shortest wavelengths, and only high mountain sites and air- or space-borne platforms can be used for submillimeter astronomy. From high mountains, observations are possible in three primary atmospheric windows, at wavelengths of about 0.35, 0.45, and 0.85 mm. Radiation of 0.85 mm penetrates down to about 2 km (1.2 mi) above sea level; while on the driest nights, only about one-half of photons with wavelengths of 0.35 or 0.45 mm that enter the upper atmosphere can be detected at the 4200-m (14,000-ft) summit of Mauna Kea, Hawaii. Even higher sites have some access to a higher-frequency band at 0.2 mm, such as high peaks in the Andes.

Several submillimeter telescopes operate on high mountain tops, including the 15-m (49-ft) James Clerk Maxwell Telescope (JCMT; *see* illustration) and the 10.4-m (34-ft) Caltech Submillimeter Observatory (CSO) telescope, both on Mauna Kea, and the 12-m (39-ft) German–European Southern Observatory ESO–Swedish Atacama Pathfinder Experiment (APEX) telescope at Chajnantor in Chile. These telescopes look like satellite antennas, but their surfaces are smoother and precisely shaped to an accuracy of order 10 μm. These ground-based telescopes are subject to more atmospheric absorption than airborne and space telescopes; but because they have larger apertures, they can collect more radiation, boosting their sensitivity. *See* TELESCOPE.

James Clerk Maxwell Telescope (JCMT), 15 m (49 ft) in diameter, situated at Mauna Kea, Hawaii.

Air- and space-borne observatories are essential. The 3.5-m-aperture (11.5-ft) *Herschel Space Observatory* is scheduled to launch in 2009, and the Stratospheric Observatory for Infrared Astronomy (SOFIA), a 2.5-m-aperture (8.2-ft) telescope in a 747 aircraft, is under development. Like the former Kuiper Airborne Observatory (KAO), SOFIA will fly at up to 14 km (9 mi) altitude. Stratospheric balloons, flying at altitudes of about 40 km (25 mi), can also be used to carry submillimeter telescopes; an example is the detailed image of the cosmic microwave background determined by the BOOMERANG experiment in 2000.

The signals from several submillimeter telescopes can be combined in an interferometer to provide a telescope with resolution as fine as one with an aperture as large as the greatest separation between the individual telescopes. Examples of this type of telescope are the interferometer on France's Plateau de Bure, operated by IRAM (Institut de Radio Astronomie Millimétrique); the Submillimeter Array (SMA) now operating on Mauna Kea; the Combined Array for Research in Millimeter-wave Astronomy (CARMA) in California; and the Atacama Large Millimeter/Submillimeter Array (ALMA), a United States–Europe–Japan collaboration under construction in Chile. *See* RADIO TELESCOPE.

Submillimeter observations are used by astronomers with a wide range of interests to study matter in the universe at temperatures of about 10–100 K and the cosmic microwave background radiation. Interstellar chemistry in clouds of gas and dust, the formation of protostars embedded in stellar nurseries within the Milky Way, the properties of the interstellar medium in even the most distant galaxies, and the evolution of galaxies are all studied in the submillimeter waveband. Submillimeter observations are especially important for observations of regions that are rich in dust and gas. These are opaque to visible and infrared light, but submillimeter light can still escape, allowing these regions to be studied. Andrew Blain

Submillimeter-wave technology The generation, detection, and application of radiation in the submillimeter-wave region of the electromagnetic spectrum. In practical terms, this region corresponds to wavelengths between about 30 micrometers and 3 millimeters (see illustration). At its extremes the region overlaps with the microwave and infrared, and a unique feature is that the distinct propagation methods of these adjacent regions, guided-wave and free-space propagation, are used throughout it. *See* ELECTROMAGNETIC RADIATION; INFRARED RADIATION; MICROWAVE.

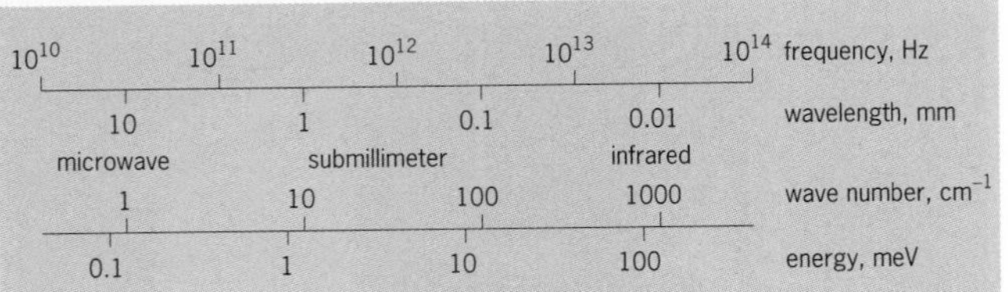

Electromagnetic spectrum in and near the submillimeter-wave region, showing its frequency, wavelength, wave number, and photon energy equivalents.

Widespread use of the submillimeter-wave region did not occur until two revolutionary developments in the late 1950s and early 1960s: (1) the practical realization of Fourier transform spectroscopy, which led to significant improvements in the quality of broad band submillimeter-wave spectroscopy; and (2) the discovery of two reasonably intense submillimeter-wave laser sources, the hydrogen cyanide and the water-vapor electrically excited lasers. *See* INFRARED SPECTROSCOPY; LASER.

Radiation sources. The main application of incoherent radiation sources is as broad band sources in Fourier transform spectrometers. The two most commonly used sources are the mercury-vapor lamp and the globar. *See* MERCURY-VAPOR LAMP; MICROWAVE TUBE.

There are several categories of coherent submillimeter-wave radiation sources: (1) nonrelativistic electron-beam tubes such as klystrons, extended interaction oscillators, orotrons, ledatrons, and carcinotrons or backward-wave oscillators, which oscillate with varying amounts of tunability and output power; (2) relativistic electron-beam devices such as gyrotrons, free-electron lasers, and synchrotrons; and (3) solid-state devices such as Gunn and IMPATT oscillators. With the exception of the free-electron laser, these are long-wave devices. The final category, gas lasers, covers the entire submillimeter-wave region. *See* GYROTRON; KLYSTRON; MICROWAVE SOLID-STATE DEVICES; SYNCHROTRON RADIATION.

Detectors. Most submillimeter-wave detectors are either thermal, rectifying, or photon detectors. The first respond to heat generated by absorbed radiation, the second by rectification of radiation-field-induced currents, the third by absorption of individual photons. They are commonly used in direct and heterodyne detection systems. Thermal detectors include room-temperature thermopile, Golay cell, and pyroelectric detectors, while more sensitive semiconductor bolometers operate at liquid-helium temperatures. Rectifying devices include metal-semiconductor and metal-metal point contacts and Schottky barrier diodes, all room-temperature devices, and Josephson-type devices operated at about 4.2 K (−452°F). Submillimeter-wave photon detectors are based on semiconductors and rely on the absorption of a photon changing the free-carrier density or mobility and hence the conductivity of the detector. *See* BOLOMETER; PYROELECTRICITY; RADIOMETRY; SUPERCONDUCTING DEVICES.

Atmospheric transparency. Between the visible and microwave regions the atmosphere is effectively opaque, with the exception of two windows in the near-infrared and three in the submillimeter-wave region, at 94, 140, and 220 GHz. The significant feature of the submillimeter-wave windows is that the levels of attenuation involved are not as seriously increased by adverse atmospheric conditions as those of the near-infrared. This makes them attractive for many applications. The attenuation associated with these windows is, nonetheless, quite high and means that ground-level low-frequency submillimeter-wave systems have a relatively limited range. This can be advantageous for some applications.

Applications. In the past, applications of radiation in the submillimeter-wave region were distributed evenly over the region and largely concerned research activities. Now, however, as a result of developments toward integrated, inexpensive transmitter and receiver systems, applications include many radar- and telecommunications-related activities centered on the 94-GHz atmospheric window. These applications are mainly military, but spin-offs into the civil area are expected.

Photon energies in the submillimeter-wave region correspond to activation energies of many physical, chemical, and bioloical phenomena, and the spectroscopic techniques of this region have been widely applied to their study. Other spectroscopic applications have concerned the development of radiometric methods to meet the requirements of fusion plasma diagnostics and astronomy. *See* SUBMILLIMETER ASTRONOMY.

The submillimeter-wave region has been extensively used for fundamental frequency metrology, relating the frequencies of submillimeter-wave gas lasers directly back to the primary frequency standard. Important consequences have been the definition of the speed of light as a fixed quantity and the redefinition of the meter as a unit derived from it. *See* LIGHT; PHYSICAL MEASUREMENT.

The submillimeter-wave region impinges on fusion plasma research in two ways. First, in machines such as tokamaks it provides a way of studying the spatial and temporal evolution of the electron temperature. The second application is electron cyclotron heating, in which electron heating by microwave radiation at the electron cyclotron frequency has been demonstrated. High-power gyrotron oscillators are generally used as the radiation sources. *See* NUCLEAR FUSION; PLASMA (PHYSICS); PLASMA DIAGNOSTICS.

In the study and applications of fast electromagnetic pulse propagation, a number of novel techniques for the submillimeter-wave region have been developed. The generation and detection of such pulses can be done with ultrashort optical pulses driving photoconductive switches, or by all-electronic systems based on nonlinear transmission lines. The submillimeter-wave radiation used in these systems is variously known as T waves or terahertz waves. *See* OPTICAL PULSES.

J. R. Birch

Subsonic flight Movement of a vehicle through the atmosphere at a speed appreciably below that of sound waves. Subsonic flight extends from zero (hovering) to a speed approximately 85% of sonic speed corresponding to the ambient temperature. At higher vehicle velocities the local velocity of air passing over the vehicle surface may exceed sonic speed, and true subsonic flight no longer exists.

Vehicle type may range from a small helicopter, which operates at all times in the lower range of the velocity scale, to an intercontinental ballistic missile, which is operative throughout this and other velocity regimes, but is in subsonic flight for only a few seconds. The design of each is affected by the same principles of subsonic aerodynamics. Subsonic flow of a fluid such as air may be subdivided into a range of velocities in which the flow may be considered incompressible without appreciable error (below a velocity of approximately 300 mi/h or 135 m/s), and a higher range in which the compressible nature of the fluid becomes significant. In both cases the viscosity of the fluid is important. The theories which apply to compressible, inviscid fluids may be used almost without modification in some low-subsonic problems, and in other cases the results offered by these theories may be modified to account for the effects of viscosity and compressibility. *See* COMPRESSIBLE FLOW; TRANSONIC FLIGHT; VISCOSITY.

A typical subsonic wing cross section (airfoil) has a rounded front portion (leading edge) and a sharp rear portion (trailing edge). Air approaching the leading edge comes to rest at some point on the leading edge, with flow above this point proceeding around the upper airfoil surface to the trailing edge, and flow below passing along the lower surface to the same point, where the flow again theoretically has zero velocity. The two points of zero local velocity are known as stagnation points. If the path from front to rear stagnation point is longer along the upper surface than along the lower surface, the mean velocity of flow along the upper surface must be greater than that along the lower surface. Thus, in accordance with the principle of conservation of energy, the mean static pressure must be less on the upper surface than on the lower surface. This pressure difference, applied to the surface area with proper regard to force direction, gives a net lifting force. Lift is defined as a force perpendicular to the direction of fluid flow relative to the body, or more clearly, perpendicular to the free-stream velocity vector. *See* AIRFOIL; BERNOULLI'S THEOREM.

The wing, as the lifting device, whether fixed, as in the airplane, or rotating, as in the helicopter, is probably the most important aerodynamic part of an aircraft. However, stability and control characteristics of the subsonic airplane depend on the complete structure. Control is the ability of the airplane to rotate about any of the three mutually perpendicular axes meeting at its center of gravity. Static stability is the tendency of the airplane to return to its original flight attitude when disturbed by a moment about any of the axes. James E. May

Substitution reaction One of a class of chemical reactions in which one atom or group (of atoms) replaces another atom or group in the structure of a molecule or ion. Usually, the new group takes the same structural position that was occupied by the group replaced.

Substitution reactions involve the attack of a reagent, which is the source of the new atom or group, on the substrate, the molecule or ion in which the replacement occurs. They involve the formation of a new bond and the breaking of an old bond. Substitution reactions are classified according to the nature of the reagent (electrophilic, nucleophilic, or radical) and according to the nature of the site of substitution (saturated carbon atom or aromatic carbon atom). *See* ELECTROPHLIC AND NUCLEOPHILIC REAGENTS.

Systematic names for substitution reactions are composed of the parts: name of group introduced + de + name of group replaced + ation, with suitable elision or change of vowels for euphony. Thus, the replacement of bromine by a methoxy group is called methoxydebromination. *See* ORGANIC REACTION MECHANISM. Joseph F. Bunnett

Subsynchronous resonance The resonance between a series-capacitor-compensated electric system and the mechanical spring-mass system of a turbine-generator at subsynchronous frequencies. Beginning about 1950, series capacitors were installed in long alternating-current transmission lines [250 km (150 mi) or more] to cancel part of the inherent inductive reactance of the line. Until 1971, up to 70% of the 60-Hz inductive reactance was canceled by series capacitors in some long lines with little concern for side effects. (If 70% of a line's inductive reactance is canceled, the line is said to have 70% series compensation.) In 1970, and again in 1971, a turbine-generator at the Mohave Power Plant in southern Nevada experienced shaft damage that required several months of repairs on each occasion. This followed switching events that placed the turbine-generator so that it was radial on a series-compensated transmission line. The shaft damage was due to torsional oscillations between the two ends of the generator-exciter shaft. Shortly after the second event, it was determined that the torsional oscillations were caused by torsional interaction, which is a type of subsynchronous resonance. There have been no reported occurrences of two other types of subsynchronous resonance, the induction generator effect and torque amplification. *See* ALTERNATING-CURRENT GENERATOR; TRANSMISSION LINES; TURBINE.

It has been clearly established that subsynchronous resonance can be controlled with the use of countermeasures, thus making it possible to benefit from the distinct advantages of series capacitors. About a dozen countermeasures have been successfully applied such that there has been no reported subsynchronous resonance event since 1971. Subsynchronous resonance countermeasures protect turbine-generator shafts from harmful torsional oscillations by one of two methods. First, the turbine-generator can be tripped when a subsynchronous resonance condition is detected. This limits the number of torsional oscillations experienced by the turbine-generator shafts. This type of countermeasure is relatively inexpensive but is not acceptable if the anticipated subsynchronous resonance conditions are expected to occur frequently. Generally, such a countermeasure will not be applied as the sole subsynchronous resonance protection if it is expected to cause a turbine-generator to be tripped more than once every 10 years. Three types of tripping countermeasures are applied: torsional motion relay, armature current relay, and the generator tripping logic scheme. The second method of protection does not involve turbine-generator tripping, but eliminates or limits harmful torsional oscillations. *See* ELECTRIC POWER SYSTEMS. Richard G. Farmer

Subtraction One of the four fundamental operations of arithmetic and algebra. Subtraction is often regarded as an operation inverse to addition, that is, if a and b are numbers, the number $a - b$ is defined as that number which added to b gives a. The more modern viewpoint eliminates subtraction completely by considering the number $a - b$ as the sum of a and that number (denoted by $-b$) which added to b gives 0. The number symbolized by $-b$ is called the inverse of b (with respect to addition). Every real number has a unique inverse (the number 0 is its own inverse). In this sense "subtraction" may be performed on objects of many different kinds, and the original numerical operation greatly extended. *See* ADDITION; ALGEBRA; DIVISION; MULTIPLICATION; NUMBER THEORY. Leonard M. Blumenthal

Subway engineering The branch of transportation engineering that deals with feasibility study, planning, design, construction, and operation of subway (underground railway) systems. In addition to providing rapid and comfortable service, subways consume less energy per passenger carried in comparison with other modes of transportation such as automobiles and buses. They have been adopted in many cities as a primary mode of transportation to reduce traffic congestion and air pollution.

Subways are designed for short trips with frequent stops, compared to above-ground, intercity railways. Many factors considered in the planning process of subway systems are quite similar to those for railway systems. Subway system planning starts with a corridor study, which includes a forecast of ridership and revenues, an estimation of construction and operational costs, and a projection of the potential benefits from land development. *See* RAILROAD ENGINEERING; TRANSPORTATION ENGINEERING.

All subway systems have three major types of structures: stations, tunnels, and depots. The most important task in planning a new subway system or a new subway line is to locate stations and depots and to determine the track alignment. Subway lines are normally located within the right-of-way of public roads and as far away as possible from private properties and sites of importance. Because stations and entrances are usually located in densely populated areas, land acquisition is often a major problem. One solution is to integrate entrances into nearby developments such as parks, department stores, and public buildings, which lessens the visual impact of the entrances and reduces their impediment to pedestrian flow.

Design of the permanent works includes structural and architectural elements and electrical and mechanical facilities. There are two types of structures: stations and tunnels. For stations, space optimization and passenger flow are important. The major elements in a typical station are rails, platform, staircases, and escalators. For handicapped passengers, provisions should be made for the movement of wheelchairs in elevators and at fare gates, and special tiles should be available to guide the blind to platforms.

In both stations and tunnels, ventilation is essential for the comfort of the passengers and for removing smoke during a fire. Sufficient staircases are required for passengers to escape from the station platform to a point of safety in case of a fire. The electrical and mechanical facilities include the rolling stock, signaling, communication, power supply, automated fare collection, and environmental control (air-conditioning) systems. Corrosion has caused problems to structures in some subways; therefore, corrosion-resistant coatings may be required. To minimize noise and vibration from running trains, floating slabs can be used under rails or building foundations in sections of routes crossing densely populated areas and in commercial districts where vibration and secondary airborne noise inside buildings are unacceptable. *See* VENTILATION.

Underground stations are normally constructed by using an open-cut method. For open cuts in soft ground, the sides of the pits are normally retained by wall members and braced using struts. The pits are fitted with decks for maintaining traffic at the surface. For new lines that pass under existing lines, it is not possible to have open cuts. In such cases, stations have to be constructed using mining methods (underground excavation). *See* TUNNEL.

Many modern subway systems are fully automated and require only a minimal staff. Train movements are monitored and regulated by computers in a control center. Therefore, engineering is limited to the function and maintenance of the electrical and mechanical facilities. The electrical and mechanical devices requiring constant care include the rolling stock, signaling, communication and broadcasting systems, power supply, elevators and escalators, automated fare collection, and environmental control systems. Also included are depot facilities, and station and tunnel service facilities. *See* ELECTRIC DISTRIBUTION SYSTEMS; RAILROAD CONTROL SYSTEMS.

Za-Chieh Moh; Richard N. Hwang

Sucrose An oligosaccharide, α-D-glucopyranosyl-β-D-fructofuranoside, also known as saccharose, cane sugar, or beet sugar. The structure is shown below. Sucrose is very soluble in water and crystallizes from the medium in the anhydrous form. The sugar occurs universally throughout the plant kingdom in fruits, seeds, flowers, and roots of plants. Honey consists principally of sucrose and its hydrolysis products. Sugarcane and sugarbeets are the chief sources for the preparation of sucrose on a large scale. Another source of commercial interest is the sap of maple trees. *See* OLIGOSACCHARIDE; SUGARBEET; SUGARCANE.

William Z. Hassid

Suctoria A small specialized subclass of the protozoan class Ciliatea whose members were long considered entirely separate from the "true" ciliates. The sole order of this subclass is Suctorida. These forms show a number of highly specialized features. Most conspicuous are their tentacles, often numerous, which serve as mouths. These multiple organelles of ingestion fasten to the pellicle of prey organisms, generally passing ciliates. By forces not entirely understood, the tentacles are used to suck out the prey's protoplasm to provide sustenance for the suctorian. Nearly all species are stalked, and the sedentary, mature forms are devoid of any external ciliature. Young larval forms are produced by both endogenous and exogenous budding. These forms bear locomotor cilia and serve, as in the case of species of the Peritrichia, for dissemination (see illustration). *See* CILIATEA; PERITRICHIA.

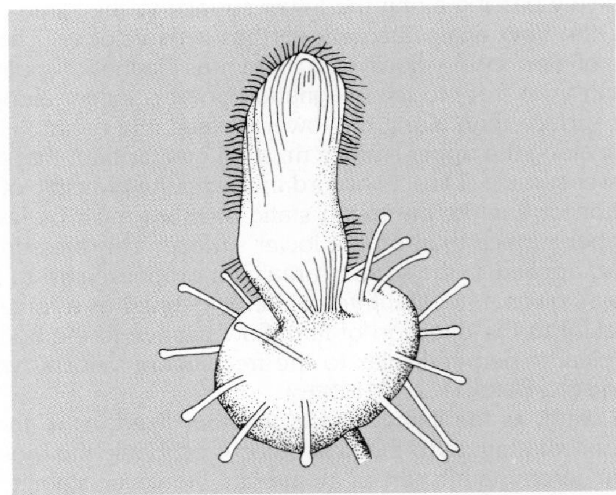

Endogenous budding in the suctorian *Podophrya*, a species which measures 10–28 micrometers.

John O. Corliss

Sudangrass An annual, warm-season grass of tropical origin said to have been grown in Egypt since early times, though its value was first recognized in Sudan only in 1909. In that same year it was introduced into the United States as a replacement for johnsongrass, which had become a noxious weed in many southern states. Sudangrass (*Sorghum bicolor* var. *sudanense*, also called *S. sudanense* and *S. vulgare* var. *sudanense*) and its hybrids are commonly used as pasture, greenchop, silage, or hay. They fill an important need in many regions of the United States, because they produce high-quality forage for cattle and sheep during the summer, when other pasture is in short supply or of low quality. Many of the varieties and hybrids produce forage until frost. *See* GRASS.

Melvin R. George

Sudden infant death syndrome (SIDS) The sudden and unexpected death of an infant less than one year of age whose death remains unexplained after a postmortem investigation, including an autopsy, death scene inspection, and a medical history. In a typical situation, a parent checks on a child who is thought to be sleeping only to find the infant has died. The largest number of infants die when they are between 2 and 4 months of age, with 95% of deaths occurring between birth and 6 months. Although sudden infant death syndrome can happen to any infant, there is a higher incidence among minority groups (American Indian/Alaskan natives and African Americans in the United States), premature infants, low-birth-weight infants, infants with little or no prenatal care, infants born to teenage mothers, infants of multiple gestations such as twins and triplets, siblings of SIDS infants, and infants who have had an apparent life-threatening event of not breathing (ALTE).

Since the year 2000, the rate of SIDS in the United States has been about 0.6 per 1000 infant live births. This means that between 2000 and 2500 deaths per year are attributed to SIDS. During the 1990s, the rate of SIDS dropped by over 50%. The reason for the decrease in the number of deaths is most likely due to public health campaigns to reduce modifiable SIDS risk factors by government and nonprofit agencies. One such modifiable risk factor is sleep position. Infants who were placed to sleep on their stomachs or sides were shown to have a 2 to 12 times greater risk of SIDS than infants who were placed to sleep on their backs. Other factors found to reduce death include the following: infants not being exposed to tobacco (prenatally or after birth), infants sleeping without heavy coverings, infants not being overheated, and infants placed in cribs kept in the mothers' rooms. *See* PUBLIC HEALTH.

A modifiable risk factor is not a *cause* of SIDS but, during the vulnerable age range, may act as a trigger for death in a child who is already compromised at birth due to some physiological defect. A triple risk model of SIDS guides most research today on sudden infant death. The three components are a subtle defect or deficiency at birth, an exogenous factor that affects the infant after birth, and a vulnerable age. The theory proposes that none of these factors alone would likely lead to death but in combination can be lethal. Research findings are pointing to several neurological and biochemical abnormalities in SIDS infants that may result from genetic factors or may be the result of prenatal insults. *See* AUTONOMIC NERVOUS SYSTEM; CONGENITAL ANOMALIES; HUMAN GENETICS; NERVOUS SYSTEM DISORDERS.

Betty McEntire; Dorothy Kelly

Sugar Usually, sucrose, the common sugar of commerce. This sugar is a disaccharide, $C_{12}H_{22}O_{11}$, which is split, as shown in the reaction below, by hydrolysis into two monosaccharides,

$$\underset{\text{Sucrose}}{C_{12}H_{22}O_{11}} + \underset{\text{Water}}{H_2O} \rightarrow \underset{\text{Glucose}}{C_6H_{12}O_6} + \underset{\text{Fructose}}{C_6H_{12}O_6}$$

or simple sugars: glucose (dextrose) and fructose (levulose). Sucrose rotates the plane of polarized light to the right, as does glucose, but fructose is so strongly levorotatory that it overcomes the effect of glucose. Thus mixtures of equal amounts of glucose and fructose are levorotatory. The hydrolytic reaction is called inversion of sugar, and the product is invert sugar or, simply, invert. *See* CARBOHYDRATE; OPTICAL ACTIVITY.

Sucrose is widely distributed in nature, having been found in all green plants that have been carefully examined for its presence. The total quantity of all sugars formed each year has been estimated at a colossal 4×10^{11} tons (3.6×10^{11} metric tons). In spite of its availability in all green plants, sucrose is obtained commercially in substantial amounts from only two plants: sugarcane, which supplies about 56% of the world total, and the sugarbeet, which provides 44%.

Cane sugar manufacture. The manufacture of cane sugar is usually done in two series of operations. First, raw sugar of about 98% purity is produced at a location adjacent to the cane fields. The raw sugar is then shipped to refineries, where a purity that is close to 100% is achieved.

Raw cane sugar. The production of raw cane sugar begins with growing the cane in tropical or subtropical areas. The cane is harvested after a season varying from 7 months in subtropical areas to 12–22 months in the tropics. The cane stalks are harvested either mechanically or by heavy hand knives. The trend is toward mechanization. The stalks are transported to a mill by oxcart, rail, or truck. *See* SUGARCANE.

At the mill the stalks are crushed and macerated between heavy grooved iron rollers while being sprayed countercurrently with water to dilute the residual juice. The expressed juice contains 95% or more of the sucrose present. The fibrous residue, or bagasse, is usually burned under the boilers, although increasing amounts are being made into paper, insulating board, and hardboard as well as furfural, which is a chemical intermediate for the synthesis of furan and tetrahydrofuran.

Cane juice. The cane juice is treated with lime to bring its pH to about 8.2. This pH prevents the inversion reaction, which is favored by heat and acid and would lower the yield of crystallizable sugar. The juice is then heated to facilitate the precipitation of impurities, which are removed by continuous filtration.

The purified juice is concentrated by multiple-stage vacuum evaporation (usually four or five stages) and, when sufficiently concentrated, is boiled to grain or seeded with sucrose crystals in a single-stage vacuum pan. Usually three successive crops of crystals are grown, cooled, and centrifuged. The final mother liquor, which is resistant to further crystallization, is called blackstrap molasses. It is used principally as a feed for cattle. Relatively small amounts are still fermented to produce industrial alcohol and rum.

Raw cane sugar refining. The refining of raw sugar begins with dissolution of the molasses which remains in a thin film on the crystals in spite of the centrifugation. This step, called affination, brings the purity from about 98 to about 99%. The crystals are dissolved in hot water and percolated through bone char columns to remove color by adsorption. The sucrose is finally concentrated by vacuum evaporation, crystallized by seeding, centrifuged, and dried.

A major step forward has been the use of bone char in a continuous countercurrent manner (Grosvenor patent). The bone char is washed, dried, and burned to remove impurities; it is reused until it wears out mechanically and is discarded as fines. Even the fines have value as fertilizer because of their high calcium phosphate content.

Bone char has been replaced in some refineries by granular carbon derived from coal or wood waste. It is used in columns and regenerated by a process similar to the one used for bone char, which unlike granular carbon is primarily calcium phosphate. Granular carbon is frequently used in combination with decolorizing resins to reduce the color of process liquors coming from the carbon columns. Pulverized activated vegetable carbon is used in a few instances to remove color from raw sugar liquors, but its use is declining. Ion-exchange resins are used to remove ash from raw sugar liquors, especially when the liquor is to be converted directly into refined liquid sugar without going through the crystallization step first. *See* ION EXCHANGE.

Beet sugar. In the United States, sugarbeets are grown under contract by farmers from seed supplied by a beet sugar company. Because sugarbeets, like the other temperate-zone crops, thrive best under crop rotation, they are not well adapted to one-crop agriculture. *See* SUGARBEET.

Beet processing. When beets are delivered to a factory, they are washed and sliced, and the slices are extracted countercurrently with hot water to remove the sucrose. The resulting solution is purified by repeatedly precipitating calcium carbonate, calcium sulfite, or both, in it. Colloidal impurities are entangled in the growing crystals of precipitate and removed by continuous filtration. The resulting solution is nearly colorless, and the sucrose is concentrated by multiple-effect vacuum evaporation. The syrup is seeded, cooled, and centrifuged, and the beet sugar crystals are washed with water and dried.

Beet molasses differs from cane molasses in having a much lower content of invert sugar. It is, therefore, relatively stable to the action of alkali and, in the United States, is usually treated with calcium oxide to yield a precipitate of calcium sucrate. This is a mixture of loose chemical aggregates of sucrose and calcium oxide which are relatively insoluble in water. The precipitate is filtered, washed, and added to the incoming crude sugar syrup, where it furnishes calcium for the precipitations of calcium carbonate and sulfite referred to above, which remove impurities. Carbon dioxide in the form of flue gas is the other reagent, and sulfur dioxide from burning sulfur is used to produce sulfite. Ion exclusion processes have been developed for recovery of sucrose from beet molasses.

Beet residue use. The beet tops and extracted slices, as well as the molasses, are valuable as feeds. More feed for cattle and other ruminants can be produced per acre-year from beets than from any other crop widely grown in the United States. This is independent of the food energy in the crystallized sucrose, which exceeds that available from any other temperate-zone plant. It is for these reasons that the densely populated countries of Europe have expanded their beet sugar production, in spite of the ready availability of cane sugar from the tropics.

The increased use of nitrogenous fertilizer has resulted in augmenting the protein content of the molasses and other beet byproducts.

Nutritional value. Sucrose has, in the past, been attacked by some nutritionists on the ground that it provides only "empty calories," without protein, minerals, or vitamins. This argument lost much of its force when it was shown that all the vitamins and minerals recommended by the National Research Council can be obtained by consuming any of a great variety of foods in amounts that yield a total of only one-half of the caloric requirements of an average person. The wide use by the public of vitamin supplements has caused some nutritionists to express an opposite worry over excessive vitamin consumption.

In addition to the charges that it provides empty calories, sugar has been blamed for such health problems as diabetes mellitus, coronary heart disease, dental caries, and obesity. These charges and the scientific evidence associated with them have been extensively reviewed by a special committee of the Federation of American Societies for Experimental Biology. A report of the committee findings, published by the Food and Drug Administration in 1976, includes the following: Sucrose is a contributor to the formation of dental caries when used at levels and in the manner prevailing in 1976; otherwise, there is no clear evidence in the available information that demonstrates a hazard to the public when sucrose is used at those levels.

Other sources of sugars. In Hawaii, the pineapple industry recovers both sucrose and citric acid from the rinds. The residue is fed to cattle. The total quantity of sucrose obtained from waste fruits is statistically negligible. On the other hand, a substantial fraction of the total sucrose consumed is that naturally present in a large number of fruits, vegetables, and nuts. *See* SUCROSE; SUGAR CROPS.

Lactose. Cow's milk, on a dry basis, is about 38% lactose or milk sugar. When milk is converted into cheese, the lactose remains in the whey, from which it may easily be isolated and purified. Lactose is a disaccharide which is split by hydrolysis into glucose and galactose. Lactose is about one-tenth as soluble in water as sucrose and it is one-sixth to one-half as sweet, depending on the concentration. Some of the major uses for lactose are as a direct-compression vehicle for the manufacture of pharmaceutical tablets and as a diluent for pharmaceuticals and synthetic sweeteners. *See* CHEESE; LACTOSE.

Starch. Starches can be hydrolyzed either by dilute acid or by enzymes. The product of acid hydrolysis varies with time and conditions but contains glucose, maltose, maltotriose, maltotetrose, and other sugars up to the dextrins. Only glucose, a monosaccharide, is readily isolated. Crystallized as the monohydrate, it is used in foods as a sweetener. *See* CORN; GLUCOSE.

Syrups high in maltose, a disaccharide, can be obtained by the action of amylases on starch. This hydrolysis has been of great importance for thousands of years in splitting starches for alcoholic fermentation. As yet, there is no large-scale production of pure maltose. *See* MALTOSE.

A major type of syrup was developed from starch hydrolyzates by the mid-1980s. An enzyme that isomerizes dextrose (glucose) was commercially developed to manufacture hydrolyzed starch syrups (corn syrups) containing varying amounts of fructose (levulose). The enzyme isomerizes dextrose into fructose theoretically to a 50-50 mixture of dextrose and fructose. In practice, the isomerization is allowed to progress until a syrup containing (on a solids basis) approximately 50% dextrose, 42% fructose, and 8% higher saccharide is achieved. Syrups containing more than 42% fructose (solid basis) can be made by subjecting the 42% fructose syrup to an ion-exclusion process that separates dextrose and fructose. The fructose-rich fraction from the separation is converted into a finished product, and the dextrose-rich fraction is returned to the isomerization process. Commercially produced syrups contain 42% or 55% fructose. The combined processes are also used to manufacture commercial fructose. The high-fructose corn syrups are competitive with sucrose and invert sugar in a variety of food products. The most widely used application for 55% high-fructose corn syrup is in soft drinks, displacing sucrose and invert sugar. *See* FRUCTOSE; STARCH.

Maple sugar. Before America was settled by Europeans, Indians were collecting and concentrating the juice of the hard maple (*Acer saccharum*), making maple syrup. The practice was copied by the new settlers, and production of maple syrup has been an industry ever since in the regions where hard maples are common, principally the northeastern United States.

The maple flavor does not exist in the sap but is developed by heating it. By additional heating at about 250°F (120°C), a flavor four or five times more intense can be developed. Maple syrup so produced is of special value for adding flavor to the less expensive products of the sucrose industry. Maple sugar is sucrose of about 95–98% purity; the characteristic flavor makes up only a small percent. Fairly satisfactory imitation maple flavors are available.

Honey. Honey is a form of relatively pure invert sugar dissolved in water to form a concentrated solution. However, honey also contains flavors derived from the nectar of the flowers from which it was obtained by the bee. Nutritionally, it is nearly equivalent in invert sugar but contains an excess of fructose over glucose. The sucrose found in the flowers is inverted by the enzyme honey invertase. Tupelo honey is remarkable in containing about twice as much fructose as glucose, and hence has little tendency to deposit glucose crystals.

The ready availability of the food energy in honey was known to athletes in ancient Greece. Only in recent times has it been discovered that, paradoxically, the energy of sucrose is still more quickly available. The flavors of various honeys run a wide gamut. That from the Mount Hymettus region, which is flavored by wild thyme, has been known and treasured since the poems of Homer.

Molasses. Virtually all molasses is distributed in the form of a concentrated viscous solution, but it can be reduced to a powder by means of spray-drying. It can then be handled without an investment by the consumer in tanks, pipes, and pumps. Unfortunately, later contact with moisture converts the dried molasses to a gummy mass. The availability of vaporproof bags (for example, those lined with polyethylene) has provided one solution to this problem. There are also various additives which, when mixed with molasses, reduce its tendency to pick up moisture. So far, dried molasses has made little headway against the practice of handling concentrated solutions.

Syrups. Syrups are relatively concentrated, somewhat viscous, solutions of various sugars, frequently in admixture to hinder crystallization. Large amounts of sucrose have been distributed in the form of high-purity syrups. These syrups consist of sucrose, invert sugar, or both. In some instances syrups include mixtures with dextrose and various corn syrups. The so-called liquid sugars are important for several reasons: the requirement for the use of syrups in a number of food processes (conversion of granular sucrose to a syrup was eliminated at the point of use); economy of handling, being moved with pumps, pipes, tank trucks, and tank cars; the high degree of sanitation inherent in the storage and distribution of sugar within the plant in a closed system, including the elimination of dust from the dumping of bags; and the availability of a combination of sugars with differing characteristics in a single product.

The quantity of sucrose distributed in syrup form has been reduced drastically because of the inroads made by high-fructose corn syrups, which have a composition somewhat comparable to that of invert sugar. These products are available only as syrups. They do not have taste characteristics identical to sucrose and invert sugar syrups, but lower prices catalyze their substitution for sucrose.

It is the aim of the sugar refiner and the beet processor to eliminate color and all flavors other than sweetness. The manufacture of table syrups, which are widely used on waffles and pancakes, aims at a broader spectrum of flavor. Corn syrup, which is somewhat lacking in sweetness, ordinarily has about 15% sucrose added. The high viscosity of the corn syrup, resulting from the content of dextrins, tends to hinder crystallization and is an advantage in the manufacture of certain candies.

Some sugar refiners manufacture so-called refiners' syrups for use in the manufacture of foods and table syrups. In addition to these syrups, a group of products generally referred to as cane syrup and edible molasses are manufactured from cane juice. Cane syrup is concentrated whole cane juice, while edible molasses is concentrated cane juice from which some sugar has been removed by crystallization. These syrups and molasses contain both sucrose and invert sugar, and have a relatively dark color and a distinctive flavor. They also contain many of the nonsugars found in sugarcane juice. They are used principally by food processors, but substantial amounts are packaged for home use. Sorghum syrup is made by extracting juice from the stalk of sweet sorghum and evaporating the juice to syrup consistency. It is a dark, pungently flavored product that is not widely distributed. *See* FOOD ENGINEERING; FOOD MANUFACTURING; SUGAR CROPS.

H. B. Hass; Charles B. Broeg

Sugar crops

Crops produced as major sources of sugar, syrup, and other sugar substances. Sugarbeet and sugarcane are crops which serve as a source of sucrose, the sugar of commerce. Sugar is a broad term applied to a large number of carbohydrates that have a more or less sweet taste. The primary sugar, glucose, is a product of photosynthesis and occurs in all green plants. Through chemical union, diverse sugars and starches are elaborated and become the major reserve food in storage organs, fruits, and the sap of plants. In most plants the sugars occur as a mixture that cannot be readily separated into the components. In the sap of some plants the sugar mixtures are condensed into syrup. The juices of sugarcane and sugarbeet are usually rich in pure sucrose. *See* CARBOHYDRATE; PHOTOSYNTHESIS; SUCROSE; SUGARBEET; SUGARCANE.

Other sugar crops are sweet sorghum, sugar maple, sugar palm, honey, chicory, and corn sugar. *See* CHICORY; MAPLE; SORGHUM.

Dewey Stewart

Sugarbeet

The plant *Beta vulgaris*, developed in modern times to fill the need for a sugar crop that could be grown in temperate climates. The sugarbeet was the source of only 5% of the world's commercial sugar in 1840, but by 1890 it supplied almost 50%, and since about 1920 this has dropped to about 40% with 60% derived from sugarcane. European countries produce most of their sugar from sugarbeet, with some countries having surplus sugar for export. *See* SUGARCANE.

The sugarbeet, fodder beet, garden beet, and leaf beet (Swiss chard) are cultivars of *B. vulgaris* and are genetically related. In addition to *B. vulgaris*, which includes all cultivated forms of beet, there are 12 species of wild beet in the Mediterranean and Middle Eastern regions. These wild beets are of interest in sugarbeet improvement as sources of disease resistance. Most varieties of sugarbeet in the United States are hybrid.

Dewey Stewart

Sugarcane

Saccharum officinarum, a member of the grass family. This crop originated in New Guinea about 15,000–8000 B.C., and was later moved by primitive peoples westward into Southeast Asia and India and eastward into Polynesia. Most current commercial varieties are interspecific hybrids involving primarily two or more of the following species: *S. officinarum*, *S. robustum*, and *S. spontaneum*.

Sugar is generally removed from sugarcane in large factories by either milling or diffusion. Milling crushes the stalks as they pass between a series of large metal rollers separating the fiber (bagasse) from the juice laden with sugar. The diffusion process separates sugar from finely cut stalks by dissolving it in hot water or hot juice. Processing of the juice is completed by clarification to remove nonsugar components, by evaporation to remove water, and by removal of molasses in high-speed centrifuges to produce centrifugal sugar. Bagasse and molasses are the principal byproducts from processing sugarcane.

Robert E. Cohen

Suicide

A death that is self-inflicted and carried out with intent to die; also, one who completes such an act. Suicide is a serious epidemiological public health problem that devastates individuals, families, and communities throughout the world. There are many risk factors for suicide and suicidal behaviors, as well as a variety of methods of suicide. Our understanding of both the methods and risk factors involved can help inform models for suicide prevention. *See* AFFECTIVE DISORDERS; PUBLIC HEALTH.

The World Health Organization has estimated that worldwide there are between 815,000 and 1,000,000 suicides annually. The rate of suicide is almost universally higher among men than women by an aggregate ratio of 3.5 to 1. For both genders, suicides increase with age, peaking among those 75 years and older. Suicide rates have been the highest in the former Soviet republics; these rates have been attributed to the political (thus sociocultural and socioeconomic) instability and significant alcohol use. Suicide ranks as one of the top three causes of death among 15- to 24-year-olds worldwide and has dramatically increased in the last few decades, particularly in Australia, New Zealand, Portugal, Spain, and Greece. *See* ALCOHOLISM.

For every completed suicide, there are an estimated 25 nonfatal attempts. Females attempt suicide three times more often than do males.

The prediction of suicide is impossible. Risk factors for suicide range from those that are historic and, thus, not open to change by interventions (perpetuating risk factors) to those current and changeable (predisposing risk factors). Offspring of parents with a history of suicide, violence, substance abuse, or significant mental disorder (perpetuating risk factors) have 5–11 times the risk of completing suicide as those without these parental histories. One in three of those who have made a prior suicide attempt (another perpetuating risk factor) will make further attempts, and 10–15% ultimately will die by suicide. Examples of predisposing risk factors include mental disorders, suicide ideation, hopelessness, social isolation, and decreased protective factors.

The modal method of suicide in the United States is firearms. Hanging is the most commonly used method worldwide. Ingestion (drug overdose) is the most frequent method used by nonfatal attempters and by females. Prevention approaches range from "case finding" models, that is, models of early detection and referral to treatment of those at risk, such as screening programs, to means restriction programs, wherein available access to lethal means is reduced. Alan L. Berman

Sulfate minerals Minerals containing the sulfate anion (SO_4^{2-}). Most metals and their ores react with sulfuric acid to give metal sulfates, and oxidation of metal sulfides produces sulfates. Gypsum is the most common sulfate mineral.

Alunite [$KAl_3(SO_4)_2(OH)_6$] occurs in white to gray rhombohedral crystals or in fine-grained, compact masses. The mineral is generally found associated with quartzand kaolinite. Alunite is used as a source of potash or for making alum. *See* ALUM; FERTILIZER; POTASSIUM.

Anglesite ($PbSO_4$) occurs in white or gray, orthorhombic, tabular or prismatic crystals or compact masses. It is a common secondary mineral, usually formed by the oxidation of galena. *See* GALENA.

Anhydrite ($CaSO_4$) occurs commonly in white and grayish granular masses, rarely in large, orthorhombic crystals. Anhydrite is an important rock-forming mineral and occurs in association with gypsum, limestone, dolomite, and salt beds. Under natural conditions anhydrite hydrates slowly, but readily, to gypsum. *See* SALINE EVAPORITES.

Barite ($BaSO_4$) is isostructural with celestite, $SrSO_4$, and anglesite, $PbSO_4$. Since barite is dense and relatively soft, its principal use is as a weighting agent in rotary well-drilling fluids. It is the major ore of barium salts, used in glass manufacture, as filler in paint, and, owing to the presence of a heavy metal and inertness, as an absorber of radiation in x-ray examination of the gastrointestinal tract.

Celestite ($SrSO_4$) occurs commonly in colorless to sky-blue, orthorhombic, tabular crystals. Celestite occurs in association with gypsum, anhydrite, salt beds, limestone, and dolomite. Celestite is the major source of strontium.

Epsomite ($MgSO_4 \cdot 7H_2O$), or epsom salt, occurs in clear, needlelike, orthorhombic crystals. Epsomite is found as a capillary coating in limestone caves and in coal or metal mine galleries.

Gypsum is characterized by the chemical formula $CaSO_4 \cdot 2H_2O$; it shows little variation from this composition. Gypsum is used for a variety of purposes, but chiefly in the manufacture of plaster of paris, in the production of wallboard, in agriculture to loosen clay-rich soils, and in the manufacture of fertilizer. *See* PLASTER OF PARIS.

Iron sulfates. These are formed by the oxidation of iron sulfides. Jarosite [$KFe_3(SO_4)_2(OH)_6$] is an ocher-yellow or brown alunite mineral having rhombohedral crystal structure. It is also known as utahite. Plumbojarosite [$PbFe_6(SO_4)_4(OH)_{12}$] is a mineral composed of basic lead iron sulfate; it is isostructural with jarosite. Edward C. T. Chao; Paul B. Moore; Marc L. Helman; Charlotte Schreiber

Sulfide and arsenide minerals Minerals based on the anions sulfide (S^{2-}) and arsenide (As^{3-}), or both. The arsenides are structurally similar and have similar properties to the sulfides. Sulfide and arsenide minerals are deposited by hydrothermal solutions. These fluids that are able to dissolve metals such as lead, zinc, and copper. Percolating ground water may also dissolve disseminated sulfides in granitic host rocks and reprecipitate new sulfides at the water table boundary by a process known as supergene enrichment. *See* HYDROTHERMAL ORE DEPOSITS; HYDROTHERMAL VENT.

Sulfide minerals are among the most important ores for several metals. These compounds are two- or three-dimensional polymers containing interconnected metal cations and sulfide S^{2-} or persulfido S_2^{2-} anions. Orange cinnabar (HgS) and yellow greenockite (CdS) are the major ores for mercury and cadmium, respectively. Molybdenite (MoS_2) is the major ore of molybdenum. Pyrites (FeS_2), also known as iron pyrites or fool's gold, are the most common sulfide minerals.

Arsenic is found native as the mineral skutterudite, but generally occurs among surface rocks combined with sulfur or metals such as manganese, iron, cobalt, nickel, or tin. The principal arsenide mineral is FeAsS (arsenopyrite, mispickel); other metal arsenide ores are $FeAs_2$ (löllingite), NiAs (nickeline), CoAsS (cobalt glance), NiAsS (gersdorffite), and $CoAs_2$ (smaltite). Naturally occurring arsenates and thioarsenates are common, and most sulfide ores contain arsenic. As_4S_4 (realgar) and As_4S_6 (orpiment) are the most important sulfur-containing minerals. The oxide, arsenolite, As_4O_6 is found as the product of the weathering of other arsenide minerals, and is also recovered from flue dusts collected during the extraction of nickel, copper, and tin from their ores. *See* MINERAL. T. B. Rauchfuss; John L. T. Waugh

Sulfide phase equilibria The chemistry of the sulfides is rather simple inasmuch as most sulfide minerals involve only two major elements, although some involve three, and a few four. Understanding of the phase relations among these minerals is important both to the geologist, whose task it is to locate and exploit ore deposits, and to the metallurgist, whose task it is to extract the metals from the ores for industrial use.

Sulfide ore deposits are the most important sources of numerous metals such as lead, zinc, copper, nickel, cobalt, and molybdenum. In addition, sulfide ores provide substantial amounts of noble metals such as platinum, gold, and silver, and of other industrially important elements such as cadmium, rhenium, and selenium. Although iron sulfides usually are the most common minerals in such deposits, most commercial iron is mined from iron oxide ores and most sulfur from elemental sulfur deposits.

Some of the more common sulfide minerals are listed in the table.

Mineral name	Chemical formula
Pyrite	FeS_2
Pyrrhotites (three or more varieties)	$Fe_{1-x}S$
Covellite	CuS
Digenite	Cu_9S_5
Chalcocite	Cu_2S
Bornite	Cu_5FeS_4
Chalcopyrite	$CuFeS_2$
Cubanite	$CuFe_2S_3$
Galena	PbS
Sphalerite and wurtzite	ZnS
Metacinnabar and cinnabar	HgS
Argentite and acanthite	Ag_2S
Molybdenite	MoS_2
Millerite	NiS
Pentlandite	$(Fe,Ni)_9S_8$

The first eight sulfide minerals listed may be plotted in the ternary system copper-iron-sulfur. Similarly, the first two and the last two minerals may be plotted in the ternary system iron-nickel-sulfur. Thus, by studying in detail the phase equilibria in two ternary systems a great deal of information can be obtained about many of the common sulfides occurring in ore deposits. However, before the ternary systems can be explored in a systematic way, the binary systems bounding the ternaries have to be studied in detail. Similarly, it is necessary that the four bounding ternary systems be fully understood before a quaternary system can be systematically investigated. Thus, it is seen that before a systematic study of, for example, the immensely important quaternary system copper-iron-nickel-sulfur can proceed, much preliminary information is required. The prerequisite data include complete knowledge of the four ternary systems Cu-Fe-S, Cu-Fe-Ni, Fe-Ni-S, and Cu-Ni-S. In turn, the phase relations in these ternary systems cannot be systematically studied before the six binary systems Fe-Cu, Fe-Ni, Cu-Ni, Fe-S, Cu-S, and Ni-S have been thoroughly explored. *See* PHASE EQUILIBRIUM.

The enormous differences in the vapor pressures over the different phases occurring in sulfide systems add complications to the diagrammatic representation. For instance, in the Fe-S system the vapor pressure over pure iron is about 10^{-25} atm (10^{-20} pascal) at 450°C (840°F), whereas that over pure sulfur is a little more than 1 atm (10^5 pascals) at the same temperature. A complete diagrammatic representation of the relations in such a system, therefore, requires coordinates for composition and temperature, as well as for pressure. In a two-component system, such as the Fe-S system, such a representation is feasible because only three coordinates are necessary. However, in ternary (where such diagrams involve four-dimensional space) and in multicomponent systems, this type of diagrammatic representation is not possible. For this reason it is customary to use composition and temperature coordinates only for the diagrammatic representation of sulfide systems. The relations as shown in such diagrams in reality represent a projection from composition-temperature-pressure space onto a two-dimensional composition-temperature plane or onto a three-dimensional prism, depending upon whether the system contains two or three components.

Pyrite or pyrrhotite, or both, occur almost ubiquitously, not only in ore deposits but in nearly all kinds of rocks. Of the binary systems mentioned above, therefore, the iron-sulfur system is of the most importance to the economic geologist. *See* PYRITE.

Gunnar Kullerud

Sulfonamide One of the group of organosulfur compounds, RSO_2NH_2. Many sulfonamides, which are the amides of sulfonic acids, have the marked ability to halt the growth of bacteria. The therapeutic drugs of this group are known as sulfa drugs. *See* ORGANOSULFUR COMPOUND.

The antibacterial spectrum of sulfonamides comprises a wide variety of gram-positive and gram-negative bacteria, including staphylococci, streptococci, meningococci, and gonococci, as well as the gangrene, tetanus, coli, dysentery, and cholera bacilli. They have only slight activity against *Mycobacterium tuberculosis*, while certain closely related sulfones are quite active against *M. leprae*. The use of sulfonmethoxine has proved to be effective in the treatment of chloroquine-resistant malaria. The relative potency of sulfonamides against the different microorganisms varies, and their action is bacteriostatic rather than bactericidal.

The antibacterial effect, both in patients and in test tube cultures, is antagonized by *p*-aminobenzoic acid (PABA) and PABA-containing natural or synthetic products, such as folic acid and procaine. Accordingly, the mode of action of sulfonamides is considered to be an antimetabolite activity, dependent upon the inhibition of enzyme systems involving the essential PABA. *See* BACILLARY DYSENTERY; CHOLERA; LEPROSY; MALARIA; MENINGOCOCCUS; STAPHYLOCOCCUS; STREPTOCOCCUS; TETANUS; TUBERCULOSIS.

Bacterial resistance has developed to all known sulfonamides, and many sulfonamide-resistant strains are encountered among the gram-positive and gram-negative bacteria. The emergence of resistance to sulfonamides, however, seems less rapid and less widespread than resistance to most antibiotics.

Sulfonamides are used today mostly as auxiliary drugs or in combination with antibiotics. In certain infectious diseases, however (for instance, in meningococcal infections and most infections of the urinary tract), sulfonamides deserve preference over antibiotics. *See* ANTIMICROBIAL AGENTS; DRUG RESISTANCE.

Rudolph L. Mayer; Emanuel Grunberg

Sulfonation and sulfation Sulfonation is a chemical reaction in which a sulfonic acid group, $—SO_3H$, is introduced into the structure of a molecule or ion in place of a hydrogen atom. Sulfation involves the attachment of the $—OSO_2OH$ group to carbon, yielding an acid sulfate, $ROSO_2OH$, or of the $—SO_4—$ group between two carbons, forming the sulfate, $ROSO_2OR$.

Sulfonation of aromatic compounds is the most important type of sulfonation. This is accomplished by treating the aromatic compound with sulfuric acid, as in the reaction below. The product of sulfonation is a sulfonic acid.

$$\text{Napthalene} + H_2SO_4 \xrightarrow{320°F \text{ or } 160°C} \beta\text{-Napthalenesulfonic acid } (SO_3H) + H_2O$$

Sulfonation may also be defined as any chemical process by which the sulfonic acid group, $—SO_2OH$, or the corresponding salt or sulfonyl halide group, for example, $—SO_2Cl$, is introduced into an organic compound. These groups may be bonded to either a carbon or a nitrogen atom. The latter compounds are designated *N*-sulfonates or sulfamates.

Most sulfonates are employed as such in acid or salt form for applications where the strongly polar hydrophilic $—SO_2OH$ group confers needed properties on a comparatively hydrophobic nonpolar organic molecule. Some sulfonates, such as methanesulfonic and toluenesulfonic acids, are used as catalysts. The major quantity of sulfonates and sulfates is both marketed and used in salt form. This category includes detergents; emulsifying, demulsifying, wetting, and solubilizing agents; lubricant additives; and rust inhibitors. *See* ORGANOSULFUR COMPOUND; SUBSTITUTION REACTION.

Joseph F. Bunnett; Robert S. Kapner

Sulfonic acid A derivative of sulfuric acid ($HOSO_2OH$) in which an OH has been replaced by a carbon group, as shown in the structure below. Sulfonic acids are strongly acidic,

$$R—S(=O)_2—OH$$

water-soluble, nonvolatile, and hygroscopic; they do not act as oxidizing agents and are typically highly stable compounds. Sulfonic acids rarely occur naturally. An exception, taurine, $NH_2CH_2CH_2SO_3H$, occurs in bile.

The aliphatic sulfonic acids are generally made by oxidation of thiols. Several have unique properties. For example, trifluoromethanesulfonic acid, CF_3SO_3H, is such a strong acid that it will protonate sulfuric acid. A compound derived from natural camphor, 10-camphorsulfonic acid, is used extensively in the optical resolution of amines.

Aromatic sulfonic acids are much more important than those of the aliphatic series. Aromatic sulfonic acids are produced by sulfonation of aromatic compounds with sulfuric acid or fuming sulfuric acid. Sulfonation of aromatic hydrocarbons is a reversible process; treatment of an aromatic sulfonic acid with superheated steam removes the $—SO_3H$ group. This process can be used in purifying aromatic hydrocarbons. Aromatic sulfonic acids and their derivatives, especially metal salts, are important industrial chemicals. *See* SULFURIC ACID.

The most extensive use of the sulfonation reaction is in the production of detergents. The most widely used synthetic detergents are sodium salts of straight-chain alkylbenzenesulfonic acids. *See* DETERGENT.

Sulfonated polymers, particularly sulfonated polystyrenes, act as ion-exchange resins which have important applications in water softening, ion-exchange chromatography, and metal separation technology. Both sulfonated polymers and simple aromatic sulfonic acids, particularly *p*-toluenesulfonic acid, are frequently used as acid catalysts in organic reactions such as esterification and hydrolysis. *See* HOMOGENEOUS CATALYSIS; ION EXCHANGE.

The sulfonic group, in either acid or salt form, is capable of making many substances water soluble, increasing their usefulness. This application is particularly significant in the dying industry, in which a majority of the dyes are complex sodium sulfonates. Many acid-base indicators are soluble due to the presence of a sodium sulfonate moiety. Some pigments used in the paint and ink industry are insoluble metal salts or complexes of sulfonic acid derivatives. Most of the brighteners used in detergents compounded for laundering are sulfonic acid derivatives of heterocyclic compounds. *See* ACID-BASE INDICATOR; DETERGENT; DYE; INK; ORGANOSULFUR COMPOUND; PAINT. Carl R. Johnson

Sulfur A chemical element, S, atomic number 16, and atomic weight 32.065. The atomic weight reflects the fact that sulfur is composed of the isotopes ^{32}S (95.1%), ^{33}S (0.74%), ^{34}S (4.2%), and ^{36}S (0.016%). The ratios of the various isotopes vary slightly but measurably according to the history of the sample. By virtue of its position in the periodic table, sulfur is classified as a main-group element. *See* PERIODIC TABLE.

The chemistry of sulfur is more complex than that of any other elemental substance, because sulfur itself exists in the largest variety of structural forms. At room temperature, all the stable forms of sulfur are molecular; that is, the individual atoms aggregate into discrete molecules, which in turn pack together to form the solid material. In contrast, other elements near sulfur in the periodic table normally exist as polymers (silicon, phosphorus, arsenic, selenium, tellurium) or as diatomic molecules (oxygen, nitrogen, chlorine). Selenium and phosphorus can exist as molecular solids, but the stable forms of these elements are polymeric.

At room temperature the most stable form of sulfur is the cyclic molecule S_8. The molecule adopts a crownlike structure, consisting of two interconnected layers of four sulfur atoms each. The S—S bond distances are 0.206 nanometer and the S—S—S bond angles are 108°. Three allotropes are known for cyclo-S_8. The most common form is orthorhombic α-sulfur, which has a density of 2.069 g/cm^3 (1.200 oz/in.^{3}) and a hardness of 2.5 on the Mohs scale. It is an excellent electrical insulator, with a room temperature conductivity of 10^{18} ohm^{-1} cm^{-1}. Sublimed sulfur and "flowers" of sulfur are generally composed of α-S_8. Sulfur is quite soluble in carbon disulfide (CS_2; 35.5/100 g or 1.23 oz/3.52 oz at 25°C or 77°F), poorly soluble in alcohols, and practically insoluble in water. At 95.3°C (203°F), sulfur changes into the monoclinic β allotrope. This form of sulfur also consists of cyclic S_8 molecules, but it has a slightly lower density at 1.94–2.01 g/cm^3 (1.12–1.16 oz/in.^{3}). A third allotrope containing S_8 is triclinic γ-sulfur. The β and γ allotropes of sulfur slowly revert to the α form at room temperature. Crystals of sulfur are yellow and have an absorption maximum in the ultraviolet at 285 nm, which shifts to higher energy as the temperature decreases. At low temperatures, S_8 is colorless. Even at room temperature, however, finely powdered sulfur can appear to be nearly white.

The best-studied system is α-S_8, which converts to the β form at 90°C (194°F), which then melts at 120°C (248°F) to give a golden yellow liquid. If this melt is quickly recooled, it refreezes at 120°C (248°F), thus indicating that it consists primarily of S_8 molecules. If the melt is maintained longer at 120°C (248°F), then the freezing point is lowered about 5°C (9°F), indicating the formation of about 5% of other rings and some polymer. At 159.4°C (318.9°F), the melt suddenly assumes a red-brown color. Over the range 159.4–195°C (318.9–383°F), the viscosity of the melt increases 10,000-fold before gradually decreasing again. This behavior is very unusual, since the viscosity of most liquids decreases with increasing temperature. The strong temperature dependence of the viscosity is due to the polymerization and eventual depolymerization of sulfur. Polymeric sulfur retains its elastomeric character even after being cooled to room temperature. There are several polymeric forms of sulfur, but all of them revert to α-S_8 after a few hours.

Sublimination of S_8 occurs when it is maintained in a vacuum at a temperature below its melting point. It vaporizes at 444.61°C (832.30°F). Below 600°C (1110°F), the predominant species in the gas are S_8 followed by S_7 and S_6. Above 720°C (1328°F), violet S_2 is the major species.

Principal inorganic compounds. Hydrogen sulfide (H_2S) is the most important compound that contains only sulfur and hydrogen. It is a gas at room temperature with a boiling point of −61.8°C (−79.2°F) and a freezing point of −82.9°C (−117°F). The low boiling point of hydrogen sulfide is attributed to the weakness of intermolecular S···H hydrogen bonding; the O···H hydrogen bond is much stronger, as evidenced by the high boiling point of water. Gaseous hydrogen sulfide is 1.19 times more dense than air, and air-H_2S mixtures are explosive. Hydrogen sulfide has a strong odor similar to that of rotten eggs; its odor is detectable at concentrations below 1 microgram/m^3. At high concentrations, H_2S has a paralyzing effect on the olfactory system, which is very hazardous because H_2S is even more toxic than carbon monoxide (CO).

The most common compound that contains only carbon and sulfur is carbon disulfide (CS_2). Carbon disulfide molecules are linear, consisting of two sulfur atoms connected to a central carbon atom. Carbon disulfide is a toxic, highly flammable, and volatile liquid that melts at −111°C (−168°F) and boils at 46°C (115°F). Commercial carbon disulfide has a strong unpleasant odor due to impurities. It is manufactured from methane and elemental sulfur and is used for the production of carbon tetrachloride, rayon, and cellophane. Structurally related to carbon disulfide is carbonyl sulfide (SCO), which forms from carbon monoxide and elemental sulfur. The chlorination of CS_2 gives Cl_3CSCl, which can be reduced by H_2S to thiophosphene, $CSCl_2$. Thiophosgene ($CSCl_2$) [boiling point 73°C or 163°F] is a planar molecule with the carbon at the center of a triangle defined by the sulfur and two chlorine atoms. Thiocyanate, the linear anion NCS^-, is prepared by the reaction of cyanide (—CN) with elemental sulfur.

Several sulfur oxides exist, but the dioxide and trioxide are of preeminent importance. Sulfur dioxide (SO_2) is a colorless gas that boils at −10.02°C (113.97°F) and freezes at −75.46°C (−103.8°F). The density of liquid sulfur dioxide at −10°C (14°F) is 1.46 g/cm^3 (0.84 oz/in.^{3}). Liquid sulfur dioxide is an excellent solvent. The sulfur dioxide molecule is bent, with an O—S—O angle of 119°.

Sulfur trioxide (SO_3) is a planar molecule that is a liquid at room temperature that exists in equilibrium with a cyclic trimeric structure known as β-SO_3. When β-SO_3, actually S_3O_9, is treated with traces of water, it converts to either of two polymeric forms referred to as γ- and α-sulfur trioxide. These are fibrous materials, proposed to have the formula $(SO_3)_xH_2$, where x is in the thousands. Sulfur trioxide is prepared by the oxidation of sulfur dioxide, although at very high temperatures this reaction reverses. Exposure of sulfur trioxide to water yields sulfuric acid (H_2SO_4); exposure of SO_3 to sulfuric acid yields disulfuric acid ($H_2S_2O_7$). *See* SULFURIC ACID.

Chlorine and sulfur react to give a family of compounds with the general formula S_xCl_2, several members of which have been obtained in pure form. The structures of these compounds are based on an atom or chain of sulfur atoms terminated with Cl. Sulfur monochloride (S_2Cl_2), also known as sulfur monochloride, is the most widely available of the series. It is a yellow oil that boils at 138°C (280°F), and reacts with chlorine in the presence of iron(III) chloride ($FeCl_3$) catalyst to give sulfur dichloride (SC_2), which is a red volatile liquid with a boiling point of 59°C (138°F).

Treatment of sulfur dichloride with sodium fluoride (NaF) gives SF_4.

Thionyl chloride ($OSCl_2$) is a colorless reactive compound with a boiling point of 76°C (169°F); it is used to convert hydroxy compounds to chlorides. Important applications include the preparation of anhydrous metal halides and alkyl halides. Sulfuryl chloride (O_2SCl_2; boiling point 69°C or 156°F) is used as a source of chlorine.

Organosulfur compounds. This family of compounds contains carbon, hydrogen, and sulfur, and it is a particularly vast area of sulfur chemistry. Thiols, also known as mercaptans, feature the linkage C—S—H. Mercaptans are foul-smelling compounds. They are the sulfur analogs of alcohols, but they are more volatile. They can be prepared by the action of hydrogen sulfide (H_2S) on olefins. Deprotonation of thiols gives thiolate anions, which form stable compounds with many heavy metals. Thiols and especially thiolates can be oxidized to form disulfides (persulfides), which have the connectivity of C—S—S—C. The organic persulfides are also related to organic polysulfides, which have chains of sulfur atoms terminated with carbon. The introduction of such mono-, di-, and polysulfide linkages is the basis of the vulcanization process, which imparts desirable mechanical properties to natural or synthetic polyolefin rubbers. This is accomplished by heating the polymer with sulfur in the presence of a zinc catalyst. *See* RUBBER.

Thioethers, also known as organic sulfides, feature the connectivity C—S—C and are often prepared from the reaction of thiolates and alkyl halides. Like mercaptans, thioethers often have strong unpleasant odors, but they are also responsible for the pleasant odors of many foods and perfumes. They are intentionally introduced at trace levels in order to impart an odor to gaseous hydrocarbon fuels. The reaction of alkyl dihalides and sodium polysulfides affords organic polysulfide polymers known as thiokols.

There are many organic sulfur oxides; prominent are sulfonic acids (RSO_3H), which are the organic derivatives of sulfuric acid. These compounds are prepared by the oxidation of thiols as well as by treatment of benzene derivatives with sulfuric acid, for example, benzene sulfonic acid. Most detergents are salts of sulfonic acids. *See* DETERGENT.

Biochemistry. Sulfur is required for life. Typical organisms contain 2% sulfur dry weight. Three amino acids contain sulfur, as do many prosthetic groups in enzymes. Some noteworthy sulfur compounds include the disulfide lipoic acid, the thioethers biotin and thiamine (vitamin B_1), and the thiol coenzyme A. Sulfide ions, S^{2-}, are found incorporated in metalloproteins and metalloenzymes such as the ferredoxins, nitrogenases, and hydrogenases. *See* AMINO ACIDS; ENZYME; PROTEIN.

Many bacterial species obtain energy by the oxidations of sulfides. Bacteria of the genus *Thiobacillus* couple the conversion of carbon dioxide (CO_2) to carbohydrates to the aerobic oxidation of mineral sulfides to sulfuric acid. This activity can be turned to good use for leaching low-grade mineral ores. Often, however, the sulfuric acid runoff (such as in mines or sewers) has negative environmental consequences. The purple and green bacteria as well as the blue-green algae are remarkable because they are photosynthetic but anaerobic; they oxidize sulfide, not water (as do most photosynthetic organisms). Depending on the species, the sulfur produced in this energy-producing pathway can accumulate inside or outside the cell wall. *See* BACTERIAL PHYSIOLOGY AND METABOLISM; PHOTOSYNTHESIS.

Minerals. Sulfide minerals are among the most important ores for several metals. These compounds are two- or three-dimensional polymers containing interconnected metal cations and sulfide S^{2-} or persulfido S_2^{2-} anions. In general, metal sulfides are darkly colored, often black, and they are not soluble in water. They can sometimes be decomposed by using strong acids, with liberation of hydrogen sulfide. Certain sulfides will also dissolve in the presence of excess sulfide or polysulfide ions.

Pyrites (FeS_2), also known as iron pyrites or fool's gold, are the most common sulfide minerals and can be obtained as very large crystals that have a golden luster. Sphalerite (zinc blende; ZnS) and galena (PbS) are major sources of zinc and lead. Orange cinnabar (HgS) and yellow greenockite (CdS) are the major ores for mercury and cadmium, respectively. Molybdenite (MoS_2) is the major ore of molybdenum.

The sulfur content of fossil fuels results from the sulfur in the ancient organisms as well as from subsequent incorporation of mineral sulfur into the hydrocarbon matrix. Gaseous fossil fuels are often contaminated with hydrogen sulfide, which is an increasingly important source of sulfur. Organic derivatives containing the C—S—C linkage are primarily responsible for the sulfur content of petroleum and coal. The so-called organic sulfur in petroleum can be removed by hydrodesulfurization catalysis, involving reaction with hydrogen over a molybdenum catalyst, to give hydrocarbons and hydrogen sulfide. T. B. Rauchfuss

Sulfuric acid A strong mineral acid with the chemical formula H_2SO_4. It is a colorless, oily liquid, sometimes called oil of vitriol or vitriolic acid. The pure acid has a density of 1.834 at 25°C (77°F) and freezes at 10.5°C (50.90°F). It is an important industrial commodity, used extensively in petroleum refining and in the manufacture of fertilizers, paints, pigments, dyes, and explosives.

Sulfuric acid is produced on a large scale by two commercial processes, the contact process and the lead-chamber process. In the contact process, sulfur dioxide, SO_2, is converted to sulfur trioxide, SO_3, by reaction with oxygen in the presence of a catalyst. Sulfuric acid is produced by the reaction of the sulfur trioxide with water. The lead-chamber process depends upon the oxidation of sulfur dioxide by nitric acid in the presence of water, the reaction being carried out in large lead rooms.

Sulfuric acid reacts vigorously with water to form several hydrates. The concentrated acid, therefore, acts as an efficient drying agent, taking up moisture from the air and even abstracting the elements of water from such compounds as sugar and starch. The concentrated acid also acts as a strong oxidizing agent. It reacts with most metals upon heating to produce sulfur dioxide. *See* SULFUR. Francis J. Johnston

Sum rules Formulas in quantum mechanics for transitions between energy levels, in which the sum of the transition strengths is expressed in a simple form. Sum rules are used to describe the properties of many physical systems, including solids, atoms, atomic nuclei, and nuclear constituents such as protons and neutrons. The sum rules are derived from quite general principles, and are useful in situations where the behavior of individual energy levels is too complex to describe by a precise quantum-mechanical theory. *See* ENERGY LEVEL (QUANTUM MECHANICS).

In general, sum rules are derived by using Heisenberg's quantum-mechanical algebra to construct operator equalities, which are then applied to particles or the energy levels of a system. *See* QUANTUM MECHANICS. George F. Bertsch

Sun The star around which the Earth revolves, and the planet's source of light and heat, hence life. The Sun is a globe of gas, 1.4×10^6 km (8.65×10^5 mi) in diameter with a mass 333,000 times the Earth, held together by its own gravity. The surface temperature of the Sun is about 6000 K (10,000°F); since solids and liquids do not exist at these temperatures, the Sun is entirely gaseous. Almost all the gas is in atomic form, although a few molecules exist in the coolest surface regions, such as sunspots.

The Sun is a typical member of the spectral class dG2, stars of surface temperature 6000 K. The d stands for a dwarf, a normal star of that class. *See* SPECTRAL TYPE.

Solar structure. The interior of the Sun can be studied only by inference from the observed properties of the entire star. The great mass of the Sun presses down on the center, requiring a gas with a central density of near 90 g/cm^3 and 2×10^7 K (3.6×10^6°F) temperature to support it. At these huge temperatures and densities, nuclear reactions take place. The radiation produced flows outward till it is radiated into space by the surface (photosphere) at 6000 K (10,000°F).

Energy production. The energy of the Sun is produced by the conversion of hydrogen into helium. For each hydrogen atom converted, one neutrino is produced. These neutrinos are detected, but less than the expected number. See NEUTRINO; SOLAR NEUTRINOS.

The material at the center of the Sun is so dense that a few millimeters are opaque, so the photons created by nuclear reactions are continually absorbed and reemitted and thus make their way to the surface by a random walk. The atoms in the center of the Sun are entirely stripped of their electrons by the high temperatures, and most of the absorption is by continuum processes, such as the scattering of light by electrons. Because there are so many absorption and emission processes along the way, it can take as long as a million years to complete the random walk to the surface.

Convection. In the outer regions of the solar interior, the temperature is low enough for ions and even neutral atoms to form and, as a result, atomic absorption becomes very important. The high opacity makes it very difficult for the radiation to continue outward, so steep temperature gradients are established that result in convective currents. Most of the outer envelope of the Sun is in such convective equilibrium. These large-scale mass motions are responsible for the complex phenomena observed at the surface. See CONVECTION (HEAT).

Radiation. Electromagnetic energy is produced by the Sun in essentially all wavelengths. However, more than 95% of the energy is concentrated in the relatively narrow band between 290 and 2500 nm and is accessible to routine observation from ground stations on Earth. The maximum radiation is in the green region, and the eyes of human beings have naturally evolved to be sensitive to this range of the spectrum. The total radiation is called the solar constant. It is not exactly constant but varies slightly ($\pm 0.1\%$) with the solar cycle. The ultraviolet flux, however, varies by substantial factors depending on the exact wavelength, and this affects the Earth's upper atmosphere. See ELECTROMAGNETIC RADIATION; SOLAR CONSTANT.

Atmosphere. Although the Sun is gaseous, it can be seen only to the point at which the density is so high that the material is opaque. This layer, the visible surface of the Sun, is termed the photosphere. Light from father down reaches the Earth by repeated absorption and emission by the atoms, but the deepest layers cannot be seen directly. The surface is actually not sharp, but the Sun is so far away that the smallest distance that can be resolved with the best telescope is about 300 km (200 mi). Since the density e-folding height (scale height) is less than 200 km (120 mi), the edge appears sharp. See PHOTOSPHERE.

Above the photosphere the atmosphere is transparent, and its density falls off much more slowly because magnetic fields support the ionized particles. The atmosphere can be seen by using a narrow-band filter or a spectrograph to pick out the isolated wavelengths absorbed by the atmospheric gases. In the upper photosphere it is cooler, and the lines are dark. If the light is imaged in the strongest lines, such as those of hydrogen, a region higher still is seen, called the chromosphere. The light from this region is dominated by the red hydrogen alpha (level $2 \rightarrow 3$ transition) line, which gives it a rosy color seen at a solar eclipse. The chromosphere is a rapidly fluctuating region of jets and waves coming up from the surface. When all the convected energy coming up from below reaches the surface, it is concentrated in the thin material and produces considerable activity. Where the magnetic field is stronger, these waves are absorbed, and raise the temperature to 7000–8000 K (12,000–14,000°F). The scale height of the chromosphere is 1000 km (600 mi) or more, so there no longer is a sharp edge. See CHROMOSPHERE; ECLIPSE.

When the Moon obscures the Sun at a total solar eclipse, the vast extended atmosphere of the Sun called the corona can be seen. The corona is a million times fainter than the photosphere, so it is visible only when seen against the dark sky of an eclipse or with very special instruments. Its density is low, but its temperature is high (more than 10^6 K or 1.8×10^6°F). The hot gas evaporating out from the corona flows steadily to the Earth and farther in the solar wind. See SOLAR CORONA; SOLAR WIND.

Coronal holes. Early coronal observations showed that the corona was occasionally absent over certain regions. In particular, at sunspot minimum it was quite weak over the poles. X-ray pictures revealed great bands of the solar surface essentially devoid of corona for many months. These proved to be regions where the local magnetic fields were connected to quite distant places, so the fields actually reached out to heights from which the solar wind could sweep the gas outward. Analysis of solar wind data showed that equatorial coronal holes were associated with high-velocity streams in the solar wind, and recurrent geomagnetic storms were associated with the return of these holes. Thus the relative intensity of the corona over sunspot regions is partly due to their strong, closed magnetic fields which trap the coronal gas.

Solar activity. There are a number of transient phenomena known collectively as solar activity. These are all connected with sunspots.

Sunspots. Sunspots were discovered around 1610. Heinrich Schwabe announced in 1843 that their number rose and fell with a 10-year period. Subsequent study of the old records revealed an 11-year period since the original discovery.

The number of sunspots peaks soon after the beginning of each cycle and decays to a minimum in 11 years. The first spots of a number cycle always occur at higher latitudes, between 20° and 35°, and the latitude of occurrence decreases as the cycle unfolds (Spörer's law). Almost no spots are observed outside the latitude range of 5–35°. The great majority are small and last a few days, but some last for two rotations. In 1908, George Ellery Hale discovered that sunspots had strong magnetic fields. Each spot group contains positive and negative magnetic polarity (monopoles are forbidden by Maxwell's laws). Hale found that the polarities were mirrored, with the same polarity generally leading in one hemisphere and following in the other. He found that with each new number cycle the lead polarity switches, so that the complete magnetic cycle lasts 22 years. But each new number cycle starts a few years before the end of the previous one, so the average duration of a half-cycle is nearly 14 years. See MAGNETISM.

The darkness of sunspots (Fig. 1) is probably due to the intense magnetic fields (3000 gauss or 0.3 tesla), which cool the surface by suppressing the normal convective energy flow from below. It takes several days for the darkening to occur.

Although the sunspot is cool, its neighborhood is the scene of the hottest and most intense activity, generally referred to as an active region. Magnetic energy is continually released there. The corona above an active region is hot and dense, roughly three times hotter and denser than in quiet regions.

Prominences. The term "prominence" is used for any cloud of cool gas in the corona, where it appears bright against the sky. Because these clouds absorb the chromospheric light and scatter it, they appear dark against the solar disk in Hα and other strong lines. In continuous light they are transparent. At the limb we see the chromospheric light they scatter against the dark sky. Since they are much denser than the corona, something must hold them up against gravity. Prominences are found only in regions of horizontal magnetic fields that support them. Thus filaments on the disk, which may last for weeks, are good markers of

Solar corona, during solar eclipse on March 29, 2006, observed from Kastellorizo, Greece. Long equatorial streamers and short polar brushes are typical of solar activity during the minimum portion of the sunspot cycle. (*Jay M. Pasachoff/Science Faction, Williams College Eclipse Expedition*)

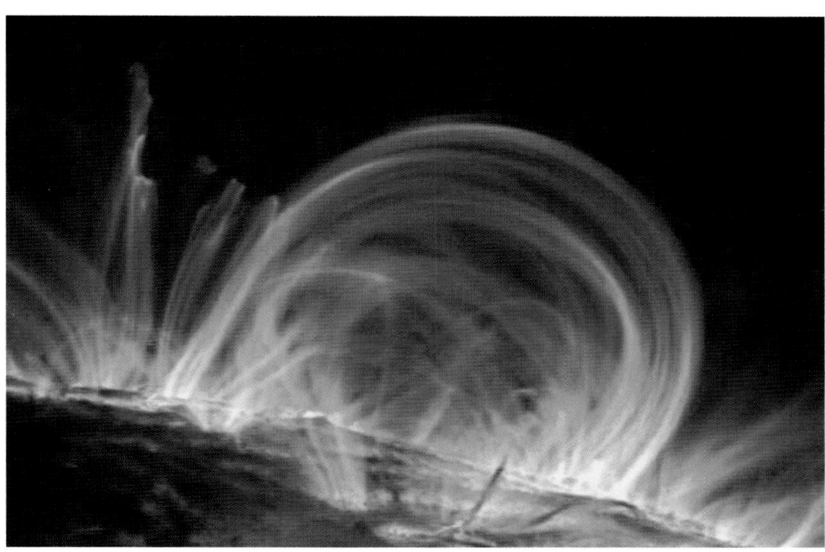

Coronal loop imaged by the *Transition Region and Coronal Explorer* (*TRACE*) spacecraft at a wavelength of 17.1 nm, characteristic of hot plasma at a temperature of 10^6 K. (*TRACE, NASA*)

TRACE 17.1-nm image of a filament eruption, measuring 75,000 mi (120,000 km) from footprint to peak, lifting off from the Sun's surface, channeled by the Sun's magnetic field. (*TRACE, NASA*)

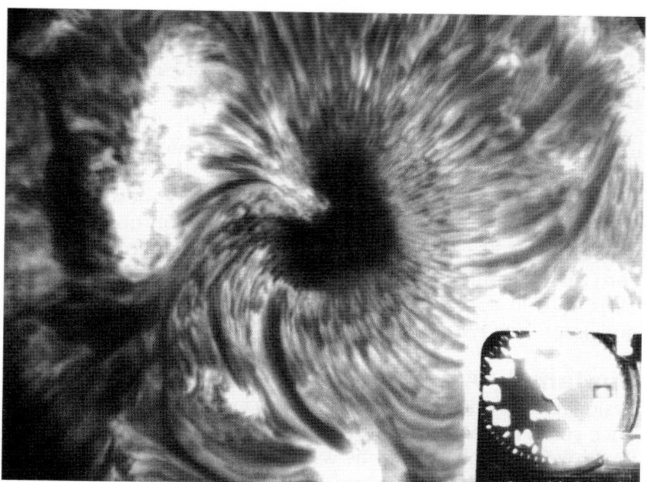

Fig. 1. Large symmetric sunspot photographed in Hα light. Clock indicates time of photograph. (*Big Bear Solar Observatory*)

the magnetic boundaries. When the magnetic structure changes, prominences become unstable and erupt, always upward. They also may be ejected by solar flares or appear as graceful loops raining from the corona after flares. Erupting prominences are probably the source of coronal mass ejections, in which a bubble of coronal material erupts outward at several hundred kilometers per second and flows out into interplanetary space.

Plages. Just as prominences occur when the magnetic field changes from one sign to the other, plages occur whenever the magnetic field is vertical and relatively strong but not strong enough to form a sunspot. They are bright regions in any strong spectrum line, because the chromosphere is heated there. In a typical active region, the preceding magnetic field is clumped in a sunspot and the following field spread out in a plage. In Hα light, the plage is seen to be connected to the sunspot by dark fibrils outlining the lines of force.

Flares. The most spectacular activity associated with sunspots is the solar flare (Fig. 2). A flare is defined as an abrupt increase in the Hα emission from the sunspot region. The brightness of the flare may be up to eight times that of the chromosphere; the rise time is seldom longer than a few minutes. The Hα brightening results from heating of the chromosphere at the foot points of the magnetic field by a tremendous energy release in the atmosphere. While flares are usually visible only in chromospheric lines, the foot points of big flares can be seen in white light. From the foot points, a cloud of hot material, up to 3×10^7 K (5.4×10^{7}°F) arises and concentrates at the arch tops. This cloud condenses out in an array of loop prominences. An active sunspot group produces a hierarchy of flares, a few big and many small ones.

Fig. 2. The great "sea horse" flare of August 7, 1972, late in the flare, photographed in the blue wing of the Hα line. The neutral line between two bright strands is crossed by an arcade of bright loop prominences raining down from the corona. (*Big Bear Solar Observatory*)

Flares are often associated with the eruption of filaments. A few minutes after the eruption begins, there is an abrupt acceleration and a storm of energetic particles is produced, heating the corona to flare brightness.

The flare produces a huge stream of solar energetic particles (SEP) as well as shock waves. A huge magnetohydrodynamic shock wave flies out at about 1000 km/s (600 mi/s) and continues into interplanetary space, often reaching the Earth. The wave produces a huge radio burst in the meter-wavelength range as it excites the coronal layers. The energetic nuclei produce gamma-ray lines from nuclear reactions as they penetrate to the photosphere. If they are sufficiently numerous, they heat the photosphere faster than it can reemit energy and a white light flare is observed, usually in the form of bright transient flashes at the foot points of the flare loops. The particles reach the Earth in a great particle storm.

Harold Zirin

Sun dog A bright spot of light that sometimes appears on either side of the Sun, the same distance above the horizon as the Sun, and separated from it by an angle of about 22° (see illustration). For higher Sun elevations, the angle increases slightly. These spots are known by many common names: sun dogs, mock suns, false suns, or the 22° parhelia. They usually show a red edge on the side closest to the Sun. On some occasions the entire spectrum of colors can be spread out in the sun-dog spot but, commonly, the red edge is followed by an orange or yellow band that merges into a diffuse white region. The effects result from the refraction by sunlight through small, flat, hexagonal-shaped ice crystals falling through the air such that their flat faces are oriented nearly horizontally. *See* HALO; METEOROLOGICAL OPTICS.

Sun dogs on either side of the Sun. Also visible are the 22° halo and the parhelic circle. (*Courtesy of Robert Greenler*)

Robert Greenler

Sundew Any plant of the genus *Drosera* (over 170 species) of the family Droseraceae. Sundews are small, herbaceous, insectivorous plants that grow on all the continents, especially Australia. Numerous glandular hairs (tentacles) on the leaf secrete a viscous fluid which traps a visiting insect. The tentacles then bend inward about the victim, bringing it into contact with the surface of the leaf where it is digested. The droplets secreted by the glands on the leaves glitter like dewdrops in the morning sunlight; hence the name sundew. *See* INSECTIVOROUS PLANTS; NEPENTHALES; SECRETORY STRUCTURES (PLANT).

Perry D. Strausbaugh; Earl L. Core

Sundial An instrument for telling time by the Sun. It is composed of a style that casts a shadow and a dial plate, which is the surface upon which hour lines are marked and upon which the shadow falls. The style lies parallel to Earth's axis. The construction of the hour lines is based on the assumption that the apparent motion of the Sun is always on the celestial equator. The most widely used form is the horizontal dial that indicates local apparent time (Sun time). Other forms of the sundial indicate local mean time, and standard time. *See* TIME. Robert N. Mayall

Sunfish A name given primarily to certain members of the freshwater family Centrarchidae, as well as to members of the marine family Molidae and the freshwater family Elassomatidae. These fishes are characterized by brilliant metallic skin coloration, especially on the cheeks and belly. *See* PERCIFORMES; TETRAODONTIFORMES.

The freshwater centrarchid sunfishes are endemic to North America. The largest centrarchid genus is *Lepomis* with 12 species. The crappies (genus *Pomoxis*) are game fish of considerable importance. Crappies are a favorite species for pond culture. They can be readily transplanted and, under favorable conditions, multiply prodigiously. The average size for these fishes is 0.3 m (1 ft), with a weight of 0.5 kg (1 lb). They feed principally on insect larvae and crustaceans; in the autumn they consume small minnows and darters. Charles B. Curtin

Sunflower *Helianthus annuus*, the most widely distributed of the 50 native North American species of this genus of the family Compositae. It is an extremely variable species, with two main divisions. The first involves wild weedy plants found along roadways and other recently disturbed areas; the second, domesticated plants grown in fields and gardens (see illustration). *See* ASTERALES; FLOWER.

Sunflower (*Helianthus annuus*). (*Photo by Andrew Bennett*)

Within the domestic type there are two categories of plants: the ornamental, which has a few branches with larger heads than the wild, and the crop type, which has only a single stem and the largest head of all sunflowers. Crop types are either oil or nonoil. Plant breeders have modified the plant for adaptation to modern, large-scale farming and have increased the oil content of the seeds. The worldwide interest in growing sunflowers as a crop has been due to the increased yield of the new commercially available oilseed hybrids. *See* AGRICULTURAL SCIENCE (PLANT); BREEDING (PLANT).

Sunflowers are grown in many countries throughout the world. Russia is the major producer, followed by Argentina, the United States, and Canada. Most of the United States oil seed production is in North Dakota, Minnesota, South Dakota, and Texas. Sunflower oil (sunoil) is a high-quality, edible vegetable oil. It is high in linoleic fatty acid and is comparable to corn oil and safflower oil in unsaturation of the fatty acids. The oil is used in cooking, salad dressing, mayonnaise, margarine, and soap. *See* FAT AND OIL (FOOD). Benjamin H. Beard

Sunlamp A special form of mercury arc discharge lamp designed to produce ultraviolet radiation. These lamps also produce some radiant energy in the visible region of the spectrum, thus having a light output as well as an ultraviolet output. The lamps are principally used for tanning. The less common uses include therapeutically producing vitamin D in the body to treat rickets and causing fluorescence or photochemical reactions.

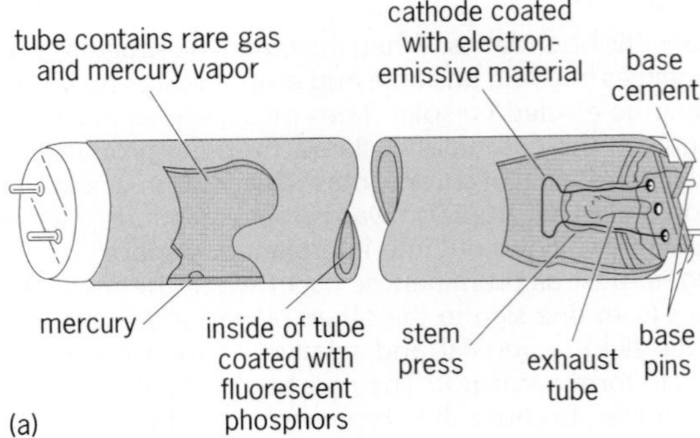

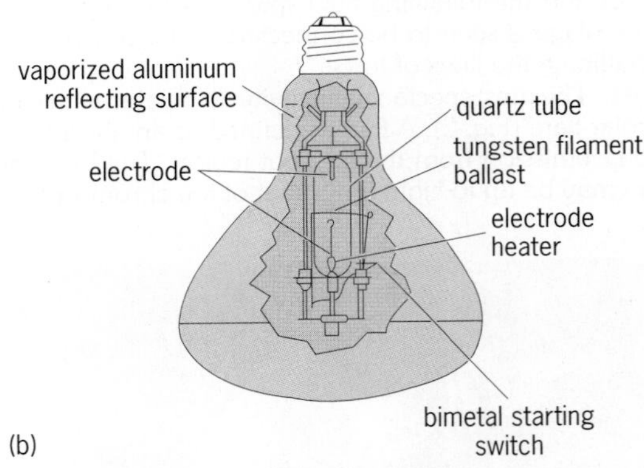

Structure of sunlamps. (*a*) FS-40. (*b*) RS.

Sunlamps have ultraviolet radiation at wavelengths above 280 nanometers. The lower limit is set by the fact that the quartz and high-silica glass used for lamp envelopes do not transmit below 280 nm. The two lamps designed for tanning are the FS-40 and the RS. The FS-40 tubular fluorescent lamp (illustration *a*) operates with a low-pressure mercury arc that causes a special chemical phosphor coating inside the tube to radiate ultraviolet energy. The RS sunlamp (illus. *b*) is a reflector unit containing a high-pressure mercury arc tube for generating ultraviolet energy plus a tungsten filament (similar to incandescent lamp filaments) in series with the arc tube to serve as a ballast. *See* MERCURY-VAPOR LAMP; ULTRAVIOLET LAMP. G. R. Peirce

Superacid An acid which has an extremely great proton-donating ability. It has proved convenient to define a superacid somewhat arbitrarily as an acid, or more generally, an acidic medium, which has a proton-donating ability equal to or greater than that of anhydrous (100%) sulfuric acid.

Superacids belong to the general class of proton or Brönsted acids. A proton acid is defined as any species which can act as a source of protons and which will therefore protonate a suitable base, as in reaction (1).

$$HA + B \rightleftharpoons BH^+ + A^- \qquad (1)$$

The strengths of acids are often compared by measuring the extent of their ionization in water, that is, the extent to which they can protonate the base water, as in reaction (2).

$$HA + H_2O \rightleftharpoons H_3O^+ + A^- \qquad (2)$$

However, all strong acids are fully ionized in dilute aqueous solution, and they therefore appear to have the same strength. Their strengths are said to be reduced or leveled to that of the hydronium ion (H_3O^+), which is the most highly acidic species that can exist in water. In any case, many of the superacids react with and are destroyed by water. For these reasons, the strengths of superacids cannot be measured by the conventional means of utilizing their aqueous solutions. The acidities of superacids can, however, be conveniently measured in terms of the Hammett acidity function. *See* ACID AND BASE.

The Hammett acidity function is a method of measuring acidity based on the determination of the ionization ratios of suitable weak bases (indicators), usually by means of the change in absorption spectrum that occurs on protonation of the base, although the nuclear magnetic resonance (NMR) spectrum has also been used. The Hammett acidity function (H_0) is defined by Eq. (3), where K_{BH+} is the dissociation constant of the acid

$$H_0 = pK_{BH+} - \log \frac{[BH^+]}{[B]} \qquad (3)$$

form of the indicator and $[BH^+]/[B]$ is the ionization ratio of the indicator. Hammett acidity function (H_0) values for a number of superacids are given in the table. In each case the value refers to the 100% (anhydrous) acid. Each of the superacids in the table is a liquid at room temperature, and each forms the basis of a solvent system. *See* IONIC EQUILIBRIUM; SOLUTION.

Ronald J. Gillespie

Hammett acidity function values for several superacids

Superacid	Formula	$-H_0$
Sulfuric acid	H_2SO_4	11.9
Chlorosulfuric acid	HSO_3Cl	13.8
Trifluoromethane sulfonic acid	HSO_3CF_3	14.0
Disulfuric acid	$H_2S_2O_7$	14.4
Fluorosulfuric acid	HSO_3F	15.1
Hydrogen fluoride	HF	15.1

Supercharger An air pump or blower in the intake system of an internal combustion engine. Its purpose is to increase the air-charge weight and therefore the power output from an engine of a given size. In an aircraft engine, the supercharger counteracts the power loss that results from the decrease of atmospheric pressure with increase of altitude. Various types of pumps and compressors may be used as superchargers, which are either mechanically driven by the engine crankshaft or powered by the engine exhaust gas. *See* COMPRESSOR; PUMP; TURBOCHARGER.

Some production automobile engines use a spiral-type supercharger, while others use a pressure-wave supercharger. However, automotive, marine, and stationary engines generally use a positive-displacement Roots blower driven from the engine crankshaft. *See* INTERNAL COMBUSTION ENGINE.

In a supercharged diesel engine, the increased air charge allows the engine to burn more fuel and produce greater power without creating excessive pressure inside the cylinder. Supercharging the diesel engine makes ignition of the fuel easier without requiring a fuel of better quality. *See* DIESEL ENGINE.

To enable a reciprocating aircraft engine to develop its rated sea-level power at altitude, a supercharger must be used to increase the pressure and weight of the intake air charge. Centrifugal compressors are generally used because of their relatively small size for a given capacity, and are driven either by a gear drive from the crankshaft or by the engine exhaust gas. With gear drive, a lower ratio is commonly used at low and medium altitudes, with a change to a higher ratio at high altitudes. *See* AIRCRAFT ENGINE.

J. A. Bolt; Donald L. Anglin

Supercomputer A computer which, among existing general-purpose computers at any given time, is superlative, often in several senses: highest computation rate, largest memory, or highest cost. Predominantly, the term refers to the fastest "number crunchers," that is, machines designed to perform numerical calculations at the highest speed that the latest electronic device technology and the state of the art of computer architecture allow.

The demand for the ability to execute arithmetic operations at the highest possible rate originated in computer applications areas collectively referred to as scientific computing. Large-scale numerical simulations of physical processes are often needed in fields such as physics, structural mechanics, meteorology, and aerodynamics. A common technique is to compute an approximate numerical solution to a set of partial differential equations which mathematically describe the physical process of interest but are too complex to be solved by formal mathematical methods. This solution is obtained by first superimposing a grid on a region of space, with a set of numerical values attached to each grid point. Large-scale scientific computations of this type often require hundreds of thousands of grid points with 10 or more values attached to each point, with 10 to 500 arithmetic operations necessary to compute each updated value, and hundreds of thousands of time steps over which the computation must be repeated before a steady-state solution is reached. *See* COMPUTATIONAL FLUID DYNAMICS; NUMERICAL ANALYSIS; SIMULATION.

Two lines of technological advancement have significantly contributed to what roughly amounts to a doubling of the fastest computers' speeds every year since the early 1950s—the steady improvement in electronic device technology and the accumulation of improvements in the architectural designs of digital computers.

Computers incorporate very large-scale integrated (VLSI) circuits with tens of millions of transistors per chip for both logic and memory components. A variety of types of integrated circuitry is used in contemporary supercomputers. Several use high-speed complementary metallic oxide semiconductor (CMOS) technology. Throughout most of the history of digital computing, supercomputers generally used the highest-performance switching circuitry available at the time—which was usually the most exotic and expensive. However, many supercomputers now use the conventional, inexpensive device technology of commodity microprocessors and rely on massive parallelism for their speed. *See* COMPUTER STORAGE TECHNOLOGY; CONCURRENT PROCESSING; INTEGRATED CIRCUITS; LOGIC CIRCUITS ; SEMICONDUCTOR MEMORIES.

Increases in computing speed which are purely due to the architectural structure of a computer can largely be attributed to the introduction of some form of parallelism into the machine's design: two or more operations which were performed one after the other in previous computers can now be performed simultaneously. *See* COMPUTER SYSTEMS ARCHITECTURE.

Pipelining is a technique which allows several operations to be in progress in the central processing unit at once. The first form of pipelining used was instruction pipelining. Since each instruction must have the same basic sequence of steps performed, namely instruction fetch, instruction decode, operand fetch, and execution, it is feasible to construct an instruction pipeline, where each of these steps happens at a separate stage of the pipeline. The efficiency of the instruction pipeline depends on the likelihood that the program being executed allows a steady stream of instructions to be fetched from contiguous locations in memory.

The central processing unit nearly always has a much faster cycle time than the memory. This implies that the central processing unit is capable of processing data items faster than a memory unit can provide them. Interleaved memory is an organization of memory units which at least partially relieves this problem.

Parallelism within arithmetic and logical circuitry has been introduced in several ways. Adders, multipliers, and dividers now operate in bit-parallel mode, while the earliest machines performed bit-serial arithmetic. Independently operating parallel functional units within the central processing unit can each perform an arithmetic operation such as add, multiply, or shift. Array processing is a form of parallelism in which the instruction execution portion of a central processing unit is replicated several times and connected to its own memory device as well as to a common instruction interpretation and control unit. In this way, a single instruction can be executed at the same time on each of several execution units, each on a different set of operands. This kind of architecture is often referred to as single-instruction stream, multiple-data stream (SIMD).

Vector processing is the term applied to a form of pipelined arithmetic units which are specialized for performing arithmetic operations on vectors, which are uniform, linear arrays of data values. It can be thought of as a type of SIMD processing, since a single instruction invokes the execution of the same operation on every element of the array. *See* COMPUTER PROGRAMMING; PROGRAMMING LANGUAGES.

A central processing unit can contain multiple sets of the instruction execution hardware for either scalar or vector instructions. The task of scheduling instructions which can correctly execute in parallel with one another is generally the responsibility of the compiler or special scheduling hardware in the central processing unit. Instruction-level parallelism is almost never visible to the application programmer.

Multiprocessing is a form of parallelism that has complete central processing units operating in parallel, each fetching and executing instructions independently from the others. This type of computer organization is called multiple-instruction stream, multiple-data stream (MIMD). *See* MULTIPROCESSING.

David W. Mizell

Superconducting devices Devices that perform functions in the superconducting state that would be difficult or impossible to perform at room temperature, or that contain components which perform such functions. The superconducting state involves a loss of electrical resistance and occurs in many metals and alloys at temperatures near absolute zero. An enormous impetus was provided by the discovery in 1986 of a new class of ceramic, high-transition-temperature (T_c) superconductors, which has resulted in a new superconducting technology at liquid nitrogen temperature. Superconducting devices may be conveniently divided into two categories: small-scale thin-film devices, and large-scale devices which employ zero-resistance superconducting windings made of type II superconducting materials. *See* SUPERCONDUCTIVITY.

Small-scale devices. A variety of thin-film devices offer higher performance than their nonsuperconducting counterparts. The prediction and discovery in the early 1960s of the Josephson effects introduced novel opportunities for ultrasensitive detectors, high-speed switching elements, and new physical standards. Niobium-based devices, patterned on silicon wafers using photolithographic techniques taken over from the semiconductor industry, have reached a high level of development, and a variety of such devices are commercially available. These devices operate at or below 4.2 K (−452°F), the temperature of liquid helium boiling under atmospheric pressure. *See* LIQUID HELIUM.

The discovery of the high-transition-temperature superconductors has enabled the operation of devices in liquid nitrogen at 77 K (−321°F). Not only is liquid nitrogen much cheaper and more readily available than liquid helium, but it also boils away much more slowly, enabling the use of simpler and more compact dewars or simpler, relatively inexpensive refrigerators. Of the new ceramic superconductors, only $YBa_2Cu_3O_{7-x}$ (YBCO) has been developed in thin-film form to the point of practical applications, and several devices are available. Intensive materials research has resulted in techniques, notably laser-ablation and radio-frequency sputtering, for the epitaxial growth of high-quality films with their crystalline planes parallel to the surface of the substrate. Most of the successful Josephson-junction devices have been formed at the interface between two grains of YBCO. These so-called grain-boundary junctions are made by depositing the film either on a bicrystal in which the two halves of the substrate have a carefully engineered in-plane misalignment of the crystal axes, or across a step-edge patterned in the substrate. *See* CRYOGENICS; GRAIN BOUNDARIES.

Two types of superconducting quantum interference device (SQUID) detect changes in magnetic flux: the dc SQUID and the rf SQUID. The dc SQUID, which operates with a dc bias current, consists of two Josephson junctions incorporated into a superconducting loop. The maximum dc supercurrent, known as the critical current, and the current-voltage (*I-V*) characteristic of the SQUID oscillate when the magnetic field applied to the device is changed. The oscillations are periodic in the magnetic flux Φ threading the loop with a period of one flux quantum, $\Phi_0 = h/2e \approx 2.07 \times 10^{-15}$ weber, where h is Planck's constant and e is the magnitude of the charge of the electron. Thus, when the SQUID is biased with a constant current, the voltage is periodic in the flux. The SQUID is almost invariably operated in a flux-locked loop. A change in the applied flux gives rise to a corresponding current in the coil that produces an equal and opposite flux in the SQUID. The SQUID is thus the null detector in a feedback circuit, and the output voltage is linearly proportional to the applied flux. *See* JOSEPHSON EFFECT; SQUID.

The rf SQUID consists of a single Josephson junction incorporated into a superconducting loop and operates with an rf bias. The SQUID is coupled to the inductor of an *LC*-resonant circuit excited at its resonant frequency, typically 30 MHz. The characteristics of rf voltage across the tank circuit versus the rf current depends on applied flux. With proper adjustment of the rf current, the amplitude of the rf voltage across the tank circuit oscillates as a function of applied flux. The rf SQUID is also usually operated in a feedback mode.

SQUIDs are mostly used in conjunction with an input circuit. For example, magnetometers are made by connecting a superconducting pickup loop to the input coil to form a flux transformer. A magnetic field applied to the pickup loop induces a persistent current in the transformer and hence a magnetic flux in the SQUID. These magnetometers have found application in geophysics, for example, in magnetotellurics. *See* GEOPHYSICAL EXPLORATION; MAGNETOMETER.

Low-transition-temperature SQUIDs are widely used to measure the magnetic susceptibility of tiny samples over a wide temperature range. Another application is a highly sensitive voltmeter, used in measurements of the Hall effect and of thermoelectricity. Low-transition-temperature SQUIDs are used as ultrasensitive detectors of nuclear magnetic and nuclear quadrupole resonance, and as transducers for gravitational-wave antennas. So-called scanning SQUIDs are used to obtain magnetic images of objects ranging from single-flux quanta trapped in superconductors to subsurface damage in two metallic sheets riveted together. *See* HALL EFFECT; MAGNETIC SUSCEPTIBILITY; NUCLEAR MAGNETIC RESONANCE (NMR); NUCLEAR QUADRUPOLE RESONANCE; THERMOELECTRICITY; VOLTMETER.

Perhaps the single largest area of application is biomagnetism, notably to image magnetic sources in the human brain or heart. In these studies an array of magnetometers or gradiometers is placed close to the subject, both generally being in a magnetically shielded room. The fluctuating magnetic signals recorded by the various channels are analyzed to locate their source. These techniques have been used, for example, to pinpoint the origin of focal epilepsy and to determine the function of the brain surrounding a tumor prior to its surgical removal. *See* BIOMAGNETISM.

The most sensitive detector available for millimeter and submillimeter electromagnetic radiation is the superconductor-insulator-superconductor (SIS) quasiparticle mixer. In this tunnel junction, usually niobium–aluminum oxide–niobium, the Josephson supercurrent is quenched and only single electron tunneling occurs. The current-voltage characteristic exhibits a very sharp onset of current at a voltage $2\Delta/e$, where Δ is the superconducting energy gap. The mixer is biased near this onset where the characteristics are highly nonlinear and used to mix the signal frequency with the frequency of a local oscillator to produce an intermediate frequency that is coupled out into a low-noise amplifier. These mixers are useful at frequencies up to about 750 GHz (wavelengths down to 400 micrometers). Such receivers are of great importance in radio astronomy, notably for airborne, balloon-based, or high-altitude, ground-based telescopes operating above most of the atmospheric water vapor. *See* RADIO ASTRONOMY.

The advent of high-transition-temperature superconductors stimulated major efforts to develop passive radio-frequency and microwave components that take advantage of the low electrical losses offered by these materials compared with normal conductors in liquid nitrogen. The implementation of thin-film YBCO receiver coils has improved the signal-to-noise ratio of nuclear magnetic resonance (NMR) spectrometers by a factor of 3 compared to that achievable with conventional coils. This improvement enables the data acquisition time to be reduced by an order of magnitude. These coils also have potential applications in low-frequency magnetic resonance imaging (MRI). High-transition-temperature bandpass filters have application in cellular communications. *See* ELECTRIC FILTER; MOBILE RADIO.

John Clarke

Large-scale devices. Large-scale applications of superconductivity comprise medical, energy, transportation, high-energy physics, and other miscellaneous applications such as high-gradient magnetic separation. When strong magnetic fields are needed, superconducting magnets offer several advantages over conventional copper or aluminum electromagnets. Most important is lower electric power costs because once the system is energized only the refrigeration requires power input, generally only 5–10% that of an equivalent-field resistive magnet. Relatively high magnetic fields achievable in unusual configurations and in smaller total volumes reduce the costs of expensive force-containment structures. *See* MAGNET.

Niobium-titanium (NbTi) has been used most widely for large-scale applications, followed by the A15 compounds, which include niobium-tin (Nb_3Sn), niobium-aluminum (Nb-Al), niobium-germanium (Nb-Ge), and vanadium-gallium (Va_3Ga). Niobium-germanium held the record for the highest critical field (23 K; −418.5°F) until the announcement of high-temperature ceramic superconductors. *See* A15 PHASES.

Significant advances have been made in high-temperature superconducting wire development. Small coils have been wound that operate at 20 K (−410°F). Current leads are in limited commercial use. Considerable development remains necessary to use these materials in very large applications.

MRI dominates superconducting magnet systems applications. Most of the MRI systems are in use in hospitals and clinics, and incorporate superconducting magnets. *See* MEDICAL IMAGING.

Some of the largest-scale superconducting magnet systems are those considered for energy-related applications. These include magnetic confinement fusion, superconducting magnetic energy storage, magnetohydrodynamic electrical power generation, and superconducting generators. *See* MAGNETOHYDRODYNAMIC POWER GENERATOR; NUCLEAR FUSION.

In superconducting magnetic energy storage superconducting magnets are charged during off-peak hours when electricity demand is low, and then discharged to add electricity to the grid at times of peak demand. The largest systems would require large land areas, for example, an 1100-m-diameter (3600-ft) site for a 5000-MWh system. However, intermediate-size systems are viable. A 6-T peak-field solenoidal magnet system designed for the Alaskan power network stores 1800 megajoules (0.5 MWh). High-purity-aluminum-stabilized niobium-titanium alloy conductor carrying 16 kiloamperes current is used for the magnet winding.

Superconducting magnets have potential applications for transportation, such as magnetically levitated vehicles. In addition, superconducting magnets are used in particle accelerators and particle detectors. *See* MAGNETIC LEVITATION; PARTICLE ACCELERATOR; PARTICLE DETECTOR.

Alberta M. Larsen

Superconductivity A phenomenon occurring in many electrical conductors, in which the electrons responsible for conduction undergo a collective transition into an ordered state with many unique and remarkable properties. These include the vanishing of resistance to the flow of electric current, the appearance of a large diamagnetism and other unusual magnetic effects, substantial alteration of many thermal properties, and the occurrence of quantum effects otherwise observable only at the atomic and subatomic level.

Superconductivity was discovered by H. Kamerlingh Onnes in Leiden in 1911 while studying the temperature dependence of the electrical resistance of mercury within a few degrees of absolute zero. He observed that the resistance dropped sharply to an unmeasurably small value at a temperature of 4.2 K (−452°F). The temperature at which the transition occurs is called the transition or critical temperature, T_c. The vanishingly small resistance (very high conductivity) below T_c suggested the name given the phenomenon.

In 1933 W. Meissner and R. Ochsenfeld discovered that a metal cooled into the superconducting state in a moderate magnetic field expels the field from its interior. This discovery demonstrated that superconductivity involves more than simply very high or infinite electrical conductivity, remarkable as that alone is. *See* MEISSNER EFFECT.

In 1957, J. Bardeen, L. N. Cooper, and J. R. Schrieffer reported the first successful microscopic theory of superconductivity. The Bardeen-Cooper-Schrieffer (BCS) theory describes how the electrons in a conductor form the ordered superconducting state. The BCS theory still stands as the basic explanation of superconductivity, even though extensive theoretical work has embellished it.

There are a number of practical applications of superconductivity. Powerful superconducting electromagnets guide elementary particles in particle accelerators, and they also provide the magnetic field needed for magnetic resonance imaging. Ultrasensitive superconducting circuits are used in medical studies of the human heart and brain and for a wide variety of physical science experiments. A completely superconducting prototype computer has even been built. *See* MEDICAL IMAGING; PARTICLE ACCELERATOR; SUPERCONDUCTING DEVICES.

Transition temperatures. It was realized from the start that practical applications of superconductivity could become much more widespread if a high-temperature superconductor, that is, one with a high T_c, could be found. For instance, the only practical way to cool superconductors with transition temperatures below 20 K (−424°F) is to use liquid helium, which boils at a temperature of 4.2 K (−452°F) and which is rather expensive. On the other hand, a superconductor with a transition temperature of 100 K (−280°F) could be cooled with liquid nitrogen, which boils at 77 K (−321°F) and which is roughly 500 times less expensive than liquid helium. Another advantage of a high-T_c material is that, since many of the other superconducting properties are proportional to T_c, such a material would have enhanced properties. In 1986 the discovery of transition temperatures possibly as high as 30 K (−406°F) was reported in a compound containing barium, lanthanum, copper, and oxygen. In 1987 a compound of yttrium, barium, copper, and oxygen was shown to be superconducting above 90 K (−298°F). In 1988 researchers showed that a bismuth, strontium, calcium, copper, and oxygen compound was superconducting below 110 K (−262°F), and transition temperatures as high as 135 K (−216°F) were found in a mercury, thallium, barium, calcium, copper, and oxygen compound.

Occurrence. Some 29 metallic elements are known to be superconductors in their normal form, and another 17 become superconducting under pressure or when prepared in the form of thin films. The number of known superconducting compounds and alloys runs into the thousands. Superconductivity is thus a rather common characteristic of metallic conductors. The phenomenon also spans an extremely large temperature range. Rhodium is the element with the lowest transition temperature (370 μK), while $Hg_{0.2}Tl_{0.8}Ca_2Ba_2Cu_3O$ is the compound with the highest (135 K or −216°F).

Despite the existence of a successful microscopic theory of superconductivity, there are no completely reliable rules for predicting whether a metal will be a superconductor. Certain trends and correlations are apparent among the known superconductors, however—some with obvious bases in the theory—and these provide empirical guidelines in the search for new superconductors. Superconductors with relatively high transition temperatures tend to be rather poor conductors in the normal state.

The ordered superconducting state appears to be incompatible with any long-range-ordered magnetic state: Usually the ferromagnetic or antiferromagnetic metals are not superconducting. The presence of nonmagnetic impurities in a superconductor usually has very little effect on the superconductivity, but the presence of impurity atoms which have localized magnetic moments can markedly depress the transition temperature even in concentrations as low as a few parts per million. *See* ANTIFERROMAGNETISM; FERROMAGNETISM.

Some semiconductors with very high densities of charge carriers are superconducting, and others such as silicon and germanium have high-pressure metallic phases which are superconducting. Many elements which are not themselves superconducting form compounds which are.

Certain organic conductors are superconducting. For instance, brominated polymeric chains of sulfur and nitrogen, known as $(SNBr_{0.4})_x$, are superconducting below 0.36 K. Other more complicated organic materials have T_c values near 10 K (−442°F). *See* ORGANIC CONDUCTOR.

Although nearly all the classes of crystal structure are represented among superconductors, certain structures appear to be especially conducive to high-temperature superconductivity. The so-called A15 structure, shared by a series of intermetallic compounds based on niobium, produced several superconductors with T_c values above 15 K (−433°F) as well as the record holder, NbGe, at 23 K (−418°F). Indeed, the robust applications of superconductivity that depend on the ability to carry high current in the presence of high magnetic fields still exclusively use two members of this class: NbTi with $T_c = 8$ K (−445°F), and Nb_3Sn with $T_c = 18.1$ K (−427°F). *See* A15 PHASES.

After 1986 the focus of superconductivity research abruptly shifted to the copper-oxide-based planar structures, due to their significantly higher transition temperatures. Basically there are three classes of these superconductors, all of which share the common feature that they contain one or more conducting planes of copper and oxygen atoms. The first class is designated by the chemical formula $La_{2-x}A_xCuO_4$, where the A atom can be barium, strontium, or calcium. Superconductivity was originally discovered in the barium-doped system, and systematic study of the substitutions of strontium, calcium, and so forth have produced transition temperatures as high as 40 K (−388°F).

The second class of copper-oxide superconductor is designated by the chemical formula $Y_1Ba_2Cu_3O_{7-\delta}$, with $\delta < 1.0$. Here, single sheets of copper and oxygen atoms straddle the rare-earth yttrium ion and chains of copper and oxygen atoms thread among the barium ions. The transition temperature, 92 K (−294°F), is quite insensitive to replacement of yttrium by many other rare-earth ions. *See* RARE-EARTH ELEMENTS.

The third class is the most complicated. These compounds contain either single thallium-oxygen layers, represented by the chemical formula $Tl_1Ca_{n-1}Ba_2Cu_nO_{2n+3}$, where n refers to the number of copper-oxygen planes, or double thallium-oxygen layers, represented by the chemical formula $Tl_2Ca_{n-1}Ba_2Cu_nO_{2n+4}$. The number of copper-oxygen planes may be varied, and as many as three planes have been included in the structure. Thallium may be replaced by bismuth, thus generating a second family of superconductors. In all of these compounds, the transition temperature appears to increase with the number of planes, but T_c decreases for larger values of n.

The spherical molecule comprising 60 carbon atoms (C_{60}), known as a buckyball, can be alloyed with various alkaline atoms which contribute electrons for conduction. By varying the number of conductors in C_{60}, it is possible to boost T_c to a maximum value of 52 K (−366°F). *See* FULLERENE.

Superconductivity was discovered in magnesium diboride (MgB_2) in January 2001 in Japan. This material may be a good alternative for some of the applications envisioned for high-T_c superconductivity, since this compound has T_c of 39 K (−389°F), is relatively easy to make, and consists of only two elements.

Magnetic properties. The existence of the Meissner-Ochsenfeld effect, the exclusion of a magnetic field from the interior of a superconductor, is direct evidence that the superconducting state is not simply one of infinite electrical conductivity. Instead, it is a true thermodynamic equilibrium state, a new phase which has lower free energy than the normal state at temperatures below the transition temperature and which somehow requires the absence of magnetic flux.

The exclusion of magnetic flux by a superconductor costs some magnetic energy. So long as this cost is less than the condensation energy gained by going from the normal to the superconducting phase, the superconductor will remain completely superconducting in an applied magnetic field. If the applied field becomes too large, the cost in magnetic energy will outweigh the gain in condensation energy, and the superconductor will become partially or totally normal. The manner in which this occurs depends on the geometry and the material of the superconductor. The geometry which produces the simplest behavior is that of a very long cylinder with field applied parallel to its axis. Two distinct types of behavior may then occur, depending on the type of superconductor—type I or type II.

Below a critical field H_c which increases as the temperature decreases below T_c, the magnetic flux is excluded from a type I superconductor, which is said to be perfectly diamagnetic. For a type II superconductor, there are two critical fields, the lower critical field H_{c1} and the upper critical field H_{c2}. In applied fields less than H_{c1}, the superconductor completely excludes the field, just as a type I superconductor does below H_c. At fields just above H_{c1}, however, flux begins to penetrate the superconductor, not in a uniform way, but as individual, isolated microscopic filaments called fluxoids or vortices. Each fluxoid consists of a normal core in which the magnetic field is large, surrounded by a superconducting region in which flows a vortex of persistent supercurrent which maintains the field in the core. *See* DIAMAGNETISM.

Thermal properties. The appearance of the superconducting state is accompanied by rather drastic changes in both the thermodynamic equilibrium and thermal transport properties of a superconductor.

The heat capacity of a superconducting material is quite different in the normal and superconducting states. In the normal state (produced at temperatures below the transition temperature by applying a magnetic field greater than the critical field), the heat capacity is determined primarily by the normal electrons (with a small contribution from the thermal vibrations of the crystal lattice) and is nearly proportional to the temperature. In zero applied magnetic field, there appears a discontinuity in the heat capacity at the transition temperature. At temperatures just below the transition temperature, the heat capacity is larger than in the normal state. It decreases more rapidly with decreasing temperature, however, and at temperatures well below the transition temperature varies exponentially as $e^{-\Delta/kT}$, where Δ is a constant and k is Boltzmann's constant. Such an exponential temperature dependence is a hallmark of a system with a gap Δ in the spectrum of allowed energy states. Heat capacity measurements provided the first indications of such a gap in superconductors, and one of the key features of the macroscopic BCS theory is its prediction of just such a gap. *See* SPECIFIC HEAT OF SOLIDS.

Ordinarily a large electrical conductivity is accompanied by a large thermal conductivity, as in the case of copper, used in electrical wiring and cooking pans. However, the thermal conductivity of a pure superconductor is less in the superconducting state than in the normal state, and at very low temperatures approaches zero. Crudely speaking, the explanation for the association of infinite electrical conductivity with vanishing thermal conductivity is that the transport of heat requires the transport of disorder (entropy). The superconducting state is one of perfect order (zero entropy), and so there is no disorder to transport and therefore no thermal conductivity. *See* ENTROPY; THERMAL CONDUCTION IN SOLIDS.

Two-fluid model. C. J. Gorter and H. B. G. Casimir introduced in 1934 a phenomenological theory of superconductivity based on the assumption that in the superconducting state there are two components of the conduction electron "fluid" (hence the name given this theory, the two-fluid model). One, called the superfluid component, is an ordered condensed state with zero entropy; hence it is incapable of transporting heat. It does not interact with the background crystal lattice, its imperfections, or the other conduction electron component and exhibits no resistance to flow. The other component, the normal component, is composed of electrons which behave exactly as they do in the normal state. It is further assumed that the superconducting transition is a reversible thermodynamic phase transition between two thermodynamically stable phases, the normal state and the superconducting state, similar to the transition between the liquid and vapor phases of any substance. The validity of this assumption is strongly supported by the existence of the Meissner-Ochsenfeld effect and by other experimental evidence. This assumption permits the application of all the powerful and general machinery of the theory of equilibrium thermodynamics. The results tie together the observed thermodynamic properties of superconductors in a very satisfying way.

Microscopic (BCS) theory. The key to the basic interaction between electrons which gives rise to superconductivity was provided by the isotope effect. It is an interaction mediated by the background crystal lattice and can crudely be pictured as follows: An electron tends to create a slight distortion of the elastic lattice as it moves, because of the Coulomb attraction between the negatively charged electron and the positively charged lattice. If the distortion persists for a brief time (the lattice may ring like a struck bell), a second passing electron will see the distortion and be affected by it. Under certain circumstances, this can give rise to a weak indirect attractive interaction between the two electrons which may more than compensate their Coulomb repulsion.

The first forward step was taken by Cooper in 1956, when he showed that two electrons with an attractive interaction can bind together to form a "bound pair" (often called a Cooper pair) if they are in the presence of a high-density fluid of other electrons, no matter how weak the interaction is. The two partners of a Cooper pair have opposite momenta and spin angular momenta. Then, in 1957, Bardeen, Cooper, and Schrieffer showed how to construct a wave function in which all of the electrons (at least, all of the important ones) are paired. Once this wave function is adjusted to minimize the free energy, it can be used as the basis for a complete microscopic theory of superconductivity.

The successes of the BCS theory and its subsequent elaborations are manifold. One of its key features is the prediction of an energy gap. Excitations called quasiparticles (which are something like normal electrons) can be created out of the superconducting ground state by breaking up pairs, but only at the expense of a minimum energy of Δ per excitation; Δ is called the gap parameter. The original BCS theory predicted that Δ is related to T_c by $\Delta = 1.76kT_c$ at $T = 0$ for all superconductors. This turns out to be nearly true, and where deviations occur they are understood in terms of modifications of the BCS theory. The manifestations of the energy gap in the low-temperature heat capacity and in electromagnetic absorption provide strong confirmation of the theory.

D. N. Langenberg; Robert J. Soulen, Jr.; Michael Osofsky

Supercontinent

Supercontinent The six major continents today are Africa, Antarctica, Australia, Eurasia, North America, and South America. Prior to the formation of the Atlantic, Indian, and Southern ocean basins over the past 180 million years by the process known as sea-floor spreading, the continents were assembled in one supercontinent called Pangea (literally "all Earth"). Pangea came together by the collision, about 300 million years ago (Ma), of two smaller masses of continental rock, Laurasia and Gondwanaland. Laurasia comprised the combined continents of ancient North America (known as Laurentia), Europe, and Asia. Africa, Antarctica, Australia, India, and South America made up Gondwanaland (this name comes from a region in southern India). The term "supercontinent" is also applied to Laurasia and Gondwanaland; hence it is used in referring to a continental mass significantly bigger than any of today's continents. A supercontinent may therefore incorporate almost all of the Earth's continental rocks, as did Pangea, but that is not implied by the word. *See* CONTINENT; CONTINENTS, EVOLUTION OF.

Laurasia, Gondwanaland, and Pangea are the earliest supercontinental entities whose former existence can be proven. Evidence of older rifted continental margins, for example surrounding Laurentia and on the Pacific margins of South America, Antarctica, and Australia, point to the existence of older supercontinents. The hypothetical Rodinia (literally "the mother of all continents") may have existed 800–1000 Ma, and Pannotia (meaning "the all-southern supercontinent") fleetingly around 550 Ma. Both are believed to have included most of the Earth's continental material. There may have been still earlier supercontinents, because large-scale continents, at least the size of southern Africa or Western Australia, existed as early as 2500 Ma at the end of Archean times. *See* ARCHEAN; CONTINENTAL MARGIN.

The amalgamation and fragmentation of supercontinents are the largest-scale manifestation of tectonic forces within the Earth.

The cause of such events is highly controversial. *See* PLATE TECTONICS. Ian W. O. Dalziel

Supercritical fields Static fields that are strong enough to cause the normal vacuum, which is devoid of real particles, to break down into a new vacuum in which real particles exist. This phenomenon has not yet been observed for electric fields, but it is predicted for these fields as well as others such as gravitational fields and the gluon field of quantum chromodynamics.

Vacuum decay in quantum electrodynamics. The original motivation for developing the new concept of a charged vacuum arose in the late 1960s in connection with attempts to understand the atomic structure of superheavy nuclei expected to be produced by heavy-ion linear accelerators. *See* PARTICLE ACCELERATOR.

The best starting point for discussing this concept is to consider the binding energy of atomic electrons as the charge Z of a heavy nucleus is increased. If the nucleus is assumed to be a point charge, the total energy E of the $1s_{1/2}$ level drops to 0 when $Z = 137$. This so-called $Z = 137$ catastrophe had been well known, but it was argued loosely that it disappears when the finite size of the nucleus is taken into account. However, in 1969 it was shown that the problem is not removed but merely postponed, and reappears around $Z = 173$. Any level $E(nj)$ can be traced down to a binding energy of twice the electronic rest mass if the nuclear charge is further increased. At the corresponding charge number, called Z_{cr}, the state dives into the negative-energy continuum of the Dirac equation (the so-called Dirac sea). The overcritical state acquires a width and is spread over the continuum. *See* ANTIMATTER; RELATIVISTIC QUANTUM THEORY.

When Z exceeds Z_{cr} a K-shell electron is bound by more than twice its rest mass, so that it becomes energetically favorable to create an electron-positron pair. The electron becomes bound in the $1s_{1/2}$ orbital and the positron escapes. The overcritical vacuum state is therefore said to be charged. *See* POSITRON.

Clearly, the charged vacuum is a new ground state of space and matter. The normal, undercritical, electrically neutral vacuum is no longer stable in overcritical fields: it decays spontaneously into the new stable but charged vacuum. Thus the standard definition of the vacuum, as a region of space without real particles, is no longer valid in very strong external fields. The vacuum is better defined as the energetically deepest and most stable state that a region of space can have while being penetrated by certain fields.

Superheavy quasimolecules. Inasmuch as the formation of a superheavy atom of $Z > 173$ is very unlikely, a new idea is necessary to test these predictions experimentally. That idea, based on the concept of nuclear molecules, was put forward in 1969: a superheavy quasimolecule forms temporarily during the slow collision of two heavy ions. It is sufficient to form the quasimolecule for a very short instant of time, comparable to the time scale for atomic processes to evolve in a heavy atom, which is typically of the order 10^{-18} to 10^{-20}. Suppose a uranium ion is shot at another uranium ion at an energy corresponding to their Coulomb barrier, and the two, moving slowly (compared to the K-shell electron velocity) on Rutherford hyperbolic trajectories, are close to each other (compared to the K-shell electron orbit radius). Then the atomic electrons move in the combined Coulomb potential of the two nuclei, thereby experiencing a field corresponding to their combined charge of 184. This happens because the ionic velocity (of the order of $c/10$) is much smaller than the orbital electron velocity (of the order of c), so that there is time for the electronic molecular orbits to be established, that is, to adjust to the varying distance between the charge centers, while the two ions are in the vicinity of each other. *See* QUASIATOM.

Giant nuclear systems. The energy spectrum for positrons created in, for example, a uranium-curium collision consists of three components: the induced, the direct, and the spontaneous, which add up to a smooth spectrum. The presence of the spontaneous component leads only to 5–10% deviations for normal nuclear collisions along Rutherford trajectories. This situation raises the question as to whether there is any way to get a clear qualitative signature for spontaneous positron production. Suppose that the two colliding ions, when they come close to each other, stick together for a certain time Δt before separating again. The longer the sticking, the better is the static approximation. For Δt very long, a very sharp line should be observed in the positron spectrum with a width corresponding to the natural lifetime of the resonant positron-emitting state. The observation of such a sharp line will indicate not only the spontaneous decay of the vacuum but also the formation of giant nuclear systems ($Z > 180$). *See* LINEWIDTH; NUCLEAR MOLECULE.

Search for spontaneous positron emission. The search for spontaneous positron emission in heavy-ion collisions began in 1976. Of special interest are peak structures in the positron energy distribution. However, the issue of spontaneous positron production in strong fields remains open. If line structures have been observed at all, they are most likely due to nuclear conversion processes. The observation of vacuum decay very much depends on the existence of sufficiently long-lived (at least 10^{-20} s) giant nuclear molecular systems. Therefore, the investigation of nuclear properties of heavy nuclei encountering heavy nuclei at the Coulomb barrier is a primary task.

Other field theories. The idea of overcriticality also has applications in other field theories, such as those of pion fields, gluon fields (quantum chromodynamics), and gravitational fields (general relativity).

A heavy meson may be modeled as an ellipsoidal bag, with a heavy quark Q and antiquark $\bar{Q}$ located at the foci of the ellipsoid. The color-electric or glue-electric field lines do not penetrate the bag surface. The Dirac equation may be solved for light quarks q and antiquarks $\bar{q}$ in this field of force. In the spherical case the potential is zero, and the solutions with different charges degenerate. As the source charges Q and $\bar{Q}$ are pulled apart, the wave functions start to localize. At a critical deformation of the bag, positive and negative energy states cross; that is, overrcriticality is reached and the color field is strong enough that the so-called perturbative vacuum inside the bag rearranges so that the wave functions are pulled to opposite sides and the color charges of the heavy quarks are completely shielded. Hence two new mesons of types $\bar{Q}q$ and $Q\bar{q}$ appear; the original meson fissions. *See* GLUONS; QUANTUM CHROMODYNAMICS; QUARKS. Walter Greiner

Supercritical fluid Any fluid at a temperature and a pressure above its critical point; also, a fluid above its critical temperature regardless of pressure. Below the critical point the fluid can coexist in both gas and liquid phases, but above the critical point there can be only one phase. Supercritical fluids are of interest because their properties are intermediate between those of gases and liquids, and are readily adjustable. *See* CRITICAL PHENOMENA; PHASE EQUILIBRIUM.

In a given supercritical fluid the thermodynamic and transport properties are a function of density, which depends strongly on the fluid's pressure and temperature. The density may be adjusted from a gaslike value of 0.1 g/ml to a liquidlike value as high as 1.2 g/ml. Furthermore, as conditions approach the critical point, the effect of temperature and pressure on density becomes much more significant. Increasing the density of supercritical carbon dioxide from 0.2 to 0.5 g/ml, for example, requires raising the pressure from 85 to 140 atm (8.6 to 14.2 megapascals) at 158°F (70°C), but at 95°F (35°C) the required change is only from 65 to 80 atm (6.6 to 8.1 MPa).

For a given fluid, the logarithm of the solubility of a solute is approximately proportional to the solvent density at constant temperature. Therefore, a small increase in pressure, which causes a large increase in the density, can raise the solubility a few orders of magnitude. While almost all of a supercritical fluid's properties vary with density, some of these properties are more like those of a liquid while others are more like those of a gas. *See* SUPERCRITICAL-FLUID CHROMATOGRAPHY.

In most supercritical-fluid applications the fluid's critical temperature is less than 392°F (200°C) and its critical pressure is less than 80 atm (8.1 MPa). High critical temperatures require operating temperatures that can damage the desired product, while high critical pressures result in excessive compression costs. In addition to these pure fluids, mixed solvents can be used to improve the solvent strength. David Dixon; Keith Johnston; Richard Lemert

Supercritical-fluid chromatography Any separation technique in which a supercritical fluid is used as the mobile phase. For any fluid, a phase diagram can be constructed to show the regions of temperature and pressure at which gases and liquids, gases and solids, and liquids and solids can coexist. For the gas-liquid equilibrium, there is a certain temperature and pressure, known as the critical temperature and pressure, below which a gas and a liquid can coexist but above which only a single phase (known as a supercritical fluid) can form. *See* CRITICAL PHENOMENA.

For example, the density of supercritical fluids is usually between 0.25 and 1.2 g/ml and is strongly pressure-dependent. Their solvent strength increases with density, so that molecules that are retained on the column can often be eluted simply by increasing the pressure under which the fluid is compressed.

Diffusion coefficients of solutes in supercritical fluids are tenfold greater than the corresponding values in liquid solvents (although about three orders of magnitude less than the corresponding values in gases). The high diffusivity of solutes in supercritical fluids decreases their resistance to mass transfer in a chromatographic column and hence allows separations to be made either very quickly or at high resolution. Molecules can be separated in a few minutes or less by using packed columns of the type used for high-performance liquid chromatography (HPLC). *See* DIFFUSION.

Despite the relatively large number of separations made using packed-column supercritical-fluid chromatography that have been described, supercritical-fluid chromatography became popular only with the commercial introduction of capillary (open-tubular) supercritical-fluid chromatographs. While supercritical-fluid chromatography may never be as widely used as gas chromatography or high-performance liquid chromatography, it provides a useful complement to these chromatographies. *See* CHEMICAL SEPARATION TECHNIQUES; CHROMATOGRAPHY; FLUIDS; GAS CHROMATOGRAPHY; LIQUID CHROMATOGRAPHY. Peter R. Griffiths

Supercritical wing A wing with special streamwise sections, or airfoils, which provide substantial delays in the onset of the adverse aerodynamic effects which usually occur at high subsonic flight speeds.

When the speed of an aircraft approaches the speed of sound, the local airflow about the airplane, particularly above the upper surface of the wing, may exceed the speed of sound. Such a condition is called supercritical flow. On previous aircraft, this supercritical flow resulted in the onset of a strong local shock wave above the upper surface of the wing (illustration *a*). This local wave caused an abrupt increase in the pressure on the surface of the wing, which may cause the surface boundary-layer flow to separate from the surface, with a resulting severe increase in the turbulence of the flow. The increased turbulence leads to a severe increase in drag and loss in lift, with a resulting decrease in flight efficiency. The severe turbulence also caused buffet or shaking of the aircraft and substantially changed its stability or flying qualities. *See* AERODYNAMIC FORCE; AERODYNAMIC WAVE DRAG; LOW-PRESSURE GAS FLOW; SHOCK WAVE; TRANSONIC FLIGHT.

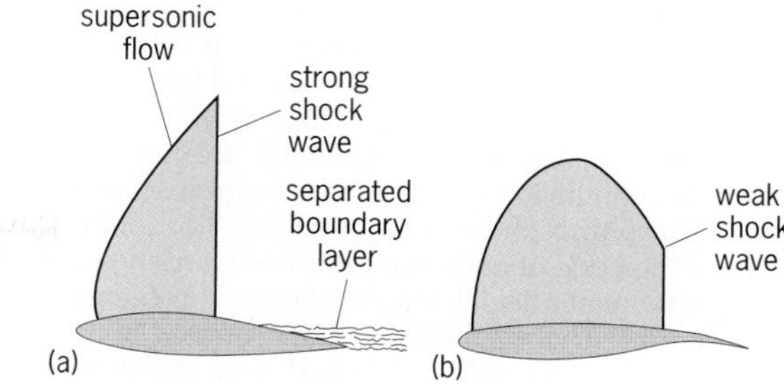

Comparison of airflow about airfoils. (*a*) Conventional airfoils, Mach number = 0.7. (*b*) Supercritical airfoils. Mach number = 0.8.

Supercritical airfoils are shaped to substantially reduce the strength of the shock wave and to delay the associated boundary-layer separation (illus. *b*). Since the airfoil shape allows efficient flight at supercritical flight speeds, a wing of such design is called a supercritical wing. *See* AIRPLANE; WING. Richard T. Whitcomb

Superfluidity The frictionless flow of liquid helium at low temperature; also, the flow of electric current without resistance in certain solids at low temperature (superconductivity).

Both helium isotopes have a superfluid transition, but the detailed properties of their superfluid states differ considerably because they obey different statistics. ^{4}He, with an intrinsic spin of 0, is subject to Bose-Einstein statistics, and ^{3}He, with a spin of $1/2$, to Fermi-Dirac statistics. There are two distinct superfluid states in ^{3}He called A and B.

The term "superfluidity" usually implies He II or the A and B phases of ^{3}He, but the basic similarity between these and the "fluid" consisting of pairs of electrons in superconductors is sufficiently strong to designate the latter as a charged superfluid. Besides flow without resistance, superfluid helium and superconducting electrons display quantized circulating flow patterns in the form of microscopic vortices. *See* BOSE-EINSTEIN STATISTICS; LIQUID HELIUM; QUANTIZED VORTICES; SECOND SOUND; SUPERCONDUCTIVITY. Laurence J. Campbell

Supergiant star A member of a class of evolved stars that occupy the top of the Hertzsprung-Russell diagram to the right of the main sequence. The absolute visual magnitudes (M_V) of supergiants range approximately between −4 and −10, and they are the largest and brightest stars. They are recognized by their spectroscopic characteristics. For example, those in class A have narrow hydrogen lines. Supergiants are subdivided into classes Ib (M_V about −5) and Ia (M_V about −7). A "hypergiant class zero" was later added near $M_V = -10$; the use of transition class Ia-0 at absolute visual magnitudes near −9 is now common. Red supergiants are the largest of all stars, and at maximum can reach diameters approaching that of the orbit of Saturn.

In an evolutionary sense, supergiants are stars above about 8 or 10 solar masses and absolute visual magnitude −6 or so that began main-sequence life hotter than class B1 or B2 and cannot evolve to become white dwarfs. As progeny of O and the hottest B stars, supergiants are exceedingly rare. Supergiant masses allow further burning of carbon and oxygen to oxygen, neon, and magnesium; thence to silicon and sulfur; and thence to iron. The iron cores collapse to produce type II (and IIB) supernovae, the condensed remnants at the centers becoming either neutron stars or, from the most massive stars, black holes. *See* ASTRONOMICAL SPECTROSCOPY; BLACK HOLE; HERTZSPRUNG-RUSSELL DIAGRAM; NEUTRON STAR; SPECTRAL TYPE; STAR; STELLAR EVOLUTION; SUPERNOVA. James B. Kaler

Supergranulation A system of convective cells, called supergranules, with diameters of about 30,000 km (20,000 mi)

and lifetimes of 1 or 2 days, which cover the visible disk of the Sun, the photosphere. Smaller convection cells called granules, which are about 1500 km (900 mi) across and have lifetimes of about 10 min, are seen through a telescope in white-light images of the photosphere, but the supergranules are not visible this way. The supergranulation is seen in Dopplergrams that measure motions in the photosphere. The gas flows horizontally from the center to the edge of each supergranule, carrying along the photospheric magnetic field that is detected by the Zeeman effect in magnetograms. The magnetism is concentrated at the edges of the supergranules, forming polygon-shaped structures called the magnetic network.

The visible disk of the Sun is the photosphere. In 1801 William Herschel noticed that the solar photosphere has a granular appearance. More than a century later, high-resolution photographs of the Sun revealed that this granulation is composed of closely packed cells having bright centers surrounded by dark lanes. The cells mark the top of the convection zone. *See* PHOTOSPHERE.

The solar granulation exhibits a nonstationary, overturning motion, a visible manifestation of the convection beneath. The bright center of each granule, or convection cell, is the highest point of a rising column of hot gas. The dark edges of each granule are the sinking cooled gas. *See* CONVECTION (HEAT).

A larger cellular pattern, called the supergranulation, is detected by observing motions in the photosphere.

In 1960, Robert Leighton developed an instrument that obtains a Doppler picture of the photosphere by using a spectroheliograph to subtract a long-wavelength image from a short-wavelength one. The resulting Dopplergrams revealed the supergranulation, a pattern of convection cells that are about 30,000 km (20,000 mi) across, or about twice the size of the Earth. Roughly 2500 supergranules are seen on the visible solar disk, each persisting for 1 or 2 days. Like the ordinary granulation, the supergranulation is caused by convection. *See* DOPPLER EFFECT; SPECTROGRAPH; SPECTROHELIOGRAPH.

The supergranular flow carries both the granules and the magnetic field across the photosphere, like leaves floating on a river, sweeping the magnetism to the edges of the supergranulation cells, where it collects and strengthens. Astronomers use an array of detectors that measure the Zeeman effect at different locations across the photosphere, producing a magnetogram that displays the strength and direction of the solar magnetic fields all across the visible solar disk. Time sequences of magnetograms show that the supergranular motions sweep small magnetic elements, called flux tubes, from the center to the edges of each supergranule. The photosphere's magnetic field is thus concentrated into a magnetic network that traces out the polygonal supergranulation pattern. *See* MAGNETOSPHERE; SOLAR MAGNETIC FIELD.

The Sun, is a volatile, ever-changing incandescent ball of gas. The granules constantly evolve and change, producing a honeycomb pattern of rising and falling gas that is in constant turmoil, completely changing on time scales of minutes and never exactly repeating itself. And the Sun'smagnetic network continually rearranges itself in response to the birth and decay of supergranules, with a time scale of 1 or 2 days. Moreover, the supergranulation undergoes oscillations and supports waves with periods of 6 to 9 days, which apparently make the supergranulation rotate faster around the Sun than the magnetic features. *See* SUN.

Kenneth R. Lang

Supergravity A theory that attempts to unify gravitation with the other fundamental interactions. The first, and only, completely successful unified theory was constructed by James Clerk Maxwell, in which the up-to-then unrelated electric and magnetic phenomena were unified in his electrodynamics. *See* FUNDAMENTAL INTERACTIONS; MAXWELL'S EQUATIONS.

Electroweak theory. The second stage of unification concerns the unification of electromagnetic and weak interactions, using Maxwell's theory as a guide. This was accomplished making use of the nonabelian gauge theories invented by C. N. Yang and R. L. Mills, and of spontaneous symmetry breaking. The symmetry of Maxwell's theory is very similar to spatial rotations about an axis, rotating the vector potentials while leaving the electric and magnetic fields unchanged. It is a local invariance because the rotations about a fixed axis can be made by different amounts at different points in space-time. Thus, Maxwell's theory is invariant under a one-parameter group of transformations U(1). In Yang-Mills theory this local invariance was generalized to theories with larger symmetry groups such as the three-dimensional rotation group $SO(3) \simeq SU(2)$ which has three parameters. The number of parameters of the local symmetry (gauge) group is also equal to the number of 4-vector potentials in the gauge theory based on that group. A detailed analysis of weak and electromagnetic forces shows that their description requires four 4-vector potentials (gauge fields), so that the gauge group must be a four-parameter group. In fact, it is the product $SU(2) \cdot U(1)$. *See* ELECTROWEAK INTERACTION; GAUGE THEORY; GROUP THEORY; SYMMETRY BREAKING.

Grand unified theories. In the third stage of unification, electroweak and strong forces are regarded as different components of a more general force which mediates the interactions of particles in a grand unified model. Strong forces are responsible for the interactions of hadrons and for keeping quarks confined inside hadrons. They are described by eight massless 4-vector potentials (gluons), the corresponding eight-parameter group being SU(3). This local symmetry is called color, and the corresponding theory quantum chromodynamics (QCD). The combination $SU(3) \cdot SU(2) \cdot SU(1)$ has strong experimental support, and has come to be known as the standard model. Thus the gauge group of any grand unified model must include the standard model as a subsymmetry. The most dramatic prediction of these theories is the decay of protons. *See* GLUONS; GRAND UNIFICATION THEORIES; PROTON; QUANTUM CHROMODYNAMICS; QUARKS; STANDARD MODEL.

Supersymmetry and supergravity theories. A still higher and more ambitious stage of unification deals with the possibility of combining grand unified and gravity theories into a superunified theory, also known as supergravity. To achieve this, use is made of the dual role played by local internal symmetry groups. On the one hand, they describe the behavior of forces. On the other hand, they classify the elementary particles (fields) of the theory into multiplets: spin-zero fields in one multiplet, spin-1/2 fields in another multiplet, and so forth, but never fermions and bosons in a single irreducible multiplet of internal symmetry. This last restriction used to be a major obstacle on the way to superunification. This is because, of all the elementary particles, only the quanta of the gravitational field (gravitons) have spin 2, so that a multiplet of elementary particles including the graviton must of necessity involve particles of different spin. But then by an internal symmetry transformation, which is by definition distinct from space-time (Lorentz) transformations, it is possible to "rotate" particles of different spin into one another, thus altering their space-time transformation properties. This apparent paradox can be circumvented if both the internal symmetry and Lorentz transformations are part of a larger (supersymmetric) transformation group which also includes the spin-changing transformations. The irreducible multiplets of such supergroups naturally contain both fermions and bosons. This is how supersymmetry makes its appearance in supergravity theories. *See* GRAVITON; LORENTZ TRANSFORMATIONS; RELATIVITY; SUPERSYMMETRY; SYMMETRY LAWS (PHYSICS).

Effective theory. If supergravity models are regarded not as fundamental theories but as effective theories describing the low-energy behavior of superstring theories, it is possible to make a strong case for their usefulness. In that case, since supergravity is no longer a fundamental theory, it is no longer crucial that supergravity satisfy very stringent physical requirements such as renormalizability. In its role as an effective theory, supergravity has been used in a number of problems in particle physics. *See* ELEMENTARY PARTICLE; QUANTUM FIELD THEORY; SUPERSTRING THEORY.

Freydoon Mansouri

Superheater A component of a steam-generating unit in which steam, after it has left the boiler drum, is heated above its saturation temperature. The amount of superheat added to the steam is influenced by the location, arrangement, and amount of superheater surface installed, as well as the rating of the boiler. The superheater may consist of one or more stages of tube banks arranged to effectively transfer heat from the products of combustion. *See* STEAM-GENERATING UNIT.

George W. Kessler

Superheavy elements A term that originally referred to those elements with spherically shaped nuclei that would reside on the "island of stability." More recently, the term "superheavy elements" has been widened to include heavy nuclei whose existence is due to the nuclear shell effect. In this widened definition, therefore, elements with atomic number $Z \geq 104$ are generally considered to be superheavy (which also coincides with the beginning of the transactinide elements).

The nuclear shell model is one of the fundamental theoretical models used to understand the internal structure of the atomic nucleus. This model successfully describes the enhanced stability or "magicity" of certain combinations of neutrons or protons. The nuclear shell model currently explains that those nuclides with proton number Z or neutron number N equal to 2, 8, 20, 28, 50, or 82, or neutrons with $N = 126$, are more stable than nuclei with even just one more or less nucleon. *See* NUCLEAR STRUCTURE.

Furthermore, this model can be used to predict the properties of unknown elements and nuclides. A key core extrapolation of the shell model is the prediction of the next doubly magic nucleus. In doubly magic nuclei, both neutrons and protons completely fill orbitals, like ^{208}Pb with $Z = 82$ and $N = 126$, the heaviest known doubly magic nuclide. Indeed, since the mid-1960s, the existence of a region of spherical, long-lived nuclei with extra stability near $Z = 114$ and $N = 184$ has been predicted. This region has been called variously the island of stability or magic island. *See* MAGIC NUMBERS.

Search motivations. This prediction of long-lived nuclides on the island of stability initiated intensive searches for these elements in many different kinds of experiments. The chance to critically test such a fundamental nuclear physics theory as the shell model, together with the prospects of discovering new superheavy elements with unmeasured nuclear properties and with unknown but certainly fascinating chemistry, has captured the imagination of many scientists working in this field.

Production of new elements. A variety of experimental techniques have been used to make new chemical elements, including heavy-ion transfer reactions, cold- or hot-fusion evaporation reactions, neutron captures, and light-ion charged particle-induced reactions. In a heavy-ion transfer reaction, typically, a heavy nucleus hits a target nucleus and transfers one or more (generally a few) nucleons (protons or neutrons). A light-ion charged particle-induced reaction involves a light nucleus (like a deuteron, proton, or ^4He nucleus) hitting a target nucleus and transferring a particle with charge such as a proton or alpha particle. Neutron capture reactions involve a neutron hitting a target nucleus and being absorbed into that nucleus. All of these techniques involve transmuting one kind of nucleus into another. *See* NUCLEAR REACTION.

Cold- and hot-fusion reactions. Modern "nuclear alchemists" working on the synthesis of superheavy elements "transmute" one element into another using particle accelerators and smashing a beam of one element into a target of another element to produce the desired element. The types of nuclear reactions that have been successfully used to produce new elements in the last decade or so are cold-fusion reactions and hot-fusion reactions. *See* PARTICLE ACCELERATOR.

Both of these types of reactions utilize doubly magic nuclei, as either target or projectile to attempt to increase the stability of the compound nucleus. Cold-fusion reactions have been successful in producing elements 104–112, and hot-fusion reactions have been used in experiments for elements 113–116 and 118.

Separation and detection techniques. Often the produced isotope decays rapidly and is produced inefficiently, so sophisticated separation and detection techniques are required for studying the reaction products. Typical experiments involve the use of post-target separators to remove the unused beam particles and unwanted reaction products from other types of reactions occurring simultaneously with the fusion-evaporation reaction of interest. Position-sensitive solid-state semiconductor detectors measure the energy and decay mode of the product of interest, called the evaporation residue. *See* PARTICLE DETECTOR.

Mark A. Stoyer

Superionic conductivity The electrical conductivity exhibited by a small group of solids with high ionic conductivity and negligible electronic conductivity. In general, ionic conductivity is due to the motion of ions, whereas the electronic conductivity results from the flow of electrons. For superionic conductors, also called fast ion conductors or solid electrolytes, the specific conductivity (σ) is usually within the range from about 10^{-3} to 10 siemens per centimeter. These values are very high for a crystalline ionic solid, but are still lower than many electronic conductors such as metals, which have typical values ranging from 10 to 10^5 S cm^{-1}. *See* CONDUCTANCE; CONDUCTION (ELECTRICITY); ELECTRICAL CONDUCTIVITY OF METALS; ELECTROLYTIC CONDUCTANCE; SOLID-STATE CHEMISTRY.

Since ionic conductivity increases with increasing temperature, many superionic conductors (such as ZrO_2) exhibit high ionic conductivity only at temperatures substantially higher than room temperature. High temperatures provide the thermal energy needed to overcome the activation energy for ion hopping (from site to site) and increase the number of defect sites needed for ion migration. For some crystalline solids (such as Li_2SO_4 and AgI), high temperatures lead to polymorphic phase transitions, which cause an abrupt increase in the ionic conductivity. *See* IONIC CRYSTALS; POLYMORPHISM (CRYSTALLOGRAPHY).

While ionic conductivity is a common property of liquid electrolyte solutions or molten salts, a typical ionic solid (such as NaCl) has negligible ionic conductivity, often below 10^{-9} S cm^{-1}. Special structural features are required in order for a crystalline solid to have superionic conductivity. One common feature is that such solids have a rigid framework with open channels or layers along which ions can migrate. Other factors that contribute to the high ionic conductivity include small ion size, low ion charge, low coordination number for mobile ions, high concentration of mobile ions, a large number of vacant sites, and high polarizability of anionic frameworks. *See* CRYSTAL DEFECTS; CRYSTAL STRUCTURE; ION.

Superionic conductors can be classified according to the type of mobile ions, the dimensionality of conduction pathways, or the structure type of the nonmobile portion of the crystal structure.

Common cationic conductors usually contain ions such as Ag^+, Na^+, Li^+, or H^+. Silver iodide and its derivatives are among the earliest studied ionic conductors. Li^+ superionic conductors are highly desirable for all-solid-state lithium batteries.

Common anionic conductors are usually oxide (O^{2-}) or fluoride (F^-) conductors, such as stabilized zirconias and PbF_2, which require high temperatures for fast ion conductivity.

In addition to crystalline inorganic solids, superionic conductivity can also be found in amorphous inorganic materials. In general, glassy electrolytes based on sulfides and halides have much higher conductivity than oxides at ambient temperatures. Many Li^+ and Ag^+ conducting glasses are known. *See* AMORPHOUS SOLID.

Superionic conductivity is also known in polymer-based solids. These polymer electrolytes have two general types: polymer-salt complexes and polyelectrolytes.

Superionic conductors are an important group of materials that have large-scale technological applications in areas such as energy storage and generation (electrolyzers, batteries, and fuel cells), gas sensors, and electrochromic devices. They are essential for the development of all-solid-state electrochemical devices, which have many advantages over those based on liquid electrolytes including ease of miniaturization and high-temperature stability. Oxygen gas sensors based on oxide conductors are being widely used to monitor automobile exhaust gases. Oxide conductors are also being studied for the construction of solid-oxide fuel cells. Sulfonated polymers are under investigation as proton conductors in polymer electrolyte membrane fuel cells. *See* BATTERY; ELECTROCHROMIC DEVICES; ENERGY STORAGE; FUEL CELL; MICROSENSOR.

Xianhui Bu; Pingyun Feng

Superluminal motion Proper motion of an astronomical object apparently exceeding the velocity of light, *c*. This phenomenon is relatively common in the nuclei of quasars, many of which exhibit systematic changes in images of their radio-frequency emission over periods of months to years. In some cases, features in the image appear to separate at a speed inferred to be more than 10 times the speed of light, given the great distance of the quasars from Earth.

Superluminal motion was one of the most exciting discoveries to emerge from a technique in radio astronomy first developed in the late 1960s and called very long baseline interferometry (VLBI). This method involves the tape recording of radio signals from large antennas at up to 10–15 locations across the Earth, and the combination of these signals in a computer to form a radio image of the quasar at extremely high resolution (less than 0.001 arcsecond). *See* QUASAR; RADIO ASTRONOMY; RADIO TELESCOPE.

Superluminal motion is seen mostly in quasars but also in some other active galactic nuclei. This rapid motion is confined to within a few tens of parsecs of the nucleus, whose power source is believed to be a massive black hole. At least 30 examples of superluminal motion are now known. Most show apparent speeds less than 10*c*, but examples of speeds above 20*c* have been found. A very few objects in the Milky Way Galaxy also show superluminal motion. An example is GRS 1915 + 105, a relativistic jet source, which emits strongly in the x-ray as well as in the radio spectrum. Because the object is within the Milky Way Galaxy, its apparent speed of 1.25*c* is detectable in less than a day. *See* GALAXY, EXTERNAL; X-RAY ASTRONOMY.

Announcement of the discovery of superluminal motion in 1972 caused widespread concern because of the apparent violation of Albert Einstein's special theory of relativity, even though the basic explanation still favored now was in fact predicted some years before the announcement. Many explanations were proposed (besides Einstein's theory being incorrect), but only the relativistic jet model has stood the test of time. Superluminal motion is explained in this model as primarily a geometric effect. A feature (perhaps a cloud of relativistic plasma) moves away from the nucleus of the quasar at high (relativistic) speed (but less than *c*) at a small angle to the line of sight to the Earth. Radio waves from the moving feature arrive only slightly later than waves from the nucleus, whereas the feature took a much longer time to reach its current position. The motion appears superluminal because the speed is calculated using this much shorter time interval. As the speed approaches *c* and the angle to the line of sight decreases, the apparent speed can be arbitrarily large. In this explanation, no material speeds faster than *c* are required, so there is no conflict with special relativity. *See* RELATIVITY.

Stephen C. Unwin

Supermassive stars Hypothetical objects with masses exceeding 60 solar masses, the mass of the largest known ordinary stars (1 solar mass equals 4.4×10^{30} lbm or 2×10^{30} kg). The term is most often used in connection with objects larger than 10^4 solar masses that might be the energy source in quasars and active galaxies. These objects probably do not exist. However, their nonexistence is one of the major assumptions that makes the case for giant black holes, rather than supermassive stars, being the central engines of quasars. *See* BLACK HOLE; QUASAR; STAR.

Harry L. Shipman

Supermultiplet A generalization of the concept of a multiplet. A multiplet is a set of quantum-mechanical states, each of which has the same value of some fundamental quantum number and differs from the other members of the set by another quantum number which takes values from a range of numbers dictated by the fundamental quantum number. The number of states in the set is called the multiplicity or dimension of the multiplet. The concept was originally introduced to describe the set of states in a nonrelativistic quantum-mechanical system with the same value of the orbital angular momentum, L, and different values of the projection of the angular momentum on an axis, M. The values that M can take are the integers between $-L$ and L, $2L + 1$ in all. This is the dimension of the multiplet. If the hamiltonian operator describing the system is rotationally invariant, all states of the multiplet have the same energy. A supermultiplet is a generalization of the concept of multiplet to the case when there are several quantum numbers that describe the quantum-mechanical states. *See* ANGULAR MOMENTUM; SYMMETRY LAWS (PHYSICS).

Both concepts, multiplet and supermultiplet, acquire a precise mathematical meaning by the use of the theory of group transformations. A multiplet is an irreducible representation of a group, G. The quantum number called fundamental in the paragraph above labels the representation of the group. The other quantum number labels the representation of a subgroup G' of G. For angular momentum, the group G is the rotation group, called special orthogonal group in three dimensions, SO(3), and its subgroup G' is the group of special orthogonal transformations in two dimensions, SO(2). A supermultiplet is a generalization to the case in which the group G is not a group of rank one but has larger rank. A group of rank one has only one quantum number to label its representations. The concept of a multiplet or supermultiplet is particularly useful in the classification of states of physical systems. *See* GROUP THEORY; QUANTUM MECHANICS; QUANTUM NUMBERS.

The term supermultiplet was first used by E. P. Wigner in 1932 in order to classify the quantum-mechanical states of light atomic nuclei. The constituents of these are protons, p, and neutrons, n. Each proton and neutron has an intrinsic spin, S, of $^1/_2$ in units of $\hbar$, which is Planck's constant divided by 2π. The projection of the intrinsic spin on an axis, S_z, is then $S_z = {}^1/_2$ or $-{}^1/_2$ (spin up or down). In addition to having the same spin, the proton and neutron have essentially the same mass but differ in that the proton is charged whereas the neutron is not. They can thus be regarded as different charge states of the same particle, a nucleon. The distinction can be made formal by introducing a quantum number called isotopic spin, T, which has the value $^1/_2$. The two charge orientations, T_z, are taken to be $^1/_2$ for the proton and $-{}^1/_2$ for the neutron. There are thus four constituents of nuclei, protons and neutrons with spin up and down, that is, $p\uparrow$, $p\downarrow$, $n\uparrow$, and $n\downarrow$. The set of transformations among these constituents forms a group called SU(4), the special unitary group in four dimensions. This is the group G for Wigner's theory. The representations of SU(4), that is, Wigner supermultiplets, are characterized by three quantum numbers (λ_1, λ_2, λ_3) with $\lambda_1 \geq \lambda_2 \geq \lambda_3$. *See* I-SPIN; NUCLEAR STRUCTURE. Francesco Iachello

Supernova The catastrophic, explosive death of a star, accompanied by the sudden, transient brightening of the star to an optical luminosity comparable to that of an entire galaxy.

A supernova shines typically for several weeks to several months with a luminosity between 2×10^8 and 5×10^9 times that of the Sun, then gradually fades away. Each explosion ejects from one to several tens of solar masses at speeds ranging from thousands to tens of thousands of kilometers per second. The total kinetic energy, 10^{44} joules (2.5×10^{28} megatons of high explosive), is about 100 times the total light output, making supernovae some of the highest-energy explosions in the universe. Unlike its fainter relative, the nova, a supernova does not recur for the same object. *See* NOVA.

Supernovas may be grouped according to either their observational characteristics or their explosion mechanism. Basically, type I supernovae have no hydrogen in their spectrum; type II supernovae do. Two mechanisms are involved: thermonuclear explosion in white dwarfs and gravitational collapse in massive stars. Type I supernovae of different subclasses can occur by either mechanism, but it is thought that most type II supernovae are powered by gravitational collapse.

During the last thousand years, there have been approximately seven supernovae visible to the unaided eye, in 1006, 1054, 1181, 1408, 1572, 1604, and 1987. SN 1006 may have been as bright as the quarter moon. The first six of these occurred in the Earth's vicinity of the Milky Way Galaxy. But the last, and only, naked-eye supernova since the invention of modern instrumentation occurred in the Large Magellanic Cloud, a small satellite galaxy of the Milky Way about 160,000 light-years away. Supernovae are discovered in other galaxies at a rate of about 150 per year. Most supernovae in the Milky Way Galaxy are obscured by dust, but various arguments suggest that about two type II supernovae per century and one type Ia every other century occur in the Milky Way Galaxy. *See* MAGELLANIC CLOUDS; MILKY WAY GALAXY.

Type Ia supernovae. Type Ia supernovae may be regarded as nature's largest thermonuclear bombs. They occur when an accreting white dwarf, composed of carbon and oxygen, grows to a mass 1.38 times that of the Sun, almost the critical mass that can be supported by electron degeneracy pressure, and ignites carbon fusion near its center. Ignition occurs when carbon fusion at the center releases energy faster than neutrinos can carry it away. Because the pressure is insensitive to the temperature, a nuclear runaway occurs. Fusion releases energy, which raises the temperature, which makes fusion go faster, but the gas cannot expand and cool. The nuclear runaway spreads in about 1 second through the star. The energy released by this nuclear burning is more than enough to completely blow the white dwarf apart with high velocity. Nothing remains—no neutron star, no black hole, and no burst of neutrino emission. *See* BINARY STAR; THERMONUCLEAR REACTION.

Type II supernovae. A typical type II supernova results from a star somewhat over 8 solar masses, on the main sequence, that spends its last years as a red supergiant burning progressively heavier fuels in its center. The radius of the star, after hydrogen has burned and the star is part way through helium burning, is roughly 500 solar radii, and its luminosity is already about 100,000 times that of the Sun. Each burning stage is shorter than the previous one. The last stage turns silicon and sulfur into a ball of roughly 1.4 solar masses of iron. Once iron has been produced, no more nuclear energy is available. *See* SUPERGIANT STAR.

A combination of instabilities now leads to the implosion of the iron core to a neutron star. When the density at the center reaches several times that of the atomic nucleus, the collapse halts and briefly springs back owing to the short-range repulsive component of the nuclear force. But the energy of this bounce is soon dissipated, and a hot young neutron star remains which, over the next few seconds, radiates away its heat and binding energy as neutrinos. *See* NEUTRINO; STRONG NUCLEAR INTERACTIONS.

The energy output in these neutrinos is enormous, about 3×10^{46} joules or 15% of the rest mass of the Sun converted to energy; rivaling the luminosity of the rest of the observable universe in light. A small fraction of these neutrinos, about 0.3%, are absorbed in reactions with neutrons and protons in the regions just outside the neutron star and deposit their energy. Even this small amount of energy is much greater than the gravitational binding of the remaining part of the star external to the newly formed neutron star. A bubble of radiation is inflated by the neutrino energy deposition, the outer boundary of which expands supersonically, driving a shock wave through the rest of the star and ejecting it with high velocity. The main energy of the explosion, though, is carried away as neutrinos. This general picture was confirmed when a neutrino burst of the predicted energy and duration was detected February 23, 1987, from the Large Magellanic Cloud in conjunction with SN 1987A. *See* NEUTRINO ASTRONOMY; SHOCK WAVE.

Nucleosynthesis. Supernovae are major element factories, responsible for producing most of the elements in nature heavier than nitrogen. The largest yields are of the more abundant elements, including oxygen, silicon, magnesium, neon, iron, and a portion of carbon, but dozens of other elements are also made. *See* NUCLEOSYNTHESIS.

Type Ia cosmological applications. Because of their brightness and the regularity of their light curves, type Ia supernovae have long been used as standard candles to survey cosmological distances. More recently it has been realized that the relatively small variation that occurs in the peak brilliance of such supernovae may be correlated with their decline rates. Use of this so-called Phillips relation allows even greater precision in distance determination. Using type Ia supernovae in this fashion reveals a surprising result. Two independent analyses show that the expansion rate of the universe is not slowing as might be expected long after the big bang, but is actually accelerating. The pull of gravity can only cause deceleration, so the acceleration is attributed to an invisible form of dark energy that enters into the cosmological equations as a repulsive term. *See* ASTROPHYSICS, HIGH-ENERGY; COSMOLOGY; STAR; STELLAR EVOLUTION; UNIVERSE; VARIABLE STAR. Stan E. Woosley

Superoxide chemistry A branch of chemistry that deals with the reactivity of the superoxide ion (O_2^-), a one-electron (e^-) adduct of molecular oxygen (dioxygen; O_2) formed by the combination of O_2 and e^-. Because 1–15% of the O_2 that is respired by mammals goes through the O_2^- oxidation state, the biochemistry and reaction chemistry of the species are important to those concerned with oxygen toxicity, carcinogenesis, and aging. Although the name superoxide has prompted many to assume an exceptional degree of reactivity for O_2^-, the use of the prefix in fact was chosen to indicate stoichiometry. Superoxide was the name given in 1934 to the newly synthesized potassium salt (KO_2) to differentiate its two-oxygens-per-metal stoichiometry from that of most other metal–oxygen compounds (NA_2O, Na_2O_2, NaOH, Fe_2O_3).

Ionic salts of superoxide (yellow-to-orange solids), which form from the reaction of dioxygen with metals such as potassium, rubidium, or cesium, are paramagnetic, with one unpaired electron per two oxygen atoms.

In 1969, by means of electron spin resonance (ESR) spectroscopy, superoxide ion was detected as a respiratory intermediate, and metalloproteins were discovered that catalyze the disproportionation of superoxide, that is, superoxide dismutases (SODs), as shown in the reaction below.

$$2O_2^- + 2H^+ \xrightarrow{SOD} O_2 + H_2O_2$$

The biological function of superoxide dismutases is believed to be the protection of living cells against the toxic effects of superoxide. The possibility that superoxide might be an important intermediate in aerobic life provided an impetus to the study of superoxide reactivity.

The most general and universal property of O_2^- is its tendency to act as a strong Brønsted base. Its strong proton affinity manifests itself in any media. Another characteristic of O_2^- is its ability to act as a moderate one-electron reducing agent. *See* ACID AND BASE; OXIDATION-REDUCTION.

In general, superoxide ion chemistry does not appear to be sufficiently robust to make superoxide ion a toxin. However, it can interact with protons, halogenated carbons, and carbonyl compounds to yield peroxy radicals that are toxic. *See* OXYGEN TOXICITY; REACTIVE INTERMEDIATES.

Superoxide does not appear to have exceptional reactivity. Nevertheless, superoxide will continue to be an interesting species for study because of the multiplicity of its chemical reactions and because of its importance as an intermediate in reactions that involve dioxygen and hydrogen peroxide. *See* BIOINORGANIC CHEMISTRY; OXYGEN. Donald T. Sawyer

Superplastic forming A process for shaping superplastic materials, a unique class of crystalline materials that exhibit exceptionally high tensile ductility. Superplastic materials may be stretched in tension to elongations typically in excess of 200% and more commonly in the range of 400–2000%. There are rare reports of higher tensile elongations reaching as much as 8000%. The high ductility is obtained only for superplastic materials and requires both the temperature and rate of deformation (strain rate) to be within a limited range. The temperature and strain rate required depend on the specific material. A variety of forming processes can be used to shape these materials; most of the processes involve the use of gas pressure to induce the deformation under isothermal conditions at the suitable elevated temperature. The tools and dies used, as well as the superplastic material, are usually heated to the forming temperature. The forming capability and complexity of configurations producible by the processing methods of superplastic forming greatly exceed those possible with conventional sheet forming methods, in which the materials typically exhibit 10–50% tensile elongation. *See* SUPERPLASTICITY.

There are a number of commercial applications of superplastic forming and combined superplastic forming and diffusion bonding, including aerospace, architectural, and ground transportation uses. Examples are turbo-fan-engine cooling-duct components, external window frames in the space shuttle, and architectural siding for buildings. *See* METAL FORMING. C. Howard Hamilton

Superplasticity The unusual ability of some metals and alloys to elongate uniformly thousands of percent at elevated temperatures, much like hot polymers or glasses. Under normal creep conditions, conventional alloys do not stretch uniformly, but form a necked-down region and then fracture after elongations of only 100% or less. The most important requirements for obtaining superplastic behavior include a very small metal grain size, a well-rounded (equiaxed) grain shape, a deformation temperature greater than one-half the melting point, and a slow deformation rate. *See* ALLOY; CREEP (MATERIALS); EUTECTICS.

Superplasticity is important to technology primarily because large amounts of deformation can be produced under low loads. Thus, conventional metal-shaping processes (for example, rolling, forging, and extrusion) can be conducted with smaller, and cheaper equipment. Nonconventional forming methods can also be used; for instance, vacuum-forming techniques, borrowed from the plastics industry, have been applied to sheet metal to form car panels, refrigerator door linings, and TV chassis parts. *See* METAL FORMING; PLASTICITY. Erwin E. Underwood

Superposition principle The principle, obeyed by many equations describing physical phenomena, that a linear combination of the solutions of the equation is also a solution.

An effect is proportional to a cause in a variety of phenomena encountered at the level of fundamental physical laws as well as in practical applications. When this is true, equations which describe such a phenomenon are known as linear, and their solutions obey the superposition principle. Thus, when $f, g, h, \cdots$, solve the linear equation, then s ($s = \alpha f + \beta g + \gamma h + \cdots$, where $\alpha, \beta, \gamma, \cdots$, are coefficients) also satisfies the same equation. *See* LINEAR ALGEBRA; LINEARITY.

For example, an electric field is proportional to the charge that generates it. Consequently, an electric force caused by a collection of charges is given by a superposition—a vector sum—of the forces caused by the individual charges. The same is true for the magnetic field and its cause—electric currents. Each of these facts is connected with the linearity of Maxwell's equations, which describe electricity and magnetism. *See* ELECTRIC FIELD; MAXWELL'S EQUATIONS.

The superposition principle is important both because it simplifies finding solutions to complicated linear problems (they can be decomposed into sums of solutions of simpler problems) and because many of the fundamental laws of physics are linear. Quantum mechanics is an especially important example of a fundamental theory in which the superposition principle is valid and of profound significance. This property has proved most useful in studying implications of quantum theory, but it is also a source of the key conundrum associated with its interpretation.

Its effects are best illustrated in the double-slit superposition experiment, in which the wave function representing a quantum object such as a photon or electron can propagate toward a detector plate through two openings (slits). As a consequence of the superposition principle, the wave will be a sum of two wave functions, each radiating from its respective slit. These two waves interfere with each other, creating a pattern of peaks and troughs, as would the waves propagating on the surface of water in an analogous experimental setting. However, while this pattern can be easily understood for the normal (for example, water or sound) waves resulting from the collective motion of vast numbers of atoms, it is harder to understand its origin in quantum mechanics, where the wave describes an individual quantum, which can be detected, as a single particle, in just one spot along the detector (for example, photographic) plate. The interference pattern will eventually emerge as a result of many such individual quanta, each of which apparently exhibits both wave (interference-pattern) and particle (one-by-one detection) characteristics. This ambivalent nature of quantum phenomena is known as the wave-particle duality. *See* INTERFERENCE OF WAVES; QUANTUM MECHANICS. Wojciech Hubert Zurek

Superposition theorem (electric networks) When there is more than one source (either voltage or current) in a linear electrical network, it is convenient to compute at any element of the network the response of voltage or of current that results from one source acting alone while all others are deactivated, and then the response resulting from another source alone, while all other sources are deactivated, and so on for all sources, and finally to compute the total response to all sources acting together by adding these individual responses. The superposition theorem states that this procedure is permissible.

Thus, if a load of constant resistance is supplied with electrical energy from a linear network containing two batteries, two generators, or one battery and one generator, it would be correct to find the current that would be supplied to the load by one source (the other being reduced to zero), then to find the current that would be supplied to the load by the second source (the first source now being reduced to zero), and finally to add the two currents so computed to find the total current that would be produced in the load by the two sources acting simultaneously.

By means of the principle of superposition, effects are added instead of causes. This principle seems so intuitively valid that there is far greater danger of applying superposition where it is incorrect than of failing to apply it where it is correct. It must be recognized that for superposition to be correct the relation between cause and effect must be linear. Hugh H. Skilling

Supersaturation A solution is at the saturation point when dissolved solute in its crystallizes from it at the same rate at which it dissolves. Under prescribed experimental conditions of temperature and pressure, a solution can contain at saturation only one fixed amount of dissolved solute. However, it is possible to prepare relatively stable solutions which contain a quantity of a dissolved solute greater than that of the saturation value provided solute phase is absent. Such solutions are said to be supersaturated. They can be prepared by changing the experimental conditions of a system so that greater solubility is obtained, perhaps by heating the solution, and then carefully returning the system to or near its original state. The addition of solute phase will immediately relieve supersaturation. Solutions in which there is no spontaneous formation of solute phase for extended periods of time are said to be metastable. There is no sharp line of demarcation between an unstable and metastable solution. The process whereby initial aggregates within a supersaturated solution develop spontaneously into particles of new stable phase is known as nucleation. The greater the degree of supersaturation, the greater will be the number of nuclei formed. *See* NUCLEATION; PHASE EQUILIBRIUM; PRECIPITATION (CHEMISTRY). Louis Gordon; Royce W. Murray

Supersonic diffuser A passive compressor (or shaped duct) in which gas enters at a velocity greater than the speed of sound, is decelerated in a contracting section, and reaches sonic speed at a throat.

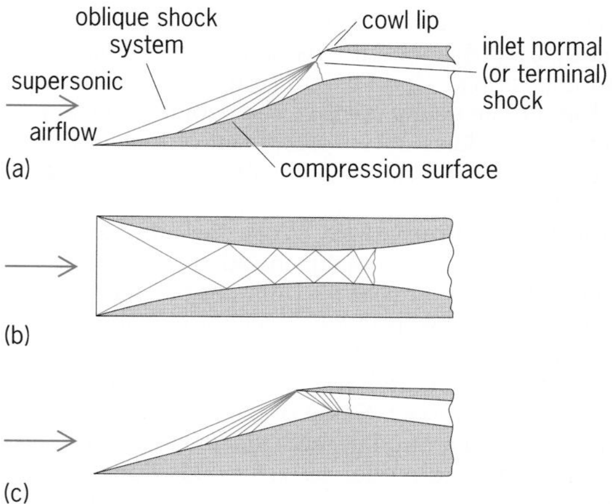

Basic supersonic compression systems. (*a*) External compression. (*b*) Internal compression. (*c*) Combination of external and internal compression.

Supersonic compression systems can be categorized to three basic types (see illustration). External-compression inlets have the supersonic diffusion taking place at or ahead of the cowl lip (or throat station) and generally employ one or more oblique waves ahead of the normal shock. Internal-compression inlets accomplish supersonic diffusion internally downstream of the cowl lip. Deceleration of the flow is produced by a number of weak reflecting waves in a gradually convergent channel. The third system is a combination of external and internal compression and appears to represent an effective compromise. James F. Connors; Leonard J. Obery

Supersonic flight Relative motion of a solid body and a gas at a velocity greater than that of sound propagation under the same conditions. The general characteristics of supersonic flight can be understood by considering the laws of propagation of a disturbance or pressure impulse, in a compressible fluid.

If the fluid is at rest, the pressure impulse propagates uniformly with the velocity of sound in all directions, the effect always acting along an ever-increasing spherical surface. If, however, the source of the impulse is placed in a uniform stream, the impulse will be carried by the stream simultaneously with its propagation at sonic velocity relative to the stream. Hence the resulting propagation is faster in the direction of the stream and slower against the stream. If the velocity of the stream past the source of disturbance is supersonic, the effect of the impulse is restricted to a cone whose vertex is the source of the impulse and whose vertex angle decreases from 90° (corresponding to Mach number equal to 1) to smaller and smaller values as the Mach number of the stream increases (see illustration). If the source of the pressure impulse travels through the air at rest, the conditions are analogous. *See* MACH NUMBER.

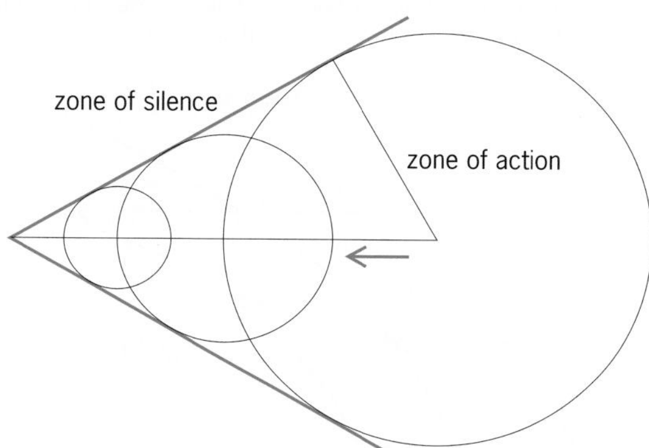

Generation of Mach wave by body at supersonic velocity; zones of action and silence are separated.

Consider the supersonic motion of a wing moving into air at rest. Because signals cannot propagate ahead of the wing, the presence of the wing has no effect on the undisturbed air until the wing passes through it. Hence there must be an abrupt change in the properties of the undisturbed air as it begins to flow over the wing. This abrupt change takes place in a shock wave which is attached to the leading edge of the wing, provided that the leading edge is sharp and the flight Mach number is sufficiently large. As the air passes through the shock wave, its pressure, temperature, and density are markedly increased.

Further aft of the leading edge, the pressure of the air is decreased as the air expands over the surface of the wing. Hence the pressure acting on the front part of the wing is higher than the ambient pressure, and the pressure acting on the rear part of the wing is lower than the ambient pressure. The pressure difference between front and rear parts produces a drag, even in the absence of skin friction and flow separation. The wing produces a system of compression and expansion waves which move with it. This phenomenon is similar to that of a speedboat moving with a velocity greater than the velocity of the surface waves. Because of this analogy, supersonic drag is called wave drag. It is peculiar to supersonic flight, and it may represent the major portion of the total drag of a body. *See* HYPERSONIC FLIGHT; SUBSONIC FLIGHT; TRANSONIC FLIGHT. John E. Scott, Jr.

Supersonic flow Fluid motion in which the Mach number M, defined as the speed of the fluid relative to the sonic speed in the same medium, is more than unity. It is, however, common to call the flow transonic when $0.8 < M < 1.4$, and hypersonic when $M > 5$. *See* MACH NUMBER.

Mach waves. A particle moving in a compressible medium, such as air, emits acoustic disturbances in the form of spherical waves. These waves propagate at the speed of sound ($M = 1$). If the particle moves at a supersonic speed, the generated waves cannot propagate upstream of the particle. The spherical waves are enveloped in a circular cone called the Mach cone. The generators of the Mach cone are called Mach lines or Mach waves.

Shock waves. When a fluid at a supersonic speed approaches an airfoil (or a high-pressure region), no information is communicated ahead of the airfoil, and the flow adjusts to the downstream conditions through a shock wave. Shock waves propagate faster than Mach waves, and the flow speed changes abruptly from supersonic to less supersonic or subsonic across the wave. Similarly, other properties change discontinuously across the wave. A Mach wave is a shock wave of minimum strength. A normal shock is a plane shock normal to the direction of flow, and an oblique shock is inclined at an angle to the direction of flow. The velocity upstream of a shock wave is always supersonic. Downstream of an oblique shock, the velocity may be subsonic resulting in a strong shock, or supersonic resulting in a weak shock. The downstream velocity component normal to any shock wave is always subsonic. There is no change in the tangential velocity component across the shock.

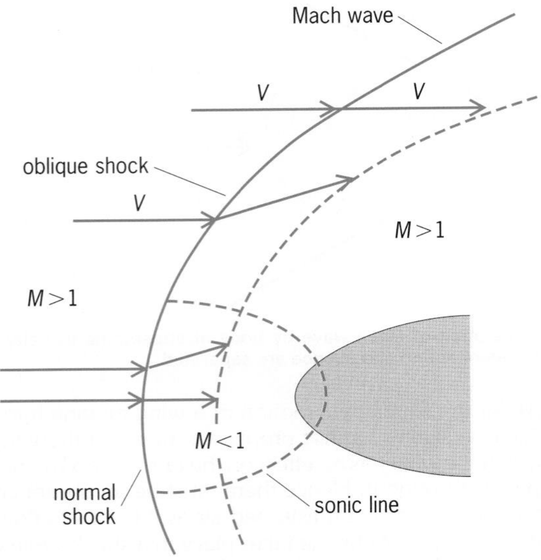

Typical normal shock, oblique shock, and Mach wave pattern in supersonic flow past a blunt body. *M* is the Mach number and *V* is the particle speed. The curved line parallel to normal and oblique shock waves indicates the end of the velocity vectors.

In a two-dimensional supersonic flow around a blunt body (see illustration), a normal shock is formed directly in front of the body, and extends around the body as a curved oblique shock. At a sufficient distance away, the flow field is unaffected by the presence of the body, and no discontinuity in velocity occurs. The shock then reduces to a Mach wave. *See* COMPRESSIBLE FLOW; FLUID FLOW; SUPERSONIC FLIGHT. Michel A. Saad

Superstring theory A proposal for a unified theory of all interactions, including gravity. At present, the strong, weak, and electromagnetic interactions are accounted for within the framework of the standard model. This model correctly describes experiments up to the highest energies performed so far, and gives a complete description of the elementary particles and their interactions down to distances of the order of 10^{-18} m. Nevertheless, it has serious limitations, and attempts to overcome them and to unify the forces of nature have been only partly successful. Moreover, these attempts have left standing fundamental difficulties in reconciling gravitation and the laws of quantum mechanics. Superstring theory represents an ambitious program to unify all of the interactions observed in nature, including gravitation, in a theory with no unexplained parameters. In other words, this theory, if successful, should be able to account for all of the particles observed in nature and their interactions. *See* ELEMENTARY PARTICLE; FUNDAMENTAL INTERACTIONS; STANDARD MODEL.

String concept. In string theory, the fundamental objects are not point particles, as in standard theories of elementary particles, but one-dimensional extended objects, the open and closed strings. In such a theory, what are usually called the elementary particles are simply particular quantum states of the string. In superstring theories, space-time is ten-dimensional (space is nine-dimensional). If such theories are to describe nature, six dimensions must be "curled up" or "compact." The main consequence of such extra dimensions is the existence of certain very massive particles. *See* SPACE-TIME.

The essential features of string theories can be understood by analogy with the strings of a musical instrument. Such strings vibrate at a characteristic frequency, as well as any integer multiple of that frequency. Each of these modes of vibration (so-called normal modes) can be excited by plucking or striking the string. In classical physics, the amplitudes of vibration of each mode can take on a continuum of values. If there were a string of atomic dimensions, subject to the laws of quantum mechanics, the energies of this quantum string could take on only discrete values, corresponding to particular quantum states. *See* QUANTUM MECHANICS; VIBRATION.

The strings of superstring theory are quite similar. The main difference is that they obey Einstein's principles of special relativity. As a result, since each quantum state has a particular energy, it has a definite mass. Thus, each state of the string behaves as a particle of definite mass. Because it is possible, in principle, to pump an arbitrarily large amount of energy into the string, the theory contains an infinity of different types of particles of arbitrarily large mass. The interactions of these particles are governed by the ways in which the strings themselves interact. To be consistent with the principles of relativity, a string can interact only by splitting into two strings or by joining together with another string to form a third string. As a result, the interactions of strings are nearly unique. This geometric picture of string interactions translates into a precise set of rules for calculating the interaction of individual string states, that is, particles. *See* RELATIVITY.

String theory landscape. While the presence of extra dimensions may seem troubling, the idea that space-time may have more than the familiar 3+1 dimensions dates back almost a century, and was taken seriously by Einstein and others. Some of these dimensions must be small, or "compactified." Superstring theory turns out to have many solutions in which several dimensions are compact. Some of these solutions look remarkably like the world around us, with the gauge bosons and matter fields of the standard model, as well as additional particles and interactions. It has come to be appreciated that the number of possible solutions is enormous. The set of possible solutions has been termed the landscape, by analogy to situations that exist in the study of proteins and certain condensed-matter systems. There is reason to believe that among these states there are many that look precisely like the world around us. Understanding this problem, and how the universe might have found itself in the observed state, is a subject of current research. It is possible that these studies may lead to predictions for phenomena at high-energy accelerators, such as the Large Hadron Collider at CERN, near Geneva, Switzerland. *See* GAUGE THEORY; PARTICLE ACCELERATOR. Michael Dine

Supersymmetry A conjectured enhanced symmetry of the laws of nature that would relate two fundamental observed classes of particles, bosons and fermions.

All particles can be classified as fermions, such as the electron and quarks, or bosons, such as the photon and graviton. A fundamental characteristic distinguishing these two classes is that they carry different quantum-mechanical spin. If the amount of spin of an elementary particle is measured in terms of the fundamental quantum unit of angular momentum—equal to Planck's constant divided by 2π—then bosons always have integer amounts of spin (that is, 0, 1, 2 . . .), while fermions have odd half-integer amounts of spin (that is, 1/2, 3/2, 5/2, . . .). *See* SPIN (QUANTUM MECHANICS).

There is seemingly a fundamental distinction between particles with differing amounts of spin. For example, bosons like to act collectively (Bose-Einstein statistics), producing such distinctive behavior as the laser, while, conversely, fermions obey the Pauli exclusion principle (and the Pauli-Dirac statistics), which disallows two identical fermions to be in the same state, and explains the stability of matter. Moreover, all the symmetries that are observed in the world relate different particles of the same spin. *See* BOSE-EINSTEIN STATISTICS; FERMI-DIRAC STATISTICS; QUANTUM STATISTICS; SYMMETRY LAWS (PHYSICS).

In contrast, supersymmetry would relate bosons and fermions. This would be a remarkable step forward in understanding the physical world. However, if supersymmetry were realized as an exact symmetry, the particles so related should have almost all their characteristics, such as mass and charge, preserved. Explicitly, any fermion of spin 1/2 should have a boson superpartner of spin 0, while any gauge boson of spin 1 should have a fermion superpartner of spin 1/2. This is apparently a disaster for the idea of supersymmetry since it predicts, for instance, that there should exist a spin-0 boson partner of the electron, the selectron, with electric charge and mass equal to that of the electron. Such a particle would be easy to detect and is certainly ruled out by very many experiments.

The crucial caveat to this negative result is the condition that supersymmetry be realized as an exact symmetry. A fundamental concept of modern physics is spontaneously broken symmetry. Physics displays many examples of symmetries that are exact symmetries of the fundamental equations describing a system, but not of their solutions. In particle physics the spontaneous breaking of a symmetry usually results in a difference in the masses of the particles related by the symmetry; the amount of breaking can be quantified by this mass difference. *See* SYMMETRY BREAKING.

If supersymmetry is broken by a large amount, then all the superpartners have masses much greater than the particles that are currently observed, and there is little hope of seeing evidence for supersymmetry. However, evidence that supersymmetry is broken by only a moderate amount comes from examination of the properties of the fundamental forces at high energy.

Of the four fundamental forces, the three excluding gravity are very similar in their basic formulation; they are all described by gauge theories, generalizations of the quantum theory of electromagnetism, and quantum electrodynamics (QED). The strength of electrical interaction between two electrons can be quantified in terms of a number, the coupling constant α_1. However, the quantity α_1 is actually not a constant, but depends on the energies at which the interaction strength is measured. In fact, the interaction strengths, α_1, α_2, and α_3, of the three forces (excluding gravity) all depend on energy, μ. The couplings $\alpha_{1,2,3}$ satisfy differential equations—renormalization group equations—that depend on the types of elementary particles that exist with mass at or below the energy scale μ and that are charged with respect to each of the three interactions. If the fundamental particles include not only the observed particles but also their superpartners, taken to have masses not greater than 1000 GeV heavier than their (observed) partners, then from the renormalization group equations, the couplings α_i are predicted to meet (unify) at a huge energy of 2×10^{16} GeV. In contrast, if either supersymmetry is not an underlying symmetry of the world, or it is very badly broken so that the superpartners are very massive, the couplings fail to unify at a single point. *See* FUNDAMENTAL INTERACTIONS; GAUGE THEORY; QUANTUM ELECTRODYNAMICS; RENORMALIZATION.

Although the unification of couplings is the most significant indication that supersymmetry is a new law of nature, there are a number of other hints in this same direction. By observing the large-scale motions of the galaxies, the average density of large volumes of the universe can be deduced, resulting in a value that is substantially greater than that directly observed in luminous matter such as stars and hot gas. Therefore, a substantial fraction of the mass of the universe must be composed of some form of nonluminous or dark matter. Remarkably, many attractive models of supersymmetry predict that the lightest of all the superpartners is a weakly interacting massive particle with just the right characteristics to be this dark matter. *See* COSMOLOGY; UNIVERSE; WEAKLY INTERACTING MASSIVE PARTICLE (WIMP).

John March-Russell

Supply chain management Beginning with the work of Ford W. Harris in 1915 on the economic order quantity model, many researchers developed a variety of mathematical models for minimizing the costs associated with holding inventories (raw materials, components, subassemblies, work in process, and finished goods) in industries and businesses. The subject dealing with these problems was initially called inventory control. These models were essentially single-decision-maker models involving one item. In those days, several reasons for holding sizable inventories were given, including economies of scale; uncertainties in demand, supply, delivery lead times, and prices; and a desire to hold buffer stocks as a cushion against unexpected swings in demand and to assure smooth production flow. *See* INVENTORY CONTROL.

Starting in the 1950s, Japanese manufacturers (in particular Toyota) initiated the just-in-time (JIT) philosophy to reduce work-in-process inventories to a minimum, and implemented it using a simple *Kanban* (which means card or ticket in Japanese) system to track the flow of in-process materials through the various operations. In the late 1960s, Toyota extended the JIT philosophy to reduce all inventories to a minimum by developing collaborative working relationships with its component suppliers and distributors with the aim of encouraging them to make and accept small and frequent JIT deliveries; providing for careful monitoring of quality and workflow; and ensuring that products were produced or received only as they were needed.

In today'sworld of rapid technological developments, frequent design changes, and shorter product cycles, carrying as little stock as possible is crucial. The more one relies on stock, the more difficult it will be to accommodate design changes. That's why the JIT philosophy is now being integrated into the overall business strategy worldwide, changing the nature of manufacturing and business dramatically. It has expanded beyond the walls of a factory or shop to include the capabilities, skills, and cooperation of its suppliers and the insights of its customers. This new expanded system is now referred to as the supply chain. Supply chain management comprises planning and processing orders; handling, transporting, and storing all materials purchased, processed, or distributed; and managing inventories in a harmonious, coordinated, and synchronized manner among all the players on the chain to build to order (to fulfill customer orders as they arise) rather than build to stock (to build up stock level to fulfill anticipated future demand).

As part of the collaboration with their suppliers, many companies are adopting the practice of vendor-managed inventories in which the company provides warehouse space to its suppliers for storing their components, to be delivered to the company as demand arises (demand pull basis).

Instead of ordering once daily, some companies order once in every planning period (maybe a shift or half-a-shift). For the purpose of this discussion, a day will be used as the planning period.

In this mode of operation, it is critical to maintain good databases on demand, production, quality, and inventory levels, and to develop Web-based interfaces containing relevant information to which all players on the supply chain have access.

The key to making this whole process run smoothly is accurate forecasting of the component demand each day (or whatever planning period is used).

Successful inventory management systems depend heavily on good demand forecasts. The purpose of forecasting is commonly misunderstood to be that of generating a single number. This misunderstanding is created because all existing demand forecasting methods output only an estimate of the expectation of future demand. So, these methods are useful only when changes in the probability distribution of demand can be captured by the value of a single parameter, the expectation. All these methods that forecast only the expected value seem very inadequate to capture all the dynamic changes occurring in the shapes of probability distributions of demand.

Another important factor today is the highly competitive environment and the rapid rate of technological change that is shortening product life cycles. Because of this constant change, multiperiod stochastic (probabilistic) inventory models based on a stable demand distribution do not seem to be appropriate for application. Single period models of the type discussed above, combined with frequent updating of the demand distribution based on the most recent data, offer the most practical value. *See* STOCHASTIC PROCESS.

Considering these arguments, it has been suggested that a better strategy is to approximate the probability distribution of demand by its empirical distribution obtained from past data, which in this context is called the discretized distribution of demand. This is the initial distribution at the time it is computed. This distribution is periodically updated using recent data. *See* OPERATIONS RESEARCH; PRODUCTION PLANNING. Katta G. Murty

Suppression (electricity) The process or technique of reducing electrical interference to acceptable levels or to situations having no adverse effect. Suppression techniques may be applied to the interference source, the intervening path, the victim or receptor, or any combination. Normal strategy for interference control is to first suppress the source, if possible, since it may disturb many victims.

For intentional transmitters, suppressing interference may include reducing or eliminating harmonic radiations, restricting the bandwidth, or restricting levels of unnecessary or excessive modulation sidebands. These are usually accomplished by radio frequency filters. *See* ELECTRIC FILTER.

For many devices involving incidental radiators, such as brush-type motors and fluorescent lights, interference suppression may require both filtering and shielding. Electrical filtering may take the form of transient surge suppressors, feed-through capacitors, electromagnetic interference (EMI) filters, ferrite absorbers, isolation transformers, or Faraday shielded isolation transformers. *See* ELECTRIC PROTECTIVE DEVICES.

Shielding to control radiation involves using metal boxes, cases, cabinet housings, or metalized plastic versions thereof. Since the interconnecting cables between equipment offer the greatest threat as an "antenna farm," the dominant suppression technique is to shield the cables, wires, or harnesses. For the best protection from electromagnetic interference, the cable shield should be designed as an extension of the box or equipment shield. Other forms of interference hardening of cables include twisting parallel wire pairs, multiple-layer shields, and absorbing jackets. Where possible, best electromagnetic compatibility performance is obtained by replacing the signal cables with fiber-optic links. *See* ELECTRICAL SHIELDING.

From a system point of view, the radiated emission propensity or pickup susceptibility of interconnected equipment is proportional to the maximum length or dimension and frequency until this length corresponds to an electrical length of about one-half wavelength. Thus, to help suppress electromagnetic interference, the corresponding frequency should not fall within the passband of victims or receptors. *See* ELECTRICAL INTERFERENCE; ELECTROMAGNETIC COMPATIBILITY. Donald R. J. White

Supramolecular chemistry A highly interdisciplinary field covering the chemical, physical, and biological features of complex chemical species held together and organized by means of intermolecular (noncovalent) bonding interactions. *See* CHEMICAL BONDING; INTERMOLECULAR FORCES.

When a substrate binds to an enzyme or a drug to its target, and when signals propagate between cells, highly selective interactions occur between the partners that control the processes. Supramolecular chemistry is concerned with the study of the basic features of these interactions and with their implementation in biological systems as well as in specially designed nonnatural ones. In addition to biochemistry, its roots extend into organic chemistry and the synthetic procedures for receptor construction, into coordination chemistry and metal ion-ligand complexes, and into physical chemistry and the experimental and theoretical studies of interactions. *See* BIOINORGANIC CHEMISTRY; ENZYME; LIGAND; PHYSICAL ORGANIC CHEMISTRY; PROTEIN.

The field started with the selective binding of alkali metal cations by natural as well as synthetic macrocyclic and macropolycyclic ligands, the crown ethers and cryptands. This led to the emergence of molecular recognition as a new domain of chemical research that, by encompassing all types of molecular components and interactions as well as both oligo and polymolecular entities, became supramolecular chemistry. It underwent rapid growth with the development of synthetic receptor molecules of numerous types for the strong and selective binding of cationic, anionic, or neutral complementary substrates of organic, inorganic, or biological nature by means of various interactions (electrostatic, hydrogen bonding, van der Waals, and donor-acceptor). Molecular recognition implies the (molecular) storage and the (supramolecular) retrieval and processing of molecular structural (geometrical and interactional) information. *See* HYDROGEN BOND; MACROCYCLIC COMPOUND; MOLECULAR RECOGNITION.

Many types of receptor molecules have been explored (crown ethers, cryptands, cyclodextrins, calixarenes, cavitands, cyclophanes, cryptophanes, and so on), and many others may be imagined for the binding of complementary substrates of chemical or biological significance. They allow, for instance, the development of substrate-specific sensors or the recognition of structural features in biomolecules (for example, nucleic acid probes, affinity cleavage reagents, and enzyme inhibitors). *See* BIOPOLYMER; CYCLOPHANE; ENZYME INHIBITION.

A major step in the development of supramolecular chemistry over the last 20 years involved the design of systems capable of spontaneously generating well-defined, supramolecular entities by self-assembly under a given set of conditions.

The information necessary for supramolecular self-assembly to take place is stored in the components, and the program that it follows operates via specific interactional algorithms based on binding patterns and molecular recognition events. Thus, rather than being preorganized, constructed entities, these systems may be considered as self-organizing, programmed supramolecular systems.

Self-assembly and self-organization have recently been implemented in numerous types of organic and inorganic systems. By clever use of metal coordination, hydrogen bonding, and donor-acceptor interactions, researchers have achieved the spontaneous formation of a variety of novel and intriguing species such as inorganic double and triple helices termed helicates, catenates, threaded entities (rotaxanes), cage compounds, grids of metal ions, and so on.

Another major development concerns the design of molecular species displaying the ability to perform self-replication, based on components containing suitable recognition groups and reactive functions. Self-recognition processes involve the spontaneous selection of the correct partner(s) in a self-assembly event—for instance, the correct ligand strand in helicate formation.

A major area of interest is the design of supramolecular devices built on photoactive, electroactive, or ionoactive components, operating respectively with photons, electrons, or ions. Thus, a variety of photonic devices based on photoinduced energy and electron transfer may be imagined. Molecular wires, ion carriers, and channels facilitate the flow of electrons and ions through membranes. Such functional entities represent entries into molecular photonics, electronics, and ionics, which deal with the storage, processing, and transfer of materials, signals, and information at the molecular and supramolecular levels. Dynamic and mechanical devices exploit the control of motion within molecular and supramolecular entities. See INORGANIC PHOTOCHEMISTRY; ION TRANSPORT; PHOTOCHEMISTRY.

The design of systems that are controlled, programmed, and functionally self-organized by means of molecular information contained in their components represents new horizons in supramolecular chemistry and provides an original approach to nanoscience and nanotechnology. In particular, the spontaneous but controlled generation of well-defined, functional supramolecular architectures of nanometric size through self-organization—supramolecular nanochemistry—represents a means of performing programmed engineering and processing of nanomaterials. It offers a powerful alternative to the demanding procedures of nanofabrication and nanomanipulation, bypassing the need for external intervention. A rich variety of architectures, properties, and processes should result from this blending of supramolecular chemistry with materials science. See NANOCHEMISTRY; NANOTECHNOLOGY. Jean-Marie Lehn

Surface (geometry) A two-dimensional geometric figure (a collection of points) in three-dimensional space. The simplest example is a plane—a flat surface. Some other common surfaces are spheres, cylinders, and cones, the names of which are also used to describe the three-dimensional geometric figures that are enclosed (or partially enclosed) by those surfaces. In a similar way, cubes, parallelepipeds, and other polyhedra are surfaces. See CUBE; POLYHEDRON; SOLID (GEOMETRY).

Any bounded plane region has a measure called the area. If a surface is approximated by polygonal regions joined at their edges, an approximation to the area of the surface is obtained by summing the areas of these regions. The area of a surface is the limit of this sum if the number of polygons increases while their areas all approach zero. See AREA; CALCULUS; INTEGRATION; PLANE GEOMETRY; POLYGON.

Methods of description. The shape of a surface can be described by several methods. The simplest is to use the commonly accepted name of the surface, such as sphere or cube. In mathematical discussions, surfaces are normally defined by one or more equations, each of which gives information about a relationship that exists between coordinates of points of the surface, using some suitable coordinate system. See COORDINATE SYSTEMS.

Some surfaces are conveniently described by explaining how they might be formed. If a curve, called the generator in three-dimensional space, is allowed to move in some manner, then each position the generator occupies during this motion is a collection of points, and the set of all such points constitutes a surface that can be said to be swept out by the generator. In particular, if the generator is a straight line, a ruled surface is formed. If the generator is a straight line and the motion is such that all positions of the generator are parallel, a cylindrical surface (or just cylinder) is formed. If the generator is a straight line and all positions of the generator have a common point of intersection, a conical surface (or just cone) is formed. A ruled surface that could be bent to lie in a plane (the bending to take place without stretching or tearing) is called a developable surface. See CONE; CYLINDER.

Dihedron. A dihedron is the surface formed by bending a plane along a line in that plane. More formally, a dihedron is the union of two half-planes that share the same boundary line.

Quadric surfaces. A surface whose implicit equation $F(x, y, z) = 0$ is second degree is a quadric surface, a three-dimensional analog of a conic section. A plane section of a quadric surface is either a conic section or one of its degenerate forms (a point, a line, parallel lines, or intersecting lines). See CONIC SECTION; QUADRIC SURFACE.

Surfaces of revolution. When a plane curve (the generator) is revolved about a line in that plane (the axis of revolution, or just axis), a surface of revolution can be said to be swept out. The resulting surface will be symmetric about the axis of revolution.

A circular cylinder (a quadric surface) is formed when the generator and the axis of revolution are distinct parallel lines. If the generator is only a segment of a line (rather than the entire line), a bounded circular cylinder is generated.

A circular cone is a quadric surface formed when a straight-line generator intersects the axis of revolution at an acute angle. The cone consists of two parts, the nappes, joined at the point of intersection, which is the vertex of the cone.

A sphere (a quadric surface) is usually defined as a collection of points in three-dimensional space at a fixed distance (the radius) from a given point (the center). However, a sphere can also be defined as the surface of revolution formed when a semicircle (or the entire circle) is revolved about its diameter.

The intersection of any plane with a sphere will be a circle (except for tangent planes). Such a circle is called, respectively, a great circle or a small circle, depending on whether or not the plane contains the center of the sphere.

If only part of a semicircle is revolved about the diameter, a part of a sphere called a zone is formed. If a semicircle is revolved about its diameter through an angle less than one revolution, the surface swept out is a lune (see illustration). See SPHERE.

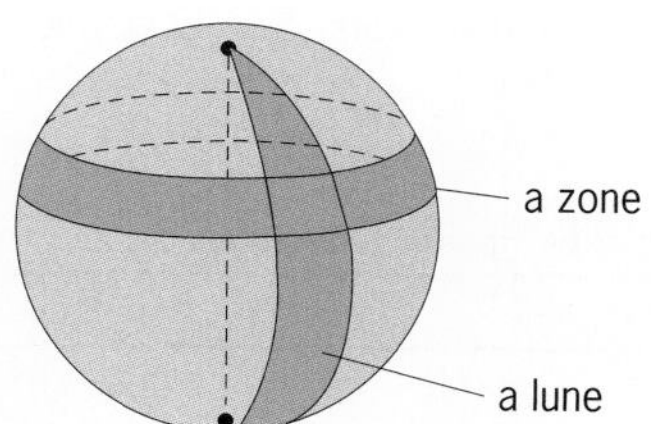

Sphere, with a zone and a lune.

A spheroid (also called an ellipsoid of revolution) is the quadric surface generated when an ellipse is revolved about either its major or minor axis. If the revolving is about the minor axis of the ellipse, the surface can be thought of as a flattened sphere, called an oblate spheroid. If the revolving is about the major axis, the surface can be thought of as a stretched sphere, called a prolate spheroid. A circular paraboloid is the quadric surface formed when a parabola is revolved about its axis. See ELLIPSE; PARABOLA.

A circular hyperboloid is the quadric surface formed when a hyperbola is revolved about either its transverse axis or its conjugate axis. The surface will be, respectively, a hyperboloid of one sheet or two sheets, depending on whether the revolving is about the conjugate axis or the transverse axis of the hyperbola. *See* HYPERBOLA.

A torus is generated when a circle is revolved about a line that does not intersect the circle. This doughnut-shaped surface has the property that not all points on the surface have the same sign of curvature. *See* DIFFERENTIAL GEOMETRY; TORUS.

Harry L. Baldwin, Jr.

Surface-acoustic-wave devices Devices that employ surface acoustic waves in the analog processing of electronic signals in the frequency range 10^7–10^9 Hz. Surface acoustic waves are mechanical vibrations that propagate along the surfaces of solids. In 1885, Lord Rayleigh discovered a type of surface acoustic wave that contains both compressional and transverse components 90° out of phase with one another. Since that time, other types of surface acoustic waves have been discovered and are an active area of intense research. A few notable examples include a wave propagating along a layer on a surface (Love wave), a wave propagating along an interface between two solids (Stoneley wave), and transverse guided waves on solids (Bleustein-Gulyaev-Shimizu waves). Love waves are shear-horizontal (SH) waves that have displacement only in a direction perpendicular to the plane of propagation.

Piezoelectric materials provide the required coupling between electrical signals and mechanical motion to generate surface acoustic waves. Crystalline piezoelectric materials, such as quartz, lithium niobate, and lithium tantalate, exhibit low attenuation and dispersion, and are therefore ideal for acoustic propagation. Surface acoustic waves in such a material are generated through a localized electric field at the surface that is created by applying voltage to an array of metal electrodes or fingers. This electrode array is known as an interdigital transducer (IDT). The IDT can also be used to detect surface waves, producing electrical output and hence an overall response. *See* PIEZOELECTRICITY; TRANSDUCER.

Surface-acoustic-wave (SAW) devices have led to a versatile technology for analog signal processing in the frequency range 10^7–10^9 Hz. The much slower propagation velocity of acoustic waves as compared to electromagnetic waves permits time delays in SAW devices, as compared to electrical delay lines, that are crucial for signal processing applications. Notable devices include band-pass filters, resonators, oscillators, pulse compression filters, fast Fourier transform processors, and more recently chemical and biological sensors. *See* ELECTRIC FILTER; OSCILLATOR; SIGNAL PROCESSING.

Chemical detection systems based on Rayleigh-wave SAW devices have largely used ST-quartz for its excellent temperature stability. To achieve specificity to a particular chemical agent, select polymer films are applied to the surface of the SAW device. Signal transduction occurs when chemicals partition into the polymer film, thereby altering the film properties, causing a change in the electrical response (that is, phase, frequency, time delay, or amplitude). Consumer application areas include mobile phones, television and satellite receivers, keyless entry systems (garage doors, cars, and so forth), and wireless applications. Commercial applications include fiber-optic communication, oscillators, local-area networks (LANs), test equipment, and chemical and biological detection systems, with military applications in radar, sonar, and advanced communications. *See* LOCAL-AREA NETWORKS; MOBILE COMMUNICATIONS; OPTICAL-FIBER COMMUNICATIONS; RADAR; SONAR; TELEVISION RECEIVER.

In a basic SAW device known as a delay line, a piezoelectric substrate has a polished upper surface on which two IDTs are deposited using photolithographic methods. The input transducer is connected, via fine bonded leads, to the electric source through an electrical matching network and source resistance. The output transducer drives the load, usually 50 ohms, through another electrical matching network. The center frequency (f_c) is governed by the Rayleigh wave velocity (V_R) on the piezoelectric substrate and the electrode width (a) of a single finger, according to Eq. (1). For SAW de-

$$f_c = \frac{V_R}{4a} \tag{1}$$

vices, the velocity of the wave depends on the properties of the piezoelectric crystal and its crystallographic orientation. Computer models have proven essential to iterate through numerous crystallographic orientations to search for the existence and type of a suitable acoustic wave. *See* DELAY LINE.

The acoustic response at frequency f for the SAW delay line can be calculated approximately by regarding each IDT as having N electrodes or sources. Through summation of sources for their amplitude and phase, the frequency response can be determined as proportional to $|(\sin x)/x|$ [that is, $|\text{sinc}(x)|$], where x is given by Eq. (2), providing a band-pass filter characteristic.

$$x = \frac{N\pi(f - f_c)}{f_c} \tag{2}$$

The electrical matching networks are normally arranged to minimize filter loss without compromising the acoustic response. The optimum number of periods, N, is inversely proportional to the

Surface-sensitive experimental techniques

Technique	Source*	Detectors	Level of information
Auger electron spectroscopy (AES)	Electrons 2–3 keV	Cylindrical mirror of retarding field	Elemental composition
X-ray photoelectron spectroscopy (XPS)	X-rays 1254 eV (Mg) 1487 eV (Al)	Hemispherical or cylindrical mirror	Elemental composition and oxidation state
Ultraviolet photoelectron spectroscopy (UPS)	UV radiation 21 eV He(I) 41 ev He(II)	Hemispherical or cylindrical mirror	Electronic properties of adsorbate and/or bulk material
Energy loss spectroscopy (ELS)	Electrons 507–1000 eV	Electron energy analyzer	Electronic structure of surface
High-resolution electron energy loss spectroscopy (HREELS)	Electrons 1–10 eV	Electron energy analyzer	Vibrational losses
Low-energy electron diffraction (LEED)	Electrons 20–500 eV	Retarding fields and phosphorescent screen	Surface structure or periodicity
Infrared spectroscopy (IRS)	Photons	Mercury-cadmium-telluride or indium antimony	Molecular identity
Optical ellipsometry	Photons	Photomultiplier	Adsorbate layer thickness
Scanning tunneling microscopy (STM)	Tunneling current	Ammeter	Substrate roughness and texture

*Mg = magnesium; Al = aluminum; He = helium.

piezoelectric coupling as is the filter band-pass width. *See* TRANSMISSION LINES.

Darren W. Branch

Surface and interfacial chemistry Chemical processes that occur at the phase boundary between gas–liquid, liquid–liquid, liquid–solid, or gas–solid interfaces.

The chemistry and physics at surfaces and interfaces govern a wide variety of technologically significant processes. Chemical reactions for the production of low-molecular-weight hydrocarbons for gasoline by the cracking and reforming of the high-molecular-weight hydrocarbons in oil are catalyzed at acidic oxide materials. Surface and interfacial chemistry are also relevant to adhesion, corrosion control, tribology (friction and wear), microelectronics, and biocompatible materials. In the last case, schemes to reduce bacterial adhesion while enhancing tissue integration are critical to the implantation of complex prosthetic devices, such as joint replacements and artificial hearts. *See* CRACKING; HETEROGENEOUS CATALYSIS; MEDICAL CHEMICAL ENGINEERING; PROSTHESIS; SURFACE PHYSICS.

Interactions with the substrate may alter the electronic structure of the adsorbate. Those interactions that lower the activation energy of a chemical reaction result in a catalytic process. Adsorption of reactants on a surface also confines the reaction to two dimensions as opposed to the three dimensions available for a homogeneous process. The two-dimentional confinement of reactants in a bimolecular event seems to drive biochemical processes with higher reaction efficiencies at proteins and lipid membranes. *See* ADSORPTION.

A limitation in the study of surfaces and interfaces rests with the low concentrations of the participants in the chemical process. Concentrations of reactants at surfaces are on the order of 10^{-10} to 10^{-8} mole/cm^2. Such low concentrations pose a sensitivity problem from the perspective of surface analysis. Experimental techniques with high sensitivity are required to examine the low concentrations of a surface species at interfaces.

Electron spectroscopy methods are widely used in the study of surfaces because of the small penetration depth of electrons through solids. This attribute makes electron spectroscopy inherently surface-sensitive, since only a few of the outermost atomic layers are accessible. The methods of electron spectroscopy used in surface studies have several common characteristics (see table). A source provides the incident radiation to the sample, which can be in the form of electrons, x-radiation, or ultraviolet radiation. Electron beams are generated from the thermionic emission of metal filaments or metal oxide pellets. The incident radiation induces an excitation at the surface of the sample, which alters the energy distribution of electrons that leave the surface. This distribution provides a diagnostic of the composition or structure of the interface. *See* ELECTRON SPECTROSCOPY.

Optical spectroscopy techniques (visible and infrared) are also useful for probing the chemical composition and molecular arrangement of surface species. Typical application configurations are the transmission and reflection (both external and internal) modes. Transmission spectroscopy relies on the passage of the probe beam through the sample. External and internal reflection spectroscopies involve the reflection of the probe beam from a medium with a lower refractive index to a medium with a higher refractive index, and from a higher to lower refractive index, respectively. The sample support must be optically transparent to the probe beam for the internal reflection mode. In both cases, the substrates are polished to a smooth, mirrorlike finish. *See* SPECTROSCOPY.

Mary M. Walczak; Marc D. Porter

Surface condenser A heat-transfer device used to condense a vapor, usually steam, by absorbing its latent heat in a cooling fluid, ordinarily water. Most surface condensers consist of a chamber containing a large number of corrosion-resisting alloy tubes through which cooling water flows. The vapor contacts the outside surface of the tubes and is condensed on them. The tubes are arranged so that the cooling water passes through the vapor space one or more times. Air coolers are normally an integral part of the condenser but may be separate and external to it. The condensate is removed by a condensate pump and the noncondensables by a vacuum pump. *See* STEAM CONDENSER; VAPOR CONDENSER.

Joseph F. Sebald

Surface hardening of steel The selective hardening of the surface layer of a steel product by one of several processes which involve changes in microstructure with or without changes in composition. Surface hardening imparts a combination of properties to the finished product not produced by bulk heat treatment alone. Among these properties are high wear resistance and good toughness or impact properties, increased resistance to failure by fatigue resulting from cyclic loading, and resistance to surface indentation by localized loads. The use of surface hardening frequently is also favored by lower costs and greater flexibility in manufacturing.

The principal surface hardening processes are: (1) carburizing, (2) the modified carburizing processes of carbonitriding, cyaniding, and liquid carburizing, (3) nitriding, (4) flame hardening and induction hardening, and (5) surface working. Carburizing introduces carbon into the surface layer of low-carbon steel parts and converts that layer into high-carbon steel, which can be quench-hardened by appropriate heat treatment. Carbonitriding, cyaniding, and liquid carburizing, in addition to supplying carbon, introduce nitrogen into the surface layer; this element permits lower case-hardening temperatures and has a beneficial effect on the subsequent heat treatment. In nitriding, only nitrogen is supplied, and reacts with special alloy elements present in the steel. Whereas the foregoing processes change the composition of the surface layer, flame hardening and induction hardening depend on a heat treatment applied selectively to the surface layer of a medium-carbon steel. Surface working by shot peening, surface rolling, or prestressing improved fatigue resistance by producing a stronger case, compressive stresses, and a smoother surface. *See* HEAT TREATMENT (METALLURGY); STEEL.

Michael B. Bever; Carl F. Floe

Surface mining A mining method used to obtain valuable minerals from the earth by first removing the overlying soil and rock (overburden) and subsequently recovering the valuable mineral. Surface mining is done until the accumulated overburden removed becomes too thick (and expensive to remove) for the economic recovery of the mineral. For mineral deposits too deep to be surface-mined, underground mining methods are used. *See* MINING.

Surface mining methods include placer mining, open gloryhole mining or milling, open-pit mining, strip mining, and quarrying.

Placer mining methods are used to recover unconsolidated mineral-rich (placer) deposits, often in streams, rivers, or their associated flood plain. Hydraulic methods, such as high-pressure water, can be used to loosen dry, unconsolidated material. The water–ore mixture then flows by gravity to an associated facility for processing into a salable product. Due to environmental concerns, this mining method is seldom used today. Another method of mining placer deposits is to use a dredge for excavating unconsolidated material underwater. Sand and gravel mining often use dredges. *See* PLACER MINING.

Open glory holes or milling pits are combined surface and underground mining operations, where a large opening or shaft is excavated vertically from the surface through a mineral-rich deposit to an underground mine. Drilling and blasting of the ore is directed downward from the surface around the shaft. The broken ore falls through the shaft and is recovered in the underground mine. These methods are typically used for precious-metal ore deposits such as gold, silver, or copper.

Open-pit mining is used to recover a large ore deposit by removing the overlying earth and rock in conjunction with removing the ore. An open-pit mine is designed in a circular or oval shape, creating a deep pit. Steep, unstable slopes are avoided since they could trigger a landslide and endanger the miners. The pit is kept dry by draining water to a sump and pumping it to a treatment facility outside the pit. The overburden and processed waste rock are placed in a landfill area. The mining operation can continue until the cost of removing the overburden exceeds the payment received for the recovered mineral. Open-pit mining typically is used at very large ore bodies, with the economic life sometimes exceeding 50 years. Open-pit mining is also called open-cut mining since the overburden is not returned to the pit and an open cut remains after mining is completed. *See* OPEN-PIT MINING.

Strip mining is the surface mining of coal. Before mining begins, ditches and ponds are constructed for containing sediment from erosion caused by rainstorms. The area is then cleared of trees, and the topsoil is removed and stockpiled. The three types of strip mining are area mining, contour mining, and mountaintop removal. *See* COAL; COAL MINING.

Surface mining of valuable stones, such as building stone and limestone, are called quarries. A typical quarry operation removes only a small amount of overburden, while most of the rock mined is sold for crushed rock. The rock is typically broken by blasting, loaded into trucks, and hauled to a processing facility, often a crusher. Crushers are mechanical devices used to break rocks into smaller sizes. Large screens are used to separate the various rock sizes. Ornamental rock quarries, such as marble, may cut the rock to dimensions requested by a purchaser. Most of the rock mined is sold, with a large opening remaining after mining, as well as a near-vertical rock slope called a highwall. To restore a quarry, rock and soil are placed against the highwall to create a stable slope, which is subsequently covered by topsoil and revegetated. *See* QUARRYING.

Modern surface mines have high reclamation standards. They restore the mined area to approximately the same contour as previously existed, revegetate with native grasses and trees, and comply with water pollution standards. Industry and environmental organizations such as the American Society of Mining and Reclamation promote reclamation practices for lands disturbed during mineral extraction. Previously unreclaimed surface mine sites are also being restored by the surface mining industry. Some are restored directly in conjunction with the mining of adjacent mineral resources. *See* LAND RECLAMATION. Thomas Gray

Surface physics The study of the structure and dynamics of atoms and their associated electron clouds in the vicinity of a surface, usually at the boundary between a solid and a low-density gas. Surface physics deals with those regions of large and rapid variations of atomic and electron density that occur in the vicinity of an interface between the two "bulk" components of a two-phase system. In conventional usage, surface physics is distinguished from interface physics by the restriction of the scope of the former to interfaces between a solid (or liquid) and a low-density gas, often at ultrahigh-vacuum pressures $p = 10^{-10}$ torr (1.33×10^{-8} newton/m^2 or 10^{-13} atm). *See* SOLID-STATE PHYSICS.

Surface physics is concerned with two separate but complementary areas of investigation into the properties of such solid-"vacuum" interfaces. Interest centers on the experimental determination and theoretical prediction of surface composition and structure (that is, the masses, charges, and positions of surface species), of the dynamics of surface atoms (such as surface diffusion and vibrational motion), and of the energetics and dynamics of electrons in the vicinity of a surface (such as electron density profiles and localized electronic surface states). As a practical matter, however, the nature and dynamics of surface species are determined experimentally by scattering and emission measurements involving particles or electromagnetic fields (or both) external to the surface itself. Thus, a second major interest in surface physics is the study of the interaction of external entities (that is, atoms, ions, electrons, electromagnetic fields, or mechanical probes) with solids at their vacuum interfaces. It is this aspect of surface physics that most clearly distinguishes it from conventional solid-state physics, because quite different scattering, emission, and local probe experiments are utilized to examine surface as opposed to bulk properties.

Techniques for characterizing the solid-vacuum interface are based on one of three simple physical mechanisms for achieving surface sensitivity. The first, which is the basis for field emission, field ionization, and scanning tunneling microscopy (STM), is the achievement of surface sensitivity by utilizing electron tunneling through the potential-energy barrier at a surface. This concept provides the basis for the development of STM to directly examine the atomic structure of surfaces by measuring with atomic resolution the tunneling current at various positions along a surface. It also has been utilized for direct determinations of the energies of individual electronic orbitals of adsorbed complexes via the measurement of the energy distributions either of emitted electrons or of Auger electrons emitted in the process of neutralizing a slow (energy $E \sim 10$ eV) external ion. *See* FIELD-EMISSION MICROSCOPY; SCANNING TUNNELING MICROSCOPE; TUNNELING IN SOLIDS.

The second mechanism for achieving surface sensitivity is the examination of the elastic scattering or emission of particles which interact strongly with the constituents of matter, for example, "low energy" ($E \lesssim 10^3$ eV) electrons, thermal atoms and molecules, or "slow" (300 eV $\lesssim E \lesssim 10^3$ eV) ions. Since such entities lose appreciable ($\Delta E \sim 10$ eV) energy in distances of the order of tenths of a nanometer, typical electron analyzers with resolutions of tenths of an electronvolt are readily capable of identifying scattering and emission processes which occur in the upper few atomic layers of a solid. This second mechanism is responsible for the surface sensitivity of photoemission, Auger electron, electron characteristic loss, low-energy electron diffraction (LEED), and ion scattering spectroscopy techniques. The strong particle-solid interaction criterion that renders these measurements surface-sensitive is precisely the opposite of that used in selecting bulk solid-state spectroscopies. In this case, weak particle-solid interactions (that is, penetrating radiation) are desired in order to sample the bulk of the specimen via, for example, x-rays, thermal neutrons, or fast ($E \gtrsim 10^4$ eV) electrons. These probes, however, can sometimes be used to study surface properties by virtue of special geometry, for example, the use of glancing-angle x-ray diffraction to determine surface atomic structure. *See* AUGER EFFECT; ELECTRON DIFFRACTION; ELECTRON SPECTROSCOPY; PHOTOEMISSION; X-RAY CRYSTALLOGRAPHY.

The third mechanism for achieving surface sensitivity is the direct measurement of the force on a probe in mechanical contact or near contact with the surface. At near contact, the van der Waals force can be measured directly by probes of suitable sensitivity. After contact is made, a variety of other forces dominate, for example, the capillary force for solid surfaces covered with thin layers of adsorbed liquid (that is, most solid surfaces in air at atmospheric pressure). When this mechanism is utilized via measuring the deflection of a sharp tip mounted on a cantilever near a surface, the experiment is referred to as atomic force microscopy (AFM) and results in maps of the force across the surface. Under suitable circumstances, atomic resolution can be achieved by this method as well as by STM. Atomic force microscopy opens the arena of microscopic surface characterization of insulating samples as well as electrochemical and biochemical interfaces at atmospheric pressure. Thus, its development is a major driving force for techniques based on surface physics. *See* INTERMOLECULAR FORCES.

Another reason for the renaissance in surface physics is the capability to generate in a vacuum chamber special surfaces that approximate the ideal of being atomically flat. These surfaces may be prepared by cycles of fast-ion bombardment, thermal outgassing, and thermal annealing for bulk samples (for example, platelets with sizes of the order of 1 cm × 1 cm × 1 mm), molecular beam epitaxy of a thin surface layer on a suitably prepared substrate, or field evaporation of etched tips for field-ion microscopes. Alternatively, the sample may be cleaved in a vacuum chamber. In such a fashion, reasonable facsimiles of uncontaminated, atomically flat solid-vacuum interfaces of many metals and semiconductors have been prepared and subsequently characterized by various spectroscopic techniques. Such characterizations must be carried out in an ultrahigh vacuum ($p \sim 10^{-8}$ N/m^2) so that the surface composition and structure are not altered by gas adsorption during the course of the measurements. *See* EPITAXIAL STRUCTURES. C. B. Duke

Surface tension The force acting in the surface of a liquid, tending to minimize the area of the surface. Surface forces, or more generally, interfacial forces, govern such phenomena as the wetting or nonwetting of solids by liquids, the capillary rise of liquids in fine tubes and wicks, and the curvature of free-liquid surfaces. The action of detergents and antifrothing agents and the flotation separation of minerals depend upon the surface tensions of liquids.

In the body of a liquid, the time-averaged force exerted on any given molecule by its neighbors is zero. Even though such a molecule may undergo diffusive displacements because of random collisions with other molecules, there exist no directed forces upon it of long duration. It is equally likely to be momentarily displaced in one direction as in any other. In the surface of a liquid, the situation is quite different; beyond the free surface, there exist no molecules to counteract the forces of attraction exerted by molecules in the interior for molecules in the surface. In consequence, molecules in the surface of a liquid experience a net attraction toward the interior of a drop. These centrally directed forces cause the droplet to assume a spherical shape, thereby minimizing both the free energy and surface area.

Liquids which wet the walls of fine capillary tubes rise to a height which depends upon the tube radius, the surface tension, the liquid density, and the contact angle between the solid and the liquid (measured through the liquid). In the illustration a liquid of a certain density is shown as having risen to a height h in a capillary whose radius is r. A balance exists between the force exerted by gravity on the mass of liquid raised in the capillary and the opposing force caused by surface tension.

Detergents, soaps, and flotation agents owe their usefulness to their ability to lower the surface tension of water, thereby stabilizing the formation of small bubbles of air. At the same time, the interfacial tension between solid particles and the liquid phase is lowered, so that the particles are more readily wetted and floated after attachment to air bubbles. *See* FLOTATION; INTERFACE OF PHASES; SURFACTANT. Norman H. Nachtrieb

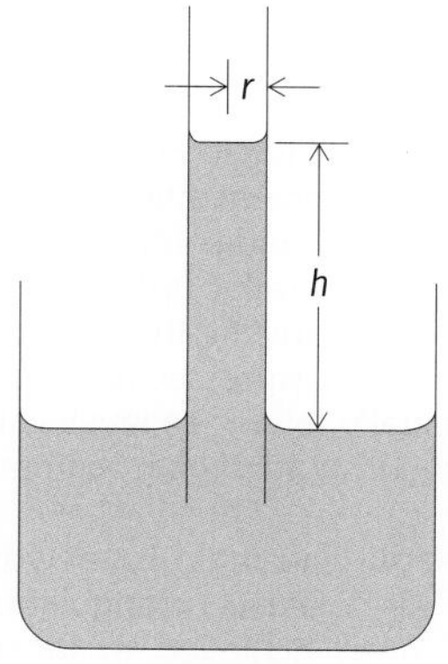

Rise of liquid in capillary tube.

Surface waves Disturbances that are propagated at a gas-liquid interface are dependent primarily upon the gravitational fluid property (surface tension and viscosity being of secondary importance). Wave motions which occur in confined fluids (either liquid or gaseous) are primarily dependent upon the elastic property of the medium.

Oscillatory waves may be generated in a rectangular channel by a simple harmonic translation of a vertical wall forming one end of the flume. A standing wave can be considered to be composed of two equal oscillatory wave trains traveling in opposite directions.

A solitary wave consists of a single crest above the original liquid surface which is neither preceded nor followed by another elevation or depression of the surface. Such a wave is generated by the translation of a vertical wall starting from an initial position at rest and coming to rest again some distance downstream. In practice, solitary waves are generated by a motion of barges in narrow waterways or by a sudden change in the rate of inflow into a river; they are therefore related to a form of flood wave. Donald R. F. Harleman

Surfactant An amphiphilic (also called amphipathic) compound that adsorbs at interfaces to form oriented monolayers and shows surface activity. An amphiphilic compound is a molecule that has a hydrophilic (polar) head and a hydrophobic (nonpolar) tail. Common synonyms for the term "surfactant" include amphiphile, surface-active agent, and tenside. If a surfactant is placed into contact with both a polar medium, such as water, and a nonpolar medium, such as an oil, one part of its molecule has an affinity for the polar medium and one part that has an affinity for the nonpolar medium. Three consequences of the amphiphilic nature of surfactants are their ability (1) to adsorb and form layers at interfaces, (2) to reduce the interfacial tension between fluids, and (3) to associate to form clusters, called micelles. *See* ADSORPTION; INTERFACE OF PHASES; MICELLE; SURFACE TENSION.

Depending on the nature of the polar (hydrophilic) part of the molecule, surfactants are classified as anionic (negatively charged), cationic (positively charged), nonionic (noncharged), or zwitterionic (able to contain either or both kinds of charge). A common anionic surfactant is sodium dodecyl sulfate, $CH_3(CH_2)_{11}SO_4^-Na^+$. A common cationic surfactant is cetyl trimethylammonium bromide, $CH_3(CH_2)_{15}N^+(CH_3)_3Br^-$, a common type of nonionic surfactant is polyoxyethylene alcohol, $C_nH_{2n+1}(OCH_2CH_2)_mOH$. An example of a zwitterionic surfactant is dodecyl betaine, $C_{12}H_{25}N^+(CH_3)_2CH_2COO^-$. A soap is a particular kind of anionic surfactant comprising any of the surface-active fatty-acid salts containing at least eight carbon atoms. Typically, the molar masses of surfactants range from a few hundreds to several thousands of grams per mole. *See* SOAP.

Of the numerous surfactant applications, the most common are emulsifying, foaming, suspending and floating, wetting, and detergency. Emulsion, foam, and suspension products span a diverse range of industries, including agriculture, food processing, cosmetics and personal care, pharmaceuticals, papermaking and deinking, water and wastewater treatment, environmental remediation, oil recovery, and minerals beneficiation. Laurier L. Schram

Surge arrester A protective device which is connected between an electrical conductor and the ground with the intent of limiting the magnitude of transient overvoltages on equipment. A lightning arrester is really a voltage-surge arrester.

The valve arrester consists of disks of zinc oxide material that exhibit low resistance at high voltage and high resistance at low voltage. By selecting an appropriate configuration of disk material, the arrester will conduct a low current of a few milliamperes at normal system voltage. When insulation levels of equipment are coordinated with the surge arrester protective characteristics, and under lightning or switching surge overvoltag conditions, the surge current is limited by the circuit; and for the magnitudes of current that can be delivered to the arrester location, the resulting voltage will be limited to controlled values, and to safe levels as well.

A typical surge arrester consists of disks of zinc oxide material sized in cross-sectional area to provide desired energy discharge capability, and in axial length proportional to the voltage capability. The disks are then placed in porcelain enclosures to provide physical support and heat removal, and sealed for isolation from contamination in the electrical environment. *See* LIGHTNING; LIGHTNING AND SURGE PROTECTION. Glenn D. Breuer

Surge suppressor A device that is designed to offer protection against voltage surges on the power line that supplies electrical energy to the sensitive components in electronic devices and systems. The device offers a limited type of protection to computers, television sets, high-fidelity equipment, and similar types of electronic systems.

A voltage surge is generally considered to be a transient wave of voltage on the power line. The amplitude of the surge may be several thousand volts, and the duration may be as short as 1 or 2 milliseconds or as long as about 100 ms. Typical effects can be damage to the electronics or loss of programs and data in computer memories. Many events can cause the surges, including lightning that strikes the power lines at a considerable distance from the home or office; necessary switching of transmission lines by the utilities; and rapid connections or disconnections of large loads, such as air conditioners and motors, from the power line, or even other appliances in the home. Lightning is perhaps the most common. *See* LIGHTNING.

The suppressor acts to limit the peak voltage applied to the electronic device to a level that normally will not cause either damage to the device or software problems in the computers. The device may include a pilot light, a fuse, a clipping circuit, resistors, and a main switch. The clipper circuit is the principal item, and the design of this portion is usually proprietary information. *See* CLIPPING CIRCUIT; FUSE (ELECTRICITY). Edwin C. Jones, Jr.

Surgery That branch of medicine which generally treats diseases by operative intervention. Surgical procedures may involve relieving mechanical obstruction of a tubular organ, such as the intestine; or removing a diseased organ, which cannot be salvaged by medical treatment, such as a gangrenous appendix or inflamed gallbladder; or removing a malignant tumor with a margin of normal tissue; or repairing an injured organ, or removing it if the organ is irreparable and its absence is compatible with survival.

The field of surgery has become increasingly specialized, primarily by organ system, so that surgical diseases of the kidney, bladder, and other components of the urinary tract are treated by surgeons called urologists; surgery of the central nervous system, including the brain and spinal cord, is done by neurosurgeons; reconstructive and cosmetic surgery is done primarily by plastic surgeons; general surgeons continue to do most abdominal surgery, some head and neck surgery, and surgery of the soft tissues of the extremities; and surgical diseases of the bones and joints are treated by orthopedic surgeons. Some specialties within surgery have also developed into specialties that are not limited to one organ system, such as surgical oncology (cancer surgery), so that cancers in most parts of the body may be treated by surgeons with special training in malignant diseases. *See* MEDICINE. Robert F. Jones

Surveillance radar Any radar equipment used by civil or military authorities for locating aircraft, ships, or ground vehicles; most commonly, in civil air-traffic management, a radar used by controllers to indicate the position of aircraft aloft or on the airport surface. Two types of air-traffic control surveillance radar are employed internationally. Ground-based conventional, or primary, radar operates by transmitting microwave pulses and detecting the resulting energy reflected from the aircraft body (airframe). Secondary surveillance radar employs cooperative radio receiver-transmitter units (transponders) on the aircraft.

Although secondary radar generally provides more reliable and complete surveillance information than does primary radar, air-traffic authorities use both. In the United States, two major examples are the Federal Aviation Administration's Airport Surveillance Radar (ASR) and Air Route Surveillance Radar (ARSR), the latter shared at some locations with the U.S. Air Force. Both normally operate unattended, their outputs fed by a landline or microwave link to their control positions. Primary radar provides backup coverage for nonequipped aircraft and for aircraft with airframe shielding. Primary radar can also detect potentially hazardous precipitation cells and potentially hostile aircraft or inadvertent intruders. Secondary radar is sometimes used without a primary radar backup for long-range, high-altitude surveillance where all aircraft are transponder-equipped, where airframe shielding is rare, and where the sensitivity disadvantage of primary radar requires costly high-power transmitters. The Federal Aviation Administration has authorized use of certain secondary radars having faster update rates and monopulse accuracy for monitoring aircraft conducting multiple parallel approaches in bad weather. *See* AIR-TRAFFIC CONTROL; MONOPULSE RADAR; RADAR. Raymond R. LaFrey

Surveying The measurement of dimensional relationships among points, lines, and physical features on or near the Earth's surface. Basically, surveying determines horizontal distances, elevation differences, directions, and angles. These basic determinations are applied further to the computation of areas and volumes and to the establishment of locations with respect to some coordinate system.

Surveying is typically used to locate and measure property lines; to lay out buildings, bridges, channels, highways, sewers, and pipelines for construction; to locate stations for launching and tracking satellites; and to obtain topographic information for mapping and charting.

Horizontal distances are usually assumed to be parallel to a common plane. Each measurement has both length and direction. Length is expressed in feet or in meters. Direction is expressed as a bearing of the azimuthal angle relationship to a reference meridian, which is the north-south direction. It can be the true meridian, a grid meridian, or some other assumed meridian. The degree-minute-second system of angular expression is standard in the United States.

Reference, or control, is a concept that applies to the positions of lines as well as to their directions. In its simplest form, the position control is an identifiable or understood point of origin for the lines of a survey. Conveniently, most coordinate sys- tems have the origin placed west and south of the area to be surveyed so that all coordinates are positive and in the northeast quadrant.

Vertical measurement adds the third dimension to an object's position. This dimension is expressed as the distance above some reference surface, usually mean sea level, called a datum. Mean sea level is determined by averaging high and low tides during a lunar month.

Horizontal control. The main framework, or control, of a survey is laid out by traverse, triangulation, or trilateration. Some success has been achieved in locating control points from Doppler measurements of passing satellites, from aerial phototriangulation, from satellites photographed against a star background, and from inertial guidance systems. In traverse, adopted for most ordinary surveying, a line or series of lines is established by directly measuring lengths and angles. In triangulation, used mainly for large areas, angles are again directly measured, but distances are computed trigonometrically. This necessitates triangular patterns of lines connecting intervisible points and starting from a baseline of known length. New baselines are measured at intervals. Trigonometric methods are also used in trilateration, but lengths, rather than angles, are measured. The development of electronic distance measurement (EDM) instruments brought trilateration into significant use.

Distance measurement. Traverse distances are usually measured with a surveyor's tape or by EDM, but also may sometimes be measured by stadia, subtense, or trig-traverse.

Whether on sloping or level ground, it is horizontal distances that must be measured. In taping, horizontal components of hillside distances are measured by raising the downhill end of the tape to the level of the uphill end. On steep ground this technique is used with shorter sections of the tape. The raised end is positioned over the ground point with the aid of a plumb bob. Where slope distances are taped along the ground, the slope angle can be measured with the clinometer. The desired horizontal distance can then be computed.

In EDM the time a signal requires to travel from an emitter to a receiver or reflector and back to the sender is converted to a distance readout. The great advantage of electronic distance measuring is its unprecedented precision, speed, and convenience. Further, if mounted directly onto a theodolite, and especially if incorporated into it and electronically coupled to it, the EDM instrument with an internal computer can in seconds measure distance (even slope distance) and direction, then compute the coordinates of the sighted point with all the accuracy required for high-order surveying.

In the stadia technique, a graduated stadia rod is held upright on a point and sighted through a transit telescope set up over another point. The distance between the two points is determined from the length of rod intercepted between two horizontal wires in the telescope.

In the subtense technique the transit angle subtended by a horizontal bar of fixed length enables computation of the transit-to-bar distance (Fig. 1). In trig-traverse the subtense bar is replaced by a measured baseline extending at a right angle from the survey line whose distance is desired. The distance calculated in either subtense or trig-traverse is automatically the horizontal distance and needs no correction.

Fig. 1. Subtense bar. (*Lockwood, Kessler, and Bartlett Inc.*)

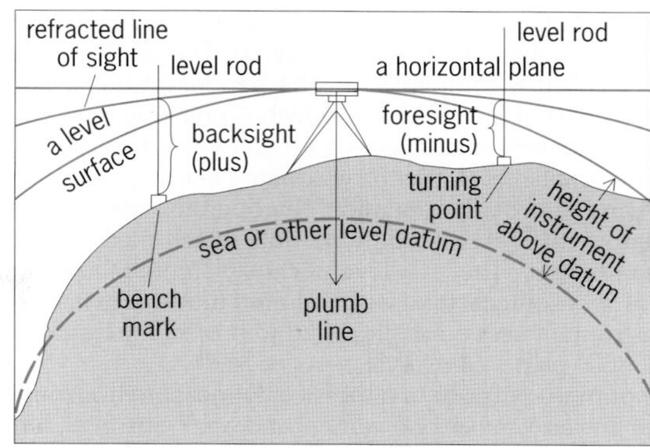

Fig. 2. Theory of differential leveling.

Angular measurement. The most common instrument for measuring angles is the transit or theodolite. It is essentially a telescope that can be rotated a measurable amount about a vertical axis and a horizontal axis. Carefully graduated metal or glass circles concentric with each axis are used to measure the angles. The transit is centered over a point with the aid of either a plumb bob suspended by a string from the vertical axis or (on some theodolites) an optical plummet, which enables the operator to sight along the instrument's vertical axis to the ground through a right-angle prism.

Elevation differences. Elevations may be measured trigonometrically in conjunction with reduction of slope measurements to horizontal distances, but the resulting elevation differences are of low precision.

Most third-order and all second- and first-order measurements are made by differential leveling, wherein a horizontal line of sight of known elevation is sighted on a graduated rod held vertically on the point being checked (Fig. 2). The transit telescope, leveled, may establish the sight line, but more often a specialized leveling instrument is used. For approximate results a hand level may be used.

Other methods of measuring elevation include trigonometric leveling which involves calculating height from measurements of horizontal, distance and vertical angle; barometric leveling, a method of determining approximate elevation difference with aid of a barometer; and airborne profiling, in which a radar altimeter on an aircraft is used to obtain ground elevations.

Astronomical observations. To determine meridian direction and geographic latitude, observations are made by a theodolite or transit on Polaris, the Sun, or other stars. Direction of the meridian (geographic north-south line) is needed for direction control purposes; latitude is needed where maps and other sources are insufficient. The simplest meridian determination is made by sighting Polaris at its elongation, as the star is rounding the easterly or westerly extremity of its apparent orbit. An angular correction is applied to the direction of sighting, which is referenced to a line on the ground. The correction value is found in an ephemeris. *See* EPHEMERIS; TOPOGRAPHIC SURVEYING AND MAPPING.

B. Austin Barry

Surveying instruments Instruments used in surveying operations to measure vertical angles, horizontal angles, and distance. Such devices were originally mechanical only, but technological advances led to mechanical-optical devices, optical-electronic devices, and finally, electronic-only devices.

Four types of levels are available: optical, automatic, electronic, and laser. An optical level is used to project a line of sight that is at a 90° angle to the direction of gravity. Both dumpy and tilting types use a precision leveling vial to orient to gravity. The dumpy type was used primarily in the United States, while the tilting type was of European origin and used in the remainder of the world. Automatic levels use a pendulum device, in place of the precision vial, for relating to gravity. The pendulum mechanism is called a compensator. The pendulum has a prism or mirror, as part of the telescope, which is precisely positioned by gravity. The electronic level has a compensator similar to that on an automatic level, but the graduated leveling staff is not observed and read by the operator. The operator has only to point the instrument at a bar-code-type staff, which then can be read by the level itself. The laser levels actually employ three different types of light sources: tube laser, infrared diode, and laser diode. The instrument uses a rotating head to project the laser beam in a level 360° plane. *See* LEVEL MEASUREMENT.

The primary purpose of a transit is to measure horizontal and vertical angles. Circles, one vertical and one horizontal, are used for these measurements. The circles are made of metal or glass and have precision graduations engraved or etched on the surface. A vernier is commonly used to improve the accuracy of the circle reading. The theodolite serves the same purpose as the transit, and they have many similar features. The major differences are that the measuring circles are constructed only of glass and are observed through magnifying optics to increase the accuracy of angular readings. The electronic theodolite uses electronic reading circles in place of the optically read ones. *See* VERNIER.

The U.S. Department of Defense installed a satellite system known as the Global Positioning System for navigation and for establishing the position of planes, ships, vehicles, and so forth. This system uses special receivers and sophisticated software to calculate the longitude and latitude of the receiver. It was discovered early in the program that the distance between two nonmoving receivers could be determined very accurately and that the distance between receivers could be many miles apart. This technology has become the standard for highly accurate control surveys, but it is not in general use because of the expense of the precision receivers, the time required for each setup, and the sophistication of the process. *See* SATELLITE NAVIGATION SYSTEMS; SURVEYING.

Kerry W. Kemper

Susceptance The imaginary part of the admittance of an alternating-current circuit.

The admittance, Y, of an alternating current circuit is a complex number given by Eq. (1). The imaginary part, B, is the

$$Y = G + jB \tag{1}$$

susceptance. The units of susceptance like those of admittance are called siemens or mhos. Susceptance may be either positive or negative. For example, the admittance of a capacitor C at frequency ω is given by Eq. (2), and so B is positive. For an inductor L, the admittance is given by Eq. (3), and so B is negative.

$$Y = jC\omega = jB \tag{2}$$

$$Y = -\frac{j}{L\omega} = jB \tag{3}$$

In general, the susceptance of a circuit may depend on the resistors as well as the capacitors and inductors. For example, the circuit in the illustration has impedance given by Eq. (4) and admittance given by Eq. (5), so that the susceptance, given by Eq. (6), depends on the resistor R as well as the inductor L.

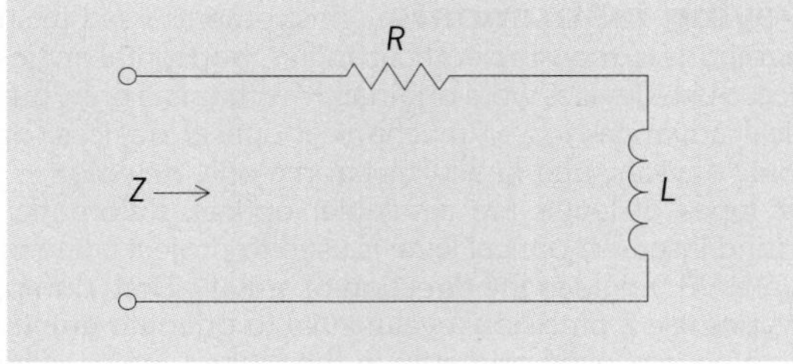

Circuit with a resistor and inductor in series.

$$Z = R + jL\omega \tag{4}$$

$$Y = \frac{1}{R + jL\omega} \tag{5}$$

$$B = \frac{-L\omega}{R^2 + L^2\omega^2} \tag{6}$$

See ADMITTANCE; ALTERNATING CURRENT; ELECTRICAL IMPEDANCE.

J. O. Scanlan

Suspension A system of small solid particles dispersed in a liquid. Suspensions are a type of colloidal dispersion. By classical definition, suspension particles have diameters (or, if not spherical, the largest dimension) between 1 nanometer and 1 micrometer. However, for practical purposes the principles of colloid science can be usefully applied to suspensions for which the particles are smaller, down to about 0.1 nm (nanoparticle suspensions), and also to those for which the largest particle dimensions are tens or even hundreds of micrometers. The particles in a suspension are large enough that they do not behave like the atoms and molecules of classical chemistry. For example, suspensions do not generally behave like true solutions, and may have undetectable freezing-point depressions. On the other hand, the particles are small enough that they do not behave like the macroscopic particles of classical physics. For example, sedimentation of a suspension'sparticles may occur extremely slowly in apparent violation of Stokes' law. *See* COLLOID; PARTICULATES; SEDIMENTATION (INDUSTRY).

In most suspensions, the solid particles are dispersed in an aqueous liquid phase (that is, water containing dissolved substances), which may be denoted solids-in-water or S/W. In some suspensions, however, the liquid phase is an oil (usually a hydrocarbon liquid), and the suspension is denoted solids-in-oil or S/O.

Suspensions can be formed by one of two basic methods. In the nucleation process, either a gas is made to condense to form solid particles or particles are precipitated from a solution. Once the fine particles have been created, they can grow to larger sizes as additional material condenses or precipitates onto the surfaces of existing particles (Ostwald ripening). The second basic suspension preparation method involves breaking-up large particles into progressively smaller ones (comminution). This process requires shearing forces, which can be provided by various devices, including cutting, rolling, crushing, or grinding mills. When the solid particles are already quite small, other devices such as propeller-style mixers or ultrasound generators can be used.

In order for dispersed particles to be reasonably stable and not aggregate or settle quickly, suspension particles need to be stabilized. The two most common means of stabilizing suspended particles are by creating electric charges or adsorbing polymers on the particles' surfaces. Some particles gain a surface electric charge due to reactions that occur when they are dispersed into water. Another way to create particle surface charges is by adsorption of chemicals such as ionic surfactants or polymers. Electrically charged particles repel each other, which can dramatically slow down the rates of both aggregation and sedimentation, making a suspension reasonably stable. *See* ADSORPTION; POLYMER; SURFACTANT.

Suspensions of colloid-sized particles are important because they feature prominently in both desirable and undesirable contexts and in a wide variety of practical disciplines, products, and industrial processes.

Suspensions commonly occur in the environment. Examples range from the suspended sediments in rivers, lakes, and oceans to bogs and quicksand. Suspensions are commonly used in many industries, such as pulp and papermaking. They are created at an early stage of processes used to separate valuable minerals by froth flotation. Suspensions are also quite common in the petroleum industry, and may occur in reservoirs, drilling fluids, production fluids, process plant streams, and tailings ponds.

Some familiar suspensions include those occurring in foods like batters, puddings, and sauces, in pharmaceuticals like cough syrups and laxatives, and in household and industrial products like inks, paints, and "liquid" waxes. Examples in personal care products include lipstick and lip balms, which are usually concentrated suspensions of solid oils in a liquid oil or in a mixture of liquid oils. Laurier L. Schramm

Swamp, marsh, and bog Wet flatlands, where mesophytic vegetation is areally more important than open water, commonly developed in filled lakes, glacial pits and potholes, or poorly drained coastal plains or floodplains. Swamp is a term usually applied to a wetland where trees and shrubs are an important part of the vegetative association, and bog implies lack of solid foundation. Some bogs consist of a thick zone of vegetation floating on water.

Unique plant associations characterize wetlands in various climates and exhibit marked zonation characteristics around the edge in response to different thicknesses of the saturated zone above the firm base of soil material. Coastal marshes covered with vegetation adapted to saline water are common on all continents. Presumably many of these had their origin in recent inundation due to post-Pleistocene rise in sea level. *See* GLACIATED TERRAIN; MANGROVE. Luna B. Leopold

Sweat gland A coiled, tubular gland found in mammals. There are two kinds, merocrine (or eccrine) and apocrine. The latter are generally associated with hair follicles (see illustration). Merocrine glands are distributed extensively over the body in the human, whereas the apocrine variety is restricted to the scalp, nipples, axilla, external auditory meatus, external genitals, and perianal areas. Apocrine sweat glands are more numerous in

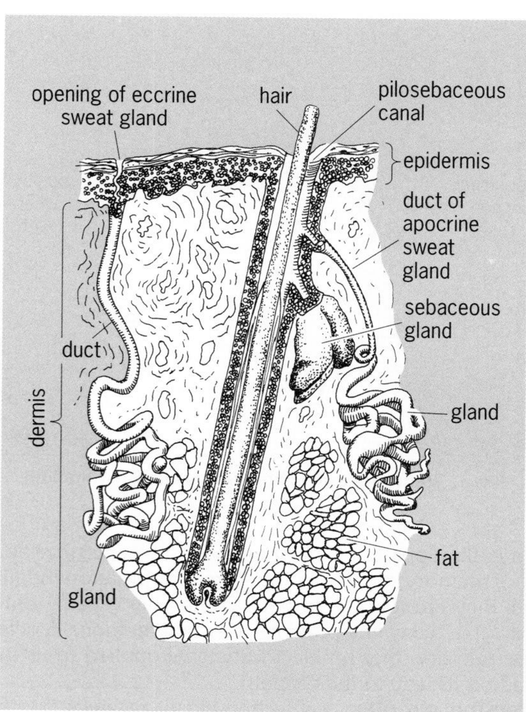

Human skin showing structure of both eccrine and apocrine sweat glands.

mammals, with the exception of the chimpanzee and human, in which the merocrine variety predominates. The mammary glands probably represent modified apocrine sweat glands which grow inward and increase in complexity. *See* GLAND; MAMMARY GLAND. Olin E. Nelsen

Sweetgum The tree *Liquidambar styraciflua*, also called redgum, a deciduous tree of the southeastern United States. It is found northward as far as southwestern Connecticut, and also grows in Central America. Sweetgum is readily distinguished by its five-lobed, or star-shaped, leaves and by the corky wings or ridges usually developed on the twigs. The erect trunk is a dark gray, but the branches are lighter in color. In winter the persistent, spiny seedballs are an excellent diagnostic feature.

Sweetgum is used for furniture, interior trim, railroad ties, cigar boxes, crates, flooring, barrels, woodenware, and wood pulp, and it is one of the most important materials for plywood manufacture. Sweetgum is one of the most desirable ornamental trees, chiefly because of its brilliant autumn coloration. *See* HAMAMELIDALES. Arthur H. Graves; Kenneth P. Davis

Swim bladder A gas-filled sac found in the body cavities of most bony fishes (Osteichthyes). The swim bladder has various functions in different fishes, acting as a float which gives the fish buoyancy, as a lung, as a hearing aid, and as a sound-producing organ. In many fishes it serves two or three of these functions, and in the African and Asiatic knifefishes (Notopteridae) it may serve all four. The swim bladder contains the same gases that make up air, but often in different proportions. R. McNeill Alexander

Swine production The science of breeding, raising, and marketing swine. This agricultural enterprise is usually located in proximity to sources of high-energy feedstuffs. Swine, as nonruminants, utilize large quantities of concentrate feeds; it is estimated that 30% of the maize grain produced is fed to swine. Swine can utilize only limited quantities of roughage in their diet.

The United States produces about one-fourth of the world's hogs. The swine production industry has changed from the status of a supplement to farm income to that of intensified units in which swine production is a major enterprise. More than 90% of all hogs are from independent producers. The others are produced by the intensified methods of integrated enterprises in confinement facilities which require highly trained biological managers capable of applying the principles of breeding, nutrition, physiology, environmental control, and economics. The primary products of swine are pork, lard, hides, and innumerable pharmaceutical by-products. Pork and lard supply about 15% of the total calories consumed as food in the United States. Pork is more successfully cured and stored than many other meats.

There are at least 85 recognized swine breeds in the world, and perhaps more than 220 "varieties" recognized as breeds in certain geographical locations. The pure breeds of major importance in the United States include Berkshire, Chester White, Duroc, Hampshire, Landrace, Poland China, spotted Poland China, and Yorkshire. In Great Britain and Europe, the Landrace and Yorkshire breeds are most prevalent. *See* BREEDING (ANIMAL). Aldon H. Jensen

Switched capacitor circuit A module consisting of a capacitor with two metal oxide semiconductor (MOS) switches connected as shown in illus. *a*. These elements in the module are easily realized as an integrated circuit on a silicon chip by using MOS technology. The switched capacitor module is approximately equivalent to a resistor, as shown in illus. *b*. The fact that resistors are relatively difficult to implement gives the switched capacitor a great advantage in integrated-circuit applications requiring resistors. Some of the advantages are that the cost is significantly reduced, the chip area needed is reduced,

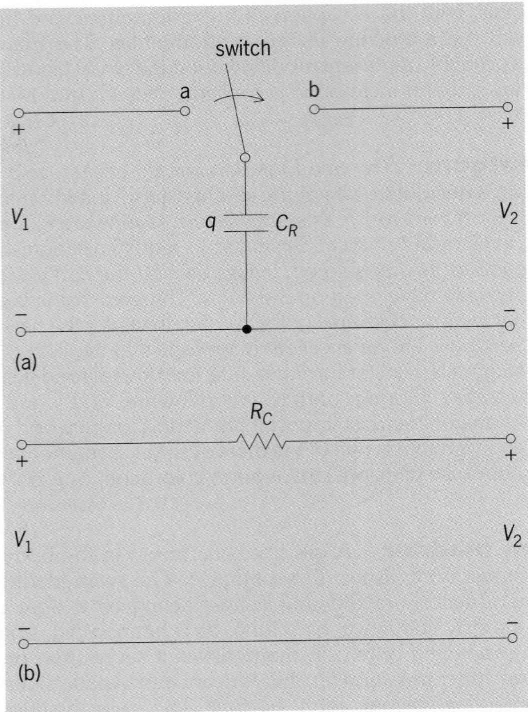

Switched capacitor, (*a*) Basic circuit, (*b*) Equivalent resistive circuit.

and precision is increased. Although the switched capacitor can be used for any analog circuit realization such as analog-to-digital or digital-to-analog converters, the most notable application has been to voice-frequency filtering. *See* ANALOG-TO-DIGITAL CONVERTER; DIGITAL-TO-ANALOG CONVERTER; ELECTRIC FILTER; INTEGRATED CIRCUITS.

M. E. Van Valkenburg

Switching circuit A constituent electric circuit of a switching or digital data-processing system which receives, stores, or manipulates information in coded form to accomplish the specified objectives of the system. Examples include digital computers, dial telephone systems, and automatic accounting and inventory systems. *See* DIGITAL COMPUTER; SWITCHING SYSTEMS (COMMUNICATIONS); SWITCHING THEORY.

Physically, switching circuits consist of conducting paths interconnecting discrete-valued electrical devices. The most generally used switching circuit devices are two-valued or binary, such as switches and relays in which manual or electromagnetic actuation opens and closes electric contacts; vacuum and gas-filled electronic tubes, and semiconductor rectifiers and transistors, which do or do not conduct current; and magnetic structures, which can be saturated in either of two directions.

The electrical conditions controlling these switching circuit devices are also generally two-valued or binary, such as open versus closed path, full voltage versus no voltage, large current versus small current, and high resistance versus low resistance. Such two-valued electrical conditions, as applied to the input of a switching circuit, represent either (1) a combination of events or situations which exist or do not exist; (2) a sequence of events or situations which occur in a certain order; or (3) both combinations and sequences of events or situations. The switching circuit responds to such inputs by delivering at its output, also in two-valued terms, new information which is functionally related to the input information.

Functional characteristics of switching circuits are defined by the logical operation and memory capabilities of the discrete devices from which they are assembled, as well as by the means used to interconnect the devices. For example, switching circuits embody such logical relationships as output X is to exist only if input A and B occur simultaneously; and output Y is to exist if either input A or input B occurs. The factor of memory, in turn, enables a switching circuit to hold or retain a given state after the condition that produced the state has passed.

Basic combinational circuits. A combinational switching circuit is one in which a particular set of input conditions always establishes the same output, irrespective of the history of the circuit.

In electronic switching circuits, so-called gates are used to perform logical functions equivalent to these series-parallel networks of switch contacts. In this sense, an electronic gate is an elementary combinational circuit. Gates do not function by physically inserting or removing metallic conduction paths between contacts of manually operated switches or remotely controlled relays. Instead, they function by control of voltage or current levels at their output.

The most commonly encountered gates are the AND and the OR gates. The AND gate produces an output only if all its inputs are concurrently present; an OR gate produces an output if any one or any combination of its inputs is present. *See* LOGIC CIRCUITS.

Basic sequential circuits. A sequential switching circuit is one whose output depends not only upon the present state of its input, but also on what its input conditions have been in the past. Sequential circuits, therefore, require memory elements.

A typical electronic memory element used in sequential circuits is a simple circuit called a flip-flop. A flip-flop consists of two amplifiers connected so that the output of one amplifier is the input of the other. A voltage pulse will set the flip-flop into one of two states, and that state remains until another voltage pulse resets the flip-flop or returns it to its original state. It can therefore be used to remember that an event has taken place.

Figure 1 is an *npn*-transistor flip-flop. When set, transistor A is conducting and transistor B is cut off. When reset, transistor B is conducting and transistor A is cut off. A positive output voltage with respect to ground may be obtained from either transistor to indicate the condition of the flip-flop. *See* TRANSISTOR.

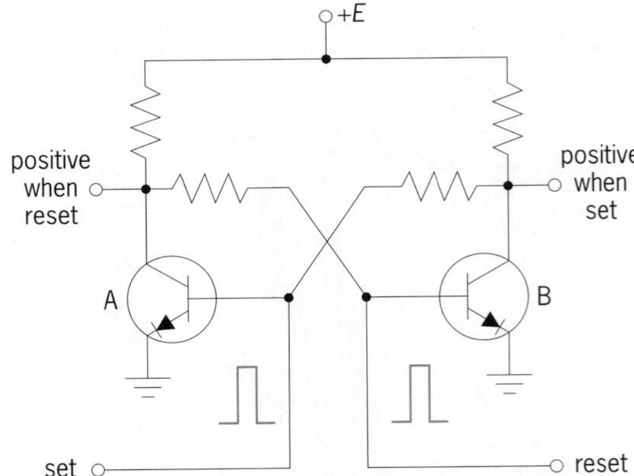

Fig. 1. Transistor switching memory element (flip-flop).

Relays, flip-flops, and similar memory elements provide static, or fixed, memory; they hold the stored information indefinitely, or until they are told to "forget," commonly called "resetting." In contrast, a delay line provides transient memory. A delay line has the property that an electrical signal applied to its input is delayed on its way to the output.

Selecting circuits. A selecting circuit receives the identity (called the address) of a particular item and selects that item from among a number of similar ones. The selectable items are often represented by terminals or leads. Selection usually involves marking the specified terminal or lead by applying to it some electrical condition, such as a voltage or current pulse, or a steady-state dc signal. By means of this electrical condition, the selected circuit is alerted, sized, or controlled.

Connecting circuits. A switching system is an aggregate of functional circuit units, some of which must sometimes be directly coupled to each other to interchange information. Figure 2 shows a simple electronic connecting circuit using AND and OR gates. In this arrangement a communication path is provided over a single link from any one of the three functional circuits A, B, C, to either the X or Y circuit by an external control circuit activating the appropriate pair of AND gates. To provide a multilead link, or to provide for other simultaneous interconnections, additional AND gates would, of course, be required. The OR gate maintains separation of the inputs at the common junction point.

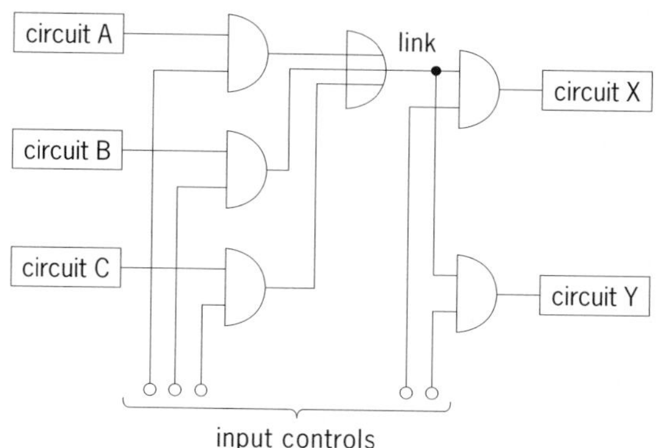

Fig. 2. Connecting circuit using AND and OR gates.

Lockout circuits. In switching systems, situations often arise where several similar circuit units are ready at the same instant to request collaboration with another type of functional circuit. Mutual interference among the requesting circuits is prevented by the lockout circuit (sometimes referred to as hunting or finding circuits). In response to concurrent inputs from a number of external circuits, a lockout circuit provides an output indication corresponding to one, and only one, of these circuits at any time.

Translating circuits. Switching systems process information in coded form; the information is generally in the form of numbers. Numerical codes are many and varied, each with its own characteristics and more or less distinct advantages for different switching circuit situations. Therefore, one of the common functional circuits in switching systems is the translating circuit, which translates information received in one code into the same information expressed in another code. These translating circuits are combinational circuits; a given input signal combination representing a code to be translated always produces the same output signals, which represent the desired code.

Register circuits. Information received by a switching system is not always used immediately. It must be stored in register circuits for future use.

In a register circuit the coded information to be stored is applied as input and retained by memory elements of the circuit, and when needed, the registered information is taken as output in the same code or in a different code. Register circuits are devised with a great variety of memory elements and have capacities to store from a few to millions of information bits. A frequently encountered form of register circuit is the shift register. This type of register has the ability to shift its stored digital information internally to positions representing higher or lower numerical values in the code employed. For example, in decimal code registration a digit may be shifted from the units to the tens position. An obvious use of such registers is in digital computers when, for example, partial multiplication products have to be lined up for addition.

Counting circuits. One of the most frequently encountered circuits in switching systems is the counting circuit whose function, in general, is to detect and count repeated current or voltage pulses which represent incoming information. Joseph A. Pecar

Switching systems (communications)

The assemblies of switching and control devices provided so that any station in a communications system may be connected as desired with any other station. A telecommunications network consists of transmission systems, switching systems, and stations. Transmission systems carry messages from an originating station to one or more distant stations. They are engineered and installed in sufficient quantities to provide a quality of service commensurate with the cost and expected benefits. To enable the transmission facilities to be shared, stations are connected to and reached through switching system nodes that are part of most telecommunications networks. Switching systems act under built-in control to direct messages toward their ultimate destination or address.

Most switching systems, known as central or end offices in the public network and as private branch exchanges (PBXs) when applied to business needs, are used to serve stations. These switching systems are at nodes that are strategically and centrally located with respect to the community of interest of the served stations. With improvements in technology, it has become practical to distribute switching nodes closer to stations. In some cases to serve stations within a premise, switching is distributed to take place at the stations themselves. A smaller number of systems serve as tandem (intermediate) switching offices for large urban areas or toll (long-distance) offices for interurban switching. These end and intermediate office functions are sometimes combined in the same switching system. *See* PRIVATE BRANCH EXCHANGE.

Switching system fundamentals. Telecommunications switching systems generally perform three basic functions: they transmit signals over the connection or over separate channels to convey the identity of the called (and sometimes the calling) address (for example, the telephone number), and alert (ring) the called station; they establish connections through a switching network for conversational use during the entire call; and they process the signal information to control and supervise the establishment and disconnection of the switching network connection.

In some data or message switching when real-time communication is not needed, the switching network is replaced by a temporary memory for the storage of messages. This type of switching is known as store-and-forward switching.

Signaling and control. The control of circuit switching systems is accomplished remotely by a specific form of data communications known as signaling. Switching systems are connected with one another by telecommunication channels known as trunks. They are connected with the served stations or terminals by lines.

In some switching systems the signals for a call directly control the switching devices over the same path for which transmission is established. For most modern switching systems the signals for identifying or addressing the called station are received by a central control that processes calls on a time-shared basis. Central controls receive and interpret signals, select and establish communication paths, and prepare signals for transmission. These signals include addresses for use at succeeding nodes or for alerting (ringing) the called station.

Most electronic controls are designed to process calls not only by complex logic but also by logic tables or a program of instructions stored in bulk electronic memory. The tabular technique is known as action translator (AT). The electronic memory is now the most accepted technique and is known as stored program control (SPC). Either type of control may be distributed among the switching devices rather than residing centrally. Microprocessors on integrated circuit chips are a popular form of distributed stored program control. *See* COMPUTER STORAGE TECHNOLOGY; INTEGRATED CIRCUITS; MICROPROCESSOR.

Common channel signaling (CCS) comprises a network of separate data communication paths used for transmitting all signaling information between offices. It became practical as a result of processor control. To reduce the number of data channels between all switching nodes, a signaling network of signal switching nodes is introduced. The switching nodes, known as signal transfer points (STPs), are fully interconnected with each other and the switching offices they serve. All links and signal transfer points are duplicated to ensure reliable operation. Each stored-program-control toll switching system connects to the two signal transfer points in its region.

Switching fabrics. Space and time division are the two basic techniques used in establishing connections. When an individual conductor path is established through a switch for the duration of a call, the system is known as space division. When the transmitted speech signals are sampled and the samples multiplexed in time so that high-speed electronic devices may be used simultaneously by several calls, the switch is known as time division.

Most switching is now automatic. The switching fabric frequently comprises two primary-secondary arrangements: first, the line link (LL) frames on which the telephone lines appear and, second, the trunk link (TL) frames on which the trunks appear. A switching entity may grow to a maximum of 60 line link and 30 trunk link frames. Each line link frame is interconnected with every trunk link frame by a network of links called junctors. Each line link frame has a basic capacity for 290 telephone lines and may be supplemented in 50-line increments to a maximum of 590 lines. The size used in a particular office depends upon the calling rate and holding time of the assigned lines.

Electronic switching. Stored program control has become the principal type of control for all types of new switching systems throughout the world, including toll, private branch, data, and Telex systems. Two types of data are stored in the memories of electronic switching systems. One type is the data associated with the progress of the call, such as the dialed address of the called line. Another type, known as the translation data, contains infrequently changing information, such as the type of service subscribed to by the calling line and the information required for routing calls to called numbers. These translation data, like the program, are stored in a memory which is easily read but protected to avoid accidental erasure. This information may be readily changed, however, to meet service needs. The flexibility of a stored program also aids in the administration and maintenance of the service so that system faults may be located quickly.

Untethered switched services. Modern mobile radio service has a considerable dependency on switching. The territory served by a radio carrier is divided into cells of varying geographical size, from microcells that might serve the floors of a business or domicile to cells several miles across that provide a space diversity for serving low-power radios.

Switching systems reach each radio cell site that detects, by signal strength, when a vehicle is about to move from one cell to another. The switching system then selects a frequency and land line for the communication to continue on another channel in a different cell without interruption. This is known as cellular mobile radio service and is used not only for voice but also facsimile and other forms of telecommunication. *See* MOBILE RADIO; TELEPHONE; TELEPHONE SERVICE. Amos E. Joel, Jr.

Sycamore American sycamore (*Platanus occidentalis*), a member of the plane tree family, known also as American plane tree, buttonball, or buttonwood, and ranging from southern Maine to Nebraska and south into Texas and northern Florida. It has the most massive trunk of any American hardwood. Characteristic are the white patches which are exposed when outer layers of the bark slough off; the simple, large, lobed leaves whose stalks completely cover the conical winter buds; and the spherical fruit heads that are always borne singly in the American species and persist throughout the winter. The tough, coarse-grained wood is difficult to work, but is useful for butchers' blocks, saddle trees, vehicles, tobacco and cigar boxes, crates, and slack cooperage. *See* ROSALES. Arthur H. Graves; Kenneth P. Davis

Sycettida An order of the subclass Calcaronea in the class Calcarea. This order comprises a rather diverse group of calcareous sponges, and includes the families Sycettidae, Heteropiidae, Grantiidae, Amphoriscidae, and Lelapiidae. Choanocytes with apical nuclei are limited to flagellated chambers and never occur lining the general spongocoel. The family Sycettidae resembles the contrasting order Leucosoleniida in lacking the true dermal membrane or cortex possessed by the other five families. The most massive skeleton (of bundles of modified triradiate spicules) is found in the family Lelapiidae. *See* CALCAREA; LEUCOSOLENIIDA. W. D. Russell-Hunter

Syenite A phaneritic (visibly crystalline) plutonic rock with granular texture composed largely of alkali feldspar (orthoclase, microcline, usually perthitic) with subordinate plagioclase (oligoclase) and dark-colored (mafic) minerals (biotite, amphibole, and pyroxene). If sodic plagioclase (oligoclase or andesine) exceeds the quantity of alkali feldspar, the rock is called monzonite. Monzonites are generally light to medium gray, but syenites are found in a wide variety of colors (gray, green, pink, red), some of which make the material ideal for use as ornamental stone. Syenite is an uncommon plutonic rock and usually occurs in relatively small bodies (dikes, sills, stocks, and small irregular plutons). *See* IGNEOUS ROCKS. Carleton A. Chapman

Symbiotic star A double star system in the late stage of stellar evolution. Since the symbiotic phase represents a brief span in the life of the binary, symbiotic stars are rare objects. The "near-official" list of symbiotic stars contains 188 safe entries; 15 of them extragalactic, and 30 suspected candidates.

Symbiotics are always associated with a nebular environment. The spectra indicate the presence of a cool M-type star with a surface temperature below 4000 K, and a hot nebula quite similar to a planetary nebula. The light output is variable on time scales of days, months, and years. *See* ASTRONOMICAL SPECTROSCOPY; LIGHT CURVES; NEBULA; PLANETARY NEBULA; VARIABLE STAR.

Evidence has accumulated that all symbiotics are double stars with periods between one year and many dozens of years. Their orbits are sufficiently wide for the two stars not to be in direct contact, and both stars lie safely within their Roche lobe. Thus, for the two stars stellar evolution proceeds over the whole main-sequence lifetime practically uninfluenced by the partner. That changes dramatically in the later stages of their evolution. *See* DOPPLER EFFECT; ECLIPSING VARIABLE STARS.

The gas temperature in the symbiotic nebula, approximately 15,000 K, is much too low to collisionally ionize the atoms in the nebula to the observed degree. The nebular gas must be radiatively ionized by a small, very hot star. Observations and model calculations show the cool star to be a red giant. For the hot star, temperatures between 50,000 and 200,000 K are found. The star's size is that of a white dwarf or of a central star in a planetary nebula, between $\sim$0.01 and $\sim$0.1 solar radius.

The symbiotic nebula has, as a rule, the same chemical composition as expected from red giants. This is taken as evidence that the red giant suffers considerable mass loss. Symbiotics began as stars with masses of the order of 5 solar masses. The more massive star evolved faster through the main sequence, and when it arrived in the red giant region, shed most of its mass through a stellar wind. That material can occasionally still be detected in the wider environment. The star has become a hot white dwarf. The originally less massive star has retained its mass and has now entered the red giant phase, with large mass loss. *See* STELLAR EVOLUTION; WHITE DWARF STAR.

A fraction of the mass lost by the red giant is captured by the white dwarf. The accretion liberates gravitational energy which can convert into radiative energy and be responsible for some of the irregular luminosity variations. When the white dwarf has accumulated a critical mass of hydrogen, thermonuclear reactions on its surface lead to an energy outburst lasting over 100 years. The energy production outburst can reach many thousand times that of the Sun. The mechanism resembles a nova explosion. However, a classical nova has higher peak energy output and shorter duration than a symbiotic nova. Whether some symbiotics can in this way accumulate sufficient mass to explode as a supernova of type Ia is still an open question. *See* CATACLYSMIC VARIABLE; NOVA; SUPERNOVA.

Observations have shown traces of the mass formerly lost by the hot star. However, as roted above, the bulk of the matter now detected as an ionized nebula is due to the present mass loss of the red giant in its stellar wind. In its active phase the white dwarf may have a stellar wind of its own. In that case the nebular environment will be strongly structured by the collision of the two winds. This leads to shock zones with temperatures of several million kelvins. The origin of bipolar gaseous jets observed in several symbiotics is not yet known. Observational evidence indicates that the white dwarf possesses a strong magnetic field which could play a major role. Harry Nussbaumer

Symbolic computing The manipulation of symbols, representing variables, functions, and other mathematical objects, and combinations of these symbols, representing formulas, equations, and expressions, according to mathematical rules.

Giving the solution to an equation using symbols instead of numbers is referred to as symbolic manipulation or symbolic computing. Often, symbolic computing refers to computations or calculations that involve the manipulation of mathematical symbols to obtain solutions. For example, the differentiation and integration of functions in calculus, manipulation of matrices in linear algebra, or determination of the function that solves a differential equation are now commonly referred to as symbolic computations. *See* CALCULUS; LINEAR ALGEBRA.

In the past, symbolic computation was done by the human mind with paper and pencil or in some similar manner. Now computers are able to do symbolic manipulations, commonly using software packages or hand-held graphing calculators. Some common software packages that can do symbolic computing are Mathematica®, Maple®, and MATLAB®. Texas Instruments and Casio are two of the most common hand-held calculators that have symbolic computing features. Microsoft Office's Excel allows Maple as an add-in so that symbolic computing can be done. Software programming environments that can do symbolic computing are usually called symbolic computing packages or computer algebra systems. *See* CALCULATORS.

Symbolic computing software packages are able do algebra and calculus in any dimension—that is, they have the ability to do scalar or vector calculus. They can do linear algebra operations. They can solve many differential equations and perform transformations on objects, like differential equations. Many packages have components that do abstract analysis. Symbolic computing packages can work with real or complex numbers and usually have some routine for dealing with complex arithmetic and its subtleties. Symbolic computing packages are capable of doing most university undergraduate mathematics and many areas of high-level research mathematics. Most of the standard algorithms used in mathematical and scientific research are incorporated in the package. Sometimes the algorithm is stored in a library inside the software package and has to be loaded into the package from a directory. *See* ALGORITHM; MATHEMATICAL SOFTWARE.

Symbolic computing software packages allow a user to do symbolic manipulations in seconds. These same manipulations can take minutes, hours, or even days for humans to do. Some of the manipulations are not humanly possible. James Sochacki

Symmetrodonta A group of extinct mammals that range from the Late Triassic to Late Cretaceous. Their fossil remains have been found in Mesozoic deposits worldwide. These small insectivorous or carnivorous mammals are the size of a shrew or mouse. They are considered to be distant relatives of the more derived therian mammals, including the extinct eupantotheres, such as dryolestids and the living placental and marsupial mammals.

Symmetrodonts are characterized by a distinctive feature: the main cusps of their cheek teeth are arranged in a symmetrical triangle. The triangular upper and lower cheek teeth fill in the gaps between the adjacent teeth of the opposite tooth row, and are specialized for crushing insects or slicing worms. However, symmetrodonts lack a basinlike heel in the lower teeth that would allow the grinding of the ingested food in the more derived living therian mammals and their kin. Symmetrodonts are also distinguished from more derived therians by lacking the angle on the mandible.

Symmetrodonts are classified as the order Symmetrodonta of the class Mammalia. Three families are notable among the diverse symmetrodonts: Kuehneotheriidae considered to be close to the ancestry of all other therian mammals; Amphidontidae; and Spalacotheriidae, which are characterized by the highly acute triangular cheek teeth and are the most diverse symmetrodont family. Zhexi Luo

Symmetry breaking A deviation from exact symmetry. According to modern physical theory the fundamental laws of physics possess a very high degree of symmetry. Several deep insights into nature arise in understanding why specific physical systems, or even the universe as a whole, exhibit less symmetry than the laws themselves.

Spontaneous symmetry breaking. This mechanism occurs in quite diverse circumstances. The most symmetrical solutions of the fundamental equations governing a given system may be unstable, so that in practice the system is found to be in a less symmetrical, but stable, state. When this occurs, the symmetry is said to have been broken spontaneously.

For example, the laws of physics are unchanged by any translation in space, but a crystalline lattice is unchanged only by special classes of translations. A crystal does retain a large amount of symmetry, for it is unchanged by those finite translations, but this falls far short of the full symmetry of the underlying laws. *See* CRYSTAL STRUCTURE.

Another example is provided by ferromagnetic materials. The spins of electrons within such materials are preferentially aligned in some particular direction, the axis of the poles of the magnet. The laws of physics governing the interactions among these spins are unchanged by any rotation in space, but the aligned configuration of spins has less symmetry. Indeed, it is left unchanged only by rotations about the polar axis. *See* FERROMAGNETISM.

In both these examples, the loss of symmetry is associated with the appearance of order. This is a general characteristic of spontaneous symmetry breaking.

Consequences. There is a cluster of important observable consequences associated with spontaneous symmetry breaking.

Nambu-Goldstone bosons are a class of low-energy excitations associated with gentle variations of the order. Thus, there is a class of excitations of the ferromagnet, the magnons, that exist as a consequence of the spontaneous symmetry breaking, and that have very low energy. Similarly, in the case of crystals, phonons are associated with gentle distortions of the lattice structure. *See* MAGNON; PHONON.

At high temperatures the energy gained by assuming an ordered structure is increasingly outweighed by the entropy loss associated with the constraints it imposes, and at some point it will no longer be favorable to have spontaneous symmetry breaking in thermal equilibrium. Changes from broken symmetry to unbroken symmetry are marked by phase transitions. For a magnet, the transition occurs at the Curie temperature. For a crystal, it is melting into a liquid or sublimation into a gas. *See* CURIE TEMPERATURE; ENTROPY; PHASE TRANSITIONS; THERMODYNAMIC PRINCIPLES.

Defects are imperfections in the ordering. The most familiar examples are domain walls in magnets. *See* CRYSTAL DEFECTS; DOMAIN (ELECTRICITY AND MAGNETISM).

In systems with long-range forces as well as spontaneous symmetry breaking, it need no longer be true that gradual changes require only a small input of energy, because even distant regions interact significantly. Thus, the Nambu-Goldstone bosons no longer have very low energies, and they are not easily excited. Conversely, the system will exhibit a special rigidity, with strong correlations between distant points. These ideas are central to modern theories of superconductivity and of particle physics (the Higgs mechanism). *See* ELECTROWEAK INTERACTION; HIGGS BOSON; SUPERCONDUCTIVITY. Frank Wilczek

Symmetry laws (physics) The physical laws which are expressions of symmetries. The term symmetry, as it is used in mathematics and the exact sciences, refers to a special property of bodies or of physical laws, namely that they are left unchanged by transformations which, in general, might have changed them. For example, the geometric form of a sphere is not changed by any rotation of the sphere around its center, and so a sphere can be said to be symmetric under rotations. Symmetry can be very powerful in constraining form. Indeed, referring to the same example, the only sort of surface which is symmetric under arbitrary rotations is a sphere.

The concept that physical laws exhibit symmetry is more subtle. A naive formulation would be that a physical law exhibits symmetry if there is some transformation of the universe that might have changed the form of the law but in reality does not. However, the comparison of different universes is generally not feasible or desirable. A more fruitful definition of the symmetry of physical law exploits locality, the principle that the behavior of a given system is only slightly affected by the behavior of other bodies far removed from it in space or time. Because of locality, it is possible to define symmetry by using transformations that do not involve the universe as a whole but only a suitably isolated portion of it. Thus the statement that the laws of physics are symmetric under rotations means that (say) astronauts in space would not be able to orient themselves—to determine a preferred direction—by experiments internal to their space station. They could do this only by referring to weak effects from distant objects, such as the light of distant stars or the small residual gravity of Earth.

Symmetries of space and time. Perhaps the most basic and profound symmetries of physical laws are symmetry under translation in time and under translations in space.

The statement that fundamental physical laws are symmetric under translation in time is equivalent to the statement that these laws do not change or evolve. Time-translation symmetry is supposed to apply, fundamentally, to simple isolated systems. Large complicated systems, and in particular the universe as a whole, do of course age and evolve. Thus in constructing the big-bang model of cosmology, it is assumed that the properties of individual electrons or protons do not change in time, although of course the state of the universe as a whole, according to the model, has changed quite drastically.

The statement that fundamental laws are symmetric under translations in space is another way of formulating the homogeneity of space. It is the statement that the laws are the same throughout the universe. It says that the astronauts in the previous example cannot infer their location by local experiments within their space station. The power of this symmetry is that it makes it possible to infer, from observations in laboratories on Earth, the behavior of matter anywhere in the universe.

The symmetry of physical law under rotations, mentioned above, embodies the isotropy of space.

In the mathematical formulation of dynamics, there is an intimate connection between symmetries and conservation laws. Symmetry under time translation implies conservation of energy; symmetry under spatial translations implies conservation of momentum; and symmetry under rotation implies conservation of angular momentum. *See* ANGULAR MOMENTUM; CONSERVATION OF ENERGY; CONSERVATION OF MOMENTUM.

The fundamental postulate of the special theory of relativity, that the laws of physics take the same form for observers moving with respect to one another at a fixed velocity, is clearly another statement about the symmetry of physical law. The idea that physical laws should be unchanged by such transformations was discussed by Galileo, who illustrated it by an observer's inability to infer motion while on a calm sea voyage in an enclosed cabin. The novelty of Einstein's theory arises from combining this velocity symmetry with a second postulate, deduced from experiments, that the speed of light is a universal constant and must take the same value for both stationary and uniformly moving observers. *See* GALILEAN TRANSFORMATIONS; LORENTZ TRANSFORMATIONS; RELATIVITY.

Discrete symmetries. Before 1956, it was believed that all physical laws obeyed an additional set of fundamental symmetries, denoted *P*, *C*, and *T*, for parity, charge conjugation, and time reversal, respectively. Experiments involving particles known as *K* mesons led to the suggestion that *P* might be violated in the weak interactions, and violations were indeed observed. This discovery led to questioning—and in some cases overthrow—of other cherished symmetry principles.

Parity, *P*, roughly speaking, transforms objects into the shapes of their mirror images. If *P* were a symmetry, the apparent behavior of the images of objects reflected in a mirror would also be the actual behavior of corresponding real objects. *See* PARITY (QUANTUM MECHANICS).

Charge conjugation, *C*, changes particles into their antiparticles. It is a purely internal transformation; that is, it does not involve space and time. If the laws of physics were symmetric under charge conjugation, the result of an experiment involving antiparticles could be inferred from the corresponding experiment involving particles.

Remarkably, by combining the transformations *P* and *C*, a result is obtained, *CP*, which is much more nearly a valid symmetry than either of its components separately. However, in 1964 it was discovered experimentally that even *CP* is not quite a valid symmetry.

Although the preceding discussion has emphasized the failure of *P*, *C*, and *CP* to be precise symmetries of physical law, both the strong force responsible for nuclear structure and reactions and the electromagnetic force responsible for atomic structure and chemistry do obey these symmetries. Only the weak force, responsible for beta radioactivity and some relatively slow decays of exotic elementary particles, violates them. Thus these symmetries, while approximate, are quite useful and powerful in nuclear and atomic physics. *See* ELECTROWEAK INTERACTION; FUNDAMENTAL INTERACTIONS; WEAK NUCLEAR INTERACTIONS.

The operation of time-reversal symmetry, *T*, involves changing the direction of motion of all particles. For example, it relates reactions of the type $A + B \rightarrow C + D$ to their reverse $C + D \rightarrow A + B$. No direct violation of *T* has been detected. *See* TIME REVERSAL INVARIANCE.

Time-reversal symmetry, even if valid, applies in a straightforward way only to elementary processes. It does not, for example, contradict the one-way character of the second law of thermodynamics, which states that entropy can only increase with time. *See* THERMODYNAMIC PRINCIPLES; TIME, ARROW OF.

Fundamental principles of quantum field theory suggest that the combined operation *PCT*, which involves simultaneously reflecting space, changing particles into antiparticles, and reversing the direction of time, must be a symmetry of physical law. Existing evidence is consistent with this prediction. *See* CPT THEOREM.

Internal symmetry. Internal symmetries, like *C*, do not involve transformations in space-time but change one type of particle into another. An important, although approximate, symmetry of this kind is isospin or i-spin symmetry. It is observed experimentally that the strong interactions of the proton and neutron are essentially the same. *See* I-SPIN.

There have been several successful predictions of the existence and properties of new particles, based on postulates of internal symmetries. Perhaps the most notable was the prediction of the mass and properties of the Ω^- baryon, based on an extension of the symmetry group SU(2) of isospin to a larger approximate SU(3) symmetry acting on strange particles as well. These symmetries were an important hint that the fundamental strong interactions are at some level universal, that is, act on all quarks in the same way, and thus paved the way toward modern quantum chromodynamics, which does implement such universality. *See* BARYON; GROUP THEORY; QUANTUM CHROMODYNAMICS; UNITARY SYMMETRY.

Much simpler mathematically than SU(2) internal symmetry, but quite profound physically, is U(1) internal symmetry. The important case of the electric charge quantum number will be considered. The action of the U(1) internal symmetry transformation with parameter λ is to multiply the wave function of a state of electric charge q by the factor $e^{i\lambda q}$. An amplitude between two states with electric charges q_r and q_s will therefore be multiplied by a factor $e^{i\lambda(q_s-q_r)}$. Since the physical predictions of quantum mechanics depend on such amplitudes, these predictions will be unchanged only if the phase factors multiplying all nonvanishing amplitudes are trivial. This will be true, in turn, only if the amplitudes between states of unequal charge vanish; that is, if charge-changing amplitudes are forbidden, which is just a backhanded way of expressing the conservation of charge. *See* QUANTUM MECHANICS; QUANTUM NUMBERS.

Localization of symmetry. The concept of local gauge invariance, which is central to the standard model of fundamental particle interactions, and in a slightly different form to general relativity, may be approached as a generalization of the U(1) internal symmetry transformation, where a parameter λ independent of space and time appears. Such a parameter goes against the spirit of locality, according to which each point in space-time has a certain independence. There is therefore reason to consider a more general symmetry, involving a space-time-dependent transformation in which the wave function is multiplied by $e^{i\lambda(x,t)q(x,t)}$, where $q(x,t)$ is the density of charge at the space-time point (x,t). These transformations are much more general than those discussed above, and invariance under them leads to much more powerful and specific consequences.

For electromagnetism, the required interactions of matter with the electromagnetic field are predicted precisely. Thus, the theory of the electromagnetic field—Maxwell's equations and quantum electrodynamics—can be said to be the unique ideal embodiment of the abstract concept of a space-time-dependent symmetry, that is, of local gauge symmetry. *See* MAXWELL'S EQUATIONS; QUANTUM ELECTRODYNAMICS.

Frank Wilczek

Symmorphosis A theory of structural design of biological organisms postulating that structure is quantitatively matched to functional demand as a result of regulated morphogenesis during growth and maintenance. Symmorphosis is a theory of economic design. In biological organisms, all functions depend on structural design, specifically on the morphometric characteristics of the organs. In general terms, the larger the structure the greater the functional capacity. A central postulate of symmorphosis, as a theory of economic design, is that differences in the functional demand on an organ require quantitative adjustments of its structural design parameters in order to match functional capacity to (maximal) functional demand.

The notion that animals, and humans, should be designed economically follows from common sense, but it is also supported by many observations. Blood vessel architecture ensures blood flow distribution with minimal energy loss. Bone structure is patterned according to stress distribution and also quantitatively adapted to total stress. With training, athletes can specifically adjust the structure of their muscles and of their cardiovascular system to higher functional demands, and these modifications are soon reversed when training is stopped.

The theory of symmorphosis postulates ideal adaptation of structural design to functional capacity. This is, however, hardly a reasonable assumption, because good engineering design of complex systems requires some redundancies as safety factors in view of imperfections in functional performance and variable boundary conditions. By using the concept of symmorphosis, such deviations from idealized economic design can be detected.

Ewald R. Weibel

Sympathetic nervous system The portion of the autonomic nervous system concerned with nonvolitional preparation of the organism for emergency situations. *See* AUTONOMIC NERVOUS SYSTEM.

The sympathetic nervous system is best understood in mammals. It consists of two neuron chains from the thoracic and lumbar regions of the spinal cord to viscera and blood vessels. The first or preganglionic neuron has its cell body in the spinal cord and sends its axon to synapse with a postganglionic sympathetic neuron, which lies either in a chain of sympathetic ganglia paralleling the spinal cord or in a sympathetic ganglion near the base of the large blood vessels vascularizing the alimentary

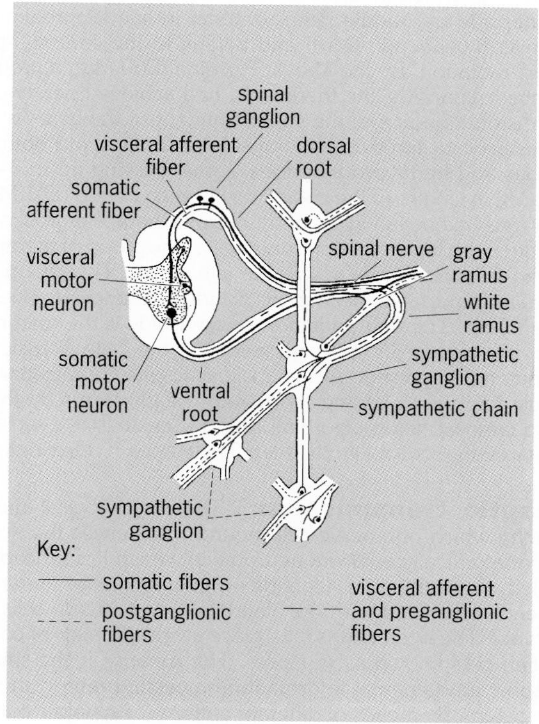

The visceral reflex arc and the sympathetic chain. (*After B. A. Houssay et al., Human Physiology, 2d ed., McGraw-Hill, 1955*)

viscera. The postganglionic axons are longer than the preganglionic axons and extend to glands or smooth muscles of viscera and blood vessels. Sensory visceral nerve fibers innervate blood vessels and viscera and carry sensory information to the spinal cord, thus providing a visceral reflex (see illustration). *See* NERVOUS SYSTEM (VERTEBRATE); PARASYMPATHETIC NERVOUS SYSTEM.
Douglas B. Webster

Sympathetic vibration The driving of a mechanical or acoustical system at its resonant frequency by energy from an adjacent system vibrating at this same frequency. Examples include the vibration of wall panels by sounds issuing from a loudspeaker, vibration of machinery components at specific frequencies as the speed of a motor increases, and the use of tuned air resonators under the bars of a xylophone to enhance the acoustic output. Increasing the damping of a vibrating system will decrease the amplitude of its sympathetic vibration but at the same time widen the band of frequencies over which it will partake of sympathetic vibration. *See* RESONANCE (ACOUSTICS AND MECHANICS); VIBRATION.
Lawrence E. Kinsler

Symphyla A class of the Myriapoda. The symphylans, like the pauropods, are tiny, pale, centipedelike creatures that inhabit humus or soil, or live under debris; in general, they live wherever there is sufficient moisture to preclude excessive water loss. They are similar to the Pauropoda and Diplopoda in being progoneate and anamorphic. Each of their mandibles, like those of millipedes, bears a movable gnathal lobe; at the same time their two pairs of maxillae are more reminiscent of the chilopods and lower insects than of the singly maxillate millipedes and pauropods. The class consists of three families to which not more than 60 species have been assigned.
Ralph E. Crabill

Synapsida The clade that includes all living and extinct species that share a closer relationship with modern mammals than with modern reptiles. The evolutionary history of synapsids is usually described in terms of three major radiations. The oldest synapsids are Middle Pennsylvanian in age (approximately 320 million years ago [Ma]) and belong to the earliest, "pelycosaur" radiation. By the Middle Permian (260 Ma), a group of advanced synapsids, the therapsids, had achieved nearly complete distribution across the supercontinent of Pangea. As with their pelycosaur forebears, therapsids diversified into both carnivorous and herbivorous species, some growing up to 3–4 m (10–13 ft) in length. At the end of the Permian Period (251 Ma), a major mass extinction killed off numerous therapsid groups and left others with far fewer representatives. One group of therapsids that did survive, however, was the cynodonts. This group is of particular importance because it includes the ancient ancestors of mammals. The final radiation of synapsids is the mammals, which first appear in the fossil record in the Late Triassic and continue to the present. All synapsid skulls are characterized by a single temporal opening, bordered in early forms by the jugal, squamosal, and postorbital bones. *See* ANIMAL EVOLUTION; MAMMALIA; PELYCOSAURIA; REPTILIA; THERAPSIDA.
Christian A. Sidor

Synaptic transmission The physiological mechanisms by which one nerve cell (neuron) influences the activity of an anatomically adjacent neuron with which it is functionally coupled. Brain function depends on interactions of nerve cells with each other and with the gland cells and muscle cells they innervate. The interactions take place at specific sites of contact between cells known as synapses. The synapse is the smallest and most fundamental information-processing unit in the nervous system. By means of different patterns of synaptic connections between neurons, synaptic circuits are constructed during development to carry out the different functional operations of the nervous system. *See* NEURON.

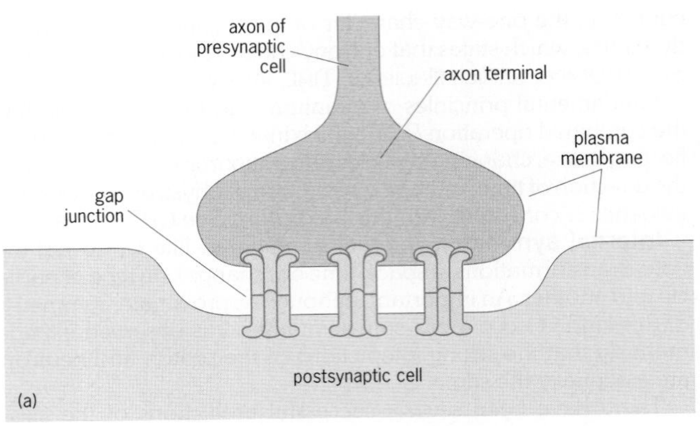

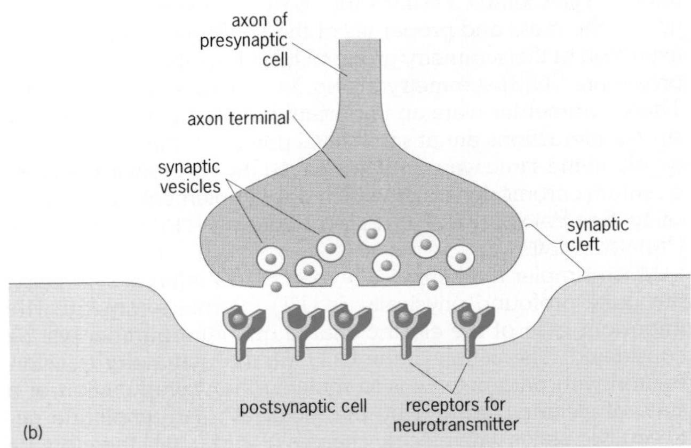

Types of synapses. (*a*) An electrical synapse, showing the plasma membranes of the presynaptic and postsynaptic cells linked by gap junctions. (*b*) A chemical synapse, showing neurotransmitters released by the presynaptic cell diffusing across the synaptic cleft and binding to receptors on the postsynaptic membrane.

The simplest type of synapse is the electrical synapse (see illustration), which consists of an area of unusually close contact between two cells packed with channels that span the two membranes and the cleft between them. Electrical synapses are also known as gap junctions. Electrical and metabolic communication between two cells is established by the components of the gap junctions. A variety of influences, including calcium ions, pH, membrane potential, neurotransmitters, and phosphorylating enzymes, may act on the channels to regulate their conductance in one direction (rectification) or both directions.

Electrical synapses are present throughout the animal kingdom. In vertebrates, they are numerous in the central nervous systems of fish as well as in certain nuclei of the mammalian brain, in regions where rapid transmission and synchronization of activity is important. Electrical synapses also interconnect glial cells in the brain. *See* BIOPOTENTIALS AND IONIC CURRENTS.

The most prevalent type of junction between nerve cells is the chemical synapse (see illustration). At chemical synapses neurotransmitters are released from the presynaptic cell, diffuse across the synaptic clefts, and bind to receptors on the postsynaptic cell. Chemical synapses are found only between nerve cells or between nerve cells and the gland cells and muscle cells that they innervate. The neuromuscular junction, that is, the junction between the axon terminals of a motoneuron and a muscle fiber, is a prototypical chemical synapse. Three basic elements constitute this synapse: a presynaptic process (in this case, the motoneuron axon terminal) containing synaptic vesicles; an end plate (a specialized site of contact between the cells); and a postsynaptic process (in this case, the muscle cell). The postsynaptic membrane contains receptors for the transmitter substance released from the presynaptic terminal.

The chemical substance that serves as the transmitter at the vertebrate neuromuscular junction is acetylcholine (ACh). Within the nerve terminal, acetylcholine is concentrated within small spherical vesicles. At rest, these vesicles undergo exocytosis at low rates, releasing their acetylcholine in quantal packets to diffuse across the cleft and bind to and activate the postsynaptic receptors. Each quantum gives rise to a small depolarization of the postsynaptic membrane. These miniature end-plate potentials are rapid, lasting only some 10 milliseconds, and small in amplitude, only some 500 microvolts, below the threshold for effecting any response in the muscle. They represent the resting secretory activity of the nerve terminal.

When an organism wants to move its muscles, electrical impulses known as action potentials are generated in the motoneurons. These are conducted along the axon and invade the terminal, causing a large depolarization of the presynaptic membrane. This opens special voltage-gated channels for calcium ions, which enter the terminal and bind to special proteins, causing exocytosis of vesicles simultaneously. Calcium ions are the crucial link between the electrical signals in the presynaptic neuron and the chemical signals sent to the postsynaptic neuron. The combined action of this acetylcholine on postsynaptic receptors sets up a large postsynaptic depolarization, which exceeds the threshold for generating an impulse in the surrounding membrane, and causes the muscle to contract. The action of acetylcholine at its receptor is terminated by an enzyme, acetylcholinesterase, which is present in the synaptic cleft and hydrolyzes the acetylcholine to acetate and choline.

In order for neurotransmission to continue, synaptic vesicles have to be regenerated. Vesicles are rapidly and efficiently reformed in the nerve terminal by endocytosis. Specific neurotransmitter transporters within the membrane then fill the synaptic vesicle with appropriate neurotransmitter. The regenerated vesicle either returns to the plasma membrane where it rejoins the releasable pool of vesicles, or remains in the nerve terminal as part of a reserve pool. *See* ACETYLCHOLINE; MUSCLE.

In the central nervous system the presynaptic process containing synaptic vesicles is most often an axon terminal and the postsynaptic process a dendrite, making an axodendritic synapse, but other relationships are also seen. The effect of transmitter on a postsynaptic cell is either excitatory or inhibitory, meaning that it either depolarizes or hyperpolarizes the membrane. Whether a transmitter has an excitatory or inhibitory effect on a cell is determined by the type of ion able to pass through the cell's receptor channels.

There are two principal types of central synapses.

Type 1 central synapses commonly release an amino acid transmitter, such as glutamate, whose action produces an excitatory postsynaptic potential. At low levels of activity the glutamate binds to the glutamate receptor and activates a relatively small conductance increase for sodium and potassium ions. Glutamate is the transmitter at many excitatory synapses throughout the central nervous system. The type 2 central synapse is usually associated with inhibitory synaptic actions. The most common inhibitory transmitter is gamma-amino butyric acid (GABA). Most inhibitory interneurons in different regions of the brain make these kinds of synapses on relay neurons in those regions. The GABA receptor is a complex channel-forming protein with several types of binding sites and several conductance states.

Other classes of synaptic transmitter substances include the biogenic amines, such as the catecholamines, norepinephrine and dopamine, and the indoleamine 5-hydroxytryptamine (also known as serotonin). *See* DOPAMINE; NORADRENERGIC SYSTEM; SEROTONIN.

A final type of transmitter substance consists of a vast array of neuropeptides. They are a diverse group and include substances that stimulate the release of hormones; those that act at synapses in pain pathways in the brain (the endogenous morphinelike substances, enkephalins and endorphins); and many of still undetermined functions. Peptides may act also indirectly, modifying the state of a receptor in its response to other transmitter substances, and they may do this in an activity-dependent manner. In view of the complexity and slow time course of many of their effects, these peptides are often referred to as neuromodulators. *See* ENDORPHINS; HORMONE.

At central synapses, rapid responses to transmitters (typically within milliseconds) are most commonly the result of direct synaptic transmission, in which the receptor itself is the ion channel. Such receptors are referred to as ionotropic. There are, however, other receptor molecules for each of the neurotransmitters, and many of these are not themselves ion channels. They are known as metabotropic receptors, and they affect neurotransmission indirectly via a set of intermediary proteins called G-proteins. Activated G-proteins have a variety of effects on synaptic processes. Some are mediated by direct interactions with ion channels. However, many other effects are mediated by activation of cellular second messenger systems involving messengers such as calcium and cyclic adenosine monophosphate (cyclic AMP). The time courses for effects caused by activation of G-protein-coupled receptors is much longer than that of ionotropic channels (milliseconds), reflecting the lifetime of the activated G-protein subunits (seconds) and second messengers (seconds to minutes). Such longer lasting signals greatly increase the complexity of chemical neurotransmission and synaptic modulation.

The synapse is a dynamic structure whose function is very dependent on its activity state. In this way the synapse is constantly adjusted for its information load. At glutamatergic synapses, high levels of input activity bring about a different transmission state. The buildup of postsynaptic depolarization relieves the normal block of a specialized glutamate receptor channel, permitting influx of calcium ions into the postsynaptic process. Since the conductance of the channel is dependent on the depolarization state of the membrane, it is said to have a voltage-gated property, in addition to being ligand-gated by its transmitter. The calcium ion acts as a second messenger to bring about a long-lasting increase in synaptic efficacy, a process known as long-term potentiation. The conjunction of increased pre- and postsynaptic activity to give an increase in synaptic efficacy is called Hebbian, and is believed to be the type of plasticity mechanism involved in learning and memory. *See* LEARNING MECHANISMS; MEMORY.

The synapse is one of the primary targets of drug actions. The first example to be identified was the arrow poison curare, which blocks neuromuscular transmission by binding to acetylcholine receptor sites. Many toxic agents have their actions on specific types of receptors; organic fluorophosphates, for example, are widely used pesticides that bind to and inactivate acetylcholinesterase. Most psychoactive drugs exert their effects at the synaptic level. *See* NERVOUS SYSTEM (VERTEBRATE); NEUROBIOLOGY.

Gordon M. Shepherd; Phyllis I. Hanson

Synbranchiformes An order of eel-like fishes in the class Actinopterygii, commonly called swamp eels. Synbranchiforms are among the most highly specialized fishes. They are characterized by a usually scaleless eel-like body lacking pelvic fins with pectoral fins present or absent; gill openings entirely ventral and often confluent across the breast; premaxillae lacking ascending processes and nonprotrusible; ectopterygoid bones enlarged; endopterygoids reduced or absent; swim bladder present or absent; and gills poorly developed, with respiration in some species accomplished in part by highly vascularized buccopharyngeal pouches. The order has no fossil record and comprises three families, 15 genera, and about 99 species, all but three of which are freshwater inhabitants. Synbranchiformes are thought to form a monophyletic group with the smegmamorphs. *See* ACTINOPTERYGII.

Swamp eels in the family Synbranchidae are the only teleosts to possess an amphistylic jaw suspension (the upper jaw is attached directly to the skull). They have a maximum total length of 150 cm (59 in.). Most species are protogynous hermaphrodites (first females, then males). Most species can breathe air, with some of them highly evolved in this capacity by having lunglike suprabranchial pouches. Swamp eels are primarily freshwater (all but three species), living in swamps, in caves, and in sluggish waters of tropical and subtropical western Africa, Asia, the Indo-Australian Archipelago, and Mexico to South America. Swamp eels are capable of overland excursions, and some species can live out of water for extended periods by burrowing in the substrate at the onset of a dry season. Herbert Boschung

Syncarida A unique superorder of malacostracan Crustacea, noted particularly for the total lack of a carapace or carapace shield. Two orders are recognized, Anaspidacea and Bathynellacea. The Stygocarididae, at one time considered a distinct order, are now assigned familial rank within the Anaspidacea.

The syncarid body plan is simple, consisting of a cephalon (head and sometimes one thoracic somite), thorax of seven or eight somites, and an abdomen of six somites and telson or five somites and pleotelson (fused sixth somite-telson). Eyes, with apposition optics, may be stalked or sessile, or sometimes absent. There are six to eight thoracic appendages (thoracopods), usually two to five abdominal appendages (pleopods), and uropods. Sexes are separate. Young hatch as immature adults and pass through several molts before reaching maturity.

Extant syncarids are specially adapted to fresh-water, often interstitial environments. Whereas the Anaspidacea are restricted to the Southern Hemisphere (Australia, Tasmania, New Zealand, South America), bathynellids have been found worldwide, except in Antarctica. *See* BATHYNELLACEA; CRUSTACEA; MALACOSTRACA. Patsy A. McLaughlin

Synchronization The process of maintaining one operation in step with another. The commonest example is the electric clock, whose motor rotates at some integral multiple or submultiple of the speed of the alternator in the power station. In television, synchronization is essential in order that the electron beams of receiver picture tubes are at exactly the same spot on the screen at each instant as is the beam in the television camera tube at the transmitter. *See* TELEVISION. John Markus

Synchronous converter A synchronous machine used to convert alternating current (ac) to direct current (dc), or vice versa. The ac-to-dc converter has been superseded by solid-state converters using silicon controlled rectifiers (SCRs) or power MOSFETs (for reasons of efficiency, lower maintenance costs, and less trouble) or by motor-generator sets. Converters are no longer manufactured, but there are converters still in use. *See* DIRECT-CURRENT GENERATOR; SYNCHRONOUS MOTOR. Loyal V. Bewley

Synchronous motor An alternating-current (ac) motor which operates at a fixed speed proportional to the frequency of the applied ac power (commonly known as the synchronous speed). A synchronous machine may operate as a generator, motor, or capacitor (reactive power source) depending only on its applied shaft torque (whether positive, negative, or zero) and its excitation. There is no fundamental difference in the theory, design, or construction of a machine intended for any of these roles, although certain design features are stressed for each of them. Operating within an electric power system, the machine may change its role depending on prevailing system conditions. For these reasons it is preferable not to set up separate theories for synchronous generators, motors, and capacitors. It is better to establish a general theory which is applicable to all three and in which the distinction between them is merely a difference in the direction of the currents and the sign of the torque angles. *See* ALTERNATING-CURRENT GENERATOR; ALTERNATING-CURRENT MOTOR; SYNCHRONOUS CAPACITOR For special types of synchronous motors *see* HYSTERESIS MOTOR; RELUCTANCE MOTOR. Richard T. Smith

Synchrotron radiation Electromagnetic radiation emitted by relativistic charged particles following a curved path in magnetic or electric fields. Synchrotron radiation is generally produced in laboratories by a beam of relativistic (moving near the speed of light) electrons or positrons in a generally circular path in a specialized synchrotron called a storage ring. A storage ring produces electromagnetic radiation with high flux, brightness, and coherent power levels. Synchrotron radiation is used for a wide variety of basic and applied research in biology, chemistry, and physics, as well as for applications in medicine and technology. *See* ELECTROMAGNETIC RADIATION; PARTICLE ACCELERATOR; RELATIVISTIC ELECTRODYNAMICS.

Electron storage rings provide radiation over a very broad range of photon energies or wavelengths. Radiation is generally produced in the infrared, visible, near-ultraviolet, vacuum-ultraviolet, soft x-ray, and hard x-ray parts of the electromagnetic spectrum, depending on the energy of the electrons in the storage ring and the strength of the magnetic field used to produce the radiation. Lasers are a much brighter source of radiation in the visible than are storage rings, so synchrotron radiation is only rarely used in the visible part of the spectrum. However, synchrotron radiation is particularly useful for producing x-ray beams.

Numerous storage ring sources of synchrotron radiation are in operation, construction, or design in many countries. The flux [photons/(second, unit bandwidth)], brightness (or brilliance) [flux/(unit source size, unit solid angle)], and coherent power (important for imaging applications and proportional to brightness) available for experiments—particularly in the vacuum-ultraviolet, soft x-ray, and hard x-ray parts of the spectrum—are steadily increasing as higher-performance storage rings are constructed. A beam from a modern storage ring has flux, brightness, and coherent power many orders of magnitude higher than is available from other sources.

Synchrotron radiation has many features (natural collimation, high intensity and brightness, broad spectral bandwidth, high polarization, pulsed time structure, small source size, and high-vacuum environment) that make it ideal for a wide variety of applications in experimental science and technology. Very powerful sources of synchrotron radiation in the ultraviolet and x-ray parts of the spectrum became available when high-energy physicists began operating electron synchrotrons in the 1950s. Although synchrotrons produce large amounts of radiation, their cyclic nature results in pulse-to-pulse variations, rendering difficult their use in many types of research. By contrast, the electron-positron storage rings developed for colliding-beam experiments starting in the 1960s offered a constant electron-beam intensity and much better stability. Beam lines to bring synchrotron radiation from a storage ring to users were constructed on both synchrotrons and storage rings to allow the radiation produced in the bending magnets of these machines to leave the ring vacuum system and reach experimental stations. In most cases the research programs were pursued on a parasitic basis, secondary to the high-energy physics programs. These parasitic light sources are now referred to as first-generation light sources.

Storage rings dedicated as synchrotron light sources are called second-generation light sources. Special arrays of magnets may be inserted into the straight sections between storage-ring bending magnets to produce beams with extended spectral range or with higher flux and brightness than is possible with the ring bending magnets. These devices, called wiggler and undulator magnets, utilize periodic transverse magnetic fields to produce transverse oscillations of the electron beam with no net deflection

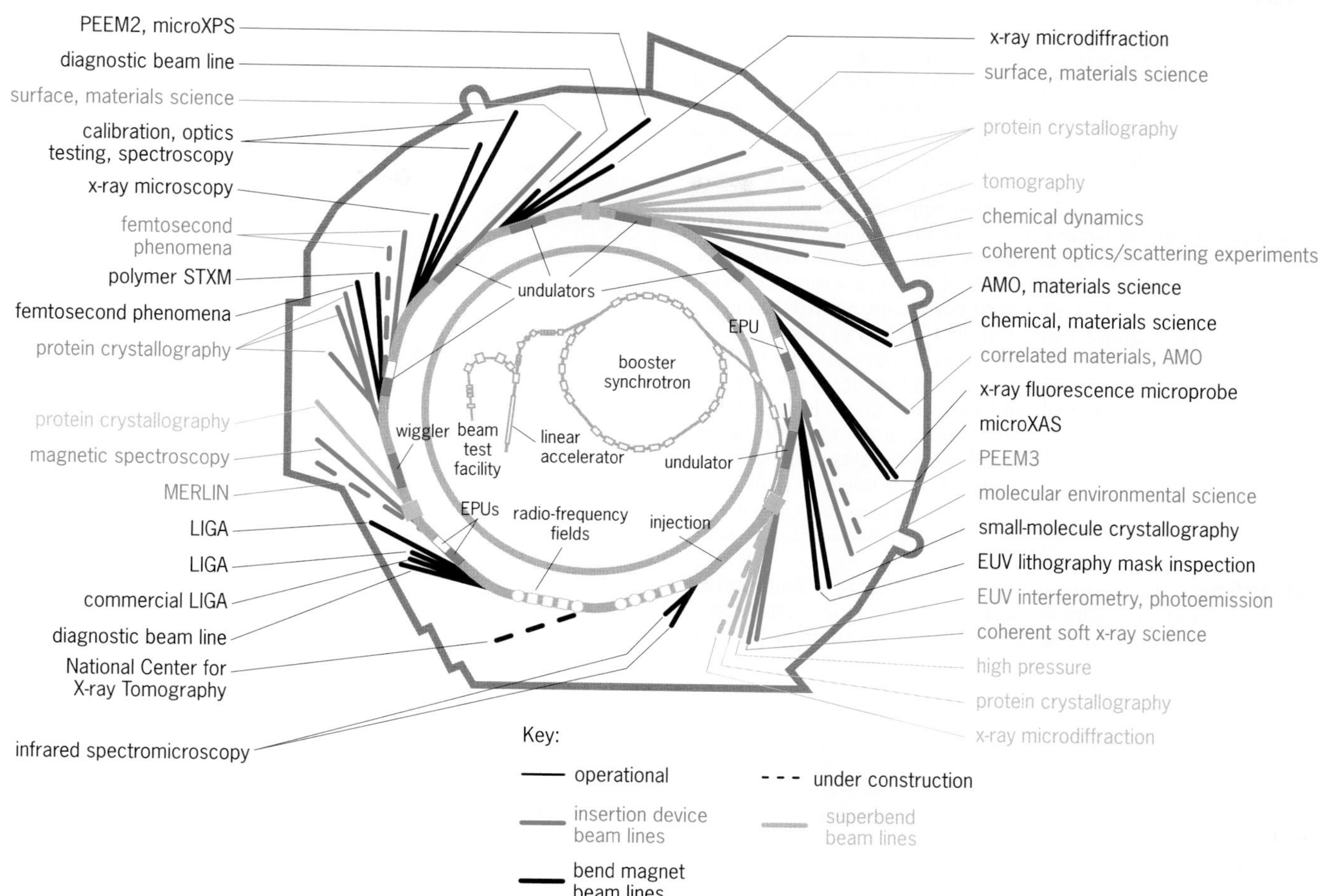

Layout of the 1.9-GeV Advanced Light Source at Lawrence Berkeley Laboratory, a low-energy, third-generation synchrotron radiation source, as of 2006. Applications of experimental stations on beam lines are indicated. EPU = elliptically polarizing undulator; PEEM = photoemission electron microscope; XPS = x-ray photoelectron spectroscopy; STXM = scanning transmission x-ray microscopy; LIGA = Lithographie, Galvanoformung, und Abformung (lithography, electroforming, and molding); AMO = atomic, molecular, and optical physics; XAS = x-ray absorption spectroscopy; EUV = extreme ultraviolet.

or displacement. They provide orders-of-magnitude increase in flux and brightness over storage-ring bending magnets, again opening up new research opportunities.

Third-generation light sources are storage rings with many straight sections for wiggler and undulator insertion device sources and with a smaller transverse size and angular divergence of the circulating electron beam. The product of the transverse size and divergence is called the emittance. The lower the electron-beam emittance, the higher the photon-beam brightness and coherent power level. With smaller horizontal emittances and with straight sections that can accommodate longer undulators, third-generation rings provide two or more orders of magnitude higher brightness and coherent power level than earlier sources. One consequence of the extraordinary brightness (brilliance) of these light sources is that the x-ray beam is partially coherent, allowing research in interferometry and speckle spectroscopy, among other uses. *See* COHERENCE; INTERFEROMETRY; SPECKLE.

A considerable number of third-generation rings are in operation. Third-generation light sources with a low-energy electron beam (typically 1–2 GeV) [see illustration] are generally optimized to produce high-brightness radiation in the vacuum ultraviolet (VUV) and soft (low-energy) x-ray spectral range, up to photon energies of about 2–3 keV. (Lower-brightness light extends to above 10 keV, and the use of superconducting bending magnets can extend the photon energy range to considerably higher values.) Third-generation light sources with a high-energy electron beam (6–8 GeV) can produce harder (higher-energy) x-rays, with energies of tens of kiloelectronvolts and above. Some of the newest storage rings that are intermediate in size, with beam energies up to 3 GeV, have even lower emittances than the first third-generation machines, and are equipped with advanced insertion-device technology with which they can also reach high brightness in the hard x-ray region.

The next logical step in the evolution of synchrotron radiation sources will be a fully coherent source, namely an x-ray laser. A fully coherent source will have a peak brightness 10 orders of magnitude greater than existing third-generation light sources. To produce a source this brilliant requires a very small and well-collimated electron beam; the best way to achieve a particle beam with the required characteristics is not with a storage ring but with a linear accelerator (linac). The beam accelerated by the linac will pass through a very long undulator (approximately 100 m or 300 ft in length) and generate x-ray beams of very high brightness through a process called self-amplified stimulated emission (SASE). Along with a 10,000-fold increase in average x-ray brightness over third-generation sources, fourth-generation free-electron laser sources will have subpicosecond pulse durations, resulting in x-ray beams with unparalleled instantaneous power levels. Several free-electron lasers are being built to produce vaccum ultraviolet and x-ray beams, such as the LCLS (Linac Coherent Light Source) at Stanford University. The unique properties of x-ray free-electron lasers, such as the instantaneous or peak brightness of a single ultrashort pulse, will surely open new fields of scientific research. *See* LASER.

The radiation produced by an electron in circular motion at low energy (speed much less than the speed of light) is weak and rather nondirectional. At relativistic energies (speed close to the speed of light), the radiated power increases markedly, and the emission pattern is folded forward into a cone with a half-opening angle in radians given approximately by mc^2/E, where mc^2 is the rest-mass energy of the electron (0.51 MeV) and E is the total energy of the electron. Thus, at electron energies of the order of 1 GeV, much of the very strong radiation produced is confined to a forward cone with an instantaneous opening angle of about 1 mrad (0.06°). At higher electron energies this cone is even smaller. The large amount of radiation produced combined with the natural collimation gives synchrotron radiation its intrinsic high brightness. Brightness is further enhanced by the small cross-sectional area of the electron beam, which is as low as 0.01 mm^2 in the third-generation storage rings.

Alfred S. Schlachter; Arthur L Robinson; Arthur Bienenstock; Dennis Mills; Gopal Shenoy; Herman Winick

Syncline In its simplest form, a geologic structure marked by the folding of originally horizontal rock layers into a systematically curved, concave upward profile geometry (illus. *a*). A syncline is convex in the direction of the oldest beds in the folded sequence, concave in the direction of the youngest beds. Although typically upright, a syncline may be overturned, recumbent, or upside down (illus. *d*). Synclines occur in all sizes, from microscopic to regional. Profile forms may be curved smoothly (illus. *a*) to sharply angular (illus. *b*). Fold tightness of a syncline, as measured by the angle at which the limbs of the syncline join, may be so gentle that the fold is barely discernible, to so tight that the limbs are virtually parallel to one another (illus. *c*). The orientation of the axis of folding is horizontal to shallowly plunging, but synclines may plunge as steeply as vertical.

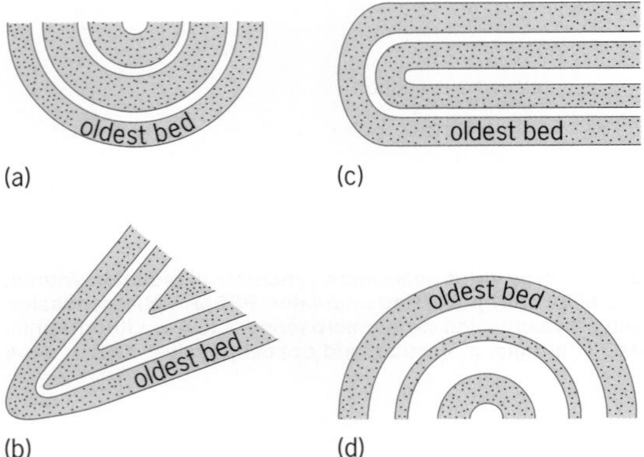

Varieties of synclines as seen in profile view. (*a*) Upright syncline with smoothly curved limbs. (*b*) Overturned, sharply angular syncline with planar limbs. (*c*) Recumbent, isoclinal syncline with parallel limbs. (*d*) Upside-down syncline, sometimes called an antiformal syncline.

Synclines are products of the layer-parallel compression that arises commonly during mountain building. The final profile form of the fold reflects the mechanical properties of the rock sequence under the temperature-pressure conditions of folding, and the percentage of shortening required by the deformation. *See* ANTICLINE; FOLD AND FOLD SYSTEMS.

George H. Davis

Synthetic aperture radar (SAR) Radar, airborne or satellite-borne, that uses special signal processing to produce high-resolution images of the surface of the Earth (or another object) while traversing a considerable flight path. The technique is somewhat like using an antenna as wide as the flight path traversed, that being the large "synthetic aperture," which would form a very narrow beam. Synthetic aperture radar is extremely valuable in both military and civil remote-sensing applications, providing surface mapping regardless of darkness or weather conditions that hamper other methods.

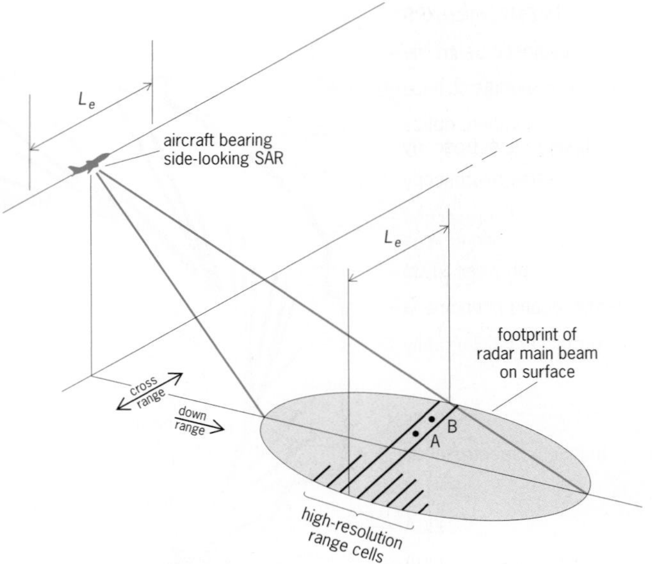

The basic idea of synthetic aperture radar (SAR); a side-looking case is illustrated. Two example scatterers, A and B, are shown in the ground scene. L_e = maximum flight path length for effective SAR processing.

Resolution is the quality of separating multiple objects clearly. In radar imaging, fine resolution is desired in both the down-range and cross-range dimensions. In radar using pulses, down-range resolution is achieved by using broad-bandwidth pulses, the equivalent of very narrow pulses, allowing the radar to sense separate echoes from objects very closely spaced in range. This technique is called pulse compression; resolution of a few nanoseconds (for example, 5 ns = 5×10^{-9} s gives about 0.75 m or 2.5 ft resolution) is readily achieved in modern radar.

Cross-range resolution is much more difficult to achieve. Generally, the width of the radar's main beam determines the cross-range, or lateral, resolution. For example, a 3° beam width resolves targets at a range of 185 km (100 nautical miles) only if they are separated laterally by more than 100 m (330 ft), not nearly enough resolution for quality imaging.

However, surface objects produce changing Doppler shifts as an airborne radar flies by. In side-looking radar (see illustration), even distant objects actually go from decreasing in range very slightly to increasing in range, producing a Doppler-time function. If the radar can sustain high-quality Doppler processing for as long as the "footprint" of the beam illuminates the scene, these Doppler histories will reveal the lateral placement of objects. In fact, if such processing can be so sustained, the cross-range resolution possible is one-half the physical width of the actual antenna being used, a few feet perhaps. Furthermore, this resolution is independent of range, quite unlike angle-based lateral resolution in conventional radar. *See* DOPPLER EFFECT.

Many synthetic aperture radars use other than just a fixed side-looking beam. Spotlighting involves steering the beam to sustain illumination for a longer time or to illuminate a designated scene at some other angle. The principles remain unchanged: fine resolution in both down-range and cross-range dimensions (achieved by pulse compression and Doppler processing, respectively) permits imaging with picture cells (pixels) of remarkably fine resolution. Many synthetic aperture radars today achieve pixels of less than 1 m (3 ft) square. Robert T. Hill

Synthetic fuel A gaseous, liquid, or solid fuel that does not occur naturally; also known as synfuel. Synthetic fuels can be made from coal, oil shale, or tar sands. Included in the category are various fuel gases, such as substitute natural gas and synthesis gas. *See* COAL; OIL SAND; OIL SHALE.

Syncrude is a synthetic crude oil, a complex mixture of hydrocarbons somewhat similar to petroleum. It is obtained from coal (liquefaction), from synthesis gas (a mixture of carbon monoxide and hydrogen), or from oil shale and tar sands. Syncrudes generally differ in composition from petroleum; for example, syncrude from coal usually contains more aromatic hydrocarbons than petroleum. Gaseous fuels can be produced from sources other than petroleum and natural gas. *See* PETROLEUM.

The most important source of synthetic crude oil is the tar sand deposit that occurs in northeastern Alberta, Canada. Tar sand is a common term for oil-impregnated sediments that can be found in almost every continent. The routes by which synthetic fuels can be prepared from coal involve either gasification or liquefaction.

Gasification can yield clean gases for combustion or synthesis gas, which has a controlled ratio of hydrogen to carbon monoxide. Catalytic conversion of synthesis gas to liquids (indirect liquefaction) can be carried out in fixed- and fluidized-bed reactors and in dilute-phase systems. Another method for producing synthetic fuels from coal involves gasification of the coal to a fuel gas that may be used as such or as a source of synthetic liquids. *See* COAL GASIFICATION; FLUIDIZATION.

Coal liquefaction is accomplished by four principal methods: direct catalytic hydrogenation, solvent extraction, pyrolysis, and indirect catalytic hydrogenation (of carbon monoxide). *See* COAL LIQUEFACTION; HYDROGENATION; PYROLYSIS; SOLVENT EXTRACTION.

Shale oil is readily produced by the thermal processing of oil shales. The basic technology is available, and commercial plants are operated in many parts of the world. James G. Speight

Syphilis A sexually transmitted infection of humans caused by *Treponema pallidum* ssp. *pallidum*, a corkscrew-shaped motile bacterium (spirochete). Due to its narrow width, *T. pallidum* cannot be seen by light microscopy but can be observed with staining procedures (silver stain or immunofluorescence) and with dark-field, phase-contrast, or electron microscopy. The organism is very sensitive to environmental conditions and to physical and chemical agents. The complete genome sequence of the *T. pallidum* Nichols strain has been determined. The nucleotide sequence of the small, circular treponemal chromosome indicates that *T. pallidum* lacks the genetic information for many of the metabolic activities found in other bacteria. Thus, this spirochete is dependent upon the host for most of its nutritional requirements. *See* BACTERIAL GENETICS; ELECTRON MICROSCOPE; IMMUNOFLUORESCENCE.

Syphilis is usually transmitted through direct sexual contact with active lesions and can also be transmitted by contact with infected blood and tissues. If untreated, syphilis progresses through various stages (primary, secondary, latent, and tertiary). Infection begins as an ulcer (chancre) and may eventually involve the cardiovascular and central nervous systems, bones, and joints. Congenital syphilis results from maternal transmission of *T. pallidum* across the placenta to the fetus. *See* SEXUALLY TRANSMITTED DISEASES.

Treponema pallidum is an obligate parasite of humans and does not have a reservoir in animals or the environment. Syphilis has a worldwide distribution. Its incidence varies widely according to geographical location, socioeconomic status, and age group. Although syphilis is controlled in most developed countries, it remains a public health problem in many developing countries. Studies have shown that syphilis is a risk factor for infection with the human immunodeficiency virus (HIV) since syphilitic lesions may act as portals of entry for the virus. There is little natural immunity to syphilis infection or reinfection.

Parenteral penicillin G is the preferred antibiotic for treatment of all stages of syphilis. Alternative antibiotics for syphilis treatment include erythromycin and tetracycline. There is currently no vaccine to prevent syphilis. However, it is anticipated that information obtained from the *T. pallidum* genome sequence will lead to further improvements in diagnostic tests for syphilis and to the eventual development of a vaccine that would prevent infection. *See* ANTIBIOTIC; PUBLIC HEALTH. Lola V. Stamm

Systellommatophora An order (or superorder) in the gastropod molluscan subclass Pulmonata containing three families of sluglike mollusks that lack any trace of a shell, have separate external male and female orifices, lack a mantle cavity, have a posterior anus and excretory pore, and bear eyes on the tops of two contractile, but not retractile, tentacles. The Rathouisiidae include about 30 species of carnivorous land slugs and range from South China and India to Queensland, Australia, in very wet tropical forests. The Veronicellidae comprise about 200 species of slugs that feed on decaying plant matter. The Onchidiidae include perhaps 250 species of intertidal to high-subtidal dwellers, reaching greatest abundance in the rocky shore areas of the tropics. They range from a few inches to nearly 4 in. (100 mm) in length. *See* PULMONATA. G. Alan Solem

System design evaluation A comprehensive and vigorous assessment of the effectiveness and cost of competing system designs, in order to choose the best candidate. System design evaluation is essential within the systems engineering process. It should be embedded appropriately within the process and then pursued continuously as system design and development progress.

The systems engineering process is a morphology for linking technology to customer needs. It is composed of three major activities: synthesis, analysis, and evaluation. Evaluation of each candidate system design is accomplished after receiving design-dependent parameter values for the candidate. It is the specific values for design-dependent parameter's that differentiate (or instance) candidate systems. Each candidate is optimized before being subjected to the design decision schema. It is here that the best candidate is sought (based on the customer's subjective evaluation). *See* SYSTEMS ENGINEERING.

To be comprehensive, system design evaluation must encompass both system effectiveness and life-cycle cost. To be rigorous, evaluation should be pursued with the aid of models and simulation. Even when expert opinion must be used in place of formal analytical approaches, a modeling structure may offer guidance. *See* MODEL THEORY; SIMULATION.

The factors used to evaluate candidate system design must be aggregated to make them apparent. Accordingly, an evaluation process must be followed which combines evaluation factors and decision making. The candidate system design that complies best with the customer's requirement is the preferred choice and may be implemented. *See* DECISION SUPPORT SYSTEM; OPERATIONS RESEARCH; OPTIMIZATION; SYSTEMS ANALYSIS. Wolter J. Fabrycky

System families Large systems that are formed from a variety of component systems: custom systems, which are newly engineered from the "ground up;" existing commercial-off-the-shelf (COTS) systems that are subsequently tailored for a particular application; and existing or legacy systems. Such related terms as systems of systems (SOS), federations of systems (FOS), federated systems of systems (F-SOS), and coalitions of systems (COS) are often used to characterize these systems. Modern systems have five characteristics, initially summarized by Mark Maier, that make one of the system family designations appropriate:

1. *Operational independence of the individual systems.* A system of systems is composed of systems that are independent and useful in their own right. If a system of systems is disassembled, the constituent systems are capable of independently performing useful operations by themselves and independently of one another.

2. *Managerial independence of the systems.* The component systems generally operate independently in order to achieve the technological, human, and organizational purposes of the individual unit that operates the system. These component systems are generally individually acquired, serve an independently useful purpose, and often maintain a continuing operational existence that is quite independent of the larger system of systems.

3. *Geographic distribution.* Geographic dispersion of the constituent systems in a system of systems is often quite large. Often, these constituent systems can readily exchange only information and knowledge with one another, and not substantial quantities of physical mass or energy.

4. *Emergent behavior.* The system of systems performs functions and carries out purposes that do not reside uniquely in any of the constituent systems. These behaviors arise as a consequence of the formation of the entire system of systems and are not the behavior of any constituent system. The principal purposes supporting engineering of these individual systems and the composite system of systems are fulfilled by these emergent behaviors.

5. *Evolutionary and adaptive development.* A system of systems is never fully formed or complete. Development of these systems is evolutionary and adaptive over time, and structures, functions, and purposes are added, removed, and modified as experience of the community with the individual systems and the composite system grows and evolves.

In a formal sense, almost anything could be regarded as a system of systems. A personal computer is a system. However, the monitor, microprocessor, disk drive, and random-access memory are also systems. Thus, should a personal computer be called a system of systems? Is the Internet a system of systems? The difference in these two examples is that the former system may be monolithic in purpose, while the latter is capable of supporting myriad communications and commerce purposes. *See* INTERNET; MICROCOMPUTER.

Often, appropriate missions exist for relatively large systems of systems in which there is a very limited amount of centralized command-and-control authority. Instead, a coalition of partners has decentralized power and authority and potentially differing perspectives of situations. It is useful to term such a system a "federation of systems" and sometimes a "coalition of systems." The participation of the federation or coalition of partners is based upon collaboration and coordination to meet the needs of the federation or coalition. *See* DISTRIBUTED SYSTEMS (CONTROL SYSTEMS); LARGE SYSTEMS CONTROL THEORY; SYSTEMS ARCHITECTURE; SYSTEMS ENGINEERING; SYSTEMS INTEGRATION. Andrew P. Sage

Systematics The comparative analysis of living and fossil species, including their discovery, description, evolutionary relationships to other species, and patterns of geographic distribution.

Systematics can be divided into four major fields. Taxonomy, often equated with systematics, is the discipline concerned with the discovery, description, and classification of organism groups, termed taxa (singular, taxon). Classification is the clustering of species into a hierarchical arrangement according to some criterion, usually an understanding of their relationships to other species. Phylogenetic analysis, an increasingly important aspect of systematics, is the discovery of the historical, evolutionary relationships among species; this pattern of relationships is termed a phylogeny. The fourth component of systematics is biogeography, the study of species' geographic distributions. Historical biogeography examines how species' distributions have changed over time in relationship to the history of landforms, ocean basins, and climate, as well as how those changes have contributed to the evolution of biotas (groups of species living together in communities and ecosystems).

Systematic data and interpretations underlie progress in all of biology. An understanding of relationships, in particular, is fundamental for interpreting comparative data across different kinds of organisms, whether those data be morphological, physiological, or biochemical. Joel Cracraft

Systems analysis The application of mathematical methods to the study of complex human physical systems. A system is an arrangement or collection of objects that operate together for a common purpose. The objects may include machines (mechanical, electronic, or robotic), humans (individuals, organizations, or societal groups), and physical and biological entities. Everything excluded from a system is considered to be part of the system's environment. A system functions within its environment. Examples of systems include the solar system, a regional ecosystem, a nation's highway system, a corporation's production system, an area's hospital system, and a missile's guidance system. A system is analyzed so as to better understand the relationships and interactions between the objects that compose it and, where possible, to develop and test strategies for managing the system and for improving its outcomes.

The term "systems analysis" is reserved for the study of systems that include the human element and behavioral relationships between the system's human element and its physical and mechanical components, if any. Examples of public policy systems are the federal government's welfare system, a state's criminal justice system, a county's educational system, a city's public safety system, and an area's waste management system. Examples of industrial systems are a manufacturer's production distribution system and an oil company's exploration, production, refining, and marketing system. Examples with physical environmental components are the atmospheric system and a water supply system. The direct transfer of systems engineering concepts to the study of a system in which the human element must be considered is restricted by limitations in the ability to comprehend and quantify human interactions. (Operations research, a related field of study, is directed toward the analysis of components of such systems. Public policy analysis is the term used for a system study of a governmental problem area.) *See* DECISION THEORY; OPERATIONS RESEARCH; SYSTEMS ENGINEERING.

Systems comprise interrelated objects, with the objects having a number of measurable attributes. A mathematical model of a system attempts to quantify the attributes and to relate the objects mathematically. The resultant model can then be used to study how the real-world system would behave as initial conditions, attribute values, and relationships are varied systematically. *See* MODEL THEORY.

The systems analysis process is an iterative one that cycles repeatedly through the following interrelated and somewhat indistinct phases: (1) problem statement, in which the system is defined in terms of its environment, goals, objectives, constraints, criteria, actors (decision makers, participants in the system, impacted constituency), and other objects and their attributes; (2) alternative designs, in which solutions are identified; (3) mathematical formulation, in which a mathematical description of the system is developed, tested, and validated; (4) evaluation of alternatives, in which the mathematical model is used to evaluate and rank the possible alternative designs by means of the criteria; and (5) selection and implementation of the most preferred solution. The process includes feedback loops in which the outcomes of each phase are reconsidered based on the analyses and outcomes of the other phases. For example, during the implementation phase, constraints may be uncovered that hinder the solution's implementation and thus cause the mathematical model to be reformulated. The analysis process continues until there is evidence that the mathematical structure is suitable; that is, it has enough validity to yield answers that are of value to the system designers or the decision maker. *See* OPTIMIZATION; SIMULATION.

As originally developed, systems analysis studies have been applied to those areas that are "hard" in that they are well defined and well structured in terms of objectives and feasible alternative systems (for example, blood-bank design, and integrated production and inventory processes). The aim of hard systems analysis is to select the best feasible alternative. In contrast, soft systems are concerned with problem areas that involve ill-defined and unstructured situations, especially those that have strong political, social, and human components. These generally involve public and private organizations (for example, design of a welfare system, and structure and impact of a corporate mission statement). The objectives of soft systems and the means to accomplish them are problematical and, in fact, a systemic view of the problem area is not assumed. The aim of soft systems analysis is to find a plan of action that accommodates the different interests of its human actors.

There is also need for further study of large-scale systems, which by definition are most complex. It is important to find ways to describe mathematically the systems that represent the totality of an industrial organization, the pollution concerns of a country and a continent, or the worldwide agricultural system. These are multicriteria problems with the solutions conflicting across criteria, individuals, and countries. The possibility that such systems may be studied in a computer-based laboratory is very promising. But this challenge must be approached cautiously, with the awareness that the methods and models employed are only abstractions to be used with due consideration of the goals of the individual and society. *See* LARGE SYSTEMS CONTROL THEORY; LINEAR SYSTEM ANALYSIS.

Saul I. Gass

Systems architecture

Systems architecture The discipline that combines system elements which, working together, create unique structural and behavioral capabilities that none could produce alone. The word "architecture" is commonly used to describe the underlying structure of networks, command-and-control systems, spacecraft, and computer hardware and software. The degree to which well-designed systems-level architectures are critical to the success of large-scale projects—or the lack thereof to failure—has been dramatically demonstrated. The explosion of technological opportunities and customer demands has driven up the size, complexity, costs, and investment risks of such projects to levels feasible for only major companies and governments. Without sound systems architectures, these projects lack the firm foundation and robust structure on which to build.

Complexity and its consequences. Systems are collections of dissimilar elements which collectively produce results not achievable by the elements separately. Their added value comes from the relationships or interfaces among the elements. (For example, open-loop and closed-loop architectures perform very differently.) But this value comes at a price: a complexity potentially too great to be handled by standard rules or rational analysis alone.

As projects have become ever more complex and multidisciplinary, new structures were needed for projects to succeed. Analytic techniques could not be used to find optimal solutions. Indeed, given the disparate perspectives of different customers, suppliers, and government agencies, unique optimal solutions generally would not exist. Instead, many possibilities might be good enough, with the choice dependent more on ancillary constraints or on the criteria for success than on detailed analysis.

Conceptual phases. As increasingly complex systems were built and used, it became clear that success or failure had been determined very early in their projects. In the early phases all the critical assumptions, constraints, choices, and priorities are made that will determine the end result. Unfortunately, no one knows in the beginning just what the final performance, cost, and schedule will be.

Systems-level architecture specifies how system-level functions and requirements are gathered together in related groups. It indicates how the subsystems are partitioned, the relationships between the subsystems, what communication exists between the subsystems, and what parameters are critical. It makes possible the setting of specifications, the analysis of alternatives at the subsystem level, the beginnings of detailed cost modeling, and the outlines of a procurement strategy.

There rarely is enough information early in the design stage for the client to decide on the relative priority of the requirements without having some idea of what the end system might be. Instead, provisional requirements and alternative system concepts have to be iterated until a satisfactory match is produced. Unavoidably, successful systems architecting in the conceptual phase becomes a joint process in which both client and architect participate heavily. In the ideal situation, the client makes the value judgments and the architect makes the technical decisions.

Systems-level architecture begins with a conceptual model, a top-level abstraction which attempts to discard features deemed not essential at the system level. Such a model is an essential tool of communication between client, architect, and builder, each viewing it from a different perspective. As the system comes into being, the model is progressively refined. *See* SOFTWARE ENGINEERING.

Eberhardt Rechtin

Systems ecology

Systems ecology The analysis of how ecosystem function is determined by the components of an ecosystem and how those components cycle, retain, or exchange energy and nutrients. Systems ecology typically involves the application of computer models that track the flow of energy and materials and predict the responses of systems to perturbations that range from fires to climate change to species extinctions. Systems ecology is closely related to mathematical ecology, with the major difference stemming from systems ecology's focus on energy and nutrient flow and its borrowing of ideas from engineering. Systems ecology is one of the few theoretical tools that can simultaneously examine a system from the level of individuals all the way up to the level of ecosystem dynamics. It is an especially valuable approach for investigating systems so large and complicated that experiments are impossible, and even observations of the entire system are impractical. In these overwhelming settings, the only approach is to break down the research into measurements of components and then assemble a system model that pieces together all components. An important contribution of ecosystem science is the recognition that there are critical ecosystem services such as cleansing of water, recycling of waste materials, production of food and fiber, and mitigation of pestilence and plagues. *See* ECOLOGICAL COMMUNITIES; ECOLOGICAL ENERGETICS; ECOLOGY; ECOSYSTEM; GLOBAL CLIMATE CHANGE; THEORETICAL ECOLOGY.

P. M. Kareiva

Systems engineering

Systems engineering A management technology involving the interactions of science, an organization, and its environment as well as the information and knowledge bases that support each. The purpose of systems engineering is to support organizations that desire improved performance. This improvement is generally obtained through the definition, development, and deployment of technological products, services, or processes that support functional objectives and fulfill needs.

Systems engineering has triple bases: a physical (natural) science basis, an organizational and social science basis, and an information science and knowledge basis. The natural science basis involves primarily matter and energy processing. The organizational and social science basis involves human, behavioral, economic, and enterprise concerns. The information science and knowledge basis is derived from the structure and organization inherent in the natural sciences and in the organizational and social sciences.

Systems engineering may also be defined as management technology to assist and support policy making, planning, decision making, and associated resource allocation or action deployment. It accomplishes this by quantitative and qualitative formulation, analysis, and interpretation of the impacts of action alternatives upon the needs perspectives, the institutional perspectives, and the value perspectives of clients to a systems engineering study. Each essential phase of a systems engineering effort—definition, development, and deployment—is associated with formulation, analysis, and interpretation. These enable systems engineers to define the needs for a system, develop the system, and deploy it in an operational setting and provide for maintenance over time, all within time and cost constraints.

Contemporary systems engineering focuses on tools, methods, and metrics, as well as on the engineering of life-cycle processes that enable appropriate use of these tools to produce trustworthy systems. There is also a focus on systems management to enable the wise determination of appropriate processes. *See* SYSTEMS INTEGRATION.

Much contemporary thought concerning innovation, productivity, and quality can be cast into a systems engineering framework. This framework can be valuably applied to systems engineering in general and information technology and software engineering in particular. The information technology revolution provides the necessary tool base that, together with knowledge management–enabled systems engineering and systems management, allows the needed process-level improvements for the development of systems of all types. The large number of ingredients necessary to accomplish needed change fit well within a systems engineering framework. Systems engineering constructs are useful not just for managing big systems engineering projects according to requirements, but for creative management of the organization itself. *See* INFORMATION SYSTEMS ENGINEERING; LARGE SYSTEMS CONTROL THEORY; QUALITY CONTROL; SYSTEMS ANALYSIS.

Andrew P. Sage

Systems integration A discipline that combines processes and procedures from systems engineering, systems management, and product development for the purpose of developing large-scale complex systems. These complex systems involve hardware and software and may be based on existing or legacy systems coupled with new requirements to add significant added functionality. Systems integration generally involves combining products of several contractors to produce the working system. Systems integration applications range from creation of complex inventory tracking systems to designing flight simulation models and reengineering large logistics systems.

Life-cycle activities. Application of systems integration processes and procedures generally follows the life cycle for systems engineering. Minimally, these systems engineering life-cycle phases are requirements definition, design and development, and operations and maintenance. For systems integration, these three phases are usually expanded to include feasibility analysis, program and project plans, logical and physical design, design compatibility and interoperability tests, reviews and evaluations, and graceful system retirement.

Primary uses. Systems integration is essential to the design and development of information systems that automate key operations for business and government. It is required for major procurements for the military services and for private businesses.

Advantages. Systems integration approaches enable early capture of design and implementation needs. The interactions and interfaces across existing system fragments and new requirements are especially critical. It is necessary that interface and intermodule interactions and relationships across components and subsystems that bring together new and existing equipment and software be articulated. The systems integration approach supports this through application of both a top-down and a bottom-up design philosophy; full compliance with audittrail needs, system-level quality assurance, and risk assessment and evaluation; and definition and documentation of all aspects of the program. It also provides a framework that incorporates appropriate systems management application to all program aspects. A principal advantage of this approach is that it disaggregates large and complex issues and problems into well-defined sequences of simpler problems and issues that are easier to understand, manage, and build. *See* INFORMATION SYSTEMS ENGINEERING; SYSTEMS ANALYSIS; SYSTEMS ENGINEERING.

James D. Palmer

Syzygy The alignment of three celestial objects within a solar system (or within any other system of objects in orbit about a star). Syzygy is most often used to refer to the alignment of the Sun, Earth, and Moon at the time of new or full moon. Alignments need not be perfect in order for syzygy to occur: because the orbital planes for any three bodies in the solar system rarely coincide, the geometric centers of three objects that are in syzygy almost never lie along the same line. *See* PHASE; SOLAR SYSTEM.

In general, syzygy occurs whenever an observer on one of the three objects would see the other two objects either in opposition or in conjunction. Opposition occurs when two objects appear 180° apart in the sky as viewed from a third object. Conjunction occurs when two objects appear near one another in the sky as seen from a third object.

Solar and lunar eclipses are dramatic results of syzygy. During a solar eclipse, when the Moon is in its new phase, the alignment of the Sun, Earth, and Moon is so nearly perfect that the Moon's shadow falls on the Earth. During a lunar eclipse, which occurs at the time of the full moon, the Moon passes through the Earth's shadow. *See* ECLIPSE.

An occultation is another type of eclipse that can occur during syzygy. For an Earth-based observer, an occultation occurs when the Moon is seen to pass in front of a planet or other member of the solar system. The occultation of a star by the Moon does not qualify as syzygy, since the star is far beyond the limits of the solar system. *See* OCCULTATION.

Harold P. Coyle

T

T Tauri star A young low-mass star characterized by variability, the presence of hydrogen emission lines, and association with dark or bright nebulae. T Tauri stars are named after a variable star in the constellation Taurus that exhibits particularly strong hydrogen emission. They were originally believed to be ordinary field stars passing through and interacting with a star-forming nebula. Their association with these regions was soon discovered to be more than coincidental, however, and they are now identified with the earliest phase of young stellar evolution, in which a star emerges from its natal molecular cloud to be detectable at visible wavelengths. T Tauri stars with very strong emission lines are designated classical T Tauri stars; their counterparts with reduced hydrogen emission are known as weak-line T Tauri stars. In addition to the original defining properties, T Tauri stars are generally accompanied by excess x-ray, ultraviolet, infrared, and millimeter-wave emission that arises from an accreting circumstellar disk of dust and gas. *See* MOLECULAR CLOUD.

The most sensitive infrared observations to date indicate that circumstellar disks are detected around virtually all classical T Tauri stars. Far fewer are observed around weak-line T Tauri stars. This is as expected by theoretical interpretations that link emission line strength and the accretion of gas from a disk onto the star. The low detection rate may indicate the presence of disks that have evolved toward the formation of a planetary system, or it may belie the existence of young stars with little or no circumstellar material to begin with. Small amounts of dust from the collisions of large boulders and planetesimals in "debris disks" with very little molecular gas are detected around some stars in the solar neighborhood. Disks with these properties are currently difficult to observe at the distance of nearby star-forming regions, but a handful of detections around weak-line T Tauri stars indicates that at least some are surrounded by transitional planetesimal disks. Taken as a whole, the properties, rate of occurrence, and inferred evolution for disks around classical and weak-line T Tauri stars provide strong evidence that planet formation is a common by-product of the star-forming process. *See* EXTRASOLAR PLANETS; INFRARED ASTRONOMY; PROTOSTAR; SOLAR SYSTEM; STELLAR EVOLUTION David Koerner

Tabulata One of two principal orders of extinct Paleozoic corals. Tabulate corals were exclusively colonial. Polyps secreted slender calcitic (calcium carbonate) tubes (corallites, ranging 0.5–20 mm in diameter, but predominantly 1–3 mm in diameter), polygonal in cross section when in contact, or cylindrical when surrounded by colonial skeletal material (coenenchyme) or not in contact (see illustration). The corallites are almost always partitioned by flat or curved, complete or incomplete plates (tabulae). Septal structures, as spines, or less commonly thin plates radially arranged in corallites, often number 12 when well developed. However, these structures are usually weakly developed or completely absent. The Tabulata is divided into six suborders—Lichenariina, Sarcinulina, Favositina, Halysitina, Heliolitina, and Auloporina—based mainly on the structural arrangement of corallites in the colony and the presence or absence of communication (mural pores or connecting tubules) between adjacent corallites. Tabulate coral colonies could take on a range of external forms. Shape was determined by the interaction of internal controls on colonial growth with prevailing environmental conditions. Tabulate corals were most abundant and diverse in temperate to warm shelf environments, particularly in biostromes and bioherms. Smaller, laminar to domal massive colonies inhabited deeper, cooler waters. They were an important component of true reefs, particularly in back-reef, reef-flat and fore-reef environments, but their limited ability of secure attachment to hard surfaces restricted their contribution to reef framework. *See* ANTHOZOA; HEXACORALLIA. Colin Scrutton

Widespread tabulate coral *Favosites*. Detail of surface showing polygonal corallites, ×1.5.

Tacan A member of the rho-theta family of air navigation systems which define an aircraft's position by its distance and bearing to a single beacon. Such systems inherently answer the navigator's questions of "in what direction" and "how far," without additional computation, and are of particular value when the beacon has to be placed on a ship, oil-drilling rig, or small island. The main weakness of such systems is that bearing errors cause spatial error to increase with distance from the beacon. Major attention is therefore given to the reduction of bearing errors.

Tacan allows the distance-measuring equipment (DME) to provide bearing service also, without the large antennas or site errors characteristic of the civil very high-frequency omnidirectional range (VOR). Range and accuracy are the same as DME (300 mi or 480 km, and 0.1 mi or 0.16 km, respectively), with a bearing accuracy of 1°. As in DME, operation is on 252 channels, spaced 1 MHz apart, 962–1213 MHz.

To provide the added bearing service, the DME transponder is first arranged to operate at constant duty cycle. This means that the number of output pulses is held constant, whether the beacon is being interrogated by one or a hundred aircraft.

The total output of the transponder is amplitude-modulated by the rotating directional antenna system (see illustration). At the center of this system is the central radiator connected to the DME transponder, just as in the conventional DME. However, rotating around this radiation at 15 revolutions per second are two concentric dielectric cylinders. The inside one, about 6 in. (15 cm) in diameter, contains a single parasitic reflector which imparts a 15-Hz amplitude modulation to the DME replies, and the outside one, about 33 in. (84 cm) in diameter, contains nine parasitic elements which impart a 135-Hz amplitude modulation. On the same rotating shaft are mounted reference pulse generators which additionally modulate the transmitter with coded pulses, once per revolution for the 15-Hz signal (called the north

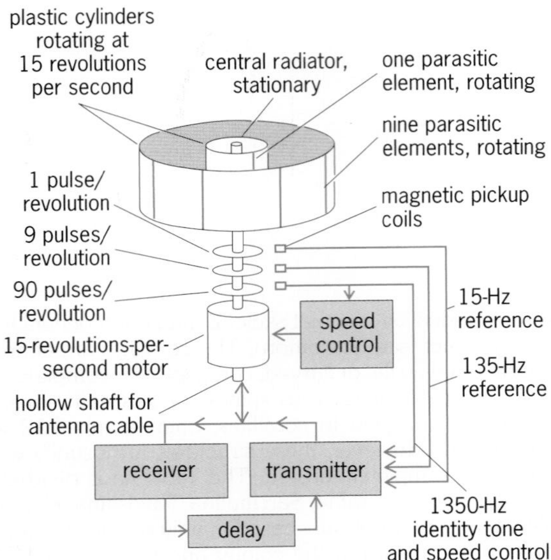

Tacan transponder with rotating directional antenna system.

reference burst) and nine times per revolution for the 135-Hz signal (called the auxiliary reference bursts).

In the airborne receiver the 15- and 135-Hz sine waves are detected and filtered and compared with the decoded reference bursts to provide a two-speed or fine-coarse bearing display whose accuracy and site freedom are superior to those of conventional VOR, yet with a ground antenna system which is small enough to mount on a ship's mast. *See* DISTANCE-MEASURING EQUIPMENT; DOPPLER VOR; ELECTRONIC NAVIGATION SYSTEMS; INSTRUMENT LANDING SYSTEM (ILS); RHO-THETA SYSTEM.

Sven H. Dodington

Tachometer An instrument that measures angular speed, as that of a rotating shaft. The measurement may be in revolutions over an independently measured time interval, as in a revolution counter, or it may be directly in revolutions per minute. The instrument may also indicate the average speed over a time interval or the instantaneous speed. Tachometers are used for direct measurement of angular speed and as elements of control systems to furnish a signal as a function of angular speed.

Alfred H. Wolferz

Tachyon A hypothetical faster-than-light particle consistent with the special theory of relativity. According to this theory, a free particle has an energy E and a momentum $\mathbf{p}$ which form a Lorentz four-vector. The length of this vector is a scalar, having the same value in all inertial reference frames. One writes, Eq. (1), where c is the speed of light and the parameter m^2 is a

$$E^2 - c^2\mathbf{p}^2 = m^2c^4 \tag{1}$$

property of the particle, independent of its momentum and energy. Three cases may be considered: m^2 may be positive, zero, or negative. The case $m^2 > 0$ applies for atoms, nuclei, and the macroscopic objects of everyday experience. The positive root m is called the rest-mass. If $m^2 = 0$, the particle is called massless. A few of these are known: the electron neutrino, the muon neutrino, the photon, and the graviton. The name tachyons (after a Greek word for swift) has been given to particles with $m^2 < 0$.

In general, the particle speed is given by Eq. (2). If $m^2 < 0$,

$$v = \frac{c\mathbf{p}}{E}c \tag{2}$$

Eq. (1) implies $E > c\mathbf{p}$ and Eq. (2) gives $v < c$. If $m^2 = 0$, then $E = c\mathbf{p}$ and $v = c$. In case $m^2 < 0$, one finds $E < c\mathbf{p}$ and $v > c$. Tachyons exist only at faster-than-light speeds. *See* ELEMENTARY PARTICLE; RELATIVITY.

Roland H. Good, Jr.

Taconite The name given to the siliceous iron formation from which the high-grade iron ores of the Lake Superior district have been derived. It consists chiefly of fine-grained silica mixed with magnetite and hematite. As the richer iron ores approach exhaustion in the United States, taconite becomes more important as a source of iron.

Cornelius S. Hurlbut, Jr.

Taeniodonta An extinct order of quadrupedal eutherian land mammals known from the early Cenozoic deposits of the Rocky Mountain intermontane basins of western North America and, based on a single tooth of *Ectoganus gliriformis*, from early Cenozoic rocks in South Carolina. The nine known genera of taeniodonts are classified into two families: (1) the medium-size (5–15 kg or 11–33 lb), relatively primitive, omnivorous conoryctids and (2) the larger (15–110 kg or 33–243 lb) and more advanced stylinodontids. Conoryctids developed enlarged canines, but the lower jaws were unspecialized and the cheek teeth were low-crowned, enamel-enclosed, and cuspidate. Stylinodontid taeniodonts developed deep massive jaws, peglike teeth that were ever-growing, large curved canines bearing enamel bands, and large laterally compressed and recurved claws on the front paws. In terms of modern analogs, an advanced stylinodontid may be thought of as an aardvark with the head of a pig. *See* EUTHERIA.

Robert M. Schoch

Taiga A zone of forest vegetation encircling the Northern Hemisphere between the arctic-subarctic tundras in the north and the steppes, hardwood forests, and prairies in the south. The chief characteristic of the taiga is the prevalence of forests dominated by conifers. The dominant trees are particular species of spruce, pine, fir, and larch. Other conifers, such as hemlock, white cedar, and juniper, occur locally, and the broad-leaved deciduous trees, birch and poplar, are common associates in the southern taiga regions. Taiga is a Siberian word, equivalent to "boreal forest." *See* TUNDRA.

The northern and southern boundaries of the taiga are determined by climatic factors, of which temperature is most important. However, aridity controls the forest-steppe boundary in central Canada and western Siberia. In the taiga the average temperature in the warmest month, July, is greater than 50°F (10°C), distinguishing it from the forest-tundra and tundra to the north; however, less than four of the summer months have averages above 50°F (10°C), in contrast to the summers of the deciduous forest further south, which are longer and warmer. Taiga winters are long, snowy, and cold—the coldest month has an average temperature below 32°F (0°C). Permafrost occurs in the northern taiga. It is important to note that climate is as significant as vegetation in defining taiga. Thus, many of the world's conifer forests, such as those of the American Pacific Northwest, are excluded from the taiga by their high precipitation and mild winters.

J. C. Ritchie

Tail assembly An assembly at the rear of an airplane, consisting of the tail cone, the horizontal tail, and one or more vertical tails.

The tail assembly, or empennage, of an airplane is normally composed of a vertical tail and a horizontal tail attached to the rear, or tail cone, of the airplane's fuselage. The vertical tail is composed of the vertical stabilizer and the rudder (see illustration). The vertical stabilizer is attached rigidly to the fuselage and is intended to provide stability about a vertical axis through the airplane's center of gravity. The rudder is attached by hinges to the rear of the vertical stabilizer and can rotate from side to side in response to pilot control input. It also contributes to stability,

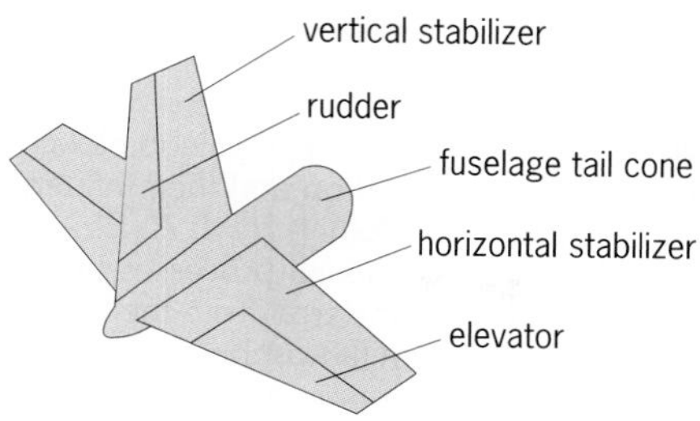

Tail assembly parts (normal configuration).

but its main function is to provide a yawing moment about the airplane's vertical (yaw) axis, thereby causing the airplane to yaw (turn) to the left or right. *See* AIRCRAFT RUDDER; FLIGHT CONTROLS; FUSELAGE; STABILIZER (AIRCRAFT).

The horizontal tail, similar to the vertical tail, is composed of the horizontal stabilizer and the elevator. The horizontal stabilizer is fixed rigidly to the fuselage and provides stability about a horizontal axis directed along the wing and through the center of gravity, and known as the pitch axis. The elevator is hinged to the rear of the horizontal stabilizer and rotates up and down as the pilot moves the control column fore and aft. The elevator also contributes to stability about the pitch axis, but its main purpose is to provide a pitching moment about the pitch axis, which causes the airplane to nose up or down. *See* ELEVATOR (AIRCRAFT).

Airplanes that operate at supersonic speeds usually have the horizontal tail swept back and in one piece that is movable and is controlled by the motion of the pilot's control stick. Such a surface is frequently called a stabilator. *See* SUPERSONIC FLIGHT.

Many airplanes employ empennages that depart from the normal configuration. *See* AIRPLANE. Barnes W. McCormick, Jr.

Takakiales An order of liverworts in the subclass Jungermanniidae, consisting of a single genus and two species. Some authors put the Takakiales in the Calobryales owing to branching from a prostrate branched stem, lack of rhizoids, copious mucilage secretion, and massive frequently scattered archegonia that lack protective envelopes. However, sporophytes have never been seen, and so it is difficult to demonstrate any meaningful relationship.

The members of the Takakiales consist of a very small gametophyte made up of a branched system of prostrate, leafless stolons and erect, radially organized branches with terete appendages. The "leafy" branches are simple or forked and have a weak central strand of narrow cells enclosed by larger cells. The leafy appendages are small and scalelike below, larger and crowded above, and variable in arrangement. The leaf cells lack oil bodies. Mucilage is secreted by simple filaments on leafy branches and by branched filaments clustered on both leafy and stoloniform branches. The archegonia are scattered and not enclosed by protective structures. *See* BRYOPHYTA; CALOBRYALES. Howard Crum

Talc A hydrated magnesium layer silicate (phyllosilicate) with composition close to $Mg_3Si_4O_{10}(OH)_2$. Talc commonly is white, but it may appear pale green or grayish depending on the amount of minor impurities. Talc has a greasy feel and pearly luster, and has been used as one of the hardness standards for rock-forming minerals with the value 1 on the Mohs scale. Because talc is soft, it can be scratched by fingernails. *See* HARDNESS SCALES.

Talc frequently occurs in magnesium-rich metamorphosed serpentinites and siliceous dolomites. Talc-rich rocks include massive soapstones, massive steatite, and foliated talc schists.

Talc is a good insulating material. It has been commonly used in industry as a raw material for ceramics, paints, plastics, cosmetics, papers, rubber, and many other applications. *See* SILICATE MINERALS. J. G. Liou

Tall oil A by-product from the pulping of pine wood by the kraft (sulfate) process. In the kraft process the wood is digested under pressure with sodium hydroxide and sodium sulfide. The volatilized gases are condensed to yield sulfate turpentine. During the pulping the alkaline liquor saponifies fats and converts the fatty and resin acids to sodium salts. Concentration of the pulping solution (black liquor) prior to recovery of the inorganic pulping chemicals allows the insoluble soaps to be skimmed from the surface. Acidification of the skimmed soap yields crude tall oil.

Crude tall oil from southern pines contains 40–60% resin acids (rosin), 40–55% fatty acids, and 5–10% neutral constituents. Abietic and dehydroabietic acids comprise over 60% of the resin acids, while oleic and linoleic acids predominate in the fatty acid fraction. Fatty acids from tall oil distillation may contain as much as 10–40% resin acids or as little as 0.5%.

Major uses of tall oil fatty acids as chemical raw materials are in coatings, resins, inks, adhesives, and soaps and detergents, and as flotation agents. Tall oil is an important source of rosin in the United States. *See* PINE TERPENE; ROSIN; WOOD CHEMICALS. Irving S. Goldstein

Tanaidacea An order of the eumalacostracans of the superorder Peracarida, derived from the genus *Tanais*. These animals have a worldwide distribution and with few exceptions are marine. They occur from the shore down to abyssal depths. They are free-living and benthonic. The order is divided into 2 suborders with 5 families, 44 genera, and about 350 species. The body is linear, more or less cylindrical or dorsoventrally depressed. Thoracic segments 1 and 2 are fused with the head, and the last abdominal segment is fused with the telson. Eight pairs of thoracic legs are present, of which the first pair are maxillipeds, the second pair chelipeds (the first pair of pereiopods), and the following six pairs pereiopods. *See* PERACARIDA. Karl Lang

Tangent A term describing a relationship of two figures (usually of the same dimension) in the neighborhood of a common point. The figures are tangent at a point P if they touch at P but do not intersect in a sufficiently small neighborhood of P. To be more precise, if P denotes a point of a curve C (see illustration), a line L is a tangent to C at P provided L is the limit

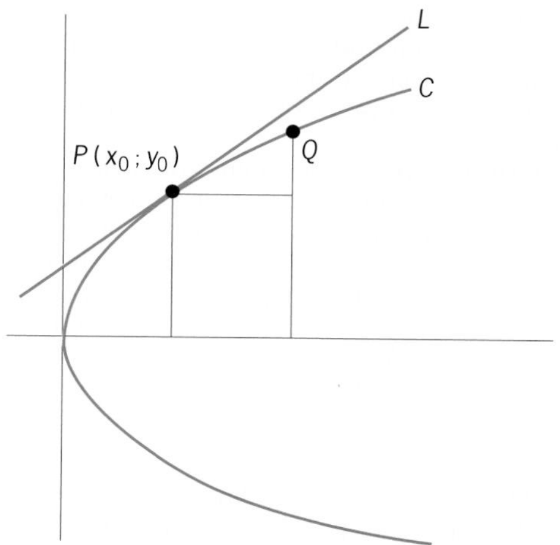

Line *L* tangent to curve *C* at *P*.

of lines joining P to a variable point Q of C, as Q approaches P along C (that is, for Q sufficiently close to P, the line PQ is arbitrarily close to L). Leonard M. Blumenthal

Tangerine A name applied to certain varieties of a variable group of loose-skinned citrus fruits belonging to the species *Citrus reticulata*. Although mandarin and tangerine are often used interchangeably to designate the whole group, tangerine is applied more strictly to those varieties (cultivars) having deep-orange or scarlet rinds, whereas the term mandarin is more properly used to include all members of this quite variable group of citrus fruits. *See* MANDARIN.

The fruits are deep orange, loose-skinned, and small to medium-sized, and possess small seeds with green cotyledons. Tangerines are easily peeled and eaten out of hand as fresh fruit. About one-third of the crop is utilized in juice, sherbets, and canned sections. Frank E. Gardner; C. Jack Hearn

Tantalum A chemical element, symbol Ta, atomic number 73, and atomic weight 180.948. It is a member of the vanadium group of the periodic table and is in the 5*d* transitional series. Oxidation states of IV, III, and II are also known. *See* PERIODIC TABLE; TRANSITION ELEMENTS.

Tantalum metal is used in the manufacture of capacitors for electronic equipment, including citizen band radios, smoke detectors, heart pacemakers, and automobiles. It is also used for heat-transfer surfaces in chemical production equipment, especially where extraordinarily corrosive conditions exist. Its chemical inertness has led to dental and surgical applications. Tantalum forms alloys with a large number of metals. Of special importance is ferrotantalum, which is added to austenitic steels to reduce intergranular corrosion.

The metal is quite inert to acid attack except by hydrofluoric acid. It is very slowly oxidized in alkaline solutions. The halogens and oxygen react with it on heating to form the oxidation-state-V halides and oxide. At high temperature it absorbs hydrogen and combines with nitrogen, phosphorus, arsenic, antimony, silicon, carbon, and boron. Tantalum also forms compounds by direct reaction with sulfur, selenium, and tellurium at elevated temperatures. Edwin M. Larsen

Tantulocarida A subclass of the Crustacea. Tantulocarids are minute ectoparasites, less than 0.01 in. (0.3 mm) in length, that infest deep-sea copepods, isopods, tanaids, and ostracods. As a result of their parasitic mode of life, adult females have lost all resemblance to crustaceans; males are free-living but nonfeeding. Infection of the host occurs at the tantalus larval stage, which is believed to occur immediately after the larva is released. The head, or cephalon, of the tantulus larva is covered by a dorsal shield that may protrude anteriorly to form a rostrum. The mechanism of attachment, the mouth tube, is located on the ventral surface with a distal mouth opening, and is surrounded by an oral disk.

Phylogenetic relations of the Tantulocarida are not clearly understood. Although tantulocarids lack the antennules and antennae characteristic of Crustacea, two clusters of sensory hairs in adult males are probably antennulary in origin. The basic tagmosis exhibited and male gonopore position suggest an affinity with some cirriped taxa. Some researchers include the Tantulocarida as a subclass of the Maxillopoda; others accept it as a distinct class. *See* CIRRIPEDIA; COPEPODA; CRUSTACEA; MAXILLOPODA.
Patsy A. McLaughlin

Taper pin A tapered self-holding pin used to connect parts together. Standard taper pins have a diametral tape $^{1}/_{4}$ in. to 12 in. (0.6 cm to 30 cm) and are driven in holes drilled and reamed to fit. The pins are made of soft steel or are cyanidehardened. They are sometimes used to connect a hub or collar to a shaft. Taper pins are frequently used to maintain the location of one surface with respect to another. A disadvantage of the taper pin is that the holes must be drilled and reamed after assembly of the connected parts; hence they are not interchangeable. *See* COTTER PIN. Paul H. Black

Taphonomy A subdiscipline of paleobiology that investigates the processes of preservation and their influence on information in the fossil record. Coined by J. A. Efremov in 1940, the term "taphonomy" involves all processes that affected the organism during its life, its transferral from the living world (biosphere) to the geological realm (lithosphere), and all physical and chemical interactions from the time of burial until collection. Besides the conspicuous characteristics of the preserved organism that can be seen easily—either morphological (external) and/or anatomical (internal) features—there are often less prominent details that record the fossil's history. Taphonomists are forensic scientists. By analyzing preserved details, paleontologists can understand an organism's mode of death or disarticulation; the biological processes that may have modified the remains before burial, including their use by hominids; the response of the organism or one of its parts to transport by animals, sediment, water, or wind; its residency time in a depositional setting before final entombment; and the alterations of tissues or skeletal parts within a wide range of chemical settings. The processes of fossilization appear to be environmentally site-specific, resulting in a mosaic of preservational traits in terrestrial and marine environments. Few fossil assemblages are exactly identical with regard to formative processes, but general patterns exist. An understanding of taphonomic assemblage features within its environmental context allows for a more accurate interpretation of the fossil record.

Once fossilized, organic remains successfully have passed through several taphonomic stages, including necrology, biostratinomy, and diagenesis. However, remains may be affected at any stage, effectively removing them from the fossilization process.

Necrology is the death or loss of a part of an organism. Death is a prerequisite for most animals, but not for plants that shed many different parts during their life cycle. Necrology may be induced by physiological means (old age, disease, or temporal or climatic shedding) or traumatic means (sudden catastrophic death or part loss in response to natural disruptions such as violent storms, volcaniclastic ash fall, mudflows, or flood events). The potential for any organic remains to successfully pass into the fossil record is dependent on the original biochemical composition. Organic remains of more resistant biochemicals or mineralized hard parts have a higher probability of fossilization than those of easily degraded compounds.

Biostratinomy focuses on the processes and interactions that follow necrology, until final burial. Biologic interactions include the effects of scavenging on carcasses, the use of discarded parts as domiciles, or borings in resistant structural parts. Abiotic processes include mechanical and physical alteration or breakdown under different transport conditions (fragmentation and rounding in river channels), orientation or concentration of resistant parts under varying hydrological regimes, or reexposure and reworking of previously buried remains in response to changing geological circumstances. This stage ends when organic remains are entombed within sediment, effectively isolating the material from the effects of biological degradation.

Diagenesis affects organic remains following burial and involves the physical (compaction) and chemical (cementation, recrystallization) changes in the sediment before, during, and after lithification (conversion of soft, unconsolidated sediments into hard rock). *See* FOSSIL; PALEOBOTANY; PALEOECOLOGY; PALEONTOLOGY. Robert A. Gastaldo

Tapir An odd-toed ungulate of the family Tapiridae. These animals, with four species, have a discontinuous distribution in South America and Asia and consequently have a theoretical importance biologically.

These animals have a prehensile, trunklike muzzle, a short tail, and small eyes. The forefoot has four toes, the hindfoot three. The female gives birth to a single young after a gestation period of 13 months. Tapirs are nocturnal, timid animals; they spend the daytime hours in dense thickets. Charles B. Curtin

Tardigrada A phylum of microscopic, bilaterally symmetrical invertebrates which are generally less than 1 mm in length. About 400 species are known. Commonly called water bears, bear animalcules, or urslets, they are worldwide in distribution and are found in all habitats.

The tardigrade body consists of an anterior prostomium and five segments. The mouth is located in the prostomium in a centroterminal position. A soft, nonchitinous cuticle surrounds the body and lines the fore- and hindgut. The cuticle may be smooth or sculptured and forms innervated cephalic appendages and spines on the trunk and legs. Four pairs of ventrolateral legs arise from the trunk and terminate in claws or other modified structures. The digestive tract is tubular and more or less lobed due to the presence of diverticular dilations. In the heterotardigrades, separate anal and genital openings occur, while in the eutardigrades there is a single anogenital opening, the cloaca. The sexes are separate and the gonads are unpaired dorsal sacs with paired gonoducts in the male and in theory also in the female. Food storage cells float in the spacious body cavity, the coelom, which lacks a parietal or visceral peritoneum in the adult. Circulatory and respiratory structures are lacking. These animals exhibit the phenomenon known as cell constancy. The number of epidermal cells is the same in all species of a genus.

Tardigrades lay eggs and development is direct. Embryonic development lasts 3–40 days, varying according to the species and surrounding temperature. During their active life of 18 months, tardigrades molt about 12 times. They are unable to feed in the 5–10 days of molting; the buccal cavity requires at least 5 days for renewal.

Tardigrada live as active forms, without encystment, only when surrounded by a pellicle of water. They are mainly herbivorous, and feed by piercing the wall of plant cells with their stylets. When the surrounding medium dries up, most tardigrades continue to live as inactive or anabiotic barrel-shaped structures called cysts without any protective cover. Desiccation begins when there is a loss of oxygen from the water. The animal responds by contraction and loss of body water. Dried eggs also survive. Moistened animals usually revive, but anabiosis and revival cannot be repeated indefinitely.

Most species are widely distributed. Dissemination may be by wind, birds, and certain terrestrial animals which transport tardigrade eggs and barrels. The tropics have few species, and Scutechiniscidae are rare on the Antarctic continent. Most tardigrades are terrestrial. They are found among lichens, liverworts, densely growing soft-leaved mosses, and also in rather hard-leaved Pottiacea and Grimmiacea. *See* Cell constancy; Eutardigrada; Heterotardigrada; Polychaeta. Eveline Marcus

Tarragon A herb of the genus *Artemesia* in the aster family (Asteraceae) that is used as a spice. Tarragon is often divided into French and Russian types. French tarragon can be distinguished by its highly aromatic leaves, low seed set or fertility, and compact growth habit. Due to its long history of vegetative propagation to preserve its fine scent characteristics, French tarragon has all but lost its ability to form viable seeds. Russian tarragon does produce seeds, but the plant is much lower in overall oil content and is not as fine-scented.

Tarragon is a perennial that grows in one season, and then dies back to the ground with frost. In spring the plant sprouts from underground rhizomes and resumes its growth. The leaves contain volatile oil with an odor similar to anise and chervil. *See* Asterales; Spice and flavoring. Seth Kirby

Tartaric acid Any of the stereoisomeric forms of 2,3-dihydroxybutanedioic acid: L(+), D(−), and meso [(**1**), (**2**), and (**3**), respectively]. L(+)-Tartaric acid is present in the juice of var-

$$
\begin{array}{ccc}
CO_2H & CO_2H & CO_2H \\
| & | & | \\
H-C-OH & HO-C-H & H-C-OH \\
| & | & | \\
HO-C-H & H-C-OH & H-C-OH \\
| & | & | \\
CO_2H & CO_2H & CO_2H \\
(1) & (2) & (3)
\end{array}
$$

ious fruits and is produced from grape juice as a by-product of the wine industry. The monopotassium salt precipitates in wine vats, and L(+)-tartaric acid is recovered from this residue. On heating in alkaline solution, the L(+) acid is converted to the racemic mixture of (**1**) and (**2**), plus a small amount of the meso acid (**3**).

Tartaric acid has played a central role in the discovery of several landmark stereochemical phenomena. In 1848, L. Pasteur isolated enantiomers (**1**) and (**2**) by mechanical separation of hemihedral crystals of the racemic mixture. He also used tartaric acid and its salts to demonstrate a distinction between the meso isomer (**3**) and the racemic mixture (**1**) + (**2**), and between enantiomers and diastereoisomers in general. The difference in properties between (**1**) [or (**2**)] and the meso form (**3**) was later a key in establishing the relative configuration of the pentose and hexose sugars.

Both L(+)- and D(−)-tartaric acid and the esters are inexpensive compounds and are used as chiral auxiliary reagents in the oxidation of alkenes to enantiomerically pure epoxides. This method employs a hydroperoxide oxidant, titanium alkoxide catalyst, and L(+)- or D(−)-tartrate, and involves chirality transfer from the tartrate to the product. *See* Asymmetric synthesis; Epoxide.

Tartaric acid has some use as an acidulant in foods and also as a chelating agent. Potassium hydrogen tartrate (cream of tartar) is an ingredient of baking powder. The potassium sodium salt, commonly called Rochelle salt, was the first compound used as a piezoelectric crystal. *See* Chelation; Piezoelectricity. James A. Moore

Taste Taste, or gustation, is one of the senses used to detect the chemical makeup of ingested food—that is, to establish its palatability and nutritional composition. Flavor is a complex amalgam of taste, olfaction (smell), and other sensations, including those generated by mechanoreceptor and thermoreceptor sensory cells in the oral cavity. Taste sensory cells respond principally to the water-soluble chemical stimuli present in food, whereas olfactory sensory cells respond to volatile (airborne) compounds. *See* Chemical senses; Sensation.

The sensory organs of gustation are termed taste buds. In humans and most other mammals, taste buds are located on the tongue in the fungiform, foliate, and circumvallate papillae and in adjacent structures of the throat. There are approximately 5000 taste buds in humans, although this number varies tremendously. Taste buds are goblet-shaped clusters of 50 to 100 long slender cells. Microvilli protrude from the apical (upper) end of sensory cells into shallow taste pores. Taste pores open onto the tongue surface and provide access to the sensory cells. Individual sensory nerve fibers branch profusely within taste buds and make contacts (synapses) with taste bud sensory cells. Taste buds also contain supporting and developing taste cells. *See* Tongue.

The basic taste qualities experienced by humans include sweet, salty, sour, and bitter. (In some species, pure water also strongly stimulates taste bud cells). A fifth taste, umami, is now recognized by many as distinct from the other qualities. Umami is a Japanese term roughly translated as "good taste" and is approximated by the English term "savory." It refers to the taste of certain amino acids such as glutamate (as in monosodium glutamate) and certain monophosphate nucleotides. These compounds occur naturally in protein-rich foods, including meat, fish, cheese, and certain vegetables.

The middorsum (middle top portion) of the tongue surface is insensitive to all tastes. Only small differences, if any, exist for the taste qualities between different parts of the tongue. No simple direct relationship exists between chemical stimuli and a particular taste quality except, perhaps, for sourness (acidity). Sourness is due to H^+ ions. The taste qualities of inorganic salts are complex, and sweet and bitter tastes are elicited by a wide variety of diverse chemicals. Carl Pfaffmann; Nirupa Chaudhari; Stephen Roper

Taurus The Bull, a large zodiacal northern constellation prominent in the evening winter sky (*see* illustration). *See* ZODIAC.

The head of the bull is marked by the Hyades, a large V-shaped open cluster, against which is projected the red giant star Aldebaran, though it is only half as far away. From the V near Aldebaran, two bright stars farther out mark the points of Taurus's long horns. Near one of those stars, Zeta, is the Crab Nebula (M1, the first object in Messier's eighteenth-century catalogue, though it had been previously discovered by John Bevis), the remnant of a supernova whose light reached Earth in AD 1054, and which is detectable all across the spectrum of gamma rays, x-rays, ultraviolet, light, infrared, and radio waves. *See* ALDEBARAN; CRAB NEBULA; HYADES; MESSIER CATALOG.

The Pleiades, the seven sisters of Greek mythology, are a star cluster riding on Taurus's back. Only six are visible to the unaided eye, though binoculars or telescopes show dozens or hundreds more stars in the cluster. The Pleiades (M45, the 45th object in Messier's catalogue) have the shape of a small dipper, and long exposures show dust surrounding many of the stars, which are hot and therefore bluish. The dust reflects the starlight toward us. *See* CONSTELLATION; PLEIADES. Jay M. Pasachoff

Tautomerism The reversible interconversion of structural isomers of organic chemical compounds. Such interconversions usually involve transfer of a proton, but anionotropic rearrangements may be reversible and so be classed as tautomeric interconversions. A cyclic system containing the grouping —CONH— is called a lactam, and the isomeric form, —COH=N—, a lactim. These terms have been extended to include the same structures in open-chain compounds when considering the shift of the hydrogen from nitrogen to oxygen.

Molecular grouping (**1**) may in certain substances exist partly or wholly as (**2**). The former constitutes the keto form and the

—COCH< (1) —COH=C< (2) >C=N— (3) —CN< (4)

latter the enol form. The existence of an enol in an acyclic system requires that a second carbonyl group (or its equivalent, for example, (**3**) be attached to the same (**4**) as an aldehyde or ketone carbonyl. Thus, ethyl acetoacetate tautomerizes demonstrably, but ethyl malonate does not. Where the enol form includes an aromatic ring such as phenol, the existence of the keto form is often not demonstrable, although in some substances

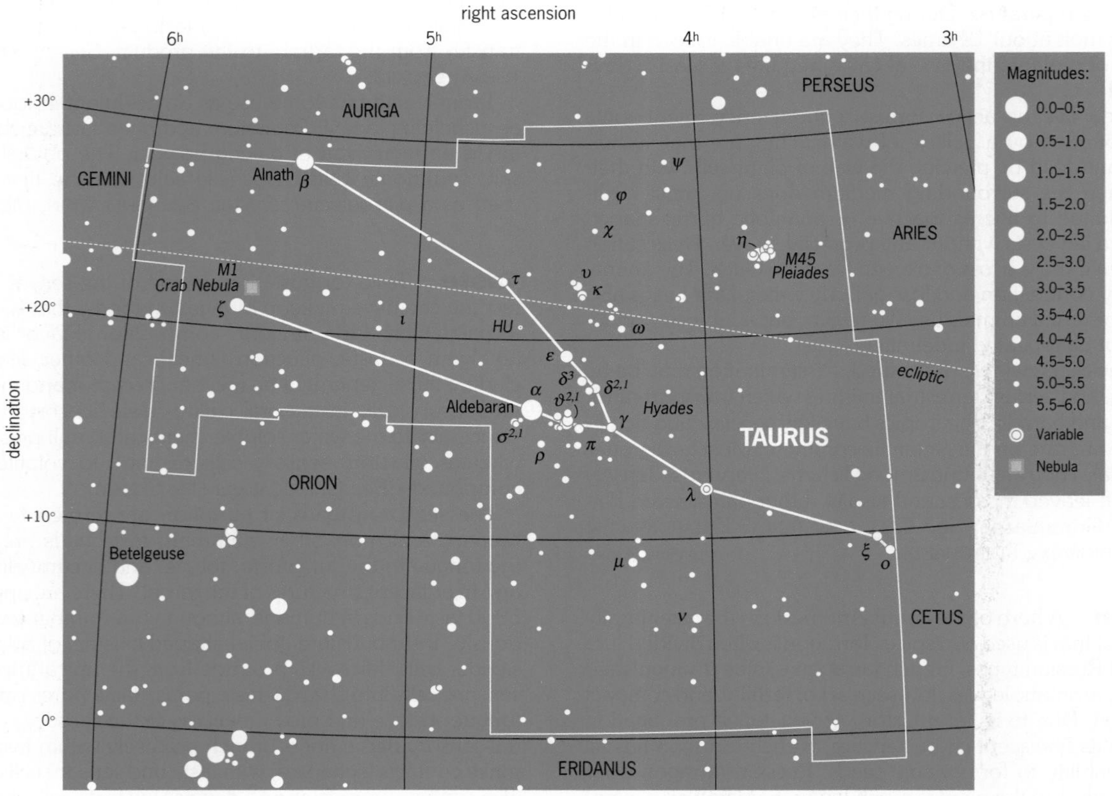

Modern boundaries of the constellation Taurus, the Bull. The celestial equator is 0° of declination, which corresponds to celestial latitude. Right ascension corresponds to celestial longitude, with each hour of right ascension representing 15° of arc. Apparent brightness of stars is shown with dot sizes to illustrate the magnitude scale, where the brightest stars in the sky are 0th magnitude or brighter and the faintest stars that can be seen with the unaided eye at a dark site are 6th magnitude. (*Wil Tirion*)

there may be either chemical or spectroscopic evidence for both forms. Closely related to keto-enol tautomerism is the prototropic interconversion of nitro and aci forms of aliphatic nitro compounds such as nitromethane.

In general, tautomeric forms exist in substances possessing functional groups which can interact additively and which are so placed that intramolecular reaction leads to a stable cyclic system. The cyclic form usually predominates (especially if it contains five or six members). *See* MOLECULAR ISOMERISM.

Wyman R. Vaughan

Taxis A mechanism of orientation by means of which an animal moves in a direction related to a source of stimulation. There exists a widely accepted terminology in which the nature of the stimulus is indicated by a prefix such as phototaxis (light), chemotaxis (chemical compounds), geotaxis (gravity), thigmotaxis (contact), rheotaxis (water current), and anemotaxis (air current). The directions toward or away from the stimulus are expressed as positive or negative, respectively. Finally, the sensory and locomotory mechanisms by means of which the orientation is achieved are denoted by a second type of prefix forming a compound noun with taxis. Positive phototropotaxis thus describes a mechanism by means of which an animal carries out a directed movement toward a source of light along a path which permits the animal's paired eyes to receive equal intensities of light throughout the movement. Otto E. Lowenstein

Taxodonta An order of bivalves in subclass Lamellibranchia whose hinge dentition is characterized by a series of numerous similar alternating teeth and sockets. Typical genera of so-called ark shells are *Arca*, *Barbatia*, and *Anadara*.

Ark-shell bivalves are probably distantly related to the true marine mussels (Mytilacea), with which they share certain features of gill ciliation and byssal organization. Formerly the name Taxodonta was used to designate a subclass of bivalves consisting of a mixed group of lamellibranch ark shells and primitive protobranchs. *See* BIVALVIA; LAMELLIBRANCHIA; MOLLUSCA; PROTOBRANCHIA. W. D. Russell-Hunter

Taxonomic categories Any one of a number of formal ranks used for organisms in a traditional Linnaean classification. Biological classifications are orderly arrangements of organisms in which the order specifies some relationship. Taxonomic classifications are usually hierarchical and comprise nested groups of organisms. The actual groups are termed taxa. In the hierarchy, a higher taxon may include one or more lower taxa, and as a result the relationships among taxa are expressed as a divergent hierarchy that is formally represented by tree diagrams. In Linnaean classifications, taxonomic categories are devices that provide structure to the hierarchy of taxa without the use of tree diagrams. By agreement, there is a hierarchy of categorical ranks for each major group of organisms, beginning with the categories of highest rank and ending with categories of lowest rank, and while it is not necessary to use all the available categories, they must be used in the correct order (see table).

Conceptually, the hierarchy of categories is different than the hierarchy of taxa. For example, the taxon Cnidaria, which is ranked as a phylum, includes the classes Anthozoa (anemones), Scyphozoa (jellyfishes), and Hydrozoa (hydras). Cnidaria is a particular and concrete group that is composed of parts. Anthozoa is part of, and included in, Cnidaria. However, categorical ranks are quite different. The category "class" is not part of, nor included in, the category "phylum." Rather, the category "class" is a shelf in the hierarchy, a roadmark of relative position. There are many animal taxa ranked as classes, but there is only one "class" in the Linnaean hierarchy. This is an important strength of the system because it provides a way to navigate through a classification while keeping track of relative hierarchical levels with only a few ranks for a great number of organisms.

Categories commonly used in botanical and zoological classifications, from highest to lowest rank

Botanical categories	Zoological categories
Divisio	Phylum
Classis	Class
Ordo	Order
Familia	Family
Genus	Genus
Species	Species

When Linnaeus invented his categories, there were only class, order, family, genus, and species. These were sufficient to serve the needs of biological diversity in the late eighteenth century, but were quite insufficient to classify the increasing number of species discovered since 1758. As a result, additional categorical levels have been created. These categories may use prefixes, such as super- and sub-, as well as new basic levels such as tribe. An example of a modern expanded botanical hierarchy of ranks between family and species is:

Familia
Subfamilia
Tribus
Subtribus
Genus
Subgenus
Sectio
Subsectio
Series
Subseries
Species

Linnaean categories are the traditional devices used to navigate the hierarchy of taxa. But categories are only conventions, and alternative logical systems, such as those used by phylogenetic systematists (cladists), are frequently used. *See* CLASSIFICATION, BIOLOGICAL; PHYLOGENY; PLANT TAXONOMY; SYSTEMATICS; TAXONOMY; ZOOLOGICAL NOMENCLATURE. Edward O. Wiley

Taxonomy The human activity of naming organisms and organizing these names according to a given criterion. In biology, taxonomy has evolved into formal rules and systems to classify organisms. It is part of the broader discipline of systematics, a discipline that covers all of comparative biology, such as comparative morphology, genomics, and biogeography. Systematists who develop formal biological classifications and publish taxonomic revisions and keys are practicing taxonomy. *See* CLASSIFICATION, BIOLOGICAL; PLANT TAXONOMY; SYSTEMATICS.

Most modern taxonomies are hierarchical classifications of groups within groups, resulting in a biological classification. In addition, modern taxonomies include detailed descriptions of the organisms and a history of the various names by which these organisms have been called. Such works are usually referred to as taxonomic revisions. In classifications and taxonomic revisions, each group of plants or animals is given a name. The group itself is a taxon (examples: *Homo sapiens*, Vertebrata). The place of the group within the taxonomy is denoted by ranking. In Linnaean taxonomies, rank is denoted by placing the taxon name with a category rank (example: class Vertebrata). Each category has a place on the hierarchy relative to other categories. Each taxon is unique, but the categories are not unique. For example, there are over 400 different families of bony fishes. This relative ranking is set by convention for each system of taxonomy. There are several systems of taxonomy, including systems for bacteria (prokaryotes) and viruses. The two most familiar are for plants and animals. *See* TAXONOMIC CATEGORIES.

One major goal of taxonomy is naming species. Each species has a two-part name, a genus name and a specific epitaph. *Homo sapiens* (humans) and *Pinus ponderosa* (ponderosa pine trees) are examples. Species names are always "set apart" from the rest of the text, usually by italics (*Homo sapiens*). The second major goal of taxonomy is to organize these species into larger groups and give these groups names. Each group of species is given a single name (example: Mammalia). Groups of species are frequently referred to as supraspecific taxa or higher taxa, with "higher" meaning that their rank category is higher than the rank of species. *See* SPECIES CONCEPT. E. O. Wiley

Tea A small tree, *Camellia* (*Thea*) *sinensis*; a preparation of its leaves dried and cured by various processes; and a beverage made from these leaves. The plant is an evergreen tree of the Theaceae family, native to southeastern Asia, and does best in a warm climate where the rainfall averages 90–200 in. (2200–5000 mm). China, Japan, Taiwan, India, Sri Lanka, and Indonesia are among the leading tea-producing countries, with China contributing about one-half of the world's supply.

Tea leaves contain caffeine, various tannins, aromatic substances attributed to an essential oil, and other materials of a minor nature, including proteins, gums, and sugars. The tannins provide the astringency, the caffeine the stimulating properties. *See* THEALES. Earl L. Core

Technetium A chemical element, Tc, atomic number 43, discovered by Carlo Perrier and Emilio Segre in 1937. They separated and isolated it from molybdenum (Mo; atomic number 42), which had been bombarded with deuterons in a cyclotron. Technetium does not occur naturally. *See* ATOMIC NUMBER; ELEMENT (CHEMISTRY).

In the periodic table, technetium is located in the middle of the second-row transition series, situated between manganese and rhenium. Because of the lanthanide contraction, the chemistry of technetium is much more like that of rhenium, its third-row congener, than it is like that of manganese. Its location in the center of the periodic table gives technetium a rich and diverse chemistry. Oxidation states -1 to $+7$ are known, and complexes with a wide variety of coordination numbers and geometries have been reported. *See* LANTHANIDE CONTRACTION; PERIODIC TABLE; RHENIUM; TRANSITION ELEMENTS.

The most readily available chemical form of technetium is the ion pertechnetate (TcO_4^-), the starting point for all of its chemistry. In its higher oxidation states ($+4$ to $+7$), technetium is dominated by oxo chemistry, which is dominated by complexes containing one or two multiply bonded oxygen (oxo) groups.

Since about 1979, inorganic chemists have been very interested in understanding and developing the fundamental chemistry of technetium, utilizing the isotope ^{99g}Tc. Most of this interest has arisen from the utility of another isotope of technetium, ^{99m}Tc (where m designates metastable), to diagnostic nuclear medicine. In fact, ^{99m}Tc in some chemical form is used in about 90% of all diagnostic scans performed in hospitals in the United States. The nuclear properties of ^{99m}Tc make it an ideal radionuclide for diagnostic imaging. This technetium isotope has a 6-h half-life, emits a 140-keV gamma ray which is ideal for detection by the gamma cameras used in hospitals, and emits no alpha or beta particles. Silvia S. Jurisson

Technology Systematic knowledge and action, usually of industrial processes but applicable to any recurrent activity. Technology is closely related to science and to engineering. Science deals with humans' understanding of the real world about them—the inherent properties of space, matter, energy, and their interactions. Engineering is the application of objective knowledge to the creation of plans, designs, and means for achieving desired objectives. Technology deals with the tools and techniques for carrying out the plans.

Robert S. Sherwood; Harold B. Maynard

Tektite A member of one of several groups of objects that are composed almost entirely of natural glass formed from the melting and rapid cooling of terrestrial rocks by the energy accompanying impacts of large extraterrestrial bodies. Tektites are dark brown to green, show laminar to highly contorted flow structure on weathered surfaces and in thin slices, are brittle with excellent conchoidal fracture, and occur in large masses but are mostly small to microscopic in size.

Five major groups of tektites are known: North American, 34,000,000 years old; Czechoslovakian (moldavites), 15,000,000 years; Ivory Coast, 1,300,000 years old; Russian (irgizites) 1,100,000 years old; and (5) Australasian, 700,000 years old. The North American, Ivory Coast, and Australasian tektites also occur as microtektites in oceanic sediment cores near the areas of their land occurrences. In the land occurrences, virtually all of the tektites are found mixed with surface gravels and recent sediments that are younger than their formation ages.

The chemical compositions of tektites differ from those of ordinary terrestrial rocks principally in that they contain less water and have a greater ratio of ferrous to ferric iron, both of which are almost certainly a result of their very-high-temperature history.

Elbert A. King

Telecast A television broadcast, involving the transmission of the picture and sound portions of the program by separate transmitters at assigned carrier frequencies within the channel assigned to a television station. A telecast is intended for reception by the general public, just as is a radio broadcast. *See* TELEVISION; TELEVISION TRANSMITTER. John Markus

Telecommunications civil defense system Warning centers and private-line telecommunications systems that carry vital emergency information and alerts to both the military and the general public. When any possible emergency threatens the security or socioeconomic structure of the United States or any political subdivision, an appropriate warning must be made in order for a reaction to occur within an appropriate period of time. Any such reactions cannot operate without a means of communication to prewarn of impending disaster or attack and to assist the civil defense and military effort after a disaster has occurred.

The Federal Emergency Management Agency has primary responsibility for financial and technical support of the two basic communications systems that meet this need during a crisis situation: the National Warning System (NAWAS) and the Emergency Broadcast System (EBS).

Operating 24 h a day, the National Warning System is a voice-grade service that consists of over 65,000 channel miles (100,000 km) connecting 2300 warning points in the United States with a national primary warning center located with the North American Air Defense (NORAD) Command. Should the primary warning center be rendered inoperative, alternate centers can cover all warning points that were dependent on the inoperative center. Similarly, modern switching facilities are available to replace any portion of the network that may become inoperative. The warning centers are located at key defense installations so that full advantage can be taken of all available information. Warning information can be sent to all points on the network in approximately 15 s. From the 2300 warning points on the network, warning information is relayed through state and local systems to more than 5000 local points in an average time of 7 min.

The mission of the Emergency Broadcast System is to provide the President of the United States with a means of communicating with the general public through the use of nongovernment broadcast stations during the period preceding, during, and following an enemy attack, natural disaster, or other national emergency. The Broadcast System consists of four multipoint private-line services routed to 31 locations at 2400 bits per second. Either of the two point-of-origin, send-receive stations (a primary location and its alternate) can transmit the activation, termination, and National Industry Advisory Council (NIAC) order messages to all other stations. Each point-of-origin send-receive station is equipped with a line controller unit, a 4420 terminal, diagnostic modems, and floppy disk storage units. All other receive-only stations are equipped with 2024 modems and a Model 43 buffered terminal which is programmed to provide an answer with the depression of a key. Activation of a switch at these locations and at the point of origin allows voice confirmation and emergency coordination. *See* RADIO BROADCASTING; TELEPHONE SERVICE.

James P. McQuillan

Teleconferencing Broadly, the various ways and means by which people communicate with one another over some distance. In a narrow sense, a teleconference is a two-way, interactive meeting, between relatively small groups of people (approximately 1 to 10 at each end), who usually use permanent teleconferencing facilities. A teleconference involves audio communication between the locations, but may also involve video or graphics. *See* TELEPHONE SERVICE; VIDEOTELEPHONY.

A teleseminar is utilized for educational purposes; it is primarily one-way communication to many destinations from one source. A teleseminar almost always uses audio communication, and may also use video and some form of graphics. A means is provided for the receiving locations to ask questions of the instructor via microphones or telephone handsets.

A telemeeting is often called an ad hoc teleconference, with the ad hoc referring to places, times, participants, and purpose. A telemeeting is similar to a teleseminar in that it is primarily a one-way communication, usually staged or prepared by video program professionals. It may be set up to order, using temporary equipment or circuits.

Computer conferencing is a method for people to communicate by using computers. The medium is quite flexible, as it can be used between just two people, between one and many people, or among many people. Basically, computer conferencing involves typing a message on a computer terminal and transmitting it to one or more destinations electronically. Sophisticated networks are required to accomplish computer conferencing between many users, or a simple data modem and telephone circuit can allow two people at a time to conference. *See* DATA COMMUNICATIONS; LOCAL-AREA NETWORKS; WIDE-AREA NETWORKS.

Currently, the most popular form of teleconferencing by far is the audio conference. Using plain speaker telephones, special speaker phones, corporate private branch exchange (PBX) systems, or special services, most business people use this form of teleconference regularly. Probably next in popularity is the Teleseminar, used for "distance learning," or formal education and training. The Internet has become a popular medium to facilitate both of these forms of teleconferencing, as high-speed connections such as digital subscriber line (DSL), cable modem, and satellite have become more prevalent. *See* COMMUNICATIONS SATELLITE; INTERNET; PRIVATE BRANCH EXCHANGE.

John J. Bleiweis

Telegraphy A method of communication employing electrical signaling impulses produced and received manually or by machines. Telegraph signals are transmitted over open wire or cable land lines, submarine cables, or radio. Telegraphy as a communication technique uses essentially a narrow frequency band and a transmission rate adapted to machine operations. *See* ELECTRICAL COMMUNICATIONS.

Early equipment devised by Samuel F. B. Morse consisted of a mechanical transmitter and receiver or register. Operators soon learned to handle messages faster using simple manual keys and audible sounders. Subsequently, telegraph transmission and reception again became mechanized. Telegraphy may also be used in other ways.

Telegraph facilities for use by the general public to transmit messages both domestically and internationally are provided by communication companies and government administrations. Special telegraph facilities include those for news services, distribution of market prices of securities and commodities, and private lines between such points as the factories and offices of a company for the exchange of messages, orders, payroll data, and inventories. Fire and police alarms are a special form of telegraphy. The armed forces have extensive fixed and mobile telegraph systems.

For manual operation, the telegraph code consists of short dot and long dash signals (see illustration). Most automatic printing telegraph circuits, including American cable operation, use a code of five equally spaced signals or units per letter or other symbol perforated into a paper tape. Machines translate automatic teleprinter code to cable code, cable code to automatic

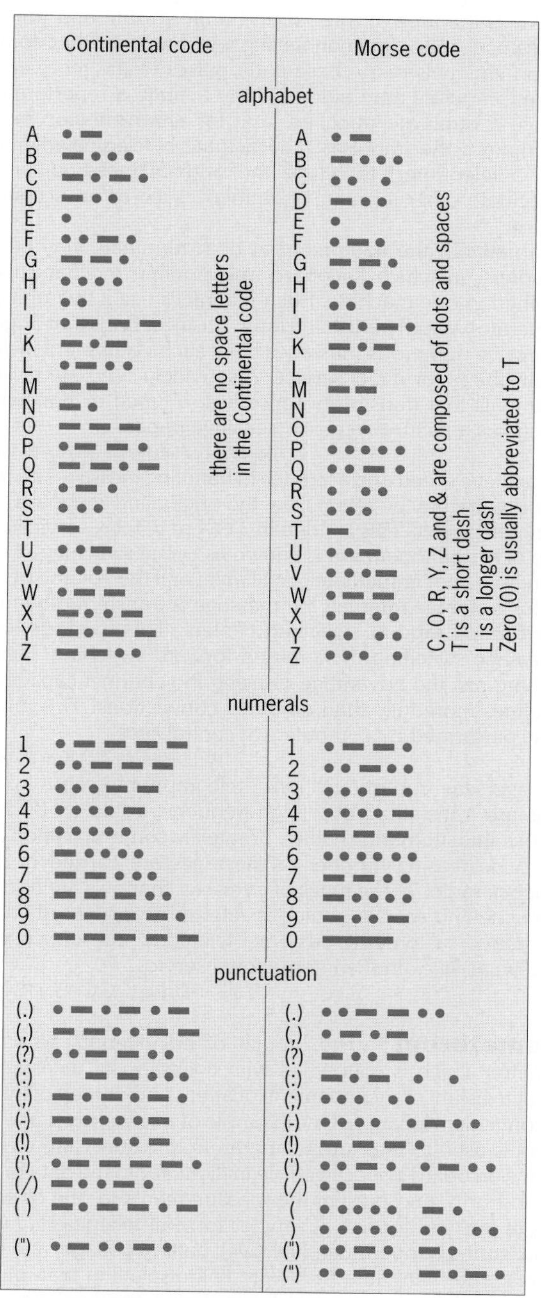

Continental code (International Morse) is commonly used for telegraph communication. Morse code (American Morse) continues in use on a few land lines in the United States and Canada.

code, or, for special applications, make other translations. For stock quotation systems and teletypesetter operation, a six-unit code is used to provide control of machine action. For data transmission or machine control, a seven- or eight-unit code may be used.

A single circuit provides transmission in only one direction at a time. For transmission in both directions simultaneously, a duplex circuit is used. Multiplex (time-division) apparatus provides two, three, or more channels operable in both directions simultaneously over a single circuit. Carrier-current techniques enable several circuits, each comprising one or more communication channels, to operate through the same wide-band wire, cable, or radio facility. *See* MULTIPLEXING.

In automatic transmission, an operator at a manual keyboard, operated like a typewriter, perforates a tape. The tape is fed through an electromechanical or photocell tape reader that drives the tape at a rapid and uniform rate and transmits the electrical code. Interconnecting wire lines or other communication channels carry these code pulses to the receiver. The received impulses may automatically actuate a reperforator to produce a duplicate punched tape for retransmission or later transcription; the impulses may actuate a teletypewriter (also called a teleprinter) to retype the original message; or they may actuate other terminal equipment, such as an accounting machine.

Alternatively, the equipment at both terminals may be teletypewriters, in which keyboard and printing mechanisms are combined in one machine. Such machines use a five-unit code but operate without perforated tape. Business firms that originate and receive numerous messages have such equipment installed at their offices for direct service. Also widely used are facsimile instruments that transmit or reproduce a typed or handwritten message as a picture on electrosensitive paper.

Grosvenor Hotchkiss; William R. Webster

To achieve speed and accuracy in handling messages en route, direct circuits may be set up from the originaling keyboard to the distant teleprinter. This is done in TWX and Telex services. This mode of system operation is known as circuit switching. If direct channels are not immediately available, or if the volume of traffic calls for a storage interval, the message is transferred by means of perforated tape at switching centers. This mode is known as message switching or store-and-forward switching. Message switching has the advantage of using the channel capacities of trunk lines more fully than do direct connections. This function is now performed by computers at central sites.

W. E. Girardin; William R. Webster

In overseas communications, telegraph messages or telegrams are transmitted over high-frequency radio (3–30 MHz), transoceanic submarine cables, or satellite communication channels. A decreasing number of messages are carried on high-frequency radio. These types of overseas channels are also used for private-line service, which is referred to as leased-channel service, and customer-to-customer teletypewriter service, which is known as Telex in the international service.

Eugene D. Becken; William R. Webster

Telemetering The branch of engineering, also called telemetry, which is concerned with collection of measurement data at a distant or inconvenient location, and display of the data at a convenient location. One example of a complex telemetering system is used to measure temperature, pressure, and electrical systems on board a space vehicle in flight, radio the data to a station on Earth, and present the measurements to one or several users in a useful format. A simpler form of telemetering would be measuring temperature, fuel level, battery voltage, and speed in a car, and then via a hard-wire link displaying that information on the instrument panel.Telemetering involves movement of data over great distances, as in the above example, or over just a few meters, as in monitoring activity on the rotating shaft of a gas turbine. It may involve less than 10 measurement points or more than 10,000.

Telemetering involves a number of separate functions: (1) generating an electrical variable which is proportional to each of several physical measurements; (2) converting each electrical variable to a proportional voltage in a common range; (3) combining all measurements into a common stream; (4) moving the combined measurements to the desired receiving location, as by radio link; (5) separating the measurements and identifying each one; (6) processing selected measurements to aid in mission analysis; (7) displaying selected measurements in a useful form for analysis; and (8) storing all measurements for future analysis.

The largest category is commonly called aerospace telemetry, used in testing developmental aircraft and in monitoring low-orbit space vehicles. Other applications include missile and rocket testing, automobile testing, and testing of other moving vehicles.

The version of telemetry commonly used in an industrial application includes supervisory control of remote stations as well as data acquisition from those stations over a bidirectional communications link. The generic term is supervisory control and data acquisition (SCADA); the technology is normally used in electrical power generation and distribution, water distribution, and other wide-area industrial applications. *See* ELECTRIC POWER SYSTEMS; WATER SUPPLY ENGINEERING.

One SCADA application involves monitoring wind direction and velocity as indicated by anemometers located on the approach and departure paths near an airport, so that air-traffic controllers can make pilots aware of dangerous differences in wind direction and velocity, known as wind shear. This type of system is operated over a radio communications link, with the appropriate anemometers being interrogated as their measurements are needed by the computer for wind analysis. Somewhat similar systems are used in oceanographic data collection and analysis, where instrumented buoys send water temperature and other data on command. *See* AERONAUTICAL METEOROLOGY; INSTRUMENTED BUOYS; WIND MEASUREMENT.

Because two-way communication with a complex and distant Earth synchronous satellite or other spacecraft presents a unique challenge, a technology called packet telemetry is in widespread use for these applications. Here, messages between the Earth station and the spacecraft are formed into groups of measurements or commands called packets to facilitate routing and indentification at each end of the link. Each packet begins with a definitive preamble and ends with an error-correcting code for data quality validation. Packet technology is defined by an international committee, the Consultative Committee for Space Data Systems (CCSDA). *See* PACKET SWITCHING; SPACE COMMUNICATIONS.

Many unique systems are in use. A special multichannel medical telemetry system is used on some ambulances to monitor and radio vital signs from a person being transported to a hospital, so that medical staff can prepare to treat the specific condition which caused the emergency. Another system is possibly the oldest user of radio telemetry, the radiosonde. A data collection and transmission system is lifted by a balloon to measure and transmit pressure, temperature, and humidity measurements from various altitudes as an aid to weather prediction. *See* BALLOON; METEOROLOGICAL INSTRUMENTATION.

O. J. Strock; Tomas C. Chavez

Teleostei The largest, youngest (first appearing in the Upper Triassic), and most successful group of the actinopterygians (rayfin fishes), and a sister group of the Amiidae (bow fins). The 23,600 species of teleosts make up more than half of all recognized species of living vertebrates, and over 96% of all living fish species. There are 494 families, of which 69 are extinct leaving 425 extant families, 43% of which have no fossil record. *See* ACTINOPTERYGII; AMIIFORMES.

Much of the evidence for teleost monophyly and relationships comes from the caudal skeleton and concomitant acquisition of a homocercal tail (upper and lower lobes symmetrical). Other characters of the teleosts include the mobile premaxilla bone, the extension of the posterior myodome (the eye muscle canal) into the basioccipital bone, and the development of the swim bladder. *See* SWIM BLADDER.

The Osteoglossomorpha, Elopomorpha, and Clupeomorpha are now generally regarded as successive clades (groups) above the level of the fossil, paraphyletic pholidophorids. The Clupeomorpha is considered to be the sister group of the Ostariophysi.

Clupeomorpha and Ostariophysi. The Clupeomorpha includes almost 80 genera and some 360 Recent species in three main families: Engraulidae (anchovies), Dussumieridae (round herrings), Clupeidae (herrings), as well as Pristigasteridae. The Ostariophysi make up nearly 75% of the fresh-water fishes of the world (over 6000 species) and include the Cypriniformes (carps, loaches, and relatives), Siluriformes (catfishes), Gymnotiformes (knife fishes, electric eel), and *Chanos*. The Clupeomorpha plus Ostariophysi are the sister group of the Euteleostei. *See* CYPRINIFORMES; OSTEOGLOSSIFORMES; PHOLIDOPHORIFORMES.

Euteleostei. The Euteleostei are by far the largest teleost taxon with more than 22,000 species arranged in some 340 families. They comprise two major lineages, the Protacanthopterygii and the Neognathi.

The Protacanthopterygii (often regarded as lower euteleosteans) include four groups: the salmonoids (salmon and allies; 66 species) and the osmeroids (smelts, salangrids, *Lepidogalaxius*; 72 species), which together are the sister group of the alepocephaloids (slickheads; 60+ species) plus the argentinoids (argentines or herring smelts; 60+ species). The Neognathi, on the other hand, are made up of the esocoids (pike and mudminnows; 10+ species) plus the sister group Neoteleostei. *See* SALMONIFORMES.

The Neoteleostei, with more than 15,319 species, comprise four main clades, the Stomiiformes, Aulopiformes, Myctophiformes, and Acanthomorpha (paracanthopterygians plus acanthopterygians). The acanthopterygians are the largest subgroup of the Euteleostei, distributed among 12 orders and 218 families. The Perciformes form the largest of these orders with 9293 species. *See* PERCIFORMES. Brian Gardiner

Teleostomi A monophyletic grade of craniates comprising Acanthodii, Actinopterygii (ray-finned fishes), and Sarcopterygii (coelacanths, lungfishes, and tetrapods). [Monophyletic refers to any form evolved from a single interbreeding population. Craniates are vertebrates distinguished by a cranium.] The Acanthodii (all fossil species) is the sister group to the actinopterygians and sarcopterygians, a relationship based on otoliths and certain details of the vertebral column and associated elements. The grade Teleostomi contains about 53,633 extant valid species, of which 26,891 are actinopterygians and 26,742 are sarcopterygians. *See* ACANTHODII; ACTINOPTERYGII; OSTEICHTHYES; SARCOPTERYGII. Herbert Boschung

Telephone An instrument containing a transmitter for converting the acoustic signals of a person's voice to electrical signals, a receiver for reconverting electrical signals to acoustic signals, and associated signaling devices (the dial and alerter) for communicating with other persons using similar instruments connected to a network. The term "telephone" also refers to the complicated system of transmission paths and switching points, called the Public Switched Network (PSN), connected to this instrument. *See* TELEPHONE SERVICE.

The transmitter is a transducer that converts acoustic energy into electric energy. The carbon transmitter was the key to practical telephony because it amplified the power of the speech signal, making it possible to communicate over distances of many miles. Current designs are based on the charged electret (a condenser microphone) or on electrodynamic principles. Both the electret and electrodynamic transmitters use transistors to provide needed power gain; they introduce less distortion than the carbon transmitter. *See* ELECTRET TRANSDUCER; MICROPHONE; TRANSDUCER; TRANSISTOR.

Transmitters in a basic set have a frequency-response range intentionally limited to 250–5000 Hz to improve voice clarity on the Public Switched Network. Sets that use another network for transmission may have a wider frequency response. Even though normal human hearing has a much broader frequency response, speech heard on the telephone resembles closely that heard by a listener a few feet from the person speaking. *See* HEARING (HUMAN).

The heart of an electret transmitter is an electrical capacitor formed by the metal on the diaphragm, a conductive coating on top of the metalized lead frame, and the plastic and air between the metal layers. The diaphragm is made of a special plastic that can be given a permanent electrostatic charge (analogous to the magnetization of a permanent magnet). As sound waves entering the sound port cause pressure changes, the diaphragm moves closer to and farther away from the metalized lead frame. This changes the value of capacitance and produces a varying electric voltage which is the analog of the impinging sound-pressure wave. The signal is amplified by the internal amplifier chip to a level that is suitable for transmission on the telephone network.

The receiver transducer operates on the relatively low power used in the telephone circuit; it converts electric energy back into acoustic energy. As in the transmitter, careful design of the relationship of the acoustical and electrical elements produces a desired response-frequency characteristic.

Historically, there were two common types of receiver units with fixed-coil windings, the ring armature receiver and the bipolar receiver. Fixed-coil receivers were designed for close coupling to the ear and were placed in a cuplike enclosure shaped to fit closely to the ear. Moving-coil designs, similar to loudspeakers, have replaced these in most telephones. This change can provide hands-free (speakerphone) capability in an ordinary handset but, because an ear-shaped cup is often not used, results in poorer performance in noisy environments. *See* LOUDSPEAKER.

Telephone sets may include two cords: a line cord that connects the instrument to tip and ring, and a handset cord that connects the telephone handset to the chassis. Cordless telephones use a two-way radio link between one or more handsets and a base unit; this replaces the handset cord. The base unit is connected through a normal line cord to the telephone network.

The cellular radio system is a form of mobile radio telephony. Cellular telephones use a two-way radio link to replace the line cord; thus tip and ring are not brought to the telephone. The telephone communicates with one of a number of base stations spread in cells throughout the service area. As the telephone user changes location, the link is automatically switched from cell to cell to maintain a good connection. Several different protocols are used by providers between the telephone and base station, so the telephone used must be chosen to be compatible with the service provider's network. Also, enhanced features, such as text messaging and transmission of photographs or video, are available in some protocols and require telephones having the necessary hardware and software to support them. *See* MOBILE COMMUNICATIONS. Richard M. Rickert

Telephone service The technology of providing many types of communications services via networks that transmit voice, data, image facsimile, and video by using both analog and digital encoding formats.

Telephone services involve three distinct sectors of components: (1) customer premises equipment (CPE), such as telephones, fax machines, personal and mainframe computers, and systems private branch exchanges (PBXs); (2) transmission systems, such as copper wires, coaxial cables, fiber-optic cables, satellites, point-to-point microwave routes, and wireless radio links, plus their associated components; and (3) switching systems that often can access associated databases, which can add new intelligent controls for the network's users. *See* COAXIAL CABLE; COMMUNICATIONS CABLE; COMMUNICATIONS SATELLITE; MICROWAVE; OPTICAL COMMUNICATIONS; OPTICAL FIBERS; PRIVATE BRANCH EXCHANGE; RADIO; SWITCHING SYSTEMS (COMMUNICATIONS); TELEPHONE.

Infrastructure. Made over a web of circuits known as a network, a connection often involves several different telephone companies or carriers. In the United States, there are more than 1300 local exchange companies (LECs) providing switched local service, plus more than 500 large and small interexchange companies (IECs) that provide switched long-distance services.

In addition, wireless mobile telephone services are provided on a city-by-city basis by cellular systems operated by two different companies in each metropolitan area. Each cellular system is connected to the wire networks of the local exchange companies and interexchange companies. *See* MOBILE COMMUNICATIONS.

Local exchange company operations of wired systems are divided geographically into 164 local access and transport areas (LATAs), each containing a number of cities and towns. Within a LATA, the telephone company operates many local switches, installed in facilities known as central offices or exchanges.

Each central office or exchange switch in connected to local telephone customer premises by a system of twisted-pair wires, coaxial cables, and fiber-optic lines called the loop plant. Direct current (dc) electricity that carries signals through the wires (or powers the lasers and photodetectors in the glass-fiber lines) is provided by a large 48-V stationary battery in the central office which is constantly recharged to maintain its power output. If a utility power outage occurs, the battery keeps the transmission system and the office operating for a number of hours, depending on the battery's size.

The transmission is usually analog to and from the customer site, especially for residences. However, in many systems a collection point between the customer and the switch, known as a subscriber loop carrier (SLC), provides a conversion interface to and from one or more digital cable facilities called T1 carriers (each with 24 channels) leading to the switch. In turn, most switches are interconnected by digital fiber-optic, coaxial, or copper-wire cables or by microwave transmission systems either directly or via intermediate facilities called tandem switches between local switches or between a group of central offices and a long-distance toll switch. *See* TELEPHONE SYSTEMS CONSTRUCTION.

Switching and transmission designs. Telephone switching machines used analog technology from 1889 to 1974, when the first all-digital switch was introduced, starting with long-distance or toll service and expanded later to central offices or local exchanges. Digital switching machines are comparable to digital computers in their components and functions. Unlike the analog electromechanical switches of the past, digital switches have no moving parts and can operate at much faster speeds. In addition, digital switches can be modified easily by use of operations systems—special software programs loaded into the switch's computer memory to provide new services or perform operational tasks such as billing, collecting and formatting traffic data from switches, monitoring the status of transmission and switching facilities, testing trunk lines between end offices, and identifying loop troubles. *See* DIGITAL COMPUTER.

In North America, digital switching and transmission are conducted within a hierarchy of multiplexing levels. The single digital telephone line, rated at 64 kilobits per second (kb/s), is known as a digital signal 0 (DS-0) level. The lowest digital network transmission level is DS-1, equivalent to 24 voice channels multiplexed by time-division multiplexing (TDM) to operate at 1.544 megabits per second (Mb/s). DS-1 is the most common digital service to customer premises. Two interim levels, now seldom used, are followed by DS-3, perhaps the most widely used high-speed level. DS-3 operates at 44.736 Mb/s, often rounded out to 45 Mb/s, the highest digital signal rate conventionally provided to customer premises by the telephone network. Of course, groups of DS-3 trunks can be connected to a customer facility if needed. Outside the United States, other digital multiplexing schemes are used. This results, for example, in a DS-1 level having 30 voice channels rather than 24 channels. *See* MULTIPLEXING AND MULTIPLE ACCESS.

A second network hierarchy appeared during the early 1990s and was gradually implemented in the world's industrial nations. Known in North America as SONET (synchronized optical network) and in the rest of the world as SDH (synchronized digital hierarchy), these standards move voice, data, and video information over a fiber network at any of eight digital transmission rates. These range from OC-1 at 51.84 Mb/s to OC-48 at 2.488 gigabits per second (Gb/s), and the hierarchy can be extended to more than 13 Gb/s. The OC designation stands for optical carrier.

Asynchronous transfer mode technology is based on cell-oriented switching and multiplexing. This is a packet-switching concept, but the packets are much shorter (always 53 bytes) and faster (for better response times) than are the packets in such applications as the global Internet and associated service networks which can have up to 4096 bytes of data in one packet. The asynchronous transfer mode cell-relay system moves cells through the network at speeds measured in megabits per second instead of kilobits. With its tremendous speed and capacity, asynchronous transfer mode technology permits simultaneous switching of data, video, voice, and image signals over cell-relay networks. *See* DATA COMMUNICATIONS; PACKET SWITCHING.

Out-of-band signaling systems. When the voice or communications information is encoded in a stream of bits, it becomes possible to use designated bits instead of analog tones for the supervisory signaling system. This simplifies the local switching equipment and allows the introduction of sophisticated data into the signaling, which significantly expands the potential range of telephone services.

Operations information now flows between switches as packet-switched signaling data via a connection that is independent of the channel being used for voice or data communications (a technique known as out-of-band signaling). Typically, a channel between two switching systems is known as a trunk. A trunk group between switches carries a number of channels whose combined signaling data are transmitted over a separate common channel as a signaling link operated at 56 kb/s. This technique is called common-channel interoffice signaling (CCIS).

In 1976, the first United States version of CCIS was introduced as a modification of an international standard, the CCITT No. 6 signaling system. The out-of-band signaling system was deployed in the long-distance (toll) network only, linking digital as well as analog switches via a network of packet-data switches called signal transfer points. Also known as Signaling System 6 (SS6), it helped to make possible numerous customer-controlled services, such as conferencing, call storage, and call forwarding.

A major advance in common-channel signaling was introduced with the CCITT's Signaling System 7 (SS7), which operates at up to 64 kb/s and carries more than 10 times as much information as SS6. In addition, Signaling System 7 is used in local exchange service areas as well as in domestic and international toll networks.

In addition to increasing network call setup efficiency, Signaling System 7 enables and enhances services such as ISDN applications; automatic call distributor (ACD); and local-area signaling services (LASS), known as caller identification. Other innovations include 800 or free-phone service, in which customers can place free long-distance calls to businesses and government offices; 900 services, in which customers pay for services from businesses; and enhanced 911 calls, in which customer names and addresses are automatically displayed in police, fire, or ambulance centers when calls are placed for emergency services. The Signaling System 7 software also can include numerous other custom calling services for residential and business customers such as call waiting and call forwarding.

Intelligent networks. The greatest potential for Signaling System 7 use is in emerging intelligent networks, both domestic and international. The intent is to provide customers with much greater control over a variety of network functions, yet to protect the network against misuse or disruption. Intelligent networks are evolving from the expanding use of digital switching and transmission, starting with the toll networks. The complexity of intelligent networks mandates the use of networked computers, programmed with advanced software.

An intelligent network allows the customer to setup and use a virtual private network as needed, and be charged only for the time that network is being used. A conventional private leased network, by contrast, reserves dedicated circuits on a full-time basis and charges for them whether they are used or not. A virtual private network enables a business to simultaneously reap the benefits of a dedicated network and the shared public-switched network by drawing on the infrastructure of intelligent networking.

Wireless. The rapid growth of cellular-service subscribers has led to congestion in some cellular systems. The solution is to introduce digital cellular systems, which can hande up to 10 times more calls in the same frequency range. However, in countries such as the United States, which already have an analog infrastructure, the FCC has mandated that any digital system must be compatible with the existing analog system. The initial digital designs, based on a technology called time-division multiple access (TDMA), provide a three-times growth factor. Another digital technology, based on code-division multiple access (CDMA), a spread-spectrum technique, could increase the growth factor to 10 or even 20. *See* RADIO SPECTRUM ALLOCATIONS; SPREAD SPECTRUM COMMUNICATION.

Personal communications service (PCS) is a sort of cellular system in which a pocket-size telephone is carried by the user. A series of small transmitter-receiver antennas operating at lower power than cellular antennas are installed throughout a city or community (mounted on lampposts and building walls, for example). All the antennas of the personal communications network (PCN) are linked to a master telephone switch that is connected to the main telephone network. John H. Davis

Telephone systems construction

Construction of the physical components of the telephone system from the central office to the end user. The telephone system can be roughly divided into three major components: outside plant, switching equipment, and interoffice facilities. The outside plant can be thought of as the local cables that bring the phone service to the subscribers' premises. The switching equipment resides in a building called a central office and serves to route the call to the correct location. The interoffice facilities connect one central office to another. When a call is initiated, the outside plant carries the signal to the local central office. The switching equipment in the local central office then determines the location to which the call needs to be routed. The call is then routed over the interoffice facilities to the central office serving the person being called, where the switching equipment then connects the call to the local outside plant. This article describes the outside plant portion of the telephone system.

Outside plant consists of cables and their supporting facilities. This cable can be aerial (supported by poles), underground in conduit, or buried directly in the ground. There are two main types of cable, metallic and optical.

Metallic cable. Modern telecommunications cable consists of copper conductors coated with a plastic color-coded insulation. The pairs are then twisted together and grouped into 25-pair complements, which are further twisted together to make a cable. The cable is then covered in a protective sheath, also called a turnplate, made of aluminum or steel. The turnplate is then covered in a sheath made of polyethylene. The center of the cable can be left as is, and is referred to as air core; or it can be flooded with a moisture-repelling compound, and is referred to as jelly-filled cable.

Fiber cable. Optical fiber consists of a glass core, coated with a glass cladding, and then coated with plastic. Light is transmitted along the core of the fiber. The interface between the core and the cladding causes the light to reflect back into the core, and in effect "bounce" back and forth as it travels along the core. Two major types of fiber are used in telephony, single-mode and multimode. In a single-mode fiber, the light can only take one path along the core. In a multimode fiber, the light can take multiple paths along the fiber. Single-mode fiber provides higher transmission speeds over longer distances. *See* COMMUNICATIONS CABLE; OPTICAL COMMUNICATIONS; OPTICAL FIBERS.

Aerial plant. Aerial plant consists of telephone cable that is suspended from structures, typically telephone poles. The majority of telephone poles are made of wood; however, alternative products are available, including fiberglass, concrete, metal, and laminated wood poles.

Telephone poles are set into the ground and connected by a steel strand that is anchored to each pole. Cable is placed by hanging rollers, called blocks, on the strand. The cable is then pulled through the blocks.

Underground plant. Underground plant consists of facilities that are placed in conduit and vaults. Vaults are typically constructed of concrete. A typical telecommunications vault is approximately 7 ft (2.5 m) deep, 6 ft (2 m) wide, and 12 ft (4 m) long. Most telecommunication vaults have two lids. The vaults are connected by conduit, which is also known as ducts. At present, conduit is typically 4-in. (10-cm) polyvinyl chloride (PVC) pipe. When the conduit system is constructed, a pulling line is installed in each duct for the installation of cable. Once a rope is in place, it is used to pull in a cable that is winched through the duct.

Buried plant. Buried plant consists of cables that are buried directly in the ground. In a conduit system, if a cable needs to be replaced or upgraded, it can be pulled out and a new one pulled in. With direct-buried cable, the entire trench must be reexcavated. Direct-buried cables are usually made specifically for being buried and are armored and jelly-filled.

Splicing. Once cable is placed in the outside plant, it must be connected by splicing. There are a number of methods used to join copper conductors. Some connectors permanently join together a single conductor. If such a connector needs to be removed, it must be cut out and a new connector used. Other connectors are designed to be pluggable. Pluggable connectors make use of the fact that cables are built in 25-pair complements.

Fiber cable can be joined using either mechanical or fusion methods. With mechanical methods, the ends of the fiber are cleaned and cleaved to create a flat clean end, and then placed in a device that holds them precisely together. The fusion method involves use of an electric arc that fuses the ends of the two fibers together.

Air pressure. If moisture gets into a copper cable, it will begin to act as a path for electricity to flow between conductors. If the cable becomes wet enough, the conductors in the cable will short and the circuit stops working. Telecommunications cables are pressurized with air to help prevent this failure.

An air system consists of dryers, air pipes, and transducers. An air dryer provides a source of pressurized dry air. The air pipe provides pressurized air to the air core of a cable. Any breach in a cable's sheath that would let water in instead forces air out. Thus, this pressure difference protects the cable against water. Transducers are mounted on the pipe panel to monitor air pressure and flow. An increase in flow or a decrease in pressure serves as an early warning that a cable may be damaged. David F. Dolch

Telescope An instrument used to collect, measure, or analyze electromagnetic radiation from distant objects. A telescope overcomes the limitations of the eye by increasing the ability to see faint objects and discern fine details. In addition, when used in conjunction with modern detectors, a telescope can "see" light that is otherwise invisible. The wavelength of the light of interest can have a profound effect on the design of a telescope. *See* ELECTROMAGNETIC RADIATION; LIGHT.

For many applications, the Earth's atmosphere limits the effectiveness of larger telescopes. The most obvious deleterious effect is image scintillation and motion, collectively known as poor seeing. Atmospheric turbulence produces an extremely rapid motion of the image resulting in a smearing. On the very best nights at ideal observing sites, the image of a star will be spread out over a 0.25-arcsecond seeing disk; on an average night, the seeing disk may be between 0.5 and 2.0 arcseconds.

The upper atmosphere glows faintly because of the constant influx of charged particles from the Sun. The combination of the finite size of the seeing disk of stars and the presence of airglow limits the telescope's ability to see faint objects. One solution is placing a large telescope in orbit above the atmosphere. In practice, the effects of air and light pollution outweigh those of airglow at most observatories in the United States. *See* AIRGLOW.

There are basically three types of optical systems in use in astronomical telescopes: refracting systems whose main optical elements are lenses which focus light by refraction; reflecting systems, whose main imaging elements are mirrors which focus light by reflection; and catadioptric systems, whose main elements are a combination of a lens and a mirror. The most notable example of the last type is the Schmidt camera.

Astronomers seldom use large telescopes for visual observations. Instead, they record their data for future study. Modern developments in photoelectric imaging devices are supplanting photographic techniques for many applications. The great advantages of detectors such as charge-coupled devices is their high sensitivity, and the images can be read out in a computer-compatible format for immediate analysis. *See* CHARGE-COUPLED DEVICES.

Light received from most astronomical objects is made up of radiation of all wavelengths. The spectral characteristics of this radiation may be extracted by special instruments called spectrographs. *See* ASTRONOMICAL SPECTROSCOPY.

As collectors of radiation from a specific direction, telescopes may be classified as focusing and nonfocusing. Nonfocusing telescopes are used for radiation with energies of x-rays and above (x-ray, gamma-ray, cosmic-ray, and neutrino telescopes). Focusing telescopes, intended for nonvisible wavelengths, are similar to optical ones (solar, radio, infrared, and ultraviolet telescopes), but they differ in the details of construction. *See* CERENKOV RADIATION; COSMIC RAYS; GAMMA-RAY ASTRONOMY; INFRARED ASTRONOMY; NEUTRINO ASTRONOMY; RADIO TELESCOPE; SUN; ULTRAVIOLET ASTRONOMY; X-RAY TELESCOPE.

The 5-m (200-in.) Hale telescope at Palomar Mountain, California, was completed in 1950. The primary mirror is 5 m in diameter with a 1.02-m (40-in.) hole in the center.

The 4-m (158-in.) Mayall reflector at the Kitt Peak National Observatory was dedicated in 1973. The prime focus has a field of view six times greater than that of the Hale reflector. An identical telescope was subsequently installed at Cerro Tololo Inter-American Observatory, in Chile.

The mirrors for these traditional large telescopes were all produced using the same general methodology. A large, thick glass mirror blank was first cast; then the top surface of the mirror was laboriously ground and polished to the requisite shape. The practical and economical limit to the size of traditional mirror designs was nearly reached by the 6-m (236-in.) telescope in the Caucasus Mountains, Russia. Newer telescopes have been designed and built that use either a number of mirrors mounted such that the light collection by them is brought to a common focus, or lightweight mirrors in computer-controlled mounts.

Gemini North (Gillette) telescope in its dome. (*Copyright Gemini Observatory/AURA/NOAO/NSF, all rights reserved*)

The Keck Telescopes on Mauna Kea, Hawaii, completed in 1993–1996, are the largest of the segmented mirror telescopes to be put into operation. The telescopes are a fairly traditional design. However, their primary mirrors are made up of 36 individual hexagonal segments mosaiced together to form single 10-m (386-in.) mirrors. Electronic sensors built into the edges of the segments monitor their relative positions, and feed the results to a computer-controlled actuator system.

In 1989, the European Southern Observatory put into operation their New Technology Telescope. The 3.58-m (141-in.) mirror was produced by a technique known as spin-casting, where molten glass is poured into a rotating mold.

Worldwide efforts are under way on a new generation of large, ground-based telescopes, using both the spin-casting method and the segmented method to produce large mirrors. The Gemini project of the National Optical Astronomy Observatories has built twin 8.1-m (319-in.) telescopes, Gemini North (now called Gillette see illustration) on Mauna Kea, Hawaii (1999), and Gemini South on Cerro Pachon in Chile (2000).

The Very Large Telescope (VLT), operated by the European Southern Observatory on Cerro Paranel, Chile, consists of four 8-m (315-in.) "unit" telescopes with spin-cast mirrors and small 1.8-m (71-in.) auxiliary telescopes. The light from the four telescopes is combined to give the equivalent light-gathering power of a 16-m (630-in.) telescope. The last of the four-unit telescopes began collecting scientific data in September 2000.

The ability of large telescopes to resolve fine detail is limited by a number of factors. Distortion due to the mirror's own weight causes problems in addition to those of atmospheric seeing. The Earth-orbiting *Hubble Space Telescope* (*HST*) with an aperture of 2.4 m (94 in.), was designed to eliminate these problems. The telescope operates in ultraviolet as well as visible light, resulting in a great improvement in resolution not only by the elimination of the aforementioned terrestrial effects but by the reduced blurring by diffraction in the ultraviolet. *See* DIFFRACTION; RESOLVING POWER (OPTICS).

Soon after the telescope was launched in 1990, it was discovered that the optical system was plagued with spherical aberration, which severely limited its spatial resolution. After space-shuttle astronauts serviced and repaired the telescope in 1993, adding what amounted to eyeglasses for the scientific instruments, the telescope exceeded its prelaunch specifications for spatial resolution. Subsequent servicing missions replaced instruments with newer technology. *See* HUBBLE SPACE TELESCOPE.

Robert D. Chapman; William M. Sinton

Telestacea An order of the cnidarian subclass Alcyonaria (Octocorallia). Telestacea are typified by *Telesto*, which forms an erect branching colony by lateral budding from the body wall of an elongated primary or axial polyp. The stolon is bandlike or membranous. Sclerites are scattered singly, partly fused, or entirely fused to form a rigid tube. *See* OCTOCORALLIA (ALCYONARIA).

Kenji Atoda

Television The electrical transmission and reception of transient visual images. Like motion pictures, television consists of a series of successive images which are registered in the brain as a continuous picture because of the persistence of vision. Each visual image impressed on the eye persists for a fraction of a second. At the television transmitter, minute portions of a scene are sampled individually for brightness and color, and the information for each portion is transmitted consecutively. At the receiver, each portion is synchronized and reproduced in its proper position and with correct brightness and color to reproduce the original scene. *See* TELEVISION RECEIVER; TELEVISION TRANSMITTER.

Fundamental principles. The technology of television is based on the conversion of light rays from still or moving scenes and pictures into electronic signals for transmission or storage, and subsequent reconversion into visual images on a screen. A similar function is provided in the production of motion picture film. However, whereas film records the brightness variations of a complete scene on a single frame in an exposure no longer than a fraction of a second, the elements of a television picture must be scanned one piece at a time. In the television system, a scene is dissected into a frame composed of a mosaic of picture elements or pixels. This process is accomplished by (1) analyzing the image with a photoelectric device in a sequence of horizontal scans from the top to the bottom of the image to produce an electric signal in which the brightness and color values of the individual picture elements are represented as voltage levels of a video waveform; (2) transmitting the values of the picture elements in sequence as voltage levels of a video signal; and (3) reproducing the image of the original scene in a video signal display of parallel scanning lines on a viewing screen. *See* TELEVISION SCANNING.

Three primary color transmission standards are in use today:

1. NTSC (National Television Systems Committee)—used in the United States, Canada, Mexico, Central America, some of South America, and Japan. In addition, NTSC is used in various countries or possessions heavily influenced by the United States.

2. PAL (Phase Alternate each Line)—used in the United Kingdom, most countries and possessions influenced by England, most European countries, and China. Variation exists in PAL systems.

3. SECAM (SEquential Color with [Avec] Memory)—used in France, countries and possessions influenced by France, the nations of the former Soviet Union, other former Soviet Bloc nations, including East Germany, and other areas influenced by Russia.

The three standards are incompatible for a variety of technical reasons, but standard converter devices are commercially available. *See* TELEVISION STANDARDS.

The band of frequencies assigned to a television station for the transmission of synchronized picture and sound signals is called a television channel. In the United States a television channel is 6 MHz wide, with the visual carrier frequency 1.25 MHz above the lower edge of the band and the aural carrier 0.25 MHz below the upper edge of the band.

Television channels in the United States are in three frequency bands. Channels 2–6 occupy 54 to 88 MHz, channels 7–13 are 174 to 216 MHz, and channels 14–69 are 470 to 806 MHz. The first two groups of channels fall in the very high frequency (VHF) band; the channels in the last group are in the ultrahigh-frequency (UHF) band. When digital television, whose stations operate at different radio frequencies, is widely implemented, the current radio frequencies used by the analog television stations will be used for other transmision services. *See* RADIO SPECTRUM ALLOCATIONS.

Digital television. Digital television (DTV) has ushered in a new era in television broadcasting. The impact of DTV is more significant than simply moving from an analog system to a digital system. DTV permits a level of flexibility wholly unattainable with analog broadcasting. An important element of this flexibility is the ability to expand system functions by building upon the technical foundations specified in the basis standards.

With NTSC, and its PAL and SECAM counterparts, the video, audio, and some limited data information are conveyed by modulating a radio-frequency (RF) carrier in such a way that a receiver of relatively simple design can decode and reassemble the various elements of the signal to produce a program consisting of video and audio, and perhaps related data (for example, closed captioning). As such, a complete program is transmitted by the broadcaster that is essentially in finished form. In the DTV system, however, additional levels of processing are required after the receiver demodulates the RF signal. The receiver processes the digital bit stream extracted from the received signal to yield a collection of program elements (video, audio, or data) that match the services the consumer has selected. This selection is made using system and service information that is also transmitted. Audio and video are delivered in digitally compressed form and must be decoded for presentation.

The development of digital television occurred over a period of many years on several continents. Three major systems emerged:

1. The Advanced Television Systems Committee (ATSC), a North American–based DTV standard organization, developed the ATSC terrestrial DTV series of standards.

2. The Digital Video Broadcasting Project (DVB), a European-based standard organization, developed the DVB series of DTV standards, which were standardized by the European Telecommunication Standard Institute (ETSI).

3. The Integrated Service Digital Broadcasting (ISDB) standards, a series of DTV standards, were developed and standardized by the Association of Radio Industries and Business (ARIB) in Japan.

The remainder of this article focuses on the ATSC DTV system. The Digital Television Standard describes a system designed to transmit high-quality video and audio and ancillary data over a single 6-MHz channel. The system can deliver about

19 megabits per second in a 6-MHz terrestrial broadcasting channel and about 38 Mbits/s in a 6-MHz cable television channel. This means that encoding high-definition video essence at raw data rates that are typically greater than 1 gigabit per second requires a bit rate reduction by about a factor of 50. To achieve this bit rate reduction, the system is designed to be efficient in utilizing available channel capacity by exploiting complex video and audio compression technology. The compression scheme optimizes the throughput of the transmission channel by representing the video, audio, and data sources with as few bits as possible while preserving the level of quality required for the given application. *See* DATA COMPRESSION.

The resolution of the displayed picture is the most basic attribute of any video production system. Generally speaking, a high-definition television (HDTV) image has approximately twice as much luminance definition horizontally and vertically as the 525-line NTSC system or the 625-line PAL and SECAM systems. The total number of luminance picture elements (pixels) in the image, therefore, is four times as great. The wider aspect ratio of the HDTV system adds even more visual information. The HDTV image is 25% wider than the conventional video image for a given image height; the ratio of image width to height in HDTV systems is 16:9, or 1.78. The conventional video image has a 4:3 aspect ratio.

As a result of these attributes, the HDTV image may be viewed more closely than is customary in conventional television systems. Full visual resolution of the detail of conventional television is available when the image is viewed at a distance equal to about six or seven times the height of the display. The HDTV image may be viewed from a distance of about three times picture height for the full detail of the scene to be resolved.

Thus, HDTV offers the potential for approximately twice the horizontal and twice the vertical resolution of current (NTSC) television. When combined with a wide-screen format (16:9 aspect ratio), this can result in considerably more visual information than conventional television. HDTV consumer television sets can have 720 or 1080 active vertical scanning lines, and are capable of decoding the transmitted 720 × 1280 and 1080 × 1920 ATSC formats and displaying them as a 16:9 aspect ratio image. These two high-definition formats can potentially provide over eight times as much picture information as delivered over broadcast NTSC. While these high-definition transmission formats will be supported by such sets, the actual delivered resolution may vary by broadcaster, by product, and by program. HDTV is normally accompanied by digital surround-sound capability.

Jerry C. Whitaker

Television camera An electrooptical system used to pick up and convert a visual image or scene into an electrical signal called video. The video may be transmitted by cable or wireless means to a suitable receiver or monitor some distance from the actual scene. It may also be recorded on a video tape recorder for playback at a later time.

A television camera may fall within one of several categories: studio (see illustration), telecine, or portable. It may also be one of several highly specialized cameras used for remote viewing of inaccessible places, such as the ocean bottom or the interior of nuclear power reactors. The camera may be capable of producing color or monochrome (black and white) pictures. Most modern cameras are entirely solid-state, including the light-sensitive element, which is composed of semiconductors called charge-coupled devices (CCDs). Inexpensive or special-purpose cameras, however, may use one or more vacuum tubes, called vidicon, with a light-sensitive surface in lieu of the charge-coupled devices. *See* CHARGE-COUPLED DEVICES.

Every camera shares certain essential elements: an optical system, one or more picture pickup devices, preamplifiers, scanning circuits, blanking and synchronizing circuits, video processing circuits, and control circuits. Color cameras also include some kind of color-encoding circuit.

Television studio camera. (*Thomson Multimedia*)

Other functions that are necessary to obtain high-quality pictures include gamma correction, aperture correction, registration, and color balance. Gamma correction is required because the pickup devices do not respond linearly to increasing light levels. It allows the camera to capture detail in the dark areas of high-contrast scenes, essentially by "stretching" the video levels in those areas. Aperture correction provides several benefits mainly related to an even overall response to scenes with more or less detail. It also helps to improve the signal-to-noise ratio of the camera's output video. Registration must be adjusted on multiple-tube cameras to ensure that the separate red, blue, and green images are precisely aligned on one another; charge-coupled-device cameras are usually registered once, at the factory. Color balance must be properly set on color cameras and must be consistent from dark scenes to bright scenes, or there will be an objectionable tint to the camera output.

Studio cameras are equipped with several ancillary systems to enhance their operation. An electronic viewfinder (actually a small television monitor) shows the camera operator what the camera is seeing, making it possible to frame and focus the picture. The tally system consists of one or more red lights that illuminate when the camera's picture is "on the line" so that production and on-camera personnel know which camera is active. Generally an intercom system is built into the camera so that the director can communicate with the camera operator. The camera itself may be mounted upon a tripod, but more often it is on a dolly and pedestal, which allows the camera to be moved around on the studio floor and raised or lowered as desired. A pan head permits the camera to be rotated to the left or right and furnishes the actual mounting plate for the camera. The lens zoom and focus controls are mounted on a panning handle convenient to the operator.

Telecine cameras are used in conjunction with film or slide projectors to televise motion pictures and still images. Many of the usual controls are automatic so as to require less operator attention.

Portable cameras usually combine all of the basic elements into one package and may be used for a multitude of purposes. They have found their way into electronic news gathering for broadcast television, and into electronic field production, where they can be used for production of broadcast programs, commercials, and educational programs. The units often have built-in microphones, videocassette recorders, and batteries for completely self-contained operation.

Cameras used in high-definition television (HDTV) are fundamentally similar in appearance and operation to previous cameras. In fact, some modern cameras are switchable to produce either a conventional output or an HDTV output. The conventional output has a 4:3 aspect ratio raster and the scan rates match the 525-horizontal-line, 59.94-Hz vertical-field-rate NTSC standard in the United States. When switched to HDTV mode, the aspect ratio becomes 16:9 and the horizontal scan rate is usually increased to either 720 progressively scanned lines or 1080 interlace-scanned lines with a 60-Hz vertical field rate. See TELEVISION; TELEVISION RECEIVER; TELEVISION STANDARDS; TELEVISION STUDIO; TELEVISION TRANSMITTER. Earl F. Arbuckle, III

Television camera tube An electron tube having a light-sensitive receptor that converts an optical image into an electrical television video signal. The tube is used in a television camera to generate a train of electrical pulses representing the light intensities present in an optical image focused on the tube. Each point of this image is interrogated in its proper turn by the tube, and an electrical impulse corresponding to the amount of light at that point of the optical image is generated by the tube. This signal represents the video or picture portion of a television signal. See TELEVISION CAMERA.

Image orthicon. The image orthicon made broadcast television practical. It was used for more than 20 years as the primary studio and field camera tube for black and white and color television programming because of its high sensitivity and its ability to handle a wide range of scene contrast and to operate at very low light levels. It is one of the most complicated camera tubes. The image orthicon is divided into an image section, a scanning section, and a multiplier section, within a single vacuum envelope. The image isocon is a further development of the image orthicon.

Photoconductive tubes. These types have a photoconductor as the light-sensitive portion. The name vidicon was applied to the first photoconductive camera tube developed by RCA. It is loosely applied to all photoconductive camera tubes, although some manufacturers adopt their own brand names. The vidicon tube is a small tube that was first developed as a closed-circuit or industrial surveillance television camera tube. The development of new photoconductors has improved its performance to the point where it is now utilized in one form or another in most television cameras. Its small size and simplicity of operation make it well suited for use in systems to be operated by relatively unskilled people.

The vidicon is a simply constructed storage type of camera tube (see illustration). The signal output is developed directly from the target of the tube and is generated by a low-velocity scanning beam from an electron gun. The target generally consists of a transparent signal electrode deposited on the faceplate of the tube and a thin layer of photoconductive material, which is deposited over the electrode. The photoconductive layer serves two purposes. It is the light-sensitive element, and it forms the storage surface for the electrical charge pattern that corresponds to the light image falling on the signal electrode.

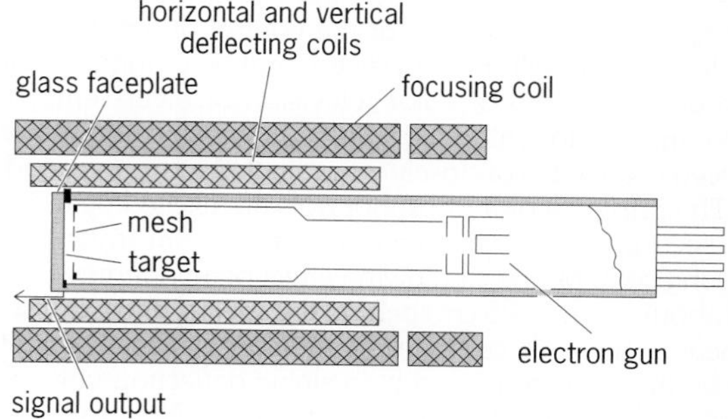

Cross section of a vidicon tube and its associated deflection and focusing coils.

Photoconductor properties determine to a large extent the performance of the different types of vidicon tubes. The first and still most widely used photoconductor is porous antimony trisulfide. The latest photoconductors are the lead oxide, selenium-arsenic-tellurium, cadmium selenide, zinc-cadmium telluride, and silicon diode arrays.

Silicon intensifier. The silicon intensifier camera tube utilizes a silicon diode target, but bombards it with a focused image of high-velocity electrons. Each high-energy electron can free thousands of electron carriers in the silicon wafer (compared to one carrier per photon of light on a silicon diode vidicon). This high amplification allows the camera to operate at light levels below that of the dark-adapted eye. The silicon intensifier tube is utilized for nighttime surveillance and other extremely low-light-level television uses in industrial, scientific, and military applications.

Solid-state imagers. These are solid-state devices in which the optical image is projected onto a large-scale integrated-circuit device which detects the light image and develops a television picture signal. Typical of these is the charge-coupled-device imager. The term charge-coupled device (CCD) refers to the action of the device which detects, stores, and then reads out an accumulated electrical charge representing the light on each portion of the image. The device detects light by absorbing it in a photoconductive substrate, such as silicon. The charge carriers generated by the light are accumulated in isolated wells on the surface of the silicon that are formed by voltages applied to an array of electrodes on top of an oxide insulator formed on the surface of the silicon. A practical CCD imager consists of a structure that forms several hundred thousand individual wells or pixels, and transfers the charges accumulated in these pixel wells out to an output amplifier in the proper sequence. See CHARGE-COUPLED DEVICES. Robert G. Neuhauser

Television networks Arrangements of communications channels, suitable for transmission of video and accompanying audio signals, which link together groups of television broadcasting stations or closed-circuit television users in different cities so that programs originating at one point can be fed simultaneously to all others.

In the United States, television network service is furnished by the long-distance or local-exchange carriers (hereafter identified as telephone companies) and satellite carriers, as well as by broadcaster-owned networks. The facilities, when provided by the telephone companies, consist of intercity channels, which interconnect the principal long-distance telephone offices in various cities, and local channels, which connect the telephone offices with the broadcasters' studios or other user locations in each city. In the terminating offices, the intercity and local channels are brought together in a television operating center (TOC) where means are provided for testing, monitoring, and connecting the channels in various patterns as required for service. See TELEPHONE SERVICE.

The principal users of television network facilities are broadcast network organizations (broadcast networks). These may be national, regional, or local in scope, and may be regular or occasional in nature and may be commercial or noncommercial. Broadcast networks typically consist of a programming entity which simultaneously feeds content over the network interconnection facilities to commonly owned and operated stations, as well to other stations (affiliates) that have a contractual relationship with the broadcast network.

Geostationary satellite antenna footprints can be made to cover large areas of the Earth, thus making them ideal for point-to-multipoint transmissions. Direct broadcast satellite systems have been placed into operation in many developed nations.

Their implementation in the United States is typical of other systems.

While all satellite television networks once used FM transmission, many operators now use digital video compression to transmit 4–10 signals over a single satellite transponder using quadrature phase-shift keying (QPSK) modulation. This changes the instantaneous amplitude of the two orthogonal carriers (I and Q) at a rate of about 38 Mbps. The received data require good forward error correction for perfect decompression to video and audio. Picture quality at four to six compressed video programs per satellite transponder is as good, or better, than FM-transmitted signals.

Over 96% of television households in the United States have access to cable television, and most of them receive all their television programming in analog format via cable. Programming usually includes the local broadcast stations and satellite services.

Because of the closed nature of cable television systems, all the frequencies between 50 and 750 MHz and higher can be used for distribution. The analog signals received at cable television headends are combined side by side on a frequency-division multiplex (FDM) basis for carriage in the cable systems. Digital satellite signals are decompressed in integrated receiver decoders, and the programs are passed through to the subscribers as analog signals. *See* MULTIPLEXING AND MULTIPLE ACCESS.

Joseph B. Glaab; Clifford M. Harrington

Television receiver The equipment used to receive the transmitted modulated radio-frequency signals and produce synchronized visual images and sound. The radio-frequency portion operates on the superheterodyne principle. *See* MODULATION; RADIO RECEIVER.

The first television receivers to be mass-produced were monochrome; that is, they provided pictures in black and white only. Later, color receivers, which produce pictures in full color as well as black and white, became available. Many television receivers now can receive stereophonic sound or alternate language in accordance with multichannel television sound standards. For basic discussion of a television system *see* TELEVISION; TELEVISION STANDARDS; STEREOPHONIC RADIO TRANSMISSION.

Early television receivers used vacuum-tube technology. Present-day receivers use solid-state technology with many functions integrated on a few chips. The only function still primarily implemented by using vacuum-tube technology is the display by the cathode-ray tube (CRT). *See* CATHODE-RAY TUBE; INTEGRATED CIRCUITS; VACUUM TUBE.

Since most broadcast television transmissions in the United States are horizontally polarized, the most basic type of television-receiving antenna is the horizontally mounted half-wave dipole. More complex antennas combine several dipole elements of various lengths, and passive reflectors may be used to achieve some degree of horizontal directivity, which increases the amplitude of the receiver signal and reduces interference from other stations. *See* ANTENNA (ELECTROMAGNETISM); POLARIZATION OF WAVES; YAGI-UDA ANTENNA.

The tuner of a television receiver selects the desired channel and converts the frequencies received to lower frequencies within the passband of the intermediate-frequency amplifier. The output from the tuner is applied to the intermediate-frequency (i-f) amplifier. The output of the intermediate-frequency amplifier consists of modulated radio-frequency signals, which when detected provide signals corresponding to the transmitted picture and sound information. *See* AMPLITUDE MODULATION; AMPLITUDE-MODULATION DETECTOR; FREQUENCY MODULATION.

Picture synchronizing information is obtained from the video signal by means of sync separation circuits. In general, sync separation circuits perform the following functions: (1) separation of the sync information from the picture information; (2) separation of the desired horizontal and vertical timing information by means of frequency selection; and (3) rejection of noise signals. *See* ELECTRICAL NOISE; TELEVISION SCANNING.

The display device for a monochrome television receiver is a cathode-ray tube, consisting of an evacuated bulb containing an electron gun and a phosphor screen, which emits light when excited by an electron beam. The intensity of the electron beam is controlled by the video signal, which is applied either to the grid or to the cathode of the electron gun. Television receivers designed to produce images in full color are necessarily more complex than those designed to produce monochrome images only. In monochrome systems, the video signal controls only the luminance of the various areas of the image. In color systems, it is necessary to control both the luminance and the chrominance of the picture elements. *See* PICTURE TUBE.

The chrominance of a color refers to those attributes which cause it to differ from a neutral (white or gray) color of the same luminance. In qualitative terms, chrominance may be regarded as those properties of a color that control the psychological sensations of hue and saturation. For color television purposes, chrominance is most frequently expressed quantitatively in terms of the amounts of two hypothetical, zero-luminance primary colors (usually designated I and Q), which must be added to or subtracted from a neutral color of a given luminance to produce the color in question. As a practical matter, color television receivers produce full-color images as additive combinations of red, green, and blue primary-color images, and it is necessary to process the luminance and chrominance information in a color signal in such a way as to make it usable by a practical reproducing device. *See* COLOR.

Color television broadcasts in the United States employ signal specifications that are fully compatible with those used for monochrome, making it possible for color programs to be received on monochrome receivers and monochrome programs to be received on color receivers. (Color pictures are produced, of course, only when color programs are viewed through color receivers.) Compatibility is achieved by encoding the color information at the transmitting end of a color television system in such a way that the transmitted signal consists essentially of a normal monochrome signal (conveying luminance information) supplemented by an additional modulated wave conveying chrominance information. Although it is added directly to the monochrome signal component before transmission, the color subcarrier signal does not cause objectionable interference, because of the use of the frequency interlace technique. Because the chrominance information involves two variables, the modulated subcarrier signal varies in both amplitude and phase, and it is necessary to employ synchronous detectors to recover the two variables. A phase reference for the special local oscillator, which provides the synchronized carriers in each color receiver, is transmitted in the form of so-called color synchronizing bursts. These are short samples of unmodulated subcarrier transmitted during the horizontal blanking periods after the horizontal sync pulses.

Special decoding circuits are necessary in a color receiver to process the luminance and chrominance information in a color signal so that it can be used for the control of a practical color cathode-ray tube utilizing red, green, and blue primary colors.

The great majority of color television receivers employ the shadow-mask color cathode-ray tube in which color images are produced in the form of closely intermingled red, green, and blue dots. The primary-color phosphor dots are excited by three separate electron beams, which are prevented from striking dots of the wrong color by the shadowing effect of an aperture mask located about $^1/_2$ in. (1.25 cm) behind the special phosphor screen. The beams in such a cathode-ray tube are deflected simultaneously by the fields produced by a single deflection yoke placed conventionally around the neck of the tube. New cathode-ray-tube designs and deflection yokes are self-converging and do not require auxiliary convergence deflection.

In addition to the same controls required for monochrome receivers (such as brightness and contrast), color receivers normally have controls for convergence, hue, and saturation. The convergence controls, considered servicing adjustments only, adjust the relative amplitudes and phases of the signal components that are added together to form the proper waveforms for the convergence yoke. The hue control usually adjusts the phase of the burst-controlled oscillator and alters all the colors in the image in a systematic manner comparable to the effect achieved when a color circle diagram is rotated in one direction or the other. The proper setting for the hue control is normally determined by observing skin tones of persons on the television screen. The saturation control, frequently labeled chroma or simply color, adjusts the gain of the chrominance circuits relative to the monochrome channel and controls the saturation or vividness of the reproduced colors. When this control is set too low, the colors are all pale or pastel, and when it is reduced to zero, the picture is seen in black and white only. Carl G. Eilers

Television scanning The process used to convert a three-dimensional image intensity into a one-dimensional television signal waveform. The image information captured by a television camera conveys color intensity (in terms of red, green, and blue primary colors) at each spatial location, with horizontal and vertical coordinates, and at each time instance. Thus, the image intensity is multidimensional, since it involves two spatial dimensions and time. It needs to be converted to a unidimensional signal so that processing, storage, communications, and display can take place.

The television scene is sampled many times per second in order to create a sequence of images (called frames). Then, within each frame, sampling is done vertically to create scan lines. Scanning proceeds sequentially, left to right and top to bottom. In a television camera, an electron beam scans across an electrically photosensitive target upon which the image is focused. At the other end of the television chain, with raster scanned displays, an electronic beam scans and lights up the picture elements in proportion to the light intensity. While it is convenient to think of all the samples of a single frame occurring at a single time (similar to the simultaneous exposure of a single frame for film), the scanning in a camera and in a display results in every sample corresponding to a different instance in time, and successive lines occur later in time. *See* TELEVISION CAMERA TUBE; TELEVISION RECEIVER.

There are two types of scanning approaches: progressive (also called sequential) and interlaced. In progressive scanning, the television scene is first sampled in time to create frames, and within each frame all the raster lines are scanned from top to bottom. Therefore, all the vertically adjacent scan lines are also temporally adjacent and are highly correlated even in the presence of rapid motion in the scene. Film can be thought of as naturally progressively scanned, since all the lines were originally exposed simultaneously, so the correlation between adjacent lines is guaranteed. Almost all computer displays (except some low-end computers) are sequentially scanned. *See* ELECTRONIC DISPLAY.

In interlaced scanning, all the odd-numbered lines in the entire frame are scanned first, and then the even numbered lines. This process produces two distinct images per frame, representing two distinct samples of the image sequence at different points in time. The set of odd-numbered lines constitute the odd field, and the even-numbered lines make up the even field. All current television systems use interlaced scanning. One principal benefit of interlaced scanning is to reduce the scan rate (or the bandwidth). This is done with a relatively high field rate (a lower field rate would cause flicker), while maintaining a high total number of scan lines in a frame (lower number of lines per frame would reduce resolution on static images). Interlace cleverly preserves the high-detail visual information and, at the same time, avoids visible large-area flicker at the display due to temporal postfiltering by the human eye.

An agreement (not a standard) on high-definition formats has been reached and is in use, enabling the transition from analog to digital television. In early 2001, there were more than 150 stations broadcasting high-definition television. The format has 1920 active pixels per line and 1080 active (out of a total of 1125) lines in a frame. The frames may be interlaced or progressive and the frame rate is 29.97 Hz. Progressive frames at a rate of 23.976 Hz are also permitted to accommodate film material. An alternative progressive-only format that provides additional temporal resolution at the expense of some spatial resolution has also been approved and is currently in use. This format has 1280 active pixels per line and 720 active (out of a total of 750) lines per frame, and the frame rates permitted are 23.976 Hz, 29.97 Hz, or 59.94 Hz. Both high-definition formats have square pixels and a 16 × 9 aspect ratio. *See* DATA COMPRESSION; IMAGE PROCESSING; TELEVISION. Arun N. Netravali

Television standards The accepted criteria for a television system, including the image aspect ratio, number of lines per frame, type of scanning, original video signal bandwidth, transmission format and bandwidth, reception, demodulation, decoding, and sound system. The implementation of high-definition television (HDTV), where the image resolution and audio fidelity are significantly higher than for conventional television, has required new standards. *See* TELEVISION.

In the early 1950s, the National Television Systems Committee (NTSC) was formed to set standards for a color television signal (in the United States) that would be fully compatible with the existing monochrome signal. Any color can be formed as a linear combination, that is, a weighted sum, of red (R), green (G), and blue (B). The NTSC standard started with three image signals—R, G, and B—and matrixed these three primary color image signals as linear combinations into one luminance signal (the conventional black-and-white video signal, often called the Y-signal) and two chrominance signals that control hue and saturation. The chrominance signals are termed the in-phase (I) signal and the quadrature (Q) signal. Although the luminance signal retained the original 4.2-MHz bandwidth, the characteristics of human color perception allowed the I-signal to be limited to 1.5 MHz and the Q-signal to only 0.5 MHz.

While NTSC color television standards are used in North America, South America, and Japan, the European International Radio Consultative Committee (CCIR) system is used in England, Germany, Italy, and Spain. The color system employed with CCIR television is called Phase Alternate Line (PAL). A modified CCIR television standard is used in France and Russia, where the color system is called SECAM (Sequential Couleur à Mémoire).

The next generation of television, HDTV, is not compatible with the previous television systems. HDTV relies on digital technology, making it more amenable with computer displays, while taking full advantage of the power and efficiency of digital signal processing (DSP). The desired characteristics of the HDTV system that would replace the NTSC system was expected to have a resolution that would approach the quality of a 35-mm film, that is, approximately twice the horizontal and twice the vertical resolution of conventional television, with a widescreen aspect ratio of 16:9. The target HDTV system standard was required to avoid interlace scanning artifacts, as well as chrominance artifacts and deliver digital multichannel audio. The end result was a 1996 FCC digital television (DTV) standard, and HDTV broadcasting commenced in the United States in 1998.

In Europe, a different transmission standard has been adopted. The European system is referred to as the Digital Video Broadcasting (DVB) standard.

The FCC expects everyone to be using new HDTV receivers by the year 2011, at which time NTSC broadcasting will cease, and all NTSC color television receivers will need to be replaced or modified with some type of converter to be able to decode and display HDTV images. Joseph L. LoCicero

Television studio A facility designed for the production of television programs, which may be broadcast live concurrently with the production or recorded for later broadcast. A television studio consists of the studio room, wherein the actual program takes place, and various support rooms, which include the control room, the equipment room, and the property room.

The studio room is where the program action occurs and is analogous to a theatrical stage. Studio rooms may be of almost any size, depending on use, but invariably provide certain facilities, such as a flexible lighting and scenery system, one or more cameras, one or more microphones for sound pickup, and a communications system to allow coordination during the program. Most studios use a lighting grid suspended from a high ceiling that allows flexible placement of the various lighting fixtures. *See* MICROPHONE; SOUND-REPRODUCING SYSTEMS; TELEVISION CAMERA.

The studio control room is the nerve center of the television production facility. The control room usually has a bank of video monitors with screens which display the output of each camera, videotape recorder, or special-effects generator, as well as a previous monitor which shows the director what the next shot will look like and a line monitor which shows the scene currently on the air. The sound engineer operates an audio mixer console which has every microphone used in the studio connected to a separate input. Other sources, such as turntables, audio tape recorders, tape cartridge machines, compact disk players, and audio hard disk recorders (or audio servers), may also be connected to the audio mixer. *See* SOUND-RECORDING STUDIO.

Earl F. Arbuckle, III

Television transmitter An electronic device that converts audio and video signals into modulated radio-frequency (rf) energy which can be radiated from an antenna and received by a television receiver. The term can also refer to the entire television transmitting plant, consisting of the transmitter proper, associated visual and aural input and monitoring equipment, transmission line, the antenna with its tower or other support structure, and the building in which the equipment is housed. In the United States, both analog NTSC (National Television Systems Committee) and digital 8-VSB transmitters are in service. The digital transmitters are used for what is termed high-definition television (HDTV).

An analog television transmitter can be thought of as two separate transmitters integrated into a common cabinet. Video information is transmitted via a visual transmitter, while audio information is transmitted via an aural transmitter. Because video and audio have different characteristics, the two transmitters differ in terms of bandwidth, modulation technique, and output power level. Nevertheless, a common transmitting antenna is generally used, and the two transmitters feed this antenna via an rf diplexer or combiner.

A digital transmitter accepts a single encoded digital bit stream that may contain video, audio, and data. In the United States, the digital terrestrial transmission standard is known as 8-VSB, which is an eight-level, vestigal sideband format. The FCC has mandated that all U.S. television stations convert to difital and terminate analog transmissions.

Television stations are licensed to operate on a particular channel, but since it takes a very wide bandwidth to transmit a television picture, these channels are allocated over a broad range of frequencies. Channels 2 through 6 are low-band very-high-frequency (VHF) channels, while channels 7 through 13 are high-band VHF channels. Channels 14 through 69 are ultrahigh-frequency (UHF) channels. Each channel is 6 MHz wide. Because of the wide range of frequencies, television transmitters are designed to work in only one of the foregoing groups, and employ specific circuits which are most efficient for the channels involved. *See* RADIO SPECTRUM ALLOCATION.

The horizontal radiation pattern of most television transmitting antennas is circular, providing equal radiated signal strength to all points of the compass. Higher-gain antennas achieve greater power in the direction of the horizon by reducing the power radiated at vertical angles above and below the horizon. Since this could result in weaker signals at some receivers close to the transmitter, beam tilt and null fill are often used to lower the angle of maximum radiated power. Because television signals travel in a "line of sight," transmitting antennas are usually placed as high as possible above ground with respect to the surrounding service area. This allows viewers to orient their receiving antennas in one direction for the best reception from all of the stations. *See* ANTENNA (ELECTROMAGNETISM).

There are two broad classes of VHF analog visual television transmitter design philosophy. The classical approach modulates the visual carrier at a moderate power level, amplifies the carrier to rated output power by means of high-power linear amplifiers, and then filters this high-power carrier to obtain the required vestigial-sideband signal. The more contemporary approach, used by nearly all transmitter manufacturers, employs modulation at a very low power level of an intermediate-frequency (i-f) signal. The required vestigial-sideband filtering is imposed on this low-level signal, generally by means of a highly stable surface-acoustic-wave filter, whereupon the signal is upconverted to the carrier frequency and amplified by linear amplifiers to rated output power. *See* AMPLIFIER; SURFACE-ACOUSTIC-WAVE DEVICES; TELEVISION. Earl F. Arbuckle, III

Tellurium A chemical element, Te, atomic number 52, and chemical atomic weight 127.60. There are eight stable isotopes of natural tellurium. Tellurium makes up approximately 10^{-9}% of the Earth's igneous rock. It is found as the free element, sometimes associated with selenium. It is more often found as the telluride sylvanite (graphic tellurium), nagyagite (black tellurium), hessite, tetradymite, altaite, coloradoite, and other silver-gold tellurides, as well as the oxide, tellurium ocher. *See* PERIODIC TABLE.

There are two important allotropic modifications of elemental tellurium, the crystalline and the amorphous forms. The crystalline form has a silver-white color and metallic appearance. This form melts at 841.6°F (449.8°C) and boils at 2534°F (1390°C). It has a specific gravity of 6.25, and a hardness of 2.5 on Mohs scale. The amorphous form (brown) has a specific gravity of 6.015. Tellurium burns in air with a blue flame, forming tellurium dioxide, TeO_2. It reacts with halogens, but not sulfur or selenium, and forms, among other products, both the dinegative telluride anion (Te^{2-}), which resembles selenide, and the tetrapositive tellurium cation (Te^{4+}) which resembles platinum(IV).

Tellurium is used primarily as an additive to steel to increase its ductility, as a brightener in electroplating baths, as an additive to catalysts for the cracking of petroleum, as a coloring material for glasses, and as an additive to lead to increase its strength and corrosion resistance. *See* SELENIUM. Stanley Kirschner

Telosporea A class of the subphylum Sporozoa. These protozoa are divided into two subclasses, the Gregarinia and Coccidia. All members of the group are either intra- or extracellular parasites, and the life cycles have both sexual and asexual phases. The spores lack a polar capsule and develop from an oocyst. The sporozoite is the usual infective stage which initiates the asexual phase in the life cycle. *See* COCCIDIA; GREGARINIA; PROTOZOA; SPOROZOA. Elery R. Becker; Norman Levine

Temnopleuroida An order of Echinacea with a camarodont lantern, smooth or sculptured test, tubercles imperforate or perforate (and usually crenulate), ambulacral plates of diademoid or echinoid type, and branchial slits which are usually shallow. There are three included families: Glyphocyphidae, Temnopleuridae, and Toxopneustidae. *See* ECHINACEA. Howard B. Fell

Temnospondyli One of the largest recognized groups of early amphibians with about 180 described genera. The Temnospondyli first appear in the fossil record in the lower Carboniferous and reach substantial diversity in the later Carboniferous and Permian. The end-Permian extinction event wiped out several families, but the surviving lineages rediversified and temnospondyls remained numerous through the Triassic before declining in the later Mesozoic, the last survivor (*Koolasuchus*) being known from the mid-Cretaceous of Australia. *See* GEOLOGIC TIME SCALE.

Earlier temnospondyls were superficially salamander-like with large flat heads, no necks, short limbs, and an undulating gait. They had short straight ribs which combined with the large skulls, suggest that they breathed by buccal pumping, ramming air into the lungs, and not using costal respiration. The skull was massively constructed and survives intact in the fossil record even when the rest of the skeleton is lost.

Carboniferous and Permian temnospondyls include a wide range of morphological types. As well as amphibious salamander-like forms, they evolved into larger superficially crocodile-like animals both amphibious and terrestrial (*Eryops*). Others were specialized aquatic forms, retaining lateral-line canals to a large size and filling eel-like niches. Most were 30 cm (1 ft) to 1 m (3.3 ft) in length, with a few forms growing to 2 m (6 ft). After the end-Permian extinction event, one subgroup of temnospondyls, the Stereospondyli, produced a new adaptive radiation of crocodile-like forms, mostly amphibious or aquatic and growing to 5 m (16.5 ft) [*Mastodonsaurus*]. The origin of the modern amphibian groups is controversial, but many workers consider the Temnospondyli to be the group from which they evolved. *See* AMPHIBIA; ANTHRACOSAURIA; ICHTHYOSTEGA; LISSAMPHIBIA. Andrew Milner

Temperature A concept related to the flow of heat from one object or region of space to another. The term refers not only to the senses of hot and cold but to numerical scales and thermometers as well. Fundamental to the concept are the absolute scale and absolute zero and the relation of absolute temperatures to atomic and molecular motions.

Thermometers do not measure a special physical quantity. They measure length (as of a mercury column) or pressure or volume (with the gas thermometer at the National Institute of Standards and Technology) or electrical voltage (with a thermocouple). The basic fact is that if a mercury column has the same length when touching two different, separated objects when the objects are placed in contact, no heat will flow from one to the other. *See* THERMOMETER.

The numbers on the thermometer scales are merely historical choices; they are not scientifically fundamental. The most widely used scales are the Fahrenheit (°F) and the Celsius (°C). The centigrade scale with 0° assigned to ice water (ice point) and 100° assigned to water boiling under one atmosphere pressure (steam point) was formerly used, but it has been succeeded by the Celsius scale, defined in a different way than the centigrade scale. However, on the Celsius scale the temperatures of the ice and steam points differ by only a few hundredths of a degree from 0° and 100°, respectively. The illustration shows how the Celsius and Fahrenheit scales compare and how they fit onto the absolute scales. *See* ICE POINT.

In 1848 William Thomson (Lord Kelvin), following ideas of Sadi Carnot, stated the concept of an absolute scale of temperature in terms of measuring amounts of heat flowing between objects. Most important, Kelvin conceived of a body which would not give up any heat and which was at an absolute zero of

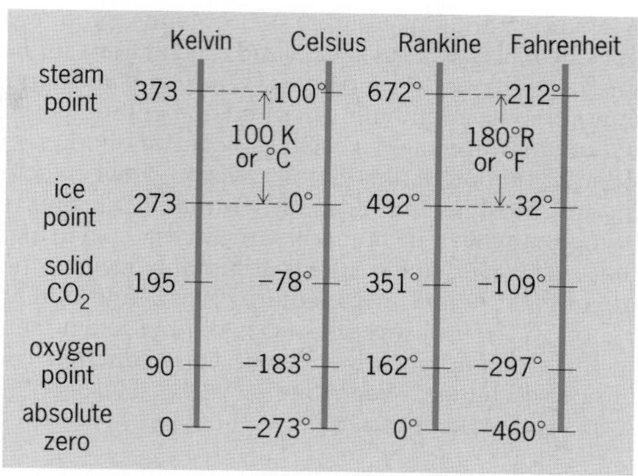

Comparisons of Kelvin, Celsius, Rankine, and Fahrenheit temperature scales. Temperatures are rounded off to nearest degree. (*After M. W. Zemansky, Temperatures Very Low and Very High, Van Nostrand, 1964*)

temperature. Experiments have shown that absolute zero corresponds to −273.15°C or −459.7°F. Two absolute scales, shown in the illustration, are the Kelvin (K) and the Rankine (°R).

In practice, absolute temperatures are measured by using low-density helium gas and dilute paramagnetic crystals, the most nearly ideal of real materials. The measurement of a single temperature with a gas or magnetic thermometer is a major scientific event done at a national standards laboratory. Only a few temperatures have been measured, including the freezing point of gold (1337.91 K or 1948.57°F), and the boiling points of sulfur (717.85 K or 832.46°F), oxygen (90.18 K or −297.35°F), and helium (4.22 K or −452.07°F). Various types of thermometers (platinum, carbon, and doped germanium resistors; thermocouples) are calibrated at these temperatures and used to measure intermediate temperatures. *See* TEMPERATURE MEASUREMENT; THERMODYNAMIC PRINCIPLES. Roland A. Hultsch

Temperature adaptation The ability of animals to survive and function at widely different temperatures as a result of specific physiological adaptations. Temperature is an all-pervasive attribute of the environment that limits the activity, distribution, and survival of animals.

Changes in temperature influence biological systems, both by determining the rate of chemical reactions and by specifying equilibria. Because temperature exerts a greater effect upon the percentage of molecules that possess sufficient energy to react (that is, to exceed the activation energy) than upon the average kinetic energy of the system, modest reductions in temperature (for example, from 77 to 59°F or from 25 to 15°C, corresponding to only a 3% reduction in average kinetic energy) produce a marked depression (two- to threefold) in reaction rate. In addition, temperature specifies the equilibria between the formation and disruption of the noncovalent (electrostatic, hydrophobic, and hydrogen-bonding) interactions that stabilize both the higher levels of protein structure and macromolecular aggregations such as biological membranes. Maintenance of an appropriate structural flexibility is a requirement for both enzyme catalysis and membrane function, yet cold temperatures constrain while warm temperatures relax the conformational flexibility of both proteins and membrane lipids, thereby perturbing biological function. *See* CELL MEMBRANES; ENZYME.

Animals are classified into two broad groups depending on the factors that determine body temperature. For ectotherms, body temperature is determined by sources of heat external to the body; levels of resting metabolism (and heat production) are low, and mechanisms for retaining heat are limited. Such animals are frequently termed poikilothermic or cold-blooded, because the body temperature often conforms to the temperature of the environment. In contrast, endotherms produce more metabolic heat and possess specialized mechanisms for heat retention. Therefore, body temperature is elevated above ambient temperature; some endotherms (termed homeotherms or warm-blooded animals) maintain a relatively constant body temperature. There is no natural taxonomic division between ecto- and endotherms. Most invertebrates, fish, amphibians, and reptiles are ectotherms, while true homeothermy is restricted to birds and mammals. However, flying insects commonly elevate the temperature of their thoracic musculature prior to and during flight (to 96°F or 36°C), and several species of tuna retain metabolic heat in their locomotory musculature via a vascular countercurrent heat exchanger. *See* ADIPOSE TISSUE; HIBERNATION AND ESTIVATION; THERMOREGULATION. Jeffrey R. Hazel

Temperature inversion The increase of air temperature with height; an atmospheric layer in which the upper portion is warmer than the lower. Such an increase is opposite, or inverse, to the usual decrease of temperature with height, or lapse rate, in the troposphere. However, above the tropopause, temperature increases with height throughout the stratosphere, decreases in the mesosphere, and increases again in the thermosphere. Thus inversion conditions prevail throughout much of the atmosphere much or all of the time, and are not unusual or abnormal. *See* AIR TEMPERATURE; ATMOSPHERE.

Inversions are created by radiative cooling of a lower layer, by subsidence heating of an upper layer, or by advection of warm air over cooler air or of cool air under warmer air. Outgoing radiation, especially at night, cools the Earth's surface, which in turn cools the lowermost air layers, creating a nocturnal surface inversion a few inches to several hundred feet thick.

Inversions effectively suppress vertical air movement, so that smokes and other atmospheric contaminants cannot rise out of the lower layer of air. California smog is trapped under an extensive subsidence inversion; surface radiation inversions, intensified by warm air advection aloft, can create serious pollution problems in valleys throughout the world; radiation and subsidence inversions, when horizontal air motion is sluggish, create widespread pollution potential, especially in autumn over North America and Europe. *See* AIR POLLUTION; SMOG. Arnold Court

Temperature measurement Measurement of the hotness of a body relative to a standard scale. The fundamental scale of temperature is the thermodynamic scale, which can be derived from any equation expressing the second law of thermodynamics. Efforts to approximate the thermodynamic scale as closely as possible depend on relating measurements of temperature-dependent physical properties of systems to thermodynamic relations expressed by statistical thermodynamic equations, thus in general linking temperature to the average kinetic energy of the measured system. Temperature-measuring devices, thermometers, are systems with properties that change with temperature in a simple, predictable, reproducible manner. *See* TEMPERATURE; THERMODYNAMIC PRINCIPLES.

In the establishment of a useful standard scale, assigned temperature values of thermodynamic equilibrium fixed points are agreed upon by an international body (General Conference of Weights and Measures), which updates the scale about once every 20 years. Thermometers for interpolating between fixed points and methods for realizing the fixed points are prescribed, providing a scheme for calibrating thermometers used in science and industry.

The scale now in use is the International Temperature Scale of 1990 (ITS-90). Its unit is the kelvin, K, arbitrarily defined as 1/273.16 of the thermodynamic temperature T of the triple point of water (where liquid, solid, and vapor coexist). For temperatures above 273.15 K, it is common to use International Celsius Temperatures, t_{90} (rather than International Kelvin Temperatures, T_{90}), having the unit degree Celsius, with symbol °C. The degree Celsius has the same magnitude as the kelvin. Temperatures, t_{90}, are defined as $t_{90}/°\text{C} = T_{90}/\text{K} - 273.15$, that is, as differences from the ice-point temperature at 273.15 K. The ice point is the state in which the liquid and solid phases of water coexist at a pressure of 1 atm (101,325 pascals). [The Fahrenheit scale, with symbol °F, still in common use in the United States, is given by $t_F/°\text{F} = (t_{90}/°\text{C} \times 1.8) + 32$, or $t_F/°\text{F} = (T_{90}/\text{K} \times 1.8) - 459.67$.] The ITS-90 is defined by 17 fixed points. *See* TRIPLE POINT.

Primary thermometers are devices which relate the thermodynamic temperature to statistical mechanical formulation. The fixed points of ITS-90 are all based on one or more types of gas thermometry or on spectral radiation pyrometry referenced to gas thermometry. Secondary thermometers are used as reference standards in the laboratory because primary thermometers are often too cumbersome. It is necessary to establish standard secondary thermometers referenced to one or more fixed points for interpolation between fixed points. Lower-order thermometers are used for most practical purposes and, when high accuracy is required, can usually be calibrated against reference standards maintained at laboratories, such as the U.S. National Institute of Standards and Technology, or against portable reference devices (sealed boiling or melting point cells). *See* GAS THERMOMETRY; LOW-TEMPERATURE THERMOMETRY; PYROMETER; THERMISTOR; THERMOCOUPLE; THERMOMETER. B. W. Mangum

Tempering The reheating of previously quenched alloy to a predetermined temperature below the critical range, holding the alloy for a specified time at that temperature, and then cooling it at a controlled rate, usually by immediate rapid quenching, to room temperature. The term is broadly applied to any process that toughens a material.

In alloys, if the composition is such that cooling produces a supersaturated solid solution, the resulting material is brittle. Heating the alloy to a temperature only high enough to allow the excess solute to precipitate out and then rapidly cooling the saturated solution fast enough to prevent further precipitation or grain growth result in a microstructure combining hardness and toughness.

With steel, the tempering must be carried out by slow heating to avoid steep temperature gradients, stress relief being one of the objectives. Properties produced by tempering depend on the temperature to which the steel is raised and on its alloy composition. For example, if hardness is to be retained, molybdenum or tungsten is used in the alloy. *See* HEAT TREATMENT (METALLURGY). Frank H. Rockett

Temporary structure (engineering) A structure erected to aid in the construction of a permanent project. Temporary structures are used to facilitate the construction of buildings, bridges, tunnels, and other above- and below-ground facilities by providing access, support, and protection for the facility under construction, as well as assuring the safety of the workers and the public. Temporary structures either are dismantled and removed when the permanent works become self-supporting or completed, or are incorporated into the finished work. Temporary structures are also used in inspection, repair, and maintenance work.

The many types of temporary structures include cofferdams; earth-retaining structures; tunneling supports; underpinning; diaphragm/slurry walls; roadway decking; construction ramps, runways, and platforms; scaffolding; shoring; falsework; concrete formwork; bracing and guying; site protection structures such as sidewalk bridges, boards, and nets for protection against falling objects, barricades and fences, and signs; and unique structures that are specially conceived, designed, and erected to aid in a specific construction operation.

These temporary works have a primary influence on the quality, safety, speed, and profitability of all construction projects.

More failures occur during construction than during the lifetimes of structures, and most of those construction failures involve temporary structures. However, codes and standards do not provide the same scrutiny as they do for permanent structures. Typical design and construction techniques and some industry practices are well established, but responsibilities and liabilities remain complex and present many contractual and legal pitfalls.

Robert T. Ratay

Tendon A cord connecting a muscle to another structure, often a bone. A tendon is a passive material, lengthening when the tension increases and shortening when it decreases. This characteristic contrasts with the active behavior of muscle. Away from its muscle, a tendon is a compact cord. At the muscle, it spreads into thin sheets called aponeuroses, which lie over and sometimes within the muscle belly. The large surface area of the aponeuroses allows the attachment of muscle fibers with a total cross-sectional area that is typically 50 times that of the tendon. *See* MUSCLE.

Tendons are living tissues that contain cells. In adult tendons, the cells occupy only a very small proportion of the volume and have a negligible effect on the mechanical properties. Like other connective tissues, tendon depends on the protein collagen for its strength and rigidity. The arrangement of the long, thin collagenous fibers is essentially longitudinal, but incorporates a characteristic waviness known as crimp. The fibers lie within a matrix of aqueous gel. Thus, tendon is a fiber-reinforced composite (like fiberglass), but its collagen is much less stiff than the glass and its matrix is very much less stiff than the resin. *See* COLLAGEN; COMPOSITE MATERIAL.

The function of tendons is to transmit force. They allow the force from the muscle to be applied in a restricted region. For example, the main muscles of the fingers are in the forearm, with tendons to the fingertips. If the hand had to accommodate these muscles, it would be too plump to be functional. Tendon extension can also be significant in the movement of a joint. For example, the tendon which flexes a human thumb joint is about 7 in. (170 mm) long. The maximum force from its muscle stretches this tendon about 0.1 in. (2.9 mm), which corresponds to rotation of the joint through an angle of about 21°. *See* JOINT (ANATOMY).

Some tendons save energy by acting as springs. In humans, the Achilles tendon reduces the energy needed for running by about 35%. This tendon is stretched during the first half of each step, storing energy which is then returned during takeoff. This elastic energy transfer involves little energy loss, whereas the equivalent work done by muscles would require metabolic energy in both stages. *See* CONNECTIVE TISSUE; MUSCULAR SYSTEM; SKELETAL SYSTEM.

Robert F. Ker

Tenrec An insectivorous mammal indigenous to Madagascar. There are 30 species in 10 genera. These animals are nocturnally active and feed on insects, worms, and mollusks. All tenrecs are essentially primitive unspecialized mammals, with poor vision. The digits are clawed, and the first digit is not opposable to the others. The body of the tenrec is covered with a mixture of hair, spines, and bristles; the tail is rudimentary; and the toes may be separate or webbed, depending on the species. The female is prolific, and has litters of 12–20 young. *See* INSECTIVORA; MAMMALIA.

Charles B. Curtin

Tensor analysis The systematic study of tensors which led to an extension and generalization of vectors, begun in 1900 by two Italian mathematicians. G. Ricci and T. Levi-Civita, following G. F. B. Riemann's proposal concerning a generalization of euclidean geometry. The principal aim of the tensor calculus (absolute differential calculus) is to construct relationships which are generally covariant in the sense that these relationships or laws remain valid in all coordinate systems. The differential equations for the geodesics in a Riemannian space are covariant expressions; they yield a description of the geodesics which is valid for all coordinate systems. On the other hand, Newton's equations of motion require a preferred coordinate system for their description, namely, one for which force is proportional to acceleration (an inertial frame of reference). Thus Albert Einstein was led to a study of Riemannian geometry and the tensor calculus in order to construct the general theory of relativity. *See* CALCULUS OF VECTORS; RIEMANNIAN GEOMETRY.

Harry Lass

Terbium Element number 65, terbium, Tb, is a very rare metallic element of the rare-earth group. Its atomic weight is 158.925, and the stable isotope ^{159}Tb makes up 100% of the naturally occurring element. *See* PERIODIC TABLE.

The common oxide, Tb_4O_7, is brown and is obtained when its salts are ignited in air. Its salts are all trivalent and white in color and, when dissolved, give colorless solutions. The higher oxides slowly decompose when treated with dilute acid to give the trivalent ions in solution. Although the metal is attacked readily at high temperatures by air, the attack is extremely slow at room temperatures. The metal has a Néel point at about 229 K and a Curie point at about 220 K. For properties of the metal *see* RARE-EARTH ELEMENTS.

Frank H. Spedding

Terebratulida An order of articulated brachiopods consisting of a group of sessile, suspension-feeding, marine, benthic, epifaunal bivalves with representatives occurring from the Early Devonian Era. It is the most diverse and abundant group of living brachiopods, which probably exhibits maximum diversity in present-day seas.

The shells are biconvex and usually smooth, although some show radial ribbed ornamentation. The valves articulate about a hinge structure and posterior edges of the valves are not coincident with the hinge axis (nonstrophic condition). The shell is calcareous and punctate. A fleshy pedicle usually attaches the animal to the substrate, but in thecidine brachiopods the ventral valve is cemented to the substrate. The tentacular feeding organ (lophophore) occupies the mantle cavity as a looped structure (the ptycholophe) in the smaller forms or as a looped and coiled structure (the plectolophe) in the larger forms. In the smaller forms the lophophore is supported by a calcareous ridge, but in the larger forms by a calcareous loop.

Three suborders are recognized: Centronellida, Terebratulidina, and Terebratellidina. *See* BRACHIOPODA; RHYNCHONELLIFORMEA.

Mark A. James

Terpene A class of natural products having a structural relationship to isoprene, as shown below. Over 5000 structurally

CH3
H2C=C(CH3)–CH=CH2

Isoprene unit

Isoprene

determined terpenes are known; many of these have also been synthesized in the laboratory. Historically terpenes have been isolated from green plants, but new compounds structurally related to isoprene continue to be isolated from other sources as well, so the class is also referred to as terpenoids, reflecting the biochemical origin without specification of the natural source. *See* ISOPRENE.

Terpenes are classified according to the number of isoprene units of which they are composed, as follows:

5	hemi	25	ses-
10	mono-	30	tri-
15	sesqui-	40	tetra-
20	di-	$(5)_n$	poly-

Although they may be named according to the systematic nomenclature and numbering systems set by the International Union of Pure and Applied Chemistry for all organic compounds, it is often easier to refer to terpenes by their common names, which usually reflect the botanical or zoological name of their source. Tomas Hudlicky

Terracing (agriculture) A method of shaping land to control erosion on slopes of rolling land used for cropping and other purposes. In early practice the land was shaped into a series of nearly level benches or steplike formations. Modern practice in terracing, however, consists of the construction of low-graded channels or levees to carry the excess rainfall from the land at nonerosive velocities. The physical principle involved is that, when water is spread in a shallow stream, its flow is retarded by the roughness of the bottom of the channel and its carrying, or erosive, power is reduced. Since direct impact of rainfall on bare land churns up the soil and the stirring effect keeps it in suspension in overland flow and rills, terracing does not prevent sheet erosion. It serves only to prevent destruction of agricultural land by gullying and must be supplemented by other erosion-control practices, such as grass rotation, cover crops, mulching, contour farming, strip cropping, and increased organic matter content. *See* EROSION; SOIL CONSERVATION.

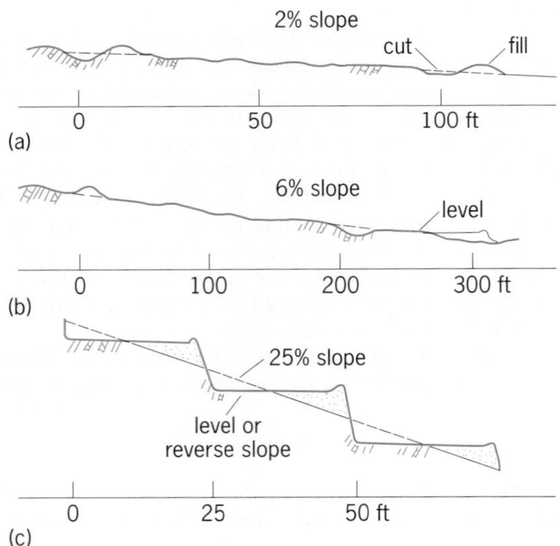

Types of terraces. (*a*) Broadbase. (*b*) Conservation bench. (*c*) Bench. 1 ft = 0.3 m. (*After Soil and Water Conservation Engineering, 2d ed., The Ferguson Foundation Agricultural Engineering Series, John Wiley and Sons, Inc., 1966*)

The two major types of terraces are the bench and the broadbase (see illustration). The bench terrace is essentially a steepland terrace and consists of an almost vertical retaining wall, called a riser, or a steep vegetative slope to hold the nearly level surface of the soil for cultivation, orchards, vineyards, or landscaping. The broadbase terrace has the distinguishing characteristic of farmability; that is, crops can be grown on this terrace and worked with modern-day machinery. These terraces are constructed either to remove or retain water and, based on their primary function, are classified either as graded or level. *See* LAND DRAINAGE (AGRICULTURE). Charles B. Ogburn

Terrain areas Subdivisions of the continental surfaces distinguished from one another on the basis of the form, roughness, and surface composition of the land. The pattern of landform differences is strongly reflected in the arrangement of such other features of the natural environment as climate, soils, and vegetation. These regional associations must be carefully considered in planning of activities as diverse as agriculture, transportation, city development, and military operations.

Eight classes of terrain are distinguished on the basis of steepness of slopes, local relief (the maximum local differences in elevation), cross-sectional form of valleys and divides, and nature of the surface material. Approximate definitions of terms used and percentage figures indicating the fraction of the world's land area occupied by each class are as follows; (1) flat plains: nearly level land, slight relief, 4%; (2) rolling and irregular plains: mostly gently sloping, low relief, 30%; (3) tablelands: upland plains broken at intervals by deep valleys or escarpments, moderate to high relief, 5%; (4) plains with hills or mountains: plains surmounted at intervals by hills or mountains of limited extent, 15%; (5) hills: mostly moderate to steeply sloping land of low to moderate relief, 8%; (6) low mountains: mostly steeply sloping, high relief, 14%; (7) high mountains: mostly steeply sloping, very high relief, 13%; and (8) ice caps: surface material, glacier ice, 11%. Edwin H. Hammond

Terrestrial coordinate system The perpendicular intersection of two curves or two lines, one relatively horizontal and the other relatively vertical, is the basis for finding and describing terrestrial location. The Earth's graticule, consisting of an imaginary grid of east-to-west-bearing lines of latitude and north-to-south bearing lines of longitude, is derived from the Earth's shape and rotation, and is rooted in spherical geometry. Plane coordinate systems, equivalent to horizontal X and vertical Y coordinates, are based upon cartesian geometry and differ from the graticule in that they have no natural origin or beginning for their grids.

The Earth, which is essentially a sphere, rotates about an axis that defines the geographic North and South poles. The poles serve as the reference points on which the system of latitude and longitude is based (see illustration). *See* LATITUDE AND LONGITUDE.

Latitude is arc distance (angular difference) from the Equator and is defined by a system of parallels, or lines that run east to west, each fully encompassing the Earth. The Equator is the parallel that bisects the Earth into the Northern and Southern hemispheres, and lies a constant 90° arc distance from both poles. As the only parallel to bisect the Earth, the Equator is considered a great circle. All other parallels are small circles (do not bisect the Earth), and are labeled by their arc distance north or south from the Equator and by the hemisphere in which they fall. Parallels are numbered from 0° at the Equator to 90° at the poles. For example, 42°S describes the parallel 42 degrees arc distance from the Equator in the Southern Hemisphere. For increased location precision, degrees of latitude and longitude are further subdivided into minutes ($1^\circ = 60'$) and seconds ($1' = 60''$). *See* EQUATOR; GREAT CIRCLE, TERRESTRIAL.

Longitude is defined by a set of imaginary curves extending between the two poles, spanning the Earth. These curves, called meridians, always point to true geographical north (or south) and converge at the poles. In the present-day system of longitude, meridians are numbered by degrees east or west of the beginning meridian, called the Prime Meridian or the Greenwich Meridian, which passes through the Royal Observatory in Greenwich, England. The Prime Meridian was assigned a longitude of 0°.

Since the Earth is fundamentally a sphere, its circumference describes a circle containing 360°, the arc distance through which the Earth rotates in 24 hours. The arc distance from the Prime

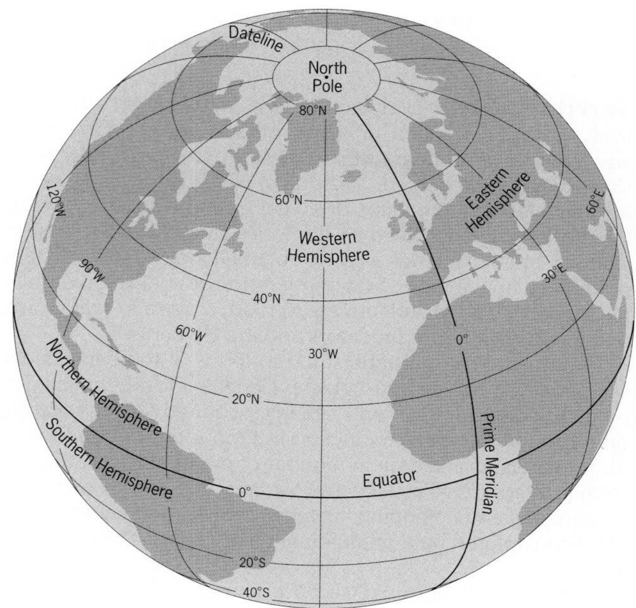

Earth's graticule. Meridians of longitude run from north to south, but are measured east or west of the Prime Meridian. Parallels run from east to west, but are measured north or south of the Equator.

Meridian describes the location of any meridian (see illustration). The 180° meridian is commonly referred to as the International Dateline. Together, the Prime Meridian and the International Dateline describe a great circle that bisects the Earth, as do all other meridian circles. The west half of the Earth, located between the Prime Meridian and the International Dateline, comprises the Western Hemisphere, and the east half on the opposite side forms the Eastern Hemisphere. Meridians within the Western Hemisphere are labeled with a W, and meridians within the Eastern Hemisphere are labeled with an E. A complete description of longitude includes an angular measurement and a hemispheric label. For example, 78°W is the meridian 78° west of the Prime Meridian. Neither the 0° meridian (Prime) nor the 180° meridian (Dateline) is given a hemispheric suffix because they divide the two hemispheres, and therefore do not belong to either one.

Coordinate system alternatives to the graticule evolved in the early twentieth century because of the complexity of using spherical geometry in determining latitude, longitude, and direction. Plane (two-dimensional) or cartesian coordinate systems presume that a relatively nonspherical Earth exists in smaller areas. Plane coordinates are superimposed upon these small areas, with coordinates being determined by the equivalent of a grid composed of a number of parallel vertical lines (X) and a complementary set of parallel horizontal lines (Y).

The State Plane Coordinate system (SPC) is used only in the United States and partitions each state into zones. Each zone has its own coordinate system. The number of zones designated in each state is determined by the size of the state. Zone boundaries follow either meridians or parallels depending on the shape of the state. All measurements are made in feet.

The Universal Transverse Mercator (UTM) system is a worldwide coordinate system in which locations are expressed using metric units. The basis for the UTM system is the Universal Transverse Mercator map projection. This projection becomes vastly distorted in polar areas above 80°, and for this reason the UTM system is confined to extend from 84°N to 80°S. The UTM system partitions the Earth into 60 north-south elongated zones, each having a width of 6° of longitude. *See* MAP PROJECTIONS.

A number of other coordinate systems are in use today. Foremost among these are the U.S. Public Land Survey System, the Universal Polar Stereographic (UPS) system, and the World Geographic Reference (GEOREF) system. Stephen Lavin

Terrestrial ecosystem A community of organisms and their environment that occurs on the land masses of continents and islands. Terrestrial ecosystems are distinguished from aquatic ecosystems by the lower availability of water and the consequent importance of water as a limiting factor. Terrestrial ecosystems are characterized by greater temperature fluctuations on both a diurnal and seasonal basis than occur in aquatic ecosystems in similar climates. The availability of light is greater in terrestrial ecosystems than in aquatic ecosystems because the atmosphere is more transparent than water. Gases are more available in terrestrial ecosystems than in aquatic ecosystems. Those gases include carbon dioxide that serves as a substrate for photosynthesis, oxygen that serves as a substrate in aerobic respiration, and nitrogen that serves as a substrate for nitrogen fixation. Terrestrial environments are segmented into a subterranean portion from which most water and ions are obtained, and an atmospheric portion from which gases are obtained and where the physical energy of light is transformed into the organic energy of carbon-carbon bonds through the process of photosynthesis.

Terrestrial ecosystems occupy 55,660,000 mi^2 (144,150,000 km^2), or 28.2%, of Earth's surface. Although they are comparatively recent in the history of life (the first terrestrial organisms appeared in the Silurian Period, about 425 million years ago) and occupy a much smaller portion of Earth's surface than marine ecosystems, terrestrial ecosystems have been a major site of adaptive radiation of both plants and animals. Major plant taxa in terrestrial ecosystems are members of the division Magnoliophyta (flowering plants), of which there are about 275,000 species, and the division Pinophyta (conifers), of which there are about 500 species. Members of the division Bryophyta (mosses and liverworts), of which there are about 24,000 species, are also important in some terrestrial ecosystems. Major animal taxa in terrestrial ecosystems include the classes Insecta (insects) with about 900,000 species, Aves (birds) with 8500 species, and Mammalia (mammals) with approximately 4100 species. *See* PLANT TAXONOMY; SYSTEMATICS; TAXONOMY.

Organisms in terrestrial ecosystems have adaptations that allow them to obtain water when the entire body is no longer bathed in that fluid, means of transporting the water from limited sites of acquisition to the rest of the body, and means of preventing the evaporation of water from body surfaces. They also have traits that provide body support in the atmosphere, a much less buoyant medium than water, and other traits that render them capable of withstanding the extremes of temperature, wind, and humidity that characterize terrestrial ecosystems. Finally, the organisms in terrestrial ecosystems have evolved many methods of transporting gametes in environments where fluid flow is much less effective as a transport medium.

The organisms in terrestrial ecosystems are integrated into a functional unit by specific, dynamic relationships due to the coupled processes of energy and chemical flow. Those relationships can be summarized by schematic diagrams of trophic webs, which place organisms according to their feeding relationships. The base of the food web is occupied by green plants, which are the only organisms capable of utilizing the energy of the Sun and inorganic nutrients obtained from the soil to produce organic molecules. Terrestrial food webs can be broken into two segments based on the status of the plant material that enters them. Grazing food webs are associated with the consumption of living plant material by herbivores. Detritus food webs are associated with the consumption of dead plant material by detritivores. The relative importance of those two types of food webs varies considerably in different types of terrestrial ecosystems. Grazing food webs are more important in grasslands, where over half of net primary productivity may be consumed by herbivores. Detritus food webs are more important in forests, where less than 5% of net primary productivity may be consumed by herbivores. *See* FOOD WEB; SOIL ECOLOGY.

There is one type of extensive terrestrial ecosystem due solely to human activities and eight types that are natural ecosystems. Those natural ecosystems reflect the variation of precipitation and temperature over Earth's surface. The smallest land areas are occupied by tundra and temperate grassland ecosystems, and the largest land area is occupied by tropical forest. The most productive ecosystems are temperate and tropical forests, and the least productive are deserts and tundras. Cultivated lands, which together with grasslands and savannas utilized for grazing are referred to as agroecosystems, are of intermediate extent and productivity. Because of both their areal extent and their high average productivity, tropical forests are the most productive of all terrestrial ecosystems, contributing 45% of total estimated net primary productivity on land. *See* DESERT; ECOLOGICAL COMMUNITIES; ECOSYSTEM; FOREST AND FORESTRY; GRASSLAND ECOSYSTEM; SAVANNA; TUNDRA.

S. J. McNaughton

Terrestrial radiation Electromagnetic radiation emitted from the Earth and its atmosphere. Terrestrial radiation, also called thermal infrared radiation or outgoing longwave radiation, is determined by the temperature and composition of the Earth's atmosphere and surface. The temperature structure of the Earth and the atmosphere is a result of numerous physical, chemical, and dynamic processes. In a one-dimensional context, the temperature structure is determined by the balance between radiative and convective processes.

The Earth's surface emits electromagnetic radiation according to the laws that govern a blackbody or a graybody. A blackbody absorbs the maximum radiation and at the same time emits that same amount of radiation so that thermodynamic equilibrium is achieved as to define a uniform temperature. A graybody is characterized by incomplete absorption and emission and is said to have emissivity less than unity. The thermal infrared emissivities from water and land surfaces are normally between 90 and 95%. It is usually assumed that the Earth's surfaces are approximately black in the analysis of infrared radiative transfer. Exceptions include snow and some sand surfaces whose emissivities are wavelength-dependent and could be less than 90%. Absorption and emission of radiation by atmospheric molecules are more complex and require a fundamental understanding of quantum mechanics. *See* ATMOSPHERE; BLACKBODY; GRAYBODY; HEAT BALANCE, TERRESTRIAL ATMOSPHERIC; HEAT RADIATION; RADIATIVE TRANSFER.

The radiant energy emitted from a number of temperatures covering the Earth and the atmosphere is measured as a function of wavenumber and wavelength. This energy is called Planck intensity (or radiance), and the units that are commonly used are denoted as watt per square meter per solid angle per wavenumber ($W/m^2 \cdot sr \cdot cm^{-1}$). Terrestrial radiation originating from the Earth-atmosphere-ocean system, as well as solar radiation reflected and scattered back to space, is measured on a daily basis by meteorological satellites. Instruments on meteorological satellites measure visible, ultraviolet, infrared, and microwave radiation. *See* ABSORPTION OF ELECTROMAGNETIC RADIATION; ELECTROMAGNETIC RADIATION; METEOROLOGICAL SATELLITES; REFLECTION OF ELECTROMAGNETIC RADIATION; SCATTERING OF ELECTROMAGNETIC RADIATION.

Each spectral region provides meteorologists and other Earth system scientists with information about atmospheric ozone, water vapor, temperature, aerosols, clouds, precipitation, lightning, and many other parameters. Measuring atmospheric radiation allows the detection of sea and land temperature, snow and ice cover, and winds at the surface of the ocean. By tracking the movement of clouds and other atmospheric features, such as aerosols and water vapor, it is possible to obtain estimates of winds above the surface. *See* SATELLITE METEOROLOGY.

K. N. Liou; Thomas H. Vonder Haar

Territoriality Behavior patterns in which an animal actively defends a space or some other resource. One major advantage of territoriality is that it gives the territory holder exclusive access to the defended resource, which is generally associated with feeding, breeding, or shelter from predators or climatic forces. Feeding and breeding territories can be mobile, such as when an animal defends a newly obtained food source or a temporarily receptive mate. Stationary territories often serve multiple functions and include access to food, a place to rear young, and a refuge site from predators and the elements.

Territoriality can be understood in terms of the benefits and costs accrued to territory holders. Benefits include time saved by foraging in a known area, energy acquired through feeding on territorial resources, reduction in time spent on the lookout for predators, or increase in number of mates attracted and offspring raised. Costs usually involve time and energy expended in patrolling and defending the territorial site, and increased risk of being captured by a predator when engaged in territorial defense.

Because territories usually include resources that are in limited supply, active defense is often necessary. Such defense frequently involves a graded series of behaviors called displays that include threatening gestures such as vocalizations, spreading of wings or gill covers, lifting and presentation of claws, head bobbing, tail and body beating, and finally, direct attack. Direct confrontation can usually be avoided by advertising the location of a territory in a way that allows potential intruders to recognize the boundaries and avoid interactions with the defender. Such advertising may involve odors that are spread with metabolic by-products, such as urine or feces in dogs, cats, or beavers, or produced specifically as territory markers, as in ants. Longer-lasting territorial marks can involve visual signals such as scrapes and rubs, as in deer and bear. *See* CHEMICAL ECOLOGY; ETHOLOGY; POPULATION ECOLOGY; REPRODUCTIVE BEHAVIOR.

Gene S. Helfman

Tertiary The older major subdivision (period) of the Cenozoic Era, extending from the Cretaceous (top of the Mesozoic Era) to the beginning of the Quaternary (younger Cenozoic Period). The term Tertiary corresponds to all the rocks and fossils formed during this period. Although the International Commission on Stratigraphy uses the terms Paleogene and Neogene (pre-Quaternary part) in place of Tertiary, it is still widely used in the geologic literature. Typical sedimentary rocks include widespread limestones, sandstones, mudstones, marls, and conglomerates deposited in both marine and terrestrial environments; igneous rocks include extrusive and intrusive volcanics as well as rocks formed deep in the Earth's crust (plutonic). *See* CRETACEOUS; FOSSIL; ROCK.

The Tertiary Period is characterized by a rapid expansion and diversification of marine and terrestrial life. In the marine realm, a major radiation of oceanic microplankton occurred following the terminal Cretaceous extinction events. This had its counterpart on land in the rapid diversification of multituberculates, marsupials, and insectivores—holdovers from the Mesozoic—and primates, rodents, and carnivores, among others, in the ecologic space vacated by the demise of the dinosaurs and other terrestrial forms. Shrubs and grasses and other flowering plants diversified in the middle Tertiary, as did marine mammals such as cetaceans (whales), which returned to the sea in the Eocene Epoch. The pinnipeds (walruses, sea lions, and seals) are derived from land carnivores, or fissipeds, and originated in the Neogene temperate waters of the North Atlantic and North Pacific. Indeed, the great diversification on land and in the sea of birds and, particularly, mammals has led to the informal designation of the Tertiary as the Age of Mammals in textbooks on historical geology.

The modern configuration of continents and oceans developed during the Cenozoic Era as a result of the continuing process known as plate tectonics. Mountain-building events (orogenies) and uplifts of large segments of the Earth's crust (epeirogenies) alternated with fluctuating transgressions and regressions of the seas over land. The middle to late Tertiary Alpine-Himalayan orogeny and the late Tertiary Cascadian orogeny led to the east-west and north-south mountain ranges, respectively, which are located in Eurasia and western North America. *See* CORDILLERAN BELT; MOUNTAIN SYSTEMS; OROGENY; PLATE TECTONICS.

W. A. Berggren

Testis The organ of sperm production. In addition, the testis (testicle) is an organ of endocrine secretion in which male hormones (androgens) are elaborated. In the higher vertebrates (reptiles, birds, and mammals), the testes are paired and either ovoid or elongated in shape. In mammals, the testes are usually ovoid or round. In many species (for example, humans) they are suspended in a pouch (scrotum) outside the main body cavity; in other species they are found in such a pouch only at the reproductive season; in still others the testicles are permanently located in the abdomen (for example, in whales and bats).

Within a firm and thick capsule of connective tissue, the tunica albuginea, the testis contains a varying number of thin but very long seminiferous tubules which are the sites of sperm formation. Essentially, these tubules are simple loops which open with both their limbs into a network of fine, slitlike canals, the rete testis. From this the sperm drains through a few, narrow ducts, the ductuli efferentes, into the epididymis, where sperm mature and are stored.

The seminiferous tubules comprise most of the testis, and in different species vary greatly in complexity. Each tubule is surrounded by a layer of thin cells which is contractile and enables the tubules to wriggle slowly. The spaces between tubules are filled with connective tissue, blood vessels, an extensive network of very thin-walled lymph vessels, and secretory cells, the interstitial cells or cells of Leydig, which secrete male hormone.

The sperm cells, spermatozoa, develop in the wall of the seminiferous tubules, either periodically, as in most vertebrates, or continually, as in humans. Most of the cells in the tubules are potential spermatozoa (spermatogenic or germ cells). Nursing cells (Sertoli cells) are interspersed at regular intervals between them. The Sertoli cells support and surround the developing spermatogenic cells and provide a specialized environment, which is absolutely necessary for normal sperm development. *See* SPERM CELL.

Spermatogenesis in the testis is the result of a balance between proliferation and differentiation, and cell degeneration or apoptosis. Apoptosis of the spermatogenic cells is largely hormonally controlled, and specifically directed apoptosis occurs in conditions of testicular damage due to environmental insults such as heat, radiation, or chemical toxicants. Recovery of spermatogenesis is possible provided the stem cells are not depleted by these processes. *See* SPERMATOGENESIS.

Mark P. Hedger; Edward C. Roosen-Runge

The functions of the testis are dependent on the secretion of gonadotropic hormones, the release of which from the pituitary gland is in turn regulated by the central nervous system. In mammals, male-hormone production resides in the Leydig cells, located in the intertubular tissue of the testes.

The principal androgenic hormone released by the testis into the bloodstream is testosterone. The testis is able to form cholesterol and to convert this via a number of pathways to testosterone. Testosterone may be further metabolized into estrogens in the testis. The production of estrogens in the male varies quite widely among species, from relatively low in humans to very high, for example in stallions and boars. Estrogens are important in the development and proper function of the ducts which drain the testis (the rete testis and ductuli efferentes), even in species with relatively low levels of estrogens. *See* ANDROGENS.

Testosterone synthesis is normally limited by the rate of pituitary gonadotrophin secretion: administration of the luteinizing hormone or of chorionic gonadotrophin results in increased testosterone synthesis and release within minutes. These hormones also stimulate growth and multiplication of Leydig cells. Hypophysectomy leads to cessation of androgen formation. *See* ADENOHYPOPHYSIS HORMONE; PITUITARY GLAND.

At the ambisexual stage of embryonic development, the testis promotes the growth of the paired Wolffian ducts and their differentiation into the epididymis, vasa deferentia, and seminal vesicles; the fetal testis also causes masculinization of the urogenital sinus, fusion of the labioscrotal folds in the midline, and development of the genital tubercle into a phallus. *See* EMBRYOLOGY.

Toward puberty, increased secretion of testosterone stimulates the growth of the penis, scrotum, and male accessory glands responsible for the formation of the seminal plasma, for example, the prostate and seminal vesicles. The hormone brings about the appearance of secondary sex characters, such as the male-type distribution of hair and body fat and lowered pitch of voice in man, the growth of the comb and wattles in birds, the clasping pads of amphibians, or the dorsal spine of certain fishes.

Unlike the ovary, the testis remains functional throughout life, with ongoing spermatogenic development. However, the efficiency of spermatogenesis falls away, and androgen levels begin to fall due to a declining Leydig cell activity. These events can lead to reduced fertility, and androgen insufficiency problems in later life in some men. *See* REPRODUCTIVE SYSTEM.

Mark P. Hedger; Hans R. Lindner

Tetanus An infectious disease, also known as lockjaw, which is caused by the toxin of *Clostridium tetani*. The bacterium may be isolated from fertile soil and the intestinal tract or fecal material of humans and other animals. Infection commonly follows dirt contamination of deep wounds or other injured tissues.

The incubation period of tetanus is usually 5–10 days, and the disease is characterized by convulsive tonic contraction of voluntary muscles. Prevention of tetanus rests on the proper, prompt surgical care of contaminated wounds and prophylactic use of antitoxin if the individual has not been protected by active immunization with toxoid. *See* IMMUNOLOGY. Leland S. McClung

Tetraodontiformes An order of the class Actinopterygii, also known as the Plectognathi. This most highly derived order of actinopterygian fishes is noted for its reduction in skeletal elements. Lost are the parietal, nasal, and infraorbital bones of the skull. Usually there are no pleural ribs; vertebrae are as few as 16 and no more than 30; and the maxillae are united with the premaxillae. The gill opening is a short vertical slit in front of the pectoral fin. Scales are usually modified to form spines, shields, or plates; a lateral line may be present or absent; and a swim bladder is present, except in the family Molidae. The 357 extant species are in three suborders (Triacanthodoidei, Balistoidei, and Tetraodontoidei), nine families, and about 101 genera. Some popular types are triggerfishes, spikefishes, filefishes, trunkfishes, cowfishes, and puffers. *See* ACTINOPTERYGII; OSTEICHTHYES; SWIM BLADDER.

Puffers are members of the family Tetraodontidae and are identified by a robust body, naked or with short prickles on sides and belly; some species with small fleshy appendages (lappets) on the sides; and maximum size of 3.5–120 cm (1.4–47.2 in.) total length, depending on the species. Puffers are capable of greatly inflating themselves with water or air when agitated. Some puffers contain the potential lethal poison tetrodotoxin, especially in the viscera and in the gonads of some during spawning season. They are chiefly marine, usually inhabiting shallow inshore waters of tropical and temperate waters of the Atlantic, Indian, and Pacific oceans; many species inhabit brackish water, but only 12 of about 162 species of puffers are limited to freshwater. *See* TOXIN. Herbert Boschung

Tetraphididae A subclass of the mosses (class Bryopsida) consisting of two families and three genera, especially characterized by growth from protonematal flaps, three-ranked leaves, and peristomes of four teeth made up of whole cells (rather than thickened parts of cells). The Tetraphididae include small acrocarpous mosses with peristome teeth in fours. The plants grow from buds produced on leaflike protonematal flaps. They are erect and simple or merely forked, with oblong-ovate leaves in three rows. The leaves usually have a single costa that ends near the apex. *See* BRYOPHYTA; BRYOPSIDA. Howard Crum

Tetraphyllidea An order of tapeworms of the subclass Cestoda. All species are intestinal parasites of elasmobranch fishes and are small in size, usually less than 2 in. (5 cm) in length. An outstanding feature of the order is the variation in the structure of the holdfast organ or scolex. All species are segmented and segments are usually shed from the body while sexually immature; these develop to sexual maturity as independent units in the host's intestine. Segment anatomy is very similar to that of Proteocephaloides. A complete life cycle is not known, but larval forms have been found in a variety of invertebrates and bony fishes. *See* CESTODA; PROTEOCEPHALOIDEA. Clark P. Read

Tetrapoda The superclass of the subphylum Vertebrata whose members typically possess limbs in contrast to the superclass, the Pisces (fishes), whose members have fins. *See* PISCES (ZOOLOGY).

The animals making up the Tetrapoda typically live part or all of their lives on land, whereas the members of the Pisces live in water. The classes of the Tetrapoda are Amphibia, Reptilia, Aves, and Mammalia. The term Tetrapoda comes from Greek words meaning "four feet," but there are tetrapods that have only two limbs or none at all, such as some amphibians and reptiles. These forms have, however, evolved from four-footed ancestors. *See* AMPHIBIA; AVES; MAMMALIA; REPTILIA. Richard G. Zweifel

Teuthoidea An order of the class Cephalopoda (subclass Coleoidea) commonly known as squids. They are characterized by 10 appendages (eight arms and two longer tentacles) around the mouth; an elongate, tapered, usually streamlined body; an internal, rod- or bladelike chitinous shell (gladius); and fins on the body. The two tentacles are strongly elastic, contractile, but not reactile into pockets as in cuttlefishes (Sepioidea). Two rows of suckers (infrequently four or six rows) occur on the arms on muscular stalks, with sucker rings that are chitinous, smooth, toothed, or modified as clawlike hooks. The muscular tentacles have terminal clubs with two rows, usually four, ranging up to many rows of suckers (and/or hooks in some families). Adults of the family Octopoteuthidae and genera *Gonatopsis* and *Lepidoteuthis* characteristically lose their tentacles. *See* SEPIOIDEA.

The Teuthoidea are divided into two suborders. The Myopsida, the nearshore, shallow-water squids, have a transparent skin (cornea) covering the two eyes, with a minute pore anteriorly, arms and tentacular clubs with suckers only, never hooks, and a single gonoduct in females, not paired. The Oegopsida, the oceanic squids, have no cornea over the eyes and no anterior pore, arms and tentacular clubs with suckers (and/or hooks in many families), and paired gonoducts in females (some exceptions).

Squids inhabit a wide variety of marine habitats, depending on the species, from very shallow grass flats, mangrove roots, lagoons, bays, and along coasts (myopsids) to the open ocean from the surface of the sea (*Ommastrephes*) to nearly 9600 ft (3000 m) in the deep sea (other oegopsids such as *Bathyteuthis*, *Neoteuthis*, and *Grimalditeuthis*). *See* CEPHALOPODA; COLEOIDA; SQUID. Clyde F. E. Roper

Textile A material made mainly of natural or synthetic fibers. Modern textile products may be prepared from a number of combinations of fibers, yards, films, sheets, foams, furs, or leather. They are found in apparel, household and commercial furnishings, vehicles, and industrial products. *See* MANUFACTURED FIBER; NATURAL FIBER.

The term fabric may be defined as a thin, flexible material made of any combination of cloth, fiber, or polymer (film, sheet, or foams); cloth as a thin, flexible material made from yarns; yarn as a continuous strand of fibers; and fiber as a fine, rodlike object in which the length is greater than 100 times the diameter. The bulk of textile products are made from cloth.

The natural progression from raw material to finished product requires: the cultivation or manufacture of fibers; the twisting of fibers into yarns (spinning); the interlacing (weaving) or interlooping (knitting) of yarns into cloth; and the finishing of cloth prior to sale.

The conversion of staple fiber into yarn (spinning) requires the following steps: picking (sorting, cleaning, and blending), carding and combing (separating and aligning), drawing (reblending), drafting (reblended fibers are drawn out into a long strand), and spinning (drafted fibers are further attenuated and twisted into yarn).

The process of weaving allows a set of yarns running in the machine direction (warp) to be interlaced with another set of yarns running across the machine (filling or weft). The weaving process involves four functions: shedding (raising the warp yarns by means of the appropriate harnesses); picking (inserting the weft yarn); battening (pushing the weft into the cloth with a reed); and taking up and letting off (winding the woven cloth onto the cloth beam and releasing more warp yarn from the warp beam.

Knit cloth is produced by interlocking one or more yarns through a series of loops. The lengthwise columns of loops are known as the wales, and the crosswise rows of loops are called courses. Filling (weft) knits are those in which the courses are composed of continuous yarns, while in warp knits the wale yarns are continuous.

Newly constructed knit or woven fabric must pass through various finishing processes to make it suitable for its intended purpose. Finishing enhances the appearance of fabric and also adds to its serviceability. Finishes can be solely mechanical, solely chemical, or a combination of the two. Those finishes, such as scouring and bleaching, which simply prepare the fabric for further use are known as general finishes. Functional finishes, such as durable press treatments, impart special characteristics to the cloth. For discussions of important finishing operations *see* BLEACHING; DYEING; TEXTILE CHEMISTRY; TEXTILE PRINTING. Ira Block

Textile chemistry The applied science of textile materials, consisting of the application of the principles of the many basic fields of chemistry to the understanding of textile materials and to their functional and esthetic modification into useful and desirable items. The study of textile chemistry begins with the knowledge of the textile fibers themselves. These are normally divided into three groups: natural, manufactured, and synthetic. *See* MANUFACTURED FIBER; NATURAL FIBER; TEXTILE.

Chemicals. The enormous number of chemicals used in textile processing may be divided broadly into two categories: those intended to remain on the fiber, and those intended to wet or clean the fiber or otherwise function in some related operation. The former includes primarily dyes and finishes. The latter group consists mainly of surface-active agents, commonly known as surfactants. *See* SURFACTANT.

Preparation. Preparation is a term applied to a group of essentially wet chemical processes having as their object the removal of all foreign matter from the fabric. This results in a clean, absorbent substrate, ready for the subsequent coloring and finishing operations.

The operations constituting preparation depend primarily on the fibers being handled. Synthetic fibers contain little or no natural impurities, so that the only materials that normally must be removed are the oils and lubricants or water-soluble sizes needed to facilitate earlier processing. This is generally accomplished by washing with water and a mild detergent capable of emulsifying the oils and waxes. On the other hand, natural fibers contain relatively high amounts of natural impurities, and in addition frequently are sized with materials presenting difficulties in removal. In the case of cotton, prolonged hot treatment with alkali, usually sodium hydroxide, and strong detergent is necessary to break down and remove the naturally occurring impurities. Special scours are necessary for cleaning such materials as wool and silk. The protein fibers are very sensitive to alkali and strong detergents; they are usually washed with mild soap or sulfated alcohols.

After other impurities are removed from the fiber, it is usually desirable to remove any coloring material. This process is known as bleaching. By far the major bleaching agent in use is hydrogen peroxide, which is efficient in color removal, while still being considered relatively controllable and safe for use. *See* BLEACHING; HYDROGEN PEROXIDE.

Mercerization. Mercerization is a special process applied only to cotton. The fabric or yarn is treated with a strong sodium hydroxide solution while being held under tension. This process causes chemical and physical changes within the fiber itself, resulting in a substantial increase in luster and smoothness of the fabric, plus important improvements in dye affinity, stabilization, tensile strength, and chemical reactivity.

Coloring. Although many textiles reach the consumer in their natural color or as a bleached white, most textiles are colored in one way or another. Coloring may be accomplished either by dyeing or printing, and the coloring materials may be either dyes or pigments.

Dyeing essentially consists of immersing the entire fabric in the solution, so that the whole fabric becomes colored. On the other hand, printing may be considered as localized dyeing. In printing, a thickened solution of dyestuff or pigment is used. This thickened solution, or paste, is applied to specific areas of the fabric by means such as engraved rollers or partially porous screens. Application of steam or heat then causes the dyestuff to migrate from the dried paste into the interior of the fiber, but only in those specific areas where it has been originally applied. *See* DYE; DYEING; TEXTILE PRINTING.

Finishing. Finishing includes a group of mechanical and chemical operations which give the fabric its ultimate feel and performance characteristics. Many desirable characteristics may be imparted to the fabric through the application of various chemical agents at this point.

Softeners are used to give a desirable hand or feel to the fabric. These chemicals are generally long fatty chains, with solubilizing groups which may be cationic, nonionic, or occasionally anionic in character. They are essentially surfactants constructed so as to contain a relatively high proportion of fatty material in the molecule. Conversely, certain types of polymeric material such as polyvinyl acetate or polymerized urea formaldehyde resins are used to impart a stiff or crisp hand to a fabric. *See* POLYVINYL RESINS; UREA-FORMALDEHYDE-TYPE RESINS.

It is in finishing that the so-called proof finishes are applied, including fire-retardant and water-repellent finishes. A fire-retardant finish is a chemical or mixture containing a high proportion of phosphorus, nitrogen, chlorine, antimony, or bromine. A truly waterproof fabric may be made by coating with rubber or vinyl, but water-repellent fabrics are produced by treating with hydrophobic materials such as waxes, silicones, or metallic soaps.

Many other types of highly specialized treatments, such as antistatic, antibacterial, or soil-repellent finishes, may be applied to fit the fabric to a particular use. David H. Abrahams

Textile microbiology

That branch of industrial microbiology concerned with textile materials. Most of the microorganisms on textiles—the fungi, actinomycetes, and bacteria—originate from air, soil, and water. Some of the microorganisms are harmful to either the fibers or the consumer. They may decompose the cellulose or protein in the fiber or affect the consumer's health. Since the minimum moisture content for microorganism development is 7%, dry storage is an effective prevention measure. Some of the microorganisms are useful, for example in the retting process, in which fibers are liberated from the stalks of such fiber plants as flax, hemp, and jute.

In principle, retting consists of a breaking down of pectic substances between the cell walls (middle lamellae) of the individual cells of the tissue surrounding the bundles. As a result, the bundles become separated from the surrounding tissue and can then easily be extracted mechanically. In water retting, the stalks are immersed in cold or warm, slowly renewed water, for from 4 days to several weeks. The active organism is *Clostridium felsineum* and related types, which break down the pectin to a mixture of organic acids (chiefly acetic and butyric), alcohols (butanol, ethanol, and methanol), carbon dioxide (CO_2), and hydrogen (H_2). In dew retting the stems are spread out in moist meadows; here the pectin decomposition is accomplished by molds and aerobic bacteria with the formation of CO_2 and H_2O. *See* INDUSTRIAL MICROBIOLOGY; TEXTILE CHEMISTRY. A. N. J. Heyn

Textile printing

The localized application of color on fabrics. In printing textiles, a thick paste of dye or pigment is applied to the fabric by appropriate mechanical means to form a design. The olor is then fixed or transferred from the paste to the fiber itself, maintaining the sharpness and integrity of the design. In a multicolor design, each color must be applied separately and in proper position relative to all other colors. Printing is one of the most complex of all textile operations. *See* DYE.

A design may be applied in three major ways: raising the design in relief on a flat surface (block printing); cutting the design below a flat surface (intaglio or engraved printing); and cutting the design through a flat metal or paper sheet (stencil or screen printing). All three methods have been used for hand printing and reciprocating printing machines. In addition, these methods have been converted into rotary action by replacing blocks or plates with cylinders. Another method of printing utilizes individual computer-controlled nozzles for each color. The nozzles are used to paint a design on the fabric.

Each printing method requires a paste with special characteristics, frequently referred to as flow characteristics. The choice of thickener is dependent not only on the type of dyestuff, but on the type of printing machine on which the printing is to be done, and frequently also on the type of fixation to be used. Most natural thickening agents are based on combinations of starch and gum. The synthetic thickening agents used are generally extremely high-molecular-weight polymers capable of developing a very high viscosity at a relatively low concentration.

The first step is the preparation of print paste, which is made by dissolving the dyestuff and combining it with a solution of the appropriate thickening agent. The fabric is then printed by any of the standard methods and then dried in order to retain a sharp printed mark.

The next operation, steaming, may be likened to a dyeing operation. Before steaming, the bulk of the dyestuff is held in a dried film of thickening agent. During the steaming operation, the printed areas absorb moisture and form a very concentrated dyebath, from which dyeing of the fiber takes place. The thickening agent prevents the dyestuff from spreading outside the area originally printed, because the printed areas act as a concentrated dyebath that exists more in the form of a gel than a solution and restricts any tendency to bleed.

Printed goods are generally washed thoroughly to remove thickening agent, chemicals, and unfixed dyestuff. Drying of the washed goods is the final operation of printing. *See* TEXTILE CHEMISTRY. David H. Abrahams

Thaliacea A small class of pelagic Tunicata especially abundant in warmer seas. This class of animals contains three orders: the Salpida, Doliolida, and Pyrosomida. Oral and atrial apertures occur at opposite ends of the body. Members of the orders Salpida and Doliolida are transparent forms, partly or wholly ringed by muscular bands. The contractions of these bands produce currents used in propulsion, feeding, and respiration. The order Pyrosomida includes species which form tubular swimming colonies and which are often highly luminescent. *See* BIOLUMINESCENCE; TUNICATA. Donald P. Abbott

Thallium A chemical element, Tl, atomic number 81, relative atomic weight of 204.38. The valence electron notation corresponding to its ground state term is $6s^2 6p^1$, which accounts for the maximum oxidation state of III in its compounds. Compounds of oxidation state I and apparent oxidation state II are also known. *See* PERIODIC TABLE.

Thallium occurs in the Earth's crust to the extent of 0.00006%, mainly as a minor constituent in iron, copper, sulfide, and selenide ores. Minerals of thallium are considered rare. Thallium compounds are extremely toxic to humans and other forms of life.

The insolubility of thallium(I) chloride, bromide, and iodide permits their preparation by direct precipitation from aqueous solution; the fluoride, on the other hand, is water-soluble. Thallium(I) chloride resembles silver chloride in its photosensitivity.

Thallium(I) oxide is a black powder which reacts with water to give a solution from which yellow thallium hydroxide can be crystallized. The hydroxide is a strong base and will take up carbon dioxide from the atmosphere.

Thallium also forms organometallic compounds of the following general classes, R_3Tl, R_2TlX, and $RTlX_2$, where R may be an alkyl or aryl group and X a halogen. *See* ORGANOMETALLIC COMPOUND. Edwin M. Larsen

Thallobionta One of the two commonly recognized subkingdoms of plants, encompassing the euglenoids and various classes of algae. In contrast to the more closely knit subkingdom Embryobionta, the Thallobionta (often also called Thallophyta) are diverse in pigmentation, food reserves, cell-wall structure, and flagellar structure. The Thallobionta are united more by the absence of certain specialized tissues or organs than by positive resemblances. They do not have the multicellular sex organs commonly found in most divisions of Embryobionta. Many of the Thallobionta are unicellular, and those which are multicellular seldom have much differentiation of tissues. None of them has tissues comparable to the xylem found in most divisions of the Embryobionta, and only some of the brown algae have tissues comparable to the phloem found in most divisions of the Embryobionta.

A large proportion of the Thallobionta are aquatic, and those which grow on dry land seldom reach appreciable size. The Thallobionta thus consist of all those plants which have not developed the special features that mark the progressive adaptation of the Embryobionta to life on dry land. *See* CHLOROPHYCOTA; CHRYSOPHYCEAE; EMBRYOBIONTA; EUGLENOPHYCEAE; PHAEOPHYCEAE; PLANT KINGDOM; RHODOPHYCEAE. Arthur Cronquist

Theales An order of flowering plants, division Magnoliophyta (Angiospermae), in the subclass Dilleniidae of the class Magnoliopsida (dicotyledons). The order consists of 18 families and nearly 3500 species. The largest families are the Clusiaceae, sometimes called Guttiferae (about 1200 species), Theaceae (about 600 species), Dipterocarpaceae (about 600 species), and Ochnaceae (about 400 species). They are mostly woody plants, less often herbaceous, with simple or occasionally compound leaves. The perianth and stamens of the flowers are attached directly to the receptacle; the calyx is arranged in a tight spiral; and the petals are usually separate from each other (see illustration). The stamens are numerous and initiated in centrifugal sequence or, less often, are few and cyclic; the pollen is nearly always binucleate. The tea plant (*Thea sinensis*), *Camellia* (in the Theaceae), and St.-John's-wort (*Hypericum*, in the Clusiaceae) are familiar members of the Theales. *See* DILLENIIDAE; FLOWER; MAGNOLIOPHYTA; PLANT KINGDOM; TEA. Arthur Cronquist; T. M. Barkley

Franklin tree (*Franklinia alatamaha*) flowers, a characteristic member of the family Theaceae in the order Theales, named in honor of Benjamin Franklin. (*Photograph by A. W. Ambler, from National Audubon Society*)

Thecanephria An order of Pogonophora, a group of elongate, tentaculated, tube-dwelling, sedentary, nonparasitic marine worms lacking a digestive system. In this order the "coelomic" space in the anterior tentacular region is horseshoe-shaped, and the excretory (osmoregulatory) portion of its ducts come close together medially near the median, adneural blood vessel. The species in this order are multitentaculate.

The order includes four families: Polybrachiidae (with seven genera), Sclerolinidae (one genus), Lamellisabellidae (two genera), and Spirobrachiidae (one genus). *See* ATHECANEPHRIA; POGONOPHORA. Edward B. Cutler

Thelodontida Sometimes called Coelolepida, this is an extinct group of Agnatha or jawless vertebrates known from the Lower Silurian to Middle Devonian of Europe, Asia, Australia, and North America. Since they had no hard skeleton, their structure and relationships are poorly known. Well-preserved specimens of *Logania* suggest a relationship to Heterostraci in their widely spaced, lateral eyes, nearly terminal mouth, pectoral flaps at the posterior end of the head, and hypocercal tail with a downwardly directed main lobe. *Phlebolepis* resembles Anaspida in general form but lacks the dorsal nostril of that group. *See* AGNATHA; HETEROSTRACI; OSTEOSTRACI; PTERASPIDOMORPHA. Robert H. Denison

Theorem A proposition arrived at by the methods of logical deduction from a set of basic postulates or axioms accepted as primitive and therefore not subject to deductive proof. So long as a theorem is part of a purely formal system, it is not meaningful to speak about the "truth" of a theorem but only about its "correctness." It becomes true when it, or its consequences, can be shown to be in accord with observable facts. *See* LOGIC.

Percy W. Bridgman; Henry Margenau

Theoretical ecology The use of models to explain patterns, suggest experiments, or make predictions in ecology. Because ecological systems are idiosyncratic, extremely complex, and variable, ecological theory faces special challenges. Unlike physics or genetics, which use fundamental laws of gravity or of inheritance, ecology has no widely accepted first-principle laws. Instead, different theories must be invoked for different questions, and the theoretical approaches are enormously varied. A central problem in ecological theory is determining what type of model to use and what to leave out of a model. The traditional approaches have relied on analytical models based on differential or difference equations; but recently the use of computer simulation has greatly increased. *See* ECOLOGICAL MODELING; ECOLOGY; ECOLOGY, APPLIED; SIMULATION.

The nature of ecological theory varies depending on the level of ecological organization on which the theory focuses. The primary levels of ecological organization are (1) physiological and biomechanical, (2) evolutionary (especially applied to behavior), (3) population, and (4) community.

At the physiological and biomechanical level, the goals of ecological theory are to understand why particular structures are present and how they work. The approaches of fluid dynamics and even civil engineering have been applied to understanding the structures of organisms, ranging from structures that allow marine organisms to feed, to physical constraints on the stems of plants.

At the behavioral evolutionary level, the goals of ecological theory are to explain and predict the different choices that individual organisms make. Underlying much of this theory is an assumption of optimality: the theories assume that evolution produces an optimal behavior, and they attempt to determine the characteristics of the optimal behavior so it can be compared with observed behavior. One area with well-developed theory is foraging behavior (where and how animals choose to feed). Another example is the use of game theory to understand the evolution of behaviors that are apparently not optimal for an individual but may instead be better for a group. *See* BEHAVIORAL ECOLOGY; GAME THEORY; OPTIMIZATION.

The population level has the longest history of ecological theory and perhaps the broadest application. The simplest models of single-species populations ignore differences among individuals and assume that the birth rates and death rates are proportional to the number of individuals in the population. If this is the case, the rate of growth is exponential, a result that goes back at least as far as Malthus's work in the 1700s. As Malthus recognized, this result produces a dilemma: exponential growth cannot continue unabated. Thus, one of the central goals of population ecology theory is to determine the forces and ecological factors that prevent exponential growth and to understand the consequences for the dynamics of ecological populations. *See* ECOLOGICAL METHODS; POPULATION ECOLOGY.

Modifications and extensions of theoretical approaches like the logistic model (which uses differential equations to explain the stability of populations) have also been used to guide the management of renewable natural resources. Here, the most basic concept is that of the maximum sustainable yield, which is the greatest level of harvest at which a population can continue to persist. *See* ADAPTIVE MANAGEMENT; MATHEMATICAL ECOLOGY.

The primary goal of ecological theory at the community level is to understand diversity at local and regional scales. Recent work has emphasized that a great deal of diversity in communities may depend on trade-offs. For example, a trade-off between competitive prowess and colonization ability is capable of explaining why so many plants persist in North American prairies. Another major concept in community theory is the role of disturbance. Understanding how disturbances (such as fires, hurricanes, or wind storms) impacts communities is crucial because humans typically alter disturbance. *See* BIODIVERSITY; ECOLOGICAL COMMUNITIES.

Alan Hastings

Theoretical physics The description of natural phenomena in mathematical form. It is impossible to separate theoretical physics from experimental physics, since a complete understanding of nature can be obtained only by the application of both theory and experiment. There are two main purposes of theoretical physics: the discovery of the fundamental laws of nature and the derivation of conclusions from these fundamental laws.

Physicists aim to reduce the number of laws to a minimum to have as far as possible a unified theory. When the laws are known, it is possible from any given initial conditions of a physical system to derive the subsequent events in the system. Sometimes, especially in quantum theory, only the probability of various events can be predicted. *See* DETERMINISM; QUANTUM MECHANICS.

The conclusions to be derived from the fundamental laws of nature may be of several different types.

1. Conclusions may be derived in order to test a given theory, particularly a new theory. An example is the derivation of the spectrum of the hydrogen atom from quantum mechanics; the verification of the predictions by accurate measurements is a good test of quantum mechanics. On rather rare occasions an experiment has been found to contradict the predictions of an existing theory, and this has then led to the discovery of important new physical laws. An example is the Michelson-Morley experiment on the constancy of the velocity of light, an experiment which led to special relativity theory. *See* ATOMIC STRUCTURE AND SPECTRA; LIGHT; RELATIVITY.

2. Theory may be required for experiments designed to determine physical constants. Most fundamental physical constants cannot be accurately measured directly. Elaborate theories may be required to deduce the constant from indirect experiments. *See* FUNDAMENTAL CONSTANTS.

3. Predictions of physical phenomena may be made in order to gain understanding of the structure of the physical world. In this category fall theories of the structure of the atom leading to an understanding of the periodic system of elements, or of the structure of the nucleus in which various models are tested (for example, shell model or collective model). In the same category fall applications of theoretical physics to other sciences, for example, to chemistry (theory of the chemical bond and of the rate of chemical reactions), astronomy (theory of planetary motion, internal constitution, and energy production of stars), or biology.

4. Engineering applications may be drawn from fundamental laws. All of engineering may be considered an application of physics, and much of it is an application of mathematical physics, such as elasticity theory, aerodynamics, electricity, and magnetism. The generation and propagation of radio waves of all frequencies are examples of application of theoretical physics to direct practice. *See* AERODYNAMICS; ELASTICITY; ELECTRICITY; MAGNETISM; RADIO-WAVE PROPAGATION.

Apart from the classification of the fields of theoretical physics according to purpose, a classification can also be made according to content. Here one may perhaps distinguish three classification principles: type of force, scale of physical phenomena, and type of phenomena. *See* MATHEMATICAL PHYSICS; PHYSICS.

Hans A. Bethe

Therapsida An order of Reptilia, subclass Synapsida, often called advanced mammallike reptiles, that flourished from the middle Permian through the Late Triassic. The group is highly diverse and subdivided into six suborders. Two of these, Eotitanosuchia and Dinocephalia, include relatively primitive mid-Permian carnivores and herbivores. A third, the Dicynodontia, made up of small to large herbivores, was abundant in the late Permian. Dicynodonts were associated with two carnivorous suborders, the Therocephalia and Gorgonopsia, which are morphologically intermediate between Eotitanosuchia and the cynodonts. Although these five developing lines are distinct, the skulls and skeletons in each became increasingly mammallike.

The trend continued among the highly diverse Cynodontia. This suborder includes a variety of carnivores, omnivores, and herbivores. The most highly derived herbivorous cynodonts were the tritylodonts of the Late Triassic and Early Jurassic. They were very mammallike.

Climatic changes during the Triassic rather than direct competition largely accounted for the decline of the therapsids and the rapid expansion of the Archosauria. Only the very mammallike therapsids, the herbivorous tritylodonts, and the minute derived first mammals survived into the Early Jurassic. *See* ARCHOSAURIA; MAMMALIA; REPTILIA; SYNAPSIDA. Everett C. Olson

Theria One of the four subclasses of the class Mammalia, including all living mammals except the monotremes. The Theria were by far the most successful of the several mammalian stocks that arose from the mammallike reptiles in the Triassic. The subclass is divided into three infraclasses: Pantotheria (no living survivors), Metatheria (marsupials), and Eutheria (placentals). Therian mammals are characterized by the distinctive structural history of the molar teeth. The fossil record shows that all the extremely varied therian molar types were derived from a common tribosphenic type in which three main cusps, arranged in a triangle on the upper molar, are opposed to a reversed triangle and basinlike heel on the lower molar. *See* MAMMALIA; THERAPSIDA. D. Dwight Davis; Frederick S. Szalay

Thermal analysis A group of analytical techniques developed to continuously monitor physical or chemical changes of a sample which occur as the temperature of a sample is increased or decreased. Thermogravimetry, differential thermal analysis, and differential scanning calorimetry are the three principal thermoanalytical methods. *See* ANALYTICAL CHEMISTRY; COMPUTER; THERMOCHEMISTRY.

The occurrence of physical or chemical changes upon heating a sample may be explained from either a kinetic or thermodynamic viewpoint. Kinetically the rate of a process may be increased by raising the temperature as shown by the Arrhenius equation (1), where A, E_a, and R represent the preexponential

$$\text{Rate} = Ae^{-E_a/RT} \qquad (1)$$

factor, activation energy, and the gas law constant, respectively. At some point the rate becomes significant and readily observable. Similarly an increase in temperature can change the Gibbs free energy [Eq. (2), where ΔG° is the Gibbs free energy, ΔH°

$$\Delta G^\circ = \Delta H^\circ - T\Delta S^\circ \qquad (2)$$

is the reaction enthalpy, and ΔS° is the entropy change for the process] to a more favorable (that is, more negative) value. In particular, ΔG° will become more negative if ΔS° is positive and the temperature is increased. In many cases a combination of these factors causes the observed physiochemical process. *See* CHEMICAL THERMODYNAMICS; KINETICS (CLASSICAL MECHANICS).

Thermogravimetry involves measuring the changes in mass of a substance, typically a solid, as it is heated. Specially designed thermobalances are required to continuously monitor sample mass during the heating process. Modern balances have a capacity of 1–1500 milligrams and can accurately detect mass changes of 0.1 microgram.

Any type of physiochemical process which involves a change in sample mass may be observed by using thermogravimetry. Mass losses are observed for dehydration, decomposition, desorption, vaporization, sublimation, pyrolysis, and chemical reactions with gaseous products. Mass increases are noted with adsorption, absorption, and chemical reactions of the sample with the atmosphere in the oven, such as the oxidation of metals.

Quantitative gravimetric analyses may be performed due to the precise measure of the mass change obtained. Rates of mass change have been used to evaluate the kinetics of a process and to estimate activation energies. Fine details of these thermograms may also be used to deduce reaction intermediates and reaction mechanisms.

Primary applications of thermogravimetry are to deduce stabilities of compounds and mixtures of elevated temperatures and to determine appropriate drying temperatures for compounds and mixtures. Evaluation of polymers, food products, and pharmaceuticals is a major application of thermogravimetry.

Differential thermal analysis involves the monitoring of the temperature difference T_D between a sample and inert reference material (such as aluminum oxide) as they are simultaneously heated, or cooled, at a predetermined rate. Multijunction thermocouples and thermistors are the most common temperature sensors used for this purpose; they are arranged in an oven. As enthalpic changes occur, T_D will be positive if the process is exothermic and negative if it is endothermic.

More physical and chemical processes may be observed using differential thermal analysis as compared to thermogravimetry. Endothermic physical processes include crystalline transitions, fusion, vaporization, sublimation, desorption, and adsorption. Endothermic physical processes include crystalline transitions, fusion, vaporization, sublimation, desorption, and adsorption. Endothermic chemical processes include dehydration, decomposition, gaseous reduction, redox reactions, and solid-state metathesis. Exothermic processes include adsorption, chemisorption, decomposition, oxidation, redox reactions, and solid-state metathesis reactions. Both solids and liquids can be studied by differential thermal analysis. Hermetically sealed capsules are often used for liquids and some solids. Other samples are studied in open or crimped pans.

Analytical applications of this technique include the identification, characterization, and quantitation of a wide variety of materials, including polymers, pharmaceuticals, metals, clays, minerals, and inorganic and organic compounds. Characteristic thermograms can be used to determine purity, heats of reaction, thermal stability, phase diagrams, catalytic properties, and radiation damage.

In differential scanning calorimetry a sample and a reference are individually heated, by separately controlled resistance heaters, at a predetermined rate. Enthalpic (heat-generating or -absorbing) processes are detected as differences in electrical energy supplied to either the sample or the reference material to maintain this heating rate. This difference in electrical energy, in milliwatts per second, of the heat flow into or out of the sample is due to the occurrence of a physical or chemical process. Modulated differential scanning calorimetry is a new method that superimposes a sine wave on the heating ramp. A significant increase in sensitivity is often observed with modulated differential scanning calorimetry.

Analytical uses of differential scanning calorimetry are very similar to those of differential thermal analysis. Usually one calibration standard is sufficient to calibrate the entire operating range of the instrument. Differential scanning calorimetry instruments are highly sensitive and may measure heat flows as small as 1 nanowatt. Differential scanning calorimetry is very useful in determining heat capacities of substances over large temperature ranges. Such evaluation has become important in polymer and biochemical studies. Small (approximately 1–10 mg) samples are used in most cases, although some instruments have been developed which use up to 1 ml of a liquid sample. *See* CALORIMETRY.

Neil D. Jespersen

Thermal conduction in solids Thermal conduction in a solid is generally measured by stating the thermal conductivity K, which is the ratio of the steady-state heat flow (heat transfer per unit area per unit time) along a long rod to the temperature gradient along the rod. Thermal conductivity varies widely among different types of solids, and depends markedly on temperature and on the purity and physical state of the solids, particularly at low temperatures.

From the kinetic theory of gases the thermal conductivity can be written as $K =$ (constant) Svl, where S is the specific heat per unit volume, v is the average particle velocity, and l is the mean free path. In solids, thermal conduction results from conduction by lattice vibrations and from conduction by electrons. In insulating materials, the conduction is by lattice waves; in pure metals, the lattice contribution is negligible and the heat conduction is primarily due to electrons. In many alloys, impure metals, and semiconductors, both conduction mechanisms contribute. *See* CONDUCTION (HEAT); KINETIC THEORY OF MATTER; LATTICE VIBRATIONS; SPECIFIC HEAT.

In superconductors at temperatures below the critical temperature, the electronic conduction is reduced; at sufficiently low temperatures, the thermal conductivity becomes entirely due to lattice waves and is similar to the form of the thermal conductivity of an insulating material. *See* SUPERCONDUCTIVITY.

Kathryn A. McCarthy

Thermal converters Devices consisting of a conductor heated by an electric current, with one or more hot junctions of a thermocouple attached to it, so that the output emf responds to the temperature rise, and hence the current. Thermal converters are used with external resistors for alternating-current (ac) and voltage measurements over wide ranges and generally form the basis for calibration of ac voltmeters and the ac ranges of instruments providing known voltages and currents.

In the most common form, the conductor is a thin straight wire less than 0.4 in. (1 cm) long, in an evacuated glass bulb, with a single thermocouple junction fastened to the midpoint by a tiny electrically insulating bead. Thermal inertia keeps the temperature of the heater wire constant at frequencies above a few hertz, so that the constant-output emf is a true measure of the root-mean-square (rms) heating value of the current. The reactance of the short wire is so small that the emf can be independent of frequency up to 10 MHz or more. An emf of 10 mV can be obtained at a rated current less than 5 mA, so that resistors of reasonable power dissipation, in series or in shunt with the heater, can provide voltage ranges up to 1000 V and current ranges up to 20 A. However, the flow of heat energy cannot be controlled precisely, so the temperature, and hence the emf, generally changes with time and other factors. Thus an ordinary thermocouple instrument, consisting of a thermal converter and a millivoltmeter to measure the emf, is accurate only to about 1–3%. *See* THERMOCOUPLE; VOLTMETER.

To overcome this, a thermal converter is normally used as an ac-dc transfer instrument (ac-dc comparator) to measure an unknown alternating current or voltage by comparison with a known nearly equal dc quantity (see illustration). By replacing the millivoltmeter with an adjustable, stable, opposing voltage V_b in series with a microvoltmeter D, very small changes in emf can be detected. The switch S is connected to the unknown ac voltage V_{ac}, and V_b is adjusted for a null (zero) reading of D. Then S is immediately connected to the dc voltage V_{dc}, which is adjusted to give a null again, without changing V_b. Thus $V_{ac} = V_{dc}$

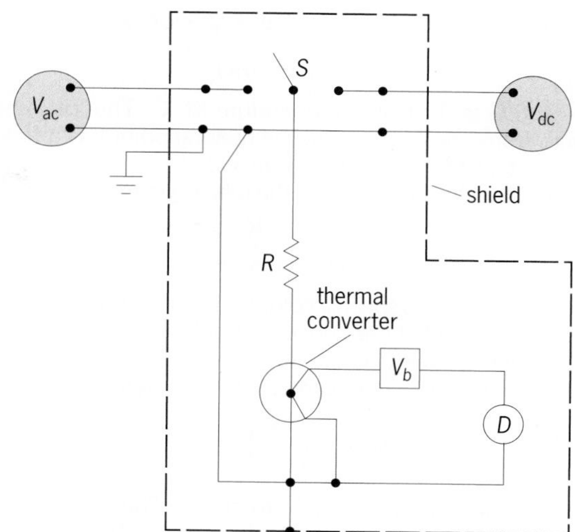

Basic circuit for ac-dc transfer measurements of ac voltages.

$(1 + d)$, where d is the ac-dc difference of the transfer instrument, which can be as small as a few parts per million (ppm).

In many commercial instruments, all of the components are conveniently packaged in the shield, shown with a broken line, and several ranges are available by taps on R. Accuracies of 0.001% are attainable at audio frequencies. *See* ELECTRICAL MEASUREMENT.

F. L. Hermach; Joseph R. Kinard

Thermal ecology The examination of the independent and interactive biotic and abiotic components of naturally heated environments. Geothermal habitats are present from sea level to the tops of volcanoes and occur as fumaroles, geysers, and hot springs. Hot springs typically possess source pools with overflow, or thermal, streams (rheotherms) or without such streams (limnotherms). Hot spring habitats have existed since life began on Earth, permitting the gradual introduction and evolution of species and communities adapted to each other and to high temperatures. Other geothermal habitats do not have distinct communities.

Hot-spring pools and streams, typified by temperatures higher than the mean annual temperature of the air at the same locality and by benthic mats of various colors, are found on all continents except Antarctica. They are located in regions of geologic activity where meteoric water circulates deep enough to become heated. The greatest densities occur in Yellowstone National Park (Northwest United States), Iceland, and New Zealand. Source waters range from 40°C (104°F) to boiling (around 100°C or 212°F depending on elevation), and may even be superheated at the point of emergence. Few hot springs have pH 5–6; most are either acidic (pH 2–4) or alkaline (pH 7–9). Conrad Wickstrom

Thermal expansion Solids, liquids, and gases all exhibit dimensional changes for changes in temperature while pressure is held constant. The molecular mechanisms at work and the methods of data presentation are quite different for the three cases.

The temperature coefficient of linear expansion α_l is defined by Eq. (1), where l is the length of the specimen, t is the tem-

$$\alpha_l = \frac{1}{l}\left(\frac{\partial l}{\partial t}\right)_{p=\text{const}} \qquad (1)$$

perature, and p is the pressure. For each solid there is a Debye characteristic temperature Θ, below which α_l is strongly dependent upon temperature and above which α_l is practically constant. Many common substances are near or above Θ at room temperature and follow approximate equation (2), where l_0 is the

$$l = l_0(1 + \alpha_l t) \tag{2}$$

length at 0°C and t is the temperature in °C. The total change in length from absolute zero to the melting point has a range of approximately 2% for most substances.

So-called perfect gases follow the relation in Eq. (3), where p is

$$\frac{pv}{T} = \frac{R}{\text{molecular weight}} \tag{3}$$

absolute pressure, v is specific volume, T is absolute temperature, and R is the so-called gas constant. Real gases often follow this equation closely. *See* GAS CONSTANT.

The coefficient of cubic expansion α_v is defined by Eq. (4),

$$\alpha_v = \frac{1}{v}\left(\frac{\partial v}{\partial t}\right)_{p=\text{const}} \tag{4}$$

and for a perfect gas this is found to be $1/T$. The behavior of real gases is largely accounted for by the van der Waals equation. *See* GAS; KINETIC THEORY OF MATTER.

For liquids, α_v is somewhat a function of pressure but is largely determined by temperature. Though α_v may often be taken as constant over a sizable range of temperature (as in the liquid expansion thermometer), generally some variation must be accounted for. For example, water contracts with temperature rise from 32 to 39°F (0 to 4°C), above which it expands at an increasing rate. *See* THERMOMETER. Ralph A. Burton

Thermal hysteresis A phenomenon in which a physical quantity depends not only on the temperature but also on the preceding thermal history. It is usual to compare the behavior of the physical quantity while heating and the behavior while cooling through the same temperature range. The illustration shows the thermal hysteresis which has been observed in the behavior of the dielectric constant of single crystals of barium titanate. On heating, the dielectric constant was observed to follow the path *ABCD*, and on cooling the path *DCEFG*. *See* DIELECTRIC CONSTANT; FERROELECTRICS.

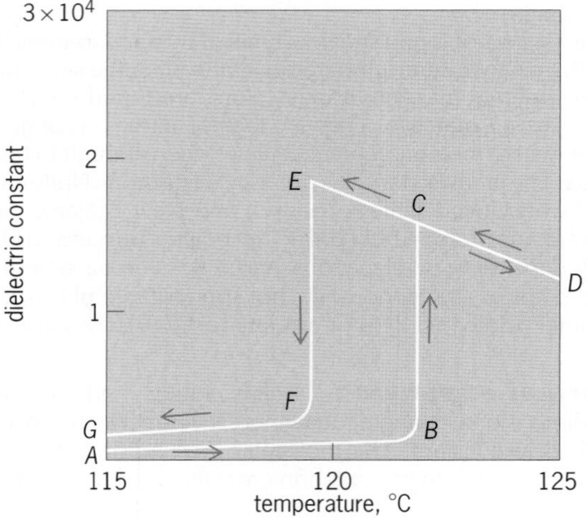

Plot of dielectric constant versus temperature for a single crystal of barium titanate. (*After M. E. Drougard and D. R. Young, Phys. Rev., 95:1152–1153, 1954*)

Perhaps the most common example of thermal hysteresis involves a phase change such as solidification from the liquid phase. In many cases these liquids can be dramatically supercooled. Elaborate precautions to eliminate impurities and outside disturbances can be instrumental in supercooling 60 to 80°C. On raising the temperature after freezing, however, the system follows a completely different path, with melting coming at the prescribed temperature for the phase change. *See* CRYSTAL; NUCLEATION; PHASE TRANSITIONS. H. B. Huntington; R. K. MacCrone

Thermal neutrons Neutrons whose energy distribution is governed primarily by the kinetic energy distribution of molecules of the material in which the neutrons are found. *See* NEUTRON.

The molecules of the material usually have a kinetic energy distribution very close to a Maxwell-Boltzmann distribution. This distribution shows a peak at an energy equal to half the product of the temperature and the Boltzmann constant. At high energies it decreases exponentially, and at low energies it is proportional to the square root of the energy. When the material is large and very weakly absorbing, the neutron energy distribution closely approaches this maxwellian. *See* BOLTZMANN STATISTICS; KINETIC THEORY OF MATTER.

The most common way of generating thermal neutrons is to allow neutrons from a source—reactor, accelerator, or spontaneous fission neutron emitter—to diffuse outward through a large block or tank of very weakly absorbing moderator. *See* REACTOR PHYSICS. Bernard I. Spinrad

Thermal stress Mechanical stress induced in a body when some or all of its parts are not free to expand or contract in response to changes in temperature. In most continuous bodies, thermal expansion or contraction cannot occur freely in all directions because of geometry, external constraints, or the existence of temperature gradients, and so stresses are produced. Such stresses caused by a temperature change are known as thermal stresses.

Problems of thermal stress arise in many practical design problems, such as those encountered in the design of steam and gas turbines, diesel engines, jet engines, rocket motors, and nuclear reactors. The high aerodynamic heating rates associated with high-speed flight present even more severe thermal-stress problems for the design of spacecraft and missiles. *See* STRESS AND STRAIN. Shih-Yuan Chen

Thermal wind The difference in the geostrophic wind between two heights in the atmosphere over a given position on Earth. It approximates the variation of the actual winds with height for large-scale and slowly changing motions of the atmosphere. Such structure in the wind field is of fundamental importance to the description of the atmosphere and to processes causing its day-to-day changes. The thermal wind embodies a basic relationship between vertical fluctuations of the horizontal wind and horizontal temperature gradients in the atmosphere. This relationship arises from the combination of the geostrophic wind law, the hydrostatic equation, and the gas law.

The geostrophic wind law applies directly to steady, straight, and unaccelerated horizontal motion and is a good approximation for large-scale and slowly changing motions in the atmosphere. The hydrostatic equation combined with the gas law relates the atmospheric pressure and temperature fields. The relationship is accurate for most atmospheric situations but not for small-scale and rapidly changing conditions such as in turbulence and thunderstorms. The equation gives the change of pressure in the vertical direction as a function of pressure and temperature. The key conclusion is that at a given level in the atmosphere the pressure change (decrease) with height is more rapid in cold air than in warm air. *See* ATMOSPHERE; GEOSTROPHIC WIND; HYDROSTATICS; TROPOSPHERE. David D. Houghton

Thermionic emission The emission of electrons into vacuum by a heated electronic conductor. In its broadest meaning, thermionic emission includes the emission of ions, but this process is quite different from that normally understood by the term. Thermionic emitters are used as cathodes in electron tubes and hence are of great technical and scientific importance. Although in principle all conductors are thermionic emitters, only a few materials satisfy the requirements set by practical applications. Of the metals, tungsten is an important practical thermionic emitter; in most electron tubes, however, the oxide-coated cathode is used to great advantage. For a detailed discussion of practical thermionic emitters. *See* VACUUM TUBE.

The thermionic emission of a material may be measured by using the material as the cathode in a vacuum tube and collecting the emitted electrons on a positive anode. If the anode is sufficiently positive relative to the cathode, space charge (a concentration of electrons near the cathode) can be avoided and all electrons emitted can be collected; the saturation thermionic current is then measured. *See* SCHOTTKY EFFECT.

The emission current density J increases rapidly with increasing temperature; this is illustrated by the following approximate values for tungsten:

T(K)	1000	2000	2500	3000
J(amperes/cm^2)	10^{-15}	10^{-3}	0.3	15

The temperature dependence of J is given by the Richardson-Duchman (or Richardson) equation below. Here A is a constant,

$$J = AT^2e^{-(\phi - kT)}$$

k is Boltzmann's constant ($=1.38 \times 10^{-23}$ joule/degree), and ϕ is the work function of the emitter. The work function has the dimensions of energy and is a few electronvolts for thermionic emitters. *See* WORK FUNCTION (ELECTRONICS). A. J. Dekker

Thermionic power generator A device for converting heat into electricity through the use of thermionic emission and no working fluid other than electric charges. An elementary thermionic generator, or thermionic converter, consists of a hot metal surface (emitter) separated from a cooler electrode (collector) by an insulator seal (see illustration). The interelectrode gap is usually a fraction of a millimeter in width. The hermetic enclosure contains a small amount of an easily ionizable gas, such as cesium vapor maintained by a liquid-cesium reservoir. In some experimental devices, the enclosure may be evacuated.

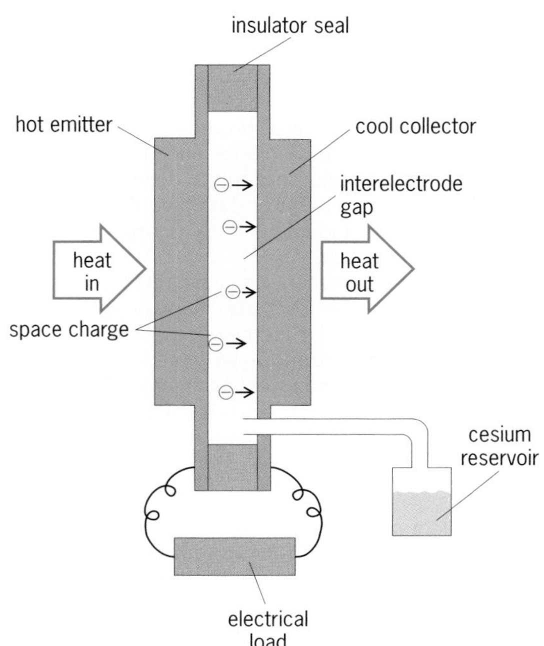

Diagram of thermionic converter.

Electrons evaporated from the emitter cross the interelectrode gap, condense on the collector, and are returned to the emitter via the external electrical load circuit. The thermionic generator is essentially a heat engine utilizing an electron gas as the working fluid. The temperature difference between the emitter and the collector drives the electron current.

Thermionic generators are characterized by high operating temperatures [typically emitter temperatures between 1600 and 2500 K (2420 and 4040°F) and collector temperatures ranging from 800 to 1100 K (980 to 1520°F)]; low output voltage (approximately 0.5 V per converter); high current density (around 5–10 A/cm^2); and high conversion efficiency (about 10–15%). These characteristics, especially the relatively high heat-rejection temperature, make the thermionic generator attractive for producing electric power in space applications with nuclear-reactor or radioisotope energy sources. The high electrode temperatures make thermionic generators also attractive as topping units for steam power plants, and for the cogeneration of electricity in combination with heat for intermediate-temperature industrial processes. Topping units increase the overall system efficiency. *See* COGENERATION; NUCLEAR BATTERY.

Elias P. Gyftopoulos; George N. Hatsopoulos

Thermionic tube An electron tube that relies upon thermally emitted electrons from a heated cathode for tube current.

Thermionic emission of electrons means emission by heat. In practical form an electrode, called the cathode because it forms the negative electrode of the tube, is heated until it emits electrons. The cathode may be either a directly heated filament or an indirectly heated surface. With a filamentary cathode, heating current is passed through the wire, which either emits electrons directly or is covered with a material that readily emits electrons. Indirectly heated cathodes have a filament, commonly called the heater, located within the cathode electrode to bring the surface of the cathode to emitting temperature. The majority of all vacuum tubes are thermionic tubes. *See* ELECTRON TUBE; GAS TUBE; THERMIONIC EMISSION; VACUUM TUBE. Leon S. Nergaard

Thermistor An electrical resistor with a relatively large negative temperature coefficient of resistance. Thermistors are useful for measuring temperature and gas flow or wind velocity. Often they are employed as bolometer elements to measure radio-frequency, microwave, and optical power. They also are used as electrical circuit components for temperature compensation, voltage regulation, circuit protection, time delay, and volume control. A common type of thermistor is a semiconducting ceramic composed of mixtures of several metal oxides. Metal electrodes or wires are attached to the ceramic material so that the thermistor resistance can be measured conveniently. The temperature coefficient of resistance is negative for these thermistors. Other types can have either negative or positive temperature coefficients. *See* BOLOMETER; ELECTRICAL RESISTIVITY; VOLTAGE REGULATOR; VOLUME CONTROL SYSTEMS.

At room temperature the resistance of a thermistor may typically change by several percent for a variation of 1° of temperature, but the resistance does not change linearly with temperature. The temperature coefficient of resistance of a thermistor is approximately equal to a constant divided by the square of the temperature in kelvins. The constant is equal to several thousand kelvins and is specified for a given thermistor and the temperature range of intended use.

The electrical and thermal properties of a thermistor depend upon the material composition, the physical dimensions, and the environment provided by the thermistor enclosure. Thermistors range in form from small beads and flakes less than 25 micrometers (10^{-3} in.) thick to disks, rods, and washers with centimeter dimensions. The small beads are often coated with glass to prevent changes in composition or encased in glass probes or cartridges to prevent damage. Beads are available with room-temperature resistances ranging from less than 100 Ω to tens of megohms, and with time constants that can be less than a second. Large disks and washers have a similar resistance range and can have time constants of minutes. *See* ELECTRIC POWER MEASUREMENT; FLOW MEASUREMENT; MICROWAVE POWER MEASUREMENT; TEMPERATURE MEASUREMENT; TIME CONSTANT; VELOCIMETER.

Thomas P. Crowley

Thermoacoustics The study of phenomena that involve both thermodynamics and acoustics. A sound wave in a gas is usually regarded as consisting of coupled pressure and displacement oscillations, but temperature oscillations accompany the pressure oscillations. When there are spatial gradients in the oscillating temperature, oscillating heat flow also occurs. The combination of these four oscillations produces a rich variety of thermoacoustic effects. *See* ACOUSTICS; OSCILLATION; SOUND; THERMODYNAMIC PRINCIPLES.

Although the oscillating heat transfer at solid boundaries does contribute significantly to the dissipation of sound in enclosures such as buildings, thermoacoustic effects are usually too small to be obviously noticeable in everyday life. However, thermoacoustic effects in intense sound waves inside suitable cavities can be harnessed to produce extremely powerful pulsating combustion, thermoacoustic refrigerators, and thermoacoustic engines.

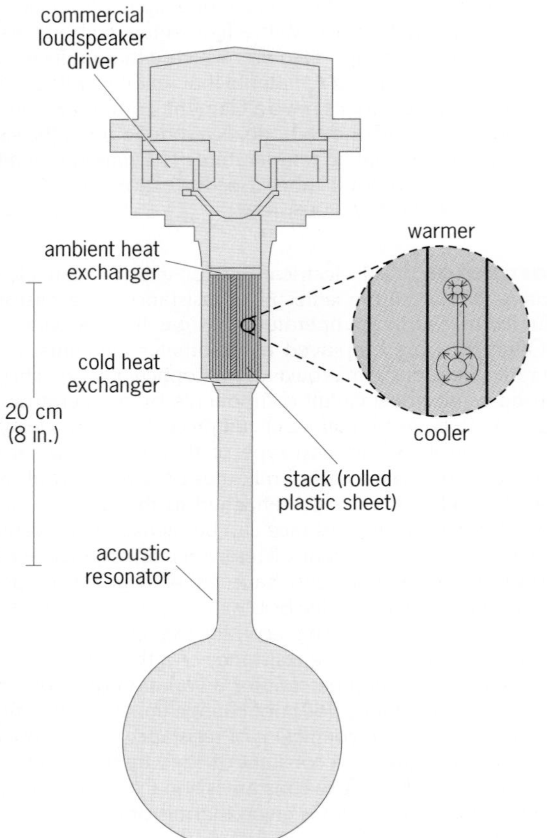

An early standing-wave thermoacoustic refrigerator that cooled to −60°C (−76°F). Heat is carried up the temperature gradient in the stack. At the right is a magnified view of the oscillating motion of a typical parcel of gas. The volume of the parcel depends on its pressure and temperature. (*After T. J. Hofler, Thermoacoustic Refrigerator Design and Performance, Ph.D. thesis, University of California at San Diego, 1996*)

Pulsating combustion. Oscillations can occur whenever combustion takes place in a cavity. In industrial equipment and residential appliances, these oscillations are sometimes encouraged in order to stir or pump the combustion ingredients, while in rocket engines such oscillations must usually be suppressed because they can damage the rocket structure. The oscillations occur spontaneously if the combustion progresses more rapidly or efficiently during the compression phase of the pressure oscillation than during the rarefaction (expansion) phase—the Rayleigh criterion. *See* COMBUSTION; GAS DYNAMICS.

Thermoacoustic refrigerators. Thermoacoustic refrigerators use acoustic power to pump heat from a low temperature to ambient temperature (see illustration). The heat-pumping mechanism takes place in the pores of a structure called a stack. As a typical parcel of the gas oscillates along a pore, it experiences changes in temperature. Most of the temperature change comes from adiabatic compression and expansion of the gas by the sound pressure, and the rest is a consequence of the local temperature of the solid wall of the pore. A thermodynamic cycle results from the coupled pressure, temperature, position, and heat oscillations. The overall effect, much as in a bucket brigade, is the net transport of heat from the cold heat exchanger to room temperature. *See* ADIABATIC PROCESS; SOUND PRESSURE; THERMODYNAMIC CYCLE; THERMODYNAMIC PROCESSES.

Thermoacoustic engines. While standing-wave thermoacoustic systems have matured only recently, Stirling engines and refrigerators have a long, rich history. New insights have resulted from applying thermoacoustics to Stirling systems, treating them as traveling-wave thermoacoustic systems in which the extrema in pressure and gas motion are approximately 90° out of phase in time. In the thermoacoustic-Stirling engine, the thermodynamic cycle is accomplished in a traveling-wave acoustic network, and acoustic power is produced from heat with an efficiency of 30%.

Greg Swift

Thermochemistry A branch of physical chemistry concerned with the absorption or evolution of heat that accompanies chemical reactions. Closely related topics are the latent heat associated with a change in phase (crystal, liquid, gas), the chemical composition of reacting systems at equilibrium, and the electrical potentials of galvanic cells. Thermodynamics provides the link among these phenomena.

A knowledge of such heat effects is important to the chemical engineer for the design and operation of chemical reactors, the determination of the heating values of fuels, the design and operation of refrigerators, the selection of heat storage systems, and the assessment of chemical hazards. Thermochemical information is used by the physiologist and biochemist to study the energetics of living organisms and to determine the calorific values of foods. Thermochemical data give the chemist an insight to the energies of, and interactions among, molecules. *See* CHEMICAL THERMODYNAMICS.

A calorimeter is an instrument for measuring the heat added to or removed from a process. There are many designs, but the following parts can generally be identified: the vessel in which the process is confined, the thermometer which measures its temperature, and the surrounding environment called the jacket. The heat associated with the process is calculated by the equation below, where T is the temperature. The quantity C, the

$$q = C[(T(\text{final}) - T(\text{initial})] - q_{ex} - w$$

energy equivalent of the calorimeter, is obtained from a separate calibration experiment. The work transferred to the process, w, is generally in the form of an electric current (as supplied to a heater, for example) or as mechanical work (as supplied to a stirrer, for example) and can be calculated from appropriate auxiliary measurements. The quantity q_{ex} is the heat exchanged between the container and its jacket during the experiment. It is calculated from the temperature gradients in the system and the measured thermal conductivities of its parts.

Two principal types of calorimeters are used to measure heats of chemical reactions. In a batch calorimeter, known quantities of reactants are placed in the vessel and the initial temperature is measured. The reaction is allowed to occur and then the final equilibrium temperature is measured. If necessary, the final contents are analyzed to determine the amount of reaction which occurred.

In a flow calorimeter, the reactants are directed to the reaction vessel in two or more steady streams. The reaction takes place quickly and the products emerge in a steady stream. The rate of heat production is calculated from the temperatures, flow velocities, and heat capacities of the incoming and outgoing streams, and the rates of work production and heat transfer to the jacket. Dividing this result by the rate of reaction gives the heat of reaction. *See* CALORIMETRY.

In the past, thermochemical quantities usually have been given in units of calories. A calorie is defined as the amount of heat needed to raise the temperature of 1 gram of water 1°C. However, since this depends on the initial temperature of the water, various calories have been defined, for example, the 15° calorie, the 20° calorie, and the mean calorie (average from 0 to 100°C). In addition, a number of dry calories have been defined. Those still used are the thermochemical calorie (exactly 4.184 joules) and the International Steam Table calorie (exactly 4.1868 J).

Randolph C. Wilhoit

Thermocouple A device in which the temperature difference between the ends of a pair of dissimilar metal wires is deduced from a measurement of the difference in the thermoelectric potentials developed along the wires. The presence of a temperature gradient in a metal or alloy leads to an electric potential gradient being set up along the temperature gradient. This thermoelectric potential gradient is proportional to the temperature gradient and varies from metal to metal. It is the fact that the thermoelectric emf is different in different metals and alloys for the same temperature gradient that allows the effect to be used for the measurement of temperature.

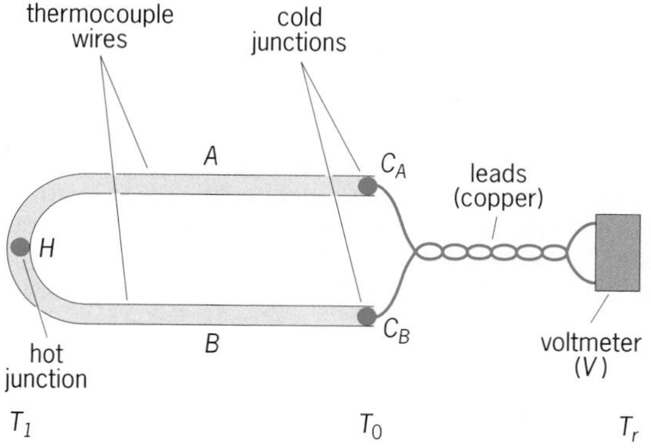

Basic circuit of a thermocouple.

The basic circuit of a thermocouple is shown in the illustration. The thermocouple wires, made of different metals or alloys A and B, are joined together at one end H, called the hot (or measuring) junction, at a temperature T_1. The other ends, C_A and C_B (the cold or reference junctions), are maintained at a constant reference temperature T_0, usually but not necessarily 32°F (0°C). From the cold junctions, wires, usually of copper, lead to a voltmeter V at room temperature T_r. Due to the thermoelectric potential gradients being different along the wires A and B, there exists a potential difference between C_A and C_B. This can be measured by the voltmeter, provided that C_A and C_B are at the same temperature and that the lead wires between C_A and V and C_B and V are identical (or that V is at the temperature T_0, which is unusual). Such a thermocouple will produce a thermoelectric emf between C_A and C_B which depends only upon the temperature difference $T_1 - T_0$. *See* TEMPERATURE MEASUREMENT; THERMOELECTRICITY.

Letter designations and compositions for standardized thermocouples*

Type designation	Materials
B	Platinum-30% rhodium/platinum-6% rhodium
E	Nickel-chromium alloy/a copper-nickel alloy
J	Iron/another slightly different copper-nickel alloy
K	Nickel-chromium alloy/nickel-aluminum alloy
R	Platinum-13% rhodium/platinum
S	Platinum-10% rhodium/platinum
T	Copper/a copper-nickel alloy

*After T. J. Quinn, *Temperature*, Academic Press, 1983.

A large number of pure metal and alloy combinations have been studied as thermocouples, and the seven most widely used are listed in the table. The thermocouples in the table together cover the temperature range from about −420°F (−250°C or 20 K) to about 3300°F (1800°C). The most accurate and reproducible are the platinum/rhodium thermocouples, types R and S, while the most widely used industrial thermocouples are probably types K, T, and E.

T. J. Quinn

Thermodynamic cycle A procedure or arrangement in which one form of energy, such as heat at an elevated temperature from combustion of a fuel, is in part converted to another form, such as mechanical energy on a shaft, and the remainder is rejected to a lower-temperature sink as low-grade heat.

A thermodynamic cycle requires, in addition to the supply of incoming energy, (1) a working substance, usually a gas or vapor; (2) a mechanism in which the processes or phases can be carried through sequentially; and (3) a thermodynamic sink to which the residual heat can be rejected. The cycle itself is a repetitive series of operations.

There is a basic pattern of processes common to power-producing cycles. There is a compression process wherein the working substance undergoes an increase in pressure and therefore density. There is an addition of thermal energy from a source such as a fossil fuel, a fissile fuel, or solar radiation. There is an expansion process during which work is done by the system on the surroundings. There is a rejection process where thermal energy is transferred to the surroundings. The algebraic sum of the energy additions and abstractions is such that some of the thermal energy is converted into mechanical work. *See* HEAT.

The basic processes of the cycle, either open or closed, are heat addition, heat rejection, expansion, and compression. These processes are always present in a cycle even though there may be differences in working substance, the individual processes, pressure ranges, temperature ranges, mechanisms, and heat transfer arrangements.

Many cyclic arrangements, using various combinations of phases but all seeking to convert heat into work, have been proposed by many investigators whose names are attached to their proposals, for example, the Diesel, Otto, Rankine, Brayton, Stirling, Ericsson, and Atkinson cycles. All proposals are not equally efficient in the conversion of heat into work. However, they may offer other advantages which have led to their practical development for various applications. *See* BRAYTON CYCLE; CARNOT CYCLE; DIESEL CYCLE; OTTO CYCLE; STIRLING ENGINE; THERMODYNAMIC PROCESSES.

Theodore Baumeister

Thermodynamic principles Laws governing the transformation of energy. Thermodynamics is the science of the transformation of energy. It differs from the dynamics of Newton by taking into account the concept of temperature, which is outside the scope of classical mechanics. In practice, thermodynamics is useful for assessing the efficiencies of heat engines (devices that transform heat into work) and refrigerators (devices that use external sources of work to transfer heat from a hot system to cooler sinks), and for discussing the spontaneity of chemical reactions (their tendency to occur naturally) and the work that they can be used to generate.

The subject of thermodynamics is founded on four generalizations of experience, which are called the laws of thermodynamics. Each law embodies a particular constraint on the properties of the world. The connection between phenomenological thermodynamics and the properties of the constituent particles of a system is established by statistical thermodynamics, also called statistical mechanics. Classical thermodynamics consists of a collection of mathematical relations between observables, and as such is independent of any underlying model of matter (in terms, for instance, of atoms). However, interpretations in terms of the statistical behavior of large assemblies of particles greatly enriches the understanding of the relations established by thermodynamics. *See* STATISTICAL MECHANICS.

Zeroth law of thermodynamics. The zeroth law of thermodynamics establishes the existence of a property called temperature. This law is based on the observation that if a system A is in thermal equilibrium with a system B (that is, no change in the properties of B take places when the two are in contact), and if system B is in thermal equilibrium with a system C, then it is invariably the case that A will be found to be in equilibrium with C if the two systems are placed in mutual contact. This law suggests that a numerical scale can be established for the common property, and if A, B, and C have the same numerical values of this property, then they will be in mutual thermal equilibrium if they were placed in contact. This property is now called the temperature. *See* TEMPERATURE.

First law of thermodynamics. The first law of thermodynamics establishes the existence of a property called the internal energy of a system. It also brings into the discussion the concept of heat.

The first law is based on the observation that a change in the state of a system can be brought about by a variety of techniques. Indeed, if attention is confined to an adiabatic system, one that is thermally insulated from its surroundings, then the work of J. P. Joule shows that same change of state is brought about by a given quantity of work regardless of the manner in which the work is done. This observation suggests that, just as the height through which a mountaineer climbs can be calculated from the difference in altitudes regardless of the path the climber takes between two fixed points, so the work, w, can be calculated from the difference between the final and initial properties of a system. The relevant property is called the internal energy, U. However, if the transformation of the system is taken along a path that is not adiabatic, a different quantity of work may be required. The difference between the work of adiabatic change and the work of nonadiabatic change is called heat, q. In general, Eq. (1) is satisfied, where ΔU is the change in internal energy

$$\Delta U = w + q \tag{1}$$

between the final and initial states of the system. *See* ADIABATIC PROCESS; ENERGY; HEAT.

The implication of this argument is that there are two modes of transferring energy between a system and its surroundings. One is by doing work; the other is by heating the system. Work and heat are modes of transferring energy. They are not forms of energy in their own right. Work is a mode of transfer that is equivalent (if not the case in actuality) to raising a weight in the surroundings. Heat is a mode of transfer that arises from a difference in temperature between the system and its surroundings. What is commonly called heat is more correctly called the thermal motion of the molecules of a system.

The first law of thermodynamics states that the internal energy of an isolated system is conserved. That is, for a system to which no energy can be transferred by the agency of work or of heat, the internal energy remains constant. This law is a cousin of the law of the conservation of energy in mechanics, but it is richer, for it implies the equivalence of heat and work for bringing about changes in the internal energy of a system (and heat is foreign to classical mechanics).

Second law of thermodynamics. The second law of thermodynamics deals with the distinction between spontaneous and nonspontaneous processes. A process is spontaneous if it occurs without needing to be driven. In other words, spontaneous changes are natural changes, like the cooling of hot metal and the free expansion of a gas. Many conceivable changes occur with the conservation of energy globally, and hence are not in conflict with the first law; but many of those changes turn out to be nonspontaneous, and hence occur only if they are driven.

The second law was formulated by Lord Kelvin and by R. Clausius in a manner relating to observation: "no cyclic engine operates without a heat sink" and "heat does not transfer spontaneously from a cool to a hotter body," respectively (see illustration). The two statements are logically equivalent in the sense that failure of one implies failure of the other. However, both may be absorbed into a single statement: the entropy of an isolated system increases when a spontaneous change occurs. The property of entropy is introduced to formulate the law quantitatively in exactly the same way that the properties of temperature and internal energy are introduced to render the zeroth and first laws quantitative and precise.

The entropy, S, of a system is a measure of the quality of the energy that it stores. The formal definition is based on Eq. (2),

$$dS = \frac{dq_{\text{reversible}}}{T} \tag{2}$$

where dS is the change in entropy of a system, dq is the energy transferred to the system as heat, T is the temperature, and the subscript "reversible" signifies that the transfer must be carried out reversibly (without entropy production other than in the system). When a given quantity of energy is transferred as heat, the change in entropy is large if the transfer occurs at a low temperature and small if the temperature is high.

This definition of entropy is illuminated by L. Boltzmann's interpretation of entropy as a measure of the disorder of a system. The connection can be appreciated qualitatively at least by noting that if the temperature is high, the transfer of a given quantity of energy as heat stimulates a relatively small additional disorder in the thermal motion of the molecules of a system; in contrast,

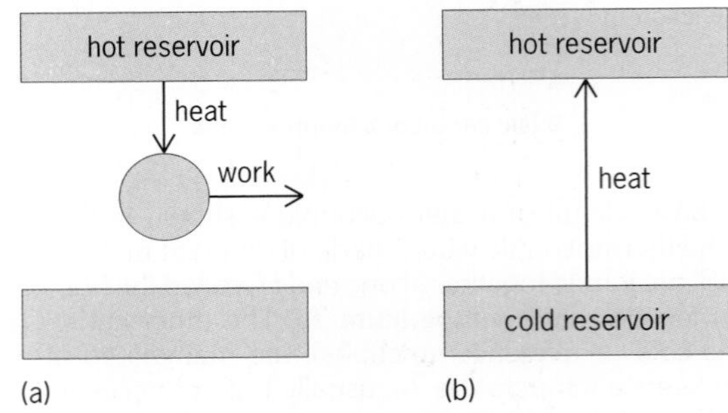

Representation of the statements of the second law of thermodynamics by (*a*) Lord Kelvin and (*b*) R. Clausius. In each case, the law states that the device shown cannot operate as shown.

if the temperature is low, the same transfer could stimulate a relatively large additional disorder.

The illumination of the second law brought about by the association of entropy and disorder is that in an isolated system the only changes that may occur are those in which there is no increase in order. Thus, energy and matter tend to disperse in disorder (that is, entropy tends to increase), and this dispersal is the driving force of spontaneous change. *See* ENTROPY; TIME, ARROW OF.

Third law of thermodynamics. The practical significance of the second law is that it limits the extent to which the internal energy may be extracted from a system as work. In order for a process to generate work, it must be spontaneous. For the process to be spontaneous, it is necessary to discard some energy as heat in a sink of lower temperature. In other words, nature in effect exacts a tax on the extraction of energy as work. There is therefore a fundamental limit on the efficiency of engines that convert heat into work.

The quantitative limit on the efficiency, ϵ, which is defined as the work produced divided by the heat absorbed from the hot source, was first derived by S. Carnot. He found that, regardless of the details of the construction of the engine, the maximum efficiency (that is, the work obtained after payment of the minimum allowable tax to ensure spontaneity) is given by Eq. (3), where

$$\epsilon = 1 - \frac{T_{\text{cold}}}{T_{\text{hot}}} \tag{3}$$

T_{hot} is the temperature of the hot source and T_{cold} is the temperature of the cold sink. The greatest efficiencies are obtained with the coldest sinks and the hottest sources, and these are the design requirements of modern power plants. *See* CARNOT CYCLE.

Perfect efficiency ($\epsilon = 1$) would be obtained if the cold sink were at absolute zero ($T_{\text{cold}} = 0$). However, the third law of thermodynamics, which is another summary of observations, asserts that absolute zero is unattainable in a finite number of steps for any process. Therefore, heat can never be completely converted into work in a heat engine. The implication of the third law in this form is that the entropy change accompanying any process approaches zero as the temperature approaches zero. That implication in turn implies that all substances tend toward the same entropy as the temperature is reduced to zero. It is therefore sensible to take the entropy of all perfect crystalline substances (substances in which there is no residual disorder arising from the location of atoms) as equal to zero. A common short statement of the third law is therefore that all perfect crystalline substances have zero entropy at absolute zero ($T = 0$). This statement is consistent with the interpretation of entropy as a measure of disorder, since at absolute zero all thermal motion has been quenched. *See* ABSOLUTE ZERO.

P. W. Atkins

Thermodynamic processes

Changes of any property of an aggregation of matter and energy, accompanied by thermal effects.

Systems and processes. To evaluate the results of a process, it is necessary to know the participants that undergo the process, and their mass and energy. A region, or a system, is selected for study, and its contents determined. This region may have both mass and energy entering or leaving during a particular change of conditions, and these mass and energy transfers may result in changes both within the system and within the surroundings which envelop the system.

To establish the exact path of a process, the initial state of the system must be determined, specifying the values of variables such as temperature, pressure, volume, and quantity of material. The number of properties required to specify the state of a system depends upon the complexity of the system. Whenever a system changes from one state to another, a process occurs.

The path of a change of state is the locus of the whole series of states through which the system passes when going from an initial to a final state. For example, suppose a gas expands to twice its volume and that its initial and final temperatures are the same. An extremely large number of paths connect these initial and final states. The detailed path must be specified if the heat or work is to be a known quantity; however, changes in the thermodynamic properties depend only on the initial and final states and not upon the path. A quantity whose change is fixed by the end states and is independent of the path is a point function or a property.

Pressure-volume-temperature diagram. Whereas the state of a system is a point function, the change of state of a system, or a process, is a path function. Various processes or methods of change of a system from one state to another may be depicted graphically as a path on a plot using thermodynamic properties as coordinates.

The variable properties most frequently and conveniently measured are pressure, volume, and temperature. If any two of these are held fixed (independent variables), the third is determined (dependent variable). To depict the relationship among these physical properties of the particular working substance, these three variables may be used as the coordinates of a three-dimensional space. The resulting surface is a graphic presentation of the equation of state for this working substance, and all possible equilibrium states of the substance lie on this *P-V-T* surface.

Because a *P-V-T* surface represents all equilibrium conditions of the working substance, any line on the surface represents a possible reversible process, or a succession of equilibrium states.

The portion of the *P-V-T* surface shown in Fig. 1 typifies most real substances; it is characterized by contraction of the substance on freezing. Going from the liquid surface to the liquid-solid surface onto the solid surface involves a decrease in both temperature and volume. Water is one of the few exceptions to this condition; it expands upon freezing, and its resultant *P-V-T* surface is somewhat modified where the solid and liquid phases abut.

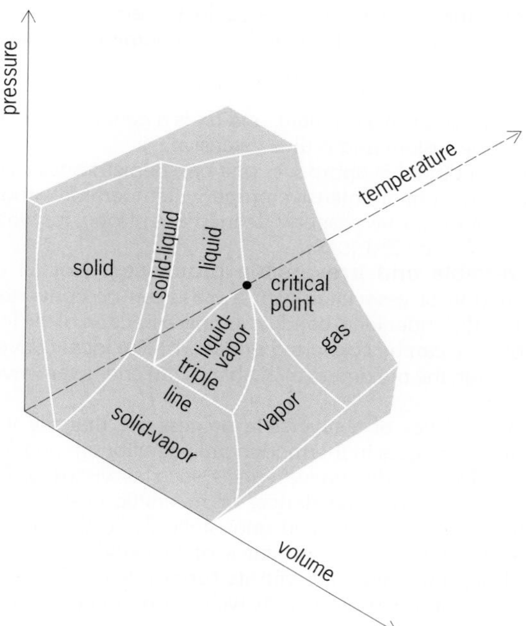

Fig. 1. Portion of pressure-volume-temperature (*P-V-T*) surface for a typical substance.

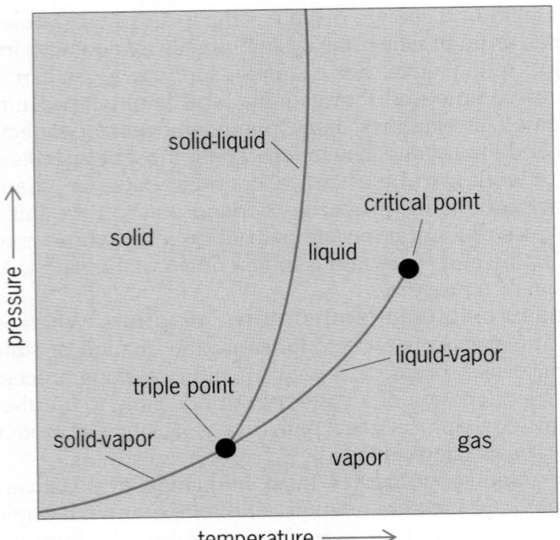

Fig. 2. Portion of equilibrium surface projected on pressure-temperature (*P-T*) plane.

One can project the three-dimensional surface onto the *P-T* plane as in Fig. 2. The triple point is the point where the three phases are in equilibrium. When the temperature exceeds the critical temperature (at the critical point), only the gaseous phase is possible.

Temperature-entropy diagram. Energy quantities may be depicted as the product of two factors: an intensive property and an extensive one. Examples of intensive properties are pressure, temperature, and magnetic field; extensive ones are volume, magnetization, and mass. Thus, in differential form, work is the product of a pressure exerted against an area which sweeps through an infinitesimal volume, as in Eq. (1). As a gas expands,

$$dW = P\,dV \qquad (1)$$

it is doing work on its environment. However, a number of different kinds of work are known. For example, one could have work of polarization of a dielectric, of magnetization, of stretching a wire, or of making new surface area. In all cases, the infinitesimal work is given by Eq. (2), where X is a generalized applied force

$$dW = X\,dx \qquad (2)$$

which is an intensive quantity, and dx is a generalized displacement of the system and is thus extensive.

By extending this approach, one can depict transferred heat as the product of an intensive property, temperature, and a distributed or extensive property defined as entropy, for which the symbol is S. *See* ENTROPY.

Reversible and irreversible processes. Not all energy contained in or associated with a mass can be converted into useful work. Under ideal conditions only a fraction of the total energy present can be converted into work. The ideal conversions which retain the maximum available useful energy are reversible processes.

Characteristics of a reversible process are that the working substance is always in thermodynamic equilibrium and the process involves no dissipative effects such as viscosity, friction, inelasticity, electrical resistance, or magnetic hysteresis. Thus, reversible processes proceed quasistatically so that the system passes through a series of states of thermodynamic equilibrium, both internally and with its surroundings. This series of states may be traversed just as well in one direction as in the other.

Actual changes of a system deviate from the idealized situation of a quasistatic process devoid of dissipative effects. The extent of the deviation from ideality is correspondingly the extent of the irreversibility of the process. *See* THERMODYNAMIC PRINCIPLES.

Philip E. Bloomfield; William A. Steele

Thermoelectric power generator A solid-state heat engine which employs the electron gas as a working fluid. It directly converts heat energy into electrical energy using the Seebeck effect. This phenomenon can be demonstrated using a thermocouple which comprises two legs (thermoelements) of dissimilar conducting materials joined at one end to form a junction. If this junction is maintained at a temperature which differs from ambient, a voltage is generated across the open ends of the thermoelements. When the circuit is completed with a load, a current flows in the circuit and power is generated. In practice the thermocouples are fabricated generally from *n*- and *p*-type semiconductors, and several hundred are connected electrically in series to form a module which is the active component of a thermoelectric generator. Provided a temperature difference is maintained across the device, it will generate electrical power. Heat is provided from a variety of sources depending on the application, and they include burning fossil fuels in terrestrial and military applications, decaying long-life isotopes in medical and deep-space applications, and waste heat. The performance of the thermoelectric generator, in terms of efficiency, output power, and economic viability, depends upon its temperature regime of operation; the materials used in the module construction; its electrical, thermal, and geometrical design; and the generator load. The power output spectrum of thermoelectric generators spans 14 orders of magnitude and ranges from nanowatt generators fabricated using integrated circuit technology to the nuclear-reactor-powered 100-kW SP-100 generator intended to provide electrical power to orbiting space stations. *See* NUCLEAR BATTERY; RADIOACTIVITY AND RADIATION APPLICATIONS; SEEBECK EFFECT; SPACE POWER SYSTEMS; SPACE PROBE; THERMOCOUPLE; THERMOELECTRICITY.

Historically the use of thermoelectric generators has been restricted to specialized applications where combinations of their desirable properties, such as the absence of moving parts, reliability, and silent operation, has outweighed their low overall conversion efficiency, typically 5%. In these applications, fuel cost or weight is a major consideration, and improving the conversion efficiency is the main research target. Improving the figure of merit has been regarded as the most important factor in increasing the conversion efficiency of a thermoelectric generator.

Selection of thermocouple material depends upon the generator's temperature regime of operation. The figures of merit of established thermoelectric materials reach maxima at different temperatures, and semiconductor compounds or alloys based on bismuth telluride, lead telluride, and silicon germanium cover the temperature ranges up to 150°C (300°F), 650°C (1200°F), and 1000°C (1830°F) respectively, with the best materials capable of generating electrical power with an efficiency of around 20%. Material research is focused on improving the figure of merit and to a lesser, though an increasing, extent, the electrical power factor.

The emergence of thermoelectrics as a technology for application in waste heat recovery has resulted in a successful search for materials with high electrical power factors and cheap materials. The rare-earth ytterbium-aluminum compound $YbAl_3$ has a power factor almost three times that of bismuth telluride, the established material for low-temperature application, while magnesium tin (MgSn) has almost the same performance as lead telluride but is available at less than a quarter of the cost. D. M. Rowe

Thermoelectricity The direct conversion of heat into electrical energy, or the reverse, in solid or liquid conductors by means of three interrelated phenomena—the Seebeck effect, the Peltier effect, and the Thomson effect—including the influence of magnetic fields upon each. The Seebeck effect concerns

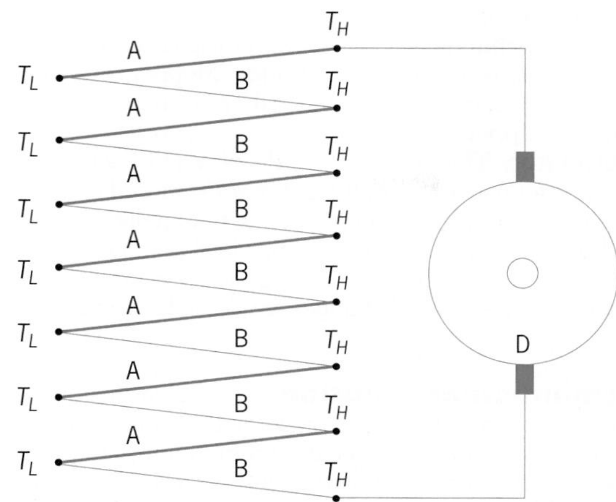

Thermopile, a battery of thermocouples connected in series. *D* is a device appropriate to the particular application; *A* and *B* are the two different conductors.

the electromotive force (emf) generated in a circuit composed of two different conductors whose junctions are maintained at different temperatures. The Peltier effect refers to the reversible heat generated at the junction between two different conductors when a current passes through the junction. The Thomson effect involves the reversible generation of heat in a single current-carrying conductor along which a temperature gradient is maintained. Specifically excluded from the definition of thermoelectricity are the phenomena of Joule heating and thermionic emission. *See* ELECTROMOTIVE FORCE (EMF); JOULE'S LAW; PELTIER EFFECT; SEEBECK EFFECT; THERMIONIC EMISSION; THOMSON EFFECT.

The three thermoelectric effects are described in terms of three coefficients: the absolute thermoelectric power (or thermopower) S, the Peltier coefficient II, and the Thomson coefficient μ, each of which is defined for a homogeneous conductor at a given temperature. These coefficients are connected by the Kelvin relations, which convert complete information about one into complete information about all three. It is therefore necessary to measure only one of the three coefficients; usally the thermopower S is chosen.

The most important practical application of thermoelectric phenomena is in the accurate measurement of temperature. The phenomenon involved is the Seebeck effect. Of less importance are the direct generation of electrical power by application of heat (also involving the Seebeck effect) and thermoelectric cooling and heating (involving the Peltier effect).

A basic system suitable for all four applications is shown schematically in the illustration. Several thermocouples are connected in series to form a thermopile, a device with increased output (for power generation or cooling and heating) or sensitivity (for temperature measurement) relative to a single thermocouple. The junctions forming one end of the thermopile are all at the same low temperature T_L, and the junctions forming the other end are at the high temperature T_H. The thermopile is connected to a device D which is different for each application. For temperature measurement, the temperature T_L is fixed, for example, by means of a bath; the temperature T_H becomes the running temperature T, which is to be measured; and the device is a potentiometer for measuring the thermoelectric emf generated by the thermopile. For power generation, the temperature T_L is fixed by connection to a heat sink; the temperature T_H is fixed at a value determined by the output of the heat source and the thermal conductance of the thermopile; and the device is whatever is to be run by the electricity that is generated. For heating or cooling, the device is a current generator that passes current through the thermopile. If the current flows in the proper direction, the junctions at T_H will heat up, and those at T_L will cool down. If T_H is fixed by connection to a heat sink, thermoelectric cooling will be provided by T_L. Alternatively, if T_L is fixed, thermoelectric heating will be provided at T_H. Such a system has the advantage that at any given location it can be converted from a cooler to a heater merely by reversing the direction of the current.

Thermoelectric power generators, heaters, or coolers made from even the best presently available materials have the disadvantages of relatively low efficiencies and concomitant high cost per unit of output. Their use has therefore been largely restricted to situations in which these disadvantages are outweighed by such advantages as small size, low maintenance due to lack of moving parts, quiet and vibration-free performance, light weight, and long life. *See* THERMOELECTRIC POWER GENERATOR. Jack Bass

Thermoluminescence The emission of light when certain solids are warmed, generally to a temperature lower than that needed to provoke visible incandescence. Two characteristics of thermoluminescence distinguish it from incandescence. First, the intensity of thermoluminescent emission does not remain constant at constant temperature, but decreases with time and eventually ceases altogether. Second, the spectrum of the thermoluminescence is highly dependent on the composition of the material and is only slightly affected by the temperature of heating. If a thermoluminescent material emits both thermoluminescence and incandescent light at some temperature of observation, the transient light emission is the thermoluminescence and the remaining steady-state emission is the incandescence. The transient nature of the thermoluminescent emission suggests that heating merely triggers the release of stored energy previously imparted to the material. Supporting this interpretation is the fact that after the thermoluminescence has been reduced to zero by heating, the sample can be made thermoluminescent again by exposure to one of a number of energy sources: x-rays and gamma rays, electron beams, nuclear particles, ultraviolet light, and, in some cases, even short-wave visible light (violet and blue). A thermoluminescent material, therefore, has a memory of its earlier exposure to an energizing source, and this memory is utilized in a number of applications. Many natural minerals are thermoluminescent, but the most efficient materials of this type are specially formulated synthetic solids (phosphors). *See* LUMINESCENCE.

In addition to special sites capable of emitting light (luminescent centers), thermoluminescent phosphors have centers that can trap electrons or holes when these are produced in the solid by ionizing radiation. The luminescent center itself is often the hole trap, and the electron is trapped at another center, although the reverse situation can also occur. In the former case, if the temperature is low and the energy required to release an electron from a trap (the trap depth) is large, electrons will remain trapped and no luminescence will occur. If, however, the temperature of the phosphor is progressively raised, electrons will receive increasing amounts of thermal energy and will have an increased probability of escape from the traps. Freed electrons may then go over to luminescent centers and recombine with holes trapped at or near these centers. The energy liberated by the recombination can excite the luminescent centers, causing them to emit light. *See* HOLE STATES IN SOLIDS; TRAPS IN SOLIDS.

Radiation dosimeters based on thermoluminescence are widely used for monitoring integrated radiation exposure in nuclear power plants, hospitals, and other installations where high-energy radiations are likely to be encountered. The key elements of the dosimeters, thermoluminescent phosphors with deep traps, can store some of the energy absorbed from these radiations for very long periods of time at normal temperatures and release it as luminescence on demand when appropriately heated. The brightness (or light sum) of the luminescence is a measure of the original radiation dose. *See* DOSIMETER.

James H. Schulman

Thermomagnetic effects Electrical and thermal phenomena occurring when a conductor or semiconductor which is carrying a thermal current (that is, is in a temperature gradient) is placed in a magnetic field. *See* SEMICONDUCTOR.

Let the temperature gradient be transverse to the magnetic field H_z, for example, along x. Then the following transverse-transverse effects are observed:

1. Ettingshausen-Nernst effect, an electric field along y.
2. Righi-Leduc effect, a temperature gradient along y.
3. An electric potential change along x, amounting to a change of thermoelectric power.
4. A temperature gradient change along x, amounting to a change of thermal resistance.

Let the temperature gradient be along H. Then changes in thermoelectric power and in thermal conductivity are observed in the direction of H.

For related phenomena *see* HALL EFFECT; MAGNETORESISTANCE.

Elihu Abrahams; Frederic Keffer

Thermometer An instrument that measures temperature. Although this broad definition includes all temperature-measuring devices, they are not all called thermometers. Other names have been generally adopted. For a discussion of two such devices *see* PYROMETER; THERMOCOUPLE. For a general discussion of temperature measurement *see* TEMPERATURE MEASUREMENT.

Liquid-in-glass thermometer. This thermometer consists of a liquid-filled glass bulb and a connecting partially filled capillary tube. When the temperature of the thermometer increases, the differential expansion between the glass and the liquid causes the liquid to rise in the capillary. A variety of liquids, such as mercury, alcohol, toluene, and pentane, and a number of different glasses are used in thermometer construction, so that various designs cover diverse ranges between about −300°F and +1200°F (−184°C and +649°C).

Bimetallic thermometer. In this thermometer the differential expansion of thin dissimilar metals, bonded together into a narrow strip and coiled into the shape of a helix or spiral, is used to actuate a pointer. In some designs the pointer is replaced with low-voltage contacts to control, through relays, operations which depend upon temperature, such as furnace controls.

Filled-system thermometer. This type of thermometer has a bourdon tube connected by a capillary tube to a hollow bulb. When the system is designed for and filled with a gas (usually nitrogen or helium) the pressure in the system substantially follows the gas law, and a temperature indication is obtained from the bourdon tube. The temperature-pressure-motion relationship is nearly linear. Atmospheric pressure effects are minimized by filling the system to a high pressure. When the system is designed for and filled with a liquid, the volume change of the liquid actuates the bourdon tube.

Vapor-pressure thermal system. This filled-system thermometer utilizes the vapor pressure of certain stable liquids to measure temperature. The useful portion of any liquid-vapor pressure curve is between approximately 15 psia (100 kilopascals absolute) and the critical pressure, that is, the vapor pressure at the critical temperature, which is the highest temperature for a particular liquid-vapor system. A nonlinear relationship exists between the temperature and the vapor pressure, so the motion of the bourdon tube is greater at the upper end of the vapor-pressure curve. Therefore, these thermal systems are normally used near the upper end of their range, and an accuracy of 1% or better can be expected.

Resistance thermometer. In this type of thermometer the change in resistance of conductors or semiconductors with temperature change is used to measure temperature. Usually, the temperature-sensitive resistance element is incorporated in a bridge network which has a reasonably constant power supply. Although a deflection circuit is occasionally used, almost all instruments of this class use a null-balance system, in which the resistance change is balanced and measured by adjusting at least one other resistance in the bridge. Metals commonly used as the sensitive element in resistance thermometers are platinum, nickel, and copper.

Thermistor. This device is made of a solid semiconductor with a high temperature coefficient of resistance. The thermistor has a high resistance, in comparison with metallic resistors, and is used as one element in a resistance bridge. Since thermistors are more sensitive to temperature changes than metallic resistors, accurate readings of small changes are possible. *See* THERMISTOR.

Howard S. Bean

Thermonuclear reaction A nuclear fusion reaction which occurs between various nuclei of the light elements when they are constituents of a gas at very high temperatures. Thermonuclear reactions, the source of energy generation in the Sun and the stable stars, are utilized in the fusion bomb. *See* HYDROGEN BOMB; NUCLEAR FUSION; STELLAR EVOLUTION; SUN.

Thermonuclear reactions occur most readily between isotopes of hydrogen (deuterium and tritium) and less readily among a few other nuclei of higher atomic number. At the temperatures and densities required to produce an appreciable rate of thermonuclear reactions, all matter is completely ionized; that is, it exists only in the plasma state. Thermonuclearer fusion reactions may then occur within such an ionized gas when the agitation energy of the stripped nuclei is sufficient to overcome their mutual electrostatic repulsions, allowing the colliding nuclei to approach each other closely enough to react. For this reason, reactions tend to occur much more readily between energy-rich nuclei of low atomic number (small charge) and particularly between those nuclei of the hot gas which have the greatest relative kinetic energy. This latter fact leads to the result that, at the lower fringe of temperatures where thermonuclear reactions may take place, the rate of reactions varies exceedingly rapidly with temperature. *See* CARBON-NITROGEN-OXYGEN CYCLES; KINETIC THEORY OF MATTER; MAGNETOHYDRODYNAMICS; NUCLEAR REACTION; PINCH EFFECT; PROTON-PROTON CHAIN.

Richard F. Post

Thermoregulation The processes by which many animals actively maintain the temperature of part or all of their body within a specified range in order to stabilize or optimize temperature-sensitive physiological processes. Body temperatures of normally active animals may range from 32 to 115°F (0 to 46°C) or more, but the tolerable range for any one species is much narrower.

Animals are commonly classified as warm-blooded or cold-blooded. When the temperature of the environment varies, the body temperature of a warm-blooded or homeothermic animal remains high and constant, while that of a cold-blooded or poikilothermic animal is low and variable. However, supposedly cold-blooded reptiles and insects, when active, may regulate body temperatures within 2–4°F (1–2°C) of a species-specific value. Supposedly warm-blooded mammals and birds may allow their temperature to drop to 37–68°F (5–20°C) during hibernation or torpor. Further, optimal temperature varies with organ, time of day, and circumstance. Thus, this classification is often misleading.

A better classification is based on the principal source of heat used for thermoregulation. Endotherms (birds, mammals) use heat generated from food energy. Ectotherms (invertebrates, fish, amphibians, reptiles) use heat from environmental sources. This classification also has limitations, however. For example, endotherms routinely use external heat sources to minimize the food cost of thermoregulation, and some ectotherms use food energy for thermoregulation. *See* PHYSIOLOGICAL ECOLOGY (ANIMAL).

Behavior is the most obviously active form of thermoregulation. Most animals are mobile, sensitive to their environment, and capable of complex behaviors. The simplest thermoregulatory behavior consists of moving to a favorable location. Operative temperature may also be altered by changing posture in one place. Lizards face the sun to minimize the area exposed to solar heating, or orient broadside to maximize it, and some ground squirrels use their tail as a sunshade. Some reptiles and amphibians also expand or contract pigmented cells in their skin to increase or decrease solar heating. *See* CHROMATOPHORE.

Evaporation is an effective means of cooling the body. Evaporation from the respiratory mucous membranes is the most common mechanism. Evaporation from the mucous membranes cools the nose during inhalation. During exhalation, water vapor condenses on the cool nasal membranes and is recovered. Evaporation can be greatly increased by exhaling from the mouth to prevent condensation. Additional increases in evaporation result from panting, that is, rapid breathing at the resonant frequency of the respiratory system. Evaporation from the eyes and the mucous membranes of the mouth and tongue is another source of cooling. Water is also evaporated from the skin of all animals, and can be varied for thermoregulation. Some desert frogs control evaporation by spreading an oily material on the skin. Reptiles, birds, and mammals have relatively impermeable skins, but evaporation can be increased by various means. The sweat glands in the skin of mammals are particularly effective and are one of the few purely thermoregulatory organs known. *See* SKIN.

Changes in circulation can be used to regulate heat flow. Countercurrent heat exchange is used to regulate heat flow to particular parts of the body while maintaining oxygen supply. Large vessels may be divided into intermingled masses of small vessels to maximize heat exchange, forming an organ called a rete. However, retes can be bypassed by alternative circulation paths to regulate heat flow. Many animals living in warm environments have a rete that regulates brain temperature by cooling the arterial blood supply to the brain with blood draining from the nasal membranes, eyes, ears, or horns. *See* CARDIOVASCULAR SYSTEM; COUNTERCURRENT EXCHANGE (BIOLOGY).

Heat exchange with the environment is limited by the fur of mammals, feathers of birds, and furlike scales or setae of insects. Erection or compression of this insulation varies heat flow. Insulation thickness varies over the body to exploit variations in local operative temperature. Thermal windows are thinly insulated areas that are either shaded (abdomen of mammals, axilla of birds and mammals) or of small size (ears, face, legs) so that solar heating is minimized.

The oxidation of foodstuffs within the metabolic pathways of the body releases as much heat as if it were burned. Basal metabolism is the energy use rate of a fasting animal at rest. Activity, digestion, and thermoregulation increase metabolism above the basal rate. Endothermy is the utilization of metabolic heat for thermoregulation. Birds and mammals are typical examples, but significant endothermy also occurs in large salt-water fish, large reptiles, and large flying insects. Endotherms regulate only the temperature of the body core, that is, the brain, heart, and lungs. The heat production of these metabolically active organs is often supplemented with heat produced in muscles. Heat produced as a by-product of activity may substitute partially for thermoregulatory heat production, and imposes no thermoregulatory energy cost. In contrast, shivering produces heat only for thermoregulation and results in an extra cost. Some animals have specialized heater organs for nonshivering thermogenesis, which is more efficient than shivering. Brown adipose tissue is a fatty tissue with a high density of mitochondria. It is found in the thorax of mammals, especially newborns and hibernators, and it warms the body core efficiently. *See* ADIPOSE TISSUE; METABOLISM.

The variety of mechanisms used in thermoregulation indicates a corresponding complexity in neural control. Temperature sensors distributed over the skin respond nearly immediately to changes in the environment and provide the major input. Nearly all parts of the central nervous system also respond to local thermal stimulation. These peripheral and central thermal inputs are integrated at a series of centers beginning in the spinal cord. This series clearly extends to the cerebral cortex, as a learning period is required before behavioral thermoregulation reaches maximum precision. Various components respond to the rate of temperature change as well as the difference between preferred and actual temperature. The neuroendocrine system then regulates metabolic heat production, the sympathetic nervous system controls blood flow, and the cerebral cortex controls behavioral thermoregulation. *See* ENDOCRINE SYSTEM (VERTEBRATE); HOMEOSTASIS; NERVOUS SYSTEM (VERTEBRATE). George Bakken

Thermosbaenacea A small order of the superorder Peracarida in the superclass Crustacea. In thermosbaenaceans, the carapace, which may cover part of the cephalic region and one to several thoracic somites, is fused only to the first thoracic somite. The carapace of females is expanded to provide a dorsally positioned brood pouch where embryos hatch as subadults. Eyes are reduced or absent. The abdomen consists of six somites and a telson; however, in *Thermosbaena*, at least, the telson and sixth somite are fused to form a pleotelson. The first pair of thoracic appendages are modified as maxillipeds, and may be sexually dimorphic; the remaining five to seven pairs provide the animals with locomotion. Sexes are separate. Thermosbaenaceans have been found principally in thermal, but occasionally in cool, fresh- or brackish-water, lakes, springs, and interstitial coastal areas, and also in cave pools, in a geographic band stretching from the Mediterranean to the Caribbean and Gulf of Mexico. *See* PERACARIDA. Patsy A. McLaughlin

Thermosphere A rarefied portion of the atmosphere, lying in a spherical shell between 50 and 300 mi (80 and 500 km) above the Earth's surface, where the temperature increases dramatically with altitude. The thermosphere responds to the variable outputs of the Sun, the ultraviolet radiation at wavelengths less than 200 nanometers, and the solar wind plasma that flows outward from the Sun and interacts with the Earth's geomagnetic field. This interaction energizes the plasma, accelerates charged particles into the thermosphere, and produces the aurora borealis and aurora australis, which are nearly circular-shaped regions of luminosity that surround the magnetic north and south poles respectively. Embedded within the thermosphere is the ionosphere, a weakly ionized plasma. *See* IONOSPHERE; MAGNETOSPHERE; PLASMA PHYSICS; SOLAR WIND.

In the thermosphere, these molecular species are subjected to intense solar ultraviolet radiation and photodissociation that gradually turns the molecular species into the atomic species oxygen, nitrogen, and hydrogen. Up to above 60 mi (100 km), atmospheric turbulence keeps the atmosphere well mixed, with the molecular concentrations dominating in the lower atmosphere. Above 60 mi, solar ultraviolet radiation most strongly dissociates molecular oxygen, and there is less mixing from atmospheric turbulence. The result is a transition area where molecular diffusion dominates and atmospheric species settle according to their molecular and atomic weights. Above 60 mi, atomic oxygen is the dominant species. *See* ATMOSPHERE.

About 60% of the solar ultraviolet energy absorbed in the thermosphere and ionosphere heats the ambient neutral gas and ionospheric plasma; 20% is radiated out of the thermosphere as airglow from excited atoms and molecules; and 20% is stored as chemical energy of the dissociated oxygen and nitrogen molecules, which is released later when recombination of the atomic species occurs. Most of the neutral gas heating that establishes the basic temperature structure of the thermosphere is derived from excess energy released by the products of ion-neutral and neutral chemical reactions occurring in the thermosphere and ionosphere. *See* AIRGLOW; ULTRAVIOLET RADIATION.

The average vertical temperature profile is determined by a balance of local solar heating by the downward conduction of molecular thermal product to the region of minimum temperature near 50 mi (80 km). For heat to be conducted downward within the thermosphere, the temperature of the thermosphere must increase with altitude. The global mean temperature increases from about 200 K (−100°F) near 50 mi to 700–1400 K (800–2100°F) above 180 mi (300 km), depending upon the intensity of solar ultraviolet radiation reaching the Earth. Above 180 mi, molecular thermal conduction occurs so fast that vertical temperature differences are largely eliminated; the isothermal temperature in the upper thermosphere is called the exosphere temperature.

As the Earth rotates, absorption of solar energy in the thermosphere undergoes a daily variation. Dayside heating causes the atmosphere to expand, and the loss of heat at night causes it to contract. This heating pattern creates pressure differences that drive a global circulation, transporting heat from the warm dayside to the cool nightside. R. G. Roble

Thermostat An instrument which directly or indirectly controls one or more sources of heating and cooling to maintain a desired temperature. To perform this function a thermostat must have a sensing element and a transducer. The sensing element measures changes in the temperature and produces a desired effect on the transducer. The transducer converts the effect produced by the sensing element into a suitable control of the device or devices which affect the temperature.

The most commonly used principles for sensing changes in temperature are (1) unequal rate of expansion of two dissimilar metals bonded together (bimetals), (2) unequal expansion of two dissimilar metals (rod and tube), (3) liquid expansion (sealed diaphragm and remote bulb or sealed bellows with or without a remote bulb), (4) saturation pressure of a liquid-vapor system (bellows), and (5) temperature-sensitive resistance element.

The most commonly used transducers are (1) switches that make or break an electric circuit, (2) potentiometer with a wiper that is moved by the sensing element, (3) electronic amplifier, and (4) pneumatic actuator.

The most common thermostat application is for room temperature control. The illustration shows a typical on-off heating-cooling room thermostat. In a typical application the thermostat controls a gas valve, oil burner control, electric heat control, cooling compressor control, or damper actuator.

Thermostats are also used extensively in safety and limit application. Thermostats are generally of the following types: insertion types that are mounted on ducts with the sensing element extending into a duct; immersion types that control a liquid in a pipe or tank with the sensing element extending into the liquid; and surface types in which the sensing element is mounted on a pipe or similar surface. *See* COMFORT HEATING; FURNACE; OIL BURNER. Nathaniel Robbins, Jr.

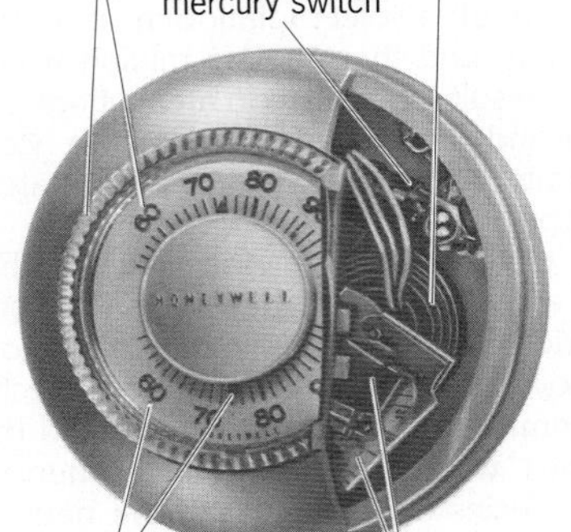

setting knob and scale
bimetal sensing element
double throw mercury switch
thermometer and scale
adjustable heat anticipator and scale

Typical heat-cool thermostat. (*Honeywell Inc.*)

Thermotherapy The treatment of disease by the local or general application of heat. The following discussion is limited to the local application of heat as an adjunct to therapeutic management of musculoskeletal and joint diseases. The most commonly used methods for this form of treatment include hot packs, hydrotherapy, radiant heat, shortwave diathermy, microwave diathermy, ultrasound, and laser therapy.

The reason so many different methods are employed is that each modality heats selectively different anatomical structures, and thus the modality selected for a given treatment is based on the temperature distribution produced in the tissues. For vigorous heat application to a given site, the location of the peak temperature produced by the modality must coincide with the site so that maximally tolerated tissue temperatures can be obtained there without burning elsewhere. Customarily, the modalities are divided into those that heat superficial tissues and those that heat deep-seated tissues. Hot packs, hydrotherapy, and radiant heat are used to heat superficial tissues. Photons of ultraviolet radiation, x-rays, and radium penetrate deeper into the tissues and produce photochemical reactions long before the temperature increases significantly. Other forms of energy used for heating deep-seated tissues include shortwave, microwave, and ultrasound. *See* ELECTROMAGNETIC RADIATION; MICROWAVE; ULTRASONICS.

The therapeutic effects produced by heating selectively by ultrasound include an increase in the extensibility of collagen tissues. Disease or injury, such as arthritis, burns, scarring, or long-term immobilization in a cast, may cause shortening of collagen tissue producing severe limitation of the range of motion at a joint. This application of heat (mostly ultrasound) is often used in conjunction with physical therapy. Heat applied by using shortwaves and microwaves may reduce muscle spasms secondary to musculoskeletal pathology. Selective heating of muscle with microwave radiation has been used to accelerate absorption of hematomas and to prepare for stretching of the contracted and stiffened muscle. Heat therapy in the form of hyperthermia has been used as an effective adjunct to cancer therapy in combination with ionizing radiation in the form of x-rays or radium therapy.

The most commonly used laser in physical therapy is the helium-neon laser. The depth of penetration and the heating effect are similar to infrared light. However, the major difference between laser light and diffuse light of the same wavelength is that, with laser light, any desirable intensity can be easily produced. *See* LASER PHOTOBIOLOGY. Justus F. Lehmann

Thévenin's theorem (electric networks) This theorem, from electric circuit theory, is also known as the Helmholtz or Helmholtz-Thévenin theorem, since H. Helmholtz stated it in an earlier form prior to M. L. Thévenin. Closely related is the Norton theorem, which will also be discussed. Laplace transform notation will be used. *See* LAPLACE TRANSFORM.

Thévenin's theorem states that at a pair of terminals a network composed of lumped, linear circuit elements may, for purposes of analysis of external circuit or terminal behavior, be replaced by a voltage source $V(s)$ in series with a single impedance $Z(s)$. The source $V(s)$ is the Laplace transform of the voltage across the pair of terminals when they are open-circuited; $Z(s)$ is the transform impedance at the two terminals with all independent sources set to zero (Fig. 1).

Norton's theorem states that a second equivalent network consists of a current source $I(s)$ in parallel with an impedance $Z(s)$.

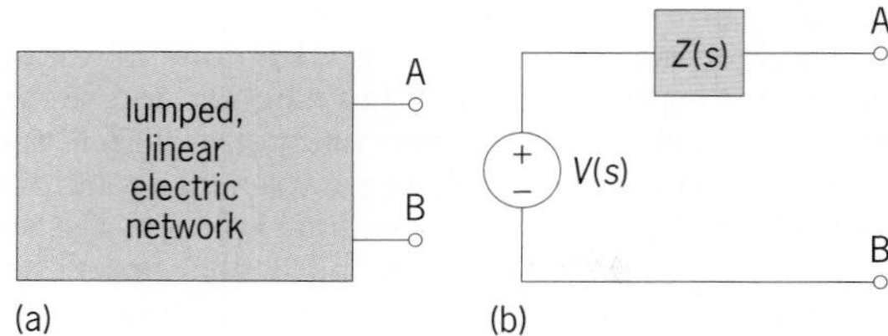

Fig. 1. Network and its Thévenin equivalent. (*a*) Original network. (*b*) Thévenin equivalent circuit.

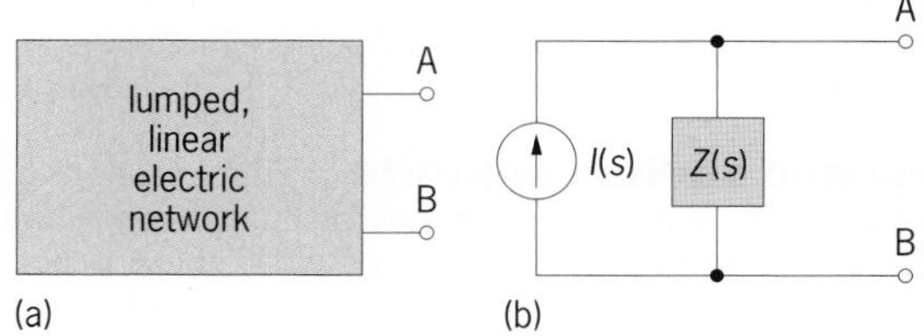

Fig. 2. Network and its Norton equivalent. (*a*) Original network. (*b*) Norton equivalent circuit.

The impedance $Z(s)$ is identical with the Thévenin impedance, and $I(s)$ is the Laplace transform of the current between the two terminals when they are short-circuited (Fig. 2).

Thévenin's and Norton's equivalent networks are related by the equation $V(s) = Z(s) \cdot I(s)$.

These theorems are useful for the study of the behavior of a load connected to a (possibly complex) system that is supplying electric power to that load. *See* SUPERPOSITION THEOREM (ELECTRIC NETWORKS). Edwin C. Jones, Jr.

Thiamine A water-soluble vitamin found in many foods; pork, liver, and whole grains are particularly rich sources. It is also known as vitamin B_1 or aneurin. The structural formula of thiamine is shown below.

```
                                     CH3
                                      |
        N = C — NH2                   C = C — CH2 — CH2OH
        |   |                        /    |
  H3C — C   C — CH2 — N                   |
        ||  ||       +   \\               |
        N — C              CH — S
```

Thiamine deficiency is known as beriberi in humans. Other species manifest the deficiency with polyneuritic conditions (which are characterized by degenerative or inflammatory lesions of several nerves simultaneously, usually symmetrical). Muscle and nerve tissues are affected by the deficiency, and poor growth is observed. People with beriberi are irritable, depressed, and weak; they often die of cardiac failure. Wernicke-Korsakoff syndrome, often observed in chronic alcoholics, is a cerebral beriberi characterized by brain lesions, liver disease, and partial paralysis, particularly of the motor nerves of the eye. As is the case in other vitamin B deficiencies, the deficiency of thiamine is commonly accompanied by insufficiencies of some of the other vitamins. *See* MALNUTRITION; VITAMIN. Stanley N. Gershoff; Donald B. McCormick

Thick-film sensor A sensor that is based on a thick-film circuit. Thick-film circuits are formed by the deposition of layers of special pastes onto an insulating substrate. The pastes are usually referred to as inks, although there is little resemblance to conventional ink. The printed pattern is fired in a manner akin to the production of pottery, to produce electrical pathways of a controlled resistance. Parts of a thick-film circuit can be made sensitive to strain or temperature. The thick-film pattern can include mounting positions for the insertion of conventional silicon devices, in which case the assembly is known as a thick-film hybrid. The process is relatively cheap, especially if large numbers of devices are produced, and the use of hybrid construction allows the sensor housing to include sophisticated signal conditioning circuits. These factors indicate that thick-film technology is likely to play an increasing role in sensor design.

The three main categories of thick-film inks are conductors, dielectrics (insulators), and resistors. Conductors are used for interconnections, such as the wiring of bridge circuits. Dielectrics are used for coating conducting surfaces (such as steel) prior to laying down thick-film patterns, for constructing thick-film capacitors, and for insulating cross-over points, where one conducting path traverses another. Resistor inks are the most interesting from the point of view of sensor design, since many thick-film materials are markedly piezoresistive. The piezoresistive properties of thick-film resistor inks can be used to form strain sensors. This approach is commonly used to manufacture pressure sensors and is exploited to produce accelerometers. *See* ACCELEROMETER; PRESSURE TRANSDUCER; STRAIN GAGE.

Piezoresistive sensors. Piezoresistive sensors are formed by placing stress-sensitive resistors on highly stressed parts of a suitable mechanical structure. The piezoresistive transducers are usually attached to cantilevers, or other beam configurations, and are connected in a Wheatstone bridge circuit. The beam may carry a seismic mass to form an accelerometer or may deform in response to an externally applied force. The stress variations in the transducer are converted into an electrical output, which is proportional to strain, by the piezoresistive effect. *See* WHEATSTONE BRIDGE.

Temperature sensors. The linear temperature coefficient of resistance possessed by certain platinum-containing conductive inks has allowed thermistors to be printed onto suitable substrates using thick-film fabrication techniques. Thick-film thermistors are very inexpensive and physically small, and have the further advantage of being more intimately bonded to the substrate than a discrete component. It has been shown that thick-film thermistors can have as good, if not better, performance than a comparable discrete component. *See* THERMISTOR.

Chemical sensors. Thick-film materials have been used for a number of chemical sensing applications, including the measurement of gas and liquid composition, acidity (pH), and humidity. A classification based on two categories seems to cover most devices: impedance-based transducers, in which the measurand causes a variation of resistance, capacitance, and so forth; and electrochemical systems, in which the sensed quantity causes a change in electrochemical potential or current. *See* ELECTRICAL IMPEDANCE; ELECTROCHEMISTRY. John D. Turner

Thickening The production of a concentrated slurry from a dilute suspension of solid particles in a liquid. In practice, a thickener also usually generates a clear liquid; therefore clarification is generally a concurrent process. Thickening and clarification are outcomes of sedimentation, and both are representative of a group of industrial processes termed mechanical separations. Although thickening may be carried out either batchwise or continuously, the latter method is more common. *See* CLARIFICATION; SEDIMENTATION (INDUSTRY).

Thickeners are especially useful when large volumes of dilute slurries must be treated, as in manufacture of cement, production of magnesium from seawater, treatment of sewage, purification of water, treatment of coal, and dressing of metallurgical ores. Vincent W. Uhl

Thinner A material used in paints and varnishes to adjust the consistency for application. Thinners are usually solvents for the vehicle used in the coating and are expected to evaporate after application. Water is used as a thinner in emulsion paints and in certain water-soluble paints such as watercolors and calcimines. Petroleum fractions are most commonly used for oil and resin coatings. Stronger solvents contain substantial amounts of aromatic hydrocarbons and may be derived from petroleum or coal tar.

Since numerous coating resins are not sufficiently soluble in hydrocarbons, other materials or mixtures must be used. These include alcohols such as denatured ethyl or isopropyl alcohols for shellac, esters such as amyl acetate for nitrocellulose, and ketones and other compounds for acrylic and vinyl resins. Chlorinated hydrocarbons are used for some materials which are otherwise hard to dissolve, but toxicity limits their usefulness. *See* PAINT AND COATINGS; SOLVENT; SURFACE COATING; TURPENTINE; VARNISH. C. R. Martinson; C. W. Sisler

Thiocyanate A compound containing the —SCN group, typically a salt or ester of thiocyanic acid (HSCN). Thiocyanates are bonded through the sulfur(s) and have the structure R—S—C≡N. They are isomeric with the isothiocyanates, R—N=C≡S, which are the sulfur analogs of isocyanates (—NCO). The thiocyanates may be viewed as structural analogs of the cyanates (—OCN), where the oxygen (O) atom is replaced by a sulfur atom.

The principal commercial derivatives of thiocyanic acid are ammonium and sodium thiocyanates. Thiocyanates and isothiocyanates have been used as insecticides and herbicides. Specifically, ammonium thiocyanate is used as an intermediate in the manufacture of herbicides and as a stabilizing agent in photography. Sodium and potassium thiocyanates are used in the manufacture of textiles and the preparation of organic thiocyanates.

In living systems, thiocyanates are the product of the detoxification of cyanide ion (CN—) by the action of 3-mercaptopyruvate sulfur transferase. In addition, thiocyanates can interfere with thyroxine synthesis in the thyroid gland and are part of a class known as goitrogenic compounds. *See* CYANIDE; SULFUR; THYROXINE. Thomas J. Meade

Thiophene An organic heterocyclic compound containing a diunsaturated ring of four carbon atoms and one sulfur atom. *See* HETEROCYCLIC COMPOUNDS.

Thiophene (**1**), methylthiophenes, and other alkylthiophenes are found in relatively small amounts in coal tar and petroleum. Thiophene accompanies benzene in the fractional distillation of coal tar. 2,5-Dithienylthiophene (**2**) has been found in the marigold plant. Biotin, a water-soluble vitamin, is a tetrahydrothiophene derivative.

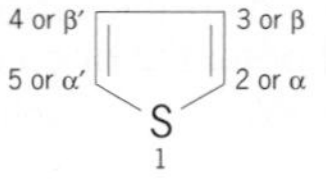

4 or β′ | 3 or β | 5 or α′ | 2 or α | S | 1

(1)

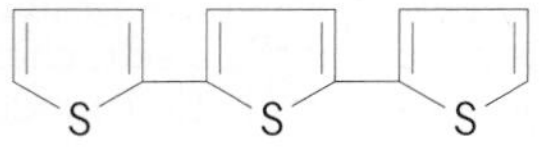

S S S

(2)

The parent compound (**1**) is nearly insoluble in water, with mp −38.2°C (−36.8°F), bp 84.2°C (183.6°F), and specific gravity (20/4) 1.0644. Thiophene is considered to be an aromatic compound. Thiophenes are stable to alkali and other nucleophilic agents, and are relatively resistant to disruption by acid. *See* AROMATIC HYDROCARBON; ORGANOSULFUR COMPOUND. Walter J. Gensler; Martin Stiles

Thiosulfate A salt containing the negative ion $S_2O_3^{2-}$. This species is an important reducing agent and may be viewed as a structural analog of the sulfate ion (SO_4^{2-}) where one of the oxygen (O) atoms has been replaced by a sulfur (S) atom. The sulfur atoms of the thiosulfate ion are not equivalent. Thiosulfate is tetrahedral, and the central sulfur is in the formal oxidation state 6+ and the terminal sulfur is in the formal oxidation state 2−.

Principal uses of thiosulfates include agricultural, photographic, and analytical applications. Ammonium thiosulfate [$(NH_4)_2S_2O_3$] is exploited for both the nitrogen and sulfur content, and it is combined with other nitrogen fertilizers such as urea. Thiosulfate ion is an excellent complexing agent for silver ions (bound through sulfur). The sodium salt and the ammonium salt are well known as the fixing agent "hypo" used in photography. The aqueous thiosulfate ion functions as a scavenger for unreacted solid silver bromide on exposed film and therefore prevents further reaction with light. In nature, thiosulfate is converted into hydrogen sulfide (H_2S) via enzymatic reduction. Hydrogen sulfide, in turn, is converted into the thiol group of cysteine by the reaction with *O*-acetylserine. *See* COORDINATION COMPLEXES; OXIDATION-REDUCTION; PHOTOGRAPHIC MATERIALS; SULFUR. Thomas J. Meade

Thirst and sodium appetite The sensations caused by dehydration, the continuing loss of fluid through the skin and lungs and in the urine and feces while there is no water intake into the body. Thirst becomes more and more insistent as dehydration worsens. Water and electrolytes are needed to replace losses, and an adequate intake of sodium as well as water is important for maintaining blood volume. Normally, the amounts of water drunk and taken in food are more than enough to maintain hydration of the body, and the usual mixed diet provides all the electrolytes required.

Deficit-induced drinking occurs when a deficit of fluid in one or both of the major fluid compartments of the body serves as a signal to increase drinking. Cellular dehydration, detected by osmoreceptors, causes thirst and vasopressin release. Hypovolemia (low blood volume), detected by volume receptors in the heart and large veins and the arterial baroreceptors, causes immediate thirst, a delayed increase in sodium appetite, activation of the renin-angiotensin system, and increased mineralocorticoid and vasopressin secretion. Increases or decreases in amounts drunk in disease may result from normal or abnormal functioning of mechanisms of thirst or sodium appetite.

Cellular dehydration. Observations using a variety of osmotic challenges have established that hyperosmotic solutions of solutes that are excluded from cells cause more drinking than equiosmolar amounts of solutes that penetrate cells. Thus, the osmotic shift of water out of the cells caused by the excluded solutes provides the critical stimulus to drinking. Continuing water loss in the absence of intake is perhaps a more significant cause of cellular dehydration than administration of an osmotic load, but the same mechanisms apply. *See* OSMOSIS.

Sharing in the overall cellular dehydration are osmoreceptors which initiate the responses of thirst and renal conservation of water. Osmoreceptors are mainly located in the hypothalamus. The nervous tissue in the hypothalamus surrounding the anterior third cerebral ventricle and, in particular, the vascular organ of the lamina terminalis also respond to osmotic stimuli. Osmoreceptors initiating thirst work in conjunction with osmoreceptors initiating antidiuretic hormone (ADH) release to restore the cellular water to its prehydration level. In addition to reducing urine loss, ADH may lower the threshold to the onset of drinking in response to cellular dehydration and other thirst stimuli. The cellular dehydration system is very sensitive, responding to changes in effective osmolality of 1–2%.

Hypovolemia. The cells of the body are bathed by sodium-rich extracellular fluid that corresponds to the aquatic environment of the unicellular organism. Loss of sodium from the extracellular fluid is inevitably accompanied by loss of water, resulting in hypovolemia with thirst followed by a delayed increase in sodium appetite. If not corrected, continuing severe sodium loss eventually leads to circulatory collapse.

Stretch receptors in the walls of blood vessels entering and leaving the heart and in the heart itself are thought to initiate hypovolemic drinking. Volume receptors in the venoatrial junctions and receptors that register atrial and ventricular pressure respond to the underfilling of the circulation with a reduction in inhibitory nerve impulses to the thirst centers, which results in increased drinking. Angiotensin II and other hormones (such as aldosterone and ADH) are also involved in this response. Arterial baroreceptors function in much the same way as the volume receptors on the low-pressure side of the circulation, exerting continuous inhibitory tone on thirst neurons. A fall in blood pressure causes increased drinking, whereas an acute rise in blood pressure inhibits drinking. The anterior third cerebral ventricle region, which is implicated in angiotensin-induced drinking, plays a crucial role in hypovolemic drinking, body fluid homeostasis, and blood pressure control.

Renin-angiotensin systems. It is believed that drinking caused by hypovolemic stimuli partly depends on the kidneys. The renal thirst factor is the proteolytic enzyme renin, which is secreted into the circulation by the kidney in response to hypovolemia. Renin cleaves an inactive decapeptide, angiotensin I, from angiotensinogen, an α_2-globulin that is synthesized in the liver and released into the circulation. Angiotensin I is converted to the physiologically active octapeptide angiotensin II during the passage of blood through the lungs. Angiotensin II is an exceptionally powerful stimulus of drinking behavior in many animals when administered systemically or into the brain. Increased activation of the renin-angiotensin system may sometimes account for pathologically increased thirst in humans. Angiotensin II also produces (1) a rise in arterial blood pressure, release of norepinephrine from sympathetic nerve endings, and secretion of adrenomedullary hormones; and (2) water and sodium retention by causing release of ADH from the posterior pituitary and stimulation of renal tubular transport of sodium through direct action on the kidney and indirectly through increased aldosterone secretion from the adrenal cortex. *See* ALDOSTERONE; KIDNEY.

Neuropharmacology. Many substances released by neurons, and in some cases by neuroglial cells, affect drinking behavior when injected into the brain and may interact with the brain and modify angiotensin-induced drinking. Substances may stimulate or inhibit drinking, or both, depending on the species and the conditions of the experiment. Acetylcholine is a particularly powerful stimulus to drinking in rats, and no inhibitory effects on drinking have been described. Histamine also seems to be mainly stimulatory. However, a lengthening list of neuroactive substances, including norepinephrine, serotonin, nitric oxide, opioids, bombesin-like peptides, tachykinins, and neuropeptide Y, may either stimulate or inhibit drinking with varying degrees of effectiveness, depending on the species or the site of injection in the brain. Natriuretic peptides, prostaglandins, and gamma-amino butyric acid seem to be exclusively inhibitory. *See* ACETYLCHOLINE; NEUROBIOLOGY; SYNAPTIC TRANSMISSION.

Many hormones also affect water or sodium intake. Relaxin stimulates water intake, and ADH (or vasopressin) lowers the threshold to thirst in some species. Vasopressin injected into the third cerebral ventricle may stimulate water intake, suggesting a possible role for vasopressinergic neurons. Increased sodium appetite in pregnancy and lactation depends partly on the conjoint action of progesterone, estrogen, adrenocorticotrophic hormone (ACTH), cortisol, corticosterone, prolactin, and oxytocin. Aldosterone and other mineralocorticoids, the stress hormones of the hypothalamo-pituitary-adrenocortical axis, corticotrophin, ACTH, and the glucocorticoids also stimulate sodium intake. *See* ENDOCRINE MECHANISMS; NEUROHYPOPHYSIS HORMONE.

The effect of many of these substances on drinking behavior shows both species and anatomical diversity. The multiplicity of effects of many of these substances makes it impossible to generalize on their role in natural thirst, but none of these substances seems to be as consistent and as universal a stimulus of increased thirst and sodium appetite as angiotensin. James T. Fitzsimons

Thomson effect A phenomenon discovered in 1854 by William Thomson, later Lord Kelvin. He found that there occurs a reversible transverse heat flow into or out of a conductor of a particular metal, the direction depending upon whether a longitudinal electric current flows from colder to warmer metal or from warmer to colder. Any temperature gradient previously existing in the conductor is thus modified if a current is turned on. The Thomson effect does not occur in a current-carrying conductor which is initially at uniform temperature. *See* THERMOELECTRICITY. John W. Stewart

Thoracica The major order of the crustacean subclass Cirripedia. The adult animals are permanently attached. The mantle is usually reinforced by calcareous plates. Six pairs of biramous cirri are present, and the abdomen is absent or represented by caudal appendages. Antennules are present in the adult, and cement glands are strongly developed. Most species are hermaphroditic. Thoracica are subdivided into three suborders: Lepadomorpha, stalked or goose barnacles; Balanomorpha, the common acorn barnacles; and Verrucomorpha, a rare group of asymmetric barnacles. *See* CIRRIPEDIA. H. G. Stubbings

Thorite A mineral, thorium silicate. The idealized chemical formula of thorite is $ThSiO_4$. All natural material departs widely from this composition owing to the partial substitution of uranium, rare earths, calcium, and iron for thorium. The specific gravity ranges between about 4.3 and 5.4. The hardness on Mohs scale is about $4^1/_2$. The color commonly is brownish yellow to brownish black and black.

Vein deposits containing thorite occur in Colorado, Idaho, and Montana. A vein deposit of monazite containing thorium is mined at Steenkampskraal near Van Rhynsdorp, Cape Province, South Africa. *See* METAMICT STATE; RADIOACTIVE MINERALS; SILICATE MINERALS; THORIUM. Clifford Frondel

Thorium A chemical element, Th, atomic number 90. Thorium is a member of the actinide series of elements. It is radioactive with a half-life of about 1.4×10^{10} years. *See* PERIODIC TABLE.

Thorium has an atomic weight of 132.04. The temperature at which pure thorium melts is not known with certainty; it is thought to be about 1750°C (3182°F). Good-quality thorium metal is relatively soft and ductile. It can be shaped readily by any of the ordinary metal-forming operations. The massive metal is silvery in color, but it tarnishes on long exposure to the atmosphere; finely divided thorium has a tendency to be pyrophoric in air.

All of the nonmetallic elements, except the rare gases, form binary compounds with thorium. With minor exceptions, thorium exhibits a valence of 4+ in all of its salts. Chemically, it has some resemblance to zirconium and hafnium. The most common soluble compound of thorium is the nitrate which, as generally prepared, appears to have the formula $Th(NO_3)_4 \cdot 4H_2O$. The common oxide of thorium is ThO_2, thoria. Thorium combines with halogens to form a variety of salts. Thorium sulfate can be obtained in the anhydrous form or as a number of hydrates. Thorium carbonates, phosphates, iodates, chlorates, chromates, molybdates, and other inorganic salts of thorium are well known. Thorium also forms salts with many organic acids, of which the water-insoluble oxalate, $Th(C_2O_4)_2 \cdot 6H_2O$, is important in preparing pure compounds of thorium.

Monazite, the most common and commercially most important thorium-bearing mineral, is widely distributed in nature. Monazite is chiefly obtained as a sand, which is separated from other sands by physical or mechanical means. *See* MONAZITE.

Thorium oxide compounds are used in the production of incandescent gas mantles. Thorium oxide has also been incorporated in tungsten metal, which is used for electric light filaments.

It is employed in catalysts for the promotion of certain organic chemical reactions and has special uses as a high-temperature ceramic material. The metal or its oxide is employed in some electronic tubes, photocells, and special welding electrodes. Thorium has important applications as an alloying agent in some structural metals. Perhaps the major use for thorium metal, outside the nuclear field, is in magnesium technology. Thorium can be converted in a nuclear reactor to uranium-233, an atomic fuel. The energy available from the world's supply of thorium has been estimated as greater than the energy available from all of the world's uranium, coal, and oil combined. *See* ACTINIDE ELEMENTS; RADIOACTIVITY. Harley A. Wilhelm

Throat The region that includes the pharynx, the larynx, and related structures. Both the nasal passages and the oral cavity open into the pharynx, which also contains the openings of the Eustachian tubes from the ears. The lower portion of the pharynx leads into the esophagus and the trachea or windpipe. The rather funnel-shaped pharynx is suspended from the base of the skull and the jaws; it is surrounded by three constrictor muscles that function primarily in swallowing. *See* EAR; PHARYNX.

The larynx, or voice box, is marked externally by the shield-shaped thyroid cartilage which forms the Adam's apple. The larynx contains the vocal cords that act as sphincters for air regulation and permit phonation. The lower end of the larynx is continuous with the trachea, a tube composed of cartilaginous rings and supporting tissues. *See* LARYNX.

The term throat is also used in a general sense to denote the front (ventral side) of the neck. Thomas S. Parsons

Thrombosis The process of forming a thrombus, which is a solid mass or plug in the living heart or vessels composed of the constituents of the blood. Thrombosis usually occurs in a diseased blood vessel, as a result of arteriosclerosis. The consequences of thrombosis include local obstruction causing both tissue death and hemorrhage. Thrombosis is a significant factor in the death of an individual affected by arteriosclerotic cardiovascular disease, malignancy, and infection. *See* HEMORRHAGE; INFARCTION.

Thrombosis is usually initiated by vascular damage and consequent platelet adhesion and clumping. The process is initiated when platelets specifically adhere to the subendothelial collagen at the points of damage to the endothelium. At the same time that the platelets begin to aggregate and release products that will further promote thrombus formation, the protein factors of the blood, which help to form the insoluble meshwork of the thrombus, become activated. This latter process is known as blood coagulation. The proteins of the coagulation system, through a series of cascading reactions, eventually reach a final common pathway to form fibrin, the insoluble protein that forms the scaffolding of the thrombus. As blood flows by the thrombus, more platelets and fibrin are deposited. Red blood cells and white blood cells become entrapped in the thrombus and are integrated into its structure. *See* FIBRINOGEN.

Once a thrombus forms, it can have one of four fates. (1) It may be digested, destroyed, and removed by proteolytic enzymes of the plasminogen-plasmin system. (2) It may continue to propagate itself and eventually occlude the vessel. (3) It may give rise to an embolus. Emboli may cause tissue damage at sites distant from the origin of the thrombus. (4) It may undergo a process known as organization. Organization helps stabilize the thrombus, and it may result in incorporation of a contracted fibrous mass into the vessel wall. *See* EMBOLISM.

Maintaining good blood flow (especially in the veins) helps prevent thrombosis. Treating hypertension and hypercholesterolemia retards atherosclerosis, which is a major cause of arterial thrombosis. Agents that interfere with platelet function, such as aspirin and fish oils, may help avoid thrombotic episodes. Anticoagulants prevent the formation of fibrin and may also be used to prevent thrombosis. If treatment can be given in the early stages of thrombosis, fibrinolytic therapy, utilizing agents that will help form plasmin, can minimize the tissue damage caused by thrombosis. *See* ARTERIOSCLEROSIS; CIRCULATION DISORDERS; PHLEBITIS. Irwin Nash; Romeo A. Vidone

Thrust The force that propels an aerospace vehicle or marine craft. Thrust is a vector quantity. Its magnitude is usually given in newtons (N) in International System (SI) units or pounds-force (lbf) in U.S. Customary Units. A newton is defined as 1 kilogram mass times an acceleration of 1 meter per second squared. One newton equals approximately 0.2248 lbf. *See* FORCE; UNITS OF MEASUREMENT.

The thrust power of a vehicle is the thrust times the velocity of the vehicle. It is expressed in joules (J) per second or watts (W) in SI units. In U.S. Customary Units thrust power is expressed in foot-pounds per second, which can be converted to horsepower by dividing by 550. *See* POWER; PROPULSION; RAMJET; RECIPROCATING AIRCRAFT ENGINE; ROCKET; TURBOJET. J. Preston Layton

Thulium A chemical element, Tm, atomic number 69, atomic weight 168.934. It is a rare metallic element belonging to the rare-earth group. The stable isotope ^{169}Tm makes up 100% of the naturally occurring element. *See* PERIODIC TABLE.

The salts of thulium possess a pale green color and the solutions have a slight greenish tint. The metal has a high vapor pressure at the melting point. When ^{169}Tm is irradiated in a nuclear reactor, $_{170}$Tm is formed. The isotope then emits strongly an 84-keV x-ray, and this material is useful in making small portable x-ray units for medical use. *See* RARE-EARTH ELEMENTS.

Frank H. Spedding

Thunder The acoustic radiation produced by thermal lightning channel processes. The lightning return stroke is a high surge of electric current that has a very short duration, depositing approximately 95% of its electrical energy during the first 20 microseconds. Spectroscopic studies have shown that the lightning channel is heated to temperatures in the 20,000–30,000 K (36,000–54,000°F) range by this process. The pressure of the hot channel exceeds 10 atm ($>10^6$ pascals). The hot, high-pressure channel expands supersonically and forms a shock wave as it pushes against the surrounding air. Because of the momentum gained in expanding, the shock wave overshoots, causing the pressure in the core of the channel to go below atmospheric pressure temporarily. The outward-propagating wave separates from the core of the channel, forming an N-shaped wave that eventually decays into an acoustic wavelet. *See* SHOCK WAVE; STORM ELECTRICITY.

The sound that is eventually heard or detected, thunder, is the sum of many individual acoustic pulses, each a remnant of a shock wave, that have propagated to the point of observation from the generating channel segments. The first sounds arrive from the nearest part of the lightning channel and the last sounds from the most distant parts.

The higher the source of the sound, the farther it can be heard. Frequently, the thunder that is heard originates in the cloud and not in the visible channel. On some occasions, the observer may hear no thunder at all; this is more frequent at night when lightning can be seen over long distances and thunder can be heard only over a limited range ($\sim$10 km or 6 mi). *See* LIGHTNING; THUNDERSTORM. Arthur A. Few

Thunderstorm A convective storm accompanied by lightning and thunder and a variety of weather such as locally heavy rainshowers, hail, high winds, sudden temperature changes, and occasionally tornadoes. The characteristic cloud is the cumulonimbus or thunderhead, a towering cloud, generally with an anvil-shaped top. A host of accessory clouds, some attached and some detached from the main cloud, are often observed in conjunction with cumulonimbus. *See* LIGHTNING; THUNDER.

Thunderstorms are manifestations of convective overturning of deep layers in the atmosphere and occur in environments in which the decrease of temperature with height (lapse rate) is sufficiently large to be conditionally unstable and the air at low levels is moist. In such an atmosphere, a rising air parcel, given sufficient lift, becomes saturated and cools less rapidly than it would if it remained unsaturated because the released latent heat of condensation partly counteracts the expansional cooling. The rising parcel reaches levels where it is warmer (by perhaps as much as 18°F or 10°C over continents) and less dense than its surroundings, and buoyancy forces accelerate the parcel upward. The rising parcel is decelerated and its vertical ascent arrested at altitudes where the lapse rate is stable, and the parcel becomes denser than its environment. The forecasting of thunderstorms thus hinges on the identification of regions where the lapse rate is unstable, low-level air parcels contain adequate moisture, and surface heating or uplift of the air is expected to be sufficient to initiate convection. *See* CONVECTIVE INSTABILITY; FRONT.

Thunderstorms are most frequent in the tropics, and rare poleward of 60° latitude. Thunderstorms are most common during late afternoon because of the diurnal influence of surface heating.

Radar is used to detect thunderstorms at ranges up to 250 mi (400 km) from the observing site. Much of present-day knowledge of thunderstorm structure has been deduced from radar studies, supplemented by visual observations from the ground and satellites, and in-place measurements from aircraft, surface observing stations, and weather balloons. *See* METEOROLOGICAL INSTRUMENTATION; RADAR METEOROLOGY; SATELLITE METEOROLOGY.

Thunderstorms are considered severe when they produce winds greater than 58 mi/h (26 m/s or 50 knots), hail larger than 3/4 in. (19 mm) in diameter, or tornadoes. While thunderstorms are generally beneficial because of their needed rains (except for occasional flash floods), severe storms have the capacity of inflicting utter devastation over narrow swaths of the countryside. Severe storms are most frequently supercells which form in environments with high convective instability and moderate-to-large vertical wind shears. The supercell may be an isolated storm or part of a squall line. *See* HAIL; SQUALL; SQUALL LINE; TORNADO.

Robert Davies-Jones

Thyme Any of a large and diverse group of plants in the genus *Thymus* utilized for their essential oil and leaves in both cooking and medicine. Hundreds of different forms, or ecotypes, of thyme are found in the Mediterranean area, where thyme occurs as a wild plant. *Thymus vulgaris*, generally considered to be the true thyme, is the most widely used and cultivated species. Both "French" and "German" thyme are varieties of this species. Most types of thyme are low-growing perennials, typically having small smooth-edged leaves that are closely spaced on stems that become woody with age.

Wild European thyme, the source of much imported material, is usually harvested only once a year, while cultivated plants in the United States are harvested mechanically up to three times a year. As with most herbs, both stems and leaves are harvested and then dehydrated. Dried stems and leaves are separated mechanically. Thyme oil is extracted from fresh material. Thyme is a widely used herb, both alone and in blends such as "fine herbs." Thyme oil is used for flavoring medicines and has strong bactericidal properties. *See* SPICE AND FLAVORING.

Seth Kirby

Thymosin A polypeptide hormone synthesized and secreted by the endodermally derived reticular cells of the thymus gland.

Thymosin exerts its actions in several loci: (1) in the thymus gland, either on precursor stem cells derived from fetal liver or from bone marrow, or on immature thymocytes, and (2) in peripheral sites, on either thymic-derived lymphoid cells or on precursor stem cells. The precursor stem cells, which are immunologically incompetent whether in the thymus or in peripheral sites, have been designated as predetermined T cells or T_0 cells, and mature through stages termed T_1 and T_2, each reflecting varying degrees of immunological competence. Thymosin promotes or accelerates the maturation of T_0 cells to T_1 cells as well as to the final stage of a T_2. In addition to this maturation influence, the hormone also increases the number of total lymphoid cells by accelerating the rate of proliferation of both immature and mature lymphocytes. *See* IMMUNITY; THYMUS GLAND.

Abraham White

Thymus gland An important central lymphoid organ in the neck or upper thorax of all vertebrates from elasmobranchs to mammals. The thymus gland is most prominent during early life. In many laboratory species of mammals and in humans it reaches its greatest relative weight at the time of birth, but its absolute weight continues to increase until the onset of puberty. Thereafter, it begins to undergo an involution and progressively decreases in size throughout adult life.

The thymic stem cells generate a large population of small lymphocytes (thymocytes) through a series of mitotic divisions. Simultaneously these dividing lymphocytes show evidence of cellular differentiation within the special thymic environment. During this division and maturation phase the developing thymocytes undergo an intrathymic migration from the peripheral cortical area to the medullary core of the organ. Some thymocytes degenerate within the organ, but many enter the circulating blood and lymph systems at various stages of maturity. A small percentage of the T lymphocyte population (5–10%) within the thymus is antigenically competent and capable of recognizing antigenic determinents on foreign cells or substances. Some of the T lymphocytes have the capacity to lyse the foreign tissue cells, while others are involved in recognizing the "foreignness" of the antigens and assisting a second subpopulation of bone-marrow-derived lymphocytes (B lymphocytes) to respond to the antigen by producing a specific antibody. These two types of immunocompetent T lymphocytes are called killer cells and helper cells, respectively. They are involved in both tissue transplantation and humoral antibody responses. On the other hand, the vast majority of the thymic lymphocytes are immunologically incompetent (90–95%). Some thymocytes are thought to give rise to the smaller pool of immunocompetent T lymphocytes, but many emigrate into the circulating blood. *See* CELLULAR IMMUNOLOGY; LYMPHATIC SYSTEM; THYMOSIN.

Charles E. Slonecker

Thyrocalcitonin A hormone, the only known secretory product of the parenchymal or C cells of the mammalian thyroid and of the ultimobranchial glands of lower forms.

In conjunction with the parathyroid hormone, thyrocalcitonin is of prime importance in regulating calcium and phosphate metabolism. Its major function is to protect the organism from the dangerous consequences of elevated blood calcium. Its sole known effect is that of inhibiting the resorption of bone. It thus produces a fall in the concentration of calcium and phosphate in the blood plasma because these two minerals are the major constituents of bone mineral and are released into the bloodstream in ionic form when bone is resorbed. *See* BONE; CALCIUM METABOLISM; PARATHYROID GLAND; PARATHYROID HORMONE.

Thyrocalcitonin also causes an increased excretion of phosphate in the urine under certain circumstances, but a question remains as to whether this is a direct effect of the hormone upon the kidney or an indirect consequence of the fall in blood calcium which occurs when the hormone inhibits bone resorption. *See* PHOSPHATE METABOLISM; THYROID GLAND. Howard Rasmussen

Thyroid gland An endocrine gland found in all vertebrates that produces, stores, and secretes the thyroid hormones. In humans, the gland is located in front of, and on either side of, the trachea (see illustration). Thyrocalcitonin, one hormone of the thyroid gland, assists in regulating serum calcium by reducing its levels. The iodine-containing hormones thyroxine and triiodothyronine regulate metabolic rate in warm-blooded animals and are essential for normal growth and development. To produce these, the thyroid gland accumulates inorganic iodides from the bloodstream and unites them with the amino acid tyrosine. This activity is regulated by thyrotropic hormone from the anterior lobe of the pituitary gland. *See* THYROID HORMONES; THYROXINE. George C. Kent

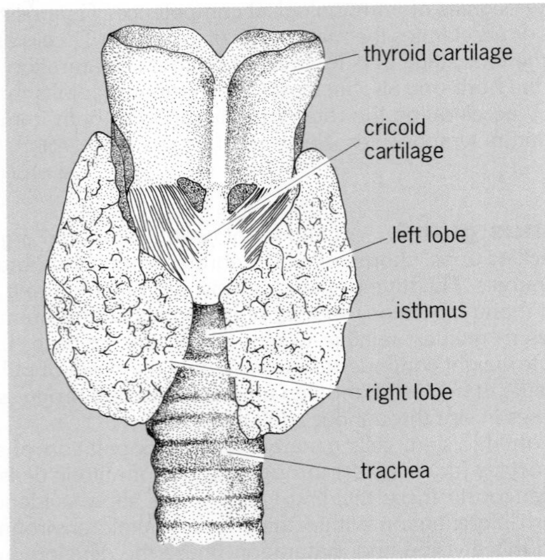

Ventral view of human thyroid gland shown in relation to trachea and larynx. (*After C. K. Weichert, Elements of Chordate Anatomy, 3d ed., McGraw-Hill, 1967*)

Thyroid gland disorders Disorders of the thyroid gland may be classified according to anatomical and functional characteristics. Those thyroid disorders that are primarily anatomical include goiter and neoplasia; those that are primarily functional result in either hyperthyroidism or hypothyroidism.

Thyroid gland enlargement, or goiter, is the most common disorder. Its classification is based upon both the anatomy and function of the gland. An enlarged but normally functioning thyroid gland is termed a nontoxic goiter. This condition affects hundreds of millions of people throughout the world in areas where the diet is deficient in iodine. In other areas it may be caused by subtle disorders in the biosynthesis of thyroid hormone. In both cases, there is compensatory enlargement of the gland that can be diffuse and symmetrical or can produce a multinodular goiter. The multinodular goiter can grow independently from pituitary gland control and produce excess thyroid hormone, causing hyperthyroidism. However, hyperthyroidism is most often the result of Graves' disease. Thyroid enlargement can also be caused by Hashimoto's thyroiditis, in which the individual's immune system develops abnormal antibodies that react with proteins in the thyroid gland. This autoimmunity can make the gland enlarge or become underactive. *See* AUTOIMMUNITY.

Tumors of the thyroid account for a small fraction of human neoplasms and an even smaller fraction of deaths due to cancer. The vast majority of thyroid neoplasms are follicular adenomas, which are benign; however, some thyroid neoplasms are malignant. Rarely, tumors arising elsewhere in the body can metastasize to the thyroid gland.

Hyperthyroidism is the clinical condition that results from excessive levels of the circulating thyroid hormones thyroxine and triiodothyronine, which are secreted by the thyroid gland. Signs and symptoms include weight loss, tachycardia (increased heart rate), heat intolerance, sweating, and tremor. Graves' disease, the most common form of hyperthyroidism, is mediated by an abnormal antibody directed to the thyroid-stimulating hormone (TSH) receptor on the surface of the thyroid cell, which stimulates secretion of thyroid hormone. Unique to Graves' disease is the associated protrusion of the eyes (exophthalmos).

Hypothyroidism is the clinical state that results from subnormal levels of circulating thyroid hormones. Manifestations in infancy and childhood include growth retardation and reduced intelligence; in adults, cold intolerance, dry skin, weight gain, constipation, and fatigue predominate. Individuals with hypothyroidism often have a slow pulse (bradycardia), puffy dry skin, thin hair, and delayed reflexes. In its most extreme form, hypothyroidism can lead to coma and death if untreated. *See* THYROID GLAND; THYROID HORMONES. Leslie J. DeGroot; David A. Ehrmann

Thyroid hormones Any of the chemical messengers produced by the thyroid gland, including thyrocalcitonin, a polypeptide, and thyroxine and triiodothyronine, which are iodinated thyronines. *See* HORMONE; THYROCALCITONIN; THYROID GLAND. M. Todd Washington; Howard Rasmussen

Thysanoptera An order of small, slender insects, commonly called thrips, having exopterygote development, sucking mouthparts, and highly modified wings. The order is a relatively small one, but individuals are often very numerous in favorable environments. *See* THRIP.

The mouthparts are conical and used for scraping, piercing, and sucking. The wings are exceptionally narrow, with few or no veins, and are bordered by long hairs. The tarsi terminate in an inflatable membranous bladder, which has remarkable adhesive properties.

The eggs of thrips are laid on the surface of twigs (suborder Tubulifera) or in small cuts made by the ovipositor (suborder Terebrantia). There are usually four nymphal stages, the last of these being quiescent and pupalike. There are from one to several generations produced in a single year. *See* INSECTA. Frank M. Carpenter

Thysanura (Zygentoma) An order of wingless insects with soft, fusiform bodies 0.12–0.8 in. (3–20 mm) long, often covered with flat scales forming diverse patterns. The mouthparts are free with dicondylous mandibles used for scraping and chewing. Antennae are long and threadlike (filiform). Visual organs may be a cluster of simple ommatidia (cylinder-shaped units of a compound eye) or lacking altogether. The abdomen terminates in three "tails": a pair of lateral cerci and a median caudal filament (telson). Females have well-developed ovipositors; males have a penis and often one or rarely two pairs of parameres.

The order Thysanura (also called Zygentoma) includes about 400 described species which most taxonomists group into three families. The Lepidotrichidae, known as forest silverfish, are well represented in Oligocene fossils but were thought to be extinct until 1959, when living specimens were discovered in northwestern California. The single extant species lives in decaying bark and rotten wood. Members of the family Nicoletiidae are typically found underground in caves, mammal burrows, or the nests of ants or termites. Some species mimic ants and steal their food. The family Lepismatidae is worldwide in distribution. It encompasses more than 300 species of fast-running insects that feed as scavengers on algae, lichens, starchy vegetable matter, fungal hyphae (mycelium), and woodland debris. This family includes all of the common species found in human dwellings, notably the silverfish, *Lepisma saccharina*, and the firebrat, *Thermobia domestica*. Silverfish (sometimes known as fishmoths) are silvery gray, active at night, and often regarded as nuisance pests. *See* APTERYGOTA; INSECTA.

John Meyer

Tick paralysis A loss of muscle function or sensation in humans or certain animals following the prolonged feeding of female ticks. Paralysis, of Landry's type, usually begins in the legs and spreads upward to involve the arms and other parts of the body. Evidence suggests that paralysis is due to a neurotoxin formed by the feeding ticks rather than the result of infection with microorganisms. *See* IXODIDES.

The disease has been reported in North America, Australia, South Africa, and occasionally in some European countries and is caused by appropriate species of indigenous ticks. In Australia, *Ixodes holocyclus* causes frequent cases in dogs, and occasional cases in humans, and paralysis has been known to progress even after removal of ticks. *Ixodes cubicundus* is associated with the disease in South Africa.

Cornelius B. Philip

Tidal bore A part of a tidal rise in a river which is so rapid that water advances as a wall often several feet high. The phenomenon is favored by a substantial tidal range and a channel which shoals and narrows rapidly upstream, but the conditions are so critical that it is not common. Although the bore is a very striking feature, the tide continues to rise after the passage of the bore. Bores may be eliminated by changing channel depth or shape. *See* RIVER TIDES; TIDE.

Tidal bore of the Petitcodiac River, Bay of Fundy, New Brunswick, Canada. Rise of water is about 4 ft (1.2 m). (*New Brunswick Travel Bureau*)

In North America three bores have been observed: at the head of the Bay of Fundy (see illustration), at the head of the Gulf of California, and at the head of Cook Inlet, Alaska. The largest known bore occurs in the Tsientang Kiang, China. At spring tides this bore is a wall of water 15 ft (4.5 m) high moving upstream at 25 ft/s (7.5 m/s).

Blair Kinsman

Tidal datum A reference elevation of the sea surface from which vertical measurements are made, such as depths of the ocean and heights of the land. The intersection of the elevation of a tidal datum with the sloping shore forms a line used as a horizontal boundary. In turn, this line is also a reference from which horizontal measurements are made for the construction of additional coastal and marine boundaries.

Since the sea surface moves up and down from infinitely small amounts to hundreds of feet over periods of less than a second to millions of years, it is necessary to stop the vertical motion in order to have a practical reference. This is accomplished by hydraulic filtering, numerical averaging, and segment definition of the record obtained from a tide gage affixed to the adjacent shore. Waves of periods up through wind waves are effectively damped by a restricting hole in the measurement well. Recorded hourly heights are averaged to determine the mean of the higher (or only) high tide of each tidal day (24.84 h), all the high tides, all the hourly heights, all the low tides, and the lower (or only) low tide. The length of the averaging segment is a specific 19 year, which averages all the tidal cycles through the regression of the Moon's nodes and the metonic cycle. [The metonic cycle is a time period of 235 lunar months (19 years); after this period the phases of the Moon occur on the same days of the same months.] But most of all, the 19-year segment is meaningful in terms of measurement capability, averaging meteorological events, and for engineering and legal interests. However, the 19-year segment must be specified and updated because of sea-level changes occurring over decades. The present tidal datum epoch is 1983 through 2001.

Tidal datums are legal entities. Because of variations in gravity, semistationary meteorological conditions, semipermanent ocean currents, changes in tidal characteristics, ocean density differences, and so forth, the sea surface (at any datum elevation) does not conform to a mathematically defined spheroid. *See* GEODESY; TIDE.

Steacy D. Hicks

Tidal power Tidal-electric power is obtained by utilizing the recurring rise and fall of coastal waters. Marginal marine basins are enclosed with dams, making it possible to create differences in the water level between the ocean and the basins. The oscillatory flow of water filling or emptying the basins is used to drive hydraulic turbines which propel electric generators.

Large amounts of electric power could be developed in the world's coastal regions having tides of sufficient range, although even if fully developed this would amount to only a small percentage of the world's potential water (hydroelectric) power. *See* ELECTRIC POWER GENERATION; TIDE.

George G. Adkins

Tidalites Sediments deposited by tidal processes. Until recently, "tidalites" referred to sediments deposited by tidal processes in both the intertidal zone (between normal low- and high-tide levels) and shallow, subtidal (permanently submerged), tide-dominated environments less than 200 m (660 ft) deep. Tidalites are now known also to occur within supratidal environments (above normal high tide and flooded only during storms or very high spring tides) and submarine canyons at depths much greater than 200 m. Common usage has drifted toward describing tidalites as ripple- and dune-scale features rather than more composite deposits such as large linear sand ridges of tidal origin present on continental shelves or point bars associated with migrating tidal channels. Both of these larger-scale features, however, would be composed of tidalites.

By identifying tidalites in either the modern or the ancient geological record, geologists are implying that they know that the sediments were deposited by tidal processes rather than by storms or waves. Tidalites are not always easy to identify with certainty, especially in the rock record. In order to do so, it is necessary to understand the basic tidal cycles that can influence sedimentation. *See* TIDE.

To recognize tidalites in the geologic record, geologists look for evidence of one or more of the following:

1. Sediment deposited by reversing currents (that is, flood–ebb cycles).

2. A stacked sequence of sediments that show a recurring change from sediments transported (and deposited) by currents at maximum current velocity to sediments deposited from suspension at minimum current velocity.

3. Stacked packages of sediments in which each package shows evidence of subaerial exposure superimposed on sediments deposited in subaqueous settings (sediments transported and deposited during flood tides and exposed during low ebb tide).

4. A sequence of sediment packages in which the thickness or accretion of successive packages of sediments varies in a systematic way, suggesting diurnal, semidiurnal, and/or neap–spring tidal cycles.

Deposits of tidalites are known from every geologic period from the modern back into the Precambrian and from depositional environments with water chemistries ranging from fresh to hypersaline. Studies of tidalites are important because geologists have used these features not only to interpret the original depositional settings of the deposits but also to calculate ancient Earth–Moon distances, interpret paleoclimates existent during deposition, and calculate sedimentation rates. *See* DEPOSITIONAL SYSTEMS AND ENVIRONMENTS; GEOLOGIC TIME SCALE; MARINE SEDIMENTS; SEDIMENTOLOGY.

Erik P. Kvale

Tide Stresses exerted in a body by the gravitational action of another, and related phenomena resulting from these stresses. Every body in the universe raises tides, to some extent, on every other. This article deals only with tides on the Earth, since these are fundamentally the same as tides on all bodies, and more specifically with variations of sea level, whatever their origin. *See* GRAVITATION; SEA-LEVEL FLUCTUATIONS.

The tide-generating forces arise from the gravitational action of Sun and Moon, the effect of the Moon being about twice as effective as that of the Sun in producing tides. The tidal effects of all other bodies on the Earth are negligible. The tidal forces act to generate stresses in all parts of the Earth and give rise to relative movements of the matter of the solid Earth, ocean, and atmosphere. In the ocean the tidal forces act to generate alternating tidal currents and displacements of the sea surface.

If the Moon attracted every point within the Earth with equal force, there would be no tide. It is the small difference in direction and magnitude of the lunar attractive force, from one point of the Earth's mass to another, which gives rise to the tidal stresses. The tide-generating force is proportional to the mass of the disturbing body (Moon) and to the inverse cube of its distance. This inverse cube law accounts for the fact that the Moon is 2.17 times as important, insofar as tides are concerned, as the Sun, although the latter's direct gravitational pull on the Earth, which is governed by an inverse-square law, is about 180 times the Moon's pull.

At most places in the ocean and along the coasts, sea level rises and falls in a regular manner. The highest level usually occurs twice in any lunar day, the times bearing a constant relationship with the Moon's meridional passage. The time between the Moon's meridional passage and the next high tide is called the lunitidal interval. The difference in level between successive high and low tides, called the range of the tide, is generally greatest near the time of full or new Moon, and smaller near the times of quadrature. The range of the tide usually exhibits a secondary variation, being greater near the time of perigee (when the Moon is closest to the Earth) and smaller at apogee (when the Moon is farthest away).

The above situation is observed at places where the tide is predominantly semidiurnal. At many other places, it is observed that one of the two maxima in any lunar day is higher than the other. This effect is known as the diurnal inequality and represents the presence of an appreciable diurnal variation. At these places, the tide is said to be of the "mixed" type. At a few places, the diurnal tide actually predominates, there generally being only one high and low tide during the lunar day.

The range of the ocean tide varies between wide limits. The highest range is encountered in the Bay of Fundy, where values exceeding 50 ft (15 m) have been observed. In some places in the Mediterranean, South Pacific, and Arctic, the tidal range never exceeds 2 ft (0.6 m).

Owing to the rotation of the Earth, there is a gyroscopic, or Coriolis, force acting perpendicularly to the motion of any water particle in motion. In the Northern Hemisphere this force is to the right of the current vector. The horizontal, or tractive, component of the tidal force generally rotates in the clockwise sense in the Northern Hemisphere. As a result of both these influences the tidal currents in the open ocean generally rotate in the clockwise sense in the Northern Hemisphere, and in the counterclockwise sense in the Southern Hemisphere. *See* CORIOLIS ACCELERATION; EARTH TIDES.

Gordon W. Groves

Tie rod A tie rod or tie bar, usually circular in cross section, is used in structural parts of machines to tie together or brace connected members, or in moving parts of machines or mechanisms it may connect arms or parts to transmit motion. In the first use the rod ends are usually a threaded fastening, while in the latter they are usually forged into an eye for a pin connection.

Paul H. Black

Tile As a structural material, a burned clay product in which the coring exceeds 25% of the gross volume; as a facing material, any thin, usually flat, square product. Structural tile used for load bearing may or may not be glazed; it may be cored horizontally or vertically. Two principal grades are manufactured: one for exposed masonry construction, and the other for unexposed construction.

As a facing, clay products are formed into thin flat, curved, or embossed pieces, which are then glazed and burned. Commonly used on surfaces subject to water splash or that require frequent cleaning, such vitreous glazed wall tile is fireproof. Unglazed tile is laid as bathroom floor. By extension, any material formed into a size comparable to clay file is called tile. Among the materials formed into tile are asphalt, cork, linoleum, vinyl, and porcelain.

Frank H. Rockett

Till The generic term for sediment deposited directly from glacier ice. Till is characteristically nonsorted and nonstratified and is deposited by lodgement or melt-out beneath a glacier or by melt-out on the surface of a glacier. The texture of till varies greatly and all tills are characterized by a wide range, of particle sizes. Till contains a variety of rock and mineral fragments which reflect the source material over which the glacier flowed. The particles in the deposit usually show a preferred orientation related to the nature and direction of the ice flow. The overall character of the fill reflects the source material, position and distance of transport, nature and position of deposition, and postdepositional changes.

W. Hilton Johnson

Tillodontia An extinct order of early Cenozoic (about 65 to 40 million years ago) quadrupedal eutherian land mammals, represented by nine known genera, from the late Paleocene to middle Eocene of North America (*Esthonyx* [*Azygonyx*], *Megalesthonyx*, *Trogosus*, and *Tillodon*), early Paleocene to late Eocene of China (*Lofochaius*, *Meiostylodon*, *Adapidium*, *Trogosus* [*Kuanchuanius*]), middle Eocene of Pakistan (*Basalina*), and the early Eocene of Europe (*Plesiesthonyx*). The tillodonts left no known descendants and were probably most closely related to the extinct order Pantodonta (another group of extinct ungulatelike mammals from the Paleocene and Eocene that in turn may be related to the early eutherian mammals known as arctocyonids).

Tillodonts were medium- to large-sized mammals (their skulls range in length from 5 to 37 cm or 2 to 15 in.) that probably fed primarily on roots and tubers in warm temperate to subtropical habitats. Tillodonts were most common in the early Eocene faunas of North America. They developed large second incisors that became rodentlike, relatively long snouts, massive skeletons, and moderately large claws. *See* ARCHAIC UNGULATE; MAMMALIA; TOOTH.

Robert M. Schoch

Time The dimension of the physical universe that orders the sequence of events at a given place; also, a designated instant in this sequence, such as the time of day, technically known as an epoch, or sometimes as an instant.

Measurement. Time measurement consists of counting the repetitions of any recurring phenomenon and possibly subdividing the interval between repetitions. Two aspects to be considered in the measurement of time are frequency, or the rate at which the recurring phenomena occur, and epoch, or the designation to be applied to each instant.

A determination of time is equivalent to the establishment of an epoch or the correction that should be applied to the reading of a clock at a specified epoch. A time interval may be measured as the duration between two known epochs or by counting from an arbitrary starting point, as is done with a stopwatch. Time units are the intervals between successive recurrences of phenomena, such as the period of rotation of the Earth or a specified number of periods of radiation derived from an atomic energy–level transition. Other units are arbitrary multiples and subdivisions of these intervals, such as the hour being 1/24 of a day, and the minute being 1/60 of an hour. *See* DAY; MONTH; TIME-INTERVAL MEASUREMENT; YEAR.

Bases. Several phenomena are used as bases with which to determine time. The phenomenon traditionally used has been the rotation of the Earth, where the counting is by days. Days are measured by observing the meridian passages of the Sun or stars and are subdivided with the aid of precision clocks. The day, however, is subject to variations in duration because of the variable rotation rate of the Earth. Thus, when a more uniform time scale is required, other bases for time must be used.

The angle measured along the celestial equator between the observer's local meridian and the vernal equinox is the measure of sidereal time. In practice, a conventionally adopted mathematical expression provides this time as a function of civil time. It is reckoned from 0 to 24 hours, each hour being subdivided into 60 sidereal minutes and the minutes into 60 sidereal seconds. Sidereal clocks are used for convenience in many astronomical observatories because a star or other object outside the solar system comes to the same place in the sky at virtually the same sidereal time.

The angle measured along the celestial equator between the observer's local meridian and the Sun is the apparent solar time. The only true indicator of local apparent solar time is a sundial. Mean solar time has been devised to eliminate the irregularities in apparent solar time that arise from the inclination of the Earth's orbit to the plane of the Sun's motion and the varying speed of the Earth in its orbit. In practice it is defined by a conventionally adopted mathematical expression. Intervals of sidereal time can be converted into intervals of mean solar time by dividing by 1.002 737 909 35. Both sidereal and solar time depend on the rotation of the Earth for their time base. *See* EQUATION OF TIME.

Historically, the mean solar time determined for the meridian of 0° longitude using astronomical observations was referred to as UT1. Currently UT1 is used only as an angle expressed in time units that depends on the Earth's rotation with respect to the celestial reference system. It is defined by a conventional mathematical expression and continuing astronomical observations. These are made at a number of observatories around the world. The International Earth Rotation and Reference System Service (IERS) receives these data and provides daily values of the difference between UT1 and civil time. *See* EARTH ROTATION AND ORBITAL MOTION.

Because the Earth has a nonuniform rate of rotation and a uniform time scale is required for many timing applications, a different definition of a second was adopted in 1967. The international agreement calls for the second to be defined as 9,192,631,770 periods of the radiation derived from an energy-level transition in the cesium atom. This second is referred to as the international or SI (International System) second and is independent of astronomical observations. International Atomic Time (TAI) is maintained by the International Bureau of Weights and Measures (BIPM) from data contributed by time-keeping laboratories around the world. *See* ATOMIC TIME.

Coordinated Universal Time (UTC) uses the SI second as its time base. However, the designation of the epoch may be changed at certain times so that UTC does not differ from UT1 by more than 0.9 s. UTC forms the basis for civil time in most countries and may sometimes be referred to unofficially as Greenwich Mean Time. The adjustments to UTC to bring this time scale into closer accord with UT1 consist of the insertion or deletion of integral seconds. These "leap seconds" may be applied preferably at 23 h 59 m 59 s of June 30 or December 31 of each year according to decisions made by the IERS. UTC differs from TAI by an integral number of atomic seconds.

Civil and standard times. Because rotational time scales are local angular measures, at any instant they vary from place to place on the Earth. Persons traveling westward around the Earth must advance their time 1 day, and those traveling eastward must retard their time 1 day in order to be in agreement with their neighbors when they return home. The International Date Line is the name given to a line where the change of date is made. It follows approximately the 180th meridian but avoids inhabited land. To avoid the inconvenience of the continuous change of mean solar time with longitude, zone time or civil time is generally used. The Earth is divided into 24 time zones, each approximately 15° wide and centered on standard longitudes of 0°, 15°, 30°, and so on. Within each of these zones the time kept is related to the mean solar time of the standard meridian. *See* INTERNATIONAL DATE LINE.

Many countries, including the United States, advance their time 1 hour, particularly during the summer months, into "daylight saving time."

Dennis D. McCarthy

Time, arrow of The uniform and unique direction associated with the apparent inevitable flow of time into the future. There appears to be a fundamental asymmetry in the universe. Herein lies a paradox, for all the laws of physics, whether they are the equations of classical mechanics, classical electromagnetism, general relativity, or quantum mechanics, are time reversible in that they admit solutions in either direction of time. This reversibility raises the question of how these fundamentally time-symmetrical equations can result in the perceived asymmetry of temporally ordered events.

The symmetry breaking of temporal order has not yet been fully explained. There are certain indications that an intrinsic asymmetry exists in temporal evolution. Thus it may be that the fundamental laws of physics are not really time symmetric and that the currently known laws are only symmetrized approximations to the truth. Indeed, the decay of the K^0 meson is not time reversible. However, it is not clear how such a rare and exotic instance of time asymmetry could emerge into the world of essentially macroscopic, electromagnetic phenomena as an everyday observable. *See* TIME REVERSAL INVARIANCE.

Another, more ubiquitous example of a fundamentally time-asymmetric process is the expansion of the universe. It has been speculated that this expansion is the true basis of time asymmetry. *See* BIG BANG THEORY; COSMOLOGY.

Alternatively, even a time-symmetrical universe will have a statistical behavior in which configurations of molecules and localizations of energy have significant probabilities of recurring only after enormously long time intervals. Indeed, such time intervals are longer than the times required for the ceaseless expansion of the universe and the evolution of its component particles. Time's arrow is destined, either by the nature of space-time or the statistics of large assemblies, to fly into the future. *See* STATISTICAL MECHANICS.

P. W. Atkins

Time constant A characteristic time that governs the approach of an exponential function to a steady-state value. When a physical quantity is varying as a decreasing exponential function of time as in Eq. (1), or as an increasing exponential function as in Eq. (2), the approach to the steady-state value achieved after a long time is governed by a characteristic time T as given in Eq. (3). This time T is called the time constant.

$$f(t) = e^{-kt} \tag{1}$$

$$f(t) = 1 - e^{-kt} \tag{2}$$

$$t = \frac{1}{k} = T \tag{3}$$

When time t is zero, $f(t)$ in Eq. (1) has the magnitude 1, and when t equals T the magnirude is $1/e$. Here e is the transcendental number whose value is approximately 2.71828, and the change in magnitude is $1 - (1/e) = 0.63212$. The function has moved 63.2% of the way to its final value. The same factor also holds for Eq. (2). *See* E (MATHEMATICS).

The initial rate of change of both the increasing and decreasing functions is equal to the maximum amplitude of the function divided by the time constant.

The concept of time constant is useful when evaluating the presence of transient phenomena. Robert L. Ramey

Time-interval measurement A determination of the duration between two instants of time (epochs). Time intervals are measured with high precision with a digital display counter. An electronic oscillator generates pulses; the count begins with a start signal and ends with a second signal. Two atomic clocks can be compared in epoch to 1 picosecond (1 ps = 10^{-12} s) by electronic interpolation. *See* ATOMIC CLOCK; DIGITAL COUNTER; OSCILLATOR; OSCILLOSCOPE.

Rapid motions can be studied at short intervals by means of a large variety of high-speed cameras, including stroboscopic, rotating film-drum, rotating mirror, streak, and image converter cameras. An electronic streak camera can separate two pulses 1 ps apart. *See* PHOTOGRAPHY; STROBOSCOPIC PHOTOGRAPHY.

Ultrashort laser pulses are used to study rapid processes caused by the interaction of photons with an atom or molecule. Pulses as short as three wavelengths of 620-nm light, with τ = 6 femtoseconds (1 fs = 10^{-15} s), have been formed. *See* LASER; LASER PHOTOCHEMISTRY; OPTICAL PULSES; ULTRAFAST MOLECULAR PROCESSES.

Radioactive decay is used to measure long time intervals, to about 5×10^9 years, concerning human history, the Earth, and the solar system. *See* GEOCHRONOMETRY; RADIOCARBON DATING.

William Markowitz

Time-of-flight spectrometer Any of a general class of instruments in which the speed of a particle is determined directly by measuring the time that it takes to travel a measured distance. By knowing the particle's mass, its energy can be calculated. If the particles are uncharged (for example, neutrons), difficulties arise because standard methods of measurement (such as deflection in electric and magnetic fields) are not possible. The time-of-flight method is a powerful alternative, suitable for both uncharged and charged particles.

The time intervals are best measured by counting the number of oscillations of a stable oscillator that occur between the instants that the particle begins and ends its journey. Oscillators operating at 100 MHz are in common use. *See* MASS SPECTROSCOPE; NEUTRON SPECTROMETRY; TIME-INTERVAL MEASUREMENT.

Frank W. K. Firk

Time-projection chamber An advanced particle detector for the study of ultra-high-energy collisions of positrons and electrons. The underlying physics of the scattering process can be studied through precise measurements of the momenta, directions, particle species, and correlations of the collision products. The time-projection chamber (TPC) provides a unique combination of capabilities for these studies and other problems in elementary particle physics by offering particle identification over a wide momentum range, and by offering high resolution of intrinsically three-dimensional spatial information for accurate event reconstruction.

The time-projection chamber concept is based on the maximum utilization of ionization information, which is deposited by high-energy charged particles traversing a gas. The ionization trail, a precise image of the particle trajectory, also contains information about the particle velocity. A strong, uniform magnetic field and a uniform electric field are generated within the time-projection chamber active volume in an exactly parallel orientation. The parallel configuration of the fields permits electrons, products of the ionization processes, to drift through the time-projection chamber gas over great distances without distortion; the parallel configuration offers a further advantage in that the diffusion of the electrons during drift can be greatly suppressed by the magnetic field, thus preserving the quality of track information. *See* PARTICLE DETECTOR. David R. Nygren

Time reversal invariance A symmetry of the fundamental (microscopic) equations of motion of a system; if it holds, the time reversal of any motion of the system is also a motion of the system. To date, only two phenomena have shown evidence (at least indirect) for violation of time reversal invariance. One is the violation of *CP* invariance observed in the decays of the neutral mesons K_L and B^0, $\overline{B}^0$. The other is the baryon asymmetry of the universe.

Time reversal invariance is not evident from casual observation of everyday phenomena. If a movie is taken of a phenomenon, the corresponding time-reversed motion can be exhibited by running the movie backward. The result is usually strange. For instance, water in the ground is not ordinarily observed to collect itself into drops and shoot up into the air. However, if the system is sufficiently well observed, the direction of time is not obvious. For instance, a movie that showed the motion of the planets (in which each of the objects that make up the system can be observed individually) would look just as right run backward or forward. The apparent irreversibility of everyday phenomena results from the combination of imprecise observation and starting from an improbable situation (a state of low entropy, to use the terminology of statistical mechanics). *See* ENTROPY; STATISTICAL MECHANICS.

If time reversal invariance holds, no particle (a physical system with a definite mass and spin) can have an electric dipole moment. Although a polar body, for example, a water (H_2O) molecule, has an electric dipole moment, this shows up only in transition matrix elements between its eigenstates of energy (and spin). No particle has been observed to have an electric dipole moment; for instance, the present experimental upper limit on the electric moments of the electron and the neutron are approximately 10^{-27} cm times e and 10^{-25} cm times e, respectively, where e is the charge of the proton. *See* DIPOLE MOMENT; ELECTRON; NEUTRON; POLAR MOLECULE; SPIN (QUANTUM MECHANICS).

Another test of time reversal invariance is to compare the cross sections for reactions which are inverse to one another. The present experimental upper limit on the relative size of the time reversal invariance–violating amplitude of such reactions is approximately 3×10^{-3}; unfortunately, this is far larger than any expected violation. *See* NUCLEAR REACTION.

A consequence of the *CPT* theorem is that violation of T (time reversal invariance) is equivalent to violation of *CP*, that is, invariance of the fundamental equations under the combined operations of charge conjugation C and space inversion P. Hence CP violation observed in the decay of the long-lived neutral K meson (K_L) and in the decay of the neutral B mesons is evidence for T violation. *See* CPT THEOREM; MESON.

According to the standard model, CP violation arises from the CKM (Cabibbo-Kobayashi-Maskawa) quark-mixing matrix. When a down-type quark (d, s, or b) absorbs a W^+ or emits a W^- boson, it becomes a linear combination of up-type quarks (u, c, t); the coefficients of these three linear combinations is called the CKM matrix. The CKM matrix gives the relative couplings of the charged weak boson to quarks. In general, all the couplings cannot be made real by any choice of the phases of the quark states; this results in violation of T. The CKM matrix consistently describes all weak-decay observations, including the CP violation seen in neutral B decay. *See* MATRIX THEORY.

An indirect but very prominent evidence for T violation is the baryon asymmetry of the universe, that is, the fact that ordinary matter contains baryons, not antibaryons. Put more quantitatively: The observed fact that there is roughly 1 baryon for every 10^9 cosmic blackbody photons means that early in the history of the universe, when the value of kT (the product of Boltzmann's constant and the thermodynamic temperature) was larger than 1 GeV (and hence the number of both baryons and antibaryons was roughly the same as the number of photons), there was an excess of baryons over antibaryons of the order of one part in 10^9. It is conceivable that this rather small baryon-number asymmetry developed in an originally baryon-number symmetric universe. This would require interactions, acting in the early universe, which violate both baryon number and T. The existence of the former would not be surprising, since particle interactions that do not conserve baryon number are always present in grand unified theories—gauge theories in which the strong and electroweak interactions, and likewise quarks and leptons, are unified. *See* COSMIC BACKGROUND RADIATION; COSMOLOGY; ELEMENTARY PARTICLE; GRAND UNIFICATION THEORIES; SYMMETRY LAWS (PHYSICS). Charles J. Goebel

Time-reversed signal processing A means for improving the performance of remote sensing and communication systems that rely on electromagnetic- or acoustic-wave propagation but must contend with wave reflections, diffraction, and scattering. In particular, for applications of remote sensing—including radar, sonar, biomedical imaging, and nondestructive evaluation—the main intent is the detection, localization, and identification of distant objects or features that either scatter or generate electromagnetic or acoustic waves. Although the disparity in objectives between military-radar and ultrasonic-imaging systems may be considerable, they have in common that random wave scattering and diffraction tend to limit the accuracy and confidence with which such systems can be used and the distances over which such systems can operate. For radar and sonar, turbulence and wave motions in the ocean or atmosphere, or rough terrestrial or ocean surfaces, may cause such random wave scattering and diffraction. In ultrasonic remote sensing, the random scattering and diffraction may be caused by grain structure within metals or by variations between and within tissues.

The basic process of time reversal can be described by four steps. First, waves are generated or scattered by an object or feature of interest and travel forward through the environment to an antenna or array of transducers. The paths that waves follow between the object and the array may be complicated and unknown. Second, the array records the signal in the usual manner. These recordings will include signal distortion from echoes, scattering, and diffraction in the environment. Third, the signal recorded by each element in the array is retransmitted from that element with the direction of time inverted; the end of the signal is transmitted first, and the start of the signal is transmitted last. In the final step, these array-transmitted time-reversed waves travel backward through the environment, retracing their paths to converge at the location where they originated. *See* ACOUSTIC SIGNAL PROCESSING; BIOMEDICAL ULTRASONICS; NONDESTRUCTIVE EVALUATION; RADAR; SONAR; ULTRASONICS; UNDERWATER SOUND.

The time-reversed waves are able to remove distortion and focus well when the transducer array is many wavelengths long, the absorption of wave energy by the environment is weak, background noise is low, and the environment changes little (or not at all) between forward and backward wave travel. Both the tight focusing and distortion-removal capabilities of time reversal are of interest in remote sensing. *See* REMOTE SENSING.

David R. Dowling

Timothy A plant, *Phleum pratense*, of the order Cyperales, long the most important hay grass for the cooler temperate humid regions. It is easily established and managed, produces seed abundantly, and grows well in mixtures with alfalfa and clover. It is a short-lived perennial, makes a loose sod, and has moderately leafy stems and a dense cylindrical inflorescence. Timothy-legume mixtures still predominate in hay and pasture seedings for crop rotations in the northern half of the United States. *See* GRASS. Howard B. Sprague

Tin A chemical element, symbol Sn, atomic number 50, atomic weight 118.71. Tin forms tin(II) or stannous (Sn^{2+}), and tin(IV) or stannic (Sn^{4+}) compounds, as well as complex salts of the stannite (M_2SnX_4) and stannate (M_2SnX_6) types. *See* PERIODIC TABLE.

Tin melts at a low temperature, is highly fluid when molten, and has a high boiling point. It is soft and pliable and is corrosion-resistant to many media. An important use of tin has been for tin-coated steel containers (tin cans) used for preserving foods and beverages. Other important uses are solder alloys, bearing metals, bronzes, pewter, and miscellaneous industrial alloys. Tin chemicals, both inorganic and organic, find extensive use in the electroplating, ceramic, plastic, and agricultural industries.

The most important tin-bearing mineral is cassiterite, SnO_2. No high-grade deposits of this mineral are known. The bulk of the world's tin ore is obtained from low-grade alluvial deposits. *See* CASSITERITE.

Two allotropic forms of tin exist: white (β) and gray (α) tin. Tin reacts with both strong acids and strong bases, but it is relatively resistant to solutions that are nearly neutral. In a wide variety of corrosive conditions, hydrogen gas is not evolved from tin and the rate of corrosion becomes controlled by the supply of oxygen or other oxidizing agents. In their absence, corrosion is negligible. A thin film of stannic oxide forms on tin upon exposure to air and provides surface protection. Salts that have an acid reaction in solution, such as aluminum chloride and ferric chloride, attack tin in the presence of oxidizers or air. Most nonaqueous liquids, such as oils, alcohols, or chlorinated hydrocarbons, have slight or no obvious effect on tin. Tin metal and the simple inorganic salts of tin are nontoxic. Some forms of organotin compounds, on the other hand, are toxic. Some important physical constants for tin are shown in the table.

Properties of tin

Property	Value
Melting point, °C	231.9
Boiling point, °C	2270
Specific gravity, α form (gray tin)	5.77
β form (white tin)	7.29
Specific heat, cal/g*, white tin at 25°C	0.053
Gray tin at 10°C	0.049

*1 cal = 4.184 joules.

Stannous oxide, SnO, is a blue-black, crystalline product which is soluble in common acids and strong alkalies. It is used in making stannous salts for plating and glass manufacture. Stannic oxide, SnO_2, is a white powder, insoluble in acids and alkalies. It is an excellent glaze opacifier, a component of pink, yellow, and maroon ceramic stains and of dielectric and refractory bodies. It is an important polishing agent for marble and decorative stones.

Stannous chloride, $SnCl_2$, is the major ingredient in the acid electrotinning electrolyte and is an intermediate for tin chemicals. Stannic chloride, $SnCl_4$, in the pentahydrate form is a white solid. It is used in the preparation of organotin compounds and chemicals to weight silk and to stabilize perfume and colors in soap. Stannous fluoride, SnF_2, a white water-soluble compound, is a toothpaste additive.

Organotin compounds are those compounds in which at least one tin-carbon bond exists, the tin usually being present in the + IV oxidation state. Organotin compounds that find applications in industry are the compounds with the general formula R_4Sn, R_3SnX, R_2SnX_2, and $RSnX_3$. R is an organic group, often methyl, butyl, octyl, or phenyl, while X is an inorganic substituent, commonly chloride, fluoride, oxide, hydroxide, carboxylate, or thiolate. *See* TIN ALLOYS. Joseph B. Long

Tin alloys Solid solutions of tin and some other metal or metals. Alloys cover a wide composition range and many applications because tin alloys readily with nearly all metals. *See* TIN; TIN METALLURGY.

Soft solders constitute one of the most widely used and indispensable series of tin-containing alloys. Common solder is an alloy of tin and lead, usually containing 20–70% tin. *See* SOLDERING.

Bronzes form an important group of structural metals. Of the true copper-tin bronzes, up to 10% tin is used in wrought phosphor bronzes, and from 5 to 10% tin in the most common cast bronzes. Many brasses, which are basically copper-zinc alloys, contain 0.75–1.0% tin for additional corrosion resistance. *See* COPPER ALLOYS.

Other useful tin alloys include babbitt metal (tin containing 4–8% each copper and antimony), an alloy with bearing applications; pewter; a tin-base alloy containing small quantifies of antimony and copper; and type metals, which are lead-base alloys containing 3–15% tin. *See* ALLOY. Bruce W. Gonser

Tin metallurgy The extraction of tin from its ores and its subsequent refining and preparation for use. Most tin concentrates are primarily cassiterite (SnO_2), the naturally occurring oxide of tin. These are comparatively easy to reduce by using carbon at high temperatures. However, this operation differs from the smelting of most common metals because retreatment of the slag is necessary to obtain efficient metal recovery. *See* CASSITERITE.

In primary smelting, carbon monoxide (CO) formed during heat-up reacts with the solid cassiterite particles to produce tin (Sn) and carbon dioxide (CO_2). As the temperature increases, silica (present in nearly all concentrates) also reacts under reducing conditions with the SnO_2 to give stannous silicate. Iron, also present as an impurity in all concentrates, reacts with the silica to form ferrous silicate ($FeSiO_3$). These silicates fuse with the added fluxes to form a liquid slag, at which point unreacted carbon from the fuel becomes the predominant reductant in reducing both stannous silicate to tin and ferrous silicate to iron. The metallic iron then reduces tin from stannous silicate, as shown in the reaction below.

$$SnSiO_3 + Fe \rightleftharpoons FeSiO_3 + Sn$$

Secondary tin from metal scrap amounts to about one-quarter of the total tin consumed in the United States. Most of this comes from tin-bearing alloys, and secondary smelters rework them into alloys and chemicals. However, additional tin of high purity is recovered from the detinning of tinplate scrap. *See* ELECTROCHEMICAL PROCESS; ELECTROMETALLURGY; HEAT TREATMENT (METALLURGY); PYROMETALLURGY, NONFERROUS; TIN. Daniel Maykuth

Tintinnida An order of the Spirotrichia whose members are conical or trumpet-shaped pelagic forms bearing shells (loricae). These protozoa are planktonic ciliates and are especially abundant in oceans, notably the Pacific. The lorica is composed of a resistant organic compound in which various foreign mineral grains are embedded; its shape may range from trumpet- or bell-form to cylindrical or subspherical.

Fossil tintinnids, representing practically the only fossilized species of ciliate protozoa known to science, are identified on the basis of the shape of the lorica in cross section as seen in randomly oriented thin sections of the rocks in which they are found. Twelve genera of fossil tintinnids have been described from limestones and cherts of the Jurassic and Cretaceous. *See* CILIOPHORA; SPIROTRICHIA. John O. Corliss; Daniel J. Jones

Tire A continuous pneumatic rubber and fabric cushion encircling and fitting onto the rim of a wheel. In modern tire building, chemicals are compounded into the rubber to help it withstand wear, heat, and aging and to produce desired changes in its characteristics. Fabric (rayon, nylon, or polyester) is used to give the tire body strength and resilience. In belted tires, additional layers of fabric (rayon, fiber glass, finely drawn steel, or aramid) are placed just under the tread rubber to increase mileage and handling. Steel wire is used in the bead that holds the tire to the rim.

A tire is made up of two basic parts: the tread, or road-contacting part, which must provide traction and resist wear and abrasion, and the body or carcass, consisting of rubberized fabric that gives the tire strength and flexibility. In compounding the rubber, large amounts of carbon black are mixed with it to improve abrasion resistance. Other substances, such as sulfur, are added to enable satisfactory processing and vulcanization. *See* RUBBER.

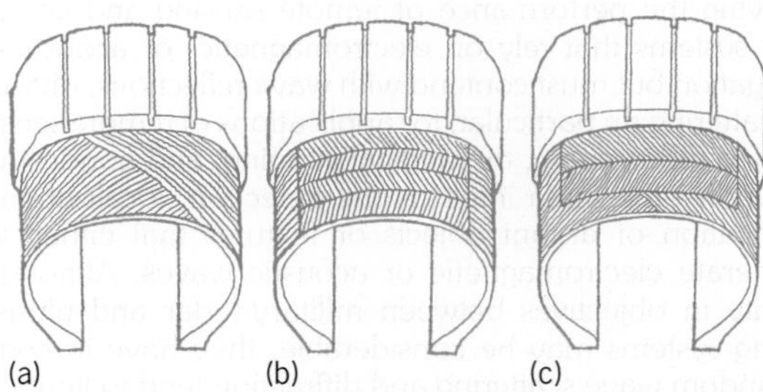

Tire construction. (*a*) Bias-ply. (*b*) Radial-ply. (*c*) Belted bias-ply. (*Goodyear Tire and Rubber Co.*)

There are three types of tires: bias, belted bias, and radial (see illustration). For bias tires, cords in the piles extend diagonally across the tire from bead to bead. The cords run in opposite directions in each successive ply, resulting in a crisscross pattern. For belted bias tires, plies are placed in a manner similar to that used in the bias-ply tire, with belts of material placed circumferentially around the tire between the plies and the tread rubber. For radial tires, cords in the plies extend transversely from bead to bead, substantially perpendicular to the direction of travel. Belts are placed circumferentially around the tire. David B. Harrison

Tissue An aggregation of cells more or less similar morphologically and functionally. The animal body is composed of four primary tissues, namely, epithelium, connective tissue (including bone, cartilage, and blood), muscle, and nervous tissue.

The process of differentiation and maturation of tissues is called histogenesis. *See* HISTOLOGY. Charles B. Curtin

Tissue culture The branch of biology in which tissues or cells of higher animals and plants are grown artificially in a controlled environment. Tissue culture is possible when cells are attached to a solid substrate, such as glass or cellophane, and if the necessary complex nutrient medium is provided. All cultures are now also grown in liquid suspension. Tissue cultures are used in the study of cell growth, multiplication, and differentiation, as well as in cancer research, hereditary mechanisms, radiation biology, all hybridization, and virus studies. *See* MICROBIOLOGICAL METHODS. Theodore T. Puck

Tissue typing A procedure involving a test or a series of tests to determine the compatibility of tissues from a prospective donor and a recipient prior to transplantation. The immunological response of a recipient to a transplant from a donor is directed against many cell-surface histocompatibility antigens controlled by genes at many different loci. However, one of these loci, the major histocompatibility complex (MHC), controls antigens that evoke the strongest immunological response. The human MHC is known as the HLA system, which stands for the first (A) human leukocyte blood group system discovered. *See* CELLULAR IMMUNOLOGY; HISTOCOMPATIBILITY.

The success of transplantation is greatly dependent on the degree of histocompatibility (identity) between the donor and recipient, which is determined by the HLA complex. When the donor and recipient have a low degree of histocompatibility, the organ is said to be mismatched, and the recipient mounts an immune response against the donor antigen. By laboratory testing, the degree of antigenic similarity between the donor and the recipient and the degree of preexisting recipient sensitization to donor antigens (and therefore preformed antibodies) can be determined. This is known as cross-matching.

Phenotyping of HLA-A, -B, and -C (ABC typing) of an individual is determined by reacting that individual's lymphocytes with a large panel of antisera directed against specific HLA antigens. The procedure is known as complement-mediated cytotoxicity assay. The person's lymphocytes are incubated with the different antisera and complement is added. Killing of the cells being tested indicates that they express the HLA antigens recognized by the particular antiserum being used. Killing of potential donor lymphocytes in the complement-mediated cytotoxicity assay is a contraindication to transplantation of tissue from that donor. *See* COMPLEMENT; IMMUNOASSAY.

In addition to its important role in organ transplantation, determination of the HLA phenotype is useful in paternity testing, forensic medicine, and the investigation of HLA-disease associations. *See* TRANSPLANTATION BIOLOGY.

M. Wayne Flye; T. Mohanakumar

Titanite A calcium, titanium silicate, $CaTiOSiO_4$, of high titanium content. Titanite is also known as sphene. Titanite is an orthosilicate (nesosilicate) with a hardness of 5–5$^1/_2$ on the Mohs scale, a distinct cleavage, a specific gravity of 3.4–3.55, and an adamantine to resinous luster. It commonly occurs as distinct wedge-shaped crystals that are usually brown in hand specimens. Titanite may also be gray, green, yellow, or black. *See* CRYSTAL STRUCTURE; HARDNESS SCALES.

Titanite is a common accessory mineral in many igneous and metamorphic rocks. It may be the principal titanium-bearing silicate mineral. It occurs in abundance in the Magnet Cove, igneous complex in Arkansas and in the intrusive alkalic-rocks of the Kola Penninsula, Russia.

The composition of titanite may diverge from pure $CaTiSiO_4$ because of a variety of chemical substitutions. Calcium ions (Ca^{2+}) can be partially replaced by strontium ions (Sr^{2+}) and rare-earth ions such as thorium (Th^{4+}) and uranium (U^{4+}). Because titanite commonly contains radioactive elements, it has been used for both uranium-lead and fission track methods of dating. *See* DATING METHODS; IGNEOUS ROCKS; METAMORPHIC ROCKS; SILICATE MINERALS; TITANIUM. John C. Drake

Titanium A chemical element, Ti, atomic number 22, and atomic weight 47.87. It occurs in the fourth group of the periodic table, and its chemistry shows many similarities to that of silicon and zirconium. As a first-row transition element, titanium has an aqueous solution chemistry, especially of the lower oxidation states, showing some resemblances to that of vanadium and chromium. *See* PERIODIC TABLE; TRANSITION ELEMENTS.

The catalytic activity of titanium complexes forms the basis of the well-known Ziegler process for the polymerization of ethylene. This type of polymerization is of great industrial interest since, with its use, high-molecular-weight polymers can be formed. In some cases, desirable special properties can be obtained by forming isotactic polymers, or polymers in which there is a uniform stereochemical relationship along the chain. *See* POLYOLEFIN RESINS.

The dioxide of titanium, TiO_2, occurs most commonly in a black or brown tetragonal form known as rutile. Less prominent naturally occurring forms are anatase and brookite (rhombohedral). Both rutile and anatase are white when pure. The dioxide may be fused with other metal oxides to yield titanates, for example, K_2TiO_3, $ZnTiO_3$, $PbTiO_3$, and $BaTiO_3$. The black basic oxide, $FeTiO_3$, occurs naturally as the mineral ilmenite; this is a principal commercial source of titanium.

Titanium dioxide is widely used as a white pigment for exterior paints because of its chemical inertness, superior covering power, opacity to damaging ultraviolet light, and self-cleaning ability. The dioxide has also been used as a whitening or opacifying agent in numerous situations, for example, as a filler in paper, a coloring agent for rubber and leather products, a pigment in ink, and a component of ceramics. It has found important use as an opacifying agent in porcelain enamels, giving a finish coat of great brilliance, hardness, and acid resistance. Rutile has also been found as brilliant, diamondlike crystals, and some artificial production of it in this form has been achieved. Because of its high dielectric constant, it has found some use in dielectrics.

The alkaline-earth titanates show some remarkable properties. The dielectric constants range from 13 for $MgTiO_3$ to several thousand for solid solutions of $SrTiO_3$ in $BaTiO_3$. Barium titanate itself has a dielectric constant of 10,000 near 120°C (248°F), its Curie point; it has a low dielectric hysteresis. These properties are associated with a stable polarized state of the material analogous to the magnetic condition of a permanent magnet, and such substances are known as ferroelectrics. In addition to the ability to retain a charged condition, barium titanate is piezoelectric and may be used as a transducer for the interconversion of sound and electrical energy. Ceramic transducers containing barium titanate compare favorably with Rochelle salt and quartz, with respect to thermal stability in the first case, and with respect to the strength of the effect and the ability to form the ceramic in various shapes, in the second case. The compound has been used both as a generator for ultrasonic vibrations and as a sound detector. *See* PIEZOELECTRICITY. Arthur W. Adamson

In addition to important uses in applications such as structural materials, pigments, and industrial catalysis, titanium has a rich coordination chemistry. The formal oxidation of titanium in molecules and ions ranges from −II to +IV. The lower oxidation states of −II and −I occur only in a few complexes containing strongly electron-withdrawing carbon monoxide ligands.

The lower oxidation states of titanium are all strongly reducing. Thus, unless specific precautions are taken, titanium complexes are typically oxidized rapidly to the +IV state. Moreover, many titanium complexes are extremely susceptible to hydrolysis. Consequently, the handling of titanium complexes normally requires oxygen- and water-free conditions. *See* COORDINATION CHEMISTRY.

L. Kieth Woo

Titanium metallurgy The winning of metallic titanium (Ti) from its ores followed by alloying and processing into forms and shapes that can be used for structural purposes.

All commercial titanium metal is produced from titanium tetrachloride ($TiCl_4$), an intermediate compound produced during the chlorination process for titantium oxide pigment. The process involves chlorination of ore concentrates; reacting TiO_2 with chlorine gas (Cl_2) and coke (carbon; C) in a fluidized-bed reactor forms impure titanium tetrachloride. For the production of acceptable metal, purification of the raw tetrachloride is required to remove other metal chlorides that would contaminate the virgin titanium.

The purified titanium tetrachloride is delivered as a liquid to the reactor vessel. In these vessels, constructed of carbon or stainless steel, the titanium tetrachloride is reacted with either magnesium (Mg) or sodium (Na) to form the pure metal called sponge, because of its porous cellular form. To avoid contamination by oxygen or nitrogen, the reaction is carried out in an argon atmosphere. The sponge, removed from the reactor pot by boring, is cleaned by acid leaching.

A mass of sponge, alloy additions, and scrap are mixed, then compressed into compacts and welded together to form a sponge electrode. This is melted by an electric arc into a water-cooled copper crucible in a vacuum or an atmosphere of purified argon. The arc progressively consumes the sponge electrode to form an ingot. Ingots up to 30,000 lb (13,600 kg) are routinely produced by using this consolidation method.

The conversion of the titanium ingot into mill products, such as forging billet, plate, sheet, and tubing, is accomplished for the most part on conventional metalworking equipment. Mills designed to roll and shape stainless or alloy steel are used with only slight modifications. For this reason titanium and its structural alloys are produced in most of the same forms and shapes as stainless steel. *See* METAL FORMING; STAINLESS STEEL.

Pure titanium is soft, weak, and extremely ductile. However, through appropriate additions of other elements, the titanium metal base is converted to an engineering material having unique characteristics, including high strength and stiffness, corrosion resistance, and usable ductility. The type and quantity of alloy addition determine the mechanical and, to some extent, the physical properties. Some common titanium alloy additions include aluminum, oxygen, nitrogen, carbon, molybdenum, chromium, iron, nickel, zirconium, and tin. *See* ALLOY; METALLURGY; TITANIUM.

Ward W. Minkler; Stanley R. Seagle

Titanium oxides Chemical compounds of the metal titanium and oxygen. The most commonly found and used titanium oxides are the titanium dioxides, TiO_2; but other oxides are known including the sesquioxide Ti_2O_3, the monoxide TiO, and nonstoichiometric phases TiO_x, with x taking values between 0.7 and 1.3. Titanium dioxide exists in three common crystalline forms under ambient conditions: rutile, anatase (also known as octahedrite), and brookite. Each polymorph contains titanium atoms surrounded by a distorted octahedron of oxygen atoms, and each form differs in the way in which the octahedral units are linked by various combinations of edge and corner sharing to give extended network structures. Rutile is considered to be the most stable form of TiO_2, since at temperatures above 500°C (930°F) both anatase and brookite are converted into rutile. *See* OXIDE; RUTILE; TITANIUM.

Both rutile and anatase are found in nature as minerals. There are also other more complex titanium-containing oxides, such as ilmenite ($FeTiO_3$) and leucoxene ($TiO_2 \cdot xFeO \cdot yH_2O$), from which the titanium dioxides may be extracted. Most of the titanium dioxide for commercial applications is extracted from titanium ores using one of two processes. In the chloride process, titanium ore is chlorinated using chorine gas and charcoal to yield volatile $TiCl_4$, which is separated by distillation from impurities and then oxidized by flame treatment to yield pure TiO_2. In the sulfate process, titanium ore is dissolved in sulfuric acid to produce a suspension of titanium oxyhydroxide from which TiO_2 is produced after filtration and firing. *See* ILMENITE.

The most widespread use of titanium dioxides is in the area of white pigments. Both anatase and rutile have exceptionally high refractive indices and therefore scatter light very effectively, offering highly effective opacity or hiding power as well as imparting whiteness and brightness to products. Rutile is most widely used since it has superior properties over anatase (a higher refractive index and a higher density, therefore offering greater opacity per gram), and has major pigmentary applications in coatings, paints, paper, inks, plastics, and rubbers, as well as in foods, cosmetics, toothpastes, and ultraviolet (UV) protection products. *See* PIGMENT (MATERIAL).

Titanium dioxides have been widely investigated as photocatalysts for a variety of applications, including catalyst materials for the production of energy by the decomposition of water into oxygen and hydrogen for solar energy conversion and the degradation of organic pollutants.

Richard Walton

Titration A quantitative analytical process that is basically volumetric. However, in high-precision titrimetry the titrant solution is sometimes delivered from a weight buret, so that the volumetric aspect is indirect. Generally, a standard solution, that is, one containing a known concentration of substance X (titrant), is progressively added to a measured volume of a solution of a substance Y (titrand) that will react with the titrant. The addition is continued until the end point is reached.

Ideally, this is the same as the equivalence point, at which an excess of neither X nor Y remains. If the stoichiometry or exact ratio in which X and Y react is known, it is possible to calculate the amount of Y in the unknown solution. *See* VOLUMETRIC ANALYSIS.

The normal requirements for the performance of a titration are: a standard titrant solution; calibrated volumetric apparatus, including burets, pipets, and volumetric flasks; and some means of detecting the end point. *See* BURET; PIPET; VOLUMETRIC FLASK.

Classification by chemical reaction. For the purposes of titrimetry, chemical reactions can be placed in three general categories: acid-base or neutralization, combination, and oxidation-reduction.

Acid-base titrations involve neutralization of an acid by titration with a base, or vice versa. However, the process is often nonspecific; in the titration of a mixture of nitric and hydrochloric acids, only the total acidity can be found without recourse to additional measurements. A salt derived from a strong base and a very weak acid can often be titrated just as if it were a base. *See* ACID AND BASE.

In titrimetry, attention is usually focused upon the combination of an ion in the titrant with one of the opposite sign in the titrand solution. Sometimes the combination may involve more than two species, some of which may be nonionic. The combinations may result in precipitation or formation of a complex. *See* COORDINATION COMPLEXES; PRECIPITATION (CHEMISTRY).

In so-called redox titrations the titrant is usually an oxidizing agent, and is used to determine a substance that can be oxidized and hence can act as a reducing agent. *See* OXIDATION-REDUCTION.

Coulometric titration. The passage of a uniform current for a measured period of time can be used to generate a known amount of a product such as a titrant. This fact is the basis of the technique known as coulometric titration. An obvious requirement is that generation shall proceed with a fixed, preferably 100%, current efficiency. The uniform current is then analogous to the concentration of an ordinary titrant solution, while the total

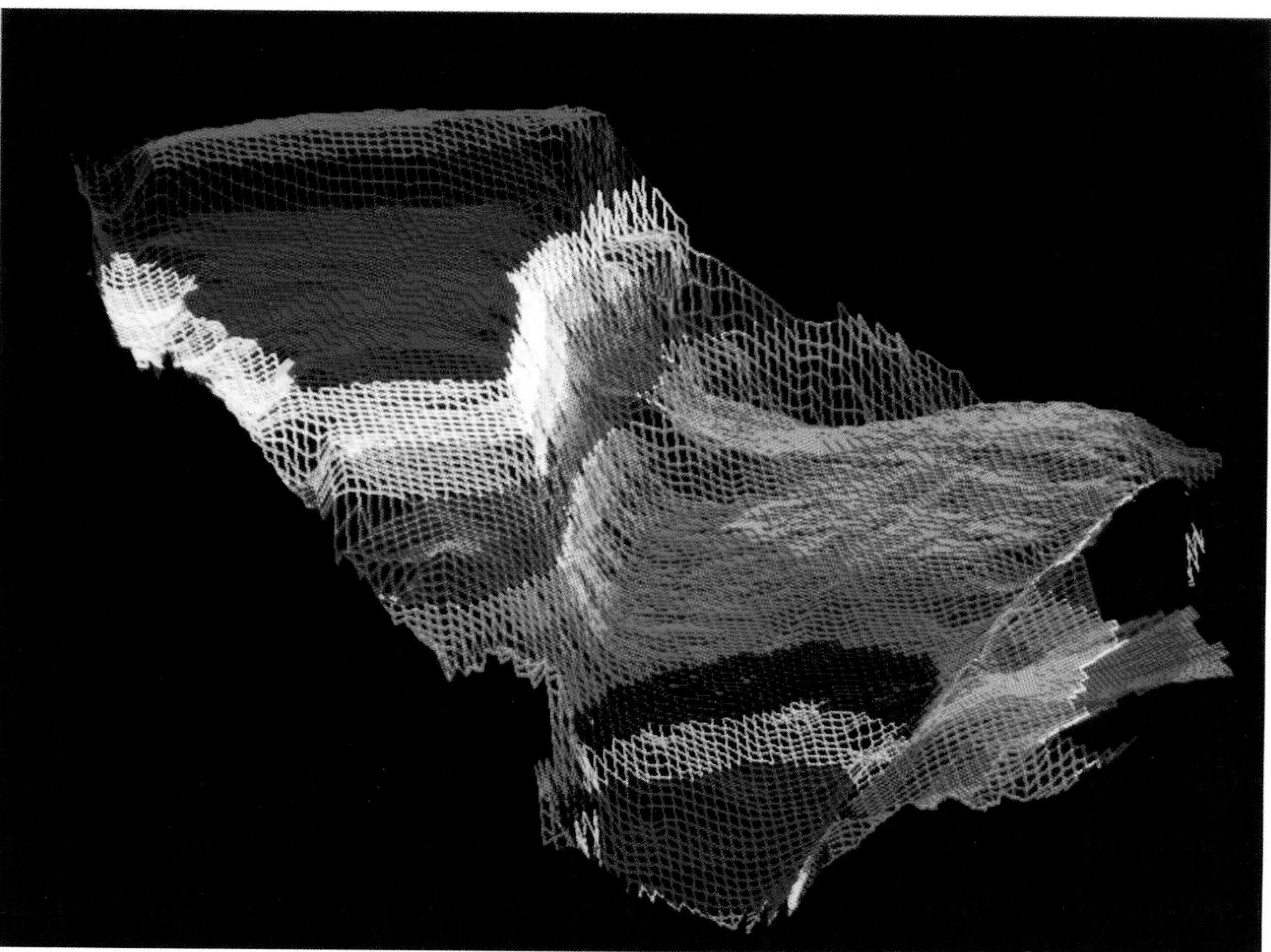

Digital Terrain Model (DTM) of the topographic map of the beach and river channel at River Mile 194 of the Colorado River in Grand Canyon (oblique view facing downstream). Each color band represents 1 meter of elevation change.

time of passage is analogous to the volume of such a solution that would be needed to reach the end point. *See* ELECTROLYSIS.

Classification by end-point techniques. The precision and accuracy with which the end point can be detected is a vital factor in all titrations. Because of its simplicity and versatility, chemical indication is quite common, especially in acid-base titrimetry.

Indicators. An acid-base indicator is a weak acid or a weak base that changes color when it is transformed from the molecular to the ionized form, or vice versa. The color change is normally intense, so that only a low concentration of indicator is needed. The working range, or visual color change, of a typical acid-base indicator is spread over about a hundredfold (~2 pH units) change in hydrogen ion concentration. Available indicators have individual working ranges that together cover the entire range of hydrogen ion concentration likely to be encountered in general acid-base titration. *See* ACID-BASE INDICATOR; HYDROGEN ION; PH.

Sometimes no suitable chemical indicator can be found for a desired titration. Possibly the concentrations involved may be so low that chemical indication functions poorly. Other situations might be the need for high precision or for the automatic arrest of the titration. Recourse is then made to some physical method of end-point detection.

Potentiometric titration. If a pH meter is used, its associated electrodes are first standardized by use of a buffer solution of known pH. By suitable choice of electrodes, potentiometric methods can also be applied to combination titrations and to oxidation-reduction titrations. The advent of modern ion-selective electrodes has greatly extended the scope of potentiometric titration and of other branches of titrimetry. *See* ELECTRODE POTENTIAL; ION-SELECTIVE MEMBRANES AND ELECTRODES.

Conductometric titration. Conductometric titration is sometimes successful when chemical indication fails. The underlying principles of conductometric titration are that the solvent and any molecular species in solution exhibit only negligible conductance; that the conductance of a dilute solution rises as the concentration of ions is increased; and that at a given concentration the hydrogen ion and the hydroxyl ion are much better conductors than any of the other ions. *See* ELECTROLYTIC CONDUCTANCE.

Spectrophotometric titration. The spectrophotometer is an optical device that responds only to radiation within a selected very narrow band of wavelengths in the visual, ultraviolet, or infrared regions of the spectrum. The response can be made both quantitative and linearly related to the concentration of a species that absorbs radiation within this band. Titrations at wavelengths within the visual region are by far the most common. *See* SPECTROPHOTOMETRIC ANALYSIS.

Amperometric titration. By use of a dropping-mercury or other suitable microelectrode, it is possible to find a region of applied electromotive force (emf) in which the current is proportional to the concentration of one or both of the reactants in a titration.

Biamperometric titration is a closely related technique. An emf that is usually small is applied across two identical microelectrodes that dip into the titrand solution. This arrangement, which involves no liquid-liquid junctions, is valuable in nonaqueous titrations, but also finds much use in aqueous titrimetry. *See* POLAROGRAPHIC ANALYSIS.

Thermometric or enthalpimetric titration. Many chemical reactions proceed with the evolution of heat. If one of these is used as the basis of a titration, the temperature first rises progressively and then remains unchanged as the titration is continued past the end point. If the reaction is endothermic, the temperature falls instead of rising. Thermometric titration is applicable to all classes of reactions. *See* THERMOCHEMISTRY.

Nonaqueous titration. This technique is used to perform titrations that give poor or no end points in water. Although applicable in principle to all classes of reactions, acid-base applications have greatly exceeded all others. Nonaqueous titrations in which the solvent is a molten salt or salt mixture are also possible.

Automatic titration. Automation is particularly valuable in routine titrations, which are usually performed repeatedly. One approach is to record the titration curve and to interpret it later. Another method is to stop titrant addition or generation automatically at, or very near to, the end point. Although a constant-delivery device is desirable, an ordinary buret with an electromagnetically controlled valve is often used.

Microcomputer control permits such refinements as the continuous adjustment of the titrant flow rate during the titration. In some cases, it is possible to automate an entire analysis, from the measurement of the sample to the final washout of the titration vessel and the printout of the result of the analysis. *See* ANALYTICAL CHEMISTRY.

John T. Stock

Tobacco The plant genus *Nicotiana*, certain species in the genus, and dried leaves of these plants are all called tobacco. Most often tobacco means a leaf product containing 1–3% of the alkaloid nicotine, which produces a narcotic effect when smoked, chewed, or snuffed. The plant *N. rustica* provides tobacco in parts of Europe, but the tobacco of world commerce is *N. tabacum*. Tobacco is American in origin. *See* SOLANALES.

Gordon S. Taylor

Toggle Any of a wide variety of mechanisms, many used to open or close electrical contacts abruptly and all characterized by the control of a large force by a small one. The basic action of a toggle mechanism is shown in illustration *a*. When $\alpha = 90^\circ$ the forces P and Q are independent of each other. Again, when $\alpha = 0^\circ$ the forces are isolated, force Q being sustained entirely by the frame, and force P serving only to hold the link in position. At $\alpha = 45^\circ$ from the symmetry $|P| = |Q|$, the mechanism serves to transfer the direction of forces to achieve equilibrium. *See* COUPLING.

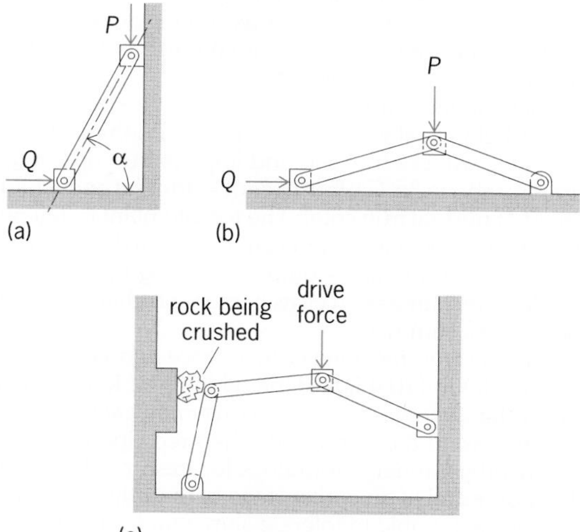

Toggle mechanism. (*a*) Simple structure. (*b*) Traditional configuration. (*c*) Typical application.

Because the simple configuration of illus. *a* requires low-friction sliders, it is impractical. A more useful structure replaces the vertical slider with a second link pinned to the frame (illus. *b*), in which case input P sets up forces in both links. A further modification (illustration *c*) replaces the other slider with a link. *See* FOUR-BAR LINKAGE; LINKAGE (MECHANISM).

Frank H. Rockett

Tolerance Amount of variation permitted or "tolerated" in the size of a machine part. Manufacturing variables make it impossible to produce a part of exact dimensions; hence the designer must be satisfied with manufactured parts that are between a maximum size and a minimum size. Tolerance is the difference between maximum and minimum limits of a basic dimension. For instance, in a shaft and hole fit, when the hole is a minimum size and the shaft is a maximum, the clearance will be the smallest, and when the hole is the maximum size and the shaft the minimum, the clearance will be the largest.

If the initial dimension placed on the drawing represents the size of the part that would be used if it could be made exactly to size, then a consideration of the operating conditions of the pair of mating surfaces shows that a variation in one direction from the ideal would be more dangerous than a variation in the opposite direction. The dimensional tolerance should be in the less dangerous direction. This method of stating tolerance is called unilateral tolerance and has largely displaced bilateral tolerance, in which variations are given from a basic line in plus and minus values. Paul H. Black

Tomato An important vegetable, belonging to the genus *Lycopersicon*, especially *L. esculentum*, that is grown for its edible fruit. The tomato was first domesticated in Mexico, and was introduced to Europe in the mid-sixteenth century. It has been grown in the United States since colonial days, but it became an important vegetable there only in the past century.

The genus *Lycopersicon* (Greek, "wolf peach") is a member of the Solanaceae, the nightshade family. *Lycopersicon esculentum*, the familiar tomato, can be hybridized with each of the eight other species of *Lycopersicon*. Tomato breeders have transferred many genes, particularly for disease resistance, from wild *Lycopersicon* species to the tomato. *See* BREEDING (PLANT).

The tomato is a herbaceous perennial, but is usually grown as an annual in temperate regions since it is killed by frost. Cultivated tomatoes are self-fertile. The fruit is a berry with 2 to 12 locules containing many seeds. Most tomato varieties have red fruit, due to the red carotenoid lycopene. Different single genes are known to produce various shades of yellow, orange, or green fruit. There is no basis for the common belief that yellow-fruited tomatoes are low in acidity.

Tomatoes prefer warm weather. Cool temperature, 10°C (50°F) and below, delays seed germination, inhibits vegetative development, reduces fruit set, and impairs fruit ripening. High temperature, above 35°C (95°F), reduces fruit set and inhibits development of normal fruit color. The tomato plant is day-neutral, flowering when grown with either short or long days. This makes it possible to grow tomatoes outdoors during the short days of winter in frost-free areas as well as in more northern areas during the long days of summer.

The tomato is the most important processed vegetable, constituting over 23 lb (10.4 kg) of the 54 lb (24.3 kg) of processed vegetables the average American consumes each year. Tomatoes for processing are harvested when red ripe and are soon sent to a nearby cannery. Tomatoes for fresh market, however, are often harvested at an earlier stage of maturity when they are still firm and better able to tolerate shipment to distant markets. Most tomatoes for fresh market are harvested by hand, but almost all of the processing tomatoes in California are harvested mechanically. *See* AGRICULTURAL MACHINERY.

The tomato is highly esteemed as a source of vitamin C, and one medium-sized tomato provides about half of the required daily allowance for adults. Tomatoes are also a significant source of vitamin A and are a good source of protein, but most of it is in the seeds. Tomato juice contains 19 amino acids, principally glutamic acid. *See* ASCORBIC ACID; VITAMIN A.

The tomato has been a favored organism for genetic studies. Over a thousand genes are known for the tomato, and several hundred of these have been located on their respective chromosomes. *See* AGRICULTURAL SCIENCE (PLANT); FOOD MANUFACTURING; PLANT PHYSIOLOGY. R. W. Robinson

Tommotian fauna The first diverse assemblages of unquestionable animal fossils at the Proterozoic-Phanerozoic transition, which marks the change from a predominantly microbial biosphere to a modern type of biosphere abundant with multicellular life. The name derives from the Early Cambrian Tommotian Stage in Siberia, where the significance of this fauna of early skeletal fossils was first realized. However, the concept goes beyond the geographical and temporal boundaries of the Tommotian Stage. It can be traced back to the low-diversity assemblages of skeletal animal fossils appearing near the end of the Neoproterozoic. *See* CAMBRIAN; FOSSIL.

Traditionally, the fauna was considered to predate the earliest trilobites; however, recent stratigraphic studies suggest that at least some of the Tommotian Stage correlates with trilobite-carrying beds elsewhere. A number of the characteristic Tommotian taxa are now also known to continue into post-Tommotian strata. The Tommotian fauna is known to have radiated from all continents, but it is particularly diverse and abundant on the Siberian Platform in Russia, in Australia, and in the belt of phosphorite-rich deposits that extends from the South China Platform through Mongolia, Kazakhstan, the Himalayas, and Iran. *See* TRILOBITA.

The Tommotian fauna has been instrumental in forming our understanding of the Cambrian explosion (dramatic evolutionary radiation of animals beginning about 545 million years ago), because its preservation is generally not dependent on extraordinary conditions and is therefore less spotty. Also, fossilized embryos of Tommotian animals make it possible to understand the complete life cycles of some of the most basal members of the metazoan evolutionary tree. *See* ANIMAL EVOLUTION.

Characteristic morphologic features of the Tommotian fauna include mineralized tubes, spicules (knoblike supporting structures), sclerites (hardened plates), and shells, often belonging to animals of unknown affinities. The minerals involved are opal (a hydrated gel of silica), apatite (calcium phosphate), and aragonite/calcite (calcium carbonate)—the same minerals that are common in animal skeletons today. Stefan Bengtson

Ton of refrigeration A rate of cooling that is equivalent to the removal of heat at 200 Btu/min (200 kilojoules/min), 12,000 Btu/h (13 megajoules/h), or 288,000 Btu/day (300 MJ/day). This unit of measure stems from the original use of ice for refrigeration. One pound of ice, in melting at 32°F (0°C), absorbs as latent heat approximately 144 Btu/lb (335 J/kg), and 1 ton (0.9 metric ton) of ice, in melting in 24 h, absorbs 288,000 Btu/day (300 MJ/day). In Europe, where the metric system is used, the equivalent cooling unit is the frigorie, which is a kilogram calorie, or 3.96 Btu. Thus 3000 frigories/h is approximately 1 ton of refrigeration. A standard ton of refrigeration is one developed at standard rating conditions of 5°F (−15°C) evaporator and 86°F (30°C) condenser temperatures, with 9°F (−13°C) liquid subcooling and 9°F (−13°C) suction superheat. *See* REFRIGERATION. Carl F. Kayan

Tone (music and acoustics) Physically, a sound that is composed of discrete frequency (or sine-wave) components; psychologically, an auditory sensation that is characterized foremost by its pitch or pitches.

The physical definition distinguishes a tone from a noise, wherein the components form a continuum of frequencies. Tones may be pure, consisting of a single frequency, or they may be complex. Complex tones, in turn, may be periodic or not periodic. Periodic complex tones repeat themselves at rapid regular intervals. They have frequency components that are harmonics—discrete frequencies that are integer multiples of a fundamental frequency. For example, the tone of an oboe consists of a fundamental frequency of 440 hertz, a second harmonic component with a frequency of 880 Hz, a third harmonic at 1320 Hz, and so on. In general, musical instruments that generate continuous sounds—the bowed strings, the brasses, and the woodwinds—create such periodic tones. Tones that are not periodic (aperiodic) have frequency components that do not fit a harmonic series. Percussive instruments such as kettledrums and bells make such aperiodic tones. *See* HARMONIC (PERIODIC PHENOMENA); MUSICAL ACOUSTICS; MUSICAL INSTRUMENTS; PERIODIC MOTION.

Pitch is a sensation of highness or lowness that is the basic element of melody. Periodic complex tones tend to have a single pitch, which listeners will match by a pure tone having a frequency equal to the fundamental frequency of the periodic complex tone. Aperiodic complex tones tend to have multiple pitches. A second psychological attribute of complex tones is tone color or timbre. Tone color is often represented by descriptive adjectives. The adjectives may be linked to the physical spectrum. Thus, a tone with strong harmonics above 1000 Hz may be called "bright." A tone with no harmonics at all above 1000 Hz may be called "dull" or "stuffy." *See* PITCH; PSYCHOACOUSTICS; SOUND.

William M. Hartmann

Tongue An organ located at the base of the oral cavity and found in all vertebrate animals. It is best developed in terrestrial vertebrates, where it takes on the functions of food procurement, food transport, and acquisition of chemosensory signals. The tongue generally is not a significant independent organ in fish, and it is secondarily reduced in organisms that feed aquatically, such as crocodilians and some turtles.

Within terrestrial vertebrates, there is considerable variability in the specific structure of the tongue, the degree of participation of the hyoid skeleton (that is, a complex of bones at the base of the tongue which supports the tongue and its muscles), and the mechanisms of movement. In birds the tongue is merely a thickened epithelium that overlies the hyoid apparatus. Movement is produced by moving various hyoid elements. In most amphibians, including both frogs and salamanders, the hyoid provides extensive support, but considerable intrinsic tongue musculature exists. In squamate reptiles (lizards and snakes) and mammals the tongue is largely independent of the hyoid apparatus and is composed entirely of muscle. The musculature is tightly packed in the tongue and is generally arranged in three mutually perpendicular planes. In the mammalian tongue the musculature is arranged into longitudinal, transverse, and vertical bundles. Organs composed entirely of muscle and lacking independent skeletal systems, termed muscular hydrostats, are widespread. One of the primary advantages of a muscular hydrostat is that bending is not restricted to movement at joints, and the highly subdivided muscular and neural systems seen in mammalian tongues in particular produce movements that are remarkably specific, complex, and diverse.

While muscular-hydrostatic movements characterize the tongue of most mammals and many lizards and snakes, many of the most spectacular tongue projectors, such as chameleon lizards and plethodontid salamanders, do not use this mechanism in protrusion. These organisms have developed separate mechanisms in which the muscular tongue is projected ballistically from the body. In both, a muscle squeezes a process of the hyoid apparatus to generate the projectile force. *See* TASTE.

Kathleen K. Smith

Tonsil Localized aggregation of diffuse and nodular lymphoid tissue found in the region where the nasal and oral cavities open into the pharynx. The tonsils are important sources of blood lymphocytes. They often become inflamed and enlarged, necessitating surgical removal. *See* TONSILLITIS.

The two palatine (faucial) tonsils are almond-shaped bodies measuring 1 by 0.5 in. (2.5 by 1.2 cm) and are embedded between folds of tissue connecting the pharynx and posterior part of the tongue with the soft palate. These are the structures commonly known as the tonsils. The lingual tonsil occupies the posterior part of the tongue surface. It is really a collection of 35–100 separate tonsillar units, each having a single crypt surrounded by lymphoid tissue. Each tonsil forms a smooth swelling about 0.08–0.16 in. (2–4 mm) in diameter. The pharyngeal tonsil (called adenoids when enlarged) occupies the roof of the nasal part of the pharynx. This tonsil may enlarge to block the nasal passage, forcing mouth breathing. *See* LYMPHATIC SYSTEM.

Theodore Snook

Tonsillitis An inflammation of the tonsil. Tonsillitis is a nonspecific term usually referring to bacterial or viral infection involving all or part of Waldeyer's ring, a collection of lymphatic tissue encircling the pharynx. It consists primarily of the tonsils (palatine tonsils), adenoids (pharyngeal tonsils), and lingual tonsils.

The complications of tonsillitis depend on which tonsil is involved. Recurrent adenoiditis with adenoid hypertrophy is frequently associated with recurrent otitis media, middle-ear fluid, and at times nasal obstruction with mouth breathing and snoring. Acute palatine tonsillitis may be complicated by peritonsillar abscess which may develop lateral to the tonsillar capsule. Removal of the adenoids is considered when there is residual middle-ear fluid. Palatine tonsils must be removed after peritonsillar abscess, but otherwise their removal depends upon the frequency of recurrent attacks of bacterial pharyngotonsillitis in relation to the patient's age. *See* TONSIL.

James A. Donaldson

Tooth One of the structures found in the mouth of most vertebrates which, in their most primitive form, were conical and were usually used for seizing, cutting up, or chewing food, or for all three of these purposes. The basic tissues that make up the vertebrate tooth are enamel, dentin, cementum, and pulp (see illustration).

Enamel is the hardest tissue in the body because of the very high concentration, about 96%, of mineral salts. The remaining 4% is water and organic matter. The enamel has no nerve supply, although it is nourished to a very slight degree from the dentin it surrounds. The fine, microscopic hexagonal rods (prisms) of apatite which make up the enamel are held together by a cementing substance.

Dentin, a very bonelike tissue, makes up the bulk of a tooth, consisting of 70% of such inorganic material as calcium and phosphorus, and 30% of water and organic matter, principally

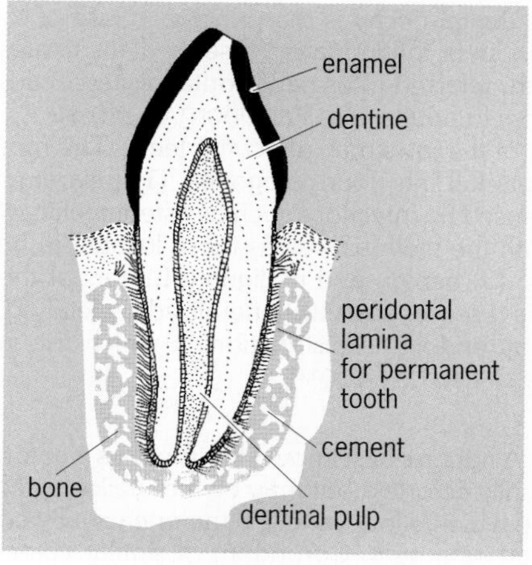

Structure of a tooth.

collagen. The rich nerve supply makes dentin a highly sensitive tissue; this sensitivity serves no obvious physiological function.

Cement is a calcified tissue, a type of modified bone less hard than dentin, which fastens the roots of teeth to the alveolus, the bony socket into which the tooth is implanted. A miscellaneous tissue, consisting of nerves, fibrous tissue, lymph, and blood vessels, known as the pulp, occupies the cavity of the tooth surrounded by dentin.

The dentition of therian mammals, at least primitively, consists of four different kinds of teeth. The incisors (I) are usually used for nipping and grasping; the canines (C) serve for stabbing or piercing; the premolars (Pm) grasp, slice, or function as additional molars; and the molars (M) do the chewing, cutting, and grinding of the food. Primitively the placentals have 40 teeth and the marsupials 50.

In therian mammals, probably because of the intricacies and vital importance of tooth occlusion, only part of the first (or "milk") dentition is replaced. This second, or permanent, dentition is made up of incisors, canines, and premolars; as a rule only one premolar is replaced in marsupials. Although the molars erupt late in development and are permanent, that is, not replaced, they are part of the first, or deciduous, dentition. *See* TOOTH DISORDERS.

Frederick S. Szalay

Tooth disorders Diseases and disturbances of the teeth and associated structures, including abnormal formation and growth of the teeth and jaws, tooth decay, inflammation of the tissues housing the roots of the teeth, and various diseases of the jaw bones.

Defective formation of dentin and enamel are referred to respectively as dentinogenesis and amelogenesis imperfecta, and may be caused by febrile illness during the period of tooth formation, or by faulty calcium or phosphorus metabolism, such as occurs in rickets. Some or all of the teeth may fail to form completely, or extra or supernumerary teeth may be present.

In most societies, dental decay, or caries, is one of the most important and common tooth disorders. Decay occurring during the adolescent years is caused by a class of microorganisms referred to as the cariogenic streptococci, of which *Streptococcus mutans* is a predominant member. However, a susceptible host and a cariogenic diet containing sucrose are also essential factors. Although the mechanism by which bacteria cause decay is not completely understood, most experts believe that cariogenic organisms, by using sucrose, and to a lesser extent other sugars, produce polymers which bind the organisms to the tooth surface and acids which cause demineralization resulting in cavity formation. Fluoride administration, either in the drinking water or by other routes, effectively reduces dental decay about 50–70%. The mechanism by which fluoride causes decreased decay rates is not understood.

Whereas dental decay is the principal cause of tooth loss in young individuals, inflammatory disease of the tissues surrounding the teeth, referred to as periodontal disease, causes most of the tooth loss in adults. *See* PERIODONTAL DISEASE.

Diseases of the jaws may affect the teeth. The most common jaw disorders fall into four categories: (1) inflammation of the jawbone caused by infections such as osteomyelitis; (2) cysts associated with the teeth as well as those located in bone sutures of the jaws; (3) benign and malignant tumors of the jaws; and (4) systemic diseases such as generalized skeletal abnormalities produced by endocrine disfunction in which the jaws are affected. *See* DENTISTRY; TOOTH.

Roy C. Page

Topaz A mineral best known for its use as a gemstone. Crystals are usually colorless but may be red, yellow, green, blue, or brown. The wine-yellow variety is the one usually cut and most highly prized as a gem. Corundum of similar color sometimes goes under the name of Oriental topaz. Citrine, a yellow variety of quartz, is the most common substitute and may be sold as quartz topaz.

Topaz is a nesosilicate with chemical composition $Al_2SiO_4(F,OH)_2$. The mineral crystallizes in the orthorhombic system and is commonly found in well-developed prismatic crystals with pyramidal terminations. It has a perfect basal cleavage which enables it to be distinguished from minerals otherwise similar in appearance. Hardness is 8 on Mohs scale; specific gravity is 3.4–3.6. *See* HARDNESS SCALES.

Fine yellow and blue crystals have come from Siberia and much of the wine-yellow gem material from Minas Gerais, Brazil. In the United States topaz has been found near Florissant, Colorado; in Thomas Range, Utah; in San Diego County, California; and near Topsham, Maine. *See* GEM; SILICATE MINERALS.

Cornelius S. Hurlbut, Jr.

Topographic surveying and mapping The measurement of surface features and configuration of an area or a region, and the graphic expression of those features. Surveying is the art and science of measurement of points on, above, or under the surface of the Earth. Topographic maps show the natural and cultural features of a piece of land. The natural features include configuration (relief), hydrography, and vegetation. The cultural features include roads, buildings, bridges, political boundaries, and the sectional breakdown of the land. Topographic maps are used by a wide variety of people, such as engineers designing a new road; backpackers finding their way into remote areas; scientists describing soil or vegetation types, wildlife habitat, or hydrology; and military personnel planning field operations. *See* CARTOGRAPHY; MAP PROJECTIONS; MAP REPRODUCTION.

Topographic maps that show natural and cultural features only in plan view are called planimetric maps, while maps that show relief are called hypsometric maps. Contour lines join points along a line of the same elevation across the ground. Contours show not only the elevation of the ground but also the geomorphic shape of features. *See* CONTOUR.

A digital terrain model (DTM) is a computer-generated grid laid over the topographic information, which can then be rotated, tilted, and vertically exaggerated to give a three-dimensional view of the ground from different perspectives, including oblique representations. This technology is an excellent presentation tool: it utilizes the advances that have been made in computer mapping and drafting software.

Prior to starting collection of data, a network of known horizontal and vertical control points must be established. The network also allows measurements made from several different locations in the same coordinate system to fit together into the same reference datum (the basis for the coordinate system). Field methods that are used to make measurements include ground surveys, geographic positioning systems, and hydrographic surveys. Photogrammetry and remote sensing techniques involve the use of photography to obtain reliable measurements. Photographs can be taken from airplanes, helicopters, and even satellites; thus the term remote sensing is applied to this technology. *See* AERIAL PHOTOGRAPH; HYDROGRAPHY; PHOTOGRAMMETRY; REMOTE SENSING; SURVEYING INSTRUMENTS.

Chris Brod

Topology The branch of mathematics that studies the qualitative properties of spaces, as opposed to the more delicate and refined geometric or analytic properties. While there are earlier results that belong to the field, the beginning of the subject as a separate branch of mathematics dates to the work of H. Poincaré during 1895–1904. The ideas and results of topology have a central place in mathematics, with connections to almost all the other areas of the subject.

The difference between topological and geometric properties is illustrated by the example of a space with three separate pieces. The exact shapes of the pieces constitute a geometric property of the space, and the study of these shapes is in the domain

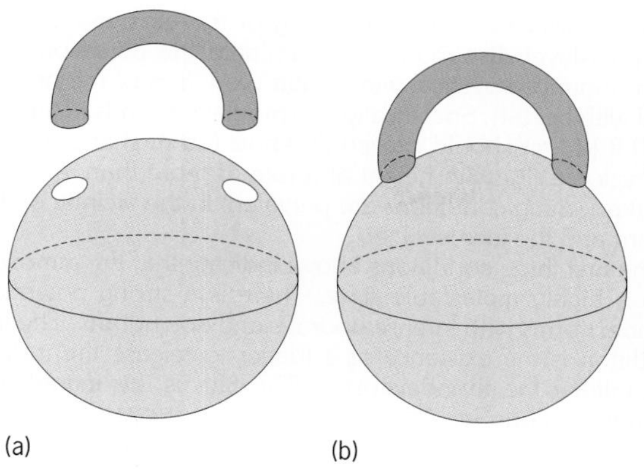

Process of adding a handle to a 2-sphere. (*a*) Cutting of holes in sphere. (*b*) Connecting of holes by a curved pipe.

of differential geometry, but the fact that the space has three separate pieces is a qualitative or topological property. As another example, if a round sphere is deformed to be pear-shaped (or even more irregularly shaped, like the surface of the Earth), then the geometric notions of distance, straight line, and angle are changed, but the topological properties of the surface are left unchanged. However, if a handle is added by cutting two holes in the sphere and connecting them by a curved pipe, then the topology of the surface is changed (see illustration).

Four major areas of topology are algebraic topology, homotopy theory, general topology, and manifold theory. Algebraic topology, the first area of modern topology to be developed, is concerned with associating algebraic invariants to geometric spaces in order to measure higher-dimensional analogs of the number of pieces of a space or the number of handles of a surface. (An algebraic invariant of a space is an algebraic object associated to the space that remains unchanged if the space is replaced by a homeomorphic space. By an algebraic object is meant either an algebraic structure, such as a group, ring, or field, or an element of an algebraic structure.) Algebraic topology has tremendous influence on other branches of mathematics, both direct (application of the invariants of algebraic topology to problems from other areas of mathematics and physics) and indirect (application in other contexts of ideas arising from algebraic topology).

As algebraic topology developed, it became clear that if one function could be continuously deformed to another (that is, if they were homotopic), then these two functions behaved in the same way as far as the invariants of algebraic topology were concerned. This led naturally to the study of invariants that remain unchanged as the maps are deformed by homotopies, that is, homotopy invariants. This study, which is an offshoot of algebraic topology, is called homotopy theory. Some of the most interesting homotopy invariants are the higher homotopy groups. These proved extremely difficult to compute, even for spaces as simple as the sphere, and are the subject of much investigation.

Early in the development of topology it was realized that the foundations of the subject needed attention. General or point-set topology studies the relationships between the basic topological properties that spaces may possess.

Before abstract topological spaces were defined, there were numerous examples of spaces arising from geometric and analytic problems. The most important of these is a class of spaces known as manifolds. Both because of their ubiquitous appearance throughout mathematics and because they possess extraordinarily rich topological properties, manifolds became one of the central objects of study in topology. The basic theme in manifold theory is to find sufficient algebraic invariants to classify, that is, to list comprehensively, all manifolds, and to give methods for evaluating these invariants in geometric cases. *See* MANIFOLD (MATHEMATICS).

John W. Morgan

Torbanite A variety of coat that resembles a carbonaceous shale in outward appearance. It is fine-grained, brown to black, and tough, and breaks with a conchoidal or subconchoidal fracture. Torbanite is synonymous with boghead coal and is related to cannel coal. It is derived from colonial algae identified with the modern species of *Botryococcus braunii* and antecedent forms. High-assay torbanite yields paraffinic oil, whereas low-assay material yields asphaltic oil. *See* COAL.

Irving A. Breger

Torch A gas-mixing and burning tool that produces a hot flame for the welding or cutting of metal. The torch usually delivers acetylene and commercially pure oxygen producing a flame temperature of 5000–6000°F (2750–3300°C), sufficient to melt the metal locally. The torch thoroughly mixes the two gases and permits adjustment and regulation of the flame. Acetylene can produce a higher flame temperature than other fuel gases. *See* ACETYLENE; WELDING AND CUTTING OF METALS.

Torches are of two types: low-pressure and high-pressure. In a low-pressure, or injector, torch, acetylene enters a mixing chamber, where it meets a jet of high-pressure oxygen. The amount of acetylene drawn into the flame is controlled by the velocity of this oxygen jet. In a high-pressure torch both gases are delivered under pressure.

A welding torch mixes the fuel and gas internally and well ahead of the flame. For cutting, the torch delivers an additional jet of pure oxygen to the center of the flame. The oxyacetylene flame produced by the internally mixed gases raises the metal to its ignition temperature. The central oxygen jet oxidizes the metal, the oxide being blown away by the velocity of the gas jet to leave a narrow slit or kerf.

Frank H. Rockett

Tornado A violently rotating, tall, narrow column of air (vortex), typically about 300 ft (100 m) in diameter, that extends to the ground from a cumulonimbus cloud. The vast majority of tornadoes rotate cyclonically (counterclockwise in the Northern Hemisphere). Of all atmospheric storms, tornadoes are the most violent. *See* CLOUD; CYCLONE.

Tornadoes are made visible by a generally sharp-edged, funnel-shaped cloud pendant from the cloud base, and a swirling cloud of dust and debris rising from the ground (see illustration). The funnel consists of small waterdroplets that form as moist air entering the tornado's partial vacuum expands and cools. The condensation funnel may not extend all the way to the ground and may be obscured by dust. Many condensation funnels exist aloft without tangible signs that the vortex is in contact with the ground; these are known as funnel clouds. Tornado funnels assume various forms: a slender smooth rope, a cone (often truncated by the ground), a thick turbulent black cloud on the ground, or multiple funnels (vortices) that revolve around the axis of the overall tornado.

Many tornadoes evolve as follows: The tornado begins outside the precipitation region as a dust whirl on the ground and a short funnel pendant from a wall cloud on the southwest side of the thunderstorm; it intensifies as the funnel lengthens downward, and attains its greatest power as the funnel reaches its greatest width and is almost vertical; then it shrinks and becomes more tilted, and finally becomes contorted and ropelike as it decays. A downdraft and curtain of rain and large hail gradually spiral from the northeast cyclonically around the tornado, which often ends its life in rain. *See* HAIL; PRECIPITATION (METEOROLOGY); THUNDERSTORM.

Most tornadoes and practically all violent ones develop from a larger-scale circulation, the mesocyclone, which is 2–6 mi (3–9 km) in diameter and forms in a particularly virulent variety of thunderstorm, the supercell. The mesocyclone forms first at midaltitudes of the storm and in time develops at low levels

The Cordell, Oklahoma, tornado of May 22, 1981, in its decay stage. (*National Severe Storms Laboratory/University of Mississippi Tornado Intercept Project*)

and may extend to high altitudes as well. The tornado forms on the southwest side (Northern Hemisphere) of the storm's main updraft, close to the downdraft, after the development of the mesocyclone at low levels. Some supercells develop up to six mesocyclones and tornadoes repeatedly over great distances at roughly 45-min intervals. Tornadoes associated with supercells are generally of the stronger variety and have larger parent cyclones. Hurricanes during and after landfall may spawn numerous tornadoes from small supercells located in their rainbands. *See* HURRICANE.

Tornadoes are classified as weak, strong, or violent, or from F0 to F5 on the Fujita (F) scale of damage intensity. Sixty-two percent of tornadoes are weak (F0 to F1). These tornadoes have maximum windspeeds less than about 50 m/s (110 mi/h) and inflict only minor damage, such as peeling back roofs, overturning mobile homes, and pushing cars into ditches. Thirty-six percent of tornadoes are strong (F2 to F3) with maximum windspeeds estimated to be 50–90 m/s (110–200 mi/h). Strong tornadoes extensively damage the roofs and walls of houses but leave some walls partially standing. They demolish mobile homes, and lift and throw cars. The remaining 2% are violent (F4 to F5), with windspeeds in excess of about 90 m/s (200 mi/h). They level houses to their foundations, strew heavy debris over hundreds of yards, and make missiles out of heavy objects such as roof sections, vehicles, utility poles, and large, nearly empty storage tanks. *See* WIND.

Tornadoes occur most often at latitudes between 20° and 60°, and they are relatively frequent in the United States, Russia, Europe, Japan, India, South Africa, Argentina, New Zealand, and parts of Australia. Violent tornadoes are confined mainly to the United States, east of the Rocky Mountains.

Essentially, there are five atmospheric conditions that set the stage for wide-spread tornado development: (1) a surface-based layer, at least 3000 ft (1 km) deep, of warm, moist air, overlain by dry air at midlevels; (2) an inversion separating the two layers, preventing deep convection until the potential for explosive overturning is established; (3) rapid decrease of temperature with height above the inversion; (4) a combination of mechanisms, such as surface heating and lifting of the air mass by a front or upper-level disturbance, to eliminate the inversion locally; (5) pronounced vertical wind shear (variation of the horizontal wind with height). Specifically, storm-relative winds in the lowest 6000 ft (2 km) should exceed 20 knots (10 m/s) and veer (turn anticyclonically) with height at a rate of more than 10°/1000 ft (30°/km). Such conditions are prevalent in the vicinity of the jet stream and the low-level jet.

The first three conditions above indicate that the atmosphere is in a highly metastable state. There is a strong potential for thunderstorms with intense updrafts and downdrafts. The fourth condition is the existence of a trigger to release the instability and initiate the thunderstorms. The fifth is the ingredient for updraft rotation. *See* AIR MASS; FRONT; JET STREAM; TEMPERATURE INVERSION.

Robert Davies-Jones

Torpediniformes An order of batoid fishes occurring in the subclass Elasmobranchii and known as the torpedo electric rays and numbfishes. Typical members of Torpediniformes are identifiable by their flat, pancakelike body sector (disc); relatively robust tail sector; smooth skin; small or obsolete eyes; distinct caudal fin; and ovoviviparous development. Of special interest is a pair of enlarged electric organs located on the disc lateral to the gill slits. These electric organs, which may constitute 17% of the total body weight, deliver shocks up to 220 volts. The voltage depends on the species, its size, and physical condition. The electric organs are thought to be used primarily for feeding and defense.

Electric rays are poor swimmers, depending primarily on the tail because the disc is rather inflexible and of little use in locomotion. They spend most of their time partially buried in sand or mud and feed on a variety of invertebrates, including crustaceans, mollusks, and worms, as well as small fishes. Members of the genus *Torpedo* are reported to reach a length of 1.8 m (6 ft) and a weight of 44 kg (100 lb). Torpediniforms occur in intertidal waters to deep waters in temperate to tropical zones of all oceans. The order comprises two families: Torpedinidae and Narcinidae. *See* BATOIDEA; ELASMOBRANCHII; ELECTRIC ORGAN (BIOLOGY).

Herbert Boschung

Torque The product of a force and its perpendicular distance to a point of turning; also called the moment of the force. Torque produces torsion and tends to produce rotation. Torque arises from a force or forces acting tangentially to a cylinder or from any force or force system acting about a point. A couple, consisting of two equal, parallel, and oppositely directed forces, produces a torque or moment about the central point. A prime mover such as a turbine exerts a twisting effort on its output shaft, measured as torque. In structures, torque appears as the sum of moments of torsional shear forces acting on a transverse section of a shaft or beam. *See* COUPLE; TORSION. Nelson S. Fisk

Torque converter A device for changing the torque-speed ratio or mechanical advantage between an input shaft and an output shaft. A pair of gears is a mechanical torque converter. A hydraulic torque converter is an automatically and continuously variable torque converter, in contrast to a gear shift, whose torque ratio is changed in steps by an external control. *See* AUTOMOTIVE TRANSMISSION; GEAR DRIVE.

A mechanical torque converter transmits power with only incidental losses; thus, the power, which is the product of torque T and rotational speed N, at input I is substantially equal to the power at output O of a mechanical torque converter, or $T_I N_I = kT_O N_O$, where k is the efficiency of the gear train. This equal-power characteristic is in contrast to that of a fluid coupling in which input and output torques are equal during steady-state operations. *See* FLUID COUPLING.

In a hydraulic torque converter, efficiency depends intimately on the angles at which the fluid enters and leaves the blades of the several parts. Because these angles change appreciably over the operating range, k varies, being by definition zero when the output is stalled, although output torque at stall may be three times engine torque for a single-stage converter and five times engine torque for a three-stage converter. Depending on its input absorption characteristics, the hydraulic torque converter tends to pull down the engine speed toward the speed at which the engine develops maximum torque when the load pulls down the converter output speed toward stall. Henry J. Wirry

Torricelli's theorem The speed of efflux of a liquid from an opening in a reservoir equals the speed that the liquid would acquire if allowed to fall from rest from the surface of the reservoir to the opening.

Torricelli, a student of Galileo, observed this relationship in 1643. In equation form, $v^2 = 2gh$, in which v is the speed of efflux, h the head (or elevation difference between reservoir surface and center line of opening if in a vertical plane), and g the acceleration due to gravity. (The equation is the same as that for a solid particle dropped a distance h in a vacuum.) The relationship can be derived from the energy equation for flow along a streamline, if energy losses are neglected. *See* FLOW MEASUREMENT. Victor L. Streeter

Torsion A straining action produced by couples that act normal to the axis of a member. Torsion is identified by a twisting deformation.

In practice, torsion is often accompanied by bending or axial thrust as in the case of line shafting driving gears or pulleys, or propeller shafts for ship propulsion. Other important examples include springs and machine mechanisms usually having circular sections, either solid or tubular. Members with noncircular sections are of interest in special applications, such as structural members subjected to unsymmetrical bending loads that twist and buckle beams. *See* SPRING (MACHINES); TORSION BAR.

When subjected only to torque, the member is in pure torsion, which produces pure shear stresses. The shear properties of materials are determined by a torsion test. *See* SHEAR; TORQUE. John B. Scalzi

Torsion bar A spring flexed by twisting about its axis. Design of a torsion bar spring is primarily based on the relationships between the torque applied in twisting the spring, the angle through which the torsion bar twists, and the physical dimensions and material (modulus of elasticity in shear) from which the torsion bar is made. The illustration shows the elements of a simple torsion bar and the important dimensions involved in its design.

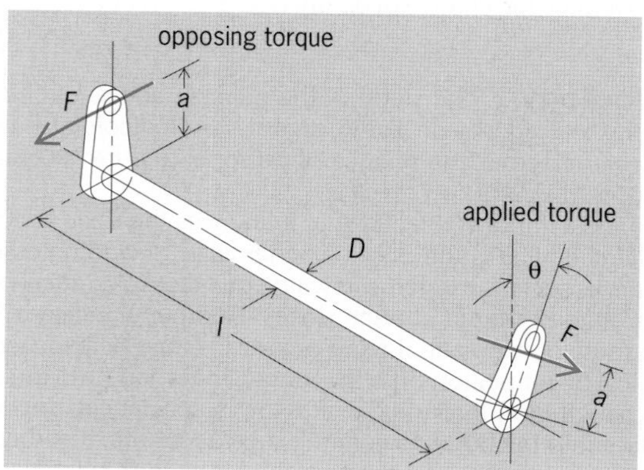

Diagram of torsion bar.

The equation below relates these dimensions. Here θ is angle

$$\theta = \frac{32Fal}{\pi D^4 G}$$

of twist in radians, F is force in pounds, a is radius arm of force in inches, l is length of torsion bar in inches, D is diameter of torsion bar in inches, and G is modulus of elasticity in shear in pounds per square inch.

Torsion bar springs are found in the spring suspension of truck and passenger car wheels, in production machines where space limitations are critical, and in high-speed mechanisms where inertia forces must be minimized. *See* SPRING (MACHINES). L. Sigfred Linderoth, Jr.

Torus A surface obtained by rotating a circle about a line that lies in its plane, but which has no points in common (see

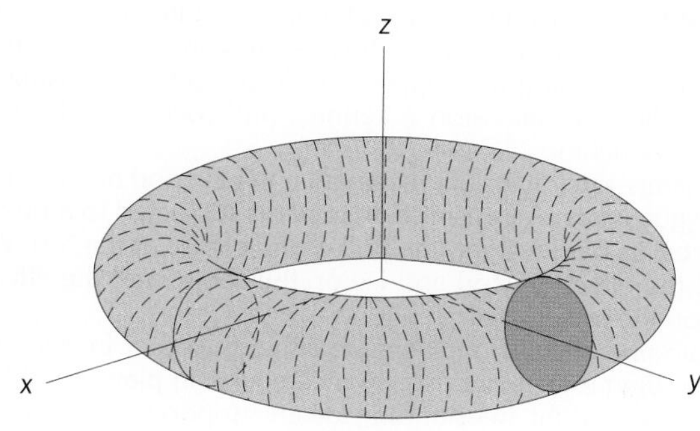

Diagram of a torus.

illustration). It is a two-dimensional manifold of genius 1 and connectivity 3. *See* MANIFOLD (MATHEMATICS); TOPOLOGY. Leonard M. Blumenthal

Tourette's syndrome A neurobehavioral disorder characterized by frequent, recurrent motor and vocal tics. The motor tics include brief, rapid, and darting movements of almost any muscle group, and can include eye blinking, eye rolling or deviations, nose wrinkling, facial grimacing, and head shaking. Some motor tics are more complex, are slow, and appear purposeful such as head turning, shoulder shrugging, touching, hopping, or twirling. Vocal tics are brief guttural sounds such as recurrent sniffing, throat clearing, coughing, and grunting or barking sounds. Complex vocal tics can be more meaningful and include verbal expressions. Tourette's syndrome has been described in nearly every country and ethnic group, with an estimated prevalence of one or two occurrences per 2000 people.

The motor and vocal tics begin in childhood, often worsen during adolescence, and tend to improve during the twenties and thirties. Symptoms increase with stress and excitement and decrease with activities that require focused effort. While the motor and vocal tics are involuntary, they can be suppressed for brief periods of time, giving the false impression that the movements and sounds are voluntary.

The pattern of inheritance is consistent with a single autosomal dominant gene whose expression is variable and dependent on the sex of the person. Tic symptoms can vary from transient tics to Tourette's syndrome and can include obsessive-compulsive symptoms. The complexity of symptoms is likely related to the various brain regions implicated in the development of Tourette's syndrome. Treatment can be targeted toward suppressing tics and the specific associated behavioral problems. Methods for tic suppression include medications that affect the brain by blocking the neurotransmitter dopamine at the site of nerve-to-nerve connections. *See* BRAIN; HUMAN GENETICS; NERVOUS SYSTEM DISORDERS. John T. Walkup; Mark A. Riddle

Tourmaline A cyclosilicate mineral family with (BO_3) triangular groups and a complex chemical composition. The general formula can be written $XY_3Al_6(OH)_4(BO_3)_3(Si_6O_{18})$, in which X = Na, Ca, and Y = Al, Fe^{3+}, Li, Mg, Mn^{2+}. The more common tourmalines are dravite (X,Y = Na,Mg), schorl (X,Y = Na,Fe), uvite (X,Y = Ca,Mg), and elbaite (X,Y = Na,Li). Fluorine commonly substitutes in the hydroxyl position. Tourmaline is a hard ($7\frac{1}{2}$ on Mohs scale), varicolored mineral which can be an important semiprecious gemstone. *See* GEM; SILICATE MINERALS.

The tourmaline crystal is polar; thus it is piezoelectric; that is, if pressure is exerted at one end, opposite electrical charges will occur at opposite poles. It is also pyroelectric, with the electrical charges developed at the ends of the polar axis on a change in temperature. Because of its piezoelectric property, tourmaline can be cut into gages to measure transient pressures. *See* PIEZOELECTRICITY; PYROELECTRICITY. Paul Brian Moore

Tower A concrete, metal, or timber structure that is relatively high for its length and width. Towers are constructed for many purposes, including the support of electric power transmission lines, radio and television antennas, and rockets and missiles prior to launching.

Transmission towers are rectangular in plan and are not steadied by guy wires. A transmission tower is subjected to a number of forces; its own weight, the pull of the cables at the top of the tower, the effect of wind and ice on the cable, and the effect of wind on the tower itself.

Radio and television towers are either guyed or freestanding. Freestanding towers are usually rectangular in plan. In addition to their own weight, freestanding towers support the weight of the antenna and accessories and the weight of ice, unless a deicing circuit is installed. Wind forces must also be carefully considered. Guyed towers are usually triangular in plan, with the main structural members, or legs, at the vertexes of the triangle. The legs are usually solid round steel bars. *See* ANTENNA (ELECTROMAGNETISM); TRANSMISSION LINES. Charles M. Antoni

Towing tank A tank of water used to determine the hydrodynamic performance of waterborne bodies such as ships and submarines, as well as torpedoes and other underwater forms. In the narrow sense, towing tanks are considered to be experimental facilities used to measure the forces, such as drag, on ship models and in turn to predict the performance of the full-scale prototype. In general, towing tanks are rectangular in planform with a uniform cross section. Different section shapes are used, ranging from rectangular to semicircular.

The principal measurements made in a towing tank are force measurements, particularly drag or resistance of a towed ship model or other body. One of two principal systems for towing a model is used in most towing tanks. The simpler system consists of a gravity dynamometer and an endless cable attached to the model. A weight provides a constant towing force (see illustration). The time to traverse a fixed distance is measured when the model reaches a constant speed, thus establishing the speed-resistance relationship for the model. This dynamometer is simple and capable of high accuracy, but is limited to the measurement of the drag force of waterborne bodies. It is used in the smaller towing tanks in which the models are generally under 6 ft (1.8 m) in length.

In larger towing tanks the model is towed by a towing carriage mounted on rails at the side of the towing tank or suspended from an overhead track system. Speed can be controlled and measured precisely on these carriages. Most carriages are equipped with a drag dynamometer as a permanent component. Jacques B. Hadler

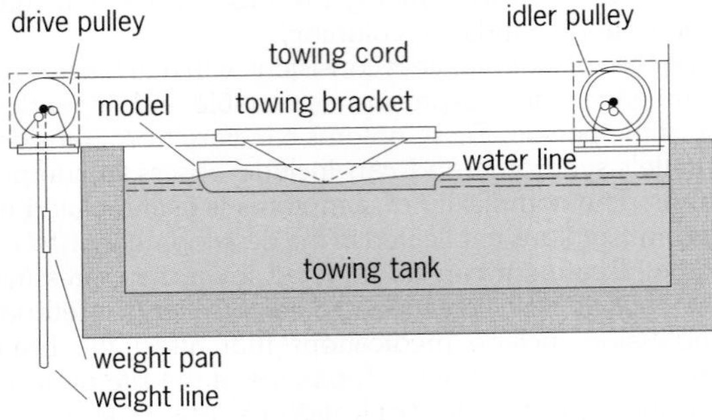

Tank with model towed by falling weight.

Townsend discharge A particular part of the voltage-current characteristic curve for a gaseous discharge device named for J. S. Townsend, who studied it about 1900. It is that part for low current where the discharge cannot be maintained by the field alone. Thus, if the agents producing the initial ionization were removed, conduction would cease. *See* ELECTRICAL CONDUCTION IN GASES. Glenn H. Miller

Toxic shock syndrome A serious, sometimes life-threatening disease usually caused by a toxin produced by some strains of the bacterium *Staphylococcus aureus*. The signs and symptoms are fever, abnormally low blood pressure, nausea and vomiting, diarrhea, muscle tenderness, and a reddish rash, followed by peeling of the skin.

Toxic shock syndrome was first reported in 1978 in seven pediatric patients. However, in 1980 hundreds of cases were reported among young women without apparent staphylococcal infections. Epidemiologists observed that the illness occurred predominantly in young women who were menstruating and were using tampons, especially those that contained so-called superabsorbent synthetic materials. A toxin [toxic shock syndrome toxin number 1 (TSST-1)] that occurs in some strains of staphylococci was later identified. These bacteria are known to proliferate in the presence of foreign particles in human infections, and it has been postulated that the tampons acted as foreign particles, allowing toxin-producing staphylococci to multiply in the vagina.

Several hundred cases of toxic shock syndrome not associated with menstruation have been reported. In these cases, which occurred in males as well as females, there was almost always an overt staphylococcal infection. Susceptibility may depend on lack of antibodies to the toxin that occur in most adults.

The toxin has been shown to occur in only about 1% of the staphylococcal strains studied. Moreover, there is some evidence that the syndrome may be caused also by other staphylococcal toxins, particularly enterotoxins. Cases of toxic shock syndrome that were caused by streptococci have been reported. A toxin distinct from TSST-1 appears involved. Persons with the symptoms of toxic shock syndrome should receive immediate medical care to reduce the chance of death. *See* STAPHYLOCOCCUS; TOXIN. Jay O. Cohen

Toxicology The study of the adverse effects of chemical and physical agents on living organisms. Toxicology has also been referred to as the science of poisons. *See* ENVIRONMENTAL TOXICOLOGY; POISON.

The most important factor that influences the toxic effect of a specific chemical is the dose. All chemicals, including essential substances such as oxygen and water, produce toxic effects when administered in large enough doses. Another significant factor is the route of exposure. Living organisms may be exposed to a chemical by inhalation (into the lungs), ingestion (into the stomach), penetration through the skin, or, in special circumstances, injection into the body. In general, substances are absorbed into the body most efficiently through the lungs so that inhalation is often the most serious route of exposure.

A third factor is the fate of the chemical after the organism is exposed. The chemical may not be absorbed at all, limiting its possible adverse effects to the site of exposure. If it is absorbed, then it may travel throughout the body and has the potential to cause toxic effects at one or more sites remote from the site of entry. The remote sites where these adverse effects occur are called target organs.

Another significant variable is the time course of the exposure. A quantity of chemical administered at one time may have an effect even though the same quantity administered in small doses over time has no effect.

In view of the importance of timing in producing adverse effects, toxicologists distinguish between two broad classes of toxicity, acute and chronic. Acute toxicity refers to effects that occur shortly after a single exposure or small number of closely spaced exposures. Chronic toxicity refers to delayed effects that occur after long-term repeated exposures.

Traditionally, the effect of most concern for acute toxicants (such as cyanide) is death. Acute toxicity is generally measured by using an assay to determine the lethal dose; rodents are given single doses and the number that have died 14 days later is recorded. The data are plotted for each dose, and the dose that is lethal for 50% of the animals (lethal dose 50 or LD_{50}) is used as the criterion for acute toxicity. *See* LETHAL DOSE 50.

Some synthetic chemicals, such as polychlorinated biphenyls (PCBs) and dichlorodiphenyltrichloroethane (DDT), exhibit their effects only after a number of repeated exposures and are considered chronic hazards, with cancer and reproductive effects being of greatest concern. To determine the dose at which chronic effects occur, rodents are exposed to daily doses of the chemical under study for long periods of time—from a few months to a lifetime. The highest dose at which no effects can be observed, the no observed effect level (NOEL), is used as a measure of chronic toxicity. *See* MUTAGENS AND CARCINOGENS.

Michael Kamrin; Robert W. Leader

Toxin Properly, a poisonous protein, especially of bacterial origin. However, nonproteinaceous poisons, such as fungal aflatoxins and plant alkaloids, are often called toxins. *See* AFLATOXIN; ALKALOID.

Bacterial exotoxins are proteins of disease-causing bacteria that are usually secreted and have deleterious effects. Several hundred are known. In some extreme cases a single toxin accounts for the principal symptoms of a disease, such as diphtheria, tetanus, and cholera. Bacteria that cause local infections with pus often produce many toxins that affect the tissues around the infection site or are distributed to remote organs by the blood. *See* CHOLERA; DIPHTHERIA; STAPHYLOCOCCUS; TETANUS.

Toxins may assist the parent bacteria to combat host defense systems, to increase the supply of certain nutrients such as iron, to invade cells or tissues, or to spread between hosts. Sometimes the damage suffered by the host organism has no obvious benefit to the bacteria. For example, botulinal neurotoxin in spoiled food may kill the person or animal that eats it long after the parent bacteria have died. In such situations it is assumed that the bacteria benefit from the toxin in some other habitat and that the damage to vertebrates is accidental. *See* FOOD POISONING.

Certain bacterial and plant toxins have the unusual ability to catalyze chemical reactions inside animal cells. Such toxins are always composed of two functionally distinct parts termed A and B, and they are often called A-B toxins. The B part binds to receptor molecules on the animal cell surface and positions the toxin upon the cell membrane. Subsequently, the enzymically active A portion of the toxin crosses the animal cell membrane and catalyzes some intracellular chemical reaction that disrupts the cell physiology or causes cell death. *See* IMMUNOLOGIC CYTOTOXICITY.

A large group of toxins breach the normal barrier to free movement of molecules across cell membranes. In sufficient concentration such cytolytic toxins cause cytolysis, a process by which soluble molecules leak out of cells, but in lower concentration they may cause less obvious damage to the cell's plasma membrane or to its internal membranes. *See* CELL MEMBRANES; CELL PERMEABILITY.

Tetanus and botulinal neurotoxins block the transmission of nerve impulses across synapses. Tetanus toxin blockage results in spastic paralysis, in which opposing muscles contract simultaneously. The botulinal neurotoxins principally paralyze neuromuscular junctions and cause flaccid paralysis.

Gram-negative bacteria, such as *Salmonella* and *Hemophilus*, have a toxic component in their cell walls known as endotoxin or lipopolysaccharide. Among other detrimental effects, endotoxins cause white blood cells to produce interleukin-1, a hormone responsible for fever, malaise, headache, muscle aches, and other nonspecific consequences of infection. The exotoxins of toxic shock syndrome and of scarlet fever induce interleukin-1 and also tumor necrosis factor, which has similar effects. *See* ENDOTOXIN; FEVER; SCARLET FEVER; TOXIC SHOCK SYNDROME.

Toxoids are toxins that have been exposed to formaldehyde or other chemicals that destroy their toxicities without impairing immunogenicity. When injected into humans, toxoids elicit specific antibodies known as antitoxins that neutralize circulating toxins. Such immunization (vaccination) is very effective for systemic toxinoses, such as diphtheria and tetanus. *See* ANTIBODY; IMMUNITY; VACCINATION.

D. Michael Gill

Toxin-antitoxin reaction A term used in serology to denote the combination of a toxic antigen with its corresponding antitoxin. If the antitoxin is derived from any species other than the horse, precipitation occurs over a wide range of reactant ratios, as in other antigen-antibody reactions. With horse antitoxin, flocculation occurs only if toxin and antitoxin are near equivalence, a twofold excess of either reactant giving soluble complexes. In most instances, the reaction results in partial or complete neutralization of the toxic activity of the antigen. *See* ANTIBODY; ANTIGEN; ANTITOXIN; NEUTRALIZATION REACTION (IMMUNOLOGY); SEROLOGY.

Henry P. Treffers

Toxoplasmea A class of the subphylum Sporozoa. The organisms are small and crescent-shaped. They move by body flection or gliding and have no flagella or pseudopodia. Characteristic structures are the two-layered pellicle and underlying longitudinal microtubules, micropyle, a conoid, paired organelles, and micronemes.

The most distinguishing characteristic of the Toxoplasmea is the unique means of reproduction. Electron microscope studies indicate that endodyogeny is the sole method. Endodyogeny is an internal budding wherein two daughter cells are produced within a mother cell which is destroyed in the process.

Only two stages are known in the life cycle of most animals. One stage, the proliferative form or trophozoite, occurs singly or in groups within host cells. The other stage, the so-called cysts, consists of a large number of organisms which, with minor differences, are structurally similar to the proliferative forms. *See* SPOROZOA; TOXOPLASMIDA.

Harley G. Sheffield

Toxoplasmida An order of the class Toxoplasmea. Four genera, *Toxoplasma*, *Besnoitia*, *Sarcocystis*, and *Encephalitozoon*, make up the order. The organisms are parasites of vertebrates. *Toxoplasma* is often found encysted in nerve tissue, *Besnoitia* in connective tissue, and *Sarcocystis* in muscle. Very little is known about *Encephalitozoon*, but the parasite has been found in the brain of rabbits. *See* TOXOPLASMEA.

Harley G. Sheffield

Trace fossils Fossilized evidence of animal behavior, also known as ichnofossils, biogenic sedimentary structures, bioerosion structures, or lebensspuren. The fossils include burrows, trails, and trackways created by animals in unconsolidated sediment (see illustration), as well as borings, gnawings, raspings, and scrapings excavated by organisms in harder materials, such as rock, shell, bone, or wood. Some workers also consider coprolites (fossilized feces), regurgitation pellets, burrow excavation pellets, rhizoliths (plant root penetration structures), and algal stromatolites to be trace fossils. *See* STROMATOLITE.

Trace fossils are important in paleontology and paleoecology, because they provide information about the presence of unpreserved soft-bodied members of the original communities, life habits of fossil organisms, evolution of certain behavior patterns through geologic time, and biostratigraphy of otherwise unfossiliferous deposits. Trace fossils also are useful in sedimentology and paleoenvironmental studies, because they are sedimentary structures that are preserved in place and are very rarely reworked and transported, as body fossils of animals and plants commonly are. This fact allows trace fossils to be regarded as reliable indicators of original conditions in the sedimentary environment. The production of trace fossils involves disruption of original stratification and sometimes results in alteration of sediment texture or composition. *See* FOSSIL; PALEOECOLOGY; PALEONTOLOGY; SEDIMENTOLOGY.

Trace fossils occur in sedimentary deposits of all ages from the late Precambrian to the Recent. Host rocks include limestone, sandstone, siltstone, shale, coal, and other sedimentary rocks. These deposits represent sedimentation in a broad spectrum of settings, ranging from subaerial (such as eolian dunes and soil horizons) to subaqueous (such as rivers, lakes, swamps, tidal flats, beaches, continental shelves, and the deep-sea floor). *See* DEPOSITIONAL SYSTEMS AND ENVIRONMENTS.

Organisms may produce fossilizable traces on the sediment surface (epigenic structures) or within the sediment (endogenic structures). Trace fossils may be preserved in full three-dimensional relief (either wholly contained within a rock or weathered out as a separate piece) or in partial relief (either as a depression or as a raised structure on a bedding plane). Simply because a trace fossil is preserved on a bedding plane does not indicate that it originally was an epigenic trace. Diagenetic alteration of sediment commonly enhances the preservation of trace fossils by differential cementation or selective mineralization. In some cases, trace fossils have been preferentially replaced by chert, dolomite, pyrite, glauconite, apatite, siderite, or other minerals. *See* DIAGENESIS.

Agrichnial farming traces (burrows produced in order to farm or trap food inside the sediment) of unknown organisms, including a double-spiral tunnel (*Spirorhaphe*) and a meshlike network of tunnels (*Paleodictyon*). Tertiary, Austria. (*Photograph by W. Häntzschel*)

The study of trace fossils is known as ichnology. The prefix "ichno-" (as in ichnofossil and ichnotaxonomy) and the suffix "-ichnia" (as in epichnia and hypichnia) commonly are employed to designate subjects relating to trace fossils. The suffix "-ichnus" commonly is attached to the ichnogenus name of many trace fossils (as in *Dimorphichnus* and *Teichichnus*).

A. A. Ekdale

Trachylina An order of jellyfish of the class Hydrozoa of the phylum Cnidaria. These jellyfish are of moderate size. They differ from other hydrozoan jellyfish in having balancing organs which develop partly from the digestive epithelium and in having only a small polyp stage or none at all. Many authorities recognize three distinct orders of trachylines—Limnomedusae, Trachymedusae, and Narcomedusae—and in this case the older term Trachylina is abandoned. *See* HYDROZOA.

Sears Crowell

Trachyte A light-colored, aphanitic (very finely crystalline) rock of volcanic origin, composed largely of alkali feldspar with minor amounts of dark-colored (mafic) minerals (biotite, hornblende, or pyroxene). If sodic plagioclase (oligoclase or andesine) exceeds the quantity of alkali feldspar, the rock is called latite. Trachyte and latite are chemically equivalent to syenite and monzonite, respectively. *See* LATITE; MONZONITE; SYENITE.

Streaked, banded, and fluidal structures due to flowage of the solidifying lava are commonly visible in many trachytes and may be detected by a parallel arrangement of tabular feldspar phenocrysts. A distinctive microscopic feature is trachytic texture in which the tiny, lath-shaped sanidine crystals of the rock matrix are in parallel arrangement and closely packed.

Trachyte is not an abundant rock, but it is widespread. It occurs as flows, tuffs, or small intrusives (dikes and sills). It may be associated with alkali rhyolite, latite, or phonolite. *See* IGNEOUS ROCKS; MAGMA; SPILITE.

Carleton A. Chapman

Tractor A wheeled, self-propelled vehicle for hauling other vehicles or equipment and for operating the towed implements; also, a crawler which runs on an endless, self-laid track and performs similar functions.

A farm tractor is a multipurpose power unit. It has a drawbar for drawing tillage tools and a power takeoff device for driving implements or operating a belt pulley. The acreage to be worked, type of crops grown, and the terrain all impose their requirements on tractor design. Accordingly, models vary in such details as power generated, weight, ground clearance, turning radius, and facilities for operating equipment. All models can, however, be grouped under four general types: four-wheel, row-crop or high-wheel, tricycle, and crawler.

Tractors are rated by the horsepower they deliver at the drawbar and at the belt. On small models, the drawbar and belt horsepower may run as low as 10 (7.5 kW); on large models the drawbar horsepower runs as high as 132 (98 kW), while belt horsepower reaches about 144 (107 kW).

The major components are engine, clutch, and transmission. These components are intimately related and designed to work in conjunction with each other to accomplish specific work. Tractor engines are relatively low-speed; their maximum horsepower is generated at crankshaft speeds in the neighborhood of 2000 rpm. These engines have one, two, three, four, six, or eight cylinders and operate on gasoline, kerosine, liquid petroleum gas, or diesel fuel. They are of the spark-ignition or diesel type, operating on the four-stroke-cycle principle, and are cooled by water or air.

Power is transmitted to the rear wheels or to all four wheels. Drive to the front wheels is mechanical or hydrostatic, its purpose being to increase drawbar pull at the will of the operator. Transmissions have 3, 4, 5, 6, 8, 10, or 12 forward speeds and one or two reverse gears. Clutchless hydraulic transmissions are also used, making it possible to shift gears while in motion. Vehicle speeds are low, ranging from slightly more than 1 mi/h (1.6 km/s) to about 18 mi/h (13 km/h) in high gear. *See* CLUTCH; TORQUE CONVERTER.

The basic design of an industrial tractor for hauling and for operating construction equipment departs little from that of a farm tractor, and differences in design of models fit the vehicle to its intended work. Because high ground clearance is not needed for industrial work, the tractor is commonly built with a lower center of gravity and is capable of traveling a few miles per hour faster than a farm tractor. If its use is confined to hauling, it may not be equipped with hydraulic power. If it is to be used for operating a scraper, backhoe, or front-end loader, its structure may be heavier and more rugged. *See* BULK-HANDLING MACHINES.

Philip H. Smith

Tractrix A plane curve for which the length of any tangent between the curve and a fixed line is constant c (see illustration). If the x axis is the fixed line, its differential equation is

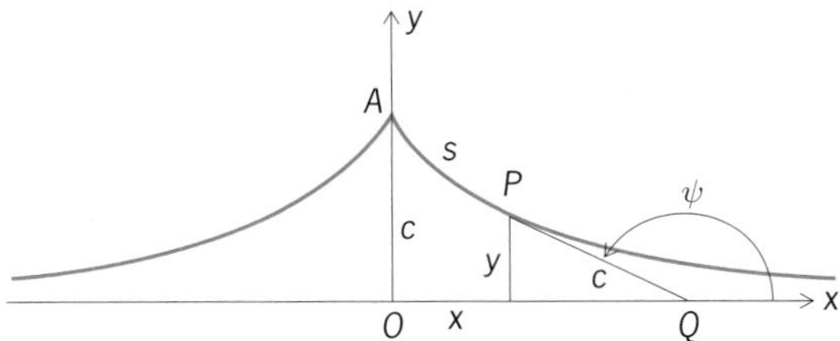

Tractrix (upper half).

Eq. (1). With ψ, the inclination of the tangent, as parameter, this yields equations (2). The tractrix has the x axis as asymptote and cusps at $(0, \pm c)$. The arc AP, measured from a cusp, is as in Eq. (3).

$$(dy/dx)^2 = y^2/(c^2 - y^2) \tag{1}$$

$$x = c \log \tan {}^1/_2|\psi| + c \cos \psi \tag{2}$$

$$y = c \sin \psi$$

$$s = -c \log \sin \psi = c \log \frac{c}{y} \tag{3}$$

Louis Brand

Traffic-control systems Systems that act to control the movement of people, goods, and vehicles in order to ensure their safe, orderly, and expeditious flow throughout the transportation system. Each of the five areas of transportation—roadways, airports and airways, railways, coastal and inland waterways, and pipelines—have unique systems of control.

Roadway traffic-control systems are intended to improve safety, increase the operational efficiency and capacity of the roadway, and contribute to the traveler's comfort and convenience. They range from simple control at isolated intersections using signs and markings, to sophisticated traffic-control centers which have the ability to react to changes in the traffic environment. Traffic-control systems are used at roadway intersections, on highways and freeways, at ramp entrances to freeways, and in monitoring and controlling wider-area transportation networks. Intelligent roadway traffic control, now known as intelligent transportation systems (ITS), is a very sophisticated form of traffic control for roadway and other areas. *See* HIGHWAY ENGINEERING.

The U.S. federal government has designated airspace as either uncontrolled or controlled. In uncontrolled airspace, pilots may conduct flights without specific authorization. In controlled airspace, pilots may be required to maintain communications with the appropriate air-traffic control facility to receive authorization and instruction on traversing, taking off from, or landing, in that controlled area. Air-traffic control systems for controlled areas may be divided loosely into en route and terminal systems. *See* AIR NAVIGATION; AIR-TRAFFIC CONTROL; AIR TRANSPORTATION.

Railroads operate high-speed freight and passenger services essentially over an exclusive right of way. Railroads use both semaphore and light signals for traffic control. Semaphores convey visual messages to train operators according to predetermined rules indicating how the train is to be operated in specified areas. Automatic block signaling prevents rear-end and head-on collisions on signal tracks. In this system, track sections are divided into blocks. Only one train is permitted to occupy a block at any time. Blocks are monitored by automatic circuitry that controls traffic signals, indicating the appropriate clear or stop signals to following or approaching trains. Similar block systems are used for subway systems. Centralized traffic-control systems may control hundreds of miles of track signals and switches. A dispatcher at a central location monitors the location of trains by means of visual displays of colored lights on a large track diagram, and can operate the switches and signals at key points from the central control console. *See* RAILROAD CONTROL SYSTEMS; RAILROAD ENGINEERING.

Vessel traffic control consists largely of marine aides that function more for informational, advisory, and guidance purposes than as positive traffic-control devices. Lighted or unlighted buoys indicate navigable areas in coastal waters and within waterways. Lightships and lighthouses with fog signals and radio beacons are placed as markers at prominent points during periods of limited visibility. Radar devices have become common, even on smaller ships. Navigation systems often employ the Differential Global Positioning System. The Vessel Traffic Service (VTS) is available in selected areas. Services may range from the provision of single advisory messages to extensive management of traffic communication and radar services. *See* BUOY; LIGHTHOUSE; MARINE NAVIGATION; NAVIGATION.

The 450,000 mi (750,000 km) of pipelines in the United States are a major part of the nation's transportation network, carrying about 25% of all intercity freight-ton mileage. The primary goods moved through pipelines are oil and oil by-products, natural gas, and fertilizers. The movement of goods in pipelines is controlled by systems of valves, pumps, and compressors. *See* PIPELINE; TRANSPORTATION ENGINEERING.

James Costantino; Donna C. Nelson

Trajectory The curve described by a body moving through space, as of a meteor through the atmosphere, a planet around the Sun, a projectile fired from a gun, or a rocket in flight. In general, the trajectory of a body in a gravitational field is a conic section—ellipse, hyperbola, or parabola—depending on the energy of motion. The trajectory of a shell or rocket fired from the ground is a portion of an ellipse with the Earth's center as one focus; however, if the altitude reached is not great, the effect of gravity is essentially constant, and the parabola is a good approximation. *See* BALLISTICS.

John P. Hagen

Tranquilizer A psychopharmacologic drug that tends to calm overexcited individuals, producing a state of "tranquility." The term tranquilizer was originally applied to two groups of drugs. Members of one group, the antianxiety drugs, were called minor tranquilizers, and members of the second group, the antipsychotic drugs, were called major tranquilizers. Although drugs in both groups may have a calming effect in appropriate doses, it is now clear that the two groups are quite different.

Antianxiety drugs, also known as anxiolytics, comprise three subgroups: propanediols, of which meprobamate (Miltown) is the best known; alcohol (ethanol) and the barbiturates, such as phenobarbital; benzodiazepines, of which diazepam (Valium) is the best known. Like all drugs, antianxiety drugs have a range of effects at different doses, including—in the low-to-high order of doses at which the effects generally occur—(1) a period of transient excitement for some persons; (2) a reduction in anxiety, the effect for which the drugs are prescribed; (3) antagonism of certain kinds of seizure activity in the brain, a so-called anticonvulsant activity; (4) a sedative or sleep-inducing effect; and (5) a generalized reduction in muscle tension, a muscle relaxant effect. *See* AFFECTIVE DISORDERS.

There are a large number of antipsychotic drugs, all of which are equally useful in managing psychotic symptoms. Because of the symptom control that antipsychotics provide, it is possible to manage the majority of psychotic patients (mostly schizophrenics) on an outpatient basis. However, the antipsychotics vary widely in potency, that is, the amount of drug that is required to produce effective symptom control. Furthermore, different side effects are associated with different potencies. The major complication of treatment with antipsychotic drugs is the array of movement disturbances that they can induce. Most of these disturbances can be controlled by concurrent medication with other drugs. *See* PSYCHOPHARMACOLOGY; PSYCHOSIS; SCHIZOPHRENIA.

Peter L. Carlton

Transamination Transfer of an amino group from one carbon chain to another without the intermediate formation of ammonia. Typically, an α-amino acid serves as the donor of the amino group and is converted to an α-keto acid, and an α-keto acid serves as the acceptor of the amino group and is converted to an α-amino acid. Transamination reactions are catalyzed by enzymes called aminotransferases (or transaminases). Most physiologically important aminotransferases have a preferred amino acid/keto acid substrate and utilize α-ketoglutarate/glutamate as the counter keto acid/amino acid. *See* AMINO ACIDS; ENZYME.

Transamination reactions play a prominent role in the synthesis of the dispensable (or nonessential) amino acids in higher organisms. Amino groups are directly transferred to the keto acids pyruvate, oxaloacetate, and α-ketoglutarate to form alanine, aspartate, and glutamate, respectively. Transamination is also involved in the pathways of de novo synthesis of serine, glycine, glutamine, asparagine, proline, and arginine. Martha H. Stipanuk

Transcription The process that occurs within a living cell in which an enzyme makes a ribonucleic acid (RNA) copy of the deoxyribonucleic acid (DNA) that contains the genetic information. The resulting RNA is often used as a template to make proteins by the cellular protein-synthesizing (translation) machinery. Transcription is an essential step in the growth and differentiation of all cells and contributes to almost every aspect of the development of an organism. Because transcription is so fundamental to life, there is significant conservation of proteins and enzymatic steps involved in transcription in cells of bacterial, archaeal, and eukaryotic origin. An RNA polymerase catalyzes synthesis of RNA from DNA by binding to a sequence termed the promoter located adjacent to or upstream of the coding portion of the linked gene. Genetic and biochemical studies have defined distinct steps in the transcriptional process. The first step of the transcription cycle is initiation, which involves melting (strand separation) of the template DNA and formation of the first chemical bond that makes the RNA. Like a revving engine, the RNA polymerase carries out several rounds of synthesis of short RNA transcripts (termed abortive transcripts) until it becomes disengaged from the promoter (promoter escape) and moves along the DNA, synthesizing RNA at an estimated rate of 1500 bases per minute. This step of the transcription cycle is called elongation, which is followed by termination and reinitiation. Transcription is a highly regulated process; each step requires participation of accessory proteins and is subject to control by transcription factors that respond to intra- and extracellular signals. *See* DEOXYRIBONUCLEIC ACID (DNA); ENZYME; GENE; PROTEIN; RIBONUCLEIC ACID (RNA).

Modulation of transcription impacts on many areas of medicine, both as a cause and as a cure of disease. Mutations in transcription factors or in gene regulatory sequences have been implicated in numerous developmental and neurological disorders. An example is Rett syndrome, which involves mutations in a protein that regulates transcription. It is now becoming clear that these mutations subtly alter the global program of gene expression in a way that kills a small but vitally important subset of neurons in the brain. Cancer may be thought of as a transcriptional disease, and many cancer-causing proteins (oncoproteins and mutated tumor suppressors) function as transcription factors. The majority of human tumor cells carry mutations that directly affect the function of two critical transcription factors, p53 and the retinoblastoma protein. In normal cells, these proteins provide a protective role by preventing damaged cells from proliferating. *See* MUTATION; ONCOGENES. Naoko Tanese; Angus C. Wilson

Transducer A device that converts variations in one energy form into corresponding variations in another, usually electrical form. Measurement transducers or input transducers may exploit a wide range of physical, chemical, or biological effects to achieve transduction, and their design principles usually revolve around high sensitivity and minimum disturbance to the measurand, that is, the quantity to be measured. Output transducers or actuators are designed to achieve some end effect, for example, opening of a valve or deflection of a control surface on an aircraft. Actuators, therefore, normally operate at high power levels. The term sensor is often used instead of transducer, but strictly a sensor does not involve energy transformation; the term should be reserved for devices such as a thermistor, which is not energy-changing but simply changes its intrinsic electrical resistance in response to changes in temperature.

Both input and output transducers, together with the instrumentation to which they are connected, may be called upon to respond to both slowly varying or dynamic signals. This means that the transducer, together with its instrumentation system, must be designed to meet such a specification. Some prior knowledge is therefore required of the type of signal to be transduced, and the bandwidth of the transducer and instrumentation system must be suitably matched to this signal.

Transducers are often described in terms of their sensitivity to input signals (responsivity). This is simply defined as the ratio of the output signal to the corresponding input signal. Once again, the responsivity of a transducer must be matched to the expected levels of signal to be transduced. *See* SENSITIVITY (ENGINEERING).

The measurement of force is very often accomplished by allowing an elastic member (spring or cantilever beam) to deflect and then measuring the deflection by using some form of displacement transducer. Transducers designed to measure acceleration are frequently based on the simple equation below, where

$$f = ma$$

f is force, m is mass, and a is acceleration. Thus, if the force due to the movement of a known mass can be measured, it is possible to derive the acceleration. Very often, the measurement technique employed uses piezoelectric, magnetostrictive, or mechanoresistive materials. Acceleration transducers or accelerometers are frequently employed for the measurement of vibration. *See* ACCELEROMETER; FORCE; MAGNETOSTRICTION.

Transducers for a wide range of chemical species are available, but probably the most widely applied is the pH transducer for the measurement of hydrogen-ion concentration. The traditional method has relied on a glass membrane electrode used to make up an electrochemical cell. *See* HYDROGEN ION; ION-SELECTIVE MEMBRANES AND ELECTRODES; PH.

Measurements of the partial pressure of oxygen (pO_2) may be accomplished by the use of a Clark oxygen cell, which comprises a gas-permeable membrane controlling the rate of arrival of oxygen molecules at a noble-metal cathode that is held at 600–800 mV potential with respect to the anode. The ensuing reduction process gives rise to a cathode current from which oxygen concentration can be derived.

Other electrochemical transducers are used in such applications as voltametry, polarography, and amperometry. Chemical transduction is also possible by adsorbing a species onto a surface and detecting its presence by mass change, electrical property change, color change, and so on. *See* ELECTROCHEMICAL TECHNIQUES; POLAROGRAPHIC ANALYSIS; TITRATION.

Measurements of the partial pressure of oxygen and the partial pressure of carbon dioxide (pCO_2) are also of particular importance in the context of blood gas analysis in medicine, and by using the Clark cell they can be performed without removing the blood from the body and noninvasively, that is, without puncturing the skin.

There have been remarkable advances in the area of biological transducers or biosensors. Examples are the ion-selective field-effect transducer (ISFET), the insulated-gate field-effect transducer (IGFET), and the chemically sensitive field-effect transducer (CHEMFET).

A smart transducer or smart sensor is a device that not only undertakes measurement but also can adapt to the environment in which it is placed. Such adaptation may range from simple changes in the characteristics of the transducer in response to changes in temperature, to more complex procedures such as adaptation of the transducer's performance to conform to overall system requirements. In integrated transducers, much of the signal processing that might previously be done remotely is brought into the transducer packaging.

The development of inexpensive fiber-optic materials for communications has led to an examination of the potential for using these devices as the basis for transduction. Two major types of devices have resulted: fiber-optic transducers for physical variables and similar devices devoted to chemical and biological determinations. The advantages of the all-optical transducer are its lack of susceptibility to electrical interference and its intrinsic safety. Small deformations of an optical-fiber waveguide cause a change in the light transmission of the fiber, and this has been exploited to produce force and pressure transducers. Alternatively, miniature transducers based on color chemistry can be fabricated at the end of a fiber and the color change can be sensed remotely. Devices of this type have been developed for measuring pH, the partial pressures of oxygen and carbon dioxide, and glucose. *See* FIBER-OPTIC SENSOR.

The most important recent technological development in the area of transducers, sensors, and actuators is micro-electromechanical systems (MEMS). There are a wide variety of MEMS devices, mostly fabricated in silicon. *See* MICRO-ELECTRO-MECHANICAL SYSTEMS (MEMS). Peter A. Payne

Transduction (bacteria) A mechanism for the transfer of genetic material between cells. The material is transferred by virus particles called bacteriophages (in the case of bacteria), or phages. The transfer method differentiates transduction from transformation. In transformation the genetic material (deoxyribonucleic acid) is extracted from the cell by chemical means or released by lysis. *See* BACTERIAL GENETICS; BACTERIOPHAGE; DEOXYRIBONUCLEIC ACID (DNA); TRANSFORMATION (BACTERIA).

The transduction mechanism has two features to distinguish it from the more usual mechanism of gene recombination, the sexual process. The most striking feature is the transfer of genetic material from cell to cell by viruses. The second feature is the fact that only a small part of the total genetic material of any one bacterial cell is carried by any particular transducing particle. However, in general transduction, all of the genetic material is distributed among different particles.

Transduction is not accomplished by all bacteriophages. It is done by some that are classified as "temperate." When such temperate bacteriophages infect sensitive bacteria, some of the bacteria respond by producing more bacteriophage particles. These bacteria donate the transducing material. Other bacteria respond to the infection by becoming more or less permanent carriers of the bacteriophage, in a kind of symbiotic relationship; these are called lysogenic bacteria. Bacteria in this latter class survive the infection, and it is among these that transduced cells are found. The proportion of bacteria in any culture that responds to infection in either manner can be influenced by the particular environment at the time of infection. *See* LYSOGENY.

Certain phages carry out a more restricted kind of transduction. They carry only a specific section of bacterial genetic material; they transduce only a few genes. Retroviruses carry out specific or restricted transduction. It has long been known that these viruses can cause the formation of tumors (oncogenesis) in animals. It is now known that these viruses exchange a small portion of their genome for a mutant cellular gene that has a role in gene regulation or replication. These viruses carrying mutant genes infect cells, causing them to be transformed into tumor cells. *See* ANIMAL VIRUS; RETROVIRUS. Norton D. Zinder

Transfer cells Plant cells characterized by the elaboration of an unlignifed, secondary cell wall to form fingerlike projections or wall ingrowths which protrude into the cytoplasm of the cell. These ingrowths are enveloped by plasma membrane, forming a wall-membrane apparatus which increases the surface-volume ratio of the cell.

The location of transfer cells within the plant provides circumstantial evidence for their involvement in solute transport. They are situated at many sites in the plant where secretion or absorption takes place. Transfer cells are most frequently associated with the conducting elements of the xylem and phloem, where they may play a role in either loading or unloading solutes from these cells. Four main categories of secretion or absorption involving transfer cells are generally recognized: (1) absorption of solutes from the external environment; (2) secretion of solutes to the external environment; (3) absorption of solutes from internal, extracytoplasmic compartments; and (4) secretion of solutes into internal, extracytoplasmic compartments. *See* PHLOEM; SECRETORY STRUCTURES (PLANT); XYLEM. R. L. Peterson

Transform fault One of the three fundamental types of boundaries between the mobile lithospheric plates that cover the surface of the Earth. Whereas spreading centers mark sites where crust is created between diverging plates, and subduction zones are where crust is destroyed between convergent plates, transform faults separate plates that are sliding past each other with neither creation nor destruction of crust. The primary tectonic feature of all transform faults is a strike-slip fault zone, a generally vertical fracture parallel to the relative motion between the two plates that it separates. Strike-slip fault zones are described as right-lateral if the far side is moving right relative to the near side or left-lateral if it is moving to the left. Not all such fault zones are plate-bounding transform faults. Small-scale strike-slip faulting is a common secondary feature of many subduction zones, especially where plate convergence is oblique, and of some spreading centers, especially those with propagating rifts; it also occurs locally deep in plate interiors. The distinguishing characteristic of a transform fault is that both ends extend to a junction with another type of plate boundary. At these junctions the divergent or convergent motion along the other boundaries is transformed into purely lateral slip. *See* EARTH CRUST; PLATE TECTONICS; SUBDUCTION ZONES.

Transform faults are most readily classified by the types of plate boundary intersected at their ends, the variety of lithosphere (oceanic or continental) they separate, and by whether they are isolated or are part of a multifault system. The common oceanic type is the ridge-ridge transform, linking two literally offset axes of a spreading center. Also common are transform faults that link the end of a spreading center to a triple junction, the meeting place of three plates and three plate boundaries. *See* LITHOSPHERE; MID-OCEANIC RIDGE.

Other types are long trench-trench transforms at the northern and southern margins of the Caribbean plate, and the combined San Andreas/Gulf of California transform, which separates the North American and Pacific plates for 1500 mi (2400 km) between triple junctions at Cape Mendocino (California) and the mouth of the Gulf of California. Strike-slip faulting in the Gulf of California (and on the northern Caribbean plate boundary) occurs along several parallel zones linked by short spreading centers, and the overall structure is more properly called a transform fault system; similar fault patterns are found at many ridge-ridge transforms. Along a few strike-slip fault zones, lithospheric plates slide quietly and almost continuously past each other by the process called aseismic creep. Much more often, frictional resistance to the sliding in the brittle crust causes the accumulation of shear stresses that are episodically or periodically relieved by sudden shifts of crustal blocks, creating earthquakes. The largest lateral shifts (slips) of the ground surface along major continental transform faults have been associated with some of the largest earthquakes on record; in 1906 the Pacific plate alongside 270 mi (450 km) of the San Andreas Fault suddenly moved an average of 15 ft (4.5 m) northwest relative to the North American plate on the other side, and the resulting magnitude-8.2 earthquake destroyed much of San Francisco. The average slip in this single event was equivalent to about 150–250 years of Pacific–North American plate motion. *See* EARTHQUAKE; FAULT AND FAULT STRUCTURES; SEISMOLOGY. Peter Lonsdale

Transformation (bacteria) The addition of deoxyribonucleic acid (DNA) to living cells, thereby changing their genetic composition and properties. The recipient bacteria are usually closely related to the donor strain. The process may occur in natural conditions, for example, in a host animal infected with two parasitic strains, and indeed it might play a part in the rapid evolution of pathogenic bacteria. There are several species of bacteria in which transformation has been achieved in the laboratory.

That bacterial transformation is true genetic transmission on a small scale, rather than controlled mutation, is demonstrated by the following characteristics: (1) A specific trait is introduced, coming always from donors bearing the trait. (2) The trait is transferred by determinant, genelike material far less complex than whole cells or nuclei, and this material, DNA, is known to be present in gene-carrying chromosomes. (3) The trait is inherited by the progeny of the changed bacteria. (4) The progeny produce, when they grow, increased amounts of DNA carrying the specific property. (5) The traits are transferred as units exactly in the patterns in which they appear or in which they are induced by mutation. (6) The DNA transmits the full potentialities of the donor strain, whether these are in an expressed or in a latent state. (7) The traits are often attributable to the presence of a specific gene-determined enzyme protein. (8) Certain groups of determinants may occur "linked" within DNA molecules, just as genes may be linked, and if so, heat denaturation, radiation, or enzyme action will inactivate or separate them just to the extent that they can damage or break apart the DNA molecules. (9) Linked determinants, while transforming a new cell, may become exchanged (recombined) between themselves and their unmarked or unselective alternate forms in such a way that they bring about genetic variation, and in a pattern indicating the existence of larger organized genetic units. *See* BACTERIAL GENETICS; GENE.

Through the application of a number of procedures prior to adding the DNA, transformation was extended first to many different bacterial species and then to eukaryotic cells. Today almost any cell type can be transformed. In some cases, tissues can be injected directly with naked DNA and transformed. However, unlike with bacteria, the naked DNA adds almost anywhere in the genome rather than recombining with its indigenous homolog. However, with special highly selective procedures, homologous recombination can be obtained. By treating embryonic stem cells and adding them to embryos that then go to term, specific and nonspecific transgenic animals can be obtained (for example, mice). *See* GENETIC ENGINEERING.

When the source of the DNA is some entity capable of independent replication, such as a virus or plasmid, the phenomenon is called transfection. If foreign DNA is then inserted into these entities, the result is recombinant DNA that can lead to transduction. *See* MOLECULAR BIOLOGY; TRANSDUCTION (BACTERIA). Rollin D. Hotchkiss; Norton D. Zinder

Transformer An electrical component used to transfer electric energy from one alternating-current (ac) circuit to another by magnetic coupling. Essentially it consists of two or more multiturn coils of insulated conducting material, so arranged that any magnetic flux linking one coil will link the others also. This creates mutual inductances between the coils. The mutual magnetic field acts to transfer energy from one input coil or primary winding to the other coils, which are then referred to as secondary windings. Under steady-state conditions, only one winding can serve as a primary. *See* COUPLED CIRCUITS; INDUCTANCE.

The transformer accomplishes one or more of the following effects between two circuits: (1) an induced voltage of different magnitude, (2) an induced current of different magnitude, (3) a difference in phase angle, (4) a difference in impedance level, and (5) a difference in voltage insulation level, either between the two circuits or to ground.

In electric power systems, transformers are used to perform a wide range of functions. Pole-type distribution transformers supply relatively small amounts of power to residences. Power transformers are used at generating stations to step up the generated voltage to high levels for transmission. The transmission voltages are then stepped down by transformers at the substations for local distribution. Instrument transformers are used to enable accurate measurements of measure voltages and currents. In other applications, audio- and video-frequency transformers must function over a broad band of frequencies. Radio-frequency transformers transfer energy in narrow frequency bands from one circuit to another. *See* INSTRUMENT TRANSFORMER.

Power transformers. Power transformers, as a class, may be defined as those designed to operate at power-system frequencies: 60 Hz in the United States and Canada, and 50 Hz in much of the rest of the world. The largest power transformers connect generators to the power grid. Since a generator, together with its driving turbine and prime energy source, is called a generating unit, such transformers are called unit transformers. The classification "distribution transformers" refers to those supplying power to the ultimate consumers. They are designed for lower power and output-voltage ratings than the other transformers in the system.

Typical configurations for single-phase transformers are shown in Fig. 1. The arrangement in Fig. 1*a* is called a shell-form transformer, while that in Fig. 1*b* is called a core-form transformer. Each of the rectangles labeled "windings" in this figure represents at least two coils. The coils may be concentric, or interleaved. In the shell form, all of the windings are on the center leg. In the core form, half of the turns of the primary winding and half

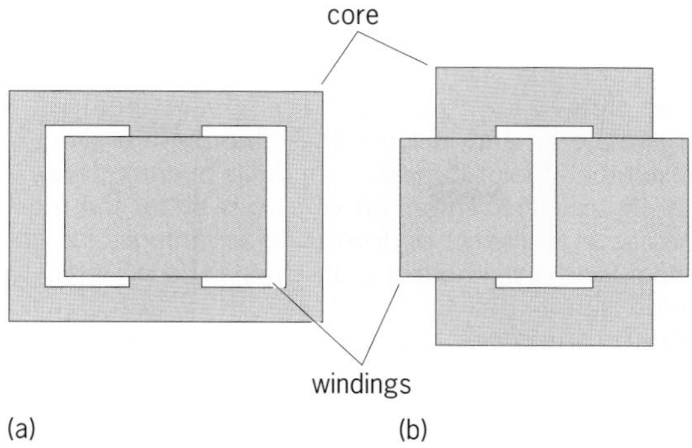

Fig. 1. Location of windings in single-phase cores. (*a*) Shell form. (*b*) Core form.

of those of the secondary are on each leg. The two halves of a given winding may be connected in series or in parallel.

Transformers operate on the basis of two fundamental physical laws: Faraday's voltage law and Ampère's law. Faraday's law states that the voltage induced in a winding by a magnetic flux linking that winding is proportional to the number of turns and the time rate of change of the flux; that is, Eq. (1) holds, where

$$e_i = N_i \frac{d\phi_i}{dt} \qquad \text{volts} \tag{1}$$

e_i is the voltage induced in a coil of N_i turns which is threaded by a flux of ϕ_i webers changing at a rate of $d\phi_i/dt$ webers per second. The ratio of the voltages induced in two windings of a transformer by the core flux is, then, given by Eq. (2). In other

$$\frac{e_1}{e_2} = \frac{N_1}{N_2}\frac{d\phi_{\text{core}}/dt}{d\phi_{\text{core}}/dt} = \frac{N_1}{N_2} \tag{2}$$

words, the voltages induced in the windings are proportional to the numbers of turns in the windings. This is the basic law of the transformer. A high-voltage winding will have many turns, and a low-voltage winding only a few. The N_1/N_2 ratio is usually called the turns ratio or transformation ratio and is designated by the symbol a, so that Eq. (3) holds.

$$\frac{e_1}{e_2} = a \tag{3}$$

See FARADAY'S LAW OF INDUCTION.

Since the flux must change to induce a voltage, steady-state voltages can be obtained only by a cyclically varying flux. This means that alternating voltages and fluxes are required for normal transformer operation, and that is the fundamental reason for ac operation of power systems. Devices operated on ac have lower losses when the voltages and fluxes are sinusoidal in form, and sinusoidal fluxes and terminal voltages will be assumed in this discussion. *See* ALTERNATING CURRENT.

Figure 2 shows the elements of a two-winding, shell-form transformer. The center leg of the core carries the full mutual flux, and each of the outside legs carries half of it. Thus the cross-sectional area of each outside leg is half of that of the center leg.

Efficiency is defined by Eq. (4). The output power is equal to

$$\eta = \frac{\text{output power}}{\text{input power}} \tag{4}$$

the input power, less the internal losses of the transformer. These losses include the ohmic (I^2R) loss in the windings, called copper loss and the core loss, called the no-load loss. The input power is thus the sum of the output power and the copper and core losses. Typical efficiency for a 20,000-kVA power transformer at full load is 99.4%, while that of a 5-kVA transformer is 94%.

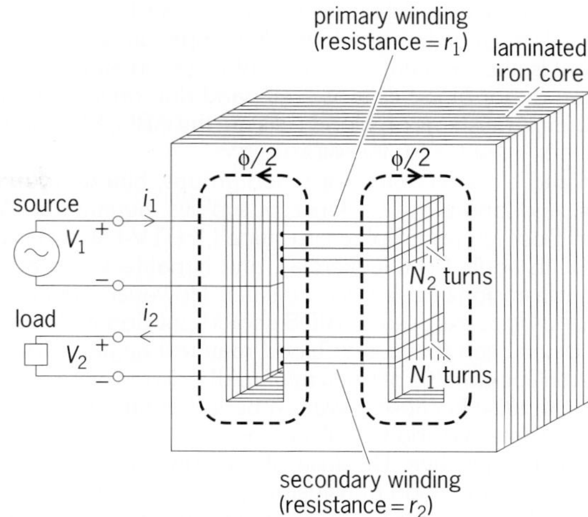

Fig. 2. Elements of a transformer.

Transformer losses in the windings and core generate heat, which must be removed to prevent deterioration of the insulation and the magnetic properties of the core. Most power transformers are contained in a tank of oil. The oil is especially formulated to provide good electrical insulation, and also serves to carry heat away from the core and windings by convection. Transformers which are designed to operate in air are called "dry-type" transformers.

George McPherson, Jr.

Audio- and radio-frequency transformers. Audio or video (broad-frequency-band) transformers are used to transfer complex signals containing energy at a large number of frequencies from one circuit to another. Radio-frequency (rf) and intermediate-frequency (i-f) transformers are used to transfer energy in narrow frequency bands from one circuit to another. Audio and video transformers are required to respond uniformly to signal voltages over a frequency range three to five or more decades wide (for example, from 10 to 100,000 Hz), and consequently must be designed so that very nearly all of the magnetic flux threading through one coil also passes through the other.

Audio and video transformers have two resonances (caused by existing stray and circuit capacitances) just as many tuned transformers do. One resonance point is near the low-signal-frequency limit; the other is near the high limit. As the coefficient of coupling in a transformer is reduced appreciably below unity by removal of core material and separation of the windings, tuning capacitors are added to provide efficient transfer of energy. The two resonant frequencies combine to one when the coupling is reduced to the value known as critical coupling, then stay relatively fixed as the coupling is further reduced.

The rF and i-f transformers use two or more inductors, loosely coupled together, to limit the band of operating frequencies. Efficient transfer of energy is obtained by resonating one or more of the inductors. By using higher than critical coupling, a wider bandwidth than that from the individual tuned circuits is obtained, while the attenuation of side frequencies is as rapid as with the individual circuits isolated from one another.

Keats A. Pullen

Transfusion The administration of blood or its components as a part of a medical treatment. There are certain fairly well-delineated indications for the use of some form of transfusion. Hemorrhage, severe burns, and certain forms of shock are perhaps the most important conditions for which blood transfusion is utilized. Other disorders in which hemotherapy may be indicated include hemophilia, leukemia, certain anemias, and rare hereditary or familial disorders in which some portion of the blood is lacking or deficient. *See* HEMATOLOGIC DISORDERS.

In order for a recipient to accept a blood transfusion, the donor blood cells must be immunologically compatible with the recipient. That is, the recipient must recognize certain molecules (antigens) on donor blood cells as "self" and not foreign. The three main antigen systems on blood cells are the ABO, Rh, and HLA. *See* BLOOD GROUPS; RH INCOMPATIBILITY.

Donated blood is tested for blood groups, blood-group antibodies, and laboratory evidence of syphilis, hepatitis B, AIDS, human T-cell lymphotrophic virus type I (HTLV-I, which is associated with adult T-cell leukemia), and hepatitis C. As a result, blood transfusions have become safer. However, persons who may have been exposed to AIDS should not donate blood, because it has been found that blood may test negative for AIDS and yet still be capable of transmitting AIDS to recipients. This situation can arise because there is a period of time during which a recently infected individual has not yet made sufficient antibody to test positive. Because of the concern that AIDS can be transmitted by blood transfusion, patients sometimes request donations from specific family members and friends. In general, such directed donations are statistically no safer than volunteer blood donation. More patients are donating their own blood (autologous blood) before elective surgery, for their own use during and after the surgery. Autologous blood is the safest blood for transfusion. *See* ACQUIRED IMMUNE DEFICIENCY SYNDROME (AIDS).

Shortly before transfusion, the blood of the donor and recipient is tested once again to make sure that the blood groups are compatible. In an emergency, these tests are abbreviated, or type O red blood cells which can be transfused safely to any individual, are used. Transfusion of blood and blood components is essential to support many patients undergoing surgery, treatment for cancer, or organ transplantation, as well as premature infants.

In addition to transmission of infections, adverse effects of blood transfusion are due to immune reactions between donor and recipient. Fever, the most frequent reaction, is caused by reaction of recipient antibody against donor white blood cells. Hives are due to allergic reactions to substances in donor plasma. Destruction of donor red cells (hemolytic reaction) occurs if the wrong type of blood is inadvertently given, or if recipient antibody is not detected prior to transfusion. *See* BLOOD. Pearl Toy

Transistor A solid-state device involved in amplifying small electrical signals and in processing of digital information. Transistors act as the key element in amplification, detection, and switching of electrical voltages and currents. They are the active electronic component in all electronic systems which convert battery power to signal power. Almost every type of transistor is produced in some form of semiconductor, often single-crystal materials, with silicon being the most prevalent. There are several different types of transistors, classified by how the internal mobile charges (electrons and holes) function. The main categories are bipolar junction transistors (BJTs) and field-effect transistors (FETs).

Single-crystal semiconductors, such as silicon from column 14 of the periodic table of chemical elements, can be produced with two different conduction species, majority and minority carriers. When made with, for example, 1 part per million of phosphorus (from column 15), the silicon is called *n*-type because it adds conduction electrons (negative charge) to form the majority carrier. When doped with boron (from column 13), it is called *p*-type because it has added positive mobile carriers called holes. For *n*-type doping, electrons are the majority carrier while holes become the minority carrier. For *p*-type doping holes are in larger numbers, hence they are the majority carriers, while electrons are the minority carriers. All transistors are made up of regions of *n*-type and *p*-type semiconducting material. *See* SEMICONDUCTOR; SINGLE CRYSTAL.

The bipolar transistor has two conducting species, electrons and holes. Field-effect transistors can be called unipolar because their main conduction is by one carrier type, the majority carrier. Therefore, field-effect transistors are either *n*-channel (majority electrons) or *p*-channel (majority holes). For the bipolar transistor, there are two forms, n^+pn and p^+np, depending on which carrier is majority and which is the minority in a given region. As a result the bipolar transistor conducts by majority as well as by minority carriers. The n^+pn version is by far the most used as it has several distinct performance advantages, as does the *n*-channel for the field-effect transistors. (The n^+ indicates that the region is more heavily doped than the other two regions.)

Bipolar transistors. Bipolar transistors have additional categories: the homojunction for one type of semiconductor (all silicon), and heterojunction for more than one (particularly silicon and silicon-germanium, $Si/Si_{1-x}Ge_x/Si$). At present the silicon homojunction, usually called the BJT, is by far the most common. However, the highest performance (frequency and speed) is a result of the heterojunction bipolar transistor (HBT).

Bipolar transistors are manufactured in several different forms, each appropriate for a particular application. They are used at high frequencies, for switching circuits, in high-power applications, and under extreme environmental stress. The bipolar junction transistor may appear in discrete form as an individually encapsulated component, in monolithic form (made in and from a common material) in integrated circuits, or as a so-called chip in a thick-film or thin-film hybrid integrated circuit. In the *pn*-junction isolated integrated-circuit n^+pn bipolar transistor, an n^+ subcollector, or buried layer, serves as a low-resistance contact which is made on the top surface (Fig. 1). *See* INTEGRATED CIRCUITS; JUNCTION TRANSISTOR.

Field-effect transistors. Majority-carrier field-effect transistors are classified as metal-oxide-semiconductor field-effect transistor (MOSFET), junction "gate" field-effect transistor (JFET), and metal "gate" on semiconductor field-effect transistor (MESFET) devices. MOSFETs are the most used in almost all computers and system applications. However, the MESFET has high-frequency applications in gallium arsenide (GaAs), and the silicon JFET has low-electrical noise performance for audio components and instruments. In general, the *n*-channel field-effect transistors are preferred because of larger electron mobilities, which translate into higher speed and frequency of operation.

An *n*-channel MOSFET (Fig. 2) has a so-called source, which supplies electrons to the channel. These electrons travel through the channel and are removed by a drain electrode into the external circuit. A gate electrode is used to produce the channel or to remove the channel; hence it acts like a gate for the electrons, either providing a channel for them to flow from the source to the drain or blocking their flow (no channel). With a large enough voltage on the gate, the channel is formed, while at a low gate voltage it is not formed and blocks the electron flow to the drain. This type of MOSFET is called enhancement mode because the gate must have sufficiently large voltages to create a channel through which the electrons can flow. Another way of saying the same idea is that the device is normally "off" in an nonconducting state until the gate enhances the channel.

In the JFET (Fig. 3), a conducting majority-carrier *n* channel exists between the source and drain. When a negative voltage is

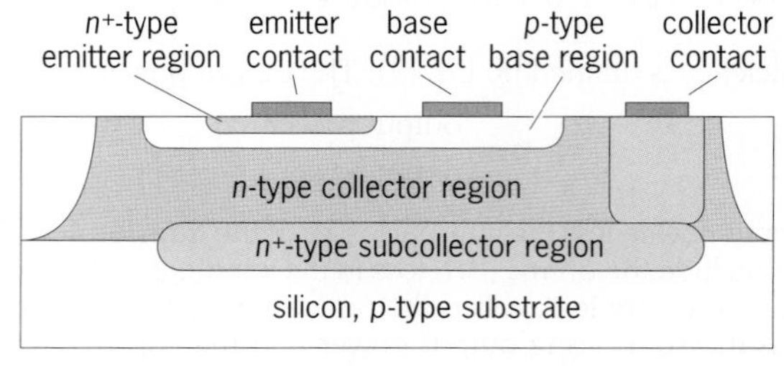

Fig. 1. Isolated n^+pn bipolar junction transistor for integrated-circuit operation.

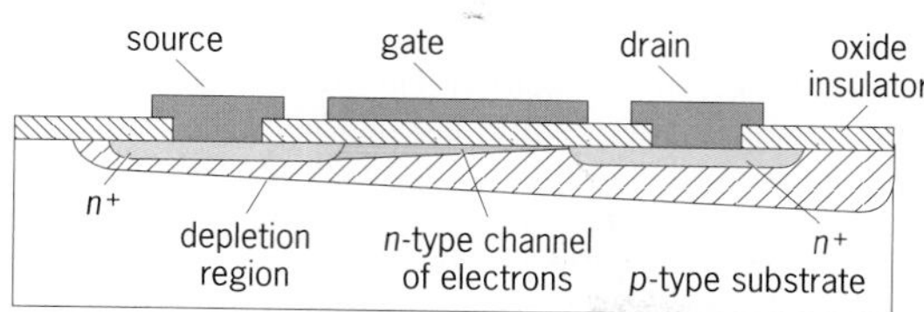

Fig. 2. An *n*-channel enhancement-mode metal-oxide-semiconductor field-effect transistor (MOSFET).

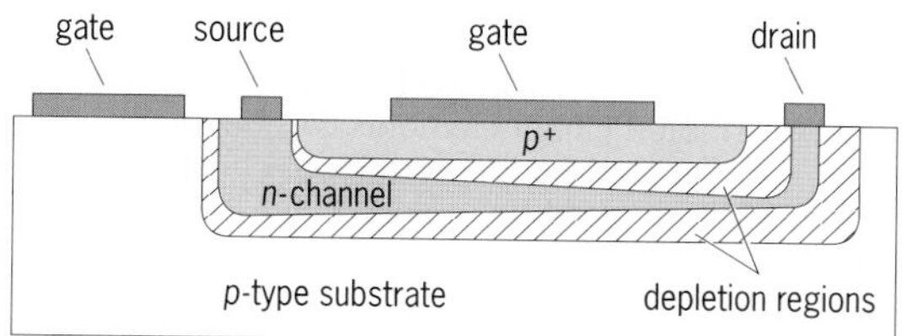

Fig. 3. An *n*-channel junction field-effect transistor (JFET).

applied to the p^+ gate, the depletion regions widen with reverse bias and begin to restrict the flow of electrons between the source and drain. At a large enough negative gate voltage (symbolized V_P), the channel pinches off.

The MESFET is quite similar to the JFET in its mode of operation. A conduction channel is reduced and finally pinched off by a metal Schottky barrier placed directly on the semiconductor. Metal on gallium arsenide is extensively used for high-frequency communications because of the large mobility of electrons, good gain, and low noise characteristics. Its cross section is similar to that of the JFET (Fig. 3), with a metal used as the gate. *See* SCHOTTKY BARRIER DIODE.

Gerold W. Neudeck

Transit (astronomy) The apparent passage of a celestial body across the apparent disk of a larger body, such as a planet across its parent star or of a satellite across its parent planet; also, the apparent passage of a celestial object or reference point across an adopted line of reference in a celestial coordinate system. Classically, the observed data were instants of internal and external tangency of the disks (contacts) at ingress and egress of the smaller body. In the modern era, data may also include the differential brightness of the two disks and the duration of any change of brightness.

Mercury and Venus are the only planets whose orbits lie between the Earth and the Sun and thus can be seen from Earth to cross the disk of the Sun. The conditions are that the planet is in inferior conjunction at the same time that it passes one of the two nodes of its orbit, thus putting it essentially in a straight line between the Earth and the Sun. Historically, transits of Mercury were observed for the purpose of getting precise positions of the planet to improve knowledge of its orbit, and transits of Venus to determine the solar parallax. In a century, there are 13 or 14 transits of Mercury. The size, shape, and orientation of the orbit of Venus causes transits to be very rare. Transits usually occur in pairs separated by 8 years, with 105.5 or 121.5 years between pairs. A pair occurs on June 8, 2004, and June 6, 2012. *See* MERCURY (PLANET); PLANET; VENUS.

Transits of the galilean satellites of Jupiter are used mainly to estimate the albedo (reflectivity) of the satellites relative to that of Jupiter. As each satellite passes in front of the planet, it casts its shadow on the planet's disk and causes the phenomenon of shadow-transit. *See* JUPITER; SATELLITE (ASTRONOMY).

Until the close of the twentieth century, passages of stars and other celestial bodies across the local meridian were observed extensively for determining precise coordinates of the stars and planets, accurate time, or the position of the observer. The instrument commonly used is often called a transit circle. This type of observation has been almost completely superseded by interferometric methods from Earth's surface and orbiting satellites, and by other astrometric observations from spacecraft. *See* ASTRONOMICAL COORDINATE SYSTEMS; ASTRONOMICAL TRANSIT INSTRUMENT.

The search for planets around solarlike stars other than the Sun yielded the first positive results in 1995. More than 100 planets have been found since, and the number is rising rapidly. They have been found by techniques using astrometry and radial velocity. The most recently developed of several techniques is to detect photometrically the minute decrease in brightness of a star as an orbiting planet crosses its face. This can occur only if the planet's orbital plane lies edge-on to the Earth. The first planetary companion detected this way was reported in 1999, orbiting the star HD 209458 in Pegasus. Even though that companion was larger than Jupiter, the technique is considered the most mature for detecting Earth-class extrasolar planets, that is, those that are 0.5–10 times the size of Earth.

Alan D. Fiala

Transition elements In broad definition, the elements of atomic numbers 21–31, 39–49, and 71–81, inclusive. A more restricted classification of the transition elements, preferred by many chemists, is limited to elements with atomic numbers 22–28, 40–46, and 72–78, inclusive. All of the elements in this classification have one or more electrons present in an unfilled *d* subshell in at least one well-known oxidation state.

All the transition elements are metals and, in general, are characterized by high densities, high melting points, and low vapor pressures. Within a given subgroup, these properties tend to increase with increasing atomic weight. Facility in the formation of metallic bonds is demonstrated also by the existence of a wide variety of alloys between different transition metals.

The transition elements include most metals of major economic importance, such as the relatively abundant iron, nickel, and zinc, on one hand, and the rarer coinage metals, copper, silver, and gold, on the other. Also included are the rare and relatively unfamiliar element, rhenium, and technetium, which is not found naturally in the terrestrial environment, but is available in small amounts as a product of nuclear fission.

In their compounds, the transition elements tend to exhibit multiple valency, the maximum valence increasing from +3 at the beginning of a series (Sc, Y, Lu) to + 8 at the fifth member (Mn, Re). One of the most characteristic features of the transition elements is the ease with which most of them form stable complex ions. Features which contribute to this ability are favorably high charge-to-radius ratios and the availability of unfilled *d* orbitals which may be used in bonding.

Most of the ions and compounds of the transition metals are colored, and many of them are paramagnetic. Both color and paramagnetism are related to the presence of unpaired electrons in the *d* subshell. Because of their ability to accept electrons in unoccupied *d* orbitals, transition elements and their compounds frequently exhibit catalytic properties.

Broadly speaking, the properties of the transition elements are intermediate between those of the so-called representative elements, in which the subshells are completely occupied by electrons (alkali metals, halogen elements), and those of the inner or *f* transition elements, in which the subshell orbitals play a much less significant role in influencing chemical properties (rare-earth elements, actinide elements). *See* ACTINIDE ELEMENTS; ATOMIC STRUCTURE AND SPECTRA; RARE-EARTH ELEMENTS.

Burris B. Cunningham

Transition point The point at which a substance changes from one state of aggregation to another. This general definition would include the melting point (transition from solid to liquid), boiling point (liquid to gas), or sublimation point (solid to gas); but in practice the term transition point is usually restricted to the transition from one solid phase to another, that is, to the temperature (for a fixed pressure, usually 1 atm or 10^5 pascals) at which a substance changes from one crystal structure to another.

Another kind of transition point is the culmination of a gradual change (for example, the loss of ferromagnetism in iron or nickel) at the lambda point, or Curie point. This behavior is typical of second-order transitions. *See* BOILING POINT; MELTING POINT; PHASE EQUILIBRIUM; SUBLIMATION; TRIPLE POINT. Robert L. Scott

Transition radiation detectors Detectors of energetic charged particles that make use of radiation emitted as the particle crosses boundaries between regions with different indices of refraction. An energetic charged particle moving through matter momentarily polarizes the material nearby. If the particle crosses a boundary where the index of refraction changes, the change in polarization gives rise to the emission of electromagnetic transition radiation. About one photon is emitted for every 100 boundaries crossed, for transitions between air and matter of ordinary density. Transition radiation is emitted even if the velocity of the particle is less than the light velocity of a given wavelength, in contrast to Cerenkov radiation. Consequently, this radiation can take place in the x-ray region of the spectrum where there is no Cerenkov radiation, because the index of refraction is less than one. *See* CERENKOV RADIATION; PARTICLE DETECTOR; REFRACTION OF WAVES. William J. Willis

Translucent medium A medium which transmits rays of light so diffused that objects cannot be seen distinctly; that is, the medium is only partially transparent. Familiar examples are various forms of glass which admit considerable light but impede vision. Inasmuch as the term translucent seems to imply seeing, usage of the term is ordinarily limited to the visible region of the spectrum. M. G. Mellon

Transmission lines A system of conductors suitable for conducting electric power or signals between two or more termini. Transmission lines take many forms in practice and have application in many disciplines. For example, they traverse the countryside, carrying telephone signals and electric power. The same transmission lines, with similar functions, may be hidden above false ceilings in urban buildings. With the need to reliably and securely transmit ever larger amounts of data, the required frequency of operation has increased from the high-frequency microwave range to the still higher frequency of light. Optical fibers are installed in data-intensive buildings and form a nationwide network. Increasing demand also requires that transmission lines handle greater values of electric power.

Transmission lines can, in some cases, be analyzed by using a fairly simple model that consists of distributed linear electrical components. Models of this type, with some permutations, can also be used to describe wave propagation in integrated circuits and along nerve fibers in animals. The study of hollow metal waveguides or optical fibers is usually based upon an analysis starting from Maxwell's equations rather than employing transmission-line models. Fundamentals and definitions that can initially be obtained from a circuit model of a transmission line carry over to waveguides, where the analysis is more complicated. *See* OPTICAL FIBERS; WAVEGUIDE.

Coaxial cables and strip lines. Two particular types of transmission lines for communication that have received considerable attention are the coaxial cable and the strip line. The coaxial cable is a flexible transmission line and typically is used to connect two electronic instruments together. In the coaxial cable, a dielectric separates a center conducting wire from a concentric conducting sleeve. The strip line is used in integrated circuits to connect, say, two transistor circuits together. A strip line also has a dielectric that separates the top conducting element from the base, which may be an electrical ground plane in the circuit. *See* COAXIAL CABLE; INTEGRATED CIRCUITS.

Circuit model. These transmission lines can be most easily analyzed in terms of electrical circuit elements consisting of distributed linear inductors and capacitors. The values of these elements are in terms of the physical dimensions of the coaxial cable and the strip line, and the permittivity of the dielectric. Each of the elements is interpreted to be measured in terms of a unit length of the element. An equivalent circuit represents either the coaxial cable or the strip line as well as other transmission lines such as two parallel wires. *See* CAPACITANCE; CAPACITOR; INDUCTANCE; INDUCTOR; PERMITTIVITY.

Losses are incorporated into the transmission-line model with the addition of a distributed resistance in series with the inductor and a distributed conductance in parallel with the capacitor. Additional distributed circuit elements can be incorporated into the model in order to describe additional effects. For example, the linear capacitors could be replaced with reverse-biased varactor diodes and the propagation of nonlinear solitons could be studied. *See* CONDUCTANCE; ELECTRICAL RESISTANCE; SOLITON; VARACTOR. Karl E. Lonngren

Power transmission lines. Electric power generating stations and distribution substations are connected by a network of power transmission lines, mostly overhead lines. Power transmitted is generally in the form of three-phase alternating current (ac) at 60 or 50 Hz. In a few instances, where a clear technical or economic advantage exists, direct-current (dc) systems may be used. As the distances over which the power must be transmitted become great and as the amount of power transmitted increases, the power lost in the transmission lines becomes an important component of the production cost of electricity, and it becomes advantageous to increase the transmission voltage. This basic consideration has led to electric power networks which use higher voltages for long-distance bulk power transfers, with several layers of underlying regional networks at progressively lower voltages which extend over shorter distances. The most common transmission voltages in use are 765, 500, 400, 220 kV, and so forth. Voltages below 69 kV are termed subtransmission or distribution voltages, and at these and lower voltages the networks may have fewer alternative supply paths (loops) or may be entirely radial in structure. *See* ALTERNATING CURRENT; DIRECT CURRENT; DIRECT-CURRENT TRANSMISSION; ELECTRIC DISTRIBUTION SYSTEMS; ELECTRIC POWER TRANSMISSION. Arun G. Phadke

Transmutation The nuclear change of one element into another, either naturally, in radioactive elements, or artificially, by bombarding an element with electrons, deuterons, or α-particles in particle accelerators or with neutrons in atomic piles.

Natural transmutation was first explained by Marie Curie about 1900 as the result of the decay of radioactive elements into others of lower atomic weight. Ernest Rutherford produced the first artificial transmutation (nitrogen into oxygen and hydrogen) in 1919. Artificial transmutation is the method of origin of the heavier, artificial transuranium elements, and also of hundreds of radioactive isotopes of most of the chemical elements in the periodic table. *See* NUCLEAR REACTION; PERIODIC TABLE. Frank H. Rockett

Transonic flight In aerodynamics, flight of a vehicle at speeds near the speed of sound. When the velocity of an airplane approaches the speed of sound (roughly 660 mi/h or 1056 km/h at 35,000 ft or 10.7 km altitude), the flight characteristics become radically different from those at subsonic speeds. The drag increases greatly, the lift at a given altitude decreases, the moments acting on the airplane change abruptly, and the vehicle may shake or buffet. Such phenomena usually persist to flight velocities somewhat above the speed of sound. These flight characteristics, as well as the speeds at which they occur, are usually referred to as transonic. The extent of the speed range of these changes depends on the form of the airplane; for configurations designed for subsonic flight they may occur at velocities of 70–110% of the speed of sound (Mach numbers of 0.7–1.1); for airplanes intended for transonic or supersonic flight they may be present only at Mach numbers of 0.95–1.05. *See* MACH NUMBER.

The transonic flight characteristics result from the development of shock waves about the airplane. Because of the accelerations of airflow over the various surfaces, the local velocities become supersonic while the airplane itself is still subsonic. (The flight speed at which such local supersonic flows first occur is called the critical speed.) Shock waves are associated with deceleration of these local supersonic flows to subsonic flight velocities. Such shock waves cause abrupt streamwise increases of pressure on the airplane surfaces. These gradients may cause a reversal and separation of the flow in the boundary layer on the wing surface in roughly the same manner as do similar pressure changes at lower subcritical speeds. *See* AERODYNAMIC FORCE; AERODYNAMIC WAVE DRAG; BOUNDARY-LAYER FLOW; SCHLIEREN PHOTOGRAPHY; SHOCK WAVE.

As for boundary-layer separation at lower speeds, the flow breakdown in this case leads to increases of drag, losses of lift, and changes of aerodynamic moments. The unsteady nature of the separated flow results in an irregular change of the aerodynamic forces acting on the airplane with resultant buffeting and shaking. As the Mach number is increased, the shock waves move aft so that at Mach numbers of about 1.0 or greater, depending on the configurations, they reach the trailing edges of the surfaces. With the shocks in these positions, the associated pressure gradients have relatively little effect on the boundary layer, and the shock-induced separation is greatly reduced. *See* SUBSONIC FLIGHT; SUPERSONIC FLIGHT. Richard T. Whitcomb

Transplantation biology The science of transferring a graft from one part of the body to another or from one individual to another. The graft may consist of an organ, tissue, or cells. If donor and recipient are the same individual, the graft is autologous. If donor and recipient are genetically identical (monozygotic), it is syngeneic. If donor and recipient are any other same-species individuals, the graft is allogeneic. If the donor and recipient are of different species, it is called xenogeneic.

In theory, virtually any tissue or organ can be transplanted. The principal technical problems have been defined and, in general, overcome. Remaining major problems concern the safety of methods used to prevent graft rejection and the procurement of adequate numbers of donor organs.

Living volunteers can donate one of a pair of organs, such as a kidney, only one of which is necessary for normal life. Volunteer donors may also be employed for large unpaired organs such as small bowel, liver, or pancreas, segments of which can be removed without impairment of function. Living donors can also provide tissues capable of regeneration; these include blood, bone marrow, and the superficial layers of the skin. In the case of a vital, unpaired organ, such as the heart, the use of cadaver donors is obligatory. In practice, with the exception of blood and bone marrow, the great majority of transplanted organs are cadaveric in origin, a necessity that presents difficult logistic problems.

Autografts are used for an increasing number of purposes. Skin autografts are important in the treatment of full-thickness skin loss due to extensive burns or other injuries. Provided that the grafts comprise only the superficial levels of the skin, the donor sites reepithelialize spontaneously within a week or two. The saphenous vein of the ankle is frequently transplanted to the heart to bypass coronary arteries obstructed by atherosclerosis. Autologous hematopoietic stem cell transplantation is sometimes used to restore blood cells to cancer patients who receive forms of chemotherapy that are lethal to their bone marrow.

The most serious problem restricting the use of allografts is immunological. Because cells in the donor graft express on their surface a number of genetically determined transplantation antigens that are not present in the recipient, allografts provoke a defensive reaction analogous to that evoked by pathogenic microorganisms. As a consequence, after a transient initial period of apparent well-being, graft function progressively deteriorates and the donor tissue is eventually destroyed. The host response, known as allograft rejection, involves a large number of immunological agents, including cytotoxic antibodies and effector lymphocytes of various types. There are a few special exemptions from rejection that apply to certain sites in the body or to certain types of graft. For example, the use of corneal allografts in restoring sight to individuals with corneal opacification succeeds because of the absence of blood vessels in the host tissue.

Successful transplantation of allografts such as kidneys and hearts currently requires suppressing the recipient's immune response to the graft without seriously impairing the immunological defense against infection. Treatment of individuals with so-called immunosuppressive drugs and other agents prevents allograft rejection for prolonged periods, if not indefinitely. Under cover of nonspecific immunosuppression, the recipient's immune system appears to undergo an adaptation to the presence of the graft, allowing the dosage of the drugs to be reduced. However, in almost all successfully transplanted individuals, drug therapy at some low dose is required indefinitely. *See* IMMUNOSUPPRESSION.

An individual's response against an allograft is directed against a large number of cell-surface transplantation antigens controlled by allelic genes at many different loci. However, in all species, one of these loci, the major histocompatibility complex (MHC), transcends all the other histocompatibility loci in terms of its genetic complexity and the strength of the antigenic response it controls. In humans, the MHC, known as the HLA (human leukocyte antigen) complex, is on the sixth chromosome; its principal loci are designated A, B, C, DR, and DQ. The allelic products of the HLA genes can be detected by serology, polymerase chain reaction technology, or microcytotoxicity assays. The ABO red cell antigens are also important because they are expressed on all tissues. *See* BLOOD GROUPS; HISTOCOMPATIBILITY.

In kidney transplants between closely related family members, the degree of HLA antigen matching can be determined very precisely, and there is a very good correlation between the number of shared HLA antigens and the survival of the graft. With grafts from unrelated donors, HLA matching is more difficult and can delay transplantation, but it may be beneficial. HLA matching is not as clearly beneficial in the case of most other solid organ grafts, and no attempt is made to HLA-match heart, lung, liver, and pancreas grafts. With few exceptions, however, most donors and recipients are matched for the expression of ABO blood group antigens.

Bone marrow transplantation presents a unique problem in its requirements for HLA matching and for immunosuppression in advance of grafting. In addition to the possibility of rejection of the graft by the recipient, by virtue of immunologically competent cells still present in the recipient, bone marrow grafts can react against the transplantation antigens of their hosts. These are known as graft-versus-host reactions, and they can be fatal. *See* IMMUNOLOGY.

The immunological events that lead to the rejection of xenografts are different from and less well understood than those responsible for allograft rejection. The small number of xenografts attempted to date have failed. In particular, xenografts are susceptible to hyperacute rejection by humans. This is due to the presence of certain glycoproteins in blood vessels of many species that are recognized by antibodies present in the blood of all humans. John P. Mordes; Rupert E. Billingham; Dale L. Greiner; Aldo A. Rossini

Transport processes The processes whereby mass, energy, or momentum are transported from one region of a material to another under the influence of composition, temperature, or velocity gradients. If a sample of a material in which the chemical composition, the temperature, or the velocity vary from point to point is isolated from its surroundings, the transport processes act so as to eventually render these quantities uniform throughout the material. The nonuni-form state required to generate these transport processes causes them to be known also as nonequilibrium processes. Associated with gradients of composition, temperature, and velocity in a material are the transport processes of diffusion, thermal conduction, and viscosity, respectively. *See* DIFFUSION IN GASES AND LIQUIDS; DIFFUSION IN SOLIDS; THERMAL CONDUCTION IN SOLIDS; VISCOSITY. W. A. Wakeham

Transportation engineering That branch of engineering related to the movements of goods and people by highway, rail, air, water, and pipeline. Special categories include urban and intermodal transportation.

Engineering for highway transportation involves planning, construction, and operation of highway systems, urban streets, roads, and bridges, as well as parking facilities. Important aspects of highway engineering include (1) overall planning of routes, financing, environmental impact evaluation, and value engineering to compare alternatives; (2) traffic engineering, which plans for the volumes of traffic to be handled, the methods to accommodate these flows, the lighting and signing of highways, and general layout; (3) pavement and roadway engineering, which involves setting of alignments, planning the cuts and fills to construct the roadway, designing the base course and pavement, and selecting the drainage system; and (4) bridge engineering, which involves the design of highway bridges, retaining walls, tunnels, and other structures. *See* HIGHWAY ENGINEERING; TRAFFIC-CONTROL SYSTEMS; VALUE ENGINEERING.

Engineering for railway transportation involves planning, construction, and operation of terminals, switchyards, loading/unloading facilities, trackage, bridges, tunnels, and traffic-control systems for freight and passenger service. For freight operations, there is an emphasis on developing more efficient systems for loading, unloading, shifting cars, and operating trains. Facilities include large marshaling yards where electronic equipment is used to control the movement of railroad cars. Also, there is a trend to developing more automated systems on trackage whereby signals and switches are set automatically by electronic devices. To accommodate transportation of containers, tunnels on older lines are being enlarged to provide for double-stack container cars. *See* RAILROAD ENGINEERING; TUNNEL.

Engineering for air transportation encompasses the planning, design, and construction of terminals, runways, and navigation aids to provide for passenger and freight service. High-capacity, long-range, fuel-efficient aircraft, such as the 440-seat Boeing 777 with a range of 7200 mi (12,000 km), are desirable. Wider use of composites and the substitution of electronic controls for mechanical devices reduce weight to improve fuel economy. Smaller planes are more efficient for shorter runs. *See* AIR NAVIGATION; AIR-TRAFFIC CONTROL; AIR TRANSPORTATION; COMPOSITE MATERIAL.

Engineering for water transportation entails the design and construction of a vast array of facilities such as canals, locks and dams, and port facilities. The transportation system ranges from shipping by barge and tugboat on inland waterways to shipping by oceangoing vessels. Although there is some transportation of passengers, such as on ferries and cruise ships, water transportation is largely devoted to freight. *See* CANAL; DAM; RIVER ENGINEERING.

Pipeline engineering embraces the design and construction of pipelines, pumping stations, and storage facilities. Pipelines are used to transport liquids such as water, gas, and petroleum products over great distances. Also, products such as pulverized coal and iron ore can be transported in a water slurry. *See* PIPELINE; STORAGE TANK.

Engineering for urban transportation concerns the design and construction of light rail systems, subways, and people-movers, as well as facilities for traditional bus systems. To enhance public acceptance of new and expanded systems, increased use is being made of computer-aided design (CAD) to visulaize alternatives for stations and facilities. Also, animated video systems are used for interactive visualization of plans. *See* COMPUTER-AIDED ENGINEERING; HARBORS AND PORTS; SUBWAY ENGINEERING.

Intermodal transportation, often referred to as containerization, entails the use of special containers to ship goods by truck, rail, or ocean vessel. Engineers must design and construct intermodal facilities for efficient operations. The containers are fabricated from steel or aluminum, and they are designed to withstand the forces from handling. The ships are constructed with a cellular grid of compartments for containers below deck, and they can accommodate one or two layers on deck as well. Advantages include savings in labor costs, less pilferage, and lower insurance costs. *See* HOISTING MACHINES; MARINE CONTAINERS; MERCHANT SHIP.

The environment and energy consumption are taken into major consideration when planning, designing, and constructing transportation facilities. Efforts to curb energy use arise from a variety of concerns, including security issues and environmental implications. Efforts to relieve congestion in urban areas through incentives to make greater use of car pooling, such as special freeway lanes, and encouraging greater use of mass transit, deserve further emphasis. Roger L. Brockenbrough

Transposons Types of transposable elements which comprise large discrete segments of deoxyribonucleic acid (DNA) capable of moving from one chromosome site to a new location. In bacteria, the transposable elements can be grouped into two classes, the insertion sequences and the transposons. The ability of transposable elements to insert into plasmid or bacterial virus (bacteriophage) which is transmissible from one organism to another allows for their rapid spread. *See* BACTERIOPHAGE; PLASMID.

The insertion sequences were first identified by their ability to induce unusual mutations in the structural gene for a protein involved in sugar metabolism. These insertion sequences are relatively small (about 500–1500 nucleotide pairs) and can only be followed by their ability to induce these mutations. Most bacterial chromosomes contain several copies of such insertion sequence elements.

The transposons are larger segments of DNA (2000–10,000 base pairs) that encode several proteins, usually one or two required for the movement of the element and often an additional protein that imparts a selective advantage to the host containing a copy of that element. The structure of many transposons suggests they may have evolved from the simpler insertion sequence elements.

All transposable elements, both the simple insertion sequence elements and the more complex transposons, have a similar structure and genetic organization. The ends of the element represent recognition sites and define the segment of DNA undergoing transposition. A short sequence present at one end of the element is repeated in an inverted fashion at the other end. These terminal inverted repeats are characteristic for each element.

Members of a widespread group of transposons, the Tn3 family, all have a similar structure and appear to move by a similar mechanism. Transposase, one protein encoded by the element, promotes the formation of intermediates called cointegrates, in which the element has been duplicated by replication. A second element-encoded protein, resolvase, completes the process by converting the cointegrates into the end products of transposition, a transposon inserted into a new site. A third protein encoded by the Tn3 element imparts resistance to the antibiotic ampicillin.

Transposons are known that encode resistances to almost all antibiotics as well as many toxic metals and chemicals. In addition, some transposons have acquired the ability to direct the synthesis of proteins that metabolize carbohydrates, petroleum, and pesticides. Other transposable elements produce enterotoxins that cause travelers to become ill from drinking water contaminated with bacteria carrying the element. The broad spectrum of activities encoded by the transposable elements demonstrates the strong selective advantage that has accompanied their evolution.

Transposable elements are not restricted to prokaryotes. Yeast as well as higher eukaryotes have DNA segments that move and cause mutations. The eukaryotic elements have much in common with their prokaryotic counterparts: the termini of the elements are composed of inverted repeats, and many of the larger elements are composed of two small insertion sequence-like regions flanking a unique central region. One class of eukaryotic virus, the ribonucleic acid (RNA) retrovirus, also has this structure and is thought to integrate into the host chromosome through a transpositionlike mechanism. *See* ANTIBIOTIC; GENE; RETROVIRUS; VIRUS.

Randall Reed

Transuranium elements Those synthetic elements with atomic numbers larger than that of uranium (atomic number 92). They are the members of the actinide series, from neptunium (atomic number 93) through lawrencium (atomic number 103), and the transactinide elements (with higher atomic numbers than 103). Of these elements, plutonium, an explosive ingredient for nuclear weapons and a fuel for nuclear power because it is fissionable, has been prepared on the largest (ton) scale, while some of the others have been produced in kilograms (neptunium, americium, curium) and in much smaller quantities (berkelium, californium, and einsteinium).

The concept of atomic weight in the sense applied to naturally occurring elements is not applicable to the transuranium elements, since the isotopic composition of any given sample depends on its source. In most cases the use of the mass number of the longest-lived isotope in combination with an evaluation of its availability has been adequate. Good choices at present are neptunium, 237; plutonium, 242; americium, 243; curium, 248; berkelium, 249; californium, 249; einsteinium, 254; fermium, 257; mendelevium, 258; nobelium, 259; lawrencium, 260; rutherfordium, 261; dubnium, 262; and seaborgium, 263. The actinide elements are chemically similar and have a strong chemical resemblance to the lanthanide, or rare-earth, elements (atomic numbers 57–71). The transactinide elements, with atomic numbers 104 and higher, appear in an expanded periodic table under the row of elements beginning with hafnium, number 72, and ending with radon, number 86. This arrangement allows prediction of the chemical properties of these elements and suggests that they will have a chemical analogy with the elements which appear immediately above them in the periodic table.

The transuranium elements up to and including fermium (atomic number 100) are produced in largest quantity through the successive capture of neutrons in nuclear reactors. The yield decreases with increasing atomic number, and the heaviest to be produced in weighable quantity is einsteinium (number 99). Many additional isotopes are produced by bombardment of heavy target isotopes with charged atomic projectiles in accelerators; beyond fermium, all elements are produced by bombardment with heavy ions.

Beyond roentgenium (atomic number 111), transactinide elements 112 and 113 have been produced, although their acceptance is pending. *See* ACTINIDE ELEMENTS; AMERICIUM; BERKELIUM; BOHRIUM; CALIFORNIUM; CURIUM; DUBNIUM; EINSTEINIUM; ROENTGENIUM; ELEMENT 112; FERMIUM; HASSIUM; LAWRENCIUM; MEITNERIUM; MENDELEVIUM; NEPTUNIUM; NOBELIUM; NUCLEAR CHEMISTRY; NUCLEAR FISSION; NUCLEAR REACTION; PERIODIC TABLE; PLUTONIUM; RARE-EARTH ELEMENT; RUTHERFORDIUM; SEABORGIUM.

Peter J. Armbruster; M. Schädel; Glenn T. Seaborg

Traps in solids Localized regions in a material that can capture and localize an electron or hole, thus preventing the electron or hole from moving through the material until supplied with sufficient thermal or optical energy. Traps in solids are associated with imperfections in the material caused by either impurities or crystal defects. *See* BAND THEORY OF SOLIDS; CRYSTAL DEFECTS; HOLE STATES IN SOLIDS.

Imperfections that behave as traps are commonly distinguished from imperfections that behave as recombination centers. If the probability for a captured electron (or hole) at the imperfection to be thermally reexcited to the conduction (or valence) band before recombination with a free hole (or free electron) is greater than the probability for such recombination, then the imperfection is said to behave like an electron (or hole) trap. If the probability for a captured electron (or hole) at the imperfection to recombine with a free hole (or free electron) is greater than the probability for being thermally reexcited to the band, the imperfection is said to behave like a recombination center. It is possible for a specific chemical or structural imperfection in the material to behave like a trap under one set of conditions of temperature and light intensity, and as a recombination center under another.

Traps play a significant role in many phenomena involving photoconductivity and luminescence. In photoconductors, for example, the presence of traps decreases the sensitivity and increases the response time. Their effect is detectable through changes in the rise and decay transients of photoconductivity and luminescence, thermally stimulated conductivity and luminescence in which the traps are filled at a low temperature and then emptied by increasing the temperature in a controlled way, electron spin responance associated with trapped electrons with unpaired spins, and a variety of techniques involving the capacitance of a semiconductor junction such as photocapacitance and deep-level transient spectroscopy. *See* ELECTRON PARAMAGNETIC RESONANCE (EPR) SPECTROSCOPY; LUMINESCENCE; PHOTOCONDUCTIVITY; THERMOLUMINESCENCE.

Richard H. Bube

Trauma Injury to tissue by physical or chemical means. Mechanical injury includes abrasions, contusions, lacerations, and incisions, as well as stab, puncture, and bullet wounds. Trauma to bones and joints results in fractures, dislocations, and sprains. Head injuries are often serious because of the complications of hemorrhage, skull fracture, or concussion.

Thermal, electrical, and chemical burns produce severe damage partly because they coagulate tissue and seal off restorative blood flow. Asphyxiation, including that caused by drowning, produces rapid damage to the brain and respiratory centers, as well as to other organs.

Frequent complications of trauma are shock, the state of collapse precipitated by peripheral circulatory failure, and also hemorrhage, infection, and improper healing. *See* HEMORRHAGE; INFECTION; SHOCK SYNDROME.

Edward G. Stuart; N. Karle Mottet

Traveling-wave tube A microwave electronic tube in which a beam of electrons interacts continuously with a wave that travels along a circuit, the interaction extending over a distance of many wavelengths. Traveling-wave tubes can provide amplification over exceedingly wide bandwidths. Typical bandwidths are 10–100% of the center frequency, with gains of 20– 60 dB. Low-noise traveling-wave tube amplifiers serve as the inputs to sensitive radars or communications receivers. High-efficiency medium-power traveling-wave tubes are the principal final amplifiers used in communication satellites, the space shuttle communications transmitter, and deep-space planetary probes and landers. High-power traveling-wave amplifiers operate as the final stages of radars, wide-band radar

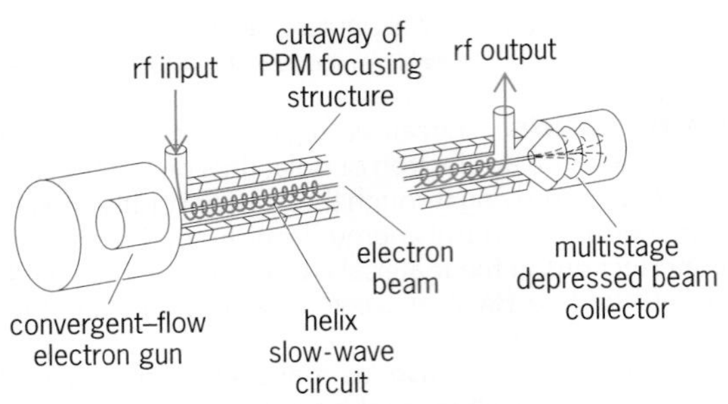

Periodic-permanent-magnet (PPM) focused traveling-wave tube.

countermeasure systems, and scatter communication transmitters. They are capable of delivering continuous-wave power levels in the kilowatt range and pulsed power levels exceeding a megawatt. *See* COMMUNICATIONS SATELLITE; ELECTRONIC WARFARE; RADAR; SPACE COMMUNICATIONS; SPACE PROBE.

In a forward-wave, traveling-wave tube amplifier (see illustration), a thermionic cathode produces the electron beam. An electron gun initially focuses the beam, and an additional focusing system retains the electron stream as a beam throughout the length of the tube until the beam is captured by the collector electrode. The microwave signal to be amplified enters the tube near the electron gun and propagates along a slow-wave circuit. The tube delivers amplified microwave energy into an external matched load connected to the end of the circuit near the collector. The slow-wave circuit serves to propagate the microwave energy along the tube at approximately the same velocity as that of the electron beam. Interaction between beam and wave is continuous along the tube with contributions adding in phase.

Velocity and current modulations of the electron beam occur because the waves and the electrons travel in near synchronism and the amplification process takes place continuously. The electron beam is injected into the helix with a velocity slightly faster than the waves. In the ensuing interaction between fields and waves, the energy lost by the average deceleration of the electrons is the source of energy for the growing waves in the circuit. Because of the continuously distributed interaction, the power in the circuit wave grows exponentially with distance along the tube. *See* MAGNETRON; MICROWAVE TUBE; OSCILLATOR.

Lester A. Roberts

Travertine A rather dense, banded limestone, sometimes moderately porous, that is formed either by evaporation about springs, as is tufa, or in caves, as stalactites, stalagmites, or dripstone. Where travertine or tufa (calcareous sinter) is deposited by hot springs, it may be the result of the loss of carbon dioxide from the waters as pressure is released upon emerging at the surface; the release of carbon dioxide lowers the solubility of calcium carbonate and it precipitates. High rates of evaporation in hot-spring pools also lead to supersaturation. Travertine formed in caves is simply the result of complete evaporation of waters containing mainly calcium carbonate. *See* LIMESTONE; STALACTITES AND STALAGMITES; TUFA.

Raymond Siever

Tree A perennial woody plant at least 20 ft (6 m) in height at maturity, having an erect stem or trunk and a well-defined crown or leaf canopy. However, no sharp lines can be drawn between trees, shrubs, and lianas (woody vines). The essence of the tree form is relatively large size, long life, and a slow approach to reproductive maturity. The difficulty of transporting water, nutrients, and storage products over long distances and high into the air against the force of gravity is a common problem of large treelike plants and one that is not shared by shrubs or herbs.

Classification. Almost all existing trees belong to the seed plants (Spermatophyta). An exception are the giant tree ferns which were more prominent in the forests of the Devonian Period and today exist only in the moist tropical regions. The Spermatophyta are divided into the Pinophyta (gymnosperms) and the flowering plants, Magnoliophyta (angiosperms). The gymnosperms bear their seed naked on modified leaves, called scales, which are usually clustered into structures called cones—for example, pine cones. By contrast the seed of angiosperms is enclosed in a ripened ovary, the fruit. *See* MAGNOLIOPHYTA; PINOPHYTA; POLYPODIALES; TREE FERNS.

The orders Cycadales, Ginkgoales, and Pinales of the Pinophyta contain trees. *Ginkgo biloba*, the ancient maidenhair tree, is the single present-day member of the Ginkgoales. The Cycadales, characteristic of dry tropical areas, contain many species which are small trees. The Pinales, found throughout the world, supply much of the wood, paper, and building products of commerce. They populate at least one-third of all existing forest and include the pines (*Pinus*), hemlocks (*Tsuga*), cedars (*Cedrus*), spruces (*Picea*), firs (*Abies*), cypress (*Cupressus*), larches (*Larix*), Douglas-fir (*Pseudotsuga*), sequoia (*Sequoia*), and other important genera. The Pinales are known in the lumber trade as softwoods and are popularly thought of as evergreens, although some (for example, larch and bald cypress) shed their leaves in the winter. *See* CEDAR; CYCADALES; CYPRESS; DOUGLAS-FIR; FIR; HEMLOCK; LARCH; PAPER; PINALES; PINE; PINOPHYTA; SEQUOIA; SPRUCE.

In contrast to the major orders of gymnosperms which contain only trees, many angiosperm families are herbaceous and include trees only as an exception. Only a few are exclusively arborescent. The major classes of the angiosperms are the Liliopsida (monocotyledons) and the Magnoliopsida (dicotyledons). The angiosperm trees, commonly thought of as broad-leaved and known as hardwoods in the lumber market, are dicotyledons. Examples of important genera are the oaks (*Quercus*), elms (*Ulmus*), maples (*Acer*), and poplars (*Populus*). *See* ELM; LILIOPSIDA; MAGNOLIOPSIDA; MAPLE; OAK; POPLAR.

The Liliopsida contain few tree species, and these are never used for wood products, except in the round as posts. Examples of monocotyledonous families are the palms (Palmae), yucca (Liliaceae), bamboos (Bambusoideae), and bananas (Musaceae). *See* BAMBOO; BANANA.

Morphology. The morphology of a tree is similar to that of other higher plants. Its major organs are the stem, or trunk and branches; the leaves; the roots; and the reproductive structures. Almost the entire bulk of a tree is nonliving. Of the trunk, branches, and roots, only the tips and a thin layer of cells just under the bark are alive. Growth occurs only in these meristematic tissues. Meristematic cells are undifferentiated and capable of repeated division. *See* FLOWER; LATERAL MERISTEM; LEAF; PLANT GROWTH; ROOT (BOTANY); STEM.

Growth. Height is a result of growth only in apical meristems at the very tips of the twigs. A nail driven into a tree will always remain at the same height, and a branch which originates from a bud at a given height will never rise higher. The crown of a tree ascends as a tree ages only by the production of new branches at the top and by the death and abscission of lower, older branches as they become progressively more shaded. New growing points originate from the division of the apical meristem and appear as buds in the axils of leaves. *See* APICAL MERISTEM; BUD; PLANT GROWTH.

In the gymnosperms and the dicotyledonous angiosperms, growth in diameter occurs by division in only a single microscopic layer, three or four cells wide, which completely encircles and sheaths the tree. This lateral meristem is the cambium. It divides to produce xylem cells (wood) on the inside toward the core of the tree and phloem cells on the outside toward the bark. In trees of the temperate regions the growth of each year is seen in cross section as a ring. *See* BARK; PHLOEM; XYLEM.

Xylem elements become rigid through the thickening and modification of their cell wall material. The tubelike xylem cells transport water and nutrients from the root through the stem to the leaves. In time the xylem toward the center of the trunk becomes impregnated with various mineral and metabolic products, and it is no longer capable of conduction. This nonfunctional xylem is called heartwood and is recognizable in some stems by its dark color. The light-colored, functional outer layer of the xylem is the sapwood. *See* WOOD ANATOMY.

The phloem tissue transports dissolved carbohydrates and other metabolic products manufactured by the leaves throughout the stem and the roots. Most of the phloem cells are thin-walled and are eventually crushed between the bark and the cambium by the pressures generated in growth. The outer bark is dead and inelastic but the inner bark contains patches of cork cambium which produce new bark. As a tree increases in circumference, the old outer bark splits and fissures develop, resulting in the rough appearance characteristic of the trunks of most large trees.

In the monocotyledons the lateral cambium does not encircle a central core, and the vascular or conducting tissue is organized in bundles scattered throughout the stem. The trunk is not wood as generally conceived although it does in fact have secondary xylem. *See* DENDROLOGY; FOREST AND FORESTRY; FOREST TIMBER RESOURCES; PLANT PHYSIOLOGY; PLANT TAXONOMY. F. Thomas Ledig

Tree diseases Diseases of both shade and forest trees have the same pathogens, but the trees differ in value, esthetics, and utility. In forests, disease is significant only when large numbers of trees are seriously affected. Diseases with such visible symptoms as leaf spots may be alarming on shade trees but hardly noticed on forest trees. Shade trees with substantial rot may be ornamentals with high value, whereas these trees would be worthless in the forest. Emphasis on disease control for the same tree species thus requires a different approach, depending on location of the tree.

Forest trees. From seed to maturity, forest trees are subject to many diseases. Annual losses of net sawtimber growth from disease (45%) are greater than from insects and fire combined. Young, succulent seedlings, especially conifers, are killed by certain soil-inhabiting fungi (damping-off). Root systems of older seedlings may be destroyed by combinations of nematodes and such fungi as *Cylindrocladium*, *Sclerotium*, and *Fusarium*. Chemical treatment of seed or soil with formulations containing nematicides and fungicides, and cultural practices unfavorable to root pathogens help to avoid these diseases.

Roots rots are caused by such fungi as *Heterobasidion* (=*Fomes*) *annosus* (mostly in conifers) and *Armillariella* (= *Armillaria*) *mellea* (mostly in hardwoods). These fungi cause heart rot in the roots and stems of large trees and also invade and kill young, vigorous ones.

In natural forests, leaf diseases are negligible, but in nurseries and plantations, fungal infections cause severe defoliation, retardation of height growth, or death. *Scirrhia acicola* causes brown spot needle blight and prevents early height growth of longleaf pine in the South; it defoliates Christmas tree plantations of Scotch pine (*Pinus sylvestris*) in northern states. Fungicides and prescribed burning are used successfully for control.

Oak wilt is a systemic disease, with the entire tree affected through its water-conducting system. The causal fungus, *Ceratocystis fagacearum*, spreads to nearby healthy trees by root grafts and to trees at longer distances by unrelated insects. The sporulating mats of the fungus develop between bark and wood, producing asexual and, sometimes, sexual spores, which are disseminated by insects. Control is possible by eradicating infected trees and by disruption of root grafts by trenching or by chemicals.

Stem rust diseases occur as cankers or galls on coniferous hosts and as minor lesions on other ones. A few, such as white pine blister rust and southern fusiform rust, are epidemic, lethal, and economically important. Others of less immediate importance (such as western gall rust) are capable of serious, widespread infection. Resistant varieties are favored for control. Other control measures include pruning out early infections and spraying nursery trees with chemicals during periods favoring needle infection.

Stem infections by numerous fungi, resulting in localized death of cambium and inner bark, range from lesions killing small stems in a year (annual) to gross stem deformities (perennial), where cankers enlarge with stem growth. Chestnut blight, first known in the United States in 1904, destroyed the American chestnut as a commercial species; and is an example of the annual lesion type. The less dramatic or devastating *Nectria* canker destroys stems of timber value, and is an example of the perennial lesion type.

All tree species, including decay-resistant ones such as redwood, are subject to ultimate disintegration by fungi. Decay fungi (Hymenomycetes) are associated with nondecay fungi (Deuteromycetes) and bacteria. These microflora enter the tree through wounds, branch stubs, and roots, and are confined to limited zones of wood by anatomical and wound-stimulated tissue barriers. The extent of decay is limited by compartmentalization of decay in trees. Trees aged beyond maturity are most often invaded by wood-rotting fungi; losses can be minimized by avoiding wounds and by shortening cutting rotations. Losses from rot are especially serious in overmature coniferous stands in the western United States, Canada, and Alaska. *See* FOREST PEST CONTROL; WOOD DEGRADATION.

Shade trees. Many shade trees are grown under conditions for which they are poorly adapted, and are subject to environmental stresses not common to forest trees. Both native and exotic trees planted out of natural habitats are predisposed to secondary pathogens following environmental stress of noninfectious origins. They are also susceptible to the same infectious diseases as forest trees. Appearance is more important than the wood produced, and individual value is higher per tree than for forest trees. Thus, disease control methods differ from those recommended for forest trees.

The most important and destructive shade tree disease known is Dutch elm disease, introduced from Europe to North America before 1930. The causal fungus, *Ophiostoma ulmi* (*Ceratocytis ulmi*), is introduced to the water-conducting system of healthy elms by the smaller European elm bark beetle (*Scolytus multistriatus*) or the American elm bark beetle (*Hylurgopinus rufipes*). One or more new and more aggressive strains of the fungus have arisen since 1970. More devastating than the original ones, they are destroying the elms in North America and Europe that survived earlier epidemics. Effective means of prevention are sanitation (destroying diseased and dying and dead elm wood); insecticidal sprays; disruption of root systems; and early pruning of new branch infections. Of much promise are resistant varieties of elm, systemic fungicides, and insect pheromones.

The most common bacterial disease of shade trees is wetwood of elm and certain other species. It is reported to be caused by a single bacterial species (*Erwinia nimipressuralis*), although causal associations of other bacteria are now suspected. The bacteria are normally present in the heartwood of mature elms and cause no disease unless they colonize sapwood by exterior wounds. Fermented sap under pressure bleeds from wounds and flows down the side of the tree. Sustained bleeding kills underlying cambial tissue. Internal gas pressure and forced spread of bacterial toxins inside living tissues of the tree can be reduced by strategic bleeding to avoid seepage into bark and cambium.

A second bacterial disease of elm, elm yellows, is caused by a mycoplasmalike organism, considered to be a unique kind of bacterium. Elm yellows is as lethal as Dutch elm disease but is more limited in distribution. The pathogen is carried by the elm leafhopper (*Scaphoideus luteolus*), which sucks phloem juice from leaf veins. Spread of disease also occurs through grafted root systems. Control measures include early destruction of infected trees, disruption of root systems, and insecticidal sprays. Injection with tetracycline and other antibiotics helps to slow the progress of the bacterium.

The most common and complex diseases of shade trees are diebacks and declines (such as maple decline). Many species show similar patterns of symptoms caused by multiple factors, but no single causal factor is known to cause any one of these diseases. Noninfectious agents of shade tree diseases are drought, soil compaction, mineral deficiency, soil pollution from waste or salt, air pollution, and so on. Trees affected experience chlorosis, premature fall coloration and abscission, tufting of new growth, dwarfing and sparseness of foliage, progressive death of terminal twigs and branches, and decline in growth. Such trees are often infested with borers and bark beetles, and infected by branch canker and root rot fungi. Noninfectious stress predisposes trees to infectious disease that is caused by different kinds of weakly parasitic fungi as secondary pathogens. *See* PLANT PATHOLOGY.

Richard J. Campana

Tree ferns Plants belonging to the families Cyatheaceae and Dicksoniaceae, whose members typically develop tall trunks crowned with leaves (fronds) which often reach some 20 ft (6.1 m) in length and 5–6 ft (1.5–1.8 m) in width. Tree ferns reach their greatest development in the rainforests and cloud forests of the mountainous tropics. *See* RAINFOREST.

Tree fern trunks may reach 65 ft (19.8 m) tall; their diameters vary from 0.4 in. (1.0 cm) to 4 ft (1.2 m). The lower trunk is often densely covered with matted adventitious roots which greatly increase its diameter. Certain specimens branch near the base of the trunk or higher up; perhaps this branching occurs in response to injury.

The degree of division of the leaves varies from simple to four or five pinnate. The leaflets (pinnae) are usually smaller toward the base of the leaf; when these basal leaflets are branched into threadlike divisions, they are called aphlebiae. The Dicksoniaceae have marginal sori terminal on the veins and protected by a bivalved indusium. The Cyatheaceae produce sori well away from the margin, usually seated at the forking of a vein or midway along a simple vein.

The typical vascular system of both families is dictyostelic. The Cyatheaceae have accessory vascular strands in the pith and cortex. Fibrous sheaths around the vascular tissue and just inside the epidermis provide mechanical support. The xylem consists of scalariform tracheids and parenchyma, the phloem of sieve tubes and parenchyma. Numerous mucilage canals are embedded in the pith and cortex. *See* STEM; TREE.

Gerald J. Gastony

Tree growth Trees grow to a larger size at maturity than other woody perennials, have a comparatively long period of development to maturity, and live a long time in the mature state. Trees, like other vascular plants, are made up of cells; growth is the result of adding more cells through cell division, and of the elongation and maturation of those cells into functional tissues. Cells and the tissues they compose differ in structure and function. Tissues may also be compared by their origin. Primary tissues are derived from apical meristems of the stem and root, whereas secondary tissues arise from secondary cambia. Two important secondary cambia in trees are the vascular cambium which produces cells which become secondary xylem (the wood of commerce) to the inside and secondary phloem tissue to the outside, and the cork cambium or phellogen which produces corky bark tissues to the outside and usually phelloderm tissue to the inside. Collectively, these secondary cambia account for most of the diameter growth in trees. *See* PLANT GROWTH; PLANT TISSUE SYSTEMS.

Photoperiod has a large influence on the growth of trees. Plant response is actually to length of the night rather than of the day. Thus, when it is said that growth is promoted by long photoperiods, in fact the plant senses short nights and so continues to grow rapidly. The glycoprotein-pigment complex phytochrome, located in the plasmalemma, is responsible for transducing the night length into the plant's physiological system. The exact nature of the biological action of the phytochrome system is not known. *See* PHOTOPERIODISM.

In vascular plants, growth generally refers to the development of organized structures. Longitudinal growth depends upon cell division in apical meristems and upon subsequent elongation of these cells. Unlike animal cells, plant cells are encased in a rigid cell wall. Growth begins by breaking of chemical bonds in the cellulosic framework of the wall, followed immediately by insertion of new cellulose molecules into the wall. This type of cell-wall growth is termed intussusception and allows the primary cell wall to grow in area while retaining the same thickness. Primary walls are laid down by immature cells during the growth process, but as growth ends, secondary walls are formed in some types of cells (for example, the secondary xylem tracheids and vessels). The secondary wall is formed by adding wall material to the inner surface of the enlarged cell. This is termed apposition and differs from the intussusception growth process of the primary cell wall. *See* APICAL MERISTEM; CELL WALLS (PLANT).

With few exceptions, periods of rapid stem elongation in trees alternate with periods of either very slow or no elongation even in environments favorable for continuous growth. Tropical trees do not exhibit seasonal growth periodicity as is commonly observed in north temperate species. Trees of temperate regions all display intermittent growth. Periods of shoot growth vary with species and genotype, but always occur during periods of favorable temperature and adequate moisture. Trees usually grow more rapidly and longer when they are young than later when they have achieved a larger size. Research suggests that the cessation of vegetative growth early in the season allows the onset of reproductive competency. *See* TREE.

William C. Carlson

Tree physiology The study of how trees grow and develop in terms of genetics; biochemistry; cellular, tissue, and organ functions; and interaction with environmental factors. While many physiological processes are similar in trees and other plants, trees possess unique physiologies that help determine their outward appearance. These physiological processes include carbon relations (photosynthesis, carbohydrate allocation), cold and drought resistance, water relations, and mineral nutrition.

Three characteristics of trees that define their physiology are longevity, height, and simultaneous reproductive and vegetative growth. Trees have physiological processes that are more adaptable than those in the more specialized annual and biennial plants. Height allows trees to successfully compete for light, but at the same time this advantage creates transport and support problems. These problems were solved by the evolution of the woody stem which combines structure and function into a very strong transport system. Simultaneous vegetative and reproductive growth in adult trees causes significant competition for carbohydrates and nutrients, resulting in decreased vegetative growth. Trees accommodate both types of growth by having cyclical reproduction: one year many flowers and seeds are produced, followed by a year or more in which few or no flowers are produced.

Carbon relations. While biochemical processes of photosynthesis and carbon assimilation and allocation are the same in trees and other plants, the conditions under which these processes occur in trees are more variable and extreme. In evergreen species, photosynthesis can occur year round as long as the air temperature remains above freezing, while some deciduous species can photosynthesize in the bark of twigs and stem during the winter.

Carbon dioxide fixed into sugars moves through the tree in the phloem and xylem to tissues of high metabolism which vary with season and development. At the onset of growth in the spring, sugars are first mobilized from storage sites, primarily in the secondary xylem (wood) and phloem (inner bark) of the woody twigs, branches, stem, and roots. The sugars, stored as starch, are used to build new leaves and twigs, and if present, flowers. Once the new leaves expand, photosynthesis begins and sugars are produced, leading to additional leaf growth. Activation of the vascular cambium occurs at the same time, producing new secondary xylem and phloem. In late spring, the leaves begin photosynthesizing at their maximum rates, creating excess sugars which are translocated down the stem to support further branch, stem, and root growth. From midsummer through fall until leaf abscission (in deciduous trees) or until temperatures drop to freezing (in evergreen trees), sugars replenish the starch used in spring growth. Root growth may be stimulated at this time by sugar availability and warm soil temperatures. Throughout the winter, starch is used for maintenance respiration, but sparingly since low temperatures keep respiration rates low. *See* PHLOEM; PHOTOSYNTHESIS; XYLEM.

In adult trees, reproductive structures (flowers in angiosperms or strobili in gymnosperms) develop along with new leaves and represent large carbohydrate sinks. Sugars are preferentially utilized at the expense of leaf, stem, and root growth. This reduces the leaf area produced, affecting the amount of sugars produced during that year, thereby reducing vegetative growth even further. The reproductive structures are present throughout the growing season until seed dispersal and continually utilize sugars that would normally go to stem and root growth.

Cold resistance. The perennial nature of trees requires them to withstand low temperatures during the winter. Trees develop resistance to freezing through a process of physiological changes beginning in late summer. A tree goes through three sequential stages to become fully cold resistant. The process involves reduced cell hydration along with increased membrane permeability. The first stage is initiated by shortening days and results in shoot growth cessation, bud formation, and metabolic changes. Trees in this stage can survive temperatures down to 23°F (−5°C). The second stage requires freezing temperatures which alter cellular molecules. Starch breakdown is stimulated, causing sugar accumulation. Trees can survive temperatures as low as −13°F (−25°C) at this stage. The last stage occurs after exposure to very low temperatures (−22 to −58°F or −30 to −50°C), which increases soluble protein concentrations that bind cellular water, preventing ice crystallization. Trees can survive temperatures below −112°F (−80°C) in this stage. A few days of warmer temperatures, however, causes trees to revert to the second stage.

Water relations. Unlike annual plants that survive drought as seeds, trees have evolved traits that allow them to avoid desiccation. These traits include using water stored in the stem, stomatal closure, and shedding of leaves to reduce transpirational area. All the leaves can be shed and the tree survives on stored starch. Another trait of some species is to produce a long tap root that reaches the water table, sometimes tens of meters from the soil surface. On a daily basis, trees must supply water to the leaves for normal physiological function. If the water potential of the leaves drops too low, the stomata close, reducing photosynthesis. To maintain high water potential, trees use water stored in their stems during the morning which is recharged during the night. *See* PLANT-WATER RELATIONS.

Transport and support. Trees have evolved a means of combining long-distance transport between the roots and foliage with support through the production of secondary xylem (wood) by the vascular cambium. In older trees the stem represents 60–85% of the aboveground biomass. However, 90% of the wood consists of dead cells. These dead cells function in transport and support of the tree. As these cells develop and mature, they lay down thick secondary walls of cellulose and lignin that provide support, and then they die with the cell lumen becoming an empty tube. The interconnecting cells provide an efficient transport system, capable of moving 106 gal (400 liters) of water per day. The living cells in the wood (ray parenchyma) are the site of starch storage in woody stems and roots. *See* PLANT TRANSPORT OF SOLUTES.

Mineral nutrition. Nutrient deficiencies are similar in trees and other plants because of the functions of these nutrients in physiological processes. Tree nutrition is unique because trees require lower concentrations, and they are able to recycle nutrients within various tissues. Trees adapt to areas which are low in nutrients by lowering physiological functions and slowing growth rates. In addition, trees allocate more carbohydrates to root production, allowing them to exploit large volumes of soil in search of limiting nutrients. Proliferation of fine roots at the organic matter-mineral soil interface where many nutrients are released from decomposing organic matter allows trees to recapture nutrients lost by leaf fall. *See* PLANT MINERAL NUTRITION; PLANT PHYSIOLOGY; TREE.

Jon D. Johnson

Trematoda A loose grouping of acoelomate, parasitic flatworms of the phylum Platyhelminthes formerly accorded class rank and containing the subclasses (or orders) Digenea, Monogenea, and Aspidobothria. These organisms commonly occur as adults in or on all vertebrate groups. They exhibit cephalization, bilateral symmetry, and well-developed anterior and ventral, or anterior and posterior, holdfast structures. The mouth is anterior, and usually a blind, forked gut occurs, as well as three muscle layers. The excretory system consists of flame cells and collecting tubules. These animals are predominantly hermaphroditic and oviparous with operculated egg capsules. The life histories of the Digenea are complex, while those of the Monogenea and Aspidobothria are simple.

Trematodes parasitize a wide variety of invertebrates and vertebrates and occupy almost every available niche within these hosts. The adaptations demanded of the worms for survival are as varied as the characteristics of the microhabitats. Over the millions of years of coevolution of the hosts and their parasites, delicate balances have, for the most part, been attained, and under normal conditions it is probable that trematodes rarely demand more than the host can supply without undue strain. It seems axiomatic that the host must survive until the parasite can again gain access to another host or until the life cycle is completed. Those parasites which cause the least disruption of the host's activities are probably the oldest as well as the most successful. Immunities are sometimes developed by the hosts. Many trematodes seem to possess such rigid requirements and such responses to particular hosts that host specificity is a phenomenon of considerable significance. Monogeneids appear more host-specific than digeneids, and aspidobothreids seem less specific than both. Trematodes are of considerable veterinary and medical importance because under certain conditions they cause debility, even death. *See* ASPIDOGASTREA; DIGENEA; MONOGENEA; PLATYHELMINTHES.

William J. Hargis, Jr.

Tremolite The name given to magnesium-rich monoclinic calcium amphibole $Ca_2Mg_5Si_8O_{22}(OH)_2$ The mineral is white to gray, but colorless in thin section. Unlike other end-member compositions of the calcium amphibole group, very pure tremolite is found in nature. Substitution of Fe for Mg is common, but pure ferrotremolite, $Ca_2Fe_5Si_8O_{22}(OH)_2$ is rare. Intermediate compositions between tremolite and ferrotremolite are referred to as actinolites, and are green in color and encompass a large number of naturally occurring calcium amphiboles. *See* AMPHIBOLE.

Barry L. Doolan

Trepostomata An extinct order of bryozoans in the class Stenolaemata. Trepostomes possess generally robust colonies, composed of tightly packed, moderately complex, long, slender, tubular or prismatic zooecia, with solid calcareous zooecial walls. Colonies show a moderately gradual transition from endozone to exozone regions, and they are exclusively free-walled. Trepostome colonies range from small and delicate to large and massive; they can be thin to thick encursting sheets; tabular, nodular, hemispherical, or globular masses; or bushlike or frondlike erect growths.

Apparently exclusively marine, the trepostomes first appeared about the start of the Middle Ordovician; they apparently share a common ancestor with cystoporates. They remained abundant through Silurian time, declined during the Devonian, and died out in the Late Triassic. *See* BRYOZOA; CYSTOPORATA; STENOLAEMATA. Roger J. Cuffey; Frank K. McKinney

Trestle A succession of towers of steel, timber, or reinforced concrete supporting the horizontal beams of a roadway, bridge, or other structure. Little distinction can be made between a trestle and a viaduct, and the terms are used interchangeably by many engineers. A viaduct is defined as a long bridge consisting of a series of short concrete or masonry spans supported on piers or towers, and is used to carry a road or railroad over a valley, a gorge, another roadway, or across an arm of the sea. *See* BRIDGE.

A trestle or a viaduct usually consists of alternate tower spans and spans between towers. For low trestles the spans may be supported on bents, each composed of two columns adequately braced in a transverse direction. A pair of bents braced longitudinally forms a tower. *See* TOWER. Charles M. Antoni

Triassic The oldest period of the Mesozoic Era, encompassing an interval between about 248 and 206 million years ago (Ma). It was named in 1848 by F. A. von Alberti for the threefold division of rocks at its type locality in central Germany, where continental redbeds and evaporites of the older Buntsandstein and younger Keuper formations are separated by marine limestones and marls of the Muschelkalk formation. *See* MESOZOIC.

As a very brief interval of geologic time (about 40 million years), the Triassic Period uniquely embraces both the final consolidation of Pangaea and the initial breakup of the landmass, which in the Middle Jurassic led to the opening of the Central Atlantic Ocean and formation of modern-day continental margins. The Triassic marks the beginning of a new cycle of ocean-basin opening through lithospheric extension and oceanic closing through subducting oceanic lithospheres along continental margins. *See* LITHOSPHERE; PLATE TECTONICS.

The most important tectonic event in the Mesozoic Era was the rifting of the Pangaea craton, which began in the Late Triassic, culminating in the Middle Jurassic with the formation of the Central Atlantic ocean basin and the proto-Atlantic continental margins. Rifting began in the Tethys region in the Early Triassic, and progressed from western Europe and the Mediterranean into the Central Atlantic off Morocco and eastern North America by the Late Triassic. As crustal extension continued throughout the Triassic, the Tethys seaway spread farther westward and inland. By that time, rifting and sea-floor spreading extended into the Gulf of Mexico, separating North and South America. *See* BASALT; CRETACEOUS; PALYNOLOGY.

Continental rift basins, passive continental margins, and ocean basins form in response to divergent stresses that extend the crust. Crustal extension, as it pertains to the Atlantic, embraces a major tectonic cycle marked by Late Triassic–Early Jurassic rifting and Middle Jurassic to Recent (Holocene) drifting. The rift stage, involving heating and stretching of the crust, was accompanied by uplift, faulting, basaltic igneous activity, and rapid filling of deep elongate rift basins. The drift stage, involving the slow cooling of the lithosphere over a broad region, was accompanied by thermal subsidence with concomitant marine transgression of the newly formed plate margin. The transition from rifting to drifting, accompanied by sea-floor spreading, is recorded by the postrift unconformity. *See* CONTINENTAL DRIFT; CONTINENTAL MARGIN; HOLOCENE; UNCONFORMITY.

Permian-to-Triassic consolidation of Pangaea in western North America led to the Sonoma orogeny (mountain building), which resulted from overthrusting and suturing of successive island-arc and microcontinent terranes to the western edge of the North American Plate. However, toward the end of the Triassic Period, as crustal extension was occurring in the Central Atlantic region, the plate moved westward, overriding the Pacific Plate along a reversed subduction zone. This created, for the remainder of the Mesozoic Era, an Andean-type plate edge with a subducting sea floor and associated deep-sea trench and magmatic arc. These effects can be studied in the Cordilleran mountain belt. *See* CORDILLERAN BELT; OROGENY.

As the epicontinental seas regressed westward, nonmarine fluvial, lacustrine, and windblown sands were deposited on the craton. Today many of these red, purple, ash-gray, and chocolate-colored beds are some of the most spectacular and colorful scenery in the American West. For example, the Painted Desert of Arizona, known for its petrified logs of conifer trees, was developed in the Chinle Formation. *See* PETRIFIED FORESTS.

Triassic faunas are also distinguished from earlier ones by newly evolved groups of plants and animals. In marine communities, molluscan stocks proliferated vigorously. Bivalves diversified greatly and took over most of the niches previously occupied by brachiopods; ammonites proliferated rapidly from a few Permian survivors. The scleractinian (modern) corals appeared, as did the shell-crushing placodont reptiles and the ichthyosaurs. In continental faunas, various groups of reptiles appeared, including crocodiles and crocodilelike forms, the mammallike reptiles, and the first true mammals, as well as dinosaurs. *See* CEPHALOPODA; CROCODYLIA; DINOSAUR; MAMMALIA; MOLLUSCA; PLACODONTIA; SCLERACTINIA.

Triassic land plants contain survivors of many Paleozoic stocks, but the gymnosperms became dominant and cycads appeared. *See* CYCADEOIDALES; PALEOZOIC; PINOPHYTA. Warren Manspeizer

Tribology The science and technology of interactive surfaces in relative motion. It incorporates various scientific and technological disciplines such as surface chemistry, fluid mechanics, materials, lubricants, contact mechanics, bearings, and lubrication systems. It is customarily divided into three branches: friction, lubrication, and wear.

Friction. This phenomenon is encountered whenever there is relative motion between contacting surfaces, and it always opposes the motion. As no mechanically prepared surfaces are perfectly smooth, when the surfaces are first brought into contact under light load, they touch only along the asperities (real area of contact). The early theories attributed friction to the interlocking of asperities; however, it is now understood that the phenomenon is far more complicated. *See* FRICTION.

Lubrication. When clean surfaces are brought into contact, their coefficient of friction decreases drastically if even a single molecular layer of a foreign substance (for example, an oxide) is introduced between the surfaces. For thicker lubricant films, the coefficient of friction can be quite small and no longer dependent on the properties of the surfaces but only on the bulk properties of the lubricant. Most common lubricants are liquids and gases, but solids such as molybdenum disulfide or graphite may be used. *See* LUBRICATION.

Wear. This is the progressive loss of substance of one body because of rubbing by another body. There are many different types of wear, including sliding wear, abrasive wear, corrosion, and surface fatigue. *See* WEAR. Andras Z. Szeri

Trichasteropsida A monospecific order of Asteroidea established for *Trichasteropsis wiessmanni*, the only Triassic asteroid known from articulated specimens. It is a small starfish with a relatively large disc and short arms. The skeleton is differentiated into marginals, actinals, and abactinals, the marginals comprising a single series of blocklike ossicles. Oral plates are large and well developed. Pedicellariae are not present. *Trichasteropsis* is the most primitive post-Paleozoic asteroid known and lies close to the latest common ancestor of all living asteroids. *See* ASTEROIDEA; ECHINODERMATA. Andrew B. Smith

Trichomycetes A polyphyletic class of Eumycota in the subdivision Zygomycotina, containing the orders Amoebidiales, Asellariales, Eccrinales, and Harpellales. These orders are grouped together because they usually exist only as commensals in the mid- and hindgut of arthropods (*Amoebidium parasiticum* can be found on the outside of its hosts).

Asexual reproduction is accomplished by the formation of trichospores (Harpellales), arthrospores (Asellariales), sporgiospores (Eccrinales), or ameboid cells, cystospores, or rigid-walled spores (Amoebidiales). Sexual reproduction (zygospore formation) is known only in Harpellales, although conjugations have been observed in Asellariales. The thallus of Amoebidiales is aseptate, but is regularly septate in the other three orders; septa with plugs in the lenticular cavities are produced by Asellariales and Harpellales. Similar septa and plugs are formed by Dimargaritales and Kickxellales (Zygomycetes). Classification is based on the type of reproduction; thallus branching pattern, complexity, and septation; and nature of the holdfast. Trichomycetes occur worldwide and may be found anywhere a suitable host exists; they inhabit a more or less equatic environment (the gut); and the spores are discharged with feces. *See* EUMYCOTA; FUNGI; ZYGOMYCETES; ZYGOMYCOTA. Gerald L. Benny

Trichoptera An aquatic order of the class Insecta commonly known as the caddis flies. The adults have two pairs of well-veined hairy wings, long antennae, and mouthparts capable of lapping only liquids. The larvae are wormlike, with distinct heads, three pairs of legs on the thorax, and a pair of hook-bearing legs at the end of the body. The pupae are delicate, with free appendages held close to the body, and have a pair of sharp mandibles, or jaws, which are used to cut an exit from the cocoon.

The Trichoptera include about 10,000 described species, divided into 34 families, and occur in practically all parts of the world. Except for a brackish-water species in New Zealand and a few moss-inhabiting species in Europe and North America, caddis flies occur only in fresh water. They abound in cold or running water relatively free from pollution. Altogether they compose a large and important segment of the biota of such habitats and of the fish feed economy. *See* INSECTA. Herbert H. Ross

Trichroism When certain optically anisotropic transparent crystals are subjected to white light, a cube of the material is found to transmit a different color through each of the three pairs of parallel faces. Such crystals are sometimes termed trichroic, and the phenomenon is called trichroism. This expression is used only rarely today since the colors in a particular crystal can appear quite different if the cube is cut with a different orientation with respect to the crystal axes. Accordingly, the term is frequently replaced by the more general term pleochroism. Even this term is being replaced by the phrase linear dichroism or circular dichroism to correspond with linear birefringence or circular birefringence. *See* BIREFRINGENCE; CRYSTAL OPTICS; DICHROISM; PLEOCHROISM. Bruce H. Billings

Tricladida An order of the Turbellaria (of the phylum Platyhelminthes) known commonly as planaria. They have a diverticulated intestine with a single anterior branch and two posterior branches separated by a plicate pharynx or pharynges. Rhabdites are numerous and, except in cave planarians, two to many eyes are present. *See* TURBELLARIA. E. Ruffin Jones

Trigger circuit An electronic circuit that generates or modifies an existing waveform to produce a pulse of short time duration with a fast-rising leading edge. This waveform, or trigger, is normally used to initiate a change of state of some relaxation device, such as a multivibrator. The most important characteristic of the waveform generated by a trigger circuit is usually the fast leading edge. The exact shape of the failing portion of the waveform often is of secondary importance, although it is important that the total duration time is not too great. A pulse generator such as a blocking oscillator may also be used and identified as a trigger circuit if it generates sufficiently short pulses. *See* BLOCKING OSCILLATOR; PULSE GENERATOR.

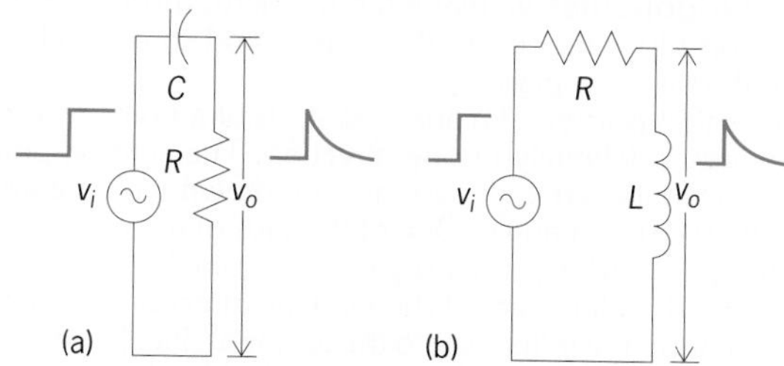

Diagrams of simple peaking circuits. (*a*) Resistance-capacitance network. (*b*) Resistance-inductance network. V_i = input voltage; V_o = output voltage.

Peaking circuits, which accent the higher-frequency components of a pulse waveform, cause sharp leading and trailing edges and are therefore used as trigger circuits. The simplest form of peaking circuits are the simple *RC* and *RL* networks shown in the illustration. If a steep wavefront of amplitude *V* is applied to either of these circuits, the output will be a sudden rise followed by an exponential decay. These circuits are often called differentiating circuits because the outputs are rough approximations of the derivative of the input waveforms, if the *RC* or *R*/*L* time constant is sufficiently small.

A circuit that is highly underdamped, or oscillatory, and is supplied with a step or pulse input is often referred to as a ringing circuit. When used in the output of a field-effect or bipolar transistor, this circuit can be used as a trigger circuit. *See* WAVE-SHAPING CIRCUITS. Glenn M. Glasford

Triglyceride (triacylglycerol) A simple lipid. Triglycerides are fatty acid triesters of the trihydroxy alcohol glycerol which are present in plant and animal tissues, particularly in the food storage depots, either as simple esters in which all the fatty acids are the same or as mixed esters in which the fatty acids are different. The triglycerides constitute the main component of natural fats and oils.

The generic formula of a triglyceride is shown below, where

$$\begin{array}{l} CH_2 - OOC - R \\ | \\ CH - OOC - R' \\ | \\ CH_2 - OOC - R'' \end{array}$$

RCO_2H, $R'CO_2H$, and $R''CO_2H$ represent molecules of either the same or different fatty acids, such as butyric or caproic (short chain), palmitic or stearic (long chain), oleic, linoleic, or linolenic (unsaturated). Saponification with alkali releases glycerol and the alkali metal salts of the fatty acids (soaps). The triglycerides in the food storage depots represent a concentrated energy source, since oxidation provides more energy than an equivalent weight of protein or carbohydrate. *See* LIPID METABOLISM; SOAP.

The physical and chemical properties of fats and oils depend on the nature of the fatty acids present. Saturated fatty acids give higher-melting fats and represent the main constituents of solid fats, for example, lard and butter. Unsaturation lowers the melting point of fatty acids and fats. Thus, in the oil of plants, unsaturated fatty acids are present in large amounts, for example, oleic acid in olive oil and linoleic and linolenic acids in linseed soil. *See* FAT AND OIL (FOOD); LIPID. Roy H. Gigg; Herbert E. Carter

Trigonometry The study of triangles and the trigonometric functions. Trigonometry has evolved from use by surveyors, engineers, and navigators to applications involving ocean tides, the rise and fall of food supplies in certain ecologies, brain-wave patterns, the analysis of alternating-current electricity, and many other phenomena of a vibratory character.

Plane trigonometry. Plane trigonometry mostly deals with the relationships among the three sides and three angles of a triangle that lies in a plane.

A ray is that portion of a line that starts at a point on the line and extends indefinitely in one direction. The starting point of a ray is called its vertex. If two rays are drawn with a common vertex, they form an angle. One of the rays of an angle is called the initial side, and the other ray is the terminal side. The angle that is formed is identified by showing the direction and amount of rotation from the initial side to the terminal side. If the rotation is in the counterclockwise direction, the angle is positive; if the rotation is clockwise, the angle is negative.

The angle formed by rotating the initial side exactly once in the counterclockwise direction until it coincides with itself (1 revolution) is said to measure 360 degrees, written 360°. Thus, one degree, 1°, is 1/360 of a revolution. One-sixtieth of a degree is called a minute, written 1′.

By using a circle of radius r, an angle can be constructed whose vertex is at the center of this circle and whose rays subtend an arc on the circle whose length equals r. Such an angle measures 1 radian. For a circle of radius r, a central angle of θ radians subtends an arc whose length s is given by Eq. (1):

$$s = r\theta \qquad (1)$$

Because a central angle of 1 revolution (360°) subtends an arc equal to the circumference of the circle ($2\pi r$), it follows that an angle of 1 revolution equals 2π radians; that is, 2π radians = 360°. *See* PLANE GEOMETRY; RADIAN MEASURE.

Trigonometric functions. A unit circle is a circle whose radius is one and whose center is at the origin of a rectangular system of coordinates. For the unit circle, Eq. (1) states that a central angle of θ radians subtends an arc whose length $s = \theta$. If t is any real number, let θ be the angle equal to t radians and P be the point on the unit circle that is also on the terminal side of θ. If $t \geq 0$, then the point P is reached by moving counterclockwise along the unit circle, starting at the point with coordinates (1,0), for a length of arc equal to t units (Fig. 1*a*). If $t < 0$, this point P is reached by moving clockwise along the unit circle beginning at (1,0), for a length of arc equal to $|t|$ units (Fig. 1*b*). Thus, to each real number t there corresponds a unique point $P = (a,b)$ on the unit circle. The coordinates of this point P are used to define the six trigonometric functions: If $\theta = t$ radians, the sine, cosine, tangent, cosecant, secant, and cotangent of θ, respectively abbreviated as sin θ, cos θ, tan θ, csc θ, sec θ, cot θ, are given by Eqs. (2), (3), and (4).

$$\sin\theta = b \qquad \cos\theta = a \qquad (2)$$

$$\text{if } a \neq 0, \quad \tan\theta = b/a, \quad \sec\theta = 1/a \qquad (3)$$

$$\text{if } b \neq 0, \quad \tan\theta = a/b, \quad \sec\theta = 1/b \qquad (4)$$

See COORDINATE SYSTEMS.

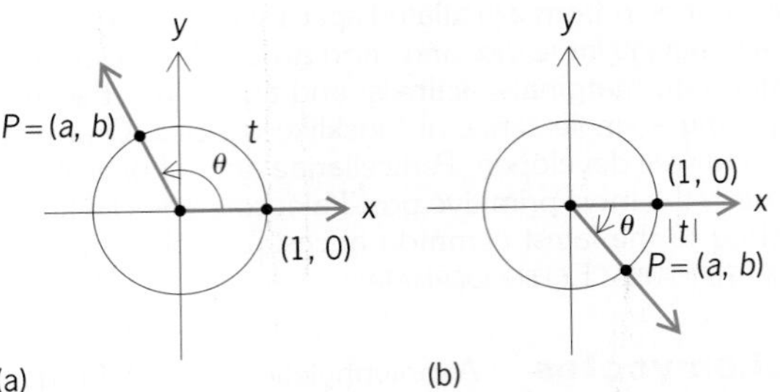

Fig. 1. Point *P* on the unit circle corresponding to $\theta = t$ radians. (*a*) $t \geq 0$: length of arc from (1,0) to *P* is *t* units. (*b*) $t < 0$: length of arc from (1,0) to *P* is $|t|$ units.

For example, for $\theta = 0$, the point (1,0) is on the terminal side of θ and is on the unit circle so that Eqs. (5) hold, with csc 0 and

$$\sin 0 = 0 \quad \cos 0 = 1 \quad \tan 0 = 0 \quad \sec 0 = 1 \qquad (5)$$

cot 0 not defined. The trigonometric functions of angles that are integral multiples of $\pi/2$ (90°) are found similarly (Table 1).

Table 1. Values of trigonometric functions at integral multiples of $\pi/4$ (90°)

θ, radians	θ	sin θ	cos θ	tan θ	csc θ	sec θ	cot θ
0	0°	0	1	0	n.d.	1	n.d.
$\pi/2$	90°	1	0	n.d.	1	n.d.	0
π	180°	0	−1	0	n.d.	−1	n.d.
$3\pi/2$	270°	−1	0	n.d.	−1	n.d.	0

*n.d. = not defined.

The coordinates of points on the unit circle that are on the terminal sides of angles that are integral multiples of $\pi/6$ (30°), $\pi/4$ (45°), and $\pi/3$ (60°) can be found. With Eqs. (2), (3), and (4), the trigonometric functions of these angles are obtained (Table 2).

Table 2. Values of trigonometric functions at integral multiples of $\pi/6$ (30°), $\pi/4$ (45°), and $\pi/3$ (60°)

θ, radians	θ	sin θ	cos θ	tan θ	csc θ	sec θ	cot θ
$\pi/6$	30°	1/2	$\sqrt{3}/2$	$\sqrt{3}/3$	2	$2\sqrt{3}/3$	$\sqrt{3}$
$\pi/4$	45°	$\sqrt{2}/2$	$\sqrt{2}/2$	1	$\sqrt{2}$	$\sqrt{2}$	1
$\pi/3$	60°	$\sqrt{3}/2$	1/2	$\sqrt{3}$	$2\sqrt{3}/3$	2	$\sqrt{3}/3$
$2\pi/3$	120°	$\sqrt{3}/2$	−1/2	$-\sqrt{3}$	$2\sqrt{3}/3$	−2	$-\sqrt{3}/3$
$3\pi/4$	135°	$\sqrt{2}/2$	$-\sqrt{2}/2$	−1	$\sqrt{2}$	$-\sqrt{2}$	−1
$5\pi/6$	150°	1/2	$-\sqrt{3}/2$	$1-\sqrt{3}/3$	2	$-2\sqrt{3}/3$	$-\sqrt{3}$
$7\pi/6$	210°	−1/2	$-\sqrt{3}/2$	$\sqrt{3}/3$	−2	$-2\sqrt{3}/3$	$\sqrt{3}$
$5\pi/4$	225°	$-\sqrt{2}/2$	$-\sqrt{2}/2$	1	$-\sqrt{2}$	$-\sqrt{2}$	1
$4\pi/3$	240°	$-\sqrt{3}/2$	−1/2	$\sqrt{3}$	$-2\sqrt{3}/3$	−2	$\sqrt{3}/3$
$5\pi/3$	300°	$-\sqrt{3}/2$	1/2	$-\sqrt{3}$	$-2\sqrt{3}/3$	2	$-\sqrt{3}/3$
$7\pi/4$	315°	$-\sqrt{2}/2$	$\sqrt{2}/2$	−1	$-\sqrt{2}$	$\sqrt{2}$	−1
$11\pi/6$	330°	−1/2	$\sqrt{3}/2$	$-\sqrt{3}/3$	−2	$2\sqrt{3}/3$	$-\sqrt{3}$

Properties of trigonometric functions. Based on Eqs. (2) and the above geometric construction (Fig. 1) for sin θ and cos θ, θ can be any angle, so the domain of the sine and cosine functions is all real numbers. In Eqs. (3), if $a = 0$, the tangent and secant functions are not defined, so the domain of these functions is all real numbers, except odd multiples of $\pi/2$ (90°). In Eqs. (4), if $b = 0$, the cotangent and cosecant functions are not defined, so the domain of these functions is all real numbers, except multiples of π (180°). Also, since $|a| \leq 1$ and $|b| \leq 1$, the range of the sine and cosine functions is −1 to 1 inclusive. Since $|b| = |\sin\theta| \leq 1$ and $|a| = |\cos\theta| \leq 1$, it follows that $|\csc\theta| = 1/|b| \geq 1$ and $|\sec\theta| = 1/|a| \geq 1$. Thus the range of the secant and cosecant functions consists of all real numbers less than or equal to -1 or greater than or equal to 1. The range of both the tangent and cotangent functions consists of all real numbers.

Equations (2), (3), and (4) also reveal the reciprocal identities, given in Eqs. (6). Two other useful identities, given in Eqs. (7), also follow.

$$\csc\theta = 1/\sin\theta \quad \sec\theta = 1/\cos\theta \quad \cot\theta = 1/\tan\theta \qquad (6)$$

$$\tan\theta = \sin\theta/\cos\theta \quad \cot\theta = \cos\theta/\sin\theta \qquad (7)$$

Since (a,b) is on the unit circle, $a^2 + b^2 = 1$, and so $(\sin\theta)^2 + (\cos\theta)^2 = 1$. This is called a pythagorean identity and is written as Eq. (8).

$$\sin^2\theta + \cos^2\theta = 1 \qquad (8)$$

Sum and difference formulas. The sum and difference formulas for the cosine function are given in Eqs. (9) and (10), and for the sine function they are given in Eqs. (11) and (12).

$$\cos(\alpha + \beta) = \cos\alpha\cos\beta - \sin\alpha\sin\beta \qquad (9)$$

$$\cos(\alpha - \beta) = \cos\alpha\cos\beta + \sin\alpha\sin\beta \qquad (10)$$

$$\sin(\alpha + \beta) = \sin\alpha\cos\beta + \cos\alpha\sin\beta \qquad (11)$$

$$\sin(\alpha - \beta) = \sin\alpha\cos\beta - \cos\alpha\sin\beta \qquad (12)$$

Inverse trigonometric functions. In the equation $x = \sin y$, if y is restricted so that $-\pi/2 \le y \le \pi/2$, then the solution of the equation for y is unique and is denoted by $y = \sin^{-1}x$ (read "y is the inverse sine of x"). Sometimes $y = \sin^{-1}x$ is written as $y = \text{Arcsin}\, x$. Thus, $y = \sin^{-1}x$ is a function whose domain is $-1 \le x \le 1$ and whose range is $-\pi/2 \le y \le \pi/2$. For example, $\sin^{-1} 1/2 = \pi/6$ and $\sin^{-1}(-1) = -\pi/2$.

Likewise in the equation $x = \cos y$, if y is restricted so that $0 \le y \le \pi$, then the solution of the equation for y is unique and is denoted by $y = \cos^{-1}x$ (read "y is the inverse cosine of x"). Thus, $y = \cos^{-1}x$ is a function whose domain is $-1 \le x \le 1$ and whose range is $0 \le y \le \pi$. Finally, in the equation $x = \tan y$, if y is restricted so that $-\pi/2 < y < \pi/2$, then the solution of the equation for y is unique and is denoted by $y = \tan^{-1}x$ (read "y is the inverse tangent of x"). Thus, $y = \tan^{-1}x$ is a function whose domain is $-\infty < x < \infty$ and whose range is $-\pi/2 < y < \pi/2$.

Solution of right triangles. The trigonometric functions can be expressed as ratios of the sides of a right triangle. Indeed, by Eqs. (6), (7), and (8), it follows that $\sin\beta = b/c$, $\cos\beta = a/c$, $\tan\beta = b/a$, and so on, where a and b are the sides adjacent to the right angle, c is the hypotenuse, and α and β are the angles opposite a and b respectively (Fig. 2). If an angle and a side or else two sides of a right triangle are known, then the remaining angles and sides can be found.

Solution of oblique triangles. If none of the angles of a right triangle is a right angle, the triangle is oblique. To solve such triangles, there are four possibilities to consider: (1) one side and two angles are given; (2) two sides and the angle opposite one of them are given; (3) two sides and the included angle are given; and (4) three sides are given. In the following discussion, the

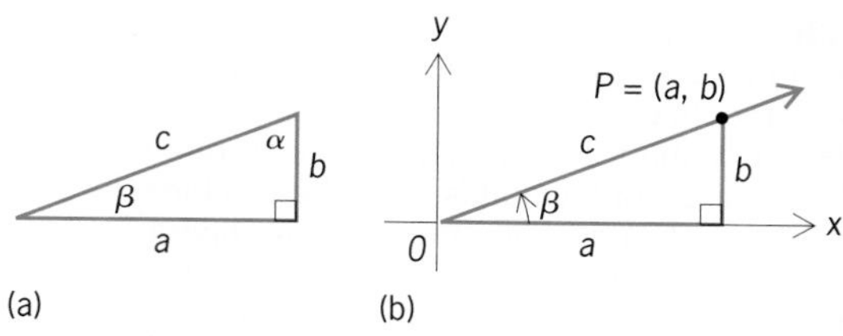

Fig. 2. Right triangle. (*a*) Labeling of sides and angles. (*b*) Relationship of sides to coordinates defining trigonometric functions.

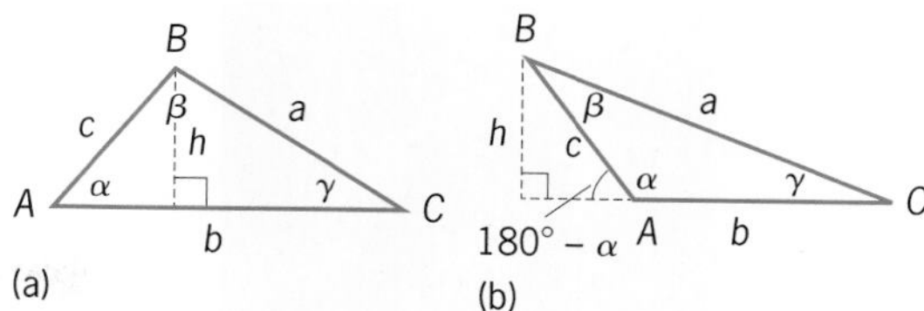

Fig. 3. Labeling of sides, angles, and vertices of oblique triangle, with altitude *h* used in proving the law of sines. (*a*) Angle α is acute. (*b*) Angle α is obtuse.

sides are labeled a, b, and c; the angles opposite these sides are α, β, and γ respectively; and the corresponding vertices are A, B, and C (Fig. 3).

The law of sines, Eq. (13), is used to solve possibilities (1)

$$\frac{\sin\alpha}{a} = \frac{\sin\beta}{b} = \frac{\sin\gamma}{c} \qquad (13)$$

and (2). The law of cosines, used to solve possibilities (3) or (4), may be stated by three equivalent formulas, Eqs. (14), (15), and (16).

$$c^2 = a^2 + b^2 - 2ab\cos\gamma \qquad (14)$$

$$b^2 = a^2 + c^2 - 2ac\cos\beta \qquad (15)$$

$$a^2 = b^2 + c^2 - 2bc\cos\alpha \qquad (16)$$

Infinite series representation. With the definition $n! = 1 \cdot 2 \cdot 3 \cdots n$, the sine and cosine functions may be represented by the two infinite series, given in Eqs. (17) and (18).

$$\sin x = \sum_{k=0}^{\infty}(-1)^k \frac{x^{2k+1}}{(2k+1)!} = x - \frac{x^3}{3!} + \frac{x^5}{5!} - \cdots \qquad (17)$$

$$\cos x = \sum_{k=0}^{\infty}(-1)^k \frac{x^{2k}}{(2k)!} = 1 - \frac{x^2}{2!} + \frac{x^4}{4!} - \cdots \qquad (18)$$

These series converge for all x. Thus, to find a value of $\sin x$ or $\cos x$, only as many terms of the series need to be used, as required to ensure required accuracy.

Michael Sullivan

Trihedron

A geometric figure bounded by three noncoplanar rays called edges that emanate from a common point called the vertex, and by the plane sectors called faces that are formed by each pair of edges (see illustration). A trihedron has

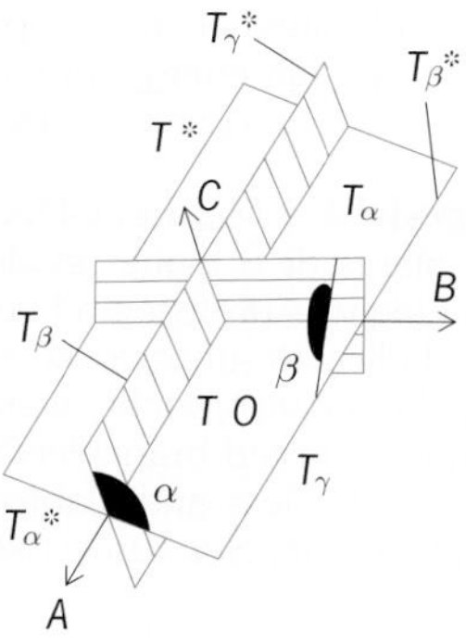

Trihedron and trihedral angles. Three planes having a common point *O* cut space into eight trihedrons, T, T^*, T_α, T_α^*, T_β, T_β^*, T_γ, and T_γ^*. *A*, *B*, and *C* are directions of common edges of pairs of planes; α, β, and γ are face angles.

three dihedrons formed by pairs of face planes, and three face angles formed by pairs of edges.

J. Sutherland Frame

Trilobita

A class of extinct Paleozoic arthropods, occurring in marine rocks of Early Cambrian through late Permian age.

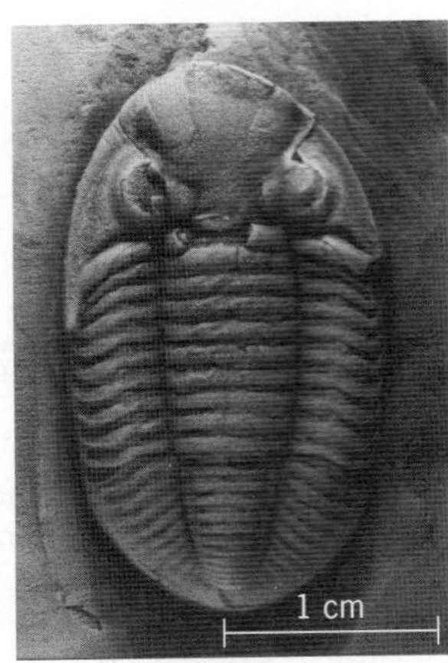

***Griffithides*, Mississippian (Indiana). Dorsal view of exoskeleton.**

Their closest living relatives are the chelicerates, including spiders, mites, and horseshoe crabs (Xiphosura). About 3000 described genera make trilobites one of the most diverse and best-known fossil groups. Species diversity peaked during the Late Cambrian and then declined more or less steadily until the Late Devonian mass extinction. Only four families survived to the Mississippian, and only one lasted until the group's Permian demise. Their dominance in most Cambrian marine settings is essential to biostratigraphic correlation of that system. *See* ARTHROPODA; CAMBRIAN; CHELICERATA; DEVONIAN; INDEX FOSSIL; PERMIAN.

Trilobites are typically represented in the fossil record by the mineralized portion of their exoskeleton, either as carcass or molt remains. The mineralized exoskeleton was confined mostly to the dorsal surface (see illustration), curved under as a rimlike doublure; a single mineralized ventral plate, the hypostome, was suspended beneath the median region of the head. The mineralized exoskeleton was composed of low magnesian calcite and a minor component of organic material. Most of the ventral exoskeleton, including the appendages, was unmineralized.

Most trilobites were benthic deposit feeders or scavengers, living on the sediment-water interface or shallow-burrowing just beneath it. Some were evidently carnivores, equipped with sharp spines and processes projecting ventrally from their appendages. The morphology and broad geographic and environmental ranges of these groups suggest they were active swimmers. Through their history, trilobites became adapted to all marine environments, from shallow high-energy shorefaces to deep-water, disaerobic habitats. Gregory D. Edgecombe; Jonathan Adrain

Trimerophytopsida Mid-Early-Devonian into Middle-Devonian vascular plants at a higher evolutionary level than Rhyniopsida. Branching was profuse and varied, dichotomous, pseudomonopodial, helical to subopposite and almost whorled, and often trifurcate. Vegetative branches were often in a tight helix, terminated by tiny recurved branchlets simulating leaf precursors. The axes were leafless and glabrous or spiny. Xylem is known only in *Psilophyton*. *See* EMBRYOBIONTA; RHYNIOPSIDA.

Harlan P. Banks

Triple point A particular temperature and pressure at which three different phases of one substance can coexist in equilibrium. In common usage these three phases are normally solid, liquid, and gas, although triple points can also occur with two solid phases and one liquid phase, with two solid phases and one gas phase, or with three solid phases.

According to the Gibbs phase rule, a three-phase situation in a one-component system has no degrees of freedom (that is, it is invariant). Consequently, a triple point occurs at a unique temperature and pressure, because any change in either variable will result in the disappearance of at least one of the three phases. Triple points are shown in the illustration of part of the phase diagram for water. *See* PHASE EQUILIBRIUM.

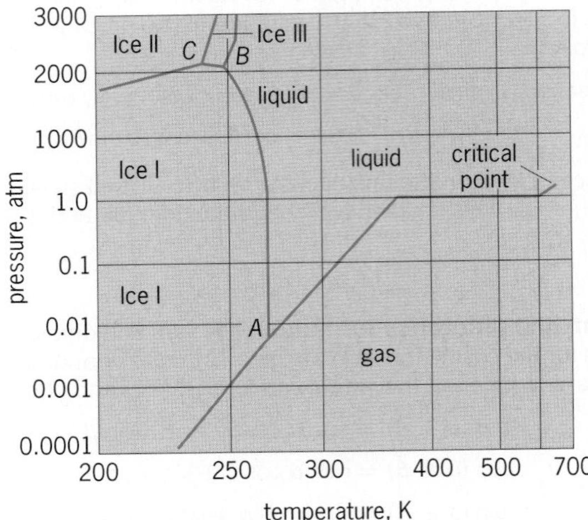

Phase diagram for water, showing gas, liquid, and several solid (ice) phases; triple points at *A*, *B*, and *C*. The pressure scale changes at 1 atm from logarithmic scale at low pressure to linear at high pressure. 1 atm = 10^2 kilopascals.

For most substances the solid-liquid-vapor triple point has a pressure less than 1 atm (10^2 kilopascals); such substances then have a liquid-vapor transition at 1 atm (normal boiling point). However, if this triple point has a pressure above 1 atm, the substance passes directly from solid to vapor at 1 atm. *See* BOILING POINT; ICE POINT; MELTING POINT; SUBLIMATION; TRANSITION POINT; VAPOR PRESSURE; WATER. Robert L. Scott

Triplet state An electronic state of a molecule that occurs when its total spin angular momentum quantum number *S* is equal to one. The triplet state is an important intermediate of organic chemistry. In addition to the wide range of triplet molecules available through photochemical excitation techniques, numerous molecules exist in stable triplet ground states, for example, oxygen molecules. *See* ATOMIC STRUCTURE AND SPECTRA; MOLECULAR STRUCTURE AND SPECTRA; SPIN (QUANTUM MECHANICS).

A triplet may result whenever a molecule possesses two electrons which are both orbitally unpaired and spin unpaired. Orbital unpairing of electrons results when a molecule absorbs a photon of visible or ultraviolet light. Direct formation of a triplet as a result of this photon absorption is a very improbable process since both the orbit and spin of the electron would have to change simultaneously. Thus, a singlet state is generally formed by absorption of light. However, quite often the lifetime of this singlet state is sufficiently long to allow the spin of one of the two electrons to invert, thereby producing a triplet. *See* MOLECULAR ORBITAL THEORY. Nicholas J. Turro

Tripylida An order of nematodes in which the cephalic cuticle is simple and not duplicated; there is no helmet. The body cuticle is smooth or sometimes superficially annulated. Cephalic sensilla follow the typical pattern in which one whorl is circumoral and the second whorl is often the combination of circlets two and three. The pouchlike amphids have apertures that are inconspicuous or transversally oval. The stoma is variable, being simple, collapsed, funnel shaped or cylindrical, and armed or unarmed. In most taxa the stoma is surrounded by esophageal tissue; that is, it is entirely esophastome. When the stoma is expanded, both the cheilostome and esophastome are evident.

Esophagi are cylindrical-conoid. Esophageal glands open anterior to the nerve ring. Males generally have three supplementary organs, more in some taxa. A gubernaculum accompanies the spicules. Caudal glands are generally present.

The two tripylid superfamilies are the Tripyloidea and Ironoidea. The characteristically well-developed cuticular annulation of the Tripyloidea is only rarely seen in other Enoplida. These nematodes are commonly found in fresh water or very moist soils; however, some are found in brackish water and marine habitats. Intestinal contents indicate that their food consists primarily of small microfauna that often include nematodes and rotifers. The Ironoidea contain species (presumably carnivorous) occurring in both fresh-water and soil habitats. *See* NEMATA (NEMATODA).

Armand R. Maggenti

Triterpene A hydrocarbon or its oxygenated analog containing 30 carbon atoms and composed of six isoprene units. Triterpenes form the largest group of terpenoids, but are classified into only a few major categories. Resins and saps contain triterpenes in the free state as well as in the form of esters and glycosides.

Biogenetically triterpenes arise by the cyclization of squalene and subsequent skeletal rearrangements. Apart from the linear squalene itself and some bicyclic, highly substituted skeletons, most triterpenes are either tetracyclic or pentacyclic compounds. The various structural classes are designated by the names of representative members. *See* ISOPRENE; TERPENE.

Tomas Hudlicky

Triticale A cereal grass plant (× *Triticosecale*) obtained from hybridization of wheat (*Triticum*) with rye (*Secale cereale*). It is a crop plant with a small-seeded cereal grain that is used for human food and livestock feed. Worldwide, triticale is slowly gaining importance as a cereal grain. The European continent dominates triticale production with 70% of the total area.

Triticale was first developed in 1876, but not until the 1960s were types developed that were suitable for cultivation. Modern varieties are called secondary triticales because they were selected after interbreeding of various triticales, including primary types. In some triticale varieties, one or more rye chromosomes have been replaced by wheat chromosomes, giving secondary-substituted triticales, as contrasted to complete triticale having all seven rye chromosomes.

Triticale is produced by deliberate hybridization of either bread wheat [*Triticum aestivum*; diploid number of chromosomes $(2n) = 42$] or durum wheat (*T. turgidum* var. *durum*; $2n = 28$) with rye ($2n = 14$), followed by the doubling of the chromosome number of the hybrid plant. Hexaploid triticale (durum wheat × rye; $2n = 42$) is a more successful crop plant than octoploid triticale. The octoploid form ($2n = 56$) is produced by hybridization of bread wheat ($2n = 42$) with rye ($2n = 14$).

Triticale is grown from seeds sown in soil by using cultivation practices similar to those of wheat or rye. Both winter-hardy and nonhardy types exist, the latter used where winters are mild or for spring sowing. Triticale tends to have a greater ability than wheat to grow in adverse environments, such as saline or acid soils or under droughty conditions.

Being a cereal grain, triticale can be used in food products made from wheat flour. Varieties tend to have large, somewhat irregularly shaped grains that produce a lower yield of milled flour than wheat. Bread and pastry products can be made very well with triticale flour. As a livestock feed, triticale grain is a good source of carbohydrate and protein.

Intense breeding and selection have made very rapid genetic improvements in triticale seed quality. The agronomic advantages and improved end-use properties of the grain of triticale over wheat make triticale an attractive option for increasing global food production. *See* BREEDING (PLANT); RYE; WHEAT.

Calvin O. Qualset

Tritium The heaviest isotope of the element hydrogen and the only one which is radioactive. Tritium occurs in very small amounts in nature but is generally prepared artificially by processes known as nuclear transmutations. It is widely used as a tracer in chemical and biological research and is a component of the so-called thermonuclear or hydrogen bomb. It is commonly represented by the symbol ^3_1H indicating that it has an atomic number of 1 and an atomic mass of 3, or by the special symbol T. For information about the other hydrogen isotopes *see* DEUTERIUM; HYDROGEN. *See also* TRANSMUTATION.

Properties of hydrogen and tritium

Property	H_2	T_2
Melting point	−259.20°C (−434.56°F)	−252.54°C (−422.57°F)
Boiling point at 1 atm (10^5 pascals)	−252.77°C (−423.00°F)	−248.12°C (−414.62°F)
Heat of vaporization	216 cal/mol (904 J/mol)	333 cal/mol (1390 J/mol)
Heat of sublimation	247 cal/mol (1030 J/mol)	393 cal/mol (1640 J/mol)

Both molecular tritium, T_2, and its counterpart hydrogen, H_2, are gases under ordinary conditions. Because of the great difference in mass, many of the properties of tritium differ substantially from those of ordinary hydrogen, as indicated in the table. Chemically, tritium behaves quite similarly to hydrogen. However, because of its larger mass, many of its reactions take place more slowly than do those of hydrogen.

The nucleus of the tritium atom, often called a triton and symbolized t, consists of a proton and two neutrons. It undergoes radioactive decay by emission of a β-particle to leave a helium nucleus of mass 3. No γ-rays are emitted in this process. The half-life for the decay is 12.26 years. When tritium is bombarded with deuterons of sufficient energy, a nuclear reaction known as fusion occurs and energy considerably greater than that of the bombarding particle is released. This reaction is one of those which supply the energy of the thermonuclear bomb. It is also of major importance in the development of controlled thermonuclear reactors. *See* HEAVY WATER; NUCLEAR FUSION; TRITON.

Louis Kaplan

Triton The nucleus of ^2_1H (tritium); it is the only known radioactive nuclide belonging to hydrogen. The triton is produced in nuclear reactors by neutron absorption in deuterium ($^2_1H + ^1_0n \rightarrow +\gamma$), and decays by β^- emission to ^3_2H with a half-life of 12.4 years. Much of the interest in producing ^3_1H arises from the fact that the fusion reaction $^3_1H + ^1_1H \rightarrow ^4_2H$ releases about 20 MeV of energy. Tritons are also used as projectiles in nuclear bombardment experiments. *See* NUCLEAR REACTION; TRITIUM.

Henry E. Duckworth

Trochodendrales An order of flowering plants, division Magnoliophyta (Angiospermae), in the Eudicotyledon. The order consists of two families, the Trochodendraceae and Tetracentraceae, each with only a single species. The group is of considerable botanical and evolutionary interest, as it is situated near the base of the advanced Eudicotyledon and links this larger group with more primitive flowering plants. Trochodendrales comprise trees of eastern and southeastern Asia with primitive (without vessels) wood. The flowers have a much reduced perianth with scarcely sealed carpels that are only slightly fused to each other. *See* EUDICOTYLEDONS; MAGNOLIOPHYTA.

K. J. Sytsma

Trogoniformes A small order of birds that contains only the family Trogonidae. Thirty-seven species are found throughout the tropics; two species reach the southern border of the

United States. The trogons and quetzals are jay-sized birds with large heads, and tails that vary from medium length and squared to very elongated and tapered. The dorsal plumage of trogons and quetzals is predominantly metallic green, with blue, violet, red, black, or gray in a few. The ventral feathers are bright red, yellow, or orange. Despite their vivid coloration, the birds are inconspicuous when sitting quietly in the forest. Sexes are dissimilar in appearance, with the males being more brightly colored. Legs are short and feet are weak, with the toes arranged in a heterodactyl fashion, with the first and second toes reverted, opposing the third and fourth toes. Flight is rapid, undulating, and brief; trogons rarely walk. The diet consists of fruit and small invertebrates, as well as insects caught in flight as the bird darts out from a perch.

Trogons are nonmigratory, arboreal, and sedentary, and they can remain on a perch for hours. The monogamous pairs nest in solitude in a hollow tree or termite nest. After the eggs have been incubated by both parents, the naked hatchlings remain in the nest and are cared for by both parents. Walter J. Bock

Trojan asteroids Asteroids located near the equilateral lagrangian stability points of a Sun-planet system (*see* illustration). As shown by J. L. Lagrange in 1772, these are two of the five stable points in the circular, restricted, three-body system, the other three points being located along a line through the two most massive bodies in the system. In 1906 Max Wolf discovered an asteroid located near the lagrangian point preceding Jupiter in its orbit. Within a year, two more were found, one of which was located near the following lagrangian point. It was quickly decided to name these asteroids after participants in the Trojan War as given in Homer's *Iliad*.

The term "Trojans" is sometimes used in a generic sense to refer to objects occupying the equilateral lagrangian points of other pairs of bodies. On June 20, 1990, D. H. Levy and H. E. Holt at Palomar Observatory discovered an asteroid, later named (5261) Eureka, occupying the following lagrangian point of the planet Mars. In 2001 the first Trojan of Neptune (2001 QR322) was discovered in the course of the Deep Ecliptic Survey, and as of December 2008 four additional Neptunian Trojans had been discovered. As of September 2006 there were 2059 known Jupiter Trojans. *See* ASTEROID. Edward F. Tedesco

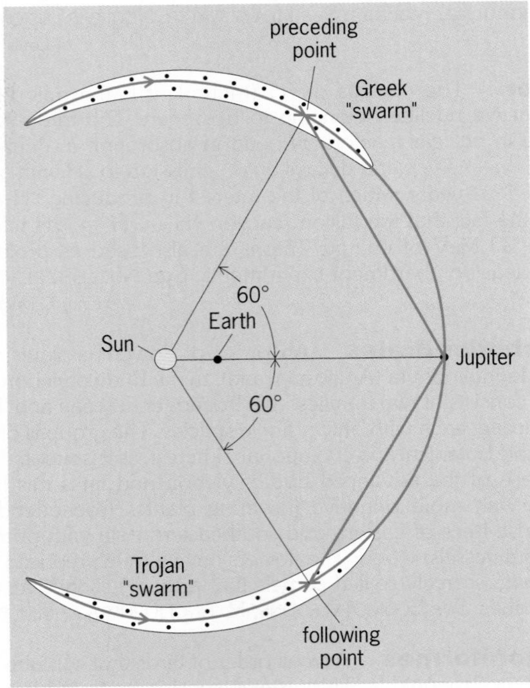

Lagrangian points and Trojan asteroids.

Trombidiformes A suborder of the Acarina (also known as Prostigmata) commonly called the trombidiform mites, more closely related to the Sarcoptiformes than to the other suborders. They are usually distinguished by presence of a respiratory system opening at or near the base of the chelicerae.

The Trombidiformes are probably the most heterogeneous group of mites, both morphologically and ecologically, varying from baglike forms with degenerate legs to the highly evolved, fully developed, parasitic forms. Economically, this group contains two families of plant-feeding mites of great importance to agriculture: the Tetranychidae (spider mites) and the Eriophyidae (bud mites or gall mites). Medically, the Trombiculidae (chiggers or red bugs) are important because the larval forms, which are parasites of vertebrates, can cause intense irritation to their hosts by their feeding. More seriously, some transmit a rickettsial disease, scrub typhus, to humans in the Far East and South Pacific regions. Edward W. Baker

Trophic ecology The study of the structure of feeding relationships among organisms in an ecosystem. Researchers focus on the interplay between feeding relationships and ecosystem attributes such as nutrient cycling, physical disturbance, or the rate of tissue production by plants and the accrual of detritus (dead organic material). Feeding or trophic relationships can be represented as a food web or as a food chain. Food webs depict trophic links between all species sampled in a habitat, whereas food chains simplify this complexity into linear arrays of interactions among trophic levels. Thus, trophic levels (for example, plants, herbivores, detritivores, and carnivores) are amalgamations of species that have similar feeding habits. (However, not all species consume prey on a single trophic level. Omnivores are species that feed on more than one trophic level.) *See* ECOLOGY; ECOSYSTEM; FOOD WEB.

The three fundamental questions in the field of trophic ecology are: (1) What is the relationship between the length of food chains and plant biomass (the relative total amount of plants at the bottom of the food chain)? (2) How do resource supply to producers (plants) and resource demand by predators determine the relative abundance of organisms at each trophic level in a food chain? (3) How long are real food chains, and what factors limit food chain length?

A central theory in ecology is that "the world is green" because carnivores prevent herbivores from grazing green plant biomass to very low levels. Trophic structure (the number of trophic levels) determines trophic dynamics (as measured by the impact of herbivores on the abundance of plants). Indirect control of plant biomass by a top predator is called a trophic cascade. Cascades have been demonstrated to varying degrees in a wide variety of systems, including lakes, streams, subtidal kelp forests, coastal shrub habitats, and old fields. In all of these systems, the removal of a top predator has been shown to precipitate dramatic reductions in the abundance of species at lower trophic levels. Food chain theory predicts a green world when food chains have odd numbers of trophic levels, but a barren world (plants suppressed by herbivores) in systems with even numbers of trophic levels.

Although predators often have strong indirect effects on plant biomass as a result of trophic cascades, both predation (a top-down force) and resource supply to producers (a bottom-up force) play strong roles in the regulation of plant biomass. The supply of inorganic nutrients at the bottom of a food chain is an important determinant of the rate at which the plant trophic level produces tissue (primary production, or productivity). However, the degree to which nutrient supply enhances plant biomass accrual depends on two factors: (1) how many herbivores are present (which in turn depends on how many trophic levels

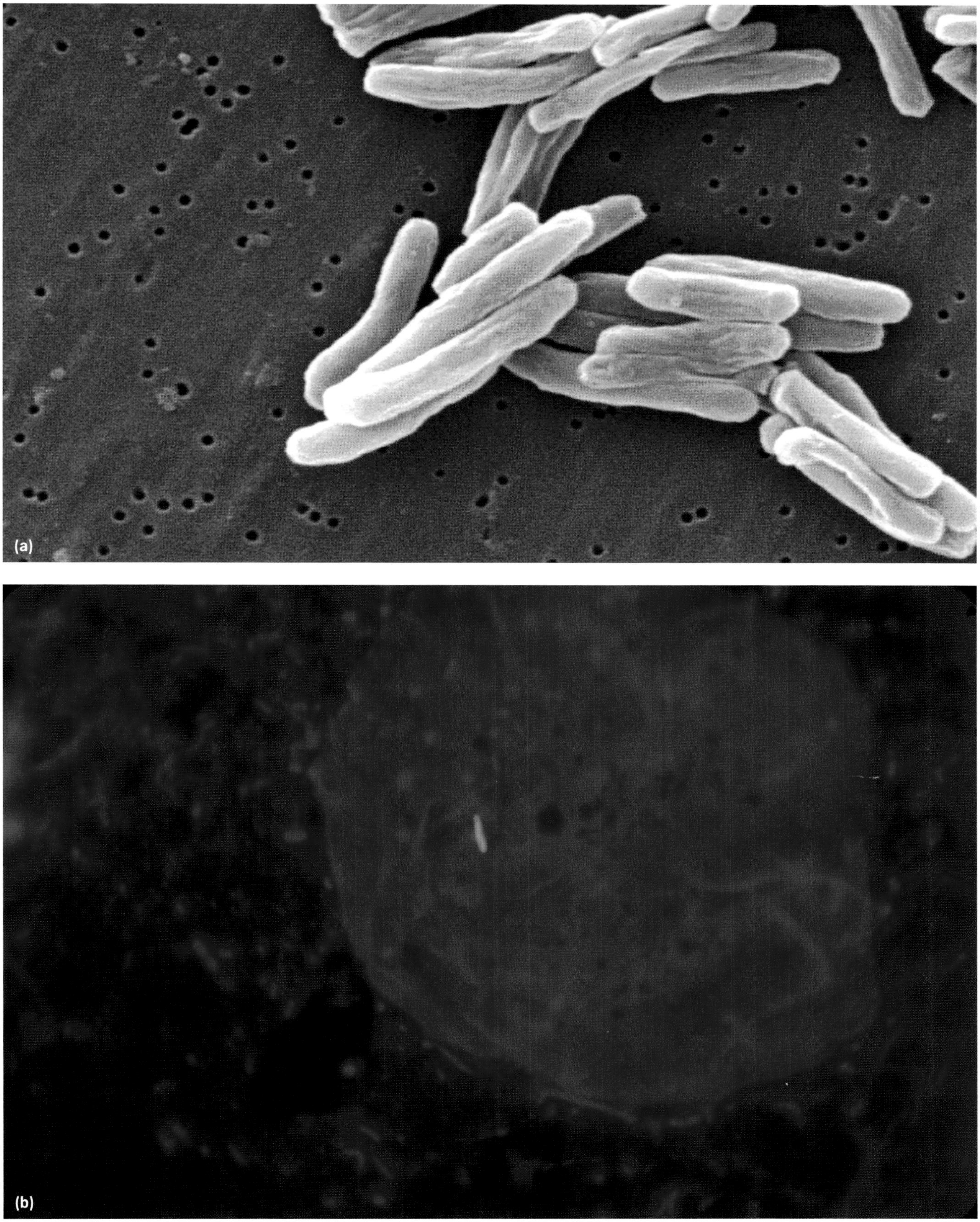

(a)

(b)

(*a*) Colorized scanning electron micrograph depicting some of the ultrastructural details seen in the cell wall configuration of a number of Gram-positive *Mycobacterium tuberculosis* bacteria. As an obligate aerobic organism, *M. tuberculosis* can only survive in an environment containing oxygen. This bacterium ranges in length between 2 and 4 μm, with a width between 0.2 and 0.5 μm (*image provided by Ray Butler and Janice Carr, Centers for Disease Control and Prevention, Public Health Image Library*). (*b*) *Mycobacterium tuberculosis* in a sputum smear stained using fluorescent auramine with acridine orange counterstain. The *M. tuberculosis* bacteria glow yellow under ultraviolet light microscopy (*image provided by Ronald W. Smithwick, Centers for Disease Control and Prevention, Public Health Image Library*).

there are in the system) and (2) the degree to which the herbivores can respond to increases in plant productivity and control plant biomass. The relative importance of top-down (demand) versus bottom-up (supply) forces is well illustrated by lake systems, in which the supply of phosphorus (bottom-up force) and the presence of piscivorous (fish-eating) fish (top-down force) have significant effects on the standing stock of phytoplankton, the plant trophic level in lake water columns. *See* BIOLOGICAL PRODUCTIVITY; BIOMASS; FRESHWATER ECOSYSTEM; LAKE; PHYTOPLANKTON.

Increases in productivity may act to lengthen food chains. However, food chain length may be limited by the efficiency at which members of each trophic level assimilate energy as it moves up the food chain; the resilience of the chain (measured as the inverse of the time required for all trophic levels to return to previous abunance levels after a disturbance); and the size of the ecosystem—small habitats are simply not large enough to support the home range or provide ample habitat for larger carnivorous species. *See* ECOLOGICAL ENERGETICS; SYSTEMS ECOLOGY; THEORETICAL ECOLOGY. John L. Sabo; Leah R. Gerber

Tropic of Cancer The parallel of latitude about $23^1/_2{}^\circ$ (23.45°) north of the Equator. The importance of this line lies in the fact that its degree of angle from the Equator is the same as the inclination of the Earth's axis from the vertical to the plane of the ecliptic. Because of this inclination of the axis and the revolution of the Earth in its orbit, the vertical overhead rays of the Sun may progress as far north as $23^1/_2{}^\circ$. At no place north of the Tropic of Cancer will the Sun, at noon, be 90° overhead.

On June 21, the summer solstice (Northern Hemisphere), the Sun is vertical above the Tropic of Cancer. On this same day the Sun is 47° above the horizon at noon at the Arctic Circle, and at the Tropic of Capricorn, only 43° above the horizon. The Tropic of Cancer is the northern boundary of the equatorial zone called the tropics, which lies between the Tropic of Cancer and Tropic of Capricorn. *See* LATITUDE AND LONGITUDE; MATHEMATICAL GEOGRAPHY; SOLSTICE. Van H. English

Tropic of Capricorn The parallel of latitude approximately $23^1/_2{}^\circ$(23.45°)south of the Equator. It was named for the constellation Capricornus (the goat), for astronomical reasons which no longer prevail.

Because the Earth, in its revolution around the Sun, has its axis inclined $23^1/_2{}^\circ$ from the vertical to the plane of the ecliptic, the Tropic of Capricorn marks the southern limit of the zenithal position of the Sun. Thus, on December 22 (Southern Hemisphere summer, but northern winter solstice) the Sun, at noon, is 90° above the horizon.

The Tropic of Capricorn is the southern boundary of the equatorial zone referred to as the tropics, which lies between the Tropic of Capricorn and the Tropic of Cancer. *See* MATHEMATICAL GEOGRAPHY; SOLSTICE; TROPIC OF CANCER. Van H. English

Tropical meteorology The study of atmospheric structure and behavior in the areas astride the Equator, roughly between 30° north and south latitude. The weather and climate of the tropics involve phenomena such as trade winds, hurricanes, intertropical convergence zones, jet streams, monsoons, and the El Niño Southern Oscillation. More energy is received from the Sun over the tropical latitudes than is lost to outer space (infrared radiation). The reverse is true at higher latitudes, poleward of 30°. The excess energy from the tropics is transported by winds to the higher latitudes, largely by vertical circulations that span roughly 30° in latitudinal extent. These circulations are known as Hadley cells.

For the most part, the oceanic tropics (the islands) experience very little change of day-to-day weather except when severe events occur. Tropical weather can be more adverse during the summer seasons of the respective hemispheres. The near equatorial belt between 5°S and 5°N is nearly always free from hurricanes and typhoons: the active belt lies outside this region over the tropics. The land areas experience considerable heating of the Earth's surface, and the summer-to-winter contrasts are somewhat larger there. For instance, the land areas of northern India experience air temperatures as high as 108°F (42°C) in the summer (near the Earth's surface), while in the winter season the temperatures remain 72°F (22°C) for many days. The diurnal range of temperature is also quite large over land areas on clear days during the summer (32°F or 18°C) as compared to winter (18°F or 10°C).

The steady northeast surface winds over the oceans of the Northern Hemisphere between 5° and 20°N and southeast winds over the corresponding latitudes of the southern oceans constitute the trade winds. Trade winds have intensities of around 5–10 knots (2.5–5 m/s). They are the equatorial branches of the anticyclonic circulation (known as the subtropical high pressure). The steadiness of wind direction is quite high in the trades. *See* WIND.

Hurricanes are also known as typhoons in the west Pacific and tropical cyclones in Indian Ocean and south Pacific. If the wind speed exceeds 65 knots (33 m/s) in a tropical storm, the storm is labeled a hurricane. A hurricane usually forms over the tropical oceans, north or south of 5° latitude from the Equator. *See* HURRICANE.

Intertropical convergence zones are located usually between 5 and 10°N latitude. They are usually oriented west to east and contain cloud clusters with rainfall of the order of 1.2–2 in. (30–50 mm) per day. The trade winds of the two hemispheres supply moisture to this precipitating system. *See* CLOUD PHYSICS; PRECIPITATION (METEOROLOGY).

A number of fast-moving air currents, known as jets, are important elements of the tropical general circulation. With speeds in excess of 30 knots (15 m/s), they are found over several regions of the troposphere. *See* ATMOSPHERIC GENERAL CIRCULATION; JET STREAM; TROPOSPHERE.

Basically the entire landmass from the west coast of Africa to Asia and extending to the date line experiences a phenomenon known as the monsoon. Monsoon circulations are driven by differential heating between relatively cold oceans and relatively warm landmasses. *See* MONSOON METEOROLOGY.

Every 2–6 years the eastern equatorial Pacific Ocean experiences a rise in sea surface temperature of about 5–9°F (3–5°). This phenomenon is known as El Niño, which is part of a larger cycle referred to as the El Niño Southern Oscillation (ENSO). The other extreme in the cycle is referred to as La Niña. El Niño has been known to affect global-scale weather. *See* EL NIÑO; MARITIME METEOROLOGY. T. N. Krishnamurti

Tropopause The boundary between the troposphere and the stratosphere in the atmosphere. The tropopause is broadly defined as the lowest level above which the lapse rate (decrease) of temperature with height becomes less than 5.8°F mi^{-1} (2°C km^{-1}). In low latitudes the tropical tropopause is at a height of 9.3–11 mi at about −135°F (15–17 km at about 180 K), and the polar tropopause between tropics and poles is at about 6.2 mi at about −63°F (10 km at about 220 K). There is a well-marked "tropopause gap" or break where the tropical and polar tropopauses overlap at 30–40° latitude. The break is in the region of the subtropical jet stream and is of major importance for the transfer of air and tracers (humidity, ozone, radioactivity) between stratosphere and troposphere. The height of the tropopause varies seasonally and also daily with the weather systems, being higher and colder over anticyclones than over depressions. *See* AIR TEMPERATURE; ATMOSPHERE; STRATOSPHERE; TROPOSPHERE. R. J. Murgatroyd

Troposphere The lowest major layer of the atmosphere. The troposphere extends from the Earth's surface to a height of 6–10 mi (10–16 km), the base of the stratosphere. It contains about four-fifths of the mass of the whole atmosphere. *See* ATMOSPHERE.

On the average, the temperature decreases steadily with height throughout this layer, with a lapse rate of about 19°F mi^{-1} (6.5°C km^{-1}), although shallow inversions (temperature increases with height) and greater lapse rates occur, particularly in the boundary layer near the Earth's surface. Appreciable water-vapor contents and clouds are almost entirely confined to the troposphere. Hence it is the seat of all important weather processes and the region where interchange by evaporation and precipitation (rain, snow, and so forth) of water substance between the surface and the atmosphere takes place. *See* CLIMATOLOGY; CLOUD PHYSICS; METEOROLOGY; WEATHER. R. J. Murgatroyd

Tropospheric scatter A term applied to propagation of radio waves caused by irregularities in the refractive index of air. The phenomenon is predominant in the lower atmosphere; little or no scattering of importance occurs above the troposphere. Tropospheric scatter propagation provides very useful communication services but also causes harmful interference. For example, it limits the geographic separation required for frequency assignments to services such as television and frequency-modulation broadcasting, very-high-frequency omnidirectional ranges (VOR), and microwave relays. It is used extensively throughout most of the world for long-distance point-to-point services, particularly where high information capacity and high reliability are required. Typical tropospheric scatter relay facilities are commonly 200–300 mi (320–480 km) apart. Some single hops in excess of 500 mi (800 km) are in regular use. High-capacity circuits carry 200–300 voice circuits simultaneously. Robert S. Kirby

Truck A motor vehicle carrying its load on its own wheels and primarily designed for the transportation of goods or cargo. A truck is similar to a passenger car in many basic aspects, but truck construction is usually heavier throughout with strengthened chassis and suspension, and lower transmission and drive-axle ratios to cope with hilly terrain. Other common truck characteristics include cargo-carrying features such as rear doors or tailgate, and a flat floor. However, there are many different kinds of trucks, often specially designed with unique features for performing a particular job. *See* AUTOMOBILE; AUTOMOTIVE ELECTRICAL SYSTEM; BUS.

A truck is rated by its gross vehicle weight (gvw), which is the combined weight of the vehicle and load. Trucks are classified as light-, medium-, or heavy-duty according to gross vehicle weight. Although a variety of models and designs are available in each category, there are two basic types of vehicles, the straight truck and the truck tractor. The straight truck has the engine and body mounted on the same chassis. The truck tractor is essentially a power unit that is the control and pulling vehicle for truck trailers such as full trailers or semitrailers. A full trailer has a front axle and one or more rear axles and is constructed so that all its own weight and that of its load rests on its own wheels. A semitrailer has one or more axles at the rear, and is constructed so that the front end and a substantial part of its own weight and that of its load rests upon another vehicle. A retractable mechanism mounted at the front end of the semitrailer is lowered to support it when the pulling vehicle is disconnected. A full trailer may be drawn by a truck, or behind a semitrailer.

The tractor-semitrailer combination permits the use of longer bodies with greater carrying capacity and better maneuverability than is possible with a straight truck. The forward positioning of the cab, the short wheelbase of the tractor, and the multiplicity of axles provide maximum payloads and operating economy in the face of restriction on overall length imposed by some states, and regulations limiting the weight carried on a single axle. Donald L. Anglin

Truss An assemblage of structural members joined at their ends to form a stable structural assembly. If all members lie in one plane, the truss is called a planar truss or a plane truss. If the members are located in three dimensions, the truss is called a space truss.

A plane truss is used like a beam, particularly for bridge and roof construction. A plane truss can support only weight or loads contained in the same plane as that containing the truss. A space truss is used like a plate or slab, particularly for long span roofs where the plan shape is square or rectangular, and is most efficient when the aspect ratio (the ratio of the length and width) does not vary above 1.5. A space truss can support weight and loads in any direction.

Because a truss can be made deeper than a beam with solid web and yet not weigh more, it is more economical for long spans and heavy loads, even though it costs more to fabricate. *See* BRIDGE; ROOF CONSTRUCTION.

The simplest truss is a triangle composed of three bars with ends pinned together. If small changes in the lengths of the bars are neglected, the relative positions of the joints do not change when loads are applied in the plane of the triangle at the apexes.

Multiple-span plane trusses (defined as statically indeterminate or redundant) and space trusses require very complex and tedious hand calculations. Modern high-speed digital computers and readily available computer programs greatly facilitate the structural analysis and design of these structures. *See* COMPUTER; STRUCTURAL ANALYSIS. Charles Thornton; I. Paul Lew

Trypanites A simple cylindrical boring common in the fossil record for the past 540 million years. *Trypanites* is an ichnogenus, which is a formal name given to a distinct trace fossil (evidence of organism behavior in the fossil record as distinct from the remains or other representations of the organism's body). *Trypanites* is thus not a type of life but a structure built by a variety of wormlike animals over time. *See* TRACE FOSSILS.

Trypanites is formally defined as a cylindrical, unbranched boring in a hard substrate (such as a rock or shell) with a length up to 50 times its width. Most *Trypanites* are only a few millimeters long, but some are known to be up to 12 cm (4.72 in.). They usually penetrate the substrate perpendicularly to its surface and remain straight, but they are sometimes found at oblique angles, and a few may curve in response to irregularities in the material they bored. Almost all *Trypanites* are found in calcareous substrates, which is evidence that the producing organisms may have used at least in part some sort of acid or other dissolving chemical to excavate the substrate. The organisms which produced *Trypanites* have varied over geological time. Most *Trypanites* were made by marine wormlike animals, with a few known from freshwater paleoenvironments.

Trypanites is important to paleontologists as one of the earliest examples of bioerosion (the erosion of hard substrates by biological actions). It is the first known macroboring (a boring easily visible to the naked eye). *See* PALEOECOLOGY. Mark A. Wilson

Trypanorhyncha An order of tapeworms of the subclass Cestoda, also known as the Tetrarhynchoidea. All are parasitic in the intestine of elasmobranch fishes. They are distinguished from all other tapeworm groups by having spiny, eversible proboscides on the head. The head also bears two or four shallow, weakly muscular suckers. A complete life history is not known for any trypanorhynchid, although larval forms have been found in the tissues of various marine invertebrates and teleost fishes. *See* CESTODA. Clark P. Read

Trypanosomatidae A family of Protozoa, order Kinetoplastida, containing flagellated parasites which change their morphology; that is, they exhibit polymorphism during their life cycles. The life cycles of the organisms may involve only an invertebrate host, or an invertebrate and a vertebrate host, or an invertebrate and a plant host. Several distinct morphological forms are recognized: trypanosomal, crithidial, leptomonad, and leishmanial. Differentiation into genera is dependent upon the host infected as well as the morphologic types involved. None of the stages possesses a mouth opening, and nutritive elements are absorbed through the surface of the body; that is, the organisms are saprozoic.

Trypanosoma is the important genus of the family Trypanosomatidae from a number of standpoints. It contains the largest number of species infecting a wide variety of hosts such as mammals, birds, fishes, amphibians, and reptiles. Although most of the species cause no damage to the hosts, there are several which produce serious diseases in humans, domesticated animals, and wild animals. The pathogenic species are prevalent in Africa.

Leishmania is the second most important genus. Three species parasitize humans but have also been found naturally infecting dogs, cats, and perhaps other lower animals. The sand fly, *Phlebotomus*, transmits the parasite from vertebrate to vertebrate.

M. M. Brooke; Horace W. Stunkard

Trypanosomiasis A potentially fatal infection caused by parasites of the genus *Trypanosoma*.

The African trypanosomes, the cause of African trypanosomiasis or African sleeping sickness, are flagellated protozoan parasites. *Trypanosoma brucei rhodesiense* and *T. b. gambiense* cause disease in humans. *Trypanosoma brucei* is restricted to domestic and wild animals. The trypanosomes are transmitted by the tsetse fly (*Glossina*), which is restricted to the African continent. The trypanosomes are taken up in a blood meal and grow and multiply within the tsetse gut. After 2–3 weeks, depending upon environmental conditions, they migrate into the salivary glands, where they become mature infective forms, and are then transmitted by the injection of infected saliva into a new host during a blood meal. The survival of the tsetse fly is dependent upon temperature and humidity, and the fly is confined by the Sahara to the north and by the colder drier areas to the south, an area about the size of the United States. Approximately 50 million people live within this endemic area, and 15,000–20,000 new human cases of African trypanosomiasis are reported annually.

In humans and other mammals the trypanosomes are extracellular. During the early stages of infection, the trypanosomes are found in the blood and lymph but not in cerebrospinal fluid. There are fever, malaise, and enlarged lymph nodes. In the absence of treatment the disease becomes chronic and the trypanosomes penetrate into the cerebrospinal fluid and the brain. The symptoms are headaches, behavioral changes, and finally the characteristic sleeping stage. Without treatment the individual sleeps more and more and finally enters a comatose stage which leads to death. Treatment is more difficult if the infection is not diagnosed until the late neurological stage.

Trypanosoma cruzi, the cause of American trypanosomiasis (Chagas' disease), is predominantly an intracellular parasite in the mammalian host. During the intracellular stage, *T. cruzi* loses its flagellum and grows predominantly in cells of the spleen, liver, lymphatic system, and cardiac, smooth, and skeletal muscle. The cells of the autonomic nervous system are also frequently invaded. *See* MEDICAL PARASITOLOGY; PARASITOLOGY.

John Richard Seed

Tsunami A set of ocean waves caused by any large, abrupt disturbance of the sea surface. If the disturbance is close to the coastline, a local tsunami can demolish coastal communities within minutes. A very large disturbance can cause local devastation and destruction thousands of miles away. Tsunami comes from the Japanese language, meaning harbor wave.

Tsunamis rank high on the scale of natural disasters. Since 1850, they have been responsible for the loss of over 420,000 lives and billions of dollars of damage to coastal structures and habitats. Most casualties were caused by local tsunamis that occur about once per year somewhere in the world. Predicting when and where the next tsunami will strike is currently impossible. Once a tsunami is generated, forecasting its arrival and impact is possible through modeling and measurement technologies.

Tsunamis are most commonly generated by earthquakes in marine and coastal regions. Major tsunamis are produced by large (greater than magnitude 7), shallow-focus (less than 30 km or 19 mi deep in the Earth) earthquakes associated with the movement of oceanic and continental plates. They frequently occur in the Pacific, where dense oceanic plates slide under the lighter continental plates. When these plates fracture, they provide a vertical movement of the sea floor that allows a quick and efficient transfer of energy from the solid earth to the ocean. Underwater landslides associated with smaller earthquakes are also capable of generating destructive tsunamis. Other large-scale disturbances of the sea surface that can generate tsunamis are explosive volcanoes and asteroid impacts. The resulting tsunami propagates as a set of waves whose energy is concentrated at wavelengths corresponding to the earth movements (~100 km or 60 mi), at wave heights determined by vertical displacement (~1 m or 3 ft), and at wave directions determined by the adjacent coastline geometry. Because each earthquake is unique, every tsunami has unique wavelengths, wave heights, and directionality. From a tsunami warning perspective, this makes forecasting tsunamis in real time daunting. *See* ASTEROID; EARTHQUAKE; OCEAN WAVES; PLATE TECTONICS; VOLCANO.

Eddie N. Bernard

Tuberculosis An infectious disease caused by the bacillus *Mycobacterium tuberculosis*. It is primarily an infection of the lungs, but any organ system is susceptible, so its manifestations may be varied. Effective therapy and methods of control and prevention of tuberculosis have been developed, but the disease remains a major cause of mortality and morbidity throughout the world. The treatment of tuberculosis has been complicated by the emergence of drug-resistant organisms, including multiple-drug-resistant tuberculosis, especially in those with HIV infection. *See* ACQUIRED IMMUNE DEFICIENCY SYNDROME (AIDS).

Mycobacterium tuberculosis is transmitted by airborne droplet nuclei produced when an individual with active disease coughs, speaks, or sneezes. When inhaled, the droplet nuclei reach the alveoli of the lung. In susceptible individuals the organisms may then multiply and spread through lymphatics to the lymph nodes, and through the bloodstream to other sites such as the lung apices, bone marrow, kidneys, and meninges.

The development of acquired immunity in 2 to 10 weeks results in a halt to bacterial multiplication. Lesions heal and the individual remains asymptomatic. Such an individual is said to have tuberculous infection without disease and will show a positive tuberculin test. The risk of developing active disease with clinical symptoms and positive cultures for the tubercle bacillus diminishes with time and may never occur, but is a lifelong risk. Only 5% of individuals with tuberculous infection progress to active disease. Progression occurs mainly in the first 2 years after infection; household contacts and the newly infected are thus at risk.

Many of the symptoms of tuberculosis, whether pulmonary disease or extrapulmonary disease, are nonspecific. Fatigue or tiredness, weight loss, fever, and loss of appetite may be present for months. A fever of unknown origin may be the sole indication of tuberculosis, or an individual may have an acute influenzalike illness. Erythema nodosum, a skin lesion, is occasionally associated with the disease.

The lung is the most common location for a focus of infection to flare into active disease with the acceleration of the growth of organisms. There may be complaints of cough, which can produce sputum containing mucus, pus, and, rarely, blood. Listening to the lungs may disclose rales or crackles and signs of pleural effusion (the escape of fluid into the lungs) or consolidation if present. In many, especially those with small infiltration, the physical examination of the chest reveals no abnormalities.

Miliary tuberculosis is a variant that results from the bloodborne dissemination of a great number of organisms resulting in the simultaneous seeding of many organ systems. The meninges, liver, bone marrow, spleen, and genitourinary system are usually involved. The term miliary refers to the lung lesions being the size of millet seeds (about 0.08 in. or 2 mm). These lung lesions are present bilaterally. Symptoms are variable.

Extrapulmonary tuberculosis is much less common than pulmonary disease. However, in individuals with AIDS, extrapulmonary tuberculosis predominates, particularly with lymph node involvement. Fluid in the lungs and lung lesions are other common manifestations of tuberculosis in AIDS. The lung is the portal of entry, and an extrapulmonary focus, seeded at the time of infection, breaks down with disease occurring.

Development of renal tuberculosis can result in symptoms of burning on urination, and blood and white cells in the urine; or the individual may be asymptomatic. The symptoms of tuberculous meningitis are nonspecific, with acute or chronic fever, headache, irritability, and malaise.

A tuberculous pleural effusion can occur without obvious lung involvement. Fever and chest pain upon breathing are common symptoms.

Bone and joint involvement results in pain and fever at the joint site. The most common complaint is a chronic arthritis usually localized to one joint. Osteomyelitis is also usually present.

Pericardial inflammation with fluid accumulation or constriction of the heart chambers secondary to pericardial scarring are two other forms of extrapulmonary disease.

The principal methods of diagnosis for pulmonary tuberculosis are the tuberculin skin test (an intracutaneous injection of purified protein derivative tuberculin is performed, and the injection site examined for reactivity), sputum smear and culture, and the chest x-ray. Culture and biopsy are important in making the diagnosis in extrapulmonary disease.

A combination of two or more drugs is used in the initial therapy of tuberculous disease. Drug combinations are used to lessen the chance of drug-resistant organisms surviving. The preferred treatment regimen for both pulmonary and extrapulmonary tuberculosis is a 6-month regimen of the antibiotics isoniazid, rifampin, and pyrazinamide given for 2 months, followed by isoniazid and rifampin for 4 months. Because of the problem of drug-resistant cases, ethambutol can be included in the initial regimen until the results of drug susceptibility studies are known. Once treatment is started, improvement occurs in almost all individuals. Any treatment failure or individual relapse is usually due to drug-resistant organisms. *See* DRUG RESISTANCE.

The community control of tuberculosis depends on the reporting of all new suspected cases so case contacts can be evaluated and treated appropriately as indicated. Individual compliance with medication is essential. Furthermore, measures to enhance compliance, such as directly observed therapy, may be necessary. *See* MYCOBACTERIAL DISEASES. George Lordi

Tubeworms The name given to marine polychaete worms (particularly to many species in the family Serpulidae) which construct permanent calcareous tubes on rocks, seaweeds, dock pilings, and ship bottoms. The individual tubes with hard walls of calcite-aragonite are firmly cemented to any hard substrate and to each other. Economically they are among the most important fouling organisms both on ship hulls (where they are second only to barnacles) and inside sea-water cooling pipes of power stations.

About 340 valid species of serpulid tubeworms have been described. The majority are truly marine, but several species thrive in brackish waters of low salinity, and one species occurs in fresh waters in Karst limestone caves. The wide geographical distribution of certain abundant species owes much to human transport on the bottoms of relatively fast ships and occurred within the last 120 years. *See* ANNELIDA; POLYCHAETA. W. D. Russell-Hunter

Tubulidentata An order of mammals containing a single living genus *Orycteropus*, the aardvark. Aardvarks occur in suitable habitats throughout sub-Saharan Africa. This order exhibits the results of an extreme adaptation for burrowing and feeding on small food items (particularly termites and ants). *See* MAMMALIA.

Aardvarks, also known as antbears, resemble a medium-sized to large pig. The body is massive with an elongate head and a piglike snout. The tough thick skin is sparsely covered with bristly hair. The ears are large and donkeylike. The short thick legs possess powerful sharp claws that are used to excavate burrows and to open the nests of termites and ants which are gathered by the aardvark's long sticky tongue. Adult aardvarks lack incisor and canine teeth. The simple peglike teeth on the sides of the jaws consist of tubular dentin covered by cement ("tubule teeth"). They lack enamel and grow continuously during the animal's life. *See* AARDVARK.

In the past, tubulidentates were often considered closely related to ungulates. However, recent mitochondrial and nuclear gene data show a close relationship to elephant-shrews, paenungulates (hyraxes, sirenians, and proboscideans), and golden moles (Chrysochloridae). All of these ecologically divergent forms probably originated in Africa. The earliest known tubulidentate (*Myorycteropus*) is from early Miocene deposits found in Kenya in East Africa. Donald W. Linzey

Tufa A spongy, porous limestone formed by precipitation from evaporating spring and river waters; also known as calcareous sinter. Calcium carbonate commonly precipitates from supersaturated waters on the leaves and stems of plants growing around the springs and pools and preserves some of their plant structures. Tufa tends to be fragile and friable. *See* LIMESTONE; TRAVERTINE. Raymond Siever

Tuff Fragmental volcanic products from explosive eruptions that are consolidated, cemented, or otherwise hardened to form solid rock. In strict scientific usage, the term "tuff" refers to consolidated volcanic ash, which by definition consists of fragments smaller than 2 mm. However, the term is also used for many pyroclastic rocks composed of fragments coarser than ash and even for pyroclastic material that has undergone limited posteruption reworking. If the thickness, temperature, and gas content of a tuff-forming pyroclastic flow are sufficiently high, the constituent fragments can become compacted and fused to form welded tuff. The term "tuff" is also used in the naming of several related types of small volcanic edifices formed by hydrovolcanic eruptions, triggered by the explosive interaction of hot magma or lava with water. *See* IGNIMBRITE; PYROCLASTIC ROCKS; VOLCANO.

Robert Tilling

Tularemia A worldwide disease caused by infection with the bacterium *Francisella tularensis*, which affects multiple animal species. Infection in humans occurs frequently from skinning infected animals bare-handed or from the bites of infected animals, ticks, deer flies. The mortality rate varies by species, but with treatment it is low. Ungulates are frequently infected but generally suffer low mortality.

Tularemia can be difficult to differentiate from other diseases because it can have multiple clinical manifestations. Nonspecific signs frequently include fever, lethargy, anorexia, and increased pulse and respiration rates. The disease can overlap geographically with plague, and both may lead to enlarged lymph nodes (buboes). However, with tularemia, the buboes are more likely to ulcerate. If tularemic infection results from inhalation of dust from contaminated soil, hay, or grain, either pneumonia or a typhoidal syndrome can occur. Rarely, the route of entry for the bacteria is the eyes, leading to the oculoglandular type of tularemia. If organisms are ingested, the oropharyngeal form can develop, characterized by abdominal pain, diarrhea, vomiting, and ulcers. *See* PLAGUE.

Tularemia is not transmitted directly from individual to individual. If the infected person or animal is untreated, blood remains infectious for 2 weeks and ulcerated lesions are infectious for a month. Deer flies (*Chrysops discalis*) are infective for 2 weeks, and ticks are infective throughout their lifetime (usually 2 years).

A number of antibacterial agents are effective against *F. tularensis*, the most effective being streptomycin. Penicillin and the sulfonamides have no therapeutic effect. *See* ANTIBIOTIC.

Millicent Eidson

Tulip tree A tree, *Liriodendron tulipifera*, also known in forestry as yellow poplar, belonging to the magnolia family, Magnoliaceae. One of the largest and most valuable hardwoods of eastern North America, it is native from southern New England and New York westward to southern Michigan, and south to Louisiana and northern Florida.

Botanically, this tree is distinguished by leaves which are squarish at the tip as if cut off, true terminal buds flattened and covered by two valvate scales, an aromatic odor resembling that of magnolia, and cone-shaped fruit which is persistent in winter. The name tulip refers to the large greenish-yellow and orange-colored flowers.

The wood of the tulip tree is light yellow to brown, hence the common name yellow poplar, which is a misnomer. It is a soft and easily worked wood, used for construction, interior finish, containers (boxes, crates, baskets), woodenware, excelsior, veneer, and sometimes for paper pulp. *See* MAGNOLIALES.

Arthur H. Graves; Kenneth P. Davis

Tumbling mill A grinding and pulverizing machine consisting of a shell or drum rotating on a horizontal axis. The material to be reduced in size is fed into one end of the mill. The mill is also charged with grinding material such as iron balls. As the mill rotates, the material and grinding balls tumble against each other, the material being broken chiefly by attrition.

Tumbling mills are variously classified as pebble, ball, or rod depending on the grinding material, and as cylindrical, conical, or tube depending on the shell shape. *See* CRUSHING AND PULVERIZING; GRINDING MILL; PEBBLE MILL.

Ralph M. Hardgrove

Tumor Literally, a swelling; in the past the term has been used in reference to any swelling of the body, no matter what the cause. However, the word is now being used almost exclusively to refer to a neoplastic mass, and the more general usage is being discarded.

A neoplastic mass or neoplasm is a pathological lesion characterized by the progressive or uncontrolled proliferation of cells. The cells involved in the neoplastic growth have an intrinsic heritable abnormality such that they are not regulated properly by normal methods. The stimulus which elicits this growth is not usually known.

It is common to divide tumors into benign or malignant. The decision as to which category a tumor should be assigned is usually based on information gained from gross or microscopic examination, or both. Benign neoplasms usually grow slowly, remain so localized that they cause little harm, and generally can be successfully and permanently removed. Malignant or cancerous neoplasms tend to grow rapidly, spread throughout the body, and recur if removed. Not all tumors which have been classified as benign are harmless to the host, and some can cause serious problems. Difficulties may occur as a result of mechanical pressure.

The cells of benign tumors are well differentiated. This means that the cells are very like the normal tissue in size, structure, and spatial relationship. The cells forming the tumor usually function normally. Cell proliferation usually is slow enough so that there is not a large number of immature cells. As the cellular mass increases in size, most benign tumors develop a fibrous capsule around them which separates them from the normal tissue. The cells of a benign tumor remain at the site of origin and do not spread throughout the body. Anaplasia (loss of differentiation) is not seen in benign tumors.

The cells of malignant tumors may be well differentiated, but most have some degree of anaplasia. Anaplastic cells tend to be larger than normal and are abnormal, even bizarre, in shape. The nuclei tend to be very large, and irregular, and they often stain darkly. Malignant tumors may be partially but never completely encapsulated. The cells of the cancer infiltrate and destroy surrounding tissue. They have the ability to metastasize; that is, cells from the primary tumor are disseminated to other regions of the body where they are able to produce secondary tumors called metastases.

In most cases the formation of a neoplasm is irreversible. It results from a permanent cellular defect which is passed on to daughter cells. Tumors should undergo medical appraisal to determine what treatment, if any, is needed.

N. Karle Mottet; Carol Quaife

Tumor suppressor genes are a class of genes which, when mutated, predispose an individual to cancer. The mutations result in the loss of function of the particular tumor suppressor protein encoded by the gene. Although this class of genes was named for its link to human cancer, it is now clear that these genes play a critical role in the normal development, growth, and proliferation of cells and organs within the human body. The protein product of many tumor suppressor genes constrains cell growth and proliferation so that these events occur in a controlled manner. Thus, these genes appear to act in a manner antagonistic to that of oncogenes, which promote cell growth and proliferation.

The retinoblastoma (RB), p53, and p16 genes are the best-understood tumor suppressors. Inactivating mutations in the RB gene have been observed in retinoblastomas, osteosarcomas (cancer of the bone), as well as cancers of the lung, breast, and bladder. The p16 mutations have been observed in cancers of the skin, lung, breast, brain, bone, bladder, kidney, esophagus, and pancreas. The tumor suppressor p53 is the most frequently mutated gene associated with the development of many different types of human cancer, including those of the breast, lung, and colon. It is also associated with the rare inherited disease, Li-Fraumeni syndrome. Affected individuals manifest an increased likelihood of breast carcinomas, soft tissue sarcoma, brain tumors, osteosarcoma, leukemia, and adrenocortical carcinoma. Like RB and p16, p53 has a role in cell cycle regulation. In addition, p53 functions in the cell's decision on whether to undergo programmed cell death (apoptosis). Deregulated cell proliferation and escape from apoptosis appear to be two common pathways leading to tumor formation. *See* CANCER (MEDICINE); GENE; MUTATION; ONCOLOGY; TUMOR SUPPRESSOR GENES; TUMOR VIRUSES.

Mark Ewen

Tumor suppressor genes A functionally heterogenous class of genes, ranging from nuclear transcription factors to cell adhesion molecules having as their common denominator that their function must be compromised during development and progression of a given tumor. The most important and most widely studied tumor suppressor is called p53. Its importance becomes evident from the fact that mutations in its gene are found in approximately 50% of all human cancers, thus constituting the most frequent alteration in a single human cancer-associated gene. It is now clear that the major function of p53 is to preserve the integrity of the genome of a cell under various conditions of cellular stress, endowing p53 with the title "guardian of the genome."

Tumorigenesis is a very complex process that is initiated by the accumulation of genetic alterations in at least two classes of genes: Activation of proto-oncogenes to oncogenes drives tumor initiation and progression by furthering uncontrolled cell growth and division, while inactivation of tumor suppressor genes leads to the elimination of important checkpoints in cellular proliferation that normally prevent uncontrolled proliferation. This implies that the products of tumor suppressor genes, the tumor suppressor proteins, exert functions that are disadvantageous to tumor development and progression. Inactivation of tumor suppressor genes usually occurs through mutations but can also occur by other means, like gene silencing by epigenetic means. As mutations in a tumor suppressor gene usually are characterized by a "loss of function," mutational inactivation of one allele of a tumor suppressor gene is not sufficient to eliminate the protective effect of the tumor suppressor, as the remaining nonmutated allele is still active. Inactivation of a tumor suppressor gene thus always requires that both of its two alleles are mutated.

Like no other tumor suppressor, the central role of p53 at the crossroad of cell cycle control, DNA repair, and apoptosis, as well as the fact that p53 is mutated in a large number of tumors, renders p53 a prominent target for applied tumor research. In contrast to wild-type p53, which is usually an unstable protein, mutant p53 accumulates in tumor cells. Detection of enhanced levels of p53 in tumor tissue is indicative for the expression of a mutant p53 and is already commonly used in tumor diagnosis. The most challenging aspect of applied p53 research is the development of therapy regimens based on the p53 status of the tumor cells. *See* CANCER (MEDICINE); CELL CYCLE; GENETICS; MUTATION.

Wolfgang Deppert

Tumor viruses Viruses associated with tumors can be classified in two broad categories depending on the nucleic acid in the viral genome and the type of strategy to induce malignant transformation.

RNA viruses. The ribonucleic acid (RNA) tumor viruses are retroviruses. When they infect cells, the viral RNA is copied into deoxyribonucleic acid (DNA) by reverse transcription and the DNA is inserted into the host genome, where it persists and can be inherited by subsequent generation of cells. Transformation of the infected cells can be traced to oncogenes that are carried by the viruses but are not necessary for viral replication. The viral oncogenes are closely similar to cellular genes, the proto-oncogenes, which code for components of the cellular machinery that regulates cell proliferation, differentiation, and death. Incorporation into a retrovirus may convert proto-oncogenes into oncogenes in two ways: the gene sequence may be altered or truncated so that it codes for proteins with abnormal activity; or the gene may be brought under the control of powerful viral regulators that cause its product to be made in excess or in inappropriate circumstances. Retroviruses may also exert similar oncogenic effects by insertional mutation when DNA copies of the viral RNA are integrated into the host-cell genome at a site close to or even within proto-oncogenes. *See* RETROVIRUS.

RNA tumor viruses cause leukemias, lymphomas, sarcomas, and carcinomas in fowl, rodents, primates, and other species. The human T-cell leukemia virus (HTLV) types I and II are endemic in Southeast Asian populations and cause adult T-cell leukemia and hairy-cell leukemia. *See* AVIAN LEUKOSIS; LEUKEMIA; LYMPHOMA; ROUS SARCOMA.

DNA viruses. DNA viruses replicate lytically and kill the infected cells. Transformation occurs in nonpermissive cells where the infection cannot proceed to viral replication. The transforming ability of DNA tumor viruses has been traced to several viral proteins that cooperate to stimulate cell proliferation, overriding some of the normal growth control mechanisms in the infected cell and its progeny. Unlike retroviral oncogenes, DNA virus oncogenes are essential components of the viral genome and have no counterpart in the normal host cells. Some of these viral proteins bind to the protein products of two key tumor suppressor genes of the host cells, the retinoblastoma gene and the p53 gene, deactivating them and thereby permitting the cell to replicate its DNA and divide. Other DNA virus oncogenes interfere with the expression of cellular genes either directly or via interaction with regulatory factors. There is often a delay of several years between initial viral infection in the natural host species and the development of cancer, indicating that, in addition to virus-induced transformation, other environmental factors and genetic accidents are involved. A specific or general impairment of the host immune responses often plays an important role.

DNA tumor viruses belong to the families of papilloma, polyoma, adeno, hepadna, and herpes viruses and produce tumors of different types in various species. DNA tumor viruses are thought to play a role in the pathogenesis of about 15–20% of human cancers. These include Burkitt's lymphoma, nasopharyngeal carcinoma, immunoblastic lymphomas in immunosuppressed individuals and a proportion of Hodgkin's lymphomas that are all associated with the Epstein-Barr virus of the herpes family; and liver carcinoma associated with chronic hepatitis B virus infection. *See* ANIMAL VIRUS; CANCER (MEDICINE); EPSTEIN-BARR VIRUS; HODGKIN'S DISEASE; INFECTIOUS PAPILLOMATOSIS; MUTATION; ONCOLOGY.

Maria G. Masucci

Tuna Certain perciform (spiny-rayed) fishes in the family Scombridae. Like all other scombrids (such as mackerels, bonitos, wahoo, and sierras), tunas have a fusiform and moderately compressed body and certain other characteristics that adapt them for sustained swimming at high speeds. The long spinous dorsal fin is depressible in a groove in the back; the pelvic fins and usually the pectoral fins are small and retractable in shallow depressions; the scales are typically small, but enlarged scales usually cover the anterior part of the body and lateral line, and form an envelopment called the corselet. The eyes protrude very little, if at all, beyond the surface of the head; the mouthparts fit snugly against the pointed head; and the gill covers fit snugly against the body. These features eliminate almost all irregularities that would cause resistance to the water. Tunas are also recognized by the finlets (independent multibranched rays, each appearing as a small fin) behind the dorsal and anal fins. The slender caudal peduncle, supported on each side by two small keels and a large median keel in between, and the lunate caudal fin are driven by powerful muscles for fast and sustained swimming. Sustained swimming depends on red muscle (comparatively thin muscle fibers containing large amounts of myoglobin and mitochondria), and the body temperature of tunas may be several degrees above water temperature. Tunas feed on a wide variety of fishes, squids, and crustaceans. *See* MUSCULAR SYSTEM; PERCIFORMES.

In American waters there are nine species of scombrids in four genera which bear the name tuna. However, there are other scombrids that are just as much tuna fishes as those that bear the name. Practically all tunas are pelagic (that is, living in the open sea) and oceanodromous (migratory in salt water), with most species being highly migratory. Most species are very important commercial food fishes as well as game fishes.

Albacore (*Thunnus alalunga*) constitute the most distinctive species of *Thunnus*, with a long paddle-shaped pectoral fin that reaches past the second dorsal fin and often to the second finlet. They are cosmopolitan in range, inhabiting the tropical and temperate waters of all oceans, including the Mediterranean Sea. The fork length is 140 cm (55 in.).

Herbert Boschung

Tundra An area supporting some vegetation beyond the northern limit of trees, between the upper limit of trees and the lower limit of perennial snow on mountains, and on the fringes of the Antarctic continent and its neighboring islands. The term is of Lapp or Russian origin, signifying treeless plains of northern regions. Biologists, and particularly plant ecologists, sometimes use the term tundra in the sense of the vegetation of the tundra landscape. Tundra has distinctive characteristics as a kind of landscape and as a biotic community, but these are expressed with great differences according to the geographic region.

Characteristically tundra has gentle topographic relief, and the cover consists of perennial plants a few inches to a few feet or a little more in height. The general appearance during the growing season is that of a grassy sward in the wetter areas, a matted spongy turf on mesic sites, and a thin or sparsely tufted lawn or lichen heath on dry sites. In winter, snow mantles most of the surface. By far, most tundra occurs where the mean annual temperature is below the freezing point of water, and perennial frost (permafrost) accumulates in the ground below the depth of annual thaw and to depths at least as great as 1600 ft (500 m). *See* PERMAFROST.

William S. Benninghoff

Tung tree The plant *Aleurites fordii*, a species of the spurge family (Euphorbiaceae). The tree, native to central and western China, is the source of tung oil. It has been grown successfully in the southern United States. The globular fruit has three to seven large, hard, rough-coated seeds containing the oil, which is expressed after the seeds have been roasted. Tung oil is used to produce a hard, quick-drying, superior varnish, which is less apt to crack than other kinds. The foliage, sap, fruit, and commercial tung meal contain a toxic saponin, which causes gastroenteritis in animals that eat it. *See* DRYING OIL; EUPHORBIALES; VARNISH.

Perry D. Strausbaugh; Earl L. Core

Tungsten A chemical element, W, atomic number 74, and atomic weight 183.84. Naturally occurring tungsten consists of five stable isotopes having the following mass numbers and relative abundances: 180 (0.14%), 182 (26.4%), 183 (14.4%), 184 (30.6%), and 186 (28.4%). Twelve radioactive isotopes ranging from 173 to 189 also have been characterized. *See* PERIODIC TABLE.

Tungsten crystallizes in a body-centered cubic structure in which the shortest interatomic distance is 274.1 picometers at 25°C (77°F). The pure metal has a lustrous, silver-white appearance. It possesses the highest melting point, lowest vapor pressure, and the highest tensile strength at elevated temperature of all metals. Some important physical properties of tungsten are compiled in the table.

Physical properties of tungsten

Property	Value
Melting point	3410 ± 20°C (6170 ± 36°F)
Boiling point	5700 ± 200°C (10,300 ± 360°F)
Density, 27°C (81°F)	19.3 g/cm^3 (11.2 oz/in.^{3})
Specific heat, 25°C (77°F)	0.032 cal/g-°C (0.13 J/g-°C)
Heat of fusion	52.2 ± 8.7 cal/g (218 ± 36 J/g)
Vapor pressure, 2027°C (3681°F)	6.4×10^{-12} atm (6.5×10^{-7} Pa)
3382°C (6120°F)	2.3×10^{-5} atm (2.3 Pa)
5470°C (9878°F)	0.53 atm (5.4×10^4 Pa)
Electrical resistivity, 27°C (81°F)	5.65 microhm-cm
1027°C (1881°F)	34.1
3027°C (5481°F)	103.3
Thermal conductivity, 27°C (81°F)	0.43 cal/cm-s-°C (1.8 J/cm-s-°C)
1027°C (1881°F)	0.27 (1.1)
Absorption cross section, 0.025-eV neutrons	18.5 ± 0.5 barns (18.5 ± 0.5 × 10^{-24} cm^2)

At room temperature tungsten is chemically resistant to water, oxygen, most acids, and aqueous alkaline solutions, but it is attacked by fluorine or a mixture of concentrated nitric and hydrofluoric acids.

Tungsten is used widely as a constituent in the alloys of other metals, since it generally enhances high-temperature strength. Several types of tool steels and some stainless steels contain tungsten. Heat-resistant alloys, also termed superalloys, are nickel-, cobalt-, or iron-based systems containing varying amounts (typically 1.5–25 wt %) of tungsten. Wear-resistant alloys having the trade name Stellites are composed mainly of cobalt, chromium, and tungsten. *See* ALLOY; HIGH-TEMPERATURE MATERIALS.

The major use of tungsten in the United States is in the production of cutting and wear-resistant materials. Tungsten carbides (representing 60% of total tungsten consumption) are used for cutting tools, mining and drilling tools, dies, bearings, and armor-piercing projectiles.

Unalloyed tungsten (25% of tungsten consumption) in the form of wire is used as filaments in incandescent and fluorescent lamps, and as heating elements for furnaces and heaters. Because of its high electron emissivity, thorium-doped (thoriated) tungsten wire is employed for direct cathode electronic filaments. Tungsten rods find use as lamp filament supports, electrical contacts, and electrodes for arc lamps.

Tungsten compounds (5% of tungsten consumption) have a number of industrial applications. Calcium and magnesium tungstates are used as phosphors in fluorescent lights and television tubes. Sodium tungstate is employed in the fireproofing of fabrics and in the preparation of tungsten-containing dyes and pigments used in paints and printing inks. Compounds such as WO_3 and WS_2 are catalysts for various chemical processes in the petroleum industry. Both WS_2 and WSe_2 are dry, high-temperature lubricants. Other applications of tungsten compounds have been made in the glass, ceramics, and tanning industries.

Miscellaneous uses of tungsten account for the remainder (2%) of the metal consumed.

Charles Kutal

Tunicata (Urochordata) A subphylum of marine animals of the Chordata. They are characterized by a perforated pharynx or branchial sac used for food collection, a dorsal notochord restricted to the tail of the larva (and the adult in one class), absence of mesodermal segmentation or a recognizable coelom, and secretion of an outer covering (the test or tunic) which contains large amounts of polysaccharides related to cellulose. Three classes are usually recognized: the sessile Ascidiacea (sea squirts or ascidians); planktonic Thaliacea (salps, doliolids, and pyrosomids); and Appendicularia, minute planktonic forms with tails living inside a specialized test or house adapted for filtering and food gathering. Approximately 2000 species of Tunicata are recognizable. The group is found in all parts of the ocean. Tunicates have little economic importance except as fouling organisms. A few species have pharmacological properties, and a few larger ascidians are used for food. *See* APPENDICULARIA (LARVACEA); ASCIDIACEA; CHORDATA; THALIACEA.

Ivan Goodbody

Tuning The process of adjusting the frequency of a vibrating system to obtain a desired result. In electronic circuits, there are a variety of frequency-determining elements. The most widely used is a combination of an inductance L (which stores energy in a magnetic field) and a capacitance C (which stores it in an electric field). The frequency of oscillation is determined by the rate of exchange of the energy between the two fields, and is inversely proportional to LC. Tuning is accomplished by adjusting the capacitor or the inductor until the desired frequency is reached. The desired frequency may be one that matches (resonates with) another frequency. Another purpose of tuning may be to match a frequency standard, as when setting an electronic watch to keep accurate time. The frequency-determining element in such watches, as well as in radio transmitters, digital computers, and other equipment requiring precise frequency adjustment, is a vibrating quartz crystal. The frequency of vibration of such crystals can be changed over a narrow range by adjusting a capacitor connected to it. *See* QUARTZ CLOCK; RESONANCE (ALTERNATING-CURRENT CIRCUITS).

Another function of tuning in electronics is the elimination of undesired signals. Filters for this purpose employ inductors and capacitors, or crystals. The filter is tuned to the frequency of the undesired vibration, causing it to be absorbed elsewhere in the circuit. *See* ELECTRIC FILTER.

Automatic tuning by electrical control is accomplished by a varactor diode. This is a capacitor whose capacitance depends on the direct-current (dc) voltage applied to it. The varactor serves as a portion of the capacitance of the tuned circuit. Its capacitance is controlled by a dc voltage applied to it by an associated circuit, the voltage and its polarity depending on the extent and direction of the mismatch between the desired frequency and the actual frequency. *See* VARACTOR. Donald G. Fink

Tuning fork A steel instrument consisting of two prongs and a handle which, when struck, emits a tone of fixed pitch. Because of their simple mechanical structure, purity of tone, and constant frequency, tuning forks are widely used as standards of frequency in musical acoustics. In its electrically driven form, a tuning fork serves to control electric circuits by producing

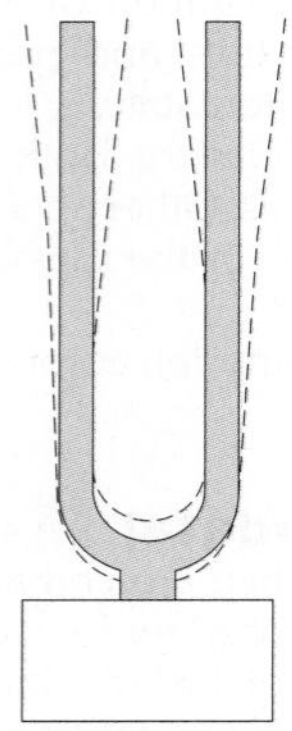

A tuning fork vibrating at its fundamental frequency.

frequency standards of high accuracy and stability. A tuning fork is essentially a transverse vibrator (see illustration). *See* VIBRATION. Lawrence E. Kinsler

Tunnel An underground space of substantial length, usually having a tubular shape. Tunnels can be either constructed or natural and are used as passageways, storage areas, carriageways, and utility ducts. They may also be used for mining, water supply, sewerage, flood prevention, and civil defense.

Tunnels are constructed in numerous ways. Shallow tunnels are usually constructed by burying sections of tunnel structures in trenches dug from the surface. This is a preferred method of tunneling as long as space is available and the operation will not cause disturbance to surface activities. Otherwise, tunnels can be constructed by boring underground. Short tunnels are usually bored manually or by using light machines. If the ground is too hard to bore, a drill-and-blast method is frequently used. For long tunnels, it is more economical and much faster to use tunneling boring machines which work on the full face (complete diameter of the opening) at the same time. In uniform massive rock formations without fissures or joints, tunnels can be bored without any temporary supports to hold up the tunnel crowns. However, temporary supports are usually required because of the presence of destabilizing fissures and joints in the rock mass. *See* DRILLING AND BORING, GEOTECHNICAL.

For tunnels to be constructed across bodies of water, an alternative to boring is to lay tunnel boxes directly on the prepared seabed. These boxes, made of either steel or reinforced concrete, are usually buried in shallow trenches dug for this purpose and covered by ballast so they will not be affected by the movement of the water. The joints between tunnel sections are made watertight by using rubber gaskets, and water is pumped out of the tunnel to make it ready for service. *See* CONCRETE; STEEL.

Za-Chieh Moh; Richard N. Hwang

Tunnel diode A two-terminal semiconductor junction device (also called the Esaki diode) which does not show rectification in the usual sense, but exhibits a negative resistance region at very low voltage in the forward-bias characteristic and a short circuit in the negative-bias direction.

This device is a version of the semiconductor *pn* junction diode which is made of a *p*-type semiconductor, containing mobile positive charges called holes (which correspond to the vacant electron sites), and an *n*-type semiconductor, containing mobile electrons (the electron has a negative charge). The densities of holes and electrons in the respective regions are made extremely high by doping a large amount of the appropriate impurities with an abrupt transition from one region to the other. In semiconductors, the conduction band for mobile electrons is separated from the valence band for mobile holes by an energy gap, which corresponds to a forbidden region. Therefore, a narrow transition layer from *n*-type to *p*-type, 5 to 15 nanometers thick, consisting of the forbidden region of the energy gap, provides a tunneling barrier. Since the tunnel diode exhibits a negative incremental resistance with a rapid response, it is capable of serving as an active element for amplification, oscillation, and switching in electronic circuits at high frequencies. The discovery of the diode, however, is probably more significant from the scientific aspect because it opened up a new field of research—tunneling in solids. *See* BAND THEORY OF SOLIDS; JUNCTION DIODE; NEGATIVE-RESISTANCE CIRCUITS; SEMICONDUCTOR; SEMICONDUCTOR DIODE; TUNNELING IN SOLIDS. Lloyd P. Hunter

Tunneling in solids A quantum-mechanical process which permits electrons to penetrate from one side to the other through an extremely thin potential barrier to electron flow. The barrier would be a forbidden region if the electron were treated as a classical particle. A two-terminal electronic device in which such a barrier exists and primarily governs the transport characteristic (current-voltage curve) is called a tunnel junction. *See* QUANTUM MECHANICS.

During the infancy of the quantum theory, L. de Broglie introduced the fundamental hypothesis that matter may be endowed with a dualistic nature—particles such as electrons, alpha particles, and so on, may also have the characteristics of waves. This hypothesis found expression in the definite form now known as the Schrödinger wave equation, whereby an electron or an alpha particle is represented by a solution to this equation. The nature of such solutions implies an ability to penetrate classically forbidden regions of negative kinetic energy and a probability of tunneling from one classically allowed region to another. The concept of tunneling, indeed, arises from this quantum-mechanical result. The subsequent experimental manifestations of this concept, such as high-field electron emission from cold metals, alpha decay, and so on, in the 1920s, can be regarded as one of the early triumphs of the quantum theory. *See* FIELD EMISSION; RADIOACTIVITY; SCHRÖDINGER'S WAVE EQUATION.

The tunnel diode (also called the Esaki diode), discovered in 1957 by L. Esaki, demonstrated the first convincing evidence of electron tunneling in solids. *See* TUNNEL DIODE.

Negative resistance phenomena can be observed in novel tunnel structures in semiconductors. Double tunnel barriers and periodic structures with a combination of semiconductors exhibit resonant tunneling and negative resistance effects. *See* SEMICONDUCTOR HETEROSTRUCTURES.

Tunneling had been considered to be a possible electron transport mechanism between metal electrodes separated by either a narrow vacuum or a thin insulating film usually made of metal oxides. In 1960, I. Giaever demonstrated that, if one or both of the metals were in a superconducting state, the current-voltage curve in such metal tunnel junctions revealed many details of that state.

In 1962, B. Josephson made a penetrating theoretical analysis of tunneling between two superconductors by treating the two superconductors and the coupling process as a single system, which would be valid if the insulating oxide were sufficiently thin, say 2 nanometers. His theory predicted the existence of a supercurrent, arising from tunneling of the bound electron pairs. This led to two startling conclusions: the dc and ac Josephson effects. The dc effect implies that a supercurrent may flow even if no voltage is applied to the junction. The ac effect implies that, at finite voltage V, there is an alternating component of the supercurrent which oscillates at a frequency of 483.6 MHz per microvolt of voltage across the junction, and is typically in the microwave range. *See* JOSEPHSON EFFECT. Leo Esaki

Tupelo A tree belonging to the genus *Nyssa* of the sour gum family, Nyssaceae. The most common species is *N. sylvatica*, variously called pepperidge, black gum, or sour gum, the authorized name being black tupelo. Tupelo grows in the easternmost third of the United States; southern Ontario, Canada; and Mexico.

The tree can be identified by the comparatively small, obovate, shiny leaves and by branches that develop at a wide angle from the axis. The fruit is a small blue-black drupe, a popular food for birds. The wood is yellow to light-brown and hard to split because of the twisted grain. Tupelo is used for boxes, baskets, and berry crates, and as backing on which veneers of rarer and more expensive woods are glued. It is also used for flooring, rollers in glass factories, hatters' blocks, and gunstocks. *See* MYRTALES. Arthur H. Graves; Kenneth P. Davis

Turbellaria A class of the phylum Platyhelminthes commonly known as the flatworms. These animals are chiefly free-living and have simple life histories. The bodies are elongate and flat to oval or circular in cross section. Their length ranges from less than 0.04 in. (1 mm) to several inches, but may exceed 20 in. (50 cm) in land planaria. Large forms are often brightly colored. This class, which numbers some 3400 described species, is ordinarily subdivided into 12 orders: Acoela, Catenulida, Haplopharyngida, Lecithoepitheliata, Macrostomida, Nemertodermatida, Neorhabdocoela, Polycladida, Prolecithophora, Proseriata, Temnocephalida, and Tricladida. Although widely distributed in fresh and salt water and moist soil, they are usually overlooked because of their generally small size, secretive habits, and inconspicuous color. *See* ACOELA; POLYCLADIDA; TRICLADIDA. E. Ruffin Jones

Turbidite A bed of sediment or sedimentary rock that was deposited from a turbidity current. The term turbidite is fundamentally genetic and interpretive in nature, rather than being a descriptive term (like common rock names). Turbidites are clastic sedimentary rocks, but they may be composed of silicic grains (quartz, feldspar, rock fragments) and therefore be a type of sandstone, or they may be composed of carbonate grains and therefore be a type of limestone. A geologist's description of a rock as a turbidite is actually an expression of an opinion that the rock was deposited by a turbidity current, rather than being a description of a particular type of rock. *See* TURBIDITY CURRENT.

No single feature of a deposit is sufficient to identify it as a turbidite and not all turbidites are marine. There are well-documented examples of modern turbidity currents and turbidites described from lakes. Although probably most turbidites were originally deposited in water of considerable depth (hundreds to thousands of meters), it is generally difficult to be specific about estimating the depth of deposition. The most that can be said is that (in most cases) there is no sign of sedimentary structures formed by the action of waves. *See* DEPOSITIONAL SYSTEMS AND ENVIRONMENTS; SEDIMENTARY ROCKS; SEDIMENTOLOGY. G. V. Middleton

Turbidity current A flow of water laden with sediment that moves downslope in an otherwise still body of water. The driving force of a turbidity current is obtained from the sediment, which renders the turbid water heavier than the clear water above. Turbidity currents occur in oceans, lakes, and reservoirs. They may be triggered by the direct inflow of turbid water, by wave action, by subaqueous slumps, or by anthropogenic activities such as dumping of mining tailings and dredging operations.

Turbidity currents are characterized by a well-defined front, also known as head, followed by a thinner layer known as the body of the current. They are members of a larger class of stratified flows known as gravity or density currents. Sediment can be entrained from or deposited on the bed, thus changing the total amount of sediment in suspension. A turbidity current must generate enough turbulence to hold its sediment in suspension. Under certain conditions, a turbidity current might erode its bed, pick up sediment, become heavier, accelerate, and pick up even more sediment, increasing its driving force in a self-reinforcing cycle akin to the formation of a snow avalanche. *See* DEPOSITIONAL SYSTEMS AND ENVIRONMENTS.

Turbidity currents constitute a major mechanism for the transport of fluvial, littoral, and shelf sediments onto the ocean floor. These flows are considered to be responsible for the scouring of submarine and sublacustrine canyons. These canyons are often of massive proportions and rival the Grand Canyon in scale. Below the mouths of most canyons, turbidity currents form vast depositional fans that have many of the features of alluvial fans built by rivers and constitute major hydrocarbon reservoirs. The sedimentary deposits created by turbidity currents, known as turbidites, are a major constituent of the geological record. *See* MARINE GEOLOGY; MARINE SEDIMENTS; SUBMARINE CANYON; TURBIDITE. Marcelo H. Garcia

Turbine A machine for generating rotary mechanical power from the energy in a stream of fluid. The energy, originally in the form of head or pressure energy, is converted to velocity energy by passing through a system of stationary and moving blades in the turbine. Changes in the magnitude and direction of the fluid velocity are made to cause tangential forces on the rotating blades, producing mechanical power via the turning rotor.

The fluids most commonly used in turbines are steam, hot air or combustion products, and water. Steam raised in fossil fuel-fired boilers or nuclear reactor systems is widely used in turbines for electrical power generation, ship propulsion, and mechanical drives. The combustion gas turbine has these applications in addition to important uses in aircraft propulsion. Water turbines are used for electrical power generation. Collectively, turbines drive over 95% of the electrical generating capacity in the world. *See* ELECTRIC POWER GENERATION; GAS TURBINE; HYDRAULIC TURBINE; STEAM TURBINE; TURBINE PROPULSION; TURBOJET.

Turbines effect the conversion of fluid to mechanical energy through the principles of impulse, reaction, or a mixture of the two. For the impulse turbine, high-pressure fluid at low velocity in the boiler is expanded through the stationary nozzle to low pressure and high velocity. The blades of the turning rotor reduce the velocity of the fluid jet at constant pressure, converting kinetic energy (velocity) to mechanical energy. *See* IMPULSE TURBINE.

For the reaction turbine, the nozzles are attached to the moving rotor. The acceleration of the fluid with respect to the nozzle causes a reaction force of opposite direction to be applied to the rotor. The combination of force and velocity in the rotor produces mechanical power. *See* REACTION TURBINE. Frederick G. Baily

Turbine engine subsystems A typical aircraft gas turbine engine is composed of an assemblage of individual turbine engine subsystems and components which function together to provide thrust to propel the aircraft and power for its onboard functioning.

Inlets. Virtually all operational aircraft gas turbine engines work with subsonic axial inlet Mach numbers at the fan or compressor inlet face. It is the function of the inlet to intercept the stream tube that contains the airstream that will be ingested by the engine at any flight speed, accommodating any power level of the engine and any flight attitude and to accelerate or diffuse that airstream with a minimum of pressure drop and distortion to the Mach number accepted by the fan or compressor or fan face.

Inlet particle separators. Helicopters can churn up dust that is sucked into the engine inlet. If allowed to enter the engine, this sand and dust will clog small air holes that supply vital cooling air to hot parts. Modern helicopter engines include inlet particle separators in which the sand and dust particles are extracted with a portion of the airflow by an ejector or a blower. Turboprop engines may include primitive separator systems to capture ingested birds, hailstones, or other foreign objects that might damage the engine. *See* HELICOPTER; TURBOPROP.

Fans and compressors. The vast majority of fans and compressors (whose function is to increase the pressure of the incoming flow of air) of aircraft gas turbine engines are axial-flow turbo-compressors. These devices consist of a series of individual stages, each comprising a rotating row (or cascade) of radially oriented airfoils (or rotor blades) followed by a stationary row of radially oriented airfoils (or stator blades). The rotating row of blades acts to increase the pressure and tangential velocity of the flow. The flow exiting the rotating row of blades is directed onto the stationary blade row which converts some of the tangential component of velocity imparted by the preceding rotor blade into a further increase in stage pressure rise.

Combustors. The combustion chamber or combustor of the aircraft gas turbine engine provides for the burning of fuel in the airflow exiting from the compressor, and for supplying the resulting stream of high-temperature, high-pressure products of combustion to the turbine. In modern engines the combustor is usually an annular chamber, in the middle of the core between the compressor exit and the turbine entry, surrounding the center shaft of the rotor. Fuel is introduced at the upstream end through injectors, which may exploit any of several mechanisms to condition the fuel for the combustion process.

Turbines. Turbines in aircraft gas turbine engines extract mechanical energy from the stream of high-pressure, high-temperature airflow exiting the combustor. There may be several turbines in series in a typical engine, each driving an individual spool of the compressor ahead of it, or a fan component, or an external load like a propeller or helicopter rotor.

Each stage of turbines in modern aircraft gas turbine engines generally includes a stationary row (or cascade) of radially oriented blades or nozzle passages which accelerate and turn the incoming flow and direct it onto the following rotating row of radially oriented airfoils (or rotor blades). *See* TURBINE.

Regenerators and recuperators. A regenerator is a heat exchanger used in a gas turbine engine to transfer heat from the waste heat in the exhaust stream to the air being discharged from the compressor, which enters the combustor. The net effect is to reduce the amount of fuel required in the combustor with a net improvement in the fuel economy of the engine. Although regenerators are common in stationary and surface vehicle gas turbines, they have not yet proven useful in aircraft propulsion because of their weight.

Exhaust nozzles. Simple jet engines and turbofan engines, which do not have an external propulsor such as a propeller, derive their thrust from the acceleration of the airflow through the turbine engine. This dictates that the energy generated in the engine as high-pressure, high-temperature airflow must be used to accelerate the airflow through an exhaust nozzle at the rear of the engine. *See* NOZZLE.

Afterburners. An afterburner may be added to low-bypass turbofans and turbojets as a means of achieving thrust augmentation by burning fuel in the exhaust of the engine before it exits the engine through the exhaust nozzle. Afterburners are extremely inefficient in fuel usage and are applied only in circumstances requiring short-duration bursts of thrust augmentation for extreme conditions such as takeoff, transonic acceleration in supersonic aircraft, and combat maneuvers in military aircraft. *See* AFTERBURNER.

Thrust reversers. Aircraft gas turbine engines may include provision for reversing the direction of flow of the airstream handled by the propulsor in order to provide braking thrust for the aircraft, primarily during its landing roll on very short runways or runways with icy surfaces. For turboprops and ultrahigh-bypass engines, this is usually accomplished by varying the pitch angle of the propeller or of the fan blading, so that the propulsor pumps the air in the direction opposite to that required for forward flight. Simple jet engines may include a target reverser, in which a barrier is deployed in the jet stream exiting from the rear of the engine to redirect the jet in a direction with a forward component of velocity.

Mixers and noise suppressors. Jet noise generally involves a broad frequency band and is dealt with by providing a highly convoluted exhaust nozzle, making a transition from the annular duct of the turbomachinery exit to a multijet configuration with an extended perimeter. The action of the nozzle's extended perimeter is to intensify the mixing and hence to shorten the length of the noise-generating shear layer between the jet and the lower-speed airflow external to the nozzle. Supersonic transport engines use low-bypass turbofans or turbojet engines with much higher jet velocities that require such suppression devices. *See* AERODYNAMIC SOUND; TURBOJET.

Thrust deflectors. To achieve increased agility and maneuverability, the low-bypass turbofans or turbojet engines of military combat aircraft may include provision for mechanically deflecting the jet nozzle from the direction of flight to a small angle upward or downward to provide a component of thrust in the direction of climb or descent. In a more extreme application, the thrust deflectors may be designed to deflect the exhaust stream to a full 90° or more to facilitate vertical takeoff and landing. *See* VERTICAL TAKEOFF AND LANDING (VTOL).

Control systems. The control system is centered on the main fuel control, which accepts a variety of signals from sensors situated around the engine that provide data indicative of the status of the engine's operation (such as measurements of key pressures and temperatures within the engine, rotative speeds of the engine's spools, actual positions of the engine's variable geometry, torque in the engine's output shaft, and rates of fuel flow to the engine's combustion systems); data indicative of the aircraft's operating conditions; and a signal from the pilot's throttle or power demand lever. The control must also provide for starting and shutdown of the engine; must protect the engine from surge, overspeed, overtemperature, and overtorque; and in the event of any failure, must provide for residual power or for safe shutdown. *See* CONTROL SYSTEMS. Fredric F. Ehrich

Turbine propulsion Propulsion of a vehicle by means of a gas turbine. Gas turbines have come to dominate most areas of common carrier aircraft propulsion, have made significant inroads into the propulsion of surface ships, and are being incorporated into military tanks.

The primary power producer common to all gas turbines used for propulsion is the core or gas generator, operating on a continuous flow of air as working fluid. The air is compressed in a rotating compressor, heated at constant pressure in a combustion chamber burning a liquid hydrocarbon fuel, and expanded through a core turbine which drives the compressor. This manifestation of the Brayton thermodynamic cycle generates a continuous flow of high-pressure, high-temperature gas which is the primary source of power for a large variety of propulsion schemes. The turbine is generally run as an open cycle; that is, the airflow is ultimately exhausted to the atmosphere rather than being recycled to the inlet. *See* BRAYTON CYCLE.

The residual energy available in the high-temperature, high-pressure airstream exiting from the core is used for propulsion in a variety of ways. For traction-propelled vehicles (buses, trucks, automobiles, military tanks, and most railroad locomotives), the core feeds a power turbine which extracts the available energy from the core exhaust and provides torque to a high-speed drive shaft as motive power for the vehicle. With a free-turbine arrangement, this power turbine is a separate shaft, driving at a speed not mechanically linked to the core speed. With a fixed turbine, this power turbine is on the same shaft as the core turbine, and must drive at the same speed as the core spool. In traction vehicles the power turbine generally drives through a transmission system which affords a constant- or a variable-speed reduction to provide the necessary torque-speed characteristics to the traction wheels.

Aircraft, ships, and high-speed land vehicles, which cannot be driven by traction, are propelled by reaction devices. Some of the ambient fluid around the vehicle (that is, the water for most ships, and the air for all other vehicles) is accelerated by some turbomachinery (a ship propeller, aircraft propeller, helicopter rotor, or a fan integrated with the core to constitute a turbofan engine). The reaction forces on this propulsion turbomachinery, induced in the process of accelerating the ambient flow, provide the propulsion thrust to the vehicle. In all these cases, motive power to the propeller or fan is provided by a power turbine extracting power from the gas generator exhaust. In the case of a jet engine, exhaust from the gas generator is accelerated through a jet nozzle, so that the reaction thrust is evolved in the gas generator rather than in an auxiliary propeller or fan. Indeed, in turboprop and turbofan engines, both forms of reaction thrust (from the stream accelerated by the propeller or fan and from the stream accelerated by the core and not fully extracted by the power turbine) are used for propulsion. *See* GAS TURBINE; TURBOFAN; TURBOJET; TURBOPROP.

Fredric F. Ehrich

Turbocharger An air compressor or supercharger on an internal combustion piston engine that is driven by the engine exhaust gas to increase or boost the amount of fuel that can be burned in the cylinder, thereby increasing engine power and performance. On an aircraft piston engine, the turbocharger allows the engine to retain its sea-level power rating at higher altitudes despite a decrease in atmospheric pressure. *See* RECIPROCATING AIRCRAFT ENGINE; SUPERCHARGER.

The turbocharger is a turbine-powered centrifugal supercharger. It consists of a radial-flow compressor and turbine mounted on a common shaft. The turbine uses the energy in the exhaust gas to drive the compressor, which draws in outside air, precompresses it, and supplies it to the cylinders at a pressure above atmospheric pressure.

Common turbocharger components include the rotor assembly, bearing housing, and compressor housing. The shaft bearings usually receive oil from the engine lubricating system. Engine coolant may circulate through the housing to aid in cooling. *See* ENGINE COOLING; INTERNAL COMBUSTION ENGINE.

Donald L. Anglin

Turbodrill A rotary tool used in drilling oil or gas wells in which the bit is rotated by a turbine motor inside the well. The principal difference between rotary and turbodrilling lies in the manner in which power is applied to the rotating bit or cutting tool. In the rotary method, the bit is attached to a drill pipe rotated through power supplied on the surface. In the turbodrill method, power is generated at the bottom of the hole by a mud-operated turbine. *See* ROTARY TOOL DRILL.

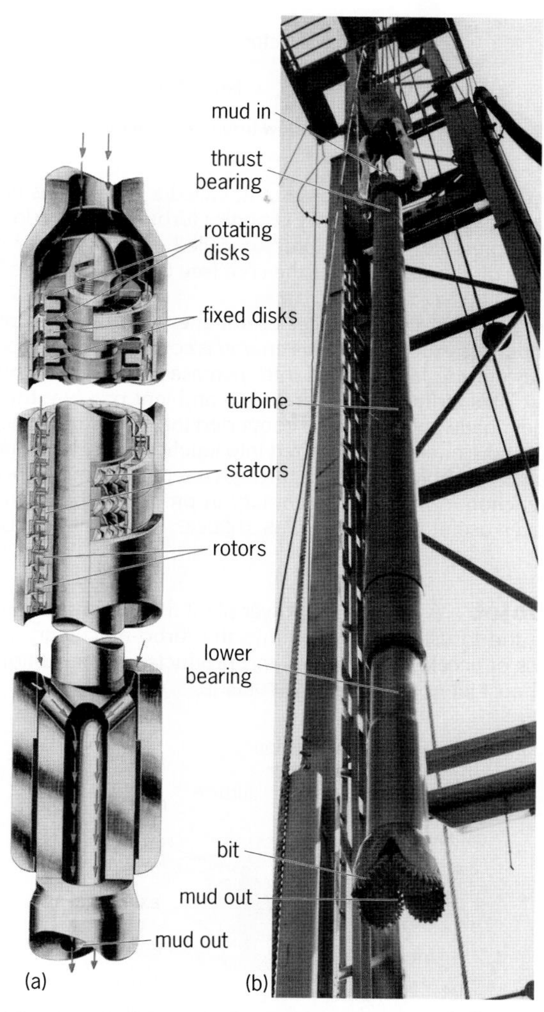

The components of a turbodrill. (*a*) Cutaway view of turbine, bearings, and bit. (*b*) Drill string suspended above hole. (*Dresser Industries*)

The turbodrill (see illustration) consists of four basic components: the upper, or thrust, bearing; the turbine; the lower bearing; and the bit. In operation, mud is pumped through the drill pipe, passing through the thrust bearing and into the turbine. In the turbine, stators attached to the body of the tool divert the mud flow onto rotors attached to the shaft. This causes the shaft, which is connected to the bit, to rotate. The mud passes through a hollow part of the shaft in the lower bearing and through the bit, as in rotary drilling, to remove cuttings, cool the bit, and perform the other functions of drilling fluid. Capacity of the mud pump, which is the power source, determines rotational speed. *See* OIL AND GAS WELL DRILLING; TURBINE.

Ade L. Ponikvar

Turbofan An air-breathing aircraft gas turbine engine with operational characteristics between those of the turbojet and the turboprop. Like the turboprop, the turbofan consists of a

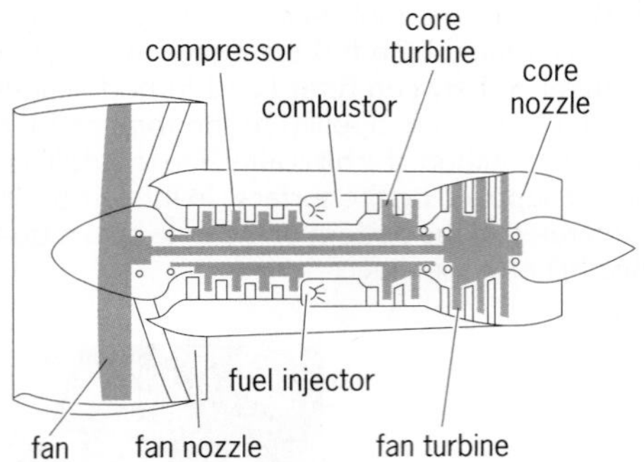

High-bypass, separate-flow turbofan configuration.

compressor-combustor-turbine unit, called a core or gas generator, and a power turbine. This power turbine drives a low- or medium-pressure-ratio compressor, called a fan, some or most of whose discharge bypasses the core (see illustration). *See* TURBOJET; TURBOPROP.

The gas generator produces useful energy in the form of hot gas under pressure. Part of this energy is converted by the power turbine and the fan it drives into increased pressure of the fan airflow. This airflow is accelerated to ambient pressure through a fan jet nozzle and is thereby converted into kinetic energy. The residual core energy is converted into kinetic energy by being accelerated to ambient pressure through a separate core jet nozzle. The reaction in the turbomachinery in producing both streams produces useful thrust. *See* GAS TURBINE; TURBINE PROPULSION.

Fredric F. Ehrich

Turbojet A gas turbine power plant used to propel aircraft, where the thrust is derived within the turbo-machinery in the process of accelerating the air and products of combustion out an exhaust jet nozzle. *See* GAS TURBINE.

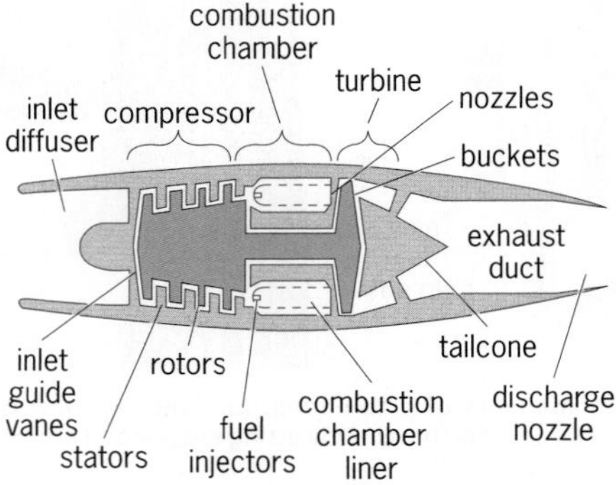

Basic turbojet engine with axial-flow components.

In its most elementary form (see illustration), the turbojet operates on the gas turbine or Brayton thermodynamic cycle. The working fluid, air drawn into the inlet of the engine, is first compressed in a turbo-compressor with a pressure ratio of typically 10:1 to 20:1. The high-pressure air then enters a combustion chamber, where a steady flow of a hydrocarbon fuel is introduced in either spray or vapor form and burned continuously at constant pressure. The exiting stream of hot high-pressure air, at an average temperature whose maximum value may range typically from 1800 to 2800°F (980 to 1540°C), is then expanded through a turbine, where energy is extracted to power the compressor. Because heat had been added to the air at high pressure, there is a surplus of energy left in the stream of combustion products that exits from the turbine and that can be harnessed for propulsion. *See* BRAYTON CYCLE; GAS TURBINE.

Turbojets have retained a small niche in the aircraft propulsion spectrum, where their simplicity and low cost are of paramount importance, such as in short-range expendable military missiles, or where their light weight may be an overriding consideration, such as for lift jets in prospective vertical takeoff and landing aircraft. *See* AIRCRAFT PROPULSION; TURBINE PROPULSION; VERTICAL TAKEOFF AND LANDING (VTOL).

Fredric F. Ehrich

Turboprop A gas turbine power plant producing shaft power to drive a propeller or propellers for aircraft propulsion. Because of its high propulsive efficiency at low flight speeds, it is the power plant of choice for short-haul and low-speed transport aircraft where the flight speeds do not exceed Mach 0.5–0.6. Developments in high-speed, highly loaded propellers have extended the range of propellers to flight speeds up to Mach 0.8–0.9, and there are prospects of these extremely efficient prop-fans assuming a much larger role in powering high-speed transport aircraft. *See* GAS TURBINE.

As with all gas turbine engines, the basic power production in the turboprop is accomplished in the gas generator or core of the engine, where a steady stream of air drawn into the engine inlet is compressed by a turbocompressor. The high-pressure air is next heated in a combustion chamber by burning a steady stream of hydrocarbon fuel injected in spray or vapor form. The hot, high-pressure air is then expanded in a turbine that is mounted on the same rotating shaft as the compressor and supplies the energy to drive the compressor. By virtue of the air having been heated at higher pressure, there is a surplus of energy in the turbine that may be extracted in additional turbine stages to drive a useful load, in this case a propeller or propellers.

A large variety of detailed variations are possible within the core. The compressor may be an axial-flow type, a centrifugal (that is, radial-flow) type, or a combination of stages of both types (that is, an axi-centrifugal compressor). In modern machines, the compressor may be split in two sections (a low-pressure unit followed by a high-pressure unit), each driven by its own turbine through concentric shafting, in order to achieve very high compression ratios otherwise impossible in a single spool. *See* AIRCRAFT PROPULSION; COMPRESSOR; PROPELLER (AIRCRAFT); TURBINE PROPULSION.

Fredric F. Ehrich

Turbulent flow A fluid motion in which velocity, pressure, and other flow quantities fluctuate irregularly in time and space. The illustration shows a slice of a water jet emerging from a circular orifice into a tank of still water. A small amount of fluorescent dye mixed in the jet makes it visible when suitably illuminated by laser light, and tags the water entering the tank. There is a small region close to the orifice where the dye concentration does not vary with position, or with time at a given position. This represents a steady laminar state. Generally in laminar motion, all variations of flow quantities, such as dye concentration, fluid velocity, and pressure, are smooth and gradual in time and space. Farther downstream, the jet undergoes a transition to a new state in which the eddy patterns are complex, and flow quantities (including vorticity) fluctuate randomly in time and three-dimensional space. This is the turbulent state. *See* JET FLOW; LAMINAR FLOW.

Turbulence occurs nearly everywhere in nature. It is characterized by the efficient dispersion and mixing of vorticity, heat, and contaminants. In flows over solid bodies such as airplane wings or turbine blades, or in confined flows through ducts and pipelines, turbulence is responsible for increased drag and heat transfer. Turbulence is therefore a subject of great engineering interest. On the other hand, as an example of collective interaction of many coupled degrees of freedom, it is also a subject at the forefront of classical physics. *See* DEGREE OF FREEDOM (MECHANICS); DIFFUSION; HEAT TRANSFER; PIPE FLOW; PIPELINE.

The illustration demonstrates the principal issues associated with turbulent flows. The first is the mechanism (or mechanisms)

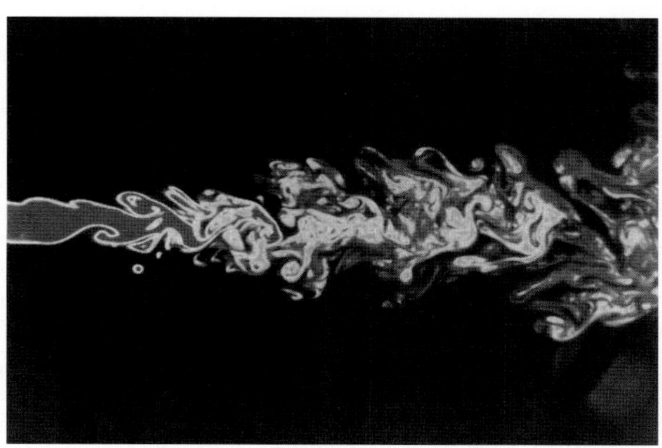

Two-dimensional image of an axisymmetric water jet, obtained by the laser-induced fluorescence technique. (*From R. R. Prasad and K. R. Sreenivasan, Measurement and interpretation of fractal dimension of the scalar interface in turbulent flows, Phys. Fluids A, 2:792–807, 1990*)

responsible for transition from the steady laminar state to the turbulent state. A second issue concerns the description of fully developed turbulence typified by the complex state far downstream of the orifice. Finally, it is of technological importance to be able to alter the flow behavior to suit particular needs. Less is known about eddy motions on the scale of centimeters and millimeters than about atomic structure on the subnanometer scale, reflecting the complexity of the turbulence problem. *See* NAVIER-STOKES EQUATION.

Origin of turbulence. A central role in determining the state of fluid motion is played by the Reynolds number. In general, a given flow undergoes a succession of instabilities with increasing Reynolds number and, at some point, turbulence appears more or less abruptly. It has long been thought that the origin of turbulence can be understood by sequentially examining the instabilities. This sequence depends on the particular flow and, in many circumstances, is sensitive to a number of details. A careful analysis of the perturbed equations of motion has resulted in a good understanding of the first two instabilities in a variety of circumstances. *See* REYNOLDS NUMBER.

Fully developed turbulence. Quite often in engineering, the detailed motion is not of interest, but only the long-time averages or means, such as the mean velocity in a boundary layer, the mean drag of an airplane or pressure loss in a pipeline, or the mean spread rate of a jet. It is therefore desirable to rewrite the Navier-Stokes equations for the mean motion. The basis for doing this is the Reynolds decomposition, which splits the overall motion into the time mean and fluctuations about the mean. These macroscopic fluctuations transport mass, momentum, and matter (in fact, by orders of magnitude more efficiently than molecular motion), and their overall effect is thus perceived to be in the form of additional transport or stress. This physical effect manifests itself as an additional stress (called the Reynolds stress) when the Navier-Stokes equations are rewritten for the mean motion (the Reynolds equations). The problem then is one of prescribing the Reynolds stress, which contains the unknown fluctuations in quadratic form. A property of turbulence is that the Reynolds stress terms are comparable to the other terms in the Reynolds equation, even when fluctuations are a small part of the overall motion. An equation for the Reynolds stress itself can be obtained by suitably manipulating the Navier-Stokes equations, but this contains third-order terms involving fluctuations, and an equation for third-order terms involves fourth-order quantities, and so forth. This is the closure problem in turbulence. The Navier-Stokes equations are themselves closed, but the presence of nonlinearity and the process of averaging result in nonclosure.

Given this situation, much of the progress in the field has been due to (1) exploratory experiments and numerical simulations of the Navier-Stokes equations at low Reynolds numbers; and (2) plausible hypotheses in conjunction with dimensional reasoning, scaling arguments, and their experimental verification.

Control of turbulent flows. Some typical objectives of flow control are the reduction of drag of an object such as an airplane wing, the suppression of combustion instabilities, and the suppression of vortex shedding behind bluff bodies. Interest in flow control has been stimulated by the discovery that some turbulent flows possess a certain degree of spatial coherence at large scales. Successful control has also been achieved through the reduction of the skin friction on a flat plate by making small longitudinal grooves, the so-called riblets, on the plate surface, imitating shark skin. *See* FLUID MECHANICS. K. R. Sreenivasan

Turmeric A dye or a spice obtained from the plant *Curcuma longa*, which belongs to the ginger family (Zingiberaceae). It is a stout perennial with short stem, tufted leaves, and short, thick rhizomes which contain the colorful condiment. As a natural dye, turmeric is orange-red or reddish brown, but it changes color in the presence of acids or bases. As a spice, turmeric has a decidedly musky odor and a pungent, bitter taste. It is an important item in curry and is used to flavor and color butter, cheese, pickles, and other food. *See* SPICE AND FLAVORING; ZINGIBERALES.

Perry D. Strausbaugh; Earl L. Core

Turn and bank indicator A combination instrument which provides an aircraft pilot with two distinct pieces of information: the aircraft's rate of turn about the vertical axis, and the relationship between this rate and the aircraft's angle of bank. It is also known as the needle and ball indicator or the turn and slip indicator. The turn needle is operated by a gyroscope and indicates the rate at which the aircraft is turning about the vertical axis in degrees per second. In a turn, gyroscopic precession causes the rotor to tilt in the direction opposite the turn with a magnitude proportional to the turn rate. A mechanical linkage converts this precession to reversed movement of a turn needle, thus indicating proper turn direction. *See* GYROSCOPE.

The bank or slip indicator is a simple inclinometer consisting of a curved glass tube containing fluid and a black ball bearing which is free to move in the fluid. The indicator is actually a balance indication, showing the relationship between the rate of turn and the angle of bank of the aircraft. *See* AIRCRAFT INSTRUMENTATION; ROTATIONAL MOTION. Grady W. Wilson

Turnbuckle A device for tightening a rod or wire rope. Its parts are a sleeve with a screwed connection at one end and a swivel at the other or, more commonly, a sleeve with screwed connections of opposite hands (left and right) at each end so that by turning the sleeve, the connected parts will be drawn together, taking up slack and producing tension (see illustration). Types of

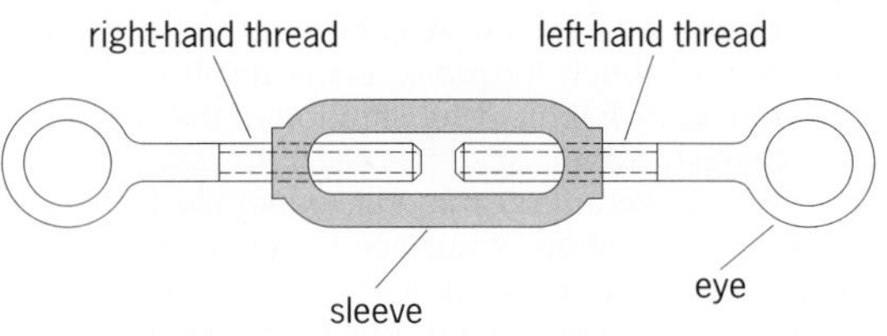

Turnbuckle with eyes.

ends are hook, eye, and clevis. The turn-buckle can be connected at any convenient place in the rod or rope, and several may be used in series if required. Paul H. Black

Turning (woodworking) The shaping of wood by rotating it in a lathe and cutting it with a chisel. The lathe consists essentially of a bed on which are mounted a headstock, a tailstock, and a tool rest (see illustration). The headstock is rotated by a motor and holds one end of the wood to be turned. The tailstock holds the other end of the wood, allowing it to rotate freely. The tool rest provides a fixed guide along which the operator can handle the chisels if the turning is by hand, or along which the tool is driven if the turning is mechanized. *See* WOODWORKING.

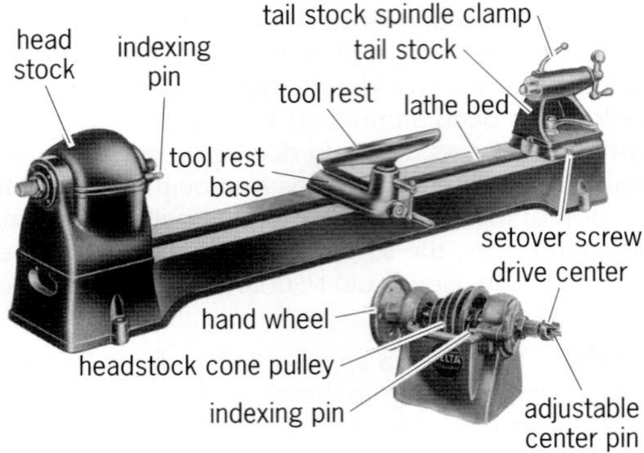

Wood-turning lathe and detail of headstock. (*Delta*)

Alan H. Tuttle

Turnip The plant *Brassica rapa*, or *B. campestris* var. *rapa*, a cool-season, hardy crucifer of Asiatic origin belonging to the order Capparales which is grown for its enlarged root and its foliage, which are eaten cooked as a vegetable. Popular white-fleshed varieties (cultivars) grown for their roots are Purple Top Globe and White Milan; Yellow Globe and Golden Ball are common yellow-fleshed varieties. Shogoin is a popular variety grown principally in the southern United States for turnip greens. Principal areas of production in the United States are in the South. *See* CAPPARALES.

H. John Carew

Turquoise A mineral of composition $CuAl_6(PO_4)_4(OH)_8 \cdot 5H_2O$ in which considerable ferrous ion (Fe^{2+}) may substitute for copper. Ferric ion (Fe^{3+}) may also substitute for part or all of the aluminum (Al), forming a complete chemical series from turquoise to chalcosiderite [$CuFe_6(PO_4)_4(OH)_8 \cdot 5H_2O$]. Turquoise with a strong sky-blue or bluish-green to apple green color is easily recognized, and such material is commonly used as a gem. Some variscite, of composition $AlPO_4 \cdot 2H_2O$ with minor chemical substitutions of Fe^{3+} and or chromium ion (Cr^{3+}) for aluminum and with a soft, clear green color, may be marketed as green turquoise. *See* GEM.

Most turquoise is massive, dense, and cryptocrystalline to fine-granular. It commonly occurs as veinlets or crusts and in stalactitic or concretionary shapes. It has a hardness on the Mohs scale of about 5 to 6 and a vitreous to waxy luster. The distinctive light blue coloration of much turquoise is the result of the presence of cuprous ion (Cu^{2+}); limited substitution of the copper by Fe^{2+} produces greenish colors. *See* HARDNESS SCALES.

Turquoise is a secondary mineral, generally formed in arid regions by the interaction of surface waters with high-alumina igneous or sedimentary rocks. It occurs most commonly as small veins and stringers traversing more or less decomposed volcanic rocks. Since the times of antiquity, turquoise of very fine quality has been produced from a deposit in Persia (now Iran) near Nishapur. It occurs also in Siberia, Turkistan, China, the Sinai Peninsula, Germany, and France.

The southwestern United States has been a major source of turquoise, especially the states of Nevada, Arizona, New Mexico, and Colorado. Extensive deposits in the Los Cerillos Mountains, near Santa Fe, New Mexico, were mined very early by Native Americans and were a major early source of gem turquoise. However, much of the gem-quality turquoise has been depleted in the Southwest.

Cornelis Klein

Twilight The period between sunset and darkness in the evening, and between darkness and sunrise in the morning. The characteristic light is caused by atmospheric scattering, which transmits sunlight to the observer for some time after sunset and before sunrise. It depends geometrically on latitude, longitude, and elevation of the observer, and on the time of year. Physically it depends also on local conditions, particularly the weather.

Gerald M. Clemence; Jay M. Pasachoff

Twinkling stars A phenomenon by which light from the stars, as it passes through fluctuations in the Earth's atmosphere, is rapidly modulated and redirected to make the starlight appear to flicker. The twinkling phenomenon affects all wavelengths that manage to penetrate the Earth's atmosphere, from the visible to the radio wavelengths. At visible wavelengths, atmospheric fluctuations are caused predominantly by temperature irregularities along the line of sight. Such irregularities introduce slight changes in the index of refraction of air, and these changes affect light waves in two ways: they modulate the intensity of the light, and they deflect the light waves in one direction and then another. At radio wavelengths, electron density irregularities in the ionosphere modulate and redirect radio waves. *See* REFRACTION OF WAVES.

The twinkling phenomenon is of utmost interest to astronomers who view the skies from ground-based telescopes. While modulation variations are present, it is the deflection of light that causes the most serious problems. The composite star image produced by a large telescope is a blurry circle that results when the randomly deflected light waves are added together in an extended time exposure. To diminish atmospheric effects, telescopes are built on high mountains, and are placed at least 30–45 m (100–150 ft) above the ground. *See* ASTRONOMICAL OBSERVATORY.

To completely remove the twinkling effects of the atmosphere, there are two alternatives. The first is to place a telescope in orbit above the atmosphere, as with the Hubble Space Telescope. The second alternative is to monitor the random deflections of the atmosphere and, within the telescope, to bend the deflected light back onto its original path. This optical technique is given the name adaptive optics. *See* ADAPTIVE OPTICS; HUBBLE SPACE TELESCOPE.

Laird A. Thompson

Twinning (crystallography) A process in which two or more crystals, or parts of crystals, assume orientations such that one may be brought to coincidence with the other by reflection across a plane or by rotation about an axis. Crystal twins represent a particularly symmetric kind of grain boundary; however, the energy of the twin boundary is much lower than that of the general grain boundary because some of the atoms in the twin interface are in the correct positions relative to each other. *See* GRAIN BOUNDARIES.

Robb M. Thomson

Twins (human) Two babies born to a mother at one birth. There are two types of twins, monozygotic and dizygotic. Members of a twin pair are called co-twins.

Controversy surrounding the definition of a twin arose with the advent of reproductive technologies enabling the simultaneous fertilization of eggs, with separate implantation. The unique "twinlike" relationships that would result between parents and cloned children (who would be genetically identical to their parents) also challenge current conceptions of twinship. Monozygotic twins are clones (genetically identical individuals derived from a single fertilized egg), but parents and cloned children would not be twins for several reasons, such as their differing prenatal and postnatal environments. *See* REPRODUCTIVE TECHNOLOGY.

Monozygotic twins. The division of a single fertilized egg (or zygote) between 1 and 14 days postconception results in monozygotic twins. They share virtually all their genes and, with very rare exception due to unusual embryological events, are of the same sex. A common assumption is that because monozygotic co-twins have a shared heredity, their behavioral or physical differences are fully explained by environmental factors. However, monozygotic twins are never exactly alike in any measured trait, and may even differ for genetic reasons.

Sometimes chromosomes fail to separate after fertilization, causing some cells to contain the normal chromosome number (46) and others to contain an abnormal number. This process, called mosaicism, results in monozygotic co-twins who differ in chromosomal constitution. There are several other intriguing variations of monozygotic twinning. Splitting of the zygote after day 7 or 8 may lead to mirror-image reversal in certain traits, such as handedness or direction of hair whorl. The timing of zygotic division has also been associated with placentation. Monozygotic twins resulting from earlier zygotic division have separate placentae and fetal membranes (chorion and amnion), while monozygotic twins resulting from later zygotic division share some or all of these structures. Should the zygote divide after 14 days, the twins may fail to separate completely. This process, known as conjoined twinning, occurs in approximately 1 monozygotic twin birth out of 200. The many varieties of conjoined twins differ as to the nature and extent of their shared anatomy. Approximately 70% of such twins are female. There do not appear to be any predisposing factors to conjoined twinning. *See* MOSAICISM.

Dizygotic twins. Dizygotic twins result when two different eggs undergo fertilization by two different spermatozoa, not necessarily at the same time. Dizygotic twins share, on average, 50% of their genes, by descent, so that the genetic relationship between dizygotic co-twins is exactly the same as that of ordinary brothers or sisters. Dizygotic twins may be of the same or opposite sex, outcomes that occur with approximately equal frequency.

There are some unusual variations of dizygotic twinning. There is the possibility of polar body twinning, whereby divisions of the ovum prior to fertilization by separate spermatozoa could result in twins whose genetic relatedness falls between that of monozygotic and dizygotic twins. Blood chimerism, another variation, refers to the presence of more than one distinct red blood cell population, derived from two zygotes, and has been explained by connections between two placentae. In humans, chimerism can occur in dizygotic twins. Superfecundation is the conception of dizygotic twins following separate fertilizations, usually within several days, in which case each co-twin could have a different father. Superfetation, which refers to multiple conceptions occurring several weeks or even one month apart, may be evidenced by delivery of full-term infants separated by weeks or months and by the birth or abortion of twin infants displaying differential developmental status. *See* FERTILIZATION (ANIMAL); OOGENESIS.

Epidemiology. According to conventional twinning rates, monozygotic twins represent approximately one-third of twins born in Caucasian populations and occur at a rate of 3–4 per 1000 births. The biological events responsible for monozygotic twinning are not well understood. It is generally agreed that monozygotic twinning occurs randomly and not as a genetically transmitted tendency. Some recent evidence from Sweden suggests an increased tendency for mothers who are monozygotic twins to bear same-sex twins themselves; further work will be needed to resolve this question.

Dizygotic twinning represents approximately two-thirds of twins born in Caucasian populations. The dizygotic twinning rate is lowest among Asian populations (2 per 1000 births), intermediate among Caucasian populations (8 per 1000 births), and highest among African populations (50 per 1000 births in parts of Nigeria). The natural twinning rate increases with maternal age, up to between 35 and 39 years, and then declines. Dizygotic twinning has also been linked to increased parity, or the number of children to which a woman has previously given birth. Mothers of dizygotic twins are also significantly taller and heavier, on average, than mothers of monozygotic twins and singletons. Dizygotic twinning appears to be genetically influenced, although the pattern of transmission within families is unknown.

Twinning rates have risen dramatically since about 1980 mainly due to advances in fertility treatments (for example, in vitro fertilization and ovulation induction), but also due to delays in child-bearing. The increase has mainly involved dizygotic twinning, in which multiple ovulation and maternal age are key factors.

Nancy L. Segal

Two-phase flow

Two-phase flow The simultaneous flow of two phases or two immiscible liquids within common boundaries. Two-phase flow has a wide range of engineering applications such as in the power generation, chemical, and oil industries. Flows of this type are important for the design of steam generators (steam-water flow), internal combustion engines, jet engines, condensers, cooling towers, extraction and distillation processes, refrigeration systems, and pipelines for transport of gas and oil mixtures.

The most important characteristic of two-phase flow is the existence of interfaces, which separate the phases, and the associated discontinuities in the properties across the phase interfaces. Because of the deformable nature of gas-liquid and liquid-liquid interfaces, a considerable number of interface configurations are possible. Consequently, the various heat and mass transfers that occur between a two-phase mixture and a surrounding surface, as well as between the two phases, depend strongly on the two-phase flow regimes. Multiphase flow, when the flow under consideration contains more than two separate phases, is a natural extension of these principles. *See* FLUID MECHANICS; INTERFACE OF PHASES.

From a fundamental point of view, two-phase flow may be classified according to the phases involved as (1) gas-solid mixture, (2) gas-liquid mixture, (3) liquid-solid mixture, and (4) two-immiscible-liquids mixture. *See* GAS; LIQUID; PHASE EQUILIBRIUM.

Industrial applications of two-phase flow include systems that convert between phases, and systems that separate or mix phases without converting them (adiabatic systems). Many of the practical cycles used to convert heat to work use a working fluid. In two or more of the components of these cycles, heat is either added to or removed from the working fluid, which may be accompanied by a phase-change process. Examples of these applications include steam generators, evaporators, and condensers, air conditioning, and refrigeration systems, and steam power plants.

In adiabatic systems, the process of phase mixing or separation occurs without heat transfer or phase change. Examples of these systems are airlift pumps, pipeline transport of gas-oil mixtures, and gas-pulverization of solid particles. *See* ADIABATIC PROCESS.

H. H. Bruun; F. Hamad; B. K. Pierscionek

Tylenchida

Tylenchida An order of nematodes in which the labial region is variable and may be distinctly set off or smoothly rounded and well developed; the hexaradiate symmetry is most often retained or discernible. The hollow stylet is the product of the cheilostome (conus, "guiding apparatus," and framework) and the esophastome (shaft and knobs). Throughout the order and its suborders the stylet may be present or absent and may or may not be adorned with knobs. The variable esophagus is most often divisible into the corpus, isthmus, and glandular posterior bulb. The corpus is divisible into the procorpus and metacorpus. The metacorpus is generally valved but may not occur in some females and males, and the absence is characteristic of some taxa.

The orifice of the dorsal esophageal gland opens either into the anterior procorpus or just anterior to the metacorporal valve. The excretory system is asymmetrical, and there is but one longitudinal collecting tubule. Females have one or two genital branches; when only one branch is present, it is anteriorly directed. Except for sex-reversed males there is only one genital branch. Males may have one (=phasmid) or more caudal papillae. The spicules are always paired and variable in shape; they may or may not be accompanied by a gubernaculum. The order comprises five superfamilies: Tylenchoidea, Criconematoidea, Aphelenchoidea, Aphelenchoidoidea, and Sphaerularoidea. *See* NEMATA (NEMATODA); PLANT PATHOLOGY. Armand R. Maggenti

Tyndall effect Visible scattering of light along the path of a beam of light as it passes through a system containing discontinuities. The luminous path of the beam of light is called a Tyndall cone. An example is shown in the illustration. In colloidal systems the brilliance of the Tyndall cone is directly dependent on the magnitude of the difference in refractive index between the particle and the medium.

The luminous light path known as the Tyndall cone or Tyndall effect. (*Courtesy of H. Steeves and R. G. Babcock*)

For systems of particles with diameters less than one-twentieth the wavelength of light, the light scattered from a polychromatic beam is predominantly blue in color and is polarized to a degree which depends on the angle between the observer and the incident beam. The blue color of tobacco smoke is an example of Tyndall blue. As particles are increased in size, the blue color of scattered light disappears and the scattered radiation appears white. If this scattered light is received through a nicol prism which is oriented to extinguish the vertically polarized scattered light, the blue color appears again in increased brilliance. This is called residual blue, and its intensity varies as the inverse eighth power of the wavelength. *See* COLLOID; SCATTERING OF ELECTROMAGNETIC RADIATION. Quentin Van Winkle

Type (printing) The intelligible images organized into readable text of various styles and sizes. Type used in printing on paper or video display is divided into four categories: foundry, machine-cast, photocomposed, and digitized type. In the first two the face of the letter is raised on one end of a piece of metal. It is from that surface, when inked, that the impression of type was made from its invention until the 1970s. In photocomposition, the type is reproduced photographically. Digitized methods assemble dots into typographic letters, lines, or pages.

Classification. Foundry type, also known as hand type, is cast as single characters. Machine-cast metal type is produced by Linotype, Intertype, Ludlow, and Monotype machines. Very few of these machines are still in operation. All but the last cast type in lines, or slugs. The Monotype—in reality two devices, a keyboard and a caster—produces individual types that are then manually set in lines of desired lengths.

The output of the photographic type machine is the image of type on film or on photosensitized paper in negative or positive form.

The term cold type is applied to text matter produced on a typewriter or laser printer and to words or lines made up of individual printed characters assembled or pasted together for photographic reproduction. Digital type uses a number of printout technologies, imaging small dots into lines (called rasters) and positioning them on pages from patterns of zeros and ones called bitmaps. Because this method is electronic, more sizes and variations (for example, condensing or expanding or other distortion) are possible.

About 6000 styles of type are in everyday use throughout the world. The most widely used method for classifying them is the serif-evolution system, based on the different shapes of the terminals or endings of letters. This provides eight classifications: Venetian, Old Style (Dutch-English and French), Transitional, Modern, Contemporary (sans serif and square serif), Scripts, Black Letter, and Decorative Letters.

Type measurements. The American point system made the unit of type measurement a point, 0.01383 in. (0.35128 mm) or nearly $^1/_{72}$ in. This system replaced the sixteenth-century practice of giving all type sizes names such as pearl, agate, nonpareil, brevier, long primer, and pica. Some names remained and have been assigned other functions. Agate, $5^1/_2$-point type or 14 lines of type to an inch, has come to be used for measuring newspaper advertising space. Publications quote small space rates by the agate line. Nonpareil is used to designate 6 points of space in and around type. The word pica is commonly used to denote a unit of space measuring 12 points. It is applied as a dimensional unit to the length of type lines and to the width and depth of page. With the advent of desktop publishing, many software programs have standardized on 72 points equaling exactly 1 inch. With these programs, the point now equals 0.01389 in., compared with the traditional point, 0.01383 in. Some of these programs allow the user to select which point system to use.

The standard height of metal type in the United States is 0.918 in. (2.3 cm), a dimension called type-high and one observed by photoengravers and electrotypers who make plates to be combined with cast type. Letterpress printing presses are adjusted to fit this standard height.

Fonts. A complete complement of letters of one size and style from A to Z, together with the arabic numerals, punctuation, reference marks and symbols, is called a font. Most fonts also include the ampersand and currency symbols and the ligatures, ff, fe, ffi, fl, ffl, æ, and œ. A letter from a different font that is included by accident in type composition is known as a wrong font. All typographic images, from letters to symbols to oriental ideographs, are called glyphs.

Type families. A family of type may be likened to the shades of a color in that it includes variations of a given type face or design. Weights are varied, from light to medium, bold, and extra bold; letters are condensed and expanded, as well as outlined, inlined, and shadowed (see illustration).

DEFGHI light
DEFGHI medium
DEFGHI bold
DEFGHI extra bold
DEFGHI condensed
DEFGHI shadowed

A family of type.

Type for desktop publishing. Most typesetting today is performed using desktop publishing (DTP), which refers to the creation of documents through the use of an inexpensive personal computer off-the-shelf software that uses page description language, and laser-based output devices that convert the digital information into images on paper and other surfaces, such as printing plates for offset lithography.

Eugene M. Ettenberg; Frank J. Romano; Thomas Destree

Type method

The nomenclatural method for providing a fixed reference point for a taxon (plural, taxa), a group of organisms. When Linnaeus, the eighteenth-century Swedish naturalist, ushered in the modern period of systematic biology through the publication of his classical work, *Systema Naturae*, he established the basis of binomial nomenclature. By means of this method the systematist who recognizes new, or supposedly new, taxa makes these known by means of a technical description and a scientific name. If the new taxon is a species, the material (specimens or parts of specimens) which the author had before him when he described it, together with such additional material as he may have gathered later, was considered typical, and such specimens were known as types. The types represented the author's notion of the new species and served as the basis of comparison. As knowledge increased, it was realized that each taxonomic name must be tied to a single specimen—the "type"—which is a nomenclatural, not a biological, type. Having a fixed reference for each name, later workers could correct taxonomic errors and modify the limits of species taxa without subsequent confusion in scientific names.

Although terms are available for more than 100 kinds of "type" specimens, these fall in two major classes: (1) nomenclatural types, the name-bearing specimens (onomatophores), which serve as reference points for nomenclatural purposes, and (2) taxonomic types which serve as standards of description and reference points for authors' concepts. The first category has been sanctioned by the International Commission on Zoological Nomenclature, the second category by taxonomic practice.

Just as types are required as nomenclatural reference points for the application of scientific names to species, so they are also necessary for the stabilization of names for higher taxa. This is especially true at the level of the genus, not only because there are more generic names than those of any other higher category but also because generic names provide roots for the names of several taxa immediately above them, as tribe, family, and so forth. As with species names, the necessity for refining early generic concepts and the retroactive application of the generic type concept to the works of previous authors have resulted in the development of a rather precise series of rules and practices governing their selection. These are specified in detail in the International Rules of Zoological Nomenclature. *See* ZOOLOGICAL NOMENCLATURE.

Walter J. Bock

Typewriter

A machine that produces printed copy, character by character, as it is operated. Its essential parts are a keyboard, a set of raised characters or a thermal print head, an inked ribbon, a platen for holding paper, and a mechanism for advancing the position at which successive characters are printed. The QWERTY keyboard (named for the sequence of letters of the top row of the alphabet worked by the left hand) was designed in the 1870s. It contains a complete alphabet, along with numbers and the symbols commonly used in various languages and technical disciplines. The manual typewriter was introduced in 1874, followed by the electrically powered typewriter in 1934. By the late 1970s, electronic typewriters offered memory capability, additional automatic functions, and greater convenience. Further advances in electronic technology led to additional capabilities, including plug-in memory and function diskettes and cartridges, visual displays, nonimpact printing, and communications adapters. Although many typewriters are still in use, computers and word processing software largely have supplanted them. *See* WORD PROCESSING.

Edward W. Gore, Jr.; Robert A. Rahenkamp

U

Ulcer A lesion on the surface of the skin or a mucous membrane characterized by a superficial loss of tissue. Ulcers are most common on the skin of the lower extremities and in the gastrointestinal tract, although they may be encountered at almost any site. The diverse causes of ulcers range from circulatory disturbances or bacterial infections to complex, multifactorial disorders. The superficial tissue sloughs, leaving a crater that extends into the underlying soft tissue, which then becomes inflamed and is subject to further injury by the original offender or secondary infection.

Peptic ulcer is the most common ulcer of the gastrointestinal tract and refers to breaks in the mucosa of the stomach or the proximal duodenum that are produced by the action of gastric secretions. It is still unknown why peptic ulcers develop. However, with rare exceptions, a person who does not secrete hydrochloric acid will not develop a peptic ulcer. Ulcers of the stomach tend to develop as a result of superficial inflammation of the stomach. These individuals tend to have normal or decreased amounts of hydrochloric acid. By contrast, most individuals with peptic ulcers of the proximal duodenum secrete excessive amounts of acid. Importantly, a bacterium, *Helicobacter pylori*, has been isolated from the stomach of most people with peptic ulcers and is thought to play a causative role. Although stress has been anecdotally related to peptic ulcers for at least a century, serious doubt has been cast upon this concept. *See* INFLAMMATION.

Ulcerative colitis is a disease of the large intestine characterized by chronic diarrhea and rectal bleeding. The disorder is common in the Western world, occurring principally in young adults. Its cause is not known, but there is some evidence for a familial predisposition to the disease.

Other ulcers of the gastrointestinal tract are caused by infectious agents. Bacterial and viral infections produce ulcers of the oral cavity. Diseases such as typhoid, tuberculosis, and bacillary dysentery and parasitic infestation with ameba lead to ulcers of the small and large intestines. Narrowing of the arteries to the legs caused by atherosclerosis, particularly in persons with diabetes mellitus, often causes ulcers of the lower extremities. *See* ARTERIOSCLEROSIS; BACILLARY DYSENTERY; DIABETES; TUBERCULOSIS.

Emanuel Rubin

Ultimobranchial bodies Small, enigmatic structures which originate as terminal outpocketings from each side of the embryonic pharynx. They occur only in vertebrates, where they are almost universal but difficult to homologize. They probably represent an expression of continued growth activity caudally, associated with pouch- or gill-forming potentialities of foregut entoderm. They are usually bilateral in mammals.

Originally the bodies were regarded as vestigial lateral, or accessory, thyroids since, because of the mechanics of growth in mammals, they were found to join with lateral lobes of the thyroid. These "bodies" are now interpreted as relatively indifferent tissue, but capable of modification by various factors. In lower vertebrates, ultimobranchial bodies do not join the thyroid but remain isolated in connective tissue between the thyroid gland and heart. Morphologically, ultimobranchial tissue can be highly variable, even within individuals of a single species. Furthermore, because of the proximity of these ultimobranchial bodies to more cephalic "true" pouches, which normally give rise to parathyroid and thymus tissue, and because foregut entoderm may possess reciprocal potentialities, this tissue on occasion may also carry induced, if not intrinsic, attributes for formation of accessory thymus or parathyroid tissue.

Since in many animals the ultimobranchial bodies have been shown to be inconsistent and to possess few specific, permanent structural attributes, they have been regarded by many as not only variable and erratic in development but also relatively unstable after birth. Despite this unusual character for endocrine tissue, it has been discovered that the ultimobranchial bodies can produce a hormone distinctly different from that of other thyroid hormones. This hormone is a polypeptide and has been labeled calcitonin. Collectively, these cells constitute an endocrine system that plays a role as a delicate and perhaps subtle calcium-regulating hormone. In response to hypercalcemia (high calcium levels of blood), these ultimobranchial cells (or "C" cells) secrete large quantities of this hormone, which rapidly lowers plasma calcium by inhibiting bone resorption, thus preventing excessive osteolysis and calcium deposition in tissues. *See* HORMONE; PARATHYROID GLAND; THYMUS GLAND.

John H. Van Dyke

Ultracentrifuge A centrifuge of high or low speed which provides convection-free conditions and which is used for quantitative measurement of sedimentation velocity or sedimentation equilibrium or for the separation of solutes in liquid solution. *See* CENTRIFUGATION.

The ultracentrifuge is used (1) to measure molecular weights of solutes and to provide data on molecular weight distributions in polydisperse systems; (2) to determine the frictional coefficients, and thereby the sizes and shapes, of solutes; and (3) to characterize and separate macromolecules on the basis of their buoyant densities in density gradients. *See* MOLECULAR WEIGHT.

The ultracentrifuge is most widely used to study high polymers, particularly proteins, nucleic acids, viruses, and other macromolecules of biological origin. However, it is also used to study solution properties of small solutes. In applications to macromolecules, the analytical ultracentrifuge, which is used for accurate determination of sedimentation velocity or equilibrium, is distinguished from the preparative ultracentrifuge, which is used to separate solutes on the basis of their sedimentation velocities or buoyant densities.

The application of a centrifugal field to a solution causes a net motion of the solute. If the solution is denser than the solvent, the motion will be away from the axis of rotation. The nonuniform concentration distribution produced in this way leads to an opposing diffusion flux tending to reestablish uniformity. In sedimentation-velocity experiments, sedimentation prevails over diffusion, and the solute sediments with finite velocity toward the bottom of the cell, although the concentration profile may be markedly influenced by diffusion. In sedimentation-equilibrium experiments, centrifugal and diffusive forces balance out, and an equilibrium concentration distribution results which may be analyzed by thermodynamic methods. *See* CENTRIFUGAL FORCE.

Victor A. Bloomfield

Ultrafast molecular processes Various types of physical and molecular changes occurring on time scales of 10^{-14} to 10^{-9} s that are studied in the field of photophysics, photochemistry, and photobiography. The time scale ranges from near the femtosecond regime (10^{-15} s) on the fast side, embodies the entire picosecond regime (10^{-12} s), and borders the nanosecond regime (10^{-9} s) on the slow side.

A typical experiment is initiated with an ultrashort pulse of energy—radiation (light) or particles (electrons). Rapid changes in the system under study are brought about by the absorption of these ultrashort energy pulses. These changes can be measured by ultrafast detection methods. Such methods are usually based on some type of linear or nonlinear spectroscopic monitoring of the system, or on optical delay lines that employ the speed of light itself to create a yardstick of time (a distance of 3 mm equals a time of 10 picoseconds). These studies are important because elementary motion, such as rotation, vibrational exchange, chemical bond breaking, and charge transfer, takes place on these ultrafast time scales. The relationship between such motions and overall physical, chemical, or biological changes can thus be observed directly. *See* CHEMICAL DYNAMICS; OPTICAL PULSES; SPECTROSCOPY.

Just as a very fast shutter speed is necessary to obtain a sharp photograph of a fast-moving object, energy pulses for the study of ultrafast molecular motions must be extremely narrow in time. The basic technique for producing ultrashort light pulses is mode-locking a laser. One way to do this is to put a nonlinear absorption medium in the laser cavity, which functions somewhat like a shutter. In the time domain, the developing light pulse, bouncing back and forth between the laser cavity reflectors, experiences an intensity-dependent loss each time it passes through the nonlinear absorption medium. Highintensity light penetrates the medium, and its intensity is allowed to build up in the laser cavity; weak-intensity light is blocked. This process shaves off the low-amplitude edges of a pulse, thus shortening it. One of the laser cavity reflectors has less than 100% reflectance at the laser wavelength, so part of the pulse energy is coupled out of the cavity each time the pulse reaches that reflector. In this way, a train of pulses is emitted from the mode-locked laser, each pulse separated from the next by the round-trip time for a pulse traveling at the speed of light in the laser cavity, about 13 nanoseconds for a 2-m (6.6-ft) laser cavity length. *See* LASER; NONLINEAR OPTICS.

In addition to the extremely narrow energy pulses in the time domain, special ultrafast detection techniques are required in order to preserve the time resolution of an experiment.

A photobiological process, which has received considerable attention by spectroscopists studying the femtosecond regime, is light-driven transmembrane proton pumping in purple bacteria (*Halobacterium halobium*). Each link in the complex chain of molecular processes, such as an ultrafast photoionization event, is "fingerprinted" by a somewhat different absorption spectrum, making it possible to sort out these events by methods of ultrafast laser spectroscopy. The necessity for use of fast primary events in nature concerns overriding unwanted competing chemical and energy loss processes. The faster the wanted process, the less is the likelihood that unwanted, energy-wasting processes will take place. *See* LASER PHOTOBIOLOGY; LASER SPECTROSCOPY.

G. Wilse Robinson; Ningyi Luo

Ultrafiltration A filtration process in which particles of colloidal size are retained by a filter medium while solvent and accompanying low-molecular-weight solutes are allowed to pass through. Ultrafilters are used (1) to separate colloid from suspending medium, (2) to separate particles of one size from particles of another size, and (3) to determine the distribution of particle sizes in colloidal systems by the use of filters of graded pore size.

Ultrafilter membranes have been prepared from various types of gel-forming substances: Unglazed porcelain has been impregnated with gels such as gelatin or silicic acid. Filter paper has been impregnated with glacial acetic acid collodions. Another type of ultrafilter membrane is made up of a thin plastic sheet containing millions of tiny pores evenly distributed over its surface. *See* COLLOID; FILTRATION.

Quentin Van Winkle

Ultralight aircraft A lightweight, single-seat aircraft with low flight speed and power, used for sport or recreation. Ultralights evolved from hang gliders. *See* GLIDER.

There have been tremendous variety of ultralight airframe configurations and control systems. Airframe types include powered weight-shift hang gliders, flying wings, canard designs, antique biplane replicas, and traditional, monoplane structures with conventional tail designs. Control systems can be either weight-shift systems, two-axis controls, or conventional three-axis controls. In weight-shift designs pilots must shift their weight, using a movable seat, to change the attitude. Two-axis designs use controls to the elevator and rudder only; there are no ailerons. Modern designs use three-axis control systems; weight-shift and two-axis control systems are no longer in production. Three-axis controls resemble those of a standard airplane and include either ailerons or spoilerons (spoiler-type systems used to make turns). Some more sophisticated ultralights are equipped with wing flaps, used to steepen approach profiles and to land at very slow airspeeds. *See* AIRFRAME; FLIGHT CONTROLS.

Typically, ultralight airframes are made of aircraft-grade aluminum tubes covered with Dacron sailcloth. Areas of stress concentration are reinforced with double-sleeving or solid aluminum components. Most ultralights are cable-braced and use aircraft-grade stainless steel cables with reinforced terminals. Wingspans average 30 ft (9 m), and glide ratios range from 7:1 to 10:1, depending primarily on gross weight and wing aspect ratio. *See* AIRFOIL; ASPECT RATIO; WING.

Ultralight engines are lightweight, two-stroke power plants with full-power values in the 28–35-hp (21–26-kW) range. They operate on a mixture of gasoline and oil, and most transmit power to the propeller via a reduction or belt drive, a simple transmission that enables the propeller to rotate at a lower, more efficient rate than the engine shaft. *See* AIRPLANE; PROPELLER (AIRCRAFT); RECIPROCATING AIRCRAFT ENGINE.

Thomas A. Horne

Ultrasonics The science of sound waves having frequencies above the audible range, that is, above about 20,000 Hz. Original workers in this field adopted the term supersonics. However, this name was also used in the study of airflow for velocities faster than the speed of sound. The present convention is to use the term ultrasonics as defined above. Since there is no marked distinction between the propagation and the uses of sound waves above and below 20,000 Hz, the division is artificial. *See* SOUND.

Ultrasonic generators and detectors. Ultrasonic transducers have two functions: transmission and reception. There may be separate transducers for each function or a single transducer for both functions. The usual types of generators and detectors for air, liquids, and solids are piezoelectric and magnetostrictive transducers. Quartz and lithium niobate ($LiNbO_3$) crystals are used to produce longitudinal and transverse waves; thin-film zinc oxide (ZnO) transducers can generate longitudinal waves at frequencies up to 96 GHz. Another class of materials used to generate ultrasonic signals is the piezoelectric ceramics. In contrast to the naturally occurring piezoelectric crystals, these ceramics have a polycrystalline structure. The most commonly produced piezoelectric ceramics are lead zirconate titanate (PZT), barium titanate ($BaTiO_3$), lead titanate ($PbTiO_3$), and lead metaniobate ($PbNb_2O_6$). Composite transducers are transducers in which the radiating or receiving element is a diced piezoelectric plate with filler between the elements. They are called "composite" to account for the two disparate elements, the piezoelectric diced into rods and the compliant adhesive filler. *See* MAGNETOSTRICTION; PIEZOELECTRICITY.

High-power ultrasound (typically 600 W) can be obtained with sonicators, consisting of a converter, horn, and tip. The converter transforms electrical energy to mechanical energy at a frequency of 20 kHz. Oscillation of piezoelectric transducers is transmitted and focused by a titanium horn that radiates energy into the liquid being treated. Horn and tip sizes are determined by the volume to be processed and the intensity desired. As the tip diameter increases, intensity or amplitude decreases.

Engineering applications. The engineering applications of ultrasonics can be divided into those dealing with low-amplitude sound waves and those dealing with high-amplitude (usually called macrosonics) waves.

Low-amplitude applications. Low-amplitude applications are in sonar (an underwater-detection apparatus), in the measurement of the elastic constants of gases, liquids, and solids by a determination of the velocity of propagation of sound waves, in the measurement of acoustic emission, and in a number of ultrasonic devices such as delay lines, mechanical filters, inspectoscopes, thickness gages, and surface-acoustic-wave devices. All these applications depend on the modifications that boundaries and imperfections in the materials cause in wave propagation properties. The attenuation and scattering of the sound in the media are important factors in determining the frequencies used and the sizes of the pieces that can be utilized or investigated. *See* SONAR.

High-amplitude applications. High-amplitude acoustic waves (macrosonic) have been used in a variety of applications involving gases, liquids, and solids. Some common applications are mentioned below.

A liquid subjected to high-amplitude acoustic waves can rupture, resulting in the formation of gas- and vapor-filled bubbles. When such a cavity collapses, extremely high pressures and temperatures are produced. The process, called cavitation, is the origin of a number of mechanical, chemical, and biological effects.

Cavitation plays an integral role in a wide range of processes such as ultrasonic cleaning and machining, catalysis of chemical reactions, disruption of cells, erosion of solids, degassing of liquids, emulsification of immiscible liquids, and dispersion of solids in liquids. Cavitation can also result in weak emission of light, called sonoluminescence. *See* CAVITATION; SONOCHEMISTRY.

One of the principal applications of ultrasonics to gases is particle agglomeration. This technique has been used in industry to collect fumes, dust, sulfuric acid mist, carbon black, and other substances.

Another industrial use of ultrasonics has been to produce alloys, such as lead-aluminum and lead-tin-zinc, that could not be produced by conventional metallurgical techniques. Shaking by ultrasonic means causes lead, tin, and zinc to mix.

Analytical uses. In addition to their engineering applications, high-frequency sound waves have been used to determine the specific types of motions that can occur in gaseous, liquid, and solid mediums. Both the velocity and attenuation of a sound wave are functions of the sound frequency. By studying the changes in these properties with changes of frequency, temperature, and pressure, indications of the motions taking place can be obtained. *See* SOUND ABSORPTION.

Medical applications. Application of ultrasonics in medicine can be generally classified as diagnostic and therapeutic. The more common of these at present is the diagnostic use of ultrasound, specifically ultrasonic imaging. *See* BIOMEDICAL ULTRASONICS; MEDICAL IMAGING; NONLINEAR ACOUSTICS.

Ultrasonic fields of sufficient amplitude can generate bioeffects in tissues. Although diagnostic ultrasound systems try to limit the potential for these effects, therapeutic levels of ultrasound have been used in medicine for a number of applications. Conventional therapeutic ultrasound is a commonly available technique used in physical therapy. High-frequency acoustic fields (typically 1 MHz) are applied through the skin to the affected area in either a continuous wave or long pulses.

Extracorporeal shock-wave lithotripsy (ESWL) disintegrates kidney stones with a high-amplitude acoustic pulse passing through the skin of the patient. The procedure eliminates the need for extensive surgery. Bioeffects are limited to the location of the stone by using highly focused fields which are targeted on the stone by imaging techniques such as ultrasound or fluoroscopy. Henry E. Bass; J. Brian Fowlkes; Veerle M. Keppens

Ultraviolet astronomy Astronomical observations carried out in the region of the electromagnetic spectrum with wavelengths from approximately 10 to 350 nanometers. Ultraviolet radiation from astronomical sources contains important diagnostic information about the composition and physical conditions of these objects. This information includes atomic absorption and emission lines of all the most abundant elements in many states of ionization. The hydrogen molecule (H_2), the most abundant molecule in the universe, has its absorption and emission lines in the far-ultraviolet. *See* ASTRONOMICAL SPECTROSCOPY; ULTRAVIOLET RADIATION.

Ultraviolet radiation with wavelengths less than 310 nm is strongly absorbed by molecules in the atmosphere of the Earth. Therefore, ultraviolet observations must be carried out by using instrumentation situated above the atmosphere. Ultraviolet astronomy began with instrumentation aboard sounding rockets. The first major ultraviolet satellite observatories to be placed in space were the United States *Orbiting Astronomical Observatories* (*OAO*s). The *International Ultraviolet Explorer* (*IUE*), launched into a geosynchronous orbit in 1978, realized the full potential of ultraviolet astronomy. *See* ROCKET ASTRONOMY.

NASA's *Extreme Ultraviolet Explorer* (*EUVE*) satellite (1992–2001) contained telescopes designed to produce images of the extreme-ultraviolet sky and spectra of bright extreme-ultraviolet sources in the wavelength range from approximately 10 to 90 nm. Most of the sources of radiation detected with the *EUVE* were stars with hot active outer atmospheres (or coronae) and hot white dwarf stars. *See* WHITE DWARF STAR.

Although it had an inauspicious beginning, the *Hubble Space Telescope*, launched in 1990, has been the centerpiece of both ultraviolet and visible astronomy. The complement of instruments aboard the *Hubble Space Telescope* has included imaging cameras operating at ultraviolet, visual, and infrared wavelengths, and spectrographs operating at ultraviolet and visual wavelengths. *See* HUBBLE SPACE TELESCOPE; SPECTROGRAPH; TELESCOPE.

The *Far-ultraviolet Spectrograph Explorer* (*FUSE*) satellite, launched in 1999, is designed to explore the universe in the 90–120-nm region of the spectrum at high spectral resolution. *Galaxy Evolution Explorer* (*GALEX*), launched in 2003, is a modest-sized ultraviolet imaging and spectroscopic survey mission that probes star formation over 80% of the age of the universe.

The important discoveries of ultraviolet astronomy span all areas of modern astronomy and astrophysics. Some of the notable discoveries in the area of solar system astronomy include new information on the upper atmospheres of the planets, including planetary aurorae and the discovery of the enormous hydrogen halos surrounding comets. In studies of the interstellar medium, ultraviolet astronomy has provided fundamental information about the molecular hydrogen content of cold interstellar clouds along with the discovery of the hot phase of the interstellar medium, which is created by the supernova explosions of stars. In stellar astronomy, ultraviolet measurements led to important insights about the processes of mass loss through stellar winds and have permitted comprehensive studies of the conditions in the outer chromospheric and coronal layers of cool stars. The *IUE*, *Hubble Space Telescope*, and *FUSE* observatories have contributed to the understanding of the nature of the hot gaseous corona surrounding the Milky Way Galaxy. Ultraviolet observations of exotic astronomical objects, including exploding stars, active galactic nuclei, and quasars, have provided new insights about the physical processes affecting the behavior of matter in extreme environments. The spectrographs aboard the *Hubble Space Telescope* have revealed the existence of large numbers of hydrogen clouds in the intergalactic medium. These intergalactic clouds may contain much more normal (baryonic) matter than exists in the known luminous galaxies and stars. The measures of the abundance of deuterium in the Milky Way Galaxy and beyond have provided important constraints on the conditions in the evolving universe when it was only several minutes old. *See* COMET; COSMOLOGY; GALAXY, EXTERNAL; INTERSTELLAR MATTER; MILKY WAY GALAXY; PLANETARY PHYSICS; QUASAR; SATELLITE (ASTRONOMY); STAR; SUPERNOVA. Blair D. Savage

Ultraviolet lamp A mercury-vapor lamp designed to produce ultraviolet radiation. Also, some fluorescent lamps and mercury-vapor lamps that produce light are used for ultraviolet effects. *See* FLUORESCENT LAMP; MERCURY-VAPOR LAMP.

Fluorescent and mercury lamps can be filtered so that visible energy is absorbed and emission is primarily in the near-ultraviolet or black-light spectrum (320–400 nanometers). The ultraviolet energy emitted is used to excite fluorescent pigments in paints, dyes, or natural materials.

Mercury-vapor lamps are sometimes designed with pressures that produce maximum radiation in the middle ultraviolet region (280–320 nm), using special glass bulbs that freely transmit this energy. One such lamp type is the sunlamp. Other lamps designed for middle-ultraviolet radiation are known as photochemical lamps. *See* SUNLAMP.

Some radiation in the 220–280-nm wavelength band has the capacity to destroy certain kinds of bacteria. Mercury lamps designed to produce energy in this region (the 253.7-nm mercury line) are electrically identical with fluorescent lamps; they differ from fluorescent lamps in the absence of a phosphor coating and in the use of glass tubes that transmit far ultraviolet. *See* ULTRAVIOLET RADIATION (BIOLOGY). Alfred Makulec

Ultraviolet radiation Electromagnetic radiation in the wavelength range 4–400 nanometers. The ultraviolet region begins at the short wavelength (violet) limit of visibility and extends to the wavelength of long x-rays. It is loosely divided into the near (400–300 nm), far (300–200 nm), and extreme (below 200 nm) ultraviolet regions (see illustration). In the extreme ultraviolet, strong absorption of the radiation by air requires the use of evacuated apparatus; hence this region is called the vacuum ultraviolet. Important phenomena associated with ultraviolet radiation include biological effects and applications, the generation of fluorescence, and chemical analysis through characteristic absorption or fluorescence. *See* ULTRAVIOLET RADIATION (BIOLOGY).

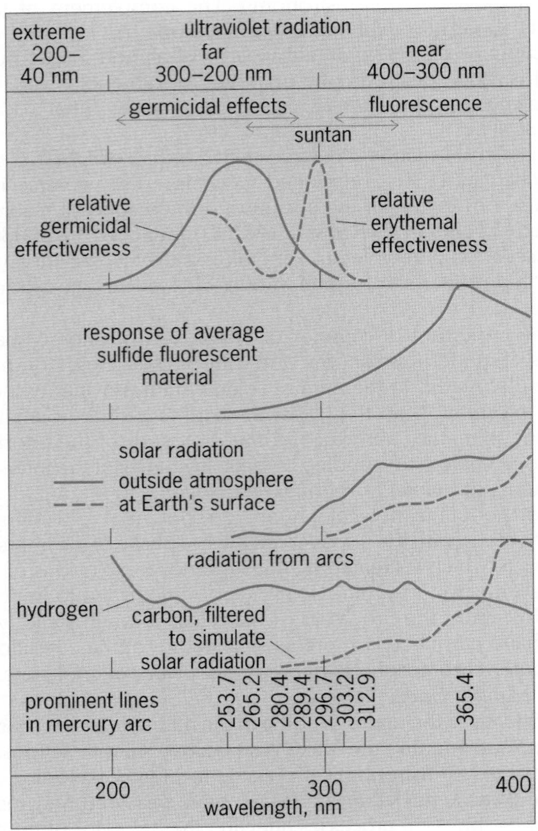

Phenomena associated with ultraviolet radiation. (*After L. R. Koller and General Electric*)

Sources of ultraviolet radiation include the Sun (although much solar ultraviolet radiation is absorbed in the atmosphere); arcs of elements such as carbon, hydrogen, and mercury; and incandescent bodies. *See* ULTRAVIOLET LAMP. Fred W. Billmeyer

Ultraviolet radiation (biology) Radiations between 200 and 300 nanometers are selectively absorbed by organic matter and produce the best-known effects of ultraviolet radiations on organisms. Ultraviolet radiations, in contrast to x-rays, do not penetrate far into larger organisms; therefore, the effects they produce are surface effects, such as sunburn and development of D vitamins from precursors present in skin or fur. The effects of ultraviolet radiations on life have, therefore, been assayed chiefly with unicellular organisms such as bacteria, yeast, and protozoans, although suspensions of cells of higher organisms, for example, eggs and blood corpuscles, have been useful as well.

Photobiological effects. Only the ultraviolet radiations which are absorbed can produce photobiological action. All life activities are shown to be affected by ultraviolet radiations, the effect depending upon the dosage. Small dosages activate unfertilized eggs of marine animals, reduce the rate of cell division, decrease the synthesis of nucleic acid, especially in the nucleus, reduce the motility of cilia and of contractile vacuoles, and sensitize cells to heat. Large doses increase the permeability of cells to various substances, inhibit most synthetic processes, produce mutations, stop division of cells, decrease the rate of respiration, and may even disrupt cells. The effect of ultraviolet radiations upon cells is invariably deleterious.

Effects on the skin. Erythema is the reddening of the skin following exposure to ultraviolet radiation of wavelength shorter than 320 nm, wavelength 296.7 nm being most effective. These radiations injure cells in the outer layer of the skin, or epidermis, liberating substances which diffuse to the inner layer of the skin, or dermis, causing enlargement of the small blood vessels. A minimal erythemal dose just induces reddening of the skin observed 10 h after exposure. A dose several times the minimal gives a sunburn, killing some cells in the epidermis after which serum and white blood cells accumulate, causing a blister. After the dried blister peels, the epidermis is temporarily thickened and pigment develops in the lower layers of the epidermis, both of these factors serving to protect against subsequent exposure to ultraviolet.

Both thickening of the epidermis and tanning may occur without blistering. Since the pigment in light-skinned races develops chiefly below the sensitive cells in the epidermis, it is not as effective as in dark-skinned races where the pigment is scattered throughout the epidermis. Consequently, the minimal erythemal dose is much higher for the dark- than for the light-skinned races.

Excessive exposure to ultraviolet radiation has been found to lead to cancer in mice, and it is claimed by some to cause cutaneous cancer in humans.

Clinical use. Ultraviolet radiations were once used extensively in the treatment of rickets, many skin diseases, tuberculosis other than pulmonary, especially skin tuberculosis (lupus vulgaris), and of many other diseases. The enthusiasm for sun bathing is, in part, a relic of the former importance of ultraviolet radiation as a clinical tool. Vitamin preparations, synthetic drugs, and antibiotics have displaced ultraviolet radiations in such therapy or are used in conjunction with the radiations.

Ultraviolet radiations alone are still employed to treat rickets in individuals sensitive to vitamin D preparations. In conjunction with chemicals, they are used in treating skin diseases, for example, psoriasis, pityriasis rosea, and sometimes acne, as well as for the rare cases of sensitivity to visible light. Ultraviolet radiations, however, are more important in research than in clinical practice. Arthur C. Giese

Ultrawideband (UWB) systems An electronic system which has either an instantaneous bandwidth of at least 500 MHz or a fractional bandwidth (ratio of the instantaneous bandwidth measured at the −10 dB points to the center frequency) of at least 20%, whichever is greater.

Development of technology. Ultrawideband technology stems from work in time-domain electromagnetics begun in 1962 by Gerald F. Ross to describe fully the transient behavior of microwave networks through their characteristic impulse response. Ross, recognizing that a linear system could be fully characterized by its response to an impulse excitation (that is, the result of a single measurement), developed real-time hardware techniques to implement such a measurement.

The tunnel diode, with its extremely wide bandwidth (tens to hundreds of gigahertz), permitted subnanosecond pulse generation essential for impulse excitation and could also be used as a sensitive thresholding device for the subsequent detection of extremely short duration waveforms. In 1962, time-domain sampling oscilloscopes based on the tunnel diode were introduced for high-speed triggering and detection, first enabling the capture and display of ultrawideband waveforms. Impulse measurement techniques were subsequently applied to the analysis of wideband radiating antenna elements, in which the impulse response is the radiated electromagnetic field. In doing so, it quickly became apparent that short-pulse radar and even communications systems could be developed using the same set of tools. *See* ANTENNA (ELECTROMAGNETISM); TUNNEL DIODE; TUNNELING IN SOLIDS.

Until the late 1980s, ultrawideband technology was alternatively referred to as baseband, carrier-free, or impulse. The term "ultrawideband" was not applied until approximately 1989.

Significance of technology. The early interest in short-pulse techniques was primarily concentrated in short-range radar applications. From first principles, the shorter a radar's pulse duration, the finer the resultant range resolution. Thus, with pulse durations measured in hundreds of picoseconds, resolutions of only a few inches (several centimeters) or less were now possible.

From the time-scaling property of the Fourier transform relationship between time and frequency domains, the shorter the transmitted pulse, the wider the instantaneous bandwidth. Thus, since short-pulse waveforms could be inexpensively produced directly at baseband, applications of ultrawideband to ground-penetrating radar and through-wall imaging followed quickly. Here, low-frequency wideband excitation translated into significantly enhanced material penetration capability. Ground-penetrating radar is used to locate buried objects, including humans in disasters such as building or mine collapses. Through-wall imaging is used for law enforcement and intelligence activities. *See* FOURIER SERIES AND TRANSFORMS; GROUND-PENETRATING RADAR; RADAR.

The early interest in ultrawideband for communications stemmed from the fact that short-pulse waveforms are extremely difficult to detect. Since the pulse bandwidth is spread over many hundreds of megahertz to gigahertz, a conventional (that is, narrowband) receiver will receive only a small slice of this spectral energy. The total received energy is directly proportional to the ratio of the intercept receiver to ultrawideband spread bandwidths.

Applications. Recent applications include an ultrawideband airborne wireless intercommunications system (AWICS), which uses the low-probability-of-intercept and multipath mitigation features of the ultrawideband waveform to provide a wireless communications capability within an aircraft or helicopter fuselage; ultrawideband wireless transceivers for the relay of voice, data, and compressed video for covert network communications; ultrawideband perimeter intrusion radars for detecting unauthorized personnel; and ultrawideband through-wall sensing radars for police and fire rescue. Robert J. Fontana

Umklapp process A concept in the theory of transport properties of solids which has to do with the interaction of three or more waves in the solid, such as lattice waves or electron waves. In a continuum, such interactions occur only among waves described by wave vectors $\mathbf{k}_1$, $\mathbf{k}_2$, and so on, such that the interference condition, given by Eq. (1), is satisfied.

$$\mathbf{k}_1 + \mathbf{k}_2 + \mathbf{k}_3 = 0 \quad (1)$$

The sign of $\mathbf{k}$ depends on whether the wave absorbs or emits energy. Since $\hbar\mathbf{k}$ is the momentum of a quantum (or particle) described by the wave, Eq. (1) corresponds to conservation of momentum. In a crystal lattice further interactions occur, satisfying Eq. (2), where $\mathbf{b}$ is any integral combination of the

$$\mathbf{k}_1 + \mathbf{k}_2 + \mathbf{k}_3 = \mathbf{b} \quad (2)$$

three inverse lattice vectors $\mathbf{b}_i$, defined by $\mathbf{a} \cdot \mathbf{b}_j = 2\pi\delta_{ij}$, the $\mathbf{a}$'s being the periodicity vectors. The group of processes described in Eq. (2) are the Umklapp processes or flip-over processes, so called because the total momentum of the initial particles or quanta is reversed. *See* CRYSTAL. Paul G. Klemens

Uncertainty principle A fundamental principle of quantum mechanics, which asserts that it is not possible to know both the position and momentum of an object with arbitrary accuracy. This contrasts with classical physics, where the position and momentum of an object can both be known exactly. In quantum mechanics, this is no longer possible, even in principle. More precisely, the indeterminacy or uncertainty principle, derived by W. Heisenberg, asserts that the product of Δx and Δp—measures of indeterminacy of a coordinate and of momentum along that coordinate—must satisfy inequality (1). The

$$\Delta x \times \Delta p \gtrsim \hbar = \frac{h}{2\pi} \quad (1)$$

Planck constant, $h \simeq 6.63\ 10^{-34}$ joule-second, is very small, which makes inequality (1) unimportant for the measurements that are carried out in everyday life. Nevertheless, the consequences of the inequality are critically important for the interactions between the elementary constituents of matter, and are reflected in many of the properties of matter that are ordinarily taken for granted. For example, the density of solids and liquids is set to a large degree by the uncertainty principle, because the sizes of atoms are determined with decisive help of inequality (1).

In classical physics, simultaneous knowledge of position and momentum can be used to predict the future trajectory of a particle. Quantum indeterminacy and the limitations it imposes force such classical notions of causality to be abandoned.

Another well-known example of indeterminacy involves energy and time, as given by inequality (2). Physically, its origins

$$\Delta E \times \Delta t \gtrsim \hbar \quad (2)$$

are somewhat different from those of inequality (1). Inequality (2) relates, for example, lifetimes of unstable states with the widths of their lines. *See* LINEWIDTH.

In quantum physics, relations similar to inequalities (1) and (2) hold for pairs of many other quantities. They demonstrate that the acquisition of the information about a quantum object cannot be usually achieved without altering its state. Much of the strangeness of quantum physics can be traced to this impossibility of separating the information about the state from the state itself. *See* QUANTUM MECHANICS. Wojciech Hubert Zurek

Unconformity In the stratigraphic sequence of the Earth's crust, a surface of erosion that cuts the underlying rocks and is overlain by sedimentary strata. The unconformity represents an interval of geologic time, called the hiatus, during which no deposition occurred and erosion removed preexisting rock. The result is a gap, in some cases encompassing millions of years, in the stratigraphic record. *See* STRATIGRAPHY.

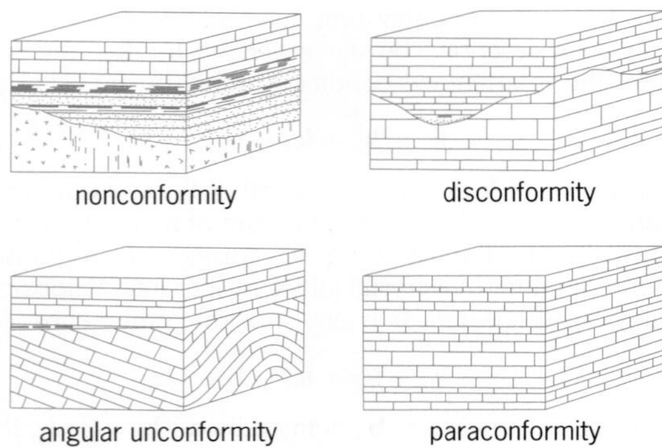

Four types of unconformity. (*After C.O. Dunbar and J. Rodgers, Principles of Stratigraphy, Wiley, 1957*)

There are four kinds of unconformable relations (see illustration): *Nonconformity*—underlying rocks are not stratified, such as massive crystalline rocks formed deep in the Earth. *Angular unconformity*—underlying rocks are stratified but were deformed before being eroded, resulting in angular discordance; this was the first type to be recognized; the term unconformity was originally used to describe the geometric relationship between the underlying and overlying bedding planes. *Disconformity*—underlying strata are undeformed and parallel to overlying strata, but separated by an evident erosion surface. *Paraconformity*—strata are parallel and the erosion surface is subtle, or even indistinguishable from a simple bedding plane. Charles W. Byers

Underground mining The extraction of ore from beneath the surface of the ground. Underground mining is also applied to deposits of industrial (nonmetallic) minerals and rocks, and underground or deep methods are used in coal mining. Some ores and industrial minerals can be recovered from beneath the ground surface by solution mining or in-place leaching using boreholes. *See* COAL MINING; SOLUTION MINING.

Underground mining is done where mineral deposits are situated beyond the economic depth of open pit mining; it is generally applied to steeply dipping or thin deposits and to disseminated or massive deposits for which the cost of removing the overburden and the maintaining of a slope angle in adjacent waste rock would be prohibitive. *See* OPEN PIT MINING.

Underground mine entries are by shaft, adit, incline, or spiral ramp (see illustration). Development workings, passageways for gaining access to the orebody from stations on individual mine levels, are called drifts if they follow the trend of the mineralization, and cross-cuts if they are driven across the mineralization. Workings on successive mine levels are connected by raises, passageways that are driven upward. Winzes are passageways that are sunk downward, generally from a lowermost mine level.

In a fully developed mine with a network of levels, sublevels, and raises for access, haulage, pumping, and ventilation, the ore is mined from excavations referred to as stopes. Pillars of unmined material are left between stopes and other workings for temporary or permanent natural support. In large-scale mining methods and in methods where an orebody and its overlying waste rock are allowed to break and cave under their own weight, the ore is extracted in large collective units called blocks, panels, or slices. *See* MINING.

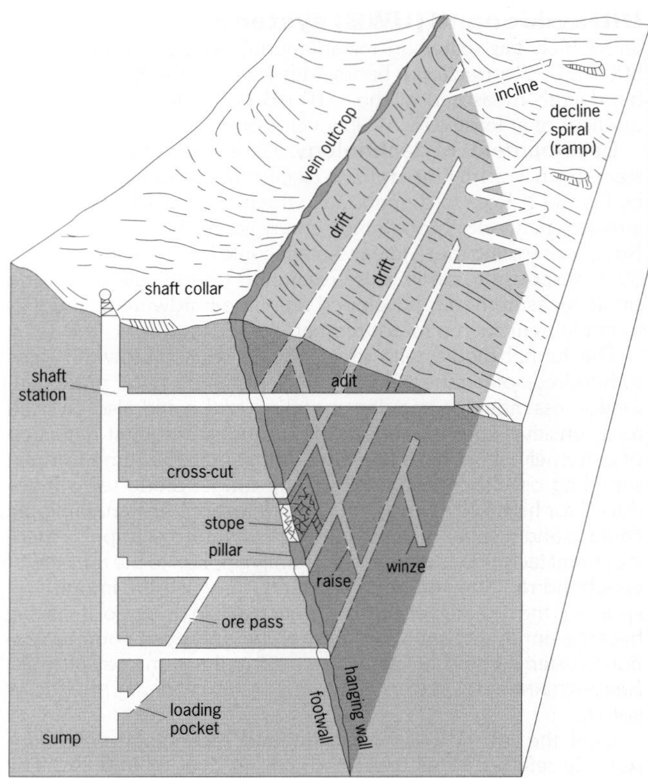

Underground mining entries and workings.

Information on the size, shape, and attitude of a deposit and information for estimating the tonnage and grade of the ore is taken from drill holes and underground exploration workings. Underground exploration workings are used for bulk and detailed sampling, rock mechanics testing, and the siting of machinery for underground drilling.

Where high topographic relief allows for an acceptable tonnage of ore above a horizontal entry site, an adit or blind tunnel is driven as a cross-cut to the deposit or as a drift following the deposit from a portal at a favorable location for the surface plant, drainage facilities, and waste disposal. In situations where the deposit lies below or at a great distance from any portal site for an adit, entry must be made from a shaft collar or from an incline or decline portal. A large mine will commonly have a main multipurpose entry and several more shafts or adits to accommodate personnel, supplies, ventilation, communication, and additional production.

A fundamental condition in the choice of mining method is the strength of the ore and wall rock. Strong ore and rock permit relatively low-cost methods with naturally supported openings or with a minimum of artificial support. Weaker ore and wall rock necessitate more costly methods requiring widespread temporary or permanent artificial support such as rock bolting. Large deposits with weak ore and weak walls that collapse readily and provide suitably broken material for extraction may be mined by low-cost caving methods. Few mineral deposits are so uniform that a single method can be used without modification in all parts of the mine. William C. Peters

Underwater demolition The controlled use of explosives to achieve specific underwater work requiring cutting, fragmenting, perforating, or pounding.

In addition to wreck removal and channel widening and deepening uses, a wide range of commercial applications has been developed for underwater demolition. In the offshore oil and gas industries, charges may be placed and fired well below the ocean floor to open fissures in the rock, cut off steel pipe, open trenches, and remove old structures. The military uses underwater demolition to remove obstacles to amphibious assault.

Special charge shapes and sizes allow very precise work to be done with explosives. While charges for military purposes are usually placed by divers, explosives used commercially can also be placed by crewed submersibles or remote-operated vehicles. *See* DIVING; EXPLOSIVE; SHIP SALVAGE. William I. Milwee

Underwater navigation The process of directing the movements of submersible vehicles, and divers, from one point to another. The development of improved submersible vehicles, coupled with advances in saturated diving, has resulted in new requirements for underwater navigation. Various methods which have proved successful include acoustic transponder systems, dead reckoning, surface-referenced navigation from a support ship, terrain map matching, homing, and various combinations of these. The choice of the navigation system depends on such factors as precision required, area to be covered, availability of surface vessels, sea state under which they are expected to operate, and for crewed vehicles the redundancy necessary for safety. *See* DIVING; SUBMARINE; UNDERWATER VEHICLE.

Acoustic transponders. The most accurate undersea navigation systems use acoustic transponders to precisely determine position with respect to points on the sea floor. A minimum of two transponders are normally dropped from a surface or submerged vehicle to establish a bottom-tethered array. The transponders can be interrogated by surface vessels, submersible vehicles, or divers. When the navigation solution is computed on the surface vessel, the submerged vehicle also carries a transponder, and the position of the submerged vehicle is computed relative to the bottom-tethered array. The location of these transponders can be determined by an accurately positioned surface vehicle that transmits a series of interrogation pulses from different locations. The location of the submersible is determined by a series of round-trip time measurements obtained by interrogating transponders at known locations.

Dead reckoning. An undersea vehicle needs sensors to show distance traveled and direction of travel to mechanize dead-reckoning estimation of position. The most promising sensor for operation near the sea floor is a Doppler sonar. Ground speed, fore-aft and athwartship, can be determined from the Doppler shift in frequency of signals returned from the sea floor. Pulse-type and continuous-wave Doppler sonars are available. The pulse type, which makes use of a frequency tracker circuit to lock on the received pulse, appears to offer the best accuracy, and it has the ability to operate to 600 ft (180 m) above the sea floor. *See* SONAR.

In addition to Doppler sonar, some specialized submersibles carry inertial dead reckoning in the form of a small, stabilized platform. The accuracy of any dead reckoning system can be improved by receiving additional position information that is more accurate than the errors that have accumulated since the last position update. This opens up opportunities to use sonar-based underwater terrain map matching or occasional Global Positioning Satellite System (GPS) fixes collected by floating an antenna to the surface as sources of these independent position updates. *See* DEAD RECKONING; INERTIAL GUIDANCE SYSTEM.

Other methods. Most submersible vehicles carry additional navigation sensors. The horizontal obstacle sonar (HOS) which is a constant-transmission frequency-modulated sonar (CTFM), is normally used to detect objects ahead of the submersible. This sonar also has a transponder channel for determining bearing and range to specially designed transponders. An altitude-depth sonar provides vertical navigation by furnishing depth and altitude off the bottom. Finally, a vertical obstacle sonar (VOS) determines heights of objects in the path of the vehicle. Both horizontal and vertical obstacle-avoidance sonars are useful in under-ice navigation. *See* HOMING; PILOTING.

An acoustic tracking system allows the monitoring and vectoring of the position of a vehicle from a surface vessel, where space and weight are not at a premium, and Loran C, the Global Positioning System, Omega, or other radio techniques can be used to determine the ship's own position in geographic coordinates. Surface tracking systems are of two types: ultrashort baseline, employing orthogonal line hydrophone arrays; and short baseline, employing four separate hydrophones. *See* MARINE NAVIGATION.

Submarines must operate over wide areas of the ocean and often under highly secure and covert conditions; therefore, navigation techniques which depend upon sonic or electromagnetic emissions, or which restrict movements of the vehicle to either near-bottom or near-surface regions, are judged to be too restrictive. The ability of inertial navigation systems to operate without frequent recourse to external position updates makes them prime candidates for submarine navigation. The need for covert and secure long-duration navigation has resulted in the evolution of a navigation configuration that uses a pair of electrostatic gyroscope navigators (ESGNs) as the primary elements of an integrated navigation subsystem. The very stable, low-drift characteristics of the electrostatic gyroscope result in an inertial navigation system with a low and highly predictable error growth. Weapons system accuracy is further enhanced by velocity derived from both gyroscopes and secure correlation velocity sonar techniques to give a direct measure of the submarine ground speed. Vertical deflection maps, which are used for vertical-axis tilt compensation of the gyroscopes, are generated from combined satellite and oceanographic surveys of the Earth's gravity field. The relatively rare position resets are selected from either Global Positioning System satellite or bathymetric sonar data. The passive electromagnetic-log, which measures vehicle velocity relative to the water and is therefore subject to ocean current errors, is used primarily to damp the gyroscope computational (Schuler) oscillations. The various data are processed in a central navigation computer which uses statistical estimation algorithms. *See* ESTIMATION THEORY.

Navigation has always been a difficult problem for divers. Some of the same sonars developed for submersible pilotage offer an answer for a diver. A horizontally swept constant-transmission frequency-modulated sonar is well matched to the excursions of saturated divers, who are limited in allowable vertical movement. By communication with the habitat, divers who stray too far or lose their orientation can be guided back to safety.

Charles J. Hrbek; Joseph A. Cestone; Emery St. George, Jr.; Richard L. Greenspan

Underwater photography The techniques involved in using photographic equipment underwater. By far the greatest percentage of underwater photography is done within sport-diving limits in the tropical oceans.

Underwater photographers are faced with specific technical challenges. Water is 600 times denser than air and is predominantly blue in color. Depth affects light and creates physiological considerations for the photographer. As a result, underwater photography requires an understanding of certain principles of light beneath the sea.

As in all photography, consideration of the variables of light transmission is crucial to underwater photography. When sunlight strikes the surface of the sea, its quality and quantity change in several ways. As light travels from air to a denser medium, such as water, the light rays are bent (refracted); one result is magnification of underwater objects by one-third as compared to viewing them in air. The magnification effect must be considered when estimating distances underwater, which is critical for both focus and exposure. Light is absorbed when it propagates through water. Variables affecting the level of light penetration include the time of day (affects the angle at which the sunlight strikes the surface of the water); cloud cover; clarity of the water; depth (light is increasingly absorbed with increasing depth); and surface conditions (if the sea is choppy, more light will be reflected off the surface and less light transmitted to the underwater scene). *See* LIGHT; PHOTOGRAPHY; REFRACTION OF WAVES.

Depth affects not only the quantity of light but also the quality of light. Once light passes from air to water, different wavelengths of its spectrum are absorbed as a function of the color of the water and depth. Even in the clearest tropical sea, water serves as a powerful cyan (blue-green) filter. Natural full-spectrum photographs can be taken only with available light in very shallow depths. In ideal daylight conditions and clear ocean water, photographic film fails to record red at about 15 ft (4.5 m) in depth. Orange disappears at 30 ft (9 m), yellow at 60 ft (18 m), green at 80 ft (24 m), and at greater depth only blue and black are recorded on film. To restore color, underwater photographers must use artificial light. *See* SEAWATER.

The water column between photographer and subject degrades both the resolution of the image and the transmission of artificial light (necessary to restore color). Therefore, the most effective underwater photos are taken as close as possible to the subject, thereby creating the need for a variety of optical tools to capture subjects of various sizes within this narrow distance limitation.

There are two types of underwater cameras—amphibious and housed. Amphibious cameras may be used either underwater or topside, although some lenses are for underwater use only (known as water contact lenses). A housed camera is a conventional above-water camera that has been protected from the damaging effects of seawater by a waterproof enclosure. The amphibious camera is protected by a series of O-rings, primarily located at the lens mount, film loading door, shutter release, and other places where controls are necessary. The O-rings make the system not only resistant to leaks but also impervious to dust or inclement weather when used above water. Stephen Frink

Deep-sea underwater photography—approximately 150 ft (35 m)—requires the design and use of special camera and lighting equipment. Watertight cases are required for both camera and light source, and they must be able to withstand the pressure generated by the sea. For each 33 ft (10 m) of depth, approximately one additional atmosphere ($\sim 10^2$ kilopascals) of pressure is exerted. At the greatest ocean depths, about 40,000 ft (12,000 m), a case must be able to withstand 17,600 lb/in.^{2} (1200 kg/cm^2). The windows for the lens and electrical seals must also be designed for such pressure to prevent water intrusion.

Auxiliary lighting is required, since daylight is absorbed in both intensity and hue. The camera must be positioned and triggered to render the desired photograph, and the great depths preclude a free-swimming human operator. Operation is often from a cable via sonar sensing equipment or from deep-diving underwater vehicles. Bottom-sensing switches can operate deep-sea cameras for photographing the sea floor, and remotely operated vehicles (ROVs) can incorporate both video and still cameras. *See* SONAR; UNDERWATER VEHICLE.

When an observer descends to great depths in a diving vehicle, the camera can assist in documentation by recording what is seen. Furthermore, the visual data will assist in accurate description of the observed phenomena. Elapsed-time photography with a motion picture camera in the sea is important in studying sedimentation deposits caused by tides, currents, and storms. Similarly, the observation of biological activity taken with the elapsed-time camera and then speeded up for viewing may reveal processes that cannot ordinarily be observed. *See* DIVING; PHOTOGRAPHY; UNDERWATER TELEVISION.

Howard E. Edgerton; Stephen Frink

Underwater sound The production, propagation, reflection, scattering, and reception of sound in seawater. The sea covers approximately 75% of the Earth's surface. In terms of exploration, visible observation of the sea is limited due to the high attenuation of light, and radar has very poor penetrability into salt water. Because of the extraordinary properties that sound has in the sea, and because of some of the inherent characteristics of the sea, acoustics is the principal means by which the sea has been explored. *See* OCEAN.

Absorption. Sound has a remarkably low loss of energy in seawater, and it is that property above all others that allows it to be used in research and other application. Absorption is the loss of energy due to internal causes, such as viscosity. Over the frequency range from about 100 Hz (cycles per second) to 100 kHz, absorption is dominated by the reactions of two molecules, magnesium sulfate ($MgSO_4$) and boric acid [$B(OH)_3$]. These molecules are normally in equilibrium with their ionic constituents. The pressure variation caused by an acoustic wave changes the ionic balance and, during the passage of the pressure-varying acoustic field, it cannot return to the same equilibrium, and energy is given up. This is called chemical relaxation. At about 65 kHz magnesium sulfate dominates absorption, and boric acid is important near 1 kHz. *See* SEAWATER; SOUND; SOUND ABSORPTION.

Sound speed. The speed of sound in seawater and its dependence on the parameters of the sea, such as temperature, salinity, and density, have an enormous effect on acoustics in the sea. Generally the environmental parameter that dominates acoustic processes in oceans is the temperature, because it varies both spatially and temporally. Solar heating of the upper ocean has one of the most important effects on sound propagation. As the temperature of the upper ocean increases, so does the sound speed. Winds mix the upper layer, giving rise to a layer of water of approximately constant temperature, below which is a region called the thermocline. Below that, most seawater reaches a constant temperature. All these layers depend on the season and the geographical location, and there is considerable local variation, depending on winds, cloud cover, atmospheric stability, and so on. Shallow water is even more variable due to tides, fresh-water mixing, and interactions with the sea floor. Major ocean currents, such as the Gulf Stream and Kuroshio, have major effects on acoustics. The cold and warm eddies that are spun off from these currents are present in abundance and significantly affect acoustic propagation. *See* GULF STREAM; KUROSHIO; OCEANOGRAPHY.

Pressure waves. The science of underwater sound is the study of pressure waves in the sea over the frequency range from a few hertz to a few megahertz. The International System (SI) units are the pascal (Pa) for pressure (equal to one newton per square meter) and the watt per square meter (W/m^2) for sound intensity (the flow of energy through a unit area normal to the direction of wave propagation). In acoustics, it is more convenient to refer to pressures, which are usually much smaller than a pascal, and the consequent intensities with a different reference, the decibel. Intensity in decibels (dB) is ten times the logarithm to the base ten of the measured intensity divided by a reference intensity. *See* DECIBEL; SOUND INTENSITY; SOUND PRESSURE.

Wave propagation. The mathematical equation that sound obeys is known as the wave equation. Its derivation is based on the mathematical statements of Newton's second law for fluids (the Navier-Stokes equation), the equation of continuity (which essentially states that when a fluid is compressed, its mass is conserved), and a law of compression, relating a change of volume to a change in pressure. By the mathematical manipulation of these three equations, and the assumption that only very small physical changes in the fluid are taking place, it is possible to obtain a single differential equation that connects the acoustic pressure changes in time to those in space by a single quantity, the square of the sound speed (c), which is usually a slowly varying function of both space and time. *See* NAVIER-STOKES EQUATION; WAVE EQUATION.

Knowing the sound speed as a function of space and time allows for the investigation of the spatial and temporal properties of sound, at least in principle. The mathematics used to find solutions to the wave equation are the same as those that are used in other fields of physics, such as optics, radar, and seismics. *See* WAVE MOTION.

In addition to knowing the speed of sound, it is necessary to know the location and nature of the sources of sound, the location and features of the sea surface, the depth to the sea floor, and, in many applications, the physical structure of the sea floor. It is not possible to know the sound speed throughout the water column or know the boundaries exactly. Thus the solutions to the wave equation are never exact representations of nature, but estimates, with an accuracy that depends on both the quality of the knowledge of the environment and the degree to which the mathematical or numerical solutions to the wave equation represent the actual physical situation.

Ambient noise. A consequence of the remarkable transmission of sound is that unwanted sounds are transmitted just as efficiently. One of the ultimate limitations to the use of underwater sound is the ability to detect a signal above the noise. In the ocean, there are four distinct categories of ambient sound: biological, oceanographic physical processes, seismic, and anthropogenic. *See* ACOUSTIC NOISE; INFRASOUND.

Scattering and reverberation. The other source of unwanted sound is reverberation. Sound that is transmitted inevitably finds something to scatter from in the water column, at the sea surface, or at the sea floor. The scatter is usually in all directions, and some of it will return to the system that processes the return signals. Sources of scattering in the water column are fish, particulates, and physical inhomogeneities. The sea surface is, under normal sea conditions, agitated by winds and has the characteristic roughness associated with the prevailing atmospheric conditions. Rough surfaces scatter sound with scattering strengths that depend on the roughness, the acoustic frequency (or wavelength), and the direction of the signal. The scattering is highly time-dependent, and needs to be studied with an appropriate statistical approach. The sea floor has inherent roughness and is usually inhomogeneous, both properties causing scatter. Although scatter degrades the performance of sonars, the characteristics of the return can be determined to enable its cancellation through signal processing or array design. Scattering can also be used to study the sea surface, the sea floor, fish types and distribution, and inhomogeneities in the water column. *See* SCATTERING LAYER; SONAR. Ralph R. Goodman

Underwater television Any type of electronic camera that is located underwater in order to collect and display images. It must be packaged in a waterproof housing. The underwater camera may be packaged with its own recording device, or it can be attached to a television that is located on a ship, in a laboratory, or at a remote site. In the latter case, the images by the camera are real-time images. An underwater television may be used for sport, ocean exploration, industrial applications, or military purposes. Common imaged subjects are animals, coral reefs, underwater shipwrecks, and underwater structures such as piers, bridges, and offshore oil platforms.

During the daytime and at minimal depths, underwater television can be used to view objects illuminated with natural sunlight. For nighttime viewing or at very deep depths, artificial lights must be used. Light in the ocean is reduced in intensity quite severely: even in the clearest natural waters, a beam of blue or green light will be reduced in intensity by approximately 67% every 230 ft (70 m). Light that is propagating through an aqueous medium such as seawater or lake water will also be selectively reduced in intensity, based on wavelength. In most situations, the extreme reds and blues will be most severely attenuated, with the region of highest clarity being the yellow to blue-green wavelengths. *See* ELECTROMAGNETIC RADIATION.

Modern advances in television cameras have opened up a host of applications for underwater viewing, such as of standard-video, charge-injection-device, silicon-intensified-target, or charge-coupled-device cameras. These cameras can be adapted for high frame rate, low light level, or underwater color imaging. Computer modeling of underwater images can be used to predict the performance of an underwater imaging system in terms of range of viewing and quality of images as a function of the clarity of the water. *See* CHARGE-COUPLED DEVICES; TELEVISION; TELEVISION CAMERA; UNDERWATER PHOTOGRAPHY. Jules S. Jaffe

Underwater vehicle A submersible work platform designed to be operated either remotely or directly. Underwater vehicles are grouped into three categories: deep submersible vehicles (DSVs), remotely operated vehicles (ROVs), and autonomous underwater vehicles (AUVs). There are also hybrid vehicles which combine two or three categories on board a single platform. Within each category of submersible there are specially adaped vehicles for specific work tasks. These can be purpose-built or modifications of standard submersibles.

There are five types of DSVs: one-atmosphere untethered vehicles; one-atmosphere tethered vehicles, including observation/work bells; atmospheric diving suits; diver lockout vehicles; and wet submersibles. While they differ mainly in configuration, source of power, and number of crew members, all carry a crew at 1-atm (10^2-kilopascal) pressure within a dry chamber. An exception is the wet submersible, where the crew is exposed to full depth pressure. The purpose of the DSV is to put the trained mind and eye to work inside the ocean. The earliest submersibles had very small viewing ports fitted into thick-walled steel hulls. In the mid-1960s, experimental work began on use of massive plastics (acrylics) as pressure hull materials. Today, submersibles with depth capabilities to 3300 ft (1000 m) are being manufactured with pressure hulls made entirely of acrylic. Essentially the hull is now one huge window.

The first ROVs were developed in the late 1950s for naval use. By the mid-1970s, they were used in the civil sector. The rapid acceptance of these submersibles is due to their relatively low cost and the fact that they do not put human life at risk when undertaking hazardous missions. However, their most important attribute is that they are less complex. By virtue of their surface-connecting umbilical cable, they can operate almost indefinitely since there is no human inside requiring life support and no batteries to be recharged. There are four types of ROV: tethered free-swimming vehicles, towed vehicles, bottom reliant vehicles, and structurally reliant vehicles.

AUVs are crewless and untethered submersibles which operate independent of direct human control. Their operations are controlled by a preprogrammed, on-board computer. They were first developed in the military where applications include such tasks as minefield location and mapping, minefield installation, submarine decoys, and covert intelligence collection. Civilian tasks include site monitoring, basic oceanographic data gathering, under-ice mapping, offshore structure and pipeline inspection, and bottom mapping. These submersibles are particularly useful where long-duration measurements and observations are to be made and where human presence is not required.

AUVs span a wide range of sizes and capabilities, related to their intended missions. Each is a mobile instrumentation platform with propulsion, sensors, and on-board intelligence desig- ned to complete sampling tasks autonomously. At the large end of the scale, transport-class platforms in the order of 10 m (33 ft) length and 10 metric ton (11 tons) weight in air have been designed for missions requiring long endurance, high speed, large payloads, or high-power sensors. At the small end of the scale, network-class platforms in the order of 1 m (3.3 ft) length and 100 kg (220 lb) weight in air address missions requiring portability, multiple platforms, adaptive spatial sampling, and sustained presence in a specific region. Vehicles can also be categorized in terms of propulsion method (propeller-driven or buoyancy-driven) or in terms of their maximum operating depth.

Hybrid vehicles are those that combine crewed vehicles, remotely operated vehicles, and divers. For example, the hybrid *DUPLUS II* can operate either as a tethered free-swimming ROV or as a 1-atm tethered crewed vehicle. This evolved to provide capability for remotely conducting those tasks for which human skills are not needed, and then to put the human at the place where those skills are required. Other hybrid examples include ROVs that can be controlled remotely from the surface or at the work site by a diver performing maintenance and repair tasks.

Don Walsh

Uninterruptible power supply A source of high-quality alternating power to sensitive electrical equipment. An ideal uninterruptible power supply (UPS) provides a critical load with power when the main supply suffers from an over- or undervoltage condition causing interruption of the supply. It thus ensures power without break for the critical load. The UPS should be reliable and highly efficient, requiring low maintenance, low cost, and light weight. Uninterruptible power supplies were originally designed to furnish reliable and high-quality continuous power to large computer systems. They are now applied to a wide variety of critical equipment, such as medical facilities, life-support systems, financial transaction data storage and computer systems, telecommunications, industrial processing, and on-line management systems. UPS ratings range from a few hundred volt-ampere (VA) up to several thousand kilovolt-ampere (kVA). Three types of UPS systems are available: static, rotary, and hybrid static/rotary. *See* CIRCUIT (ELECTRICITY); ELECTRIC POWER TRANSMISSION.

Static UPS. Static UPS are available in ratings from 100 VA to 1 MVA, powering equipment ranging from personal computers and telecommunication systems, to medium power medical systems, to high power main systems. They are the most commonly used UPS systems, since they have high efficiency, high reliability, and low distortion. The disadvantages of static UPS are their poor performance with nonlinear and unbalanced loads, and high costs of achieving very high reliability. The principal configurations of static UPS systems are line-preferred, inverter-preferred, and line-interactive.

Line-preferred UPS are also referred to as standby or off-line UPS. This type of UPS is extensively employed to support different electrical loads. A line-preferred UPS contains an ac/dc converter, a battery bank, a dc/ac converter, and static and transfer switches. There are three operating modes: normal mode, recovery mode, and backup mode. *See* ALTERNATING CURRENT; CONVERTER; DIRECT CURRENT.

The inverter-preferred configuration is also called a double-conversion UPS or on-line UPS. The system consists of an ac/dc converter (rectifier/charger), a battery unit, a dc/ac converter (inverter), and static (bypass) and transfer switches. The inverter-preferred UPS system has four operating modes: normal, recovery, stored energy, and bypass.

The line-interactive UPS is also known as the grid-interactive UPS system. There are three types: single-conversion, in-line, and double-conversion topologies.

Rotary UPS. In rotary UPS systems a motor/generator set is used for energy conversion. There are two types of rotary UPS systems: flywheel rotary and diesel rotary. A flywheel rotary UPS contains an ac motor, an ac generator, a flywheel unit, a static switch, and an optional bypass switch. The flywheel unit is the energy storage component in this topology. It provides dynamic energy for the generator and feeds the critical load through the generator when the ac line is out of preset tolerance.

The diesel rotary UPS contains a motor/generator set that is connected to a diesel engine. In normal mode, the ac line provides the power for the motor; the generator merely distributes the energy to the load. In backup mode, the ac motor is disconnected from the power line by the control unit, and the diesel engine begins to operate. The rotary UPS systems are more reliable than static UPS systems. *See* ALTERNATING-CURRENT GENERATOR; DIESEL ENGINE; FLYWHEEL; MOTOR.

Hybrid static/rotary UPS. Hybrid static/rotary UPS systems combine the features of both the static and rotary UPS systems. The main advantages of this configuration are better power quality with nonlinear loads, high reliability, frequency stability, better isolation, and low cost of maintenance are.

Applications. UPS systems provide uninterruptible and reliable power for the load including power conditioning in the system. There are two approaches to deploying UPS systems: distributed and centralized.

In the distributed approach many parallel UPS units are placed in an interconnected secure network that feeds the critical loads in a distributed system. Power circulates between the UPS units and the critical load elastically. Only one large UPS unit is employed in the centralized configuration. The large UPS provides continuous operation for the whole system. This configuration is more desirable for industrial and utility applications, due to low maintenance requirements.

Ayse E. Amac; Ali Emadi

Unit operations A structure of logic used for synthesizing and analyzing processing schemes in the chemical and allied industries, in which the basic underlying concept is that all processing schemes can be composed from and decomposed into a series of individual, or unit, steps. If a step involves a chemical change, it is called a unit process; if physical change, a unit operation. These unit operations cut across widely different processing applications, including the manufacture of chemicals, fuels, pharmaceuticals, pulp and paper, processed foods, and primary metals. The unit operations approach serves as a very powerful form of morphological analysis, which systematizes process design, and greatly reduces both the number of concepts that must be taught and the number of possibilities that should be considered in synthesizing a particular process.

Most unit operations are based mechanistically upon the fundamental transport processes of mass transfer, heat transfer, and fluid flow (momentum transfer). Unit operations based on fluid mechanics include fluid transport (such as pumping), mixing/ agitation, filtration, clarification, thickening or sedimentation, classification, and centrifugation. Operations based on heat transfer include heat exchange, condensation, evaporation, furnaces or kilns, drying, cooling towers, and freezing or thawing. Operations that are based on mass transfer include distillation, solvent extraction, leaching, absorption or desorption, adsorption, ion exchange, humidification or dehumidification, gaseous diffusion, crystallization, and thermal diffusion. Operations that are based on mechanical principles include screening, solids handling, size reduction, flotation, magnetic separation, and electrostatic precipitation. The study of transport phenomena provides a unifying and powerful basis for an understanding of the different unit operations. *See* ABSORPTION; ADSORPTION; CENTRIFUGATION; CHEMICAL ENGINEERING; CLARIFICATION; COOLING TOWER; CRYSTALLIZATION; DEHUMIDIFIER; DIFFUSION; DISTILLATION; DRYING; ELECTROSTATIC PRECIPITATOR; EVAPORATION; FILTRATION; FLOTATION; HEAT EXCHANGER; HUMIDIFICATION; ION EXCHANGE; KILN; LEACHING; MAGNETIC SEPARATION METHODS; MECHANICAL SEPARATION TECHNIQUES; MIXING; PUMP; PUMPING

MACHINERY; SEDIMENTATION (INDUSTRY); SOLIDS PUMP; SOLVENT EXTRACTION; THICKENING; TRANSPORT PROCESSES; UNIT PROCESSES.

C. Judson King

Unit processes Processes that involve making chemical changes to materials, as a result of chemical reaction taking place. For instance, in the combustion of coal, the entering and leaving materials differ from each other chemically: coal and air enter, and flue gases and residues leave the combustion chamber. Combustion is therefore a unit process. Unit processes are also referred to as chemical conversions.

Together with unit operations (physical conversions), unit processes (chemical conversions) form the basic building blocks of a chemical manufacturing process. Most chemical processes consist of a combination of various unit operations and unit processes.

The basic tools of the chemical engineer for the design, study, or improvement of a unit process are the mass balance, the energy balance, kinetic rate of reaction, and position of equilibrium (the last is included only if the reaction does not go to completion). *See* CHEMICAL ENGINEERING; UNIT OPERATIONS. W. F. Furter

Unitary symmetry A type of symmetry law, an important example of which is flavor symmetry, an approximate internal symmetry law obeyed by the strong interactions of elementary particles. According to the successful theory of strong interactions, quantum chromodynamics, flavor symmetry is the consequence of the fact that the strong force (the so-called glue force, mediated by the SU_3^{color} gauge field) is the same between any two quarks, the constituents of hadrons. It follows that if all the kinds (flavors) of quarks had the same mass, strong interactions would have the symmetry SU_N^{flavor}, where N is the number of quark flavors. As a result, if nonstrong (electromagnetic and weak) interactions were neglected, then hadrons would occur as degenerate (all having the same mass) multiplets (irreducible representations) of the group SU_N. *See* COLOR (QUANTUM MECHANICS); FLAVOR; HADRON; QUANTUM CHROMODYNAMICS; QUARKS.

An example of unitary symmetry is the spin independence of electric forces on slowly moving electrons. The electron has spin $^1/_2$ and thus two spin states, referred to as spin-up and spin-down. Denoting the wave functions of these two states by $|u\rangle$ and $|d\rangle$, the spin-independent physical properties (energy eigenvalues, charge density, and so on) of a system of electrons such as those in an atom, in which the forces on electrons are nearly spin independent, are unchanged by the replacements shown in Eqs. (1),

$$\begin{aligned} |u\rangle &\rightarrow \alpha|u\rangle + \beta|d\rangle \\ |d\rangle &\rightarrow -\beta^*|u\rangle + \alpha^*|d\rangle \\ |\alpha|^2 &+ |\beta|^2 = 1 \end{aligned} \tag{1}$$

where α and β are complex numbers. These transformations form a group known as SU_2. They correspond to rotations of space; α and β can be expressed in terms of the three numbers which describe the rotation. It is easily seen that many-electron states decompose under the transformation; for example, using Eqs. (1), the two-electron state $(|u, d\rangle - |d, u\rangle)/\sqrt{2}$ is unchanged, and the three remaining two-electron states, $|u, u\rangle$, $|d, d\rangle$, and $(|u, d\rangle + |d, u\rangle)/\sqrt{2}$, transform to linear combinations of themselves. This is the decomposition into singlet and triplet spin states, that is, into total spin $S = 0$ and 1, respectively; the nonmixing between them is equivalent to the invariance of S to rotation. *See* ANGULAR MOMENTUM; ELECTRON SPIN; SPIN (QUANTUM MECHANICS).

The first flavor symmetry encountered in particle physics was the near equivalence of the proton and neutron in nuclear physics. This is called charge independence, *i*- spin (isotopic spin) invariance, or (in terms of the quark model) flavor SU_2 symmetry. The nucleons, proton and neutron, both have spin $^1/_2$ and have nearly the same mass (differing by only 0.14%), and the nuclear force between any two nucleons—whether two protons, a proton and a neutron, or two neutrons—is nearly the same. This is just like the approximate spin independence of the force between two electrons and is describable in the same way, as an SU_2 symmetry, Eqs. (1). Hence one speaks of the proton and neutron as being the two *i*-spin states of a nucleon, a particle with *i*-spin $^1/_2$.

The *i*-spin symmetry extends to all the strongly interacting particles, both baryons and mesons, all of which have *i*-spin as one of their properties. The universality is explained in the quark model of hadrons by the fact that the two least massive quarks, the up (u) and down (d) quarks, are similar (analogous to the similarity of proton and neutron). The masses of the u and d quarks are estimated at roughly 3 MeV and 6 MeV, respectively. Despite this mass ratio of 2 the u and d really do behave similarly in a hadron, because they have there a zero-point energy of the order of 200 or 300 MeV, and so the effect of their masses on their dynamics is very small. *See* I-SPIN.

The third lightest quark is the strange quark (s), with mass about 0.15 GeV. If the s quark were as light as a u or d quark, then these three kinds would be equivalent (as far as the strong interactions were concerned), and any flavor- independent physical properties (energy eigenvalue, for instance) of a state of quarks would be unchanged by replacements similar to the SU_2 transformation shown in Eqs. (1) but with a third state, $|s\rangle$, involved in addition to $|u\rangle$ and $|d\rangle$, and with the complex coefficients specified by eight (rather than three) real numbers. These symmetry transformations form a group known as SU_3 [SU_N denotes the group of $N \times N$ unimodular (determinant = 1) unitary matrices]. The triplet of states $|u\rangle$, $|d\rangle$, and $|s\rangle$, equivalent under the SU_3 symmetry, is analogous to the doublet $|u\rangle$ and $|d\rangle$, equivalent under the SU_2 (*i*-spin) symmetry. Baryons form SU_3 multiplets if the quarks in the baryon are u, d, or s, namely the octet and the decuplet. Because the s quark is significantly heavier than the u and d quarks, the SU_3 (u, d, s) symmetry is only approximate; the masses of the hadrons belonging to an SU_3^{flavor} multiplet differ from one another by the order of 100 or 200 MeV. *See* STRANGE PARTICLES.

The lightest of the three so-called heavy quarks, the charmed quark (c), has a mass of about 1.25 GeV. Since this large mass badly breaks the SU_4 flavor symmetry that would hold if u, d, s, and c all had the same mass, this symmetry is not a useful starting point to relate properties of the multiplet members. The remaining heavy quarks, the bottom (b) and top (t) quarks, are heavier still with masses of about 4.5 and 175 GeV, respectively, and so SU_5 and SU_6 flavor symmetries are even more pointless to consider. In fact the top quark is so massive that it decays (weakly) before it has a chance to be a constituent of any well-defined hadron, its mean life being less than 10^{-24} s. *See* CHARM; J/PSI PARTICLE; UPSILON PARTICLES.

There are unitary symmetries that are not purely internal in the strict sense, namely those in which it is supposed that the interaction between the fundamental particles is spin-independent (as for nonrelativistic electrons interacting with electrostatic forces). *See* ELEMENTARY PARTICLE. Charles J. Goebel

Units of measurement Values, quantities, or magnitudes in terms of which other such are expressed. Units are grouped into systems, suitable for use in the measurement of physical quantities and in the convenient statement of laws relating physical quantities. A quantity is a measurable attribute of phenomena or matter.

A given physical quantity A, such as length, time, or energy, is the product of a numerical value or measure $\{A\}$ and a unit $[A]$. Thus Eq. (1) holds.

$$A = \{A\}[A] \tag{1}$$

The unit $[A]$ can be chosen arbitrarily, but it is desirable to define units in such a way that they are derived from a few base units by equations without numerical factors other than unity, and that the equations between numerical values of quantities have exactly the same form as the equations between the quantities. For example, the kinetic energy E of a body is given in terms of its mass M and speed V by Eq. (2), where $E = \{E\}[E]$,

$$E = \tfrac{1}{2}MV^2 \tag{2}$$

$M = \{M\}[M]$, $V = \{V\}[V]$, and $\tfrac{1}{2}$ is called a definitional factor and is dimensionless. If the units of E, M, and V are defined in such a way that Eq. (3) holds, then the equation between the numerical values is Eq. (4). A system of units defined in this way

$$[E] = [M][V]^2 \tag{3}$$

$$\{E\} = \tfrac{1}{2}\{M\}\{V\}^2 \tag{4}$$

is called a coherent system. It is constructed by defining the units of a few base quantities independently; these are called base units. The units of all other quantities are defined by equations similar to Eq. (3) with no numerical factors other than unity, and are called derived units.

In 1960 the General Conference on Weights and Measures (CGPM) gave official status to a single practical system, the International System of Units, abbreviated SI in all languages. The system is a modernized version of the metric system. The SI, as subsequently extended, includes seven base units and twenty-two derived units with special names. These derived units, and others without special names, are derived from the base units in a coherent manner. A set of prefixes is used to form decimal multiples and submultiples of the SI units. Certain units which are not part of the SI but which are widely used or are useful in specialized fields have been accepted for use with the SI or for temporary use in those fields. *See* METRIC SYSTEM.

Geometrical units. Units of plane angle and solid angle are purely geometrical. The SI units of plane and solid angle are regarded as dimensionless derived units.

Plane angle units. The radian (rad), the SI unit of plane angle, is the plane angle between two radii of a circle which cut off on the circumference an arc equal in length to the radius. Since the circumference of a circle is 2π times the radius, the complete angle about a point is 2π rad. *See* RADIAN MEASURE.

The degree and its decimal submultiples can be used with the SI when the radian is not a convenient unit. By definition, 2π rad $= 360^\circ$. The minute $[1' = (1/60)^\circ]$ and the second $[1'' = (1/60)']$ can also be used.

Steradian. The steradian (sr), the SI unit of solid angle, is the solid angle which, having its vertex at the center of a sphere, cuts off an area on the surface of the sphere equal to that of a square with sides of length equal to the radius of the sphere.

Mechanical units. In mechanics, it is convenient to have three base quantities, and two of these are generally chosen to be length and time. Systems of mechanical units may be classified as absolute systems, in which the third base quantity is mass, and gravitational systems, in which the third base quantity is force.

Two absolute systems of metric units are commonly employed, each named for its base units of length, mass, and time: the mks (meter-kilogram-second) absolute system, and the cgs (centimeter-gram-second) absolute system. The mks absolute system is the mechanical portion of the SI. A coherent absolute system of British units is based on the foot, the pound (1 lb $\cong$ 0.4536 kg), and the second.

Gravitational systems, in which the base quantities are length, force, and time, have been frequently employed by engineers, and are therefore sometimes called technical systems.

Length units. The meter (m) is the SI base unit of length. The use of special names for decimal submultiples of the meter should be avoided, and units formed by attaching appropriate SI prefixes to the meter should be used instead.

The angstrom (Å) is equal to 10^{-10} m. Although it has been accepted for temporary use with the SI, it is preferable to replace this unit with the nanometer, using the relation 1 Å = 0.1 nm.

The nautical mile (nmi), equal to 1852 m, has been accepted for temporary use with the SI in navigation. *See* NAVIGATION.

The foot (ft) is, as discussed above, the unit of length in the British systems of units, and it is also in customary use in the United States. Since 1959 the foot has been defined as exactly 0.3048 m. The yard (yd) is defined as exactly 3 ft or 0.9144 m. *See* LENGTH; MEASURE.

Relative measurements of x-ray wavelengths formerly could be made to a higher accuracy than absolute measurements. Before 1965, most x-ray wavelengths were expressed in terms of the X-unit, which is approximately 10^{-13} m. The X-unit was superseded by the A* unit, which is based on the tungsten $K\alpha_1$ line as a standard. The peak of this line is defined as exactly 0.2090100 A*. X-ray wavelength tables have been published in terms of this unit. At the time the A* unit was defined, it was thought to equal 10^{-10} m (the angstrom unit, Å) to within 5 parts per million, but the A* unit has since been determined to be about 15 parts per million larger than 10^{-10} m.

Special units whose values are obtained experimentally are used in astronomy. For their definitions *See* ASTRONOMICAL UNIT; LIGHT-YEAR; PARSEC.

Area units. The square meter (m^2), the SI unit of area, is the area of a square with sides of length 1 m. Other area units are defined by forming squares of various length units in the same manner. *See* AREA.

Cross sections, which measure the probability of interaction between an atomic nucleus, atom, or molecule and an incident particle, have the dimensions of area, and the appropriate SI unit for expressing them is therefore the square meter. The barn (b), a unit of cross section equal to 10^{-28} m^2, has been accepted for temporary use with the SI.

Units of volume. The cubic meter (m^3), the SI unit of volume, is the volume of a cube with sides of length 1 m. Other units of volume are defined by forming cubes of various length units in the same manner. The liter (symbol L in the United States) is equal to 1 cubic decimeter (1 dm^3), or equivalently to 10^{-3} m^3. It has been accepted for use with the SI for measuring volumes of liquids and gases.

Time units. The second (s) is the SI base unit of time. However, other units of time in customary use, such as the minute (1 min = 60 s), hour (1 h = 60 min), and day (1 d = 24 h), are acceptable for use with the SI. *See* TIME.

Frequency units. The hertz (Hz), the SI unit of frequency, is equal to 1 cycle per second. A periodic oscillation has a frequency of n hertz if it goes through n cycles in 1 s. *See* FREQUENCY (WAVE MOTION).

Speed and velocity units. The meter per second (m/s), the SI unit of speed or velocity, is the magnitude of the constant velocity at which a body traverses 1 m in 1 s. Other speed and velocity units are defined by dividing a unit of length by a unit of time in the same manner. *See* SPEED; VELOCITY.

The knot (kn) is equal to 1 nautical mile per hour (1 nmi/h); it has been accepted for temporary use with the SI.

Acceleration units. The meter per second squared (m/s^2), the SI unit of acceleration, is the acceleration of a body whose velocity changes by 1 m/s in 1 s. Other units of acceleration are defined by dividing a unit of velocity by a unit of time in the same manner. *See* ACCELERATION.

The gal or galileo (symbol Gal) is equal to 1 cm/s^2, or equivalently to 10^{-2} m/s^2.

Mass units. The kilogram (kg), the SI base unit of mass, is the only SI unit whose name, for historical reasons, contains a prefix. Names of decimal multiples and submultiples of the kilogram are formed by attaching prefixes to the word gram (g). The metric ton (t), which is equal to 10^3 kg or 1 megagram (Mg), is permitted in commercial usage of the SI.

The pound (lb), the unit of mass in British absolute system, is also in customary use in the United States. In 1959 the pound was defined to be exactly 0.45359237 kg.

The slug is the unit of mass in the British gravitational system. By definition 1 pound force (lbf) acting on a body of mass 1 slug produces an acceleration of 1 foot per second squared (1 ft/s^2). The slug is equal to approximately 32.174 lb or 14.594 kg. *See* MASS.

Force units. The newton (N), the SI unit of force, is the force which imparts an acceleration of 1 meter per second squared (1 m/s^2) to a body having a mass of 1 kg.

The dyne, the cgs absolute unit of force, is the force which imparts an acceleration of 1 centimeter per second squared (1 cm/s^2) to a body having a mass of 1 g.

The unit of force in the British absolute system is the poundal (pdl), the force which imparts an acceleration of 1 foot per second squared (1 ft/s^2) when applied to a body of mass 1 lb. One poundal is approximately 0.13825 N.

The units of force in the mks gravitational, cgs gravitational, and British gravitational systems are the forces which impart an acceleration equal to the standard acceleration of gravity, $g_n = 9.80665$ m/s$^2 \cong 32.174$ ft/s^2, when applied to bodies having masses of 1 kg, 1 g, and 1 lb, respectively. These units are named the kilogram force (kgf), gram force (gf), and pound force (lbf), respectively. Unfortunately, these units have also been called simply the kilogram, gram, and pound, giving rise to confusion with the mass units of the same name. *See* FORCE.

Pressure and stress units. The pascal (Pa), the SI unit of pressure and stress, is the pressure or stress of 1 newton per square meter (N/m^2). Other units of pressure can also be formed by dividing various units of force by various units of area, such as the pound force per square inch (lbf/in.^{2}, frequently abbreviated psi).

Pressure has been frequently expressed in terms of the bar and its decimal submultiples, where 1 bar $= 10^6$ dynes/cm$^2 = 10^5$ Pa. Pressures are also frequently expressed in terms of the height of a column of either mercury or water which the pressure will support.

Two other units which have been frequently used for measuring pressure are the standard atmosphere and the torr. The standard atmosphere (atm) is exactly 101,325 Pa, which is approximately the average value of atmospheric pressure at sea level. The torr is exactly 1/760 atmosphere, or approximately 133.322 Pa. To within 1 part per million, the torr equal to the pressure of a column of mercury of height 1 millimeter (1 mmHg) at a temperature of 0°C when the acceleration due to gravity has the standard value $g_n = 9.80665$ m/s^2. *See* ATMOSPHERE; PRESSURE; PRESSURE MEASUREMENT.

Energy and work units. The joule (J), the SI unit of energy or work, is the work done by a force of magnitude 1 newton when the point at which the force is applied is displaced 1 m in the direction of the force. Thus, joule is a short name for newton-meter (N-m) of energy or work. *See* ENERGY; WORK.

Units of energy or work in other systems are defined by forming the product of a unit of force and a unit of length in precisely the same manner as in the definition of the joule. Thus, the erg, the cgs absolute unit of energy or work, is the product of 1 dyne and 1 cm.

The foot-poundal (ft-pdl), the British absolute unit of energy or work, is the product of 1 poundal and 1 foot. The foot-pound, or, more properly, the foot-pound force (ft-lbf), the British gravitational unit of energy or work, is the product of 1 lbf and 1 ft.

Sometimes energy is measured in units which are products of a unit of power and a unit of time. Since 1 watt (W) of power equals 1 joule per second (1 J/s), as discussed below, the joule is equivalent to 1 watt-second (1 W · s). In electrical power applications, energy is frequently measured in kilowatthours (kWh), where 1 kWh $= (10^3$ W) (3600 s) $= 3.6 \times 10^6$ J.

The calorie was originally defined as the quantity of heat required to raise the temperature of 1 g of air-free water 1°C under a constant pressure of 1 atm. However, the magnitude of the calorie, so defined, depends on the place on the Celsius temperature scale at which the measurement is made. The International (Steam) Table calorie is defined as exactly 4.1868 J; this is the type of calorie most frequently used in mechanical engineering. The thermochemical calorie, which has been used in thermochemistry in preference to the other types of calorie, is exactly 4.184 J.

The British thermal unit (Btu) was originally defined as the quantity of heat required to raise the temperature of 1 lb of air-free water 1°F under a constant pressure of 1 atm. The International Table Btu is approximately 1055.056 J, and the thermochemical Btu is approximately 1054.350 J.

Power units. The watt (W), the SI unit of power, is the power which gives rise to the production of energy at the rate of 1 joule per second (1 J/s). Other units of power can be defined by forming the ratio of a unit of energy to a unit of time in the same manner. *See* POWER.

The horsepower (hp) is equal to exactly 550 ft · lbf/s, or approximately 745.700 W.

Torque units. The newton-meter (N · m), the SI unit of torque, is the magnitude of the torque produced by a force of 1 newton acting at a perpendicular distance of 1 m from a specified axis of rotation. The joule should never be used as a synonym for this unit.

Units of torque in other systems are defined by forming the product of a unit of force and a unit of length in precisely the same manner as in the definition of the newton-meter.

The foot-poundal (ft · pdl), the British absolute unit of torque, is the product of 1 poundal and 1 foot. The foot-pound (ft · lbf), the British gravitational unit of torque, is the product of 1 lbf and 1 ft. These units are sometimes called the poundal-foot (pdl · ft) and pound-foot (lbf · ft) to distinguish them from the units of energy or work. *See* TORQUE.

Electrical units. For a general discussion of electrical units, including the SI or mks system, three cgs systems [electrostatic system of units (esu), electromagnetic system of units (emu), and gaussian system], and definitions of the SI units ampere (A), volt (V), ohm (Ω), coulomb (C), farad (F), henry (H), weber (Wb), and tesla (T) *see* ELECTRICAL UNITS AND STANDARDS.

This section discusses some additional SI units and some units in the cgs electromagnetic system which are frequently encountered in scientific literature in spite of the fact that their use has been discouraged.

Siemens. The siemens (S), the SI unit of electrical conductance, is the electrical conductance of a conductor in which a current of 1 ampere is produced by an electric potential difference of 1 volt. *See* CONDUCTANCE; ELECTRICAL RESISTANCE.

The siemens was formerly called the mho () to illustrate the fact the unit is the reciprocal of the ohm.

Abampere. The abampere (abA), the cgs electromagnetic unit of current, is that current which, if maintained in two straight, parallel conductors of infinite length, of negligible circular cross section, and placed 1 cm apart in vacuum, would produce between these conductors a force equal to 2 dynes per centimeter of length. The abampere is equal to exactly 10 A.

Abvolt. The abvolt (abV), the cgs electromagnetic unit of electrical potential difference and electromotive force, is the difference of electrical potential between two points of a conductor carrying a constant current of 1 abA, when the power dissipated between these points is equal to 1 erg per second. Then 1 abV $= 10^{-8}$ V.

Maxwell. The maxwell (Mx), the cgs electromagnetic unit of magnetic flux, is the magnetic flux which, linking a circuit of one turn, produces in it an electromotive force of 1 abV as it is reduced to zero in 1 s. Then 1 maxwell $= 10^{-8}$ weber.

Units of magnetic flux density. The gauss (Gs), the cgs electromagnetic unit of magnetic flux density (also called magnetic induction), is a magnetic flux density of 1 maxwell per square centimeter (1 Mx/cm^2). Then 1 gauss = 10^{-4} tesla.

Units of magnetic field strength. The SI unit of magnetic field strength is 1 ampere per meter (1 A/m), which is the magnetic field strength at a distance of 1 m from a straight conductor of infinite length and negligible circular cross section which carries a current of 2π A. This definition is based on the definition, in the SI, of the magnetic field strength H_{SI}. At a distance r from a long straight conductor carrying I, H_{SI} is given by Eq. (5). The

$$2\pi r H_{SI} = I \tag{5}$$

left-hand side of this equation is the line integral of H_{SI} around a circular path, all of whose points are at distance r from the conductor. Substituting $r = 1$ m, and $I = 2\pi$ A in this equation gives $H_{SI} = 1$ A/m.

The oersted (Oe), the cgs electromagnetic unit of magnetic field strength, is the magnetic field strength at a distance of 1 cm from a straight conductor of infinite length and negligible circular cross section which carries a current of 0.5 abA.

Units of magnetic potential and mmf. The ampere serves as the SI unit of magnetic potential difference and magnetomotive force (mmf), as well as the unit of current. In the SI, the magnetomotive force around a closed path equals the current passing through a surface enclosed by the path. Thus, 1 A is the magnetomotive force around a closed path when a current of 1 A passes through an enclosed surface.

The gilbert (Gb), the cgs electromagnetic unit of magnetic potential difference and magnetomotive force, is the magnetomotive force around a closed path enclosing a surface through which flows a current of $(1/4\pi)$ abA.

Photometric units. Photometric units involve a new base quantity, luminous intensity. For the definition of the candela (cd), the SI unit of luminous intensity, *see* PHOTOMETRY; PHYSICAL MEASUREMENT.

For a general discussion of photometric units, including units of illuminance (illumination) and luminance, and in particular the SI units lux (lx) and candela per square meter (cd/m^2), *see* ILLUMINATION. *See also* LUMINANCE.

Lumen. The lumen (lm), the SI unit of luminous flux, is the luminous flux emitted within a unit solid angle (1 steradian) by a point source having a uniform intensity of 1 candela. *See* LUMINOUS FLUX.

Luminous energy units. The lumen-second (lm·s), the SI unit of luminous energy (also called quantity of light), is the luminous energy radiated or received over a period of 1 s by a luminous flux of 1 lumen. This unit is also called the talbot. *See* LUMINOUS ENERGY.

Radiation units. Certain quantities and units are used particularly in the area of ionizing radiation. The special units curie, roentgen, rad, and rem, which were previously adopted for use in this area, are not coherent with the SI, but their temporary use with the SI has been approved while the transition to SI units takes place.

Activity units. The becquerel (Bq), the SI unit of activity (radioactive disintegration rate), is the activity of a radionuclide decaying at the rate of one spontaneous nuclear transition per second. Thus 1 Bq = 1 s^{-1}.

The curie (Ci), the special unit of activity, is equal to 3.7×10^{10} Bq. *See* RADIOACTIVITY.

Exposure units. The SI unit of exposure to ionizing radiation, 1 coulomb per kilogram (1 C/kg), is the amount of electromagnetic radiation (x-radiation or gamma radiation) which in 1 kg of pure dry air produces ion pairs carrying 1 coulomb of charge of either sign. (The ionization arising from the absorption of bremsstrahlung emitted by electrons is not to be included in measuring the charge.) *See* BREMSSTRAHLUNG.

The roentgen (R), the special unit of exposure, is equal to 2.58×10^{-4} C/kg.

Absorbed dose units. The gray (Gy), the SI unit of absorbed dose, is the absorbed dose when the energy per unit mass imparted to matter by ionizing radiation is 1 joule per kilogram (1 J/kg).

The rad (rd), the special unit of absorbed dose, is equal to 10^{-2} Gy.

Dose equivalent units. Different types of radiation cause slightly different effects in biological tissue. For this reason, a weighted absorbed dose called the dose equivalent is used in comparing the effects of radiation on living systems. The dose equivalent is the product of the absorbed dose and various dimensionless modifying factors.

The sievert (Sv), the SI unit of dose equivalent, is the dose equivalent when the absorbed dose of ionizing radiation multiplied by the stipulated dimensionless factors is 1 joule per kilogram (1 J/kg).

The rem, the special unit of dose equivalent, is equal to 10^{-2} Sv.

Other units. Logarithmic measures may be used with the SI. *See* DECIBEL; NEPER; PH; VOLUME UNIT (VU).

For units of loudness (sone, phon) *see* LOUDNESS. For units of sound absorption by surfaces *see* ARCHITECTURAL ACOUSTICS. For units of pitch *see* PITCH. For quantities and units pertaining to sound transmission *see* SOUND.

For temperature scales *see* TEMPERATURE.

For units measuring quantities in chemistry *see* ATOMIC MASS UNIT; CONCENTRATION SCALES; ELECTROCHEMICAL EQUIVALENT; GRAM-MOLECULAR WEIGHT; MOLE (CHEMISTRY); MOLECULAR WEIGHT; RELATIVE ATOMIC MASS; RELATIVE MOLECULAR MASS.

For quantities and units pertaining to nuclear reactors *see* REACTOR PHYSICS.

For units of information content *see* BIT; INFORMATION THEORY.

Jonathan F. Weil

Universal joint A linkage that transmits rotation between two shafts whose axes are coplanar but not coinciding. The universal joint is used in almost every class of machinery: machine tools, instruments, control devices, and, most familiarly, automobiles.

A simple universal joint, known in English-speaking countries as Hooke's joint and in continental Europe as a Cardan joint, is shown in the illustration. It consists of two yokes attached to their respective shafts and connected by means of a spider. The angle between the shafts may have any value up to approximately 35°, if angular velocity is moderate when the angle is large. Although one shaft must make a single revolution for each revolution of the second shaft, the instantaneous angular displacement of the first shaft is the same as that of the second shaft only at the end of each 90° of shaft rotation. Thus, only at four positions during each revolution is angular velocity of both shafts the same.

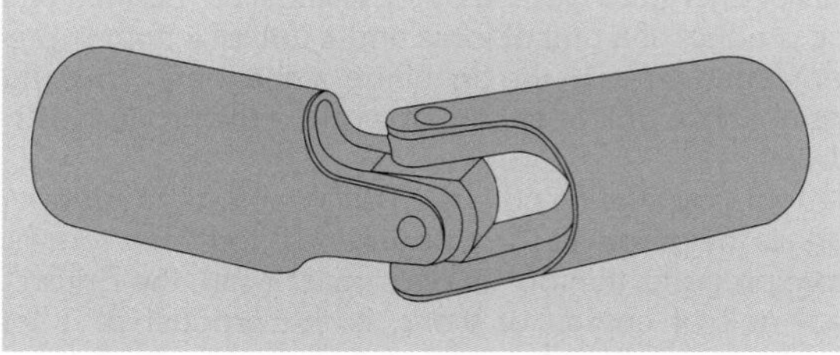

Simple universal joint. ***(After C. W. Ham, E. J. Crane, and W. L. Rogers, Mechanics of Machinery, McGraw-Hill, 1958)***

The variation in angular displacement and angular velocity between driving and driven shafts, which is objectionable in many mechanisms, can be eliminated by using two Hooke's joints, with an intermediate shaft. This arrangement is conventional for an automobile drive shaft. The axes of the driving and driven shafts need not intersect; however, it is necessary that the axes of the two yokes attached to the intermediate shaft lie respectively in planes containing the axes of adjoining shafts. See FOUR-BAR LINKAGE.

Douglas P. Adams

Universal motor A series motor built to operate on either alternating current (ac) or direct current (dc). It is normally designed for capacities less than 1 hp (0.75 kW). It is usually operated at high speed, 3500 revolutions per minutes (rpm) loaded and 8000 to 10,000 revolutions per minute unloaded. For lower speeds, reduction gears are often employed, as in the case of electric hand drills or food mixers. As in all series motors, the rotor speed increases as the load decreases and the no-load speed is limited only by friction and windage. See ALTERNATING-CURRENT MOTOR; DIRECT-CURRENT MOTOR.

Irving L. Kosow

Universe The universe comprises everything in existence, including all matter and energy, and the enormous volume which contains them. The observable universe currently spans about 2.6×10^{23} km (1.6×10^{23} mi), and contains 2.4×10^{52} kg (5.2×10^{23} lb) of matter, yielding an average density of a few atoms per cubic meter. Most of the universe, then, is empty space; the matter is distributed thinly throughout, forming objects and structures at a variety of different sizes.

Constituents. This article will start the cosmic survey with the more familiar objects, following a sequence of increasing size. A number of lesser-known and less tangible entities will complete the survey.

Baryonic matter. Most of the matter encountered in everyday life is in the form of atoms. An atom consists of a positively charged nucleus of protons and neutrons, surrounded by clouds or shells of negatively charged electrons. The protons and neutrons are responsible for most of the mass of the atom. Since both protons and neutrons belong to a class of subatomic particles known as baryons, this ordinary form of atomic matter is called baryonic matter in astronomical contexts. See BARYON.

A large fraction of the visible matter elsewhere in the universe—planets, stars, nebulae, galaxies—is also baryonic in nature, but the relative proportions of the chemical elements are very different from here on Earth. Hydrogen is by far the most abundant element in the universe, representing 75% of the total baryonic mass. Helium is also plentiful at about 23% of the total mass. All the heavier elements make up the remaining 2%.

Stars and stellar evolution. The most numerous components of the nearby universe, as seen in the night sky, are the stars. These pinpricks of light are actually objects much like the Sun, which appear faint due to their extreme distance from Earth. Stars are enormous balls of hot gas (primarily hydrogen and helium) held together by their own gravitation. They are powered by nuclear reactions deep in their interiors, where temperatures and pressures are high enough to fuse hydrogen atoms together into helium, releasing energy in the process. See CARBON-NITROGEN-OXYGEN CYCLES; NUCLEAR FUSION; PROTON-PROTON CHAIN.

Once a star has used up all of its core hydrogen, it must change its internal structure to utilize new fuel sources. A low-mass star (less than a few solar masses) will burn helium for a time, but as the helium fuel is exhausted and its internal furnace wanes, the outer layers of the star are ejected into space, and the core will gradually shrink into a tiny, dense ember, a white dwarf, glowing only from its residual heat.

Higher-mass stars will burn helium, then carbon, and then a succession of even heavier elements, each for a progressively shorter time, until fusable material of any sort abruptly runs out and the star collapses catastrophically. Much of the interior mass of the star is compacted into an ultradense core; the outer layers rebound off this core and explode into space, forming a supernova, which shines for several weeks at a billion times the luminosity of the Sun. The stellar core is usually left behind as a neutron star, a small, rapidly rotating body consisting almost entirely of neutrons. Powerful magnetic fields on the surface of a neutron star produce radio waves which appear to blink on and off as the star spins. A neutron star of this sort is called a pulsar. In the most extreme cases, the stellar core is compressed so far that it collapses to an infinitesimal point, forming a black hole. See BLACK HOLE; NEUTRON STAR; NUCLEOSYNTHESIS; PULSAR; STAR; STELLAR EVOLUTION; SUPERNOVA.

Solar system. The Sun is accompanied by a number of smaller objects of various sizes and compositions. The Sun's gravitational domain extends out to almost half a parsec (1 parsec equals 3.1×10^{13} km or 1.9×10^{13} mi), but its planetary system lies much closer, within about 5×10^{9} km (3×10^{9} mi) of the center. See PLANET; SOLAR SYSTEM.

Extrasolar planets. In recent years, it has become evident that the Sun is not the only star with a planetary system. Over 100 nearby stars, most of them similar in type to the Sun, exhibit subtle periodic shifts in their motion which are most likely due to the gravitational influence of small orbiting companions.

Interstellar material. The "empty" space between the stars actually contains significant amounts of matter—some of it distributed continuously, some of it concentrated in enormous dark clouds—collectively known as the interstellar medium. Also scattered throughout space are denser clouds of molecular hydrogen (H_2), carbon monoxide (CO), and more complex molecules, and are sprinkled with the heavier elements. These clouds often are the site of star formation, when external gravitational disturbances or shock waves trigger the collapse of portions of the clouds. See INTERSTELLAR MATTER; MOLECULAR CLOUD.

Galaxies. Stars and interstellar matter are not distributed uniformly throughout the universe but cluster together in vast units known as galaxies, each containing from 10^{7} to 10^{12} stars. Galaxies are categorized into three types—spirals, ellipticals, and irregulars—with numerous subclasses based on size and structure. The Milky Way Galaxy, the galaxy containing the Earth, is a typical spiral galaxy—a flattened disk of some 10^{11} stars. See GALAXY, EXTERNAL.

Quasars. Supermassive black holes, similar to the stellar-sized black holes which form during the supernova explosion of a massive star but containing millions or billions of solar masses instead of only a few, are thought to exist at the centers of many galaxies, including the Milky Way. As stellar or interstellar matter is drawn into the black hole, it forms an accretion disk around the central point, and is heated by friction to temperatures of millions of kelvins, so that it glows brightly with a luminosity equivalent to 10^{12} Suns. At the far reaches of the universe, these galaxy cores appear as faint star like points of light, and were thus labeled quasistellar objects or quasars when first discovered. See QUASAR.

Groups, clusters, and large-scale structure. Galaxies themselves are usually bound together in groups (up to about 50 galaxies) and clusters (50 to thousands of galaxies) spanning regions 2–10 Mpc in diameter. Clusters accumulate into yet larger entities called superclusters. Groups and clusters seem to be concentrated in thin sheets, surrounding enormous voids with very few galaxies. Superclusters sit at the edges and vertices where surfaces intersect.

Antimatter. All subatomic particles have oppositely charged antiparticle counterparts. When a particle and its corresponding antiparticle collide, they annihilate, converting all their mass to energy. Both matter and antimatter are expected to have been formed in the early universe, but clearly not in precisely equal amounts, since the observable universe is predominantly matter. See ANTIMATTER.

Nonbaryonic particles. Although most observable matter is baryonic in nature, several kinds of nonbaryonic matter have been experimentally observed or have been proposed on the basis of theory. According to current astronomical measurements, nonbaryonic matter accounts for approximately 23% of the total density of the universe.

Neutrinos are electrically neutral, very low mass particles that are generated in nuclear reactions. They were once thought to be completely massless entities, but theoretical and experimental evidence now suggests that they have a tiny but nonzero mass. *See* NEUTRINO.

Other nonbaryonic particle species might also exist. Some of them could be much heavier and slower than neutrinos, but like neutrinos, would interact only weakly with baryonic matter. The most likely such candidate is the so-called weakly interacting massive particle (WIMP). *See* WEAKLY INTERACTING MASSIVE PARTICLE (WIMP).

Dark matter. A large fraction of the mass of the universe is in some form which cannot be seen but which is evident from its gravitational effect on the motions of bright objects, such as stars and galaxies. This dark matter, or "missing mass," is present on many different scales, from galaxies to the universe as a whole.

Energy. Energy is a physical entity as real as matter, but somewhat less tangible, which makes it more difficult to categorize easily. Like matter, energy comes in many different forms, which can be readily transformed from one to another. Energy can also be converted to and from matter.

Research suggests that, on cosmological scales, gravitational attraction is countered by a previously undetected form of energy, known as dark energy or the cosmological constant, which causes space to expand. Since 1998, observations of distant supernovae and the cosmic background radiation have led astronomers to the conclusion that the expansion of the universe is accelerating. According to these data, the dark energy comprises 73% of the total density of mass and energy in the universe. The actual nature of the dark energy is still a mystery. *See* COSMOLOGICAL CONSTANT; DARK ENERGY; RELATIVITY.

The traditional four basic forces of nature—gravity, electromagnetism, and the strong and weak nuclear forces—may in fact be different facets of a single fundamental force. Particle physics experiments show that under conditions of extremely high energy, the weak nuclear force and electromagnetism merge into a single "electroweak" force. Theoretical efforts are being made to devise a grand unified theory under which the strong force and gravity are also incorporated. Such a unified force may have existed during the moments following the big bang and then fragmented into separate forces as the universe expanded and cooled. *See* ELECTROWEAK INTERACTION; FUNDAMENTAL INTERACTIONS; GRAND UNIFICATION THEORIES; STANDARD MODEL.

Origin, evolution, and fate. The universe is a dynamically evolving system. By closely studying the current distribution and motion of matter and energy, and collecting the "fossil light" from distant objects, scientists have constructed a fairly consistent picture of the creation of the universe, the big bang theory, which explains the observations fairly well.

Big bang. The observable universe originated 13.7 billion years ago in a fiery cataclysm termed the big bang. This was not an explosion of compressed matter and energy into a previously empty space. Instead, space itself, as well as everything contained therein, sprang from a single point of an infinite density and temperature, and grew to the volume observed today. As it expanded, it cooled, allowing familiar forms of matter to condense from the high-energy "soup" of energy and subatomic particles that constituted the very early universe.

Cosmological redshifts. Distinct spectral features, characteristic of the different chemical elements that are present, appear at well-established locations in the spectrum of an astronomical object, but the entire spectrum is shifted toward shorter wavelengths (bluewards) for an object approaching the observer, or toward longer wavelengths (redwards) for a receding object. The amount of this Doppler shift is closely related to the actual relative speed of the object. *See* DOPPLER EFFECT.

With the exception of a few nearby galaxies, all galaxies exhibit redshifts of varying amounts, implying that they are all moving away from the Earth. Edwin Hubble extended this work by determining the distances to these galaxies, and found that recession velocity was directly proportional to distance—that the more distant galaxies were moving away faster. The constant of proportionality relating speed and distance now bears Hubble's name, and an accurate determination of this Hubble constant (H_0) has been one of the central pursuits of modern astronomy. Recent studies seem to be converging towards a value of $H_0 = 72$ km/(s)(Mpc).

If space itself is expanding uniformly in all directions, and the galaxies are being carried along in this general expansion, then any point in the universe would see all other points moving away. The greater the distance between two points, the more space exists between them, and the faster this distance increases. Hubble's law is therefore a consequence of the uniform expansion of space, which causes more distant galaxies to exhibit larger redshifts because they are receding from the observer faster.

If space is expanding uniformly in all directions, then in the past the universe was smaller. At some point in the past, all matter and energy may have existed in a single point of infinite temperature and density. *See* HUBBLE CONSTANT; REDSHIFT.

Cosmic background radiation. If this picture is correct, and the current universe was spawned from a primordial fireball, then the sky should be filled with a residual afterglow from the era when the matter in the universe was hot and emitting strongly, in much the same way that the surfaces of stars shine today. This microwave background was first detected in 1965. The *Cosmic Background Explorer* (*COBE*) satellite mission in 1990 that the spectrum was measured its spectrum over a wide range of wavelengths, establishing the cosmic background radiation temperature at 2.726 K.

Since the cosmic background radiation is an intrinsic characteristic of the universe, observers expect to see it reaching the Earth uniformly from all directions in space, and the *COBE* measurements confirmed that this is indeed the case, once the Doppler effect from the Earth's motion through space is subtracted. Subsequent observations of the cosmic background radiation by highly sensitive microwave telescopes, including the *Wilkinson Microwave Anisotropy Probe* (*WMAP*), have revealed faint ripples in the temperature distribution of the background radiation. These patterns in the cosmic background radiation are a result of the very first clumps of matter that accreted in the early universe, as regions with slightly different densities emitted slightly hotter or cooler radiation. By measuring the amplitude and angular size of these temperature variations, researchers have been able to deduce the age, total density, and early history of the universe with unprecedented precision. *See* COSMIC BACKGROUND RADIATION; WILKINSON MICROWAVE ANISOTROPY PROBE.

Evolution of the universe. The best models for the beginning of the universe start at 10^{-43} second after the big bang itself. At 10^{-42} s, the universe was 10^{45} times smaller than it is today (approximately 6×10^{-18} cm or 2×10^{-18} in.), and had a mean temperature of 10^{30} K. At such temperatures, matter as presently known cannot exist, because the energies are so enormous that even the protons and neutrons themselves are torn apart into separate sub-subatomic particles called quarks. The universe was a featureless mixture of subatomic particles and high-energy photons. *See* LEPTON; QUANTUM GRAVITATION; QUARKS; SUPERGRAVITY; SUPERSYMMETRY.

At 10^{-34} s, when the temperature had dropped to 3×10^{26} K, heavier exotic particles like magnetic monopoles could have emerged. Around this time, the rapidly enlarging structure of space may have undergone an era of even faster expansion, driven by the energy of space itself. This era of hyperfast inflation helps to explain features of the present-day observable universe, such as the uniformity of the cosmic background radiation over the entire sky, and the way in which the average mass density of the universe is high enough for structures like stars and galaxies to form, but not so large that it would immediately recollapse upon itself. *See* INFLATIONARY UNIVERSE COSMOLOGY; MAGNETIC MONOPOLES.

When the temperature had dropped further, to about 3×10^{12} K, protons and neutrons condensed out of the quark mixture. It was still much too hot for electrons to join them and form atoms, but more complex atomic nuclei were created in a fashion similar to stellar core fusion. This era of big bang nucleosynthesis established the initial composition of the universe, about three-quarters hydrogen, one-quarter helium, and a smattering of lithium, from which all subsequent stellar and supernova nucleosynthesis has proceeded.

Temperatures were still too high for electrons to join these nuclei to make complete atoms until a time about 370,000 years after the big bang. Up to this point, the universe was relatively opaque, since unattached electrons are very good at absorbing light and other electromagnetic radiation. Once the temperature fell below 3000 K, however, protons and electrons could combine into neutral hydrogen atoms, and the universe suddenly became transparent to most wavelengths. The cosmic background radiation formed at this point and has had little interaction with matter since. The formation of structure—the development of clumpiness in the universe, which is seen today as stars, galaxies, and clusters—started slightly prior to matter-radiation decoupling.

Ultimate fate of the universe. On larger scales, the universe will probably continue to expand indefinitely. It was once thought that if the average cosmic density of matter were high enough, the mutual gravitational attraction would be sufficiently strong to gradually slow the expansion to a standstill, and then cause the universe to contract, eventually collapsing in a "Big Crunch" (or "Gnab Gib"). According to the current tally of baryonic and dark matter, however, there is not enough gravitational mass in the universe to overcome the current rate of expansion. With the added repulsive force of the dark energy, the expansion rate will increase, instead of decrease. Objects such as stars, galaxies, and clusters, which are kept together by their own self-gravity, will remain largely intact, but the enormous voids separating clusters and superclusters from each other will grow ever larger. Eventually, over time scales of trillions of years, the universe will grow empty and dim, illuminated by only a few last stellar remnants, cooling slowly toward absolute zero. *See* BIG BANG THEORY; COSMOLOGY.

Bradford B. Behr

Upper-atmosphere dynamics The motion of the atmosphere above 50 km (30 mi). The predominant dynamical phenomena of the upper atmosphere are quite different from those encountered in the lower atmosphere. Among those encountered in the lower atmosphere are cyclones, anticyclones, tropical hurricanes, thunderstorms and shower clouds, tornadoes, and dust devils. Even the largest of these phenomena do not penetrate far into the upper atmosphere. Above an altitude of about 50 km (30 mi), the predominant dynamical phenomena are internal gravity waves, tides, sound waves (including infrasonic), turbulence, and large-scale circulation.

Except under meteorological conditions characterized by convection, the atmosphere is stable against small vertical displacements of small air parcels; this results from buoyancy forces that tend to restore displaced air parcels to their original levels. An air parcel therefore tends to oscillate around its undisturbed position at a frequency known as the Brunt-Vaisala frequency ω_B. If pressure waves are generated in the atmosphere with frequencies much greater than ω_B, they propagate as sound waves. For frequencies much less than ω_B, the waves propagate as internal gravity waves; in this case, the restoring forces for the wave motion are provided primarily by buoyancy (that is, gravity) rather than by compression.

Tides are internal gravity waves of particular frequencies. The term tidal usually implies that the exciting force is gravitational attraction by the Moon or Sun. However, it is conventional in the case of atmospheric tides to include also those waves that are excited by solar heating. One is therefore concerned with three separate excitation functions—lunar gravitation, solar gravitation, and solar heating.

Tidal wind patterns in the upper atmosphere generate electrical currents in the ionosphere through a dynamo action. These in turn give rise to diurnal variations in the geomagnetic field that can be observed at the Earth's surface. *See* ATMOSPHERIC TIDES; GEOMAGNETIC VARIATIONS; IONOSPHERE.

Sound waves generated in the lower atmosphere may propagate upward; to maintain continuity of energy flow, the waves might be expected to grow in relative amplitude as they move into the more rarefied upper atmosphere. However, higher temperatures in the upper atmosphere refract most of the energy back toward the Earth's surface, giving rise to the phenomenon known as anomalous propagation. This involves the redirection of upward-moving sound waves back to the surface beyond the point where the source can be heard by waves propagating along the surface. Infrasonic waves with periods from 20 to 80 s have been observed occasionally with detectors at the Earth's surface in connection with auroral activity. *See* SOUND.

There is clear visual evidence of turbulence in the upper atmosphere; this evidence is obtained by examination of vapor trails released from rockets or of long-persisting meteor trails. The source of the turbulence is not clear. The atmosphere is thermodynamically stable against vertical displacements throughout the region above the troposphere, and work has to be done against buoyancy forces in order to produce and maintain turbulence. The only apparent source of energy is internal gravity waves, either tidal or of random period.

There are prevailing patterns of atmospheric circulation in the upper atmosphere, but they are very different from those that occur in the lower atmosphere, which are associated with weather systems and have complicated structures resulting from growth of instabilities. The upper atmospheric large-scale wind systems are mainly diurnal in nature and global in scale.

The main heat source that is responsible for the upper atmospheric circulation is ultraviolet radiation from the Sun, radiation that is mainly absorbed at altitudes between 100 and 200 km (60 and 120 mi). The atmosphere is not a good infrared radiator in this altitude region, so the temperature rises rapidly with altitude, providing a temperature gradient of such a magnitude that molecular conduction transfers the absorbed heat downward to altitudes below 100 km (60 mi) where the atmosphere does have the capability of radiating the energy back to space. Above about 300 km (180 mi), the temperature becomes roughly constant with altitude because very little energy is absorbed there (the gas is exceedingly rarefied) and the thermal conductivity is good enough under these circumstances to virtually eliminate vertical temperature gradients. *See* SOLAR RADIATION.

The region of rising temperature above 80 km (48 mi) is known as the thermosphere. The exosphere is that region of the atmosphere that is so rarefied that for many purposes collisions between molecules can be neglected; it is roughly the region above 500 km (300 mi).

At high latitudes, the ionosphere moves in response to electric fields imposed as a consequence of interactions between the Earth's magnetic field and the solar wind. The imposed electric field causes the ionosphere to drift in a generally antisunward direction over the polar caps (regions at higher latitudes than the auroral zone, or magnetic latitudes greater than about 68°), with a return circulation (that is, generally sunward in direction) just outside the polar caps. *See* ATMOSPHERIC GENERAL CIRCULATION; SOLAR WIND; WIND.

Ultraviolet radiation of suitable wavelengths can photodissociate atmospheric molecules—something of great importance in the upper atmosphere. It is even important in the stratosphere, where ozone is formed as a result of absorption by molecular oxygen of ultraviolet radiation, the important wavelengths being below 242 nanometers. *See* OZONE; STRATOSPHERE.

Francis S. Johnson

Upsilon particles A family of elementary particles whose first three members were discovered in 1977. The upsilon mesons, Υ, are the heaviest known vector mesons, with masses greater than 10 times that of the proton. They are bound states of a heavy quark and its antiquark. The quarks which bind to form the upsilons carry a new quantum number called beauty or bottomness, and they are called *b*-quarks or *b*. The mass of the *b*-quark is around 5 GeV. The anti-*b*-quark or *b* carries antibeauty, and therefore the upsilons carry no beauty and are often called hidden beauty states. Direct proof of the existence of the *b*-quark was obtained by observing the existence of *B* mesons which consist of a *b*-quark bound to a lighter quark. Thirteen *bb*-mesons have also been observed thus far. *See* ELEMENTARY PARTICLE; MESON; QUARKS.

Paolo Franzini

Upwelling The phenomenon or process involving the ascending motion of water in the ocean. Vertical motions are an integral part of ocean circulation, but they are a thousand to a million times smaller than the horizontal currents. Vertical motions are inhibited by the density stratification of the ocean because with increasing depth, as the temperature decreases, the density increases, and energy must be expended to displace water vertically. The ocean is also stratified in other properties; for example, nutrient concentration generally increases with depth. Thus even weak vertical flow may cause a significant effect by advecting nutrients to a new level. *See* GEOPHYSICAL FLUID DYNAMICS.

There are two important upwelling processes. One is the slow upwelling of cold abyssal water, occurring over large areas of the world ocean, to compensate for the formation and sinking of this deep water in limited polar regions. The other is the upwelling of subsurface water into the euphotic (sunlit) zone to compensate for a horizontal divergence of the flow in the surface layer, usually caused by winds. *See* OCEAN CIRCULATION.

Robert L. Smith

Uranium A chemical element, symbol U, atomic number 92, atomic weight 238.03. The melting point is 1132°C (2070°F) and the boiling point is 3818°C (6904°F). Uranium is one of the actinide series. *See* ACTINIDE ELEMENTS; PERIODIC TABLE.

Uranium in nature is a mixture of three isotopes: ^{234}U, ^{235}U, and ^{238}U. Uranium is believed to be concentrated largely in the Earth's crust, where the average concentration is 4 parts per million (ppm). The total uranium content of the Earth's crust to a depth of 15 mi (25 km) is calculated to be 2.2×10^{17} lb (10^{17} kg); the oceans may contain 2.2×10^{13} lb (10^{13} kg) of uranium. Several hundred uranium-containing minerals have been identified, but only a few are of commercial interest. *See* RADIOACTIVE MINERALS; URANINITE.

Because of the great importance of the fissile isotope ^{235}U, rather sophisticated industrial methods for its separation from the natural isotope mixture have been devised. The gaseous diffusion process, which in the United States is operated in three large plants (at Oak Ridge, Tennessee; Paducah, Kentucky; and Portsmouth, Ohio) has been the established industrial process. Other processes applied to the separation of uranium include the centrifuge process, in which gaseous uranium hexafluoride is separated in centrifuge cascades, the liquid thermal diffusion process, the separation nozzle, and laser excitation. *See* ISOTOPE (STABLE) SEPARATION.

Uranium is a very dense, strongly electropositive, reactive metal; it is ductile and malleable, but a poor conductor of electricity. Many uranium alloys are of great interest in nuclear technology because the pure metal is chemically active and anisotropic and has poor mechanical properties. However, cylindrical rods of pure uranium coated with silicon and canned in aluminum tubes (slugs) are used in production reactors. Uranium alloys can also be useful in diluting enriched uranium for reactors and in providing liquid fuels. Uranium depleted of the fissile isotope ^{235}U has been used in shielded containers for storage and transport of radioactive materials. *See* NUCLEAR FUELS; NUCLEAR REACTOR.

Uranium reacts with nearly all nonmetallic elements and their binary compounds. Uranium dissolves in hydrochloric acid and nitric acid, but nonoxidizing acids, such as sulfuric, phosphoric, or hydrofluoric acid, react very slowly. Uranium metal is inert to alkalies, but addition of peroxide causes formation of water-soluble peruranates. *See* URANIUM METALLURGY.

Uranium reacts reversibly with hydrogen to form UH_3 at 250°C (482°F). Correspondingly, the hydrogen isotopes form uranium deuteride, UD_3, and uranium tritide, UT_3. The uranium-oxygen system is extremely complicated. Uranium monoxide, UO, is a gaseous species which is not stable below 1800°C (3270°F). In the range UO_2 to UO_3, a large number of phases exist. The uranium halides constitute an important group of compounds. Uranium tetrafluoride is an intermediate in the preparation of the metal and the hexafluoride. Uranium hexafluoride, which is the most volatile uranium compound, is used in the isotope separation of ^{235}U and ^{238}U. The halides react with oxygen at elevated temperatures to form uranyl compounds and ultimately U_3O_8.

Fritz Weigel

Uranium metallurgy The processing treatments for the production of uranium concentrates and the recovery of pure uranium compounds, as well as the conversion chemistry for producing uranium metal and the processes employed for preparing uranium alloys.

The procedures to recover uranium from its ores are numerous, because of the great variety in the nature of uranium minerals and associated materials and the wide range of concentration in the naturally occurring ores. Recovery of uranium requires chemical processing; however, preliminary treatment of the ore may involve a roasting operation, a physical or chemical concentration step, or a combination of these. In general, one of two leaching treatments—acid leaching and carbonate leaching—is used as the initial step in chemical concentration. The choice depends on the nature of the ore, which largely determines the efficiency and the cost of the process employed. *See* LEACHING.

The concentrate, whether obtained by chemical or physical means, is treated chemically to give a uranyl nitrate solution that can be further purified by solvent extraction. The impurities remain in the aqueous phase, while the uranium is extracted into the organic phase. *See* SOLVENT EXTRACTION.

Uranium metal can be obtained from its halides by fused-salt electrolysis or by reduction with more reactive metals. The reaction of UO_2 with calcium yields metal of fair quality. The largest tonnages of good-quality uranium have been produced by metallothermic reduction of finely divided UF4 with calcium or magnesium in steel bombs lined with fused dolomitic oxide (Ames process). The charge, consisting of an intimate mixture of UF_4 with the reductant metal in granular form together with a suitable booster (usually calcium plus iodine), is placed into the reduction bomb, the lid is bolted down, and the bomb is heated to ignition temperature. The shape of the metal ingot (also referred to as a biscuit) depends on the shape of the reduction bomb.

Uranium alloys are prepared by fusing the components together. All procedures following conventional metallurgical techniques, however, may have to be carried out inside inert-gas glove boxes because many alloys are attacked by oxygen or moisture. *See* NUCLEAR FUELS; NUCLEAR FUELS REPROCESSING; URANIUM.

F. Weigel

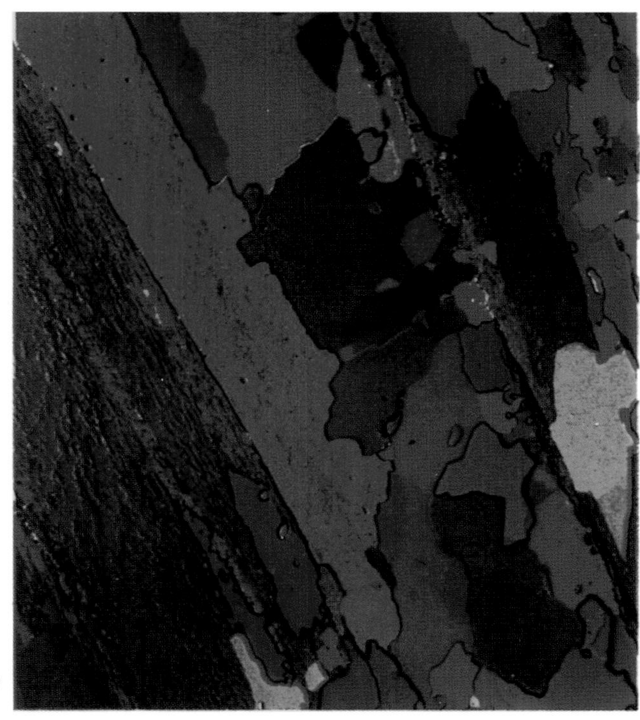

(a)

Photomicrographs of aluminum-uranium alloy. Samples have been ground and polished by hand, then polished and anodized electrolytically. This produces a very thin oxide film, whose thickness is dependent on the substrata orientation. The grain size, location and amount of different phases, and other identifying characteristics of the microstructure are revealed by the color variations. (*a*) As cast. Dark discontinuous lines are UAl_4. (*b*) Cold-worked. The UAl_4 seen in *a* has been broken up rather severely, showing a highly deformed microstructure. (*c*) Recrystallized. Fine grains have nucleated after the cold-worked alloy has been heated above the recrystallization temperature. All ×75. (*USAEC — Union Carbide Corp., Oak Ridge National Laboratory*)

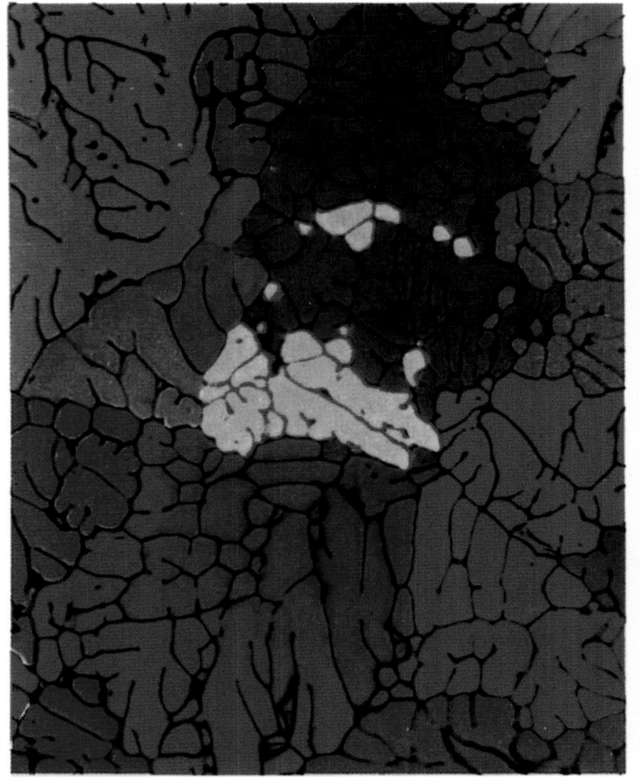

(b)

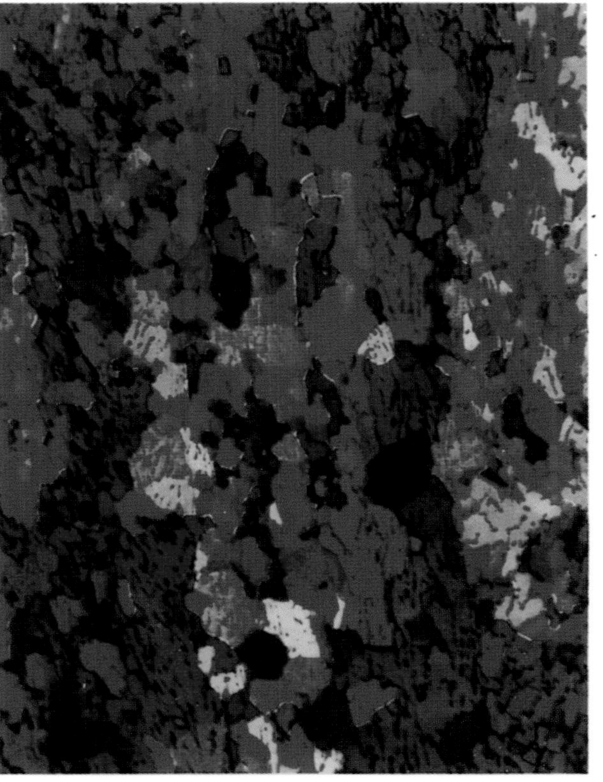

(c)

Uranus The first planet to be discovered with the telescope and the seventh in the order of distance from the Sun. It was found accidentally by W. Herschel in England on March 13, 1781. *See* PLANET.

The obliquity (inclination of rotational axis to orbit plane) of Uranus is 98°, exceeded only by the obliquities of Pluto and Venus. This means that the axis is almost in the plane of the orbit, and thus the seasons on Uranus are very unusual. During summer in one hemisphere, the pole points almost directly toward the Sun while the other hemisphere is in total darkness. Forty-two years later, the situation is reversed. *See* PLUTO; VENUS.

The linear equatorial diameter is 31,763 mi (51,118 km). The mass is 14.5 times the mass of the Earth. The corresponding mean density is greater than that of Saturn even though Uranus is smaller than Saturn. This means that Uranus is richer than Saturn (or Jupiter) in elements heavier than hydrogen and helium but not nearly so rich in these elements as Earth. The same is true for Neptune. *See* JUPITER; NEPTUNE; SATURN.

The temperature that Uranus would assume in simple equilibrium with incident solar radiation at this distance is about 55 K (−360°F). This is essentially identical with the value found by direct measurement. There is thus no evidence for the existence of an internal energy source as observed for Jupiter, Saturn, and Neptune.

Through the telescope, Uranus appears as a small, slightly elliptical blue-green disk. This appearance is confirmed in the nearly featureless pictures of the planet obtained by *Voyager 2*. Uranus owes its characteristic aquamarine color to the relatively high proportion of methane in its atmosphere. The methane abundance is 20 to 30 times the amount corresponding to a solar distribution of the elements, in agreement with the proportion of heavy elements in the planet deduced from its average density. In striking contrast, the proportion of helium to hydrogen, the two most abundant gases in the planet's atmosphere, is essentially equal to that observed in the Sun. Like Jupiter and Saturn, Uranus exhibits zonal winds that are parallel to the equator, despite the unusual orientation of the planet's rotational axis.

Before the *Voyager 2* encounter, only five satellites of Uranus had been discovered. They form a remarkably regular system with low orbital eccentricities and inclinations close to the plane of the planet's equator. The *Voyager* cameras found 11 additional small satellites, including two that may serve as gravitational "shepherds" for the outermost ring, and two were subsequently identified. Like Jupiter, Saturn, and Neptune, Uranus also has more distant, irregular satellites (moons that move in orbits with high inclinations or eccentricities). As of August 2008, nine of these unusual moons were definitely established, all discovered since 1997.

A system of 10 narrow rings can be observed around Uranus. The mean width of the widest is only 36 mi (58 km) which is still six to seven times larger than that of the next widest rings.

Tobias C. Owen

Urban climatology The branch of climatology concerned with urban areas. These locales produce significant changes in the surface of the Earth and the quality of the air. In turn, surface climate in the vicinity of urban sites is altered. The era of urbanization on a worldwide scale has been accompanied by unintentional, measurable changes in city climate. *See* CLIMATOLOGY.

The process of urbanization changes the physical surroundings and induces alterations in the energy, moisture, and motion regime near the surface. Most of these alterations may be traced to causal factors such as air pollution; anthropogenic heat; surface waterproofing; thermal properties of the surface materials; and morphology of the surface and its specific three-dimensional geometry—building spacing, height, orientation, vegetative layering, and the overall dimensions and geography of these elements. Other factors that must be considered are relief, nearness to water bodies, size of the city, population density, and land-use distributions.

In general, cities are warmer than their surroundings, as documented over a century ago. They are islands or spots on the broader, more rural surrounding land. Thus, cities produce a heat island effect on the spatial distribution of temperatures. The timing of a maximum heat island is followed by a lag shortly after sundown, as urban surfaces, which absorbed and stored daytime heat, retain heat and affect the overlying air. Meantime, rural areas cool at a rapid rate.

A number of energy processes are altered to create warming, and various features lead to those alterations. City size, the morphology of the city, land-use configuration, and the geographic setting (such as relief, elevation, and regional climate) dictate the intensity of the heat island, its geographic extent, its orientation, and its persistence through time. Individual causes for heat island formation are related to city geometry, air pollution, surface materials, and anthropogenic heat emission. There are two atmosphere layers in an urban environment, besides the planetary boundary layer outside and extending well above the city: (1) The urban boundary layer is due to the spatially integrated heat and moisture exchanges between the city and its overlying air. (2) The surface of the city corresponds to the level of the urban canopy layer. Fluxes across this plane comprise those from individual units, such as roofs, canyon tops, trees, lawns, and roads, integrated over larger land-use divisions (for example, suburbs).

Anthony J. Brazel

Urea A colorless crystalline compound, formula CH_4N_2O, melting point 132.7°C (270.9°F). Urea is also known as carbamide and carbonyl diamide, and has numerous trade names as well. It is highly soluble in water and is odorless in its purest state, although most samples of even high purity have an ammonia odor. The diamide of carbonic acid, urea has the structure below.

$$H_2N - \overset{\overset{\displaystyle O}{\|}}{C} - NH_2$$

Urea occurs in nature as the major nitrogen-containing end product of protein metabolism by mammals, which excrete urea in the urine. The adult human body discharges almost 50 g (1.8 oz) of urea daily. Urea was first isolated in 1773 by G. F. Rouelle. By preparing urea from potassium cyanate (KCNO) and ammonium sulfate (NH_4SO_4) in 1828, F. Wöhler achieved a milestone, the first synthesis of an organic molecule from inorganic starting materials, and thus heralded the modern science of organic chemistry. *See* NITROGEN; PROTEIN METABOLISM.

Because of its high nitrogen content (46.65% by weight), urea is a popular fertilizer. About three-fourths of the urea produced commercially is used for this purpose. After application to soil, usually as a solution in water, urea gradually undergoes hydrolysis to ammonia (or ammonium ion) and carbonate (or carbon dioxide). Another major use of urea is as an ingredient for the production of urea-formaldehyde resins, extremely effective adhesives used for laminating plywood and in manufacturing particle board, and the basis for such plastics as malamine. *See* FERTILIZER; UREA-FORMALDEHYDE RESINS.

Other uses of urea include its utilization in medicine as a diuretic. In the past, it was used to reduce intracranial and intraocular pressure, and as a topical antiseptic. It is still used for these purposes, to some extent, in veterinary medicine and animal husbandry, where it also finds application as a protein feed supplement for cattle and sheep. Urea has been used to brown baked goods such as pretzels. It is a stabilizer for nitrocellulose explosives because of its ability to neutralize the nitric acid that is formed from, and accelerates, the decomposition of the nitrocellulose. Urea was once used for flameproofing fabrics. Mixed with barium hydroxide, urea is applied to limestone monuments to slow erosion by acid rain and acidic pollutants.

Robert D. Walkup

Urea-formaldehyde resins The condensation products obtained by the reaction of urea or melamine with formaldehyde. Resinous condensation products of formaldehyde with other nitrogen-containing compounds, for example, aniline and amides, also belong to this group of resins but have gained only limited utility.

The urea- and melamine-formaldehyde resins (amino resins) possess an excellent combination of physical properties and can be easily fabricated in a variety of colors. They are widely used as adhesives, laminating resins, molding compounds, paper and textile finishes, and surface coatings.

The aniline- and sulfonamide-formaldehyde resins are produced from other compounds containing $—NH_2$ groups that condense with formaldehyde to form methylol derivatives which are capable of further reaction. These resins have received limited attention except for aniline and *p*-toluenesulfonamide. Because of their resistance to the absorption of water, the aniline-formaldehyde resins have been used in electrical applications, such as insulation or panels, where their natural brown color is not objectionable. The sulfonamide-formaldehyde polymers are less colored than the aniline-type resins and have been employed in surface coatings. *See* CONDENSATION REACTION; FORMALDEHYDE; POLYMERIZATION; UREA.

John A. Manson

Urediniomycetes (rust) A class of fungi known as plant rusts that belong to the division (phylum) Basidiomycota. In nature, all 7000 species are obligate parasites of many vascular plant species. They cause diseases known as rust on numerous cultivated crops, and many trees are also attacked. Each rust species infects one or just a few closely related plant host species. Rust fungi occur on all continents except Antarctica.

The body of a rust fungus consists of numerous microscopic, threadlike, branching hyphae that grow inside the tissues and between the cells of the host plant. Specialized feeding structures, haustoria, grow from these hyphae into the host cells of the plant and provide nourishment for the hyphae. Hyphal cells are binucleate. Depending upon the rust species and environmental conditions, up to six different kinds of spore-producing structures, the sori, may be produced by one rust species. These sori may be powdery, waxy, or crustlike, and whitish, yellow, orange, brown, or blackish.

Rust teliospores that are produced in sori called telia are essential elements for classification. Teliospores consist of one or more specialized probasidial cells. During the development of each of these cells, nuclei fuse and meiosis occurs, resulting in a septate metabasidium from which four haploid meiospores (basidiospores) are formed. Basidiospores are forcibly ejected from the tips of their tiny stalks and are wind-disseminated. After landing on a susceptible host, under proper conditions basidiospores produce new infections. In cold regions, a rust may survive the winter as dormant, thick-walled, dark-colored teliospores. In many rust species, teliospores mature and produce basidiospores with no or little dormancy, especially in the tropics.

Although most rust species require only one host species to complete their life cycles (autoecious rusts), some of the best-known rusts require two taxonomically unrelated hosts (heteroecious rusts).

Rust diseases are controlled most effectively by breeding resistant host varieties. Fungicides are used for some rusts. In the case of some heteroecious rusts, eradication of one of the hosts (the noneconomic one) has been successful. Some rust species are used to aid in biological control of weeds. *See* BASIDIOMYCOTA; EUMYCOTA; FUNGI; PLANT PATHOLOGY.

Joe F. Hennen

Uric acid An end product of purine metabolism in humans and higher primates. Uric acid is excreted as such in the urine, and it is very poorly soluble in aqueous solutions, causing it to crystallize when concentrations of the compound in the urine are abnormally high. This leads to crystalluria (excretion of crystals in urine), hematuria (blood in the urine), infection, or urinary tract stones. Other mammals do not experience these problems, as they express an enzyme, uricase, which catalyzes the conversion of uric acid to allantoin, which is highly soluble. A nonprimate exception is the Dalmatian dog, in which high rates of excretion of uric acid are a consequence of defective renal tubular reabsorption; thus, the uric acid is excreted before it can be oxidized. In birds and reptiles, uric acid is an excretory end product of the metabolism of proteins. *See* PURINE.

Uric acid may be found in large amounts in blood (hyperuricemia) due to rapid nucleic acid turnover (for example, in leukemia), and especially due to the acute lysis of tumor cells (for example, in successful chemotherapy). In this situation, uric acid crystallization in the kidney can cause kidney failure. Uric acid concentration may also rise in the blood because of diminished renal excretion, the situation in most patients with gout; in renal disease; under conditions such as lactic acidemia or ketoacidosis, in which organic acids such as lactic acid compete with uric acid for tubular reabsorption; and in response to certain diuretics. Hyperuricemia also results from overproduction of purines through the de novo pathway, as occurs in Lesch-Nyhan disease. Hyperuricemia that results from diminished renal excretion, but not from renal failure, can be treated with drugs that cause increased excretion of uric acid. *See* GOUT; KIDNEY DISORDERS; METABOLIC DISORDERS; METABOLISM; PROTEIN METABOLISM.

William Nyhan

Uridine diphosphoglucose (UDPG) A compound in which α-glucopyranose is esterified, at carbon atom 1, with the terminal phosphate group of uridine-5′-pyrophosphate (that is, uridine diphosphate, UDP). On very mild acid hydrolysis, glucose and UDP are liberated. Uridine diphosphoglucose occurs in animal, plant, and microbial cells and is synthesized enzymatically from uridine triphosphate (UTP) and α-glucose-1-phosphate. This compound functions as a key in the transformation of glucose to other sugars.

In general, UDP-glucose is a prominent member of a family of compounds composed of sugars activated as derivatives of nucleoside diphosphates. In these active forms they can be used directly as glycosyl donors, or they may be transformed to other more complex sugars before their utilization for the biosynthesis of both simple saccharides and complex polysaccharides. *See* CARBOHYDRATE METABOLISM; NUCLEIC ACID.

Jack L. Strominger

Urinalysis Laboratory examination of urine. Urine is a filtrate of the blood and is produced in the kidneys. It is a reflection of the metabolic activity of the body; conditions that affect the normal homeostatic mechanisms are often revealed by a careful analysis of the composition of the urine. Modern routine urinalysis can be divided into two basic procedures: macroscopic and chemical examination, and microscopic analysis.

Macroscopic and chemical examination. Macroscopic examination includes noting the color and clarity of the urine. Normal urine is pale yellow or straw colored and is usually transparent or clear; abnormal urine may vary greatly in color and may show varying degrees of cloudiness. The specific gravity of urine, that is, the ratio of the weight of a volume of urine to that of the same volume of water, is measured routinely. Specific gravity is an indicator of the kidney's ability to concentrate or dilute the urine and thus of renal tubular function. The normal range is 1.003–1.032.

Most routine chemical urinalysis is now carried out by dipstick testing, which involves the use of plastic strips, or dipsticks, bearing pads embedded with chemical reactants and color indicators. The reaction of each pad represents a separate chemical test for a specific product in the urine. Dipstick testing includes the following categories: protein, glucose, ketone bodies, blood, and bile. *See* PROTEIN.

Microscopic examination. The urine normally contains a wide variety of formed elements that can be identified by using a light microscope. Together these elements form the urinary sediment.

Cells in the urine originate in the bloodstream or in the epithelium lining the urinary tract. The main types of epithelial cells are renal tubular cells, transitional (urothelial) cells, and squamous cells. All three types occur in relatively small numbers in the normal sediment. Blood cells occur normally in the urine in small numbers, and consist mainly of polymorphonuclear neutrophils (a type of granular leukocyte) and red blood cells.

Casts, proteinaceous products of the kidney, are of major importance when present in increased numbers or seen in abnormal forms, because they usually indicate intrinsic renal disease. Casts are cylindrical and are named on the basis of their microscopic appearance and the cells they contain.

Normally, urine is sterile, and the urinary sediment should not contain microorganisms. However, in patients with serious urinary tract infections, microorganisms are usually present in the urine in considerably greater numbers. *See* CLINICAL PATHOLOGY.

Mucus is frequently found in urine sediment and has no known pathologic significance. A wide variety of crystals appear in the urine; their presence may be normal or may indicate an abnormal state. *See* KIDNEY; KIDNEY DISORDERS.

Meryl H. Haber

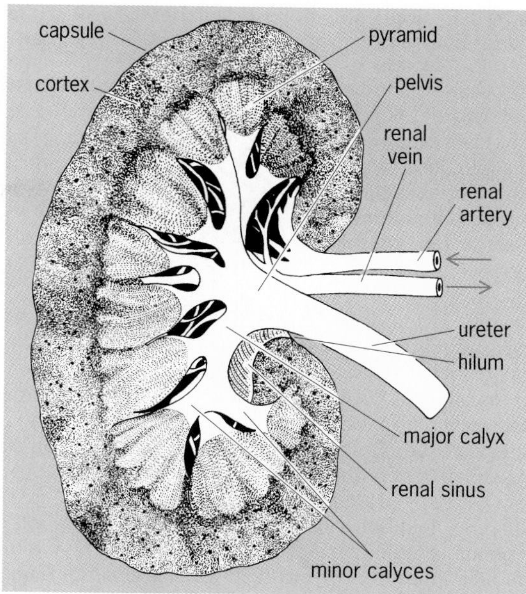

Fig. 1. Sagittal section of a human metanephric kidney (semidiagrammatic). (*After C. K. Weichert and W. Presch, Elements of Chordate Anatomy, 4th ed., McGraw-Hill, 1975*)

Urinary bladder

Urinary bladder A distensible, muscular sac in most vertebrates which serves as a reservoir for urine. Snakes, crocodilians, birds (with the exception of the ostrich), most lizards, and a few fish lack a urinary bladder. In these organisms, urine empties directly into the cloaca. The development of the urinary system is intimately associated with the development of the reproductive system. Three general types of urinary bladder are recognized among the vertebrates: tubal, cloacal, and allantoic. *See* URINE.

Most fish possess tubal bladders, that is, enlargements of the mesonephric ducts. The cloacal bladder is found in monotremes, amphibians, and some dipnoans. There is no direct connection between the excretory ducts and this type of bladder. The bladder is an outpouching or diverticulum of the cloacal wall. The cloacal opening is closed by a sphincter muscle and the urine which seeps into the cloaca from the excretory ducts is forced into the bilobed bladder.

The allantoic bladder is derived from the ventral wall of the cloaca and possibly the allantoic diverticulum. The role of the allantois in the formation of this type of bladder, which is found in most mammals, the turtles, and those lizards which have a bladder, is questioned by some embryologists. *See* ALLANTOIS.

The mammalian bladder is lined with a special epithelium composed of transitional cells. The muscular layer is composed of vertical, horizontal, and oblique fibers. The bladder drains through the urethra, the opening being controlled by a sphincter. Innervation is by the hypogastric sympathetic plexus and partly by parasympathetic fibers from the second and third sacral nerves. Stimulation of the parasympathetic causes the bladder muscle to contract and relaxes the internal sphincter. Micturition is a reflex act which is initiated voluntarily except in children. *See* PARASYMPATHETIC NERVOUS SYSTEM; SYMPATHETIC NERVOUS SYSTEM; URINARY SYSTEM.

Charles B. Curtin

Urinary system

Urinary system The urinary system consists of the kidneys, urinary ducts, and bladder. Similarities are not particularly evident among the many and varied types of excretory organs found among vertebrates. The variations that are encountered are undoubtedly related to problems with which vertebrates have had to cope in adapting to different environmental conditions.

Kidneys. In reptiles, birds, and mammals three types of kidneys are usually recognized: the pronephros, mesonephros, and metanephros. These appear in succession during embryonic development, but only the metanephros persists in the adult.

The metanephric kidneys of reptiles lie in the posterior part of the abdominal cavity, usually in the pelvic region. They are small, compact, and often markedly lobulated. The posterior portion on each side is somewhat narrower. In some lizards the hind parts may even fuse. The degree of symmetry varies.

The kidneys of birds are situated in the pelvic region of the body cavity; their posterior ends are usually joined. They are lobulated structures with short ureters which open independently into the cloaca.

A rather typical mammalian metanephric kidney (Fig. 1) is a compact, bean-shaped organ attached to the dorsal body wall outside the peritoneum. The ureter leaves the medial side at a depression, the hilum. At this point a renal vein also leaves the kidney and a renal artery and nerves enter it. The kidneys of mammals are markedly lobulated in the embryo, and in many forms this condition is retained throughout life. *See* KIDNEY.

Charles K. Weichert

Urinary bladder. At or near the posterior ends of the nephric ducts there frequently is a reservoir for urine. This is the urinary bladder. Actually there are two basic varieties of bladders in vertebrates. One is found in fishes in which the reservoir is no more than an enlargement of the posterior end of each urinary duct. Frequently the urinary ducts are conjoined and a small bladder is formed by expansion of the common duct. The far more common type of bladder is that exhibited by tetrapods. This is a sac which originates embryonically as an outgrowth from the ventral side of the cloaca. Present in all embryonic life, it is exhibited differentially in adults. All amphibians retain the bladder, but it is lacking in snakes, crocodilians, and most lizards; birds, also, with the exception of the ostrich, lack a bladder. It is present in all mammals. *See* URINARY BLADDER.

Theodore W. Torrey

Physiology. Urine is produced by individual renal nephron units which are fundamentally similar from fish to mammals (Fig. 2); however, the basic structural and functional pattern of these nephrons varies among representatives of the vertebrate classes in accordance with changing environmental demands. Kidneys serve the general function of maintaining the chemical and physical constancy of blood and other body fluids. The most striking modifications are associated particularly with the relative amounts of water made available to the animal. Alterations in

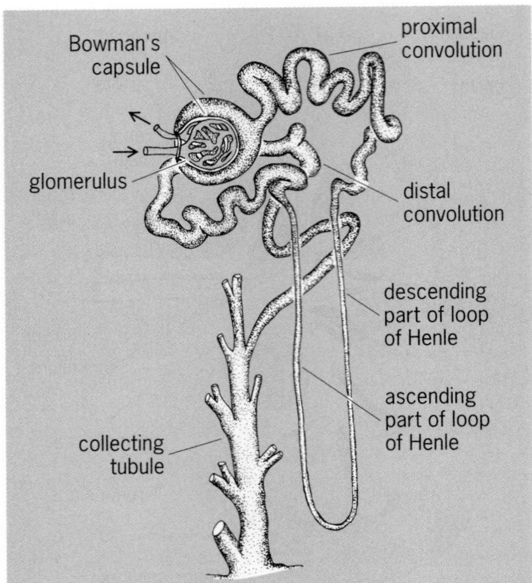

Fig. 2. A mammalian metanephric tubule, showing the renal corpuscle and secretory and collecting portions. (*After C. K. Weichert and W. Presch, Elements of Chordate Anatomy, 4th ed., McGraw-Hill, 1975*)

degrees of glomerular development, in the structural complexity of renal tubules, and in the architectural disposition of the various nephrons in relation to one another within the kidneys may all represent adaptations made either to conserve or eliminate water.

Regulation of volume. Except for the primitive marine cyclostome *Myxine*, all modern vertebrates, whether marine, fresh-water, or terrestrial, have concentrations of salt in their blood only one-third or one-half that of seawater. The early development of the glomerulus can be viewed as a device responding to the need for regulating the volume of body fluids. Hence, in a hypotonic fresh-water environment the osmotic influx of water through gills and other permeable body surfaces would be kept in balance by a simple autoregulatory system whereby a rising volume of blood results in increased hydrostatic pressure which in turn elevates the rate of glomerular filtration. Similar devices are found in fresh-water invertebrates where water may be pumped out either as the result of work done by the heart, contractile vacuoles, or cilia found in such specialized "kidneys" as flame bulbs, solenocytes, or nephridia that extract excess water from the body cavity rather than from the circulatory system. Hence, these structures which maintain a constant water content for the invertebrate animal by balancing osmotic influx with hydrostatic output have the same basic parameters as those in vertebrates that regulate the formation of lymph across the endothelial walls of capillaries. *See* OSMOREGULATORY MECHANISMS.

Electrolyte balance. A system that regulates volume by producing an ultrafiltrate of blood plasma must conserve inorganic ions and other essential plasma constituents. The salt-conserving operation appears to be a primary function of the renal tubules which encapsulate the glomerulus. As the filtrate passes along their length toward the exterior, inorganic electrolytes are extracted from them through highly specific active cellular resorptive processes which restore plasma constituents to the circulatory system.

Movement of water. Concentration gradients of water are attained across cells of renal tubules by water following the active movement of salt or other solute. Where water is free to follow the active resorption of sodium and covering anions, as in the proximal tubule, an isosmotic condition prevails. Where water is not free to follow salt, as in the distal segment in the absence of antidiuretic hormone, a hypotonic tubular fluid results.

Nitrogenous end products. Of the major categories of organic foodstuffs, end products of carbohydrate and lipid metabolism are easily eliminated mainly in the form of carbon dioxide and water. Proteins, however, are more difficult to eliminate because the primary derivative of their metabolism, ammonia, is a relatively toxic compound. For animals living in an aquatic environment ammonia can be eliminated rapidly by simple diffusion through the gills. However, when ammonia is not free to diffuse into an effectively limitless aquatic environment, its toxicity presents a problem, particularly to embryos of terrestrial forms that develop wholly within tightly encapsulated eggshells or cases. For these forms the detoxication of ammonia is an indispensable requirement for survival. During evolution of the vertebrates two energy-dependent biosynthetic pathways arose which incorporated potentially toxic ammonia into urea and uric acid molecules, respectively. Both of these compounds are relatively harmless, even in high concentrations, but the former needs a relatively large amount of water to ensure its elimination, and uric acid requires a specific energy-demanding tubular secretory process to ensure its efficient excretion. *See* UREA; URIC ACID.

Urine concentration. The unique functional feature of the mammalian kidney is its ability to concentrate urine. Human urine can have four times the osmotic concentration of plasma, and some desert rats that survive on a diet of seeds without drinking any water have urine/plasma concentration ratios as high as 17. More aquatic forms such as the beaver have correspondingly poor concentrating ability.

The concentration operation depends on the existence of a decreasing gradient of solute concentration that extends from the tips of the papillae in the inner medulla of the kidney outward toward the cortex. The high concentration of medullary solute is achieved by a double hairpin countercurrent multiplier system which is powered by the active removal of salt from urine while it traverses the ascending limb of Henle's loop (Fig. 2). The salt is redelivered to the tip of the medulla after it has diffused back into the descending limb of Henle's loop. In this way a hypertonic condition is established in fluid surrounding the terminations of the collecting ducts. Urine is concentrated by an entirely passive process as water leaves the lumen of collecting ducts to come into equilibrium with the hypertonic fluid surrounding its terminations. *See* URINE; UROGENITAL SYSTEM. Roy P. Forster

Urinary tract disorders Diseases affecting the outflow tract of the urinary system from the renal pelvis to the distal portion of the urethra. Urinary tract disorders can occur in the renal pelvis, ureters, bladder, or urethra.

Renal pelvis. The renal pelvis, the structure that collects and carries urine from the kidney to the ureter, is subject to a number of disorders. Anomalies of the renal pelvis are common. The pelvis may be subdivided to form a duplicated system, which may join a single ureter or empty into separate ureters. By itself such an anomaly is harmless, but if there is associated obstruction or vesicoureteral reflux (abnormal flow of urine from the bladder back into the kidney), there may be symptoms of pain, associated infection, and possibly destruction of renal tissue.

Calculi typically arise in the renal papillae usually below the epithelial layer. The cause is usually due to a metabolic defect, which results in excessive excretion of oxalate, urate, calcium, and/or cystine.

Fibrous bands and aberrant vessels associated with the kidneys are usually blamed for obstruction when no other intrinsic cause can be found for ureteropelvic obstruction. Tumors, either intrinsic or extrinsic, and periureteral fibrosis (fibrosis around ureter) are also capable of blocking the flow of urine from the renal pelvis.

Acute infection of the kidney, known as pyelonephritis, produces fever and flank pain. If there are no underlying causes (such as stone, obstruction, or tumor), the disease can be treated with antibiotics and has no serious aftereffects.

Tumors of the pelvis are rare; when they do occur they are usually transitional-cell carcinomas with varying degrees of differentiation. They can cause obstruction and hematuria. They are often associated with similar tumors elsewhere in the urinary system.

Ureters. These long thin tubes that carry urine from the renal pelvis to the bladder are subject to the same diseases as the renal pelvis. Duplication of the ureters can occur at any point, from a dual pelvis and fusion to form one ureter, to complete duplication and separate implantation into the bladder. The anomaly is generally harmless unless obstruction or reflux is present, which can contribute to developing infections.

One of the most common causes of excruciating abdominal and flank pain, often radiating to the groin or testicle, is the passage of small calculi. They originate from the renal pelvis and travel toward the bladder. If the stone becomes lodged along the course from the kidney to the urethra, intervention is required to remove the stone. Again, obstruction of the urinary system can lead to renal damage and loss of renal function. Occasionally, the ureter is obstructed by external pressure from retroperitoneal fibrosis (development of a fibrous mass in the back of the abdomen behind the abdominal lining), tumor, or aortic aneurysm.

If the normal valvelike mechanism that prevents urine in the bladder from going back toward the kidney is defective, vesicoureteral reflux results. Reflux can occur during bladder filling or emptying. When associated with an infection, vesicoureteral reflux can lead to renal injury.

Bladder. The urinary bladder is subject to anomalies, obstruction, inflammation, calculi, fistulae, and tumors. As the receptacle for urine from the kidneys, it is variable in shape and form depending upon the amount of urine it contains.

Bladder exstrophy is an uncommon congenital anomaly due to failure of both the lower abdominal wall and anterior bladder wall to close during development. The bladder appears directly through the abdominal wall defect.

Interference with complete emptying of the bladder may be secondary to various factors. In males it is often due to prostatic enlargement, and in females it is typically due to a cystocele, which is a bulging of the bladder into the vaginal vault. Other causes of obstruction include calculi; tumors; constriction secondary to surgical instrumentation or gonococcal infection; or congenital valves. A neurogenic bladder (which occurs when the normal balance of bladder wall contraction and sphincter relaxation is altered and the bladder cannot completely empty) usually follows spinal cord injury or disease and can lead to inadequate bladder emptying.

Obstruction from any cause leads to changes in the bladder wall. The bladder wall can weaken in areas and bulge outward, causing the formation of a diverticulum. Other effects of obstruction include increased susceptibility to infection, stone formation, and backpressure on the kidneys and ureters.

Cystitis or inflammation of the urinary bladder is a common affliction and is characterized by dysuria (painful urination), frequency, and urgency. The organisms responsible are primarily bacterial, including *Escherichia coli*, pseudomonas, and other gram-negative organisms, although yeast and fungi can cause urinary tract infections.

Abnormal connections (fistulas) between the bladder and the female urethra or vagina are not unusual. These can be due to pelvic or transvaginal surgery, childbirth, or radiation. Fistulas can result in incontinence and chronic urinary tract infection.

The bladder is frequently the site of tumors, which are almost always of transitional-cell origin. Although benign lesions can exist in the bladder, bladder cancer needs to be ruled out.

Female urethra. This structure is not often diseased. In older women a painful raspberrylike growth of inflammatory tissue may appear at the external urethral opening and is known as a caruncle. The urethra can also be a site of transitional-cell and squamous cell malignancy.

Male urethra. The male urethra is approximately 20 cm in length and is responsible for draining urine from the bladder and acting as a conduit for the delivery of semen. The first 3 cm is surrounded by the prostate gland and is termed the prostatic urethra. The prostate gland may develop cancer, and options are available to treat this condition based on stage of the tumor or extent of disease at time of detection.

The prostatic urethra is prone to obstruction. The most common cause is a noncancerous growth of the prostate called benign prostatic hyperplasia. This growth is a natural process associated with aging and can occlude the prostatic urethra, causing progressive difficulty in urinating. Individuals may complain of straining with urination, weak stream, incomplete bladder emptying, nocturia (excessive nighttime urination), frequency, and/or double voiding. *See* CANCER (MEDICINE); KIDNEY DISORDERS; REPRODUCTIVE DISORDERS; TUMOR; URINARY SYSTEM.

Roy N. Barnett; Katherine S. Rhee; Martin I. Resnick

Urine An aqueous solution of organic and inorganic substances, mostly waste products of metabolism. The kidneys maintain the internal milieu of the body by excreting these waste products and adjusting the loss of water and electrolytes to keep the body fluids relatively constant in amount and composition. The urine normally is clear and has a specific gravity of 1.017–1.020, depending upon the amount of fluid ingested, perspiration, and diet. The increase in specific gravity above that of water is due to the presence of dissolved solids, about 60% of which are organic substances such as urea, uric acid, creatinine, and ammonia; and 40% of which are inorganic substances such as sodium, chloride, calcium, potassium, phosphates, and sulfates. Its reaction is usually acid (pH 6) but this too varies with the diet. It usually has a faint yellow color due to a urochrome pigment, but the color varies depending upon the degree of concentration, and the ingestion of certain foods (for example, rhubarb) or cathartics. It usually has a characteristic aromatic odor, the cause of which is not known. *See* KIDNEY; UREA; URIC ACID; URINALYSIS; URINARY SYSTEM.

Reuben Straus

Urodela One of the orders of the class Amphibia, also known as the Caudata. The members of this order are the tailed amphibians, or salamanders, and are distinguished superficially from the frogs and toads (order Anura) by the possession of a tail, and from the caecilians (order Apoda) by the possession of limbs. Salamanders resemble lizards in that members of both groups normally have a relatively elongate body, four limbs, and a tail. The similarity is, however, only superficial. The most obvious external difference, the moist glandular skin of the amphibian and dry scaly skin of the reptile, is underlain by the numerous characters that distinguish reptiles from amphibians. *See* ANURA.

The vast majority of salamanders are well under a foot (0.3 m) in length. The largest is the giant salamander of Japan and China, which may attain a length of 5 ft (1.5 m). A close relative, the hellbender, which lives in streams in the eastern United States, grows to over 2 ft (0.6 m) in length.

Four well-developed limbs are typically present in salamanders, but the legs of the mud eel (*Amphiuma*) of the southeastern United States are very tiny appendages and of no use in locomotion. The sirens (*Siren* and *Pseudobranchus*), also aquatic salamanders of the same region, have undergone even further degeneration of the limbs and retain only tiny forelimbs. Salamanders with normal limbs usually have four toes on the front feet and five on the rear, though a few have only four on all feet. A feature of the Urodela not shared with the Anura is the ability to regenerate limbs.

The moist and highly vascularized skin of a salamander serves as an organ of respiration, and in some forms shares with the buccal region virtually the whole burden. In one entire family of salamanders, the Plethodontidae, lungs are not present. Even in those forms with lungs and gills, dermal respiration plays a very important part. The tympanum is absent in salamanders and the middle ear is degenerate. Hence these animals are deaf to airborne sounds. Undoubtedly correlated with this is the reduction of voice; salamanders are mute or produce only slight squeaking or, rarely, barking sounds when annoyed. However, salamanders can detect sounds carried through ground or water.

Salamanders live in a variety of aquatic or moist habitats. Many forms are wholly aquatic, living in streams, rivers, lakes, and ponds. Others live on land most of the year but must return to water to breed. The most advanced species live out their lives in moist places on land or, in the case of some tropical species, in trees, and never go in the water.

Producing the young alive rather than by laying eggs is very rare among salamanders. The eggs of salamanders may be deposited in water or in moist places on land. Presumably, aquatic breeding is the more primitive. Fertilization is external in most of the Cryptobranchoidea, internal by means of spermatophores in the remaining suborders. *See* NEOTENY.

The greatest concentration of families and genera of salamanders is found in the eastern United States, and these animals are one of the few major groups of vertebrate animals that is distinctly nontropical. No salamander occurs south of the Equator in the Old World, and only 18 species of the evolutionarily advanced family Plethodontidae are found as far south as South America in the New World.

The Urodela are readily arranged in seven families, but authorities differ as to the relationships among these families and their classification into suborders. Three to five suborders may be recognized, and one of these is thought by some to merit ordinal rank, equivalent to the Urodela. About 300 species of salamanders are known to be living today, and these are distributed among about 54 genera. *See* AMPHIBIA. Richard G. Zweifel

Urogenital system The combined structures composing the urinary and genital, or reproductive, organs of vertebrates. The terms urinogenital or genitourinary system are equally applicable and just as frequently used. During embryonic development, the urinary and reproductive systems are closely interrelated, as their ducts arise in associated mesodermal regions. Common passages are associated with both these systems in various vertebrates, such as the single orifice for emission of urine and sperm. *See* KIDNEY; REPRODUCTIVE SYSTEM; URINARY SYSTEM. Charles B. Curtin

Urokinase An enzyme that is a plasminogen activator. Urokinase cleaves the plasma protein plasminogen, forming the active enzyme plasmin, which subsequently degrades fibrin. Thus urokinase is an essential component in the fibrinolytic clot-dissolving system in the human body. *See* FIBRINOGEN.

Urokinase is found in human urine and in much lower concentrations in human plasma. In the body, urokinase is produced by kidney cells, and its presence in the urine promotes the dissolution of any blood clots in the urine-collecting system of the kidneys or the bladder. Urokinase is also produced by a variety of tumor cells, and it is thought to be involved in the formation of tumor metastases. Urokinase can be inhibited by plasma inhibitor proteins. The presence of cell receptors and inhibitors suggests that the regulation of urokinase function is complex.

Urokinase is employed in clinical medicine in the treatment of venous blood clots (thrombophlebitis and pulmonary emboli), acute myocardial infarction, and arterial blood clots in the legs and arms. It is also used to prevent the accumulation of blood clots in intravenous catheters used to administer long-term chemotherapy. *See* BLOOD; ENZYME; PLASMIN. Philip C. Comp

Uropygi An order of arachnids, the tailed whip scorpions, comprising about 70 species from tropical and warm temperate Asia and the Americas. Most are dark reddish-brown, of medium to giant size, 0.7–2.6 in. (18–65 mm), the largest one being *Mastigoproctus giganteus* of the southern United States and Mexico. The elongate, flattened body bears in front a pair of greatly thickened, raptorial pedipalps set with many sharp spines and used to hold and crush insect prey. The first pair of legs is elongated and modified into feelers. The abdomen terminates in a slender, many jointed, whiplike flagellum. The uropygids are harmless, nocturnal creatures without poison glands that live in dark places and burrow into the soil. When disturbed, they expel a volatile liquid, with the strong odor of acetic acid, from a gland at the base of the tail. This accounts for the name "vinegaroon" given by many Americans to these much-feared animals. *See* ARACHNIDA. Willis J. Gertsch

Uropygial gland A relatively large, compact, bilobed, secretary organ located at the base of the tail (uropygium) of most birds having a keeled sternum. It is known also as the preen, oil, or scent gland. This is the only true skin gland possessed by this class of vertebrates.

The glandular secretion, predominantly oily and sometimes of offensive odor (musk-duck, hoopoe, petrel) is discharged through an orifice at the tip of a nipplelike protuberance often encircled by short, bristly feathers. The act of preening induces a flow of secretion from the nipple which is transferred by the beak to the body plumage. In water fowl there is some evidence that this oily secretion assists in maintaining the water-repellent quality of the feathers, either directly or by preserving their physical structure. *See* FEATHER; SCENT GLAND. Mary E. Rawles

Ursa Major The Great Bear, a circumpolar constellation for observers at mid-northern latitudes or higher latitudes (see illustration). Part of this constellation is the prominent asterism (star pattern) known as the Big Dipper, whose bowl of four stars is linked to a tail of three more, perhaps representing a long bear's tail pulled out when the bear was swung by it to throw it into the heavens. In England, the Big Dipper is more usually known as Charles's Wain (Wagon) or the Plough.

The stars of the Dipper were labeled by Johan Bayer (1601) in alphabetical order with Greek letters, contrary to his usual practice of assigning Greek letters in order of brightness. The two

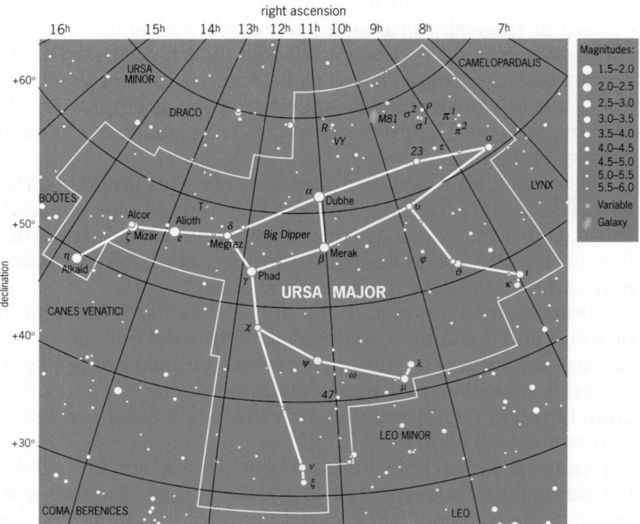

Modern boundaries of the constellation Ursa Major, the Great Bear. The celestial equator is 0° of declination, which corresponds to celestial latitude. Right ascension corresponds to celestial longitude, with each hour of right ascension representing 15° of arc. Apparent brightness of stars is shown with dot sizes to illustrate the magnitude scale, where the brightest stars in the sky are 0th magnitude or brighter and the faintest stars that can be seen with the unaided eye at a dark site are 6th magnitude. (*Wil Tirion*)

stars at the head of the dipper, alpha and beta Ursae Majoris, point toward Polaris, a star that is now within 1° of the north celestial pole. Thus these pointers are very useful for formal and informal navigation. *See* POLARIS.

In the handle of the Big Dipper, the middle star Alcor has a fainter companion, Mizar, that is separately visible to the unaided eye. On inspection with a telescope, each is a double star. *See* MIZAR.

M81 is a prominent spiral galaxy in the northern part of Ursa Major, thus making it visible year-round from mid-northern latitudes. *See* CONSTELLATION; GALAXY, EXTERNAL; MESSIER CATALOG.

Jay M. Pasachoff

Ursa Minor The Little Bear, a circumpolar constellation from northern latitudes (see illustration). Part of this constellation is the prominent asterism (star pattern) known as the Little Dipper, whose stars are fainter than the corresponding stars of the Big Dipper. At the end of the handle of the Little Dipper is Polaris, at magnitude 2.0 not a particularly bright star—only the 49th brightest in the sky—but one that happens in the present epoch to be close to (almost 1° away from) the north celestial pole, the extension of the Earth's axis of rotation into the sky. Thus the Little Dipper and Polaris are useful for formal and informal navigation. *See* CONSTELLATION; POLARIS.

Jay M. Pasachoff

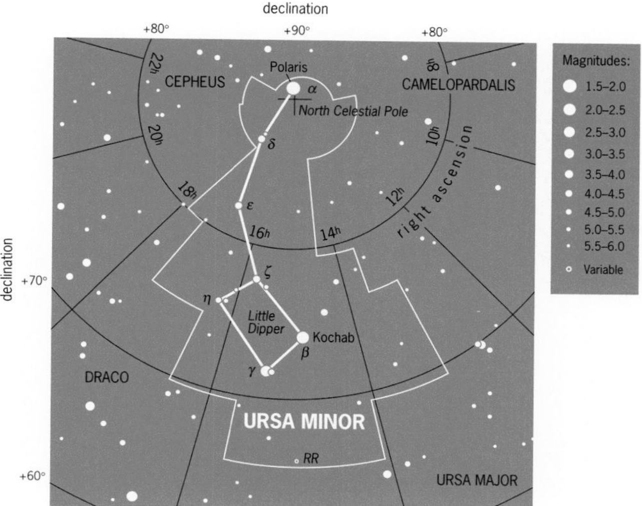

Modern boundaries of the constellation Ursa Minor, the Little Bear. The celestial equator is 0° of declination, which corresponds to celestial latitude. Right ascension corresponds to celestial longitude, with each hour of right ascension representing 15° of arc. Apparent brightness of stars is shown with dot sizes to illustrate the magnitude scale, where the brightest stars in the sky are 0th magnitude or brighter and the faintest stars that can be seen with the unaided eye at a dark site are 6th magnitude. (*Wil Tirion*)

Ustilaginomycetes (smut) A class of the subdivision Basidiomycotina. These microscopic plant-parasitic fungi are commonly known as smut fungi. Many species may have a shorter or longer saprophytic life cycle, capable of yeastlike reproduction. Several nonparasitic yeasts have been identified as closely related to the smuts. About 1500 true Ustilaginales species are known.

The body (mycelium) of the smut fungi consists of hyphae that are thin, transparent, branched, septate, and binucleate. Usually parasitic, the fungus grows in the host tissues and gives little, if any, evidence of its presence before spore formation sets in. The smut spores (teliospores or ustilospores) are the organs of dispersion and resistance, and are thick walled, pigmented, and variously ornamented. They are formed in great number in the sori, which consist of host tissues, spore masses, and sometimes modified fungal cells or tissues. The smut fungi develop variable sori in different organs (such as the roots, stems, leaves, inflorescences, flowers, anthers, and seeds) filled by powdery or agglutinated, usually colored (black, brown, violet, or yellow) spore masses. The spores may develop singly, in pairs, or in aggregates of spore balls. Spore germination results in a basidium (promycelium or ustidium) which gives rise to hyaline basidiospores. *See* BASIDIOMYCOTA; EUMYCOTA; FUNGI.

Many smut fungi produce severe losses of cereals and other cultivated plants.

Kálmán Vánky

Uterine disorders Neoplastic, functional, congenital, and inflammatory diseases of the uterus and cervix. Uterine and cervical disorders are common gynecological complaints, which can be categorized in terms of increased uterine size, abnormal bleeding, and infection.

The most common cause of increased uterine size is the presence of fibroids (leiomyoma). These benign smooth-muscle tumors are the most common tumors in women of reproductive age, occurring in approximately one-third of women. Problems arise when the fibroid presses on nearby organs such as the urinary tract. Fibroids may also cause infertility or pregnancy loss. Rarely, fibroids become malignant.

Abnormal uterine bleeding, especially in postmenopausal women, must always be evaluated. In girls and young women it is usually caused by irregular ovulation or hormonal imbalances. In women of reproductive age, problems of pregnancy must always be excluded such as miscarriage or ectopic pregnancy. Other common causes include polyps of the uterus or cervix. The most serious reason to seek medical attention for abnormal uterine bleeding is to exclude the possibility of cancer. *See* MENSTRUATION.

Uterine infections are not extremely common but are associated with trauma, childbirth, or miscarriage. Cervical infections and inflammations (cervicitis) are common. Specific agents of cervicitis include staphylococci, streptococci, gonococci, syphilis, chancroid, tuberculosis, chlamydia, and mycoplasma; appropriate antibiotics should be chosen to treat such infections. *See* REPRODUCTIVE SYSTEM DISORDERS; STAPHYLOCOCCUS; STREPTOCOCCUS; SYPHILIS; TUBERCULOSIS; UTERUS.

Machelle M. Seibel

Uterus The hollow, muscular womb, being an enlarged portion of the oviduct in the adult female. An adult human uterus, before pregnancy, measures 3 × 2 × 1 in. (7.5 × 5 × 2.5 cm) in size and has the shape of an inverted, flattened pear. The paired Fallopian tubes enter the uterus at its upper corners; the lower, narrowed portion, the cervix, projects into the vagina (see illustration). Normally the uterus is tilted slightly forward and lies behind the urinary bladder.

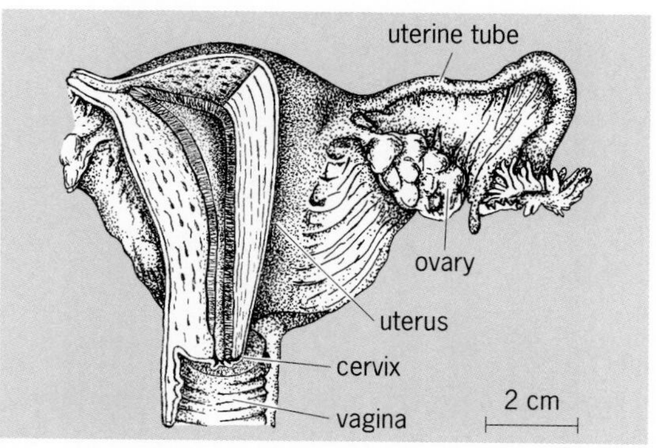

Human uterus and associated structures. (*After L. B. Arey, Developmental Anatomy, 7th ed., Saunders, 1965*)

The lining, or mucosa, responds to hormonal stimulation, growing in thickness with a tremendous increase in blood vessels during the first part of the menstrual cycle. If fertilization does not occur, the thickened vascular lining is sloughed off, producing the menstrual flow at the end of the cycle, and a new menstrual cycle begins with growth of the mucosa. When pregnancy occurs, the mucosa continues to thicken and forms an intimate connection with the implanted and enlarging placenta. *See* Menstruation; Pregnancy; Reproductive system.

Walter Bock

V

Vaccination Active immunization against a variety of microorganisms or their components, with the ultimate goal of protecting the host against subsequent challenge by the naturally occurring infectious agent. The terms vaccine and vaccination were originally used only in connection with Edward Jenner's method for preventing smallpox, introduced in 1796. In 1881 Louis Pasteur proposed that these terms should be used to describe any prophylactic immunization. Vaccination now refers to active immunization against a variety of bacteria, viruses, and parasites (for example, malaria and trypanosomes). *See* SMALLPOX.

Implicit within Jenner's method of vaccinating against smallpox was the recognition of immunologic cross-reactivity together with the notion that protection can be obtained through active immunization with a different, but related, live virus. It was not until the 1880s that the next immunizing agents, vaccines against rabies and anthrax, were introduced by Pasteur. Two facts of his experiments on rabies vaccines are particularly noteworthy.

First, Pasteur found that serial passage of the rabies agent in rabbits resulted in a weakening of its virulence in dogs. During multiple passages in an animal or in tissue culture cells, mutations accumulate as the virus adapts to its new environment. These mutations adversely affect virus reproduction in the natural host, resulting in lessened virulence. Only as the molecular basis for virulence has begun to be elucidated by modern biologists has it become possible to deliberately remove the genes promoting virulence so as to produce attenuated viruses.

Second, Pasteur demonstrated that rabies virus retained immunogenicity even after its infectivity was inactivated by formalin and other chemicals, thereby providing the paradigm for one class of noninfectious virus vaccine, the "killed"-virus vaccine.

Attenuated-live and inactivated vaccines are the two broad classifications for vaccines. Anti-idiotype antibody vaccines and deoxyribonucleic acid (DNA) vaccines represent innovations in inactivated vaccines. Recombinant-hybrid viruses are novel members of the live-virus vaccine class recently produced by genetic engineering.

Because attenuated-live-virus vaccines reproduce in the recipient, they provoke both a broader and more intense range of antibodies and T-lymphocyte-associated immune responses than noninfectious vaccines. Live-virus vaccines have been administered subdermally (vaccinia), subcutaneously (measles), intramuscularly (pseudorabies virus), intranasally (infectious bovine rhinotracheitis), orally (trivalent Sabin poliovirus), or by oropharyngeal aerosols (influenza). Combinations of vaccines have also been used. Live-virus vaccines administered through a natural route of infection often induce local immunity, which is a decided advantage. However, in the past, attenuated-live virus vaccines have been associated with several problems, such as reversion to virulence, natural spread to contacts, contaminating viruses, lability, and viral interference. *See* ANIMAL VIRUS; VIRULENCE; VIRUS CLASSIFICATION; VIRUS INTERFERENCE.

Noninfectious vaccines include inactivated killed vaccines, subunit vaccines, synthetic peptide and biosynthetic polypeptide vaccines, oral transgenic plant vaccines, anti-idiotype antibody vaccines, DNA vaccines, and polysaccharide-protein conjugate vaccines. With most noninfectious vaccines a suitable formulation is essential to provide the optimal antigen delivery for maximal stimulation of protective immune responses. Development of new adjuvant (a substance that enhances the potency of the antigen) and vector systems is pivotal to produce practical molecular vaccines. *See* ANTIBODY; ANTIGEN; IMMUNITY. Saul Kit

Vacuole An intracellular compartment, bounded by a single membrane bilayer, which functions as a primary site of protein and metabolite degradation and recycling in animals, but serves additional complex functions in fungi and plants (see illustration). Scientists who study vacuoles also define them as the terminal product of the secretory pathway. The secretory pathway functions to transport protein and metabolite-containing membrane vesicles from sites of synthesis or uptake to the vacuole. *See* CELL MEMBRANES; CELL METABOLISM; GOLGI APPARATUS; SECRETION.

In animals, a lytic vacuole known as the lysosome typically functions to process macromolecules. Such macromolecules can be targeted to the lysosome from sites of synthesis. For example, proteins that assemble incorrectly in the endoplasmic reticulum (ER) can be degraded in the lysosome and their constituent amino acids recycled. Proteins that can serve as nutrients are also targeted to the lysosome from the cell surface. An important process for the recycling of cytoplasm in eukaryotic cells is autophagy, in which molecules or organelles are encapsulated in membrane vesicles that fuse with the lysosome. *See* ENDOCYTOSIS; ENDOPLASMIC RETICULUM; LYSOSOME.

In the mammalian immune system, macrophages and neutrophils take up particles and pathogens in the process of phagocytosis, during which the pathogen is eventually digested in the lysosome. A number of diseases in humans can be caused when intracellular pathogens evade destruction in the lysosome. *See* PHAGOCYTOSIS.

In fungi, vacuoles can serve functions not found in animals. Besides a lytic function, they serve in the storage of ions as

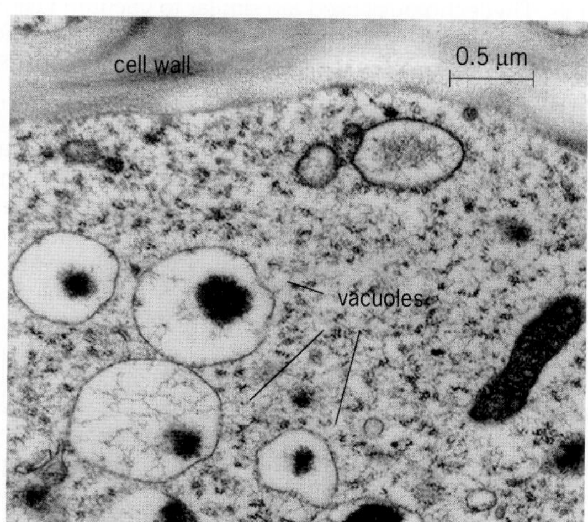

Electron micrograph of a barley root tip cell, showing multiple vacuoles within the cytoplasm.

well as amino acids for protein synthesis. In yeast, vacuoles can also function in the destruction and recycling of cellular organelles, such as peroxisomes, which help protect the cell from toxic oxygen-containing molecules. The process of peroxisome digestion by vacuoles is known as pexophagy.

The most complex vacuoles are found in plants. Some contain hydrolytic enzymes and store ions similar to those found in lysosomes, whereas others serve a role in storing pigments which impart color to flowers to attract pollinators. Specialized ER-derived vacuoles in plant seeds, known as protein bodies, function in the storage of proteins called prolamines that are common in the endosperm of cereals. Upon germination, the proteins are degraded and used as a source of amino acids and nitrogen for the growing plant. Toxins, such as alkaloids, are stored in vacuoles in parts of the plant, such as the leaves, which are subject to frequent herbivory. Scientists have learned that plants produce and store in their vacuoles a vast array of unique chemicals which may, in addition to their natural functions, have medicinal value. *See* PLANT CELL.

Another unique function that vacuoles serve in plants is in cell growth. As a consequence of the accumulation of ions, metabolites, and water, plant vacuoles are under considerable internal osmotic pressure. The vacuolar membrane in plants, known as the tonoplast, as well as the cell itself would burst under this pressure if not for the rigid wall that surrounds the cells. The resulting turgor pressure provides mechanical stability to plant stems. Loss of osmotic pressure in the vacuole due to a lack of water results in plant wilting. The osmotic pressure of the vacuoles also provides the driving force that allows plants to grow by enlarging their cell volume. Enzymes reduce the rigidity of the cell wall, which permits cell expansion under the force of turgor. This is a fundamental process in plants and explains why vacuoles can occupy as much as 95% of the volume of some cells. *See* CELL WALLS (PLANT).

Glenn R. Hicks; Natasha Raikhel

Vacuum fusion A technique of analytical chemistry for determining the oxygen, hydrogen, and sometimes nitrogen content of metals.

The metal sample is either fused or dissolved in a bath, or flux, of a second metal in a heated graphite crucible supported inside an evacuated glass or quartz vessel. Oxygen is released from the metal as carbon monoxide by reaction of oxides or dissolved oxygen with carbon from the graphite crucible at high temperature. Metal nitrides dissociate to form elemental nitrogen. Hydrogen is evolved as elemental hydrogen. The mixture of carbon monoxide, nitrogen, and hydrogen is analyzed to determine individual component concentrations by various techniques.

Inert gas fusion has been developed for determination of gases in metals. The techniques used are similar to vacuum fusion but substitute inert gases for the vacuum environment. *See* VACUUM METALLURGY.

Frank C. Benner

Vacuum measurement The determination of a gas pressure that is less in magnitude than the pressure of the atmosphere. This low pressure can be expressed in terms of the height in millimeters of a column of mercury which the given pressure (vacuum) will support, referenced to zero pressure. The height of the column of mercury which the pressure will support may also be expressed in micrometers. The unit most commonly used is the torr, equal to 1 mm (0.03937 in.) of mercury (mmHg). Less common units of measurement are fractions of an atmosphere and direct measure of force per unit area. The unit of pressure in the International System (SI) is the pascal (Pa), equal to 1 newton per square meter (1 torr = 133.322 Pa). Atmospheric pressure is sometimes used as a reference. The pressure of the standard atmosphere is 29.92 in. or 760 mm of mercury (101,325 Pa or 14.696 lbf/in.^{2}).

Pressures above 1 torr can be easily measured by familiar pressure gages, such as liquid-column gages, diaphragm-pressure gages, bellows gages, and bourdon-spring gages. At pressures below 1 torr, mechanical effects such as hysteresis, ambient errors, and vibration make these gages impractical. *See* MANOMETER; PRESSURE MEASUREMENT.

Pressures below 1 torr are best measured by gages which infer the pressure from the measurement of some other property of the gas, such as thermal conductivity or ionization. The thermocouple gage, in combination with a hot- or cold-cathode gage (ionization type), is the most widely used method of vacuum measurement today. *See* IONIZATION GAGE.

Other gages used to measure vacuum in the range of 1 torr or below are the McLeod gage, the Pirani gage, and the Knudsen gage. The McLeod gage is used as an absolute standard of vacuum measurement in the $10–10^{-4}$ torr ($10^3–10^{-2}$ Pa) range. *See* MCLEOD GAGE.

The Knudsen gage is used to measure very low pressures. It measures pressure in terms of the net rate of transfer of momentum (force) by molecules between two surfaces maintained at different temperatures (cold and hot plates) and separated by a distance smaller than the mean free path of the gas molecules.

Richard Comeau

Vacuum metallurgy The making, shaping, and treating of metals, alloys, and intermetallic and refractory metal compounds in a gaseous environment where the composition and partial pressures of the various components of the gas phase are carefully controlled. In many instances, this environment is a vacuum ranging from subatmospheric to ultrahigh vacuum (less than 760 torr or 101 kilopascals to 10^{-12} torr or 10^{-10} pascal). In other cases, reactive gases are deliberately added to the environment to produce the desired reactions, such as in reactive evaporation and sputtering processes and chemical vapor depositon. The processes in vacuum metallurgy involve liquid/solid, vapor/solid, and vapor/liquid/solid transitions. In addition, they include testing of metals in controlled environments.

There are three basic reasons for vacuum processing of metals: elimination of contamination from the processing environment, reduction of the level of impurities in the product, and deposition with a minimum of impurities. Contamination from the processing environment includes the container for the metal and the gas phase surrounding the metal. In the vacuum process, impurities, particularly oxygen, nitrogen, hydrogen, and carbon, are released from the molten metal and pumped away; and metals, alloys, and compounds are deposited with a minimum of entrained impurities. There are numerous and varied application areas for vacuum metallurgy including special areas of extractive metallurgy, melting processes, casting of shaped products, degassing of molten steel, heat treatment, surface treatment, vapor deposition, space processing, and joining processes. *See* ARC WELDING; STEEL MANUFACTURE; WELDING AND CUTTING OF METALS.

Rointan F. Bunshah

Vacuum pump A device that reduces the pressure of a gas (usually air) in a container. When gas in a closed container is lowered from atmospheric pressure, the operation constitutes an increase in vacuum in this container. *See* PRESSURE.

Vacuum pumps are evaluated for the degree of vacuum they can attain and for how much gas they can pump in a unit of time. In practice, where high vacuum is required, two or more different types of pumps are used in series.

In the rotary oil-seal pump (*see* illustration), gas is sucked into chamber A through the opening intake port by the rotor. A sliding vane partitions chamber A from chamber B. The compressed gas that has been moved from position A to position B is pushed out of the exit port through the valve, which prevents the gas from flowing back. The valve and the rotor contact point are oil-sealed. Since each revolution sweeps out a fixed volume, it is called a constant-displacement pump. Other mechanical pumps are the rotary blower pump, which operates by the propelling

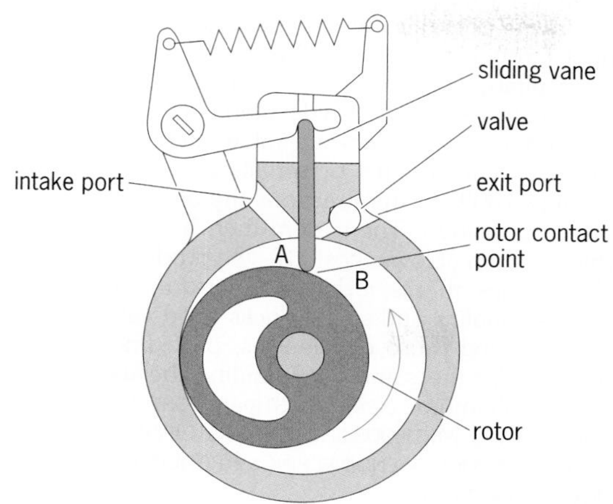

Chief components of a typical mechanical pump, the rotary oil-seal pump.

action of one or more rapidly rotating lobelike vanes, and the molecular drag pump, which operates at very high speeds, as much as 16,000 rpm. Pumping is accomplished by imparting a high momentum to the gas molecules by the impingement of the rapidly rotating body.

The water aspirator is an ejector pump. When water is forced under pressure through the jet nozzle, it will force the gas in the inlet chamber to go through the diffuser, thus lowering the pressure in the inlet chamber. When high-pressure steam is used instead of water, it is called a steam ejector.

In addition, a number of pumps have been developed which meet special pumping requirements. The ion pump operates electronically. Electrons that are generated by a high voltage applied to an anode and a cathode are spiraled into a long orbit by a high-intensity magnetic field. These electrons colliding with gas molecules ionize the molecules, imparting a positive charge to them. These are attracted to, and are collected on, the cathode. Thus a pumping action takes place. Sorption pumping is the removal of gases by adsorbing and absorbing them on a granular sorbent material such as a molecular sieve held in a metal container. Cryogenic pumping is accomplished by condensing gases on surfaces that are at extremely low temperatures.

Edward S. Barnitz

Vacuum tube An electron device in which use is made of the electrostatically or magnetically controlled flow of electrons in an evacuated space, or of the phenomena accompanying the electron emission, acceleration, and collection associated with the flow. In common usage, completely evacuated tubes are called hard tubes; tubes containing a small amount of deliberately added gas or vapor are called soft tubes, but they are usually included in the family of vacuum tubes.

Vacuum tubes are used as active and passive electrical circuit elements. In suitable circuitry they rectify, detect, gate, amplify, and oscillate. They can also scan and display images; generate heat, light, and radiation; ionize and accelerate atoms and molecules; and measure high vacuum.

Most vacuum tubes employ the Edison effect, in which a unidirectional flow of current takes place through a vacuum by thermionic emission of electrons from a hot surface (the cathode). The electrons are collected by a nearby positively charged electrode (the anode or plate). Two-electrode Edison-effect tubes were first employed as radio-frequency detectors, but the enormous potential of the device was not realized until it was shown that the current could be controlled with large amplification ratios and great sensitivity by small electrical charges on a third, meshlike electrode (the grid) interposed between cathode and anode. From the resulting three-element tube (the triode) came all the characteristic inventions of the modern electronic age. *See* ELECTRON EMISSION; ELECTRON MOTION IN VACUUM; THERMIONIC EMISSION.

Later refinements included additional grids by which certain problems inherent in triodes were overcome, and combination tubes were developed in which several functions were combined within one vacuum envelope. Elaborations of the basic idea include camera and picture tubes and magnetrons, klystrons, and traveling-wave tubes for microwave radar. *See* KLYSTRON; MAGNETRON; MICROWAVE TUBE; PICTURE TUBE; RADAR; TELEVISION CAMERA TUBE; TRAVELING-WAVE TUBE.

Originally, vacuum tubes were enclosed in cylindrical glass structures. Through the walls or base of the tube were sealed the metal electrodes which emitted, controlled, or collected the electrons traversing the empty space within. Up through the 1960s, vast numbers of small tubes were used in radio and television broadcast receivers. They also found their way into industrial instruments and controls, and into the early electronic computers, but their size, fragility, inordinate power consumption, and short lifetimes made them unpopular for these more demanding purposes. Since that time, such vacuum tubes have been almost entirely replaced by the much smaller, cheaper, and longer-lived solid-state devices. Only a few standard receiving-tube types still remain in production, and are used where their electrical characteristics are particularly apt, or where environmental or circuit conditions would destroy solid-state devices. A significant exception is the cathode-ray tube (CRT), which is still made in great numbers for consumer applications. Though bulky, complex, and fragile, it is highly versatile, and so remains in favor as a display device for television, computers, and instruments of all kinds. *See* CATHODE-RAY TUBE; ELECTRONIC DISPLAY.

Other vacuum tubes in widespread use are high-power transmitting and microwave tubes; x-ray and accelerator tubes; mass spectrometer, electron microscope, night-vision, and surface-analytical tubes; and electron beam guns (partial tubes) for evaporators and welders. Since there is no prospect that any of these types will be replaced by solid-state devices, research and development on them continues, and new types occasionally appear. *See* ELECTRON MICROSCOPE; IMAGE TUBE (ASTRONOMY); ION SOURCES; LIGHT AMPLIFIER; MASS SPECTROSCOPE; PARTICLE ACCELERATOR; PHOTOMULTIPLIER; PHOTOTUBE; SURFACE PHYSICS; WELDING AND CUTTING OF MATERIALS; X-RAY TUBE.

Nicholas Reinhardt

Vaginal disorders Inflammations, trauma, tumors, congenital defects, and functional disorders of the vagina, the muscular membranous canal from the vulvar opening to the uterine cervix. The most common vaginal disorder is vaginitis, an inflammation of varying degree, characterized by itching (pruritus), typically accompanied by a discharge. Among women of reproductive age, the three most common types of vaginitis are caused by the protozoon *Trichomonas*, the bacterium *Gardnerella*, or the monilial fungus. Women with new or multiple sexual partners should also have vaginal cultures taken for chlamydia and gonorrhea. Recurrent monilia infections could be a clue to diabetes or human immunodeficiency virus (HIV) infection. *See* ACQUIRED IMMUNE DEFICIENCY SYNDROME (AIDS); GONORRHEA.

The most important primary lesion of the vagina is carcinoma, but its incidence is low. Cancer of the cervix is much more common, and in advanced cases the cancer may extend into the vagina. *See* CANCER (MEDICINE); REPRODUCTIVE SYSTEM DISORDERS.

Machelle M. Seibel

Valence A term commonly used by chemists to characterize the combining power of an element for other elements, as measured by the number of bonds to other atoms which one atom of the given element forms upon chemical combination. The term also has come to signify the theory of all the physical and chemical properties of molecules that specially depend on molecular electronic structure.

Thus, in water, H_2O, the valence of each hydrogen atom is 1; the valence of oxygen, 2. In methane, CH_4, the valence of hydrogen again is 1; of carbon, 4. In NaCl and CCl_4 the valence of chlorine is 1, and in CH_2 the valence of carbon is 2. *See* CHEMICAL BONDING.

Most of the simple facts of valence (though certainly not all) follow from the postulate that atoms combine in such a way as to seek closed-shell or inert-gas structures (rule of eight) by the transfer of electrons between them or the sharing of a pair of electrons between them. Many molecular structures may be obtained by inspection using these rules; letting a dot represent an electron:

$$\begin{array}{cccc} \mathrm{H} & \mathrm{H} & & \\ \cdot\cdot & \cdot\cdot & & \cdot\cdot \\ :\underset{\cdot\cdot}{\mathrm{O}}:\mathrm{H} & \mathrm{H}:\underset{\cdot\cdot}{\mathrm{C}}:\mathrm{H} & [\mathrm{Na}]^{+} & \left[:\underset{\cdot\cdot}{\mathrm{F}}:\right] \\ & \mathrm{H} & & \end{array}$$

In these electron-dot symbols, the electrons in the K shell are not included for atoms after He, nor are the electrons in the K and L shells for atoms following Ne.

As generally used and here defined, the word valence is ambiguous. Before a value can be assigned to the valence of an atom in a molecule, the electronic structure of the molecule must be exactly known, and this structure must be describable simply in terms of simple bonds. In practice neither of these conditions is ever precisely fulfilled. A term not so ambiguous is oxidation number or valence number. Oxidation numbers are useful for the balancing of oxidation-reduction equations, but they are not related simply to ordinary valences. Thus the valence of carbon in CH_4, $CHCl_3$, and CCl_4 is 4; oxidation numbers of carbon in these three substances are -4, $+2$, and $+4$. *See* CHEMICAL STRUCTURES; ELECTRONEGATIVITY; MOLECULAR ORBITAL THEORY; OXIDATION-REDUCTION.
Robert G. Parr

Valence band The highest electronic energy band in a semiconductor or insulator which can be filled with electrons. The electrons in the valence band correspond to the valence electrons of the constituent atoms. In a semiconductor or insulator, at sufficiently low temperatures, the valence band is completely filled and the conduction band is empty of electrons. Some of the high energy levels in the valence band may become vacant as a result of thermal excitation of electrons to higher energy bands or as a result of the presence of impurities. The net effect of the valence band is then equivalent to that of a few particles which are equal in number and similar in motion to the missing electrons but each of which carries a positive electronic charge. These "particles" are referred to as holes. *See* BAND THEORY OF SOLIDS; CONDUCTION BAND; ELECTRIC INSULATOR; HOLE STATES IN SOLIDS; SEMICONDUCTOR; VALENCE.
H. Y. Fan

Value engineering A thinking system (also called value management or value analysis) used to develop decision criteria when it is important to secure as much as possible of what is wanted from each unit of the resource used. The resource may be money, time, material, labor, space, energy, and so on. The system is unique in that it effectively uses both knowledge and creativity, and provides step-by-step techniques for maximizing the benefits from both. It promotes development of alternatives suitable for the future as well as the present. This is accomplished by identifying and studying each function that is wanted by the customer or user, then applying knowledge and creativity to achieve the desired function. Resources are converted into costs to achieve direct, meaningful comparisons. By using the methods of value engineering, 15 to 40% reduction in the required resources often results.

Value engineering has applications in five broad areas: in design, purchase, and manufacture of products; in administrative groups, private or public, where the task is to achieve accomplishment through people; in all areas of social service work, such as hospitals, insurance services, or colleges; in architectural design and construction; and in development as well as research.

The system is used to improve value in either or both of two situations: (1) The product or service as used or as planned may provide 100% of the functions the user wants, but lower costs may be needed. The system then holds those functions but achieves them at lower cost. (2) The product or service may have deficiencies, that is, it does not perform the desired functions or lacks quality, and so also lacks good value. The system aims at correcting those deficiencies, providing the functions wanted, while at the same time holding the use of resources (costs) at a minimum. *See* INDUSTRIAL COST CONTROL; INDUSTRIAL ENGINEERING; METHODS ENGINEERING; OPERATIONS RESEARCH; OPTIMIZATION; PROCESS ENGINEERING; PRODUCTION ENGINEERING; PRODUCTION PLANNING.
Lawrence D. Miles

Valve A flow-control device. Valves are used to regulate the flow of fluids in piping systems and machinery. In machinery the flow phenomenon is frequently of a pulsating or intermittent character and the valve, with its associated gear, contributes a timing feature.

The valves commonly used in piping systems are gate valves (Fig. 1), usually operated closed or wide open and seldom used for throttling; globe valves, frequently fitted with a renewable disk and adaptable to throttling operations; check valves, for automatically limiting flow in a piping system to a single direction; and plug cocks, for operation in the open or closed position by turning the plug through 90° and with a shearing action to clear foreign matter from the seat. Safety and relief valves are automatic protective devices for the relief of excess pressure. *See* SAFETY VALVE.

For hydraulic turbines and hydroelectric systems, valves and gates control water flow for (1) regulation of power output at sustained efficiency and with minimum wastage of water, and (2) safety under the inertial flow conditions of large masses of water.

To control the kinematics of the cycle, steam-engine valves range from simple D-slide and piston valves to multiported types. Many types of reversing gear have been perfected which use the same slide valve or piston valve for both forward and backward rotation of an engine, as in railroad and marine service. *See* STEAM ENGINE.

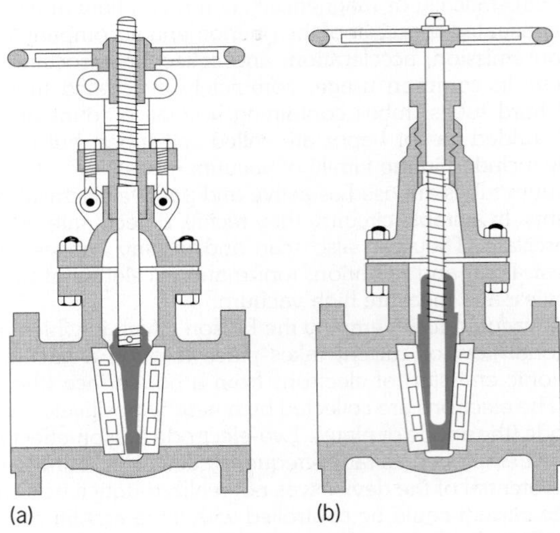

Fig. 1. Gate valves with disk gates shown in color. (*a*) Rising threaded stem shows when valve is open. (*b*) Nonrising stem valve requires less overhead.

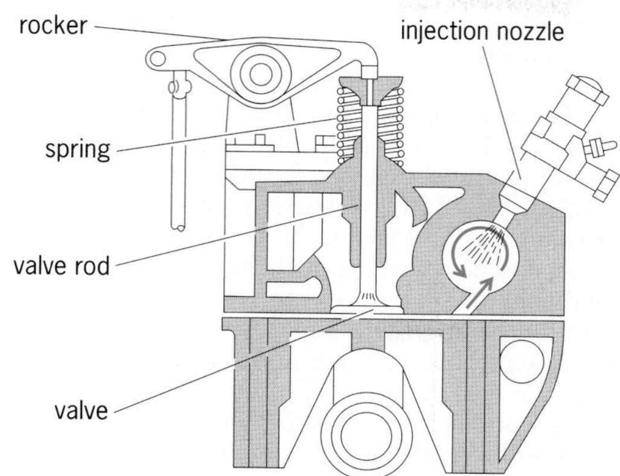

Fig. 2. **Poppet valve for internal combustion engine. (*After T. Baumeister, ed., Marks' Standard Handbook for Mechanical Engineers, 8th ed., McGraw-Hill, 1978*)**

Poppet valves are used almost exclusively in internal combustion reciprocating engines because of the demands for tightness with high operating pressures and temperatures (Fig. 2). Two-cycle engines utilize ports, alternately covered and uncovered by the main piston, for inlet or exhaust. *See* CAM MECHANISM; INTERNAL COMBUSTION ENGINE; VALVE TRAIN.

In compressors, valves are usually automatic, operating by pressure difference on the two sides of a movable, springloaded member and without any mechanical linkage to the moving parts of the compressor mechanism. Like those for compressors, pump valves are usually of the automatic type operating by pressure difference. Theodore Baumeister

Valve train The valves and valve-operating mechanism by which an internal combustion engine takes air or fuel-air mixture into the cylinders and discharges combustion products to the exhaust. *See* VALVE.

Mechanically, an internal combustion engine is a reciprocating pump, able to draw in a certain amount of air per minute. Since the fuel takes up little space but needs air with which to combine, the power output of an engine is limited by its air-pumping capacity. The flow through the engine should be restricted as little as possible. This is the first requirement for valves. The second is that the valves close off the cylinder firmly during the compression and power strokes. *See* INTERNAL COMBUSTION ENGINE.

In most four-stroke engines the valves are the inward-opening poppet type, with the valve head ground to fit a conical seat in the cylinder block or cylinder head. The valve is streamlined and as large as possible to give maximum flow, yet of low inertia so that it follows the prescribed motion at high engine speed.

Engine valves are usually opened by cams that rotate as part of a camshaft, which may be located in the cylinder block or cylinder head. Riding on each cam is a cam follower or valve lifter, which may have a flat or slightly convex surface or a roller, in contact with the cam. The valve is opened by force applied to the end of the valve stem. A valve rotator may be used to rotate the valve slightly as it opens. In engines with the camshaft and valves in the cylinder head, the cam may operate the valve directly through a cup-type cam follower. To ensure tight closing of the valve even after the valve stem lengthens from thermal expansion, the valve train is adjusted to provide some clearance when the follower is on the low part of the cam. *See* CAM MECHANISM. Donald L. Anglin

Vampyromorpha An order of coleoid cephalopods that contains only one species, *Vampyroteuthis infernalis*. It is characterized by a flat, broad, leaflike, chitinous internal shell (gladius); eight arms around the mouth connected by a deep web; no tentacles, but two small sensory filaments that retract into pockets between the bases of the first and second arms; fingerlike cirri and a single row of suckers with chitinous rings along the arms; one pair of paddle-shaped fins on the body in adults (two pairs at the juvenile stage); very dark maroon, mostly black pigmentation; and photophores on the body, head, and arms. Vampyromorphs superficially resemble cirrate octopods, but they bear features that show distinct teuthoid (squid) relationships, and in many characteristics are intermediate between teuthoids and octopods.

Vampyroteuthis infernalis is a gelatinous bathypelagic (very deep, midwater living) species that inhabits worldwide tropical, subtropical, and temperate waters. It occurs mostly at depths of 1600–3800 ft (500–1200 m), occasionally to 4800 ft (1500 m). *See* CEPHALOPODA; COLEOIDEA; OCTOPODA. Clyde F. E. Roper

Van Allen radiation The high-energy, charged particles that are trapped into orbits by the geomagnetic field, forming radiation belts that surround the Earth. The belts consist primarily of electrons and protons and extend from a few hundred kilometers above the Earth to a distance of about 8 R_e (R_e = radius of Earth = 6371 km = 3959 mi). James Van Allen and coworkers discovered them in 1958 using radiation detectors carried on satellites *Explorer 1* and *3*, and they are often referred to as the Van Allen belts.

A charged particle under the influence of the geomagnetic field follows a trajectory that can be conveniently described as a superposition of three separate motions. The first motion, produced by the magnetic force acting at right angles to both the particle velocity and the magnetic field, is a rapid spiral about magnetic field lines. As the spiraling particle moves along the field line toward either the North Pole or South Pole, the increase in magnetic field strength causes the particle to be reflected so that it bounces between the Earth's two hemispheres. Superimposed on the spiral and bounce motions is a slow east-west drift; electrons drift eastward and protons or heavier ions drift westward. (The resulting current, called the ring current, acts to decrease the strength of the Earth's (surface) northward magnetic field at low latitudes.) Thus, individual trapped particles move completely around the Earth in a complicated pattern, their motion being constrained to lie on magnetic shells.

The spatial structure of trapped radiation shows two maxima: an inner radiation belt centered at about $1.5R_e$ and an outer belt centered at 4–$5R_e$. In the inner radiation belt, the most penetrating particles are protons with energies extending to several hundred megaelectronvolts (MeV). However, the flux of high-energy protons decreases rapidly with increasing distance from the Earth and becomes insignificant beyond $4R_e$. Electrons and low-energy protons with energies up to a few MeV occur throughout the stable trapping region. The electron energies extend to several MeV, and in the outer radiation belt electrons are the most penetrating component.

The intensity, energy spectrum, and spatial distribution of particles within the radiation belts vary with time. The most dramatic variations are associated with magnetic storms. The changes are most pronounced for particles in the outer belt where the magnetic variations are the largest. During a magnetic storm, the electron flux in the outer belt may increase by an order of magnitude or more.

It is believed that most of the very high energy protons (>50 MeV) in the inner belt result from the spontaneous decay of high-energy neutrons, produced by collisions of cosmic rays with atmospheric atoms. However, the vast majority of the radiation belt populations are ions and electrons that originate from either the atmosphere or the solar wind and are accelerated by processes only partly understood.

The radiation belts are just one feature of the space plasma environment. For example, plasma from the Sun is continually impinging on the Earth's magnetic field, resulting in a "cavity" known as the magnetosphere.

Because the conditions that lead to the formation of Earth's radiation belts are so general, it is believed that any planet or moon that has a large enough magnetic field will also have radiation belts. Jupiter, Saturn, Uranus, and Neptune have strong magnetic fields and very large, intense radiation belts analogous to those of the Earth.

The term "space weather" describes the conditions in space that affect Earth and its technological systems. It is a consequence of the behavior of the Sun, and the interaction of the solar wind with the Earth's magnetic field. The fluxes of electrons and protons trapped in the radiation belts can injure both personnel and equipment on board spacecraft if the vehicle is exposed to these energetic particles for a sufficiently long time. Because of the spatial structure of the belts, the degree of damage will be strongly dependent on the position in space and hence on the orbit of the vehicle. In the inner belt region, high-energy protons can pass through several centimeters of aluminum structure and injure components in the interior of the spacecraft. In most other regions of the radiation belts, the trapped particles are less penetrating, and damage is confined to exposed equipment such as solar cells. Pete Riley; Martin Walt

Van der Waals equation An equation of state of gases and liquids proposed by J. D. van der Waals in 1873 that takes into account the nonzero size of molecules and the attractive forces between them. He expressed the pressure p as a function of the absolute temperature T and the molar volume $V_m = V/n$, where n is the number of moles of gas molecules in a volume V (see equation below). Here $R = 8.3145$ J K^{-1}

$$p = \frac{RT}{V_m - b} - \frac{a}{V_m^2}$$

mol^{-1} is the universal gas constant, and a and b are parameters that depend on the nature of the gas. Parameter a is a measure of the strength of the attractive forces between the molecules, and b is approximately equal to four times the volume of the molecules in one mole, if those molecules can be represented as elastic spheres. The equation has no rigorous theoretical basis for real molecular systems, but is important because it was the first to take reasonable account of molecular attractions and repulsions, and to emphasize the fact that the intermolecular forces acted in the same way in both gases and liquids. It is accurate enough to account for the fact that all gases have a critical temperature T_c above which they cannot be condensed to a liquid. The expression that follows from this equation is $T_c = 8a/27\,Rb$. *See* CRITICAL PHENOMENA; GAS; INTERMOLECULAR FORCES; LIQUID.

In a gas mixture, the parameters a and b are taken to be quadratic functions of the mole fractions of the components since they are supposed to arise from the interaction of the molecules in pairs. The resulting equation for a binary mixture accounts in a qualitative but surprisingly complete way for the many kinds of gas-gas, gas-liquid, and liquid-liquid phase equilibria that have been observed in mixtures. *See* PHASE EQUILIBRIUM.

The equation is too simple to represent quantitatively the behavior of real gases, and so the parameters a and b cannot be determined uniquely; their values depend on the ranges of density and temperature used in their determination. For this reason, the equation now has little practical value, but it remains important for its historical interest and for the concepts that led to its derivation. *See* THERMODYNAMIC PRINCIPLES. J. S. Rowlinson

Vanadium A chemical element, V, atomic number 23 and atomic weight 50.941. Natural deposits contain two isotopes, ^{50}V (0.24%), which is weakly radioactive, and ^{51}V (99.76%). Commercially important as an oxidation catalyst, vanadium also is used in the production of alloy steel and ceramics and as a coloring agent. Studies have demonstrated the biological occurrence of vanadium, especially in marine species; in mammals, vanadium has a pronounced effect on heart muscle contraction and renal function. *See* PERIODIC TABLE; TRANSITION ELEMENTS.

Very pure vanadium is difficult to prepare because the metal is highly reactive at temperatures above the melting point of its oxide (663°C or 1225°F) from which it is produced. Vanadium is a bright white metal that is soft and ductile. It has a melting point of 1890°C (3434°F), a boiling point of 3380°C (6116°F), and a density of 6.11 g/cm^3 (3.53 oz/in.^{3}) at 18.7°C (65.7°F). The thermal and electrical conductivity of vanadium is superior to that of titanium.

At room temperature, the metal is resistant to corrosion by oxygen, salt water, alkalies, and nonoxidizing acids, the exception being hydrogen fluoride (HF). Vanadium cannot withstand the oxidizing conditions presented by nitric acid or aqua regia. At elevated temperatures it will combine with most nonmetals to form oxides, nitrides, carbides, arsenides, and other such compounds.

The physical properties of vanadium are very sensitive to interstitial impurities. The strength varies from 30,000 lb/in.^{2} (200 megapascals) in the purest form to 80,000 lb/in.^{2} (550 MPa) in the commercial grade. The melting point is markedly altered by small impurities; vanadium containing 10% carbon has a melting point of 2700°C (4892°F).

Vanadium has a low fission neutron cross section. This property combined with the metal's excellent retention of strength at elevated temperatures has made its use in atomic energy applications attractive.

Carbon and alloy steels consume more than half the vanadium produced in the United States. Many plate, structural, bar, and pipe steels contain vanadium to enhance strength and toughness. The basis for the unique properties of these carbon alloy steels is the formation of vanadium carbide. These carbides are extremely hard and wear-resistant; they do not coalesce readily, but maintain a state of fine dispersion. Many large steel forgings contain vanadium in the range 0.05–0.15%; here vanadium acts as a grain refiner, and also improves the mechanical properties of the forgings. Tool steels are another large class of vanadium-containing steels; vanadium ensures the retention of hardness and cutting ability at the elevated temperatures generated by the rapid cutting of metals.

The production of ferrovanadium, an iron alloy, is very important since the primary commercial use of vanadium is in steel. Ferrovanadium is produced by aluminum or silicon reduction of V_2O_5 in the presence of iron in an electric arc furnace. The commonly practiced aluminum reduction is exothermic, so that little additional heat from the arc is required. Silicon processing requires a two-stage reduction to achieve efficient operation. *See* FERROALLOY; STEEL MANUFACTURE.

Vanadium compounds, especially V_2O_5 and NH_4VO_3, are excellent oxidation catalysts in the chemical industry. Processes that employ such catalysts include the manufacture of polyamides, such as nylon; sulfuric acid production by the contact process; phthalic and maleic anhydride syntheses; and various oxidations of organic compounds such as the conversion of anthracene to anthraquinone, ethanol to acetaldehyde, and sugar to oxalic acid. Vanadium pentoxide is used as a mordant in dyeing and printing fabrics and in producing aniline black for the dye industry. Vanadium compounds are used in the ceramics industry for glazes and enamels. A wide range of colors can be obtained with combinations of vanadium oxide, zirconia, silica, lead, tin, zinc, cadmium, and selenium. *See* CERAMICS; DYEING; MORDANT.

Vanadium has long been recognized as an essential element in biological systems; however, the role of the metal often is obscure. Tunicates accumulate vanadium to levels 1 million times greater than the surrounding seawater. This vanadium was once thought to act as an oxygen carrier but now is believed to be an oxidation catalyst that repairs damage to the polymeric, protective tunic of these animals. The first vanadium-dependent enzyme, vanadium bromoperoxidase, was isolated from brown, red, and green marine algae (for example, *Ascophyllum nodosum*); this enzyme catalyzes the bromination of a variety of organic molecules by using hydrogen peroxide and bromide. This activity may be the source of many important brominated compounds that potentially may be used as antifungal and antineoplastic agents. *See* ENZYME.

A variety of physiological effects in mammalian systems have been reported, the most significant being in cardiovascular and renal function. Vanadate causes constriction of veins in the kidney and can alter the retention and excretion of sodium and chloride ions. Cardiac effects of vanadium are species-specific, with observed increases (rabbit and rat) and decreases (guinea pig and cat) in heart muscle contractility. Because vanadate is a potent inhibitor of Na,K-ATPase in the laboratory, it has been suggested that this is the site of the metal's action. However, the physiology of vanadate is probably more complicated, since vanadium behaves as a hormone mimic by elevating intracellular calcium ion (Ca^{2+}) levels in a process that is poorly understood. *See* BIOINORGANIC CHEMISTRY.

Vincent L. Pecoraro

Vanilla A choice flavoring obtained from a climbing orchid, *Vanilla fragrans*, a native of tropical American forests. Its fruits are pods called vanilla beans. These are picked at the proper time before they have fully matured.

Vanillin (4-hydroxy-3-methoxybenzaldehyde) is the principal component of vanilla, although other components contribute to the distinctive flavor of the extract compared to synthetic vanilla. When they are harvested, the beans contain no free vanillin; it develops during the curing period from glucosides that break down during the fermentation and sweating of the beans. The sweating process consists of alternately drying the beans in sunlight and bunching them so that they heat and ferment.

Perry D. Strausbaugh; Earl L. Core

Vapor condenser A heat-transfer device that reduces a thermodynamic fluid from its vapor phase to its liquid phase. The vapor condenser extracts the latent heat of vaporization from the vapor, as a higher-temperature heat source, by absorption in a heat-receiving fluid of lower temperature. The vapor to be condensed may be wet, saturated, or superheated. The heat receiver is usually water but may be a fluid such as air, a process liquid, or a gas. When the condensing of vapor is primarily used to add heat to the heat-receiving fluid, the condensing device is called a heater and is not within the normal classification of a condenser.

Condensers may be divided into two major classes according to use: those used as part of a processing system (process condensers) and those used for serving engines or turbines in a steam power plant cycle (power cycle condensers). Condensers may be further classified according to mode of operation as surface condensers or as contact condensers. *See* CONTACT CONDENSER; SURFACE CONDENSER.

Condensers are required, almost without exception, to condense impure vapors, that is, vapors containing air or other noncondensable gases. Because most condensers operate at subatmospheric pressures, air leaking into the apparatus or system becomes a common cause for vapor contamination, and a variety of designs have been developed to reduce such problems. Accumulation of noncondensable gases seriously affects heat transfer, and means must be provided to direct them to a suitable outlet. Most surface and contact condensers are arranged with a separate zone of heat-transfer surface within the condenser and located at the outlet end of the vapor flow path for efficient removal of the noncondensable gases through dehumidification.

Joseph F. Sebald

Vapor cycle A thermodynamic cycle, operating as a heat engine or a heat pump, during which the working substance is in, or passes through, the vapor state. A vapor is a substance at

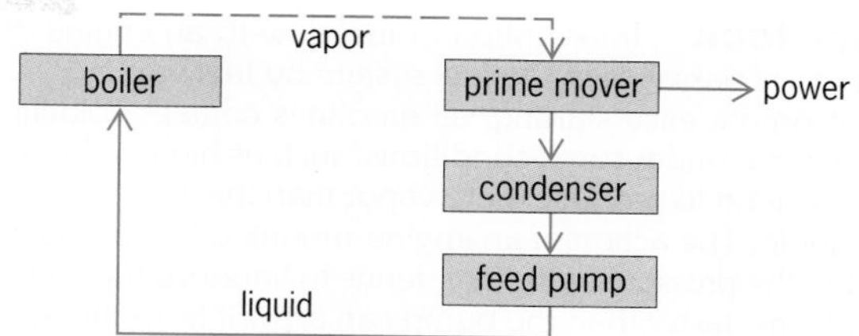

Fig. 1. Rudimentary steam power plant flow diagram.

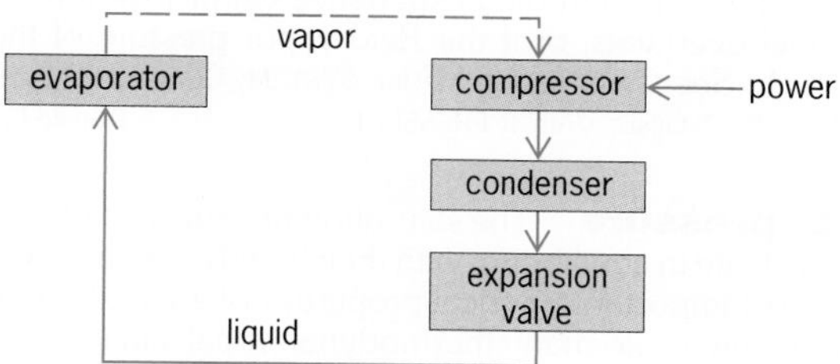

Fig. 2. Rudimentary vapor-compression refrigeration plant flow diagram.

or near its condensation point. It may be wet, dry, or slightly superheated. One hundred percent dryness is an exactly definable condition which is only transiently encountered in practice. *See* HEAT PUMP; THERMODYNAMIC CYCLE.

A steam power plant operates on a vapor cycle where steam is generated by boiling water at high pressure, expanding it in a prime mover, exhausting it to a condenser, where it is reduced to the liquid state at low pressure, and then returning the water by a pump to the boiler (Fig. 1).

In the customary vapor-compression refrigeration plant, the process is essentially reversed with the refrigerant evaporating at low temperature and pressure, being compressed to high pressure, condensed at elevated temperature, and returned as liquid refrigerant through an expansion valve to the evaporating coil (Fig. 2). *See* REFRIGERATION.

The Carnot cycle, between any two temperatures, gives the limit for the efficiency of the conversion of heat into work. This efficiency is independent of the properties of the working fluid. The Rankine cycle is more realistic in describing the ideal performance of steam power plants and vapor-compression refrigeration systems. *See* CARNOT CYCLE; RANKINE CYCLE.

Theodore Baumeister

Vapor lamp A source of radiant energy excited by a supply of electricity which creates a current of ionized gas between electrodes in an enclosure that contains the arc while permitting transmission of the radiant energy. Gaseous-discharge lamps or vapor lamps are given various names relating to the element responsible for the majority of the radiation (mercury, sodium metal-halide, xenon), to the physical attribute of the lamp (short-arc, high-pressure), or, in the case of fluorescent lamps, to the way a phosphor on the bulb wall fluoresces as a result of the lamp's low-pressure mercury-vapor excitation. *See* ARC DISCHARGE; ELECTRICAL CONDUCTION IN GASES.

Gaseous-discharge lamps are broadly used throughout the world because the conversion of electric energy to radiant energy in a gaseous discharge provides radiation in narrow bands within the range of visible light in which the rods and cones of the eye are most sensitive. These light sources have high efficiency in conversion of electricity to light.

For discussions of common types of vapor lamps *see* FLUORESCENT LAMP; MERCURY-VAPOR LAMP; NEON GLOW LAMP; SODIUM-VAPOR LAMP.

T. F. Neubecker

Vapor lock Interruption of fuel flow to an engine due to blockage of passages in the fuel system by fuel vapor.

To promote easy starting, all gasolines contain volatile constituents that under some conditions, such as high ambient temperature, tend to produce more vapor than the fuel-system vents can handle. The action of an engine-mounted fuel pump, in decreasing the pressure at its inlet, tends to vaporize the fuel. If the vapor forms faster than the pump can draw it from the fuel line, the flow of fuel to a carburetor is effectively stopped and the engine stalls. Vapor lock is much less likely on a fuel-injected engine with an electric pump in the fuel tank. However, an engine with port fuel injection may experience vapor lock if the injector or fuel overheats, or if the Reid vapor pressure of the fuel is too high. *See* CARBURETOR; FUEL SYSTEM; GASOLINE; INTERNAL COMBUSTION ENGINE; VAPOR PRESSURE. Donald L. Anglin

Vapor pressure The saturation pressures exerted by vapors which are in equilibrium with their liquid or solid forms. One of the most important physical properties of a liquid, the vapor pressure, enters into many thermodynamic calculations and underlies several methods for the determination of the molecular weights of substances dissolved in liquids. For a discussion of the vapor pressure relationships of solids *see* SUBLIMATION. *See* MOLECULAR WEIGHT; SOLUTION.

If a liquid is introduced into an evacuated vessel at a given temperature, some of the liquid will vaporize, and the pressure of the vapor will attain a maximum value which is termed the vapor pressure of the liquid at that temperature. Although the quantity of liquid remaining does not diminish thereafter, the process of evaporation does not cease. A dynamic equilibrium is established, in which molecules escape from the liquid phase and return from the vapor phase at equal rates. *See* EVAPORATION.

It is important to make a distinction between the vapor pressure of a liquid, as described above, and the pressure of a vapor. The vapor pressure of a pure liquid is a unique and characteristic property of the liquid and depends only upon the temperature. A gas or vapor may, on the other hand, exert any pressure within reason, depending upon the volume to which it is confined, provided it is not in contact with its liquid phase. *See* PHASE EQUILIBRIUM. Norman H. Nachtrieb

Varactor A solid-state device which has a capacitance that varies with the voltage applied across it. The name varactor is a contraction of the words variable and reactor. Typically the device consists of a reverse-biased *pn* junction that has been doped to maximize the change in capacitive reactance for a given change in the applied bias voltage. The device has two primary applications: frequency-tuning of radio-frequency circuits including frequency-modulation (FM) transmitters and solid-state receivers, and nonlinear frequency conversion in parametric oscillators and amplifiers. *See* CAPACITANCE; FREQUENCY MODULATOR; PARAMETRIC AMPLIFIER; REACTANCE.

A *pn* junction in reverse bias has two adjacent microscopic space-charge or depletion regions which function like the plates of a capacitor. These depletion regions get larger as the applied reverse-bias voltage is increased; however, the increase in the width of the depletion region is not linear with bias voltage, but instead is sublinear, the exact nature of the relationship depending upon the doping profile in the *pn* junction. For example, in a *pn* junction with constant doping density, the depletion region width varies as the square root of the applied reverse-bias voltage. Because the capacitance of the device is proportional to the width of the depletion region, the nonlinear relationship between bias voltage and depletion width results in a nonlinear voltage-capacitance relationship as well. The *pn* junction doping profile is adjusted by the device designer to obtain the desired capacitive nonlinearity.

The frequency response of the varactor is governed by the relationship between the series linear resistance of the diode and its nonlinear capacitance. The highest frequency for which the device will function properly is that at which the capacitive reactance (the reciprocal of the product of nonlinear capacitance and frequency) is equal to the series resistance of the device. Thus, designing a varactor for maximum frequency response involves choosing a doping density high enough for a small series resistance but low enough so that the capacitance of the device is small.

Varactors, as well as other solid-state devices, possess the advantage that they are compact and robust, permitting their use in hostile environments, as well as improving the reliability of the circuits in which they are employed. *See* JUNCTION DIODE; MICROWAVE SOLID-STATE DEVICES; SEMICONDUCTOR; SEMICONDUCTOR DIODE. David R. Andersen

Variable (mathematics) A symbol x is usually defined to be a variable if it may denote any member of a set of objects. A variable is discrete or continuous according to whether its range (the set) is discrete (for example, a subset of the natural numbers) or continuous (for example, all real numbers between two real numbers), respectively. *See* ANALYTIC GEOMETRY; NUMBER THEORY; PARAMETRIC EQUATION. Leonard Blumenthal

Variable star A star with a detectable change in brightness that is often accompanied by other physical changes. Every star varies in brightness sometime during its life, usually in the early stages (while it is forming) or in the late stages (close to its death). Therefore, variability provides important clues about the evolution and nature of stars. Depending upon the type of star, variability in brightness can provide information about its size, radius, mass, temperature, luminosity, internal and external structure, composition, and distance from the Earth. Hundreds of thousands of stars are known to vary in brightness. Variable stars can be divided into two major types: extrinsic and intrinsic variables.

The extrinsic variables are those stars in which the variability in brightness occurs because of the occultation of one star by another object, or by the rotation of a star that has dark or bright spots on its surface, similar to sunspots. *See* BINARY STAR; ECLIPSING VARIABLE STARS.

Intrinsic variables are those stars in which the variability of brightness occurs because of physical change in or on the star itself. These stars are divided into two classes: pulsating and eruptive (*see* table).

Pulsating variables. Periodic pulsation (contraction and expansion) of the star and its outer layer results in the variation in brightness, as well as variations in the star's temperature, spectrum, and radius, for pulsating variables, Cepheids are rare, highly luminous (supergiant), yellow pulsating stars. They vary with periods from 1 to 100 days, and have a range of variation from 0.1 to 2 magnitudes.

Cepheids show an important correlation between their period of variation and their relative brightness, in that those stars with longer periods are also brighter. Due to this important correlation, cepheids have been used to measure distances to nearby galaxies and in the determination of the distance scale of the universe.

A similar group of variable stars, W Virginis stars, are found in globular star clusters and the corona of the Milky Way Galaxy. These stars are bluer (and thus hotter) and older (population II) than the cepheids. They have periods from 10 to 30 days, and obey a similar period–luminosity relationship. They are sometimes called type II cepheids to distinguish them from the classical type I cepheids. *See* CEPHEIDS.

RR Lyrae are the second most common type of variable in the Galaxy. They have periods of less than 1 day, and have a small range of variation, from 0.5 to 1.5 magnitudes. They are

Representative types of intrinsic variables

Type	Range of period, days	Amplitude, magnitude	Spectra and luminosity	Example	Number known
Pulsating					
RR Lyrae	<1	<2	A to F (blue), giants	RR Lyrae	7382
Cepheids (types I and II)	1–70	0.1–2	F to G (yellow), supergiants	δ Cephei	1028
Long-period					
Mira	80–1000	2.5–6+	M (red), giants	o Ceti (Mira)	7390
Semiregular	30–1000	1–2	M (red), giants and supergiants	Z Ursae Majoris	5479
RV Tauri	30–150	Up to 3	G to K (yellow and red), supergiants	RV Tauri	134
Irregular	Irregular	Up to several magnitudes	All types		5884
Others					707
Eruptive					
Supernovae	?	15 or more		CM Tauri (Crab Nebula)	7
Novae	Centuries?	7–16	O to A, subdwarfs	GK Persei	388
Recurrent novae	20 years?	7–9		RS Ophiuchi	8
Dwarf novae (U Geminorum and Z Camelopardalis)	10–600	2–6	A to F, subdwarfs	SS Cygni, Z Camelopardalis	420
Flare	?	Up to 6	M, dwarfs	UV Ceti	1780
Nebular	Rapid and erratic	Up to a few magnitudes	B to M, main sequence and subgiants	T Tauri	1698
R Coronae Borealis		1–9	F to K, supergiants	R Coronae Borealis	42
Others					380

particularly numerous in globular clusters, and thus are sometimes referred to as cluster variables. All RR Lyrae variables have the same intrinsic luminosity, of magnitude 0.5. Therefore these variables do not follow the period-luminosity relationship. Thus, if an RR Lyrae star can be identified in a star cluster and its apparent magnitude determined, the above equation and the star's known absolute magnitude can be used to obtain the distance to the cluster.

Long-period pulsating variable stars are the most abundant of all variables in the Milky Way Galaxy. They are red, cool, giant or supergiant stars with spectral class M or R, S, or C carbon types. Long-period variables are old stars which have evolved from the main sequence and are in the late stages of their evolution. *See* STELLAR EVOLUTION.

The prototype of long-period variables is Ceti (Mira), which has given its name to those long-period variables that have a range of variation of 2.5 magnitudes and more. Mira variables have periods ranging from 100 to 1000 days. Although the change in brightness is periodic, the brightness of individual cycles may not be the same, some cycles being much brighter or fainter than others. *See* MIRA.

Red giant and supergiant stars with less regularity of variation, shorter periods, and smaller ranges of variation, less than 2.5 magnitudes, are called semiregular variables. Other red variable stars that do not exhibit any regularity in their brightness change are called irregular stars.

Rare, very luminous, yellow supergiant stars that generally show alternating shallow and deep fadings are called RV Tauri stars. These pulsating variables vary by 2 to 3 magnitudes within 30 to 150 days.

Eruptive variables. Eruptive variables are those stars that have one or more eruptions—the ejection of matter into space—in their lifetime.

The most spectacular type of stellar explosion is a supernova, wherein a star's luminosity suddenly increases to tens of thousands of times the original brightness and the star outshines the total brightness of its galaxy. The gigantic explosion may be due to the gravitational collapse of a very hot and massive star that has exhausted the energy available from nuclear reactions; the collapse then creates enormous energy to blast outward the layers surrounding the stellar core. Alternatively, the explosion may occur in a close binary system in which one of the components is a massive white dwarf, an Earth-sized, very compact, old star, that explodes when it receives too much material from the other component. *See* SUPERNOVA.

The best known of the eruptive variables are stars that brighten by 7 to 16 magnitudes (less than 0.0010 of that of a supernova) within about a day. They stay at maximum brightness for a few days or weeks and then slowly fade. The word "nova," meaning new, was used for these stars. Actually a nova is not a new star, but an already existing star that due to the eruption has become very luminous and thus visible. *See* NOVA.

Recurrent novae are stars that about every 20 years or more have eruptions during which the system brightens by 7 to 9 magnitudes.

Dwarf novae have smaller-scale eruptions, in which the star brightens by 2 to 6 magnitudes within a day, stays bright 1 to 2 weeks, and then fades to the original brightness.

R Coronae Borealis stars, instead of brightening through eruptions, irregularly decrease in brightness every 2 to 3 years by 1 to 9 magnitudes. These bright, highly luminous supergiant stars are rich in carbon and poor in hydrogen in composition. The decrease in brightness is caused by the veiling of the star by thick carbon clouds expelled to the star's atmosphere. *See* CATACLYSMIC VARIABLE; STAR.

Janet Akyüz Mattei; Arne Henden

Variational methods (physics)

Variational methods (physics) Methods based on the principle that, among all possible configurations or histories of a physical system, the system realizes the one that minimizes some specified quantity. Variational methods are used in physics both for theory construction and for calculational purposes.

The earliest use of a variational principle for physics is Fermat's principle in optics, which states that when a light ray traverses a medium with nonuniform index of refraction its path is such as to minimize its travel time. An integral expresses the time that the light takes to travel from one point to another along a particular path, and an application of the calculus of variations to this integral makes it possible to determine the particular path for which the travel time is a minimum. This problem is mathematically identical to the variational principle that determines a geodesic, the path of shortest distance, in a given geometry. In that form, the same principle determines the world lines of all objects in the general theory of relativity. *See* RELATIVITY; RIEMANNIAN GEOMETRY.

Similarly, in mechanics, Hamilton's principle for the action is defined for any system of point particles by an integral (called the action) that extends over an arbitrarily prescribed path Γ in configuration space. Hamilton's principle asserts that the trajectories of all the particles are determined by the requirement that Γ be such that, for given initial and final times, the action is a minimum; for this reason it is also called the principle of least action. If the calculus of variations is applied to implement this principle, the corresponding Euler-Lagrange equations are obtained. These are the lagrangian equations of motion, that is, Newton's equations of motion in lagrangian form. *See* ACTION; HAMILTON'S PRINCIPLE ; LAGRANGIAN FUNCTION; LEAST-ACTION PRINCIPLE.

The principle of least action has been generalized to systems with infinitely many degrees of freedom, that is, fields. A Lagrange density function is then defined, which is a function of the fields and their time derivatives at any given point in space and time. For any field theory, only the Lagrange density needs to be given; the field equations are then derivable as the corresponding Euler-Lagrange equations. A similar technique makes it possible to derive the Schrödinger equation and the Dirac equation in quantum mechanics from specific Lagrange density functions. *See* QUANTUM MECHANICS; QUANTUM THEORY OF MATTER; RELATIVISTIC QUANTUM THEORY.

This method has great procedural advantages. For example, it facilitates a check of whether the theory satisfies certain invariance principles (such as relativistic invariance or rotational invariance) by simply ascertaining whether the Lagrange density satisfies them. The corresponding conservation laws can also be derived directly from the lagrangian. *See* CONSERVATION LAWS (PHYSICS); QUANTUM FIELD THEORY; SYMMETRY LAWS (PHYSICS).

The variational method also plays an important role in quantum-mechanical calculations. For the computation of needed quantities in terms of functions that result from the solution of differential equations, it is always of great advantage to use formulas that have the special form required to make them stationary with respect to small variations of the input functions in the vicinity of the unknown, exact solutions. *See* MINIMAL PRINCIPLES.

Roger G. Newton

Varistor Any two-terminal solid-state device in which the electric current I increases considerably faster than the voltage V. This nonlinear effect may occur over all, or only part, of the current-voltage characteristic. It is generally specified as $I \propto V^n$, where n is a number ranging from 3 to 35 depending on the type of varistor. The main use of varistors is to protect electrical and electronic equipment against high-voltage surges by shunting them to ground. *See* ELECTRIC PROTECTIVE DEVICES.

One type of varistor comprises a sintered compact of silicon carbide particles with electrical terminals at each end. It has symmetrical characteristics (the same for either polarity of voltage) with n ranging from 3 to 7. These devices are capable of application to very high power levels, for example, lightning arresters. *See* LIGHTNING AND SURGE PROTECTION.

Another symmetrical device, the metal-oxide varistor, is made of a ceramiclike material comprising zinc oxide grains and a complex amorphous intergranular material. It has a high resistance (about 10^9 ohms) at low voltage due to the high resistance of the intergranular phase, which becomes nonlinearly conducting in its control range (100–1000 V) with $n > 25$.

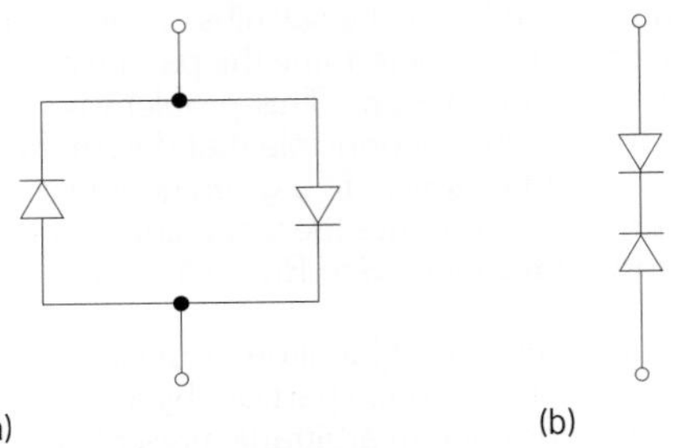

Symmetrical rectifier varistors. (*a*) Low voltage. (*b*) High voltage.

Semiconductor rectifiers, of either the *pn*-junction or Schottky barrier (hot carrier) types, are commonly utilized for varistors. A single rectifier has a nonsymmetrical characteristic which makes it useful as a low-voltage varistor when biased in the low-resistance (forward) polarity, and as a high-voltage varistor when biased in the high-resistance (reverse) polarity. Symmetrical rectifier varistors are made by utilizing two rectifiers connected with opposing polarity, in parallel (illus. *a*) for low-voltage operation and in series (illus. *b*) for high-voltage use. For the high-voltage semiconductor varistor, n is approximately 35 in its control range, which can be designed to be anywhere from a few volts to several hundred. *See* SEMICONDUCTOR RECTIFIER.

I. A. Lesk

Varnish A transparent surface coating which is applied as a liquid and then changes to a hard solid. Varnishes are solutions of resinous materials in a solvent, and dry by the evaporation of the solvent or by a chemical reaction, either with oxygen from the air or by some other means, including absorption of atmospheric moisture.

Spirit varnishes are those in which the evaporation of solvent is the only drying process; the solvent is usually alcohol, although the term is used for similar coatings made with other solvents. Shellac varnish, made by dissolving shellac in alcohol, is the most common of this type. Oleoresinous varnishes are made by treating a drying oil with a resin, usually with heat, and dissolving the reaction product in a solvent, usually a petroleum fraction; drying results from the evaporation of the solvent, followed by polymerization of the drying oil portion, a reaction which is accelerated by metallic driers added to the varnish. For a discussion of the mechanism of this drying action. *See* DRIER (PAINT); SHELLAC; SOLVENT.

Varnish coatings on wood are used to protect against abrasion, staining, and weather and to reduce the penetration of water and other materials without obscuring the grain or changing the color materially. Varnishes are used on masonry to reduce the penetration of moisture and the damage from freezing. Paper is coated with varnish to resist moisture and keep printing from being damaged. *See* PAINT AND COATINGS; PRINTING.

C. R. Martinson; C. N. Sisler

Varnish tree The plant *Toxicodendron vernicifluum* (previously known as *Rhus vernicifera*), also called lacquer tree, a member of the sumac family (Anacardiaceae). It is a native of China, but has long been cultivated in Japan. When the bark is cut, it exudes a milky juice which darkens and thickens on exposure. This is the lacquer long used in China and Japan. When properly applied, the thin transparent film becomes a varnish of extreme hardness. *See* LACQUER; SAPINDALES.

Perry D. Strausbaugh; Earl L. Core

Varve Any of a variety of distinct sediment laminations or beds deposited within the span of a single year. They are formed commonly in saline or fresh-water lakes, but examples from marine environments are known as well. Usually, varves occur in repetitive series and thus comprise vertical sequences of annual cyclic deposits. Varves range in thickness from less than a millimeter (0.04 in.) to over a meter (3 ft), but typically are a few millimeters or centimeters thick.

The classic varves are found in glacial lake sediments formed during the Pleistocene ice ages. These glacial varves occur typically as couplets of light-colored silt or sand and dark clay. The relatively coarse silt-sand layers are formed during the warm summer months when meltwater inflows and sediment yields to the lake are large. During winter when meltwater inflow is greatly reduced or stopped, the fine-grained clay settles slowly to the lake bottom to deposit the fine, dark winter layer of the varve. Similar varved sediments are found commonly in modern glacier-fed lakes that undergo large seasonal variations in inflow. *See* GLACIATED TERRAIN; PLEISTOCENE.

Varves can be used as tools for correlation as well as for chronological reconstructions. In addition to their use in dating sedimentary deposits, varves have been used to investigate sedimentation rates, cyclic deposition, climate variations, glacial histories, and as standards of comparison for other dating techniques. Varves have been identified occasionally in ancient sedimentary rocks. *See* DATING METHODS; GLACIAL EPOCH; MARINE SEDIMENTS; SEDIMENTARY ROCKS; SEDIMENTOLOGY. Norman D. Smith

Vascular disorders Those disorders that involve the arteries, veins, and lymphatics.

Arteries. Diseases of the large and medium-sized arteries are the major cause of morbidity and mortality in the Western world.

Arteriosclerotic vascular disease, also known as arteriosclerosis, affects large and medium arteries and is particularly common in the arteries supplying the heart, those supplying the brain, and those supplying the lower extremities. Progressive narrowing and finally total occlusion of these arteries lead to the development of angina pectoris, myocardial infarction, stroke, and vascular insufficiency in the limbs. Although the cause of the disease is not known, certain factors appear to predispose to its development. These include cigarette smoking, hypertension, elevated serum cholesterol, and diabetes mellitus. There can also be a genetic predisposition to the disorder. *See* ARTERIOSCLEROSIS; DIABETES; HYPERTENSION; INFARCTION.

Much less common are diseases that lead to arteritis, an inflammation of the large or medium arteries, with subsequent occlusion or rupture of the artery. The causes include bacterial infection, syphilis, allergic disorders, and hypersensitivity states. *See* SYPHILIS.

Several generalized disorders can involve the small arteries. These include scleroderma, periarteritis nodosa, lupus erythematosus, rheumatoid arthritis, and dermatomyositis. The involvement can produce areas of cell death (necrosis) and ulceration of the skin, particularly of the limbs. This appears to occur because of the occlusion of the involved small arteries, with consequent loss of blood supply. Such involvement often is preceded by a type of hypersensitivity to cold, a condition known as Raynaud's syndrome. *See* CONNECTIVE TISSUE DISEASE.

Veins. The most common disorder of the venous circulation is varicose veins. It develops because of the loss of function of the valves in the superficial veins of the limbs. The two most commonly involved veins are the greater and lesser saphenous. In this condition, the veins are enlarged and often tortuous. If the deep veins are not involved, the varicose veins are termed primary. Varicose veins tend to cluster in families and are more common in women than men. Primary varicose veins may be cosmetically unpleasing, but they rarely cause serious symptoms in the involved limbs. A more serious problem develops when the varicose veins are associated with destruction of valves in the deep veins, secondary to deep-vein thrombosis. When varicose veins develop because of problems within the deep venous system, they are called secondary varicose veins.

When patients undergo major surgery or are confined to bed because of a serious medical illness, thrombosis of the veins of the leg can occur. The thrombosis occurs most commonly in the deep veins of the calf but can progress and extend proximally to involve the popliteal, femoral, and iliac veins. *See* THROMBOSIS.

Lymphatics. When the drainage function is disturbed by obstruction of the lymphatics by tumors, parasites, or surgical excision, serious swelling (edema) can occur. This condition is called lymphedema. The pattern of swelling involves not only the lower leg but the foot and toes as well. Obstruction of the lymphatics can also be congenital or can appear unexpectedly at the time of the onset of the menses. Regardless of its cause, lymphedema is a permanent condition. *See* EDEMA; LYMPHATIC SYSTEM.

Arteriovenous communications. If there are anomalous communications between the arteries and veins proximal to the capillaries, blood is shunted away from the capillary bed. Such shunting can be congenital or acquired.

The congenital form can occur at almost any site in the vascular system. When these communications are present, they may appear as prominent veins that are much larger and more numerous than normal. These are commonly referred to as hemangiomas. The extent to which they cause problems depends upon their location, their size, and numbers of feeding communications that are present. If a foreign body such as a bullet or knife passes between a major artery and vein, an artificial communication can develop that results in the shunting of a large volume of blood. When the communication is between very large vessels such as the abdominal aorta and inferior vena cava, the amount of blood shunted may be so large that heart failure occurs. *See* CARDIOVASCULAR SYSTEM. D. Eugene Strandness, Jr.

Vector (mathematics) A mathematical element used to describe quantities having magnitude and direction. It can be represented by a directed line segment. Many physical quantities, for example, velocity, acceleration, and force, are vectors. Vectors are widely used in mathematical physics. *See* CALCULUS OF VECTORS. Marvin Yelles

Vector methods (physics) Methods that make use of the behavior of physical quantities under coordinate transformations.

From the point of view of physics, the most appropriate definition of a vector in three-dimensional space is a quantity that has three components which transform under rotations of the coordinate system like the coordinates of a point in space. What characterizes rotations is thatthe distance from the origin—

$$\sqrt{x_1^2 + x_2^2 + x_3^2}$$

of all points $\mathbf{x}$ with cartesian coordinates x_i, $i = 1, 2, 3$—remains unchanged. Specifically, if the rotation takes the x_i to new coordinates x_i', given by Eqs. (1) [in which $\det\{a_{ij}\}$ is the determinant

$$\begin{gathered} x_i' = \sum_{j=1}^{3} a_{ij} x_j \qquad i = 1, 2, 3 \\ \sum_{k=1}^{3} a_{ik} a_{jk} = \begin{cases} 1 & \text{if } i = j, \\ 0 & \text{if } i \neq j, \end{cases} \qquad \det\{a_{ij}\} = 1 \end{gathered} \tag{1}$$

of the matrix $\{a_{ij}\}$], then the three quantities V_i, $i = 1, 2, 3$, form the components of a vector $\mathbf{V}$, if in the new coordinate system the transformed coordinates are given by Eq. (2).

$$V_i' = \sum_{j=1}^{3} a_{ij} V_j \qquad i = 1, 2, 3 \tag{2}$$

If the coordinate transformation of Eqs. (1) is such that $|a_{ij}| = 1$ for $i = j$ and $a_{ij} = 0$ for $i \neq j$, but such that $\det\{a_{ij}\} = -1$ rather than $+1$ as in Eqs. (1), then it describes a reflection, in which a right-handed coordinate system is replaced by a left-handed one. If, for such a transformation, Eq. (2) also holds, then $\mathbf{V}$ is called a polar vector, whereas if the components of $\mathbf{V}$ do not change sign, it is called an axial vector or pseudovector. The vector $\mathbf{V}$ can also be looked upon as a quantity with a direction, with the magnitude

$$\sqrt{V_1^2 + V_2^2 + V_3^2}$$

pointing from the origin of the coordinate system to the point in space with the cartesian coordinates (V_1, V_2, V_3).

A quantity that remains invariant under a rotation of the coordinate system is called a scalar. The importance of vectors and scalars in physics derives from the assumed isotropy of the universe, which implies that all general physical laws should have the same form in any two coordinate systems that differ only by a rotation. It is therefore useful to classify physical quantities according to their transformation properties under coordinate rotations. Examples of scalars include the mass of an object, its electric charge, its volume, its surface area, the energy of a system, and its temperature. Other quantities have a direction and thus are vectors, such as the force exerted on a body, its velocity, its acceleration, its angular momentum, and the electric and magnetic fields. Since the sum of two scalars is a scalar, and the sum of two vectors is a vector, it is important in the formulation of physical laws not to mix quantities that have different transformation properties under coordinate rotations. The sum of a vector and a scalar has no simple transformation properties; a law that equated a vector to a scalar would have different forms in different coordinate systems and would thus not be acceptable. Roger G. Newton

Vega The fifth brightest of all stars, and the third brightest in the northern sky. It will be the north polar star in about 12,000 years. In moving through the Milky Way Galaxy, the Sun is generally heading toward the position now occupied by Vega. At a distance of 7.8 parsecs (25.3 light-years, or 2.4×10^{14} km, or 1.49×10^{14} mi), Vega, or α Lyrae, is the prototypical star of spectral class A0V, indicating that it has an effective surface temperature of 9600 K (16,800°F) and derives its energy from the thermonuclear burning of hydrogen in a stable core region. Stars of this class have main-sequence lifetimes of about 5×10^8 years, 20 times shorter than the Sun. In comparison with the Sun, Vega is approximately 2.9 times larger in diameter, 2.5 times more massive, and 60 times more luminous. *See* PRECESSION OF EQUINOXES; SPECTRAL TYPE.

Vega emits far more radiation at infrared wavelengths than would be expected. This radiation originates from a shell or disk of particles with a temperature of 100 K (−280°F) surrounding Vega out to a distance of 1.3×10^{10} km (8×10^9 mi), twice the radius of the solar system. *See* INFRARED ASTRONOMY; STAR. Harold A. McAlister

Vegetable ivory The seed of the tagua palm (*Phytelephas macrocarpa*) of tropical America. Each drupelike fruit contains six to nine bony seeds. The extremely hard endosperm of the seed is used as a substitute for ivory. Vegetable ivory can be carved and tooled to make buttons, chessmen, knobs, inlays, and various ornamental articles. *See* ARECALES. Perry D. Strausbaugh; Earl L. Core

Vegetation and ecosystem mapping The graphic portrayal of spatial distributions of vegetation, ecosystems, or their characteristics. Vegetation is one of the most conspicuous and characteristic features of the landscape and has long been a convenient way to distinguish different regions; maps of ecosystems and biomes have been mainly vegetation maps. As pressure on the Earth's natural resources grows and as natural ecosystems are increasingly disturbed, degraded, and in some cases replaced completely, the mapping of vegetation and ecosystems, at all scales and by various methods, has become more important. *See* BIOME; ECOSYSTEM.

Three approaches have arisen for mapping general vegetation patterns, (1) based on vegetation structure or gross physiognomy, (2) based on correlated environmental patterns, and (3) based on important floristic taxa. The environmental approach provides the least information about the actual vegetation but succeeds in covering regions where the vegetation is poorly understood. Most modern classification systems use a combination of physiognomic and floristic characters. *See* PLANT GEOGRAPHY.

Mapping has expanded to involve other aspects of vegetation and ecosystems as well as new methodologies for map production. Functional processes such as primary production, decomposition rates, and climatic correlates (such as evapotranspiration) have been estimated for enough sites so that world maps can be generated. Structural aspects of ecosystems, such as total standing biomass or potential litter accumulations, are also being estimated and mapped. Quantitative maps of these processes or accumulations can be analyzed geographically to provide first estimates of important aspects of world biogeochemical budgets and resource potentials.

Computer-produced maps, using Geographic Information Systems (GIS), often coupled directly with predictive models, remote-sensing capabilities, and other techniques, have also revolutionized vegetation and ecosystem mapping. This gives scientists a powerful tool for modeling and predicting the outcome from global climate change, in that feedback from the world's vegetation can be accounted for. Before computer technology exploded in the early 1980s, the spatial scale and related resolution or grain of vegetation and ecosystem mapping was limited by the static nature of hard-copy maps. The advent of GIS technology enabled the analysis of digital maps at any spatial scale, and the only limitation was the resolution at which the data were originally mapped. In addition, GIS software is used for sophisticated spatial analyses on maps, and this was virtually impossible before. *See* CARTOGRAPHY; CLIMATE MODELING; GEOGRAPHIC INFORMATION SYSTEMS; REMOTE SENSING. Blake E. Feist; Elgene O. Box

Velocimeter An instrument that measures the velocity of a flowing liquid or gas, usually understood to be the average velocity of fluid within a relatively small measuring volume. Velocimeters are distinct from flow meters, which measure spatially averaged velocity, or flow rate, across the cross section of a duct or channel. Depending on the nature of the flow and the objective of the measurement, a velocimeter may be used to measure a time-dependent flow velocity or a time-averaged value. Another characteristic of a velocimeter is its spatial resolution, namely the size of its measuring volume. Some velocimeters measure flow velocity in both magnitude and direction, whereas others only resolve a velocity component in a particular direction or plane. The term anemometer is sometimes used as synonymous to velocimeter, although, strictly speaking, the former term should apply only to instruments that measure air velocity. *See* FLOW MEASUREMENT.

Velocimetry is an essential activity in fluid mechanics research laboratories and in flow-related industrial and environmental applications. The operation of different velocimeters is based on a variety of principles.

Pressure tubes and multihole probes. Pressure tubes relate flow velocity to measurable pressure differences in the flow. Pressure tubes are thin hollow tubes inserted into a stream to measure its velocity V, static pressure P, or total pressure P_o. In idealized flow, these three parameters are related to each other through Bernoulli's equation below, where ρ is the fluid density

$$P_o = P + \tfrac{1}{2}\rho V^2$$

and the combination $\tfrac{1}{2}\rho V^2$ is called the dynamic pressure. A Pitot tube, also called total or impact tube, is a simple tube with an open end facing the flow and the other end connected through flexible tubing to a manometer or pressure transducer. A Pitot tube would ideally read the average total pressure across its face. A static tube is also hollow, but, instead, has a sealed, rounded nose facing the flow and a number of small holes on its side. It reads approximately the static pressure in the free flow. A coaxial combination of these tubes, called the Pitot-static tube, reads directly the dynamic pressure as $P_o - P$, from which the flow velocity can be easily calculated with the use of Bernoulli's equation. *See* BERNOULLI'S THEOREM; MANOMETER; PITOT TUBE; PRESSURE MEASUREMENT; PRESSURE TRANSDUCER.

Thermal anemometers. Thermal anemometers relate flow velocity to the rate of convective heat transfer from a heated sensor. In thermal anemometry, the sensor is a fine, short metallic element, inserted in the flow and supplied with an electric circuit, such that the generated Joule heat increases its temperature above that of the surrounding flow. For metallic elements, the resistance also rises above its value for the unheated sensor. If the flow velocity rises, then more heat is removed from the sensor, so that its temperature tends to decrease. The voltage across the sensor is related to the flow velocity and temperature. With proper calibration, and if the flow temperature is maintained equal to that during calibration or is measured by other means, the flow velocity may be determined from the measured voltage. *See* RESISTANCE MEASUREMENT.

The hot-wire anemometer (HWA) has a cylindrical sensor made of platinum-plated tungsten or platinum alloys. A single-sensor hot-wire probe has an insulated body containing two parallel metallic needles (prongs), at the free tips of which the sensor is mounted and through which current is flowing.

Hot films have sensors that are thin (about 1 μm thick) films of nickel, deposited on a ceramic substrate and coated by a quartz insulation, so that they are insulated from the flowing fluid and can be used in tap water and other conducting liquids. Because of their sturdier construction, they are less sensitive than hot wires to flow impurities and aerodynamic loading.

Laser Doppler velocimeter. Laser Doppler velocimeters relate flow velocity to the frequency shift of laser light scattered by fine particles transported by the flow. The laser Doppler velocimeter (LDV), alternately referred to as a laser Doppler anemometer (LDA), utilizes the Doppler phenomenon, namely the fact that the frequency of waves scattered by a moving object is different from the frequency of incident waves, with the difference (Doppler frequency) being proportional to the component of the velocity of the object along the bisector of the angle between the directions of propagation of the incident and scattered waves. For practical reasons, an optical Doppler system must use a laser beam. *See* DOPPLER EFFECT; LASER.

Ultrasonic velocimeters. The operation of the ultrasonic Doppler velocimeter is based on the same physical principle as laser Doppler velocimetry, but utilizes ultrasound rather than light. The sound is produced by a transmitter, which also alternates as a receiver, collecting scattered sound as it returns toward it. The average velocity of particles scattering sound along the path of the beam is determined from the difference in the frequencies of the transmitted and received signals. By measuring the time delay between a transmitted and a received sound pulse, it also becomes possible to determine the location of the particle, from which it is possible to obtain not only the average velocity, but also an instantaneous velocity profile along the beam path.

Particle image velocimeter. Particle tracking is the direct method of determining flow velocity by timing the displacement of markers that move with the flow. In the simplest case of readily visible, slowly moving, isolated particles, it is sufficient to simply employ a ruler and a stopwatch, but for the method to be applied to complex flows with high resolution, it is usually necessary to capture consecutive particle positions in rapid succession and to be able to separate and identify the images of closely spaced particles. This need has led to the development of digital particle image velocimetry (DPIV or PIV).

Whereas hot-wire anemometry and laser-Doppler velocimetry measure an average flow velocity within the measuring volume, thus being point-measuring methods, particle image velocimetry provides the velocity distribution within an area.

Cup and propeller anemometers. Cup and propeller anemometers measure flow velocity via a force or torque applied on a mechanical element. The cup anemometer is a familiar device, frequently installed on towers near airports and in meteorological stations for measuring wind speed. It consists of three or four hollow hemispherical cups mounted at some radial distance from a rotating shaft and such that all the concave sides face in the same circumferential direction.

A related device is the propeller anemometer, which actually operates like a wind turbine. It consists of a small axial propeller mounted on a rotating shaft, whose speed of rotation is proportional to the flow velocity parallel to its axis. Stavros Tavoularis

Velocity

Velocity The time rate of change of position of a body in a particular direction. Linear velocity is velocity along a straight line, and its magnitude is commonly measured in such units as meters per second (m/s), feet per second (ft/s), and miles per hour (mi/h). Since both a magnitude and a direction are implied in a measurement of velocity, velocity is a directed or vector quantity, and to specify a velocity completely, the direction must always be given. The magnitude only is called the speed. *See* SPEED.

A body need not move in a straight line path to possess linear velocity. When a body is constrained to move along a curved path, it possesses at any point an instantaneous linear velocity in the direction of the tangent to the curve at that point. The average value of the linear velocity is defined as the ratio of the displacement to the elapsed time interval during which the displacement took place.

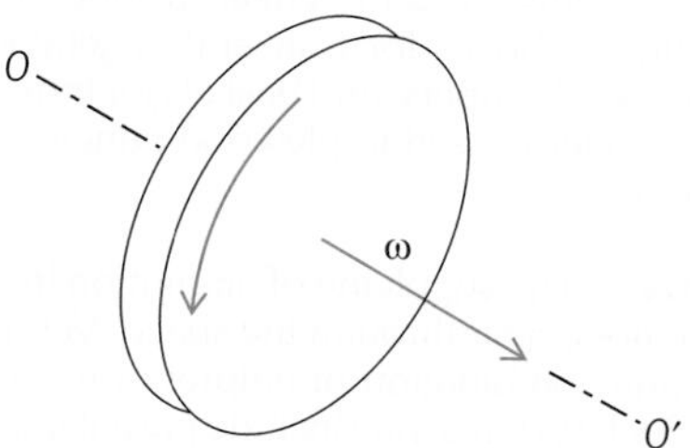

Angular velocity shown as an axial vector. Axis of rotation is *OO′*.

The representation of angular velocity ω as a vector is shown in the illustration. The vector is taken along the axis of spin. Its length is proportional to the angular speed and its direction is that in which a right-hand screw would move. If a body rotates simultaneously about two or more rectangular axes, the resultant angular velocity is the vector sum of the individual angular velocities. Rogers D. Rusk

Velocity analysis

Velocity analysis A technique for the determination of the velocities of the parts of a machine or mechanism. Graphical and analytical analyses of plane mechanisms are the two main methods that are used. Visualization is an inherent part of the graphical analysis and generally gives a better physical feel for the problem than most purely analytical methods. Analytical methods, however, are necessary for computer analyses. *See* VELOCITY.

In a high-speed machine it is important that the inertia forces be determined. This requires an acceleration analysis of the machine, and the first step in an acceleration analysis usually is a velocity analysis. In the analysis of some machines, the velocity of a particular point in the machine may itself be the important thing to be determined in the analysis—for example, the cutting speed and return speed of the cutting tool in a shaper, or the shuttle velocity in a textile machine.

In the graphical method of velocity analysis, the geometry of the mechanism is known in each phase of its motion cycle from the drawing of the mechanism in each position. In an analytical velocity analysis, the geometry is usually determined analytically; that is, a position analysis of the mechanism is performed by using trigonometry, vector mathematics, or some other analytical method as the first step in the velocity analysis.

One analytical method for the position analysis of mechanisms, developed by M. A. Chace, makes use of vector mathematics. It consists essentially of solving vector triangles containing two unknowns. Computer subprograms for Chace solutions to the plane vector equation and for all the various vector operations have been written, thus making the position and velocity analyses (as well as the acceleration analysis) of mechanisms by computer using Chace's method quite convenient. Computer analyses are particularly useful when complex mechanisms must be analyzed in many positions of their cycle of motion. *See* STRAIGHT-LINE MECHANISM; VECTOR (MATHEMATICS).

James C. Wolford

Veneer A thin sheet of wood of uniform thickness produced by peeling, slicing, or sawing. Depending on the manner of production and the portion of wood from which a veneer is made, the grain may be flat, vertical, or biased. Most veneer is rotary-cut from a bolt of wood, called a flitch, centered in chucks of a lathe. A nose bar bears against the flitch parallel to the center line of the lathe and a knife, also extending nearly the length of the lathe, peels off the veneer. Knives near the ends of the flitch cut the edges of the veneer.

Veneers cut from selected hardwoods, burls, crotches, and stumps are used for facings on furniture and the interior decoration of buildings. Other uses of veneer include plate separators in storage batteries, boxes for fruit and vegetables, drums for cheeses, and crates, hampers, and baskets for transportation and storage. Most veneer is used in plywood panels. *See* PLYWOOD; WOOD PRODUCTS.

Frank H. Rockett

Ventilation The supplying of air motion in a space by circulation or by moving air through the space. Ventilation may be produced by any combination of natural or mechanical supply and exhaust. Such systems may include partial treatment such as heating, humidity control, filtering or purification, and, in some cases, evaporative cooling. More complete treatment of the air is generally called air conditioning. *See* AIR CONDITIONING.

Natural ventilation may be provided by wind force, convection, or a combination of the two. Although largely supplanted by mechanical ventilation and air conditioning, natural ventilation still is widely used in homes, schools, and commercial and industrial buildings.

Mechanical supply ventilation may be of the central type consisting of a central fan system with distributing ducts serving a large space or a number of spaces, or of the unitary type with little or no ductwork, serving a single space or a portion of large space. Outside air connections are generally provided for all ducted systems. Outside air is needed in controlled quantities to remove odors and to replace air exhausted from the various building spaces and equipment.

Exhaust ventilation is required to remove odors, fumes, dust, and heat from an enclosed occupied space. Such exhaust may be of the natural variety or may be mechanical by means of roof or wall exhaust fans or mechanical exhaust systems. The mechanical systems may have minimal ductwork or none at all, or may be provided with extensive ductwork which is used to collect localized hot air, gases, fumes, or dust from process operations. Where it is possible to do so, the process operations are enclosed or hooded to provide maximum collection efficiency with the minimum requirement of exhaust air.

John H. Clarke

Venus The second planet in distance from the Sun. Venus has been called Earth's twin because it is similar to Earth in gross characteristics such as mass, radius, and density. In other ways Venus is apparently different. Its atmospheric mass is almost a hundred times that of the Earth; its atmosphere is mostly carbon dioxide instead of nitrogen and oxygen; an extensive cloud layer of concentrated sulfuric acid is present; its surface temperature is an unbearable 867°F (464 K); and it rotates with a period of 243 days, and from east to west, clockwise as we look down from the month, in the opposite sense of most other planets. Some of these differences are due more to alternate evolutionary paths of the two planets than to totally different initial conditions.

To the naked eye, Venus is the brightest starlike object in the sky. It is usually visible during the night either soon after sunset or close to sunrise. It can sometimes be seen during the daytime.

The light seen coming from Venus is entirely due to sunlight that is reflected from a dense cloud layer whose top is located about 45 mi (70 km) above the surface and whose bottom lies within 30 mi (50 km) of the surface. In contrast to the Earth's approximately 50% cloud cover, the clouds of Venus are present over the entire planet. The clouds of Venus consist of a large number of tiny particles, about 1 micrometer in size, that are made of a water solution of concentrated sulfuric acid.

By far, the chief gas species of Venus's atmosphere is carbon dioxide, which makes up more than 96% of the atmospheric molecules, while nitrogen accounts for almost all the remainder. In contrast to the dominance of carbon dioxide in Venus's atmosphere, the Earth's atmosphere consists mostly of nitrogen and oxygen, with carbon dioxide being present at a level of only 340 ppm. In part, this difference may stem more from temperature differences than from intrinsic differences. Over the lifetime of the Earth, an amount of carbon dioxide comparable to that in Venus's atmosphere was vented out of the Earth's hot interior, but almost all of it participated with rain in dissolving land rocks, and rivers carried the dissolved rock and carbon dioxide into the oceans, where they subsequently precipitated to form carbonate rocks, such as limestone. Venus's surface is much too hot for oceans of water to be present, and hence its atmosphere has been able to retain essentially all of the carbon dioxide vented from its interior. *See* ATMOSPHERIC CHEMISTRY.

The atmospheric temperature has a relatively cool value of −10°F (250 K) near the top of the cloud layer, which is at a pressure of about 1/20 that at the Earth's surface. The temperature gradually increases with decreasing altitude until it reaches about 850°F (454 K) at the surface, where the pressure is 90 times that at the Earth's surface. The high value of Venus's surface temperature is not due to its being closer to the Sun than the Earth. Because its cloud layer reflects to space about 75% of the incident sunlight, Venus actually absorbs less solar energy than does the Earth. Rather, the high temperature is the result of a very efficient greenhouse effect that allows a small but significant fraction of the incident sunlight to penetrate to the surface (about 2.5%), but prevents all except a negligible fraction of the heat generated by the surface from escaping directly to space. *See* GREENHOUSE EFFECT.

The similar mean densities of Venus and the Earth imply that Venus is made of rocks similar to those that make up the Earth. Venus's interior may be qualitatively similar to that of the Earth in having a central iron core, a middle mantle made of rocks rich in silicon, oxygen, iron, and magnesium, and a thin outer crust containing rocks enriched in silicon in comparison with the rocks of the mantle. However, in contrast to the situation for the Earth, Venus's core may now be either entirely solid or entirely liquid, which could account for the absence of a detectable magnetic field. *See* EARTH INTERIOR.

James B. Pollack; B. Stephen Saunders

Venus has been more intensely explored by spacecraft than any other planet. United States and Soviet spacecraft provided data on the composition of the surface and atmosphere, and the dynamics of the upper atmosphere. The Soviet *Venera* landers transmitted back the first images from the surface of another planet. The United States *Magellan* mapped nearly 98% of the surface. *See* SPACE PROBE.

Magellan's mapping mission revealed a unique global volcanic and tectonic style on Venus. Broad volcanic plains make up about 85% of the surface of Venus. The rest is tectonically deformed, higher-standing terrains with complex systems of folds and faults. Regional tectonism is evident in the widespread compressional and extensional deformation of much of the surface material. Venus apparently has had a dynamic mantle that has driven crustal warping, which may be ongoing. However, while various regions of the planet show evidence of motion, no evidence of Earth-like plate tectonics has been found. *See* PLATE TECTONICS.

Long, narrow troughs are seen in many areas where the crust has ruptured; these linear rift zones are associated with extensive broad, domical rises and shield-volcano complexes. The planet also has some unexplained surface features, including long channels meandering across the plains. Also seen are oval to circular volcanic-tectonic structures called coronae that range in diameter from 60 to 1300 mi (100 to 2100 km). Impact craters are much less abundant on Venus than on the Moon or Mars, and currently active volcanism has not been detected on Venus. *See* PLANET.

R. Stephen Saunders

Venus' flytrap *Dionaea muscipula*, an insectivorous plant of North and South Carolina (see illustration). The two

Stages *a–e* in capture and digestion of fly by leaf of Venus' flytrap. (*General Biological Supply House*)

halves of a leaf blade can move as if they were hinged along the midrib and, swinging upward and inward, the two surfaces come together. Any insect alighting on a leaf triggers this sensitive motor mechanism, and is caught between the closing halves of the leaf blade. In this trap, the insect is slowly digested by enzymes secreted by cells in the leaf. *See* INSECTIVOROUS PLANTS; NEPENTHALES; SECRETORY STRUCTURES (PLANT).

Perry D. Strausbaugh; Earl L. Core

Vermiculite A group of minerals common in some soils and clays and belonging to the family of minerals called layer silicates. Species within the vermiculite group are denoted as either dioctahedral vermiculite or trioctahedral vermiculite, with two or three octahedral cation sites occupied per formula unit, respectively. Trioctahedral vermiculite is a 2:1 layer silicate with a fundamental unit similar to that of mica. An octahedral sheet forms the basis of the layer and is sandwiched between two opposing tetrahedral sheets. *See* MICA.

Perfect basal cleavage develops by the layerlike structure. Density varies but is near 2.4 g/cm^3; hardness on Mohs scale is near 1.5; and luster is pearly. Thin sheets deform easily and may be yellow to brown. *See* HARDNESS SCALES.

Heat-treated and expanded vermiculite is used as an insulator in construction. Mixed with plaster and cement, vermiculite is used to make lightweight versions of these materials. Vermiculite is also useful as an absorbent for some environmentally hazardous liquids. *See* CLAY MINERALS; SILICATE MINERALS.

Stephen Guggenheim

Vernalization The induction in plants of the competence or ripeness to flower by the influence of cold, that is, temperatures below the optimal temperature for growth. Vernalization thus concerns the first of the three phases of flower formation in plants. In the second stage, for which a certain photoperiod frequently is required, flowers are initiated. In the third stage flowers are unfolded. *See* FLOWER; PHOTOPERIODISM; PLANT GROWTH.

Klaus Napp-Zinn

Vernier A short, auxiliary scale placed along the main instrument scale to permit accurate fractional reading of the least main division of the main scale. The auxiliary, or vernier, scale is graduated in one or both directions from the fiducial (index) mark in numbered divisions which are fractionally shorter (in a direct vernier) or longer (in a retrograde vernier) than those on the main scale (see illustration). The position of the fiducial

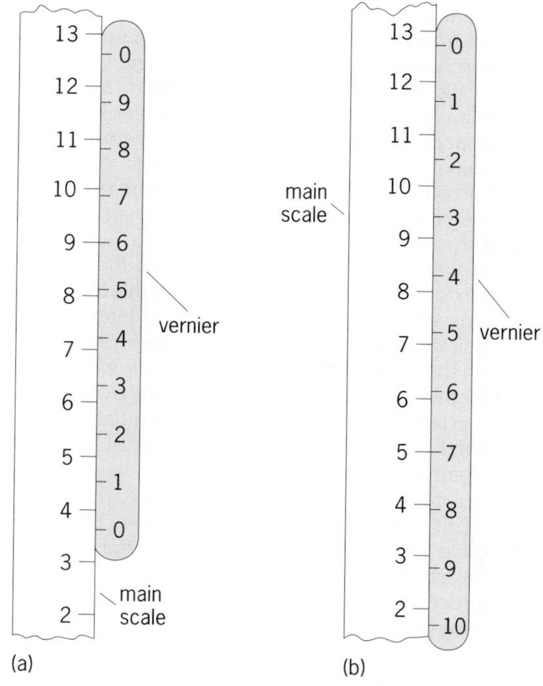

Two types of vernier scales. (*a*) Direct (reading 3.6). (*b*) Retrograde (reading 12.7)

mark (the zero mark of the vernier scale) between divisions on the main scale is indicated by the number of the graduation on the vernier scale which lines up exactly with a graduation on the main scale.

William A. Wildhack

Vertebra The basic unit of the vertebral column. Collectively, the vertebrae surround and protect the spinal cord and provide some type of axial support for the body. The stresses that the vertebral column must meet change somewhat from one end of the animal to the other, and also differ greatly between aquatic and terrestrial vertebrates because the problems of support and locomotion in these media are quite different.

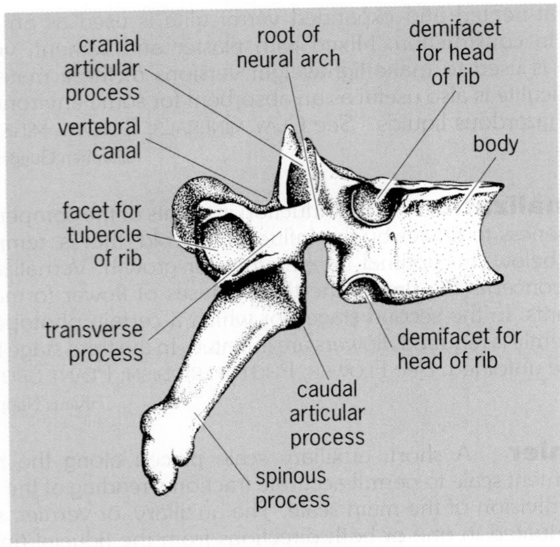

Lateral view of a human thoracic vertebra. Anterior is toward the right. (*After C. M. Goss, ed., Gray's Anatomy of the Human Body, 28th ed., Lea and Febiger, 1964*)

Accordingly, vertebral structure varies widely; yet all vertebrae have many features in common. *See* SPINE.

A vertebra from the thoracic region of a mammal illustrates the basic morphology well (see illustration). The ventral portion consists of a disc-shaped mass of bone known as the body or centrum. An arch of bone, the neural arch, extends dorsally from the centrum and encompasses a space, the vertebral canal, in which the spinal cord lies. The bases, or roots, of the arch are narrower than other parts so that clefts, the intervertebral foramina, lie between the arch bases of successive vertebrae. Spinal nerves pass through these foramina.

Certain muscles and ligaments attach onto the spinous process, which extends dorsally from the top of the arch, and onto a pair of transverse processes, which extend laterally from the arch. One pair of articular processes, or zygapophyses, extends forward from the neural arch, and another pair extends posteriorly. Articular processes of successive vertebrae overlap and help to hold the vertebrae together. The centra of adjacent vertebrae are also joined together by invertebral discs of fibrocartilage. Numerous ligaments interlace the vertebrae.

Ribs articulate onto the thoracic vertebrae. Typically, each mammalian rib has two articular surfaces—a terminal head and a tubercle situated a short distance distal to the head. The tubercle articulates with a facet located on the end of the transverse process; the head usually articulates intervertebrally onto the intervertebral disc and adjacent parts of the bordering centra.

In mammals different parts of the column are clearly specialized to subserve certain functions in addition to their general supportive role. The head moves independently of the trunk, and a distinct neck region, consisting of cervical vertebrae, is present. With few exceptions all mammals from a shrew to a giraffe have seven cervical vertebrae. Well-developed ribs, which play an important role in respiratory movements, articulate with the anterior trunk or thoracic vertebrae of mammals. The number of thoracic vertebrae varies between species but is on the order of 11 (bat) to 18 or 20 (horse). Humans have 12. Lumbar vertebrae occupy the posterior part of the mammal trunk region. They are characterized by relatively large transverse processes to which certain of the powerful back muscles attach. Again the number of lumbar vertebrae varies between species but is on the order of 5 (bat) to 8 (whale). Humans have 5.

Correlated with the greater efficiency of terrestrial locomotion and the need for strong support for the powerful hindlegs, the number of sacral vertebrae increases during evolution from the single one of amphibians. Reptiles usually have two, and most mammals have three which are fused together, along with their embryonic rib rudiments, into a solid complex of bone called the sacrum. Humans, which are bipeds, have 5, and certain of the powerful hoofed mammals, for example, the horse, also have 5. Birds, whose hindlegs act as shock absorbers upon landing, have between 10 and 23 vertebrae fused together in their synsacrum.

The tail and caudal musculature no longer play an important role in the locomotion of most mammals (the Cetacea being a conspicuous exception), and the tail is greatly reduced in size. The spinal cord of mammals ends within the lumbar region, and only a few spinal nerves continue through the vertebrae canal into the tail. Caudal vertebrae are small and become progressively incomplete as one moves distally along the tail until only centra are left. Tail length, and hence the number of caudal vertebrae, vary widely. Some opossums have as many as 35, and humans, in which the tail is absent as an external structure, have only 3 to 5 caudal vertebrae. These form an internal coccyx to which certain anal muscles attach. Warren F. Walker, Jr.

Vertebrata The largest natural group (clade, monophyletic group) of chordate animals. Like all chordates, vertebrates have a notochord, but it is largely replaced by vertebrae (initially neural arches that surround the nerve chord) during development. Together with the living hagfishes (Hyperotreti; Myxiniiformes) and many fossil groups, vertebrates are members of the Craniata. One major theme of craniate evolution is the increasing elaboration of the brain, its cranial nerves, and many sensory organs, such as the organs of smell (olfactory), hearing (inner ear), and feel (sensory organs of the lateral line system). A unique innovation is the evolution of neural crest cells, cells that begin development along the neural crest of the embryo and which then migrate to many parts of the body, giving rise to or contributing to the formation of such structures as the cranium and the branchial arches (a remnant of which remains in humans as parts of the larynx). Vertebrates have a long fossil history. Earliest fossils of jawless vertebrates are known from the late Cambrian (490 million years before present, mybp). Jawed vertebrate fossils are known from the Ordovician (450 mybp) and this group began to predominate in the Devonian (about 360 mybp). *See* CHORDATA; NERVOUS SYSTEM (VERTEBRATE).

Formal vertebrate classifications of the past emphasized overall distinctiveness and grade of organization. Modern (evolutionary phylogenetic) classifications are based on common ancestry and genealogy. Thus, traditional classifications have been highly modified. *See* SYSTEMATICS.

The most basal living members of Vertebrata are the lampreys (Hyperoatia). Lampreys are jawless fishes whose eel-like body resembles hagfishes, but whose predatory, round, and toothed mouth is unique among vertebrates. Lampreys and many fossil jawless vertebrates that flourished in the Paleozoic have a number of unique evolutionary innovations. Major evolutionary innovations characterized the jawed vertebrates, Gnathostomata ("jawed mouth"). In addition to jaws, gnathostomes have paired fins or the homologue of paired fins, legs. They also have a mineralized exoskeleton that surrounds certain cartilages of the head and fins and consists of true bone (as in human leg bones) or calcified cartilage (as in many sharks).

A current classification of living vertebrates is shown below.

Subphylum Chordata
  Infraphylum Craniata
    Superclass Hyperotreti (hagfishes)
      Vertebrata
    Class Hyperoatia (lampreys)
      Gnathostomata

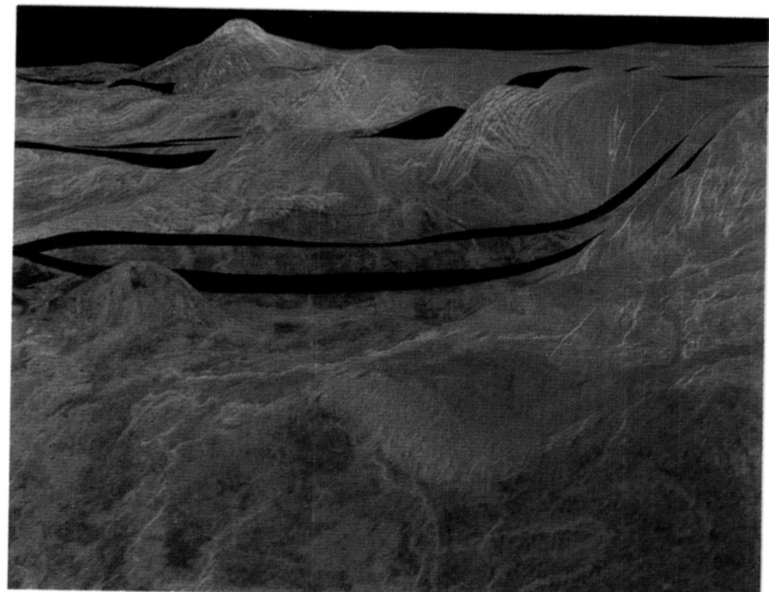

(a)

(b)

Images of the surface of Venus from the *Magellan* radar-mapping mission. Color has been added to simulate the appearance of the Venus surface. (*a*) Perspective view of the southern boundary of Lakshmi Planum, a plateau 1.5–2.5 mi (2.5–4.0 km) high. (*b*) Perspective view of a portion of Western Eistla Regio, with a rift valley in the foreground and the volcanoes Gula Mons (right) and Sif Mons (left) on the horizon. (*c*) Sapas Mons, one of several large volcanic edifices that compose the broad equatorial rise known as Atla Regio.

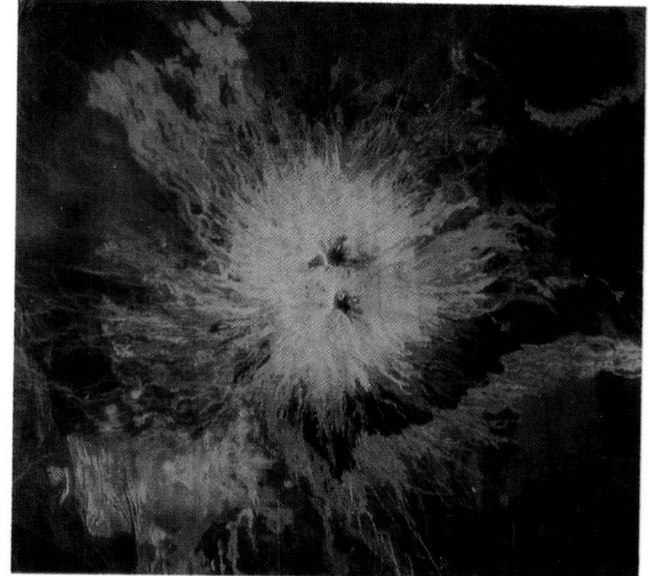

(c)

Infraclass Chondrichthyes (chimeras, sharks, rays)
Osteichthyes
Subclass Actinopterygii (bony fishes, sturgeons, trouts, basses, tunas, etc.)
Sarcopterygii (lungfishes, coelacanths, amphibians, reptiles, birds, mammals)

E. O. Wiley

Vertebrate brain (evolution) A highly complex organ consisting of sensory and motor systems that constitutes part of the nervous system. Virtually all of the brain systems that are found in mammals occur in birds, reptiles, and amphibians, as well as in fishes and sharks. These systems have become more complex and sophisticated as the adaptive requirements of the animals changed. Occasionally, new sensory systems arose, most likely as specializations of existing systems; however, some of these disappeared as animals left the aquatic world. Some sensory systems arose and declined several times in different lineages. In spite of the many changes in brain structure, some of them quite dramatic, the evolution of the nervous system, from the earliest vertebrates through to those of today, has been relatively conservative. See NERVOUS SYSTEM (INVERTEBRATE); NERVOUS SYSTEM (VERTEBRATE).

Central nervous system. The nervous system consists of two main divisions: the central nervous system, which is made up of the brain and the spinal cord, and the peripheral nervous system. The peripheral nervous system consists of the nerves that bring information from the outside world via the sensory systems, and the nerves that carry information from the body's interior to the spinal cord and brain. These nerves also convey commands from the brain and spinal cord to the external muscles that move the skeleton, as well as to various internal organs and glands. See CENTRAL NERVOUS SYSTEM.

The brain consists of a variety of systems, some of which are sensory and deal with the acquisition of information from the internal and external environments. Other systems are motor and are involved with the movement of the skeletal muscles; the muscles of the internal organs, such as the heart and the digestive and respiratory systems; and the secretions of certain glands, such as the salivary and tear glands. The bulk of the brain, however, is composed of systems that are integrative and organize, coordinate, and direct the activities of the sensory and motor systems. These integrative systems regulate such processes as sleep and wakefulness, attention, the coordination of various muscle groups, emotion, social behavior, learning, memory, thinking, planning, and other aspects of mental life. Social behavior itself is highly complex and includes such interactions between individuals as courtship, mating, parental care, and the organization and structure of groups of individuals. See BRAIN; MOTOR SYSTEMS.

Development. The nervous system develops in the embryo as a hollow tube. The remnants of this hollow tube in the adult are known as the brain ventricles and are filled with the cerebrospinal fluid. Two types of overall brain organization are found among vertebrates based on the relationship of the neurons to these ventricles. In brains with laminar organization, the neurons have not migrated very far from the layer immediately surrounding the hollow ventricular core of the brain. This type of organization is typical of amphibia and is also found in the brains of some sharks, fishes (especially those in which their skeletons are partly or mostly made of cartilage rather than bone), and lampreys. In contrast, in the brains of those vertebrates with the elaborated type of organization, the neurons have migrated from the zone around the ventricles to occupy nearly all of the interior of the brain. This type of organization is typical of reptiles, mammals, birds, fishes (with fully bony skeletons), as well as skates, rays, some sharks, and hagfishes. See NEURON.

In general, brain size or weight varies in proportion to the size of the body. In some species, however, brain size or weight is greater than would be expected for that body weight, for example, in humans, chimpanzees, and porpoises. Birds have brains that are comparable to those of mammals of equivalent body weight. In fact, crows have brain weights that would be expected of a small primate of equivalent body weight. In contrast to birds and mammals, amphibians, reptiles, and bony fishes have relatively small brains for their body weights, as do jawless fishes.

Subdivisions of brain. The brain is divided into a hindbrain, a midbrain, and a forebrain.

Hindbrain. The hindbrain is a region that contains nerve endings that receive information from the outside world and from the body interior; these are known as sensory cranial nerves. The neuron groups upon which they terminate are known as sensory cranial nuclei. Also found in the hindbrain are motor nerves that control internal and skeletal muscles and glands, which are called motor cranial nerves; the neuron groups from which they originate are known as motor cranial nerve nuclei.

Many animals possess senses that humans do not possess. One such is the lateral line sense, which derives from receptors located in the lateral line organ which can easily be seen on most bony fishes as a thin, horizontal line running the length of the body from behind the gill opening to the tail. Other lateral line organs can be found on the head and jaws. These organs contain mechanoreceptors that respond to low-frequency pressure waves that might be produced by other fishes nearby or the bow wave of a fast-swimming predator about to strike. Lateral line systems and a special region of the hindbrain dedicated to lateral line sense are found in fishes and sharks, jawless fishes, and bony fishes of various sorts.

Electroreception is another way of dealing with a murky environment. Scientists have described two types of electroreception: active and passive. The receptors are also located in the lateral line canals and sometimes on the skin. Animals with passive electroreception, such as sharks and rays, platypuses, and echidna, can detect the presence of the very weak electric fields that are generated around a living body, which they then follow to capture their prey. Animals with active electroreception generate stronger electric fields around themselves using specialized electric organs. By detecting changes in these electric fields, they can derive a picture of their environment. Electrosensory cranial nerves terminate in a region of the hindbrain known as the electrosensory area. A second group of active electrosensory fishes are capable of generating electric fields so powerful they can stun a prey or an enemy. Among these are the electric eel, the electric catfish, and an electric shark (the torpedo). These animals also use their low-level electric fields to detect objects and creatures in the environment. See ELECTRIC ORGAN (BIOLOGY).

Not only did the hindbrain change in response to sensory evolution, but it also underwent major motor transformations; for example, motor-neuron groups involved in swallowing, chewing, and salivating evolved as a consequence of the transition to land and the loss of the water column to carry food from the opening of the mouth into the throat.

The hindbrain also contains two important coordinating or integrating systems: the cerebellum and the reticular formation. The functions of the cerebellum are varied; they include the integration of a sense of balance with aspects of movement and motor learning and motor memory, as well as playing an important role in electrosensory reception.

The reticular formation coordinates the functions of various muscle groups. For example, the actions of the jaws and tongue must be coordinated so that an animal does not eat its own tongue while eating its meal. It also coordinates the motor-neuron groups that control the air column that enters and leaves the mouth and throat, which produces the various vocalizations of land animals, including speech. The reticular formation also is involved in sleep, wakefulness, and attention.

Midbrain. The midbrain contains the motor cranial nerves that move the eyes. It also contains neuron groups that are organized to form maps of visual space, auditory space, and the body. These maps are coordinated with each other such that a sudden, unexpected sound will cause the head and eyes to move to the precise region of space from which the sound originated. In those animals that make extensive use of sound localization, such as owls and bats, the map areas of the midbrain are very highly developed. In addition, certain snakes, such as rattlesnakes and boa constrictors, have infrared detectors on the snout or under the eyes that can sense the minute heat from a small animal's body at a distance of 1 m (3 ft) or more. The midbrains of these animals also have infrared maps that are in register with the auditory, visual, and body maps to permit the animal to correlate all the necessary information to make a successful strike on prey in virtually total darkness.

Forebrain. The forebrain is a very complex region that consists of the thalamus, the hypothalamus, the epithalamus, and the cerebrum or telencephalon. In addition, the forebrain contains the limbic system, which has components in all regions of the forebrain as well as continuing into the midbrain. The thalamus processes and regulates a large quantity of the information that enters and emanates from the forebrain. As the cerebrum increases in size and complexity in land animals, the thalamus increases accordingly. The hypothalamus regulates autonomic functions as well as behaviors such as feeding, drinking, courtship and reproduction, parental, territoriality, and emotional, which it controls in conjunction with the limbic system. The hypothalamus also regulates the endocrine system. The size and complexity of the hypothalamus, relative to the rest of the brain, is greatest in fishes and sharks; it declines considerably in proportion to the rest of the brain in land animals. The epithalamus contains the pineal gland, which is involved in various biological rhythms that depend on daylight, including seasonal changes. In some animals, such as certain reptiles, the pineal takes on the form of an eye, located on the top of the head and known as the parietal eye. This eye has a lens and a primitive retina that capture light and transmit information, such as the amount of daylight, to the hypothalamus. The epithalamus, like the hypothalamus, is relatively smaller in the brains of land animals.

The greatest evolutionary expansion of the forebrain is seen in the cerebrum. The cerebrum consists of an outer layer, the pallium, and a series of deep structures, known as the subpallium. The subpallium is composed of the corpus striatum, the amygdala, and the hippocampus. The outer layer of the cerebrum in mammals is known as the cerebral cortex. Considerable debate surrounds the evolutionary relationship between the cerebral cortex of mammals and the pallium and subpallium of nonmammals. Most specialists, however, seem to agree that the mammalian cortex arose from the pallium and certain regions of the subpallium. The cerebrum is relatively small in animals with laminar brains and larger in those with complex brains. Scientists have only begun to catalog the many complicated behavioral functions of the cerebrum. Among them appear to be memory, thinking and reasoning, and planning. With the advent of life on the land, the cerebrum underwent an extreme degree of elaboration in reptiles and birds and especially in mammals. *See* ENDOCRINE SYSTEM (VERTEBRATE).

William Hodos

Vertical takeoff and landing (VTOL) A flight technique in which an aircraft rises directly into the air and settles vertically onto the ground. Such aircraft do not need runways but can operate from a small pad or, in some cases, from an unprepared site. The helicopter was the first aircraft that could hover and take off and land vertically, and is now the most widely used VTOL concept. *See* HELICOPTER.

The helicopter is ideally suited for hovering flight, but in cruise its rotor must move essentially edgewise through the air, causing vibration, high drag, and large power losses. The aerodynamic efficiency of the helicopter in cruising flight is only about one-quarter of that of a good conventional airplane. The success of the helicopter in spite of these deficiencies started a wide-ranging study of aircraft concepts that could take off like a helicopter and cruise like an airplane. The term VTOL is usually used to designate the aircraft other than the helicopter that can take off and land vertically. The term V/STOL indicates an aircraft that can take off vertically when necessary, but can also use a short running takeoff, when space is available, to lift a greater load. *See* SHORT TAKEOFF AND LANDING (STOL).

The tiltrotor is closest to the conventional helicopter and relies heavily on helicopter technology. The rotor disks are horizontal in VTOL operation and are tilted 90° to act as propellers in cruising flight. Such aircraft can have cruise efficiencies at least twice those of the helicopter, making them especially useful for helicopter missions where greater range, speed, and time on station are desired.

Numerous advanced rotorcraft concepts are being investigated in research programs. These include designs that take off and land as a rotor system and fix the blades to act as wings for cruise flight. Other concepts feature tiltrotor technology for vertical flight, but fold the blades and employ other propulsion modes for cruise.

The vectored-thrust concept features an engine with four rotating nozzles that deflect the thrust from horizontal for conventional flight to vertical for VTOL operation. This activity led to the very successful British Harrier aircraft, which is considered to be a V/STOL aircraft because its normal mode of operation is to use a short takeoff run, when space is available, to greatly increase its payload and range capability.

With the success of the Harrier, design studies and technology development programs have been directed at expanding the flight envelope of this class of aircraft to include supersonic capability. The term STOVL (short takeoff vertical landing) is used to define this supersonic fighter-attack aircraft since the large benefits in payload and range of a short takeoff will be factored into the basic design.

W. P. Nelms

Vessel traffic service A marine traffic management system, operated by a competent authority, for improving the safety and efficiency of vessel traffic and protecting the environment within designated service areas. A vessel traffic service (VTS) has the ability and the authority to interact with marine traffic and respond to developing traffic situations. It improves order and predictability on the waterway throughout its service area by using surveillance, communications, and standard operating procedures to collect, evaluate, and share information to support professional mariners' navigation decision making. Opportunities for human error are reduced through the provision of traffic information, traffic organization, and navigation-assistance services. Captains, mates, and marine pilots use the delivered vessel traffic services to direct and control the maneuvering of ships, tugs with tows, and passenger ferries.

Functions and capabilities vary among the world's more than 300 VTS systems, but their basic operational function is to provide three services: information, navigational assistance, and traffic organization. A VTS also provides support to allied services including customs and law enforcement, firefighting and emergency responders, and commercial maritime interests in the port. A modern VTS system uses a variety of sensors to collect information and assemble a "traffic image" of all vessel movements, events, and conditions in its area of responsibility. These sensors may include surface radar, closed-circuit television cameras, wind speed and visibility monitors, tide and current gauges, and simple visual overlooks. The newest tool available to vessel traffic service is the automatic identification system (AIS). This consists of a ship-borne device that automatically reports each ship's position, identification, destination, and other information needed for the shore-based authority to conduct traffic management. All information is gathered from these sensors and sent to the vessel traffic center using microwave or high-capacity communications lines. There, it is presented to the vessel traffic service operators on a sophisticated, composite display so they may monitor the complete traffic situation and relay relevant information to the participating mariners. Information is typically sent to the mariner using dedicated voice radio channels. However, more vessel traffic services are coming to rely on data transfer using digitized messages instead of ship-to-shore voice communications. Information of general interest to all vessels may be sent at regular broadcast intervals. Information or advice directed at a specific ship is sent when the ship requests information or when the vessel traffic service feels the ship needs it.

In the United States, the Coast Guard has been named in law as the competent authority to operate vessel traffic services. As such, the vessel traffic service is an extension of the Coast Guard's authority to regulate marine traffic and direct the movement of vessels. This arrangement is consistent with other VTS operations throughout the world.

In most areas around the world, participation in a vessel traffic service is mandatory for commercial vessels greater than 270 metric tons (300 tons) or greater than 20 m (65 ft) in length. Passenger ferries are usually required to participate. Inland or harbor craft, such as tugs, may be excluded. Small fishing vessels and recreational craft rarely take part in the vessel traffic service. In some cases, these types of craft may be prohibited from participating to reduce the communications load on the system. In other areas, these craft may be allowed to participate as needed.

During the course of the vessel's transit through the area, the vessel traffic service will monitor its progress and adherence to the sailing plan. The vessel traffic service may occasionally provide information to the vessel relevant to opposing traffic, channel (fairway) conditions, weather, aids to navigation, or anything else that might affect the transit or influence the sailing plan. The vessel may wish to contact the vessel traffic service during its transit for updated information or to report that it must deviate from the agreed-upon sailing plan. Upon completing its transit, a vessel might make a final call to the vessel traffic service reporting "all secure" or to inform the vessel traffic service of its plans. *See* ELECTRONIC NAVIGATION SYSTEMS; PILOTING; RADAR.

J. Michael Sollosi

Vestibular system The system that subserves the bodily functions of balance and equilibrium. It accomplishes this by assessing head and body movement and position in space, generating a neural code representing this information, and distributing this code to appropriate sites located throughout the central nervous system. Vestibular function is largely reflex and unconscious in nature.

The centrifugal flow of information begins at sensory hair cells located within the peripheral vestibular labyrinth. These hair cells synapse chemically with primary vestibular afferent nerve fibers, causing them to fire with a frequency code of action potentials that include the parameters of head motion and position. These vestibular afferents, in turn, enter the brain and terminate within the vestibular nuclei and cerebellum. Information carried by the firing patterns of these afferents is combined within these central structures with incoming sensory information from the visual, somatosensory, cognitive, and visceral systems to compute a central representation of head and body position in space. This representation is called the gravito-inortial vector and is an important quantity that the central nervous system employs to achieve balance and equilibrium. *See* BRAIN; NERVOUS SYSTEM (VERTEBRATE); POSTURAL EQUILIBRIUM; REFLEX.

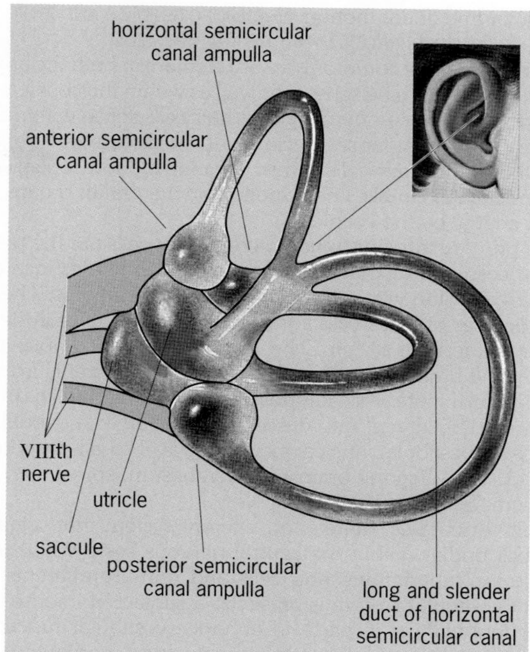

Fig. 1. The vestibular labyrinth is located within the inner ear. It communicates with the brain via the VIIIth nerve. Each of the three semicircular canals has an ampulla and a long and slender duct. The utricle primarily senses motion in an earth-parallel plane, while the saccule primarily senses motion and gravity in an earth-perpendicular plane.

The vestibular labyrinth is housed within the petrous portion of the temporal bone of the skull along with the cochlea, the organ of hearing (Fig. 1). The receptor element or primary motion sensor within the labyrinth is the hair cell (Fig. 2). Hair cells respond to bending of their apical sensory hairs by changing the electrical potential across their cell membranes. These changes are called receptor potentials, and the apical surface of the hair cell thus functions as a mechanical-to-electrical transducer. The frequency of the resulting action potentials in the VIIIth cranial (vestibulocochlear) nerve encodes the parameters

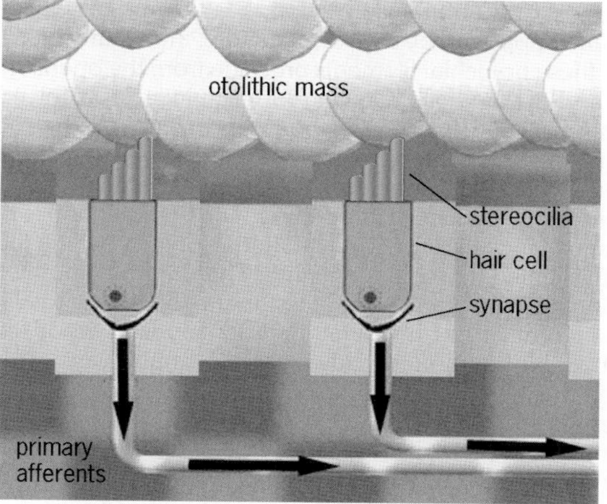

Fig. 2. Otolithic macula at rest. The arrows within the primary afferents or VIIIth nerve fibers indicate spontaneous activity in these fibers in the absence of motion of the otolithic mass relative to the hair cell stereocilia. (The otolithic membrane is not illustrated for clarity.)

of angular and linear motion. *See* BIOPOTENTIALS AND IONIC CURRENTS; EAR (VERTEBRATE); SYNAPTIC TRANSMISSION.

Hair cells are the common sensory element in both the angular and linear labyrinthine sensors as well as within the cochlea. The particular frequency of energy that hair cells sense within these diverse end organs arises because of the accessory structures surrounding the hair cells. Thus, angular motion is sensed by the semicircular canals, linear motion by the otolith organs, and sound energy by the cochlea.

The primary afferents innervated by hair cells are the peripheral processes of bipolar neurons having cell bodies located in Scarpa's ganglion within the internal auditory meatus. The central processes of these cells contact neurons in the brainstem of the central nervous system. The vestibular nuclei complex is defined as the brainstem region where primary afferents from the labyrinth terminate. It is composed of four main nuclei: the superior, medial, lateral, and descending nuclei. The axonal projections of vestibular nuclear neurons travel to all parts of the neuraxis, including the brainstem, cerebellum, spinal cord, and cerebrum. *See* MOTOR SYSTEMS.

In all vertebrates, there is an efferent system that originates from cell bodies within the central nervous system and terminates upon labyrinthine hair cells and primary afferents. The efferent vestibular system is presently a subject of intense study but undoubtedly is in place to enhance vestibular function. It is interesting that evolution felt it necessary to modify incoming vestibular information before it could enter the central nervous system. Stephen M. Highstein

Vestimentifera A group of benthic marine annelid worms that is restricted to habitats rich in sulfide (for example, hydrothermal vents and sulfide seeps). As adults they lack a mouth, gut, and anus, and are nourished by internal symbionts. All vestimentiferans live in tubes of varying hardness and rigidity. Tube material is secreted by internal glands and is a mixture of chitin and protein. *See* HYDROTHERMAL VENT.

In the past, vestimentiferans had been considered to be members of the phylum Pogonophora as well as of the phylum Annelida, and it even had been suggested that they are a phylum in their own right. They are now considered to be a part of the annelid family Siboglinidae along with other members of the former phylum Pogonophora and the bone-eating worms *Osedax*. Currently there are 10 genera containing a total of 15 species: *Alaysia*, *Arcovestia*, *Escarpia*, *Ridgeia*, *Riftia*, *Lamellibrachia*, *Oasisia*, *Paraescarpia*, *Seepiophila*, and *Tevnia*. *See* ANNELIDA.

Vestimentiferans have four regions along their length. The anteriormost, the obturacular region, has a mass of vascularized filaments supported by a paired structure, the obturaculum. Normally these branchial filaments protrude from the opening of the tube and act as a gill, allowing for exchange of dissolved substances between the worm's body and seawater. The second region, the vestimentum, is muscular and serves to maintain the plume of branchial filaments in the open seawater by pressure on the inner tube surface at its opening. The vestimentum is the site of many glands that contribute material for lengthening the tube and thickening it near the opening. The third region, the trunk, is a single segment and constitutes about 75% of the total length of the worm. It has a pair of large longitudinal blood vessels. The bulk of the trunk is made up of the trophosome, an organ containing masses of sulfide-oxidizing bacteria. Gonads are also present in the trunk. Much of the body wall of the trunk is lined internally by longitudinal muscles that allow the worm to withdraw into its tube for protection from predators. The fourth and most posterior region, the opisthosoma, is made up of many segments, each segment in the anterior portion bearing a row of small hooks that can be set into the inner surface of the tube. This provides an anchor against which the body of the worm retracts when the longitudinal muscles of the trunk contract during withdrawal into the tube. Meredith L. Jones; Greg Rouse

Vesuvianite A sorosilicate mineral of complex composition crystallizing in the tetragonal system; also known by the name idocrase. Crystals, frequently well formed, are usually prismatic with pyramidal terminations. It commonly occurs in columnar aggregates but may be granular or massive. The luster is vitreous to resinous; the color is usually green or brown but may be yellow, blue, or red. Hardness is $6^1/_2$ on Mohs scale; specific gravity is 3.35–3.45. *See* HARDNESS SCALES.

The composition of vesuvianite is expressed by the formula $Ca_{10}Al_4(Mg,Fe)_2Si_9O_{34}(OH)_4$. Magnesium and ferrous iron are present in varying amounts, and boron or fluorine is found in some varieties. Beryllium has been reported in small amounts.

Vesuvianite is found characteristically in crystalline limestones resulting from contact metamorphism. It is there associated with other contact minerals such as garnet, diopside, wollastonite, and tourmaline. Noted localities are Zermatt, Switzerland; Christiansand, Norway; River Vilui, Siberia; and Chiapas, Mexico. In the United States it is found in Sanford, Maine; Franklin, New Jersey; Amity, New York; and at many contact metamorphic deposits in western states. A compact green variety resembling jade is found in California and is called californite. *See* SILICATE MINERALS. Cornelius S. Hurlbut, Jr.

Vetch Any of a group of plants which are mostly annual and perennial legumes with weak viny stems often terminating in tendrils. There are about 150 species in the temperate zones of four continents. The leaves are compound with many leaflets. Vetches are used mainly for green manure, cover crops, hay, and pasture and silage. The seeds are used as concentrate in animal feeds, and some vetches are used as a vegetable for human consumption. Cool temperatures promote best development. In general, vetches are more tolerant of acidic soil conditions than are most legume crops, but they have a relatively high requirement for phosphorus. Identification of vetches is difficult until pods and seeds develop. *See* LEAF; LEGUME.

By the 1960s, some 35 vetch species and subspecies had been found in the United States. Of these 35 species, 16 are native to the United States.

The vetch complex *Vicia villosa* is the most widely grown. The subspecies of this complex (*V. villosa* ssp. *villosa*), commonly known as hairy vetch, is the most winter-hardy and is mostly adapted to the eastern and southern United States, where it is grown as a winter annual or biannual. A second subspecies of this group is woollypod or smooth vetch (*V. villosa* ssp. *varia*). It is less winter-hardy than the hairy vetch type and is adapted to the Pacific Coast states.

The vetch complex *V. sativa* is the second most important species in use in the United States. Common vetch (*V. sativa* ssp. *sativa*) is used throughout the United States. Narrow-leaf or blackpod vetch (*V. sativa* ssp. *nigra*) occurs mostly as a weed in waste places in the United States. Underground vetch (*V. sativa* ssp. *amphicarpa*) produces pods above and below the ground and is being studied for revegetation on marginal lands of the Far East.

Commonly known as the faba bean, horsebean, or broadbean, *V. faba* produces coarse, upright plants with large leaves and pods. It is one of only several vetches that are important sources of food for human beings.

Purple vetch (*V. bengalensis*) possesses poor winter hardiness; however, it is useful in the California rice-growing area, where it is used as a soil-improving crop in rotation with the rice (see illustration). Hungarian vetch (*V. pannonica*), a more winter-hardy type from Central Europe, is grown in the Pacific Northwest for forage, green manure, and seed.

Crown vetch, *Coronilla varia*, is a long-lived, winter-hardy perennial legume, but it is not a true vetch. It spreads by seeds and rhizomes to form a dense, weed-free, erosion-resisting ground cover. Its greatest use is for erosion and weed control on unmowed slopes of highways, industrial developments, and

Purple vetch in flower. (*USDA*)

military installations. Its forage value is being studied. *See* COVER CROPS; SOIL CONSERVATION.

Fred V. Grau; Walter Graves

Vibration The term used to describe a continuing periodic change in the magnitude of a displacement with respect to a specified central reference. The periodic motion may range from the simple to-and-fro oscillations of a pendulum, through the more complicated vibrations of a steel plate when struck with a hammer, to the extremely complicated vibrations of large structures such as an automobile on a rough road. Vibrations are also experienced by atoms, molecules, and nuclei. *See* PENDULUM.

A mechanical system must possess the properties of mass and stiffness or their equivalents in order to be capable of self-supported free vibration. Stiffness implies that an alteration in the normal configuration of the system will result in a restoring force tending to return it to this configuration. Mass or inertia implies that the velocity imparted to the system in being restored to its normal configuration will cause it to overshoot this configuration. It is in consequence of the interplay of mass and stiffness that periodic vibrations in mechanical systems are possible.

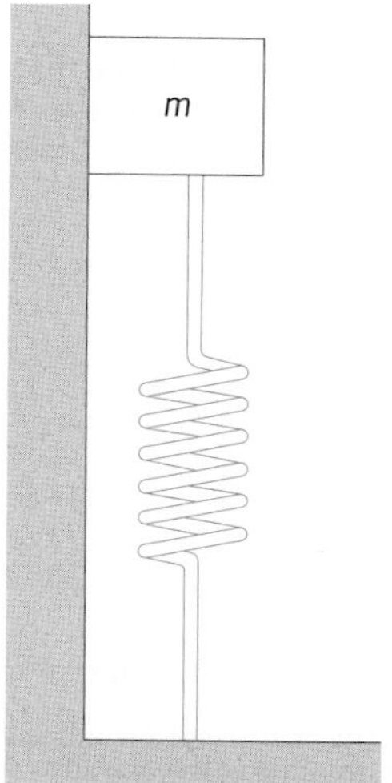

Fig. 1. Simple oscillator.

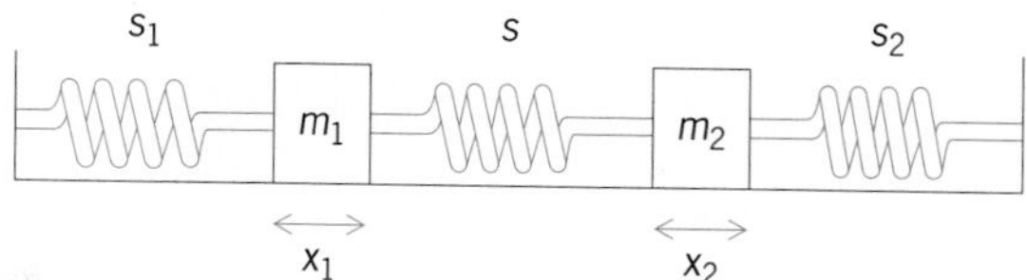

Fig. 2. Simple oscillator with two degrees of freedom. Masses m_1 and m_2, with displacements x_1 and x_2, are connected by springs s_1, s, and s_2.

Mechanical vibration is the term used to describe the continuing periodic motion of a solid body at any frequency. When the rate of vibration of the solid body ranges between 20 and 20,000 hertz (Hz), it may also be referred to as an acoustic vibration, for if these vibrations are transmitted to a human ear they will produce the sensation of sound. The vibration of such a solid body in contact with a fluid medium such as air or water induces the molecules of the medium to vibrate in a similar fashion and thereby transmit energy in the form of an acoustic wave. Finally, when such an acoustic wave impinges on a material body, it forces the latter into a similar acoustic vibration. In the case of the human ear it produces the sensation of sound. *See* MECHANICAL VIBRATION; SOUND.

Systems with one degree of freedom are those for which one space coordinate alone is sufficient to specify the system's displacement from its normal configuration. An idealized example known as a simple oscillator consists of a point mass m fastened to one end of a massless spring and constrained to move back and forth in a line about its undisturbed position (Fig. 1). Although no actual acoustic vibrator is identical with this idealized example, the actual behavior of many vibrating systems when vibrating at low frequencies is similar and may be specified by giving values of a single space coordinate.

When the restoring force of the spring of a simple oscillator on its mass is directly proportional to the displacement of the latter from its normal position, the system vibrates in a sinusoidal manner called simple harmonic motion. This motion is identical with the projection of uniform circular motion on a diameter of a circle. *See* HARMONIC MOTION.

When two simple vibrating systems are interconnected by a flexible connection, the combined system has two degrees of freedom (Fig. 2). Such a system has two normal modes of vibration of two frequencies. Both of these frequencies differ from the respective natural frequencies of the individual uncoupled oscillators.

A vibrating system is said to have several degrees of freedom if many space coordinates are required to describe its motion. One example is n masses $m_1, m_2, \ldots, m_n$ constrained to move in a line and interconnected by $(n - 1)$ coupling springs with additional terminal springs leading from m_1 and m_n to rigid supports. This system has n normal modes of vibration, each of a distinct frequency. *See* DAMPING; VIBRATION DAMPING; VIBRATION ISOLATION.

Lawrence E. Kinsler

Vibration damping The processes and techniques used for converting the mechanical vibrational energy of solids into heat energy. While vibration damping is helpful under conditions of resonance, it may be detrimental in many instances to a system at frequencies above the resonant point. This is due to the fact that the relative motion between the base of the vibration isolator and the mounted body tends to become smaller as the isolator becomes more efficient at the higher frequencies. With damping present, the force transmitted by the elastic element is unable to overcome the damping force; this leads to a resulting increase in transmissibility. *See* DAMPING; VIBRATION; VIBRATION ISOLATION.

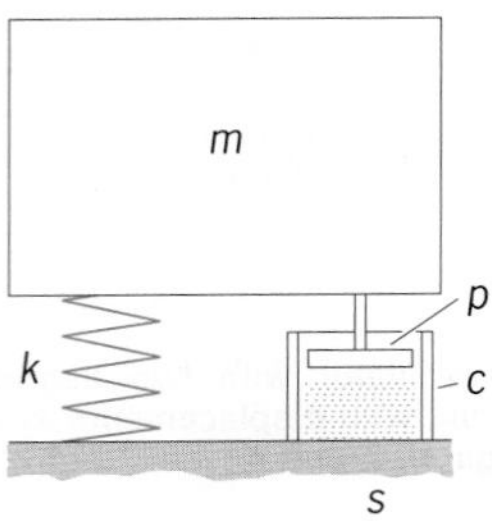

Fig. 1. Automobile shock absorber.

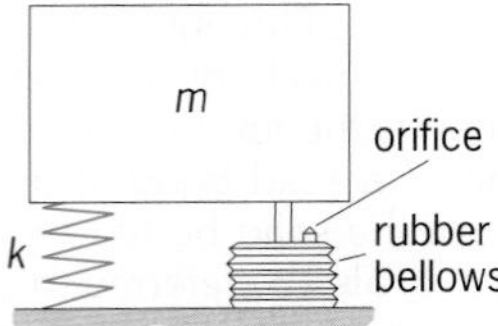

Fig. 2. System employing viscous damping with air.

All metal springs which include structural members such as brackets and shelves have some damping. However, such damping is insufficient for vibration isolators and must be augmented by special damping devices. See SPRING (MACHINES).

Several different types of damping devices have been developed and used successfully. Probably the most familiar is that used on automobiles, which, although known as a shock absorber, is in reality a damper, and functions as a limiter to the spring system of spring constant k. The system is shown in Fig. 1. A piston p is attached to the body m and is arranged to move vertically through the liquid in a cylinder c which is secured to the support s. As the piston moves, the force required to cause the liquid to flow from one side of the piston to the other is approximately proportional to the velocity of the piston in the cylinder. This type of damping is known as viscous damping. The damping force is controlled by the viscosity of the liquid and by the size of the orifice in the piston. See SHOCK ABSORBER.

Some of the disadvantages of viscous damping may be overcome by using air instead of liquid as the damping medium. Air, being compressible, will add to the effective spring force with large displacements. If the air is housed within a flexible bellows, damping will be attainable horizontally as well as vertically. Such a system is illustrated in Fig. 2. This type of damping has proved very effective in vibration isolators.

Damping forces may be generated by causing one dry member to slide on another. This is known as dry friction or coulomb damping. Friction damping is used in several commercially available isolators because it provides a simple means to control the damping forces.

Magnetic damping is attainable as a result of the electric current induced in a conductor moving through a magnetic field. The damping force can be made proportional to the velocity of the conductor moving through the field. See ELECTROMAGNETIC INDUCTION.

K. W. Johnson

Vibration isolation The isolation, in structures, of those vibrations or motions that are classified as mechanical vibration. Vibration isolation involves the control of the supporting structure, the placement and arrangement of isolators, and control of the internal construction of the equipment to be protected.

The simplest kind of mechanical vibration has the waveform of sinusoidal motion. Vibrations in structures, although generally more complex in waveform, exist wherever movement takes place. Such movement may be caused, for example, by the engine in an automobile, by engines or wind buffeting in aircraft, or by a punch press in a building. Delicate electronic equipment and precision instruments must normally be isolated from these motions if accurate measurements are to be obtained. See MECHANICAL VIBRATION; VIBRATION.

Vibration, in most cases, may be effectively isolated by placing a resilient medium, or vibration isolator, between the source of vibration and its surrounding area to reduce the magnitude of the force transmitted from a structure to its support or, alternatively, to reduce the magnitude of motion transmitted from a vibrating support to the structure. Isolating vibration at its source is commonly termed active or source isolation; isolating an instrument from its surroundings is known as passive isolation.

The vibration isolators may be positioned and arranged in many different ways, all variations of three basic types, each of which requires a definite amount of space: (1) isolators attached underneath equipment, known as an underneath mounting system; (2) isolators located in the plane of the center of gravity of the equipment, known as a center-of-gravity system; (3) mountings arranged four on each side in the plane of the radius of gyration, known as a double side-mounted system or radius-of-gyration system. See CENTER OF GRAVITY; RADIUS OF GYRATION.

K. W. Johnson

Vibration machine A device for subjecting a system to controlled and reproducible mechanical vibration. Vibration machines, commonly called shake tables, are widely used in vibration measurement and analysis. There are three types of vibration machines in general use. These are the mechanical direct-drive type, the mechanical reaction type, and the electrodynamic type. Other types, such as hydraulic excitation devices, resonant systems, piezoelectric vibration generators used for instrument calibration, and machines for testing packages, have only limited or specialized applications. See MECHANICAL VIBRATION; VIBRATION.

K. W. Johnson

Vibration pickup An electromechanical transducer capable of converting mechanical vibrations into electrical voltages. Depending upon their sensing element and output characteristics, such pickups are referred to as accelerometers, velocity pickups, or displacement pickups.

The accelerometer consists essentially of a mass which is seismically supported with respect to a surrounding case by means of a spring and guided to prevent motions other than those along the seismic direction of support. The mass exerts a force on the spring's support which is directly proportional to the acceleration being measured. This, in turn, is converted into an electrical voltage by means of stresses produced in a piezoelectric crystal. See ACCELEROMETER.

The velocity pickup generates a voltage proportional to the relative velocity between two principal elements of the pickup, the two elements usually being a coil of wire and a source of magnetic field (see illustration).

The displacement pickup is a device that generates an output voltage which is directly proportional to the relative displacement between two elements of the instrument. These pickups are

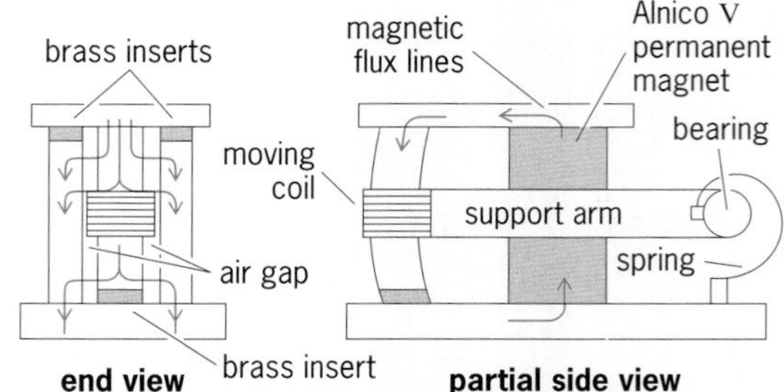

Velocity pickup. The coil swings on one end of an arm which is supported by bearings at the opposite end. The case follows the motion of the structure to which it is attached. (*MB Manufacturing Co.*)

similar in construction and behavior to velocity pickups. The only essential difference is the use of a frequency-weighting network, required to make them direct-reading. *See* VIBRATION.

Lawrence E. Kinsler

Vibrotaction The response of tactile nerve endings to varying forces on the skin and to oscillatory motion of the skin. Grasping, holding, and tactile exploration of an object are part of everyday experience. The performance of all these activities is dependent on the dynamic response of specialized nerve endings located in the skin.

Knowledge of the neural and psychophysical processes involved in vibrotaction is necessary to develop effective tactile communication systems, such as vibrotactile pagers, and multipin vibrotactile stimulators used to produce images on an extended skin surface. Potential users range from business people to the visually and hearing impaired: potential applications range from activities involving visual and aural sensory saturation (such as pilots in high stress situations) to those requiring sensory stimulation (for example, virtual reality).

Small electrical impulses, or action potentials, generated by a single nerve ending can be recorded by inserting a microelectrode into the arm of an alert human subject and positioning the electrode tip in a nerve fiber. Studies of the action potentials produced when the skin is locally depressed at a fingertip have identified activity in four networks of distinctive nerve endings lying within 1–2 mm of the skin surface. These nerve endings, which transform skin motion into neural signals, consist of physically and functionally different mechanoreceptors.

Three types of mechanoreceptor in the fingertip have been implicated in vibrotaction. One of these is a population of slowly adapting mechanoreceptors (SA type I, or SAI); their action potentials persist for some time after a transient skin indentation. The two others are populations of rapidly adapting receptors (FAI and FAII types). Type I receptors respond only to skin indentation within a few millimeters of the nerve ending, whereas type II receptors also respond to stimuli at a greater distance from the nerve ending. A fourth mechanoreceptor type appears to respond primarily to skin stretch (denoted SAII). *See* CUTANEOUS SENSATION; MECHANORECEPTORS.

Anthony J. Brammer

Vicuna A rare animal whose fiber makes the world's most costly and most exquisite cloth, surpassing all others in fineness and beauty. It is found in an almost inaccessible area of the Andes Mountains.

A single animal yields only approximately $^1/_4$ lb (0.11 kg) of hair; thus 40 animals are required to provide enough hair for the average coat. The fiber of the vicuna is the softest and most delicate of the known animal fibers; yet it is strong for its weight, is resilient, and has a marked degree of elasticity and surface cohesion. *See* ALPACA; CAMEL'S HAIR; CASHMERE; LLAMA; MOHAIR; NATURAL FIBER; WOOL.

M. David Potter

Video amplifier A low-pass amplifier having a bandwidth in the range from 2 to 100 MHz. Typical applications are in television receivers, cathode-ray-tube computer terminals, and pulse amplifiers. The function of a video amplifier is to amplify a signal containing high-frequency components without introducing distortion.

Modern video amplifiers use specially designed integrated circuits. With one chip and an external resistor to control the voltage gain, it is possible to make a video amplifier with a bandwidth between 50 and 100 MHz having voltage gains ranging from 20 to 500. *See* AMPLIFIER; INTEGRATED CIRCUITS.

Harold F. Klock

Video disk A medium used to record, distribute, and play video information. The video disk exists in three major forms: the Digital Video Disk (DVD; sometimes also called the Digital Versatile Disk); the Laser Video disk (sometimes called the Laser Vision disk); and the Video CD. *See* COMPACT DISK.

The advantages of DVD and Laser Video disks over videocassette tapes include very high video quality (because the signal-to-noise ratio is very high); use of a full television frame on playback; multiple audio channels, digital and analog; archival quality, since there is no disk deterioration over time; the possibility of interactivity, whereby cues, branches, and so forth can be programmed in; a freeze-frame option with no loss of quality; the ability to interface and interact with computers using digital cues; the availability of rapid access to any point in the recording; the possibility of auto-stop, whereby a frame is preselected for freezing; frame-by-frame viewing, providing slow motion with no loss of quality; and programmability with certain systems. The Video CD does not possess these advantages, but is a lower-cost option with limited image quality and minimal interactivity.

All three types of video disks have basic elements in common, notably a solid-state laser a fraction of a square millimeter in size that emits a few milliwatts of coherent light focused to a micrometer-sized spot. This laser spot becomes essentially an optical stylus that performs various functions such as tracking, focusing, and reading the total integrated light. *See* LASER.

The laser spot follows a track on the rotating disk that is either the data itself or a groove. It does this by sensing the symmetry of the laser light reflected back to a split segmented detector; if the spot is not symmetric, the position of the head is jogged until it is back to a perfectly registered spot on the detector. The jogging movement is created by a small motor, often a voice coil motor, that is driven by an electronic servo system that "closes the loop" in an iterative manner. The laser spot and the disk are kept in focus through a similar servo-based controller that uses the image of the laser on the detector.

By adding up the total integrated light on the detector, it is possible to read the total light returned from the disk. While the laser light travels over the disk, differences in the intensity of the light, affected by the disk features that represent the data, are detected. These disk features can be pits in the surface (using the phase contrast of light coming off the disk) or the absorbed light in a dye or other material. This change in intensity is detected and the pulse train decoded by the system to produce a series of data bits that are then reconstructed to form a video image. This principle is used by all the above video disk applications. *See* OPTICAL RECORDING.

David H. Davies

Video games Entertainment systems in which a computer is used to drive a video display and interact with players using a variety of input devices. Video games can be divided into arcade systems, home computers, and game consoles. The distinction between a home computer and a game console is that a computer can be used for a variety of other applications such as word processing and Internet access while a game console is specifically designed for entertainment purposes.

Arcade systems are typically built for a particular game or set of related games. Everything from the choice of controls to the design on the sides and face of the unit are geared toward the game itself. The display is typically larger than that on a home computer, providing a greater sense of immersion. In some cases, specialized optics are used to further increase the sense of breadth and depth. In some racing simulations, for example, multiple screens are arranged side by side in an arc around the player in order to give an even greater feeling of being "in" the game. A combination of conventional arcade games and specialized high-end systems involve virtual reality gear, large rear-projection screens, motion platforms, and sophisticated input devices. *See* SIMULATION; VIRTUAL REALITY.

A game console typically does not include its own display, but is hooked up to a television set. This provides only limited resolution. The input devices on game consoles are limited to simple multipurpose controllers that are included with the console. Most consoles come with two controllers in order to support two-player competitive games. Consoles are much less expensive than the average home computer system and often have specialized hardware for fast graphics and high-quality sound. The games are distributed either as cartridges ("carts") or compact disks (CDs) that contain the game logic programmed into read-only memory (ROM). *See* COMPACT DISK.

Home computer systems have grown steadily more powerful to the point where they are more than capable of serving as game machines. Since they use an actual computer monitor rather than relying on a television set, the resolution and overall image quality is much higher than that of game consoles and often rivals or exceeds that of the arcade systems. A variety of controllers are available for the personal computer, ranging from simple analog joysticks to sophisticated input devices with full-force feedback. Production, packaging, and distribution costs for computer games are similar to those for game consoles that use CD-ROM. However, widespread Internet access allows game companies to make their games available for download. *See* COMPUTER GRAPHICS ; COMPUTER STORAGE TECHNOLOGY; INTERNET; MICROPROCESSOR.

Regardless of the platform (home computer, game console, arcade system), there are a number of basic genres of game. Some of these genres are more common on one type of platform than another, mostly due to technical limitations. Action games are popular on all platforms. Adventure games are mostly popular on home computers, though some game consoles do support them. Simulation games usually involve operating some sort of vehicle, such as an aircraft or a high-performance race car. Simulators can also be found on all popular platforms. Strategy games are based on planning and anticipating future events. There are two subcategories: war games and system simulators. War games are often based on reenacting historic battles of land, sea, or air. System simulators work by simulating a system such as a city, an ant colony, or an entire planet. Strategy games are found mostly on home computers and are not seen in arcade systems.

Bernie Roehl

Video microscopy The use of a high-quality video camera or other fast camera (such as a charge-coupled device) attached to a research-quality light microscope for the purpose of real-time or high-speed imaging of samples on a microscope stage. These images are recorded at regular intervals (often at "video rate" of 30 images per second), and the time-lapse sequence can be played back in the form of a movie. The term "video microscopy" originally referred to microscope imaging using true video (30 frames per second) but now generally refers to rapid time-lapse imaging techniques. Video microscopy is used frequently to image small structures that move rapidly within cells as well as movement of whole cells. This motion can be quantitated, and in the case of fluorescence microscopy, changes in fluorescent intensity (reflecting the local chemical environment of the fluorescent molecule or the number of fluorescent molecules) can be quantitated as well. *See* CAMERA; CELL MOTILITY; CHARGE-COUPLED DEVICES; FLUORESCENCE; FLUORESCENCE MICROSCOPE; IMAGE PROCESSING; MICROSCOPE.

George M. Langford; John F. Presley

Videotelephony A means of simultaneous, two-way communication comprising both audio and video elements. Participants in a video telephone call can both see and hear each other in real time. Videotelephony is a subset of teleconferencing, broadly defined as the various ways and means by which people communicate with one another over some distance. Initially conceived as an extension to the telephone, videotelephony is now possible using computers with network connections. *See* TELECONFERENCING.

Small residential video telephones, computer-based desktop video telephones, and small videoconferencing setups have been introduced to fulfill diverse needs. One such commercially available residential videophone is about as big as a typical office desk telephone with a small flip-up screen that has an eyeball camera above it. Although it will work with several standards, this phone is primarily designed for use over Integrated Services Digital Network (ISDN) lines in which a residence gets three circuits; one circuit is used for control and the other two for voice and video. *See* INTEGRATED SERVICES DIGITAL NETWORK (ISDN).

An example of a computer-based desktop videophone consists of a PCI (Peripheral Component Interconnect) video/audio CODEC board to add to a personal computer, a composite color camera, audio peripherals, and visual collaboration software.

Videotelephony software has been developed and made widely available that permits real-time collaboration and conferencing, including multipoint and point-to-point conferencing. Multipoint means, for example, that three people in three different locations could have a video telephone conference call in which each could see and hear the others. In addition to the basic audio and video capabilities, such software provides several other features such as a whiteboard, background file transfer, program sharing, and remote desktop sharing.

John Bleiweis

Vinegar A food condiment containing mainly acetic acid that is produced by the bacterial oxidative fermentation of various ethanolic solutions. Vinegar is one of the oldest fermented foods used by humans. Babylonian records indicate that vinegars prepared from wines and beer were widely used as early as 5000 B.C. *See* ACETIC ACID; WINE.

The names for the different vinegars are based on the substrates from which they are made. These include the juices from different fruits, starchy vegetables, cereals, and distilled ethanol (ethyl alcohol). In the United States, white distilled vinegar makes up over 80% of the annual production. Much of the white distilled vinegar is made from denatured synthetic ethanol. Nitrogen compounds, minerals, and other nutrients must be added to the alcohol medium to support the growth and metabolism of the acetic acid bacteria. Most other vinegars are made from ethanolic solutions generated by a yeast fermentation. Cider vinegar is produced from yeast-fermented apple juice, wine vinegar (such as Balsamic vinegar) from fermented grape juice. Other juices such as pineapple, orange, and pear, as well as sugarcane syrup and molasses, can serve as fermentation substrates. Vinegar is also produced from starchy vegetables such as potatoes and from cereals such as barley, corn, and rice. Malt vinegar is made from an infusion of barley malt and other cereals in which the enzymes in the malt have converted the starch to fermentable sugars. *See* ENZYME; ETHYL ALCOHOL.

Vinegar sold in the United States must contain at least 4 g of acetic acid per 100 ml of solution (40-grain vinegar). Other federal specifications describe permitted color, odor, presence of trace metals, and alcohol content. Nondistilled vinegars possess distinctive colors and flavors that reflect the properties of the original substrate.

Bacteria belonging to the genus *Acetobacter* are primarily responsible for the vinegar fermentation. Four species are recognized; they are distinguished on the basis of nutrient requirements, production of brown pigments, and tolerance to ethanol. All acetic acid bacteria require oxygen for growth and metabolism. *See* INDUSTRIAL MICROBIOLOGY.

A variety of methods are used for the production of vinegar. The simplest fermentations are slow and inefficient but are relatively foolproof and utilize inexpensive equipment. Submerged fermenters produce 120–150-grain vinegar in minimal time, but they are expensive and system failures can occur. *See* FERMENTATION.

Don F. Splittstoesser

Violales An order of flowering plants, division Magnoliophyta (Angiospermae), in the subclass Dilleniidae of the class Magnoliopsida (dicotyledons). The order consists of 24 families and more than 5000 species, including most of the families that have been referred to the Englerian order Parietales. The

Downy yellow violet (*Viola pubescens*). (*Photograph by Elsie M. Rodgers, National Audubon Society*)

largest families are the Flacourtiaceae (about 800 species), Violaceae (about 800 species), Cucurbitaceae (about 900 species), Begoniaceae (about 1000 species), and Passifloraceae (about 650 species).

The most characteristic feature of this morphologically heterogeneous order is the unilocular, compound ovary with mostly parietal placentation. When the stamens are numerous they are initiated in centrifugal sequence. The more primitive members of the order, such as some of the Flacourtiaceae, are trees with stipulate, alternate leaves; perfect, polypetalous flowers with numerous centrifugal stamens; a compound pistil with free styles and parietal placentation; and seeds with a well-developed endosperm. Tendencies toward unisexuality, reduction in the number of stamens, fusion of filaments, development of a corona, reduction in the number of carpels, fusion of styles, and loss of endosperm from the seed can all be seen in the Flacourtiaceae. These are also some of the prominent characteristics used in combination to define many of the other families of the order. *See* FLOWER; LEAF; SEED.

Cucumbers and melons (species of *Cucumis*, in the Cucurbitaceae), pumpkins and squashes (species of *Cucurbita*), begonias, and violets (*Viola*; see illustration) are familiar members of the Violales. *See* CANTALOUPE; CUCUMBER; DILLENIIDAE; HONEY DEW MELON; MAGNOLIOPHYTA; MAGNOLIOPSIDA; MUSKMELON; PERSIAN MELON; PLANT TAXONOMY; PUMPKIN; SQUASH; WATERMELON.

Arthur Cronquist; T. M. Barkley

Viral inclusion bodies Abnormal structures formed in the cytoplasm or the nucleus of the host cell (or both) during the multiplication of some viruses. These structures vary from virus to virus and may represent aggregations of mature virus particles but more often are areas where viral synthesis takes place. These areas of viral synthesis have an altered appearance when treated with typical diagnostic stains.

The presence of inclusion bodies is often important in diagnosis. Brain tissue taken from suspected cases of rabies is examined for the presence of cytoplasmic inclusions known as Negri bodies. In human herpes simplex infection, intranuclear inclusions are found that are surrounded by a clear halo. These inclusions are sites of replication of the virus, or "virus factories." *See* HERPES; RABIES; VIRUS.

Marcia M. Pierce

Virgo The Virgin, a southern zodiacal constellation, is visible in the summer evening sky from midnorthern latitudes. It contains the nearest cluster of galaxies to us; the Milky Way Galaxy is in an outer part of the cluster. *See* VIRGO CLUSTER; ZODIAC.

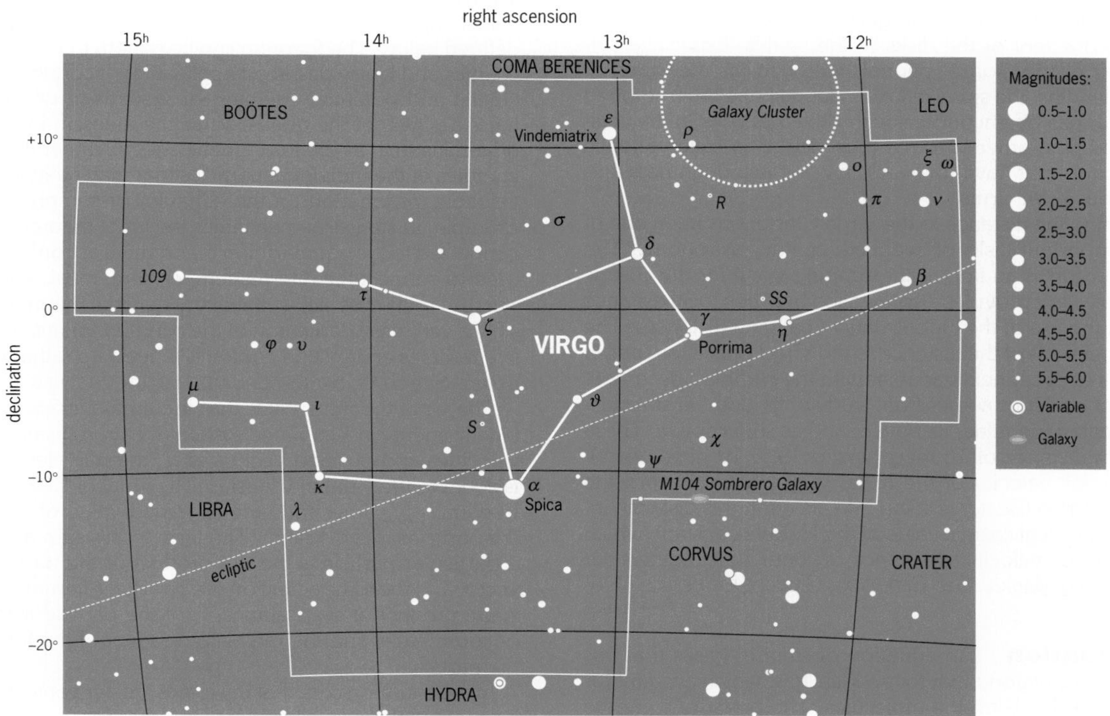

Modern boundaries of the constellation Virgo, the Virgin. The celestial equator is 0° of declination, which corresponds to celestial latitude. Right ascension corresponds to celestial longitude, with each hour of right ascension representing 15° of arc. Apparent brightness of stars is shown with dot sizes to illustrate the magnitude scale, where the brightest stars in the sky are 0th magnitude or brighter and the faintest stars that can be seen with the unaided eye at a dark site are 6th magnitude. (*Wil Tirion*)

Spica, the brightest star in Virgo, at first magnitude is the 16th brightest star in the sky. It is a B-star, a hot star on the main sequence of the Hertzsprung-Russell diagram, and glows blue-white. It is slightly variable in brightness. *See* CONSTELLATION; SPICA.

Jay M. Pasachoff

Virgo Cluster The nearest large cluster of galaxies and one of the best studied. It dominates the distribution of bright galaxies in the sky, forming a conspicuous clump in the distribution of easily visible elliptical and spiral galaxies. The clump lies mostly in the constellation of Virgo but also extends into neighboring constellations, especially Coma Berenices. The well-known spirals M58, M61, and M100, for example, are members of the Virgo Cluster. *See* CONSTELLATION; UNIVERSE; VIRGO.

The first exhaustive exploration of the Virgo Cluster was the Du Pont Telescope survey, carried out in the 1980s by A. Sandage and B. Binggeli. Their survey showed that the cluster has the appearance of two superimposed clusters. The two concentrations are approximately 5° apart; one is centered near the very luminous elliptical galaxy M87 and the other near M49, another giant elliptical galaxy. They lie along a nearly north–south line. The spiral galaxies are much less concentrated toward these central peaks than are either the giant ellipticals or the dwarf ellipticals. Furthermore, the spirals are more common, relative to other types, in the southern concentration than in the northern. It is also found that the dwarf elliptical galaxies are less concentrated toward the center of the entire complex than are any of the other types, possibly the result of mass segregation, with the more massive galaxies, under the effects of gravity and encounters, having fallen toward the center of mass more than the low-mass ones.

The Virgo Cluster is an example of a mixed-population cluster, with both early-type (elliptical and S0) and late-type (spiral and irregular) galaxies present. As the nearest large cluster, the Virgo Cluster provides a good testing ground for theories of how the dense cluster environment affects the properties of galaxies. The first peculiarity of the Virgo galaxies to be noticed was the rather limited extent of the disks of the spirals. Especially evident in their distribution of neutral hydrogen, as measured at radio wavelengths, the spiral galaxies clearly have been stripped of their outer gas by encounters with other galaxies in the cluster. A related phenomenon is the presence of so-called anemic galaxies, spirals that have unusually low surface brightness and inconspicuous spiral arms.

Determining the distance to the Virgo Cluster has been one of the more important tasks of twentieth-century astronomy. The Virgo Cluster forms an important stepping-stone for the cosmic distance scale, as it provides a connection between local distance criteria and the use of the Hubble law of gauging distances. The best measurements of the distance to the Virgo Cluster are based on the study of its cepheid variables with the *Hubble Space Telescope*. Other evidence comes from study of its globular star clusters, its planetary nebulae, its novae, and its supernovae. These give a mean distance of 16 megaparsecs (5×10^7 light-years). When the infall velocity of the Milky Way Galaxy toward the center of the Virgo Cluster is taken into account, this distance indicates that the local cosmic value of the Hubble constant, which relates expansion velocity to distance, is about 70 kilometers per second per megaparsec. *See* CEPHEIDS; GALAXY, EXTERNAL; HUBBLE CONSTANT.

Paul Hodge

Virial equation An equation of state of gases that has additional terms beyond that for an ideal gas, which account for the interactions between the molecules. The pressure p can be expressed in terms of the molar volume $V_m = V/n$ (where n is the number of moles of gas molecules in a volume V), the absolute temperature T, and the universal gas constant $R = 8.3145$ J K^{-1} mol^{-1} or, in more commonly used practical units, 0.082058 L atm K^{-1} mol^{-1} [Eq. (1)]. In the equation the virial coefficients

$$\frac{V_m}{RT} = 1 + \frac{B_2(T)}{V_m} + \frac{B_3(T)}{V_m^2} + \frac{B_4(T)}{V_m^3} + \cdots \tag{1}$$

$B_n(T)$ are functions only of the temperature and depend on the nature of the gas. In an ideal gas, in which all interactions between the molecules can be neglected because V_m is sufficiently large, only the first term, unity, survives on the right-hand side. *See* GAS.

The equation is important because there are rigorous relations between the coefficients B_2, B_3, and so on, as well as the interactions of the molecules in pairs, triplets, and so forth. It provides a valuable route to a knowledge of the intermolecular forces. Thus if the intermolecular energy of a pair of molecules at a separation r is $u(r)$, then the second virial coefficient can be expressed as Eq. (2). Here $N_A = 6.0221 \times 10^{23}$ mol^{-1} is the Avogadro constant,

$$B_2(T) = 2\pi N_A \int_0^\infty (1 - e^{-u(r)/kT}) r^2\, dr \tag{2}$$

and $k = R/N_A$. In a gas mixture, B_n is a polynomial of order n in the mole fractions of the components. *See* INTERMOLECULAR FORCES.

The virial equation is useful in practice because it represents the pressure accurately at low and moderate gas densities, for example, up to about 4 mol L^{-1} for nitrogen at room temperature, which corresponds to a pressure of about 100 atm (10 MPa). It is not useful at very high densities, where the series may diverge, and is inapplicable to liquids. It can be rearranged to give the ratio pV_m/RT as an expansion in powers of the pressure instead of the density, which is equally useful empirically, but the coefficients of the pressure expansion are not usually called virial coefficients, and lack any simple relation to the intermolecular forces or, in a mixture, to the composition of the gas. *See* VAN DER WAALS EQUATION.

J. S. Rowlinson

Virial theorem A theorem in classical mechanics which relates the kinetic energy of a system to the virial of Clausius, as defined below. The theorem can be generalized to quantum mechanics and has widespread application. It connects the average kinetic and potential energies for systems in which the potential is a power of the radius. Since the theorem involves integral quantities such as the total kinetic energy, rather than the kinetic energies of the individual particles that may be involved, it gives valuable information on the behavior of complex systems. For example, in statistical mechanics the virial theorem is intimately connected to the equipartition theorem; in astrophysics it may be used to connect the internal temperature, mass, and radius of a star and to discuss stellar stability. The virial theorem makes possible a very easy derivation of the counterintuitive result that as a star radiates energy and contracts it heats up rather than cooling down. *See* STAR; STATISTICAL MECHANICS; STELLAR EVOLUTION.

The virial theorem states that the time-averaged value of the kinetic energy in a confined system (that is, a system in which the velocities and position vectors of all the particles remain finite) is equal to the virial of Clausius. The virial of Clausius is defined to equal $-1/2$ times the time-averaged value of a sum over all the particles in the system. The term in this sum associated with a particular particle is the dot product of the particle's position vector and the force acting on the particle. Alternatively, this term is the product of the distance, r, of the particle from the origin of coordinates and the radial component of the force acting on the particle.

In the common case that the forces are derivable from a power-law potential, V, proportional to r^k, where k is a constant, the virial is just $-k/2$ times the potential energy. Thus, in this case the virial theorem simply states that the kinetic energy is $k/2$ times the potential energy. For a system connected by Hooke's-lawsprings, $k = 2$, and the average kinetic and potential energies are equal. For $k = 1$, that is, for gravitational or Coulomb forces, the potential energy is minus twice the kinetic energy. *See* COULOMB'S LAW; GRAVITATION; HARMONIC MOTION.

Albert G. Petschek

Viroids Small infectious nucleic acid particles containing independently replicating single-stranded circular ribonucleic acids (RNAs) of 246–399 nucleotides, which are known to infect certain monocot and dicot plants. Viroids are the smallest known disease agents, with an estimated molecular weight of as little as $1.0–1.3 \times 10^4$, in marked contrast to the conventional plant virus genomes, which have molecular weights of approximately 2×10^6. Conventional viruses are made up of nucleic acid encapsulated in protein (capsid), whereas viroids are uniquely characterized by the absence of a capsid. In spite of their small size, viroid RNAs can replicate and produce characteristic disease syndromes when introduced into cells. Although the viroids thus far identified are associated with plants, based on the unusual properties of the infectious agents they may also be found to affect other forms of life, including humans. It is possible that in instances where a viral etiology of disease has been assumed but where no causal virus has been identified, viroid infectious agents are involved.

Up to 27 different viroid species have been described from widely separated geographical locations. Viroids are classified in two families. The first family, Pospiviroidae, contains 24 viroids which have a central conserved nucleotide region which contain the same nucleotide sequences. Some examples include potato spindle tuber viroid (PSTV), citrus exocortis viroid (CEV), chrysanthemum stunt viroid (CSV), and cucumber pale fruit viroid (CPFV). The second family, Avsunviroidae, contains three viroids, which do not have a central conserved region: avocado sunblotch viroid (ASV), peach latent mosaic viroid (PLMV), and chrysanthemum chlorotic mottle viroid (ChCMV).

Diseases caused by viroids are agriculturally relevant, affecting economically important plants and an assortment of herbaceous and woody plants, such as potato, tomato, cucumber, hop, citrus, grapevine, coconut palm, fruit trees (apple, pear, plum, peach, avocado), and ornamentals (chrysanthemum and coleus). Viroid infections in some plant species produce profound disease symptoms ranging from stunting and leaf epinasty (downward bending) to plant death, whereas infections in other species produce few detectable symptoms compared with uninoculated control plants. Viroids generally have a restricted host range, although several viroids can infect the same hosts and cause similar symptoms. Good controls are not available for diseases caused by these small infectious agents other than indexing procedures to provide viroid-free propagules. *See* PLANT VIRUSES AND VIROIDS; VIRUS.

R. K. Horst

Virtual acoustics The stimulation of the complex acoustic field experienced by a listener within an environment. The technology is also known as three-dimensional sound and auralization. Going beyond the simple left-right volume adjustment of normal stereo techniques, the goal is to process sounds so that they appear to come from particular locations in three-dimensional space. Although loudspeaker systems have been developed, much of the work in the field focuses on using headphones for playback and is the outgrowth of earlier analog techniques. For example, in binaural recording, the sound of an orchestra playing classical music is recorded through small microphones in the two imitation ear canals of an artificial or dummy head placed in the audience of a concert hall. When the recorded piece is played back over headphones, the listener passively experiences the illusion of hearing the violins on the left and the cellos on the right, along with all the associated echoes, resonances, and ambience of the original environment. Techniques use digital signal processing to synthesize the acoustical properties that people use to localize a sound source in space. Thus, they provide the flexibility of a kind of digital dummy head, allowing a more active experience in which a listener can both design and move around or interact with a simulated acoustic environment in real time. *See* BINAURAL SOUND.

The success of virtual acoustics is critically dependent on whether the acoustical cues used by humans to locate sounds have been adequately synthesized. There may be many cumulative effects on the sound as it makes its way to the eardrum, but all of these effects can be expressed as a single filtering operation much like the effects of a graphic equalizer in a stereo system. The exact nature of this filter can be measured by an experiment in which an impulse (a single, very short sound pulse or click) is produced by a loudspeaker at a particular location. The acoustic shaping by the two ears is then measured by recording the outputs of small probe microphones placed inside the ear canals of the individual or an artificial head. If the measurement of the two ears occurs simultaneously, the responses, when taken together as a pair of filters, include an estimate of the interaural differences as well. Thus, this technique makes it possible to measure all of the relevant spatial cues together for a given source location, for a given listener, and in a given room or environment. *See* EAR; EQUALIZER; HEARING; PSYCHOACOUSTICS; SOUND.

Elizabeth M. Wenzel

Virtual manufacturing The modeling of manufacturing systems using audiovisual or other sensory features to simulate or design alternatives for an actual manufacturing environment, or the prototyping and manufacture of a proposed product using computers. The motivation for virtual manufacturing is to enhance people's ability to predict potential problems and inefficiencies in product functionality and manufacturability before real manufacturing occurs. *See* MANUFACTURING ENGINEERING; MODEL THEORY.

The concepts underlying virtual manufacturing include virtual reality, high-speed networking and software interfaces, agile manufacturing, and rapid prototyping.

Virtual reality is broadly defined as the ability to create and interact in cyberspace, that is, a simulated space that represents an environment very similar to the actual environment. The subset of virtual reality that is used in virtual manufacturing is commonly known as virtual environment. The perceived visual space is three-dimensional rather than two-dimensional, the human-machine interface is multimodal, and the user is immersed in the computer-generated environment; the screen separating the user and the computer becomes invisible to the user. The virtual environment for virtual manufacturing is simulated through immersion in computer graphics coupled with an acoustic interface, domain-independent interacting devices such as wands, and domain-specific devices such as steering and brakes for cars or earthmovers or instrument clusters for airplanes. *See* VIRTUAL REALITY.

High-speed networking and software interfaces are concerned with computer-aided-design (CAD) model portability among systems, trade-offs of high-detail models versus real-time interaction and display, rapid prototyping, collaborative design using virtual reality over distance, use of the Web for small- or medium-business virtual manufacturing, use of qualitative information (illumination, sound levels, ease of supervision, handicap accessibility) to design manufacturing systems, use of intelligent and autonomous agents in virtual environments, and the validity of virtual reality versus reality.

Agile manufacturing integrates an organization's people and technologies through innovative management and organization, knowledgeable and empowered people, and flexible and intelligent technologies. Virtual manufacturing provides a model for making rapid changes in products and processes based on customer requirements, and an agile manufacturing system attempts to implement it.

Rapid prototyping is an area in which virtual manufacturing has made an impact in processes such as stereolithography, selective laser sintering, and fused deposition modeling. A CAD drawing of a part is processed to create a layered file of the part. The part is built one layer at a time, precisely depositing layer upon layer of material.

Global virtual manufacturing extends the definition of virtual manufacturing to include the use of the Internet and intranets (global communications networks) for virtual component sourcing, and the use of virtual collaborative design and testing environments by multiple organizations or sites. Pat Banerjee

Virtual reality A form of human-computer interaction in which a real or imaginary environment is simulated and users interact with and manipulate that world. Users travel within the simulated world by moving toward where they want to be, and interact with things in that world by grasping and manipulating simulated objects. In the most successful virtual environments, users feel that they are truly present in the simulated world and that their experience in the virtual world matches what they would experience in the environment being simulated. This sensation is referred to as engagement, immersion, or presence, and it is this quality that distinguishes virtual reality from other forms of human-computer interaction. *See* HUMAN-COMPUTER INTERACTION.

When a user interacts with a virtual environment, the computer-generated graphics display must be updated with each turn of the head or movement of the hand. The virtual environment must be able to generate and display realistic-looking views of the simulated world quickly enough that the interaction feels responsive and natural. *See* COMPUTER GRAPHICS.

Hardware. Virtual reality relies on a variety of specialized input and output devices to achieve this sense of natural interaction.

The most important of the input devices used in a virtual environment, a tracker is capable of reporting its location in space and its orientation. Tracking devices can be optical, magnetic, or acoustic. A tracker is sometimes combined with a traditional computer input device, such as a mouse or a joystick. *See* COMPUTER PERIPHERAL DEVICES.

An attempt to provide a truly natural input device, the data glove is outfitted with sensors that can read the angle of each of the finger joints in the hand. Wearing such a glove, users can interact with the virtual world through hand gestures, such as pointing or making a fist. *See* FIBER-OPTIC SENSOR; STRAIN GAGE.

The real-world visual experience is approximated in virtual environments by using stereoscopic displays. Two views of the simulated world are generated, one for each eye, and a stereoscopic display device is used to show the correct view to each eye.

Applications. Virtual reality can be applied in a variety of ways. In scientific and engineering research, virtual environments are used to visually explore whatever physical world phenomenon is under study. Training personnel for work in dangerous environments or with expensive equipment is best done through simulation. Airplane pilots, for example, train in flight simulators. Virtual reality can enable medical personnel to practice new surgical procedures on simulated individuals. As a form of entertainment, virtual reality is a highly engaging way to experience imaginary worlds and to play games. Virtual reality also provides a way to experiment with prototype designs for new products. *See* AIRCRAFT DESIGN; COMPUTER-AIDED DESIGN AND MANUFACTURING. M. Pauline Baker

Virtual work principle The principle stating that the total virtual work done by all the forces acting on a system in static equilibrium is zero for a set of infinitesimal virtual displacements from equilibrium. The infinitesimal displacements are called virtual because they need not be obtained by a displacement that actually occurs in the system. The virtual work is the work done by the virtual displacements, which can be arbitrary, provided they are consistent with the constraints of the system. *See* CONSTRAINT.

The principle of virtual work is equivalent to the conditions for static equilibrium of a rigid body expressed in terms of the total forces and torques. That is, the principle of virtual work can be derived from these conditions, and conversely. *See* EQUILIBRIUM OF FORCES; STATICS.

One advantage of the principle of virtual work is that it can serve as a basis for all of statics. In the solution of problems the principle of virtual work is often useful for eliminating the need for consideration of the forces of constraint, since these forces often are perpendicular to the virtual displacements and consequently do no work. Paul W. Schmidt

Virulence The ability of a microorganism to cause disease. Virulence and pathogenicity are often used interchangeably, but virulence may also be used to indicate the degree of pathogenicity. Scientific understanding of the underlying mechanisms of virulence has increased rapidly due to the application of the techniques of biochemistry, genetics, molecular biology, and immunology. Bacterial virulence is better understood than that of other infectious agents.

Virulence is often multifactorial, involving a complex interplay between the parasite and the host. Various host factors, including age, sex, nutritional status, genetic constitution, and the status of the immune system, affect the outcome of the parasite-host interaction. Hosts with depressed immune systems, such as transplant and cancer patients, are susceptible to microorganisms not normally pathogenic in healthy hosts. Such microorganisms are referred to as opportunistic pathogens. The attribute of virulence is present in only a small portion of the total population of microorganisms, most of which are harmless or even beneficial to humans and other animals. *See* OPPORTUNISTIC INFECTIONS.

The spread of an infectious disease usually involves the adherence of the invading pathogen to a body surface. Next, the pathogen multiplies in host tissues, resisting or evading various nonspecific host defense systems. Actual disease symptoms are from damage to host tissues caused either directly or indirectly by the microorganism's components or products.

Most genetic information in bacteria is carried in the chromosome. However, genetic information is also carried on plasmids, which are independently replicating structures much smaller than the chromosome. Plasmids may provide bacteria with additional virulence-related capabilities (such as pilus formation, iron transport systems, toxin production, and antibiotic resistance). In some bacteria, several virulence determinants are regulated by a single genetic locus. *See* BACTERIA; CELLULAR IMMUNOLOGY; PLASMID; VIRUS. Brian Wilkinson

Virus Any of the elementary agents that possess some of the properties of living systems, such as having a genome and being able to adapt to changing environments. However, viruses are not functionally active outside their host cells. Viruses share three characteristics: (1) their simple, acellular organization consisting of a nucleic acid genome surrounded by a protective protein shell, which may itself be enclosed within an envelope that includes a membrane; (2) the presence of either DNA or RNA, but not both; and (3) their inability to reproduce independent of host cells. In essence, viruses are nucleic acid molecules, that is, genomes that can enter cells, replicate in them, and encode proteins capable of forming protective shells around them. It is preferable to refer to them as functionally active or inactive rather than living or dead.

Viruses are recognized as significant causes of disease in animals and plants. Many of the most important diseases that afflict humankind, including poliomyelitis, hepatitis, influenza, the common cold, measles, mumps, chickenpox, herpes, rubella, hemorrhagic fevers, encephalitis, and the acquired immunodeficiency syndrome (AIDS), are caused by viruses. Viruses also cause diseases in livestock (foot-and-mouth disease, avian influenza, hog cholera, distemper) and plants (tobacco mosaic disease, tomato ring spot, rice dwarf disease) that are of great

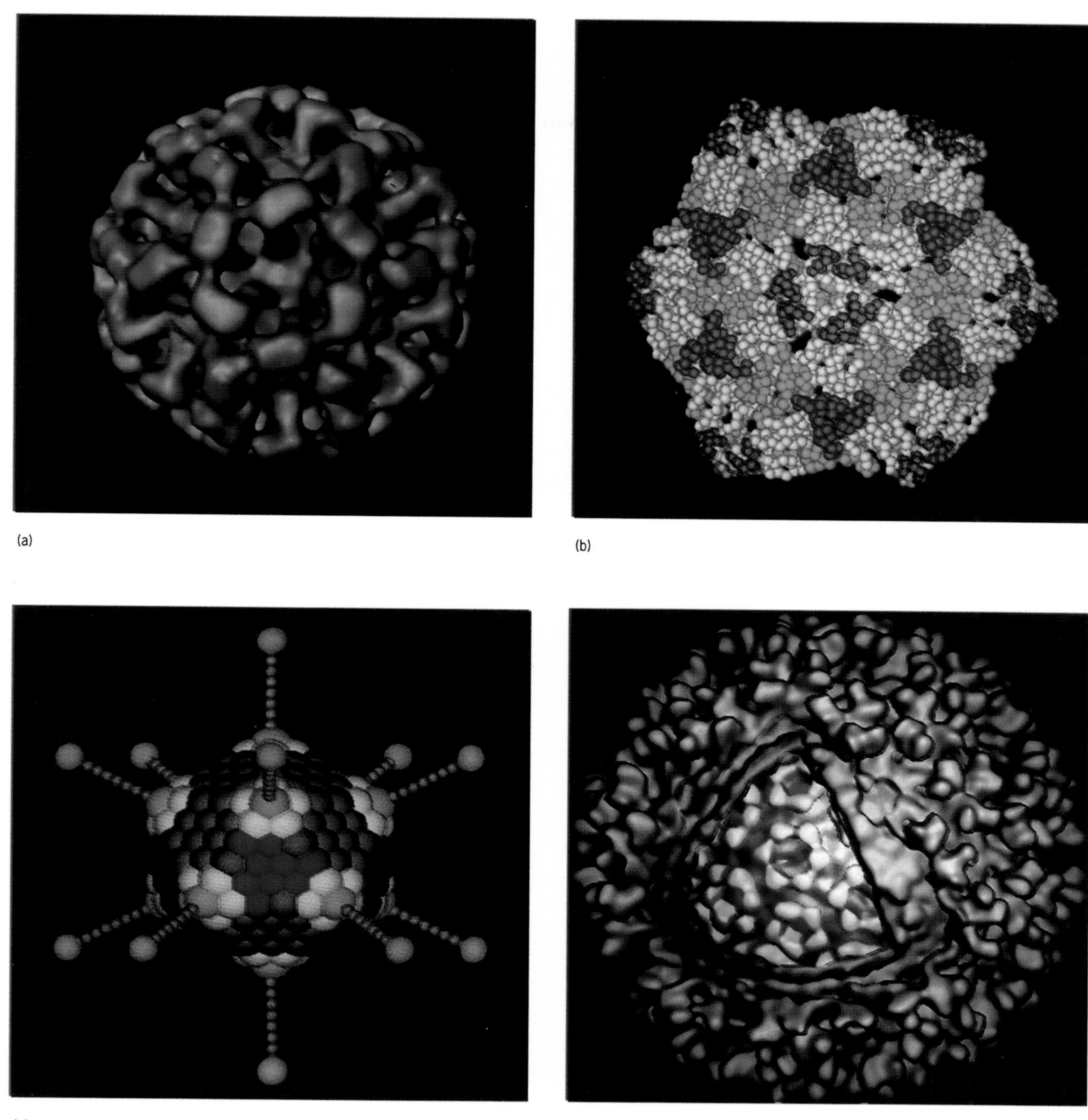

(a) (b) (c) (d)

Computer-generated representations of viruses. The protein coats of the viruses are shown in three dimensions. (*a*) Norwalk virus. (*b*) Front view: Southern Bean Mosaic Virus. (*c*) Adenovirus. (*d*) Ross River Virus. (*All courtesy of Richard J. Feldmann, except d, courtesy of B. V. V. Prasad, R. Rothnagel, and A. L. Shaw*)

economic importance. *See* ACQUIRED IMMUNE DEFICIENCY SYNDROME (AIDS); PLANT VIRUSES AND VIROIDS.

Viruses are also the simplest model systems for studying basic problems in biology. Their genomes are often no more than one-millionth the size of, for example, the human genome; yet the principles that govern the behavior of viral genes are the same as those that control the behavior of human genes. Viruses thus afford unrivaled opportunities for studying mechanisms that control the replication and expression of genetic material. *See* HUMAN GENOME.

Virus particles range in size from about 10 to 400 nanometers in diameter. Although viruses differ widely in shape and size (see illustration), they are constructed according to certain common principles. Basically, viruses consist of nucleic acid and protein arranged in a structure known as the nucleocapsid. The nucleic acid is the genome containing the information necessary for virus multiplication and survival; the protein is arranged around the genome in the form of a layer or shell that is termed the capsid. Some viruses consist only of a naked nucleocapsid, while in others the nucleocapsid is surrounded by a membrane on the outside of which "spikes" composed of glycoproteins may be attached; this is termed the envelope. The complete virus particle is known as the virion, a term that denotes both intactness of structure and the property of infectiousness.

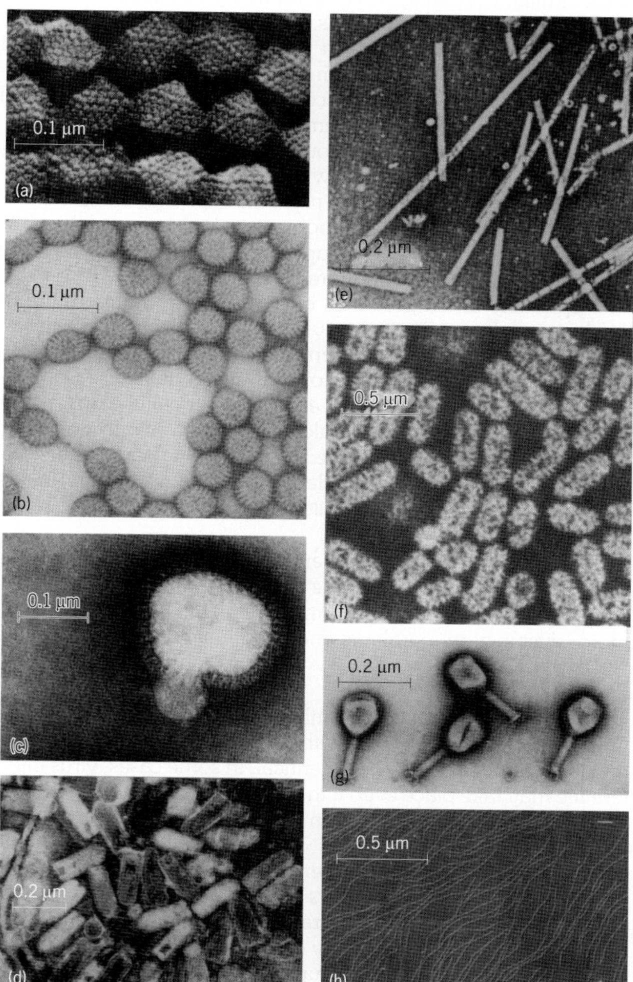

Electron micrographs of highly purified preparations of some viruses. (*a*) Adenovirus. (*b*) Rotavirus. (*c*) Influenza virus (*courtesy of George Laser*). (*d*) Vesicular stomatitis virus. (*e*) Tobacco mosaic virus. (*f*) Alfalfa mosaic virus. (*g*) T4 bacteriophage. (*h*) M13 bacteriophage.

Viral genomes are astonishingly diverse. Some consist of deoxyribonucleic acid (DNA), others of ribonucleic acid (RNA). Some are double-stranded, others single-stranded. Some are linear, others circular. Some consist of one molecule, others of several (up to 12). Their size also varies within wide limits: they range from 3000 to 280,000 base pairs if double-stranded, and from 5000 to 27,000 nucleotides if single-stranded. *See* DEOXYRIBONUCLEIC ACID (DNA); RIBONUCLEIC ACID (RNA); VIRUS CLASSIFICATION.

Viral genomes encode three types of genetic information. First, they encode the structural proteins of virus particles. Second, most viral genomes can be neither replicated nor decoded (that is, transcribed into messenger RNA) by host-cell enzymes. Most viruses, therefore, encode enzymes capable of transcribing their genomes into messenger RNA molecules that are then translated by host-cell ribosomes, as well as nucleic acid polymerases capable of replicating their genomes. Many viruses also encode nonstructural proteins with catalytic and other functions necessary for virus particle maturation and morphogenesis. Third, many viruses encode proteins that interact with components of host-cell defense mechanisms against invading infectious agents. The more successful these proteins are in neutralizing these defenses, the more virulent viruses are and the more severe the resulting disease.

The two most commonly observed virus–cell interactions are the lytic interaction, which results in virus multiplication and lysis of the host cell, and the transforming interaction, which results in the integration of the viral genome into the host genome and the permanent transformation or alteration of the host cell with respect to morphology, growth habit, and the manner in which it interacts with other cells. Transformed animal and plant cells are also capable of multiplying; they often grow into tumors, and the viruses that cause such transformation are known as tumor viruses. *See* CANCER (MEDICINE); ONCOLOGY; RETROVIRUS; TUMOR VIRUSES.

Because viruses enter host cells and make use of host-cell enzymes and constituents to reproduce, development of drugs to treat viral infections seemed a remote possibility for many years. A drug that would block virus reproduction would likely be toxic to the host. However, inhibitors of virus-specific enzymes and life-cycle stages have now been discovered. Most antiviral drugs in use today disrupt either viral nucleic acid synthesis or specific stages in the virus life cycle. Viral nucleic acid synthesis is almost always carried out by virus-encoded enzymes that do not exist in uninfected cells and are therefore excellent targets for antiviral chemotherapy. Numerous chemical compounds have been described that inhibit the multiplication of viruses. Only a few, however, inhibit virus multiplication efficiently in the body without undesirable side effects. Another viral function that has been targeted is the cleavage of polyproteins, precursors of structural proteins, to their functional components by virus-encoded proteases; examples include the HIV protease inhibitors used for treatment in AIDS patients. *See* CHEMOTHERAPY AND OTHER ANTINEOPLASTIC DRUGS; CYTOMEGALOVIRUS INFECTION; HERPES; INFLUENZA; RESPIRATORY SYNCYTIAL VIRUS; VIRUS CHEMOPROPHYLAXIS.

Antiviral agents on which much interest is focused are the interferons. Interferons are cytokines or lymphokines that regulate cellular genes concerned with cell division and the functioning of the immune system. Their formation is strongly induced by virus infection; they provide the first line of defense against viral infections until antibodies begin to form. Interferons interfere with the multiplication of viruses by preventing the translation of early viral messenger RNAs. As a result, viral capsid proteins cannot be formed and no viral progeny results. Interferon is clinically useful in the treatment of hepatitis B and C infection.

By far the most effective means of preventing viral diseases is through mobilization of the immune system by vaccines. There are two types of antiviral vaccines: inactivated and attenuated active. Most of the antiviral vaccines currently in use are of the latter kind. The principle of antiviral vaccines is that inactivated virulent or active attenuated virus particles cause the formation of antibodies that neutralize a virulent virus when it invades the body. *See* ANIMAL VIRUS; VACCINATION; VIRUS, DEFECTIVE.

W. K. Joklik; Marcia M. Pierce

Virus, defective A virus that by mutation has lost the ability to be replicated in the host cell without the aid of a helper virus. The virus particles (virions) contain all the viral structural components; they can attach, penetrate, and release their nucleic acid (RNA; DNA) within the host cell. However, since the mutation has destroyed an essential function, new virions will not be made unless the cell was simultaneously infected with the helper virus, which can provide the missing function. Only then will the cell produce a mixed population of new helper and defective viruses. Occasionally, when their nucleic acids become integrated in the DNA of the host cell, defective viruses persist in nature by propagation from mother cell to daughter cell. *See* ANIMAL VIRUS; MUTATION; VIRULENCE.

The most important group of defective viruses are deletion mutants. They are derived from their homologous nondefective (wild-type) virus through errors in the nucleic acid replication that result in the deletion of a fragment in the newly synthesized molecules. The defective nucleic acid must be capable of self-replication, at least in the presence of the wild-type virus, and must combine with other viral components to form a particle in order to exit the cell.

The defective RNA tumor viruses are deletion mutants. Mammalian and most avian sarcoma viruses require a nondefective leukemia virus as a helper virus. Usually the specificity for a certain type of host cell exhibited by the defective virion depends on the helper virus, indicating that one of the virion surface proteins has been furnished by the helper virus gene. These proteins are involved in interactions with cellular surface receptors, and thus determine whether a cell can serve as a host for viral infection. *See* ONCOLOGY; TUMOR VIRUSES. M. E. Reichmann

Virus chemoprophylaxis The treatment of viral disease by chemical drugs has been approached experimentally through the inhibition of intracellular metabolic processes that otherwise lead to the synthesis of viral constituents. Compounds have also been tested for their capacity to inhibit adsorption, penetration, or release of infectious virus from the host cell. In order to be considered for chemoprophylaxis, a compound must have a greater specificity for inhibiting virus-directed reactions than for normal host cell reactions. Although some success has been achieved, there are no inhibitors available that can be used in the routine treatment of viral infections in humans. *See* CHEMOTHERAPY AND OTHER ANTINEOPLASTIC DRUGS; VIRUS. Joseph L. Melnick

Virus classification There is no evidence that viruses possess a common ancestor or are in any way phylogenetically related. Nevertheless, classification along the lines of the Linnean system into families, genera, and species has been partially successful. Based on the organisms they infect, the first broad division is into vertebrate; algae, fungi, yeast, and protozoan; invertebrate; plant; and bacterial viruses. (However, viral families may fall into more than one of these classes.) Within these classes, other criteria for subdivision are used. Among these are general morphology: envelope or the lack of it; nature of the genome (deoxyribonucleic acid, DNA, or ribonucleic acid, RNA); structure of the genome [single-stranded (ss) or double-stranded (ds), linear or circular, fragmented or nonfragmented]; mechanisms of gene expression and virus replication (positive- or negative-strand RNA); serological relationship; host and tissue susceptibility; and pathology (symptoms, type of disease).

Families are sometimes subdivided into subfamilies; the suffix -virinae may then be used. The subgroups of a family or subfamily are equivalent to the genera of the Linnean classification. A number of plant viruses are placed into groups rather than families. Bacterial viruses are also known as bacteriophages or phages. They may be tailed or nontailed. *See* ANIMAL VIRUS; BACTERIOPHAGE; PLANT VIRUSES AND VIROIDS; VIRUS.

Marcia M. Pierce; M. E. Reichmann

Virus infection, latent, persistent, slow Initially inapparent (covert) viral infections in which an equilibrium between the virus and the host has been established. Pathology may occur periodically, or chronically at a low level. If manifested in degenerative diseases of the central nervous system, these infections may result in death. The basis of these infections is an uncharacteristically low level of viral replication which may persist throughout the normal lifetime of the host, and which may follow the recovery from a more severe bout with the virus. The acute illness may have elicited production of antibodies, and an equilibrium between virus neutralization and antibody generation is established which may produce life-long immunity.

Persistent, slow, and latent infections are not easily differentiated. Persistent infections are usually accompanied by detectable formation of antibodies without host pathology. (However, in the case of measles the persistent infection may change into a slow infection culminating in the degenerative brain disease termed subacute sclerosing panencephalitis, or SSPE.) Slow infections, while similar, eventually develop pathology. In latent infections, demonstrable presence of virus and symptoms of disease may be absent for a long time, with periodic outbreaks precipitated usually at a time when the immune system has been compromised in some way. This may occur through a variety of external influences, both physical and emotional.

Several mechanisms are thought to account for the generation of latent, persistent, slow infections. Attenuation of the virulence of a virus can take place not only by immunological neutralization but also by spontaneous mutation of the virus itself. The most important group of mutants pertinent to these phenomena are the temperature-sensitive mutants, which replicate optimally at temperatures below the body temperature of the host, and very inefficiently at the elevated temperatures. The presence of temperature-sensitive mutants has been demonstrated in persistent infections of foot-and-mouth disease, coxsackie, sindbis, influenza, and equine encephalitis viruses. Another reason for diminished virulence in persistent infections may be due to the generation of defective interfering particles. Defective interfering particles are deletion mutants which are unable to replicate on their own. *See* VIRULENCE; VIRUS, DEFECTIVE; VIRUS INTERFERENCE.

The tissue system most frequently affected in slow infections is the central nervous system. The involvement of the central nervous system in slow infections manifests itself by disorientation, motor dysfunction, paralysis, and eventually death. *See* ANIMAL VIRUS; VIRUS. M. E. Reichmann

Virus interference Inhibition of the replication of a virus by a previous infection with another virus. The two viruses may be unrelated, related, or identical. In some cases, virus interference may take place even if the first virus was inactivated. The term mutual exclusion has been applied to this phenomenon in bacterial viruses.

Several mechanisms of interference can be distinguished: (1) Inactivation of cell receptors by one virus may prevent subsequent adsorption and penetration by another virus. (2) The first virus may inhibit or modify cellular enzymes or proteins required for replication of the superinfecting virus. (3) The first virus may generate destructive enzymes or induce the cell to synthesize protective substances which prevent superinfection. (4) The first virus may generate defective interfering particles or mutants which may inhibit the replication of the infecting virus by competing with it for a protein (or enzyme) available in limited quantities; this type of viral interference has been called autointerference, and depends on a greater replicative efficiency of the defective interfering particles or mutants, compared to the infecting virus. *See* ANIMAL VIRUS; VIRULENCE; VIRUS; VIRUS, DEFECTIVE.

M. E. Reichmann

Viscosity The material property that measures a fluid's resistance to flowing. For example, water flows from a tilted jar more quickly and easily than honey does. Honey is more viscous than water, so although gravity creates nearly the same stresses in honey and water, the more viscous fluid flows more slowly.

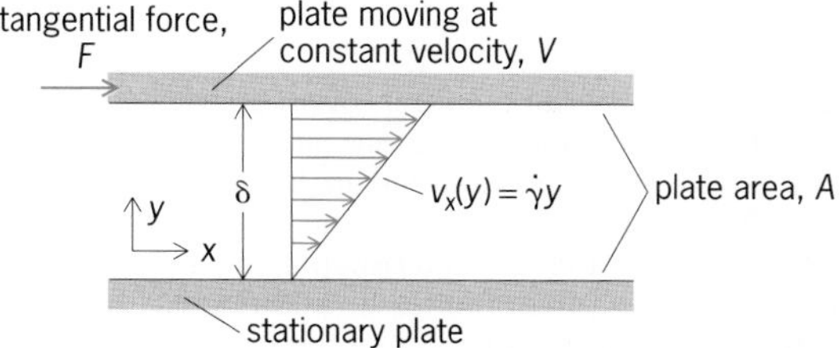

Planar Couette flow. v_x = fluid velocity at distance y above the stationary plate, $\dot{\gamma}$ = velocity gradient or shear rate, δ = distance between plates.

The viscosity can be measured where the fluid of interest is sheared between two flat plates which are parallel to one another (see illustration). This is known as planar Couette flow. The shear flow created between the plates has the linear velocity profile given by Eq. (1), where v_x is the velocity parallel to the plates

$$v_x = \dot{\gamma} y \tag{1}$$

at a perpendicular distance y above the stationary plate. The coefficient $\dot{\gamma}$, called the velocity gradient or shear rate, is given by V/δ, where δ is the distance between the plates. The shear stress is the ratio of the tangential force F needed to maintain the moving plate at a constant velocity V to the plate area A. It is expected that the shear stress increases with increasing shear rate but that the ratio of these two quantities depends only on the fluid between the plates. This ratio is used to define the shear viscosity, η, as in Eq. (2). The shear viscosity may depend on temperature, pressure, and shear rate.

$$\eta \equiv \frac{\text{shear stress}}{\text{shear rate}} = \frac{F/A}{V/\delta} \tag{2}$$

Isaac Newton is credited with first suggesting a model for the viscous property of fluids in 1687. Newton proposed that the resistance to flow caused by viscosity is proportional to the velocity at which the parts of the fluid are being separated from one another because of the flow. G. G. Stokes gave the first mathematical formulation of Newton's law of viscosity in 1845. Although Newton's law of viscosity is an empirical idealization, many fluids, such as low-molecular-weight liquids and dilute gases, are well characterized by it over a large range of conditions. However, many other fluids, such as polymer solution and melts, blood, ink, liquid crystals, and colloidal suspensions, are not described well by Newton's law. Such fluids are referred to as non-newtonian.

For planar Couette flow, the one-dimensional form of Newton's law of viscosity is given mathematically by Eq. (3),

$$\tau_{yx} = \mu \frac{dv_x}{dy} = \mu\dot{\gamma} \tag{3}$$

where τ_{yx} is the shear stress, and μ, a function of temperature and pressure, is the coefficient of viscosity or simply the viscosity. Therefore, by comparing Eqs. (2) and (3) the shear viscosity is equal to the coefficient of viscosity (that is, $\eta = \mu$) for a newtonian fluid. Because of this relation the shear viscosity is also often referred to as the viscosity. However, it should be clear that the two quantities are not equivalent; μ is a newtonian-model parameter, which varies only with temperature and pressure, while η is a more general material property which may vary nonlinearly with shear rate. *See* FLUID FLOW; NEWTONIAN FLUID.

From Eqs. (2) and (3), the units of viscosity are given by force per area per inverse time. If in planar Couette flow, for example, 1 dyne of tangential force is applied for every 1 cm^2 area of plate to create a velocity gradient of 1 s^{-1}, then the fluid between the plates has a viscosity of 1 poise (=1 dyne · s/cm^2). Several viscosity units are in common use (see table). Comparison of the viscosities of different fluids demonstrates some general trends. For example, the viscosity of gases is generally much less than that of liquids. Whereas gases tend to become more viscous as temperature is increased, the opposite is true of liquids. Other data also show that increasing pressure tends to increase the viscosity of dense gases, but pressure has only a small effect on the viscosity of dilute gases and liquids.

Whereas dilute gas molecules interact primarily in pairs as they collide, molecules in the liquid phase are in continuous interaction with many neighboring molecules. The concepts of average velocity and mean free path have little meaning for liquids. It is clear, however, that increasing temperature increases the mobility of molecules, thus allowing neighboring molecules to more easily overcome energy barriers and slip past one another. Such arguments lead to an exponential relation for the dependence of viscosity on temperature. *See* GAS; LIQUID.

Many non-newtonian fluids not only exhibit a viscosity which depends on shear rate (pseudoplastic or dilantant) but also exhibit elastic properties. These viscoelastic fluids require a large number of strain-rate-dependent material properties in addition to the shear viscosity to characterize them. The situation can become more complex when the material properties are time dependent (thixotropic or rheopectic). Fluids that are nonhomogeneous or nonisotropic require even more sophisticated analysis. The field of rheology attempts to deal with these complexities. *See* RHEOLOGY.

Lewis E. Wedgewood

Vision The sense of sight, which perceives the form, color, size, movement, and distance of objects. Of all the senses, vision provides the most detailed and extensive information about the environment. In the higher animals, especially the birds and primates, the eyes and the visual areas of the central nervous

Viscosity conversions

Unit	poise	cp	Pa · s	$lb_m/(ft \cdot s)$	$lb_f \cdot s/ft^2$
1 poise*	1	100	0.1	6.72×10^{-2}	2.089×10^{-3}
1 centipoise	0.01	1	0.001	6.72×10^{-4}	2.089×10^{-5}
1 pascal-second†	10	1000	1	0.672	2.089×10^{-2}
1 $lb_m/(ft \cdot s)$	14.88	1488	1.488	1	3.108×10^{-2}
1 $lb_f \cdot s/ft^2$	478.8	4.788×10^4	47.88	32.17	1

*1 poise = 1 dyne · s/cm^2 = 1 g/(cm · s).
†1 Pa · s = 1 kg/(m · s).

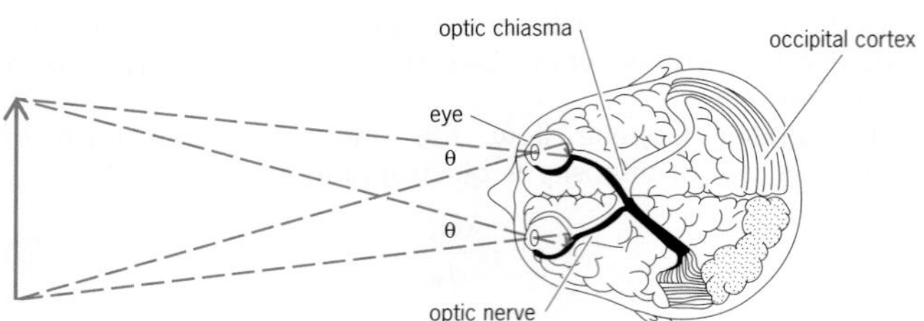

Fig. 1. Diagram showing the eyes and visual projection system. The visual angle θ is measured in degrees.

Characteristics of human vision

Characteristic	Scotopic vision	Photopic vision
Photochemical substance	Rhodopsin	Cone pigments
Receptor cells	Rods	Cones
Speed of adaptation	Slow (30 min or more)	Rapid (8 min or less)
Color discrimination	No	Yes
Region of retina	Periphery	Center
Spatial summation	Much	Little
Visual acuity	Low	High
Number of receptors per eye	120,000,000	7,000,000
Cortical representation	Small	Large
Spectral sensitivity peak	505 nm	555 nm

system have developed a size and complexity far beyond the other sensory systems.

Visual stimuli are typically rays of light entering the eyes and forming images on the retina at the back of the eyeball (Fig. 1). Human vision is most sensitive for light comprising the visible spectrum in the range 380–720 nanometers in wavelength. In general, light stimuli can be measured by physical means with respect to their energy, dominant wavelength, and spectral purity. These three physical aspects of the light are closely related to the perceived brightness, hue, and saturation, respectively. *See* COLOR; LIGHT.

Anatomical basis for vision. The anatomical structures involved in vision include the eyes, optic nerves and tracts, optic thalamus, primary visual cortex, and higher visual areas of the brain. The eyes are motor organs as well as sensory; that is, each eye can turn directly toward an object to inspect it. The two eyes are coordinated in their inspection of objects, and they are able to converge for near objects and diverge for far ones. Each eye can also regulate the shape of its crystalline lens to focus the rays from the object and to form a sharp image on the retina. Furthermore, the eyes can regulate the amount of light reaching the sensitive cells on the retina by contracting and expanding the pupil of the iris. These motor responses of the eyes are examples of involuntary action that is controlled by various reflex pathways within the brain. *See* EYE (VERTEBRATE).

The process of seeing begins when light passes through the eye and is absorbed by the photoreceptors of the retina. These cells are activated by the light in such a way that electrical potentials are generated. These potentials serve to generate nerve responses in various successive neural cells in the vicinity of excitation. Impulses emerge from the eye in the form of repetitive discharges in the fibers of the optic nerve, which do not mirror exactly the excitation of the photoreceptors by light. Complex interactions within the retina serve to enhance certain responses and to suppress others. Furthermore, each eye contains more than a hundred times as many photoreceptors as optic nerve fibers. Thus it would appear that much of the integrative action of the visual system has already occurred within the retina before the brain has had a chance to act.

The optic nerves from the two eyes traverse the optic chiasma. Figure 1 shows that the fibers from the inner (nasal) half of each retina cross over to the opposite side, while those from the outer (temporal) half do not cross over but remain on the same side. The effect of this arrangement is that the right visual field, which stimulates the left half of each retina, activates the left half of the thalamus and visual cortex. Conversely the left visual field affects the right half of the brain. This situation is therefore similar to that of other sensory and motor projection systems in which the left side of the body is represented by the right side of the brain and vice versa.

The visual cortex includes a projection area in the occipital lobe of each hemisphere. Here there appears to be a point-for-point correspondence between the retina of each eye and the cortex. Thus the cortex contains a "map" or projection area, each point of which represents a point in visual space as seen by each eye. Other important features of an object such as its color, motion, orientation, and shape are simultaneously perceived. The two retinal maps are merged to form the cortical projection area. This allows the separate images from the two eyes to interact with each other in stereoscopic vision, binocular color mixture, and other phenomena. In addition to the projection areas on the right and left halves of the cortex, there are visual association areas and other brain regions that are involved in vision. Complex visual acts, such as form recognition, movement perception, and reading, are believed to depend on widespread cortical activity beyond that of the projection areas. *See* BRAIN.

Scotopic and photopic vision. Night animals have eyes that are specialized for seeing with a minimum of light. This type of vision is called scotopic. Day animals have predominantly photopic vision. They require much more light for seeing, but their daytime vision is specialized for quick and accurate perception of fine details of color, form, and texture, and location of objects. Color vision, when it is present, is also a property of the photopic system. Human vision is duplex; humans are in the fortunate position of having both photopic and scotopic vision. Some of the chief characteristics of human scotopic and photopic vision are enumerated in the table.

Scotopic vision occurs when the rod receptors of the eye are stimulated by light. The outer limbs of the rods contain a photosensitive substance known as visual purple or rhodopsin. This substance is bleached away by the action of strong light so that the scotopic system is virtually blind in the daytime. In darkness, however, the rhodopsin is regenerated by restorative reactions based on the transport of vitamin A to the retina by the blood. One experiences a temporary blindness upon walking indoors on a bright day, especially into a dark room. As the eyes become accustomed to the dim light the scotopic system gradually begins to function. This process is known as dark adaptation. Complete dark adaptation is a slow process during which the rhodopsin is restored in the rods. A 10,000-fold increase in sensitivity of is often found to occur during a half-hour period of dark adaptation. By this time some of the rod receptors are so sensitive that only one photon is necessary to trigger each rod into action. Faulty dark adaptation or night blindness is found in persons who lack rod receptors or have a dietary deficiency in vitamin A. This scotopic vision is colorless or achromatic. *See* VITAMIN A.

Normal photopic vision has the characteristics enumerated in the table. Emphasis is placed on the fovea centralis, a small region at the very center of the retina of each eye.

Foveal vision is achieved by looking directly at objects in the daytime. The image of an object falls within a region almost exclusively populated by cone receptors, closely packed together in the central fovea, each of which is provided with a series of specialized nerve cells that process the incoming pattern of stimulation and convey it to the cortical projection area. In this way the cortex is supplied with superbly detailed information about any pattern of light that falls within the fovea centralis.

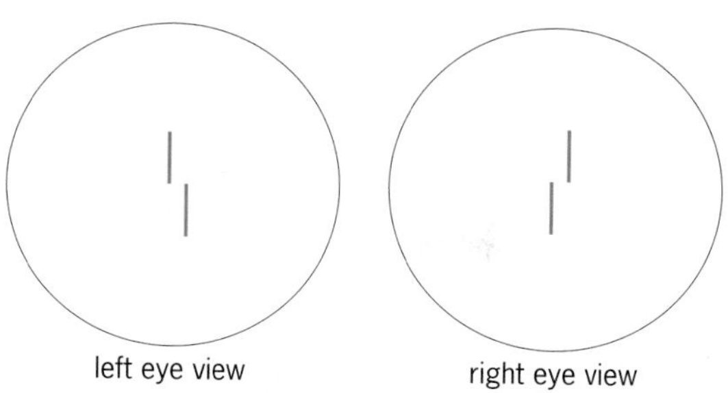

Fig. 2. Vernier and stereoscopic discriminations of space.

Peripheral vision takes place outside the fovea centralis. Vision extends out to more than 90° from center, so that one can detect moving objects approaching from either side. This extreme peripheral vision is comparable to night vision in that it is devoid of sharpness and color.

There is a simple anatomical explanation for the clarity of foveal vision as compared with peripheral vision. The cones become less and less numerous in the retinal zones that are more and more remote from the fovea. In the extreme periphery there are scarcely any, and even the rods are more sparsely distributed. Furthermore, the plentiful neural connections from the foveal cones are replaced in the periphery by network connections in which hundreds of receptors may activate a single optic nerve fiber. This mass action is favorable for the detection of large or dim stimuli in the periphery or at night, but it is unfavorable for visual acuity (the ability to see fine details of an object) or color vision, both of which require the brain to differentiate between signals arriving from closely adjacent cone receptors.

Space and time perception. Vernier and stereoscopic discrimination are elementary forms of space perception. Here, the eye is required to judge the relative position of one object in relation to another (Fig. 2). The left eye, for example, sees the lower line as displaced slightly to the right of the upper. This is known as vernier discrimination. The eye is able to distinguish fantastically small displacements of this kind, a few seconds of arc under favorable conditions. If the right eye is presented with similar lines that are oppositely displaced, then the images for the two eyes appear fused into one and the subject sees the lower line as nearer than the upper. This is the principle of the stereoscope. Again it is true that displacements of a few seconds of arc are clearly seen, this time as changes in distance. The distance judgment is made not at the level of the retina but at the cortex where the spatial patterns from the separate eyes are fused together. The fineness of vernier and stereoscopic discrimination transcends that of the retinal mosaic and suggests that some averaging mechanism must be operating in space or time or both.

The spatial aspects of the visual field are also of interest. Good acuity is restricted to a narrowly defined region at the center of the visual field. Farther out, in the peripheral regions, area and intensity are reciprocally related for all small sizes of stimulus field. A stimulus patch of unit area, for example, looks the same as a patch of twice the same area and half the luminance. This high degree of areal summation is achieved by the convergence of hundreds of rod receptors upon each optic nerve fiber. It is the basis for the ability of the dark-adapted eye to detect large objects even on a dark night.

In daytime vision, spatial inhibition, rather than summation, is most noticeable. The phenomenon of simultaneous contrast is present at a border between fields of different color or luminance. This has the effect of heightening contours and making forms more noticeable against their background.

The temporal characteristics of vision are revealed by studying the responses of the eye to various temporal patterns of stimulation. When a light is first turned on, there is a vigorous burst of nerve impulses that travel from the eye to the brain. Continued illumination results in fewer and fewer impulses as the eye adapts itself to the given level of illumination. Turning the light off elicits another strong neural response. The strength of a visual stimulus depends upon its duration as well as its intensity. Below a certain critical duration, the product of duration and intensity is found to be constant for threshold stimulation. A flash of light lasting only a few milliseconds may stimulate the eye quite strongly, providing its luminance is sufficiently high. A light of twice of the original duration will be as detectable as the first if it is given half the original luminance.

Voluntary eye movements enable the eyes to roam over the surface of an object of inspection. In reading, for example, the eyes typically make four to seven fixational pauses along each line of print, with short jerky motions between pauses. An individual's vision typically takes place during the pauses, so that one's awareness of the whole object is the result of integrating these separate impressions over time.

A flickering light is one that is going on and off (or undergoing lesser changes in intensity) as a function of time. At a sufficiently high flash rate (called the critical frequency of fusion, cff) the eye fails to detect the flicker, and the light pulses seem to fuse to form a steady light that cannot be distinguished from a continuous light that has the same total energy per unit of time. As the flash rate is reduced below the cff, flicker becomes noticeable, and at very low rates the light may appear more conspicuous than flashes occurring at higher frequency. The cff is often used clinically to indicate a person's visual function as influenced by drugs, fatigue, or disease. *See* COLOR VISION; PERCEPTION.

Lorrin A. Riggs; Charles E. Sternheim

Visual debugging Visualization of computer program state and program execution to facilitate understanding and, if necessary, alteration of the program. Debuggers are universal tools for understanding what is going on when a program is executed. Using a debugger, one can execute the program in a specific environment, stop the program under specific conditions, and examine or alter the content of the program variables or pointers. Traditional command-line oriented debuggers allowed only a simple textual representation of the program variables (program state).

Textual representation did not change even when modern debuggers came with a graphical user interface. Although variable names became accessible by means of menus, the variable values were still presented as text, including structural information, such as pointers and references. Likewise, the program execution is available only as a series of isolated program stops. (Pointers are variables that contain the "addresses" of other variables.) Compared to traditional debuggers, the techniques of visual debugging allow quicker exploration and understanding of what is going on in a program. *See* COMPUTER PROGRAMMING; PROGRAMMING LANGUAGES; SOFTWARE; SOFTWARE ENGINEERING.

The GNU Data Display Debugger (DDD), for example, is a graphical front-end to a command-line debugger, providing menus and other graphical interfaces that eventually translate into debugger commands. As a unique feature, DDD allows the visualization of data structures as graphs. The concept is simple: Double-clicking on a variable shows its value as an isolated graph node. By double-clicking on a pointer, the dereferenced value or the variable pointed is to shown as another graph node, with an edge relating pointer and dereferenced value. By subsequent double-clicking on pointers, the programmer can unfold the entire data structure.

If a pointer points to a value that is already displayed (for example, in a circular list), no new node is created; instead, the edge is drawn to the existing value. Using this alias recognition, the programmer can quickly identify data structures that are referenced by multiple pointers.

In principle, DDD can render arbitrary data structures by means of nodes and edges. However, the programmer must choose what to unfold, as the screen size quickly limits the number of variables displayed. Nonetheless, DDD is one of the most popular debugging tools under Unix and Linux. Andreas Zeller

Visual impairment Abnormal visual acuity. The term is used to describe visual acuity substantially less than normal. The World Health Organization defines visual impairment as acuity less than 20/60 (normal being 20/20). The legal definition of blindness in the United States is visual acuity of 20/200 or worse (or severely restricted peripheral vision). The World Health Organization defines blindness as visual acuity worse than 20/400. Visual impairment and blindness increase substantially with age. The major causes of blindness differ substantially by race. Cataracts, which involve opacification of the normally clear lens, and diabetic retinopathy, which is an accumulation of fluid or the growth of abnormal blood vessels in the retina (most commonly in insulin-dependent diabetics), are important causes of blindness. A large proportion of people having blindness caused by cataracts can be treated by surgery.

A growing proportion of blindness caused by diabetic retinopathy can be prevented by laser surgery.

Other important causes of blindness are glaucoma and macular degeneration. Early detection of glaucoma requires routine, careful screening and examination, since many patients remain asymptomatic until much of the optic nerve is destroyed. Macular degeneration involves atrophy of that portion of the retina responsible for fine (reading) vision. People with macular degeneration rarely suffer the total blindness of advanced glaucoma or diabetic retinopathy. *See* DIABETES.

Visual problems are far more common in developing countries, where there are limited resources for dealing with problems that are otherwise readily treated, such as cataract, or prevented, such as trachoma or xerophthalmia ("blinding malnutrition" caused by vitamin A deficiency). *See* CATARACT; GLAUCOMA; VISION. Alfred Sommer

Vitamin An organic compound required in very small amounts for the normal functioning of the body and obtained mainly from foods. Vitamins are present in food in minute quantities compared to the other utilizable components of the diet, namely, proteins, fats, carbohydrates, and minerals.

Synthetic and natural vitamins usually have the same biological value. Different vitamins, which are often not related to each other chemically or functionally, are conventionally divided into a fat-soluble group (vitamins A, D, E, and K) and a water-soluble group [vitamin C (ascorbic acid) and the various B vitamins: thiamine, vitamin B, riboflavin, vitamin B_2, vitamin B_6, niacin, folic acid, vitamin B_{12}, biotin, and pantothenic acid]. The vitamins, particularly the water-soluble ones, occur almost universally throughout the animal and plant kingdoms individual articles on each vitamin.

The B vitamins function as coenzymes that catalyze many of the anabolic and catabolic reactions of living organisms necessary for the production of energy; the synthesis of tissue components, hormones, and chemical regulators; and the detoxification and degradation of waste products and toxins. On the other hand, vitamin C and the fat-soluble vitamins do not function as coenzymes. Vitamins C and E and β-carotene (a precursor of vitamin A) act as antioxidants, helping to prevent tissue injury from free-radical reactions. In addition, vitamin C functions as a cofactor in hydroxylation reactions. Vitamin D has hormonelike activity in calcium metabolism; vitamin A plays a critical role in night vision, growth, and maintaining normal differentiation of epithelial tissue; and vitamin K has a unique posttranscriptional role in the formation of active blood-clotting factors. *See* ANTIOXIDANT; CAROTENOID; COENZYME; NUTRITION. Lawrence J. Machlin

Vitamin A A pale-yellow alcohol, soluble in fat but not in water. In pure form, it is readily destroyed by oxidation and light, which may cause losses during storage. Vitamin A is found in all animal tissues, although it is particularly concentrated in the liver. There are two different dietary sources for the vitamin: animal sources which contain vitamin A itself, mostly in the form of retinyl esters, and plant sources which contain carotenoids that are converted to vitamin A in animal tissues such as the absorptive cells in the intestine. The most vitamin A–enriched animal food source is fish liver oil. Plant carotenoids are found in green and yellow fruits and vegetables such as carrots, apricots, asparagus, broccoli, and green leafy vegetables. *See* CAROTENOID.

In vitamin A deficiency, the epithelial tissues of many organs are affected. Growth failure occurs, and young animals can suffer from neurological symptoms resulting from pressures on the central nervous system. Vitamin A deficiency is also strongly associated with depressed immune function and higher morbidity and mortality due to infectious diseases such as diarrhea, measles, and respiratory infections. A severe manifestation of vitamin A deficiency is night blindness and inflammation of the eyes (xerophthalmia), followed by irreversible blindness.

The symptoms seen in vitamin A deficiency reflect the multiple roles of this compound in animals. These roles are fulfilled by two compounds that are synthesized from vitamin A in the body: vitamin A aldehyde (retinaldehyde), which is critical for vision, and vitamin A acid (retinoic acid), which controls many physiological functions in both the embryo and the adult. *See* VISION.

Studies of many mammalian species suggested that approximately 20 IU (6 μg) of vitamin A per kilogram of body weight will support growth and prevent symptoms of deficiency. The current intake recommendations of vitamin A in the United States is 3 mg/day and about 1 mg/day in the European Union. *See* VITAMIN. Stanley N. Gershoff; Noa Noy

Vitamin B_6 A vitamin which exists as three chemically related and water-soluble forms found in food; pyridoxine, pyridoxal, and pyridoxamine. All three forms have equal activity for animals and yeast.

Vitamin B_6 deficiency is accompanied by poor growth, dermatitis, microcytic anemia, epileptiform convulsions, and kidney and adrenal lesions. There is evidence that some women in the third trimester of pregnancy may have a special requirement for vitamin B_6 in that its administration often relieves the nausea of pregnancy. Some types of human dermatitis respond to local application of this vitamin.

It is difficult to set requirements for vitamin B_6, since no single set of assay conditions or criteria has received universal acceptance. Adults probably require about 1.5–2 milligrams per day, and a dietary intake of 0.4 milligrams per day would probably be satisfactory for most infants. *See* VITAMIN. Stanley N. Gershoff

Vitamin B_{12} A group of closely related polypyrrole compounds, with the structure shown, containing trivalent cobalt; it is often called cobalamin, The vitamin is a dark-red crystalline compound; in aqueous solution and at room temperature it is most stable at pH 4–7.

Vitamin B_{12} deficiency in animals is characterized primarily by anemia and neuropathy. In humans, this deficiency is called pernicious anemia. People suffering from this disease lack a factor secreted in gastric juice which, by affecting absorption directly and by protecting vitamin B_{12} from intestinal destruction, enables the vitamin to be absorbed. *See* ANEMIA.

Requirements for vitamin B_{12} are increased by reproduction or hyperthyroidism. Of the known vitamins B_{12} is the most active biologically. A daily injection of 1 microgram of vitamin B_{12} will prevent the recurrence of symptoms in people with pernicious anemia. For normal people a diet containing 3–5 μg per day (providing 1–1.5 μg absorbed) will satisfy vitamin B_{12} requirements. *See* VITAMIN. Stanley N. Gershoff

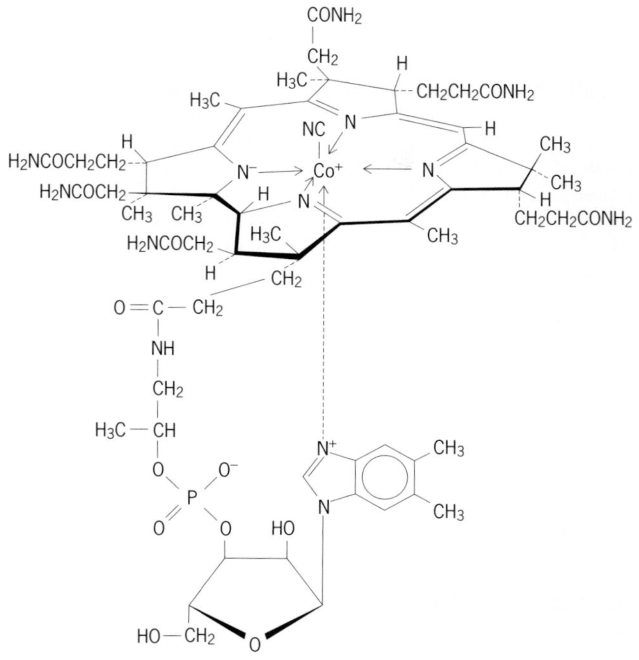

Vitamin D Either of two fat-soluble sterol-like compounds, ergocalciferol (vitamin D_2) and activated cholecalciferol (vitamin D_3). Vitamin D_2 is formed from the irradiation of ergosterol, a plant sterol. However, vitamin D_3 is normally manufactured in the skin, where ultraviolet light activates the compound 7-dehydrocholesterol. Vitamins D_3 and D_2 are about equal in activity in all mammals except New World monkeys and birds, in which vitamin D_2 is approximately one-tenth as active as vitamin D_3. *See* VITAMIN.

Vitamin D as acquired from the diet or produced in the skin is biologically inactive. It must be metabolized by the liver to produce 25-hydroxyvitamin D_3. However, this compound is also biologically inactive under physiological circumstances and must be activated by the kidney to produce the final vitamin D hormone, 1,25-dihydroxyvitamin D_3. This hormonal form of vitamin D plays an essential role in stimulating intestinal absorption of calcium and phosphorus, in the mobilization of calcium from bone, and in renal reabsorption of calcium. The function of vitamin D has been expanded beyond regulating plasma calcium and phosphorus levels, and hence healing the diseases of rickets and osteomalacia. It is now known that the vitamin D hormone controls parathyroid gland growth and production of the parathyroid hormone. It is an immunomodulator. Vitamin D hormone also appears to play a role in the regulation of insulin production or secretion. Finally, it is required for female reproduction. These new sites of action of vitamin D are under intense investigation. *See* HORMONE.

Vitamin D is largely absent from the food supply. It is found in large amounts in fish liver oils; cod liver oil has long been known to be an important source of vitamin D. Fortified foods are the major dietary source of vitamin D, but the major overall source is the production of vitamin D in skin by exposure to sunlight or ultraviolet irradiation. In winter months at temperate latitudes, insufficient amounts of vitamin D are produced in skin, and unless it is replaced by a dietary source, danger of insufficiency exists.

A deficiency of vitamin D in growing animals results in the disease rickets. A similar disease, osteomalacia, occurs in adult animals. By far the most serious disorder of vitamin D deficiency is the low-blood calcium levels which result in convulsions known as hypocalcemic tetany. Moderate deficiency of vitamin D may contribute to osteoporosis, especially in the elderly. *See* BONE; OSTEOPOROSIS.

The recommended daily requirement for vitamin D_3 is 10 micrograms or 400 international units (IU). Higher requirements are reported for the elderly and for rapidly growing adolescents: 20 μg or 800 IU per day. It is possible that the average requirement is lower than 10 μg per day. The exact absolute requirement has never been determined. Hector F. Deluca

Vitamin E A group of compounds, α, β, γ, and δ tocopherols, that have a chromanol ring and phytyl side chain and are widely distributed in nature, especially in edible vegetable oils (wheat germ, sunflower, cottonseed, safflower, canola, soybean, and corn oil). Unprocessed grains, nuts, and vegetables are other sources. When used as supplements, there are two vitamin E products available: natural source RRR α-tocopherol and synthetic all-rac-α-tocopherol. The latter is a mixture of eight different stereoisomers, of which only one is RRR α-tocopherol. *See* VITAMIN.

The terms "vitamin E" and "α-tocopherol" are frequently used interchangeably in human nutrition, but it is imperative to distinguish between supplements of RRR α-tocopherol and of synthetic α-tocopherol because their biological activity is different. Since the major function of vitamin E is to serve as a chain-breaking antioxidant, protecting cell membranes against free-radical damage, the most potent form of the vitamin should be used as a supplement. Although gastrointestinal absorption of all forms of vitamin E is equivalent, the subsequent physiological steps are sharply in favor of the RRR form. This action is mediated by a cellular liver transfer protein that is specific for the RRR form of α-tocopherol. It maintains the plasma level by selectively choosing the RRR form and recycling it into plasma lipoproteins for distribution of the vitamin to every tissue and organ in the body.

When deficiencies of the vitamin occur in humans as a consequence of acquired malabsorption or genetic abnormalities of lipoproteins or of the transfer protein, the major symptoms that develop are in the nervous system. Ataxia (lack of muscular coordination) and other neurologic symptoms result in severe incoordination and subsequent musculoskeletal changes.

The recommended daily allowance for vitamin E is 15 milligrams, which is present in the usual Western diet. Herbert J. Kayden

Vitamin K A group of compounds derived from 2-methyl-1,4-naphthoquinone that prevent bleeding in mammals and birds. Vitamin K_1 (phylloquinone) is produced by green plants; a related form, vitamin K_2 (menaquinone), is produced by intestinal bacteria. Chemically synthesized forms include vitamin K_1, K_2, and menadione (vitamin K_3). All of these compounds are fat-soluble liquids at room temperature that become biologically inactive when exposed to light or alkali. *See* VITAMIN.

Vitamin K_1 (phylloquinone) is a photosynthesis cofactor in plants, and various forms of vitamin K_2 (menaquinones) participate in energy transfer reactions in bacteria. Mammals and many other animals need vitamin K hydroquinone (the active form) as a cofactor for the specific synthesis of the amino acid γ-carboxyglutamate (Gla) in certain proteins, which enable the proteins to bind calcium and phospholipids with high specificity. Gla is an essential part of coagulation factors and of other proteins that regulate blood clotting (hence the disruption of blood clotting and internal bleeding associated with vitamin K deficiency). *See* AMINO ACIDS; BLOOD; COENZYME; PROTEIN.

Humans depend on continuous vitamin K supplies, since storage is minimal. Good dietary sources of vitamin K_1 are green vegetables and fruits; certain fermented Asian foods, especially natto, have a high vitamin K_2 content.

Intestinal bacteria produce vitamin K_2, and under most circumstances enough is absorbed to prevent bleeding. However, spontaneous bleeding occurs if both dietary intake and production by intestinal bacteria are persistently low. Other health risks related to inadequate vitamin K intake may include accelerated loss of bone minerals and hardening of arteries. Martin Kohlmeier

Vivianite A mineral of the vivianite group, other important members of which are annabergite and erythrite.

Vivianite is a hydrated ferrous phosphate, $Fe_3(PO_4)_2 \cdot 8H_2O$; usually ferric iron is present as the result of oxidation. It crystallizes in the monoclinic system, with crystals generally prismatic. Vivianite also occurs in earthy form and as globular and encrusting masses of fibrous structure. Crystals are colorless and transparent when fresh. Oxidation changes the color progressively to pale blue, greenish blue, dark blue, or bluish black. Wayne R. Lowell

Vocal cords The pair of elastic, fibered bands inside the human larynx. The cords are covered with a mucous membrane and pass horizontally backward from the thyroid cartilage (Adam's apple) to insert on the smaller, paired arytenoid cartilages at the back of the larynx. The vocal cords act as sphincters for air regulation and may be vibrated to produce sounds. Separation, approximation, and alteration of tension are produced by action of laryngeal muscles acting on the pivoting arytenoids. Vibration of the cords produces fundamental sounds and overtones. These can be modified by the strength of the air current, the size and shape of the glottis (the opening between the cords), and tension in the cords. *See* LARYNX; SPEECH. Walter Bock

Voice over IP A technology that transports voice using data packets instead of circuit switching, which the traditional public switched telephone network (PSTN) uses. Voice over IP (VoIP), using packet technology, allows for more efficient transport of voice while providing the quality of service and reliability of PSTN.

Comparison of VoIP and PSTN. The present public switched telephone network transfers voice by converting speech received from the telephone into 64-kbps (kilobits per second) digital data and transporting it in a timeslot (limited time interval) that is periodically inserted in a higher-capacity signal, a procedure referred to as time-division multiplexing (TDM). Two timeslots, one for each direction, are allocated for each phone call. These timeslots are set up by the signaling function of the public switched telephone network and kept in use for the duration of the call. The timeslots are switched in and out of use based on the calls taking place at any point in time, and are referred to as being circuit-switched. *See* MULTIPLEXING AND MULTIPLE ACCESS; TELEPHONE SERVICE.

By contrast, voice over IP uses IP packets to carry speech over a data network. Only those packets that contain speech need to be transported, thereby allowing voice over IP to improve bandwidth efficiency by transporting packets only in the direction of the call participant who is listening. To further improve efficiency, voice over IP uses speech-compression algorithms to reduce speech from 64 to 2.4–8 kbps. *See* DATA COMMUNICATIONS; DATA COMPRESSION; PACKET SWITCHING.

The IP network does not allocate specific timeslots for a particular user. This allows the IP network to take advantage of silent periods of a normal call by not sending any packets, further increasing its efficiency.

While the public switched telephone network was developed for voice and backfitted for data, IP was developed for data and is being backfitted for voice. The ubiquity of IP allows the convergence of new services dependent on voice and data. Voice over IP is being expanded to support applications such as videophone, video conferencing, and whiteboard conferencing. *See* TELECONFERENCING; VIDEOTELEPHONY.

VoIP-PSTN interoperation. The public switched telephone network can communicate with the IP network, transforming an IP call to a public switched telephone network call, and vice versa. The packet voice gateway or media gateway converts time-division multiplexing used on the public switched telephone network to and from voice being carried in real-time protocol (RTP) packets in the Internet network. *See* INTERNET.

Security. The public switched telephone network is relatively secure, since the network resources are dedicated for the duration of a call. Packet networks or the Internet are much less secure with hackers being able to attack IP networks. In order to build reliability into a voice over IP network, service providers must be able to increase its security level to the level of the public switched telephone network. Numerous tools have been developed for helping to secure IP networks such as firewalls, intrusion detection and prevention elements, and virus scanners. IP Security (IPSec), Secure Sockets Layer (SSL), and voice over IP–specific protocols are also being developed to increase IP network security. Daniel Heer

Voice response The generation of synthetic speech signals in order to convey information to listeners, usually based upon a verbal or textual request by the users. This speech synthesis typically employs a computer program and requires access to storage of portions of speech previously spoken by humans. The naturalness of the synthetic voice depends on several factors, including the vocabulary of words to pronounce, the amount of stored speech, and the complexity of the synthesis programs. The most basic voice response simply plays back appropriate short verbal responses, which are only copies of human speech signals stored using digital sampling technology. The most universal systems are capable of transforming any given text into comprehensible speech for a given language. These latter systems so far exist for only 20 or so of the world's major languages, and are flawed in producing speech that, while usually intelligible, sounds unnatural.

Voice response is also known as text-to-speech synthesis (TTS) because the task usually has as input a textual message (to be spoken by the machine). The text could be in tabular form (for example, reading aloud a set of numbers), or, more typically, formatted as normal sentences. Speech synthesizers are much more flexible and universal than their speech-recognition counterparts, for which human talkers must significantly constrain their verbal input to the machines in order to achieve accurate recognition. In TTS, a computer database usually determines the text to be synthetically spoken, following an automatic analysis of each user request. The user may pose the request in response to a menu of inquiries (for example, by an automated telephone dialogue, by pushing a sequence of handset keys, or by a series of brief verbal responses). Thus, the term "voice response" is used to describe the synthetic speech as an output to a user inquiry. The value of such a synthetic voice is the capability of efficiently receiving information from a computer without needing a computer screen or printer. Given the prevalence of telephones, as well as the difficulty of reading small computer screens on many portable computer devices, voice response is a convenient way to get data. *See* SPEECH RECOGNITION.

The simplest approach to voice response is to digitally sample natural speech and output the samples later as needed. A common Nyquist sampling rate is 10,000 samples per second, which preserves sound frequencies up to almost 5 kHz, allowing quite natural speech. High-frequency energy in fricative sounds is severely attenuated (but less so than on telephone lines), but this usually has little impact on intelligibility. Straightforward sampling requires 12 bits per sample, which requires memory at 120 kbits/s. Such high data rates are prohibitive except for

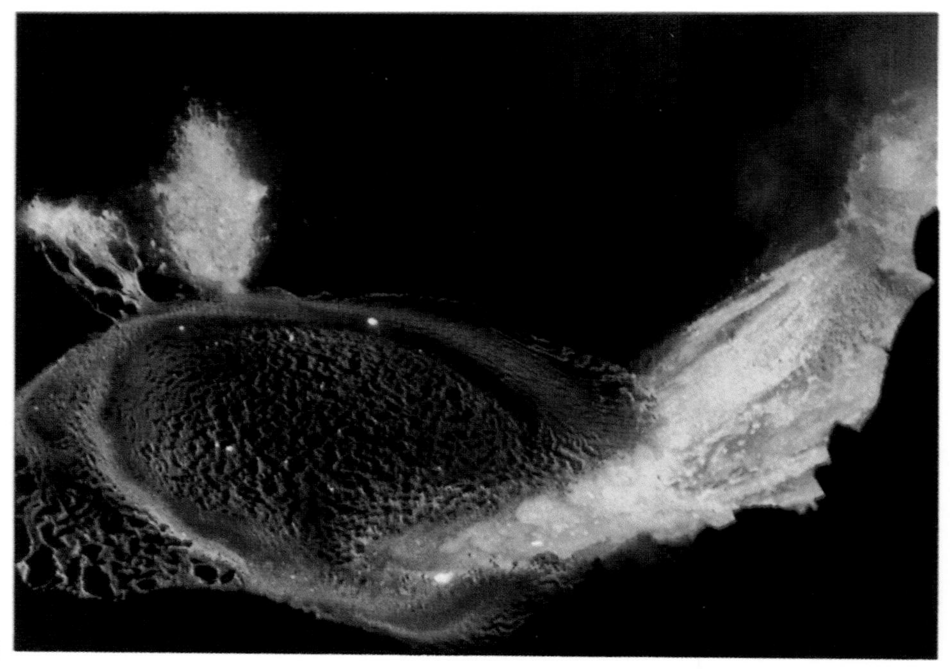

(a)

Volcanic activity. *(a)* Lava cascade from a nearby vent pouring into a crater. *(b)* View through skylights (collapsed roof of lava tube) of underground lava rivers. *(c)* Lava flowing down a steep wall. *(d)* Lava fountains from a central vent. *(e)* View through skylights of lava stalactites on roof of active lava tubes. *(f)* Curtain of fire: lines of fountains from fissures. *(g)* Lava cascading down crater wall. *(h)* Surface flow (left to right) of a lava lake. *(Part f courtesy of Robin T. Holcomb, USGS; other parts, R. I. Tilling, USGS)*

(b) 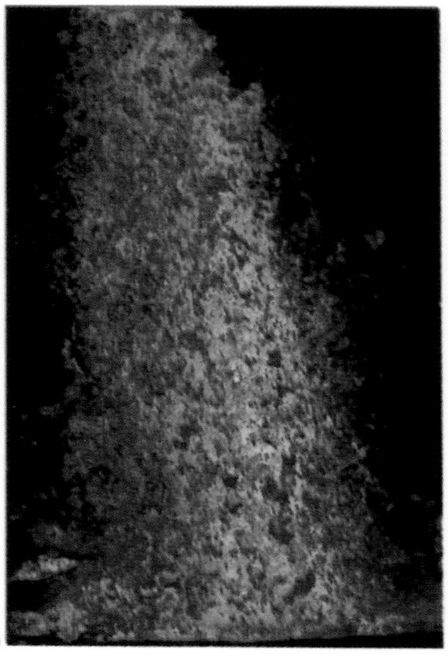 (d) 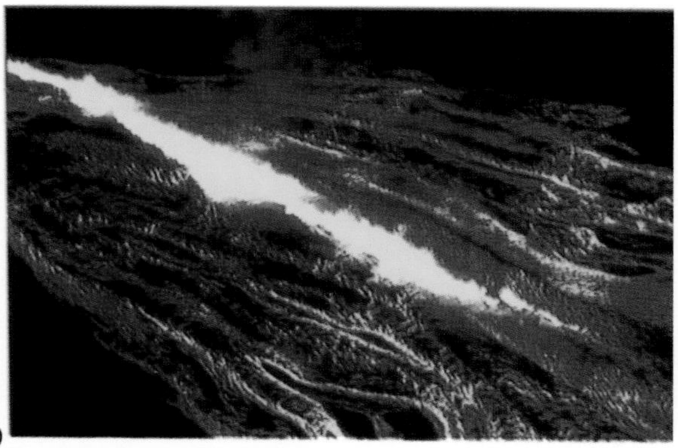 (f) 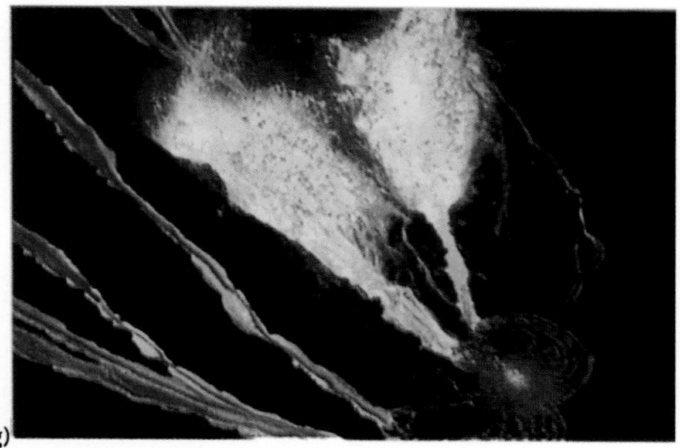 (g)  (c) 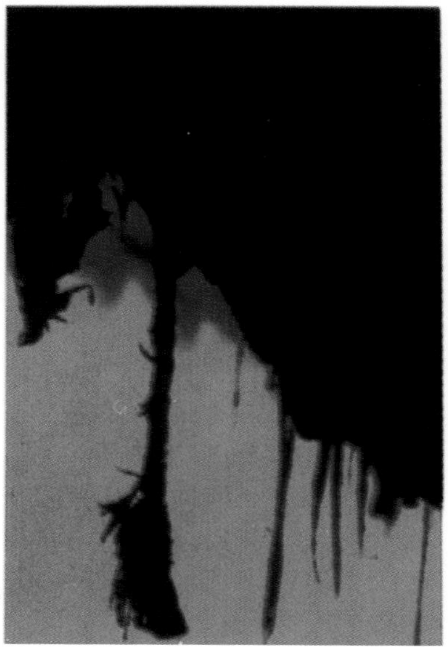 (e) 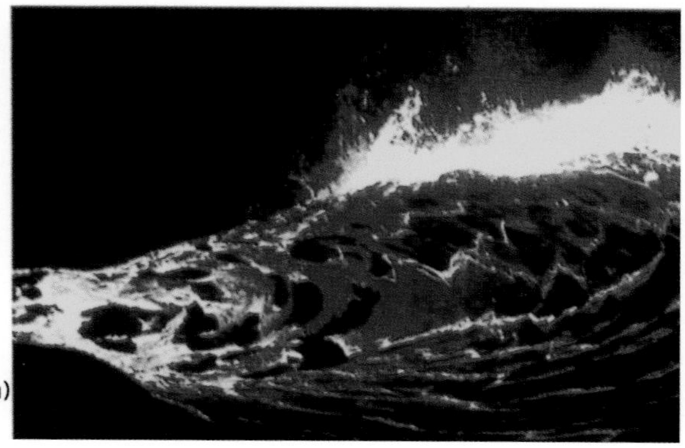 (h)

applications with very small vocabularies. Even in cases with more limited bandwidth (for example, 8000 samples per second in telephone applications) and more advanced coding schemes, the straightforward playback approach is unacceptable for general TTS. Despite rapidly decreasing costs for computer memory, it will remain impossible to store all the necessary speech signals except for applications with very restricted vocabulary needs. *See* COMPACT DISK; DATA COMPRESSION; INFORMATION THEORY; PULSE MODULATION.

A voice response system which minimizes memory needs generates synthetic speech from sequences of brief basic sounds and has great flexibility. Since most languages have only 30–40 phonemes (distinct linguistic sounds), storing units of such size and number is trivial. However, the spectral features of these short concatenated sounds (lasting 50–200 ms) must be adjusted at their frequent boundaries to avoid severely discontinuous speech. Normal pronunciation of each phoneme in an utterance depends heavily on its phonetic context (for example, on neighboring phonemes, intonation, and speaking rate). The adjustment process and the need to calculate an appropriate intonation for each context lead to complicated synthesizers with correspondingly less natural output speech.

Current synthesizers usually compromise between the extremes of minimizing storage and complexity. One approach is to store thousands of speech units of varying size, which can be automatically extracted from natural speech. In contrast to automatic speech recognition, where segmentation of speech into pertinent units is very difficult, TTS training exploits prior knowledge of the text (the training speaker reads a furnished text).

Synthesizers that accept general text as input need a linguistic processor to convert the text into phonetic symbols in order to access the appropriate stored speech units. One task is to convert letters into phonemes. This may be as simple as a table look-up: a computer dictionary with an entry for each word in the chosen language, noting its pronunciation (including syllable stress), syntactic category, and possibly some semantic information. Many systems also have language-dependent rules, which examine the context of each letter in a word to determine how it is pronounced; for example, the letter [p] in English is pronounced /p/, except before the letter [h] (for example, in "telephone"; however, it has normal pronunciation in "cupholder"). English needs hundreds of such rules. TTS often employs these rules as a backup procedure to handle new words, foreign words, and typographical mistakes (that is, cases not in the dictionary). *See* PHONETICS.

The problem of determining an appropriate intonation for each input text continues to confound TTS. In simple voice response, the stored units are large (for example, phrases), and pitch and intensity are usually stored explicitly with the spectral parameters or implicitly in the signals of waveform synthesizers. However, when smaller units are concatenated, the synthetic speech sounds unnatural unless the intonation is adjusted for context. Intonation varies significantly among languages. Although automatic statistical methods show some promise, intonation analysis has mostly been manual.

Simple voice-response systems work equally well for all languages since they just play back previously stored speech units. For general TTS, however, major synthesizer components are highly language-dependent. The front end of TTS systems, dealing with letter-to-phoneme rules, the relationship between text and intonation, and different sets of phonemes, is language-dependent. The back end, representing simulation of the vocal track via digital filters, is relatively invariant across languages. Even languages with sounds (for example, clicks) other than the usual pulmonic egressives require only simple modifications.

Commercial synthesizers are widely available for about 10 languages. They often combine software, memory, and processing chips, and range from expensive systems providing close-to-natural speech to inexpensive personal computer programs. General digital signal processing chips are widely used for TTS. Current microprocessors can easily handle the speeds for synthesis, and indeed synthesizers exist entirely in software. Memory requirements can still be a concern, especially for some of the newer waveform concatenation systems. *See* MICROPROCESSOR; SPEECH.

Douglas O'Shaughnessy

Volatilization The process of converting a chemical substance from a liquid or solid state to a gaseous or vapor state. Other terms used to describe the same process are vaporization, distillation, and sublimation. A substance can often be separated from another by volatilization and can then be recovered by condensation of the vapor. The substance can be made to volatilize more rapidly either by heating to increase its vapor pressure or by removal of the vapor using a stream of inert gas or a vacuum pump. Chemical reactions are sometimes utilized to produce volatile products. Volatilization methods are generally characterized by great simplicity and ease of operation, except when high temperatures or highly corrosion-resistant materials are needed. *See* CHEMICAL SEPARATION TECHNIQUES; DISTILLATION; SUBLIMATION; VAPOR PRESSURE.

Louis Gordon; Royce W. Murray

Volcanic glass A natural glass formed by rapid cooling of magma. Magmas typically comprise crystals and bubbles of gas within a silicate liquid. On slow cooling, the liquid portion of the magma usually crystallizes, but if cooling is sufficiently rapid, it may convert to glass—an amorphous, metastable solid that lacks the long-range microscopic order characteristic of crystalline solids. *See* LAVA; MAGMA.

Silica-rich, rhyolitic magmas frequently quench to glass during explosive eruptions and make up the bulk of the solid material in many pyroclastic deposits (usually as shards, pumice lumps, and other fragments); but they also can erupt quiescently to form massive glassy rocks (known as obsidian, the most common source of volcanic glass on land) even in the slowly cooled interiors of flows tens of meters thick. In contrast, more basic, basaltic glasses (sometimes known as tachylite) are less common and rarely form in more than small quantities unless rapidly cooled in a volcanic eruption. Pele's hair is an example of basaltic glass formed in this way. *See* BASALT; OBSIDIAN; RHYOLITE.

Edward M. Stolper

Volcano A mountain or hill, generally steep-sided, formed by accumulation of magma (molten rock with associated gas and crystals) erupted through openings or volcanic vents in the Earth's crust; the term volcano also refers to the vent itself. During the evolution of a long-lived volcano, a permanent shift in the locus of principal vent activity can produce a satellitic volcanic accumulation as large as or larger than the parent volcano, in effect forming a new volcano on the flanks of the old.

Planetary exploration has revealed dramatic evidence of volcanoes and their products on the Earth's Moon, Mars, Mercury, Venus, and the moons of Jupiter, Neptune, and Uranus on a scale much more vast than on Earth. However, only the products and landforms of terrestrial volcanic activity are described here. *See* MARS; MERCURY (PLANET); MOON; NEPTUNE; URANUS; VENUS; VOLCANOLOGY.

Volcanic vents. Volcanic vents, channelways for magma to ascend toward the surface, can be grouped into two general types: fissure and central (pipelike). Magma consolidating below the surface in fissures or pipes forms a variety of igneous bodies, but magma breaking the surface produces fissure or pipe eruptions. Fissures, most of them less than 10 ft (3 m) wide, may form in the summit region of a volcano, on its flanks, or near its base; central vents tend to be restricted to the summit area of a volcano. For some volcanoes or volcanic regions, swarms of fissure vents are clustered in swaths called rift zones.

Volcanic products. Magma erupted onto the Earth's surface is called lava. If the lava is chilled and solidifies quickly, it forms volcanic glass; slower rates of chilling result in greater crystallization before complete solidification. Lava may accrete near the vent to form various minor structures or may pour out in streams called lava flows, which may travel many tens of miles from the vents. During more violent eruption, lava torn into fragments and hurled into the air is called pyroclastic (fire-broken materials). *See* CRYSTALLIZATION; LAVA; MAGMA; PYROCLASTIC ROCKS; VOLCANIC GLASS.

Volcanic gases. Violent volcanic explosions may throw dust and aerosols high into the stratosphere, where it may drift across the surface of the globe for many thousands of miles. Most of the solid particles in the volcanic cloud settle out within a few days, and nearly all settle out within a few weeks, but the gaseous aerosols (principally sulfuric acid droplets) may remain suspended in the stratosphere for several years. Such stratospheric clouds of volcanic aerosols, if sufficiently voluminous and long-lived, can have an impact on global climate. *See* ACID RAIN; AEROSOL; AIR POLLUTION.

In general, water vapor is the most abundant constituent in volcanic gases; the water is mostly of meteoric (atmospheric) origin, but in some volcanoes can have a significant magmatic or juvenile component. Excluding water vapor, the most abundant gases are the various species of carbon, sulfur, hydrogen, chlorine, and fluorine.

Mudflows are common on steep-side volcanoes where poorly indurated or nonwelded pyroclastic material is abundant. Probably by far the most common cause, however, is simply heavy rain saturating a thick cover of loose, unstable pyroclastic material on the steep slope of the volcano, transforming the material into a mobile, water-saturated "mud," which can rush downslope at a speed as great as 50–55 mi (80–90 km) per hour. Such a dense, fast-moving mass can be highly destructive, sweeping up everything loose in its path.

Volcanic landforms. Much of the Earth's solid surface, on land and below the sea, has been shaped by volcanic activity. Landscape features of volcanic origin may be either positive (constructional) forms, the result of accumulation of volcanic materials, or negative forms, the result of the lack of accumulation or collapse.

Not all volcanoes show a graceful, symmetrical cone shape, such as that exemplified by Mount Fuji, Japan. Most volcanoes, especially those near tectonic plate boundaries, are more irregular, though of grossly conical shape. Such volcanoes, called stratovolcanoes or composite volcanoes, typically erupt explosively and are composed dominantly of andesitic, relatively viscous and short lava flows, interlayered with beds of ash and cinder that thin away from the principal vents. Volcanoes constructed primarily of fluid basaltic lava flows, which may spread great distances from the vents, typically are gentle-sloped, broadly upward convex structures. Such shield volcanoes, classic examples of which are Mauna Loa volcano, Hawaii, tend to form in oceanic intraplate regions and are associated with hot-spot volcanism. The shape and size of a volcano can vary widely between the simple forms of composite and shield volcanoes, depending on magma viscosity, eruptive style (explosive versus nonexplosive), migration of vent locations, duration and complexity of eruptive history, and posteruption modifications.

Some of the largest volcanic edifices are not shaped like the composite or shield volcanoes. In certain regions of the world, voluminous extrusions of very fluid basaltic lava from dispersed fissure swarms have built broad, nearly flat-topped accumulations. These voluminous outpourings of lava are known as flood basalts or plateau basalts. *See* BASALT.

Submarine volcanism. Deep submarine volcanism occurs along the spreading ridges that zigzag for thousands of miles across the ocean floor, and it is exposed above sea level only in Iceland. Because of the logistical difficulties in making direct observations posed by the great ocean depths, no deep submarine volcanic activity has been actually observed during eruption. However, evidence that deep-sea eruptions are happening is clearly indicated by (1) seismic and acoustic monitoring networks; (2) the presence of deep-ocean floor hydrothermal vents; (3) episodic hydrothermal discharges, measured and mapped as thermal and geochemical anomalies in the ocean water; and (4) the detection of new lava flows in certain segments of the oceanic ridge system. *See* HYDROTHERMAL VENT; MID-OCEANIC RIDGE.

Volcanic eruptions in shallow water are very similar in character to those on land but, on average, are probably somewhat more explosive, owing to heating of water and resultant violent generation of supercritical steam. Much of the ocean basin appears to be floored by basaltic lava. *See* OCEANIC ISLANDS.

Fumaroles and hot springs. Vents at which volcanic gases issue without lava or after the eruption are known as fumaroles. They are found on active volcanoes during and between eruptions and on dormant volcanoes, persisting long after the volcano itself has become inactive. Fumaroles grade into hot springs and geysers. The water of most, if not all, hot springs is predominantly of meteoric origin, and is not water liberated from magma. Some hot springs are of volcanic origin and the water may contain volcanic gases. *See* GEYSER.

Distribution of volcanoes. Over 500 active volcanoes are known on the Earth, mostly along or near the boundaries of the dozen or so lithospheric plates that compose the Earth's solid surface. Lithospheric plates show three distinct types of boundaries: divergent or spreading margins—adjacent plates are pulling apart; convergent margins (subduction zones)—plates are moving toward each other and one is being destroyed; and transform margins—one plate is sliding horizontally past another. All these types of plate motion are well demonstrated in the Circum-Pacific region, in which many active volcanoes form the so-called Ring of Fire. Some volcanoes, however, are not associated with plate boundaries, and many of these so-called intraplate volcanoes form roughly linear chains in the interior parts of the oceanic plates, for example, the Hawaiian-Emperor, Austral, Society, and Line archipelagoes in the Pacific Basin. Intraplate volcanism also has resulted in voluminous outpourings of fluid lava to form extensive plateau basalts, or of more viscous and siliceous pyroclastic products to form ash flow plains.

Robert I. Tilling

Volcanology The scientific study of volcanic phenomena, especially the processes, products, and hazards associated with active or potentially active volcanoes. It focuses on eruptive activity that has occurred within the past 10,000 years of the Earth's history, particularly eruptions during recorded history. Strictly speaking, it emphasizes the surface eruption of magmas and related gases, and the structures, deposits, and other effects produced thereby. Broadly speaking, however, volcanology includes all studies germane to the generation, storage, and transport of magma, because the surface eruption of magma represents the culmination of diverse physicochemical processes at depth. This article considers the activity of erupting volcanoes and the nature of erupting lavas. For a discussion of the distribution of volcanoes and the surface structures and deposits produced by them, *see* PLATE TECTONICS; VOLCANO

On average, about 50 to 60 volcanoes worldwide are active each year. About half of these constitute continuing activity that began the previous year, and the remainder are new eruptions. Analysis of historic records indicates that eruptions comparable in size to that of Mount St. Helens or El Chichón tend to occur about once or twice per decade, and larger eruptions such as Pinatubo about once per one or two centuries. On a global basis, eruptions the size of that at Nevado del Ruiz in November 1985 are orders of magnitude more frequent.

Generalized relationships between magma composition, relative viscosity, and common eruptive characteristics

Magma composition	Relative viscosity	Common eruptive characteristics
Basaltic	Fluidal	Lava fountains, flows, and pools
Andesitic	Less fluidal	Lava flows, explosive ejecta, ashfalls, and pyroclastic flows
Dacitic-rhyolitic	Viscous	Explosive ejecta, ashfalls, pyroclastic flows, and lava domes

Modern volcanology perhaps began with the founding of well-instrumented observations at Asama Volcano (Japan) in 1911 and at Kilauea Volcano (Hawaii) in 1912. The Hawaiian Volcano Observatory, located on Kilauea's caldera rim, began to conduct systematic and continuous monitoring of seismic activity preceding, accompanying, and following eruptions, as well as other geological, geophysical, and geochemical observations and investigations.

The eruptive characteristics, products, and resulting landforms of a volcano are determined predominantly by the composition and physical properties of the magmas involved in the volcanic processes (see table). Formed by partial melting of existing solid rock in the Earth's lower crust or upper mantle, the discrete blebs of magma consist of liquid rock (silicate melt) and dissolved gases. Driven by buoyancy, the magma blebs, which are lighter than the surrounding rock, coalesce as they rise toward the surface to form larger masses. *See* IGNEOUS ROCKS; LITHOSPHERE; MAGMA.

Magma consists of three phases: liquid, solid, and gas. Volcanic gases generally are predominantly water; other gases include various compounds of carbon, sulfur, hydrogen, chlorine, and fluorine. All volcanic gases also contain minor amounts of nitrogen, argon, and other inert gases, largely the result of atmospheric contamination at or near the surface.

Temperatures of erupting magmas have been measured in lava flows and lakes, pyroclastic deposits, and volcanic vents by means of infrared sensors, optical pyrometers, and thermocouples. Reasonably good and consistent measurements have been obtained for basaltic magmas erupted from Kilauea and Mauna Loa volcanoes, Hawaii, and a few other volcanoes. Measured temperatures typically range between 2100 and 2200°F (1150 and 1200°C), and many measurements in cooling Hawaiian lava lakes indicate that the basalt becomes completely solid at about 1800°F (980°C). *See* GEOLOGIC THERMOMETRY.

The character of a volcanic eruption is determined largely by the viscosity of the liquid phase of the erupting magma and the abundance and condition of the gas it contains. Viscosity is in turn affected by such factors as the chemical composition and temperature of the liquid, the load of suspended solid crystals and xenoliths, the abundance of gas, and the degree of vesiculation. The subsequent violent expansion during eruption shreds the frothy liquid into tiny fragments, generating explosive showers of volcanic ash and dust, accompanied by some larger blocks (volcanic "bombs"); or it may produce an outpouring of a fluidized slurry of gas, semisolid bits of magma froth, and entrained blocks to form high-velocity pyroclastic flows, surges, and glowing avalanches. *See* PYROCLASTIC ROCKS; VISCOSITY.

Types of eruptions customarily are designated by the name of a volcano or volcanic area that is characterized by that sort of activity, even though all volcanoes show different modes of eruptive activity on occasion and even at different times during a single eruption.

Eruptions of the most fluid lava, in which relatively small amounts of gas escape freely with little explosion, are designated Hawaiian eruptions. Most of the lava is extruded as successive, thin flows that travel many miles from their vents. An occasional feature of Hawaiian activity is the lava lake, a pool of liquid lava with convectional circulation that occupies a preexisting shallow depression or pit crater. *See* LAVA.

Strombolian eruptions are somewhat more explosive eruptions of lava, with greater viscosity, and produce a larger proportion of pyroclastic material. Many of the volcanic bombs and lapilli assume rounded or drawn-out forms during flight, but commonly are sufficiently solid to retain these shapes on impact.

Generally still more explosive are the vulcanian type of eruptions. Angular blocks of viscous or solid lava are hurled out, commonly accompanied by voluminous clouds of ash but with little or no lava flow.

Peléean eruptions are characterized by the heaping up of viscous lava over and around the vent to form a steep-sided hill or volcanic dome. Explosions, or collapses of portions of the dome, may result in glowing avalanches (nuées ardentes).

Plinian eruptions are paroxysmal eruptions of great violence—named after Pliny the Elder, who was killed in A.D. 79 while observing the eruption of Vesuvius—and are characterized by voluminous explosive ejections of pumice and by ash flows. The copious expulsion of viscous siliceous magma commonly is accompanied by collapse of the summit of the volcano, forming a caldera, or by collapse of the broader region, forming a volcano-tectonic depression. *See* CALDERA.

A major component of the science of volcanology is the systematic and, preferably, continuous monitoring of active and potentially active volcanoes. Scientific observations and measurements—of the visible and invisible changes in a volcano and its surroundings—between eruptions are as important, perhaps even more crucial, than during eruptions. Measurable phenomena important in volcano monitoring include earthquakes; ground movements; variations in gas compositions; and deviations in local gravity, electrical, and magnetic fields. These phenomena reflect pressure and stresses induced by subsurface magma movements and or pressurization of the hydrothermal envelope surrounding the magma reservoir. The monitoring of volcanic seismicity and ground deformations before, during, and following eruptions has provided the most useful and reliable information. *See* EARTHQUAKE; SEISMOLOGY.

Volcanoes are in effect windows into the Earth's interior; thus research in volcanology, in contributing to an improved understanding of volcanic phenomena, provides special insights into the chemical and physical processes operative at depth. However, volcanology also serves an immediate role in the mitigation of volcanic and related hydrologic hazards (mudflows, floods, and so on). Progress toward hazards mitigation can best be advanced by a combined approach. One aspect is the preparation of comprehensive volcanic hazards assessments of all active and potentially active volcanoes, including a volcanic risk map for use by government officials in regional and local land-use planning to avoid high-density development in high-risk areas. The other component involves improvement of predictive capability by upgrading volcano-monitoring methods and facilities to adequately study more of the most dangerous volcanoes. An improved capability for eruption forecasts and predictions would permit timely warnings of impending activity, and give emergency-response officials more lead time for preparation of contingency plans and orderly evacuation, if necessary. Robert I. Tilling

Volt-ampere The apparent-power index of drive level for sinusoidal alternating-current loads. A circuit branch with E volts

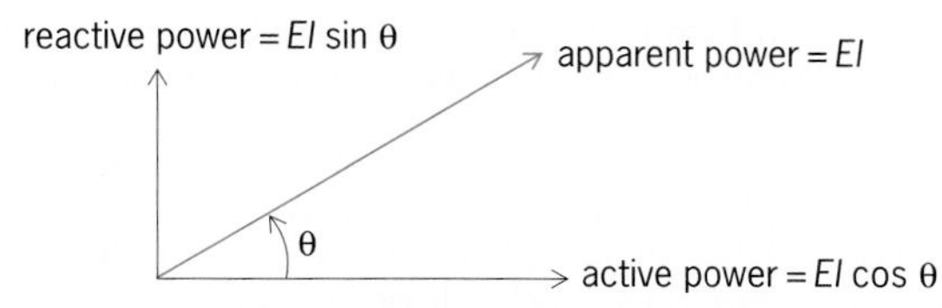

Apparent, active, and reactive power.

across its two terminals, carrying *I* amperes from one to the other, is said to be receiving *EI* volt-amperes of apparent power, whatever may be the phase lag θ of current behind voltage. If such a load is driven through a transformer having negligible losses, the volt-amperes into the transformer primary is the same number as the volt-amperes into the load.

Practical computation of real power, reactive power, and apparent power is usually done with complex-number algebra using the geometrical diagram of the illustration. Mark G. Foster

Voltage amplifier An electronic circuit whose function is to accept an input voltage and produce a magnified, accurate replica of this voltage as an output voltage. The voltage gain of the amplifier is the amplitude ratio of the output voltage to the input voltage. Often, electronic amplifiers designed to operate in different environments are categorized by criteria other than their voltage gain, even though they are voltage amplifiers in fact. Many specialized circuits are designed to provide voltage amplification. *See* AUDIO AMPLIFIER; CASCODE AMPLIFIER; VIDEO AMPLIFIER.

Voltage amplifiers are distinguished from other categories of amplifiers whose ability to amplify voltages, or lack thereof, is of secondary importance. Amplifiers in other categories usually are designed to deliver power gain (power amplifiers, including push-pull amplifiers) or to isolate one part of a circuit from another (buffers and emitter followers). Power amplifiers may or may not have voltage gain, while buffers and emitter followers generally produce power gain without a corresponding voltage gain. *See* BUFFERS (ELECTRONICS); EMITTER FOLLOWER; POWER AMPLIFIER; PUSH-PULL AMPLIFIER.

Transistor amplifiers, such as the junction field-effect transistor (JFET) or the bipolar junction transistor (BJT) amplifier, will not operate properly without proper gate (JFET) or base (BJT) bias voltages applied in series with the signal voltage. These bias circuits can be modeled as ideal voltage sources. The bias and signal voltages are chosen so that the total input voltage—bias plus signal—will not cut off or saturate the amplifier for any value in the range of the input signal voltage. In addition to a bias voltage source, well-designed bipolar transistor amplifiers require negative feedback at dc to protect the transistor from thermal runaway. *See* BIAS (ELECTRONICS).

To obtain high gain, cascades of single amplifier circuits are used, usually with a coupling network, actually a simple filter, inserted between the stages of amplification. One such filter is a high-pass network formed by a coupling capacitor, the output resistances of the driving stage, and the input resistance of the driven stage. Since dc voltages are blocked by the capacitor, this ac coupling permits independently setting dc bias voltages for each amplifier stage in the cascade. The coupling network also rejects signals with ac frequency components below a cutoff. The capacitor must be sufficiently large not to attenuate any of the frequencies that are to be amplified. If dc is to be amplified, a direct-coupled amplifier is required, and the design is somewhat more complicated since dc bias voltages on each transistor now cannot be set independently. *See* DIRECT-COUPLED AMPLIFIER; ELECTRIC FILTER.

The amplifiers discussed above are called single-ended amplifiers, since their input and output voltages are referred to a common reference point which by convention is called ground. These single-ended circuits, while satisfactory for most noncritical applications, have several weaknesses which degrade their performance in high-gain, weak-signal applications. Their unbalanced construction and their use of a common ground point for return currents makes them susceptible to noise pickup.

To minimize noise on sensitive signal lines, special balanced differential amplifier circuits are often used in critical amplifier applications. Differential amplifiers are designed to have equal impedances to ground for each side of the signal line and to have an output voltage proportional to the difference of the voltages from each signal line to ground. This symmetry cancels common-mode noise voltages, voltages which tend to appear on each of the signal lines as equal voltages to ground. Proper circuit design, with attention to the symmetry of the input circuit construction, can ensure that the majority of undesired noise pickup will be common-mode noise and, hence, will be attenuated by the differential amplifier. *See* DIFFERENTIAL AMPLIFIER; INSTRUMENTATION AMPLIFIER.

In cases where a voltage amplifier is required for some special purpose, operational amplifiers are often used to fill the need. The operational amplifier is an integrated circuit containing a cascade of differential amplifier stages, usually followed by a push-pull amplifier acting as a buffer. The differential voltage gain of the operational amplifier is very high, about 100,000 at low frequencies, while its input impedance is in the megohm range and its output impedance is usually under 100 ohms. The amplifier is designed to be used in a negative-feedback configuration, where the desired gain is controlled by a resistive voltage divider feeding a fraction of the output voltage to the inverting input of the operational amplifier.

With needed amplification built into many integrated circuits and with the availability of operational amplifiers for special-purpose amplification needs, there is seldom a need to design and build a voltage amplifier from discrete components. *See* AMPLIFIER; OPERATIONAL AMPLIFIER. Philip V. Lopresti

Voltage measurement Determination of the difference in electrostatic potential between two points. The unit of voltage in the International System of Units (SI) is the volt, defined as the potential difference between two points of a conducting wire carrying a constant current of 1 ampere when the power dissipated between these two points is equal to 1 watt.

Direct-current voltage measurement. The chief types of instruments for measuring direct-current (constant) voltage are potentiometers, resistive voltage dividers, pointer instruments, and electronic voltmeters.

The most fundamental dc voltage measurements from 0 to a little over 10 V can now be made by direct comparison against Josephson systems. At a slightly lower accuracy level and in the range 0 to 2 V, precision potentiometers are used in conjunction with very low-noise electronic amplifiers or photo-coupled galvanometer detectors. Potentiometers are capable of self-calibration, since only linearity is important, and can give accurate measurements down to a few nanovolts. When electronic amplifiers are used, it may often be more convenient to measure small residual unbalance voltages, rather than to seek an exact balance. *See* AMPLIFIER; GALVANOMETER; JOSEPHSON EFFECT.

Voltage measurements of voltages above 2 V are made by using resistive dividers. These are tapped chains of wire-wound resistors, often immersed in oil, which can be self-calibrated for linearity by using a buildup method. Instruments for use up to 1 kV, with tappings typically in a binary or binary-coded decimal series from 1 V, are known as volt ratio boxes, and normally provide uncertainties down to a few parts per million. Another configuration allows the equalization of a string of resistors, all operating at their appropriate power level, by means of an internal bridge. The use of series-parallel arrangements can provide certain easily adjusted ratios.

Higher voltages can be measured by extending such chains, but as the voltage increases above about 15 kV, increasing attention must be paid to avoid any sharp edges or corners, which could give rise to corona discharges or breakdown. High-voltage dividers for use up to 100 kV with an uncertainty of about 1 in 10^5, and to 1 MV with an uncertainty of about 1 in 10^4, have been made. *See* CORONA DISCHARGE; ELECTRICAL BREAKDOWN.

For most of the twentieth century the principal dc indicating voltmeters have been moving-coil milliammeters, usually giving full-scale deflection with a current between 20 microamperes and 1 milliampere and provided with a suitable series resistor.

Many of these will certainly continue to be used for many years, giving an uncertainty of about 1% of full-scale deflection.

The digital voltmeter has become the principal means used for voltage measurement at all levels of accuracy, even beyond one part in 10^7, and at all voltages up to 1 kV. Essentially, digital voltmeters consist of a power supply, which may be fed by either mains or batteries; a voltage reference, usually provided by a Zener diode; an analog-to-digital converter; and a digital display system. This design provides measurement over a basic range from zero to a few volts, or up to 20 V. Additional lower ranges may be provided by amplifiers, and higher ranges by resistive attenuators. The accuracy on the basic range is limited to that of the analog-to-digital converter. *See* ANALOG-TO-DIGITAL CONVERTER; ELECTRONIC POWER SUPPLY.

Most modern digital voltmeters use an analog-to-digital converter based on a version of the charge balance principle. In such converters the charge accumulated from the input signal during a fixed time by an integrator is balanced by a reference current of opposite polarity. This current is applied for the time necessary to reach charge balance, which is proportional to the input signal. The time is measured by counting clock pulses, suitably scaled and displayed. Microprocessors are used extensively in these instruments. *See* MICROPROCESSOR.

Alternating-current voltage measurements. Since the working standards of voltage are of the direct-current type, all ac measurements have to be referred to dc through transfer devices or conversion systems. A variety of techniques can be used to convert an ac signal into a dc equivalent automatically. All multimeters and most ac meters make use of ac-dc conversion to provide ac ranges. These are usually based on electronic circuits. Rectifiers provide the most simple example. *See* MULTIMETER.

In a commonly used system, the signal to be measured is applied, through a relay contact, to a thermal converter. In order to improve sensitivity, a modified single-junction thermal converter may be used in which there are two or three elements in a single package, each with its own thermocouple. The output of the thermal converter is measured by a very sensitive, high-resolution analog-to-digital converter, and the digital value memorized. When a measurement is required, the relay is operated, and the thermal converter receives its input, through a different relay contact, from a dc power supply, the amplitude of which is controlled by a digital and analog feedback loop in order to bring the analog-to-digital converter output back to the memorized level. The dc signal is a converted value of the ac input and can be measured. Modern versions of this type of instrument make use of microprocessors to control the conversion process, enhance the speed of operation, and include corrections for some of the errors in the device and range-setting components.

As in the dc case, digital voltmeters are now probably the instruments in widest use for ac voltage measurement. The simplest use diode rectification of the ac to provide a dc signal, which is then amplified and displayed as in dc instruments. This provides a signal proportional to the rectified mean. For most purposes an arithmetic adjustment is made, and the root-mean-square value of a sinusoidal voltage that would give the same signal is displayed. Several application-specific analog integrated circuits have been developed for use in instruments that are required to respond to the root-mean-square value of the ac input. More refined circuits, based on the logarithmic properties of transistors or the Gilbert analog multiplier circuit, have been developed for use in precision instruments. The best design, in which changes in the gain of the conversion circuit are automatically compensated, achieves errors less than 10 ppm at low and audio frequencies.

Sampling digital voltmeters are also used, in which the applied voltage is switched for a time very short compared with the period of the signal into a sample-and-hold circuit, of which the essential element is a small capacitor. The voltage retained can then be digitized without any need for haste. At low frequencies this approach offers high accuracy and great versatility, since the voltages can be processed or analyzed as desired. At higher frequencies, for example, in the microwave region, it also makes possible the presentation and processing of fast voltage waveforms using conventional circuits. *See* OSCILLOSCOPE.

Voltage measurements at radio frequencies are made by the use of rectifier instruments at frequencies up to a few hundred megahertz, single-junction converters at frequencies up to 500 MHz, or matched bolometers or calorimeters. At these higher frequencies the use of a voltage at a point must be linked to information regarding the transmission system in which it is measured, and most instruments effectively measure the power in a matched transmission line, usually of 50 ohms characteristic impedance, and deduce the voltage from it. *See* BOLOMETER; MICROWAVE MEASUREMENTS; MICROWAVE POWER MEASUREMENT; TRANSMISSION LINES.

Pulse voltage measurements are made most simply by transferring the pulse waveform to an oscilloscope, the deflection sensitivity of which can be calibrated by using low-frequency sine waves or dc. Digital sampling techniques may also be used. *See* ELECTRICAL MEASUREMENTS; VOLTMETER. R. B. D. Knight

Voltage-multiplier circuit A circuit which produces a dc output voltage that is a fixed multiple of the input dc voltage or the peak voltage of an ac input waveform. This fixed multiple is approximately an integer ($\pm 2, \pm 3, \pm 4, \ldots$). Voltage multipliers are used to produce a high dc voltage where modest load current is required. The circuit can be implemented with diodes and capacitors or with switched capacitors.

A diode passes conventional current in only one direction. In the diode-capacitor voltage doubler, this current through one of the diodes charges one of the capacitors to approximately the peak value of the input voltage. The series combination of the input source and the capacitor generates a voltage across the diode that varies from approximately 0 to -2 times the peak input voltage. The diode is reverse biased except for short intervals to replace charge drained from the capacitor to supply the load current or leakage. A second diode and capacitor form another peak detecting circuit with the voltage across the first diode as the input. This results in a negative dc output voltage equal to twice the peak of the input waveform. The polarity of the output can be reversed by simply reversing both diodes. Another version of the voltage doubler produces two outputs, one equal to the peak input voltage and the other to its negative, using the same components. This is realized by connecting a positive and a negative peak detector across the input. *See* CAPACITOR; DIODE.

A typical voltage-multiplier circuit that uses switched capacitors is a dc-to-dc converter that generates +10-V and −10-V dc sources from a single +5-V dc source. A common application for this circuit is to use a +5-V dc power supply to power integrated circuits which require positive and negative voltages. Such switched-capacitor applications are well suited to integrated circuits because the transistor switches and the associated control circuitry take up very little space and the added cost is negligible. *See* ELECTRONIC POWER SUPPLY; INTEGRATED CIRCUITS; SWITCHED CAPACITOR CIRCUIT; TRANSISTOR. Norman G. Dillman

Voltage regulation The change in voltage magnitude that occurs when the load (at a specified power factor) is reduced from the rated or nominal value to zero, with no intentional manual readjustment of any voltage control, expressed in percent of nominal full-load voltage. Voltage regulation is a convenient measure of the sensitivity of a device to changes in loading. *See* GENERATOR; TRANSFORMER; VOLTAGE REGULATOR. Paul M. Anderson

Voltage regulator A device or circuit that maintains a load voltage nearly constant over a range of variations of input

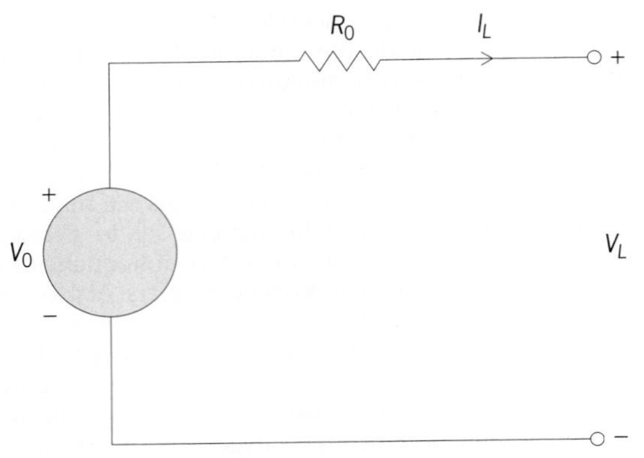

Equivalent circuit of a basic dc power supply.

voltage and load current. Voltage regulators are used wherever the unregulated voltage would vary more than can be tolerated by the electrical equipment using that voltage. Alternating-current distribution feeders use regulators to keep the voltage supplied to the user within a prescribed range. Electronic equipment often has voltage regulators in dc power supplies.

Electronic regulator. A dc power supply is an essential component in any electronic system. The illustration shows a Thévenin equivalent circuit of a power supply voltage source having an open-circuit (no-load) voltage V_0 and output resistance R_0. When a load with resistance R_L is connected to the output of this supply, a load current $I_L = V_L/R_L$ flows through R_0, resulting in a drop in load voltage V_L as given by Eq. (1).

$$V_L = V_0 - I_L R_0 \tag{1}$$

A basic power supply as modeled above exhibits two undersirable characteristics. The first is that the load voltage V_L decreases with increasing load current I_L. This effect can be severe in supplies with large effective output resistances. Such supplies are said to have poor load regulation. The second problem is that V_L depends directly on the source V_0. In practice, V_0 might be derived from a relatively inaccurate source such as the ac line voltage (after suitable rectification and filtering) or a battery. Fluctuations in V_0 reflect directly onto the voltage experienced by the load. In this case, the supply is said to have poor line regulation. An ideal power supply would exhibit perfect load regulation (V_L independent of I_L) and perfect line regulation (V_L independent of V_0) and would provide an output voltage of the form of Eq. (2), where V_{ref} is a well-defined reference voltage and k is

$$V_L = kV_{ref} \tag{2}$$

a constant scaling factor. The task of an electronic voltage regulator is to provide an output voltage characteristic that closely approximates the ideal of Eq. (2), given an unregulated supply voltage as an input.

Feedback network. Typically, electronic voltage regulators employ a feedback network, where a high-gain amplifier compares a fraction of the load voltage V_L/k with a constant reference V_{ref}. Any difference between these two voltages is amplified and used to control a series pass device in a manner whereby this difference is minimized. For an ideal amplifier with zero offset and infinite voltage gain, the difference is reduced to zero and the ideal relationship of Eq. (2) is realized. *See* FEEDBACK CIRCUIT.

The wide range of applications for electronic voltage regulators has led to the development of these circuits in fully monolothic integrated circuit technology, where all or most of the required circuit components are realized on a single chip of silicon. Offering various output current and voltage ratings, and output voltages of either positive or negative polarity, several commercial regulator integrated circuits are now available to suit the requirements of most applications. The designs of these regulators have matured and have become rather sophisticated. In addition to implementation of the high-gain feedback amplifier, the series pass element, and an accurate voltage reference, all on a single silicon die, built-in protection against overload conditions (such as output short circuits and excessive operating temperature) is now standard. Novel circuit-design, processing, and packaging techniques have been developed and implemented to achieve increased accuracy, temperature stability, efficiency, reliability, and power-handling capability, while reducing package size and cost. *See* INTEGRATED CIRCUITS. Ashok P. Nedungadi

Power-system regulator. Voltage regulators are used on distribution feeders to maintain voltage constant, irrespective of changes in either load current or supply voltage. Voltage variations must be minimized for the efficient operation of industrial equipment and for the satisfactory functioning of domestic appliances, television in particular. Voltage is controlled at the system generators, but this alone is inadequate because each generator supplies many feeders of diverse impedance and load characteristics. Regulators are applied either in substations to control voltage on a bus or individual feeder or on the line to reregulate the outlying portions of the system. These regulators are variable autotransformers with the primary connected across the line. The secondary, in which an adjustable voltage is induced, is connected in series with the line to boost or buck the voltage. *See* AUTOTRANSFORMER; ELECTRIC DISTRIBUTION SYSTEMS; ELECTRIC POWER SUBSTATION. D. D. MacCarthy

Voltage regulators are used on rotating machines in power generation applications to automatically control the field excitation so as to maintain a desired machine output voltage. Rotating machines, both small (down to 1 kW) and large (up to 1,000,000 kW), are the predominant means of power generation throughout the world, and voltage regulators of varying design and sophistication are employed on most of them. Even ac generators (or alternators) in automotive applications employ voltage regulators utilizing similar principles. *See* DIRECT-CURRENT GENERATOR; ELECTRIC POWER GENERATION; ELECTRIC ROTATING MACHINERY; GENERATOR. J. D. Hurley

Voltmeter

An instrument for the measurement of the electric potential difference between two conductors. Many different kinds of instruments are available to suit different purposes. Voltages of the order of picovolts (10^{-12} V) to megavolts (10^6 V) can be measured. Frequencies from zero (dc) to many megahertz and accuracies in the range from a fraction of part per million (ppm) to a few percent may be covered. *See* ELECTRICAL UNITS AND STANDARDS; VOLTAGE MEASUREMENT.

Analog voltmeters. Where no great accuracy is required, a voltage may be indicated by a mechanical displacement of a pointer against a scale. There is a wide variety of principles on which instruments of this type can be based. The d'Arsonval movement (see illustration) is one of the most popular constructions. This is basically a current-sensing instrument and is used in conjunction with a suitable resistance in series to measure voltage. A further variant, taut-band suspension, uses a pair of resilient strips under tension to carry the current to the coil, locate it, and provide the rotational restoring force. *See* AMMETER; MULTIMETER.

The permanent-magnet, moving-coil instrument is very sensitive, but by its nature is responsive only to the average value of the current flowing through the coil. It is therefore unsuitable for ac. A rectifier circuit can be used in order to combine the sensitivity of the movement with ac response. A transformer can be used to reduce the nonlinearity that results from the forward voltage drop of the diode rectifiers, at the expense of current drain. *See* RECTIFIER; TRANSFORMER.

Electronic voltmeters. The movements so far described require energy from the signal being measured to cause the deflection. The resulting current is liable to modify the voltage at the measurement point. To reduce this loading effect, active circuits

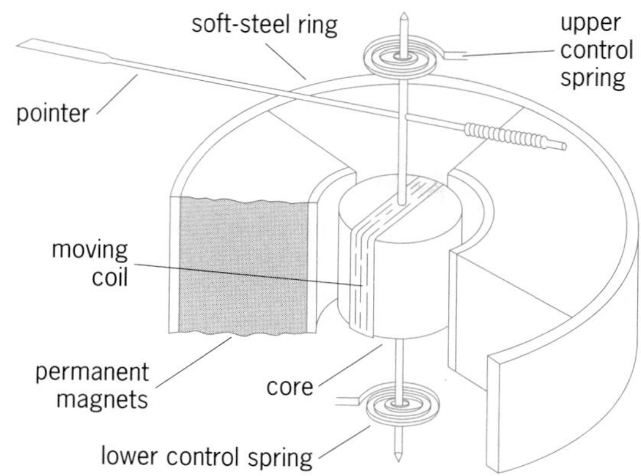

D'Arsonval moving-coil instrument. (*General Electric Co.*)

are often used between the input terminals and the indicating movement. Once an independent source of power is available, electronic circuits can be used to provide other features, including a variety of kinds of signal processing and digital presentation of the results.

Digital voltmeters. Digital voltmeters (DVMs) are now the preferred instruments for ac and dc measurements at all levels of accuracy and at all voltages up to 1 kV. Essentially a digital voltmeter consists of a voltage reference, usually provided by a Zener diode, an analog-to-digital converter and digital display system, and a power supply, which may be derived from either the mains or a battery. The basic range of the instrument provides measurement from zero to 10 or 20 V. Additional lower ranges may be provided by amplifiers, whose gain is stabilized by precision resistors. These electronic input amplifiers often provide a very high input impedance, perhaps exceeding 10^{10} Ω. Since this impedance is obtained by active means, a much lower impedance may be found when the instrument is switched off. Higher voltage ranges are provided by the use of resistive attenuators, usually limited to a value of 10 MΩ by economic restraints. The best accuracy is always obtained on the basic range, where it is limited to that of the analog-to-digital converter. *See* AMPLIFIER; ANALOG-TO-DIGITAL CONVERTER; ELECTRONIC POWER SUPPLY; ZENER DIODE.

Sampling voltmeters. A sampling voltmeter is an instrument that uses sampling techniques and has advantages at very low frequencies, that is, below 1 Hz, and also at very high frequencies, where conventional measuring circuits become difficult or even impossible. Low-frequency sampling instruments achieve uncertainties as small as 50 ppm with 10-V signals; high-frequency instruments can achieve a few percent with frequencies as high as 12 GHz and amplitudes as small as 1 mV. Measurements are generally of rectified-mean or root-mean-square voltage. Modern digital sampling voltmeters may also be capable of calculating and displaying voltages or energy density as a function of frequency. Sampling voltmeters, like conventional voltmeters, may use scale and pointer meters, graphic recorders, cathode-ray tubes, or digital indicators for readout of measured quantities. *See* WAVEFORM DETERMINATION.

R. B. D. Knight

Volume control systems

Systems that maintain proper audio signal levels in applications such as sound recording, public address systems, and broadcasting. Two types of electronic devices, compressors and limiters, perform this operation automatically without the need for human intervention, but differ in the way that they perform.

A compressor slowly varies its gain so as to maintain its output volume level at some constant average value. When used to process the audio signal from a microphone, a compressor equalizes the loudness of speech from different talkers. It compresses the dynamic range of the signal. A limiter controls the peak levels of an audio signal. Limiters are used to prevent overmodulation of transmitters in broadcast facilities, to prevent peak clipping in public address system audio amplifiers, and to prevent overload of audio recorders. A compressor followed by a limiter can be used to dramatically enhance the loudness of an audio signal without an increase in the peak level. This is a common practice in broadcasting to increase signal range and to maximize program loudness.

The input of a compressor or a limiter is applied to a variable-gain amplifier. A gain-control circuit monitors the level of the output and generates a voltage which controls the amplifier gain. Manual controls at the input and output allow setting the levels for proper interface to external devices. *See* AUTOMATIC GAIN CONTROL (AGC); GAIN; LIMITER CIRCUIT; RADIO BROADCASTING; SOUND RECORDING; SOUND-REINFORCEMENT SYSTEM; SOUND-REPRODUCING SYSTEM.

W. Marshall Leach, Jr.

Volume unit (vu)

A unit used to measure the strength of electrical waves produced by microphones. In sound recording, broadcasting, and public address systems, microphones convert acoustical signals into electrical waves that are nonperiodic and cannot be simply described in terms of voltage, current, power, or frequency. To provide a practical means of assigning a numerical value to the strength of such waves, the concept of "volume" is used. The volume of an audio program wave is the magnitude of the wave as measured with a standard volume indicator. This is a meter calibrated to read in volume units (vu, pronounced vee-you).

William M. Leach, Jr.

Volumetric efficiency

In describing an engine or gas compressor, the ratio of volume of working substance actually admitted, measured at specified temperature and pressure, to the full piston displacement volume. For a liquid-fuel engine, such as a diesel engine, volumetric efficiency is the ratio of volume of air drawn into a cylinder to the piston displacement. For a gas-fuel engine, such as a gasoline engine with carburetor, throttle body, or port injection, volumetric efficiency is based on the charge of fuel and air drawn into the cylinder. *See* COMPRESSOR; ENGINE.

Volumetric efficiency of naturally aspirated automobile and aircraft reciprocating engines may be 85–90% at rated speed. Supercharging or turbocharging increases volumetric efficiency, giving values over 100%. Air compressors, refrigerator compressors, and dry vacuum pumps are generally specified for a volumetric efficiency of 60–90%. *See* INTERNAL COMBUSTION ENGINE; RECIPROCATING AIRCRAFT ENGINE.

Donald L. Anglin

Volvocales

A large order of green algae (Chlorophyceae) comprising all forms that normally are flagellate and motile. In zoological classification, it is called Volvocida, in the class Phytomastigophora. The cells are solitary or united into colonies of definite structure (coenobia), often with morphological and functional differentiation among the component cells. Some taxonomists place the unicellular forms in a separate order, Chlamydomonadales. Unicells that have volvocalean cytological features but are nonflagellate and sedentary in their vegetative phase are considered here to constitute the order Tetrasporales. Or these sedentary forms may be retained in the Chlamydomonadales or Volvocales. *See* CHLOROPHYCEAE.

Of the unicellular forms, *Chlamydomona*, which is presumed to be similar to ancestral Volvocales, has a very small cell body (usually 7–40 micrometers in greatest dimension) that is spherical, ellipsoid, or pyriform. The wall (or theca), which consists primarily of glycoproteins, is usually smooth, but may have blunt protuberances. The cell bears two smooth flagella of equal length at its apex, with contractile vacuoles and an eye-spot at the anterior end. Chloroplasts contain one or more pyrenoids. As in all Volvocales, the cells are uninucleate. Other genera with chlamydomonad features are distinguished on the basis of flagellar number, presence or absence of photosynthetic pigments, presence or absence of pyrenoids, and cell shape. Together with *Chlamydomonas* they constitute the family Chlamydomonadaceae. Two other families of unicellular Volvocales are recognized: Dunaliellaceae, in which the unicells are naked; and Phacotaceae, in which the unicells are surrounded by a lorica—a rigid, hyaline or dark brown, wall-like structure, which is often impregnated with compounds of calcium, iron, or manganese and which may be sculptured. Unicellular Volvocales are ubiquitous in fresh, brackish, and coastal marine waters and in such terrestrial habitats as soil, snow, and ice. They are especially abundant in organically enriched bodies of fresh water.

Most colonial (coenobial) Volvocales are placed in the family Volvocaceae. In this family, coenobia are composed of *Chlamydomonas*-like biflagellate cells embedded in a firm gelatinous matrix. Coenobia have the form of a slightly curved plate or are ellipsoid to spherical. The component cells are arranged according to a predetermined pattern. Asexual reproduction is by repeated bipartition of the protoplast of an ordinary or specialized cell to form a new coenobium (autocoenobium) that is a miniature of the parent. Sexual reproduction forms a phylogenetic series from isogamy through anisogamy to oogamy. The Volvocaceae are abundant in organically enriched bodies of fresh water, especially temporary pools.

Paul C. Silva; Richard L. Moe

Volvocida An order of the class Phytamastigophorea. The protozoans, also known as the Phytomonadida, are grass-green, but a few are colorless (*Polytoma*). Individual cells may be as small as 8 micrometers. They closely relate the flagellates to the algae. They have one, two, four, or eight flagella; plural flagella are usually equal. The group is large and about one-fourth of the approximately 100 genera form colonies, with flagella only in reproductive cells. Cell walls are of cellulose and often they are thick. Chromatophores contain the same chlorophylls as higher plants. Pyrenoids have the usual form commonly found with starch as the reserve material. *See* CILIA AND FLAGELLA; PHYTAMASTIGOPHOREA.

James B. Lackey

VOR (VHF omnidirectional range) A short-range air navigation aid, which provides azimuth aid by visual means of cockpit instruments. A VOR system provides properly equipped aircraft with bearing information relative to the VOR station and magnetic north. The VOR system is used for landing, terminal, and en route guidance. It also gives virtually static-free regular weather broadcasts, special flight instructions, and voice and code station identification. The VOR service operates in the very high frequency (VHF) band between 108 and 118 MHz, sharing alternate channels with the localizer in the instrument landing system. Typically, VOR stations are co-located with a distance measuring equipment (DME) system or a tactical are navigation (TACAN) system. The combined systems are referred to as VOR/DME or VORTAC stations and provide both azimuth and distance information. *See* DISTANCE-MEASURING EQUIPMENT; INSTRUMENT LANDING SYSTEM (ILS); TACAN.

The VOR operates on the principle that the phase difference between two signals can be employed as a means of determining azimuth if one of the signals maintains a fixed phase through 360°, so it can be used as a reference, while the other is made to vary as a direct function of azimuth. The phase difference between these two signals will then equal the azimuth of the aircraft. In practice, two demodulated 30-Hz signals are used. These are called the reference-phase and variable-phase signals. *See* ELECTRONIC NAVIGATION SYSTEMS.

There are two types of VOR equipment. One employs a four-loop antenna array and is called conventional VOR. The other, Doppler VOR, having a 50-loop or larger antenna array located around a single-loop carrier antenna, is based on the Doppler principle and is designed for installation at locations which present especially difficult siting problems due to multipath conditions. Although the two types differ in design, from the standpoint of air navigation they function in essentially the same manner and can be received by the same equipment. Both types are used to define route intersections and most domestic airways. *See* DOPPLER EFFECT; DOPPLER VOR.

VOR/DME beacons provide air navigation services throughout the world. However, the emergence of satellite navigation systems has motivated civil aviation administrations to reconsider their continued support of these services. In the United States, the Federal Aviation Administration (FAA) plans a phased reduction (but not elimination) of VOR/DME services in the U.S. National Airspace beginning in 2010 based on the anticipated decrease in the use of VOR/DME. *See* SATELLITE NAVIGATION SYSTEMS.

William I. Thompson III; Robert B. Flint; Richard L. Greenspan

Vortex In common usage, a fluid motion dominated by rotation about an isolated curved line in space, as in a tornado, a whirlpool, a hurricane, or a similar natural phenomenon. The importance of vortices is due to two characteristics: general fluid flows can be represented by a superposition of vortices; and vortices, once created, have a persistence that increases as the effects of viscosity are reduced. The aerodynamic lift forces and most other contributors to the forces and moments on aircraft and other bodies moving through fluids do not exist in the absence of vortices. *See* AERODYNAMIC FORCE; HURRICANE; TORNADO; WATERSPOUT.

The strength of rotation is measured by a vector called the vorticity, ω, defined as the curl of the velocity vector. A flow devoid of vorticity is known as irrotational flow or potential flow, since the velocity vector can be expressed as the gradient of a scalar function, the velocity potential. The spatial distribution of the vorticity vector provides a precise characterization of the rotation effects in fluids, and the nature of what subjectively and popularly would be called a vortex. *See* CALCULUS OF VECTORS; LAPLACE'S IRROTATIONAL MOTION.

The vorticity vector field can be constructed by measuring the instantaneous angular velocity of small masses of fluid. The vorticity vector is twice the local angular velocity vector. Starting at any arbitrary point in the fluid, a line, called a vortex line, can be drawn everywhere parallel to the vorticity vector.

A bundle of vortex lines defines a tubular region of space, called a vortex tube, with a boundary surface that no vortex line crosses.

Two simple rules follow from the definitions: (1) a vortex tube must either close on itself or end on a boundary of the fluid (including extending to "infinity" if the fluid is imagined to fill all space); and (2) at every cross section of a given vortex tube, the area integral of the normal vorticity has the same value at any given instant. The area integral is, by Stokes' theorem, equal to a line integral around the periphery of the tube, namely, the line integral of the velocity component parallel to the direction of the closed curve defining the line integral. This quantity is also known as the circulation around the line, so at an instant of time a vortex tube has a unique value of the circulation applying to

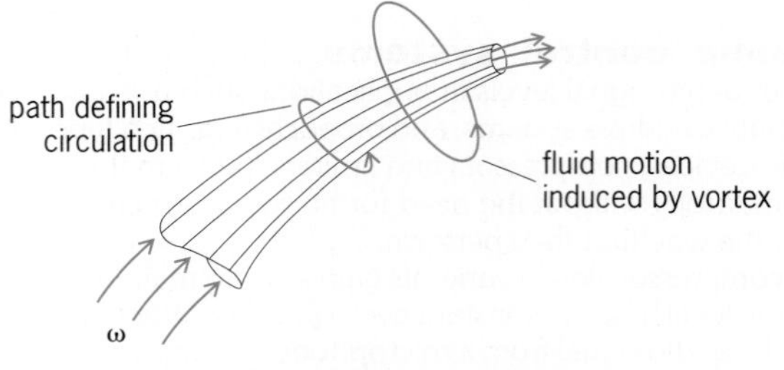

Vortex tube; ω is the vorticity.

all cross sections (see illustration). The circulation is a measure of the speed at which fluid circulates around the path.

Vortex lines confined to a layer rather than a tube describe fluid motion of a different character. This is most easily visualized when the direction of the vorticity does not vary, so all of the vortex lines are straight and parallel. Assuming the vorticity has zero magnitude outside the layer, this vortex layer represents a flow with a different speed and direction on either side of the layer. Such a change in speed occurs at the edge of wakes produced by wind passing over an obstacle. Reducing the thickness of this layer of vorticity to zero leads to an idealization known as a vortex sheet, a surface in space across which there is a finite jump in velocity tangent to the surface. Vortex sheets have a tendency to roll up because of self-induction. Aerodynamicists approximate flows in the wake of aircraft wings by vortex sheets, and the roll-up of this sheet leads to a pair of concentrated vortices trailing behind. The lift force on an aircraft depends on the formation of this trailing vortex system, and its strength is proportional to the weight of the aircraft. Sidney Leibovich

Vorticity A vector proportional to the local angular velocity of a fluid flow. The vorticity, $\vec{\omega}$, is a derived quantity in fluid mechanics, defined, for a flow field with velocity $\vec{u}$, by Eq. (1). As the curl of the velocity vector, the vorticity is a vector

$$\vec{\omega} = \nabla \times \vec{u} \quad (1)$$

with the dimensions of both a frequency and an angular velocity $[\vec{\omega}] = [s^{-1}]$. The component of vorticity along a particular axis is related to the rate of rotation of the fluid about the axis. For this reason, flows for which $\vec{\omega} = 0$ are described as irrotational. *See* CALCULUS OF VECTORS; DIMENSIONAL ANALYSIS; VELOCITY.

Circulation. Closely related to vorticity is the fluid circulation, Γ, defined, for any closed contour, C, in a fluid,by Eq. (2). In

$$\Gamma = \oint_C \vec{u} \cdot d\vec{l} = \iint_S \vec{\omega} \cdot \vec{n}\, dS \quad (2)$$

this definition, $\oint_C$ indicates the conventional counterclockwise contour integral around the contour C, $\vec{l}$ is a unit vector tangent to the contour, S is an arbitrary curved surface bounded by the contour C, and $\vec{n}$ is a unit vector normal to the surface. [The equality of the two integrals in Eq. (2) may be deduced by the application of Stokes' theorem.] The circulation is thus a scalar quantity equal to the integrated component of vorticity normal to the surface around which Γ is taken. Circulation is important because the Kutta-Joukowski law of aerodynamics states that the lift generated by a two-dimensional airfoil is $L = \rho U \Gamma$. In this expression, ρ is the fluid density, U is the free-stream velocity, and Γ is the bound circulation of the airfoil, defined conventionally as the negative of the definition above. *See* AERODYNAMICS; AIRFOIL; STOKES' THEOREM; SUBSONIC FLIGHT.

Vortex line and vortex tube. A vortex line is defined as a line that is everywhere tangent to the local vorticity vector (analogous to a streamline). A series of adjacent vortex lines is referred to as a vortex tube. The first Helmholtz vortex law states that at any instant in time the circulation about all loops taken around the exterior of a vortex tube is the same. Thus, vortex tubes must either form loops entirely within a fluid or terminate at some fluid boundary. *See* VORTEX.

Kelvin's theorem. Kelvin's theorem considers how the circulation Γ around a material loop in a fluid (a loop that moves with the fluid) varies in time. Starting with the Navier-Stokes equations, Lord Kelvin showed that if (1) the fluid is inviscid along the loop, (2) the fluid is subject only to potential body forces, and (3) the fluid pressure is a function of density alone, then the rate of change of Γ is 0. In other words, the circulation around a material loop is time-independent. Kelvin's theorem may also be stated slightly differently: subject to the above three constraints, vortex lines are material lines, convected with the local fluid velocity. *See* KELVIN'S CIRCULATION THEOREM.

Generation. Kelvin's theorem can tell what happens when vorticity is already present in a flow, but it sheds no light on how vorticity is generated. To answer this question, it is useful to consider situations for which Kelvin's theorem is inapplicable: flow with viscosity, with nonpotential body forces, and for which the pressure is not solely a function of the density.

The action of viscosity has two effects on vorticity. One effect of viscosity is to cause the diffusion of vorticity in a fluid. A second effect of viscosity is the generation of vorticity at a wall where there is a pressure gradient at the wall. *See* VISCOSITY.

A common example of a nonpotential body force is the Coriolis force, which is present in a rotating frame of reference. This force generates vorticity in a fluid, and is a major cause of the large-scale circulation in the atmosphere and oceans. *See* CORIOLIS ACCELERATION.

There are many flows for which the pressure may not be solely a function of the density (so-called baroclinic flows), such as the flow of gas with heat addition and the flow of water with salinity variations. Pressure gradients in such flows generate vorticity. This source of vorticity is called baroclinic torque, and is important in atmospheric flow, buoyancy-driven flow, and oceanographic flow. *See* BAROCLINIC FIELD; DYNAMIC INSTABILITY; DYNAMIC METEOROLOGY. Sheldon I. Green

Wake flow The flow downstream of a body immersed in a stream, or the flow behind a body propagating through a fluid otherwise at rest. Wakes are narrow elongated regions, aligned with the flow direction and filled with large and small eddies. The wake eddies of a bridge pier immersed in a river stream, or of a ship propelled through the water, are often visible on the surface. On windy days, similar wakes form downstream of smoke stacks or other structures, but such eddies in the air are not visible unless some smoke or dust is entrained in them.

Turbulence in the wake of bluff bodies consists of all sizes of eddies, which interact with each other in their unruly motion. Yet, out of this chaos emerges some organization, whereby large groups of eddies form a well-ordered sequence of vortices. These vortices are rolling and moving downstream much like the waves on the surface of the ocean, and for this reason they are often referred to as rollers. The sense of rotation of these vortices alternates, and their spacing is quite regular. As a result, they can drive a structure that they encounter, or they can exert on the body that created them a force alternating in sign with the same frequency as that of the formation of the vortices. Such rollers can be experienced when following in the wake of a large truck. *See* FLUID MECHANICS; KÁRMÁN VORTEX STREET; TURBULENT FLOW.

Wakes are sustained for very large distances downstream of a body. Ship wakes retain their turbulent character for miles behind a vessel and can be detected by special satellites hours after their generation. Similarly, condensation in the wake of aircraft sometimes looks like narrow braided clouds, traversing the sky.

Demetri P. Telionis

Wall construction Methods for constructing walls for buildings. Walls are constructed in different forms and of various materials to serve several functions. Exterior walls protect the building interior from external environmental effects such as heat and cold, sunlight, ultraviolet radiation, rain and snow, and sound, while containing desirable interior environmental conditions. Walls are also designed to provide resistance to passage of fire for some defined period of time, such as a one-hour wall. Walls often contain doors and windows, which provide for controlled passage of environmental factors and people through the wall line.

Walls are designed to be strong enough to safely resist the horizontal and vertical forces imposed upon them, as defined by building codes. Such loads include wind forces, self-weight, possibly the weights of walls and floors from above, the effects of expansion and contraction as generated by temperature and humidity variations as well as by certain impacts, and the wear and tear of interior occupancy. *See* LOADS, DYNAMIC; LOADS, TRANSVERSE.

Modern building walls may be designed to serve as either bearing walls or curtain walls or as a combination of both in response to the design requirements of the building as a whole. Both types may appear similar when complete, but their sequence of construction is usually different.

Bearing-wall construction may be masonry, cast-in-place or precast reinforced concrete, studs and sheathing, and composite types. The design loads in bearing walls are the vertical loading from above, plus horizontal loads, both perpendicular and parallel to the wall plane. Bearing walls must be erected before supported building components above can be erected.

Curtain-wall construction takes several forms, including lighter versions of those used for bearing walls. These walls can also comprise assemblies of corrugated metal sheets, glass panels, or ceramic-coated metal panels, each laterally supported by light subframing members. The curtain wall can be erected after the building frame is completed, since it receives vertical support by spandrel beams, or relieving angles, at the wall line.

Masonry walls are a traditional, common, and durable form of wall construction used in both bearing and curtain walls. They are designed in accordance with building codes and are constructed by individual placement of bricks, blocks of stone, cinder concrete, cut stone, or combinations of these. The units are bonded together by mortar. *See* CONCRETE; MASONRY; MORTAR.

Reinforced concrete walls are used for both strength and esthetic purposes. Such walls may be cast in place or precast, and they may be bearing or curtain walls. Some precast concrete walls are constructed of tee-shaped or rectangular prestressed concrete beams, which are more commonly used for floor or roof deck construction. They are placed vertically, side by side, and caulked at adjacent edges. *See* CONCRETE BEAM; REINFORCED CONCRETE.

Stud and sheathing walls are a light type of wall construction, commonly used in residential or other light construction where they usually serve as light bearing walls. They usually consist of wood sheathing nailed to wood or steel studs, usually with the dimensions 2 × 4 in. (5 × 10 cm) or 2 × 6 in. (5 × 15 cm), and spaced at 16 in. (40 cm) or 24 in. (60 cm) on center—all common building module dimensions. The interior sides of the studs are usually covered with an attached facing material. This is often sheetrock, which is a sandwich of gypsum between cardboard facings. Composite walls are essentially a more substantial form of stud walls. They are constructed of cementitious materials, such as weatherproof sheetrock or precast concrete as an exterior sheathing, and sheetrock as an interior surface finish. *See* GYPSUM; PRECAST CONCRETE.

Prefabricated walls are commonly used for curtain-wall construction and are frequently known as prefab walls. Prefabricated walls are usually made of corrugated steel or aluminum sheets, although they are sometimes constructed of fiber-reinforced plastic sheets, fastened to light horizontal beams (girts) spaced several feet apart. Prefab walls are often made of sandwich construction: outside corrugated sheets, an inside liner of flat or corrugated sheet, and an enclosed insulation are fastened together by screws to form a thin, effective sandwich wall. These usually have tongue-and-groove vertical edges to permit sealed joints when the units are erected at the building site by being fastened to framing girts.

Glass, metal, or ceramic-coated metal panel walls are a common type of curtain wall used in high-rise construction. They are typically assembled as a sandwich by using glass, formed metal, or ceramic-coated metal sheets on the outside, and some form of liner, including possibly masonry, on the inside; insulation is enclosed.

Tilt-up walls are sometimes used for construction efficiency. Here, a wall of any of the various types is fabricated in a horizontal position at ground level, and it is then tilted up and connected at its edges to adjacent tilt-up wall sections. Interior partitions are a lighter form of wall used to separate interior areas in buildings. They are usually nonbearing, constructed as thinner versions of some of the standard wall types; and they are often designed for some resistance to fire and sound. Retaining walls are used as exterior walls of basements to resist outside soil pressure. They are usually of reinforced concrete; however, where the basement depth or exterior soil height is low, the wall may be constructed as a masonry wall. *See* BUILDINGS; RETAINING WALL; STRUCTURAL MATERIALS. Milton Alpern

Walnut This name is applied to about a dozen species of large deciduous trees widely distributed over temperate North and South America, southeastern Europe, and central and eastern Asia. The genus (*Juglans*) is characterized by pinnately compound aromatic leaves and chambered or laminate pith. The staminate (male) flowers are borne in unbranched catkins on the previous season's growth, and the pistillate (female) flowers are terminal on the current season's shoots. The shells of the nuts of most species are deeply furrowed or sculptured.

Two species, the black walnut (*J. nigra*) and the Persian or English walnut (*J. regia*), are of primary importance for their timber and nuts. The butternut finds local use in the northeastern United States. The other species are sparingly used as shade trees, as grafting stocks, and as sources of nuts. *See* JUGLANDALES. Laurence H. MacDaniels

Warm-air heating system In a general sense, a heating system which circulates warm air. Under this definition both a parlor stove and a steam blast coil circulate warm air. Strictly speaking, however, a warm-air system is one containing a direct-fired furnace surrounded by a bonnet through which air circulates to be heated.

When air circulation is obtained by natural gravity action, the system is referred to as a gravity warm-air system. If positive air circulation is provided by means of a centrifugal fan (referred to in the industry as a blower), the system is referred to as a forced-air heating system.

Direct-fired furnaces are available for burning of solid, liquid, or gaseous fuels. Furnaces have also been designed which have air circulating over electrical resistance heaters. A completely equipped furnace-blower package consists of furnace, burner, bonnet, blower, filter, and accessories. The furnace shell is usually of welded steel. The burner supplies a positively metered rate of fuel and a proportionate amount of air for combustion. A casing, or jacket, encloses the furnace and provides a passage for the air to be circulated over the heated furnace shell. The casing is insulated and contains openings to which return-air and warm-air ducts can be attached. The blower circulates air against static pressure. The air filter removes dust particles from the circulating air. *See* OIL BURNER.

The complete forced-air heating system consists of the furnace-blower package unit; the return-air intake, or grille, together with return-air ducts leading from the grille to the return-air plenum chamber at the furnace; and the supply trunk duct and branch ducts leading to the registers located in the different spaces to be heated. *See* COMFORT HEATING. Seichi Konzo

Washer A flattened, ring-shaped device used to improve the tightness of a screw fastener. Three types of washer are in common use: plain, spring-lock, and antiturn (tooth-lock washers). Standard plain washers are used to protect a part from damage or to provide for a wider distribution of the load. Because a plain washer will not prevent a nut from turning, a locking-type washer should be used to prevent a bolt or nut from loosening under vibration (see illustration). Lock washers create a continuous pressure between the parts and the fastener. The antiturn-type washers may be externally serrated, internally serrated, or both. The bent teeth bite into the bearing surface to prevent the nut from turning and the fastening from loosening under vibration. *See* SCREW FASTENER. Warren J. Luzadder

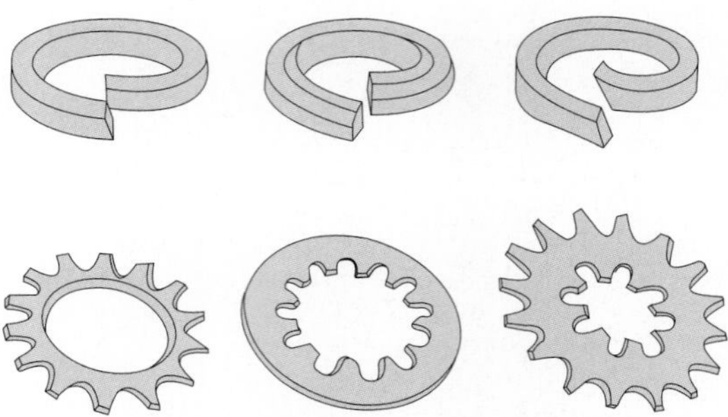

Lock washers. (*After W. J. Luzadder, Fundamentals of Engineering Drawing, 6th ed., Prentice-Hall, 1971*)

Wastewater reuse The use of treated wastewater effluent for beneficial purposes such as irrigation, industrial processes, and municipal water supply. Several developments have prompted wastewater reuse, including shortages of freshwater, stringent requirements for wastewater effluent quality, and advancements in treatment technology.

Wastewater reuse is classified as nonpotable and potable. Nonpotable uses of wastewater effluent include irrigation, industrial process or cooling water, and recreational impoundments (such as lakes or reservoirs for public use). Potable uses of high-quality wastewater effluent can be achieved by sending it back to water supply reservoirs, to potable ground-water aquifers, or directly to water treatment plants.

As the shortage of freshwater becomes more serious and the quality of wastewater effluent improves, reuse of municipal wastewater will become one of the most attractive water resource management alternatives. Advances in health risk assessment will make treated wastewater a widely acceptable freshwater source. The field of water and wastewater engineering is undergoing rapid changes. It is expected that, through proper source control and treatment, water and wastewater will be indistinguishable. *See* WATER SUPPLY ENGINEERING. Clark C. K. Liu

Watch A portable instrument that measures time. All modern timepieces, operate by using a part that swings back and forth in equal intervals called an oscillator (for example, a pendulum,

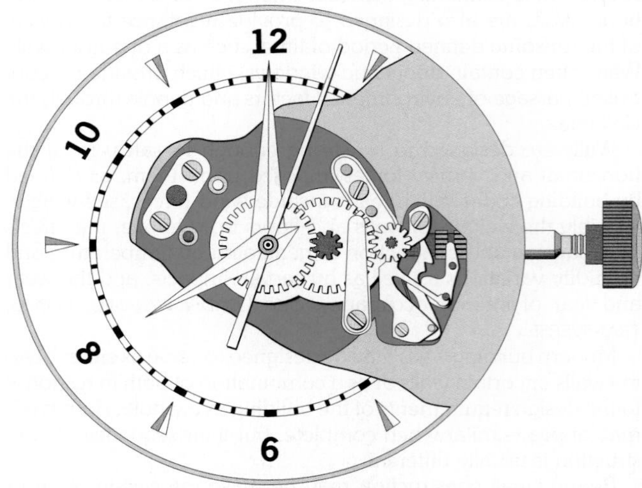

Cutaway view of classic mechanical watch.

balance wheel, tuning fork, or quartz crystal); a power source that energizes the swings (for example, a mainspring, battery, or capacitors); and a system that counts the swings (*see* illustration). The earliest watches consisted of a "swinger" (basically a balance wheel with an attached hairspring) as an oscillator; a mainspring as a power source; and a dial and hands to count the swings. This type of model is still manufactured and is comparatively expensive. Some are called chronometers and are precise timekeepers. But most watches today are quartz watches. *See* CLOCK (MECHANICAL); TUNING FORK.

In a quartz watch, the quartz crystal requires an intermediary to reduce the oscillations to activate the setting system for dial and hands. This is accomplished with a stepping motor, which is energized by the integrated circuit (microchip). *See* STEPPING MOTOR.

After the quartz watch with its stepping motor, the solid-state quartz, a watch that displayed the time with numbers (digits) rather than with hands and a dial was invented. In a solid-state watch, the time display is formed strictly electronically and inert matter forms the digits on the display screen. The integrated circuit controls the voltages and currents that activate a liquid crystal display. *See* ELECTRONIC DISPLAY; LIQUID CRYSTALS. Benjamin Matz

Water The chemical compound with two atoms of hydrogen and one atom of oxygen in each of its molecules. It is formed by the direct reaction (1) of hydrogen with oxygen. The other

$$2H_2 + O_2 \rightarrow 2H_2O \quad (1)$$

compound of hydrogen and oxygen, hydrogen peroxide, readily decomposes to form water, reaction (2). Water also is formed in

$$2H_2O_2 \rightarrow 2H_2O + O_2 \quad (2)$$

the combustion of hydrogen-containing compounds, in the pyrolysis of hydrates, and in animal metabolism. Some properties of water are given in the table.

Gaseous state. Water vapor consists of water molecules which move nearly independently of each other. The relative positions of the atoms in a water molecule are shown in the illustration. The dotted circles show the effective sizes of the isolated atoms. The atoms are held together in the molecule by chemical bonds which are very polar, the hydrogen end of each bond being electrically positive relative to the oxygen. When two molecules near each other are suitably oriented, the positive hydrogen of one molecule attracts the negative oxygen of the other, and while in this orientation, the repulsion of the like charges is comparatively small. The net attraction is strong enough to hold the molecules together in many circumstances and is called a hydrogen bond. *See* CHEMICAL BONDING; ELECTRONEGATIVITY; GAS; VALENCE.

Solid state. Ordinary ice consists of water molecules joined together by hydrogen bonds in a regular arrangement. It appears that there is considerable empty space between the molecules. This unusual feature is a result of the strong and directional hydrogen bonds taking precedence over all other intermolecular

Properties of water

Property	Value
Freezing point	0°C (32°F)
Density of ice, 0°C (32°F)	0.92 g/cm^3 (0.53 oz/in.^{3})
Density of water, 0°C (32°F)	1.00 g/cm^3 (0.578 oz/in.^{3})
Heat of fusion	80 cal/g (335 J/g)
Boiling point	100°C (212°F)
Heat of vaporization	540 cal/g (2260 J/g)
Critical temperature	347°C (657°F)
Critical pressure	217 atm (22.0 MPa)
Specific electrical conductivity at 25°C	1×10^{-7}/ohm-cm
Dielectric constant, 25°C (77°F)	78

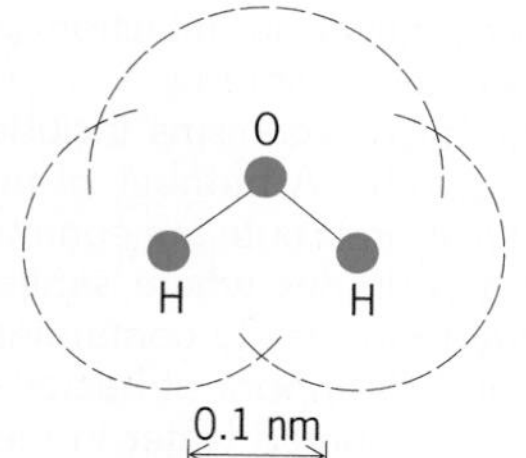

The water molecule.

forces in determining the structure of the crystal. *See* CRYSTAL; HYDROGEN BOND.

Liquid state. The molecules in liquid water also are held together by hydrogen bonds. When ice melts, many of the hydrogen bonds are broken, and those that remain are not numerous enough to keep the molecules in a regular arrangement. As water is heated from 0°C (32°F), it contracts until 4°C (39°F) is reached and then begins the expansion which is normally associated with increasing temperature. This phenomenon and the increase in density when ice melts both result from a breaking down of the open, hydrogen-bonded structure as the temperature is raised. *See* LIQUID.

Properties. The electrical conductivity of water is at least 1,000,000 times larger than that of most other nonmetallic liquids at room temperature. The current in this case is carried by ions produced by the dissociation of water according to reaction (3).

$$H_2O \rightleftharpoons H^+ + OH^- \quad (3)$$

Water is an excellent solvent for many substances, but particularly for those which dissociate to form ions. Its principal scientific and industrial use as a solvent is to furnish a medium for purifying such substances and for carrying out reactions between them. *See* SOLUTION; SOLVENT.

Water is not a strong oxidizing agent, although it may enhance the oxidizing action of other oxidizing agents, notably oxygen. Water is an even poorer reducing agent than oxidizing agent. One of the few substances that it reduces rapidly is fluorine. *See* OXIDATION-REDUCTION.

Substances with strong acidic or basic character react with water. For example, calcium oxide, a basic oxide, reacts in a process called the slaking of lime, reaction (4). Another type of sub-

$$CaO + H_2O \rightarrow Ca(OH)_2 \quad (4)$$

stance with strong acidic character is an acid chloride. An example and its reaction with water is boron trichloride, reaction (5).

$$BCl_3 + 3H_2O \rightarrow H_3BO_3 + 3HCl \quad (5)$$

This is termed hydrolysis, as is the reaction of an ester with water. *See* HYDROLYSIS.

Water also reacts with a variety of substances to form solid compounds in which the water molecule is intact, but in which it becomes a part of the structure of the solid. Such compounds are called hydrates. *See* HYDRATE.

For various aspects of water, its uses, and occurrence *see* HEAVY WATER; HYDROGEN; HYDROLOGY; IRRIGATION (AGRICULTURE); NUTRITION; OXYGEN; PLANT-WATER RELATIONS; PRECIPITATION (METEOROLOGY); SEAWATER; TRIPLE POINT; VAPOR PRESSURE; WATER ANALYSIS; WATER CONSERVATION; WATER MICROBIOLOGY; WATER POLLUTION; WATER PURIFICATION; WATER SOFTENING; WATER SUPPLY ENGINEERING; WATER TABLE; WATER TREATMENT; WATERPOWER.

Harold L. Friedman

Water-borne disease Disease acquired by drinking water contaminated at its source or in the distribution system, or by direct contact with environmental and recreational waters.

Water-borne disease results from infection with pathogenic microorganisms or chemical poisoning.

These pathogenic microorganisms include viruses, bacteria, protozoans, and helminths. A number of microbial pathogens transmitted by the fecal-oral route are commonly acquired from water in developing countries where sanitation is poor. Viral pathogens transmitted via fecally contaminated water include hepatitis viruses A and E. Important bacterial pathogens transmitted via fecally contaminated water in the developing world are *Vibrio cholerae*, enterotoxigenic *Escherichia coli*, *Shigella*, and *Salmonella enterica* serotype Typhi. Water-borne protozoan pathogens in the developing world include *Giardia lamblia* and *Entamoeba histolytica*. The major water-borne helminthic infection is schistosomiasis; however, transmission is not fecal-oral. Another water-borne helmenthic infection is dracunculiasis (guinea worm infection).

In developed countries, fecal contamination of drinking water supplies is less likely. However, there have been outbreaks of diseases such as shigellosis and giardiasis associated with lapses in proper water treatment, such as cross-contamination of wastewater systems and potable water supplies. Animals are therefore more likely to play a role in water-borne disease in developed countries. Bacterial pathogens acquired from animal feces such as nontyphoid *S. enterica*, *Campylobacter jejuni*, and *E. coli* serotype O157:H7 have caused outbreaks of water-borne disease in developed countries where water is not properly chlorinated. Hikers frequently acquire *G. lamblia* infections from drinking untreated lake and stream water. *Giardia lamblia* may have animal reservoirs and can persist in the environment. A recently recognized pathogen apparently resistant to standard chlorination and filtration practices is the protozoan *Cryptosporidium parvum*. This organism is found in the feces of farm animals and may enter water supplies through agricultural runoff.

Chemical poisoning of drinking water supplies causes disease in both developing and developed countries. Lead, copper, and cadmium have been frequently involved. *See* CHOLERA; ESCHERICHIA; MEDICAL PARASITOLOGY; SCHISTOSOMIASIS.

Steve L. Moseley

Water conservation The protection, development, and efficient management of water resources for beneficial purposes. Nearly every human activity—from agriculture to transportation to daily living—relies on water resources and affects the availability and quality of those resources. Water resource development has played a role in flood control, agricultural production, industrial and energy development, fish and wildlife resource management, navigation, and a host of other activities. As a result of these impacts, natural hydrologic features have changed through time, pollution has decreased the quality of remaining water resources, and global climate change may affect the distribution of water in the future. *See* HYDROLOGY.

Water availability varies substantially between geographic regions, but it is also affected strongly by the population of the region. Asia, for example, has an extremely large total runoff but the lowest per-capita water availability. In addition, nearly 40% of the world's population lives in areas that experience severe to moderate water stress. Thus the combination of water and population distribution has resulted in a large difference in per-capita water use between countries.

Worldwide, nearly 4000 km^3 of water is withdrawn every year from surface and ground waters. This is a sixfold increase from the levels withdrawn in 1900 (since which time population has increased four times). Agriculture accounts for the greatest proportion of water use, with about two-thirds of water withdrawals and 85% of water consumption. It also accounts for a great proportion of the increase in water use, with irrigated cropland more than doubling globally since 1960. However, in Europe and North America particularly, industry consumes a large proportion of available water; industrial uses for water are anticipated to grow on other continents as well.

Land development has substantially affected the distribution of water resources. It is estimated that one-half of the natural wetlands in the world have been lost in the last century. In some areas, such as California, wetland loss is estimated to be greater than 90%. The vast majority of wetlands loss has been associated with agricultural development, but urban and industrial changes have reduced wetlands as well. River channels have also been altered to enhance irrigation, navigation, power production, and a variety of other human activities.

Ground-water resources have been depleted in the last century, with many aquifers or artesian sources being depleted more rapidly than they can be recharged. This is called ground-water overdraft. In the United States, ground-water overdraft is a serious problem in the High Plains from Nebraska to Texas and in parts of California and Arizona. *See* GROUND-WATER HYDROLOGY.

Streams have traditionally served for waste disposal. Towns and cities, industries, and mines provide thousands of pollution sources. Pollution dilution requires large amounts of water. Treatment at the source is safer and less wasteful than flushing untreated or poorly treated wastes downstream. However, sufficient flows must be released to permit the streams to dilute, assimilate, and carry away the treated effluents. *See* WATER POLLUTION.

The availability of freshwater is also likely to be affected by global climate change. There is substantial evidence that global temperatures have risen and will continue to rise. Although the precise effects of this temperature risk on water distribution are challenging to predict, most models of climate change do anticipate increased global precipitation. It is likely that some areas, particularly those at mid to high latitudes, will become wetter, but the increased precipitation will be more seasonal than current patterns. Other areas are likely to receive less precipitation than they do currently. In addition, many models predict increases in the intensity and frequency of severe droughts and floods in at least some regions. These changes will affect natural stream flow patterns, soil moisture, ground-water recharge, and thus the timing and intensity of human demands for freshwater supplies. *See* GLOBAL CLIMATE CHANGE.

Land management vitally influences the distribution and character of runoff. Inadequate vegetation or surface organic matter; compaction of farm, ranch, or forest soils by heavy vehicles; frequent crop-harvesting operations; repeated burning; or excessive trampling by livestock or wild ungulates all expose the soil to the destructive energy of rainfall or rapid snowmelt. On such lands little water enters the soil, soil particles are dislodged and quickly washed into watercourses, and gullies may form. *See* LAND-USE PLANNING; SOIL CONSERVATION.

There are a variety of measures that can be taken to reduce water consumption. In the United States, for example, per-capita water usage dropped 20% from 1980 to 1995. In many cases, improvements to existing systems would contribute to additional water savings. In the United States, an average of 15% of the water in public supply systems (for cities with populations greater than 10,000) is unaccounted for, and presumably lost.

Improvements can also be achieved by changing industrial and agricultural practices. Agricultural water consumption has an estimated overall water use efficiency of 40%. More effective use of water in agricultural systems can be achieved, for example, with more efficient delivery methods such as drip irrigation. More accurate assessment of soil and plant moisture can allow targeted delivery of water at appropriate times. In industrial settings, recycling and more efficient water use has tremendous potential to reduce water consumption. Overall, industrial water usage dropped by 30% in California between 1980 and 1990, with some sectors achieving even greater reductions. Japan has achieved a 25% reduction in industrial water use since the

1970s. Additional potential to reduce this usage still exists even in locations where many conservation measures are already in place.

Residential water consumption can also be reduced through conservation measures. High-efficiency, low-flow toilets can reduce the water required to flush by 70% or more. Additional savings are possible with efficient faucet fixtures and appliances.

Water conservation in the United States faces a number of institutional as well as technological challenges. States must administer the regulatory provisions of their pollution-control laws, develop water quality standards and waste-treatment requirements, and supervise construction and maintenance standards of public service water systems. Some states can also regulate ground-water use to prevent serious overdrafts. Artesian wells may have to be capped, permits may be required for drilling new wells, or reasonable use may have to be demonstrated. Federal responsibilities consist largely of financial support or other stimulation of state and local water management. Federal legislation permits court action on suits involving interstate streams where states fail to take corrective action following persistent failure of a community or industry to comply with minimum waste-treatment requirements.

The watershed control approach to planning, development, and management rests on the established interdependence of water, land, and people. Coordination of structures and land-use practices is sought to prevent erosion, promote infiltration, and retard high flows (to prevent flooding). The Natural Resources Conservation and Forest Services of the Department of Agriculture administer the program. The Natural Resources Conservation Service cooperates with other federal and state agencies and operates primarily through the more than 2000 soil conservation districts.

Because watersheds often span political boundaries, many efforts to conserve and manage water require cooperation between states and countries. Many countries currently have international treaties addressing water allocation and utilization. In 1997, the United Nations adopted the Convention on the Law of the Non-navigational Uses of International Watercourses, which includes an obligation not to cause significant harm to other watercourse states, as well as provisions for dispute resolution. In addition, in 1996 the Global Water Partnership and the World Water Council were formed for the purpose of addressing ongoing international water concerns.

Bernard Frank; Michelle McClure; George Press

The increasing utilization of the continental shelf for oil drilling and transport, siting of nuclear power plants, and various types of planned and inadvertent waste disposal, as well as for food and recreation, requires careful management of human activities in this ecosystem. Nearshore waters are presently subject to both atmospheric and coastal input of pollutants in the form of heavy metals, synthetic chemicals, petroleum hydrocarbons, radionuclides, and other urban wastes. Overfishing is an additional human-induced stress. Physical transport of pollutants, their modification by the coastal food web, and demonstration of transfer to humans are sequential problems of increasing complexity on the continental shelf.

One approach to quantitatively assess the above pollutant impacts is to construct simulation models of the coastal food web in a systems analysis of the continental shelf. Models of physical transport of pollutants have been the most successful, for example, as in studies of beach fouling by oil. Incorporation of additional biological and chemical terms in a simulation model, however, requires dosage response functions of the natural organisms to each class of pollutants, as well as a quantitative description of the "normal" food web interactions of the continental shelf. *See* ECOLOGICAL MODELING; FOOD WEB.

In addition to toxic materials introduced by oil spills, sewage, and agricultural and industrial run-off, coastal waters are vulnerable to thermal pollution. Thermal pollution is caused by the discharge of hot water from power plants or factories and from desalination plants. A large power installation may pump in 10^6 gal/min (63 m^3/s) of seawater to act as a coolant and discharge it at a temperature approximately 18°F (10°C) above that of the ambient water. In a shallow bay with restricted tidal flow, the rise in temperature can cause gross alterations to the natural ecology. Federal standards prohibit heating of coastal waters by more than 0.9°F (0.5°C). *See* THERMAL ECOLOGY.

Finally, dredging waters to fill wetlands for house lots, parking lots, or industrial sites destroys the marshes that provide sanctuary for waterfowl and for the young of estuarine fishes. As the bay bottom is torn up, the loosened sediments shift about with the current and settle in thick masses on the bottom, suffocating animals and plants. In this way, the marshes are eliminated and the adjoining bays are degraded as aquatic life zones. The northeast Atlantic states have lost 45,000 acres (182 km^2) of coastal wetlands in only 10 years, and San Francisco Bay has been nearly half obliterated by filling. Dredging to remove sand and gravel has the same disruptive effects as dredging for landfill or other purposes, whether the sand and gravel are sold for profit or used to replenish beach sand eroded away by storms. The dredging of boat channels adds to the siltation problem, and disposal of dredge spoils is being regulated in coastal areas. John J. Walsh

Water desalination The removal of dissolved minerals (including salts) from seawater or brackish water. This may occur naturally as part of the hydrologic cycle, or as an engineered process. Engineered water desalination processes, which produce potable water from seawater or brackish water, have become important because many regions throughout the world suffer from water shortages caused by the uneven distribution of the natural water supply and by human use. *See* WATER SUPPLY ENGINEERING.

Seawater, brackish water, and fresh water have different levels of salinity, which is often expressed by the total dissolved solids (TDS) concentration. Seawater has a TDS concentration of about 35,000 mg/L, and brackish water has a TDS concentration of 1000–10,000 mg/L. Water is considered fresh when its TDS concentration is below 500 mg/L, which is the secondary (voluntary) drinking water standard for the United States. Salinity is also expressed by the water's chloride concentration, which is about half of its TDS concentration. *See* SEAWATER.

Water desalination processes separate feed water into two streams: a fresh-water stream with a TDS concentration much less than that of the feed water, and a brine stream with a TDS concentration higher than that of the feed water.

Distillation is a process that turns seawater into vapor by boiling, and then condenses the vapor to produce fresh water. Boiling water is an energy-intensive operation, requiring about 4.2 kilojoules of energy (or latent heat) to raise the temperature of 1 kg of water by 1°C. After water reaches its boiling point, another 2257 kJ of energy (or the heat of vaporization) is required to convert it to vapor. The boiling point depends on ambient atmospheric pressure—at lower pressure, the boiling point of water is lower. Therefore, keeping water boiling can be accomplished either by providing a constant energy supply or by reducing the ambient atmospheric pressure. *See* DISTILLATION.

Reverse osmosis, the process that causes water in a salt solution to move through a semipermeable membrane to the fresh-water side, is accomplished by applying pressure in excess of the natural osmotic pressure to the salt solution. The operational pressure of reverse osmosis for seawater desalination is much higher than that for brackish water, as the osmotic pressure of seawater at a TDS concentration of 35,000 mg/L is about 2700 kPa (400 psi), while the osmotic pressure of brackish water at a TDS concentration of 3000 mg/L is only about 230 kPa (30 psi).

Salts dissociate into positively and negatively charged ions in water. The electrodialysis process uses semipermeable and ion-specific membranes, which allow the passage of either positively or negatively charged ions while blocking the passage of the oppositely charged ions. An electrodialysis membrane unit consists of a number of cell pairs bound together with electrodes on the outside. These cells contain an anion exchange membrane and cation exchange membrane. Feed water passes simultaneously in parallel paths through all of the cells, separating the product (water) and ion concentrate. *See* DIALYSIS; ION EXCHANGE. Clark C. K. Liu; Jae-Woo Park

Water hammer The propagation in a liquid of an acoustic wave that is caused by a rapid change in fluid velocity. Such relatively sudden changes in the liquid velocity are due to events such as the operation of pumps or valves in pipelines, the collapse of vapor bubbles within the liquid, underwater explosions, or the impact of water following the rapid expulsion of air from a vent or a partially open valve. Alternative terms such as pressure transients, pressure surge, hydraulic transients, and hydraulic shock are often employed. Although the physics and mathematical characterization of water hammer and underwater acoustics (employed in sonar) are identical, underwater sound is always associated with very small pressure changes compared to the potential of moderate to very large pressure differences associated with water hammer. *See* CAVITATION; SOUND; UNDERWATER SOUND.

A pressure change Δp is always associated with the rapid velocity change ΔV across a water hammer wave, as formulated from the basic physics of mass and momentum conservation by the Joukowsky equation, $\Delta p = -\rho a \Delta V$. Here ρ is the liquid mass density and a is the sonic velocity of the pressure wave in the fluid medium. In a pipe, this velocity depends on the ratio of the bulk modulus of the liquid to the elastic modulus of the pipe wall, and on the ratio of the inside diameter of the pipe to the wall thickness. In water in a very rigid pipe or in a tank, or even the sea, the acoustic velocity is approximately 1440 m/s (4720 ft/s), a value many times that of any liquid velocity.

Liquid-handling systems are designed so that water hammer does not result from sudden closure, but is limited to more gradual flow changes initiated by valves or other devices. The dramatic pressure rise (or drop) results can be significantly reduced by reflections of the original wave from pipe-area changes, tanks, reservoirs, and so forth. Although the Joukowsky equation applies across every wavelet, the effect of complete valve closure over a period of time greater than a minimum critical time can be quite beneficial. This critical time is the time required for an acoustic wave to propagate twice the distance along the pipe from the point of wave creation to the location of the first pipe-area change. *See* HYDRODYNAMICS; PIPE FLOW. C. Samuel Martin

Water-jet cutting The use of high-pressure water jets, which may contain abrasive powder, for cutting and removing materials. For example, water accelerated up to twice the speed of sound [343 m/s (1125 ft/s) at 20°C (68°F)] can penetrate and cut rock in a few seconds.

Among the methods of cutting metal and nonmetallic materials, pure and abrasive water-jet cutting techniques have a distinct advantage because of their versatility and speed. They can cut all materials, including hard-to-machine materials such as superalloy, Kevlar, and boron carbide. They can also easily cut aerospace materials such as graphite composite and titanium, and brittle materials such as advanced ceramics, granite, marble, and glass (see illustration). The pure water jet technique with a relatively high flow rate (2–10 gal/min or 8–38 L/min) is widely used by the construction industry for applications such as road repair and tunnel boring. It is also used by the food industry to cut candy and chocolate bars, meats, vegetables, and fruits. Precision enhancement of the technology has led to medical applications, including orthopedic surgery for hip joint replacement. Other biomedical applications include removing clots from blood vessels and a micro water jet for corneal surgery.

Abrasive water-jet cutting of 0.5-in.-thick (12.5-mm) titanium at a pressure of 45,000 lb/in.^{2} (310 MPa). (*Flow International Corp.*)

The advantages of pure and abrasive water-jet cutting are (1) absence of thermal distortion and work hardening; (2) noncontact during cutting, thus eliminating tool wear and contact force; and (3) omnidirectional cutting, allowing the cutting of complex shapes and contours.

Although the use of the water-jet system is rapidly growing, the technique has some drawbacks and limitations. Water-jet technology has not yet developed fully for high-tolerance and -precision machining. The initial capital investment for the system, including the motion-control equipment and operating costs, is relatively high. The noise level (80 adjusted decibels) is somewhat high, but the system can be specially designed to isolate the noise source. *See* MACHINABILITY OF METALS.

The water-jet pump and its delivery system are designed to produce a high-velocity jet stream within a relatively short trajectory distance, since the kinetic energy of the water and abrasive particles is directly proportional to the square of the jet velocity. In abrasive jet cutting applications, the abrasives entrained in the jet stream usually attain approximately 80% of the water-droplet velocity at the nozzle tip. The jet cuts the material by a rapid erosion process, when its force exceeds the compressive strength of the material. Since the area eroded by the abrasive is also swept by the water stream, the heat generated during the cutting is dissipated immediately, resulting in a small rise in temperature (less than 90°F or 50°C) in the workpiece. Therefore, no thermal distortion or work hardening is associated with water-jet cutting. The cutting by rapid erosion also significantly reduces the actual force exerted on the material, enabling the water jet to cut fragile or deformable materials such as glass and honeycomb structures. *See* JET FLOW; METAL, MECHANICAL PROPERTIES OF; SHEAR. Thomas J. Kim

Water pollution A change in the chemical, physical, biological, and radiological quality of water that is injurious to its existing, intended, or potential uses (for example, boating, waterskiing, swimming, the consumption of fish, and the health of aquatic organisms and ecosystems). The term “water pollution” generally refers to human-induced (anthropogenic) changes to water quality. Thus, the discharge of toxic chemicals from a pipe or the release of livestock waste into a nearby water body is considered pollution. Conversely, nutrients that originate from animals in the wild or toxins that originate from natural processes are not considered pollution.

The contamination of ground water, rivers, lakes, wetlands, estuaries, and oceans can threaten the health of humans and aquatic life. Sources of water pollution are generally divided into two categories. The first is point-source pollution, in which contaminants are discharged from a discrete location. Sewage outfalls and oil spills are examples of point-source pollution. The second category is non-point-source or diffuse pollution, referring to all of the other discharges that deliver contaminants to water bodies. Acid rain and unconfined runoff from agricultural or urban areas are examples of non-point-source pollution. The principal contaminants of water include toxic chemicals, nutrients and biodegradable organics, and bacterial and viral pathogens.

Water pollution can threaten human health when pollutants enter the body via skin exposure or through the direct consumption of contaminated food or drinking water. Priority pollutants, including dichlorodiphenyl trichloroethane (DDT) and polychlorinated biphenyls (PCBs), persist in the natural environment and bioaccumulate in the tissues of aquatic organisms. These persistent organic pollutants are transferred up the food chain (in a process called biomagnification), and they can reach levels of concern in fish species that are eaten by humans. Finally, bacteria and viral pathogens can pose a public health risk for those who drink contaminated water or eat raw shellfish from polluted water bodies. *See* ENVIRONMENTAL TOXICOLOGY; FOOD WEB.

Contaminants have a significant impact on aquatic ecosystems. for example, enrichment of water bodies with nutrients (principally nitrogen and phosphorus) can result in the growth of algae and other aquatic plants that shade or clog streams. If wastewater containing biodegradable organic matter is discharged into a stream with inadequate dissolved oxygen, the water downstream of the point of discharge will become anaerobic and will be turbid and dark. Settleable solids, if present, will be deposited on the streambed, and anaerobic decomposition will occur. Over the reach of stream where the dissolved-oxygen concentration is zero, a zone of putrefaction will occur with the production of hydrogen sulfide, ammonia, and other odorous gases. Because many fish species require a minimum of 4–5 mg of dissolved oxygen per liter of water, they will be unable to survive in this portion of the stream.

Direct exposures to toxic chemicals is also a health concern for individual aquatic plants and animals. Chemicals (e.g., pesticides) are frequently transported to lakes and rivers via runoff, and they can have unintended and harmful effects on aquatic life. Toxic chemicals have been shown to reduce the growth, survival, reproductive output, and disease resistance of exposed organisms. These effects can have important consequences for the viability of aquatic populations and communities. *See* INSECTICIDE.

Wastewater discharges are most commonly controlled through effluent standards and discharge permits. Under this system, discharge permits are issued with limits on the quantity and quality of effluents. Water-quality standards are sets of qualitative and quantitative criteria designed to maintain or enhance the quality of receiving waters. Receiving waters are divided into several classes depending on their uses, existing or intended, with different sets of criteria designed to protect uses such as drinking water supply, bathing, boating, fresh-water and shellfish harvesting, and outdoor sports for seawater. For toxic compounds, chemical-specific or whole-effluent toxicity studies are used to develop standards and criteria. In the chemical-specific approach, individual criteria are used for each toxic chemical detected in the wastewater. Criteria can be developed to protect aquatic life against acute and chronic effects and to safeguard humans against deleterious health effects, including cancer. In the whole-effluent approach, toxicity or bioassay tests are used to determine the concentration at which the wastewater induces acute or chronic toxicity effects. *See* HAZARDOUS WASTE; SEWAGE DISPOSAL; SEWAGE TREATMENT.

Nathaniel Scholz; George Tchobanoglous

Water resources The Earth's water supply and its natural distribution. Although water is a renewable resource, which is continually being replaced by precipitation, it is not evenly distributed and is scarce in many areas.

World water*

Location	Surface area, km^2	Water volume, 10^3 km^3	Percent of total
World ocean	362,000,000	1,400,000	95.96
Mixed layer		50,000	
Thermocline		460,000	
Abyssal		890,000	
Glacial ice	18,000,000	43,400	2.97
Ground water		15,300	1.05
Lakes, fresh	855,000	125	0.009
Rivers		1.7	0.0001
Soil moisture		65	0.0045
Atmosphere†		15.5	0.001
Biosphere		2	0.0006
TOTAL	510,000,000	1,459,000	100

*SOURCE: E. K. Berner and R. A. Berner, *Global Environment: Water, Air and Geochemical Cycles*, Prentice Hall, 1996; National Research Council (NRC), *Global Change in the Geosphere–Biosphere*, National Academy Press, 1986; R. L. Nace, Terrestrial water, in AccessScience@McGraw-Hill, http://www.accessscience.com.DOI 10.1036/1097-8542.685800, last modified July 15, 2002 (surface area).
†As liquid water.

Water is stored on the Earth's surface in a number of places called reservoirs (see table). The main fresh-water resources available for humans on the Earth's surface are ground water and lake and river water, which together only constitute about 1.1% of the Earth's total water. *See* GROUND-WATER HYDROLOGY; SURFACE WATER.

Water does not permanently remain in any one reservoir on the Earth but is continually in motion through the hydrologic or water cycle. Residence time is the length of time water spends in a particular reservoir before being removed through the water cycle. The water cycle is driven by solar energy. Water is evaporated from the oceans to the atmosphere in the form of water vapor gas (a flux of 435,400 km^3/year). Part of this water vapor cools and is precipitated back to the oceans in the form of rain (398,000 km^3/year). Some water vapor (37,400 km^3/year) is transported through the atmosphere to the continents where it joins water vapor evaporated from the land (69,600 km^3/year). The vapor cools and precipitates out as rain or snow (107,000 km^3/year). Much of this precipitation runs along the Earth's surface to rivers, which flow to the oceans. A small amount of precipitation trickles down to join the ground water.

The uses of water worldwide are 70% for agriculture, 10% for domestic purposes such as drinking water, and 20% for industry (more than half of which is used for hydropower). Countries that have scarce water include a number in the belt of low precipitation (20–30° N and S latitude) such as the northern tier of Africa (Mauritania, Algeria, Morocco, Libya, Niger, and Egypt) and the Middle East (Saudi Arabia, Palestine, Syria, and Jordan). Worldwide there are 500 million people in countries with scarce water.

Water for human consumption is unsafe in many places, particularly in the developing countries. It is estimated that as much as 80% of diseases in developing countries are water-related, and 1.7 million people, often children, die from these diseases mainly in Africa and southeast Asia. Typical diseases are diarrhea, cholera, typhoid, and malaria. The main problem is that unsafe disposal of human and animal waste contaminates water for domestic use and irrigation.

More than 50% of the water used by industry (20% of the total) is used for hydropower plants. These plants provide one-fifth of the world's electricity. Hydropower is relatively clean and nonpolluting and is renewable. Dams used for hydropower generation also store water resources for agricultural irrigation, flood prevention, and domestic use.

Agriculture uses 70% of water worldwide, primarily for irrigation. About 65% of irrigation water is "consumed" in distribution and application and by crops and not available for reuse. Irrigation can be wasteful of water and can lead to salt buildup in soils if the soil is poorly drained. Agricultural and lawn runoff often cause overfertilization of water from nitrate and phosphate, causing algal blooms and loss of oxygen in bottom water of rivers, lakes, and estuaries. There have also been problems with agricultural pesticides polluting ground and surface water.

Desalinization of salt water currently supplies only about 0.1% of fresh water. It is expensive since it requires a lot of energy. Thus, it is used primarily for drinking water in water-poor areas. *See* WATER DESALINATION. Elizabeth K. Berner

Water softening The process of removing divalent cations, usually calcium or magnesium, from water. When a sample of water contains more than 120 mg of these ions per liter (0.016 oz/gal), expressed in terms of calcium carbonate ($CaCO_3$), it is generally classified as a hard water. Hard waters are frequently unsuitable for many industrial and domestic purposes because of their soap-destroying power and tendency to form scale in equipment such as boilers, pipelines, and engine jackets. Therefore it is necessary to treat the water either to remove or to alter the constituents for it to be fit for the proposed use.

The principal water-softening processes are precipitation, cation exchange, electrical methods, or combinations of these. The factors to be considered in the choice of a softening process include the raw-water quality, the end use of softened water, the cost of softening chemicals, and the ways and costs of disposing of waste streams. *See* ION EXCHANGE; WATER TREATMENT. Yi Hua Ma

Water supply engineering A branch of civil engineering concerned with the development of sources of supply, transmission, distribution, and treatment of water. The term is used most frequently in regard to municipal water works, but applies also to water systems for industry, irrigation, and other purposes.

Water obtained from subsurface sources, such as sands and gravels and porous or fractured rocks, is called ground water. Ground water flows toward points of discharge in river valleys and, in some areas, along the seacoast. The flow takes place in water-bearing strata known as aquifers. In an unconfined stratum the water table is the top or surface of the ground water. It may be within a few inches of the ground surface or hundreds of feet below. *See* AQUIFER; GROUND-WATER HYDROLOGY; WATER TABLE.

Wells are vertical openings, excavated or drilled, from the ground surface to a water-bearing stratum or aquifer. Pumping a well lowers the water level in it, which in turn forces water to flow from the aquifer. Thick, permeable aquifers may yield several million gallons daily with a drawdown (lowering) of only a few feet. Thin aquifers, or impermeable aquifers, may require several times as much drawdown for the same yields, and frequently yield only small supplies.

Dug wells, several feet in diameter, are frequently used to reach shallow aquifers, particularly for small domestic and farm supplies. They furnish small quantities of water, even if the soils penetrated are relatively impervious. Large-capacity dug wells or caisson wells, in coarse sand and gravel, are used frequently for municipal supplies. Drilled wells are sometimes several thousand feet deep.

The distance between wells must be sufficient to avoid harmful interference when the wells are pumped. In general, economical well spacing varies directly with the quantity of water to be pumped, and inversely with the permeability and thickness of the aquifer. It may range from a few feet to a mile or more.

Specially designed pumps, of small diameter to fit inside well casings, are used in all well installations, except in flowing artesian wells or where the water level in the well is high enough for direct suction lift by a pump on the surface (about 15 ft or 5 m maximum). Well pumps are set some distance below the water level, so that they are submerged even after the drawdown is established. *See* ARTESIAN SYSTEMS; WELL.

Natural sources, such as rivers and lakes, and impounding reservoirs are sources of surface water. Water is withdrawn from rivers, lakes, and reservoirs through intakes. The simplest intakes are pipes extending from the shore into deep water, with or without a simple crib and screen over the outer end. Intakes for large municipal supplies may consist of large conduits or tunnels extending to elaborate cribs of wood or masonry containing screens, gates, and operating mechanisms. Intakes in reservoirs are frequently built as integral parts of the dam and may have multiple ports at several levels to permit selection of the best water. *See* DAM; RESERVOIR; SURFACE WATER.

The water from the source must be transmitted to the community or area to be served and distributed to the individual customers. The major supply conduits, or feeders, from the source to the distribution system are called mains or aqueducts. The oldest and simplest type of aqueducts, especially for transmitting large quantities of water, are canals. Canals are used where they can be built economically to follow the hydraulic gradient or slope of the flowing water. If the soil is suitable, the canals are excavated with sloping sides and are not lined. Otherwise, concrete or asphalt linings are used. Gravity canals are carried across streams or other low places by wooden or steel flumes, or under the streams by pressure pipes known as inverted siphons. Tunnels are used to transmit water through ridges or hills; tunnels may follow the hydraulic grade line and flow by gravity or may be built below the grade line to operate under considerable pressure. Pipelines are a common type of transmission main, especially for moderate supplies not requiring large aqueducts or canals. *See* CANAL; PIPELINE; TUNNEL.

Included in the distribution system are the network of smaller mains branching off from the transmission mains, the house services and meters, the fire hydrants, and the distribution storage reservoirs. The network is composed of transmission or feeder mains, usually 12 in. (30 cm) or more in diameter, and lateral mains along each street, or in some cities along alleys between the streets. The mains are installed in grids so that lateral mains can be fed from both ends where possible. Valves at intersections of mains permit a leaking or damaged section of pipe to be shut off with minimum interruption of water service to adjacent areas.

Distribution reservoirs are used to supplement the source of supply and transmission system during peak demands, and to provide water during a temporary failure of the supply system. Ground storage reservoirs, if on high ground, can feed the distribution system by gravity, but otherwise it is necessary to pump water from the reservoir into the distribution system. Circular steel tanks and basins built of earth embankments, concrete, or rock masonry are used. Elevated storage reservoirs are tanks on towers, or high cylindrical standpipes resting on the ground. Storage reservoirs are built high enough so that the reservoir will maintain adequate pressure in the distribution system at all times. Elevated tanks are usually of steel plate, mounted on steel towers, but wood is sometimes used for industrial and temporary installations.

Pumps are required wherever the source of supply is not high enough to provide gravity flow and adequate pressure in the distribution system. The pumps may be high or low head depending upon the topography and pressures required. Booster pumps are installed on pipelines to increase the pressure and discharge, and adjacent to ground storage tanks for pumping water into distribution systems. Pumping stations usually include two or more pumps, each of sufficient capacity to meet demands when one unit is down for repairs or maintenance. The station must also include piping and valves arranged so that a break can be isolated quickly without cutting the whole station out of service. Richard Hazen

Drinking water comes from surface and ground-water sources. Surface waters normally contain suspended matter, pathogenic organisms, and organic substances. Ground water normally contains dissolved minerals and gases. Both require treatment. Conventional water treatment processes include pretreatment, aeration, rapid mix, coagulation and flocculation, sedimentation, filtration, disinfection, and other unit processes to meet specific requirements. *See* FILTRATION; SEDIMENTATION (INDUSTRY); WATER TREATMENT.

Aeration (air or oxygen into water) and air stripping (water into air) primarily are used to remove dissolved gases, such as hydrogen sulfide which causes taste and odor, and to oxidize iron and manganese. Robert A. Corbitt

Water table The upper surface of the zone of saturation in permeable rocks not confined by impermeable rocks. It may also be defined as the surface underground at which the water is at atmospheric pressure. Saturated rock may extend a little above this level, but the water in it is held up above the water table by capillarity and is under less than atmospheric pressure; therefore, it is the lower part of the capillary fringe and is not free to flow into a well by gravity. Below the water table, water is free to move under the influence of gravity.

The position of the water table is shown by the level at which water stands in wells penetrating an unconfined water-bearing formation. Where a well penetrates only impermeable material, there is no water table and the well is dry. But if the well passes through impermeable rock into water-bearing material whose hydrostatic head is higher than the level of the bottom of the impermeable rock, water will rise approximately to the level it would have assumed if the whole column of rock penetrated had been permeable. This is called artesian water. *See* ARTESIAN SYSTEMS; GROUND-WATER HYDROLOGY; WELL. Albert N. Sayre; Ray K. Linsley

Water treatment Physical and chemical processes for making water suitable for human consumption and other purposes. Drinking water must be pathogenically safe, free from toxic or harmful chemical or substances, and comparatively free of turbidity, color, and taste- or odor-producing substances. Excessive hardness and high concentration of dissolved solids are also undesirable, particularly for boiler feedwater and industrial purposes. The treatment processes commonly used are sedimentation, coagulation, filtration, disinfection, softening, and aeration.

Sedimentation. Silt, clay, and other fine material settle if the water is allowed to stand or flow at low velocity. Sedimentation occurs naturally in reservoirs and is accomplished in treatment plants using basins or settling tanks. Basic sedimentation does not remove extremely fine or colloidal material within a reasonable time; therefore, it is used principally as a preliminary process to other treatment methods or following the coagulation–flocculation process.

Coagulation and flocculation. The chemical process (coagulation) of neutralizing the chemical and electrostatic forces that repel particles is followed by the physical process (flocculation) of encouraging particle collisions and agglomerating particles for successful removal by gravity sedimentation. Thus, extremely fine particles and colloidal material are combined into larger masses, called floc, which are large enough to settle in basins and to be caught on the surface of filters. Waters high in organic material and iron may coagulate naturally with gentle mixing. The term "coagulation" is usually applied to chemical coagulation, in which iron or aluminum salts are added to the water to form insoluble hydroxide floc. The floc is a feathery, absorbent substance to which color-producing colloids, bacteria, fine particles, and other substances become attached and are removed from the water.

Filtration. Suspended solids, colloidal material, bacteria, and other organisms are filtered out by passing the water through a bed of sand or pulverized coal, or through a matrix of fibrous material supported on a perforated core. Filtration of turbid or highly colored water usually follows sedimentation or coagulation and sedimentation. Soluble materials, such as salts and metals in ionic form, are not removed by filtration. *See* FILTRATION.

Disinfection. There are several water treatment methods designed to kill harmful microorganisms, particularly pathogenic bacteria, spores, viruses, and protozoa. The application of chlorine or chlorine compounds is the most common method; however, its use is under increasing scrutiny. Emerging disinfection methods include the use of ultraviolet light and ozone, while boiling is used as a household emergency measure. *See* CHLORINE.

Water softening. Water softening is the process of removing the "hardness" caused by calcium and magnesium salts. Municipal water softening is common where the natural water has a hardness in excess of 150 parts per million. Two methods are used: (1) The water is treated with lime and soda ash to precipitate the calcium and magnesium as carbonate and hydroxide, after which the water is filtered. (2) The water is passed through a porous cation exchanger which has the ability of substituting sodium ions in the exchange medium for calcium and magnesium in the water. The exchange medium may be a natural sand known as zeolite or may be manufactured from organic resins. It must be recharged periodically by backwashing with brine. *See* ION EXCHANGE; WATER SOFTENING; ZEOLITE.

Aeration. Aeration is a process of exposing water to air by dividing the water into small drops, by forcing air through the water, or by a combination of both. The first method uses jets, fountains, waterfalls, and riffles; in the second, compressed air is admitted to the bottom of a tank through perforated pipes or porous plates; in the third, drops of water are met by a stream of air produced by a fan.

Aeration is used to add oxygen to water and to remove carbon dioxide, hydrogen sulfide, and taste-producing gases or vapors. Aeration is also used in iron-removal plants to oxidize the iron ahead of the sedimentation or filtration processes. *See* WATER POLLUTION. Karl G. Linden; Richard Hazen

Water-tube boiler A steam boiler in which water circulates within tubes and heat is applied from outside the tubes. The outstanding feature of the water-tube boiler is the use of small tubes [usually 1–3 in. (2.5–7.5 cm) outside diameter] exposed to the products of combustion and connected to steam and water drums which are shielded from these high-temperature gases.

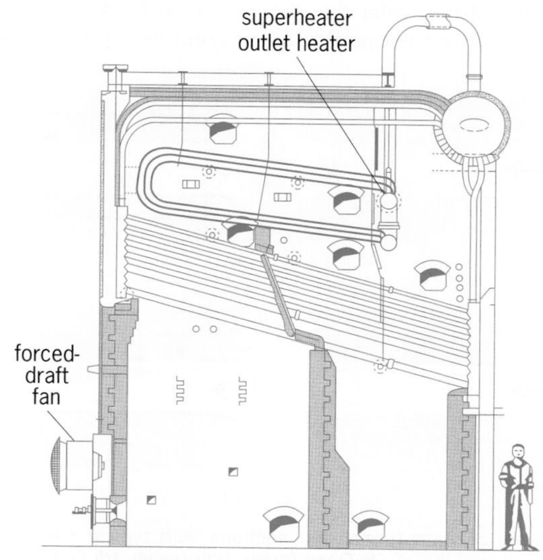

Diagram of a straight-tube-type boiler.

Thus, possible failure of boiler parts exposed to direct heat transfer is restricted to the small-diameter tubes and, in the event of failure, the energy released is reduced and explosion hazards are minimized.

There are many types of water-tube boilers but, in general, they can be grouped into two categories: the straight-tube (see illustration) and the bent-tube types. In essence, both types consist of banks of parallel tubes which are connected to, or by, headers or drums. Most modern water-tube boilers utilize a water-cooled surface in the furnace, and this surface is an integral part of the boiler's circulatory system. Further, modern water-tube boilers generally incorporate the use of superheaters, economizers, or air heaters to utilize more efficiently the heat from the fuel and to provide steam at a high potential for useful work in an engine or turbine. *See* BOILER; STEAM; STEAM-GENERATING FURNACE; STEAM-GENERATING UNIT. George W. Kessler

Water tunnel (research and testing) A hydrodynamic facility used for research, test, and evaluation, comprising a well-guided and controlled stream of water in which items for test are placed. The water tunnel is in many ways similar in appearance, arrangement, and operation to a subsonic wind tunnel. It is related to and complementary to the towing tank, in which the test item, usually a scale model of a ship or ship component, is towed through stationary water and evaluated through observation and measurement. In a water tunnel the test item is held stationary while the water is circulated around it. Many water tunnels are capable of operation with variable internal pressure to simulate the phenomenon of cavitation. *See* CAVITATION; TOWING TANK; WIND TUNNEL.

Water tunnels may be classified, in part, by the type of test section used. The most common section is the closed throat (illus. *a*) in which the test section flow has solid boundaries. The advantage of this arrangement is its simplicity and efficiency, but the model must be small relative to the tunnel cross section to avoid large wall effects. Small wall effects are theoretically correctable. In an open-throat test section (illus. *b*) the water jet passes through a water-filled chamber of larger diameter. This minimizes wall effects, and many tunnels dedicated to propeller testing use such an arrangement. When very low test section static pressures are required, or for fully cavitating flows, a free jet (illus. *c*) in which the water jet passes through an air-filled chamber is useful. However, capture of the free jet and removal of excess entrained air prior to recirculation is not easily achieved over a broad range of test conditions. To study cavity flows on surface-piercing components or hydrofoils which operate near a free surface, a free-surface tunnel (illus. *d*) is required in which three sides of the water flow are bounded by solid walls and the upper surface is open to air at controlled pressure levels.

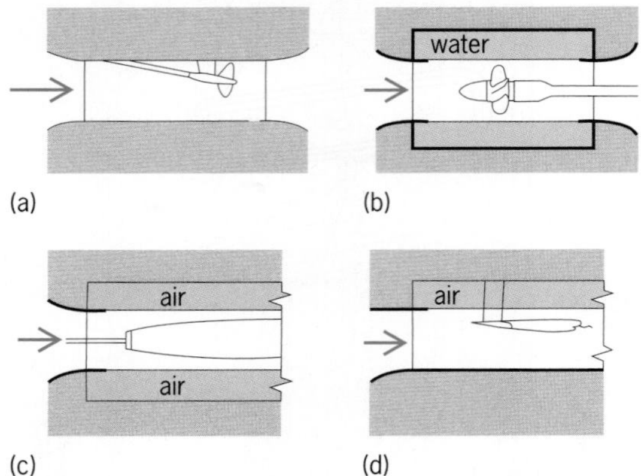

Types of water-tunnel test sections with typical models. (*a*) Closed throat. (*b*) Open throat. (*c*) Free jet. (*d*) Free surface.

This arrangement is often referred to as a water channel. *See* HYDROFOIL CRAFT; OPEN CHANNEL.

Another distinction among water tunnel types is whether or not they recirculate the flow. If the tunnel is nonrecirculating, water may "blow down" from a pressurized or elevated water tank or water may be diverted from a continuous source such as a waterfall or dam on a river. The blow-down tunnel has a limited test period proportional to the size of the storage tank. All water tunnels make use of transparent test-section viewing windows.

Water tunnels are used to investigate the dynamics, hydrodynamics, and cavitation of submerged and semisubmerged bodies such as propellers, ships, submarines, torpedoes, and hydrofoils and of turbomachinery. They are also indispensable to research on general flow phenomena in liquids. An application of great importance has been the acoustic characterization of propellers under both cavitating and noncavitating conditions. *See* HYDRODYNAMICS; PROPELLER (MARINE CRAFT).

Richard Stone Rothblum; Robert J. Etter; Steven L. Ceccio

Watermelon The edible fruit of *Citrullus lanatus*, of the family Cucurbitaceae. The plant is an annual prostrate vine with multiple stems.

The numerous cultivars of watermelon are highly diverse in fruit size (5–85 lb; 2.3–38.3 kg), shape (round, oval, oblong-cylindrical), rind color (very light to very dark green and often striped or mottled), flesh color (red, pink, orange, yellow, white), and seed size and color. The flesh contains 6–12% sugar, depending upon variety and condition of growth, with 8% sugar being acceptable on most markets.

Watermelon juice of the sweet cultivars can be reduced to edible sugar and syrup. Watermelon seeds are relished as food in some Near East countries and in China. Fresh watermelon flesh has a unique melting quality that has proved impossible to preserve in palatable form through any processing technique. The flesh consists of water (91%), fiber, and sugar, and little else of obvious nutritional value, but may have some as yet unproved health-promoting value. It is said to have diuretic properties, and both frozen concentrate and canned juice have been available for the treatment of nephritis. The seeds also are said to contain substances effective in the control of hypertension. C. F. Andrus

Waterspout An intense columnar vortex (not necessarily containing a funnel-shaped cloud) of small horizontal extent, over water. Typical visible vortex diameters are of the order of 33 ft (10 m), but a few large waterspouts may exceed 330 ft (100 m) across. In the case of Florida waterspouts, only rarely does the visible funnel extend from parent cloudbase to sea surface. Like the tornado, most of the visible funnel is condensate. Therefore, the extension of the funnel cloud downward depends upon the distribution of ambient water vapor, ambient temperature, and pressure drop due to the vortex circulation strength. These vortices are most frequently observed during the warm season in the oceanic tropics and subtropics.

All waterspouts undergo a regular life cycle composed of five discrete but overlapping stages. (1) The dark-spot stage signifies a complete vortex column extending from cloud-base to sea surface. (2) The spiral-pattern stage is characterized by development of alternating dark- and light-colored bands spiraling around the dark spot on the sea surface. (3) The spray ring (incipient spray vortex) stage is characterized by a concentrated spray ring around the dark spot, with a lengthening funnel cloud above. (4) The mature waterspout stage (see illustration) is characterized by a spray vortex of maximum intensity and organization. (5) The decay stage occurs when the waterspout dissipates (often abruptly).

Waterspouts and tornadoes are qualitatively similar, differing only in certain quantitative aspects: tornadoes are

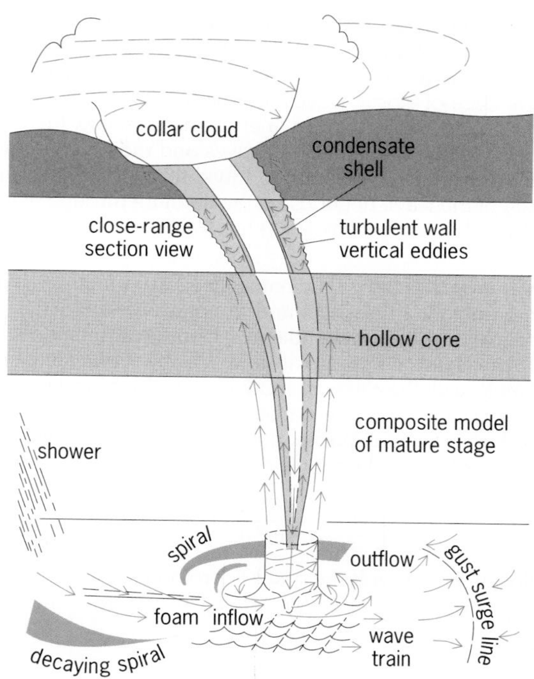

Composite schematic model of a mature waterspout. For scaling reference, the maximum funnel diameters in this stage, just below the collar cloud, range from 10 to 460 ft (3 to 140 m).

usually more intense, move faster, and have longer lifetimes—especially maxi-tornadoes. Tornadoes are associated with intense, baroclinic (frontal), synoptic-scale disturbances with attendant strong vertical wind shear, while waterspouts are associated with weak, quasibarotropic disturbances (weak thermal gradients) and consequent weak vertical wind shear. *See* TORNADO; WIND.

Joseph H. Golden

Watt balance An electromechanical apparatus for establishing the watt as an SI electrical unit. Prior to 1990, voltage units in use in the United States, France, the United Kingdom, and Russia differed from each other by up to 9 parts per million (ppm). A different kind of apparatus at the National Physical Laboratory in the United Kingdom and the National Institute for Science and Technology in the United States enabled the electrical units to be put on a sound SI basis. The accuracy in deriving the SI unit of voltage in this way was considered to be better than 0.2 ppm. Apparatuses are being developed in metrological laboratories with the objective of defining the kilogram in terms of fundamental physical constants instead of having to rely on a carefully preserved artifact. To do this, the accuracy achieved by the apparatus will have to be of the order of 0.01 ppm. *See* CURRENT BALANCE; ELECTRICAL UNITS AND STANDARDS; PHYSICAL MEASUREMENT.

Working principle. A conductor whose total length is l is wound into a horizontal circular coil and placed in a horizontal magnetic flux B which is everywhere perpendicular to the coil. If a current I flows in the conductor, the consequent vertical force can be opposed by a mass M which is subject to the earth's gravitational acceleration, g. That is, Eq. (1) is satisfied.

$$Mg = BIl \qquad (1)$$

If, in a separate measurement, the coil is moved vertically with velocity u, a voltage V will appear across it and will be given by Eq. (2).

$$V = Blu \qquad (2)$$

Combining Eqs. (1) and (2) eliminates the product Bl, which is too difficult to measure with the required accuracy, and gives Eq. (3). This equation relates the product VI to Mgu watts as

$$VI = Mgu \qquad (3)$$

measured directly in terms of the SI units of mass, length, and time. By defining voltage and current units in terms of the Josephson effect constant, $2e/h$, and the quantum Hall effect constant, h/e^2, the mass unit, the kilogram, would be redefined in terms of a defined value of Planck's constant, h. *See* ELECTROMAGNETIC INDUCTION; FARADAY'S LAW OF INDUCTION; HALL EFFECT; JOSEPHSON EFFECT; MAGNETISM.

Bryan P. Kibble

Watt-hour meter An electrical energy meter, that is, an electricity meter that measures and registers the integral, with respect to time, of the power in the circuit in which it is connected. This instrument can be considered as having two parts: a transducer, which converts the power into a mechanical or electrical signal, and a counter, which integrates and displays the value of the total energy that has passed through the meter. Either or both of these parts can be based on mechanical or electronic principles.

In its wholly mechanical form the transducer is an electric motor designed so that its torque is proportional to the electric power in the circuit. The motor spindle carries a conducting disk that rotates between the poles of one or more strong permanent magnets. These provide a braking torque that is proportional to the disk rotational speed, so the motor runs at a rate that accurately represents the circuit power. The integrating register is simply connected to the motor through a gear train that gives the required movement of the dials in relation to the passage of electrical energy. *See* MOTOR.

Mechanical meters can measure either dc or ac energy. Some form of commutator motor similar to those with a shunt field winding is commonly used for dc energy measurement. It is most convenient for the field to carry the circuit current, while the armature is fed with a signal from the circuit voltage. The Ferraris, or induction-type, meter is used for ac energy measurement. The stator carries two windings. An ordinary energy meter will easily achieve an accuracy of 2% over a wide range of loads; precision models may reach 0.1%. Ferraris meters are in very wide use and measure the consumption of the vast majority of domestic and industrial users of electric power throughout the world.

Electronic meters have an electronic watt transducer, which is a solid-state circuit that performs the multiplication of current and voltage signals, and delivers an output in the form of a pulse train at a rate proportional to power. The simplest solid-state watt-hour meter is completed by adding an electronic register to record the energy consumed. Precision electronic energy meters can give errors less than 0.005%. Electronic instruments are available in which six registers are provided, to record consumption at four different times of day and two levels of maximum demand. *See* TRANSDUCER; VOLTMETER.

Intelligent, or smart, meters can provide a wide variety of load and tariff-control functions, as well as remote reading of energy consumption.

Electronic and mechanical techniques can be combined in a variety of ways. Signals from a mechanical transducer may be used to operate electronic registers in order to obtain the advantages of the facilities that they can provide. A mechanical impulse register may be used in conjunction with an electronic transducer, where it is considered important to maintain a record without the need for batteries or nonvolatile memory elements.

Watt-hour meters that operate at potentials of 100–250 V and currents up to 100 A are widely manufactured. At higher levels, voltage or current transformers are used to reduce the signals handled by the meter to more convenient values, frequently 110 V and 5 A. By this means it is possible to carry out energy metering at any level required, including hundreds of kilovolts and tens of kiloamperes. *See* ELECTRIC POWER MEASUREMENT; ELECTRICAL ENERGY MEASUREMENT; INSTRUMENT TRANSFORMER; TRANSFORMER. R. B. D. Knight

Wattmeter An instrument that measures electric power. *See* ELECTRIC POWER MEASUREMENT.

A variety of wattmeters are available to measure the power in ac circuits. They are generally classified by names descriptive of their operating principles. Determination of power in dc circuits is almost always done by separate measurements of voltage and current. However, some of the instruments described will also function in dc circuits, if desired.

Probably the most useful instrument in the measurement of ac power at commercial frequencies is the indicating (deflecting) electrodynamic wattmeter. It is similar in principle to the double-coil dc ammeter or voltmeter in that it depends on the interaction of the fields of two sets of coils, one fixed and the other movable. The moving coil is suspended, or pivoted, so that it is free to rotate through a limited angle about an axis perpendicular to that of the fixed coils. As a single-phase wattmeter, the moving (potential) coil, usually constructed of fine wire, carries a current proportional to the voltage applied to the measured circuit, and the fixed (current) coils carry the load current. This arrangement of coils is due to the practical necessity of designing current coils of relatively heavy conductors to carry large values of current. The potential coil can be lighter because the operating current is limited to low values. *See* AMMETER; VOLTMETER.

A thermal converter consists of a resistive heater in close thermal contact with one or more thermocouples. When current flows through the heater, the temperature rises. Thermocouples give an output voltage proportional to the temperature difference between their junctions, in this case proportional to the square of the current, and so make suitable transducers for the construction of thermal wattmeters. *See* ELECTRICAL RESISTANCE; THERMAL CONVERTERS; THERMOCOUPLE; THERMOELECTRICITY.

The electrostatic force between two conductors is proportional to the product of the square of the potential difference between them and the rate of change of capacitance with displacement. A differential electrostatic instrument may therefore be used to construct a quarter-squares wattmeter. In spite of the problems of matching the capacitance changes of the two elements and the small forces available, electrostatic wattmeters were used as standards for many years.

Digital wattmeters combine the advantages of electronic signal processing and a high-resolution, easily read display. Electrical readout of the measurement is also possible. A variety of electronic techniques for carrying out the necessary multiplication of the signals representing the current and voltage have been used. Usually the electronic multiplier is an analog system which gives as its output a voltage proportional to the power indication required. This voltage is then converted into digital form in one of the standard ways. Many of the multipliers were originally developed for use in analog computers. *See* ANALOG COMPUTER; ELECTRONIC DISPLAY.

The instruments described are designed for single-phase power measurement. In polyphase circuits, the total power is the algebraic sum of the power in each phase. This summation is assisted by simple modifications of single-phase instruments. *See* ALTERNATING CURRENT. R. B. D. Knight

Wave (physics) The general term applied to the description of a disturbance which propagates from one point in a medium to other points without giving the medium as a whole any permanent displacement.

Waves are generally described in terms of their amplitude, and how the amplitude varies with both space and time. The actual description of the wave amplitude involves a solution of the wave equation and the particular boundary conditions for the case being studied. *See* WAVE EQUATION; WAVE MOTION.

Acoustic waves, or sound waves, are a particular kind of the general class of elastic waves. Elastic waves are propagated in media having two properties, inertia and elasticity. Electromagnetic waves (for example, light waves and radio waves) are not elastic waves and therefore can travel through a vacuum. The velocity of the wave depends on the medium through which the wave travels. *See* ELECTROMAGNETIC WAVE. William J. Galloway

Wave equation The name given to certain partial differential equations in classical and quantum physics which relate the spatial and time dependence of physical functions. In this article the classical and quantum wave equations are discussed separately, with the classical equations first for historical reasons.

In classical physics the name wave equation is given to the linear, homogeneous partial differential equations which have the form of Eq. (1). Here v is a parameter with the dimensions

$$\left[\nabla^2 - \frac{1}{v^2}\frac{\partial^2}{\partial t^2}\right] f(\mathbf{r},t) = 0 \qquad (1)$$

of velocity; $\mathbf{r}$ represents the space coordinates x, y, z; t is the time; and ∇^2 is Laplace's operator defined by Eq. (2). The function

$$\nabla^2 = \frac{\partial^2}{\partial x^2} + \frac{\partial^2}{\partial y^2} + \frac{\partial^2}{\partial z^2} \qquad (2)$$

$f(\mathbf{r},t)$ is a physical observable; that is, it can be measured and consequently must be a real function.

The simplest example of a wave equation in classical physics is that governing the transverse motion of a string under tension and constrained to move in a plane.

A second type of classical physical situation in which the wave equation (1) supplies a mathematical description of the physical reality is the propagation of pressure waves in a fluid medium. Such waves are called acoustical waves, the propagation of sound being an example. A third example of a classical physical situation in which Eq. (1) gives a description of the phenomena is afforded by electromagnetic waves. In a region of space in which the charge and current densities are zero, Maxwell's equations for the photon lead to the wave equations (3). Here $\mathbf{E}$ is

$$\left[\nabla^2 - \frac{1}{c^2}\frac{\partial^2}{\partial t^2}\right] E(\mathbf{r},t) = 0$$
$$\left[\nabla^2 - \frac{1}{c^2}\frac{\partial^2}{\partial t^2}\right] B(\mathbf{r},t) = 0 \qquad (3)$$

the electric field strength and $\mathbf{B}$ is the magnetic flux density; they are both vectors in ordinary space. The parameter c is the speed of light in vacuum. *See* ELECTROMAGNETIC RADIATION; MAXWELL'S EQUATIONS.

The nonrelativistic Schrödinger equation is an example of a quantum wave equation. Relativistic quantum-mechanical wave equations include the Schrödinger-Klein-Gordon equation and the Dirac equation. *See* QUANTUM MECHANICS; RELATIVISTIC QUANTUM THEORY. David L. Weaver

Wave mechanics The modern theory of matter holding that elementary particles (such as electrons, protons, and neutrons) have wavelike properties. In 1924 L. de Broglie postulated that the wave-particle duality which had been demonstrated for electromagnetic radiation also was a property of the elementary particles making up the atoms and molecules forming ordinary matter. In particular, de Broglie postulated that a particle has an associated wavelength obeying the same relation as was found to hold for photons, namely: the wavelength equals Planck's constant divided by the particle's momentum (as customarily defined in elementary mechanics). This hypothesis was verified in 1927 in an experiment in which a beam of electrons having known momentum is diffracted by a crystal into special directions. Such diffraction seems understandable only on the hypothesis that the electrons are waves. Furthermore, the wavelength of the electrons in the incident beam, computed via the same formula as was used to derive x-ray wavelengths in x-ray diffraction experiments, agreed precisely with the de Broglie relation. *See* ELECTRON DIFFRACTION; X-RAY DIFFRACTION.

Subsequent experiments have confirmed that not merely electrons but material particles in general, such as neutrons and neutral sodium atoms, manifest the wave-particle duality and obey the de Broglie relation. The de Broglie relation and the qualitative wave-particle duality concept have been incorporated into the highly successful modern theory of quantum mechanics. *See* ATOM OPTICS; DE BROGLIE WAVELENGTH; INTERFERENCE OF WAVES; QUANTUM MECHANICS.

Edward Gerjuoy

Wave motion

Wave motion The process by which a disturbance at one point in space is propagated to another point more remote from the source with no net transport of the material of the medium itself. For example, sound is a form of wave motion; wind is not. Wave motion can occur only in a medium in which energy can be stored in both kinetic and potential form. In a mechanical medium, kinetic energy results from inertia and is stored in the velocity of the molecules, while potential energy results from elasticity and is stored in the displacement of the molecules.

Media. In a free traveling wave (as distinguished from a stationary or standing wave) one part of the medium disturbs an adjacent part, thereby imparting energy to it. This portion of the medium, in turn, disturbs another part, thereby causing a flow of energy in a given direction away from the source. More technically, wave propagation is the result of kinetic energy at one point being transferred into potential energy at an adjacent point, and vice versa. The rate of travel of the disturbance, or velocity of propagation, is determined by the constants of the medium. A stationary wave is the combination of two waves of the same frequency and strength traveling in opposite directions so that no net transfer of energy away from the source takes place. A standing wave is the same but with the returning wave (toward the source) being of lesser intensity than the outwardly traveling wave so that a net transfer of energy away from the source does take place.

Wave motion can occur in a vacuum (electromagnetic waves), in gases (sound waves), in liquids (hydrodynamic waves), and in solids (vibration waves). Electromagnetic waves can also travel in gases, liquids, and solids provided that the electrical conductivity of the medium is not perfect or that the imaginary part of the dielectric constant is not infinitely great. By current usage, elastic waves propagated in gases, liquids, and solids, regardless of whether one can hear them or not, are called acoustic waves.

Fundamental relations. A wave is commonly referred to in terms of either its wavelength or its frequency. In any type of wave motion, these two quantities are related to a third quantity, velocity of propagation, by the simple relation $f\lambda = c$, where $f =$ frequency, $\lambda =$ wavelength, and $c =$ velocity of propagation. The period T is the reciprocal of the frequency, and the amplitude A is the maximum magnitude taken on by the variable of the wave at a given point in space. It is a basic property of wave motion that the frequency of a wave remains constant under all circumstances except for a relative motion between the source of the wave and the observer. The velocity of propagation is dependent on the properties of the medium (and, sometimes, also on the frequency) and the wavelength will vary with the velocity in accordance with the equation above.

Electromagnetic waves. The media in which electromagnetic waves travel possess no elasticity or inertia, but rather the ability to store energy in the electric and magnetic fields. J. C. Maxwell recognized in about 1863 that the basic equations governing these fields could be combined to yield an equation resembling the wave equation for mechanical wave motion. Thus he predicted the existence of electromagnetic waves which had not been suspected theretofore. Later, electromagnetic waves proved to be identical with light waves. *See* ELECTROMAGNETIC RADIATION; LIGHT; MAXWELL'S EQUATIONS; WAVE EQUATION.

Leo L. Beranek

Wave optics

Wave optics The branch of optics which treats of light (or electromagnetic radiation in general) with explicit recognition of its wave nature. The counterpart to wave optics is ray optics or geometrical optics, which does not assume any wave character but treats the propagation of light as a straight-line phenomenon except for changes of direction induced by reflection or refraction. *See* ELECTROMAGNETIC RADIATION; GEOMETRICAL OPTICS; OPTICS.

Any optical phenomenon which is correctly describable in terms of geometrical optics can also be correctly described in terms of wave optics. However, the many phenomena of interference, diffraction, and polarization are incontrovertible evidence of the wave nature of light, and geometrical optics often gives an incomplete or incorrect description of the behavior of light in an optical system. This is especially true if changes of refractive index occur within a space which is of the order of several wavelengths of the light. *See* DIFFRACTION; INTERFERENCE OF WAVES; POLARIZED LIGHT.

Richard C. Lord

Wave-shaping circuits

Wave-shaping circuits Electronic circuits used to create or modify specified time-varying electrical voltage or current waveforms using combinations of active electronic devices, such as transistors or analog or digital integrated circuits, and resistors, capacitors, and inductors. Most wave-shaping circuits are used to generate periodic waveforms. *See* INTEGRATED CIRCUITS; TRANSISTOR.

The common periodic waveforms include the square wave, the sine and rectified sine waves, the sawtooth and triangular waves, and the periodic arbitrary wave. The arbitrary wave can be made to conform to any shape during the duration of one period. This shape then is followed for each successive cycle. *See* FUNCTION GENERATOR; WAVEFORM.

A number of traditional electronic and electromechanical circuits are used to generate these waveforms. Sine-wave generators and *LC*, *RC*, and beat-frequency oscillators are used to generate sine waves; rectifiers, consisting of diode combinations interposed between sine-wave sources and resistive loads, produce rectified sine waves; multivibrators can generate square waves; electronic integrating circuits operating on square waves create triangular waves; and electronic relaxation oscillators can produce sawtooth waves. *See* ALTERNATING CURRENT; DIODE; MULTIVIBRATOR; OPERATIONAL AMPLIFIER; OSCILLATOR; RECTIFIER.

In many applications, generation of these standard waveforms is now implemented using digital circuits. Digital logic or microprocessors generate a sequence of numbers which represent the desired waveform mathematically. These numerical values then are converted to continuous-time waveforms by passing them through a digital-to-analog converter. Digital waveform generation methods have the ability to generate waveforms of arbitrary shape, a capability lacking in the traditional approaches. *See* CIRCUIT (ELECTRONICS); DIGITAL-TO-ANALOG CONVERTER; LOGIC CIRCUITS; MICROPROCESSOR.

Philip V. Lopresti

Waveform

Waveform The pictorial representation of the form or shape of a wave, obtained by plotting the amplitude of the wave with respect to time. There are an infinite number of possible waveforms (see illustration). One such waveform is the square wave, in which a quantity such as voltage alternately assumes two discrete values during repeating periods of time. Other waveforms of particular interest in electronics are the sine wave and rectified sine wave, the sawtooth wave and triangular wave, and the arbitrary wave—a recurrent waveform which takes on an

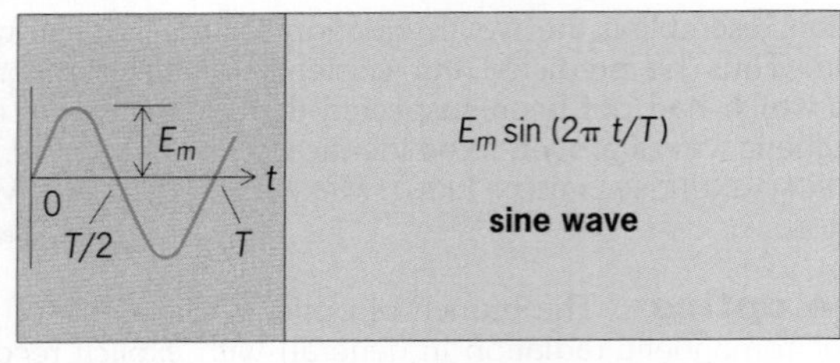

$$\frac{4E_m}{\pi}\sum^{\infty}\frac{1}{n}\sin(2\pi nt/T)$$
$$n = 1, 3, 5, \ldots$$
square wave

$$\frac{E_m}{2}-\frac{E_m}{\pi}\sum^{\infty}\frac{1}{n}\sin(2\pi nt/T)$$
$$n = 1, 2, 3, \ldots$$
sawtooth wave

$$\frac{8E_m}{\pi^2}\sum^{\infty}(-1^{(n-1)/2})\frac{1}{n^2}\sin(2\pi nt/T)$$
$$n = 1, 3, 5, \ldots$$
triangular wave

$$\frac{E_m}{\pi}\left[1+\frac{\pi}{2}\sin(2\pi t/T) - 2\sum^{\infty}\frac{1}{4n^2-1}\cos(4\pi nt/T)\right]$$
$$n = 1, 2, 3, \ldots$$
half-wave rectified sine wave

$$\frac{2E_m}{\pi}\left[1-2\sum^{\infty}\frac{1}{4n^2-1}\cos(2\pi nt/T)\right]$$
$$n = 1, 2, 3, \ldots$$
full-wave rectified sine wave

Common electrical waveforms.

arbitrary shape over one complete cycle; this shape is then repeated in successive cycles.

Each of these waveforms has a shape which repeats periodically in time. It is possible to characterize any of them mathematically by a Fourier series, a weighted sum of terms consisting of the basic periodic trigonometric functions: sines and cosines. A periodic waveform thus can be represented as a constant or dc term, plus a sum of harmonically related sine and cosine terms where the sine and cosine frequencies are integral multiples of the fundamental frequency. The Fourier series is given beside each waveform in the illustration as a function of time t, where E_m is the maximum value of the wave and T is the period. *See* FOURIER SERIES AND TRANSFORMS; NONSINUSOIDAL WAVEFORM; WAVE-SHAPING CIRCUITS. Philip V. Lopresti

Waveform determination A waveform describes the variation of a quantity with respect to time. The necessary measurements to determine a waveform are normally carried out and presented in one of two ways: the amplitude may be presented as a function of time (time domain), or an analysis may be given of the relative amplitudes and phases of the frequency components (frequency domain). Although the simplest instruments measure and display the information in the same domain, it is possible to convert the results from the time domain to the frequency domain and vice verse by mathematical processing.

Waveforms may be divided into two classes, depending on whether the signal is repeated at regular intervals or represents a unique event. The former signal is defined as a periodic or continuous wave, the latter as an aperiodic signal or transient. *See* ELECTRIC TRANSIENT; NONSINUSOIDAL WAVEFORM; WAVEFORM.

The oscilloscope is an example of an instrument that measures and displays directly in the time domain, by deflecting an electron beam in a vertical direction in accordance with the signal while scanning at a uniform rate in the horizontal direction. The position of the beam is revealed by a fluorescent screen. *See* GRAPHIC RECORDING INSTRUMENTS; OSCILLOSCOPE.

Several methods may be used to obtain the spectral content of a waveform. In the simplest, the signal is applied to a filter that is manually tuned in turn to each frequency that is expected to be present. In order to automate the measurement, the tuning of a filter may be varied by a linear, logarithmic, or other sweep and the resulting output displayed. However, any transient signal that occurs at a frequency that is different from the filter frequency will be missed. This is an important restriction that limits the technique to continuous waveforms. Specific instruments that display the frequency components of the signal are often called spectrum analyzers or harmonic analyzers. *See* ELECTRIC FILTER; HETERODYNE PRINCIPLE; SPECTRUM ANALYZER.

Many modern instruments use techniques in which the signal to be measured is sampled and digitized. The key elements of digital oscilloscopes are a high-speed analog-to-digital converter, sufficient high-speed memory to store the results, and a display. Because the signal is continuously digitized and stored in memory, these instruments are ideal for measuring transient waveforms. As the data can be read out of digital storage at any convenient speed, low-bandwidth display circuits are adequate. *See* ANALOG-TO-DIGITAL CONVERTER.

Once the data have been collected in digital form, they can be processed in many different ways. The application of a discrete Fourier transformation (DFT) to the amplitude data enables the information to be presented in the frequency domain. In this way, harmonic distortion that was invisible on a directly displayed waveform can be made obvious. The fast Fourier transform (FFT) is a method by which the arithmetic involved is simplified and thus accelerated. Wavelet transforms are another powerful technique that can be used to analyze waveforms. These are well suited to showing localized intermittent periodicities such as natural events and transients. *See* FOURIER SERIES AND TRANSFORMS; SPEECH RECOGNITION; WAVELETS.

Often, waveforms contain too much detailed information for easy comparison. Instead, they can be represented by simple parameters, such as the pulse duration or transition duration. The Institute of Electrical and Electronic Engineers has written a standard which sets out definitions of all these parameters and how they should be measured. The transition durations (also commonly known as rise time and fall time) are measured between the 10% and the 90% pulse amplitude reference levels. The pulse duration (also commonly known as full width at half

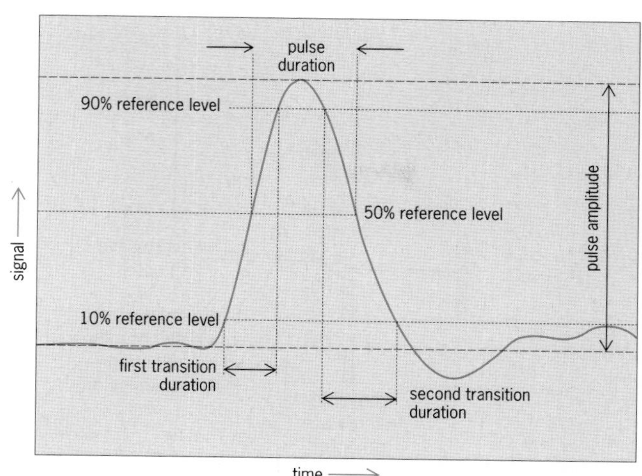

Typical pulse waveform showing the definitions of the transition duration and pulse duration parameters.

maximum or FWHM) is measured at the 50% reference level. The pulse duration and transition durations are the key parameters often used to characterize a pulse waveform (see illustration).

David A Humphreys

Waveform generator A term usually applied to an instrument that generates voltage waves using digital technology. These digital instruments have largely supplanted the analog instruments that traditionally served in this role. Digital instruments have a number of important advantages over analog ones: they are more flexible and can be used to generate waveforms that are beyond the capabilities of analog instruments, they produce more accurate replicas of the desired mathematical waveforms, and their output is more stable and predictable over time. *See* FUNCTION GENERATOR; SIGNAL GENERATOR; WAVE-SHAPING CIRCUTS.

The basic components of these instruments are: a digital signal memory that contains numeric samples of the desired signal waveform, a sequencing system that accesses this memory in a predetermined pattern, a high-accuracy digital-to-analog converter that transforms the contents of the accessed memory location to proportional analog voltages, a low-pass filter to smooth the output of the converter, and a digital clock—the timing device that sets the memory access rate. *See* DIGITAL-TO-ANALOG CONVERTER; ELECTRIC FILTER; SEMICONDUCTOR MEMORIES.

The two most common approaches to implementing these instruments are direct digital synthesis (DDS) and arbitrary waveform generation (AWG). Each has its own advantages and each plays an important role. Direct digital synthesis typically is used to generate high-accuracy versions of standard analog signal waveforms: sine, triangular, and square waves. Instruments that employ arbitrary waveform generation are much more flexible and can generate a repetitive signal waveform that takes on values limited only by the word size of each memory cell and by the number of memory locations in the instrument. Instruments that use direct digital synthesis can be programmed to change their output from one form to another essentially instantaneously, while devices that use arbitrary waveform generation are limited by the time it takes to replace their signal memory contents with new signal patterns.

Philip V. Lopresti

Waveguide A device that constrains or guides the propagation of electromagnetic radiation along a path defined by the physical construction of the guide. Electromagnetic waves may propagate in space, as radio waves, but for many purposes waves need to be guided with minimum loss from the generating point to a point of application. Several guiding systems are of importance, including two-conductor transmission lines, various forms of striplines used in microwave integrated circuits, hollow-pipe waveguides, and dielectric waveguides. Hollow-pipe guides are used primarily in the microwave region of the spectrum, dielectric guides primarily in the optical region. *See* COAXIAL CABLE; ELECTROMAGNETIC WAVE TRANSMISSION; MICROWAVE; TRANSMISSION LINES.

Hollow-pipe waveguides consist of a dielectric region, usually air, surrounded by a closed good conductor such as silver, copper, aluminum, or brass. The cross-sectional shape is usually rectangular, but may be circular or of a variety of other shapes. Voltage and current concepts, so useful for transmission lines, are not so useful for waveguides. Distributions of electric and magnetic fields, obtained from Maxwell's equations, are needed. *See* MAXWELL'S EQUATIONS.

Solution of the field equations shows that there is an infinite number of modes for these guides, where a mode is a solution that maintains its transverse pattern but attenuates and shifts in phase as it propagates along the guide. These modal patterns may be likened to the resonant modes of a drumhead, with low-order modes having only a few transverse variations and high-order modes having many.

All modes in these guides have cutoff properties. Below a certain critical frequency, the mode will not propagate but only attenuates if excited; above the cutoff frequency, though, it may propagate with a finite phase velocity and small attenuation (zero attenuation for perfect dielectric and conductor). Higher-order modes usually have higher cutoff frequencies.

If the guiding pipe is perfectly conducting, the modes may be divided into transverse magnetic (TM) and transverse electric (TE) types. For the former, the magnetic field is confined to the transverse plane, while the electric field is so confined for the latter. These classifications remain useful for the practical metallic conductors used for such guides.

A variety of circuit elements is required for exciting desired modes, filtering, coupling to passive and active elements, and other necessary networking functions. Excitation of a particular mode may be by probes, along the direction of the mode's electric field, by loops normal to magnetic field lines, or by the charge streams of an active vacuum-tube or semiconductor device placed within the guide.

Other important waveguide elements are the directional coupler and isolator. In the directional coupler, there is coupling to an auxiliary guide in such a way that the output of one of its ports is proportional to the wave traveling in the forward direction, and the output of the other is proportional to the reverse wave. The isolator makes use of the nonreciprocal properties of ferrites with an applied magnetic field to pass the forward-traveling wave of the guide but to eliminate the reflected wave. *See* DIRECTIONAL COUPLER.

A dielectric waveguide consists of one dielectric material, called the core, surrounded by a different dielectric, called the cladding. The permittivity (dielectric constant), or refractive index, of the core is larger than that of the cladding, and under proper conditions electromagnetic energy is confined largely to the core through the phenomenon of total reflection at the boundary between the two dielectrics. *See* PERMITTIVITY; REFLECTION OF ELECTROMAGNETIC RADIATION; REFRACTION OF WAVES.

Early dielectric guides were so lossy that they could be used only over short distances. Dielectric light pipes found surgical and laboratory use. In 1969 silica fibers were developed with attenuations of 32 dB/mi (20 dB/km), low enough to be of use for optical communication applications. Since then, further improvements have reduced losses to as low as 0.3 dB/mi (0.2 dB/km). Fiber guides are now the basis for a worldwide optical communication network. *See* FIBER-OPTICS IMAGING.

Planar, rectangular, and thin-film forms of dielectric guides are also important in guiding optical energy from one device to another in optoelectronic and integrated optic devices, for example, from a semiconductor laser to an electrooptic modulator on a gallium arsenide substrate. Because of the simpler geometry, the planar forms will be used to explain the principle. *See* INTEGRATED OPTICS; LASER; OPTICAL MODULATORS.

By far the most important dielectric guide at present is the optical fiber used for optical communications. Here the round core is surrounded by a cladding of slightly lower refractive index. The combination is surrounded by a protective jacket to prevent corrosion and give added strength, but this jacket plays no role in the optical guiding. *See* OPTICAL FIBERS. John R. Whinnery

Wavelength The distance between two points on a wave which have the same value and the same rate of change of the value of a parameter, for example, electric intensity, characterizing the wave. The wavelength, usually designated by the Greek letter λ, is equal to the speed of propagation c of the wave divided

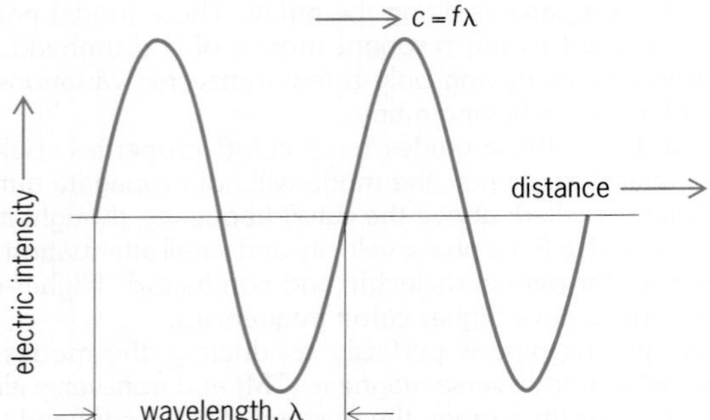

Wavelength λ and related quantities.

by the frequency of vibration f; that is, $\lambda = c/f$ (see illustration). *See* WAVE (PHYSICS); WAVE MOTION. William J. Galloway

Wavelength measurement Determination of the distance between successive wavefronts of equal phase of a wave. This article discusses wavelength measurement of electromagnetic waves in the radio, microwave, infrared, and optical regions. *See* ELECTROMAGNETIC RADIATION; WAVE MOTION; WAVELENGTH.

From the relation $\lambda = c/f$ between wavelength λ, speed c, and frequency f, the wavelength of a wave motion can be calculated if the speed is known and its frequency is measured. The ease and accuracy of electronic counting and timing make frequency measurement the most precise of all physical measurements. This method of wavelength determination is thus one of the most accurate, but only if the speed (phase velocity) is known. In free space the speed of an electromagnetic wave c_0 is, through the 1983 definition of the meter, fixed at exactly 299,792,458 m/s (≈186,282.397 mi/s), or roughly 300,000 km/s. Unless otherwise specified, it is general practice to quote the wavelength of an electromagnetic wave as the free-space value λ_0, given by the equation below.

$$\lambda_0 = \frac{c_0}{f}$$

See FREQUENCY MEASUREMENT; PHASE VELOCITY.

Radio and microwave regions. The presence of any dielectric material (such as air) or any magnetic matter with a permeability greater then unity will cause the wave to travel at a velocity lower than its free-space value. The speed is also altered if the waves pass through an aperture, are focused by a lens or mirror, or are constrained by a waveguide or transmission line. In such cases it may be more appropriate to measure the wavelength directly. In the pioneering experiments on radio waves,

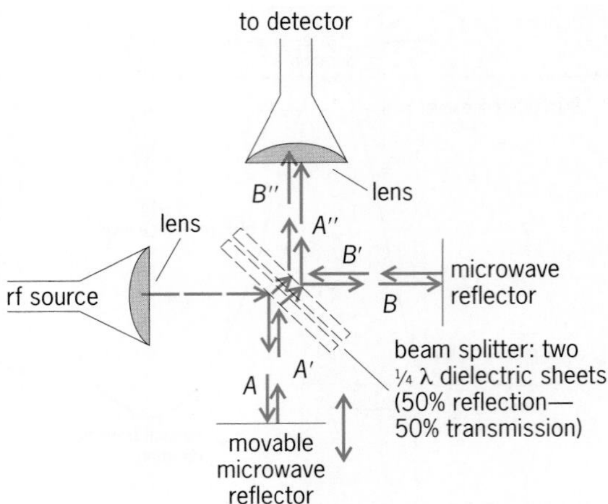

Wavelength measurement by the Michelson interferometer used at millimeter wavelengths.

it was found that standing waves existed in space whenever reflections occurred and that these provided a convenient means of measuring the wavelength. It thus became the convention to characterize waves by their wavelength rather than by their frequency, as is now more commonly the case. Specifying the frequency is preferred because, unlike the wavelength, it is independent of the speed of propagation and does not change as the wave moves from one medium to another. *See* ELECTROMAGNETIC WAVE TRANSMISSION; TRANSMISSION LINES; WAVEGUIDE.

In the microwave region the wavelengths are sufficiently short that it is convenient to measure them by using interferometer techniques directly analogous to those used with light.

In a typical interferometer used in the millimeter wavelength range (see illustration), a microwave beam is directed at a beamsplitter, which splits the beam into two parts, A and B, by partial reflection. The A beam is reflected to a movable reflector and reflected again as A'. The beamsplitter transmits part of this as A''. The transmitted part B of the original beam is reflected by a fixed microwave reflector as B'. This is partially reflected by the beamsplitter as B''. The beams A'' and B'' combine to form standing waves, which are then detected. Movement of the movable reflector causes the position of the standing wave to move, which causes the detected signal to pass through successive points of maximum and minimum amplitude. The distance between points of successive maxima or minima is one-half wavelength. This distance may be determined from the motion of the movable reflector. *See* INTERFEROMETRY.

Infrared and optical regions. The free-space wavelengths of monochromatic visible and infrared radiations can be derived directly from the above equation if their corresponding frequencies are known. Air wavelengths can be determined by dividing the vacuum wavelength by the refractive index of air. The optical frequencies of a number of standards have been precisely determined by measurement of their frequency referenced to the cesium-133 primary frequency standard, using femtosecond combs. The most accurate standards published to date are based on transitions in single trapped ions and have an uncertainty within a factor of 3 or better of the cesium standard. These optical standards are used for the realization of the unit of length, but are also expected to become secondary representations of the second. *See* FREQUENCY COMB; FREQUENCY MEASUREMENT; LENGTH; LIGHT; PHYSICAL MEASUREMENT; WAVELENGTH STANDARDS.

Wavelength values to an accuracy of 1 part in 10^5 can be determined with a spectrometer, spectrograph, or monochromator, in which a prism or diffraction grating is used as a dispersive element. Each wavelength forms a line image of the entrance slit at a particular angle. An unknown wavelength can be determined by interpolation with the pattern formed by a lamp emitting the tabulated characteristic wavelengths of a particular element. *See* DIFFRACTION GRATING; OPTICAL PRISM; SPECTROSCOPY.

The most precise wavelength measurements use an interferometer to compare the unknown wavelength λ_1 with a standard wavelength λ_2. Usually either the two-beam Michelson form or the multiple-beam Fabry-Perot form of interferometer is used.

When a number of wavelengths are mixed together in the input to a moving-carriage two-beam interferometer, the output signal is the summation of the many separate sine-wave signals having different periods. A Fourier analysis of this composite signal enables the separate wavelengths to be identified. This Fourier transform method is particularly useful for the measurement of complex spectra in the infrared. *See* FOURIER SERIES AND TRANSFORMS; INFRARED SPECTROSCOPY.

W. R. C. Rowley; Geoffrey P. Barwood

Wavelets The elementary building blocks in a mathematical tool for analyzing functions. The functions can be very diverse; examples are solutions of a differential equation, and one- and two-dimensional signals. The tool itself, the wavelet transform, is the result of a synthesis of ideas from many different fields, ranging from pure mathematics to quantum physics and electrical engineering.

In many practical applications, it is desirable to extract frequency information from a signal—in particular, which frequencies are present and their respective importance. An example is the decomposition into spectral lines in spectroscopy. The tool that is generally used to achieve this is the Fourier transform. Many applications, however, concern nonstationary signals, in which the makeup of the different frequency components is constantly shifting. An example is music, where this shifting nature has been recognized for centuries by the standard notation, which tells a musician which note (frequency information) to play when and how long (time information). For signals of this nature, a time-frequency representation is needed. *See* FOURIER SERIES AND TRANSFORMS.

There exist many different mathematical tools leading to a time-frequency representation of a given signal, each with its own strengths and weaknesses. The wavelet transform is such a time-frequency analysis tool. Its strength lies in its ability to deal well with transient high-frequency phenomena, such as sudden peaks or discontinuities, as well as with the smoother portions of the signal. (An example is a crack in the sound from a damaged record, or the attack at the start of a music note.) The wavelet transform is less well adapted to harmonically oscillating parts in the signal, for which Fourier-type methods are more indicated.

Applications of wavelets include various forms of data compression (such as for images and fingerprints), data analysis (nuclear magnetic resonance, radar, seismograms, and sound), and numerical analysis (fast solvers for partial differential equations). *See* DIFFERENTIAL EQUATION; INTEGRAL TRANSFORM; NUMERICAL ANALYSIS.

Ingrid Daubechies

Wavellite A hydrated phosphate of aluminum mineral with composition $Al_3(OH)_3(PO_4)_2 \cdot 5H_2O$, in which small amounts of fluorine and iron may substitute for the hydroxyl group (OH) and aluminum (Al), respectively. Wavellite crystallizes in the orthorhombic system. The crystals are stout to long prismatic, but are rare. Wavellite commonly occurs as globular aggregates of fibrous structure and as encrusting and stalactitic masses. Wavellite crystals range in color from colorless and white to different shades of blue, green, yellow, brown, and black. Wavellite is found at many places in Europe and North America; it is also abundant in tin veins at Llallagua, Bolivia. *See* PHOSPHATE MINERALS.

Wayne R. Lowell

Wax, animal and vegetable Any of the substances containing esters of higher fatty acids and long-chain monohydric alcohols. From a practical standpoint, this definition is inadequate because it allows liquids such as sperm whale oil and jojoba oil to be called waxes, and it fails to indicate the complexity of waxes. While waxes do contain wax esters, they are seldom if ever pure. They are usually mixtures that may contain high-molecular-weight acids, alcohols, esters, ketones, hydrocarbons, sterols, diesters, hydroxyacids, and so forth, as well as the wax esters. *See* ESTER.

Practical wax formulators use physical, rather than chemical, properties to define waxes. Waxes must: be a solid at 68°F (20°C); be crystalline; melt above 140°F (40°C) without decomposition; have relatively low viscosity above the melting point; have consistency and solubility properties that are strongly dependent upon temperature; and be capable of being polished under slight pressure.

Wax sources are abundant. They have been isolated from the outer layers of bacteria, the roots, stems, leaves, fruit, and flowers of plants, the exudates of insects, the skin and hair of some animals, and the bodies of certain marine and land animals. Carnauba wax is extracted from an exudate on the leaves of the carnauba palm (*Copernicia prunifera*). Candelilla wax is obtained from a coating on the stem of *Euphorbia antisyphilitica*, a leafless desert shrub. Beeswax is an exudate of the honeybee. Wool wax is obtained from sheep wool. Many other waxes have been of commercial importance from time to time, but their availability or cost has forced them from the market. *See* CARNAUBA WAX; FAT AND OIL.

Howard M. Hickman

Wax, petroleum A substance produced primarily from the dewaxing of lubricating-oil fractions of petroleum. It may be of either the crystalline or microcrystalline type. Petroleum wax has a wide variety of uses. It is used to coat paper products, to blend with other waxes for the manufacture of candles, in the manufacture of electrical equipment and many polishes for home and industry, and as a source material for oxidized products. The softer waxes, such as petroleum jelly, after proper purification, are being used as medicinal products. *See* DEWAXING OF PETROLEUM; PETROLEUM PROCESSING; PETROLEUM PRODUCTS.

Wayne E. Kuhn

Weak nuclear interactions Fundamental interactions of nature that play a significant role in elementary-particle and nuclear physics, and are distinguished from other such interactions by special properties such as participation of all the fundamental fermions and failure to conserve parity and to respect particle–antiparticle symmetry. According to present understanding, the four fundamental forces of nature are the gravitational, the electromagnetic, the strong, and the weak. The weak force has short range (less than 10^{-17} m) and at low energies is very feeble compared to the strong and electromagnetic forces, but it can be distinguished from the latter by its special character. For example, all of matter (with the possible exception of dark matter) appears to consist of certain basic constituents, the quarks and leptons, collectively called the fundamental fermions. While only quarks participate in strong interactions, and only the quarks and the charged leptons ($e^{\pm}$, $\mu^{\pm}$, $\tau^{\pm}$) participate in electromagnetic interactions, all of the fundamental fermions, including neutrinos, engage in weak interactions. Moreover, while the strong and electromagnetic interactions respect spatial inversion symmetry (parity) and are also particle–antiparticle (charge conjugation) symmetric, the weak interaction violates these two symmetries. *See* FUNDAMENTAL INTERACTIONS; LEPTON; PARITY (QUANTUM MECHANICS); QUARKS; SYMMETRY LAWS (PHYSICS).

Weak interactions are classified as charged or neutral, depending on whether or not a particle participating in a weak reaction suffers a change of electric charge of one electronic unit. Observed charged weak interactions include nuclear beta decay and electron capture, muon capture on nuclei, decays of the μ and τ

leptons and the $\pi^{\pm}$ and K mesons, the slow hyperon decays, decays of c- and b-quarked mesons and baryons, and the decay of the top quark. Charged neutrino–nucleon and neutrino–lepton scattering reactions, and the phenomenon of neutrino oscillations, are additional manifestations of the charged weak interaction. Neutral weak interactions include but are not limited to neutral neutrino–nucleon and neutrino–lepton scattering as well as the weak electron–nucleon reaction. *See* BARYON; ELEMENTARY PARTICLE; HYPERON; MESON; NEUTRAL CURRENTS.

In 1934 E. Fermi proposed a theory that gave a good phenomenological account of many aspects of nuclear beta decay, especially after its generalization by G. Gamow and E. Teller in 1936, and anticipated a number of future developments. The theory was further generalized in 1958, following the discovery of parity nonconservation in the weak interactions. However, Fermi's theory is not renormalizable and it leads to difficulties (breakdown of unitarity, that is, nonconservation of probability) at high energies. *See* RENORMALIZATION.

The solution to these defects was found in a theory uniting weak and electromagnetic interactions, and it was based on a combination of subtle ideas. First is the notion that the theory must be invariant under local gauge transformations, which are phase transformations of a quantum field that can vary in an arbitrary manner from one space–time point to another. Quantum electrodynamics is such a theory. *See* GAUGE THEORY.

The second important idea is spontaneous symmetry breaking. This means that one has a quantum field theory possessing a certain symmetry not shared by the ground state of the system. S. Glashow, S. Weinberg (1967) and independently A. Salam (1968) constructed a successful gauge theory combining weak and electromagnetic interactions, and this theory was proved to be renormalizable by G. 't Hooft in 1971. In 1974, the Glashow-Weinberg-Salam theory was combined with a new gauge theory of strong interactions, called quantum chromodynamics, and the synthesis became known as the standard model. *See* ELECTROWEAK INTERACTION; HIGGS BOSON; QUANTUM CHROMODYNAMICS; STANDARD MODEL; SYMMETRY BREAKING. Eugene D. Commins

Weakly interacting massive particle (WIMP)

A hypothetical elementary particle that might make up most of the matter in the universe, and that is also predicted to exist in supersymmetry theory. Most matter is detected only through its gravitational effects; this "dark matter" has not been observed to emit, absorb, or reflect light of any wavelength. The total amount of dark matter appears to be approximately ten times as great as all the ordinary matter in the universe, and about one hundred times as great as all the visible matter. The nature of the dark matter is not yet known, although many experiments are under way to try to discover it directly or indirectly. *See* COSMOLOGY; UNIVERSE.

Almost all the currently available data in elementary particle physics can be accounted for by a theory called the standard model, in which matter is made of quarks (the building blocks of protons and neutrons) and leptons (including electrons and neutrinos), while the strong, weak, and electromagnetic forces are transmitted by particles like the photon (the carrier of electromagnetic forces). However, the standard model does not predict the existence of any particle—say, X—that could be the dark matter. Most efforts to go beyond the standard model of particle physics have been based on the idea of supersymmetry, and most versions of supersymmetry predict that there will be a stable weakly interacting massive particle (WIMP) that would be a natural candidate for the X particles. Dark matter made of WIMPs would be "cold" dark matter (CDM), and a version of CDM theory has become the standard theory of structure formation in cosmology. *See* ELEMENTARY PARTICLE; STANDARD MODEL; SUPERSYMMETRY.

There is now abundant evidence for dark matter around galaxies and clusters of galaxies, and on larger scales in the universe. Gas and satellites at large distances from galaxies have orbital velocities similar to those at smaller distances from the center, which indicates that most of the mass in the galaxy must not be near the center, where most stars are, but in a roughly spherical dark matter halo that extends to perhaps ten times the optical size of the galaxy and has a mass at least ten times that of all the stars. Confirmation of the existence of such dark-matter halos has come from gravitational lensing observations, showing that light from more distant galaxies is bent by the gravity of nearer galaxies. *See* CELESTIAL MECHANICS; GALAXY, EXTERNAL; GRAVITATIONAL LENS; MILKY WAY GALAXY.

There is also much evidence for dark matter in clusters of galaxies. The astronomer Fritz Zwicky pointed out in 1933 that the galaxies in one nearby cluster were moving at such high speeds that they would not be held together gravitationally unless there was much more mass than was indicated by the light from their stars. This same was subsequently found to be true of other clusters. Later, similar conclusions were reached from x-ray observations and gravitational lensing observations of clusters. *See* X-RAY ASTRONOMY.

Supersymmetry is the hypothesis that there is a relationship between the two known classes of particles, bosons and fermions. According to supersymmetry, for every kind of boson in the universe, there must also be a corresponding fermion with the same electric charge and very similar interactions with other particles. Since these hypothetical sypersymmetric partner particles have not been discovered yet, if supersymmetry is right their masses must be too large for them to have been produced at current particle accelerators. Thus far, the evidence for supersymmetry is only indirect, but if the theory is right many supersymmetric partner particles should be produced at accelerators such as the Large Hadron Collider (LHC) being built in Geneva, Switzerland. *See* PARTICLE ACCELERATOR; QUANTUM STATISTICS.

Efforts to detect WIMPs directly are based on detecting their scattering from nuclei. WIMPs can also be detected indirectly, for example by looking for particles coming from their annihilation. WIMPs are also expected to be produced at accelerators such as the LHC from rapid decays of heavier supersymmetric partner particles, and this could be where they are discovered first if they are not seen before that in direct or indirect search experiments. Failure to see supersymmetric particles at LHC energies would mean that current ideas about supersymmetry are wrong. Joel R. Primack

Wear

The removal of material from a solid surface as a result of sliding action. It constitutes the main reason why the artifacts of society (automobiles, washing machines, tape recorders, cameras, clothing) become useless and have to be replaced. There are a few uses of the wear phenomenon, but in the great majority of cases wear is a nuisance, and a tremendous expenditure of human and material resources is required to overcome the effects.

Adhesive wear is the only universal form of wear, and in many sliding systems it is also the most important. It arises from the fact that, during sliding, regions of adhesive bonding, called junctions, form between the sliding surfaces. If one of these junctions does not break along its original interface, then a chunk from one of the sliding surfaces will have been transferred to the other surface. In this way, an adhesive wear particle will have been formed. Initially adhering to the other surface, adhesive particles soon become loose and can disappear from the sliding system. *See* FRICTION.

Abrasive wear is produced by a hard, sharp surface sliding against a softer one and digging out a groove. The abrasive agent may be one of the surfaces (such as a file), or it may be a third component (such as sand particles in a bearing abrading material from each surface). Abrasive wear coefficients are large compared to adhesive ones. Thus, the introduction of abrasive particles into a sliding system can greatly increase the wear rate;

automobiles, for example, have air and oil filters to catch abrasive particles before they can produce damage.

Corrosive wear arises when a sliding surface is in a corrosive environment, and the sliding action continuously removes the protective corrosion product, thus exposing fresh surface to further corrosive attack. *See* CORROSION.

Surface fatigue wear occurs as result of the formation and growth of cracks. It is the main form of wear of rolling devices such as ball bearings, wheels on rails, and gears. During continued rolling, a crack forms at or just below the surface and gradually grows until a large particle is lifted right out of the surface.

Most manifestations of wear are highly objectionable, but the phenomenon does have a few uses. Thus, a number of systems for recording information (pencil and paper, chalk and blackboard) operate via a wear mechanism. Some methods of preparing solid surfaces (filling, sandpapering, sandblasting) also make use of wear. *See* ABRASIVE. Ernest Rabinowicz

Weasel The common name for 14 species belonging to the family Mustelidae. These carnivores are native to every continent except Australia and Antarctica. They are also absent from Madagascar and most oceanic islands.

Weasels have a long slender body and short legs. Most are brown or reddish-brown on the back and whitish or yellowish on the underside. During winter in northern areas, the fur of some species, such as the long-tailed (*Mustela frenata*) and short-tailed (*M. erminea*) weasels, changes entirely to white except for the tip of the tail, which may remain black. The limbs possess five digits with nonretractile, curved claws. Ears are short and rounded.

Weasels are very quick and their movements are difficult to follow. When running, the back is strongly arched at each bound and the tail is held straight out behind or at an angle. Weasels have keen senses of smell, sight, and hearing and are tireless hunters. They are voracious feeders, catching animals their own size or larger for food. They feed primarily on mice, rats, shrews, and moles, although earthworms, insects, frogs, lizards, rabbits, squirrels, snakes, and small birds may also be eaten. Weasels sever the prey's spinal column with one or two strong, swift bites at the base of the skull. The weasel's slim body and short legs enable it to hunt in the narrow openings of stone walls, under logs and rocks, and in rodent burrows. Although chiefly ground-dwelling animals, many weasels are excellent tree climbers and are able to raid birds' nests. Weasels are more active at night than during the day, and they remain active in winter. They inhabit farmland as well as woodlands and swamps. *See* CARNIVORA. Donald W. Linzey

Weather The state of the atmosphere, as determined by the simultaneous occurrence of several meteorological phenomena at a geographical locality or over broad areas of the Earth. When such a collection of weather elements is part of an interrelated physical structure of the atmosphere, it is termed a weather system, and includes phenomena at all elevations above the ground. More popularly, weather refers to a certain state of the atmosphere as it affects humans' activities on the Earth's surface. In this sense, it is often taken to include such related phenomena as waves at sea and floods on land.

A weather element is any individual physical feature of the atmosphere. At a given locality, at least seven such elements may be observed at any one time. These are clouds, precipitation, temperature, humidity, wind, pressure, and visibility. Each of these principal elements is divided into many subtypes. *See* WEATHER MAP.

The various forms of precipitation are included by international agreement among the hydrometeors, which comprise all the visible features in the atmosphere, besides clouds, that are due to water in its various forms. For convenience in processing weather data and information, this definition is made to include some phenomena not due to water, such as dust and smoke. Some of the more common hydrometeors include rain, snow, fog, hail, dew, and frost. *See* PRECIPITATION (METEOROLOGY).

Certain optical and electrical phenomena have long been observed among the weather elements. These include lightning, aurora, solar or lunar corona, and halo. *See* AIR MASS; ATMOSPHERE; AURORA; CLOUD; FRONT; LIGHTNING; METEOROLOGY; STORM; WEATHER OBSERVATIONS; WIND. Philip F. Clapp

Weather forecasting and prediction Processes for formulating and disseminating information about future weather conditions based upon the collection and analysis of meteorological observations. Weather forecasts may be classified according to the space and time scale of the predicted phenomena. Atmospheric fluctuations with a length of less than 100 m (330 ft) and a period of less than 100 s are considered to be turbulent. The study of atmospheric turbulence is called micrometeorology; it is of importance for understanding the diffusion of air pollutants and other aspects of the climate near the ground. Standard meteorological observations are made with sampling techniques that filter out the influence of turbulence. Common terminology distinguishes among three classes of phenomena with a scale that is larger than the turbulent microscale: the mesoscale, synoptic scale, and planetary scale. *See* MICROMETEOROLOGY.

The mesoscale includes all moist convection phenomena, ranging from individual cloud cells up to the convective cloud complexes associated with prefrontal squall lines, tropical storms, and the intertropical convergence zone. Also included among mesoscale phenomena are the sea breeze, mountain valley circulations, and the detailed structure of frontal inversions. Most mesoscale phenomena have time periods less than 12 h. The prediction of mesoscale phenomena is an area of active research. Most forecasting methods depend upon empirical rules or the short-range extrapolation of current observations, particularly those provided by radar and geostationary satellites. Forecasts are usually couched in probabilistic terms to reflect the sporadic character of the phenomena. Since many mesoscale phenomena pose serious threats to life and property, it is the practice to issue advisories of potential occurrence significantly in advance. These "watch" advisories encourage the public to attain a degree of readiness appropriate to the potential hazard. Once the phenomenon is considered to be imminent, the advisory is changed to a "warning," with the expectation that the public will take immediate action to prevent the loss of life. *See* MESOMETEOROLOGY; SQUALL.

The next-largest scale of weather events is called the synoptic scale, because the network of meteorological stations making simultaneous, or synoptic, observations serves to define the phenomena. The migratory storm systems of the extratropics are synoptic-scale events, as are the undulating wind currents of the upper-air circulation which accompany the storms. The storms are associated with barometric minima, variously called lows, depressions, or cyclones. The synoptic method of forecasting consists of the simultaneous collection of weather observations, and the plotting and analysis of these data on geographical maps. An experienced analyst, having studied several of these maps in chronological succession, can follow the movement and intensification of weather systems and forecast their positions. This forecasting technique requires the regular and frequent use of large networks of data. *See* WEATHER MAP.

Planetary-scale phenomena are persistent, quasistationary perturbations of the global circulation of the air with horizontal dimensions comparable to the radius of the Earth. These dominant features of the general circulation appear to be correlated with the major orographic features of the globe and with the latent and sensible heat sources provided by the oceans. They tend to control the paths followed by the synoptic-scale storms, and to draw upon the synoptic transients for an additional source of heat and momentum. *See* ATMOSPHERE; METEOROLOGICAL INSTRUMENTATION; WEATHER OBSERVATIONS. John P. Gerrity

Numerical weather prediction is the prediction of weather phenomena by the numerical solution of the equations governing the motion and changes of condition of the atmosphere. Numerical weather prediction techniques, in addition to being applied to short-range weather prediction, are used in such research studies as air-pollutant transport and the effects of greenhouse gases on global climate change. *See* AIR POLLUTION; GREENHOUSE EFFECT; JET STREAM; UPPER-ATMOSPHERE DYNAMICS.

The first operational numerical weather prediction model consisted of only one layer, and therefore it could model only the temporal variation of the mean vertical structure of the atmosphere. Computers now permit the development of multilevel (usually about 10–20) models that could resolve the vertical variation of the wind, temperature, and moisture. These multilevel models predict the fundamental meteorological variables for large scales of motion. Global models with horizontal resolutions as fine as 125 mi (200 km) are being used by weather services in several countries. Global numerical weather prediction models require the most powerful computers to complete a 10-day forecast in a reasonable amount of time.

Research models similar to global models could be applied for climate studies by running for much longer time periods. The extension of numerical predictions to long time intervals (many years) requires a more accurate numerical representation of the energy transfer and turbulent dissipative processes within the atmosphere and at the air-earth boundary, as well as greatly augmented computing-machine speeds and capacities.

Long-term simulations of climate models have yielded simulations of mean circulations that strongly resemble those of the atmosphere. These simulations have been useful in explaining the principal features of the Earth's climate, even though it is impossible to predict the daily fluctuations of weather for extended periods. Climate models have also been used successfully to explain paleoclimatic variations, and are being applied to predict future changes in the climate induced by changes in the atmospheric composition or characteristics of the Earth's surface due to human activities. *See* CLIMATE HISTORY; CLIMATE MODIFICATION; PALEOCLIMATOLOGY. Richard A. Anthes

Surface meteorological observations are routinely collected from a vast continental data network, with the majority of these observations obtained from the middle latitudes of both hemispheres. Commercial ships of opportunity, military vessels, and moored and drifting buoys provide similar in-place measurements from oceanic regions. Information on winds, pressure, temperature, and moisture throughout the troposphere and into the stratosphere is routinely collected from (1) balloon-borne instrumentation packages (radiosonde observations) and commercial and military aircraft which sample the free atmosphere directly; (2) ground-based remote-sensing instrumentation such as wind profilers (vertically pointing Doppler radars), the National Weather Service Doppler radar network, and lidars; and (3) special sensors deployed on board polar orbiting or geostationary satellites. The remotely sensed observations obtained from meteorological satellites have been especially helpful in providing crucial measurements of areally and vertically averaged temperature, moisture, and winds in data-sparse (mostly oceanic) regions of the world. Such measurements are necessary to accommodate modern numerical weather prediction practices and to enable forecasters to continuously monitor global storm (such as hurricane) activity. *See* LIDAR; METEOROLOGICAL INSTRUMENTATION; RADAR METEOROLOGY.

Forecast products and forecast skill are classified as longer term (greater than 2 weeks) and shorter term. These varying skill levels reflect the fact that existing numerical prediction models such as the medium-range forecast have become very good at making large-scale circulation and temperature forecasts, but are less successful in making weather forecasts. An example is the prediction of precipitation amount and type given the occurrence of precipitation and convection. Each of these forecasts is progressively more difficult because of the increasing importance of mesoscale processes to the overall skill of the forecast. *See* PRECIPITATION (METEOROLOGY). Lance F. Bosart

Nowcasting is a form of very short range weather forecasting. The term nowcasting is sometimes used loosely to refer to any area-specific forecast for the period up to 12 h ahead that is based on very detailed observational data. However, nowcasting should probably be defined more restrictively as the detailed description of the current weather along with forecasts obtained by extrapolation up to about 2 h ahead. Useful extrapolation forecasts can be obtained for longer periods in many situations, but in some weather situations the accuracy of extrapolation forecasts diminishes quickly with time as a result of the development or decay of the weather systems. *See* WEATHER. Keith A. Browning

Forecasts of time averages of atmospheric variables, for example, sea surface temperature, where the lead time for the prediction is more than 2 weeks, are termed long-range or extended-range climate predictions. Extended-range predictions of monthly and seasonal average temperature and precipitation are known as climate outlooks. The accuracy of long-range outlooks has always been modest because the predictions must encompass a large number of possible outcomes, while the observed single event against which the outlook is verified includes the noise created by the specific synoptic disturbances that actually occur and that are unpredictable on monthly and seasonal time scales. According to some estimates of potential predictability, the noise is generally larger than the signal in middle latitudes.

Edward A. O'Lenic

Weather map A map or a series of maps that is used to depict the evolution and life cycle of atmospheric phenomena at selected times at the surface and in the free atmosphere. Weather maps are used for the analysis and display of in-place observational measurements and computer-generated analysis and forecast fields derived from weather and climate prediction models by research and operational meteorologists, government research laboratories, and commercial firms. Similar analyses derived from sophisticated computer forecast models are displayed in map form for forecast periods of 10–14 days in advance to provide guidance for human weather forecasters. *See* METEOROLOGICAL INSTRUMENTATION; WEATHER OBSERVATIONS.

Rapid advances in computer technology and visualization techniques, as well as the continued explosive growth of the Internet distribution of global weather observations, satellite and radar imagery, and model analysis and forecast fields, have revolutionized how weather, climate, and forecast data and information can be conveyed to both the general public and sophisticated users in the public and commercial sectors. People and organizations with access to the Internet can access weather and climate information in a variety of digital or map forms in support of a wind range of professional and personal activities. *See* CLIMATOLOGY; METEOROLOGICAL SATELLITES; METEOROLOGY; RADAR METEOROLOGY; WEATHER FORECASTING AND PREDICTION.

Lance F. Bosart

Weather modification Human influence on the weather and, ultimately, climate. This can be either intentional, as with cloud seeding to clear fog from airports or to increase precipitation, or unintentional, as with air pollution, which increases aerosol concentrations and reduces sunlight. Weather is considered to be the day-to-day variations of the environment—temperature, cloudiness, relative humidity, wind-speed, visibility, and precipitation. Climate, on the other hand, reflects the average and extremes of these variables, changing on a seasonal basis. Weather change may lead to climate change, which is assessed over a period of years. *See* AIR POLLUTION; CLIMATE HISTORY; CLOUD PHYSICS.

Specific processes of weather modification are as follows: (1) Change of precipitation intensity and distribution result from changes in the colloidal stability of clouds. For example, seeding of supercooled water clouds with dry ice (solid carbon dioxide, CO_2) or silver iodide (AgI) leads to ice crystal growth and fallout; layer clouds may dissipate, convective clouds may grow. (2) Radiation change results from changes of aerosol or clouds (deliberately with a smoke screen, or unintentionally with air pollution from combustion), from changes in the gaseous constituents of the atmosphere (as with carbon dioxide from fossil fuel combustion), and from changes in the ability of surfaces to reflect or scatter back sunlight (as replacing farmland by houses.) (3) Change of wind regime results from change in surface roughness and heat input, for example, replacing forests with farmland. *See* PRECIPITATION (METEOROLOGY). John Hallett

Weather observations The measuring, recording, and transmitting of data of the variable elements of weather. In the United States the National Weather Service (NWS), a division of the National Oceanic and Atmospheric Administration (NOAA), has as one of its primary responsibilities the acquisition of meteorological information. The data are sent by various communication methods to the National Meteorological Center.

At the Center, the raw data are fed into large computers that are programmed to plot, analyze, and process the data and also to make prognostic weather charts. The processed data and the forecast guidance are then distributed by special National Weather Service systems and conventional telecommunications to field offices, other government agencies, and private meteorologists. They in turn prepare forecasts and warnings based on both processed and raw data. *See* WEATHER MAP.

A wide variety of meteorological data are required to satisfy the needs of meteorologists, climatologists, and users in marine activities, forestry, agriculture, aviation, and other fields. This has led to a dual surface-observation program: the Synoptic Weather Program and the Basic Observations Program. *See* AERONAUTICAL METEOROLOGY; AGRICULTURAL METEOROLOGY; INDUSTRIAL METEOROLOGY.

The Synoptic Weather Program is designed to assist in the preparation of forecasts and to provide data for international exchange. Worldwide surface observations are taken at standard times [0000, 0600, 1200, and 1800 Universal Time Coordinated (UTC)] and sent in synoptic code.

The Basic Observations Program routinely provides meteorological data every hour. Special observations are taken at any intervening time to report significant weather events or changes. Observation sites are located primarily at airports; a few are in urban centers. At these sites, human observers report the weather elements.

Present weather consists of a number of hydrometers, such as liquid or frozen precipitation, fog, thunderstorms, showers, and tornadoes, and of lithometers, such as haze, dust, smog, dust devils, and blowing sand. The amount of cloudiness is also reported. *See* FOG; METEOROLOGICAL OPTICS; PRECIPITATION (METEOROLOGY); SMOG; THUNDERSTORM; TORNADO.

Pressure measurements are read from either a mercury or precision aneroid barometer located at the station. A microbarograph provides a continuous record of the pressure, from which changes in specific intervals of time are reported. Pressure changes are frequently quite helpful in short-range prediction of weather events. *See* AIR PRESSURE.

Temperature and humidity are measured by a hygrothermometer, located near the center of the runway complex at many airport stations. The readings are transmitted to the observation site. The temperature dial indicator is equipped with pointers to determine maximum and minimum temperature extremes. *See* HUMIDITY; HYGROMETER; TEMPERATURE MEASUREMENT.

Wind speed and direction measurements are telemetered into most airport stations. The equipment, consisting of an anemometer and a wind vane, is located near the center of the runway complex at participating airports; elsewhere it is placed in an unsheltered area. *See* WIND MEASUREMENT.

Various types of clouds and their heights are reported. The lowest height of opaque clouds covering half or more of the sky is known as the ceiling, and is normally measured by a ceilometer at first-order stations. *See* CLOUD.

Upper-air observations have been made by the National Weather Service with radiosondes. The radiosonde is a small, expendable instrument package that is suspended below a 6-ft-diameter (2-m) balloon filled with hydrogen or helium. As the radiosonde is carried aloft, sensors on it measure profiles of pressure, temperature, and relative humidity. By tracking the position of the radiosonde in flight with a radio direction finder or radio navigation system, such as Loran or the Global Positioning System (GPS), information on wind speed and direction aloft is also obtained.

Understanding and accurately predicting changes in the atmosphere requires adequate observations of the upper atmosphere. Radiosonde observations, plus routine aircraft reports, radar, and satellite observations, provide meteorologists with a three-dimensional picture of the atmosphere. *See* LORAN; METEOROLOGICAL INSTRUMENTATION; SATELLITE NAVIGATION SYSTEMS; WEATHER OBSERVATIONS; WIND MEASUREMENT.

Weather radars distributed throughout the United States are used to observe precipitation within a radius of about 250 nmi (460 km), and associated wind fields (utilizing the Doppler principle) within about 125 nmi (230 km). The primary component of this set of weather radars is known as NEXRAD (Next Generation Weather Radar). These radars provide information on rainfall intensity, likelihood of tornadoes or severe thunderstorms, projected paths of individual storms (both ambient and within-storm wind fields), and heights of storms for short-range (up to 3 h) forecasts and warnings. *See* DOPPLER RADAR; RADAR METEOROLOGY.

Geostationary weather satellites near 22,000 mi (36,000 km) above the Earth transmit pictures depicting the cloud cover over vast expanses of the hemisphere. Using still photographs and animated images, the meteorologist can determine, among other things, areas of potentially severe weather and the motion of clouds and fog. In addition, the satellite does an outstanding job of tracking hurricanes over the ocean where few other observations are taken. *See* HURRICANE; METEOROLOGICAL SATELLITES.

Ground-based lightning detection systems detect the electromagnetic wave that emanates from the lightning path as the lightning strikes the ground. Lightning information has proven to be operationally valuable to a wide variety of users and as a supplement to other observing systems, particularly radar and satellites. *See* LIGHTNING; LIGHTNING AND SURGE PROTECTION; MESOMETEOROLOGY; METEOROLOGY; WEATHER FORECASTING AND PREDICTION. Frederick S. Zbar; Ronald L. Lavoie

Weathering processes The response of geologic materials to the environment (physical, chemical, and biological) at or near the Earth's surface. This response typically results in a reduction in size of the weathering materials; some may become as tiny as ions in solution.

The agents and energies that activate weathering processes and the products resulting therefrom have been classified traditionally as physical and chemical in type. In classic physical weathering, rock materials are broken by action of mechanical forces into smaller fragments without change in chemical composition, whereas in chemical weathering the process is characterized by change in chemical composition. In practice, however, the two processes commonly overlap.

Specific agents of weathering may be recognized and correlated with the types of effects they produce. Important agents of weathering are water in all surface occurrences (rain, soil and ground water, streams, and ocean); the atmosphere

(H_2O, O_2, CO_2, wind); temperature (ambient and changing, especially at the freezing point of water); insolation (on large bare surfaces); ice (in soil and glaciers); gravity; plants (bacteria and macroforms); animals (micro and macro, including humans). Human modifications of otherwise geologic weathering that have increased exponentially during recent centuries include construction, tillage, lumbering, use of fire, chemically active industry (fumes, liquid, and solid effluents), and manipulation of geologic water systems.

Products of physical weathering include jointed (horizontal and vertical) rock masses, disintegrated granules, frost-riven soil and surface rock, and rock and soil flows. Products of chemical weathering include the soil, and the clays used in making ceramic structural products, whitewares, refractories, various fillers and coating of paper, portland cement, absorbents, and vanadium. These are the relatively insoluble products of weathering; characteristically they occur in clays, siltstones, and shales. Sand-size particles resulting from both physical and chemical weathering may accumulate as sandstones.

After precipitation, the relatively soluble products of chemical weathering give rise to products and rocks such as limestone, gypsum, rock salt, silica, and phosphate and potassium compounds useful as fertilizers.

Walter D. Keller

Wedge A piece of resistant material whose two major surfaces make an acute angle. It is closely related to the inclined plane and is used to multiply the applied force and to change the direction in which it acts (see illustration). *See* INCLINED PLANE.

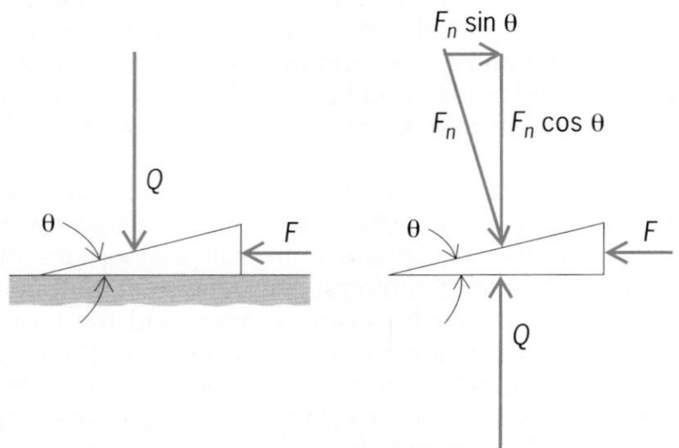

Forces acting on a wedge.

Force F is the smaller applied force and Q is the larger force to be exerted. In the absence of friction, forces must act normal to their surfaces; thus the actual force on the inclined surface is not Q but a larger force F_n. Summing up forces in the horizontal and vertical directions gives Eqs. (1).

$$\begin{aligned} F_n \sin\theta - F &= 0 \\ Q - F_n \cos\theta &= 0 \end{aligned} \tag{1}$$

Combining the expressions for F and Q and solving for F gives Eq. (2).

$$F = Q \tan\theta \tag{2}$$

If angle θ is small, the reaction of Q against F is exceeded by the friction between the face of the wedge and the adjacent body on which it rests. Thus the wedge tends to remain in position even when loaded by a large force Q. *See* SIMPLE MACHINE.

Richard M. Phelan

Weeds Unwanted plants or plants whose negative values outweigh the positive values in a given situation. Weeds impact growers each year in reduced yield and quality of agricultural products. Especially in tropical areas, irrigation systems have become unusable because of clogging with weeds. Weeds can harbor deleterious disease organisms and insects that harm crops and livestock. In addition, weeds can cause allergic reactions and serious skin problems (poison ivy), break up pavement, slow or stop water flow in municipal water supplies, interfere with power lines, cause fire hazards around buildings and along railroad tracks, and produce poisonous plant parts. *See* ALLERGY.

The most serious weeds are those that succeed in invading new areas and surviving at the expense of other plants by monopolizing light, nutrients, and water or by releasing chemicals detrimental to the growth of surrounding vegetation (allelopathy). In the plant kingdom, dozens of species have been shown to release allelopathic chemicals from roots, leaves, and stems. *See* ALLELOPATHY.

Classification. Weeds can be classified as summer annuals, which germinate in the spring, set seed, and die in the fall (crabgrass); winter annuals, which germinate in the fall, set seed, and die in the spring (common chickweek); biennials, which germinate one year, overwinter, set seed, and die the following summer (wild carrot); simple perennials, which live for several years but spread only by seed (dandelion); and creeping perennials, which live for several years and can spread both by seed and by underground roots or rhizomes (field bindweed).

Control methods. Hand pulling, fire, flooding, and tillage are useful for controlling weeds. Insects and pathogens have also been introduced to control certain weed species. Techniques such as herbicides, computerization of spray and tillage technology, remote sensing for weed mapping and identification, and laser treatment have also been explored. Herbicides have resulted in large improvements in the availability and quality of food. They have increased the feasibility of no-till agriculture, leading to significant reductions in soil erosion. They commonly kill weeds by disrupting a physiological process that is not present in animals. Some, however, are moderately high in toxicity and must be used carefully. Rare individual weeds that are genetically resistant to the herbicide have flourished and reproduced, leading to populations of weeds resistant to that herbicide. *See* HERBICIDE.

Arnold P. Appleby

Biological control. Biological control involves the use of natural enemies (parasites, pathogens, and predators) to control pest populations. In the case of weed pests, the primary natural enemy groups utilized are arthropods, fungal pathogens, and vertebrates. The two major approaches are classical and inundative biological control. Biocontrol can be a highly effective and cost-efficient means of controlling weeds without the use of chemical herbicides.

Classical biocontrol (also termed the inoculative or importation method) is based on the principle of population regulation by natural enemies. Most naturalized weeds leave behind their natural enemies when they colonize new areas, and so can increase to significant densities. Classical biocontrol involves the importation of natural enemies, usually from the area of origin of the weed (and preferably from a part of its native range that is a good climatic match with the intended control area), and their field release. Imported biocontrol agents must be host specific to the target weed.

Inundative control is the mass production and periodic release of large numbers of biocontrol agents to achieve controlling densities. It can be used where existing populations of agents are lacking or where existing populations that are not self-sustaining at high, controlling densities can be augmented. A chief advantage of this method is that it can be integrated with conventional fanning practices on cultivated croplands.

Arthropods, especially insects, are heavily utilized as imported biocontrol agents in uncultivated environments such as grasslands and aquatic systems.

Rusts (Uredinales) are the fungal group most frequently employed as imported agents in classical biocontrol programs. The rust *Puccinia chondrillina*, imported from Italy, controlled the narrow-leaf form of skeletonweed (*Chondrilla juncea*) in Australia. Formulations of spores of foreign and endemic pathogens can be used as mycoherbicides in inundative applications.

Vertebrate animal agents, such as goats, typically do not possess a high degree of host-plant specificity, and their feeding has to be carefully managed to focus it on the target weeds. *See* ARTHROPODA; FUNGI.

Charles E. Turner

Weight The gravitational weight of a body is the force with which the Earth attracts the body. By extension, the term is also used for the attraction of the Sun or a planet on a nearby body. This force is proportional to the body's mass and depends on the location. Because the distance from the surface to the center of the Earth decreases at higher latitudes, and because the centrifugal force of the Earth's rotation is greatest at the Equator, the observed weight of a body is smallest at the Equator and largest at the poles. The difference is sizable, about 1 part in 300. At a given location, the weight of a body is highest at the surface of the Earth. Weight is measured by several procedures. *See* BALANCE; MASS; WEIGHT MEASUREMENT.

Howard S. Bean

Weightlessness A condition induced by the effective lack of resistance to gravitational force on an object or organism, sometimes known as free fall.

Newton proposed the law of universal gravitation, which states that two bodies of matter in the universe attract each other with a force that is directly proportional to the product of their masses and inversely proportional to the square of the distance between their centers. According to this law, even a small increase in the distance between bodies will produce a large decrease in the gravitational force, since the force decreases with the square of the distance. As a body moves from the Earth's surface to a location an infinite distance from the Earth, the gravitational force approaches zero and the body approaches weightlessness. In the true sense, a body can be weightless only when it is an infinite distance from all other objects.

Weightlessness is also defined as a condition in which no acceleration, whether of gravity or any other force, can be detected by an object or organism within the system in question. According to Albert Einstein's principle of equivalence, there is no way to distinguish between the forces of gravitational fields and the forces due to inertial motion. When a gravitational force on a body is opposed by an equal and opposite inertial force, a weightless state is produced. This is based on the fact that the mass that determines the gravitational force of a body is the same as the mass related to the acceleration produced by an inertial force of any kind. These inertial forces have no external physical origin, but are the consequences of an accelerated state of motion. Because of inertia, a moving object always tends to follow a straight line. When a person swings a bucket by the handle in a large circle, he or she feels a pull on the hand, because inertial force (also called centrifugal force in this case) tends to keep the bucket moving in a straight line, while the bucket holder exerts a counterforce constraining the bucket to move along the circle. A similar situation exists in a spaceship orbiting the Earth 200 mi (320 km) above the Earth's surface, where the gravitational field is only slightly weaker than at sea level. The ship is in a state of free fall and therefore weightless. However, while it is falling or in effect being pulled toward the Earth by the Earth's gravitational attractive force, the inertial force of the moving ship is directed radially outward from the Earth. Consequently, the ship follows an orbital path. *See* GRAVITATION; GRAVITY; INERTIA; NEWTON'S LAWS OF MOTION; SPACE FLIGHT; SPACE PROCESSING.

Thora Waters Halstead

Welded joint The joining of two or more metallic components by introducing fused metal (welding rod) into a fillet between the components or by raising the temperature of their surfaces or edges to the fusion temperature and applying pressure (flash welding).

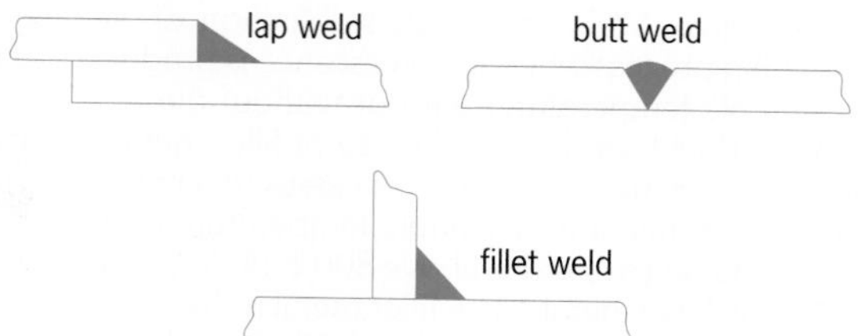

Fig. 1. Three types of welded joints.

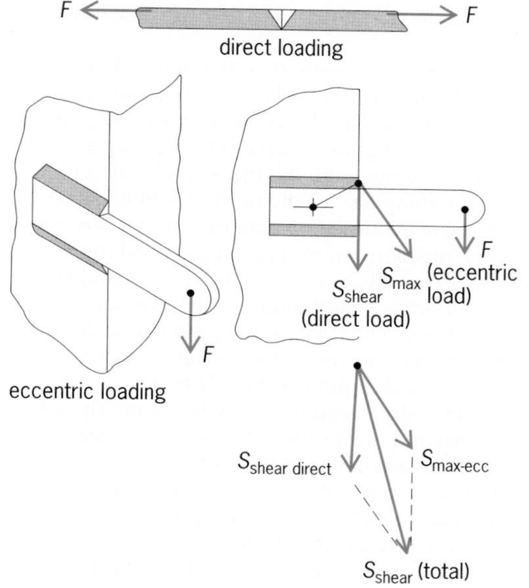

Fig. 2. Loading forces on a welded joint.

Figure 1 shows three types of welded joints. In a lap weld, the edges of a plate are lapped one over the other and the edge of one is welded to the surface of the other. In a butt weld, the edge of one plate is brought in line with the edge of a second plate and the joint is filled with welding metal or the two edges are resistance-heated and pressed together to fuse. For a fillet weld, the edge of one plate is brought against the surface of another not in the same plane and welding metal is fused in the corner between the two plates, thus forming a fillet. The joint can be welded on one or both sides.

Because welded joints are usually exposed to a complex stress pattern as a result of the high temperature gradients present when the weld is made, it is customary to design joints by use of arbitrary and simplified equations and generous safety factors. The force F of direct loading, and consequently the stress S, is applied directly along or across a weld. The stress-force equation is then simply $F = SA$, in which A is the area of the plane of failure (Fig. 2). For eccentric loading, the force F causes longitudinal and transverse forces of varying magnitudes along the weld. *See* RIVETED JOINT; STRUCTURAL CONNECTIONS; WELDING AND CUTTING OF METALS.

L. Sigfred Linderoth, Jr.

Welding and cutting of materials Processes based on heat to join and sever metals. Welding and cutting are grouped together because, in many manufacturing operations, severing precedes welding and involves the same production personnel. Welding is one of the joining processes, others being riveting, bolting, gluing, and adhesive bonding. *See* BOLTED JOINT; WELDED JOINT.

The American Welding Society's definition of welding is "a metal-joining process wherein coalescence is produced by heating to suitable temperatures with or without the application of pressure, and with or without the use of filler metal." Brazing is defined as "a group of welding processes wherein coalescence is produced by heating to suitable temperature and by using a filler metal, having a liquidus above 800°F (427°C) and below the solidus of the base metals. The filler metal is distributed between the closely fitted surfaces of the joint by capillary attraction." Soldering is similar in principle, except that the melting point of solder is below 800°F (427°C). The adhesion of solder depends not so much on alloying as on its keying into small irregularities in the surfaces to be joined. *See* JOINT (STRUCTURES); BRAZING; SOLDERING.

Cutting is one of the severing and material-shaping processes, some others being sawing, drilling, and planning. Thermal cutting is defined as a group of cutting processes wherein the severing or removing of metals is effected by melting or by the chemical reaction of oxygen with the metal at elevated temperatures. Welding and cutting are widely used in building ships, machinery, boilers, spacevehicles, structures, atomic reactors, aircraft, railroad cars, missiles, automobiles, buses and trailers, and pressure vessels, as well as in constructing piping and storage tanks of steel, stainless steel, aluminum, nickel, copper, lead, titanium, tantalum, and their alloys. For many products, welding is the only joining process that achieves the desired economy and properties, particularly leak-tightness. *See* TORCH.

Nearly all industrial welding involves fusion. The edges or surfaces to be welded are brought to the molten state. The liquid metal bridges the gap between the parts. After the source of welding heat has been removed, the liquid solidifies, thus joining or welding the parts together. The principal sources of heat for fusion welding are electric arc, electric resistance, flame, laser, and electron-beam. *See* ARC WELDING; LASER WELDING; RESISTANCE WELDING.

Mel M. Schwartz

Well An artificial excavation made to extract water, oil, gas, brine, or other fluid substance from the earth. Most wells are of the drilled type. Dug wells are almost obsolete, because of the greater speed of drilling and the greater efficiency of drilled wells. *See* ARTESIAN SYSTEMS; OIL AND GAS WELL DRILLING.

Drilled wells, commonly 2–36 in. (5–90 cm) in diameter, usually are fitted with a steel tube or casing inserted in the drilled hole to the desired depth. Where the water-bearing formation is competent to stand without support, the casing is set, or finished, at the top of solid rock. Where there is danger of caving, as in sand or gravel, the casing is carrried below the top of the water-bearing bed, and a perforated pipe or screen extends below the casing to the bottom of the hole. The construction includes a considerable period of pumping, surging, or other treatment (called well development), during which the finer particles of the formation are drawn into the well and removed. This process substantially increases the initial yield of the well.

Most wells of large capacity are equipped with pumps of the deep-well turbine type to lift the water to the surface. When a well is pumped, the pressure head at the well is lowered and a hydraulic gradient toward the well is established which causes water to flow toward the well. This lowering of head is called drawdown. *See* GROUND-WATER HYDROLOGY; PUMPING MACHINERY.

Albert N. Sayre; Ray K. Linsley

Well logging The technique of making measurements in drill holes with probes designed to measure the physical and chemical properties of rocks and their contained fluids. Much information can be obtained from samples of rock brought to the surface in cores or bit cuttings, or from other clues while drilling, such as penetration rate; however, the greatest amount of information comes from well logs. *See* BOREHOLE LOGGING.

Well logs result from a probe lowered into the borehole at the end of an insulated cable. The resulting measurements are recorded graphically or digitally as a function of depth. These records are known as geophysical well logs, petrophysical logs, or more commonly well logs, or simply logs.

Although the most common uses of logs are for correlation of geological strata and location of hydrocarbon zones, there are many other important subsurface parameters that need to be detected or measured. Also, different borehole and formation conditions can require different tools to measure the same basic property. In petroleum engineering, logs are used to: identify potential reservoir rock; determine bed thickness; determine porosity; estimate permeability; locate hydrocarbons; estimate water salinity; quantify amount of hydrocarbons; estimate type and rate of fluid production; estimate formation pressure; identify fracture zones; measure borehole inclination and azimuth; measure hole diameter; aid in setting casing; evaluate quality of cement bonding; locate entry, rate, and type of fluid into borehole; and trace material injected into formations (such as artificial fractures). In geology and geophysics, they are used to: correlate between wells; locate faults; determine dip and strike of beds; identify lithology; deduce environmental deposition of sediments; determine thermal and pressure gradients; create synthetic seismograms; calibrate seismic amplitude anomalies to help identify hydrocarbons from surface geophysics; calibrate seismic with velocity surveys; and calibrate gravity surveys with borehole gravity meter. Other applications include: locating fresh-water aquifers; locating solid minerals; and studying soil and rock conditions for foundations of large structures.

Electrical devices employ instruments that generate data based on electrical measurements. The spontaneous potential, usually recorded along with resistivity curves, is a simple but valuable aid to help geologists correlate from one well to another and to assist them in inferring the depositional environment of the sediments. It defines permeable zones from surrounding non-permeable shales and can be used to estimate the salinity of formation water. Resistivity is one of the most important physical properties to record. A sand filled with salt water will conduct electricity much easier (and therefore has a lower resistivity) than a sand in which most of the salt water is replaced with a nonconducting oil or gas. The primary purpose of resistivity measurements is to determine hydrocarbon saturation.

Quantitative interpretation of hydrocarbon saturation requires a knowledge of porosity. One of the most widely used porosity logs is the acoustic log, also known as a sonic log, acoustilog, or velocity log. The acoustical log measures the shortest time for sound waves to travel through 1 ft (or 1 m) of formation. This log is used for many other purposes, including identification of lithology, prediction of pressures, and assisting with geophysical interpretation. In addition, information from the amplitude of the acoustic wave aids in detection of fractures and gas zones, determination of mechanical properties of rocks, and analysis of cement bond behind pipe.

Nuclear devices are instruments which generate data based on measurements of nuclear particles. In most cases the gamma-ray log may be considered as a shale log; that is, clays or shaly rocks will give a higher radioactive count than clean sands or carbonates. Like the spontaneous potential, this provides a good correlation curve, defines bed thicknesses, and aids interpretation of environmental deposition. Almost all the naturally occurring radioactivity that is detected by the gamma-ray log comes from three elements: potassium, thorium, and uranium.

An important porosity tool is the density log, which emits a beam of gamma rays into the rock and these rays interact with electrons in the formation through Compton scattering. The tool provides a bulk density measurement. Another important device for determination of porosity is the neutron log. These logs respond primarily to hydrogen atoms. Therefore, in clean or shale-free formations the neutron log reflects the amount of liquid-filled porosity. A tool developed to distinguish hydrocarbon from salt water behind casing is the pulsed neutron capture tool which employs a neutron generator that repeatedly emits pulses of high-energy neutrons. After these neutrons are slowed down to the thermal state, they are captured by nuclei of the various atoms surrounding the tool. With each capture a corresponding emission of gamma rays occurs.

Production logging tools are designed to locate fluid movement behind the casing and into the well bore, to detect the type of fluid, and to determine the flow rate. Production logging tools include various types of flow meters, fluid density devices, sensitive thermometers, radioactive tracer devices, noise loggers, and capacitance logs. In addition, these tools usually record some correlation curve such as gamma-ray or collar locator to tie the production curves to the formation and casing. Closely related to production tools are logging devices that either magnetically or mechanically inspect the condition of the casing in the borehole.

Small but rugged and powerful computers designed for field operation record logs on magnetic tapes as they are run and mathematically manipulate them in the logging trucks at the well site. Detailed computations are made quickly on a foot by foot basis to give information on fluid and rock properties. The digital information from practically every curve can be incorporated into some type of analysis or interpretation program.

Richard E. Wyman

Wellpoint systems A method of keeping an excavated area dry by intercepting the flow of groundwater with pipe wells located around the excavation area. Intercepting the flow before it reaches the excavated area also improves the stability of the edge of the excavation, permitting steeper bank slopes and often eliminating the need for supporting or shoring the banks. *See* CONSTRUCTION METHODS; GROUND-WATER HYDROLOGY; WELL.

Wellpoint systems are most effective in coarse-grained soils, such as gravel or sand. They are not effective in fine soils, such as silts and clays, where the small size of the pores between grains restricts the flow of water.

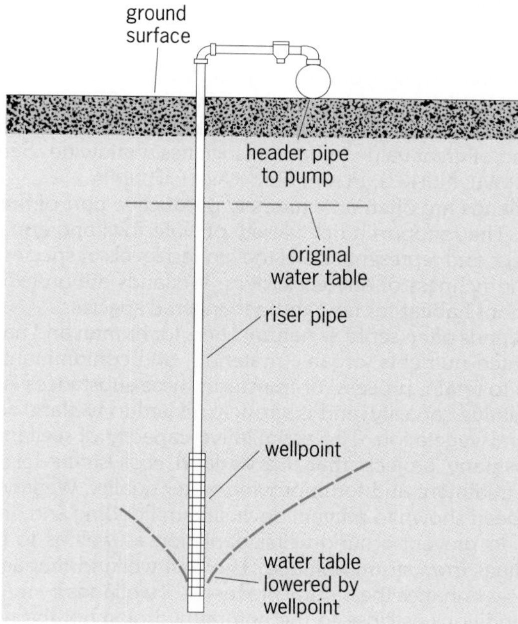

Components of a wellpoint system.

The basic components of a wellpoint system are the wellpoint, the riser pipe, the header pipe or manifold, and the pump (see illustration). The wellpoint consists of a perforated pipe equipped with a ball valve to regulate the flow of water, a screen to prevent the entry of sand during pumping, and a jetting tip. The steel riser pipe brings the groundwater to the surface, where it is collected by the horizontal manifold pipe or header pipe. The pumps are located above the water table and collect the water from the header pipes for discharge away from the excavation area. *See* PUMP; PUMPING MACHINERY.

William Hershleder

Welwitschiales An order of the class Cycadopsida having one species, *Welwitschia mirabilis*. This plant is native to the very arid deserts of southwestern Africa. Its appearance is bizarre—it has a very short, unbranched, woody stem (sometimes to 3 ft or 1 m in diameter), which is cushion- or saucer-shaped and tapers quickly to a long taproot. There are only two leaves, and these persist throughout the life of the plant. The leaf is broadly strap-shaped, as wide as the stem, firm, and leathery, and gradually splits lengthwise between the veins. The leaf (of indefinite growth) develops from a meristem at this point of connection to the stem. The species is dioecious; the cones are borne on branched axes originating between the crown of the stem and the base of the leaf. The order is known only from the one living species and not from any fossils. *See* CYCADOPSIDA; PINOPHYTA; PLANT KINGDOM.

Thomas A. Zanoni

Wentzel-Kramers-Brillouin method A special technique for obtaining an approximation to the solutions of the one-dimensional time-independent Schrödinger equation, valid when the wavelength of the solution varies slowly with position. It is named after G. Wentzel, H. A. Kramers, and L. Brillouin, who independently in 1926 contributed to its understanding in the quantum-mechanical application. It is also called the WKB method, BWK method, the classical approximation, the quasi-classical approximation, and the phase integral method. *See* QUANTUM MECHANICS; SCHRÖDINGER'S WAVE EQUATION.

Roland H. Good, Jr.

West Indies An archipelago, including the Bahamas, the Greater Antilles (including Cuba, Jamaica, Hispaniola—the Dominican Republic and Haiti—and Puerto Rico), the Lesser Antilles, and other islands, curving 2500 mi (4000 km) from Yucatan Peninsula and southeastern Florida to northern Venezuela and enclosing the Caribbean Sea. Situated between latitude 10° and 27°N and longitude 59° and 85°W, in the zone of the northeast trade winds, the West Indies have a subtropical and predominantly oceanic climate, with even warmth and steady breezes. Temperatures vary little from season to season, ranging from means of 80–85°F (27–29°C) in July to 70–78°F (21–26°C) in January at sea level. Freezing is unknown, and the hottest temperatures rarely exceed 90°F (32°C). Precipitation ranges from a low of 25–50 in. (64–127 cm) a year on low-lying islands and drier coasts up to 300 in. (7.6 m) on the highest peaks, which are almost perpetually cloud-capped. At lower elevations, rainfall is erratic from year to year and from season to season, but reaches a maximum in the summer and fall, when the northeast trades are replaced by light, variable winds. This is also the season of hurricanes, destructive tropical cyclones which sweep west and northwest across the Caribbean, sparing only the southernmost islands. The winter months are generally dry, and there is frequently a shorter dry season in July or August. *See* CARIBBEAN SEA; TROPICAL METEOROLOGY.

The West Indian flora is chiefly derived from Central and South America, but there are a number of endemic species, notably palms; many mainland plants failed to colonize the islands, which are floristically poor. The effect of isolation and small size is evident in the meager character of West Indian fauna. Animal species are limited; there are few mammals and no large ones, except for domesticated animals and, especially on Hispaniola, feral cattle, goats, pigs, and horses.

David Lowenthal

West Nile Virus An arbovirus, first identified in the West Nile area of Uganda in the early 1930s, that has been an increasing threat in North America since the late 1990s. Arboviruses (*ar*thropod-*bo*rne viruses) are carried by arthropods and transmitted to the host through the bite of the insect. West Nile virus (WNV) found its way to North America in 1999, possibly through international travel, the importation of infected birds or mosquitoes, or migration of infected birds. Severe infection with WNV can result in viral encephalitis, a dangerous and sometimes fatal inflammation of the brain. Since 1999, WNV infections in humans, birds, and mosquitoes have been reported from all states except Hawaii, Alaska, and Oregon. The incidence of WNV encephalitis in the United States is seasonal, with peaks corresponding to the times of the year during which adult mosquitoes are active.

In order to be effectively transmitted, arboviruses must have the ability to infect vertebrates (such as birds or humans) as well as the invertebrate vector. The virus must then be able to grow in the vertebrate's bloodstream and multiply (this is known as a "viremia"), remaining in the blood long enough for infection of a second invertebrate vector. Finally, it must infect the second invertebrate's salivary gland to allow transmission to other vertebrate hosts. For WNV, this cycle occurs when the *Culex* mosquito bites an infected bird, taking up the virus and becoming infected in its turn, after which the mosquito bites another bird and continues the transmission cycle.

Humans are usually "dead-end hosts," meaning they cannot spread WNV back to the vector because they do not maintain a persistent viremia. It is now recognized, however, that the virus can be transmitted by blood transfusions and organ transplants. Despite this development, the reservoirs for this virus are birds, not humans. Reservoirs are susceptible hosts that allow the reinfection of other arthropods. Therefore, there is no transmission of WNV from human to mosquito. WNV has been detected in at least 138 species of birds. Although birds infected with WNV may become ill or die (especially crows and jays), most survive infection.

Upon entering the host, WNV travels through the host's plasma to tissue sites for which it has a "tropism," or preference. The initial symptoms include fever, chills, headache, and backaches caused by the resulting viremia. Most viral infections do not progress beyond this stage; however, a secondary viremia may occur, which can result in infection of organs such as the brain, liver, and the skin. This can lead to fatality in the infected host. Typically, fatal encephalitis occurs in the elderly and in younger people with underlying illnesses, such as diabetes, emphysema, acquired immune deficiency syndrome (AIDS), and chronic diseases of the liver, kidneys, or heart. Only one in five people infected with WNV show symptoms, and less than 1 in 100 develop serious complications from the infection.

Prevention and control of arbovirus disease is best accomplished through a comprehensive mosquito management program; personal, household, and community prevention through public education; and targeted prevention programs aimed at educating high-risk groups. Treatment for WNV encephalitis and fever is largely supportive: possible hospitalization, administration of intravenous fluids, respiratory support (ventilation), and prevention of secondary infections. There is no specific antiviral therapy for WNV infection. No vaccine is available for use to prevent WNV disease in humans. *See* ARBOVIRAL ENCEPHALITIDES; MOSQUITO; VIRUS; ZOONOSES. Marcia M. Pierce

Wetlands Ecosystems that form transitional areas between terrestrial and aquatic components of a landscape. Typically they are shallow-water to intermittently flooded ecosystems, which results in their unique combination of hydrology, soils, and vegetation. Examples of wetlands include swamps, fresh- and salt-water marshes, bogs, fens, playas, vernal pools and ponds, floodplains, organic and mineral soil flats, and tundra. As transitional elements in the landscape, wetlands often develop at the interface between drier uplands such as forests and farmlands, and deep-water aquatic systems such as lakes, rivers, estuaries, and oceans. Thus, wetland ecosystems are characterized by the presence of water that flows over, ponds on the surface of, or saturates the soil for at least some portion of the year.

Wetland soils can be either mineral (composed of varying percentages of sand, silt, or clay) or organic (containing 12–20% organic matter). Through their texture, structure, and landscape position, soils control the rate of water movement into and through the soil profile (the vertical succession of soil layers). Retention of water and organic carbon in the soil environment controls biogeochemical reactions that facilitate the functioning of wetland soils. *See* BIOGEOCHEMISTRY.

Vegetated wetlands are dominated by plant species, called hydrophytes, that are adapted to live in water or under saturated soil conditions. Adaptations that allow plants to survive in a water-logged environment include morphological features, such as pneumatophores (the "knees," or exposed roots, of the bald cypress), buttressed tree trunks, shallow root systems, floating leaves, hypertrophied lenticels, inflated plant parts, and adventitious roots. Physiological adaptations also allow plants to survive in a wetland environment. These include the ability of plants to transfer oxygen from the root system into the soil immediately surrounding the root (rhizosphere oxidation); the reduction or elimination of ethanol accumulation due to low concentrations of alcohol dehydrogenase; and the ability to concentrate malate (a nontoxic metabolite) instead of ethanol in the root system. *See* ROOT (BOTANY).

Wetlands differ with respect to their origin, position in the landscape, and hydrologic and biotic characteristics. For example, work has focused on the hydrology as well as the geomorphic position of wetlands in the landscape. This hydrogeomorphic approach recognizes and uses the fundamental physical properties that define wetland ecosystems to distinguish among classes of wetlands that occur in riverine, depressional, estuarine or lake fringe, mineral or organic soil flats, and slope environments.

The extent of wetlands in the world is estimated to be $2–3 \times 10^6$ mi^2 ($5–8 \times 10^6$ km^2), or about 4–6% of the Earth's land surface. Wetlands are found on every continent except Antarctica and in every clime from the tropics to the frozen tundra. Rice paddies, which comprise another 500,000–600,000 mi^2 ($1.3–1.5 \times 10^6$ km^2), can be considered as a type of domesticated wetland of great value to human societies worldwide. *See* BOG; MANGROVE; MUSKEG; PLAYA; SALT MARSH; TUNDRA.

Wetlands are often an extremely productive part of the landscape. They support a rich variety of waterfowl and aquatic organisms, and represent one of the highest levels of species diversity and richness of any ecosystem. Wetlands are an extremely important habitat for rare and endangered species.

Wetlands often serve as natural filters for human and naturally generated nutrients, organic materials, and contaminants. The ability to retain, process, or transform these substances is called assimilative capacity, and is strongly related to wetland soil texture and vegetation. The assimilative capacity of wetlands has led to many projects that use wetland ecosystems for wastewater treatment and for improving water quality. Wetlands also have been shown to prevent downstream flooding and, in some cases, to prevent ground-water depletion as well as to protect shorelines from storm damage. The best wetland management practices enhance the natural processes of wetlands by maintaining conditions as close to the natural hydrology of the wetland as possible. *See* GROUND-WATER HYDROLOGY.

The world's wetlands are becoming a threatened landscape. Loss of wetlands worldwide currently is estimated at 50%. Wetland loss results primarily from habitat destruction, alteration of wetland hydrology, and landscape fragmentation. Global warming may soon be added to this list, although the exact loss of coastal wetlands due to sea-level rise is not well documented. Worldwide, destruction of wetland ecosystems primarily has been through the conversion of wetlands to agricultural land.

Hydrologic modifications that destroy, alter, and degrade wetland systems include the construction of dams and water diversions, ground-water extraction, and the artificial manipulation of the amount, timing, and periodicity of water delivery. The primary impact of landscape fragmentation on wetland ecosystems is the disruption and degradation of wildlife migratory corridors, reducing the connectivity of wildlife habitats and rendering wetland habitats too small, too degraded, or otherwise irreversibly altered to support the critical life stages of plants and animals.

The heavy losses of wetlands in the world, coupled with the recognized values of these systems, have led to a number of policy initiatives at both the national and international levels.

Wetland restoration usually refers to the rehabilitation of degraded or hydrologically altered wetlands, often involving the reestablishment of vegetation. Wetland enhancement generally refers to the targeted restoration of one or a set of ecosystem functions over others, for example, the focused restoration of a breeding habitat for rare, threatened, or endangered amphibians. Wetland creation refers to the construction of wetlands where they did not exist before. Created wetlands are also called constructed or artificial wetlands. Restoring, enhancing, or creating a wetland requires a comprehensive understanding of hydrology and ecology, as well as engineering skills. See DAM; ECOSYSTEM; ESTUARINE OCEANOGRAPHY; HYDROLOGY; RIVER ENGINEERING; RESTORATION ECOLOGY.

William J. Mitsch; Peggy L. Fiedler; Lyndon C. Lee; Scott R. Stewart

Wheat A food grain crop. Wheat is the most widely grown food crop in the world, and is increasing in production. It ranks first in world crop production and is the national food staple of 43 countries. At least one-third of the world's population depends on wheat as its main staple. The principal food use of wheat is as bread, either leavened or unleavened. The United States is second to Russia in total production, but the average yield per acre in the United States is about twice that of Russia. Other major wheat-producing countries in the world are Canada, China, India, France, Argentina, and Australia.

Wheat is best adapted to a cool dry climate, but is grown in a wide range of soils and climates. Much of the world's wheat is seeded in the fall season and, after being dormant or growing very slowly during winter, it makes rapid growth in the spring and develops grain for harvest in early summer.

Wheat for milling is classified according to hardness, color, and best use. In the United States, there are seven official market classes of which the following five are the most important: (1) hard red winter, for bread; (2) hard red spring, for bread and rolls; (3) soft red winter, for cake and pastries; (4) white, for bread, breakfast foods, and pastries; and (5) durum, for macaroni products.

The wheat inflorescence is a spike bearing sessile spikelets arranged alternately on a zigzag rachis. Two, three, or more florets may develop in each spikelet and bear grains. The grain may be white, red (brown), or purple, and it may be hard or soft in texture. Size of the grain or caryopsis may be large, as in durum, or very small, as in shot wheat (*Triticum sphaerococcum*). Wheats vary in plant height and in the ability to produce tillers. The stems are usually hollow. The wheat grain is composed of the endosperm and embryo enclosed by bran layers. The endosperm portion is principally starch and is therefore used as energy food. Wheat is also an important protein source, especially for those people who use wheat as their main staple. See SEED.

Botanically, wheat is a member of the grass family to which rice, barley, corn, and several other cereal grain crops also belong. The *Triticum* genus includes a wide range of wheat forms. Taxonomic studies place the goat grasses (*Aegilops*) and wheat (*Triticum*) in one genus, *Triticum*. Wheat has been crossed with rye (*Secale*) and with *Agropyron* (a grass). New forms, called *Triticale*, have been derived from crossing rye and wheat followed by doubling the chromosomes in the hybrid. See TRITICALE.

Most countries in which wheat is grown have wheat breeding programs in which the objective is to develop more productive and more stable varieties (cultivars). Many methods are combined in these programs, but in nearly all of them specially selected parent types are crossbred followed by pure-line selection among the progeny to develop new combinations of merit. Varieties and genetic types from all over the world become candidate parents to provide the desired recombinations of good quality, winter and drought hardiness, straw strength, yield, and disease resistance. Wheats must be bred for specific milling processes and to provide quality end-use products. Many new varieties have complex pedigrees. See BREEDING (PLANT); GRAIN CROPS.

Louis P. Reitz

Milling of wheat has evolved from rudimentary crushing or cracking to sophisticated separation and refining. The main purpose of milling is isolation of the starch-protein matrix, that is, separation of the endosperm from the high-fiber bran and high-lipid germ. Under optimal conditions, milling yields a high-quality, uniformly colored flour with a relatively stable shelf-life. The flours of hard wheats (11 to 13% protein) develop strong gluten complexes during mixing and are therefore suitable for making bread. Whole soft wheats (9 to 11% protein) yield flours that are used primarily for cakes, cookies, and pastries. Durum wheat is used to produce a relatively coarse flour, semolina, used for manufacture of pasta products. See FOOD MANUFACTURING.

Mark A. Uebersax

Wheatstone bridge An extensively used electrical network of four resistances R_{1-4} (or impedances Z_{1-4} if alternating currents are involved). This network (see illustration), was first described by S. H. Christie in 1833, only 7 years after Georg S. Ohm discovered the relationship between voltage and current. Since 1843, when Charles Wheatstone called attention to Christie's work, Wheatstone's name has been associated with this network. See ELECTRICAL IMPEDANCE; ELECTRICAL RESISTANCE; OHM'S LAW.

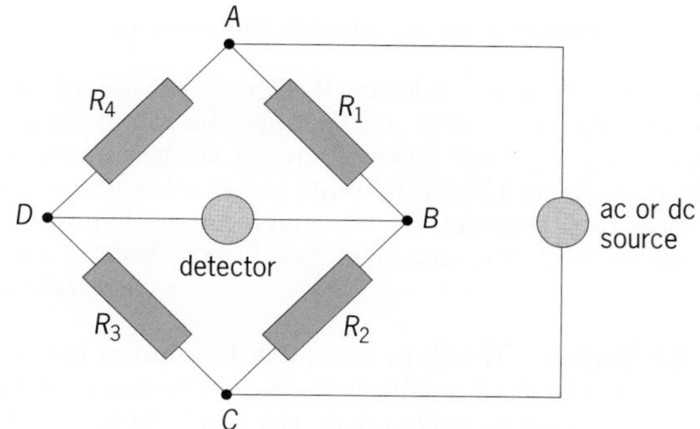

Wheatstone bridge circuit.

The potentials at nodes B and D with respect to node C (or A), depend only on the ratios R_1/R_2 and R_4/R_3 respectively, and some or all of the values of R_{1-4} can be adjusted until these potentials are equal. A detector connected between B and D will then indicate zero current or voltage and Eq. (1) will be satisfied [or Eq. (2) for alternating-current impedances].

$$R_1/R_2 = R_4/R_3 \quad (1)$$

$$Z_1/Z_2 = Z_4/Z_3 \quad (2)$$

The network possesses the following useful properties:

1. Obtaining the detector indication of zero current or voltage (which is often termed "balancing the bridge") and the consequent validity of Eq. (1) or Eq. (2) is independent of the internal resistance (or impedance) of the source or detector.

2. The same result is obtained if the positions of source and detector in the network are interchanged. This is a simple example of reciprocity in linear systems. *See* RECIPROCITY PRINCIPLE.

3. From Eq. (1) or Eq. (2), the value of one of the four resistances or impedances can be calculated if one of the others and the ratio of the remaining two are known.

4. The balanced network will eliminate unwanted electrical signals (interference and pickup) between the source and circuitry connected to *B* and *D*.

The response of the detector to a deviation from the condition of Eq. (1) or Eq. (2) depends on the values of $R_{1\text{–}4}$ or $Z_{1\text{–}4}$ as well as the internal resistances or impedances of both source and detector. This response can be calculated by network analysis. *See* NETWORK THEORY; RESISTANCE MEASUREMENT. Bryan P. Kibble

Wheel and axle A wheel and its axle or, more generally, two wheels with different diameters or a wheel and drum, as in a windlass, rigidly connected together so that they rotate as a unit on a common axis. The principle of operation is the same as that of the lever in that, for static equilibrium, the summation of torques about the axis of rotation equals zero. Where flexible members, such as ropes, have been firmly attached to a wheel and drum (see illustration) and the machine is mounted on frictionless bearings, $F_1R_1 - F_2R_2 = 0$.

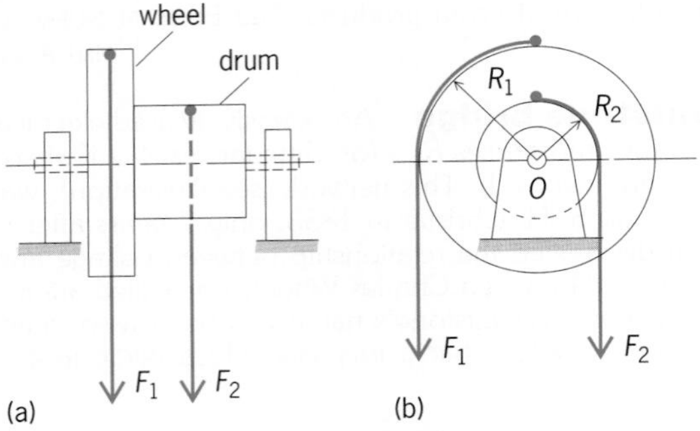

Wheel and axle. (*a*) Side view. (*b*) Front view.

The main difference between the lever and the wheel and axle is that the wheel and axle permits the forces to operate through a much greater distance. In the illustration the wheel and drum could be allowed to rotate any number of revolutions if the ropes were wrapped the required number of times around each before they were attached. *See* FORCE; SIMPLE MACHINE. Richard M. Phelan

Wheel base The distance in the direction of travel from front to rear wheels of a vehicle, measured between centers of ground contact under each wheel. For a vehicle with two rear axles, the rear measuring point is on the ground midway between rear axles. Tread of a vehicle is the distance perpendicular to the direction of travel between front wheels, or between rear wheels, measured from centers of ground contact. Frank H. Rockett

White dwarf star The smallest kind of ordinary star, about the size of the Earth. In most white dwarf stars, an amount of matter equal to 60% of the Sun's mass is compressed into this small volume. One quarter liter (about 1 cup) of white dwarf material has a mass of 600 tons. Several thousand white dwarf stars have been discovered.

Some of the other properties of white dwarf stars reach extreme values. The hottest known stars are either white dwarfs or stars that are just about to become white dwarfs, and have temperatures of a few hundred thousand kelvins. A few white dwarf stars have very strong magnetic fields, with field strengths exceeding 10^8 gauss (10^4 teslas), many times stronger than can be generated in laboratories. Some white dwarf stars rotate every few minutes, and some others rotate so slowly that no spin has been detected over several decades. The coolest white dwarf stars emit less energy per second than any other type of visible star. *See* HIGH MAGNETIC FIELDS.

White dwarfs are the final stages in the life cycles of low-mass stars like the Sun, the most common types of stars. At present, the Sun is on the main sequence, fusing hydrogen to form helium in its core. Five billion years from now, the hydrogen in the center of the Sun will run out, and the Sun will become a red giant star as it turns to other nuclear reactions to provide its internal heat. Stars like the Sun will become only hot enough inside that they will fuse helium nuclei to form a mixture of carbon and oxygen. They will then reach the end of the nuclear fusion road, and will no longer be able to generate their own sources of energy. At this point in their life cycle, they will be on the way to becoming white dwarf stars, the most common types of stellar remnants. *See* NUCLEOSYNTHESIS; STELLAR EVOLUTION.

Just before a star becomes a white dwarf star, it will shed its outer layers as its stellar wind strengthens at the very end of the red giant stage of its life cycle. In many and perhaps all cases, the expelled gas will become visible as a glowing gas cloud called a planetary nebula. *See* PLANETARY NEBULA.

A star becomes a bona fide white dwarf when it contracts to a state where heat pressure plays a very small role in determining its internal structure. The material in white dwarf stars is a form of matter called degenerate matter. The electron clouds in degenerate matter actually touch each other, and this touching gives white dwarf matter a resistance to compression that balances the force of gravity.

White dwarf stars have no nuclear energy sources, so they simply cool. The hottest star to have compacted itself to the white dwarf state has a temperature of 190,000 K (350,000°F), while the coolest white dwarf stars have temperatures near 3800 K (6400°F), still hotter than the coolest main-sequence stars. The internal structure of these two stars is virtually the same.

However, the surface layers of white dwarfs are much more varied than is the case of other types of stars. For example, white dwarf stars come in two basic flavors, hydrogen-rich stars, whose hydrogen/helium ratios can exceed 10^7, and helium-rich stars, whose hydrogen/helium ratio can be less than 10^{-7}. (In virtually every other type of star, the hydrogen/helium ratio is near its cosmic value of approximately 10:1.) At least part of this extraordinary variation (over 14 orders of magnitude) probably comes from subtle differences in the way that white dwarfs form in the late stages of red giant life cycles. Harry L. Shipman

Whooping cough An acute infection of the tracheobronchial tree caused by *Bordetella pertussis*, a bacteria species exclusive to infected humans. The disease (also known as pertussis) follows a prolonged course beginning with a runny nose, and finally develops into violent coughing, followed by a slow period of recovery. The coughing stage can last 2–4 weeks, with a whooping sound created by an exhausted individual rapidly breathing in through a narrowed glottis after a series of wrenching coughs. The classical disease occurs in children 1–5 years of age, but in immunized populations infants are at greatest risk and adults with attenuated (and unrecognized) disease constitute a major source of transmission to others. *Bordetella pertussis* is highly infectious, particularly following face-to-face contact with an individual who is coughing. The disease is caused by structural components and extracellular toxins elaborated by *B. pertussis*. Multiple virulence factors produced by the organism play important roles at various stages of pertussis.

A vaccine produced from whole *B. pertussis* cells and combined with diphtheria and tetanus toxoids has been used throughout the world for routine childhood immunization. Concern over vaccine morbidity has caused immunization rates to decline in some developed countries. These drops in immunization rates have often been followed by widespread outbreaks of disease, including deaths. Considerable effort has been directed toward the development of a vaccine which would minimize side effects but maintain efficacy. A new acellular vaccine is available and has fewer side effects than the whole-cell vaccine. *See* DIPHTHERIA; TETANUS; VACCINATION.

Although *B. pertussis* is susceptible to many antibiotics, their use has little effect once the disease reaches the coughing stage. Erythromycin is effective in preventing spread to close contacts and in the early stage. Kenneth J. Ryan

Wide-area networks Communication networks that are regional, nationwide, or worldwide in geographic area, with a minimum distance typical of that between major metropolitan areas. Smaller networks include metropolitan and local-area networks. A communication network provides common transmission, multiplexing, and switching functions that enable users to transport data between many sources and many destinations. Under ideal circumstances, the data that arrive at the destination are identical to the data that were sent. The rate of arrival of bits at any point in the network is said to be the data rate at that point and is typically measured in bits per second. These bits may come from one source or from a multiplicity of sources. The capacity of a network to transmit at a cerain data rate is known as its bandwidth. *See* LOCAL-AREA NETWORKS.

There are several fundamental attributes and concepts that facilitate the accurate transmission of data within and between digital networks. To communicate between computers, a set of rules, formats, and delivery procedures known as protocols must be established.

Part of the communications protocol allows for definition of where the packets of digital data are to be routed. Each packet of data contains the unique address of a computer or other network as its destination. The routing of the data is known as packet switching since the nodes in the network can switch the packet to various transmission paths. Networks are interconnected by means of routers. *See* PACKET SWITCHING.

Another part of the communications protocol allows for including error detection and correction information in the data packets. The destination computer or network will verify the data in the packet utilizing the error control data, such as a checksum. Protocols are also used to implement flow control. This allows the receiving computer or network to communicate back to the sender when it can or cannot receive additional data. *See* DIGITAL COMPUTER.

Certain protocols have become standards for a majority of wide-area networks. Asynchronous Transfer Mode (ATM) is a protocol used in business-to-business (B2B) communications when high data rates are needed. Typically, one ATM port supports 45 megabits per second (Mbps). Frame Relay is another business-to-business protocol. The advantage of Frame Relay is that the data rate can be scaled to the individual company's needs. The third protocol, and the one having the most worldwide impact on both business and personal communications, is the Internet Protocol, or IP. *See* INTEGRATED SERVICES DIGITAL NETWORK (ISDN).

Wide-area networks may operate on a mix of transmission media for either fixed or mobile applications. Fixed applications mean that the receiver of digital data is stationary. Examples of wireline transmission media for fixed applications are fiber-optic cable, copper wire, and coaxial cable. For copper, a technique known as digital subscriber line (DSL) allows for transmission in excess of 1 Mbps over regular phone lines. In fiber, a technique known as wave division multiplexing (WDM) allows the simultaneous transmission of different streams of digital data over each spectral component of the light wave. This allows bundles of fiber-optic cable to transport billions of bits (gigabits) and even trillions of bits (terabits) of data per second. *See* COMMUNICATIONS CABLE; OPTICAL COMMUNICATIONS; OPTICAL FIBERS; TRANSMISSION LINES.

Wide-area networks also operate over a variety of wireless media. Wireless media can support either fixed or mobile applications. Another common distinction in wireless networks is whether it is point-to-point or point-to-multipoint. In point-to-point the originating transmission has one receiver, whereas in point-to-multipoint the originating transmission has multiple receivers. Examples of wireless media include radio-wave, microwave, cellular, and satellite. *See* COMMUNICATIONS SATELLITE; MOBILE RADIO; RADIO SPECTRUM ALLOCATIONS; RADIO-WAVE PROPAGATION.

There are many different types of content transmitted over WANs. Examples of content are data, voice, video, audio, paging messages, and fax. By virtue of the ability to digitize all of this content, the major difference in transmission requirements is the bandwidth, or capacity, required to transmit digital packets of any type of content. *See* DATA COMMUNICATIONS; ELECTRICAL COMMUNICATIONS; FACSIMILE.

In order to utilize the bandwidth capacity of WANs more efficiently, digital compression techniques are now used in many applications. Fundamentally, digital compression reduces the amount of bits in data packets by removing repetitive strings of bits and replacing them with shorter packets that numerically describe the amount of repetitive data. *See* DATA COMPRESSION.

With massive amounts of information being transmitted over WANs using both public and private infrastructure, data security is increasingly important. In order to secure digital data transmitted over networks, encryption techniques have been developed. Encryption of digital data involves using hardware and software to manipulate the bits in a data packet, making it unrecognizable and unusable to anyone not authorized to use the data. *See* COMMUNICATIONS SCRAMBLING; COMPUTER SECURITY; CRYPTOGRAPHY.

The Internet, using the IP, has become by far the most ubiquitous WAN in the world. Internet users are able to transmit video clips, electronic mail (e-mail), telephone calls (called Voice Over IP), to digitized x-rays over the Internet. There are three basic variations of IP networks. First is the overall Internet itself which encompasses all personal and business users of the Internet. The second type of IP-based networks is intranets. An intranet is usually deployed within a specific organization or company. A company intranet may be used to manage human resources and financial processes and keep employees updated on company news. The third type of IP-based networks is extranets. Typically, extranets are used to link multiple organizations or companies for some common business purpose. In both intranets and extranets, a technology known as a firewall is employed to prevent unauthorized access to the network or unauthorized URL (Uniform Resources Locator) from the network. The URL is the basic unique address or location for any Web site or other Internet service. *See* ELECTRONIC MAIL; INTERNET. Richard L. Jensen

Wiedemann-Franz law An empirical law of physics which states that the ratio of the thermal conductivity of a metal to its electrical conductivity is a constant times the absolute temperature, as given by the equation below. Here K_c is the thermal

$$K_c = L_0 \sigma T$$

conductivity due to the conduction electrons, σ is the electrical conductivity, T is the absolute temperature, and L_0 is known as the Lorentz number. The Wiedemann-Franz law provides an important check on theories of electrical and thermal conductivity. *See* CONDUCTION (HEAT); THERMAL CONDUCTION IN SOLIDS. Frank J. Blatt

Wilkinson Microwave Anisotropy Probe A spacecraft which is precisely measuring the cosmic microwave background (CMB) radiation. The *Wilkinson Microwave*

Anisotropy Probe (*WMAP*) is a space mission of the National Aeronautics and Space Administration (NASA) that has put fundamental theories of the nature of the universe to a precise test. Since August 2001, *WMAP* has continually surveyed the full sky, mapping out tiny differences in the temperature of the cosmic microwave background radiation, which is the radiant heat from the big bang. A fossil remnant of the hot big bang, the cosmic microwave background permeates the universe and is seen today with an average temperature of only 2.725 kelvins. Tiny variations about this average temperature were first discovered by NASA's *Cosmic Background Explorer* (*COBE*) mission. *WMAP* followed up on the *COBE* results by characterizing the detailed statistical nature of the temperature variations of the cosmic microwave background (called anisotropy), revealing a wealth of detail about the global properties of the universe. *See* BIG BANG THEORY.

The *WMAP* science requirements dictated that the relative cosmic microwave background temperature be measured accurately over the full sky. The overriding design requirement was to control systematic errors that would otherwise contaminate the measurements. To achieve this, *WMAP* uses differential microwave radiometers that are polarization-sensitive to measure temperature differences between pairs of spots on the sky.

WMAP observes the sky from an orbit about the second Sun–Earth Lagrangian point (L2), 1.5×10^6 km (0.9×10^6 mi) from Earth in the direction opposite to that of the Sun. L2 is about four times farther from Earth than the Moon. The observatory can always point away from the Sun, Earth, and Moon while maintaining an unobstructed view of deep space. *WMAP* observes the full sky every 6 months. *See* CELESTIAL MECHANICS.

The first results from the *WMAP* mission were reported on February 11, 2003. This included the most detailed full-sky "baby picture" of the universe taken so far. The *WMAP* data were compared and combined with other diverse cosmic measurements, and a new unified and precise understanding of the universe emerged.

The universe is 13.7×10^9 years old, with a margin of error of about 1%. The cosmic microwave signal in the *WMAP* map is from 379,000 years after the big bang. The expansion rate of the universe (the Hubble constant) value is $H_0 = 71$ (km/s)/Mpc, with a margin of error less than 5%. The contents of the universe include 4% baryons, 23% cold dark matter, and 73% dark energy. *See* COSMOLOGY; DARK ENERGY; DARK MATTER; HUBBLE CONSTANT; UNIVERSE.

The polarized data also provide new evidence for inflation, the rapid expansion of the universe a fraction of a second after its birth. The inflationary theory predicts that density is very close to the critical density, producing a flat (Euclidean) universe. *See* COSMIC BACKGROUND RADIATION; INFLATIONARY UNIVERSE COSMOLOGY.

Charles L. Bennett

Willemite A rare nesosilicate mineral, composition Zn_2SiO_4, crystallizing in the hexagonal system. It is usually massive or granular with a vitreous luster; crystals are rare. The mineral may be variously colored, most commonly green, red, or brown. Hardness is $5^1/_2$ on Mohs scale; specific gravity is 3.9–4.2. Willemite forms a valuable ore of zinc at Franklin, New Jersey. At this famous zinc deposit Willemite fluoresces a yellow-green. *See* SILICATE MINERALS.

Cornelius S. Hurlbut, Jr.

Willow A deciduous tree and shrub of the genus *Salix*, order Salicales, common along streams and in wet places in the United States, Europe, and China. The twigs are often yellow-green and bear alternate leaves which are characteristically long, narrow, and pointed, usually with fine teeth along the margins. Flowers occur in catkins. The fruit contains several silky seeds. *See* SALICALES.

Willow lumber is used for fuel and in making charcoal, excelsior, ball bats, boxes, crates, boats, waterwheels, and wicker furniture. The tough, pliable shoots of many species are used to make baskets; the bark of other species is used for tanning. Willows are of great value in checking soil erosion. A few species are ornamental shade trees.

James F. Ferry

Wind The motion of air relative to the Earth's surface. The term usually refers to horizontal air motion, as distinguished from vertical motion, and to air motion averaged over a chosen period of 1–3 min. Micrometeorological circulations (air motion over periods of the order of a few seconds) and others small enough in extent to be obscured by this averaging are thereby eliminated.

The direct effects of wind near the surface of the Earth are manifested by soil erosion, the character of vegetation, damage to structures, and the production of waves on water surfaces. At higher levels wind directly affects aircraft, missile and rocket operations, and dispersion of industrial pollutants, radioactive products of nuclear explosions, dust, volcanic debris, and other material. Directly or indirectly, wind is responsible for the production and transport of clouds and precipitation and for the transport of cold and warm air masses from one region to another. *See* ATMOSPHERIC GENERAL CIRCULATION; WIND MEASUREMENT.

Cyclonic and anticyclonic circulation are each a portion of the pattern of airflow within which the streamlines (which indicate the pattern of wind direction at any instant) are curved so as to indicate rotation of air about some central point of the cyclone or anticyclone. The rotation is considered cyclonic if it is in the same sense as the rotation of the surface of the Earth about the local vertical, and is considered anticyclonic if in the opposite sense. Thus, in a cyclonic circulation, the streamlines indicate counterclockwise (clockwise for anticylonic) rotation of air about a central point on the Northern Hemisphere or clockwise (counterclockwise for anticyclonic) rotation about a point on the Southern Hemisphere. When the streamlines close completely about the central point, the pattern is denoted respectively a cyclone or an anticyclone. Since the gradient wind represents a good approximation to the actual wind, the center of a cyclone tends strongly to be a point of minimum atmospheric pressure on a horizontal surface. Thus the terms cyclone, low-pressure area, or low are often used to denote essentially the same phenomenon. *See* GRADIENT WIND.

Convergent or divergent patterns are said to occur in areas in which the (horizontal) wind flow and distribution of air density is such as to produce a net accumulation or depletion, respectively, of mass of air. The horizontal mass divergence or convergence is intimately related to the vertical component of motion. For example, since local temporal rates of change of air density are relatively small, there must be a net vertical export of mass from a volume in which horizontal mass convergence is taking place. Only thus can the total mass of air within the volume remain approximately constant.

The horizontal mass divergence or convergence is closely related to the circulation. In a convergent wind pattern the circulation of the air tends to become more cyclonic; in a divergent wind pattern the circulation of the air tends to become more anticyclonic. A convergent surface wind field is typical of fronts. As the warm and cold currents impinge at the front, the warm air tends to rise over the cold air, producing the typical frontal band of cloudiness and precipitation. *See* FRONT.

Zonal surface winds patterns result from a longitudinal averaging of the surface circulation. This averaging typically reveals a zone of weak variable winds near the Equator (the doldrums) flanked by northeasterly trade winds in the Northern Hemisphere and southeasterly trade winds in the Southern Hemisphere, extending poleward in each instance to about latitude 30°. The doldrum belt, particularly at places and times at which it is so narrow that the trade winds from the two hemispheres impinge upon it quite sharply, is designated the intertropical convergence zone, or ITCZ. The resulting convergent wind field is associated with abundant cloudiness and locally heavy rainfall. *See* MONSOON METEOROLOGY.

Local winds commonly represent modifications by local topography of a circulation of large scale. They are often capricious and violent in nature and are sometimes characterized by extremely low relative humidity. Examples are the mistral which blows down the Rhone Valley in the south of France, the bora which blows down the gorges leading to the coast of the Adriatic Sea, the foehn winds which blow down the Alpine valleys, the williwaws which are characteristic of the fiords of the Alaskan coast and the Aleutian Islands, and the chinook which is observed on the eastern slopes of the Rocky Mountains. *See* CHINOOK.

Frederick Sanders; Howard Bluestein

Wind measurement The determination of three parameters: the size of an air sample, its speed, and its direction of motion. Air movement or wind is a vector that is specified by speed and direction; meteorological convention indicates wind direction is the direction from which the wind blows (for example, a southeast wind blows toward the northwest). Anemometers measure wind speed, while wind vanes indicate direction. On average, the wind blows horizontally over flat terrain; however, gusts, thermals, cloud outflows, and many other conditions have associated with them significant short-term vertical wind components. While research wind instruments typically measure both horizontal and vertical air movement, operational and personal wind sensors measure only the horizontal component. *See* ANEMOMETER.

There are many types of wind measurement instruments. In situ devices measure characteristics of air in contact with the instrument; often they are referred to as immersion sensors because they are immersed in the fluid (air) they measure. Remote wind sensors make measurements without physical contact with the portion of the atmosphere measured. Active remote sensors emit electromagnetic (for example, light or radio waves) or sound waves into the atmosphere and measure the amount and nature of the electromagnetic or acoustic power returned from the atmosphere. *See* LIDAR; METEOROLOGICAL INSTRUMENTATION; METEOROLOGICAL RADAR; WIND.

Walter F. Dabberdt

Wind power The extraction of kinetic energy from the wind and conversion of it into a useful type of energy: thermal, mechanical, or electrical. Wind power has been used for centuries.

It has been estimated that the total wind power in the atmosphere averages about 3.6×10^{12} kW, which is an annual energy of about 107,000 quads (1 quad = 2.931×10^{11} kWh). Only a fraction of this wind energy can be extracted, estimated to be a maximum of 4000 quads per year. According to what is commonly known as the Betz limit, a maximum of 59% of this power can be extracted by a wind machine. Practical machines actually extract from 5 to 45% of the available power. Because the available wind power varies with the cube of wind speed, it is very important to find areas with high average wind speeds to locate wind machines. *See* WIND.

Most research on wind power has been concerned with producing electricity. Wind power is a renewable energy source that has virtually no environmental problems. However, wind power has limitations. Wind machines are expensive and can be located only where there is adequate wind. These high-wind areas may not be easily accessible or near existing high-voltage lines for transmitting the wind-generated energy. Another disadvantage occurs because the demand for electricity varies with time, and electricity production must follow the demand cycle. Since wind power varies randomly, it may not be available when needed. The storage of electrical energy is difficult and expensive, so that wind power must be used in parallel with some other type of generator or with nonelectrical storage. Wind power teamed with hydroelectric generators is attractive because the water can be used for energy storage, and operation with underground compressed-air storage is another option. *See* ELECTRIC POWER GENERATION; ENERGY SOURCES; ENERGY STORAGE.

The most common type of wind turbine for producing electricity has a horizontal axis, with two or more aerodynamic blades mounted on the horizontal shaft. With a horizontal-axis machine, the blade tips can travel at several times the wind speed, which results in a high efficiency. The blade shape is designed by using the same aerodynamic theory as for aircraft. *See* PROPELLER (AIRCRAFT); TURBINE.

Gary Thomann

Wind rose A diagram in which statistical information concerning the direction and speed of the wind at a particular location may be conveniently summarized. In the standard wind rose a line segment is drawn in each of perhaps eight compass directions from a common origin (see illustration). The length of

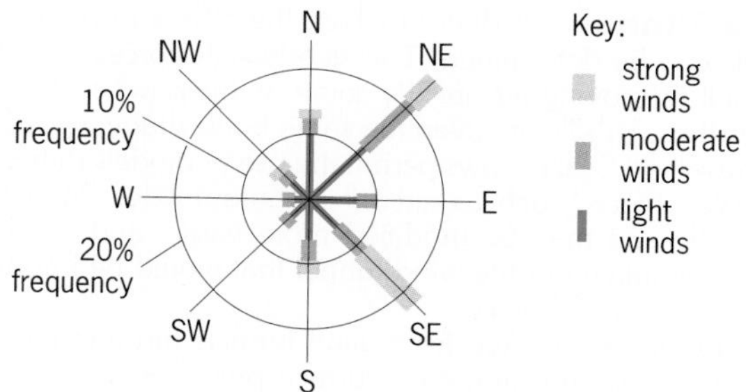

Standard wind rose.

a particular segment is proportional to the frequency with which winds blow from that direction. Parts of a given segment are given various thicknesses, indicating frequencies of occurrence of various classes of wind speed from the given direction. *See* WIND MEASUREMENT.

Frederick Sanders

Wind stress The drag or tangential force per unit area exerted on the surface of the Earth by the adjacent layer of moving air. Erosion of ground surfaces and the production of waves on water surfaces are manifestations of wind stress. Surface wind stress determines the exchange of momentum between the Earth and the atmosphere and exerts a strong influence on the typical variation of wind through the lowest kilometer of the atmosphere. Estimated values of the surface wind stress range up to several dynes per square centimeter (0.1 pascal), depending on the nature of the surface and the character of the adjacent airflow. *See* METEOROLOGY.

Significant stresses arise within the lower atmosphere because of the strong shear of the wind between the slowly moving air near the ground and the more rapidly moving air a kilometer above and because of the turbulent nature of the airflow in this region. The turbulent eddies referred to here have characteristic dimensions ranging up to a few hundreds of meters.

Wind pressure is the force exerted by the wind per unit area of solid surface exposed normal to the wind direction and is also known as dynamic pressure. In contrast to shearing stresses, the wind pressure arises from the difference in pressure between the windward and lee sides of the exposed surface. Wind pressure thus represents a substantial force when the wind speed is high. *See* WIND.

Frederick Sanders

The drag or tangential force of the wind on the sea is expressed in units of dynes per square centimeter or micronewtons per square meter but is normally taken to represent the mean drag over an undefined area, perhaps several kilometers square, containing many waves. It is usually related to an appropriate time and space average of the wind near the sea surface (at 10 m above the mean level, for example).

The drag coefficient over the sea is an important quantity in both meteorology and oceanography since it relates the wind speed to the drag, which generates ocean waves, drives the ocean currents, and sets the scale of the atmospheric turbulence that transfers water vapor and heat from the ocean to the atmosphere to provide the energy for clouds and weather systems.

The drag coefficient of the sea surface depends on the wave field and on the turbulent structure of the flow in the air and the water. Present knowledge of the complicated fluid mechanics involved is not sufficient to allow theoretical calculation of it. *See* ATMOSPHERIC GENERAL CIRCULATION; MARITIME METEOROLOGY; OCEAN CIRCULATION; OCEAN WAVES.

There is substantial agreement that the drag of the wind on the sea is small relative to that of a fixed soil surface with the same geometry. It is largely independent of the fetch and so seems to depends less on the larger waves than on the short waves and ripples. Surface-active agents, which affect the shortest waves, may therefore be important. Henry Charnock

Wind tunnel A duct in which the effects of airflow past objects can be determined. The steady-state forces on a body held still in moving air are the same as those when the body moves through still air, given the same body shape, speed, and air properties. Scaling laws permit the use of models rather than full-scale objects, such as aircraft or automobiles. Models are less costly and may be modified more easily, and conditions may be simulated in the wind tunnel that would be impossible or dangerous in full scale.

Most data are secured from wind tunnels through measurement of forces and moments, surface pressures, changes produced in the airstream by the model, local temperatures, and motions of dynamically scaled models, and by visual studies.

A balance system separates and measures the six components of the total force. The three forces taken parallel and perpendicular to a flight path are lift, drag, and side force. The three moments about these axes are yawing moment, rolling moment, and pitching moment, respectively.

Surface pressures are measured by connecting orifices flush with the model surface to pressure-measuring devices. Local air load, total surface load, moment about a control surface hinge line, boundary-layer characteristics, and local Mach number may be obtained from pressure data.

Measurements of stream changes produced by the model may be interpreted in terms of forces and moments on the model. In two-dimensional tunnels, where an aircraft model spans the tunnel, it is possible to determine the lift and center of pressure by measuring the pressure changes on the floor and ceiling of the tunnel. The parasite drag of a wing section may be determined by measuring the total pressure of the air which has passed over the model and calculating its loss of momentum.

Measurements of surface temperatures indicate the rate of heat transfer or define the amount of cooling that may be necessary.

In elastically and dynamically scaled models used for flutter testing, measurements of amplitude and frequency of motion are made by using accelerometers and strain gages in the structure. In free-flight models, such as bomb or missile drop tests, data are frequently obtained photographically.

At low speeds, smoke and tufts are often used to show flow direction. A mixture of lampblack and kerosine painted on the model shows the surface streamlines. A suspension of talcum powder and a detergent in water is also used.

For aircraft at velocities near or above the speed of sound, some flow features may be made visible by optical devices. *See* INTERFEROMETRY; SCHLIEREN PHOTOGRAPHY; SHADOWGRAPH.

The V/STOL wind tunnel is a newer development of low-speed wind tunnels having a large very-low-speed section to permit testing of aircraft designed for vertical or short takeoff and landing (V/STOL) while operating in the region between vertical flight and cruising flight. Robert G. Joppa

Windings in electric machinery Windings can be classified in two groups: armature windings and field windings. The armature winding is the main current-carrying winding in which the electromotive force (emf) or counter-emf of rotation is induced. The current in the armature winding is known as the armature current. The field winding produces the magnetic field in the machine. The current in the field winding is known as the field or exciting current. *See* ELECTRIC ROTATING MACHINERY; GENERATOR; MOTOR.

The location of the winding depends upon the type of machine. The armature windings of dc motors and generators are located on the rotor, since they must operate in conjunction with the commutator, and the field windings are mounted on stator field poles. *See* DIRECT-CURRENT GENERATOR; DIRECT-CURRENT MOTOR.

Alternating-current synchronous motors and generators are normally constructed with the armature winding on the stator and the field winding on the rotor. There is no clear distinction between the armature and field windings of ac induction motors or generators. One winding may carry the main current of the machine and also establish the magnetic field. It is customary to use the terms stator winding and rotor winding to identify induction motor windings. The word armature, when used with induction motors, applies to the winding connected to the power source (usually the stator). *See* ALTERNATING-CURRENT GENERATOR; ALTERNATING-CURRENT MOTOR; SYNCHRONOUS MOTOR. Arthur R. Eckels

Wine Alcoholic beverages made by fermentation of the juice of fruits or berries, essentially grape juice. Wines from other materials are always required to show their source on the label, for example, apple wine, berry wine, and cherry wine. California produces the largest percentage of the grape wine made in the United States, and well over a hundred varieties of the cultivated grape (*Vitis vinifera*) are used.

Classification of wine depends on the color, relative sweetness, alcoholic content, presence of carbon dioxide, the variety of grape, and the region where the grapes are grown. Wines may be red or white. The terms dry and sweet refer to the relative sugar content of a wine. Table wines contain less than 14% alcohol by volume, and dessert wines contain over 14%, usually 20%, alcohol. The higher alcohol content of dessert wines is obtained by the addition of brandy, which is called fortification. Sparkling wines such as champagne and sparkling Burgundy contain carbon dioxide.

Grapes are harvested and go through a crusher-stemmer, which crushes the berries, but not the seeds, and removes the stems. The must, still containing seeds and skins, is pumped to fermentation tanks where it receives sulfur dioxide gas. This controls, to a large extent, the growth of bacteria and wild yeasts, whereas the various strains of wine yeast are adapted to this amount of SO_2. A pure yeast starter is then added. In the production of white wines, the juice is separated from the skins and seeds at an early stage of fermentation. *See* FERMENTATION.

After the initial fermentation, the wine is transferred to the storage cellar for completion of fermentation, clarification, aging, stabilization, and bottling. These operations are called cellar practices. The sediment of yeast and other insoluble matter is called lees. The new wine is drawn off, or racked, from the lees to avoid picking up undesirable flavors from the lees. Aging is then continued. Wines are often chilled to precipitate excess cream of tartar, the potassium acid tartrate from the grape, which otherwise might precipitate upon chilling of bottled wine. Champagne is made by allowing a secondary fermentation to occur with a special flocculating type of yeast. either in bottles, by the original French procedure, or in bulk, using pressure tanks followed by bottling. The carbon dioxide, CO_2, of sparkling wines is produced by yeast fermentation. *See* ALCOHOLIC BEVERAGES. Emil M. Mrak; Herman J. Phaff

Wing The planar surface of a heavier-than-air body whose primary purpose is to produce lift during flight. A wing is composed of a series of airfoil-shaped components joined together to form one integral structure. On an airplane, the wing also provides storage space for the fuel and undercarriage, and is often used as the structural anchor point for the propulsion systems. The wing also includes important flight-control surfaces, such as the aileron, flaps, slats, and spoilers.

Wing planform definitions. Several key geometric parameters are used to define the wing planform (see illustration). The chord of a wing section (c) is defined as the straight-line distance between the leading and trailing edges. Other important parameters are the leading-edge sweep angle (Λ), the root chord (c_r), the tip chord (c_t), the span (b), the planform area (S), the taper ratio (c_t/c_r), the mean geometric chord ($c_g = S/b$), and the aspect ratio ($A_R = b/c_g = b^2/S$). The wing angle of attack (α_r) is defined as the angle between the root chord line and the oncoming airflow, known as the free-stream velocity (V_∞). A wing will have a particular angle of attack at which no net lift is produced, and this is known as the wing zero-lift angle (α_{L0}).

The speed of sound in air (a) is measured by the Mach number and defined as $M_\infty = V_\infty/a$. If $M_\infty < 1.0$, the flow is classified as subsonic, while, if $M_\infty > 1.0$, the flow is supersonic. The speed of sound decreases slowly with altitude, and has a value at sea level of 340.3 m/s (1116.45 ft/s or 661.5 knots). During subsonic flight, an airfoil produces lift by forcing the air to follow a path over the upper surface which has a larger amount of curvature than that over the lower surface. This flow field causes the air to have a lower static pressure on the upper surface relative to the flow underneath the wing. At each wing tip the lower surface high-pressure air is free to move onto the upper surface, and, as the wing moves forward, two corkscrew flow structures are trailed behind the airplane. These structures are referred to as the wing-tip vortices, and on a humid day they often become visible as water condenses out in the low-pressure vortex core to form two vapor trails behind the airplane. The wing-tip vortices induce a downward flow, known as the downwash (w, measured in the same units as the free-stream velocity), not only behind the wing, but also ahead of the wing.

Downwash gives rise to a drag component known as the induced drag (or drag due to lift), which increases as the wing aspect ratio decreases. At low angles of attack, wings which have a large aspect ratio produce more lift for a lower induced drag than wings of low aspect ratio. Thus, high-aspect-ratio wings have a higher lift-to-drag ratio and are often referred to as being aerodynamically efficient.

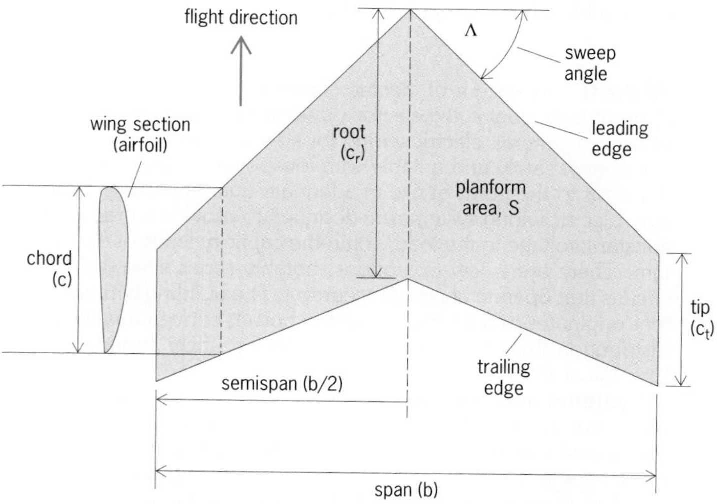

Wing planform definitions.

On many aircraft, the leading edge of the wing is swept back by an angle Λ degrees (see illustration). The effect of sweep relies on the fact that it is only the airspeed normal to the leading edge which controls the variation of air velocity over the upper and lower surfaces of the wing. Thus, even though the free-stream Mach number (M_∞) may be equal to unity, the airspeed normal to the leading edge ($M_\infty \cos \Lambda$) may be small enough to avoid increased drag due to shock-wave formation over the wing. Typical values of sweepback for modern transport aircraft range from 25 to 37°.

Wing high-lift devices, specifically referred to as slats and flaps, are deployed during take-off and landing when high lift is required at slow flight speeds. However, they are not used during cruise, since they greatly increase the drag on the airplane. When deployed, these control surfaces increase both the camber and chord of the cruise airfoil section. Increasing the camber increases the curvature of the air as it flows over the airfoil, which further decreases the upper-surface static air pressure (over the cruise configuration) and, thus, increases wing lift. The increased chord increases the wing planform area, which also increases wing lift. After touchdown, spoilers are deployed, which increase the drag on the airplane and quickly reduce the lift, thus applying the weight of the airplane on the braking wheels. Spoilers can also be used individually during high-speed flight to reduce lift on one wing only and thus roll the aircraft, and, when used together, to act as an airbrake to slow the aircraft quickly. *See* AIRFOIL; AIRPLANE; WING STRUCTURE. Andrew J. Niven

Winged bean A plant (*Psophocarpus tetragonolobus*), also known as four-cornered bean, asparagus pea, goa bean, and manila bean, in the family Leguminosae. It is a climbing perennial that is usually grown as an annual. It has been suggested that it originated either in East Africa or Southeast Asia, but there is more evidence to support an African origin. However, Southeast Asia and the highlands of Papua New Guinea represent two foci of its domestication.

Traditionally, winged bean is grown as a backyard vegetable in Southeast Asia and a few islands of the Pacific. It is grown as a field crop in Burma and Papua New Guinea. However, between 1980 and 1990 winged bean was introduced throughout the tropical world.

Almost all parts of this plant are edible and are rich sources of protein. The green pods, tubers, and young leaves can be used as vegetables, and the flowers can be added to salads. The dry seeds are similar to soybeans and can be used for extracting edible oil, feeding animals, and making milk and traditional Southeast Asian foods such as tempeh, tofu, and miso. Flour from the winged bean can also be used as a protein supplement in bread making.

A number of diseases and insect pests may limit winged bean yield. The most widespread and damaging disease appears to be false rust or orange gall (caused by *Synchytrium psophocarpi*). *See* BREEDING (PLANT); PLANT PATHOLOGY; ROSALES. Tanveer N. Khan

Wire A thread or slender rod of metal. Wire is usually circular in cross section and is flexible. If it is of such a diameter or composition that it is fairly stiff, it is termed rod. The wire may be of several small twisted or woven strands, but if used for lifting or in a structure, it is classed as cable. Wire may be used structurally in tension, as in a suspension bridge, or as an electrical conductor, as in a power line. The working of metal into wire greatly increases its tensile strength. Thus, a cable of stranded small-diameter wires is stronger as well as more flexible than a corresponding solid rod. *See* MAGNET WIRE. Frank H. Rockett

Wire drawing The reduction of the diameter of a metal rod or wire by pulling it through a die. The working region of dies are typically conical (see illustration). The tensile stress on

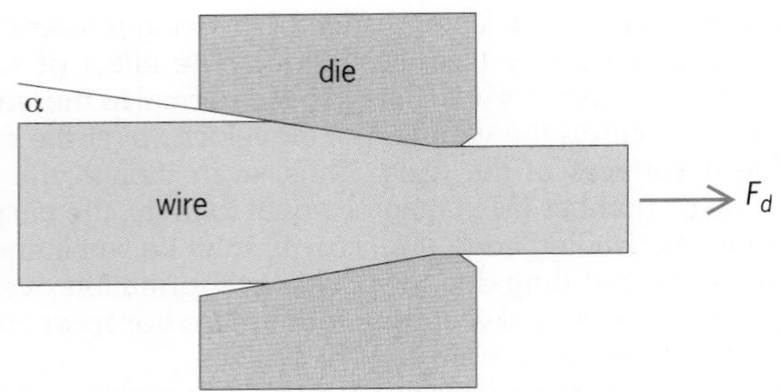

Wire being drawn through a die.

the drawn wire, that is, the drawing stress, must be less than the wire's yield strength. Otherwise the drawn section will yield and fail without pulling the undrawn wire though the die. Because of this limitation on the drawing stress, there is a maximum reduction that can be achieved in a single drawing pass. After large drawing reductions, wires or rods develop crystallographic textures or preferred orientations of grains. The textures are characteristic of the crystal structure of the metal. *See* ALLOY; CRYSTAL STRUCTURE; METAL; METAL, MECHANICAL PROPERTIES OF; METALLURGY. William F. Hosford

Wireless fidelity (Wi-Fi) In general, the wireless local area network (LAN) technology based on the Institute of Electrical and Electronics Engineers (IEEE) 802.11 standard. It enables computing devices to wirelessly exchange data with each other or with a wired network over a distance of up to about 300 ft (90 m), in a normal office environment, using unlicensed portions of the radio-frequency spectrum. Strictly speaking, Wi-Fi® technology refers to wireless LAN technologies that have passed interoperability tests designed by the Wi-Fi Alliance, an industry organization of vendors of 802.11 wireless LAN products. *See* LOCAL-AREA NETWORKS.

The IEEE 802.11 standard consists of medium access control (MAC) specifications and physical layer (PHY) specifications. The MAC specifications define how a wireless LAN entity exchanges data with others using a shared wireless medium. The PHY specifications define the wireless signals that carry the exchanged data between wireless LAN entities and the wireless channels over which the wireless signals are transmitted. IEEE 802.11 was introduced in 1997 with a MAC specification and three PHY specifications. It has been amended with many extensions, including a MAC security specification 802.11i (2004) and three high-speed PHY specifications: 802.11b for 2.4 GHz (1999), 802.11a for 5 GHz (1999), and 802.11g for 2.4 GHz (2003). *See* RADIO SPECTRUM ALLOCATION.

IEEE 802.11 defines two networking components, a station and an access point. A station has a MAC layer and a PHY layer; it is responsible for sending or receiving data to or from another station over the wireless medium, and does not relay data for any other station. An access point is a special station whose main task is to relay data for stations associated with it. *See* DATA COMMUNICATIONS.

IEEE 802.11 MAC. The IEEE 802.11 MAC provides three functions: fair access to the shared wireless medium with optional quality of service, reliable delivery of data frames over the wireless medium, and security protection.

Fair access. The basic medium access mechanism employed by the IEEE 802.11 MAC is called carrier sense multiple access with collision avoidance (CSMA/CA). CSMA/CA is a wait-before-send protocol; that is, after the shared wireless medium becomes available, a station must wait a required period plus a back-off time before it can transmit a data frame, provided that the wireless medium is still available at the end of back off. The back-off time is randomly selected by the station within a specified time period, called a contention window. CSMA/CA ensures that stations with the same contention window size and the same required waiting period have an equal probability to seize the shared wireless medium to transmit.

To serve more complicated quality-of-service requirements, new medium access methods are developed in the 802.11e draft standard. In order to respond to market demand, the Wi-Fi Alliance started an interoperability test in 2004 for one of these methods, called EDCA in the 802.11e draft standard, or Wi-Fi multimedia (WMM) by the Wi-Fi Alliance. WMM provides differentiated services for four traffic classes. They are voice (highest priority), video, best-effort data, and background data (lowest priority).

Reliable delivery. Since an 802.11 wireless LAN operates in unlicensed spectrum, the wireless medium could be very noisy due to the existence of other types of radio devices operating on the same band and not conforming to CSMA/CA. Thus, data frames can be frequently corrupted. In order to ensure reliable delivery of data frames over the noisy wireless medium, the IEEE 802.11 MAC employs a data frame exchange protocol for transmission of every data frame.

Security protection. Since anyone in the range of a wireless LAN can receive data frames transmitted over the wireless medium and can send data frames to stations or access points on the wireless LAN, authentication and data encryption must be implemented for access control and privacy. Initially, the IEEE 802.11 MAC defined the wired equivalent privacy (WEP) protocol to serve these needs. Unfortunately, WEP has significant security flaws. In order to replace WEP, the 802.11i standard has been developed. It specifies two authentication methods: preshared key (PSK) authentication (for simple wireless LAN configurations) and the IEEE 802.1x/extensible authentication protocol (EAP) authentication (for centralized access control on large wireless LANs); and two encryption methods: temporal key integrity protocol (TKIP) with message integrity check (MIC) and advanced encryption standard (AES). *See* COMPUTER SECURITY; CRYPTOGRAPHY.

IEEE 802.11 PHY. As of 2005 three 802.11 PHY had been put in use: 802.11b, 802.11a, and 802.11g, with 802.11b and 802.11g having a large installed base.

The 802.11b standard defines a direct sequence spread spectrum (DSSS) radio PHY for the 2.4-GHz industrial, scientific, and medical (ISM) band. The major interfering sources for 802.11b-based wireless LAN are microwave ovens, 2.4-GHz cordless phones, and Bluetooth® devices.

The 802.11a standard defines an orthogonal frequency division multiplexing (OFDM) radio PHY for the 5-GHz unlicensed national information infrastructure (U-NII) band.

The 802.11g standard defines an OFDM radio PHY that supports the same high data rates offered by 802.11a but remains compatible with the popular 802.11b. Hui Luo

Wiring A system of electric conductors, components, and apparatus for conveying electric power from source to the point of use. In general, electric wiring for light and power must convey energy safely and reliably with low power losses, and must deliver it to the point of use in adequate quantity at rated voltage. Electric wiring systems are designed to provide a practically constant voltage to the load within the capacity limits of the system. There are a few exceptions, notably series street-lighting circuits that operate at constant current. The building wiring system originates at a source of electric power, conventionally the distribution lines or network of an electric utility system. *See* ELECTRICAL CODES.

Systems and service. Wiring systems are generally three-phase to conform to the supply systems. Energy is transformed to the desired voltage levels by a bank of three single-phase transformers or a single three-phase transformer. The transformers may be connected in either a delta or Y configuration.

Service provided at the primary voltage of the utility distribution system, typically 13,800 or 4160 volts, is termed primary

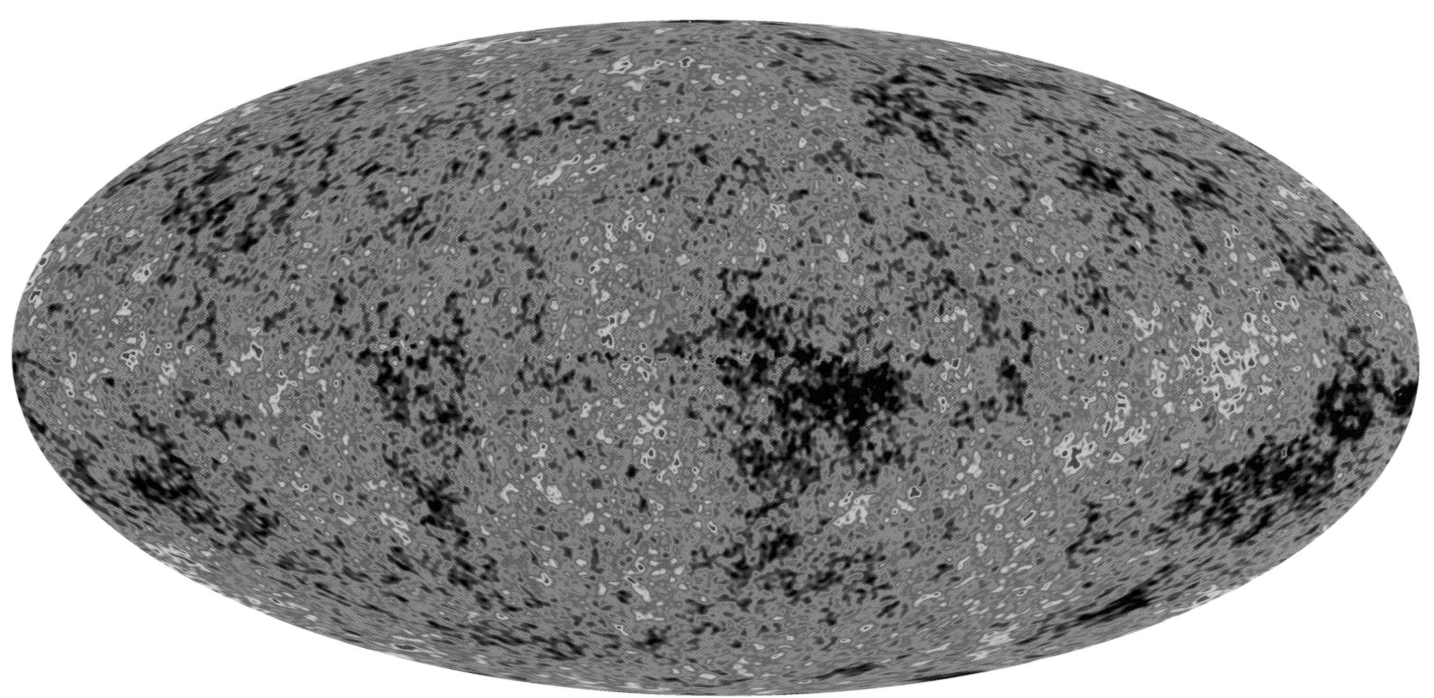

Detailed all-sky picture of the infant universe from three years of data from the *Wilkinson Microwave Anisotropy Probe* (*WMAP*). The image reveals the 13.7×10^9-year-old temperature fluctuations (shown as color differences) that corresponded to the seeds that grew to become galaxies. The signal from our Milky Way Galaxy was subtracted using multifrequency data. The image shows a temperature range of ±200 microkelvin. (*NASA/WMAP Science Team*)

service. Service provided at secondary or utilization voltage, typically 120/208 or 277/480 volts, is called secondary service.

Service at primary voltage levels is often provided for large industrial, commercial, and institutional buildings, where the higher voltage can be used to advantage for power distribution within the buildings. Where primary service is provided, power is distributed at primary voltage from the main switchboard through feeders to load-center substations installed at appropriate locations throughout the building.

Most secondary services in the United States are 120/ 208 volts, three-phase, four-wire, or 120/240 volts, single-phase, three-wire serving both light and power. For relatively large buildings where the loads are predominantly fluorescent lighting and power (as for air conditioning), the service is often 277/480 volts, three-phase, four-wire, supplying 480 volts for power and 277 volts, phase-to-neutral, for the lighting fixtures.

From the service entrance, power is carried in feeders to the main switchboard, then to distribution panelboards. Smaller feeders extend from the distribution panelboards to light and power panelboards. Branch circuits then carry power to the outlets serving the various lighting fixtures, plug receptacles, motors, or other utilization equipment.

Methods. Methods of wiring in common use for light and power circuits are as follows: (1) insulated wires and cables in metallic or nonmetallic raceways; (2) nonmetallic sheathed cables; (3) metallic armored cables; (4) busways; (5) copper-jacketed, mineral-insulated cables; (6) aluminum-sheathed cables; (7) nonmetallic sheathed and armored cables in cable support systems; and (8) open insulated wiring on solid insulators (knob and tube).

The selection of the wiring method or methods is governed by a variety of considerations, which usually include code rules limiting the use of certain types of wiring materials; suitability for structural and environmental conditions; installation (exposed or concealed); accessibility for changes and alterations; and costs.

Circuit design. The design of a particular wiring system is developed by considering the various loads, establishing the branch-circuit and feeder requirements, and then determining the service-entrance requirements. Outlets for lighting fixtures, motors, portable appliances, and other utilization devices are indicated on the building plans and the load requirement of each outlet noted in watts or horsepower. Lighting fixtures and plug receptacles are then grouped on branch circuits and connections to the lighting panelboard indicated.

Lighting branch circuits may be loaded to 80% of circuit capacity. However, there is a reasonable probability that the lighting equipment will be added or replaced at some future time by equipment of higher output and greater load. Therefore, in modern practice, lighting branch circuits are loaded only to about 50% capacity. Lighting branch circuits are usually rated at 20 A. Smaller 15-A branch circuits are used mostly in residences.

Joseph F. McPartland; Brian J. McPartland

Wiring diagram A drawing illustrating electrical and mechanical relationships between parts of a component between which electrical wiring must be connected.

A wiring diagram is distinguished from an electrical schematic in that the arrangement of the schematic bears no necessary relationship to the mechanical arrangement of the electrical elements in the component. The wiring diagram provides an accurate picture of how the wiring on the components and between components should appear in order that the electrical wiring technician can install the wiring in the manner that will best contribute to the optimum performance of the device.

Wiring diagrams also include such information as type of wire, color coding, methods of wire termination, and methods of wire and cable clamping. *See* SCHEMATIC DRAWING. Robert W. Mann

Witherite The mineral form of barium carbonate. Witherite has orthorhombic symmetry and the aragonite structure type. Crystals, often twinned, may appear hexagonal in outline. It may be white or gray with yellow, brown, or green tints. Its hardness is 3.5 and its specific gravity 4.3.

Witherite may be found in veins with barite and galena. It is found in many places in Europe, and large crystals occur at Rosiclare, Illinois. *See* BARIUM; CARBONATE MINERALS; HARDNESS SCALES.

Robert I. Harker

Wolf-Rayet star A type of hot, luminous star that is distinguished by its extremely dense and fast wind. The spectacularly bright, discrete bands of atomic emission from these winds greatly facilitated their discovery with the aid of a visual spectroscope by the French astronomers Charles Wolf and Georges Rayet at the Paris Observatory in 1867. *See* ASTRONOMICAL SPECTROSCOPY.

The Wolf-Rayet phenomenon includes a typical phase in the advanced evolution of a massive star (about 20–100 times the Sun's mass at birth), or sometimes (15% of cases) among lower-mass stars in the planetary nebula stage. The first fusion process in the core of a massive star involves the conversion of hydrogen into helium via the carbon-nitrogen-oxygen (CNO) cycle, in which helium and nitrogen are enhanced at the expense of the initially abundant hydrogen and traces of carbon and oxygen. When such fusion products are visible in the winds, a WN-type Wolf-Rayet star is seen, whose spectrum is dominated by Doppler-broadened atomic lines of helium and nitrogen in various stages of ionization. Later, when the second fusion process occurs, helium is converted mainly into carbon and oxygen, with nitrogen being virtually destroyed. A WC-type Wolf-Rayet star is then seen with lines mainly of carbon and helium. A brief oxygen-rich phase may occur (WO) toward the end of the helium-burning phase, after which all subsequent exotic nuclear-burning phases are so rapid that they remain hidden by the much more slowly changing stellar surface. At that point, it is believed that the Wolf-Rayet star will explode as a supernova, resulting in the collapse to a black hole in most cases and an associated gamma-ray burst in very rare instances. *See* BLACK HOLE; DOPPLER EFFECT; NUCLEOSYNTHESIS; PLANETARY NEBULA; SUPERNOVA.

Beneath the dense winds that often hide the stellar surface, massive Wolf-Rayet stars have surface temperatures ranging from 30,000 to 150,000 (54,000 to 270,000°F), radii of 1 to 15 solar units, and luminosities of 10^5 to 10^6 times that of the Sun. They are losing matter at a rate that is typically 10^9 times that of the Sun'swind, at speeds ranging from 1000 to 3000 km/s (600 to 1800 mi/s). Although massive Wolf-Rayet stars appear to be rare (about 250 are known so far in the Milky Way Galaxy, out of an estimated total population of 1000–2000), all massive stars likely pass through a Wolf-Rayet stage toward the end of their relatively short lives. *See* SOLAR WIND; STAR; STELLAR EVOLUTION.

Anthony F. J. Moffat

Wolframite A mineral with composition $(Fe,Mn)WO_4$, intermediate between ferberite and huebnerite, which form a complete solid solution series. Wolframite occurs commonly in short, brownish-black, monoclinic, prismatic, bladed crystals. It is probably the most important tungsten mineral. China is the major producer of wolframite. Tungsten minerals of the wolframite series occur in many areas of the western United States; the major producing district is Boulder and northern Gilpin counties in Colorado. *See* HUEBNERITE. Edward C. T. Chao

Wollastonite A mineral inosilicate with composition $CaSiO_3$. Commonly it is massive, or in cleavable to fibrous aggregates. Hardness is 5–5$^1/_2$ on Mohs scale; specific gravity is 2.85. On the cleavages the luster is pearly or silky; the color is white to gray.

Wollastonite is found in large masses in the Black Forest of Germany; Brittany, France; Chiapas, Mexico; and Willsboro, New York, where it is mined as a ceramic material. *See* SILICATE MINERALS. Cornelius S. Hurlbut, Jr.

Wolverine The largest member of the family Mustelidae, to which weasels, mink, fishers, martens, ferrets, badgers, and otters belong. The powerfully built wolverine (*Gulo gulo*) has an overall shaggy appearance. The heavy body is covered with long guard hairs that overlie a coarse, dense, kinky, woolly underfur. The general coloration is blackish brown. A light brown band extends along each side of the body from shoulder to rump before merging at the base of the tail. Most wolverines have a light silver facial mask. The large head is broad and somewhat rounded with small, wide-set eyes and short, rounded, well-furred ears. The legs, feet, and most of the relatively short, shaggy tail are dark. The legs are massive, stocky, and powerful, and the feet are large. The claws are pale and sharply curved. A pair of anal scent glands are present which secrete a yellowish-brown fluid. Adults have a head and body length of 650–1050 mm (25–41 in.), a tail length of 170–260 mm (6.5–10 in.), and weigh 7–32 kg (15–70 lb).

The wolverine has a circumpolar distribution, occurring in the tundra and taiga zones throughout northern Europe, Asia, and North America. It is uncommon in the United States and most of Canada and is very rare in Scandinavia. It is active all year and may be found in forests, mountains, or open plains. Wolverines are primarily terrestrial, but they can also climb trees and are excellent swimmers. These mammals have a keen sense of smell, but they lack keen senses of vision and hearing. Wolverines are solitary wanderers except during the breeding season. Food consists of carrion (reindeer, red deer, moose, caribou), eggs of ground-nesting birds, lemmings, and berries. *See* BADGER; CARNIVORA; FERRET; FISHER; MARTEN; MINK; OTTER; WEASEL.

Donald W. Linzey

Wood anatomy Wood is composed mostly of hollow, elongated, spindle-shaped cells that are arranged parallel to each other along the trunk of a tree. The characteristics of these fibrous cells and their arrangement affect strength properties, appearance, resistance to penetration by water and chemicals, resistance to decay, and many other properties.

Just under the bark of a tree is a thin layer of cells, not visible to the naked eye, called the cambium. Here cells divide and eventually differentiate to form bark tissue to the outside of the cambium and wood or xylem tissue to the inside. This newly formed wood (termed sapwood) contains many living cells and conducts sap upward in the tree. Eventually, the inner sapwood cells become inactive and are transformed into heartwood. This transformation is often accompanied by the formation of extractives that darken the wood, make it less porous, and sometimes provide more resistance to decay. The center of the trunk is the pith, the soft tissue about which the first wood growth takes place in the newly formed twigs. *See* STEM.

In temperate climates, trees often produce distinct growth layers. These increments are called growth rings or annual rings when associated with yearly growth; many tropical trees, however, lack growth rings. These rings vary in width according to environmental conditions.

Many mechanical properties of wood, such as bending strength, crushing strength, and hardness, depend upon the density of wood; the heavier woods are generally stronger. Wood density is determined largely by the relative thickness of the cell wall and the proportions of thick- and thin-walled cells present. *See* WOOD PROPERTIES.

In hardwoods (for example, oak or maple), these three major planes along which wood may be cut are known commonly as end-grain, quarter-sawed (edge-grain) and plain-sawed (flat-grain) surfaces (see illustration).

Hardwoods have specialized structures called vessels for conducting sap upward. Vessels are a series of relatively large cells with open ends, set one above the other and continuing as open passages for long distances. In most hardwoods, the ends of the individual cells are entirely open; in others, they are separated by a grating. On the end grain, vessels appear as holes and are termed pores. The size, shape, and arrangement of pores vary considerably between species, but are relatively constant within a species.

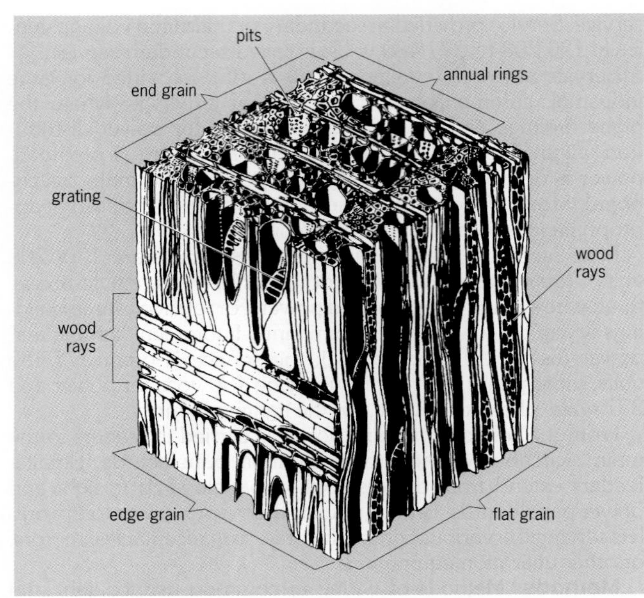

Structure of a typical hardwood. (*USDA*)

Most smaller cells on the end grain are wood fibers which are the strength-giving elements of hardwoods. They usually have small cavities and relatively thick walls. Thin places or pits in the walls of the wood fibers and vessels allow sap to pass from one cavity to another. Wood rays are strips of short horizontal cells that extend in a radial direction. Their function is food storage and lateral conduction. *See* PARENCHYMA; SECRETORY STRUCTURES (PLANT).

The rectangular units that make up the end grain of softwood are sections through long vertical cells called tracheids or fibers. Because softwoods do not contain vessel cells, the tracheids serve the dual function of transporting sap vertically and giving strength to the wood. The wood rays store and distribute sap horizontally.

The principal compound in mature wood cells is cellulose, a polysaccharide of repeating glucose molecules which may reach 4 μm in length. These cellulose molecules are arranged in an orderly manner into structures about 10–25 nm wide called microfibrils. The microfibrils wind together like strands in a cable to form macrofibrils that measure about 0.5 μm in width and may reach 4 μm in length. These cables are as strong as an equivalent thickness of steel.

This framework of cellulose macrofibrils is cross-linked with hemicelluloses, pectins, and lignin. Lignin, the second most abundant polymer found in plants, gives the cell wall rigidity and the substance that cements the cells together. *See* CELL WALLS (PLANT); CELLULOSE; LIGNIN; PECTIN; PLANT ANATOMY; TREE.

Regis B. Miller

Wood chemicals Substances derived from wood. Woody plants comprise the greatest part of the organic materials produced by photosynthesis on a renewable basis, and were the precursors of the fossil coal deposits. The derivation of chemicals from wood is carried out wherever technical utility and economic conditions have combined to make it feasible.

Wood is a mixture of three natural polymers—cellulose, hemicelluloses, and lignin—in an approximate abundance of 50:25:25. In addition to these polymeric cell wall components which make up the major portion of the wood, different species contain varying amounts and kinds of extraneous materials called extractives. The nature of the chemicals derived from wood depends on the wood component involved. *See* CELLULOSE; HEMICELLULOSE.

Chemicals derived from wood include: bark products, cellulose, cellulose esters, cellulose ethers, charcoal, dimethyl sulfoxide, ethyl alcohol, fatty acids, furfural, hemicellulose extracts, kraft lignin, lignin sulfonates, pine oil, rayons, rosin, sugars, tall oil, turpentine, and vanillin. Most of these are either direct products or by-products of wood pulping, in which the lignin that cements the wood fibers together and stiffens them is dissolved away from the cellulose. High-purity chemical cellulose or dissolving pulp is the starting material for such polymeric cellulose derivatives as viscose rayon and cellophane, cellulose esters such as the acetate and butyrate for fiber, film, and molding applications, and cellulose ethers such as carboxymethylcellulose, ethylcellulose, and hydroxyethylcellulose for use as gums. *See* CELLOPHANE; DIMETHYL SULFOXIDE; ETHYL ALCOHOL; FURFURAL; ROSIN; TURPENTINE; VANILLA; WAX, ANIMAL AND VEGETABLE; WOOD PRODUCTS.

Irving S. Goldstein

Wood composites Wood-based composites and a combination of two or more elements held together by a matrix. By this definition, what we call "solid wood" is a composite. Solid wood is a three-dimensional composite composed of cellulose, hemicelluloses, and lignin polymers with smaller amounts of inorganics and extractives held together in a lignin matrix. Among the advantages of developing wood composites from solid wood are the ability to: (1) use trees or tree parts that are too small for lumber or veneer production (includes thinning, limbs, brush, and other forest biomass), (2) use underutilized wood species, (3) use mixed wood species, (4) remove defects, (5) create more uniform materials, (6) develop materials that are stronger than the original solid wood, (7) make materials of different shapes, and (8) produce both large-volume low-value and low-volume value-added materials. A beneficial by-product is the removal of biomass from overcrowded forests, especially in the western United States, to improve forest health and to reduce the fuel load that increases the threat of catastrophic fires.

Wood composites can be classified in several ways: by density, element type, process, or application. Classification by specific gravity includes high-, medium-, and low-density composites. Classification by element type includes strandboard, waferboard, chipboard, flakeboard, particleboard, and fiberboard. Classification by process includes wet or dry production processes. Classification by application would include products such as insulation board, underlayment composites, and structural and core composites. The basic adhesives most commonly used for wood composites are formaldehyde, urea, melamine, phenol, resorcinol, and isocyanates. Despite the apparent simplicity in terms of families of chemicals, the formulations are highly complex mixtures of chemicals and additives depending on the specific application. *See* WOOD ENGINEERING DESIGN; WOOD PRODUCTS; WOOD PROPERTIES.

Roger M. Rowell

Wood degradation Decay of the components of wood. Despite its highly integrated matrix of cellulose, hemicellulose, and lignin, which gives wood superior strength properties and a marked resistance to chemical and microbial attack, a variety of organisms and processes are capable of degrading wood. The decay process is a continuum, often involving a number of organisms over many years. Wood degrading agents are both biotic and abiotic, and include heat, strong acids or bases, organic chemicals, mechanical wear, and sunlight (uv degradation).

Abiotic degradation. Heat degrades both cellulose and hemicellulose, reducing strength and causing the wood to darken. At temperatures above 451°F (219°C), combustion occurs. Strong acids eventually degrade the carbohydrate portion of wood, reducing its strength. Strong bases attack the lignin, leaving the wood appearance bleached and white. Other chemicals, such as concentrated organics or salt solutions, can also disrupt the lignocellulosic matrix, reducing material properties of the wood. Sunlight, primarily through the action of ultraviolet light, also degrades wood through the creation of free radicals which then degrade the wood polymers. Mechanical wear of wood can occur in a variety of environments.

Biotic degradation. Biotic damage can occur from a variety of agents, including bacteria, fungi, insects, marine borers, and birds and animals. Birds and animals generally cause mechanical damage in isolated instances. All biotic agents have four basic requirements: adequate temperature (32–104°F or 0–40°C) with most optima between 77–90°F (25–32°C), oxygen (or other suitable terminal election acceptor), water, and a food source. Water is a critical element for biotic decay agents: it serves as reactant in degradative reactions, a medium for diffusion of enzymes into wood and degradative products back to the organism, and a wood swelling agent.

Bacteria are not major degraders of wood products, but they can damage pit membranes, thereby increasing permeability, and some are capable of cell wall degradation. *See* BACTERIA.

Fungi are among the most important wood-degrading organisms because they play an important role in terrestrial carbon cycling. Wood-degrading fungi can be classified as molds, stainers, soft rotters, brown rotters, and white rotters on the basis of the attack patterns. Molds, stainers, and soft rotters are members of the ascomycetes and the deuteromycetes (Fungi Imperfecti). *See* ASCOMYCOTA; DEUTEROMYCOTA; FUNGI.

A number of insects have evolved to attack wood, including termites (Isoptera), beetles (Coleoptera), and bees and ants (Hymenoptera). Termites are the most important wood-degrading insects in most environments, and their activity causes severe economic losses. *See* COLEOPTERA; HYMENOPTERA; ISOPTERA.

In saline environments, marine borers can cause significant wood losses. Three groups of marine borers—shipworms, pholads, and gribbles (*Limnoria*)—cause most wood damage in these areas. *See* BORING BIVALVES.

Wood protection. Protecting wood from degradation can take a number of forms. By far the simplest method is to employ designs which limit wood exposure to moisture. In some cases, however, water exclusion is not possible and alternative methods must be employed. The simplest of these methods is the use of heartwood from naturally durable species. Decay- or insect-resistant species include redwood (*Sequoia sempervirens*), western red cedar (*Thuja plicata*), and ekki (*Lophira alata*), while marine-borer-resistant heartwoods include greenheart (*Ocotea rodiaei*) and ekki. Most marine-borer-resistant woods contain high levels of silica which discourages marine borer attack, while species resistant to terrestrial decay agents often contain toxic phenolics. Wood can also be protected from degradation by spraying, dipping, soaking, or pressure treatment with preservatives. *See* WOOD PROPERTIES.

Jeffrey J. Morrell

Wood engineering design The process of creating products, components, and structural systems with wood and wood-based materials. Wood engineering design applies concepts of engineering in the design of systems and products that must carry loads and perform in a safe and serviceable fashion. Examples include structural systems such as buildings or electric power transmission structures, components such as trusses or prefabricated stressed-skin panels, and products such as furniture or pallets and containers. The design process considers the shape, size, physical and mechanical properties of the materials, type and size of the connections, and the type of system response needed to resist both stationary and moving (dynamic) loads and to function satisfactorily in the end-use environment. *See* ENGINEERING DESIGN; STRUCTURAL DESIGN.

Wood is used in both light frame structures and heavy timber structures. Light frame structures consist of many relatively small wood elements such as lumber covered with a sheathing material such as plywood. The lumber and sheathing are connected to act together as a system in resisting loads; an example is a residential house wood floor system where the plywood is nailed to lumber bending members or joists. In this system, no one joist is heavily loaded because the sheathing spreads the load out over many joists. Service factors such as deflections or vibration often govern the design of floor systems rather than strength. Light frame systems are often designed as diaphragms or shear walls to resist lateral forces resulting from wind or earthquake. *See* FLOOR CONSTRUCTION; WOOD PROCESSING.

In heavy timber construction, such as bridges or industrial buildings, there is less reliance on system action and, in general, large beams or columns carry more load transmitted through decking or panel assemblies. Strength, rather than deflection, often governs the selection of member size and connections. There are many variants of wood construction using poles, wood shells, folded plates, prefabricated panels, logs, and combinations with other materials. *See* WOOD PRODUCTS.

Thomas E. McLain

Wood processing Peeling, slicing, sawing, and chemically altering hardwoods and softwoods to form finished products such as boards or veneer; particles or chips for making paper, particle, or fiber products; and fuel. *See* PAPER; VENEER.

A high percentage of the weight of freshly cut or green wood is water. Green wood contains free water in the cell cavities and bound water in the cell walls. When all the free water has been extracted and before any of the bound water has been removed, the wood is said to be at the fiber saturation point. As the moisture content falls below the fiber saturation point, the bound water leaves the cell walls and the wood shrinks. During the drying process, differential shrinkage can cause internal stresses in the wood. If not controlled, this can result in defects such as cracks, splits, and warp. Below the fiber saturation point, wood takes on and gives off water molecules depending on the relative humidity of the air around it and swells and shrinks accordingly.

Wood is machined to bring it to a specific size and shape for fastening, gluing, or finishing. With the exception of lasers, which have a limited application at this time, all machining is based on a sharpened wedge that is used to sever wood fibers. Tools for sawing, boring holes, planing, and shaping, as well as the particles in sandpaper, use some version of the sharpened wedge. *See* WOOD ENGINEERING.

Wood is ground to fibers for hardboard, medium-density fiberboard, and paper products. It is sliced and flaked for particleboard products, including wafer boards and oriented strand boards. Whether made from waste products (sawdust, planer shavings, slabs, edgings) or roundwood, the individual particles generally exhibit the anisotropy and hygroscopicity of larger pieces of wood. The negative effects of these properties are minimized to the degree that the three wood directions (longitudinal, tangential, and radial) are distributed more or less randomly. *See* WOOD PRODUCTS.

Charles J. Gatchell

Wood products Materials developed from use of the hard fibrous substance (wood) which makes up the greater part of the trunks and limbs of trees.

Solid wood products include lumber, veneer and plywood, furniture, poles, piling, mine timbers, and posts; and composite wood products such as laminated timbers, insulation board, hard-board, and particle board. *See* PLYWOOD.

Fiber wood products can be referred to as those which develop initially from the various processes for pulping wood. All are intended to separate the cellulose fibers one from another in relatively pure form to be recombined into layers of pulp, paper sheets, or paperboards. *See* PAPER.

Chemical wood products result from the chemical modification or conversion of cellulose, lignin, and extractives. Chief among these products are textile fibers such as rayon, and many cellulose plastics products such as cellophane, nitrocellulose, photographic film, telephone parts, and plastic housewares and toys. *See* CELLULOSE; TEXTILE; WOOD CHEMICALS.

Willard S. Bromley

Wood properties Physical and mechanical characteristics of wood which are controlled by specific anatomy, moisture content, and to a lesser extent, mineral and extractive content. The properties are also influenced by wood's directional nature, which results in markedly different properties in the longitudinal, tangential, and radial directions or axes. Wood properties within a species vary greatly from tree to tree and within a single axis.

Physical properties. The physical properties (other than appearance) are moisture content, shrinkage, density, permeability, and thermal and electrical properties.

Moisture content is a major factor in the processing of wood because it influences all physical and mechanical properties, and durability and performance during use. Normal in-use moisture content of processed wood that has been dried ranges 8–13%. Moisture content for wood is expressed on either a fractional or percentage basis. Moisture content is defined as the ratio of the mass of water contained in the wood to the mass of the same sample of dry wood.

Shrinkage occurs when wood loses moisture below the fiber saturation point. Above that point, wood is dimensionally stable. The amount of the shrinkage depends on its direction relative to grain orientation and the amount of moisture lost below the fiber saturation point. Wood shrinks significantly more in the radial and tangential directions than in the longitudinal direction.

The density of wood is determined by the amount of cell wall substance and the volume of voids caused by the cell cavities (lumens) of the fibers. Density can vary widely across a growth or annual ring. The percentage of earlywood and latewood in each growth ring determines the overall density of a wood sample.

Permeability is a measure of the flow characteristics of a liquid or gas through wood as a result of the total pressure gradient. Permeability is influenced by the anatomy of the wood cells, the direction of flow (radial, tangential, and longitudinal), and the properties of the fluid being measured. Permeability is also affected by the species, by whether the wood is sapwood or heartwood, and by the chemical and physical properties of the fluid.

The primary thermal properties of wood are conductivity, specific heat, and coefficient of thermal expansion. The conductivity of wood is determined by density, moisture content, and direction of conduction. Thermal conductivity in the transverse directions (radial and tangential) is approximately equal. Conductivity in the longitudinal direction is greater than in the transverse directions. For most processing operations, the dominant heating direction is transverse. Thermal conductivity is important to wood processing because heating—whether for drying, curing, pressing, or conditioning—is an integral step. Specific heat of wood is dependent on moisture content and, to less extent, on temperature. *See* SPECIFIC HEAT.

Dry wood is an excellent insulator. By measuring wood's electrical resistance, electrical moisture meters accurately determine the moisture content of wood in the 5–25% range. Two other electrical properties of interest are the dielectric constant and the dielectric power factor for alternating current. These dielectric properties are dependent on density, moisture content, frequency of current, grain orientation, and temperature. The power factor is a measure of the stored energy that is converted to heat.

Mechanical properties. The mechanical properties of wood include elastic, strength, and vibration characteristics. These properties are dependent upon species, grain orientation, moisture content, loading rate, and size and location of natural characteristics such as knots.

Wood is both an elastic and plastic material. Elasticity manifests itself during loading and at moisture contents and temperatures that occur in most service uses of wood. The elastic stiffness or modulus of elasticity of wood is dependent on grain orientation, moisture content, species, temperature, and rate of loading. The stiffness of wood in the longitudinal (fiber) direction is utilized in the manufacture of composite products such as oriented strand board, in which the grain or fiber direction is controlled. *See* ELASTICITY.

The strength of wood, like its elastic properties, is dependent upon rate of loading, species, moisture content, orientation, temperature, size and location of natural characteristics such as knots, and specimen size. The strength of individual wood fibers in the longitudinal direction can be significantly greater than that of larger samples with their complex anatomy and many defects. As with stiffness, the excellent strength characteristics of wood in the direction of the fiber can be maximized during the manufacture of wood composites by controlling fiber alignment.

Damping and sound velocity are two primary vibration phenomena of interest in structural applications. Damping occurs when internal friction dissipates mechanical energy as heat. The velocity of a sound wave through wood can be used to estimate mechanical stiffness and strength: the higher the velocity, the higher the stiffness and strength. Like other properties of wood, the velocity of sound along the three principal axes differs. Sound velocity in the longitudinal direction is two to four times greater than in the transverse directions. *See* WOOD ANATOMY.

James B. Wilson

Woodward-Hoffmann rule A concept which can predict or explain the stereochemistry of certain types of reactions in organic chemistry. It is also described as the conservation of orbital symmetry, and is named for its developers, R. B. Woodward and Roald Hoffmann. The rule applies to a limited group of reactions, called pericyclic, which are characterized by being more or less concerted (that is, one-step, without a distinct intermediate between reactants and products) and having a cyclic arrangement of the reacting atoms of the molecule in the transition state. Most pericyclic reactions fall into one of three major classes: electrocyclic, cycloaddition, or sigmatropic. *See* STEREOCHEMISTRY.

David L. Dalrymple

Woodworking The shaping and assembling of wood and wood products into finished articles such as mold patterns, furniture, window sashes and frames, and boats. The pronounced grain of wood requires modifications in the working techniques when cutting with the grain and when cutting across it. Five principal woodworking operations are sawing, planing, steam bending, gluing, and finishing. To shape round pieces, wood is worked on a lathe. *See* TURNING (WOODWORKING).

Wood is sawed by cutting or splitting its fibers by the continuous action of a series of uniformly spaced teeth alternately staggered to move in closely parallel work planes. Action of the cutting teeth produces a path or kerf of uniform width through the workpiece from which the fibers have been severed and removed. Sawing across the grain or cell structure of the wood is called crosscutting. Cutting parallel with the grain of the piece is referred to as ripping. Saw teeth are bent alternately to the left and right to provide clearance for the blade. Some blades include straight raker teeth for cleaning fibers from the cut.

Flat or uniformly contoured surfaces of wood are roughed down, smoothed, or made level by the shaving and cutting action of a wide-edged blade or blades. Planing may be accomplished either manually or by power-operated tools.

Wooden members are bent or formed to a desired shape by pressure after they have been softened or plasticized by heat and moisture. If thick pieces of wood are to be bent to a permanent shape without breaking, some form of softening or plasticizing such as steaming is necessary. When a piece of wood is bent, its outer or convex side is actually stretched in tension while its concave side is simultaneously compressed. Actually, plasticized wood can be stretched but little. It can, however, be compressed a considerable amount. When a piece of plasticized wood is successfully bent, the deformation is chiefly compression distributed almost uniformly over the curved portion. Curvature results from many minute folds, wrinkles, and slippages in the compressed area.

Wood pieces may be fastened together by the adhesive qualities of a substance that sets or hardens into a permanent bond. Adhesives for wood are of two principal types, synthetic and natural-origin. The term glue was first applied to bonding materials of natural origin, while adhesive has been used to describe those of synthetic composition. *See* ADHESIVE.

The finishing operation is the preparation and sealing or covering of a surface with a suitable substance in order to preserve it or to give it a desired appearance. The preparation and conditioning of a surface may include cleaning, sanding, use of steel wool, removing or covering nails and screws, gluing or fastening loose pieces, filling cracks and holes with putty or crack filler, shellacking, and dusting. An inconsequential item with a painted surface does not require the thorough surface preparation that a piece of fine furniture does. The quality of surface conditioning directly affects the end result. *See* WOOD PROPERTIES.

Alan H. Tuttle

Wool A textile fiber made of the undercoat of various animals, especially sheep. Wool provides warmth and physical comfort that cotton and linen fabrics cannot give. These qualities, combined with its soft resiliency, make wool a necessity for apparel and for rugs and blankets.

Sheep are generally shorn of their fleeces in the spring, but the time of shearing varies in different parts of the world. Each fleece contains different grades, or sorts, of wool, and the raw stock must be carefully graded and segregated according to length, diameter, and quality of fiber. Wool from different parts of the body of the lamb differs greatly. The shoulders and sides generally yield the best quality of wool, because the fibers from those parts are longer, softer, and finer.

Wool's major physical characteristics include fiber diameter (fineness or grade), staple length, and clean wool yield. Also significant are soundness, color, luster, and content of vegetable matter. Grade refers specifically to mean fiber diameter and its variability. Fiber diameter is the most important manufacturing characteristic. Fleeces are commercially graded visually through observation and handling by men of long experience in the industry. Degree of crimp and relative softness of the fleece are important deciding factors employed by the graders.

The inherent advantages of wool have been exploited, and its limitations as a textile fiber have been overcome by the application of technology to manufacturing processes. The use of the insecticide dieldrin as a dye renders wool mothproof for life. Permanent pleats have been imparted to garments which are shrinkproofed and can be home laundered. Each such technological advance enables wool to hold its competitive place in the field of textile manufacture and use. *See* TEXTILE.

Thomas D. Watkins

Word processing The use of a computer and specialized software to write, edit, format, print, and save text. In addition to these basic capabilities, the latest word processors enable users to perform a variety of advanced functions. Although the advanced features vary among the many word processing applications, most of the latest software facilitates the exchange of information between different computer applications, allows easy access to the World Wide Web for page editing and linking, and enables groups of writers to work together on a common project. *See* COMPUTER; SOFTWARE; WORLD WIDE WEB.

Writing is accomplished by using the computer's typewriterlike keyboard. The characters appear on the computer screen as they are typed. A finite number of characters can be typed across the computer screen. The word processor "knows" when the user has reached this limit and automatically moves the cursor to the next line for uninterrupted typing. The position on the computer screen where a character can be typed is marked by a blinking cursor. The cursor can be positioned anywhere on the screen by using the mouse, or the keys marked with arrows on the keyboard. *See* COMPUTER PERIPHERAL DEVICES; ELECTRONIC DISPLAY.

In addition to writing, the latest word processors provide tools to create and insert drawings anywhere in the document. Typical features allow users to draw lines, rectangles, circles, and arrowheads, and to add text.

Editing allows users to correct typographical errors, add new sentences or paragraphs, move entire blocks of text to a different location, delete portions of the document, copy text and paste it somewhere else in the document, or insert text or graphics from an entirely different document. Most word processing programs can automatically correct many basic typographical errors, such as misspelled words, two successive capital letters in a word, and failure to capitalize the first letter of the names of days and of the first word in a sentence. Some other helpful editing tools commonly found in word processors include an automatic spelling checker, a thesaurus, and a grammar checker.

Formatting enables users to define the appearance of the elements in a document, such as the font and type size of all headings and text, the left, right, top, and bottom margins of each page, and the space before and after sentences and paragraphs. Most word processors allow all the elements in a document to be formatted at once. This is accomplished by applying a "style."

Word processors are approaching the formatting power of full-featured desktop publishing applications. The formatted page can be viewed on the computer screen exactly as it will be printed. This is referred to as "what you see is what you get" (WYSIWYG).

The latest word processors have many features for allowing groups of people to work together on the same document. For instance, multiple versions of a document can be saved to a single file for version control; access levels can be assigned so that only a select group of people can make changes to a document; edits can be marked with the date, time, and editor's name; and text colors can be assigned to differentiate editors. In addition, some word processors have editing features that include highlighting text, drawing lines through text to represent deleted text, and using red underscoring to identify changed text. Carlos Quiroga

Work In physics, the term work refers to the transference of energy that occurs when a force is applied to a body that is moving in such a way that the force has a component in the direction of the body's motion. Thus work is done on a weight that is being lifted, or on a spring that is being stretched or compressed, or on a gas that is undergoing compression in a cylinder.

When the force acting on a moving body is constant in magnitude and direction, the amount of work done is defined as the product of just two factors: the component of the force in the direction of motion, and the distance moved by the point of application of the force. Thus the defining equation for work W is given below, where f and s are the magnitudes of the force

$$W = f \cos\phi \cdot s$$

and displacement, respectively, and ϕ is the angle between these two vector quantities (see illustration). Because $f \cos \phi \cdot s = f \cdot s \cos\phi$, work may be defined alternatively as the product of the force and the component of the displacement in the direction of the force.

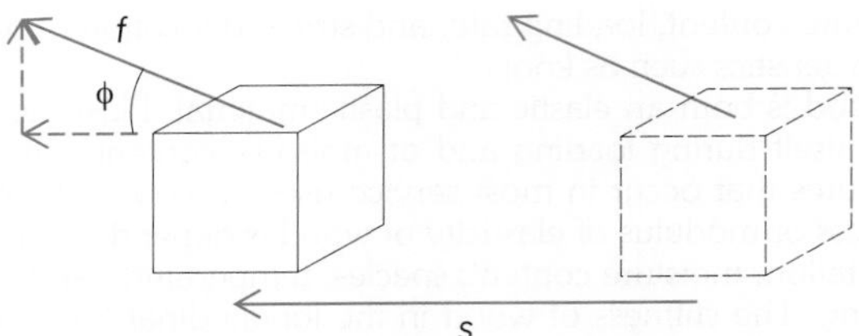

Work of constant force f is $fs \cos \phi$.

The work done is positive in sign whenever the force or any component of it is in the same direction as the displacement; one then says that work is being done *by* the agent exerting the force and *on* the moving body. The work is said to be negative whenever the direction of the force or force component is opposite to that of the displacement; then work is said to be done *on* the agent and *by* the moving body. From the point of view of energy, an agent doing positive work is losing energy to the body on which the work is done, and one doing negative work is gaining energy from that body.

The work principle, which is a generalization from experiments on many types of machines, asserts that, during any given time, the work of the forces applied to the machine is equal to the work of the forces resisting the motion of the machine, whether these resisting forces arise from gravity, friction, molecular interactions, or inertia.

The work done by any conservative force, such as a gravitational, elastic, or electrostatic force, during a displacement of a body from one point to another has the important property of being path-independent: Its value depends only on the initial and final positions of the body, not upon the path traversed between these two positions. On the other hand, the work done by any nonconservative force, such as friction due to air, depends on the path followed and not alone on the initial and final positions, for the direction of such a force varies with the path, being at every point of the path tangential to it. *See* ENERGY; FORCE. Leo Nedelsky

Work function (electronics) A quantity with the dimensions of energy which determines the thermionic emission of a solid at a given temperature. For metals, the work function may also be determined by measuring the photoemission as a function of the frequency of the incident electromagnetic radiation: the work function is then equal to the minimum (threshold) frequency for which electon emission is observed times Planck's constant h ($= 6.63 \times 10^{-34}$ joule second). The work function of a solid is usually expressed in electronvolts.

The work function of metals varies from one crystal plane to another and also varies slightly with temperature. For a metal, the work function has a simple interpretation. At absolute zero, the energy of the most energetic electrons in a metal is referred to as the Fermi energy; the work function of a metal is then equal to the energy required to raise an electron with the Fermi energy to the energy level corresponding to an electron at rest in vacuum. The work function of a semiconductor or an insulator has the same interpretation, but in these materials the Fermi level is in general not occupied by electrons and thus has a more abstract meaning. *See* FIELD EMISSION; PHOTOEMISSION; THERMIONIC EMISSION. Adrianus J. Dekker

Work function (thermodynamics) The thermodynamic function better known as the Helmholtz energy, $A = U - TS$, where U is the internal energy, T is the thermodynamic (absolute) temperature, and S is the entropy of the system. At constant temperature, the change in work function is equal to the maximum work that can be done by a system ($\Delta A = w_{max}$). *See* FREE ENERGY. P. W. Atkins

Work measurement The determination of a set of parameters associated with a task. There are four reasons, common to most organizations whether profit seeking or not, why time, effort, and money are spent to measure the amount of time a job takes. These are cost accounting, evaluation of alternatives, acceptable day's work, and scheduling. The fifth, pay by results, is used only by a minority of organizations.

There are three common ways to determine time per job: stopwatch time study (sequential observations), occurrence sampling (nonsequential observations), and standard data.

Stopwatch time study can be used for almost any existing job. It is reasonable in cost and gives reasonable accuracy. However, it does require the worker to be rated. Once the initial cost of standard data system has been incurred, standard data may be the lowest-cost, most accurate, and most accepted technique.

Occurrence sampling is also called work sampling or ratio-delay sampling. If time study is a "movie," then occurrence sampling is a "series of snapshots." The primary advantage of this approach may be that occurrence sampling standards are obtained from data gathered over a relatively long time period, so the sample is likely to be representative of the universe. That is, the time from the study is likely to be representative of the long-run performance of the worker.

Three levels of detail for standard time systems

Micro system (typical component time range from 0.01 to 1 s; MTM nomenclature)

Element	*Code*	*Time*
Reach	R10C	12.9 TMU
Grasp	G4B	9.1
Move	M10B	12.2
Position	P1SE	5.6
Release	RL1	2.0

Elemental system (typical component time range from 1 to 1000 s)

Element	*Time*
Get equipment	1.5 min
Polish shoes	3.5
Put equipment away	2.0

Macro system (typical component times vary upward from 1000 s)

Element	*Time*
Load truck	2.5 h
Drive truck 200 km	4.0
Unload truck	3.4

27.8 TMU D 1 s; 1 s D 0.036 TMU.
SOURCE: S. A. Konz, *Work Design*, published by Grid, 4666 Indianola Avenue, Columbus, OH 43214, 1979.

Reuse of previous times (standard data) is an alternative to measuring new times for an operation. There are three levels of detail: micro, elemental, and macro (see table). Micro-level systems have times of the smallest component ranging from about 0.01 to 1 s. Components usually come from a predetermined time system such as methods-time-measurement (MTM) or Work-Factor. Elemental level systems have the time of the smallest component, ranging from about 1 to 1000 s. Components come from time study or micro-level combinations. Macro-level systems have times ranging upward from about 1000 s. Components come from elemental-level combinations, from time studies, and from occurrence sampling. *See* METHODS ENGINEERING; PERFORMANCE RATING; PRODUCTIVITY. Stephan A. Konz

Work standardization The establishment of uniformity of technical procedures, administrative procedures, working conditions, tools, equipment, workplace arrangements, operation and motion sequences, materials, quality requirements, and similar factors which affect the performance of work. It involves the concepts of design standardization applied to the performance of jobs or operations in industry or business. *See* DESIGN STANDARDS; WORK MEASUREMENT.

Work standardization is part of methods engineering and, where it is practiced, usually precedes the setting of time standards. The objectives of work standardizations are lower costs, greater productivity, improved quality of workmanship, greater safety, and quicker and better development of skills among workers. *See* METHODS ENGINEERING; PRODUCTIVITY.

One of the best known of the more formal techniques of work standardization is group technology. This is the careful description of a heterogeneous lot of machine or other piece parts with a view to discovering as many common features in materials and dimensions as can be identified. It is then possible to start a rather large lot of a basic part through the production process, doing the common operations on all of them. Any changes or additional operations required to produce the final different parts can then be made at a later stage. The economy is realized in being able to do the identical jobs at one time.

There has been considerable progress in computerized systems to facilitate group technology. The techniques are closely linked to computer-aided design and to formalized codes that permit the detailed description of many operations. *See* COMPUTER-AIDED DESIGN AND MANUFACTURING. John E. Ullman

World Wide Web A part of the Internet that contains linked text, image, sound, and video documents. Before the World Wide Web (WWW), information retrieval on the Internet was text-based and required that users know basic UNIX commands. The World Wide Web has gained popularity largely because of its ease of use (point-and-click graphical interface) and multimedia capabilities, as well as its convenient access to other types of Internet services (such as e-mail, Telnet, and Usenet). *See* INTERNET.

Improvements in networking technology, the falling cost of computer hardware and networking equipment, and increased bandwidth have helped the Web to contain richer content. The Web is the fastest medium for transferring information and has universal reach (crossing geographical and time boundaries). It is also easy to access information from millions of Web sites using search engines (systems that collect and index Web pages, and store searchable lists of these pages). The Web's unified networking protocols make its use seamless, transparent, and portable. As the Web has evolved, it has incorporated complementary new technologies for developing online commerce and video on demand, to name a few.

Individual documents are called Web pages, and a collection of related documents is called a Web site. All Web documents are assigned a unique Internet address called a Uniform Resource Locator (URL) by which they can be accessed by all Web browsers. A URL (such as http://www.hq.nasa.gov/ office/procurement/index.html) identifies the communication protocol used by the site (http), its location [domain name or server (www.hq.nasa.gov)], the path to the server (office/ procurement), and the type of document (html).

The language used to create and link documents is called Hypertext Markup Language (HTML). Markup is the process of adding information to a document that is not part of the content but identifies the structure or elements. Markup languages are not new. HTML is based on the Standard Generalized Markup Language (SGML).

Though the initial format for creating a Web site was pure HTML, new and extended HTML has the ability to include programming language scripts such as common gateway interface (CGI), active server page (ASP), and Java server page (JSP), which can be used to create dynamic and interactive

Web pages as opposed to just static HTML text. Dynamic Web pages allow users to create forms for transactions and data collection; perform searches on a database or on a particular Web site; create counters and track the domain names of visitors; customize Web pages to meet individual user preferences; create Web pages on the fly; and create interactive Web sites.

XML, developed by the World Wide Web Consortium, is another derivative of SGML and is rapidly becoming the standard information protocol for all commercial software such as office tools, messaging, and distributed databases. XML is a flexible way to create common information formats and share both the format and the data on the World Wide Web, intranets, and other Web-based services. Alladi Venkatesh

Wrought iron As defined by the American Society for Testing and Materials, "a ferrous material, aggregated from a solidifying mass of pasty particles of highly refined metallic iron, with which, without subsequent fusion, is incorporated a minutely and uniformly distributed quantity of slag." This slag is a ferrous silicate resulting from the oxidizing reactions of refining, and it varies in amount from 1 to 3% in various types of final product. It is in purely mechanical association with the iron base metal, as contrasted with the alloying relationships of the metalloids present in steel. *See* STEEL.

A distinguishing characteristic of wrought iron is a fragmented or irregular fracture, as contrasted with a fibrous or crystalline type in steel. Metallographic analysis shows that this results from the fiberlike slag inclusions. Although wrought iron once held competitive merit in certain uses, notably where corrosion- and shock-resistance were important, very little wrought iron is produced at present. *See* IRON ALLOYS. James Aston; John F. Wallace

Wulfenite A mineral consisting of lead molybdate, $PbMoO_4$. Wulfenite occurs commonly in yellow, orange, red, and grayish-white crystals, with a luster from adamantine to resinous. Wulfenite may also be massive or granular. Its fracture is uneven. Its hardness is 2.7–3 on Mohs scale and its specific gravity 6.5–7. Its streak is white.

Wulfenite is found in numerous localities in the western and southwestern United States. Brilliant orange tabular wulfenite crystals up to 2 in. (5 cm) in size have been found from the Red Cloud and Hamburg mines in Yuma County, Arizona. *See* MOLYBDENUM. Edward C. T. Chao

Wurtzilite A black, infusible carbonaceous substance occurring in Uinta County, Utah. It is insoluble in carbon disulfide, has a density of about 1.05, and consists of 79–80% carbon, 10.5–12.5% hydrogen, 4–6% sulfur, 1.8–2.2% nitrogen, and some oxygen. Wurtzilite is derived from shale beds deposited near the close of Eocene (Green River) time. The material was introduced into the calcareous shale beds as a fluid after which it polymerized to form nodules or veins. *See* ASPHALT AND ASPHALTITE; ELATERITE; IMPSONITE. Irving A. Breger

X-ray astronomy The study of x-ray emission from extrasolar sources, including virtually all types of astronomical objects from stars to galaxies and quasars. The x-ray region of the electromagnetic spectrum extends from wavelengths of about 10 picometers to a few tens of nanometers, with shorter wavelengths corresponding to higher-energy photons (1 nm corresponds to about 1000 eV). X-ray astronomy is traditionally divided into broad bands—soft and hard—depending on the energy of the radiation being studied. Observations in the soft band (below about 10 keV) must be carried out above the atmosphere from rockets or satellites, while hard x-ray observations can be made at high altitudes achievable by balloons. *See* ELECTROMAGNETIC RADIATION; ROCKET ASTRONOMY; SATELLITE ASTRONOMY; X-RAYS.

Most observations have been in the soft band. Sky surveys, particularly the *ROSAT* all-sky survey, have located tens of thousands of sources at a sensitivity of about 1/100,000 of the strength of the brightest source, Sco X-1. Some of these sources are concentrated along the galactic equator. Such a concentration corresponds to objects in the Milky Way Galaxy, particularly in the disk, which contains most of the galactic stars and the spiral arms. Other x-ray sources are uniformly spread over the sky, associated mainly with extragalactic objects, such as individual galaxies, clusters of galaxies, and quasars. *See* MILKY WAY GALAXY.

Galactic sources. Galactic x-ray sources have been identified with different types of unusual objects. The Crab Nebula, which was the first nonsolar x-ray source to be identified with a specific celestial object, is a supernova remnant left over from the explosive death of a star. It contains a rapidly rotating neutron star at its center as well as a nebula consisting of hot gas and energetic particles. About 10% of the x-ray emission from the Crab is pulsed radiation from the neutron star; the rest is extended emission associated with the nebula. Many other galactic supernova remnants have been detected as x-ray sources, and there have also been detections of supernova remnants in the nearest neighboring galaxies, especially in the Large Magellanic Cloud. *See* CRAB NEBULA; NEUTRON STAR; PULSAR; SUPERNOVA.

While normal stars radiate some energy in the x-ray band, most of their luminosity is output in the visible region of the spectrum. X-ray surveys have discovered a class of objects where the majority of energy is radiated in the x-ray portion of the electromagnetic spectrum. These are known as x-ray stars. With accurate locations, it has been possible to identify them with binary star systems. The combination of x-ray and optical data leads to the conclusion that these systems usually consist of a relatively normal visible star and one subluminous star in a gravitationally bound orbit about each other. The x-ray star is often a collapsed object, such as a white dwarf, neutron star, or even a black hole. Such compact sources form the majority of galactic emitters.

The theoretical model for x-ray emission in these systems consists of matter being transferred from the normal star to the compact object. This process, known as accretion, usually leads to the creation of a disk of infalling material spiraling down toward the surface of the compact star. During infall the matter reaches very high temperatures as gravitational energy is converted to heat, which is then radiated as x-rays. It is believed that similar processes are involved regardless of whether the compact object is a white dwarf, neutron star, or black hole. However, different types of detailed behavior are expected for each of these objects. Rotating neutron stars lead to x-ray pulsars. For white dwarfs, there appear to be fluctuations in x-ray intensity on time scales of hours to days that might be indicative of changes at the surface of the star where accreted material is collecting, and then flashing in a burst of thermonuclear energy release. In the case of black hole candidates, there are variations in x-ray intensity at extremely short time scales that indicate very small regions of x-ray emission.

Another type of time behavior is represented by the so-called bursters. These sources emit at some constant level, with occasional short bursts or flares of increased brightness. These outbursts may well be instabilities in the x-ray emission processes associated with this type of source. *See* BINARY STAR; BLACK HOLE; CATACLYSMIC VARIABLE; STELLAR EVOLUTION; WHITE DWARF STAR.

Extragalactic sources. Among the more interesting types of objects detected have been apparent normal galaxies, galaxies with active nuclei, radio galaxies, clusters of galaxies, and quasars.

In active nuclei galaxies (those with strong optical emission lines and nonthermal continuum spectra), x-ray emission is usually orders of magnitude in excess of that from normal galaxies. In most cases the emission comes from the galactic nucleus, and must be confined to a relatively small region on the basis of variability and lack of structure at the current observational limits of a few seconds of arc. Many astrophysicists believe that galactic nuclei are the sites of massive black holes, at least 10^6–10^9 times the mass of the Sun, and that the radiation from these objects is due to gravitational energy released by infalling material. The broad range of properties, such as x-ray luminosity, may reflect the size of the black hole and availability of infalling matter. Quasars may represent the extreme case of this mechanism. Many of the observable properties depend on the viewing angle of the observer as well as the size of the nuclear black hole. *See* QUASAR.

Clusters of galaxies are collections of hundreds to thousands of individual galaxies which form a gravitationally bound system. The space between galaxies in such clusters has been found to contain hot (approximately 10^7–10^8 K) tenuous gas which glows in x-rays, and whose mass is equal to (or exceeds) the mass of the visible galaxies. *See* ASTROPHYSICS, HIGH-ENERGY; GALAXY, EXTERNAL.

Stephen S. Murray

X-ray crystallography The study of crystal structure by x-ray diffraction techniques. For the experimental aspects of x-ray diffraction *see* X-RAY DIFFRACTION

Structurally, a crystal is a three-dimensional periodic arrangement in space of atoms, groups of atoms, or molecules. If the periodicity of this pattern extends throughout a given piece of material, one speaks of a single crystal. The exact structure of any given crystal is determined if the locations of all atoms making up the three-dimensional periodic pattern called the unit cell are known. The very close and periodic arrangement of the atoms in a crystal permits it to act as a diffraction grating for x-rays. *See* CRYSTALLOGRAPHY.

Lawrence F. Dahl

X-ray diffraction The scattering of x-rays by matter with accompanying variation in intensity in different directions due to interference effects. X-ray diffraction is one of the most important tools of solid-state chemistry, since it constitutes a powerful and readily available method for determining atomic arrangements in matter. X-ray diffraction methods depend upon the fact that x-ray wavelengths of the order of 1 nanometer are readily available and that this is the order of magnitude of atomic dimensions. When an x-ray beam falls on matter, scattered x-radiation is produced by all the atoms. These scattered waves spread out spherically from all the atoms in the sample, and the interference effects of the scattered radiation from the different atoms cause the intensity of the scattered radiation to exhibit maxima and minima in various directions. *See* DIFFRACTION.

Uses. Some of the uses of x-ray diffraction are: (1) differentiation between crystalline and amorphous materials; (2) determination of the structure of crystalline materials (crystal axes, size and shape of the unit cell, positions of the atoms in the unit cell); (3) determination of electron distribution within the atoms, and throughout the unit cell; (4) determination of the orientation of single crystals; (5) determination of the texture of polygrained materials; (6) identification of crystalline phases and measurement of the relative proportions; (7) measurement of limits of solid solubility, and determination of phase diagrams; (8) measurement of strain and small grain size; (9) measurement of various kinds of randomness, disorder, and imperfections in crystals; and (10) determination of radial distribution functions for amorphous solids and liquids.

Techniques. The techniques employed in the study of crystalline substances, gases, and liquids are discussed below.

Laue method. The Laue pattern uses polychromatic x-rays provided by the continuous spectrum from an x-ray tube operated at 35–50 kV. The different diffracted beams have different wavelengths, and their directions are determined solely by the orientations of the set of planes with Miller indices *hkl*. Transmission Laue patterns were once used for structure determinations, but their many disadvantages have made them practically obsolete. On the other hand, the back-reflection Laue pattern is used a great deal in the study of the orientation of crystals.

Rotating crystal method. The original rotating crystal method was employed in the Bragg spectrometer. A sufficiently monochromatic beam, of wavelength of the order of 1 Å, is collimated by a system of slits and then falls on the large extended face of a single crystal as shown by Fig. 1. The Bragg spectrometer has been used extensively in obtaining quantitative measurements of the integrated intensity from planes parallel to the face of the crystal. The chamber is set at the correct 2θ angle with a slit so wide that all of the radiation reflected from the crystal can enter and be measured. The crystal is turned at constant angular speed through the Bragg law position, and the total diffracted energy received by the ionization chamber during this process is measured. Similar readings with the chamber set on either side of the peak give a background correction.

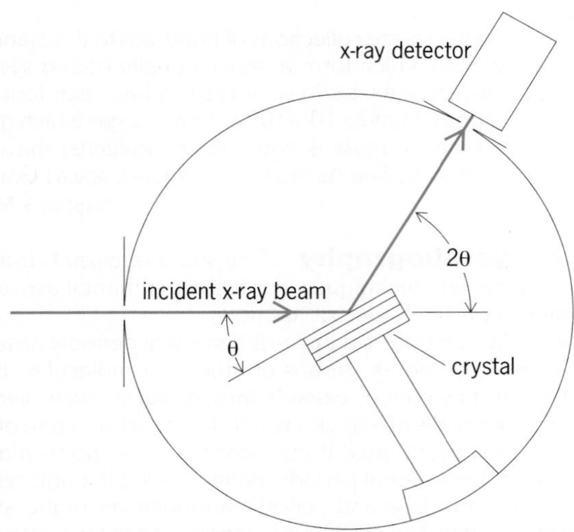

Fig. 1. Schematic of Bragg spectrometer.

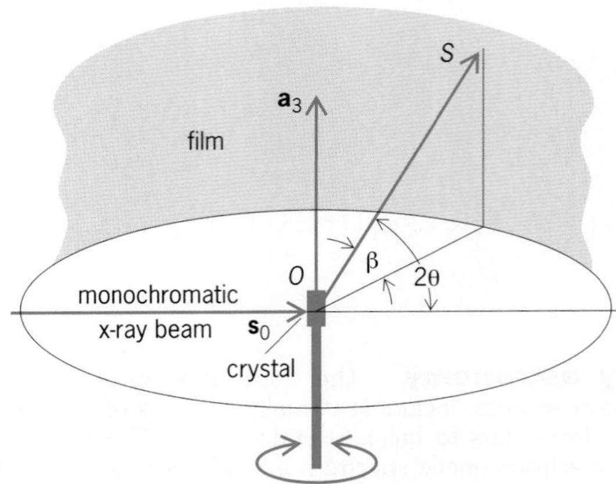

Fig. 2. Schematic of rotation camera.

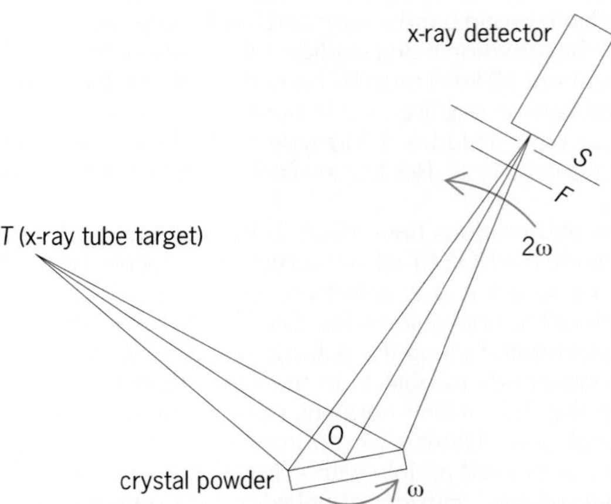

Fig. 3. Schematic representation of the Geiger counter diffractometer for powder samples.

The rotation camera, which is frequently used for structure determinations, is illustrated in Fig. 2. The monochromatic primary beam $\mathbf{s}_0$ falls on a small single crystal at *O*. The crystal is mounted with one of its axes (say, $\mathbf{a}_3$) vertical, and it rotates with constant velocity about the vertical axis during the exposure. The various diffracted beams are registered on a cylindrical film concentric with the axis of rotation.

Powder method. The powder method involves the diffraction of a collimated monochromatic beam from a sample containing an enormous number of tiny crystals having random orientation. Powder pattern studies are made with Geiger counter, or proportional counter, diffractometers. The apparatus is shown schematically in Fig. 3. X-rays diverging from a target at *T* fall on the sample at *O*, the sample being a flat-faced briquet of powder. Diffracted radiation from the sample passes through the receiving slit at s and enters the Geiger counter. During the operation the sample turns at angular velocity ω and the counter at 2ω. The distances *TO* and *OS* are made equal to satisfy approximate focusing conditions. A filter *F* before the receiving slit gives the effect of a sufficiently monochromatic beam. A chart recording of the amplified output of the Geiger counter gives directly a plot of intensity versus scattering angle 2θ. Bertram E. Warren

Gases. Gases and liquids are found to give rise to x-ray diffraction patterns characterized by one or more halos or interference rings which are usually somewhat diffuse. These diffraction patterns, which are similar to those for glasses and amorphous solids, are due to interference effects depending both upon the electronic distribution of each of the individual atoms or molecules and upon their relative positions in the system.

For monatomic gases the only appreciable interference effects giving rise to a distribution of scattered intensities are those produced by the electronic distribution about each nucleus. These interference effects giving rise to so-called coherent intensities are the result of the interference of the individual waves scattered by electrons in different parts of the atom. The electronic distribution of an atom is described in terms of a characteristic atomic scattering factor which is defined as the ratio of the resultant amplitude scattered by an atom to the amplitude that a free electron would scatter under the same conditions.

Liquids. One cannot, as in the cases of dilute gases and crystalline solids, derive unambiguous, detailed descriptions of liquid structures from diffraction data. Nevertheless, diffraction studies of liquids do provide most useful information. Instead of comparing the experimental intensity distributions with theoretical distributions computed for various models, the experimental results are usually provided in the form of a radial distribution function which specifies the density of atoms or electrons as a function of the radial distance from any reference atom or electron in the system without any prior assumptions about the structure. From the radial distribution function one can obtain (1) the average interatomic distances most frequently occurring in the structure corresponding to the positions of the first, second, and possibly third nearest neighbors; (2) the distribution of distances; and (3) the average coordination number for each interatomic distance.

Lawrence F. Dahl

X-ray fluorescence analysis A nondestructive physical method used for chemical elemental analysis of materials in the solid or liquid state. The specimen is irradiated by photons or charged particles of sufficient energy to cause its elements to emit (fluoresce) their characteristic x-ray line spectra. The detection system allows the determination of the energies of the emitted lines and their intensities. Elements in the specimen are identified by their spectral line energies or wavelengths for qualitative analysis, and the intensities are related to their concentrations for quantitative analysis. Computers are widely used in this field, both for automated data collection and for reducing the x-ray data to weight-percent and atomic-percent chemical composition or area-related mass (of films). *See* FLUORESCENCE.

The materials to be analyzed may be solids, powders, liquids, or thin foils and films. The crystalline state normally has no effect on the analysis, nor has the state of chemical bonding, except for very light elements. All elements above atomic number 12 can be routinely analyzed in a concentration range from 0.1 to 100 wt %. Special techniques are required for the analysis of elements with lower atomic numbers (4–11) or of lower concentrations, and for trace analysis. The counting times required for analysis range from a few seconds to several minutes per element, depending upon specimen characteristics and required accuracy; but they may be much longer for trace analysis and thin films. The results are in good agreement with wet chemical and other methods of analysis. The method is generally nondestructive for most inorganic materials in that a suitably prepared specimen is not altered by the analytical process.

There are two general methods for producing x-ray spectra for fluorescence analysis excitation by photons and excitation by charged particles. The most common method is to expose the specimen to the entire spectrum emitted from a standard x-ray tube. It is sometimes modified by using a secondary target material (or monochromator) outside the x-ray tube to excite fluorescence. This has the advantage of selecting the most efficient energy close to the absorption edge of the element to be analyzed and reducing or not exciting other interfering elements, but the intensity is reduced by two or three orders of magnitude. Further alternatives are radioactive sources and synchrotron radiation.

The other method, used in electron microscopes and the electron microprobe, uses an electron beam directly on the specimen, and each element generates its own x-ray spectrum, under electron bombardment, as in an x-ray tube. *See* ELECTRON MICROSCOPE.

The output signals from a detector are fed into the analyzer, where the photon counts are stored in memory locations (1024–8192 channels are generally used) that are related to the energies of these photons. This also allows visual observation on a cathode-ray-tube screen of the accumulated spectrum and of the simultaneous counting process. Analyzers are usually provided with cursor markers to easily identify the peaks in the spectrum. Computer memories can be used for storage of the spectral counts, thus providing efficient access to computer routines for further data evaluation.

The electron microprobe is widely used for elemental analysis of small areas. An electron beam of 1 micrometer (or smaller) is used, and the x-ray spectrum is analyzed with a focusing (curved) crystal spectrometer or with an energy dispersive solid-state detector. Usually two or three spectrometers are used to cover different spectral regions. Light elements down to beryllium, boron, and carbon can be detected. An important use of the method is in point-to-point analysis with a few cubic micrometers of spatial resolution. *X-Y* plots of any element can be made by moving the specimen to determine the elemental distribution. *See* ELECTRON-PROBE MICROANALYSIS.

William Parrish; Michael Mantler

X-ray microscope A technique and an instrument or combination of instruments which utilize x-radiation for chemical analysis and for magnification of 100–1000 diameters. The resolution possible is about 0.25 micrometer. The contrast in the x-ray microscopic image is caused by the varying x-ray attenuation in the specimen. The advantage of x-ray microscopy is that it yields quantitative chemical information, besides structural information, about objects, including those which are opaque to light. It is a reliable ultramicrochemical analytical technique by which amounts of elements and weights of samples as small as 10^{-12} to 10^{-14} g can be analyzed with an error of only a few percent. *See* MICRORADIOGRAPHY.

There are four general principles of image formation in x-ray microscopy: (1) contact microradiography (illus. *a*), (2) projection x-ray microscopy (illus. *b*), (3) reflection x-ray microscopy (illus. *c*), and (4) x-ray image spectrography (illus. *d*).

In contact microradiography the thin specimen is placed in close contact with an extremely fine-grained photographic emulsion which has a resolution of more than 25, 000 lines/in. (1000 lines/mm), and radiographed with x-rays of suitable wavelength. Thus an absorption image in scale 1:1 is obtained, and this image is subsequently viewed in a light microscope. The maximal resolution is that of the optical microscope (0.25 μm), but the image has more information, which can be obtained by examining the microradiogram in the electron microscopes. *See* ELECTRON MICROSCOPE.

Projection x-ray microscopy, or x-ray shadow microscopy, is based on the possibility of producing an extremely fine x-ray focal spot. This is achieved by an electronic lens system similar to that in the electron microscope. The fine focal spot is produced on a very thin metal foil which serves as a transmission target. The x-rays are generated on the target by the impact of the electrons. The sample is placed near the target, and the primary magnification depends on the ratios of the distances from focal spot to sample and sample to film. Resolution is of the same order as the size of the focal spot; the best value is about 0.1 μm in favorable objects.

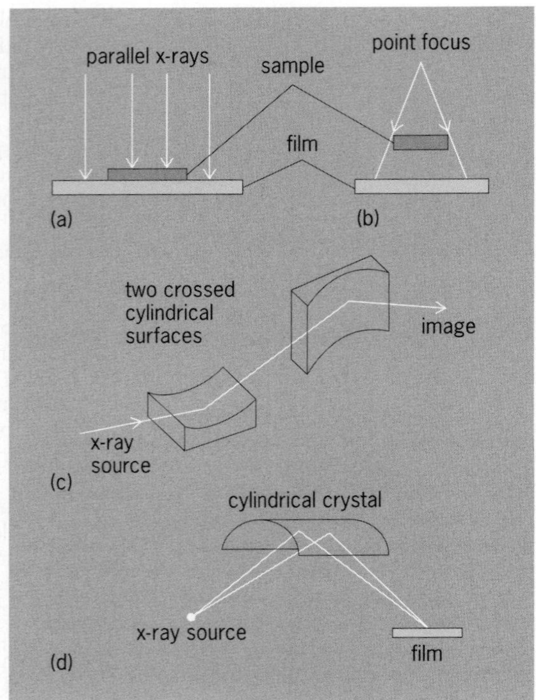

Principles for x-ray microscopy. (*a*) Contact microradiography; (*b*) projection x-ray microscopy; (*c*) reflection x-ray microscopy; and (*d*) x-ray image spectrography.

The method of reflection x-ray microscopy is based on the fact that the refractive index for x-rays in solids is a very small amount less than 1. Thus at grazing incidence (that is, incidence at very small angles), the x-rays are totally reflected, and if the reflecting surface is made cylindrical, there will be a focusing action in one dimension. By crossing two such surfaces a true image formation can be obtained, although with some astigmatism, which can be corrected by giving the surfaces a complicated optical shape. The resolution by this procedure is about 0.5–1 μm.

X-ray image spectrography utilizes Bragg reflections in a cylindrically bent crystal and produces slightly enlarged emission images; this technique is best classified as a micromodification of x-ray fluorescence analysis. The resolution is about 50 μm. *See* X-RAY DIFFRACTION; X-RAY FLUORESCENCE ANALYSIS.

In biology, x-ray microscopy has been utilized for the quantitative determination of the dry weight, water content, and elementary composition of many tissues, especially mineralized tissues. Arne Engstrom

X-ray optics By analogy with the science of optics, those aspects of x-ray physics in which x-rays exhibit properties similar to those of light waves. X-ray optics may also be defined as the science of manipulating x-rays with instruments analogous to those used in visible-light optics. These instruments employ optical elements such as mirrors to focus and deflect x-rays, zone plates to form images, and diffraction gratings to analyze x-rays into their spectral components. X-ray optics is important in many fields, including x-ray astronomy, biology, medical research, thermonuclear fusion, and x-ray microlithography. It is essential to the construction of instruments that manipulate and analyze x-rays from synchrotrons and particle storage rings for synchrotron radiation research. *See* GEOMETRICAL OPTICS; OPTICS; PHYSICAL OPTICS; X-RAYS.

When W. C. Roentgen discovered x-rays in 1895, he unsuccessfully attempted to reflect, refract, and focus them with mirrors, prisms, and lenses of various materials. The reason for his lack of success became evident after it was established that x-rays are electromagnetic waves of very short wavelength for which the refractive index of all materials is smaller than unity by a only a small decrement. In addition, x-rays are absorbed by materials. The refractive index can be written as a complex quantity, as in Eq. (1), where $1 - \delta$ represents the real part, n, of the

$$\tilde{n} = 1 - \delta - i\beta \qquad (1)$$

refractive index and β is the absorption index. These quantities are strongly dependent on the wavelength of the x-rays and the material. X-rays of wavelength about 0.1 nanometer or less are called hard x-rays and are relatively penetrating, while x-rays of wavelength 1–10 nm are less penetrating and are called soft x-rays. Radiation in the wavelength range 10–50 nm, called the extreme-ultraviolet (EUV) region, is very strongly absorbed by most materials. Values of δ remain very small throughout the x-ray and extreme-ultraviolet regions with the consequence that radiation is very weakly refracted by any material. Thus lenses for x-rays would have to be very strongly curved and very thick to achieve an appreciable focusing effect. However, because the absorption index, β, is so high in comparison, such thick lenses would absorb most of the incident radiation, making such lenses impractical. *See* ABSORPTION OF ELECTROMAGNETIC RADIATION; REFRACTION OF WAVES; ULTRAVIOLET RADIATION.

If radiation is incident normally (that is, perpendicular) to a surface between two media of differing refractive index, the fraction of the energy that is reflected is $^1/_4(\delta^2 + \beta^2)$. This is clearly impractically small for a normal-incidence mirror for x-rays. However, useful mirrors can be constructed by using the principle of total reflection. If electromagnetic waves are incident on the boundary between one material of refractive index n_1 and another of lower refractive index n_2, there exists an angle of incidence I_c, called the critical angle, given by Eq. (2). If the angle of incidence

$$\sin I_c = \frac{n_2}{n_1} \qquad (2)$$

(the angle of incident radiation with respect to the normal to the surface) is greater than this critical angle, all the wave energy is reflected back into the first medium. This phenomenon can be seen when looking upward into an aquarium tank; objects in the tank are reflected in the surface of the water, which acts as a perfect mirror. An analogous situation occurs for x-rays. Since the refractive index for all materials is slightly less than 1, x-rays incident from vacuum (or air) on a polished surface of, say, a metal encounter a lower refractive index and there exists a critical angle given by $\sin I_c = 1 - \delta$. Since δ is very small, I_c is very close to 90°. In this case the angle of incidence is customarily measured from the tangent to the surface rather than from the normal, and the angle $\theta_c = 90° - I_c$ is termed the angle of glancing (or grazing) incidence. This angle is typically in the range 0.1–1.0°. *See* REFLECTION OF ELECTROMAGNETIC RADIATION.

Although the reflectivity of surfaces at glancing angles greater than the critical angle is very small, this reflectivity can be enhanced by depositing a stack of ultrathin films having alternately high and low values of δ on the surface. The individual thicknesses of these films is adjusted so that the reflections from each interface add in phase at the top of the stack in exact analogy to the multilayer mirrors used for visible light. However, whereas visible multilayers require film thicknesses of hundreds of nanometers, in the x-ray region the thickness of each film must be between 1 and 100 nm. Such ultrathin films can be made by a variety of vacuum deposition methods, commonly sputtering and evaporation. The response of these artificial multilayers is strongly wavelength-selective. *See* X-RAY DIFFRACTION.

As a coating for glancing-incidence optics, multilayers allow a mirror to be used at a shorter wavelength (higher x-ray energy) for a given glancing angle, increasing the projected area and thus the collection efficiency of the mirror. At wavelengths longer than 3 or 4 nm, multilayer mirrors can be used to make normal-incidence mirrors of relatively high reflecting power. For example, stacks consisting of alternating layers of molybdenum and silicon can have reflectivities as high as 65% at wavelengths of 13 nm and longer. These mirrors have been used to construct optical systems that are exact analogs of mirror optics used for visible light. For example, normal-incidence x-ray telescopes have photographed the Sun's hot outer atmosphere at wavelengths of around 18 nm. Multilayer optics at a wavelength of 13.5 nm can be used to perform x-ray microlithography by the projection method to print features of dimensions less than 100 nm.

Crystals are natural multilayer structures and thus can reflect x-rays. Many crystals can be bent elastically (mica, quartz, silicon) or plastically (lithium fluoride) to make x-ray focusing reflectors. These are used in devices such as x-ray spectrometers, electron-beam microprobes, and diffraction cameras to focus the radiation from a small source or specimen on a film or detector. Until the advent of image-forming optics based on mirrors and zone plates, the subject of x-ray diffraction by crystals was called x-ray optics. *See* X-RAY CRYSTALLOGRAPHY; X-RAY SPECTROMETRY.

Zone plates are diffraction devices that focus x-rays and form images. They are diffracting masks consisting of concentric circular zones of equal area, and are alternately transparent and opaque to x-rays. Whereas mirrors and lenses focus radiation by adjusting the phase at each point of the wavefront, zone plates act by blocking out those regions of the wavefront whose phase is more than a half-period different from that at the plate center. Thus a zone plate acts as a kind of x-ray lens. Zone-plate microscopy is the most promising candidate method for x-ray microscopy of biological specimens. *See* DIFFRACTION.

James H. Underwood

X-ray powder methods Physical techniques used for the identification of substances, and for other types of analyses, principally for crystalline materials in the solid state. In these techniques, a monochromatic beam of x-rays is directed onto a polycrystalline (powder) specimen, producing a diffraction pattern that is recorded on film or with a diffractometer. This x-ray pattern is a fundamental and unique property resulting from the atomic arrangement of the diffracting substance. Different substances have different atomic arrangements or crystal structures, and hence no two chemically distinct substances give identical diffraction patterns. Identification may be made by comparing the pattern of the unknown substance with patterns of known substances in a manner analogous to the identification of people by their fingerprints. The analytical information is different from that obtained by chemical or spectrographic analysis. X-ray identification of chemical compounds indicates the constituent elements and shows how they are combined.

The x-ray powder method is widely used in fundamental and applied research; for instance, it is used in the analysis of raw materials and finished products, in phase-diagram investigations, in following the course of solid-state chemical reactions, and in the study of minerals, ores, rocks, metals, chemicals, and many other types of material. The use of x-ray powder diffraction methods to determine the actual atomic arrangement, which has been important in the study of chemical bonds, crystal physics, and crystal chemistry, is described in related articles. *See* X-RAY CRYSTALLOGRAPHY; X-RAY DIFFRACTION.

There are many types of powder diffractometer available ranging from simple laboratory instruments to versatile and complex instruments using a synchrotron source. Specialized instruments allow recording of diffraction patterns under nonambient conditions, including variable temperature, pressure, and atmosphere. Completely automated equipment for x-ray analysis is available. Most laboratory instruments consist of a high-voltage generator which provides stabilized voltage for the x-ray tube, so that the x-ray source intensity varies by less than 1%. A diffractometer goniometer is mounted on a table in front of the x-ray tube window. Electronic circuits use an x-ray detector to convert the diffracted x-ray photons to measurable voltage pulses, and to record the diffraction data.

Ron Jenkins

X-ray spectrometry A rapid and economical technique for quantitative analysis of the elemental composition of specimens. It differs from x-ray diffraction, whose purpose is the identification of crystalline compounds. It differs from spectrometry in the visible region of the spectrum in that the x-ray photons have energies of thousands of electronvolts and come from tightly bound inner-shell electrons in the atoms, whereas visible photons come from the outer electrons and have energies of only a few electronvolts.

In x-ray spectrometry the irradiation of a sample by high-energy electrons, protons, or photons ionizes some of the atoms, which then emit characteristic x-rays whose wavelength depends on the atomic number of the element, and whose intensity is related to the concentration of that element. Generally speaking, the characteristic x-ray lines are independent of the physical state (solid or liquid) and of the type of compound (valence) in which an element is present, because the x-ray emission comes from inner, well-shielded electrons in the atom. The illustration shows the removal of one of the innermost, *K*-shell, electrons by a high-energy photon. The photon energy must be greater than the binding energy of the electron; the difference in energy appears as the kinetic energy of the ejected electron. The *K* ionized atom is unstable, and one of the *L*- or *M*-shell electrons drops into the *K*-shell vacancy. As this transition occurs, a characteristic x-ray photon is emitted with an energy equal to the difference in energy between the *K* and the *L* (or *M*) shell, or an additional electron, called an Auger electron, is ejected from the atom. Either the x-rays or the Auger electrons may be used for analysis. *See* AUGER EFFECT.

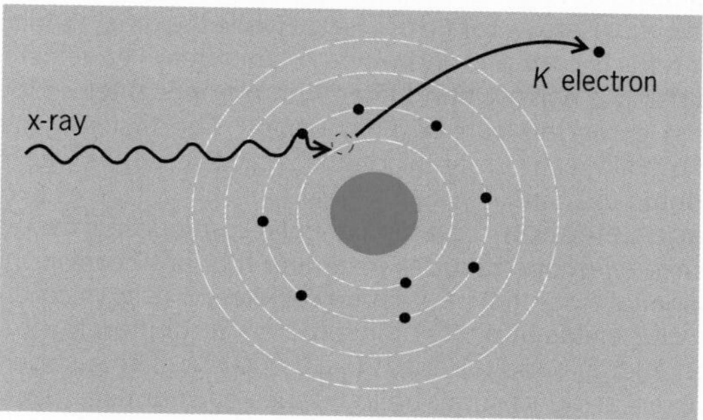

Removal of a *K* election from an atom by a primary x-ray photon. (*After L. S. Birks, X-Ray Spectrochemical Analysis, 2d ed., Wiley-Interscience, 1969*)

X-ray spectrometry generally does not require any separation of elements before measuring, because the x-ray lines are easily resolved. However, preconcentration methods are sometimes useful as a means for improving the limit of detection. One limitation of x-ray spectrometry is the progressive difficulty of measurement below atomic number 11. *See* SPECTROSCOPY.

L. S. Birks

X-ray telescope An instrument designed to collect and detect x-rays emitted from a source outside the Earth's atmosphere and to resolve the x-rays into an image. Absorption by the atmosphere requires that x-ray telescopes be carried to high altitudes. Balloons are used for detection systems designed for higher-energy (hard) x-ray observations, whereas rockets and satellites are required for softer x-ray detectors. *See* X-RAY ASTRONOMY; X-RAYS.

Image formation. An image-forming telescope lens for x-ray wavelengths can be based either on the phenomenon of total external reflection at a surface where the index of refraction changes (grazing-incidence telescope) or on the principles of constructive interference (multilayer telescope).

In the case of x-rays, the index of refraction in matter is slightly less than unity. By application of Snell's law, the condition for total external reflection is that the radiation be incident at small grazing angles, less than a critical angle of about 1°, to the reflecting surface. Based on these properties, x-ray mirrors have been constructed which focus an image in two dimensions. Various configurations of surfaces are possible. *See* REFLECTION OF ELECTROMAGNETIC RADIATION; REFRACTION OF WAVES.

A second type of x-ray telescope is based on the principles of constructive interference in extremely thin layers of material deposited on a mirror surface. Unlike the grazing-incidence telescope, multilayer telescopes do not require the x-rays to strike at shallow angles in order to be reflected. Instead these mirrors are similar to normal optical telescope mirrors where the incoming radiation strikes the mirror at nearly normal incidence to be reflected and focused. *See* INTERFERENCE OF WAVES.

The advantage of a multilayer mirror is that all of the mirror area is used in collecting the x-ray radiation, whereas the grazing-incidence telescopes have only a small projected area of the actual mirror surface collecting radiation. There is an offsetting disadvantage, as the multilayer mirror reflects x-rays only within a very narrow range of energy while the grazing-incidence mirror reflects over a broad range of energies. The effect is similar to using a narrow-band filter with an optical telescope, and in some cases this can be very useful. The first astronomical use of multilayer, normal-incidence x-ray telescopes was in photographing the Sun.

Image detection. The high angular resolution of the grazing-incidence telescope requires a camera that has correspondingly good spatial resolution. One type of detector uses microchannel plates and yields about 20 micrometers resolution for x-rays in the soft energy band (about 100–10,000 eV). The microchannel plate is an array of small hollow tubes or channels (about 10–15 μm in diameter) which are processed to have high secondary electron yield from their inner walls.

The charge-coupled device (CCD) has been developed for x-ray imaging applications. This solid-state detector consists of microscopic silicon picture elements (pixels) in which electronic charges produced by the passage of an x-ray photon are collected. Typical charge-coupled devices have pixels that are 15–25 μm on a side, and there are up to 4096×4096 pixels on one such device. The charge-coupled device not only records the position of an event but can also yield information on the energy of the photon. *See* CHARGE-COUPLED DEVICES.

A position-sensitive proportional counter is a gas-filled counter in which x-rays are photoelectrically absorbed, yielding an electron which is detected by the ionization it produces in the gas. By operation of the counter in the proportional mode and use of planes of wires to localize the electrical signals, the position and amplitude of each event can be recorded. These detectors generally have lower spatial resolution (about 200 μm) than do microchannel plate or charge-coupled-device detectors, but they can be made larger in size and provide better energy resolution than microchannel plate detectors. *See* IONIZATION CHAMBER.

Another class of detectors consists of very low-temperature devices that detect the heat deposited in an absorber when an x-ray is stopped. These devices promise improved energy resolution and efficiency over a broad range of energies. *See* TELESCOPE.

Stephen S. Murray

X-ray tube An electronic device used for the generation of x-rays. X-rays are produced when electrons that have been accelerated by a large potential are stopped in a target, which is usually made of metal. The x-rays radiate in all directions from the spot on the target where the electrons hit. X-rays are produced by two mechanisms in a metal target: bremsstrahlung (or braking radiation in German), which produces a continuous energy spectrum of x-rays; and characteristic x-rays, which are produced by the creation and subsequent filling of a vacancy in an inner shell of the target atoms. *See* BREMSSTRAHLUNG; X-RAYS.

In gas tubes, electrons are freed from a cold cathode by positive ion bombardment. For the existence of the positive ions a certain gas pressure is required without which the tube will allow no current to pass.

Many designs of gas tubes have been built for useful application, particularly in the medical field. Metals, such as platinum and tungsten, are placed in the path of the electron beam to serve as the target. Concave metal cathodes are used to focus the electrons on a small area of the metal target and increase the sharpness of the resulting shadows on the fluorescent screen or the photographic film.

The operational difficulties and erratic behavior of gas x-ray tubes are inherently associated with the gas itself and the positive ion bombardment that takes place during operation. The high-vacuum x-ray tube eliminates these difficulties by using other means of emitting electrons from the cathode. One typical form of a modern commercial hot-cathode high-vacuum x-ray tube is built with a liquid-cooled, copper-backed tungsten target, which operates over a wide range of energy ratings and is capable of rectifying its own current.

The inherent advantages of the hot-cathode high-vacuum tube over the gas tube are (1) flexibility—the voltage and current may be varied independently; (2) stability—this permits more accurate reproducibility of results; (3) small size; (4) operation—it can be operated directly from a transformer, making possible a very simple unit; and (5) long life.

Many special forms of x-ray tubes have been built for application in medicine, industry, and fundamental science. They vary from a few centimeters to many meters in length. X-ray tubes are used as an aid in medical diagnosis and in therapeutic treatment. They are used in the nondestructive testing of materials throughout industry. In both fundamental and applied science they are widely used in crystal-diffraction work, chemical analysis by x-ray spectra and absorption, and research on atomic structure. *See* MEDICAL IMAGING; NONDESTRUCTIVE EVALUATION; RADIOGRAPHY; RADIOLOGY; X-RAY CRYSTALLOGRAPHY; X-RAY DIFFRACTION; X-RAY FLUORESCENCE ANALYSIS; X-RAY POWDER METHODS.

Ernest E. Charlton

X-rays X-rays, or Roentgen rays, are electromagnetic waves; they are the same as visible light, except that they have shorter wavelengths (higher photon energies). Thus x-rays, visible light, ultraviolet, infrared, microwaves, and radio waves are all electromagnetic radiation in different wavelength (energy) spectral regions. X-rays are generated when fast-moving electrons slow down and stop in matter, when an inner-shell vacancy in an atom is filled by another electron, and when electrons moving at relativistic speeds (speeds near the speed of light) change their direction of motion in space. *See* ELECTROMAGNETIC RADIATION.

X-rays were discovered by W. C. Roentgen in 1895. Other scientists studying x-rays found the essential experimental conditions to prove that x-rays can be polarized, diffracted by crystals, refracted in prisms and in crystals, reflected by mirrors, and diffracted by ruled gratings. *See* X-RAY OPTICS.

The wavelength range of x-rays in the electromagnetic spectrum, as excited in x-ray tubes by the bombardment of a target by electrons under a high accelerating potential, overlaps the ultraviolet range on the order of 100 nanometers on the long-wavelength side, and the shortest-wavelength limit moves downward as voltages increase. An accelerating potential of 10^9 V, now readily generated, produces a λ of 10^{-6} nm. An average wavelength used in research is 0.1 nm, or about 1/6000 the wavelength of yellow light. *See* ULTRAVIOLET RADIATION; X-RAY TUBE.

X-rays (and visible light) can be considered as an electromagnetic wave. X-rays can also be considered as discontinuous bundles of energy, or quanta, in accordance with the laws first enunciated by M. Planck and extended by A. Einstein early in the twentieth century. In diffraction, refraction, polarization, and interference phenomena, x-rays, together with all other electromagnetic radiation, appear to act as waves and λ has a real significance. There is duality, meaning that light and x-rays have both wave and particle properties, although these are generally not observed at the same time in a given experiment. Beams of electrons and neutrons also have wave properties, and are diffracted in appropriate media. In other phenomena—such as the appearance of sharp spectral lines, a definite short-wavelength limit λ_0 of the continuous "white" spectrum [defined by $\lambda_0 = hc/eV$, where h is Planck's constant, c the velocity of electromagnetic radiation (including light and x-rays), e the charge of electron, and V the accelerating voltage], the shift in wavelength of x-rays scattered by electrons in atoms (Compton effect), and the photoelectric effect—the energy is propagated and transferred in quanta (called photons) defined by values of $h\upsilon$, where the frequency υ is c/λ. *See* COMPTON EFFECT; ELECTRON DIFFRACTION; NEUTRON DIFFRACTION; PHOTOEMISSION; PHOTON; QUANTUM MECHANICS.

X-rays are a valuable probe of matter, as they interact selectively with electrons; electrons in matter account for most of the significant properties of matter (excepting nuclear properties). Medical uses abound, as x-rays can penetrate matter, and are selectively absorbed by atoms containing many electrons—elements with a high atomic number. X-ray microscopes and microprobes have been developed which allow imaging with high spatial resolution while obtaining contrast with different elemental discrimination, distinguishing chemical bonds, and, by use of circularly polarized x-rays, the orientation of magnetization. X-rays are used in protein crystallography to determine the spatial structure of proteins, viruses, and other objects that can be made into crystals but that otherwise cannot be seen. *See* COMPUTERIZED TOMOGRAPHY; HISTORADIOGRAPHY; MEDICAL IMAGING; MICRORADIOGRAPHY; NONDESTRUCTIVE EVALUATION; RADIOGRAPHY; RADIOLOGY; X-RAY CRYSTALLOGRAPHY; X-RAY FLUORESCENCE ANALYSIS; X-RAY MICROSCOPE; X-RAY POWDER METHODS.

George L. Clark; Alfred S. Schlachter

Xanthophyceae A class of plants, comprising the yellow-green algae, in the chlorophyll *a–c* phylectic line (Chromophycota). Alternate names are Tribophyceae, derived from *Tribonema*, a filamentous member of the class, and Heterokontae, referring to the presence of two kinds of flagella on each motile cell. The absence of starch is an important characteristic in distinguishing yellow-green algae from green algae, which they may resemble superficially in form and color. *See* ALGAE; CHROMOPHYCOTA.

Motile cells have two unequal flagella borne apically or laterally. One (usually the longer) is directed forward and is pleuronematic (hairy); the other is directed backward and is acronematic (smooth). The compound zoospore (synzoospore) of *Vaucheria* is exceptional in having numerous pairs of slightly unequal smooth flagella. There are usually two chloroplasts in each motile cell, the ventral one with a reddish refractile eyespot at its anterior end. One or two contractile vacuoles occur at the anterior end of a motile cell. Cells in most Xanthophyceae are uninucleate, but in some genera they are multinucleate. The mitochondria have tubular cristae. Each vegetative cell contains one to many yellow-green, laminate or discoid, parietal chloroplasts.

Most Xanthophyceae occur in fresh water, especially soft water. Frequently, they are epiphytic on aquatic plants. The chief exception is *Vaucheria*, which has fresh-water, brackish-water, and marine species. Xanthophyceae also occur in and on soil and mud, in snow and ice, and on tree trunks and damp walls.

Paul C. Silva; Richard L. Moe

Xenacanthidae A family of Paleozoic elasmobranch sharks that are characterized by teeth with large divergent lateral cusps (diplodont crowns), an elongate dorsal fin preceded by a distinctive spine, and paired fins with an unusually well-developed skeletal axis. The xenacanths are one of the most clearly defined and stable groups to emerge from the early radiation of cartilaginous fishes (Chondrichthyes), with body fossils known from the early Carboniferous (*Diplodoselache*) through the late Triassic (*Xenacanthus*) [see illustration]. Numerous articulated specimens have been discovered, predominantly from the upper Carboniferous and Permian of Europe and North America, wherein xenacanths are a classic feature of faunas associated with brackish and freshwater lagoons and coal swamps. These xenacanths include the most frequently depicted genera, such as *Orthacanthus* and *Pleuracanthus*, with eellike body forms of up to 3 m (10 ft) or more in length.

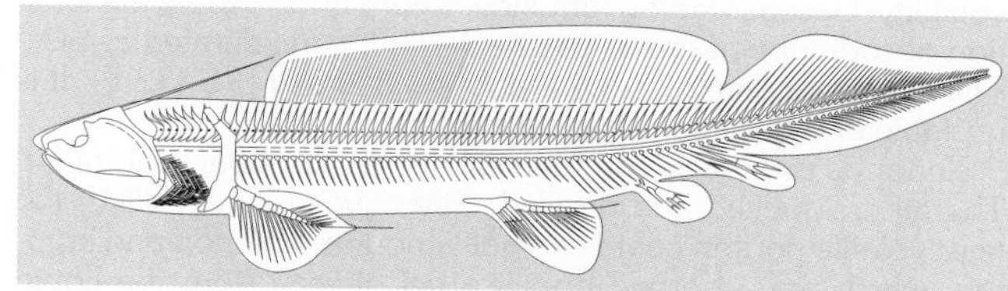

***Xenacanthus meisenheimensis*. (After U. H. J. Heidtke, 2003)**

The exceptional abundance and quality of xenacanth fossil material, relative to other Paleozoic chondrichthyans, has allowed their skeletal anatomy to be studied in unusual detail. The fossil record of xenacanth teeth exceeds that of the less resilient cartilaginous skeleton. Diplodont teeth are among the very earliest examples of sharklike teeth known, and include *Leonodus* from the Early Devonian of Europe. *See* CHONDRICHTHYES; ELASMOBRANCHII.

Michael Coates

Xenolith A rock fragment enclosed in another rock, and of varying degrees of foreigness. Cognate xenoliths, for example, are pieces of rock that are genetically related to the host rock that contains them, such as pieces of a border zone in the interior of the same body. Included blocks of unrelated rocks are more deserving of the xenolith label. Such foreign rocks help establish the once molten condition of invading magma capable of incorporating and mixing an assemblage of unrelated rock inclusions. *See* LAVA; MAGMA.

Xenoliths tend to react with the enclosing magma, so that their constituent minerals become like those in equilibrium with the melt. Reaction is rarely complete, however. Even completely equilibrated xenoliths may be conspicuous because the equilibration process does not require either the texture or the proportions of the minerals in the xenolith to match those in the enclosing rock. *See* PLUTON.

Xenoliths may be angular to round, millimeters to meters in diameter, aligned or haphazard, and sharply or gradationally bounded. Xenoliths are present in most bodies of igneous rock. *See* IGNEOUS ROCKS.

Alfred T. Anderson, Jr.

Xenon A chemical element, Xe, atomic number 54, and atomic weight 131.29. It is a member of the family of noble gases, group 18 in the periodic table. Xenon is colorless, odorless, and tasteless; it is a gas under ordinary conditions (see table). *See* INERT GASES; PERIODIC TABLE.

Xenon is the only one of the nonradioactive noble gases which forms chemical compounds that are stable at room temperature. Xenon also forms weakly bonded clathrates with such substances as water, hydroquinone, and phenol. *See* CLATHRATE COMPOUNDS.

The three fluorides, XeF_2, XeF_4, and XeF_6, are thermodynamically stable compounds at room temperature, and they may be

Physical properties of xenon

Property	Value
	131.29
Melting point (triple point)	−111.8°C (−169.2°F)
Boiling point at 1 atm pressure	−108.1°C (−162.6°F)
Gas density at 0°C and 1 atm pressure, g/liter	5.8971
Liquid density at its boiling point, g/ml	3.057
Solubility in water at 20°C, ml xenon (STP) per 1000 g water at 1 atm partial pressure of xenon	108.1

prepared simply by heating mixtures of xenon and fluorine at 300–400°C (570–750°F).

The reaction of XeF_6 with water gives $XeOF_4$; if the reaction is allowed to continue, XeO_3 is formed. XeO_3 is a colorless, odorless, and dangerously explosive white solid of low volatility. Gaseous xenon tetroxide, XeO_4, is formed by the reaction of sodium perxenate, Na_4XeO_6, with concentrated H_2SO_4. The vapor pressure of XeO_4 is about 3.3 kPa at 0°C (32°F). It is unstable and has a tendency to explode.

Xenon is produced commercially in an air-separation plant. The air is liquefied and distilled. The oxygen is redistilled; the least volatile portion contains small amounts of xenon and krypton, which are adsorbed on silica gel directly from the liquid oxygen. The crude xenon and krypton thus obtained are separated and further purified by distillation and selective absorption, at controlled low temperatures, on activated carbon. Remaining impurities are removed by passing the xenon over hot titanium, which reacts with all but the inert gases. *See* AIR SEPARATION.

Xenon is used to fill a type of flashbulb used in photography and called an electronic speed light. These bulbs produce a white light with a good balance of all the colors in the visible spectrum, and can be used 10,000 times or more before burning out.

A xenon-filled arc lamp gives a light intensity approaching that of the carbon arc; it is particularly valuable in projecting motion pictures. *See* VAPOR LAMP.

An important development in high-energy physics was the detection of nuclear radiation, such as gamma rays and mesons, by bubble chambers, in which a liquid is kept at a temperature just above its boiling point. Nucleation by the radiation results in bubble formation along the path of the particle. The tracks made by the particles are then photographed. Liquid xenon is one of the liquids used in these bubble chambers.

Xenon is used to fill neutron counters, x-ray counters, gas-filled thyratrons, and ionization chambers for cosmic rays; it is also used in high-pressure arc lamps to produce ultraviolet radiation.

Between 3 and 5% of the fissions in a nuclear reactor using uranium as fuel lead to the formation of xenon-135.

Arthur W. Francis

Xenophyophorida An order of Protozoa in the subclass Granuloreticulosia. The group includes deep-sea forms which are multinucleate at maturity. They develop as discoid to fan-shaped or algalike branching forms (illus. *a* and *b*) covered with a hyaline organic layer and sometimes measuring 2 mm or more overall. The aggregate contains many tubes which vary in diameter and form netlike or branching patterns. The lack of detailed information makes the taxonomic status of these organisms somewhat uncertain. The genera included are *Psammetta, Stanomma*, and *Stannophyllum*. *See* GRANULORETICULOSIA.

Richard P. Hall

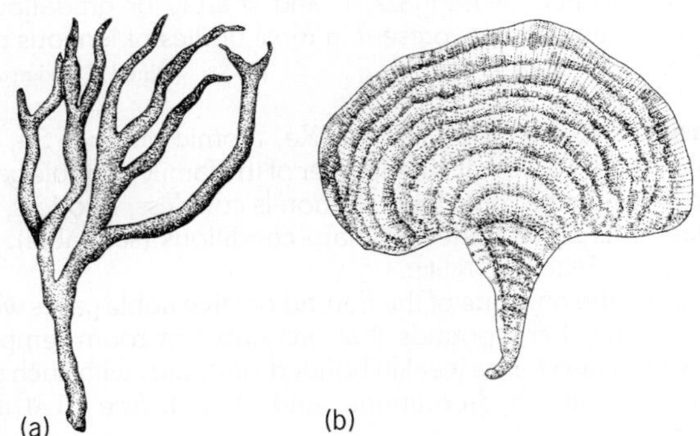

Xenophyophorids. ***(a)* Stanomma dendroides. *(b)* Stannophyllum zonarium.**

Xenoturbella A simple marine worm whose natural habitat is in soft mud bottoms at a depth of around 60 m (200 ft) off the coasts of Sweden and Norway. *Xenoturbella* was first discovered in 1915 by the Swedish biologist Sixten Bock, but was not described until 30 years later by Einar Westblad. Due to its simplicity of form, Westblad initially classified *Xenoturbella* as a primitive turbellarian, a class of ciliated nonparasitic flatworms, and named it accordingly (*Xenoturbella* = strange flatworm). However, recent analyses using deoxyribonucleic acid (DNA) have found *Xenoturbella* to be a primitive member of the deuterostomes, the group of animals that includes the vertebrates.

Xenoturbella is a small worm up to 3 cm (1.2 in.) in length with a mouth opening into a gut cavity. It has no anus and lacks defined excretory structures, body cavities, and reproductive organs. It has a gravity-sensing organ called a statocyst and two sensory grooves with a dense underlying nerve plexus. *Xenoturbella* has neither brain nor condensed nerve cord; its nervous system consists simply of a nerve net (a diffuse network of neurons) under the epidermis. Externally, it is completely covered in hairlike structures called cilia, which it uses to glide along the muddy seafloor. The reproduction and embryology of Xenoturbella are still a mystery. *See* ANIMAL EVOLUTION; DEUTEROSTOMIA.

Sarah J. Bourlat

Xenungulata An order of large, primitive, digitigrade, hoofed, tapirlike mammals with relatively short, slender limbs and five-toed feet with broad, flat phalanges. They are restricted to the Paleocene deposits of Brazil and Argentina. Only one genus (*Carodnia*) constitutes the single family (Carodniidae) in the order. The dentition is complete with strong, procumbent, chisel-shaped incisors, strong sharp-pointed canines, and low-crowned cheek teeth with bilophodont molars. *See* DENTITION; EUTHERIA.

The affinities of the Xenungulata remain uncertain. Affinities with the Dinocerata are strongly supported by the dental characteristics, but the structure of the tarsus suggests that the xenungulates had common ancestry with the Pyrotheria, to which they were tentatively assigned originally. *See* DINOCERATA; PYROTHERIA.

Gideon T. James; Everett C. Olson

Xiphosurida An order (xiphosurids, or horseshoe crabs) of the class Xiphosura (phylum Arthropoda). These "sword-tailed" arthropods are a classic example of a "living fossils" group that has seemingly changed little over a period of many millions of years. Xiphosurids have three distinct body divisions: a prominent horseshoe-shaped, fused head and thorax, called the prosoma; a fused "abdomen," or opisthosoma; and a spike-like terminal segment, called the telson. Xiphosurida is a low-diversity group that has a moderately good fossil record in some paleoenvironments, especially marginal-marine environments. Present-day xiphosurid species are tolerant of wide ranges of salinity, from marine to brackish water, and occasionally even venture into freshwater; ancient species may have had similar salinity tolerances. *See* LIVING FOSSILS; MEROSTOMATA.

The order Xiphosurida is a monophyletic group of arachnomorph arthropods, a large group that includes nektaspids, trilobites, eurypterids, arachnids, and synziphosurines. Two xiphosurid suborders are recognized: the more primitive Bellinurina, all of which are now extinct, and the more derived Limulina, which includes all the four living species of the group. The most primitive group of xiphosurids (suborder Bellinurina) probably evolved from a stem-group xiphosuran, called a synziphosurine, during the Late Devonian (about 365 million years ago), and became extinct in the Permian Period.

Xiphosurids include 13 genera known only from fossils, two genera known only from living species, and one genus that is known from both fossil and living species. There are four extant species of limulines: *Limulus polyphemus*, which occurs on the eastern coast of North America, and three species distributed in the Indo-Pacific region (*Tachypleus tridentatus, T. gigas*, and *Carcinoscorpius rotundicauda*).

The carapace (test or shell) is composed of chitin, strengthened internally in the adults by bridgelike struts within the shell margins. Sensory structures on the carapace include chemoreceptors for taste on the bases of the legs and a pair of simple eyes, or ocelli. The multilensed compound eyes form mosaic images and are sensitive to polarized light. The organ systems, except the peripheral blood vessels and nerves, digestive, and reproductive glands, which branch extensively into the lateral portions of the prosoma, are mainly within the prominent axial region. This region contains the musculature of the paired appendages, hinge, and telson, the digestive tube, the coxal (excretory) gland, and the respiratory, central circulatory, and nervous structures. The dorsally situated tubular heart lies within a sinus that receives venous blood, the hemolymph, from the book gills (the respiratory structures) and the body. The central nervous system and the major nerve branches are sheathed by arteries. A plate of "cartilaginous" chitin, the endocranium, forms a roof over the brain.

Carl N. Shuster, Jr.; Loren E. Babcock

Xylem The principal water-conducting tissue and the chief supporting system of higher plants. This tissue and the associated phloem constitute the vascular system of vascular plants. Xylem is composed of various kinds of cells, living or nonliving. The structure of these cells differs in their functions, but characteristically all have a rigid and enduring cell wall that is well preserved in fossils.

In terms of their functions, the kinds of cells in xylem are those related principally to conduction and support, tracheids; to conduction, vessel members; to support, fibers; and to food storage, parenchyma. Vessel members and tracheids are often called tracheary elements. The cells in each of the four categories vary widely in structure. *See* PARENCHYMA.

Xylem tissues arise in later stages of embryo development of a given plant and are added to by differentiation of cells derived from the apical meristems of roots and stems. Growth and differentiation of tissues derived from the apical meristem provide the primary body of the plant, and the xylem tissues formed in it are called primary. Secondary xylem, when present, is produced by the vascular cambium. *See* LATERAL MERISTEM.

In the trade, softwood is a name for xylem of gymnosperms (conifers) and hardwood for xylem of angiosperms. The terms do not refer to actual hardness of the wood. Woods of gymnosperms are generally composed only of tracheids, wood parenchyma, and small rays, but differ in detail. Resin ducts are present in many softwoods. Woods of angiosperms show extreme variation in both vertical and horizontal systems, but with few exceptions have vessels.

Vernon I. Cheadle; Katherine Esau

Y

Y-delta transformations

Y-delta transformations Relationships between electrically equivalent networks with three terminals, one being connected internally by a Y configuration and the other being connected internally by a Δ configuration. These relationships are also known as star-delta transformations. More complicated relationships can also be derived between a many-armed star and the equivalent many-sided mesh.

If a network connects three terminals with one another, it is called a three-terminal network. The simplest configurations of three-terminal networks are the Y or star (illus. *a*) and the Δ or mesh (illus. *b*).

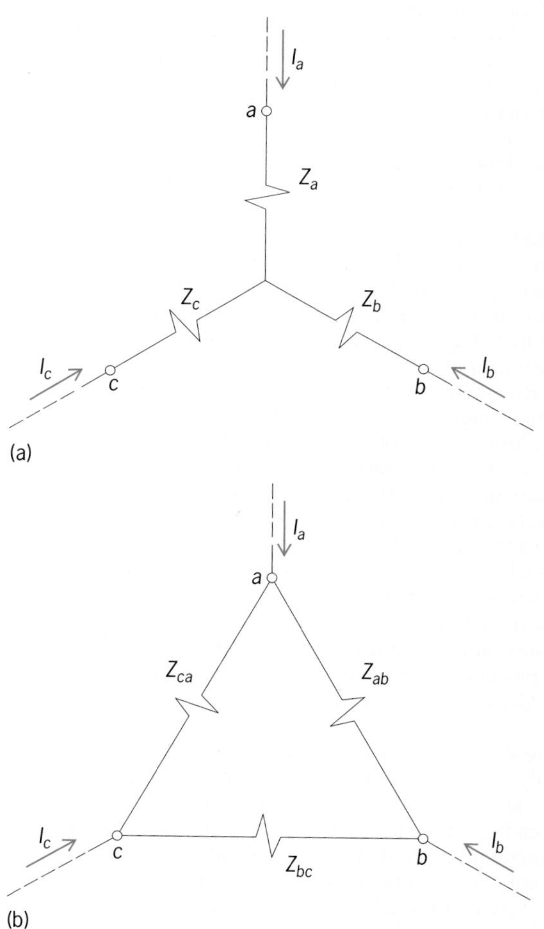

Three-terminal networks: (*a*) Y or star, (*b*) Δ or mesh.

If a given three-terminal network, passive and linear, has a Y configuration, it is possible to determine an equivalent Δ network that could be substituted for the Y without changing the relations of voltage and current at the network terminals, or elsewhere external to the network. Similarly, if a Δ network is given, an equivalent Y network can be found. Impedances of equivalent networks are usually functions of frequency, and realization of these impedances by use of physically possible elements is usually limited to a single frequency.

Hugh H. Skilling

Yagi-Uda antenna

Yagi-Uda antenna An antenna in which the gain of a single dipole element is enhanced by placing a reflector element behind the dipole (the driver) and one or more director elements in front of it (see illustration). It was invented in 1926 by H. Yagi and S. Uda in Japan. The gain is slightly increased by the reflector and further enhanced by the first director element. Additional director elements further increase the gain and improve the front-to-back ratio, up to a point of diminishing returns. This type of antenna has traditionally been used for local television reception. Its variants have found applications in the more modern communication systems at higher frequencies and smaller sizes, and have even been adapted to printed-circuit techniques in some applications. The same electromagnetic induction principle used in such linear elements can be applied to loop and disk elements as well with similar results. *See* Directivity; Electromagnetic induction; Gain; Printed circuit board.

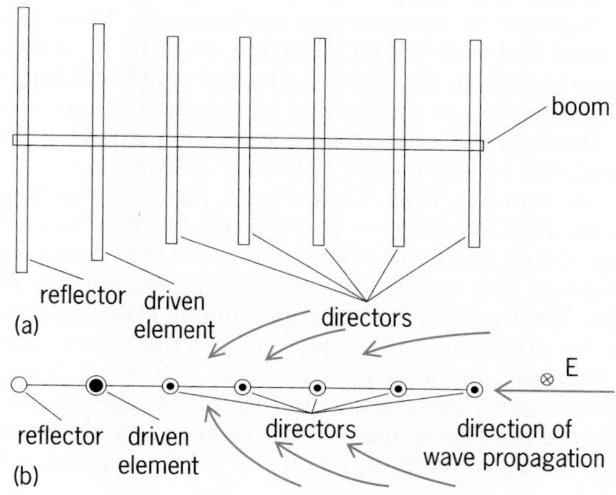

Yagi-Uda antenna. Typical radiation patterns are shown. (*a*) Side view with radiation pattern in the E plane (the plane containing the electric field). (*b*) Top view with radiation pattern in the H plane (the plane containing the magnetic field).

Since these antennas can be made highly directive with good radiation efficiency, they have found new applications and new manufacturing techniques with miniaturization. They can be printed on microwave circuit substrates with high dielectric constants, which reduces their size even further. The parasitic electromagnetic coupling demonstrated in the Yagi-Uda antenna has been adapted to many new types of miniaturized antennas applicable to mobile communication devices in wide use, and will be used in future wireless Internet devices. *See* Antenna (electromagnetism); Internet; Microwave solid-state devices; Mobile communications.

Jamal S. Izadian

Yak

Yak A member of the mammalian cattle family Bovidae in the order Artiodactyla. Yaks (*Bos grunniens*) have been domesticated for at least several hundred years. They are powerful yet

Adult male domestic yak (*Bos grunniens grunniens*). (*Copyright © Brent Huffman, 2000*)

docile. Above 2000 m (6600 ft), they are the most useful domestic mammal. They are used for meat and milk production, as a mount, and as a beast of burden. The hair is used for making cloth and tents, the dung is used as fuel, and the tail serves as a flybrush.

The yak, a relative of the bison, is a blackish-brown bovid with large black horns that grow out from the sides of the head and curve upward (see illustration). The horns, which are present in both sexes, are composed of bony cores attached to the frontal bones of the skull and hard sheaths of horny material. The massive body is covered by long guard hairs that almost reach the ground and a dense underfur of soft hair. The shoulders are high and humped, with a broad, drooping head. The limbs are short and sturdy with broad hooves and large dewclaws as an adaptation to mountainous environments. Males may attain a head and body length up to 325 cm (10.8 ft), a shoulder height up to 200 cm (6.6 ft), and a weight of 305–820 kg (670–1805 lb). Females are smaller and may weigh only one-third as much. Domesticated animals are considerably smaller than wild yaks and may be red, brown, or black in coloration.

Yaks are found in Tibet, China, India, Kashmir, and Nepal at elevations of 4000–6000 m (12,800–19,200 ft). They inhabit desolate steppe regions of alpine tundra and cold desert regions with swamps and moors. They are sure-footed and are expert climbers. They spend a portion of the summer in the lower plains, but as the temperature rises they move to the higher plateaus. They swim and bathe in icy cold lakes and rivers. Yaks feed on grasses, herbs, mosses, and lichens. Since vegetation is very sparse, yaks must travel great distances from one grazing site to another. *See* ARTIODACTYLA; BISON. Donald W. Linzey

Yaw indicator An aircraft instrument that provides a measure of the sideslip or yaw of the aircraft, defined as the angular measure of the relative wind in relation to a vertical plane passing through the longitudinal center of the aircraft. This device is not common to production aircraft, being found in only a few vertical takeoff and landing (VTOL) aircraft which operate in the low-speed regime. The measurement of yaw is accomplished either by a balanced vane which aligns with the wind, or through a set of pressure sensors that determine the difference in static pressure on each side of the vertical plane and hence the sideslip. The device provides equivalent information in the lateral-directional (yaw and roll) axis to the angle of attack in the longitudinal (pitch) axis. *See* AIRCRAFT INSTRUMENTATION; VERTICAL TAKEOFF AND LANDING (VTOL). Grady W. Wilson

Yaws An infectious disease of humans caused by the spirochete *Treponema pertenue*. It is also known as frambesia and is largely confined to the tropics. Usually yaws is contracted in childhood by direct contact or from small flies feeding in succession on infected lesions and open wounds. No race or age possesses natural immunity. Thomas B. Turner

Year The period of the Earth's revolution around the Sun. This period can be measured in several ways.

The most commonly used type of year, known as the tropical year, is the mean period for the Sun to go around the sky from one vernal equinox to the next. The tropical year varies slightly over time. Tropical year 2000 was approximately 365.242190 mean solar days, or 365 d 5 h 48 m 45.2 s. Seasons repeat with intervals of a tropical year.

The sidereal year is the average period of the Earth's revolution around the Sun until the Earth-Sun line returns to the same direction in space as measured by the star background. A sidereal year is approximately 365.25636 mean solar days long, or 365 d 6 h 9 m 10 s.

The anomalistic year is the average period of the Earth's return to perihelion in its orbit around the Sun. An anomalistic year is approximately 365.25964 days, or 365 d 6 h 13 m 53 s.

An eclipse year is the duration between alternate eclipse seasons, measured as the interval between the Moon passing alternate nodes. An eclipse year lasts only 346.62005 days, or 346 d 14 h 52 m 52 s.

A Gaussian year, defined as the period derived from Kepler's law for 1 astronomical unit, is 365.25690 days, or 365 d 6 h 9 m 56 s. *See* ASTRONOMICAL UNIT; KEPLER'S LAWS.

The term "year" is also used for the period of the orbit of other planets, both inside the solar system and around other stars. For example, a Jupiter year is 5.2 Earth years long. One of the planets around a distant star orbits in only 2 Earth days, so its year is only 2 Earth days long. *See* TIME. Jay M. Pasachoff

Yeast A collective name for those fungi which possess, under normal conditions of growth, a vegetative body (thallus) consisting, at least in part, of simple, single cells. The cells making up the thallus occur in pairs, in groups of three, or in straight or branched chains consisting of as many as 12 or more cells. Vegetative reproduction is characterized by budding or fission. Sexual reproduction also occurs in yeast, and is differentiated from that of other fungi by sexual states that are not enclosed in a fruiting body. Yeasts are a phylogenetically diverse group of organisms that occur in two divisions of fungi (Ascomycotina and Basidiomycotina) and 100 genera. The 700 or more species that have been described possibly represent only 1% of the species in nature, so the majority of the yeasts have yet to be discovered. Yeast plays a large part in industrial fermentation processes such as the production of industrial enzymes and chemicals, food products, industrial ethanol, and malt beverage and wine; in diseases of humans, animals and plants; in food spoilage; and as a model of molecular genetics. *See* DISTILLED SPIRITS; FOOD MICROBIOLOGY; GENETIC ENGINEERING; MALT BEVERAGE; MEDICAL MYCOLOGY; WINE.

The shape and size of the individual cells of some species vary slightly, but in other species the cell morphology is extremely heterogeneous. The shape of yeast cells may be spherical, globose, ellipsoidal, elongate to cylindrical with rounded ends, more or less rectangular, pear-shaped, apiculate or lemon-shaped, ogival or pointed at one end, or tetrahedral. The diameter of a spherical cell may vary from 2 to 10 micrometers. The length of cylindrical cells is often 20–30 μm and, in some cases, even greater.

The asexual multiplication of yeast cells occurs by a budding process, by the formation of cross walls or fission, and sometimes by a combination of these two processes. Yeast buds are sometimes called blastospores or blastoconidia. When yeast reproduces by a fission mechanism, the resulting cells are termed arthrospores or arthroconidia.

Yeasts are categorized into two groups, based on their methods of sexual reproduction: the ascomycetous (Division Ascomycotina) and basidiomycetous (Division Basidiomycotina) yeasts.

The sexual spores of the ascomycetous yeasts are termed ascospores, which are formed in simple structures, often a vegetative cell. Such asci are called naked asci because of the absence of an ascocarp, which is a more complex fruiting body found in the higher Ascomycetes. If the vegetative cells are diploid, a cell may transform directly into an ascus after the 2*n* nucleus undergoes a reduction or meiotic division. *See* ASCOMYCOTA.

Certain yeasts have been shown to be heterothallic; that is, sporulation occurs when strains of opposite mating type (usually indicated by "a" and α) are mixed on sporulation media. However, some strains may be homothallic (self-fertile), and reduction division and karyogamy (fusion of two haploid nuclei) take place during formation of the sexual spore. Yeasts that produce sporogenous cells represent the teleomorphic form of the life cycle. In cases, in which sexual cycles are unknown, the yeast represents the asexual or anamorphic form. A species of yeast may be originally discovered in the anamorphic form and named accordingly; subsequently, the sexual state may be found and a name applied to represent the teleomorph. Consequently, the anamorphic and teleomorphic names will differ.

Basidiospores and teliospores are the sexual spores that are produced in the three classes of basidiomycetous yeasts: Urediniomycetes, Hymenomycetes, and Ustilaginomycetes. Sexual reproduction and life cycle in these yeasts is typical of other basidiomycetes in that it can include both unifactorial (bipolar) and bifactorial (tetrapolar) mating systems. *See* BASIDIOMYCOTA.

Some yeasts have the ability to carry out an alcoholic fermentation. Other yeasts lack this property. In addition to the fermentative type of metabolism, fermentative yeasts as a rule have a respiratory type of metabolism, whereas nonfermentative yeasts have only a respiratory, or oxidative, metabolism. Both reactions produce energy, with respiration producing by far the most, which is used in part for synthetic reactions, such as assimilation and growth. Part is lost as heat. In addition, small or sometimes large amounts of by-products are formed, including organic acids, esters, aldehydes, glycerol, and higher alcohols. When a fermenting yeast culture is aerated, fermentation is suppressed and respiration increases. This phenomenon is called the Pasteur effect. *See* FERMENTATION.

Yeasts are ubiquitous in nature. They exist on plants and animals; in waters, sediments, and soils; and in terrestrial, aquatic, and marine habitats. Yeasts require oxygen for growth and reproduction; therefore they do not inhabit anaerobic environments such as anoxic sediments. Many species have highly specific habitats, whereas others are found on a variety of substrates in nature.

Jack W. Fell; Herman J. Phaff

Yellow fever An acute, febrile, mosquito-borne viral disease characterized in severe cases by jaundice, albuminuria, and hemorrhage. Inapparent infections also occur.

The agent is a flavivirus, an arbovirus of group B. The virus multiplies in mosquitoes, which remain infectious for life. After the mosquito ingests a virus-containing blood meal, an interval of 12–18 days (called the extrinsic incubation period) is required for it to become infectious. *See* ANIMAL VIRUS; ARBOVIRAL ENCEPHALITIDES.

The virus enters the body through a mosquito bite and multiplies in lymph nodes, circulates in the blood, and localizes in the liver, spleen, kidney, bone marrow, and lymph glands. The severity of the disease and the major signs and symptoms which appear depend upon where the virus localizes and how much cell destruction occurs. The incubation period is 3–6 days. At the onset, the individual has fever, chills, headache, and backache, followed by nausea and vomiting. A short period of remission often follows. On about the fourth day, the period of intoxication begins with a slow pulse relative to a high fever and moderate jaundice. In severe cases, there are high levels of protein in the urine, and manifestations of bleeding appear; the vomit may be black with altered blood; and there is an abnormally low number of lymphocytes in the blood. When the disease progresses to the severe stage (black vomit and jaundice), the mortality rate is high. However, the infection may be mild and go unrecognized. Diagnosis is made by isolation of the virus from the serum obtained from an individual as early as possible in the disease, or by the rise in serum antibody. *See* ANTIBODY; COMPLEMENT-FIXATION TEST; NEUTRALIZATION REACTION (IMMUNOLOGY).

There are two major epidemiological cycles of yellow fever: classical or urban epidemic yellow fever, and sylvan or jungle yellow fever. Urban yellow fever involves person-to-person transmission by *Aedes aegypti* mosquitoes in the Western Hemisphere and West Africa. This mosquito breeds in the accumulations of water that accompany human settlement. Jungle yellow fever is primarily a disease of monkeys. In South America and Africa, it is transmitted from monkey to monkey by arboreal mosquitoes (*Haemagogus* and *Aedes* species) that inhabit the moist forest canopy. The infection in animals ranges from severe to inapparent. Persons who come in contact with these mosquitoes in the forest can become infected. Jungle yellow fever may also occur when an infected monkey visits a human habitation and is bitten by *A. aegypti*, which then transmits the virus to a human.

Vigorous mosquito abatement programs have virtually eliminated urban yellow fever. However, with the speed of modern air travel, the threat of a yellow fever outbreak exists where *A. aegypti* is present. An excellent attenuated live-virus vaccine is available. *See* VACCINATION.

Joseph L. Melnick

Yersinia A genus of bacteria in the Enterobacteriaceae family. The bacteria appear as gram-negative rods and share many physiological properties with related *Escherichia coli*. Of the 11 species of *Yersinia, Y. pestis, Y. enterocolitica*, and *Y. pseudotuberculosis* are etiological agents of human disease. *Yersinia pestis* causes flea-borne bubonic plague (the black death), an extraordinarily acute process believed to have killed over 200 million people during human history. Enteropathogenic *Y. pseudotuberculosis* and *Y. enterocolitica* typically cause mild chronic enteric infections. The remaining species either promote primary infection of fish (*Y. ruckeri*) or exist as secondary invaders or inhabitants of natural environments (*Y. aldovae, Y. bercovieri, Y. frederiksenii, Y. intermedia, Y. kristensenii, Y. mollaretii*, and *Y. rohdei*). *See* MEDICAL BACTERIOLOGY; PLAGUE.

Robert R. Brubaker

Yew A genus of evergreen trees and shrubs, *Taxus*, with a fruit containing a single seed surrounded by a scarlet, fleshy, cuplike envelope (aril). The leaves are flat and acicular (needle-shaped), green below, with stalks extending downward on the stem. The only native American species of commercial importance is the Pacific yew (*T. brevifolia*), a medium-sized tree of the Pacific Coast and northern Rocky Mountain regions. Although it is not a common tree, its wood is sometimes used for poles, paddles, bows, and small cabinetwork.

The English yew (*T. baccata*), native in Europe, North Africa, and northern Asia, and the Japanese yew (*T. cuspidata*) are much cultivated in the United States as evergreen ornamentals. *See* FOREST AND FORESTRY; TREE.

Arthur H. Graves; Kenneth P. Davis

Ylide A variety of organic compounds that contain two adjacent atoms bearing formal positive and negative charges, and in which both atoms have full octets of electrons. Heteroatoms most commonly utilized as the positive atom are phosphorus, nitrogen, sulfur, selenium and oxygen, and the negative atom usually involves carbon, nitrogen, oxygen, or sulfur. The ylide may be in an alicyclic or cyclic environment, and in the latter the ylide function may be endocyclic or exocyclic. Ylides most useful in organic synthesis are those containing phosphorus or sulfur and an adjacent carbanion.

A phosphorus ylide is known as a phosphorane. Ylides of this type are highly reactive. Sulfur-containing ylides (π-sulfuranes) are of two types, sulfonium ylides and oxosulfonium ylides, which differ in having an oxygen atom attached to the sulfur in the latter. *See* REACTIVE INTERMEDIATES.

The ylide function may also be incorporated into heterocyclic systems. For example, the meso-ionic sydnone molecule contains an azomethine imine ylide, and the mesomeric betaine derived from 3-hydroxypyridine contains an azomethine ylide. These ylides are often referred to as masked ylides.

Kevin T. Potts

Yolk sac An extraembryonic membrane which extends through the umbilicus in vertebrates. In some elasmobranchs, birds, and reptiles, it is laden with yolk which serves as the nutritive source of embryonic development.

In mammals, as in birds, the yolk sac generally develops from extraembryonic splanchnopleure, and extends beneath the developing embryo. A blood vessel network develops in the mammalian yolk sac lining. Though these blood vessels are empty, they play an important role in absorbing nourishing food and oxygen from the mother. Thus, although the yolk sac in higher mammals may be considered an evolutionary vestige from its yolky-egged ancestors, it still serves important functions in the young embryo. As the embryo ages, the yolk sac shrinks in size, and the allantois takes over the role of nutrition. *See* ALLANTOIS.

Steven B. Oppenheimer

Young's modulus A constant designated E, the ratio of stress to corresponding strain when the material behaves elastically. Young's modulus is represented by the slope $E = \Delta S/\Delta \varepsilon$ of the initial straight segment of the stress-strain diagram. More correctly, E is a measure of stiffness, having the same units as stress: pounds per square inch or pascals. When stress and strain are not directly proportional, E may be represented as the slope of the tangent or the slope of the secant connecting two points on the stress-strain curve. The modulus is then designated as tangent modulus or secant modulus at stated values of stress. The modulus of elasticity applying specifically to tension is called Young's modulus. *See* ELASTICITY; HOOKE'S LAW; STRESS AND STRAIN.

W. J. Krefeld; W. G. Bowman

Ytterbium A chemical element, Yb, atomic number 70, and atomic weight 173.05. Ytterbium is a metal element of the rare-earth group. There are 7 naturally occurring stable isotopes. *See* PERIODIC TABLE.

The common oxide, Yb_2O_3, is colorless and dissolves readily in acids to form colorless solutions of trivalent salts which are paramagnetic. Ytterbium also forms a series of divalent compounds. The divalent salts are soluble in water but react very slowly with water to liberate hydrogen.

The metal is best prepared by distillation. It is a silvery soft metal which corrodes slowly in air and resembles the calcium-strontium-barium series more than the rare-earth series. For a discussion of the properties of the metal and its salts *see* RARE-EARTH ELEMENTS.

Frank H. Spedding

Yttrium A chemical element, Y, atomic number 39, and atomic weight 88.906. Yttrium resembles the rare-earth elements closely. The stable isotope ^{89}Y constitutes 100% of the natural element, which is always found associated with the rare earths and is frequently classified as one. *See* PERIODIC TABLE.

Yttrium metal absorbs hydrogen, and in alloys up to a composition of YH_2 they resemble metals very closely. In fact, in certain composition ranges, the alloy is a better conductor of electricity than the pure metal.

Yttrium forms the matrix for the europium-activated yttrium phosphors which emit a brilliant, clear-red light when excited by electrons. The television industry uses these phosphors in manufacturing television screens.

Yttrium is used commercially in the metal industry for alloy purposes and as a "getter" to remove oxygen and nonmetallic impurities in other metals. For properties of the metal and its salts *see* RARE-EARTH ELEMENTS.

Frank H. Spedding

Z

Z transform The preferred operational-calculus tool for analysis and design of discrete-time systems. (It should not be confused with the z transformation.) The role of the z transform with regard to discrete-time systems is similar to that of the Laplace transform for continuous systems. In fact, the Laplace transform is a specialized case of the z transform. The z transform is by far the more insightful tool, and the Laplace transform is just the limiting case of the z transform in a practical as well as a conceptual way. *See* CONTROL SYSTEMS; DIGITAL FILTER; LAPLACE TRANSFORM; LINEAR SYSTEM ANALYSIS.

It is useful to consider a band-limited real continuous signal, $x(t)$, with no significant amount of energy above a frequency, f_c. This signal is sampled at uniformly spaced intervals of time, $T, 2T, 3T, \ldots, nT, \ldots$, where the sampling interval, T, and the sampling frequency, f_s, are reciprocals of one another, and where f_s is greater than $2f_c$, a condition necessary for unambiguous interpretation of the sampled signal. If the common shorthand notation $x(nT) = x_n$ is used, the definition of the z transform of $x(t)$ is given by the equation

$$X(z) = x_0 + x_1 z^{-1} + x_2 z^{-2} + x_3 z^{-3} + \cdots$$

The coefficient of z^{-p} is therefore the value of the pth sample of the time signal. This gives a great deal of physical insight to the use of the z transform, a feature not shared by the Laplace transform. The Laplace transform can be found by evaluating the limit of the z transform as T approaches zero. *See* ANALOG-TO-DIGITAL CONVERTER; INFORMATION THEORY. Stanley A. White

Zebra One of three species of striped, horselike mammals belonging to the family Equidae and found wild in Africa. Zebras are classified in the order Perissodactyla (odd-toed ungulates), a group of mammals in which the middle toe is functional and the second and fourth digits are vestigial. All equids walk on the tips of their toes (unguligrade locomotion). Zebras stand 1.2 to 1.5 m (4–5 ft) high at the withers. Adults weigh 106–202 kg (235–450 lb).

Zebras differ from all other members of the horse family because of their startling color pattern. They have parallel black or dark brown stripes on a whitish background arranged in exact designs. These stripes run all over the body, meeting diagonally down the sides of the head. Researchers have rejected the hypotheses that stripes serve as camouflage for zebras, that they visually confuse predators and pests, or that they assist in regulating body temperature through heat absorption. Instead, it is thought that stripes facilitate group cohesion and socialization.

The plains, or common, zebra (*Equus burchellii*) inhabits the grasslands, light woodlands, open scrub, and savannahs of East Africa south of the Sahara Desert. The largest population (over 250,000) occurs in the Serengeti of Tanzania. The mountain zebra (*E. zebra*) is confined to the mountainous grasslands of southwest Africa. It has narrower stripes than the plains zebra. Grevy's zebra (*E. grevyi*) is the largest living wild equine, reaching a height of 1.5 m (5 ft) at the withers. It inhabits subdesert steppe and arid bushy grasslands in Ethiopia, Somalia, and northern Kenya. Zebra numbers continue to decline, especially on unprotected lands where humans bring settlement, crops, and livestock ranching. Natural predators include lions and hyaenas. *See* AFRICA; EQUIDAE; PERISSODACTYLA. Donald W. Linzey

Zebu A breed of humped, domestic cattle native to India and belonging to the family Bovidae in the order Artiodactyla. Also known as Brahman, they are probably descended from the wild auroch (*Bos taurus*), which is considered to be the ancestor of domestic cattle.

Zebus (*Bos indicus*) have been domesticated in Asia for approximately 6500 years. They were imported by breeders into the United States in 1849 and have been crossed with local breeds of beef cattle to produce strains better adapted to the hot humid Gulf states. The breed has also contributed to beef production though cross breeding with European cattle.

Brahman cattle have a very distinctive appearance with a large, fleshy, fatty hump (or sometimes double hump) over the shoulders, a rounded forehead, loose skin under the throat (dewlap), large drooping ears, and white legs (see illustration).

Gray Brahman bull (*Bos indicus*). (*Photo: Salinas Ranch, Carmine, Texas*)

The hump is formed from two overdeveloped muscles plus fatty tissue and probably represents an energy reserve for emergencies. These cattle are generally light to medium gray in color, but domestic animals may also be red, brown, or black. Brahman cattle have short hair and well-developed sweat glands that enable them to withstand heat and humidity, ticks, and insects. Zebus are protected as sacred cows by the Hindus in India, who allow them to roam freely through the streets and villages. In India, they are used as draft animals as well as for milk. *See* ARTIODACTYLA; BEEF CATTLE PRODUCTION; BOVIDAE. Donald W. Linzey

Zeeman effect A splitting of spectral lines when the light source being studied is placed in a magnetic field. Discovered by P. Zeeman in 1896, the effect furnishes information of prime importance in the analysis of spectra. Each kind of spectral term has its characteristic mode of splitting, and the types of terms are most definitely identified by this property, Furthermore, the effect allows an evaluation of the ratio of charge to mass of the electron and an evaluation of its precise magnetic moment.

The normal Zeeman effect is a splitting into two or three lines, depending on the direction of observation, as shown in the illustration. The light of these components is polarized in ways indicated in the illustration. The normal effect is observed for all lines belonging to singlet systems, those for which the spin quantum number $S = 0$. The change of frequency of the shifted components can be evaluated on classical electromagnetic principles.

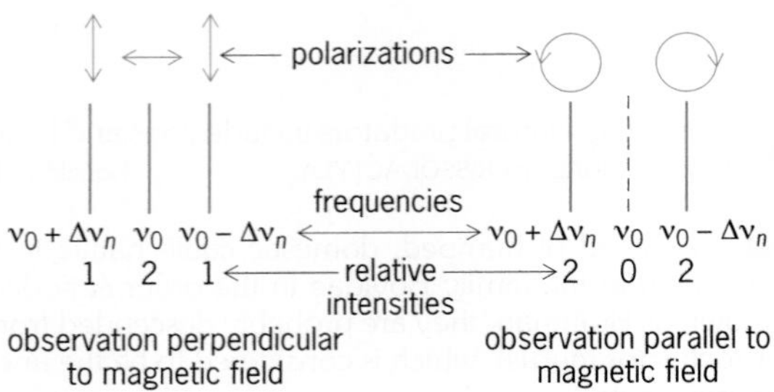

Triplet observed in normal Zeeman effect. ν_0 = unshifted frequency; $\Delta\nu_n$ = frequency shift.

The anomalous Zeeman effect is a more complicated type of line splitting, so named because it did not agree with the predictions of classical theory. It occurs for any spectral line arising from a combination of terms of multiplicity greater than one. Since multiplicity in spectral lines is caused by the presence of a resultant spin vector S of the electrons, the anomalous effect must be attributed to a nonclassical magnetic behavior of the electron spin.

The quadratic Zeeman effect, which depends on the square of the field strength, is of two kinds. The first results from second-order terms, and the second from the diamagnetic reaction of the electron when revolving in large orbits.

The inverse Zeeman effect is the Zeeman effect of absorption lines. It is closely related to the Faraday effect, the rotation of plane-polarized light by matter situated in a magnetic field. *See* FARADAY EFFECT.

The Zeeman effect in molecules is, in general, so small as to be unobservable, even for molecules which have a permanent magnetic moment. An exception occurs for some light molecules where the magnetic moment is coupled so lightly to the frame of the molecule that it can orient itself freely in the magnetic field just as for atoms.

A clear Zeeman effect also can be observed in many crystals with sharp spectrum lines in absorption or fluorescence. Such crystals are found particularly among rare-earth salts.

The magnetic moment of the nucleus causes a Zeeman splitting in atomic spectra which is of an order of magnitude a thousand times smaller than the ordinary Zeeman effect. This Zeeman effect of the hyperfine structure usually is modified by a nuclear Paschen-Back effect. *See* PASCHEN-BACK EFFECT.

F. A. Jenkins; G. H. Dieke; W. W. Watson

Zeiformes The order of teleost fishes collectively called dories. The order has representatives in all seas of the world from the tropics to Antarctica; however, they are most common in tropical and temperate zones, in waters of moderate depth to at least 1800 m. Most species are mesopelagic or bathypelagic. Structurally intermediary to the beryciform and the perciform fishes, zeiforms are characterized by a body varying from moderately elongate to very deep and compressed; jaws usually greatly protractile; colors predominantly silvery or silver gray; absence of an orbitosphenoid bone; a simple posttemporal bone rigidly united to the skull; a pelvic fin, each with one spine or none and from two to ten soft rays; a first dorsal fin with five to ten spines, usually strong; and an anal fin with one to four spines. Most species are of small size and of minor economic importance.

The order consists of five families: Oreosomatidae, Parazenidae, Zeniontidae (formerly Macrurocyttidae), Grammicolepididae, and Zeidae.

Herbert Boschung

Zener diode A two-terminal semiconductor junction device with a very sharp voltage breakdown as reverse bias is applied. The device is used to provide a voltage reference. It is named after C. Zener, who first proposed electronic tunneling as a mechanism of electrical breakdown in insulators. *See* SEMICONDUCTOR; TUNNELING IN SOLIDS.

A classic circuit to define a very stable current uses an operational amplifier and three stable resistors (see illustration). The voltage across the Zener itself defines a higher level from which the current is drawn. Thus, a stable noise-free Zener defines its own stable noise-free current. *See* OPERATIONAL AMPLIFIER.

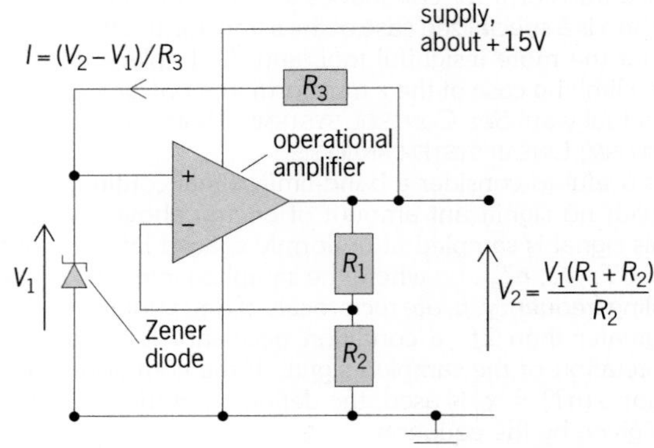

Zener diode circuit to define a very stable current, I, using an operational amplifier and three stable resistors, R_1, R_2, and R_3. The voltage V_1 across the Zener itself defines a higher level V_2 from which the current I is drawn.

The effect of temperature on the breakdown voltage can be nulled by having a second forward-biased junction, which has a small negative temperature coefficient, in series with the Zener junction. Such a device is called compensated Zener and has a breakdown voltage of 6.2 V rather than the normal 5.6 V (for the smallest possible temperature coefficient). Alternatively, a Zener junction can be part of an integrated circuit which adds a whole temperature controller to keep the silicon substrate at a constant temperature. For the very best performance, only four components are integrated into the silicon: the Zener, a heater resistor, a temperature-sensing transistor, and a current-sensing transistor. A separate selected dual operational amplifier then completes the current-and-temperature-control circuit. Such a circuit sets the chip temperature at, say 122°F (50°C), and the junction condition is then largely independent of ambient temperature. *See* INTEGRATED CIRCUITS.

The compact, robust Zener diode, with the circuits described to set its current and temperature, makes a fine portable voltage standard. This is used to disseminate the voltage level from national or accredited calibration laboratories to industry and to research laboratories. *See* JUNCTION DIODE; SEMICONDUCTOR DIODE; VOLTAGE MEASUREMENT; VOLTAGE REGULATOR.

Peter J. Spreadbury

Zenith The point in the sky directly above an observer. If the observer is at a pole, then the observer's zenith is a celestial pole; but for observers at midlatitudes the zenith is a point in the sky that corresponds to a changing right ascension but a constant declination as the sky rotates overhead. The point directly below the observer is the observer's nadir, and is 180° in longitude and in latitude from the observer's zenith.

The zenith distance of an object in the sky is the angle across the sky from the zenith to the object. The zenith distance is 90°

minus the altitude of an object above the horizon. *See* ASTRONOMICAL COORDINATE SYSTEMS; CELESTIAL SPHERE. Jay M. Pasachoff

Zeolite Any mineral belonging to the zeolite family of minerals and synthetic compounds characterized by an aluminosilicate tetrahedral framework, ion-exchangeable large cations, and loosely held water molecules permitting reversible dehydration. The general formula can be expressed as $X_y^{1+,2+}Al_x^{3+}Si_{1-x}^{4+}O_2 \cdot nH_2O$. The amount of large cations (X) present is conditioned by the aluminum/silicon (Al:Si) ratio and the formal charge of these large cations. Typical large cations are the alkalies and alkaline earths such as sodium (Na^+), potassium (K^+), calcium (Ca^{2+}), strontium (Sr^{2+}), and barium (Ba^{2+}). The large cations, coordinated by framework oxygens and water molecules, reside in large cavities in the crystal structure; these cavities and channels may even permit the selective passage of organic molecules. Thus, zeolites are extensively studied from theoretical and technical standpoints because of their potential and actual use as molecular sieves, catalysts, and water softeners. *See* MOLECULAR SIEVE; SILICATE MINERALS; WATER SOFTENING.

Zeolites are low-temperature and low-pressure minerals and commonly occur as late minerals in amygdaloidal basalts, as devitrification products, as authigenic minerals in sandstones and other sediments, and as alteration products of feldspars and nepheline. Phillipsite and laumontite occur extensively in sediments on the ocean floor. Stilbite, heulandite, analcime, chabazite, and scolecite are common as large crystals in vesicles and cavities in various basalts.

Zeolites are usually white, but often may be colored pink, brown, red, yellow, or green by inclusions; the hardness is moderate (3–5) and the specific gravity low (2.0–2.5) because of their rather open framework structures. Paul B. Moore; Norman Herron

The porous, yet crystalline nature of zeolites has been exploited commercially in three main areas—as sorbents, as cation-exchange materials, and as catalysts.

Most aluminosilicate zeolites have an extremely high (yet reversible) affinity for water and are widely used as desiccants. The discrimination between molecules on the basis of their size as compared to the zeolite pore size is the basis of several extremely important separation processes in a procedure known as molecular sieving. *See* CHEMICAL SEPARATION TECHNIQUES.

The major commodity use of zeolites takes advantage of their cation-exchange ability and leads to their use in low-phosphate detergents. Mineral zeolites have found utility in agricultural and wastewater treatment applications, where they either ion-exchange harmful metal ions out of the stream being treated or absorb ammonia by reacting, as their acidic form, to produce the absorbed ammonium ion. *See* ION EXCHANGE.

The most important and expanding area of application for zeolites is as heterogeneous catalysts. By combining the properties of excellent thermal stability (>800°C or 1500°F), a pore size of molecular dimensions, and the ready introduction of a wide assortment of cations via ion exchange, numerous very selective catalysts can be prepared. *See* HETEROGENEOUS CATALYSIS.

Norman Herron

Zero In mathematics, the concept zero is used in two ways: as a number and as a value of a variable. The positional system of number notation, developed first by the Babylonians (about 500 B.C.) with the base 60, and a millennium later by the Hindus and the Chinese with the base 10, required for greater clarity a special marker of the empty, nonoccupied position.

The zero as a number, however, is a new concept, introduced by the Hindus and Chinese about the same time (6th century). Brahmagupta (born A.D. 598) remarked that the number 0 has special properties: $a \pm 0 = a$, and $a \cdot 0 = 0$, where a may be any number (integer).

In a modern way, zero can be called the identity element of the infinite Abelian additive group of integers. If in an integral domain a product is equal to zero, then at least one factor of the product is zero. In the second concept zero is the value of a variable for which a function is equal to zero. Hans Rademacher; Emil Grosswald

Zinc A chemical element, Zn, atomic number 30, and atomic weight 65.38. Zinc is a malleable, ductile, gray metal. Because of chemical similarities among zinc, cadmium, and mercury, these three metals are classed together in a transition-elements subgroup of the periodic table. *See* METAL; PERIODIC TABLE; TRANSITION ELEMENTS.

Fifteen isotopes of zinc are known, of which five are stable, having atomic masses of 64, 66, 67, 68, and 70. About half of ordinary zinc occurs as the isotope of atomic mass 64. The half-lives of the radioactive isotopes range from 88 s for ^{61}Zn to 244 days for ^{65}Zn.

Zinc is a fairly active metal chemically. It can be ignited with some difficulty to give a blue-green flame in air and to discharge clouds of zinc oxide smoke. Zinc ranks above hydrogen in the electrochemical series, so that metallic zinc in an acidic solution will react to liberate hydrogen gas as the zinc passes into solution to form dipositively charged zinc ions, Zn^{2+}. This reaction is slow with very pure zinc, but the presence of small amounts of impurities, addition of a trace of copper sulfate, or contact between the zinc surface and such metals as nickel or platinum facilitates formation of gaseous hydrogen and speeds the reaction. The combination of zinc and dilute acid is often used to generate small quantities of hydrogen in the laboratory. Zinc also dissolves in strongly alkaline solutions, such as sodium hydroxide, to liberate hydrogen and form dinegatively charged tetrahydroxozincate ions, $Zn(OH)_4^{2-}$, sometimes written as ZnO_2^{2-} in the formulas of the zincate compounds. Zinc also dissolves in solutions of ammonia or ammonium salts. The common soluble zinc compounds undergo to some extent the process of hydrolysis, which makes their solutions slightly acidic. The ion Zn^{2+} is colorless, so that the relatively few zinc compounds that are not colorless in large crystals, or white as powders, receive their color through the influence of the other constituents. Some of the atomic and ionic properties of zinc are shown in the table. *See* ELECTROCHEMICAL SERIES; HYDROLYSIS.

Atomic and ionic properties of zinc

Property	Value
Electronic configuration	$1s^2, 2s^2, 2p^6, 3s^2, 3p^6, 3d^{10}, 4s^2$
Ionization potentials	
1st electron loss	9.39 eV
2d electron loss	17.9 eV
Ionic radius, Zn^{2+}	0.072 nm
Covalent radius (tetrahedral)	0.131 nm
Oxidation potentials	$Zn \rightleftharpoons Zn^{2+} + 2e^-, E^\circ = 0.76$ V $Zn + 4OH^- \rightleftharpoons ZnO_2^{2-} + 2H_2O + 2e^-, E^\circ = 1.22$ V

Zinc also forms many coordination compounds. The zincates are actually coordination compounds, or complexes, in which hydroxide ions, OH^-, are bound to the zinc ions. Ammonia, NH_3, forms complexes with zinc, such as the typical tetrammine zinc ion, $[Zn(NH_3)_4]^{2+}$. Zinc cyanide, usually given the simple formula $Zn(CN)_2$, is a coordination compound in which many alternating zinc and cyanide ions are three-dimensionally bound together in a very large molecule. This compound is still widely used in zinc plating, but concern over environmental pollution has led to increasing use of zinc chloride plating baths. In most coordination compounds of zinc, the fundamental structural unit is a central zinc ion surrounded by four coordinated groups arranged spatially at the corners of a regular tetrahedron. *See* COORDINATION CHEMISTRY; COORDINATION COMPLEXES.

Pure, freshly polished zinc is bluish-white, lustrous, and moderately hard (2.5 on Mohs scale). Moist air brings about a superficial tarnish to give the metal its usual grayish color. Pure zinc is malleable and ductile enough to be rolled or drawn, but small amounts of other metals present as contaminants may render it brittle. Malleability of even pure zinc is improved by heating zinc to 100–150°C (212–300°F). If heated zinc is mechanically worked; it does not embrittle on cooling. Zinc melts at 420°C (788°F) and boils at 907°C (1665°F). Its density is 7.13 times that of water, so that 1 ft^3 (0.028 m^3) of zinc weighs 445 lb (200 kg).

As a conductor of heat and of electricity, zinc ranks fairly high. However, its electrical resistivity (5.92 microhm-cm at 20°C or 68°F) is almost four times that of silver, the best conductor. As a conductor of heat, zinc is likewise only about one-fourth as efficient as silver. At 0.91 K zinc is an electrical superconductor. Pure zinc is not ferromagnetic, but the alloy compound $ZrZn_2$ displays ferromagnetism below 35 K.

The most important uses of zinc are in its alloys and as a protective coating on other metals. Coating iron or steel with zinc is known as galvanizing, and it may be done by immersing the article in melted zinc (hot-dip process), depositing zinc electrolytically onto the article in a plating bath (electrogalvanizing), exposing the article to powdered zinc near its melting point (sherardizing), or spraying the article with melted zinc (metallizing). The mere physical presence of the zinc coat prevents corrosion of iron, and even if breaks in the coat expose portions of the iron, the greater chemical activity of the zinc causes it to be consumed in preference to the iron. Adding small amounts of other metals to galvanizing baths has been found to improve the adhesion and weathering qualities of the coating. *See* ELECTROPLATING OF METALS; GALVANIZING.

Even such nonstructural materials as cardboard can be zinc-coated by low-temperature flame spraying. Other important uses of zinc are in brass and zinc die-casting alloys, in zinc sheet and strip, in electrical dry cells, in making certain zinc compounds, and as a reducing agent in chemical preparations.

A so-called tumble-plating process coats small metal parts by applying zinc powder to them with an adhesive, then tumbling them with glass beads to roll out the powder into a continuous coat of zinc. Rechargeable nickel-zinc batteries offer higher energy densities than conventional dry cells. Foamed zinc metal has been suggested for use in lightweight structures such as aircraft and spacecraft. Some other uses of zinc are in dry cells, roofing, lithographic plates, fuses, organ pipes, and wire coatings. Zinc dust, a flammable material when dry, is used in fireworks and as a chemical catalyst and reducing agent. Radioactive ^{65}Zn is used medically in the study of metabolism of zinc, and also in determining rates of wear for zinc-containing alloys. *See* METAL COATINGS; ZINC ALLOYS. William E. Cooley

Zinc is believed to be needed for normal growth and development of all living species, including humans; actually, life without zinc would be impossible. Zinc is a common element that is present in virtually every type of human food, and zinc deficiency is therefore not considered to be a common problem in humans. Zinc is a trace element; that is, it is present in biological fluids at a concentration below 1 ppm, and only a small amount (normally $<$25 mg) is required in the daily diet. (The recommended daily allowance for zinc is 15 mg/day for adults and 10 mg/day for growing children.) It is relatively nontoxic, without noticeable side effects at intake levels of up to 10 times the normal daily requirement. James F. Riordan; Kenneth H. Falchuk

Zinc alloys Combinations of zinc with one or more other metals. If zinc is the primary constituent of the alloy, it is a zinc-base alloy. Zinc also is commonly used in varying degrees as an alloying component with other base metals, such as copper, aluminum, and magnesium. A familiar example of the latter is the association of varying amounts of zinc (up to 45%) with copper to produce brass. *See* BRASS; COPPER ALLOYS.

Zinc-base alloys have two major uses: for casting and for wrought applications. Casting includes both die casting and gravity casting, which differs from die casting primarily in that no pressure is applied, except the force of gravity, in forcing the molten metal into the mold. *See* METAL CASTING.

The great bulk of zinc die casting is carried out with two alloys, commonly identified as alloys 3 and 5. A third alloy has been added and is commonly designated as alloy 7. The entire group in the United States is often spoken of by its original trade name as Zamak alloys, and in England as the Mazak alloys.

Aluminum, the major alloying constituent in alloys 3 and 5, is added in amounts of about 4%, which has proved to be the optimum composition from the standpoint of strength, ductility, and stability. The aluminum content also sharply reduces the rate of attack of molten zinc on iron containers. Aluminum also avoids "soldering," or sticking of the casting to the die. This permits the casting of these zinc alloys in the more productive hot-chamber (plunger) type of die-casting machine. *See* ALUMINUM.

Alloys 3 and 5 differ primarily in their copper content. Copper additions increase strength and hardness and improve corrosion resistance. The copper content of alloys is held to a specified maximum of 1.25% (in alloy 5) to avoid making the alloy unstable through aging and to avoid reducing the impact strength to a very low value.

Alloy 7 is considered a modification of alloy 3. It has a lower magnesium content, and iron, lead, tin, and cadmium are also held to lower levels. A small amount of nickel is added. When the alloys are compared, alloy 7 exhibits improved casting properties, making it easier to secure high-quality surfaces for finishing (mostly chromium plating) and to obtain higher production rates.

In the wrought zinc area, numerous compositions and alloys are used, depending on ultimate product requirements. Alloying metals can be used to improve various properties, such as stiffness, for special applications. The zinc-copper-titanium alloy has become the dominant wrought-zinc alloy for applications demanding superior performance. *See* ZINC; ZINC METALLURGY.

Adolph L. Ponikvar

Zinc metallurgy Separating and extracting zinc from ores, refining it, and preparing it into usable forms. Ordinarily, after being mined, ores must first be separated into a concentrated mineral and a waste rock. This concentrate then is reduced to the metal in a metallurgical works. Finally, the metal may be further refined and alloyed to commercially usable form. Commonly, these three extractive metallurgical operations are conducted at separate locations and are broadly categorized as concentrating, smelting, and refining, respectively.

Zinc sulfide ores are usually beneficiated (concentrated or ore-dressed) adjacent to the mine site. First, the ore must be crushed and ground in order to free the mineral lattices from those of the waste rock (gangue). Next, the finely divided ore is mixed into a slurry with water and the mineral and gangue particles are separated utilizing the effect of gravity. The most common means of separation is the froth flotation process. *See* FLOTATION; ORE DRESSING.

The second operation, smelting, involves first the chemical or physical changes required in the source material to prepare it into a crude zinc oxide form and a specified particle size (depending upon the particular smelting method contemplated). Such process steps variously entail the operations of roasting, sintering, or pyroconcentration. The other basic part of smelting is the reduction step, wherein the zinc is reduced from its oxide to its elemental form. Several very successful and varied processes accomplish this function, including horizontal retort, vertical retort, electrothermic furnace, and blast furnace (all of which are pyrometallurgical and use carbon as a reducing agent). *See* PYROMETALLURGY; SINTERING.

Purity of zinc produced by the various smelting processes is largely dependent upon controls and operating procedures practiced during the preparation steps. The quality produced is variously suitable for hot-dip galvanizing, continuous-line galvanizing, and in some cases for brass manufacture and rolled (wrought) zinc; however, for the sizable usage in die-casting alloys, output from this type of smelter must undergo a refining step. Fractional distillation in reflux refining columns is the leading method of upgrading the lower-purity zinc metal. *See* ZINC ALLOYS.

Carl H. Cotterill

Zincite A mineral with composition ZnO (zinc oxide). It crystallizes in the hexagonal system with a wurtzite-type structure. Thus its principal axis is polar and different forms appear at top and bottom of crystals. Such crystals are rare and the mineral is usually massive. Its hardness is 4 and its specific gravity 5.6. The mineral has a subadamantine luster and a deep-red to orange-yellow color. Zincite is rare except at the zinc deposits at Franklin and Sterling Hill, N.J. There, associated with franklinite and willemite, it is mined as a valuable ore of zinc. *See* FRANKLINITE; WILLEMITE; ZINC.

Cornelius S. Hurlbut, Jr.

Zingiberales An order of flowering plants, division Magnoliophyta, in the monocots, consisting of eight families and about 1800 species. Zingiberales (also known as Scitamineae or Scitaminales) are morphologically well defined and clearly circumscribed in DNA sequence analyses. The largest families are Zingiberaceae (about 1000 species), Marantaceae (about 400 species), Costaceae (about 150 species), and Heliconiaceae (about 100 species). Zingiberales are most closely related to Commelinales.

Zingiberales are herbs or scarcely branched trees or shrubs with pinnately veined leaves and irregular flowers that have well-differentiated sepals and petals, an inferior ovary, septal or septa-derived nectaries, and usually either one or five functional stamens. The economic crops, ginger (*Zingiber officinale*) in Zingiberaceae and banana (*Musa*) in Musaceae, and the ornamentals, bird of paradise flower (*Strelitzia*) in Strelitziaceae and *Canna* in Cannaceae, are familiar members of Zingiberales. *See* BANANA; COMMELINALES; FLOWER; FRUIT; GINGER; LILIOPSIDA; MAGNOLIOPHYTA.

Michael F. Fay

Zingiberidae A subclass of Liliopsida (monocotyledons) of the division Magnoliophyta (Angiospermae), the flowering plants, containing two orders (Bromeliales and Zingiberales), with nine families and about 3800 species. The subclass has been associated with the Commelinidae and the Liliidae, sharing some features with each but aberrant in either, and it differs from both in usually having four or more supporting cells around each stomate. The flowers are usually epigynous, usually bisexual, and often irregular, and they are usually showy and adapted to pollination by insects or other animals. The pistil consists of three united carpels, often with septal nectaries opening at the top of the ovary. The seeds usually have well-developed, mealy or starchy endosperm. *See* BROMELIALES; LILIOPSIDA; MAGNOLIOPHYTA; PLANT KINGDOM; SECRETORY STRUCTURES (PLANT); ZINGIBERALES.

T. M. Barkley

Zircon A mineral with the idealized composition $ZrSiO_4$, one of the chief sources of the element zirconium. Structurally, zircon is a nesosilicate, with isolated SiO_4 groups. *See* SILICATE MINERALS; ZIRCONIUM.

Zircon often occurs as well-formed crystals. The color is variable, usually brown to reddish brown, but also colorless, pale yellowish, green, or blue. The transparent colorless or tinted varieties are popular gemstones. Hardness is $7^1/_2$ on Mohs scale; specific gravity is 4.7, decreasing in metamict types. *See* METAMICT STATE.

Because of its chemical and physical stability, zircon resists weathering and accumulates in residual deposits and in beach and river sands, from which it has been obtained commercially in Florida and in India, Brazil, and other countries. *See* HEAVY MINERALS.

Clifford Frondel

Zirconium A chemical element, Zr, atomic number 40, atomic weight 91.22. Its naturally occurring isotopes are 90, 91, 92, 94, and 96. Zirconium is one of the more abundant elements, and is widely distributed in the Earth's crust. Being very reactive chemically, it is found only in the combined state. Under most conditions, it bonds with oxygen in preference to any other element, and it occurs in the Earth's crust only as the oxide, ZrO_2, baddeleyite, or as part of a complex of oxides as in zircon, elpidite, and eudialyte. Zircon is commercially the most important ore. Zirconium and hafnium are practically indistinguishable in chemical properties, and occur only together. *See* HAFNIUM; PERIODIC TABLE; ZIRCON.

Most of the zirconium used has been as compounds for the ceramic industry: refractories, glazes, enamels, foundry mold and core washes, abrasive grits, and components of electrical ceramics, The incorporation of zirconium oxide in glass significantly increases its resistance to alkali. The use of zirconium metal is almost entirely for cladding uranium fuel elements for nuclear power plants. Another significant use has been in photo flashbulbs.

Zirconium is a lustrous, silvery metal, with a density of 6.5 g/cm^3 at 20°C (68°F). It melts at about 1850°C (3362°F). Estimates of the boiling point from appropriate data have commonly been of the order of 3600°C (6500°F), but observations suggest about 8600°C (15,500°F). The free energies of formation of its compounds indicate that zirconium should react with any nonmetal, other than the inert gases, at ordinary temperatures. In practice, the metal is found to be nonreactive near room temperature because of an invisible, impervious oxide film on its surface. The film renders the metal passive, and it remains bright and shiny in ordinary air indefinitely. At elevated temperatures it is very reactive to the non-metallic elements and many of the metallic elements, forming either solid solutions or compounds.

Zirconium generally has normal covalency of 4, and commonly exhibits coordinate covalencies of 5, 6, 7, and 8. Zirconium is at oxidation number 4 in nearly all of its compounds, Halides in which its oxidation numbers are 3 and 2 have been prepared. While zirconium is often part of cationic or anionic complexes, there is no definite evidence for a monatomic zirconium ion in any of its compounds.

Most handling and testing of zirconium compounds have indicated no toxicity. There has generally been no ill consequence of contact of zirconium compounds with the unabraded skin. However, some individuals appear to have allergic sensitivity to zirconium compounds, characteristically manifested by appearance of nonmalignant granulomas. Inhalation of sprays containing some zirconium compounds and of metallic zirconium dusts have had inflammatory effects.

Warren B. Blumenthal

Zoanthidea An order of the cnidarian subclass Hexacorallia. Sometimes known popularly as "mat anemones" and technically also as Zoanthiniaria (the name Zoantharia is an alternative for subclass Hexacorallia), this order contains relatively few species. *See* ANTHOZOA; CNIDARIA; HEXACORALLIA.

The column of one of these sedentary, anemonelike anthozoans is typically encrusted with sand grains, sponge spicules, tests of foraminiferans, and other detritus that invades the mesoglea, thus acting as a sort of adventitious skeleton. Zoanthid musculature is poorly developed. The basal end of most is somewhat or deeply embedded in a mass of tissue that connects members of the colony; in solitary zoanthids, a pedal disk is absent. The simple, unbranched tentacles are arrayed in two cycles. There are two suborders: Brachycnemina and Macrocnemina. Members of Brachycnemina are abundant in warm, shallow waters, including on coral reefs, and typically harbor zooxanthellae; they comprise the families Neozoanthidae, Sphenopidae, and Zoanthidae. The Macrocnemina is characteristic of the deep sea, where colonies of some species encrust objects such as sea whips, and others associate with hermit crabs, forming a carcinoecium around the crustacean's abdomen; the families of this group are Epizoanthidae, Gerardiidae, and Parazoanthidae. Like all cnidarians, zoanthids are carnivorous. Some zoanthids are hermaphroditic (individuals are both male and female) and some gonochoric (individuals are either male or female); most broadcast their gametes but some brood. Daughter polyps of a colony arise by budding from the stolon or the polyp base, or by longitudinal fission; transverse fission occurs in the solitary zoanthid *Sphenopus*. Daphne G. Fautin

Zodiac The band of sky through which the Sun, Moon, and planets apparently move in the course of the year. The Babylonians, about 2500 years ago, divided the zodiac into 12 parts, which correspond to constellations. These zodiacal constellations, in order around the sky, are Aries, the Ram; Taurus, the Bull; Gemini, the Twins; Cancer, the Crab; Leo, the Lion; Virgo, the Virgin; Libra, the Scales; Scorpius, the Scorpion; Sagittarius, the Archer, Capricornus, the Sea Goat; Aquarius, the Water Carrier; and Pisces, the Fish. These constellations are based on Greek myths.

Because of the precession of the equinoxes, the positions of the constellations in the sky have drifted from the dates of the year with which they were associated thousands of years ago. Thus the popular astrological "signs" of the zodiac are not actually those that currently correspond to the sky. Though the vernal equinox is often called the first point of Aries, precession has moved it into Pisces, not far from Aquarius. *See* PRECESSION OF EQUINOXES.

The Sun actually passes through parts of 13 constellations, as currently defined. Also, if the zodiac is defined as the region within latitudes $\pm 8^\circ$, which accommodates the eight planets through Neptune, it contains all or part of 24 constellations. *See* CONSTELLATION; ECLIPTIC. Jay M. Pasachoff

Zodiacal light A diffuse, night-sky luminosity easily seen at low to middle geographic latitudes in the absence of moonlight. It is caused by sunlight scattered and absorbed by interplanetary (solar system) dust particles. Zodiacal light extends over the entire sky, but it is brightest toward the Sun and in the zodiacal band. It is best seen in the west after evening twilight and in the east before morning twilight when the ecliptic is close to the vertical. In the Northern Hemisphere this corresponds to spring evenings and autumn mornings. *See* ECLIPTIC; INTERPLANETARY MATTER.

The density of particles responsible for the zodiacal light falls off somewhat faster than $1/R$, where R is heliocentric distance. The particles are primarily in the size range of tens to hundreds of micrometers in diameter, and are believed to originate primarily from comets. *See* COMET.

The visual zodiacal light brightness decreases monotonically with elongation (angular distance from the Sun) to a relatively flat minimum in the ecliptic at 120 to 140°, after which it gradually increases out to the antisolar point at 180°. The enhanced brightness near the antisolar point is called the Gegenschein or counterglow. It is barely above the visible threshold and is described by visual observers as being oval in appearance, 6 by 10° or larger, with the long axis in the ecliptic. Space observations have shown the Gegenschein to be an intrinsic part of the zodiacal light; that is, the zodiacal dust particles have an increased scattering efficiency near backscattering.

Although the Earth's atmosphere complicates attempts to observe zodiacal light closer than approximately 30° to the Sun, eclipse and space observations have shown the brightness to increase smoothly all the way in to the F-corona. The F-corona or inner zodiacal light and the "primary" zodiacal light seen at larger elongations come primarily from dust particles located relatively close to the Sun. *See* SOLAR CORONA; SUN. J. L. Weinberg

Zone refining One of a number of techniques used in the preparation of high-purity materials. The technique is capable of producing very low impurity levels, namely, parts per million or

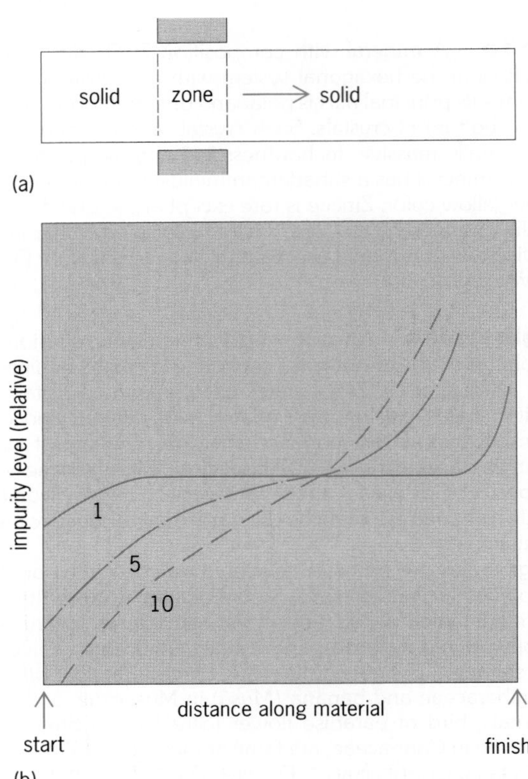

Zone refining. (*a*) Passage of a molten zone along the material to be purified. (*b*) Effect of 1, 5, and 10 zone passes on the impurity distribution along the material.

less in a wide range of materials, including metals, alloys, intermetallic compounds, semiconductors, and inorganic and organic chemical compounds. In principle, zone refining takes advantage of the fact that the solubility level of an impurity is different in the liquid and solid phases of the material being purified; it is therefore possible to segregate or redistribute an impurity within the material of interest. In practice, a narrow molten zone is moved slowly along the complete length of the specimen in order to bring about the impurity segregation. *See* SOLUTION; SOLVENT.

Impurity atoms either raise or lower the melting point of the host material. There is also a difference in the concentration of the impurity in the liquid phase and in the solid phase when the liquid and solid exist together in equilibrium. In zone refining, advantage is taken of this difference, and the impurity atoms are gradually segregated to one end of the starting material. To do this, a molten zone is passed from one end of the impure material to the other, as in illus. *a*, and the process is repeated several times. The end to which the impurities are segregated depends on whether the impurity raises or lowers the melting point of the pure material; a lowering of the melting point is more common, in which case impurities are moved in the direction of travel of the molten zone. The effect of multiple zone passes (in the same direction) on impurity content along the material is illustrated in illus. *b*. *See* METALLURGY. Alan Lawley

Zooarcheology Zooarcheology or archeozoology is the study of animal remains from archeological sites. Most such remains derive from people's meals. In other words, zooarcheology is essentially the study of ancient garbage, mainly bones and teeth of mammals, as well as birds, fish, mollusks, and even insects. Zooarcheology helps to provide a more complete picture of people's environment and way of life, especially their economy, as reflected in the relationship between people and animals 100,000 or even just 100 years ago. It is a multidisciplinary endeavor requiring knowledge of anatomy and biometry as well as an appreciation of the archeological questions that need to be addressed. Unlike most paleontological collections, zooarcheological collections are usually well dated and comprise large numbers of bones. They provide excellent opportunities to study microevolution. However, much remains to be explored in this relatively new science. *See* ANTHROPOLOGY; ARCHEOLOGY; FOSSIL HUMANS; ZOOLOGY. Simon J. M. Davis

Zoogeography The subdivision of the science of biogeography that is concerned with the detailed description of the distribution of animals and how their past distribution has produced present-day patterns. Scientists in this field attempt to formulate theories that explain the present distributions as elucidated by geography, physiography, climate, ecological correlates (especially vegetation), geological history, the canons of evolutionary theory, and an understanding of the evolutionary relationships of the particular animals under study.

The field of zoogeography is based upon five observations and two conclusions. The observations are as follows. (1) Each species and higher group of animals has a discrete nonrandom distribution in space and time (for example, the gorilla occurs only in two forest areas in Africa). (2) Different geographical regions have an assemblage of distinctive animals that coexist (for example, the fauna of Africa south of the Sahara with its monkeys, pigs, and antelopes is totally different from the fauna of Australia with its platypuses, kangaroos, and wombats). (3) These differences (and similarities) cannot be explained by the amount of distance between the regions or by the area of the region alone [for example, the fauna of Europe and eastern Asia is strikingly similar although separated by 6900 mi (11,500 km) of land, while the faunas of Borneo and New Guinea are extremely different although separated by a tenth of that distance across land and water]. (4) Faunas strikingly different from those found today previously occurred in all geographical regions (for example, dinosaurs existed over much of the world in the Cretaceous). (5) Faunas resembling those found today or their antecedents previously occurred, sometimes at sites far distant from their current range (for example, the subtropical-warm temperate fauna of Eocene Wyoming, including many fresh-water fishes, salamander, and turtle groups, is now restricted to the southeastern United States).

The conclusions are as follows. (1) There are recognizible recurrent patterns of animal distribution. (2) These patterns represent faunas composed of species and higher groups that have evolved through time in association with one another.

Two rather different approaches have dominated the study of zoogeography since the beginning of the nineteenth century: ecological and historical. Ecological zoogeography attempts to explain current distribution patterns principally in terms of the ecological requirements of animals, with particular emphasis on environmental parameters, physiological tolerances, ecological roles, and adaptations. The space and time scales in this approach are narrow, and emphasis is upon the statics and dynamics of current or very recent events. Historical zoogeography recognizes that each major geographical area has a different assemblage of species, that certain systematic groups of organisms tend to cluster geographically, and that the interaction of geography, climate, and evolutionary processes over a long time span is responsible for the patterns or general tracks. Emphasis in this approach is upon the statics and dynamics of major geographical and geological events ranging across vast areas and substantial time intervals of up to millions of years. The approach is based on concordant evolutionary association of diverse groups through time. *See* ECOLOGY. Jay M. Savage

Zoological nomenclature The system of naming animals that was adopted by zoologists and detailed in the International Code of Zoological Nomenclature, which applies to both living and extinct animals. The present system is founded on the 10th edition of C. Linnaeus's *Systema Naturae* (1758) and has evolved through international agreements culminating in the Code adopted in 1985. The primary objective of the Code is to promote the stability of the names of taxa (groups of organisms) by providing rules concerning name usage and the activity of naming new taxa. The rules are binding for taxa ranked at certain levels and nonbinding on taxa ranked at other levels. *See* SYSTEMATICS.

Zoological nomenclature is built around four basic features. (1) The correct names of certain taxa are either unique or unique combinations. (2) These names are formed and treated as Latin names and are universally applicable, regardless of the native language of the zoologist. (3) The Code for animals is separate and independent from similar codes for plants and bacteria. (4) No provisions of the Code are meant to restrict the intellectual freedom of individual scientists to pursue their own research.

There are four common reasons why nomenclature may change. (1) New species are found that were once considered parts of other species. (2) Taxonomic revisions may uncover older names or mistakes in identification of types. (3) Taxa may be combined, creating homonyms that require replacement. (4) Concepts of the relationships of animals change. Stability is subservient to progress in understanding animal diversity.

The articles in the Code are directed toward the names of taxa at three levels. The family group includes taxa ranked as at the family and tribe levels (including super- and subfamilies). The genus group includes taxa below subtribe and above species. The species group includes taxa ranked as species or subspecies. Taxa above the family group level are not specifically treated, and their formation and use are not strictly regulated. For each group, provisions are made that are either binding or recommended.

Binominal nomenclature. The basis for naming animals is binominal nomenclature, that is, a system of two-part names. The first name of each species is formed from the generic name, and the second is a trivial name, or species epithet. The two names agree in gender unless the specific epithet is a patronym (named for a person). The combination must be unique; no other animal can have the same binominal. The formal name of a species also includes the author, so the formal name for humans is *Homo sapiens* Linnaeus. The genus, as a higher taxon, may have one or many species, each with a different epithet. *Homo* includes *H. sapiens, H. erectus, H. habilis*, and so forth. One feature of the Linnaean binominal system is that species epithets can be used over and over again, so long as they are used in different genera. *Tyrannosaurus rex* is a large dinosaur, and *Percina rex* is a small fresh-water fish. The epithet *rex* is not the species name of *Percina rex* because all species names are binominal in form. It is recommended that names of genera and species be set in a different typeface from normal text; italics is conventional. Different names for the same species are termed synonyms, and the senior synonym is usually correct (principle of priority). Modern species descriptions are accompanied by a description that attempts to show how the species is different from others, and the designation of one or more type specimens.

Higher taxa. All higher taxonomic names have one part (uninominal) and are plural. Names of taxa of the family and genus groups must be unique. The names of genera are in Latin or latinized, are displayed in italics, and may be used alone. The names of the family group are formed by a root and an ending specific to a particular hierarchical level: family Hominidae (root + idae), subfamily Homininae (root + inae). The endings of superfamilies (root + oidea) and tribes (root + ini) are recommended but not mandated. The endings of taxa higher than the family group vary. For example, orders of fishes are formed by adding -iformes to the root (Salmoniformes), while in insects the ending is usually -ptera (Coleoptera). *See* CLASSIFICATION, BIOLOGICAL; TAXONOMIC CATEGORIES. E. O. Wiley

Zoology The science that deals with knowledge of animal life. With the great growth of information about animals, zoology has been much subdivided. Some major fields are anatomy, which deals with gross and microscopic structure; physiology, with living processes in animals; embryology, with development of new individuals; genetics, with heredity and variation; parasitology, with animals living in or on others; natural history, with life and behavior in nature; ecology, with the relation of animals to their environments; evolution, with the origin and differentiation of animal life; and taxonomy, with the classification of animals. *See* ANATOMY, REGIONAL; DEVELOPMENTAL BIOLOGY; GENETICS; PARASITOLOGY; PHYLOGENY; PLANT EVOLUTION; TAXONOMY. Tracy I. Storer

Zoomastigophorea A class of protozoans of the subphylum Sarcomastigophora. Zoomastigophorea, also known as Zoomastigina, are flagellates. Some are simple, some are specialized; some have pseudopodia besides flagella. One group engulfs solid food at any body point; another shows localized ingestive areas. All are colorless. None produce starch or paramylum, and lipids and glycogen are assimilation products. Cells are naked or have delicate membranes. Colony formation is common. Colonies differ greatly in form and may be amorphous, linear, spherical, arboroid, or plane. Flagella vary from none to many.

Nine orders are included in this class: Choanoflagellida, Bicosoecida, Rhizomastigida, Kinetoplastida, Retortomonadida, Diplomonadida, Oxymonadida, Trichomonadida, and Hypermastigida. *See* PROTOZOA; SARCOMASTIGOPHORA. James B. Lackey

Zoonoses Diseases that are transmitted from animals to humans. Zoonotic agents are found among all major categories of microbes, including bacteria, fungi, parasites, and viruses. Although anyone can acquire a zoonotic infection, young children, older adults, and immunocompromised individuals are most at risk. Animals can also serve as reservoirs for zoonotic agents; that is, they are a source of the microorganism that allows transmission to the human population. Domesticated animals can often serve as a bridge for zoonotic agents that are commonly found in feral animal populations. However, sometimes animals serve only as indicators that an organism is present in the environment. In these cases, they become infected from contaminated soil, water, or air, as do humans. Additionally, there are diseases, called anthroponoses, that are transmitted from humans to animals.

Important zoonotic agents that can invade the nervous system in humans include *Toxoplasma gondii* and *Cryptococcus neoformans. Bordetella bronchiseptica* is a secondary bacterial invader that occurs commonly in the respiratory tracts of dogs, cats, and pigs, and less commonly in horses. Agents causing primary gastrointestinal disease that have been documented to be transmitted to humans from an animal source include *Salmonella, Campylobacter, Mycobacteria, Cryptosporidia*, and *Giardia*. Agents that have been implicated as zoonotic agents include *Entamoeba, Isospora, and Microsporidia*.

Some parasitic diseases of animals have zoonotic potential. The dog heartworm, *Dirofilaria immitis*, is endemic in many areas of North America and the rest of the world, requiring daily chemical prophylaxis in pet dogs during the mosquito season. Humans can be accidentally infected by *D. immitis* microfilaria-carrying mosquito vectors due to the high prevalence of *D. immitis*, although the disease is rare. *Toxocara canis* and *T. cati* are round worms of the dog and cat. Humans acquire infection with these parasites by ingesting embryonated eggs shed in the feces of animals.

Bartonella spp., including *B. henselae* (formerly *Rochalimaea*), a member of the rickettsia family, have been implicated as causative agents in cat scratch disease and bacillary angiomatosis (vascular tumors of the skin, internal organs, bone marrow, and lymph nodes). In humans, cat scratch disease is characterized by persistent regional lymphadenopathy in the lymphatic drainage area of the site of a recent cat bite or cat scratch. *See* BARTONELLOSIS; CAT SCRATCH DISEASE.

Pasteurella multocida is an important pathogen in animal bites and a documented life-threatening pathogen in immunocompromised patients. Transmission occurs via traumatic wounds such as bites and scratches, by the respiratory route via aerosol, and by ingestion. *See* PASTEURELLA.

Overall, the risk of contracting zoonotic diseases by susceptible individuals is likely to be small, but physicians and veterinarians can decrease this risk by educating patients on the transmission and prevention of these diseases. *See* ANIMAL VIRUS; MEDICAL BACTERIOLOGY; MEDICAL PARASITOLOGY. Linda S. Mansfield

Zooplankton Animals that inhabit the water column of oceans and lakes and lack the means to counteract transport currents. Zooplankton inhabit all layers of these water bodies to the greatest depths sampled, and constitute a major link between primary production and higher trophic levels in aquatic ecosystems. Many zooplankton are capable of strong swimming movements and may migrate vertically from tens to hundreds of meters; others have limited mobility and depend more on water turbulence to stay afloat. All zooplankton, however, lack the ability to maintain their position against the movement of large water masses.

Zooplankton can be divided into various operational categories. One means of classification is based on developmental stages and divides animals into meroplankton and holoplankton. Meroplanktonic forms spend only part of their life cycles as plankton and include larvae of benthic worms, mollusks, crustaceans, echinoderms, coral, and even insects, as well as the eggs and larvae of many fishes. Holoplankton spend essentially their whole existence in the water column. Examples are chaetognaths, pteropods, larvaceans, siphonophores, and many copepods. Nearly every major taxonomic group of animals has either meroplanktonic or holoplanktonic members.

Size is another basis of grouping all plankton. A commonly accepted size classification scheme includes the groupings: picoplankton (<2 micrometers), nanoplankton (2–20 μm), microplankton (20–200 μm), mesoplankton (0.2–20 mm), macroplankton (20–200 mm), and megaplankton (>200 mm).

The classic description of the trophic dynamics of plankton is a food chain consisting of algae grazed by crustacean zooplankton which are in turn ingested by fishes. This model may hold true to a degree in some environments such as upwelling areas, but it masks the complexity of most natural food webs. Zooplankton have an essential role in linking trophic levels, but several intermediate zooplankton consumers can exist between the primary producers (phytoplankton) and fish. Thus, food webs with multiple links to different organisms indicate the versatility of food choice and energy transfer and are a more realistic description of the planktonic trophic interactions.

Size is of major importance in planktonic food webs. Most zooplankton tend to feed on organisms that have a body size smaller than their own. However, factors other than size also modify feeding interactions. Some phytoplankton are noxious and are avoided by zooplankton, and others are ingested but not digested. Furthermore, zooplankton frequently assume different feeding habits as they grow from larval to adult form. They may ingest bacteria or phytoplankton at one stage of their life cycle and become raptorial feeders later. Other zooplankton are primarily herbivorous but also ingest heterotrophic protists and can opportunistically become carnivorous. Consequently, omnivory, which is considered rare in terrestrial systems, is a relatively common trophic strategy in the plankton. In all food webs, some individuals die without being consumed and are utilized by scavengers and ultimately by decomposers (bacteria and fungi). *See* ECOLOGY; ECOSYSTEM; MARINE ECOLOGY; PHYTOPLANKTON. Robert W. Sanders

Zoraptera An order of minute insects superficially resembling termites (Isoptera) or booklice (Psocoptera). Although species nest in decaying wood, they are of no economic importance. Interest in the group principally derives from their unique biology, anatomy, and evolutionary relationship to other insects.

Zorapterans are generally less than 3 mm (0.11 in.) in total body length and can be distinguished by their characteristic two-segmented tarsi; unsegmented cerci; nine-segmented antennae; dehiscent, paddle-shaped wings with uniquely reduced pattern of venation; greatly enlarged hind femora bearing rows of stiff spines along their undersurfaces; asymmetrical male genitalia; and vestigial ovipositor. Individuals of each species occur in two morphs: eyed and winged forms that shed their wings after dispersal (becoming what are called "dealates"), or blind and wingless forms that predominate in colonies (see illustration). Development progresses through a series of nymphal stages. The relatively soft integument is pale in nymphal stages and typically reddish brown in adults.

Zoraptera are gregarious, living in small colonies of 15–120 individuals in crevices or under bark of moist, decaying logs. Species prefer rotting wood that has decomposed to the point that logs can be easily torn through by hand or with an ordinary garden tool. Colonies harbor in naturally formed spaces in logs that are not reached by light. Species feed principally on fungal hyphae and spores but can also be generalist scavengers or predators, victimizing nematodes, mites, or other tiny arthropods.

Zorapterans are notable for having only 32 modern species and six fossil species known, these classified into two genera (*Zorotypus* and *Xenozorotypus*) in a single family. Once believed to be related to termites or booklice, current hypotheses associate the zorapterans with webspinners (Embiodea, Embiidina, Embioptera), as the two groups share a unique pattern of musculature in the hind legs, a reduced anal region in the wings, occurrence of wingless morphs, and a gregarious biology. Zoraptera are principally distributed pantropically, with only four species occurring north of the Tropic of Cancer: two species in North America and two in Tibet. Fossil records for Zoraptera are all from ambers formed in warm, tropical paleoclimates. *See* INSECTA; ISOPTERA; PSOCOPTERA. Michael S. Engel

Adult female of *Zorotypus hubbardi* (blind and wingless morph). (*Modified from D. Grimaldi and M. S. Engel, Evolution of the Insects, Cambridge University Press, 2005; photo © M. S. Engel*)

Zosterophyllopsida An extinct class of the division Lycophyta, comprising primitive vascular land plants that evolved from the rhyniophytes and reached a relatively brief peak in the late Early Devonian, when they were an ecologically important and widely distributed element of the exclusively herbaceous terrestrial flora.

The Zosterophyllopsida consists of approximately 18 genera, although few of these include species that have been fully reconstructed from their disarticulated organs. All genera exhibited rhizomatous growth. The rhizomes produced adventitious roots and upright aerial branches, which were typically densely crowded. The group is united by the key lycophyte characters of exarch protoxylem maturation and laterally borne, vascularized, kidney-shaped sporangia with multilayered walls and complete distal dehiscence.

The most primitive generally accepted lycophyte is *Zosterophyllum*. *Sawdonia* and *Gosslingia* are more advanced zosterophyllopsids, possessing characters that are absent from the lycopsids. Although briefly successful as dense stands on floodplains and channel margins, these more sophisticated zosterophyllopsids were virtually extinct by the end of the Devonian. *See* LYCOPHYTA; LYCOPSIDA. Richard M. Bateman; William A. DiMichele

Zygnematales A large order of green algae (Chlorophyceae) that is characterized by the lack of flagellate cells. Sexual reproduction is effected by the fusion of ameboid or passive gametes in conjugation tubes, which give rise to the alternate name Conjugales. Biochemical and ultrastructural features indicate that the Zygnematales are more closely related to charophytes than they are to most other green algae. Members of this order are essentially restricted to fresh-water and subaerial habitats. The order Zygnematales is divided into two suborders, Zygnematineae and Desmidiineae. Paul C. Silva; Richard L. Moe

Zygomycetes A class of terrestrial fungi in the phylum Zygomycota, reproducing sexually with a zygospore and asexually by sporangiospore or conidium. Zygomycetes can be distinguished by the production of sporangia (sacs) that undergo internal cleavage to produce multiple sporangiospores as opposed to the production of conidiospores as observed in most other groups of fungi (Ascomycota and Basidiomycota). A notable exception is the order Entomophthorales, which reproduces asexually by conidium formation. Zygomycetes are all microscopic with the exception of *Endogone* and relatives that produce a zygospore-containing structure called a sporocarp. Sexual reproduction results in the formation of a thick-walled zygosporangium following the conjugation of morphologically relatively similar, undifferentiated gametangia (single-cell structures functioning as gametes). *See* EUMYCOTA; FUNGI; ZYGOMYCOTA.

The Zygomycetes comprise six orders: Dimargaritales, Endogonales, Entomophthorales, Kickxellales, Mucorales, Zoopagales. These orders span a great diversity of ecological habitat. The Dimargaritales and Zoopagales contain many species which parasitize other fungi (mycoparasites). The Entomophthorales are specialized as pathogens primarily of insects. The Kickxellales contain both mycoparasitic species and saprobic species (especially found on soil and animal dung). Endogonales are unique in that some species associate with plant roots as ectomycorrhizae in what is likely to be a mutualistic relationship. The Mucorales is the largest order containing the most commonly encountered species. These species are primarily saprobic (live on decaying organic matter) in dung and soil but are notorious for spoilage of fruits during storage. The Mucorales include some of the fastest-growing fungi. Timothy James

Zygomycota A phylum of filamentous, microscopic fungi that undergo sexual reproduction by the formation of a zygospore. The Zygomycota are often referred by the common names "pin molds" or "sugar molds" and are ubiquitous fungi though rarely observed by humans because of the diminutive size of their reproductive structures. The Zygomycota have a large diversity of ecological roles: as saprophytes on various substrates, especially sugary substrates and dung, as parasites of animals or plants, as commensualistic (both nonpathogenic and nonbeneficial) associates of arthropods, as parasites of other fungi, and as mycorrhizal symbionts with plant roots. *See* MYCORRHIZAE.

As with other fungi, members of the Zygomycota reproduce through spores. Asexual reproduction appears to be the dominant mode of reproduction in most species; sexual reproduction may be rare, or lacking altogether. Asexual reproduction occurs through the production of sporangiospores produced by the internal cleavage of saclike structures termed sporangia. Sporangia are usually borne at the end of a stalk (the sporangiophore) and are dispersed by air or water following the dehiscence (bursting open) of the sporangia wall. Entomophthorales are an exception and reproduce through the production of forcibly discharged conidia.

The Zygomycota may be divided into two classes: the Zygomycetes and the Trichomycetes. The Trichomycetes are distinguished from the Zygomycetes by the production of a specialized spore, the trichospore, which is a sporangiospore with an appendage to aid in aquatic dispersal. The Trichomycetes comprise those Zygomycota which have a commensual relationship with arthropods. *See* EUMYCOTA; FUNGI; ZYGOMYCETES.

Timothy James

Zygophyllales An order of flowering plants in the eurosid I group of the rosid eudicots. This order comprises just two small families of distant relationship (as evidenced by DNA sequence studies) to all other rosid eudicots: Krameriaceae and Zygophyllaceae. Krameriaceae (25 species) are photosynthetic parasitic plants ("hemiparasites") found from the southwestern United States to South America, whereas Zygophyllaceae (240 species) are free-living and widely distributed in arid to semiarid zones throughout the warmer regions of the world. Some species of the latter family produce especially hard woods such as lignumvitate (*Bulnesia* and *Guaiacum*), and a few others produce medicinal compounds, such as the creosote bush (*Larrea*). *See* FLOWER; LIGNUMVITAE; MAGNOLIOPHYTA; MAGNOLIOPSIDA; POLYGALALES; ROSIDAE; SAPINDALES. Mark W. Chase

Appendix

Contributors

Index

Bibliographies

ACOUSTICS

Berg, R. E., and D. G. Stork, *The Physics of Sound,* 3d ed., 2004.

Blackstock, D. T., *Fundamentals of Physical Acoustics,* 2000.

Crocker, M. J. (ed.)**,** *Handbook of Acoustics,* 1998.

Hall, D. E., *Musical Acoustics,* 3d ed., 2001.

Kinsler, L. E., et al., *Fundamentals of Acoustics,* 4th ed., 2000.

Levy, M., H. E. Bass, and R. R. Stern (eds.)**,** *Handbook of Elastic Properties of Solids, Liquids, and Gases* (4-vol. set), 2000.

Mason, W. P., R. N. Thurston, and A. D. Pierce (eds.)**,** *Physical Acoustics: Principles and Methods,* vols. 1–25, 1964–1999.

Mechel, F. P., *Formulas of Acoustics,* 2d ed., 2008.

Pierce, A. D., *Acoustics: An Introduction to Its Physical Principles and Applications,* 1981, reprint 1989.

Rossing, T. D., R. F. Moore, and P. A. Wheeler, *The Science of Sound,* 3d ed., 2001.

Speaks, C. E., *Introduction to Sound Acoustics for Hearing and Speech Sciences,* 3d ed., 1999.

Urick, R. J., *Principles of Underwater Sound,* 3d ed., 1983, reprint 1996.

Journal of the Acoustical Society of America**,** American Institute of Physics, monthly.

AERODYNAMICS

Anderson, J. D., Jr., *Fundamentals of Aerodynamics,* 4th ed., 2006.

Bertin, J. J., and R. M. Cummings, *Aerodynamics for Engineers,* 5th ed., 2008.

Houghton, E. L., and P. W. Carpenter, *Aerodynamics for Engineering Students,* 5th ed., 2001.

Kuethe, A. M., and C. Y. Chow, *Foundations of Aerodynamics: Bases of Aerodynamic Design,* 5th ed., 1997.

Smith, H. C., *The Illustrated Guide to Aerodynamics,* 2d ed., 1992.

Wegener, P. P., *What Makes Airplanes Fly?: History, Science and Applications of Aerodynamics,* 2d ed., 1997.

AERONAUTICAL ENGINEERING

Anderson, J. D., *Aircraft Performance and Design,* 1999.

Anderson, J. D., *Fundamentals of Aerodynamics,* 4th ed., 2006.

Anderson, J. D., Jr., *Introduction to Flight,* 6th ed., 2008.

Barnard R. H., and D. R. Philpott, *Aircraft Flight,* 2d ed., 1995.

Bertin, J. J., and R. M. Cummings, *Aerodynamics for Engineers,* 5th ed., 2008.

Davies, M. (ed.)**,** *The Standard Handbook for Aeronautical and Astronautical Engineers,* 2003.

Etkin, B., and L. D. Reid, *Dynamics of Flight: Stability and Control,* 3d ed., 1996.

Hancock, G. J., *An Introduction to the Flight Dynamics of Rigid Aeroplanes,* 1995.

Kermode, A. C., R. H. Barnard, and D. R. Philpott, *Mechanics of Flight,* 11th ed., 1996.

Kroes, M. J., and J. R. Rardon, *Aircraft: Basic Science,* 7th ed., 1993.

Megson, T. H. G., *Aircraft Structures for Engineering Students,* 4th ed., 2007.

Schmidt, L. V., *Introduction to Aircraft Flight Dynamics,* 1998.

Wagtendonk, W. J., *Principles of Helicopter Flight,* 2d ed., 2007.

AIAA Journal**,** American Institute of Aeronautics and Astronautics, monthly.

Aviation and Aerospace Almanac**,** Aviation Daily and Aerospace Daily, yearly.

Aviation Week and Space Technology**,** weekly.

AGRICULTURE

Adams, C. R., and M. P. Early, *Principles of Horticulture,* 4th ed., 2004.
Barnes, R. F., *Forages, Volume II: The Science of Grassland Agriculture,* 6th ed., 2007.
Blakely, J., and D. H. Bade, *The Science of Animal Husbandry,* 6th ed., 1993.
Ensminger, M. E. (ed.)**,** *Animal Science,* 9th ed., 1991.
Finch, H. J. S., A. M. Samuel, and G. P. F. Lane, *Lockhart and Wiseman's Crop Husbandry,* 8th ed., 2002.
Gliessman, S. R., *Agroecosystem Sustainability: Developing Practical Strategies,* 2001.
Martin, J. H., et al., *Principles of Field Crop Production,* 4th ed., 1986.
Miller, D. A., *Forage Crops,* 1984.
Sparks, D. L. (ed.)**,** *Advances in Agronomy,* vols. 47–98, 1992–2008.

Agronomy Journal, American Society of Agronomy, bimonthly.
Journal of Agricultural Science (England), bimonthly.
Journal of Animal Science, American Society of Animal Science, monthly.

ALGEBRA

Birkhoff, G., and S. MacLane, *A Survey of Modern Algebra,* 5th ed., 1996.
Bittinger, M. L., *Introductory Algebra,* 10th ed., 2006.
Blitzer, R. F., *College Algebra,* 4th ed., 2006.
Durbin, J. R., *Modern Algebra: An Introduction,* 5th ed., 2005.
Fraleigh, J. B., *A First Course in Abstract Algebra,* 7th ed., 2002.
Hungerford, T. W., *Algebra* (Graduate Texts in Mathematics), 1974, reprint 2003.
Jacobson, N., *Basic Algebra,* vol. 1, 2d ed., 1985, vol. 2, 2d ed., 1989.
Kaufman, J. E., and K. L. Schwitters, *Algebra for College Students,* 8th ed., 2006.
Lang, S., *Algebra,* 3d ed., 1993, reprint 2002.
Lial, M. L., J. Hornsby, and T. McGinnis, *Beginning Algebra,* 10th ed., 2007.
Siever, N., *Intermediate Algebra,* 1990.
Streeter, J., D. Hutchinson, and L. Hoelzle, *Beginning Algebra,* 4th ed., 1998.
Sullivan, M., *College Algebra,* 8th ed., 2007.

Journal of Algebra, semimonthly.
Journal of Pure and Applied Algebra, approximately 18 issues per year.

ANALYTICAL CHEMISTRY

Bard, A. J., and I. Rubenstein (eds.)**,** *Electroanalytical Chemistry: A Series of Advances,* vol. 22, 2003.
Bockris, J. O., et al. (eds.)**,** *Modern Aspects of Electrochemistry,* vol. 34, 2001.
Christian, G. D., *Analytical Chemistry,* 6th ed., 2003.
Crow, D. R., *Principles and Applications of Electrochemistry,* 4th ed., 1994.
Haines, P. J. (ed.)**,** *Principles of Thermal Analysis and Calorimetry,* 2002.
Harris, D. C., *Quantitative Chemical Analysis,* 7th ed., 2006.
Patnaik, P., *Dean's Analytical Chemistry Handbook,* 2d ed., 2004.
Settle, F. A. (ed.)**,** *Handbook of Instrumental Techniques for Analytical Chemistry,* 1997.
Skoog, D. A., F. J. Holler, and S. R. Crouch, *Principles of Instrumental Analysis,* 6th ed., 2006.
Skoog, D. A., et al., *Fundamentals of Analytical Chemistry,* 8th ed., 2003.
Wang, J., *Analytical Electrochemistry,* 3d ed., 2006.

American Laboratory, monthly.
Analytical Chemistry, American Chemical Society, semimonthly.
Journal of the Electrochemical Society, monthly.
Trends in Analytical Chemistry, Elsevier, monthly.

ANATOMY

Marieb, E. N., *Human Anatomy and Physiology,* 7th ed., 2006.
Martini, F. H., *Fundamentals of Anatomy and Physiology,* 7th ed., 2005.
Schlossberg, L., and G. D. Zuidema, *The Johns Hopkins Atlas of Human Functional Anatomy,* 4th ed., 1997.
Senisi Scott, A., and E. Fong, *Body Structures and Functions,* 11th ed., 2009.
Standring, S., *Gray's Anatomy,* 40th ed., 2009.
Tortora, G., *Principles of Anatomy and Physiology,* 11th ed., 2006.

Anatomical Record, American Association of Anatomists, monthly.
Journal of Anatomy (England), monthly.

ANTHROPOLOGY AND ARCHEOLOGY

Campbell, B. G., T. D. Loy, and K. Cruz-Uribe, *Humankind Emerging,* 9th ed., 2005.
Ember, C. R., M. Ember, and P. Peregrine, *Anthropology,* 12th ed., 2006.
Fagan, B., *World Prehistory: A Brief Introduction,* 7th ed., 2008.
Fagan, B. M., *People of the Earth,* 12th ed., 2006.
Jurmain, R., et al., *Introduction to Physical Anthropology,* 9th ed., 2003.
Klein, R. G., and B. Edgar, *The Dawn of Human Culture,* 2002.
Renfrew, C., and P. Bahn, *Archaeology: Theories, Methods and Practice,* 4th ed., 2004.
Strier, K. B., *Primate Behavioral Ecology,* 3d ed., 2006.
Thomas, D. H., and R. L. Kelly, *Archaeology,* 5th ed., 2009.
Whitten, P., and D. E. Hunter, *Anthropology: Contemporary Perspectives,* 8th ed., 2000.

Evolutionary Anthropology**,** bimonthly.

ARCHITECTURAL ENGINEERING

Allen, E., *How Buildings Work: The Natural Order of Architecture,* 3d ed., 2005.
Ambrose, J. E., *Building Structures,* 2d ed., 1993.
Ching, F. D. K., *Building Construction Illustrated,* 4th ed., 2008.
Guthrie, P., *The Architect's Portable Handbook,* 3d ed., 2003.
Harris, C. M. (ed.)**,** *Dictionary of Architecture and Construction,* 4th ed., 2005.
Roth L. M., *Understanding Architecture: Its Elements, History, and Meaning,* 2d ed., 2006.

Architectural Record, McGraw-Hill, monthly.
Journal of Architectural Engineering, American Society of Civil Engineers, quarterly.

ASTRONOMY

Beatty, J. K., C. C. Petersen, and A. L. Chaikin (eds.), *The New Solar System,* 4th ed., 1999.
Chaisson, E., and S. McMillan, *Astronomy Today,* 6th ed., 2007.
Foukal, P. V., *Solar Astrophysics,* 2d ed., 2004.
Fraknoi, A., D. Morrison, and S. Wolff, *Voyages Through the Universe,* 3d ed., 2004.
Golub, L., and J. M. Pasachoff, *Nearest Star: The Surprising Science of Our Sun,* 2001.
Kutner, M., *Astronomy: A Physical Perspective,* 2d ed., 2003.
Pasachoff, J. M., *Astronomy: From the Earth to the Universe,* 6th ed., 2002.
Pasachoff, J. M., and A. Filippenko, *The Cosmos: Astronomy in the New Millennium,* 3d ed., 2006.
Seeds, M., *Foundations of Astronomy,* 10th ed., 2008.
Zeilik, M., *Astronomy: The Evolving Universe,* 9th ed., 2002.

Annual Review of Astronomy and Astrophysics, annually.
Astronomical Calendar, Universal Workshop, yearly.
Astronomical Journal, American Astronomical Society, monthly.
Astronomy, monthly.
Astrophysical Journal, American Astronomical Society, three issues per month.
Mercury, Astronomical Society of the Pacific, monthly.
Monthly Notices of the Royal Astronomical Society, three issues per month.
Observers' Handbook, Royal Astronomical Society of Canada, annually.
The Planetary Report, The Planetary Society, bimonthly.
Sky & Telescope, monthly.

ATOMIC PHYSICS

Adam, S., *Frontiers: Twentieth Century Physics,* 2000.
Beiser, A., *Concepts of Modern Physics,* 6th ed., 2003.
Brehm, J. J., and W. J. Mullin, *Introduction to the Structure of Matter: A Course in Modern Physics,* 1989.
Cowan, R. D., *The Theory of Atomic Structure and Spectra,* 1981.
Haken, H., H. C. Wolf, and W. D. Brewer, *The Physics of Atoms and Quanta: Introduction to Experiments and Theory,* 7th ed., 2005.
Ohanian, H. C., *Modern Physics,* 2d ed., 1996.
Serway, R. A., C. J. Moses, and C. A. Moyer, *Modern Physics,* 3d ed., 2004.
Thornton, S. T., and A. Rex, *Modern Physics for Scientists and Engineers,* 3d ed., 2005.
Yang, F., and J. H. Hamilton, *Modern Atomic and Nuclear Physics,* 1996.

Physical Review A, American Physical Society, monthly.

AUTOMOTIVE ENGINEERING

Braess, H.-H., and U. Seiffert (eds.), *Handbook of Automotive Engineering,* 2005.
Robert Bosch GmbH, *Automotive Handbook,* 7th ed., 2007.
Duffy, J. E., *Modern Automotive Technology,* 2003.

Automotive Engineering, Society of Automotive Engineers, International (CSAE), monthly.

BIOCHEMISTRY

Alberts, B., et al., *Molecular Biology of the Cell,* 5th ed., 2007.
Devlin, T. M. (ed.), *Textbook of Biochemistry,* 6th ed., 2005.
Elliott, W. H., and D. C. Elliott, *Biochemistry and Molecular Biology,* 3d ed., 2005.
Horton, H. R., et al., *Principles of Biochemistry,* 4th ed., 2005.
Murray, R. K., et al., *Harper's Illustrated Biochemistry,* 27th ed., 2006.
Nelson, D., and M. Cox, *Lehninger Principles of Biochemistry,* 5th ed., 2008.
Stipanuk, M. H., *Biochemical and Physiological Aspects of Human Nutrition,* 2000.
Stryer, L., *Biochemistry Extended,* 2002.
Van Holde, K. E., C. Johnson, and P. Ho, *Principles of Physical Biochemistry,* 2d ed., 2005.
Voet, D. J., J. G. Voet, and C. Pratt, *Fundamentals of Biochemistry: Life at the Molecular Level,* 3d ed., 2008.

Annual Review of Biochemistry, annually.
Archives of Biochemistry and Biophysics, biweekly.
Biochemical Journal, biweekly.
Biochimica et Biophysica Acta, 100+ issues per year.
FEBS Journal, semimonthly.
Journal of Biological Chemistry, American Society of Biological Chemists, weekly.
Trends in Biochemical Sciences, monthly.

BIOMEDICAL ENGINEERING

Bronzino, J. D. (ed.), *The Biomedical Engineering Handbook,* 3d ed., 2006.
Domach, M. F., *Introduction to Biomedical Engineering,* 2004.
Dyro, J. F. (ed.), *Clinical Engineering Handbook,* 2004.
Khandpur, R. S., *Biomedical Instrumentation,* 2005.
Kutz, M. (ed.), *Standard Handbook of Biomedical Engineering & Design,* 2003.

Annals of Biomedical Engineering, monthly.
Annual Review of Biomedical Engineering, annually.
IEEE Transactions on Biomedical Engineering, monthly.

BIOPHYSICS

Austin, R. H., and S. Chan, *Biophysics for Physicists,* 2003.
Cerdonio, M., and R. W. Noble, *Introductory Biophysics,* 1988.
Kreighbaum, E., and K. Barthels, *Biomechanics: A Qualitative Approach for Studying Human Movements,* 4th ed., 1995.
Sybesma, C. (ed.), *Biophysics: An Introduction,* 1989.
Valenta, J. (ed.), *Biomechanics,* 1993.
Wainwright, S. A., et al., *Mechanical Design in Organisms,* 1982.

Annual Review of Biophysics and Bioengineering, annually.
Biophysical Journal, monthly.

CALCULUS

Berkey, D. D., and P. Blanchard, *Calculus of One Variable,* 3d ed., 1997.
Courant, R., and F. John, *Introduction to Calculus and Analysis,* 2 vols., 1989.
Grossman, S. I., and R. B. Lane, *Calculus,* 1993.
Kaplan, W., *Advanced Calculus,* 5th ed., 2002.
Lang, S., *A First Course in Calculus,* 5th ed., 1986.
Larson, R., B. H. Edwards, and R. P. Hostetler, *Calculus with Analytic Geometry,* 8th ed., 2005.
Salas, S. L., E. Hille, and G. Etgen, *Calculus,* 9th ed., 2003.
Taylor, A. E., and W. R. Mann, *Advanced Calculus,* 3d ed., 1982.
Thomas, G. B., and R. L. Finney, *Calculus and Analytic Geometry,* 9th ed., 1996.

CELESTIAL MECHANICS

See Astronomy.

CELL AND MOLECULAR BIOLOGY

Alberts, B., et al., *Molecular Biology of the Cell,* 5th ed., 2007.
Becker, W. M., *The World of the Cell,* 7th ed., 2008.
Calladine, C. R., et al., *Understanding DNA: The Molecule and How It Works,* 3d ed., 2004.
Lodish, H., et al., *Molecular Cell Biology,* 6th ed., 2007.
Pollard, T., *Cell Biology,* 2d ed., 2002.
Watson, J. D., et al., *Molecular Biology of the Gene,* 6th ed., 2007.

Cell, biweekly.
Current Opinion in Cell Biology, bimonthly.
EMBO Journal, European Molecular Biology Organization, semimonthly.
Journal of Cell Biology, biweekly.
Journal of Molecular Biology, weekly.
Molecular and Cellular Biology, American Society for Microbiology, semimonthly.
Molecular Biology, bimonthly.
Molecular Biology of the Cell, monthly.
Nature Cell Biology, monthly.
Trends in Cell Biology, monthly.

CHEMICAL ENGINEERING

Bailey, J. E., and D. F. Ollis, *Biochemical Engineering Fundamentals,* 2d ed., 1986.
Espenson, J. H., *Chemical Kinetics and Reaction Mechanisms,* 2d ed., 1995.
Fogler, H. S., *Elements of Chemical Reaction Engineering,* 4th ed., 2005.
Furter, W. F. (ed.)**,** *A Century of Chemical Engineering,* 1981.
Green, D. W. and, R. H. Perry, *Perry's Chemical Engineers' Handbook,* 8th ed., 2007.
Himmelblau, D. M., and J. B. Riggs, *Basic Principles and Calculations in Chemical Engineering,* 7th ed., 2003.
Humphrey, J. L., and G. E. Keller II, *Separation Process Technology,* 1997.
Kirk-Othmer Encyclopedia of Chemical Technology, 5th ed., 27 vols., 2007.
McCabe, W. L., J. Smith, and P. Harriott, *Unit Operations of Chemical Engineering,* 7th ed., 2005.
Miller, R. W., *Flow Measurement Engineering Handbook,* 3d ed., 1996.
Schweitzer, P. A., *Handbook of Separation Techniques for Chemical Engineers,* 3d ed., 1997.

Biotechnology Progress, American Institute of Chemical Engineers, bimonthly.
Chemical and Engineering News, American Chemical Society, weekly.
Chemical Engineering, monthly.
Chemical Engineering Progress, American Institute of Chemical Engineers, monthly.
Industrial & Engineering Chemistry Research, American Chemical Society, biweekly.

CIVIL ENGINEERING

Ambrose, J., *Simplified Design of Concrete Structures,* 8th ed., 2007.
Chen, W. F. and E. M. Lui (eds.)**,** *Handbook of Structural Engineering,* 2d ed., 2005.
Chopra, A. K., *Dynamics of Structures: Theory and Applications to Earthquake Engineering,* 3d ed., 2006.
Mamlouk, M. S., and J, P. Zaniewski, *Materials for Civil and Construction Engineers,* 2d ed., 2005.
Merritt, F. S., M. K. Loftin, and J. T. Ricketts (eds.)**,** *Standard Handbook for Civil Engineering,* 5th ed., 2003.
Paz, M., and W. Leigh, *Structural Dynamics: Theory and Computation,* 5th ed., 2006.
Scott, J. S., *Dictionary of Civil Engineering,* 4th ed., 1993.
Toneas, D. E., and J. J. Zhao, *Bridge Engineering,* 2d ed., 2006.

Civil Engineering, American Society of Civil Engineers, monthly.

CLASSICAL MECHANICS

Barger, V. D., and M. G. Olsson, *Classical Mechanics: A Modern Perspective,* 2d ed., 1995.
Goldstein, H., C. P. Poole, and J. L. Safko, *Classical Mechanics,* 3d ed., 2002.
Greiner, W., *Classical Mechanics: Point Particles and Relativity,* 2004.
Halliday, D., R. Resnick, and J. Walker, *Fundamentals of Physics,* 8th ed., 2007.
José, J. V., and E. J. Saletan, *Classical Dynamics, A Contemporary Approach,* 1998.
Kibble, T. W. B., and F. H. Berkshire, *Classical Mechanics,* 5th ed., 2004.
Morin, D., *Introduction to Classical Mechanics,* 2008.
Sussman, G. J., and J. Wisdom, *Structure and Interpretation of Classical Mechanics,* 2001.
Taylor, J. R., *Classical Mechanics,* 2005.

Archive for Rational Mechanics and Analysis, 16 times a year.
Quarterly Journal of Mechanics and Applied Mathematics, quarterly.

CLIMATOLOGY AND METEOROLOGY

Ahrens, C. D., *Meteorology Today,* 9th ed., 2008.
Finlayson-Pitts, B. J., and J. N. Pitts, Jr., *Chemistry of the Upper and Lower Atmosphere: Theory, Experiments, and Applications,* 1999.
Holton, J. R., *An Introduction to Dynamic Meteorology,* 4th ed., 2004.
Houghton, J., *Global Warming: The Complete Briefing,* 3d ed., 2004.
Intergovernmental Panel on Climate Change, *Climate Change 2007: Impacts, Adaptation and Vulnerability,* 2008.
Liou, K. N., *An Introduction to Atmospheric Radiation,* 2d ed., 2002.
Lutgens, F. K., et al., *The Atmosphere: An Introduction to Meteorology,* 10th ed., 2006.
Pruppacher, H. R., and J. D. Klett, *Microphysics of Atmospheric Clouds and Precipitation,* 2d ed., 1997.
Wallace, J. M., and P. V. Hobbs, *Atmospheric Science,* 2d ed., 2006.
Wang, P. K., *Ice Microdynamics,* 2002.

***Bulletin of the American Meteorological Society*,** American Meteorological Society, monthly.
***Journal of the Atmospheric Sciences*,** American Meteorological Society, semimonthly.
***Journal of Climate*,** American Meteorological Society, monthly.
***Journal of Applied Meteorology*,** American Meteorological Society, monthly.

COMMUNICATIONS

See Data communications; Radio communication; Telecommunications.

COMPUTER SCIENCE

Abelson, H., G. Sussman, and J. Sussman, *Sturcture and Interpretation of Computer Science,* 2d ed., 1997.
Hennessy, J. L., and D. A. Patterson, *Computer Organization and Design,* 4th ed., 2006.
Horowitz, E., S. Sahni, and S. Rajasekaran, *Computer Algorithms,* 2d ed., 2007.
Long, L., and N. Long, *Computers: Information Technology in Perspective,* 11th ed., 2003.
Maxfield, C., *Bebop to Boolean Boogie,* 3d ed., 2008.
Null, L., and J. Lobur, *The Essentials of Computer Organization and Architecture,* 2d ed., 2006.
Patterson, D. A., and J. L. Hennessy, *Computer Organization and Design,* 3d ed., 2007.
Ralston, A., and E. D. Reilly (eds.)**,** *Encyclopedia of Computer Science and Engineering,* 4th ed., 2003.
Sebesta, R., *Concepts of Programming Languages,* 8th ed., 2007.
Sommerville, I., *Software Engineering,* 8th ed., 2006.
Tanenbaum, A., *Modern Operating Systems,* 3d ed., 2007.
Tjaden, B. C., *Fundamentals of Secure Computing Systems,* 2002.
Tocci, R. J., and F. J. Ambrosio, *Microprocessors and Microcomputers: Hardware and Software,* 6th ed., 2002.

***Communications of the ACM*,** monthly.
Computer and Control Abstracts (Science Abstracts, Section C), INSPEC, Institution of Electrical Engineers, monthly.
***IEEE Spectrum*,** Institute of Electrical and Electronics Engineers, Inc., monthly.
***Journal of the ACM*,** bimonthly.

CONSERVATION

Chiras, D. D., J. P. Reganold, and O. S. Owen, *Natural Resource Conservation: Management for a Sustainable Future,* 9th ed., 2004.
Dasmann, R. F., *Environmental Conservation,* 6th ed., 2004.
Frankham, R., J. D. Ballou, and D. A. Briscoe, *A Primer of Conservation Genetics,* 2004.
Hambler, C., *Conservation* (Studies in Biology), 2004.
Meffe, G. K., and C. R. Carroll, *Principles of Conservation Biology,* 3d ed., 2005.
Primack, R. B., *Essentials of Conservation Biology,* 2004.

***Audubon Magazine*,** Audubon Society, bimonthly.
***Conservation Biology*,** bimonthly.
***New York State Conservationist*,** New York State Department of Environmental Conservation, bimonthly.

CONTROL SYSTEMS

Clark, R. N., *Control System Dynamics,* 1996.
Close, C. M., D. K. Frederick, and J. C. Newell, *Modeling and Analysis of Dynamic Systems,* 3d ed., 2002.
D'Azzo, J. J., C. H. Houpis, and S. N. Sheldon, *Linear Control System Analysis and Design,* 5th ed., 2003.
Dorf, R. C., and R. H. Bishop, *Modern Control Systems,* 11th ed., 2007.
Kuo, B. C., and F. Golnaraghi, *Automatic Control Systems,* 8th ed., 2002.
Lewis, F. L., and V. L. Symos, *Optimal Control,* 2d ed., 1995.
Nise, N. S., *Control Systems Engineering,* 5th ed., 2007.
Phillips, C. L., and R. Harbor, *Feedback Control Systems,* 4th ed., 1999.
Raven, F., *Automatic Control Engineering,* McGraw-Hill Series in Mechanical Engineering, 5th ed., 1994.

Computer and Control Abstracts (Science Abstracts, Section C0), INSPEC, Institution of Electrical Engineers, monthly.
***Control Engineering*,** monthly.

COSMOLOGY

Kirshner, R. P., *The Extravagant Universe,* 2002.
Lemonick, M., *Echo of the Big Bang,* 2003.
Rees, M., *Our Cosmic Habitat,* 2001.
Rowan-Robinson, M., *Cosmology,* 4th ed., 2003.
Steinhardt, P. J., and N. Turok, *Endless Universe: Beyond the Big Bang,* 2007.
Stern, S. A., *Our Universe: The Thrill of Extragalactic Exploration as Told by Leading Experts,* 2001.
Weinberg, S., *Cosmology,* 2008.

CRYSTALLOGRAPHY

Borchardt-Ott, W., *Crystallography,* 2d ed., 1995.
Clegg, W. (ed.)**,** *Crystal Structure Analysis: Principles and Practice,* 2001.

De Graef, M., and M. E. McHenry, *Structure of Materials: An Introduction to Crystallography, Diffraction and Symmetry,* 2007.
Giacovazzo, C. (ed.), *Fundamentals of Crystallography,* 2d ed., 2002.
Glusker, J. P., M. Lewis, and M. Rossi, *Crystal Structure Analysis for Chemists and Biologists,* 1994.
Guinier, A., *X-Ray Diffraction,* 1963, reprint 1994.
Hammond, C., *The Basics of Crystallography and Diffraction* (International Union on Crystallography Texts on Crystallography), 2d ed., 2001.
Klug, H. P., and L. E. Alexander, *X-Ray Diffraction Procedures,* 2d ed., 1974.
Rhodes, G., *Crystallography Made Clear: A Guide for Users of Macromolecular Models,* 2006.

Acta Crystallographica, Sections A, B, and D, bimonthly; Section C, monthly.

DATA COMMUNICATIONS

Bertsekas, D. P., and R. Gallager, *Data Networks,* 2d ed., 1991.
De Prycker, M., *Asynchronous Transfer Mode: Solution for Broadband ISDN,* 3d ed., 1995.
Forouzan, B. A., *Data Communications and Networking,* 4th ed., 2006.
Ginsburg, D., *ATM Solutions for Enterprise Internetworking,* 2d ed., 1999.
Halsall, F., *Data Communications, Computer Networks and Open Systems,* 4th ed., 1996.
Kurose, J. F., and K. W. Ross, *Computer Networking: A Top-Down Approach Featuring the Internet,* 3d ed., 2005.
Stallings, W., *Data and Computer Communications,* 8th ed., 2006.
Tanenbaum, A. S., *Computer Networks,* 4th ed., 2003.
White, C. M., *Data Communications and Computer Networks: A Business Users Approach,* 5th ed., 2008.

Data Communications, 18 issues per year.

DEVELOPMENTAL BIOLOGY

Beysens, D., G. Forgacs, and F. Gaill, *Interplay of Genetic and Physical Processes in the Development of Biological Form,* 1995.
Browder, L. W., C. A. Erickson, and W. R. Jeffrey, *Developmental Biology,* 4th ed., 2001.
Carlson, B. M., *Human Embryology and Developmental Biology,* 3d ed., 2004.
Gilbert, S. F., *Developmental Biology,* 8th ed., 2006.
Hall, B. K., *Evolutionary Developmental Biology,* 1998.
Oppenheimer, S. B., *Introduction to Embryonic Development,* 4th ed., 2004.
Slack, J. M. W., *Essential Developmental Biology,* 2d ed., 2005.
Vander Zanden, J. W., et al., *Human Development,* 8th ed., 2006.

Current Opinion in Genetics and Development, bimonthly.
Development Genes and Evolution, monthly.
Developmental Biology, semimonthly.
Excerpta Medica, Section 21: Developmental Biology and Teratology, monthly.
Genes and Development, Cold Spring Harbor Laboratory Press, semimonthly.

ECOLOGY

Bolen, E. G., and W. L. Robinson, *Wildlife Ecology and Management,* 6th ed., 2008.
Conner, J. K., and D. L. Hartl, *A Primer of Ecological Genetics,* 2004.
Gotelli, N. J., *A Primer of Ecology,* 4th ed., 2008.
Molles, M. C., Jr., *Ecology: Concepts and Applications,* 4th ed., 2008.
Pianka, E. R., *Evolutionary Ecology,* 6th ed., 1999.
Ricklefs, R. E., and G. Miller, *Ecology,* 4th ed., 1999.
Roughgarden, J., *Primer of Ecological Theory,* 1998.
Smith, R. L., and T. Smith, *Ecology and Field Biology,* 6th ed., 2001.
Walker, L. R., and R. del Moral, *Primary Succession and Ecosystem Rehabilitation* (Cambridge Studies in Ecology), 2003.

Annual Review of Ecology, Evolution, and Systematics, annually.
Ecological Monographs, Ecological Society of America, quarterly.
Ecology, Ecological Society of America, 6 issues per year.
Journal of Environmental Sciences, Institute of Environmental Sciences, bimonthly.

ELECTRICAL POWER ENGINEERING

Chapman, S. J., *Electric Machinery Fundamentals* (McGraw-Hill Series in Electrical and Computer Engineering), 4th ed., 2005.
Dorf, R. C., *The Electrical Engineering Handbook,* 6 vols., 2d ed., 2006.
El-Hawary, M. E., *Electrical Energy Systems,* 2000.
El-Hawary, M. E., *Principles of Electric Machines with Power Electronic Applications,* 2002.
Fink, D. G., and H. W. Beaty (eds.), *Standard Handbook of Electrical Engineering,* 15th ed., 2006.
Fitzgerald, A. E., C. Kingsley, and S. Umans, *Electric Machinery,* 6th ed., 2003.
Glover, J. D., M. S. Sarma, and T. Overbye, *Power System Analysis and Design,* 4th ed., 2007.
Grainger, J. J., and W. D. Stevenson, *Power System Analysis,* 1994.
Guru, B. S., and H. R. Hiziroğlu, *Electric Machinery and Transformers,* 3d ed., 2001.
Hambley, A. R., *Electrical Engineering: Principles and Applications,* 4th ed., 2007.
Richardson, D. V., and A. J. Caisse, Jr., *Rotating Electric Machinery and Transformer Technology,* 4th ed., 1996.
Rizzoni, G., *Fundamentals of Electrical Engineering,* 2008.
Saadat, H., *Power System Analysis,* 2d ed., 2004.

C&M (Electrical Construction and Maintenance), monthly.
Electric Power and Energy Magazine, IEEE Power Engineering Society, bimonthly.
Electrical and Electronics Abstracts (Science Abstracts, Section B), INSPEC, Institution of Electrical Engineers, monthly.

Electrical World, monthly.
IEEE Spectrum, Institute of Electrical and Electronics Engineers, Inc., monthly.

ELECTRICITY AND MAGNETISM

Fowler, R. J., *Electricity: Principles and Applications,* 7th ed., 2007.
Griffiths, D. J., *Introduction to Electrodynamics,* 3d ed., 1999.
Jackson, J. D., *Classical Electrodynamics,* 3d ed., 1999.
Panofsky, W. K. H., and M. Phillips, *Classical Electricity and Magnetism,* 2d ed., 1962, reprint, 2005.
Purcell, E. M., *Electricity and Magnetism,* Berkeley Physics Course, vol. 2, 2d ed., 1985.
Reitz, J. R., F. J. Milford, and R. W. Christy, *Foundations of Electromagnetic Theory,* 4th ed., 2008.
Schwartz, M., *Principles of Electrodynamics,* reprint, 1987.

Journal of Magnetism and Magnetic Materials, 3 issues per month.

ELECTROMAGNETIC FIELDS AND RADIATION

Cheng, D. K., *Field and Wave Electromagnetics,* 2d ed., 1989.
Guru, B., and H. Hiziroğlu, *Electromagnetic Field Theory Fundamentals,* 2d ed., 2004.
Inan, U. S., and A. S. Inan, *Engineering Electromagnetics,* 1999.
Kraus, J. D., and D. Fleisch, *Electromagnetics with Applications,* 5th ed., 1999.
Lorrain, P., D. R. Corson, and F. Lorrain, *Electromagnetic Fields and Waves,* 3d ed., 1988.
Paul, C. R., K. W. Whites, and S. A. Nasar, *Introduction to Electromagnetic Fields,* 3d ed., 1997.
Popović, Z., and B. D. Popović, *Introductory Electromagnetics,* 2000.
Sadiku, M. N. O., *Elements of Electromagnetics* (Oxford Series in Electrical and Computer Engineering), 4th ed., 2007.
Smith, G. S., *An Introduction to Classical Electromagnetic Radiation,* 1997.

ELECTRONIC CIRCUITS

Bogart, T. F., J. S. Beasley, and J. Rico, *Electronic Devices and Circuits,* 6th ed., 2003.
Boylestadt, R., and L. Nashelsky, *Electronic Devices and Circuit Theory,* 9th ed., 2005.
Burns, S. G., and P. R., Bond, *Principles of Electronic Circuits,* 2d ed., 1997.
Fortney, L. R., *Principles of Electronics: Analog and Digital,* 2005.
Grey, P. R., et al., *Analysis and Design of Analog Integrated Circuits,* 4th ed., 2001.
Malvino, A. P., and D. J. Bates, *Electronic Principles,* 7th ed., 2006.
Millman, J., and A. Grabel, *Microelectronics* (McGraw-Hill Series in Electrical Engineering), 2d ed., 1987.
Paynter, R. T., *Introductory Electronic Devices and Circuits* (Electron Flow Version), 7th ed., 2005.
Rabaey, J. M., A. Chandrakasan, and B. Nikolic, *Digital Integrated Circuits,* 2d ed., 2002.
Reed, D. G., *The AARL Handbook for Radio Communications,* Association for Amateur Radio, annually.
Sedra, A. S., and K. C. Smith, *Microelectronic Circuits* (Oxford Series in Electrical Engineering), 5th ed., 2007.

Electrical and Electronics Abstracts (Science Abstracts, Section B), INSPEC, Institution of Electrical Engineers, monthly.
Electronic Design, biweekly.
Electronics, biweekly.
IEEE Spectrum, Institute of Electrical and Electronics Engineers, Inc., monthly.

ELECTRONICS

See Electronic circuits; Physical electronics.

ELEMENTARY PARTICLES

Aitchison, I. J. R., and A. J. G. Hey, *Gauge Theories in Particle Physics,* 3d ed., 2 vols., 2003.
Allday, J., *Quarks, Leptons and the Big Bang,* 2d ed., 2002.
Barnett, R. M., H. Mühry, and H. R. Quinn, *The Charm of Strange Quarks: Mysteries and Revolutions of Particle Physics,* 2000.
Close, F. E., *Particle Physics: A Very Short Introduction,* 2004.
Cottingham, W. A., and D. A. Greenwood, *An Introduction to the Standard Model of Particle Physics,* 2d ed., 2007.
Coughlan, G. D., J. E. Dodd, and B. M. Gripaios, *The Ideas of Particle Physics: An Introduction for Scientists,* 3d ed., 2006.
Griffiths, D., *Introduction to Elementary Particles,* 2d ed., 2008.
Halzen, F., and A. D. Martin, *Quarks and Leptons: An Introductory Course in Modern Particle Physics,* 1984.
Kane, G., *Modern Elementary Particle Physics: The Fundamental Particles and Forces?,* updated ed., 1993.
Schumm, B. A., *Deep Down Things: The Breathtaking Beauty of Particle Physics,* 2004.
Weinberg, S., *The Discovery of Subatomic Particles,* revised, 2d ed., 2003.

Nuclear Physics, Section B, 81 issues per year.
Physical Review D, American Physical Society, semimonthly.

EMBRYOLOGY

See Developmental biology.

ENGINEERING

See Aeronautical engineering; Architectural engineering; Automotive engineering; Biomedical engineering; Chemical engineering, Civil engineering; Electrical power engineering; Environmental engineering; Food engineering; Genetic engineering; Industrial engineering; Marine engineering; Mechanical engineering; Metallurgical engineering; Mining engineering; Nuclear engineering; Petroleum engineering; Production engineering; Systems engineering; Transportation engineering.

ENVIRONMENTAL ENGINEERING

American Water Works Association, et al., *Water Treatment Plant Design,* 4th ed., 2004.
Corbitt, R. A. (ed.), *Standard Handbook of Environmental Engineering,* 2d ed., 1999.
Freeman, H. M. (ed.), *Standard Handbook of Hazardous Waste Treatment and Disposal,* 2d ed., 1997.
Lee, C. C., and S. D. Lin, *Handbook of Environmental Engineering Calculations,* 2d ed., 2007.
Manahan, S. E., *Environmental Chemistry,* 8th ed., 2004.
Metcalf and Eddy, Inc., *Wastewater Engineering: Treatment, Disposal, and Reuse,* 4th ed., 2002.
Pichtel, J., *Waste Management Practices: Municipal, Hazardous, and Industrial,* 2005.
Salvato, J. A., N. L. Nemerow, and F. J. Agardy, *Environmental Engineering,* 5th ed., 2003.
Vallero, D. A., *Environmental Contaminants: Assessment and Control,* 2004.

The Diplomate, American Academy of Environmental Engineers, quarterly.
Environmental Engineering Science, 10 issues.
Journal of Environmental Engineering, monthly.
Journal of Environmental Sciences, Institute of Environmental Engineers, bimonthly.

EVOLUTION

Bell, G., *Selection: The Mechanisms of Evolution,* 1996.
Futuyma, D. J., *Evolutionary Biology,* 3d ed., 1998.
Hall, B. K., *Evolutionary Developmental Biology,* 2d ed., 1998.
Hall, B. K., and B. Hallgrímsson, *Strickberger's Evolution,* 4th ed., 2008.
Levinton, J. S., *Genetics, Paleontology, and Macroevolution,* 2d ed., 2001.
Li, W.-H., and D. Graur, *Fundamentals of Molecular Evolution,* 2d ed., 2000.
Williams, G. C., *Adaptation and Natural Selection: A Critique of Some Current Evolutionary Thought,* 1996.
Zimmer, C., *Evolution: The Triumph of an Idea,* 2001.

Evolution, Society for the Study of Evolution, monthly.
Journal of Molecular Evolution, monthly.

FLUID MECHANICS

Cengel, Y. A., and J. M. Cimbala, *Fluid Mechanics,* 2005.
Crowe, C. T., et al., *Engineering Fluid Mechanics,* 9th ed., 2008.
Currie, I. G., *Fundamental Mechanics of Fluids,* 3d ed., 2003.
Douglas, J. F., et al., *Fluid Mechanics,* 5th ed., 2006.
Finnemore, J. E., and J. B. Franzini, *Fluid Mechanics with Engineering Applications,* 10th ed., 2001.
Fox, R. W., A. T. McDonald, and P. J. Pritchard, *Introduction to Fluid Mechanics,* 7th ed., 2008.
Kindu, P. K., and I. M. Cohen, *Fluid Mechanics,* 4th ed., 2008.
Mott, R. L., *Applied Fluid Mechanics,* 6th ed., 2005.
Munson, B. R., et al., *Fundamentals of Fluid Mechanics,* 6th ed., 2009.
Shames, I. H., *Mechanics of Fluids,* 4th ed., 2003.
Streeter, V. L., E. B. Wylie, and K. W. Bedford, *Fluid Mechanics,* 9th ed., 1997.
White, F. M., *Fluid Mechanics,* 6th ed., 2006.

Applied Scientific Research, Central National Organization for Applied Scientific Research in the Netherlands, quarterly.
Physics of Fluids, American Institute of Physics, monthly.

FOOD ENGINEERING

Heldman, D. R., and D. B. Lund, *Handbook of Food Engineering,* 2d ed., 2006.
Jay, J. M., M. J. Loessner, and D. A. Golden, *Modern Food Microbiology,* 7th ed., 2006.
Lee, D. S., K. L. Yam, and L. Piergiovanni, *Food Packaging Science and Technology,* 2008.
Singh, R. P., and D. R. Heldman, *Introduction to Food Engineering,* 4th ed., 2008.

Food Chemistry, Elsevier, monthly.
Food Technology, Institute of Food Technologists, monthly.
Journal of Agricultural and Food Chemistry, American Chemical Society, biweekly.

FORENSIC SCIENCE

Fisher, B. A. J., *Techniques of Crime Scene Investigation,* 7th ed., 2004.
Nordby, J. J., *Forensic Science: An Introduction to Scientific and Investigative Techniques,* 2005.
Saferstein, R. E., *Criminalistics: An Introduction to Forensic Science,* 9th ed., 2006.

Forensic Science Communications, FBI, quarterly.
Forensic Science International, Elsevier, semimonthly.
Journal of Analytical Toxicology, Society of Forensic Toxicologists, 8 issues per year.
Journal of Forensic Identification, International Association for Identification, bimonthly.
Journal of Forensic Sciences, American Academy of Forensic Sciences, bimonthly.
Science and Justice, Forensic Science Society, U.K., quarterly.

FORESTRY

Anderson, D., and I. I. Holland (eds.), *Forests and Forestry,* 4th ed., 1997.
Davis, L. S., and K. Norman Johnson, *Forest Management,* 4th ed., 2000.
Landsberg, J. J., and S. T. Gower, *Applications of Physiological Ecology to Forest Management,* 1996.
Matthews, J. D., *Silvicultural Systems,* 1991.
Sharpe, G. W., and C. Hendes, *Introduction to Forestry,* 6th ed., 1995.
Sharpe, G., J. Hendee, and W. Sharpe, *Introduction to Forest and Renewable Resources,* 2003.
Shugart, H. H., *A Theory of Forest Dynamics: The Ecological Implications of Forest Succession Models,* 2003.

Forest Products Journal, Forest Products Society, monthly.
Forest Science, Society of American Foresters, bimonthly.

Journal of Forestry, Society of American Foresters, 8 issues per year.

GENERAL CHEMISTRY

Brady, J. E., and F. Senese, *Chemistry: The Study of Matter and Its Changes,* 5th ed., 2007.
Brown, T. E., et al., *Chemistry: The Central Science,* 11th ed., 2008.
Chang, R., *Chemistry,* 9th ed., 2006.
Corwin, C. H., *Introductory Chemistry: Concepts & Connections,* 5th ed., 2007.
Speight, J. G. (ed.), *Lange's Handbook of Chemistry,* 16th ed., 2004.
Oxtoby, D. W., et al., *Principles of Modern Chemistry,* 6th ed., 2007.
Timberlake, K. C., *Chemistry: An Introduction to General, Organic, and Biological Chemistry,* 10th ed., 2008.
Zumdahl, S. S., and S. Zumdahl, *Chemistry,* 7th ed., 2006.

GENETIC ENGINEERING

Bourgaize, D., T. R. Jewell, and R. G. Buiser, *Biotechnology: Demystifying the Concepts,* 1999.
Nicholl, D. S. T., *An Introduction to Genetic Engineering,* 3d ed., 2008.
Primrose, S., and R. Twyman, *Principles of Gene Manipulation and Genomics,* 7th ed., 2006.
Setlow, J. K. (ed.), *Genetic Engineering: Principles and Methods,* 2002.
Singer, M., and P. Berg, *Exploring Genetic Mechanisms,* 1997.
Watson, J. D., et al., *Recombinant DNA,* 3d ed., 2007.
Weaver, R. F., *Molecular Biology,* 4th ed., 2007.

Biotechnology Journal, monthly.
Biotechnology Progress, bimonthly.
Genetic Engineering and Biotechnology News, biweekly.
Journal of Biotechnology, semimonthly.
Nature Biotechnology, monthly.

GENETICS

Dobzhansky, T., *Dobzhansky's Genetics of Natural Populations,* 2003.
Dobzhansky, T., *Genetics and the Origin of Species,* 1982.
Hall, J. C., J. C. Dunlap, and T. Friedman, *Advances in Genetics,* vol. 56, 2006.
Hartwell, L., et al., *Genetics: From Genes to Genomes,* 3d ed., 2006.
King, R. C., and W. D. Stansfield, *Dictionary of Genetics,* 7th ed., 2006.
Klug, W. S., and M. R. Cummings, *Concepts of Genetics,* 9th ed., 2008.
Scriver, C. R., et al., *The Metabolic Bases of Inherited Disease,* 8th ed., 2001.
Starr, C., and R. Taggart, *Cell Biology and Genetics,* 11th ed., 2007.
Vogel, F., and A. G. Motulsky, *Human Genetics: Problems and Approaches,* 1997.
Watson, J. D., and A. Berry, *DNA: The Secret of Life,* 2003.
Watson, J. D., et al., *Molecular Biology of the Gene,* 6th ed., 2008.

American Journal of Human Genetics, monthly.
Annual Review of Genetics, annually.
Annual Review of Genomics and Human Genetics, annually.
Gene, semimonthly.
Genetics, Genetics Society of America, monthly.
Journal of Heredity, bimonthly.
Molecular Genetics and Genomics, monthly.

GEOCHEMISTRY

Albarède, F., *Geochemistry: An Introduction,* 2003.
Chester, R., *Marine Geochemistry,* 2d ed., 2003.
Faure, G., *Principles and Applications of Inorganic Geochemistry,* 2d ed., 1997.
Hoefs, J., *Stable Isotope Geochemistry,* 5th ed., 2004.
Killops, S. D., and J. K. Vanessa, *An Introduction to Organic Geochemistry,* 2d ed., 2005.

Chemical Geology, Elsevier, monthly.
Geochimica et Cosmochimica Acta, Elsevier, semimonthly.

GEOGRAPHY

DeMers, M. N., *Fundamentals of Geographic Information Systems,* 4th ed., 2008.
Gaile, G. L., and C. J. Willmott (eds.), *Geography in America at the Dawn of the 21st Century,* 2004.
Gould, P., *Becoming a Geographer (Space, Place, and Society),* 2000.
McKnight, T. L., and D. Hess, *Physical Geography: A Landscape Appreciation,* 9th ed., 2007.
Muehrcke, P. C., *Map Use: Reading, Analysis and Interpretation,* 5th ed., 2005.
Robinson, A. H., et al., *Elements of Cartography,* 6th ed., 1995.
Strahler, A. H., and A. N. Strahler, *Introducing Physical Geography,* 4th ed., 2005.

Focus, American Geographical Society, quarterly.
Geographical Review, American Geographical Society, quarterly.
The Professional Geographer, Association of American Geographers, quarterly.

GEOLOGY

Condie, K. C., and R. E. Sloon, *Origin and Evolution of Earth: Principles of Historical Geology,* 1998.
Grotzinger, J., et al., *Understanding Earth,* 5th ed., 2006.
Lutgens, F. K., E. J. Tarbuck, and D. Tasa, *Essentials of Geology,* 10th ed., 2008.
Miall, A. D., *The Geology of Stratigraphic Sequences,* 1996.
Plummer, C., D. Carlson, and D McGeary, *Physical Geology,* 11th ed., 2006.
Press, F., and R. Siever, *Understanding Earth,* 3d ed., 2000.
Skinner, B. J., and S. C. Porter, *Dynamic Earth: An Introduction to Physical Geology,* 5th ed., 2003.
Tarbuck, E. J., F. K. Lutgens, and D. Tasa, *The Earth: An Introduction to Physical Geology,* 8th ed., 2004.

Annual Reviews of Earth and Planetary Sciences, annually.
Episodes, International Union of Geological Sciences, quarterly.
Earth, American Geological Institute, monthly.
GSA Bulletin, Geological Society of America, monthly.
Journal of Geology, University of Chicago Press, bimonthly.

GEOMETRY

Baldwin, H. L., Jr., *Essential Geometry,* 1993.
Coxeter, H. S. M., *Non-Euclidean Geometry,* 6th ed., 1998.
Coxeter, H. S. M., *Projective Geometry,* 2d ed., 1974, reprint, 2003.
do Carmo, M. P., *Riemannian Geometry,* 2d ed., 1992.
Greenberg, M., *Euclidean and Non-Euclidean Geometries: Development and History,* 4th ed., 2007.
Gustafson, R. D., and P. D. Frisk, *Elementary Geometry,* 2d ed., 1991.
Holliday-Darr, K., *Applied Descriptive Geometry,* 2d ed., 1998.
Jacobs, H. R. (ed.)**,** *Geometry: Seeing, Doing, Understanding,* 3d ed., 2004.
Jurgensen, R. C., R. G. Brown, and J. W. Jurgensen, *Geometry,* 2000.
Lang, S., and G. Murrow, *Geometry,* 2d ed., 1988, reprint, 2008.
O'Daffer, P. G., and S. R. Clemens, *Geometry: An Investigative Approach,* rev. 2d ed., 1992.
O'Neil, B., *Elementary Differential Geometry,* rev. 2d ed., 2006.
Reid, M., *Undergraduate Algebraic Geometry,* 1988.
Smith, K. E., et al., *An Invitation to Algebraic Geometry,* 2000.

GEOMORPHOLOGY

Bloom, A. L., *Geomorphology,: A Systematic Analysis of Late Cenozoic Landforms,* 3d ed., 2004.
Easterbrook, D. J., *Surface Processes and Landforms,* 2d ed., 1998.
Goudie, A., *The Changing Earth: Geomorphological Processes and Time,* 1995.
Leopold, L. B., M. G. Wolmon, and J. P. Miller, *Fluvial Process in Geomorphology,* 1995.
Ritter, D. F., R. C. Kochel, and J. R. Miller, *Process Geomorphology,* 4th ed., 2006.
Thomas, D. S. (ed.)**,** *Arid Zone Geomorphology: Process, Form and Change in Drylands,* 2d ed., 1999.

Earth Surface Processes and Landforms, Wiley, 14 issues.

GEOPHYSICS

Bertotti, B., and P. Farinella, *Physics of the Earth and the Solar System, Dynamics, Evolution, Space, Navigation, Space-time Structure* (Geophysics and Astrophysics Monographs), 1990.
Fowler, C. M., *The Solid Earth: An Introduction to Global Geophysics,* 2d ed., 2004.
Shearer, P., *Introduction to Seismology,* 1999.
Telford, W. M., L. P. Geldart, and R. E. Sheriff, *Applied Geophysics,* 2d ed., 1990.
Turcotte, D. L., and G. Schubert, *Geodynamics,* 2d ed., 2001.

Eos, American Geophysical Union, weekly.
Journal of Geophysical Research (issued in three parts), American Geophysical Union, monthly.
Reviews in Geophysics, American Geophysical Union, quarterly.

GRAPHIC ARTS

Adams, J. M., and P. A. Dolin, *Printing Technology,* 5th ed., 2001.
Bennett, P. K., *The Handbook for Digital Printing and Variable-Data Printing,* 2006.
Field, G. G., *Color and Its Reproduction,* 3d ed., 2004.
Levenson, H. R., *Understanding Graphic Communication: Selected Readings,* 2000.
Romano, F. (ed.), *Pocket Pal: A Graphic Arts Production Handbook,* 20th ed., 2007.
Ryan, W. E., and T. E. Conover, *Graphic Communications Today,* 4th ed., 2003.

HYDROLOGY

Berner, E. K., and R. A. Berner, *Global Environment: Water, Air and Geochemical Cycles,* 1996.
Black, P. E., *Watershed Hydrology,* 2d ed., 1996.
Dunne, T., and L. B. Leopold, *Water in Environmental Planning,* 1995.
Gupta, R. S. S., *Hydrology and Hydraulic Systems,* 2d ed., 2001.
Linsley, R. K., et al., *Hydrology for Engineers,* 3d ed. (McGraw-Hill Series in Water Resources & Environmental Engineering), 1982.
Todd, D. K., and L. W. Mays, *Groundwater Hydrology,* 3d ed., 2004.
Ward, A. D., and S. W. Trimble, *Environmental Hydrology,* 2d ed., 2003.

Hydrological Sciences Journal, International Association of Hydrological Sciences (U.K.), quarterly.
Journal of Hydrology, Elsevier, monthly.

ILLUMINATION

Coaton, J. R., and M. Marsden (eds.)**,** *Lamps and Lighting,* 4th ed., 1997.
Descottes, H., *Ultimate Lighting Design,* 2005.
Fetters, J. L., *Applied Illumination Engineering,* 2008.
Grosslight, J., *Light, Light, Light: Effective use of Daylight and Electric Lighting in Residential and Commercial Spaces,* 3d ed., 1998.
Helms, R., and M. C. Belcher, *Lighting for Energy Efficient Luminous Environment,* 1991.
Karlen, M., and J. Benya, *Lighting Design Basics,* 2004.
Lindsey, J. L., *Applied Illumination Engineering,* 2d ed., 1997.
Murdoch, J. B., *Illumination Engineering: From Edison's Lamp to the Laser,* 1985, reprint 1994.
Rea, M. (ed.)**,** *IESNA Lighting Handbook,* Illuminating Engineering Society of North America, 9th ed., 2000.
Thumann, A., and E. Woodroof, *Lighting Efficiency: Technology and Applications,* 2008.

LD & A (Lighting Design & Application), Illuminating Engineering Society, monthly.

IMMUNOLOGY

Coico, R., and G. Sunshine, *Immunology: A Short Course,* 6th ed., 2008.
Kindt, T. J., B. A. Osborne, and R. A. Goldsby, *Kuby Immunology,* 6th ed., 2006.
Murphy, K. M., P. Travers, and M. Walport, *Janeway's Immunobiology,* 7th ed., 2007.
Parham, P., *The Immune System,* 2d ed., 2004.
Parslow, T. G., et al., *Medical Immunology,* 10th ed., 2001.
Paul, W. E. (ed.), *Fundamental Immunology,* 6th ed., 2008.
Sell, S., and E. Max, *Immunology, Immunopathology, and Immunity,* 6th ed., 2001.

Advances in Immunology, quarterly.
Annual Review of Immunology, annually.
Immunobiology, 10 issues per year.
Immunology (England), monthly.
Immunology Letters, 14 issues per year.
Journal of Immunology, American Association of Immunologists, semimonthly.
Trends in Immunology, monthly.

INDUSTRIAL ENGINEERING

Aft, L. S., *Work Measurement and Methods Improvement (Engineering Design and Automation),* 2000.
Besterfield, D. H., *Quality Control,* 8th ed., 2008.
Chopra, S. and P. Meindl, *Supply Chain Management,* 3d ed., 2006.
Groover, M. P., *Automation, Production Systems, and Computer-Aided Manufacturing,* 3d ed., 2007.
Hobbs, D. P., *Lean Manufacturing Implementation: A Complete Execution Manual for Any Size Manufacturer,* 2003.
Salvendy, G. (ed.), *Handbook of Industrial Engineering: Technology and Operations Management,* 2001.

INFORMATION THEORY

Blahut, R. E., *Digital Transmission of Information,* 1990.
Cover, T. M., and J. A. Thomas, *Elements of Information Theory,* 2d ed., 2006.
Golomb, S. W., R. E. Peile, and R. A. Scholtz, *Basic Concepts in Information Theory and Coding,* 1994.
MacKay, J. C., *Information Theory, Inference, and Learning Algorithms,* 2003.
Roman, S., *Coding and Information Theory* (Graduate Texts in Mathematics), 1992.
Usher, M. J., *Information Theory for Information Technologists* (Computer Science Series), 1984.
Van Der Lubbe, J. C. A., and H. J. Hoeve, *Information Theory,* 1997.
Vickery, B. C., and A. Vickery, *Information Sciences Theory and Practice* (Guides to Information Sources), 3d ed., 2004.
Yeung, R. W., *A First Course in Information Theory,* 2002.

IEEE Transactions on Information Theory, Institute of Electrical and Electronics Engineers, Information Theory Group, bimonthly.
Information Sciences, 36 issues per year.

INORGANIC CHEMISTRY

Bowser, J. R., *Inorganic Chemistry,* 1993.
Cotton, F. A., and G. Wilkinson, *Advanced Inorganic Chemistry,* 6th ed., 1999.
Douglas, B. E., *Concepts & Models of Inorganic Chemistry,* 2d ed., 1993.
Greenwood, N. N., and A. Earnshaw, *Chemistry of the Elements,* 2d ed., 1997.
Housecroft, C. E., and A. G. Sharp, *Inorganic Chemistry,* 2001.
Hueehy, J. E., E. A. Keiter, and R. L. Keiter, *Inorganic Chemistry: Principles of Structure & Reactivity,* 4th ed., 1993.
Lee, J. D., *Concise Inorganic Chemistry,* 5th ed., 1996.
Mackay, K. M., et al., *Introduction to Modern Inorganic Chemistry,* 6th ed., 2002.
Miessler, G. L., and D. A. Tarr, *Inorganic Chemistry,* 3d ed., 2004.
Mingos, D. M. P., *Essential Trends in Inorganic Chemistry,* 1998.
Porterfield, W. W., *Inorganic Chemistry: A Unified Approach,* 2d ed., 1993.
Rayner-Canham, G., and T. Overton, *Descriptive Inorganic Chemistry,* 4th ed., 2006.
Shriver, D., and P. Atkins, *Inorganic Chemistry,* 4th ed., 2006.
Wulfsberg, G., *Inorganic Chemistry,* 2000.

Inorganic Chemistry, American Chemical Society.
Inorganica Chimica Acta, Elsevier.
Dalton Transactions, Royal Society of Chemistry.

INVERTEBRATE ZOOLOGY

See Zoology.

LASERS

Hecht, J., *Understanding Lasers: An Entry-Level Guide* (IEEE Press Understanding Science & Technology Series), 3d ed., 2008.
Meschede, D., *Optics Light and Lasers,* 2d ed. 2007.
Milonni, P. W., and J. H. Eberly, *Lasers* (Wiley Series in Pure and Applied Optics), 1988.
Saleh, B. E. A., and M. C. Teich, *Fundamentals of Photonics,* 2d ed., 2007.
Shimoda, K., *Introduction to Laser Physics,* 2d ed., 1991.
Siegman, A. E., *Lasers,* 1986.
Silfvast, W. T., *Laser Fundamentals,* 2d ed., 2004.
Svelto, O., et al., *Principles of Lasers,* 4th ed., 1998.
Verdeyen, J. T., *Laser Electronics,* 3d ed., 1994.
Yariv, A., *Photonics: Optical Electronics in Modern Communications,* 6th ed., 2006.
Young, M., *Optics and Lasers: Including Fibers and Optical Waveguides* (Advanced Texts in Physics), 5th rev. ed., 2000.

Applied Optics, Optical Society of America, American Institute of Physics, 36 issues per year.
Applied Physics B, Deutsche Physikalische Gesellschaft, 12 issues per year.
Journal of the Optical Society of America, A, B, American Institute of Physics, monthly.

LOW-TEMPERATURE PHYSICS

Annett, J. F., *Superconductivity, Superfluids and Condensates,* 2004.
Betts, D. S., et al., *Introduction to Millikelvin Technology* (Cambridge Studies in Low Temperature Physics), 1989, reprint, 2005.
Brewer, D. F., et al. (eds.)**,** *Progress in Low Temperature Physics,* vols. 1–16, 1956–2008.
Dahl, P. F., *Superconductivity: Its Historical Roots and Development from Mercury to the Ceramic Oxides,* 1992, reprint 2008.
De Gennes, P. G., *Superconductivity of Metals and Alloys* (Advanced Book Classics), 1999.
Flynn, T. M., *Cryogenic Engineering,* 2d ed., 2005.
Pobell, F., *Matter and Methods at Low Temperatures,* 2d ed., 1996.
Poole, C. P., et al., *Superconductivity,* 2d ed., 2007.
Richardson, R. C., and E. N. Smith (eds.)**,** *Experimental Techniques in Condensed Matter Physics at Low Temperatures* (Advanced Book Classics), 1998.
Schrieffer, J. R., *Theory of Superconductivity* (Advanced Book Classics), 1999.
Tilley, D. R., and J. Tilley, *Superfluidity and Superconductivity,* 3d ed., 1990.
Timmerhaus, K. D., et al. (eds.)**,** *Advances in Cryogenic Engineering,* vols. 1–54, 1960–2008.
Tinkham, M., *Introduction to Superconductivity* (Dover Books on Physics), 2d ed., 2004.
Wilks, J., and D. S. Betts, *An Introduction to Liquid Helium,* 2d reprint ed.,1990.

Cryogenics, monthly.
Journal of Low Temperature Physics, 24 issues per year.

MAGNETISM

See Electricity and magnetism.

MARINE ENGINEERING

Blank, D. A., and A. E. Bock (eds.)**,** *Introduction to Naval Engineering,* 2d ed., 1986, paperback, 2005.
Carlton, J., *Marine Propellers and Propulsion,* 2d ed., 2007.
Harrington, R. L. (ed.)**,** *Marine Engineering,* rev. ed., 1992.
Hunt, E. C., et al., *Modern Marine Engineer's Manual,* 3d ed., 1999.
McGeorge, H. D., *Marine Auxiliary Machinery,* 7th ed., 1995.
Taylor, D. A., *Introduction to Marine Engineering,* 2d rev. ed., 1996.

Journal of Ship Research, Society of Naval Architects and Marine Engineers, quarterly.
Marine Engineers Review, Institute of Marine Engineers, monthly.
Marine Technology, SNAME NEWS, Society of Naval Architects and Marine Engineers, quarterly.
Transactions of the Society of Naval Architects and Marine Engineers, annually.

MATERIALS SCIENCE

Ashby, M. F., *Materials Selection in Mechanical Design,* 3d ed., 2005.
Callister, W. D. Jr., *Materials Science and Engineering: An Introduction,* 7th ed., 2006.
de Podesta, M., *Understanding the Properties of Matter,* 2d ed., 2002.
Interrante, L. V., and M. J. Hampden-Smith (eds.)**,** *Chemistry of Advanced Materials: An Overview,* 1997.
Jones, I. P., *Materials Science for Electrical and Electronic Engineers,* 2001.
Shackelford, J. F., *Introduction to Materials Science for Engineers,* 6th ed., 2005.
Smith, W. F., *Foundations of Materials Science and Engineering,* 3d ed., 2005.
White, M. A., *Properties of Materials,* 1999.

Advanced Materials, Wiley-VHC, monthly.
MRS Bulletin, Materials Research Society, monthly.
Materials Today, Elsevier, monthly.
Nature Materials, Nature Publishing Group, monthly.
Nano Letters, American Chemical Society, monthly.

MATHEMATICS

Barker, V. C., R. N. Aufmann, and J. Lockwood, *Essential Mathematics with Applications,* 7th ed., 2005.
Keedy, M. L., M. Bittinger, and W. B. Rudolph, *Essential Mathematics,* 7th ed., 1992.
Korn, G. A., and T. M. Korn, *Mathematics Handbook for Scientists and Engineers: Definitions, Theorems, and Formulas for Reference and Review,* 2d rev. ed., 1968, reprint, 2000.
Kramer, E. E., *The Nature and Growth of Modern Mathematics,* 1970, reprint, 1983.
Maki, D. P., and M. Thompson, *Finite Mathematics,* 5th ed., 2005.
Mathematical Society of Japan, *Encyclopedic Dictionary of Mathematics,* 2d ed., 2 vols., edited by K. Ito, 1993.
Research and Education Association Staff, *The Essentials of Numerical Analysis,* no. 1 and 2, rev. ed., 1989.
Setek, W. M., and M. A. Gallo, *Fundamentals of Mathematics,* 10th ed., 2004.
Shapiro, S., *Thinking About Mathematics: The Philosophy of Mathematics,* 2000.
Sullivan, M., *Finite Mathematics: An Applied Approach,* 10th ed., 2007.

Advances in Mathematics, 16 issues per year.
American Journal of Mathematics, bimonthly.
American Mathematical Monthly, Mathematical Association of America, 10 issues per year.
Annals of Mathematics, bimonthly.

MECHANICAL ENGINEERING

Avallone, E. A., T. Baumeister III, and A. Sadegh (eds.)**,** *Marks' Standard Handbook for Mechanical Engineers,* 11th ed., 2006.
Budynas, R. G., and J. K. Nisbett, *Shigley's Mechanical Engineering Design,* 8th ed., 2007.
Hicks, T. G., *Handbook of Mechanical Engineering Calculations,* 2d ed., 2006.
Nayler, G. H. F., *Dictionary of Mechanical Engineering,* 4th ed., 1996.
Rothbart, H. A., *Mechanical Design Handbook,* 2d ed., 2006.
Shigley, J. E., C. R. Mischke, and R. Budynas, *Mechanical Engineering Design,* 7th ed., 2003.

The American Society of Mechanical Engineers, (ASME), Publisher 24 Journals on various topics in mechanical engineering.

MECHANICS

See Classical mechanics; Fluid mechanics; Quantum mechanics; Statistical mechanics.

MEDICAL MICROBIOLOGY

Brooks, G. F., et al., *Jawetz, Melnick & Adelberg's Medical Microbiology,* 24th ed., 2007.
Goering, R., et al., *Mims' Medical Microbiology,* 4th ed., 2008.
Levinson, W., *Medical Microbiology and Immunology,* 10th ed., 2008.
Ryan, K. J., and C. G. Ray, *Sherris Medical Microbiology: An Introduction to Infectious Diseases,* 4th ed., 2003.

Annual Review of Microbiology, annually.
Antimicrobial Agents and Chemotherapy, monthly.
Journal of Clinical Microbiology, monthly.
Journal of Medical Microbiology, monthly.

MEDICINE AND PATHOLOGY

DeVita, V. T., Jr., T. S. Lawrence, and S. A. Rosenberg (eds.)**,** *DeVita, Hellman, and Rosenberg's Cancer: Principles & Practice of Oncology,* 8th ed., 2004.
Fauci, A., et al. (eds.)**,** *Harrison's Principles of Internal Medicine,* 17th ed., 2008.
Johnson, L. R., and J. H. Byrne (eds.)**,** *Essential Medical Physiology,* 3d ed., 2003.
Kronenberg, H. M., et al., *Williams Textbook of Endocrinology,* 11th ed., 2008.
Kumar, V., et al., *Robbins Basic Pathology,* 8th ed., 2007.
Kumar, V., N. Fausto, and A. Abbas, *Robbins & Cotran Pathologic Basis of Disease,* 7th ed., 2004.
Rosai, J., *Rosai and Ackerman's Surgical Pathology,* 9th ed., 2004.
Tamparo, C., and M. A. Lewis, *Diseases of the Human Body,* 4th ed., 2005.
Weinberg, R. A., *The Biology of Cancer,* 2006.

American Journal of Cardiology, semimonthly.
American Journal of Medicine, monthly.
American Journal of Pathology, monthly.
American Journal of Surgery, monthly.
Annals of Internal Medicine, American College of Physicians, semimonthly.
Cancer, American Cancer Society, semimonthly.
Cardiovascular News, monthly.
Circulation, American Heart Association, weekly.
Contemporary Surgery, monthly.
Endocrinology, monthly.
Human Pathology, monthly.
JAMA: Journal of the American Medical Association, weekly.
Journal of Nuclear Medicine, monthly.
Modern Medicine, monthly.
New England Journal of Medicine, weekly.

METALLURGICAL ENGINEERING

Abbaschian, R., and R. E. Reed-Hill, *Physical Metallurgy Principles,* 4th ed., 2008.
Dieter, G., *Mechanical Metallurgy,* 1986.
Neely, J. E., and T. J. Bertone, *Practical Metallurgy and Materials of Industry,* 6th ed., 2002.
Sinha, A., *Physical Metallurgy Handbook,* 2002.
Vander Voort, G. F. (ed.)**,** *ASM Handbook Volume 9: Metallography and Microstructures,* 2004.

Iron & Steel Technology, The Association for Iron & Steel Technology (AIST), monthly.
JOM: The Member Journal of The Minerals, Metals & Materials Society, monthly.
Metallurgical and Materials Transactions A, The Minerals, Metals & Materials Society, monthly.
Metallurgical and Materials Transactions B, The Minerals, Metals & Materials Society, bimonthly.
Mineral Processing and Extractive Metallurgy: Transactions of the Institute of Mining and Metallurgy, Section C, several times per year (irregular).

METEOROLOGY

See Climatology and meteorology.

MICROBIOLOGY

Garrity, G., et al. (eds.)**,** *Bergey's Manual of Systematic Bacteriology,* vols. 1 and 2, 2d ed., 2001 and 2005.
Lederberg, J., *Encyclopedia of Microbiology,* 2d ed., 2000.
Madigan, M. T., et al., *Brock Biology of Microorganisms,* 12th ed., 2009.
Nester, E., et al., *Microbiology: A Human Perspective,* 4th ed., 2004.
Tortora, G. J., C. L. Case, and B. R. Funke, *Microbiology: An Introduction,* 9th ed., 2006.
Willey, J., L. Sherwood, and C. Woolverton, *Prescott, Harley, and Klein's Microbiology,* 7th ed., 2007.

Advances in Applied Microbiology, annually.
Annual Reviews of Microbiology, annually.
Applied and Environmental Microbiology, semimonthly.
International Journal of Systematic and Evolutionary Microbiology, monthly.
Journal of Bacteriology, biweekly.
Microbiology, monthly.
Microbiology and Molecular Biology Reviews, quarterly.

MICROSCOPY

Bradbury, S., *An Introduction to the Optical Microscope,* vol. 1., 1989.
Briggs, A., *An Introduction to Scanning Acoustic Microscopy* (Microscopy Handbooks no. 12), 1986.
Duke, P. J., and A. G. Michette (eds.)**,** *Modern Microscopy: Techniques and Applications,* 1990.
Goldstein, J. I., et al., *Scanning Electron Microscopy and X-Ray Microanalysis,* 3d ed., 2003.
Johari, O., et al. (eds.)**,** *Scanning Electron Microscopy,* 1983.

Mayer, O., and D. J. Mayer, *Clinical Wide-Field Specular Microscopy,* 1984.
Shotton, D. M. (ed.)**,** *Electronic Light Microscopy: The Principles and Practice of Video-enhanced, Contrast Digital Intensified Fluorescence, and Confocal, and Scanning Light Microscopy.* 1993.
Slayter, E. M., and H. S. Slayter, *Light and Electron Microscopy,* 1992.
Wilson, T. C., and J. R. Sheppard, *Theory and Practice of Scanning Optical Microscopy,* 1984.
Wilson, T. (ed.)**,** *Confocal Microscopy,* 1990.
Wischnitzer, S., *Introduction to Electron Microscopy,* 3d ed., 1981.

Micron, quarterly.
Microscope, Cutter Laboratories, 10 issues per year.
The Microscope, McCrone Research Institute, quarterly.

MINERALOGY

Klein, C. and B. Dutrow, *Manual of Mineral Science,* 23d ed., 2007.
Howie, R. A., J. Zussman, and W. A. Deer, *An Introduction to the Rock-Forming Minerals,* 2d ed., 1992.
Mann, S., *Biomineralization,* 2002.
Nesse, W. D., and D. J. Schulze, *Introduction to Mineralogy and An Atlas of Minerals in Thin Section,* 2004.

American Mineralogist, Mineralogical Society of America, monthly.
Mineralogical Magazine, Mineralogical Society of Great Britain, monthly.
Physics and Chemistry of Minerals, Springer-Verlag, monthly.

MINING ENGINEERING

Guilbert, J. M., and C. F. Park, *The Geology of Ore Deposits,* 2007.
Hartman, H. L., et al., *SME Mining Engineering Handbook,* 2d rev. ed., 1992.
Hartman, H. L., and J. Mutmansky, *Introductory Mining Engineering,* 2d ed., 2002.
Kennedy, B. A. (ed.)**,** *Surface Mining,* 2d ed., 1990.

Mining Engineering, Society of Mining Engineers (of AIME), monthly.

MOLECULAR BIOLOGY

See Cell and molecular biology.

MOLECULAR PHYSICS

Atkins, P. W., *Atkins' Molecules,* 2d ed., 2003.
Atkins, P. W., and R. S. Friedman, *Molecular Quantum Mechanics,* 4th ed., 2005.
Banwell, C. N., and E. M. McCash, *Fundamentals of Molecular Spectroscopy,* 4th ed., 1994.
Bernath, P. F., *Spectra of Atoms and Molecules* (Topics in Physical Chemistry), 2d ed., 2005.
Bransden, B. H., and C. J. Joachain, *Physics of Atoms and Molecules,* 2d ed., 2003.
Demtröder, W., *An Introduction to Atomic and Molecular Physics,* 2005.
Erkoç, Ş., and T. Uzer, *Lecture Notes on Atomic and Molecular Physics,* 1996.
Richards, W. G., and P. R. Scott, *Structure and Spectra of Molecules,* 1985.
Struve, W. S., *Fundamentals of Molecular Spectroscopy,* 1989.
Svanberg, S., *Atomic and Molecular Spectroscopy: Basic Aspects and Practical Applications,* 4th ed., 2004.

Journal of Molecular Structure, 48 issues per year.
Molecular Physics, 18 issues per year.

MYCOLOGY

Ainsworth, G. C., et al., *Ainsworth and Bisby's Dictionary of Fungi,* 9th ed., 2001.
An, Z. (ed.)**,** *Handbook of Industrial Mycology,* 2005.
Burnett, J. H., *Fungal Populations and Species* (Life Science), 2003.
Carlile, M. J., et al., *The Fungi,* 2d ed., 2001.
Ulloa, M., and R. T. Hanlin, *Illustrated Dictionary of Mycology,* 2000.
Watanabe, T., *Pictorial Atlas of Soil and Seed Fungi: Morphologies of Cultured Fungi and Key to Species,* 2d ed., 2002.

FEMS Yeast Research, 8 issues per year.
Fungal Biology Reviews, quarterly.
Fungal Genetics and Biology, monthly.
Medical Mycology, 8 issues per year.
Mycological Research, British Mycological Society, monthly.

NAVAL ARCHITECTURE

Benford, H., *Naval Architecture for Non-Naval Architects,* 1991.
Eyres, D. J., *Ship Construction,* 6th ed., 2007.
Gillmer, T. C., and B. Johnson, *Introduction to Naval Architecture,* 1982.
Lewis, E. V. (ed.)**,** *Principles of Naval Architecture,* 3 vols., 1988.
Taylor, D. A., *Merchant Ship Construction,* 4th ed., 1998.
Tupper, E. C., *Introduction to Naval Architecture,* 4th ed., 2004.
Zubaly, R. B., *Applied Naval Architecture,* 2d ed., 1996.

Naval Architect, Royal Institution of Naval Architects, 10 issues per year.
Transactions of the Society of Naval Architects and Marine Engineers, annually.
U. S. Naval Institute Proceedings, monthly.

NAVIGATION

Clausing, D. J., *Aviator's Guide to Navigation,* 4th ed., 2007.
Cunliffe, T., *Coastal and Offshore Navigation,* 3d ed., 2009.
Cutler, T. J., *Dutton's Nautical Navigation,* 15th ed., 2003.
Hobbs, R. R., *Marine Navigation: Piloting and Celestial Electronic Navigation,* 4th ed., 1998.

Kaplan, E. D., and C. J. Hegarty (eds.), *Understanding GPS: Principles and Applications,* 2d ed., 2006.
Karl, J., *Celestial Navigation in the GPS Age,* 2007.
Kayton, M., and W. R. Fried, *Avionics Navigation Systems,* 2d ed., 1997.
Misra, P., and P. Enge, *Global Positioning System,* 2001.
Parkinson, B., and J. Spilker (eds.), *Global Positioning System: Theory and Applications* (vols. 1 and 2), 1996.
Tetley, L., and D. Calcutt, *Electronic Navigation Systems,* 3d ed., 2001.

Navigation, Journal of the Institute of Navigation, quarterly.

NEUROPSYCHOLOGY

Beaumont, J. G., P. Kenealy, and M. J. Rogers, *The Blackwell Dictionary of Neuropsychology,* 1999.
Gazzaniga, M. S., R. B. Ivry, and G. R. Mangun, *Cognitive Neuroscience: The Biology of the Mind,* 3d ed., 2008.
Heilman, K., and E. Valenstein, *Clinical Neuropsychology* (Medicine), 2003.
Kolb, B., and I. Q. Whishaw, *Fundamentals of Neuropsychology,* 6th ed., 2008.
Lezak, M. D., *Neuropsychological Assessment,* 2004.
Strauss, E., E. M. S. Sherman, and O. Spreen, *A Compendium of Neuropsychological Tests: Administration, Norms, and Commentary,* 3d ed., 2006.

Neuropsychology, bimonthly.

NEUROSCIENCE

Bear, M., B. Connors, and M. Paradiso, *Neuroscience: Exploring the Brain,* 3d ed., 2006.
Cooper, J. R., F. E. Bloom, and R. H. Roth, *Biochemical Basis of Neuropharmacology,* 8th ed., 2002.
Kandel, E. R., J. H. Schwartz, and T. M. Jessell (eds.), *Principles of Neural Science,* 4th ed., 2000.
Levitan, I. B., and L. Kaczmarek, *The Neuron,* 3d ed., 2001.
Nestler, E. J., S. E. Hyman, and R. C. Malenka, *Molecular Basis of Neuropharmacology,* 2d ed., 2008.
Nicholls, J. G., et al., *From Neuron to Brain,* 4th ed., 2001.
Purves, D., et al. (eds.), *Neuroscience,* 4th ed., 2007.
Shepherd, G. M. (ed.), *Synaptic Organization of the Brain,* 5th ed., 2003.

Annual Review of Neuroscience, annually.
Journal of Neuroscience, Society for Neuroscience, weekly.
Nature Neuroscience, monthly.
Neuron, semimonthly.
Trends in Neurosciences, monthly.

NUCLEAR ENGINEERING

Almenas, K., and R. Lee, *Nuclear Engineering: An Introduction,* 1992.
Bodansky, D., *Nuclear Energy: Principles, Practices, Prospects,* 2d ed., 2004.
Foster, A. R., and R. L. Wright, Jr., *Basic Nuclear Engineering* (Allyn and Bacon Series in Engineering), 4th ed., 1983.
Glasstone, S., and W. H. Jordan, *Nuclear Power and Its Environmental Effects,* 1980.
Glasstone, S., and A. Sesonske, *Nuclear Reactor Engineering,* 2 vols., 4th ed., 1994.
Knief, R. A., *Nuclear Engineering: Theory and Technology of Commercial Nuclear Power* (SCPP), 2d ed., 1992.
Lamarsh, J. R., and A. J. Barrata, *Introduction to Nuclear Engineering,* 3d ed., 2001.
Lewins, J., and M. Becker (eds.), *Advances in Nuclear Science & Technology,* vols. 1–26, 1962–1999.
Ligou, J. P., and S. Mitter, *Elements of Nuclear Engineering,* rev ed., 1987.
Murray, R. L., *Nuclear Energy: An Introduction to the Concepts, Systems, and Applications of Nuclear Processes,* 6th ed., 2008.

IEEE Transactions on Nuclear Science, Institute of Electrical and Electronics Engineers, Nuclear and Plasma Sciences Society, monthly.
Nuclear Science and Engineering, American Nuclear Society, 9 issues per year.

NUCLEAR PHYSICS

Basdevant, J.-L., J. Rich, and M. Spiro, *Fundamentals in Nuclear Physics,* 2005.
Burcham, W. E., and M. Jobes, *Nuclear and Particle Physics,* 1995.
Cottingham, W. N., and D. A. Greenwood, *An Introduction to Nuclear Physics,* 2d ed., 2001.
Das, A., and T. Ferbel, *Introduction to Nuclear and Particle Physics,* 2d ed., 2003.
Eisenberg, J. M., and W. Greiner, *Nuclear Theory,* vol. 1, 3d ed., 1988; vol. 2, 3d ed., 1988; and vol. 3, 2d ed., 1976.
Enge, H. A., and R. P. Redwine, *Introduction to Nuclear Physics,* 2d ed., 1999.
Krane, K. S., *Introductory Nuclear Physics,* 1988.
Lilley, J. S., *Nuclear Physics: Principles and Applications,* 2001.
Walecka, J. D., *Theoretical Nuclear and Subnuclear Physics,* 2d ed., 2004.
Wong, S. S. M., *Introductory Nuclear Physics,* 2d ed., 1998.

Annual Review of Nuclear and Particle Science, annually.
Nuclear Physics, Section A, 68 issues per year.
Physical Review C, American Physical Society, monthly.

OCEANOGRAPHY

Colling, A., *Ocean Circulation,* 2d ed., 2001.
Garrison, T. S., *Oceanography: An Invitation to Marine Science,* 6th ed., 2007.
Knauss, J. A., *An Introduction to Physical Oceanography,* 2d ed., 1996.
Lalli, C. M., *Biological Oceanography,* 2d ed., 1997.
Pilson, M. E. Q., *An Introduction to the Chemistry of the Sea,* 1998.
Thurman, H. V., and A. P. Trujillo, *Essentials of Oceanography,* 9th ed., 2007.

Journal of Physical Oceanography, American Meteorological Society, monthly.
Oceanus, Woods Hole Oceanographic Institute, semiannually.

OPTICS

Born, M., and E. Wolf, *Principles of Optics,* 7th ed., 1999.
Hecht, E., *Optics,* 4th ed., 2001.
Jenkins, F. A., and H. E. White, *Fundamentals of Optics,* 4th ed., 1976, reprint, 2001.
Meyer-Arendt, J., *Introduction to Classical and Modern Optics,* 4th ed., 1994.
Optical Society of America, *Handbook of Optics,* 2d ed., 4 vols., 1995–2001.
Pedrotti, F. J., L. M. Pedrotti, and L. S., Pedrotti, *Introduction to Optics,* 3d ed., 2006.
Smith, W. J., *Modern Optical Engineering,* 4th ed., 2008.
Yu, F. T. S., and X. Yang, *Introduction to Optical Engineering,* 1997.

Applied Optics, Optical Society of America, American Institute of Physics, 36 issues per year.
Journal of the Optical Society of America, A, B, American Institute of Physics, monthly.

ORGANIC CHEMISTRY

Carey, F. A., and R. J. Sundberg, *Advanced Organic Chemistry: Structure and Mechanisms,* 5th ed., 2 vols., 2008.
Carey, F. A., *Organic Chemistry,* 7th ed., 2007.
Crabtree, R. H., *The Organometallic Chemistry of the Transition Metals,* 4th ed., 2005.
Ege, S. N., *Organic Chemistry: Structure and Reactivity,* 5th ed., 2003.
Eliel, E., and S. H. Wilen, *Stereochemistry of Organic Compounds,* 1994.
Fleming, I., *Frontier Orbitals in Organic Chemical Reactions,* 2d ed., 2008.
Holum, J. R., *Fundamentals of General, Organic and Biological Chemistry,* 6th ed., 1997.
Juaristi, E., *Introduction to Stereochemistry and Conformational Analysis,* 1991.
March, J., and M. B. Smith, *March's Advanced Organic Chemistry: Reactions, Mechanisms, and Structure,* 6th ed., 2007.
McMurry, J. E., *Organic Chemistry,* 7th ed., 2007.
Morrison, R. T., and R. N. Boyd, *Organic Chemistry,* 7th ed., 2003.
Nicolaou, K. C., and S. A. Snyder, *Classics in Total Synthesis II: More Targets, Strategies, Methods,* 2003.
Science of Synthesis: Houben-Weyl Methods of Molecular Transformations, 48 vols., 2001–2008.
Smith, M. B., *Organic Synthesis,* 2d ed., 2001.
Traynham, J. G., *Organic Nomenclature: A Programmed Introduction,* 6th ed., 2008.
Trost, B. M. (ed.), *Comprehensive Organic Synthesis: Selectivity, Strategy & Efficiency in Modern Organic Chemistry,* 9 vols., 1991.
Von Zelewsky, A., *Stereochemistry of Coordination Compounds,* 1996.

European Journal of Organic Chemistry, Wiley-VCH, 36 issues per year.
Journal of Organic Chemistry, American Chemical Society, biweekly.
Journal of Organometallic Chemistry, Elsevier (Switzerland), 48 issues per year.
Organic Letters, American Chemical Society.
Synlett: Accounts and Rapid Communications in Synthetic Organic Chemistry, Thieme, 20 issues per year.
Synthesis: Journal of Synthetic Organic Chemistry, Thieme, 24 issues per year.
Tetrahedron, Elsevier, weekly.
Tetrahedron Letters, Elsevier, weekly.

PALEONTOLOGY

Benton, M. J., *Vertebrate Palaeontology,* 3d ed., 2005.
Briggs, D., D. Erwin, and F. J. Collier, *The Fossils of the Burgess Shale,* 1995.
Briggs, D. E. G., and P. R. Crowther, *Palaeobiology II,* 2002.
Carroll, R., *Vertebrate Paleontology and Evolution,* 1987.
Clack, J. A., *Gaining Ground: The Origin and Early Evolution of Tetrapods,* 2002.
Colbert, E. H., M. Morales, and E. C. Minkoff, *Colbert's Evolution of the Vertebrates: A History of the Backboned Animals through Time,* 5th ed., 2001.
Gould, S. J., *The Book of Life: An Illustrated History of the Evolution of Life on Earth,* 2d ed., 2001.
Grimaldi, D., and M. S. Engel, *Evolution of the Insects,* 2005.
Jackson, J. B. C., S. Lidgard, and F. K. McKinney, *Evolutionary Patterns: Growth, Form, and Tempo in the Fossil Record,* 2001.
Knoll, A. H., *Life on a Young Planet: The First Three Million Years of Evolution on Earth,* 2003.
Prothero, D. R., *Bringing Fossils to Life: An Introduction to Paleobiology,* 2003.
Weishampel, D. B., P. Dodson, and H. Osmolska (eds.), *The Dinosauria,* 2d ed., 2004.
Willis, K., and J. McElwain, *The Evolution of Plants,* 2002.

Journal of Paleontology, bimonthly.
Journal of Vertebrate Paleontology, Society of Vertebrate Paleontology, quarterly.
Palaeontology, Palaeontological Association, bimonthly.
Paleobiology, quarterly.

PETROLEUM ENGINEERING

Baker, R., *A Primer of Oilwell Drilling,* 6th ed., 2000.
Gluyas, J., and R. Swarbrick, *Petroleum Geoscience,* 2003.
Hyne, N. J., *Nontechnical Guide to Petroleum Geology, Exploration, Drilling and Production,* 2d ed., 2001.
Luthi, S. M., *Geological Well Logs: Their Use in Reservoir Modeling,* 2001.
Lyons, W. C., *Standard Handbook of Petroleum and Natural Gas Engineering,* 2d ed., 2004.

Hydrocarbon Processing, Gulf Publishing Company, monthly.
Journal of Petroleum Technology, Society of Petroleum Engineers, monthly.

PETROLOGY

Best, M. G., *Igneous and Metamorphic Petrology,* 2d ed., 2002.
Blatt, H., R. Tracy, and B. Owens, *Petrology: Igneous, Sedimentary, and Metamorphic,* 3d ed., 2005.
Boggs, S. Jr., *Petrology of Sedimentary Rocks,* 2003.

Bucher, K., and M. Frey, *Petrogenesis of Metamorphic Rocks,* 7th ed., 2002.

Raymond, L. A., *Petrology: The Study of Igneous, Sedimentary and Metamorphic Rocks,* 2d ed., 2007.

J. D. Winter, *Introduction to Igneous and Metamorphic Petrology,* 2001.

Contributions to Mineralogy and Petrology, Springer-Verlag, monthly.

Journal of Petrology (England), quarterly.

PHARMACOLOGY

Brunton, L., J. Lazo, and K. Parker, *Goodman & Gilman's The Pharmacological Basis of Therapeutics,* 11th ed., 2005.

Chabner, B. A., and D. L. Longo (eds.)**,** *Cancer Chemotherapy and Biotherapy: Principles and Practice,* 2005.

Katzung, B. G. (ed.)**,** *Basic and Clinical Pharmacology,* 10th ed., 2006.

Licinio, J., and M.-L. Wong, *Pharmacogenomics: The Search for Individualized Therapies,* 2002.

Neal, M. J., *Medical Pharmacology at a Glance,* 5th ed., 2005.

Physicians' Desk Reference, 63d ed., 2009.

Biochemical Pharmacology, semimonthly.

Excerpta Medica, Section 30: Clinical and Experimental Pharmacology, 30 issues per year.

Pharmacy Today, monthly.

PHYSICAL CHEMISTRY

Adamson, A. W., and A. P. Gast, *Physical Chemistry of Surfaces,* 6th ed., 1997.

Adamson, A. W., and P. D. Fleischauer, (eds.)**,** *Concepts of Inorganic Photochemistry,* 1975, reprint.

Atkins, P. W., *Concepts in Physical Chemistry,* 1995.

Baggott, J., *Beyond Measure: Modern Physics, Philosophy, and the Meaning of Quantum Theory,* 2004.

Berry, R. S., et al., *Physical Chemistry,* 2d ed., 2002.

Bruce, P. G. (ed.), *Solid State Electrochemistry,* 1997.

Burdett, J. K., *Chemical Bonding in Solids,* 1995.

Calais, J.-L., *Quantum Chemistry Workbook: Basic Concepts and Procedures in the Theory of the Electronic Structure of Matter,* 1994.

Cheetham, A. K., and P. Day, *Solid-State Chemistry,* vol. 1: Techniques, 1990, vol. 2: Compounds, 1992.

Denbigh, K. G., *The Principles of Chemical Equilibrium,* 4th ed., 1981.

Hoffman, R., *Solids and Surfaces: A Chemist's View of Bonding in Extended Structures,* 1989.

Horspool, W. M., and P.-S. Song (eds.)**,** *Handbook of Organic Photochemistry and Photobiology,* 2d ed., 2003.

Houston, P. L., *Chemical Kinetics and Reaction Dynamics,* 2001.

Klotz, I. M., and R. M. Rosenberg, *Chemical Thermodynamics: Basic Theory and Methods,* 7th ed., 2008.

Kopecky, J., *Organic Photochemistry: A Visual Approach,* 1992.

Ladd, M. F. C., *Structure and Bonding in Solid State Chemistry,* 1979.

Laidler, K. L., and J. Keith, *Chemical Kinetics,* 3d ed., 1987.

Levine, I. N., *Physical Chemistry,* 5th ed., 2001.

Levine, R. D., *Molecular Reaction Dynamics,* 2005.

Lowe, J. P., K. Petevgen *Quantum Chemistry,* 3d ed., 2005.

McQuarrie, D. A., *Quantum Chemistry,* 2d ed., 2007.

McQuarrie, D. A., and J. D. Simon, *Physical Chemistry: A Molecular Approach,* 1997.

Meites, L., *An Introduction to Chemical Equilibrium and Kinetics,* 1981.

Moore, J. W., and R. G. Pearson, *Kinetics and Mechanism,* 3d ed., 1981.

Rabek, J. F., *Photochemistry and Photophysics,* vol. 4, 1991.

Schmalzried, H., *Solid State Reactions,* 2d ed., 1981.

Servos, J. W., *Physical Chemistry from Ostwald to Pauling: The Making of a Science in America,* 1996.

Silbey, R. J., et al., *Physical Chemistry,* 4th ed., 2004.

Smart, L. E., and E. A. Moore, *Solid State Chemistry: An Introduction,* 3d ed., 2005.

Smith, E. B., *Basic Chemical Thermodynamics,* 5th ed., 2004.

Smith, V. H., Jr., et al., *Applied Quantum Chemistry,* 1986.

Szabo, A., and N. S. Ostlund, *Modern Quantum Chemistry: Introduction to Advanced Electronic Structure Theory,* 1996.

Turro, N. J., *Modern Molecular Photochemistry,* 1981, reprint 1991.

West, A. R., *Basic Solid State Chemistry,* 2d ed., 1999.

International Journal of Quantum Chemistry, 15 issues per year.

Journal of Chemical Physics, American Institute of Physics, weekly.

Journal of Physical Chemistry, American Chemical Society, biweekly.

Photochemistry and Photobiology (text in English, French, or German), American Society for Photobiology, bimonthly.

PHYSICAL ELECTRONICS

Enderlein, R., and N. J. M. Horing, *Fundamentals of Semiconductor Physics and Devices,* 1997.

Ferendeci, A. M., *Physical Foundations of Solid State and Electron Devices,* 1991.

Ghandi, S. K., *VLSI Fabrication Principles: Silicon and Gallium Arsenide,* 2d ed., 1994.

Grasserbauer, M., and H. W. Werner (eds.)**,** *Analysis of Microelectronic Materials and Devices,* 1991.

Mishra, U. K., and J. Singh, *Semiconductor Device Physics and Design,* 2008.

Neaman, D., *Semiconductor Physics and Devices,* 3d ed., 2003.

Ng, K. K., *Complete Guide to Semiconductor Devices,* 2d ed., 2002.

Pierret, R. F., *Semiconductor Device Fundamentals,* 2d ed., 1996.

Seeger, K., *Semiconductor Physics: An Introduction,* 9th ed., 2004.

Sze, S. M., and K. K. Ng, *Physics of Semiconductor Devices,* 3d ed., 2007.

Yu, P. Y., and M. Cardona, *Fundamentals of Semiconductors,* 3d ed., 2001.

Electrical and Electronics Abstracts (Science Abstracts, Section B), INSPEC, Institution of Electrical Engineers, monthly.

IEEE Spectrum, Institute of Electrical and Electronics Engineers, Inc., monthly.

IEEE Transactions on Electron Devices, Institute of Electrical and Electronics Engineers, Electron Devices Society, monthly.

Sold State Technology, monthly.

PHYSICS

Beiser, A., *Concepts of Modern Physics,* 6th ed., 2002.
Brown, L., A. Pais, and B. Pippard (eds.)**,** *Twentieth Century Physics,* 3 vols., 2d ed., 1995.
Bueche, F., *Principles of Physics* (McGraw-Hill Schaum's Outline Series in Science), rev. 6th ed., 1994.
Feynman, R., *The Character of Physical Law (Messenger Lectures, 1964),* 2001.
Feynman, R. P., R. B. Leighton, and M. Sands, *The Feynman Lectures on Physics including Feynman's Tipson Physics: The Definitive and Extended Edition,* 2d ed., 2005.
Halliday, D., R. Resnick, and J. Walker, *Fundamentals of Physics, Extended Edition,* 8th ed., 2007.
Lightman, A. P., *Great Ideas in Physics,* 3d ed., 2000.
Sachs, M., *Concepts of Modern Physics: The Haifa Lectures,* 2007.
Serway, R. A., and C. Vuille, *College Physics,* 8th ed., 2008.
Young, H. D., and R. A. Freedman, *Sears and Zemansky's University Physics, With Modern Physics,* 12th ed., 2007.

American Journal of Physics, American Institute of Physics, monthly.
Annals of Physics, 18 issues per year.
Journal of Applied Physics, American Institute of Physics, monthly.
Journal of Physics, A, B, D, and G, Institute of Physics, Section G, monthly; other sections, semimonthly.
Physical Review, American Physical Society, American Institute of Physics, Section B, 4 issues per month; Section D, semimonthly; Sections A, C, E, monthly.
Physical Review Letters, American Physical Society, American Institute of Physics, weekly.
Physics Abstracts (Science Abstracts, Section A), INSPEC, Institution of Electrical Engineers, biweekly.
Physics Letters, Section A, weekly; Section B, weekly.
Physics Reports, 90 issues per year.
Physics Today, American Institute of Physics, monthly.
Proceedings of the Royal Society of London, Series A: Mathematical, Physical and Engineering Sciences, monthly.
Reviews of Modern Physics, American Physical Society, American Institute of Physics, quarterly.

PHYSIOLOGY

Costanzo, L. S., *Physiology,* 4th ed., 2006.
Ganong, W. A., *Review of Medical Physiology,* 22d ed., 2005.
Guyton, A. C., and J. E. Hall, *Textbook of Medical Physiology,* 11th ed., 2005.
Koeppen, B. M., and B. A. Stanton, *Berne and Levy Physiology,* 6th ed., 2008.
Malvin, R. L., et al., *Concepts of Human Physiology,* 1997.
Porterfield, S. P., and B. A. White, *Endocrine Physiology,* 3d ed., 2007.
Prosser, C. L., *Comparative Animal Physiology,* 4th ed., 1991.
Randall, D., et al., *Eckert Animal Physiology: Mechanisms and Adaptations,* 6th ed., 2009.
Rhoades, R. A., and R. G. Pflanzer, *Human Physiology,* 4th ed., 2002.
Senisi Scott, A., and E. Fong, *Body Structures and Functions,* 11th ed., 2009.
Sherwood, L., *Human Physiology: From Cells to Systems,* 6th ed., 2006.
Sherwood, L., H. Klandorf, and P. H. Yancy, *Animal Physiology,* 2005.
Sperelakis, N., *Cell Physiology Source Book: A Molecular Approach,* 2001.

American Journal of Physiology, American Physiological Society, monthly.
Annual Review of Physiology, annually.
Journal of Comparative Physiology, monthly.
Journal of General Physiology, monthly.
Journal of Physiology (England), semimonthly.
Physiological Reviews, American Physiological Society, quarterly.

PLANT ANATOMY

Bowes, B. G., *A Colour Atlas of Plant Structure,* 1999.
Dickison, W. C., *Integrative Plant Anatomy,* 2000.
Evert, R. F., and S. E. Eichhorn, *Esau's Plant Anatomy,* 3d ed., 2005.
Fahn, A., *Plant Anatomy,* 4th ed., 1990.
Mauseth, J. D., *Plant Anatomy,* 1988.
Nabors, M., *Introduction to Botany,* 2004.
Pearson, L. C., *The Diversity and Evolution of Plants,* 1995.
Raven, P. H., R. F. Evert, and S. E. Eichhorn, *Biology of Plants,* 1999.
Rudall, P., *Anatomy of Flowering Plants: An Introduction to Structure and Development,* 3d ed., 2007.

PLANT PATHOLOGY

Agrios, G. N., *Plant Pathology,* 5th ed., 2005.
Dickinson, M. J., *Molecular Plant Pathology,* 2003.
Lucas, G. B., C. L. Campbell, and L. T. Lucas, *Introduction to Plant Diseases: Identification and Management,* 2d ed., 1992.
Lucas, J. A., *Plant Pathology and Plant Pathogens,* 1998.
Matthews, R. E., et al., *Plant Virology,* 4th ed., 2001.
Sinclair, W. A., and H. Lyon, *Diseases of Trees and Shrubs,* 2d ed., 2005.
Strange, R. N., *Introduction of Plant Pathology,* 2003.
Talbot, N., *Plant-Pathogen Interactions,* 2004.
Vidhyasekaran, P., *Concise Encyclopedia of Plant Pathology,* 2004.
Zamir, K. P. (ed.)**,** *Fungal Disease Resistance in Plants: Biochemistry, Molecular Biology, and Genetic Engineering,* 2004.

Annual Review of Phytopathology, annually.
Plant Disease, American Phytopathological Society, monthly.

PLANT PHYSIOLOGY

Blankenship, R. E., *Molecular Mechanisms of Photosynthesis,* 2002.
Chrispeels, M. J., and D. E. Sadava, *Plants, Genes, and Crop Biotechnology,* 2d ed., 2003.
Davies, P. J. (ed.)**,** *Plant Hormones: Biosynthesis, Signal Transduction, Action!,* 2004.
Epstein, E., and A. J. Bloom, *Mineral Nutrition of Plants: Principles and Perspectives,* 2d ed., 2005.
Govindjee, et al. (eds.)**,** *Discoveries in Photosynthesis* (Advances in Photosynthesis and Respiration), 2005.

Hedden, P., and S. G. Thomas (eds.), *Plant Hormone Signaling,* 2006.
Leyser, O., and S. Day, *Mechanisms in Plant Development,* 2003.
Shäfer, E., and F. Nagy (eds.), *Photomorphogenesis in Plants,* 3d ed., 2006.
Taiz, L., and E. Zeiger, *Plant Physiology,* 4th ed., 2006.

Current Opinion in Plant Biology, bimonthly.
The Plant Cell, American Society of Plant Biologists, monthly.
Plant, Cell and Environment, monthly.
Plant Journal, bimonthly.
Plant Physiology, American Society of Plant Biologists, monthly.
Planta, monthly.
Trends in Plant Science, monthly.

PLASMA PHYSICS

Bellan, P. M., *Fundamentals of Plasma Physics,* 2006.
Chen, F. F., *Introduction to Plasma Physics and Controlled Fusion,* 2d ed., 1984, corrected 2d printing, 2006.
Dendy, R. O. (ed.), *Plasma Physics: An Introductory Course,* new edited ed., 1993.
Freidberg, J., *Plasma Physics and Fusion Energy,* 2007.
Goldston, R. J., and P. H. Rutherford, *Introduction to Plasma Physics* (Plasma Physics Series), 1995.
Hutchinson, I. H., *Principles of Plasma Diagnostics,* 2d ed., 2002.
Leontovich, M. A., B. B. Kadomstev, and V. D. Shafranov (eds.), *Reviews of Plasma Physics,* vols. 1–24, 1965–2008.
Parks, G. K., *Physics of Space Plasmas,* 2d ed., 2004.
Sturrock, P. A., *Plasma Physics: An Introduction to the Theory of Astrophysical, Geophysical, and Laboratory Plasmas,* new edited ed., 1994.
Wesson, J., *Tokamaks* (The International Series on Monographs on Physics), 3d ed., 2004.

IEEE Transactions on Plasma Science, Institute of Electrical and Electronics Engineers, Plasma Sciences Society, bimonthly.
Plasma Physics and Controlled Fusion, Institute of Physics, monthly.

POLYMER CHEMISTRY

Allcock, H., et al., *Contemporary Polymer Chemistry,* 3d ed., 2003.
Allen, G., and J. C. Bevington (eds.), *Comprehensive Polymer Science: The Synthesis, Characterization, Reactions and Applications of Polymers: Polymer Characterization,* 1990.
Elias, H. G., *An Introduction to Polymer Science,* 1997.
Feldman, D., and A. Barbalata, *Synthetic Polymers: Technology, Properties, and Applications,* 1996.
Fried, J., and J. Hunger, *Polymer Science and Technology,* 2d ed., 2003.
Kirk-Othmer Encyclopedia of Chemical Technology, 5th ed., 27 vols., 2007.
Mark, H. F. (ed.), *Encyclopedia of Polymer Science and Technology,* 3d ed., 12 vols., 2004.
Morawetz, H., *Polymers: The Origins and Growth of a Science,* 1985.
Odian, G., *Principles of Polymerization,* 4th ed., 2004.
Seymour, R. B., and C. E. Carraher, Jr., *Seymour/Carraher's Polymer Chemistry,* 7th ed., 2007.
Stevens, M. P., *Polymer Chemistry: An Introduction,* 3d ed., 1998.
Young, R. J., and P. Lovell, *Introduction to Polymers,* 2d ed., 2000.

Macromolecules, American Chemical Society.
Polymer, Elsevier.
Progress in Polymer Science, Elsevier.

PRODUCTION ENGINEERING

Anderson, D. M., *Design for Manufacturability & Concurrent Engineering,* 2008.
Black, J T., and S. L. Hunter, *Lean Manufacturing Systems and Cell Design,* 2003.
Heizer, J., and B. Render, *Principles of Operations Management,* 6th ed., 2005.
Kalpakjian, S., and S. R. Schmid, *Manufacturing Engineering and Technology,* 5th ed., 2005.
Montgomery, D., *Introduction to Statistical Quality Control,* 5th ed., 2005.
Tompkins, J. A., and J. A. White, *Facilities Planning,* 3d ed., 2002.
White, J. A. (ed.), *Production Handbook,* 1987.
Yang, K., and B. S. El-Haik, *Design for Six Sigma: A Roadmap for Product Development,* 2003.

PROPULSION

Cumpsty, N., *Jet Propulsion,* 2d ed., 2003.
El-Sayed, A. F., *Aircraft Propulsion and Gas Turbine Engines,* 2008.
Farokhi, S., *Aircraft Propulsion,* 2008.
Hill, P., and C. Peterson, *Mechanics and Thermodynamics of Propulsion,* 2d ed., 1992.
Hünecke, K., *Jet Engines,* 1997.
Kerrebrock, J. L., *Aircraft Engines and Gas Turbines,* 2d ed., 1992.
Kroes, M. J., and T. W. Wild, *Aircraft Powerplants,* 7th ed., 1994.
Mattingly, J. D., *Elements of Propulsion: Gas Turbines and Rockets,* 2006.
Sutton, G. P., and O. Biblarz, *Rocket Propulsion Elements: An Introduction to the Engineering of Rockets,* 7th ed., 2001.
Treager, I. E., *Aircraft Gas Turbine Engine Technology,* 3d ed., 1995.

PSYCHIATRY

American Psychiatric Association, *Diagnostic and Statistical Manual of Mental Disorders,* 4th ed., 2000.
Coccaro, E. F. (ed.), *Aggression: Psychiatric Assessment and Treatment,* 2003.
First, M. B., and A. Tasman, *Clinical Guide to the Diagnosis and Treatment of Mental Disorders,* 2006.
Freud, S., et al., *An Outline of Psychoanalysis,* 2003.
Lambert, M. J., *Bergin and Garfield's Handbook of Psychotherapy and Behavior Change,* 2004.
Tasman, A., et al., *Psychiatry,* 3d ed., 2008.
Waldinger, R. J., *Psychiatry for Medical Students,* 3d ed., 1997.

American Journal of Psychiatry, American Psychiatric Association, monthly.
Psychiatric News, American Psychiatric Association, semimonthly.

PSYCHOLOGY

Aronson, E., *The Social Animal,* 10th ed., 2007.
Carlson, N. R., *Physiology of Behavior,* 9th ed., 2006.
Gerow, J. R., and K. Bordens, *Psychology: An Introduction,* 8th ed., 2004.
Gerow, J., K. Bordens, and E. Blanch-Payne, *General Psychology: With Spotlights on Diversity,* 2007.
Gleitman, H., D. Reisberg, and J. Gross, *Psychology,* 7th ed., 2007.
Kalat, J. W., *Introduction to Psychology,* 8th ed., 2007.

American Psychologist, American Psychological Association, 9 issues per year.
Annual Review of Psychology, annually.

QUANTUM MECHANICS

Cohen-Tannoudji, C., B. Diu, and F. Laloe, *Quantum Mechanics,* 2 vols., 1977.
Dirac, P. A., *Principles of Quantum Mechanics,* 4th ed., 1958, reprint, 1982.
Eisberg, R., and R. Resnick, *Quantum Physics Of Atoms, Molecules, Solids, Nuclei, and Particles,* 2d ed., 1985.
Gottfried, K., and T. M. Yan, *Quantum Mechanics: Fundamentals,* 2d ed., 2004.
Griffiths, D. J., *Introduction to Quantum Mechanics,* 2d ed., 2004.
Liboff, R. L., *Introductory Quantum Mechanics,* 4th ed., 2002.
Mandl, F., *Quantum Mechanics,* 1992.
McMahon, D., *Quantum Mechanics Demystified,* 2006.
Miller, D. A. B., *Quantum Mechanics for Scientists and Engineers,* 2008.
Park, D., *Introduction to the Quantum Theory* (International Series in Pure and Applied Physics), 3d ed., 1992.
Shankar, R., *Principles of Quantum Mechanics,* 2d ed., 1994.
Zettili, N., *Quantum Mechanics: Concepts and Applications,* 2001.

RADAR

Barton, D. K., and S. A. Leonov, *Radar Technology Encyclopedia,* 1997.
Cole, H., *Understanding Radar,* 2d ed., 1992.
Edde, B., *Radar: Principles, Technology, Applications,* 1992.
Raemer, H. R., *Radar Systems Principles,* 1997.
Raghavan, S., *Radar Meteorology,* 2003.
Richards, M. A., *Fundamentals of Radar Signal Processing,* 2005.
Skolnik, M. I., *Introduction to Radar Systems,* 3d ed., 2001.
Skolnik, M. I. (ed.)**,** *Radar Handbook,* 3d ed., 2008.
Stimson, G. W., *Introduction to Airborne Radar,* 2d ed., 1998.
Toomay, J. C., and P. J. Hannen, *Radar Principles for the Non-Specialist,* 3d ed., 2004.

IEE Proceedings: Radar, Sonar and Navigation, Institution of Electrical Engineers, semimonthly.

RADIO COMMUNICATIONS

American Radio Relay League, *The ARRL Handbook for Radio Communications,* annually.
Carson, R. S., *Radio Communications Concepts: Analog,* 1990.
Jacobs, G. (ed.)**,** *World Radio TV Handbook,* annually.
Proakis, J. G., and M. Salehi, *Communication Systems Engineering,* 2d ed., 2001.
Rappaport, T. S., *Wireless Communications: Principles and Practice,* 2d ed., 2002.
Roddy, D., *Satellite Communications,* 4th ed., 2006.
Sabin, W., and E. Schoenike (eds.)**,** *Single-Sideband Systems and Circuits,* 2d ed., 1995.
Steele, R., and L. Hanzo, *Mobile Radio Communications,* 2d ed, 1999.
Torrieri, D., *Principles of Spread-Spectrum Communication Systems,* 2005.
Valkenburg, M., and W. Middleton (eds.)**,** *Reference Data for Engineers: Radio, Electronics, Computer, and Communications,* 9th ed., 2002.
Williams, E. A. (ed.)**,** *NAB Engineering Handbook,* 10th ed., 2007.

IEEE Transactions on Broadcasting, Institute of Electrical and Electronics Engineers, Broadcast Technology Society, quarterly.

RADIOCHEMISTRY

Adloff, J. P., and R. Guillaumont, *Fundamentals of Radiochemistry,* 1993.
Arnikar, H. J., *Essentials of Nuclear Chemistry,* 2d ed., 1987.
Choppin, G. R., J.-O. Liljenzin, and J. Rydberg, *Radiochemistry and Nuclear Chemistry: Nuclear Chemistry, Theory and Applications,* 2d ed., 1995.

Journal of Labelled Compounds and Radiopharmaceuticals, Wiley, monthly.
Radiation Physics and Chemistry, Elsevier, monthly.

RELATIVITY

Bohm, D., *The Special Theory of Relativity,* 1965, reprint, 1996.
Carroll, S. M., *Spacetime and Geometry: An Introduction to General Relativity,* 2003.
Hartle, J. B., *Gravity: An Introduction to Einstein's General Relativity,* 2003.
Jagerman, L., *The Mathematics of Relativity for the Rest of US,* 2001.
McMahon, D., *Relativity Demystified,* 2006.
Misner, C., K. Thorne, and J. Wheeler, *Gravitation,* 1973.
Rindler, W., *Relativity: Special, General, and Cosmological,* 2d ed., 2006.
Schutz, B., *Gravity From the Ground Up,* 2003.
Schwarz, P. M., and J. H. Schwarz, *Special Relativity: From Einstein to Strings,* 2004.
Stephani, H., *Relativity: An Introduction to Special and General Relativity,* 3d ed., 2004.
Taylor, E. F., and J. A. Wheeler, *Spacetime Physics: Introduction to Special Relativity,* 2d ed., 1992.
Wald, R., *General Relativity,* 1984.

Will, C. M., *Was Einstein Right? Putting General Relativity to the Test,* 2d rev. ed., 1993.

General Relativity and Gravitation, International Committee on General Relativity and Gravitation, monthly.

SOIL

Bowles, J. E., *Engineering Properties of Soils and Their Measurement,* 4th ed., 1992.
Brady, N. C., and R. R. Weil, *The Nature and Properties of Soils,* 13th ed., 2001.
Crunkilton, J. R., et al., *The Earth and Agriscience,* 1995.
Daniels, R. B., and R. D. Hammer, *Soil Geomorphology,* 1992.
Dowdy, R. H. (ed.)**,** *Chemistry in the Soil Environment,* 1981.
Foster, A. B., and D. A. Bosworth, *Approved Practices in Soil Conservation,* 5th ed., 1982.
Foth, H. D., *Fundamentals of Soil Science,* 8th ed., 1990.
Hanks, R. J., *Applied Soil Physics: Soil Water and Temperature Applications,* 2d ed., 1992.
Morgan, R. P. C., *Soil Erosion and Conservation,* 2d ed., 1995.
Plaster, E. J., *Soil Science and Management,* 3d ed., 1997.

Agronomy Journal, American Society of Agronomy, bimonthly.

SOLID-STATE PHYSICS

Chaikin, P. M., and T. C. Lubensky, *Principles of Condensed Matter Physics,* 1995, paperback 2000.
Ehrenreich, H., et al. (eds.)**,** *Solid State Physics,* vols. 1–60, 1955–2006.
Grosso, G., and G. P. Paravicini, *Solid State Physics,* 2000.
Hook, J. R., and H. E. Hall, *Solid State Physics,* 2d ed., 1991, corrected reprint, 2000.
Ibach, H., and H. Luth, *Solid State Physics: An Introduction to Principles of Materials Science* (Advanced Texts in Physics), 3d ed., 2003.
Kittel, C., *Introduction to Solid State Physics,* 8th ed., 2005.
Marder, M. P., *Condensed Matter Physics,* 2000.
Myers, H. P., *Introductory Solid State Physics,* 2d ed., 1997.
Omar, M. A., *Elementary Solid State Physics,* revised printing, 1993.
Snoko, D. W., *Solid State Physics: Essential Concepts,* 2008.
Tanner, B. K., *Introduction to the Physics of Electrons in Solids,* 1995.

Advances a Solid State Physics, annually.
Applied Physics A, Deutsche Physikalische Gesellschaft, 12 issues per year.
Journal of Physics: Condensed Matter, Institute of Physics, 50 issues per year.
Journal of Physics and Chemistry of Solids, monthly.
Physica B, 60 issues per year.
Physical Review B, American Institute of Physics, 4 issues per month.
Solid State Communications, 48 issues per year.

SPACE TECHNOLOGY

Boden, D. S., and W. J. Larson, *Cost-Effective Space Mission Operations,* 2d ed., 2006.
Brown, C. D., *Elements of Spacecraft Design* (AIAA Education Series), 2002.
Griffin, M. D., and J. B. French, *Space Vehicle Design* (AIAA Education Series), 2004.
Pisacane, V. L. (ed.)**,** *Fundamentals of Space Systems* (The John Hopkins University/Applied Physics Laboratory Series in Science & Engineering), 2d ed., 2005.
Roddy, D., *Satellite Communications,* 4th ed., 2006.
Sellers, J. J., et al., *Understanding Space: An Introduction to Astronautics,* 3d ed., 2007.
Tribble, A. C., *The Space Environment: Implications for Spacecraft Design,* rev. expanded ed., 2003.
Wertz, J. R., and W. J. Larson (eds.)**,** *Space Mission Analysis and Design,* 3d ed., 1999.

Aerospace America, American Institute of Aeronautics and Astronautics, monthly.
Journal of Spacecraft and Rockets, American Institute of Aeronautics and Astronautics, bimonthly.

SPECTROSCOPY

Andrews, D. L., and A. A. Demidov (eds.)**,** *An Introduction to Laser Spectroscopy,* 2d ed., 2002.
Bernath, P. F., *Spectra of Atoms and Molecules* (Topics in Physical Chemistry), 2d ed., 2005.
Briggs, D., and M. P. Seah (eds.)**,** *Practical Surface Analysis: Auger and X-Ray Photoelectron Spectroscopy,* 2d ed., 1990.
De Hoffmann, E., and V. Stroobant, *Mass Spectrometry: Principles and Applications,* 3d ed., 2007.
Demtröder, W., *Laser Spectroscopy,* 4th ed., 2 vols., 2008.
Dickson, D. P. E., and F. J. Berry (eds.)**,** *Mössbauer Spectroscopy,* 1986.
Ferraro, J. R., K. Nakamoto, and C. W. Brown, *Introductory Raman Spectroscopy,* 2d ed., 2003.
Graybeal, J. D., *Molecular Spectroscopy,* rev. ed., 1993.
Hollas, J. M., *Modern Spectroscopy,* 4th ed., 2004.
Hüfner, S., *Photoelectron Spectroscopy,* 3d ed., 2003.
Jacobsen, N. E., *NMR Spectroscopy Explained,* 2007.
Schmidt, V., et. al., *Electron Spectrometry of Atoms Using Synchrotron Radiation,* 1997.
Stuart, B. H., *Infrared Spectroscopy: Fundamentals and Applications,* 2004.
Tsuji, K., J. Injuk, and R. Van Grieken (eds.)**,** *X-Ray Spectrometry: Recent Technological Advances,* 2004.

Applied Spectroscopy, Society for Applied Spectroscopy, monthly.
Journal of Magnetic Resonance Spectroscopy, 4 copies per year.
Spectrochimica Acta, Parts A and B, monthly.

STATISTICAL MECHANICS

Bowley, R., and M. Sánchez, *Introductory Statistical Mechanics,* 2d ed., 1999.
Garrod, C., *Statistical Mechanics and Thermodynamics,* 1995.

Greiner, W., L. Neise, and H. Stöcker, *Thermodynamics and Statistical Mechanics,* 1995.
Kubo, R., *Statistical Mechanics,* 2d ed., 1988.
McQuarrie, D. A., *Statistical Mechanics,* 2000.
Pathria, R. K., *Statistical Mechanics,* 2d ed., 1996.
Ter Harr, D., *Elements of Statistical Mechanics,* 3d ed., 1995.
Van Vliet, C. M., *Equilibrium and Non-equilibrium Statistical Mechanics,* 2008.
Zwanzig, R., *Nonequilibrium Statistical Mechanics,* 2001.

Journal of Statistical Physics, 12 issues per year.
Physica A, 52 issues per year.

STATISTICS AND PROBABILITY

Coolidge, F. L., *Statistics: A Gentle Introduction,* 2d ed., 2006
DeGroot, M. H., and M. J. Schervish, *Probability and Statistics,* 3d ed., 2001.
Freedman, D., R. Pisani, and R. Purves, *Statistics,* 4th ed., 2007.
Larsen, R. J., and M. L. Marx, *Introduction to Mathematical Statistics and Its Applications,* 4th ed., 2005.
Moore, D. S., G. P. McCabe, and B. Craig, *Introduction to the Practice of Statistics,* 6th ed., 2007.
Naiman, A., R. Rosenfeld, and G. Zirkel, *Understanding Statistics,* 4th ed., 1996.
Rohatgi, V. K., and A. K. M. Ehsanes Saleh, *An Introduction to Probability and Statistics,* 2001.
Ross, S. M., *A First Course in Probability,* 8th ed., 2009.
Sanders, D. H., and R. K. Smidt, *Statistics: A First Course,* 6th ed., 1999.
Tabachnick, B. G., and L. S. Fidell, *Using Multivariate Statistics,* 5th ed., 2006.
Triola, M. F., *Elementary Statistics,* 10th ed., 2007.
Weiss, N. A., *Introductory Statistics,* 8th ed., 2007.

Annals of Statistics, Institute of Mathematical Statistics, bimonthly.
***Journal of the American Statistical Association* (JASA)**, quarterly.

SYSTEMATICS

Avise, J. C., *Phylogeography: The History and Formation of Species,* 2000.
Brooks, D. R., and D. A. McLennan, *The Nature of Diversity: An Evolutionary Voyage of Discovery,* 2002.
Ereshefsky, M., *The Poverty of the Linnaean Hierarchy: A Philosophical Study of Biological Taxonomy,* 2001.
Felsenstein, J., *Inferring Phylogenies,* 2004.
Gaston, K. J., and J. I. Spicer, *Biodiversity: An Introduction,* 2004.
Hillis, D. M., C. Moritz, and B. K. Mable (eds.)**,** *Molecular Systematics,* 2d ed., 1996.
Lomolino, M. V., B. R. Riddle, and J. H. Brown, *Biogeography,* 3d ed., 2006.
Schuh, R. T., *Biological Systematics: Principles and Applications,* 2000.
Wilson, R. A. (ed.)**,** *Species: New Interdisciplinary Essays,* 1999.

Cladistics, Willi Hennig Society, bimonthly.
Systematic Biology, Society of Systematic Biologists, bimonthly.

SYSTEMS ENGINEERING

Blanchard, B. S., *System Engineering Management,* 4th ed, 2008.
Booher, H. (ed.)**,** *Handbook of Human Systems Integration,* 2003.
Haimes, Y. Y., *Risk Modeling, Assessment, and Management,* 3d ed., 2008.
Kossiakoff, A., and W. N. Sweet, *Systems Engineering: Principles and Practice,* 2003.
Rouse, W. B., *Essential Challenges of Strategic Management,* 2001.
Sage, A. P., and W. B. Rouse (eds.)**,** *Handbook of Systems Engineering and Management,* 1999.
Sage, A. P., and J. E. Armstrong, *Introduction to Systems Engineering,* 2000.
Sheridan, T. B., *Humans and Automation: System Design and Research Issues,* 2002.

TELECOMMUNICATIONS

Clayton, J., *McGraw-Hill Illustrated Telecom Dictionary,* 4th ed., 2002.
Freeman, R. L., *Telecommunication System Engineering,* 4th ed., 2004.
Frenzel, L. E., *Principles of Electronic Communication Systems,* Student Edition, 3d ed., 2007.
Goleniewski, L., and K. W. Jarrett, *Telecommunications Essentials,* 2d ed., 2006.
Green, J. H., *The Irwin Handbook of Telecommunications,* 5th ed., 2005.
Haykin, S., and M. Moher, *An Introduction to Digital and Analog Communications,* 2d ed., 2007.
Horak, R., *Telecommunications and Data Communications Handbook,* 2007.
Lathi, B. P., *Modern Digital and Analog Communication Systems* (Oxford Series in Electrical and Computer Engineering), 3d ed., 1998.
Miller, G. M., and J. Beasley, *Modern Electronic Communication,* 9th ed., 2007.
Newton, H., *Newton's Telecom Dictionary,* published annually.
Proakis, J. G., and D. K. Manolakis, *Digital Signal Processing,* 4th ed., 2006.
Proakis, J. G., and M. Salehi, *Digital Communications,* 5th ed., 2008.
Shrader, R. L., *Electronic Communications,* 6th ed., 1991.

Bell Labs Technical Journal, John Wiley, quarterly.
IEEE Spectrum, Institute of Electrical and Electronics Engineers, Inc. monthly.

TELEPHONY

Bellamy, J. C., *Digital Telephony* (Wiley Series in Telecommunications and Signal Processing), 3d ed., 2000.
Bigelow, S. J., J. J. Carr, and S. Winder, *Understanding Telephone Electronics,* 4th ed., 2001.
Collins, D., *Carrier Grade Voice over IP,* 2d ed., 2003.
Minoli, D., and E. Minoli, *Delivering Voice over IP Networks,* 2d ed., 2002.
Noll, A. M., *Introduction to Telephones and Telephone Systems* (Artech House Telecommunications Library), 3d ed., 1998.

TELEVISION

Boston, J., *DTV Survival Guide,* 2000.
Ciciora, W., et al., *Modern Cable Television Technology,* 2d ed., 2004.
Poynton, C., *Digital Video and HDTV: Algorithms and Interfaces,* 2003.
Robin, M. and M. Poulin, *Digital Television Fundamentals,* 2000.
Symes, P., *Digital Video Compression,* 2003.
Whitaker, J., *Mastering Digital Television: The Complete Guide to DTV Conversion,* 2006.
Whitaker, J., and B. K. Benson, *Standard Handbook of Video and Television Engineering,* 4th ed., 2003.
Williams, E. A. (ed.)**,** *NAB Engineering Handbook,* 10th ed., 2007.

TEXTILES AND FIBERS

Collier, B. J., and P. G. Tortora, *Understanding Textiles,* 7th ed., 2008.
Kadolph, S. J., and A. L. Langford, *Textiles,* 10th ed., 2006.
Lewin, M., *Handbook of Fiber Chemistry,* 3d ed., 2006.

***Textile Research Journal*,** Textile Research Institute, monthly.

THEORETICAL PHYSICS

Arfken, G. B., and H. J. Weber (eds.)**,** *Mathematical Methods for Physicists,* 6th ed., 2005.
Courant, R., and D. Hilbert, *Methods of Mathematical Physics,* 2 vols., 1989.
Koks, D., *Explorations in Mathematical Physics: The Concepts Behind an Elegant Language,* 2006.
Lam, L. (ed.)**,** *Nonlinear Physics for Beginners: Fractals, Chaos, Solitons, Pattern Formation, Cellular Automata, and Complex Systems,* 1998.
Landau, L. D., et al., *Course of Theoretical Physics,* 9 vols., 1959–1987.
Reed, M., and B. Simon, *Methods of Modern Mathematical Physics,* 4 vols., 1975–1980.
Thirring, W., *Classical Mathematical Physics,* 3d ed., 1997, paper, 2003.
Thirring, W., *Quantum Mathematical Physics,* 2d ed., 2002, rev. printing, 2003.

***Communications in Mathematical Physics*,** 18 issues per year.
***Journal of Mathematical Physics*,** American Institute of Physics, monthly.

THERMODYNAMICS AND HEAT

Black, W. Z., and J. G. Hartley, *Thermodynamics* (English/SI Version), 3d ed., 1997.
Cengel, Y. A., and M. A. Boles, *Thermodynamics: An Engineering Approach,* 6th ed., 2007.
Gokcen, N. A., and R. G. Reddy, *Thermodynamics,* 2d ed., 1996.
Holman, J. P., *Thermodynamics,* 4th ed., 1988.
Moran, M. J., and H. N. Shapiro, *Fundamentals of Engineering Thermodynamics,* 6th ed., 2007.
Sandler, S. I., *Chemical, Biochemical, and Engineering Thermodynamics,* 4th ed., 2006.
Sonntag, R. E., G. J. Van Wylen, and C. Borgnakke, *Fundamentals of Thermodynamics,* 7th ed., 2008.
Wark, K., and D. E. Richards, *Thermodynamics,* 6th ed., 1999.
Wood, S. E., and R. Battino, *Thermodynamics of Chemical Systems,* 1990.
Zemansky, M. W., and R. H. Dittman, *Heat and Thermodynamics,* 7th ed., 1997.

TRANSPORTATION ENGINEERING

Brockenbrough, R. L., and K. J. Boedecker (eds.)**,** *Highway Engineering Handbook,* 2d ed., 2003.
Chowdhury, M. A., and A. W. Sadek, *Fundamentals of Intelligent Transportation Systems Planning,* 2003.
Garber, N. J., and L. A. Hoel, *Traffic and Highway Engineering,* 4th ed., 2008.
Khisty, C. J., and K. Lall, *Transportation Engineering: An Introduction,* 3d ed., 2002.
Mannering, F. L., W. P. Kilareski, and S. S. Washburn, *Principles of Highway Engineering and Traffic Analysis,* 4th ed., 2008.
Papacostas, C. S., and P. D. Prevedouros, *Transportation Engineering,* 3d ed., 2000.
Wright, P. H., and K. Dixon, *Highway Engineering,* 7th ed., 2003.

***Journal of Transportation Engineering*,** American Society of Civil Engineering, bimonthly.
***Transportation Journal*,** American Society of Traffic and Logistics, quarterly.

TRIGONOMETRY

Aufmann, R. N., V. C. Barker, and R. D. Nation, *College Trigonometry,* 6th ed., 2007.
Baley, J. D., and G. Sarell, *Trigonometry,* rev. 3d ed., 2002.
Drooyan, I., and C. C. Carico, *Trigonometry: An Analytical Approach,* 6th ed., 1991.
Heineman, E. R., and J. D. Tarwater, *Plane Trigonometry,* 7th ed., 1993.
Hungerford, T. W., and R. Mercer, *Trigonometry,* 1992.
Smith, K. J., *Trigonometry for College Students,* 7th ed., 1998.
Sullivan, M., *Trigonometry: A Unit Circle Approach,* 8th ed., 2007.

VERTEBRATE ZOOLOGY

See Zoology.

VETERINARY MEDICINE

Coetzer, J. A. W., and R. C. Tustin (eds.)**,** *Infectious Diseases of Livestock,* 2d ed., 2005.
Ettinger, S. J., and E. C. Feldman, *Textbook of Veterinary Internal Medicine,* 6th ed., 2004.
Fenner, W. R., *Quick Reference to Veterinary Medicine,* 3d ed., 2000.
Kahn, C. M. (ed.)**,** *The Merck Veterinary Manual,* 9th ed., 2005.
Salmon, M. D. (ed.)**,** *Animal Disease Surveillance and Survey Systems,* 2003.
Timoney, J. F., *Hagan and Bruner's Microbiology and Infectious Diseases of Domestic Animals,* 8th ed., 1988.

Williams, E. S., and I. K. Barker (eds.)**,** *Infectious Diseases of Wild Mammals,* 3d ed., 2001.

VIROLOGY

Cann, A. J., *Principles of Molecular Virology,* 4th ed., 2005.
Dimmock, N., A. Easton, and K. Leppard, *Introduction to Modern Virology,* 6th ed., 2007.
Flint, S. J., et al. (eds.)**,** *Principles of Virology,* 3d ed., 2009.
ICTVdB, The Universal Virus Database of the International Committee on Taxonomy of Viruses, maintained by C. Büchen-Osmond [http://www.ncbi.nlm.nih.gov/ICTVdB/ICTVdBintro.htm].
Knipe, D. M., et al. (eds.)**,** *Fields Virology,* 2006.
Matthews, R. E., et al., *Matthews' Plant Virology,* 4th ed., 2001.
Nathanson, N. (ed.)**,** *Viral Pathogenesis and Immunity,* 2d ed., 2007.
Richman, D. D., R. J. Whitley, and F. G. Hayden, *Clinical Virology,* 2d ed., 2002.
Voyles, B. A., *Biology of the Viruses,* 2d ed., 2002.
Wagner, E. K., et al. (eds.)**,** *Basic Virology,* 3d ed., 2007.

Advances in Virus Research, irregularly.
Journal of Medical Virology, monthly.
Journal of Virology, American Society for Microbiology, semimonthly.

ZOOLOGY

Alcock, J., *Animal Behavior: An Evolutionary Approach,* 8th ed., 2005.
Barton, M., *Bond's Biology of Fishes,* 3d ed., 2006.
Brusca, R. C., and G. J. Brusca, *Invertebrates,* 2d ed., 2003.
Grzimek, B. (ed.)**,** *Grzimek's Encyclopedia of Mammals,* 2d ed., 1989.
Hickman, C. P., et al., *Integrated Principles of Zoology,* 14th ed., 2008.
Hildebrand, M., and G. Goslow, *Analysis of Vertebrate Structure,* 5th ed., 2001.
Kardong, K. V., *Vertebrates,* 5th ed., 2008.
Linzey, D. W., *Vertebrate Biology,* 2001.
McFarland, D., *Animal Behavior, Psychobiology, Ethology, and Evolution,* 3d ed., 1998.
Miller, S. A., and J. P. Harley, *Zoology,* 7th ed., 2007.
Pearse, V., et al., *Living Invertebrates,* 1987.
Pough, F., et al., *Vertebrate Life,* 8th ed., 2008.
Ruppert, E. E., R. S. Fox, and R. D. Barnes, *Invertebrate Zoology,* 7th ed., 2003.
Vaughan, T. A., J. M. Ryan, and N. J. Czaplewski, *Mammalogy,* 4th ed., 2000.

Integrative and Comparative Biology, 6 issues per year.
Journal of Zoology, monthly.

Measurement systems

U.S. Customary System and the metric system

Over the past 200 years or so, scientists and engineers have used two major systems of units in measurement. These are commonly called the U.S. Customary System (inherited from the British Imperial System) and the metric system (developed at the time of the French Revolution).

In the U.S. Customary System the units yard and pound with their divisions, such as the inch, and multiples, such as the ton, are basic. The metric system has been adopted for general use by most countries. It is used nearly everywhere for precise measurements in science. The meter and kilogram with their multiples, such as the kilometer, and fractions, such as the gram, are basic to the metric system. Until the second half of the twentieth century, most of the base units in these systems were defined in terms of specific physical artifacts, such as the International Prototype Kilogram, a platinum-iridium cyclinder maintained in Sèvres, France.

In the U.S. Cutomary System, units of the same kind are related almost at random. For example, there are the units of length, the inch, yard, and mile. In the metric system the relationships between units of the same kind are strictly decimal (millimeter, meter, and kilometer).

However, in technical writing there is no uniformity within each of these two systems as to the choice of units for the same quantities. For example, the hour or the second, the foot or the inch, and the centimeter or the millimeter could be chosen as the unit of measurement for the quantities time and length.

International System

To simplify matters and to make communication more understandable, an internationally accepted system of units came into use in 1960. This is termed the International System of Units, which is abbreviated SI in all languages (from the French Système International d'Unités).

SI is fundamentally metric. The base units are derived from scientific formulas or natural constants. Their definitions have been modified from time to time in order to allow for their more accurate realization as techniques of measurement have evolved. In several instances, artifact standards have been replaced by physically invariant quantities, such as atomic transition frequencies and fundamental physical constants. For example, the meter in the SI is defined as the length of the path traveled by light in vacuum during a time interval of 1/299 792 458 of a second. The effect of this definition, which was adopted in 1983, is to fix the speed of light at exactly 299 792 458 meters per second.

The second in the SI is defined as the duration of 9 192 631 770 periods of the radiation corresponding to the transition between two hyperfine levels of the ground state of the cesium-133 atom.

Interestingly, the kilogram, the SI unit of mass, is still the mass of the kilogram kept at Sèvres. However, it is probable that eventually the unit will be redefined in terms of atomic mass or through a definition that will fix the value of the Planck constant h.

Although the SI is increasingly used by scientists and engineers, there are some other units in everyday use which will probably remain, for example, minute, hour, day, degree (angle), and liter. The point should be made, however, that these terms will not be employed in a scientific context if the SI is fully adopted.

Because of their extremely common use among scientists, several units are still permitted in conjunction with SI units, for example, the electronvolt, rad, roentgen, barn, and curie. In time their usage might be phased out.

One futher point is that in October 1967 the Thirteenth General Conference of Weights and Measures decided to name the SI unit of thermodynamic temperature "kelvin" (symbol K) instead of "degree Kelvin" (symbol °K). For example, the notation is 273 K and not 273°K.

The base units and derived units of the SI are shown in **Tables 1** and **2**.

In the SI the prefixes differ from a unit in steps of 10^3. A list of prefix terms, symbols, and their factors is given in **Table 3**. Some examples of the use of these prefixes follow:

$$1000 \text{ m} = 1 \text{ kilometer} = 1 \text{ km}$$
$$1000 \text{ V} = 1 \text{ kilovolt} = 1 \text{ kV}$$
$$1\,000\,000\ \Omega = 1 \text{ megohm} = 1 \text{ M}\Omega$$
$$0.000\,000\,001 \text{ s} = 1 \text{ nanosecond} = 1 \text{ ns}$$

Only one prefix is to be employed for a unit. For example:

$$1000 \text{ kg} = 1 \text{ Mg} \quad \text{not } 1 \text{ kkg}$$
$$10^{-9} \text{ s} = 1 \text{ ns} \quad \text{not } 1 \text{ m}\mu\text{s}$$
$$1\,000\,000 \text{ m} = 1 \text{ Mm} \quad \text{not } 1 \text{ kkm}$$

Also, when a unit is raised to a power, the power applies to the whole unit including the prefix. For example:

$$\text{km}^2 = (\text{km})^2 = (1000 \text{ m})^2 = 10^6 \text{ m}^2$$
$$\text{not } 1000 \text{ m}^2$$

Some common units defined in terms of SI units are given in **Table 4** (the definitions in the fourth column are exact).

Conversion factors for the measurement systems

This Encyclopedia has retained the U.S. Customary and metric systems, but has incorporated SI units in many cases. Conversion factors between the three measurement systems are given in **Table 5** for some prevalent units; in each of the subtables the user proceeds as follows:

To convert a quantity expressed in a unit in the left-hand column to the equivalent in a unit in the top row of a subtable, multiply the quantity by the factor common to both units. For example, to convert 7 ft to the equivalent in meters, go to subtable A, "Units of length." and find 1 ft in the left-hand column and m in the top row. The conversion factor common to these units is 0.3048. Therefore, 7 ft = 7 × 0.3048 = 2.1336 m.

The conversion factors have been carried out to seven significant figures, as derived from the fundamental constants and the definitions of the units. However, this does not mean that the factors are always known to that accuracy. Numbers followed by ellipses are to be continued indefinitely with repetition of the same pattern of digits. Factors written with fewer than seven significant digits are exact values. Numbers followed by an asterisk are definitions of the relation between the two units.

TABLE 1. Base units of the International System

Quantity	Name of unit	Unit symbol
length	meter	m
mass	kilogram	kg
time	second	s
electric current	ampere	A
temperature	kelvin	K
luminous intensity	candela	cd
amount of substance	mole	mol

TABLE 2. Derived units of the International System*

Quantity	Name of unit	Unit symbol, or unit expressed in terms of other SI units	Unit expressed in terms of SI base units
plane angle	radian	rad	$m/m = 1$
solid angle	steradian	sr	$m^2/m^2 = 1$
area	square meter		m^2
volume	cubic meter		m^3
frequency	hertz	Hz	s^{-1}
density	kilogram per cubic meter		kg/m^3
velocity	meter per second		m/s
angular velocity	radian per second	rad/s	$m/(m \cdot s) = s^{-1}$
acceleration	meter per second squared		m/s^2
angular acceleration	radian per second squared	rad/s^2	$m/(m \cdot s^2) = s^{-2}$
volumetric flow rate	cubic meter per second		m^3/s
force	newton	N	$kg \cdot m/s^2$
surface tension	newton per meter, joule per square meter	$N/m, J/m^2$	kg/s^2
pressure	pascal, newton per square meter	$Pa, N/m^2$	$kg/(m \cdot s^2)$
viscosity, dynamic	pascal-second, newton-second per square meter	$Pa \cdot s, N \cdot s/m^2$	$kg/(m \cdot s)$
viscosity, kinematic	meter squared per second		m^2/s
work, torque, energy, quantity of heat	joule, newton-meter, watt-second	$J, N \cdot m, W \cdot s$	$kg \cdot m^2/s^2$
power, heat flux	watt, joule per second	W, J/s	$kg \cdot m^2/s^3$
heat flux density	watt per square meter	W/m^2	kg/s^3
volumetric heat release rate	watt per cubic meter	W/m^3	$kg/(m \cdot s^3)$
heat transfer coefficient	watt per square meter kelvin	$W/(m^2 \cdot K)$	$kg/(s^3 \cdot K)$
heat capacity (specific)	joule per kilogram kelvin	$J/kg \cdot K)$	$m^2/(s^2 \cdot K)$
capacity rate	watt per kelvin	W/K	$kg \cdot m^2/(s^3 \cdot K)$
thermal conductivity	watt per meter kelvin	$W/(m \cdot K), \frac{J \cdot m}{s \cdot m^2 \cdot K}$	$kg \cdot m/(s^3 \cdot K)$
quantity of electricity	coulomb	C	$A \cdot s$
electromotive force	volt	V, W/A	$kg \cdot m^2/(A \cdot s^3)$
electric field strength	volt per meter	V/m	$kg \cdot m/(A \cdot s^3)$
electric resistance	ohm	Ω, V/A	$kg \cdot m^2/(A^2 \cdot s^3)$
electric conductance	siemens	S, A/V	$A^2 \cdot s^3/(kg \cdot m^2)$
electric conductivity	ampere per volt meter	$A/(V \cdot m)$	$A^2 \cdot s^3/(kg \cdot m^3)$
electric capacitance	farad	$F, A \cdot s/V$	$A^2 \cdot s^4/(kg \cdot m^2)$
magnetic flux	weber	$Wb, V \cdot s$	$kg \cdot m^2/A \cdot s^2)$
inductance	henry	$H, V \cdot s/A$	$kg \cdot m^2/(A^2 \cdot s^2)$
magnetic permeability	henry per meter	H/m	$kg \cdot m/(A^2 \cdot s^2)$
magnetic flux density	tesla, weber per square meter	$T, Wb/m^2$	$kg/(A \cdot s^2)$
magnetic field strength	ampere per meter		A/m
magnetomotive force	ampere		A
luminous flux	lumen	$lm, cd \cdot sr$	$cd \cdot m^2/m^2 = cd$
luminance	candela per square meter		cd/m^2
illumination	lux, lumen per square meter	$lx, lm/m^2, cd \cdot sr/m^2$	$cd \cdot m^2/m^4 = cd/m^2$
activity (of radionuclides)	becquerel	Bq	s^{-1}
absorbed dose	gray	Gy, J/kg	m^2/s^2
dose equivalent	sievert	Sv, J/Kg	m^2/s^2
catalytic activity	katal	kat	mol/s

*The degree Celsius (°C), listed in Table 4, is also a derived unit of the International System.

TABLE 3. Prefixes for units in the International System

Prefix	Symbol	Power	Example	Prefix	Symbol	Power	Example
yotta	Y	10^{24}		deci	d	10^{-1}	
zetta	Z	10^{21}		centi	c	10^{-2}	centimeter (cm)
exa	E	10^{18}		milli	m	10^{-3}	milligram (mg)
peta	P	10^{15}		micro	μ	10^{-6}	microgram (μg)
tera	T	10^{12}	terawatt (TW)	nano	n	10^{-9}	nanosecond (ns)
giga	G	10^{9}	gigawatt (GW)	pico	p	10^{-15}	picofarad (pF)
mega	M	10^{6}	megahertz (MHz)	femto	f	10^{-15}	femtosecond (fs)
kilo	k	10^{3}	kilometer (km)	atto	a	10^{-18}	
hecto	h	10^{2}		zepto	z	10^{-21}	
deka	da	10^{1}		yocto	y	10^{-24}	

TABLE 4. Some common units defined in terms of SI units

Quantity	Name of unit	Unit symbol	Definition of unit
length	inch	in.	2.54×10^{-2} m
mass	pound (avoirdupois)	lb	0.45359237 kg
force	kilogram-force	kgf	9.80665 N
pressure	atmosphere	atm	101325 Pa
pressure	torr	torr	(101325/760) Pa
pressure	conventional millimeter of mercury*	mmHg	$13.5951 \times 980.665 \times 10^{-2}$ Pa
energy	kilowatt-hour	kWh	3.6×10^{6} J
energy	thermochemical calorie	cal	4.184 J
energy	international steam table calorie	cal_{IT}	4.1868 J
thermodynamic temperature (*T*)	degree Rankine	°R	(5/9) K
customary temperature (*t*)	degree Celsius	°C	$t(°C) = T(K) - 273.15$
customary temperature (*t*)	degree Fahrenheit	°F	$t(°F) = [1.8 \times t(°C)] + 32 = T(°R) - 459.67$
radioactivity	curie	Ci	3.7×10^{10} Bq
energy†	electronvolt	eV	$eV = 1.60218 \times 10^{-19}$ J
mass†	unified atomic mass unit	u	$u = 1.66054 \times 10^{-27}$ kg

*The conventional millimeter of mercury, symbol mmHg (not mm Hg), is the pressure exerted by a column exactly 1 mm high of a fluid of density exactly 13.5951 g cm^{-3} in a place where the gravitational acceleration is exactly 980.665 cm · s^{-2}. The mmHg differs from the torr by less than 2×10^{-7} torr.

†These units defined in terms of the best available experimental values of certain physical constants may be converted to SI units. The factors for conversion of these units are subject of change in the light of new experimental measurements of the constants involved.

TABLE 5. Conversion factors for the U.S. Customary System, metric system, and International System

A. Units of length

Units	cm	m	in.	ft	yd	mi
1 cm	= 1	0.01*	0.3937008	0.03280840	0.01093613	6.213712×10^{-6}
1 m	= 100.	1	39.37008	3.280840	1.093613	6.213712×10^{-4}
1 in.	= 2.54*	0.0254	1	0.08333333...	0.02777777...	1.578283×10^{-5}
1 ft	= 30.48	0.3048	12.*	1	0.3333333...	$1.893939... \times 10^{-4}$
1 yd	= 91.44	0.9144	36.	3.*	1	$5.681818... \times 10^{-4}$
1 mi	= 1.609344×10^{5}	1.609344×10^{3}	6.336×10^{4}	5280.*	1760.	1

B. Units of area

Units	cm^2	m^2	$in.^2$	ft^2	yd^2	mi^2
1 cm^2	= 1	10^{-4}*	0.1550003	1.076391×10^{-3}	1.195990×10^{-4}	3.861022×10^{-11}
1 m^2	= 10^{4}	1	1550.003	10.76391	1.195990	3.861022×10^{-7}
1 $in.^2$	= 6.4516*	6.4516×10^{-4}	1	$6.944444... \times 10^{-3}$	7.716049×10^{-4}	2.490977×10^{-10}
1 ft^2	= 929.0304	0.09290304	144.*	1	0.1111111...	3.587007×10^{-8}
1 yd^2	= 8361.273	0.8361273	1296.	9.*	1	3.228306×10^{-7}
1 mi^2	= 2.589988×10^{10}	2.589988×10^{6}	4.014490×10^{9}	2.78784×10^{7}*	3.0976×10^{6}	1

C. Units of volume

Units	m^3	cm^3	liter	$in.^3$	ft^3	qt	gal
1 m^3	= 1	10^{6}	10^{3}	6.102374×10^{4}	35.31467	1.056688×10^{3}	264.1721
1 cm^3	= 10^{-6}	1	10^{-3}	0.06102374	3.531467×10^{-5}	1.056688×10^{-3}	2.641721×10^{-4}
1 liter	= 10^{-3}	1000.*	1	61.02374	0.03531467	1.056688	0.2641721
1 $in.^3$	= 1.638706×10^{-5}	16.38706*	0.01638706	1	5.787037×10^{-4}	0.01731602	4.329004×10^{-3}
1 ft^3	= 2.831685×10^{-2}	28316.85	28.31685	1728.*	1	29.92208	7.480519
1 qt	= 9.463529×10^{-4}	946.3529	0.9463529	57.75	0.0342014	1	0.25
1 gal (U.S.)	= 3.785412×10^{-3}	3785.412	3.785412	231.*	0.1336806	4.*	1

D. Units of mass

Units	g	kg	oz	lb	metric ton	ton
1 g	= 1	10^{-3}	0.03527396	2.204623×10^{-3}	10^{-6}	1.102311×10^{-6}
1 kg	= 1000.	1	35.27396	2.204623	10^{-3}	1.102311×10^{-3}
1 oz (avdp)	= 28.34952	0.02834952	1	0.0625	2.834952×10^{-5}	3.125×10^{-5}
1 lb (avdp)	= 453.5924	0.4535924	16.*	1	4.535924×10^{-4}	$5. \times 10^{-4}$
1 metric ton	= 10^{6}	1000.*	35273.96	2204.623	1	1.102311
1 ton	= 907184.7	907.1847	32000.	2000.*	0.9071847	1

*Usage of the asterisk is defined in the accompanying text.

TABLE 5. Conversion factors for the U.S. Customary System, metric system, and International System (cont.)

E. Units of density

Units	$g \cdot cm^{-3}$	$g \cdot L^{-1}$, $kg \cdot m^{-3}$	$oz \cdot in.^{-3}$	$lb \cdot in.^{-3}$	$lb \cdot ft^{-3}$	$lb \cdot gal^{-1}$
1 $g \cdot cm^{-3}$	= 1	1000.	0.5780365	0.03612728	62.42795	8.345403
1 $g \cdot L^{-1}$, $kg \cdot m^{-3}$	= 10^{-3}	1	5.780365×10^{-4}	3.612728×10^{-5}	0.06242795	8.345403×10^{-3}
1 $oz \cdot in.^{-3}$	= 1.729994	1729.994	1	0.0625	108.	14.4375
1 $lb \cdot in.^{-3}$	= 27.67991	27679.91	16.	1	1728.	231.
1 $lb \cdot ft^{-3}$	= 0.01601847	16.01847	9.259259×10^{-3}	5.787037×10^{-4}	1	0.1336806
1 $lb \cdot gal^{-1}$	= 0.1198264	119.8264	6.926407×10^{-2}	4.329004×10^{-3}	7.480519	1

F. Units of pressure

Units	Pa, $N \cdot m^{-2}$	$dyn \cdot cm^{-2}$	bar	atm	$kgf \cdot cm^{-2}$	mmHg (torr)	in. Hg	$lbf \cdot in.^{-2}$
1 Pa, 1 $N \cdot m^{-2}$	= 1	10	10^{-5}	9.869233×10^{-6}	1.019716×10^{-5}	7.500617×10^{-3}	2.952999×10^{-4}	1.450377×10^{-4}
1 $dyn \cdot cm^{-2}$	= 0.1	1	10^{-6}	9.869233×10^{-7}	1.019716×10^{-6}	7.500617×10^{-4}	2.952999×10^{-5}	1.450377×10^{-5}
1 bar	= 10^{5}*	10^{6}	1	0.9869233	1.019716	750.0617	29.52999	14.50377
1 atm	= 101325*	1.01325×10^{6}	1.01325	1	1.033227	760.	29.92126	14.69595
1 $kgf \cdot cm^{-2}$	= 98066.5	980665	0.980665	0.9678411	1	735.5592	28.95903	14.22334
1 mmHg (torr)	= 133.3224	1333.224	1.333224×10^{-3}	1.315789×10^{-3}	1.3595510×10^{-3}	1	0.03937008	0.01933678
1 in. Hg	= 3386.388	33863.88	0.03386388	0.03342105	0.03453155	25.4	1	0.4911541
1 $lbf \cdot in.^{-2}$	= 6894.757	68947.57	0.06894757	0.06804596	0.07030696	51.71493	2.036021	1

G. Units of energy

Units	g mass (energy equiv)	J	eV	cal	cal_{IT}	Btu_{IT}	kWh	hp-h	ft-lbf	$ft^3 \cdot lbf \cdot in.^{-2}$	liter-atm
1 g mass (energy equiv)	= 1	8.987552×10^{13}	5.609589×10^{32}	2.148076×10^{13}	2.146640×10^{13}	8.518555×10^{10}	2.496542×10^{7}	3.347918×10^{7}	6.628878×10^{13}	4.603388×10^{11}	8.870024×10^{11}
1 J	= 1.112650×10^{-14}	1	6.241509×10^{18}	0.2390057	0.2388459	9.478172×10^{-4}	$2.777777... \times 10^{-7}$	3.725062×10^{-7}	0.7375622	5.121960×10^{-3}	9.869233×10^{-3}
1 eV	= 1.782662×10^{-33}	1.602177×10^{-19}	1	3.829294×10^{-20}	3.826733×10^{-20}	1.518570×10^{-22}	4.450490×10^{-26}	5.968206×10^{-26}	1.181705×10^{-19}	8.206283×10^{-22}	1.581225×10^{-21}
1 cal	= 4.655328×10^{-14}	4.184*	2.611448×10^{19}	1	0.9993312	3.965667×10^{-3}	$1.162222... \times 10^{-6}$	1.558562×10^{-6}	3.085960	2.143028×10^{-2}	0.04129287
1 cal_{IT}	= 4.658443×10^{-14}	4.1868*	2.613195×10^{19}	1.000669	1	3.968321×10^{-3}	1.163×10^{-6}	1.559609×10^{-6}	3.088025	2.144462×10^{-2}	0.04132050
1 Btu_{IT}	= 1.173908×10^{-11}	1055.056	6.585141×10^{21}	252.1644	251.9958	1	2.930711×10^{-4}	3.930148×10^{-4}	778.1693	5.403953	10.41259
1 kWh	= 4.005540×10^{-8}	3600000.*	2.246943×10^{25}	860420.7	859845.2	3412.142	1	1.341022	2655224.	18349.06	35529.24
1 hp-h	= 2.986931×10^{-8}	2684520.	1.675545×10^{25}	641615.6	641186.5	2544.33	0.7456998	1	1980000.*	13750.	26494.15
1 ft-lbf	= 1.508551×10^{-14}	1.355818	8.462351×10^{18}	0.3240483	0.3238315	1.285067×10^{-3}	3.766161×10^{-7}	$5.050505... \times 10^{-7}$	1	$6.944444... \times 10^{-3}$	0.01338088
1 ft^3 lbf $\cdot$ $in.^{-2}$	= 2.172313×10^{-12}	195.2378	1.218578×10^{21}	46.66295.	46.63174	0.1850497	5.423272×10^{-5}	$7.272727... \times 10^{-5}$	144.*	1	1.926847
1 liter-atm	= 1.127393×10^{-12}	101.325	6.3242109×10^{20}	24.21726	24.20106	0.09603757	2.814583×10^{-5}	3.774419×10^{-5}	74.73349	0.5189825	1

*Usage of the asterisk is defined in the accompanying text.

Units of temperature in measurement systems

Temperature is a basic physical quantity. It is a measure of the thermal energy of random motion of particles in a system. As such it has been chosen as one of the base quantities in the SI. It is to be treated as are the units of length, mass, time, electric current, and luminous intensity. In the SI the unit of length is the meter, the unit of time the second, and so on. The question arises as to the choice of the unit of temperature in the SI.

In the past it was customary to refer to scales of temperature, for example, the Celsius and Fahrenheit scales. On the Celsius scale, 0 designates the freezing point (ice point) and 100 the boiling point (steam point) of water. Corresponding numbers on the Fahrenheit scale are 32 and 212. There are 100 units between the ice point and steam point on the Celsius scale, and 180 units between these points in the Fahrenheit system.

By measuring the volume changes of a gas within the 100-unit interval for the ice point and steam point of water on the Celsius scale, it was found that a numerical value could be assigned for a basic unit of temperature. Careful measurement of this ice-steam interval in a gas thermometer determined that the ice point of water should be assigned the value of 273.15 kelvins. The unit of temperature was thus called the kelvin with the symbol K. Further experiments led to the decision to define the kelvin in the SI along the same lines but in terms of the triple point of water. This is the temperature and pressure at which ice, liquid water, and water vapor coexist at equilibrium. The triple point was chosen because it was a more reproducible value than the ice point.

This change led to the SI definition of temperature in terms of the temperature of the triple point of water, which contains exactly 273.16 kelvins.

It follows that the Celsius temperature (°C) is an intermediate scale. It is useful in defining Kelvin temperature in the SI. Celsius temperature (t) is related to Kelvin temperature (K) as follows:

$$t_{\text{ice point}} = 0°\text{C}$$
$$t_{\text{steam point}} = 100°\text{C}$$
$$0\,\text{K} = -273.15°\text{C}$$

A summary of the conventions in the SI as proposed in the Thirteenth General Conference of Weights and Measures pertaining to temperature units is given below.

1. The unit of SI temperature is the kelvin, symbol K.
2. The word "scale" is not to be used except in terms of measurement of temperature between certain fixed points on the Celsius scale.
3. The terms "thermodynamic scale" or "absolute scale" are not to be used to describe temperature. The degree sign is to be eliminated with the symbol K.
4. When Celsius temperatures are used (°C), it is understood that the temperature unit is the kelvin.

Not all scientists and engineers have adopted the SI of temperature terminology. Furthermore, many engineers in the United States still use the Fahrenheit system in discussing practical engineering systems.

In converting Fahrenheit (°F) to Celsius (°C) the following formula applies.

$$°\text{C} = \frac{°\text{F} - 32°}{1.8}$$

In converting Celsius to Fahrenheit the following formula can be used.

$$°\text{F} = (°\text{C} \times 1.8) + 32°$$

In changing from Celsius terminology (t) to kelvin units (K) the following formula can be used.

$$\text{K} = \text{t} + 273.15$$

Chemistry

Symbols for the chemical elements

The mass number, atomic number, number of atoms, and ionic charge of an element are often indicated by means of four indices placed around the symbol. The positions occupied are left upper index, mass number; left lower index, atomic number; right upper index, ionic charge; and right lower index, number of atoms of an element in a molecule or formula unit of a given species: for example, ${}^{12}_{6}C$, Ca^{2+}, O_2, and Al_2O_3. The atomic number, which is redundant, is omitted in most cases: that is, ${}^{12}_{6}C$ can be written as ${}^{12}C$.

Ionic charge is indicated by a plus or minus superscript following the symbol of the ion; for multiple charges an arabic superscript numeral precedes the plus or minus sign, for example, Na^+, NO_3^-, Ca^{2+}, PO_4^{3+}.

An alphabetical list of the elements, their symbols, and their atomic numbers is shown in **Table 6**.

Chemical nomenclature

The International Union of Pure and Applied Chemistry (IUPAC) has established definitive rules for chemical nomenclature. Chemical species are identified in the Dictionary by a systematic name, frequently accompanied by a formula. Occasionally common names are used.

For inorganic compounds, systematic names of compounds are formed by identifying the constituents and their proportions in a specific order, for example, dinitrogen oxide (N_2O). Also accepted by IUPAC is Stock's system, in which the proportions of the constituents are indicated indirectly, and roman numerals are used to represent the oxidation number or stoichiometric valence of an element, for example, iron(II) chloride ($FeCl_2$). Complex compounds are also named according to rules specified by IUPAC; an example is potassium oxodichloroimidophosphate, $K[POCl_2(NH)]$. Examples of accepted trivial names are diborane (B_2H_6), silane (SiH_4), and ammonia (NH_3).

There also are definitive rules for naming organic compounds. Because of the infinite variety of disciplines and industrial applications involving organic compounds, the rules encompass different types of names. Sometimes a single compound can correctly be identified by a number of names; for example, chloral hydrate is also known as 2,2,2-trichloro-1,1-ethanediol and trichloroacetaldehyde monohydrate.

TABLE 6. The chemical elements

Name	Symbol	At. no.	Name	Symbol	At. no.	Name	Symbol	At. no.
Actinium	Ac	89	Gold	Au	79	Promethium	Pm	61
Aluminum	Al	13	Hafnium	Hf	72	Protactinium	Pa	91
Americium	Am	95	Hassium	Hs	108	Radium	Ra	88
Antimony	Sb	51	Helium	He	2	Radon	Rn	86
Argon	Ar	18	Holmium	Ho	67	Rhenium	Re	75
Arsenic	As	33	Hydrogen	H	1	Rhodium	Rh	45
Astatine	At	85	Indium	In	49	Roentgenium	Rg	111
Barium	Ba	56	Iodine	I	53	Rubidium	Rb	37
Berkelium	Bk	97	Iridium	Ir	77	Ruthenium	Ru	44
Beryllium	Be	4	Iron	Fe	26	Rutherfordium	Rf	104
Bismuth	Bi	83	Krypton	Kr	36	Samarium	Sm	62
Bohrium	Bh	107	Lanthanum	La	57	Scandium	Sc	21
Boron	B	5	Lawrencium	Lr	103	Seaborgium	Sg	106
Bromine	Br	35	Lead	Pb	82	Selenium	Se	34
Cadmium	Cd	48	Lithium	Li	3	Silicon	Si	14
Calcium	Ca	20	Lutetium	Lu	71	Silver	Ag	47
Californium	Cf	98	Magnesium	Mg	12	Sodium	Na	11
Carbon	C	6	Manganese	Mn	25	Strontium	Sr	38
Cerium	Ce	58	Meitnerium	Mt	109	Sulfur	S	16
Cesium	Cs	55	Mendelevium	Md	101	Tantalum	Ta	73
Chlorine	Cl	17	Mercury	Hg	80	Technetium	Tc	43
Chromium	Cr	24	Molybdenum	Mo	42	Tellurium	Te	52
Cobalt	Co	27	Neodymium	Nd	60	Terbium	Tb	65
Copper	Cu	29	Neon	Ne	10	Thallium	Tl	81
Curium	Cm	96	Neptunium	Np	93	Thorium	Th	90
Darmstadtium	Ds	110	Nickel	Ni	28	Thulium	Tm	69
Dubnium	Db	105	Niobium	Nb	41	Tin	Sn	50
Dysprosium	Dy	66	Nitrogen	N	7	Titanium	Ti	22
Einsteinium	Es	99	Nobelium	No	102	Tungsten	W	74
Element 112*		112	Osmium	Os	76	Uranium	U	92
Erbium	Er	68	Oxygen	O	8	Vanadium	V	23
Europium	Eu	63	Palladium	Pd	46	Xenon	Xe	54
Fermium	Fm	100	Phosphorus	P	15	Ytterbium	Yb	70
Fluorine	F	9	Platinum	Pt	78	Yttrium	Y	39
Francium	Fr	87	Plutonium	Pu	94	Zinc	Zn	30
Gadolinium	Gd	64	Polonium	Po	84	Zirconium	Zr	40
Gallium	Ga	31	Potassium	K	19			
Germanium	Ge	32	Praseodymium	Pr	59			

*This element does not have an official name or symbol.

Symbols in scientific writing

Symbols commonly encountered in scientific writing are listed in **Table 7**. Symbols following the ellipses and separated by commas are alternatives that are used only when there is some reason for not using the symbol given first. The letters of the Greek alphabet, frequently used to represent terms, are shown in **Table 8**.

Some frequently encountered symbols for particles and quanta are as follows:

neutron	n	pion	π
proton	p	muon	μ
deuteron	d	electron	e
triton	t	neutrino	ν
alpha particle	α	photon	γ

The meaning of abbreviated notations for nuclear reactions should be the following:

initial nuclide (incoming particle(s) or quanta, outgoing particle(s) or quanta) final nuclide

Some examples are:

$^{14}N(\alpha, p)^{17}O$ $\quad$ $^{59}Co(n, \gamma)^{60}Co$

$^{23}Na(\gamma, 3n)^{20}Na$ $\quad$ $^{31}P(\gamma, pn)^{29}Si$

TABLE 7. Commonly used symbols in scientific writing

Space, time, mass, and related quantities

length l
height h
radius r
diameter d
path, length of arc s
plane angle $\alpha, \beta, \gamma, \theta, \phi, \psi$
solid angle ω
area A, S
volume $V \ldots v$
specific volume v
wavelength λ
wavenumber σ, ν
time t
period or other characteristic interval T, τ
frequency ν, f
angular frequency ($2\pi\nu$) ω
velocity $v \ldots u, w$
angular velocity ω
acceleration a
acceleration of free fall g
mass m
moment of inertia I
density ρ
relative density d

Molecular and related quantities

molecular mass m
molar mass M
Avogadro's number N_0, L, N
number of molecules N
number of moles n
mole fraction $x \ldots X, y$
molality m
concentration c
molar concentration of substance B c_B, [B], c(B)
molecular concentration C
partition function Q
statistical weight $g \ldots p$
symmetry number σ
characteristic temperature θ
diameter of molecule $\sigma \ldots D$
mean free path l
diffusion coefficient D
osmotic pressure Π
surface concentration Γ

Mechanical and related quantities

force F
force due to gravity (weight) $G \ldots W$
moment of force M
power P
pressure p, P
traction σ
shear stress τ
modulus of elasticity E
shear modulus G
compressibility κ
compression modulus ($1/\kappa$) K
viscosity η
fluidity ϕ
kinematic viscosity ν
friction coefficient f
surface tension $\gamma \ldots \sigma$
angle of contact θ

Thermodynamic and related quantities

temperature $\theta \ldots t$
temperature, absolute T
gas constant R
Boltzmann constant k
heat q, Q
work w, A
energy (Gibbs ε) $E \ldots U$
entropy (Gibbs η) S
*Helmholtz free energy (Gibbs Ψ) A
enthalpy (Gibbs χ) H
*Gibbs function (ζ) $G \ldots F$
heat capacity C
specific heat at constant pressure c_p
specific heat at constant volume c_v
ratio of specific heats c_p/c_v γ, κ
chemical potential μ
activity, absolute λ
activity, relative a
activity coefficient f, γ
osmotic coefficient g, ϕ
thermal conductivity γ
Joule-Thomson coefficient μ

*The terms for the Helmholtz and Gibbs energies were modified by action of the IUPAC Council. Montreal, August 1961, as follows:
Helmholtz energy (Gibbs $\Psi = E_2TS$): A Gibbs energy (Gibbs $\zeta = H-TS$): G

TABLE 7. Commonly used symbols in scientific writing (cont.)

Chemical reactions

stoichiometric number of molecules (negative for reactants, positive for products) ν
standard equation of chemical reaction $\Sigma \nu_B B = 0$
affinity $(-\Sigma \nu_B \mu_B)$ of a reaction A
equilibrium constant K
equilibrium quotient or equilibrium product (of molalities) Q
extent of reaction $(dn_B = \nu_B d\xi)$ ξ
degree of reaction (e.g., degree of dissociation) α
rate constant k
collision number (collisions per unit volume and unit time) Z
rate constant corresponding to the rate Z z
rate of reaction $\nu \ldots r, s, J$

Light

Planck's constant h
Planck's constant divided by 2π $\hbar$
quantity of light q
radiant power, flux of light (dQ/dt) Φ
illumination $(d\Phi/dS)$ E
luminance L,B
luminous emittance H
absorption factor (fraction of incident radiant power which is absorbed) α
reflection factor (fraction of incident radiant power which is reflected) ρ
transmission factor (fraction of incident radiant power which is transmitted) τ
transmittance $(T = I/I_0)$ T
absorption (extinction) coefficient $[\kappa/c = \ln (1/T)]$ κ
absorbance (extinction) $[A = \log (1/T)]$ $A \ldots E$
absorptivity (specific absorbance) [decadic absorption or extinction coefficient] a
molar absorptivity (molar decadic absorption or extinction coefficient) $[\varepsilon/c = A]$ ϵ
refraction index n
refractivity r
angle of optical rotation α

Electricity and magnetism

elementary change e
quantity of electricity Q
change density p
surface change density σ
electric current $I \ldots i$
electric current density J
electric potential V
electric field strength E
electric displacement D
electrokinetic potential ζ
capacitance C
permittivity (dielectric constant) ϵ
dielectric polarization P
dipole moment μ
electric polarizability of a molecule α, γ
magnetic field strength H
magnetic induction B
magnetic permeability μ
magnetization M
magnetic susceptibility χ
resistance R
resistivity ρ
self inductance L
mutual inductance M, L_{12}
reactance X
impedance Z
admittance Y

Electrochemistry

Faraday's constant (the faraday) F
charge number of an ion, plus or minus z
degree of electrolytic dissociation α
ionic strength $I \ldots \mu$
electrolytic conductivity (specific conductance) κ
equivalent or molar conductance of electrolyte or ion A
transport number t, T
electromotive force E
overpotential η

TABLE 8. Greek alphabet

Upper and lower cases	Name	Upper and lower cases	Name
Α α	Alpha	Ν ν	Nu
Β β	Beta	Ξ ξ	Xi
Γ ξ	Gamma	Ο ο	Omicron
Δ δ ∂	Delta	Π π	Pi
Ε ϵ ε	Epsilon	Ρ ρ	Rho
Ζ ζ	Zeta	Σ σ	Sigma
Η η	Eta	Τ τ	Tau
Θ θ	Theta	Υ υ	Upsilon
Ι ι	Iota	Φ ϕ φ	Phi
Κ κ	Kappa	Χ χ	Chi
Λ λ	Lambda	Ψ ψ	Psi
Μ μ	Mu	Ω ω	Omega

Mathematics

Mathematical signs and symbols

Symbol	Definition	Symbol	Definition	Symbol	Definition
$+$	plus (sign of addition)	$\ddagger$ $\neq$	not equal to	$\partial u/\partial x$	partial derivative of u with respect to x
$+$	positive	$\rightarrow$ $\doteq$	approaches	$\int$	integral of
$-$	minus (sign of subtraction)	$\propto$	varies as	$\int_b^a$	integral of, between limits a and b
$-$	negative	∞	infinity	$\oint$	line integral around a closed path
$\pm$ $(\mp)$	plus or minus (minus or plus)	$\sqrt{\ }$	square root of	Σ	(sigma) summation of
$\times$	times, by (multiplication sign)	$\sqrt[3]{\ }$	cube root of	$f(x)$, $F(x)$	functions of x
$\cdot$	multiplied by	$\therefore$	therefore	∇	del or nabla, vector differential operator
$\div$	sign of division	$\parallel$	parallel to	∇^2	Laplacian operator
/	divided by	( ) [ ] { }	parentheses, brackets and braces; quantities enclosed by them to be taken together in multiplying dividing, etc.	$\pounds$	Laplace operational symbol
:	ratio sign, divided by, is to	$\overline{AB}$	length of line from A to B	4!	factorial $4 = 1 \times 2 \times 3 \times 4$
::	equals, as (proportion)	π	(pi) = 3.14159 +	$\lvert x\rvert$	absolute value of x
$<$	less than	$^\circ$	degrees	$\dot{x}$	first derivative of x with respect to time
$>$	greater than	$'$	minutes	$\ddot{x}$	second derivative of x with respect to time
$\ll$	much less than	$''$	seconds	$\mathbf{A} \times \mathbf{B}$	vector product; magnitude of **A** times magnitude of **B** times sine of the angle from **A** to **B**; $AB \sin \overline{AB}$
$\gg$	much greater than	$\angle$	angle	$\mathbf{A} \cdot \mathbf{B}$	scalar product of **A** and **B**; magnitude of **A** times magnitude of **B** times cosine of the angle from **A** to **B**; $AB \cos \overline{AB}$
$=$	equals	dx	differential of x		
$\equiv$	identical with	Δ	(delta) difference		
$\sim$	similar to	Δx	increment of x		
$\approx$	approximately equals				
$\cong$	approximately equals, congruent				
$\leq$	equal to or less than				
$\geq$	equal to or greater than				

Mathematical notation

Mathematical logic

p, q, $P(x)$	Sentences, propositional functions, propositions
$\neg p$, $\sim p$, non p, Np	Negation, read "not p" ($\neq$: read "not equal")
$p \vee q$, $p + q$, Apq	Disjunction, read "p or q," "p, q," or both
$p \wedge q$, $p \cdot q$, $p\&q$, Kpq	Conjunction, read "p and q"
$p \rightarrow q$, $p \supset q$, $p \Rightarrow q$, Cpq	Implication, read "p implies q" or "if p then q"
$p \leftrightarrow q$, $p \equiv q$, $p \Leftrightarrow q$, Epq, p iff q	Equivalence, read "p is equivalent to q" or "p if and only if q"
n.a.s.c.	Read "necessary and sufficient condition"
( ), [ ], { }, . . .	Parentheses
$\forall$, $\forall$, Σ	Universal quantifier, read "for all" or "for every"
$\exists$, $\exists$, Π	Existential quantifier, read "there is a" or "there exists"
$\vdash$	Assertion sign ($p \vdash q$: read "q follows from p"; $\vdash p$: read "p is or follows from an axiom," or "p is a tautology"
0, 1	Truth, falsity (values)
$=$	Identity
$\overset{\text{Df}}{=}$, $\overset{\text{df}}{=}$, $=_{\text{df}}$, $\equiv$	Definitional identity
■	"End of proof"; "QED"

Set theory, relations, functions

X, Y	Sets
$x \in X$	x is a member of the set X
$x \notin X$	x is not a member of X
$A \subset X$, $A \subseteq X$	Set A is contained in set X
$A \not\subset X$, $A \nsubseteq X$	A is not contained in X
$X \cup Y$, $X + Y$	Union of sets X and Y
$X \cap Y$, $X \cdot Y$	Intersection of sets X and Y
$+$, $\dot{+}$, $\bigcirc$	Symmetric difference of sets
$\cup X_i$, ΣX	Union of all the sets X_i
$\cap X_i$, ΠX_i	Intersection of all the sets X_i
$\varnothing$, 0, Λ	Null set, empty set
X', $\complement X$, CX	Complement of the set X
$X - Y$, $X \backslash Y$	Difference of sets X and Y
$\hat{x}(P(x))$, $\{x \mid P(x)\}$, $\{x{:}P(x)\}$	The set of all x with the property P
(x,y,z), (x, y, z)	Ordered set of elements x, y, and z; to be distinguished from (x,z,y), for example
$\{x,y,z\}$	Unordered set, the set whose elements are x, y, z, and no others
$\{a_1, a_2, \ldots, a_n\}$, $\{a_i\}_{i=1,2,\ldots,n}$, $\{a_i\}_{i=1}^{n}$	The set whose members are a_i, where i is any whole number from 1 to n
$\{a_1, a_2, \ldots\}$, $\{a_i\}_{i=1,2,\ldots}$, $\{a_i\}_{i=1}^{\infty}$	The set whose members are a_i, where i is any positive whole number
$X \times Y$	Cartesian product, set of all (x,y) such that $x \in X$, $y \in Y$

$\{a_i\}_{i \in I}$	The set whose elements are a_i, where $i \in I$
xRy, $R\{x,y\}$	Relation
$\equiv, \cong, \sim, \simeq$	Equivalence relations, for example, congruence
$\geqq, \geq, >, \varepsilon, \gg, \leqq, \leq, <$	Transitive relations, for example, numerical order
$f{:}X \to Y$, $X \xrightarrow{f} Y$, $X \to Y$, $f \in Y^X$	Function, mapping, transformation
f^{-1}, $\overset{-1}{f}$, $X \xleftarrow{f^{-1}} Y$	Inverse mapping
$g \circ f$	Composite functions: $(g \circ f)(x) = g(f(x))$
$f(X)$	Image of X by f
$f^{-1}(X)$	Inverse-image set, counter image
1-1, one-one	Read "one-to-one correspondence"
$\begin{matrix} X & \xrightarrow{f} & Y \\ \phi\downarrow & & \downarrow\psi \\ W & \xrightarrow{g} & Z \end{matrix}$	Diagram: the diagram is commutative in case $\psi \circ f = g \circ \phi$
$f\|A$	Partial mapping, restriction of function f to set A
$\overset{=}{X}$, card X, $\|X\|$	Cardinal of the set A
$\aleph_0$, d	Denumerable infinity
$\mathfrak{c}$, c, $2^{\aleph_0}$	Power of continuum
ω	Order type of the set of positive integers
σ-	Read "countably"

Number, numerical functions

1.4; 1,4; 1·4	Read "one and four-tenths"
1(1)2(10)100	Read "from 1 to 20 in intervals of 1, and from 20 to 100 in intervals of 10"
const	Constant
$A \geqq 0$	The number A is nonnegative, or the matrix A is positive definite, or the matrix A has nonnegative entries
$x\|y$	Read "x divides y"
$x \equiv y \bmod p$	Read "x congruent to y modulo p"
$a_0 + \frac{1}{a_1} + \frac{1}{a_2} + \cdots$, $a_0 + \frac{1\|}{\|a_1} + \cdots$	Continued fractions
$[a,b]$	Closed interval
$[a,b)$, $[a,b[$	Half-open interval (open at the right)
(a,b), $]a,b[$	Open interval
$[a,\infty)$, $[a,\to[$	Interval closed at the left, infinite to the right
$(-\infty,\infty)$, $[\leftarrow,\to[$	Set of all real numbers
$\max_{x \in X} f(x)$, $\max\{f(x) \| x \in X\}$	Maximum of $f(x)$ when x is in the set X
min	Minimum
sup, l.u.b.	Supremum, least upper bound
inf, g.l.b.	Infimum, greatest lower bound
$\lim_{x \to a} f(x) = b$, $\lim_{x=a} f(x) = b$, $f(x) \to b$ as $x \to a$	b is the limit of $f(x)$ as x approaches a
$\lim_{x \to a-} f(x)$, $\lim_{x=a-0} f(x)$, $f(a-)$	Limit of $f(x)$ as x approaches a from the left
lim sup, $\overline{\lim}$	Limit superior
lim inf, $\underline{\lim}$	Limit inferior
l.i.m.	Limit in the mean
$z = x + iy = re^{i\theta}$, $\zeta = \xi + i\eta$, $w = u + iv = \rho e^{i\phi}$	Complex variables
z^*	Complex conjugate
Re, $\Re$	Real part
Im, $\Im$	Imaginary part
arg	Argument
$\frac{\partial(u,v)}{\partial(x,y)}$, $\frac{D(u,v)}{D(x,y)}$	Jacobian, functional determinant
$\int_E f(x)\, d\mu(x)$	Integral (for example, Lebesgue integral) of function f over set E with respect to measure μ
$f(n) \sim \log n$ as $n \to \infty$	$f(n)/\log n$ approaches 1 as $n \to \infty$
$f(n) = O(\log n)$ as $n \to \infty$	$f(n)/\log n$ is bounded as $n \to \infty$
$f(n) = o(\log n)$	$f(n)/\log n$ approaches zero
$f(x) \nearrow b$, $f(x) \uparrow b$	$f(x)$ increases, approaching the limit b
$f(x) \downarrow b$, $f(x) \searrow b$	$f(x)$ decreases, approaching the limit b

a.e., p.p.	Almost everywhere
ess sup	Essential supremum
C^0, $C^0(X)$, $C(X)$	Space of continuous functions
C^k, $C^k[a,b]$	The class of functions having continuous kth derivative (on $[a,b]$)
C'	Same as C^1
Lip_α, Lip α	Lipschitz class of functions
L^p, L_p, $L^p[a,b]$	Space of functions having integrable absolute pth power (on $[a,b]$)
L'	Same as L^1
(C,α), (C,p)	Cesàro summability

Special functions

$[x]$	The integral part of x
$\binom{n}{k}$, ${}^nC_k$, ${}_nC_k$	Binomial coefficient $n!/k!(n-k)!$
$\left(\frac{n}{p}\right)$	Legendre symbol
e^x, exp x	Exponential function
sinh x, cosh x, tanh x	Hyperbolic functions
sn x, cn x, dn x	Jacobi elliptic functions
$\wp(x)$	Weierstrass elliptic function
$\Gamma(x)$	Gamma function
$J_\nu(x)$	Bessel function
$\chi_X(x)$	Characteristic function of the set X: $\chi_X(x) = 1$ in case $x \in X$, otherwise $\chi_X(x) = 0$
sgn x	Signum: sgn 0 = 0, while sgn $x = x/\|x\|$ for $x \neq 0$
$\delta(x)$	Dirac delta function

Algebra, tensors, operators

$+, \cdot, \times, \circ, \mathrm{T}, \tau$	Laws of composition in algebraic systems
e, 0	Identity, unit, neutral element (of an additive system)
e, I, I	Identity, unit, neutral element (of a general algebraic system)
e, e, E, P	Idempotent
a^{-1}	Inverse of a
Hom(M,N)	Group of all homomorphisms of M into N
G/H	Factor group, group of cosets
$[K{:}k]$	Dimension of K over k
$\oplus$, +	Direct sum
$\otimes$	Tensor product, Kronecker product
$\wedge$	Exterior product, Grassmann product
$\vec{x}$, $\mathbf{x}$, $\underset{\to}{x}$, $\mathfrak{x}$, x	Vector
$\vec{x} \cdot \vec{y}$, $\mathbf{x} \cdot \mathbf{y}$, $(\mathfrak{x}, \mathfrak{y})$	Inner product, scalar product, dot product
$\mathbf{x} \times \mathbf{y}$, $[\mathfrak{x},\mathfrak{y}]$, $\mathbf{x} \wedge \mathbf{y}$	Outer product, vector product, cross product
$\|x\|$, $\|x\|$, $\\|x\\|$, $\\|x\\|_p$	Norm of the vector x
Ax, xA	The image of x under the transformation A
δ_{ij}	Kronecker delta: $\delta_{ii} = 1$, while $\delta_{ij} = 0$ for $i \neq j$
A', tA, A^t, ${}^tA$	Transpose of the matrix A
A^*, $\bar{A}$	Adjoint, Hermitian conjugate of A
tr A, Sp A	Trace of the matrix A
det A, $\|A\|$	Determinant of the matrix A
$\Delta^n f(x)$, $\Delta_k{}^n f$, $\underset{h}{\Delta^n} f(x)$	Finite differences
$[x_0,x_1]$, $[x_0,x_1,x_2]$, $\underset{x_1}{\Delta} u_{x_0}$, $[x_0,x_1]_f$	Divided differences
∇f, grad f	Read "gradient of f"
$\nabla \cdot$ v, div v	Read "divergence of **v**"
$\nabla \times$ v, curl v, rot **v**	Read "curl of **v**"
∇^2, Δ, div grad	Laplacian
$[X,Y]$	Poisson bracket, or commutator, or Lie product
GL(n,R)	Full linear group of degree n over field R
O(n,R)	Full orthogonal group
SO(n,R), O$^+(n,R)$	Special orthogonal group

Topology

E^n	Euclidean n space
S^n	n sphere
$\rho(p,q)$, $d(p,q)$	Metric, distance (between points p and q)
$\bar{X}$, X^-, cl X, X^c	Closure of the set X
FrX, fr$X$, ∂X, bdry X	Frontier, boundary of X
int X, $\mathring{X}$	Interior of X
T_2 space	Hausdorff space
F_σ	Union of countably many closed sets
G_δ	Intersection of countably many open sets
dim X	Dimensionality, dimension of X
$\pi_1(X)$	Fundamental group of the space X
$\pi_n(X)$, $\pi_n(X,A)$	Homotopy groups
$H_n(X)$, $H_n(X,A;G)$, $H_*(X)$	Homology groups
$H^n(X)$, $H^n(X,A;G)$, $H^*(X)$	Cohomology groups

Probability and statistics

X, Y	Random variables
$P(X \leqq 2)$, $\Pr(X \leqq 2)$	Probability that $X \leqq 2$
$P(X \leqq 2 \mid Y \geqq 1)$	Conditional probability
$E(X)$, $E(X)$	Expectation of X
$E(X \mid Y \geqq 1)$	Conditional expectation
c.d.f.	Cumulative distribution function
p.d.f.	Probability density function
c.f.	Characteristic function
$\bar{x}$	Mean (especially, sample mean)
σ, s.d.	Standard deviation
σ^2, Var, var	Variance
μ_1, μ_2, μ_3, μ_i, μ_{ij}	Moments of a distribution
ρ	Coefficient of correlation
$\rho_{12 \cdot 34}$	Partial correlation coefficient

Fundamental constants

Recommended values (2006) of selected fundamental constants of physics and chemistry[a]

Quantity	Symbol	Numerical value[b]	Unit[c]	Relative uncertainty (standard deviation)
UNIVERSAL CONSTANTS				
Speed of light in vacuum	c, c_0	299 792 458	m s^{-1}	(exact)
Magnetic constant	μ_0	$4\pi \times 10^{-7}$	N A^{-2}	
		$= 12.566\,370\,614\ldots \times 10^{-7}$	N A^{-2}	(exact)
Electric constant, $1/\mu_0 c^2$	ε_0	$8.854\,187\,817\ldots \times 10^{-12}$	F m^{-1}	(exact)
Characteristic impedance of vacuum, $\sqrt{\mu_0/\varepsilon_0} = \mu_0 c$	Z_0	$376.730\,313\,461\ldots$	Ω	(exact)
Newtonian constant of gravitation	G	$6.674\,28(67) \times 10^{-11}$	m^3 kg^{-1} s^{-2}	1.0×10^{-4}
	$G/\hbar c$	$6.708\,81(67) \times 10^{-39}$	(GeV/c^2)$^{-2}$	1.0×10^{-4}
Planck constant	h	$6.626\,068\,96(33) \times 10^{-34}$	J s	5.0×10^{-8}
in eV s		$4.135\,667\,33(10) \times 10^{-15}$	eV s	2.5×10^{-8}
hc in eV m		$1.239\,841\,875(31) \times 10^{-6}$	eV m	2.5×10^{-8}
$h/2\pi$	$\hbar$	$1.054\,571\,628(53) \times 10^{-34}$	J s	5.0×10^{-8}
in eV s		$6.582\,118\,99(16) \times 10^{-16}$	eV s	2.5×10^{-8}
$\hbar c$ in eV m		$197.326\,9631(49) \times 10^{-9}$	eV m	2.5×10^{-8}
Planck mass, $(\hbar c/G)^{1/2}$	m_P	$2.176\,44(11) \times 10^{-8}$	kg	5.0×10^{-5}
Planck length, $\hbar/m_P c = (\hbar G/c^3)^{1/2}$	l_P	$1.616\,252(81) \times 10^{-35}$	m	5.0×10^{-5}
Planck time, $l_P/c = (\hbar G/c^5)^{1/2}$	t_P	$5.391\,24(27) \times 10^{-44}$	s	5.0×10^{-5}
ELECTROMAGNETIC CONSTANTS				
Elementary charge	e	$1.602\,176\,487(40) \times 10^{-19}$	C	2.5×10^{-8}
Magnetic flux quantum, $h/2e$	Φ_0	$2.067\,833\,667(52) \times 10^{-15}$	Wb	2.5×10^{-8}
Josephson constant[d], $2e/h$	K_J	$483\,597.891(12) \times 10^9$	Hz V^{-1}	2.5×10^{-8}
von Klitzing constant[e], $h/e^2 = \mu_0 c/2\alpha$	R_K	$25\,812.807\,557(18)$	Ω	6.8×10^{-10}
Bohr magneton, $e\hbar/2m_e$	μ_B	$927.400\,915(23) \times 10^{-26}$	J T^{-1}	2.5×10^{-8}
in eV T^{-1}		$5.788\,381\,7555(79) \times 10^{-5}$	eV T^{-1}	1.4×10^{-9}
	μ_B/h	$13.996\,246\,04(35) \times 10^9$	Hz T^{-1}	2.5×10^{-8}
	μ_B/hc	$46.686\,4515(12)$	m^{-1} T^{-1}	2.5×10^{-8}
	$\mu_B k$	$0.671\,7131(12)$	K T^{-1}	1.7×10^{-6}
Nuclear magneton, $e\hbar\,2m_p$	μ_N	$5.050\,783\,24(13) \times 10^{-27}$	J T^{-1}	2.5×10^{-8}
in eV T^{-1}		$3.152\,451\,2326(45) \times 10^{-8}$	eV T^{-1}	1.4×10^{-9}
	μ_N/h	$7.622\,593\,84(19)$	MHz T^{-1}	2.5×10^{-8}
	μ_N/hc	$2.542\,623\,616(64) \times 10^{-2}$	m^{-1}T^{-1}	2.5×10^{-8}
	μ_N/k	$3.658\,2637(64) \times 10^{-4}$	K T^{-1}	1.7×10^{-6}
ATOMIC AND NUCLEAR CONSTANTS				
General				
Fine-structure constant, $e^2/4\pi\varepsilon_0\hbar c$	α	$7.297\,352\,5376(50) \times 10^{-3}$		6.8×10^{-10}
Inverse fine-structure constant	α^{-1}	$137.035\,999\,679(94)$		6.8×10^{-10}
Rydberg constant, $\alpha^2 m_e c/2h$	R_∞	$10\,973\,731.568\,527(73)$	m^{-1}	6.6×10^{-12}
	$R_\infty c$	$3.289\,841\,960\,361(22) \times 10^{15}$	Hz	6.6×10^{-12}
$R_\infty hc$ in eV		$13.605\,691\,93(34)$	eV	2.5×10^{-8}
Bohr radius, $\alpha/4\pi R_\infty = 4\pi\varepsilon_0\hbar^2/m_e e^2$	a_0	$0.529\,177\,208\,59(36) \times 10^{-10}$	m	6.8×10^{-10}
Hartree energy, $e^2/4\pi\varepsilon_0 a_0 = 2R_\infty hc = \alpha^2 m_e c^2$	E_h	$4.359\,743\,94(22) \times 10^{-18}$	J	5.0×10^{-8}
in eV		$27.211\,383\,86(68)$	eV	2.5×10^{-8}
Electroweak				
Fermi coupling constant[f]	$G_F/(\hbar c)^3$	$1.166\,37(1) \times 10^{-5}$	GeV^{-2}	8.6×10^{-6}

[a]Footnotes are at table end.

Recommended values (2006) of selected fundamental constants of physics and chemistry[a] (cont.)

Quantity	Symbol	Numerical value[b]	Unit[c]	Relative uncertainty (standard deviation)
		ATOMIC AND NUCLEAR CONSTANTS (cont.)		
		Electron, e$^-$		
Electron mass	m_e	$9.109\,382\,15(45) \times 10^{-31}$	kg	5.0×10^{-8}
in u		$5.485\,799\,0943(23) \times 10^{-4}$	u	4.2×10^{-10}
Energy equivalent in MeV	m_ec^2	$0.510\,998\,910(13)$	MeV	2.5×10^{-8}
Electron charge to mass quotient	$-e/m_e$	$-1.758\,820\,150(44) \times 10^{11}$	C kg^{-1}	2.5×10^{-8}
Compton wavelength, h/m_ec	λ_C	$2.426\,310\,2175(33) \times 10^{-12}$	m	1.4×10^{-9}
$\lambda_c/2\pi = \alpha a_0 = \alpha^2/4\pi R_\infty$	λ_C	$386.159\,264\,59(53) \times 10^{-15}$	m	1.4×10^{-9}
Classical electron radius, $\alpha^2 a_0$	r_e	$2.817\,940\,2894(58) \times 10^{-15}$	m	2.1×10^{-9}
Thomson cross section, $(8\pi/3)r_e^2$	σ_e	$0.665\,245\,8558(27) \times 10^{-28}$	m^2	4.1×10^{-9}
Electron magnetic moment	μ_e	$-928.476\,377(23) \times 10^{-26}$	J T^{-1}	2.5×10^{-8}
Electron magnetic moment anomaly, $\lvert\mu_e\rvert/\mu_B - 1$	a_e	$1.159\,652\,181\,11(74) \times 10^{-3}$		6.4×10^{-10}
Electron gyromagnetic ratio, $2\lvert\mu_e\rvert/\hbar$	γ_e	$1.760\,859\,770(44) \times 10^{11}$	s^{-1} T^{-1}	2.5×10^{-8}
	$\gamma_e/2\pi$	$28\,024.95\,364(70)$	MHz T^{-1}	2.5×10^{-8}
		Muon, μ^-		
Muon mass	m_μ	$1.883\,531\,30(11) \times 10^{-28}$	kg	5.6×10^{-8}
in u		$0.113\,428\,9256(29)$	u	2.5×10^{-8}
Energy equivalent in MeV	$m_\mu c^2$	$105.658\,3668(38)$	MeV	3.6×10^{-8}
Muon-electron mass ratio	m_μ/m_e	$206.768\,2823(52)$		2.5×10^{-8}
Muon Compton wavelength, $h/m_\mu c$	$\lambda_{C,\mu}$	$11.734\,441\,04(30) \times 10^{-15}$	m	2.5×10^{-8}
$\lambda_{C,\mu}/2\pi$	$\lambda_{C,\mu}$	$1.867\,594\,295(47) \times 10^{-15}$	m	2.5×10^{-8}
Muon magnetic moment	μ_μ	$-4.490\,447\,86(16) \times 10^{-26}$	J T^{-1}	3.6×10^{-8}
Muon magnetic moment anomaly, $\lvert\mu_\mu\rvert/(e\hbar/2m_\mu) - 1$	a_μ	$1.165\,920\,69(60) \times 10^{-3}$		5.2×10^{-7}
		Tau, τ^-		
Tau mass[g]	m_τ	$3.167\,77(52) \times 10^{-27}$	kg	1.6×10^{-4}
in u		$1.907\,68(31)$	u	1.6×10^{-4}
Energy equivalent in Mev	$m_\tau c^2$	$1776.99(29)$	MeV	1.6×10^{-4}
Tau-electron mass ratio	m_τ/m_e	$3\,477.48(57)$		1.6×10^{-4}
		Proton, p		
Proton mass	m_p	$1.672\,621\,637(83) \times 10^{-27}$	kg	5.0×10^{-8}
in u		$1.007\,276\,466\,77(10)$	u	1.0×10^{-10}
Energy equivalent in MeV	m_pc^2	$938.272\,013(23)$	MeV	2.5×10^{-8}
Proton-electron mass ratio	m_p/m_e	$1836.152\,672\,47(80)$		4.3×10^{-10}
Proton charge to mass quotient	e/m_p	$9.578\,833\,92(24) \times 10^7$	C kg^{-1}	2.5×10^{-8}
Proton Compton wavelength, h/m_pc	$\lambda_{C,p}$	$1.321\,409\,8446(19) \times 10^{-15}$	m	1.4×10^{-9}
$\lambda_{C,p}/2\pi$	$\lambda_{C,p}$	$0.210\,308\,908\,61(30) \times 10^{-15}$	m	1.4×10^{-9}
Proton magnetic moment	μ_p	$1.410\,606\,662(37) \times 10^{-26}$	J T^{-1}	2.7×10^{-8}
to nuclear magneton ratio	μ_p/μ_N	$2.792\,847\,356(23)$		8.2×10^{-9}
Shielded proton magnetic moment[h]	μ'_p	$1.410\,570\,419(38) \times 10^{-26}$	J T^{-1}	2.7×10^{-8}
to nuclear magneton ratio	μ'_p/μ_N	$2.792\,775\,598(30)$		1.1×10^{-8}
Proton gyromagnetic ratio, $2\mu_p/\hbar$	γ_p	$2.675\,222\,099(70) \times 10^8$	s^{-1} T^{-1}	2.6×10^{-8}
	$\gamma_p/2\pi$	$42.577\,4821(11)$	MHz T^{-1}	2.6×10^{-8}
Shielded proton gyromagnetic ratio[h], $2\mu'_p/\hbar$	γ'_p	$2.675\,153\,362(73) \times 10^8$	s^{-1} T^{-1}	2.7×10^{-8}
	$\gamma'_p/2\pi$	$42.576\,3881(12)$	MHz T^{-1}	2.7×10^{-8}
		Neutron, n		
Neutron mass	m_n	$1.674\,927\,211(84) \times 10^{-27}$	kg	5.0×10^{-8}
in u		$1.008\,664\,915\,97(43)$	u	4.3×10^{-10}
Energy equivalent in MeV	m_nc^2	$939.565\,346(23)$	MeV	2.5×10^{-8}
Neutron magnetic moment	μ_n	$-0.966\,236\,41(23) \times 10^{-26}$	J T^{-1}	2.4×10^{-7}
to nuclear magneton ratio	μ_n/μ_N	$-1.913\,042\,73(45)$		2.4×10^{-7}
		Deuteron, d		
Deuteron mass	m_d	$3.343\,583\,20(17) \times 10^{-27}$	kg	5.0×10^{-8}
in u		$2.013\,553\,212\,724(78)$	u	3.9×10^{-11}
Energy equivalent in MeV	m_dc^2	$1875.612\,793(47)$	MeV	2.5×10^{-8}
Deuteron magnetic moment	μ_d	$0.433\,073\,465(11) \times 10^{-26}$	J T^{-1}	2.6×10^{-8}
to nuclear magneton ratio	μ_d/μ_N	$0.857\,438\,2308(72)$		8.4×10^{-9}

[a]Footnotes are at table end.

Recommended values (2006) of selected fundamental constants of physics and chemistry[a] (cont.)

Quantity	Symbol	Numerical value[b]	Unit[c]	Relative uncertainty (standard deviation)
		PHYSICOCHEMICAL CONSTANTS		
Avogadro constant	N_A, L	$6.022\ 141\ 79(30) \times 10^{23}$	mol^{-1}	5.0×10^{-8}
Atomic mass constant, $m_u = \frac{1}{12}m(^{12}C) = 1$ u $= 10^{-3}$ kg mol$^{-1}/N_A$	m_u	$1.660\ 538\ 782(83) \times 10^{-27}$	kg	5.0×10^{-8}
Energy equivalent in MeV	m_uc^2	931.494 028(23)	MeV	2.5×10^{-8}
Faraday constant[i], N_Ae	F	96 485.3399(24)	C mol^{-1}	2.5×10^{-8}
Molar Planck constant	N_Ah	$3.990\ 312\ 6821(57) \times 10^{-10}$	J s mol^{-1}	1.4×10^{-9}
	N_Ahc	0.119 626 564 72(17)	J m mol^{-1}	1.4×10^{-9}
Molar gas constant	R	8.314 472(15)	J mol^{-1} K^{-1}	1.7×10^{-6}
Boltzmann constant, R/N_A	k	$1.380\ 6504(24) \times 10^{-23}$	J K^{-1}	1.7×10^{-6}
in eV K^{-1}		$8.617\ 343(15) \times 10^{-5}$	eV K^{-1}	1.7×10^{-6}
	k/h	$2.083\ 6644(36) \times 10^{10}$	Hz K^{-1}	1.7×10^{-6}
	k/hc	69.503 56(12)	$m^{-1}K^{-1}$	1.7×10^{-6}
k^{-1} in K eV^{-1}		11 604.505(20)	K eV^{-1}	1.7×10^{-6}
Molar volume of ideal gas, RT/p, for $T = 273.15$ K, $p = 101.325$ kPa	V_m	$22.413\ 996(39) \times 10^{-3}$	m^3 mol^{-1}	1.7×10^{-6}
Loschmidt constant, N_A/V_m	n_o	$2.686\ 7774(47) \times 10^{25}$	m^{-3}	1.7×10^{-6}
Stefan-Boltzmann constant, $(\pi^2/60)k^4/\hbar^3c^2$	σ	$5.670\ 400(40) \times 10^{-8}$	W m^{-2} K^{-4}	7.0×10^{-6}
First radiation constant, $2\pi hc^2$	c_1	$3.741\ 771\ 18(19) \times 10^{-16}$	W m^2	5.0×10^{-8}
Second radiation constant, hc/k	c_2	$1.438\ 7752(25) \times 10^{-2}$	m K	1.7×10^{-6}
Wien displacement law constant, $b = \lambda_{max} T = c_2/4.965\ 114\ 231\ldots$	b	$2.897\ 7685(51) \times 10^{-3}$	m K	1.7×10^{-6}
		NON-SI UNITS ACCEPTED FOR USE WITH THE SI		
Electronvolt: (e/C) J	eV	$1.602\ 176\ 487(40) \times 10^{-19}$	J	2.5×10^{-8}
(Unified) atomic mass unit: 1 u $= m_u = \frac{1}{12}m(^{12}C)$	u	$1.660\ 538\ 782(83) \times 10^{-27}$	kg	5.0×10^{-8}

[a]This table presents a selection of the values of the fundamental constants recommended by the Committee on Data for Science and Technology (CODATA). These "2006 CODATA recommended values" form a self-consistent set based on the data available through December 31, 2006, and are generally recognized for use in all fields of science and technology. There is a detailed description of the data and analysis that led to these results in P. J. Mohr, B. N. Taylor, and D. B. Newell, CODATA recommended values of the fundamental constants: 2006, in the *Reviews of Modern Physics,* vol. 80, no. 2, pp. 633–730, 2008, and the *Journal of Physical and Chemical Reference Data,* vol. 37, no. 3, pp. 1187–1284, 2008. The 2006 adjustment was carried out under the auspices of the CODATA Task Group on Fundamental Constants. The recommended values are available on the World Wide Web at http://physics.nist.gov/cuu/Constants/index.html.

[b]The digits in parentheses represent one-standard-deviation uncertainties in the final two digits of the quoted values.

[c]A = ampere, C = coulomb, F = farad, Hz = hertz, J = joule, K = kelvin, kg = kilogram, m = meter, mol = mole, N = newton, Pa = pascal, s = second, T = tesla, W = watt, Wb = weber, Ω = ohm, eV = electronvolt, u = (unified) atomic mass unit. Prefixes: k = 10^3, M = 10^6, G = 10^9.

[d]The conventional value of the Josephson constant, adopted internationally for realizing representations of the volt using the Josephson effect, is K_{J-90} = 483 597.9 GHz V^{-1}.

[e]The conventional value of the von Klitzing constant, adopted internationally for realizing representations of the ohm using the quantum Hall effect, is R_{K-90} = 25 812.807 Ω.

[f]Value recommended by the Particle Data Group in the 2006 Review of Particle Physics (W.-M. Yao et al., *Journal of Physics G: Nuclear and Particle Physics,* vol. 33, no. 1, pp. 1–1232, 2006). The same value is given in the 2008 Review of Particle Physics [C. Amsler et al., *Physics Letters,* vol. B667, nos. 1–5, pp. 1–1340, 2008 (http://pdg.lbl.gov)].

[g]This and all other values involving m_τ are based on the value of $m_\tau c^2$ in MeV recommended by the Particle Data Group in the 2006 Review of Particle Physics (*ibid.*), but with a standard uncertainty of 0.29 MeV rather than the quoted uncertainty of −0.26 MeV, + 0.29 MeV. The value of $m_\tau c^2$ recommended in the 2008 Review of Particle Physics (*ibid.*) is 1776.84 ± 0.17 MeV (relative uncertainty = 9.6 ¥ 10^{-5}).

[h]Based on nuclear magnetic resonance (NMR) frequency of protons in a sphere of pure water (H_2O) at 25°C surrounded by vaccum.

[i]The numerical value of F to be used in coulometric chemical measurements is 96 485.3401(48) [relative uncertainty = 5.0×10^{-8}] when the relevant current is measured in terms of representations of the volt and ohm based on the Josephson and quantum Hall effects and the conventional values of the Josephson and von Klitzing constants, K_{J-90} and R_{K-90}.

The fundamental particles[a]

Gauge bosons $J_C^P = 1_-^-$ Self-conjugate except $\overline{W^+} = W^-$

Name	Symbol	Charge[b]	Mass and width, GeV	Couplings
Photon	γ	0	0	$A \Rightarrow \gamma A$
Gluon[c]	g	0	0	$A \Rightarrow gA'$
Weak bosons				
Charged[d]	$W^\pm$	±1	80.4, 2.1	$d \Rightarrow W^+u$
Neutral[e]	Z^0	0	91.2, 2.5	$A \Rightarrow Z^0A$

Fermions[f] $J = 1/2$ All have distinct antiparticles, except perhaps the neutrinos

Name	Charge[b]	Symbol and mass, GeV		Symbol and mass, GeV		Symbol and mass, GeV	
Leptons							
Neutrinos	0	ν_e	$<2 \times 10^{-9}$	ν_μ	<.0002	ν_τ	<.02
Charged leptons[g]	−1	e	.00051	μ	.106[h]	τ	1.777[h]
Quarks[c]							
Up type	$2/3$	u	.0015–.004	c	1.2	t	175[i]
Down type	$-1/3$	d	.004–.008	s	.10	b	4.5

[a] The graviton, with $J_C^P = 2_+^+$, has been omitted, since it plays no role in high-energy particle physics.

[b] In units of the proton charge.

[c] The gluon is a color SU_3 octet (8); each quark is a color triplet (3). These colored particles are confined constituents of hadrons; they do not appear as free particles, which is indicated by placing their entries in boxes.

[d] The branching ratios (%) of the decay modes of the W^+ are:

$u\bar{d}$, $c\bar{s}$	34 each
$\nu_e e^+$, $\nu_\mu \mu^+$, $\nu_\tau \tau^+$	11 each

[e] The branching ratios (%) of the decay modes of the Z^0 are:

$d\bar{d}$, $s\bar{s}$, $b\bar{b}$	15.6 each
$u\bar{u}$, $c\bar{c}$	11.6 each
$\nu_e\bar{\nu}_e$, $\nu_\mu\bar{\nu}_\mu$, $\nu_\tau\bar{\nu}_\tau$	6.7 each
e^+e^-, $\mu^+\mu^-$, $\tau^+\tau^-$	3.4 each

[f] The three known families (generations) of fermions are displayed in three columns.

[g] Any further charged leptons have mass greater than 100 GeV.

[h] The μ and τ leptons are unstable, with the following mean life and principal decay modes (branching ratios in %):

μ	$\tau_\mu = 2.2 \times 10^{-6}$ s	$e\bar{\nu}_e\nu_\mu$ 100
τ	$\tau_\tau = 2.9 \times 10^{-13}$ s	$\mu\bar{\nu}_\mu\nu_\tau$ 17, $e\bar{\nu}_e\nu_\tau$ 18, (hadrons) $\bar{\nu}_\tau$ 65

[i] The t quark has a width ≈ 2 GeV, with dominant decay to Wb.

Geologic time scale and related aspects

Eon	Era	Period (Interval length, $\times 10^6$ years)	Epoch	Beginning of epoch, $\times 10^6$ years ago	Forms of life	Other features
Phanerozoic	Cenozoic "Age of Mammals"	Quaternary (1.8)	Recent	0.01	Modern humans	Extensive glaciation during portions of Quaternary
			Pleistocene	1.8	Early humans	
		Tertiary (63.7)	Pliocene	5.3	Large carnivores	
			Miocene	23.8	Whales, apes, grazing animals	Formation of most present mountain ranges
			Oligocene	33.9	Large browsing animals	
			Eocene	55.8	Rise of modern floras	
			Paleocene	65.5	First placental mammals	
	Mesozoic "Age of Reptiles"	Cretaceous (80)	Upper (Base Santonian)		Last of dinosaurs	Widespread seas over much of present-day Europe, North Africa, and North America
			Middle (Base Albian)		Last of ammonites	
			Lower	145.5	Rise of flowering plants	
		Jurassic (54.1)	Upper (Base Callovian)		Toothed birds	Beginning of formation of Atlantic Ocean
			Middle (Base Bajocian)		Flying reptiles	
			Lower	199.6	First primitive mammals	
		Triassic (51.4)	Upper		Rise of dinosaurs	
			Middle		Rise of ammonites	
			Lower	251	Rise of cycads	
	Paleozoic "Age of Invertebrates"	Permian (48)	Upper (Base Ochoan)		Primitive reptiles	Mountain building during late Paleozoic
			Middle (Base Guadalupian)		Last of trilobites	
			Lower	299	Glossopteris flora	
		Carboniferous[a]: Pennsylvanian (19.1)	Upper		Spread of amphibians	Widespread coal swamps
			Middle		Great coal forests	
			Lower	318.1	Climax of spore-forming plants	
		Carboniferous[a]: Mississippian (41.1)	Upper		Abundant sharks	
			Middle		Climax of crinoids and blastoids	
			Lower	359.2		
		Devonian (56.8)	Upper		First forests	
			Middle		Rise of ferns	
			Lower	416	Earliest known amphibians	
		Silurian (27.7)	Upper		Appearance of land plants	
			Middle		First known scorpions	
			Lower	443.7	Expansion of brachiopods and corals	
		Ordovician (44.6)	Upper		Appearance of primitive fishes	
			Middle		Climax of trilobites	
			Lower	488.3	Rise of cephalopods	
		Cambrian (53.7)	Upper (Base Croixian)		Abundant trilobites	
			Middle		Many kinds of shelled invertebrates	
			Lower	542		
Proterozoic	Precambrian		Neoproterozoic (357[b])	1000	Marine algae, wormlike organisms, other simple forms	Extensive glaciation Oldest dated rocks (Greenland)
			Mesoproterozoic (700[b])	1600		
			Paleoproterozoic (900[b])	~2500		
Archean[c]			Late (500[b])	2800	Blue-green algae (cyanobacteria)	Abundant dark sediments Formation of earliest known rocks
			Middle (400[b])	3200		
			Early (>400[b])	>3600	Bacteria	

[a]The period known as Carboniferous in Europe is divided into two periods in North American usage.
[b]Length of epoch, $\times 10^6$ years.
[c]The beginning of the Archean corresponds to the age of the oldest preserved continental crust.

Biographical Listing

Abbe, Ernst (1840–1905), German physicist. Developed optical instruments, such as an apochromatic objective and a crystal refractometer.

Abel, Frederick Augustus (1827–1902), English chemist. Expert on the chemistry of explosives; originated the Abel test for determination of the flash point of petroleum.

Abel, John Jacob (1857–1938), American pharmacologist and physiologist. Isolated epinephrine, and insulin in crystal form.

Abel, Niels Henrik (1802–1829), Norwegian mathematician. Contributed to the theory of elliptical functions.

Abell, George Ogden (1927–1983), American astronomer. Research in problems relating to organization, structure, and distribution of galaxies; observational cosmology; and planetary nebulae.

Abney, William de Wiveleslie (1843–1920), English photographic chemist and physicist. Photographed the infrared solar spectrum.

Abrikosov, Alexei A. (1928–), Russian-born American physicist. Applied the Ginzburg-Landau theory to explain the behavior of type II superconductors. Nobel Prize, 2003.

Adams, John Couch (1819–1892), English astronomer. Discovered, independently of U. J. J. Leverrier, Neptune.

Adanson, Michel (1727–1806), French naturalist. Classified plants in his *Les Familles Naturelles des Plantes.*

Addison, Thomas (1793–1860), English physician. Identified pernicious anemia and Addison's disease of the adrenal cortex.

Adler, Alfred (1870–1937), Austrian psychiatrist and psychologist. Founded the school of individual psychology.

Adrian of Cambridge, Edgar Douglas Adrian, Baron (1889–1977), English physiologist. Investigated physiology of nervous system; showed that change in electric potential in electroencephalograph is due to electrical activity of cortex; Nobel Prize, 1932.

Afzelius, Adam (1750–1837), Swedish botanist. Founded the Linnaean Institute.

Agassiz, Jean Louis Rudolphe (1807–1873), Swiss-born American naturalist. Wrote books on ichthyology, especially relating to classification.

Agnesi, Maria Gaetana (1718–1799), Italian mathematician. Author of *Instituzioni Analitiche*, a complete treatment of algebra and analysis; shared in the discovery of a cubic curve ("witch of Agnesi").

Agre, Peter (1949–), American medical doctor and scientist. Discovered water channels in cell membranes; Nobel Prize, 2003.

Agricola, Georgius, real name Georg Bauer (1494–1555), German physician and mineralogist. Known as the father of systematic mineralogy.

Ahlfors, Lars Valerian (1907–1996), Finnish-American mathematician. Did research on covering surfaces related to Riemann surfaces of inverse functions of entire and meromorphic functions; opened up new fields of analysis; Fields Medal, 1936.

Airy, George Biddell (1801–1892), English astronomer. Discovered inequality in the motions of Venus and the Earth; determined the mass of the Earth.

Aitken, John (1839–1919), Scottish physicist. Studied dust particles in the atmosphere, known as Aitken nuclei.

Aitken, Robert Grant (1864–1951), American astronomer. Discovered more than 3000 binary stars.

Alder, Kurt (1902–1958), German chemist. Codeveloper of the Diels-Alder reaction for diene synthesis; contributed to stereochemistry; Nobel Prize, 1950.

Alembert, Jean le Rond d' (1717–1783), French mathematician. Developed d'Alembert's principle and the calculus of partial differences.

Alferov, Zhores Ivanovich (1930–), Russian physicist and electronics engineer. Developed semiconductor heterostructures used in high-speed and opto-electronics, including fast transistors, laser diodes, and light-emitting diodes; Nobel Prize, 2000.

Alfvén, Hannes Olof Gösta (1908–1995), Swedish physicist. Studies in magnetohydrodynamics, planetary physics, antiferromagnetism, and ferrimagnetism; Nobel Prize, 1970.

Alhazen (965–1038), Arab mathematician and astronomer. Provided the first accounts of atmospheric refraction and reflection from concave surfaces; constructed spherical and parabolic mirrors.

al-Khwarizmi (780–?850), Arab mathematician. Wrote treatises on arithmetic and algebra, which were important in the mathematical knowledge of medieval Europe.

Allen, Edgar (1892–1943), American biologist. Discovered estrogen; investigated hormonal mechanisms controlling female reproductive cycle.

Allen, Willard Myron (1904–1993), American physician. With G. W. Corner, discovered progesterone, and proved it necessary for development of embryo in early pregnancy; with O. Wintersteiner, synthesized crystalline progesterone.

Altman, Sidney (1939–), American chemist. Discovered an unusual enzyme that contains ribonucleic acid (RNA) in addition to a protein, leading to the discovery that RNA molecules have catalytic properties similar to those of enzymes; Nobel Prize, 1989.

Alvarez, Luis Walter (1911–1988), American physicist. Pioneer in building liquid hydrogen bubble chambers, and in developing measurement devices and computer systems to analyze data from these chambers; discovered large numbers of short-lived elementary particles; Nobel Prize, 1968.

Amagat, Émile (1841–1915), French physicist. Investigated relationship of pressure, density, and temperature in gases and liquids, particularly at high pressure.

Amici, Giovanni Battista (1786–1863), Italian astronomer, optician, and naturalist. Invented the Amici microscope; designed parabolic mirrors for reflecting telescopes.

Ampère, André Marie (1775–1836), French physicist and mathematician. Founder of electrodynamics; formulated Ampère's law; invented the astatic needle.

Anaximander (611–547 B.C.), Greek astronomer and mathematician. Reputed inventor of geographical maps; formulated the concept of the universe as infinite (apeiron).

Anderson, Carl David (1905–1991), American physicist. Discovered the meson in cosmic rays; discovered the positron; Nobel Prize, 1936.

Anderson, Philip Warren (1923–), American physicist. Demonstrated existence of electronic localization in disordered solids, and of localized magnetism in metals; Nobel Prize, 1977.

Andrade, Edward Neville da Costa (1887–1971), English physicist. Discovered Andrade's creep law and a law governing variation of viscosity of liquids with temperature.

Andrews, Roy Chapman (1884–1960), American naturalist. Discovered many plant and animal fossils.

Anfinsen, Christian Boehmer (1916–1995). American biochemist. Discovered how three-dimensional structures of ribonuclease and other proteins are formed; Nobel Prize, 1972.

Angström, Anders Jonas (1814–1874), Swedish physicist. Mapped the solar spectrum; discovered hydrogen in the solar atmosphere.

Apollonius of Perga (247–205 B.C.), Greek mathematician. Wrote about conic sections; coined the terms parabola, ellipse, and hyperbola.

Appleton, Edward Victor (1892–1965), English physicist. Demonstrated the existence of the ionosphere and discovered its region known as the Appleton layer; contributed to the development of radar; Nobel Prize, 1947.

Arago, Dominique François (1786–1853), French astronomer and physicist. Discovered the magnetic properties of nonferrous materials, and the production of magnetism by electricity.

Arber, Werner (1929–), Swiss molecular biologist. Determined the molecular mechanism of host-controlled restriction modification of bacterial viruses and discovered the restriction enzymes; Nobel Prize, 1978.

Archimedes (287–212 B.C.), Greek physicist and mathematician. Formulated Archimedes' principle; invented the compound pulley and Archimedes' screw.

Argand, Jean Robert (1768–1822), Swiss mathematician. Developed the Argand diagram.

Argelander, Friedrich Wilhelm August (1799–1875), German astronomer. Prepared a star catalog; introduced decimal division of stellar magnitudes.

Aristarchus of Samos (310–250 B.C.), Greek astronomer. Invented the hemispherical sundial; determined the movement of the Earth around the stationary Sun; added a correction factor of 1/1623 of a day to the length of the year.

Aristotle (384–322 B.C.), Greek philosopher. Exponent of the methodology and division of sciences; contributed to physics, astronomy, meteorology, psychology, and biology.

Arkwright, Richard (1732–1792). English inventor. Developed the first practical mechanized spinning frame, utilizing rollers.

Armstrong, Edwin Howard (1890–1954). American electronics engineer. Prominent for his contributions to radio technology, including invention of the superheterodyne receiver and frequency modulation (FM) radio transmission.

Arrhenius, Svante August (1859–1927), Swedish physicist and chemist. Developed theory of electrolytic dissociation; investigated osmosis and viscosity of solutions; Nobel Prize, 1903.

Arsonval, Jacques Arsène d' (1851–1940), French physicist and physiologist. Pioneered in electrotherapy; invented d'Arsonval galvanometer.

Aston, Francis William (1877–1945), English physicist and chemist. Discovered isotopes in nonradioactive elements by using the mass spectrograph he invented; Nobel Prize, 1922.

Atiyah, Michael Francis (1929–), British mathematician. Work centered on the interaction between geometry and analysis; developed K theory in collaboration with F. Hirzebruch; with I. M. Singer, proved the index theorem concerning elliptic differential operators on compact differentiable manifolds, which was later seen to have applications to theoretical physics; Fields Medal, 1966; Abel Prize, 2004.

Atwood, George (1746–1807), English mathematician. Invented the Atwood machine.

Audubon, John James (1785–1851), Haitian-born American ornithologist and artist. Made drawings and paintings of birds and animals.

Auger, Pierre Victor (1899–1993), French physicist. Discovered the Auger effect.

Avicenna (979–1037), Arab physician. Wrote the medical text *Canon Medicinae*.

Avogadro, Amedeo (1776–1856), Italian physicist. Formulated Avogadro's law.

Axel, Richard (1946–), American molecular biologist. Recognized along with Linda Buck for pioneering research of the olfactory system, which led to the discovery of the genes and proteins involved in the transmission of olfactory information. Later independent work clarified the cellular and molecular mechanisms underlying the olfactory system. Nobel Prize, 2004.

Axelrod, Julius (1912–2004), American biochemist and pharmacologist. Showed that many drugs act by modifying storage of neurotransmitters at nerve terminals; made discoveries concerning metabolism, and mechanisms for formation and inactivation of norepinephrine; Nobel Prize, 1970.

Ayrton, William Edward (1847–1908), English physicist and electrical engineer. Invented the ammeter, voltmeter, and other electrical measuring instruments.

Baade, Walter (1893–1960), German-born American astronomer. Formulated concept of stellar populations; increased distance scale of universe by factor of 2.

Babbage, Charles (1792–1851), English mathematician. Devised a primitive computer to calculate and print mathematical and astronomical tables.

Babcock, Harold Delos (1882–1968) and **Horace Welcome** (1912–2003), American astronomers. Invented the Babcock magnetograph, observed weak solar magnetic fields, and discovered stellar magnetic fields.

Babcock, Stephen Moulton (1843–1931), American agricultural chemist. Pioneer in nutrition; devised the Babcock test to measure fat content in milk.

Babinet, Jacques (1794–1872), French physicist. Invented a polariscope and a goniometer.

Back, Ernst E. A. (1881–1959), German physicist. Developed improved spectrographs; made spectroscopic observations leading to Paschen-Back effect.

Badger, Richard McLean (1896–1974), American physical chemist and spectroscopist. Studied structures of polyatomic molecules; formulated Badger's rule concerning molecular bonds.

Baekeland, Leo Hendrik (1864–1944), Belgian-born, American chemist. Invented the phenol-formaldehyde polymer, Bakelite, the first commercial synthetic polymer.

Baer, Karl Ernst von (1792–1876), Estonian embryologist. Discovered the mammalian ovum and the notochord; developed the theory of embryonic germ layers.

Baeyer, Johann Friedrich Wilhelm Adolf von (1835–1917), German chemist. Synthesized indigo and hydroaromatic compounds; Nobel Prize, 1905.

Baily, Francis (1774–1844), English astronomer. A founder of the Royal Astronomical Society; first observed phenomenon of Baily's beads.

Baire, René Louis (1874–1932), French mathematician. Contributed to theory of functions of real variables; introduced concept of Baire functions.

Baker, Alan (1939–), British mathematician. Proved a generalization of the Gelfond-Schneider theorem; from this work, generated transcendental numbers not previously identified and solved problems in the theory of Diophantine equations; Fields Medal, 1970.

Balfour, Francis Maitlant (1851–1882), English biologist. Founder of comparative embryology.

Balmer, Johann Jakob (1825–1898), Swiss physicist. Expressed the mathematical formula for frequencies of hydrogen lines in the visible spectrum.

Baltimore, David (1938–), American virologist. Investigated interaction between ribonucleic acid tumor viruses and genetic material; independently of H. M. Temin, discovered reverse transcriptase; Nobel Prize, 1975.

Banach, Stefan (1892–1945), Polish mathematician. Laid foundations of contemporary functional analysis; introduced concept of Banach space and discovered its fundamental properties.

Bang, Bernhard Laurits Frederik (1848–1932), Danish veterinarian. Discovered method of eradicating bovine tuberculosis; discovered *Brucella abortus*, the agent of contagious abortion (Bang's disease) and brucellosis.

Banting, Frederick Grant (1891–1941), Canadian physician. With J. J. R. Macleod and C. H. Best, discovered insulin and its role in diabetes; Nobel Prize. 1923.

Bárány, Robert (1876–1936), Austrian physician. Developed new methods of diagnosing ear diseases; Nobel Prize, 1914.

Bardeen, John (1908–1991), American physicist. With L. N. Cooper and J. R. Schrieffer, formulated a theory of superconductivity; invented the transistor; Nobel Prize, 1956 and 1972.

Barkhausen, Heinrich Georg (1881–1956), German electronic engineer and physicist. Contributed to theory and application of electron tubes; with K. Kurz, developed Barkhausen-Kurz oscillator; discovered Barkhausen effect.

Barkla, Charles Glover (1877–1944), English physicist. Described characteristics of x-rays and other short-wave emissions of elements.

Barnard, Edward Emerson (1857–1923), American astronomer. Discovered 16 comets, the fifth satellite of Jupiter, and dark nebulae; contributed to celestial photography.

Barnett, Samuel Jackson (1873–1956), American physicist. Discovered Barnett effect and used it to measure the gyromagnetic ratio of ferromagnetic materials; gave experimental proof of existence of ionosphere.

Barr, Murray Llewellyn (1908–1995), Canadian anatomist. Discovered the Barr body on the X chromosome of the human female.

Barré-Sinoussi, Françoise (1947–), French virologist. Prominent for research on retroviruses. Co-discoverer of the human immunodeficiency virus (HIV); Nobel Prize, 2008.

Bartholin, Kaspar (1655–1738), Danish physician. Discovered Bartholin's glands of the vagina and a sublingual duct.

Bartholin, Thomas (1616–1680), Danish physician. Discovered lymphatic glands; described the lymphatic system.

Bartlett, James Holly (1904–2000), American physicist. Introduced concept of Bartlett force; did research on nuclear shell model, electrochemical potentiostat, and restricted three-body problem.

Barton, Derek Harold Richard (1918–1998), British chemist. Developed and expanded concept of conformation to include large molecules with complex ring systems; Nobel Prize, 1969.

Basov, Nicolai Gennediyevich (1922–2001), Soviet physicist. Conducted fundamental studies in quantum electronics; with A. M. Prokhorov, developed quantum optical generators; Nobel Prize, 1964.

Bassham, James Alan (1922–), American chemist. Helped to elucidate basic photosynthetic carbon cycle.

Bates, Henry Walter (1825–1892), English naturalist. Discovered Batesian mimicry among butterflies and moths.

Baudot, Èmile (1845–1903), French engineer. Invented an improved telegraph transmitter.

Baumé, Antoine (1728–1804), French chemist. Invented a graduated hydrometer which utilizes the Baumé scale.

Bautz, Laura Patricia (1940–), American astronomer. Collaborated with W. W. Morgan in developing Bautz-Morgan classification of galaxy clusters.

Bayer, Johann (1572–1624), German astronomer. Charted 12 constellations; first to use Greek letters to designate the order of brightness of stars in a constellation.

Bayes, Thomas (1702–1761), English mathematician. Formulated a basis for statistical inference.

Bayliss, William Maddock (1860–1924), English physiologist. Did research on electrophysiology of heart action; discovered the hormone secretin.

Beadle, George Wells (1903–1989), American geneticist. With E. L. Tatum, proved that genes affect heredity by controlling cell chemistry; Nobel Prize, 1958.

Beams, Jesse Wakefield (1898–1977), American physicist. Developed vacuum-type ultracentrifuges, used in purification and molecular weight determination of large-molecular-weight substances, isotope separation, and determination of the gravitational constant.

Beattle, James Alexander (1895–1981), American chemist and physicist. Studied ionic theory and thermodynamics; with P. W. Bridgman, proposed Beattie and Bridgman equation for gases.

Beaufort, Francis (1774–1857), English hydrographer. Devised scale of wind velocity.

Beaumont, William (1785–1853), American physician. Did pioneering studies of digestion and gastric juices.

Béchamp, Pierre Jacques Antoine (1816–1908), French chemist. Discovered a method of preparing aniline.

Beckmann, Ernst Otto (1853–1923), German chemist. Discovered Beckmann molecular transformation; invented the Beckmann thermometer.

Becquerel, Antoine César (1788–1878), French physicist. Pioneer in electrochemistry; first to extract metals from ore by electrolysis.

Becquerel, Antoine Henri (1852–1908), French physicist. A discoverer of radioactivity in uranium.

Bednorz, Johannes Georg (1950–), German physicist. With K. A. Müller, discovered high-temperature superconductivity in copper oxide ceramic materials; Nobel Prize, 1987.

Beebe, Charles William (1877–1962), American naturalist. Pioneer in deep-sea exploration; made ornithological collections.

Beer, August (1825–1863), German physicist. Discovered Beer's law of light absorption.

Behring, Emil Adolph von (1854–1917), German bacteriologist. Produced diphtheria and tetanus antitoxins; Nobel Prize, 1901.

Békésy, Georg von (1889–1972), Hungarian-born American physicist. Studied hearing processes, especially inner-ear mechanics; Nobel Prize, 1961.

Bell, Alexander Graham (1847–1922), Scottish-born American inventor. Invented the telephone, pho-tophone, graphophone, and one of the earliest gramophones.

Bell, Charles (1774–1842), Scottish anatomist. Discovered that sensory and motor nerves are anatomically and functionally distinct.

Bellman, Richard Ernest (1920–1984), American mathematician. Research in analytic number theory, differential equations, stochastic processes, dynamic programming, and mathematical biosciences; discovered Bellman's principle of optimality.

Benacerraf, Baruj (1920–), American immunologist. Discovered immune-response (Ir) genes that control specific immune responses to thymus-dependent antigens; Nobel Prize, 1980.

Benioff, Hugo (1899–1968), American geophysicist. Investigated earthquakes, particularly through instrumental seismology.

Bentham, George (1800–1884), English botanist. With J. Hooker, wrote *Genera Plantarum*.

Berg, Paul (1926–), American biochemist. Investigated the biochemistry of deoxyribonucleic acid (DNA) and designed a technique for gene splicing; Nobel Prize, 1980.

Bergey, David Hendricks (1860–1937), American bacteriologist. Authority on classification of bacteria.

Bergius, Friedrich (1884–1949), Polish-born German chemist. Developed Bergius process for hydrogenation of coal to a petroleumlike oil; Nobel Prize, 1931.

Bergström, Sune Carl (1916–2004), Swedish biochemist and medical scientist. Studied the metabolism of unsaturated fatty acids and determined the chemical structure of prostaglandins; Nobel Prize, 1982.

Bernard, Claude (1813–1878), French physiologist. Studied digestion; discovered that glycogen is produced by the liver.

Berners-Lee, Tim (1955–), British-born physicist and software engineer. Proposed the World Wide Web, and then invented hypertext markup language (HTML), HyperText Transfer Protocol (HTTP), the Internet addressing scheme (Universal Resource Locator or URL), and the first Web browser.

Bernoulli, Daniel (1700–1782), Swiss mathematician born in the Netherlands. Founder of mathematical physics; worked on hydrodynamics and differential equations; formulated the Bernoulli equation.

Bernoulli, Jacques or Jacob (1654–1705), Swiss mathematician. Contributed to mathematics of curves, calculus, and probability; developed the Bernoulli number.

Bernoulli, Jean or Johann (1667–1748), Swiss mathematician. A founder of calculus of variations; contributed to exponential calculus, complex numbers, geodesies, and trigonometry.

Berthelot, Pierre Eugène Marcellin (1827–1907), French chemist. Founder of thermochemistry; first to synthesize organic compounds; demonstrated nitrogen fixation.

Bertrand, Joseph Louis Francois (1822–1900), French mathematician. Contributed to analysis, differential geometry, and probability theory.

Berzelius, Jons Jakob (1779–1848), Swedish chemist. Discovered the elements cerium, selenium, thorium, and silicon; developed a system for classification and nomenclature of compounds.

Bessel, Friedrich Wilhelm (1784–1846), German astronomer and mathematician. Developed Bessel functions; determined parallax of the star 61 Cygni; postulated existence of Neptune and dark stars.

Bessemer, Henry (1813–1898), English engineer. Invented the Bessemer process, the first method for manufacturing steel on a large scale.

Best, Charles Herbert (1899–1978), Canadian physiologist and medical researcher. Associated with F. G. Banting and J. J. R. Macleod in the discovery of insulin.

Bethe, Hans Albrecht (1906–2005), German-born American physicist. Formulated the theory of energy production in stars; research on nuclear physics; Nobel Prize, 1967.

Bhabha, Homi Jehangir (1909–1966), Indian physicist. With W. Heitler, developed theory of cascade showers of cosmic rays; observed slowing of decay rate of high-velocity mesons.

Bianchi, Luigi (1856–1928), Italian mathematician. Contributed to differential geometry and study of noneuclidean geometries; discovered Bianchi identity.

Bichat, Marie François Xavier (1771–1802), French anatomist and physiologist. Founder of animal histology; originated the term "tissues," and distinguished 21 types in his particular scheme.

Bieberbach, Ludwig (1886–1982), German mathematician. Research on complex function theory, differential equations, geometry, and algebra; postulated Bieberbach's conjecture.

Bienaymé, Irénée Jules (1796–1878), French mathematician. Studied calculus of probabilities and its application to financial science; with P. L. Chebyshev, discovered Bienaymé-Chebyshev inequality.

Billet, Felix (1808–1882). French physicist, Invented Billet split lens.

Binet, Alfred (1857–1911), French psychologist. Investigated development and measurement of intelligence.

Binnig, Gerd (1947–), German physicist. With H. Rohrer, developed scanning tunneling microscope; Nobel Prize, 1986.

Biot, Jean Baptiste (1774–1862), French mathematician and physicist. Discovered circular polarization of light; invented a polariscope; with D. Brewster, discovered biaxial crystals; helped formulate Biot-Savart law.

Birkhoff, George David (1884–1944), American mathematician. Investigated differential equations, dynamical systems, ergodic theory, mechanics of fluids, and foundations of relativity and quantum mechanics.

Bishop, John Michael (1936–), American virologist and biochemist. With H. Varmus, researched the genetic basis of human cancers; their work led to the identification of over 50 cellular genes that can become oncogenes; Nobel Prize, 1989.

Bjerknes, Vilhelm Fremann Doren (1862–1951), Norwegian physicist. Research on electric waves; originated the polar-front theory in meteorology.

Black, James (1924–), British pharmacologist. Developed the first beta blocker drug, propranolol; also credited with the discovery of another important class of drugs, the H2 antagonists; Nobel Prize, 1988.

Black, Joseph (1728–1799) Scottish physicist and chemist. Established the concepts of latent heat and specific heat, and discovered carbon dioxide.

Blackett, Patrick Maynard Stuart (1897–1974), English physicist. Built an improved cloud chamber used to photograph tracks of a nuclear disintegration and of a cosmic-ray shower; discovered the positron; Nobel Prize, 1948.

Blobel, Günter (1936–), German-born American cell and molecular biologist. Discovered that proteins carry signals that help direct their movement among the organelles of the cell; Nobel Prize, 1999.

Bloch, Felix (1905–1983), Swiss-born American physicist. Discovered a technique for studying magnetism of atomic nuclei in normal matter; Nobel Prize, 1952.

Bloch, Konrad Emil (1912–2000), German-born American biochemist. Traced the transformations of fat and carbohydrate metabolites to cholesterol; Nobel Prize, 1964.

Blodgett, Katharine Burr (1898–1979), American chemical physicist. Studied surface science, and is best know for her work with Irving Langmuir and the development of the Langmuir-Blodgett film.

Bloembergen, Nicolaas (1920–), Netherlands-born American physicist. Contributed to development of maser; made extensive contributions to theoretical and experimental development of nonlinear optics; Nobel Prize, 1981.

Blumberg, Baruch Samuel (1925–), American physician and biologist. Research leading to a test for hepatitis viruses in blood and to an experimental hepatitis vaccine; Nobel Prize, 1976.

Boas, Franz (1858–1942), American anthropologist. Pioneered in physical anthropology; contributed to stratigraphic archeology in Mexico; emphasized importance of linguistic analysis.

Bobillier, Étienne (1798–1840), French mathematician and physicist. Contributed to geometry and statics; discovered Bobillier's law.

Bode, Johann Elert (1747–1826), German astronomer. Prepared a celestial atlas showing about 17,000 stars; formulated Bode's law.

Boerhaave, Hermann (1668–1738), Dutch physician. A great teacher at the University of Leyden; wrote the physiology textbook *Institutiones Medicae*.

Bohm, David (1917–1992), American-born British physicist. Research in quantum theory and new modes of description in physics.

Bohr, Aage (1922–), Danish physicist. With B. R. Mottelson, developed theory which unifies shell and liquid-drop models of the atomic nucleus, and which explains nonspherical nuclei; Nobel Prize, 1975.

Bohr, Niels (1885–1962), Danish physicist. Devised an atomic model; codeveloped the quantum theory, applying it to atomic structure in Bohr's theory; Nobel Prize, 1922.

Boltzmann, Ludwig Eduard (1844–1906), Austrian physicist. An authority on the kinetic theory of gases; demonstrated the Stefan-Boltzmann law of blackbody radiation, Boltzmann's law of energy, and the Boltzmann constant.

Bolyai, János (1802–1860), Hungarian mathematician. Independently of K. F. Gauss and N. I. Lobachevski, originated a system of noneuclidean geometry.

Bolzano, Bernard (1781–1848), Czechoslovakian philosopher, logician, and mathematician. Contributed to theory of real functions; proved Bolzano's theorem and Bolzano-Weierstrass theorem.

Bombieri, Enrico (1940–), Italian mathematician. Made major contributions to the study of prime numbers, partial differential equations and minimal surfaces, univalent functions and the local Bieberbach conjecture, and functions of several complex variables; Fields Medal, 1974.

Bond, George Phillips (1825–1865), American astronomer. Introduced concept of Bond albedo; pioneered astronomical photography.

Boole, George (1815–1864), English mathematician and logician. Developed new system of mathematical logic, which is known as Boolean algebra.

Borcherds, Richard Ewan (1959–), British mathematician. Worked in algebra and geometry; proved the moonshine conjecture, which relates the so-called monster group and elliptical curves, using methods borrowed from string theory in theoretical physics; Fields Medal, 1998.

Borda, Jean Charles (1733–1799), French physicist and mathematician. Introduced Borda mouthpiece; developed instruments for navigation, geodesy, and determination of weights and measures.

Bordet, Jules Jean Baptiste Vincent (1870–1961), Belgian physiologist. Made discoveries in immunology; with O. Gengou, developed the technique of the complement fixation reaction; Nobel Prize. 1919.

Borel, Félix Edouard Émile (1871–1956), French mathematician. Work in infinitesimal calculus and the calculus of probabilities.

Born, Max (1882–1970), German-born British theoretical physicist. Pioneered in the development of quantum mechanics; Nobel Prize, 1954.

Bosch, Carl (1874–1940), German chemist. Developed chemical high-pressure methods, and the Haber-Bosch process for ammonia synthesis; Nobel Prize, 1931.

Bose, Jagadis Chandra (1858–1937), Indian plant physiologist and physicist. Founded the Bose Research Institute in Calcutta; investigated photosynthesis, "nervous mechanism" of plants, and other plant subjects.

Bose, Satyendra Nath (1894–1974), Indian physicist. Originated Bose-Einstein statistics to describe photons.

Bothe, Walter (1891–1957), German physicist. Devised the coincidence method for the investigation of nuclear reactions and cosmic radiation; Nobel Prize, 1954.

Bouguer, Pierre (1698–1758), French geodesist, hydrographer, and physicist. Laid foundations of photometry; discovered Bouguer-Lambert law of light intensity.

Bourdon, Eugène (1808–1884), French inventor. Invented Bourdon pressure gage.

Bourgain, Jean (1954–), Belgian mathematician. Worked in several areas of mathematical analysis, including the geometry of Banach spaces, convexity in high dimensions, harmonic analysis, ergodic theory, and nonlinear partial differential equations; Fields Medal, 1994.

Boussinesq, Joseph Valentin (1842–1929), French mathematical physicist. Research in hydrodynamics; introduced Boussinesq approximation.

Bovet, Daniel (1907–1992), Swiss-born Italian pharmacologist. Research on synthetic compounds that inhibit the action of the vascular system and skeletal muscles; Nobel Prize, 1957.

Bowditch, Nathaniel (1773–1838), American navigator and mathematician. Produced navigation guide; translated and improved P. S. de Laplace's *Mécanique céleste*.

Bowen, Norman Levi (1887–1956), Canadian-born American geologist. Studied physical chemistry of geological processes and phase equilibria of silicates; discovered significance of reaction principle in petrogenesis.

Boyer, Paul D. (1918–), American chemist. Made major contributions toward elucidating the enzymatic mechanism underlying the synthesis of adenosine triphosphate (ATP) by proposing the binding change mechanism; Nobel Prize, 1997.

Boyle, Robert (1627–1691), British physicist and chemist. Conducted experiments on properties of the air pump; his law concerning gases is named for him; advanced the atomistic theory of matter.

Bradley, James (1693–1762), English astronomer. Discovered aberration of light due to the Earth's motion, and nutation of the Earth's axis; prepared astronomical tables and star catalogs.

Bragg, William Henry (1862–1942), English physicist. Codeveloper, with W. L. Bragg, of the x-ray spectrometer; used x-ray diffraction to determine crystal structure; Nobel Prize, 1915.

Bragg, William Lawrence (1890–1971), British physicist. With W. H. Bragg, developed x-ray analysis of the atomic arrangement in crystalline structures; Nobel Prize, 1915.

Brahe, Tycho or Tyge (1546–1601), Danish astronomer. Made painstaking observations of the planetary system; wrote *Astronomiae Instauratae Progymnasmata*.

Brattain, Walter Houser (1902–1987), American physicist. Investigated properties of semiconductors; research on surface properties of solids; Nobel Prize, 1956.

Braun, Karl Ferdinand (1850–1918), German physicist. Research on cathode rays and wireless telegraphy; Nobel Prize, 1909.

Bravais, Auguste (1811–1863), French physicist. Studied relationship between crystal form and structure; derived Bravais lattices.

Breit, Gregory (1899–1981), Russian-born American physicist. Research on quantum theory, quantum electrodynamics, hyperfine structure, and ionosphere.

Brenner, Sydney (1927–), South African-born British molecular biologist. Established the nematode *Caenorhabditis elegans* as an experimental model system for studying genetic regulation of cell division, cell specialization, and organ development in multicellular animals. Also credited with the identification of messenger ribonucleic acid (mRNA). Nobel Prize, 2002.

Breuer, Josef (1842–1925), Austrian neurologist. Evolved abreaction method for treatment of neuroses.

Brewster, David (1781–1868), Scottish physicist. Formulated Brewster's law on polarization of light; codiscoverer, with J. B. Biot, of biaxial crystals.

Brianchon, Charles Julien (1783–1864), French mathematician. Proved Brianchon's theorem.

Bridgman, Percy Williams (1882–1961), American physicist. Worked in high-pressure physics and thermodynamics of liquids; Nobel Prize, 1946.

Briggs, Henry (1561–1631), English mathematician. Prepared logarithmic tables (later known as common logarithms); devised sophisticated interpolation techniques.

Bright, Richard (1789–1858), English physician. Made biochemical study of disease; researched Bright's disease of the kidneys.

Brillouin, Leon (1889–1969), French physicist. With G. Wentzel and H. A. Kramers, developed Wentzel-Kramers-Brillouin method; originated concept of Brillouin zones.

Brillouin, Louis Marcel (1854–1948), French physicist. Work on crystal structure, viscosity of liquids and gases, radiotelegraphy, and relativity.

Brinell, Johann August (1849–1925), Swedish engineer. Invented the Brinell machine to measure the hardness of alloys and metals in terms of the Brinell number.

Brockhouse, Bertram Neville (1918–2003), Canadian physicist. Developed slow neutron spectroscopy technique for studying dynamics of atoms in solids and liquids; Nobel Prize, 1994.

Broglie, Louis Victor de (1892–1987), French physicist. Worked in nuclear physics; first to link wave and corpuscular theory; Nobel Prize, 1929.

Bromwich, Thomas John I'Anson(1875–1929), English mathematician. Showed how the Heaviside calculus could be developed in a manner acceptable to pure mathematicians through use of contour integrals.

Brönsted, Johannes Nicolaus (1879–1947), Danish chemist. Researched kinetic properties of ions, catalysis, and nitramide; formulated the Brönsted theory of acid-base reactions.

Brown, Herbert Charles (1912–2004), British-born American chemist. Developed methods for chemical synthesis of diborane and organoboranes; Nobel Prize, 1979.

Brown, Michael S. (1941–), American biochemist and geneticist. With Joseph L. Goldstein, discovered low-density lipoprotein receptors and their function in cholesterol metabolism; Nobel Prize, 1985.

Browning, John Moses (1855–1926), American inventor. Invented the Browning machine gun.

Brun, Viggo (1885–1978), Norwegian mathematician. Worked in number theory, introducing what is now known as the Brun's sieve, which made some progress in the resolution of such problems as Goldbach's conjecture and the twin prime problem.

Brunauer, Stephen (1903–1986), Hungarian-born American chemist. Contributed to surface and colloid chemistry; with P. H. Emmett and E. Teller, developed Brunauer-Emmett-Teller equation for surface area determinations.

Brunel, Isambard Kingdom (1806–1859), English engineer. Constructed great bridges in England; designed important steamships.

Buchner, Eduard (1860–1917), German chemist. Studied alcoholic fermentation of sucrose; Nobel Prize, 1907.

Buck, Linda (1947–), American molecular biologist. Recognized along with Richard Axel for pioneering research of the olfactory system, which led to the discovery of the large gene family encoding the different odorant receptors located on the membrane of olfactory receptor cells. Later independent work clarified the olfactory system, from the molecular level to the organization of the cells. Nobel Prize, 2004.

Buckingham, Edgar (1867–1940), American physicist. Worked on thermodynamics and dimensional analysis; derived Buckingham's π theorem.

Buffon, Georges Louis Leclerc, Comte de (1707–1788), French naturalist. Compiled *Histoire Narurelle*, a monumental work on natural history.

Bullen, Keith Edward (1906–1976), Australian applied mathematician. Carried out mathematical studies of earthquake waves; with H. Jeffreys, prepared the Jeffreys-Bullen tables on seismic travel time.

Bunsen, Robert Wilhelm (1811–1899), German chemist. Discovered, with G. R. Kirchhoff, spectrum analysis; invented the Bunsen burner, Bunsen cell, and Bunsen ice calorimeter; formulated law of reciprocity with H. E. Roscoe.

Burbank, Luther (1849–1926), American horticulturist. Experimented on crossing and in-breeding of plant varieties.

Burnet, Frank Macfarlane (1899–1985), Australian immunologist. With P. B. Medawar, studied the body's tolerance of antigenic substances; Nobel Prize, 1960.

Bush, Vannevar (1890–1974), American electrical engineer. Originated the concept of hypertext.

Butenandt, Adolph Friedrich Johann (1903–1995), German chemist. Researched sex hormones; Nobel Prize (declined), 1939.

Buys-Ballot, Christoph Hendrik Didericus (1817–1890), Dutch meteorologist. Devised a system of storm signals; formulated Buys-Ballot's law for determination of wind direction.

Byron, Augusta Ada, Countess of Lovelace (1815–1852), English mathematician. Daughter of the poet Lord Byron; wrote the first program for Charles Babbage's analytical engine; often described as the first computer programmer.

Cailletet, Louis Paul (1832–1913), French chemist. Researched liquefaction of gases; first to obtain liquid oxygen, hydrogen, nitrogen, and air.

Callendar, Hugh Longbourne (1863–1930), English physicist and engineer. Developed platinum resistance thermometer and continuous-flow calorimeter.

Callow, John Michael (1867–1940), English-born American mining engineer and metallurgist. Invented Callow flotation cell and Callow screen.

Calvin, Melvin (1911–1997), American chemist. With J. A. Bassham, traced the path of carbon in photosynthesis; Nobel Prize, 1961.

Cannizzaro, Stanislao (1826–1910), Italian chemist. Promulgated Avogadro's work as related to atomic weights; discovered Cannizzaro's reaction in organic chemistry.

Cantor, Georg (1845–1918), Russian-born German mathematician. Founded set theory; introduced fundamental concepts in topology; worked on the theory and representations of real numbers.

Capecchi, Mario R. (1937–), American molecular geneticist (born in Italy). Made pioneering contributions to the development of gene targeting in mouse embryo–derived stem cells, which permits the creation of mice with mutations in any desired gene; Nobel Prize, 2007.

Carathéodory, Constantin (1873–1950), German mathematician. Developed calculus of variations for curves with corners; introduced Carathéodory outer measure; gave mathematical formulation of second law of thermodynamics (Carathéodory's principle).

Cardano, Geronimo, or Jerome Cardan (1501–1576), Italian physician and mathematician. Wrote on algebra, medicine, and astronomy; invented the Cardan shaft.

Carleson, Lennart (1928–), Swedish mathematician. Fundamental contributions in harmonic analysis and the theory of smooth dynamical systems; Abel Prize, 2006.

Carlsson, Arvid (1923–), Swedish pharmacologist. Did research on dopamine, leading to the discovery of its role as a key neurotransmitter in the brain and in the control of movement; the development of L-dopa, a precursor of dopamine, into a drug to treat Parkinson's disease; the elucidation of the mode of action of antipsychotic drugs, which affect synaptic transmission by blocking dopamine receptors; did work (along with that of Paul Greengard and Eric Kandel) leading to the elucidation of the molecular mechanisms involved in slow synaptic transmission in the nervous system; Nobel Prize, 2000.

Carnot, Nicolas Léonard Sadi (1796–1832), French physicist. Formulated Carnot's theorems in thermodynamics.

Carrel, Alexis (1873–1944), French surgeon and biologist. Worked on transplanting organs, suturing blood vessels, treating deep wounds, and prolonging tissue life; Nobel Prize, 1912.

Carrington, Richard Christopher (1826–1875), English astronomer. Investigated motions of sunspots.

Carver, George Washington (1864–1943), American botanist. Did research on industrial uses of the peanut.

Cassegrain, N. (17th century), French physician. Designed the Cassegrain reflecting telescope.

Cassini, Jean Dominique (1625–1712), Italian-born French astronomer. Director of the Paris Observatory; discovered four new satellites of Saturn and the Cassini division in Saturn's ring; conducted pendulum experiments related to the shape of the Earth.

Castigliano, Carlo Alberto (1847–1884), Italian structural engineer. Proved Castigliano's theorem.

Cauchy, Augustin Louis, Baron (1789–1857), French mathematician. Wrote extensively on wave propagation, calculus, and elasticity.

Cavendish, Henry (1731–1810), English physicist and chemist. Determined the density of the Earth and the composition of the atmosphere; studied properties of carbon dioxide and hydrogen.

Cayley, Arthur (1821–1895), English mathematician. Proposed the theory of matrices; developed the theory of invariants and covariants; worked on quantics and the theory of groups.

Cech, Thomas R. (1947–), American chemist. By studying the single-cell organisms *Tetrahymena* and *Thermophila*, discovered that molecules of ribonucleic acid (RNA) have catalytic properties similar to those of enzymes; Nobel Prize, 1989.

Celsius, Anders (1701–1744), Swedish astronomer. Constructed the thermometer using the Celsius (centigrade) scale.

Cerenkov, Pavel Alexeyevich (1904–1990), Soviet physicist. Discovered the Cerenkov effect of radiation; devised the Cerenkov counter for particle detection; Nobel Prize, 1958.

Cerf, Vinton G. (1943–), American computer scientist. Co-invented (with Robert E. Kahn)

Transmission Control Protocol/Internet Protocol (TCP/IP).

Cesalpino, Andrea (1519–1603), Italian physician and botanist. First to attempt to classify plants according to characteristics of fruit and seed in his *De Plantis*.

Cesàro, Ernesto (1859–1906), Italian mathematician. Formulated an intrinsic geometry; introduced the Cesàro summation.

Ceva, Giovanni (1647?–1734), Italian mathematician. Formulated a theorem on the concurrency of straight lines passing through the vertices of a triangle.

Chadwick, James (1891–1974), English physicist. Established experimentally the existence of the neutron; Nobel Prize, 1935.

Chain, Ernst Boris (1906–1979), German-born British biochemist. With H. W. Florey, worked on the chemical structure of penicillin and its first clinical trials; Nobel Prize, 1945.

Chalfie, Martin (1947–), American neurobiologist. Work in the areas of nerve cell development and function. Pioneering research in the application of green fluorescent protein (GFP) as a genetic tag in biological phenomena; Nobel Prize, 2008.

Chamberlain, Owen (1920–2006), American physicist. With E. G. Segrè, demonstrated the existence of the antiproton; Nobel Prize. 1959.

Chamberlin, Thomas Chrowder (1843–1928), American geologist. Studied the fundamental geology of the solar system.

Chandler, Seth Carlo (1846–1913), American astronomer. Discovered the Chandler wobble.

Chandrasekhar, Subrahmanyan (1910–1995), Indian astrophysicist. Developed a theory of white dwarf stars; Nobel Prize, 1983.

Chaplygin, Sergel Alekseevich (1869–1942), Russian physicist, engineer, and mathematician. Made contributions to fluid mechanics, particularly aerodynamics.

Chapman, Sydney (1888–1970), English mathematician and physicist. Discovered (independently of D. Enskog) gaseous thermal diffusion; studied the daily variations of the geomagnetic field and magnetic storms.

Chaptal, Jean Antoine Claude, Comte de Chanteloup (1756–1832), French chemist. Wrote on technical chemistry; introduced the metric system after the Revolution.

Charcot, Jean Martin (1825–1893), French neurologist. Director of the Salpetrière clinic, where he made systematic clinical studies of chronic nervous disorders, including cerebrospinal disease.

Charles, Jacques Alexandre César (1746–1823), French physicist, chemist, and inventor. Formulated Charles' law, relating gas volume to pressure.

Charpak, Georges (1924–), French physicist. Invented the multiwire proportional chamber, used as a detector in high-energy physics experiments; Nobel Prize, 1992.

Chauvin, Yves (1930–), French chemist. Made fundamental contributions in the field of homogeneous catalysis, in particular to the development of the metathesis method in organic synthesis; also created several important petrochemical industrial processes; Nobel Prize, 2005.

Chebyshev, Pafnuti Lvovich (1821–1894), Russian mathematician. Research on convergence of Taylor series, prime numbers, probability theory, quadratic forms, and integral theory.

Chebyshev, Pafnuti Lvovich (1821–1894), Russian mathematician. Research on convergence of Taylor series, prime numbers, probability theory, quadratic forms, and integral theory.

Chladini, Ernst Florenz Friedrich (1756–1827), German physicist. Discovered Chladini's figures and used them to study vibrations of solid plates.

Christoffel, Elwin Bruno (1829–1900), Swiss mathematician. Worked in higher analysis, geometry, mathematical physics, and geodesy.

Chu, Ching-Wu (1941–), Chinese-born American physicist. Discovered superconductivity at temperatures over 90 K (−298°F) in yttrium-barium-copper-oxygen compounds.

Chu, Steven (1948–), American physicist. Developed a system of opposed laser beams to cool atoms to extremely low temperatures, and a magnetooptical trap to capture them; Nobel Prize, 1997.

Ciechanover, Aaron (1947–), Israeli medical doctor and scientist. Discovered ubiquitin-mediated protein degradation; Nobel Prize, 2004.

Clairaut, Alexis Claude (1713–1765), French mathematician. Studied the shape of the Earth, evolving Clairaut's theorem; made astronomical calculations concerning Halley's comet.

Claisen, Ludwig (1851–1930), German organic chemist. Developed Claisen condensation; contributed to understanding of tautomerism; worked on rearrangement of allyl aryl ethers into phenols.

Clapeyron, Benoit Paul Émile (1799–1864), French engineer. Developed N. L. S. Carnot's concept of a universal function of temperature.

Clarke, Alexander Ross (1828–1914), British geodesist. Worked on the triangulation of the British Isles; proposed Clarke ellipsoids as geodetic standards.

Claude, Albert (1899–1983), American cytologist, born in Luxembourg. Pioneered in applying electron microscopy to cell studies and in using centrifuge to separate cell components; Nobel Prize. 1974.

Clausius, Rudolf Julius Emmanuel (1822–1888), German physicist. A founder of thermodynamics: worked out the Clausius-Clapeyron equation for the universal temperature function.

Clebsch, Rudolf Friedrich Alfred (1833–1872), German mathematician. Contributions to theory of invariants and algebraic geometry.

Cockcroft, John Douglas (1897–1967), English physicist. With E. T. S. Walton, split nuclei by bombarding them with accelerated protons.

Cohen, Paul Joseph (1934–2007), American mathematician. Proved that the axiom of choice is independent of the other axioms of set theory, and that the continuum hypothesis is independent of the axiom of choice, a result with profound implications for the foundations of mathematics; Fields Medal, 1966.

Cohen, Stanley (1922–), American biochemist. With R. Levi-Montalcini, made landmark studies of nerve growth factor and its functions; Nobel Prize, 1986.

Cohen-Tannoudji, Claude (1933–), French physicist. Helped develop methods to cool and trap atoms with laser light, and explained how atoms could be cooled to temperatures lower than the previously calculated theoretical limits; Nobel Prize, 1997.

Cohn, Ferdinand Julius (1828–1898), German botanist. A founder of bacteriology; did research in plant pathology; first to classify bacteria according to genus and species.

Cole, Kenneth Stewart (1900–1984), American biophysicist. Research on structure and function of living cell membranes and nerve membranes in particular, concentrating on electrical approach; with brother, R. H. Cole, introduced Cole-Cole plot of dielectric behavior.

Cole, Robert Hugh (1914–1990), American chemist and physicist. Research on dielectric properties of matter and intermolecular forces; with brother, K. S. Cole, introduced Cole-Cole plot of dielectric behavior.

Collins, Samuel Cornette (1898–1984), American engineer. Invented Collins helium liquefier.

Compton, Arthur Holly (1892–1962), American physicist. Discovered the Compton effect of x-rays; studied cosmic rays; helped develop the atomic bomb; Nobel Prize, 1927.

Conant, James Bryant (1893–1978), American chemist. Researched free radicals, hemoglobin, and chlorophyll; contributed to atomic energy development.

Condon, Edward Uhler (1902–1974). American physicist. Contributed to the Franck-Condon principle, by extending and giving quantum-mechanical treatment to J. Franck's concept of nuclear motion to molecules in transition from one energy level to another.

Connes, Alain (1947–), French mathematician. Did fundamental work on the theory and application of operator algebras, particularly von Neumann algebras; Fields Medal, 1982.

Coolidge, William David (1873–1975), American physicist. Invented Coolidge tube; discovered method for making tungsten strong and ductile.

Cooper, Leon N. (1930–), American physicist. Showed that electrons could form Cooper pairs; with J. R. Schrieffer and J. Bardeen. formulated a theory of superconductivity; Nobel Prize, 1972.

Copernicus, Nicolaus (1473–1543), Polish (or Prussian) astronomer. Proposed the Copernican system, with the Sun as the center of planetary orbits.

Corey, Elias James (1928–), American chemist. Developed theories and methods of organic chemical synthesis that have made possible the production of a wide variety of complex biologically active substances and useful chemicals; Nobel Prize, 1990.

Cori, Carl Ferdinand (1896–1984), and **Cori, Gerty Theresa Radnitz** (1896–1957). Czechoslovakian-born American biochemists. Discovered the enzymatic mechanism of glucose-glycogen interconversion and the effects of hormones on this mechanism; Nobel Prize, 1947.

Coriolis, Gaspard Gustave de (1792–1843), French physicist. Contributed to theoretical and applied mechanics; clarified and supplied concepts of work and kinetic energy; derived Coriolis acceleration.

Cormack, Allan MacLeod (1924–1998), American physicist. Contributed to the development of computerized axial tomography; Nobel Prize, 1979.

Cornell, Eric Allin (1961–), American physicist. With C. E. Wieman, succeeded for the first time in producing Bose-Einstein condensates in a dilute gas of alkali (rubidium) atoms, and carried out fundamental studies of their properties, including studies of collective excitations and vortex formation in condensates; Nobel Prize, 2001.

Corner, George Washington (1889–1981), American medical biologist. Contributed to understanding of anatomical details of menstrual cycle and functions of estrogen and progesterone.

Cornforth, John Warcup (1917–), Australian-born British chemist. Investigated stereochemistry of enzyme-catalyzed reactions; Nobel Prize, 1975.

Cornu, Marie Alfred (1841–1902), French physicist. Used Cornu spiral for determination of intensities in interference phenomena.

Coster, Dirk (1889–1950), Dutch physicist. Work in x-ray spectroscopy; with G. von Hevesy, discovered hafnium; with R. Kronig, discovered Coster-Kronig transitions.

Cottreli, Frederick Gardner (1877–1948), American chemist. Invented the Cottrell process for precipitation of particles from gas; researched nitrogen fixation, liquefaction of gases, and recovery of helium.

Coulomb, Charles Augustin de (1736–1806), French physicist. Formulated Coulomb's law of electric charges.

Courant, Richard (1888–1972), German-born American mathematician. Research in geometric function theory, differential equations of mathematical physics, and transition by limiting processes from finite difference equations to differential equations.

Cournand, André Frédéric (1895–1988), French-born American physician. Studied normal and abnormal human cardiovascular and pulmonary functions; Nobel Prize, 1956.

Cowan, Clyde L., Jr. (1919–1974), American physicist. With F. Reines, made first detection of the neutrino, a fundamental particle.

Crafts, James Mason (1839–1917), American chemist. With C. Friedel, discovered the Freidel-Crafts reaction, wherein anhydrous aluminum chloride acts as a catalyst.

Cram, Donald J. (1919–2001), American chemist. Expanded the field of crown ether chemistry by using crown ethers to synthesize structures that mimic the action of biological molecules; Nobel Prize, 1987.

Cramer, Gabriel (1704–1752), Swiss mathematician. Contributed to Cramer's rule for solving linear equations.

Crick, Francis Harry Compton (1916–2004), English molecular biologist. With J. D. Watson, proposed a double-helix structure for the deoxyribo-nucleic acid molecule; Nobel Prize, 1962.

Cronin, James Watson (1931–), American physicist. Collaborated with V. L. Fitch on experiment showing that the principle of time-reversal invariance is violated in the decay of neutral *K* mesons; Nobel Prize, 1980.

Cronstedt, Axel Fredrik, Baron (1722–1765), Swedish mineralogist. Discovered nickel; developed a chemical classification system for minerals.

Crookes, William (1832–1919), English physicist and chemist. Invented Crookes tube to study electrical discharges in high vacuum, and a radiometer; discovered thallium.

Crutzen, Paul J. (1933–), Dutch chemist. Demonstrated that nitrogen oxides react catalytically with ozone, thus accelerating the rate of reduction of the ozone layer; Nobel Prize, 1995.

Curie, Marie, born Marya Sklodowska (1867–1934), Polish physical chemist in France. Explored nature of radioactivity; codiscoverer of radium, and first to separate polonium; Nobel Prize, 1903 and 1911.

Curie, Pierre (1859–1906), French chemist and physicist. Codiscoverer of radium; formulated the Curie point, relating magnetic properties and temperature; discovered the piezoelectric effect; Nobel Prize, 1903.

Curl, Robert F., Jr. (1933–), American chemist. Made major contributions toward the discovery of fullerenes; Nobel Prize, 1996.

Cushing, Harvey (1869–1930), American surgeon. Innovator in neurosurgical techniques; research on function and diseases of the pituitary gland.

Cuvier, Georges Léopold Chrétien Frédéric Dagobert, Baron (1769–1838), French naturalist. Made a detailed classification of the animal kingdom; wrote on comparative anatomy.

Daguerre, Louis Jacques Mandé (1787–1851), French inventor. Invented the daguerreotype photographic process.

Dale, Henry Hallett (1875–1968), British pharmacologist and physiologist. Isolated acetylcholine and recognized its effect to be similar to that brought about by parasympathetic nerves; Nobel Prize, 1936.

Dalén, Nils Gustaf (1869–1937), Swedish physicist. Invented automatic gas lighting for unsupervised lighthouses and railroad signals; Nobel Prize, 1912.

Dalitz, Richard Henry (1925–2006), Australian-born British theoretical physicist. Research on properties of mesons and baryons and nuclear interactions of the lambda hyperon; proposed models for elementary particles; introduced the Dalitz plot.

Dalton, John (1766–1844), English chemist and physicist. Proposed the atomic theory of chemical reactions; developed the law of partial pressures of gases; studied color-blindness.

Dam, Carl Peter Henrik (1895–1976), Danish biochemist and nutritionist. Discovered vitamin K and studied its role in human hemorrhagic disease; Nobel Prize, 1943.

Danckwerts, Peter Victor (1916–1984), British chemical engineer. Proposed the surface-renewal model of liquids.

Daniell, John Frederic (1790–1845), English physicist and chemist. Invented the Daniell cell.

Darwin, Charles Robert (1809–1882), English naturalist. Proposed far-reaching theory of evolution of species and theory of natural selection in his *Origin of Species*.

Darwin, George Howard (1845–1912), English mathematician and astronomer. Applied detailed dynamical analysis to cosmological and geological problems.

Dausset, Jean (1916–), French biologist and medical scientist. Studied antigen in human leukocytes and their role in transplant acceptance or rejections; Nobel Prize, 1980.

Davis, Raymond, Jr. (1914–2006), American physicist. Developed pioneering experiments to detect solar neutrinos, and observed deficit in their number. Nobel Prize, 2002.

Davisson, Clinton Joseph (1881–1958), American physicist. Studied magnetism, radiant energy, and electricity; independent of G. P. Thomson, discovered electron diffraction by crystals; Nobel Prize, 1937.

Davy, Humphry (1778–1829), English chemist. Discovered potassium and sodium; invented the Davy safety lamp for use in coal mines; proposed theoretical explanations of electrolysis and voltaic action.

Debye, Peter Joseph William (1884–1966), American physical chemist born in the Netherlands. Worked on dipole moments and the diffraction of x-rays in gases; formulated Debye-Hückel theory on the behavior of strong electrolytes; Nobel Prize, 1936.

Dedekind, Julius Wilhelm Richard (1831–1916), German mathematician. Worked in number theory and analysis, particularly with algebraic integers, algebraic functions, and ideals; defined real numbers by Dedekind cuts.

de Duve, Christian René (1917–), Belgian biochemist and cytologist. Refined centrifuge technique for studying cell components; discovered lysosomes; Nobel Prize, 1974.

De Forest, Lee (1873–1961), American inventor. Pioneer in radio technology; invented audio amplifier and the four-electrode valve; incorporated the grid into the thermionic valve.

de Gennes, Pierre-Gilles (1932–2007), French physicist. Applied physical principles to the study of complex systems, including liquid crystals and polymers; Nobel Prize, 1991.

de Haas, Wander Johannes (1878–1960), Dutch physicist. Demonstrated the Einstein-de Haas effect; worked on production of extremely low temperatures by adiabatic demagnetization; with P. Van Alphen, discovered de Haas-Van Alphen effect.

Dehmelt, Hans Georg (1922–), German-born American physicist. Developed the Penning trap, which uses magnetic and electric fields to hold ions in a small volume; used the traps to isolate a single electron and carry out extremely accurate measurements of atomic properties; Nobel Prize, 1989.

Deisenhofer, Johann (1943–), German chemist. With R. Huber and M. Hartmut, elucidated the structure of a bacterial protein that performs photosynthesis; Nobel Prize, 1988.

Delbruck, Max (1906–1981), German-born American biologist. Pioneered in molecular biology; research on bacterial viruses; Nobel Prize, 1969.

Deligne, Pierre René (1944–), Belgian mathematician. Solved three conjectures of A. Weil concerning generalizations of the Riemann hypothesis to finite fields—work which brought together algebraic geometry and algebraic number theory; Fields Medal, 1978.

Demoivre, Abraham (1667–1754), French-born English mathematician. Originated two theorems on expansions of trigonometrical expansions; proposed methods to approximate functions of large numbers, and the concept of the normal distribution curve.

De Morgan, Augustus (1806–1871), English mathematician. Wrote textbooks and treatises on arithmetic, algebra, and trigonometry; formulated the De Morgan theorem.

Desargues, Gérard (1593–1662), French mathematician. A founder of modern geometry; proposed the theory of involution and transversals.

Descartes, René (1596–1650), French mathematician. Originated cartesian, or coordinate, geometry.

de Sitter, Willem (1872–1934), Dutch astronomer. Worked on the application of Einstein's theory to the problems of the universe; computed the size of the universe.

De Vries, Hugo (1848–1935), Dutch botanist. Formulated the mutation theory of evolution.

Dewar, James (1842–1923), British chemist. Made pioneering studies of matter at low temperatures; first to liquefy hydrogen; invented the Dewar vacuum flask.

Dick, George Frederick (1881–1967), American physician and bacteriologist. Isolated scarlet fever streptococci, developed scarlet fever streptococcus antitoxin, and developed Dick test.

Dicke, Robert Henry (1916–1997), American physicist. Developed new relativistic theory of gravitation with C. Brans; investigated cosmic blackbody radiation; worked on development of radar.

Diels, Otto Paul Hermann (1876–1954), German chemist. Codiscoverer of the Diels-Alder reaction (diene synthesis); worked on sterol chemistry; discovered carbon suboxide; Nobel Prize, 1950.

Diesel, Rudolf (1858–1913), German inventor. Designed and built the diesel engine.

Diocles (2d century B.C.), Greek mathematician. Contributions to theory of conics and geometry.

Diophantus of Alexandria (3d century), Greek mathematician. Known as the father of algebra; the first to use conventional algebraic notation.

Dirac, Paul Adrien Maurice (1902–1984), English physicist. Worked in quantum mechanics; his theory of negative-energy holes predicted existence of the positron; Nobel Prize, 1933.

Dirichlet, Peter Gustave Lejeuné (1805–1859), German mathematician. Applied higher analysis to the theory of numbers; work on definite integrals.

Djerassi, Carl (1923–), Austrian-born, American chemist. Synthesized the first oral contraceptive.

Dobzhansky, Theodosius (1900–1975), Russian-born American biologist. Elucidated the mechanisms of heredity and variation through studies of *Drosophila*.

Doherty, Peter C. (1940–), Australian-born American immunologist. Collaborated with R. M. Zinkerngagel in the discovery of specificity of cell-mediated immune defense; Nobel Prize, 1996.

Dolsy, Edward Adelbert (1893–1986), American biochemist. Isolated pure crystalline compounds important to human health, such as sex hormones and vitamins; Nobel Prize, 1943.

Domagk, Gerhard (1895–1964), German biochemist. Discovered sulfamidocrysoidin, the first synthetic microbial of broad clinical usefulness. Nobel Prize (declined), 1939.

Donaldson, Simon Kirwan (1957–), British mathematician. Worked on topology of four-manifolds and showed that there exist exotic four-spaces, that is, four-dimensional differentiable manifolds that are topologically but not differentiably equivalent to the standard Euclidean four-space; Fields Medal, 1986.

Donati, Giovanni Battista (1826–1873), Italian astronomer. Studied stellar spectra; discovered six comets.

Donnan, Frederick George (1870–1956), Irish chemist born in Ceylon. Research in chemical kinetics; originated the Donnan theory of membrane equilibrium.

Doppler, Christian Johann (1803–1853), Austrian physicist and mathematician. Formulated Doppler's principle, relating the frequency of wave motion to velocity; described the Doppler effect.

Douglas, Jesse (1897–1965), American mathematician. Solved the Plateau problem (the problem of determining the existence of a minimal surface with a given space curve as its boundary) about the same time as T. Radó and studied generalizations of this problem; Fields Medal, 1936.

Drake, Frank Donald (1930–), American astronomer. Research on solar system, 21-centimeter radio line, search for extraterrestrial intelligence (introduced Drake equation), and radio telescope development.

Draper, Henry (1837–1882), American scientist. Research in spectroscopy; the Draper catalog is named in his honor.

Drinfeld, Vladimir Gershonovich (1954–), Ukrainian mathematician. Worked in algebraic geometry, number theory, and the theory of quantum groups; proved a special case of the Langlands conjecture; Fields Medal, 1990.

Drude, Paul Karl Ludwig (1863–1906), German physicist. Attempted to correlate and account for optical, electrical, thermal, and chemical properties of substances; developed theory of properties of metals based on free electrons treated as a gas.

Duane, William (1872–1935), American physicist and radiologist. Developed treatment of cancer by radioisotopes and x-rays; with F. L. Hunt, discovered Duane-Hunt law of x-rays.

Du Bois-Reymond, Emil (1818–1896), German physiologist. Pioneer work in electrical properties of living tissues, especially nerves.

DuBridge, Lee Alvin (1901–1994), American physicist. Developed Fowler-DuBridge theory of photoelectric emission.

Ducrey, Augusto (1860–1940), Italian dermatologist. Discovered *Hemophilus ducreyi*, the agent of chancroid.

Dufay, Charles François de Cisternay (1698–1739), French chemist. Discovered positive and negative types of electricity.

Dulbecco, Renato (1914–), Italian-born American virologist. Developed techniques for studying animal viruses; investigated interaction between deoxyribonucleic acid tumor viruses and genetic material; Nobel Prize, 1975.

Dulong, Pierre Louis (1785–1838), French chemist and physicist. With A. T. Petit, formulated the law of the constancy of atomic heats; developed the Dulong formula for heat value of fuels.

Dumas, Jean Baptiste André (1800–1884), French chemist. Research on organic compounds; determined many atomic weights.

Dürer, Albrecht (1471–1528), German painter, graphic artist, and mathematician. Work on scientific perspective and mathematical proportion.

du Vigneaud, Vincent (1901–1978), American biochemist. Synthesized a polypeptide hormone, oxytocin: worked on other biologically important sulfur compounds; Nobel Prize, 1955.

Eccles, John Carew (1903–1997), Australian physiologist. Elucidated the action of nerve impulses across zones of close contact between nerve cells; Nobel Prize, 1963.

Eddington, Arthur Stanley (1882–1944), English astronomer and writer. Theoretical research on stellar movements and internal makeup of stars; wrote on theory of relativity.

Edelman, Gerald Maurice (1929–), American biochemist. Worked to determine chemical structure of immunoglobulins; Nobel Prize, 1972.

Edison, Thomas Alva (1847–1931), American electrician and inventor. Invented the gramophone, carbon transmitter for the telephone, incandescent electric lamp, moving pictures, and the diplex method of telegraphy.

Ehrenfest, Paul (1880–1933), Austrian-born Dutch theoretical physicist. Contributed to statistical mechanics and quantum mechanics; developed Ehrenfest's principle; proved Ehrenfest's theorem.

Ehrlich, Paul (1854–1915), German bacteriologist. Research in chemotherapy, notably the discovery of Salvarsan for treatment of syphilis; pioneered in the study of hematology and immunity; Nobel Prize, 1908.

Eigen, Manfred (1927–), German chemist. Devised relaxation techniques to study high-speed chemical reactions; Nobel Prize, 1967.

Eijkman, Christiaan (1858–1930), Dutch physician. Studied dietary deficiency disease, in particular beriberi; Nobel Prize, 1929.

Einstein, Albert (1879–1955), German-born American physicist. Proposed the theory of relativity; extended the application of quantum theory; Nobel Prize, 1921.

Einthoven, Willem (1860–1927), Dutch physiologist born in Java. Used the string galvanometer to record electrical activity of the heart, thereby inventing the electrocardiograph; Nobel Prize, 1924.

Elion, Gertrude Belle (1918–1999), American biochemist. With G. H. Hitchings, pioneered research that led them to the development of drugs for the treatment of leukemia, malaria, gout, herpes, bacterial and fungal infections, and autoimmune diseases and organ-transplant rejection; Nobel Prize, 1988.

Elsasser, Walter Maurice (1904–1991), German-born American geophysicist. Formulated the dynamo theory of the Earth's permanent terrestrial magnetic force.

Elster, Johann Philipp Ludwig Julius (1854–1920), German experimental physicist. With H. F. Geitel, studied atmospheric electricity, radioactivity, and photoelectricity, and invented photocell.

Emmett, Paul Hugh (1900–1985), American chemist. Worked on catalysts for ammonia synthesis and the water-gas conversion reaction; with S. Brunauer and E. Teller, formulated Brunauer-Emmett-Teller equation for surface area determinations.

Encke, Johann Franz (1791–1865), German astronomer. Determined period of revolution of Encke's comet, discovered by J. L. Pons; measured the distance of the Sun from the Earth.

Enders, John Franklin (1897–1985), American microbiologist. With F. C. Robbins and T. H. Weller, discovered the capacity of poliomyelitis virus to grow in various tissue cultures; Nobel Prize, 1954.

Enskog, David (1884–1947), Swedish physicist. With S. Chapman, developed the Chapman-Enskog theory for solving the Boltzmann transport equation.

Eötvös, Roland, Baron (1848–1919), Hungarian physicist. Research on gravitation and terrestrial magnetism; formulated a law which relates surface tension to temperature of liquids; designed the Eötvös torsion balance.

Erasistratus (3d century B.C.), Greek physician and anatomist. Founder of physiology; distinguished between the cerebrum and cerebellum, and sensory and motor nerves.

Eratosthenes (3d century B.C.), Greek astronomer. Suggested an extra day in the calendar every fourth year; made a determination of the size of the Earth; measured obliquity of the ecliptic.

Erdös, Paul (1913–1996), Hungarian mathematician. Did wide-ranging work in algebra, analysis, combinatorial theory, geometry, topology, number theory, and graph theory; with A. Selberg, gave an elementary proof of the prime number theorem.

Erlanger, Joseph (1874–1965), American physiologist. With H. S. Gasser, created new methods for amplifying and recording electrical impulses in nerves; research on the function of the synapse; Nobel Prize, 1944.

Erlenmeyer, Richard August Carl Emil (1825–1909), German organic chemist. Research on synthesis and constitution of aliphatic compounds; introduced modern structural notation; invented Erlenmeyer flask.

Ernst, Richard R. (1933–). Swiss chemist. Developed methods that transformed nuclear magnetic resonance (NMR) spectroscopy from a tool with a narrow application to a key analytical technique in chemistry as well as many other fields: Nobel Prize, 1991.

Ertl, Gerhard (1936–), German physicist and professor of chemistry. Made fundamental contributions to the understanding of chemical reactions taking place on surfaces, thereby helping lay the foundation of modern surface chemistry; Nobel Prize, 2007.

Esaki, Leo (1925–), Japanese physicist. Discovered a new negative-resistance characteristic in semiconductor *pn* junctions, leading to the discovery of the tunnel, or Esaki. diode; Nobel Prize, 1973.

Euclid (ca. 330-ca. 275 B.C.), Greek mathematician. Wrote geometry textbooks; euclidean geometry is named after him.

Eudoxus of Cnidus (ca. 408-ca. 355 B.C.). Greek astronomer and mathematician. Expounded theory of motion of planets based on homocentric spheres; developed theory of proportions and methods of measuring areas and volumes of geometrical figures.

Euler, Leonhard (1707–1783), Swiss mathematician. Contributed to algebraic series and differential and integral calculus; realized the significance of coefficients (Euler numbers) of certain trigonometrical expansions.

Euler-Chelpin, Hans Karl August Simon von (1873–1964), German-Swedish chemist. Research on enzyme action and fermentation of sugars; Nobel Prize, 1929.

Evans, Sir Martin J. (1941–), British geneticist. Made pioneering contributions to the principles of gene targeting, in particular methods for culturing embryonic stem cells that would allow DNA modifications to be inherited; Nobel Prize, 2007.

Ewald, Paul Peter (1888–1985), German-born. American physicist. Developed dynamic theory of x-ray interference in crystals.

Ewing, William Maurice (1906–1974), American geophysicist. Made fundamental contributions to seismology, geodesy, oceanography, and submarine geology.

Eyring, Henry (1901–1981), Mexican-born American chemist. Pioneered in the application of quantum and statistical mechanics to chemistry; conceived the theory of absolute reaction rates and the significant structures theory of liquids.

Fabricius, Hieronymus, or Girolamo Fabrizio (ca. 1533–1619), Italian anatomist. Made painstaking descriptions of valves in veins; did comparative research in animal embryology.

Fabry, Charles (1867–1945), French physicist. With A. Pérot, invented Fabry-Perot interferometer; experimentally verified Doppler broadening and Doppler effect.

Fahrenheit, Gabriel Daniel (1686–1736), German physicist. Constructed thermometers; invented the Fahrenheit temperature scale.

Fallopio, Gabriele (1523–1562), Italian anatomist. Discovered Fallopian tubes; gave first clear description of organs of inner and middle ear.

Faltings, Gerd (1954–), German mathematician. Used methods of arithmetic algebraic geometry to prove the Mordell conjecture, which states that there are only finitely many rational points on a curve of genus greater than 1; this was a step toward the later proof of Fermat's last theorem by A. Wiles; Fields Medal, 1986.

Fanning, John Thomas (1837–1911), American civil engineer. Designed water works and water supply systems.

Faraday, Michael (1791–1867), English chemist and physicist. Discovered electromagnetic induction; formulated two laws of electrolysis; invented the dynamo.

Fechner, Gustav Theodor (1801–1887), German psychologist. Founded psychophysics; developed the Fechner law concerning intensity of sensation produced by a stimulus.

Fefferman, Charles Louis (1949–), American mathematician. Worked in Fourier analysis, partial differential equations, and the theory of functions of several complex variables; discovered the dual of the Hardy space H^1; Fields Medal, 1978.

Feit, Walter (1930–2004), Austrian-born American mathematician. Worked in group theory; with J. G. Thompson, proved that all noncyclic finite simple groups have even order.

Fenn, John B. (1917–), American chemist. Developed soft desorption ionization method for mass spectrometric analyses of biological macromolecules; Nobel Prize, 2002.

Fermat, Pierre de (1601–1665), French mathematician. Founder of the modern theory of numbers; originated Fermat's last theorem, and Fermat's principle in optics.

Fermi, Enrico (1901–1954), Italian-born American physicist. Research on producing radioactive isotopes by neutron bombardment; directed construction of the first atomic pile; Nobel Prize, 1938.

Fert, Albert (1938–), French physicist. Co-discoverer of the phenomenon of giant magnetoresistance (GMR); numerous contributions to the field of spin electronics; Nobel Prize, 2007.

Feynman, Richard Phillips (1918–1988), American physicist. Proposed a theory to eliminate difficulties that had arisen in the study of the interaction of electrons, positrons, and radiation; Nobel Prize, 1965.

Fibiger, Johannes Andreas Grib (1867–1928), Danish pathologist. First to produce cancer experimentally; Nobel Prize, 1926.

Finsen, Niels Ryberg (1860–1904), Danish physician. Originated ultraviolet light therapy for certain diseases; Nobel Prize, 1903.

Fire, Andrew Z. (1959–), American molecular geneticist. Co-discoverer of RNA interference, gene silencing by double-stranded RNA; Nobel Prize, 2006.

Fischer, Edmond H. (1920–), American biochemist. With E. G. Krebs, discovered phosphorylation processes that play a critical role in cell-protein regulation; they isolated the first protein kinase, a class of enzymes that transfer phosphate from adenosinetriphosphate to proteins; Nobel Prize, 1992.

Fischer, Emil Hermann (1852–1919), German chemist. Synthesized many natural substances, including purines, D-glucose and other sugars, and the first nucleotide; studied polypeptides and proteins; Nobel Prize, 1902.

Fischer, Ernst Otto (1918–2007), German chemist. Studied how metals and organic molecules combine to form unique molecules with sandwichlike structures; Nobel Prize, 1973.

Fischer, Hans (1881–1945), German organic chemist. Investigated and synthesized pyrrole pigments; studied structure of chlorophylls; Nobel Prize, 1930.

Fisher, Ronald Aylmer (1890–1962), English geneticist and statistician. Developed statistical techniques for analysis of variance, and for use and validation of small samples; developed theory of the evolution of dominance.

Fitch, Val Logsdon (1923–), American physicist. Collaborated with J. W. Cronin on experiment showing that the principle of time-reversal invariance is violated in the decay of neutral K mesons; Nobel Prize, 1980.

FitzGerald, George Francis (1851–1901), Irish physicist. Proposed Lorentz-FitzGerald contraction, relating to a material moving through an electromagnetic field.

Fizeau, Armand Hippolyte Louis (1819–1896), French physicist. First to accurately measure the velocity of light; conducted experiments on the velocity of electricity, use of light wavelength to measure length, and measurement of diameter of stars through the method of interference.

Flamsteed, John (1646–1719), English astronomer. Made a trustworthy catalog of stars; invented conical projection in mapmaking.

Fleming, Alexander (1881–1955), British bacteriologist. Discovered lysozyme and penicillin; Nobel Prize, 1945.

Fleming, John Ambrose (1849–1945), English electrical engineer. Invented the thermionic valve; contributed to widespread application of electric lighting and heating.

Florey, Howard Walter (1898–1968), British pathologist born in Australia. Contributed, with E. B. Chain, to development of penicillin as a chemotherapeutic agent; Nobel Prize, 1945.

Flory, Paul John (1910–1985), American physical chemist. Developed analytic techniques to explore properties and molecular structures of long-chain molecules; Nobel Prize, 1974.

Fock (Fok), Vladimir Alexandrovitch (1898–1974), Soviet theoretical physicist. Contributions to quantum electrodynamics, quantum field theory, electromagnetic diffraction and propagation, and general relativity; developed Hartree-Fock approximation of wave functions.

Forbush, Scott Ellsworth (1904–1984), American geophysicist. Discovered the worldwide decrease in cosmic-ray intensity associated with some magnetic storms.

Forssmann, Werner Theodor Otto (1904–1979), German physician. Developed the technique of cardiac catheterization; Nobel Prize, 1956.

Foucault, Jean Bernard Léon (1819–1868), French physicist. Accurately determined the velocity of light; constructed the Foucault pendulum and the Foucault prism; determined experimentally the rotation of the Earth.

Fourier, Jean Baptiste Joseph, Baron (1768–1830), French geometrician and physicist. Proposed the Fourier series on arbitrary functions; formulated the law of heat propagation.

Fowler, Ralph Howard (1889–1944), English physicist. Applied statistical mechanics to matter at high temperatures and high pressures; explained structure of white dwarf stars; with E. A. Guggenheim, R. F. Peierls, and others, developed Ising model.

Fowler, William Alfred (1911–1995), American physicist. Fundamental contributions to understanding of nuclear reactions that generate the energy of stars and synthesize the elements of the universe; Nobel Prize, 1983.

Franck, James (1882–1964), German physicist. With G. Hertz, studied energy transfer in collisions of molecules; formulated Franck-Condon principle of transition from one energy state to another; Nobel Prize, 1925.

Frank, Ilya Milkhallovich (1908–1990), Soviet physicist. With I. Y. Tamm, proposed a theoretical interpretation of Cerenkov radiation; Nobel Prize, 1958.

Franklin, Benjamin (1706–1790), American physicist, oceanographer, meteorologist, and inventor. Formulated a theory of general electrical "action"; introduced principle of conservation of charge; showed that lightning is an electrical phenomenon; invented lighting rod.

Fraunhofer, Joseph von (1787–1826), German optician and physicist. First to study the dark lines in the solar spectrum (Fraunhofer lines); invented a heliometer; improved the spectroscope.

Fredholm, Eric Ivar (1866–1927), Swedish mathematician. Developed the theory of integral equations (Fredholm equations).

Freedman, Michael Hartley (1951–), American mathematician. Developed new methods for topological analysis of four-manifolds and applied them to prove the Poincaré conjecture for dimension 4; Fields Medal, 1986.

Frenet, Jean Frédéric (1816–1900), French mathematician. Helped to develop Frenet-Serret formulas.

Frenkel, Yakov Ilyich (1894–1954), Soviet physicist. Pioneered in modern atomic theory of solids; developed quantum-mechanical explanations for electron mean free path in metals, and for paramagnetism and ferromagnetism; postulated excitons, Frenkel excitons, and Frenkel defects.

Fresnel, Augustin Jean (1788–1827), French physicist. Investigated effects (Fresnel's fringes) due to the interference of light; developed a wave theory of light; originated Fresnel's reflection formula.

Freud, Sigmund (1856–1939), Austrian psychoanalyst. Founder of psychoanalysis, with emphasis on dream interpretation and free association; developed a theory of personality involving id, ego, and superego, and stressing importance of the libido.

Friedel, Charles (1832–1899), French chemist and mineralogist. With J. M. Crafts, described the Friedel-Crafts reaction; work on artificial production of minerals; studied crystals, ketones, and aldehydes.

Friedman, Jerome Isaac (1930–), American physicist. Collaborated in experiments that demonstrated that protons, neutrons, and similar particles are made up of quarks; Nobel Prize, 1990.

Frisch, Karl von (1886–1982), Austrian zoologist. Discovered means by which bees communicate information about the distance and direction of food; Nobel Prize, 1973.

Frobenius, Georg Ferdinand (1849–1917), German mathematician. Developed representation theory of finite groups and method for solving linear homogeneous ordinary differential equations.

Froude, William (1810–1879), English engineer. Discovered the Froude law of comparison, concerning the towing of an object in a liquid.

Fubini, Guido (1879–1943), Italian mathematician. Worked in algebra, analysis, and differential projective geometry; proved Fubini's theorem.

Fukui, Kenichi (1918–1998), Japanese chemist. Developed frontier orbital theory, a quantum-mechanical model useful in prediction of the combinative properties of molecules; Nobel Prize, 1981.

Fuller, R. Buckminster (1895–1983), American engineer and architect. Designed geodesic dome; the carbon molecular form C_{60} was named buckminsterfullerene because of its structured resemblance to the geodesic dome, and the name fullerene was given to any closed-cage molecule containing an even number of carbon atoms.

Furchgott, Robert F. (1916–), American pharmacologist. Discovered endothelium-derived relaxing factor (EDRF), a signaling molecule in the cardiovascular system that makes vascular smooth muscle cells relax (Louis J. Ignarro, working independently and with Furchgott, later concluded that EDRF was nitric oxide); Nobel Prize, 1998.

Gabor, Dennis (1900–1979), Hungarian-born British physicist and engineer. Invented holography; Nobel Prize, 1971.

Gajdusek, Daniel Carleton (1923–2008), American physician and virologist. Discovered causal virus and transmission mechanism of kuru; Nobel Prize, 1976.

Galen (2d century), Greek physician and medical writer. Wrote treatises long used as textbooks; experimented on animal nervous systems; made anatomical descriptions of structure and functions of body parts.

Galileo Galilei (1564–1642), Italian astronomer. First to use the telescope for observational purposes; made many discoveries related to the planets and the Sun; did theoretical work on classical physics.

Gallo, Robert (1937–), American virologist. Fundamental research on acquired immunodeficiency syndrome (AIDS) including the determination that human immunodeficiency virus (HIV) is the causative agent.

Galois, Evariste (1811–1832), French mathematician. Developed Galois theory of polynomials.

Gamow, George (1904–1968), Russian-born American physicist. Made theoretical contributions to nuclear physics, astronomy, and biology; with E. Teller, formulated the selection rule for beta emission: proposed theoretically the genetic code.

Garvey, Gerald Thomas (1935–), American physicist. Research in experimental nuclear physics, particularly nuclear reactions, isobaric spin studies, and weak interactions in nuclear systems.

Gasser, Herbert Spencer (1888–1963), American physiologist. With J. Erlanger, provided a new method for recording electrical impulses of nerves; studied functions of nerve fibers; Nobel Prize, 1944.

Gatterman, Friedrich August Ludwig (1860–1920), German chemist. Originated Gatterman-Koch synthesis of aldehydes; isolated and analyzed nitrogen trichloride; synthesized aromatic carboxylic acids, thionaphthalene, and thioanilide.

Gauss, Karl Friedrich (1777–1855), German mathematician, astronomer, and physicist. Formulated the Gauss theorem in the mathematics of electricity; made many contributions to pure and applied mathematics; determined orbits of planets and comets from observational data.

Gay-Lussac, Joseph Louis (1778–1850), French chemist and physicist. Discovered the law of expansion of gases by heat, and the law of combining volumes of gases; studied chemistry of iodine and cyanogen.

Geiger, Hans Wilhelm (1882–1945), German physicist and inventor. Invented the Geiger counter to detect alpha particles; investigated properties of alpha particles, cosmic rays, and artificial radiation.

Geissler, Johann Heinrich Wilhelm (1815–1879), German instrument maker. Developed Geissler pump and Geissler tube.

Geitel, Hans Friedrich (1855–1923), German experimental physicist. With J. Elster, studied atmospheric electricity, radioactivity, and photoelectricity, and invented photocell.

Gell-Mann, Murray (1929–), American physicist. Proposed law of conservation of strangeness; used unitary symmetry to classify and explain elementary particles; postulated concept of quarks; Nobel Prize, 1969.

Gesner, Konrad von (1516–1565), Swiss naturalist. Wrote *Historia Animalium*, beginning zoology as a science.

Giacconi, Riccardo (1931–), Italian-born American astrophysicist. Pioneered the field of x-ray astronomy, working out the principles of an x-ray telescope and leading the development of the early x-ray telescopes. Nobel Prize, 2002.

Giaever, Ivar (1929–), Norwegian-born American physicist. Discovered that current-voltage characteristics of an electron tunneling across a thin insulating film separating two metals, one or both of which is in a superconducting state, can be used to obtain electron density of states of superconductors; Nobel Prize, 1973.

Giauque, William Francis (1895–1982), Canadian-born American chemist. Developed adiabatic demagnetization technique for production of extremely low temperatures; collaborated in discovery of isotopes of oxygen; Nobel Prize, 1949.

Gibbs, Josiah Willard (1839–1903), American mathematician and physicist. Made a mathematical treatment of chemical subjects, notably thermodynamics; worked on statistical mechanics, leading to the basis for the phase rule of heterogeneous equilibria.

Gilbert, Walter (1932–), American biochemist. Developed methods for determining nucleotide sequence (independently of F. Sanger), advancing the technology of DNA recombination; Nobel Prize, 1980.

Gilman, Alfred G. (1941–), American pharmacologist. With M. Rodbell, discovered G-proteins and their role in cellular signal transduction; Nobel Prize, 1994.

Ginzburg, Vitaly Lazarevich (1916–), Soviet physicist. Developed Ginzburg-Landau and Ginzburg-London theories of superconductivity; Nobel Prize, 2003.

Giorgi, Giovanni (1871–1950), Italian electrical engineer, physicist, and mathematician. Developed the meter-kilogram-second-ampere system of units.

Glaser, Donald Arthur (1926–), American physicist. Invented the bubble chamber for detecting the paths of high-energy atomic particles; Nobel Prize, 1960.

Glashow, Sheldon Lee (1932–), American physicist. Contributed to development of theory uniting electromagnetism and weak nuclear interactions; postulated existence of charmed particles; Nobel Prize, 1979.

Glauber, Johann Rudolf (1604–1670), German chemist. Discovered Glauber's salt (sodium sulfate) and hydrochloric acid; conducted experiments on compounds of mercury, arsenic, and antimony.

Gmelin, Leopold (1788–1853), German chemist. Wrote *Handbuch der Chemie*, first systematic treatment of chemical knowledge; devised Gmelin's test for presence of bile pigments; studied cyanides.

Goethals, George Washington (1858–1928), American civil and military engineer. As chairman and chief engineer of the Isthmian Canal Commission he directed the completion of the construction of the Panama Canal.

Goldbach, Christian (1690–1764), German-born Russian mathematician. Research on number theory and analysis; proposed the Goldbach conjecture.

Goldhaber, Maurice (1911–), Austrian-born American physicist. With J. Chadwick, discovered photodisintegration and disintegration of light elements by slow neutrons; with L. Grodzins and A. W. Sunyar, discovered that the neutrino has left-handed spin.

Goldstein, Joseph L. (1940–), American biochemist and geneticist. With M. S. Brown, discovered low-density lipoprotein receptors and their function in cholesterol metabolism; Nobel Prize, 1985.

Golgi, Camillo (1843–1926), Italian physician. Pioneered in the study of histology of the nervous system; discovered the Golgi bodies; Nobel Prize, 1906.

Gordan, Paul Albert (1837–1912), German mathematician. Worked on the theory of invariants and on solutions of algebraic equations and their groups of substitutions.

Gordon, Walter (1893–1940), German-born Swedish physicist. Contributed to relativistic quantum theory; with O. B. Klein, originated the Klein-Gordon equation.

Goudsmit, Samuel Abraham (1902–1978), Dutch-born American physicist. With G. E. Uhlenbeck, discovered electron spin.

Gowers, William Timothy (1963–), British mathematician. Contributed to functional analysis, in particular, to the theory of Banach spaces, using methods of combinatorial theory; Fields Medal, 1998.

Gram, Hans Christian Joachim (1853–1938), Danish physician. Developed the Gram method for staining and differentiating bacteria.

Gram, Jorgen Pedersen (1850–1916), Danish mathematician. Worked in number theory and analysis; with E. Schmidt, originated Gram-Schmidt process for obtaining orthogonal set of vectors.

Granit, Rangar Arthur (1900–1991), Finnish-born Swedish physiologist. Research on vision and on motor control by afferent neurons; Nobel Prize, 1967.

Grashof, Franz (1826–1893), German mechanical engineer. Applied mathematics and physics to engineering problems; derived fundamental equations in the theory of elasticity; introduced Grashof number.

Gray, Henry (1825–1861), English anatomist. Wrote *Anatomy of the Human Body*.

Green, George (1793–1841), English mathematician. Worked in analysis; derived Green's theorem and Green's identities; introduced Green's function.

Greengard, Paul (1925–), American neurobiologist. Discovered the mechanism by which dopamine and other chemical neurotransmitters (such as norepinephrine and serotonin) affect the nervous system—work contributing to the elucidation of the molecular mechanisms involved in slow synaptic transmission in the nervous system; Nobel Prize, 2000.

Gregory, James (1638–1675), Scottish geometer. Provided first proof of the theorem of calculus; gave first description of the reflecting telescope; discovered the series from which π can be calculated.

Grignard, François Auguste Victor (1871–1935), French chemist. Discovered organomagnesium compounds, or Grignard reagents, useful in synthesis of organic and organometallic compounds; Nobel Prize, 1912.

Gross, David J. (1941–), American physicist. Discovered asymptotic freedom in the strong interactions (in collaboration with F. Wilczek, and independent of H. D. Politzer). Nobel Prize, 2004.

Grothendieck, Alexander (1928–), German-born French mathematician. Made fundamental advances in algebraic geometry and related fields, providing unifying themes in geometry, number theory, topology, and complex analysis; introduced the idea of K theory; revolutionized homological algebra; developed theory of schemes, allowing conjectures in number theory to be solved; Fields Medal, 1966.

Grubbs, Robert H. (1942–), American chemist. Has made fundamental contributions to the design and synthesis of catalysts for use in organic and polymer synthesis, in particular to the metathesis method; Nobel Prize, 2005.

Grünberg, Peter (1939–), German physicist (born in former Czechoslovakia). Co-discoverer of the phenomenon of giant magnetoresistance (GMR); Nobel Prize, 2007.

Gruneisen, Eduard (1877–1949), German physicist. Formulated laws relating specific heat and other properties of solids.

Guillaume, Charles Édouard (1861–1938), Swiss-born French physicist. Studied nickel-steel alloys and invented Invar; Nobel Prize, 1920.

Guillemin, Ernst Adolph (1898–1970), American electrical engineer. Worked on network analysis and synthesis problems; invented Guillemin line; developed network to produce loran pulses.

Guillemin, Roger Charles Louis (1924–), French-born American physiologist. With A. V. Schally, isolated and analyzed peptide hormones secreted in hypothalamic region of brain which control anterior pituitary hormone secretion; Nobel Prize, 1977.

Gullstrand, Allvar (1862–1930), Swedish ophthalmologist. Discovered intracapsular accommodation of the eye lens; improved techniques for studying eye structure; Nobel Prize, 1911.

Gunn, John Battiscombe (1928–), Egyptian-born American physicist. Discovered Gunn effect and used it to develop Gunn oscillator.

Gunter, Edmund (1581–1626), English mathematician and astronomer. Invented Gunter's chain used in surveying, and the logarithmic scale (Gunter's scale) which is the principle of the slide rule.

Gutenberg, Johann (ca. 1397–1468), German inventor. Invented the movable-type printing press.

Haar, Alfred (1885–1933), Hungarian mathematician. Studied orthogonal systems of functions, complex functions, partial differential equations, and calculus of variations; introduced the Haar measure on groups.

Haber, Fritz (1868–1934), German chemist. Developed the Haber-Bosch process for synthesis of ammonia; made electrochemical studies; Nobel Prize, 1918.

Hadamard, Jacques (1865–1963), French mathematician. Proved theorem on the asymptotic behavior of the function giving the number of prime numbers less than a given number; introduced concept of "the problem correctly posed" in the solution of partial differential equations.

Haeckel, Ernst Heinrich (1834–1919), German biologist. Studied various invertebrates; classified animals as uni- and multicellular organisms; proposed the theory of recapitulation: ontogeny repeats phylogeny.

Hagen, Carl Ernst Bessel (1851–1923), German physicist. With H. Rubens, conducted experiments confirming Maxwell's electromagnetic theory of light, permitting determination of electrical conductivity of metals by optical measurements alone.

Hagen, Gotthilf Heinrich Ludwig (1797–1884), German hydraulic engineer. Discovered Hagen-Poiseuille law independently of J. L. M. Poiseuille; directed construction of dikes, harbor installations, and dune fortifications.

Hahn, Hans (1879–1934), Austrian mathematician. With S. Banach, proved the Hahn-Banach theorem of linear functionals.

Hahn, Otto (1879–1968), German chemist. With L. Meitner and F. Strassman, discovered that fission of heavy nuclei was possible by irradiation with neutrons; discovered protactinium with Meitner; Nobel Prize, 1944.

Hahnemann, Christian Friedrich Samuel (1775–1843), German physician. Founder of homeopathy.

Haldane, John Bourdon Sanderson (1892–1964), British geneticist and physiologist. Pioneered in mathematical treatment of population genetics; studied respiration in humans; wrote about enzymes.

Haldane, John Scott (1860–1936), British physiologist. Research on respiration, particularly the effects of high and low atmospheric pressures.

Hale, George Ellery (1868–1938), American astronomer. Established the Mount Wilson and Mount Palomar observatories; proved existence of magnetic fields in sunspots; invented the spectroheliograph.

Hall, Charles Martin (1863–1914), American commercial chemist. Discovered Hall process for extracting aluminum.

Hall, Edwin Herbert (1855–1938), American physicist. Discovered Hall effect and conducted studies of this and other galvanomagnetic and thermomagnetic effects.

Hall, John L. (1934–), American physicist. Responsible for a number of major innovations in laser technology, in particular in laser-based precision spectroscopy, including the optical frequency comb technique; Nobel Prize, 2005.

Halley, Edmond (1656–1742), English astronomer. Published a southern star catalog; computed orbits of 24 comets; discovered Halley's comet.

Hamel, Georg Karl Wilhelm (1877–1954), German mathematician. Research on analysis and applied mathematics; introduced Hamel basis of vectors.

Hamilton, William Rowan (1805–1865), Irish mathematician and mathematical physicist. Discovered quaternions; developed mathematical theories encompassing wave and particle optics and mechanics; introduced Hamilton's principle and a form of the Hamilton-Jacobi theory.

Hankel, Hermann (1839–1873), German mathematician. Studied complex and hypercomplex numbers, theory of functions, and Hankel functions; proved that no hypercomplex number system can satisfy ail the laws of ordinary arithmetic.

Hänsch, Theodor W. (1941–), German physicist. Made fundamental contributions to the development of laser-based precision spectroscopy, including the optical frequency comb technique; Nobel Prize, 2005.

Hansen, Armauer (1841–1912), Norwegian bacteriologist. Discovered the bacillus of leprosy, or Hansen's disease.

Harden, Arthur (1865–1940), English chemist. Research on enzymes and alcoholic fermentation; Nobel Prize, 1929.

Hardy, Godfrey Harold (1877–1947), English mathematician. With S. Ramanujan, discovered formula for number of ways of writing a positive integer as the sum of positive integers; proved that the Riemann zeta function has an infinite number of zeros with real part equal to 1/2.

Harker, David (1906–1991), American crystallographer. Completed development of Patterson-Harker method of x-ray diffraction analysis of crystal structure.

Hartley, Ralph Vinton Lyon (1888–1970), American electrical engineer. Invented Hartley oscillator; developed Hartley principle in information theory.

Hartline, Haldan Keffer (1903–1983), American biophysicist. Elucidated cellular electrical activity in the eye and optic nerve; Nobel Prize, 1967.

Hartmann, Johannes Franz (1865–1936), German astronomer. Derived Hartmann dispersion formula relating index of refraction and wavelengths; devised Hartmann test for telescope mirrors; gave first observational proof of interstellar matter.

Hartree, Douglas Rayner (1897–1958). English mathematician and mathematical physicist. Developed methods of numerical analysis which made it possible to apply Hartree method to calculation of atomic wave functions.

Hartwell, Leland H. (1939–), American geneticist. Discovered more than 100 genes that control the cell cycle and introduced the concept of cell cycle checkpoints, ordered groups of genes and protein which halt progress through the cell cycle if DNA is damaged, allowing time for DNA repair; Nobel Prize, 2001.

Harvey, William (1578–1657), English anatomist and physician. Described the true circulation of blood and the action of the heart.

Hassel, Odd (1897–1981), Norwegian chemist. Developed concept of conformation by studying three-dimensional structure of cyclohexane molecule, and explaining the orientation of attached atoms or functional groups: Nobel Prize, 1969.

Hauptman, Herbert A. (1917–), American chemist. With J. Karle, developed computer-aided mathematical techniques for use in x-ray crystallography to determine three-dimensional structures of molecules: Nobel Prize, 1985.

Hausdorff, Felix (1868–1942), German mathematician. Founded and advanced general topology and the general theory of metric spaces.

Haüy, Réné Just, Abbé (1743–1822), French mineralogist. Formulated the geometrical law of crystallization; pioneer in the science of crystallography.

Havers, Clopton (1665–1702), English osteologist. Provided the first full discussion of Haversian lamellae and Haversian canals.

Haworth, Walter Norman (1883–1950), English chemist. Synthesized ascorbic acid; studied carbohydrates, including the structure of sugars; Nobel Prize, 1937.

Heaviside, Oliver (1850–1925), English physicist. Proposed the Heaviside layer in the upper atmosphere.

Heeger, Alan J. (1936–), American physicist. Discovered and developed conductive polymers with Alan MacDiarmid and Hideki Shirakawa; Nobel Prize, 2000.

Hefner-Alteneck, Friedrich Franz von (1845–1904), German engineer. Invented Hefner candle as a standard of luminous intensity.

Heine, Heinrich Eduard (1821–1881), German mathematician. Formulated concept of uniform continuity.

Heisenberg, Werner (1901–1976), German physicist. Founder of quantum mechanics; studied structure of the atom and the Zeeman effect; formulated the principle of indeterminancy in nuclear physics; Nobel Prize, 1932.

Heitler, Watler Heinrich (1904–1981), German-born Swiss theoretical physicist. Developed Heitler-London covalence theory of chemical bonding.

Helmholtz, Hermann Ludwig Ferdinand von (1821–1894), German physicist. anatomist, and physiologist. Physiological research on the nervous system and the human eye and ear, and theoretical work on conservation of force in physics; invented the ophthalmoscope.

Hench, Philip Showalter (1896–1965), American physiologist. Discovered that ACTH and cortisone could be used to treat rheumatoid arthritis; Nobel Prize, 1950.

Henle, Friedrich Gustav Jacob (1809–1885), German pathologist and anatomist. Wrote *Handbuch der Rationellen Pathologie*, integrating the study of physiology and pathology; discovered looped portion of the kidney tubules, and the epithelium.

Henry, Joseph (1797–1878), American physicist. Studied electromagnetic induction, solar phenomena, meteorology, and acoustics.

Henry, William (1775–1836), English chemist and physician. Formulated Henry's law of solubility of gases in liquid.

Henyey, Louis George (1910–1970), American astronomer. Research on reflection nebulae, interstellar matter, stellar atmospheres and evolution, and optical design.

Hermite, Charles (1822–1901), French mathematician. First to solve a fifth-degree equation; investigated *e*, the base of natural logarithms.

Hero of Alexandria (3d century or earlier), Greek mathematician. Wrote on the geometry of plane and solid figures, mechanics, and simple machines; showed that the angle of incidence equals the angle of reflection.

Héroult, Paul Louis Toussaint (1863–1914), French metallurgist. Designed the Héroult furnace for electric steel; developed the Héroult process for aluminum extraction.

Herschbach, Dudley R. (1932–), American chemist. With Y. T. Lee, developed crossed molecular-beam technique for tracing chemical reactions; Nobel Prize, 1986.

Herschel, John Frederick William (1792–1871), English mathematician, physicist, and astronomer. Discovered many nebulae and clusters; pioneered in celestial photography.

Herschel, William or Friedrich Wilhelm (1738–1822), German-born English astronomer. Discovered Uranus, two of its satellites, and two satellites of Saturn; discovered the Sun's intrinsic motion; proposed the concept of the form of the Milky Way.

Hershey, Alfred Day (1908–1997), American biologist. With M. Chase, experimented with bacteriophage, confirming earlier indications that the material basis of heredity is contained in nucleic acids; Nobel Prize, 1969.

Hershko, Avram (1937–), Israeli medical doctor and scientist. Discovered ubiquitin-mediated protein degradation; Nobel Prize, 2004.

Hertz, Gustav (1887–1975), German physicist. With J. Franck, studied effects of electron impacts on atoms; Nobel Prize, 1925.

Hertz, Heinrich Rudolph (1857–1894), German physicist. Discovered Hertzian waves in the ether; proved experimentally Maxwell's theories of electricity and magnetism.

Hertzsprung, Ejnar (1873–1967), Danish astronomer. Discovered method of spectroscopic parallax for measuring stellar distances; determined relation between color and luminosity of stars; with H. N. Russell, developed Hertsprung-Russell diagram.

Herzberg, Gerhard (1904–1999), German-born Canadian physicist. Determined electronic structure and geometry of diatomic and polyatomic molecules, particularly free radicals; Nobel Prize, 1971.

Hess, Victor Franz (1883–1964), Austrian physicist. Studied alpha particles from radium; discovered cosmic rays; Nobel Prize, 1936.

Hess, Walter Rudolf (1881–1973), Swiss physiologist. Discovered the organizer function of the middle brain in coordinating activity of internal organs; developed technique of using electrodes to stimulate localized brain areas; Nobel Prize, 1949.

Hevelius or Hewel, Johannes (1611–1687), Danzig-born astronomer. Discovered four comets; charted surface of the Moon; cataloged numerous stars.

Hevesy, George von (1886–1966), Hungarian chemist. Experimented with radioisotope indication, leading to the technique of isotope tracing of biological and chemical processes; Nobel Prize, 1943.

Hevroský, Jaroslav (1890–1967), Czechoslovakian physical chemist. Developed the technique of polarographic analysis; Nobel Prize, 1959.

Hewish, Antony (1924–), British astronomer. Pioneered in discovery of pulsars, by means of radio telescopes; Nobel Prize, 1974.

Heymans, Corneille (1892–1968), French-Belgian physiologist. Investigated the carotid sinus in connection with the mechanism of breathing; Nobel Prize, 1938.

Hilbert, David (1862–1943), German mathematician. Contributed to theory of numbers and theory of invariants; applied integral equations to physical problems.

Hill, Archibald Vivian (1886–1977), English physiologist. Worked on heat loss and oxygen consumption in muscle contraction; Nobel Prize, 1922.

Hinshelwood, Cyril Norman (1897–1967), British chemist. Elucidated chain reaction and chain branching mechanisms; Nobel Prize, 1956.

Hippias of Elis (5th century B.C.), Greek philosopher and mathematician. Discovered quadratrix.

Hipparchus of Rhodes (fl. 130 B.C.), Greek astronomer. Calculated the inclination of the ecliptic and the precession of the equinoxes; invented trigonometry; made the first star catalog.

Hippocrates (460?–377 B.C.), Greek physician. Known as the father of medicine; writings attributed to him contain clinical observations of diseases, descriptions of surgical practice, and the Hippocratic doctrine of the four humors.

Hironaka, Heisuke (1931–), Japanese-American mathematician. Worked in algebraic geometry; solved the problem of the resolution of singularities on an algebraic variety for algebraic varieties of any dimension over a field of characteristic 0, generalizing work of O. Zariski; Fields Medal, 1970.

Hirzebruch, Friedrich Ernst Peter (1927–), German mathematician. Collaborated with M. F. Atiyah in the development of K theory.

Hitchings, George Herbert (1905–1998), American biochemist. With G. B. Elion, pioneered research that led them to the development of drugs for the treatment of leukemia, malaria, gout, herpes, bacterial and fungal infections, and autoimmune diseases and organ-transplant rejection; Nobel Prize, 1988.

Hittorf, Johann Wilhelm (1824–1914), German physicist. Described effects of Hittorf rays in vacuum tubes; studied electrolysis, and electrical discharge in rarefied gases with the Hittorf tube.

Hodgkin, Alan Lloyd (1914–1998), British biophysicist. With A. Huxley, devised a system of mathematical equations describing the nerve impulse; presented evidence for the sodium theory of nervous conduction; Nobel Prize, 1963.

Hodgkin, Dorothy Crowfoot (1910–1994), Egyptian-born British chemist. Determined the structure of the vitamin B_{12} molecule through x-ray crystallographic analysis; Nobel Prize, 1964.

Hodgkin, Thomas (1798–1866), English physician. First to describe Hodgkin's disease, a glandular disorder.

Hoffmann, Roald (1937–), Polish-born American chemist. Developed methods for predicting whether a chemical reaction is possible based on molecular orbital models; Nobel Prize, 1981.

Hofmann, August Wilhelm von (1818–1892), German chemist. Studied reactions of derivatives; developed the Hofmann reaction for preparing primary amines.

Hofmeister, Wilhelm Friedrich Benedict (1824–1877), German botanist. Did fundamental work on plant embryology; explained the alternating life cycles of mosses and ferns.

Hofstadter, Robert (1915–1990), American physicist. Investigated the properties and behavior of the proton and neutron; determined the size and shape of many nuclei; discovered the construction scheme of fundamental atomic nuclei; Nobel Prize, 1961.

Hölder, Otto Ludwig (1859–1937), German mathematician. Contributed to analysis and group theory; introduced Hölder condition; proved Hölder inequality and Jordan-Hölder theorem.

Holley, Robert William (1922–1993), American biochemist. With coworkers, made first determination of a nucleotide sequence of a nucleic acid; Nobel Prize, 1968.

Holmberg, Erik Bertil (1908–2000), Swedish astronomer. Investigations of galaxies, especially photometry of galaxies.

Hooke, Robert (1635–1703), English inventor. Invented the compound microscope, wheel barometer, universal (Hooke's) joint, and the reflecting telescope; formulated theories on light and on the motion of the Earth.

Hooker, Joseph Dalton (1817–1911), English botanist. With G. Bentham, wrote *Genera Plantarum*; prepared works on the flora of New Zealand, Antarctica, and India.

Hopkins, Frederick Gowland (1861–1947), English biochemist. Discovered the amino acid tryptophan and the tripeptide glutathione; did experimental work leading to the discovery of vitamins; Nobel Prize, 1929.

Hopper, Grace (1906–1992), American mathematician and computer scientist. Pioneered in the development of computer software, including the invention of the first compiler. Credited with having discovered the first computer "bug," a moth that literally had to be removed from the wiring of an early computer.

Hörmander, Lars (1931–), Swedish mathematician. Worked on partial differential equations; in particular, made major contributions to the general theory of linear differential operators; Fields Medal, 1962.

Horvitz, H. Robert (1947–), American geneticist and neurobiologist. Identified and characterized the genes controlling programmed cell death in the nematode *Caenorhabditis elegans*. Showed that these genes interact with each other in cell death and correspond to existing genes in humans. Nobel Prize, 2002.

Houdry, Eugene J. (1892–1962), French-born American engineer. Devised catalytic method of producing oil.

Hounsfield, Godfrey Newbold (1919–2004), British electronics engineer. Invented computerized axial tomography; Nobel Prize, 1979.

Houssay, Bernardo Alberto (1887–1971), Argentine physiologist. Research on the functions and effects of the hypophysis, including its relationship to carbohydrate metabolism; Nobel Prize, 1947.

Howell, William Henry (1860–1945), American physiologist. Discovered heparin; isolated thrombin and thromboplastin; discovered Howell-Jolly bodies; proved that blood platelets are formed in lungs.

Hubble, Edwin Powell (1889–1953), American astronomer. Studied nebulae; formulated Hubble's law of extragalactic nebulae.

Hubel, David Hunter (1926–), American neurobiologist. Contributed to the study of the processing of visual information in the brain; Nobel Prize, 1981.

Huber, Robert (1937–), German chemist. With J. Deisenhofer and M. Hartmut, elucidated the structure of a bacterial protein that performs photosynthesis; Nobel Prize, 1988.

Hückel, Erich (1896–1980), German chemist. With P. J. W. Debye, formulated Debye-Hückel theory of strong electrolytes; devised theoretical explanation of electron properties of aromatic hydrocarbons.

Huggins, Charles Brenton (1901–1997), Canadian-born American surgeon and cancer researcher. Developed treatment of cancers using endocrinologic methods; Nobel Prize, 1966.

Hughes, David Edward (1831–1900), English-born American inventor. Invented the Hughes electromagnet, a printing telegraph, microphone, and induction balance.

Hugoniot, Pierre Henry (1851–1887), French physicist. Developed theory of shock waves.

Hulse, Russell A. (1950–), American astronomer and physicist. With J. H. Taylor, discovered the binary pulsar and studied it to observe phenomena predicted by general relativity; worked in plasma physics; Nobel Prize, 1993.

Humboldt, Friedrich Heinrich Alexander, Baron von (1769–1859), German naturalist. Founder of physical geography; made scientific explorations of South America and Central Asia.

Hume-Rothery, William (1899–1968), English metallurgist and chemist. Discovered Hume-Rothery rule concerning electron compounds.

Hunt, Franklin Livingston (1883–1973), American physicist. Research on x-ray spectroscopy; with W. Duane, discovered and applied Duane-Hunt law.

Hunt, R. Timothy (1943–), British biologist who discovered cyclins, proteins that regulate cyclin-dependent kinase activity, and found that their periodic degradation is an important general control mechanism of the cell cycle; Nobel Prize, 2001.

Hurwitz, Adolf (1859–1919), German-born Swiss mathematician. Worked on modular functions, number theory, Riemann surfaces, complex function theory, and analytic number theory; formulated condition satisfied by Hurwitz polynomials.

Huxley, Andrew Fielding (1917–), British physiologist. With A. L. Hodgkin, discovered the ionic mechanism involved in excitation in the cell membrane of peripheral nerves; Nobel Prize, 1963.

Huygens, Christiaan (1629–1695), Dutch mathematician, physicist, and astronomer. Discovered Saturn's rings; contributed to dynamics and optics; proposed the wave theory of light.

Hylleraas, Egil Andersen (1898–1965), Norwegian physicist. Applied quantum theory to helium atom, negative hydrogen ion, and other atoms, molecules, and crystals; developed variational method and other methods for mathematical solution of quantum-mechanical problems.

Ignarro, Louis J. (1941–), American pharmacologist. Working independently and with Robert Furchgott, he concluded that endothelium-derived relaxing factor was identical to nitric oxide; Nobel Prize, 1998.

Ingenhousz, Jan (1730–1799), Dutch physician and naturalist. Demonstrated the cycle of photosynthesis in plants.

Ising, Ernest (1900–1998), German-born American physicist. Introduced Ising model of ferromagnetic material; research in solid-state physics and ferromagnetism.

Itō, Kiyosi (1915–), Japanese mathematician. Studied stochastic processes; introduced Itō's integral and Itō's formula.

Jacob, François (1920–), French biologist. Discovered episomes, a class of genetic elements; with J. Monod, proposed the concepts of messenger ribonucleic acid and of the operon; Nobel Prize, 1965.

Jacobi, Karl Gustav Jacob (1804–1851), German mathematician. Worked on elliptic functions and differential equations; developed the theory of determinants.

Jacquard, Joseph Marie (1752–1834), French inventor. Designed and built the Jacquard loom for figured weaving.

Jaeger, Frans Maurits (1877–1945), Dutch crystallographer and physical chemist. Measured physical properties of molten salts and silicates at extremely high temperatures.

James, William (1842–1910), American psychologist and writer. Coformulator of James-Lange theory that emotions are the perception of physiological changes.

Janet, Pierre Marie Félix (1859–1947), French psychologist. Studied hysteria, obsession, and neurosis; wrote a textbook on the theory of hysteria.

Jeans, James Hopwood (1877–1946), English physicist and astronomer. Worked in stellar dynamics; proposed the tidal theory of the origin of planets.

Jenner, Edward (1749–1823), English physician. Discovered vaccination.

Jensen, J. Hans D. (1906–1973), German physicist. With M. G. Mayer, formulated the nuclear shell model; Nobel Prize, 1963.

Jerne, Niels Kaj (1911–1994), Swiss immunologist. Formulated three important theories involving the immune system: how the body produces specific antibodies, how the immune system develops and matures, and how the various interrelated aspects of the immune response are coordinated by the body; Nobel Prize, 1984.

Johnson, Harold Lester (1921–1980), American astronomer. Research in astrophysics, astronomical photometry, infrared astronomy, applications of electronics to astronomy, and Fourier transform spectroscopy; collaborated with W. W. Morgan in developing Johnson-Morgan system of stellar magnitudes.

Joliot-Curie, Irène (1897–1956), French physicist. With M. Curie, discovered projection of atomic nuclei by neutrons; with J. F. Joliot-Curie, discovered artificial radiation; Nobel Prize, 1935.

Joliot-Curie, Jean Frédéric (1900–1958), French physicist. With I. Joliot-Curie, produced an artificial radioactive substance by bombarding boron with fast alpha particles; Nobel Prize, 1935.

Jones, Vaughn Frederick Randal (1952–), New Zealand-American mathematician. Proved an index theorem for von Neumann algebras, discovered a relationship between these algebras and geometric topology, and discovered a new polynomial invariant for knots; this work provided a connecting link for widely separated areas of mathematics and physics; Fields Medal, 1990.

Jordan, Camille (1838–1921), French mathematician. Discovered many fundamental results in group theory; gave a proof of the Jordan curve theorem (later shown to be incorrect).

Jordan, Pascual (1902–1980), German physicist. Contributed to formulation of quantum mechanics; introduced Jordan algebra in an attempt to generalize quantum mechanics.

Josephson, Brian David (1940–), British physicist. Predicted the Josephson effect concerning electron pairs; Nobel Prize, 1973.

Joukowski, Nikolai Jegorowitch (1847–1921), Russian applied mathematician and aerodynamicist. Helped to introduce concept of Kutta-Joukowski airfoil and to prove Kutta-Joukowski theorem.

Joule, James Prescott (1818–1889), English physicist. Formulated a mechanical theory of heat; demonstrated Joule-Thomson effect relating to the fall in temperature of a gas; first to estimate the velocity of a gas molecule.

Jung, Carl Gustav (1875–1961), Swiss psychologist and psychiatrist. Evolved a theory of complexes; founded the analytical school of psychoanalysis and psychotherapy.

Jussieu, Antoine Laurent de (1748–1836), French botanist. Wrote *Genera Plantarum*, the basis of modern natural botanical classification.

Kahn, Robert E. (1938–), American electrical engineer. Co-invented (with Vinton G. Cerf) Transmission Control Protocol/Internet Protocol (TCP/IP).

Kalman, Rudolf Emil (1930–), Hungarian born American mathematician and electrical engineer. Worked on mathematical theory of control systems; developed the Kalman filter; introduced concepts of controllability and observability.

Kaluza, Theodor Franz Eduard (1885–1954), German mathematical physicist. Developed theory which attempted to unify gravitation and electromagnetism.

Kamerlingh Onnes, Heike (1853–1926), Dutch physicist. Research on cryogenics, critical phenomena, and low temperatures; discovered the phenomenon of superconductivity; Nobel Prize, 1913.

Kandel, Eric R. (1929–), Austrian-born American neurobiologist. Discovered that protein phosphorylation plays an important role in the molecular mechanisms underlying learning and memory formation; his work contributed to the elucidation of the molecular mechanisms involved in slow synaptic transmission in the nervous system; Nobel Prize, 2000.

Kapitza, Pjotr Leonidovich (1894–1984), Russian physicist. Studied magnetism and low temperature; designed hydrogen and helium liquefaction plants; Nobel Prize, 1978.

Kapteyn, Jacobus Cornelius (1851–1922), Dutch astronomer. Studied the proper motion of stars; with P. J. van Rhijn, evolved a theory of the universe.

Karle, Jerome (1918–), American crystallographer. With H. A. Hauptman, developed computer-aided mathematical techniques for use in x-ray crystallography to determine three-dimensional structures of molecules; Nobel Prize, 1985.

Karrer, Paul (1889–1971), Swiss chemist. Pioneering research on vitamins A and B_2 and on the flavins and carotenoids; Nobel Prize, 1937.

Kastler, Alfred (1902–1984), French physicist. Developed a double-resonance method to study energy levels of atoms in excited states; Nobel Prize, 1966.

Kater, Henry (1777–1835), English geodesist. Developed Kater's reversible pendulum to obtain accurate values of acceleration of gravity.

Katz, Bernard (1911–2003), German-born British physiologist. Made discoveries concerning mechanism for release of transmitter substances at nerve-muscle junction; Nobel Prize, 1970.

Keesom, Willem Hendrik (1876–1956), Dutch physicist. Worked in low-temperature physics; first to solidify helium; studied molecular structure of liquids and compressed gases.

Kekulé von Stradonitz, Friedrich August (1829–1896), German chemist. A founder of structural organic chemistry; made theoretical proposal of the structure of benzene.

Kelvin, William Thomson, 1st Baron (1824–1907), British mathematician and physicist. Invented the Kelvin balance; formulated Kelvin's laws concerning electric cables; contributed to thermodynamics.

Kendall, Edward Calvin (1886–1972), American biochemist. Chemical investigation of the adrenal cortex, leading to the isolation of crystalline cortical hormones, especially cortisone; Nobel Prize, 1950.

Kendall, Henry Way (1926–1999), American physicist. Collaborated in experiments that demonstrated that protons, neutrons, and similar particles are made up of quarks; Nobel Prize, 1990.

Kendrew, John Cowdery (1917–1997), British molecular biologist. First to successfully determine the structure of a protein; Nobel Prize, 1962.

Kennelly, Arthur Edwin (1861–1939), American electrical engineer. Discovered the ionized layer in the atmosphere, independently of O. Heaviside; proposed the theory of alternating currents.

Kepler, Johannes (1571–1630), German astronomer. Proposed Kepler's three laws of planetary motion; worked in optics.

Kerr, John (1824–1907), Scottish physicist. Discovered the Kerr magnetooptic effect.

Ketterle, Wolfgang (1957–), German physicist. Produced Bose-Einstein condensates in a dilute gas of sodium atoms (independent of the work of E. A. Cornell and C. E. Wieman) and carried out fundamental studies of their properties, including the production of interference patterns and atom lasers; Nobel Prize, 2001.

Khorana, Har Gobind (1922–), Indian-born American biochemist. Synthesized complicated nucleic acids: proved that genetic code consists of nonoverlapping triplets of bases without gaps between triplets; Nobel Prize, 1968.

Kilby, Jack St. Clair (1923–2005), American physicist, electronics engineer, and inventor. Participated in the invention of the integrated circuit; Nobel Prize, 2000.

Kirchhoff, Gustav Robert (1824–1887), German physicist. With R. W. Bunsen, discovered method of spectrum analysis; formulated Kirchhoff's law of electric currents and electromotive forces in a network.

Kirkwood, Daniel (1814–1895), American astronomer. Discovered Kirkwood gaps; studied nature, origin, and evolution of solar system, particularly role of asteroids, comets, meteors, and meteorites.

Kitasato, Shibasaburo (1852–1931), Japanese bacteriologist. Independently of A. E. J. Yersin, discovered the bacillus of bubonic plague; isolated bacilli of symptomatic anthrax, dysentery, and tetanus.

Klebs, Edwin (1834–1913), German pathologist. Described diphtheria (Klebs-Löffler) bacillus; studied bacteriology of malaria, anthrax, and tuberculosis.

Klein, Christian Felix (1849–1925), German mathematician. Contributed to function theory, noneuclidean geometry, group therapy, and applied mathematics; introduced Klein bottle.

Klein, Oskar Benjamin (1894–1977), Swedish physicist. Codeveloper of Klein-Gordon equation. Klein-Nishina formula, and Klein-Rydberg method; proposed theory of overall structure of the universe.

Kleinrock, Leonard (1934–), American electrical engineer. Developed the basic principles of packet switching for communication networks, the underlying technology of the Internet.

Klitzing, Klaus von (1943–), German physicist. Discovered quantum Hall effect; Nobel Prize, 1985.

Klug, Aaron (1926–), South African-born British biochemist. Developed crystallographic electron microscopy and elucidated biologically important nucleic acid-protein complexes; Nobel Prize, 1982.

Knowles, William S. (1917–), American chemist. Developed chirally catalyzed hydrogenation reactions; Nobel Prize, 2001.

Knudsen, Martin Hans Christian (1871–1949), Danish physicist and hydrographer. Studied flow and diffusion of gases at low pressure; developed Knudsen cell and Knudsen gage; developed methods to measure the properties of seawater.

Kobayashi, Makoto (1944–), Japanese physicist. Fundamental contributions to the theory of CP violation. Co-discoverer of the origin of the broken symmetry leading to the prediction of the existence of at least three families of quarks in nature; Nobel Prize, 2008.

Koch, Robert (1843–1910), German physician and bacteriologist. Studied cholera, tuberculosis, and bubonic plague; showed a specific bacillus to be the cause of anthrax; discovered the tubercle bacillus; Nobel Prize, 1905.

Kocher, Emil Theodor (1841–1917), Swiss surgeon. Studied the functions and malfunctions of the thyroid gland; Nobel Prize, 1909.

Kodaira, Kunihiko (1915–1997), Japanese mathematician. Did research on harmonic integrals and harmonic forms with applications to Kählerian and more specifically algebraic varieties; demonstrated, by sheaf cohomology, that such varieties are Hodge manifolds; Fields Medal, 1954.

Köhler, Georges J. F. (1946–1995), Swiss immunologist. With C. Milstein, discovered a laboratory technique for producing monoclonal antibodies, highly uniform immune bodies that are selective in responding to target substances; Nobel Prize, 1984.

Kohn, Walter (1923–), Austrian-born American physicist. Developed density-functional theory, which solves equations for electron density rather than positions of individual electrons; it is one of the developments that has significantly sped up computational quantum chemistry; Nobel Prize, 1998.

Kolbe, Adolf Wilhelm Hermann (1818–1884), German chemist. Contributed to the synthesis concept of compound formation, doing much to eliminate the division of chemistry into two branches: organic and inorganic.

Kolmogorov, Andrei Nikolaevich (1903–1987), Soviet mathematician. Formulated set-theoretic basis of probability theory.

Kontsevich, Maxim (1964–), Russian mathematician and mathematical physicist. Worked in algebraic geometry, algebraic topology, string theory, and quantum field theory; demonstrated equivalence of two models of quantum gravitation; discovered an invariant for classifying knots; Fields Medal, 1998.

Kornberg, Arthur (1918–2007), American biochemist. Discovered deoxyribonucleic acid polymerase, providing the first rational enzymatic mechanism for the replication of genetic material of the cell; Nobel Prize, 1959.

Kornberg, Roger (1947–), American biochemist. Made fundamental contributions to the understanding of the molecular basis of eukaryotic transcription; Nobel Prize, 2006.

Korteweg, Diederik Johannes (1848–1941), Dutch mathematician. Work in applied mathematics, mechanics, and hydrodynamics; with G. de Vries, proposed equation of wave motion with soliton solution.

Koshiba, Masatoshi (1926–), Japanese physicist. Developed large water detectors (Kamiokande and Super Kamiokande) that observed solar and atmospheric neutrinos and neutrinos from Supernova 1987A, and found evidence for neutrino oscillations. Nobel Prize, 2002.

Kossel, Albrecht (1853–1927), German chemist. Investigated the chemistry of cells and of proteins; Nobel Prize, 1910.

Krafft-Ebing, Richard, Baron von (1840–1902), German neurologist. Authority on psychological disorders and their forensic implications; wrote *Psychopathia Sexualis*, a collection of case histories.

Kramers, Hendrik Anthony (1894–1952), Dutch physicist. Developed quantum theory of dispersion, establishing Kramers-Kronig relation; with G. Wentzel and L. Brillouin, developed Wentzel-Kramers-Brillouin method.

Krebs, Edwin Gerhard (1918–), American biochemist. With E. H. Fischer, discovered phosphorylation processes that play a critical role in cell-protein regulation; they isolated the first protein kinase, a class of enzymes that transfer phosphate from adenosine triphosphate to proteins; Nobel Prize, 1992.

Krebs, Hans Adolf (1900–1981), German-born British biochemist. Elucidated metabolic pathways, including the tricarboxytic acid cycle; Nobel Prize, 1953.

Kroemer, Herbert (1928–), German-American physicist and electronics engineer. Developed semiconductor heterostructures used in high-speed and opto-electronics, including fast transistors, laser diodes, and light-emitting diodes; Nobel Prize, 2000.

Krogh, Schack August Steenberg (1874–1949), Danish physiologist. Discovered the regulation of the vasomotor mechanism of capillaries; devised the nitrous oxide method for measuring human circulation; Nobel Prize, 1920.

Kronecker, Leopold (1823–1891), German mathematician. Contributed to theory of elliptical functions, algebra, and number theory, and attempted to unify these disciplines; attempted to base all mathematics on integers and finite processes.

Kroto, Harold W. (1939–), British chemist. Studied long-chain molecules found in radioastronomy by using microwave spectroscopy, which ultimately led to the discovery of fullerenes; Nobel Prize, 1996.

Kruskal, Martin David (1925–2006), American mathematician and physicist. Research in plasma physics, asymptotic phenomena, relativity, and minimal surfaces.

Kuhn, Richard (1900–1967), German chemist. Research on the structures and synthesis of vitamins and carotenoids; Nobel Prize, 1938 (declined).

Kundt, August Adolph (1839–1894), German physicist. Used Kundt tube to determine speed of sound in gases; determined ratio of specific heats of monatomic gases; with W. K. Röntgen, demonstrated Faraday effect in gases.

Kurchatov, Igor Vasilievich (1903–1960), Soviet physicist. Discovered nuclear isomers; studied nuclear reactions; developed nuclear weapons and nuclear power.

Kusch, Polykarp (1911–1993), German-born American physicist. Precisely determined the magnetic moment of the electron; Nobel Prize, 1955.

Kutta, Wilhelm Martin (1867–1944), German applied mathematician. Helped introduce concept of Kutta-Joukowski airfoil, prove Kutta-Joukowski theorem, and develop Runge-Kutta method.

Kwolek, Stephanie (1923–), American polymer chemist. Developed Kevlar, poly(*p*-phenylene terephthalamide), a lightweight synthetic fiber that is stronger than steel.

Lacaille, Nicolas Louis de (1713–1762), French astronomer and geodesist, Determined positions of nearly 10,000 stars in southern skies; measured lunar and solar parallax; showed that Earth has equatorial bulge.

Lafforgue, Laurent (1966–), French mathematician. Proved the global Langlands correspondence for function fields, a major advance toward the realization the Langlands program, which deals with deep connections between number theory, analysis, and group representation. Fields Medal, 2002.

Lagrange, Joseph Louis, Count (1736–1813), French geometer and astronomer. Invented the calculus of variations; studied the mathematics of sound; wrote *Mécanique Analytique*, concerning statics and dynamics.

Laguerre, Édmond Nicolas (1834–1886), French mathematician. Discovered Laguerre's differential equations and Laguerre polynomials.

Lamarck, Jean Baptiste Pierre Antoine de Monet, Chevailer de (1744–1829), French naturalist. Proposed theory that changes in animal and plant structure are caused by changes in environment; classified animals into vertebrates and invertebrates.

Lamb, Willis Eugene, Jr. (1913–2008), American physicist. Made precise atomic measurements leading to a new understanding of the theory of electron interactions and electromagnetic radiation; Nobel Prize, 1955.

Lambert, Johann Heinrich (1728–1777), German physicist. Formulated the Lambert theorem concerning the illumination of a surface.

Lamé, Gabriel (1795–1870), French mathematician, physicist, and engineer. Introduced curvilinear coordinates and applied them to differential equations, elasticity, thermodynamics, and number theory.

Landau, Lev Davydovich (1908–1968), Soviet physicist. Made theoretical explanation of the nature and properties of liquid helium; investigated condensed matter; Nobel Prize, 1962.

Landé, Alfred (1888–1975), German-born American physicist. Introduced Landé g factor; discovered Landé interval rule and Landé Γ-permanence rule.

Landsteiner, Karl (1868–1943), Austrian-born American pathologist. Discovered human blood groups and factors M and N; with A. S. Weiner, discovered the Rh factor; Nobel Prize, 1930.

Lange, Carl Georg (1834–1900), Danish physician and psychologist. With W. James, proposed the James-Lange theory of emotion.

Langerhans, Paul (1847–1888), German pathologist and anatomist. Studied human and animal microscopical anatomy, particularly structures of skin and pancreas; discovered islets of Langerhans.

Langevin, Paul (1872–1946), French physicist. Developed quantitative theories of paramagnetism and diamagnetism; helped to elucidate the theory of relativity; contributed to the development of sonar.

Langley, Samuel Pierpont (1834–1906), American astronomer and airplane pioneer. Studied infrared solar spectrum; constructed in 1896 the first mechanical heavier-than-air machine to fly.

Langmuir, Irving (1881–1957), American chemist. With G. N. Lewis, proposed the Lewis-Langmuir atomic theory; studied surface chemistry and thermionic emission; Nobel Prize, 1932.

Laplace, Pierre Simon, Marquis de (1749–1827), French astronomer and mathematician. Contributed to celestial mechanics, especially to the study of the Moon, Saturn, and Jupiter; formulated the theory of probability; discovered the Laplace differential equation.

Larmor, Joseph (1857–1942), British physicist. Developed electron theory which fused electromagnetic and optical concepts; introduced Larmor precession and derived Larmor formula.

Latimer, Louis Howard (1848–1928), American inventor. A collaborator of Alexander Graham Bell and Thomas Edison, this self-educated son of an escaped slave is best known for key inventions in the field of electric lighting.

Laue, Max Theodor Felix von (1879–1960), German physicist. Proposed the theory of x-ray diffraction by crystals; developed the Laue method of investigating crystal structure; Nobel Prize, 1914.

Laughlin, Robert Betts (1950–), American physicist. Provided a theoretical explanation for the fractional quantum Hall effect by showing how electrons acting together in strong magnetic fields can form new types of quasiparticles with charges that are fractions of electron charges; Nobel Prize, 1998.

Laurent, Pierre Alphonse (1813–1854), French mathematician and physicist. Introduced Laurent series; research on wave theory of light.

Lauterbur, Paul C. (1929–2007), American chemist. Discovered that it was possible to create a two-dimensional image of the body with magnetic resonance by introducing gradations in the external magnetic field, providing the basis for the development of magnetic resonance as a useful medical imaging technique. Nobel Prize, 2003.

Laveran, Charles Louis Alphonse (1845–1922), French physician. Discovered the malaria parasite; researched sleeping sickness; Nobel Prize, 1907.

Lavoisier, Antoine Laurent (1743–1794), French chemist. The founder of modern chemistry; studied combustion and respiration; published a table of the elements.

Lawrence, Ernest Orlando (1901–1958), American physicist. Discovery, development, and use of the cyclotron; Nobel Prize, 1939.

Lax, Peter D. (1926–), American mathematician (born in Hungary). Fundamental contributions in the field of partial differential equations; significant contributions in other fields including mathematical physics; author of numerous textbooks; Abel Prize, 2005.

Lebesgue, Henry Léon (1875–1941), French mathematician. Developed theory of measure and integration; studied trigonometric series.

Le Chatelier, Henry Louis (1850–1936), French chemist and metallurgist. Research on cement chemistry, gas combustion, blast furnace reactions, chemical equilibria, alloy properties, and chemistry and metallurgy of iron and steel; formulated Le Chatelier's principle.

Leclanché, Georges (1839–1882), French chemist and electrician. Invented the Leclanché galvanic cell.

Lederberg, Joshua (1925–2008), American geneticist. With E. L. Tatum, discovered genetic recombination in bacteria and organization of genetic material; Nobel Prize, 1958.

Lederman, Leon Max (1922–), American physicist. Collaborated in experiment that demonstrated the existence of two types of neutrino; led an experiment that discovered the upsilon particle; Nobel Prize, 1988.

Lee, David Morris (1931–), American physicist. With D. D. Osheroff and R. C, Richardson, discovered superfluidity in helium-3; Nobel Prize, 1996.

Lee, Tsung-Dao (1926–), Chinese-born American physicist. With C. N. Yang disproved the parity principle; worked on statistical mechanics, astrophysics, nuclear and subnuclear physics, and field theory; Nobel Prize, 1957.

Lee, Yuan Tseh (1936–), American chemist. With Dudley R. Herschbach, developed crossed molecular-beam technique for tracing chemical reactions; Nobel Prize, 1986.

Legendre, Adrien Marie (1752–1833), French mathematician. Worked on elliptic functions, the theory of numbers, and the method of least squares.

Leggett, Anthony J. (1938–), British and American physicist. Developed a theory explaining the complex behavior of superfluid helium-3. Nobel Prize, 2003.

Lehn, Jean-Marie (1939–), French chemist. Studied crown ethers and developed the synthesis of related structures known as cryptands; Nobel Prize, 1987.

Leibniz, Gottfried Wilhelm, Baron von (1646–1716), German mathematician. Contributed to the development of differential calculus.

Leloir, Luis Federico (1906–1987), French-born Argentine biochemist. Discovered sugar nucleotides and their role in carbohydrate biosynthesis; Nobel Prize, 1970.

Lenard, Phillipp Eduard Anton (1862–1947), Hungarian-born German physicist. Studied cathode rays outside the discharge tube; worked on photoelectricity; Nobel Prize, 1905.

L'Enfant, Pierre Charles (1754–1825), French engineer. Designed Washington, D.C.

Lennard-Jones, John Edward (1894–1954), English physicist and chemist. Proposed Lennard-Jones potential for interatomic forces; contributed to quantum theory of molecular structure and statistical mechanics of liquids, gases, and surfaces.

Lenz, Heinrich Friedrich Emil (1804–1865), German physicist. Formulated Lenz's law governing induced current.

Leverrier, Urbain Jean Joseph (1811–1877), French astronomer. Studied Mercury, Uranus, and Neptune; credited, with J. C. Adams in England, as discoverer of Neptune.

Levi-Civita, Tullio (1873–1941), Italian mathematician and mathematical physicist. With G. Ricci-Curbastro, developed tensor analysis; introduced concept of parallelism in curved spaces.

Levi-Montalcini, Rita (1909–), Italian biologist. With S. Cohen, made landmark studies of nerve growth factor and its functions; Nobel Prize, 1986.

Lewis, Edward B. (1918–2004), American geneticist. Shared in discoveries showing the genetic involvement of early embryonic development with C. Nusselin-Volhard and E. F. Wieschaus; Nobel Prize, 1995.

Lewis, Gilbert Newton (1875–1946), American chemist. Collaborated in developing the Lewis-Langmuir atomic theory; worked on the electronic theory of valency and chemical thermodynamics.

Leydig, Franz von (1821–1908), German histologist and anatomist. Founder of comparative histology; promoted use of microscope in anatomical study; described cells in testes believed to secrete male hormones.

l'Hospital (l'Hôpital), Guillaume Françöis Antoine de, Marquis de Sainte-Mesme, Compte d'Entremont (1661–1704), French mathematician. Wrote first textbook on differential calculus, which gives l'Hospital's rule.

Libby, Willard Frank (1908–1980), American chemist. Developed the method of radiocarbon dating; Nobel Prize, 1960.

Lie, Marius Sophus (1842–1899), Norwegian mathematician. Originated the theory of tangential transformations.

Liebig, Justus, Baron von (1803–1873), German chemist. Discovered chloroform and chloral; founded agricultural chemistry; invented the Liebig condenser.

Linnaeus, Carolus, real name Carl von Linné (1707–1778), Swedish botanist. Developed the Linnaean system of biological classification.

Lions, Pierre-Louis (1956–), French mathematician. Made important contributions to the theory of nonlinear partial differential equations; Fields Medal, 1994.

Liouville, Joseph (1809–1882), French mathematician. Proved existence of transcendental functions; developed concept of geodesic curvature; originated theory of doubly periodic functions.

Lipmann, Fritz Albert (1899–1986), German-born American biochemist. Formulated general rules for the biotechnology of energy transmission; discovered coenzyme A; Nobel Prize, 1953.

Lippmann, Gabriel (1845–1921), French physicist. Produced the first colored photograph of the light spectrum; invented the Lippmann capillary electrometer; Nobel Prize, 1908.

Lipscomb, William Nunn, Jr. (1919–), American physical chemist. Studied structure and bonding of boranes, providing insight into nature of chemical bonding; Nobel Prize, 1976.

Lissajous, Jules Antoine (1822–1880), French physicist. Invented the vibration microscope, involving Lissajous figures.

Lister, Joseph, 1st Baron (1827–1912), English surgeon. Introduced antiseptics to surgery; pioneered in bacteriology.

Littlewood, John Endensor (1885–1977), British mathematician. Work on diophantine approximation, Tauberian theorems, Fourier series and associated function theory, the zeta function, additive number theory, and inequalities.

Littrow, Joseph Johann von (1781–1840), Austrian astronomer. Studied light refraction; worked on telescope construction.

Lloyd, Humphrey (1800–1881), Irish physicist. Discovered Lloyd's mirror interference; verified W. R. Hamilton's prediction of conical refraction.

Lobachevski, Nikola Ivanovich (1793–1856), Russian mathematician. Originated the first comprehensive system of noneuclidean geometry.

Loewi, Otto (1873–1961), German pharmacologist. Investigated nerve impulses; proved the role of ace-tylcholine in nerve impulse transmission; Nobel Prize, 1936.

Löffler, Friedrich August Johannes (1852–1915). German bacteriologist. Isolated the diphtheria (Klebs-Löffler) bacillus; developed protective serum against foot-and-mouth disease.

London, Fritz (1900–1954), German-born American physicist. Developed, with W. Heitler, theory of covalent bonding; with H. London, theory of superconductivity; and theory of superfluidity.

London, Heinz (1907–1970), German-born English physicist. Research on electrodynamic and ther-modynamic behavior of superconductors and properties of superfluid helium.

Lorentz, Hendrik Antoon (1853–1928), Dutch physicist. Proposed the electron theory to explain electromagnetic properties of materials; proposed the Lorentz-FitzGerald contraction and the Lorentz transformation, contributing to the theory of relativity; studied Zeeman effect; Nobel Prize, 1902.

Lorenz, Konrad Zacharias (1903–1989), Austrian zoologist. Pioneered in study of animal behavior patterns; discovered imprinting in birds; Nobel Prize, 1973.

Loschmidt, Johann Joseph (1821–1895), Austrian physicist and chemist, born in Bohemia. Worked on graphical and structural molecular formulas; attempted to estimate size of air molecules and number of air molecules per unit volume.

Lowry, Thomas Martin (1874–1936), British chemist. Simultaneously with Johannes Brønsted, defined acid-base theory in terms of proton transfer—the Brønsted-Lowry theory (of acids and bases).

Lummer, Otto Richard (1860–1925), German physicist. Codeveloper of Lummer-Brodhun sight box and Lummer-Gehrcke plate; constructed an improved bolometer.

Luria, Salvador Edward (1912–1991), Italian-born American biologist. Devised fluctuation test to demonstrate and study spontaneous mutations in bacteria and viruses; Nobel Prize, 1969.

Lwoff, André Michael (1902–1994), French biologist. Explained the phenomenon of lysogeny in bacteria; Nobel Prize, 1965.

Lyapunov, Aleksandr Mikhailovich (1857–1918), Soviet mathematician and physicist. Determined in what cases linear approximations can be used to solve the problem of stability of a mechanical system with a finite number of degrees of freedom; proved existence of various figures of equilibrium for a rotating liquid.

Lyell, Charles (1797–1875), British geologist. Wrote *Principles of Geology*, refuting catastrophic theory of geological changes.

Lyman, Theodore (1874–1954), American physicist. Observed ultraviolet spectra; clarified nature of Lyman ghosts; discovered Lyman series.

Lynen, Feodor (1911–1979), German biochemist. Research on the formation of the cholesterol molecule; discovered chemistry of biotin; Nobel Prize, 1964.

Lyot, Bernard Ferdinand (1879–1952), French astronomer. Invented the coronagraph and developed monochromatic filters that greatly extended knowledge of the solar corona.

MacDiarmid, Alan G. (1927–2007), New Zealand-born American chemist. Discovered and developed conductive polymers with Hideki Shirakawa and Alan Heeger; Nobel Prize, 2000.

Mach, Ernst (1838–1916), Austrian physicist. Research on supersonic flight, leading to Mach angle and Mach number; studied airflow over objects at high speeds.

MacKinnon, Roderick (1956–), American medical doctor and scientist. Determined structure and mechanism of ion channels in cell membranes; Nobel Prize, 2003.

Maclaurin, Colin (1698–1746), Scottish mathematician. Systematized and developed Newton's calculus; introduced Maclaurin series and Maclaurin-Cauchy test.

Macleod, John James Rickard (1876–1935), Scottish physiologist. Shared in discovery of insulin with F. G. Banting and C. H. Best; Nobel Prize, 1923.

Magnus, Heinrich Gustav (1802–1872), German physicist and chemist. Made first quantitative analysis of blood gases; showed that arterial blood has higher oxygen content than venous blood; discovered Magnus effect.

Majorana, Ettore (1906–1938), Italian physicist. Studied properties of elementary particles; postulated Majorana force.

Maksutov, Dmitry Dmitrievich (1896–1964), Soviet physicist and astronomer. Developed general theory of aplanatic optical systems; developed Maksutov system.

Malpighi, Marcello (1628–1694), Italian anatomist. Discovered the capillaries; made microscopic studies in embryology; discovered the Malpighian layer of the epidermis and the Malpighian corpuscles in the kidney.

Malus, Étienne Louis (1775–1812), French engineer and physicist. Formulated Malus' cosine-squared law concerning polarized light and Malus' law of rays.

Mandelstam, Stanley (1928–), American physicist, born in South Africa. Research in theoretical physics of elementary particles; introduced Mandelstam plane and Mandelstam representation.

Mansfield, Peter (1933–), British physicist. Developed mathematical techniques for capturing, analyzing, and processing magnetic resonance signals more efficiently, making it possible to produce three-dimensional images of internal organs. Nobel Prize, 2003.

Marconi, Guglielmo, Marquis (1874–1937), Italian electrician and inventor. Developed commercial wireless telegraphy; Nobel Prize, 1909.

Margulis, Gregori Aleksandrovich (1946–), Russian mathematician. Worked in combinatorics, differential geometry, ergodic theory, dynamical systems, and Lie groups; developed innovative analysis of the structure of Lie groups, describing their discrete subgroups; Fields Medal, 1978.

Mariotte, Edmé (?-1684), French physicist and physiologist. Discovered blind spot; studied circulation of sap in plants, collisions of bodies, properties of air, refraction and color of light, hydrostatics, hydraulics, and meteorology.

Marcus, Rudolph Arthur (1923–), Canadian-born American chemist. Developed the mathematical analysis for electron transfer reactions in chemical systems; Nobel Prize, 1992.

Mark, Herman Francis (1895–1992), Austrian-born American chemist. Elucidated molecular structures of natural and synthetic polymers; developed theory of polymerization; studied relation between structure and properties of macromolecular systems.

Markov, Andrei Andreevich (1856–1922), Russian mathematician. Formulated rigorous proofs of law of large numbers and central-limit theorem; introduced Markov chain.

Marshall, Barry J. (1951–), Australian microbiologist. Co-discoverer of the bacterium *Helicobacter pylori* and its role in gastritis and peptic ulcer disease; Nobel Prize, 2005.

Martin, Archer John Porter (1910–2002), English chemist. With R. L. M. Synge, developed partition chromatography; Nobel Prize, 1952.

Mascheroni, Lorenzo (1750–1800), Italian mathematician. Calculated Euler's constant; proved that all plane construction problems that can be solved with a ruler and compass can also be solved with a compass alone.

Maskawa, Toshihide (1940–), Japanese physicist. Fundamental contributions to the theory of CP violation. Co-discoverer of the origin of the broken symmetry leading to the prediction of the existence of at least three families of quarks in nature; Nobel Prize, 2008.

Mather, John C. (1946–), American physicist. Coordinated the Cosmic Background Explorer satellite project and had principal responsibility for the experiment demonstrating the blackbody form of the microwave background radiation; Nobel Prize, 2006.

Mathieu, Emile Leonard (1835–1890), French mathematician and physicist. Worked on solution of partial differential equations; research in celestial and analytical mechanics; studied Mathieu equation and introduced Mathieu functions.

Matthias, Bernd Teo (1919–1980), German-born American physicist. Tested metals and alloys for superconductivity; developed empirical rules to predict new superconducting materials.

Maunder, Edward Walter (1851–1928), British astronomer. Observations of the sun, sunspots and eclipses.

Maupertuis, Pierre Louis Moreau de (1698–1759), French mathematician and astronomer. Discovered the principle of least action: mathematical writings on the properties of curves.

Maxwell, James Clerk (1831–1879), Scottish physicist. Formulated the electromagnetic theory of light and the Maxwell distribution of molecular velocities of gases; invented the Maxwell disk concerning color vision.

Mayall, Nicholas Ulrich (1906–1993), American astronomer. Research on nebulae, globular star clusters, and external galaxies.

Mayer, Julius Robert von (1814–1878), German physicist. Discovered the principle of conservation of energy.

Mayer, Maria Goeppert (1906–1972), German-born American nuclear physicist. With J. H. D. Jensen, discovered nuclear shell structure; Nobel Prize, 1963.

McClintock, Barbara (1902–1992), American geneticist. Discovered mobile genetic elements known as jumping genes; Nobel Prize, 1983.

McCoy, Elijah (1844–1929), Canadian-born American inventor. Educated in Scotland as a mechanical engineer, this son of former slaves is best known for inventing automatic lubricating systems for steam engines and industrial machines, greatly advancing the Industrial Revolution; the popularity of his products is believed to have led to the American expression "the real McCoy."

McMillan, Edwin Mattison (1907–1991), American physicist. Discovered element 93 (neptunium), which led to the creation of element 94 (plutonium); conceived the theory of phase stability; Nobel Prize, 1951.

McMullen, Curtis Tracy (1958–), American mathematician. Worked in hyperbolic geometry and in complex dynamics, also known as chaos theory; Fields Medal, 1998.

Meckel, Johann Friedrich (1781–1833), German anatomist and embryologist. Gave first comprehensive description of birth defects; described Meckel's cartilage; discovered Meckel's diverticulum.

Medawar, Peter Brian (1915–1987), Brazilian-born British biologist and medical scientist. Discovered acquired immunological tolerance; Nobel Prize, 1960.

Meissner, Alexander (1883–1958), Austrian-born German radio engineer. Helped develop improved electrical insulators and continuous-wave transmission; invented Meissner oscillator.

Meissner, Walther (1882–1974), German physicist. Research in low-temperature physics; discovered Meissner effect.

Meitner, Lise (1878–1968), German physicist. With O. Hahn, discovered protactinium; found evidence of four other radioactive elements; with Hahn and F. Strassmann, accomplished fission of uranium.

Mello, Craig C. (1960–), American molecular geneticist. Co-discoverer of RNA interference, gene silencing by double-stranded RNA; Nobel Prize, 2006.

Mendel, Gregor Johann (1822–1884), Austrian botanist. Formulated Mendel's laws of heredity, the foundation of genetics.

Mendeleev, Dmitri Ivanovich (1834–1907), Russian chemist. Formulated Mendeleev's periodic law and table of the elements; research on interatomic and intermolecular forces.

Menelaus of Alexandria (1st century), Greek mathe matician and astronomer. Founded spherical trigonometry.

Mercator, Gerhardus, real name Gerhard Kremer (1512–1594), Flemish geographer. Created a chart of the world (Mercator projection); made surveying instruments.

Mercer, John (1791–1866), English chemist. Invented the mercerizing process for cotton.

Merrifield, Robert Bruce (1921–2006), American biochemist. Developed methods of protein synthesis, including solid-phase peptide synthesis that produces proteins by assembling amino acids sequentially into peptide chains; Nobel Prize, 1984.

Mersenne, Marin (1588–1648), French physicist. Showed that pitch is proportional to frequency and calculated frequencies of musical notes; discovered Mersenne's law for vibrating strings, and similar relations for wind and percussion instruments.

Messier, Charles (1730–1817), French astronomer. Credited with discovering 21 comets; compiled a catalog of nebulae.

Metchnikoff, Élie (1845–1916), Russian-born French zoologist and bacteriologist. Work on cholera and immunology; Nobel Prize, 1908.

Meusnier de la Place, Jean Baptiste Marie Charles (1754–1793), French mathematician, physicist, and chemist. Derived Meusnier's theorem of curvature of surface curve; with A. L. Lavoisier, did research on analysis and synthesis of water.

Meyerhof, Otto Fritz (1884–1951), German physiologist. Studied the glycogen-lactic acid cycle of muscles; Nobel Prize, 1923.

Michaelis, Leonor (1875–1949), German-born American biochemist. Developed theory of kinetics of enzyme-catalyzed reactions.

Michel, Hartmut (1948–), German chemist. With J. Deisenhofer and R. Huber, elucidated the structure of a bacterial protein that performs photosynthesis; Nobel Prize, 1988.

Michelson, Albert Abraham (1852–1931), American physicist. Experimented on the velocity of light with S. Newcomb; invented the Michelson interferometer; performed, with E. W. Morley, an experiment to determine the Earth's motion through the ether; Nobel Prize, 1907.

Mie, Gustav (1868–1957), German physicist. Carried out rigorous electrodynamic calculation of Mie scattering; attempted to formulate theory of matter.

Miller, William Hallowes (1801–1880), British crystallographer and mineralogist. Introduced Miller indices for identifying crystallographic planes.

Millikan, Robert Andrews (1868–1953), American physicist. Determined an accurate value for Planck's constant; originated the "oil drop" experiment to measure electronic charge; work on x-rays and cosmic rays; Nobel Prize, 1923.

Milnor, John Willard (1931–), American mathematician. Proved that a seven-dimensional sphere can have several differential structures, opening up the new field of differential topology; contributions to algebraic K theory, differential geometry, and algebraic topology; Fields Medal, 1962.

Milstein, César (1927–2002), British immunologist. With Georges J. F. Köhler, discovered a laboratory technique for producing monoclonal antibodies, highly uniform immune bodies that are selective in responding to target substances; Nobel Prize, 1984.

Minkowski, Hermann (1864–1909), Russian-born German mathematician. Studied the mathematical basis of relativity, notably the concept of the space-time continuum.

Minot, George Richards (1885–1950), American physician. With W. P. Murphy, first to recognize the value of liver therapy for pernicious anemia; studied arthritis, cancer, and vitamin B deficiency; Nobel Prize, 1934.

Mitchell, Peter (1920–1992), British chemist. Explained how plant and animal cells store and transfer energy by creating protonic gradients in the oxidative and photosynthetic phosphorylation processes; Nobel Prize, 1978.

Möbius, August Ferdinand (1790–1868), German mathematician and astronomer. Founder of topology; developed the Möbius strip.

Mohl, Hugo von (1805–1872), German botanist. Worked on the anatomy and physiology of higher plant forms; discovered protoplasm.

Mohorovičić, Andrija (1857–1936), Yugoslav meteorologist and seismoiogist. Discovered Mohorovičić seismic discontinuity.

Mohr, Carl Friedrich (1806–1879), German chemist. Developed titration procedures, including use of Mohr's salt.

Mohr, Christian Otto (1835–1918), German civil engineer. Studied stresses and strains of bodies, and failure of materials; introduced Mohr's stress circle.

Mohs, Friedrich (1773–1839), German mineralogist. Developed Mohs scale of hardness.

Moissan, Ferdinand Frédéric Henri (1852–1907), French chemist. First to isolate fluorine; invented an electric furnace and used it to produce synthetic metal compounds and samples of less common metals; Nobel Prize, 1906.

Molina, Mario J. (1943–), American chemist. Demonstrated that chemically inert chlorofluorocarbon (CFC) could be transported up to the ozone layer and could react with ultraviolet light and deplete the ozone layer; Nobel Prize, 1995.

Mollier, Richard (1863–1935), German physicist and engineer. Presented properties of thermodynamic media in form of charts and diagrams; introduced concept of enthalpy and Mollier diagram.

Moniz, Antonio Egas (1874–1955), Portuguese neurosurgeon. Developed cerebral angiography; introduced the prefrontal lobotomy; Nobel Prize, 1949.

Monod, Jacques (1910–1976), French biologist. With F. Jacob, proposed the concepts of messenger ribonucleic acid and of the operon; Nobel Prize, 1965.

Montagnier, Luc (1932–), French virologist. Fundamental contributions on the nature of viruses, their effects on host organisms, and defense mechanisms. Co-discoverer of the human immunodeficiency virus (HIV); Nobel Prize, 2008.

Moody, Lewis Ferry (1880–1953), American hydraulic engineer. Made improvements in hydraulic turbines, pumps, and accessories.

Moore, Stanford (1913–1982), American biochemist. With W. H. Stein, developed technique for determining amino acid sequence in proteins, and applied it to ribonuclease; Nobel Prize, 1972.

Mordell, Louis Joel (1888–1972), American-born British mathematician. Worked in number theory; proved the finite basis theorem concerning the finite generation of the group of rational points on an elliptic curve; conjectured that there are only finitely many rational points on a curve of genus greater than 1 (the Mordell conjecture).

Morgagni, Giovanni Battista (1682–1771), Italian anatomist. Founded pathological anatomy; first to describe liver cirrhosis.

Morgan, Thomas Hunt (1866–1945), American geneticist, embryologist, and zoologist. Proposed the chromosome theory of heredity; Nobel Prize, 1933.

Morgan, William Wilson (1906–1994), American astronomer. Developed methods for investigating more precisely the structure of the Milky Way Galaxy and of other galaxies; collaborated in developing the Johnson-Morgan system of stellar magnitudes (with H. L. Johnson) and

Bautz-Morgan classification of galaxy clusters (with L. P. Bautz).

Mori, Shigefumi (1951–), Japanese mathematician. Worked in algebraic geometry, particularly on the classification of algebraic varieties of dimension three; Fields Medal, 1990.

Morley, Edward Williams (1838–1923), American chemist and physicist. Associated with A. A. Michelson in an experiment on ether drift; research on variations of atmospheric oxygen content.

Morse, Samuel Finley Breese (1791–1872), American inventor. Invented the receiving and sending instruments for the telegraph, and a code for sending messages.

Moseley, Henry Gwyn Jeffries (1887–1915), English physicist. Discovered Moseley's law for frequency of x-ray spectral lines.

Mössbauer, Rudolf Ludwig (1929–), German physicist. Discovered the property of recoilless resonance absorption, the ability of some nuclei to emit and absorb gamma rays without energy loss; Nobel Prize, 1961.

Mossotti, Ottaviano Fabrizio (1791–1863), Italian physicist. Developed theory of dielectrics, from which he derived the Clausius-Mossotti equation.

Mott, Nevill Francis (1905–1996), British physicist. Applied quantum mechanics to study of charged particle scattering; with R. W. Gurney, developed Gurney-Mott theory of photographic process; introduced fundamental concepts elucidating electronic properties of disordered materials; Nobel Prize, 1977.

Mottelson, Ben Roy (1926–), American-born Danish physicist. With A. Bohr, developed theory which unifies shell and liquid-drop models of atomic nucleus, and which explains nonspherical nuclei; Nobel Prize, 1975.

Muller, Hermann Joseph (1890–1967), American geneticist. Studied genetic mutation rates under natural and artificial conditions; discovered the effect of x-rays on mutation rate; Nobel Prize, 1946.

Müller, Johannes Peter (1801–1858), German physiologist and anatomist. Proposed the principle of specific nerve energies, concerning stimuli to sense organs; discovered the Müllerian duct, an early embryonic structure.

Müller, Karl Alexander (1927–), Swiss physicist. With J. G. Bednorz, discovered high-temperature superconductivity in copper oxide ceramic materials; Nobel Prize, 1987.

Müller, Paul Hermann (1899–1965), Swiss chemist. Discovered the insecticidal properties of DDT; Nobel Prize, 1948.

Mulliken, Robert Sanderson (1896–1986), American chemist. Applied principles of quantum mechanics to study of chemical bonding; with F. Hund, systematized electronic states of molecules in terms of molecular orbitals; Nobel Prize, 1966.

Mullis, Kary B. (1944–), American chemist. Invented the polymerase chain reaction (PCR) method used for studying DNA molecules; Nobel Prize, 1993.

Mumford, David Bryant (1937–), American mathematician. Worked in algebraic geometry, especially on problems of the existence and structure of varieties of moduli and on the theory of algebraic surfaces; Fields Medal, 1974.

Murad, Ferid (1936–), American pharmacologist. Analyzed the action of nitroglycerin and related vasodilating compounds, leading to the discovery that they release nitric oxide, which relaxes smooth muscle cells; his work, along with the research of Robert F. Furchgott and Louis J. Ignarro, led to the discovery of nitric oxide as a signaling molecule in the cardiovascular system; Nobel Prize, 1998.

Murchison, Roderick Impey (1792–1871), British geologist. Studied the order of rock formations in Great Britain; with A. Sedgwick, differentiated the Silurian and Devonian.

Murphy, William Parry (1892–1987), American physician. With G. R. Minot, first to suggest liver diet as a treatment for pernicious anemia; Nobel Prize, 1934.

Murray, Joseph (1919–), American physician. Performed the first successful transplant of a human organ, a kidney; with E. D. Thomas, helped define and then overcome the immunological mechanisms behind organ rejection; Nobel Prize, 1990.

Nambu, Yoichiro (1921–), American physicist (born in Japan). Work in the areas of theoretical physics, particle physics, and field theory, with particular interest in the problem of mass hierarchy of particles. Discoverer of the mechanism of spontaneous broken symmetry in subatomic physics; Nobel Prize, 2008.

Napier or Neper, John, Laird of Merchiston (1550–1617), Scottish mathematician. Invented the theory of logarithms and developed methods to compute them.

Nathans, Daniel (1928–1999), American biologist. Pioneered in the use of restriction enzymes to study the structure and functions of deoxyribonucleic acid (DNA) molecules; Nobel Prize, 1978.

Natta, Giulio (1903–1979), Italian chemist. Discovered stereospecific polymerization, making possible the production of new classes of macromolecules from inexpensive raw materials; Nobel Prize, 1963.

Navier, Claude Louis Marie Henri (1785–1836), French physicist and engineer. Studied analytical mechanics and its application to strength of materials, machines, and motion of solid and liquid bodies; formulated Navier-Stokes equations.

Néel, Louis Eugène Félix (1904–2000), French physicist. Proposed the theory of behavior of antiferromagnetic and other ferrimagnetic materials in which the cyrstal lattice is divided into one or more sublattices; Nobel Prize, 1970.

Neher, Erwin (1944–), German biophysicist. With B. Sakmann, using the "patch clamp" technique they developed, showed how individual ion channels control the passage of charged ions into and out of cells; Nobel Prize, 1991.

Nernst, Hermann Walther (1864–1941), German chemist. Proposed the heat theorem (third law of thermodynamics); determined the specific heat of solids at low temperatures; proposed the chain reaction theory in photochemistry; Nobel Prize, 1920.

Neumann, Carl Gottfried (1832–1925), German mathematician. Believed to be founder of logarithmic potentials; developed the potential theory.

Newcomb, Simon (1835–1909), American astronomer. With A. A. Michelson, determined the velocity of light; studied the motions of the Moon and planets.

Newton, Isaac (1642–1727), English mathematician. Proposed a dynamical theory of gravitation; discovered three basic laws of motion which are the foundation of practical mechanics; made discoveries in optics and mathematics.

Neyman, Jerzy (1894–1981), Russian-born American statistician. Developed methodology for making decisions based only on results of experiments or observations subject to chance.

Nicholson, Seth Barnes (1891–1963), American astronomer. Discovered four satellites of Jupiter; with E. Petit, invented a thermocouple to measure surface temperature of planets.

Nicol, William (1763–1851), Scottish physicist. Invented the Nicol prism for investigating the polarization of light.

Nicolle, Charles Jules Henri (1866–1936), French physician. Discovered the louse to be the transmission vector of typhus; Nobel Prize, 1928.

Nicomedes (3d century B.C.), Greek mathematician. Discovered the conchoid.

Nirenberg, Marshall Warren (1927–), American biochemist. Pioneered in deciphering genetic code; Nobel Prize, 1968.

Nishina, Yoshio (1890–1951), Japanese physicist. Pioneer in study of cosmic rays; with O. B. Klein, originated the Klein-Nishina formula.

Nobel, Alfred Bernhard (1833–1896), Swedish chemist and engineer. Invented dynamite and a blasting gelatin containing nitroglycerin; established the annual Nobel prizes.

Noguchi, Hideyo (1876–1928), Japanese bacteriologist. First to produce pure cultures of syphilis spirochetes; discovered the parasite of yellow fever.

Norrish, Ronald George Wreyford (1897–1978), British physical chemist. With G. Porter and colleagues, developed methods of flash photolysis and kinetic spectroscopy for the study of very fast reactions; Nobel Prize, 1967.

Northrop, John Howard (1891–1987), American biochemist. Isolated several enzymes and proved them to be proteins; isolated the first bacterial virus; established the chemical nature of enzymes and viruses; Nobel Prize, 1946.

Novikov, Sergi Petrovich (1938–), Russian mathematician. Worked in algebraic topology; proved the topological invariance of the Pontryagin classes of a differentiable manifold; studied the cohomology and homotopy of Thom spaces; Fields Medal, 1970.

Noyce, Robert Norton (1927–1990), American physicist, electronics engineer, and inventor. Participated in the invention of the integrated circuit.

Noyori, Ryoji (1938–), Japanese chemist. Developed chirally catalyzed hydrogenation reactions; Nobel Prize, 2001.

Nurse, Paul M. (1949–), British biologist. Identified, cloned, and characterized a key regulator of the cell cycle, cyclin-dependent kinase; Nobel Prize, 2001.

Nusselt, Ernst Kraft Wilhelm (1882–1957), German mechanical engineer and physicist. Used dimensional analysis to derive functional form of solutions to equations for heat flux in a flowing fluid.

Nüsslein-Volhard, Christiane (1942–), German developmental biologist. Using *Drosophila*, she and Eric Wieschaus identified and classified a small number of genes that are important in determining the body plan and the formation of body segments; their work, along with that of American development biologist Edward B. Lewis, led to the discovery of important genetic mechanisms which control early embryonic development; Nobel Prize, 1995.

Nyquist, Harry (1889–1976), Swedish-born American physicist and engineer. Discovered conditions necessary to keep feedback control circuits stable; determined Nyquist rate for communications channels.

Ochoa, Severe (1905–1993), Spanish-born American biochemist. Discovered a bacterial enzyme that synthesizes ribonucleic acid from nucleoside diphosphates; first to synthesize a ribonucleic acid; Nobel Prize, 1959.

Ockham, William of (ca. 1284–1347), English philosopher and theologian. Developed nominalist school of thought; postulated Ockham's razor.

Oersted, Hans Christian (1777–1851), Danish physicist, chemist, and electromagnetist. Discovered a fundamental principle of electromagnetism: a magnetic needle turns at right angles to an electric current.

Ohm, Georg Simon (1787–1854), German physicist. Discovered Ohm's law relating electrical resistance to voltage and current.

Okounkov, Andrei (1969–), Russian mathematician working in the United States. Major contributions bridging probability, representation theory, and algebraic geometry; Fields Medal, 2006.

Olah, George A. (1927–), American chemist. Made fundamental contributions in carbocation chemistry; Nobel Prize, 1994.

Olbers, Heinrich Wilhelm Matthias (1758–1840), German astronomer. Devised new method for computing cometary orbits; proposed Glbers' paradox.

Onsager, Lars (1903–1976), Norwegian-born American chemist. Laid the foundation of irreversible thermodynamics; contributed to theories of dielectrics, electrolytes, and cooperative phenomena; Nobel Prize, 1968.

Oort, Jan Hendrik (1900–1992), Dutch astronomer. Research on the structure and dynamics of the galactic system; investigated the origin of comets.

Oppenheimer, J. Robert (1904–1967), American physicist. Research on nuclear disintegration, quantum theory, cosmic rays, and relativity; directed production of the atomic bomb.

Orr, John Boyd, Baron (1880–1971). Scottish physiologist and nutritionist. Work on animal nutrition; pioneer in science of human nutrition; Nobel Peace Prize, 1949.

Osheroff, Douglas D. (1945–), American physicist. With D. M. Lee and R. C. Richardson, discovered superfluidity in helium-3; Nobel Prize, 1996.

Ostwald, Friedrich Wilhelm (1853–1932), German chemist born in Latvia. Researches on affinity and mass action; discovered the Ostwald dilution law; worked on the catalytic oxidation of ammonia; Nobel Prize, 1909.

Otto, Nikolaus August (1832–1891), German inventor. Built the first four-stroke internal combustion engine.

Paget, James (1814–1899), English surgeon and pathologist. Studied pathology of tumors and bone and joint diseases; described osteitis deformans (Paget's disease).

Palade, George Emil (1912–), Rumanian born American cytologist. Applied electron microscope and centrifuge techniques to study of ultrastructure of cells; discovered ribosomes; Nobel Prize, 1974.

Papanicolaou, George Nicholas (1883–1962), Greek-born American cytologist and anatomist. Developed the Papanicolaou test for diagnosis of uterine cervical and endometrial cancer.

Pappus, Alexandrinus (ca. 3d–4th century), Greek mathematician. Wrote *Mathematical Collection*, an account of Greek geometry; formulated Pappus' theorems.

Paracelsus, Philippus Aureolus, real name Theophrastus Bombastus von Hohenheim (1493–1541), Swiss physician. Emphasized use of chemicals in medicine; advocated that diseases are specific and require specific remedies.

Paré, Ambroise (1509–1590), French surgeon. Advocated the treatment of wounds by tying arteries with ligatures rather than by cauterization; proposed improvements in operating methods.

Parkinson, James (1755–1824), English physician and paleontologist. Described parkinsonism.

Parseval des Chenes, Marc Antoine (1755–1836), French mathematician. Introduced an equation from which the theorem now known as Parseval's theorem is derived.

Pascal, Blaise (1623–1662), French mathematician and physicist. Contributed to the geometry of conies; formulated Pascal's law, relating to the pressure of a liquid at rest; applied Pascal's triangle to the calculation of probabilities.

Paschen, Louis Carl Heinrich Friedrich (1865–1947), German physicist. Established Paschen's law; with E. Back, discovered Paschen-Back effect; verified predictions of relativistic fine structure made by Bohr-Sommerfeld theory.

Pasteur, Louis (1822–1895), French biologist. Founder of microbiology; discovered the role of bacteria in fermentation; discovered anaerobic bacteria; developed the pasteurization process; demonstrated the efficacy of vaccination, especially for rabies.

Patterson, Arthur Lindo (1902–1966), New Zealand-born American physicist and crystallographer. Developed Patterson-Harker method of x-ray diffraction analysis of crystal structure.

Paul, Wolfgang (1913–1993), German physicist. Invented the Paul trap, which uses radio-frequency radiation to hold ions in a small volume; Nobel Prize, 1989.

Pauli, Wolfgang (1900–1958), Austrian-born American physicist. Worked on quantum theory; formulated the Pauli exclusion principle; contributed to matrix mechanics; Nobel Prize, 1945.

Pauling, Linus Carl (1901–1994), American chemist. Applied quantum theory to chemistry; research on molecular structure and chemical bonds; contributed to electrochemical theory of valency; Nobel Prize, 1954; Nobel Peace Prize, 1963.

Pavlov, Ivan Petrovich (1849–1936), Russian pathologist. Discovered the nerve fibers affecting heart action and the secretory nerves of the pancreas; research on the physiology of digestive glands; studied conditioned reflexes; Nobel Prize, 1904.

Peano, Giuseppe (1858–1932), Italian mathematician. Pioneer in symbolic logic and foundations of mathematics; promoted axiomatic method in mathematics; formulated postulates for natural numbers.

Pearl, Raymond (1879–1940), American biologist and statistician. Applied statistics to the study of population changes; introduced logistic curve describing population growth.

Pearson, Karl (1857–1936), English applied mathematician, statistician, and biometrician. Pioneered in application of statistics to biology; introduced chi-square test.

Pedersen, Charles J. (1904–1989), American chemist. Developed the synthesis of cyclic polyethers known as crown ethers; Nobel Prize, 1987.

Peierls, Rudolf Ernst (1907–1995). German-born British physicist. Developed theory of heat conduction in nonmetallic crystals; with O. R. Frisch, calculated critical mass of uranium-235.

Peirce, Charles Santiago Sanders (1839–1914), American mathematician, logician, and physicist. Laid foundation for logical analysis of mathematics; contributed to probability theory.

Pelletier, Pierre Joseph (1788–1842), French chemist. Discovered quinine, strychnine, and other alkaloids.

Peltier, Jean Charles Athanase (1785–1845), French physicist. Discovered the Peltier effect in thermoelectricity.

Penrose, Roger (1931–), British mathematician and physicist. Developed twistor theory of space-time geometry; studied singularities in classical general relativity theory.

Penzias, Arno A. (1933–), American astrophysicist. With R. W. Wilson, discovered cosmic background radiation, confirming the big bang theory of the origin of the universe; Nobel Prize, 1978.

Perelman, Grigory (1966–), Russian mathematician. Major contributions to geometry and the understanding of the analytical and geometric structure of the Ricci flow. Proved the Thurston Geometrization Conjecture and recognized for his proof of the Poincaré Conjecture; Fields Medal (declined), 2006.

Perl, Martin L. (1927–), American physicist. Discovered the tau lepton, a fundamental particle; Nobel Prize, 1995.

Pérot, Jean Baptiste Gaspard Gustav Alfred (1863–1925), French physicist. With C. Fabry, developed Fabry-Pérot interferometer.

Perrin, Jean Baptiste (1870–1942), French physicist. Research on the particle nature of cathode rays; found values for Avogadro's number, thereby proving the existence of molecules; Nobel Prize, 1926.

Perutz, Max Ferdinand (1914–2002), Austrian-born British crystallographer and molecular biologist. Worked on the structure of hemoglobin; introduced the method of isomorphous replacement with heavy atoms into protein crystallography; Nobel Prize, 1962.

Petit, Alexis Thérèse (1791–1820), French physicist. With P. L. Dulong, formulated the law of constancy of atomic heats; devised methods for determining thermal expansion and specific heats of solids.

Pettit, Edison (1890–1962), American astronomer. Studied the Sun and formulated laws alleged to govern the movement of prominences; constructed the interference polarizing monochromator; with S. B. Nicholson, devised a sensitive thermocouple to measure the surface temperatures of planets.

Pfaff, Johann Friedrich (1765–1825), German mathematician. Developed theory of Pfaffian differential equations, which is basic to general solution of partial differential equations.

Pfeiffer, Richard Friedrich Johann (1858–1945), German bacteriologist. Discovered Pfeiffer's bacillus in influenza; described Pfeiffer's reaction for determination of cholera.

Phillips, William Daniel (1948–), American physicist. Developed a method of slowing and trapping atoms in an atomic beam by using an opposed laser beam and a magnetic trap, and cooled these atoms to temperatures lower than the previously calculated theoretical limits: Nobel Prize, 1997.

Piaget, Jean (1896–1980), Swiss psychologist. Elucidated development of cognitive functions in the child.

Picard, Charles Émile (1856–1941), French mathematician. Formulated Picard's theorem relating to functions.

Piccard, Auguste (1884–1962), Swiss physicist. Conducted a data-collecting exploration of the stratosphere in an airtight gondola of a balloon; constructed and tested a bathysphere for deep-sea exploration.

Pickering, Edward Charles (1846–1919), American astronomer. Invented the meridian photometer; pioneered in stellar spectroscopy.

Pierce, George Washington (1872–1956). American physicist and electronic engineer. Developed theoretical basis of electrical communications: developed Pierce oscillator; with A. E. Kennelly, discovered concept of motional impedance.

Pitzer, Kenneth Sanborn (1914–1997), American chemist. Pioneered in the development of useful approximations which made possible the calculation of chemical thermodynamic properties of broad classes of chemical substances.

Planck, Max Karl Ernst Ludwig (1858–1947), German physicist. Presented the quantum theory; introduced Planck's constant, or quantum of action.

Planté, Gaston (1834–1889), French physicist. Constructed a storage battery, the first primitive accumulator.

Plateau, Joseph Antoine Ferdinand (1801–1883), Belgian physicist. Experimented with soapy films bounded by wires, noting that the surfaces formed were minimal surfaces; from this he formulated the Plateau problem (the problem of determining the existence of a minimal surface with a given space curve as its boundary).

Podolsky, Boris (1896–1966), Russian-born American physicist. Collaborated in formulation of Einstein-Podolsky-Rosen paradox; research on quantum electrodynamics.

Poggendorff, Johann Christian (1796–1877), German physicist. Introduced the small mirror on a suspended system to magnify small deflections of a light beam; invented the galvanometer.

Poincaré, Jules Henri (1854–1912), French mathematician. Worked on the theory of functions, on differential equations, and on the theory of orbits in astronomy.

Poinsot, Louis (1777–1859), French mathematician. Originated theory of couples.

Poiseuille, Jean Léonard Marie (1797–1869), French physiologist and physicist. Studied physiology of arterial circulation; invented improved methods for measuring blood pressure; discovered Hagen-Poiseuille law independently of G. H. L. Hagen.

Poisson, Siméon Denis (1781–1840), French mathematician. Worked on mathematical physics; contributed to the wave theory of light; formulated the Poisson ratio concerning the elasticity of materials.

Politzer, H. David (1949–), Discovered asymptotic freedom in the strong interactions (independent of D. J. Gross and F. Wilczek). Nobel Prize, 2004.

Polanyi, John C. (1929–), Canadian chemist. Studied chemiluminescence, a phenomenon in which the energy states of excited molecules are revealed by their emission of light; Nobel Prize, 1986.

Pomeranchuk, Isaak Yakolevich (1913–1966), Soviet physicist. Showed that energy of cosmic-ray electrons reaching the atmosphere is limited by their radiation in Earth's magnetic field; proved the Pomeranchuk theorem for scattering cross sections.

Pons, Jean Louis (1761–1831), French astronomer. Discovered 37 comets, including Encke's comet.

Pople, John A. (1925–2004), British-born chemist. Developed computational methods in quantum chemistry; Nobel Prize, 1998.

Porro, Ignazio (1801–1875), Italian topographer, geodesist, and physicist. Invented optical surveying instruments, Porro prism erecting system, and modern prism binoculars.

Porter, George (1920–2002), British chemist. With R. G. W. Norrish, developed the technique of flash photolysis to initiate and record very fast chemical reactions; Nobel Prize, 1967.

Porter, Rodney Robert (1917–1985), British biochemist. Research to determine chemical structure of immunoglobulins; Nobel Prize, 1972.

Powell, Cecil Frank (1903–1969), British physicist. Made practical the use of photographic emulsions in nuclear research; with G. P. S. Occhialini and others, discovered and investigated production of pions from cosmic radiation in the Earth's atmosphere; Nobel Prize, 1950.

Poynting, John Henry (1852–1914), English physicist. Determined the constant of gravitation and explained why a comet's tail points away from the Sun.

Prandtl, Ludwig (1875–1953), German physicist. Contributed to fluid mechanics, particularly aerodynamics; introduced concept of boundary layer.

Pregl, Fritz (1869–1930), Austrian chemist. Developed microchemical methods of analysis; Nobel Prize, 1923.

Prelog, Vladimir (1906–1998), Yugoslavian-born Swiss chemist. Investigated stereochemistry of organic molecules and reactions; Nobel Prize, 1975.

Prevost, Pierre (1751–1839), Swiss physicist. Developed theory of exchanges, explaining nature of heat.

Priestley, Joseph (1733–1804), English chemist and physicist. Discovered oxygen, ammonia, oxides of nitrogen, hydrochloric acid gas, nitrogen, carbon monoxide, and sulfur dioxide.

Prigogine, Ilya (1917–2003), Soviet-born Belgian chemist. Contributed to nonequilibrium thermodynamics, particularly the theory of dissipative structures; Nobel Prize, 1977.

Prokhorov, Aleksandr Mikhailovich (1916–2002), Soviet physicist. With N. G. Basov, devised a new method for amplifying electromagnetic radiation; Nobel Prize, 1964.

Prout, William (1785–1850), English physician and chemist. Formulated Prout's hypothesis concerning atomic weights.

Prusiner, Stanley B. (1942–), American neurologist. Discovered prions, a new biological agent of infection; Nobel Prize, 1997.

Ptolemy (2d century), Greco-Egyptian astronomer, geographer, and geometer at Alexandria. Proposed the Ptolemaic system, with the Earth as the center of the universe.

Pupin, Michael (1858–1935), Yogoslavian-born American physicist and electrical engineer. Developed inductance coils for telephone lines; contributed to x-ray fluoroscopy, design of radio transmitters, and network theory.

Purcell, Edward Mills (1912–1997), American physicist. Developed the method of nuclear resonance absorption; Nobel Prize, 1952.

Purkinje, Johannes Evangelista (1787–1869), Czech physiologist. Discovered the Purkinje effect in eye physiology and Purkinje cells in the cerebral cortex.

Pythagoras (6th century B.C.), Greek mathematician. Originated a system of geometry, including the Pythagorean theorem.

Quételet, Lambert Adolphe Jacques (1796–1874), Belgian statistician. Did pioneer work on statistics; applied the calculus of probabilities to sociological studies.

Quillen, Daniel Gray (1940–), American mathematician. Developed algebraic K-theory, an extension of ideas of A. Grothendieck to commutative rings, which employed geometric and topological methods and ideas to formulate and solve major problems in algebra, particularly ring theory and module theory; Fields Medal, 1978.

Rabi, Isidor Isaac (1898–1988), Austrian-born American physicist. Research on neutrons, magnetism, quantum mechanics, and nuclear physics; Nobel Prize, 1944.

Radô, Tibor (1895–1965), Hungarian-American mathematician. Solved the Plateau problem (the problem of determining the existence of a minimal surface with a given space curve as its boundary) about the same time as J. Douglas.

Radon, Johann (1887–1956), Bobemian-born Austrian mathematician. Work in calculus of variations and integration theory.

Rainwater, Leo James (1917–1986), American physicist. Suggested that shell-model potentials of certain atomic nuclei are not spherical but are deformed into spheroids, and proposed mechanism for this distortion; Nobel Prize, 1975.

Raman, Chandrasekhara Venkata (1888–1970), Indian physicist. Research on diffraction and oscillation; discovered the Raman effect; Nobel Prize, 1930.

Ramón y Cajal, Santiago (1852–1934), Spanish histologist. Isolated the neuron and made discoveries concerning nerve cells in gray matter and the spinal cord; Nobel Prize, 1906.

Ramsay, William (1852–1916), British chemist. With J. W. S. Rayleigh, discovered argon; with M. W. Travers, discovered neon, krypton, and xenon; Nobel Prize, 1904.

Ramsden, Jesse (1735–1800), English mathematical-instrument maker. Invented an eyepiece containing cross-wires as a measuring scale; introduced equatorial mounting for telescopes.

Ramsey, Norman Foster (1915–), American physicist. Invented an accurate method of measuring differences between atomic energy levels that formed the basis for the cesium atomic clock; worked on the hydrogen maser; Nobel Prize, 1989.

Rankine, William John Macquorn (1820–1872), Scottish civil engineer. Contributed to thermodynamics and theories of elasticity and waves; wrote textbooks on the steam engine and civil engineering.

Raoult, François Marie (1830–1901), French chemist. Formulated Raoult's law concerning vapor pressure of a solution.

Rathke, Martin Heinrich (1793–1860), German biologist. Discovered gill slits and gill arches in embryo birds and mammals, and Rathke's pocket in developing vertebrates.

Ray or Wray, John (1627?–1706), English naturalist. Identified the difference between mono- and dicotyledons; arranged plants according to their natural form, the foundation of the natural system of classification.

Rayleigh, John William Strutt, 3d Baron (1842–1919), English physicist. Worked on the theory of sound and on physical optics; with W. Ramsay, discovered argon; Nobel Prize, 1904.

Réaumur, René Antoine Ferchault de (1683–1757), French entomologist. Worked in biology and metallurgy; invented the Réaumur thermometer scale.

Regge, Tullio (1931–), Italian physicist. Played a role in introducing the idea of complex angular momenta into elementary particle physics.

Regiomontanus, real name Johann Müller (1436–1476), German astronomer. Erected the first European observatory, in 1471 in Nürnberg; produced mathematical tables.

Reichstein, Tadeus (1897–1996), Polish-born Swiss organic chemist. Isolated about 30 of the 40 substances produced by the adrenal cortex; synthesized and described the structure and properties of many of these substances; Nobel Prize, 1950.

Reines, Frederick (1918–1998), American physicist. With C. L. Cowan, made first detection of the neutrino, a fundamental particle; Nobel Prize, 1995.

Reynolds, Osborne (1842–1912), British engineer and physicist. Demonstrated streamline and turbulent flow in pipes, and showed that transition between them occurs at a critical velocity determined by Reynolds' number; introduced Reynolds' analogy.

Riccati, Jacopo Francesco (1676–1754), Italian mathematician. Research on analysis, particularly differential equations, and geometry.

Ricci-Curbastro, Gregorio (1853–1924), Italian mathematician and mathematical physicist. Developed theory of tensor analysis, providing mathematical foundation for general relativity.

Richards, Dickinson Woodruff (1895–1973), American physician. With A. F. Cournand, uitlized the technique of cardiac catheterization and proved its value as a diagnostic tool; Nobel Prize, 1956.

Richards, Theodore William (1868–1928), American chemist. Worked on atomic weights; experimentally confirmed the existence of isotopes of lead from uranium and thorium; Nobel Prize, 1914.

Richardson, Owen Willans (1879–1959), English physicist. Studied the emission of electricity from hot bodies and the electron theory of matter; Nobel Prize, 1928.

Richardson, Robert Coleman (1937–), American physicist. With D. M. Lee and D. D. Osheroff, discovered superfluidity in helium-3; Nobel Prize, 1996.

Richet, Charles Robert (1850–1935), French physiologist. Studied serum therapy and discovered anaphylaxis; Nobel Prize, 1913.

Richter, Burton (1931–), American physicist. Independently of S. C. C. Ting, discovered a new heavy elementary particle, which he named the psi particle; Nobel Prize, 1976.

Richter, Jeremias Benjamin (1762–1807), German chemist. Discovered the law of equivalent proportions.

Riemann, Georg Friedrich Bernhard (1826–1866), German mathematician. Originated Riemannian geometry, a noneuclidean system.

Riesz, Frigyes or Frederic (1880–1956), Hungarian mathematician. Did research on abstract and general theories related to mathematical analysis, particularly functional analysis; independently of E. Fischer, discovered Riesz-Fisher theorem.

Righi, Augusto (1850–1920), Italian physicist. Discovered magnetic hysteresis and Righi-Leduc effect, independently of S. A. Leduc; demonstrated that microwaves have all properties characteristic of light waves.

Ritchey, George Wills (1864–1945), American astronomer. Made important astronomical observations, particularly on the Andromeda nebula; with H. Chrétien, developed Ritchey-Chrétien optics.

Ritz, Walter (1878–1909), Swiss-born German physicist. Introduced Ritz combination principle; developed Ritz method for numerical solution of boundary-value problems.

Robbins, Frederick Chapman (1916–2003), American microbiologist. Discovered that poliomyelitis virus can be grown in various human tissue cultures; Nobel Prize, 1954.

Roberts, Richard J. (1943–), British geneticist. Independently of P. Sharp, discovered split genes; Nobel Prize, 1993.

Robinson, Robert (1886–1975), English chemist. Worked on plant pigments, alkaloids, and phenanthrene derivatives; Nobel Prize, 1947.

Roche, Edouard Adelbert (1820–1883), French physicist, mathematician, and meteorologist. Studied the internal structure and free-surface form of the celestial bodies; applied results to study of cosmogonic hypotheses.

Rodbell, Martin (1925–1998), American pharmacologist. With A. Gilman, discovered G-proteins and the role of these proteins in cellular signal transduction; Nobel Prize, 1994.

Roebling, John Augustus (1806–1869), Prussian-born American engineer. Pioneering designer of bridges, in particular suspension bridges including the Brooklyn Bridge in New York.

Rohrer, Heinrich (1933–), Swiss physicist. With G. Binnig, developed scanning tunneling microscope; Nobel Prize, 1986.

Rolle, Michel (1652–1719), French mathematician. Worked on Diophantine analysis and algebra of equations.

Röntgen, Wilhelm Konrad (1845–1923), German physicist. Discovered x-rays; Nobel Prize, 1901.

Roscoe, Henry Enfield (1833–1915), English chemist. With R. W. Bunsen, evolved the law of reciprocity and invented the actinometer; first to isolate metallic vanadium.

Rose, Irwin (1926–), American biochemist. Discovered ubiquitin-mediated protein degradation; Nobel Prize, 2004.

Rosen, Nathan (1909–1995), American-born Israeli physicist. Collaborated in formulation of Einstein-Podolsky-Rosen paradox; research on general relativity and gravitational waves.

Ross, Ronald (1857–1932), British physician. Proved that malaria is transmitted by the female *Anopheles* mosquito; Nobel Prize, 1902.

Rossby, Carl Gustaf Arvid (1898–1957), Swedish-born American meteorologist. Formulated theories of large-scale air movements; derived the Rossby formula, relating speed of propagation of perturbations to airflow and wavelengths of perturbations; devised the Rossby diagram, used to plot air mass properties.

Roth, Klaus Friedrich (1925–), German-born British mathematician. Solved a problem previously studied by A. Thue and C. Siegel concerning the approximation to algebraic numbers by rational numbers; proved that a sequence with no three numbers in arithmetic progression has zero density; Fields Medal, 1958.

Rous, Francis Peyton (1879–1970), American physician and virologist. Produced cancer in chickens by inoculating them with filterable virus procured from tissue of chickens with tumors; Nobel Prize, 1966.

Routh, Edward John (1831–1907), British mathematical physicist. Made contributions to classical mechanics, including procedure for eliminating cyclic coordinates from equations of motion.

Roux, Pierre Paul Emile (1853–1933), French physician and bacteriologist. Helped develop modern serum therapeutics, especially concerning diphtheria.

Rowland, F. Sherwood (1927–), American chemist. Demonstrated that chemically inert chlorofluorocarbon (CFC) could be transported up to the ozone layer and could react with ultraviolet light and deplete the ozone layer; Nobel Prize, 1995.

Rowland, Henry Augustus (1848–1901), American physicist. Developed the Rowland grating in spectroscopy; studied electromagnetism and heat.

Rubbia, Carlo (1934–), Italian physicist. Principal architect of experiment that first detected intermediate vector bosons, an important step in confirming theory uniting electromagnetic and weak nuclear interactions; Nobel Prize, 1984.

Rubens, Heinrich (1865–1922), German physicist. With E. B. Hagen, conducted electromagnetic experiments; built new types of galvanometer and bolometer.

Rumford, Benjamin Thompson, Count (1753–1814), British physicist. Carried out research on heat.

Runge, Carl David Tolme (1856–1927), German mathematician and physicist. Research on theoretical and experimental spectroscopy, particularly data reduction and development of series formulas; developed methods for numerical and graphical computation, including Runge-Kutta method.

Ruska, Ernst (1906–1988), German electronic engineer. Developed the electron microscope; Nobel Prize, 1986.

Russell Bertrand Arthur William (1872–1970), English mathematician and philosopher. With A. N. Whitehead, pioneered in study of mathematical logic.

Russell, Henry Norris (1877–1957), American astronomer and physicist. Analyzed eclipsing binary stars: with E. Hertzsprung, introduced Hertzsprung-Russell diagram; determined abundance of chemical elements in solar atmosphere; with F. A. Saunders, devised theory of Russell-Saunders coupling.

Rutherford, Ernest, 1st Baron (1871–1937), British physicist. Discovered alpha, beta, and gamma rays; suggested the divisible nuclear atom; effected the transmutation of an atom: Nobel Prize, 1908.

Ružička, Leopold (1887–1976), Swiss chemist born in Croatia. Research on many-membered rings and higher terpenes (including male sex hormones); Nobel Prize, 1939.

Rydberg, Johannes Robert (1854–1919), Swedish physicist. Developed a formula for series of spectral lines, involving Rydberg's constant.

Ryle, Martin (1918–1984), British astronomer. Devised aperture synthesis method in radiotelescopy; designed equipment and made observations in radio astronomy; Nobel Prize, 1974,

Sabatier, Paul (1854–1941), French chemist. Discovered, with J. B. Senderens, the process for catalytic hydrogenation of oils to solid fat; Nobel Prize, 1912.

Sabin, Albert Bruce (1906–1993), Polish-born American physician and virologist. Studied nature, mode of transmission, and epidemiology of human poliomyelitis; developed oral polio virus vaccine.

Sabine, Edward (1788–1883), British physicist and astronomer. Headed a magnetic survey of the world which discovered a connection between sunspots and terrestrial magnetic disturbances.

Sabine, Wallace Clement Ware (1868–1919), American physicist. Pioneered in architectural acoustics; discovered law determining reverberation time in acoustics.

Sachs, Julius von (1832–1897), German botanist. Studied the connection between sunlight and chlorophyll; worked on heliotropism and geotropism.

Saha, Meghnad (1894–1956), Indian physicist. Developed theory for degree of ionization of hot gases, a basic component of modern astrophysics.

Sakmann, Bert (1942–), German physiologist. With E. Neher, using the "patch clamp" technique they developed, showed how individual ion channels control the passage of charged ions into and out of cells: Nobel Prize, 1991.

Salam, Abdus (1926–1996), Pakistani physicist. Independently of S. Weinberg, developed theory uniting two of the basic forces of nature, electromagnetism and the weak nuclear interactions; Nobel Prize, 1979.

Salk, Jonas Edward (1914–1995), American physician. Produced killed-virus vaccine effective in preventing poliomyelitis.

Salpeter, Edwin Ernest (1924–), Austrian-born American physicist. Research in quantum theory of atoms, quantum electrodynamics, nuclear theory, energy production of stars, and theoretical astrophysics; with H. A. Bethe, introduced Bethe-Salpeter equation.

Samuelsson, Bengt Ingemar (1934–), Swedish biochemist and medical scientist. Studied prostaglandin metabolism and the formation of prostaglandin from arachidonic acid; Nobel Prize, 1982.

Sanctorius, real name Santorio Santorio (1561–1636), Italian physician. Invented the clinical thermometer; experimented with metabolism.

Sanger, Frederick (1918–), English chemist. Determined the exact order of amino acids in insulin; first to establish amino acid sequence for a protein; developed methods for determining nucleotide sequences (independently of W. Gilbert), advancing the technology of DNA recombination; Nobel Prizes. 1958 and 1980.

Savart, Félix (1791–1841), French physicist. Helped formulate the Biot-Savart law in electromagnetism.

Schally, Andrew Victor (1926–), Polish-born American physiologist. With R. Guillemin, isolated and analyzed peptide hormones secreted in hypothalmic region of brain which control anterior pituitary hormone secretion; Nobel Prize, 1977.

Schawlow, Arthur Leonard (1921–1999), American physicist. Contributed to invention of laser; made numerous contributions to laser spectroscopy, particularly the development of Doppler-free spectroscopy; Nobel Prize, 1981.

Scheele, Karl Wihelm (1742–1786), Swedish chemist. Made many discoveries, including oxygen (independently of J. Priestley), chlorine, and glycerin; synthesized many organic acids.

Schiaparelli, Glovanni Virginio (1835–1910), Italian astronomer. Discovered the connection between comets and meteorites, and the "canals" of Mars.

Schiff, Hugo Josef (1834–1915). German-born Italian organic chemist. Discovered Schiff bases; devised Schiff test; devised an improved nitrometer.

Schmidt, Bernhard Voldemar (1879–1935), Estonian-born German astronomer. Invented Schmidt system for astronomical telescopes.

Schmidt, Erhard (1876–1959), German mathematician. Extended D. Hilbert's work on integral equations; formalized and developed concept of Hilbert space.

Schoenflies, Arthur Moritz (1853–1928), German mathematician and crystallographer. Classified the 230 crystallographic space groups.

Schottky, Walter (1886–1976), Swiss-born German physicist. Discovered Schottky effect; invented screen grid and tetrode; developed Schottky theory of semiconductor-metal junctions.

Schrieffer, John Robert (1931–), American physicist. With J. Bardeen and L. N. Cooper, formulated a theory of superconductivity: Nobel Prize, 1972.

Schrock, Richard R. (1945–), American chemist. Has made fundamental contributions to the field of organic synthesis, in particular through the development of catalysts enabling the metathesis method; Nobel Prize, 2005.

Schrödinger, Erwin (1887–1961), German physicist. Proposed concept of atomic structure based on wave mechanics; contributed to quantum theory and color theory; Nobel Prize, 1933.

Schur, Issai (1875–1941), Russian-born German mathematician. Contributed to representation theory of groups; research on group theory, matrices, algebraic equations, and number theory.

Schwartz, Laurent (1915–2002), French mathematician. Developed the theory of distributions, which provided an abstract and rigorous mathematical foundation for methods of formal calculation such as the Dirac delta function, and greatly extended their range of application; Fields Medal, 1950.

Schwartz, Melvin (1932–2006), American physicist. Collaborated in an experiment that demonstrated the existence of two types of neutrino; Nobel Prize, 1988.

Schwarz, Hermann Amandus (1843–1921), German mathematician. Introduced Schwarz-reflection principle and Schwarz's lemma while proving the Riemann mapping theorem.

Schwarzschild, Karl (1873–1916), German astronomer. Developed photographic methods for measuring brightness of stars; discovered Schwarzschild solution of equations of general relativity.

Schwarzschild, Martin (1912–1997), German-born American astronomer. Numerical studies of the internal structure and evolution of stars; astronomical observations with balloon-borne telescopes.

Schwinger, Julian Seymour (1918–1994), American physicist. Made fundamental contributions to the quantum theory of radiation; worked out the mathematical formalism of interaction between charged particles and an electromagnetic field; Nobel Prize, 1965.

Seaborg, Glenn Theodore (1912–1999), American chemist. Synthesized and identified eight transuranium elements and over a hundred isotopes; Nobel Prize, 1951.

Secchi, Pietro Angelo (1818–1878), Italian astronomer. Originated the spectroscopic survey of the heavens; made the first classification of stars according to spectral type.

Sedgwick, Adam (1785–1873), English geologist. With R. I. Murchison, established the Devonian system.

Seebeck, Thomas Johann (1770–1831), German physicist. Investigated thermoelectricity and invented the thermocouple.

Segrè, Emilio Gino (1905–1989), Italian-born American physicist. Codiscovered the elements technetium, astatine, and plutonium, slow neutrons, and the antiproton; Nobel Prize, 1959.

Seidel, Philipp Ludwig von (1821–1896), German astronomer and mathematician. Developed theory of aberrations; made first accurate photometric measurements of stars and planets, and evaluated them with probability theory.

Selberg, Atle (1917–2007), Norwegian-American mathematician. Worked on generalizations of the sieve methods of V. Brun; proved major results on zeros of the Riemann zeta function; with P. Erdös, gave an elementary proof of the prime number theorem, with a generalization to numbers in an arbitrary arithmetic progression; Fields Medal, 1950.

Semenov, Nikolai Nikolaevich (1896–1986), Soviet chemist. Elucidated the mechanisms of chemical reactions, especially the chain mechanism; Nobel Prize, 1956.

Senderens, Jean Baptiste (1856–1937), French chemist. With P. Sabatier, discovered hydrolysis of oils by catalysis.

Serber, Robert (1909–1997), American physicist. Laid foundations of orbit theory of high-energy particle accelerators; introduced Serber potential to describe nuclear forces.

Serre, Jean-Paul (1926–), French mathematician. Applied spectral sequences to discover fundamental connections between the homology groups and homotopy groups of a space and to prove important results on the homotopy groups of spheres; reformulated and extended some of the main results of complex variable theory in terms of sheaves; Fields Medal, 1954; Abel Prize, 2003.

Serret, Joseph Alfred (1819–1885), French mathematician and astronomer. Helped develop Frenet-Serret formulas in the theory of space curves.

Servetus, Michael (1511–1553), Spanish physician. Discovered the pulmonary circulation and the purification of the blood by the lungs.

Shannon, Claude Elwood (1916–2001), American mathematician. Developed mathematical theory of communication, making use of analogy between concepts of entropy and information.

Shapley, Harlow (1885–1972), American astronomer. Worked on a theory to explain cepheid variables; made an estimate of the size of the universe: provided a description of the universe's form of construction.

Sharp, Phillip A. (1944–), American geneticist. Independently of R. Roberts, discovered split genes; Nobel Prize, 1993.

Sharpless, K. Barry (1941–), American chemist. Developed chirally catalyzed oxidation reactions; Nobel Prize, 2001.

Sherrington, Charles Scott (1861–1952), English physiologist. Studied the neuron and its function and other aspects of the nervous system; Nobel Prize, 1932.

Shimomura, Osamu (1928–), American biochemist (born in Japan). Pioneering research in the field of bioluminescence; discoverer of green fluorescent protein (GFP) in the marine jellyfish *aequorea*; characterized GFP and its companion protein aequorin; Nobel Prize, 2008.

Shirakawa, Hideki (1936–), Japanese polymer scientist. Discovered and developed conductive polymers with Alan Heeger and Alan MacDiarmid; Nobel Prize, 2000.

Shockley, William (1910–1989), English-born American physicist. Discovered the transistor effect for electronic amplification by means of solid-state semiconductors; Nobel Prize, 1956.

Shor, Peter (1959–), American mathematician. Worked in combinatorial analysis and the theory of quantum computing; developed a computational method for factorizing large numbers on quantum computers, which, theoretically, could be used to break many of the coding systems currently employed.

Shubnikov, Aleksei Vasilevich (1887–1970), Soviet crystallographer. Classified Shubnikov groups; developed techniques for growing crystals, including synthetic rubies used in lasers.

Shull, Clifford G. (1915–2001), American physicist. Developed the neutron diffraction technique for studying the atomic structure of solids and liquids; Nobel Prize, 1994.

Siegbahn, Kai Manne Börje (1918–2007), Swedish physicist. Pioneered the development of high-resolution electron spectroscopy; Nobel Prize, 1981.

Siegbahn, Karl Manne Georg (1886–1978), Swedish physicist. Studied x-ray spectroscopy, in which he discovered the M series; Nobel Prize, 1924.

Siegel, Carl Ludwig (1896–1981), German mathematician. Worked on number theory, functions of one or several complex variables, and differential equations.

Siemens, Ernst Werner von (1816–1892), German engineer and electrician. Developed telegraphy and self-acting dynamo.

Siemens, William or Karl Wilhelm (1823–1883), German inventor in London. Made many inventions, including a differential governor, bathometer, dynamometer, and electric furnace.

Sikorsky, Igor Ivan (1889–1972), Ukrainian-born American aeronautical engineer. Accomplishments include the first multi-engine airplane and the first practical helicopter.

Simpson, Thomas (1710–1761), English mathematician. Formulated the Simpson rule for finding the area of a figure, given only a limited number of data.

Singer, Isadore Manual (1924–), American mathematician. Worked in global analysis, especially the theory of elliptic operators and their applications to topology and geometry, and in mathematical physics; collaborated with M. F. Atiyah in proving the index theorem; Abel Prize, 2004.

Skou, Jens C. (1918–), Dutch chemist. Discovered an ion-transporting enzyme—sodium, potassium-stimulated adenosine triphosphatase (Na^+, K^+-ATPase)—maintaining the balance of sodium and potassium ions in the living cell; Nobel Prize, 1997.

Slater, John Clarke (1900–1976), American physicist. Introduced Slater determinant describing many-electron systems; developed theory of magnetrons.

Smale, Stephen (1930–), American mathematician. Worked in differential topology, differential equations, and dynamical systems; proved that the sphere can be turned inside out and that the generalized Poincaré conjecture is valid for dimensions greater than 4; discovered strange attractors that lead to chaotic dynamical systems; Fields Medal, 1966.

Smalley, Richard E. (1943–2005), American chemist. Designed and built a special laser-supersonic cluster beam apparatus that was used in the discovery of fullerenes; Nobel Prize, 1996.

Smith, Hamilton Othanel (1931–), American geneticist. Isolated a restriction enzyme that cleaves deoxyribonucleic acid (DNA) molecules at a specific site; Nobel Prize, 1978.

Smith, Michael (1927–2000), Canadian chemist. Made fundamental contribution toward oligonucleotide-based, site-directed mutagenesis and its development for protein studies within DNA-based chemistry; Nobel Prize, 1993.

Smith, Robert (1689–1768), English physicist. Developed a particulate theory of light; developed geometric propositions for computing properties of optical systems; derived a special case of the Smith-Helmholtz law.

Smithies, Oliver (1925–), American geneticist (born in the United Kingdom). Made pioneering contributions to the development of gene targeting; co-discoverer of a technique enabling introduction of DNA material into cells mimicking a natural process known as homologous DNA recombination; Nobel Prize, 2007.

Smoot, George F. (1945–), American physicist. Principally responsible for the discovery of anisotropy (small variations in temperature in varying directions) in the cosmic microwave background radiation, which points to the beginnings of the aggregation of matter in the early universe; Nobel Prize, 2006.

Snell, George Davis (1903–1996), American immunogeneticist. Demonstrated the x-ray induction of mutational changes in a mammal; contributed to the study of immunological systems and to the development of transplant immunology; Nobel Prize, 1980.

Snell, Willebrod van Roijen (1591–1626), Dutch mathematician. Formulated Snell laws concerning angles of incidence and refraction; conceived the idea of measuring the Earth by triangulation.

Soddy, Frederick (1877–1956), English chemist. With E. Rutherford, developed theory of atomic disintegration of radioactive substances; research on isotopes; Nobel Prize, 1921.

Solvay, Ernest (1838–1922), Belgian industrial chemist. Developed the Solvay process for production of sodium carbonate.

Sommerfeld, Arnold (1868–1951), German physicist. Developed quantum theory, especially in its application to sprectral lines and the Bohr atomic model.

Sörensen, Sören Peter Lauritz (1868–1939), Danish biochemist. Did pioneer work on hydrogen ion concentration; invented the symbol pH.

Spemann, Hans (1869–1941), German zoologist. Studied embryonic development and discovered the organizer function of certain tissues; Nobel Prize, 1935.

Sperry, Roger Wolcott (1913–1994), American neuroscientist. Discovered the functional split between the left and right hemispheres of the brain; Nobel Prize, 1981.

Spörer, Gustav Friedrich Wilhelm (1822–1895), German astronomer. Observations of the sun and sunspots.

Stanley, Wendell Meredith (1904–1971), American biochemist. Discovered that a virus is a nucleoprotein and can be crystallized; Nobel Prize, 1946.

Stanton, Thomas Ernest (1865–1931), English engineer. Studied surface friction of fluids; built wind tunnel for wind velocity investigations; studied strength of materials, heat transmission, and lubrication.

Stark, Johannes (1874–1957), German physicist. Studied radiation and atomic theory; discovered the Stark effect on spectrum lines and the Doppler effect in canal rays; Nobel Prize, 1919.

Staudinger, Hermann (1881–1965), German chemist. Conceived and elaborated the explanation of phenomenon of polymerization; Nobel Prize, 1953.

Steenrod, Norman Earl (1910–1971), American mathematician. Worked in topology; introduced Steenrod algebra.

Stefan, Josef (1835–1893), Austrian physicist. Originated Stefan's (or Stefan-Boltzmann) law of blackbody radiation; proposed theory of diffusion of gases; studied gas conductivity.

Stein, William Howard (1911–1980), American biochemist. With S. Moore, developed technique for determining amino acid sequence in proteins, and applied it to ribonuclease; Nobel Prize, 1972.

Steinberger, Jack (1931–), German-born American physicist. Collaborated in an experiment that demonstrated the existence of two types of neutrino; Nobel Prize, 1988.

Steinmetz, Charles Proteus (1865–1923), German-born American electrical engineer. Developed complex number technique for analyzing alternating-current circuits; made numerous electrical inventions; applied mathematical methods to solution of electrical engineering problems.

Stern, Otto (1888–1969), German-born American physicist. Developed the molecular beam method and used it to prove directly the existence of the magnetic moment of atoms and nuclei and to measure their magnitudes; Nobel Prize, 1943.

Stieltjes, Thomas Jan (1856–1894), Dutch-born French mathematician. Developed analytic theory of continued fractions, and Stieltjes integral as a tool for their study.

Stirling, James (1692–1770), British mathematician. Discovered Stirling's formula and Stirling's interpolation formula.

Stokes, George Gabriel (1819–1903), British mathematician and physicist. Originated the idea of determining the chemical composition of the Sun and stars from their spectra; studied double refraction and electromagnetic waves.

Stone, Marshall Harvey (1903–1989), American mathematician. Studied structural aspects of mathematical situations having origins in classic problems of analysis, geometry, and logic.

Störmer, Carl Fredrik Mülertz (1874–1957), Norwegian mathematician and geophysicist. Studied atmospheric phenomena; discovered the Stormer cone concerning cosmic rays.

Störmer, Horst Ludwig (1949–), German-American physicist. With D. C. Tsui, discovered the fractional quantum Hall effect, a manifestation of a new form of quantum fluid with fractionally charged excitations; Nobel Prize, 1998.

Strassman, Fritz (1902–1980), German chemist. With O. Hahn and L. Meitner, discovered nuclear fission; research on uranium and thorium isotopes.

Strömgren, Bengt Georg Daniel (1908–1987), Swedish-born American astronomer. Developed theory of neubulae consisting of hydrogen ionized by hot stars.

Struve, Friedrich Georg Wilhelm von (1793–1864), German-born Russian astronomer. Authority on double stars and nebulae; one of the first to measure a stellar parallax.

Struve, Otto Wilhelm von (1819–1905), Russian astronomer. Discovered some 500 new double stars; calculated the constant of precession.

Sturgeon, William (1783–1850), English electrician and inventor. Constructed the first useful electromagnet and the first moving-coil galvanometer.

Sturm, Jacques Charles François (1803–1855), French mathematician. Formulated the Sturm theorems, concerning real roots of an equation.

Suhl, Harry (1922–), German-born American physicist. Discovered Suhl effect; invented Suhl amplifier; studied resonance in magnetic materials, superconductivity, and general theory of magnetism.

Sulston, John E. (1942–), British geneticist. Using the nematode *Caenorhabditis elegans* as a model system, discovered that specific cells in the cell lineage always die through programmed cell death. Described the visible steps in the cellular death process and demonstrated the first mutations of genes participating in programmed cell death. Nobel Prize, 2002.

Sumner, James Batcheller (1887–1955), American biochemist. First to isolate an enzyme in pure, crystalline form and characterize it as a protein; Nobel Prize, 1946.

Sutherland, Earl Wilbur, Jr. (1915–1974), American physiologist. Uncovered intermediary role of cyclic adenylic acid in the mechanism of hormone control over human metabolic activities; Nobel Prize, 1971.

Svedberg, Theodor (1884–1971), Swedish chemist. An authority on colloid chemistry (dispersed phase); developed a centrifuge for colloidal particles and protein molecules; Nobel Prize, 1926.

Sydenham, Thomas (1624–1689), English physician. Gave classic descriptions of gout, venereal disease, fevers, hysteria, and Sydenham's chorea.

Synge, Richard Laurence Millington (1914–1994), English chemist. Developed partition chromatography with A. J. P. Martin; Nobel Prize, 1952.

Szent-Györgyi, Albert von Nagyrapolt (1893–1986), Hungarian biochemist. Isolated vitamin C; research on combustion processes in plant and animal tissues, muscular contraction, and cell division; Nobel Prize, 1937.

Talbot, William Henry Fox (1800–1877), English inventor and mathematician. Invented the calotype photographic process.

Tamm, Igor Yevgenevich (1895–1971), Soviet physicist. With I. M. Frank, formulated the mathematical theory explaining the physical origin and properties of Cerenkov radiation; Nobel Prize, 1958.

Tanaka, Koichi (1959–), Japanese engineer. Developed soft desorption ionization method for mass spectrometric analyses of biological macromolecules; Nobel Prize, 2002.

Tao, Terence (1975–), Australian mathematician working in the United States. Major contributions to the fields of harmonic analysis, partial differential equations, combinatorics, and number theory; Fields Medal, 2006.

Tatum, Edward Lawrie (1909–1975), American biochemist and geneticist. Researched the relation of genes to biochemical reactions in bacterial, yeast, and mold cells; with G. W. Beadle, discovered the phenomenon of genetic recombination in bacteria; Nobel Prize, 1958.

Taube, Henry (1915–2005), American chemist. Elucidated the mechanisms of electron transfer reactions, especially in metal complexes; Nobel Prize, 1983.

Taylor, Brook (1685–1731), English mathematician. Formulated Taylor's theorem and worked on mathematics of physical problems.

Taylor, Geoffrey Ingram (1886–1975), British mathematician. Work in theoretical hydrodynamics, particularly turbulence and effect of rotation on fluid flow.

Taylor, Joseph Hooten, Jr. (1941–), American astronomer. With R. A. Hulse, discovered a binary pulsar and studied it to observe phenomena predicted by general relativity; Nobel Prize, 1993.

Taylor, Richard Edward (1929–), American physicist. Collaborated in experiments that demonstrated that protons, neutrons, and similar particles are made up of quarks; Nobel Prize, 1990.

Taylor, Richard Lawrence (1962–), British mathematician. Assisted A. Wiles in the proof of Fermat's last theorem; collaborated with C. Breuil, B. Conrad, and F. Diamond in extending this work by proving the Taniyama-Shimura conjecture on elliptic curves.

Teisserenc de Bort, Léon Philippe (1855–1913), French meteorologist. Discovered the stratosphere.

Teller, Edward (1908–2003), Hungarian-born American physicist. With associates, developed the concept which led to the construction of the first hydrogen bomb; with G. Gamow, proposed the Gamow-Teller interaction and Gamow-Teller selection rules.

Temin, Howard Martin (1934–1994), American virologist. Proposed that genetic information is transferred from ribonucleic acid tumor viruses to deoxyribonucleic acid; independently of D. Baltimore, discovered reverse transcriptase; Nobel Prize, 1975.

Tesla, Nikola (1856–1943), American inventor born in what is now Croatia of Serbian parents. Invented the induction motor, a high-frequency electric coil; improved design of dynamos, transformers, and electric bulbs.

Thales (ca. 640-ca. 546 B.C.), Greek mathematician and astronomer. First to scientifically predict an eclipse of the Sun; discovered static electricity; credited with formulating several theorems.

Theiler, Max (1899–1972), South African physician and virologist. Developed a vaccine to prevent human yellow fever; Nobel Prize, 1951.

Theorell, Axel Hugo Teodor (1903–1982), Swedish biochemist. Made discoveries concerning the nature and mode of action of oxidative enzymes; Nobel Prize, 1955.

Thiele, F. K. Johannes (1865–1918), German chemist. Research on nitrogen compounds and the theory of unsaturated organic molecules.

Thom, René (1923–2002), French mathematician. Invented and developed the theory of cobordism in algebraic topology, a classification of manifolds that used homotopy theory in a fundamental way; developed catastrophe theory; Fields Medal, 1958.

Thomas, Edward Donnall (1920–), American physician. Performed the first successful transfer of bone marrow from one individual to another; with J. Murray, helped define and then overcome the immunological mechanisms behind organ rejection; Nobel Prize, 1990.

Thomas, Llewellyn Hilleth (1903–1992), English-born American physicist. Discovered Thomas precession; with E. Fermi, developed Thomas-Fermi atomic model; developed basic theory for Thomas cyclotron.

Thompson, John Griggs (1932–), American-British mathematician. Worked on theory of finite groups; with W. Feit, proved that all noncyclic finite simple groups have even order; determined the finite simple groups whose proper subgroups are solvable (minimal simple groups); Fields Medal, 1970. Nobel prize, 2008.

Thomson, George Paget (1892–1975), English physicist. Discovered, independently of C. J. Davisson, the diffraction of electrons by crystals; Nobel Prize, 1937.

Thomson, Joseph John (1856–1940), English physicist. Discovered that cathode rays consist of negatively charged particles, or electrons; Nobel Prize, 1906.

't Hooft, Gerardus (1946–), Dutch physicist. With M. J. G. Veltman, elucidated the quantum structure of the electroweak interactions, placing the theory of these interactions and similar particle physics theories on a firmer mathematical foundation by showing how they may be used for precise calculations of physical quantities; Nobel Prize, 1999.

Thouless, David James (1934–), British physicist. Studied many-body problem and its applications to nuclear and condensed matter physics, including phase transitions in superfluid helium films and electrons in disordered systems.

Thue, Axel (1863–1922), Norwegian mathematician. Worked on number theory, especially algebraic numbers and Diophantine equations.

Thurston, William Paul (1946–), American mathematician. Advanced the study of topology in two and three dimensions, showing relationships between analysis, topology, and geometry; suggested that a very large class of closed three-manifolds carry a hyperbolic structure; Fields Medal, 1982.

Timoshenko, Stephen P. (1878–1972), Ukrainian-born engineer who worked in several countries including Russia. Ukraine, the United States, and Germany. Considered a pioneer in engineering mechanics, whose contributions included beam theory as well as numerous standard textbooks. The first recipient of the Timoshenko Medal in applied mechanics of the American Society of Mechanical Engineers.

Tinbergen, Nikolaas (1907–1988), Dutch-born British zoologist. Pioneered in study of social behavior of animals and their responses to complex stimuli; conducted experimental studies of the effects of selection pressures and evolutionary response to them; Nobel Prize, 1973.

Ting, Samuel Chao Chung (1936–), American physicist. Independently of B. Richter, discovered a new heavy elementary particle, which he named the J particle; Nobel Prize, 1976.

Tiselius, Arne Wilhelm Kaurin (1902–1971), Swedish biochemist. Research on electrophoresis and absorption analysis; made discoveries concerning the complex nature of serum proteins; Nobel Prize, 1948.

Tits, Jacques (1930–), French mathematician (born in Belgium). Fundamental contributions to the field of group theory; Abel Prize, 2008.

Todd of Trumpington, Alexander Robertus Todd, Baron (1907–1997), British chemist. Worked on the structure and synthesis of nucleotides, and nucleotide coenzymes, and the related problem of phosphorylation; Nobel Prize, 1957.

Tomlinson, Ray (1941–), American computer engineer. Invented electronic mail, including the use of @ in the address.

Tomonaga, Sin-Itiro (1906–1979), Japanese physicist. Showed the modern theory of quantum electrodynamics to be quantitatively consistent with observed physical phenomena; Nobel Prize, 1965.

Tonegawa, Susumu (1939–), Japanese immunologist. Discovered how a limited number of genes are capable of producing a vast number of diverse antibodies, each designed for a specific invading foreign substance; Nobel Prize, 1987.

Torricelli, Evangelista (1608–1647), Italian physicist. Invented the mercury barometer.

Townes, Charles Hard (1915–), American physicist. Invented the maser; Nobel Prize, 1964.

Townsend, John Sealy Edward (1868–1957), British physicist. Developed collision theory of ionization of gases in an electric field.

Travers, Morris William (1872–1961), English chemist. Discovered, with W. Ramsay, krypton, xenon, and neon; investigated low-temperature phenomena.

Ts'ai Lun (fl. 105), Chinese inventor. Invented paper.

Tsien, Roger Y. (1952–), American cell biologist. Pioneering research on the visualization of cellular phenomena through the design and application of fluorescent and photolabile molecules. Recognized for contributions to the understanding of how green fluorescent protein (GFP) fluoresces; Nobel Prize, 2008.

Tsui, Daniel Chee (1939–), Chinese-born American physicist. With H. L. Störmer, discovered the fractional quantum Hall effect, a manifestation of a new form of quantum fluid with fractionally charged excitations; Nobel Prize, 1998.

Turing, Alan Mathison (1912–1954), English mathematician. Developed concept of Turing machine; research on mathematical logic, group theory, and computer technology.

Tychonoff (Tikhonov), Andrei Nikolaevich (1906–1993), Soviet mathematician and geophysicist. Proved Tychonoff theorem in topology and introduced concept of Tychonoff space.

Tyndall, John (1820–1893), British physicist. Studied temperature waves in metals and diathermancy of gases; discovered the effect of atmospheric density on sound transmission.

Uhlenbeck, George Eugene (1900–1988), Javanese-born American physicist. With S. Goudsmit, developed hypothesis of electron spin.

Urey, Harold Clayton (1893–1981), American chemist. Isolated heavy water and thus discovered the heavy isotope of hydrogen; Nobel Prize, 1934.

Urysohn, Pavel Samuilovich (1898–1924), Soviet mathematician. Proved Urysohn's lemma in topology.

Van Allen, James Alfred (1914–2006), American physicist. Discovered that the Earth is circled by two high-energy radiation belts, leading to major revisions in concepts of the Earth's atmosphere and magnetic field.

Van de Graaff, Robert Jemison (1901–1967), American physicist. Contributed to the development of the direct particle accelerator and invented the electrostatic belt generator.

van der Meer, Simon (1925–), Dutch physicist. Devised method to ensure frequent and efficient collision of accelerated protons and antiprotons in the superproton synchrotron at CERN, contributing to discovery of intermediate vector bosons; Nobel Prize, 1984.

Vandermonde, Alexandre Théophile (1735–1796), French mathematician. Gave first logical exposition of theory of determinants; developed methods to test solvability of algebraic equations.

van der Waals, Johannes Diderik (1837–1923), Dutch physicist. Formulated van der Waals equation; investigated van der Waals forces, concerning intermolecular attraction; Nobel Prize, 1910.

Vane, John Robert (1927–2004), English pharmacologist. Discovered prostaglandin X (prostacyclin) and the role of aspirinlike drugs as blocking agents in the prostaglandin synthesis; Nobel Prize, 1982.

van Rhijn, Pieter Johannes (1886–1960), Dutch astrophysicist. With J. C. Kapetyn, evolved a theory of the universe.

van't Hoff, Jacobus Hendricus (1852–1911), Dutch chemist. Pioneered in the study of stereochemistry; studied reaction rates, thermodynamics applied to chemistry, and the theory of dilute solutions; Nobel Prize, 1901.

Van Vleck, Jan Hasbrouck (1899–1980), American mathematical physicist. Pioneer in the development of the modern quantum-mechanical theory of magnetism; Nobel Prize, 1977.

Varadhan, Srinivasa S. R. (1940–), American mathematician (born in India). Fundamental contributions to probability theory; created a unified theory of large deviations; also recognized for his work on diffusion proceeses processes and gradient models; Abel Prize, 2007.

Varmus, Harold (1939–), American researcher in molecular virology and oncogenesis. With J. M. Bishop, researched the genetic basis of human cancers; their work led to the identification of over 50 cellular genes that can become oncogenes; Nobel Prize, 1989.

Vauquelin, Louis Nicola (1763–1829), French chemist. Discovered chromium and its compounds and beryllium compounds.

Vega, George, Baron von (1756–1802), Austrian mathematician. Prepared logarithmic tables.

Veltman, Martinus J. G. (1931–), Dutch physicist. With G. 't Hooft, elucidated the quantum structure of the electroweak interactions, placing the theory of these interactions and similar particle physics theories on a firmer mathematical foundation by showing how they may be used for precise calculations of physical quantities; Nobel Prize, 1999.

Verdet, Marcel Emile (1824–1866), French physicist. Determined dependence of Faraday effect on magnetic field strength, wavelength of the light, and index of refraction of the material.

Vernier, Pierre (1580–1637), French technician. Invented the Vernier scale.

Vesalius, Andreas (1514–1564), Belgian anatomist in Italy. Known as the father of modern anatomy; corrected many of Galen's mistaken doctrines.

Viète, François, or Franciscus Vieta (1540–1603), French mathematician. Worked on the solution of equations up to the fourth degree and laid the foundation of modern algebra.

Virtanen, Artturi Ilmari (1895–1973), Finnish biochemist. Research on problems of human nutrition and agriculture; investigated acidity (pH) and biological nitrogen fixation; Nobel Prize, 1945.

Voevodsky, Vladimir (1966–), Russian mathematician. Developed a new cohomology theory for algebraic varieties, which represented an important advance in number theory and algebraic geometry. Fields Medal, 2002.

Vogel, Hermann Wilhelm (1834–1898), German photochemist. Invented the orthochromatic photographic plate and designed a photometer.

Voigt, Woldemar (1850–1919), German physicist. Introduced transformation equations (later known as Lorentz transformations).

Volta, Alessandro, Count (1745–1827), Italian physicist. Invented the voltaic pile; developed the theory of current electricity.

Volterra, Vito (1860–1940), Italian mathematician. Developed method of solving Volterra equations; pioneered in developing functional analysis.

Von Braun, Wernher (1912–1977), German-born American rocket engineer. Directed development of the German V-2 and Wassefall missiles; instrumental in launch of *Explorer 1*, first American artificial satellite; supervised development of Saturn rockets for the Apollo program.

Von Euler, Ulf Svante (1905–1983), Swedish physiologist. Identified norepinephrine as neurotransmitter of sympathetic nervous system; isolated and characterized norepinephrine storage granules in nerves; Nobel Prize, 1970.

von Kármán, Theodore (1881–1963), American aerodynamicist. Theoretical contributions to aerodynamics; formulated von Kármán's theory of vortex streets, an early step in the mathematical treatment of turbulent motion.

von Neumann, John (1903–1954), Hungarian-born American mathematician. Research in logic, theory of quantum mechanics, theory of high-speed computing machines, and mathematical theory of games and strategy.

Wagner von Jauregg (Wagner-Jauregg), Julius (1857–1940), Austrian psychiatrist. Developed use of malarial infection to treat general paresis; Nobel Prize, 1927.

Waksman, Selman Abraham (1888–1973), Russian-born American bacteriologist. Isolated the antibiotic streptomycin; Nobel Prize, 1952.

Wald, George (1906–1997), American biologist and biochemist. Discovered the role of vitamin A in vision; Nobel Prize, 1967.

Walker, John E. (1941–), British chemist. Made major contributions toward elucidating the enzymatic mechanism underlying the synthesis of adenosine triphosphate (ATP) by clarifying the structural conditions of the enzyme; Nobel Prize, 1997.

Wallace, Alfred Russel (1823–1913), English naturalist. Originated, independently of C. Darwin, theory of natural selection; postulated Wallace's line regarding geographical distribution of animals.

Wallach, Otto (1847–1931), German chemist. Research on essential oils and the terpenes; Nobel Prize, 1910.

Wallis, John (1616–1703), English mathematician. Worked on algebraic curves, interpolation, evaluation of integrals, infinite series, mechanics, and algebra; derived Wallis product and Wallis formulas.

Walton, Ernest Thomas Sinton (1903–1995), British physicist. With J. D. Cockcroft, devised high-voltage apparatus capable of producing fast atomic particles with energies up to 700,000 electronvolts; showed the capability of these particles to disintegrate many light elements; Nobel Prize, 1951.

Wannier, Gregory Hugh (1911–1983), Swiss-born American physicist. Developed harmonization of localized and nonlocalized descriptions of electrons in solids.

Warburg, Otto Heinrich (1883–1970), German physiologist. Worked on chemistry of respiration and on cancer; Nobel Prize, 1931.

Warren, J. Robin (1937–), Australian pathologist. Co-discoverer of the bacterium *Helicobacter pylori* and its role in gastritis and peptic ulcer disease; Nobel Prize, 2005.

Wassermann, August von (1866–1925), German physician. Discovered the Wassermann test for the detection of syphilis.

Watson, James Dewey (1928–), American biochemist. With F. H. C. Crick, determined the double-helix structure of deoxyribonucleic acid; Nobel Prize, 1962.

Watson, John Broadus (1878–1958), American psychologist. Founded the behaviorist school of psychology.

Watt, James (1736–1819), Scottish inventor. Improved the steam engine, making it a commercial success.

Weber, Ernst Heinrich (1795–1878), German anatomist and physiologist. Discovered Weberian apparatus; applied hydrodynamics to study of blood circulation; discovered inhibitory power of vagus nerve; proposed Weber's law of stimuli.

Weber, Heinrich (1842–1913), German mathematician. Introduced Weber differential equation; demonstrated Abel theorem in its most general form; proved that absolute Abelian fields are cyclotomic.

Weber, Wilhelm Eduard (1804–1891), German physicist. Devised instruments for measurement of electrical and magnetic quantities; formulated absolute electrical and magnetic units.

Wegener, Alfred Lothar (1880–1930), German geologist. Presented the idea of continental drift.

Welerstrass, Karl Theodor (1815–1897), German mathematician. Worked on the theory of functions and on the calculus of variations.

Weil, Adolf (1848–1916), German physician. Gave classic description of Weil's disease.

Weil, André (1906–1998), French-born American mathematician. Laid the foundations for abstract algebraic geometry and the modern theory of algebraic varieties, starting a rapid advance in both algebraic geometry and number theory.

Weinberg, Steven (1933–), American physicist. Independently of A. Salam, developed theory uniting two of the basic forces of nature, electromagnetism and the weak nuclear interactions; Nobel Prize, 1979.

Weismann, August (1834–1914), German biologist. Contributed to the theory of heredity, which he attributed to variations in "germ-plasm."

Weiss, Pierre (1865–1940), French physicist. Developed phenomenological theory of ferromagnetism.

Weizsäcker, Carl Friedrich von (1912–2007), German physicist. Helped develop method for calculating bremsstrahlung in high-energy collisions; developed a theory of origin of solar system.

Weller, Thomas Huckle (1915–2008), American virologist and parasitologist. Isolated the virus of chickenpox and herpes zoster and proved the common etiology of the two diseases; first to propagate German measles virus; Nobel Prize, 1954.

Wentzel, Gregor (1898–1978), German-born American physicist. Helped develop Wentzel-Kramers-Brillouin method; research on theory of atomic spectra, wave mechanics, quantum electrodynamics, meson field theories, and statistical mechanics of many-body problems, especially superconductivity.

Werner, Alfred (1866–1919), Swiss chemist. Formulated the coordination theory of valency; Nobel Prize, 1913.

Werner, Wendelin (1968–), French mathematician (born in Germany). Research in mathematical physics and probability theory; honored for contributions to the development of stochastic Loewner evolution, the geometry of two-dimensional Brownian motion, and conformal field theory; Fields Medal, 2006.

Westinghouse, George (1846–1914). American engineer and industrialist. Important contributions to railroad technology, such as the invention of the air brake, and in the generation and distribution of electricity.

Weyl, Hermann (1885–1955), German-born American mathematician and mathematical physicist. Basic research on group representations and Riemann surfaces.

Wheatstone, Charles (1802–1875), English physicist and inventor. Conducted experiments on sound; invented Wheatstone's bridge, an instrument for comparing electrical resistances.

Wheeler, John Archibald (1911–2008), American physicist. Introduced the concepts of the scattering matrix and resonating group structure into nuclear physics; with N. Bohr, elucidated the mechanism of nuclear fission and predicted the fissibility of plutonium.

Whewell, William (1794–1866), British astronomer. Work on the tides, and on the history and philosophy of science.

Whipple, George Hoyt (1878–1976), American pathologist. Studied anemia and liver treatment; Nobel Prize, 1934.

Whitehead, Alfred North (1861–1947), English mathematician, physicist, and philosopher. With B. Russell, pioneered in mathematical logic and foundations of mathematics.

Whittaker, Edmund Taylor (1873–1956), British mathematician and physicist. Studied special functions of mathematical physics and equations satisfied by them, particularly Whittaker's differential equation; found general integral representation for harmonic functions; made major contributions to analytical dynamics.

Wiedemann, Gustave Heinrich (1826–1899), German physicist and physical chemist. With R. Franz, discovered Wiedemann-Franz law of thermal conductivity of metals; discovered Wiedemann effect.

Wieland, Heinrich (1877–1957), German chemist. Studied bile acids, chlorophyll, and hemoglobin; Nobel Prize, 1927.

Wieman, Carl Edwin (1951–), American physicist. With E. A. Cornell, succeeded in producing Bose-Einstein condensates in a dilute gas of alkali (rubidium) atoms, and carried out fundamental studies of their properties, including studies of collective excitations and vortex formation in condensates; Nobel Prize, 2001.

Wien, Wilhelm (1864–1928), German physicist. Formulated the two Wien laws pertaining to radiation from blackbodies; Nobel Prize, 1911.

Wiener, Norbert (1894–1964), American mathematician. Formulated a mathematical theory of Brownian motion; founded science of cybernetics.

Wieschaus, Eric F. (1947–), American developmental biologist. Using *Drosophila*, he and the German developmental biologist Christiane Nüsslein-Volhard identified and classified a small number of genes that are important in determining the body plan and the formation of body segments; their work, along with that of American development biologist Edward B. Lewis, led to the discovery of important genetic mechanisms which control early embryonic development; Nobel Prize, 1995.

Wiesel, Torsten Nils (1924–), Swedish physiologist. Contributed to the study of the processing of visual information in the brain; Nobel Prize, 1981.

Wigner, Eugene Paul (1902–1995). Hungarian-born American mathematical physicist. With G. Breit, worked out the Breit-Wigner formula for resonant nuclear reactions; proposed the Wigner theorem of conservation of the angular momentum of electron spin; Nobel Prize, 1963.

Wilczek, Frank (1951–), Discovered asymptotic freedom in the strong interactions (in collaboration with D. J. Gross, and independent of H. D. Politzer). Nobel Prize, 2004.

Wiles, Andrew John (1953–), British mathematician. Proved Fermat's last theorem with assistance of R. Taylor.

Wilkins, Maurice Hugh Frederick (1916–2004), English biophysicist born in New Zealand. Made x-ray diffraction studies that contributed to the structural determination of deoxyribonucleic acid; Nobel Prize, 1962.

Wilkinson, Geoffrey (1921–1996), British chemist. Research to determine how metals and organic molecules combine to form unique molecules which have sandwichlike structures; Nobel Prize, 1973.

Williamson, William Crawford (1816–1895), English naturalist. Laid the foundation for paleobotany and showed the importance of plant life forms in coal.

Willstätter, Richard (1872–1942), German chemist. Worked on plant pigments; investigated alkaloids and their derivatives; Nobel Prize, 1915.

Wilson, Charles Thomson Rees (1869–1959), British physicist. Worked on ionization; originated the cloud chamber method of studying ionized particles; Nobel Prize, 1927.

Wilson, Kenneth Geddes (1936–), American physicist. Used renormalization group theory to analyze critical phenomena in the behavior of matter at phase transitions; Nobel Prize, 1982.

Wilson, Robert Woodrow (1936–), American astrophysicist. With A. A. Penzias, discovered cosmic background radiation, confirming the big bang theory of the origin of the universe; Nobel Prize, 1978.

Windaus, Adolf (1876–1959), German chemist. Worked on sterols; discovered that ultraviolet light activates ergosterol and gives vitamin D_2; Nobel Prize, 1928.

Witten, Edward (1951–), American mathematical physicist. Applied advanced mathematical tools to theoretical physics, particularly quantum field theory, supersymmetry, and string theory; his physical insights were the basis for major developments in mathematics; Fields Medal, 1990.

Wittig, Georg (1897–1987), German chemist. Work on the linking of carbon and phosphorus (Wittig reaction) made it possible to synthesize new types of compounds, including metal-organic complex compounds; Nobel Prize, 1979.

Wöhler, Friedrich (1800–1882), German chemist. First to synthesize an organic compound, urea.

Wolf, Maximilian Franz Joseph Cornelius (1863–1932), German astronomer. Invented the photographic method of discovering asteroids.

Wollaston, William Hyde (1766–1828), English chemist and physicist. Discovered the lines in the solar spectrum; discovered palladium and rhodium; invented the Wollaston lens.

Woodward, Robert Burns (1917–1979), American chemist. Contributed to the development of total synthesis of complex natural products, and structural determination of several complex natural molecules, later confirmed by total synthesis; Nobei Prize, 1965.

Wright, Almroth Edward (1861–1947), British physician and pathologist. Studied parasitic disease; introduced inoculation against typhoid.

Wright, Wilbur (1867–1912) **and Orville** (1871–1948), American pioneers in aviation. Built the first successful airplane, which each flew at Kitty Hawk, North Carolina, on Dec. 17, 1903.

Wundt, Wilhelm Max (1832–1920), German physiologist and psychologist. Founded the first laboratory for experimental psychology.

Wurtz, Charles Adolphe (1817–1884), French chemist. Discovered methyl and ethyl amines; evolved the Wurtz reaction for synthesis of hydrocarbons.

Wüthrich, Kurt (1938–), Swiss chemist. Developed nuclear magnetic resonance spectroscopy for determining the three-dimensional structure of biological macromolecules in solution; Nobel Prize, 2002.

Yalow, Rosalyn Sussman (1921–), American medical physicist. Developed a radioimmunoassay technique to detect and measure minute levels of substances such as hormones in the body; Nobel Prize, 1977.

Yang, Chen Ning (1922–), Chinese-born American physicist. With T. Lee, disproved the law of conservation of parity for weak interactions; Nobel Prize, 1957.

Yau, Shing-Tung (1949–), Chinese-born American mathematician. Worked in differential geometry and partial differential equations; solved the Calabi conjecture in algebraic geometry and the positive mass conjecture of general relativity theory; Fields Medal, 1982.

Yersin, Alexandre Émile Jean (1863–1943), Swiss bacteriologist. Discovered the bubonic plague bacillus in Hong Kong, working independently of S. Kitasato, and developed a serum for it.

Yoccoz, Jean-Christophe (1957–), French mathematician. Worked on the theory of dynamical systems; Fields Medal, 1994.

Young, Thomas (1773–1829), English physicist and physician. Discovered the effect of the ciliary muscle on the shape of the eye lens (the mechanism of accommodation).

Yukawa, Hideki (1907–1981), Japanese physicist. Postulated the existence of a new fundamental particle, the meson; Nobel Prize, 1949.

Zariski, Oscar (1899–1986), Russian-born American mathematician. Worked in algebraic geometry, in paricular, on local uniformization and reduction of singularities of algebraic varieties.

Zeeman, Pieter (1865–1943), Dutch physicist. Discovered the Zeeman effect in magnetooptics; Nobel Prize, 1902.

Zelmanov, Efim Isaakovich (1955–), Russian mathematician. Contributed to the theory of Jordan algebras and the theory of Lie algebras; solved the restricted Burnside problem, one of the fundamental questions in group theory; Fields Medal, 1994.

Zener, Clarence Melvin (1905–1993), American physicist. Proposed mechanism of Zener breakdown.

Zeno of Elea (ca. 490–425 B.C.), Greek philosopher and mathematician. Formulated a group of paradoxes, important for their stimulation of philosophical and mathematical thought, which appear to deny the possibility of motion.

Zernike, Fritz (1888–1966), Dutch physicist. Developed the phase-contrast microscope, making possible the first microscopic examination of the internal structure of living cells; Nobel Prize, 1953.

Zewail, Ahmed H. (1946–), Egyptian-born American chemist. Studies the transition states of chemical reactions using femtosecond spectroscopy; Nobel Prize, 1999.

Ziegler, Karl (1898–1973), German organic chemist. Developed a low-pressure process for production of polyethylene; Nobel Prize, 1963.

Zinkerngagel, Rolf M. (1944–), Swiss-born American immunologist and physician. Shared in the discovery of the specificity of cell-mediated immune defense with P. C. Doherty; Nobel Prize, 1996.

Zinsser, Hans (1878–1940), American bacteriologist. Developed methods of immunization against typhus.

Zsigmondy, Richard (1865–1929), German chemist. Studied colloidal solutions; introduced the ul-tramicroscope; Nobel Prize, 1925.

zur Hausen, Harald (1936–), German virologist. Pioneering research on the links between viral infections and the development of specific types of cancer in humans, in particular on human papilloma virus causing cervical cancer; Nobel Prize, 2008.

Zworykin, Vladimir Kosma (1889–1982), Russian-born American physicist. Pioneer in the development of television and the electron microscope.

Contributor Affiliations

The list below comprises an alphabetical sequence, according to surname, of all contributors to the Encyclopedia. A brief affiliation is provided for each author.

A

Abbott, Dr. Donald P. Deceased; formerly, Professor of Biology, Hopkins Marine Station, Stanford University. THALIACEA.

Able, Dr. Kenneth P. Professor Emeritus, Department of Biological Sciences, State University of New York, Albany. MIGRATORY BEHAVIOR.

Abraham, Dr. Bernard M. Deceased; formerly, Solid State Science Division, Argonne National Laboratory, Illinois. HELIUM.

Abrahams, David H. Dexter Chemical Corporation, New York. DYE; DYEING; TEXTILE CHEMISTRY; TEXTILE PRINTING.

Abrahams, Prof. Elihu. Director, Center for Materials Theory, Department of Physics and Astronomy, Rutgers University, New Jersey. ANTIFERROMAGNETISM; CURIE-WEISS LAW; DIAMAGNETISM; FERROMAGNETISM; LARMOR PRECESSION; MAGNETIC SUSCEPTIBILITY; PARAMAGNETISM; QUANTUM THEORY OF MATTER.

Abrahamson, Prof. Warren G., II. David Burpee Professor of Plant Genetics, Department of Biology, Bucknell University, Lewisburg, Pennsylvania. PLANT-ANIMAL INTERACTIONS.

Abramovitch, Dr. Rudolph A. H. Professor, Emeritus, Department of Chemistry and Geology, Clemson University. HETEROCYCLIC COMPOUNDS.

Abzug, Dr. Charles. Department of Computer Science James Madison University, Harrisonburg, Virginia. COMPUTER ARCHITECTURE.

Abzug, Dr. Malcolm J. Manager, Controls Design & Analysis Department, TRW Systems Group, Redondo Beach, California. AILERON; ELEVATOR (AIRCRAFT); STABILIZER (AIRCRAFT).

Achey, Prof. Phillip M. Professor Emeritus, College of Agricultural and Life Sciences, University of Florida. RADIATION BIOLOGY.

Adair, Dr. Robert K. Professor Emeritus of Physics at Yale University. I-SPIN.

Adams, Dr. Chris J. Unilever Research, Merseyside, England. IODINE.

Adams, Prof. Douglas P. Department of Mechanical Engineering, Massachusetts Institute of Technology. CAM MECHANISM; PANTOGRAPH; SLIDER-CRANK MECHANISM; UNIVERSAL JOINT.

Adams, Dr. Elijah. Deceased; formerly, Department of Biological Chemistry, School of Medicine, University of Maryland. POISON.

Adams, Prof. Maurice W. Department of Crop and Soil Sciences, Michigan State University. BEAN.

Adams, Dr. Michael. Information Systems, Statistics, and Operations Management, The University of Alabama, Tuscaloosa. STATISTICAL PROCESS CONTROL.

Adams, Dr. Raymond D. Bullard Professor Emeritus of Neuropathology, Harvard Medical School. ANALGESIC; BARBITURATES; other articles.

Adamson, Prof. Arthur W. Department of Chemistry, University of Southern California. TITANIUM.

Addicott, Prof. Frederick T. Department of Botany, University of California, Davis. ABSCISSION.

Addison, Cyril C. Professor of Inorganic Chemistry, University of Nottingham, England. NITROGEN OXIDES.

Adelberg, Dr. Edward A. Professor Emeritus, Genetics, Yale University. AMINO ACIDS.

Adelman, Dr. William J., Jr. Laboratory of Biophysics, Department of Health and Human Services, National Institutes of Health, Bethesda, Maryland. BIOPOTENTIALS AND IONIC CURRENTS.

Ader, Prof. Robert. Distinguished University Professor at the University of Rochester School of Medicine. PSYCHONEUROIMMUNOLOGY.

Adewumi, Prof. Michael A. Professor, Petroleum and Natural Gas Engineering, Pennsylvania State University. NATURAL GAS.

Adkins, George G. S. Chief, Department of River Basins, Bureau of Power, Federal Power Commission, Washington, DC. TIDAL POWER.

Adkins, Prof. Gregory. Department of Physics and Astronomy, Franklin and Marshall College, Lancaster, Pennsylvania. POSITRONIUM.

Adler, Dr. David. Deceased; formerly, Department of Electrical Engineering, Massachusetts Institute of Technology. GLASS SWITCH.

Adler, Dr. Ronald J. W. W. Hansen Experimental Physics Laboratory, Stanford University, California. RELATIVITY; REST MASS.

Adrain, Dr. Jonathan. Department of Geosciences, University of Iowa, Iowa City. TRILOBITA.

Agarwal, Dr. Gyan C. Department of Electrical Engineering, University of Illinois. MEDICAL CONTROL SYSTEMS.

Aguiar, Jose L. Farm Advisor, University of California Cooperative Extension, Riverside County, Indio. DATE.

Aguirre, Dr. Anthony N. Institute for Advanced Study, Princeton University, Princeton, New Jersey. GRAVITATIONAL REDSHIFT; REDSHIFT.

Ahearn, Prof. Gregory A. Department of Zoology, University of Hawaii, Manoa. ABSORPTION (BIOLOGY).

Ahlberg, Dr. Per Erik. Department of Physiology and Developmental Biology, Uppsala University, Sweden. OSTEOLEPIFORMES; PANDERICHTHYS.

Ahlstrom, Dr. Elbert H. Deceased; formerly, Fishery-Oceanography Center, U.S. Bureau of Commercial Fisheries, La Jolla, California. MONOGONONTA; SEISONACEA.

Ahmadjian, Dr. Vernon. Emeriti Faculty, Clark University, Worcester, Massachusetts. LICHENS.

Ahuja, Prof. Narendra. Department of Electrical and Computer Engineering, Beckman Institute and Coordinated Science Laboratory, University of Illinois, Urbana. COMPUTER VISION.

Ajzenberg-Selove, Prof. Fay. Emeritus, Physics Department, University of Pennsylvania. NUCLEAR SPECTRA.

Akasofu, Prof. S. I. Department of Geophysics, University of Alaska. AURORA; MAGNETOSPHERE.

Akers, Dr. Donald L. School of Medicine, Tulane University. ANEURYSM.

Akesson, Dr. Norman B. Professor Emeritus, Department of Agricultural

Engineering, University of California, Davis. AGRICULTURAL AIRCRAFT.

Akiba, Prof. Kin-ya. Hiroshima University, Department of Chemistry, Faculty of Science, Kagamiyama, Higashi-Hiroshima, Japan. HYPERVALENT COMPOUNDS.

Albaugh, Reuben. Extension Animal Scientist, University of California, Davis. AGRICULTURAL SCIENCE (ANIMAL).

Albert, Dr. Henry J. Manager, Physics and Metallurgy Laboratory, Research and Development Department, Engelhard Industries Division, Engelhard Minerals and Chemicals Corporation, Newark, New Jersey. PLATINUM.

Albright, Dr. Louis D. Department of Agricultural and Biological Engineering, Cornell University, Ithaca, New York. GREENHOUSE TECHNOLOGY.

Albright, Dr. Lyle F. Department of Chemical Engineering, Purdue University. HALOGENATION; NITRATION.

Aldridge, Dr. R. J. Department of Geology, University of Nottingham, England. CONODONT.

Aldwinkle, Prof. H. S. Department of Plant Pathology, Cornell University. APPLE.

Alexander, Dr. David. Director, Disaster Management Center, Cranfield University, England. ENVIRONMENTAL MANAGEMENT.

Alexander, Dr. Gene. Laboratory of Neurosciences, National Institutes of Aging and Health, Bethesda, Maryland. ALZHEIMER'S DISEASE.

Alexander, Prof. R. McNeill. Department of Biology, University of Leeds, England. ADAPTATION (BIOLOGY); SWIM BLADDER.

Alexander, Prof. Vera. School of Fisheries and Ocean Sciences, University of Alaska-Fairbanks. BERING SEA.

Alexeff, Prof. Igor. Laboratory for Information Technologies, University of Tennessee, Knoxville. ARC DISCHARGE.

Alger, Prof. Philip L. Deceased; formerly, Consulting Professor of Electrical Engineering, Rensselaer Polytechnic Institute. MAGNET WIRE.

Algermissen, S. T. U.S. Geological Survey, Department of the Interior, Denver, Colorado. SEISMIC RISK.

Allan, Roger. Executive Editor, "Electronic Design Magazine," Rochelle Park, New Jersey. EMBEDDED SYSTEMS.

Alldredge, Alice L. Department of Biological Sciences, University of California, Santa Barbara. APPENDICULARIA (LARVACEA).

Allegretti, Thomas A. President and CEO, The American Waterways Operators, Arlington, Virginia. INLAND WATERWAYS TRANSPORTATION.

Allen, Dr. John A. U.M.B.S.M. Marine Station, Isle of Cumbrae, Scotland. PROTOBRANCHIA; SEPTIBRANCHIA.

Allen, Prof. Philip B. Department of Physics and Astronomy, Stony Brook University, New York. ELECTRICAL CONDUCTIVITY OF METALS.

Allen, Prof. Phillip E. School of Electrical and Computer Engineering, Georgia Institute of Technology. DIFFERENTIAL AMPLIFIER.

Allen, Dr. Robert Day. Deceased; formerly, Department of Biological Science, Dartmouth College. CYTOPLASM.

Allenby, Dr. Braden R. Vice President, AT&T Environment, Health and Safety, Basking Ridge, New Jersey. INDUSTRIAL ECOLOGY.

Aller, Prof. Lawrence H. Deceased; formerly, Department of Physics and Astronomy, University of California, Los Angeles. ASTRONOMICAL SPECTROSCOPY.

Alley, Prof. Charles L. Emeritus, Department of Electrical and Computer Engineering, University of Utah. FREQUENCY-MODULATION DETECTOR; FREQUENCY MODULATOR; FREQUENCY MULTIPLIER.

Allfrey, Vincent G. Department of Cell Biology, Rockefeller University, New York. NUCLEOPROTEIN.

Allport, Dr. H. Burnham. Project Manager, Carbon Products Division, Union Carbide Corporation, Cleveland, Ohio. ACTIVATED CARBON.

Allred, Dr. Albert L. Department of Chemistry, Northwestern University. ELECTRONEGATIVITY.

Almgren, Dr. Frederick R. Deceased; formerly, Department of Mathematics, Princeton University. MINIMAL SURFACES.

Almond, Dr. Peter R. Louisville, Kentucky. ISOTOPIC IRRADIATION.

Aloia, Dr. John F. Osteoporsois Diagnosois, Treatment, and Research Center, Winthrop-University Hospital, Mineola, New York. OSTEOPOROSIS.

Alpern, Milton. Manager, Special Projects Division, Shah Associates, P.C., Bellmore, New York. FLOOR CONSTRUCTION; ROOF CONSTRUCTION; WALL CONSTRUCTION.

Altamuro, Vincent M. President, VMA, Inc., Toms River, New Jersey. PARETO'S LAW.

Altman, Dr. David. Division Vice President and Technical Director, United Technology Center, Sunnyvale, California. METAL-BASE FUEL.

Alvim, Dr. Paulo de T. Centro de Pesquisas do Cacau (CEPEC), Itabuna, Bahia, Brazil. CACAO.

Alvord, Dr. Ellsworth C. Department of Neuropathology, Harborview Medical Center, Seattle, Washington. NERVOUS SYSTEM DISORDERS.

Amac, Ayse E. Kocaeli University, Turkey. UNINTERRUPTIBLE POWER SUPPLY.

Amar, Dr. Marcelo. National Heart, Lung and Blood Institute, National Institutes of Health, Bethesda, Maryland. ABC LIPID TRANSPORTERS.

Ammons, Dr. Nelle. Retired; formerly, Department of Biology, West Virginia University. BUD; DECIDUOUS PLANTS; EVERGREEN PLANTS; STEM.

Amoreira, Celine. Ph. D. student, University of Zurich, Switzerland. MOLECULAR MECHANICS.

Anagnostakis, Dr. Sandra L. Department of Pathology & Ecology, Connecticut Agriculture Experimental Station, New Haven. CHESTNUT.

Anders, James E., Sr. President, Hydraulics Associates, Bethlehem, Pennsylvania. HYDRAULIC ACCUMULATOR.

Andersen, Prof. David R. Department of Electrical and Computer Engineering, University of Iowa, Iowa City. ELECTRONICS; NONLINEAR OPTICAL DEVICES; OPTICAL BISTABILITY; VARACTOR.

Anderson, Prof. Alan P. Department of Electronic and Electrical Engineering, University of Sheffield, England. HOLOGRAPHY.

Anderson, Prof. Alfred T., Jr. Department of Geophysical Sciences, University of Chicago. ANORTHOSITE; MAGMA; PETROGRAPHY; other articles.

Anderson, Prof. Dana Z. Joint Institute for Laboratory Astrophysics, University of Colorado. NONLINEAR OPTICS.

Anderson, Harry R. H. R. Anderson and Associates, Inc., Consulting Engineers, Eugene, Oregon. RADIO BROADCASTING.

Anderson, Dr. Jason S. Assistant Professor, Anatomy, Faculty of Veterinary Medicine, University of Calgary, Canada. LEPOSPONDYLI.

Anderson, Dr. John D. Senior Research Scientist, Jet Propulsion Laboratory. ASTRONOMICAL UNIT.

Anderson, Dr. John D., Jr. Curator of Aerodynamics, National Air and Space Museum, Smithsonian Institution, Washington, DC and Glenn L. Martin Distinguished Professor of Aerospace Engineering, Department of Aerospace Engineering, University of Maryland, College Park. AERODYNAMIC FORCES; AERODYNAMICS; AIRFOIL; ISENTROPIC FLOW; MACH NUMBER.

Anderson, Dr. John G. Seismological Lab, University of Nevada, Reno. SEISMIC RISK.

Anderson, Lorinda K. Assistant Professor, Department of Biology, Colorado State University, Fort Collins. CHROMOSOME.

Anderson, Dr. O. Roger. Department of Biology, Lamont-Doherty Geological Laboratories, Palisades, New York. ACANTHAREA; ACANTHOMETRIDA; ACTINOPODEA; ARCELLINIDA; RADIOLARIA.

Anderson, Dr. Paul. Power Math Associates, Inc., San Diego, California. ALTERNATING-CURRENT GENERATOR; SHORT CIRCUIT; SPEED REGULATION; VOLTAGE REGULATION.

Anderson, Dr. Peter. Department of Materials Science and Engineering, Ohio State University, Columbus. PLASTIC DEFORMATION OF METALS.

Anderson, Prof. Richard G. W. Department of Cell Biology and Neuroscience, University of Texas Southwestern Medical Center, Dallas. ENDOCYTOSIS.

Andors, Dr. Allison V. Department of Ornithology, American Museum of Natural History, New York. GASTORNITHIFORMES.

Andreasen, Dr. Christian. Bureau Hydrographique International, Monaco. HYDROGRAPHY.

Andrews, Prof. Lester S. Department of Chemistry, University of Virginia, Charlottesville. MATRIX ISOLATION.

Andrews, Dr. Peter. Department of Palaeontology, British Museum (Natural History), London, England. APES.

Andrew, Dr. Warren. Chairperson, Department of Anatomy, School of Medicine, Indiana University. LYMPHATIC SYSTEM.

Andriole, Dr. Stephen J. Department of Information Systems and Systems Engineering, George Mason University, Fairfax, Virginia. INFORMATION SYSTEMS ENGINEERING.

Andrus, C. F. Agricultural Consultant; formerly, U.S. Department of Agriculture, Charleston, South Carolina. WATERMELON.

Angelakos, Dr. Diogenes J. Deceased; formerly, College of Engineering, University of California, Berkeley. ANTENNA (ELECTROMAGNETISM); CAVITY RESONATOR.

Anglin, Donald L. Consultant, Automotive and Technical Writing, Honolulu, Hawaii. AIR-CUSHION VEHICLE; AUTOMOBILE; ENGINE COOLING; FUEL INJECTION; TURBOCHARGER; other articles.

Angrisano, Rear Admiral Guiseppe. Bureau Hydrographique International, Monaco. HYDROGRAPHY.

Angus, James W. Retired; formerly Intercontinental Dynamics Corporation, Englewood, New Jersey. ALTIMETER; GLIDE-PATH INDICATOR; other articles.

Anscombe, Prof. Francis J. Deceased; formerly, Department of Statistics, Yale University. QUEUEING THEORY.

Anthes, Prof. Richard A. President, University Corporation for Atmospheric Research, Boulder, Colorado. WEATHER FORECASTING AND PREDICTION.

Antoni, Dr. Charles M. Retired; formerly, Department of Civil Engineering, Syracuse University. STRUCTURAL MATERIALS; TOWER; other articles.

Antonides, Lloyd E. International Minerals Division, U.S. Bureau of Mines, Washington, D.C. MINING.

Antsaklis, Prof. Panos. Department of Electrical Engineering, University of Notre Dame, Indiana. MULTIVARIABLE CONTROL.

Anwyl, Robert D. Retired; formerly, Photographic Consultant, Eastman Kodak Company, Rochester, New York. PHOTOGRAPHIC MATERIALS; PHOTOGRAPHY.

Apel, Dr. John R. APEL Associates, Rockville, Maryland. SEAWATER.

Apfel, Prof. Robert E. Deceased; formerly, Yale University, Department of Mechanical Engineering. ACOUSTIC LEVITATION; ACOUSTIC RADIATION PRESSURE.

Apker, Dr. L. Deceased; formerly, General Electric Research Laboratory, Schenectady, New York. PHOTOEMISSION.

Appel, Dr. Max J. James A. Baker Institute for Animal Health, Cornell University, Ithaca, New York. CANINE DISTEMPER.

Appelman, Dr. Evan H. Chemistry Division, Argonne National Laboratory, Argonne, Illinois. XENON.

Appelquist, Dr. Thomas. Department of Physics, Yale University, New Haven, Connecticut. COLOR (QUANTUM MECHANICS).

Appenzeller, Dr. Otto. Department of Neurology, University of New Mexico School of Medicine. SPORTS MEDICINE.

Appleby, Prof. A. John. Center for Electrochemical Systems and Hydrogen Research, Texas Engineering Experiment Station, Texas A&M University System. FUEL CELL.

Appleby, Dr. Arnold P. Professor Emeritus of Crop Science, Oregon State University, Corvallis. WEEDS.

Applegate, Charles E. Retired; formerly, Consulting Engineer, Weston, Massachusetts. CABLE TESTING (ELECTRICITY); KELVIN BRIDGE.

Arbuckle, Earl F., III. Vice President, Engineering, Fox Television Stations, New York, New York. TELEVISION CAMERA; TELEVISION STUDIO; TELEVISION TRANSMITTER.

Arbuckle, P. Douglas. NASA, Langley Research Center, Hampton, Virginia. FLIGHT CONTROLS.

Arbuckle, Prof. Wendell S. Deceased; formerly, Department of Dairy Science, University of Maryland. ICE CREAM.

Archer, Mary D. Grantchester, Cambridge, United Kingdom. ELECTROMOTIVE FORCE (CELLS).

Archibald, Dr. J. David. Department of Biology, San Diego State University, California. ARCHAIC UNGULATE.

Ardema, Dr. Mark D. Research Scientist, National Aeronautics and Space Administration, Ames Research Center, Moffett Field, California. AIRSHIP.

Arden, Nancy. Adjunct Assistant Professor, Department of Epidemiology and Biostatistics, Texas A&M. INFLUENZA.

Arey, Prof. Leslie B. Department of Cell Biology and Anatomy, Northwestern University Medical School, Chicago, Illinois. ORAL GLANDS; PALATE.

Armbruster, Prof. Peter J. Kernphysik II, Gesellschaft für Schwerionenforschung mbH, Darmstadt, Germany. BOHRIUM; DARMSTADTIUM; ELEMENT 112; HASSIUM; MEITNERIUM; ROENTGENIUM; TRANSURANIUM ELEMENTS.

Armstrong, Dr. Richard L. Institute of Arctic and Alpine Research, University of Colorado. AVALANCHE.

Arndt, Dr. Kenneth A. Department of Dermatology, Beth Israel Hospital, Boston, Massachusetts. DERMATITIS.

Arney, Cyril. Engineering & Technology, Marathon Oil Company, Houston, Texas. OIL AND GAS, OFFSHORE.

Arnold, Dr. Chester A. Professor of Botany and Geology and Curator of Paleobotany, University of Michigan. ARCHAEOPTERIDALES; CAYTONIALES.

Arnsdorf, Dr. Morton. Section of Cardiology, The University of Chicago, Pritzker School of Medicine, Illinois. CARDIAC ELECTROPHYSIOLOGY.

Aron, Dr. Walter. Consultant, Menlo Park, California. ELECTRICITY.

Aronow, Dr. Wilbert S. MD, Department of Medicine, New York Medical College, Valhalla, New York. GANGRENE.

Arthur, Dr. Michael A. Graduate School of Oceanography, University of Rhode Island. SAPROPEL.

Ash, Sidney. Albuquerque, New Mexico. PETRIFIED FORESTS.

Ashkin, Dr. Arthur. Retired; formerly, Bell Laboratories, Lucent Technologies, Holmdel, New Jersey. RADIATION PRESSURE.

Ashton, Dr. Mark S. Yale School of Forestry and Environmental Studies, New Haven, Connecticut. SILVICULTURE.

Asik, Dr. Joseph R. Scientific Research Laboratories, Ford Motor Company, Dearborn, Michigan. IGNITION SYSTEM; SPARK KNOCK.

Askey, Dr. Richard A. Mathematics Research Center, University of Wisconsin. HYPERGEOMETRIC FUNCTIONS; LEGENDRE FUNCTIONS; SPECIAL FUNCTIONS.

Aspnes, Dr. David E. Department of Physics, North Carolina State University, Raleigh. CRYSTAL ABSORPTION SPECTRA; ELLIPSOMETRY.

Aston, Dr. James. Deceased; formerly, Consulting Metallurgist, A. M. Beyers Company. WROUGHT IRON.

Aström, Prof. Karl Johan. Department of Automatic Control, Lund Institute of Technology, Lund, Sweden. ADAPTIVE CONTROL.

Athenstaedt, Dr. Herbert. Director, Institut für Molekular-Physikalische Physiologie, Hanover, Germany. PYROELECTRICITY (BIOLOGY).

Atkins, Dr. Elisha. School of Medicine, Yale University. FEVER.

Atkins, Prof. Peter W. Department of Chemistry, Lincoln College/Oxford University, United Kingdom. CHEMICAL ENERGY; DALTON'S LAW; EVAPORATION; GAS CONSTANT; MOLECULE; PHYSICAL CHEMISTRY; THERMODYNAMIC PRINCIPLES; WORK FUNCTION (THERMODYNAMICS).

Atkinson, Prof. Gordon. Department of Chemistry, University of Oklahoma. HYDROGEN ION.

Atkinson, Dr. Robert d'E. Deceased; formerly, Department of Astronomy, Indiana University. CLOCK (MECHANICAL).

Atlas, Dr. Ronald M. Department of Biology, University of Louisville. MICROBIAL ECOLOGY.

Atoda, Dr. Kenji. Professor of Zoology, Tohoku University, Japan. ACTINIARIA; GORGONACEA; HELIOPORACEA; SCLERACTINIA; other articles.

Atwood, Prof. Jerry. Department of Chemistry, University of Missouri-Columbia. CLATHRATE COMPOUNDS.

Atwood, Dr. Ronald L. Foote Mineral Company, Kings Mountain, North Carolina. FLOTATION.

Au, Dr. Whitlow W. L. Marine Mammal Research Program, Hawaii Institute of Marine Biology, Kailua. BIOACOUSTICS, ANIMAL.

Auclair, Dr. W. Retired; formerly, Marine Biological Laboratories, Woods Hole, Massachusetts. CENTRIFUGATION.

Audus, Dr. Leslie J. Retired; formerly, Professor of Botany, Bedford College, University of London, England. PLANT MOVEMENTS.

Ausich, Dr. William I. Professor of Geological Sciences, Ohio State University. CAMERATA.

Austrian, Dr. Robert. Deceased; formerly, Division of Infectious Disease, University of Pennsylvania Health System, Philadelphia. STREPTOCOCCUS PNEUMONIAE (PNEUMOCOCCUS).

Avallone, Prof. Eugene A. Formerly, Department of Mechanical Engineering, City University of New York. BOILER FEEDWATER.

Averbach, Prof. Benjamin L. Department of Materials Science and Engineering, Massachusetts Institute of Technology. HEAT TREATMENT (METALLURGY).

Aviszus, Dr. Katja. Integrated Department of Immunology, National Jewish Medical and Research Center, Denver, Colorado. IMMUNOGLOBULIN GENES.

Awan, Dr. Shakil A. Division of Enabling Metrology, National Physical Laboratory, Teddington, Middlesex, United Kingdom. BRIDGE CIRCUIT; CAPACITANCE MEASUREMENT.

Axley, Milton J. Department of Health and Human Services, National Heart, Lung and Blood Institute, Bethesda, Maryland. SELENIUM.

Ayers, Arthur R. Assistant Professor, Cellular and Development Biology, Harvard University. PHYTOALEXINS.

Ayers, Dr. Walter B. Bureau of Economic Geology, University of Texas, Austin. LIGNITE.

Azarnoff, Dr. Daniel L. President, Searle Research and Development, G. D. Searle & Co., Skokie, Illinois. ASPARTAME.

Aziz, Dr. Michael J. Division of Applied Sciences, Harvard University. CRYSTAL GROWTH.

B

Baab, Dr. Karen. Department of Anatomical Sciences, Health Sciences Center, Stony Brook University, Stony Brook, New York. FOSSIL HUMANS.

Baars, Dr. Bernard J. Wright Institute, Berkeley, California. CONSCIOUSNESS.

Babcock, Dr. Loren E. Department of Geological Sciences, Ohio State University, Columbus. CONULARIIDA; PERMINERALIZATION; PHANEROZOIC; XIPHOSURIDA.

Bachinski, Dr. Sharon W. Department of Geology, University of New Brunswick, Fredericton, Canada. LAMPROPHYRE.

Back, Dr. Thomas G. Department of Chemistry, Faculty of Science, University of Calgary, Alberta, Canada. ORGANO-SELENIUM COMPOUND.

Backer, Dr. Donald C. Department of Astronomy, University of California. HOROLOGY; PULSAR.

Bada, Dr. Jeffrey L. Scripps Institution of Oceanography, University of California, La Jolla. AMINO ACID DATING.

Badgley, Dr. Peter C. Earth Science Division, Office of the Naval Reserve, Arlington, Virginia. REMOTE SENSING.

Baer, Prof. Charles J. Department of Mechanical Engineering, University of Kansas. DESCRIPTIVE GEOMETRY; DRAFTING; ENGINEERING DRAWING; PICTORIAL DRAWING.

Baer, Dr. Donald R. Jackson Laboratory, Organic Chemical Department, E. I. du Pont de Nemours and Company, Wilmington, Delaware. DYE.

Baer, Dr. Norbert S. Conservation Center of the Institute of Fine Arts, New York University. ART CONSERVATION CHEMISTRY.

Baghdachi, Dr. Jamil. Northville, Michigan. PAINT AND COATINGS.

Bagley, Dr. Brian G. Bell Laboratories, Murray Hill, New Jersey. AMORPHOUS SOLID.

Bagnall, Dr. Kenneth W. Research Chemist, Atomic Energy Research Establishment, Harwell, England. POLONIUM.

Bahcall, Dr. John N. Deceased; formerly, School of Natural Sciences, Institute for Advanced Studies, Princeton, New Jersey. SOLAR NEUTRINOS.

Bahr, Dr. Janice M. Professor of Animal Science and Physiology, Department of Animal Sciences, College of Agriculture, University of Illinois, Urbana-Champaign. MENSTRUATION.

Bai, Dr. Er-Wei. Department of Electrical & Computer Engineering, University of Iowa, Iowa City. FORCED OSCILLATION.

Bailar, Prof. John C., Jr. Deceased; formerly, School of Chemical Sciences, University of Illinois, Urbana-Champaign. CHROMIUM; STOICHIOMETRY.

Bailey, A. Earle. Deceased; formerly, Superintendent of Electrical Science, National Physical Laboratory, London, England. DIPOLE; ELECTRICAL UNITS AND STANDARDS; ELECTROMOTIVE FORCE (EMF); other articles.

Bailey, Prof. A. G. Department of Electrical Engineering, University of Southampton, England. ELECTROSTATICS; STATIC ELECTRICITY.

Bailey, Prof. Catherine H. Department of Horticulture, Rutgers University. NECTARINE.

Bailey, Prof. Milton E. Department of Food Science and Nutrition, College of Agriculture, University of Missouri. MAILLARD REACTION.

Bailey, Dr. Reeve M. Deceased; formerly, Professor Emeritus of Biological Sciences, and Curator Emeritus, Division of Fishes, Museum of Zoology, University of Michigan. ACIPENSERIFORMES; PISCES (ZOOLOGY); other articles.

Bailey, Dr. S. W. Department of Geology, University of Wisconsin. CHLORITE.

Baily, Frederick G. Turbine Application Engineering, Large Steam Turbine-Generator Department, General Electric Company, Schenectady, New York. STEAM TURBINE; TURBINE.

Baker, Edward W. Entomology Research Division, Agricultural Research Service, U.S. Department of Agriculture. TROMBIDIFORMES.

Baker, Dr. M. Pauline. National Center for Supercomputing Applications, University of Illinois, Urbana. VIRTUAL REALITY.

Baker, Dr. Paul A. Department of Geology, Duke University. DOLOMITE ROCK.

Baker, Dr. Victor R. Department of Geosciences, University of Arizona. ESCARPMENT; FLOODPLAIN; MORPHOLOGY; REGOLITH; RIVER—in part.

Bakken, Prof. George S. Center Director, Center for Biodiversity Studies, Indiana State University, Terre Haute. THERMOREGULATION.

Balakrishnan, Dr. V. K. Department of Mathematics, University of Maine. GRAPHIC METHODS.

Balch, Dr. Alan L. Department of Chemistry, University of California, Davis. IRIDIUM; RHODIUM.

Balderes, Dr. Theodore. Redondo Beach, California. FINITE ELEMENT METHOD.

Baldi, Dr. Bruce G. Plant Hormone Laboratory, U.S. Department of Agriculture, Beltsville, Maryland. AUXIN.

Baldridge, Dr. Kim. University of Zurich, Organic Chemistry Institute, Switzerland. MOLECULAR MECHANICS.

Baldridge, Dr. W. Scott. Earth and Space Science Division, Los Alamos National Laboratory, Los Alamos, New Mexico. CALDERA.

Baldwin, Harry L., Jr. Department of Mathematics, San Diego City College, San Diego, California. PLANE GEOMETRY; POLYHEDRON; SOLID (GEOMETRY); SURFACE (GEOMETRY).

Bales, Robert A. Center for Advanced Aviation Development, MITRE Corporation, McLean, Virginia. AIRPORT SURFACE DETECTION EQUIPMENT.

Balick, Prof. Bruce. Department of Astronomy, University of Washington, Seattle. PLANETARY NEBULA.

Baliga, Prof. B. Jayant. Department of Electrical and Computer Engineering, North Carolina State University. POWER INTEGRATED CIRCUITS.

Ball, Dr. John E. D. Telecommunications Consultant, Vienna, Virginia. CLOSED CAPTION TELEVISION.

Ballard, Prof. William W. Department of Biological Sciences, Dartmouth College. ESOPHAGUS.

Ballou, Prof. Clinton E. Department of Molecular and Cell Biology, University of California, Berkeley. INOSITOL.

Bally, Dr. John. Center of Astrophysics and Space Astronomy, University of Colorado, Boulder. MOLECULAR CLOUD; ORION NEBULA.

Baltay, Prof. Charles. Chair and Higgins Professor, Physics Department, Yale University, New Haven, Connecticut. ELECTROWEAK INTERACTION; NEUTRAL CURRENTS.

Balzani, Prof. Vincenzo. Departimento de Chimica, Universitá di Bologna, Italy. MOLECULAR MACHINE.

Bamburg, Dr. James R. Department of Biochemistry and Molecular Biology, Colorado State University. NEUROBIOLOGY.

Banerjee, Dr. Pat. Associate Professor, Mechanical Engineering, University of Illinois, Chicago. VIRTUAL MANUFACTURING.

Banks, Dr. Harlan P. Emeritus Professor of Botany, Division of Biological Sciences, Cornell University. HYENIALES; PSILOPHYTALES; RHYNIOPSIDA; TRIMEROPHYTOPSIDA; other articles.

Banks, Dr. John E. Interdisciplinary Arts & Sciences, University of Washington, Tacoma. AGROECOSYSTEM.

Banks, Prof. William P. Chairperson, Department of Psychology, Pomona College, Claremont, California. PSYCHOLOGY.

Banner, Dr. Donald W. Banner, Birch, McKie & Beckett, Washington, D.C. PATENT.

Barash, Prof. David P. Department of Psychology, University of Washington, Seattle. SOCIOBIOLOGY.

Barber, Richard T. Duke University Marine Laboratory, Beaufort, North Carolina. SEAWATER FERTILITY.

Barbosa, Dr. Pedro. Department of Entomology, College of Life Sciences, University of Maryland. INSECT CONTROL, BIOLOGICAL.

Barbour, Dr. Neil. The Charles Stark Draper Laboratory, Inc., Cambridge, Massachusetts. GYROSCOPE; INERTIAL NAVIGATION SYSTEM.

Barcelona, Dr. Michael J. Department of Civil and Environmental Engineering, University of Michigan, Ann Arbor. GROUND-WATER HYDROLOGY.

Barclay, Dr. Harriet G. Professor of Botany, Department of Life Sciences, University of Tulsa. PARAMO; PUNA.

Bard, Dr. Allen J. Department of Chemistry, College of Natural Sciences, University of Texas, Austin. ELECTROLYSIS; OVERVOLTAGE.

Barer, Prof. Robert. Department of Human Biology and Anatomy, University of Sheffield, England. PHASE-CONTRAST MICROSCOPE.

Barger, Prof. Vernon D. Department of Physics, University of Wisconsin, Madison. NUTATION (ASTRONOMY AND MECHANICS); PRECESSION; RIGID-BODY DYNAMICS.

Barghoorn, Prof. Elso S. Department of Biology and Botanical Museum, Harvard University. NOEGGERATHIALES; PETRIFACTION.

Barkalow, Dr. David G. Department of Biochemistry, Purdue University. CELLULOSE.

Barkan, Prof. Christopher P. L. Newmark Civil Engineering Laboratory, University of Illinois at Urbana-Champaign. RAILROAD ENGINEERING.

Barker, Prof. Daniel S. Department of Geological Sciences, University of Texas, Austin. CARBONATITE; KIMBERLITE.

Barker, Dr. Fred. U.S. Geological Survey, Department of the Interior, Denver, Colorado. GRANITE.

Barker, Dr. Horace A. Department of Biochemistry, University of California. BACTERIAL PHYSIOLOGY AND METABOLISM.

Barker, Dr. Joseph W. Retired; formerly, Chairperson of the Board, Research Corporation, New York. ENGINEERING.

Barker, Dr. R. W. J. Department of Electrical and Electronic Engineering, Trent Polytechnic, Peterborough, Ontario, Canada. ELECTROMETER.

Barkley, Prof. Theodore M. Division of Biology, Kansas State University. ALISMATIDAE; CAMPANULALES; DILLENIALES; MAGNOLIOPHYITA; PLANT PHYLOGENY; other articles.

Barksdale, Dr. Lane. Department of Microbiology, School of Medicine, New York University Medical Center. ACTINOPHAGE; BACTERIOPHAGE; LYSOGENY.

Barnard, Dr. J. Laurens. Deceased; formerly, Curator of Crustacea, Smithsonian Institution, Washington, D.C. AMPHIPODA.

Barnes, Dr. Ivan. Department of Physics, U.S. Geological Survey, Water Resources Division, Menlo Park, California. SPRING (HYDROLOGY).

Barnes, Dr. Robert D. Department of Biology, Gettysburg College, Pennsylvania. EYE (INVERTEBRATE).

Barnes, Dr. Robert K. Research and Development Department, Chemicals and Plastics Operations Division, Union Carbide Corporation, South Charleston, West Virginia. ETHYLENE; ETHYLENE GLYCOL; OZONOLYSIS; POLYETHYLENE GLYCOL.

Barnes, Dr. Ted. Physics Division, Oak Ridge National Laboratory, Oak Ridge, Tennessee. STRANGE PARTICLES.

Barnett, Dr. Roy N. Department of Medicine, Section of Gastroenterology, Norwalk Hospital, Connecticut. URINARY TRACT DISORDERS.

Barnitz, Edward S. Consultant in Vacuum Engineering, Rochester, New York. VACUUM PUMP.

Barr, Dr. David. Department of Entomology, Royal Ontario Museum, Toronto, Canada. ACARI.

Barr, Dr. Donald J. S. Principal Research Scientist, Centre for Land and Biological Resources Research, Ottawa, Ontario, Canada. OOMYCOTA.

Barrett, Dr. A. G. M. Department of Chemistry, Northwestern University. NITROPARAFFIN.

Barrett, Dr. Paul M. Department of Palaeontology, The Natural History Museum, London, United Kingdom. DINOSAURIA; ORNITHISCHIA; SAURISCHIA.

Barron, Prof. Eric. Director, Earth System Science Center, Pennsylvania State University, University Park. PALEOCLIMATOLOGY.

Barronet, Dr. Alain. Centre de Recherche sur les Méchanismes de la Croissance Cristalline, Marseille, France. CHRYSOTILE.

Barry, Prof. B. Austin. Department of Civil Engineering, Manhattan College. SURVEYING.

Barry, Dr. Michelle M. Department of Medical Microbiology and Immunology, University of Alberta, Edmonton, Canada. CYTOLYSIS.

Bars, Dr. Itzhak. Department of Physics, Yale University. RENORMALIZATION.

Bar-Sela, Dr. Shlomo. Department of Medicine, Allergy-Immunology Section, Medical College of Wisconsin. ANAPHYLAXIS.

Barth, Prof. T. F. W. Deceased; formerly, Geologic Museum, Oslo, Norway. METAMORPHIC ROCKS; other articles.

Bartlett, Prof. Robert W. Dean, College of Mines, University of Idaho, Moscow. SOLUTION MINING.

Barton, Dr. Jennifer Kehlet. Biomedical Engineering Program, University of Arizona, Tucson. OPTICAL COHERENCE TOMOGRAPHY.

Barton, Dr. Mark D. Department of Earth and Space Sciences, University of California, Los Angeles. BERYL; BERYLLIUM MINERALS.

Barton, Dr. Richard L. Lyndon B. Johnson Space Center, National Aeronautics and Space Administration, Houston, Texas. ATMOSPHERIC ENTRY.

Barwood, Dr. Geoffrey P. National Physical Laboratory, Teddington, Middlesex, United Kingdom. WAVELENGTH MEASUREMENT.

Bash, Dr. Jack. University–National Oceanographic System Office, University of Rhode Island, Saunderstown. OCEANOGRAPHIC VESSELS.

Basolo, Prof. Fred. Department of Chemistry, Northwestern University.

COORDINATION CHEMISTRY; COORDINATION COMPLEXES.

Bass, Prof. Henry E. National Center for Physical Acoustics, University of Mississippi. SOUND ABSORPTION; ULTRASONICS.

Bass, Prof. Jack. Department of Physics and Astronomy, Michigan State University. THERMOELECTRICITY.

Bassemir, Dr. Robert. Graphic Arts Laboratories, Sun Chemical Corporation, Carlstadt, New Jersey. INK; PRINTING.

Bassett, Dr. C. Andrew L. Department of Orthopaedic Surgery, Bioelectric Research Center, Columbia University, Riverdale. ELECTROTHERAPY.

Basso, Dr. F. National Heart Lung and Blood Institute, National Institutes of Health, Bethesda, Maryland. ABC LIPID TRANSPORTERS.

Bateman, Dr. Richard M. Department of Geology, Royal Museum of Scotland, Edinburgh. ASTEROXYLALES; ISOETALES; LEPIDODENDRALES; LYCOPHYTA; LYCOPODIALES; LYCOPSIDA; PROTOLEPIDODENDRALES; SELAGINELLALES; ZOSTEROPHYLLOPSIDA.

Bates, Dr. Robert L. Department of Geology, Ohio State University. STONE AND STONE PRODUCTS.

Batterman, Dr. Scott. Consultants Associates, Inc., Cherry Hill, New Jersey. FORENSIC ENGINEERING.

Batterman, Dr. Steven. Consultants Associates, Inc., Cherry Hill, New Jersey. FORENSIC ENGINEERING.

Battles, Byron W. The Battles Group, LLC, Silver Spring, Maryland. INTERCOMMUNICATING SYSTEMS (INTERCOMS).

Bauer, Prof. E. Department of Physics, Arizona State University, Tempe. ELECTRON DIFFRACTION.

Bauer, Prof. Simon H. Department of Chemistry, Baker Laboratory, Cornell University, Ithaca, New York. SHOCK TUBE.

Baumeister, Prof. Theodore. Deceased; formerly, Consulting Engineer; Stevens Professor of Mechanical Engineering, Emeritus, Columbia University; Editor in Chief, *Standard Handbook for Mechanical Engineers.* BOILER; CARNOT CYCLE; HYDRAULIC TURBINE; MEAN EFFECTIVE PRESSURE; REFRIGERATION CYCLE; other articles.

Baumgarten, Dr. Alexander. Department of Laboratory Medicine, School of Medicine, Yale University. AGGLUTINATION REACTION; ALPHA FETOPROTEIN; IMMUNOASSAY; IMMUNOELECTROPHORESIS; other articles.

Baumiller, Dr. Tomasz. Department of Geological Sciences, University of Michigan, Ann Arbor. CRINOIDEA; INADUNATA.

Baver, Prof. Leonard D. Retired; formerly, Agronomy Department, Ohio State University. SORGHUM.

Bayfield, Dr. James E. Department of Physics, University of Pittsburgh. ELECTRON CONFIGURATION; MOLECULAR BEAMS; RYDBERG ATOM.

Baylis, Dr. Matthew. University of Liverpool, United Kingdom. SCRAPIE.

Beall, Christine. Consulting Architect, Columbus, Texas. MASONRY; PRECAST CONCRETE.

Beam, Dr. Charles F. Department of Chemistry, College of Charleston, South Carolina. ACETAL; EPOXIDE.

Beams, Dr. Jesse W. Deceased; formerly, Department of Physics, University of Virginia. ELECTRODYNAMIC INSTRUMENT; ELECTROMAGNETIC FIELD.

Bean, Howard S. Deceased; formerly, Consultant on Fluid Metering, Liquids and Gases, Sedona, Arizona. HYDROMETER; THERMOMETER; WEIGHT.

Beard, Dr. Benjamin H. Agronomy and Range Science, Agricultural Research Western Region, U.S. Department of Agriculture, Davis, California. SUNFLOWER.

Beatty, J. Kelly. Executive Editor, Sky and Telescope Magazine, Editor, Night Sky Magazine, Cambridge, Massachusetts. MARS; PLANET.

Beaty, H. Wayne. Consultant, Fairfax, Virginia. CONDUCTOR (ELECTRICITY).

Beaumont, Prof. Ross A. Department of Mathematics, University of Washington. DETERMINANT; EQUATIONS, THEORY OF; LINEAR SYSTEMS OF EQUATIONS; other articles.

Beaver, Dr. Wallace W. Deceased; formerly, Director of Research and Development, Brush Beryllium Company, Cleveland, Ohio. BERYLLIUM ALLOYS; BERYLLIUM METALLURGY.

Beck, Dr. Charles B. Professor of Botany, University of Michigan. ARCHAEOPTERIDALES; GINKGOALES; GINKGOOPSIDA.

Beck, Prof. Curt W. Department of Chemistry, Amber Research Laboratory, Vassar College, Poughkeepsie, New York. ARCHEOLOGICAL CHEMISTRY.

Beck, James. Boeing Defense and Space Group, Rocketdyne Division, Boeing North American Inc., Canoga Park, California. AEROSPIKE ENGINE.

Beck, Karl M. Abbott Laboratories, North Chicago, Illinois. SACCHARIN.

Becken, Eugene D. Retired; formerly, President, RCA Global Communications, Inc., New York. TELEGRAPHY.

Becker, Dr. Elery R. Deceased; formerly, Division of Life Sciences, Arizona State University. COCCIDIA; GREGARINIA.

Becker, Dr. Henry A. Department of Chemical Engineering, Queen's University, Kingston, Ontario, Canada. FLAME.

Becker, Dr. Philippe. AT&T Bell Laboratories, Murray Hill, New Jersey. OPTICAL PULSES.

Bederson, Dr. Benjamin. Department of Physics, New York University. ATOMIC NUMBER; ATOMIC PHYSICS; MOLECULAR PHYSICS.

Bedford, Prof. Richard D. Chairperson, Department of Geography, University of Waikato, Hamilton, New Zealand. PACIFIC ISLANDS.

Beer, Dr. Steven V. Department of Plant Pathology, College of Agriculture and Life Sciences, Cornell University. APPLE.

Behr, Dr. Bradford B. NRC Research Associate, Naval Research Laboratory/U.S. Naval Observatory, Gaithersburg, Maryland. UNIVERSE.

Belfit, Dr. Robert W., Jr. Manager, Quality Standards, Quality Assurance Department, Dow Chemical Company, Midland, Michigan. CHLORINE.

Bell, Morton A. Associated Air Products, Inc., San Antonio, Texas. AIR FILTER.

Bell, Prof. Norman R. Department of Electrical Engineering, North Carolina State University. INDUCTION COIL.

Bell, Robin E. Lamont-Doherty Geological Observatory of Columbia University, Palisades, New York. ISOSTASY.

Bell, Dr. William E. Department of Pediatrics, University of Iowa. MENINGITIS.

Belrose, Dr. John S. Department of Communications, Communication Research Centre, Ottawa, Ontario, Canada. FREQUENCY-MODULATION RADIO.

BeMiller, Prof. James N. Director, Whistler Center for Carbohydrate Research, Purdue University. CORN.

Bemis, Dr. David A. Director, Clinical Bacteriology and Mycology Services, University of Tennessee, Knoxville. KENNEL COUGH.

Bemis, Dr. William E. Zoology Department, University of Massachusetts. DIPNOI.

Bender, Prof. Myron L. Deceased; formerly, Department of Chemistry, Northwestern University. CYCLOAMYLOSE.

Bender, Prof. Paul J. Department of Chemistry, University of Wisconsin. FREE ENERGY; FUGACITY; INTERNAL ENERGY.

Benedict, Dr. Manson. Institute Professor Emeritus, Department of Nuclear Engineering, Massachusetts Institute of Technology. NUCLEAR CHEMICAL ENGINEERING.

Bengston, Dr. Stefan. Senior Curator (Invertebrate Fossils), Department of Palaeozoology, Swedish Museum of Natural History, Stockholm, Sweden. FOSSIL; TOMMOTIAN FAUNA.

Benito, Dr. Jesús. Department of Animal Biology I—Zoology of Invertebrates, University Complutense, Madrid, Spain. ENTEROPNEUSTA; HEMICHORDATA; PTEROBRANCHIA.

Benner, Dr. Frank C. Assistant Director, Research and Development Department, Norton Company, Worcester, Massachusetts. VACUUM FUSION.

Bennett, Prof. Albert F. Department of Ecology and Evolutionary Biology, University of California, Irvine. SKIN.

Bennett, Prof. Charles. Department of Physics and Astronomy, The Johns Hopkins University, Baltimore, Maryland. WILKINSON MICROWAVE ANISOTROPY PROBE.

Bennett, Prof. Joan W. Department of Cell and Molecular Biology, Tulane University, New Orleans, Louisiana. AFLATOXIN.

Bennett, Prof. Michael V. L. Department of Neuroscience, Albert Einstein College of Medicine, New York. ELECTRIC ORGAN (BIOLOGY).

Bennett, Dr. Mike B. Department of Anatomical Sciences, University of Queensland, Australia. SKELETAL SYSTEM.

Bennett, Dr. Orus L. GRC Panels Unlimited, Leeds, Alabama. LAND RECLAMATION.

Bennett, Dr. Ralph E. Director, Science Information Department, Squibb Institute for Medical Research, New Brunswick, New Jersey. RIBOFLAVIN.

Benninghoff, Dr. William S. Department of Botany, University of Michigan. TUNDRA.

Benny, Dr. Gerald L. Department of Plant Pathology, University of Florida, Gainesville. TRICHOMYCETES; ZYGOMYCETES.

Benson, Prof. Frank A. Department of Electronic and Electrical Engineering, University of Sheffield, England. SKIN EFFECT (ELECTRICITY).

Bent, Prof. Brian E. Department of Chemistry, Columbia University, New York. ADSORPTION.

Beranek, Dr. Leo L. Retired; formerly, Chief Scientist, Bolt Beranek and Newman, Inc., Cambridge, Massachusetts. WAVE MOTION.

Beresford, Dr. William A. Department of Anatomy, West Virginia University. CARTILAGE.

Berg, Dr. Clyde. Clyde Berg and Associates, Long Beach, California. SOLIDS PUMP.

Berger, Dr. France B. Kearfott Division, Singer Company, Little Falls, New Jersey. DOPPLER RADAR.

Berger, Harold. Industrial Quality, Inc., Gaithersburg, Maryland. RADIOGRAPHY.

Berger, Dr. Richard D. Department of Plant Pathology, University of Florida, Gainesville. PLANT PATHOLOGY.

Berger, Prof. Wolfgang H. Geological Research Division, Scripps Institution of Oceanography, La Jolla, California. PALEOCEANOGRAPHY.

Berggren, Dr. William A. Department of Geological Sciences, Rutgers University, Piscataway, New Jersey. EOCENE; MIOCENE; TERTIARY.

Bergh, Dr. Berthold O. Department of Plant Sciences, University of California, Riverside. AVOCADO.

Berglund, Dr. C. Neil. President, Northwest Technology Group, Vancouver, Washington. INTEGRATED CIRCUITS.

Bergmann, Peter G. Deceased; formerly, Department of Physics, Syracuse University, Syracuse, New York. SPACE-TIME.

Berkowitz, Dr. Gerald A. Institute for Photobiology of Cells and Organelles, Brandeis University. PHOTOSYNTHESIS.

Berman, Dr. Lanny. Executive Director, American Association of Suicidology, Washington, D.C. SUICIDE.

Bernard, Dr. E. N. National Oceanic and Atmospheric Administration, Pacific Marine Environmental Laboratory, Seattle, Washington. TSUNAMI.

Bernath, Prof. Peter F. Department of Chemistry, University of Waterloo, Ontario, Canada. MOLECULAR STRUCTURE AND SPECTRA.

Berner, Dr. Elizabeth. Geology and Geophysics Department, Yale University, New Haven, Connecticut. BIOGEOCHEMISTRY; WATER RESOURCES.

Berner, Prof. Lewis. Acting Director, Division of Biological Sciences, University of Florida. EPHEMEROPTERA; PLECOPTERA.

Berner, Dr. Robert A. Department of Geology and Geophysics, Yale University. BIOGEOCHEMISTRY.

Bernson, Dr. Laurence. Environmental Management Department, AT&T Bell Laboratories, Murray Hill, New Jersey. GAS AND ATMOSPHERE ANALYSIS.

Bernstein, Dr. I. M. Department of Metallurgy and Materials Science, Carnegie-Mellon University. EMBRITTLEMENT.

Berry, Dr. Spencer J. Department of Biology, Wesleyan University. CELL LINEAGE; GAMETOGENESIS; MATERNAL INFLUENCE; OOGENESIS; SPERMATOGENESIS.

Berry, Dr. William B. N. Department of Geology and Geophysics, University of California, Berkeley. GRAPTOLITHINA; ORDOVICIAN; SILURIAN.

Bersted, Dr. Bruce. Amoco Polymers, Inc., Alpharetta, Georgia. POLYOLEFIN RESINS.

Bertsch, Dr. George F. Department of Physics, Michigan State University. SUM RULES.

Bethe, Prof. Hans A. Deceased; formerly, Floyd R. Newman Laboratory of Nuclear Studies, Cornell University; Nobelist. THEORETICAL PHYSICS.

Betts, Dr. Russell. Argonne National Laboratory, Argonne, Illinois. NUCLEAR ISOMERISM.

Beutler, Dr. Bruce. Howard Hughes Medical Institute, University of Texas, Dallas. CACHEXIA.

Bever, Prof. Michael B. Deceased; formerly, Department of Materials Science and Engineering, Massachusetts Institute of Technology. METALLURGY; SURFACE HARDENING OF STEEL.

Bewley, Dr. Loyal V. Deceased; formerly, Dean, College of Engineering, Lehigh University. ELECTRIC TRANSIENT; SYNCHRONOUS CONVERTER.

Beyer, Dr. Ann L. Department of Microbiology, University of Virginia. RIBONUCLEIC ACID (RNA).

Bhansali, Dr. Kirit J. Metallurgy Division, National Institute of Standards and Technology, Washington, D.C. ALLOY STRUCTURES.

Bichsel, Prof. Hans. Emeritus, Center for Experimental Nuclear Physics and Astrophysics, University of Washington. CHARGED PARTICLE BEAMS.

Bidanset, Dr. Jesse H. Department of Toxicology, St. John's University. FORENSIC TOXICOLOGY.

Biederman, Dr. Joseph. Pediatric Psychopharmacological Unit, Massachusetts General Hospital, Boston. ATTENTION DEFICIT HYPERACTIVITY DISORDER.

Biegel, Prof. John E. Department of Industrial Engineering, Syracuse University. PRODUCTION PLANNING.

Bieler, David B. Department of Geology and Geophysics, Centenary College of Louisiana, Shreveport. ROCK CLEAVAGE.

Bienenstock, Dr. Arthur. Stanford Linear Accelerator Center, Stanford University. SYNCHROTRON RADIATION.

Biermann, Dr. Christopher. Department of Forest Products, Forest Research Laboratory, Oregon State University, Corvallis. CELLOPHANE; PAPER.

Bigeleisen, Prof. Jacob. Department of Chemistry, University of Rochester. DEUTERIUM.

Bignami, Dr. Giovanni F. Director, Centre d'Etude Spatiale des Rayonnements (CESR), Toulouse, France. GEMINGA.

Billet, Prof. Michael. Head, Fluid Dynamics Department, Applied Research Laboratory, Pennsylvania State University, State College. CAVITATION.

Billingham, Dr. Rupert E. Department of Biology, University of Texas Health and Science Center, Dallas. TRANSPLANTATION BIOLOGY.

Billings, Dr. Bruce H. Deceased; formerly, Special Assistant to the Ambassador for Science and Technology, Embassy of the United States of America, Taipei, Taiwan. BIREFRINGENCE; CRYSTAL OPTICS; DICHROISM; INTERFERENCE OF WAVES; POLARIZED LIGHT; other articles.

Billington, Dr. Douglas S. Deceased; formerly, Senior Staff Advisor for Materials Science, Metals and Ceramics Division, Oak Ridge National Laboratory, Oak Ridge, Tennessee. RADIATION DAMAGE TO MATERIALS.

Billmeyer, Dr. Fred W. Retired; formerly, Department of Chemistry, Rensselaer Polytechnic Institute. ULTRAVIOLET RADIATION.

Bing, Prof. R. H. Department of Mathematics, University of Texas, Austin. SET THEORY.

Bingham, Prof. Carrol. Department of Physics, University of Tennessee, Knoxville. ALPHA PARTICLES.

Birch, Dr. J. R. Division of Electrical Science, National Physical Laboratory, Teddington, Middlesex, England. SUBMILLIMETER-WAVE TECHNOLOGY.

Bird, Dr. Thomas D. University of Washington, School of Medicine, Seattle. MUSCULAR DYSTROPHY.

Birdsall, Prof. Theodore G. Emeritus, Department of Electrical Engineering and

Computer Science, College of Engineering, University of Michigan, Ann Arbor. ACOUSTIC SIGNAL PROCESSING.

Birkenheuer, Dr. Daniel L. Lakewood, Colorado. SATELLITE METEOROLOGY.

Birkhoff, Prof. Garrett. Deceased; formerly, Department of Mathematics, Harvard University. BOOLEAN ALGEBRA; LATTICE (MATHEMATICS).

Birks, Dr. L. S. Head, X-Ray Optics Branch, Radiation Technology Division, Naval Research Laboratory, Washington, D.C. X RAY SPECTROMETRY.

Birnstiel, Dr. Charles. Consulting Engineer, Forest Hills, New York. ARCH; BOLT; BOLTED JOINT.

Biscaye, Dr. Pierre E. Lamont-Doherty Geological Observatory, Palisades, New York. MARINE SEDIMENTS.

Bisno, Dr. A. L. Veterans Administration Medical Center, Miami, Florida. RHEUMATIC FEVER.

Black, Dr. Harold S. Deceased; formerly, Communications Consultant, Summit, New Jersey. AMPLITUDE MODULATION.

Black, Prof. Paul H. Department of Mechanical Engineering, Ohio University. ALLOWANCE; CLEVIS PIN; HARMONIC SPEED CHANGER; PRESSURE SEAL; other articles.

Blackburn, J. Lewis. Deceased; formerly, Consulting Engineer, Bothell, Washington. ELECTRIC PROTECTIVE DEVICES; INSTRUMENT TRANSFORMER; RELAY.

Blackwell, Dr. Meredith. Department of Plant Biology, Louisiana State University, Baton Rouge. EUMYCOTA.

Blad, Prof. Blaine. Head, Department of Agricultural Meteorology, Institute of Agriculture and Natural Resources, University of Nebraska, Lincoln. AGRICULTURAL METEOROLOGY; BIOMETEOROLOGY.

Blades, Dr. Brooke. Department of Anthropology, University of Maryland, College Park. PREHISTORIC TECHNOLOGY.

Blain, Dr. Andrew. Department of Astronomy, California Institute of Technology, Pasadena. SUBMILLIMETER ASTRONOMY.

Blake, Dr. Daniel B. Department of Geology, University of Illinois, Urbana. PAXILLOSIDA; PLATYASTERIDA.

Blake, Dr. Wilson. Hayden Lake, Idaho. ROCK BURST.

Blaner, Dr. William S. Hammer Health Sciences Center, Columbia University, New York. RETINOID RECEPTOR.

Blankenship, Dr. James E. Department of Neuroscience, University of Texas Medical Branch, Galveston. NERVOUS SYSTEM (INVERTEBRATE).

Blaser, Prof. H. Weston. Department of Botany, University of Washington. BARK; PITH.

Blass, Dr. Elliott M. Department of Psychology, Johns Hopkins University. HUNGER; THIRST.

Blatt, Prof. Frank J. Department of Physics, Michigan State University. MATTHIESSEN'S RULE; WIEDEMANN-FRANZ LAW.

Blatt, Dr. Harvey. Hebrew University of Jerusalem, Givat Ram, Jerusalem, Israel. ENVIRONMENTAL GEOLOGY.

Blatter, Dr. Dawnika L. Berkeley Geochronology Center, California. DACITE.

Bleiweis, John J. Consultant, Great Falls, Virginia. TELE-CONFERENCING; VIDEOTELEPHONY.

Blencoe, Dr. James G. Chemistry Division, Oak Ridge National Laboratory, Oak Ridge, Tennessee. NEPHELINE.

Blitz, Dr. Leo. Radio Astronomy Laboratory, University of California, Berkeley. MILKY WAY GALAXY.

Blobel, Dr. Günter B. Department of Cell Biology, Rockefeller University, New York; Nobelist. ENDOPLASMIC RETICULUM.

Bloch, Dr. Spencer J. Department of Mathematics, University of Chicago. ALGEBRAIC GEOMETRY.

Block, Dr. Ira. Department of Textiles—Consumer Economics, University of Maryland. MANUFACTURED FIBER; NATURAL FIBER; TEXTILE.

Blom, Dr. Henning. Subdepartment of Evolutionary Organismal Biology, Department of Physiology and Developmental Biology, Uppsala, Sweden. ANASPIDA.

Blondel, Dr. Christophe. Laboratoire Aimé-Cotton, C.N.R.S., Orsay, France. PHOTODETACHMENT MICROSCOPY.

Bloomer, Dr. Sherman H. Department of Geosciences, Oregon State University, Corvallis. SUBDUCTION ZONES.

Bloomfield, Prof. Philip E. Bryn Mawr, Pennsylvania. ADIABATIC PROCESS; ISENTROPIC PROCESS; ISOMETRIC PROCESS; THERMODYNAMIC PROCESSES.

Bloomfield, Dr. Victor A. Department of Biochemistry, University of Minnesota. ULTRACENTRIFUGE.

Blough, Prof. Donald S. Department of Psychology, Brown University. BEHAVIORAL PSYCHOPHYSICS.

Blouin, Dr. Edmour F. Department of Veterinary Pathobiology, Oklahoma State University, Stillwater. ANAPLASMOSIS.

Bluestein, Prof. Howard B. Department of Meteorology, University of Oklahoma. ATMOSPHERIC WAVES, UPPER SYNOPTIC; BAROCLINIC FIELD; CHINOOK; JET STREAM; WIND.

Blum, Dr. Frank D. Department of Chemistry, University of Missouri-Rolla. COLLOID.

Blumenthal, Prof. Leonard M. Deceased; formerly, Defoe Distinguished Professor of Mathematics, University of Missouri. ADDITION; ANALYTIC GEOMETRY; CIRCLE; TANGENT; other articles.

Blumenthal, Warren B. Blumenthal-Zirconium, North Tonawanda, New York. HAFNIUM; ZIRCONIUM.

Boast, Dr. Warren B. Deceased; formerly, Anson Marston Distinguished Professor Emeritus of Electrical Engineering, Iowa State University. ILLUMINATION; LIGHT PANEL; LUMINAIRE.

Bobart, George F. Manager, Induction Heating and Ultrasonic Cleaning Department, Westinghouse Electric Corp., Skyesville, Maryland. DIELECTRIC HEATING; INDUCTION HEATING.

Bobbitt, Dr. James M. Department of Chemistry, University of Connecticut. INDOLE.

Bochner, Dr. Salomon. Deceased; formerly, Department of Mathematics, Rice University. E (MATHEMATICS); EXPONENT; LOGARITHM; ROOT (MATHEMATICS); SERIES. MATHEMATICS.

Bock, Prof. Jane H. University of Colorado, Boulder. FORENSIC BOTANY.

Bock, Dr. Walter J. Professor of Evolutionary Biology, Department of Biological Sciences, Columbia University. AVES; CHARADRIIFORMES; CHORDATA; FALCONIFORMES; GALLIFORMES; GRUIFORMES; NEORNITHES; PASSERIFORMES; PELECANIFORMES; PICIFORMES; PODICIPEDIFORMES; PSITTACIFORMES; RATITES; STRIGIFORMES, other articles.

Boell, Dr. Edgar J. Osborn Memorial Laboratory, Yale University. ANIMAL GROWTH.

Boero, Dr. Ferdinando. Professor of Zoology, Universitá di Lecce, Dipartmento di Biologia, Stazione di Biologia Marina, Lecce, Italy. ONTOGENY.

Boggess, Prof. William R. Tree Ring Research Laboratory, University of Arizona. DENDROCHRONOLOGY.

Boggs, Dr. Paul T. Computational Sciences and Mathematics Research Department, Sandia National Laboratories, Livermore, California. NONLINEAR PROGRAMMING.

Boggs, Prof. Steven A. Electrical Insulation Research Center, Institute of Materials Science, University of Connecticut, Storrs. ELECTRICAL INSULATION.

Bogin, Dr. Barry Allan. Department of Behavioral Sciences, University of Michigan-Dearborn. HUMAN BIOLOGICAL VARIATION.

Bogucki, Dr. Peter. Department of Anthropology, Princeton University, Princeton, New Jersey. NEOLITHIC.

Bohach, Dr. Greg. Department of Microbiology, Molecular Biology, and Biochemistry, University of Idaho, Moscow. STAPHYLOCOCCUS.

Böhme, Prof. M. Department of Chemistry, Friedrich-Wilhelms University, Germany. CYCLOPHANE.

Bøhn, Dr. Jan Helge. Department of Mechanical Engineering, Virginia Tech, Blacksburg. COMPUTER-AIDED ENGINEERING.

Boke, Dr. Norman H. George Lynn Cross Research Professor of Botany, University of Oklahoma. EPIDERMIS (PLANT); SCLERENCHYMA.

Bolender, Dr. Nicole-Fr. Department of Radiology, Group Health Central, Seattle, Washington. COMPUTERIZED TOMOGRAPHY.

Boles, Prof. James R. Department of Geological Sciences, University of California, Santa Barbara. AUTHIGENIC MINERALS.

Bolin, Dr. Steven R. Diagnostic Center for Population and Animal Health, Michigan State University, Lansing. BOVINE VIRAL DIARRHEA.

Bolt, Prof. J. A. Department of Mechanical Engineering, University of Michigan. CARBURETOR; FUEL SYSTEM; GAS TURBINE; SUPERCHARGER.

Bolton, Dr. James R. Department of Chemistry, University of Western Ontario, London, Canada. FREE RADICAL.

Bolz, Dr. Ray E. Formerly, Leonard S. Case Professor and Dean of Engineering, Case Western Reserve University. NUTATION (ASTRONOMY AND MECHANICS).

Bond, Dr. Gerard. Lamont-Doherty Geological Observatory, Palisades, New York. BASIN.

Bondos, Dr. Sarah E. Department of Biochemistry and Cell Biology, Rice University, Houston, Texas. PROTEIN FOLDING.

Bondybey, Dr. Vladimir E. Bell Laboratories, Murray Hill, New Jersey. JAHN-TELLER EFFECT; RENNER-TELLER EFFECT.

Bonner, Dr. Walter D., Jr. Department of Biochemistry and Biophysics, School of Medicine, University of Pennsylvania. CELL, SPECTRAL ANALYSIS OF.

Boothroyd, Dr. Carl W. Department of Plant Pathology, New York State College of Agriculture and. Life Sciences, Cornell University, Ithaca, New York. PLANT PATHOLOGY.

Borello, Sebastian R. Central Research Laboratories, Texas Instruments, Inc., Dallas. PHOTOCONDUCTIVE CELL.

Bormann, Dr. Bernard T. Department of Forest Sciences, Oregon State University, Corvallis. FOREST ECOSYSTEM; FOREST SOIL.

Borovsky, Dr. Joseph. Los Alamos National Laboratory, Los Alamos, New Mexico. PLASMA (PHYSICS).

Borowiecki, Prof. Barbara Z. Department of Geography, University of Wisconsin, Milwaukee. NORTH AMERICA.

Borst, Henry V. Henry V. Borst & Associates, Wayne, Pennsylvania. PROPELLER (AIRCRAFT).

Bortman, Dr. Marci L. Waste Management Institute, Marine Sciences Research Center, State University of New York, Stony Brook. RECYCLING TECHNOLOGY.

Bosart, Prof. Lance F. Department of Atmospheric Sciences, State University of New York, Albany. WEATHER FORECASTING AND PREDICTION; WEATHER MAP.

Boschung, Dr. Herbert T. Professor Emeritus, Department of Biological Sciences, University of Alabama. ANGUILLIFORMES; ATHERINIFORMES; BASSES; BATOIDEA; CHIMAERIFORMES; DIPNOI; EEL; GADIFORMES; GASTEROSTEIFORMES; GNATHOSTOMATA; HERRINGS; LAMNIFORMES; MYXINIFORMES; NOTACANTHOIDEI; OSTEICHTHYES; PERCIFORMES; RAJIFORMES; SALMONIFORMES; STICKLEBACK; TELEOSTOMI; TUNA; ZEIFORMES, other articles.

Bose, Prof. Anjan. School of Electrical Engineering and Computer Science, Washington State University, Pullman. ELECTRIC POWER SYSTEMS.

Bosh, Dr. Amanda S. Lowell Observatory, Flagstaff, Arizona. OCCULTATION.

Boss, Dr. Alan P. Carnegie Institution of Washington, Department of Terrestrial Magnetism, Washington, D.C. SOLAR SYSTEM.

Bothwell, Dr. Mark. Biochemical Sciences Laboratory, Princeton University. GROWTH FACTOR.

Botset, Holbrook G. Professor Emeritus of Petroleum Engineering, University of Pittsburgh. PHOTOCLINOMETER.

Bouchard, Prof. Thomas J. Minnesota Center for Twin and Adoption Research, Department of Psychology, University of Minnesota. BEHAVIOR GENETICS.

Bouillon, Dr. Jean. Universitá di Lecce, Dipartmento di Biologia, Italy. ONTOGENY.

Bouldin, Dr. Donald W. Electrical and Computer Engineering, University of Tennessee. CYBERNETICS.

Bourlat, Dr. Sarah J. Department of Biology, University College London, United Kingdom. XENOTURBELLA.

Bouwkamp, Dr. John C. Department of Horticulture, College of Agriculture, University of Maryland. POTATO, SWEET.

Bowler, Dr. Bruce E. Division of Chemistry and Chemical Engineering, California Institute of Technology. ELECTRON-TRANSFER REACTION in part.

Bowman, Richard G. Deceased; formerly, Technical Assistant to the President, Republic Aviation Corp., Farmingdale, New York. AIRCRAFT.

Bowman, Rush A. Consulting Engineer, Memphis, Tennessee. GAGE; OPTICAL FLAT.

Bowman, Dr. Thomas E. Division of Crustacea, Smithsonian Institution, Washington, D.C. HYPERIDEA; PERACARIDA.

Bowman, Waldo G. Deceased; formerly, Editor, "Engineering News Record.". HARDNESS SCALES; STRUCTURAL STEEL; YOUNG'S MODULUS; other articles.

Box, Dr. Elgene O. Department of Geography, University of Georgia. PLANT GEOGRAPHY; VEGETATION AND ECOSYSTEM MAPPING.

Boyd, Dr. F. R. Geophysical Laboratory, Washington, DC. DIOPSIDE.

Boyd, Dr. Richard H. Department of Chemical Engineering, University of Utah. ACID AND BASE.

Boyden, Prof. Edward A. Deceased; formerly, Department of Biological Structure, School of Medicine, University of Washington. GALLBLADDER.

Boynton, Dr. Robert M. Department of Psychology, University of California, San Diego. COLOR.

Boynton, Dr. Robert S. National Lime Association, Washington, D.C. LIME (INDUSTRY).

Bozarth, Dr. Robert F. Department of Life Science, Indiana State University, Terre Haute. FUNGAL VIRUS.

Braaten, Prof. David A. Department of Geography, University of Kansas, Lawrence. GREENLAND.

Bracewell, Dr. Ronald N. Professor of Electrical Engineering, Emeritus, Stanford University, Stanford, California. FOURIER SERIES AND TRANSFORMS.

Bradley, Dr. Robert L., Jr. Department of Food Science, College of Agricultural and Life Sciences, University of Wisconsin. BUTTER; MILK.

Bradley, Dr. S. Gaylen. Research Affairs, Penn State College of Medicine, Hershey. ACTINOMYCETES.

Bradley, Prof. William F. Deceased; formerly, Department of Chemical Engineering, University of Texas, Austin. SEPIOLITE.

Bradley, Prof. William W. Department of Civil Engineering and Institute of Colloid and Surface Science, Clarkson College of Technology, Potsdam, New York. METAL COATINGS.

Bradner, Mead. Deceased; formerly, Technical Director, Foxboro Company, Foxboro, Massachusetts. FLOW MEASUREMENT.

Bradsher, Prof. Charles K. Department of Chemistry, Duke University. FRIEDEL-CRAFTS REACTION; POLYNUCLEAR HYDROCARBON; STYRENE; other articles.

Bragg, Prof. Michael B. Department of Aeronautical and Astronautical Engineering, University of Illinois at Urbana-Champaign. AIRCRAFT ICING.

Brainard, Dr. George C. Department of Neurology, Thomas Jefferson University, Jefferson Medical College, Philadelphia, Pennsylvania. PINEAL GLAND.

Bralla, J. R. Casey. Consultant, Mechanical Engineering, Martinez, Georgia. COMPUTER NUMERICAL CONTROL; MACHINING.

Brammer, Dr. Anthony J. Institute for Microstructural Sciences, National Research Council of Canada, Ottawa, Ontario. VIBROTACTION.

Branch, Dr. Darren V. Sandia National Laboratories, Biosensor and Nanomaterials Department, Albuquerque, New Mexico. SURFACE-ACOUSTIC-WAVE DEVICES.

Brand, Dr. Louis. Deceased; formerly, M. D. Anderson Professor of Mathematics, University of Houston. HYPERBOLIC FUNCTION; PARAMETER; PROGRESSION (MATHEMATICS); other articles.

Brandt, Dr. G. Donald. Professor Emeritus, City University of New York. STRUCTURAL ANALYSIS; STRUCTURAL PLATE; STRUCTURAL STEEL.

Brandt, Prof. John C. Department of Physics and Astronomy, Institute for Astrophysics, University of New Mexico, Albuquerque. COMET; HALLEY'S COMET.

Brandt, Dr. Philip W. Department of Anatomy, College of Physicians and Surgeons, Columbia University. PHAGOCYTOSIS.

Bransden, Tania. School of Computer Science and Software Engineering, Monash University, Clayton, Victoria, Australia. COMPLEXITY THEORY.

Brassell, Prof. Simon C. Professor of Geological Sciences, Biogeochemical Laboratories, Indiana University, Bloomington. ORGANIC GEOCHEMISTRY.

Brasseur, Dr. Guy P. Director, Atmospheric Chemistry Division, National Center for Atmospheric Research, Boulder, Colorado. STRATOSPHERIC OZONE.

Braun, Dr. Hans. CERN, Geneva, Switzerland. PARTICLE ACCELERATOR.

Brazel, Prof. Anthony J. Department of Geography, Laboratory of Climatology, Arizona State University, Tempe. URBAN CLIMATOLOGY.

Breed, Dr. Michael D. Environmental Population and Organismic Biology, University of Colorado. SOCIAL HIERARCHY.

Breen, Dr. Patrick J. Department of Horticulture, Oregon State University. BLACKBERRY.

Breger, Dr. Irving A. Deceased; formerly, Office of Energy Resources, U.S. Geological Survey, Reston, Virginia. BITUMEN; IMPSONITE; WURTZILITE; other articles.

Breier, Dr. Alan. Department of Psychiatry, Yale University School of Medicine; Connecticut Mental Health Center, New Haven. NORADRENERGIC SYSTEM.

Breman, Dr. Joel D. Division of Parasitic Diseases, Centers for Disease Control, Department of Health and Human Services, Atlanta, Georgia. SMALLPOX.

Breslin, Dr. Vincent T. Waste Management Institute, Marine Sciences Research Center, State University of New York, Stony Brook. RECYCLING TECHNOLOGY.

Breslow, Dr. David S. David S. Breslow Associates, Wilmington, Delaware. AZIDE; COPOLYMER.

Bressler, Dr. Peter E. Endocrine & Diabetes Associates of Texas, Dallas. DIABETES.

Brett, Dr. Carlton. Department of Geology, University of Cincinnati, Ohio. INDEX FOSSIL; PALEOECOLOGY.

Breuer, Glenn D. Consulting Engineer; formerly, Systems Development and Engineering Department, General Electric Company, Schenectady, New York. LIGHTNING AND SURGE PROTECTION; SURGE ARRESTER.

Brewer, Dr. Charles P. Shell Development Corp., Emeryville, California. HYDROCRACKING.

Brewer, S. Theodore. Deceased; formerly, Head, Undersea and Lightwave Engineering Department, Bell Laboratories, Holmdel, New Jersey. COAXIAL CABLE; COMMUNICATIONS CABLE.

Brézin, Dr. Edouard. Laboratoire de Physique Theorique, Ecole Normale Supérieur, Paris, France. RANDOM MATRICES.

Brian, Dr. M. V. Institute of Terrestrial Ecology, Furzebrook Research Station, Wareham, England. SOCIAL INSECTS.

Brickley, Robert L. Marketing Manager, Westinghouse Electric Corp., Springfield, Massachusetts. ANALYSIS OF VARIANCE.

Bridges, Prof. Patricia S. Deceased; formerly, Department of Anthropology, Queens Collge and CUNY Graduate Center, Flushing, New York. BIOARCHEOLOGY.

Bridgman, Prof. Percy W. Deceased; formerly, Harvard University; Nobelist. EMPIRICAL METHOD; HYPOTHESIS; POSTULATE; other articles.

Brierley, Dr. Corale. Brierley Consultancy LLC, Highlands Ranch, Colorado. BIOLEACHING.

Briggs, Dr. Derek E. G. Department of Geology, University of Nottingham, England. CONODONT.

Briggs, Dr. John D. Department of Entomology, Ohio State University. INSECT DISEASES.

Briggs, M. Michael. Missile Systems Division, Integrated Systems Inc., Santa Clara, California. BALLISTIC MISSILE.

Briggs, Winslow. Emeritus Professor, Carnegie Instituion of Washington. PHOTOMORPHOGENESIS.

Brimacombe, Dr. R. Max-Planck Institut für Molekulare Genetik, Berlin, Germany. RIBOSOMES.

Brindley, Dr. George W. Deceased; formerly, Department of Geosciences, Pennsylvania State University. APOPHYLLITE; PYROPHYLLITE.

Brintzinger, Dr. Hans. Fachbereich Biologie-Chemie, Universität Konstanz, Germany. NITROGEN COMPLEXES.

Brisbin, Dr. I. Lehr. Savannah River Ecology Laboratory, University of Georgia. RADIOECOLOGY.

Briskman, Robert. Technical Executive, Sirius Satellite Radio, New York. SATELLITE RADIO.

Broadus, Dr. Arthur E. Associate Professor of Medicine, Department of Medicine, Yale University. CALCIUM METABOLISM; PARATHYROID HORMONE.

Brock, Dr. Fred V. Department of Meteorology, University of Oklahoma. PRECIPITATION MEASUREMENT.

Brockenbrough, Dr. Roger L. R. L. Brockenbrough Associates, Inc., Pittsburgh, Pennsylvania. TRANSPORTATION ENGINEERING.

Brockhouse, Dr. Bertram N. Institute for Materials Research, McMaster University, Hamilton, Ontario, Canada; Nobelist. SLOW NEUTRON SPECTROSCOPY.

Brockway, Prof. Lawrence O. Deceased; formerly, Department of Chemistry, University of Michigan. ELECTRON DIFFRACTION.

Brod, Dr. Chris. Glen Canyon Environmental Studies, Bureau of Reclamation, Flagstaff, Arizona. TOPOGRAPHIC SURVEYING AND MAPPING.

Broderson, Prof. Klaus. Professor of Inorganic and Analytic Chemistry, University of Erlangen-Nurnberg, Germany. MERCURY (ELEMENT).

Brodsky, Dr. Barbara. Department of Biochemistry, UMDNJ–Robert Wood Johnson Medical School, Piscataway, New Jersey. COLLAGEN.

Brodsky, Dr. Frances M. Becton Dickinson Monoclonal Research Center, Inc., Mountain View, California. MONOCLONAL ANTIBODIES.

Brodsky, William A. Mount Sinai School of Medicine of the City University of New York; Institute for Medical Research and Studies, New York. OSMOREGULATORY MECHANISMS.

Broeg, Dr. Charles B. Consultant, Short Hills, New Jersey. SUGAR.

Brofman, Dr. Peter J. Senior Engineer, General Technology Division, IBM Corp., Hopewell Junction, New York. ELECTRONIC PACKAGING.

Brogden, Dr. Kim A. Professor, College of Dentistry, University of Iowa, Iowa City. POLYMICROBIAL DISEASES.

Bromer, Dr. William W. Research Advisor, Eli Lilly and Company, Indianapolis, Indiana. GLUCAGON.

Bromley, Prof. D. Allan. Deceased; formerly, Sterling Professor of the Sciences and Dean of Engineering, Yale University. HYPERCHARGE; NUCLEAR PHYSICS; other articles.

Bromley, Willard S. Consulting Forester and Association Consultant, New Rochelle, New York. FOREST FIRE; FOREST PLANTING AND SEEDING.

Brongersma, Prof. Mark L. Department of Materials Science and Engineering, Geballe Laboratory for Advanced Materials, Stanford University, California. PLASMON.

Bronstein, Dr. Judith. Department of Ecology and Evolutionary Biology, University of Arizona, Tucson. MUTUALISM.

Brook, Dr. Marx. Department of Physics, New Mexico Institute of Mining and Technology, Socorro. SFERICS.

Brook, Rollins. Bolt Beranek and Newman, Inc., Canoga Park, California. STEREOPHONIC SOUND.

Brooke, Dr. M. M. Deceased; formerly, Deputy Chief, Licensure and Development Branch, Centers for Disease Control, U.S. Public Health Service, Atlanta, Georgia. KINETOPLASTIDA; TRYPANOSOMATIDAE.

Brooks, David J. Pacific Northwest Research Station, Forestry Sciences Laboratory, U.S. Department of Agriculture Forest Service, Corvallis, Oregon. FOREST TIMBER RESOURCES.

Brooks, Dr. Frank D. Department of Physics, University of Capetown, Rondebosch, South Africa. SCINTILLATION COUNTER.

Brooks, Prof. Philip R. Department of Chemistry, Wise School of Natural Sciences, Rice University. CHEMICAL DYNAMICS; CHEMICAL KINETICS.

Brouwer, Prof. Dirk. Deceased; formerly, Director, Observatory, Yale University. GRAVITY.

Browder, Dr. James Steve. Department of Physics, Division of Science and Mathematics, Jacksonville University, Jacksonville, Florida. OPTICAL MATERIALS.

Brown, Dr. Erik Thorson. Large Lakes Observatory, University of Minnesota, Duluth. COSMOGENIC NUCLIDE.

Brown, Dr. Frank. Department of Geology and Geophysics, University of Utah, Salt Lake City. DATING METHODS.

Brown, Prof. Frederick C. Everett, Washington. COLOR CENTERS.

Brown, Dr. Glenn H. Professor of Chemistry and Director of the Liquid Crystal Institute, Kent State University. LIQUID CRYSTALS.

Brown, Prof. Herbert C. Department of Chemistry, Purdue University; Nobelist. HYDROBORATION.

Brown, Michael D. Department of Biochemistry and Molecular Biology, Colorado State University. NEUROBIOLOGY.

Brown, Dr. Perry. Dean, School of Forestry, Montana Forest and Conservation Experiment Station, University of Montana, Missoula. FOREST RECREATION.

Brown, Prof. Steven. Department of Chemistry, University of Delaware, Newark. CHEMOMETRICS.

Brown, Dr. Steven S. National Oceanic and Atmospheric Administration Aeronomy Laboratory, Boulder, Colorado, and Cooperative Institute for Research in the Environmental Sciences, University of Colorado, Boulder. CHEMICAL KINETICS.

Brown, Prof. Terrence D. Forest Research Laboratory, Oregon State University, Corvallis. WOOD PRODUCTS.

Brown, Dr. Thomas G. The Institute of Optics, University of Rochester, New York. OPTICAL FIBERS.

Brown, Dr. William F. Faculty of Medicine, Electromyogram Laboratory, University Hospital of Western Ontario, London, Canada. ELECTROMYOGRAPHY.

Brown, Dr. William L., Jr. Emeritus Professor of Entomology, Cornell University. ARCHAEOGNATHA; DIPLURA; ENDOPTERYGOTA; EXOPTERYGOTA; PROTURA; PTERYGOTA; STREPSIPTERA.

Brownell, Dr. Gordon L. Head, Physics Research Laboratory, Massachusetts General Hospital, Boston. RADIOACTIVE TRACER.

Browning, Dr. Keith A. Retired; formerly, Department of Meteorology, University of Reading, United Kingdom. WEATHER FORECASTING AND PREDICTION.

Brownlee, Prof. Donald E., II. Department of Astronomy, University of Washington, Seattle. COSMIC SPHERULES; MICROMETEORITE.

Brubaker, Prof. Robert R. Department of Microbiology, Michigan State University, East Lansing. YERSINIA.

Bruhn, Dr. Hjalmar D. Agricultural Engineering Department, University of Wisconsin. AGRICULTURAL SOIL AND CROP PRACTICES.

Brummett, Prof. Anna Ruth. Associate Professor of Biology and Associate Dean, Oberlin College. FETAL MEMBRANE.

Brunelle, Daniel. Chief Technologiest, Performance Polymers, GE Global Research, Niskayuna, New York. POLYESTER RESINS.

Brunette, Stephen A. Colorado Springs, Colorado. FORENSIC EVIDENCE.

Brüning, Dr. Oliver. CERN, Geneva, Switzerland. PARTICLE ACCELERATOR.

Brunner, Dr. J. Robert. Department of Food Science and Human Nutrition, Michigan State University. CASEIN.

Brunski, Dr. John B. Department of Biomedical Engineering, Rensselaer Polytechnic Institute. ALLOY.

Brusca, Dr. Richard C. Arizona-Sonora Desert Museum, Tucson. PERACARIDA.

Brush, Dr. Alan H. Department of Biology, University of Connecticut. FEATHER.

Bruun, Dr. Hans H. Department of Mechanical and Medical Engineering, University of Bradford, United Kingdom. TWO-PHASE FLOW.

Bryan, Prof. Leslie A. Director, Institute of Aviation, University of Illinois. AVIATION.

Brylawski, Prof. Thomas. Deceased; formerly, Department of Mathematics, University of North Carolina, Chapel Hill. COMBINATORIAL THEORY.

Bryson, Prof. Arthur E., Jr. Emeritus, Department of Aeronautics and Astronautics, Stanford University. D'ALEMBERT'S PARADOX; OPTIMAL CONTROL THEORY; other articles.

Bu, Prof. Xianhui. Department of Chemistry, University of California, Riverside. SUPERIONIC CONDUCTIVITY.

Bube, Richard H. Professor of Electrical Engineering, Department of Material Sciences, Stanford University. PHOTOCONDUCTIVITY; TRAPS IN SOLIDS.

Bucher, Dr. Martin A. Department of Physics, Princeton University. GAUGE THEORY.

Buchroeder, Dr. Richard A. Optical Design Service, Tucson, Arizona. DISPERSION (RADIATION).

Buck, Dr. Richard P. Department of Chemistry, University of North Carolina. ION-SELECTIVE MEMBRANES AND ELECTRODES.

Buede, Dr. Dennis. Department of Systems Engineering, George Mason University, Fairfax, Virginia. FUNCTIONAL ANALYSIS AND MODELING (ENGINEERING).

Buell, Dr. Heinz. Plessey Electronic Systems Corporation, Wayne, New Jersey. GYROCOMPASS.

Buffetaut, Dr. Eric. Centre National de la Recherche Scientifique, Paris, France. GASTORNITHIFORMES.

Bukowski, Richard W. U.S. Department of Commerce, Washington, D.C. FIRE DETECTOR; SMOKE.

Bulinski, Eugene C. Graphic Arts Technical Foundation, Roswell, Georgia. PRINTING.

Bullington, Kenneth. Deceased; formerly, Bell Telephone Laboratories, Holmdel, New Jersey. RADIO-WAVE PROPAGATION.

Bunge, Prof. Mario. Foundations and Philosophy of Science Unit, McGill University, Montreal, Canada. PHYSICAL LAW; SCIENTIFIC METHODS.

Bunnett, Prof. Joseph F. Department of Chemistry, University of California, Santa Cruz. ELECTROPHILIC AND NUCLEOPHILIC REAGENTS; SUBSTITUTION REACTION; SULFONATION AND SULFATION.

Bunshah, Dr. Rointan F. Department of Materials Science and Engineering, School of Engineering and Applied Science, University of California, Los Angeles. PHYSICAL VAPOR DEPOSITION; VACUUM METALLURGY.

Burdett, Prof. Jeremy K. Department of Chemistry, University of Chicago. MOLECULAR ORBITAL THEORY.

Burdick, Dr. Donald. Richard B. Russell Agricultural Research Center, Athens, Georgia. ANIMAL FEEDS.

Bureau, William H. Deceased; formerly, P.H. Glatfer Company, Spring Grove, Pennsylvania. PRINTING.

Burg, Dr. William R. W. R. Burg and Associates, Inc., Lilburn, Georgia. GAS AND ATMOSPHERE ANALYSIS.

Burgdorfer, Dr. Willy. Rocky Mountain Laboratories, National Institutes of Health, Department of Health and Human Services. BORRELIA; LYME DISEASE; RELAPSING FEVER.

Burghard, Ronald A. Retired; formerly, Intel Corporation, Hillsboro, Oregon. INTEGRATED CIRCUITS.

Burke, Dr. James D. Retired-on-call, Technical Staff, Spacecraft Systems Engineering, Jet Propulsion Laboratory, California Institute of Technology, Pusadena. MOON.

Burke, Prof. Kevin C. Department of Geosciences, University of Houston, Texas. EARTH SCIENCES.

Burkhalter, Dr. Alan. Department of Pharmacology, School of Medicine, University of California, San Francisco. ANTIHISTAMINE.

Burkhart, Dr. Harold E. Department of Forestry, College of Forestry and Wildlife Resources, Virginia Polytechnic Institute and State University, Blacksburg. FOREST MEASUREMENT.

Burmeister, Prof. John. Department of Chemistry and Biochemistry, University of

Delaware, Newark. COORDINATION CHEMISTRY; COORDINATION COMPLEXES.

Burnett, Peter G. Petroleum Engineer, Chicago, Illinois. OIL AND GAS STORAGE.

Burnham, Prof. Charles W. Department of Earth and Planetary Sciences, Harvard University, Cambridge, Massachusetts. PYROXENE; PYROXENOID.

Burns, Dr. C. Patrick. Professor, Division of Hematology, Oncology, and Blood and Marrow Transplantation, Department of Medicine and The University of Iowa Cancer Center, University of Iowa, Iowa City. CHEMOTHERAPY AND OTHER ANTINEOPLASTIC DRUGS.

Burns, Roger G. Deceased; formerly, Department of Earth Science, Massachusetts Institute of Technology. MANGANESE NODULES.

Burns, Prof. Stanley G. Department of Electrical Engineering and Computer Engineering, Iowa State University. RIPPLE VOLTAGE; SEMICONDUCTOR RECTIFIER.

Burns, Dr. Virginia Mee. Cambridge, Massachusetts. MANGANESE NODULES.

Bursey, Prof. Maurice M. Durham, North Carolina. MASS SPECTROMETRY; SPECTROSCOPY.

Bursky, David. Electronic Design, San Jose, California. SEMICONDUCTOR MEMORIES.

Burstyn, Dr. Judith N. Department of Chemistry, University of Wisconsin, Madison. NITRIC OXIDE.

Burton, Dr. Glenn W. Research Geneticist, Coastal Plain Experiment Station, Agricultural Research Service, U.S. Department of Agriculture, Tifton, Georgia. BERMUDA GRASS; GRASS CROPS.

Burton, Prof. Ralph A. Chairperson, Department of Mechanical Engineering and Astronautical Sciences, Northwestern University. ICE POINT; THERMAL EXPANSION.

Burwell, Dr. Robert L., Jr. Department of Chemistry, Northwestern University. CATALYSIS; HETEROGENEOUS CATALYSIS.

Busch, Prof. Daryle H. Department of Chemistry, Ohio State University. STRUCTURAL CHEMISTRY.

Busfield, Dr. James. Department of Materials, Queen Mary, University of London, United Kingdom. RUBBER.

Bushnell, Dennis M. Langley Research Center, National Aeronautics and Space Administration, Hampton, Virginia. STREAMLINING.

Busta, Dr. Francis F. Department of Food Science and Human Nutrition, University of Florida. PASTEURIZATION.

Butler, Dr. G. W. Olin Rocket Research Company, Redmond, Washington. ELECTROTHERMAL PROPULSION.

Butler, Dr. Percy M. Egham, Surrey, United Kingdom. INSECTIVORA.

Butler, Dr. R. Paul. Staff Scientist, Carnegie Institute of Washington, Terrestrial Magnetism, Washington, D.C. EXTRASOLAR PLANETS; PLANET.

Butler, Dr. Robert W. Department of Psychology, Child Development and Rehabilitation Center, Oregon Health Sciences University, Portland. POSTTRAUMATIC STRESS DISORDER.

Byers, Dr. Charles W. Department of Geology, University of Wisconsin. BLACK SHALE; DIASTEM; UNCONFORMITY.

Byrd, Dr. Robert. Department of Pediatrics, UC Davis Medical Center, Sacramento, California. AUTISM.

Byrnes, Dr. Dennis V. Jet Propulsion Laboratory, Pasadena, California. SPACE NAVIGATION AND GUIDANCE.

C

Cable, Prof. Raymond M. Department of Biological Sciences, Purdue University. CERCARIA; DIGENEA.

Cabrera, Dr. Blas. Department of Physics, Stanford University. MAGNETIC MONOPOLES.

Caceres, Dr. Cesar A. Institute for Technology in Health Care, Washington, D.C. MEDICAL INFORMATION SYSTEMS.

Cadenhead, Prof. D. Allen. Department of Chemistry, University at Buffalo, New York. MONOMOLECULAR FILM.

Cady, Dr. Gilbert H. Deceased; formerly, Consulting Coal Geologist, Urbana, Illinois. PEAT.

Cahn, Dr. John W. National Institutes of Standards and Technology, Gaithersburg, Maryland. QUASICRYSTAL.

Cain, Dr. Joseph C. Department of Geology, Florida State University, Tallahassee. GEOMAGNETISM.

Cain, Dr. William S. John B. Pierce Foundation Laboratories, New Haven, Connecticut. OLFACTION.

Caine, Dr. Edsel A. Associate Professor of Marine Sciences, University of South Carolina. CAPRELLIDEA.

Caine, Nel. Institute of Arctic and Alpine Research, University of Colorado at Boulder. DENUDATION.

Calabrese, Dr. Ronald L. Department of Biology, Emory University. HEART (INVERTEBRATE).

Calderwood, Dr. James H. Department of Electronic Engineering, University College, Galway, Ireland. DIELECTRIC MATERIALS; DIELECTRIC MEASUREMENTS; PERMITTIVITY.

Caldwell, Prof. C. Denise. Department of Physics, Yale University. MOLECULAR WEIGHT.

Caldwell, Dr. Robert. Department of Physics and Astronomy Dartmouth College, Hanover, New Hampshire. COSMOLOGICAL CONSTANT; DARK ENERGY.

Callaham, Dr. Michael. Director, Emergency Medical Services, School of Medicine, University of California, San Francisco. EMERGENCY MEDICINE.

Callaway, Prof. Joseph. Deceased; formerly, Department of Physics and Astronomy, Louisiana State University, and Agricultural and Mechanical College, Baton Rouge. BAND THEORY OF SOLIDS; BRILLOUIN ZONE; HOLE STATES IN SOLIDS; KRONIG-PENNEY MODEL.

Callister, William D. Adjunt Professor, Department of Metallurgical Engineering, The University of Utah. PLASTICITY.

Calnek, Dr. Bruce W. Chairperson, Department of Avian and Aquatic Animal Medicine, College of Veterinary Medicine, Cornell University. AVIAN LEUKOSIS.

Cambel, Dr. Ali B. Professor Emeritus of Engineering and Applied Science, George Washington University, Washington, D.C. GAS DYNAMICS.

Camin, Dr. Joseph H. Department of Entomology, University of Kansas. OPILIOACARIFORMES.

Camp, Dr. Michael J. Director, State Crime Laboratory, Milwaukee, Wisconsin. FORENSIC CHEMISTRY.

Campana, Dr. Richard J. Department of Botany and Plant Pathology, University of Maine. TREE DISEASES.

Campanella, Dr. Angelo J. Campanella Associates, Hilliard, Ohio. SOUND ABSORPTION; SOUND-RECORDING STUDIO.

Campanella, Dr. S. Joseph. Retired; formerly, COMSAT Laboratories, Clarksburg, Maryland. DATA COMPRESSION.

Campbell, Dr. Andrew C. Department of Zoology and Comparative Physiology, Queen Mary College, University of London, England. ASTEROIDEA; ECHINODERMATA; EUECHINOIDEA.

Campbell, Dr. Eleanor. Max-Born-Institut, Berlin, Germany. ATOM CLUSTER.

Campbell, Dr. Howard W. National Fisheries and Wildlife Laboratory, Gainesville, Florida. ALLIGATOR; CROCODILE.

Campbell, Dr. John S. Department of Biological Sciences, University of Lethbridge, Alberta, Canada. ALTITUDINAL VEGETATION ZONES.

Campbell, Dr. Laurence J. Low Temperature Physics Group, Los Alamos Scientific Laboratory, Los Alamos, New Mexico. SUPERFLUIDITY.

Campbell, Dr. Wallace H. U.S. Geological Survey, Denver, Colorado. GEOMAGNETIC VARIATIONS.

Campbell, Dr. William B. Professor and Chairman, Department of Pharmacology and Toxicology, Medical College of Wisconsin, Milwaukee. EICOSANOIDS.

Canfield, Dr. Robert E. Department of Medicine, College of Physicians and Surgeons, Columbia University. LYSOZYME.

Cannell, Prof. David S. Department of Physics, University of California, Santa Barbara. SHADOWGRAPH.

Cannon, Prof. Robert H., Jr. Chairperson, Department of Aeronautics and Astronautics, Stanford University. GYROCOMPASS; SCHULER PENDULUM.

Canter, Prof. Gerald J. Department of Communicative Disorders, Northwestern University. AGNOSIA; APRAXIA.

Capellen, Dr. Jennings. Rare Earth Information Center, Iowa State University. SCANDIUM—coauthored.

Capitaine, Dr. Nicole. Observatoire de Paris, France. ASTRONOMICAL CATALOGS.

Capra, Dr. J. Donald. Department of Microbiology, Southwestern Medical School, University of Texas, Dallas. IMMUNOGLOBULIN.

Capstick, Dr. Simon. Department of Phyics, Florida State University, Tallahassee. BARYON.

Carberry, Dr. James J. Department of Chemical Engineering, University of Notre Dame. CHEMICAL REACTOR.

Carbone, Dr. Richard E. Director, Atmospheric Technology Division, National Center for Atmospheric Research, Boulder, Colorado. METEOROLOGICAL INSTRUMENTATION.

Cardullo, Dr. Frank. Watson School, Engineering Technology, State University of New York. AIRCRAFT TESTING.

Carew, Dr. H. John. Professor of Horticulture, Michigan State University. BEET; CARROT; EGGPLANT; GARLIC; PUMPKIN; other articles.

Carey, Dr. Paul D. MRC Unit, Department of Psychiatry, University of Stellenbosch Medical School, South Africa. ANXIETY DISORDERS.

Carey, Prof. Tom. Department of Computing and Information Science, College of Physical Science, University of Guelph, Ontario, Canada. COMPUTER; HUMAN-COMPUTER INTERACTION.

Carey, Prof. William M. Boston University, Department of Aerospace and Mechanical Engineering, Boston, Massachusetts. SOUND PRESSURE.

Carlson, Dr. Don M. Division of Biological Sciences, University of California, Davis. OLIGOSACCHARIDE.

Carlson, Dr. Jack H. Rhone Meriux, Inc., Athens, Georgia. FELINE PANLEUKOPENIA.

Carlson, Dr. Sandra J. Department of Geology, University of California, Davis. BRACHIOPODA; INARTICULATA; RHYNCHO-NELLIFORMEA.

Carlson, William C. Weyerhaeuser Company, Hot Springs, Arkansas. TREE GROWTH.

Carlson, Dr. William D. Department of Geological Sciences, University of Texas, Austin. ENSTATITE.

Carlsson, Prof. Anders E. Department of Physics, Washington University, St. Louis, Missouri. COHESION (PHYSICS).

Carlsson, Dr. D. J. National Research Council of Canada, Ottawa, Ontario. PHOTODEGRADATION.

Carlton, Dr. Peter L. Department of Psychiatry, University of Medicine and Dentistry of New Jersey, Rutgers Medical School. PSYCHOPHARMACOLOGY; TRANQUILIZER.

Carney, Prof. Bruce W. Samuel Baron Professor of Astronomy, University of North Carolina, Chapel Hill. STAR CLUSTERS.

Carpenter, Dr. C. J. Bath, Avon, England. MAGNETISM.

Carpenter, Dr. Frank M. Department of Biology, Harvard University, Cambridge. NEUROPTERA; THYSANOPTERA.

Carpenter, Dr. Stephen, R. Department of Biological Sciences, University of Notre Dame. TROPHIC ECOLOGY.

Carraher, Prof. Charles E., Jr. Department of Chemistry, Florida Atlantic University, Boca Raton. POLYMER.

Carroll, Dr. C. Ronald. Institute of Ecology, University of Georgia, Athens. AGROECOSYSTEM.

Carroll, Dr. John M. Formerly, Associate Professor of Computer Science, University of Western Ontario, London, Canada. CONTINUOUS-WAVE RADAR; ELECTRONIC LISTENING DEVICES; MOVING-TARGET INDICATION.

Carroll, Robert G., Jr. Manager of Technical Development, Mirafi, Inc., Charlotte, North Carolina. GEOSYNTHETIC.

Carroll, Dr. Robert L. Redpath Museum, McGill University, Montreal, Quebec, Canada. MESOSAURIA; MICROSAURIA; TEMNOSPONDYLI; other articles.

Carruthers, Dr. Marvin H. Department of Chemistry, University of Colorado. OLIGONUCLEOTIDE.

Carson, Dr. John W. President, Jenike & Johanson Inc., North Billerica, Massachusetts. FLOW OF GRANULAR SOLIDS.

Carson, Robert W. Formerly, Managing Editor, "Product Engineering," New York. CLADDING.

Carter, Archie N. Buan, Carter & Associates, Inc., Minneapolis, Minnesota. PAVEMENT; SUBWAY ENGINEERING.

Carter, Dr. G. R. Professor Emeritus, Department of Pathobiology, Virginia-Maryland Regional College of Veterinary Medicine, Virginia Polytechnic Institute and State University, Blacksburg. PASTEURELLOSIS.

Carter, Prof. Herbert E. Vice Chancellor for Academic Affairs, University of Illinois. TRIGLYCERIDE (TRIGLYCEROL).

Casasent, Prof. David. Department of Electrical and Computer Engineering, Carnegie-Mellon University. OPTICAL INFORMATION SYSTEMS.

Casey, Dr. H. C., Jr. Department of Electrical Engineering, Duke University. SEMICONDUCTOR HETEROSTRUCTURES.

Cash, Dr. William Benjamin. Department of Biology, Maryville College, Tennessee. ALLIGATOR; AMPHIBIA; ANURA; APODA; CROCODILE; CROCODYLIA; REPTILIA; SKINK; SQUAMATA.

Cassens, Prof. Linda J. Department of Plant Sciences, College of Agriculture, University of Arizona. GUAYULE.

Castro, Dr. Joaquin H. Pratt and Whitney Space Propulsion, West Palm Beach, Florida. SCRAMJET.

Cataliotti, Dr. Francesco Saverio. Department of Physics, University of Catania, Italy, and European Laboratory for Nonlinear Spectroscopy (LENS), Sesto Fiorentino, Italy. LASER SPECTROSCOPY.

Catania, Prof. A. Charles. Department of Psychology, University of Maryland. INSTRUMENTAL CONDITIONING.

Catchpole, John P. Admiralty Materials Laboratory, Dorset, England. DIMENSIONLESS GROUPS.

Caton, Prof. Jerald A. Department of Mechanical Engineering Texas A and M University, College Station. COGENERATION SYSTEMS.

Caughlan, Dr. Georgeanne R. Deceased; formerly, Department of Physics, Montana State University, Bozeman. CARBON-NITROGEN-OXYGEN CYCLES.

Cavender, Ted M. Museum of Zoology, University of Michigan. ASPIDORHYNCHIFORMES; PALAEONISCIFORMES; PYCNODONTIFORMES.

Cavey, Michael J. Department of Biology, University of Calgary, Alberta, Canada. EMBRYOGENESIS; EMBRYOLOGY.

Ceccio, Prof. Steven. Department of Mechanical Engineering, University of Michigan, Ann Arbor. WATER TUNNEL (RESEARCH AND TESTING).

Celander, Dr. Daniel W. Department of Microbiology, University of Illinois, Urbana. RIBOZYME.

Celler, Dr. George K. AT&T Bell Laboratories, Murray Hill, New Jersey. LASER-SOLID INTERACTIONS.

Cestone, Joseph A. Consultant, Navigation, Sumner, Maryland. UNDERWATER NAVIGATION.

Chace, Dr. Fenner A., Jr. Senior Zoologist, Department of Invertebrate Zoology, Smithsonian Institution, Washington, D.C. DECAPODA (CRUSTACEA).

Chambers, Dr. James P. Research Scientist and Research Assistant Professor of Mechanical Engineering, University of Mississippi, Oxford. SOUND ABSORPTION.

Chambron, Dr. Jean-Claude. Laboratoire de Chimie Organo-Minérale, Institut le Bel, Université Louis Pasteur, Strasbourg, France. CATENANES.

Chameides, Prof. William L. School of Earth and Atmospheric Sciences, Georgia Institute of Technology, Altanta. SMOG.

Chan, Dr. Moses H. W. Department of Physics, Pennsylvania State University. CRITICAL PHENOMENA.

Chandler, Dr. David W. Sandia National Laboratories, Livermore, California. LASER PHOTOCHEMISTRY.

Chang, Prof. Howard H. Department of Civil and Environmental Engineering, San Diego State University, San Diego, California. RIVER ENGINEERING.

Chang, Prof. Kang-tsung. Department of Geography, University of Idaho, Moscow. GREAT CIRCLE, TERRESTRIAL.

Chang, Dr. Luke L. Y. Department of Geology,University of Maryland, College Park. CERUSSITE; MAGNESITE.

Chang, Dr. M. C. Worcester Foundation for Experimental Biology, Shrewsbury, Massachusetts. REPRODUCTIVE SYSTEM.

Chang, Prof. S. S. L. Department of Electrical Sciences, State University of New York, Stony Brook. STOCHASTIC CONTROL THEORY.

Chao, Dr. Edward C. T. U.S. Geological Survey, Department of the Interior, Reston, Virginia. ALUNITE; ANHYDRITE; COESITE; SCHEELITE; STISHOVITE; WOLFRAMITE; other articles.

Chaplin, Prof. Robin. Department of Chemical Engineering University of New Brunswick, Canada. LEVEL MEASUREMENT.

Chapman, Dr. Carleton A. Department of Geology, University of Illinois. DACITE; DOLERITE; GRANODIORITE; IGNEOUS ROCKS; NEPHELINITE; other articles.

Chapman, Dr. Clark R. Southwest Research Institute, Boulder, Colorado. ASTEROID.

Chapman, Dr. Kenneth. Division of Respiratory Medicine, Toronto Western Hospital, Toronto, Ontario, Canada. OXIMETRY.

Chapman, Dr. Richard A. Central Research Laboratory, Texas Instruments, Inc., Dallas. PHOTOELECTRIC DEVICES.

Chapman, Dr. Robert D. Retired; Columbia, Maryland. TELESCOPE.

Chapman, Prof. Russell L. Chairperson, Department of Botany, Louisiana State University, Baton Rouge. CHAROPHYCEAE.

Chappell, Dr. Charles F. University Corporation for Atmospheric Research, COMET, Boulder, Colorado. STORM DETECTION.

Charig, Dr. Alan J. Chief Curator of Fossil Amphibians, Reptiles and Birds, British Museum, London, England. LEPIDOSAURIA; SPHENODONTA.

Charlton, Dr. Ernest E. Retired; formerly, Research Engineer, Schenectady, New York. X-RAY TUBE.

Charney, Dr. Dennis S. Department of Psychiatry, West Haven Virginia Medical Center, Yale University School of Medicine, New Haven, Connecticut. NORADRENERGIC SYSTEM.

Charnock, Dr. Henry. Department of Oceanography, University of Southampton, England. WIND STRESS.

Charrier, Dr. J. M. Department of Chemical Engineering, McGill University, Montreal, Quebec, Canada. SPINNING (TEXTILES).

Chase, Dr. Mark W. Head, Molecular Systematics Section, Jodrell Laboratory, Botanic Gardens, Kew, United Kingdom. DICOTYLEDONS; EUDICOTYLEDONS; LILIALES; MALPIGHIALES; MAGNOLIALES; MONOCOTYLEDONS; PLANT PHYLOGENY; ROSALES; SAXIFRAGALES; other articles.

Chaudhari, Dr. Nirupa. Department of Physiology and Biophysics, University of Miami School of Medicine, Miami, Florida. TASTE.

Chaudhari, Praveen. Thomas J. Watson Research Center, IBM, Yorktown Heights, New York. MAGNETIC THIN FILMS.

Chavez, Tomas C. Las Cruces, New Mexico. TELEMETERING.

Cheadle, Prof. Vernon I. Chancellor, University of California, Santa Barbara. LATERAL MERISTEM; PERICYCLE; XYLEM.

Chellis, Robert D. Deceased; formerly, Structural Engineer, Wellesley Hills, Massachusetts. CAISSON FOUNDATION.

Chen, Dr. Brian. University of Texas. PLATE GIRDER.

Chen, Dr. Chong-maw. Department of Life Sciences, Biological Sciences/Biomedical Research Institute, University of Wisconsin, Parkside. CYTOKININS.

Chen, Dr. Mo-Shing. Department of Electrical Engineering, University of Texas, Arlington. ELECTRIC HEATING.

Chen, Prof. Peter P. Director of the Institute of Computer and Information Systems Research, Louisiana State University. COMPUTER STORAGE TECHNOLOGY.

Chen, Dr. Robert T. Medical Epidemiologist, Infant Immunization Section, Centers for Disease Control, Surveillance Investigations and Research Branch, Atlanta, Georgia. DIPHTHERIA.

Chen, Dr. Shih-Yuan. Department of Environmental Sciences, Rand Corp., Santa Monica, California. AEROTHER-MODYNAMICS; HYPERSONIC FLIGHT; THERMAL STRESS.

Chenault, Dr. Keith. Department of Chemistry, University of Delaware, Newark. BIOORGANIC CHEMISTRY.

Chenault, Roy L. Retired; formerly, Chief Research Engineer, Oilwell Division, United States Steel Corporation, Dallas, Texas. OIL AND GAS FIELD EXPLOITATION.

Cheng, Dr. T. C. Department of Electrical Engineering, University of Southern California. IMPULSE GENERATOR; INSULATION RESISTANCE TESTING.

Cherlin, Dr. Gregory L. Mathematical Sciences Research Institute and Rutgers University, Berkeley, California. LOGIC.

Chern, Prof. S. S. Deceased; formerly, Mathematical Sciences Research Institute, University of California, Berkeley. RIEMANNIAN GEOMETRY.

Chernosky, Prof. J. V. Department of Geological Science, University of Maine. ANTHOPHYLLITE.

Chickering, Dr. Arthur M. Museum of Comparative Zoology, Harvard University. ARANEAE.

Chien, Dr. Shu. Institute for Biomedical Engineering, University of California, La Jolla, California. BIOMECHANICS.

Chin, Dr. Gilbert Y. Deceased; formerly, AT&T Bell Telephone Laboratories, Murray Hill, New Jersey. FERRITE; FERRITE DEVICES.

Chinitz, Dr. Wallace. Department of Mechanical Engineering, The Cooper Union, New York. CHEMICAL FUEL; ROTARY ENGINE.

Chiodo, Prof. Louis A. Chairperson, Department of Pharmacology, Texas Tech University Health Sciences Center, Lubbock. DOPAMINE.

Chipkin, Dr. Richard E. Department of Pharmacology, Schering Corporation, Bloomfield, New Jersey. AMPHETAMINE.

Chisholm, Dr. Malcolm. Department of Chemistry, Indiana University. METAL CLUSTER COMPOUNDS.

Chojnowski, Prof. Julian. The Centre of Molecular and Macromolecular Studies, Department of Heteroorganic Polymers, Poland. SILICONE RESINS.

Chokroverty, Prof. Sudhansu. Chief, Neurology Service, VA Medical Center, Lyons, New Jersey. SLEEP DISORDERS.

Chomel, Dr. Bruno B. Department of Population Health and Reproduction, School of Veterinary Medicine, University of California, Davis. CAT SCRATCH DISEASE.

Choueiri, Dr. Edgar Y. Director, Electric Propulsion and Plasma Dynamics Laboratory, Engineering Quadrangle, Princeton University. PLASMA PROPULSION.

Chow, Prof. Wen L. Emeritus, Department of Mechanical Engineering, College of Engineering and Computer Science, Florida Atlantic University, Boca Raton. NOZZLE.

Choy, Dr. Patrick C. Department of Biochemistry, University of Manitoba Faculty of Medicine, Winnipeg, Manitoba, Canada. PHOSPHOLIPID.

Christ, Dr. Charles L. Physicist, U.S. Geological Survey. BORACITE.

Christensen, Dr. Bruce M. Assistant Professor, Department of Veterinary Science, University of Wisconsin, Madison. MOSQUITO.

Christensen, Dr. Martha. Department of Botany and Plant Pathology, Oregon State University, Corvallis. FUNGAL ECOLOGY.

Christenson, Dr. Todd R. Photonics and Microfabrication Department, Sandia National Laboratories, Albuquerque, New Mexico. MICRO-ELECTRO-MECHANICAL SYSTEMS (MEMS).

Christiansen, Dr. Kenneth. Professor Emeritus of Biology, Department of Biology, Grinnell College, Iowa. COLLEMBOLA; PARAINSECTA.

Christiansen, Dr. R. L. Petroleum Engineering Department, Colorado School of Mines, Golden. PETROLEUM ENHANCED RECOVERY.

Christie, Dr. John M. Department of Geology, University of California. Los Angeles. PETROFABRIC ANALYSIS; STRUCTURAL PETROLOGY.

Christie, R. W. Hardesty & Hanover, LLP, Consulting Engineers, New York. BRIDGE.

Christie-Blick, Dr. Nicholas. Professor of Geology, Lamont-Doherty Earth Observatory, Columbia University, Palisades, New York. FORMATION; STRATIGRAPHY.

Christophorou, Dr. L. G. Head, Atomic, Molecular and High Voltage Physics Group,

Oak Ridge National Laboratory, Oak Ridge, Tennessee. BREAKDOWN POTENTIAL.

Christou, Dr. George. Department of Chemistry, Indiana University. MANGANESE.

Christy, Dr. Nicholas. College of Physicians and Surgeons, Columbia University. ADRENAL GLAND DISORDERS; PITUITARY GLAND DISORDERS.

Chromey, Prof. Frederick. Department of Physics and Astronomy, Vassar College, Poughkeepsie, New York. ASTRONOMICAL IMAGING.

Chu, Prof. Benjamin. Department of Chemistry, State University of New York, Stony Brook. OPALESCENCE.

Chu, Prof. Steven. Departments of Physics and Molecular and Cellular Biology, University of California, Berkeley; Director, Lawrence Berkeley National Laboratory; United States Secretary of Energy—designate; Nobelist. PARTICLE TRAP.

Chu, Dr. T. Ming. Roswell Park Memorial Institute, Buffalo, New York. ONCOFETAL ANTIGENS.

Churchill, Dr. Steven. Department of Anthropology, University of New Mexico, Albuquerque. NEANDERTALS.

Cicerone, Dr. Ralph J. Director, Atmospheric Chemistry and Aeronomy Division, National Center for Atmospheric Research, Boulder, Colorado. HALOGEN ELEMENTS.

Ciechanover, Dr. Aaron J. The Bruce Rappaport Faculty of Medicine, Technion-Israel Institute of Technology, Israel; Nobelist. PROTEIN; PROTEIN DEGRADATION.

Ciegler, Dr. Alex. Southern Regional Research Center, USDA Science and Education Administration, New Orleans, Louisiana. MYCOTOXIN.

Clack, Dr. Jennifer A. Department of Zoology, University of Cambridge, United Kingdom. ANTHRACOSAURIA; ICHTHYOSTEGA.

Clapp, Philip F. National Weather Service, National Oceanic and Atmospheric Administration, Washington, D.C. WEATHER.

Claridge, Dr. Elmond L. Retired; formerly, Director of Graduate Program in Petroleum Engineering, Chemical Engineering Department, University of Houston Central Campus, Houston. PETROLEUM; PETROLEUM RESERVOIR ENGINEERING.

Clark, Dr. David A. Department of Medicine, McMaster University, Hamilton, Ontario, Canada. CYTOKINE.

Clark, Prof. Francis. Department of Microbiology, University of Illinois. ANTISEPTIC.

Clark, Prof. George L. Emeritus Research Professor of Analytical Chemistry, University of Illinois. RADIOGRAPHY; X-RAYS.

Clark, Prof. John F. Director, Graduate Studies, and Professor, Space Systems, Spaceport Graduate Center, Florida Institute of Technology, Satellite Beach. SATELLITE (SPACECRAFT); SPACE COMMUNICATIONS; SPACE SHUTTLE.

Clark, Dr. Nancy Barnes. Department of Physiology and Neurobiology, University of Connecticut. ENDOCRINE SYSTEM (VERTEBRATE); PARATHYROID GLAND.

Clark, Dr. Noel A. Department of Physics and Astrophysics, University of Colorado. COLLOIDAL CRYSTALS.

Clark, Richard F. Division of Physics, National Research Council of Canada, Ottawa, Ontario. MICROWAVE MEASUREMENTS.

Clark, Wallis H., Jr. Director, Aquaculture Program, University of California, Davis. AQUACULTURE.

Clarke, Dr. J. Harold. Horticulturist, Clarke Nursery, Long Beach, Washington. BLUEBERRY; GOOSEBERRY; PINEAPPLE; POMEGRANATE; STRAWBERRY.

Clarke, Prof. John. Department of Physics, University of California, Berkeley. SQUID; SUPERCONDUCTING DEVICES.

Clarke, John H. Engineering Design Consultant, Air Conditioning and Power, Films-Packaging Division, Union Carbide Corporation, Chicago, Illinois. VENTILATION.

Clemence, Dr. Gerald M. Deceased; formerly, Observatory, Yale University. CALENDAR; TWILIGHT; other articles.

Clemens, Dr. William A. Department of Paleontology, University of California, Berkeley. DOCODONTA; EUTRICONODONTA (TRICONODONTA).

Clements, Dr. Leo P. Deceased; formerly, Emeritus Professor of Anatomy, School of Medicine, Creighton University, Omaha, Nebraska. RESPIRATORY SYSTEM.

Cleve, Dr. Hartwig. Institute for Anthropology and Human Genetics, University of Munich, Germany. BLOOD.

Cleveland, Prof. L. F. Department of Electrical Engineering, Northeastern University, Boston, Massachusetts. DIRECT-CURRENT MOTOR.

Cleveland, Dr. Robin. Department of Aerospace and Mechanical Engineering, Boston University, Boston, Massachusetts. SONIC BOOM.

Cline, Dr. Douglas. Nuclear Structure Laboratory, University of Rochester, New York. COULOMB EXCITATION.

Close, Prof. Frank. Department of Theoretical Physics, Oxford University, Oxford, United Kingdom. ELEMENTARY PARTICLE; MESON.

Cloud, Dr. Preston. Department of Geology, University of California, Santa Barbara. REEF.

Clowes, Dr. F. A. L. Department of Botany, Oxford University, England. CHIMERA.

Clydesdale, Dr. Fergus M. Department of Food Science and Nutrition, University of Massachusetts. FOOD MANUFACTURING.

Coakley, Bernard J. Lamont-Doherty Geological Observatory, Columbia University, Palisades, New York. ISOSTASY.

Coates, Dr. Michael. Department of Organismal Biology and Anatomy, University of Chicago, Illinois. CLADOSELACHE; XENACANTHIDAE.

Coates, Dr. Peter B. Retired; formerly, Division of Electrical Science, National Physical Laboratory, Teddington, Middlesex, United Kingdom. FREQUENCY MEASUREMENT.

Cochran, Dr. James R. Lamont-Doherty Earth Observatory, Columbia University, Palisades, New York. RED SEA.

Cocks, Dr. George. Los Alamos National Laboratory, Los Alamos, New Mexico. CHEMICAL MICROSCOPY.

Cocks, Dr. L. Robin M. Department of Palaeontology, The Natural History Museum, London, United Kingdom. PENTAMERIDA.

Coddington, Dr. Jonathan A. National Museum of Natural History, Smithsonian Institution, Washington, D.C. SCHIZOMIDA.

Coffin, Dr. Allison. Research Associate, Aquatic Bioacoustics Lab, University of Maryland, College Park. SENSORY CELL REGENERATION.

Cohen, Dr. Alan S. Director, Thorndike Memorial Laboratory, Boston City Hospital, Boston, Massachusetts. AMYLOIDOSIS.

Cohen, Dr. David. National Magnetic Laboratory, Massachusetts Institute of Technology. BIOMAGNETISM.

Cohen, Dr. E. Richard. Retired; formerly, Distinguished Fellow, Rockwell International Science Center, Thousand Oaks, California. FUNDAMENTAL CONSTANTS.

Cohen, Dr. Gene. National Institute on Aging, National Institutes of Health, Bethesda, Maryland. AGING.

Cohen, Dr. J. John. Department of Immunology, University of Colorado School of Medicine, Denver. ANTIBODY; FOOD ALLERGY; IMMUNOPATHOLOGY; RADIOIMMUNOASSAY; SEROLOGY.

Cohen, Dr. Jay O. Deceased; formerly, Microbiologist, Department of Health and Human Services, Public Health Service, Centers for Disease Control, Atlanta, Georgia. TOXIC SHOCK SYNDROME.

Cohen, Dr. Jerome. Department of Psychiatry, Medical School, Northwestern University. ELECTROENCEPHALOGRAPHY.

Cohen, Dr. Jerry D. Plant Hormone Laboratory, U.S. Department of Agriculture, Beltsville, Maryland. AUXIN.

Cohen, Dr. Lisa M. Department of Dermatology, Beth Israel Hospital, Boston, Massachusetts. DERMATITIS.

Cohen, Dr. Steven A. Senior Research Associate, INEOS Technologies, Naperville, Illinois. POLYOLEFIN RESINS.

Cohen, Dr. Zoe. College of Nursing, University of Arizona, Tucson. ANTIGEN; ONCOFETAL ANTIGENS.

Colbertson, Dr. Joseph O. Retired; formerly, Leader, Industrial Crops Investigations, U.S. Department of Agriculture, Beltsville, Maryland. FLAX.

Colby, Dr. Kenneth Mark. Department of Psychiatry and Behavioral Sciences, School of Medicine, University of California, Los Angeles. PARANOIA.

Cole, Dr. Edward H. Department of Immunopathology, Scripps Clinic and Research Foundation, La Jolla, California. IMMUNE COMPLEX DISEASE.

Cole, Dr. Michael. Norris Cotton Cancer Center, Dartmouth-Hitchcock Medical Center, Lebanon, New Hampshire. ONCOGENES.

Coleman, Dr. David C. Department of Entomology, University of Georgia, Athens. BIOLOGICAL PRODUCTIVITY; SOIL ECOLOGY.

Coleman, Dr. Robert E. Staff Scientist, Sugar Crops Production, National Program Staff, BARC-WEST, Beltsville, Maryland. SUGARCANE.

Collier, R. J. Bell Telephone Laboratories, Murray Hill, New Jersey. MAGNETRON.

Collins, Dr. Dean R. Director, CCD Technology Laboratory, Central Research Laboratories, Texas Instruments, Inc., Dallas. LIGHT AMPLIFIER.

Collins, Dr. Francis. Department of Health and Human Services, National Human Genome Research Institute, Bethesda, Maryland. HUMAN GENOME.

Collins, Dr. Steven J. Department of Clinical Neurosciences, St. Vincent's Hospital, Melbourne, and Department of Pathology, University of Melbourne, Australia. PRION DISEASE.

Collinson, Dr. Charles W. Department of Geology, University of Illinois. INDEX FOSSIL.

Colwell, Dr. Rita R. Department of Microbiology, Division of Agricultural and Life Sciences, University of Maryland. NUMERICAL TAXONOMY.

Comeau, Richard. Foxboro Company, Foxboro, Massachusetts. MCLEOD GAGE; VACUUM MEASUREMENT.

Comer, Prof. Jonathan. Deparment of Geography, Oklahoma State University. MATHEMATICAL GEOGRAPHY.

Comings, Dr. Edward W. University of Petroleum and Minerals, Dhahran, Saudi Arabia. HIGH-PRESSURE PROCESSES.

Cominsky, Prof. Lynn R. Department of Physics and Astronomy Sonoma State University, Rohnert Park, California. GAMMA-RAY BURSTS.

Commins, Prof. Eugene D. Department of Physics, University of California, Berkeley. HELICITY (QUANTUM MECHANICS); LEPTON; WEAK NUCLEAR INTERACTIONS.

Comp, Dr. Philip C. Director, Thrombosis and Coagulation Laboratory, Oklahoma Memorial Hospital, University of Oklahoma. SEROLOGY; UROKINASE.

Compton, Robert D. Electro-Optical Systems Design, Milton S. Kiver Publications, Inc., Chicago, Illinois. OPTICAL ISOLATOR.

Condie, Dr. Kent C. Department of Earth and Environmental Sciences, New Mexico Tech, Socorro. ARCHEAN.

Confer, Prof. Anthony W. Head, Department of Veterinary Pathology, Oklahoma State University, Stillwater. SHIPPING FEVER.

Connors, Dr. James F. Deceased; formerly, Propulsion Aerodynamics Division, National Aeronautics and Space Administration. SUPERSONIC DIFFUSER.

Conrad, Prof. Albert G. College of Engineering, University of California, Santa Barbara. INDUCTION MOTOR.

Constabel, Dr. Fred. Plant Biotechnology Institute, National Research Council, Saskatoon, Saskatchewan, Canada. BREEDING (PLANT).

Constanzo, Thomas. Capital Construction, Waterways Development, Port Authority of New York and New Jersey, New York. HARBORS AND PORTS.

Converse, Dr. Richard H. Department of Botany, Oregon State University. CURRANT.

Conway Morris, Dr. Simon. Department of Earth Sciences, The Open University, Milton Keynes, England. ANNELIDA; BURGESS SHALE.

Conwell, Dr. Esther M. Department of Chemistry, University of Rochester, New York. CONDUCTION (ELECTRICITY).

Cook, Dr. Charles G. Research Geneticist, Agricultural Research Service, U.S. Department of Agriculture, Weslaco, Texas. KENAF.

Cook, Dr. Frank. Research Assistant Professor, Department of Veterinary Science, Gluck Equine Research Center, University of Kentucky, Lexington. EQUINE INFECTIOUS ANEMIA.

Cook, Dr. Gerald. Department of Electrical and Computer Engineering, George Mason University, Fairfax, Virginia. GUIDANCE SYSTEMS; STOCHASTIC CONTROL THEORY.

Cook, Philip H. Plastics Department, Texas Division, Dow Chemical U.S.A., Freeport, Texas. GLYCEROL.

Cooley, Prof. Hollis R. Professor Emeritus of Mathematics, New York University. BINOMIAL THEOREM; ROOT-MEAN-SQUARE.

Cooley, Dr. William E. Deceased; formerly, Winton Hill Technical Center, Proctor and Gamble Company, Cincinnati, Ohio. GOLD; NICKEL; SILVER; ZINC.

Coombs, Dr. Sheryl. Professor of Hearing Sciences, Parmly Hearing Institute, Loyola University, Chicago, Illinois. LATERAL LINE SYSTEM.

Cooper, Dr. Anthony R. Senior Staff Scientist, Research and Development, Lockheed Missiles and Space Company, Inc., Palo Alto, California. GEL PERMEATION CHROMATOGRAPHY.

Cooper, Dr. Arthur W. Department of Botany, North Carolina State University. LIFE ZONES; PLANTS, LIFE FORMS OF.

Cooper, Prof. W. Charles. Department of Metallurgical Engineering, Queen's University, Kingston, Ontario, Canada. HYDROMETALLURGY; PYROMETALLURGY, NONFERROUS.

Coote, Dr. Michelle. Research School of Chemistry, Australian National University. COPOLYMER.

Copeland, Prof. Miles A. Faculty of Engineering, Carleton University, Ottawa, Ontario, Canada. ELECTRICAL INSTABILITY.

Coppens, Dr. Alan B. Associate Professor of Physics, U.S. Navy Post Graduate School, Pacific Grove, California. SOUND.

Coppersmith, Dr. Don. Thomas J. Watson Research Center, IBM Corporation, Yorktown Heights, New York. CRYPTOGRAPHY.

Corben, Dr. Herbert C. Deceased; formerly, Department of Physics, Harvey Mudd College, Claremont, California. CENTRODE; KINEMATICS; MOTION.

Corbitt, Robert A. Associate, Metcalf & Eddy, Inc., Atlanta, Georgia. ENVIRONMENTAL ENGINEERING; WATER SUPPLY ENGINEERING.

Core, Prof. Earl L. Department of Biology, West Virginia University. ALLSPICE; BOTANICAL GARDENS; BREADFRUIT; CAMPHOR TREE; INSECTIVOROUS PLANTS; VANILLA; other articles.

Corliss, Dr. John O. Department of Zoology, University of Maryland. APOSTOMATIDA; CHONOTRICHIDA; EUKARYOTAE; HETEROTRICHIDA; other articles.

Costantino, Dr. James. Intelligent Transportation Society of America, Washington, D.C. RAILROAD CONTROL SYSTEMS; TRAFFIC CONTROL SYSTEMS.

Costanzo, Thomas. Senior Harbor Planner, Port Department, Port Authority of New York and New Jersey, New York. HARBORS AND PORTS.

Costerton, Dr. J. William F. Department of Biology, University of Calgary, Alberta, Canada. BIOFILM.

Cotterill, Carl H. Chief, Division of Statistical and Technical Services, Bureau of Mines, U.S. Department of the Interior. ZINC METALLURGY.

Couch, Prof. Houston B. Department of Plant Pathology and Physiology, College of Agriculture and Life Sciences, Virginia Polytechnic Institute and State University. GRASS CROPS.

Couch, Dr. Lee. Albuquerque Technical-Vocational Institute, Albuquerque, New Mexico. BIOLOGY.

Couch, Dr. Richard B. Professor of Naval Architecture and Marine Engineering, Ship Hydrodynamics Laboratory, University of Michigan. MARINE ENGINEERING; NAVAL ARCHITECTURE.

Couch, Dr. Robert B. Department of Microbiology and Immunology, Baylor College of Medicine, Houston, Texas. COMMON COLD.

Couclelis, Prof. Helen. Department of Geography, University of California, Santa Barbara. GEOGRAPHIC INFORMATION SYSTEMS.

Coulombe, Dr. Mary J. Director, Timber Access and Supply, American Forest and Paper Association, Washington, D.C. FOREST AND FORESTRY.

Coulter, Earl E. Manager, Mechanical Technology, Babcock and Wilcox Company, Barberton, Ohio. STEAM SEPARATOR.

Coursen, D. Linn. Dynatec Explosives Consultants, Inc., Espanola, New Mexico. EXPLOSIVE; PYROTECHNICS.

Court, Dr. Arnold. Retired; formerly, Department of Climatology, California State University. ISOTHERMAL CHART; PRECIPITATION (METEOROLOGY); TEMPERATURE INVERSION.

Couto, Dr. Julia. Instructor of Computer Science, University of Georgia, Athens. DATA REDUCTION.

Cowan, Dr. Thomas E. General Atomics—Photonics Division, San Diego, California. ELECTRON-POSITRON PAIR PRODUCTION; QUASIATOM.

Cowles, Prof. Philip B. Professor Emeritus of Microbiology, School of Medicine, Yale University. COLIPHAGE; LYTIC INFECTION.

Cox, Dr. Gertrude M. Professor Emeritus, Institute of Statistics, North Carolina State University. EXPERIMENT.

Cox, Prof. Michael M. Department of Biochemistry, University of Wisconsin, Madison. GENE AMPLIFICATION.

Coxeter, Prof. H. S. M. Deceased; formerly, Department of Mathematics, University of Toronto, Canada. REGULAR POLYTOPES.

Coyle, Dr. Harold P. Harvard College Observatory, Harvard-Smithsonian Center for Astrophysics, Cambridge, Massachusetts. SEASONS; SYZYGY.

Coyne, Dr. Michael J. Senior Veterinarian Technical Services, SmithKline Beecham Animal Health, Exton, Pennsylvania. CANINE PARVOVIRUS INFECTION.

Cozzens, Christine S. Biology Laboratory, Harvard University. HEART (INVERTEBRATE).

Crabill, Ralph E. Retired; formerly, Curator and Supervisor of the Division of Myriapoda and Arachnida, Department of Entomology, National Museum of Natural History, Smithsonian Institution, Washington, D.C. CHILOPODA; PAUROPODA; SYMPHYLA.

Crabtree, Dr. Pamela. Department of Anthropology, New York University. DOMESTICATION (ANTHROPOLOGY).

Crabtree, Prof. Robert H. Chemistry Department, Yale University, New Haven, Connecticut. C-H ACTIVATION.

Cracraft, Dr. Joel. Department of Ornithology, American Museum of Natural History, New York. PHYLOGENY; SYSTEMATICS.

Crafts, Dr. Alden S. Professor Emeritus, Department of Botany, University of California, Davis. DEFOLIANT AND DESICCANT.

Craig, Dr. Homer. Deceased; formerly, Department of Mathematics, University of Texas, Austin. DYADIC; GAUSS' THEOREM; GRADIENT OF A SCALAR; GREEN'S THEOREM; other articles.

Craine, Dr. Eric R. Western Research Company, Inc., Tucson, Arizona. ASTRONOMICAL PHOTOGRAPHY.

Cramer, Prof. C. O. Department of Agricultural Engineering, University of Wisconsin, Madison. AGRICULTURAL BUILDINGS.

Cramer, Friedrich D. Max-Planck Institut für Experimentelle Medizin, Göttingen, Germany. CLATHRATE COMPOUNDS.

Cramer, Dr. William S. Ship Acoustics Department, Naval Ship Research and Development Center, Washington, D.C. DIRECTIVITY.

Crane, Julian C. Professor of Pomology, Agricultural Experimental Station, University of California, Davis. PISTACHIO.

Crane, Prof. Robert. New London, New Hampshire. RADIO-WAVE PROPAGATION.

Cranwell, Dr. Lucy M. Department of Geosciences, University of Arizona. POLLEN.

Crawford, Prof. Michael H. Chief, Division of Cardiology, University of New Mexico School of Medicine, Albuquerque. ECHOCARDIOGRAPHY.

Cremer, Sheldon E. Department of Chemistry, Marquette University. ORGANOPHOSPHORUS COMPOUND.

Creutz, Dr. Carol. Department of Chemistry, Brookhaven National Laboratory, Associated University Inc., Upton, New York. OSMIUM; RUTHENIUM.

Crittenden, Dr. Charles V. Silver Spring, Maryland. AIR.

Crocker, Dr. Burton B. Monsanto Company, St. Louis, Missouri. KILN.

Crocker, Prof. Malcolm J. Chairperson, Department of Mechanical Engineering, Auburn University. ACOUSTIC NOISE.

Cronin, Dr. Thomas. Department of Biological Sciences, University of Maryland, Baltimore. EYE (INVERTEBRATE).

Cronquist, Dr. Arthur. Deceased; formerly, Senior Curator, New York Botanical Garden, New York. ALISAMATALES; BOTANY; EMBRYOBIONTA; PLANT TAXONOMY; other articles.

Cross, Dr. Chester E. Agricultural Experiment Station, University of Massachusetts. CRANBERRY.

Crossley, Dr. F. R. E. Department of Mechanical Engineering, University of Massachusetts. EFFICIENCY.

Crouse, William H. Deceased; formerly, Consulting Editor, Automotive Books, McGraw-Hill, Inc., New York. JOINT (STRUCTURES).

Crow, Dr. James F. Department of Zoology, University of Wisconsin. HARDY-WEINBERG FORMULA; HETEROSIS.

Crowe, Prof. Clayton T. Department of Mechanical and Materials Engineering, Washington State University, Pullman. PARTICLE FLOW.

Crowe, Dr. John H. Department of Zoology, University of California, Davis. CRYPTOBIOSIS.

Crowe, Dr. Lois M. Department of Zoology, University of California, Davis. CRYPTOBIOSIS.

Crowell, Dr. Sears. Department of Zoology, Indiana University. HYDROIDA; HYDROZOA; MILLEPORINA; SIPHONOPHORA; STYLASTERINA; TRACHYLINA.

Crowley, Dr. Thomas P. National Institute of Standards and Technology, Boulder, Colorado. THERMISTOR.

Crowson, Dr. Roy A. Department of Zoology, University of Glasgow, Scotland. COLEOPTERA.

Cruikshank, Dr. Michael J. Marine Minerals Technical Center, University of Hawaii, Honolulu. MARINE MINING.

Crum, Dr. Howard. Herbarium, University of Michigan. ANDREAEOPSIDA; BRYALES; BRYOPSIDA; HOOKERIALES; MARCHANTIIDAE; SCHISTOSTEGALES; other articles.

Csanady, Dr. G. T. Department of Physical Oceanography, Old Dominion University, Norfolk, Virginia. CONTINENTAL MARGIN.

Cuckler, Lee E. Interdivisional Coordinator, Aeronautical and Instrument Division, Robertshaw Controls Company, Anaheim, California. MOISTURE-CONTENT MEASUREMENT.

Cuffey, Dr. Roger J. Department of Geosciences, The Pennsylvania State University, University Park. CHEILOSTOMATA; PHYLACTOLAEMATA; GYMNOLAEMATA; TREPOSTOMATA; other articles.

Culotta, Prof. Valeria Cizewski. Associate Professor, Johns Hopkins University School of Public Health, Department of Environmental Health Sciences, Baltimore, Maryland. METALLOCHAPERONES.

Cummings, Dr. Michael R. Department of Biological Sciences, University of Illinois at Urbana-Champaign. DOMINANCE; MENDELISM.

Cummings, Prof. Peter T. Department of Chemical Engineering, University of Tennessee, Knoxville. MOLECULAR SIMULATION.

Cummins, Dr. Denise Dellarosa. Department of Psychology, University of Arizona. COGNITION.

Cummins, Prof. Harold. Professor Emeritus, School of Medicine, Tulane University. EPIDERMAL RIDGES; INTEGUMENTARY PATTERNS.

Cunningham, Prof. Burris B. Deceased; formerly, Radiation Laboratory, University of California, Berkeley. TRANSITION ELEMENTS.

Curkendall, Dr. David W. NASA Jet Propulsion Laboratory, Pasadena, California. SPACE NAVIGATION AND GUIDANCE.

Curray, Dr. Joseph R. Scripps Institution of Oceanography, University of California, La Jolla. GULF OF CALIFORNIA.

Curtin, Dr. Charles B. Department of Biology, Creighton University, Omaha, Nebraska. CAMEL; CARIBOU; CHAMELEON; HIPPOPOTAMUS; MACKEREL; other articles.

Curtin, Dr. Denis J. XTAR, LLC, Rockville, Maryland. COMMUNICATIONS SATELLITE.

Curtiss, Prof. Charles F. Department of Chemistry, University of Wisconsin. GAS.

Curtiss, Leon F. Formerly, National Bureau of Standards. RADIOACTIVITY STANDARDS.

Cushman, Dr. Pauline K. College of Integrated Science and Technology, James Madison University, Harrisonburg, Virginia. MULTIMEDIA TECHNOLOGY.

Cushman-Roisin, Dr. Benoit. Thayer School of Engineering, Dartmouth College, Hanover, New Hampshire. GEOPHYSICAL FLUID DYNAMICS.

Cutler, Dr. Edward B. Deceased; formerly, Museum of Comparative Zoology, Harvard University. POGONOPHORA (SIBOGLINIDAE); SIPUNCULA.

Cutter, Dr. Elizabeth G. Department of Cryptogamic Botany, University of Manchester, England. APICAL MERISTEM.

Czamanske, Dr. Gerald K. U.S. Geological Survey, Menlo Park, California. LAYERED INTRUSION.

Czarnik, Dr. Anthony W. Senior Director, Chemistry, IRORI Quantum Microchemistry, La Jolla, California. COMBINATORIAL CHEMISTRY.

D

Dabberdt, Dr. Walter F. Atmospheric Technology Division, National Center for Atmospheric Research, Boulder, Colorado. WIND MEASUREMENT.

Dabbs, Dr. David J. Pathologist, Department of Pathology, University of Washington. HERPES.

Dagle, Dr. John. Department of Biochemistry, University of Iowa, Iowa City. OLIGONUCLEOTIDE.

Dahl, Lawrence F. Professor of Chemistry, University of Wisconsin. X-RAY CRYSTALLOGRAPHY; X-RAY DIFFRACTION.

Dahl, Dr. Leslie R. Consultant, General Electric Company, San Jose, California. ELECTROMAGNETIC PUMPS.

Dahl, Dr. Peter S. Department of Geology, Kent State University, Kent, Ohio. EPIDOTE.

Daiber, Dr. Franklin C. College of Marine Studies, University of Delaware. SALT MARSH.

Daigle, Dr. Gilles A. National Research Council, Institute for Microstructural Sciences, Ottawa, Ontario, Canada. ATMOSPHERIC ACOUSTICS.

Dale, Dr. James B. Veteran Affairs Medical Center, Memphis, Tennessee. RHEUMATIC FEVER.

Dales, Dr. R. Phillips. Department of Zoology, Bedford College, London, England. FEEDING MECHANISMS (INVERTEBRATE).

Dalitz, Prof. Richard H. Department of Theoretical Physics, Oxford University, Oxford, England. DALITZ PLOT; HYPERON.

Dallas, Prof. William J. College of Optical Sciences, University of Arizona, Tucson. HOLOGRAPHY.

Dalrymple, Dr. David L. Nicolet Instrument Corporation, Madison, Wisconsin. WOODWARD-HOFFMANN RULE.

Dalton, Dr. David R. Department of Chemistry, College of Arts and Sciences, Temple University, Philadelphia, Pennsylvania. STRYCHNINE ALKALOIDS.

Dalziel, Dr. Ian W. O. Department of Geological Sciences, University of Texas at Austin. SUPERCONTINENT.

Damberger, Dr. Heinz H. Illinois State Geological Survey, Coal Section, Champaign. COAL.

Damian, Dr. Raymond T. Department of Zoology, Franklin College of Arts and Sciences, University of Georgia, Athens. PARASITOLOGY.

Damman, Prof. Antoni W. H. Department of Biology, University of Connecticut. BOG.

Dan, Dr. Bruce. American Medical Association, Chicago, Illinois. HOSPITAL INFECTIONS; INFECTIOUS DISEASE.

Danby, Dr. J. M. Anthony. Emeritus, Department of Mathematics, North Carolina State University. GRAVITATION.

Daney, Dr. David E. Los Alamos National Laboratory, Los Alamos, New Mexico. CRYOGENICS.

Daniel, James R. Department of Biochemistry, Purdue University. PECTIN.

Daniels, Dr. David J. ERA Technology, Leatherhead, Surrey, England. GROUND-PENETRATING RADAR.

Daniels, Dr. Farrington. Deceased; formerly, Professor Emeritus, Solar Energy Laboratory, University of Wisconsin. CHAIN REACTION (CHEMISTRY); HALF-LIFE.

Daniels, Prof. James M. Department of Physics, University of Toronto, Ontario, Canada. DYNAMIC NUCLEAR POLARIZATION.

D'Arcangelo, Prof. Amelio M. Department of Naval Architecture, University of Michigan, Ann Arbor. MERCHANT SHIP.

Darragh, Dr. Peter J. CSIRO, Wembley, Australia. OPAL.

Darwin, Dr. C. J. Experimental Psychology, University of Sussex, England. SPEECH PERCEPTION.

Date, Prof. H. Acoustics Department, Kyusha Institute of Design, Fukuoka, Japan. OPTICAL RECORDING.

Daubechies, Dr. Ingrid. AT&T Bell Laboratories, Murray Hill, New Jersey. WAVELETS.

Davidhazy, Dr. Andrew. Rochester Institute of Technology, New York. CAMERA.

Davidson, Dr. Jaime A. Endocrine & Diabetes Associates of Texas, Dallas, Texas. DIABETES.

Davies, David H. DataPlay Inc., Boulder, Colorado. VIDEO DISK.

Davies, Prof. Glyn A. O. Department of Aeronautics, Imperial College of Science, Technology and Medicine, London, England. AIRFRAME.

Davies, Dr. John A. Institute for Materials Research, McMaster University, Hamilton, Ontario, Canada. CHANNELING IN SOLIDS.

Davies, Keith. Manager, Propulsion Group, General Electric Astrospace, Princeton, New Jersey. SPACECRAFT PROPULSION.

Davies, Dr. Peter J. Department of Botany, Cornell University. PLANT GROWTH; PLANT HORMONES; PLANT PHYSIOLOGY.

Davies-Jones, Dr. Robert P. Meteorologist, National Severe Storms Laboratory, U.S. Department of Commerce, Norman, Oklahoma. THUNDERSTORM; TORNADO.

Davis, Dr. Andrew M. Enrico Fermi Institute, University of Chicago, Illinois. COSMOCHEMISTRY; ELEMENTS, COSMIC ABUNDANCE OF.

Davis, Dr. Bernard D. Deceased; formerly, Bacterial Physiology Unit, Harvard Medical School. CULTURE.

Davis, D. Dwight. Deceased; formerly, Curator of Vertebrate Anatomy, Chicago Museum of Natural History, Chicago, Illinois. ALLOTHERIA; METATHERIA; PROTOTHERIA.

Davis, Dr. Donald R. Research Scientist Associate, College of Natural Sciences, Clayton Foundation Biochemical Institute, University of Texas, Austin. NUTRITION.

Davis, Dr. George H. Department of Geological Sciences, University of Arizona. FAULT AND FAULT STRUCTURES; MOUNTAIN SYSTEMS; STRUCTURAL GEOLOGY; SYNCLINE.

Davis, Prof. Gregory A. Department of Geological Sciences, University of Southern California. CORDILLERAN BELT.

Davis, Dr. Jack. Deceased; formerly, Research and Development, Brigh Star Industries Inc., Clifton, New Jersey. NUCLEAR BATTERY.

Davis, Dr. John H. AT&T Bell Laboratories, Holmdel, New Jersey. TELEPHONE SERVICE.

Davis, Dr. Joseph R. ARCO Oil and Gas Company, Dallas, Texas. SHALE.

Davis, Prof. Kenneth P. Deceased; formerly, School of Forestry, Yale University. ARBORVITAE; BALSA; CYPRESS; FIR; MULBERRY; SWEETGUM; other articles.

Davis, Prof. Richard A., Jr. Department of Geology, University of South Florida. BARRIER ISLANDS; COASTAL LANDFORMS; COASTAL PLAIN; CONTINENTAL MARGIN; STRAND LINE.

Davis, Dr. Simon J. M. Ancient Monuments Laboratory, Saville Row, London. ZOOARCHEOLOGY.

Dawood, Dr. M. Yusoff. Department of Obstetrics and Gynecology, University of Illinois College of Medicine, Chicago. NEUROHYPOPHYSIS HORMONE.

Dawson, Dr. David. West Roxbury Veterans Affairs Medical Center, West Roxbury, Massachusetts. CARPAL TUNNEL SYNDROME.

Dawson, Prof. Kerry J. Environmental Horticulture, University of California, Davis. LANDSCAPE ARCHITECTURE.

Dawson, Dr. Michael N. Division of Biological Sciences, Section of Evolution & Ecology, University of California, Davis. SCYPHOZOA.

Dawson, Dr. Todd E. Ecology and Systematics, Cornell University, Ithaca, New York. PHYSIOLOGICAL ECOLOGY (PLANT).

Day, Dr. William H. Manager, Advanced Industrial Programs, Pratt & Whitney, East Hartford, Connecticut. GAS TURBINE.

Dayton, Benjamin B. Consulting Physicist, East Flatrock, North Carolina. RAREFIED GAS FLOW.

de Boor, Dr. Carl. Mathematical Research Center, University of Wisconsin, Milwaukee. NUMERICAL ANALYSIS.

De Cicco, John. Department of Mathematics, Illinois Institute of Technology, Chicago. NONEUCLIDEAN GEOMETRY.

de Freitas, Prof. Duarte Mota. Department of Chemistry, Loyola University, Chicago, Illinois. LITHIUM; MAGNESIUM; POTASSIUM; SODIUM.

De Jong, Dr. Theodore M. Pomology Department, University of California, Davis. PEACH.

de la Cruz, Dr. Felix. Chief, Mental Retardation and Developmental Disabilities Branch, National Institute of Child Health and Human Development, Bethesda, Maryland. PHENYLKETONURIA.

de Levie, Prof. Robert. Department of Chemistry, Georgetown University, Washington, DC. ELECTRODE POTENTIAL.

De Micheli, Dr. Giovanni. Professor and Director, Integrated Systems Centre,Swiss Federal Institute of Technology, Lausanne. COMPUTER-AIDED CIRCUIT DESIGN.

de Montmollin, Fernand. Retired; formerly, Hoffmann-LaRocbe, Inc. THIAMINE.

de Neufville, Dr. John P. President, Voltaix, Inc., North Branch, New Jersey. PHOTOVOLTAIC EFFECT.

de Weck, Dr. A. L. Institut für Klinische Immunologic Inselspital, Universität Bern, Switzerland. ALLERGY.

Deacon, Margaret. Department of Oceanography, University of Southampton, United Kingdom. OCEANOGRAPHY.

Deacon, T. A. Deceased; formerly, Division of Electrical Science, National Physical Laboratory, Teddington, Middlesex, England. INDUCTIVE VOLTAGE DIVIDER.

Deakins, Dr. Ralph R. Director, Computer-Aided Systems, Rockwell International Corporation, El Segundo, California. LANDING GEAR.

Dean, Dr. Thomas S. Architectural Engineer, Lawrence, Kansas. ARCHITECTURAL ENGINEERING.

Dean, Dr. Walter E. U.S. Geological Survey, Denver Federal Center, Denver, Colorado. MARL.

DeCarli, Dr. Paul S. Poulter Laboratories, Stanford Research Institute, Menlo Park, California. HIGH-PRESSURE PHYSICS.

DeClaris, Prof. Nicholas. Fulton, Maryland. NEURAL NETWORK.

DeFacio, Prof. Brian. Department of Physics and Astronomy, University of Missouri, Columbia. ANGULAR MOMENTUM; BEAT; CATASTROPHE THEORY; INVERSE SCATTERING THEORY; STATICS.

Degerman, Dr. Richard. Department of Psychology, University of California, Irvine. MULTIDIMENSIONAL SCALING.

DeGroot, Dr. Leslie J. Department of Medicine, Section of Endocrinology, University of Chicago Medical Center. THYROID GLAND DISORDERS.

Dehmelt, Dr. Hans. Department of Physics, University of Washington; Nobelist. NUCLEAR QUADRUPOLE RESONANCE.

Dekker, Dr. Adrianus J. Professor of Solid State Physics, University of Groningen, Netherlands. ELECTRON EMISSION; SCHOTTKY EFFECT; THERMIONIC EMISSION; WORK FUNCTIONS (ELECTRONICS).

Del Fava, Dr. Raymond. Chairperson, Public Information Committee, American College of Radiology, Reston, Virginia. MEDICAL IMAGING.

Delahay, Prof. Paul. Formerly, Department of Chemistry, New York University. DECOMPOSITION POTENTIAL; ELECTROCHEMICAL EQUIVALENT.

Deland, Dr. Raymond J. Retired; formerly, Department of Meteorology and Oceanography, New York University. AIR PRESSURE; AIR TEMPERATURE.

Delevoryas, Prof. Theodore. Department of Botany, University of Texas, Austin. CYCADEOIDALES.

DeLia, Paul. L-3 Communications Corporation, New York. ELECTRONIC WARFARE; JAMMING.

Dell'Arciprete, Dr. Patricia. Central Nacional Patagónico (CONICET), Chubut, Argentina. FISHERIES ECOLOGY.

Delong, Dr. Dwight M. College of Biological Sciences, Ohio State University. ANOPLURA; INSECTA.

Delson, Prof. Eric. Professor of Anthropology, Lehman College and the Graduate School, City University of New York, and Department of Vertebrate Paleontology, American Museum of Natural History, New York. FOSSIL HUMANS; FOSSIL PRIMATES; MONKEY.

DeLuca, Dr. Hector F. Department of Biochemistry, University of Wisconsin, Madison. VITAMIN D.

Dempsey, Prof. Mary E. Department of Biochemistry, University of Minnesota Medical School, Minneapolis. STEROID; STEROL.

Den Hartog, Prof. J. P. Retired; formerly, Department of Mechanical Engineering, Massachusetts Institute of Technology. MECHANICAL VIBRATION.

Dengler, Dr. Nancy G. Department of Botany, University of Toronto, Ontario, Canada. LEAF.

Denison, Dr. Robert H. Deceased; formerly, Curator of Fossil Fishes, Field Museum of Natural History, Chicago, Illinois. ACANTHODII; AGNATHA; HETEROSTRACI; OSTRACODERM; other articles.

Denniston, Dr. Carter L. Department of Medical Genetics, University of Wisconsin, Madison. HUMAN GENETICS.

Denny, Robert W., Jr. Denny & Associates, Oxon Hill, Maryland. FREQUENCY-MODULATION RADIO.

Deno, Richard A. Professor of Pharmacognosy, University of Michigan. PHARMACOGNOSY; PHARMACY.

Dent, Prof. William A. Emeritus, Department of Physics and Astronomy, University of Massachusetts, Amherst. QUASAR.

DePaolo, Dr. Donald J. Department of Geology and Geophysics, University of California, Berkeley. CHEMOSTRATIGRAPHY.

DePlasco, Mark A. Air Force Flight Standard Agency, Washington, D.C. PRECISION APPROACH RADAR (PAR).

Deppert, Dr. Wolfgang. Heinrich-Pette-Institut for Experimental Virology and Immunology, University of Hamburg, Germany. TUMOR SUPPRESSOR GENES.

Derge, Prof. Gerhard. Department of Metallurgy and Material Science, Carnegie-Mellon University. FERROALLOY.

Dershowitz, Dr. Nachum. Department of Computer Science, Illinois Institute of Technology, Chicago. CALENDAR.

Destree, Thomas M. Editor in Chief, Graphic Arts Technical Foundation, Sewickley, Pennsylvania. BOOK MANUFACTURE; PHOTOCOPYING PROCESSES; TYPE (PRINTING); PHOTOGRAPHY.

DeVaney, Fred D. Consulting Metallurgist, Duluth, Minnesota. MAGNETIC SEPARATION METHODS.

DeVay, Prof. James E. Department of Plant Pathology, University of California, Davis. AGRICULTURAL SOIL AND CROP PRACTICES.

Devillez, Dr. Edward J. Department of Zoology, Miami University, Oxford, Ohio. DIGESTION (INVERTEBRATE).

DeVolpi, Dr. A. Argonne National Laboratory, Argonne, Illinois. ATOMIC BOMB; CRITICAL MASS; HYDROGEN BOMB.

DeVore, Prof. George W. Department of Geology, Florida State University. AMPHIBOLITE; PIGEONITE; SPODUMENE.

DeVries, Dr. Arthur L. Department of Physiology and Biophysics, University of Illinois, Urbana-Champaign. ANTIFREEZE (BIOLOGY).

Dewsbury, Dr. Donald A. Department of Psychology, University of Florida, Gainesville. REPRODUCTIVE BEHAVIOR.

Dexter, Dr. Theodore H. Research Supervisor, Inorganic Chemistry, Hooker Chemical Corporation Research Center, Niagara Falls, New York. HYDRAZINE.

Dhir, Prof. Vijay K. Department of Mechanical and Aerospace Engineering, University of California, Los Angeles. BOILING; CONDENSATION.

di Cicco, Dennis. Associate Editor, *Sky & Telescope*, Sky Publishing Corporation,

Cambridge, Massachusetts. EQUATION OF TIME.

Di Fiore, Dr. Anthony. Department of Anthropology, New York University. SOCIOBIOLOGY.

Di Giulio, Edmund M. Cinema Products Corporation, Los Angeles, California. CINEMATOGRAPHY.

Diamond, Dr. Seymour. Director, Diamond Headache Clinic, Ltd., Chicago, Illinois. HEADACHE.

Diaz, Alejandro. Department of Pediatrics, Weill Medical College of Cornell University, New York, New York. ALDOSTERONE.

DiBenedetto, Michael F. Avionics Engineering Center, Ohio University, Albany. MICROWAVE LANDING SYSTEM (MLS).

DiBiase, Dr. David. Department of Geography, Pennsylvania State University, University Park. CARTOGRAPHY.

Dick, Prof. B. Gale. Department of Physics, University of Utah. IONIC CRYSTALS; MADELUNG CONSTANT.

Dick, Dr. Henry J. B. Senior Scientist, Department of Geology and Geophysics, Woods Hole Oceanographic Institution, Woods Hole, Massachusetts. DUNITE.

Dick, Dr. Steven. U.S. Naval Observatory, Washington, D.C. CHRONOMETER.

Dickinson, Prof. Robert E. Department of Atmospheric Physics, University of Arizona, Tucson. CLIMATE MODELING.

Dickinson, Dr. W. R. Department of Geosciences, University of Arizona, Tucson. SANDSTONE.

Dieke, Prof. G. H. Deceased; formerly, Chairperson, Department of Physics, Johns Hopkins University. FARADAY EFFECT; MAGNETOOPTICS; ZEEMAN EFFECT.

Diekema, Dr. Daniel James. Department of Pathology, Division of Medical Microbiology, University of Iowa College of Medicine, Iowa City. DRUG RESISTANCE.

Dietrich, Prof. Gunter O. Deceased; formerly, Institute of Oceanography, University of Kiel, Germany. ATLANTIC OCEAN.

Dietrich-Buchecker, Dr. C. O. Laboratoire de Chimie Organo-Min'erale, Institut le Bel, Université Louis Pasteur, Strasbourg, France. CATENANES.

Dill, Dr. Robert F. Scripps Institute of Oceanography and Department of Geology, San Diego State University. DIVING.

Dille, Dr. James M. Deceased; formerly, Department of Pharmacology, School of Medicine, University of Washington. NARCOTIC.

Dillingham, Dr. Mark S. National Institute for Medical Research, London, United Kingdom. DNA HELICASE.

Dillman, Prof. Norman G. Department of Electrical and Computer Engineering, Kansas State University, Manhattan. ELECTRONIC POWER SUPPLY; VOLTAGE-MULTIPLIER CIRCUIT.

Dillon, J. F., Jr. Bell Telephone Laboratories, Murray Hill, New Jersey. CURIE TEMPERATURE.

Dilworth, Dr. S. M. MRC Laboratory of Molecular Biology, Cambridge, England. POLYOMA VIRUS.

DiMichele, Dr. William A. Department of Paleobotany, Smithsonian Institution, Washington, D.C. ASTEROXYLALES; ISOETALES; LEPIDODENDRALES; LYCOPHYTA; LYCOPODIALES; LYCOPSIDA; PROTOLEPIDODENDRALES; SELAGINELLALES; ZOSTEROPHYLLOPSIDA.

Dimitman, Dr. J. E. Agronomicas de Guatemala. BANANA.

Dimmel, Prof. Donald R. Research Associate, Wood Sciences, Chemical Sciences Division, Institute of Paper Chemistry, Appleton, Wisconsin. LIGNIN.

Dine, Prof. Michael. Santa Cruz Institute for Particle Physics, University of California, Santa Cruz. SUPERSTRING THEORY.

Dingle, Prof. Hugh. Department of Entomology, College of Agricultural and Environmental Sciences, University of California, Davis. BEHAVIORAL ECOLOGY.

Dinsmore, Robertson P. International Ice Patrol, Woods Hole Oceanographic Institution, Woods Hole, Massachusetts. ICEBERG.

DiTallo, Cynthia. United Technologies, Otis Elevator Company, Farmington, Connecticut. ELEVATOR.

Dittmar, Dr. Katharina. Department of Integrative Biology, Brigham Young University, Provo, Utah. SIPHONAPTERA.

Dix, Dr. Cyril H. Bishop Monkton, North Yorkshire, England. RESISTANCE MEASUREMENT.

Dixon, Prof. David. Chemistry Department, South Dakota School of Mines. SUPERCRITICAL FLUID.

Dixon, Prof. John C. Department of Geography, University of Arkansas, Fayetteville. LAKE.

Doane, Dr. J. William. Liquid Crystal Institute, Kent State University, Kent, Ohio. LIQUID CRYSTAL.

Dobrowolski, Dr. J. A. Research Emeritus, Institute for Microstructural Sciences, National Research Council of Canada, Ottawa, Ontario, Canada. INTERFERENCE FILTERS.

Dobson, Dr. Gerald R. Department of Chemistry, North Texas State University, Denton. METAL CARBONYL.

Dodds, Robert H. Deceased; formerly, Personnel Manager, Gibbs and Hill, Inc., New York. CONTOUR; PHOTOGRAMMETRY.

Dodington, Sven H. Deceased; formerly, International Telephone and Telegraph, New York. DISTANCE-MEASURING EQUIPMENT; RADIO RANGE; RHO-THETA SYSTEM; other articles.

Dogramaci, Prof. Ali. Department of Industrial and Management Engineering, Columbia University. INVENTORY CONTROL.

Doherty, Dr. Michael F. Head, Department of Chemical Engineering, University of Massachusetts. AZEOTROPIC DISTILLATION.

Dolch, David F. Verizon, Teaneck, New Jersey. TELEPHONE SYSTEMS CONSTRUCTION.

Dolphin, Dr. David. Department of Chemistry, University of British Columbia, Vancouver, Canada. PORPHYRIN.

Domning, Dr. Daryl P. Department of Anatomy, Howard University, Washington, D.C. SIRENIA.

Donahoo, Dr. William T. Division of Endocrinology, Metabolism and Diabetes, University of Colorado Health Sciences Center, Denver. LEPTIN.

Donaldson, Dr. Alex I. Pirbright Laboratory, Pirbright, Surrey, United Kingdom. FOOT-AND-MOUTH DISEASE.

Donaldson, Dr. James A. Professor and Chairperson, Department of Otolaryngology, School of Medicine, University of Washington. TONSILLITIS.

Donovan, Dr. Nowell. Department of Geology, Texas Christian University, Forth Worth. CALICHE.

Doob, Dr. Joseph L. Deceased; formerly, Department of Mathematics, University of Illinois, Urbana-Champaign. STOCHASTIC PROCESS.

Doolan, Barry L. Department of Geology, Memorial University of Newfoundland, St. John's, Canada. TREMOLITE.

Dooley, Dr. H. D. International Council for the Exploration of the Sea, Copenhagen, Denmark. NORTH SEA.

Doornbos, Dr. Durk J. Deceased; formerly, Institute for Geophysics, University of Oslo, Norway. EARTH INTERIOR.

Dorf, Richard C. Professor Emeritus, Department of Electrical and Computer Engineering, University of California, Davis. CONTROL SYSTEMS.

Dorfman, Dr. Ralph I. Senior Vice President and Director, Syntex Research Center, Palo Alto, California. ESTROGEN; PROGESTERONE.

Dorsey, Prof. John G. Department of Chemistry, Florida State University, Tallahassee. LIQUID CHROMATOGRAPHY.

Doscher, Dr. Todd M. Deceased; formerly, Department of Petroleum Engineering, University of Southern California. GAS FIELD AND GAS WELL; OIL AND GAS WELL COMPLETION; OIL AND GAS WELL DRILLING; PETROLEUM ENGINEERING; PETROLEUM ENHANCED RECOVERY; other articles.

Doshi, Dr. Mahendra. Progress in Paper Recycling, Appleton, Wisconsin. RECYCLING TECHNOLOGY.

Dossel, Dr. William E. Professor and Chairperson, Department of Anatomy, School of Medicine, Creighton University. PARATHYROID GLAND.

Dott, Prof. Robert H., Jr. Department of Geology and Geophysics; University of Wisconsin, Madison. CRATON.

Doudoroff, Dr. Michael. Deceased; formerly, Department of Bacteriology, University of California, Berkeley. ADENOSINE TRIPHOSPHATE (ATP); CARBOHYDRATE METABOLISM; NICOTINAMIDE ADENINE DINUCLEOTIDE (NAD); other articles.

Douglas, Donald W., Jr. Deceased; formerly, Corporate Vice President—Administration, McDonnell Douglas Corporation, St. Louis, Missouri; Chairperson and President, Douglas Aircraft Company of Canada, Toronto, Ontario. AIRCRAFT TESTING.

Douglass, Prof. Raymond D. Department of Mathematics, Massachusetts Institute of Technology. NOMOGRAPH.

Dover, Dr. Karl. Department of Physics, Brookhaven National Laboratory, Upton, New York. DELTA RESONANCE.

Doviak, Dr. Richard J. National Severe Storms Laboratory–NOAA, Norman, Oklahoma. PRECIPITATION MEASUREMENT.

Dowling, Prof. David R. Department of Mechanical Engineering, University of Michigan, Ann Arbor. MATCHED-FIELD PROCESSING; TIME-REVERSED SIGNAL PROCESSING.

Downey, Dr. Joe. Dow Chemical Company, Midland, Michigan. CHLORINE.

Downey, Dr. R. K. Agricultural Research Station, Saskatoon, Saskatchewan, Canada. RAPE.

Doyle, Prof. James A. Department of Biological Sciences, University of California, Davis. PLANT EVOLUTION.

Doyle, Prof. Michael P. Director, Center for Food Safety and Quality Enhancement, University of Georgia, Griffin. FOOD POISONING.

Drake, A. E. Centre for Electromagnetic and Time Metrology , National Physical Laboratory, Teddington, Middlesex, England. EDDY CURRENT; MAGNETIC INSTRUMENTS.

Drake, Prof. John C. Department of Geology, University of Vermont, Burlington. SILICA MINERALS; SMITHSONITE; SOLID SOLUTION; TITANITE.

Drake, Dr. John W. Laboratory of Genetics, National Institute of Environmental Health Sciences, Research Triangle Park, North Carolina. MUTATION.

Drake, Dr. Michael V. Department of Ophthalmology, School of Medicine, University of California, San Francisco. EYE DISORDERS; EYEGLASSES AND CONTACT LENSES.

Draper, Dr. Clifton W. Laser Studies Group, Western Electric Company, Princeton, New Jersey. LASER ALLOYING.

Dreschel, Dr. Thomas W. Biologist/Chemist, Bionetics Corporation, Environmental Monitoring and Research Program, Kennedy Space Center, Florida. HYDROPONICS.

Dresden, Prof. Max. Deceased; formerly, Institute for Theoretical Physics, State University of New York, Stony Brook. BOLTZMANN CONSTANT; BOSE-EINSTEIN STATISTICS; KINETIC THEORY OF MATTER; QUANTUM STATISTICS; STATISTICAL MECHANICS; other articles.

Drexler, Dr. Erick. Los Altos, California. NANOTECHNOLOGY.

Dropping, Barry. Hewlett-Packard, Santa Clara, California. FREQUENCY COUNTER.

D'Souza, Dr. Cyril. Department of Psychiatry, VA Medical Center, Yale University, New Haven, Connecticut. PSYCHOTOMIMETIC DRUGS.

Duan, Dr. Lian. Senior Bridge Engineer, California DOT, Sacramento. HIGHWAY BRIDGE.

Dubiel, Dr. Russell F. Research Geologist, United States Geological Survey. MESOZOIC.

DuBois, Prof. Robert D. Department of Physics, University of Missouri, Rolla. POSITRON.

Duby, Prof. Paul F. Department of Mineral Engineering, Henry Krumb School of Mines, Columbia University. ELECTROMETALLURGY.

Duckworth, Dr. Henry E. Professor Emeritus of Physics, Department of Physics, University of Manitoba, Winnipeg, Canada. ATOMIC NUCLEUS; ATOMIC NUMBER; DEUTERON; MASS NUMBER; RADIOISOTOPE; other articles.

Duffield, Prof. Roger C. Director of Graduate Studies, Department of Mechanical and Aerospace Engineering, University of Missouri. ACCELEROMETER.

Dufrêne, Dr. Yves Frédéric. Université Catholique de Louvain, Belgium. ATOMIC FORCE MICROSCOPY.

Dugan, Dr. Frank. USDA/ARS, Washington State University, Pullman. AGONOMYCETES; BLASTOMYCETES; FUNGI; HYPHOMYCETES.

Duguay, Dr. Michel A. Bell Laboratories, Holmdel, New Jersey. ELECTROOPTICS; KERR EFFECT.

Duke, Dr. Charles B. Vice President and Senior Research Fellow, Xerox Research and Technology, Webster, New York. SURFACE PHYSICS.

Duke, Kenneth L. Department of Anatomy, Duke University. OVARY.

Dunbar, Prof. Carl O. Deceased; formerly, Yale University. GASTROLITH.

Duncan, Prof. William G. Department of Agronomy, University of Kentucky, Lexington. CORN.

Duncombe, Dr. Raynor L. Department of Aerospace Engineering, University of Texas, Austin. APSIDES; ORBITAL MOTION; PERTURBATION (ASTRONOMY); other articles.

Dunlop, Prof. David J. Department of Physics, University of Toronto, Ontario, Canada. ROCK MAGNETISM.

Dunn, Dr. Alison M. School of Biology, University of Leeds, United Kingdom. SEXUAL DIMORPHISM.

Dunn, Prof. Bruce. Department of Materials Science & Engineering, University of California, Los Angeles. SOL-GEL PROCESS.

Dunn, Prof. Floyd. Department of Electrical and Computer Engineering, University of Illinois, Urbana-Champaign. MEDICAL ULTRASONIC TOMOGRAPHY.

Dunning, Dr. John B., Jr. Department of Forestry and Natural Resources, Purdue University. POPULATION DISPERSAL.

DuPont, Dr. Herbert L. Medical School, University of Texas, Houston. INFANT DIARRHEA.

Duppen, Dr. Piet Van. Instituut voor Kern-en Stralingsfysica, Celestijnenlaan, Leuven, Belgium. SPALLATION REACTION.

Duren, Prof. Peter L. Department of Mathematics, University of Michigan. CONFORMAL MAPPING.

Durst, Prof. Richard A. Director, Analytical Chemistry Laboratories, New York State Agricultural Experimental Station, Cornell University, Geneva. PH.

Dutor, Dr. J. Thomas, Jr. Museum of Natural History, U. S. Department of the Interior, Washington, D.C. PALEOZOIC.

Dutton, John. Dean of the College of Earth and Mineral Sciences, Office of the Dean, Pennsylvania State University, University Park. CHAOS.

Dyer, Dr. K. R. Institute of Oceanographic Sciences, Somerset, England. ESTUARINE OCEANOGRAPHY.

Dyke, Dr. Gareth. Department of Zoology, University College Dublin, Ireland. HESPERORNITHIFORMES; ICHTHYORNITHIFORMES.

Dym, Prof. Clive L. Department of Civil Engineering, University of Massachusetts. MECHANICAL VIBRATION.

Dymon, Prof. Ute. Department of Geography, Kent State University, Ohio. GLOBE (EARTH).

Dziuba, Ronald F. Senior Member, Electrical Measurements Laboratory, National Institute of Standards and Technology, Gaithersburg, Maryland. GALVANOMETER; POTENTIOMETER.

E

Eargle, Dr. John M. JME Consulting Corporation, Los Angeles, California. FIDELITY.

Easterbrook, Prof. Don J. Department of Geology, Western Washington University. EOLIAN LANDFORMS.

Eberl, Dr. Dennis D. U.S. Geological Survey, Denver Federal Center, Denver, Colorado. CLAY; CLAY MINERALS; HALLOYSITE; ILLITE; KAOLINITE.

Eberts, Prof. Ray. School of Industrial Engineering, Purdue University, West Lafayette, Indiana. PRODUCT USABILITY.

Eckel, Dr. Robert H. Division of Endocrinology, Metabolism and Diabetes, University of Colorado Health Sciences Center, Denver. LEPTIN.

Eckels, Dr. Arthur R. Department of Electrical Engineering, North Carolina State University, Raleigh. COMMUTATOR; DYNAMIC BRAKING; WINDINGS IN ELECTRIC MACHINERY; other articles.

Eckenfelder, Dr. W. Wesley, Jr. Eckenfelder, Inc., Nashville, Tennessee. INDUSTRIAL WASTEWATER TREATMENT.

Eckenwalder, Dr. James E. Department of Botany, University of Toronto, Toronto, Ontario, Canada. CYCADOPSIDA; GNETALES.

Economides, Dr. Michael. Chemical Engineering Department, University of Houston, Texas. LIQUEFIED NATURAL GAS (LNG).

Edelman, Robert I. Corporate Communications, Eastman Kodak Company, Rochester, New York. PHOTOCOPYING PROCESSES.

Edelson, Dr. Marshall. Department of Psychiatry, Yale University. PSYCHOANALYSIS.

Edgecombe, Gregory D. American Museum of Natural History, New York. TRILOBITA.

Edgerton, Dr. Harold E. Deceased; formerly, Institute Professor, Emeritus, Massachusetts Institute of Technology. STROBOSCOPIC PHOTOGRAPHY; UNDERWATER PHOTOGRAPHY.

Edmonds, Dr. Robert L. College of Forest Resources, University of Washington, Seattle. EUCALYPTUS.

Edwards, Alan F. Freeport-McMoran Inc., New Orleans, Louisiana. SOLUTION MINING.

Edwards, Dr. Helen T. Fermi National Accelerator Laboratory, Batavia, Illinois, and Deutches ElektronenSynchrotron, Hamburg, Germany. PARTICLE ACCELERATOR.

Edwards, Prof. John O. Professor Emeritus, Department of Chemistry, Brown University, Providence, Rhode Island. IRON.

Edwards, Dr. Peter A. Department of Biological Chemistry, University of California School of Medicine, Los Angeles. CHOLESTEROL.

Edwards, Russell K. Argonne National Laboratory, Argonne, Illinois. OXIDE.

Edwards, Dr. Steven. Department of Virology, Central Veterinary Laboratory-Weybridge, United Kingdom. CLASSICAL SWINE FEVER.

Eeckhaut, Dr. Igor. Marine Biology, University of Mons-Hainaut, Belgium. MYZOSTOMIDA; POLYCHAETA.

Eesley, Dr. Gary L. GM Research and Development Center, Warren, Michigan. RELAXATION TIME OF ELECTRONS.

Egan, Dr. Patrick O. Lawrence Livermore National Laboratory, California. MUONIUM.

Ehernberger, L. J. NASA Dryden Flight Research Facility, Edwards, California. CLEAR-AIR TURBULENCE.

Ehrich, Dr. Fredric F. Retired; formerly, General Electric Company, Aircraft Engine Business Group, Lynn, Massachusetts. AIRCRAFT ENGINE PERFORMANCE; AIRCRAFT PROPULSION; TURBINE ENGINE SUBSYSTEMS; TURBINE PROPULSION; TURBOFAN; TURBOJET.

Ehrlich, Dr. Paul R. Professor of Biological Sciences and Curator of the Entomological Collections, Stanford University. LEPIDOPTERA.

Ehrlich, Dr. Stanley L. Middletown, Rhode Island. SONOBUOY.

Ehrmann, Dr. David A. Department of Medicine, Section of Endocrinology, University of Chicago Medical Center, Chicago, Illinois. THYROID GLAND DISORDERS.

Eidson, Dr. Millicent. Veterinarian, Santa Fe, New Mexico. TULAREMIA.

Eilers, Carl G. Manager, Electronics Systems Research and Development, Zenith Radio Corporation, Glenview, Illinois. TELEVISION RECEIVER.

Einaudi, Prof. Marco T. Department of Applied Earth Science, Stanford University. GREISEN; MALACHITE; METASOMATISM; SKARN.

Eirich, Dr. F. R. Chemistry/Life Sciences Department, Polytechnic University, New York. DONNAN EQUILIBRIUM; ISOLECTRIC POINT.

Eisenbud, Dr. Merrill. Deceased; formerly, Institute of Environmental Medicine, New York University Medical Center. ENVIRONMENTAL RADIOACTIVITY.

Eisler, Dr. Thomas J. Systems Planning and Development Staff, National Office of Charting and Geodetic Services, National Oceanic and Atmospheric Administration, Rockville, Maryland. COMPUTERIZED TOMOGRAPHY.

Ekdale, Dr. Allan A. Department of Geology and Geophysics, College of Mines and Earth Sciences, University of Utah. TRACE FOSSILS.

El-Arroudi, Dr. Khalil. Department of Electrical and Computer Engineering, McGill University, Montreal, Québec, Canada. REACTIVE POWER.

Eldeib, Dr. Hany K. Department of Electrical and Computer Engineering, George Mason University, Fairfax, Virginia. LINEAR SYSTEM ANALYSIS.

Eldredge, Dr. Niles. American Museum of Natural History, New York. LIVING FOSSILS; MEROSTOMATA.

Elgomayel, Prof. Joseph. School of Industrial Engineering, Purdue University. LIMITS AND FITS.

Eliel, Prof. Ernest L. Department of Chemistry, University of North Carolina, Raleigh. MOLECULAR ISOMERISM; PROCHIRALITY.

Eliseev, Dr. Alexey V. Assistant Professor, Department of Medicinal Chemistry, State University of New York at Buffalo. COMBINATORIAL CHEMISTRY.

Ellett, Dr. C. Wayne. Professor Emeritus, Department of Plant Pathology, Ohio State University. PLANT PATHOLOGY; WHEAT.

Ellis, Dr. A. J. Director, Chemistry Division, Department of Scientific and Industrial Research, Petone, New Zealand. GEOTHERMAL POWER.

Ellis, Dr. David E. Research and Development Department, Continental Oil Company, Ponca City, Oklahoma. SCAPOLITE.

Ellis, Dr. Gordon W. Department of Biology, School of Art and Sciences, University of Pennsylvania, Leidy Laboratory of Biology, Philadelphia. POLARIZED LIGHT MICROSCOPE.

Ellis, Dr. Walther R., Jr. Division of Chemistry and Chemical Engineering, California Institute of Technology. ELECTRON-TRANSFER REACTION.

El-Mansy, Dr. Youssef A. Retired; formerly, Director of Logic Technology Development, Intel Corporation, Aloha, Oregon. INTEGRATED CIRCUITS.

Elston, Dr. Stuart B. Department of Physics, University of Tennessee, Knoxville. CHANNEL ELECTRON MULTIPLIER.

Eltoweissy, Dr. Mohamed Y. Department of Computer Science, James Madison University, Harrisonburg, Virginia. DISTRIBUTED SYSTEMS (COMPUTERS).

Ely, F. G. Retired; formerly, Research and Development Division, Babcock and Wilcox Company. RAW WATER.

Emadi, Dr. Ali. Electrical and Computer Engineering Department, Illinois Institute of Techology, Chicago. UNINTERRUPTIBLE POWER SUPPLY.

Emanuel, Dr. Kerry Andrew. Department of Earth, Atmospheric and Planetary Sciences, Massachusetts Institute of Technology. CONVECTIVE INSTABILITY; DYNAMIC INSTABILITY.

Embury, Dr. Stephen H. Chief, Hematology Division of the Medical Service, San Francisco General Hospital, San Francisco, California. SICKLE CELL DISEASE.

Emerson, Lewis P. Engineer in Charge of Flow Measurements, Foxboro Company, Foxboro, Massachusetts. FLOW MEASUREMENT.

Emerson, Dr. Steven R. Department of Oceanography, University of Washington. ANOXIC ZONES.

Emig, Christian. Directeur de Recherches au C.N.R.S., Centre d'Océanologie de Marseille, France. KUTORGINATA; LINGULIDA; OBOLELLIDA; PHORONIDA.

Emin, Prof. David. Department of Physics and Astronomy, University of New Mexico, Albuquerque. POLARON.

Encleson, Stephen C. Technical Editor, Corporate Communications, Dow Chemical Company, Midland, Michigan. MAGNESIUM.

Endicott, Prof. John F. Department of Chemistry, Wayne State University. INORGANIC PHOTOCHEMISTRY.

Engel, Prof. Michael S. Curator of Fossil Insects and Professor of Ecology and Evolutionary Biology, University of Kansas Natural History Museum. ZORAPTERA.

Engels, Dr. William R. Department of Genetics, University of Wisconsin, Madison. HYBRID DYSGENESIS.

English, Prof. Van H. Department of Geography, Dartmouth College. Equator; International date line; North Pole; Tropic of Cancer; other articles.

Engstrom, Dr. Arne. Head, Department of Medical Physics, Karolinska Institute, Stockholm, Sweden. X-ray microscope.

Engstrom, Dr. Ralph W. Electro Optics and Devices, RCA Laboratories, Lancaster, Pennsylvania. Photomultiplier.

Enlander, Dr. Derek. New York, New York. Chronic fatigue immune dysfunction syndrome.

Ensign, Dr. J. C. Department of Bacteriology, University of Wisconsin. Bacteriology.

Enslow, Prof. Philip H., Jr. School of Information and Computer Science, Georgia Institute of Technology. Multiaccess computer.

Epperly, W. R. Exxon Research and Engineering, Annandale, New Jersey. Coal gasification; Coal liquefaction.

Epstein, Dr. Charles J. Professor of Pediatrics, University of California, San Francisco. Down syndrome.

Epstein, Dr. W. W. Department of Chemistry, University of Utah. Dimethyl sulfoxide.

Erasmus, Dr. Baltus J. Head, Onderstepoort Vaccine Factory, South Africa. African horsesickness.

Erb, Dr. K. A. Oak Ridge National Laboratory, Oak Ridge, Tennessee. Scattering experiments (nuclei).

Erdelyl, Prof. A. Deceased; formerly, Professor of Mathematics, University of Edinburgh, Scotland. Elliptic function and integral; Spherical harmonics.

Erdman, Dr. Arthur G. Department of Mechanical Engineering, University of Minnesota. Belt drive.

Erickson, Dr. Robert P. Department of Pediatrics, University of Arizona Health Science Center, Tucson. Developmental genetics.

Ernst, Dr. Ralph A. Extension Poultry Specialist, Avian Sciences, University of California, Davis. Poultry production.

Ernster, Dr. Virginia L. Department of Epidemiology and International Health, School of Medicine, University of California, San Francisco. Epidemiology.

Esaki, Leo. Thomas J. Watson Research Center, International Business Machines Corporation, Yorktown Heights, New York; Nobelist. Tunneling in solids.

Esau, Prof. Katherine. Deceased; formerly, Department of Botany, University of California, Santa Barbara. Plant anatomy; Plant organs; Plant tissue systems; Secretory structures (plant); Xylem.

Espenak, Dr. Fred. NASA/Goddard Space Flight Center, Greenbelt, Maryland. Eclipse.

Essene, Prof. Eric J. Department of Geological Sciences, University of Michigan, Ann Arbor. Siderite.

Estes, Dr. Richard. Department of Biology, Boston University. Lissamphibia.

Ethridge, Prof. Frank G. Department of Earth Resources, Colorado State University. Diagenesis.

Ettenberg, Eugene M. Formerly, Lecturer in Graphic Arts, New York. Type (printing).

Etter, Robert J. David W. Taylor Naval Ship Research and Development Center, Bethesda, Maryland. Water tunnel (research and testing).

Evans, Dr. C. E. College of Veterinary Medicine, North Carolina State University, Raleigh. Digestive system.

Evans, Doris L. Deceased; formerly, Research and Development Division, Corning Glass Works, Corning, New York. Crystal structure; Crystallography; Polymorphism (crystallography).

Evans, Dr. Doyle J. Chief, Bacterial Enteropathogens Laboratory, Veterans Affairs Medical Center, Houston, Texas. Escherichia; Helicobacter; Klebsiella.

Evans, Prof. Francis C. Division of Biological Sciences, University of Michigan. Population dispersion.

Evans, Ginger S. Carter & Burgess, Arington, Virginia. Airport engineering.

Evans, Dr. J. Warren. Equine Consultant, Woodland, California. Horse production.

Evans, Dr. Nancy Remage. Harvard-Smithsonian Center for Astrophysics, Cambridge, Massachusetts. Pleiades.

Evans, Prof. Robley D. Deceased; formerly, Department of Physics, Massachusetts Institute of Technology. Half-life.

Evans, Dr. Susan E. Department of Anatomy and Developmental Biology, University College London, United Kingdom. Diapsida; Lepidosauria; Prolacertiformes; Rhynchocephalia.

Evenson, Dr. Kenneth M. Time and Frequency Division, National Institute of Standards and Technology, Boulder, Colorado. Light.

Evenson, Prof. Paul. Bartol Research Institute, University of Delaware, Newark. Cosmic rays.

Everetts, Dr. John, Jr. Professor of Architectural Engineering, Pennsylvania State University. Dehumidifier; Psychrometrics.

Ewalt, Dr. Karla. Senior Staff Scientist, The Scripps Research Institute, La Jolla, California. Genetic code; Protein; Protein engineering.

Ewen, Dr. Mark E. Dana-Farber Cancer Institute, Boston, Massachusetts. Tumor.

Ewing, Prof. Rodney. Department of Earth and Planetary Sciences, University of New Mexico, Albuquerque. Metamict state; Pleochroic halos; Radioactive minerals.

Eyde, Dr. Richard H. Department of Botany, Smithsonian Institution, Washington, D.C. Flower.

Eyring, Dr. Edward M. Department of Chemistry, University of Utah, Salt Lake City. Chemical kinetics.

Ezzell, Dr. John W. Chief, Special Pathogens Branch, U.S. Army Medical Research Institute, Fort Detrick, Maryland. Anthrax.

F

Fabrycky, Dr. Wolter J. Department of Industrial and Systems Engineering, Virginia Polytechnic Institute and State University. System design evaluation.

Failla, Gioacchino. Deceased; formerly, Argonne National Laboratory, Argonne, Illinois. Nuclear radiation (biology).

Fairbank, Dr. Henry A. Department of Physics, Duke University. Second sound.

Falchuck, Dr. Kenneth A. Center for Biochemical and Biophysical Sciences and Medicine, Harvard Medical School. Zinc.

Faller, Prof. John W. Department of Chemistry, Yale University, New Haven, Connecticut. Asymmetric synthesis; Fluxional compounds.

Fan, Prof. Hsu Y. Department of Physics, Purdue University. Acceptor atom; Conduction; Donor atom; Semiconductor; Valence band.

Fano, Prof. U. Deceased; formerly, James Franck Institute, University of Chicago. Molecular structure and spectra; Multipole radiation.

Fanta, Prof. Paul E. Department of Chemistry, Illinois Institute of Technology. Diels-Alder reaction; Grignard reaction; Ketone; Nitrile; other articles.

Farmer, Dr. David M. Ocean Science and Productivity Division, Institute of Ocean Science, Sidney, British Columbia, Canada. Seiche.

Farmer, Richard G. Department of Electrical Engineering, Arizona State University, Tempe. Subsynchronous resonance.

Farrel, Dr. Nicholas. Department of Chemistry, Virginia Commonwealth University, Richmond. Platinum.

Farrell, Joseph. Retired; formerly, President, American Waterways Operators, Inc., Arlington, Virginia. Inland waterways transportation.

Farrell, R. F. General Manager, Glidden Coatings and Resins, Division of SCM Corporation, Charlotte, North Carolina. Metal coatings.

Faure, Dr. Gunter. Department of Geological Sciences, Ohio State University. Elements, geochemical distribution of; Radioisotope geochemistry; Rock age determination.

Fautin, Dr. Daphne G. Natural History Museum, Department of Ecology and Evolutionary Biology, University of Kansas, Lawrence. Anthozoa;

CNIDARIA; CORALLIMORPHARIA; HEXACORALLIA; OCTOCORALLIA (ALCYONARIA).

Faw, Prof. Richard E. Head, Department of Nuclear Engineering, Kansas State University. RADIATION SHIELDING.

Fawcett, Prof. Eric. Department of Physics, University of Toronto, Ontario, Canada. SPIN-DENSITY WAVE.

Fay, Prof. Michael F. Jodrell Laboratory, Royal Botanic Gardens, Kew, Richmond, Surrey, United Kingdom. EUDICOTYLEDONS; PLANT KINGDOM; PROTEALES; SOLANALES; ZINGIBERALES; other articles.

Fay, Dr. Richard R. Parmly Hearing Institute & Department of Psychology, Loyola University of Chicago, Illinois. HEARING (VERTEBRATE).

Fayyad, Dr. Usama. Microsoft Research, Redmond, Washington. DATA MINING.

Feathers, Dr. James. University of Washington, Seattle. ARCHEOLOGICAL CHRONOLOGY.

Feduccia, Dr. Alan. Department of Biology, University of North Carolina, Chapel Hill. AVES.

Fehr, Dr. Walter R. Department of Agronomy, Iowa State University of Science and Technology. SOYBEAN.

Feigenbaum, Prof. Mitchell J. Toyota Professor, Rockefeller University, New York. PERIOD DOUBLING.

Fein, Dr. Jay S. Division of Atmospheric Sciences, National Science Foundation, Washington, D.C. MONSOON METEOROLOGY.

Feist, Dr. Blake E. Northwest Fisheries Science Center, Environmental Conservation Division, Seattle, Washington. VEGETATION AND ECOSYSTEM MAPPING.

Feld, Prof. Michael. Director, George R. Harrison Spectroscopy Laboratory, Massachusetts Institute of Technology, Cambridge. RAMAN EFFECT.

Feldberg, Dr. Steven W. Department of Applied Science, Brookhaven National Laboratory, Upton, New York. ELECTROMOTIVE FORCE (CELLS).

Feldman, Dr. Barry J. Naval Research Laboratories, Washington, D.C. OPTICAL PHASE CONJUGATION.

Feldman, Leonard. Electronics Laboratories, Great Neck, New York. AUDIO AMPLIFIER.

Feldman, Dr. Leonard C. AT&T Bell Laboratories, Murray Hill, New Jersey. AUGER EFFECT.

Feldman, Dr. Lewis J. Department of Botany, University of California, Berkeley. ROOT (BOTANY).

Feldmann, Dr. Rodney M. Department of Geology, Kent State University, Kent, Ohio. CRUSTACEA.

Fell, Prof. Howard B. Deceased; formerly, Professor Emeritus, Museum of Comparative Zoology, Harvard University. BLASTOIDEA; other articles.

Fell, Dr. Jack W. University of Miami, Rosenstiel School of Marine and Atmospheric Science, Marine Biology, and Fisheries, Key Biscayne, Florida. YEAST.

Feller, Prof. William. Deceased; formerly, Department of Mathematics, Princeton University. DISTRIBUTION (PROBABILITY); PROBABILITY.

Fellman, Dr. Jerome D. Department of Geography, University of Illinois. ASIA.

Felver, Prof. Richard I. Carnegie Institute of Technology. PRODUCT DESIGN.

Fendler, Dr. Janos H. Department of Chemistry, Center for Research in Membranes and Colloid Science, Syracuse University. MEMBRANE MIMETIC CHEMISTRY.

Feng, Dr. Pingyun. Department of Chemistry, University of California at Riverside. SUPERIONIC CONDUCTIVITY.

Fenton, Wayne A. Department of Genetics, Yale University School of Medicine, New Haven, Connecticut. MOLECULAR CHAPERONE.

Ferguson, Prof. J. Homer. Department of Biological Science, University of Idaho. DIVING; DIVING ANIMALS.

Ferguson, Dr. Louise. Extension Pomologist, Kearney Agricultural Center, University of California, Parlier. NUT CROP CULTURE; PERSIMMON.

Fernando, Dr. Quintus. Department of Chemistry, University of Arizona. COULOMETER.

Ferrari, Frank. Department of Invertebrate Zoology, National Museum of Natural History, Smithsonian Institution, Washington, D.C. MISOPHRIOIDA; MONSTRILLOIDA; MORMONILLOIDA.

Ferrell, Dr. James E., Jr. Department of Molecular Pharmacology, Stanford University, School of Medicine, Stanford, California. CELL CYCLE.

Ferris, Prof. Gordon F. Deceased; formerly, Natural History Museum, Stanford University. ANOPLURA.

Ferris, Howard. Department of Nematology, University of California, Davis. STICHOSOMIDA.

Ferry, Dr. James F. Department of Biology, Madison College, Harrisonburg, Virginia. EGG (FOWL); WILLOW.

Feshbach, Dr. Herman. Deceased; formerly, Department of Physics, Massachusetts Institute of Technology. PERTURBATION (MATHEMATICS); STRONG NUCLEAR INTERACTIONS.

Fetz, Dr. Eberhard E. Department of Physiology and Biophysics, School of Medicine, University of Washington. MOTOR SYSTEMS.

Few, Prof. Arthur. Department of Space Physics and Astronomy, Rice University. THUNDER.

Fiala, Dr. Alan D. Retired; formerly, U.S. Naval Observatory, Washington, D.C. SIDEREAL TIME; TRANSIT (ASTRONOMY).

Fiedler, Dr. Peggy L. Piedmont, California. RESTORATION ECOLOGY; WETLANDS.

Field, Joseph H. Benfield Corporation, Pittsburgh, Pennsylvania. CHARCOAL; FISCHER-TROPSCH PROCESS.

Fielding, Dr. Christopher. Cardiovascular Research Institute, University of California, San Francisco. LIPID RAFTS (MEMBRANES).

Fields, Paul R. Senior Chemist, Argonne National Laboratory, Argonne, Illinois. NOBELIUM.

Fields, Thomas H. Physics Division, Argonne National Laboratory, Illinois. PROTON.

Fincham, Prof. John R. S. Head of Department, Department of Genetics, University of Cambridge, England. COMPLEMENTATION (GENETICS); RECOMBINATION (GENETICS).

Findlay, Dr. David M. Department of Orthopedics and Trauma, University of Adelaide, Royal Adelaide Hospital, Adelaide, Australia. BONE.

Finegold, Dr. Sydney M. Los Angeles, California. ANAEROBIC INFECTION; BOTULISM.

Fingerman, Dr. Milton. Department of Biology, Tulane University. ENDOCRINE SYSTEM (INVERTEBRATE); NEUROSECRETION.

Fink, Donald G. Deceased; formerly, Director Emeritus, IEEE; Editor in Chief, *Electronics Engineers' Handbook*, McGraw-Hill, Inc., New York. TELEVISION STANDARDS; TUNING.

Fink, Prof. Jordan N. Chief, Allergy-Immunology Section, Department of Medicine, Medical College of Wisconsin, Milwaukee. ANAPHYLAXIS.

Finkelstein, Prof. L. Department of Physics, City University School of Engineering, London, England. INSTRUMENT SCIENCE.

Finkelstein, Dr. Richard. Department of Microbiology, University of Missouri School of Medicine, Columbia. CHOLERA.

Finks, Dr. Robert M. Department of Geology, Queens College, City University of New York. AMPHIDISCOSA; CALCAREA.

Finn, Prof. M. G. Department of Chemistry, The Scripps Research Institute, La Jolla, California. CLICK CHEMISTRY.

Finotello, Dr. Daniele. Liquid Crystal Institute, Kent State University, Kent, Ohio. LIQUID CRYSTALS.

Firebaugh, Dr. Millard S. Electric Boat Corporation, Groton, Connecticut. SUBMARINE.

Firk, Dr. Frank W. K. Department of Physics, Yale University. TIME-OF-FLIGHT SPECTROMETERS.

Firor, Dr. John W. Director, Advanced Study Program, National Center for Atmospheric Research, Boulder, Colorado. CLIMATE MODIFICATION.

Fischer, Prof. Alfred G. Department of Earth Sciences, University of Southern California. SEDIMENTARY ROCKS.

Fischer, Harold. General Motors Corporation, Flint, Michigan. DIFFERENTIAL.

Fischer, Henry W. Hardesty & Hanover, New York. BRIDGE.

Fischer, Prof. J. E. Department of Materials Science and Engineering, University of Pennsylvania, Philadelphia. INTERCALATION COMPOUNDS.

Fisher, Dr. Jack B. Chair of Botanical Sciences, Fairchild Tropical Garden, Research Center, Miami, Florida. STEM.

Fisher, Dr. Robert A. Los Alamos National Laboratory, Los Alamos, New Mexico. OPTICAL PHASE CONJUGATION.

Fisher, Dr. Robert L. Geological Research Division, Scripps Institution of Oceanography, La Jolla, California. DEEP-SEA TRENCH.

Fisher, Prof. Russell A. Department of Physics, Northwestern University. DEGREE OF FREEDOM (MECHANICS); DYNAMICS; EQUATIONS OF MOTION.

Fisher, Dr. William L. Department of Geological Science, University of Texas, Austin. DEPOSITIONAL SYSTEMS AND ENVIRONMENTS; PETROLEUM GEOLOGY.

Fisk, Prof. Nelson S. Department of Civil Engineering, Columbia University. CENTER OF GRAVITY; CENTROIDS (MATHEMATICS); EQUILIBRIUM OF FORCES; RESULTANT OF FORCES; TORQUE; other articles.

Fiske, Capt. Richard P. Retired; formerly, Director of Ocean Engineering/Supervisor of Salvage and Diving, U.S. Navy. SHIP SALVAGE.

Fitch, Prof. Val L. Department of Physics, Princeton University, Princeton, New Jersey; Nobelist. CPT THEOREM; FLAVOR.

FitzGerald, Prof. Garret A. Department of Pharmacology, Institute for Translational Medicine & Therapeutics, University of Pennsylvania School of Medicine, Philadelphia. ISOPROSTANES.

Fitzsimons, James T. Physiological Laboratory, University of Cambridge, United Kingdom. THIRST AND SODIUM APPETITE.

Flathers, George W., II. MITRE Corporation, McLean, Virginia. GROUND PROXIMITY WARNING SYSTEM.

Fleischer, Dr. R. L. Physical Science Branch, General Physics Laboratory, General Electric Company, Schenectady, New York. FISSION TRACK DATING.

Fleminger, Dr. Abraham. Research Biologist, Scripps Institution of Oceanography, La Jolla, California. BRANCHIURA.

Flessa, Dr. Karl Walter. Division of Earth Sciences, National Science Foundation, Washington, D.C. EXTINCTION (BIOLOGY).

Flexser, Dr. Leo A. Retired; formerly, Vice President of Chemical Production, Hoffmann-LaRoche, Inc., Nutley, New Jersey. RIBOFLAVIN.

Flint, Dr. Richard F. Department of Geology and Geophysics, Yale University. FIORD.

Flint, Robert B. VORTAC Consultant, Havana, Illinois. VOR (VHF OMNIDIRECTIONAL RANGE).

Floe, Dr. Carl F. Professor Emeritus, Department of Metallurgy, Massachusetts Institute of Technology. SURFACE HARDENING OF STEEL.

Flowers, Dr. T. J. School of Biological Sciences, University of Sussex, United Kingdom. PLANTS OF SALINE ENVIRONMENTS.

Flye, Prof. M. Wayne. Surgery, Microbiology and Immunology, Washington University School of Medicine, St. Louis, Missouri. TISSUE TYPING.

Flynn, Dr. Thomas M. Consultant, Cryogenic Engineering, Boulder, Colorado. LIQUEFACTION OF GASES.

Focht, Prof. Dennis D. Department of Soil and Environmental Sciences, University of California, Riverside. BIODEGRADATION.

Fogel, Dr. David B. Natural Selection, Inc., La Jolla, California. EVOLUTIONARY COMPUTATION.

Foley, Dr. Janet E. School of Veterinary Medicine, Center for Companion Animal Health, University of California, Davis. FELINE INFECTIOUS PERITONITIS.

Foner, Dr. Simon. Deceased; formerly, Associate Professor, Francis Bitter Natonal Magnet Lab, Massachusetts Institute of Technology, Cambridge. MAGNET.

Fontana, Dr. Robert J. Multispectral Solutions, Inc., Germantown, Maryland. ULTRAWIDEBAND (UWB) SYSTEMS.

Foon, Dr. Kenneth A. Roswell Park Memorial Institute, Buffalo, New York. LEUKEMIA.

Foote, Dr. Kenneth G. Woods Hole Oceanographic Institution, Massachusetts. PARAMETRIC ARRAY.

Ford, Prof. Derek C. Department of Geography, McMaster University, Hamilton, Ontario, Canada. CAVE; KARST TOPOGRAPHY.

Ford, Prof. E. B. Deceased; formerly, Genetics Laboratory, Department of Zoology, Oxford University, Oxford, England. POLYMORPHISM (GENETICS).

Ford, Prof. Joseph. Regents' Professor, School of Physics, College of Sciences and Liberal Studies, Georgia Institute of Technology. CHAOS.

Ford, Dr. Karen Grady. Department of Biology, College of Charleston, South Carolina. PLANT PROPAGATION.

Forey, Dr. Peter. Department of Paleontology, The Natural History Museum, London, United Kingdom. JAWLESS VERTEBRATES; PALAEONISCIFORMES.

Forman, Dr. Barry Marc. The City of Hope National Medical Center, Beckman Research Institute, Duartes, California. NUCLEAR HORMONE RECEPTORS.

Forster, Dr. Denis. Monsanto Company, St. Louis, Missouri. HOMOGENEOUS CATALYSIS.

Forster, Dr. Roy P. Department of Biological Sciences, Dartmouth College, Hanover, New Hampshire. URINARY SYSTEM.

Fortney, Dr. William. Department of Clinical Sciences, College of Veterinary Medicine, Kansas State University, Manhattan. HEARTWORMS.

Fortune, Dr. W. Brooks. Group Vice President, Science and Manufacturing, Eli Lilly Company, Indianapolis, Indiana. PHARMACEUTICALS TESTING.

Foster, Prof. Jackson W. Deceased; formerly, Department of Microbiology, University of Texas at Austin. BACTERIAL PHYSIOLOGY AND METABOLISM; CULTURE.

Foster, Dr. Mark G. Retired; formerly, Department of Electrical Engineering, School of Engineering and Applied Science, University of Virginia. VOLT-AMPERE.

Foster, Prof. Merrill W. Department of Geology, Bradley University, Peoria, Illinois. OBOLELLIDA; PATERINIDA; other articles.

Foster, Dr. Robert J. New York Orthopaedic Hospital Research Laboratory, Columbia-Presbyterian Medical Center, Columbia University, New York. JOINT (ANATOMY).

Foster, Prof. Robert J. Department of Geology and Geophysics, San Jose State University. EARTH.

Fountain, Dr. David M. Department of Geology and Geophysics, University of Wyoming. MOHO (MOHOROVIČIĆ DISCONTINUITY).

Fourtner, Dr. Charles R. Department of Biological Science, State University of New York, Buffalo. BLATTARIA (BLATTODEA).

Fowler, Prof. W. Beall. Department of Physics, Lehigh University. CRYSTAL DEFECTS.

Fowlkes, Dr. J. Brian. Senior Associate Research Scientist, Basic Radiological Sciences Division, Department of Radiology, University of Michigan Health System, Ann Arbor. ULTRASONICS.

Fox, Dr. Christopher. Department of Computer Science, James Madison University, Harrisonburg, Virginia. ABSTRACT DATA TYPE.

Fox, Dr. Irving. School of Tropical Medicine, School of Medicine, University of Puerto Rico. SIPHONAPTERA.

Fox, Dr. Marye Anne. Department of Chemistry, University of Texas, Austin. PHOTOCHEMISTRY.

Fox, Prof. Patrick J. School of Civil Engineering, Purdue University, West Lafayette, Indiana. COFFERDAM.

Fox, Dr. Richard. Department of Biological Sciences, University of Alberta, Edmonton, Canada. MULTITUBERCULATA.

Fox, Prof. Robert M. Department of Electrical Engineering, College of Engineering, University of Florida. ELECTRICAL MODEL.

Fox, Dr. William K. Standard Havens Research Corporation, Kansas City, Missouri. ENERGY SOURCES.

Fradkin, Dr. David M. Department of Physics, Wayne State University. PERTURBATION (QUANTUM MECHANICS); RUNGE VECTOR.

Frailey, Dr. Dennis J. Texas Instruments, Inc., Austin. COMPUTER SYSTEMS ARCHITECTURE.

Frakes, Dr. William B. Virginia Tech, Northern Virginia Center, Falls Church. SOFTWARE ENGINEERING.

Frame, Prof. J. Sutherland. Professor Emeritus of Mathematics, Michigan State University. CIRCLE; EUCLIDEAN GEOMETRY; PRISM; PYTHAGOREAN THEOREM; other articles.

Franc, Dr. Jean-Marie. Laboratoire d'Histologie et Biologie Tissulaire, Université Claude Bernard, Villeurbanne, France. CTENOPHORA.

France, Dr. Diane L. Fort Collins, Colorado. FORENSIC ANTHROPOLOGY.

Francis, Dr. Arthur W. Union Carbide Corporation, Tarry-town, New York. ARGON; CRYOBIOLOGY; HYPERBARIC OXYGEN CHAMBER; INERT GASES; OXYGEN; other articles.

Francis, Dr. Charles A. Department of Agronomy, Institute of Agriculture and Natural Resources, University of Nebraska. AGRONOMY; MULTIPLE CROPPING.

Frank, Dr. Bernard. Deceased; formerly, Professor of Watershed Management, Colorado State University, Fort Collins. WATER CONSERVATION.

Franklin, David. Program Manager, Nuclear Systems and Materials, Electric Power Research Institute, Palo Alto, California. NUCLEAR FUELS.

Franklin, Dr. Philip. Deceased; formerly, Professor of Mathematics, Massachusetts Institute of Technology. SERIES.

Franzese, Dr. Kenneth. Research and Development, Bright Star Industries, Inc., Clifton, New Jersey. NUCLEAR BATTERY.

Franzini, Dr. Paolo. Dipartimento di Fisica, Universita di Roma I, Italy. UPSILON PARTICLES.

Freedman, Dr. Wendy L. Director, Observatories of the Carnegie Institution of Washington, Pasadena, California. CEPHEIDS; HUBBLE CONSTANT.

Freeman, Dr. A. E. Department of Animal Science, Iowa State University of Science and Technology, Ames. BREEDING (ANIMAL).

Freeman, Prof. Jeremiah P. Department of Chemistry, University of Notre Dame. CARBONYL; MESOIONIC COMPOUND.

Freeman, Prof. Ralph L. Retired; formerly, Department of Mechanical Engineering, Iowa State College. COINING; EXTRUSION; SPINNING (METALS).

Freeman, Dr. Reino S. Professor of Parasitology, University of Toronto, Ontario, Canada. SPARGANOSIS.

Freeman, Prof. Walter J. Department of Molecular and Cell Biology, University of California, Berkeley. BRAIN.

Freire, Prof. Ernesto. Director, Biocalorimetry Center, Johns Hopkins University, Baltimore, Maryland. BIOCALORIMETRY.

Freiser, Prof. Henry. Department of Chemistry, University of Arizona, Tucson. GRAM-EQUIVALENT WEIGHT; HYDROLYSIS.

Freivalds, Prof. Andris. Department of Industrial & Manufacturing Engineering, Penn State University. METHODS ENGINEERING.

Fremed, Raymond F. Burson-Marsteller Associates, New York. HEAT EXCHANGER.

French, Dr. J. B. Department of Physics, University of Rochester, New York. SELECTION RULES (PHYSICS).

Fretter, Prof. William B. Deceased; formerly, Department of Physics, University of California, Berkeley. CERENKOV RADIATION; GEIGER-MÜLLER COUNTER.

Freudenstein, Dr. Ferdinand. Department of Mechanical Engineering, Columbia University, New York. FOUR-BAR LINKAGE; MECHANISM.

Freundt, Dr. E. A. Institute for Medical Microbiology, Aarhus University, Aarhus, Denmark. MYCOPLASMAS.

Frey, Dr. Kenneth J. Department of Agronomy, Iowa State University. OATS.

Fribourg, Prof. Henry A. Department of Plant and Soil Science, University of Tennessee, Knoxville. FESCUE.

Fridovich, Dr. Irwin. Department of Biochemistry, Duke University Medical Center. OXYGEN TOXICITY.

Fried, Dr. George H. Brooklyn College, City University of New York. LIVER.

Fried, Dr. Sherman. Chemistry Division, Argonne National Laboratory, Argonne, Illinois. ACTINIUM; RHENIUM.

Fried, Walter R. Deceased; formerly, Navigation Systems Consultant, Hughes Aircraft Company, Santa Ana, California. ELECTRONIC NAVIGATION SYSTEMS.

Friedlander, Prof. Sheldon K. Parson Professor of Chemical and Biomolecular Engineering Department, University of California, Los Angeles. AEROSOL.

Friedman, Dr. Harold L. Department of Chemistry, State University of New York, Stony Brook. WATER.

Friedman, Lewis. Brookhaven National Laboratory, Upton, New York. ISOTOPE DILUTION TECHNIQUES.

Friedman, Dr. Norman. New York, New York. ANTISUBMARINE WARFARE.

Frink, Dr. Stephen. Photographic Inc., Key Largo, Florida. UNDERWATER PHOTOGRAPHY.

Frisancho, Dr. A. Roberto. Center for Human Growth and Development, University of Michigan, Ann Arbor. ANTHROPOMETRY.

Fritsch, Prof. J. Michael. Department of Meteorology, College of Earth and Mineral Sciences, Pennsylvania State University. MESOMETEOROLOGY.

Fritschel, Dr. Peter K. Kavli Institute for Astrophysics and Space Research, Massachusetts Institute of Technology, Cambridge. GRAVITATIONAL RADIATION; LIGO (LASER INTERFEROMETER GRAVITATIONAL-WAVE OBSERVATORY).

Froes, Dr. F. H. Sam. Director, Institute for Materials and Advanced Processes, University of Idaho, Moscow. MECHANICAL ALLOYING.

Fröhlich, Dr. Claus. Physikalisch-Meteorologisches Observatorium Davos, World Radiation Center, Davos Dorf, Switzerland. SOLAR RADIATION.

Fromkin, Dr. Victoria A. Dean, Vice Chancellor, Graduate Programs, Department of Linguistics, University of California, Los Angeles. LINGUISTICS.

Frondel, Prof. Clifford. Retired; formerly, Department of Geological Sciences, Harvard University. QUARTZ; SILICATE MINERALS; other articles.

Fruchter, Dr. Benjamin. Department of Educational Psychology, University of Texas, Austin. FACTOR ANALYSIS.

Frumhoff, Dr. Peter. Department of Entomology, University of California, Davis. BEHAVIORAL ECOLOGY.

Fry, Prof. Albert J. Department of Chemistry, Wesleyan University, Middletown, Connecticut. ELECTROCHEMICAL TECHNIQUES; ELECTROCHEMISTRY.

Fry, Prof. Edward S. Department of Physics, Texas A&M University, College Town. HIDDEN VARIABLES.

Frye, Eugene O. Retired; formerly, Technical Staff, Collins Division, Rockwell International, Cedar Rapids, Iowa. HOMING.

Fryer, Dr. Geoffrey. Freshwater Biological Association, Cumbria, United Kingdom. ANOMOPODA; ANOSTRACA; BRANCHIOPODA; HAPLOPODA; LAEVICAUDATA; other articles.

Fryrear, Dr. Donald W. Research Leader, Agricultural Research Services, U.S. Department of Agriculture, Big Spring, Texas. AGRICULTURAL SOIL AND CROP PRACTICES; SOIL.

Fuchs, Prof. Ronald. Department of Physics, Iowa State University. BOLTZMANN TRANSPORT EQUATION.

Fudenberg, Dr. K. Hugh. Section of Hematology and Immunology, San Francisco Medical Center, University of California. GLOBULIN.

Fulford, Prof. George D. Department of Chemical Engineering, University of Waterloo, Ontario, Canada. DIMENSIONLESS GROUPS.

Fulks, J. R. Retired; formerly, National Weather Service, Chicago, Illinois. DEW POINT; HUMIDITY; PRECIPITATION (METEOROLOGY); RAIN SHADOW.

Fuller, Prof. Dudley D. Retired; formerly, Department of Mechanical Engineering, Columbia University. ANTIFRICTION BEARING.

Fuller, Dr. Mark Roy. Patuxent Wildlife Reserve Center, Laurel, Maryland. BIOTELEMETRY.

Fuller, Dr. Thomas C. Retired; formerly, Supervisor, Botany and Seed Laboratories,

California Department of Food and Agriculture, Sacramento. POISONOUS PLANTS.

Fulton, Dr. James W. Monsanto Company, St. Louis, Missouri. DEHYDROGENATION.

Fung, Prof. Y. C. Department of Applied Mechanics, University of California, San Diego. BIORHEOLOGY.

Furter, Dr. William F. Dean of Graduate Studies and Research, Royal Military College of Canada, Kingston, Ontario. ABSORPTION; CHEMICAL CONVERSION; CHEMICAL ENGINEERING; CHEMICAL PROCESS INDUSTRY; UNIT PROCESSES.

Futral, Prof. J. G. Deceased; formerly, Department of Agricultural Engineering, Georgia Agricultural Experiment Station, Griffin. FERTILIZING.

Futuyma, Dr. Douglas J. Section of Ecology and Systematics, Cornell University. ORGANIC EVOLUTION.

G

Gaasch, Dr. William H. Director of Cardiovascular Research, Section of Cardiovascular Diseases, Labey Clinic, Burlington, Massachusetts. HEART DISORDERS.

Gad-el-Hak, Prof. Mohamed. Department of Mechanical Engineering, Virginia Commonwealth University, Richmond. ADAPTIVE WINGS.

Gaffney, Dr. Eugene S. Department of Vertebrate Paleontology, American Museum of Natural History, New York. CHELONIA.

Gagliardo, Reginald S. Vice President, Engineering, Burns & Roe Enterprises, Inc., Oradell, New Jersey. POWER PLANT.

Gagne, Dr. Raymond J. Systematic Entomology Laboratory, U.S. Department of Agriculture, Washington, D.C. DIPTERA.

Gaines, Alan M. Division of Earth Science, National Science Foundation, Washington, D.C. ANKERITE; CARBONATE MINERALS; DOLOMITE.

Gaines, Dr. George L. Research and Development Center, General Electric Company, Schenectady, New York. MONOMOLECULAR FILM.

Galambos, Prof. Theodore V. Emeritus Professor, Department of Civil Engineering, University of Minnesota, Minneapolis. BEAM; STRUCTURE (ENGINEERING).

Gale, Peter A. Chief Naval Architect, John J. McMullen Associates, Inc., Arlington, Virginia. NAVAL ARCHITECTURE; SHIP DESIGN.

Galiana, Prof. Francisco D. Department of Electrical and Computer Engineering, McGill University, Montreal, Québec, Canada. REACTIVE POWER.

Galil, Prof. Noah I. Department of Civil and Environmental Engineering, Technion Institute of Technology, Israel. INDUSTRIAL WASTEWATER TREATMENT.

Gallant, Dr. Donald M. Department of Psychiatry and Neurology, Tulane University School of Medicine. SEDATIVE.

Galler, Prof. Bernard A. Ann Arbor, Michigan. COMPUTER.

Galloway, Dr. William J. Bolt, Beranek and Newman, Inc., Canoga Park, California. ECHO; FREQUENCY (WAVE MOTION); PHASE (PERIODIC PHENOMENA); SINE WAVE; other articles.

Galston, Dr. Arthur W. MCD Biology Department, Yale University, New Haven, Connecticut. PLANT REPRODUCTION.

Gamble, Dr. Joe D. National Aeronautics and Space Administration, Lyndon B. Johnson Space Center, Houston, Texas. ATMOSPHERIC ENTRY.

Gamble, Dr. John. Department of Geology, Victoria University, Wellington, New Zealand. ANDESITE.

Garcia, Dr. John. Department of Psychology, University of California, Los Angeles. CONDITIONED REFLEX.

Garcia, Prof. Marcelo. Department of Civil Engineering, University of Illinois, Urbana. TURBIDITY CURRENT.

Gardiner, Dr. Brian. Linnean Society of London, Burlington House, Piccadilly, London, United Kingdom. TELEOSTEI.

Gardiner, Dr. Kathleen. Eleanor Roosevelt Institute for Cancer Research, Denver, Colorado. ELECTROPHORESIS.

Gardner, Prof. Chester S. Department of Electrical Engineering and Computer Science, University of Illinois, Urbana. LIDAR.

Gardner, Dr. Frank E. Horticulturist, Agricultural Research Service, U.S. Department of Agriculture, Orlando, Florida. KUMQUAT; MANDARIN; TANGERINE.

Gardner, Dr. Julian W. Director of Nanotechnology Centre, Department of Engineering, University of Warwick, Coventry, United Kingdom. MICROSENSOR.

Gardner, Dr. Walter H. Department of Agronomy and Soils, Washington State University, Pullman. SOIL.

Garetz, Dr. Bruce A. Department of Chemistry, Polytechnic University, New York. BOND ANGLE AND DISTANCE.

Garfield, Dr. Sol L. Department of Psychology, Washington University, St. Louis, Missouri. PSYCHOTHERAPY.

Garland, Dr. John. Professor, Department of Forest Engineering, Oregon State University, Corvallis. LOGGING.

Garman, Philip. Connecticut Agricultural Experiment Station, New Haven. ODONATA.

Garrett, Prof. Alfred B. Department of Chemistry, Ohio State University. CHEMISTRY; ELEMENT (CHEMISTRY); SALT (CHEMISTRY).

Garrett, Dr. Roger E. Department of Agricultural Engineering, University of California, Davis. AGRICULTURAL ENGINEERING.

Gartner, Mr. Jeffrey W. Water Resources Discipline, U. S. Geological Survey, Tucson, Arizona. RIVER TIDES.

Gasaway, Donald C. E.A.R. Division, Cabot Corporation, San Antonio, Texas. EAR PROTECTORS.

Gass, Dr. Saul I. Potomac, Maryland. SYSTEMS ANALYSIS.

Gastaldo, Dr. Robert A. Department of Geology, Auburn University, Auburn, Alabama. TAPHONOMY.

Gastony, Dr. Gerald J. Department of Botany, Indiana University. TREE FERNS.

Gatchell, Charles J. U.S. Department of Agriculture Forest Service, Forestry Sciences Laboratory, Princeton, West Virginia. WOOD PROCESSING.

Gaudin, Dr. Tim. Department of Biological and Environmental Science, University of Tennessee, Chattanooga. EDENTATA.

Gaunard, Dr. Guillermo C. Naval Surface Warfare Center, Carderock Division, West Bethesda, Maryland. ACOUSTIC MINE.

Gauss, Dr. D. Max-Planck Institut für Experimentelle Medizin, Göttingen, Germany. CLATHRATE COMPOUNDS.

Gauss, F. Stephen. Retired; formerly, Director, Astrometry Department, U.S. Naval Observatory, Washington, D.C. ASTRONOMICAL TRANSIT INSTRUMENT.

Gaxiola, Roberto. Department of Plant Sciences, University of Connecticut, Storrs. PLANTS OF SALINE ENVIRONMENTS.

Gaylord, Prof. Charles N. Chairperson, Department of Civil Engineering, University of Virginia. ARCH.

Gazzaniga, Prof. Michael S. Professor of Psychology and Social Sciences in Medicine, State University of New York, Stony Brook. HEMISPHERIC LATERALITY.

Geiger, Prof. Randall L. Department of Electrical and Computer Engineering, Iowa State University. CAPACITANCE MULTIPLICATION; ELECTRONIC CIRCUITS.

Geist, Jon. Radiometric Physics Division, National Institute of Standards and Technology, Washington, D.C. PHOTOMETRY; RADIOMETRY.

Gelfand, Prof. Erwin W. Chief, Division of Immunology and Rheumatology, Hospital for Sick Children, Toronto, Ontario, Canada. IMMUNOLOGICAL ONTOGENY.

Gemmell, Dr. Donald S. Physics Division, Argonne National Laboratory, Argonne, Illinois. COULOMB EXPLOSION.

Genito, Dennis. Pasture Systems and Watershed Management Research Laboratory, U.S. Department of Agriculture–Agricultural Research Service, University Park, Pennsylvania. ACID RAIN.

Gensler, Prof. Walter J. Deceased; formerly, Department of Chemistry, Boston University, Boston, Massachusetts. FURAN; PYRIDINE; other articles.

Geobel, Prof. Charles J. Department of Physics, University of Wisonsin, Madison. DISPERSION RELATIONS.

George, Dr. John W. Department of Chemistry, University of Massachusetts. SELENIUM.

George, Dr. Kathleen F. Union Carbide Chemicals and Plastics Company, Inc., South Charleston, West Virginia. ANTIFREEZE MIXTURE.

George, Dr. Melvin R. Agronomy and Range Science Extension, University of California, Davis. SUDANGRASS.

Georgi, Dr. Todd A. Department of Biology, Doane College, Crete, Nebraska. ADIPOSE TISSUE.

Georgorapadakou, Dr. Nafsika H. Department of Infectious Diseases, DuPont Research Laboratories, Wilmington, Delaware. ANTIBIOTIC.

Gepts, Dr. Paul. Department of Agronomy and Range Science, University of California, Davis. BEAN; COWPEA.

Gerber, Leah R. National Center for Ecological Analysis and Synthesis, Santa Barbara, California. TROPHIC ECOLOGY.

Gergely, Dr. John. Department of Muscle Research, Boston Biomedical Research Institute, Boston, Massachusetts. MUSCLE PROTEINS.

Gerjuoy, Dr. Edward. Department of Physics, University of Pittsburgh. DE BROGLIE WAVELENGTH; EXCITED STATE; MATRIX MECHANICS; QUANTUM MECHANICS; other articles.

Gerlach, Dr. Arch C. Chief Geographer, U.S. Geological Survey, Department of the Interior. GLOBE (EARTH).

German, Edward R. U.S. Army Criminal Investigation Laboratory, Forest Park, Georgia. FINGERPRINT.

Gerrity, Dr. Joseph P. National Meteorological Center, National Oceanic and Atmospheric Administration, Camp Springs, Maryland. WEATHER FORECASTING AND PREDICTION.

Gersh, Prof. Isidore. Department of Animal Biology, School of Veterinary Medicine, University of Pennsylvania. BLOOD; CONNECTIVE TISSUE; EPITHELIUM; HISTOLOGY.

Gershoff, Prof. Stanley N. Department of Nutrition, Harvard School of Public Health, Boston, Massachusetts. PARA-AMINOBENZOIC ACID; BIOTIN; NIACIN; VITAMIN D; VITAMIN K; other articles.

Gertler, Dr. Janos. Department of Electrical and Computer Engineering, George Mason University, Fairfax, Virginia. FAULT ANALYSIS; REMOTE-CONTROL SYSTEM.

Gertsch, Dr. Willis J. American Museum of Natural History, New York. PALPIGRADI; UROPYGI.

Getman, Dr. Thomas D. Department of Chemistry, Northern Michigan University. CARBORANE.

Ghiorso, Dr. Albert. Department of Chemistry, Lawrence Berkeley Laboratory, University of California, Berkeley. DUBNIUM; LAWRENCIUM; RUTHERFORDIUM.

Giallorenzi, Dr. Thomas G. Superintendent, Optical Science Division, Department of the Navy, Naval Research Laboratory, Washington, D.C. FIBER-OPTIC SENSOR.

Gibeling, Dr. Jeffery C. Department of Materials Science and Engineering, Stanford University, Palo Alto, California. CREEP (MATERIALS).

Giblon, Robert P. President, George G. Sharp, Inc., New York, New York. MARINE BOILER; MARINE REFRIGERATION.

Giedt, Prof. Warren H. Professor Emeritus, Department of Mechanical Engineering, University of California, Davis. CONDUCTION (HEAT); CONVECTION (HEAT).

Giese, Prof. Arthur C. Department of Biological Sciences, Stanford University. INFRARED RADIATION (BIOLOGY); ULTRAVIOLET RADIATION (BIOLOGY).

Giese, Raymond. Department of Mechanical Engineering, University of Minnesota. BELT DRIVE.

Gigg, Roy H. Chemistry Division, National Institute for Medical Research, London, England. LIPID; TRIGLYCERIDE (TRIACYLGLYCEROL).

Giguère, Paul A. Deceased; formerly, Emeritus Professor, St. Petersburg Beach, Florida. HYDROGEN PEROXIDE.

Gilardi, Dr. Gerald L. Microbiology Laboratory, North General Hospital, New York. MORAXELLA; PSEUDOMONAS.

Gilbert, Barrie. Analog Devices Inc., Northwest Laboratories, Beaverton, Oregon. INSTRUMENTATION AMPLIFIER.

Gilbert, Dr. Edgar N. AT&T Bell Telephone Laboratories, Murray Hill, New Jersey. INFORMATION THEORY.

Gill, Dr. D. Michael. Department of Molecular Biology and Microbiology, Tufts University. TOXIN.

Gilleo, Dr. Ken. ET-Trends LLC, Warwick, Rhode Island. PRINTED CIRCUIT BOARD.

Gillespie, Dr. Ronald J. Department of Chemistry, McMaster University, Hamilton, Ontario, Canada. SUPERACID.

Gillies, Dr. C. B. School of Biological Sciences, University of Sydney, Australia. CROSSING-OVER (GENETICS).

Gilloteaux, Dr. Jacques. Professor of Cell Biology and Anatomy, Lake Erie College of Osteopathic Medicine, Erie, Pennsylvania. GALLBLADDER.

Gingrich, Prof. Newell S. Department of Physics, University of Missouri. CLASSICAL MECHANICS.

Ginsburg, Dr. Robert N. Rosenstiel School of Marine and Atmospheric Science, University of Miami. STROMATOLITE.

Girardin, W. E. President and Director, Western Union Telegraph Company, Upper Saddle River, New Jersey. TELEGRAPHY.

Giribet, Dr. Gonzalo. Department of Invertebrate Zoology, Harvard University, Cambridge, Massachusetts. METAZOA.

Girvin, Prof. Steven M. Department of Physics, Indiana University, Bloomington. HALL EFFECT.

Glaab, Joseph B. Retired; formerly, General Instrument Corporation, Hatboro, Pennsylvania. TELEVISION NETWORKS.

Glantz, Dr. Michael. National Center for Atmospheric Research, Boulder, Colorado. EL NIÑO.

Glasford, Prof. Glenn M. Department of Electrical Engineering and Computer Science, Syracuse University. COMPARATOR; DELAY LINE; ELECTRONIC SWITCH; FUNCTION GENERATOR; PULSE GENERATOR; other articles.

Glass, Dr. Robert L. Editor-in-Chief, *Journal of Systems and Software*, and Editor/Publisher, *The Software Practitioner*, Bloomington, Indiana. SOFTWARE.

Glass, Dr. Roger I. Chief, Viral Gastroenteritis Unit, Center for Infectious Diseases, Centers for Disease Control and Prevention, Atlanta, Georgia. EPIDEMIC VIRAL GASTROENTERITIS.

Glasscock, Dwight L. Deceased; formerly, Harza Engineering Company, Chicago, Illinois. PUMPED STORAGE.

Gledhill, Dr. S. J. Harpenden Herts, United Kingdom. SPECTRUM ANALYZER.

Gleim, Paul S. Manager, Advanced Products, Texas Instruments, Inc., Dallas. GERMANIUM.

Glick, Dr. Benjamin. University of Chicago, Cummings Life Science Center, Chicago, Illinois. GOLGI APPARATUS.

Glickman, Dr. Todd S. Burlington, Massachusetts. INDUSTRIAL METEOROLOGY.

Glowatz, Dr. Michael, Jr. Environmental Health and Safety Department, AT&T Bell Laboratories, Murray Hill, New Jersey. GAS AND ATMOSPHERE ANALYSIS.

Gnade, Dr. Bruce E. Defense Advanced Research Projects Agency, Arlington, Virginia. FLAT-PANEL DISPLAY DEVICE.

Gnanou, Dr. Yves. Laboratoire de Chimie des Polymères Organiques, Université Bordeaux, Pessac, France. CONTROLLED/LIVING RADICAL POLYMERIZATION.

Goddu, Dr. Robert F. Manager, Fibers and Film Research Division, Hercules, Inc., Wilmington, Delaware. POLARIMETRIC ANALYSIS; REFRACTOMETRIC ANALYSIS.

Godwin, John P. Gretna Green Associates, Los Angeles, California. SATELLITE TELEVISION BROADCASTING.

Goebel, Prof. Charles J. Emeritus, Department of Physics, University of Wisconsin. BREMSSTRAHLUNG; DISPERSION RELATIONS; PARITY (QUANTUM MECHANICS); RELATIVISTIC QUANTUM THEORY; TIME REVERSAL INVARIANCE; UNITARY SYMMETRY; other articles.

Goetschalckx, Dr. Marc. School of Industrial and Systems Engineering, Georgia Institute of Technology, Atlanta. FACILITIES PLANNING AND DESIGN.

Goff, Dr. Fraser. Los Alamos National Laboratories, Los Alamos, New Mexico. PNEUMATOLYSIS.

Goff, Dr. M. Lee. Department of Entomology, University of Hawaii at Manoa, Honolulu. FORENSIC ENTOMOLOGY.

Goguen, Dr. Jay. NASA Jet Propulsion Laboratory, Pasadena, California. ALBEDO.

Gokel, Dr. George W. Molecular Biology & Pharmacology, Washington University School of Medicine, St. Louis, Missouri. MACROCYCLIC COMPOUND.

Gold, Dr. Ronald. Hospital for Sick Children, Toronto, Ontario, Canada. MENINGOCOCCUS.

Goldberg, Dr. Alfred L. Department of Cell Biology, Harvard Medical School, Boston, Massachusetts. PROTEASOME.

Goldberg, Prof. David E. Department of General Engineering, University of Illinois, Urbana. GENETIC ALGORITHMS.

Goldberg, Dr. Ira J. Department of Medicine, Columbia College of Physicians and Surgeons, Division of Metabolism and Nutrition, New York. LIPOPROTEIN.

Golden, Dr. Joseph. National Weather Service, National Oceanic and Atmospheric Administration, Silver Spring, Maryland. TORNADO; WATERSPOUT.

Goldfarb, Dr. Marjorie S. School of Oceanography, University of Washington. HYDROTHERMAL VENT.

Goldhaber, Dr. Gerson. Lawrence Berkeley Laboratory, University of California, Berkeley. GOLDHABER TRIANGLE.

Goldich, Dr. Samuel S. Department of Geology, Colorado School of Mines, Golden. LATERITE.

Goldman, Dr. Marvin. Emeritus Professor, Department of Radiological Sciences, University of California, Davis. BACKGROUND COUNT; HEALTH PHYSICS; RADIOACTIVE FALLOUT.

Goldstein, Dr. D. J. Department of Biomedical Science, University of Sheffield, England. INTERFERENCE MICROSCOPE.

Goldstein, Dr. Irving S. Department of Wood and Paper Science, North Carolina State University, Raleigh. ROSIN; TALL OIL; WOOD CHEMICALS.

Golley, Dr. Frank B. Institute of Ecology, University of Georgia. ECOLOGY, APPLIED.

Golub, Dr. Leon. Harvard-Smithsonian Center for Astrophysics, Cambridge, Massachusetts. ROCKET ASTRONOMY; SOLAR CORONA.

Gomer, Prof. Robert. James Franck Institute, University of Chicago. FIELD EMISSION.

Gönen, Prof. Turan. Department of Electrical & Electronic Engineering, California State University, Sacramento. ELECTRIC DISTRIBUTION SYSTEMS.

Gonser, Dr. Bruce W. Battelle Memorial Institute, Columbus, Ohio. TIN ALLOYS.

Gonzani, Dr. Tshai. Palo Alto, California. DOSIMETER.

Good, Prof. Irving J. Department of Statistics, Virginia Polytechnic Institute and State University, Virginia. MONTE CARLO METHOD.

Good, Dr. Roland H., Jr. Retired; formerly, Department of Physics, Pennsylvania State University, University Park. TACHYON; WENTZEL-KRAMERS-BRILLOUIN METHOD.

Goodall, Dr. McChesney. Professor of Pharmacology and Professor of Surgery and Physiology, University of Texas Medical Branch, Galveston. DECOMPRESSION ILLNESS.

Goodbody, Dr. Ivan. Department of Zoology, University of the West Indies, Jamaica. ASCIDIACEA; TUNICATA (UROCHORDATA).

Goodchild, Dr. C. G. Candler Professor of Biology, Emory University. PLATYHELMINTHES.

Goodenough, Prof. John B. Center for Materials Science and Engineering, University of Texas, Austin. ELECTRIC INSULATOR.

Goodglass, Dr. Harold. Director, Psychology Research, Veterans Administration Hospital, Boston, Massachusetts. APHASIA.

Goodheart, Prof. Clarence F. Department of Electrical Engineering, Union College, Schenectady, New York. CIRCUIT (ELECTRICITY); OPEN CIRCUIT; PARALLEL CIRCUIT.

Gooding, Dr. James L. Director of Meteorology, Duke Energy International, Houston, Texas. METEORITE.

Goodison, Dr. Barry E. Superintendent, Hydrometeorological Impact and Development Section, Canadian Climate Centre, Atmospheric Environment Service, Downsview, Ontario. SNOW SURVEYING.

Goodkind, Prof. John M. Department of Physics, University of California, San Diego. QUANTUM SOLIDS.

Goodman, Prof. Bernard. Department of Physics, University of Cincinnati. CAYLEY-KLEIN PARAMETERS; EULER ANGLES; FRAME OF REFERENCE; MECHANICS; other articles.

Goodman, Dr. Joseph W. Retired; formerly, Stanford Electronic Laboratory, Stanford University. HOLOGRAPHY.

Goodman, Dr. Matthew S. Bell Communications Research, Inc., Morristown, New Jersey. DATA COMMUNICATIONS.

Goodman, Dr. Morris. Department of Anatomy and Cell Biology, Wayne State University School of Medicine, Detroit, Michigan. MOLECULAR ANTHROPOLOGY.

Goodman, Dr. Ralph R. Naval Research Laboratory, Stennis Space Center, Mississippi. SOFAR; UNDERWATER SOUND.

Goodman, Dr. Richard A. Division of Public Health, Emory University. EPIDEMIC.

Goodman, Sidney H. Manager, Materials Products Department, Technology Support Division, Hughes Aircraft Company, Culver City, California; Senior Lecturer, Department of Chemical Engineering, University of Southern California. PLASTICS PROCESSING.

Goodman, Dr. Steven R. Chair, Department of Cell Biology and Neuroscience, University of South Alabama College of Medicine, Mobile, Alabama. CELL BIOLOGY.

Goodwin, John P. Principal, Gretna Green Associates, Los Angeles, California. SATELLITE TELEVISION BROADCASTING.

Goolish, Dr. Edward M. NASA Astrobiology Institute, NASA Ames Research Center, Moffett Field, California. ASTROBIOLOGY.

Goosby, Dr. Eric P. AIDS Education and Training Centers Program, U.S. Public Health Service, Rockville, Maryland. ACQUIRED IMMUNE DEFICIENCY SYNDROME (AIDS).

Gordon, Dr. Arnold L. Lamont-Doherty Earth Observatory of Columbia University, Palisades, New York. ANTARCTIC OCEAN.

Gordon, Dr. Gary D. Aerospace Consultant, Washington Grove, Maryland. COMMUNICATIONS SATELLITE; SPACE TECHNOLOGY.

Gordon, Dr. James P. Rumson, New Jersey. MASER.

Gordon, Prof. Louis. Deceased; formerly, Case Institute of Technology. NUCLEATION; PRECIPITATION (CHEMISTRY); SUPERSATURATION.

Gordon, Dr. William E. Associate Professor of Physical Chemistry, Pennsylvania State University; and Consultant. EXPLOSIVE FORMING; FIRE EXTINGUISHER.

Gore, Edward W. International Business Machines, Lexington, Kentucky. TYPEWRITER.

Gosling, Dr. William. Consultant, Radio Communications, Rode, Bath, England. RADIO RECEIVER.

Gossard, Prof. Arthur C. Department of Materials, University of California, Santa Barbara. QUANTIZED ELECTRONIC STRUCTURE (QUEST).

Gotelli, Dr. Nick. Department of Biology, University of Vermont, Burlington. ISLAND BIOGEOGRAPHY.

Gots, Dr. Joseph S. Department of Microbiology, School of Medicine, University of Pennsylvania. ALLOSTERIC ENZYME.

Gould, Dr. Richard A. Department of Anthropology, University of Hawaii, Manoa. ANTHROPOLOGY.

Gould, Dr. Stephen J. Deceased; formerly, Professor of Geology, Museum of Comparative Zoology, Harvard University, Cambridge, Massachusetts. PALEONTOLOGY.

Goulding, Dr. Fred S. Lawrence Berkeley National Laboratory, Berkeley, California. PARTICLE DETECTOR.

Govindjee, Dr. Department of Botany and Department of Physiology and Biophysics, University of Illinois, Urbana-Champaign. CHLOROPHYLL; PHOTOSYNTHESIS.

Grabowski, Dr. Martha R. Center for Industrial Innovation, Department of Decision Sciences and Engineering Systems, Rensselaer Polytechnic Institute, Troy, New York. MARINE NAVIGATION.

Grace, Dr. Neville. Biotechnology Division, Department of Scientific and Industrial Research, Palmerston North, New Zealand. GEOPHAGIA.

Graham, Dr. Jeffrey B. Scripps Institution of Oceanography, La Jolla, California. HIBERNATION AND ESTIVATION.

Grande, Dr. Lance. Field Museum of Natural History, Chicago, Illinois. ACTINOPTERYGII.

Gränicher, Prof. H. Laboratory of Solid State Physics, Swiss Federal Institute of Technology, Zurich, Switzerland. PIEZOELECTRICITY; PYROELECTRICITY.

Grasshoff, Dr. K. Institut für Meereskund an der Universität Kiel, Germany. BALTIC SEA.

Grau, Dr. Fred V. Consulting Agronomist, College Park, Maryland. VETCH.

Graves, Dr. Arthur H. Deceased; formerly, Consultant in Genetics, Connecticut Agricultural Experiment Station. ARBORVITAE; BASSWOOD; ELM; HICKORY; LARCH; PINE; other articles.

Graves, Prof. Lawrence M. Professor Emeritus of Mathematics, University of Chicago, Illinois. INFINITY; INTEGRATION; REAL VARIABLE.

Graves, Walter L. San Diego, California. VETCH.

Gray, Dr. Ernest. Variety Club Children's Hospital, Department of Pediatrics, University of Minnesota. SCARLET FEVER.

Gray, G. Ronald. Director, Syncrude Canada, Ltd., Edmonton, Alberta, Canada. OIL SAND.

Gray, Dr. Gerald J. Vice President, Forest Policy Center, American Forests, Washington, D.C. FORESTRY, URBAN.

Gray, Dr. Harry B. Division of Chemistry and Chemical Engineering, California Institute of Technology. ELECTRON-TRANSFER REACTION.

Gray, Dr. Michael W. Professor of Biochemistry and Molecular Biology/Fellow, Program in Evolutionary Biology, Canadian Institute for Advanced Research, Dalhousie University, Halifax, Nova Scotia, Canada. RIBONUCLEIC ACID (RNA).

Gray, Dr. Peter. Andrey Avinoff Professor of Biology, University of Pittsburgh. MICROTECHNIQUE.

Gray, Thomas. GAI Consultants Inc., Monroeville, Pennsylvania. SURFACE MINING.

Green, Prof. David G. School of Computer Science and Software Engineering, Monash University, Clayton, Victoria, Australia. COMPLEXITY THEORY.

Green, Dr. David M. Graduate Research, Department of Psychology, University of Florida, Gainesville. PITCH.

Green, Dr. Paul. Harvard-Smithsonian Center for Astrophysics, Cambridge, Massachusetts. CARBON STAR.

Green, Dr. Richard F. Large Binocular Telescope Observatory, University of Arizona, Tucson. LIGHT CURVES.

Green, Prof. Sheldon I. Department of Mechanical Engineering, University of British Columbia, Vancouver, Canada. VORTICITY.

Greenberg, Prof. David M. Department of Biochemistry, School of Medicine, University of California, San Francisco. PROTEIN METABOLISM.

Greenberger, Prof. Daniel M. Department of Physics, City College of the City University of New York. NEUTRON OPTICS.

Greene, Prof. David C. Department of Geology and Geography, Denison University, Granville, Ohio. GREAT BASIN.

Greenfield, Dr. Irving G. Unidel Professor of Engineering, Department of Mechanical Engineering, University of Delaware. METAL.

Greenfield, Dr. Lazar J. Department of Surgery, University of Michigan Hospitals, Ann Arbor. PHLEBITIS.

Greenler, Dr. Robert. Department of Physics, University of Wisconsin. HALO; METEOROLOGICAL OPTICS; MIRAGE; RAINBOW; SUN DOG.

Greensfelder, Dr. Bernard S. Deceased; formerly, Director of Oil Research, Shell Development Company, Emeryville, California. AROMATIZATION.

Greenspan, Prof. Ehud. Department of Nuclear Engineering, University of California, Berkeley. NUCLEAR ENGINEERING.

Greenspan, Martin. National Institute of Standards and Technology. DAMPING; LISSAJOUS FIGURES; MECHANICAL IMPEDANCE.

Greenspan, Dr. Richard L. The Charles Stark Draper Laboratory, Cambridge, Massachusetts. AIR-TRAFFIC CONTROL; DIRECTION-FINDING EQUIPMENT; DOPPLER VOR; ELECTRONIC NAVIGATION SYSTEMS; MARINE NAVIGATION; UNDERWATER NAVIGATION; VOR (VHF OMNIDIRECTIONAL RANGE).

Greenstein, Prof. Jesse L. Department of Astronomy, California Institute of Technology. PARSEC; other articles.

Gregory, Dr. Joseph T. Curator of Lower Vertebrates, Museum of Paleontology, University of California, Berkeley. CROCODYLIA; PTEROSAURIA.

Greiner, Dr. Dale L. Professor of Medicine, University of Massachusetts Medical School, Worcester. TRANSPLANTATION BIOLOGY.

Greiner, Dr. Walter. Institut für Theoretische Physik, Universität Frankfurt, Germany. RADIOACTIVITY; SUPERCRITICAL FIELDS.

Grest, Dr. Gary S. Exxon Research and Engineering Company, Annandale, New Jersey. GLASS TRANSITION.

Grey, Dr. David S. Formerly, Aerospace Corporation, El Segundo, California. REFLECTING MICROSCOPE.

Grey, Jerry. President, Greyrad Corporation, Princeton, New Jersey. PROPULSION.

Grierson, Dr. William. Institute of Food and Agricultural Sciences, University of Florida. CITRON.

Griffen, Prof. Dana T. Department of Geology, Brigham Young University, Provo, Utah. OLIGOCLASE.

Griffin, Dr. James. Graduate School of Oceanography, University of Rhode Island, Charleston. OCEANOGRAPHIC VESSELS.

Griffiths, Dr. A. J. F. Department of Botany, University of British Columbia, Vancouver, Canada. FUNGAL GENETICS.

Griffiths, Prof. Peter R. Chairperson, College of Letters and Science, Department of Chemistry, University of Idaho. SUPERCRITICAL FLUID CHROMATOGRAPHY.

Grigg, Prof. Neil S. Head, Department of Civil Engineering, Colorado State University, Fort Collins. RESERVOIR.

Grilo, Dr. Carlos M. Director of Psychology, Yale Psychiatric Institute, New Haven, Connecticut. EATING DISORDERS.

Grim, Dr. Ralph E. Department of Geology, University of Illinois, Urbana-Champaign. BENTONITE; GLAUCONITE; MONTMORILLONITE.

Grimm, Robert A. AC Spark Plug Division, General Motors Corporation, Flint, Michigan. SPEEDOMETER.

Grine, Dr. Frederick E. Department of Anthropology, State University of New York, Stony Brook. AUSTRALOPITHECINE.

Grissino-Mayer, Dr. Henri D. Department of Physics, Astronomy, & Geosciences, Valdosta State University, Georgia. DENDROCHRONOLOGY.

Grivell, Dr. Les A. Afdeling Moleculaire Biologie, Laboratorium voor Biochemie, Universiteit van Amsterdam, Netherlands. MITOCHONDRIA.

Grodzins, Dr. Lee. Physics Department, Massachusetts Institute of Technology. PROTON-INDUCED X-RAY EMISSION (PIXE).

Grogan, Dr. R. G. Department of Plant Pathology, University of California, Davis. LETTUCE.

Groggins, P. H. Deceased; formerly, Chemical Division, Food Machinery and Chemical Corporation. SULFONATION AND SULFATION.

Gronstol, Prof. Hallstein. Department of Large Animal Clinical Sciences, Norwegian College of Veterinary Medicine, Oslo, Norway. LISTERIOSIS.

Groody, Dr. E. Patrick. Integrated Genetics, Framingham, Massachusetts. NUCLEOTIDE.

Gross, Prof. David. Director, Kavli Institute for Theoretical Physics, University of California, Santa Barbara; Nobelist. QUANTUM FIELD THEORY.

Gross, Prof. Jonathan L. Department of Mathematics, Columbia University. GRAPH THEORY.

Gross, Marjorie K. Newport, North Carolina. ABACUS.

Gross, William H. Technical Editor, Corporate Communications, Dow Chemical Company, Midland, Michigan. MAGNESIUM.

Grossman, Dr. Lawrence. Department of Geophysical Sciences, University of Chicago. BYTOWNITE; MICA; SODALITE; other articles.

Grosswold, Prof. Emil. Deceased; formerly, Department of Mathematics, Temple University. ZERO.

Grotzinger, Dr. John. Department of Earth, Atmosphere, and Space Sciences, Massachusetts Institute of Technology. PROTEROZOIC.

Grove, Dr. Ralph F., Jr. Department of Computer Science, James Madison University, Harrisonburg, Virginia. CELLULAR AUTOMATA.

Groves, Dr. Gordon W. Instituto de Geofisica, Torre de Ciencias, Ciudad Universitaria, Mexico. TIDE.

Gruen, Dr. Dieter M. Argonne National Laboratory, Argonne, Illinois. MAGNETOCHEMISTRY.

Grunberg, Dr. Emanuel. Director, Department of Chemotherapy, Hoffmann-LaRoche, Inc., Nutley, New Jersey. SULFONAMIDE.

Gruver, Prof. William A. School of Engineering Science, Simon Fraser University, Burnaby, British Columbia, Canada. ROBOTICS.

Grys-Rubenstein, Prof. Ellen. Interdisciplinary Arts & Sciences, University of Washington, Bothell. BIOLOGICAL PRODUCTIVITY.

Gschneidner, Prof. Karl A., Jr. Ames Laboratory, Iowa State University. SCANDIUM.

Gudo, Michael. Forschungsinstitut Senckenberg, Germany. HETEROCORALLIA.

Guggenheim, Prof. Stephen. Department of Geological Sciences, University of Illinois, Chicago. LEPIDOLITE; MUSCOVITE; PHLOGOPITE; VERMICULITE.

Guido, Prof. Louis. Associate Professor, Electrical and Computer Engineering, Virginia Tech, Blacksburg. LIGHT-EMITTING DIODE.

Guidry, Dr. Mark R. Fairchild Camera and Instrument Corporation, Mountain View, California. CHARGE-COUPLED DEVICES.

Gunckel, Prof. James E. Department of Botany, Rutgers University. STEM.

Gundersen, Prof. Martin. Electrical Engineering Department, University of Southern California, Los Angeles. GAS TUBE.

Gunkler, Dr. Albert A. Chief Process Engineer, Midland Division, Dow Chemical Company, Midland, Michigan. HALOGEN ELEMENTS.

Gunning, Dr. Brian E. S. Department of Botany, Queen's University, Belfast, Northern Ireland. CELL PLASTIDS.

Gunning, Dr. Robert C. Department of Mathematics, Princeton University, Princeton, New Jersey. COMPLEX NUMBERS.

Gupta, Dr. Madan M. Intelligent System Research Laboratory, College of Engineering, University of Saskatchewan, Saskatoon, Canada. FUZZY SETS AND SYSTEMS.

Gupta, Dr. Y. M. Department of Physics, Shock Dynamics Laboratory, Washington State University. SHOCK WAVE.

Gür, Dr. Turgut M. Senior Research Associate, Department of Materials Science and Engineering, Stanford University. IONIC CRYSTALS.

Gurevich, Michael. Masonary Consultants, Fair Lawn, New Jersey. BRICK.

Gurney, Dr. Ashley B. Systematic Entomology Laboratory, Entomology Research Division, U.S. Department of Agriculture. PSOCOPTERA; ZORAPTERA.

Gustafson, Dr. David A. Department of Computing and Information Science, Kansas State University, Manhattan. ARTIFICIAL INTELLIGENCE; SOFTWARE METRIC; SOFTWARE TESTING AND INSPECTION.

Gustafson, Dr. Steven M. University of Nottingham, United Kingdom. ARTIFICIAL INTELLIGENCE.

Guthmiller, Dr. Janet M. College of Dentistry, The University of Iowa, Iowa City. POLYMICROBIAL DISEASES.

Gutzwiller, Dr. Martin C. T. J. Watson Research Center, IBM Corporation, Yorktown Heights, New York. CORRESPONDENCE PRINCIPLE; NONLINEAR PHYSICS.

Guy, Dr. Arthur W. Bioelectromagnetism Research Laboratory, School for Medicine and College of Engineering, University of Washington, Seattle. ELECTROTHERAPY.

Guzinski, Dr. Gay M. Chief, Section of Benign Gynecology, Department of Obstetrics and Gynecology, School of Medicine, University of Maryland. MENOPAUSE; REPRODUCTIVE SYSTEM DISORDERS.

Gwynne, Dr. Darryl T. Department of Biology, University of Toronto at Mississauga, Ontario, Canada. BLATTARIA (BLATTODEA); GRYLLOBLATTODEA; ORTHOPTERA.

Gyftopoulos, Dr. Elias P. Thermo Electron Corporation, Waltham, Massachusetts. THERMIONIC POWER GENERATOR.

H

Haack, Prof. Joel K. Head, Department of Mathematics, University of Northern Iowa, Cedar Falls. ABSTRACT ALGEBRA.

Haber, Bernard. Partner, Hardesty & Hanover, LLP, Consulting Engineer. BRIDGE.

Haber, Dr. Meryl H. Director, Pathology Department, Rush Medical College, Rush Presbyterian, Chicago, Illinois. URINALYSIS.

Hackl, Dr. Ralph P. Research Director, Placer Dome, Inc., Vancouver, British Columbia, Canada. BIOLEACHING.

Hader, Rodney N. Secretary, American Chemical Society, Washington, D.C. AGRICULTURAL CHEMISTRY.

Hadler, Jacques B. Webb Institute of Naval Architecture, Glen Cove, New York. PROPELLER (MARINE CRAFT); TOWING TANK.

Hadley, Prof. Elbert H. Retired; formerly, Department of Chemistry and Biochemistry, Southern Illinois University. BENZOIC ACID; LACTATE; other articles.

Hadni, A. Laboratoire Infrarouge Lointain, Université de Nancy I, France. PYROELECTRICITY.

Haenlein, Prof. George F. W. Animal Science Department, University of Delaware. GOAT PRODUCTION.

Hagan, Dr. Maura. National Center for Atmospheric Research, Boulder, Colorado. MESOSPHERE.

Hagan, Dr. William J., Jr. Department of Chemistry, College of St. Rose, Albany, New York. PREBIOTIC ORGANIC SYNTHESIS.

Hagar, Prof. Charles F. Department of Physics and Astronomy, San Francisco State University. PLANETARIUM.

Hagen, Dr. John P. Deceased; formerly, Department of Astronomy, Pennsylvania State University. BALLISTICS; TRAJECTORY.

Hagist, Prof. Warren M. Retired; formerly, Department of Mechanical Engineering, University of Rhode Island. HYDRAULICS; HYDROSTATICS.

Hagner, A. F. Retired; formerly, Department of Geology, University of Illinois, Urbana-Champaign. ORE AND MINERAL DEPOSITS.

Hagood, Mel A. Irrigated Agriculture Research and Extension Center, Washington State University. IRRIGATION (AGRICULTURE).

Hague, Wilbur. Oxy Metal Industries Corporation, Warren, Michigan. CADMIUM.

Hahn, Dr. John F. Department of Psychology, University of Virginia. PSYCHOPHYSICAL METHODS; SENSATION.

Haight, Dr. Gilbert P., Jr. Professor Emeritus, Department of Chemistry, University of Illinois. MOLYBDENUM.

Haile, Dr. William. Swales Aerospace, Inc., Beltsville, Maryland. SPACECRAFT STRUCTURE.

Haines, John E. Deceased; formerly, Vice President, Minneapolis-Honeywell Regulator Company, Minneapolis, Minnesota. HUMIDISTAT.

Haines, Prof. Malcolm G. Blackett Laboratory, Imperial College of Science and Technology, London, England. MAGNETOHYDRODYNAMICS.

Haines, Roger W. Professional Engineer, Laguna Hills, California. HEATING SYSTEM.

Hainline, Bernard C. Systems Technology, Boeing Commercial Airplane Company, Seattle, Washington. AIRCRAFT INSTRUMENTATION.

Halbedel, Dr. Elaine M. Corralitos Observatory, Las Cruces, New Mexico. JUPITER.

Halbouty, Dr. Michel T. Consulting Geologist and Petroleum Engineer, Houston, Texas. NATURAL GAS.

Haldane, Dr. Frederick D. M. Department of Physics, Princeton University. EXCHANGE INTERACTION.

Halde, Dr. Carlyn. Department of Microbiology and Immunology, School of Medicine, University of California, San Francisco. MEDICAL MYCOLOGY.

Hale, Dr. Mason E., Jr. Deceased; formerly, Smithsonian Institution, Washington, D.C. CALICIALES; HYSTERIALES (LICHENIZED); LECANORALES; other articles.

Hale, Dr. Thomas. Department of Enteric Infections, Walter Reed Army Institute of Research, Washington, D.C. BACILLARY DYSENTERY.

Hales, Prof. Milton R. School of Medicine, West Virginia University. GALLBLADDER DISORDERS.

Halkias, Prof. Christos C. Chair of Electronics, National Technical University, Athens, Greece. BIAS (ELECTRONICS) IMPEDANCE MATCHING; other articles.

Hall, Brian. Department of Biology, Dalhousie University, Canada. GERM LAYERS.

Hall, Dr. Carl W. Dean, College of Engineering, Washington State University, Pullman. AGRICULTURAL SOIL AND CROP PRACTICES; DAIRY MACHINERY.

Hall, Prof. Dennis G. Director, Institute of Optics, University of Rochester, New York. OPTICAL GUIDED WAVES.

Hall, Dr. Hamilton. Professor, Department of Surgery, University of Toronto, Ontario; Executive Director, Canadian Spine Society. SCIATICA.

Hall, Prof. Lawrence O. Department of Computer Science and Engineering, University of South Florida, Tampa. COMPUTATIONAL INTELLIGENCE.

Hall, Dr. Michael. National Physical Laboratory, Teddington, Middlesex, United Kingdom. DEMAGNETIZATION.

Hall, Dr. Richard P. Deceased; formerly, Professor of Zoology, University of California, Los Angeles. ACANTHOPHRACTIDA; HELIOZOIA; PROTOZOA; other articles.

Hall, Prof. Robert. Director, SE Asia Research Group, Department of Geology, University of London, Egham, Surrey, United Kingdom. EAST INDIES.

Hallam, Prof. Anthony. School of Earth and Environmental Sciences, University of Birmingham, United Kingdom. JURASSIC.

Haller, Prof. Gary L. Professor of Engineering and Applied Science, Jonathan Edwards College, Yale University, New Haven, Connecticut. CATALYSIS; HETEROGENEOUS CATALYSIS.

Hallet, Dr. John. Director, Atmospheric Ice Physics Laboratory, Desert Research Institute, Atmospheric Sciences Center, University of Nevada. CLOUD; FOG; SNOW; WEATHER MODIFICATION.

Halliday, Dr. Alex. Department of Geological Sciences, University of Michigan, Ann Arbor. LEAD ISOTOPES (GEOCHEMISTRY).

Halpin, Prof. Daniel W. School of Civil Engineering, Purdue University, West Lafayette, Indiana. CONSTRUCTION ENGINEERING.

Halstead, Dr. Thora Waters. American Society for Gravitational and Space Biology, Arlington, Virginia. AEROSPACE MEDICINE; WEIGHTLESSNESS.

Halvorson, Dr. Harlyn O. Department of Biochemistry, University of Minnesota. BACTERIA.

Hamad, Dr. Falk. Department of Mechanical and Medical Engineering, University of Bradford, United Kingdom. TWO-PHASE FLOW.

Hamer, Dr. Walter J. Institute for Basic Standards, National Institute of Standards and Technology. ELECTRODE; ELECTROMOTIVE FORCE (CELLS).

Hamers, Prof. Robert. Department of Chemistry, University of Wisconsin-Madison. DIAMOND THIN FILMS.

Hamilton, Prof. C. Howard. Department of Mechanical and Materials Engineering, Washington State University, Pullman. SUPERPLASTIC FORMING.

Hamilton, Dr. Howard L. Department of Biology, University of Virginia. ANIMAL REPRODUCTION; COELOM; ESTRUS; METAMERES; NEURULATION; OVUM; other articles.

Hamilton, Dr. James. Department of Physiology and Biophysics, Boston University School of Medicine, Massachusetts. ALBUMIN.

Hamilton, John C. Department of Physics and Astronomy, University of Hawai'i–Hilo. OBSERVATORY, ASTRONOMICAL.

Hamilton, Prof. Joseph H. Department of Physics and Astronomy, Vanderbilt University, Nashville, Tennessee. EXOTIC NUCLEI; RADIOACTIVITY.

Hamilton, Dr. Mark F. Department of Mechanical Engineering, University of Texas, Austin. NONLINEAR ACOUSTICS.

Hamilton, Dr. R. I. Research Branch, Agriculture Canada, Vancouver, British Columbia. PLANT PATHOLOGY; PLANT VIRUSES AND VIROIDS.

Hammar, Dr. Samuel P. Diagnostic Specialties Laboratory, Bremerton, Washington. CANCER; LIVER DISORDERS; LYMPHOMA; PNEUMONIA; RESPIRATORY SYSTEM DISORDERS.

Hammer, Edward E. Retired; formerly, Fluorescent and High Intensity Systems Department, General Electric Company, Cleveland, Ohio. FLUORESCENT LAMP.

Hammond, Prof. Edwin H. Retired; formerly, Department of Geography, University of Tennessee. HILL AND MOUNTAIN TERRAIN; MOUNTAIN; PLAINS; PLATEAU; TERRAIN AREAS.

Hand, Dr. Cadet. Bodega Marine Laboratory, University of California, Bodega Bay. SEA ANEMONE.

Handler, Dr. Philip. Deceased; formerly, National Academy of Sciences, Department of Biochemistry. NITROGEN EXCRETION.

Handsfield, Dr. H. Hunter. Director, Sexually Transmitted Disease Control Program, Seattle-King County Public Health Department, Harborview Medical Center, Seattle, Washington. SEXUALLY TRANSMITTED DISEASES.

Haney, Dr. Robert L. Department of Meteorology, Naval Postgraduate School, Monterey, California. MARITIME METEOROLOGY.

Hang, Prof. C. C. Centre for Intelligent Control, Department of Electrical Engineering, National University of Singapore. EXPERT CONTROL SYSTEM.

Hankley, Prof. William. Department of Computing and Information Science, Kansas State University, Manhattan. SOFTWARE METRIC.

Hanlin, Dr. Richard T. Department of Plant Pathology, University of Georgia, Athens. ASCOMYCOTA; HEMIASCOMYCETES.

Hanna, Dr. Michael. Assistant Professor of Biology, Rensselaer Polytechnic Institute. CELL ADHESION.

Hanna, Dr. Steven R. Sigma Research Corporation, Concord, Massachusetts. AIR POLLUTION.

Hanor, Jeffrey S. Department of Geology, Louisiana State University. HYDROSPHERE.

Hänsch, Prof. Theodore W. Sektion Physik, Ludwig-Maximillians-Universitat Munchen, Germany; Nobelist. LASER SPECTROSCOPY; RYDBERG CONSTANT.

Hansel, Dr. Ardith K. Illinois State Geological Survey, Champaign. HOLOCENE.

Hansen, Dr. Arthur G. Retired; formerly, President, Georgia Institute of Technology. STREAMLINING.

Hansen, Dr. Everett M. Department of Botany and Plant Pathology, Oregon State University, Corvallis. FOREST ECOSYSTEM.

Hansen, Dr. John T. Department of Neurobiology and Anatomy, University of Rochester Medical Center. CAROTID BODY.

Hansen, Dr. Katherine J. Department of Earth Sciences, Montana State University, Bozeman. ALPINE VEGETATION.

Hanson, Prof. Allen L. Formerly, Department of Chemistry, Saint Olaf College, Northfield, Minnesota. ACID-BASE INDICATOR; CONCENTRATION SCALES; DETERGENT; PARAFFIN; PHENACETIN; other articles.

Hanson, Dr. Phyllis. Department of Cell Biology and Physiology, Washington University School of Medicine, St. Louis, Missouri. SYNAPTIC TRANSMISSION.

Happer, Prof. William. Department of Physics, Columbia University. MICROWAVE SPECTROSCOPY.

Haq, Dr. Bilal U. Bethesda, Maryland. CENOZOIC; OLIGOCENE; PALEOCENE; PLEISTOCENE; PLIOCENE.

Harborne, Dr. Jeffrey B. Department of Botany, University of Reading, England. PLANT PIGMENT.

Hardesty, Egbert R. Retired; formerly, Hardesty & Hanover, New York. BRIDGE.

Hardgrove, Ralph M. Sales Engineer, Stock Equipment Company, Cleveland, Ohio.

CRUSHING AND PULVERIZING; GAS FURNACE; GRINDING MILL; PEBBLE MILL; ROLL MILL; TUMBLING MILL.

Hardy, Dr. Ernest E. Department of Natural Resources, Cornell University. LAND-USE CLASSES.

Hardy, Prof. John. Cyclotron Institute, Texas A&M University, College Station. BETA PARTICLES.

Hardy, Dr. John W. Lexington, Massachusetts. ADAPTIVE OPTICS.

Hare, Prof. Jonathan. School of Chemistry and Molecular Physics, University of Sussex, Falmer, Brighton, England. FULLERENE.

Hargis, Dr. William J., Jr. Department of Marine Science, University of Virginia. ASPIDOGASTREA; MONOGENEA; TREMATODA.

Harker, Dr. Robert I. Department of Geology, University of Pennsylvania, Philadelphia. CALCITE; CERUSSITE; WITHERITE; other articles.

Harleman, Dr. Donald R. F. Deceased; Department of Civil Engineering, Massachusetts Institute of Technology. HYDRAULIC JUMP; SURFACE WAVES.

Harley, Dr. John P. Department of Biological Sciences, Eastern Kentucky University, Richmond. ACANTHAREA; ACANTHOMETRIDA; ARCHAEA; LEPTOSPIROSIS; MICROBIOTA (HUMAN).

Harman, Prof. Jay R. Department of Geography, Michigan State University, East Lansing. ECOTONE.

Harper, Prof. Michael J. K. Chief, Reproductive Biology Division, Scientific Coordinator, Center for Research, Department of Obstetrics and Gynecology, University of Texas-Health Science Center, San Antonio. REPRODUCTIVE SYSTEM.

Harrigan, Prof. Kevin. TeleLearning Network of Centres of Excellence, Department of Computer Science, University of Waterloo, Ontario, Canada. HUMAN-COMPUTER INTERACTION.

Harriman, Prof. Anthony. Laboratorie de Photochimie, Ecole Europée͏ne de Chimie, Polyméres et Matériaux, Strasbourg, France. PHOTOLYSIS.

Harrington, Clifford M. Shaw Pittman, McLean, Virginia. TELEVISION NETWORKS.

Harrington, Dr. Robert S. Deceased; formerly, U.S. Naval Observatory, Washington, D.C. CELESTIAL MECHANICS; PRECESSION OF EQUINOXES.

Harris, Dr. Cyril M. Professor of Electrical Engineering and Architecture, Department of Electrical Engineering, Columbia University. ANECHOIC CHAMBER; REFLECTION OF SOUND; REVERBERATION.

Harris, Dr. Forest K. Retired; formerly, Electrical Measurements Laboratory, National Institute of Standards and Technology, Gaithersburg, Maryland. GALVANOMETER; POTENTIOMETER.

Harris, Prof. J. Donald. Deceased; formerly, U.S. Navy Submarine Medical Laboratory, Groton, Connecticut. EARPHONES.

Harris, Dr. J. Milton. Founder and Scientific Officer, Nektar Therapeutics, Huntsville, Alabama. POLY(ETHYLENE GLYCOL).

Harris, Lawrence A. Oak Ridge National Laboratory, Oak Ridge, Tennessee. OPTICAL MICROSCOPE.

Harris, Prof. P. J. F. J.J. Thomson Physical Laboratory, University of Reading, United Kingdom. CARBON NANOTUBES.

Harris, Prof. Richard W. Department of Environmental Horticulture, University of California, Davis. ARBORICULTURE.

Harrison, Dr. David B. Public Relations Department, Goodyear Tire and Rubber Company, Akron, Ohio. TIRE.

Harrison, Prof. Edward R. Deceased; formerly, Department of Physics and Astronomy, University of Massachusetts, Amherst. OLBERS' PARADOX.

Harrison, Dr. George R. Deceased; formerly, Dean Emeritus, School of Science, Massachusetts Institute of Technology. DEWAR FLASK; LINE SPECTRUM; INCANDESCENCE; RESOLVING POWER (OPTICS).

Harrison, Prof. John P. Department of Physics, Queen's University, Ontario, Canada. KAPITZA RESISTANCE.

Harrison, Dr. Marcia. Department of Biology, Washington University, St. Louis, Missouri. APICAL DOMINANCE.

Harrison, Prof. Terry. Department of Anthropology, New York University, New York. FOSSIL APES; PILTDOWN MAN; PRIMATES.

Harrison, Prof. Walter A. Emeritus, Department of Applied Physics, Stanford University. FERMI SURFACE; FREE-ELECTRON THEORY OF METALS; SOLID-STATE PHYSICS.

Hart, Prof. Gerald W. Chair, Department of Biochemistry and Molecular Genetics, University of Alabama, Birmingham. GLYCOLIPID.

Hart, Dr. Harold. Professor Emeritus, Department of Chemistry, Michigan State University. AROMATIC HYDROCARBON.

Hart, Dr. Ronald W. Director, Department of Chemistry, National Center for Toxicological Research, Jefferson, Arizona. BIOASSAY.

Hartley, Dr. A. M. Department of Chemistry and Chemical Engineering, University of Illinois, Urbana-Champaign. BUFFERS (CHEMISTRY).

Hartman, Dr. Olga. Deceased; formerly, Allan Hancock Foundation, University of Southern California. ANNELIDA—in part; ERRANTIA; MYZOSTOMARIA; POLYCHAETA; SEDENTARIA.

Hartman, Dr. Standish C. Department of Chemistry, Boston University. PURINE; PYRIMIDINE:.

Hartman, Dr. Willard D. Associate Professor of Biology and Curator in Invertebrate Zoology, Peabody Museum of Natural History, Yale University. AMPHIDISCOSA; CALCAREA; PARAZOA; PORIFERA; RETICULOSA; other articles.

Hartmann, Dr. Dennis L. Department of Atmospheric Sciences, University of Washington, Seattle. CLIMATOLOGY.

Hartmann, Prof. Dieter H. Department of Physics and Astronomy, Clemson University, Clemson, South Carolina. NUCLEOSYNTHESIS.

Hartmann, Prof. Hudson T. Pomology Extension, University of California, Davis. OLIVE.

Hartmann, Dr. William M. Department of Physics, Michigan State University, East Lansing. TONE (MUSIC AND ACOUSTICS).

Hartnett, Dr. Richard J. Department of Engineering, U.S. Coast Guard Academy, New London, Connecticut. HYPERBOLIC NAVIGATION SYSTEM.

Hartstein, Dr. Jack. Eye Surgeons and Physicians of St. Louis, Ltd., Chesterfield, Missouri. CATARACT; GLAUCOMA.

Hartung, Dr. Walter H. Deceased; formerly, Professor of Pharmaceutical Chemistry, Medical College of Virginia. PHARMACEUTICAL CHEMISTRY.

Hartwig, Dr. William H. Deceased; formerly, Department of Electrical Engineering, University of Texas. ELECTRIC ENERGY MEASUREMENT.

Harvey, Dr. John A. Oak Ridge National Laboratory, Tennessee. NEUTRON SPECTROMETRY.

Harvey, Dr. Paul. Department of Zoology, University of Oxford, United Kingdom. ALLOMETRY.

Harvey, Dr. Philip D. Professor of Psychiatry, Mount Sinai School of Medicine, New York. SCHIZOPHRENIA.

Hasey, Dr. Janine K. Cooperative Extension, University of California, Yuba City. KIWIFRUIT.

Hasiotis, Dr. Stephen T. Assistant Professor, University of Kansas. MESOZOIC.

Hasler, Dr. Arthur D. Laboratory of Limnology, University of Wisconsin. EUTROPHICATION.

Hass, Dr. H. B. Director of Chemical Research, Kellogg Company, Piscataway, New Jersey. SUGAR.

Hassialis, Menelaos D. Krumb School of Mines, Columbia University. ORE DRESSING.

Hassibi, Dr. Babak. Information Systems Laboratory, Stanford University, Stanford, California. ADAPTIVE SIGNAL PROCESSING.

Hassid, Prof. William Z. Deceased; formerly, Department of Biochemistry, University of California, Berkeley. CARBOHYDRATE; GLYCOGEN; INSULIN; POLYSACCHARIDE; other articles.

Hastings, Dr. Alan. Division of Environmental Studies, University of California at Davis. THEORETICAL ECOLOGY.

Hastings, Dr. J. Woodland. Biological Laboratories, Department of Cellular and Developmental Biology, Harvard University. BIOLUMINESCENCE.

Hatsopoulos, Dr. George N. Thermo Electron Corporation, Waltham, Massachusetts. THERMIONIC POWER GENERATOR.

Hauck, Dr. Scott. Department of Electrical Engineering, University of Washington, Seattle. FIELD-PROGRAMMABLE GATE ARRAYS.

Haumont, Prof. S. Department of Histology, Institute of Anatomy, Louvain, Belgium. PANCREAS.

Hausner, Henry H. Consulting Engineer, New York. CERMET.

Haut, Dr. Arthur. Professor of Medicine, Division of Hematology and Oncology, University of Arkansas for Medical Sciences. ANEMIA.

Havens, Dr. Donald P. Filter Products, Rockwell International Corporation, Costa Mesa, California. ELECTRIC FILTER.

Hawkins, Dr. W. Lincoln. Bell Laboratories, Murray Hill, New Jersey. STABILIZER (CHEMISTRY).

Hawksworth, Dr. Frank G. Retired; USDA Forest Services. PLANT PATHOLOGY.

Hawley, Dr. R. Scott. Department of Genetics, Section of Molecular and Cellular Biology, University of California at Davis. MEIOSIS.

Hawthorne, Dr. Frank C. Department of Geological Sciences, University of Manitoba, Winnipeg, Canada. AMPHIBOLE; BORATE MINERALS; COLEMANITE; HORNBLENDE; STAUROLITE.

Hay, Prof. George W. Department of Chemistry, Queen's University, Ontario, Canada. GLYCOSIDE.

Hayes, Dr. Dennis E. Lamont-Doherty Geological Observatory, Palisades, New York. INDIAN OCEAN; MARINE GEOLOGY; SOUTHEAST ASIAN WATERS.

Hayes, Prof. John P. Department of Electrical Engineering and Computer Science, University of Michigan, Ann Arbor. DIGITAL COMPUTER.

Hayes, Dr. Michael P. Electrical and Computer Engineering Building University of Canterbury, New Zealand. SONAR.

Hayes, Dr. Thomas L. Donner Laboratory, University of California, Berkeley. SCANNING ELECTRON MICROSCOPE.

Hayes, William C. Retired; formerly, Editor in Chief, *Electrical World*, McGraw-Hill, Inc. New York. ELECTRIC POWER SYSTEMS.

Hayes-Roth, Dr. Frederick. Executive Vice President, Technology, Teknowledge, Inc., Palo Alto, California. EXPERT SYSTEMS.

Hayflick, Prof. Leonard. Center for Gerontological Studies, University of Florida. CELL SENESCENCE.

Haynes, David O. Deceased; formerly, Consulting Industrial Engineer, Tucson, Arizona. DERRICK.

Haynes, Dr. Richard. Pacific Northwest Forest Research Station, U.S. Department of Agriculture Forest Service, Portland, Oregon. FOREST TIMBER RESOURCES.

Hazel, Prof. Jeffrey R. Department of Zoology, Arizona State University. TEMPERATURE ADAPTATION.

Hazen, Dr. David C. Department Chairman, Aerospace Engineering, Embry-Riddle Aeronautical University, Daytona Beach, Florida. AIRPLANE; KITE.

Hazen, Richard. Deceased; formerly, Hazen and Sawyer, Consulting Engineers, New York. WATER SUPPLY ENGINEERING; WATER TREATMENT.

Hazen, Ronald M. Retired; formerly, Consultant, Indianapolis, Indiana. AIRCRAFT ENGINE.

Heaney, Dr. Peter J. Department of Geological and Geophysical Sciences, Princeton University. QUARTZ.

Hearmon, R. F. S. Formerly, Timber Mechanics Section, Forest Products Research Laboratory, Princes Risborough, Bucks, England. ELASTICITY; PLASTICITY.

Hearn, Dr. C. Jack. Research Geneticist, Horticultural Research Laboratory, U.S. Department of Agriculture, Orlando, Florida. TANGERINE.

Heath, Michelle. Canadian Energy Research Institute, Calgary, Alberta, Canada. ALTERNATIVE FUELS FOR VEHICLES.

Hecht, Dr. Hans H. Chairperson, Department of Medicine, University of Chicago. CARDIAC ELECTROPHYSIOLOGY.

Heckel, Prof. Philip H. Department of Geology, University of Iowa, Iowa City. CYCLOTHEM; DEVONIAN.

Hedger, Dr. Mark. Institute of Reproduction and Development, Monash University, Clayton, Australia. TESTIS.

Hedgpeth, Dr. Joel W. Marine Science Center, Oregon State University. PALAEOISOPUS; PYCNOGONIDA.

Hedlin, Dr. Michael A. H. Laboratory for Atmospheric Acoustics, Institute of Geophysics and Planetary Physics, University of California, San Diego. INFRASOUND.

Heer, Daniel. Lucent-Bell Laboratories, Westford, Massachusetts. VOICE OVER IP.

Hein, Dr. James R. U.S. Geological Survey, Branch of Pacific Marine Geology, Menlo Park, California. CHERT.

Heiney, Prof. Paul A. Department of Physics, University of Pennsylvania, Philadelphia. CRYSTAL.

Heinrich, Dr. E. William. Deceased; formerly, Department of Geology and Mineralogy, University of Michigan. ROCK; other articles.

Heinrikson, Dr. Robert L. Department of Biochemistry, University of Chicago. RIBONUCLEASE.

Heinzel, Dr. Frederick P. Department of Medicine, Division of Infectious Diseases, University of California, School of Medicine, San Francisco. CHICKENPOX AND SHINGLES.

Held, Dr. Gilbert. Director, 4-Degree Consulting, Macon, Georgia. MODEM.

Heldman, Dr. Dennis R. Executive Vice President, Scientific Affairs, National Food Processors Association, Washington, D.C. FOOD ENGINEERING.

Helfman, Dr. Gene S. Department of Zoology, University of Georgia. TERRITORIALITY.

Helgert, Prof. Hermann J. Department of Electrical and Computer Engineering, George Washington University, Washington, D.C. FREQUENCY MODULATION; MODULATION; MULTIPLEXING AND MULTIPLE ACCESS; PHASE MODULATION; PULSE MODULATION; SIDEBAND; SIGNAL-TO-NOISE RATIO.

Hellmers, Dr. Henry. Botany Department, Duke University. PHYTOTRONICS.

Helman, Marc L. Department of Geological Sciences, University of Durham, England. GYPSUM; HALITE.

Helmers, Dr. Carl J. Vice President, Applied Automation, Inc., Bartlesville, Oklahoma. CARBON BLACK.

Henden, Dr. Arne A. Director, American Association of Variable Star Observers, AAVSO, Cambridge, Massachusetts. MIRA; VARIABLE STAR.

Hendricks, Mr. Gregor W. R.R. Donnelly & Son, Inc., Chicago, Illinois. BOOK MANUFACTURE.

Henker, Kevin J. Department of Materials Science and Engineering, Stanford University. CREEP (MATERIALS).

Hennen, Dr. J. F. The Arthur Herbarium, Department of Botany and Plant Pathology, Purdue University, West Lafayette, Indiana. UREDINIOMYCETES (RUST).

Henriquez, Oscar. California State University of Long Beach, Department of Civil Engineering, Long Beach, California. BRIDGE.

Henry, Dr. Charles H. Bell Laboratories, Murray Hill, New Jersey. ELECTRON-HOLE RECOMBINATION.

Henry, Dr. Leo G. ESD/EMI/TLP Consultants, Fremont, California. ELECTROSTATIC DISCHARGE TESTING.

Hepburn, Capt. Richard D. Noesis Inc., Arlington, Virginia. DRY-DOCKING.

Herakovitch, Prof. Carl T. Civil Engineering and Applied Mechanics, University of Virginia, Charlottesville. COMPOSITE LAMINATES.

Herber, Prof. Rolfe H. Department of Chemistry, Rutgers University, New Jersey. MÖSSBAUER EFFECT.

Herberman, Dr. Ronald B. Chief, Biological Development Branch, National Cancer Institute, Frederick Cancer Research Facility, Frederick, Maryland. IMMUNOLOGIC CYTOTOXICITY.

Herbst, Dr. Jan F. Physics Department, General Motors Research Laboratories, Warren, Michigan. CORBINO DISK; DOMAIN (CRYSTALLOGRAPHY); FERROMAGNETISM; MAGNETORESISTANCE.

Hermach, Dr. Francis L. National Institute of Standards and Technology,

Gaithersburg, Maryland. THERMAL CONVERTERS.

Herman, C. J. Consulting Engineer, Advanced Processes and Products, Insulating Materials Department, General Electric Company, Schenectady, New York. MAGNET WIRE—coauthored.

Herring, Dr. Thomas A. Department of Earth, Atmospheric, & Planetary Sciences, Massachusetts Institute of Technology, Cambridge. GEODESY.

Herron, Dr. Norman. Central Research and Development, DuPont Research, Wilmington, Delaware. ZEOLITE.

Hersey, Dr. John B. Deceased; formerly, Office of Naval Research, U.S. Department of the Navy. SCATTERING LAYER.

Hersh, Dr. Herbert N. National Research Council, Commission of Engineering and Technical Systems, National Materials Advisory Board, Washington, D.C. CATHODOLUMINESCENCE.

Hershkowitz, Prof. Noah. Department of Nuclear Engineering, University of Wisconsin, Madison. PLASMA (PHYSICS).

Hershleder, William. Consulting Construction Engineer, New York. CONSTRUCTION ENGINEERING; CONSTRUCTION METHODS; WELLPOINT SYSTEMS.

Herz, Prof. Carl S. Department of Mathematics, McGill University, Montreal, Quebec, Canada. ORTHOGONAL POLYNOMIALS.

Herzberger, Max J. Deceased; formerly, Consulting Professor, Department of Physics, Louisiana State University. CHROMATIC ABERRATION; DIOPTER; FOCAL LENGTH; OPTICAL IMAGE; OPTICAL PRISM; other articles.

Herzog, Prof. Gregory F. Department of Chemistry, Rutgers University, Piscataway, New Jersey. COSMOGENIC NUCLIDE.

Hess, Dr. Ronald A. Department of Mechanical and Aeronautical Engineering, University of California, Davis. AUTOMATION; ENVIRONMENTAL TEST; REMOTE MANIPULATORS.

Hesseltine, Dr. C. W. Fermentation Laboratory, Agricultural Research Service, U.S. Department of Agriculture, Peoria, Illinois. FOOD FERMENTATION.

Hester, Dr. Jeff. Department of Physics and Astronomy, Arizona State University, Tempe. CRAB NEBULA.

Hewins, Prof. Roger H. Department of Geological Sciences Wright Geological Laboratory, Rutgers University, Piscataway, New Jersey. METEORITE.

Heyn, Dr. A. N. J. Department of Biological Sciences, Louisiana State University. TEXTILE MICROBIOLOGY.

Heyne, Prof. E. G. Professor of Plant Breeding, Department of Agronomy, Kansas State University. GRAIN CROPS.

Heyneman, Dr. Donald. Chair, Health and Medical Sciences Program, University of California, Berkeley. MALARIA; MEDICAL PARASITOLOGY.

Heyworth, Dr. Martin. Scientific Director, AIDS Research Center, Veterans Administration Medical Center, San Francisco, California. DIARRHEA.

Hickman, Dr. Howard M. Section Manager, Research and Development, Sherex Chemical Company, Dublin, Ohio. FAT AND OIL; WAX, ANIMAL AND VEGETABLE.

Hicks, Dr. Bruce B. Atmospheric Turbulence and Diffusion Division, National Oceanic and Atmospheric Administration, Oak Ridge, Tennessee. DEW.

Hicks, Dr. Glenn R. Associate Director Plant Biotechnology, Exelixis Pharmaceuticals, South San Francisco, California. VACUOLE.

Hicks, Philip E. President, Hicks & Associates, Consulting Industrial Engineers, Orlando, Florida. PERFORMANCE RATING.

Hicks, Tyler G. Formerly, Publisher, Professional and Reference Book Division, McGraw-Hill, Inc., New York. COMPRESSOR.

Hieftje, Dr. Gary M. Department of Chemistry, Indiana University. ATOMIC SPECTROMETRY.

Higgins, Dr. Robert P. Curator, Department of Invertebrate Zoology, National Museum of Natural History, Smithsonian Institution, Washington, D.C. KINORHYNCHA.

Highstein, Dr. Stephen M. Washington University, St. Louis, Missouri. VESTIBULAR SYSTEM.

Hildebrand, Dr. Joel H. Deceased; formerly, Department of Chemistry, University of California, Berkeley. PHYSICAL SCIENCE; SOLUTION.

Hill, Armin J. Dean, College of Physical and Engineering Science, Brigham Young University. OPTICAL PROJECTION SYSTEMS.

Hill, Dr. Lawrence S. Department of Management, California State University. GANTT CHART.

Hill, Robert T. Teacher and Consultant in Radar, Chula Vista, California, and Easton, Maryland. AIRBORNE RADAR; MONOPULSE RADAR; RADAR; RADOME; SYNTHETIC APERTURE RADAR (SAR).

Hill, Dr. Rolla B., Jr. Professor and Chairperson, Department of Pathology, Upstate Medical Center, Syracuse, New York. JAUNDICE.

Hillger, Dr. Donald W. Regional and Mesoscale Meteorology Branch (RAMMB), National Environmental Satellite, Data, and Information Service (NESDIS), National Oceanic and Atomospheric Administration (NOAA), Ft. Collins, Colorado, and U.S. Metric Association Webmaster. METRIC SYSTEM; PHYSICAL MEASUREMENT.

Hilliard, Dr. Ron. Optomechanics Research, Inc., Vail, Arizona. SPECTROGRAPH.

Hillson, Dr. Simon. Institute of Archeology, University College of London, United Kingdom. DENTAL ANTHROPOLOGY.

Hines, Dr. M. E. Vice President, Microwave Associates Inc., Burlington, Massachusetts. PARAMETRIC AMPLIFIER.

Hingorani, Dr. Narain G. Vice President and Director, Electrical Systems Division, Electric Power Research Institute, Palo Alto, California. ELECTRIC POWER SUBSTATION; STATIC VAR COMPENSATOR.

Hinrichs, Thomas C. Vice President, Magma Power, San Diego, California. GEOTHERMAL POWER.

Hinsch, Dr. Gertrude W. Department of Biology, University of South Florida. INVERTEBRATE EMBRYOLOGY; SPERM CELL.

Hinsinger, Dr. Philippe. UMR Rhizosphère & Symbiose, France. RHIZOSPHERE.

Hintz, Dr. Kenneth J. Department of Electrical and Computer Engineering, George Mason University, Fairfax, Virginia. PROGRAMMABLE CONTROLLERS; SAMPLED-DATA CONTROL SYSTEM.

Hipel, Prof. Keith W. Department of Systems Design Engineering and Statistical Actuarial Science, University of Waterloo, Ontario, Canada. RISK ASSESSMENT AND MANAGEMENT.

Hirschberg, Prof. Daniel S. Information and Computer Science, University of California, Irvine. DATA COMPRESSION.

Hirschfelder, Dr. J. O. Department of Chemistry, University of Wisconsin. GAS.

Hitchman, Dr. Matthew H. Department of Meteorology, University of Wisconsin, Madison. STRATOSPHERE.

Ho, Prof. Monto. Department of Infectious Diseases and Microbiology, University of Pittsburgh, Pennsylvania. CYTOMEGALOVIRUS INFECTION.

Hodgdon, Dr. James. Naval Health Research Center, San Diego, California. HYPOTHERMIA.

Hodge, Prof. Paul. Professor Emeritus, Department of Astronomy, University of Washington, Seattle. ANDROMEDA GALAXY; GALAXY, EXTERNAL; HYADES; MAGELLANIC CLOUDS; STAR CLOUDS; VIRGO CLUSTER.

Hodge, Prof. Philip. Chair of Polymer Chemistry, Department of Chemistry, University of Manchester, England. POLYMER-SUPPORTED REACTION.

Hodges, Prof. Kip. Department of Earth, Atmospheric, and Planetary Sciences, Massachusetts Institute of Technology, Cambridge. OROGENY.

Hodgin, David M. Spectra Associates Inc., Cedar Rapids, Iowa. SINGLE SIDEBAND.

Hodgkins, P. Douglas. Federal Aviation Administration, Technical Support, Washington, D.C. AUTOMATIC LANDING SYSTEM.

Hodos, Dr. William. Department of Psychology, University of Maryland, College Park. VERTEBRATE BRAIN (EVOLUTION).

Hoering, Dr. Thomas C. Geophysical Laboratory, Carnegie Institution, Washington, D.C. PALEOBIOCHEMISTRY.

Hoff, Prof. C. Clayton. Department of Biology, University of New Mexico. PSEUDOSCORPIONIDA; RICINULEIDA; SOLIFUGAE.

Hoffman, Dr. Carol D. Institute of Ecology, University of Georgia, Athens. AGROECOSYSTEM.

Hoffman, Dr. Paul. Department of Earth and Planetary Sciences, Harvard University. CONTINENTS, EVOLUTION OF.

Hoffman, Dr. Paul S. Department of Microbiology and Immunology, Faculty of Medicine, Dalhousie University, Halifax, Nova Scotia, Canada. LEGIONNAIRES' DISEASE.

Hoffman, Dr. Ralph E. Yale Psychiatric Institute, School of Medicine, Yale University. PSYCHOSIS.

Hoffman, Richard A. Martin Marietta Company, Denver, Colorado. SEAPLANE.

Hoffman, Prof. Richard L. Department of Biology, Radford College, Radford, Virginia. DIPLOPODA.

Hofstadter, Dr. Robert. Deceased; formerly, Department of Physics, Stanford University; Nobelist. SCINTILLATION COUNTER.

Hogan, Dr. Brigid. Mill Laboratories, Imperial Cancer Research Fund, London, England. CHIMERA.

Hogan, Dr. Robert. Department of Psychology, Johns Hopkins University. PERSONALITY THEORY.

Holbrey, Dr. John D. University of Alabama, Center for Green Manufacturing. IONIC LIQUIDS.

Holdaway, Dr. M. J. Department of Geological Sciences, Southern Methodist University. ANDALUSITE; KYANITE.

Holdeman, Dr. Louis B. Communications Satellite Corporation Laboratories, Clarksburg, Maryland. JOSEPHSON EFFECT.

Holder, Dr. Nigel. Anatomy and Human Biology Group, Biomedical Sciences Division, Kings College, University of London, England. PATTERN FORMATION (BIOLOGY).

Holland, Dr. Timothy J. B. Department of Mineralogy and Petrology, University of Cambridge, England. ECLOGITE; OMPHACITE.

Holloway, Prof. A. Gordon L. Department of Mechanical Engineering, University of New Brunswick, Canada. CENTER OF PRESSURE; NOZZLE.

Hollowell, Eugene A. Consultant in Agronomy, Port Republic, Maryland. CLOVER.

Hollweg, Prof. Joseph V. Department of Physics, Space Science Center, University of New Hampshire. ALFVÉN WAVES.

Holmes, Dr. Jack K. Holmes Associates, Los Angeles, California. SPREAD SPECTRUM COMMUNICATION.

Holmes, Dr. Mark. Department of Neurology, University of Washington, Seattle, Washington. SEIZURE DISORDERS.

Holmes, Dr. Robert W. Department of Biological Sciences, University of California, Santa Barbara. PHYTOPLANKTON.

Holst, Per A. Department of Electrical Engineering and Computer Sciences, School of Science and Technology, Stavanger College, Stavanger, Norway. ANALOG COMPUTER.

Holt, Dr. Donald L. Monsanto Company, St. Louis, Missouri. HYDROFORMYLATION; HYDROLYTIC PROCESSES.

Holt, Dr. Perry C. Department of Biology, Virginia Polytechnic Institute. OLIGOCHAETA.

Holt, Dr. Robert D. Museum of Natural History, University of Kansas. POPULATION ECOLOGY; THEORETICAL ECOLOGY.

Holthuis, Dr. Lipke B. Curator, Division of Crustacea, Rijksmuseum van Natuurlijke Historie, Leiden, Netherlands. STOMATOPODA.

Holton, Dr. Gerald. Department of Physics, Harvard University. EMPIRICAL METHOD; PHYSICAL THEORY; SCIENCE.

Holton, Prof. James R. Department of Atmospheric Sciences, University of Washington, Seattle. CYCLONE; DYNAMIC METEOROLOGY.

Holtzman, Dr. Eric. Deceased; formerly, Department of Biological Sciences, Columbia University. LYSOSOME.

Homann, Prof. Peter H. Institute of Molecular Biophysics, Florida State University. PIGMENTATION.

Hong, Dr. Richard. UVM Genetics Laboratory, Burlington, Vermont. IMMUNOLOGICAL DEFICIENCY.

Hood, Dr. Laura. Defenders of Wildlife, Washington, D.C. ENDANGERED SPECIES.

Hood, Dr. Leroy. President, Institute for Systems Biology, Seattle, Washington. GENETIC MAPPING; GENOMICS.

Hooke, Dr. Roger LeB. Deer Isle, Maine. GLACIOLOGY.

Hoover, J. Edgar. Deceased; formerly, Director, Federal Bureau of Investigation, U.S. Department of Justice. FINGERPRINT.

Horan, Dr. Michael. National Heart, Lung and Blood Institute, National Institutes of Health, Bethesda, Maryland. HYPERTENSION.

Horen, Dr. Daniel J. Nuclear Division, Oak Ridge National Laboratory, Oak Ridge, Tennessee. ANALOG STATES; ISOTOPE.

Horn, Dr. David J. Department of Entomology, Ohio State University. DERMAPTERA; ENTOMOLOGY, ECONOMIC.

Horne, Thomas A. Aircraft Owners and Pilots Association, Frederick, Maryland. ULTRALIGHT AIRCRAFT.

Horowitz, Stanley H. Consulting Electrical Engineer, Columbus, Ohio. RELAY.

Horsfall, Dr. James G. Director, Connecticut Agricultural Experiment Station, New Haven. FUNGISTAT AND FUNGICIDE.

Horst, Dr. R. Kenneth. Professor Emeritus, Plant Pathology, Cornell University, Ithaca, New York. VIROIDS.

Horstadius, Prof. Sven. Professor Emeritus of Zoology, University of Uppsala, Sweden. NEURAL CREST.

Horton, Prof. Derek. Department of Chemistry, American University, Washington, D.C. AMINO SUGAR.

Horton, G. A. Manager, Engineering Laboratories, Westinghouse Electric Corporation, Cleveland, Ohio. LUMINOUS EFFICACY; LUMINOUS EFFICIENCY; PHOTOMETER.

Horton, Dr. William A. Department of Pediatrics, Medical School, Health Science Center at Houston, University of Texas. DWARFISM AND GIGANTISM.

Horwich, Dr. Arthur. Department of Genetics, Yale School of Medicine, New Haven, Connecticut. MOLECULAR CHAPERONE.

Hosford, Prof. William F. Department of Materials Science and Engineering, College of Engineering, University of Michigan, Ann Arbor. WIRE DRAWING.

Hotchkiss, Grosvenor. Retired; formerly, Western Union Telegraph Company, New York. TELEGRAPHY.

Hotchkiss, Dr. Rollin D. Department of Genetics, Rockefeller University, New York. TRANSFORMATION (BACTERIA).

Houde, Prof. Edward D. University of Maryland, Center for Environmental Science, Chesapeake Biology Lab, Solomons, Maryland. MARINE FISHERIES.

Hough, Dr. L. F. Research Professor in Pomology, Department of Horticulture and Forestry, Rutgers University. NECTARINE.

Houghton, Dr. David D. Department of Meteorology, University of Wisconsin, Madison. THERMAL WIND.

Hougie, Prof. Cecil. School of Medicine, University of California, San Diego. HEMATOLOGIC DISORDERS; HEMOPHILIA.

Howard, Dr. Walter E. Department of Wildlife and Fisheries Biology, University of California, Davis. RODENTICIDE.

Howe, Prof. Carl E. Deceased; formerly, Professor Emeritus of Physics, Oberlin College. ACCELERATION; CENTRIFUGAL FORCE; ROTATIONAL MOTION.

Howell, Dr. Benjamin F., Jr. Department of Geological Sciences, Pennsylvania State University. GEOPHYSICS.

Howell, William E. NASA Langley Research Center, Hampton, Virginia. INERTIAL GUIDANCE SYSTEM.

Howells, Dr. T. A. Director of Continuing Education, Institute of Paper Chemistry, Appleton, Wisconsin. FLAMEPROOFING.

Howells, Dr. W. W. Department of Anthropology, Peabody Museum, Harvard University. PILTDOWN MAN.

Hoxton, Prof. Lewellyn G. Deceased; formerly, Professor Emeritus of Physics, University of Virginia. JOULE'S LAW.

Hrazdina, Dr. Geza. Department of Food Science and Technology, Institute of Food Science, New York State Agricultural Experiment Station, Cornell University, Geneva. FLAVONOID.

Hrbek, Dr. Charles J. Charles Stark Draper Laboratory, Cambridge, Massachusetts. UNDERWATER NAVIGATION.

Hrones, Dr. John A. Provost of Science and Technology, Case Western Reserve University. REGULATOR.

Hsieh, Dr. Samuel C. Department of Computer Science, Southwestern

Oklahoma State University, Weatherfort. ALGORITHM; COMPUTER PROGRAMMING.

Hu, Dr. A. S. L. Department of Biochemistry, University of Kentucky. BIOCHEMISTRY.

Hua, Dr. Hong. Director, 3D Visualization and Imaging System Lab, College of Optical Sciences, University of Arizona, Tucson. POLARIZED LIGHT.

Hubbard, Dr. Dennis. Geology Department, Oberlin College, Ohio. REEF.

Hubert, Prof. John F. Department of Geosciences, University of Massachusetts, Amherst. PLAYA.

Hudlicky, Dr. Tomas. Department of Chemistry, Illinois Institute of Technology, Chicago. TERPENE; TRITERPENE.

Hudson, Dr. Hugh S. Space Sciences Laboratory, University of California, Berkeley. SOLAR CONSTANT.

Hudson, Dr. Ralph P. Retired; formerly, Bureau International des Poids et Mesures, Sevres, France. ADIABATIC DEMAGNETIZATION; LOW-TEMPERATURE THERMOMETRY.

Huebler, Dr. Jack. Senior Vice President, Institute of Gas Technology, IIT Center, Chicago, Illinois. FUEL GAS.

Huelsman, Prof. Lawrence P. Department of Electrical and Computer Engineering, College of Engineering and Mines, University of Arizona. ELECTRIC FILTER; EQUIVALENT CIRCUIT.

Huffman, E. O. Consultant, Fertilizer Research and Development, Florence, Alabama. PRILLING.

Hufnagel, Prof. Todd C. John Hopkins University, Baltimore, Maryland. METALLIC GLASSES.

Hughes, Dr. John M. Associate Dean, College of Arts and Science, Department of Geology, Miami University, Oxford, Ohio. MONAZITE.

Huizenga, Dr. John R. Nuclear Structure Research Laboratory, University of Rochester. DEEP INELASTIC COLLISIONS; NUCLEAR FISSION.

Hulbary, Dr. Robert L. Department of Botany, University of Iowa. COLLENCHYMA; PARENCHYMA.

Hulet, Prof. Randall G. Physics and Astronomy Department, Rice University, Houston, Texas. ATOMIC FERMI GAS.

Hull, Dr. McAllister H., Jr. Department of Physics, University of New Mexico. LEAST-SQUARES METHOD; MAGNETON; RADIATION; other articles.

Hultsch, Dr. Ronald A. Department of Physics, University of Missouri, Columbia. TEMPERATURE.

Humes, Dr. Arthur G. Professor Emeritus, Department of Biology, Boston University. NEMERTEA.

Hummon, Dr. William D. Department of Zoological and Biomedical Sciences, Ohio University. CHAETONOTIDA; GASTROTRICHA; MACRODASYIDA.

Humphreys, Dr. David A. National Physical Laboratory, Teddington, Middlesex, United Kingdom. WAVEFORM DETERMINATION.

Humphreys, Dr. William F. Western Australian Museum, Perth, Western Australia. ECOLOGICAL ENERGETICS.

Hungate, Prof. Robert E. Department of Bacteriology, University of California, Davis. BACTERIA.

Hunsperger, Prof. Robert G. Department of Electrical and Computer Engineering, University of Delaware, Newark. INTEGRATED OPTICS.

Hunt, Dr. John M. Department of Chemistry, Woods Hole Oceanographic Institution, Woods Hole, Massachusetts. PETROLEUM ENGINEERING.

Hunt, Dr. Kelly. Division of Surgical Oncology, Comprehensive Breast Center, University of California, Los Angeles. BREAST DISORDERS.

Hunter, Prof. Lloyd P. Professor Emeritus, Department of Electrical Engineering, University of Rochester. CONTROLLED RECTIFIER; JUNCTION DIODE; other articles.

Hunter, Dr. R. H. F. School of Agriculture, University of Edinburgh, Scotland. REPRODUCTIVE SYSTEM.

Huntington, Prof. H. B. Department of Physics, Rensselaer Polytechnic Institute. HYSTERESIS.

Hurlbut, Prof. Cornelius S., Jr. Professor Emeritus of Mineralogy, Department of Geological Sciences, Harvard University. ANDALUSITE; CASSITERITE; EMERY; HALOGEN MINERALS; NICCOLITE; ORPIMENT; other articles.

Hurley, Joseph D. Westinghouse Electric Corporation, Orlando, Florida. VOLTAGE REGULATOR.

Hurll, John. Measurements Standards Laboratory, Marconi Instruments Ltd., Stevenage, Hertfordshire, England. PHASE-ANGLE MEASUREMENT.

Hurst, Dr. G. Samuel. Adjunct Professor, University of Tennessee, Knoxville. RESONANCE IONIZATION SPECTROSCOPY.

Hurst, Prof. James K. Department of Biochemistry and Biophysics, Washington State University. HYPOHALOUS ACID; NITRIC OXIDE; PEROXYNITRITE.

Hutcheon, Dr. Ian D. Deputy Director, Glenn T. Seaborg Institute, Chemical Biology and Nuclear Science Division, Analytical and Nuclear Chemistry Division, Lawrence Livermore National Laboratory, Livermore, California. COSMOCHEMISTRY.

Hutchinson, Prof. Ian H. Massachusetts Institute of Technology, Plasma Science and Fusion Center, Cambridge, Massachusetts. PLASMA DIAGNOSTICS.

Hutchison, Dr. Keith. Department of Biology, Brandeis University. BACTERIA.

Hwang, Dr. Patrick. Rockwell International, Cedar Rapids, Iowa. DIRECTION-FINDING EQUIPMENT.

Hwang, Dr. Richard N. Moh and Associates, Inc., Taipei, Taiwan. SUBWAY ENGINEERING; TUNNEL.

Hyde, Dr. Earl K. Lawrence Berkeley Laboratory, University of California, Berkeley. ASTATINE; FRANCIUM; RADON.

I

Iachello, Dr. Franco. Department of Physics, Yale University. SUPERMULTIPLET.

Ibrahim, Dr. Aly M. Agricultural Research Service, Western Region, U.S. Department of Agriculture, Salinas, California. ARTICHOKE.

Illis, Alexander. Director, Process Technology, International Nickel Company of Canada, Ltd., Toronto, Ontario. NICKEL METALLURGY.

Imbembo, Prof. Anthony L. Chairperson, Department of Surgery, University of Maryland Hospital, Baltimore. APPENDICITIS; SURGERY.

Ingels, Chuck A. University of California Cooperative Extension, Sacramento. COVER CROPS.

Ingersoll, Dr. Raymond V. Department of Earth and Space Science, University of California at Los Angeles. PROVENANCE (GEOLOGY).

Ingerson, Dr. Earl. Deceased; formerly, Department of Geology, University of Texas, Austin. GEOLOGIC THERMOMETRY.

Ingram, Dr. B. Lynn. Department of Geology and Geophysics, University of California, Berkeley. CHEMOSTRATIGRAPHY.

Inman, Dr. Douglas L. La Jolla, California. NEAR-SHORE PROCESSES.

Inman, Dr. Robert D. Professor of Medicine and Immunology, University of Toronto, Canada. CONNECTIVE TISSUE DISEASE.

Irish, Vivian F. MCD Biology Department, Yale University, New Haven, Connecticut. PLANT REPRODUCTION.

Ishida, Dr. Nakao. Department of Bacteriology, Tohaku University School of Medicine, Sendai, Japan. SENDAI VIRUS.

Ishihara, Prof. Teruo. Department of Mechanical and Aerospace Engineering and Materials Science, University of Minnesota. ACCELEROMETER.

Isley, Dr. Duane. Department of Botany, Iowa State University. LEGUME.

Israel, Dr. Yedy. Head, Biochemical Research, Addiction Research Foundation, Toronto, Ontario, Canada. ALCOHOLISM.

Issel, Dr. Charles J. College of Agriculture, Veterinary Science, Gluck Equine Research Center, University of Kentucky, Lexington. EQUINE INFECTIOUS ANEMIA.

Itano, Dr. Wayne M. Time and Frequency Division, National Institute of Standards and Technology, Boulder, Colorado. ATOMIC CLOCK.

Iverson, Dr. Brent L. Department of Chemistry and Biochemistry, University of Texas, Austin. CATALYTIC ANTIBODIES.

Ives, Dr. Anthony R. Department of Zoology, University of Wisconsin, Madison. PREDATOR-PREY INTERACTIONS.

Iwai, Dr. Kazuhiro. Department of Molecular Cell Biology, Graduate School of Medicine, Osaka City University, Japan. PROTEIN DEGRADATION.

Iyengar, Dr. Ravi. Professor and Chairman, Department of Pharmacology, Mount Sinai School of Medicine, New York. SIGNAL TRANSDUCTION.

Izadian, Dr. Jamal S. AntennaEM Communication, LLC, San Jose, California. YAGI-UDA ANTENNA.

J

Jackiw, Prof. Roman. Department of Physics, Massachusetts Institute of Technology. QUANTUM ANOMALIES.

Jackson, Donald C. Division of Biology and Medicine, Brown University. PH REGULATION (BIOLOGY).

Jackson, G. A. Deputy Division Manager, ERA Technology Ltd., Surrey, England. ANECHOIC CHAMBER.

Jackson, Dr. Kenneth A. AT&T Bell Telephone Laboratories, Murray Hill, New Jersey. CRYSTAL WHISKERS.

Jackson, Dr. M. P. A. Senior Research Scientist, Bureau of Economic Geology, University of Texas, Austin. DIAPIR; SALT DOME.

Jackson, Dr. Roscoe G., II. Department of Geological Science, Northwestern University. HOLOCENE.

Jackson, Dr. William D. President, HMJ Corporation, Chevy Chase, Maryland. MAGNETOHYDRODYNAMIC POWER GENERATOR.

Jacob, Prof. Daniel J. Department of Earth and Planetary Sciences, Harvard University. ATMOSPHERIC CHEMISTRY.

Jacobs, Prof. Stephen F. Emeritus, College of Optical Sciences, University of Arizona. LASER; OPTICAL ISOLATOR.

Jacobson, Prof. Nathan. Deceased; formerly, Department of Mathematics, Yale University. LINEAR ALGEBRA; RING THEORY.

Jacobson, Dr. Robert. Chief of the Clinical Branch, Department of Health and Human Services, Gillis W. Long Hansen's Disease Center, Carville, Louisiana. LEPROSY.

Jaep, William F. Central Research Department, Experimental Station, E. I. Du Pont de Nemours and Company, Wilmington, Delaware. ENTROPY.

Jaffe, Dr. Jules. Marine Physical Laboratory, Scripps Institution of Oceanography, La Jolla, California. UNDERWATER TELEVISION.

Jaffe, Prof. Robert L. Department of Physics, Massachusetts Institute of Technology, Cambridge. NUCLEON.

Jahns, Dr. Richard H. Department of Earth Sciences, Stanford University. PEGMATITE.

Jain, Dr. Subodh K. Department of Agronomy and Range Science, University of California, Davis. AMARANTH.

Jakobi, Prof. Hans. Professor of Animal Physiology, Department of Zoology, University of Parana, Brazil. BATHYNELLACEA; SYNCARIDA.

James, Prof. Arthur M. Department of Chemistry, Bedford College, London, England. CELL-SURFACE IONIZATION.

James, Dr. Goideon T. Department of Earth Sciences, East Texas State University. DESMOSTYLIA; PYROTHERIA; XENUNGULATA.

James, John W. Vice President, Research, McDonnell and Miller, Inc., Chicago. STEAM HEATING.

James, Dr. Mark A. Corriebruach House, Perthshire, United Kingdom. LINGULIDA; RHYNCHONELLIDA; TEREBRATULIDA.

James, Dr. Richard W. Head, Forecasting Branch, Oceanographic Prediction Division, U.S. Naval Oceanographic Office. SHIP ROUTING.

James, Dr. Timothy. Department of Biology, Duke University, Durham, North Carolina. ZYGOMYCETES; ZYGOMYCOTA.

Jamshidi, Dr. Mohamed. Director, CAD Laboratory for Systems and Robotics, Department of Electrical and Computer Engineering, University of New Mexico. LARGE SYSTEMS CONTROL THEORY.

Janick, Prof. Jules. Department of Horticulture, Purdue University. AGRICULTURE; HORTICULTURAL CROPS.

Janis, Dr. Christine. Department of Ecology and Evolutionary Biology, Brown University, Providence, Rhode Island. ARTIODACTYLA.

Janna, Prof. William S. Department of Mechanical Engineering, Memphis State University, Memphis, Tennessee. CHOKED FLOW; PIPE FLOW.

Jannasch, Dr. Holger W. Senior Scientist, Woods Hole Oceanographic Institution, Woods Hole, Massachusetts. MARINE MICROBIOLOGY.

Japikse, Dr. David. President, Concepts ETI, Inc., Norwich, Vermont. DIFFUSER.

Japuntich, Daniel A. Occupational Health and Safety Products Division, 3M Company, St. Paul, Minnesota. RESPIRATOR.

Jarvis, J. Gordon. Corporate Communications, Eastman Kodak Company, Rochester, New York. PHOTOCOPYING PROCESSES.

Jarvis, Dr. John F. Robotics Systems Research Department, Bell Laboratories, Holmdel, New Jersey. INTELLIGENT MACHINE.

Jaswal, Prof. Sitaram S. Behlen Laboratory of Physics, University of Nebraska. PHASE TRANSITIONS.

Jeffries, Dr. R. P. S. Department of Paleontology, British Museum of Natural History, London, England. CHALCICHORDATES.

Jegla, Dr. Dorothy E. Department of Biology, Kenyon College, Gambier, Ohio. FATE MAPS (EMBRYOLOGY).

Jen, Dr. C. K. Supervisor, Microwave Physics Group, Applied Physics Laboratory, Johns Hopkins University. MICROWAVE OPTICS.

Jenike, Andrew W. Jenike & Johanson, Inc., North Billerica, Massachusetts. FLOW OF GRANULAR SOLIDS.

Jenkins, Prof. Francis A. Deceased; formerly, Department of Physics, University of California, Berkeley. DIFFRACTION; FINE STRUCTURE (SPECTRAL LINES); HUYGEN'S PRINCIPLE; PASCHEN-BACK EFFECT; RESOLVING POWER (OPTICS); STARK EFFECT; other articles.

Jenkins, Dr. Julian L. Director of Research, Bell Helicopter Company, Fort Worth, Texas. HELICOPTER.

Jenkins, Dr. Ron. International Centre for Diffraction Data, Newtown. X-RAY POWDER METHODS.

Jennings, Dr. Donald A. Time and Frequency Division, National Institute of Standards and Technology, Boulder, Colorado. LENGTH.

Jennings, Dr. J. B. Department of Pure and Applied Zoology, University of Leeds, England. ACOELA; ANOPLA; BDELLOIDEA; ENOPLA; HETERONEMERTINI; ROTIFERA; other articles.

Jensen, Dr. Aldon H. Department of Animal Science, University of Illinois, Urbana-Champaign. SWINE PRODUCTION.

Jensen, Prof. Arthur R. School of Education, University of California, Berkeley. INTELLIGENCE.

Jensen, Dr. Howard B. Research Supervisor, Laramie Energy Research Center, Energy Research and Development Administration, Laramie, Wyoming. OIL SHALE.

Jensen, Richard L. Sirius Satellite Radio, New York. WIDE-AREA NETWORKS.

Jerome, Dr. D. Laboratoire de Physique des Solides, Associé au C.N.R.S., Université de Paris—Sud, France. ORGANIC CONDUCTOR.

Jespersen, Dr. Neil D. Department of Chemistry, St. John's University, New York. THERMAL ANALYSIS.

Joel, Amos E., Jr. Deceased; formerly, AT&T Bell Laboratories, Holmdel, New Jersey. INTEGRATED SERVICES DIGITAL NETWORK (ISDN); SWITCHING SYSTEMS (COMMUNICATIONS).

Joenje, Dr. Hans. Department of Clinical Genetics and Human Genetics, Free University Medical Center, Amsterdam, the Netherlands. COMPLEMENTATION (GENETICS).

Johannsen, Dr. Ing Gunnar. Laboratory for Systems Engineering and Human-Machine Systems, University of Kassel, Germany. PROCESS CONTROL.

Johansen, Dr. Kjell. Department of Zoophysiology, University of Aarhus, Denmark. RESPIRATION.

Johnson, Prof. Carl R. Department of Chemistry, Wayne State University. SULFONIC ACID.

Johnson, Dr. Claudia. Department of Geological Sciences, University of Indiana, Bloomington. BIVALVE, RUDIST.

Johnson, Dr. Donald Lee. Department of Geography, University of Illinois, Urbana. CONTINENT.

Johnson, Dr. Edward. Department of Cell Biology, Rockefeller University, New York, New York. NUCLEIC ACID.

Johnson, Dr. Ellis L. Thomas J. Watson Research Center, IBM Corporation, Yorktown Heights, New York. LINEAR PROGRAMMING.

Johnson, Dr. Francis S. Acting President, University of Texas, Dallas. IONOSPHERE; UPPER-ATMOSPHERE DYNAMICS.

Johnson, Dr. Giles. School of Biological Sciences, University of Manchester, United Kingdom. ENVIRONMENT.

Johnson, Prof. J. G. Deceased; formerly, Department of Geology, Oregon State University. DEVONIAN.

Johnson, Dr. James E. Department of Chemistry, Texas Women's University, Denton. OXIME.

Johnson, Dr. Jon D. Department of Forestry, University of Florida, Gainesville. TREE PHYSIOLOGY.

Johnson, K. W. President, Vibra-Grip Corporation, Jamestown, Ohio. SHOCK ISOLATION; VIBRATION DAMPING; VIBRATION ISOLATION; VIBRATION MACHINE.

Johnson, Dr. Paul E. School of Management, University of Minnesota. PROBLEM SOLVING (PSYCHOLOGY).

Johnson, Philip C. Research and Development Department, Chemicals and Plastics Operations Division, Union Carbide Corporation, South Charleston, West Virginia. POLYOL.

Johnson, Prof. Ronald R. Department of Animal Science, Institute of Agriculture, University of Tennessee. AGRICULTURAL SCIENCE (ANIMAL).

Johnson, Dr. W. Hilton. Department of Geology, University of Illinois, Urbana-Champaign. FLUVIAL EROSION LANDFORMS; other articles.

Johnson, Prof. Walter C. Department of Electrical Engineering, Princeton University. ELECTROMAGNETIC WAVE TRANSMISSION.

Johnsson, Dr. Mark J. California Coastal Commission, San Francisco. SAND.

Johnston, Prof. Francis J. Department of Chemistry, University of Georgia, Athens. AMMONIUM SALT; CYANIDE; HYDRATION; METALLOID; MORDANT; NITRIC ACID; OSMOSIS; other articles.

Johnston, Dr. James D. Senior Research Advisor, Pioneering Research, Ethyl Corporation, Baton Rouge, Louisiana. LEAD.

Johnston, Prof. Keith. Department of Chemical Engineering, University of Texas, Austin. SUPERCRITICAL FLUID.

Johnston, Dr. Mark. Department of Genetics, Washington University Medical Center, School of Medicine, St. Louis, Missouri. GENETICS.

Joklik, Dr. W. K. Department of Microbiology and Immunology, Duke University Medical Center. VIRUS.

Jones, Prof. B. Frank, Jr. Department of Mathematics, Weiss School of Natural Sciences, Rice University. MEASURE THEORY.

Jones, Dr. Clive G. Institute of Ecosystem Studies, New York Botanical Garden, Millbrook, New York. CHEMICAL ECOLOGY.

Jones, Dr. Daniel J. Department of Earth Sciences, California State College, Bakersfield. PROTISTA; SILICOFLAGELLATA; TINTINNIDA.

Jones, Dr. David R. Department of Zoology, University of British Columbia, Vancouver, Canada. CARDIOVASCULAR SYSTEM; CIRCULATION.

Jones, Dr. E. Peter. Bedford Institute of Oceanography, Dartmouth, Nova Scotia, Canada. ARCTIC OCEAN.

Jones, Dr. E. Ruffin. Department of Zoology, University of Florida. POLYCLADIDA; TRICLADIDA; TURBELLARIA.

Jones, Edwin L., Jr. Forensic scientist, Ventura County Sheriff's Crime Laboratory, Ventura, California. FORENSIC MICROSCOPY.

Jones, Prof. Edwin R. Department of Physics and Astronomy, University of South Carolina, Columbia. CENTER OF MASS; LENGTH; MASS; RADIUS OF GYRATION.

Jones, Dr. Hugh D. School of Biological Sciences, University of Manchester, United Kingdom. BURROWING ANIMALS.

Jones, Dr. J. B. Department of Plant Pathology, University of Florida. PLANT PATHOLOGY.

Jones, Dr. Jeremy B. Department of Biological Sciences, University of Nevada, Las Vegas. ECOSYSTEM.

Jones, Prof. Lawrence W. The Harrison M. Randall Laboratory, University of Michigan, Ann Arbor. QUARKS.

Jones, Dr. Meredith L. National Museum of Natural History, Smithsonian Institution, Washington, D.C. VESTIMENTIFERA.

Jones, Dr. Morton E. Director, Physical Sciences Research Laboratory, Texas Instruments, Inc., Dallas. FERRIMAGNETIC GARNETS.

Jones, Dr. Norman S. Senior Lecturer, Marine Biological Station, University of Liverpool, England. CUMACEA.

Jones, Dr. Owen C. Superintendent, Division of Electrical Science, National Physical Laboratory, Teddington, England. ELECTRICAL MEASUREMENTS.

Jones, Dr. Patricia M. Department of Mechanical and Industrial Engineering, University of Illinois, Urbana-Champaign. HUMAN-MACHINE SYSTEMS.

Jones, Prof. R. Alan. School of Chemical Sciences, University of East Anglia, England. PHASE-TRANSFER CATALYSIS.

Jones, Dr. R. Gareth. Division of Electrical Science, National Physical Laboratory, Teddington, Middlesex, United Kingdom. DIELECTRIC MEASUREMENTS; ELECTRICAL MEASUREMENTS; RESISTANCE MEASUREMENT; CALIBRATION.

Jones, Prof. Richard T. Chairperson, Department of Biochemistry, Oregon Health Sciences University. HEMOGLOBIN.

Jones, Dr. Robert F. Professor of Surgery, School of Medicine, University of Washington, SURGERY.

Jones, Prof. William D. Department of Chemistry, University of Rochester, New York. ALKANE.

Jong, Dr. S. C. Director, Microbiology, American Type Culture Collection, Manassas, Virginia. FUNGAL BIOTECHNOLOGY; MYCOLOGY.

Joppa, Dr. Robert G. Deceased; formerly, Department of Aeronautics, University of Washington. WIND TUNNEL.

Jordan, Dr. Elke. Department of Health & Human Services, National Human Genome Research Institute, Bethesda, Maryland. HUMAN GENOME.

Jory, Howard R. Varian Associates, Palo Alto, California. GYROTRON.

Joseph, Earl C. St. Paul, Minnesota. REAL-TIME SYSTEMS.

Joseph, Leonard M. Lev Zetlin Associates, New York. STRUCTURAL DESIGN.

Joseph, Dr. Simpson. Department of Chemistry and Biochemistry, University of California, La Jolla. RIBOSOMES.

Jost, Dr. Alfred. Laboratoire de Physiologie du Developpement, College de France, Paris. REPRODUCTIVE SYSTEM.

Juang, Dr. Biing Hwang. Lucent Technologies, Bell Laboratories, Murray Hill, New Jersey. SPEECH RECOGNITION.

Judd, Dr. W. John. Department of Pathology, University of Michigan Medical Center, Ann Arbor. BLOOD GROUPS.

Judd, Dr. William R. School of Civil Engineering, Purdue University. ENGINEERING GEOLOGY; ROCK MECHANICS.

Juranek, Dr. Dennis D. Centers for Disease Control and Protection, Atlanta, Georgia. GIARDIASIS.

Jurenka, Dr. Russell A. Department of Entomology, Iowa State University of Science and Technology, Ames. INSECT PHYSIOLOGY.

Jurisson, Dr. Silvia S. Department of Chemistry, University of Missouri, Columbia. TECHNETIUM.

Just, Prof. Evan. Department of Mining and Geology, Stanford University. MINING.

Juvet, Prof. Richard S., Jr. Department of Chemistry, Arizona State University. GAS CHROMATOGRAPHY.

K

Kado, Prof. Clarence I. Department of Plant Pathology, University of California, Davis. CROWN GALL.

Kaesler, Dr. Roger L. Department of Geology, University of Kansas, Lawrence. MYODOCOPIDA; OSTRACODA; PALEOCOPA.

Kahn, Dr. Bernd. Environmental Resources Center, Georgia Institute of Technology, Atlanta. LOW-LEVEL COUNTING.

Kahn, Dr. Robert. President, Corporation for National Research Initiatives, Reston, Virginia. WIDE-AREA NETWORKS.

Kailath, Prof. Thomas. Information Systems Laboratory, Stanford University. ADAPTIVE SIGNAL PROCESSING.

Kaler, Prof. James B. Department of Astronomy, College of Liberal Arts and Sciences, University of Illinois, Urbana-Champaign. COLOR INDEX; DWARF STAR; GIANT STAR; SPECTRAL TYPE; STAR; STELLAR EVOLUTION; other articles.

Kalhammer, Dr. Fritz. Electric Power Research Institute, Palo Alto, California. ENERGY STORAGE.

Kallen, Dr. Bengt. Department of Embryology, University of Lund, Sweden. NERVOUS SYSTEM (VERTEBRATE).

Kallio, Dr. R. E. School of Life Sciences, University of Illinois, Urbana-Champaign. BACTERIAL PHYSIOLOGY AND METABOLISM; CULTURE.

Kalpakjian, Prof. Serope. Mechanical and Aerospace Engineering Department, Illinois Institute of Technology. DRAWING OF METAL; FORGING; METAL FORMING; METAL ROLLING; SHEET-METAL FORMING.

Kalthoff, Dr. Klaus. School of Biological Sciences, The University of Texas at Austin. CLEAVAGE (DEVELOPMENTAL BIOLOGY); FERTILIZATION (ANIMAL).

Kaminow, Dr. Ivan P. Lucent Technologies, Holmdel, New Jersey. OPTICAL MODULATORS.

Kaminsky, Prof. Walter. Institute of Technical and Macromolecular Chemistry, University of Hamburg, Germany. SINGLE-SITE CATALYSTS (POLYMER SCIENCE).

Kamrin, Dr. Michael A. Center for Environmental Toxicology, Department of Pathology, Michigan State University. TOXICOLOGY.

Kan-Mitchell, Dr. June. School of Medicine, University of Southern California. IMMUNOTHERAPY.

Kana, Dr. Todd M. Horn Point Laboratory, Cambridge, Maryland. PHYCOBILIN.

Kaner, Prof. Richard B. Department of Chemistry and Biochemistry, University of California, Los Angeles. PERIODIC TABLE.

Kanury, Dr. A. Murty. Senior Mechanical Engineer, Stanford Research Center, Menlo Park, California. FIRE TECHNOLOGY.

Kanzig, Prof. Werner. Department of Physics, Massachusetts Institute of Technology. FERROELECTRICS.

Kaplan, Dr. Louis. Senior Chemist, Argonne National Laboratory, Argonne, Illinois. HYDROGEN; TRITIUM.

Kaplan, Robert A. Vice President in Charge of Engineering, Automatic Burner Corporation, Chicago, Illinois. OIL BURNER.

Kaplan, Dr. Simon. Department of Computer Science, University of Queensland, Brisbane, Australia. SOFTWARE.

Kaplan, Prof. Thomas A. Department of Physics and Astronomy, Michigan State University, East Lansing. MAGNON.

Kapner, Robert S. Department of Chemical Engineering, The Cooper Union, New York. SULFONATION AND SULFATION.

Kaprelian, Edward K. Deceased; formerly, Vice President and Technical Director, Keuffel and Esser Company, Morristown, New Jersey. GUNSIGHTS; PERISCOPE; RANGEFINDER (OPTICS).

Kapur, Prof. Kallash C. Director, Industrial Engineering, University of Washington, Seattle. RELIABILITY, AVAILABILITY, AND MAINTAINABILITY.

Karady, Prof. George G. Department of Electrical Engineering, Arizona State University. BUS BAR; ELECTRIC POWER TRANSMISSION; FUSE (ELECTRICITY).

Kareiva, Prof. Peter. Department of Zoology, University of Washington, Seattle. DISEASE ECOLOGY; ECOLOGICAL MODELING; SYSTEMS ECOLOGY.

Karlson, Dr. Peter. Institut für Physiologische Chemie, Philipps-Universität, Marburg, Germany. ECDYSONE.

Karner, Dr. Garry D. Lamont-Doherty Earth Observatory, Palisades, New York. LITHOSPHERE.

Karni, Prof. Shlomo. Department of Electrical and Computer Engineering, University of New Mexico. ELECTRICAL LOADING; SHUNTING.

Karovska, Dr. Margarita. Harvard-Smithsonian Center for Astrophysics, Cambridge, Massachusetts. SPECKLE.

Karp, Dr. Alan H. IBM Scientific Center, Palo Alto, California. RADIATIVE TRANSFER.

Kartalopoulos, Dr. Stamatios V. Williams Professor in Telecommunications Networking, School of Electrical and Computer Engineering, University of Oklahoma, Tulsa. OPTICAL COMMUNICATIONS.

Kastner, Dr. Miriam. Geosciences Research Division, Scripps Institution of Oceanography, University of California, La Jolla. HYDRATE.

Katz, Dr. Jack L. Department of Psychiatry, Montefiore Hospital and Medical Center, New York. ANOREXIA NERVOSA.

Katz, Dr. Joseph J. Chemistry Division, Argonne National Laboratory, Argonne, Illinois. HEAVY WATER; HYDROGEN FLUORIDE; INORGANIC CHEMISTRY.

Katz, Dr. Leonard A. Health Care Plan Medical Center, Buffalo, New York. GASTROINTESTINAL TRACT DISORDERS; INFLAMMATORY BOWEL DISEASE.

Kauffman, Prof. Erle. Department of Geological Sciences, Indiana University, Bloomington. BIVALVE, RUDIST.

Kaufmann, R. H. Deceased; formerly, Consultant, Schenectady, New York. GROUNDING.

Kayan, Prof. Carl F. Deceased; formerly, Department of Mechanical Engineering, School of Engineering, Columbia University. DRY ICE; REFRIGERATION; REFRIGERATOR; TON OF REFRIGERATION.

Kayden, Dr. Herbert J. Department of Medicine, New York University School of Medicine, New York. VITAMIN E.

Kaye, Prof. Paul H. Science and Technology Research Institute University of Hertfordshire, United Kingdom. LIGHT-SCATTERING TECHNIQUES.

Kays, Prof. John M. Animal Industries Department, University of Connecticut. MULE.

Kazmierczak, Dr. James J. Bureau of Public Health, Madison, Wisconsin. LYME DISEASE.

Keddy, Prof. Rex J. Schonland Research Center for Nuclear Sciences, University of Witwatersrand, Johannesburg, South Africa. CRYSTAL COUNTER.

Keefe, Dr. Richard S. E. Department of Psychiatry, Duke University Medical Center, Durham, North Carolina. SCHIZOPHRENIA.

Keeton, Dr. William T. Deceased; formerly, Professor and Chairperson, Section of Neurobiology and Behavior, Langmuir Laboratory, Division of Biological Sciences, Cornell University. MAGNETIC RECEPTION (BIOLOGY).

Keffer, Prof. Frederic. Deceased; formerly, Department of Physics, University of Pittsburgh. ANTIFERRIMAGNETISM; CURIE-WEISS LAW; DIAMAGNETISM; LANGEVIN FUNCTION; LARMOR PRECESSION; PARAMAGNETISM.

Keh, Dr. Charlene. Department of Chemistry, Tulane University, New Orleans, Louisiana. ATOM ECONOMY.

Keil, Dr. Stephen L. Chief, Solar Research Branch, National Optical Astronomy Observatories, National Solar Observatory, Air Force Geophysics Laboratory, Sunspot, New Mexico. PHOTOSPHERE.

Keiser, Dr. Bernhard E. President, Keiser Engineering, Inc., Vienna, Virginia. ELECTROMAGNETIC COMPATIBILITY.

Keller, George V. Department of Geophysics, Colorado School of Mines. ROCK, ELECTRICAL PROPERTIES OF.

Keller, Dr. Harold W. Botanical Research Institute of Texas, Inc., Fort Worth. MYXOMYCOTA.

Keller, Dr. Joseph M. Department of Physics, Iowa State University. HARMONIC MOTION; HARMONIC OSCILLATOR; OSCILLATION; PERIODIC MOTION.

Keller, Prof. Walter D. Department of Geology, University of Missouri. WEATHERING PROCESSES.

Keller, Dr. William E. Los Alamos National Laboratory, Los Alamos, New Mexico. BOLOMETER.

Kellerman, Dr. Kenneth. National Radio Astronomy Observatory Charlottesville, Virginia. RADIO ASTRONOMY; RADIO TELESCOPE.

Kelley, Prof. Michael C. Engineering and Theory Center, Cornell University. Ionosphere.

Kelley, Prof. Robert M. Department of Chemical Engineering, North Carolina State University, Raleigh. Biochemical engineering.

Kellogg, Dr. Douglas S. Centers for Disease Control and Protection, Atlanta, Georgia. Granuloma inguinale.

Kells, Prof. Lyman M. Deceased; formerly, Professor Emeritus, U.S. Naval Academy. Radian measure.

Kelly, Dr. Dorothy. Littleton Regional Hospital, New Hampshire. Sudden infant death syndrome (SIDS).

Kelly, Dr. Robert. Chemical and Biochemical Engineering, North Carolina State University, Raleigh. Biochemical engineering.

Kemeny, Dr. John M. Department of Mining and Geological Engineering, The University of Arizona, Tucson. Rock mechanics.

Kemper, Kerry. Surveying Division, Pentax Corporation, Englewood, Colorado. Surveying instruments.

Kempers, Dr. Roger D. Department of Obstetrics and Gynecology, Mayo Foundation, Mayo Graduate School of Medicine, Rochester, Minnesota. Infertility.

Kemppainen, Dr. Robert J. Department of Anatomy, Physiology, and Pharmacology, Auburn University, Auburn, Alabama. Adrenal gland; Epinephrine.

Kennedy, Prof. John D. School of Chemistry, University of Leeds, United Kingdom. Borane.

Kenning, Dr. David B. B. Department of Engineering Science, Oxford, United Kingdom. Heat balance.

Kenny, Dr. Stephen T. Assistant Agronomist, Irrigated Agriculture Research and Extension Center, Washington State University, Prosser. Hop.

Kensler, Dr. Charles J. Senior Vice President, Life Science Division, Arthur D. Little, Inc., Cambridge, Massachusetts. Pharmacology.

Kent, Dr. George C., Jr. Professor Emeritus, Louisiana State University; Author's Services, Inc., Baton Rouge, Louisiana. Pineal body; Thyroid gland.

Keppens, Dr. Veerle M. National Center for Physical Acoustics, University of Mississippi, Oxford. Ultrasonics.

Ker, Dr. Robert F. Department of Pure and Applied Biology, University of Leeds, United Kingdom. Tendon.

Kerr, Dr. Marilyn S. Department of Biology, Syracuse University. Crab.

Kerst, Prof. Donald W. Deceased; formerly, Department of Physics, University of Wisconsin. Betatron.

Kerstein, Prof. Morris. D. Director, Graduate and Postgraduate Medical Education Programs, Department of Surgery, Hahnemann University, Philadelphia, Pennsylvania. Aneurysm.

Kessler, Prof. Edwin. Formerly, Director, National Severe Storms Laboratory, Norman, Oklahoma. Air mass; Air pressure; Air temperature; Coriolis acceleration; Weather.

Kessler, George W. Vice President, Engineering and Technology, Power Generation Division, Babcock and Wilcox Company, Barberton, Ohio. Air heater; Boiler economizer; Buhrstone mill; Cyclone furnace; Steam temperature control; Water-tube boiler; other articles.

Kester, Dr. Dale E. Department of Pomology, College of Agriculture, University of California, Davis. Almond.

Ketterle, Prof. Wolfgang. Department of Physics, Massachusetts Institute of Technology, Cambridge; Nobelist. Atom laser; Bose-Einstein condensation.

Ketterson, Dr. J. B. Department of Physics, Northwestern University. De Haas-van Alphen effect.

Keyomarsi, Dr. Khandan. Wadsworth Center, Albany, New York. Cell division.

Khan, Dr. Tanveer N. Senior Plant Breeder, Division of Plant Industry, Western Australian Department of Agriculture, South Perth. Winged bean.

Kharasch, Prof. Norman. Department of Biomedical Chemistry, University of Southern California. Mercaptan.

Khochfar, Sadegh. Astrophysics, Somerville College, University of Oxford, United Kingdom. Black hole; Galaxy formation and evolution; Gravitational collapse.

Kibble, Dr. Bryan P. Formerly, Division of Electrical Science, National Physical Laboratory, Teddington, Middlesex, England. Capacitance measurement; Current balance; Frequency comb; Inductance measurement; Inductive voltage divider; Interference-free circuitry; Watt balance; other articles.

Kiefer, Dr. Falk. Department of Psychiatry, University Hospital of Hamburg, Germany. Alcoholism.

Kik, Dr. Pieter G. College of Optics and Photonics, University of Central Florida, Orlando. Plasmon.

Kilgore, Dr. Lee A. Deceased; formerly, Westinghouse Electric Corporation, East Pittsburgh, Pennsylvania. Alternating-current generator.

Killion, Dr. Mead C. Etymotic Research, Elk Grove Village, Illinois. Hearing aid.

Kim, Dr. C. S. National Heart Lung and Blood Institutes, National Institutes of Health, Bethesda, Maryland. ABC lipid transporters.

Kim, Prof. Chung W. Department of Physics, Johns Hopkins University. Grand unification theories.

Kim, Dr. David. Associate Professor, Department of Industrial and Manufacturing Engineering, Oregon State University. Production systems.

Kim, Dr. Thomas J. Department of Mechanical Engineering and Applied Mechanics, University of Rhode Island, Kingston. Water-jet cutting.

Kimbark, Edward W. Deceased; formerly, Head, Systems Analysis Group, Bonneville Power Administration, Portland, Oregon. Neper.

Kimbrell, James E. Brashear LP, A division of L-3 Communications Corporation, Pittsburgh, Pennsylvania. Optical tracking systems.

Kimbrough, Dr. J. W. Department of Plant Pathology, University of Florida, Gainesville. Plant pathology.

Kinard, Joseph R. Consultant, Darnestown, Maryland. Thermal converters.

Kindstedt, Dr. Paul S. Agricultural Experiment Station, Department of Animal Sciences, University of Vermont, College of Agriculture and Life Sciences, Burlington. Rennin.

King, Dr. Bryan H. Professor of Psychiatry and Pediatrics, Department of Psychiatry, Dartmouth Hitchcock Medical Center, Lebanon, New Hampshire. Mental retardation.

King, Dr. C. Judson. Department of Chemical Engineering, University of California, Berkeley. Distillation; Unit operations.

King, Dr. Donald West, Jr. Department of Pathology, University of Chicago. Metabolic disorders.

King, Dr. Elbert A. Department of Geology, University of Houston. Tektite.

King, Dr. Gregory R. Chevron Overseas Petroleum Inc., California. Coalbed methane.

Kinoshita, Prof. Toichiro. Laboratory of Nuclear Studies, Cornell University, Ithaca, New York. Electron spin; Quantum electrodynamics.

Kinsler, Lawrence E. Deceased; formerly, Professor of Physics, U.S. Naval Postgraduate School, Monterey, California. Resonance (acoustics and mechanics); Sympathetic vibration; Tuning fork; Vibration; Vibration pickup.

Kinsman, Blair. Deceased; formerly, Marine Science Research Center, State University of New York, Stony Brook. River tides; Tidal bore.

Kipouros, Dr. Georges. Dalhousie University, Halifax, Canada. Fused-salt solution.

Kirby, Dr. H. W. Mound Facility, Monsanto Research Corporation, Miamisburg, Ohio. Protactinium.

Kirby, Robert S. Research Laboratories, National Oceanic and Atmospheric Administration, U.S. Department of Commerce, Boulder, Colorado. Tropospheric scatter.

Kirby, Dr. Seth. Dixon Spice Farms, Dixon, California. Coriander; Cumin; Dill; Fennel; Marjoram; Sage; other articles.

Kirk, Prof. Wiley P. Department of Physics, Texas A&M University. KONDO EFFECT.

Kirmani, Prof. Syed N. U. A. Department of Mathematics, University of Northern Iowa, Cedar Falls. BAYESIAN STATISTICS.

Kirschner, Dr. Stanley. Department of Chemistry, Wayne State University. TELLURIUM.

Kirshner, Prof. Robert. Harvard-Smithsonian Center for Astrophysis, Cambridge, Massachusetts. ACCELERATING UNIVERSE.

Kissel, Prof. David E. Department of Crop and Soil Sciences, University of Georgia, Athens. SOIL FERTILITY.

Kit, Dr. Saul. Novagene Inc., Houston, Texas. VACCINATION.

Klass, Dr. Klaus-Dieter. Staatliche Naturhistorische, Sammlungen, Dresden, Germany. MANTOPHASMATODEA.

Kleier, Dr. Daniel. Department of Chemistry, Drexel University Philadelphia, Pennsylvania. COMPUTATIONAL CHEMISTRY.

Klein, Dr. Barbara P. Professor Emeritus, University of Illinois at Urbana-Champaign. FOOD.

Klein, Dr. Cornelis. Department of Earth and Planetary Sciences, University of New Mexico, Albuquerque. CHABAZITE; KERNITE; MINERAL; OPAL; TURQUOISE.

Klein, Dr. David H. Department of Chemistry, Hope College, Holland, Michigan. NUCLEATION.

Klein, Dr. Jan. Max-Planck Institut für Biologie, Abteilung Immungenetik, Tubingen, Germany. IMMUNOGENETICS.

Klein, Prof. Richard G. Department of Anthropology, University of Chicago. PALEOLITHIC.

Klein, Prof. Richard M. Department of Botany, University of Vermont. BONSAI; CELL WALLS (PLANT).

Kleiner, Dr. Mendel. Chalmers University of Technology, Applied Acoustics, Gotheberg, Sweden. SOUND FIELD ENHANCEMENT; SOUND-REINFORCEMENT SYSTEM.

Kleinpell, Dr. Robert M. Department of Paleontology, University of California, Berkeley. CENOZOIC; other articles.

Kleinschmidt, Mr. Peter. Division of Enabling Metrology, National Physical Laboratory, Teddington, Middlesex, United Kingdom. CURRENT COMPARATOR.

Kleinsmith, Prof. Lewis J. Division of Biological Sciences, University of Michigan. BIOLOGICAL SPECIFICITY.

Klemens, Dr. Paul G. Department of Physics, University of Connecticut. LATTICE VIBRATIONS; PHONON; UMKLAPP PROCESS.

Klepper, Dr. David Lloyd. Klepper Marshall King Associates, Inc., White Plains, New York. SOUND-REINFORCEMENT SYSTEM.

Kleppner, Prof. Daniel. Department of Physics, Massachusetts Institute of Technology, Cambridge. PHOTON.

Klich, Dr. Maren. Microbiologist, Food and Feed Safety Research, U.S. Department of Agriculture–Agricultural Research Service, New Orleans, Louisiana. MYCOTOXIN.

Klick, Dr. Clifford C. Deceased; formerly, Superintendent, Solid State Division, U.S. Naval Research Laboratory. FRANCK-CONDON PRINCIPLE; LUMINESCENCE; PHOSPHORESCENCE; other articles.

Kliewer, Dr. K. L. Argonne National Laboratory, Argonne, Illinois. PERPETUAL MOTION.

Klinger, Dr. Terry. Friday Harbor Laboratories, Friday Harbor, Washington. MARINE CONSERVATION.

Kloch, Dr. Harold F. Department of Electrical and Computer Engineering, Ohio University. FEEDBACK CIRCUIT; PHASE INVERTER; PHASE-SHIFT AMPLIFIER; POWER AMPLIFIER; other articles.

Klutke, Dr. Georgia-Ann. Associate Professor/Director, Institute for Manufacturing Systems,Texas A&M University, College Station. MAINTENANCE, INDUSTRIAL AND PRODUCTION.

Klymkowsky, Dr. Michael. Department of Molecular, Cellular, and Developmental Biology, University of Colorado, Boulder. CELL ADHESION.

Knapp, Prof. Anthony W. Department of Mathematics, State University of New York, Stony Brook. LIE GROUP; MANIFOLD (MATHEMATICS).

Knapp, Prof. Gillian R. Department of Astrophysical Sciences, Princeton University, New Jersey. SLOAN DIGITAL SKY SURVEY.

Knauss, Dr. John A. Graduate School of Oceanography, Consulting Editor for Oceanography, University of Rhode Island, Narragansett. OCEAN; OCEAN CIRCULATION.

Knebelman, Prof. Morris S. Professor of Mathematics, Bucknell University. COORDINATE SYSTEMS; DIFFERENTIAL GEOMETRY.

Kneipp, Dr. Katrin. Wellman Center for Photomedicine, Harvard-MIT Division of Health Sciences and Technology, Boston, Massachusetts. RAMAN EFFECT.

Knight, Dr. R. B. D. Retired; formerly, Division of Electrical Science, National Physical Laboratory, Teddington, Middlesex, England. CURRENT MEASUREMENT; MULTIMETER; RESISTOR; VOLTAGE MEASUREMENT; other articles.

Knotek, Dr. M. L. Sandia National Laboratories, Albuquerque, New Mexico. DESORPTION.

Knowles, P. F. Department of Agronomy and Range Science, University of California, Davis. SAFFLOWER.

Knowlton, Dr. Robert. Department of Biological Sciences, George Washington University, Washington, D.C. DECAPODA (CRUSTACEA); GASTROPODA; OLIGOCHAETA.

Kobayashi, Dr. George. Washington University School of Medicine, St. Louis, Missouri. FUNGAL INFECTIONS.

Kobilinsky, Prof. Lawrence. Department of Forensic Sciences, John Jay College of Criminal Justice, City University of New York. REFLECTING MICROSCOPE.

Kocan, Prof. Katherine M. Department of Veterinary Pathology, Oklahoma State University, Stillwater. HEARTWATER DISEASE.

Koch, Prof. Arthur L. Department of Biology, Indiana University. BACTERIAL GROWTH.

Koch, Prof. Peter M. Department of Physics and Astronomy, State University of New York, Stony Brook. ATOM; CHAOS; ISOTOPE SHIFT.

Koerker, Frederick W. Retired; formerly, Technical Expert, Dow Chemical Company, Midland, Michigan. CHLORINE.

Koerner, Dr. David. Department of Physics and Astronomy Northern Arizona University, Flagstaff. HERBIG-HARO OBJECTS; PROTOSTAR; T TAURI STAR.

Kogel, Dr. Jessica Elzea. Thiele Kaolin Company, Sandersville, Georgia. KAOLINITE.

Kogelnik, Dr. Herwig. Crawford Hill Laboratory, Bell Laboratories, Holmdel, New Jersey. REFLECTION OF ELECTROMAGNETIC RADIATION.

Kohl, Arthur L. Woodland Hills, California. HUMIDIFICATION; STRIPPING (CHEMICAL ENGINEERING).

Kohlmeier, Dr. Martin. Department of Nutrition, University of North Carolina at Chapel Hill. VITAMIN K.

Kohls, Dr. Glen M. Retired; formerly, Sanitarian Director, U.S. Public Health Service, Rocky Mountain Laboratory, Hamilton, Montana. IXODIDES.

Kohnhorst, Lloyd L. Retired; formerly, Senior Research Specialist, Rockwell International, Columbus, Ohio. AUTOPILOT.

Koller, Dr. Noemie. Department of Physics, Rutgers University. NUCLEAR MOMENTS.

Kolvoord, Dr. Robert A. College of Integrated Science and Technology, James Madison University, Harrisonburg, Virginia. MULTIMEDIA TECHNOLOGY.

Kominz, Michelle. Lamont-Doherty Geological Observatory, Palisades, New York. BASIN.

Koneman, Dr. Elmer W. Department of Pathology, Denver Veterans Affairs Hospital, University of Colorado Medical School, Denver. CLINICAL PATHOLOGY.

Konigsberg, Dr. William H. Department of Molecular Biophysics, Yale University. IMMUNOCHEMISTRY.

Konopka, Dr. Allan E. Department of Biological Sciences, Purdue University. CYANOBACTERIA.

Konz, Dr. Stephan A. Department of Industrial Engineering, Kansas State University. WORK MEASUREMENT.

Konzo, Prof. Seichi. Department of Mechanical Engineering, University of Illinois. AIR REGISTER; WARM-AIR HEATING SYSTEM.

Koopman, Dr. Karl F. Department of Mammalogy, American Museum of Natural History, New York. CHIROPTERA.

Koral, Dr. Richard L. Visiting Associate Professor, Pratt School of Architecture; Coordinator, Apartment House Institute,

New York City Community College. AIR CONDITIONING; AIR COOLING; DISTRICT HEATING; HUMIDITY CONTROL; RADIANT HEATING; other articles.

Koren, Dr. Gideon. Division of Clinical Pharmacology/Toxicology, The Hospital for Sick Children, University of Western Ontario, University of Toronto, Canada. FETAL ALCOHOL SPECTRUM DISORDER.

Korn, Prof. Robert. Biology Department, Bellarmine University, Louisville, Kentucky. CHIMERA.

Korpel, Prof. Adrian. Department of Electrical and Computer Engineering, University of Iowa. ACOUSTOOPTICS.

Korth, Dr. William W. Rochester Institute of Vertebrate Paleontology, Penfield, New York. RODENTIA.

Korus, Dr. Roger A. Chairperson, Department of Chemical Engineering, College of Engineering, Buchanan Engineering Laboratory, University of Idaho. INDUSTRIAL MICROBIOLOGY.

Kosikowski, Prof. Frank V. Department of Food Science, Cornell University. CHEESE.

Kosow, Dr. Irving L. Series Editor, Electrical Engineering Technology, John Wiley & Sons, Inc., Marietta, Georgia. REPULSION MOTOR; UNIVERSAL MOTOR.

Kosterlitz, Prof. John M. Department of Physics, Brown University. CRYSTAL DEFECTS.

Kotter, Dr. F. Ralph. National Institute of Standards and Technology. CAPACITANCE MEASUREMENT.

Kovar, Dr. Dennis G. Hahn-Meitner-Institut für Kernforschung, Berlin, Germany. NUCLEAR RADIATION; NUCLEAR REACTION.

Kowalczykowski, Dr. Stephen. Professor of Microbiology and of Molecular & Cellular Biology, University of California, Davis. DNA HELICASE.

Kowalewski, Dr. Michael. Department of Geosciences, Virginia Polytechnic University, Blacksburg. OICHNUS; TIME-AVERAGING (PALEONTOLOGY).

Kowalski, Prof. Bruce. Department of Chemistry, University of Washington, Seattle. CHEMOMETRICS.

Kozloff, Dr. Eugene. Friday Harbor Laboratories, University of Washington. DICYEMIDA (RHOMBOZOA).

Kramar, Dr. Ernst. Standard Elektrik Lorenz Aktiengesell-schaft, Stuttgart, Germany. DOPPLER VOR.

Kramer, Frank. Retired; formerly, General Foods, Briarcliff Manor, New York. COFFEE.

Krasan, Dr. Graham P. Lecturer in Pediatrics and Communicable Diseases, University of Michigan Medical School, Ann Arbor. HAEMOPHILUS.

Krauss, Dr. George. Colorado School of Mines, Golden. STEEL.

Krauss, Dr. Lawrence M. Department of Physics, Case Western Reserve University, Cleveland, Ohio. INFLATIONARY UNIVERSE COSMOLOGY.

Kreek, Dr. Mary Jeanne. Department of Neuroscience, Rockefeller University. ADDICTIVE DISORDERS.

Krefeld, Prof. William J. Deceased; formerly, Professor of Civil Engineering, Columbia University. HARDNESS SCALES; HOOKE'S LAW; RING; YOUNG'S MODULUS.

Kreilik, Dr. Robert. Department of Chemistry, University of Rochester, New York. SPIN LABEL.

Kretschmann, Dr. David E. Forest Products Laboratory, USDA Forest Service, Madison, Wisconsin. LUMBER.

Kreutel, Dr. Randall W. Mechanical Engineering Manager, Atlanta Instrumentation Division, Atlanta, Georgia. ELECTRICAL NOISE; ELECTRICAL NOISE GENERATOR.

Krider, Prof. E. Philip. Director, Institute of Atmospheric Physics, University of Arizona. LIGHTNING.

Krishnamurti, Dr. T. N. Department of Meteorology, Florida State University. TROPICAL METEOROLOGY.

Kristensen, Dr. Reinhardt Mobjerg. Zoological Museum, University of Copenhagen, Denmark. CYCLIOPHORA.

Kristofferson, Karl E. Deceased; formerly, National Aeronautics and Space Administration, John F. Kennedy Space Center, Florida. LAUNCH COMPLEX.

Krol, Dr. Edward. Department of Computer Science, University of Illinois, Urbana. ELECTRONIC MAIL.

Krombein, Dr. Karl V. Chairperson, Department of Entomology, National Museum of Natural History, Smithsonian Institution, Washington, D.C. HYMENOPTERA.

Kropinski, Dr. Mary Catherine. Department of Mathematics, Simon Fraser University, Burnaby, British Columbia, Canada. CREEPING FLOW.

Kroto, Prof. Harold W. The School of Chemistry and Molecular Physics, University of Sussex, Falmer, Brighton, England; Nobelist. FULLERENE.

Krumdieck, Dr. S. P. Mechanical Engineering, University of Canterbury, New Zealand. CHEMICAL VAPOR DEPOSITION.

Kruse, Dr. Paul W. Consultant, Infrared Technology, Edina, Minnesota. OPTICAL DETECTORS.

Krusius, Prof. Matti. Low Temperature Laboratory, Helsinki University of Technology, Helsinki, Finland. MAGNETIC THERMOMETER.

Krygier, Dr. John. Department of Geology and Geography Ohio Wesleyan University, Delaware. CONTOUR.

Kryter, Dr. Karl D. Director, Stanford Research Institute, Menlo Park, California. BEL; DECIBEL.

Krzysztofowicz, Dr. Roman. Department of Systems Engineering, University of Virginia. DECISION ANALYSIS.

Kuchner, Dr. Marc J. Exoplanets and Stellar Astrophysics Laboratory, NASA/Goddard Space Flight Center, Greenbelt, Maryland. INTERPLANETARY MATTER.

Kudo, Prof. Albert M. Department of Earth and Planetary Sciences, University of New Mexico, Albuquerque. GABBRO.

Kuenen, Dr. J. Gijs. Laboratory of Microbiology, Delft University of Technology, Netherlands. CHEMOSTAT.

Kuhn, Dr. Wayne E. Professional Engineer, Portland, Oregon. WAX, PETROLEUM.

Kuis, Wietse. Department of Molecular and Experimental Medicine, Research Institute of Scripps Clinic, La Jolla, California. NEUROIMMUNOLOGY.

Kulkarni, Sudhir S. Corporate Research Center, UOP Inc., Des Plaines, Illinois. MEMBRANE DISTILLATION; MEMBRANE SEPARATIONS.

Kullerud, Dr. Gunnar. Department of Geosciences, Purdue University. SULFIDE PHASE EQUILIBRIA.

Kumar, Dr. Nitin. Massachusetts Institute of Technology, Cambridge. SPIDER SILK.

Kuntz, Glenn. Regional Director, ETI of North America, Inc., Washington, D.C. POLYCHLORINATED BIPHENYLS.

Kunz, Dr. Kaiser S. Research Professor of Physics, New Mexico State University. CURVE FITTING; EXTRAPOLATION; INTERPOLATION.

Kunzendorf, Dr. Robert G. Department of Psychology, University of Lowell, Massachusetts. CONSCIOUSNESS.

Kuo, Dr. Benjamin C. Formerly, Department of Electrical Engineering, University of Illinois, Champaign. STEPPING MOTOR.

Kuo, Prof. John T. Department of Mining Engineering, Columbia University. EARTH TIDES.

Kupper, Dr. Walter E. Manager, Technical Marketing Services, Mettler Instrument Corporation, Hightstown, New Jersey. BALANCE.

Kusch, Prof. Polykarp. Deceased; formerly, Department of Physics, University of Texas, Dallas; Nobelist. ATOMIC BEAMS.

Kushner, Dr. D. J. Department of Microbiology, University of Toronto, Ontario, Canada. HALOPHILISM (MICROBIOLOGY).

Kutal, Dr. Charles. Department of Chemistry, University of Georgia. TUNGSTEN.

Kutzbach, Prof. John E. Department of Meteorology, University of Wisconsin. CLIMATE HISTORY.

Kvale, Dr. Erik P. Indiana Geological Survey, Indiana University, Bloomington. TIDALITES.

Kwok, Dr. Sun. Faculty of Science, The University of Hong Kong, Pokfulam. NEBULA.

L

Laarhoven, Prof. William H. Department of Organic Chemistry, University of Nijmegen, Netherlands. ORGANIC PHOTOCHEMISTRY.

Labandeira, Dr. Conrad C. Department of Paleobiology, National Museum of Natural History, Smithsonian Institution, Washington, D.C. INSECTA.

LaBerge, Dr. Stephen. Department of Psychology, Stanford University, Stanford, California. SLEEP AND DREAMING.

Labi, Prof. Samuel. School of Civil Engineering, Purdue University, West Lafayette, Indiana. HIGHWAY ENGINEERING.

Labotka, Dr. Theodore C. Department of Geological Sciences, College of Liberal Arts, University of Tennessee. CHLORITOID.

Labuza, Prof. T. P. Department of Food Science and Nutrition, University of Minnesota. FOOD MANUFACTURING.

Lacis, Dr. Andrew A. NASA Goddard Institute for Space Studies, New York. GREENHOUSE EFFECT.

Lackey, Dr. James B. Emeritus Professor, Department of Environmental Engineering, University of Florida, Gainesville. BICOSOECIDA; CHLOROMONADIDA; DIPLOMONADIDA; EUGLENIDA; SARCOMASTIGOPHORA; VOLVOCIDA; other articles.

Ladd, Dr. Harry S. Retired; formerly, National Museum, Smithsonian Institution, Washington, D.C. ATOLL.

LaFrey, Raymond R. Retired; formerly, Lincoln Laboratory, Massachusetts Institute of Technology, Lexington. AIRCRAFT COLLISION AVOIDANCE SYSTEM; SURVEILLANCE RADAR.

Lagowski, Dr. J. J. Department of Chemistry, University of Texas, Austin. SOLVENT.

Lagunoff, Dr. David. Department of Pathology, School of Medicine, University of Washington. INFLAMMATION.

Lahiri, Prof. Sukhamay. Department of Physiology, School of Medicine, University of Pennsylvania. AORTIC BODY.

Laidler, Dr. James J. Director, Chemical Technology Division, Argonne National Laboratory, Illinois. NUCLEAR FUELS REPROCESSING.

Laitinen, Prof. Herbert A. Deceased; formerly, Department of Chemistry, University of Florida, Gainesville. ANALYTICAL CHEMISTRY; ELECTROCHEMISTRY; ELECTROLYTIC CONDUCTANCE.

Lam, Dr. Stephen W. Center for Excellence for Document Analysis and Recognition, University of Buffalo, Amherst, New York. CHARACTER RECOGNITION.

Lamb, Prof. Dennis. Department of Meteorology, Pennsylvania State University. ACID RAIN.

Lamb, Prof. Richard C. Space Radiation Laboratory, California Institute of Technology, Pasadena. CERENKOV RADIATION.

Lambert, Dr. Jack L. Department of Chemistry, Kansas State University. QUALITATIVE CHEMICAL ANALYSIS.

Lambert, Dr. Samuel. Air Force Research Laboratory, Munition Directorate, Eglin Air Force Base, Florida. AIR ARMAMENT.

Lambert, Dr. W. David. Louisiana School of Math, Science, and the Arts, Natchitoches. PROBOSCIDEA.

Lamotte, Dr. Clifford E. Botany Department, Iowa State University. PLANT PROPAGATION.

Lamotte, Dr. Robert. Department of Anesthesiology, School of Medicine, Yale University. CUTANEOUS SENSATION; PROPRIOCEPTION; SOMESTHESIS.

Lan, Dr. Esther H. Department of Materials Science and Engineering, University of California, Los Angeles. SOL-GEL PROCESS.

Land, Dr. Cecil E. Sandia National Laboratories, Albuquerque, New Mexico. PHOTOFERROELECTRIC IMAGING.

Landes, Dr. Hugh S. Department of Electrical Engineering, University of Virginia. RECIPROCITY PRINCIPLE.

Landsberg, Dr. H. E. Deceased; formerly, Institute for Physical Science and Technology, University of Maryland, College Park. BIOMETEOROLOGY.

Lane, Dr. John C. Research and Development Department, Research Laboratories, Ethyl Corporation, Ferndale, Michigan. OCTANE NUMBER.

Lane, Dr. Malcolm G. Department of Computer Science, James Madison University, Harrisonburg, Virginia. COMPUTER PERIPHERAL DEVICES.

Lane, Dr. Meredith A. Director, University of Kansas Herbarium, Lawrence. PLANT; SPECIES CONCEPT.

Lane, Dr. N. Gary. Deceased; formerly, Professor of Paleontology, Indiana University. CAMERATA.

Lanford, Dr. Pamela. Life Sciences, University of Maryland. SENSORY CELL REGENERATION.

Lanford, Prof. William A. Department of Physics, State University of New York, Albany. IONIZATION CHAMBER.

Lang, Dr. Anton. Department of Botany, MSU-DOE Plant Research Laboratory, Michigan State University. PLANT PHYSIOLOGY.

Lang, Dr. Karl. Director, Department of Invertebrates, Swedish State Museum of Natural History, Stockholm. HARPACTOIDA; TANAIDACEA.

Lang, Prof. Kenneth. Department of Physics and Astronomy, Tufts University, Medford, Massachusetts. SUPERGRANULATION.

Langenberg, Prof. D. N. Department of Physics, University of Pennsylvania. SUPERCONDUCTIVITY.

Langenheim, Dr. Jean H. Professor Emeritus, Biology Department, University of California, Santa Cruz. AMBER.

Langford, Prof. George M. Department of Biological Sciences, Dartmouth University. VIDEO MICROSCOPY.

Lansana, Prof. Florence. Department of Geology and Geography, Hunter College, New York. AFRICA.

Lapedes, Daniel N. Deceased; formerly, Editor in Chief, *McGraw-Hill Encyclopedia of Science and Technology*, McGraw-Hill, Inc., New York. AMYLASE; INFECTION; LACTASE; PATHOGEN.

Laport, Edmund A. Director, Communications Engineering, RCA Corp., Princeton, New Jersey. AMPLITUDE-MODULATION RADIO.

Lappin-Scott, Hilary. Department of Biology, University of Calgary, Alberta, Canada. BIOFILM.

Lapple, Charles E. Deceased; formerly, Consultant, Fluid and Particle Technology, Air Pollution and Chemical Engineering, Los Altos, California. DUST AND MIST COLLECTION; PARTICULATES; PIPELINE DESIGN.

Larimer, Dr. J. L. Department of Zoology, University of Texas, Austin. REFLEX.

Larive, Dr. Cynthia K. Department of Chemistry, University of Kansas, Lawrence. NUCLEAR MAGNETIC RESONANCE (NMR).

Larsen, Dr. Alberta M. Technical Division, Next Linear Collider Department, Stanford Linear Accelerator Center, Stanford University. SUPERCONDUCTING DEVICES.

Larsen, Dr. Clark Spencer. Distinguished Professor and Chair, Ohio State University, Columbus. BIOARCHEOLOGY.

Larsen, Prof. Edwin M. Department of Chemistry, University of Wisconsin. GALLIUM; INDIUM; NIOBIUM; TANTALUM; THALLIUM.

Larsen, Dr. R. Paul. Superintendent and Horticulturist, Tree Fruit Research Center, Washington State University. CHERRY; FRUIT, TREE; PLUM.

Larsen, Dr. Robert G. Assistant to the Vice President, Shell Development Company, Emeryville, California. LUBRICANT.

Larsen, Dr. Sandra A. Centers for Disease Control and Prevention, Sexually Transmitted Disease Laboratory Program, Atlanta, Georgia. GONORRHEA; SYPHILIS.

Larson, Dr. Allan. Department of Biology, Washington University, St. Louis, Missouri. MACROEVOLUTION.

Larson, Dr. Roy A. Department of Horticulture, North Carolina State University. FLORICULTURE.

Lass, Dr. Harry. Research Specialist, Jet Propulsion Laboratory, California Institute of Technology, Pasadena. CALCULUS OF VECTORS; TENSOR ANALYSIS.

Lassila, Prof. Kenneth E. Department of Physics, Iowa State University, Ames. PROPAGATOR (FIELD THEORY); QUANTUM (PHYSICS); QUANTUM NUMBERS.

Last, Prof. John M. Faculty of Health Sciences, School of Medicine/Epidemiology and Community Medicine, University of Ottawa, Ontario, Canada. PUBLIC HEALTH.

Latanision, Prof. R. M. H. H. Ublig Corrosion Laboratory, Department of Materials Science and Engineering, Massachusetts Institute of Technology, Cambridge. CORROSION.

Latchman, Prof. Haniph A. Department of Electrical and Computer Engineering, University of Florida, Gainesville. POWER LINE COMMUNICATION.

Latham, Dr. David W. Harvard-Smithsonian Center for Astrophysics, Cambridge, Massachusetts. ALPHA CENTAURI; ARCTURUS; MIZAR; POLARIS; PROCYON; RIGEL; SIRIUS; SPICA.

Lathi, Prof. B. P. Department of Electrical and Electronic Engineering, California State University, Sacramento. BANDWIDTH REQUIREMENTS (COMMUNICATIONS).

Lattman, Dr. Laurence H. Dean of Mines, University of Utah, Salt Lake City. AERIAL PHOTOGRAPHY.

Lauffer, Dr. Max A. Department of Biophysics and Microbiology, University of Pittsburgh. BIOPHYSICS.

Laurent, Dr. Philippe. Service d'Astrophysique, Centre d'Etudes de Saclay, France. GAMMA-RAY ASTRONOMY.

Laves, Dr. Fritz H. Deceased; formerly, Eidg. Technische Hochschule, Institut für Kristallographie und Petrographie, Zurich, Switzerland. ANORTHOCLASE.

Lavin, Prof. Stephen. Department of Geography, University of Nebraska, Lincoln. TERRESTRIAL COORDINATE SYSTEM.

Lavoie, Dr. Ronald L. Chief, Program Requirements and Development Division, U.S. Department of Commerce, NOAA, Silver Spring, Maryland. WEATHER OBSERVATIONS.

Lavrentovich, Prof. Oleg. Liquid Crystals Institute and Chemical Physics Interdisciplinary Program, Kent State University, Ohio. LIQUID CRYSTALS.

Lawley, Dr. Alan. Department of Materials Engineering. Drexel University, Philadelphia, Pennsylvania. POWDER METALLURGY; ZONE REFINING.

Lawrence, Dr. J. Dennis. Lawrence Livermore National Laboratory, University of California, Livermore. PLANE CURVE.

Layer, David H. Director, Advanced Engineering, Science and Technology Department, National Association of Broadcasters, Washington, D.C. FREQUENCY-MODULATION RADIO; RADIO; STEREOPHONIC RADIO TRANSMISSION; RADIO BROADCASTING NETWORK.

Layton, Dr. J. Preston. Deceased; formerly, RCA AstroElectronics, Conceptual Design, Princeton, New Jersey. SPECIFIC FUEL CONSUMPTION; SPECIFIC IMPULSE; THRUST.

Layzer, Prof. David. Professor Emeritus; Department of Astronomy, Harvard University. GRAVITATIONAL REDSHIFT; REDSHIFT.

Lazarides, Dr. Elias. Department of Biology, California Institute of Technology, Pasadena. CYTOSKELETON.

Lazaruk, Kenneth, H. Shea & Gould, New York. CIVIL ENGINEERING.

Leach, Dr. Richard E. Fellow, Reproductive Endocrinology and Fertility, Department of Obstetrics and Gynecology, Mayo Graduate School of Medicine, Rochester, Minnesota. INFERTILITY.

Leach, Prof. William Marshall, Jr. School of Electrical and Computer Engineering, Georgia Institute of Technology, Atlanta. LOUDSPEAKER; VOLUME CONTROL SYSTEMS; VOLUME UNIT (UV).

Leadbetter, Prof. E. R. Department of Biology, Amherst College. BACTERIAL PHYSIOLOGY AND METABOLISM.

Leader, Dr. Robert W. Center for Environmental Toxicology, Department of Pathology, Michigan State University, East Lansing. TOXICOLOGY.

Leaist, Dr. Derek G. Department of Chemistry, University of Western Ontario, London, Ontario, Canada. DIFFUSION.

Learned, Prof. John G. Department of Physics, University of Hawaii, Honolulu. NEUTRINO ASTRONOMY.

Lebowitz, Dr. Michael. College of Public Health, University of Arizona, Tucson. AIR POLLUTION, INDOOR.

LeBuhn, Dr. Gretchen. Department of Biological Sciences, Florida State University, Tallahassee. POPULATION VIABILITY.

Ledig, Dr. F. Thomas. Senior Scientist, Institute of Forest Genetics, U.S. Department of Agriculture, Berkeley, California. FOREST GENETICS; TREE.

Lee, Prof. David M. Department of Physics, Cornell University; Nobelist. ABSOLUTE ZERO.

Lee, Dorothy B. Lyndon B. Johnson Space Center, National Aeronautics and Space Administration, Houston, Texas. ATMOSPHERIC ENTRY.

Lee, Dr. J. T. Cooperative Institute for Mesoscale Meterological Studies, National Oceanic and Atmospheric Administration, University of Oklahoma. AERONAUTICAL METEOROLOGY.

Lee, Lin-Nan. Communications Satellite Corporation, Clarksburg, Maryland. COMMUNICATIONS SCRAMBLING.

Lee, Lyndon C. L. C. Lee & Associates, Inc., Seattle, Washington. WETLANDS.

Lee, Dr. Roberto. Monsanto Company, St. Louis, Missouri. HYDROGENATION.

Lee, Prof. Steven B. Director, Forensic Science, Justice Studies Department, San Jose University, California. FORENSIC BIOLOGY.

Lee, Sunguk. Department of Electrical and Computer Engineering, University of Florida, Gainesville. POWER LINE COMMUNICATION.

Lee, Prof. T. H. Centre for Intelligent Control, Department of Electrical Engineering, National University of Singapore. EXPERT CONTROL SYSTEM.

Lee, Dr. Thomas H. Deceased; formerly, Strategic Planning Operation, General Electric Company, Fairfield, Connecticut. BLOWOUT COIL; CIRCUIT BREAKER; ELECTRIC CONTACT; ELECTRIC SWITCH.

Lee, Dr. W. John. Professor and L. F. Peterson Chair, Department of Petroleum Engineering, Texas A&M University, College Station. PETROLEUM RESERVOIR ENGINEERING.

Leete, Prof. Edward. Deceased; formerly, Department of Chemistry, University of Minnesota. NICOTINE ALKALOIDS.

Lefkowitz, Prof. Irving. Cleveland Heights, Ohio. MULTILEVEL CONTROL THEORY.

Leggett, Prof. Anthony. Department of Physics, University of Illinois, Urbana; Nobelist. QUANTUM THEORY OF MEASUREMENT.

Lehiste, Prof. Ilse. Department of Linguistics, Ohio State University. PHONETICS.

Lehmann, Prof. Justus F. Department of Rehabilitation Medicine, School of Medicine, University of Washington, Seattle. THERMOTHERAPY.

Lehn, Prof. Jean-Marie. Collège de France, Paris, and Institut de Science et Ingénierie Supramoléculaires–Université Louis Pasteur, Strasbourg, France; Nobelist. SUPRAMOLECULAR CHEMISTRY.

Lehner, Dr. Paul E. Chief Scientist, Information Systems and Technology Division, MITRE Corporation, McLean, Virginia. AUTOMATED DECISION MAKING.

Lehto, Dr. Mark R. School of Industrial Engineering, Purdue University, West Lafayette, Indiana. INDUSTRIAL HEALTH AND SAFETY.

Leibovich, Prof. Sidney. Samuel B. Eckert Professor of Mechanical and Aerospace Engineering, Cornell University, Ithaca, New York. VORTEX.

Leigh, Prof. Hoyle. Department of Psychiatry, University of California, San Francisco. PSYCHOSOMATIC DISORDERS.

Leland, Dr. Henry. Department of Psychology, Ohio State University, Columbus. DOWN SYNDROME.

Leland, Dr. Jonathan K. IGEN, Inc., Rockville, Maryland. CHEMILUMINESCENCE.

Lemert, Prof. Richard. Chemical Engineering Department, University of Toledo, Ohio. SUPERCRITICAL FLUID.

Lemont, Prof. Harvey. Chairperson, Department of Podiatric Medicine, Pennsylvania College of Podiatric Medicine, Philadelphia. FOOT DISORDERS.

Lenel, Dr. F. V. Professor Emeritus, Department of Materials Engineering, Rensselaer Polytechnic Institute. SINTERING.

Lengauer, Dr. Christoph. School of Medicine, Johns Hopkins University, Baltimore, Maryland. DNA METHYLATION.

Le Noble, Prof. William J. Department of Chemistry, State University of New York, Stony Brook. PHYSICAL ORGANIC CHEMISTRY; RESONANCE (MOLECULAR STRUCTURE).

Lentz, Dr. Steven R. Professor of Internal Medicine, University of Iowa, Iowa City. HOMOCYSTEINE.

Leon, Prof. Roberto T. Roswell, Georgia. COMPOSITE BEAM.

Leonard, Dr. Edward F. Chemical Engineering Department, Columbia University. PROCESS ENGINEERING.

Leontis, Dr. Thomas E. Magnesium Research Center, Batelle Columbus Laboratories, Columbus, Ohio. MAGNESIUM ALLOYS.

Leopold, Dr. Luna B. Department of Geology and Geophysics, University of California, Berkeley. STREAM TRANSPORT AND DEPOSITION; SWAMP, MARSH, AND BOG.

Lerner, Dr. I. Michael. Department of Genetics, University of California, Berkeley. GENETIC HOMEOSTASIS.

Lerner-Lam, Dr. Art. Lamont-Doherty Geological Observatory of Columbia University, Palisades, New York. SEISMOLOGY.

Lesins, Dr. Glen. Department of Physics & Atmospheric Science, Dalhousie University, Halifax, Nova Scotia. ATMOSPHERE.

Lesk, Dr. I. A. Vice President, Technical Staff, Motorola Inc., Phoenix, Arizona. VARISTOR.

Lesso, Dr. William G. Department of Mechanical Engineering. University of Texas, Austin. OPERATIONS RESEARCH.

Leucken, John J. Monsanto Company, Akron, Ohio. RUBBER.

Levasheff, V. V. American Potash and Chemical Corporation, Whittier, California. BORON.

Levenson, Dr. Harvey. Graphic Communication, California Polytechnic State University, San Luis Obispo, California. PRINTING.

Leventis, Dr. Nicholas. Department of Chemistry, University of Missouri, Rolla. ELECTROCHROMIC DEVICES.

Levi, Prof. Herbert W. Curator in Arachnology, Museum of Comparative Zoology, Agassiz Museum, Harvard University, Cambridge, Massachusetts. AMBLYPYGI; ARACHNIDA; OPILIONES; PHALANGIDA.

Levin, Prof. Simon A. Professor of Applied Mathematics and Ecology, Division of Biological Sciences, Cornell University, New York. MATHEMATICAL ECOLOGY.

Levine, Prof. Ira N. Department of Chemistry, Brooklyn College. ACTIVITY (THERMODYNAMICS); AVOGADRO NUMBER; BOILING POINT; LE CHATELIER'S PRINCIPLE.

Levine, Dr. Michael. Department of Psychology, The University of Illinois at Chicago. SENSATION.

Levine, Dr. Norman D. College of Veterinary Medicine, University of Illinois, Urbana. ARCHIGREGARINIDA; EUCOCCIDA; NEOGREGARINIDA; TELOSPOREA; other articles.

Levine, Seymour. Retired; formerly, Northrop Corporation, Hawthorne, California. STAR TRACKER.

Levine, T. E. Chevron Research Company, Richmond, California. OXIDATION PROCESS.

Levine, Prof. William S. Department of Electrical and Computer Engineering, University of Maryland, College Park. HYBRID CONTROL.

Levitan, Dr. Max. Department of Cell Biology, Mount Sinai School of Medicine, New York. SEX DETERMINATION.

Levy, Dr. David H. Jarnac Observatory, Vail, Arizona. MESSIER CATALOG.

Lew, Dr. I. Paul. Thornton-Tomasetti, Engineers, New York. CANTILEVER; GEODESIC DOME; TRUSS.

Lewandowski, Prof. John J. Department of Materials Science and Engineering, Case Western Reserve University, Cleveland, Ohio. BRITTLENESS.

Lewandowski, Dr. Scott M. Department of Computer Science, Brown University, Providence, Rhode Island. CLIENT-SERVER SYSTEM.

Lewis, Dr. Alcinda C. Institute of Ecosystem Studies, New York Botanical Garden, Millbrook. CHEMICAL ECOLOGY.

Lewis, Dr. D. J. Department of Mathematics, University of Michigan. FIELD THEORY (MATHEMATICS).

Lewis, Prof. Edward B. Department of Biology, California Institute of Technology, Pasadena; Nobelist. ALLELE.

Lewis, Dr. Frank L. School of Electrical Engineering, Georgia Institute of Technology, Atlanta. OPTIMAL CONTROL (LINEAR SYSTEMS).

Lewis, Dr. James K. Department of Animal Science, South Dakota State University, Brookings. BLUESTEM GRASS.

Lewis, Dr. John H. Pratt & Whitney, East Hartford, Connecticut. GAS TURBINE.

Lewis, Prof. M. J. Department of Food Science and Technology, University of California, Davis. MALT BEVERAGE.

Lewis, Dr. Richard S. Department of Psychology, Pomona College, Claremont, California. AMNESIA.

Ley, Dr. Willy. Deceased; formerly, Fairleigh Dickinson University. BALLISTICS.

Lhermitte, Dr. Roger M. Professor of Physical Meterology, University of Miami School of Marine and Atmospheric Science. DOPPLER RADAR.

Li, Prof. Chao-Jun. Canada Research Chair in Green Chemistry, Department of Chemistry, McGill University, Québec, Canada. ATOM ECONOMY.

Li, Dr. Ching Chun. Graduate School of Public Health, University of Pittsburgh. BIOMETRICS; HUMAN GENETICS; POPULATION GENETICS.

Li, Prof. Choh Hao. Deceased; formerly, Laboratory of Molecular Endocrinology, University of California. ADENOHYPOPHYSIS HORMONE; ALDOSTERONE; EPINEPHRINE; PANCREAS; other articles.

Li, Dr. Norman N. President, NL Chemical Technology, Inc., Mount Prospect, Illinois. ION EXCHANGE; MEMBRANE DISTILLATION; MEMBRANE SEPARATIONS.

Lichtenberg, Dr. Don B. Department of Physics, Indiana University, Bloomington. MAGNETON.

Liddicoat, Richard T., Jr. Gemological Institute of America, Los Angeles, California. AGATE; CARAT; GEM; JADE; PEARL; other articles.

Lieberman, Prof. A. Department of Landscape Architecture, Cornell University, Ithaca, New York. LAND-USE CLASSES.

Lieberman, Dr. Bruce. Department of Geology, University of Kansas, Lawrence. OLENELLINA.

Liebman, Dr. Joel F. Department of Chemistry and Biochemistry, University of Maryland, Baltimore. HYDRIDE.

Liedholm, George E. Retired; formerly, Department Head, Petroleum Processing, Shell Development Company, Emeryville, California. ISOMERIZATION.

Lieth, Dr. Helmut. Department of Botany, University of North Carolina. SAVANNA.

Likens, Dr. Gene E. Director, Institute of Ecosystem Studies, Millbrook, New York. MEROMICTIC LAKE.

Liley, Prof. Peter E. Lafayette, Indiana. REFRIGERATION; REFRIGERATION CYCLE.

Limpel, Eugene J. Consulting Electrical Engineer, Eagle River, Wisconsin. FLASH WELDING; SPOT WELDING.

Lin, Prof. Shao-Chi. Emeritus, Department of Mechanical and Aerospace Engineering, University of California, San Diego. SHOCK WAVE.

Lin, Dr. Yu-Ju. Computer and Information Science, Charleston Southern University, South Carolina. POWER LINE COMMUNICATION.

Linde, Dr. Ronald K. President, Envirodyne, Inc., Los Angeles, California. HIGH-PRESSURE PHYSICS.

Linden, Prof. Karl G. Department of Civil and Environmental Engineering, Duke University, Durham, North Carolina. WATER SUPPLY ENGINEERING; WATER TREATMENT.

Lindenmaier, Dr. Werner A. Hoffmann-LaRoche, Inc., Nutley, New Jersey. ASCORBIC ACID.

Linderoth, Prof. L. Sigfred, Jr. Department of Mechanical Engineering, Duke University. FLYWHEEL; SHOCK ABSORBER; SPRING (MACHINES); other articles.

Lindner, Dr. Hans R. Department of Biodynamics, Weizmann Institute of Science, Rehovot, Israel. TESTIS.

Lindroth, Dr. Eva. Atomic Physics, Stockholm University, Sweden. ATOMIC THEORY.

Lindsay, Prof. R. Bruce. Deceased; formerly, Hazard Professor of Physics, Emeritus, Brown University. ACOUSTICS; SOUND ABSORPTION.

Lindsey, J. S. Liquid Carbonic Corporation, Chicago, Illinois. CARBON DIOXIDE.

Lindstrom, Dr. Kirk W. Los Altos, California. FIBER-OPTIC CIRCUIT.

Lindzen, Prof. Richard S. Center for Meteorology and Physical Oceanography,

Massachusetts Institute of Technology, Cambridge. ATMOSPHERIC TIDES.

Lineback, Dr. David R. Department of Food Science, North Carolina State University. FOOD MANUFACTURING; STARCH.

Lineberger, Prof. Carl. Joint Institute for Laboratory Astrophysics, National Institute for Standards and Technology, University of Colorado. ELECTRON AFFINITY.

Lines, Malcolm E. Bell Telephone Laboratories, Murray Hill, New Jersey. MAGNETIC FERROELECTRICS.

Linscum, Prof. Emmanuel. University of Missouri-Columbia. PHOTOMORPHOGENESIS.

Linskey, Dr. Jeffrey. Department of Astrophysics, University of Colorado, Boulder. CHROMOSPHERE.

Linsley, Prof. Ray K. Department of Civil Engineering, Stanford University. AQUIFER; GEYSER; WATER TABLE; other articles.

Linzey, Dr. Donald. Wytheville Community College, Virginia. AARDVARK; KANGAROO; MONGOOSE; PORCUPINE; SQUIRREL; other articles.

Liou, Dr. J. G. Department of Earth and Planetary Sciences, Tokyo Institute of Technology, Japan. CALCITE; GLAUCOPHANE; OPHIOLITE; TALC.

Liou, Prof. Kuo-Nan. Department Chair, Department of Atmospheric Sciences, University of California, Los Angeles. HEAT BALANCE, TERRESTRIAL ATMOSPHERIC; TERRESTRIAL RADIATION.

Lipkens, Dr. Bart. Assistant Professor, Department of Mechanical Engineering, Virginia Commonwealth University, Richmond. SIREN (ACOUSTICS).

Lipkowitz, Prof. Kenneth B. Department of Chemistry, Indiana-Purdue University, Indianapolis. MOLECULAR MECHANICS.

Lipner, Prof. Harry. Department of Biological Science, Florida State University. LACTATION.

Lipschutz, Prof. Michael E. Department of Chemistry, Purdue University, West Lafayette, Indiana. METEORITE.

Lipscomb, Prof. Diana. Associate Dean for Faculty and Research Professor Biological Sciences, George Washington University, Washington, D.C. EUKARYOTAE.

Lipshutz, Prof. Bruce H. Department of Chemistry, University of California, Santa Barbara. ORGANOMETALLIC COMPOUND.

Lister, Prof. C. J. A. W. Wright Nuclear Structure Laboratory, Physics Department, Yale University, New Haven, Connecticut. MAGIC NUMBERS; NUCLEAR STRUCTURE.

Litherland, Prof. Albert E. Director, Isotrace Laboratory, Department of Physics, University of Toronto, Ontario, Canada. ACCELERATOR MASS SPECTROMETRY.

Little, Elbert L., Jr. Dendrologist, Forest Science, U.S. Department of Agriculture. ALDER; DENDROLOGY; DOUGLAS-FIR; EUCALYPTUS.

Little, Dr. James N. Research Division, Waters Associates, Milford, Massachusetts. POLARIMETRIC ANALYSIS; REFRACTOMETRIC ANALYSIS.

Liu, Benjamin Y. H. Department of Mechanical Engineering, University of Minnesota. RESPIRATOR.

Liu, Prof. Clark C. K. Department of Civil Engineering, University of Hawaii at Manoa. WATER DESALINATION; WASTEWATER REUSE.

Liu, Dr. S. H. Solid State Division, Oak Ridge National Laboratory, Oak Ridge, Tennessee. DENSITY MATRIX; FRACTALS.

Liu, Dr. Tanzhuo. Lamont-Doherty Earth Observatory of Columbia University, Palisades, New York. ROCK VARNISH.

Livingston, Gideon E. Food Science Associates, Rye, New York. VANILLA.

Livingston, Dr. William C. National Solar Observatory, National Optical Astronomy Observatories,Tucson, Arizona. FRAUNHOFER LINES.

Lochhead, Dr. John H. London, England. CEPHALOCARIDA; SPELAEOGRIPHACEA.

LoCicero, Prof. Joseph L. Illinois Institute of Technology, Electrical and Computer Engineering Department, Chicago. TELEVISION STANDARDS.

Lockhart, Prof. Frank J. Department of Chemical Engineering, University of Southern California, Los Angeles. COUNTERCURRENT TRANSFER OPERATIONS.

Loeb, John L. Deceased; formerly, ACLS Corporation, Silver Spring, Maryland. AUTOMATIC LANDING SYSTEM.

Loeblich, Dr. Alfred R., Jr. Department of Earth and Space Sciences, University of California, Los Angeles. NUMMULITES.

Loeblich, Prof. Helen Tappan. Department of Earth and Space Sciences, University of California, Los Angeles. FORAMINIFERIDA.

Loewenstein, Dr. Werner R. College of Physicians and Surgeons, Columbia University. MECHANORECEPTORS.

Loftus, Dr. Douglas J. Laboratory of Cell Biology, National Cancer Institute, Bethesda, Maryland. CELLULAR IMMUNOLOGY.

Loftus, Prof. Elizabeth F. Department of Psychology, University of Washington, Seattle. MEMORY.

Logan, Prof. Earl, Jr. Department of Aerospace and Mechanical Engineering, Arizona State University, Tempe. IMPULSE TURBINE.

Lohse, Prof. Detlef. Faculty of Applied Physics, University of Twente, Enschede, The Netherlands. SONOLUMINESCENCE.

Loneragan, Prof. J. F. School of Environmental and Life Science, Murdoch University, Perth, Western Australia. PLANT MINERAL NUTRITION.

Long, Prof. Franklin A. Formerly, Department of Chemistry, Cornell University. ACID AND BASE.

Long, Joseph B. Tin Research Institute, Inc., Columbus, Ohio. TIN.

Longacre, Alan. Fluor Engineers and Constructors, Irvine, California. FERTILIZER.

Longmoore, Dr. Kenneth. Department of Chemistry, University of Massachusetts. QUALITATIVE CHEMICAL ANALYSIS.

Longo, Dr. Dan L. Director, National Cancer Institute, Frederick Cancer Research Facility, Maryland. ONCOLOGY.

Lonngren, Prof. Karl E. Department of Electrical and Computer Engineering, University of Iowa, Iowa City. ARC DISCHARGE; TRANSMISSION LINES.

Lonsdale, Dr. Carol J. Center for Astrophysics and Space Sciences, University of California, San Diego. STARBURST GALAXY.

Lonsdale, Dr. Peter. Scripps Institution of Oceanography, La Jolla, California. MARINE SEDIMENTS; TRANSFORM FAULT.

Loope, Dr. David B. Department of Geology, University of Nebraska Medical Center. DUNE.

Lopiano, Dr. Michael. Chairperson, Public Information Committee, American College of Radiology, Reston, Virginia. MEDICAL IMAGING.

Lopresti, Dr. Philip V. Retired; formerly, Engineering Research Center, AT&T Bell Laboratories, Princeton, New Jersey. BREADBOARDING; CHARACTERISTIC CURVE; COMPUTER-AIDED CIRCUIT DESIGN; ELECTRICAL MODEL; LOGIC CIRCUITS; SENSE AMPLIFIER; SIGMA-DELTA CONVERTER; VOLTAGE AMPLIFIER; WAVE-SHAPING CIRCUITS; WAVEFORM; WAVEFORM GENERATOR.

Lorand, Dr. L. Department of Biochemistry and Molecular Biology, Northwestern University. FIBRINOGEN.

Lord, Prof. Richard C. Deceased; formerly, Department of Chemistry, Massachusetts Institute of Technology. OPTICS; PHYSICAL OPTICS; WAVE OPTICS.

Lordi, George M. Department of Medicine, College of Medicine and Dentistry, Newark, New Jersey. MYCOBACTERIAL DISEASES; TUBERCULOSIS.

Lorenz, Prof. Oscar A. Department of Vegetable Crops, University of California, Davis. CANTALOUPE; HONEY DEW MELON; PERSIAN MELON.

Lorimer, Dr. Duncan. Department of Physics, West Virginia University, Morgantown. PULSAR.

Lotz, Dr. Martin. Department of Molecular and Experimental Medicine, Research Institute of Scripps Clinic, La Jolla, California. NEUROIMMUNOLOGY.

Loughlin, James E. Chemical Processes Pilot Plant, Textile Research Center, Texas Tech University. DYEING.

Love, Dr. Jeffrey J. Institute of of Geophysics & Planetary Physics, University of California, San Diego. GEODYNAMO.

Love, Dr. Lonnie. Oak Ridge National Laboratory, Tennessee. FERROFLUID.

Love, Dr. Susan M. Division of Surgical Oncology, Comprehensive Breast Center, University of California, Los Angeles. BREAST DISORDERS.

Lovell, Dr. Mark A. Director, Diagnostic Molecular Pathology Laboratory, Department of Pathology, University of Virginia Health Sciences Center, Charlottesville. MOLECULAR PATHOLOGY.

Lowan, Dr. Arnold N. Deceased; formerly, Chairperson, Department of Physics, Yeshiva University. E (MATHEMATICS); EXPONENT; LOGARITHM; ROOT (MATHEMATICS).

Lowe, Dr. Christopher R. Institute of Biotechnology, Cambridge University, Cambridge, England. BIOELECTRONICS.

Lowe, Dr. David A. Department of Fermentation Development, Industrial Division, Bristol-Myers Squibb Company, Syracuse, New York. PENICILLIN.

Lowell, Dr. Cadance A. Department of Biology, Central State University, Wilberforce, Ohio. SEED.

Lowell, Prof. Wayne R. Deceased; formerly, Department of Geology, Indiana University. MONAZITE; PYROMORPHITE; other articles.

Lowenstein, Prof. Otto E. Department of Zoology, University of Birmingham, England. BIOLOGICAL EQUILIBRIUM; TAXIS.

Lowenthal, David. Department of Geography, University College, London, England. WEST INDIES.

Lowry, Dr. Brian J. Department of Chemical Engineering, University of New Brunswick, Fredericton, Canada. DENSITY; HYGROMETER.

Luban, Prof. Naomi L. C. Department of Laboratory Medicine, Transfusion Medicine, Children's Hospital, George Washington University Medical Center, Washington, D.C. RH INCOMPATIBILITY.

Luborsky, Dr. Fred E. Research and Development Center, General Electric Company, Schenectady, New York. MAGNETIC MATERIALS.

Lucas, Dr. Spencer G. New Mexico Museum of Natural History, Albuquerque. PANTODONTA.

Luckenbach, Edward C. Senior Engineering Associate, Exxon Research and Engineering Company, Florham Park, New Jersey. CRACKING.

Ludlam, Prof. Frank H. Deceased; formerly, Department of Meteorology, Imperial College, London, England. CLOUD.

Ludvik, Dr. George F. Insecticide Application Research, Agricultural Research and Development Department, Monsanto Company, St. Louis, Missouri. INSECTICIDE; PESTICIDE.

Ludwig, Prof. Reinhold. Department of Electrical and Computer Engineering, Worcester Polytechnic Institute, Massachusetts. RADIO-FREQUENCY AMPLIFIER; SMITH CHART.

Luebbers, Dr. Ralph H. Department of Chemical Engineering, University of Missouri. HEAT TRANSFER.

Lui, Dr. Eric M. Department of Civil amd Environmental Engineering, Syracuse University, New York. STRUCTURAL STABILITY.

Lund, Dr. Robert E. Director of Extractive Research and Development, Zinc Smelting Division, St. Joseph Minerals Corporation, Monaca, Pennsylvania. CADMIUM METALLURGY.

Lunte, Prof. Craig E. Department of Chemistry, University of Kansas, Lawrence. MICRODIALYSIS SAMPLING.

Lunte, Dr. Susan M. Associate Director, Center for Bioanalytical Research, University of Kansas, Lawrence. ELECTROPHORESIS.

Luo, Dr. Hui. Broadcom Corporation, Matawan, New Jersey. WIRELESS FIDELITY (WI-FI).

Luo, Ningyl. Subpicosecond and Quantum Radiation Laboratory, Texas Tech University, Lubbock. ULTRAFAST MOLECULAR PROCESSES.

Luo, Dr. Zhexi. Section of Vertebrate Fossils, Carnegie Museum of Natural History, Pittsburgh, Pennsylvania. SYMMETRODONTA.

Lusted, Dr. Lee B. Department of Radiology, University of Chicago, Illinois. RADIOLOGY.

Luthin, Prof. James N. Department of Civil Engineering, University of California, Davis. LAND DRAINAGE (AGRICULTURE).

Lutter, Dr. Leonard. Molecular Biology Research Program, Henry Ford Hospital, Detroit, Michigan. DEOXYRIBONUCLEIC ACID (DNA).

Lüttge, Dr. Ulrich. Institut für Botanik, Technische Hochschule Darmstadt, Germany. ION TRANSPORT.

Luu, Dr. Jane. Lincoln Laboratory, Massachusetts Institute of Technology, Lexington. KUIPER BELT.

Luzadder, Prof. Warren J. Department of Engineering Graphics, Purdue University. NUT (ENGINEERING); SCREW THREADS; WASHER.

Lykken, Dr. David T. Department of Psychiatry, University of Minnesota. ELECTRODERMAL RESPONSE; LIE DETECTOR.

Lyman, Prof. Charles P. Deceased; formerly, Associate Curator of Mammals, Museum of Comparative Zoology at Harvard, and Professor of Anatomy, Harvard Medical School, Boston, Massachusetts. HIBERNATION AND ESTIVATION.

Lyman, Dr. John. Deceased; formerly, Department of Oceanography, University of North Carolina. MEDITERRANEAN SEA.

Lynch, Anna. Royal Botanic Gardens Kew, Jodrell Laboratory, United Kingdom. PLANT TISSUE SYSTEMS.

Lynds, Dr. Beverly T. Center for Astrophysics and Space Astronomy, University of Colorado. GLOBULE.

Lyon, Dr. W. S. Analytical Division, Oak Ridge National Laboratory, Oak Ridge, Tennessee. ACTIVATION ANALYSIS.

Lyon, Dr. Waldo. Arctic Submarine Laboratory, Naval Ocean Systems Center, San Diego, California. SEA ICE.

Lyons, Walter. Consultant in Telecommunications, Flushing, New York. AUTOMATIC FREQUENCY CONTROL (AFC); AUTOMATIC GAIN CONTROL (AGC).

Lyons, Dr. Walter A. Certified Consulting Meteorologist, Forensic Meteorology Associates, Fort Collins, Colorado. SEA BREEZE.

M

Ma, Dr. Yi Hua. Department of Chemical Engineering, Worcester Polytechnic Institute. WATER SOFTENING.

MacCarthy, Donnell D. Manager, Voltage Regulator Engineering, Voltage Regulator Product Section, General Electric Company, Pittsfield, Massachusetts. VOLTAGE REGULATOR.

MacCoull, Nell. Deceased; formerly, Lecturer in Mechanical Engineering, Columbia University. COMPRESSION RATIO; INTERNAL COMBUSTION ENGINE.

MacCrone, Prof. R. K. Department of Physics, Rensselaer Polytechnic Institute. HYSTERESIS; THERMAL HYSTERESIS.

MacDaniels, Dr. Laurence H. Professor Emeritus of Floriculture and Ornamental Horticulture, Cornell University. BRAZIL NUT; CASHEW; MACADAMIA NUT; other articles.

MacDonald, Dr. Daniel G. Department of Estuarine and Ocean Science, University of Massachusetts-Dartmouth, New Bedford. ESTUARINE OCEANOGRAPHY.

MacDonald, Dr. Gorden A. Deceased; formerly, Institute of Geophysics, University of Hawaii. PUMICE.

MacDonald, Dr. J. M. Small Animal Clinic, Auburn University, Auburn, Alabama. DERMATITIS.

MacDonald, Ruby I. Department of Biochemistry, Molecular and Cell Biology, Northwestern University. LIPOSOMES.

Macdoughall, Prof. J. Douglas. Geological Research Division, Scripps Institution of Oceanography, La Jolla, California. ELEMENTS, GEOCHEMICAL DISTRIBUTION OF; GEOCHEMISTRY.

MacDowell, Dr. Samuel W. Department of Physics, Yale University. QUANTUM GRAVITATION.

Macher, Dr. Abe M. Medical Consultant, AIDS Education and Training Centers Program, U.S. Public Health Service, Rockville, Maryland. ACQUIRED IMMUNE DEFICIENCY SYNDROME (AIDS).

Machiels, Albert. Nuclear Engineer, Nuclear Radiation Laboratory, University of Illinois, Urbana-Champaign. NUCLEAR FUELS.

Machlin, Dr. Laurence J. Department of Vitamins and Clinical Nutrition, Hoffmann-LaRoche, Inc., Nutley, New Jersey. VITAMIN.

Mack, Dr. Chris A. KLA-Tencor Corporation, Austin, Texas. MICROLITHOGRAPHY.

Mackay, Dr. Ian R. Center for Molecular Biology and Medicine, Monash University, Clayton, Victoria, Australia. AUTOIMMUNITY.

MacKenzie, Prof. Innes K. Department of Physics, University of Guelph, Ontario, Canada. COMPTON EFFECT.

Mackey, Prof. George W. Deceased; formerly, Department of Mathematics, Harvard University. OPERATOR THEORY.

Mackey, Dr. Philip J. Noranda Technology Centre, Pointe Claire, Quebec, Canada. PYROMETALLURGY; PYROMETALLURGY, NONFERROUS.

Mackhann, Prof. Guy Mead. Neurology Department, Johns Hopkins Hospital, Baltimore, Maryland. MULTIPLE SCLEROSIS; NERVOUS SYSTEM DISORDERS; PARKINSON'S DISEASE.

Madanshetty, Prof. Sameer I. Department of Aerospace and Mechanical Engineering, Boston University, Massachusetts. ACOUSTIC INTERFEROMETER; ACOUSTIC RADIOMETER.

Maddalon, Dal V. NASA Langley Research Center, Hampton, Virginia. AIR TRANSPORTATION.

Maderson, Dr. Paul F. Department of Biology, Brooklyn College. HAIR; SCALE (ZOOLOGY).

Madestau, L. Chemicals and Plastics Division, Union Carbide Corporation, Bound Brook, New Jersey. FORMALDEHYDE.

Madin, Dr. Laurence P. Associate Scientist, Biology Department, Woods Hole Oceanographic Institution, Woods Hole, Massachusetts. BEROIDA; CESTIDA; CYDIPPIDA; LOBATA; PLATYCTENIDA.

Madison, Dr. John H. Environmental Horticulture Department, University of California, Davis. BLUEGRASS.

Madison, Dr. Vincent. Department of Medicinal Chemistry, School of Pharmacy, University of Illinois. OPTICAL ACTIVITY.

Madore, Dr. Barry F. NASA/IPAC Extragalactic Database, Jet Propulsion Laboratory, California Institute of Technology, Pasadena, California. CEPHEIDS.

Magee, Paul T. Department of Microbiology, School of Medicine, Yale University. AMINO ACIDS.

Maggenti, Dr. Armand R. Division of Nematology, University of California, Davis. NEMATA (NEMATODA); other articles.

Maher, Dr. Barbara A. Department of Geography, Lancaster University, United Kingdom. LOESS.

Mahmoud, Dr. Aly A. School of Engineering, Indiana-Purdue University, Fort Wayne. INDUCTIVE COORDINATION.

Mahoney, Dr. Lee R. Department of Chemistry, Ford Motor Company, Dearborn, Michigan. INHIBITOR (CHEMISTRY).

Maisey, Dr. John. Curator, Division of Paleontology, American Museum of Natural History, New York, New York. CHIMAERIFORMES; OSTEICHTHYES; SELACHII.

Major, Dr. Jack. Department of Botany, University of California, Davis. CHAPARRAL.

Makulec, Alfred. Product Planning and Application Specialist, Large Lamp Department, General Electric Company, Cleveland, Ohio. NEON GLOW LAMP; ULTRAVIOLET LAMP.

Malacinski, Dr. George M. Department of Biology, Indiana University. AXENIC CULTURE; EMBRYONIC INDUCTION.

Maldonado-Moll, Dr. José F. Special Assistant for Science and Technology, Sistema Universitario de la Fundation Educative Ana G. Mendez, Puerto Rico Junior College, Rio Piedras. COENUROSIS; SCHISTOSOMIASIS.

Malik, Prof. Om P. Department of Electrical and Computer Engineering, The University of Calgary, Alberta, Canada. ELECTRIC POWER SYSTEMS.

Malin, Dr. David F. Anglo-Australian Observatory, Epping, New South Wales, Australia. ASTRONOMICAL PHOTOGRAPHY.

Malm, John G. Argonne National Laboratory, Argonne, Illinois. KRYPTON; XENON.

Mamantov, Dr. Gleb. Chemistry Department, University of Tennessee. FUSED-SALT SOLUTION.

Mamay, Dr. Sergius H. U.S. Geological Survey, National Museum, Washington, D.C. SPHENOPHYLLALES.

Mandle, Gary. Product Manager, Broadcast and Professional Division, Sony Electronics Corporation, Park Ridge, New Jersey. PICTURE TUBE.

Mandler, Dr. George. Department of Psychology, University of California, San Diego. AGGRESSION; EMOTION.

Mangion, Charles. Technology Laboratory, Science and Technology Division, Systems Group, TRW, Inc., Redondo Beach, California. HYDRAULIC ACTUATOR.

Manitius, Prof. Andre Z. Department of Electrical and Computer Engineering, George Mason University, Fairfax, Virginia. ESTIMATION THEORY.

Mann, Prof. Alfred K. Department of Physics, University of Pennsylvania. HADRON.

Mann, Dr. Kenneth H. Fisheries Research Board of Canada, Dartmouth, Nova Scotia. ARHYNCHOBDELLAE; HIRUDINEA; RHYNCHOBDELLAE.

Mann, Prof. Robert W. Department of Mechanical Engineering, Massachusetts Institute of Technology. ENGINEERING DESIGN; ENVIRONMENTAL TEST; LAYOUT DRAWING; WIRING DIAGRAM; other articles.

Mann, Prof. Stephen. School of Chemistry, University of Bath, United Kingdom. NANOCHEMISTRY.

Mann, William R. Rockwell International Corporation, New Port Beach, California. ELECTRONIC TEST EQUIPMENT.

Mannhelm, Prof. Walter. Med. Zentrum für Hygiene, Universität Marburg, Germany. ACTINOBACILLUS; FRANCISELLA; PASTEURELLA.

Manning, Dr. James M. Department of Biochemistry, Rockefeller University, New York. ALBUMIN; PEPTIDE; PROTEIN.

Manning, Prof. Kenneth V. Emeritus, Pennsylvania State University. AMPÉRE'S LAW; BIOT-SAVART LAW; EDDY CURRENT; HELMHOLTZ COILS; other articles.

Manning, Dr. Paul D. V. Retired; formerly, Professor of Chemical Engineering, California Institute of Technology. MONOSODIUM GLUTAMATE.

Mansfield, Dr. Linda S. College of Veterinary Medicine, Michigan State University, East Lansing. ZOONOSES.

Mansfield, Steve. American Radio Relay League, Inc., Newington, Connecticut. AMATEUR RADIO.

Manson, Dr. John A. Deceased; formerly, Department of Chemistry, Lehigh University. HETEROCYCLIC POLYMER; POLYMERIZATION; other articles.

Manson, Prof. Steven. Department of Physics and Astronomy, Georgia State University, Atlanta. ANGULAR CORRELATIONS.

Mansouri, Dr. Freydoon. Deceased; formerly, Department of Physics, University of Cincinnati, Ohio. SUPERGRAVITY.

Mantell, Dr. Charles L. Consulting Chemical Engineer, Electrochemical Engineering, Process and Plant Design, Manhasset, New York. ELECTROCHEMICAL PROCESS.

Mantler, Dr. Michael. Institut fur Angewandte und Technische Physik,Techische Universitat Wien, Austria. X-RAY FLUORESCENCE ANALYSIS.

Manwell, Dr. Reginald D. Department of Zoology, College of Liberal Arts, Syracuse University. HAEMOSPORINA; HAPLOSPOREA; HAPLOSPORIDA; SARCOSPORIDA.

Maple, M. Brian. Department of Physics, University of California, San Diego. A15 PHASES; ANTIFERROMAGNETISM; CHEVREL PHASES.

March-Russell, Dr. John. Rudolf Peirls Center for Theoretical Physics, Department of Physics, University of Oxford, United Kingdom. EIGENFUNCTION; EIGENVALUE (QUANTUM MECHANICS); ENERGY LEVEL (QUANTUM MECHANICS); SPINOR; SUPERSYMMETRY.

Marchalonis, Dr. John J. Chairperson, Department of Biochemistry, Medical University of South Carolina. IMMUNOLOGICAL PHYLOGENY.

Marchand, Prof. Alan P. College of Arts and Sciences, Department of Chemistry, University of North Texas, Denton. PERICYCLIC REACTION; TAUTOMERISM.

Marchis, Dr. Franck. Department of Astronomy , University of California-Berkeley. BINARY ASTEROID.

Marcus, Prof. Ernesto. Deceased; formerly, Department of Zoology, University of São Paulo, Brazil. EUTARDIGRADA.

Marcus, Eveline. Department of Zoology, University of São Paulo, Brazil. EUTARDIGRADA; HETEROTARDIGRADA; TARDIGRADA.

Marcus, Prof. Marvin. Department of Computer Science, University of California, Santa Barbara. MATRIX THEORY.

Marden, Dr. Albert. Department of Mathematics, University of Minnesota, Minneapolis. RIEMANN SURFACE.

Marek, Dr. John M. Independent Mining Consultants, Inc., Tucson, Arizona. OPEN-PIT MINING.

Margai, Prof. Florence Lansana. Department of Geography, State University of New York at Binghamton. AFRICA.

Margenau, Prof. Henry. Deceased; formerly, Department of Physics, Yale University. HYPOTHESIS; POSTULATE; THEOREM.

Margolis, Prof. Simeon. Department of Endocrinology and Metabolism, Johns Hopkins University, Baltimore, Maryland. ELECTRODIAGNOSIS.

Margollash, Dr. E. Department of Biochemistry, Northwestern University. CYTOCHROME.

Margrave, Prof. John L. Dean of Research, Rice University. HIGH-TEMPERATURE CHEMISTRY.

Mariño, Prof. Miguel A. Hydrology Program, University of California, Davis. HYDROLOGY.

Marino, Dr. Paul. Department of Biology, College of Charleston, South Carolina. ECOLOGICAL COMPETITION.

Maris, Prof. Humphrey J. Department of Physics, Brown University, Providence, Rhode Island. QUANTUM ACOUSTICS.

Mark, Prof. Harry B., Jr. Department of Chemistry, University of Cincinnati, Ohio. KINETIC METHODS OF ANALYSIS.

Mark, Dr. J. Carson. Member of the Advisory Committee, Reactor Safeguards of U.S. Nuclear Regulatory Commission, Los Alamos, New Mexico. NUCLEAR EXPLOSION.

Market, Clement L. Department of Biology, Yale University, New Haven, Connecticut. EMBRYONIC DIFFERENTIATION.

Marko, Prof. Hans. Lehrstuhl für Nachrichtentechnik, Institut für Informationstechnik, Technischen Universität München, Germany. INFORMATION THEORY (BIOLOGY).

Markovitz, Dr. Hershel. Department of Chemistry, Carnegie-Mellon University, Pittsburgh, Pennsylvania. RHEOLOGY.

Markowitz, Dr. William. Deceased; formerly, Department of Physics, Nova University, Fort Lauderdale, Florida; Editor, *Geophysical Surveys.* TIME-INTERVAL MEASUREMENT.

Marks, Dr. Jonathan. Department of Sociology and Anthropology, University of North Carolina at Charlotte. HUMAN VARIATION.

Marks, Lawrence E. Pierce Foundation, New Haven, Connecticut. AUDIOMETRY; LOUDNESS; PSYCHOACOUSTICS.

Marks, Dr. Tobin J. Department of Chemistry, Northwestern University. ORGANOACTINIDES; STRUCTURAL CHEMISTRY.

Markus, John. Deceased; formerly, Consultant, Sunnyvale, California. ANTIRESONANCE; other articles.

Marler, Dr. Peter R. Field Research Center, Millbrook, New York. ETHOLOGY.

Marmur, Prof. Julius. Department of Biochemistry, Albert Einstein College of Medicine of Yeshiva University, New York. GENE.

Marra, Dr. John. Doherty Senior Scholar, Lamont-Doherty Earth Observatory, Palisades, New York. MARINE BIOLOGICAL SAMPLING.

Marsden, Dr. Joan R. Department of Zoology, McGill University, Montreal, Quebec, Canada. PHORONIDA.

Marsh, Don. Media Services Representative, Portland Cement Association, Skokie, Illinois. CEMENT; GROUT.

Marsh, Prof. Kenneth. Department of Chemical and Process Engineering, University of Canterbury, Christchurch, New Zealand. CALORIMETRY.

Marsh, Rex E. Department of Wildlife, Fish & Conservation Biology, University of California, Davis. RODENTICIDE.

Marshak, Dr. Harvey. National Institute of Standards and Technology. NUCLEAR ORIENTATION.

Marshall, Dr. H. G. Crops Research Division, Agricultural Research Service and Department of Agronomy, Pennsylvania State University. BUCKWHEAT.

Marshall, Dr. Norman B. Retired; formerly, Senior Principal Scientific Officer, British Museum of Natural History, London, England. DEEP-SEA FAUNA.

Marshall, Peter L. Independent Consultant, Granville, Ohio. AIRCRAFT DESIGN.

Marshall, Prof. Robert T. Department of Food Science and Nutrition, College of Agriculture, University of Missouri. ICE CREAM.

Marshall, Prof. William R., Jr. Director, UIR Program, University of Wisconsin, Madison. DESICCANT; DRYING.

Marsland, Charles R. Chief Technical Consultant, Handy and Harman, Fairfield, Connecticut. GOLD ALLOYS.

Marston, Prof. Philip L. Department of Physics and Astronomy, Washington State University, Pullman. ACOUSTIC LEVITATION; ACOUSTIC RADIATION PRESSURE; FREQUENCY (WAVE MOTION).

Martell, Dr. A. E. Department of Chemistry, Texas A&M University. CHELATION.

Martin, Dr. Arthur W. Department of Zoology, University of Washington. EXCRETION.

Martin, Prof. C. Samuel. School of Civil Engineering, Georgia Institute of Technology, Atlanta. WATER HAMMER.

Martin, Dr. Daniel W. Acoustics Consultant, Cincinnati, Ohio. MUSICAL INSTRUMENTS.

Martin, Franklin W. Mayaguez Institute of Tropical Agriculture, Mayaguez, Puerto Rico. CASSAVA.

Martin, Prof. James P. Department of Soils and Plant Nutrition, College of Biological and Agricultural Sciences, University of California, Riverside. HUMUS.

Martin, Prof. Peter G. Department of Astronomy and Astrophysics, University of Toronto, and Canadian Institute for Theoretical Astrophysics, Toronto, Ontario, Canada. INTERSTELLAR MATTER.

Martin, Prof. R. D. Anthropologisches Institut, Universitat Zurich-Irchel, Zurich, Switzerland. GESTATION PERIOD.

Martin, Prof. Richard J. Division of Neonatology, Rainbow Babies and Childrens Hospital, University Hospitals of Cleveland, Ohio. INFANT RESPIRATORY DISTRESS SYNDROME.

Martin, Stanley B. Stanford Research Center, Menlo Park, California. FIRE TECHNOLOGY.

Martin, Dr. Thomas. Sektion Mammalogie, Forschungsinstitut Senckenberg, Germany. MULTITUBERCULATA.

Martínez-Carrera, Dr. D. College of Postgraduates in Agricultural Sciences, Mushroom Biotechnology, Puebla, Mexico. MUSHROOM.

Martino, Dr. Steve. Yale Psychiatric Center, Yale University School of Medicine, New Haven, Connecticut. ALCOHOLISM.

Martinson, C. R. Corporate Engineering Department, Monsanto Company, St. Louis, Missouri. DRIER (PAINT); LACQUER; PAINT; VARNISH; other articles.

Maruyama, Prof. Shige. Department of Earth and Planetary Sciences, Tokyo Institute of Technology, Japan. GLAUCOPHANE; OPHIOLITE.

Marvier, Dr. Michelle A. Department of Biology, Santa Clara University, California. BIODIVERSITY.

Marx, Dr. Preston A. California Primate Research Center, University of California, Davis. RETROVIRUS.

Marzilli, Dr. Luigi G. Department of Chemistry, Emory University, Atlanta, Georgia. COBALT.

Marzilli, Dr. Patricia A. Department of Chemistry, Emory University, Atlanta, Georgia. COBALT.

Mascarenhas, Prof. Joseph P. Department of Biological Sciences, State University of New York, Albany. FERTILIZATION; GAMETOGENESIS.

Mashhoon, Prof. Bahram. Department of Physics and Astronomy, College of Arts and Sciences, University of Missouri—Columbia. GRAVITATION.

Mason, Dr. Basil John. Programme Director, Centre for Environmental Technology, Imperial College of Science and Technology, London, England. CLOUD PHYSICS.

Massey, Dr. Philip. Lowell Observatory, Flagstaff, Arizona. ASTRONOMICAL SPECTROSCOPY.

Massey, Dr. Vincent. Department of Biological Chemistry, University of Michigan, Ann Arbor. BIOLOGICAL OXIDATION.

Massingill, Dr. John. Texas State University-San Marcos. DRYING OIL.

Massion, Dr. Jean. Laboratoire de Neurobiologie et Mouvements, Centre de Recherche en Neurosciences Cognitives, Marseille, France. POSTURAL EQUILIBRIUM.

Mast, Dr. Roy C. Miami Valley Laboratories, Procter & Gamble Company, Cincinnati, Ohio. SOAP.

Mast, Dr. T. Douglas. Department of Biomedical Engineering, University of Cincinnati. HELMHOLTZ RESONATOR.

Masters, Dr. Colin L. Head of the National Creutzfeldt-Jakob Disease Registry, Department of Pathology, University of Melbourne, Victoria, Australia. PRION DISEASE.

Masucci, Dr. Maria G. Microbiology and Tumor Biology Center, Karolinska Institute, Stockholm, Sweden. TUMOR VIRUSES.

Mata-Toledo, Dr. Ramon A. Professor of Computer Science, James Madison University, Harrisonburg, Virginia. AUTOMATA THEORY; BIT; DATABASE MANAGEMENT SYSTEM; LANGUAGE THEORY; METADATA; NUMBERING SYSTEMS; NUMERICAL REPRESENTATION (COMPUTERS); other articles.

Matijevic, Prof. Egon. Chairperson, Department of Chemistry, Clarkson College of Technology, Potsdam, New York. COLLOID.

Mattei, Dr. Janet Akyüz. Deceased; formerly, Director, American Association of Variable Star Observers, Cambridge, Massachusetts. MIRA; VARIABLE STAR.

Matthews, Dr. Kathleen S. Professor of Biochemistry and Cell Biology and Dean of Natural Sciences, Rice University, Houston, Texas. PROTEIN FOLDING.

Matyas, Stephen M. Personal Path Systems, Inc., New Jersey. CRYPTOGRAPHY.

Matz, Benjamin. Deceased; formerly, Fellow, American Watchmaker-Clockmakers Institute, and Manager of Technical Information, Bulova Corporation, New York. WATCH.

Matz, Dr. Samuel A. President, Pan-Tech International, Inc., McAllen, Texas. FOOD MANUFACTURING.

Matzner, Dr. Christopher D. Department of Astronomy and Astrophysics, University of Toronto, Ontario, Canada. INTERSTELLAR MATTER.

Mauchline, Dr. John. Scottish Association for Marine Science, Argyll, Scotland. EUPHAUSIACEA; MYSIDACEA.

Mauk, Dr. Craighton S. Research Associate, Department of Horticultural Science, Mountain Horticultural Crops Research and Extension Center, School of Agriculture and Life Sciences, North Carolina State University, Fletcher. DORMANCY.

Maul, Prof. George A. Director, Division of Marine and Environmental Systems, Florida Institute of Technology, Melbourne, Florida. GULF OF MEXICO; INTRA-AMERICAS SEA.

Maurer, Dr. Brian A. Department of Fisheries and Wildlife, Michigan State University, East Lansing. BIOGEOGRAPHY.

Mausner, Dr. Leonard F. Radionuclide/Radiopharmaceutical Research, Brookhaven National Laboratory, Upton, New York. RADIOCHEMICAL LABORATORY.

May, Prof. Bella J. Department of Physical Therapy, Medical College of Georgia, Augusta. PROSTHESIS.

May, F. H. Consultant, Kerr-McGee Corporation, Whittier, California. BORON.

May, James E. Lockheed-California Company, Burbank. SUBSONIC FLIGHT.

May, Marvin B. Chief Scientist, Pennsylvania State University Applied Research Laboratory, Navigation Research and Development Center, Warminster, Pennsylvania. ELECTRONIC NAVIGATION SYSTEMS.

Mayall, Robert N. Deceased; formerly, Director, Planning and Research Associates, Boston, Massachusetts. SUNDIAL.

Mayer, Dr. Rudolph L. Deceased; formerly, Leonard Wood Memorial Hospital, Washington, D.C. SULFONAMIDE.

Maykuth, Daniel. Tin Research Institute, Inc., Columbus, Ohio. TIN METALLURGY.

Maynard, Dr. Harold B. Retired; formerly, Maynard Research Council, Inc., Pittsburgh, Pennsylvania. TECHNOLOGY.

Mays, Rolland L. Materials System Division, Union Carbride Corporation, Tarrytown, New York. MOLECULAR SIEVE.

Mazzo, Anthony. Urban Substructures, Inc., East Elmhurst, New York. PILE FOUNDATION.

Mazzone, Dr. Theodore. University of Illinois at Chicago. HORMONE.

McAlister, Prof. Harold A. Director, Center for High Angular Resolution Astronomy, Georgia State University, Atlanta. ALDEBARAN; ANTARES; BETELGEUSE; VEGA.

McAteer, J. Davitt. U.S. Department of Labor, Mine Safety and Health Administration, Arlington, Virginia. MINING.

McBride, Dr. Earle F. Department of Geology, University of Texas, Austin. GRAYWACKE.

McCabe, Dr. Allyssa. Eliot-Pearson Department of Child Study, Tufts University, Medford, Massachusetts. DEVELOPMENTAL PSYCHOLOGY.

McCabe, Dr. Warren L. Department of Chemical Engineering, North Carolina State University, Raleigh. MECHANICAL CLASSIFICATION; SCREENING.

McCann, Dr. Mary B. Nutrition Program, Center for Disease Control, Health Services and Mental Health Administration, U.S. Department of Health, Education and Welfare, Rockville, Maryland. COENZYME; METABOLISM.

McCarthy, Dr. Dennis D. Time Service Division, U.S. Naval Observatory, Washington, D.C. EARTH ROTATION AND ORBITAL MOTION, LATITUDE AND LONGITUDE; QUARTZ CLOCK; TIME.

McCarthy, Dr. John. National Center for Atmospheric Research, Boulder, Colorado. AERONAUTICAL METEOROLOGY.

McCarthy, Dr. Kathryn A. Department of Physics, Tufts University, Medford, Massachusetts. THERMAL CONDUCTION IN SOLIDS.

McCartney, Dr. Kevin. Northern Maine Museum of Science, University of Maine at Presque Isle. SILICOFLAGELLATA.

McCarty, Dr. John E. Structure Staff Supervisor, Boeing Company, Renton, Washington. FUSELAGE.

McClung, Prof. Leland S. Department of Biology, Indiana University, Bloomington. BLACKLEG; GANGRENE; TETANUS.

McClure, Dr. Michelle. Northwest Fisheries Science Center, Seattle, Washington. WATER CONSERVATION.

McConnaughey, Dr. Bayard H. Department of Biology, University of Oregon, Eugene. DICYEMIDA; MESOZOA; ORTHONECTIDA.

McConnell, Kenneth R. Director of Technical Operations, Panafax Corp., Melville, New York. FACSIMILE.

McCormick, Prof. Barnes W., Jr. Department of Aerospace Engineering, Pennsylvania State University. AIRCRAFT RUDDER; ELEVON; TAIL ASSEMBLY.

McCormick, Dr. Donald. Department of Biochemistry, Emory University School of Medicine, Atlanta, Georgia. THIAMINE.

McCormick, Dr. Thomas R. Department of Biomedical History, School of Medicine, University of Washington. DEATH.

McCoy, Prof. Barry M. Division of Engineering and Applied Physics, Harvard University, Cambridge, Massachusetts. ISING MODEL.

McCrory, Dr. Robert L. Director, Laboratory for Laser Energetics, College of Engineering and Applied Sciences, University of Rochester. OPTICAL PULSES.

McDaniel, Dr. Carl N. Department of Biology, Rensselaer Polytechnic Institute. DEVELOPMENTAL BIOLOGY.

McDaniel, Dr. J. Gregory. Associate Professor, Department of Aerospace and Mechanical Engineering, Boston University, Massachusetts. ACOUSTIC IMPEDANCE.

McDonald, Prof. Alan T. School of Mechanical Engineering, Purdue University, West Lafayette, Indiana. DYNAMIC SIMILARITY.

McDonald, Prof. John J. Retired; formerly, Head, Textile Engineering Department, University of Lowell, Massachusetts. BLEACHING.

McDowell, Dr. Robin S. Senior Chief Scientist, Chemical Structure and Dynamics, Molecular Science Research Center, Battelle Pacific Northwest Laboratories, Richland, Washington. INFRARED SPECTROSCOPY.

McEntire, Dr. Betty. American SIDS Institute, Marietta, Georgia. SUDDEN INFANT DEATH SYNDROME (SIDS).

McEvilly, Dr. Thomas V. Department of Geology, University of California, Berkeley. SEISMOGRAPHIC INSTRUMENTATION.

McEwen, Dr. Bruce. Harold and Margaret Milliken Hatch Laboratory of Neuroendocrinology, Rockefeller University, New York. STRESS (PSYCHOLOGY).

McEwen, Dr. Robert S. Deceased; formerly, Professor Emeritus, Oberlin College. RESPIRATORY SYSTEM.

McFarland, Forest R. Deceased; formerly, Engineering Consultant, Flint, Michigan. GEAR LOADING.

McFarland, Dr. Richard H. Director Emeritus, Avionics Engineering Center, Ohio University, Athens. INSTRUMENT LANDING SYSTEM (ILS).

McGannon, Harold E. Technical Editor, Research and Technology Divison, United States Steel Corporation. Pittsburgh, Pennsylvania. IRON ALLOYS.

McGaugh, Dr. James L. Department of Psychobiology, University of California, Irvine. ELECTROCONVULSIVE THERAPY.

McGillis, Donald. Department of Electrical and Computer Engineering, McGill University, Montreal, Québec, Canada. REACTIVE POWER.

McGinley, Dr. John A. Chief, Forecast Research Group, National Oceanic and Atmospheric Administration, Environmental Research Laboratory, Boulder, Colorado. SATELLITE METEOROLOGY.

McGinnis, Prof. William. Department of Molecular Biophysics and Biochemistry, Yale University, New Haven, Connecticut. HOMEOTIC (HOX) GENES.

McGinnles, William G. Professor of Dendrochronology and Arid Land Studies, University of Arizona. OASIS.

McGovern, Dr. Katharine. Wright Institute, Berkeley, California. CONSCIOUSNESS.

McGrain, Preston. Assistant State Geologist, Kentucky Geological Survey, University of Kentucky. OILFIELD WATERS.

McGuire, Ann B. Aquaculture Program, University of California, Davis. AQUACULTURE.

McIntire, Prof. Larry V. Chairperson, Institute of Biosciences and Bioengineering, Rice University, Houston, Texas. BIOMEDICAL CHEMICAL ENGINEERING.

McIntosh, Prof. J. Richard. Department of Molecular, Cellular and Developmental Biology, Laboratory for High-Voltage Electron Microscopy, University of Colorado. CENTROSOME.

McIntyre, Dr. Andrew. Lamont-Doherty Geological Observatory, Palisades, New York. COCCOLITHOPHORIDA.

McIver, Richard D. Geochem Research Inc., Houston, Texas. ORGANIC GEOCHEMISTRY.

McKenna, Dr. Malcolm C. Frick Curator, American Museum of Natural History, New York. CONDYLARTHRA; DERMOPTERA; EMBRITHOPODA; MULTITUBERCULATA; PANTOTHERIA; other articles.

McKenzie, Dr. James M. Sandia National Laboratories, Albuquerque, New Mexico. JUNCTION DETECTOR.

McKesson, Chris B. Alion Science and Technology, Alexandria, Virginia. FERRY; HYDROFOIL CRAFT.

McKinley, Prof. Gareth. Department of Mechanical Engineering, Massachusetts Institute of Technology, Cambridge. SPIDER SILK.

McKinney, Dr. Chester M. Retired; formerly, Applied Research Laboratories, University of Texas, Austin. SONAR.

McKinney, Prof. Frank K. Department of Geology, Appalachian State University, Boone, North Carolina. BRYOZOA; CHEILOSTOMATA; CTENOSTOMATA; STENOLAEMATA; TREPOSTOMATA; other articles.

McKnight, Prof. Tom. Department of Geography, University of Caifornia, Los Angeles. ANTARCTIC CIRCLE.

McLain, Prof. Thomas E. Head, Department of Forest Products, Oregon State University, Corvallis. WOOD ENGINEERING DESIGN.

McLaughlin, Dr. Patsy A. Shannon Point Marine Center, Western Washington University, Anacortes. ARTHROPODA; MAXILLOPODA; OSTRACODA; PODOCOPA; SIPHONOSTOMATOIDA; other articles.

McLelland, Dr. James. Professor of Geology, Emeritus, Colgate University. ANORTHOSITE.

McMahon, Dr. John F. Deceased; formerly, New York State Technical Service Program, Alferd. ABRASIVE; GLASS; PLASTER; POTTERY; other articles.

McNamara, Dr. Ken J. Senior Curator, Invertebrate Paleontology, Western Australian Museum, Perth, Western Australia. HETEROCHRONY.

McNamara, Dr. Peter. Department of Medical Microbiology, University of Wisconsin, Madison. INFECTIOUS DISEASE.

McNaughton, Prof. Samuel J. Department of Biology, Syracuse University. ECOLOGY; TERRESTRIAL ECOSYSTEM.

McNellis, Dr. Donald. Center for Research for Mothers and Children, National Institute of Child Health and Human Development, National Institutes of Health, Bethesda, Maryland. PRENATAL DIAGNOSIS.

McPartland, Brian J. Consultant, Tappan, New York. BRANCH CIRCUIT; ELECTRICAL CODES; ELECTRICAL CONNECTOR; GROUNDING; WIRING.

McPartland, Joseph F. Consultant, Tenafly, New Jersey. BRANCH CIRCUIT; ELECTRICAL CODES; ELECTRICAL CONNECTOR; WIRING.

McPherson, Prof. George, Jr. Department of Electrical Engineering, School of Engineering, University of Missouri. ARMATURE REACTION; ELECTRICAL DEGREE; MOTOR; RELUCTANCE MOTOR; TRANSFORMER; other articles.

McQuillan, James P. AT&T Federal Systems, Washington, D.C. TELECOMMUNICATIONS CIVIL DEFENSE SYSTEM.

McTaggart, Prof. W. Donald. International Programs Office, Arizona State University, Tempe. ASIA.

McVey, Charles I. Fluorescent and High Intensity Systems Department, General Electric Company, Twinsburg, Ohio. SODIUM-VAPOR LAMP.

McWilliams, Prof. Michael. Geological and Environmental Sciences, Stanford University, California. MAGNETOMETER; PALEOMAGNETISM.

Mead, Mr. Lance. Brandon, Vermont. QUARRYING.

Meade, Dr. Thomas J. Beckman Institute, California Institute of Technology, Pasadena. CYANATE; ELECTRON-TRANSFER REACTION; HALIDE; HYDROXIDE; LIGAND; THIOCYANATE; THIOSULFATE.

Meadows, Kevin. Library and Information Science, Drexel University, Philadelphia, Pennsylvania. SMART CARD.

Mebus, Dr. Charles A. Laboratory Chief, Foreign Animal Disease Diagnostic Laboratory, U.S. Department of Agriculture, Greenport, New York. RIFT VALLEY FEVER.

Meeuse, Dr. Bastiaan J. D. Department of Botany, University of Washington. POLLINATION.

Meggers, Dr. William F. Deceased; formerly, National Bureau of Standards. BAND SPECTRUM; SPECTRUM.

Mehra, Prof. Raman K. President, Scientific Systems, Inc., Cambridge, Massachusetts. ESTIMATION THEORY.

Mehrer, Prof. Helmut. Institut für Metallforschung, Münster, Germany. DIFFUSION.

Meijer, Prof. Paul H. E. Department of Physics, Catholic University of America, Washington, D.C. DIPOLE-DIPOLE INTERACTION.

Meijer, Roelof J. Assistant Director of Research, Philips Research Laboratories, Endhoven, the Netherlands. STIRLING ENGINE.

Meinke, Dr. David. Department of Botany and Microbiology, Oklahoma State University. EMBRYOGENESIS; EMBRYOLOGY.

Meinwald, Dr. J. Department of Chemistry, Cornell University, Ithaca, New York. PHEROMONES.

Meisel, Prof. David D. Emeritus, Department of Physics and Astronomy, State University of New York, Geneseo. METEOR.

Meisel, Dr. Jerome. Department of Electrical and Computer Engineering, Wayne State University, Detroit, Michigan. ELECTROMAGNET; SOLENOID (ELECTRICITY).

Meissner, Dr. Hans W. Department of Physics, Stevens Institute of Technology, Hoboken, New Jersey. MEISSNER EFFECT.

Meixner, Dr. Josef. Director, Institute of Theoretical Physics, Aachen, Germany. BESSEL FUNCTIONS; GAMMA FUNCTION.

Melander, John. Portland Cement Association, Skokie, Illinois. MORTAR.

Mellichamp, Prof. Duncan A. Department of Chemical and Nuclear Engineering, University of California, Santa Barbara. DIGITAL CONTROL.

Mellon, Prof. M. G. Department of Chemistry, Purdue University, West Lafayette, Indiana. TRANSLUCENT MEDIUM.

Melnick, Dr. Joseph L. Department of Virology and Epidemiology, Baylor College of Medicine. ARBOVIRAL ENCEPHALITIDES; EMBRYONATED EGG CULTURE; ENTEROVIRUS; EPSTEIN-BARR VIRUS; other articles.

Melosh, Prof. H. Jay. Lunar and Planetary Laboratory, University of Arizona, Tucson. METEORITE.

Meltzer, Dr. David J. Department of Anthropology, Southern Methodist University, Dallas, Texas. PALEOINDIAN.

Mendonca, Dr. Aubrey F. Department of Food Science and Human Nutrition, Iowa State University, Ames. FOOD PRESERVATION.

Menkes, Joshua. Science and Technology Divisions, Institute for Defense Analysis, Arlington, Virginia. GAS DYNAMICS.

Menzel, Prof. E. Roland. Director, Center for Forensic Studies, Texas Tech University, Lubbock, Texas. FORENSIC PHYSICS.

Menzies, Prof. John. Departments of Earth Sciences and Geography, Brock University, St. Catherines, Ontario, Canada. GLACIAL GEOLOGY AND LANDFORMS; GLACIAL HISTORY.

Menzies, Dr. Robert J. Department of Zoology, Duke University, Durham, North Carolina. ISOPODA.

Merold, John. M. Rosenblatt and Son, Inc., Arlington, Virginia. MARINE REFRIGERATION.

Merritt, Frederick S. Deceased; formerly, Consulting Engineer, West Palm Beach, Florida. CONCRETE; CONCRETE BEAM; CONCRETE COLUMN; CONCRETE SLAB; PRESTRESSED CONCRETE; REINFORCED CONCRETE.

Merwin, Earl J. McCormick & Company, Hunt Valley, Maryland. SPICE AND FLAVORING.

Merwin, R. F. Erie Manufacturing Company, Erie, Pennsylvania. CONVEYOR.

Mescher, Prof. Anthony L. Medical Sciences Program, Indiana University, Bloomington. REGENERATIVE BIOLOGY.

Messenger, Dr. George C. Vice President and Technical Director, Messenger & Associates, Las Vegas, Nevada. RADIATION HARDENING.

Messenger, Dr. John B. Department of Zoology, University of Sheffield, United Kingdom. CHROMATOPHORE.

Messerschmitt, Prof. David. Department of Electrical Engineering and Computer Science, University of California, Berkeley. EQUALIZER.

Messina, Dr. Paula. Department of Geology, San José State University, California. KÖPPEN CLIMATE CLASSIFICATION SYSTEM.

Metzger, Dr. James D. Research Physiologist, U.S. Department of Agriculture, Agricultural Research Service, Biosciences Research Laboratory, Fargo, Indiana. GIBBERELLIN.

Meyer, Dr. Bernard S. Department of Botany, Ohio State University. PLANT–WATER RELATIONS.

Meyer, Prof. Bradley S. Department of Physics and Astronomy, Clemson University, South Carolina. NUCLEOSYNTHESIS.

Meyer, Dr. Carl H. Information Security Consultant, Kingston, New York. CRYPTOGRAPHY.

Meyer, Prof. Christian. Columbia University, New York. CONCRETE.

Meyer, Dr. David B. Department of Anatomy, Wayne State University, Detroit, Michigan. EYE (VERTEBRATE).

Meyer, Prof. Henry O. A. Department of Earth and Atmospheric Sciences, Purdue University, West Lafayette, Indiana. DIAMOND.

Meyer, Dr. John. Department of Entomology, North Carolina State University, Raleigh. THYSANURA (ZYGENTOMA).

Meyer, Prof. John F. Computing Research Laboratory, Department of Electrical Engineering and Computer Science, College of Engineering, University of Michigan. FAULT-TOLERANT SYSTEMS.

Meyer, Dr. Judy L. Institute of Ecology, University of Georgia. FOREST ECOSYSTEM.

Meyers, Dr. Gary. Oceans and Climate Program Leader, CSIRO Marine and Atmospheric Research, Australia. SOUTHEAST ASIAN WATERS.

Miall, Dr. Andrew D. Department of Geology, University of Toronto, Ontario, Canada. FACIES (GEOLOGY); FLUVIAL SEDIMENTS; SEQUENCE STRATIGRAPHY.

Michael, Prof. John W. Director, Department of Prosthetics and Orthotics, Duke University Medical Center. PROSTHESIS.

Michalec, James R. American Electric Power Company, Columbus, Ohio. STEAM ELECTRIC GENERATOR.

Middleton, Dr. Gerald V. Department of Geology, McMaster University, Hamilton, Ontario, Canada. DEPOSITIONAL SYSTEMS AND ENVIRONMENTS; SEDIMENTOLOGY; TURBIDITE.

Middleton, Dr. Roy. Department of Physics, University of Pennsylvania. ION SOURCES.

Migliari, Dr. Simon. Astronomical Institute "Anton Pannekoek," University of Amsterdam, Netherlands. SS 433.

Mikos, Prof. Antonios G. Director of John W. Cox Laboratory of Biomedical Engineering, Department of Chemical Engineering, Rice University, Houston, Texas. BIOMEDICAL CHEMICAL ENGINEERING.

Miles, Lawrence D. Engineering Consultant, Easton, Maryland. VALUE ENGINEERING.

Miles, Thomas K. Retired; formerly, Shell Development Co., Houston, Texas. ASPHALT AND ASPHALTITE.

Mill, Dr. George S. Research Chemist, Shell Oil Company, New York. ELECTROSTATIC PRECIPITATOR.

Miller, Barry. Consultant, Santa Fe, New Mexico. ELECTRONIC WARFARE.

Miller, Dr. Clyde A. Falls Church, Virginia. AIR-TRAFFIC CONTROL.

Miller, Dr. David A. B. AT&T Bell Laboratories, Holmdel, New Jersey. ELECTROOPTICS; STARK EFFECT.

Miller, Dr. Dorothy A. Wayne State University School of Medicine, Center for Molecular Medicine and Genetics, Detroit, Michigan. CHROMOSOME; CHROMOSOME ABERRATION.

Miller, Dr. Frederick R. Department of Soil and Crop Sciences, Texas A&M University. SORGHUM.

Miller, Dr. Glenn H. Weapons Effects Division, Sandia National Laboratories, Albuquerque, New Mexico. CATHODE RAYS; CORONA DISCHARGE; IONIZATION POTENTIAL; other articles.

Miller, Prof. Gordon. Department of Chemistry, Iowa State University, Ames. MAGNETOCHEMISTRY.

Miller, Prof. J. Creighton, Jr. Texas Agricultural Experiment Station, Department of Horticultural Science, College of Agriculture, Texas A&M University. POTATO, IRISH.

Miller, Dr. Joel S. Department of Chemistry, University of Utah, Salt Lake City. FERRICYANIDE AND FERROCYANIDE; PRUSSIAN BLUE.

Miller, Dr. Julian C. Department of Horticulture, Louisiana State University. POTATO, SWEET.

Miller, Prof. Maynard M. Department of Geology, Michigan State University; and Director, Foundation for Glacial and Environmental Research, Seattle, Washington. ICE FIELD; SNOW LINE; SNOWFIELD AND NÉVÉ.

Miller, Prof. Orlando J. Wayne State University School of Medicine, Center for Molecular Medicine and Genetics, Detroit, Michigan. CHROMOSOME ABERRATION; SOMATIC CELL GENETICS.

Miller, R. A. Engineering Department, Babcock and Wilcox, Barberton, Ohio. REHEATING.

Miller, Dr. Regis B. Project Leader, Center for Wood Anatomy Research, Forest Products Laboratory, U.S. Department of Agriculture, Forest Service, Madison, Wisconsin. WOOD ANATOMY.

Miller, Dr. Shelby A. Argonne National Laboratory, Argonne, Illinois. CLARIFICATION; FILTRATION; LEACHING; MECHANICAL SEPARATION TECHNIQUES.

Miller, Terry A. AT&T Bell Laboratories, Murray Hill, New Jersey. JAHN-TELLER EFFECT; RENNER-TELLER EFFECT.

Miller, Thomas E. Department of Biological Sciences, Florida State

University, Tallahassee. POPULATION VIABILITY.

Miller, Dr. William H. Department of Chemistry, University of California, Berkeley. CHEMICAL DYNAMICS.

Milligan, Dr. W. O. Robert A. Welch Foundation, Houston, Texas. ELECTROSTATIC PRECIPITATOR.

Millman, Dr. Jacob. Department of Electrical Engineering, Columbia University, New York. SIGNAL GENERATOR.

Mills, Dr. Dennis. Advanced Photon Source, Argonne National Laboratory, Argonne, Illinois. SYNCHROTRON RADIATION.

Mills, Prof. Ian M. Department of Chemistry, University of Reading, United Kingdom. GRAM-MOLECULAR WEIGHT; MOLE (CHEMISTRY).

Mills, Thomas B. National Semiconductor Corporation, Santa Clara, California. PHASE-LOCKED LOOPS.

Milner, Dr. Andrew. School of Biological and Chemical Sciences Birkbeck College, London, United Kingdom. TEMNOSPONDYLI.

Milner, Dr. Angela C. Department of Palaeontology, The Natural History Museum, London, United Kingdom. ARCHAEOPTERYX.

Milsom, Dr. William K. Department of Zoology, University of British Columbia, Vancouver, Canada. CHEMORECEPTION.

Milstein, Prof. Laurence B. Department of Electrical and Computer Engineering, University of California, San Diego. AMPLITUDE MODULATION; ANGLE MODULATION; CARRIER (COMMUNICATIONS).

Milster, Dr. Tom D. Optical Data Storage Center, Optical Sciences Center, University of Arizona, Tucson. OPTICAL RECORDING.

Milwee, William I. Milwee Associates, Inc., Portland, Oregon. UNDERWATER DEMOLITION.

Minelli, Dr. Alessandro. Dipartimento di Biologia, Università degli Studi di Padova, Italy. CHILOPODA.

Minifie, Bernard W. Knechtel Laboratories, Bristol, England. COCOA POWDER AND CHOCOLATE.

Minkler, Ward W. Transitional Metals, Sewickley, Pennsylvania. TITANIUM METALLURGY.

Misra, Dr. Pratap. Lincoln Laboratory, Massachusetts Instititute of Technology, Lexington. SATELLITE NAVIGATION SYSTEMS.

Mitchell, Dr. Charles E. Department of Geology, State University of New York at Buffalo. GRAPTOLITHINA.

Mitchell, Dr. Jeffery C. SPACEHAB, Orlando, Florida. SPACE SHUTTLE; SPACE TECHNOLOGY; SPACE POWER SYSTEMS.

Mitchell, Dr. Kim W. Solar Energy Research Institute, Golden, Colorado. SOLAR CELL.

Mitchell, Dr. Malcolm S. Comprehensive Cancer Center, Kenneth Norris, Jr. Cancer Research Institute, University of Southern California, Los Angeles. IMMUNOTHERAPY.

Mitchell, Dr. Todd. Department of Atmospheric Sciences, University of Washington, Seattle. EL NINO.

Mitchison, Dr. N. A. Department of Zoology, University College, London, England. ACQUIRED IMMUNOLOGICAL TOLERANCE.

Mitsch, Prof. William J. Graduate Program in Environmental Science, School of Natural Resources, Ohio State University, Columbus. WETLANDS.

Mitterer, Dr. Richard M. Department of Geosciences, University of Texas, Dallas. BIOSPHERE.

Miyawaki, Dr. Ritsuro. Department of Geology, National Science Museum, Tokyo, Japan. RARE-EARTH MINERALS.

Mizell, Dr. David W. Boeing Shared Services Group, Applied Research and Technology, Seattle, Washington. SUPERCOMPUTER.

Moder, Dr. Joseph J. Chairperson, Department of Management Science, University of Miami, Florida. PERT.

Modesto, Dr. Sean. Department of Biology, Cape Breton University, Nova Scotia, Canada. ANAPSIDA.

Moe, Dr. Richard L. Department of Botany, University of California, Berkeley. AGAR; ALGAE; FUCALES; RHODOPHYCEAE; other articles.

Moennig, Dr. Volker. Institute of Virology, School of Veterinary Medicine, Germany. CLASSICAL SWINE FEVER.

Moffat, Prof. Anthony F. J. Département de Physique, Université de Montréal, Québec, Canada. WOLF-RAYET STAR.

Moffett, Dr. Mark B. Naval Underwater Systems Center, New London Laboratory, New London, Connecticut. PARAMETRIC ARRAYS.

Moh, Dr. Za-Chieh. Moh & Associates, Inc., Oriental Technopolises Tower, Taipei, Taiwan. SUBWAY ENGINEERING; TUNNEL.

Mohanakumar, Dr. T. Department of Surgery, Washington University School of Medicine, St. Louis, Missouri. TISSUE TYPING.

Mohr, Prof. David W. Department of Geology, Texas A&M University, College Station, Texas. GNEISS.

Mohr, Prof. Peter. Department of Physics, Yale University, New Haven, Connecticut. FEYNMAN DIAGRAM.

Moll, Dr. Gary A. Vice President, Urban Forest Center, American Forests, Washington, D.C. FORESTRY, URBAN.

Møllar , Prof. Aage R. Department of Neurological Surgery, University of Pittsburgh. HEARING (HUMAN).

Moment, Dr. Gairdner B. Department of Biological Sciences, Goucher College, Towson, Maryland. CELL CONSTANCY.

Montville, Dr. Thomas J. Department of Food Science, Cook College, Rutgers University. FOOD MICROBIOLOGY.

Mood, Dr. Alexander M. Professor Emeritus, Administration and Policy Analyst, University of California, Irvine. STATISTICS.

Moody, Capt. Alton B. Deceased; formerly, Navigation Consultant, San Diego, California. BUOY; CELESTIAL NAVIGATION; MAGNETIC COMPASS; NAVIGATION; PILOTING; SEXTANT; other articles.

Mooi, Dr. Rich. California Academy of Sciences, San Francisco. IRREGULARIA.

Mook, Dr. Douglas G. Department of Psychology, University of Virginia, Charlottesville. MOTIVATION.

Mooney, Dr. Walter D. U.S. Geological Survey, Menlo Park, California. EARTH CRUST.

Moore, Dr. Alice E. Sloan-Kettering Institute for Cancer Research, New York. ROUS SARCOMA.

Moore, Prof. Clyde H., Jr. Director, Applied Carbonate Research Program, Department of Geology, Agricultural and Mechanical College, Louisiana State University, Baton Rouge. LIMESTONE.

Moore, Donald V. Department of Microbiology, Southwestern Medical School, University of Texas, Dallas. ACANTHOCEPHALA; ARCHIACANTHOCEPHALA; EOACANTHOCEPHALA; PALAEACANTHOCEPHALA.

Moore, Dr. James Alexander. Department of Chemistry, University of Delaware. CHEMISTRY; ESSENTIAL OILS; MENTHOL; ORGANIC CHEMISTRY; ORGANIC NOMENCLATURE; ORGANOSULFUR COMPOUND; other articles.

Moore, Dr. Jeffrey S. Department of Chemistry, University of Illinois, Urbana-Champaign. DENDRITIC MACROMOLECULE.

Moore, Dr. Paul B. Department of the Geophysical Sciences, University of Chicago, Illinois. FLUORITE; MAGNETITE; MINERALOGY; other articles.

Moore, Dr. Richard K. Remote Sensing Laboratory, University of Kansas Center for Researching, Inc., Lawrence. ALTIMETER.

Moore, Theodore T. Research Engineer, Solvay Advanced Polymers, Alpharetta, Georgia. POLYSULFONE RESINS.

Moorer, Major, Daniel F., Jr. Ball Aerospace & Technologies, Boulder, Colorado. MILITARY SATELLITES.

Mordes, Dr. J. P. Department of Medicine, University of Massachusetts Medical School, Worcester. TRANSPLANTATION BIOLOGY.

Morel, Dr. Pierre. Director, NASA Office of Earth Science, Washington, D.C. METEOROLOGICAL SATELLITES.

Moreno, Dr. Giuliana. Laboratoire de Biophysique, Museum National d'Histoire Naturelle, Paris, France. LASER PHOTOBIOLOGY.

Moretti Ojemann, Linda. Department of Neurology, University of Washington, Seattle, Washington. SEIZURE DISORDERS.

Morgan, Dr. Alexander P. Senior Staff Research Scientist, Mathematics Department, General Motors Research Laboratories, Warren, Michigan. POLYNOMIAL SYSTEMS OF EQUATIONS.

Morgan, Dr. Anne H. Deceased; formerly, Mount Holyoke College. EPHEMEROPTERA.

Morgan, Prof. John W. Department of Mathematics, Columbia University, New York. TOPOLOGY.

Morgan, Dr. Karl Z. Deceased; formerly, Neely Professor, School of Nuclear Engineering, Georgia Institute of Technology, Atlanta. HEALTH PHYSICS; RADIOACTIVITY STANDARDS; other articles.

Morgan-Richards, Dr. Mary. Allan Wilson Centre for Molecular Ecology and Evolution, Massey University, New Zealand. PHASMATODEA.

Moritz, Prof. Helmut. Technische Universität Graz, Abteilung für Physikalische Geodäsie, Graz, Austria. EARTH, GRAVITY FIELD OF.

Morrell, Dr. Jeffrey J. Department of Forest Products, Oregon State University, Corvallis. WOOD DEGRADATION.

Morris, Prof. Glenn K. Department of Biology, University of Toronto at Mississauga, Ontario, Canada. GRYLLOBLATTODEA.

Morris, Prof. J. W., Jr. Department of Materials Sciences and Mineral Engineering, University of California, Berkeley. SOLDERING.

Morris, Prof. Philip J. Department of Aerospace Engineering, Pennsylvania State University, University Park. AERODYNAMIC SOUND.

Morris, Dr. Simon Conway. Department of Earth Sciences, University of Cambridge, England. ANNELIDA; BURGESS SHALE.

Morris, Dr. William F. Department of Zoology, Duke University, Durham, North Carolina. HERBIVORY.

Morrison, Dr. David C. Associate Director, Basic Research Programs, Department of Microbiology, University of Kansas Medical Center, Kansas City. ENDOTOXIN.

Morrison, Prof. George H. Analytical Facility of Materials Science Center, Cornell University. CHEMICAL SEPARATION TECHNIQUES.

Morrison, Dr. Kenneth J. Cooperative Extension Service, Washington State University. LENTIL; PEA.

Morse, Dr. Robert W. Associate Director and Dean of Oceanographic Studies, Woods Hole Oceanographic Institution, Woods Hole, Massachusetts. ACOUSTIC MINE; ECHO SOUNDER; SOFAR; SONAR.

Morse, Dr. Roger A. Department of Entomology, Cornell University. BEEKEEPING.

Morse, Dr. Roy E. Food Science Department, Cook College, Rutgers University. SALT (FOOD).

Morse, Dr. Stephen A. Director, Division of Sexually Transmitted Diseases Laboratory Research, Centers for Disease Control and Protection, Atlanta, Georgia. GONORRHEA.

Mortenson, Leonard E. Department of Biological Sciences, Purdue University, West Lafayette, Indiana. NITROGEN CYCLE.

Morton, Dr. Andrew. Department of Geology and Petroleum Geology, University of Aberdeen, Kings College, Aberdeen, United Kingdom. HEAVY MINERALS.

Moscowitz, Dr. Albert. Department of Chemistry, University of Minnesota. COTTON EFFECT.

Moseley, Dr. Steve L. Department of Microbiology, University of Washington School of Medicine, Seattle. ESCHERICHIA; MEDICAL BACTERIOLOGY; WATER-BORNE DISEASE.

Moshell, Dr. Alan. Director, Skin Diseases Program, National Institute of Arthritis and Musculoskeletal and Skin Diseases, National Institutes of Health, Bethesda, Maryland. SKIN DISORDERS.

Moss, Timothy J. Ph.D. Program, University of Iowa, Ames. NUCLEOPROTEIN.

Mossman, Prof. Harland W. Deceased; formerly, Professor, Department of Anatomy, University of Wisconsin. PLACENTATION.

Motani, Dr. R. Geology Department, University of California, Davis. ICHTHYOPTERYGIA.

Motekaitis, Dr. R. J. Department of Chemistry, Texas A&M University. CHELATION.

Mottershead, Dr. Allen. Physical Science Department, Cypress College, Cypress, California. CAPACITANCE.

Mottet, Dr. N. Karle. Professor of Pathology and Director of Hospital Pathology, University Hospital, University of Washington, Seattle. FETAL ALCOHOL SPECTRUM DISORDER; other articles.

Moulton, Prof. William G. National High Magnetic Field Laboratory, Tallahassee, Florida. HIGH MAGNETIC FIELDS.

Mouroulis, Dr. Pantazis. Jet Propulsion Laboratory, Pasadena, California. DIFFRACTION GRATING.

Mow, Dr. Van C. New York Orthopaedic Hospital Research Laboratory, Columbia-Presbyterian Medical Center, Columbia University, New York. JOINT (ANATOMY).

Mower, Dr. Richard L. Department of Botany, University of California, Berkeley. ERGOT AND ERGOTISM.

Moyer, Dr. Kenneth E. Department of Psychology, Carnegie-Mellon University. INSTINCTIVE BEHAVIOR.

Moyer, Dr. William W. Applied Research Laboratory, Pennsylvania State University, University Park. CALCULATORS.

Mrak, Dr. Emil M. Deceased; formerly, Office of Chancellor Emeritus, University of California, Davis. DISTILLED SPIRITS; ETHYL ALCOHOL.

Mudrick, Prof. Stephen E. Department of Atmospheric Science, University of Missouri. FRONT.

Mueller, Prof. Berndt. Department of Physics, Duke University, Durham, North Carolina. QUARK-GLUON PLASMA.

Mueller, Dr. Werner E.G. Institut fur Physiologische Chemie, Universitat Mainz, Germany. INTRON.

Muffler, Dr. L. J. Patrick. Geologist, Branch of Field Geochemistry and Petrology, U.S. Geological Survey, Department of the Interior, Menlo Park, California. GEOTHERMAL POWER.

Muller, Dr. Pamela Hallock. Professor, College of Marine Science, University of South Florida, St. Petersburg. NUMMULITES.

Mulligan, Dr. Pamela K. Formerly, Department of Biochemistry, University of North Carolina. GENETIC ENGINEERING; PROTEINS, EVOLUTION OF.

Mulliken, Dr. Robert S. Deceased; formerly, Institute of Molecular Biophysics, Florida State University; Nobelist. MOLECULAR STRUCTURE AND SPECTRA.

Munger, Dr. H. M. Department of Plant Breeding and Biometry, New York State College of Agriculture and Life Sciences, Cornell University, Ithaca, New York. SQUASH.

Munson, Prof. Bruce R. Department of Aerospace Engineering and Mechanics, Iowa State University, Ames. ARCHIMEDES PRINCIPLE.

Munz, Dr. R. J. Department of Chemical Engineering, McGill University, Montreal, Quebec, Canada. ARC HEATING; ELECTRIC FURNACE.

Murgatroyd, Dr. R. J. Meteorological Office, Bracknell, England. TROPOPAUSE; TROPOSPHERE.

Murphy, Dr. Eric J. School of Medicine and Health Sciences, University of North Dakota, Grand Forks. INOSITOL.

Murphy, Dr. Frederick A. Centers for Disease Control, Department of Health and Human Services, Tucker, Georgia. EXOTIC VIRAL DISEASES.

Murphy, Dr. Robert L. Division of Infectious Diseases, Northwestern University Medical School, Chicago, Illinois. OPPORTUNISTIC INFECTIONS.

Murphy, Dr. Terence M. Department of Anesthesiology, School of Medicine, University of Washington. PAIN.

Murray, Prof. Haydn H. Department of Geological Sciences, Indiana University, Bloomington. BAUXITE; CLAY, COMMERCIAL; FULLER'S EARTH.

Murray, Dr. James W. School of Oceanography, University of Washington, Seattle. BLACK SEA.

Murray, Dr. Joanne L. Alloy Phase Diagrams Data Center, U.S. Department of Commerce. ALLOY STRUCTURES.

Murray, Prof. M. John. Department of Medicine, University of Minnesota Medical School, Minneapolis. IRON METABOLISM.

Murray, Dr. Merritt J. Formerly, A. M. Todd Company, Kalamazoo, Michigan. PEPPERMINT; SPEARMINT.

Murray, Prof. R. G. E. Department of Microbiology and Immunology, Health Sciences Centre, University of Western Ontario, London, Canada. BACTERIAL TAXONOMY.

Murray, Dr. R. W. Department of Earth Sciences, Boston University, Massachusetts. SILICEOUS SEDIMENT.

Murray, Dr. Roger K. Department of Chemistry and Biochemistry, University of Delaware, Newark. CAGE HYDROCARBON.

Murray, Dr. Royce W. Department of Chemistry, University of North Carolina. PRECIPITATION (CHEMISTRY); SUPERSATURATION; VOLATILIZATION.

Murray, Dr. Stephen S. Center for Astrophysics, Harvard College Observatory, Smithsonian Astrophysical Observatory, Cambridge, Massachusetts. X-RAY ASTRONOMY; X-RAY TELESCOPE.

Murray, Dr. Thomas P. Applied Research Laboratory, United States Steel Corporation, Monroeville, Pennsylvania. PYROMETER.

Murray, Dr. William. Teknowledge Corporation, Palo Alto, California. EXPERT SYSTEMS.

Musacchia, Dr. X. J. Department of Physiology and Space Science Research Center, University of Missouri. HIBERNATION AND ESTIVATION; HYPOTHERMIA.

Musgrave, R. B. Consultant. AGRICULTURAL SOIL AND CROP PRACTICES.

Mushlitz, R. D. Ore Department, American Smelting and Refining Company, New York. SILVER METALLURGY.

Muther, Richard. Executive Director, Richard Muther & Associates, Inc., Kansas City, Missouri. INDUSTRIAL FACILITIES.

Mutter, Dr. Carolyn Z. Associate Research Scientist, Lamont-Doherty Earth Observatory, Columbia University, Palisades, New York. MOHO (MOHOROVIČIĆ DISCONTINUITY).

Mutter, Dr. John. Lamont-Doherty Geological Observatory, Columbia University, Multichannel Seismics Group, Palisades, New York. SEISMOLOGY.

Muus, Dr. Bent J. Research Biologist, Danish Institute for Fishery and Marine Research, Charlottenlund, Denmark. NEMATOMORPHA.

Myers, Prof. A. L. Department of Chemical Engineering, University of Pennsylvania, Philadelphia. ADSORPTION OPERATIONS.

Myers, Phil E. Tanks and Pressure Vessels, Chevron Research and Technology Company, Richmond, California. STORAGE TANK.

Myklebust, Prof. Arvid. Retired; formerly, Mechanical Engineering Department, Virginia Polytechnic Institute and State University, Blacksburg. COMPUTER-AIDED ENGINEERING.

Mynlieff, Dr. Michelle. Department of Biology, Marquette University, Milwaukee. NORADRENERGIC SYSTEM.

Myose, Prof. Roy Y. Department of Aerospace Engineering Wichita State University, Kansas. DRONE.

N

Nacamulli, Laurette. IGEN, Inc., Rockville, Maryland. CHEMILUMINESCENCE.

Nachtrieb, Prof. Norman H. Deceased; formerly, Chairperson, Department of Chemistry, University of Chicago. EVAPORATION; LIQUID; SUBLIMATION; SURFACE TENSION; VAPOR PRESSURE.

Nachtsheim, Dr. Philip R. NASA Ames Research Center, Moffett Field, California. NOSE CONE.

Nagel, David J. Superintendent, Condensed Matter and Radiation, Sciences Division, Naval Research Laboratory, Department of the Navy, Washington, D.C. MICRORADIOGRAPHY.

Nagel, Suzanne R. Retired; formerly, Director, Manufacturing Process Research and Development, Engineering Research Center, Lucent Technologies, Princeton, New Jersey. OPTICAL FIBERS.

Nagel, Dr. Walter R. NASA Goddard Space Flight Center, Greenbelt, Maryland. BALLOON.

Namias, Jerome. Deceased; formerly, Scripps Institution of Oceanography, La Jolla, California. DROUGHT.

Napp-Zinn, Dr. Klaus. Botanical Institute, Cologne, Germany. VERNALIZATION.

Nash, Dr. David J. School of the Environment, University of Brighton, United Kingdom. KALAHARI DESERT.

Nash, Dr. Irwin. School of Medicine, Yale University. EDEMA; HEMORRHAGE; INFARCTION; THROMBOSIS.

Natland, Prof. James H. Rosentiel School of Marine and Atmospheric Science, University of Miami, Florida. EARTH CRUST.

Naumann, Dr. Robert J. Associate Vice President for Research/Consortium for Materials Development in Space, University of Alabama, Huntsville. SPACE PROCESSING.

Navrotsky, Dr. Alexandra. Department of Geological and Geophysical Sciences, Princeton University, New Jersey. PEROVSKITE.

Naylor, Dr. Arch. Department of Electrical Engineering, University of Michigan, Ann Arbor. FLEXIBLE MANUFACTURING SYSTEM.

Naylor, Dr. Ernest. Reader in Zoology, University College of Swansea, Wales. VALVIFERA.

Nealson, Dr. Kenneth H. Center for Great Lakes Studies, University of Wisconsin. BACTERIAL LUMINESCENCE.

Nedelsky, Prof. Leo. Deceased; formerly, Department of Physical Science, University of Chicago. CONSERVATION OF ENERGY; CONSERVATION OF MASS; other articles.

Nedungadi, Dr. Ashok P. Integrated Electronics Facility, Schlumberger Well Services, Houston, Texas. LINEARITY; VOLTAGE REGULATOR.

Nelms, Dr. W. P. Deputy Chief, Aircraft Technology Division, National Aeronautics and Space Administration, Ames Research Center, Moffett Field, California. SHORT TAKEOFF AND LANDING (STOL); VERTICAL TAKEOFF AND LANDING (VTOL).

Nelsen, Dr. Olin E. Department of Biology, University of Pennsylvania, Philadelphia. GLAND; LACRIMAL GLAND; PHOTOPHORE GLAND; POISON GLAND; SCENT GLAND; other articles.

Nelson, Dr. Curtis L. Department of Agronomy, University of Missouri. ABACA; AGRICULTURAL SCIENCE (PLANT); COIR; FIBER CROPS; FLAX; LINEN; other articles.

Nelson, Dr. Donna C. Intelligent Transport Society of America, Washington, D.C. RAILROAD CONTROL SYSTEMS; TRAFFIC CONTROL SYSTEMS.

Nelson, Elton G. Fiber Specialist, Special Collection, National Agricultural Library; and U.S. Department of Agriculture, Beltsville, Maryland. RAMIE; SISAL; other articles.

Nelson, Prof. Frederick E. Department of Geography and Planning, Rutgers University, New Brunswick, New Jersey. PERMAFROST.

Nelson, Dr. George C. Department of Mathematics, University of Iowa, Iowa City. RECURSIVE FUNCTION.

Nelson, Prof. Philip. A. Institute of Sound and Vibration Research, University of Southampton, United Kingdom. ACTIVE SOUND CONTROL.

Nelson, Dr. Richard B. Chief Engineer, Varian Associates, Palo Alto, California. KLYSTRON.

Nelson, Dr. T. J. Bell Communications Research, Red Bank, New Jersey. MAXWELL'S EQUATIONS.

Nergaard, Dr. Leon S. Director, Microwave Research Laboratory, RCA Laboratories, Princeton, New Jersey. THERMIONIC TUBE.

Nestle, Dr. Marion. Chairperson, Department of Home Economics and Nutrition, Division of Education, New York University. MALNUTRITION; OBESITY.

Netravali, Dr. Arun N. President, Bell Laboratories, Murray Hill, New Jersey. TELEVISION SCANNING.

Neu, Dr. Harold C. Department of Medicine, Columbia Presbyterian Medical Center, New York. CEPHALOSPORINS.

Neubecker, T. F. Manager, Engineering, High Intensity and Quartz Lamp Department, General Electric Company, Twinsburg, Ohio. MERCURY-VAPOR LAMP; VAPOR LAMP.

Neudeck, Prof. Gerald W. Department of Electrical Engineering, Purdue University, West Lafayette, Indiana. TRANSISTOR.

Neuhauser, Robert G. Electronic Components, Solid State Division, RCA Corp., Lancaster, Pennsylvania. TELEVISION CAMERA TUBE.

Nevins, Dr. James L. Charles Stark Draper Laboratory, Cambridge, Massachusetts. MANUFACTURING ENGINEERING.

New, Dr. Maria. Department of Pediatrics, Mount Sinai School of Medicine, New York. ALDOSTERONE.

New, Dr. Ronald. Systems Planning and Development Staff, Office of Charting and Geodetic Services, U.S. Department of Commerce, National Oceanic and Atmospheric Administration, Rockville, Maryland. COMPUTERIZED TOMOGRAPHY.

Newcomb, Prof. Robert W. Microsystems Laboratory, Electrical Engineering Department, University of Maryland. CIRCUIT (ELECTRONICS).

Newell, Dr. Homer E. National Aeronautics and Space Administration, Washington, D.C. SCIENTIFIC AND APPLICATIONS SATELLITES.

Newhouse, Alan R. Retired; formerly, U.S. Department of Energy. SHIP NUCLEAR PROPULSION.

Newton, Dr. Chester W. National Center for Atmospheric Research, Boulder, Colorado. ATMOSPHERIC GENERAL CIRCULATION; SQUALL; STORM; other articles.

Newton, Dr. R. C. Department of the Geophysical Sciences, University of Chicago, Illinois. GRANULITE; METAMORPHIC ROCKS; METAMORPHISM.

Newton, Prof. Roger G. Department of Physics, Indiana University, Bloomington. MATHEMATICAL PHYSICS; SCATTERING MATRIX; VARIATIONAL METHODS (PHYSICS); VECTOR METHODS (PHYSICS).

Ng, Dr. Tai P. New Ground Resources Ltd., Alberta, Canada. SEISMIC EXPLORATION FOR OIL AND GAS.

Nichols, Dr. Mark L. Deceased; formerly, U.S. Department of Agriculture. AGRICULTURAL SOIL AND CROP PRACTICES.

Nicodemus, Dr. Fred E. Physicist, Optical Radiation Section, Heat Division, Institute for Basic Standards, National Institute of Standards and Technology, Gaithersburg, Maryland. RADIANCE.

Nicolson, Dr. Daniel H. Department of Botany, Smithsonian Institution, Washington, D.C. PLANT NOMENCLATURE.

Niebel, Benjamin W. Professor Emeritus of Industrial Engineering, Pennsylvania State University; and Industrial Engineering Consultant, State College, Pennsylvania. METHODS ENGINEERING; PROCESS ENGINEERING.

Nier, Prof. Alfred O. School of Physics and Astronomy, University of Minnesota. MASS SPECTROSCOPE.

Niesz, Dr. Dale E. Director, Center for Ceramics Research, College of Engineering, Rutgers University, New Brunswick, New Jersey. CERAMICS.

Nieto, Dr. Alberto S. Department of Geology, University of Illinois, Urbana. LANDSLIDE; MASS WASTING.

Nigrelli, Dr. Ross F. Osborn Laboratories of Marine Sciences, New York Aquarium, New York. ACTINOMYXIDA; CNIDOSPORA; HELICOSPORIDA; MICROSPORIDEA; MYXOSPORIDA; MYXOSPORIDEA.

Nippes, Dr. Ernest F. Department of Metallurgical Engineering, Rensselaer Polytechnic Institute. RESISTANCE WELDING.

Niven, Dr. Andrew. Department of Mechanical and Aeronautical Engineering, University of Limerick, Ireland. WING.

Niven, Charles F., Jr. Director of Research Center, Del Monte Research Center, Walnut Creek, California. STERILIZATION.

Nix, Prof. William D. Associate Chairperson, Department of Materials Science and Engineering, Stanford University. CREEP (MATERIALS).

Noback, Dr. Charles R. Department of Anatomy, Columbia University, New York. MUSCULAR SYSTEM; NERVOUS SYSTEM (VERTEBRATE).

Nobel, Prof. Park S. Professor Emeritus, Department of Biology, University of California, Los Angeles. CACTUS.

Noguchi, Dr. Thomas T. Pasadena, California. FORENSIC MEDICINE.

Nolan, Dr. R. P. Environmental Sciences Laboratory, Brooklyn College, New York. ASBESTOS.

Norberg, Dr. Ulla M. Department of Zoomorphology, Zoological Institute, University of Goteborg, Sweden. FLIGHT.

Norcross, Prof. Neil L. College of Veterinary Medicine, Cornell University, Ithaca, New York. MASTITIS.

Nordman, Prof. James E. Department of Electrical and Computer Engineering, University of Wisconsin, Madison. CONTACT POTENTIAL DIFFERENCE; SCHOTTKY BARRIER DIODE.

Norris, Prof. David O. University of Colorado, Boulder. FORENSIC BOTANY.

Norris, Prof. H. Thomas. Chairperson, Department of Clinical Pathology and Diagnostic Medicine, East Carolina University, School of Medicine, Greenville, North Carolina. PANCREAS DISORDERS.

North, Dr. Gerald R. Director, Climate System Research Program, Department of Meteorology, Texas A&M University, College Station. CLIMATE MODELING; CLIMATE MODIFICATION.

Northcutt, Dr. R. Glenn. Department of Anatomy, Case Western Reserve University, Cleveland, Ohio. NERVOUS SYSTEM (VERTEBRATE).

Norton, Dr. Jeanette. Department of Plants, Soils and Biometeorology, Utah State University, Logan. SOIL MICROBIOLOGY.

Novikoff, Dr. Phyllis. Department of Pathology, Albert Einstein College of Medicine, Bronx, New York. CYTOCHEMISTRY.

Novotny, Prof. Donald W. Department of Electrical and Computer Engineering, University of Wisconsin, Madison. ALTERNATING-CURRENT MOTOR; ELECTRIC POWER MEASUREMENT; GENERATOR; other articles.

Novotny, Prof. Milos V. Department of Chemistry, Indiana University, Bloomington. CHROMATOGRAPHY.

Nowak, Dr. Peter. Department of Rural Sociology, University of Wisconsin. PRECISION AGRICULTURE.

Noy, Dr. Noa. Division of Nutritional Sciences, Cornell University, Ithaca, New York. VITAMIN A.

Noyce, Dr. David A. Civil and Environmental Engineering, University of Wisconsin-Madison. TRANSPORTATION ENGINEERING.

Noyes, Richard M. Department of Chemistry, College of Arts and Sciences, University of Oregon. OSCILLATORY REACTION.

Noyes, Dr. Robert W. Harvard College Observatory, Cambridge, Massachusetts. HELIOSEISMOLOGY.

Nuccitelli, Prof. Richard. Zoology Department, University of California, Davis. CELL POLARITY (BIOLOGY).

Nulman, Dr. Irena. The Hospital for Sick Children, Toronto, Ontario, Canada. FETAL ALCOHOL SPECTRUM DISORDER.

Nussbaumer, Prof. Harry. Institut für Astronomie, Zürich, Switzerland. SYMBIOTIC STAR.

Nygren, Dr. David R. Physics Division, Lawrence Berkeley National Laboratory, University of California, Berkeley. TIME-PROJECTION CHAMBER.

Nygren, Stephen. Bell Telephone Laboratories, Reading, Pennsylvania. SEMICONDUCTOR DIODE.

Nyhan, Dr. William. Department of Pediatrics, University of California, San Diego. HUMAN GENETICS; URIC ACID.

Nyman, Dr. Matthew W. Senior Research Associate, Department of Earth and Planetary Sciences, University of New Mexico, Albuquerque. HORNFELS; PHYLLITE.

O

Obery, Leonard J. Chief, Plans and Programs Offices, National Aeronautics and Space Administration, Cleveland, Ohio. SUPERSONIC DIFFUSER.

O'Brien, Prof. Michael. Department of Agricultural Engineering, University of California, Davis. AGRICULTURAL MACHINERY.

O'Connell, Prof. John P. Department of Chemical Engineering, University of Florida, Gainesvlle. FLUIDS.

O'Connor, Dr. Donald J. Environmental Engineering and Science Program, Manhattan College, New York. STREAM POLLUTION.

Odom, Dr. Robert W. Charles Evans Associates, Redwood City, California. SECONDARY ION MASS SPECTROMETRY (SIMS).

O'Donnell, Franklin. Jet Propulsion Laboratory, California Institute of Technology, Pasadena, California. SPACE PROBE.

Ogasawara, Prof. Yoshihide. Department of Earth Sciences, Waseda University, Tokyo, Japan. OPHIOLITE.

Ogburn, Charles B. Cooperative Extension Service, Auburn University, Alabama. TERRACING (AGRICULTURE).

Ogle, Dr. Kenneth N. Deceased; formerly, Section of Biophysics, Mayo Clinic, Rochester, Minnesota. STEREOSCOPY.

Ogle, Dr. Michael. Techinical Dicrector, Material Handling Industrty of America, Charlotte, North Carolina. MATERIALS HANDLING.

Ogren, Ingmar. Tofs Corporation, Veddoe, Sweden. MODELING LANGUAGES.

O'Hagan, Dr. David. Department of Chemistry, University of Durham, United Kingdom. BIOSYNTHESIS.

Ohlendorf, Dr. Douglas H. Genex Corporation, Gaithersburg, Maryland. OPERON.

Ojo, Prof. Kanji. Department of Materials Science and Engineering, University of California, Los Angeles. ACOUSTIC EMISSION.

Okandan, Dr. Ender. Department of Petroleum and Natural Gas Engineering, Middle East Technical University, Ankara, Turkey. PETROLEUM.

Olcott, John W. Editor and Publisher, *Business & Commercial Aviation* (McGraw-Hill), Aerospace and Defense Group, White Plains, New York. GENERAL AVIATION.

Old, Bruce S. Senior Vice President, Arthur D. Little, Inc., Cambridge, Massachusetts. PRESSURIZED BLAST FURNACE.

Oldenburger, Prof. Rufus. Deceased; formerly, Director, Automatic Control Center, Purdue University, West Lafayette, Indiana. GOVERNOR; LINKAGE (MECHANISM).

O'Leary, Dr. John E. Department of Forest Engineering, Oregon State University, Corrallis. FOREST HARVEST AND ENGINEERING.

O'Leary, Dr. Nuala. Dana-Farber Cancer Center, Harvard Medical School, Boston, Massachusetts. CELL DIVISION.

O'Lenic, Dr. Edward A. Chief, Climate Operations Branch, Climate Prediction Center, Maryland. WEATHER FORECASTING AND PREDICTION.

Oliver, Prof. John E. Department of Geography, Geology, and Anthropology, Indiana State University, Terre Haute. PHYSICAL GEOGRAPHY.

Olmstead, Prof. Richard. Department of Botany, University of Washington, Seattle. LAMIALES.

Olness, Dr. John W. Physics Department, Brookhaven National Laboratory, Upton, New York. GAMMA RAYS.

Olson, Prof. Edward C. Department of Astronomy, University of Illinois, Urbana. BINARY STAR.

Olson, Dr. Everett C. Deceased; formerly, Department of Biology, University of California, Los Angeles. ANAPSIDA; CAPTORHINIDA; PELYCOSAURIA; SAUROPTERYGIA; other articles.

Olson, Dr. Harry F. Deceased; formerly, Staff Vice President, Acoustical and Electromechanical Research, RCA Laboratories, Princeton, New Jersey. BINAURAL SOUND SYSTEM.

Olson, Dr. Norman F. Department of Food Science, University of Wisconsin. CHEESE.

Onishi, Steven. Fairchild Imaging, Milpitas, California. CHARGE-COUPLED DEVICES.

Oosthuizen, Dr. Patrick H. Department of Mechanical Engineering, Queen's University, Kingston, Ontario, Canada. COMPRESSIBLE FLOW.

Oppenheimer, Dr. Ben R. Department of Astrophysics, American Museum of Natural History, New York. BROWN DWARF.

Oppenheimer, Dr. Steven B. Department of Biology, California State University, Northridge. YOLK SAC.

O'Rahilly, Dr. Ronan. Carnegie Embryological Laboratories, University of California, Davis. EYE (VERTEBRATE).

Orians, Dr. Gordon. Department of Zoology, University of Washington, Seattle. HUMAN ECOLOGY.

Oris, Dr. James T. Department of Zoology, Miami University, Oxford, Ohio. ENVIRONMENTAL TOXICOLOGY.

Orkin, Dr. Frederick K. University of California, San Francisco, The Medical Center at the University of California School of Medicine. ANESTHESIA.

Orr-Weaver, Dr. Terry. Whitehead Institute for Biomedical Research, Cambridge, Massachusetts. POLYPLOIDY.

Orrego, Dr. Christian. Department of Biology, Brandeis University, Waltham, Massachusetts. BACTERIA.

Orville, Dr. Harold D. Institute of Atmospheric Sciences, South Dakota School of Mines and Technology, Rapid City. HAIL.

Orville, Prof. Richard E. Cooperative Institute for Applied Meteorological Studies, Texas A&M University, College Station. LIGHTNING.

Osberg, Prof. Philip H. Department of Geological Sciences, University of Maine, Orono. ANTICLINE; FOLD AND FOLD SYSTEMS; GRABEN; HORST; MASSIF.

Osborn, Prof. Jeffrey M. Professor and Chair of Biology, Division of Science, Truman State University, Kirksville, Missouri. PALYNOLOGY.

O'Shaughnessy, Dr. Douglas. INRS-Telecom, Place Bonaventure, Montreal, Quebea, Canada. VOICE RESPONSE.

Osofsky, Dr. Michael. Naval Research laboratory Washington, D.C. SUPERCONDUCTIVITY.

Osswald, Prof. Tim. University of Wiscosin-Madison. POLYMER COMPOSITE.

Oster, Prof. Gerald. Mount Sinai School of Medicine, City University of New York. MOIRÉ PATTERN.

Ostergaard, Dr. Hanne L. Department of Medical Microbiology and Immunology, University of Alberta, Edmonton, Canada. CYTOLYSIS.

Ostro, Dr. Steven J. Jet Propulsion Laboratory, California Institute of Technology, Pasadena. RADAR ASTRONOMY.

Ostroff, Dr. Robert. Yale New Haven Psychiatric Hospital, Connecticut. ELECTROCONVULSIVE THERAPY.

Ostrom, Dr. John H. Division of Vertebrate Paleontology, Peabody Museum of Natural History, Yale University, New Haven, Connecticut. ORNITHISCHIA; SAURISCHIA.

Othmer, Prof. Donald F. Department of Chemical Engineering, Polytechnic Institute of Brooklyn, New York. ALCOHOL FUEL; ALKALI; FLUIDIZATION.

Otis, Dr. Arthur B. Department of Physiology, College of Medicine, University of Florida, Gainesville. SPIROMETRY.

Otten, Prof. Ernst W. Institut fur Physik, Johannes Gutenberg-Universitat Mainz, Germany. DOPPLER EFFECT.

Otto, Dr. Gilbert F. Department of Zoology, University of Maryland, College Park. SPIRURIDA.

Otto, Dr. Norman C. Ford Motor Research Laboratory, Dearborn, Michigan. CATALYTIC CONVERTER.

Over, Dr. Jeff. Department of Geological Sciences, State University of New York, Geneseo. DEVONIAN.

Overhauser, Dr. Albert W. Department of Physics, Purdue University, West Lafayette, Indiana. BLOCH THEOREM; CHARGE-DENSITY WAVE.

Ovshinsky, Stanford R. President, Energy Conversion Devices, Inc., Troy, Michigan. GLASS SWITCH.

Owen, Dr. Denis F. Department of Biology, Oxford Polytechnic, Oxford, England. DESERT.

Owen, Prof. Guillermo. Department of Mathematical Sciences, Rice University, Houston, Texas. FINITE MATHEMATICS.

Owen, Dr. Tobias C. Institute for Astronomy, University of Hawaii, Honolulu. NEPTUNE; PLUTO; SATELLITE (ASTRONOMY); SATURN; URANUS.

P

Pacella, Gary. United States Army Armament Research Development and Engineering Center, Picatinny Arsenal, New Jersey. ARMY ARMAMENT.

Packard, Prof. Richard E. Department of Physics, University of California, Berkeley. QUANTIZED VORTICES.

Packer, Dr. Lester. Department of Molecular and Cell Biology, University of California, Berkeley. ANTIOXIDANT.

Padian, Prof. Kevin. Museum of Paleontology, University of California, Berkeley. DINOSAURIA; PTEROSAURIA.

Page, Dr. Roy C. Director, Research Center in Oral Biology, School of Medicine, University of Washington, Seattle. DENTISTRY; MOUTH DISORDERS; PERIODONTAL DISEASE; TOOTH DISORDERS.

Pake, Dr. George E. Vice President, Xerox Corporation, and General Manager, Xerox Palo Alto Research Center, Palo Alto, California. FORCE.

Palermo, Dr. David S. Associate Dean, Research and Graduate Studies, Department of Psychology, College of Liberal Arts, Pennsylvania State University. PSYCHOLINGUISTICS.

Palevsky, Dr. Gerald. Consulting Professional Engineer, Hastings-on-Hudson, New York. CIVIL ENGINEERING; SEPTIC TANK; SEWAGE; SEWAGE COLLECTION SYSTEMS; SEWAGE SOLIDS; SEWAGE TREATMENT.

Palmer, Dr. Allison R. Institute for Cambrian Studies, Boulder, Colorado. CAMBRIAN; GEOLOGICAL TIME SCALE.

Palmer, Prof. Darwin L. Chief, Veterans Administration Hospital, Medical Center, Department of Medicine, Infectious Disease Division, University of New Mexico School of Medicine. PLAGUE.

Palmer, Prof. Fred. Lighting Research Center, Osram Sylvania, Beverly, Massachusetts. ATOMIC MASS.

Palmer, Dr. James D. School of Information Technology and Engineering, George Mason University, Fairfax, Virginia. PROTOTYPE; SYSTEMS ENGINEERING; SYSTEMS INTEGRATION.

Palmer, Dr. Melissa. Plainview, New York. CIRRHOSIS.

Palmerlee, Prof. Albert S. Deceased; formerly, Professor of Engineering Drawing, School of Engineering and Architecture, University of Kansas. DESCRIPTIVE GEOMETRY.

Palzkill, Dr. David A. Department of Plant Sciences, University of Arizona. JOJOBA.

Pandolfi, Dr. John M. Department of Paleobiology, National Museum of Natural History, Smithsonian Institute, Washington, D.C. SCLERACTINIA.

Pang, Prof. Stella. Department of Electrical Engineering and Computer Science, University of Michigan, Ann Arbor. ION IMPLANTATION.

Pangborn, Dr. John B. Director, Alternative Energy Systems Research, Institute of Gas Technology, Chicago, Illinois. AIRCRAFT FUEL.

Panish, Dr. Morton B. Head, Materials Science Research Department, AT&T Bell Laboratories, Murray Hill, New Jersey. CRYSTAL GROWTH.

Pant, Dr. H. K. Department of Environmental & Geological Sciences, Lehman College, City University of New York, Bronx. PHOSPHATASE.

Pantell, Richard. Electrical Engineering Department, Stanford University. LASER.

Panton, Prof. Ronald L. Mechanical Engineering Department, The University of Texas at Austin. KELVIN'S CIRCULATION THEOREM.

Papageorgiou, George. Department of Botany, University of Illinois, Urbana-Champaign. CHLOROPHYLL.

Papendick, Dr. Robert I. Research Leader, U.S. Department of Agriculture, Agricultural Research Service, Washington State University. AGRICULTURE SOIL AND CROP PRACTICES; FERTILIZER.

Papitashvili, Dr. Vladimir. Space Physics Research Laboratory, University of Michigan, Ann Arbor. GEOMAGNETIC VARIATIONS.

Pardee, Dr. Arthur B. Dana-Farber Center, Harvard Medical School, Boston, Massachusetts. CELL DIVISION.

Parente, Michael. MP Technologies, Hendersenville, North Carolina. COMMUNICATIONS SYSTEMS PROTECTION.

Paresce, Dr. Francesco. European Southern Observatory, Garching, Germany, and Istituto Nazionale di Astrofisica, Rome, Italy. BLUE STRAGGLER STAR.

Park, Prof. David A. Department of Physics, Williams College. CLASSICAL FIELD THEORY; DETERMINISM; PLANCK'S CONSTANT.

Park, Prof. Jae-Woo. Department of Civil Engineering, University of Hawaii at Manoa. WATER DESALINATION.

Parker, Dr. Ingrid M. Department of Biology, Earth & Marine Sciences, University of California, Santa Cruz. INVASION ECOLOGY.

Parker, Dr. Joel. Southwest Research Institute, Boulder, Colorado. CERES.

Parker, Dr. Maurice H. Consultant, Aeronautics, Hampton, Virginia. AIRCRAFT TESTING.

Parker, Sybil P. Retired; formerly, Editor in Chief, *McGraw-Hill Encyclopedia of Science and Technology,* The McGraw-Hill Companies, New York. AMBERGRIS; ENDOCRINOLOGY; HERMAPHRODITISM; STATOBLASTS.

Parma, Dr. Ana. International Pacific Halibut Commission, Seattle, Washington. ADAPTIVE MANAGEMENT.

Parmentier, Prof. E. Mark. Department of Geological Sciences, Brown University, Providence, Rhode Island. GEODYNAMICS.

Parmesan, Dr. Camille. National Center for Ecological Analysis and Synthesis, Santa Barbara, California. GLOBAL CLIMATE CHANGE.

Parpia, Prof. Jeevak M. Department of Physics, Cornell University. ABSOLUTE ZERO.

Parr, Dr. J. F. Biological Waste Management and Organic Resources Laboratory, Beltsville, Maryland. AGRICULTURAL SOIL AND CROP PRACTICES; FERTILIZER.

Parr, Dr. Robert G. Department of Chemistry, University of North Carolina. CHEMICAL BONDING; DELOCALIZATION; VALENCE.

Parrett, Prof. Douglas F. Department of Animal Science, University of Illinois, Urbana. BEEF CATTLE PRODUCTION.

Parrish, Dr. Judith Totman. Department of Geosciences, University of Arizona. PALEOCLIMATOLOGY.

Parrish, Dr. William. IBM Almaden Research Laboratory, San Jose, California. X-RAY FLUORESCENCE ANALYSIS.

Parsons, Dr. Thomas S. Ramsay Wright Zoological Laboratories, University of Toronto, Ontario, Canada. ANATOMY, REGIONAL; HEART (VERTEBRATE); LUNG; MENINGES; PERITONEUM; other articles.

Partridge, Dr. Edward G. Consultant, Redondo Beach, California. RUBBER.

Parvainen, Dr. Pekka. Polar Image, Turku, Finland. MIDNIGHT SUN.

Pasachoff, Prof. Jay M. Director, Hopkins Observatory, Williams College, Williamstown, Massachusetts. ASTRONOMICAL ATLASES; ASTRONOMY; BIG BANG THEORY; CONSTELLATION; ECLIPSE; EQUINOX; SCHMIDT CAMERA; articles on constellations; other articles.

Pascual, Dr. Miguel. Centro Nacional Patagónico, Chubut, Argentina. FISHERIES ECOLOGY.

Pasternak, Dr. Gavril W. Attending Neurologist, Department of Neurology, Memorial Hospital, Sloan-Kettering Institute Cancer Center, New York. ENDORPHINS.

Pasto, Prof. Daniel J. Department of Chemistry, University of Notre Dame. CONFORMATIONAL ANALYSIS.

Patel, Dr. C. K. N. Bell Laboratories, Murray Hill, New Jersey. PHOTOACOUSTIC SPECTROSCOPY.

Patrick, Norman W. JVC Digital Technologies Center, Escondido, California. CATHODE-RAY TUBE.

Patterson, Dr. David J. Department of Zoology, University of Bristol, England. ACTINOPHRYIDA; CENTROHELIDA; DESMOTHORACIDA.

Patterson, Dr. Eric G. Applied Research Laboratory, Pennsylvania State University, State College. COMPUTATIONAL FLUID DYNAMICS; KELVIN'S MINIMUM ENERGY THEOREM.

Patterson, Prof. Michael M. Director of Research Affairs, College of Osteopathic Medicine, Ohio University. LEARNING MECHANISMS.

Patton, Prof. Donald J. Crawfordville, Florida. SOIL CONSERVATION.

Patton, Dr. Harry D. Chairperson, Department of Physiology, Biophysics, School of Medicine, University of Washington. EFFECTOR SYSTEMS.

Patton, Peter C. Minnesota Superscomputer Institute, University of Minnesota. MULTIPROCESSING.

Patzkowsky, Dr. Mark E. Department of Geosciences, Pennsylvania State University, University Park. ORTHIDA; RHYNCHONELLIDA; SPIRIFERIDA; STROPHOMENIDA.

Paul, Dr. Burton. Department of Mechanical Engineering and Applied Mechanics, College of Engineering and Applied Science, University of Pennsylvania. SHAFT BALANCING.

Pawson, Dr. David L. National Museum of Natural History, Smithsonian Institution, Washington, D.C. ECHINODERMATA; ECHINOIDEA.

Payne, Prof. Joseph N. School of Education, University of Michigan, Ann Arbor. ARITHMETIC; PERCENT; SIGNIFICANT FIGURES.

Payne, Prof. Peter A. Retired; formerly, Vice President for Academic Development, Department of Instrumentation and Analytical Science, University of Manchester, Institute of Science and Technology, Manchester, England. TRANSDUCER.

Peacey, Dr. John G. Centre de Recherche Noranda, Pointe-Claire, Quebec, Canada. COPPER METALLURGY.

Peacor, Dr. Donald R. Department of Geology, University of Michigan. FELDSPATHOID.

Pecar, Dr. Joseph A. Joseph A. Pecar & Associates, Inc., Potomac, Maryland. SWITCHING CIRCUIT.

Pechenik, Dr. Jan. Department of Biology, Tufts University, Medford, Massachusetts. METAMORPHOSIS.

Peck, Dr. Stewart B. Department of Biology, Carleton University, Ottawa, Ontario, Canada. ONYCHOPHORA.

Pecora, Prof. Robert. Department of Chemistry, Stanford University, California. QUASIELASTIC LIGHT SCATTERING.

Pecoraro, Dr. Vincent. Department of Chemistry, University of Michigan. VANADIUM.

Pedersen, Prof. Niels C. School of Veterinary Medicine, University of California, Davis. FELINE LEUKEMIA.

Peetz, Prof. Ralf. Chemistry Department, Center for Engineered Polymeric Materials, CUNY/College of Staten Island, New York. POLYMERIZATION.

Pegg, Prof. David J. Department of Physics and Astronomy, University of Tennessee, Knoxville. NEGATIVE ION; PHOTOIONIZATION.

Pehlke, Dr. Robert D. Department of Materials Science and Engineering, University of Michigan. STEEL MANUFACTURE.

Peirce, Dr. G. R. Engineer, Champaign, Illinois. ARC LAMP; INCANDESCENT LAMP; INFRARED LAMP; LAMP; SUNLAMP.

Pelletier, Dr. S. William. Director, Institute for Natural Product Research, University of Georgia, Athens. ALKALOID.

Pelosi, Dr. James J. Manager, Integrated Requirements, ISS Payloads Office, NASA Liaison to the Italian Space Agency, Johnson Space Center, Houston, Texas. SPACE STATION.

Peltier, Dr. W. R. Department of Physics, University of Toronto, Ontario, Canada. EARTH, CONVECTION IN.

Pennak, Dr. Robert W. Department of Biology, University of Colorado. MYSTACOCARIDA.

Penneman, Dr. Robert A. Los Alamos National Laboratory, Los Alamos, New Mexico. AMERICIUM; NEPTUNIUM.

Penner-Hahn, Prof. James E. Department of Chemistry, University of Michigan, Ann Arbor. EXTENDED X-RAY ABSORPTION FINE STRUCTURE (EXAFS).

Penrose, Prof. Roger. Mathematical Institute, University of Oxford, England. POLARIZATION OF WAVES.

Pepper, Prof. Darrell W. Department of Mechanical Engineering, University of Nevada, Las Vegas. ENVIRONMENTAL FLUID MECHANICS.

Perch, Michael. Koppers Company, Inc., Monroeville, Pennsylvania. COKE.

Perl, Prof. Martin L. Stanford Linear Accelerator Center, Stanford, California; Nobelist. LEPTON.

Perlmann, Dr. Gertrude E. Deceased; formerly, Rockefeller University, New York. PROTEIN.

Perrin, Arthur M. Deceased; formerly, President, National Conveyors Company, Inc., Fairfax, New Jersey. BULK-HANDLING MACHINES; CONVEYOR; HOISTING MACHINES; INDUSTRIAL TRUCKS; MONORAIL; other articles.

Perry, Astor. Extension Agronomy Specialist, North Carolina State University. PEANUT.

Perry, Stephen F. Exxon Research, Florham Park, New Jersey. DEWAXING OF PETROLEUM.

Perry, Tekla S. Associate Editor, *Spectrum,* Institute of Electrical and Electronics Engineers, New York. VIDEO GAMES.

Perryman, Dr. Michael A. C. Space Science Department, European Space Agency, ESTEC (SCI-SA), Noordwijk, Netherlands. HERTZSPRUNG-RUSSELL DIAGRAM.

Peskin, Prof. Michael E. Stanford Linear Accelerator Center, Stanford, California. STANDARD MODEL.

Pessin, Dr. Jeffrey E. Department of Pharmacological Sciences, SUNY-Stony Brook, New York. GLUCOSE.

Peter, Dr. Patrick. Institut d'Astrophysique de Paris, France. COSMIC STRINGS.

Peters, Prof. Philip C. Deceased; formerly, Department of Physics, University of Washington. BLACK HOLE; GRAVITATIONAL COLLAPSE.

Peters, Prof. Randall D. Department of Physics, Mercer University, Macon, Georgia. ANHARMONIC OSCILLATOR; CHAOS.

Peters, Dr. William C. Professor Emeritus, Mining and Geological Engineering, University of Arizona. MINING; PROSPECTING; UNDERGROUND MINING.

Petershack, Dr. Victor D. Rexnord, Inc., Milwaukee, Wisconsin. CHAIN DRIVE.

Peterson, Dr. Benjamin B. Department of Engineering, United States Coast Guard Academy, New London, Connecticut. HYPERBOLIC NAVIGATION SYSTEM; LORAN.

Peterson, Dr. Gary L. Senior Optical Engineer, Breault Research Organization, Tucson, Arizona. GHOST IMAGE (OPTICS).

Peterson, Dr. Hart deCourdes. Cornell Medical College, New York. CEREBRAL PALSY.

Peterson, Prof. James F. Department of Geography and Planning, Southwest Texas State University, San Marcos. GEOGRAPHY.

Peterson, Dr. Kirk D. Texas Instruments Inc., Plano, Texas. BRIDGE CIRCUIT; GAIN.

Peterson, Dr. R. L. Department of Botany and Genetics, University of Guelph, Ontario, Canada. TRANSFER CELLS.

Petschek, Prof. Albert G. Department of Physics, New Mexico Institute of Mining and Technology, Socorro. VIRIAL THEOREM.

Peyghambarian, Dr. Nasser. University of Arizona, College of Optical Sciences. LASER.

Pfaffmann, Dr. Carl. Department of Psychology, Rockefeller University, New York. CHEMICAL SENSES; TASTE.

Pfeffer, Dr. Robert A. U.S. Army, Nuclear and Chemical Agency, Springfield, Virginia. ELECTROMAGNETIC PULSE (EMP).

Pfeifer, Prof. Peter. Department of Physics, University of Missouri, Columbia. QUANTUM COMPUTATION.

Pfennig, Prof. Norbert. Institut für Mikrobiologie, Göttingen Universität, Germany. BACTERIA.

Pfleeger, Dr. Charles P. ARCA Systems, Inc., Vienna, Virginia. COMPUTER SECURITY.

Phadke, Prof. Arun G. Distinguished Professor Emeritus, Virginia Polytechnic Institute and State University, Blacksburg. TRANSMISSION LINES.

Phaff, Prof. Herman J. Department of Food Science and Technology, College of Agricultural and Environmental Sciences, University of California, Davis. DISTILLED SPIRITS; ETHYL ALCOHOL; YEAST.

Phaneuf, Prof. Ronald A. Chair, Department of Physics, University of Nevada, Reno. SCATTERING EXPERIMENTS (ATOMS AND MOLECULES).

Phelan, Prof. Richard M. Department of Mechanical Systems and Design, Cornell University. BLOCK AND TACKLE; HYDRAULIC PRESS; INCLINED PLANE; MECHANICAL ADVANTAGE; WHEEL AND AXLE; other articles.

Philander, Prof. S. George. Program in Atmospheric and Oceanic Sciences, Princeton University. EQUATORIAL CURRENTS.

Philip, Dr. Cornelius B. Department of Entomology, California Academy of Sciences, San Francisco. TICK PARALYSIS; other articles.

Philipson, William R. Department of Botany, University of Canterbury, Christchurch, New Zealand. PRIMARY VASCULAR SYSTEM (PLANT).

Phillips, Dr. James A. CTR Division Office, Los Alamos National Laboratory, Los Alamos, New Mexico. PINCH EFFECT.

Phillips, Mike. Software Engineering Institute, Pittsburgh, Pennsylvania. CAPABILITY MATURITY MODELING.

Phillips, Prof. Tom L. Department of Botany, University of Illinois. COENOPTERIDALES.

Phillips, Dr. William D. National Institute of Standards and Technology, Gaithersburg, Maryland; Nobelist. LASER COOLING.

Phillips, William H. NASA Langley Research Center, Hampton, Virginia. FLIGHT CONTROLS.

Pickles, Dr. Andrew J. Caltech Opitcal Observatories, California Institute of Technology, Pasadena. SCHMIDT CAMERA.

Pielou, E. C. Department of Biology, University of Lethbridge, Alberta, Canada. SPECIATION.

Pience, Roger. Detection Technologies, South Portland, Maine. CABLE TELEVISION SYSTEM.

Pierce, Prof. Allan D. Department of Aerospace and Mechanical Engineering. Boston University. ACOUSTIC INTERFEROMETER; ACOUSTIC RADIOMETER; FUZZY-STRUCTURE ACOUSTICS.

Pierce, Prof. E. Lowe. Department of Zoology, University of Florida. CHAETOGNATHA.

Pierce, Dr. F. J. Department of Crop and Soil Sciences, Michigan State University, East Lansing. PRECISION AGRICULTURE.

Pierce, Dr. Marcia. Department of Biological Sciences, Eastern Kentucky University, Richmond. PROKARYOTAE; VIRAL INCLUSION BODIES; VIRUS; VIRUS CLASSIFICATION; WEST NILE VIRUS.

Pierre, Dr. Donald A. Department of Electrical Engineering, College of Engineering, Montana State University. OPTIMIZATION.

Pike, Dr. Andrew C. Neotronics Scientific Ltd., United Kingdom. MICROSENSOR.

Pike-Tay, Dr. Anne. Department of Anthropology, Vassar College, Poughkeepsie, New York. PALEOLITHIC.

Pilbrow, Prof. John R. Head, Department of Physics, Monash University, Clayton, Victoria, Australia. ELECTRON PARAMAGNETIC RESONANCE (EPR) SPECTROSCOPY.

Pilger, Dr. John. Department of Biology, Agnes Scott College, Decatur, Georgia. ECHIURA.

Pimm, Dr. Stuart. Department of Zoology, University of Tennessee, Knoxville. FOOD WEB.

Pinder, Prof. George F. Professor of Civil and Environmental Engineering, Professor of Mathematics, University of Vermont, Burlington. GROUND-WATER HYDROLOGY.

Pine, Dr. Daniel S. College of Physicians and Surgeons of Columbia University and the New York State Psychiatric Institute, New York. ANXIETY DISORDERS.

Pinedo, Prof. Michael. Statistics and Operations Research Department, Stern School of Business, New York University, New York. SCHEDULING.

Pinkava, Dr. Donald J. Director, ASU Herbarium, Department of Botany, Arizona State University. PLANT KEYS.

Pinkel, Dr. Benjamin. Consulting Engineering, Santa Monica, California. AFTERBURNER; PULSE JET.

Pinker, Prof. Rachel T. Department of Meteorology, University of Maryland. INSOLATION.

Pionke, Dr. Harry B. Pasture Systems and Watershed Management Research Laboratory, U.S. Department of Agriculture–Agricultural Research Service, University Park, Pennsylvania. ACID RAIN.

Pipkin, Prof. Francis M. Department of Physics, Harvard University. LINEWIDTH.

Piraino, Dr. Stefano. Università di Lecce, Dipartimento di Biologia, Italy. ONTOGENY.

Pitchford, Dr. Leanne C. Sandia National Laboratories, Albuquerque, New Mexico. GAS DISCHARGE.

Pitman, Walter C. III. Lamont-Doherty Earth Observatory, Columbia University, Palisades, New York. PLATE TECTONICS.

Pitts, Dr. Ronald. Space Telescope Science Institute, Baltimore, Maryland. PLEIADES.

Plapp, Dr. Bryce. Department of Biochemistry, University of Iowa. ENZYME INHIBITION.

Platt, Dr. Donald. Micro Aerospace Solutions, Inc., Melbourne, Florida. SPACE COMMUNICATIONS.

Plaut, Dr. Gerhard W. E. Professor Emeritus of Biochemistry, Temple University School of Medicine, Philadelphia, Pennsylvania. CITRIC ACID CYCLE.

Pleasants, Dr. Julian R. International Committee on Laboratory Animals, Lobund Laboratory, University of Notre Dame. GNOTOBIOTICS.

Plecnik, Prof. Joseph M. Department of Civil Engineering, California State University of Long Beach. BRIDGE.

Plesset, Dr. Michael. Aerospace Corporation, Los Angeles, California. COMPUTER STORAGE TECHNOLOGY.

Ploem, Dr. Johan S. Laboratorium voor Cytochemie en Cytometrie, Rijksuniversiteit Leiden, Wassenaarseweg, Netherlands. FLUORESCENCE MICROSCOPE.

Plotnick, Dr. Roy E. Department of Geological Sciences, College of Liberal Arts and Sciences, University of Illinois. EURYPTERIDA.

Plueddemann, Dr. Edwin P. Dow Corning Corporation, Midland, Michigan. SILICON.

Pocius, Dr. Alphonsus. 3M Corporate Research Materials Laboratory, St. Paul, Minnesota. ADHESIVE; ADHESIVE BONDING.

Pohl, Dr. D. W. IBM Research Division, Zurich Research Laboratory, Ruschlikon, Switzerland. OPTICAL MICROSCOPE.

Pohl, Prof. Robert O. Laboratory of Atomic and Solid State Physics, Cornell University. RADIOACTIVE WASTE MANAGEMENT; SPECIFIC HEAT.

Pohlmann, Prof. Kenneth C. School of Music, University of Miami, Coral Gables. COMPACT DISK.

Polchinski, Prof. Joseph. Institute for Theoretical Physics, University of California, Santa Barbara. DUALITY (PHYSICS); RENORMALIZATION.

Polhamus, Edward C. Langley Research Center, National Aeronautics and Space Administration, Hampton, Virginia. STRAKE.

Polking, Dr. John C. Department of Mathematics, Rice University. DIFFERENTIAL EQUATION; INTEGRAL EQUATION.

Pollack, Dr. Henry N. Professor of Geophysics, Department of Geological Sciences, University of Michigan. EARTH, HEAT FLOW IN.

Pollack, Dr. James B. Deceased; formerly, Ames Research Center, National Aeronautics and Space Administration, Moffett Field, California. VENUS.

Pollard, Dr. Thomas. Department of Cell Biology and Anatomy, Johns Hopkins University School of Medicine, Baltimore, Maryland. CELL ORGANIZATION.

Pollard, Dr. William G. Oak Ridge Associated Universities, Oak Ridge, Tennessee. PHYSICS.

Polley, Dr. Margaret J. Formerly, Department of Medicine, Cornell Medical Center, New York. ANTIBODY; ANTIGEN; HETEROPHILE ANTIGEN; IMMUNOLOGY; ISOANTIGEN.

Polly, Dr. P. David. Department of Anatomy, Queen Mary & Westfield College, London, United Kingdom. CREODONTA.

Pomykalski, Dr. James J. Assistant Professor, Integrated Science and Technology and Computer Science Programs, James Madison University, Harrisonburg, Virginia. SIMULATION.

Ponikvar, Ade L. Editor, Production and Drilling. *Petroleum Week,* New York. TURBODRILL.

Ponikvar, Adolph L. Manager, Technical Publications, International Lead Zinc Research Organization, Inc., New York. LEAD METALLURGY.

Pont-Lezica, Dr. Rafael. Department of Biology, Washington University, St. Louis, Missouri. LECTINS.

Pontecorvo, Dr. Guido. Imperial Cancer Research Fund Laboratories, London, England. LINKAGE (GENETICS).

Poole, Prof. Colin F. Department of Chemistry, Wayne State University, Detroit, Michigan. CHROMATOGRAPHY; ION EXCHANGE.

Poore, Dr. Gary C. B. Senior Curator, Museum Victoria, Australia. SPELAEOGRIPHACEA.

Pope, Dr. Joseph. Pope Engineering Company, Newton Centre, Massachusetts. NOISE MEASUREMENT.

Pope, Michael. Pope, Evans and Robbins, Consulting Engineers, New York. FLUIDIZED-BED COMBUSTION.

Popp, Dr. Robert K. Department of Geology and Geophysics, Texas A&M University. CUMMINGTONITE.

Popper, Dr. Arthur N. Director, Neuroscience and Cognitive Science Program, Department of Biology, University of Maryland, College Park. EAR (VERTEBRATE).

Popper, Dr. Daniel M. Deceased; formerly, Department of Astronomy, University of California, Los Angeles. MASS-LUMINOSITY RELATION.

Porges, Dr. Steven. Department of Human Development, University of Maryland, College Park. AUTONOMIC NERVOUS SYSTEM.

Porra, Dr. Robert J. CSIRO, Canberra, Australia. CHLOROPHYLL.

Porter, Dr. Mark D. Department of Chemistry, Iowa State University of Science and Technology. SURFACE AND INTERFACIAL CHEMISTRY.

Porter, Dr. Richard D. Chair, Department of Mathematics, Northeastern University, Boston, Massachusetts. FIBER BUNDLE.

Porter, Dr. Warren. Department of Zoology, University of Wisconsin, Madison. PHYSIOLOGICAL ECOLOGY (ANIMAL).

Posada, German. Research Associate, Department of Psychology, State University of New York, Stony Brook. MATERNAL BEHAVIOR.

Post, Dr. Richard F. Lawrence Livermore National Laboratory, Livermore, California. LAWSON CRITERION; NUCLEAR FUSION; THERMONUCLEAR REACTION.

Postgate, Prof. John R. Department of Microbiology, University of Sussex, England. NITROGEN FIXATION.

Postow, Dr. Elliot. Special Studies Section, Division of Research Grants, National Institutes of Health, Bethesda, Maryland. BIOELECTROMAGNETICS.

Potter, Dr. Kathleen N. Department of Microbiology, University of Texas Southwestern Medical Center, Dallas. IMMUNOGLOBULIN.

Potter, Prof. M. David. School of Business, San Francisco State College. CAMEL'S HAIR; CASHMERE; PIÑA; SILK; VICUNA.

Potter, Prof. Norman N. Department of Food Science, Cornell University. FOOD PRESERVATION; FOOD SCIENCE.

Potts, Dr. Kevin T. Department of Chemistry, Rensselaer Polytechnic Institute. YLIDE.

Potvin, Prof. Jean. Department of Physics, Parks College of Engineering, Aviation, and Technology, St. Louis, Missouri. PARACHUTE.

Poulson, Dr. Donald F. Department of Biology, Yale University. LETHAL GENE.

Povh, Dr. Bogdan. Max-Planck Institut für Kernphysik, Saupfercheckweg, Germany. HADRONIC ATOM; HYPERNUCLEI.

Powell, Jeffrey R. Department of Biology, Yale University. POLYMORPHISM (GENETICS).

Powell, Dr. Martha J. Department of Biological Sciences, University of Alabama, Tuscaloosa. CHYTRIDIOMYCOTA; HYPHOCHYTRIOMYCOTA; PLASMODIOPHOROMYCOTA.

Powell, Robert C. Consultant, Physics, Gaithersburg, Maryland. BARRETTER.

Powell, Dr. Wayne B. Vice President for Research and Associate Dean of the Graduate College, Oklahoma State University, Stillwater. ALGEBRA.

Pozar, Prof. David M. Department of Electrical and Computer Engineering, University of Massachusetts, Amherst. MICROWAVE FILTER.

Pozos, Dr. Robert S. Sustained Performance Program Manager, Department of the Navy, Naval Health Research Center, Department of Defense, San Diego, California. HYPOTHERMIA.

Praston, Dr. Lincoln F. Research Scientist, Institute of Arctic and Alpine Research, Department of Geology, University of Colorado, Boulder. SEA-FLOOR IMAGING.

Prendini, Dr. Lorenzo. American Museum of Natural History, Division of Invertebrate Zoology, New York. SCORPIONES.

Prescott, David M. Department of Molecular, Cellular and Developmental Biology, Boulder, Colorado. CELL (BIOLOGY); CELL BIOLOGY.

Presley, Dr. John F. National Institute of Child Health and Human Development, Bethesda, Maryland. GREEN FLUORESCENT PROTEIN; VIDEO MICROSCOPY.

Press, Dr. George. Northwest Fisheries Science Center, Seattle, Washington. WATER CONSERVATION.

Price, Dr. James F. Department of Physical Oceanography, Woods Hole Oceanographic Institution, Massachusetts. OCEAN CIRCULATION.

Price, Dr. P. Buford, Jr. Lawrence Berkeley National Laboratory, University of California, Berkeley. FISSION TRACK DATING—coauthored; PARTICLE TRACK ETCHING.

Price, Dr. Peter W. Department of Entomology, Northern Arizona University. GUILD.

Priester, Gayle B. Consulting Engineer, Baltimore, Maryland. COMFORT HEATING.

Primack, Prof. Joel R. Physics Department, University of California, Santa Barbara. WEAKLY INTERACTING MASSIVE PARTICLE (WIMP).

Prime, Mr. Jamison S. Office of Engineering and Technology, Federal Communications Commission, Washington, D.C. RADIO SPECTRUM ALLOCATION.

Princen, L. H. Associate Center Director, North Regional Research Center, Department of Agriculture, Peoria, Illinois. RENEWABLE RESOURCES.

Prinn, Prof. Ronald G. Department of Earth, Atmosphere and Planetary Sciences, Massachusetts Institute of Technology, Cambridge. STRATOSPHERIC OZONE.

Prisbrey, Prof. Keith A. Department of Metallurgical and Mining Engineering, College of Mines and Earth Resources, University of Idaho, Moscow. GOLD METALLURGY.

Pritchard, Prof. David E. Department of Physics, Massachusetts Institute of Technology, Cambridge. ATOM OPTICS; ATOMIC MASS.

Pritchard, Tram C. Lockheed Missiles and Space Company, Sunnydale, California. DRAFTING.

Pritchett, Wilson S. Senior Project Engineer, Noller Control Systems Inc., Richmond, California. INDUCTOR; NONSINUSOIDAL WAVEFORM; SATURABLE REACTOR.

Pritsker, Prof. A. Alan B. School of Industrial Engineering, Purdue University. GERT.

Proctor, Mary C. Department of Surgery, University of Michigan Hospitals, Ann Arbor. PHLEBITIS.

Prospero, Dr. Joseph M. Department of Marine and Atmospheric Chemistry, University of Miami. DUST STORM.

Prothero, Dr. Donald R. Chair, Department of Geology, Occidental College, Los Angeles, California. DRYOLESTIDA; MAMMALIA; PERISSODACTYLA.

Pryor, Dr. William A. Department of Chemistry, Louisiana State University. AUTOXIDATION; BENZOYL PEROXIDE; PEROXIDE.

Puck, Dr. Theodore T. Department of Biophysics, University of Colorado Medical Center. TISSUE CULTURE.

Pugel, Andrew. Department of Civil Engineering, California State University at Long Beach, California. BRIDGE.

Pullen, Prof. Keats A., Jr. Ballistic Research Laboratories, Aberdeen Proving Ground, Maryland. TRANSFORMER.

Punzo, Dr. Fred. Department of Biology, University of Tampa, Florida. SOLIFUGAE.

Purchon, Dr. Richard D. Department of Zoology, Chelsea College, University of London, England. PTERIOMORPHIA.

Purohit, Dr. Milind. Department of Physics and Astronomy, University of South Carolina, Columbia. DEEP INELASTIC COLLISIONS.

Purvis, Dr. Andy. Department of Zoology, University of Oxford, United Kingdom. ALLOMETRY.

Putnam, Prof. Russell C. Professor Emeritus, Case Institute of Technology. CANDLEPOWER; LUMINANCE; LUMINOUS ENERGY; LUMINOUS FLUX; LUMINOUS INTENSITY.

Putney, Dr. James W. National Institute of Environmental Health Sciences, Research Triangle Park, North Carolina. CALCIUM METABOLISM.

Putz, Dr. Thomas J. Deceased; formerly, Westinghouse Electric Corporation, Philadelphia, Pennsylvania. GAS TURBINE.

Pyeritz, Dr. Reed E. Departments of Medicine and Genetics, University of Pennsylvania, Philadelphia. SEX-LINKED INHERITANCE.

Q

Quaife, Carol. Department of Pathology, University of Washington. DISEASE; PATHOLOGY; TUMOR.

Qualset, Dr. Calvin O. Director, Genetic Resources Conservation Program, University of California, Davis. TRITICALE.

Quarles, Dr. John M. Department of Microbial and Molecular Pathogenesis, College of Medicine, Texas A&M University, College Station. INFLUENZA.

Que, Prof. Lawrence Jr. Center for Metals in Biocatalysis, University of Minnesota, Minneapolis. BIOINORGANIC CHEMISTRY.

Quéré, Dr. David. Physique de la Matière Condensée, Collège de France, Paris. FLUID COATING.

Quigg, Dr. Chris. Theoretical Physics Department, Fermi National Accelerator Laboratory, Batavia, Illinois. QUANTUM CHROMODYNAMICS.

Quimby, Dr. Edith H. Deceased; formerly, Department of Radiology, College of Physicians and Surgeons, Columbia University. NUCLEAR RADIATION (BIOLOGY).

Quinn, Dr. T. J. Bureau International des Poids et Mesures, Pavillion de Breteuil, Sevres, France. THERMOCOUPLE.

Quirin, Edward J. Consulting Engineer, Besier, Gibble & Quirin, Old Saybrook, Connecticut. COFFERDAM; other articles.

Quiroga, Dr. Carlos. Pacific Technical Documentation, Windsor, California. WORD PROCESSING.

Quiroz, Dr. Roderick S. Retired; formerly, Research Meteorologist, Upper Air Branch, National Oceanic and Atmospheric Administration, U.S. Department of Commerce. MESOSPHERE.

R

Raab, Dr. Frederick H. President and Owner, Green Mountain Radio Research Company, Winooski, Vermont. RADIO TRANSMITTER.

Raad, Prof. Peter E. Mechanical Engineering Department, School of Engineering and Applied Science, Southern Methodist University, Dallas, Texas. POTENTIAL FLOW.

Rabinowicz, Prof. Ernest. Department of Mechanical Engineering, Massachusetts Institute of Technology. FRICTION; WEAR.

Rabinowitz, Dr. Mario. Formerly, Senior Scientist, Electrical Systems Division, Electric Power Research Institute, Palo Alto, California. ELECTRICAL INSULATION.

Rachubinski, Dr. Richard A. Department of Cell Biology and Anatomy, University of Alberta, Edmonton, Alberta, Canada. PEROXISOME.

Rademacher, Prof. Hans. Deceased; formerly, Professor Emeritus of Mathematics, University of Pennsylvania. ZERO.

Radforth, Dr. Norman W. Muskeg Research Institute, University of New Brunswick, Fredericton, Canada. MUSKEG.

Radke, Dr. Rodney O. Agricultural Product Research Laboratory, Monsanto Company, St. Louis, Missouri. HERBICIDE.

Radok, Dr. Uwe. Environmental Research (CIRES), University of Colorado. POLAR METEOROLOGY.

Rae, Dr. Peter M. M. Molecular Diagnostics, Inc., West Haven, Connecticut. EXON; INTRON.

Rafferty, Vincent F. Siemens Enterprise Networks, Reston, Virginia. INTERCOMMUNICATING SYSTEMS (INTERCOMS); KEY TELEPHONE SYSTEM.

Ragnarsson, Dr. Kristjan T. Department of Rehabilitation Medicine, Mount Sinai School of Medicine, New York. CONCUSSION.

Rahenkamp, Robert A. International Business Machines, Lexington, Kentucky. TYPEWRITER.

Rahn, Frank J. Nuclear Power Division, Electric Power Research Institute, Palo Alto, California. NUCLEAR POWER; NUCLEAR REACTOR.

Rahn, Prof. Perry H. Depatment of Geology and Geological Engineering, South Dakota School of Mines and Technology, Rapid City. ENGINEERING GEOLOGY.

Raikhel, Dr. Natasha V. Department of Energy, Michigan State University, East Lansing. VACUOLE.

Rajan, Prof. Krishna. Materials Engineering Department, Rensselaer Polytechnic Institute, Troy, New York. EPITAXIAL STRUCTURES.

Raju, Prof. G. V. S. Director, Division of Engineering, University of Texas, San Antonio. BLOCK DIAGRAM.

Rakovan, Prof. John. Department of Geology, Miami University, Oxford, Ohio. MONAZITE.

Ramasethu, Dr. Jayashnee. Department of Pediatric & Neonatology, Georgetown University Hospital, Washington, D.C. RH INCOMPATIBILITY.

Ramberg, Dr. Edward G. Deceased; formerly, RCA Laboratories, Princeton, New Jersey. CHILD-LANGMUIR LAW; ELECTRON LENS; ELECTRON MICROSCOPE; MAGNETIC LENS; other articles.

Ramdas, Prof. Anant K. Department of Physics, Purude University, West Lafayette, Indiana. ACCEPTOR ATOM.

Ramey, Prof. Robert L. Department of Electrical Engineering, University of Virginia. SERIES CIRCUIT; TIME CONSTANT.

Ramsey, Dr. Norman F. Department of Physics, Harvard University; Nobelist. NEGATIVE TEMPERATURE.

Rana, Dr. Farhan. Department of Electrical and Computer Engineering, Cornell University, Ithaca, New York. QUANTUM ELECTRONICS.

Randerson, Dr. Peter. Department of Applied Biology, University of Wales, Institute of Science and Technology, Cardiff. ECOLOGICAL SUCCESSION.

Rao, Dr. V. N. M. Dupont Central Research and Development, Wilmington, Delaware. FLUORINE; HALOGENATED HYDROCARBON; FLUOROCARBON.

Raper, Prof. John R. Deceased; formerly, Professor of Botany, Harvard University. FUNGI.

Rapp, Prof. George, Jr. Dean, College of Letters and Science, University of Minnesota. ARCHEOLOGICAL CHRONOLOGY.

Raps, Prof. Shirley. Professor and Chair, Department of Biological Sciences, Hunter College of CUNY. PLANT PIGMENT.

Raski, Dr. Dewey J. Department of Nematology, University of California, Davis. NEMATICIDE.

Rasmussen, Dr. Andrew K. Entomology Program, Florida A & M University, Tallahassee. PLECOPTERA.

Rasmussen, Dr. D. Tab. Department of Anthropology, Washington University, St. Louis, Missouri. HYRACOIDEA.

Rasmussen, Dr. Howard. Department of Biochemistry, University of Pennsylvania. THYROCALCITONIN; THYROID HORMONE.

Rasmussen, Dr. Norman C. Department of Nuclear Engineering, Massachusetts Institute of Technology. CHAIN REACTION (PHYSICS).

Rasmusson, Dr. Eugene M. Geophysical Fluid Dynamics Laboratory, Environmental Science Services Administration, Princeton, New Jersey. HYDROMETEOROLOGY.

Ratay, Dr. Robert T. Consulting Engineer, Manhasset, New York. BEAM COLUMN; COLUMN; TEMPORARY STRUCTURE (ENGINEERING).

Ratnoff, Dr. Oscar D. Department of Medicine, Division of Hematology/ Oncology, Case Western Research University. BLOOD; PLASMIN.

Rau, Prof. A. R. P. Department of Physics, Louisiana State University, Baton Rouge. RESONANCE (QUANTUM MECHANICS).

Rau, Dr. Michael C. EZ Communications, Fairfax, Virginia. AMPLITUDE-MODULATION RADIO; RADIO.

Rauber, Prof. Robert M. Department of Atmospheric Science, University of Illinois at Urbana-Champaign, Urbana. RADAR METEOROLOGY.

Rauchfuss, Dr. T. B. School of Chemical Sciences, University of Illinois, Urbana. SULFUR.

Raum, Dr. Donald. Center for Blood Research, Boston, Massachusetts. PRECIPITIN; PRECIPITIN TEST; SEROLOGY.

Ravishankara, Dr. A. R. National Oceanic and Atmospheric Administration, Aeronomy Laboratory, Boulder, Colorado. CHEMICAL KINETICS.

Rawles, Dr. Mary E. Deceased; formerly, Department of Embryology, Carnegie Institution, Washington, D.C. UROPYGIAL GLAND.

Rawlinson, Dr. Nicholas. Research School of Earth Sciences, The Australian National University, Canberra. ASTHENOSPHERE.

Raymond, Prof. Loren. Department of Geology, Appalachian State University, Boone, North Carolina. IGNEOUS ROCKS.

Rayner, Dr. G. H. Consultant, Teddington, Middlesex, England. CAPACITANCE MEASUREMENT.

Raz, Prof. Tzvi. Faculty of Management, Tel Aviv University, Israel. CONTROL CHART.

Read, Prof. Clark P. Deceased; formerly, Department of Biology, Rice University. AMPHILINIDEA; EUCESTODA; GYROCOTYLIDEA; PSEUDOPHYLLIDEA; TRYPANORHYNCHA; other articles.

Reay, Dr. David A. International Research & Development Company, Ltd., Fossway, Newcastle-upon-Tyne, England. HEAT PIPE.

Rebbi, Claudio. CERN, Geneva, Switzerland. SOLITON.

Rebek, Prof. Julius, Jr. Department of Chemistry, Massachusetts Institute of Technology. MOLECULAR RECOGNITION.

Rebuck, Dr. A. S. Head, Division of Respiratory Medicine, Toronto Western Hospital, Toronto, Ontario, Canada. OXIMETRY.

Recht, Dr. Abram. Joint Center for Radiation Therapy, Boston, Massachusetts. RADIATION THERAPY.

Rechtin, Dr. Eberhardt. Retired; formerly, Department of Industrial and Systems Engineering, School of Engineering, University of Southern California. SYSTEMS ARCHITECTURE.

Reddish, Dr. George F. Deceased; formerly, St. Louis College of Pharmacy and Allied Sciences. ANTISEPTIC.

Redhead, Dr. Scott A. Mycologist, Biosystematics Research Centre, Agriculture Canada, Research Branch, Central Experimental Farm, Ottawa, Ontario, Canada. BASIDIOMYCOTA; GASTEROMYCETES; HYMENOMYCETES.

Reed, Prof. Mark A. Professor of Electrical Engineering and Applied Physics, Yale University, New Haven, Connecticut. MOLECULAR ELECTRONICS.

Reed, Dr. Randall. Department of Molecular Biology and Genetics, School of Medicine, Johns Hopkins University. TRANSPOSONS.

Reed-Hill, Dr. Robert E. Department of Metallurgical Science and Engineering, University of Florida. ALLOY.

Reeder, Prof. Richard J. Department of Geosciences, State University of New York, Stony Brook. ARAGONITE; CARBONATE MINERALS.

Rees, Dr. Bryant E. Department of Biology, Fresno State College. MECHANORECEPTORS; MECOPTERA.

Rees, Jack. Corporate Services Staff, Exxon Research and Engineering Company, Florham Park, New Jersey. OIL ANALYSIS.

Reeve, Prof. John. Emeritus, Department of Electrical Engineering, Faculty of Engineering, University of Waterloo, Ontario, Canada. COMMUTATION; CONVERTER; DIRECT CURRENT TRANSMISSION.

Reeve, Dr. Roger M. Western Regional Research Laboratory, U.S. Department of Agriculture, Albany, California. SEED.

Reganold, Dr. John P. Land Management and Water Conservation Research Unit, U.S. Department of Agriculture, Pullman, Washington. AGRICULTURAL SOIL AND CROP PRACTICES.

Regnery, Dr. Russell. Supervisory Research Microbiologist, Viral and Rickettsial Zoonoses Branch, Department of Health and Human Services, Centers for Disease Control and Prevention, Atlanta, Georgia. RICKETTSIOSES.

Reich, Prof. Herbert J. Deceased; formerly, Groveland, Massachusetts. DIODE.

Reichman, Lee B. New Jersey Medical School, National Tuberculosis Center, Newark, New Jersey. MYCOBACTERIAL DISEASES.

Reichmann, Dr. M. E. Department of Microbiology, University of Illinois. ANIMAL VIRUS; COXSACKIEVIRUS; DEPENDOVIRUS; PARAMYXOVIRUS; RHINOVIRUS; VIRUS INTERFERENCE; other articles.

Reid, Dr. David A. Agricultural Research Service, U.S. Department of Agriculture, Tucson, Arizona. BARLEY.

Reid, Prof. Evans B. Chairperson, Department of Chemistry, Colby College, Waterville, Maine. CARBOXYLIC ACID.

Reid, Dr. Walter V. World Resources Institute, Washington, DC. CONSERVATION OF RESOURCES.

Reiling, Dr. Gilbert H. Lighting Research and Technical Services Operation, General Electric Company, Cleveland, Ohio. METAL HALIDE LAMP.

Reilly, John C., Jr. Retired; formerly, Naval Historical Center, U.S. Department of the Navy. NAVAL ARMAMENT.

Reingold, Dr. Edward M. Chairman, Department of Computer Science, Illinois Institute of Technology, Chicago. CALENDAR.

Reinhardt, Dr. Nicholas. Consultant, Lexington, Massachusetts. VACUUM TUBE.

Reintjes, John F. U.S. Naval Research Lab, Washington, D. C. NONLINEAR OPTICS.

Reischman, Placidus G. Head, Department of Biology, St. Martin's College, Olympia, Washington. RHIZOCEPHALA.

Reiter, Dr. Elmar R. Department of Atmospheric Science, Colorado State University. MOUNTAIN METEOROLOGY.

Reitz, Dr. Louis P. Retired; formerly, Crops Research Division, Agricultural Research Service, U.S. Department of Agriculture, Beltsville, Maryland. CEREAL; WHEAT.

Releford, Dr. John. Department of Anthropology, State University of New York, College at Oneonta. HUMAN GENETICS.

Remde, Dr. Harry F. Chief, Basic Physics Research Section, Johns-Manville Research and Engineering Center, Manville, New Jersey. HEAT INSULATION.

Remley, Frederick M., Jr. Retired; formerly, Center for Information Technology Integration, University of Michigan. CLOSED-CIRCUIT TELEVISION.

Resnick, Dr. Martin. University Hospitals of Cleveland, Ohio. URINARY TRACT DISORDERS.

Retallack, Dr. Gregory J. Department of Geological Sciences, College of Arts and Sciences, University of Oregon. PALEOSOL.

Rettinger, Michael. Deceased; formerly, Consultant on Acoustics, Encino, California. SOUND-RECORDING STUDIO.

Reynolds, Gardner M. Dames & Moore, Los Angeles, California. FOUNDATIONS.

Reynolds, Mark F. Graduate Research Assistant, Department of Chemistry, University of Wisconsin, Madison. NITRIC OXIDE.

Rhee, Dr. Katherine. Case Medical Center, Cleveland, Ohio. URINARY TRACT DISORDERS.

Rhoads, Prof. Samuel. Honolulu Community College, Hawaii. CELESTIAL SPHERE.

Ribbe, Dr. Paul H. Department of Geological Sciences, Virginia Polytechnic Institute. ANORTHITE; FELDSPAR; LABRADORITE; MICROCLINE; other articles.

Rice, Dr. Mary E. Associate Curator, Division of Worms, Smithsonian Institution, Washington, D.C. ECHIURIDA; PRIAPULIDA.

Rice, Dr. Michael A. Department of Fisheries, Animal and Veterinary Science, University of Rhode Island, Kingston. BIVALVIA.

Rich, Dr. Alexander. Department of Biology, Massachusetts Institute of Technology. DEOXYRIBONUCLEIC ACID (DNA).

Rich, Prof. Arthur. Deceased; formerly, Department of Physics, University of Michigan. ELECTRON SPIN.

Rich, Dr. Saul. Connecticut Agricultural Experiment Station, New Haven. FUNGISTAT AND FUNGICIDE; PLANT PATHOLOGY.

Richards, Dr. Frank F. Department of Medicine, Yale University. IMMUNOCHEMISTRY.

Richards, Prof. Larry G. Center for Computer-Aided Engineering, Department of Mechanical and Aerospace Engineering, University of Virginia, Charlottesville. COMPUTER-AIDED DESIGN AND MANUFACTURING.

Richards, Dr. Oscar W. College of Optometry, Pacific University, Forest Grove, Oregon. OPTICAL MICROSCOPE.

Richardson, Dr. Philip T. Department of Physical Oceanography, Woods Hole Oceanographic Institution, Woods Hole, Massachusetts. ATLANTIC OCEAN; GULF STREAM.

Richter, Reinhard H. Consultant, Freeport, Texas. KETENE.

Rickert, Richard M. Retired; formerly, AT&T Consumer Products Laboratory, Indianapolis, Indiana. TELEPHONE.

Ricksecker, Ralph E. Director of Metallurgy, Chase Brass and Copper Company, Cleveland, Ohio. COPPER ALLOYS.

Riddle, Dr. Mark A. Department of Psychiatry and Behavioral Sciences, Johns Hopkins Medical Institutions, Baltimore, Maryland. TOURETTE'S SYNDROME.

Riedel, Dr. Rupert J. Department of Zoology, University of North Carolina. GNATHOSTOMULIDA.

Rieder, Dr. Conly L. Senior Research Scientist and Professor, Division of Molecular Medicine, Lab of Cell Regulation, Wadsworth Center, New York State Department of Health, Albany. MITOSIS.

Riedl, Dr. William R. Scripps Institution of Oceanography, University of California, La Jolla. RADIOLARIA.

Rieger, Prof. Reinhard. Institute of Zoology and Limnology, University of Innsbruck, Austria. MACROSTOMORPHA.

Rieke, Dr. George. Steward Observatory, University of Arizona, Tucson. SPITZER SPACE TELESCOPE.

Rieppel, Dr. Olivier C. Department of Geology, The Field Museum, Chicago, Illinois. PLACODONTIA.

Ries, Dr. Harold C. Stanford Research Institute, Menlo Park, California. KEROSINE.

Rigby, Dr. J. Keith. Department of Geology, Brigham Young University. HEXACTINELLIDA.

Riggs, Dr. Austen F. Department of Zoology, University of Texas, Austin. RESPIRATORY PIGMENTS (INVERTEBRATE).

Riggs, Dr. Lorrin A. Department of Psychology, Brown University. COLOR VISION; VISION.

Riley, Frank. Senior Vice President, Bodine Assembly Systems, Bridgeport, Connecticut. ASSEMBLY MACHINES.

Riley, Dr. Pete. Science Applications International Corporation, San Diego, California. SOLAR WIND; VAN ALLEN RADIATION.

Riley, Dr. Ralph. Plant Breeding Institute, Cambridge, England. BREEDING (PLANT).

Riley, Dr. Reed F. Quantum Electronics Group, General Telephone and Electronics, Bayside, New York. ALKALINE-EARTH METALS; BARIUM; CALCIUM; STRONTIUM.

Rillema, Prof. James A. Professor of Physiology, Department of Physiology, School of Medicine, Wayne State University. MAMMARY GLAND.

Rindler, Dr. Michael. Associate Professor of Cell Biology, New York University School of Medicine, New York. SECRETION.

Riordan, Prof. James. Center for Biochemical and Biophysical Sciences and Medicine, Harvard Medical School, Boston, Massachusetts. ZINC.

Risebrough, Dr. Robert W. Research Ecologist, Bodega Marine Laboratory, University of California, Berkeley. PESTICIDE.

Riser, Dr. Stephan C. School of Oceanography, University of Washington, Seattle. SEA OF OKHOTSK.

Risser, Dr. Paul G. Vice President for Research, University of New Mexico. BIOME; GRASSLAND ECOSYSTEM.

Ritchey, Dr. H. W. President, Thiokol Chemical Corporation, Ogden, Utah. PROPELLANT.

Ritchie, Dr. J. C. Department of Biology, York University, Toronto, Ontario, Canada. TAIGA.

Ritchile, Dr. Rufus H. Oak Ridge National Laboratory, Oak Ridge, Tennessee. ELECTRON WAKE.

Rizack, Dr. Martin A. Department of Medicine, Rockefeller University, New York. LIPID METABOLISM.

Robb, D. D. D. D. Robb and Associates, Consulting Engineers, Salina, Kansas. DIRECT CURRENT.

Robbins, Dr. John A. Great Lakes Environmental Research Laboratory, National Oceanic and Atmospheric Administration, Ann Arbor, Michigan. LEAD ISOTOPES (GEOCHEMISTRY).

Robbins, Nathaniel, Jr. Director of Engineering, Residential Division, Honeywell, Inc., Minneapolis, Minnesota. THERMOSTAT.

Roberts, Dr. Andrew. Department of Oceanography, University of Southampton, Southampton Oceanography Centre, United Kingdom. MAGNETIC REVERSALS.

Roberts, J. K. Retired; formerly, Research and Development Department, Standard Oil Company, Chicago, Illinois. NAPHTHA; PETROLATUM.

Roberts, Dr. Lester. DNA Sequencing and Technology Center, Stanford University, Palo Alto, California. TRAVELING-WAVE TUBE.

Roberts, Prof. Louis D. Department of Physics, University of North Carolina. HYPERFINE STRUCTURE.

Roberts, Dr. Richard. Cold Spring Harbor Laboratories, Cold Spring Harbor, New York; Nobelist. RESTRICTION ENZYME.

Robertson, Prof. Burtis L. Retired; formerly, Professor of Electrical Engineering, University of California, Berkeley. INDUCTOR; NONSINUSOIDAL WAVEFORM; POWER FACTOR; Q (ELECTRICITY).

Robertson, Dr. J. David. Chairperson and James B. Duke Professor of Neurobiology, Duke University Medical Center. CELL MEMBRANES.

Robertson, Dr. James S. Medical Department, Medical Physics Division, Brookhaven National Laboratory, Upton, New York. RADIOISOTOPE (BIOLOGY).

Robertson, Lemuel C. Newport News Shipbuilding and Dry Dock Company, Newport News, Virginia. DRY-DOCKING.

Robinovitch, Dr. Murray. Department of Oral Biology, University of Washington School of Dentistry, Seattle. DENTAL CARIES.

Robinson, Dr. Arthur. Advanced Light Source, Lawrence Berkeley National Laboratory, California. SYNCHROTRON RADIATION.

Robinson, Prof. Arthur H. Retired; formerly, Department of Geography, University of Wisconsin. MAP PROJECTIONS; TERRESTRIAL COORDINATE SYSTEM.

Robinson, Dr. G. Wilse. Department of Chemistry, Texas Tech University. ULTRAFAST MOLECULAR PROCESSES.

Robinson, Dr. R. W. Department of Seed and Vegetable Sciences, New York State Agricultural Experiment Station, Cornell University. TOMATO.

Robinson, W. V. Processor Technology Research Group, AT&T Bell Laboratories, Whippany, New Jersey. LOGIC CIRCUITS.

Roble, Dr. Raymond G. National Center for Atmospheric Research, Boulder, Colorado. THERMOSPHERE.

Rochow, Prof. Eugene G. Retired; formerly, Department of Chemistry, Harvard University. ELECTROCHEMICAL SERIES.

Rock, Dr. Irvin. Department of Psychology, University of California, Berkeley. PERCEPTION.

Rock, Prof. Peter A. Department of Chemistry, University of California, Davis. CHEMICAL EQUILIBRIUM; CHEMICAL THERMODYNAMICS.

Rockett, Frank H. Engineering Consultant, Charlottesville, Virginia. AIR BRAKE; BOYLE'S LAW; DERRICK; ENGINE; HOROLOGY; PLYWOOD; TEMPERING; other articles.

Rodden, Dr. William P. Consulting Engineer, La Canada, Flintridge, California. AEROELASTICITY; FLUTTER (AERONAUTICS).

Rodier, Dr. Patricia M. Department of Anatomy, University of Virginia. BEHAVIORAL TOXICOLOGY.

Rodman, Dr. James. Program Director for Systematic Biology, National Science Foundation, Arlington, Virginia. CAPPARALES.

Roe, Mr. K. Keith. Chairman and President, Burns & Roe Enterprises, Inc., New Jersey. POWER PLANT.

Roeder, Dr. Peter. FAO Headquarters, Agriculture Department, Animal Production and Health Division, Italy. RINDERPEST.

Roehl, Dr. Bernie. Department of Electrical and Computer Engineering, University of Waterloo, Ontario, Canada. VIDEO GAMES.

Rogallo, Francis M. Retired; formerly, NASA Langley Research Center, Hampton, Virginia. GLIDER.

Rogers, Prof. Peter. George W. Woodruff School of Mechanical Engineering, Georgia Institute of Technology. HYDROPHONE.

Rogers, Dr. Robin D. Department of Chemistry, University of Alabama, Tuscaloosa. IONIC LIQUIDS.

Rogowski, Prof. Augustus R. Department of Mechanical Engineering, Massachusetts Institute of Technology. COMBUSTION

CHAMBER; ENGINE COOLING; HYDRAULIC VALVE LIFTER; other articles.

Rohde, Dr. Klaus. School of Biological Sciences, Division of Zoology, University of New England, Australia. GYROCOTYLIDEA; PLATYHELMINTHES.

Rohde, Dr. R. A. Department of Plant Pathology, University of Massachusetts, Amherst. PLANT PATHOLOGY.

Rohlf, Prof. James W. Department of Physics, Boston University. INTERMEDIATE VECTOR BOSON.

Rohrlich, Prof. F. Department of Physics, Syracuse University. POTENTIALS.

Rokach, Dr. Joshua. Claude Pepper Institute and Department of Chemistry, Florida Institute of Technology, Melbourne. ISOPROSTANES.

Roller, Dr. Duane E. Deceased; formerly, Harvey Mudd College, Claremont, California. CONSERVATION OF ENERGY.

Romano, Frank. Consultant, Graphic Arts, Salem, New Hampshire. TYPE (PRINTING).

Romer, Prof. Alfred S. Deceased; formerly, Museum of Comparative Zoology, Harvard University. AISTOPODA; ANTHRACOSAURIA; CLADOSELACHII; XENCANTHIDA; other articles.

Romick, Dr. Gerald J. Applied Physics Laboratory, Johns Hopkins University, Laurel, Maryland. AERONOMY.

Roosen-Runge, Dr. Edward C. Department of Biological Structure, School of Medicine, University of Washington. TESTIS.

Roper, Dr. Clyde F. E. Division of Molluscs, Smithsonian Institution, U.S. National Museum, Washington, D.C. CEPHALOPODA; DECAPODA (MOLLUSCA); OCTOPODA; SQUID; TEUTHOIDEA; other articles.

Roper, Prof. J. Alan. Head, Department of Genetics, University of Sheffield, England. PARASEXUAL CYCLE.

Roper, Dr. Stephen D. University of Miami School of Medicine, Department of Physiology and Biophysics, Miami, Florida. TASTE.

Rosa, Dr. Patricia. National Institutes of Health and Allergies, Rocky Mountain Laboratories, Hamilton, Montana. BORRELIA.

Rose, Dr. William Ingersoll, Jr. Department of Geology and Geological Engineering, Michigan Technological University. DIORITE; PETROLOGY; PORPHYRY; RHYOLITE; SPILITE.

Rosen, Prof. Barry. Department of Biological Chemistry, University of Maryland. CHEMIOSMOSIS.

Rosen, Dr. Fred S. President, Center for Blood Research, Department of Pediatrics, Harvard Medical School, Children's Hospital Medical Center, Boston, Massachusetts. COMPLEMENT.

Rosen, Dr. Kenneth H. AT&T Laboratories, Holmdel, New Jersey. NUMBER THEORY.

Rosen, Dr. Robert. Department of Physiology and Biophysics, Faculty of Medicine, Dalhousie University, Halifax, Nova Scotia, Canada. MATHEMATICAL BIOLOGY.

Rosenbaum, Prof. Fred J. Deceased; formerly, Department of Electrical Engineering, Washington University, St. Louis, Missouri. GYRATOR.

Rosenberg, Dr. Eugene. Department of Microbiology, George S. Wise Faculty of Life Sciences, Tel Aviv University, Israel. PETROLEUM MICROBIOLOGY.

Rosenberg, Leon T. Senior Consultant, Generator Design, Generation, Installation, and Service, Allis-Chalmers Manufacturing Company, Milwaukee, Wisconsin. CORE LOSS.

Rosendahl, Dr. Bruce. Department of Geology, Duke University. RIFT VALLEY.

Roskam, Dr. Jan. Department of Aerospace Engineering, University of Kansas, Lawrence. STABILITY AUGMENTATION; SUBSONIC FLIGHT.

Roskoski, Prof. Robert, Jr. Blue Ridge Institute for Medical Research, Horse Shoe, North Carolina. ADENOSINE TRIPHOSPHATE (ATP); BIOLOGICAL OXIDATION; ENERGY METABOLISM; PROTEIN KINASE; PROTEIN METABOLISM.

Ross, Prof. Charles A. Department of Geology, Western Washington University. CARBONIFEROUS; FUSULINACEA; PENNSYLVANIAN; PERMIAN.

Ross, Dr. Edward S. California Academy of Sciences, San Francisco. EMBIOPTERA.

Ross, Herbert H. Illinois State Natural History Survey, Urbana. TRICHOPTERA.

Ross, Dr. J. Central Science Laboratory, Surrey, United Kingdom. INFECTIOUS MYXOMATOSIS; INFECTIOUS PAPILLOMATOSIS.

Ross, Dr. June R. P. Department of Biology, Western Washington University. BRYOZOA; HEDERELLIDA; CARBONIFEROUS; PENNSYLVANIAN; PERMIAN.

Ross, Dr. Malcolm. Washington, D.C. ASBESTOS.

Ross, Dr. Robert S. Vice President, Director of Engineering, Environmental Structures Inc., Cleveland, Ohio. BLIMP.

Ross, Dr. Sydney. Department of Chemistry, Rensselaer Polytechnic Institute. FILM; FOAM.

Rossby, Prof. H. Thomas. Graduate School of Oceanography, University of Rhode Island, Kingston. INSTRUMENTED BUOYS.

Rossing, Dr. Thomas D. Department of Physics, Northern Illinois University, DeKalb. MUSICAL ACOUSTICS.

Rossini, Dr. Aldo A. Professor of Medicine, Director, Division of Diabetes, University of Massachusetts Medical School, Worcester. TRANSPLANTATION BIOLOGY.

Rossmeissl, Dr. Paul G. Senior Research Scientist, American Institute for. HUMAN-FACTORS ENGINEERING.

Rostoker, Dr. W. Department of Materials Engineering, University of Illinois. PROSTHESIS.

Roth, Willard. Engineer, Sunbeam Equipment Corporation, Meadville, Pennsylvania. RESISTANCE HEATING.

Rothblum, Richard Stone. Retired; formerly, David W. Taylor Naval Ship Research and Development Center, Bethesda, Maryland. WATER TUNNEL (RESEARCH AND TESTING).

Rothschild, Dr. Bruce M. Youngstown, Ohio. PALEOPATHOLOGY.

Roundhill, Dr. D. Max. Department of Chemistry, Tulane University. PALLADIUM.

Rouse, Dr. Greg. South Australian Museum, Adelaide, Australia. VESTIMENTIFERA.

Rowe, Prof. D. M. NEDO Centre for Thermoelectric Engineering, Division of Electronic Engineering, University of Wales, Cardiff, United Kingdom. THERMOELECTRIC POWER GENERATOR.

Rowe, Prof. Harrison E. Department of Electrical Engineering and Computer Science, Stevens Institute of Technology, Hoboken, New Jersey. HETERODYNE PRINCIPLE.

Rowell, Prof. A. J. Department of Geology, University of Kansas. ARTICULATA (BRACHIOPODA); BRACHIOPODA; INARTICULATA.

Rowell, Dr. Roger. Forest Products Laboratory, University of Wisconsin-Madison. PLYWOOD; WOOD COMPOSITES.

Rowen, Prof. Alan L. Webb Institute of Naval Architecture, Glen Cove, New York. MARINE ENGINE; MARINE ENGINEERING; MARINE MACHINERY.

Rowen, Dr. Lee. Senior Research Scientist, Institute for Systems Biology, Seattle, Washington. GENOMICS.

Rowley, Dr. W. R. C. Department of Trade and Industry, Division of Mechanical and Optical Metrology, National Physical Laboratory, Teddington, Middlesex, England. WAVELENGTH MEASUREMENT.

Rowlinson, Prof. J. S. Physical Chemistry Laboratory, University of Oxford, England. VAN DER WAALS EQUATION; VIRIAL EQUATION.

Rownd, Dr. Robert H. Department of Molecular Biology, Northwestern University Medical School, Chicago, Illinois. PLASMID.

Roy, Natalie U. Director of Recycling & Legislative Affairs, Glass Packaging Institute, Washington, D.C. RECYCLING TECHNOLOGY.

Rozeanu, Prof. L. Department of Material Science, Technion-Israel Institute of Technology, Haifa. NUCLEAR BATTERY.

Rozen, Dr. Jerome C., Jr. Department of Entomology, American Museum of Natural History, New York. ARTHROPODA.

Rubin, Dr. Emanuel. Chairperson, Department of Pathology, Jefferson Medical College, Philadelphia, Pennsylvania. ULCER.

Rubinstein, Sue R. Marketing Program Services Group, The Aries Group, Rockville, Maryland. KEY TELEPHONE SYSTEM.

Rubis, Prof. David D. Department of Plant Sciences, College of Agriculture, University of Arizona. GUAYULE.

Ruch, Prof. Fritz. Department of General Botany, Swiss Federal Institute of Technology, Zurich, Switzerland. DICHROISM (BIOLOGY).

Ruch, Prof. Theodore C. Director, Regional Primate Research Center; and Professor,

Department of Physiology and Biophysics, University of Washington. EFFECTOR SYSTEMS.

Ruddat, Dr. Manfred. Department of Ecology and Evolution, University of Chicago, Illinois. ALLELOPATHY.

Rudel, Dr. Lawrence. Professor of Pathology and Biochemistry, Section Head for Lipid Sciences, Department of Pathology, Wake Forest University School of Medicine, Winston-Salem, North Carolina. CHOLESTEROL.

Rudnick, Dr. Dorothea. Department of Biology, Albertus Magnus College, Yale University. DEVELOPMENTAL BIOLOGY.

Ruffing, Jaime. AMSTA-AR-ASP, U.S. Army Armament Research Development and Engineering Center, Picatinny Arsenal, New Jersey. ARMY ARMAMENT.

Rulfs, Dr. Charles L. Retired; formerly, Department of Chemistry, University of Michigan. QUANTITATIVE CHEMICAL ANALYSIS; REAGENT CHEMICALS; SOLUBILIZING OF SAMPLES.

Rumble, Dr. Douglas, III. Petrologist, Geophysical Laboratory, Carnegie Institution, Washington, D.C. METAMORPHISM.

Rumsey, Prof. Francis. Institute of Sound Recording, University of Surrey, United Kingdom. STEREOPHONIC SOUND.

Rumsey, Dr. James W. Department of Agricultural Engineering, University of California, Davis. AGRICULTURAL MACHINERY.

Runcorn, Dr. Stanley K. Deceased; formerly, Department of Physics, University of Newcastle upon Tyne, England. ROCK MAGNETISM.

Rundman, Prof. Karl. Department of Metallurgical Engineering, Michigan Technological University. METAL CASTING.

Runnegar, Dr. Bruce. Department of Earth and Space Sciences, University of California, Los Angeles. CRYPTOSOMA; EDIACARAN BIOTA; MONOPLACOPHORA.

Ruoff, Dr. Kathryn L. Francis Blake Bacteriology Laboratories, Department of Microbiology and Molecular Genetics, Massachusetts General Hospital, Harvard Medical School, Boston, Massachusetts. STREPTOCOCCUS.

Rupp, Arthur F. Oak Ridge National Laboratory, Oak Ridge, Tennessee. RADIOISOTOPE.

Rusk, Dr. Rogers D. Deceased; formerly, Mount Holyoke College. ACCELERATION; RECTILINEAR MOTION; RELATIVE MOTION; VELOCITY; other articles.

Russell, Dr. Allen S. Vice President, Alcoa Laboratories, Aluminum Company of America, Pittsburgh, Pennsylvania. ALUMINUM; ALUMINUM ALLOYS.

Russell, Dr. Howard W. Deceased; formerly, Technical Director, Battelle Memorial Institute, Columbus, Ohio. DEWAR FLASK; INCANDESCENCE.

Russell-Hunter, Prof. W. D. Department of Biology, Syracuse University. CHITON; GASTROPODA; LIMPET; MOLLUSCA; OCTOPUS; TAXODONTA; other articles.

Rust, Dr. Brian R. Department of Geology, University of Ottawa, Ontario, Canada. BRECCIA; GRAVEL.

Rust, Dr. David M. The Johns Hopkins University, Applied Physics Laboratory, Laurel, Maryland. POLARIMETRY.

Rutger, J. N. Department of Agronomy, University of California. RICE.

Ruthroff, Clyde L. Consultant, Holmdel, New Jersey. MICROWAVE; PASSIVE RADAR.

Rutkow, Dr. Ira M. Hernia Center, Freehold, New Jersey. HERNIA.

Rutledge, Dr. Steven A. Department of Atmospheric Science, Colorado State University, Fort Collins. METEOROLOGICAL RADAR.

Ryan, Prof. James J. Professor Emeritus of Mechanical Engineering, University of Minnesota. BUSHING; SCREW JACK; SHAFTING.

Ryan, Dr. Kenneth J. Department of Pathology, University of Arizona Medical Center, Tucson. WHOOPING COUGH.

Ryge, Dr. Peter. Vice President for Engineering, Ancore Corporation, Santa Clara, California. DOSIMETER.

Ryland, Prof. John S. Head of Biology, School of Natural Resources, University of the South Pacific, Laucata Bay, Suva, Fiji. BRYOZOA; ENTOPROCTA.

S

Saad, Prof. Michael L. Department of Mechanical Engineering, Santa Clara University, Santa Clara, California. SUPERSONIC FLOW.

Saaty, Prof. Thomas L. Graduate School of Business, University of Pittsburgh. ANALYTIC HIERARCHY.

Sabo, Dr. John L. National Center for Ecological Analysis and Synthesis, Santa Barbara, California. TROPHIC ECOLOGY.

Sadoway, Prof. Donald R. Department of Materials Science & Engineering, Massachusetts Institute of Technology, Cambridge. BATTERY.

Sadwick, Dr. Laurence P. Electrical Engineering Department, University of Utah, Salt Lake City. MICROWAVE SOLID-STATE DEVICES.

Saenger, Dr. Peter. Centre for Coastal Management, University of New England, Northern Rivers, Lismore, Australia. MANGROVE.

Saenger, Dr. Wolfram. Max-Planck Institut für Experimentelle Medizin, Gottingen, Germany. CLATHRATE COMPOUNDS.

Safko, Prof. John. Department of Physics and Astronomy, University of South Carolina, Columbia. CANONICAL COORDINATES AND TRANSFORMATIONS; CHRONOGRAPH; GRAVITATION; HAMILTON-JACOBI THEORY; HAMILTON'S EQUATIONS OF MOTION; HAMILTON'S PRINCIPLE.

Sage, Prof. Andrew P. Founding Dean Emeritus and First American Bank Professor, University Professor, School of Information Technology and Engineering, George Mason University, Fairfax, Virginia. DECISION SUPPORT SYSTEM; DECISION THEORY; INFORMATION TECHNOLOGY; REENGINEERING; SYSTEM FAMILIES; SYSTEMS ENGINEERING.

St. Geme, Dr. Joseph W. Associate Professor of Pediatrics and Molecular Microbiology, Department of Pediatrics, Washington University School of Medicine, St. Louis, Missouri. HAEMOPHILUS.

St. George, Emery, Jr. Retired; formerly, Charles Stark Draper Laboratory, Inc., Cambridge, Massachusetts. UNDERWATER NAVIGATION.

St. Jean, Dr. Joseph, Jr. Department of Geology, University of North Carolina. SPHAERACTINOIDEA; SPONGIOMORPHIDA; STROMATOPOROIDEA.

St. Pierre, Prof. George R. Department of Metallurgical Engineering, Ohio State University. IRON METALLURGY.

Saito, Dr. Tsunemasa. Lamont-Doherty Geological Observatory, Palisades, New York. MICROPALEONTOLOGY.

Šajgalik, Dr. Pavol. Institute of Organic Chemistry, Slovak Academy of Sciences, Bratislava, Slovak Republic. COMPOSITE MATERIAL.

Salam, Prof. Abdus. Deceased; formerly, Director, International Center for Theoretical Physics, International Atomic Energy Agency, Trieste, Italy; Nobelist. FUNDAMENTAL INTERACTIONS.

Salazar, Prof. Deborah A. Department of Geography, Oklahoma State University, Stillwater. SOUTH AMERICA.

Salek, Stanley. National Association of Broadcasters, Washington, D.C. AMPLITUDE-MODULATION RADIO.

Salet, Christian. Laboratoire de Biophysique, Muséum National d'Histoire Naturelle, Paris, France. LASER PHOTOBIOLOGY.

Salisbury, Dr. Alan B. Chairman, Learning Tree International, Reston, Virginia. INFORMATION MANAGEMENT.

Salisbury, Dr. Frank B. Plant Science Department, Utah State University, Salt Lake City. PHOTOPERIODISM.

Salmon, Vincent. Acoustical Consultant, Menlo Park, California. SOUND-REPRODUCING SYSTEMS.

Salon, Dr. Sheppard. Department of Electric Power Engineering, School of Engineering, Rensselaer Polytechnic Institute. ELECTRIC ROTATING MACHINERY.

Salton, Prof. Milton R. J. Department of Microbiology, New York University Medical Center. BACTERIAL PHYSIOLOGY AND METABOLISM.

Saltzman, Dr. W. Mark. Department of Chemical Engineering, G.W.C. Whitting School of Engineering, Johns Hopkins University, Baltimore, Maryland. DRUG DELIVERY SYSTEMS.

Salutsky, Dr. Murrell L. Vice President, Research, Dearborn Chemical Division, W. R. Grace and Company, Lake Zurich, Illinois. RADIUM.

Samios, Dr. Nicholas P. Brookhaven National Laboratory, Upton, New York. CHARM.

Samson, Prof. Perry. Department of Atmospheric, Oceanic and Space Sciences, University of Michigan, Ann Arbor, Michigan. AIR POLLUTION.

Samter, Dr. Max. Institute of Allergy and Clinical Immunology, Grant Hospital, Chicago, Illinois. ASPIRIN.

San Martin, Robert L. Deputy Assistant Secretary for Renewable Energy, Department of Energy, Washington, D.C. SOLAR ENERGY.

Sanchez, Dr. Anthony. Research Officer, Department of Health and Human Services, Centers for Disease Control and Prevention, Atlanta, Georgia. EBOLA VIRUS.

Sanchez-Sinencio. Dr. Edgar. Department of Electrical Engineering, Texas A & M University. SENSITIVITY (ENGINEERING).

Sandberg, Dr. Charles A. U.S. Geological Survey, Denver, Colorado. MISSISSIPPIAN.

Sandberg, Dr. Philip A. Department of Geology, University of Illinois at Urbana-Champaign. OOLITE.

Sanders, Dr. Fredrick. Department of Meteorology, Massachusetts Institute of Technology. BAROCLINIC FIELD; CHINOOK; ISOPYCNIC; GEOSTROPHIC WIND; ISENTROPIC SURFACES; WIND STRESS; other articles.

Sanders, Dr. Robert W. Department of Zoology, Temple University, Philadelphia. ZOOPLANKTON.

Sanders, Prof. Tom H., Jr. Department of Materials Engineering, Purdue University. ALUMINUM ALLOYS.

Sandler, Prof. Stanley I. Department of Chemical Engineering, University of Delaware. ISOTHERMAL PROCESS; PHASE RULE; POLYTROPIC PROCESS.

Sankey, Dr. Bruce M. Senior Research Associate, Research Department, Imperial Oil Limited, Petroleum Products Division, Sarnia, Ontario, Canada. EXTRACTION.

Sankey, E. W. Marion Power Shovel Division, Dresser Industries, Inc., Marion, Ohio. EXCAVATOR; POWER SHOVEL.

Sarachik, Prof. Edward S. Department of Atmospheric Sciences, University of Washington, Seattle. CLIMATE MODELING; EL NIÑO.

Sasian, Prof. Jose M. Optical Sciences Center, University of Arizona, Tucson. OPTICAL PRISM.

Sass, Prof. Stephen L. Department of Materials Science and Engineering, Cornell University. GRAIN BOUNDARIES.

Sathish, Dr. Shamachary. University of Dayton, Research Institute, Ohio. ACOUSTIC MICROSCOPE.

Satir, Dr. Peter. Chairperson, Department of Anatomy and Structural Biology, Albert Einstein College of Medicine, Yeshiva University, New York. CILIA AND FLAGELLA.

Sauer, Fred. Technical Service Department, Pfizer, Inc., Groton, Connecticut. CITRIC ACID.

Sauer, Prof. Helmut. Department of Biology, Texas A&M University. CELL DIFFERENTIATION.

Saul, Dr. Everett L. Director of Scientific Liaison, Products Division, Bristol-Myers Company. CAMPHOR.

Saunders, Dr. Gary W. Department of Biology, University of New Brunswick, Fredericton, New Brunswick, Canada. PLANT PHYLOGENY.

Saunders, Dr. R. Stephen. Solar System Division, Science Mission Directorate, NASA Headquarters, Washington, D.C. VENUS.

Sauvage, Dr. Jean-Pierre. Laboratoire de Chimie Organo-Minérale, Institut le Bel, Université Louis Pasteur, Strasbourg, France. CATENANES.

Savage, Prof. Blair D. Washburn Observatory, Department of Astronomy, University of Wisconsin. INTERSTELLAR EXTINCTION; ULTRAVIOLET ASTRONOMY.

Savage, Dr. Donald E. Department of Paleontology, University of California, Berkeley. EDENTATA; PHOLIDOTA; TAENIODONTA; TILLODONTIA.

Savage, Dr. George M. Director, Product Research and Development, Upjohn International, Inc., Kalamazoo, Michigan. ANTIMICROBIAL AGENTS.

Savage, Dr. Jay M. Department of Biology, University of Miami. AMNIOTA; CHELONIA; CROCODYLIA; SQUAMATA; ZOOGEOGRAPHY.

Sawhney, Dr. V. K. Department of Biology, University of Saskatchewan, Saskatoon, Canada. PLANT MORPHOGENESIS.

Sawyer, Dr. Donald. Distinguished Professor, Department of Chemistry, Texas A&M University. SUPEROXIDE CHEMISTRY.

Sayeed, Prof. Ali H. Department of Electrical Engineering, University of California, Los Angeles. ADAPTIVE SIGNAL PROCESSING.

Sayler, Prof. John H. Electrical Engineering and Computer Science Department, University of Michigan, Ann Arbor. COMPUTER.

Sayre, Dr. Albert N. Deceased; formerly, Consulting Groundwater Geologist, Behre Dolbear and Company. ARTESIAN SYSTEMS. GEYSER; WATER TABLE.

Sayre, Prof. Lawrence M. Department of Chemistry, Case Western Reserve University, Cleveland, Ohio. OXYGEN.

Scalzi, Dr. John B. Consultant, Arlington, Virginia. BRITTLENESS; LOADS, TRANSVERSE; STRENGTH OF MATERIALS; STRUCTURAL DEFLECTIONS; other articles.

Scanes, Dr. Colin G. Mississippi State University. ANIMAL GROWTH.

Scanlan, Prof. J. O. Department of Electrical and Electronic Engineering, University College, Dublin, Ireland. ADMITTANCE; CONDUCTANCE; ELECTRICAL IMPEDANCE; IMMITTANCE; other articles.

Scawthorn, Prof. Charles. Kyoto University, Japan. EARTHQUAKE ENGINEERING.

Schacher, Dr. Juerg. Laboratory for High Energy Physics, University of Bern, Switzerland. PIONIUM.

Schachter, Dr. Julius. Department of Laboratory Medicine, San Francisco General Hospital, University of California. CHLAMYDIA.

Schädel, Dr. Matthias. Gesellschaft für Schwerionenforschung, mbH, Germany. ELEMENT 112; TRANSURANIUM ELEMENTS.

Schaeberle, Dr. Cheryl. Department of Anatomy and Cellular Biology, Tufts University School of Medicine, Boston, Massachusetts. FETAL MEMBRANE.

Schaefer, Prof. Henry F., III. Graham Perdue Professor of Chemistry, Director, Center for Computational Chemistry, School of Chemical Sciences, University of Georgia, Athens. QUANTUM CHEMISTRY.

Schaeffer, Dr. Bobb. Chairperson and Curator, American Museum of Natural History, New York. ELASMOBRANCHII.

Schärfe, Dr. Joachim. Fisheries Technology Service, Fishery Industries Division, Food and Agriculture Organization of the United Nations, Rome, Italy. MARINE FISHERIES.

Schaumann, Prof. Rolf. Department of Electrical and Computer Engineering, Portland State University, Oregon. AMPLIFIER; BUFFERS (ELECTRONICS); EMITTER FOLLOWER.

Schawlow, Prof. Arthur L. Deceased; formerly, Department of Physics, Stanford University; Nobelist. LASER.

Scheeline, Dr. Alexander. School of Chemical Sciences, University of Illinois, Urbana-Champaign. QUANTITATIVE CHEMICAL ANALYSIS.

Scheer, Prof. Bradley T. Department of Biology, University of Oregon. ENDOCRINE MECHANISMS; ION TRANSPORT; NITROGEN EXCRETION; other articles.

Schelleng, John C. Retired; formerly, Bell Telephone Laboratories. CORNER REFLECTOR ANTENNA.

Scheltema, Dr. Amelie H. Woods Hole Oceanographic Institution, Woods Hole, Massachusetts. APLACOPHORA; CHAETODERMOMORPHA; NEOMENIOMORPHA.

Scher, Dr. Robert M. Annandale, Virginia. SHIP POWERING, MANEUVERING, AND SEAKEEPING.

Scherer, Dr. George W. Department of Civil and Environmental Engineering, Princeton University, New Jersey. GEL.

Schery, Dr. Robert W. Director, Lawn Institute, Maryville, Ohio. LAWN AND TURF GRASSES.

Schetz, Prof. Joseph A. Department of Aerospace and Ocean Engineering, Virginia Polytechnic Institute and State University, Blacksburg. JET FLOW.

Schierwater, Prof. Bernd. ITZ, Division of Ecology and Evolution, Germany. PLACOZOA.

Schilb, Dr. T. P. Mount Sinai School of Medicine, City University of New York, and Institute for Medical Research and Studies, New York. OSMOREGULATORY MECHANISMS.

Schimmel, Dr. Curt. San Ramon, California. OPERATING SYSTEM.

Schimmel, Dr. Paul. The Scripps Research Institute, La Jolla, California. GENETIC CODE; PROTEIN.

Schindler, Dr. James E. Department of Biological Sciences, College of Sciences, Clemson University. LIMNOLOGY.

Schlachter, Dr. Alfred S. Advanced Light Source, Lawrence Berkeley National Laboratory, California. ELECTRON; ELECTRON CAPTURE; ELECTRONVOLT; PARTICLE DETECTOR; PHOTOIONIZATION; SYNCHROTRON RADIATION; X-RAY TUBE; X-RAYS.

Schlecht, Dr. Matthew F. Section Research Chemist, E. I. du Pont de Nemours & Company, Inc., Agricultural Products Department, Newark, Delaware. CONJUGATION AND HYPERCONJUGATION.

Schlitt, Dr. William Joseph. Process Manager/Kvaerner Metals, Davy Nonferrous Division, San Ramon, California. SOLUTION MINING.

Schmid, Dr. Rudolf. Department of Botany, University of California, Berkeley. FRUIT.

Schmidl, Dr. M. K. Group Leader, Department of Food Science and Nutrition, University of Minnesota. FOOD MANUFACTURING.

Schmidlin, Dr. Francis. National Aeronautics and Space Administration, Observational Science Branch, Laboratory for Oceans, Goddard Space Flight Center, Wallops Flights Facility, Wallops Island, Virginia. METEOROLOGICAL ROCKET.

Schmidt, Dr. Glen H. Dairy Science Department, Ohio State University. DAIRY CATTLE PRODUCTION.

Schmidt, Dr. Paul W. Deceased; formerly, Department of Physics, University of Missouri. COLLISION (PHYSICS); CONSERVATION OF MOMENTUM; D'ALEMBERT'S PRINCIPLE; POWER; other articles.

Schmidt, Dr. Thomas. Department of Physiology and Biophysics, University of Iowa, Iowa City. ANDROGENS.

Schmidt-Neilson, Dr. Bodil M. Department of Zoology, Duke University. KIDNEY.

Schmitt, Dr. Raymond. Woods Hole Oceanographic Institution, Woods Hole, Massachusetts. DOUBLE DIFFUSION.

Schmitt, Dr. Waldo L. Deceased; formerly, Zoologist Emeritus, Department of Zoology, Smithsonian Institution, Washington, D.C. CRUSTACEA.

Schnabel, Dr. Ronald. Pasture Systems and Watershed Management Research Laboratory, U.S. Department of Agriculture–Agricultural Research Service, University Park, Pennsylvania. ACID RAIN.

Schneider, Dr. Thomas R. Electric Power Research Institute, Palo Alto, California. ENERGY STORAGE.

Schnitzer, Dr. Morris. Principal Research Scientist, Land Resource Research Centre, Central Experimental Farm, Ottawa, Ontario, Canada. HUMUS.

Schoch, Dr. Robert. Attleboro, Massachusetts. TAENIODONTA; TILLODONTIA.

Schoeffler, Dr. James D. Department of Computer and Information Science, Cleveland State University. DISTRIBUTED SYSTEMS (CONTROL SYSTEMS).

Scholle, Prof. Peter A. Department of Geological Sciences, Southern Methodist University, Dallas, Texas. CHALK.

Scholz, Dr. Christopher A. Department of Earth Sciences, Syracuse University, New York. RIFT VALLEY.

Scholz, Christopher H. Lamont-Doherty Earth Observatory of Columbia University, Palisades, New York. EARTHQUAKE.

Scholz, Dr. Nathaniel. Northwest Fisheries Science Center, Environmental Conservation Division, Seattle, Washington. WATER POLLUTION.

Schönberg, Dr. Christine. Carl von Ossietzky Universität Oldenburg, Germany. BIOERODING SPONGES.

Schooler, Dr. Jonathan. Learning, Research, and Development Center, University of Pittsburgh. MEMORY.

Schooley, Dr. James F. Retired; formerly, Cheif, Division of Temperature Measurements, National Institute of Standards and Technology, Gaithersburg, Maryland. GAS THERMOMETRY.

Schoonmaker, G. R. Deceased; formerly, Vice President, Production-Exploration, Marathon Oil Company, Findlay, Ohio. OIL AND GAS, OFFSHORE.

Schotland, Richard M. Institute of Atmospheric Physics, University of Arizona. PSYCHROMETER.

Schotte, Marilyn. National Museum of Natural History, Invertebrate Zoology, Smithsonian Institution, Washington, D.C. ISOPODA.

Schowalter, Prof. Timothy D. Department of Entomology, Oregon State University, Corvallis. FOREST ECOSYSTEM.

Schram, Prof. Frederick R. Department of Biology, University of Washington, Seattle. ARTHROPODA; CIRRIPEDIA.

Schramm, Dr. Laurier L. Saskatchewan Research Council, Canada. EMULSION; SURFACTANT; SUSPENSION.

Schreiber, Prof. B. Charlotte. Department of Earth and Environmental Sciences, Queens College, New York. GYPSUM; HALITE.

Schreibman, Dr. Martin P. Department of Biology, Brooklyn College, City University of New York. PITUITARY GLAND.

Schreiner, Dr. Rodney. Department of Chemistry, University of Wisconsin-Madison. METHANE.

Schroeder, Dr. Charles A. Department of Botanical Sciences, University of California, Los Angeles. BANANA.

Schroeder, Prof. Peter A. Department of Physics and Astronomy, Michigan State University. ELECTRICAL RESISTANCE; ELECTRICAL RESISTIVITY.

Schubert, Dr. Jack. Radiation Health, Graduate School of Public Health, University of Pittsburgh. BERYLLIUM.

Schuchardt, Dr. Lee F. Merck, Sharp, and Dohme, West Point, Pennsylvania. BIOLOGICALS.

Schugar, Dr. Harvey. Department of Chemistry, Wright and Rieman Laboratories, New Brunswick, Rutgers University. COPPER.

Schuler, Mr. Robert H. RADIATION CHEMISTRY.

Schuller, Dr. Ivan K. Physics Department, University of California, La Jolla. ARTIFICIALLY LAYERED STRUCTURES.

Schulman, Dr. James H. U.S. Naval Research Laboratory. ELECTROLUMINESCENCE; FLUORESCENCE; FRANCK-CONDON PRINCIPLE; LUMINESCENCE; PHOSPHORESCENCE; THERMOLUMINESCENCE; other articles.

Schulte, Dr. Daniel. CERN, Geneva, Switzerland. PARTICLE ACCELERATOR.

Schultz, Dr. Cheryl. National Center for Ecological Analysis and Synthesis, University of California, Santa Barbara. LAND-USE PLANNING.

Schultz, Dr. Jack. Gypsy Moth Research Center, Department of Entomology, College of Agriculture, Pesticide Research Laboratory and Graduate Study Center, Pennsylvania State University. PLANT COMMUNICATION.

Schultz, Robert M. General Manager, William Langer Jewel Bearing Plant, Bulova Watch Company, Inc., Rolla, North Dakota. JEWEL BEARING.

Schultze, Dr. Hans-Peter. Institut für Paläontologie, Museum für Naturkunde, Humboldt-Universitat zu Berlin, Germany. COELACANTHIFORMES.

Schulze, Prof. Anja. Department of Marine Biology, Texas A&M University at Galveston. POGONOPHORA (SIBOGLINIDAE); SIPUNCULA.

Schülzgen, Axel. University of Arizona, College of Optical Sciences. LASER.

Schurig, Dr. Gerhardt. Virginia-Maryland Regional College of Veterinary Medicine, Virginia Tech, Center for Molecular Medicine and Infectious Disease, Blacksburg, Virginia. BRUCELLOSIS.

Schuster, Prof. Frederick L. Department of Biology, Brooklyn College. AMEBA.

Schwab, Robert W. Boeing Company, Commercial Airplane Group, Seattle, Washington. AIRCRAFT COMPASS SYSTEM.

Schwartz, Dr. Karlene V. Department of Biology, University of Massachusetts, Boston. ANIMAL KINGDOM; CLASSIFICATION, BIOLOGICAL.

Schwartz, Mel. Materials Consultant, United Technologies Corporation, Stratford, Connecticut. BRAZING; COMPOSITE MATERIAL; MAGNESIUM ALLOYS; METAL MATRIX COMPOSITE; SHAPE MEMORY ALLOYS; SILICON; STRUCTURAL MATERIALS; WELDING AND CUTTING OF MATERIAL.

Schwartz, Dr. Morton K. Chairperson, Department of Clinical Chemistry, Memorial Sloan-Kettering Cancer Center, New York. BILIRUBIN; PHOSPHATE METABOLISM; SEROTONIN; URIC ACID.

Schweizer, Ernest. President, Schweizer Aircraft Corporation, Elmira, New York. GLIDER.

Scotese, Dr. Christopher R. Department of Geology, University of Texas, Arlington. PALEOGEOGRAPHY.

Scott, Prof. Alwyn. Program in Applied Mathematics, University of Arizona, Tucson. SOLITON.

Scott, Dave. Industrial Engineering Manager, Madulator Computer Systems, Inc., Fort Lauderdale, Florida. MEASURED DAYWORK.

Scott, Dr. Edward R. D. University of Hawai'i at Manoa, Hawai'i Institute of Geophysics and Planetology, Honolulu. METEORITE.

Scott, Prof. J. F. Department of Physics, University of Colorado. SCATTERING OF ELECTROMAGNETIC RADIATION.

Scott, Dr. John E., Jr. Department of Aerospace Engineering and Engineering Physics, University of Virginia. SUPERSONIC FLIGHT.

Scott, Dr. John P. Department of Psychology, Bowling Green State University, Ohio. SOCIAL MAMMALS.

Scott, Dr. Richard A. U.S. Geological Survey, Department of the Interior, Denver, Colorado. FOSSIL SEEDS AND FRUITS.

Scott, Prof. Robert L. Department of Chemistry, University of California, Los Angeles. DELIQUESCENCE; EFFLORESCENCE; MELTING POINT; PHASE EQUILIBRIUM; other articles.

Scouten, Dr. Charles W. Stoelting Company, Oakwood Centre, Wood Dale, Illinois. MICROMANIPULATION.

Scrutton, Dr. Colin. Department of Earth Sciences, University of Durham, United Kingdom. RUGOSA; TABULATA.

Seaborg, Dr. Glenn T. Deceased; formerly, Lawrence Berkeley National Laboratory, University of California, Berkeley; Nobelist. ACTINIDE ELEMENTS; CURIUM; FERMIUM; SEABORGIUM; TRANSURANIUM ELEMENTS; other articles.

Seagle, Stanley. Vice President, Research and Technical Development, RMI Company, Niles, Ohio. TITANIUM METALLURGY.

Searles, Dr. Robert. Rheumatology Division, University of New Mexico School of Medicine. ARTHRITIS; GOUT; JOINT DISORDERS; RHEUMATISM.

Searson, Dr. Peter C. H. H. Uhlig Corrosion Laboratory, Department of Materials Science and Engineering, Massachusetts Institute of Technology. CORROSION.

Sebald, Joseph F. Deceased; formerly, Consulting Engineer and President, Heat Power Products Corporation, Bloomfield, New Jersey. CONTACT CONDENSER; COOLING TOWER; STEAM CONDENSER; STEAM JET EJECTOR; SURFACE CONDENSER; VAPOR CONDENSER.

Seed, Prof. John Richard. Department of Epidemiology, School of Public Health, University of North Carolina, Chapel Hill. TRYPANOSOMIASIS.

Segal, Dr. Nancy L. Department of Psychology, California State University, Fullerton. TWINS (HUMAN).

Segeler, C. George. Director, Technical Service, Dave Sage, Inc., New York. DEGREE-DAY.

Sehgal, Dr. Amita. Howard Hughes Medical Institute, Department of Neuroscience, University of Pennsylvania School of Medicine, Philadelphia. BIOLOGICAL CLOCKS.

Seibel, Dr. Machelle M. Faulkner Center for Reproductive Medicine, Boston, Massachusetts. REPRODUCTIVE TECHNOLOGY; UTERINE DISORDERS; VAGINAL DISORDERS.

Seidelmann, Dr. P. Kenneth. Department of Astronomy, University of Virginia, Charlottesville. ALMANAC; ASTROMETRY; ATOMIC TIME; CELESTIAL REFERENCE SYSTEM; DYNAMICAL TIME; ROCHE LIMIT.

Seitz, Dr. W. Rudolf. Department of Chemistry, College of Engineering and Physical Sciences, University of New Hampshire. LUMINESCENCE ANALYSIS.

Selamet, Prof. Ahmet. Center for Automotive Research, Ohio State University, Columbus. MUFFLER.

Selby, Prof. Michael J. Department of Earth Science, University of Waikato, New Zealand. NEW ZEALAND.

Seliskar, Dr. Denise M. Associate Research Scientist, College of Marine Studies, University of Delaware. DUNE VEGETATION.

Sell, Dr. Heinz G. Metals Development Section, Westinghouse Lamp Divisions, Bloomfield, New Jersey. BLACKBODY; EMISSIVITY; GRAYBODY; HEAT RADIATION; PLANCK'S RADIATION LAW.

Sellars, Dr. John R. Manager, Engineering Mechanics Operations, TRW, Inc., Redondo Beach, California. AERONAUTICAL ENGINEERING.

Sellin, Dr. Ivan A. Department of Physics, University of Tennessee, Knoxville. ATOMIC STRUCTURE AND SPECTRA.

Sellmyer, Prof. D. J. Behlen Laboratory of Physics, University of Nebraska. PHASE TRANSITIONS.

Selverstone, Dr. Jane. Department of Earth & Planetary Sciences, Hoffman Laboratory, Harvard University. FACIES (GEOLOGY).

Seshadri, Ms. Sridhar. Operations Management Department, Stern School of Businees, New York University. SCHEDULING.

Sessions, Dr. John. Department of Forest Engineering, Oregon State University, Corvallis. LOGGING.

Sessler, Prof. Gerhard. Institut für Nachrichtentechnik, Technische Universität Darmstadt, Germany. ELECTRET; ELECTRET TRANSDUCER; MICROPHONE.

Sethares, Dr. William A. Department of Electrical and Computer Engineering, University of Wisconsin-Madison. SCALE (MUSIC).

Shack, Prof. Roland V. Optics Science Center, University of Arizona. ABERRATION (OPTICS); GEOMETRICAL OPTICS; OPTICAL SURFACES.

Shackelford, Dr. James F. Department of Biomedical Engineering, University of California, Davis. MATERIALS SCIENCE AND ENGINEERING.

Shaffer, Dr. Albert L. Clinical Professor of Medicine, Brigham and Women's Hospital, Boston, Massachusetts. ASTHMA.

Shakhashiri, Prof. Bassam Z. Department of Chemistry, University of Wisconsin-Madison. METHANE.

Sham, Prof. Lu Jeu. Department of Physics, University of California, San Diego. EXCITON.

Shands, Dr. H. L. Agronomist, DeKalb Soft Wheat Research Center, DeKalb AgRes, Inc., West Lafayette, Indiana. RYE.

Shanmugam, Dr. G. Department of Geology, The University of Texas at Arlington. DEEP-MARINE SEDIMENTS; SUBMARINE CANYON.

Shannon, L. R. The New York Times, New York. COMPUTER PERIPHERAL DEVICES.

Shannon, Prof. Robert R. Optical Sciences Center, University of Arizona. BINOCULARS; CONFORMAL OPTICS; MIRROR OPTICS.

Shapere, Prof. Dudley. Emeritus, Reynolds Professor of Philosophy and History of Sciences, Wake Forest University, Winston-Salem, North Carolina. CAUSALITY; MATTER (PHYSICS).

Shapiro, Charles. Graphic Arts Technical Foundation, Pittsburgh, Pennsylvania. PRINTING.

Shapiro, Hymin. Research and Development Department, Ethyl Corporation, Baton Rouge, Louisiana. LEAD; METAL CARBONYL.

Sharkey, Thomas D. Department of Botany, University of Wisconsin–Madison. PHOTORESPIRATION.

Sharon, Prof. Nathan. Department of Membrane Research and Biophysics, Weizmann Institute of Science, Rehovot, Israel. GLYCOPROTEIN.

Sharpe, Ralph H. Department of Fruit Crops, University of Florida. PECAN.

Shaw, Brian R. BHP Petroleum (Americas) Inc., Houston, Texas. PETROLEUM ENGINEERING.

Shaw, Dr. Cheng-mei. Laboratory of Neuropathology, Department of Pathology, University of Washington School of Medicine. NERVOUS SYSTEM DISORDERS.

Shaw, Dr. Michael. International Association of Refrigerated Warehouses, Bethesda, Maryland. COLD STORAGE.

Shaw, Prof. Peter B. Department of Physics, Pennsylvania State University. GREEN'S FUNCTION.

Shearer, Dr. Charles K. Institute of Meteorites, Department of Earth and Planetary Sciences, University of New Mexico, Albuquerque. AMBLYGONITE.

Shechtman, Dan. Department of Materials Engineering, Israel Institute of Technology, Haifa. QUASICRYSTAL.

Shedlock, Dr. Kaye M. U.S. Geological Survey, Administrative Officer, Denver, Colorado. EARTHQUAKE.

Sheeley, Dr. Neil R., Jr. Naval Research Laboratory, Washington, DC. SOLAR MAGNETIC FIELD.

Sheff, Dr. Irving. Chemistry Division, Argonne National Laboratory, Argonne, Illinois. FLUORINE.

Sheffer, Dr. Albert L. Clinical Professor of Medicine, Brigham and Women's Hospital, Boston, Massachusetts. ASTHMA.

Sheffield, Dr. Harley G. National Institute of Allergy and Infectious Diseases, Bethesda, Maryland. TOXOPLASMEA; TOXOPLASMIDA.

Sheingold, Daniel H. Manager of Technical Marketing, Analog Devices, Inc., Norwood, Massachusetts. ANALOG-TO-DIGITAL CONVERTER; DIGITAL-TO-ANALOG CONVERTER.

Shelfer, Prof. Katherine. College of Information Science and Technology, Drexel University, Philadelphia, Pennsylvania. SMART CARD.

Shenoy, Dr. Gopal. Senior Scientific Director, The Advanced Photon Source, Argonne National Laboratory, Illinois. SYNCHROTRON RADIATION.

Shephard, Dr. Bruce D. Diplomates, American College of Obstetricians and Gynecologists, Tampa Obstetrics and Gynecology Associates, Tampa, Florida. PREGNANCY; PREGNANCY DISORDERS.

Shepherd, Dr. Gordon M. Yale University Medical School. SYNAPTIC TRANSMISSION.

Shergold, Dr. John H. Chairman, International Subcommission on Cambrian Stratigraphy, Masseret, France. CAMBRIAN.

Sheridan, Prof. Thomas B. Department of Mechanical Engineering, Massachusetts Institute of Technology. HUMAN-FACTORS ENGINEERING.

Sheriff, Prof. Robert E. Department of Geoscience, University of Houston. GEOPHYSICAL EXPLORATION; SEISMIC STRATIGRAPHY.

Sherman, Dr. David M. Department of Earth Sciences, University of Bristol, United Kingdom. MINERAL; MINERALOGY; OXIDE AND HYDROXIDE MINERALS.

Sherwood, Robert S. Deceased; formerly, Manager, Engineering, Steam Turbine Division, Worthington Corporation, Harrison, New York. MACHINERY; MECHANICAL ENGINEERING; TECHNOLOGY.

Sheviak, Dr. Charles J. Senior Scientist and Curator of Botany, New York State Museum, Albany. ORCHID.

Shiino, Dr. Sueo M. Faculty of Fisheries, Prefectural University of Mie, Japan. EPICARIDEA.

Shilling, Prof. John J. College of Computing, Georgia Institute of Technology, Atlanta. OBJECT-ORIENTED PROGRAMMING.

Shimek, Dr. Ronald L. Reef Safe Enterprises, Inc., Wilsall, Montana. SCAPHOPODA.

Shin, Prof. Y. S. Department of Mechanical Engineering, Naval Postgraduate School, Monterey, California. CLUTCH; COUPLING; PRESSURE VESSEL.

Shipman, Prof. Harry L. Department of Physics and Astronomy, University of Delaware, Newark. GRAVITATIONAL LENS; HORIZON; NEUTRON STAR; SUPERMASSIVE STARS; WHITE DWARF STAR.

Shirley, Prof. David A. Deceased; formerly, Department of Chemistry, University of Tennessee. ACETONE; QUINONE.

Shivakumar, Dr. B. Oak Ridge National Laboratory, Oak Ridge, Tennessee. DYNAMIC NUCLEAR POLARIZATION.

Shivaram, Dr. Bellave S. Department of Physics, John S. Beams Laboratory, University of Virginia. LIQUID HELIUM.

Shive, Prof. William. Department of Chemistry and Clayton Foundation Biochemical Institute, University of Texas, Austin. ENZYME INHIBITION.

Shively, Dr. Richard B. AT&T Laboratories, Whippany, New Jersey. LOGIC CIRCUITS.

Shoemaker, Dr. William C. Martin Luther King, Jr./Charles R. Drew Medical Center, Los Angeles, California. CRITICAL CARE MEDICINE.

Shore, Prof. Steven N. Chair, Department of Physics and Astronomy, Indiana University, South Bend. STELLAR POPULATION.

Shores, Prof. David. Department of Materials Science, University of Minnesota, Minneapolis. CORROSION.

Shortman, Dr. Ken. Head, Lymphocyte Differentiation Unit, Walter and Eliza Hall Institute of Medical Research, Melbourne, Australia. CLINICAL IMMUNOLOGY.

Shortreed, Dr. John. Director, Institute for Risk Research, Waterloo, Canada. RISK ASSESSMENT AND MANAGEMENT.

Shostak, Dr. Seth. SETI Insititute, Mountain View, California. EXTRATERRESTRIAL INTELLIGENCE.

Shprintzen, Dr. Robert J. Director, Center for Craniofacial Disorders, Montifiore Medical Center, Albert Einstein College of Medicine, New York. CLEFT LIP AND CLEFT PALATE.

Shuman, E. C. Regional Professional Engineer, State College, Pennsylvania. AIR CONDITIONING.

Shuman, Dr. Frederick G. Director, National Meteorological Center, National Oceanic and Atmospheric Administration, Washington, D.C. ISOBAR (METEOROLOGY).

Shung, Dr. K. Kirk. Bioengineering Program, Pennsylvania State University, University Park. BIOMEDICAL ULTRASONICS.

Shuster, Dr. Carl N., Jr. Retired; formerly, Ecologist, Bureau of Water Hygiene, U.S. Public Health Service, Cincinnati, Ohio. XIPHOSURIDA.

Sidor, Dr. Christian. Department of Biology, University of Washington, Seattle. PELYCOSAURIA; SYNAPSIDA.

Siebein, Dr. Gary W. Department of Architecture, University of Florida, Gainesville. ARCHITECTURAL ACOUSTICS.

Siegbahn, Dr. Hans. Institute of Physics, Uppsala University, Sweden. ELECTRON SPECTROSCOPY.

Siegbahn, Prof. Kai. Institute of Physics, Uppsala University, Sweden; Nobelist. ELECTRON SPECTROSCOPY.

Siegel, Prof. Benjamin M. Department of Applied Physics, Cornell University. ELECTRON MICROSCOPE.

Siegel, Dr. Richard W. Materials Science Division, Argonne National Laboratory, Argonne, Illinois. NANOSTRUCTURE.

Siegmund, C. W. Exxon Research and Engineering Company, Linden, New Jersey. FUEL OIL.

Siegmund, Dr. Walter P. Director, Research and Development, Schott Fiber Optics Inc., Southbridge, Massachusetts. FIBER-OPTICS IMAGING.

Sieh, Prof. Kerry. Seismological Laboratory, California Institute of Technology, Pasadena. PALEOSEISMOLOGY.

Siekevitz, Dr. Philip. Department of Cell Biology, Rockefeller University, New York. CELL METABOLISM.

Sienko, Michell J. Deceased; formerly, Department of Chemistry, Cornell University. NONSTOICHIOMETRIC COMPOUNDS.

Siever, Dr. Raymond. Department of Geological Sciences, University of Michigan. ARENACEOUS ROCKS; CONGLOMERATE; STALACTITES AND STALAGMITES; TRAVERTINE; other articles.

Siggia, Prof. Sidney. Deceased; formerly, Department of Chemistry, University of Massachusetts. QUALITATIVE CHEMICAL ANALYSIS.

Signorini, Dr. Sergio R. Division of Ocean Sciences, National Science Foundation, Arlington, Virginia. STORM SURGE.

Šiljak, Prof. Dragoslav D. School of Engineering, Santa Clara University, California. NONLINEAR CONTROL THEORY.

Silk, Dr. Joseph. Department of Astrophysics, University of Oxford, United Kingdom. BLACK HOLE; GALAXY FORMATION AND EVOLUTION; GRAVITATIONAL COLLAPSE.

Silman, Robert. Robert Silman Associates, P.C., New York. BUILDINGS.

Silva, Dr. Paul C. Department of Botany, University of California, Berkeley. AGAR; ALGAE; BRYOPSIDALES; CHLOROPHYCEAE; CHLOROPHYCOTA; OEDOGONIALES; other articles.

Silva-Tulla, Dr. Francisco. Consulting Engineer, Geotechnics, Lexington, Massachusetts. SOIL MECHANICS.

Silver, Dr. Edward A. Consultant. AGRICULTURAL SOIL AND CROP PRACTICES.

Silverman, Dr. Joseph. Department of Chemical Engineering, University of Maryland. RADIOACTIVITY AND RADIATION APPLICATIONS.

Simberloff, Prof. Daniel. Department of Biological Science, Florida State University. ECOLOGICAL COMMUNITIES.

Simmons, Dr. James A. Walter S. Hunter Laboratory of Psychology, Brown University. ECHOLOCATION.

Simmons, Dr. Nancy. Department of Mammalogy, American Museum of Natural History, New York. EUTHERIA.

Simmons, Prof. William H. Division of Biochemistry, Loyola University, Chicago, Stritch School of Medicine, Maywood, Illinois. SOMATOSTATIN.

Simon, Prof. Brian. IBM Professor of Mathematics and Theoretical Physics, California Institute of Technology, Pasadena. FEYNMAN INTEGRAL.

Simon, Dr. Steven B. Department of Geophysical Sciences, University of Chicago, Illinois. COSMOCHEMISTRY; MICA.

Simonsen, Dr. John. Wood Science and Engineering, College of Forestry, Oregon State University, Corvallis. RECYCLING TECHNOLOGY.

Simonson, Dr. Roy W. Retired; formerly, Director, Soil Classification and Correlation, U.S. Department of Agriculture, Hyattsville, Maryland. PEDOLOGY; SOIL.

Simpson, Dr. Carol. Department of Earth and Planetary Sciences, Johns Hopkins University. MYLONITE.

Simpson, Dr. Joanne. Head, Severe Storms Branch, Goddard Space Flight Center, National Aeronautics and Space Administration, Greenbelt, Maryland. HURRICANE.

Simpson, Dr. Robert H. Retired; formerly, Director, Experimental Meteorology Laboratory, National Weather Service, Miami, Florida. HURRICANE.

Simpson, Prof. Stephen G. Department of Chemistry, Massachusetts Institute of Technology. GRAVIMETRIC ANALYSIS.

Sims, Dr. Chester T. Consultant, Department of Materials Engineering, Rensselaer Polytechnic Institute. HIGH-TEMPERATURE MATERIALS.

Sims, Dr. Gerald K. Microbiologist, USDA-ARS, University of Illinois-Urbana. SOIL DEGRADATION.

Sincerbox, Prof. Glenn T. Optical Data Storage Center, Optical Sciences Center, University of Arizona, Tucson. OPTICAL RECORDING.

Sinclair, Malcolm W. Procurement Executive, Ministry of Defence, Royal Signals and Radar Establishment, Great Malvern, Worcestershire, England. MICROWAVE NOISE STANDARDS.

Singer, Dr. S. F. Department of Environmental Sciences, University of Virginia. SPACE.

Singer, Dr. Stanley. Director, ATHENEX Research Associates. PROPELLANT.

Singleton, John D. Supervisor of Programming, Radio Division, National Broadcasting Company, New York. RADIO.

Sinha, Prof. Kumares C. Professor and Head, Transportation and Infrastructure Systems Engineering, Purdue University, West Lafayette, Indiana. HIGHWAY ENGINEERING.

Sinha, Prof. P. K. Department of Engineering, University of Reading, Whiteknights, Reading, England. MAGNETIC LEVITATION.

Sinton, Dr. William M. Department of Astronomy, University of Hawaii, Manoa. TELESCOPE.

Sircar, Dr. S. Senior Research Associate, Air Products and Chemicals, Inc., Allentown, Pennsylvania. ADSORPTION OPERATIONS.

Sirignano, Prof. William A. Dean of Engineering, University of California, Irvine. SPRAY FLOW.

Sisler, C. W. Corporate Engineering Department, Monsanto Company, St. Louis, Missouri. DRIER (PAINT); LACQUER; PAINT; THINNER; VARNISH; other articles.

Sisler, Dr. Harry H. Department of Chemistry, University of Florida. AMMINE; AMMONIA; NITROGEN.

Sisler, Dr. Hugh D. Department of Botany, Division of Agricultural and Life Sciences, University of Maryland. FUNGISTAT AND FUNGICIDE.

Sistare, George H., Jr. Retired; formerly, Consultant, Handy and Harman, Fairfield, Connecticut. GOLD ALLOYS.

Sittig, Marshall. Assistant Director, Office of Research, Project Administration, Princeton University. ALKALI METALS; CESIUM; LITHIUM; POTASSIUM; RUBIDIUM; SODIUM.

Sittner, Dr. W. R. President, Electro Medical Systems, Inc., Englewood, Colorado. PHOTODIODE; PHOTOTRANSISTOR.

Skalak, Dr. Richard. Deceased; formerly, Founding Director, Institute for Mechanics and Materials, University of California, San Diego. BIOMECHANICS.

Skelly, Dr. David. School of Forestry and Environmental Studies, Yale University, New Haven, Connecticut. LANDSCAPE ECOLOGY.

Skilling, Prof. Hugh H. Deceased; formerly, Department of Electrical Engineering, Stanford University. ALTERNATING CURRENT; COUPLED CIRCUITS; NETWORK THEORY; SUPERPOSITION THEOREM (ELECTRIC NETWORKS); Y-DELTA TRANSFORMATIONS; other articles.

Skinner, A. Douglas. Head of Measurement Standards Laboratory, Marconi Instruments, Ltd., Stevanage, England. OHMMETER.

Skinner, Prof. Brian J. Department of Geology and Geophysics, Yale University, New Haven, Connecticut. EARTH GEOLOGY.

Skinner, E. N. Assistant to the Director, Product Research and Development, International Nickel Company, Inc., New York. NICKEL ALLOYS.

Skinner, Dr. Harry B. Department of Orthopedic Surgery, School of Medicine, University of California, San Francisco. BONE DISORDERS.

Skotland, Dr. C. B. Irrigated Agriculture Research and Extension Center, Washington State University. HOP.

Skrutskie, Dr. Michael F. Department of Physics and Astronomy, University of Massachusetts, Amherst. INFRARED ASTRONOMY.

Slabaugh, Dr. Wendell H. Deceased; formerly, Department of Chemistry, Oregon State University. INTERFACE OF PHASES.

Slack, Dr. J. M. W. Imperial Cancer Research Fund, Mill Hill Laboratories, London, England. FATE MAPS (EMBRYOLOGY).

Slaton, Prof. Milton R. J. Department of Microbiology, New York University Medical Center. BACTERIAL PHYSIOLOGY AND METABOLISM.

Slichter, Dr. Charles P. Department of Physics, University of Illinois, Urbana-Champaign. MAGNETIC RELAXATION; MAGNETIC RESONANCE.

Slichter, Prof. Louis B. Deceased; formerly, Institute of Geophysics, University of California, Los Angeles. EARTH TIDES.

Slocum, Dr. Donald W. Argonne National Laboratory, Argonne, Illinois. METALLOCENES.

Slonecker, Dr. Charles E. Associate Professor of Anatomy, University of British Columbia, Vancouver, Canada. THYMUS GLAND.

Sluder, Dr. Greenfield. Senior Scientist, Worcester Foundation for Experimental Biology, Shrewsbury, Massachusetts. CENTRIOLE.

Smartt, Dr. Raymond N. Scientist Emeritus, National Solar Observatory, Sunspot, New Mexico. CORONAGRAPH; SPECTROHELIOGRAPH.

Smith, Dr. Andrew B. Department of Paleontology, British Museum of Natural History, London, England. DIASTEROIDA; HOLASTEROIDA; NEOLAMPADOIDA; NOTOMYOTIDA; PHYMOSOMATOIDA; other articles.

Smith, Dr. David M. Department of Forestry and Environmental Studies, Yale University. SILVICULTURE.

Smith, Gaylord D. Director, High Nickel Alloys and Precious Metals, Product Development and Research Division, International Nickel Company, New York. NICKEL ALLOYS.

Smith, Dr. Guy D. Geological Institute, Krijgslaan, Ghent, Belgium. SOIL.

Smith, Prof. Harlan J. Deceased; formerly, Department of Astronomy, University of Texas, Austin. ABERRATION OF LIGHT.

Smith, Dr. Jerome A. Marine Physics Laboratory, Scripps Institution of Oceanography, University of California, La Jolla. LANGMUIR CIRCULATION.

Smith, Dr. John A. IBM Corporation, Lexington, Kentucky. STRAIGHT-LINE MECHANISM.

Smith, John Ward. Deceased; formerly, Research Supervisor, Laramie Energy Research Center, Energy Research and Development Administration, Laramie, Wyoming. OIL SHALE.

Smith, Dr. Jonathan W. AT&T Bell Laboratories, North Andover, Massachusetts. CROSSTALK.

Smith, Prof. Julian C. Department of Chemical Engineering, Cornell University. CENTRIFUGATION.

Smith, Dr. Julius O., III. Center for Computer Research in Music and Acoustics, Department of Music, Stanford University, Stanford, California. MUSICAL INSTRUMENTS.

Smith, Dr. Kathleen K. Department of Biological Anthropology and Anatomy, Duke University Medical Center, Durham, North Carolina. TONGUE.

Smith, Dr. Matthijs J. Imperial Cancer Research Fund Laboratories, London, England. GENETIC MAPPING.

Smith, Dr. Michael S. Leader, Nuclear Astrophysics Research Group, Physics Division, Oak Ridge National Laboratory, Tennessee. RADIOACTIVE BEAMS.

Smith, Prof. Norman D. Department of Geosciences, University of Nebraska, Lincoln. VARVE.

Smith, Prof. Paul G. Department of Vegetable Crops, College of Agricultural and Environmental Science, University of California, Davis. PEPPER.

Smith, Philip H. Consultant, Sherman, Connecticut. TRACTOR.

Smith, Dr. Richard T. Consultant, San Antonio, Texas. SYNCHRONOUS MOTOR.

Smith, Dr. Robert L. College of Oceanic and Atmospheric Sciences, Oregon State University, Corvallis. UPWELLING.

Smith, Prof. Robert Leroy. Retired; formerly, Associate Professor of Architecture, University of Illinois, Urbana. INCANDESCENT LAMP; INFRARED LAMP; LAMP.

Smith, Dr. Steven B. Department of Geophysical Sciences, University of Chicago. MICA.

Smith, Dr. T. W. Department of Orthopaedics, Northern General Hospital, United Kingdom. FOOT DISORDERS.

Smith, Dr. Vincent S. The Natural History Museum, London, United Kingdom. PHTHIRAPTERA.

Smith, Dr. Warren E. Science Applications International Corporation, Tucson, Arizona. IMAGE PROCESSING.

Smithson, Dr. T. R. Head of School, Science Maths and Information Technology, Cambridge Regional College, United Kingdom. AMPHIBIA.

Smullyan, Dr. Raymond M. Oscar Ewing Professor of Philosophy Emeritus, Indiana University, Bloomington, Indiana. GÖDEL'S THEOREM.

Smythe, Dr. William R. Deceased; formerly, Department of Physics, California Institute of Technology. DISPLACEMENT CURRENT; ELECTROMAGNETIC WAVE; INVERSE-SQUARE LAW; POYNTING'S VECTOR; other articles.

Snavely, Dr. Deanne. Center for Photochemical Sciences, Bowling Green State University, Bowling Green, Ohio. LASER PHOTOCHEMISTRY.

Snelgrove, Prof. Martin. Department of Electronics, Carleton University, Ontario, Canada. INTEGRATED-CIRCUIT FILTER; NEGATIVE-RESISTANCE CIRCUITS; OSCILLATOR; PUSH-PULL AMPLIFIER.

Snell, Dr. Arthur H. Deceased; formerly, Associate Director, Oak Ridge National Laboratory, Oak Ridge, Tennessee. DELAYED NEUTRON; NEUTRON.

Snell, Dr. Christopher. Department of Sport Sciences, University of the Pacific, Stockton, California. CHRONIC FATIGUE IMMUNE DYSFUNCTION SYNDROME.

Snook, Dr. Theodore. Department of Anatomy, School of Medicine, University of North Dakota. TONSIL.

Snover, Dr. Kurt A. Research Professor, Nuclear Physics Laboratory, Department of Physics, University of Washington, Seattle. GIANT NUCLEAR RESONANCES.

Snow, William W. Consulting Engineer, William W. Snow Associates, Inc., Woodside, New York. ELECTROPOLISHING; EXCITATION; FREQUENCY DIVIDER.

Sochacki, Dr. James. Department of Mathematics and Statistics, James Madison University, Harrisonburg, Virginia. MATHEMATICAL SOFTWARE; SYMBOLIC COMPUTING.

Soderblom, Dr. David. Space Telescope Science Institute, Baltimore, Maryland. STELLAR ROTATION.

Sohon, Dr. Harry. Deceased; formerly, Moore School, University of Pennsylvania. POWER-FACTOR METER.

Soini, Dr. Ylermi. Department of Pathology, University of Oulu, Finland. APOPTOSIS.

Sojka, Prof. Paul E. Maurice J. Zucrow Laboratories, School of Mechanical Engineering, Purdue University, West Lafayette, Indiana. ATOMIZATION.

Sokol, Dr. Pamela A. Department of Microbiology and Infectious Diseases, University of Calgary Health Sciences Center, Calgary, Alberta, Canada. SIDEROPHORES.

Solberg, Dr. Myron. Department of Food Science, Rutgers University. PEPSIN.

Solem, Dr. G. Alan. Curator and Head, Division of Invertebrates, Field Museum of Natural History, Chicago, Illinois. PULMONATA; STYLOMMATOPHORA; SYSTELLOMMATOPHORA.

Sollosi, J. Michael. Commandant, (G-MWV), U.S. Coast Guard, Washington, D.C. VESSEL TRAFFIC SERVICE.

Solomon, Dr. James C. Associate Curator, Missouri Botanical Garden, St. Louis. HERBARIUM.

Somlo, Dr. Peter I. CSIRO Division of Applied Physics, Lindfield, New South Wales, Australia. RADIO-FREQUENCY IMPEDANCE MEASUREMENTS.

Somlyo, Dr. Andrew Paul. Department of Physiology, School of Medicine, University of Virginia. ELECTRON-PROBE MICROANALYSIS.

Sommer, Dr. Alfred. Director, International Center for Epidemiology and Preventive Ophthalmology, Dana Center, Wilmer Institute, Johns Hopkins School of Hygiene and Public Health, Baltimore, Maryland. VISUAL IMPAIRMENT.

Sommer, Dr. Alfred H. Thermo Electron Corporation, Waltham, Massachusetts. SECONDARY EMISSION.

Song, Dr. Pill-Soon. Chairperson, Department of Chemistry, University of Nebraska, Lincoln. CAROTENOID.

Soost, Dr. R. K. Department of Plant Science, University of California, Riverside. LEMON; LIME (BOTANY); ORANGE.

Sorensen, Prof. Robert M. Department of Civil Engineering, Lehigh University, Bethlehem, Pennsylvania. COASTAL ENGINEERING; RETAINING WALL; REVETMENT.

Sorenson, Prof. S. Department of Physics, University of Tennessee, Knoxville. RELATIVISTIC HEAVY-ION COLLISIONS.

Souders, Dr. Mott. Deceased; formerly, Director, Oil Development, Shell Oil Company, Emeryville, California. AROMATIZATION; DISTILLATION COLUMN; LIQUEFIED PETROLEUM GAS (LPG).

Soulen, Dr. Robert J., Jr. Naval Research Laboratory, Washington, D.C. SUPERCONDUCTIVITY.

South, Dr. David B. School of Forestry, Auburn University, Alabama. REFORESTATION.

Southwick, Dr. Stephen M. Department of Pomology, University of California. APRICOT.

Southwick, Dr. Steven. Department of Psychiatry and Connecticut Mental Health Center, School of Medicine, Yale University. NORADRENERGIC SYSTEM.

Southwood, Prof. T. R. E. Department of Zoology, Oxford University, Oxford, England. ECOLOGICAL METHODS.

Sovonick-Dunford, Dr. S. Associate Professor of Biological Sciences, University of Cincinnati, Ohio. PLANT TRANSPORT OF SOLUTES.

Spackman, Dr. William. Department of Geology, Pennsylvania State University. COAL PALEOBOTANY.

Spalding, Jay. U.S. Coast Guard R&D Center, Groton, Connecticut. MARINE NAVIGATION.

Spear, Dr. Scott K. Staff Scientist, The University of Alabama, Center for Green Manufacturing, Tuscaloosa. IONIC LIQUIDS.

Spearman, Dr. M. Leroy. Distinguished Research Associate, NASA Langley Research Center, Hampton, Virginia. GUIDED MISSILE.

Spector, Dr. David L. Senior Staff Scientist, Cold Spring Harbor Laboratory, New York. CELL NUCLEUS.

Spedding, Prof. Frank H. Deceased; formerly, Ames Laboratory, Energy Research and Development Administration, Iowa State University. CERIUM; EUROPIUM; HOLMIUM; LANTHANIDE CONTRACTION; RARE-EARTH ELEMENTS; YTTRIUM; other articles.

Speer, Dr. J. Alexander. Department of Marine, Earth, and Atmospheric Sciences, School of Physical and Mathematical Sciences, North Carolina State University. CONTACT AUREOLE.

Speight, Dr. James G. CD&W, Inc., Laramie, Wyoming. ALKYLATION (PETROLEUM); DIESEL FUEL; FOSSIL FUEL; PETROLEUM PROCESSING AND REFINING; REFORMING PROCESSES; other articles.

Spencer, Dr. R. C. Health Protection Agency South West Regional Laboratory, United Kingdom. ANTHRAX.

Spergel, Dr. David N. Department of Astrophysical Sciences, Princeton Unversity, New Jersey. COSMOLOGY; DARK MATTER.

Spiegel, Dr. Allen M. National Institute of Diabetes and Kidney Diseases, National Institutes of Health, Bethesda, Maryland. PARATHYROID GLAND DISORDERS.

Spiegel, Dr. Carol A. Director, Clinical Microbiology, Department of Pathology and Laboratory Medicine, University of Wisconsin Hospital and Clinics, Madison. CLINICAL MICROBIOLOGY.

Spies, Dr. Maria. Section of Microbiology, University of California, Davis. DNA HELICASE.

Spinrad, Dr. Bernard I. Retired; formerly, Department of Nuclear Engineering, Oregon State University. REACTOR PHYSICS; THERMAL NEUTRONS.

Splittstoesser, Dr. Don F. New York State Agricultural Experiment Station and Department of Food Science and Technology, Cornell University. VINEGAR.

Sposito, Prof. Garrison. College of Natural Resources, Division of Ecosystem Sciences, Environmental Geochemistry Group, University of California, Berkeley. SOIL CHEMISTRY.

Sprague, Prof. G. F. Department of Agronomy, University of Illinois, Urbana. CORN.

Sprague, Dr. Howard B. Agricultural Consultant, Washington, D.C. BROMEGRASS; DALLIS GRASS; FARM CROPS; REDTOP GRASS; other articles.

Spratt, Dr. Nelson T., Jr. Department of Zoology, University of Minnesota. ALLANTOIS; AMNION; CHORION; GASTRULATION; GERM LAYERS.

Spreadbury, Dr. Peter J. Department of Engineering, University of Cambridge, United Kingdom. ZENER DIODE.

Sprinkle, Dr. James. Department of Geological Sciences, University of Texas, Austin. CARPOIDS; EOCRINOIDEA; HOMALOZOA; PARABLASTOIDEA; PARACRINOIDEA; other articles.

Spurgin, Robert A. President, Spurgin & Associates, Irvine, California. MEDICAL WASTE.

Sreenivasan, Prof. K. R. Engineering and Applied Sciences, Yale University. TURBULENT FLOW.

Srihari, Dr. Sargur N. Center of Excellence for Document Analysis & Recognition, University of Buffalo, Amherst, New York. CHARACTER RECOGNITION.

Stack, Dr. Stephen. Department of Biology, Colorado State University, Fort Collins. CHROMOSOME.

Stadtman, Dr. Theressa C. Department of Health and Human Services, National Heart, Lung and Blood Institute, National Institutes of Health, Bethesda, Maryland. SELENIUM.

Stage, Dr. Steven A. Baton Rouge, Louisiana. MICROMETEOROLOGY.

Stahnke, Dr. Herbert L. Director, Poisonous Animal Research Laboratory, Arizona State University. SCORPIONES.

Stall, Prof. Robert E. Plant Pathology Department, Institute of Food and Agricultural Sciences, University of Florida. PLANT PATHOLOGY.

Stamm, Dr. Lola. Department of Epidemiology, School of Public Health, University of North Carolina, Chapel Hill. SYPHILIS.

Standish, Dr. E. Myles. Jet Propulsion Laboratory, Pasadena, California. EPHEMERIS; PRECESSION OF EQUINOXES.

Stanford, Prof. John L. Department of Physics and Astronomy, Iowa State University, Ames. MIDDLE-ATMOSPHERE DYNAMICS.

Staniford, Ferris C., Jr. W. L. Badger Associates, Inc., Consulting Engineers, Ann Arbor, Michigan. EVAPORATOR.

Stanley, Dr. Bruce A. Department of Cellular and Molecular Physiology, College of Medicine, Pennsylvania State University, Hershey. SERUM.

Stanton, Prof. Anthony. Carnegie Mellon University, Pittsburgh, Pennsylvania. CINEMATOGRAPHY.

Stapczynski, Dr. J. Stephan. Phoenix, Arizona. SHOCK SYNDROME.

Starr, Dr. Eugene C. Bonneville Power Administration, U.S. Department of the Interior, Portland, Oregon. ELECTRIC POWER GENERATION.

Staud, Dr. Roland. Division of Rheumatology and Clinical Immunology, University of Florida, Gainesville. FIBROMYALGIA SYNDROME.

Stauffer, Dr. Randy C. Halogens Research Laboratory, Dow Chemical Company, Midland, Michigan. BROMINE.

Stearn, Colin. Ontario, Canada. SPHAERACTINOIDEA; SPONGIOMORPHIDA; STROMATOPOROIDEA.

Stebbins, Prof. G. Ledyard. Department of Genetics, College of Agriculture, University of California, Davis. POLYPOIDY.

Steel, David L. Otis Elevator Company, Farmington, Connecticut. ESCALATOR.

Steele, William A. Department of Chemistry, Pennsylvania State University. ENTHALPY; ISOBARIC PROCESS; THERMODYNAMIC PROCESSES. HEAT.

Stefanovic, Dr. Victor R. Electrical Engineering Department, University of Missouri. REACTOR (ELECTRICITY).

Stehle, Dr. Philip. Department of Physics, University of Pittsburgh. CANONICAL COORDINATES AND TRANSFORMATIONS; HAMILTON-JACOBI THEORY; HAMILTON'S EQUATIONS OF MOTION; LAGRANGE'S EQUATIONS; LEAST-ACTION PRINCIPLE; other articles.

Stein, Prof. Dale F. Retired; Formerly President Emeritus, Michigan Technological University, Houghton. RECYCLING TECHNOLOGY.

Stein, Prof. Daniel J. MRC Unit, Department of Psychiatry, University of Stellenbosch Medical School, South Africa. ANXIETY DISORDERS.

Stein, Dr. Gary S. Department of Biochemistry and Molecular Biology, J. Hillis Miller Health Center, College of Medicine, University of Florida. MOLECULAR BIOLOGY.

Stein, Dr. Janet L. Department of Biochemistry and Molecular Biology, J. Hillis Miller Health Center, College of Medicine, University of Florida. MOLECULAR BIOLOGY.

Stein, Dr. Murray B. Anxiety & Traumatic Stress Disorders Research Program, University of California, La Jolla. PHOBIA.

Steinberg, Dr. Aephraim M. Department of Physics, University of California, Berkeley. COHERENCE.

Steinberg, Ellis P. Senior Scientist, Argonne National Laboratory, Argonne, Illinois. NUCLEAR CHEMISTRY.

Steinet, Emil F. Retired; formerly, Arc Welding Division, Westinghouse Electric Corporation. ARC WELDING.

Stephenson, Dr. Edward J. Indiana University Cyclotron Facility, Bloomington. CHARGE SYMMETRY.

Stephenson, Prof. F. William. Associate Dean for Research and Graduate Studies, Department of Electrical Engineering, College of Engineering, Virginia Polytechnic Institute and State University. DISTORTION (ELECTRONIC CIRCUITS).

Stephenson, Dr. R. J. Department of Physics, Wooster College, Wooster, Ohio. ACCELERATION; CENTRIFUGAL FORCE; CENTRIPETAL FORCE; ROTATIONAL MOTION.

Steponkus, Prof. Peter L. Department of Agronomy, Cornell University. COLD HARDINESS (PLANT).

Stern, Prof. Fred. Iowa Institute of Hydraulic Research, University of Iowa, Iowa City. COMPUTATIONAL FLUID DYNAMICS; KELVIN'S MINIMUM-ENERGY THEOREM.

Stern, Dr. Robert J. Department of Geosciences, University of Texas at Dallas. SUBDUCTION ZONES.

Sternheim, Dr. Charles E. Director of Undergraduate Studies, Department of Psychology, University of Maryland, College Park. VISION.

Sterrer, Dr. Wolfgang. Curator, Bermuda Aquarium, Museum and Zoo, Bermuda. GNATHOSTOMULIDA.

Stevens, Dr. C. E. College of Veterinary Medicine, North Carolina State University, Raleigh. DIGESTIVE SYSTEM.

Stevens, Dr. Calvin H. Department of Geology, San Jose State University. CNIDARIA.

Stevens, Dr. Mary Betty. Division of Rheumatology, Johns Hopkins University School of Medicine, Baltimore, Maryland. CONNECTIVE TISSUE DISEASE.

Stevenson, Dr. Edward C. Deceased; formerly, Department of Electrical Engineering, School of Engineering and Applied Sciences, University of Virginia. POWER-FACTOR METER.

Stewart, Dr. David B. Experimental Geochemistry and Mineralogy Branch, U.S. Geological Survey, Department of the Interior. SILICATE PHASE EQUILIBRIA.

Stewart, Dewey. Crops Research Division, Agricultural Research Service, U.S. Department of Agriculture, Beltsville, Maryland. SUGAR CROPS; SUGARBEET.

Stewart, Dr. John W. Department of Physics, University of Virginia. CURRENT DENSITY; DIMENSIONAL ANALYSIS; ELECTRODYNAMICS; REFRACTION OF WAVES; other articles.

Stewart, Scott R. L. C. Lee & Associates, Inc., Seattle, Washington. WETLANDS.

Steyn, Dr. Julian J. Energy Resources International, Inc., Washington, D.C. ISOTOPE SEPARATION.

Stidd, Prof. Benton M. Department of Biological Science, Western Illinois University, Macomb, Illinois. MARATTIALES.

Stiefel, Dr. Edward I. Scientific Advisor, Exxon Research and Engineering Co., Annandale, New Jersey. MOLYBDENUM.

Stiles, Dr. Martin. Department of Chemistry, University of Kentucky. PHENOL; other articles.

Stillman, Dr. Gregory E. Department of Electrical Engineering, University of Illinois, Urbana. PHOTOVOLTAIC CELL.

Stipanuk, Dr. Martha H. Professor, Division of Nutritional Sciences, Cornell University, Ithaca, New York. TRANSAMINATION.

Stirton, Dr. Robert I. Retired; formerly, Vice President, Roger Williams Technical and Economic Services, Inc., Berkeley, California. PHENOL.

Stock, Dr. John T. Professor Emeritus, Department of Chemistry, University of Connecticut. TITRATION.

Stocum, Dr. David L. Department of Biology, Indiana University-Purdue University, Indianapolis, Indiana. MORPHOGENESIS.

Stoloff, Dr. Norman S. Department of Materials Engineering, Rensselaer Polytechnic Institute. EUTECTICS; METAL, MECHANICAL PROPERTIES OF.

Stolper, Dr. Edward M. Chairman, Division of Geological and Planetary Sciences, California Institute of Technology. METEORITE; VOLCANIC GLASS.

Stone, Prof. A. Douglas. Department of Applied Physics, Yale University. MESOSCOPIC PHYSICS.

Stone, Prof. Anthony J. Department of Chemistry, University of Cambridge, United Kingdom. INTERMOLECULAR FORCES.

Stone, Dr. Jeffrey K. Department of Botany and Plant Pathology, Oregon State University, Corvallis. FUNGAL ECOLOGY.

Stone, Prof. Kenneth G. Deceased; formerly, Department of Chemistry, Michigan State University. REAGENT CHEMICALS.

Stone, Dr. Nelson N. Urology Department, Mount Sinai School of Medicine, Elmhurst, New York. PROSTATE GLAND DISORDERS.

Storch, Dr. Henry H. Deceased; formerly, Assistant Professor of Chemistry, New York University. DESTRUCTIVE DISTILLATION.

Storer, Dr. Tracy I. Deceased; formerly, Department of Zoology, University of California, Davis. ANIMAL SYMMETRY; PROTANDRY; ZOOLOGY; other articles.

Storey, Prof. William B. Department of Botany and Plant Sciences, University of California, Riverside. FIG.

Storm, Dr. Eric. Lawrence Livermore National Laboratory; Livermore, California. NUCLEAR FUSION.

Stott, Dr. Jeffrey L. Department of Veterinary Microbiology and Immunology, School of Veterinary Medicine, University of California, Davis. BLUETONGUE.

Stout, Gary J. AT&T Submarine Systems, Inc., Morristown, New Jersey. SUBMARINE CABLE.

Stoutemyer, Dr. Vernon T. Department of Agriculture Sciences, University of California, Los Angeles. ORNAMENTAL PLANTS.

Stoyer, Dr. Mark A. Lawrence Livermore National Laboratory, California. SUPERHEAVY ELEMENTS.

Strandberg, Dr. James O. Agricultural Research Extension Center, Sanford, Florida. CABBAGE; CELERY; TURNIP.

Strandness, Dr. D. Eugene, Jr. Department of Surgery, University of Washington School of Medicine, Seattle. VASCULAR DISORDERS.

Stranks, Prof. Donald R. Department of Physical and Inorganic Chemistry, University of Adelaide, Australia. RADIOCHEMISTRY.

Straus, Dr. Reuben. Providence Medical Center, Portland, Oregon. SERUM; URINE.

Strausbaugh, Prof. Carl A. Kimberly Research & Extension Center, University of Idaho, Kimberly. SOIL STERILIZATION.

Strausbaugh, Dr. Perry D. Deceased; formerly, Professor of Botany, West Virginia University. ALLSPICE; BLACK PEPPER; CINNAMON; DIGITALIS; POISON IVY; other articles.

Streeter, Prof. Victor L. Department of Civil Engineering, University of Michigan. BUOYANCY; JET FLOW; TORRICELLI'S THEOREM; other articles.

Striker, Dr. Gary. National Institute of Diabetes, Digestive and Kidney Diseases, National Institutes of Health, Bethesda, Maryland. KIDNEY DISORDERS.

Strock, Dr. O. J. Consultant, Sarasota, Florida. TELEMETERING.

Strohl, Robert, J. Senior Research Scientist, Batelle Laboratories, Columbus, Ohio. DRONE; MILITARY AIRCRAFT.

Stroke, Dr. George W. Department of Electrical Sciences, and Head, Electro-Optical Sciences Center, State University of New York, Stony Brook. LIGHT.

Strom, Prof. Robert G. Department of Planetary Sciences, Lunar and Planetary Laboratory, University of Arizona. MERCURY (PLANET).

Strominger, Dr. Jack L. Department of Biochemistry and Molecular Biology, Harvard University. URIDINE DIPHOSPHOGLUCOSE (UDPG).

Stroock, Dr. Abraham D. Department of Chemical and Biomolecular Engineering, Cornell University, Ithaca, New York. MICROFLUIDICS.

Strunz, Dr. Kai. Department of Electrical Engineering, University of Washington, Seattle. ELECTRIC TRANSIENT.

Stuart, Dr. Alastair M. Department of Zoology, University of Massachusetts. ISOPTERA.

Stuart, Dr. Edward G. Deceased; formerly, West Virginia School of Medicine. BURN; other articles.

Stubbings, Dr. H. G. Poole, England. ACROTHORACICA; THORACICA; other articles.

Stuckas, Kenneth J. Project Engineering, Engineering Consultation Services, Jacksonville, Florida. RECIPROCATING AIRCRAFT ENGINE.

Studenny, Dr. John. CMC Electronics Inc., Montreal, Quebec, Canada. AIR NAVIGATION.

Studlbarg, Dr. Michael S. Director, Chest Clinic, Department of Medicine, Pulmonary Division, University of California School of Medicine, San Francisco. EMPHYSEMA.

Stumm, Dr. Erwin Charles. Department of Geology and Mineralogy, Museum of Paleontology, University of Michigan. RUGOSA.

Stunkard, Dr. Horace W. Research Associate, American Museum of Natural History, New York. CEPHALOBAENIDA; TRYPANOSOMATIDAE; other articles.

Sturges, Mr. Thomas B., III. Pennsylvania Drilling Company, McKees Rock. DRILLING, GEOTECHNICAL.

Sturges, Prof. Wilton. Department of Oceanography, Florida State University. OCEAN CIRCULATION.

Sturm, Prof. Terry W. School of Civil and Environmental Engineering, Georgia Institute of Technology, Atlanta. OPEN CHANNEL.

Sturtevant, Prof. Alfred H. Thomas Hunt Morgan Professor of Biology, Emeritus, California Institute of Technology. GENE; GENETICS.

Sudds, Dr. Richard. Department of Biological Sciences, State University of New York, Plattsburgh. PEDICULOSIS.

Sues, Dr. Hans-Dieter. National Museum of Natural History, Smithsonian Institute, Washington, D.C. ARCHOSAURIA.

Sulkin, Mr. Allan. TEQConsult Group, Hackensack, New Jersey. PRIVATE BRANCH EXCHANGE.

Sullivan, Prof. Michael J., Jr. Department of Mathematics, Chicago State University, Chicago, Illinois. TRIGONOMETRY.

Sumi, Dr. S. Mark. Associate Professor of Neuropathology, School of Medicine, University of Washington. MUSCULAR DYSTROPHY; MUSCULAR SYSTEM DISORDERS; MYASTHENIA GRAVIS.

Summers, Dr. B. A. James A. Baker Institute for Animal Health, Cornell University, Ithaca, New York. CANINE DISTEMPER.

Sumners, Dr. Thomas E. Plant Pathologist, Agricultural Research Service, U.S. Department of Agriculture. JUTE; SISAL.

Sumrall, Dr. Colin. Department of Earth and Planetary Sciences, University of Tennessee, Knoxville. RHOMBIFERA.

Suppe, Dr. John. Department of Geological Sciences, Princeton University. BLUESCHIST.

Surles, Dr. Terry. Limnetics, Inc., Milwaukee, Wisconisn. INTERHALOGEN COMPOUNDS.

Suryanaryana, Prof. N. V. Department of Mechanical Engineering and Engineering Mechanics, College of Engineering, Michigan Technological University. HEAT PUMP.

Suslick, Prof. Kenneth S. Department of Chemistry, University of Illinois, Urbana-Champaign. SONOCHEMISTRY.

Sutherland, J. R. Power Transformer Department, General Electric Company, Pittsfield, Massachusetts. AUTOTRANSFORMER.

Sutherland, Dr. S. S. Senior Research Officer, Animal Health Laboratories, Department of Agriculture, Western Australia. STREPTOTHRICOSIS.

Suttner, Dr. Lee J. Department of Geology, Indiana University. SANDSTONE.

Sutton, Dr. B. C. Assistant Director, International Mycological Institute, United Kingdom. AGONOMYCETES; BLASTOMYCETES; COELOMYCETES; DEUTEROMYCOTA; HYPHOMYCETES.

Sutton, George P. Consulting Engineer, Danville, California. INTERPLANETARY PROPULSION; ROCKET; ROCKET PROPULSION; ROCKET STAGING.

Suzuki, Dr. Yoichiro. Kamioka Observatory, Institute for Cosmic Ray Research, University of Tokyo, Japan. NEUTRINO.

Swanson, Barry. Department of Food Science and Human Nutrition, Washington State University. FOOD ENGINEERING.

Swanson, Dr. Eric. Vice President, Technology, Crystal Semiconductor Corporation, Austin, Texas. CHOPPING.

Swanson, Dr. M. L. Department of Physics and Astronomy, University of North Carolina. CHANNELING IN SOLIDS.

Swanson, Dr. R. Lawrence. Director, Waste Reduction and Management Institute, Marine Sciences Research Center, State University of New York, Stony Brook. RECYCLING TECHNOLOGY.

Swearingen, Dr. Will. NASA-Montana State University TechLink, Bozeman, Montana. DESERTIFICATION.

Sweat, F. W. Department of Chemistry, University of Utah. DIMETHYL SULFOXIDE.

Sweet, Dr. Kathleen M. Department of Aviation Technology, Purdue University, West Lafayette. Indiana. AVIATION SECURITY.

Swets, Dr. John A. Bolt Beranek and Newman, Inc., Cambridge, Massachusetts. SIGNAL DETECTION THEORY.

Swider, Dr. William, Jr. Space Physics Division, Air Force Geophysics Laboratory, Hanscom Air Force Base, Bedford, Massachusetts. ALKALI EMISSIONS.

Swift, Dr. Gregory W. Condensed Matter and Thermal Physics, Los Alamos National Laboratory, New Mexico. THERMOACOUSTICS.

Switzer, Dr. George. Curator, Department of Mineral Sciences, National Museum of Natural History, Smithsonian Institution, Washington, D.C. NITER; NITRATE MINERALS; SODA NITER.

Sydenham, Dr. Peter H. Rostrevor, South Australia. GRAPHIC RECORDING INSTRUMENTS.

Sykes, Megan. Transplantation Biology Research Center, Massachusetts. ACQUIRED IMMUNOLOGICAL TOLERANCE.

Symko, Orest G. Department of Physics, University of Utah. LOW-TEMPERATURE PHYSICS.

Symons, Dr. Frank. Applied Research Laboratory, Pennsylvania State University. ACOUSTIC TORPEDO.

Symons, Dr. Robert S. Litton Electrical Devices Division, San Carlos, California. ELECTRON TUBE.

Sytsma, Prof. Kenneth J. Department of Botany, University of Wisconsin at Madison. CORNALES; FAGALES; GERANIALES; TROCHODENDRALES.

Szabad, Dr. Janos. Institute of Genetics, Biological Research Center, Hungarian Academy of Sciences, Szeged. MOSAICISM.

Szalay, Dr. Frederick S. Department of Vertebrate Paleontology, American Museum of Natural History, New York. DENTITION; METATHERIA; TOOTH; other articles.

Szeri, Dr. Andras Z. Professor and Chairperson, Department of Mechanical Engineering, University of Delaware, Newark. LUBRICATION; TRIBOLOGY.

Szkody, Dr. Paula. Department of Astronomy, University of Washington, Seattle. CATACLYSMIC VARIABLE.

Sztul, Dr. Elizabeth. Department of Cell Biology, University of Alabama at Birmingham. CELL MEMBRANES.

T

Tabor, Paul. Soil Conservation Service, Athens, Georgia. KUDZU; LEGUME FORAGES; LESPEDEZA; LUPINE.

Taggart, Douglas. Overlook Systems Technologies, Vienna, Virginia. SHIP ROUTING.

Tait, Dr. David L. Department of Obstetrics and Gynecology, Vanderbilt University Medical Center, Nashville, Tennessee. OVARIAN DISORDERS.

Talbot, Prof. Lawrence. Department of Mechanical Engineering, University of California, Berkeley. GAS DYNAMICS; KNUDSEN NUMBER.

Tan, Dr. K. K. Centre for Intelligent Control, Department of Electrical Engineering, National University of Singapore. EXPERT CONTROL SYSTEM.

Tanese, Dr. Naoko. Department of Microbiology, New York University School of Medicine, New York. TRANSCRIPTION.

Tang, Prof. K. Y. Deceased; formerly Department of Electrical Engineering, Ohio State University. KIRCHHOFF'S LAWS OF ELECTRIC CIRCUITS.

Tanihata, Dr. Isao. Accelerator Facility, Institute of Physical and Chemical Research, Saritama, Japan. CHARGED PARTICLE BEAMS.

Tannas, Lawrence E., Jr. Tannas Electronics, Orange, California. ELECTRONIC DISPLAY.

Tansu, Dr. Nelson. Center for Optical Technologies, Department of Electrical and Computer Engineering, Rossin College of Engineering and Applied Science, Lehigh University, Bethlehem, Pennsylvania. PHOTOEMISSION.

Tao, Dr. B. Y. School of Agricultural and Biological Engineering, Purdue University, West Lafayette, Indiana. BIOMASS.

Tapper, Prof. Nigel J. Department of Geography and Environmental Science,

Monash University, Melbourne, Australia. AUSTRALIA.

Tarhule, Dr. Aondover. Department of Geography, University of Oklahoma, Norman. SAHEL.

Tarling, Dr. D. H. Department of Geophysics and Planetary Physics, University of New Castle, England. CONTINENTAL DRIFT.

Taton, Dr. Daniel. Laboratoire de Chimie des Polymères Organiques, ENSCPB, Université Bordeaux, Pessac, France. CONTROLLED/LIVING RADICAL POLYMERIZATION.

Tattersall, Dr. Ian. Curator of Physical Anthropology, American Museum of Natural History, New York. FOSSIL PRIMATES.

Taube, Dr. Henry. Department of Chemistry, Stanford University; Nobelist. OXIDATION-REDUCTION.

Tavoularis, Prof. Stavros. Department of Mechanical Engineering, University of Ottawa, Ontario, Canada. VELOCIMETER.

Taylor, Prof. Angus E. Deceased; formerly, President's Office, University of California, Berkeley. CALCULUS; DIFFERENTIATION; PARAMETRIC EQUATION; PARTIAL DIFFERENTIATION.

Taylor, Dr. Gordon S. Connecticut Agricultural Experiment Station, Windsor. TOBACCO.

Taylor, Prof. J. Herbert. Institute of Molecular Biophysics, Florida State University. AUTORADIOGRAPHY; HISTORADIOGRAPHY; NUCLEAR RADIATION (BIOLOGY).

Taylor, Prof. Jean E. Department of Mathematics, Rutgers University, New Brunswick, New Jersey. CALCULUS OF VARIATIONS.

Taylor, Prof. John W. Department of Plant Biology, University of California, Berkeley. FUNGI.

Taylor, Prof. R. E. Department of Anthropology, Radiocarbon Laboratory, University of California, Riverside. RADIOCARBON DATING.

Taylor, Dr. Robert J. Consultant, Microscopy, Falls Church, Virginia. CONFOCAL MICROSCOPY.

Taylor, Dr. Stuart Ros. Department of Geology, Australian National University, Canberra, Australia. HADEAN.

Taylor, Prof. Thomas N. Department of Biology, University of Kansas, Lawrence. COAL BALLS; PALEOBOTANY; PLANT-ANIMAL INTERACTIONS; PLEUROMIALES; PTERIDOSPERMS; SPHENOPHYTA.

Tchobanoglous, Prof. George. Department of Civil Engineering, University of California, Davis. WATER POLLUTION.

Teaguarden, Dr. Dennis E. Professor Emeritus of Forestry, University of California, Berkeley. FOREST MANAGEMENT.

Tease, Samuel C. Pennsalt Chemicals Corporation, Philadelphia, Pennsylvania. LYOPHILIZATION.

Tedesco, Dr. Edward F. Space Science Center, University of New Hampshire, Durham. TROJAN ASTEROIDS.

Tedford, Dr. Richard H. Department of Vertebrate Paleontology, American Museum of Natural History, New York. CARNIVORA; LITOPTERNA; MARSUPIALIA; NOTOUNGULATA; TUBULIDENTATA.

Teichert, Dr. Curt. Department of Geological Sciences, University of Rochester. NAUTILOIDEA.

Teitz, Dr. Carol C. Sports Medicine, University of Washington, Seattle. PERFORMING ARTS MEDICINE.

Teleki, Dr. Geza. Department of Geology, George Washington University. EUROPE.

Telionis, Prof. Demetri P. Department of Engineering Sciences & Mechanics, Virginia Polytechnic Institute and State University, Blacksburg, Virginia. KÁRMÁN VORTEX STREET; WAKE FLOW.

Teller, Dr. Aaron J. Teller Environmental Systems, Worcester, Massachusetts. GAS ABSORPTION OPERATIONS.

Temin, Dr. Howard M. McArdle Laboratory for Cancer Research, Department of Oncology, University of Wisconsin Medical School; Nobelist. REVERSE TRANSCRIPTASE.

Teng, Lee C. Fermi National Accelerator Laboratory, Batavia, Illinois. RELATIVISTIC ELECTRODYNAMICS.

Tennant, Dr. Raymond W. Chief, Cellular and Genetic Toxicology Branch, Department of Health and Human Resources, National Institutes of Health, National Institute of Environmental Health Sciences, Research Triangle Park, North Carolina. MUTAGENS AND CARCINOGENS.

Tenne-Sens, Andrej. Retired; formerly, Communications Research Center, Ottawa, Ontario, Canada. ENERGY; LINEAR MOMENTUM; MOMENT OF INERTIA; MOMENTUM.

Tenney, Dr. Gerold H. American Society for Nondestructive Testing, Inc., Los Alamos, New Mexico. RADIOGRAPHY.

Terrill, Dr. Clair E. Chief, Sheep and Fur Animal Research Branch, U.S. Department of Agriculture, Beltsville, Maryland. MOHAIR; SHEEP.

Teuber, Dr. Larry R. Department of Agronomy and Range Science, Agricultural Experiment Station, College of Agricultural and Environmental Sciences, University of California, Davis. ALFALFA.

Thalos, Dr. Mariam. Department of Philosophy, University of Utah, Salt Lake City. PHILOSOPHY OF SCIENCE.

Thase, Dr. Michael E. Professor of Psychiatry, University of Pittsburgh, School of Medicine, Pittsburgh, Pennsylvania. AFFECTIVE DISORDERS.

Theis, Dr. Douglas J. Computer Systems Department, Aerospace Corporation, Los Angeles, California. COMPUTER STORAGE TECHNOLOGY.

Thewissen, Dr. J. G. M. Anatomy Department, Northeastern Ohio Universities College of Medicine, Rootstown, Ohio. CETACEA.

Thoen, Prof. Charles O. College of Veterinary Medicine, Iowa State University, Ames. JOHNE'S DISEASE.

Thoma, Roy E. Oak Ridge National Laboratory, Oak Ridge, Tennessee. FUSED-SALT PHASE EQUILIBRIA.

Thomann, Dr. Gary C. Power Technologies, Inc., Schenectady, New York. WIND POWER.

Thomas, Dr. David A. Department of Materials Science and Engineering, Lehigh University, Bethlehem, Pennsylvania. METALLOGRAPHY.

Thomas, Dr. David Hurst. American Museum of Natural History, New York. ARCHEOLOGY.

Thomas, Prof. Helmuth. Department of Oceanography, Dalhousie University, Nova Scotia, Canada. ANOXIC ZONES.

Thomas, Dr. J. Kerry. Department of Chemistry, University of Notre Dame. MICELLE; SURFACTANT.

Thomas, Prof. Marlin U. Head, School of Industrial Engineering, Purdue University, West Lafayette, Indiana. INDUSTRIAL ENGINEERING; PRODUCT QUALITY.

Thompson, Prof. Gerald L. Professor of Applied Mathematics and Industrial Administration, Graduate School of Industrial Administration, Carnegie-Mellon University. GAME THEORY.

Thompson, Dr. Laird A. Department of Astronomy, College of Liberal Arts and Sciences, University of Illinois. TWINKLING STARS.

Thompson, Dr. Margaret. Departments of Medical Genetics and Pediatrics, University of Toronto and Hospital for Sick Children, Canada. HUMAN GENETICS.

Thompson, Dr. Stanley G. Senior Staff Member, Lawrence Berkeley Laboratory, University of California, Berkeley. EINSTEINIUM.

Thompson, Dr. T. E. Department of Zoology, University of Bristol, England. NUDIBRANCHIA; OPISTHOBRANCHIA; SACOGLOSSA.

Thompson, Tom. Hollis, New Hampshire. MICROCOMPUTER; MICROPROCESSOR.

Thompson, William I., III. Senior Systems Engineer, Science Applications International Corporation, Falls Church, Virginia. VOR (VHF OMNIDIRECTIONAL RANGE).

Thomson, Dr. Keith S. President, Academy of Natural Sciences, Philadelphia, Pennsylvania. ANIMAL EVOLUTION.

Thomson, Prof. Robb M. Chairperson, Department of Materials Science, State University of New York, Stony Brook. ALLOY STRUCTURES; TWINNING (CRYSTALLOGRAPHY).

Thorn, Robert J. Chemistry Division, Argonne National Laboratory, Argonne, Illinois. SOLID-STATE CHEMISTRY.

Thornton, Dr. Charles H. Thornton-Tomasetti, Engineers, New York. TRUSS.

Thorogood, Dr. Robert M. Department of Chemical Engineering, North Carolina State University, Raleigh. AIR SEPARATION.

Thorp, Prof. H. Holden. Department of Chemistry, University of North Carolina, Chapel Hill. ELECTROLYTE; HYDROXYL; ION.

Tichy, Prof. Jiri. Graduate Program in Acoustics, Pennsylvania State University. SOUND INTENSITY.

Tikofsky, Dr. Ronald S. Department of Radiology, Medical College of Wisconsin, Milwaukee. SPEECH DISORDERS.

Tilling, Dr. Robert I. Geologist, Branch of Igneous and Geothermal Processes, U.S. Geological Survey, Menlo Park, California. IGNIMBRITE; LAVA; PYROCLASTIC ROCKS; TUFF; VOLCANO; VOLCANOLOGY.

Tilton, Dr. Richard C. Department of Laboratory Medicine, University of Connecticut. CLINICAL MICROBIOLOGY.

Timmerman, Nancy. Boston, Massachusetts. AIRPORT NOISE.

Timoney, Dr. John F. University of Kentucky College of Agriculture, Department of Veterinary Science, Gluck Equine Research Center. ERYSIPELOTHRIX; GLANDERS; STRANGLES.

Ting, Dr. Irwin P. Department of Botany and Plant Sciences, University of California, Riverside. PLANT METABOLISM.

Ting, Prof. Samuel C. C. Laboratory for Nuclear Science, Massachusetts Institute of Technology; Nobelist. J/PSI PARTICLE.

Tinsley, Dr. Brian A. Center for Space Science, Department of Physics, University of Texas, Dallas. AIRGLOW.

Tirion, Dr. Wil. Capelle a/d Ijssel, the Netherlands. ANDROMEDA; AQUARIUS; articles on other constellations.

Tokuta, Dr. Alade. Department of Mathematics and Computer Science, North Carolina Central University, Durham. COMPUTER GRAPHICS; DATA STRUCTURE.

Tolman, Dr. William B. Department of Chemistry, University of Minnesota. IRON.

Tolstoy, Dr. Maya. Oceanography, Lamont Doherty Earth Observatory, Palisades, New York. MOHO (MOHOROVIČIČ DISCONTINUITY).

Tomasetti, Richard L. Senior Vice President, Lev Zetlin Associates, Inc., New York. STRUCTURAL DESIGN.

Tompkins, Prof. Charles B. Department of Mathematics, University of California, Los Angeles. CALCULUS OF VARIATIONS.

Tonkay, Prof. Gregory L. Department of Industrial Engineering, Lehigh University, Bethlehem, Pennsylvania. PRODUCTIVITY.

Tonndorf, Prof. Juergen. Deceased; formerly, Department of Otolaryngology, College of Physicians and Surgeons, Columbia University. PHYSIOLOGICAL ACOUSTICS.

Toole, Dr. Floyd. Oak Park, California. SOUND-REPRODUCING SYSTEMS.

Torchilin, Dr. Vladimir. Department of Pharmaceutical Sciences, Northeastern University, Boston, Massachusetts. DRUG DELIVERY SYSTEMS.

Torres, Dr. Alfonso. Head, Diagnostic Services Section, National Veterinary Services Laboratory, U.S. Department of Agriculture, Greenport, New York. RINDERPEST.

Torres, Monica S. Department of Plant Biology and Pathology, Rutgers University, New Brunswick, New Jersey. ENDOPHYTIC FUNGI.

Torrey, Dr. Theodore W. Professor Emeritus, Department of Zoology, Indiana University. URINARY SYSTEM.

Toth, Dr. Nicholas. Department of Anthropology, Indiana University, Bloomington. PREHISTORIC TECHNOLOGY.

Townes, Prof. Charles H. Department of Physics, University of California, Berkeley; Nobelist. MASER.

Toy, Dr. Pearl. San Francisco General Hospital, San Francisco, California. TRANSFUSION.

Toy, Dr. T. J. Department of Geography, University of Denver, Colorado. EROSION.

Träbert, Prof. Elmar. Experimentalphysik III, Ruhr-Universität Bochum, Germany. BEAM-FOIL SPECTROSCOPY.

Tracey, Dr. Joshua I., Jr. U.S. Geological Survey, Smithsonian Institution, Washington, D.C. ATOLL.

Tracy, Dr. Robert J. Department of Geological Sciences, Virginia Polytechnic Institute and State University, Blacksburg. AUGITE; ORTHORHOMBIC PYROXENE.

Trappe, Dr. James M. Department of Forest Science, Oregon State University, Corvallis. MYCORRHIZAE.

Trautman, W. Dean. Manager, Laboratory Administration, Research and Development, Brush Beryllium Company, Cleveland, Ohio. BERYLLIUM ALLOYS; BERYLLIUM METALLURGY.

Treffers, Dr. Henry P. Professor of Pathology, School of Medicine, Yale University. ANTIGEN-ANTIBODY REACTION; COMPLEMENT-FIXATION TEST; IMMUNITY; NEUTRALIZING ANTIBODY; other articles.

Treger, Dr. Darly. Strategic Analysis Inc., Arlington, Virginia. SPINTRONICS.

Treglia, Prof. Anthony. Department of Environmental Control Technology, New York City College of Technology, Brooklyn. RADIANT HEATING.

Treiman, Prof. Sam B. Deceased; formerly, Department of Physics, Princeton University, Princeton, New Jersey. SCHRÖDINGER WAVE EQUATION.

Trendall, Dr. Alec. Adjunct Professor, Department of Applied Physics, Curtin University of Technology, Western Australia. BANDED IRON FORMATION.

Treybi, Robert E. Deceased; formerly, Department of Chemical Engineering, New York University. SOLVENT EXTRACTION.

Trimble, Dr. Virginia. Department of Physics, University of California, Irvine. NOVA.

Trinkaus, Dr. Erik. Department of Anthropology, Washington University, St. Louis, Missouri. EARLY MODERN HUMANS; NEANDERTALS.

Trivett, Dr. David H. George W. Woodruff School of Mechanical Engineering, Georgia Institute of Technology. HYDROPHONE.

Trivett, Dr. Mary L. Department of Environmental and Plant Biology, Ohio University, Athens. CORDAITALES.

Trolinger, Dr. James D. Vice President, Spectron Development Laboratories, Costa Mesa, California. SCHLIEREN PHOTOGRAPHY.

Trowsdale, Dr. John. Professor of Immunology, Department of Pathology, Cambridge University. HISTOCOMPATIBILITY; IMMUNITY.

Truant, J. P. Advance Medical & Research Center, Inc., Pontiac, Michigan. STAIN (MICROBIOLOGY).

Truax, Capt. Robert C. Retired; formerly, Truax Engineering Company, Saratoga, California. ASTRONAUTICAL ENGINEERING; ASTRONAUTICS.

Trueman, Dr. John W. H. School of Botany and Zoology, Faculty of Science, The Australian National University. ODONATA.

Trump, Benjamin F. Professor and Chairperson, Department of Pathology, University of Maryland Medical Center. DEATH.

Tsaur, Dr. Bor-Yeu. Lincoln Laboratories, Lexington, Massachusetts. ION BEAM MIXING.

Tsong, Dr. Tien-Tzou. Distinguished Research Fellow, Institute of Physics, Academica Sinica, Taipei, Taiwan. FIELD-EMISSION MICROSCOPY.

Tsuchiya, Prof. Henry M. Department of Chemical Engineering, University of Minnesota. DEXTRAN.

Tucker, Dr. David H. Agriculture Research and Education Center, University of Florida. GRAPEFRUIT.

Tucker, Dr. Wallace. Harvard-Smithsonian Center for Astrophysics, Cambridge, Massachusetts. ASTROPHYSICS, HIGH-ENERGY; CHANDRA X-RAY OBSERVATORY.

Tuite, Dr. Paul. Department of Neurology, University of Minnesota, Minneapolis. PARKINSON'S DISEASE.

Tukey, Dr. Harold B. Deceased; formerly, Professor Emeritus, Department of Horticulture, Michigan State University. CRABAPPLE; QUINCE.

Tully, Dr. R. Brent. Institute for Astronomy, University of Hawaii at Manoa. LOCAL GROUP.

Tur, Dr. Moshe. School of Engineering, Tel Aviv University, Israel. DIRECTIONAL COUPLER.

Turk, Prof. J. L. Department of Pathology, Royal College of Surgeons of England, London. IMMUNOSUPPRESSION.

Turley, Dr. J. A. Marathon Oil Company, Littleton, Colorado. OIL AND GAS, OFFSHORE.

Turnbull, Prof. David. Department of Applied Physics, Harvard University. CRYSTAL GROWTH; SINGLE CRYSTAL.

Turner, Dr. Barry E. National Radio Astronomy Observatory Charlottesville, Virginia. INTERSTELLAR MATTER.

Turner, Dr. Charles E. Agricultural Research Service, U.S. Department of Agriculture, Albany, California. WEEDS.

Turner, Dr. Edwin L. Peyton Hall, Princeton University Observatory. GALAXY, EXTERNAL.

Turner, Prof. J. R. G. Department of Genetics, University of Leeds, United Kingdom. PROTECTIVE COLORATION.

Turner, Joyce B. Peyton Hall, Princeton University Observatory. GALAXY, EXTERNAL.

Turner, Megan B. The University of Alabama, Center for Green Manufacturing. IONIC LIQUIDS.

Turner, Dr. Monica G. Environmental Sciences Division, Oak Ridge National Laboratory, Oak Ridge, Tennessee. LANDSCAPE ECOLOGY.

Turner, Dr. Noel H. Chemistry Division, Naval Research Laboratory, Washington, D.C. AUGER ELECTRON SPECTROSCOPY.

Turner, Dr. Ruth D. Museum of Comparative Zoology, Harvard University. BORING BIVALVES.

Turner, Dr. Thomas B. Professor of Microbiology, Dean Emeritus, School of Medicine, Johns Hopkins University. LEPTOSPIROSIS; SYPHILIS; YAWS.

Turner, Prof. John D. Transport Research Laboratory, Crowthorne, Berks, United Kingdom. THICK-FILM SENSOR.

Turng, Prof. Lih-Sheng. Mechanical Engineering, University of Wisconsin-Madison. PLASTICS PROCESSING.

Turro, Nicholas J. Department of Chemistry, Columbia University. TRIPLET STATE.

Turturro, Dr. Angelo. National Center for Toxicological Research, Department of Health and Human Services, Jefferson, Arkansas. BIOASSAY.

Tuttle, Alan H. Deceased; formerly, Vocational Division, State University of New York Agricultural and Technical Institute, Wellsville. WOODWORKING; other articles.

Tuttle, Prof. Russell H. Department of Anthropology, University of Chicago. PHYSICAL ANTHROPOLOGY.

Tyler, Prof. Albert. Deceased; formerly, Division of Biology, California Institute of Technology. ANIMAL REPRODUCTION; ESTRUS; OVUM.

Tyler, Nathan. Glass Packaging Institute, Washington, D.C. RECYCLING TECHNOLOGY.

Tyrer, Dr. Louise B. Medical Affairs Coordinator, Planned Parenthood Federation of America, New York. BIRTH CONTROL.

U

Uchida, Dr. Tohru. Professor of Zoology, Hokkaido University, Japan. CORONATAE; CUBOZOA; RHIZOSTOMEAE; SCYPHOZOA; SEMAEOSTOMEAE; STAUROMEDUSAE.

Uebersax, Dr. Mark A. Department of Food Sciences and Human Nutrition, Michigan State University. BARLEY; BUCK-WHEAT; OATS; RICE; WHEAT; other articles.

Uhl, Dr. Vincent W. Department of Chemical Engineering, University of Virginia. MIXING; SEDIMENTATION (INDUSTRY); THICKENING.

Uhler, Dr. Michael D. Department of Biological Chemistry, University of Michigan, Ann Arbor. CYCLIC NUCLEOTIDES.

Uhrhammer, Dr. Robert A. Seismological Laboratory, University of California, Berkeley. SEISMOGRAPHIC INSTRUMENTATION.

Ullman, Dr. John E. Department of Marketing Management, Hofstra University, Hempstead, New York. DESIGN STANDARDS; WORK STANDARDIZATION.

Ullrich, Dr. Henry. Guilford, Connecticut. POLYURETHANE RESINS.

Ullrich, Dr. Robert C. Department of Botany, University of Vermont. FUNGAL GENETICS.

Underwood, Dr. Ervin E. Department of Metallurgy, Georgia Institute of Technology. SUPERPLASTICITY.

Underwood, Dr. James H. Center for X-Ray Optics, Lawrence Berkeley National Laboratory, University of California, Berkeley. X-RAY OPTICS.

Unwin, Dr. Stephen C. Deputy Project Scientist, Space Interferometry Mission, Jet Propulsion Laboratory, Pasadena, California. SUPERLUMINAL MOTION.

Uozumi, Prof. Yasuhiro. Laboratory of Complex Catalysis, Institute for Molecular Science, Okazaki, Japan. COMBINATORIAL SYNTHESIS.

Upton, Prof. Arthur. New York, New York. RADIATION INJURY (BIOLOGY).

Uzsoy, Prof. Reha M. School of Industrial Engineering, Purdue University, West Lafayette, Indiana. PRODUCTION ENGINEERING.

V

Vacelet, Dr. Jean. Centre d'Oceanologie de Marseille, France. CALCAREA.

Vaena, Dr. Daniel A. Assistant Professor, Division of Hematology, Oncology, and Blood and Marrow Transplantation, Department of Medicine and The University of Iowa Cancer Center, University of Iowa, Iowa City; and The Iowa City Veterans Administration Medical Center. CHEMOTHERAPY AND OTHER ANTINEOPLASTIC DRUGS.

Vaickus, Dr. Louis. Roswell Park Memorial Institute, Buffalo, New York. LEUKEMIA.

Valdmanis, Janis A. Bell Laboratories, Murray Hill, New Jersey. ELECTROOPTICS.

Valenti, Dr. Jeff A. National Optical Astronomy Observatories, Tucson, Arizona. STELLAR MAGNETIC FIELD.

Valentine, Prof. James W. Department of Integrative Biology, University of California, Berkeley. ANIMAL EVOLUTION; BILATERIA.

Valenzuela, Dr. Reinaldo A. Wireless Communications Research Department, Bell Labs, Lucent Technologies, Holmdel, New Jersey. MOBILE COMMUNICATIONS.

Vali, Dr. Gabor. Department of Atmospheric Science, University of Wyoming. FROST.

Vallee, Dr. Richard. Principal Scientist, Worcester Foundation for Experimental Biology, Shrewsbury, Massachusetts. CELL MOTILITY.

Vallero, Dr. Daniel A. Adjunct Professor of Engineering Ethics Pratt School of Engineering, Duke University, Durham, North Carolina. HAZARDOUS WASTE; HAZARDOUS-WASTE ENGINEERING.

Van Allen, Dr. Margot I. Staff Geneticist, Genetics Division, Hospital for Sick Children, Toronto, Ontario, Canada. CONGENITAL ANOMALIES.

van Altena, Prof. William. Department of Astronomy, Yale University, New Haven, Connecticut. PARALLAX (ASTRONOMY).

Van Andel, Prof. Tjeerd H. Department of Geology, Stanford University, California. MARINE SEDIMENTS.

van Belle, Dr. Gerard. Michelson Science Center, California Institute of Technology, Pasadena. MASS-LUMINOSITY RELATION.

Van der Pluijm, Prof. Ben. Department of Geological Sciences, University of Michigan, Ann Arbor. SCHIST; SLATE.

Van der Spiegel, Prof. Jan. Moore School of Electrical Engineering, University of Pennsylvania, Philadelphia. GATE CIRCUIT.

Van Dyke, Dr. John H. Department of Anatomy, Hahnemann Medical College, Philadelphia, Pennsylvania. ULTIMOBRANCHIAL BODIES.

Van Etten, Dr. James P. Navigation Consultant, Hawley, Pennsylvania. LORAN.

Van Fleet, Dr. D. S. Department of Botany, University of Georgia. CORTEX (PLANT); ENDODERMIS; HYPODERMIS.

Van Houten, Prof. Franklyn B. Department of Geosciences, Princeton University, New Jersey. REDBEDS.

Van Niel, Dr. Cornelis B. Deceased; formerly, Hopkins Marine Station, Pacific Grove, California. FERMENTATION.

Van Schmus, Prof. W. Randall. Department of Geology, University of Kansas. PLUTON; PRECAMBRIAN.

Van Valkenburg, Dr. M. E. Deceased; formerly, Department of Electrical Engineering, University of Illinois. SWITCHED CAPACITOR CIRCUIT.

Van Vlack, Prof, Lawrence H. Department of Materials Engineering, University of Michigan. ALLOY.

Van Wagtendonk, Dr. Willem J. Veterans Administration Hospital, Miami, Florida. ERGOSTEROL.

Van Wazer, Dr. John R. Department of Chemistry, Vanderbilt University. PHOSPHORUS.

Van Winkle, Prof. Quentin. Department of Chemistry, Ohio State University. DIALYSIS; ELECTROKINETIC PHENOMENA; STREAMING POTENTIAL; TYNDALL EFFECT; ULTRAFILTRATION.

Vanden Heuvel, Dr. John P. Center for Molecular Toxicology and Carcinogenesis, Pennsylvania State University, University Park. PEROXISOME PROLIFERATOR-ACTIVATED RECEPTOR.

Vanderkool, Dr. Jane M. Department of Medicine, School of Medicine, University of Pennsylvania. CELL, SPECTRAL ANALYSIS OF.

Vanderzee, Dr. Cecil E. Deceased; formerly, Department of Chemistry, University of Nebraska. CHEMICAL EQUILIBRIUM.

Vane-Wright, Richard I. Biogeography and Conservation Laboratory, Department of Entomology, The Natural History Museum, London, United Kingdom. LEPIDOPTERA.

Vanick, James S. Consultant, Foundry-Castings, Sea Girt, New Jersey. CAST IRON.

Vànky, Dr. Kàlmàn. Botanisches Institut, Universitat Tubingen, Germany. USTILAGINOMYCETES (SMUT).

VanPeteghem, Dr. Peter M. Deceased; formerly, Electrical Engineering Department, Texas A&M University. CURRENT SOURCES AND MIRRORS; OPERATIONAL AMPLIFIER.

Vaughan, Prof. Wyman R. Head, Department of Chemistry, University of Connecticut. TAUTOMERISM.

Vaya, Dr. Jacob. Migal-Gailee Technological Center, Rosh-Pina, Israel. ANTIOXIDANT.

Veale, Dr. Chris A. Zeneca, Inc., Wilmington, Delaware. IONOPHORE.

Veblen, Prof. David R. Chairman, Department of Earth and Planetary Sciences, Johns Hopkins University. BIOPYRIBOLE.

Vellozzi, Dr. Joseph. Consulting Engineer, Ardsley, New York. CHIMNEY.

Venditti, Dr. Ferdinand, Jr. Head, Section of Cardiovascular Diseases, Lahey Clinic, Burlington, Massachusetts. HEART DISORDERS.

Vendrasco, Michael. Department of Earth and Space Sciences, University of California, Los Angeles. POLYPLACOPHORA.

Venkatesh, Prof. Alladi. Graduate School of Management, University of California, Irvine. INTERNET; WORLD WIDE WEB.

Verma, Prof. Pramode K. Director, Telecommunications Systems Program, The University of Oklahoma-Tulsa. PACKET SWITCHING.

Vessey, Dr. Judith A. Department of Physiological Nursing. University of California School of Nursing, San Francisco. NURSING.

Vestal, Prof. Paul A. A. G. Bush Science Center and Department of Biology, Rollins College, Winter Park, Florida. EQUISETALES; HEPATICOPSIDA; POLYPODIOPSIDA; other articles.

Veverka, Prof. Joseph. Laboratory for Planetary Studies, Cornell University. ALBEDO.

Vickers, Douglas B. Annapolis, Maryland. MICROWAVE LANDING SYSTEM (MLS).

Vickers, Dr. Mary Lynne. Department of Veterinary Science, Livestock Disease Diagnostic Center, University of Kentucky, Lexington. NEWCASTLE DISEASE.

Vidone, Dr. Romeo A. Chairperson, Department of Pathology, Hospital of St. Raphael, New Haven, Connecticut. CIRCULATION DISORDERS; EDEMA; EMBOLISM; HEMORRHAGE; INFARCTION; THROMBOSIS.

Viegas, Dr. Tacey X. Senior Director, Nektar Therapeutics, Huntsville, Alabama. POLY(ETHYLENE GLYCOL).

Vierstra, Dr. Richard. Horticulture Department, University of Wisconsin. PHYTOCHROME.

Villchur, Edgar. Foundation for Hearing Aid Research, Woodstock, New York. HEARING AID.

Villiger, Peter. Research Associate, Department of Molecular and Experimental Medicine, Research Institute of Scripps Clinic, La Jolla, California. NEUROIMMUNOLOGY.

Vinson, Prof. S. B. Department of Entomology, Texas A&M University. ICHNEUMON.

Viro, Dr. Felix. Kind & Knox, Sioux City, Iowa. GELATIN.

Vitek, Dr. Charles. Medical Epidemiologist, HIV Vaccine Section, Division of HIV/AIDS Prevention, Centers for Disease Control and Prevention, Altanta, Georgia. DIPHTHERIA.

Voelker, Dr. Richard P. Science and Technology Corporation, Columbia, Maryland. ICEBREAKER.

Vogel, Dr. Friedrich. Institute for Anthropology and Human Genetics, Im Neuenheimer Feld, Germany. HUMAN GENETICS.

Vogel, Prof. Steven. Department of Zoology, Duke University, Durham, North Carolina. HOMEOSTASIS.

Vogt, Dr. Peter R. Port Republic, Maryland. MID-OCEANIC RIDGE.

Vogtle, Dr. Fritz. Department of Chemistry, Friedrich Wilhelms University, Bonn, Germany. CYCLOPHANE.

Voight, Dr. Janet R. Department of Zoology, The Field Museum, Chicago, Illinois. CEPHALOPODA; COLEOIDEA.

Voit, Dr. Mark. Department of Physics & Astronomy, Michigan State University, East Lansing. HUBBLE SPACE TELESCOPE.

Volk, Dr. Herbert F. Carbon Products Division, Union Carbide Corporation, Parma, Ohio. GRAPHITE.

Volk, Dr. Wesley A. Department of Microbiology, School of Medicine, University of Virginia. BACTERIAL PHYSIOLOGY AND METABOLISM.

Von Doenhoff, Albert E. Deceased; formerly, Staff Scientist, National Aeronautics and Space Administration. ASPECT RATIO.

Von Dorn, William G. Scripps Institution of Oceanography, La Jolla, California. TSUNAMI.

Von Graevenitz, Alexander W. C. Department of Medical Microbiology, University of Zurich, Switzerland. AEROMONAS; KLEBSIELLA; other articles.

von Huene, Dr. Roland. Camino, California. PLATE TECTONICS.

von Puttkamer, Dr. Jesco. NASA Headquarters, Office of Space Flight, Washington, D.C. SPACE FLIGHT.

Vonder Haar, Prof. Thomas H. Department of Atmospheric Sciences, Colorado State University, Fort Collins. TERRESTRIAL RADIATION.

Vonnegut, Dr. Bernard. Department of Atmospheric Science, State University of New York, Albany. ATMOSPHERIC ELECTRICITY.

Vos, Dr. Willem L. Geophysical Laboratory and Center for High Pressure Research, Carnegie Institution, Washington, D.C. HELIUM.

Voss, Prof. Edward W., Jr. Department of Microbiology, University of Illinois, Urbana. MICROBIOLOGY.

Vulovic, Rod. Vice President, Ship Management Services, Sealand Service, Inc., Charlotte, North Carolina. MARINE CONTAINERS.

Vye, David. Ansoft Product Marketing Manager, New-buryport, Massachusetts. COMPUTER-AIDED CIRCUIT DESIGN.

W

Wachtel, Dr. Thomas L. Director, Trauma Service, Good Samaritan Medical Center, Phoenix, Arizona. BURN.

Waddington, Prof. Thomas C. Deceased; formerly, Department of Chemistry, University of Durham, England. AVOGADRO'S LAW; COMBINING VOLUMES, LAW OF; COMPOUND (CHEMISTRY); EQUIVALENT WEIGHT; RELATIVE MOLECULAR MASS; other articles.

Wade, Dr. Charles Edwin. NASA Ames Research Center, Moffett Field, California. SPACE BIOLOGY.

Waggoner, Dr. Ben. Department of Biology, University of Central Arkansas, Conway. EDIACARAN BIOTA.

Wagner, Dr. E. National Heart Lung and Blood Institutes, National Institutes of Health, Bethesda, Maryland. ABC LIPID TRANSPORTERS.

Wagner, Dr. Frank. Technical Center, Celanese Chemical Company, Corpus Christi, Texas. ACETIC ACID; ATROPINE; CAFFEINE; HEPARIN; other articles.

Wagner, Prof. Henry N., Jr. Director, Divisions of Nuclear Medicine and Radiation Health Sciences, Johns Hopkins

Medical Institutions, Baltimore, Maryland. NUCLEAR MEDICINE; RADIOACTIVE TRACER.

Wagner, Dr. Richard S. AT&T Telephone Laboratories, Murray Hill, New Jersey. CRYSTAL WHISKERS.

Wagner, Dr. Warren H., Jr. Department of Botany, University of Michigan. TAXONOMY.

Wahl, Prof. Floyd M. Professor and Chairperson, Department of Geology, University of Florida. BENTONITE; GLAUCONITE; MONTMORILLONITE.

Waidelich, Dr. Donald L. Department of Electrical Engineering, University of Missouri. RECTIFIER.

Wainwright, Howard W. Coal Research Center, U.S. Bureau of Mines, Morgantown, West Virginia. COAL CHEMICALS; DESTRUCTIVE DISTILLATION.

Waisel, Prof. Y. Department of Botany, Tel Aviv University, Israel. PERIDERM.

Wakeham, Dr. W. A. Department of Chemical Engineering and Chemical Technology, Imperial College of Science and Technology, London, England. TRANSPORT PROCESSES.

Walbot, Dr. Virginia. Department of Biological Sciences, Stanford University. CELL LINEAGE.

Walczak, Mary M. Department of Chemistry, Iowa State University of Science and Technology. SURFACE AND INTERFACIAL CHEMISTRY.

Wald, Prof. Robert M. Department of Physics, University of Chicago, Illinois. CLOCK PARADOX; LORENTZ TRANSFORMATION; RELATIVITY.

Waldmann, Katherine S. National Institutes of Health, Department of Health and Human Services, Bethesda, Maryland. IMMUNOSUPPRESSION.

Waldmann, Dr. Thomas A. Chief, Metabolism Branch, National Institutes of Health, Department of Health and Human Services, Bethesda, Maryland. IMMUNOSUPPRESSION.

Waldron, Dr. Robert D. Director, Research Enterprises, Scottsdale, Arizona. DIPOLE MOMENT; ELECTRIC SUSCEPTIBILITY; ELECTROSTRICTION; POLAR MOLECULE; POLARIZATION OF DIELECTRICS.

Walker, Laurence R. Deceased; formerly, Bell Laboratories, Murray Hill, New Jersey. FERRIMAGNETISM.

Walker, Michelle Y. Department of Genetics, University of California, Davis, California. MEIOSIS.

Walker, Dr. R. M. Department of Physics, Washington University, St. Louis, Missouri. FISSION TRACK DATING.

Walker, Dr. Warren F., Jr. Emeritus Professor, Department of Biology, Oberlin College, Ohio. MUSCULAR SYSTEM; VERTEBRA.

Walkley, Dr. S. U. Department of Neuroscience, Rose F. Kennedy Center for Research in Mental Retardation and Human Development, Albert Einstein College of Medicine, Bronx, New York. LYSOSOME.

Walkup, Dr. John T. Yale University School of Medicine, New Haven, Connecticut. TOURETTE'S SYNDROME.

Walkup, Dr. Robert D. Department of Chemistry and Biochemistry, Texas Tech University. ORGANIC SYNTHESIS; UREA.

Wallace, Dr. John F. Department of Metallurgy, Case Western Reserve University. WROUGHT IRON.

Wallace, Dr. John M. Department of Atmospheric Sciences, University of Washington, Seattle. METEOROLOGY.

Wallrath, Dr. Lori L. Ph.D., Department of Biochemistry, University of Iowa. NUCLEOPROTEIN.

Walsh, Dr. Don. President, International Maritime, Inc., Myrtle Point, Oregon. UNDERWATER VEHICLES.

Walsh, John J. Brookhaven National Laboratories, Upton, New York. WATER CONSERVATION.

Walsh, Prof. Joseph L. Deceased; formerly, Department of Mathematics, University of Maryland. LAPLACE'S DIFFERENTIAL EQUATION.

Walsh, Michael A. Department of Biology, Utah State University. PHLOEM.

Walsh, Dr. Peter J. Department of Physics, Fairleigh Dickinson University. BLACKBODY; EMISSIVITY; GRAYBODY; HEAT RADIATION; PLANCK'S RADIATION LAW.

Walsh, Dr. Walter M. Bell Laboratories, Murray Hill, New Jersey. CYCLOTRON RESONANCE EXPERIMENTS.

Walstedt, Dr. R. E. Bell Laboratories, Murray Hill, New Jersey. SPIN GLASS.

Walt, Prof. Alexander J. Department of Surgery, Harper Hospital, Wayne State University, Detroit, Michigan. MAMMOGRAPHY.

Walt, Dr. Martin. Research and Development Lockheed Missiles and Space Company, Palo Alto, California. VAN ALLEN RADIATION.

Walton, Dr. Peter D. Department of Plant Science, University of Alberta, Edmonton, Canada. FORAGE CROPS.

Walton, Dr. Richard. Department of Chemistry, The Open University, Milton Keynes, United Kingdom. TITANIUM OXIDES.

Walz, Arthur H., Jr. United States Committee on Large Dams, Corps of Engineers, Washington, D.C. DAM.

Wamser, Dr. Carl C. Department of Chemistry, Portland State University, Portland, Oregon. ORGANIC REACTION MECHANISM; REACTIVE INTERMEDIATES; REARRANGEMENT REACTION.

Wang, Dr. Changchang. Visiting Scholar, Department of Electrophysics, University of Southern California, Los Angeles. IMPULSE GENERATOR.

Wang, Chen. Department of Pathology and Laboratory Medicine, Mount Sinai Hospital, Ontario, Canada. STEM CELLS.

Wang, Prof. K. K. Sibley School of Mechanical and Aerospace Engineering, Cornell University. INERTIA WELDING.

Wang, Dr. William S.-Y. Department of Electronic Engineering, City University of Hong Kong, Kowloon. SPEECH.

Wang, Dr. Xunhua. Department of Computer Science, James Madison University, Harrisonburg, Virginia. CRYPTOGRAPHY.

Ward, Dr. David. Atomic Energy of Canada Limited, Chalk River Nuclear Laboratories, Ontario, Canada. GAMMA-RAY DETECTORS.

Ward, Jeffrey. American Radio, Newington, Connecticut. STEREOPHONIC RADIO TRANSMISSION.

Ware, Dr. Bennie R. Department of Chemistry, Syracuse University. ELECTROPHORESIS.

Warf, Dr. James C. Department of Chemistry, University of Southern California. HYDRIDO COMPLEXES; METAL HYDRIDES.

Warner, Frank L. Retired; formerly, Procurement Executive, Ministry of Defence, Royal Signal and Radar Establishment, Great Malvern, England. ATTENUATION (ELECTRICITY).

Warner, Dr. Robert M. Department of Horticulture, University of Hawaii, Manoa. MANGO.

Warren, Prof. Bertram E. Department of Physics, Massachusetts Institute of Technology. X-RAY DIFFRACTION.

Warren, Dr. John J. K. Resources Pty. Ltd., Mitcham, Australia. SALINE EVAPORITES.

Warren, Dr. Mial E. Sandia National Laboratories, Albuquerque, New Mexico. MICRO-OPTO-ELECTRO-MECHANICAL SYSTEMS (MOEMS); MICRO-OPTO-MECHANICAL SYSTEMS (MOMS).

Wartik, Prof. Thomas. Department of Chemistry, Pennsylvania State University. BORANE.

Washington, Dr. M. Todd. Department of Biochemistry, University of Iowa, Iowa City. CYTOCHROME; DOPAMINE; GLUCAGON; GLUTATHIONE; RIBONUCLEASE; STEROID; THYROID HORMONES.

Wasserman, Dr. Zelda R. Department of Central Research and Development, E. I. du Pont de Nemours and Company, Wilmington, Delaware. COMPUTATIONAL CHEMISTRY.

Waters, Dr. Everett. Department of Psychology, State University of New York, Stony Brook. MATERNAL BEHAVIOR.

Waters, Dr. John. Department of Geology, Appalachian State University, Boone, North Carolina. BLASTOIDEA.

Watkins, Dr. Thomas D. Chabot College, Hayward, California. WOOL.

Watson, Dr. Robert. Department of Pharmacological Sciences, SUNY-Stony Brook, New York. GLUCOSE.

Watson, Dr. W. W. Professor Emeritus of Physics, Yale University. BAND SPECTRUM; DIFFRACTION; HUYGENS' PRINCIPLE; MAGNETOOPTICS; STARK EFFECT; other articles.

Watterson, Dr. Ray L. Department of Zoology, University of Illinois, Urbana. BLASTULATION; CLEAVAGE (DEVELOPMENTAL BIOLOGY).

Watts, Dr. Anthony B. Department of Earth Sciences, University of Oxford, England. HOT SPOTS (GEOLOGY).

Watts, Michael. National Aeronautics and Space Administration, Ames Research Center, Moffett Field, California. AIRCRAFT TESTING.

Waugh, Prof John L. T. Department of Chemistry, University of Hawaii. ANTIMONY; ARSENIC.

Wawszkiewicz, Dr. Edward J. Department of Microbiology and Immunology, University of Illinois College of Medicine, Chicago. WINE.

Way, George H. Retired; formerly, Vice President of Research and Test Development, Association of American Railroads, Washington, D.C. LOCOMOTIVE; RAILROAD ENGINEERING.

Weaver, Prof. David L. Department of Physics, Tufts University. WAVE EQUATION.

Weaver, Prof. David S. Emeritus, Department of Mechanical Engineering, McMaster University, Hamilton, Ontario, Canada. FLOW-INDUCED VIBRATION.

Weaver, Dr. E. Eugene. Research Scientist, Product Development Group, Ford Motor Company, Dearborn, Michigan. CARBON; CYANAMIDE; IRON; other articles.

Weaver, James L. Lockheed Missiles and Space Company, Palo Alto, California. PHOTOTUBE.

Weaver, Prof. Robert J. Department of Viticulture and Enology, College of Agricultural and Environmental Sciences, University of California, Davis. GRAPE.

Webb, Prof. J. E. Professor of Zoology, University College, Ibadan, Nigeria. CEPHALOCHORDATA.

Webb, Dr. Jacqueline. Villa Nova, Pennsylvania. LATERAL LINE SYSTEM.

Webb, Dr. Richard A. IBM Research Division, Thomas J. Watson Research Center, IBM Corporation, Yorktown Heights, New York. AHARONOV-BOHM EFFECT.

Webb, Willis L. Deceased; formerly, Chief, Meteorological Satellite TA, Atmospheric Sciences Laboratory, U.S. Army Electronic Command, White Sands, New Mexico. METEOROLOGICAL ROCKET.

Webber, Bonnie. Chair of Intelligent Systems, Division of Informatics, University of Edinburgh, Scotland. NATURAL LANGUAGE PROCESSING.

Webber, Dr. James H. Chief Engineer, U.S. Navy, Naval Sea System Command, Washington, D.C. SUBMARINE.

Weber, Erwin L. Deceased; formerly, Trust Department, National Bank of Commerce, Seattle, Washington. HOT-WATER HEATING SYSTEM; PANEL HEATING AND COOLING; RADIANT HEATING; RADIATOR.

Weber, Harold C. Chemical Engineer, Boston, Massachusetts. ENTHALPY; HEAT; HEAT CAPACITY.

Webster, Dr. Douglas B. Department of Otorhinolaryngology and Biocommunication, Louisiana State University Medical Center. CENTRAL NERVOUS SYSTEM; CRANIAL NERVE; EAR (VERTEBRATE); NERVE; PARASYMPATHETIC NERVOUS SYSTEM; other articles.

Webster, Owen W. DuPont Experimental Station, Wilmington, Delaware. CYANOCARBON.

Webster, William R. Assistant Vice President, Transmission Systems Operations, Western Union Telegraph Company, Upper Saddle River, New Jersey. TELEGRAPHY.

Wedell, Dr. Nina. School of Biology, University of Leeds, United Kingdoom. SEXUAL DIMORPHISM.

Wedgewood, Dr. Lewis E. Department of Chemical Engineering University of Illinois at Chicago. VISCOSITY.

Weeks, Dr. Daniel. Department of Biochemistry, University of Iowa, Iowa City. DEOXYRIBONUCLEIC ACID (DNA); OLIGONUCLEOTIDE.

Wefer, Dr. Gerold. Department of Geosciences, Universität Bremen, Germany. PALEOCEANOGRAPHY.

Wegner, Dr. Harvey E. Deceased; formerly, Physics Division, Brookhaven National Laboratory, Upton, New York. PARTICLE ACCELERATOR.

Wei, Dr. James. Department of Chemical Engineering, Princeton, University, New Jersey. CHEMICAL ENGINEERING.

Weibel, Dr. Ewald R. Vice Chairperson and Secretary, Maurice E. Muller Foundation, Bern, Switzerland. LARYNX DISORDERS; SINUS; SYMMORPHOSIS.

Weibell, Dr. Fred J. Biomedical Engineering Society, Culver City, California. BIOMEDICAL ENGINEERING.

Weichert, Dr. Charles K. Professor of Zoology and Dean, College of Arts and Sciences, University of Cincinnati. CARDIOVASCULAR SYSTEM; REPRODUCTIVE SYSTEM; URINARY SYSTEM.

Weigel, Dr. Fritz. Institut für Anorganische Chemie, Universität München, Germany. PLUTONIUM; URANIUM; URANIUM METALLURGY.

Weil, Dr. Andrew T. College of Medicine, University of Arizona Health Sciences Center, Tucson, Arizona. COCA; COCAINE.

Weil, Jonathan F. Senior Staff Editor, *McGraw-Hill Encyclopedia of Science and Technology*, The McGraw-Hill Companies, New York. ATOMIC MASS UNIT; SPACE FLIGHT; UNITS OF MEASUREMENT.

Weil, Robert T., Jr. Deceased; formerly, Dean, School of Engineering, Manhattan College. DIRECT-CURRENT GENERATOR; KIRCHHOFF'S LAWS OF ELECTRIC CIRCUITS; MAGNETO.

Weil, Prof. Rolf. Professor of Metallurgy, Stevens Institute of Technology, Hoboken, New Jersey. ELECTROLESS PLATING; ELECTROPLATING OF METALS.

Weinberg, Dr. Jerry L. Technology Applications Group, Snellville, Georgia. ZODIACAL LIGHT.

Weinberger, Dr. Peter S. Swales Aerospace, Inc., Beltsville, Maryland. SPACECRAFT STRUCTURE.

Weinstein, Dr. Brant M. Laboratory of Molecular Genetics, National Institutes of Health, Bethesda, Maryland. ANGIOGENESIS.

Weismer, Prof. Gary G. Department of Communicative Disorders, University of Wisconsin–Madison. ACOUSTIC PHONETICS.

Weiss, Dr. Allison. Department of Molecular Genetics, Biochemistry, and Microbiology, University of Cincinnati, Ohio. BORDETELLA.

Weiss, Prof. Gerald. Department of Electrical Engineering, Polytechnic Institute of New York. DIFFERENTIAL TRANSFORMER; SERVOMECHANISM.

Weiss, Dr. Theodore J. Technical Manager, Food Department, Hunt-Wesson Foods, Inc., Fullerton, California. FAT AND OIL (FOOD); MARGARINE.

Weitz, Dr. Martin. Sektion Physik, Ludwig-Maximillians-Universitat, Munchen, Germany. RYDBERG CONSTANT.

Weitzenhoffer, Dr. André M. Veterans Administration Hospital, Oklahoma City, Oklahoma. HYPNOSIS.

Wejsa, James. AMSTA-AR-FSA-M, United States Army Armament Research Development and Engineering Center, Picatinny Arsenal, New Jersey. ARMY ARMAMENT.

Welch, Dr. David B. Mark. The Marine Biological Laboratory, Woods Hole, Massachusetts. BDELLOIDEA; ROTIFERA.

Welham, Dr. Chris J. Druck Ltd., United Kingdom. MICROSENSOR.

Welhener, Dr. Herb. Independent Mining Consultants, Inc., Tucson, Arizona. OPEN-PIT MINING.

Weller, Prof. Gunter. Director, Center for Global Change and Arctic System Research, University of Alaska, Fairbanks. ANTARCTICA.

Weller, Prof. Robert A. Department of Physics, Vanderbilt University, Nashville, Tennessee. ION-SOLID INTERACTIONS; MOLECULAR ADHESION.

Wellner, Dr. Daniel. Department of Biochemistry, Cornell University Medical College. ENZYME.

Wellstead, Dr. Carl F. Department of Biology, West Virginia Institute of Technology, Montgomery. AISTOPODA; LEPOSPONDYLI; NECTRIDEA.

Wendlandt, Prof. Wesley W. Department of Chemistry, Indiana University. CHEMICAL SYMBOLS AND FORMULAS.

Wenkert, Ernest. Department of Chemistry, University of California, San Diego. STERIC EFFECT (CHEMISTRY).

Wenner, Prof. Adrian M. Department of Biological Sciences, University of California, Santa Barbara. ANIMAL COMMUNICATION.

Wentorf, Robert H., Jr. General Electric Research Laboratory, Schenectady, New York. HIGH-PRESSURE CHEMISTRY.

Wentworth, Prof. R. A. D. Department of Chemistry, Indiana University. LIGAND FIELD THEORY.

Wenzel, Dr. Elizabeth M. Ames Research Center, National Aeronautics and Space Administration, Moffett Field, California. VIRTUAL ACOUSTICS.

Wenzel, Prof. John. Ceramics Department, Rutgers University, Piscataway, New Jersey. GLASS.

West, Dr. John B. Department of Medicine, University of California, San Diego. RESPIRATION.

West, Dr. William. Eastman Kodak Company, Rochester, New York. ABSORPTION OF ELECTROMAGNETIC RADIATION; OPTICAL PUMPING.

Westbrook, Dr. Jack H. Brookline Technologies, Ballston Spa, New York. INTERMETALLIC COMPOUNDS.

Westfall, Dr. Richard S. Deceased; formerly, Department of History and Philosophy of Science, Indiana University. KEPLER'S LAWS.

Westheimer, Prof. F. H. Chemistry Department, Harvard University, Cambridge. PHOTOAFFINITY LABELING.

Wetzel, Dr. Richard. College of William and Mary, Virginia Institute of Marine Science, School of Marine Science, Gloucester Point, Virginia. MARINE ECOLOGY.

Wetzel, Dr. Robert G. Department of Biological Science, University of Alabama, Tuscaloosa. FRESHWATER ECOSYSTEM.

Wever, Ernest G. Department of Psychology, Princeton University. PHONORECEPTION.

Whaley, Dr. W. Gordon. Cell Research Institute, University of Texas, Austin. PLANT CELL.

Wheeler, Dr. Chris. Institute of Biomedical and Life Sciences, Department of Biochemistry and Molecular Biology, University of Glasgow, Scotland, United Kingdom. NITROGEN-FIXING TREES.

Wheeler, Dr. Harry E. Department of Plant Pathology, University of Kentucky. PATHOTOXIN.

Whelan, Dr. Jean K. Senior Research Specialist, Department of Chemistry, Woods Hole Oceanographic Institution, Woods Hole, Massachusetts. KEROGEN.

Whikehart, J. William. Technical Fellow, Visteon Corporation, Milford, Michigan. RADIO RECEIVER; SIGNAL PROCESSING.

Whinnery, Prof. John R. Department of Electrical Engineering and Computer Science, College of Engineering, University of California, Berkeley. MICROWAVE; WAVEGUIDE.

Whippo, Craig. Department of Biology, Indiana University, Bloomington. PLANT MOVEMENTS.

Whistler, Prof. Roy L. Department of Biochemistry, Purdue University. CELLULOSE; GUM; HEMICELLULOSE; PECTIN; other articles.

Whitaker, Dr. Jerry. Vice President, Standards Development, American Television Systems Committee (ATSC), Morgan Hill, California. TELEVISION.

Whitaker, Dr. John O., Jr. Department of Life Sciences, Indiana State University, Terre Haute. ARMADILLO; MICE AND RATS; MOLE (ZOOLOGY); MUSKRAT.

Whitcomb, Dr. Richard T. Retired; formerly, NASA Langley Research Center, Langley Field, Virginia. AERODYNAMIC WAVE DRAG; SUPERCRITICAL WING; TRANSONIC FLIGHT.

White, Dr. Abraham. Consulting Professor of Biochemistry, Stanford University School of Medicine; and Distinguished Scientist, Institute of Biological Sciences, Syntex Research, Palo Alto, California. THYMOSIN.

White, Dr. C. Langdon. Retired; formerly, Professor of Geography, Stanford University. SOUTH AMERICA.

White, Prof. Colin. Department of Epidemiology and Public Health, Yale University. EFFECTIVE DOSE 50.

White, Donald R. J. International Training Centre, Don White Consultants, Inc., Gainesville, Virginia. ELECTRICAL INTERFERENCE; ELECTRICAL SHIELDING; SUPPRESSION (ELECTRICITY).

White, Prof. Frank M. Emeritus, Department of Mechanical Engineering and Applied Mechanics, University of Rhode Island. BERNOULLI'S THEOREM; BOUNDARY FLOW; FLUID MECHANICS; INCOMPRESSIBLE FLOW; LAMINAR FLOW; REYNOLDS NUMBER; other articles.

White, Prof. James F., Jr. Department of Plant Biology and Pathology, Rutgers University, New Brunswick, New Jersey. ENDOPHYTIC FUNGI.

White, Dr. Lawrence R. Technical Director, "Popular Photography," New York. CAMERA.

White, Prof. Mary Anne. Department of Chemistry, Dalhousie University, Halifax, Nova Scotia, Canada. HEAT STORAGE SYSTEMS.

White, Dr. Morris. Howard Hughes Medical Institute, Children's Hospital, Boston, Massachusetts. INSULIN.

White, Dr. K. Preston, Jr. Department of Systems and Information Engineering, School of Engineering and Applied Science, University of Virginia. COMPUTER-AIDED DESIGN AND MANUFACTURING; MODEL THEORY.

White, Prof. Robert M. Head, Department of Electrical and Computer Engineering, Carnegie-Mellon University, Pittsburgh, Pennsylvania. MAGNETIC RECORDING.

White, Dr. Sidney E. Department of Geology and Mineralogy, Ohio State University. QUATERNARY.

White, Dr. Stanley A. SPACE Corporation, San Clemente, California. AMPLITUDE-MODULATION DETECTOR; AMPLITUDE MODULATOR; DIGITAL FILTER; INTERMEDIATE-FREQUENCY AMPLIFIER; LIMITER CIRCUIT; PHASE MODULATOR; PHASE-MODULATION DETECTOR; PULSE DEMODULATOR; PULSE MODULATOR; Z TRANSFORM.

White, Dr. Stephanie. Northop Grumman, Bethpage, New York. COMPUTER-BASED SYSTEMS.

White, Dr. Tim. School of Forest Resources, University of Florida, Gainesville. FOREST GENETICS.

Whitehead, Prof. W. Dexter. Department of Physics, Center for Advanced Studies, University of Virginia. MEAN FREE PATH.

Whitmore, Dr. T. C. Department of Geography, University of Cambridge, United Kingdom. RAINFOREST.

Whitney, Eugene C. Consulting Engineer, Pittsburgh, Pennsylvania. HYDROELECTRIC GENERATOR.

Whittingham, Dr. M. Stanley. Director, Materials Research Center, State University of New York, Binghamton. SOLID-STATE BATTERY.

Wickelgran, Dr. Wayne A. Department of Psychology, University of Oregon. INFORMATION PROCESSING (PSYCHOLOGY).

Wickramasinghe, Dr. H. Kumar. International Business Machines, Yorktown Heights, New York. SCANNING TUNNELING MICROSCOPE.

Wicks, Dr. Fred J. Department of Mineralogy, Royal Ontario Museum, Ontario, Canada. SERPENTINE; SERPENTINITE.

Wickstrom, Dr. Conrad. Department of Biological Sciences, Kent State University. THERMAL ECOLOGY.

Widder, Prof. David V. Department of Mathematics, Harvard University. INTEGRAL TRANSFORM; LAPLACE TRANSFORM.

Widom, Dr. Jennifer. IBM Almaden Research Center, San Jose, California. CONCURRENT PROCESSING.

Wiegand, Dr. Clyde E. Deceased; formerly, Lawrence Berkeley Laboratory, University of California, Berkeley. HADRONIC ATOM.

Wiegert, Dr. Richard G. Department of Zoology, Franklin College of Arts and Sciences, University of Georgia. RADIOECOLOGY.

Wiens, Prof. Herold J. Deceased; formerly, Department of Geography, Yale University. ASIA.

Wienski, Robert M. District Manager, ISDN Architecture Planning, Bell Communications Research, Red Bank, New Jersey. INTEGRATED SERVICES DIGITAL NETWORK (ISDN); SWITCHING SYSTEMS (COMMUNICATIONS).

Wiesman, Clarence K. Director, Research and Development Division, John Sexton and Company, Chicago, Illinois. ANIMAL FEEDS.

Wiest, Dr. John M. Department of Chemical Engineering, University of Alabama, Tuscaloosa. NON-NEWTONIAN FLUID.

Wightman, Prof. Arthur S. Department of Physics, Joseph Henry Laboratories, Princeton University. GROUP THEORY.

Wignall, Dr. Paul. University of Leeds, United Kingdom. BLACK SHALE.

Wijesekera, T. Department of Chemistry, University of British Columbia, Vancouver, Canada. PORPHYRIN.

Wikle, Prof. Thomas A. Department of Geography, Oklahoma State University, Stillwater. HUDSON BAY; MAP DESIGN; MAP PROJECTIONS.

Wilbur, Donald A., Sr. Professor Emeritus of Entomology, Kansas State University. FUMIGANT.

Wilbur, Prof. Paul J. Department of Mechanical Engineering, Colorado State University. ION PROPULSION.

Wilcox, Dr. H. State University College of Forestry, Syracuse University. PERIDERM.

Wilcox, Dr. Kimerly J. General College, University of Minnesota. BEHAVIOR GENETICS.

Wilcox, Prof. William R. Department of Chemical Engineering, Clarkson College of Technology, Potsdam, New York. CRYSTALLIZATION.

Wilczek, Prof. Frank. Herman Feshbach Professor of Physics, Department of Physics, Massachusetts Institute of Technology, Cambridge, Massachusetts; Nobelist. ANYONS; CONSERVATION LAWS (PHYSICS); GEOMETRIC PHASE; GROUP THEORY; SYMMETRY BREAKING; SYMMETRY LAWS (PHYSICS).

Wilder, Dr. George J. Department of Biological Sciences, University of Illinois. INFLORESCENCE.

Wildhack, William A. Deceased; formerly, Consultant, and Associate Director, Institute for Basic Standards, National Bureau of Standards. VERNIER.

Wilen, Dr. Samuel H. Deceased; formerly, Department of Chemistry, City College of the City University of New York. RACEMIZATION; STEREOCHEMISTRY.

Wiles, Dr. D. M. National Research Council of Canada, Ottawa, Ontario. PHOTODEGRADATION.

Wiley, Prof. Edward O., III. Natural History Museum, University of Kansas, Lawrence. CLASSIFICATION, BIOLOGICAL; SPECIES CONCEPT; TAXONOMIC CATEGORIES; VERTEBRATA; ZOOLOGICAL NOMENCLATURE.

Wilhelm, Harley A. Retired; formerly, Energy and Mineral Resources Research Institute and Ames Laboratory of U.S. Department of Energy, and Professor Emeritus, Iowa State University. THORIUM.

Wilhoit, Dr. Randolph C. Thermodynamics Research Center, Texas A&M University. THERMOCHEMISTRY.

Wilkening, Dr. M. Department of Physics, New Mexico Institute of Mining and Technology. RADON.

Wilkins, Dr. Jon F. Santa Fe Institute, New Mexico. MATERNAL INFLUENCE.

Wilkinson, Dr. Brian J. Department of Biological Sciences, Illinois State University. VIRULENCE.

Wilkinson, Dr. Michael K. Associate Director, Solid State Division, Oak Ridge National Laboratory, Oak Ridge, Tennessee. NEUTRON DIFFRACTION.

Willett, Prof. Hilda P. Director of Graduate Studies, Department of Microbiology, Duke University Medical Center, Durham, North Carolina. BARTONELLOSIS; CLOSTRIDIUM; MICROBIOTA (HUMAN).

Willett, Prof. Hurd C. Department of Meteorology, Massachusetts Institute of Technology. AIR MASS.

Willey, Dr. Robert B. Department of Biological Sciences, University of Illinois. ORTHOPTERA.

Williams, Dr. D. B. Department of Materials Science and Engineering, Lehigh University, Bethlehem, Pennsylvania. METALLOGRAPHY.

Williams, Dean N. LEAD ALLOYS.

Williams, Prof. Dudley. Department of Physics, Kansas State University. DIMENSIONS (MECHANICS); MINIMAL PRINCIPLES; NEWTON'S LAWS OF MOTION; other articles.

Williams, Dr. Earl G. Structural Acoustics, Naval Research Laboratory, Washington, D.C. ACOUSTICAL HOLOGRAPHY.

Williams, Prof. Earle. Center for Global Change Science, Massachusetts Institute of Technology, Cambridge. STORM ELECTRICITY.

Williams, Prof. Gary A. Department of Physics, University of California, Los Angeles. LOW-TEMPERATURE ACOUSTICS.

Williams, Dr. Jack M. Chemistry Division, Argonne National Laboratory, Argonne, Illinois. HYDROGEN BOND; STRUCTURAL CHEMISTRY.

Williams, Dr. Kathleen M. Extension Pomologist, Tree Fruit Research and Extension Center, Washington State University, Wenatchee. PEAR.

Williams, Dr. Marvin C. Department of Biology, University of Nebraska, Kearney. FUNGI.

Williams, Dr. Peter. Department of Chemistry, Arizona State University. SPUTTERING.

Williams, Prof. Todd D. Department of Chemistry, University of Kansas, Lawrence. MASS SPECTROMETRY.

Williams, Prof. W. A. Department of Agronomy and Range Science, University of California, Davis. AGRICULTURAL SCIENCE (ANIMAL).

Willis, Homer B. Consultant, Bethesda, Maryland. RIVER ENGINEERING.

Willis, Dr. William J. Physics Division, European Organization for Nuclear Research, Geneva, Switzerland. TRANSITION RADIATION DETECTORS.

Wilson, Dr. B. F. Department of Forestry and Wildlife Management, University of Massachusetts. ROOT (BOTANY).

Wilson, Dr. Curtis B. Department of Immunopathology, Scripps Clinic and Research Foundation, La Jolla, California. IMMUNE COMPLEX DISEASE.

Wilson, Prof. George S. Department of Chemistry, University of Kansas, Lawrence. ANALYTICAL CHEMISTRY; BIOSENSOR.

Wilson, Grady W. Deputy Director and Chief Test Pilot, Raspet Flight Research Laboratory, Mississippi State University. RATE-OF-CLIMB INDICATOR; TURN AND BANK INDICATOR.

Wilson, Dr. Irwin B. Department of Chemistry and Biochemistry, University of Colorado. ACETYLCHOLINE.

Wilson, Dr. James B. Department of Forest Products, Oregon State University, Corvallis. WOOD PROPERTIES.

Wilson, Dr. James Lee. Department of Geology and Mineralogy, University of Michigan. BIOHERM; BIOSTROME.

Wilson, Dr. Mark A. Department of Geology, The College of Wooster, Ohio. TRYPANITES.

Wilson, Dr. Mark V. H. Department of Biological Sciences & Laboratory for Vertebrate Paleontology, University of Alberta, Canada. ACANTHODII; HETEROSTRACI.

Wilson, Dr. Therese. Biological Laboratories, Harvard University. CHEMILUMINESCENCE.

Wilt, Dr. Fred H. Department of Zoology, University of California, Berkeley. HEMATOPOIESIS.

Wimbush, Prof. Mark. Graduate School of Oceanography, University of Rhode Island, Narragansett. KUROSHIO; OCEAN WAVES.

Winch, Prof. Ralph P. Deceased; formerly, Department of Physics, Williams College. CAPACITANCE; COULOMB'S LAW; ELECTRIC CHARGE; ELECTRIC FIELD; LINES OF FORCE; other articles.

Wing, Prof. Robert F. Emeritus, Department of Astronomy, Ohio State University, Columbus. MAGNITUDE (ASTRONOMY).

Winick, Dr. Herman. Stanford Linear Accelerator Center, Stanford Synchrotron Radiation Laboratory, Stanford, California. SYNCHROTRON RADIATION.

Winkler, Dr. Malcolm E. Department of Microbiology and Molecular Genetics, University of Texas, Houston. GENE.

Winograd, Prof. Nicholas. Department of Chemistry, Pennsylvania State University, University Park. COMBINATORIAL CHEMISTRY.

Winston, Dr. Judith E. Assistant Curator, Department of Invertebrates, American Museum of Natural History, New York. LOPHOPHORE.

Winter, Prof. John D. Geology Department, Whitman College, Walla Walla, Washington. METAMORPHISM.

Winter, Dr. John M., Jr. Department of Materials Science and Engineering, John Hopkins University, Baltimore, Maryland. NONDESTRUCTIVE EVALUATION.

Winter, Dr. Nancy L. Bedford, Massachusetts. GLOBE (EARTH).

Winter, Prof. Rolf G. Deceased; formerly, Department of Physics, College of William and Mary. COHERENCE.

Winter, Dr. Thomas C. U.S. Geological Survey, Department of the Interior, Denver, Colorado. STREAM GAGING.

Winterer, Dr. Edward L. Geological Research Division, Scripps Institution of Oceanography, La Jolla, California. OCEANIC ISLANDS; SEAMOUNT AND GUYOT.

Wirry, Henry J. Product Manager, Hydraulic Drives, Twin Disc, Inc., Rockford, Illinois. FLUID COUPLING; TORQUE CONVERTER.

Wise, Edmund M. Consultant; formerly, Assistant to Vice President of Research, International Nickel Company, Inc., New York. SILVER ALLOYS.

Wisnosky, Dennis F. Wizdom Systems, Inc., Naperville, Illinois. COMPUTER-INTEGRATED MANUFACTURING.

Wittmann, H. G. Max-Planck Institut für Molekulare Genetik, Berlin. RIBOSOMES.

Wittmers, Dr. L. E. School of Medicine, University of Minnesota. HYPOTHERMIA.

Wnek, Prof. Gary E. Chair, Department of Chemical Engineering, Virginia Commonwealth University, Richmond, Virginia. BIOPOLYMER; BRANCHED POLYMER; ORGANIC CONDUCTOR.

Wohl, Prof. Ellen. Department of Geosciences, Colorado State University, Fort Collins. FLOODPLAIN.

Wold, Dr. Marc. Department of Biochemistry, University of Iowa College of Medicine, Iowa City. DEOXYRIBONUCLEIC ACID (DNA).

Wolf, Dr. Don P. Oregon Regional Primate Research Center, Oregon Health and Science University, Beaverton. CLONING.

Wolf, Dr. Peter. IBM Zurich Research Laboratory, Zurich, Switzerland. CRYOTRON.

Wolf, Dr. Werner P. Department of Physics, Yale University. SCHOTTKY ANOMALY.

Wolfe, Dr. Ralph S. Department of Microbiology, University of Illinois, Urbana. METHANOGENESIS (BACTERIA).

Wolfe, Dr. William L. Optical Sciences Center, University of Arizona. COLOR FILTER; INFRARED RADIATION.

Wolferz, Alfred H. Retired; formerly, Chief, Tachometer Systems Development, Weston Instruments Division, Daystrom Inc., Newark, New Jersey. TACHOMETER.

Wolford, Dr. James C. Mechanical Engineering Department, University of Nebraska. VELOCITY ANALYSIS.

Wollersheim, Prof. David. Department of Mechanical and Aerospace Engineering, University of Missouri. BALLISTICS.

Wollnik, Prof. H. Physikalisches Institut, Justus Liebig-Universität, Giessen, Germany. CHARGED PARTICLE OPTICS.

Wolovich, Prof. William A. Emeritus, Division of Engineering, Brown University. MULTIVARIABLE CONTROL.

Wonders, Dr. William C. Department of Geography, University of Alberta, Edmonton, Canada. ARCTIC AND SUBARCTIC ISLANDS.

Wones, Dr. David R. Department of Geological Sciences, Virginia Polytechnic Institute and State University. BIOTITE.

Wong, Dr. George S. K. Acoustical Standards, Institute for National Measurement Standards, Ottawa, Ontario, Canada. ACOUSTIC INTERFEROMETER.

Woo, Dr. L. Keith. Associate Chair, Department of Chemistry, Iowa State University of Science and Technology. TITANIUM.

Wood, Dr. Albert E. Retired; formerly, Department of Biology, Amherst College. LAGOMORPHA; RODENTIA.

Wood, Dr. B. J. Department of Geophysical Sciences, University of Chicago. GARNET.

Wood, Dr. Chris. Electric Power Research Institute, Palo Alto, Calif. DECONTAMINATION OF RADIOACTIVE MATERIALS.

Wood, Dr. Frank Bradshaw. Deceased; formerly, Department of Physics and Astronomy, University of Florida. ECLIPSING VARIABLE STARS.

Wood, Dr. H. William. Consultant, Aerospace, Roaring Gap, North Carolina. SPACECRAFT GROUND INSTRUMENTATION.

Wood, Dr. Rachel. Department of Earth Sciences, University of Cambridge, United Kingdom. ARCHAEOCYATHA.

Woodall, James E. Division Manufacturing Manager, Westinghouse Electric Corporation, Pittsburgh, Pennsylvania. PILOT PRODUCTION.

Woodall, Prof. Jerry M. Yale University, Department of Electrical Engineering, New Haven, Connecticut. LIGHT-EMITTING DIODE.

Woodcock, C. L. F. Professor of Zoology, University of Massachussetts. NUCLEOSOME.

Woods, Dr. Jon P. Department of Medical Microbiology and Immunology, University of Wisconsin Medical School, Madison. MEDICAL MYCOLOGY.

Woodward, Prof. John B., III. Department of Naval Architecture and Marine Engineering, University of Michigan. BOAT PROPULSION.

Woody, Dr. R. W. Department of Biochemistry, Colorado State University. OPTICAL ROTATORY DISPERSION.

Woon, Prof. L. C. Centre for Intelligent Control, Department of Electrical Engineering, National University of Singapore. EXPERT CONTROL SYSTEM.

Woosley, Prof. Stan E. Lick Observatory, Santa Cruz, California. SUPERNOVA.

Woram, John M. Woram Audio Associates, Rockville Centre, New York. SOUND RECORDING.

Wray, Dr. John L. Marathon Oil Company, Littleton, Colorado. CORALLINALES; GIRVANELLA; HALIMEDA; SOLENOPORA.

Wright, Prof. Edward. Department of Physics and Astronomy, University of California, Los Angeles. COSMIC BACKGROUND RADIATION.

Wright, Elliott F. Retired; formerly, Consulting Engineer, Advanced Products Division, Studebaker-Worthington Corporation, Harrison, New Jersey. CENTRIFUGAL PUMP; DISPLACEMENT PUMP; PUMP; PUMPING MACHINERY.

Wright, James A. Strategic Marketing Manager, Motorola, Inc., Boynton Beach, Florida. RADIO PAGING SYSTEMS.

Wright, Maynard A., P. E. Acterna, Citrus Heights, California. BANDWIDTH REQUIREMENTS (COMMUNICATIONS).

Wright, Michael E. National Semiconductor Corporation, Santa Clara, California. DIGITAL COUNTER.

Wright, Prof. Raymond. Department of Civil & Environmental Engineering, University of Rhode Island, Kingston. FROUDE NUMBER.

Wright, Dr. Rita P. Department of Anthropology, New York University. PREHISTORIC TECHNOLOGY.

Wu, Prof. Tai Tsun. Division of Engineering and Applied Physics, Harvard University. ISING MODEL.

Wüest, Dr. Alfred Johny. Applied Aquatic Ecology (APEC), Kastanienbaum, Switzerland. SEICHE.

Wulff, Dr. Verner J. Associate Director of Research, Masonic Medical Research Laboratory, Utica, New York. PHOTORECEPTION.

Wurtele, Prof. Morton G. Department of Atmospheric Sciences, University of California, Los Angeles. CLEAR-AIR TURBULENCE.

Wurtman, Dr. Richard. Department of Brain and Cognitive Science, Massachusetts Institute of Technology, Cambridge. SEROTONIN.

Wyant, Prof. James C. Optical Sciences Center, University of Arizona. INTERFEROMETRY; SPECKLE.

Wygodzinsky, Dr. Pedro W. Department of Entomology, American Museum of Natural History, New York. DIPLURA; HEMIPTERA; THYSANURA (ZYGENTOMA).

Wyllie, Peter J. Chairperson, Department of Geophysical Sciences, University of Chicago. OLIVINE.

Wyman, Dr. Richard E. Canadian Hunter Exploration Ltd., Calgary, Alberta, Canada. WELL LOGGING.

Wyrtki, Dr. Klaus. UH Sea Level Center, University of Hawaii, Honolulu. INDIAN OCEAN; SOUTHEAST ASIAN WATERS.

Y

Yagi, Prof. Takehiko. Institute for Solid State Physics, University of Tokyo, Japan. HIGH-PRESSURE MINERAL SYNTHESIS.

Yalow, Dr. Rosalyn S. Distinguished Professor at Large, Albert Einstein College of Medicine, Yeshiva University, New York; Nobelist. RADIOIMMUNOASSAY.

Yamaguchi, M. Professor, Department of Vegetable Crops, University of California, Davis. ORIENTAL VEGETABLES.

Yang, Hongjun. IGEN, Inc., Rockville, Maryland. CHEMILUMINESCENCE.

Yang, Prof. Yeong-Bin. National Taiwan University, Taipei. STRUCTURAL MECHANICS.

Yates, Dr. S. F. Allied Signal Research Center, Inc., Des Plaines, Illinois. ION EXCHANGE.

Yeatman, Dr. Harry C. Department of Zoology, University of the South, Sewanee, Tennessee. CALANOIDA; COPEPODA; CYCLOPOIDA.

Yegulalp, Prof. Tuncel M. Henry Krumb School of Mines, Columbia University, New York. COAL MINING.

Yell, Dr. Ralph W. Retired; formerly, National Physical Laboratory, Teddington, Middlesex, England. MICROWAVE FREE-FIELD STANDARDS.

Yelles, Marvin. Formerly, Editor, *McGraw-Hill Encyclopedia of Science and Technology*, The McGraw-Hill Companies, New York. VECTOR (MATHEMATICS).

Yellott, John I. Deceased; formerly, Yellott Engineering Associates, Inc., Phoenix, Arizona. SOLAR HEATING AND COOLING.

Yeomans, Dr. Donald K. Jet Propulsion Laboratory, California Institute of Technology. ASTRONOMICAL COORDINATE SYSTEMS.

Yin, Prof. Frank C. P. Department of Biomedical Engineering, Washington University, St. Louis, Missouri. BIOMECHANICS.

Yochelson, Dr. Ellis L. U.S. Geological Survey, Department of the Interior, Washington, D.C. CONULARIDA.

Yokelson, Dr. Howard B. Research and Development Department, Amoco Chemical Company, Naperville, Illinois. ORGANOSILICON COMPOUND.

Yosim, Dr. Samuel J. Atomics International Division, North American Rockwell, Canoga Park, California. BISMUTH.

Yost, Prof. Don M. Formerly, California Institute of Technology. QUATERNIONS.

Yost, Dr. William A. Director, Parmly Hearing Institute, Loyola University, Chicago, Illinois. MASKING OF SOUND.

Young, Dr. Anne B. Department of Neurology, University of Michigan. HUNTINGTON'S DISEASE.

Young, Dr. David A. Department of Entomology, North Carolina State University. HOMOPTERA.

Young, Edward M. Deceased; formerly, Associate Editor, *Engineering News-Record*, McGraw-Hill, Inc., New York. CONSTRUCTION EQUIPMENT.

Young, Dr. Eliot. Department of Space Studies, Southwest Research Institute, Boulder, Colorado. PLANET.

Young, Prof. Grant M. Department of Earth Sciences, University of Western Ontario, London, Ontario, Canada. PRECAMBRIAN.

Young, Prof. N. J. Department of Mathematics and Statistics, University of Lancaster, England. HILBERT SPACE.

Young, Dr. Robert W. Associate Editor, *Journal of the Acoustical Society of America*, San Diego, California. MODE OF VIBRATION; MUSICAL INSTRUMENTS; other articles.

Young, Prof. Thomas F. Deceased; formerly, Department of Chemistry, University of Chicago. IONIC EQUILIBRIUM; PK; SOLUBILITY PRODUCT CONSTANT.

Youngberg, Dr. Harold. Department of Agronomy, Oregon State University. AGRICULTURAL SCIENCE (PLANT).

Youngs, Dr. Curtis R. Animal Science Department, Iowa State University, Ames. BREEDING (ANIMAL).

Younker, Dr. Leland W. Earth Science Department, Lawrence Livermore National Laboratory, University of California. RHYOLITE.

Yu, Dr. Robert K. Department of Biochemistry & Molecular Biophysics, Virginia Commonwealth University School of Medicine, Richmond. SPHINGOLIPID.

Yuen, Prof. Horace P. Department of Electrical Engineering and Computer Science, Northwestern University. SQUEEZED QUANTUM STATES.

Yungul, Dr. Sulhi H. Chevron Resources Company, San Francisco, California. GEOELECTRICITY.

Z

Zaitlin, Dr. Milton. Associate Director, Biotechnology Program, Plant Pathology, Cornell University. BIOTECHNOLOGY.

Zander, Andrew T. Director, Process Engineering, Genitope, Inc., California. ANALYTICAL CHEMISTRY; EMISSION SPECTROCHEMICAL ANALYSIS.

Zangerl, Dr. Rainer. Retired; formerly, Chief Curator, Department of Geology, Field Museum of Natural History, Chicago, Illinois. PLACODONTIA; SAUROPTERYGIA.

Zanoni, Thomas A. Taxonomist, Jardín Botánico Nacional, Santo Domingo, Dominican Republic; and Honorary Research Associate, New York Botanical Garden. CYCADALES; EPHEDRALES; PINOPSIDA; other articles.

Zapffe, Carl A. Consultant, Baltimore, Maryland. STAINLESS STEEL.

Zbar, Dr. Frederick. Chevy Chase, Maryland. WEATHER OBSERVATIONS.

Zebroski, Dr. Edwin L. Los Altos, California. NUCLEAR FUEL CYCLE.

Zeevaart, Dr. Jan Adriaan D. Department of Energy Plant Research Laboratory, Department of Botany, Michigan State University. ABSCISIC ACID.

Zeilik, Prof. Michael. Department of Physics and Astronomy, University of New Mexico, Albuquerque. ARCHEOASTRONOMY.

Zeilinger, Prof. Anton. University of Vienna and Institute of Quantum Optics & Quantum Information, Austrian Academy of Sciences. QUANTUM TELEPORTATION.

Zelkowitz, Prof. Marvin V. Department of Computer Science, University of Maryland, College Park. PROGRAMMING LANGUAGES.

Zeller, Dr. Andreas. Universität Passau, Lehrstuhl für Software-Systeme, Passau, Germany. VISUAL DEBUGGING.

Zeller, Prof. Michael E. Physics Department, Yale University, New Haven, Connecticut. ANTIMATTER.

Zentmyer, Dr. George A. Department of Plant Pathology, University of California, Riverside. AVOCADO; CACAO.

Zhang, Dr. Ruyuan. Department of Geological and Environmental Sciences, Stanford University, Stanford, California. GLAUCOPHANE.

Zhao, Prof. Wei. Department of Computer Science, Texas A&M University, College Station. LOCAL-AREA NETWORKS.

Zhou, Dr. Zhonghe. Institute of Vertebrate Paleontology and Paleoanthropology, Chinese Academy of Sciences, Beijing, China. CONFUCIUSORNITHIDAE; ENANTIORNITHES.

Ziemer, Prof. Rodger E. Chairperson, Department of Electrical and Computer Engineering, University of Colorado, Colorado Springs. DIFFERENCE EQUATION; LINEAR SYSTEMS ANALYSIS.

Zifcak, John H. Foxboro Company, Foxboro, Massachussets. BAROMETER; PRESSURE MEASUREMENT; STRAIN GAGE.

Zimmerman, Dr. John R. Department of Mechanical Engineering, Pennsylvania State University. GEAR; GEAR TRAIN; PLANETARY GEAR TRAIN; PULLEY; ROLLING CONTACT.

Zimmerman, Dr. Leroy H. Retired; formerly, U.S. Department of Agriculture, Tucson, Arizona. CASTOR PLANT.

Zinder, Dr. Norton D. Rockefeller University, New York. TRANSDUCTION (BACTERIA); TRANSFORMATION (BACTERIA).

Zink, Dr. Frank W. Department of Vegetable Crops, University of California, Davis. MUSKMELON.

Zinner, Prof. Ernst K. McDonnell Center for the Space Sciences, Washington University, St. Louis, Missouri. METEORITE.

Zirin, Dr. Harold. Professor Emeritus, Department of Physics, California Institute of Technology. SUN.

Zirkle, Dr. Raymond E. Deceased; formerly, Committee on Biophysics, University of Chicago. LINEAR ENERGY TRANSFER (BIOLOGY).

Zissis, Dr. George J. Erim International, Ann Arbor, Michigan. INFRARED IMAGING DEVICES.

Zlotnik, Dr. I. Microbiological Research Establishment, Wilts, England. SCRAPIE.

Zohar, Dr. Joseph. Deputy Director, Beer Sheva Mental Health Center, Beer Sheva, Israel. OBSESSIVE-COMPULSIVE DISORDER.

Zubaly, Prof. Robert B. Institute of Naval Architecture, Glen Cove, New York. MERCHANT SHIP.

Zubay, Dr. Geoffrey. Sherman Fairchild Center for the Life Sciences, Department of

Biological Sciences, Columbia University. BACTERIAL GENETICS.

Zuk, Dr. William. School of Architecture, University of Virginia. PHOTOELASTICITY.

Zuman, Prof. Petr. Department of Chemistry, Clarkson University, Potsdam, New York. POLAROGRAPHIC ANALYSIS; REFERENCE ELECTRODE.

Zumberge, Dr. Mark A. Institute of Geophysics and Planetary Physics, Scripps Institution of Oceanography, La Jolla, California. GRAVITY METER.

Zurek, Dr. Wojciech H. Los Alamos National Laboratory, Los Alamos, New Mexico. SUPERPOSITION PRINCIPLE; UNCERTAINTY PRINCIPLE.

Zweifel, Dr. Richard G. Curator, Department of Herpetology, American Museum of Natural History, New York. AMPHIBIA; ANURA; APODA; TETRAPODA; URODELA.

Zwerman, Prof. Paul J. Department of Agronomy, Cornell University. COVER CROPS.

Index

Page numbers with an asterisk indicate an article title. Entries on pages 1–1367 are found in Volume 1; those on pages 1369–2590 are found in Volume 2.

A

B

C

E

G

J

L

M

N

O

P

R

S

U

V

Z

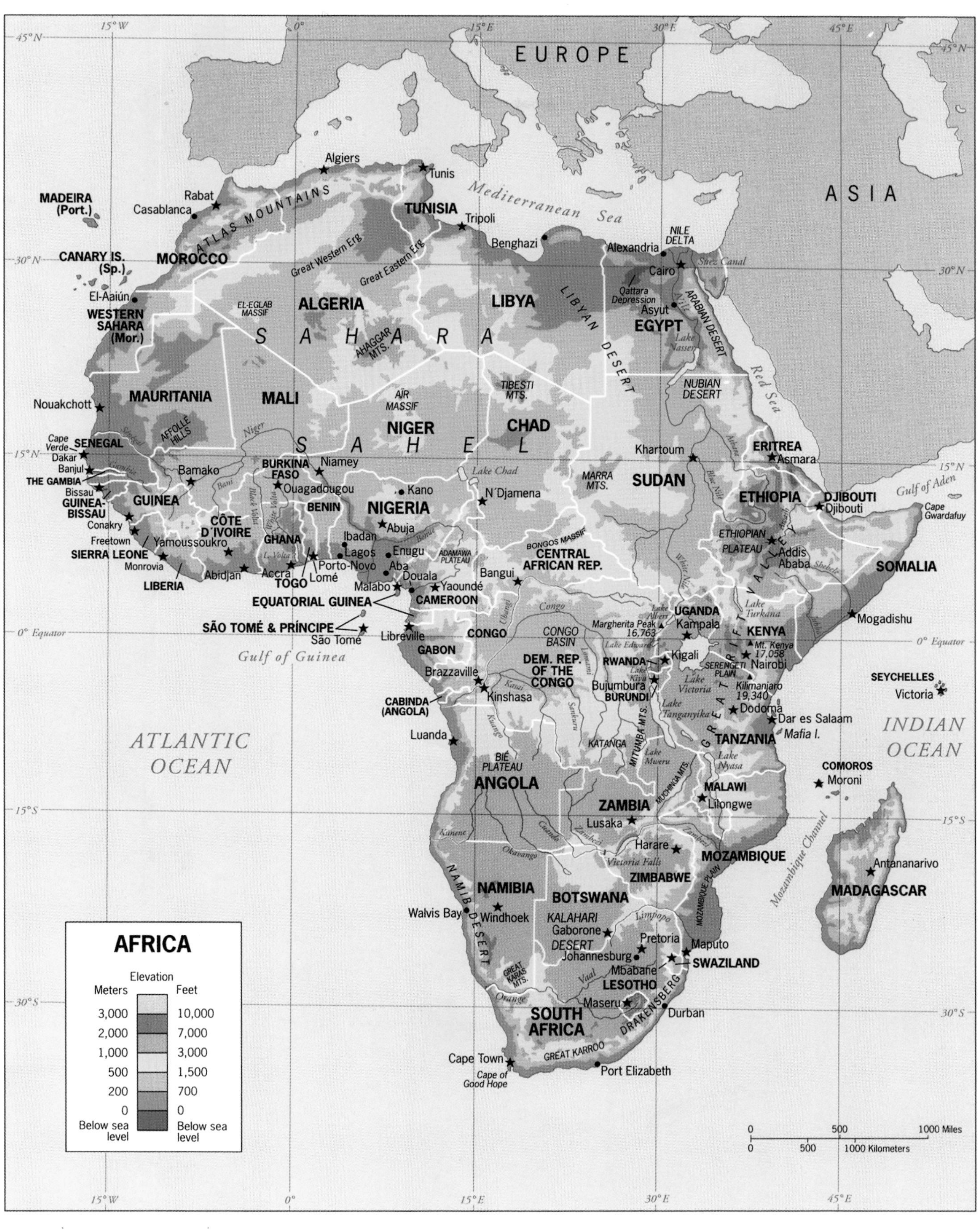

AFRICA
Elevation
Meters
Feet
3,000
10,000
2,000
7,000
1,000
3,000
500
1,500
200
700
0
0
Below sea level
Below sea level
0
500
1000 Miles
0
500
1000 Kilometers
15° W
0°
15° E
30° E
45° E
45° N
30° N
15° N
0° Equator
15° S
30° S
EUROPE
ASIA
Mediterranean Sea
Red Sea
Gulf of Aden
ATLANTIC OCEAN
INDIAN OCEAN
Gulf of Guinea
Mozambique Channel
Suez Canal
MADEIRA (Port.)
CANARY IS. (Sp.)
Algiers
Tunis
Rabat
Casablanca
TUNISIA
Tripoli
Benghazi
Alexandria
NILE DELTA
Cairo
ATLAS MOUNTAINS
MOROCCO
Great Western Erg
Great Eastern Erg
El-Aaiún
WESTERN SAHARA (Mor.)
EL-EGLAB MASSIF
ALGERIA
LIBYA
LIBYAN DESERT
Qattara Depression
Asyut
EGYPT
ARABIAN DESERT
Nile
Lake Nasser
S A H A R A
AHAGGAR MTS.
MAURITANIA
Nouakchott
MALI
AÏR MASSIF
TIBESTI MTS.
NUBIAN DESERT
NIGER
CHAD
AFFOLLÉ HILLS
Niger
Senegal
Cape Verde
SENEGAL
Dakar
S A H E L
Khartoum
ERITREA
Asmara
Banjul
THE GAMBIA
Bissau
GUINEA-BISSAU
Bamako
BURKINA FASO
Niamey
Ouagadougou
Lake Chad
N'Djamena
MARRA MTS.
SUDAN
ETHIOPIA
DJIBOUTI
Djibouti
Cape Gwardafuy
GUINEA
Conakry
Freetown
SIERRA LEONE
Monrovia
LIBERIA
CÔTE D'IVOIRE
Yamoussoukro
Abidjan
GHANA
Accra
TOGO
Lomé
BENIN
Porto-Novo
Lagos
Ibadan
NIGERIA
Kano
Abuja
Enugu
Aba
Benue
Black Volta
White Volta
L. Volta
ADAMAWA PLATEAU
BONGOS MASSIF
CENTRAL AFRICAN REP.
Bangui
ETHIOPIAN PLATEAU
Addis Ababa
Blue Nile
White Nile
Atbara
Shebele
SOMALIA
Malabo
Douala
Yaoundé
CAMEROON
EQUATORIAL GUINEA
SÃO TOMÉ & PRÍNCIPE
São Tomé
Libreville
GABON
CONGO
Ubangi
Congo
CONGO BASIN
Lake Albert
Margherita Peak 16,763
UGANDA
Kampala
Lake Turkana
Mogadishu
KENYA
Mt. Kenya 17,058
Nairobi
Lake Edward
RWANDA
Kigali
Lake Kivu
Bujumbura
BURUNDI
Lake Victoria
SERENGETI PLAIN
Kilimanjaro 19,340
DEM. REP. OF THE CONGO
Lomami
Kasai
Brazzaville
Kinshasa
CABINDA (ANGOLA)
SEYCHELLES
Victoria
Lake Tanganyika
Dodoma
Dar es Salaam
Mafia I.
TANZANIA
GREAT RIFT VALLEY
Sankuru
Kwango
MITUMBA MTS.
KATANGA
Lake Mweru
Luanda
BIÉ PLATEAU
ANGOLA
Lake Nyasa
MUCHINGA MTS.
COMOROS
Moroni
MALAWI
Lilongwe
ZAMBIA
Lusaka
Zambezi
Cuando
Kunene
Okavango
Victoria Falls
Harare
ZIMBABWE
MOZAMBIQUE
Antananarivo
MADAGASCAR
NAMIB DESERT
NAMIBIA
Walvis Bay
Windhoek
BOTSWANA
KALAHARI DESERT
Gaborone
Limpopo
MOZAMBIQUE PLAIN
Pretoria
Maputo
Johannesburg
Mbabane
SWAZILAND
GREAT KARAS MTS.
Vaal
LESOTHO
Maseru
Orange
DRAKENSBERG
Durban
SOUTH AFRICA
GREAT KARROO
Cape Town
Cape of Good Hope
Port Elizabeth

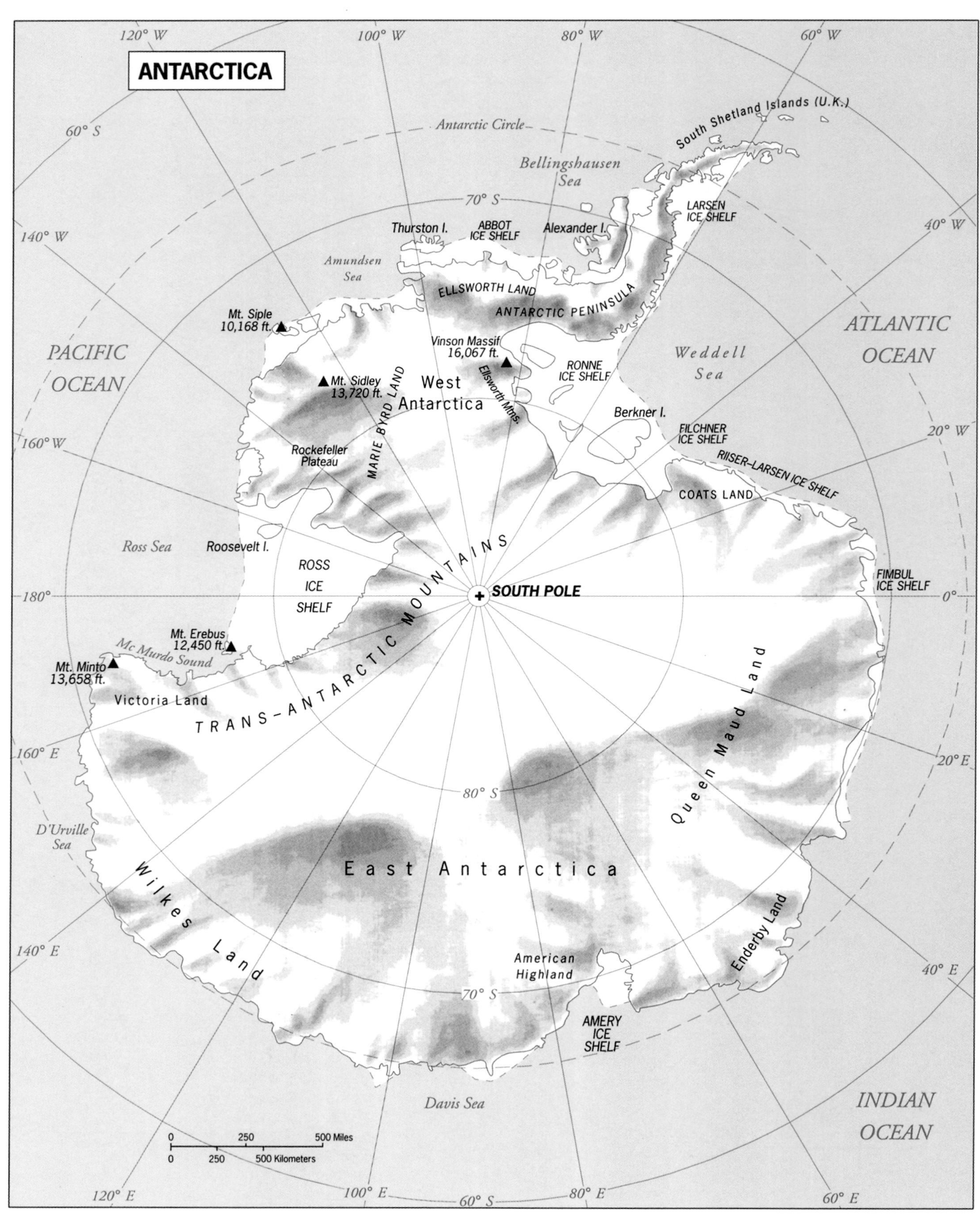

ANTARCTICA
120° W
100° W
80° W
60° W
60° S
Antarctic Circle
South Shetland Islands (U.K.)
Bellingshausen Sea
70° S
LARSEN ICE SHELF
Thurston I.
ABBOT ICE SHELF
Alexander I.
40° W
140° W
Amundsen Sea
ELLSWORTH LAND
ANTARCTIC PENINSULA
Mt. Siple 10,168 ft.
ATLANTIC OCEAN
PACIFIC OCEAN
Vinson Massif 16,067 ft.
RONNE ICE SHELF
Weddell Sea
Mt. Sidley 13,720 ft.
MARIE BYRD LAND
West Antarctica
Ellsworth Mtns.
Berkner I.
FILCHNER ICE SHELF
20° W
160° W
Rockefeller Plateau
RIISER-LARSEN ICE SHELF
COATS LAND
Ross Sea
Roosevelt I.
ROSS ICE SHELF
TRANS-ANTARCTIC MOUNTAINS
FIMBUL ICE SHELF
180°
SOUTH POLE
0°
Mt. Erebus 12,450 ft.
Mc Murdo Sound
Mt. Minto 13,658 ft.
Queen Maud Land
Victoria Land
160° E
20° E
80° S
D'Urville Sea
East Antarctica
Wilkes Land
Enderby Land
140° E
American Highland
40° E
70° S
AMERY ICE SHELF
Davis Sea
INDIAN OCEAN
0 250 500 Miles
0 250 500 Kilometers
120° E
100° E
60° S
80° E
60° E

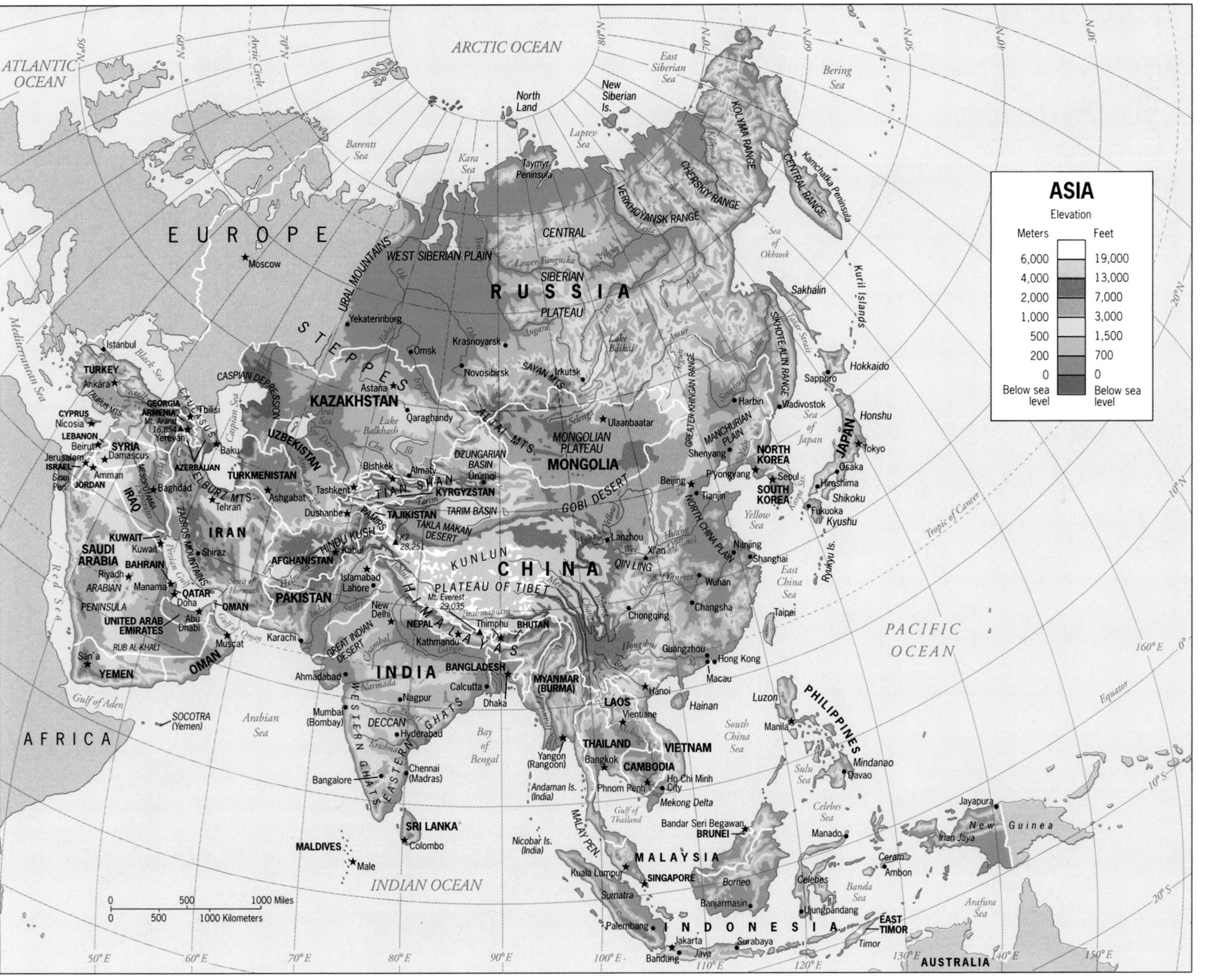

ASIA
Elevation
Meters
Feet
6,000 19,000
4,000 13,000
2,000 7,000
1,000 3,000
500 1,500
200 700
0 0
Below sea level
Below sea level
0 500 1000 Miles
0 500 1000 Kilometers
ARCTIC OCEAN
ATLANTIC OCEAN
PACIFIC OCEAN
INDIAN OCEAN
EUROPE
AFRICA
AUSTRALIA
RUSSIA
CHINA
MONGOLIA
KAZAKHSTAN
UZBEKISTAN
TURKMENISTAN
KYRGYZSTAN
TAJIKISTAN
AFGHANISTAN
PAKISTAN
INDIA
NEPAL
BHUTAN
BANGLADESH
MYANMAR (BURMA)
LAOS
THAILAND
VIETNAM
CAMBODIA
MALAYSIA
SINGAPORE
BRUNEI
INDONESIA
EAST TIMOR
PHILIPPINES
JAPAN
NORTH KOREA
SOUTH KOREA
SRI LANKA
MALDIVES
IRAN
IRAQ
SYRIA
TURKEY
CYPRUS
LEBANON
ISRAEL
JORDAN
GEORGIA
ARMENIA
AZERBAIJAN
KUWAIT
SAUDI ARABIA
BAHRAIN
QATAR
UNITED ARAB EMIRATES
OMAN
YEMEN
Moscow
Istanbul
Ankara
Nicosia
Beirut
Damascus
Jerusalem
Amman
Tbilisi
Yerevan
Baku
Baghdad
Tehran
Kuwait
Riyadh
Manama
Doha
Abu Dhabi
Muscat
San'a
Ashgabat
Tashkent
Dushanbe
Bishkek
Almaty
Astana
Qaraghandy
Yekaterinburg
Omsk
Novosibirsk
Krasnoyarsk
Irkutsk
Ulaanbaatar
Ürümqi
Kabul
Islamabad
Lahore
Karachi
New Delhi
Kathmandu
Thimphu
Dhaka
Calcutta
Ahmadabad
Mumbai (Bombay)
Nagpur
Hyderabad
Bangalore
Chennai (Madras)
Colombo
Male
Yangon (Rangoon)
Bangkok
Vientiane
Hanoi
Phnom Penh
Ho Chi Minh City
Kuala Lumpur
Bandar Seri Begawan
Jakarta
Bandung
Palembang
Surabaya
Banjarmasin
Ujungpandang
Manado
Ambon
Jayapura
Manila
Davao
Beijing
Tianjin
Shenyang
Harbin
Pyongyang
Seoul
Lanzhou
Xi'an
Chongqing
Wuhan
Changsha
Nanjing
Shanghai
Guangzhou
Hong Kong
Macau
Taipei
Vladivostok
Sapporo
Tokyo
Osaka
Hiroshima
Fukuoka
URAL MOUNTAINS
WEST SIBERIAN PLAIN
CENTRAL SIBERIAN PLATEAU
STEPPES
CASPIAN DEPRESSION
CAUCASUS
Mt. Ararat 16,854
TAURUS MTS.
MESOPOTAMIA
ELBURZ MTS.
ZAGROS MOUNTAINS
ARABIAN PENINSULA
RUB AL-KHALI
HINDU KUSH
PAMIRS
TIAN SHAN
ALTAI MTS.
SAYAN MTS.
DZUNGARIAN BASIN
TARIM BASIN
TAKLA MAKAN DESERT
K2 28,251
KUNLUN
PLATEAU OF TIBET
Mt. Everest 29,035
HIMALAYAS
GREAT INDIAN DESERT
DECCAN
WESTERN GHATS
EASTERN GHATS
MONGOLIAN PLATEAU
GOBI DESERT
QIN LING
NORTH CHINA PLAIN
MANCHURIAN PLAIN
GREATER KHINGAN RANGE
SIKHOTE-ALIN RANGE
VERKHOYANSK RANGE
CHERSKIY RANGE
KOLYMA RANGE
CENTRAL RANGE
Kamchatka Peninsula
Taymyr Peninsula
North Land
New Siberian Is.
Sakhalin
Kuril Islands
Hokkaido
Honshu
Shikoku
Kyushu
Ryukyu Is.
Luzon
Mindanao
Hainan
Borneo
Celebes
Sumatra
Java
Timor
Ceram
New Guinea
Irian Jaya
Andaman Is. (India)
Nicobar Is. (India)
SOCOTRA (Yemen)
MALAY PEN.
Mekong Delta
Barents Sea
Kara Sea
Laptev Sea
East Siberian Sea
Bering Sea
Sea of Okhotsk
Sea of Japan
Yellow Sea
East China Sea
South China Sea
Sulu Sea
Celebes Sea
Banda Sea
Arafura Sea
Gulf of Thailand
Bay of Bengal
Arabian Sea
Gulf of Oman
Strait of Hormuz
Persian Gulf
Gulf of Aden
Red Sea
Mediterranean Sea
Black Sea
Caspian Sea
Aral Sea
Lake Balkhash
Lake Baikal
Tatar Strait
Korea Str.
Arctic Circle
Tropic of Cancer
Equator
Ob
Yenisey
Lena
Lower Tunguska
Angara
Amur
Irtysh
Tobol
Ili
Chu
Syr Dar'ya
Amu Dar'ya
Tigris
Euphrates
Indus
Sutlej
Ganges
Brahmaputra
Narmada
Godavari
Krishna
Chambal
Mekong
Irrawaddy
Yangtze
Huang (Yellow)
Selenga
Vilyuy
Aldan
Kolyma
50°N
60°N
70°N
80°N
70°N
60°N
50°N
40°N
30°N
20°N
10°N
0°
10°S
20°S
50°E
60°E
70°E
80°E
90°E
100°E
110°E
120°E
130°E
140°E
150°E
160°E

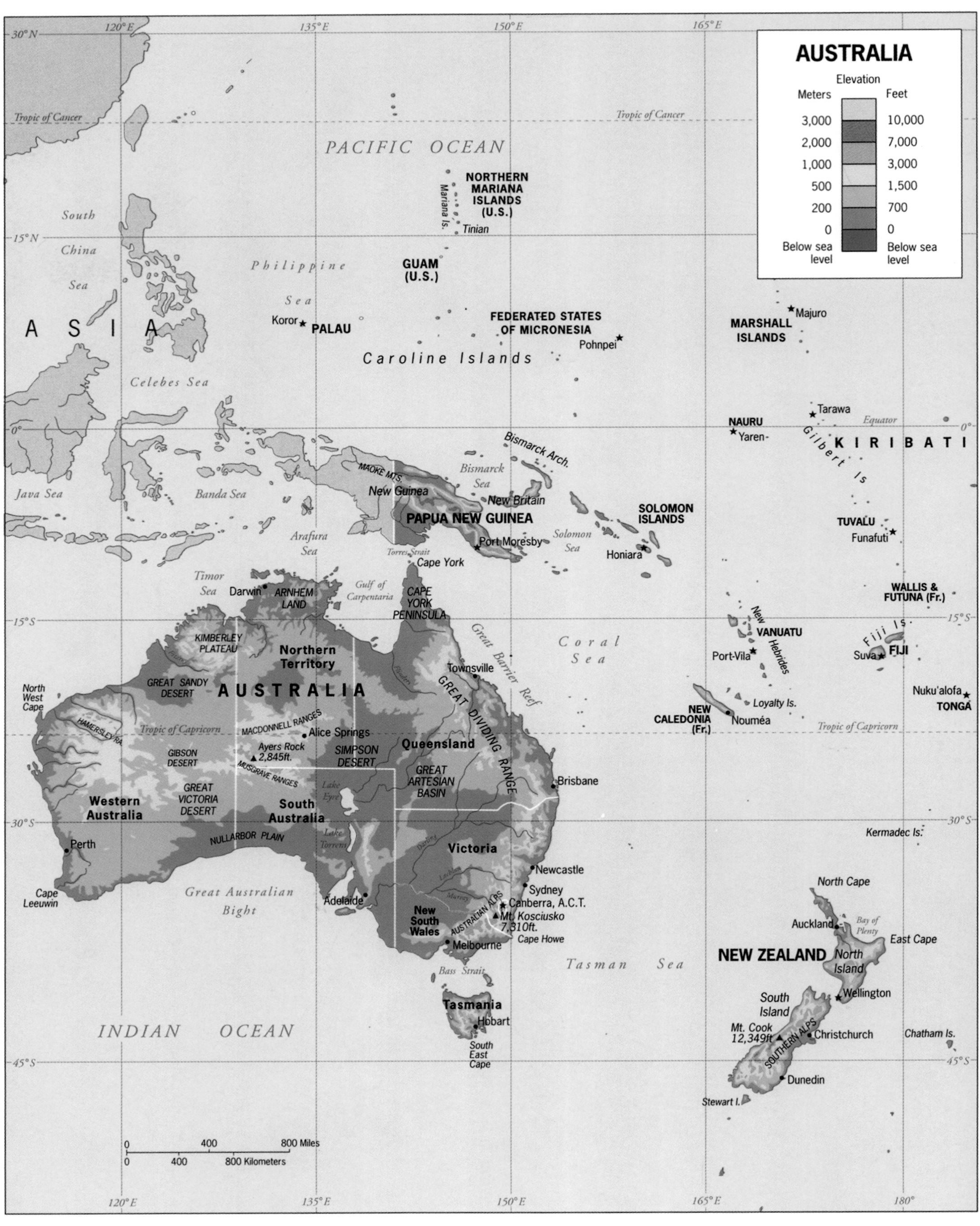
AUSTRALIA
Elevation
Meters
Feet
3,000
10,000
2,000
7,000
1,000
3,000
500
1,500
200
700
0
0
Below sea level
Below sea level
ASIA
PACIFIC OCEAN
South China Sea
Philippine Sea
Celebes Sea
Java Sea
Banda Sea
Arafura Sea
Timor Sea
NORTHERN MARIANA ISLANDS (U.S.)
Mariana Is.
Tinian
GUAM (U.S.)
Koror
PALAU
FEDERATED STATES OF MICRONESIA
Pohnpei
Caroline Islands
MARSHALL ISLANDS
Majuro
Tarawa
Equator
NAURU
Yaren
Gilbert Is
KIRIBATI
Bismarck Arch.
Bismarck Sea
MAOKE MTS.
New Guinea
New Britain
PAPUA NEW GUINEA
Port Moresby
Solomon Sea
SOLOMON ISLANDS
Honiara
TUVALU
Funafuti
Torres Strait
Cape York
Gulf of Carpentaria
Darwin
ARNHEM LAND
CAPE YORK PENINSULA
WALLIS & FUTUNA (Fr.)
New Hebrides
VANUATU
Port-Vila
Fiji Is.
Suva
FIJI
Great Barrier Reef
Coral Sea
KIMBERLEY PLATEAU
Northern Territory
Townsville
AUSTRALIA
GREAT SANDY DESERT
North West Cape
HAMERSLEY RA.
Tropic of Capricorn
MACDONNELL RANGES
Alice Springs
Ayers Rock 2,845ft.
GREAT DIVIDING RANGE
Nuku'alofa
TONGA
NEW CALEDONIA (Fr.)
Nouméa
Loyalty Is.
GIBSON DESERT
MUSGRAVE RANGES
SIMPSON DESERT
Queensland
GREAT ARTESIAN BASIN
Brisbane
Lake Eyre
Western Australia
GREAT VICTORIA DESERT
South Australia
NULLARBOR PLAIN
Lake Torrens
Kermadec Is.
Perth
Victoria
Darling
Newcastle
Sydney
Canberra, A.C.T.
Mt. Kosciusko 7,310ft.
AUSTRALIAN ALPS
Murray
Lachlan
Cape Leeuwin
Great Australian Bight
Adelaide
New South Wales
Melbourne
Cape Howe
North Cape
Auckland
Bay of Plenty
East Cape
NEW ZEALAND
North Island
Bass Strait
Tasman Sea
Wellington
Tasmania
Hobart
South East Cape
South Island
Mt. Cook 12,349ft
SOUTHERN ALPS
Christchurch
Chatham Is.
INDIAN OCEAN
Dunedin
Stewart I.
0
400
800 Miles
0
400
800 Kilometers
30° N
15° N
0°
15° S
30° S
45° S
120° E
135° E
150° E
165° E
180°
Tropic of Cancer

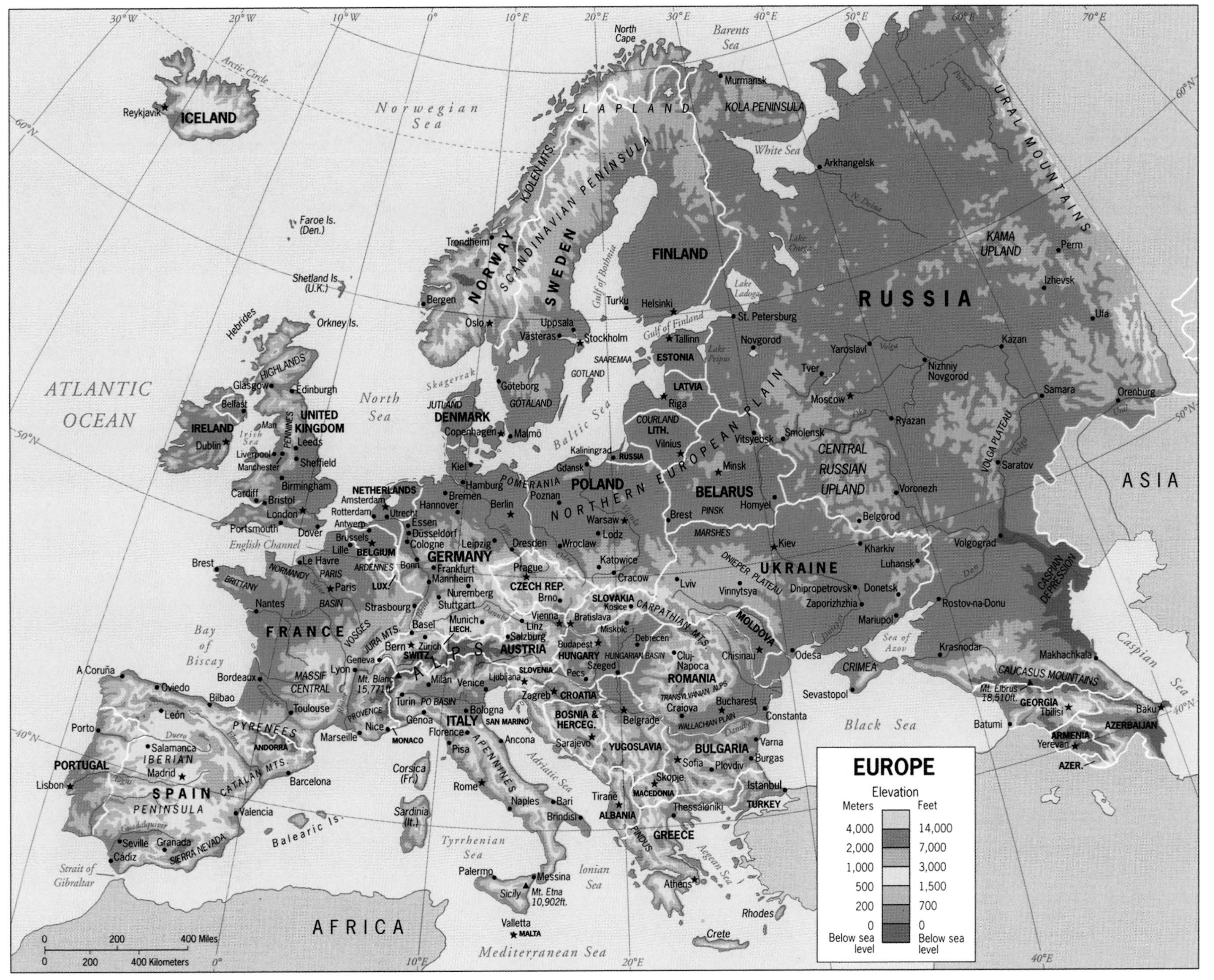

EUROPE
Elevation
Meters
Feet
4,000
14,000
2,000
7,000
1,000
3,000
500
1,500
200
700
0
0
Below sea level
Below sea level
0
200
400 Miles
0
200
400 Kilometers
ATLANTIC OCEAN
Norwegian Sea
North Sea
Barents Sea
White Sea
Baltic Sea
Bay of Biscay
Mediterranean Sea
Tyrrhenian Sea
Ionian Sea
Adriatic Sea
Aegean Sea
Black Sea
Sea of Azov
Caspian Sea
English Channel
Irish Sea
Skagerrak
Gulf of Bothnia
Gulf of Finland
Strait of Gibraltar
Arctic Circle
AFRICA
ASIA
ICELAND
Reykjavík
Faroe Is. (Den.)
Shetland Is. (U.K.)
Orkney Is.
Hebrides
IRELAND
Dublin
Belfast
UNITED KINGDOM
HIGHLANDS
PENNINES
Man
Glasgow
Edinburgh
Leeds
Liverpool
Manchester
Sheffield
Birmingham
Cardiff
Bristol
London
Portsmouth
Dover
NORWAY
Bergen
Trondheim
Oslo
KJOLEN MTS.
SCANDINAVIAN PENINSULA
LAPLAND
North Cape
SWEDEN
Uppsala
Västeras
Stockholm
Göteborg
GOTALAND
GOTLAND
SAAREMAA
Malmö
FINLAND
Turku
Helsinki
KOLA PENINSULA
Murmansk
DENMARK
JUTLAND
Copenhagen
ESTONIA
Tallinn
LATVIA
Riga
COURLAND
LITH.
Vilnius
RUSSIA
Kaliningrad
RUSSIA
Arkhangelsk
N. Dvina
Pechora
Lake Onega
Lake Ladoga
Lake Peipus
St. Petersburg
Novgorod
Tver
Yaroslavl
Moscow
Nizhniy Novgorod
Ryazan
Smolensk
Kazan
Samara
Saratov
Volgograd
Voronezh
Belgorod
Perm
Izhevsk
Ufa
Orenburg
Ural
Volga
Oka
Don
KAMA UPLAND
URAL MOUNTAINS
VOLGA PLATEAU
CENTRAL RUSSIAN UPLAND
CASPIAN DEPRESSION
NORTHERN EUROPEAN PLAIN
BELARUS
Minsk
Vitsyebsk
Homyel
Brest
PINSK MARSHES
UKRAINE
Kiev
Kharkiv
Luhansk
Donetsk
Dnipropetrovsk
Zaporizhzhia
Mariupol
Vinnytsya
Lviv
Odesa
Sevastopol
CRIMEA
Dnieper
DNIEPER PLATEAU
Rostov-na-Donu
Krasnodar
Makhachkala
CAUCASUS MOUNTAINS
Mt. Elbrus 18,510ft.
GEORGIA
Tbilisi
Batumi
ARMENIA
Yerevan
AZERBAIJAN
Baku
AZER.
MOLDOVA
Chisinau
POLAND
Gdansk
POMERANIA
Poznan
Warsaw
Lodz
Wroclaw
Katowice
Cracow
Vistula
Oder
NETHERLANDS
Amsterdam
Rotterdam
Utrecht
Antwerp
Brussels
BELGIUM
ARDENNES
LUX.
GERMANY
Kiel
Hamburg
Bremen
Hannover
Berlin
Essen
Düsseldorf
Cologne
Bonn
Leipzig
Dresden
Frankfurt
Mannheim
Nuremberg
Stuttgart
Munich
Elbe
Rhine
Danube
CZECH REP.
Prague
Brno
SLOVAKIA
Kosice
Bratislava
AUSTRIA
Vienna
Linz
Salzburg
LIECH.
SWITZ.
Basel
Zürich
Bern
Geneva
JURA MTS.
ALPS
Mt. Blanc 15,771ft.
FRANCE
Brest
BRITTANY
NORMANDY
Le Havre
Lille
PARIS BASIN
Paris
Seine
Loire
Nantes
Strasbourg
VOSGES
Lyon
MASSIF CENTRAL
Bordeaux
Garonne
Toulouse
PROVENCE
Marseille
Nice
Rhône
MONACO
Corsica (Fr.)
Sardinia (It.)
ITALY
Turin
Milan
Venice
PO BASIN
Po
Genoa
Bologna
SAN MARINO
Florence
Pisa
Ancona
APENNINES
Rome
Naples
Bari
Brindisi
Palermo
Messina
Sicily
Mt. Etna 10,902ft.
Valletta
MALTA
SLOVENIA
Ljubljana
CROATIA
Zagreb
BOSNIA & HERCEG.
Sarajevo
HUNGARY
Budapest
Miskolc
Debrecen
Pecs
Szeged
HUNGARIAN BASIN
CARPATHIAN MTS.
ROMANIA
Cluj-Napoca
TRANSYLVANIAN ALPS
Craiova
Bucharest
WALLACHIAN PLAIN
Constanta
YUGOSLAVIA
Belgrade
BULGARIA
Sofia
Plovdiv
Varna
Burgas
MACEDONIA
Skopje
ALBANIA
Tiranë
GREECE
PINDUS
Thessaloniki
Athens
Crete
Rhodes
TURKEY
Istanbul
SPAIN
IBERIAN PENINSULA
A Coruña
Oviedo
Bilbao
León
Salamanca
Duero
Ebro
Tagus
Madrid
Guadalquivir
Seville
Granada
Cádiz
SIERRA NEVADA
Valencia
Barcelona
CATALAN MTS.
PYRENEES
ANDORRA
Balearic Is.
PORTUGAL
Porto
Lisbon
30°W
20°W
10°W
0°
10°E
20°E
30°E
40°E
50°E
60°E
70°E
60°N
50°N
40°N

NORTH AMERICA
Elevation
Meters
Feet
3,000
10,000
2,000
7,000
1,000
3,000
500
1,500
200
700
0
0
Below sea level
Below sea level
0
500
1000 Miles
0
500
1000 Kilometers
ARCTIC OCEAN
ASIA
Chukchi Sea
Bering Sea
Bering Strait
St. Lawrence I.
SEWARD PEN.
Nome
Barrow
Pt. Barrow
BROOKS RANGE
Beaufort Sea
Nunivak I.
Alaska
Yukon
Bethel
Mt. McKinley 20,320
Fairbanks
ALASKA RANGE
Anchorage
Kodiak I.
Valdez
Dawson
Yukon Terr.
MACKENZIE MTS.
Gulf of Alaska
Whitehorse
Watson Lake
Juneau
Alexander Arch.
Sitka
Ketchikan
Prince Rupert
Queen Charlotte Is.
COAST MOUNTAINS
Inuvik
Mackenzie
Great Bear Lake
Northwest Territories
Yellowknife
Great Slave Lake
Ft. Nelson
Peace
Lake Athabasca
Ft. McMurray
Dawson Creek
Prince George
Alberta
Edmonton
British Columbia
Port Hardy
Vancouver I.
Kamloops
Vancouver
Victoria
Calgary
Lethbridge
ROCKY MOUNTAINS
Saskatchewan
La Ronge
Prince Albert
Saskatoon
Regina
Manitoba
Thompson
Flin Flon
Churchill
Lake Winnipeg
Winnipeg
Sachs Harbour
Banks I.
Victoria I.
Cambridge Bay
BOOTHIA PEN.
Melville I.
Queen Elizabeth Is.
Resolute
Devon I.
Ellesmere Island
Alert
Qaanaaq
Greenland (Kalaallit Nunaat) (Denmark)
Baffin Bay
Pond Inlet
Baffin Island
Davis Strait
Nuuk
Arctic Circle
Pangnirtung
Iqaluit
Nunavut
Southampton I.
Hudson Strait
Rankin Inlet
CANADA
Hudson Bay
UNGAVA PEN.
Labrador Sea
Newfoundland
Schefferville
Goose Bay
Labrador
Labrador City
Newfoundland
St. John's
Sept-Îles
Quebec
Gulf of St. Lawrence
P.E.I.
Sydney
Charlottetown
N.B.
Halifax
Fredericton
Nova Scotia
C. Sable
James Bay
Moosonee
Chibougamau
Chicoutimi
CANADIAN SHIELD
Ontario
Timmins
Val-d'Or
Québec
Thunder Bay
Sudbury
Montréal
Ottawa
Maine
Bangor
Portland
Vt.
N.H.
Boston
Mass.
Cape Cod
R.I.
Conn.
New York
Toronto
Buffalo
Michigan
L. Superior
L. Huron
L. Michigan
L. Ontario
L. Erie
PACIFIC OCEAN
Seattle
Spokane
Wash.
Portland
Eugene
Oregon
CASCADES
Great Falls
Montana
Butte
Billings
Idaho
Boise
Snake
Eureka
C. Mendocino
Pocatello
Great Salt Lake
Salt Lake City
Reno
Nevada
SIERRA NEVADA
Sacramento
San Francisco
San Jose
California
Casper
Wyoming
Cheyenne
GREAT PLAINS
Utah
Colorado
Denver
Las Vegas
Santa Barbara
Los Angeles
San Diego
Tijuana
Mexicali
Arizona
Phoenix
Tucson
Albuquerque
New Mexico
El Paso
North Dakota
Bismarck
Fargo
South Dakota
Minnesota
Minneapolis
Wisconsin
Sioux Falls
Milwaukee
Detroit
Nebraska
Omaha
Iowa
Des Moines
Chicago
Ill.
Ind.
Indianapolis
Ohio
Columbus
Cleveland
Pittsburgh
Pa.
Philadelphia
N.J.
Md.
Del.
Baltimore
Washington, D.C.
W.Va.
Va.
Richmond
Cincinnati
Louisville
Kentucky
Kansas
Kansas City
St. Louis
Missouri
Wichita
UNITED STATES
Amarillo
Okla.
Oklahoma City
Arkansas
Little Rock
Memphis
Tennessee
Nashville
Birmingham
Atlanta
Georgia
S.C.
Charleston
N.C.
Raleigh
Charlotte
APPALACHIAN MTS.
Lubbock
Dallas
Texas
Miss.
Jackson
Alabama
La.
COASTAL PLAINS
Savannah
Jacksonville
Florida
Tampa
St. Petersburg
Miami
Austin
San Antonio
Houston
New Orleans
Mobile
ATLANTIC OCEAN
Rio Grande
Ciudad Juárez
Hermosillo
Chihuahua
BAJA CALIFORNIA
Gulf of California
Ciudad Obregón
Nuevo Laredo
Brownsville
Matamoros
La Paz
Torreón
Monterrey
SIERRA MADRE OCCIDENTAL
SIERRA MADRE ORIENTAL
Durango
C. San Lucas
Mazatlán
MEXICO
Tampico
Aguascalientes
León
Guadalajara
Mexico City
Puebla
Veracruz
Acapulco
Oaxaca
Villahermosa
Campeche
Mérida
Yucatán Peninsula
Gulf of Mexico
Tropic of Cancer
BAHAMAS
Nassau
Havana
CUBA
Greater Antilles
HAITI
Port-au-Prince
DOMINICAN REP.
Santo Domingo
Puerto Rico (U.S.)
JAMAICA
Kingston
Caribbean Sea
Belmopan
BELIZE
GUATEMALA
Guatemala City
HONDURAS
Tegucigalpa
San Salvador
EL SALVADOR
NICARAGUA
Managua
San José
COSTA RICA
Panama City
PANAMA
SOUTH AMERICA
PACIFIC OCEAN
180°
160° W
140° W
120° W
80° N
70° N
60° N
50° N
40° N
30° N
20° N
10° N
110° W
100° W
90° W
80° W
70° W
60° W
40° W
20° W

SOUTH AMERICA
Elevation
Meters
Feet
3,000
10,000
2,000
7,000
1,000
3,000
500
1,500
200
700
0
0
Below sea level
Below sea level
ATLANTIC OCEAN
PACIFIC OCEAN
CENTRAL AMERICA
Caribbean Sea
Pt. Gallinas
Cristóbal Colón 19,020 ft.
Barranquilla
Cartagena
Maracaibo
Caracas
Barquisimeto
Valencia
VENEZUELA
Ciudad Guayana
CORD. DE MÉRIDA
LLANOS
GUIANA HIGHLANDS
Georgetown
Paramaribo
Cayenne
GUYANA
SURINAME
FRENCH GUIANA (Fr.)
Gulf of Panama
Medellín
Bogotá
Cali
CORD. OCCIDENTAL
CORD. CENTRAL
CORDILLERA ORIENTAL
COLOMBIA
Quito
Mt. Cotopaxi 19,347 ft.
ECUADOR
Guayaquil
Pt. Pariñas
Iquitos
Equator
Marajó Island
Belém
Manaus
SELVAS
AMAZON BASIN
Fortaleza
Cape São Roque
Natal
Recife
Maceió
Trujillo
Mt. Huascarán 22,205 ft.
PERU
ANDES
Lima
Cuzco
BRAZIL
BOLIVIA
Lake Titicaca
Mt. Illimani 21,201 ft.
La Paz
Arequipa
MATO GROSSO PLATEAU
BRAZILIAN HIGHLANDS
Brasília
Goiânia
Salvador
SERRA DO ESPINHAÇO
Santa Cruz
Mt. Sajama 21,391 ft.
ALTIPLANO
Lake Poopó
Sucre
Belo Horizonte
Mt. Licancábur 19,455 ft.
PARAGUAY
Antofagasta
ATACAMA DESERT
Mt. Llullaillaco 22,057 ft.
GRAN CHACO
Asunción
Campinas
Rio de Janeiro
São Paulo
Tropic of Capricorn
Curitiba
SERRA DO MAR
San Miguel de Tucumán
CHILE
ARGENTINA
Pôrto Alegre
Patos Lagoon
Mt. Mercedario 22,211 ft.
Mt. Aconcagua 22,834 ft.
Córdoba
Rosario
Valparaíso
Santiago
URUGUAY
Mirim Lagoon
Buenos Aires
Montevideo
La Plata
Río de la Plata
PAMPAS
Concepción
Mar del Plata
Bahía Blanca
San Matías Gulf
Valdés Peninsula
Chiloé Island
Chonos Arch.
Gulf of San Jorge
Mt. San Valentín 13,314 ft.
Cape Tres Puntas
PATAGONIA
Wellington Island
FALKLAND ISLANDS (U.K.)
Stanley
Strait of Magellan
Tierra del Fuego
Punta Arenas
Cape Horn
0 300 600 Miles
0 300 600 Kilometers